Comprehensive Organic Functional Group Transformations

Comprehensive Organic Functional Group Transformations

Editors-in-Chief

Alan R. Katritzky, FRS
University of Florida, Gainesville, FL, USA

Otto Meth-Cohn
University of Sunderland, UK

Charles W. Rees, FRS
Imperial College of Science, Technology and Medicine, London, UK

Volume 1

SYNTHESIS: CARBON WITH NO ATTACHED HETEROATOMS

Volume Editor

Stanley M. Roberts
University of Exeter, UK

PERGAMON

UK	Elsevier Science Ltd., The Boulevard, Langford Lane, Kidlington, Oxford OX5 1GB, UK
USA	Elsevier Science Inc., 660 White Plains Road, Tarrytown, New York 10591-5153, USA
JAPAN	Elsevier Science Japan, Tsunashima Building Annex, 3-20-12 Yushima, Bunkyo-ku, Tokyo 113, Japan

First edition 1995

Library of Congress Cataloging in Publication Data
Comprehensive organic functional group transformations / editors-in-chief, Alan R. Katritzky, Otto Meth-Cohn, Charles W. Rees. — 1st ed.
p. cm.
Includes indexes.
Contents: v. 1. Synthesis: carbon with no attached heteroatoms / edited by Stanley M. Roberts — v. 2. Synthesis: carbon with one heteroatom attached by a single bond / edited by Steven V. Ley — v. 3. Synthesis: carbon with one heteroatom attached by a multiple bond / edited by Gerald Pattenden — v. 4. Synthesis: carbon with two heteroatoms, each attached by a single bond / edited by Gordon W. Kirby — v. 5. Synthesis: carbon with two attached heteroatoms with at least one carbon-to-heteroatom multiple link / edited by Christopher J. Moody — v. 6. Synthesis: carbon with three or four attached heteroatoms / edited by Thomas L. Gilchrist — v. 7. Indexes.
1. Organic compounds—Synthesis. I. Katritzky, Alan R. II. Meth-Cohn, Otto. III. Rees, Charles W.
QD262.C534 1995
547'.2—dc20 95–31088

British Library Cataloguing in Publication Data
A catalogue record for this book is available from the British Library.

ISBN 0–08–040604–1 (set : alk. paper)
ISBN 0–08–042322–1 (Volume 1)

∞™ The paper used in this publication meets the minimum requirements of the American National Standard for Information Sciences—Permanence of Paper for Printed Library Materials, ANSI Z39.48–1984.

Printed and bound in Great Britain by Cambridge University Press, Cambridge, UK.

Contents

Preface

Some years ago the three of us met in a London club reviewing an ongoing publishing venture in Organic Synthesis. The conversation drifted to a consideration of volumes on the synthesis of key functional groups. No doubt the good wine helped since we actually broached the idea of a work on the synthesis of *all* functional groups. Would it be useful? Definitely. Would it be feasible? How would it be organized? Where do you start? We recognized that functionality was based on the coordination and heteroatom attachment of a carbon atom. But putting together a complete framework seemed particularly daunting. Two of us became very interested in the fascinating bouquet of the Muscat de Beaumes de Venise.

At our next dinner together Alan announced that he had solved the problems posed last time—problems that Charles and I hoped he had forgotten! He brought out a remarkable matrix analysis of *all* functional groups, analysed rigorously and logically. Even unknown functions were covered. Although we were all very impressed, the practicalities of the idea still seemed daunting. Those who know Alan's terrier instincts will appreciate that he would not give up such a challenge so easily. Our twice yearly club get-togethers, occasionally with friends from Pergamon, refined our thinking. Alan's cosmic vision was tempered by Charles's intuitive realism and fully supported by the publishers.

Another major problem remained: how to reduce our thinking into a practical handbook for authors—a dismaying task for three busy chemists. We settled on a seven-volume work and the indomitable ARK produced a rough breakdown to fit such a format. Putting flesh on these bones became feasible during a fortuitous three-month break between jobs by myself, and the largest handbook ever assembled by Pergamon (120 pages) was written and page allocations agreed—even for little or unknown functional groups. Sample chapters were commissioned and finally proved very encouraging, despite our first chosen topic uncovering virtually no known examples!

Contracts were defined and agreed, volume editors approached, and potential authors considered during a pleasant preconference stay in Grasmere. Following the sale of Pergamon to Elsevier Science Ltd there was a lull in the project but soon *Comprehensive Organic Functional Group Transformations* was back on track, and everyone adhered to a very businesslike timetable.

OTTO METH-COHN
Sunderland

CHARLES W. REES
London

ALAN R. KATRITZKY
Florida

Contributors to Volume 1

Dr A. Armstrong
School of Chemistry, University of Bath, Claverton Down, Bath BA2 7AY, UK

Dr T. N. Birkinshaw
Astra Charnwood, Bakewell Road, Loughborough, Leicestershire LE11 ORH, UK

Dr C. Bruneau
Laboratoire de Chimie de Coordination Organique, Université de Rennes 1, URA CNRS 415, Campus de Beaulieu, F-35042 Rennes Cedex, France

Dr A.J. Clark
Department of Chemistry, University of Warwick, Coventry CV4 7AL, UK

Dr I. Coldham
Department of Chemistry, University of Exeter, Stocker Road, Exeter EX4 4QD, UK

Dr J. M. Dickinson
Department of Chemistry, Manchester Metropolitan University, John Dalton Building, Chester Street, Manchester M1 5GD, UK

Professor P. H. Dixneuf
Laboratoire de Chimie de Coordination Organique, Université de Rennes 1, URA CNRS 415, Campus de Beaulieu, F-35042 Rennes Cedex, France

Dr G. L. Edwards
Department of Organic Chemistry, Heffron Building, Room 415, The University of New South Wales, Barker Street, Gate 14, Kensington, Sydney, NSW 2052, Australia

Dr M. Furber
Astra Charnwood, Bakewell Road, Loughborough, Leicestershire LE11 ORH, UK

Dr I. Gosney
Department of Chemistry, University of Edinburgh, King's Buildings, West Mains Road, Edinburgh EH9 3JJ, UK

Dr M. A. Hayes
Department of Natural Products Chemistry, Xenova Group plc, 545 Ipswich Road, Slough, Berkshire SL1 4EQ, UK

Dr J. Howarth
School of Chemical Sciences, Dublin City University, Glasnevin, Dublin 9, Republic of Ireland

Dr K. Jones
Department of Chemistry, King's College, The Strand, London WC2R 2LS, UK

Dr M. J. Kelly
Department of Chemistry, University of Exeter, Stocker Road, Exeter EX4 4QD, UK

Dr D. Lloyd
Department of Chemistry, University of St. Andrews, St. Andrews, Fife KY16 9ST, UK

Dr A. R. Maguire
Department of Chemistry, University College Cork, College Road, Cork, Republic of Ireland

Dr H. L. McFarland
Department of Chemistry, University of Liverpool, PO Box 147, Liverpool L69 3BX, UK

Dr H. McNabb
Department of Chemistry, University of Edinburgh, King's Buildings, West Mains Road, Edinburgh EH9 3JJ, UK

Dr A. P. Millar
Department of Chemistry, University of Liverpool, PO Box 147, Liverpool L69 3BX, UK

Dr P. J. Murphy
Department of Chemistry, University of North Wales, Bangor, Gwynedd LL57 2UW, UK

Dr P. C. B. Page
Department of Chemistry, University of Liverpool, PO Box 147, Liverpool L69 3BX, UK

Dr J. M. Percy
School of Chemistry, University of Birmingham, Edgbaston, Birmingham B15 2TT, UK

Dr C. M. Rayner
School of Chemistry, University of Leeds, Leeds LS2 9JT, UK

Dr A. C. Regan
Department of Chemistry, University of Manchester, Oxford Road, Manchester M13 9PL, UK

Dr P. G. Steel
Department of Chemistry, University of Durham, Science Laboratories, South Road, Durham DH1 3LE, UK

Dr A. G. Sutherland
School of Applied Chemistry, Faculty of Science, Computing and Engineering, University of North London, 166–220 Holloway Road, London N7 8DB, UK

Dr P. C. Taylor
Department of Chemistry, University of Warwick, Coventry CV4 7AL, UK

Dr J. O. Williams
Department of Chemistry, University of Exeter, Stocker Road, Exeter EX4 4QD, UK and White Knight Chemicals Ltd., Orchard House, Higher Blagdon, Paignton, Devon PQ3 3YS, UK

Contents of All Volumes

Volume 3 Synthesis: Carbon with One Heteroatom Attached by a Multiple Bond

Volume 4 Synthesis: Carbon with Two Heteroatoms, Each Attached by a Single Bond

Introduction

OBJECTIVES, SCOPE, AND COVERAGE

Comprehensive Organic Functional Group Transformations (COFGT) aims to present the vast subject of organic synthesis in terms of the introduction and interconversion of functional groups. All organic structures can be considered as skeletal frameworks of carbon atoms to which functional groups are attached[a]; it is the latter which are mainly responsible for chemical reactivity and which are highlighted in COFGT. All known functional groups fit a logical and comprehensive pattern and this forms the basis for the detailed list of contents. The format of the present work was designed with the intention to cover systematically all the possible arrangements of atoms around a carbon, including those which are quite unfamiliar. The work also considers the possibility of as yet unknown functional groups which may be constructed in the future and prove to be important; thus COFGT also indicates what is not known and so points the way to new research areas.

The philosophy of the present work has been to rationalize this enormous subject within as logical and formal a framework as possible, in a scholarly and critical fashion. COFGT is designed to provide the first point of entry to the literature for synthetic organic chemists, together with an unrivalled source for anyone interested in less common, obscure, or unknown functional groups.

All functional groups are viewed as being carbon based (even if the group contains no carbon). Thus, a nitro compound is considered from the standpoint of the immediately attached carbon atom, whether di- (*sp*), tri- (sp^2), or tetracoordinated(sp^3). The work is organized on the basis of formation or rupture of bonds to a carbon atom and it is the nature of the carbon atom left after the transformation that determines the classification of the overall sequence. Several key criteria have been used to organize the work and to minimize overlap. These are, in order of priority:

1. the number of attached heteroatoms;
2. the coordination of the carbon atom involved in the functional group;
3. the nature of the immediately attached heteroatom(s); and
4. the Latest Placement Principle.

These four key principles have been used to determine the content of each volume, and to develop the detailed chapter breakdown within each volume.

Thus, according to the number of attached heteroatoms:

Volume 1 deals with synthetic reactions which result in the alteration of bonding at carbon atoms which are left with *no* attached heteroatoms.

Volume 2 deals with syntheses which result in carbon atoms attached to *one* heteroatom by a single bond.

Volume 3 deals with syntheses which result in carbon atoms attached to *one* heteroatom by a double or by a triple bond.

Volume 4 deals with syntheses which result in carbon atoms attached to *two* heteroatoms, each by a single bond.

Volume 5 deals with syntheses which result in carbon atoms attached to *two* heteroatoms by one single and one double bond, or by two double bonds, or by one single and one triple bond.

Volume 6 deals with syntheses which result in carbon atoms attached to *three* or *four* heteroatoms.

Volume 7 comprises the author and subject indexes.

Certain key principles apply to all the volumes because all functional groups are viewed as carbon based (e.g. a nitro group is either alkyl-, vinyl-, aryl-, or alkynyl-); these are:

(a) Volumes are subdivided according to the *coordination* of the carbon atom which is the product of the reaction, i.e., tetra- coming before tri- before di- before monocoordinated carbon functions.

---

[a] The major exception to this lies in heterocyclic compounds, where the cyclic heteroatoms are more logically considered as part of the framework. The subject of heterocycles has been treated elsewhere in the companion work *Comprehensive Heterocyclic Chemistry* published in 1984 with a second edition to be published in 1996.

In Volumes 1 and 6, reactions producing four-coordinated carbon are considered first, followed by three- and then two-coordinated carbon. The other volumes contain a more limited range of coordination types (Volumes 2 and 4 only four-coordinated, Volumes 3 and 5 only two- or three-coordinated). Each type of coordination is allocated a separate section in each volume.

(b) Attached *heteroatoms* are discussed in the following order of priority:

Halogens—F, Cl, Br, I
Chalcogens—O, S, Se, Te
Nitrogen—N
Other group 15 elements—P, As, Sb, Bi
Metalloids—B, Si, Ge
Main group metals—Sn, Pb, Al, Ga, In, Tl, Be, Mg, Ca, Sr, Ba, Li, Na, K, Rb, Cs
Transition metals—Cu, Ag, Au, Zn, Cd, Hg, Ti, Zr, Hf, Cr, Mo, W, Mn, Fe, Co, Ni, Pd, Pt, and others.

Higher coordination of heteroatoms is treated after lower. Thus, in sections dealing with iodo compounds, monocoordinate (e.g. iodides) are discussed before dicoordinate (e.g. iodoxyls) and tricoordinate functions.

(c) The *Latest Placement Principle* (or Last Position Principle) is used to avoid undue overlap in the work. Thus, the carbon attached to the heteroatom is discussed at the last possible position in the above prioritizing of heteroatoms. Examples of its application are noted later. On this basis, for example, when both C—C and C—H bonds are formed the reaction will appear in the latest chapter (i.e. Chapters 1.04–1.10 rather than in the earlier Chapters 1.01–1.03), and when both C—C and C═C bonds are formed, this will be found in the later Chapter 1.17. The Latest Placement Principle is particularly important in determining where to find electrocyclic reactions in Volume 1. If C—H, ═C—C, and C═C bonds are all formed in a reaction, then the latest appropriate chapter will deal with the reaction. Only if a change in heterofunction occurs is the reaction left to a later volume.

Exceptions to the above principles are rare. However the reactions of heteroarenes are mentioned along with those of arenes. If on reduction no change in the heterofunction occurs (e.g. in going from thiophene to tetrahydrothiophene or from pyridine to 2,3,4,5-tetrahydropyridine) the reaction is found in Volume 1. However, when the function changes (e.g. pyridine to piperidine), the conversion is considered in Volume 2. Conversion of methyl phenyl sulfone into methyl cyclohexyl sulfone appears in Volume 2, whereas the formation of cyclohexyl methyl ketone is treated in Volume 1, since the coordination of the carbon atom to the heteroatoms is changed in the first but not in the second hydrogenation.

Some further exceptions to the rigorous ordering of the work have been made for the purpose of easy reference. Thus, in Volume 1, a special chapter on ions, radicals, and carbenes is added: this chapter is limited to the treatment of species capable of more than a transitory existence. Throughout the work aspects of the Latest Placement Principle are occasionally ignored for reasons of clarity. Thus metal ligands that are incidental to the chemistry under discussion are not considered when prioritizing. Also, references to aromatic substituents, some of which involve a heteroatom (e.g. pyridyl, thienyl, etc.) but are incidental to the chemistry being described, are not viewed as changing the priority.

Within each section, we have endeavored to explain the influence of important secondary effects such as inclusion in a ring, degree of strain, degree of substitution, various types of activation, influence of stereochemistry, and so on, on the transformation under consideration. General synthetic methods are treated before specific methods.

Transient intermediates, as such, do not fall within the scope of this work. Although there is clearly no sharp division, we have attempted to restrict coverage of radicals, etc., to more stable, longer lived species. It is the aim of this work to consider all organic functional groups provided that the molecules which incorporate them, though they may be unstable, can have a finite lifetime and chemistry. The whole work deals with the generation and transformation of functional groups, *not* of molecules such as CO_2, COS, CS_2, ClCN, etc. Such simple carbon derivatives are not treated unless a further carbon is attached (e.g. RN═C═O).

VOLUME 1 SYNTHESIS: CARBON WITH NO ATTACHED HETEROATOMS

Volume 1 deals solely with the formation of nonheteroatom functional groups and as such is different in style to the remaining volumes.

In addition to the general principles, Volume 1 is further organized as follows:

1. By the type of bond formed (i.e. C—H before C—C).
2. By the type of reaction involved (i.e. substitution, then addition, then rearrangement). With C=C bond formation the order is addition, elimination, condensation, then electrocyclic and other methods. One rearrangement chapter only is devoted to each of the Parts I and II.
3. In Parts II and III the treatment of formation of ions, radicals, and carbenes is added at the end of the section dealing solely with those species with a significant rather than a transient lifetime.

In Volume 1, the heteroatom sequence is a secondary feature since only remote heteroatom functions are involved in the products: but the standard order pertains in reactants that contain heteroatoms (see, e.g. Chapters 1.01 and 1.02).

All the major structural influences that are treated throughout this work apply equally (or perhaps more importantly) in Volume 1. Thus the effects of conjugation, remote substituents, rings, stereochemistry, strain, kinetic or thermodynamic factors, solvation, primary, secondary and tertiary nature, etc., are mentioned whenever relevant.

VOLUME 2 SYNTHESIS: CARBON WITH ONE HETEROATOM ATTACHED BY A SINGLE BOND

Volume 2 is arranged in three parts: I, II and III, dealing respectively with sp^3, sp^2, and sp carbon linked to the heteroatom. In each chapter we have endeavored to explain important effects due to such features as the primary, secondary, tertiary nature, ring effects, strain activation, effect of beta, gamma, and more remote functionality, stereochemical effects, and so on. Methods that are common to a larger group are dealt with at their first appearance and suitably cross-referenced.

Volumes 2–6 all deal with the synthesis of functions involving at least one heteroatom. To avoid major overlap we have applied the Latest Placement Principle; that is, the chemistry is discussed at the last possible position based on the prioritization of the carbon attached to the heteroatom. Thus the compound CH_3ONH_2 is treated under "Alkyl Chalcogenides" in the subsection "Functions Based on the RON-Unit" (i.e. 2.02.6). However, CH_3ONHCH_3 appears under "Alkyl Nitrogen Compounds" (2.06.2.3) since the Latest Placement Principle prevails. Also, dialkyl ethers appear in Part I of Volume 2 (Functions Linked by a Single Bond to an sp^3 Carbon Atom), while alkyl aryl ethers appear in Part II of Volume 2 (Functions Linked by a Single Bond to an sp^2 Carbon Atom). Exceptions to the rule are:

(a) When a fully unsaturated heterocyclic substituent (e.g. thienyl, pyridyl, etc.) is used as an example of an aryl group, the ring heteroatom(s) is (are) not taken into account (e.g. 2-methoxypyridine should strictly appear in Volume 6, but is covered in Volume 2 along with 3- and 4-methoxypyridine).

(b) Carbon-based metal ligands that are incidental to the synthesis under discussion (e.g. carbonyls, cyclopentadienyls, etc.) are not taken into consideration.

VOLUME 3 SYNTHESIS: CARBON WITH ONE HETEROATOM ATTACHED BY A MULTIPLE BOND

Volume 3 follows the logical development indicated in Volume 2. Thus, according to the Last Placement Principle, the imines, RCH=N—R, appear in Volume 3 rather than in Volume 2 (where functions singly bonded to carbon are treated). Furthermore, acetophenone, $PhCOCH_3$, is treated under α,β-unsaturated ketones (3.05) rather than saturated ketones (3.04). Chloronitroacrylonitriles would appear under the section "α,β-Vinylic Nitriles with Nitrogen-based Substituents" (3.19.2.7), not under the related earlier section dealing with halo-substituents (3.19.2.3).

VOLUME 4 SYNTHESIS: CARBON WITH TWO HETEROATOMS, EACH ATTACHED BY A SINGLE BOND

Volume 4 is in three parts. Part I deals with tetracoordinated carbon bearing two heteroatoms, Part II with tricoordinated carbon bearing two heteroatoms, and Part III (a brief chapter) with stabilized radicals, ions, and the like bearing two heteroatoms. The material is arranged according to the Latest Placement Principle: thus, the synthesis of $CHBr_2CHI_2$ would appear in the section dealing with diiodo, not dibromo functions (i.e. in 4.01.5, not 4.01.4), and the synthesis of CF_3CH-BrCl is discussed in Volume 6 (carbons bearing three heteroatoms), rather than in Volume 4.

VOLUME 5 SYNTHESIS: CARBON WITH TWO ATTACHED HETEROATOMS WITH AT LEAST ONE CARBON-TO-HETEROATOM MULTIPLE BOND

Volume 5 is in three parts. Part I deals with functions with one doubly bonded and one singly bonded heteroatom, Part II with functions containing two doubly bonded heteroatoms and Part III with one triply bonded and one singly bonded heteroatom. Part I constitutes the bulk of Volume 5.

The arrangement of the chemistry in each part follows the same logical sequence. The multiply bonded heteroatom is focused on first and then the other heteroatom in a secondary classification, both following the priority rules already described. Each section excludes the coverage of the previous sections. Thus, all carbonyl derivatives will appear in Chapters 5.01–5.10 but not in Chapters 5.11, *et seq.*

According to the Latest Placement Principle structure RC(O)OC(S)R is discussed in the chapter dealing with carbons bearing a doubly bonded sulfur and singly bonded oxygen (5.12.3), *not* in that dealing with doubly and singly bonded oxygen (5.04.1). Another effect of the Latest Placement Principle is that the amides RCONMePh are discussed under *N*-arylalkanoamides (5.06.2.4), rather than *N*-alkylalkanoamides (5.06.2.2). Again, exceptions are made to the latest placement rules for: (a) hetaryl rings used as examples of aryl substituents which are not viewed as functional groups. Thus, 2-methylimidazole is not considered as an example of an amidine function and 2-methoxypyridine is not an example of a doubly bonded nitrogen, singly bonded oxygen function; (b) metal ligands that are incidental to the organic chemistry under discussion are not viewed as functions in priority considerations.

VOLUME 6 SYNTHESIS: CARBON WITH THREE OR FOUR ATTACHED HETEROATOMS

Volume 6 is in four parts. Part I deals with tetracoordinate carbons bearing three heteroatoms. Part II covers tetracoordinate compounds bearing four heteroatoms, i.e. substituted methanes, and Part III deals with tricoordinate systems bearing three heteroatoms, i.e. where one heteroatom is attached by a double bond. Part IV is brief and deals with stabilized radicals and ions. Not surprisingly, the coverage of Volume 6 is very large—and also shows that many gaps in the development of organic chemistry still exist.

The organization within the three sections not only follows the same broad logic developed in the previous volumes, but also has a structure unique to the multiheteroatom volume. According to the Latest Placement Principle $CF_3C(NR_2)_3$ appears in the section dealing with carbons bearing three nitrogens (6.05.1.1), not that dealing with carbons bearing three halogens (6.01.2), while $(CF_3CH_2O)_2CO$ appears in Part III, not in Part I.

In the chapter dealing with iminocarbonyl functions in Part III, the substituents on nitrogen are discussed in each appropriate subsection in the order outlined above. Thus, the RN= group would be first considered with R = H, then alkyl, alkenyl, aryl and hetaryl, alkynyl and then heteroatom substituents in the usual order.

In each relevant section, we have endeavored to explain the influence of important secondary effects on the synthesis such as structure (primary, secondary, etc.), ring effects, strain, activation, stereochemistry, remote substituent effects, etc.

The arrangement of the chemistry in each of Parts I–III follows a similar pattern. Thus, each section commences with functions containing at least one halogen. This section deals with all combinations of halogen with other heteroatoms in the described order. The next section deals with functions containing at least one chalcogen in combination with any other heteroatoms except halogens. Subsequent sections each exclude the previous title heteroatom functions.

VOLUME 7 INDEXES

Subject Indexes are included in each of Volumes 1–6 and Cumulative Subject and Author Indexes appear in Volume 7. Most entries in the Subject Index consist of two or three lines: the first line is the entry itself (e.g. Lactones) and the second line is descriptive of that entry (e.g. reduction); in many cases more detail is given (e.g. with 9-BBN).

REFERENCES

The references are handled by the system previously used successfully in *Comprehensive Heterocyclic Chemistry*. In this system reference numbers appear neither in the text, nor as footnotes, nor at the end of chapters. Instead, each time a reference is cited in the text there appears (in parentheses) a two-letter code assigned to the journal being cited, which is preceded by the year (tens and units only for twentieth-century references) and followed by the page number. For example: "It was shown ⟨80TL1327⟩ that . . .". In this phrase, "80" refers to 1980, "TL" to *Tetrahedron Letters*, and "1327" to the page number. For those journals which are published in parts, or which have more than one volume number per year, the appropriate part of the volume is indicated, e.g. as in ⟨73JCS(P2)1594⟩ or ⟨78JOM(162)611⟩, where the first example refers to *J. Chem. Soc., Perkin Trans 2*, 1973, page 1594, and the second to *J. Organomet. Chem.*, 1978, volume 162, page 611.

This reference system is adopted because it is far more useful to the reader than the conventional "superscript number" system. It enables readers to go directly to the literature reference cited, without first having to consult the bibliography at the end of each chapter.

References to the last century quote the year in full. Books have a prefix "B-" and if they are commonly quoted (e.g. *Organic Reactions*) they will have a code. Otherwise, as with uncommon journals, they are given a miscellaneous code (MI) and numbered arbitrarily *abb1*, *abb2*, etc., where *abb* refers to the volume and chapter number and *1*, *2*, etc., are assigned sequentially. Patents are assigned appropriate three-letter codes.

The references are given in full at the end of each volume. They include *Chemical Abstract* references when these are likely to help; in particular, they are given for all patents, and for less accessible sources such as journals in languages other than English, French, or German, company reports, obscure books, and theses.

1.01 One or More CH Bond(s) Formed by Substitution: Reduction of C—Halogen and C—Chalcogen Bonds

ALAN G. SUTHERLAND
University of North London, UK

1.01.1 REDUCTION OF C—HALOGEN BONDS TO CH

1.01.1.1 General Methods

A large majority of the methods available for the reduction of alkyl halides to the corresponding alkanes, often referred to as the process of hydrogenolysis in earlier works, can be loosely grouped into four categories: catalytic hydrogenation, low-valent metal reduction, metal hydride nucleophilic displacement and radical substitution.

Generally, the observed reactivity is in the order I > Br > Cl > > F for all four categories, matching the order of bond strengths (C—I 53 kcal mol^{-1}, C—Br 67 kcal mol^{-1}, C—Cl 81 kcal mol^{-1}, C—F 109 kcal mol^{-1}). Benzylic and, to a lesser extent, allylic halides also tend to be more reactive than similar alkyl halides. The order of reactivity between primary, secondary and tertiary alkyl halides tends to be dependent on the reagent in use. α-Halo carbonyl compounds are particularly prone to reduction by these methods, especially in the case of low-valent metal reductions ⟨83OR163⟩, although competing carbonyl group reductions may occur with some procedures.

Until the late-1970s the first three categories appeared more commonly, as reflected in a contemporary review ⟨80S425⟩, but radical reduction procedures have dominated since.

Catalytic hydrogenation methods have been reviewed ⟨B-85MI 101-01⟩. Practical difficulties can be encountered in utilising these procedures, owing to catalyst poisoning by the hydrogen halide evolved, particularly in the reduction of alkyl fluorides ⟨80S425⟩.

Palladium-on-carbon is employed most regularly as the catalyst in these reductions. Early reports ⟨46JA261⟩ suggested that only activated systems such as ethyl bromoacetate or benzyl chloride could be reduced while, for example, primary alkyl bromides were inert, even under high hydrogen pressure. However, many examples have been reported since which demonstrate a wider reactivity, concomitant with high chemoselectivity (Scheme 1) ⟨79T885, 84CAR(130)125⟩. Isolated examples of reductions of benzylic fluorides ⟨63JA1609⟩ and even a secondary alkyl fluoride (Scheme 1) ⟨71LA(748)123⟩ have been reported.

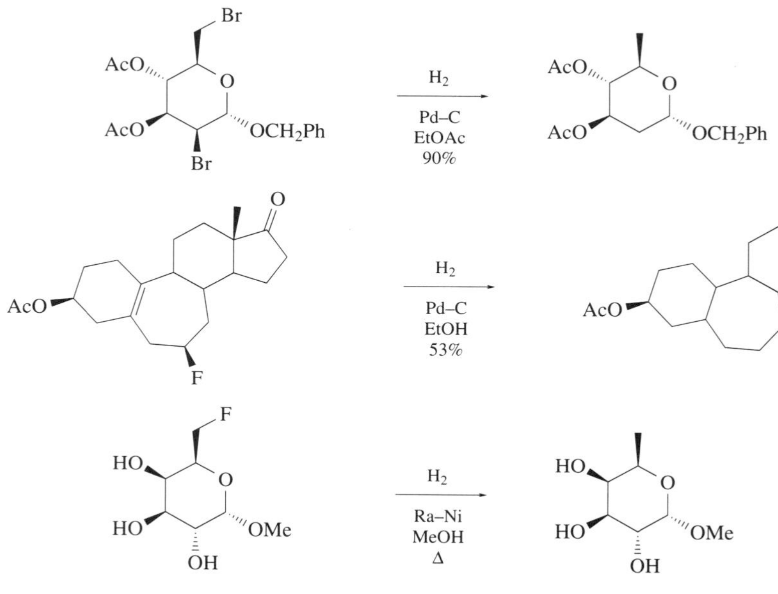

Scheme 1

α-Haloketones are readily dehalogenated with no competing carbonyl reduction ⟨58JOC1938, 78JA1786⟩, while the possibility of utilising transfer hydrogenation has also been highlighted ⟨85JOC3408⟩.

Raney-nickel has been shown to catalyse the reduction of a series of alkyl bromides and iodides in addition to tertiary and benzylic chlorides ⟨59CB1700⟩, while the range can be extended to primary alkyl fluorides under more forcing conditions (Scheme 1) ⟨60JCS298⟩. Platinum oxide has been reported to catalyse the reduction of a benzylic fluoride under relatively mild conditions ⟨65CJC1689⟩.

As indicated above, low-valent metal-based procedures appear to be the method of choice for the reduction of α-halocarbonyl compounds ⟨83OR163⟩. The metal employed most commonly in this context is zinc, although the use of iron pentacarbonyl ⟨79JOC641⟩ and of samarium(II) iodide ⟨86JOC1135⟩ has been exemplified.

The zinc reductions are generally performed in the presence of a proton source, typically acetic acid ⟨72JOC2363⟩ or ammonium chloride ⟨78BCJ2745, 78JA1765⟩, and give high yields for chloro- ⟨79JA4003, 85JOC3957⟩, bromo- ⟨78JA1786⟩ and iodocompounds (Scheme 2) ⟨65LA(681)196⟩. The chemoselectivity of these processes is high, notably alkenes which are reduced preferentially or competitively in catalytic hydrogenation procedures remain intact ⟨78JA1786⟩.

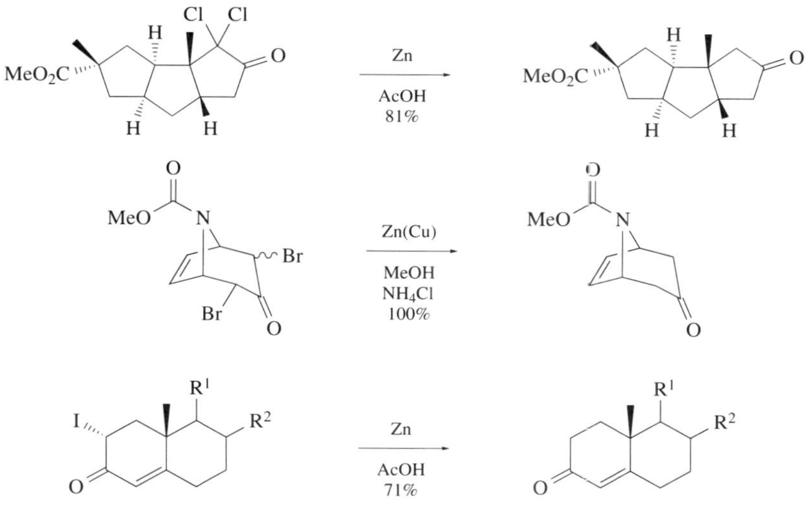

Scheme 2

Alkyl halides which are not adjacent to a carbonyl group can also be reduced by low-valent metals. Here, lithium or sodium are generally used in conjunction with an alcohol acting as proton source. This procedure is often used to reduce polycyclic alkyl halides (Scheme 3) ⟨86JA1265, 87JA7230, 90S649⟩.

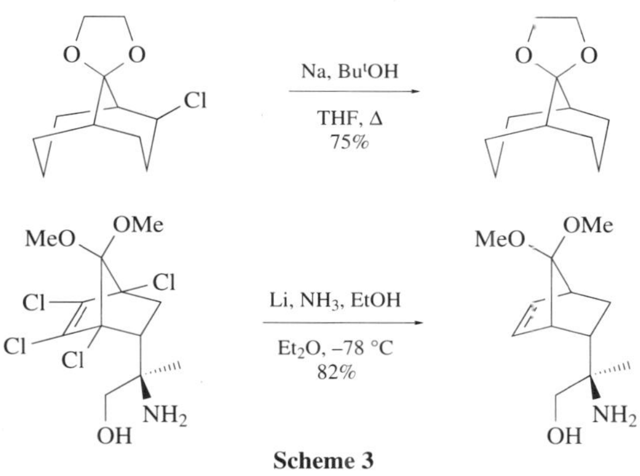

Scheme 3

Alkyl fluorides are not reduced by the above procedures. However, potassium has been reported to effect a high yielding reduction of a range of fluorinated steroids in the presence of an excess of crown ether ⟨81TL2583⟩.

Despite the obvious analogy of generating a Grignard reagent from an alkyl halide, followed by reacting with a proton source, there has been little use of magnesium in the context of a direct reduction ⟨63PCS219⟩. Samarium(II) iodide causes the reduction of primary iodides and bromides, but primary chlorides are unreactive, and both benzylic and allylic systems undergo Wurtz coupling ⟨80JA2693⟩.

The utility of a wide range of metal hydrides in the reduction of carbon–halogen bonds has been compared ⟨80JOC849⟩. Lithium aluminium hydride has long been known to reduce primary or secondary alkyl iodides and bromides and also benzylic chlorides ⟨49JA1675⟩. Benzylic fluorides have also been shown to be prone to reduction ⟨63JA1609⟩, particularly in the presence of an *ortho* or *para* electron-donating group ⟨71CC465⟩, while primary fluorides may be reduced after the addition of aluminium chloride ⟨64CJC572, 64JOC2870⟩. More recently, the simple expedient of employing a clear solution of lithium aluminium hydride in THF, rather than the more typical slurry, has been shown to markedly improve the reduction process such that secondary alkyl chlorides are reduced readily at room temperature ⟨82JOC276⟩. Alternatively, DIGLYME has been shown to be an excellent solvent for these reactions ⟨80JOC2550⟩. The reduction can also be promoted by the addition of nickel(II) chloride or cobalt(II) chloride ⟨78JOC1263⟩. The main drawback to the use of lithium aluminium hydride in this context is the poor chemoselectivity obtained, particularly through the competing reduction of carbonyl groups (Scheme 4) ⟨49JA1675, 64CJC572, 71CC465, 71LA(748)123, 87TL3883⟩.

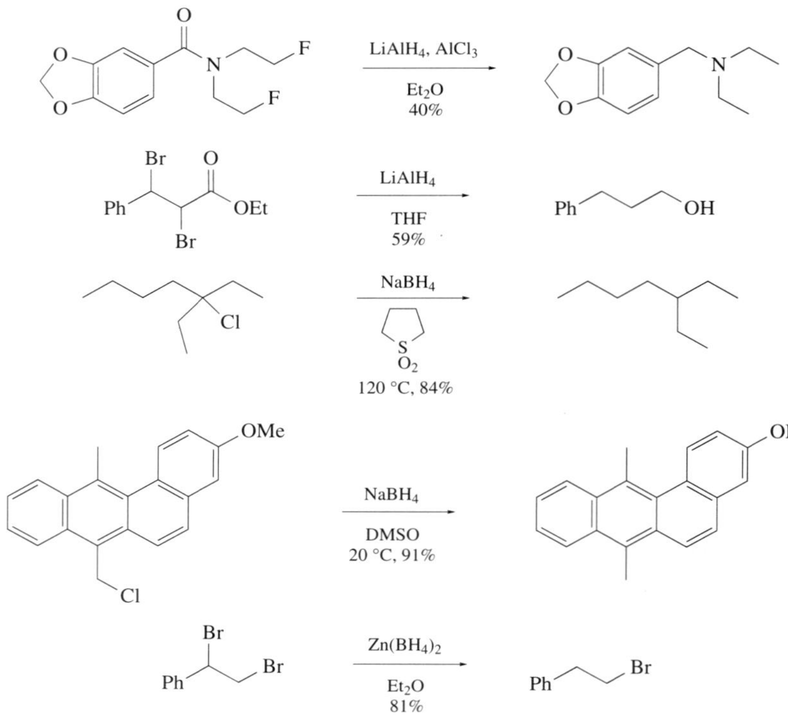

Scheme 4

Sodium borohydride has been shown to reduce alkyl chlorides, bromides and iodides but not fluorides. Initially, two-phase solvent systems were utilised ⟨66JA1473⟩, but polar aprotic solvents such as DMSO ⟨69JOC3923⟩, sulfolane ⟨71JOC1568⟩ and hexamethylphosphoramide (HMPA) ⟨78JOC2259⟩ are now used almost exclusively, although the use of phase transfer catalysis has been recommended ⟨81JOC3909⟩. There is evidence to suggest that these reactions proceed via an elimination–alkene hydroboration mechanism ⟨70CC338⟩. Tertiary alkyl chlorides, which often prove resistant to reduction by other methods, are converted cleanly, although elevated temperatures are required ⟨71JOC1568⟩. More activated systems are converted at room temperature (Scheme 4) ⟨86JOC3502⟩.

Sodium borohydride shows greater chemoselectivity than lithium aluminium hydride in that esters remain intact ⟨78JOC2259⟩. Under similar conditions, sodium cyanoborohydride has been shown to reduce alkyl bromides in the presence of ketones, aldehydes and benzylic epoxides ⟨77JOC82⟩.

Zinc borohydride offers unusual reactivity, reducing only tertiary and benzylic halides (Scheme 4) ⟨83AG(E)562⟩.

Lithium triethylborohydride reduces primary alkyl fluorides ⟨83ACS(B)141⟩ and both primary and secondary alkyl iodides, bromides and chlorides. Tertiary bromides undergo elimination ⟨83JOC3085⟩.

Radical reduction processes have been dominated by the use of organostannanes, principally tributyltin hydride ⟨87S665⟩, although organosilanes have attracted more recent attention ⟨92ACR188⟩. Primary, secondary and tertiary alkyl chlorides, bromides and iodides are all reduced; however, alkyl fluorides do not react.

The popularity of the tributyltin hydride-mediated reduction arises from the high chemoselectivity of the reaction. Thus, acetals ⟨76CB3598⟩, benzyl ethers, esters ⟨84CAR(130)125⟩, lactones ⟨92TL6673⟩, lactams ⟨79TL4631, 91JCS(P1)656⟩, carbamates ⟨84CAR(130)103⟩, ketones ⟨83AJC2243⟩, α-acyloxy-carbonyls ⟨83LA705⟩, nucleosides ⟨83S304⟩, epoxides ⟨82JOC5041⟩ and some alkenes ⟨77CB1823⟩ represent some of the functional groups which are not reduced under the same conditions (Scheme 5). Azides, converted to amines, are a notable exception here ⟨81LA1215⟩. Alkenyl halides, which contain a carbon–carbon multiple bond removed by five atoms from the halogen, undergo radical dehalogenation, but with concomitant five-membered ring formation ⟨87S665⟩.

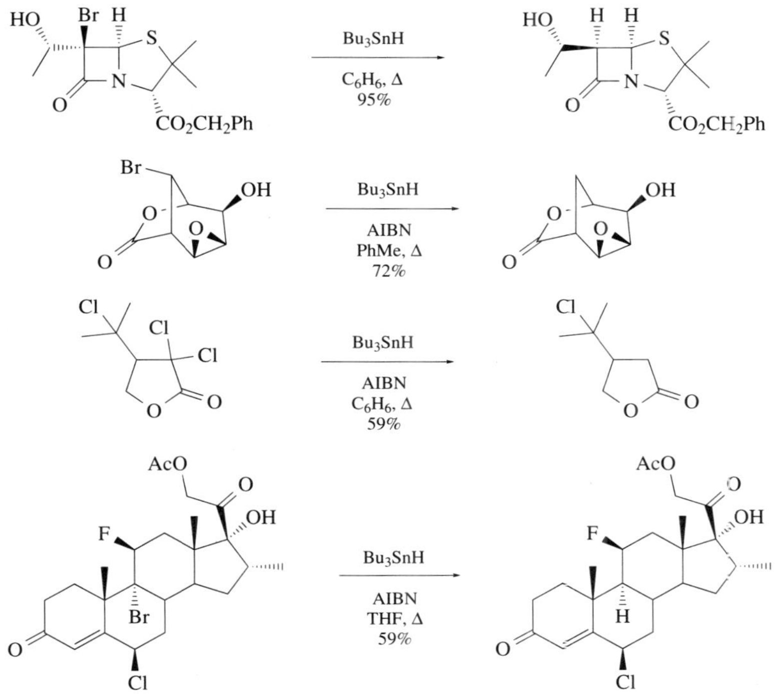

Scheme 5

The marked difference in rates of radical formation between different halogens and between the same halogen in different environments ⟨B-83MI 101-01⟩ can also be readily exploited in highly selective dehalogenations (Scheme 5) ⟨79JOC151, 84S949⟩.

The difficulties which can be encountered in removing the organotin residues from the product have led to the development of carbon–halogen bond reduction procedures which are catalytic in tin ⟨75JOC2554, 87JOC473⟩ but, in common with the use of a water-soluble tin hydride reagent ⟨90TL2957⟩, these have yet to be exploited widely.

Tris(trimethylsilyl) hydride reduces a similar range of alkyl halides as tributyltin hydride with similar efficiency ⟨88JOC3641, 91JOC678⟩. Attempts to use this relatively esoteric reagent in a catalytic process were partially successful ⟨89TL2733⟩, but simpler silanes may also prove equally useful in stoichiometric processes ⟨92JOC2427, 92JOC3405⟩. The readily available triethylsilane also exhibits utility in this context when used in conjunction with a thiol 'polarity reversal catalyst' ⟨91JCS(P1)103⟩.

α-Halocarbonyl compounds undergo a further general reaction unique to their class. Treatment with a nucleophile, typically iodide, in the presence of a Brönsted ⟨80TL3195⟩ or Lewis acid ⟨71TL137, 79S59, 80JOC3531, 83JOC3667, 86S570⟩ results in formation of a halogen–halogen bond with cleavage of the halogen–carbon bond and consequent reduction of the substrate (Scheme 6).

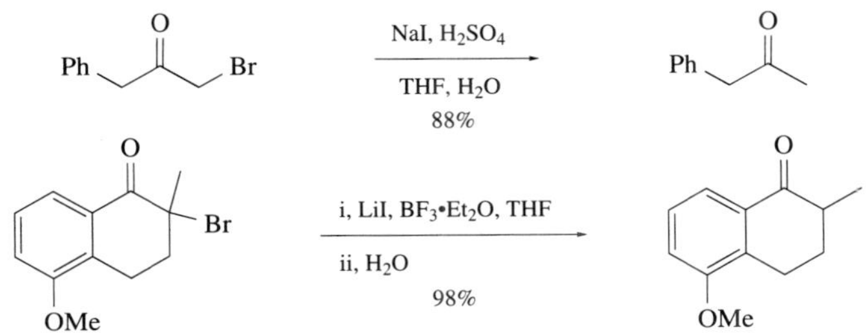

Scheme 6

1.01.1.2 Reduction of Fluoroalkanes

Fluoroalkanes are the most difficult class of alkyl halides to reduce by any given procedure. Indeed, no examples of successful radical substitution-based methods have been reported. A recurrent problem encountered is that other groups tend to be reduced preferentially ⟨B-76MI 101-01⟩.

Catalytic hydrogenation procedures tend to suffer from the catalyst poisoning effect of the fluoride ion that is liberated in the reaction ⟨80S425⟩. Nonetheless these reactions can be performed under sufficiently forcing conditions. Experiments in the mid 1950s using simple fluoroalkanes at elevated temperatures in the gas phase indicated that secondary fluorides are reduced more readily than primary under palladium-on-carbon catalysis ⟨56JPC1454⟩. The same catalyst has been shown to reduce more complex secondary fluorides (Scheme 1) ⟨71LA(748)123⟩ and benzylic fluorides ⟨63JA1609⟩ under less forcing conditions, while an example exists of the reduction of a primary fluoride at ambient temperature and elevated pressure (Equation (1)) ⟨92JA741⟩.

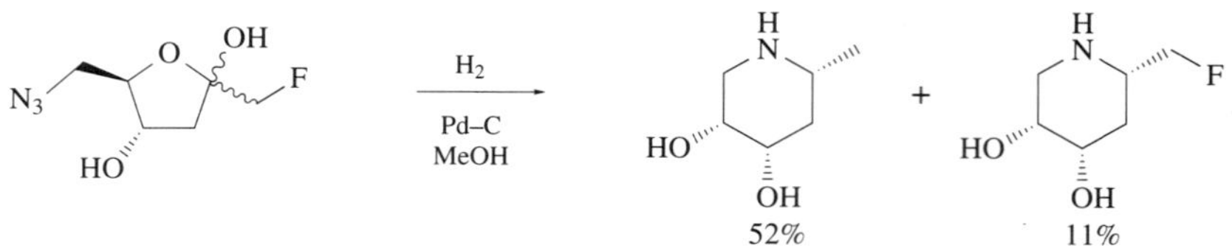

(1)

Raney-nickel has been employed to reduce a primary fluoride (Scheme 1) ⟨60JCS298⟩, while platinum oxide catalysed the reduction of a benzylic fluoride under mild conditions ⟨65CJC1689⟩.

Low-valent metal-mediated reductions have seen little use. Benzylic fluorides have been shown to be resistant to both zinc and sodium amalgam ⟨63JA1609⟩, while elimination reactions compete when magnesium is employed ⟨63PCS219⟩. The most successful of these methods is the use of potassium in the presence of 18-crown-6 which was shown to reduce a range of primary, secondary and tertiary fluorides in high yield ⟨81TL2583⟩.

Lithium aluminium hydride can sometimes be used to reduce alkyl fluorides, particularly when used in conjunction with aluminium chloride, but high yields from such reactions are rare ⟨64CJC572, 64JOC2870, 71LA(748)123⟩. Many benzylic systems are either inert or reduced in low yield ⟨49JA1767, 63JA1609⟩ but the presence of an amino- or hydroxy- functional group in the *ortho* or *para* position causes high yielding conversions to occur, possibly via an elimination–addition mechanism ⟨71CC465⟩.

Sodium borohydride ⟨81JOC3909⟩ and sodium cyanoborohydride ⟨77JOC82⟩ are apparently not powerful enough reducing agents to cleave C—F bonds, but lithium triethylborohydride does reduce primary fluorides, although the reduction of secondary or hindered primary systems is slow ⟨83ACS(B)141⟩.

Electrochemical techniques have been used to reduce α-fluorocarbonyl ⟨57JA1546, 68CC934⟩ and benzylic systems ⟨70CC738⟩.

1.01.1.3 Reduction of Chloroalkanes

Reports of the catalytic hydrogenation of chloroalkanes are relatively rare. Palladium-on-carbon catalyses the reduction of benzyl chlorides using either a hydrogen atmosphere ⟨46JA261⟩ or transfer hydrogenation ⟨85JOC3408⟩. Primary and secondary chloroalkanes do not undergo hydrogenation in the presence of Raney-nickel, but this catalyst does mediate the reduction of tertiary and benzylic systems, as well as geminal and vicinal dichloroalkanes ⟨59CB1700⟩.

Primary, secondary and tertiary chloroalkanes can be reduced by a variety of low-valent metals, typically in the presence of an alcohol proton source. Examples include lithium with ethanol ⟨90S649⟩ or *t*-butanol ⟨60CI(L)405, 87JOC4784⟩, sodium with *t*-butanol ⟨68JA7271, 86JA1265⟩, potassium ⟨81TL2583⟩, magnesium with isopropanol ⟨63PCS219⟩ and samarium(II) iodide ⟨80JA2693⟩. Wurtz coupling can occur with these procedures in the reduction of benzylic and allylic systems ⟨80JA2693⟩.

α-Chlorocarbonyl compounds are dechlorinated cleanly by zinc in acetic acid. These reaction conditions are employed most commonly in the reduction of cyclic ketone systems, for example, with four- ⟨72JOC2363, 83JA2435⟩ and five-membered rings (Scheme 2) ⟨79JA4003, 85JOC3957⟩, but are also compatible with lactone ⟨83JOC860⟩, lactam and carboxylic acid moieties (Scheme 7) ⟨87TL5339⟩. The combination of samarium(II) iodide and methanol has been shown to be of use in this context ⟨86JOC1135⟩.

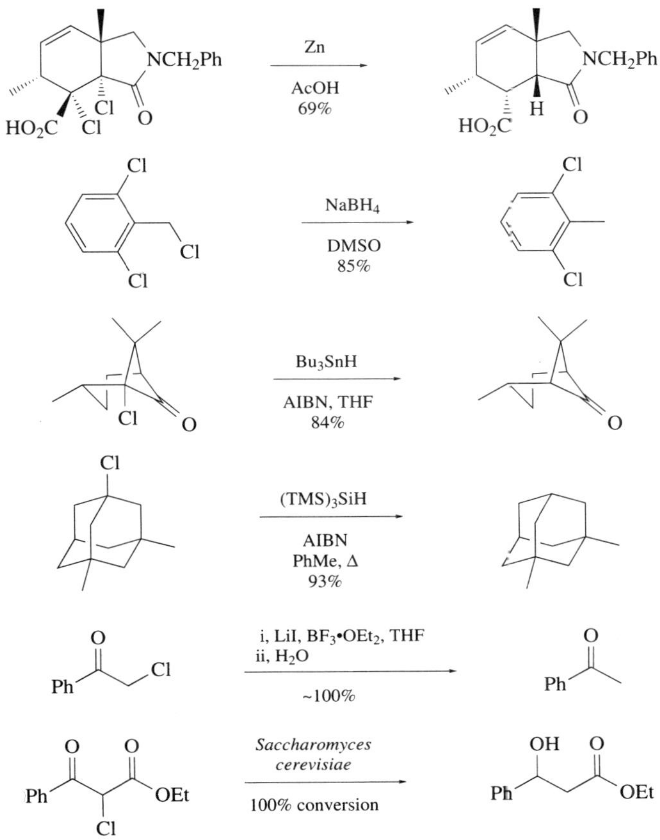

Scheme 7

Early reports ⟨49JA1675⟩, which indicated that lithium aluminium hydride reduced only activated chloroalkanes, have been countered more recently ⟨82JOC276⟩ by demonstrations that relatively unreactive systems, such as secondary chloroalkanes, can be reduced cleanly at ambient temperature by homogenous solutions of the reagent in THF. The use of cobalt(II) and nickel(II) chloride has also been shown to promote these reductions ⟨78JOC1263⟩, and even a tertiary neopentyl chloride has been reduced ⟨87TL3883⟩.

Sodium borohydride also reduces chloroalkanes ⟨66JA1473⟩. Generally, these reactions are performed in polar aprotic solvents such as DMSO ⟨69JOC3923, 86JOC3502⟩, sulfolane ⟨71JOC1568⟩ or HMPA ⟨78JOC2259⟩, and often at elevated temperatures, but still tend to offer greater chemoselectivity than lithium aluminium hydride (Scheme 7) ⟨81JOC3909, 78JOC2259⟩. Other borohydride systems of use in this context include zinc borohydride (tertiary systems) ⟨83AG(E)562⟩, sodium cyanoborohydride ⟨77JOC82⟩ and lithium triethylborohydride ⟨82JOC2590, 83JOC3085⟩.

Primary, secondary ⟨76CB3598, 81LA1215, 84S949⟩ and tertiary ⟨80TL287⟩ chloroalkanes are all reduced in high yield by tributyltin hydride by a radical mechanism. The power of this approach is illustrated by an example where low-valent metal procedures failed and ionic metal hydrides would offer no chemoselectivity (Scheme 7) ⟨87JOC307⟩.

A similar range of substrates undergo radical reduction using silyl hydrides. Tris(trimethylsilyl)silane has seen most use, with the reaction being initiated by the use of light, peroxides ⟨88JOC3641⟩ or azobisisobutyronitrile (AIBN) (Scheme 7) ⟨91JOC678⟩. The use of bis(trimethylsilyl)methylsilane ⟨92JOC3405⟩, triethylsilane in the presence of a thiol catalyst ⟨91JCS(P1)103⟩ and, although with lower efficiency, tris(alkylthio)silanes ⟨92JOC2427⟩ have also been demonstrated in this reduction.

α-Chlorocarbonyl compounds can be reduced by iodide ion in the presence of a Lewis acid. The reaction generally proceeds in high yields with both ketones and esters ⟨86S570⟩, utilising a range of Lewis acids ⟨79S59, 80JOC3531, 86S570⟩, unless the system is particularly hindered (Scheme 7) ⟨71TL137⟩.

A biological reductive dechlorination has also been reported ⟨92TL7337⟩. Treatment of ethyl α-chlorobenzoylacetate with bakers' yeast at low concentrations (1 g l^{-1}) results in reduction of both the carbonyl group and the C—Cl bond (Scheme 7). At higher concentrations, the carbonyl group is reduced but the chloro moiety remains intact, suggesting that the dechlorination is an enzyme-catalysed process and not the result of a fortuitous noncatalysed chlorination of a constituent of the microorganism.

1.01.1.4 Reduction of Bromoalkanes

Although it was suggested initially that only activated (e.g., benzylic and α-carbonyl) bromoalkane systems were reduced by hydrogenation over palladium-on-carbon ⟨46JA261⟩, it has now been demonstrated that a full range of these compounds can be reduced by this procedure (Scheme 1) ⟨58JOC1938, 79TL885, 84CAR(130)125⟩. This approach can fail if other functional groups, that are prone to hydrogenation, are present ⟨78JA1786, 84CAR(130)125⟩. Allylic bromides may be reduced by such procedures, but concomitant alkene migration can occur ⟨55JA5960⟩.

Catalytic hydrogenation can also be performed over Raney-nickel ⟨59CB1700⟩ or using a homogenous palladium(0) catalyst in conjunction with transfer hydrogenation techniques ⟨86JOC734⟩.

There are remarkably few examples of the reduction of simple bromoalkanes using a low-valent metal, and the procedure appears low yielding ⟨87JA7230⟩. Samarium(II) iodide does reduce primary alkyl bromides in high yield, but allylic and benzylic substrates undergo Wurtz coupling ⟨80JA2693⟩.

In common with the chloroalkane analogues described above, zinc cleanly reduces α-bromocarbonyl compounds (Scheme 2) ⟨78BCJ2745, 78JA1786⟩ and the diastereoselectivity of the protonation step in this reaction has been investigated ⟨59JA3634, 59JA3644⟩. These substrates can also be reduced with samarium(II) iodide in high yield (Scheme 8) ⟨86JOC1135⟩ and, in moderate yield, with iron pentacarbonyl ⟨79JOC641⟩.

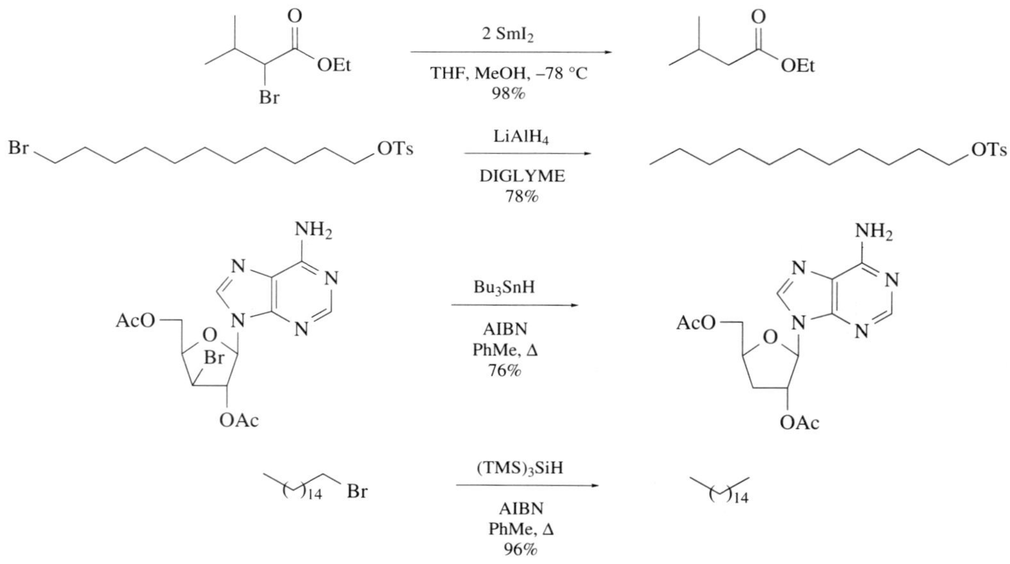

Scheme 8

Lithium aluminium hydride reduces most bromoalkanes ⟨49JA1675⟩. Benzylic and primary systems are the most reactive, but secondary and tertiary systems can also be reduced, particularly in the presence of nickel(II) or cobalt(II) chloride ⟨78JOC1263⟩ or when a homogenous solution of lithium aluminium hydride in THF is used ⟨82JOC276⟩. The carbon–bromine bond can be reduced by lithium aluminium hydride in the presence of other reactive centres when DIGLYME is used as the solvent (Scheme 8) ⟨80JOC2550⟩.

Sodium borohydride reduces primary and secondary bromoalkanes in high yield and with good chemoselectivity. These reactions are usually carried out in polar aprotic solvents, typically DMSO

⟨69JOC3923⟩, sulfolane or HMPA ⟨78JOC2259⟩, although two-phase systems involving an aqueous layer have been used successfully in conjunction with phase transfer catalysis ⟨81JOC3909⟩. Zinc borohydride shows complementary activity, reducing only tertiary and benzylic bromoalkanes, even in the presence of primary alkyl bromides (Scheme 4) ⟨83AG(E)562⟩.

Sodium cyanoborohydride in HMPA offers very high chemoselectivity. Bromine–carbon bonds can be reduced by this system in the presence of a variety of functional groups, including carboxylic acids and esters, ketones, nitriles and epoxides ⟨77JOC82⟩.

Other metal hydride systems have also been used to reduce bromoalkanes, for example, lithium triethylborohydride ⟨83JOC3085⟩ and potassium hydride ⟨87JOC4299⟩, but these reagents have yet to see widespread use.

The most popular method for the reduction of bromoalkanes is the use of tributyltin hydride in a radical substitution procedure. Essentially, all classes of bromoalkane can be reduced, usually in high yield, with excellent chemoselectivity, thus sulfides, β-lactams and benzyl ethers ⟨79TL4631⟩, γ-lactams ⟨91JCS(P1)656⟩, epoxides and lactones (Scheme 5) ⟨82JOC5041⟩, alkyl fluorides, alkyl chlorides, ketones and enones (Scheme 5) ⟨79JOC151⟩, ethers ⟨72S483⟩, α-acyloxyketones ⟨83LA705⟩, alkenes ⟨77CB1823⟩, strained ⟨83TL1147⟩ and rigid cycloalkanes ⟨83AJC2243⟩, carbohydrates ⟨84CAR(130)125, 92JA10138⟩ and nucleosides ⟨83S304⟩ are all unaffected by typical reaction conditions (Scheme 8).

The main difficulty encountered with these procedures is the removal of tin residues from the product ⟨87S665⟩. Attempts to circumvent this by developing methodology which is catalytic in tin (and stoichiometric in sodium borohydride) have achieved some success ⟨75JOC2554, 87JOC473⟩ but see little contemporary use.

Silyl hydrides have again been shown to be of use in place of tributyltin hydride. Tris(trimethylsilyl)silane has seen the most investigation (Scheme 8) ⟨88JOC3641, 91JOC678⟩, including its use in catalytic amounts ⟨89TL2733⟩, but other silanes have also been shown to work well ⟨91JCS(P1)103, 92JOC2427, 92JOC3405⟩.

The carbon–bromine bond of α-bromoketones can be reduced by treatment with iodide ion in the presence of an acid. Although sulphuric acid has been used in this context (Scheme 6) ⟨80TL3195⟩, the reaction is usually driven by a Lewis acid (Scheme 6) ⟨71TL137, 79S59⟩. When trimethylsilyl chloride is employed, the silyl enol ether intermediate is usually not isolated, and is hydrolysed *in situ* (Scheme 9) ⟨80JOC3531, 83JOC3667⟩.

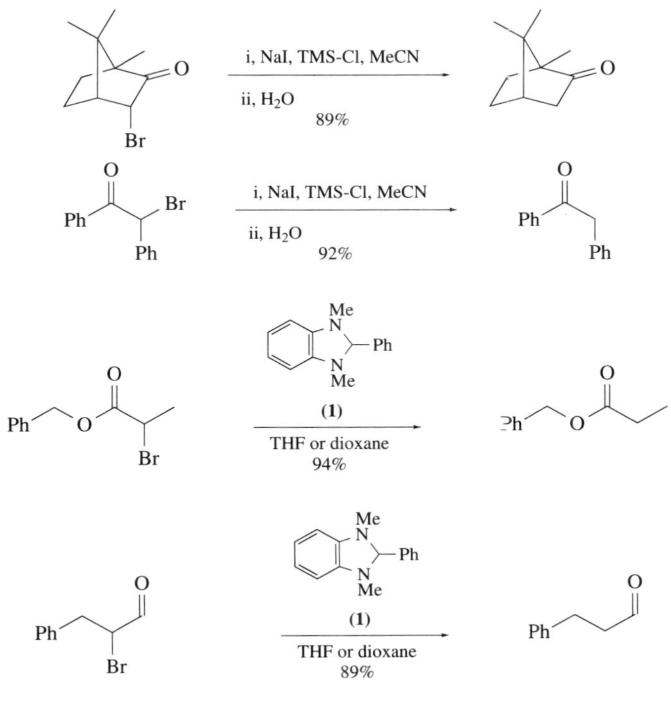

Scheme 9

The dihydrobenzimidazole (**1**) has been shown to effect an unusual, but high yielding, reduction of α-bromocarbonyl compounds ⟨86JOC5400⟩. The process, which presumably involves hydride transfer assisted by the nitrogen lone pairs, is highly chemoselective (Scheme 9).

1.01.1.5 Reduction of Iodoalkanes

Iodoalkanes undergo facile reduction by all the general methods available. Perhaps as a consequence of this, with there being little need to develop new methods or refine old ones, there are relatively few investigations of these reductions reported in the literature.

As would therefore be expected from the precedents described above, iodoalkanes can be reduced by catalytic hydrogenation over palladium ⟨80S425⟩ or Raney-nickel ⟨59CB1700⟩.

Samarium(II) iodide is a particularly efficient low-valent metal reducing agent for simple alkyl iodides ⟨80JA2693⟩, while the 'classic' combination of zinc in acetic acid cleanly reduces α-iodoketones (Scheme 2) ⟨65LA(681)196⟩.

Again, as might be predicted, lithium aluminium hydride reduces iodoalkanes readily ⟨49JA1675, 82JOC276⟩. The rate of the reaction can be promoted by the use of solvents such as GLYME or DIGLYME ⟨80JOC2550⟩ or the addition of nickel(II) or cobalt(II) salts ⟨78JOC1263⟩, but yields are generally high regardless of the method used.

Similarly, other metal hydride reagents (such as sodium borohydride ⟨69JOC3923, 81JOC3909⟩, sodium cyanoborohydride ⟨77JOC82⟩ and lithium triethylborohydride ⟨83JOC3085⟩) capable of reducing bromoalkanes also work well with iodoalkanes, while offering the possibility of greater chemoselectivity than that available with lithium aluminium hydride.

Chemoselectivity is also the strength of the radical substitution of iodoalkanes using tributyltin hydride, where high yielding reductions can be performed in the presence of a wide range of other sensitive functional groups (Scheme 10) ⟨82JFC(20)313, 84CAR(130)103, 91JOC6206, 92TL6673, 92TL7429⟩. Methodology for performing these reductions in the presence of catalytic amounts of the reducing agent, together with an inexpensive coreductant, is recorded ⟨75JOC2554⟩.

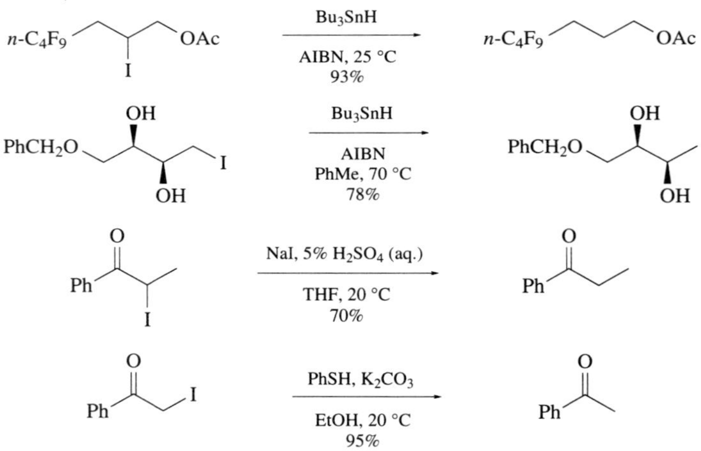

Scheme 10

As with chloro- and bromoalkanes, such radical-based reductions can also be performed with a range of silanes in place of tributyltin hydride with no loss in yield or efficiency ⟨88JOC3641, 89TL2733, 91JCS(P1)103, 91JOC678, 92JOC2427, 92JOC3405⟩.

α-Iodoketones can be reduced by a range of procedures distinct from those described above (Scheme 10). Thus treatment of aryl α-iodoketones with iodide in the presence of a Lewis ⟨71TL137⟩ or Brönsted ⟨80TL3195⟩ acid has been shown to provide a clean reduction. A closely related process is the reaction with benzene thiolate or benzene selenolate ion, which gives high yields for the reduction of both aryl and alkyl ketones ⟨81JOC2596⟩. Curiously, this reaction fails for the corresponding α-bromoketones, when thioester formation occurs. An isolated example of the reduction of an α-iodoketone in the presence of an α-bromoketone moiety using sodium bisulfite has also been reported ⟨50JA362⟩.

1.01.1.6 Reduction of Hypervalent Haloalkanes

There have been few reports of the reduction of a hypervalent haloalkane to the corresponding alkane. The esoteric nature of this reaction is highlighted by the fact that the only alkanes that have been formed by these processes are all *β*-dicarbonyl compounds.

When an ethanolic solution of phenyldimedonyliodone is refluxed, decomposition occurs to give dimedone as a minor component in a complex mixture of products (Equation (2)) ⟨70BCJ2506⟩. Cleaner reductions of the same substrate have been performed with either aryl thiols ⟨78JOC2676⟩ or sodium bisulfite ⟨77JOU1033⟩ as the reducing agent (Equation (2)), and the latter method has been shown to be applicable to a wider range of substrates.

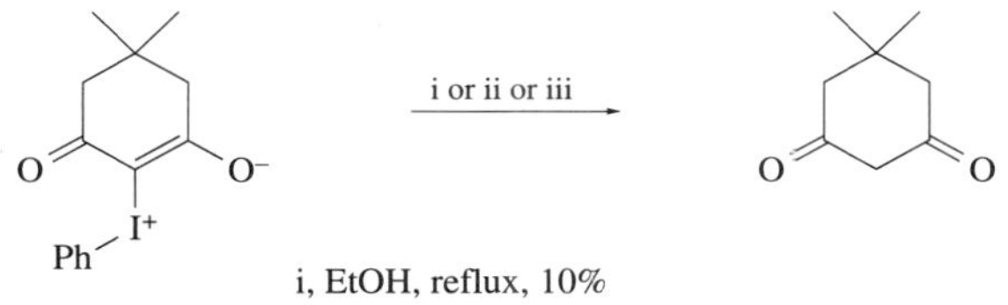

(2)

i, EtOH, reflux, 10%
ii, 4-ClC_6H_4SH, CH_2Cl_2, 86%
iii, $NaHSO_3$, H_2O, 90%

1.01.2 REDUCTION OF C—OXYGEN BONDS TO CH

1.01.2.1 General Methods

It is difficult to identify general methods for the reduction of carbon–oxygen single bond systems, particularly due to the dramatic variation in reactivity of the different structures which come into this class.

While benzylic systems can be reduced by catalytic hydrogenation (Scheme 11) ⟨53OR(7)263, 82SC983, 93T8433⟩, most alcohols (and simple derivatives thereof) are largely inert to direct reduction. These species are transformed to alkanes by conversion to a more reactive substrate, usually one of the range of thionoethers or a sulfonate, then reduction of this intermediate by a radical substitution or nucleophilic displacement procedure, respectively (Scheme 12) ⟨75JCS(P1)1574, 77JOC82⟩.

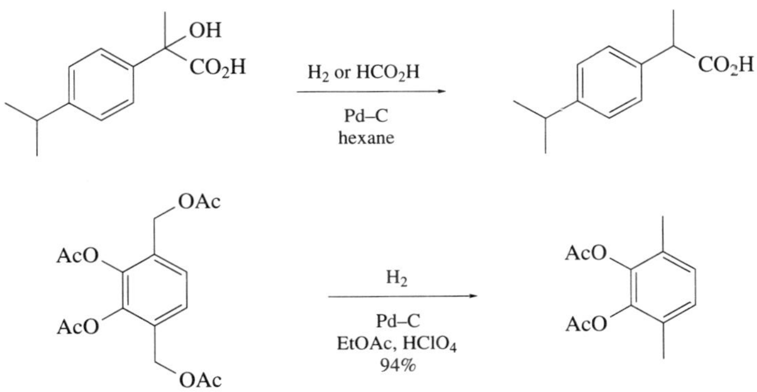

Scheme 11

Most aldehydes and ketones can be reduced to methyl and methylene compounds, respectively, by treatment with hydrazine and base, the Wolff–Kischner reduction ⟨48OR(4)378⟩. The Huang–Minlon modification of this procedure is usually employed ⟨46JA2487⟩, and high yields are often obtained (Scheme 13) ⟨49JA3301, 87JOC3205⟩.

Benzylic aldehydes and ketones can also be reduced, via the benzylic alcohol intermediate, to the corresponding alkane by catalytic hydrogenation. Again, high yields can be obtained, and the procedure is tolerant of many other functional groups (Scheme 14) ⟨80TL2637, 82JOC4304⟩.

Aldehydes and ketones can also be deoxygenated by conversion to a dithioacetal derivative, then treatment with Raney-nickel. This approach is discussed in Section 1.01.3.5.

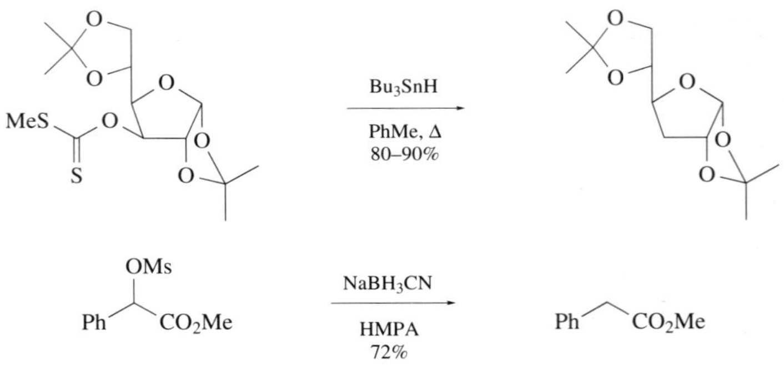

Scheme 12

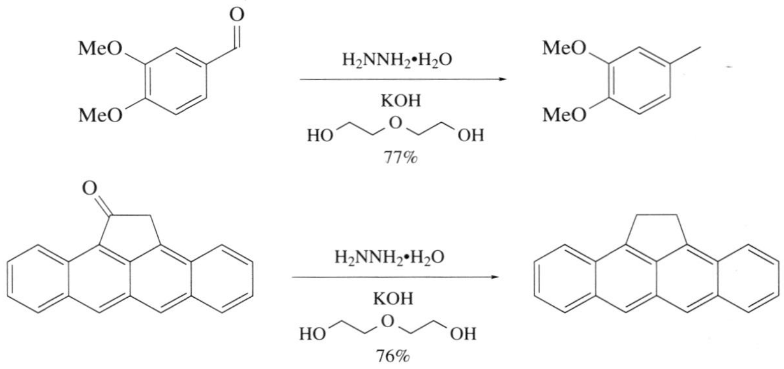

Scheme 13

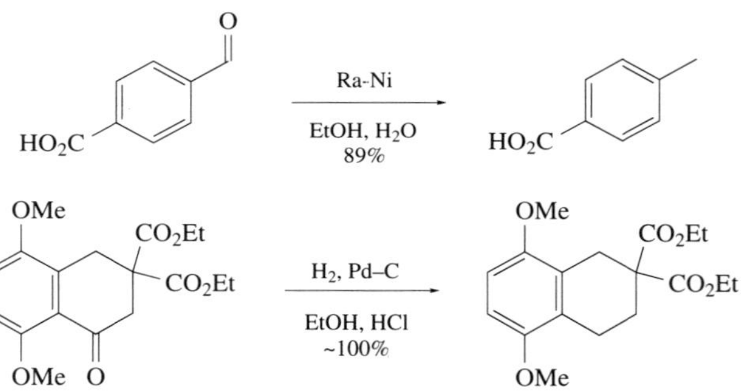

Scheme 14

1.01.2.2 Reduction of C—OX Bonds

1.01.2.2.1 Reduction of C—OH bonds

As indicated above, the C—O bond of alcohols is not particularly prone to reduction, and only systems activated through a neighbouring group can be transformed readily.

Catalytic hydrogenation over palladium-on-carbon is an effective method for the reduction of benzylic alcohols ⟨46JA261, 53OR(7)263⟩, and it has been demonstrated that transfer hydrogenation is as least as effective as the use of a hydrogen atmosphere (Schemes 11 and 15) ⟨93T8433⟩. Tertiary alcohols can be hydrogenated under platinum(IV) oxide catalysis in trifluoroacetic acid (when the intermediacy of an alkene is probable) ⟨64JOC2325⟩ or in the presence of Raney-nickel ⟨88JOC432⟩. Selective reduction of the alcohol in the presence of a primary bromoalkane functional group has been achieved by the latter method. Allylic alcohols can be converted to alkenes by hydrogenation in the presence of the hydridopentacyanocobaltate anion, although double bond migration can occur and tertiary allylic alcohols are reported not to react ⟨90TL4101⟩.

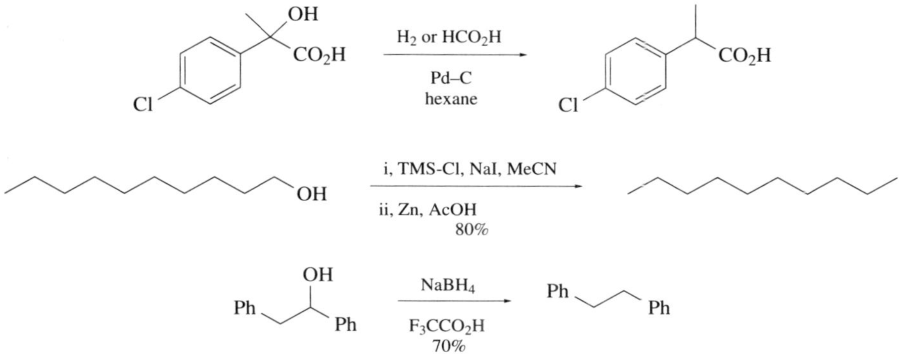

Scheme 15

By analogy with α-halocarbonyl compounds, it might be expected that the alcohol moiety of the corresponding α-hydroxycarbonyl series should be reduced by the action of low-valent metals. While this is the case, it would seem that here the combination of zinc in acetic acid ⟨54JCS3045⟩ is less popular than samarium(II) iodide ⟨86JOC1135, 87JA4424⟩.

The use of lithium in ammoniacal THF containing ammonium chloride has been shown to reduce benzylic alcohols without concomitant Birch reduction ⟨75JOC3151⟩, while zinc has been employed in the latter stage of a two-stage, one-pot process to reduce primary, secondary and benzylic alcohols through the intermediacy of the corresponding iodoalkane (Scheme 15) ⟨81S32⟩.

The combination of an acid and a hydride source, often termed ionic hydrogenation ⟨74S633⟩, reduces alcohols where the intermediate carbonium ion is relatively stable (i.e., tertiary and benzylic alcohols). Initially, triethylsilane was used as the hydride donor in the presence of trifluoroacetic acid ⟨69JOC4, 71JOC758⟩, but it has since been demonstrated that the use of sodium borohydride with the same acid is highly effective (Scheme 15) ⟨77S172⟩. The use of zinc iodide, rather than a protic acid, together with sodium cyanoborohydride, also reduces tertiary and benzylic alcohols ⟨86JOC3038⟩, while the combination of diisobutylaluminium hydride (dibal-H) with aluminium bromide reduces benzylic alcohols in high yield, and has been shown to provide a rare method for a direct, if low-yielding, reduction of a secondary alcohol ⟨92JOC2143⟩.

1.01.2.2.2 Reduction of C—O—C bonds

(i) Reduction of oxiranes

There are, not unexpectedly, few examples of the reduction of both carbon–oxygen bonds of an oxirane (epoxide). The use of titanocene has met with some success, although yields are variable, being poor for highly substituted epoxides ⟨74JA5290⟩. Triethylsilane, in combination with boron trifluoride, also performs this reduction (Scheme 16) ⟨79TL849⟩, but rearrangement of the intermediate carbonium ion can also occur.

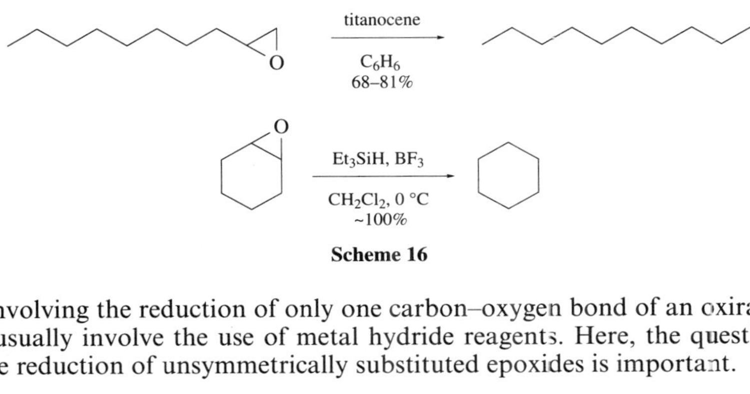

Scheme 16

Procedures involving the reduction of only one carbon–oxygen bond of an oxirane are relatively common, and usually involve the use of metal hydride reagents. Here, the question of the regioselectivity of the reduction of unsymmetrically substituted epoxides is important.

In aliphatic substituted systems where one oxirane carbon atom has fewer substituents (i.e., 2-alkyl, 2,2-dialkyl or 2,2,3-trialkyloxiranes), reduction with highly nucleophilic metal hydride reagents (such as lithium aluminium hydride ⟨92JOC1618, 92T9427⟩, lithium triethylborohydride ⟨80JOC1⟩ and dibal-H ⟨92JOC1618⟩) occurs at the less substituted position (Scheme 17). In contrast, reducing agents with Lewis acid character ⟨92JOC1618⟩, or used in the presence of an acid ⟨81JOC5214, 92JCS(P1)1881⟩ tend to deliver hydride to the more substituted carbon centre (Scheme 17).

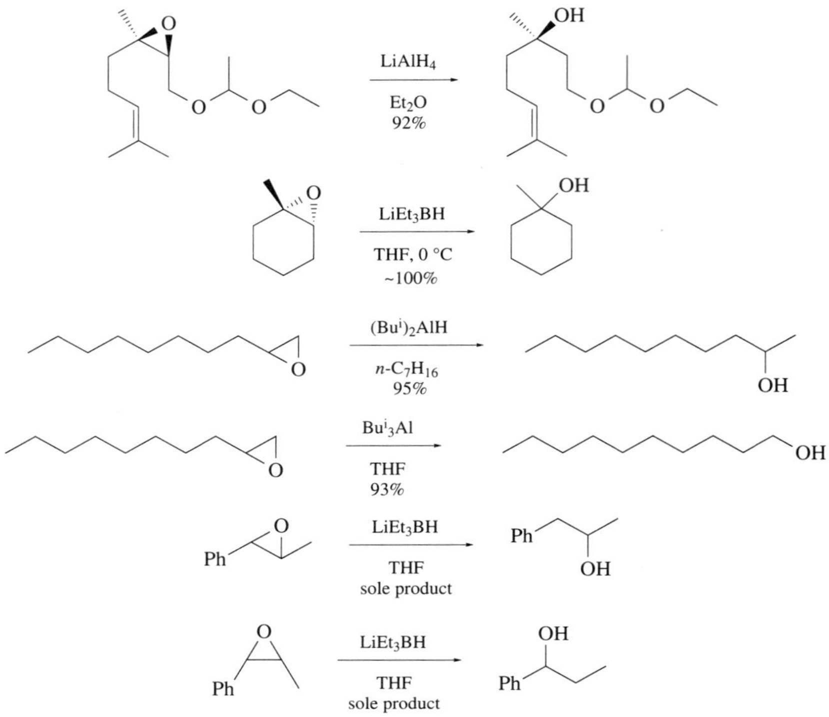

Scheme 17

When the more substituted carbon of the oxirane is also benzylic, selective delivery of hydride to that centre can be achieved by a range of reagents including either borane ⟨68CC1549⟩ or sodium cyanoborohydride ⟨81JOC5214⟩ in combination with boron trifluoride, and also triisobutylaluminium ⟨92JOC1618⟩. Reduction of these compounds at the less substituted carbon atom can still be performed using reagents such as lithium aluminium hydride, but the regioselectivity of this process is not always as clear cut as with the nonbenzylic systems ⟨92JOC1618, 68JA2927⟩.

Reductions of 2,3-disubstituted oxiranes tend to give mixtures of products, even when one substituent is aromatic, and the geometry of the oxirane can have a dramatic effect on the outcome of the reaction (Scheme 17) ⟨83JOC3091⟩.

Reduction takes place at the allylic carbon–oxygen bond when oxiranes with vinyl substituents are treated with reagents such as lithium aluminium hydride ⟨85JA7978, 87JOC4898⟩ or dibal-H ⟨82CC1292⟩.

The popularity of the Sharpless asymmetric epoxidation procedure has led to the study of the reduction of glycidols (hydroxymethyloxiranes) as an approach to 1,2- and 1,3-dihydroxylated systems. When the glycidol alcohol moiety is primary, selective reductions to give 1,3-diols can be performed using sodium bis(methoxyethoxy)aluminium dihydride (Red-Al) (Scheme 18) ⟨82CC1292, 82JOC1378⟩. Reductions to the 1,2-dihydroxy system are more difficult, unless the site of reduction is benzylic ⟨82TL3597⟩. The most effective systems are dibal-H ⟨82TL2719, 82TL3597⟩ and lithium borohydride/titanium tetraisopropoxide (Scheme 18) ⟨86TL4343⟩. Glycidols with a secondary alcohol and terminal oxirane can be reduced to the vicinal diol using lithium aluminium hydride ⟨85JOC5678⟩.

Oxiranes with a carbon–oxygen bond α to an aromatic ring or carbonyl group can be reduced selectively using low-valent metal techniques. Samarium(II) iodide ⟨86JOC2596, 87TL4437⟩, lithium and potassium ⟨86AG(E)653⟩ have all been used with success.

Catalytic hydrogenation methodology can be employed to reduce certain oxiranes, for example, those with vinyl ⟨89JA6280⟩ or carbonyl ⟨91CC535⟩ substituents.

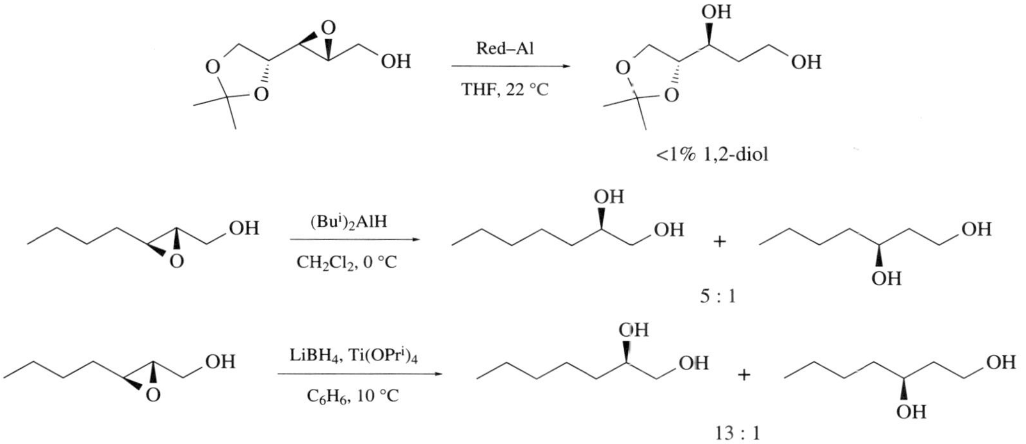

Scheme 18

(ii) Reduction of other ethers

Ethers share a resistance to reductive cleavage similar to the corresponding alcohols.

Although benzyl ethers are frequently used as alcohol protecting groups in synthesis and are subsequently cleaved by catalytic hydrogenation, the alkane product, toluene, is of no synthetic interest. Hydrogenation of cyclic benzylic ethers, where both alcohol and alkane are part of the required product, have been reported and work well ⟨48JA1490⟩.

Allylic ethers can be reduced to the corresponding alkenes by the combination of a homogenous palladium(0) catalyst and lithium triethylborohydride in high yields with little or no double bond migration ⟨82JOC4380⟩ while γ-alkoxy-α,β-unsaturated esters can be reduced to the corresponding fully saturated ester by the action of magnesium in methanol (Equation (3)) ⟨93JCS(P1)9⟩. Alkyl methyl ethers can also be reduced by low valent metal methods after *in situ* conversion to the corresponding alkyl iodide ⟨81S32⟩, while α-alkoxyketones are reduced cleanly by samarium(II) iodide ⟨86JOC1135⟩.

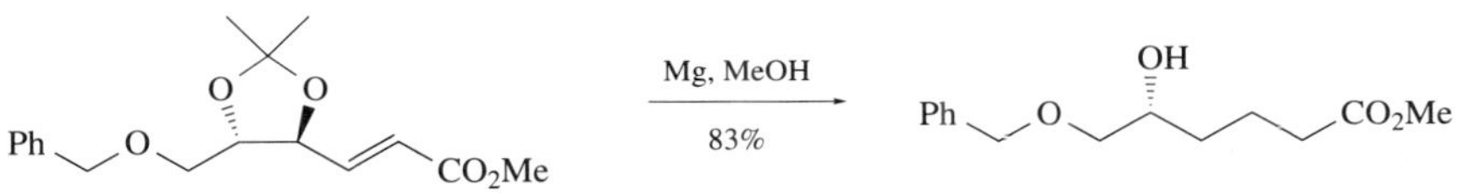

(3)

(iii) Reduction of esters

The reduction of the alkyl carbon–oxygen bond of esters can be performed in preference to the reduction at the carbonyl carbon atom in some specific situations.

In common with benzyl ethers, benzyl esters are often used as protecting groups in organic synthesis, and are subsequently removed by catalytic hydrogenation. There are, however, examples of this reaction with more complex aromatic systems, where the alkane formed is the required product of the reaction (Schemes 11 and 19) ⟨83JCS(P1)87, 82SC983⟩.

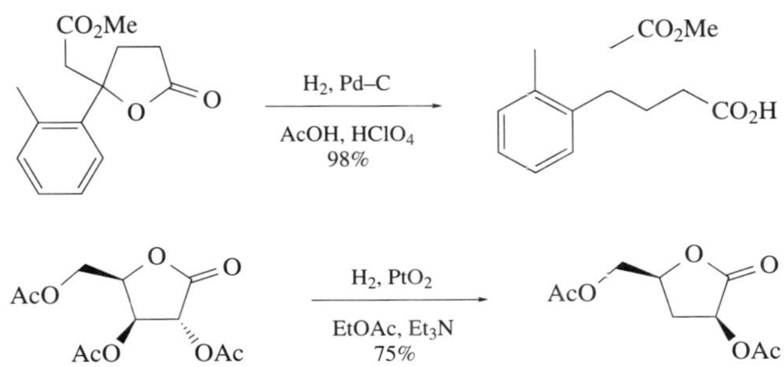

Scheme 19

The β-acetoxy group of α,β-diacetoxybutanolides can be reduced by catalytic hydrogenation over Raney-nickel in the presence of base (Scheme 19) ⟨93OS(72)48⟩, via an alkene. If platinum(IV) oxide or palladium-on-carbon is used as catalyst then reduction of the α-acetoxy group also occurs ⟨91TL4963⟩.

α-Acetoxyketones can be reduced using zinc in acetic acid ⟨55JA4367, 57JA5540⟩ or samarium(II) iodide ⟨86JOC1135⟩. The latter procedure is particularly selective, notably the reduction can be performed in the presence of a primary iodide. A vinylogous example of this type of reduction has been performed with magnesium in methanol as the reducing system, although alkene reduction also occurred under these conditions ⟨93JCS(P1)9⟩.

Sterically hindered esters are reduced by the action of lithium in ethylamine. Less hindered systems undergo oxygen–carbonyl bond cleavage under these conditions, but this selectivity can be exploited beneficially ⟨81JCS(P1)1501⟩. The similar conditions of lithium in ammoniacal THF have been employed to reduce benzylic esters selectively in the presence of alkenes ⟨87JOC4878⟩.

(iv) Reduction of thionoethers

One of the most important methods for the deoxygenation of alcohols is the radical chain reduction of a thionoether derivative, typically using tributyltin hydride as the reducing agent. This process, often termed the Barton–McCombie reaction ⟨75JCS(P1)1574⟩, has been the subject of extensive reviews ⟨83T2609, 87S665, B-89MI 101-01⟩.

A range of thionoethers has been used in this procedure. While thioesters were utilised in the original paper in this area ⟨75JCS(P1)1574⟩, these intermediates now see little use and S-methyl dithiocarbonate (xanthate) ⟨75JCS(P1)1574, 93TL2733⟩, phenylthiocarbonate ⟨87JOC3706, 90SL705⟩ and imidazoylthiocarbonyl ⟨81JOC4843, 93TA625⟩ derivatives are employed now.

These reductions are used most commonly in the deoxygenation of secondary, often cyclic, alcohols and are high yielding, notably in the presence of acid or base sensitive functionality (Schemes 12 and 20) ⟨75JCS(P1)1574, 87JOC3706, 93TA625⟩. The procedure is used only rarely for primary alcohols, and can be low yielding ⟨81JOC4843, 91JCS(P1)103⟩, although the use of high boiling solvents can be beneficial ⟨81S743⟩. Tertiary alcohols can be deoxygenated cleanly if elimination of the thionoether can be avoided, and a revised methodology for the synthesis of these intermediates has been reported ⟨93TL2733⟩.

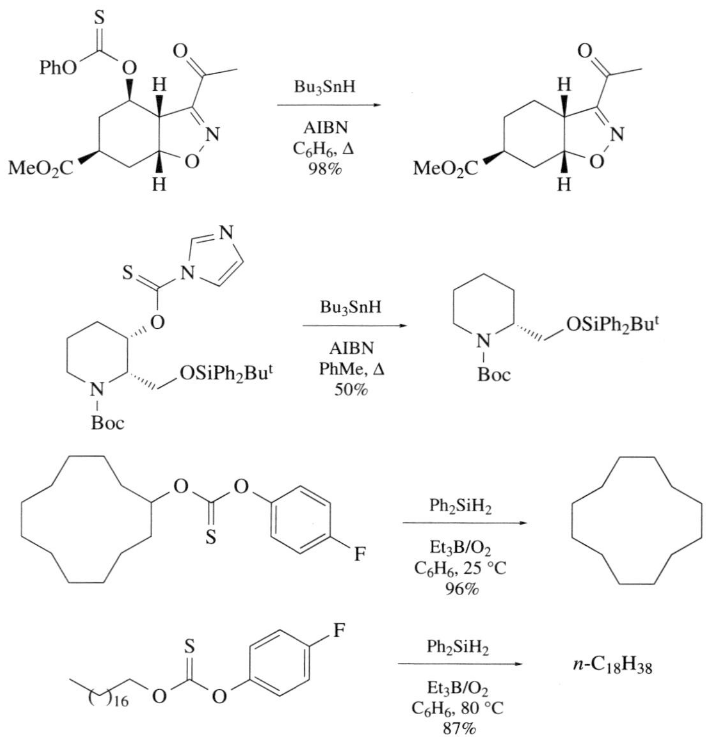

Scheme 20

The practical difficulty of removing unwanted tin-containing residues from the reaction product can be avoided by the use of polystyrene-supported tin hydrides ⟨92SL801⟩.

More recently, the use of organosilanes to reduce thionoethers to alkanes has seen a rise in popularity matching the analogous usage in the reduction of haloalkanes ⟨92ACR188⟩. The most useful of these reagents seem to be tris(trimethylsilyl)silane ⟨90SL705, 91JOC678⟩ and diphenylsilane ⟨90TL4681⟩. The combination of the latter reagent with the triethylborane/oxygen initiator system and *p*-fluorophenyl thiocarbonyl substrates seems particularly effective (Scheme 20) ⟨90TL4681⟩.

A number of other silanes have been reported to perform these reductions in high yields but have yet to see widespread use ⟨92JOC2427, 92JOC3405⟩. The least esoteric of these, triethylsilane, can be used in the presence of a thiol catalyst ⟨91JCS(P1)103⟩ or in excess with relatively large amounts of initiator in what has been shown not to be a conventional chain reaction process ⟨91TL7187, 93JOC249⟩.

1.01.2.2.3 Reduction of C—O—heteroatom bonds

The reduction of alcohols is often achieved by conversion to a group more prone to nucleophilic displacement, usually a (*p*-toluene or benzene) sulfonate, then treatment with a suitable metal hydride reagent.

Thus highly nucleophilic reagents such as lithium aluminium hydride ⟨78JOC1263⟩ and lithium triethylborohydride ⟨83JOC3085⟩ have been used to effect efficient reductions of such esters. The chemoselectivity of such reactions can be poor ⟨82AJC1895⟩, but choice of solvent can be important (Scheme 21) ⟨80JOC2550⟩.

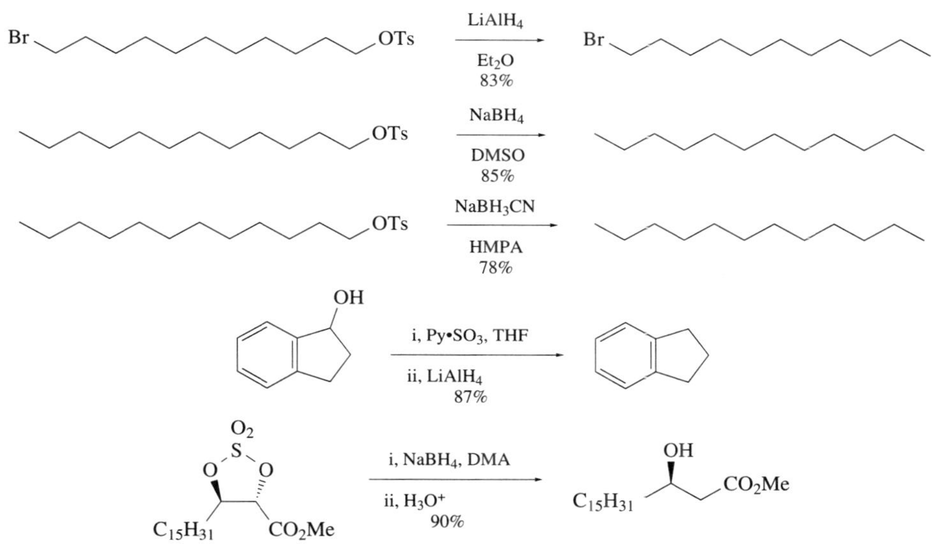

Scheme 21

The use of less reactive reagents such as sodium borohydride ⟨69JOC3923, 78JOC2259⟩ or sodium cyanoborohydride ⟨77JOC82⟩ still allows the reduction of sulfonates, but with a consequent increase in chemoselectivity (Scheme 21).

Sulfonates can also be reduced electrochemically ⟨79TL2157⟩ or, in the case of primary ⟨80JA2693⟩ or α-carbonyl ⟨86JOC1135⟩ systems, by reaction with samarium(II) iodide in the presence of a proton source.

Sulfates can also be reduced via nucleophilic displacement by a metal hydride reagent, but only benzylic, allylic or α-carbonyl systems seem to have sufficient reactivity (Scheme 21) ⟨69JOC3667, 86TL4813, 88JA7538⟩.

Phosphates and phosphoramidates can be reduced by lithium in the presence of ethylamine and *t*-butanol in a procedure that appears to be efficient for primary, secondary and tertiary systems but has few proponents ⟨72JA5098, 86JA3835⟩.

The reduction of the carbon–oxygen bond of relatively activated silyl ethers (allylic, benzylic and α-carbonyl systems) has been reported in a few specific instances ⟨81S32, 82JOC4380, 86JOC1135⟩. These procedures may be of value in particular circumstances, but would not be one of the first methods of choice if starting from the corresponding alcohol.

1.01.2.3 Reduction of C=O Bonds to CH_2

1.01.2.3.1 Reduction of aldehydes

There is a surprisingly small number of reports of the direct reduction of aldehydes to methyl groups compared to those devoted to the analogous conversion of ketones to methylene moieties, which may be more a reflection of the relative synthetic utilities of the procedures than of ease of reaction.

The Clemmensen reduction, using zinc in the presence of an acid, illustrates this point. While the original paper on this procedure reported the reductions of heptanal and benzaldehyde in moderate yield ⟨13CB1837⟩, very few aldehyde reductions have appeared subsequently ⟨42OR(1)155, 59AG726⟩, and the reaction is now used almost exclusively for the deoxygenation of ketones ⟨75OR(22)401⟩.

The Wolff–Kischner reduction, in which the carbonyl compound is treated with hydrazine and a base ⟨48OR(4)378⟩, has seen rather more use in the reduction of aldehydes than the Clemmensen procedure. The Huang–Minlon modification of the reaction, where (tri- or) diethylene glycol is used as solvent, and volatile by-products are removed by distillation as the reaction temperature is raised from 180 °C to 200 °C ⟨46JA2487⟩, is now employed generally, and aryl and aliphatic aldehydes which are tolerant of base and these elevated temperatures can be reduced in high yields (Schemes 13 and 22) ⟨49JA3301, 84JA6702⟩.

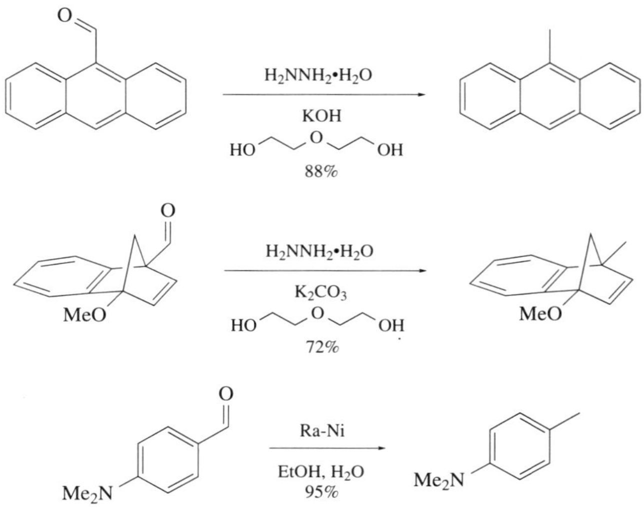

Scheme 22

The variant of the Wolff–Kischner reduction in which the tosylhydrazone of the aldehyde is formed, then converted to the methyl compound by treatment with a metal hydride reagent also works well. Lithium aluminium hydride is used in the conversion of aromatic aldehydes ⟨63T1127⟩ but tends to give alkenic by-products with aliphatic systems when sodium borohydride becomes the reagent of choice ⟨64CI(L)153⟩. Carbohydrate aldehydes have also been reduced by this approach, using potassium borohydride, in moderate yields ⟨64CI(L)1689⟩.

Benzylic aldehydes can be reduced directly to the corresponding toluene derivative by catalytic hydrogenation ⟨53OR(7)263⟩. Systems with electron-donating substituents on the ring tend to react more quickly, particularly when transfer hydrogenation techniques are used ⟨88TL3741⟩, but high yields can be obtained in many cases, especially when Raney-nickel is employed (Schemes 14 and 22) ⟨80TL2637⟩.

An aldehyde unit attached to an electron-rich aromatic system can be reduced directly to the methyl group using Red-Al ⟨69TL1739⟩ or sodium cyanoborohydride in the presence of zinc iodide ⟨86JOC3038⟩. An aliphatic aldehyde has been reduced quantitatively by the combination of triethylsilane and boron trifluoride ⟨79TL849⟩, while titanocene has been reported to reduce aliphatic, but not aromatic, aldehydes ⟨74JA5290⟩.

1.01.2.3.2 Reduction of ketones

The Clemmensen reduction, in which the substrate is reduced by the action of zinc in the presence of an acid, is one of the longest established methods for ketone deoxygenation ⟨75OR401⟩. The reaction does not work well for diketones owing to pinacol coupling, a process which can also be a significant side reaction, in an intramolecular sense, for monoketones ⟨46JA2487⟩, and the alkene unit of α,β-unsaturated ketones is reduced in tandem with the carbonyl group ⟨72BCJ264⟩. Nonetheless, high yields of alkanes can be obtained for this procedure, particularly when the reaction is carried out under anhydrous conditions (Scheme 23) ⟨72BCJ264⟩. A variation in the procedure, using base rather than acid, has also shown promise ⟨87JOC3205⟩.

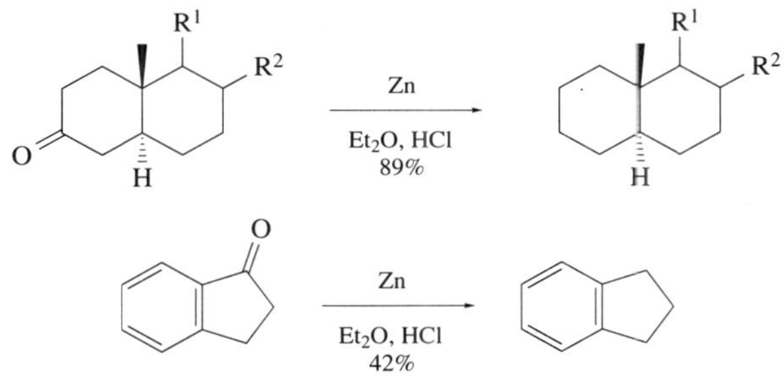

Scheme 23

The Wolff–Kischner reduction, in which the ketone is treated with hydrazine and a base, is as well established a procedure as the Clemmensen reduction ⟨48OR(4)378⟩. It has been demonstrated that, if the intermediate hydrazone is isolated first, the reaction can be carried out in DMSO at room temperature ⟨62JA1734⟩. However, most practitioners use the procedure at elevated temperature (180–200 °C) with *in situ* hydrazone formation (termed the Huang–Minlon modification ⟨46JA2487⟩), and are generally rewarded with high yields (Schemes 13 and 24) ⟨46JA2487, 82JOC2590, 84JA6690, 87JOC3205⟩.

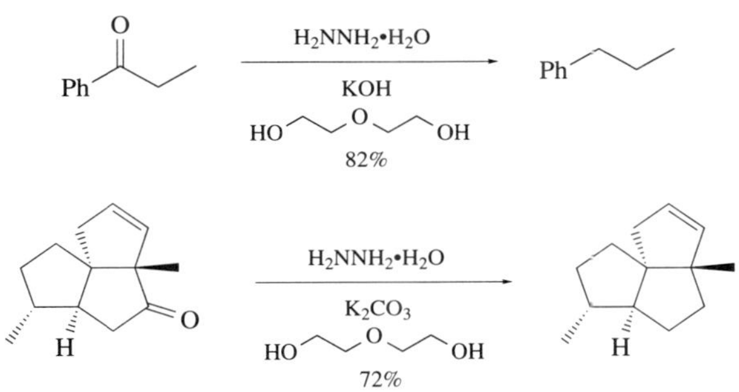

Scheme 24

Hindered ketones are not reduced by these procedures ⟨49JA3301⟩ and, sometimes, pyrazoline formation can compete with the reduction of α,β-unsaturated ketones ⟨87JOC3205⟩.

A useful variation of the Wolff–Kischner procedure involves forming the tosylhydrazone, and treating this intermediate with a reducing agent such as sodium borohydride ⟨64CI(L)153, 87TL4759⟩, sodium cyanoborohydride ⟨85JOC2607⟩ or catecholborane ⟨75JOC1834, 85JA5732⟩. These reactions are performed at much lower temperatures than the Huang–Minlon type procedures and give comparable yields (Scheme 25) ⟨75JOC1834, 87TL4759⟩.

Aromatic ketones can be reduced, via the benzyl alcohol, to arylalkanes ⟨53OR(7)263⟩. Both Raney-nickel and palladium-on-carbon are effective catalysts for this procedure, and the reactions can be carried out under a hydrogen atmosphere or using transfer hydrogenation techniques ⟨88TL3741⟩. These reactions are generally high yielding, although some functional groups (e.g., chloro- or nitroarenes) are reduced preferentially (Schemes 14 and 26) ⟨58JOC133, 80TL2637, 82JOC4304, 82S940, 83CJC2415⟩. Aliphatic ketones are normally reduced to the corresponding alcohol by catalytic hydrogenation but it has been reported that, under strongly acidic conditions and platinum(IV) oxide catalysis, conversion to the alkane (without the intermediacy of the alcohol) can occur ⟨67TL2329⟩.

The combination of sodium borohydride and trifluoroacetic acid has been shown to reduce a range of diaryl ⟨78S763⟩ and aryl alkyl ketones (Scheme 27) ⟨85JOC5451⟩ in high yield, unless a strong electron withdrawing group is present ⟨78S763⟩. This range of ketones can also be reduced efficiently by the action of hydride reagents in conjunction with Lewis acids, for example, lithium

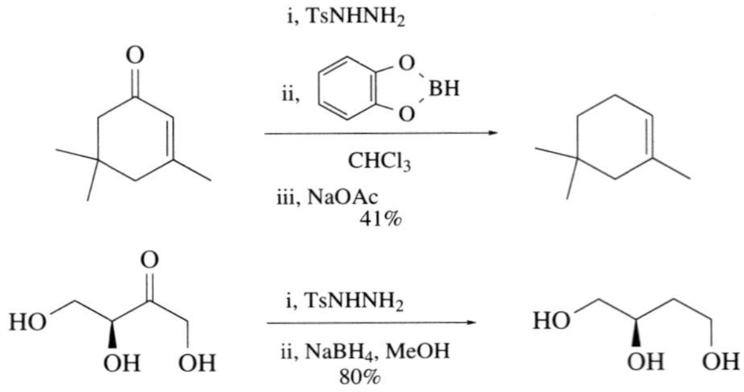

Scheme 25

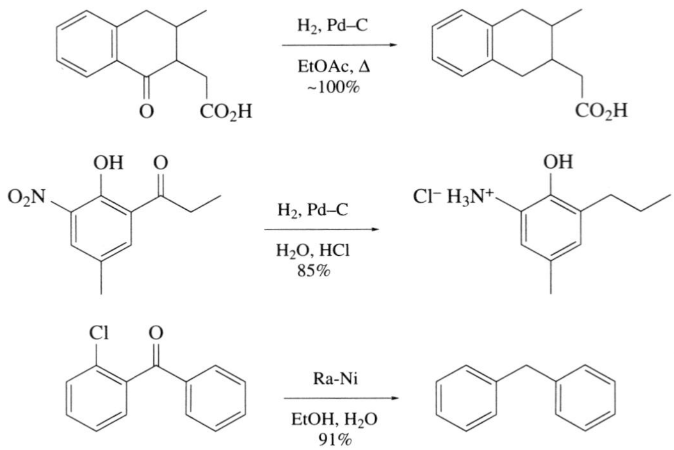

Scheme 26

aluminum hydride ⟨58JA2896, 87TL2937⟩ or sodium borohydride (Scheme 27) ⟨87S736⟩ with aluminum chloride, dibal-H with aluminum bromide ⟨92JOC2143⟩, sodium cyanoborohydride with zinc iodide ⟨86JOC3038⟩, and triethylsilane with boron trifluoride ⟨81OS(60)108⟩ have all been used to good effect. This last combination has also been shown to reduce α,β-unsaturated ketones without alkene reduction ⟨87JOC1984⟩.

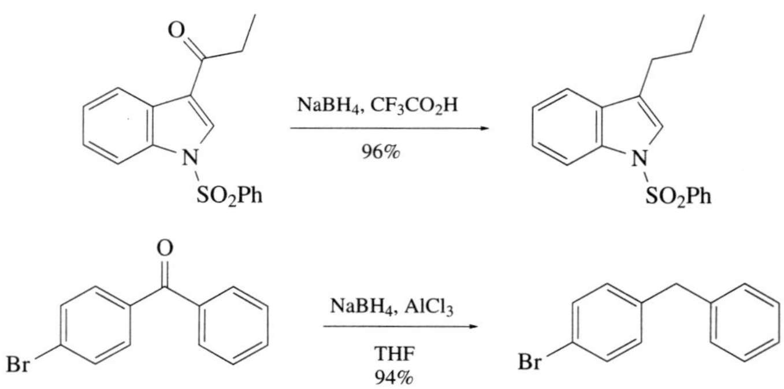

Scheme 27

1.01.2.4 Reduction of (C=O)X to CH_3

The sole example of the direct hydrogenation of a carboxylic acid derivative to a methyl group by catalytic hydrogenation is the quantitative transformation of benzoic acid to toluene under rhenium(II) oxide catalysis ⟨63JOC2347⟩. Aliphatic acids are converted to alcohols under the same conditions.

A series of benzoic acids and esters (with electron-donating substituents on the ring) has been converted to the corresponding toluene derivatives by reduction with Red-Al in refluxing xylene ⟨69TL1739⟩.

Aliphatic esters and lactones undergo this type of reduction upon treatment with titanocene, but the reduction stops at the benzylic alcohol stage with aromatic systems ⟨74JA5290⟩.

The most clearly delineated method for the full reduction of benzoic acids involves the sequential, one-pot treatment with excess trichlorosilane, then a tertiary amine, then hydroxide in aqueous methanol ⟨70JA3232⟩. The first step is thought to cause anhydride formation. In the presence of an amine, the excess silane converts the anhydride to two moles of benzyl silane, which is cleaved to give the toluene derivative and a silanol. The reaction is highly chemoselective, aryl bromides, and even ester groups remain intact (Scheme 28) ⟨70JA3232, 73JOC3660⟩. The procedure can be extended to the reduction of aryl esters by preliminary treatment with trimethylsilyl iodide and iodine, forming the trimethylsilyl ester, followed by the reduction sequence ⟨79JOC2185⟩.

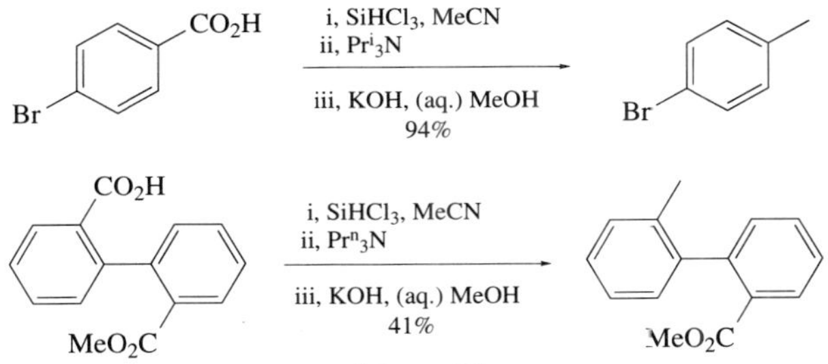

Scheme 28

1.01.2.5 Reduction of C(OX)$_n$ Systems

Many methods are available for the reduction of *ortho*-esters to acetals, or of acetals to ethers ⟨B-84MI 101-01, B-89MI 101-01⟩. However, complete reduction to the alkane oxidation state is rare.

Some simple benzylic acetals have been reduced to the corresponding toluene derivative by catalytic hydrogenation ⟨53OR(7)263⟩. Very occasionally this methodology is used to deprotect benzylidene derivatives of carbohydrates ⟨28CB1750, 63JOC1395⟩.

1.01.3 REDUCTION OF C—SULFUR, C—SELENIUM AND C—TELLURIUM BONDS TO CH

1.01.3.1 General Methods

The most general method for reduction to alkanes in this area is by treatment with Raney-nickel ⟨62CRV347, 62OR(12)356⟩. Thiols ⟨50JCS3000⟩, sulfides ⟨92TL5567⟩, sulfoxides ⟨86CC1688⟩, dithioacetals ⟨50JA4296⟩ and selenium systems ⟨76TL2643⟩ are all reduced cleanly and chemoselectively by this procedure (Scheme 29).

Radical chain reactions using stannanes are efficient for the reduction of selenium and tellurium systems (Scheme 30) ⟨80JA4438⟩ but generally do not work well for the complete reduction of sulfur analogues.

1.01.3.2 Reduction of C—SX Bonds

Historically, the most important method for the reduction of alkyl thiols is Raney-nickel ⟨62CR347, 62OR356⟩, which provides excellent chemoselectivity with consequent high yields (Schemes 29 and 31) ⟨50JCS3000, 51JCS556⟩.

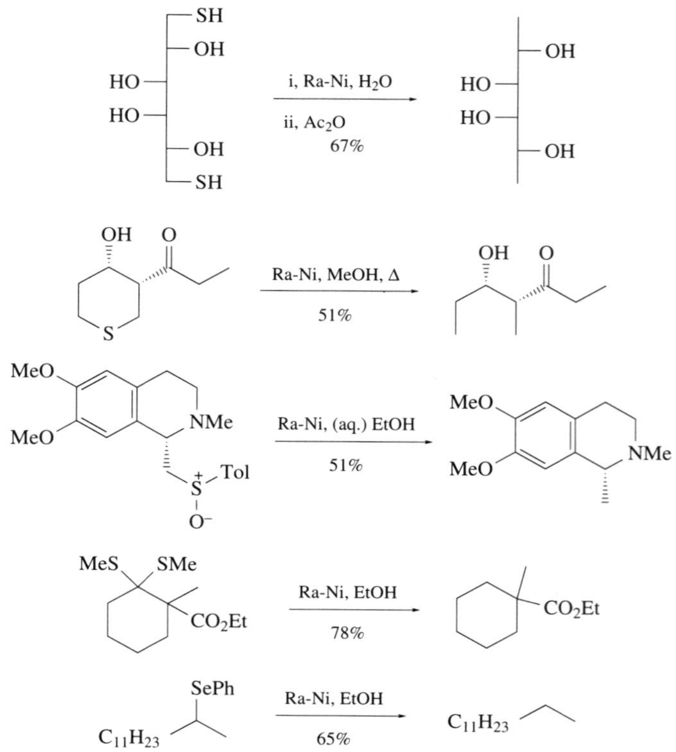

Scheme 29

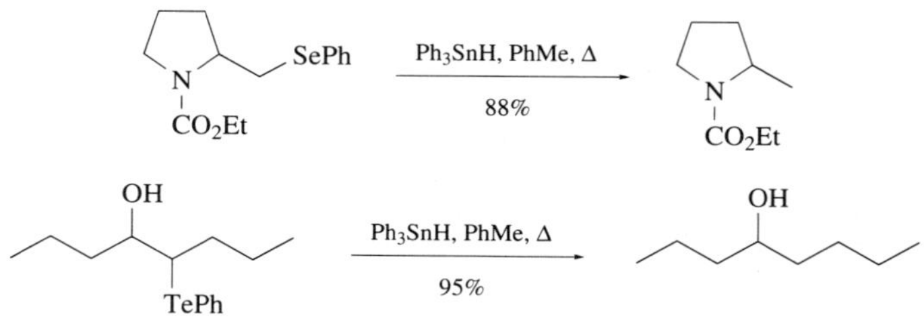

Scheme 30

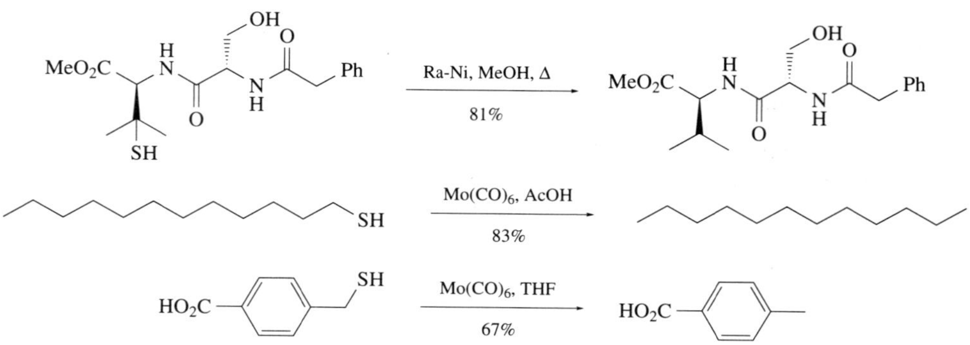

Scheme 31

More recently the use of homogenous transition metal reagents, such as molybdenum hexacarbonyl (Scheme 31) ⟨80CC169, 85JOC5413⟩, and catalysts ⟨89JOC4474⟩ have become popular ⟨90S89⟩ while stannane-mediated radical reductions have also proved successful ⟨82JA2046⟩.

In common with thiols, the desulfurisation of dialkyl or aryl alkyl sulfides is often performed by treatment with Raney-nickel ⟨62CRV347, 62OR(12)356⟩. The mild conditions afforded by this procedure are particularly useful when the stereochemical integrity of labile chiral centres elsewhere in the molecule is important (Scheme 32) ⟨92TL5567, 92TL6367⟩.

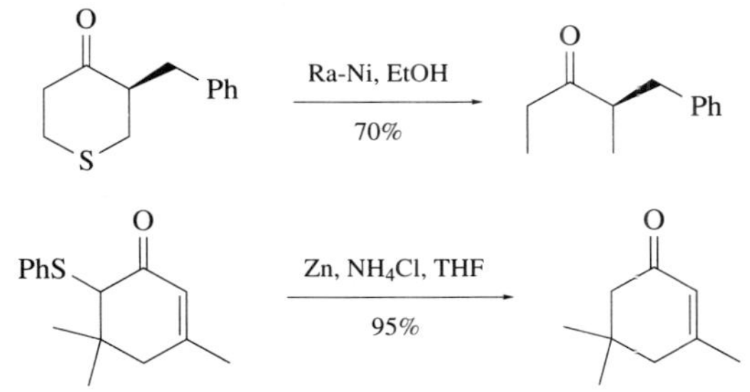

Scheme 32

Homogenous transition metal catalysts and reagents have also been used in this context ⟨90S89⟩, but seem to have a less general reactivity, with most reported reductions being of relatively activated (i.e., benzylic and α-carbonyl) sulfides ⟨85JOC5413, 89JOC4474⟩.

Sulfides that are in the α-position of a carbonyl group can be reduced by a range of methods which includes the use of Raney-nickel ⟨85JOC2589⟩, but also low-valent metals, such as samarium(II) iodide and zinc, in the presence of a proton source (Scheme 32) ⟨86JOC1135, 87JOC2317⟩. Allylic sulfides are also prone to reduction by low-valent metals ⟨80JOC4097⟩, or by lithium triethylborohydride in the presence of a palladium(0) catalyst ⟨82JOC4380⟩.

There are relatively few examples of the desulfurisation of alkyl sulfoxides, but here again Raney-nickel appears to be the reagent of choice (Schemes 29 and 33) ⟨51JA1528, 62OR(12)356, 86CC1688⟩. Sulfoxides α to a carbonyl group are reduced conveniently by low-valent metal systems such as aluminium amalgam (Scheme 34) ⟨81JA2886⟩, samarium(II) iodide ⟨86JOC1135⟩ and zinc ⟨87JOC2317⟩. Benzylic sulfoxides have been reduced by lithium aluminium hydride in the presence of a homogenous nickel catalyst ⟨89JOC4474⟩.

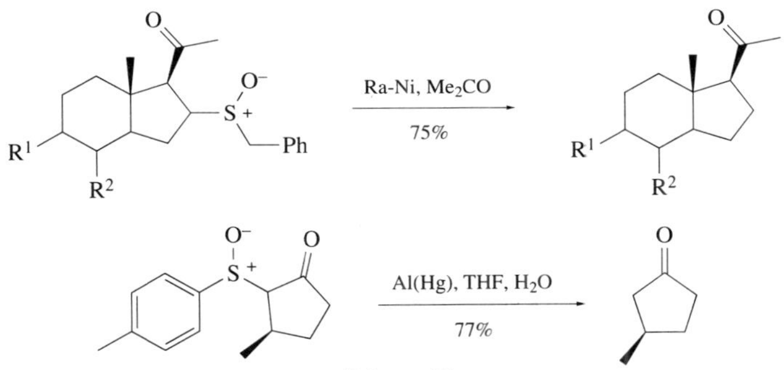

Scheme 33

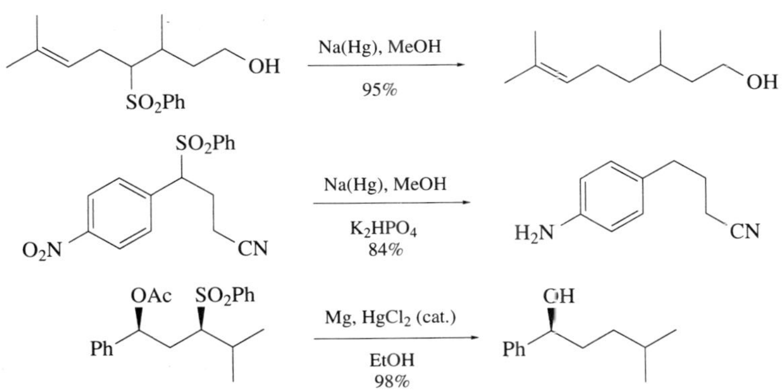

Scheme 34

The reduction of sulfones to alkanes has been reviewed ⟨B-93MI 101-01⟩. Most of these reactions have been performed with low-valent metal reducing reagents, most commonly sodium amalgam (Scheme 34) ⟨76BSF513, 76BSF525, 81S55⟩ (sodium in liquid ammonia has also been employed ⟨86JOC858⟩). More recently, other low-valent metal reagents, such as lithium in ethylamine ⟨92JOC4598⟩, aluminium amalgam ⟨84JOC1246⟩ and samarium(II) iodide ⟨86JOC1135, 91TL1949⟩, have been employed and, of these, the use of magnesium in ethanol with a catalytic amount of mercury(II) chloride seems particularly effective (Scheme 34) ⟨93TL4541⟩.

Potassium, dispersed ultrasonically in toluene, has been used to reduce only one carbon–sulfur bond of cyclic alkyl sulfones in high yields ⟨85TL4495, 91TL3551⟩. Unsymmetrical systems are reduced at the more substituted position.

Benzylic sulfones can be reduced by dibal-H ⟨84RTC220⟩ or by lithium aluminium hydride under nickel catalysis ⟨89JOC4474⟩, while allylic sulfones have been reduced by lithium triethylborohydride in the presence of a palladium catalyst ⟨82JOC4380⟩.

Sulfoximines have been reduced by Raney-nickel ⟨82JOC1193⟩.

1.01.3.3 Reduction of C=S to CH_2

There is little evidence to suggest that Raney-nickel is an effective agent for the desulfurisation of thioaldehydes or thioketones ⟨62CRV347⟩.

Biarylthioketones have been reduced by zinc in aqueous acid ⟨66CB1393⟩, diphosphorus tetraiodide then sodium bisulfite ⟨85BCJ2421⟩, a nicotinamide derivative in the presence of a dimercurated benzene derivative (Scheme 35) ⟨85JA6121⟩ and the tetrahydridocarbonylferrate anion (Scheme 35) ⟨75JOC2694⟩. The last of these methods is the only one shown to be applicable to aliphatic thioketones.

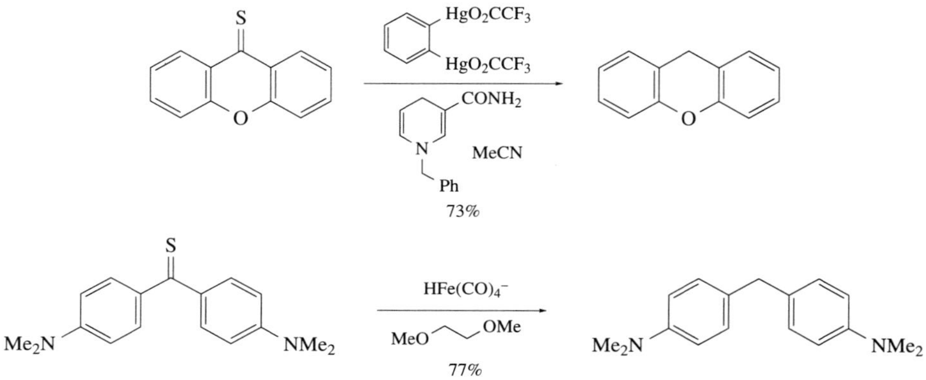

Scheme 35

1.01.3.4 Reduction of C(=S)X to CH_3

Toluenes are formed in the reduction of methyl dithiobenzoates by zinc, but only as minor components of a complex mixture of products in which the corresponding methyl benzyl sulfide predominates (Equation (4)) ⟨66CB1393⟩.

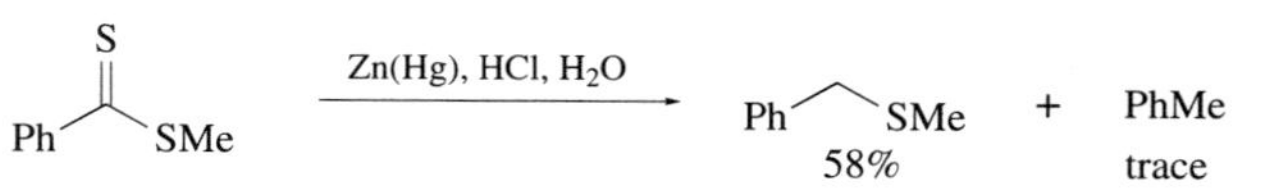

(4)

1.01.3.5 Reduction of $C(SX)_n$ Systems

One of the most important methods of deoxygenation of a ketone is by conversion to a dithioacetal derivative, which is then desulfurised by treatment with Raney-nickel (Schemes 29 and 36) ⟨49JA2870, 50JA4296, 85JOC2607⟩.

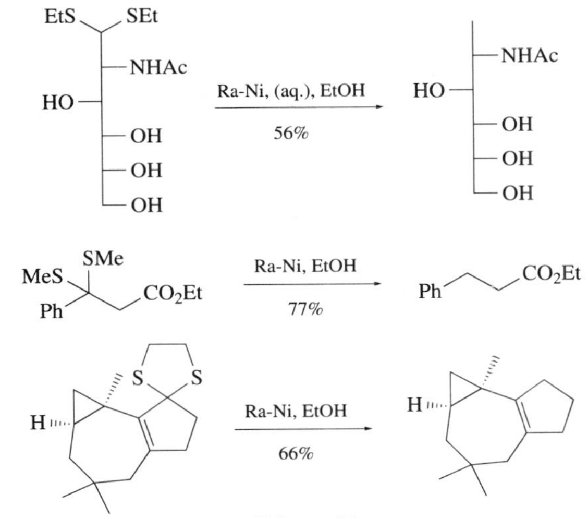

Scheme 36

Although the method involves two separate reactions, certain advantages are presented in comparison to protocols such as the Wolff–Kischner and Clemmensen reductions. Thus enone systems are cleanly reduced without alkene reduction (Scheme 36) ⟨87JA3025, 87JOC3346⟩, and dicarbonyl compounds can be reduced to a monocarbonyl derivative if selective dithioacetal formation can be obtained (e.g., an aldehyde can be reduced in the presence of a ketone ⟨86CC451⟩).

Dithioacetals in which both sulfur atoms are in the sulfone oxidation state can be reduced to the alkane by treatment with magnesium in methanol (Scheme 37) ⟨38JOC1823, 88T6355⟩.

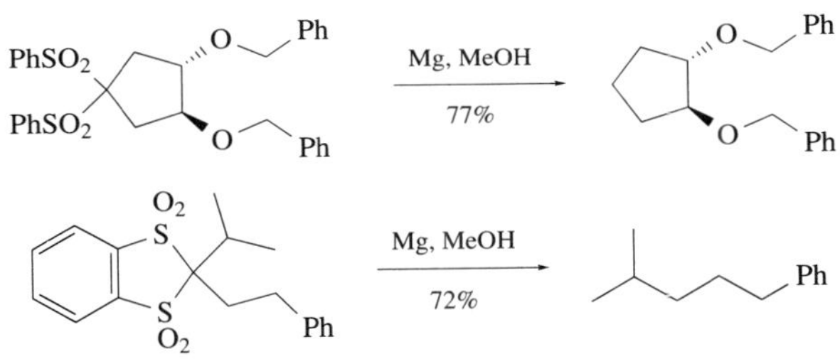

Scheme 37

1.01.3.6 Reduction of C—Se Systems

Although Raney-nickel can be used to reduce selenides and diselenoacetals (Schemes 29 and 38) ⟨76TL2643, 81T4097⟩, these compounds are more commonly reduced (in higher yields) by radical reduction procedures using stannanes (Schemes 29 and 38) ⟨80JA4438, 81TL1623, 81T4097⟩ or silanes ⟨92JOC2427, 92JOC3405⟩.

Selenides have also been reduced by lithium in ethylamine ⟨76TL2643⟩ and nickel boride ⟨84CC1417⟩. Allylic systems can be reduced by lithium triethylborohydride under palladium(0) catalysis ⟨82JOC4380⟩.

1.01.3.7 Reduction of C—Te Systems

Tellurides can be reduced in a radical chain process by triphenylstannane (Schemes 30 and 39) ⟨80JA4438⟩. Telluride dichlorides are also reduced under these conditions at faster rates despite the reduction proceeding via the corresponding telluride (Scheme 39) ⟨80JA4438⟩.

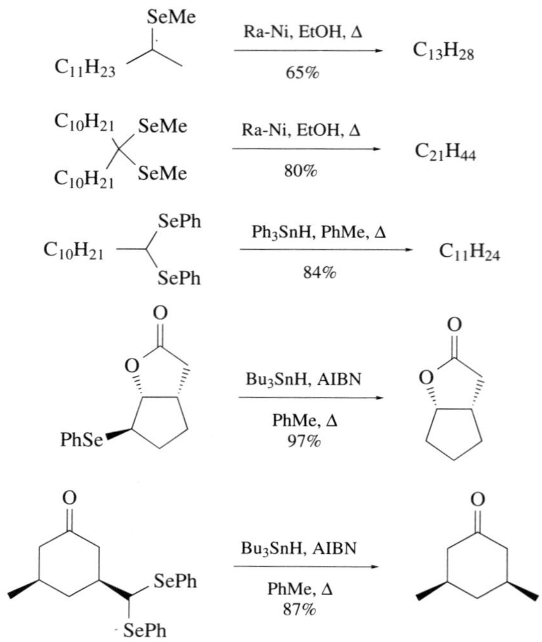

Scheme 38

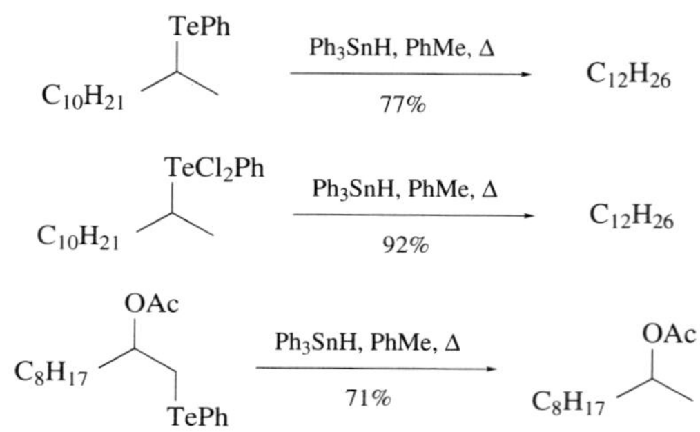

Scheme 39

1.02
One or More CH Bond(s) Formed by Substitution: Reduction of Carbon–Nitrogen, –Phosphorus, –Arsenic, –Antimony, –Bismuth, –Carbon, –Silicon, –Germanium, –Boron, and –Metal Bonds

JOSHUA HOWARTH
Dublin City University, Republic of Ireland

1.02.1 THE REDUCTION OF CARBON–NITROGEN BONDS TO CARBON–HYDROGEN BONDS

1.02.1.1 Reduction of Carbon–Nitrogen Single Bonds

Reductive deaminations have been the subject of many reviews, the last of which appeared in 1982 ⟨B-82MI 102-01⟩. There have been several other earlier reviews on reductive deamination ⟨B-56MI 102-01, 66JCE1021, B-68MI 102-01⟩. The following sections are categorized with respect to the type of nitrogen moiety lost in the deamination.

1.02.1.1.1 Loss of sulfonamide anion

The mechanism for such a reduction is essentially a nucleophilic displacement of a good leaving group by a nucleophilic hydride (Equation (1)). Sodium borohydride in polar aprotic solvents (hexamethylphosphoramide (HMPA), DMSO) furnishes a convenient and effective source of nucleophilic hydride which may be used for the reductive displacement of disulfonimides. The incorporation of two powerfully electron-withdrawing groups, principally sulfonyl ⟨78JOC2259⟩, or carbonyl ⟨73JOC1348⟩ moieties, on the nitrogen atom, can be sufficient to stabilize the departing anion. Since non-basic anions are produced when sulfonimide is used as the stabilizer of the anion, several synthetic applications have been developed using this method ⟨78JOC2259⟩. An excellent procedure exists for the production of disulfonimides from readily obtainable sulfonamides ⟨74JOC3525⟩. This procedure is, however, only applicable for unhindered primary amines. To date, no high-yielding method of producing disulfonimides from hindered amines has been found. It appears that sulfonamides resist addition of a second sulfonyl moiety in even moderately hindered environments. Research has suggested that the steric requirement of the disulfonimide surpasses that of the trimethylammonium ion ⟨66LA(691)15, 75JA6873⟩. The range of compounds investigated in this type of reduction spans several substituted aniline and straight chain primary amine derivatives as well as a number of cyclic amines (Scheme 1).

$$R^1N(SO_2R^2)_2 \xrightarrow[\text{HMPA}]{BH_4^-} R^1H + {}^-N(SO_2R^2)_2 \qquad (1)$$

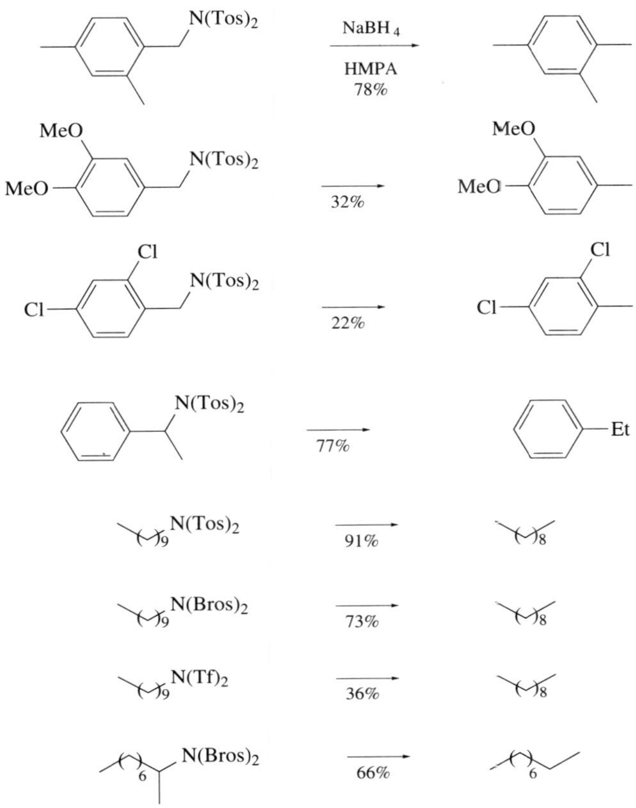

Bros = *p*-bromobenzene sulfonyl

Scheme 1

1.02.1.1.2 Loss of succinimide

The development of a monomethylation procedure for aromatic amines ⟨73JOC1348⟩ serendipitously led to a method where certain succinimides derived from Mannich bases may be reductively deaminated using sodium borohydride (Equation (2)). Fortunately, functionalities such as ester, nitrile, or amide do not interfere with the reaction. There is circumstantial evidence to suggest that the reaction proceeds via a stabilized nitrogen anion, similar to the case of the sulfonimide reductions (Scheme 2).

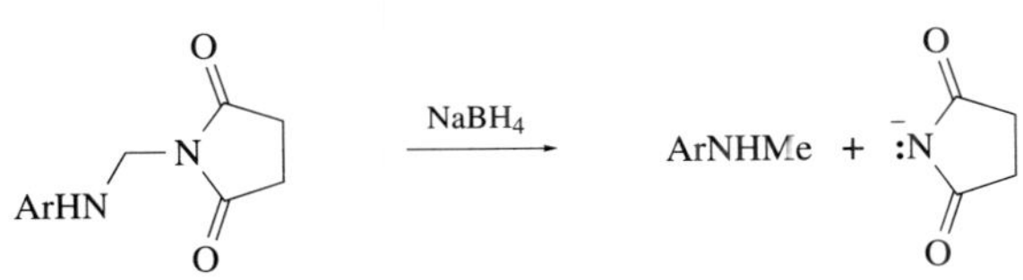

(2)

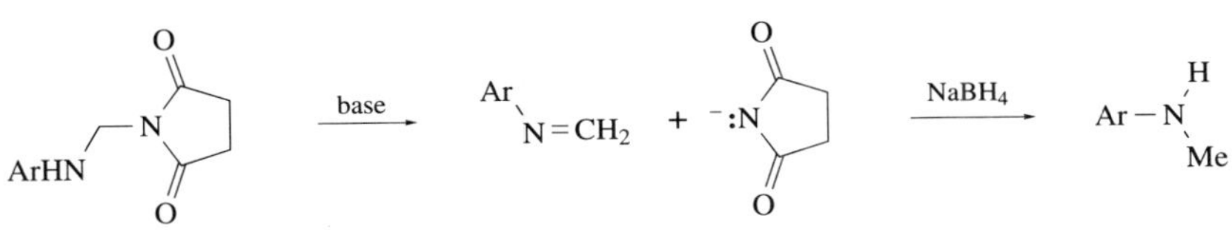

Scheme 2

1.02.1.1.3 Loss of pyridine derivatives

This method of reduction has been thoroughly investigated ⟨79CC300, 79JCS(P1)442, 80JCS(P1)2554, 80T679, 84AG(E)420, 84JCS(P1)1671⟩. Highly substituted pyrilium salts are used for a two-step conversion of the amino group in amines into numerous other functionalities. The first step involves the conversion of the pyrilium salt and the amine into a *N*-substituted pyridinium salt. This salt is then reacted with a nucleophile. In the case of reduction this nucleophile is a hydride donor (Scheme 3). The amines used can be categorized into three different groups, depending on the substituted pyrilium salt needed to perform the conversion.

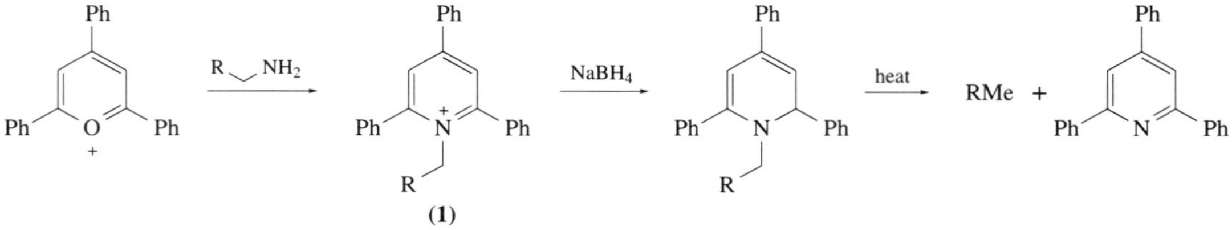

Scheme 3

The first group consists of *N*-allyl, *N*-benzyl, and *N*-heteroarylmethyl amines. The *N*-substituted 2,4,6-triphenylpyridinium salts (**1**) are reduced by sodium borohydride in good yields to the 1,2-dihydro derivatives. For these substituents the dihydropyridines decompose smoothly at around 200 °C to give 2,4,6-triphenylpyridine and the product in which the amino group in the original compound has been replaced by a hydrogen atom (Table 1).

Table 1 Yields of deaminated products RCH_3 from the thermolysis of 1-(CH_2R)-2,4,6-triphenyl-1,2-dihydropyridines.

R	*Yield* (%)
CH_2=CH	81
Ph	82
2-Furyl	82
2-Pyridyl	75
4-Pyridyl	77

NMR studies have suggested that this type of reaction could possibly involve a radical mechanism (Scheme 4) ⟨79CC300⟩, although unpublished work has suggested that the *N*-allyl type substituent may undergo deamination via an electrocyclic course (Scheme 5).

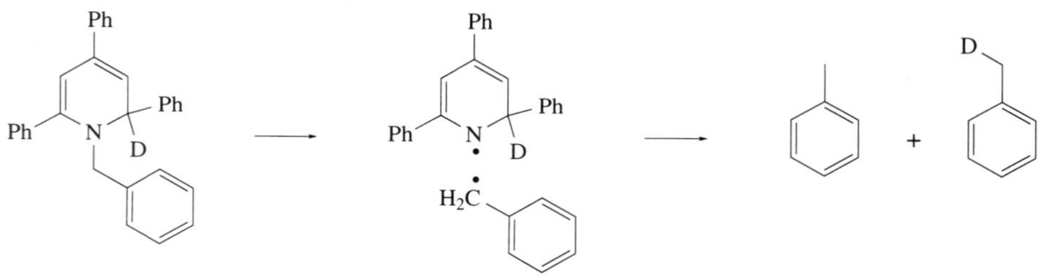

Scheme 4

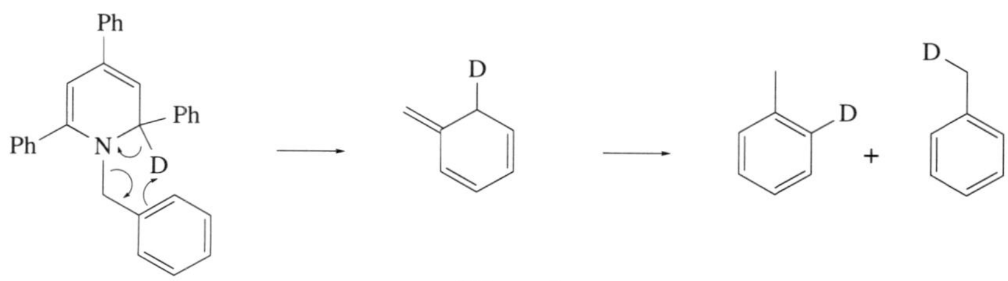

Scheme 5

The second group consists of primary alkylamines. These amines react readily with 2,3,5,6-tetraphenylpyrilium cations (**2**) to yield the corresponding pyridinium salts. Steric hindrance directs attack by the sodium borohydride to the *para* position and hence 1,4-dihydropyridines are produced (Scheme 6). These decompose at 180–200 °C to give the corresponding alkane (Table 2).

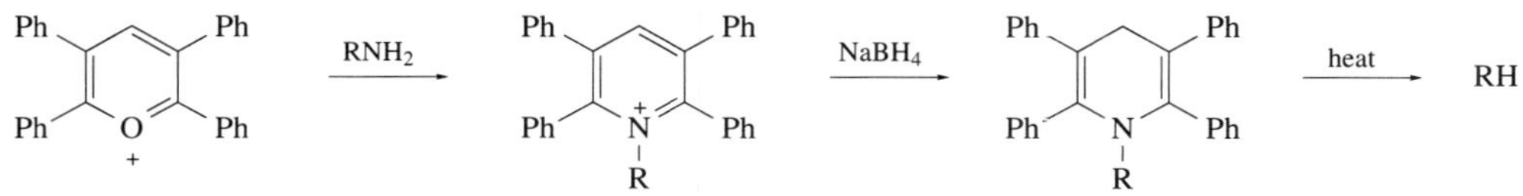

(**2**)

Scheme 6

Table 2 Yields of deaminated products RH from the thermolysis of 1-substituted-2,3,5,6-tetraphenyl-1,4-dihydropyridines.

R	*Yield* (%)
n-C_6H_{13}	58
n-C_8H_{17}	88
Ph	54
$PHCH_2$	44
$Ph(CH_2)_2$	64
p-$ClC_6H_4CH_2$	62

Primary arylamines make up the third and final group. The method described above for alkylamines can be applied to arylamines for their conversion to the corresponding hydrocarbons, but the temperature required for decomposition is high, 300 °C, and the yields using this method vary considerably. If *N*-aryl-2,4-diphenyl-5,6-dihydrobenzo[*h*]quinolinium fluorides (**3**) are formed and heated to 160 °C the corresponding arenes are produced in yields of around 60% (Table 3) (Equation (3)). The mechanism has not been resolved, but is thought to be of the radical type similar to that shown in Scheme 4 ⟨80T679⟩.

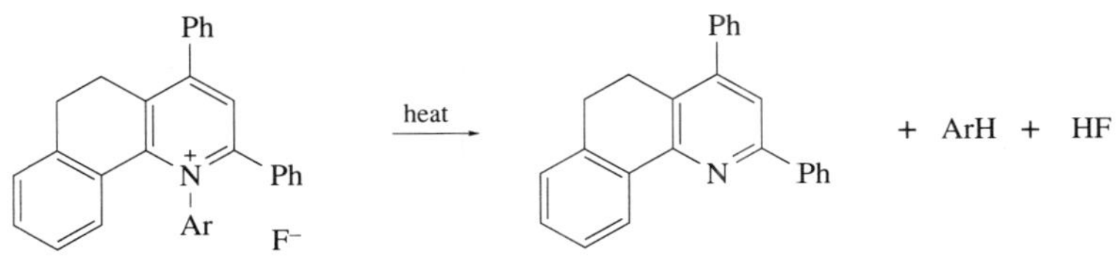

(**3**)

(3)

Table 3 Reduction deamination of primary arylamines via *N*-(Ar)-2,4-diphenyl-5,6-dihydrobenzo[*h*]quinolinium fluorides.

N-Substituent (Ar)	*Yield* (%)
C_6H_5	60
p-Cl—C_6H_5	57
p-Br—C_6H_5	62

1.02.1.1.4 Loss of amines

Many examples are found where the leaving group involved in the reduction of an amine is also an amine. Amine salts of alkyl or aryl systems can be reduced to the corresponding alkyl or aryl moiety in a number of ways, including catalytic hydrogenation, metal reductions, and various hydrides. All these methods have been extensively reviewed ⟨B-56MI 102-01, 66JCE398, B-68MI 102-01, 73S744, B-85MI 102-01⟩. In addition electrolytic reductions are known and have also been reviewed ⟨B-68MI 102-01⟩.

Catalytic hydrogenation using palladium on carbon and formic acid ⟨80JOC4926⟩ has been found to reduce amines attached to molecules containing carbon–carbon double bonds and hydroxyl functionality. Essentially the corresponding alkene is formed with little rearrangement of the carbon–carbon double bond (Table 4).

Table 4 Catalytic hydrogenation using palladium on carbon and formic acid.

Amine	*Major product*	*Yield* (%)
E-MCH(CH_3)CH=CHCH$_2$CH(OCH_3)$_2$	*Z*-3- and 4-hexenal dimethyl acetal	61
E-(Pip)CH(CH_3)CH=CHCH$_2$CH(OCH_3)$_2$	*Z*-3- and 4-hexenal dimethyl acetal	67
E-(CH_3Pip)CH(CH_3)CH=CHCH$_2$CH(OCH_3)$_2$ I	*Z*-3- and 4-hexenal dimethyl acetal	63
E-(Pip)CH$_2$C(CH_3)=CHCH$_2$C(CH_3)$_2$OH	(CH_2)$_2$C=CHCH$_2$C(CH_3)$_2$OH	59
E-(Pip)CH$_2$C(CH_3)=CHCH$_2$Ph	(CH_2)$_2$C=CHCH$_2$Ph	60
E-(Pip)C(CH_3)$_2$CH=CHCH$_2$CH(OCH_3)$_2$	(CH_3)$_2$CHCH=CHCH$_2$CH(OCH_3)$_2$	31
	(CH_3)$_2$C=CH(CH_2)CH(OCH_3)$_2$	31

M = morpholino, Pip = piperidino.

Reduction of non-allylic systems appears not to have been attempted using this method. However, reduction of a range of amines using a platinum halide on silica has been investigated ⟨84JOC2875⟩. This method can reduce amines to the alkanes or arenes and ammonia (Table 5). In some cases the temperature needed for the reduction to occur is as high as 225 °C. The reaction system will also reduce nitrogen heterocycles to arenes or alkanes and ammonia (Scheme 7).

Table 5 Reduction of amines using platinum halides on silica.

Amine	*Product*	*Yield* (%)
n-Hexylamine	*n*-Hexane	99
Cyclohexylamine	Cyclohexane	99
1-Adamantane amine	No reaction	99
1-(Methylamino)adamantane	1-Methylladamantane	99
Piperidine	*n*-Pentane	99
Quinuclidine	3-Ethylpentane	88
Aniline	Cyclohexane	99
Diphenylamine	Cyclohexane	99

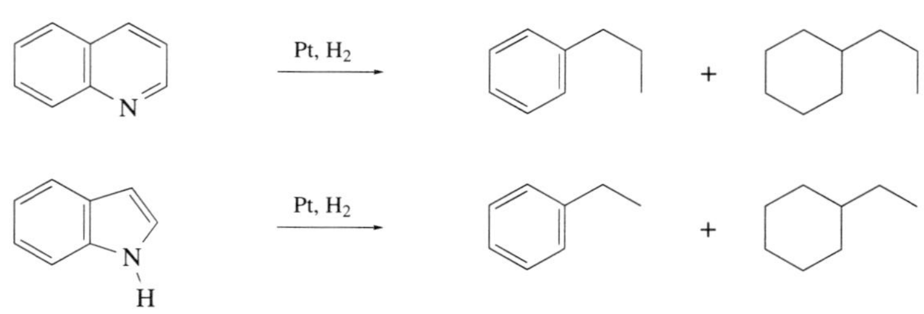

Scheme 7

In all, this is a good method for defunctionalizing simple amines to their corresponding hydrocarbons. The problem of reduction of other functionality in more complex systems has not been fully investigated.

The use of other hydrides such as tri-*n*-butyltin hydride ⟨88SC1207⟩, aluminum hydrides ⟨81T2797⟩, and borohydrides ⟨78CC1089⟩ in the reduction of amines has also been researched in some detail. Tri-*n*-butyltin hydride was used to reduce a series of similar compounds as shown in Scheme 8 ⟨88SC1207⟩. This appears to be the only examination of this reagent as an amine reductant. Temperatures of 200 °C were required to effect the reaction; the method can also be used to reduce oxides and salts of the amine (Scheme 9).

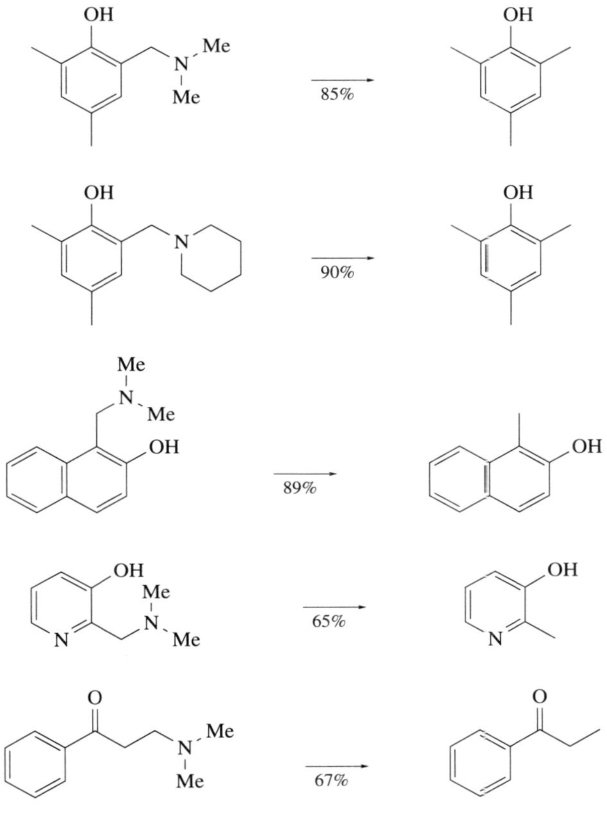

Scheme 8

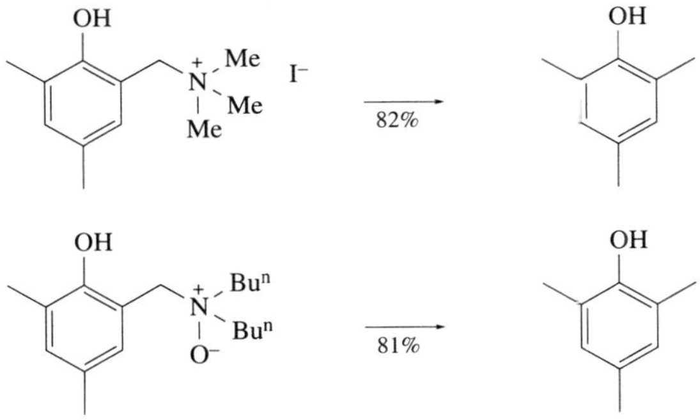

Scheme 9

The mild and selective reagent, sodium cyanoborohydride, has been used very successfully in the high-yield reduction of quaternary ammonium salts (Scheme 10) ⟨78CC1089⟩. The reduction occurs in the presence of halogen, ester, nitrile and nitro functionality. Several benzenoid derivatives have been reduced to the parent benzenoid compound (Table 6). However, the drawback with this method is the simultaneous use of two highly toxic compounds (HMPA and dimethylsulphate) in the reaction.

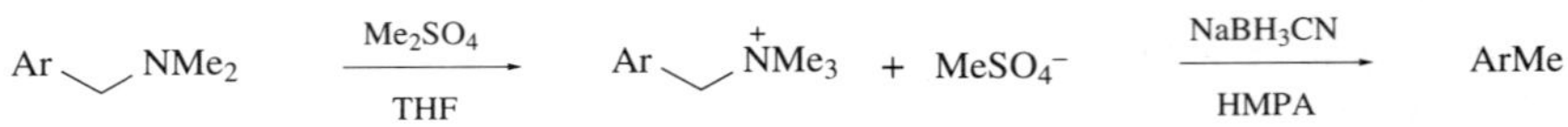

Scheme 10

Table 6 The reduction of benzenoid tertiary amine derivatives using Me_2SO_4 followed by $NaBH_3CN$.

Amine	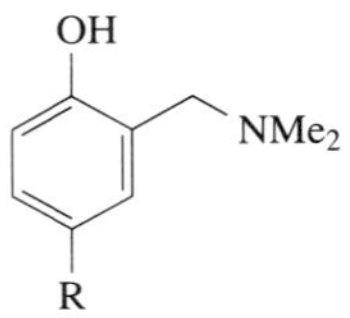					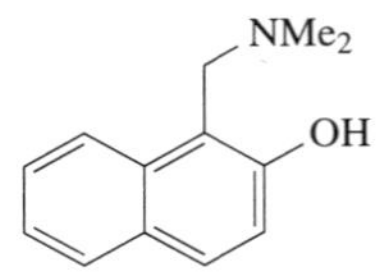	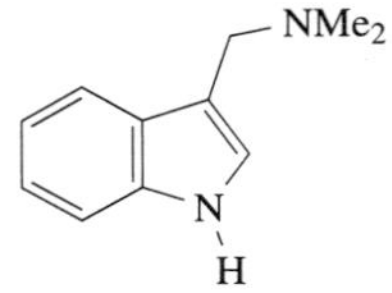
Yield of methyl product (%)	R = H	Cl	CO_2Et	CH_2CN	NO_2		
	71	83	91	72	90	74	77

In general amides can be reduced to the corresponding aldehyde. To obtain the aldehyde by cleavage of the amide it is necessary to maintain the amide in excess. The reagent commonly used for this process is an aluminum hydride, usually lithium aluminum hydride, but variations have been used ⟨B-85MI 102-01⟩.

A nice demonstration of the selectivity of hydride reagents is clearly shown by the use of lithium aluminum hydride ⟨81T2797⟩ on an amide containing an epoxide (**4**). The amide was reduced to the corresponding aldehyde without any unwanted reduction of the epoxide unit (Scheme 11).

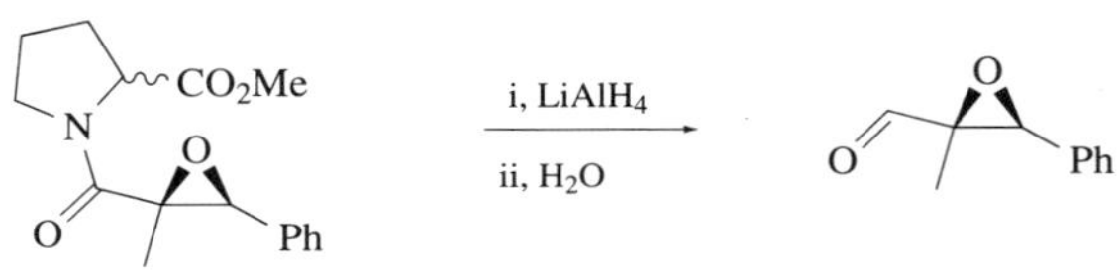

(4)

Scheme 11

Early methods of reducing amines to the corresponding alkane commonly revolved around the use of metal reductions such as zinc and acetic acid at reflux temperature ⟨58JA1654, 75BCJ1892⟩ or Raney nickel hydrogenolysis at 60–100 atm and at 80 °C (Scheme 12) ⟨53JA1128⟩. Some methods used sodium methoxide at 180 °C ⟨51JA2718⟩. These are all harsh conditions, and since more subtle and selective methods for functional group interconversions have been developed, the tendency to use these harsher methods has become minimal, although tin chloride has been used recently (Scheme 13) ⟨83SC677⟩. One of the methods that is useful in the reduction of amines is the cleavage of tetraalkyl ammonium halides using sodium in liquid ammonia. This method has been carried out on a large number of substrates and results in good yields in almost every case (Equation (4)) ⟨59JA4850⟩. This is a mild way of reducing amines, useful as long as there is no other functionality in the molecule which dissolving metal reductions can affect.

$$Bu^n{}_4N^+\ Br^- \xrightarrow[79\%]{Na/NH_3,\ 78\ ^\circ C} Bu^nH + Bu^n{}_3N + NaBr + NaNH_2 \qquad (4)$$

An example of an electrolytic reduction is shown in Equation (5). This involves the reduction of α-amino acid esters to the corresponding β-keto esters in good yields ⟨73JOC2731⟩.

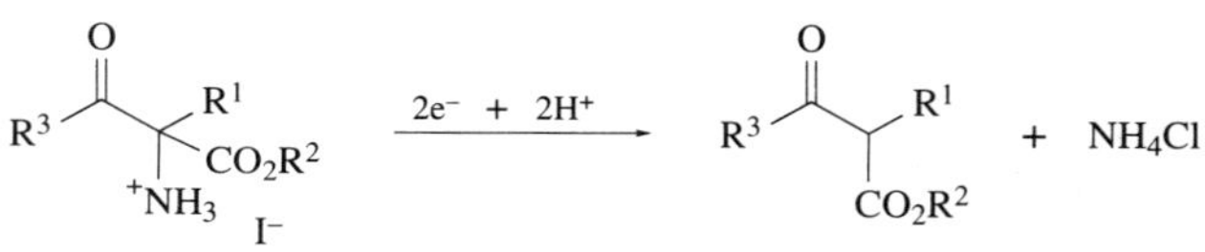

(5)

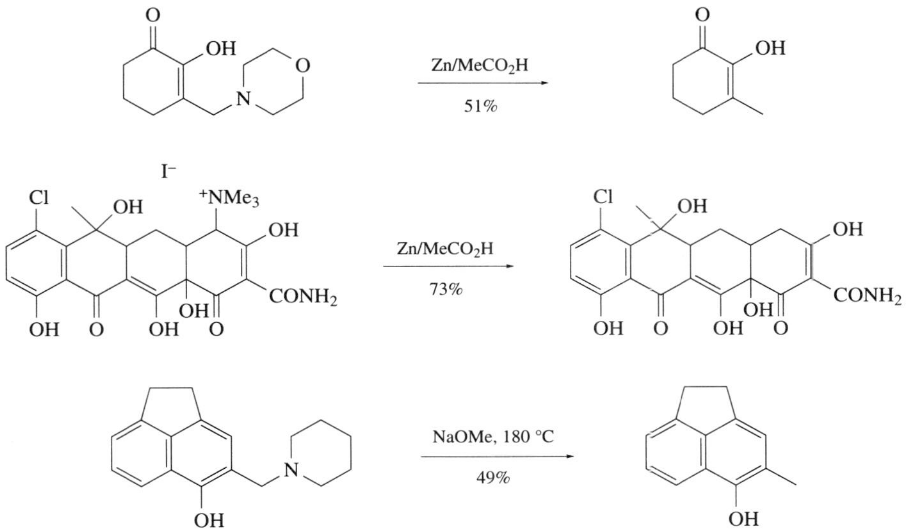

Scheme 12

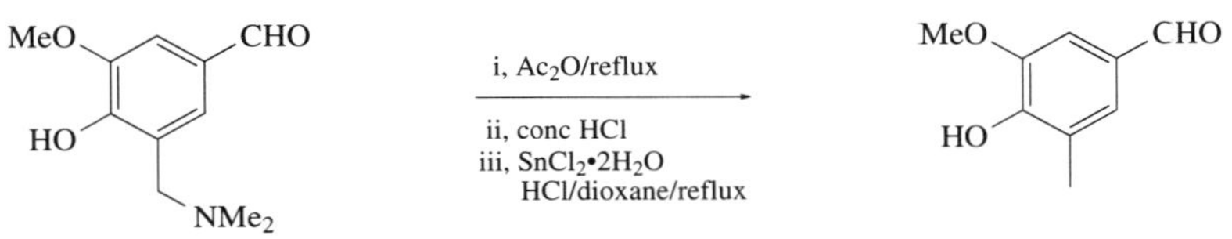

Scheme 13

1.02.1.1.5 Loss of nitrogen gas

Amine reductions which result in the formation of a nitrogen molecule are, in most cases, believed to proceed via a diimide intermediate.

An example of this strategy is the reaction of primary amines with difluoroamine (Scheme 14). This conversion and its mechanism have been previously reviewed ⟨63JOC1108, 66JCE398, B-68MI 102-01, 70MI 102-01, 71ACR193, B-72MI 102-01⟩.

$$RNH_2 + HNF_2 \longrightarrow \left[R\text{-}N{=}NH \right] \longrightarrow RH + N_2$$

Scheme 14

The first reported procedure for reductive deamination of amines to form alkanes via the diimide, involved the reaction of hydroxylamine-*O*-sulfonic acid (Scheme 15) or chloramine with arylsulphonamides. Similar reductions are observed when alkylhydrazines are oxidized to alkyldiimides ⟨63JA1108⟩. A similar reaction of primary amines with hydroxylamine-*O*-sulfonic acid has been reported (Equation (6)) ⟨78JA341, B-82MI 102-01⟩. This is a direct, general and selective method of deamination (Table 7).

$$RNH_2 \xrightarrow{ArSO_2Cl} RNHSO_2Ar \xrightarrow[NH_2X]{OH^-} RH + ArSO_3H + N_2$$

$X = OSO_3H$ or Cl

Scheme 15

$$RNH_2 \xrightarrow[HO^-,\ 0\,°C]{NH_2OSO_3H} RH + SO_4^{2-} + N_2 \qquad (6)$$

Table 7 Reduction of primary amines using hydroxylamine-*O*-sulfonic acid.

Substrate	*Product*	*Yield* (%)
2-Aminobenzoic acid	Benzoic acid	72
2-Amino-3-methylbenzoic acid	3-Methylbenzoic acid	26
Benzylamine	Toluene	65
3-Aminopropionic acid	Propionic acid	50
Alanine	Propionic acid	50
L-Alanyl-L-phenylalanine	*N*-Propionyl-L-phenylalanine	51

As can be seen from Table 7, several amines can be reduced by this method in yields of 26–72%. Carbonyl and amide groups seem not to seriously interfere with the reduction of 2-aminobenzoic acid to benzoic acid or 2-amino-3-methylbenzoic acid to 3-methyl-benzoic acid. Diazoketones may be reduced to methyl ketones by hydrogen iodide (Equation (7)) ⟨68AJC1375⟩. Since it is possible to obtain diazoketones from the corresponding amine, this is a useful type of deamination.

$$RCOCHN_2 + 2\,HI \longrightarrow RCOMe + N_2 + I_2 \qquad (7)$$

Aromatic deaminations may be accomplished in either two-step or by one-step reactions. In the latter case the diazonium ion is produced first. This is then followed by reductive dediazonisation. Hypophosphorus acid tends to be the most common reducing agent used (Equation (8)), although a number of others have been reported ⟨44OR(2)262, 52JA3074, 63RCR59, 77JOC2431, B-85MI 102-01⟩.

$$ArNH_2 \xrightarrow[\text{ii, } H_3PO_2]{\text{i, } HNO_2} ArH + N_2 \qquad (8)$$

Using the one-step procedure, the arylamines are treated with alkyl nitrites in solvents such as ethers or DMF ⟨73JCS(P1)541, 77JOC2431⟩. Variations of this method ⟨77JOC2431⟩ have resulted in good yields for deamination. Evidence exists for a free radical mechanism for this reaction (Scheme 16) ⟨77JOC2431⟩. The alkyl nitrite method of deamination has been applied to the conversion of the adenine derivative (**5**) to the corresponding purine (**6**) (Equation (9)) ⟨79TL1181⟩.

$$ArNH_2 + RONO \longrightarrow Ar\text{–}N{=}NOR + H_2O$$

$$Ar\text{–}N{=}NOR \longrightarrow ArN_2\bullet + RO\bullet$$

$$ArN_2\bullet \longrightarrow Ar\bullet + N_2$$

$$\text{Solvent} + Ar\bullet \longrightarrow ArH + \text{Solvent}\bullet$$

Scheme 16

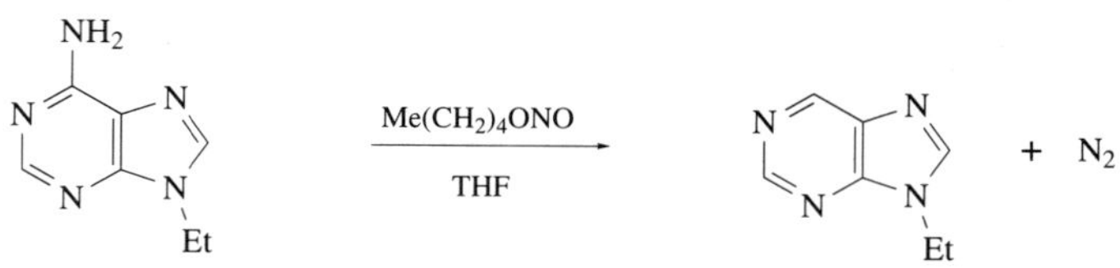

(**5**) (**6**) (9)

1.02.1.1.6 Loss of nitro groups

Much research has been carried out on the reduction of nitroalkanes to the corresponding alkane. The dominant method for achieving this conversion involves the use of trialkyltin hydrides, in particular tri-*n*-butyltin hydride ⟨81TL1705, 82JOC1047, 83CC875, 83JA4017, 83TL3477, 84HCA261, 85JOC3692, 85S269, 85T4013, 86JOC2832, 87JOC5053, 87S821⟩. The reaction is extremely versatile and only presents problems if other groups are present in the molecule which are susceptible to radical attack. The examples shown in Scheme 17 illustrate the versatility of this procedure.

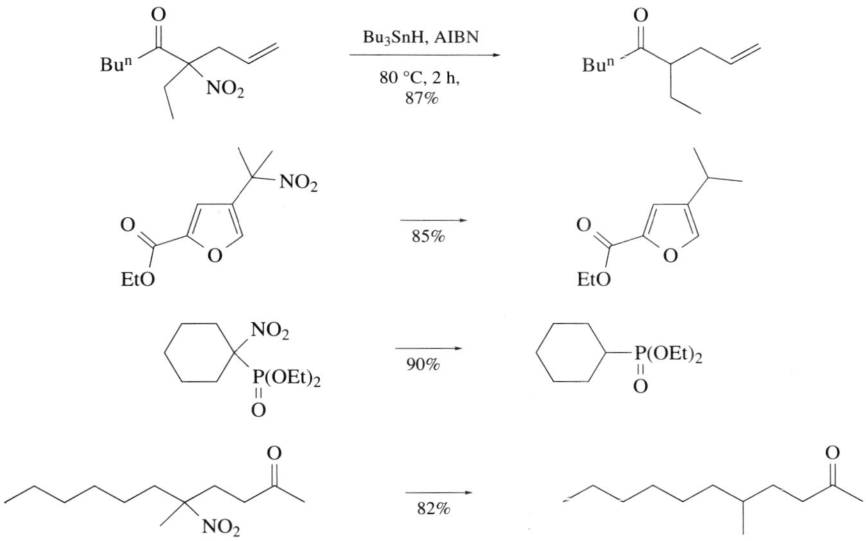

Scheme 17

Other hydrides such as sodium hydrogentelluride have been used to remove the nitro group. It is claimed ⟨85BCJ1067⟩ that this reagent overcomes some of the drawbacks associated with other methods, such as high temperatures, chromatographic separation problems and long reaction times. The reagent is successful for a range of substrates and gives high yields (Scheme 18).

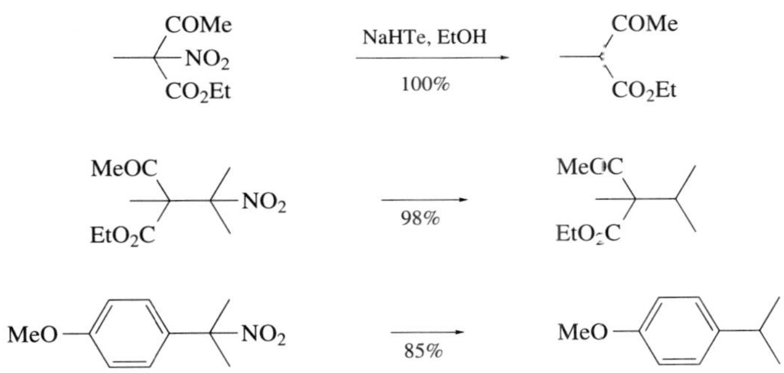

Scheme 18

Lithium aluminum hydride has also been used as a reductant (Equation (10)) ⟨83S137⟩. It is interesting to note that although lithium aluminum hydride can be used to reduce carbon–nitrogen double bonds (see Section 1.02.1.2.1) this does not occur in the reaction illustrated in Equation (10).

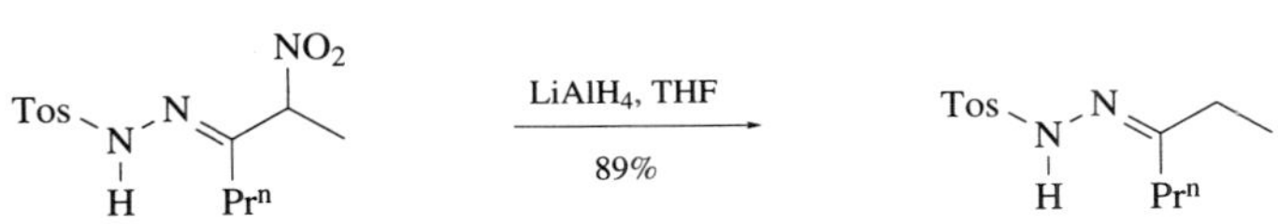

(10)

Other methods for reduction of the nitro group to the corresponding alkane include using the sodium salt of methanethiol in dipolar aprotic solvents ⟨85BCJ1067⟩, potassium hydroxide in ethylene glycol ⟨68JOC1421⟩, and *N*-benzyl-1,4-dihydronicotinamide ⟨84TL4417⟩.

1.02.1.1.7 Loss of cyanide

There are two common methods currently found in the literature for deamination via the isonitrile. One of the procedures uses tri-*n*-butyltin hydride as the reductant ⟨68JA4182, 79CC345, 79TL2291, 80S68⟩. This radical-induced deamination of isonitriles gives high yields and the reaction is tolerant of a large number of functional groups. An example of this tolerance is given below (Equation (11)). The mechanism is thought to be that shown in Scheme 19.

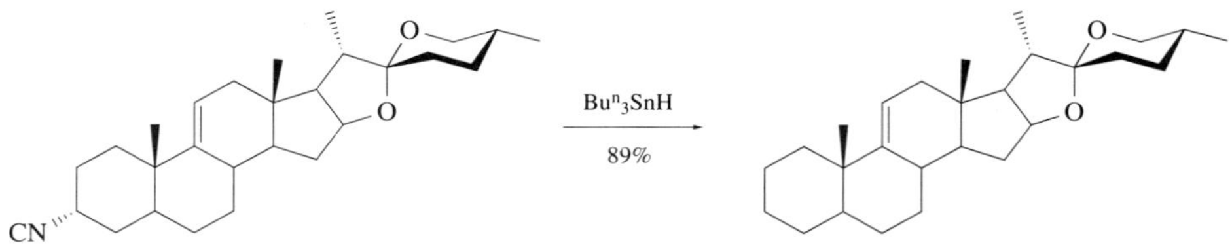

(11)

$$R^1_3SnH \longrightarrow R^1_3Sn\bullet$$

$$R^1_3Sn\bullet + R^2N{=}CH_2 \longrightarrow R^2N{=}\dot{C}SnR^1_3$$

$$R^2N{=}\dot{C}SnR^1_3 \xrightarrow{\beta\text{-scission}} R^2\bullet + R^1_3SnCN$$

$$R^2\bullet + R^1_3SnH \longrightarrow R^1_3Sn\bullet + R^2H$$

Scheme 19

The second method utilizes a dissolving metal reduction, the mechanism of which (Scheme 20) and a typical reaction (Equation (12)) are shown.

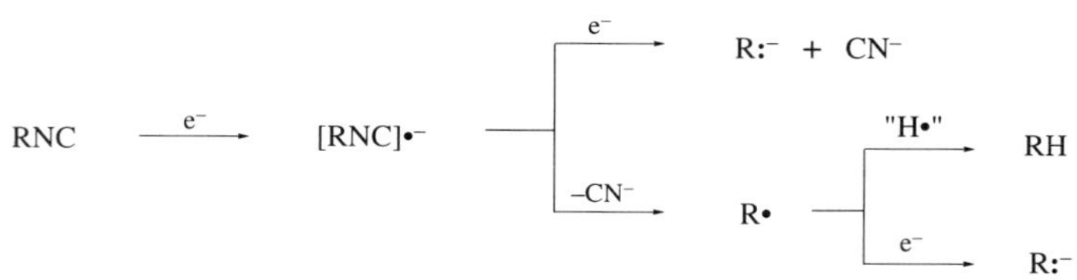

Scheme 20

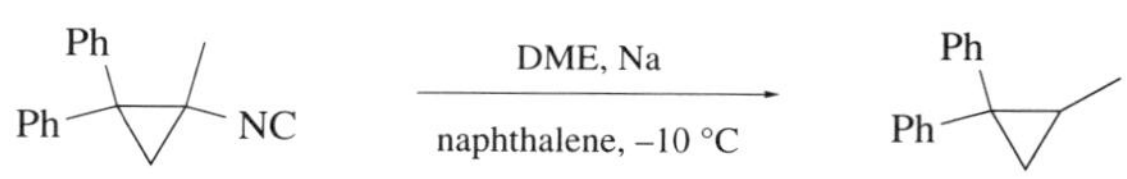

(12)

Several isonitriles have been reduced in this way using alkali or alkali earth metals in liquid ammonia (Equation (13)) ⟨61CB1157, 66HCA1145, 78JOC2396⟩. The yields are extremely good and examples are listed in Table 8.

$$RNC + 2M + NH_3 \longrightarrow RH + MCN + MNH_2 \quad (13)$$

1.02.1.2 Reduction of Carbon–Nitrogen Double Bonds to Methylene and Methyl Groups

This section considers the reduction of systems (C=N)X to methylene or methyl groups, where X is carbon, hydrogen or a heteroatom. The reduction of carbon–nitrogen double bonds can be divided into three distinct groups of substrates. Firstly there are the arylhydrazones derived from an aldehyde or ketone, with the general formula C=N—NR^{1}R^2, where R^1 is an arylsulfonyl moiety.

Table 8 Reduction of isonitriles (RNC) with M/NH_3.

R	*Metal (M)*	*Yield (%)*
Bu^t	Na	93
C_6H_{11}	Li	96
n-$C_{12}H_{25}$	Na	88
$PhCH_2$	Na	58
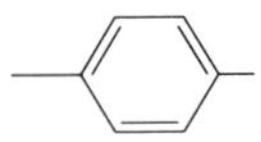	Na	57

The next group consists of those hydrazones, $C{=}N{-}NR^1R^2$ where R^1 or R^2 are anything but arylsulfonyl groups. The last group contains the molecules with the imino group $C{=}NR^1R^2$ where R^1 or R^2 can be alkyl, aryl, acyl or arylsulfonyl.

1.02.1.2.1 Reduction of $C{=}N{-}NR^1R^2$ systems where R^1 is an aryl sulfonyl group

The conversion of alkyl-arylsulphonyl compounds to alkyl compounds was investigated in a very deliberate fashion. It was known that alkyl-*p*-toluenesulfonyl hydrazides could undergo thermal decomposition, in protic or aprotic solvents, with or without basic catalysis, to nitrogen, *p*-toluenesulfinic acid and alkanes ⟨63JA1108⟩. The intermediate was thought to be an alkyl-diimide (Scheme 21) ⟨36JCS584, 63JA1108⟩.

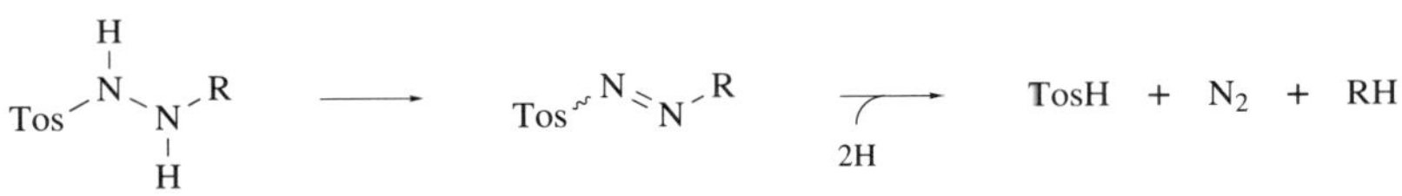

Scheme 21

Alkyl-*p*-toluenesulphonyl hydrazones, on the other hand, decompose with basic reagents to give alkenes ⟨52JCS4735, 65JA935, 66T487⟩, but via a common intermediate, diimide. So it was reasoned, if the carbon–nitrogen double bond of a *p*-toluenesulfonyl hydrazone could be reduced to a carbon–nitrogen single bond, then the decomposition of the substituted *p*-toluenesulfonyl hydrazide so formed should lead to the formation of a methylene or methyl group (Scheme 22). A whole series of *p*-toluenesulfonyl hydrazones was subjected to reaction with lithium aluminum hydride or sodium borohydride ⟨63T1127, 66T487⟩. These included *p*-toluenesulfonyl hydrazones derived from both ketones and aldehydes, the former giving a methylene group on reduction and the latter a methyl group (Table 9) ⟨63T1127⟩.

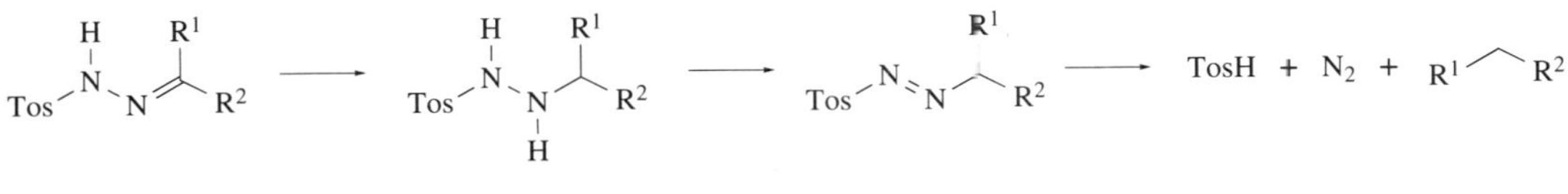

Scheme 22

The reaction using lithium aluminum hydride as the reducing agent has since been used successfully in some natural product syntheses. A particular example is the synthesis of (*Z*)-9-tricosene (**7**), the sex pheromone of the domestic housefly ⟨90JOC5159⟩. A typical reaction is given in Scheme 23 and further examples are given in Table 10.

Other reducing agents have been used to effect this reaction. One such reductant is a zinc modified cyanoborohydride reagent which has been put to use in the deoxygenation of aldehydes and ketones, using the *p*-toluenesulfonyl hydrazone in methanol and a 2:1 ratio of cyanoborohydride to zinc chloride ⟨85JOC1927⟩. The yields are in excess of 85% (Table 11). The same method was used to produce disubstituted cyclohexanes for research into aliphatic liquid crystals (Equation (14)) ⟨88CB1039⟩.

Table 9 Reduction of some *p*-toluenesulfonyl hydrazones with lithium aluminum hydride or sodium borohydride.

p-Toluenesulfonyl hydrazone of	*Product*	*Yield (%)*
5α-Cholestan-3-one	Cholestane	70
Coprostan-3-one	Coprostane	65
3β-Acetoxy-5α-cholestan-7-one	3β-Hydroxy-5α-cholestane	65
Hecogenin	Tigogenin	50
Indol-3-aldehyde	Skatole	50
α-Naphthaldehyde	α-Methylnaphthalene	70

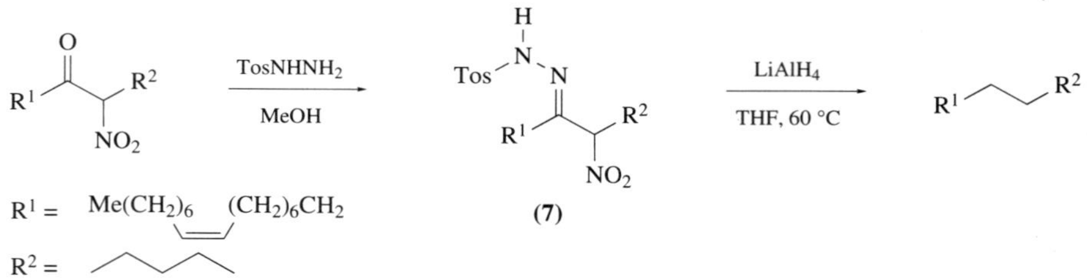

Scheme 23

Table 10 Reduction of more complex *p*-toluenesulfonyl hydrazones.

p-Toluenesulfonyl hydrazone of	*R*	*Yield of alkane (%)*
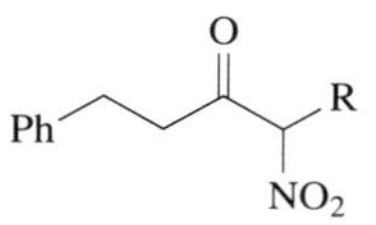	Me	70
	n-C_8H_{17}	61
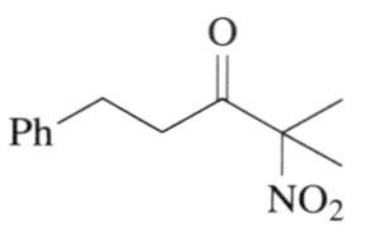		67
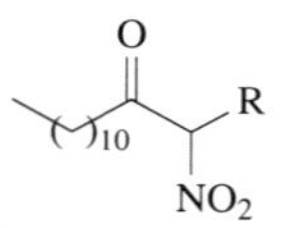	Me	65
	n-C_8H_{17}	68

Table 11 Reduction of hydrazones using cyanoborohydride and zinc chloride.

p-Toluenesulfonyl hydrazone of	*Product*	*Yield (%)*
2-Undecanone	Undecane	78
4-*t*-Butylcyclohexanone	4-*t*-Butylcyclohexane	85
Cholestan-3-one	Cholestane	94

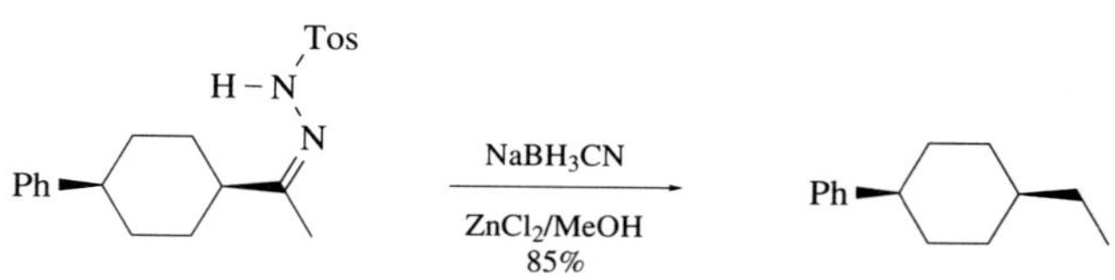

(14)

Catechol borane has also been used as a reducing agent for *p*-toluenesulfonyl hydrazone systems ⟨75JOC1834, 76JOC574, 77S124, 79SC275⟩. The metalborohydride, bis(triphenylphosphine)copper(I) tetrahydroborate can facilitate the reduction of a wide range of ketone and aldehyde hydrazones although yields for the latter are not particularly high (Table 12) ⟨80TL4031⟩.

Table 12 Reduction of hydrazones using bis(triphenylphosphine) copper(I) tetrahydroborate.

p-*Toluenesulfonyl hydrazone of*	*Product*	*Yield* (%)
Cholestanone	Cholestane	70
Adamantone	Adamantane	84
4-*t*-Butylcyclohexanone	4-*t*-Butylcyclohexane	74
Nonan-5-one	Nonane	70
Cyclohexanone	Cyclohexane	68
Cyclopentanone	Cyclopentane	57
N-Methylpiperid-4-one	*N*-Methylpiperidine	17
Nonanal	Nonane	56
Palmitaldehyde	Hexadecane	48
4-Methoxybenzaldehyde	4-Methylanisole	20
4-Nitrobenzaldehyde	4-Nitrotoluene	10

Using the *p*-toluenesulfonyl hydrazone (**8**) of a ketone adjacent to a sulfide a reaction involving the 1,2-transposition of a carbonyl group has been carried out. This reaction involves a conversion of a carbon–nitrogen double bond to a methylene unit via a vinyl sulphide, formed by the action of *n*-butyllithium on the hydrazone (Scheme 24) ⟨78CC414⟩.

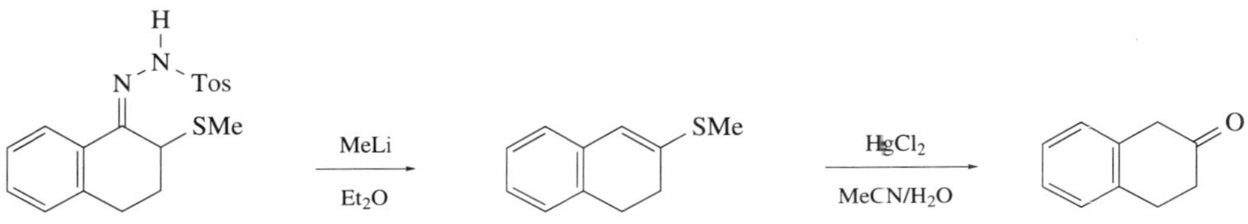

Scheme 24

1.02.1.2.2 *Reduction of C=N—NR¹R² systems where R¹ or R² is alkyl, acyl, aryl or hydrogen*

This second type of reduction of carbon–nitrogen double bonds is dominated by the Wolff–Kishner reaction (Scheme 25) which was originally discovered in 1912 ⟨48OR(4)378⟩. A ketone is reacted with hydrazine and a strong base. A hydrazone is formed *in situ*, and reacts with the strong base to form an alkane and nitrogen ⟨48OR(4)378⟩. The original reaction has almost entirely been replaced by the Huang–Minlon modification ⟨46JA2487, 49JA3301⟩ in which the reaction is carried out in diethylene glycol at reflux. The method is not suitable for α,β-unsaturated aldehydes or ketones, which give pyrazolines under these conditions. Sterically hindered ketones are resistant to reduction but can be reduced by vigorous treatment with anhydrous hydrazine ⟨55JCS2044⟩. The reaction has been used successfully on a more complex molecule (**9**) (Equation (15)) ⟨52JCS2127⟩.

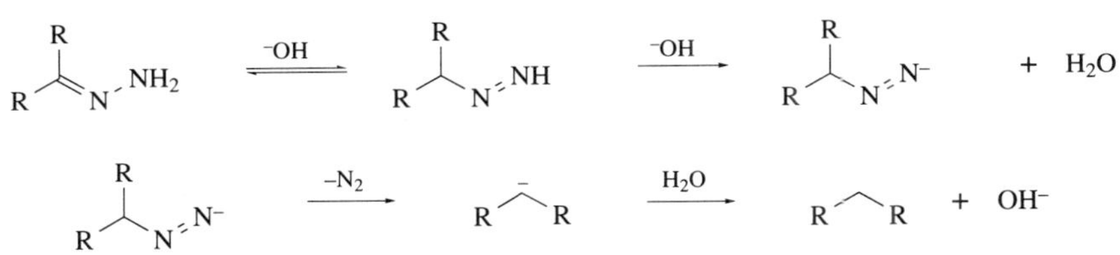

Scheme 25

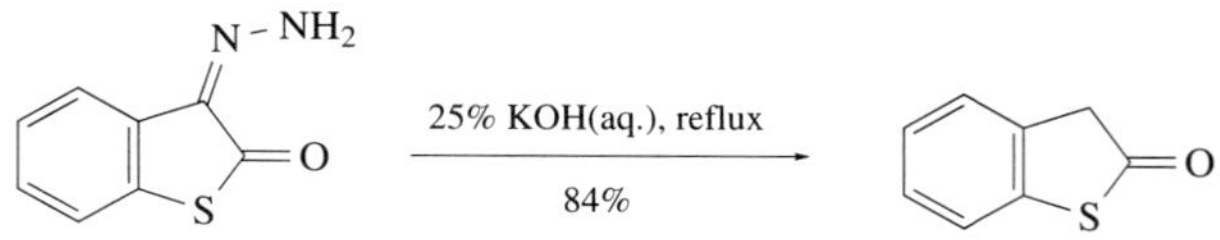

(9)

(15)

Several *N*-unsubstituted hydrazones of aldehydes have been subjected to treatment with potassium hydroxide. The product of the reaction is the methyl analogue of the original aldehyde ⟨43CB1252⟩. The yields ranged from 78% to 86% (Table 13).

Table 13 Reduction of hydrazones using potassium hydroxide.

Hydrazone of	*Product*	*Yield* (%)
Benzaldehyde	Toluene	79
2-Chlorobenzaldehyde	2-Chlorotoluene	82
2,6-Dichlorobenzaldehyde	2,6-Dichlorotoluene	80
2-Aminobenzaldehyde	2-Aminotoluene	66
2-Hydroxybenzaldehyde	2-Hydroxytoluene	86
Anthracenealdehyde-(9)	9-Methylanthracene	78
Pyrenealdehyde-(3)	3-Methylpyrene	84

The same general conditions can be applied to acylhydrazones or semicarbazones as in the conversion sequence shown in Equation (16) ⟨83JOC3866⟩. In this case the reaction conditions for reduction of the acylhydrazone (**10**) were 2-methoxymethanol and sodium hydroxide at reflux temperature for 30 minutes. The conditions are claimed to be milder than usual.

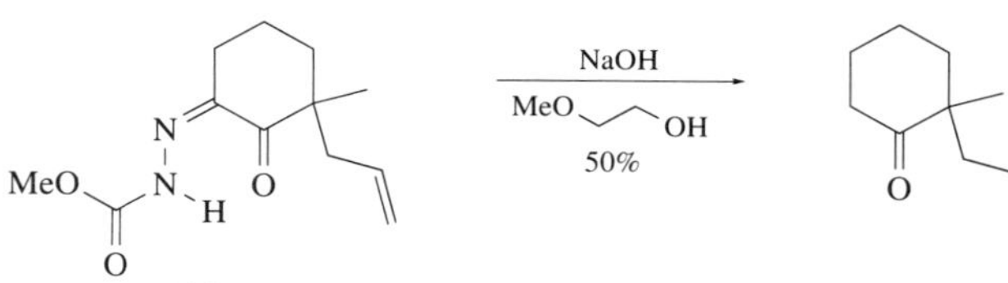

(10)

(16)

Several heterocyclic molecules, pyrrole, furan, and thiophene have been transformed from their 2-formyl derivatives to the corresponding methyl analogue by initial conversion to the semicarbazone followed by reduction using a strong base (Scheme 26) ⟨51JA4033, 56JOC918, 76T829⟩. The yields of the methyl derivatives were substantially higher than in previous methods and the reactions more straightforward. Another semicarbazone which has been reduced to the corresponding alkane is the cyclic compound (**11**) (Equation (17)) ⟨59CB916⟩.

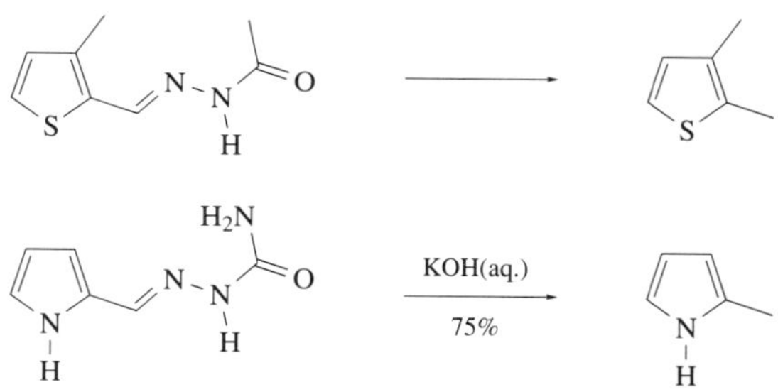

Scheme 26

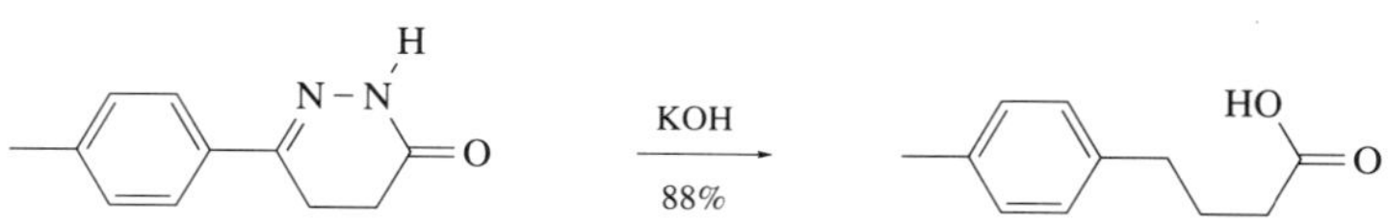

(11)

(17)

A further modification of this procedure, involving the slow addition of hydrazones of aldehydes and ketones to a solution of sublimed potassium *t*-butoxide in anhydrous DMSO at 25°C, affords a mild method of reduction (Equation (18)) ⟨62JA1734⟩.

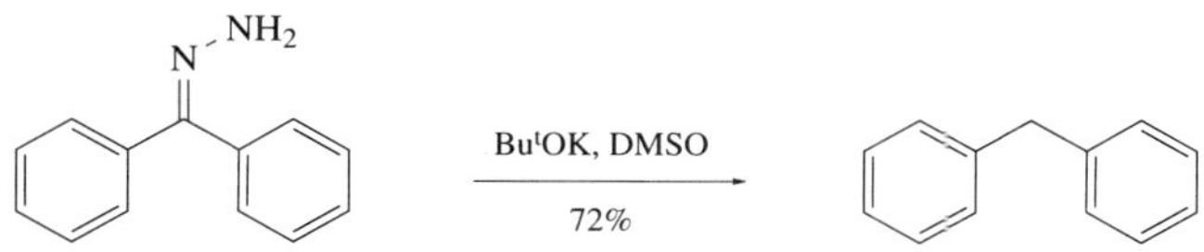

(18)

There are two other distinct methods of reducing a hydrazone. In at least two instances ⟨71JOC737, 87JCS(P1)2371⟩ palladium on carbon with hydrogen has been used as the reductant (Equation (19)). As Table 14 shows, the reaction is extremely effective, and it can be applicable even when there is a range of other functionalities in the molecule ⟨87JCS(P1)2371⟩.

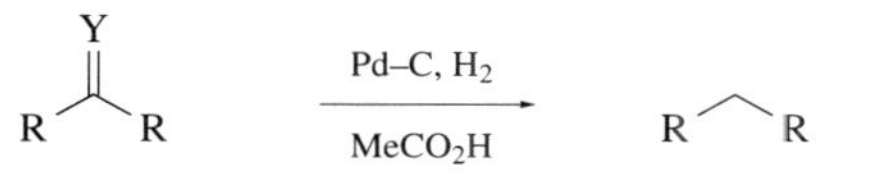

(19)

Table 14 Reduction of hydrazones to methylene using palladium on carbon and hydrogen.

	Yield (%) of alkane when Y =		
	$NNHC_6H_3(NO_2)_2$	*$NNHCONH_2$*	*$NNHC_6H_5$*
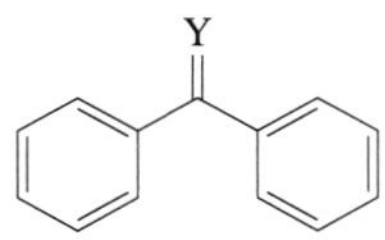	92	93	96
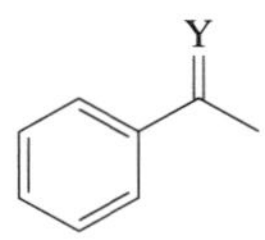	81	97	94
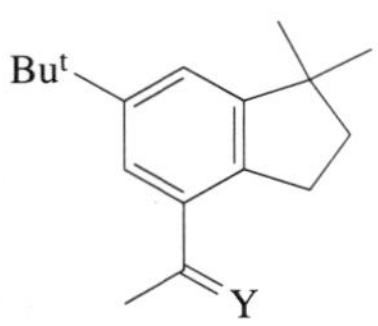	84	85	
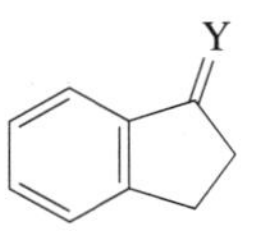	71		
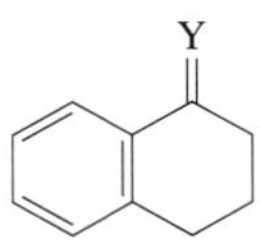	95		

An alternative method of reduction is with lithium aluminum hydride (Equation (20)). Unlike most other hydrazone reductions, the reaction proceeds at room temperature under non-acidic conditions, thus furnishing a mild pathway to the methylene group ⟨91TL1691⟩. Several substrates have been examined (Table 15) and the reaction gives moderate (44%) to good (81%) yields. The reaction utilizes an *N*-aminoaziridine to form the requisite hydrazone. This method has also been used to reduce arylhydrazones derived from aldehydes to the corresponding methyl compound ⟨58CB2383⟩; alkyl substituted benzylamines were produced (Equation (21)).

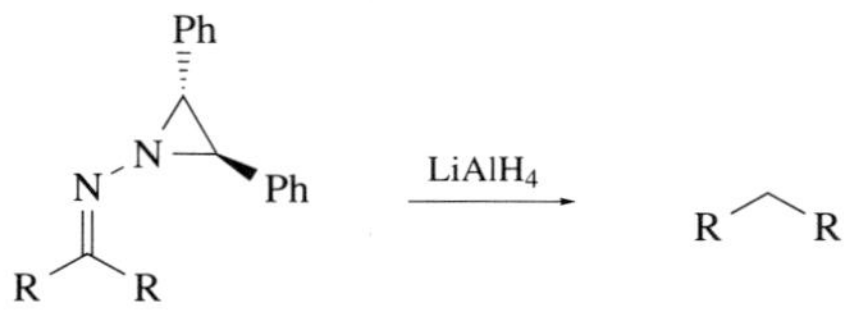

(20)

Table 15 Reduction using lithium aluminum hydride.

Hydrazone of	*Yield* (%)
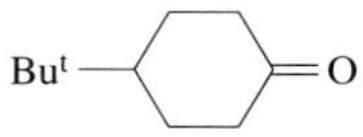	81
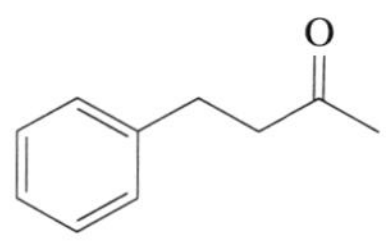	59
	48
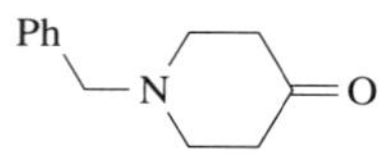	44

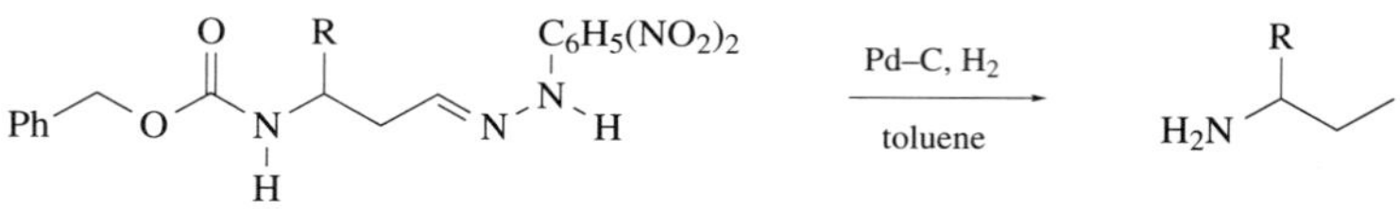

(21)

R = Ph or p-MeOC$_6$H$_5$

1.02.1.2.3 Reduction of diazo systems

Diazocompounds can be reduced to the corresponding methylene compounds using 5% palladium on charcoal and hydrogen in an autoclave at room temperature for one hour (Equation (22)) ⟨85JCS(P1)493⟩. A better reducing system (in terms of yields), developed by the same research team, involves the use of $Rh_2(OAc)_4$ (Equation (23)). However this reagent also catalyses a rearrangement which converts the neighbouring hydroxyl group to a carbonyl group ⟨79CC959⟩.

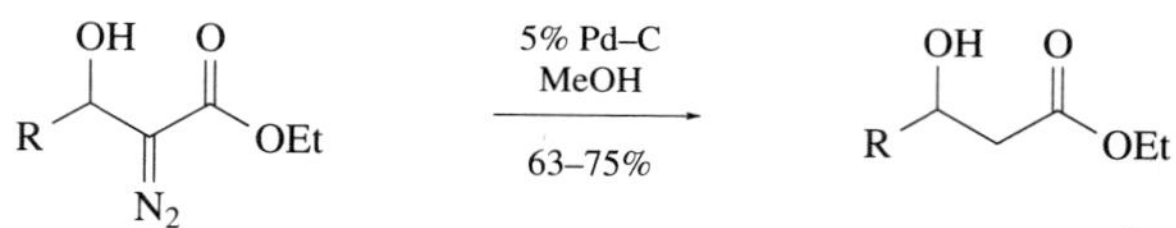

(22)

R = Me, Pr^i, n-C$_7$H$_{15}$, Ph, 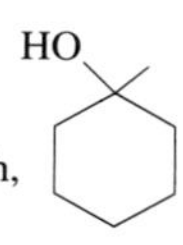

or 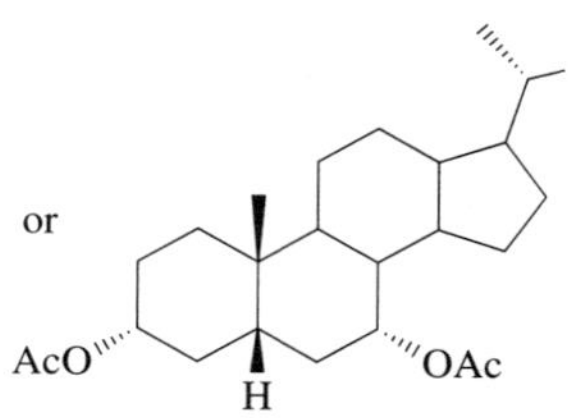

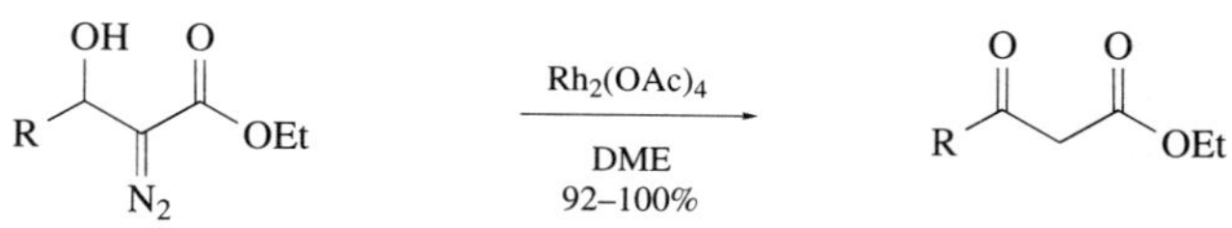

(23)

R = Pr^i, Pr^n, Ph$_2$CH, CH$_2$=CH, PhCH=CH or Ph

Several diazoketones have been reduced to the corresponding methyl ketone ⟨54JCS3617, 81CPB3489, 90SL465⟩. Many of the conversions are carried out by means of the standard Wolff method as shown in Scheme 27 ⟨43JA1516⟩. The reduction of diazoketones has been used to produce variants on antimalarial drugs ⟨46JA1813⟩.

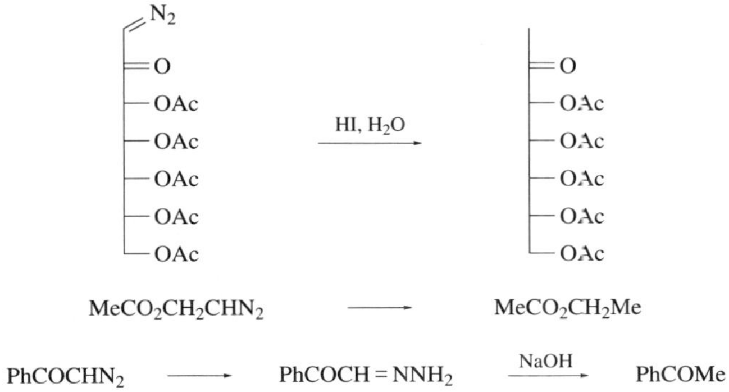

Scheme 27

1.02.1.2.4 Reduction of N*-alkylimine type systems*

Several methods for the deamination of the *N*-alkylimine type of system are found. The first is just an extension of the Haung–Minlon adaptation of the Wolff–Kishner reduction mentioned in Section 1.02.1.2.2 ⟨46JA2487⟩, an example of which is shown in Equation (24) ⟨66JCS(C)425⟩.

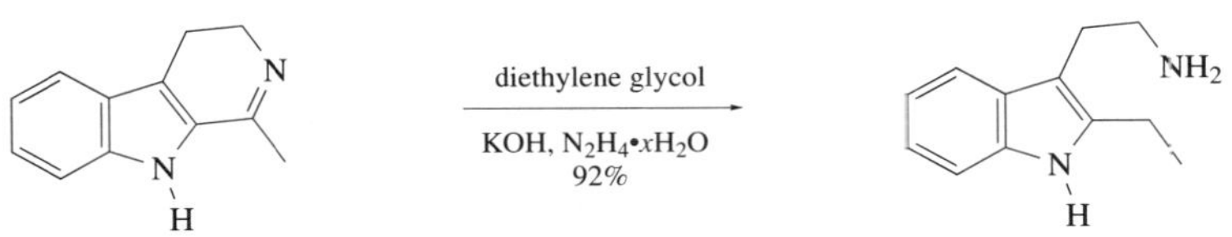

(24)

The second is by use of hydrides. Several amidines have been reduced using hydride reagents, such as lithium aluminum hydride (Table 16) ⟨87S719⟩ and lithium borohydride (Equation (25)) ⟨70KGS495⟩.

Table 16 Reduction using lithium aluminum hydride with variation of substituent R for Equation (25).

R	*Yield* (%)
Ph	50
4-$MeOC_6H_5$	72
Cyclopentyl	50
Cyclohexyl	66
Et	14
$CH_3CH(C_2H_5)$	56
Cyclopropyl	45
Cycloheptyl	44

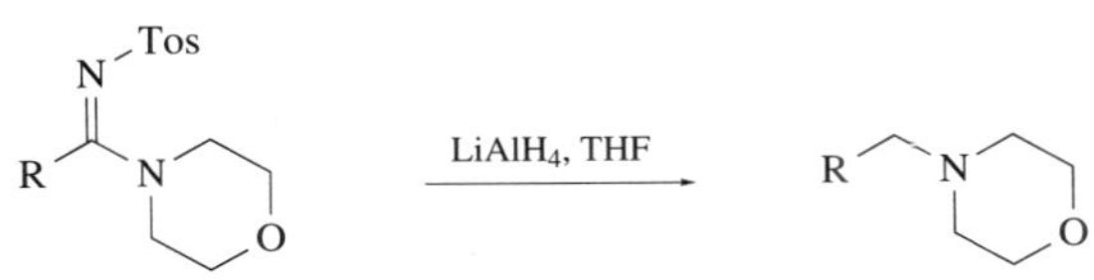

(25)

Sodium borohydride in ethanol has also been used in the reduction of diarylimines to the corresponding diarylmethanes (Equation (26)). It is the *o*-hydroxydiarylimines, such as **(12)**, that produce the diarylmethanes and a mechanism has been proposed ⟨85IJC(B)59⟩.

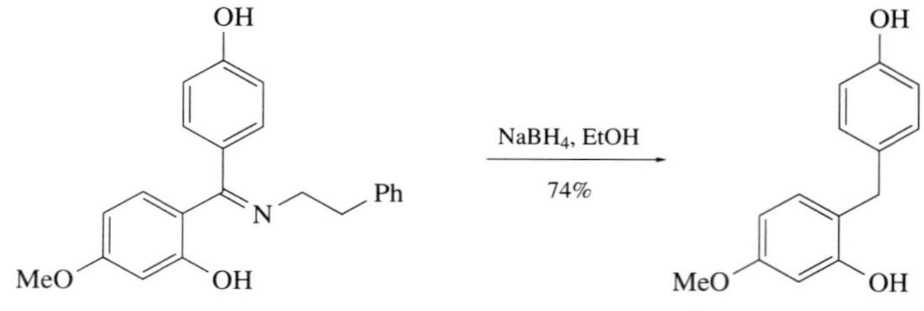

(26)

(12)

It has been shown that arylimines can be used to produce methyl substituted resorcinols ⟨43HCA800⟩ using a palladium catalyst. However the conditions employed, 120 °C for five hours and with the hydrogen pressure at 20 atm, are rather severe (Equation (27)).

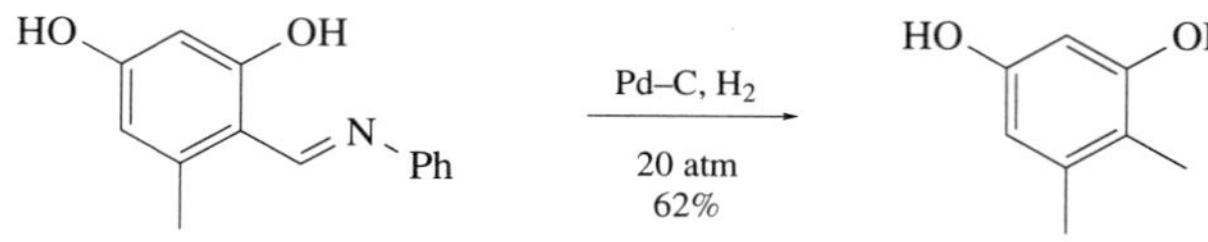

(27)

One isolated case of an imine being reduced to the corresponding methylene compound by the action of hydrogen sulfide in DMF at 20–65 °C has been reported (Equation (28)) ⟨69M724⟩.

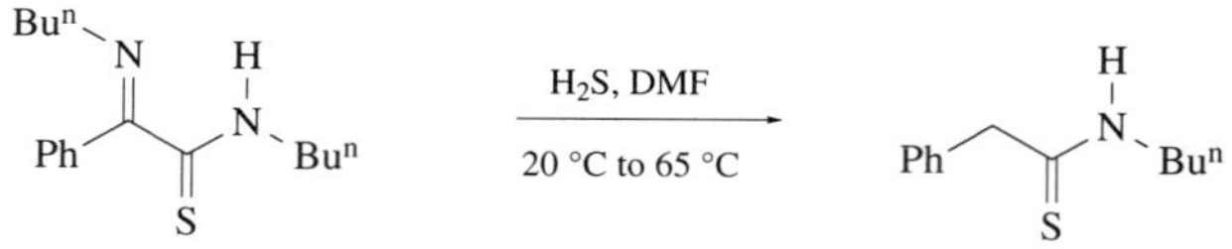

(28)

A decalin derivative was produced as shown in Scheme 28 ⟨75JOC3686⟩. The imine was produced from hydrazine and potasium hydroxide in triethyleneglycol. The reaction was left for one hour at 60 °C and then for nine hours at 210 °C (whereby a distillate was collected) then a further 12 hours at 210 °C. This procedure yielded the desired product.

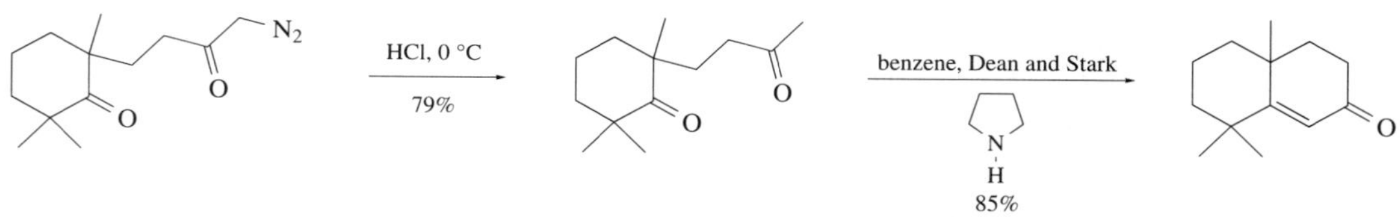

Scheme 28

The simple methylimine of benzaldehyde shown in Equation (29) was reduced to the *p*-methoxytoluene using sodium telluride ⟨88TL2571⟩. Some amine was also produced.

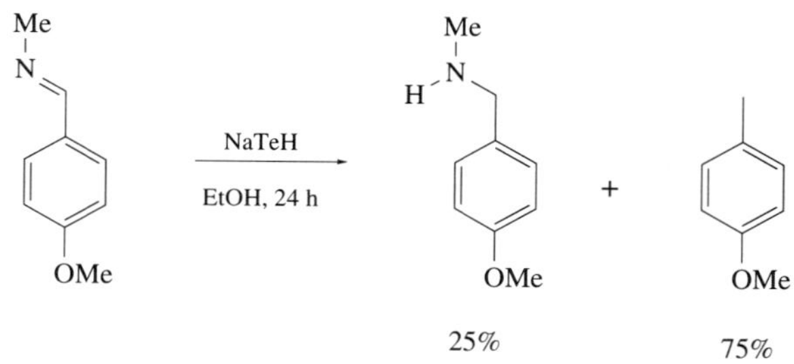

(29)

1.02.1.3 Reduction of Carbon–Nitrogen Triple Bonds to the Methyl Group

Three ways of reducing the nitrile group to a methyl group are known. Reduction using a source of hydrogen and a catalyst, which may either consist of palladium or Raney-nickel, can be used.

Associated with this method are the procedures which use metal catalysts supported by oxides such as silica. Finally the reduction can be achieved using a hydride reagent.

1.02.1.3.1 Reduction using palladium catalysts

The first example in this category was the result of an accidental discovery. The real aim of the experiment ⟨87T3817⟩ was to debenzylate the 3-indolecarbonitrile using palladium on carbon and hydrogen (Equation (30)). The reaction was left for one week; the only product formed was the 5,6-dimethoxy-2,3-dimethyl-1-(phenylmethyl)-1*H*-indole (85%).

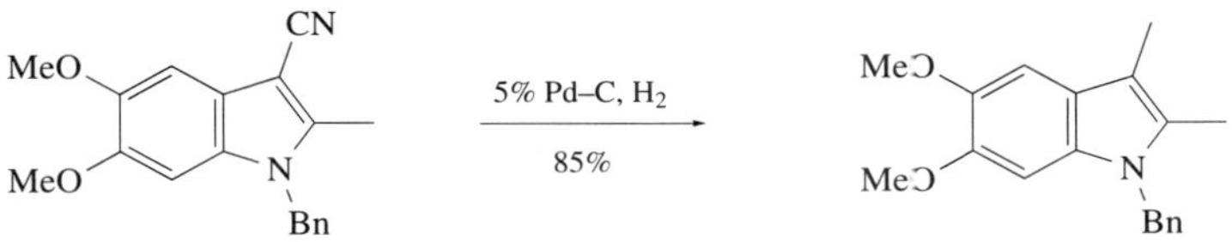

(30)

An earlier investigation ⟨66CB227⟩ examined the use of terpene hydrogen donors with palladium on carbon catalysts (Equation (31)). It was observed that aromatic nitriles were readily reduced to the corresponding methyl compounds. With the exception of three examples, all of the 21 nitriles tested could be reduced to the methyl analogue in under two hours, with half the reactions giving yields between 85% and 90%, the others giving yields in the range 70% to 85%. Temperatures needed for the reaction varied from 90–315 °C.

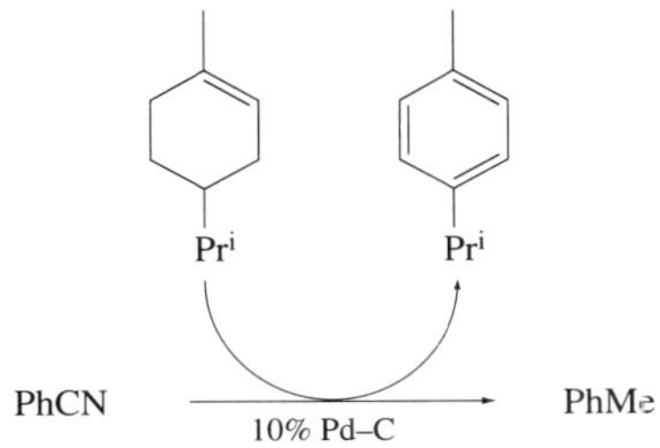

(31)

Ammonium formate has also been used as the source of hydrogen, using palladium on carbon as the catalyst (Equation (32)) ⟨82S1036⟩.

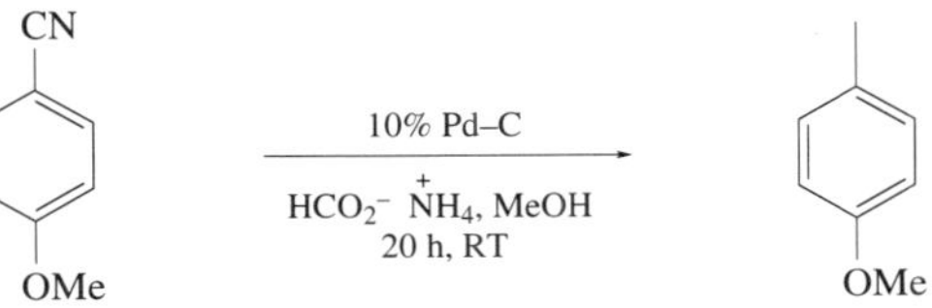

(32)

The reduction of several nitriles using Raney-nickel has been achieved by two separate methods. The first uses Raney-nickel in ethanol ⟨87S846⟩. Here the nitrile, ethanol and Raney-nickel are simply mixed and heated to reflux temperature for a period of five hours (Equation (33)). Yields of around 70% were achieved.

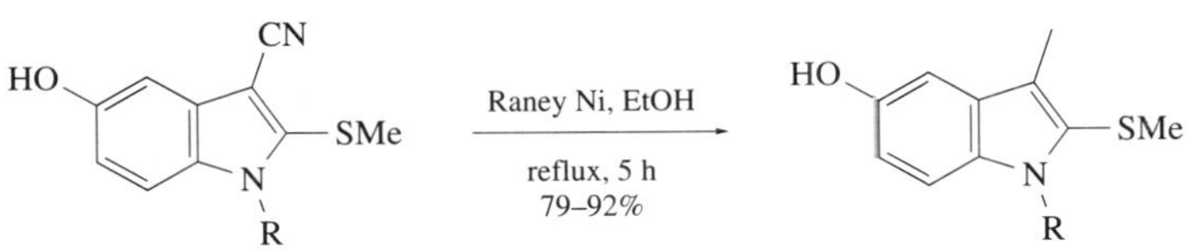

(33)

R = Ph, 4-ClC_6H_5, 4-$MeOC_6H_5$, $C_6H_5CH_2$

The second uses Raney-nickel and hydrogen as the reductant. The 2-cyano-4-methylamino-6-hydroxy-*s*-triazine (**13**) with a catalytic amount of Raney-nickel and sodium acetate in acetic anhydride, was hydrogenated at 2.6 atm and 50 °C for one hour ⟨62JA3744⟩ (Equation (34)). The product, 2-methyl-4-methylamino-6-hydroxy-*s*-triazine (**14**) was obtained in 66% yield.

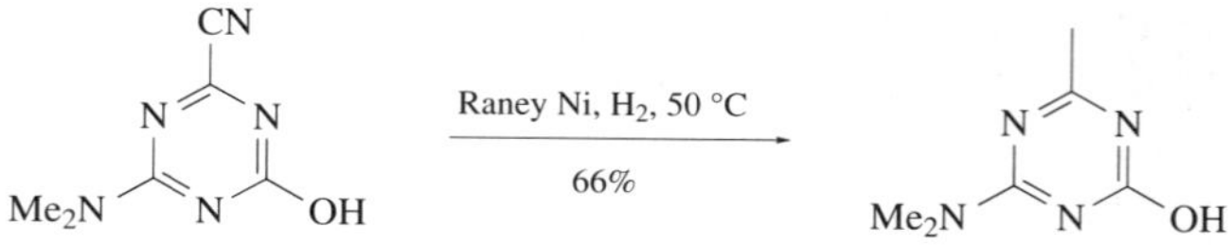

(13) **(14)** (34)

1.02.1.3.2 Reduction using oxide supported catalysts

In one of the first investigations into the use of metals and oxides to reduce nitriles to the corresponding methyl analogue ⟨59BCJ861⟩ *p*-amino benzonitrile (**15**) was heated under nitrogen to 330 °C in the presence of NiO on a Cu–SiO_2 support, to produce *p*-aminotoluene (**16**). Another example, using the same catalyst, involved the reduction of cinnamonitrile (**17**) to *β*-methylstyrene (**18**) (Scheme 29). Some years later several nitriles were reduced over a platinum-on-silica catalyst ⟨84JOC2875⟩. Although the yields for this method were, in most cases, greater than 94% (Table 17), there are two major drawbacks to its use. First, it requires a special flow apparatus, which for a one-off experiment would be prohibitive, and second the catalyst requires a high percentage of platinum, thus making the method costly. Another flow method was used to reduce cyano-adamantane (Equation (35)) ⟨79AG(E)939⟩. In this case the support was aluminum oxide and the metal was nickel. As in the previous example the yield was high (99%).

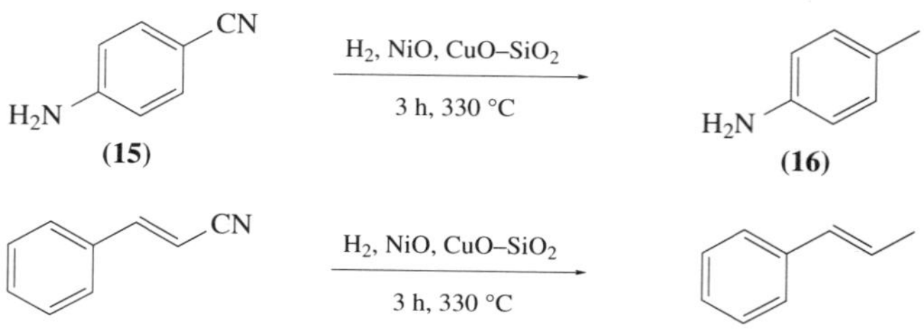

(15) **(16)** **(17)** **(18)**

Scheme 29

Table 17 Reduction of nitriles over platinum on silica catalysts with hydrogen.

Nitrile	*Product*	*Yield (%)*
Decanenitrile	Decane	98
Cyclohexane-carbonitrile	Methylcyclohexane	99
1-Cyanoadamantane	1-Methyladamantane	99
1-Adamantane acetonitrile	1-Ethyladamatane	94
Benzonitrile	Methylcyclohexane	75

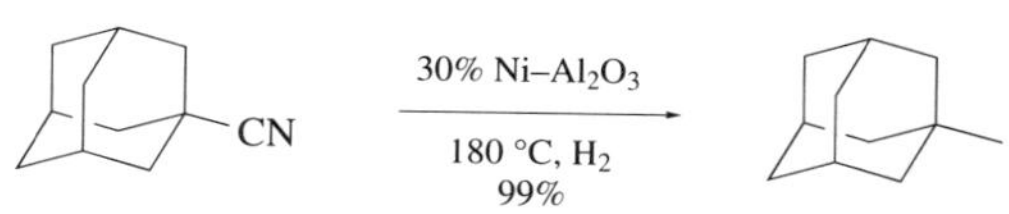

(35)

1.02.1.3.3 Reduction using hydride reagents

The use of hydrides for the reduction of nitriles to the methyl analogue seems to have been scarcely investigated to any degree. Sodium borohydride in acetonitrile was used to reduce *p*-cyanotoluene to the corresponding xylene (Equation (36)) ⟨78JCS(P1)1133⟩.

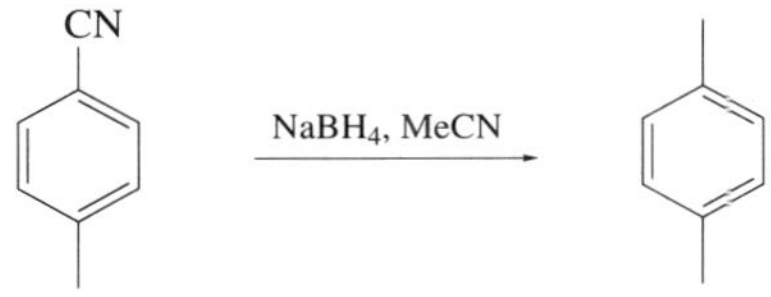

(36)

Sodium bis(2-methoxyethoxy)aluminum hydride was used to reduce the nitrile (**19**) to its methyl analogue in 82% yield (Equation (37)). The reaction seems to have been performed purely to obtain NMR data ⟨86T2429⟩. In an elegant synthesis of the natural product laurenene (**20**) (Equation (38)) ⟨88JOC477⟩, a nitrile was reduced to a methyl group in 65% yield using diisobutylaluminum hydride (dibal-H) in dry benzene over a period of one hour.

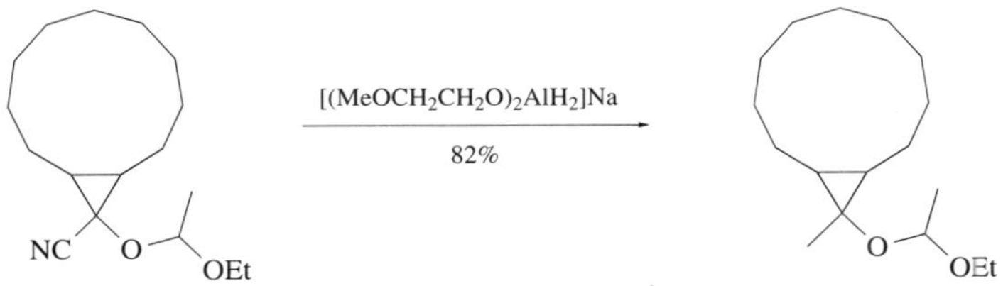

(**19**)

(37)

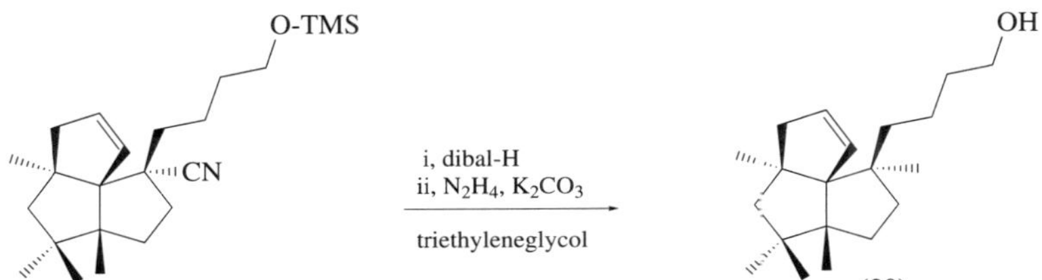

(**20**)

(38)

1.02.2 REDUCTION OF CARBON–PHOSPHORUS, –ANTIMONY AND –BISMUTH BONDS TO CARBON–HYDROGEN BONDS

1.02.2.1 Reduction of Carbon–Phosphorus Bonds

The reduction of carbon–phosphorus bonds to carbon–hydrogen single bonds or a methylene group can be segregated into two separate approaches. The first involves the use of a phosphorus moiety as an incidental unit in the overall reaction scheme. In other words the cleavage of the carbon–phosphorus bond leaves a molecule that contains no phosphorus and which is either the desired product or a substrate to be used further in the reaction pathway. The second and more common case is where the resultant molecule containing phosphorus, after carbon–phosphorus bond cleavage, is the desired product.

1.02.2.1.1 Carbon–phosphorus bond cleavage to give a desired product not containing phosphorus

(i) Cleavage by strong bases

One of the earliest methods for converting phosphorus containing compounds to the corresponding hydrocarbon moiety involved the use of hydroxides or alkoxides. It had been reported that quaternary phosphonium halides were decomposed by hydroxide to hydrocarbon and phosphine oxide with cleavage of the most electronegative group ⟨29JCS2342⟩. A further investigation into this reaction involved the phosphonium bromide (**21**) (Equation (39)) ⟨70JA2139⟩. Cleavage of the allylphosphonium bromide with hydroxide or *t*-butoxide resulted in a 3:1 mixture of the two isomers (**22**) and (**23**) (Equation (39)). There was however significant competition between allyl and phenyl as the leaving group.

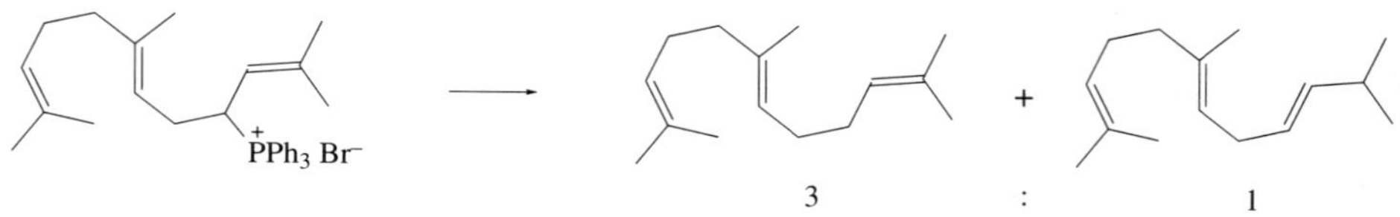

(21) **(22)** **(23)** (39)

Alkaline induced cleavages of benzyl[2-(*N*,*N*-dimethylamino)phenyl]diphenylphosphonium bromide (**24**) and benzyl[4-(*N*,*N*-dimethylamino)phenyl]diphenylphosphonium bromide (**25**) have been studied (Scheme 30). The former undergoes decomposition to give *N*,*N*-dimethylaniline 96.5% and benzene 3.5%, while the latter gives toluene (100% yield) in a reaction that is approximately 1000 times slower ⟨87JOC4829⟩.

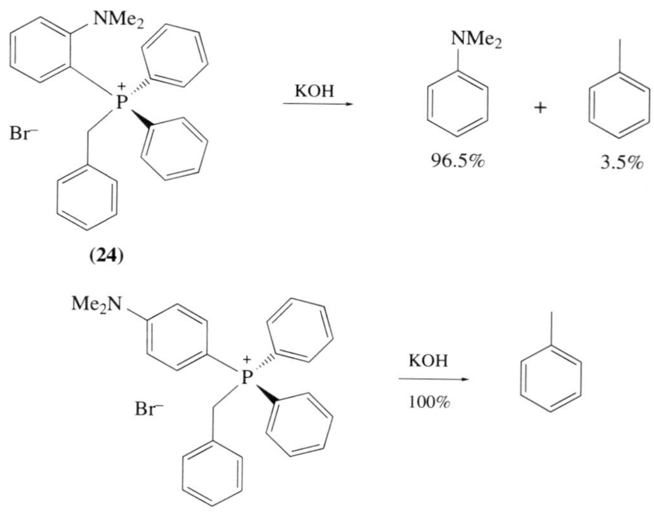

Scheme 30

An equally interesting divergence was seen between benzylbis[2-(*N*,*N*-dimethylamino)phenyl] phenylphenylphosphonium bromide and benzylbis[4-(*N*,*N*-dimethylamino)phenyl]phenylphosphonium bromide; the first compound gives $PhNMe_2$ in 94.2% yield and benzene in 1.2% yield, with no toluene being detected, while the second reaction yields only toluene in quantitative yield. An attempt at rationalizing the differences has been made, but no concrete conclusions were drawn.

Investigations have been carried out into the reaction of vinyl phosphonium salts ⟨79TL349⟩. Conditions for preferential benzene or alkene elimination were studied in detail. As shown in Scheme 31, sodium hydroxide in a water/ethanol solvent gives the alkene, whilst the same hydroxide in a water/benzene mixture produces benzene.

The mechanisms for simple alkaline decomposition of acyclic quaternary phosphonium salts is

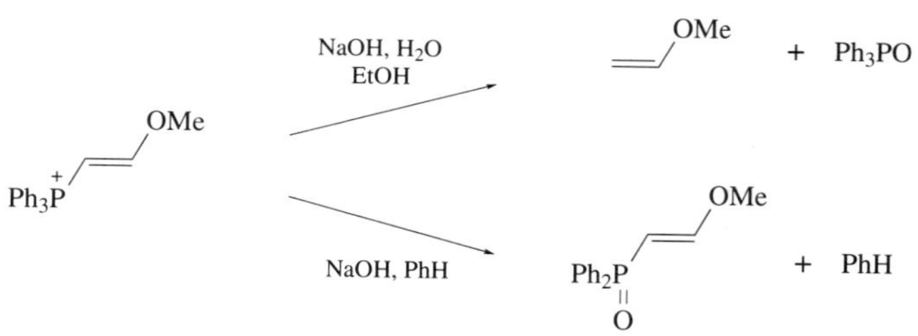

Scheme 31

well understood and documented ⟨B-73MI 102-01, 78CJC1933, 78JA7312, 83JCS(P2)1923, 86TL1209⟩. The basic mechanism is detailed in Scheme 32.

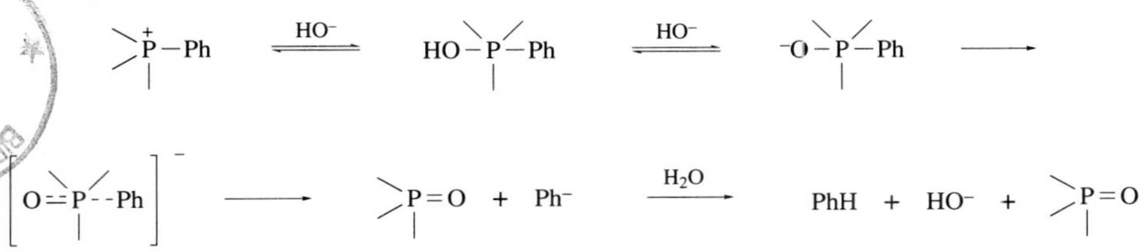

Scheme 32

Other bases can be used to cleave carbon–phosphorus bonds. The tetraphosphine allene (**26**) in Scheme 33 can have a phosphine unit cleaved off using *n*-butyllithium as the base. This results in the lithium allene anion adduct being formed, which can then react with electrophiles. The tetraphosphine allene is simply dissolved in THF at $-70\,^{\circ}\mathrm{C}$ and *n*-butyllithium is added. The electrophile is then introduced to this reaction mixture (Scheme 33) ⟨87CB1403⟩. A similar cleavage has been observed using a lithium/amine mixture ⟨55JA3230⟩.

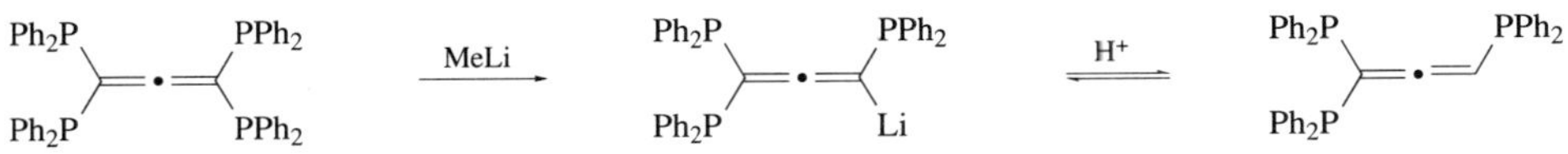

(**26**)

Scheme 33

(ii) Cleavage by hydrides

Lithium aluminum hydride has also been used for carbon–phosphorus bond cleavage ⟨57JA3567, 70JA2139⟩. When lithium aluminum hydride was used for the reduction of the phosphonium bromide (**21**) (Equation (39)) only (**23**) was produced. Lithium aluminum hydride has also been used in another reaction which is formally a reduction of a phosphorus compound, as illustrated in Equation (40) ⟨57JA3567⟩.

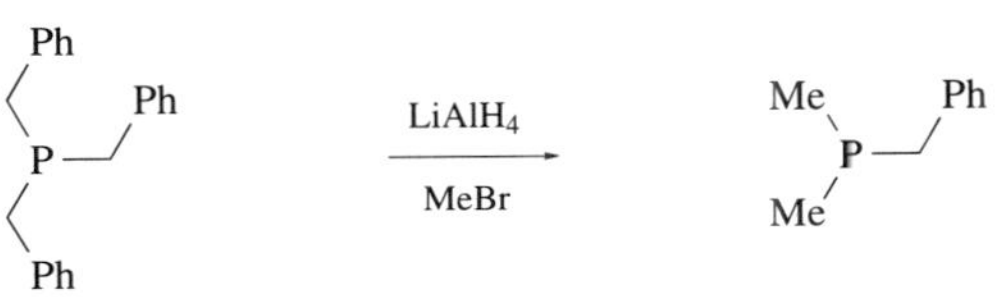

(40)

(iii) Cleavage by metal-based reductants

Acyl ylides have been reduced successfully using aluminum amalgams ⟨64JA1639, 82JOC4963⟩. Of several compounds tested, all gave greater than 81% yields. The reductions were conducted typically at room temperature, with an excess of the aluminum amalgam and with a periodic addition of acid (Scheme 34). This is an excellent method for the preparation of *β*-keto esters.

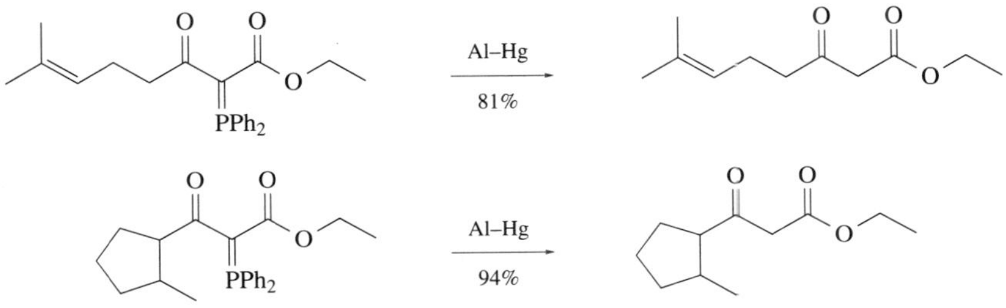

Scheme 34

Other metal based reductions of phosphoranes are found. One such reduction uses a zinc-acid mix as the reagent ⟨55JA3230, B-79MI 102-01⟩. An important industrial process which relies on the cleavage of carbon–phosphorus bonds ⟨84OM(3)923⟩ is the rhodium-promoted hydroformylation process which produces benzaldehydes and phenols. Scheme 35 outlines the general procedure.

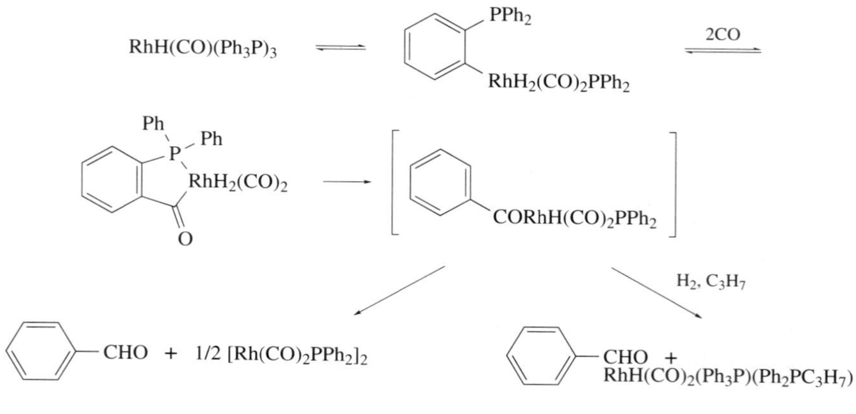

Scheme 35

1.02.2.1.2 Carbon–phosphorus bond cleavage to give a desired product containing phosphorus

Essentially this section can be sub-divided into two parts. In the first part loss of a group attached to a tertiary phosphine forms a phosphorus nucleophile or reagent. The second group is where units are lost from the phosphorus atom to form another more unusual phosphorus species.

(i) Formation of phosphorus nucleophiles and reagents

There are many methods for the formation of tertiary phosphines, Ph_2PR, where R is alkyl or aryl ⟨76MI 102-01⟩. Most of these methods cannot be utilized conveniently because they have difficult working procedures or the yields from the reactions are poor. One method exists ⟨76MI 102-01⟩, however, that is undoubtedly the most convenient, both in terms of the availability of starting materials and of the simplicity of the practical procedure. It is the reaction of lithium diphenylphosphide with an alkyl or aryl halide. This method is generally applicable to a wide range of phosphines as is shown in Scheme 36. It offers high yields, especially when the phenyllithium produced on cleavage of triphenylphosphine by lithium metal is selectively destroyed by means of *t*-butyl chloride.

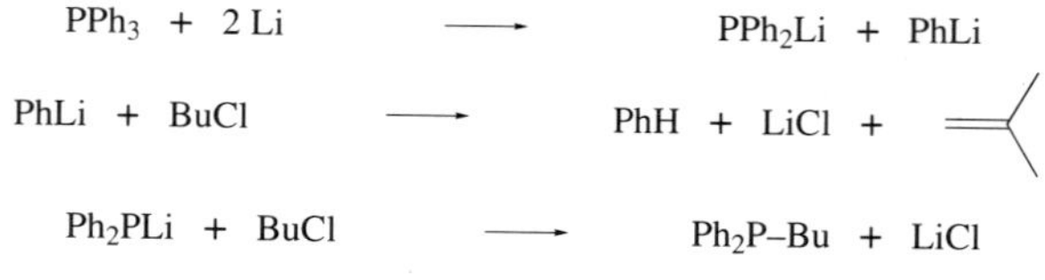

Scheme 36

Diphosphines of the class $RPhP(CH_2)_nPPhR$ ($n = 2$–6), have been prepared from $Ph_2P(CH_2)_nPPh_2$ with alkali metals followed by alkylation (Scheme 37). It was stressed that ultrasound irradiation should be carried out at the reductive cleavage stage, so as to assure the purity of the final products and give good yields, in excess of 80% in most cases ⟨85JOC4329⟩.

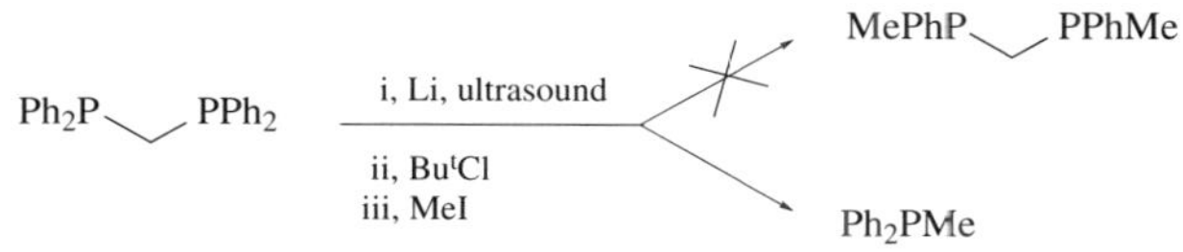

Scheme 37

However not only triphenylphosphine carbon–phosphorus bonds can be cleaved using alkali metals. There has been significant work on 1,1′-diphosphometallocene chemistry (Equation (41)) which is involved in producing 1*H*-phosphol-1-yl anions. In doing so, it is often a carbon–phosphorus bond which is broken to give an anion and hydrocarbon ⟨80JA994⟩. Various procedures for the creation of this type of anion have been reported. These include using lithium in THF at room temperature ⟨71JA6205, 71T5523⟩, potassium in boiling THF, dioxane or toluene and sodium in boiling THF ⟨74MI 102-01⟩. An example of a typical reaction ⟨80JA994⟩ is given in Equation (42).

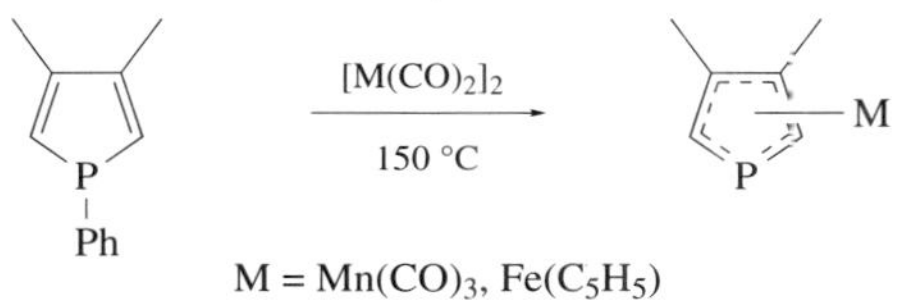

(41)

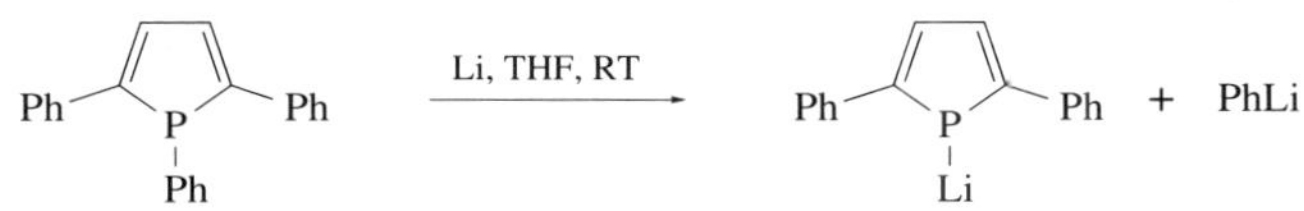

(42)

(ii) Reduction of carbon–phosphorus bonds in order to study properties of the reaction rates and chirality about phosphorus

It was mentioned in Section 1.02.2.1.1.(i) that the mechanisms for simple alkaline decomposition of acyclic quaternary phosphonium salts is well understood and documented and references to this literature were given in that section. The discussion of chirality about phosphorus in cleavage reactions has also been extensively documented ⟨75CB803, 91COS(8)858⟩.

If the rate of reaction is being studied, then both those products containing phosphorus, and more importantly for this discussion, those not containing phosphorus are of interest. However, these studies rarely produce reactions that are of synthetic value, and as such they will not be dealt with here.

1.02.2.2 Reduction of Carbon–Arsenic Bonds

When compared with phosphorus, the reduction of carbon–arsenic bonds to carbon–hydrogen bonds has been severely neglected. There are very few examples of this type of reaction in the literature.

As with certain phosphorus compounds (Section 1.02.2.1.1.(i)), it has been shown that an alkyl group on arsenic can be removed by addition of butyllithium ⟨88JA4346⟩. The example given in Scheme 38 was carried out in an attempt to produce optically active arsenic containing macrocycles.

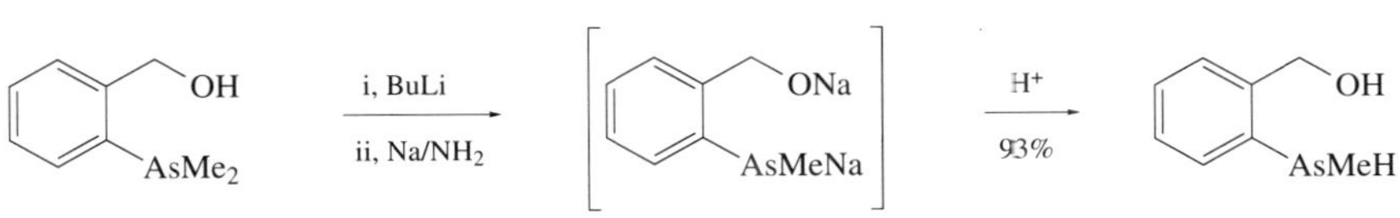

Scheme 38

A further example of reduction of a carbon–arsenic bond is where arsonium ylides of the type shown in Scheme 39 are treated with hydrogen chloride in dichloromethane at room temperature to form *β*-keto esters or *β*-keto ketones ⟨86MI 102-01⟩.

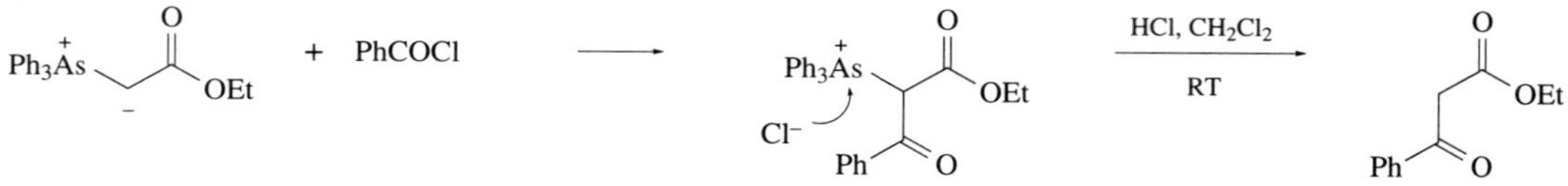

Scheme 39

1.02.2.3 Reduction of Carbon–Antimony Bonds

It is also possible to form ylides containing antimony. These are formed on reaction of a diazo compound with triphenylantimony. The resulting ylides are called stilbonium ylides. Until 1986 only one of these ylides existed (**27**) (Equation (43)) ⟨67CI(L)787⟩.

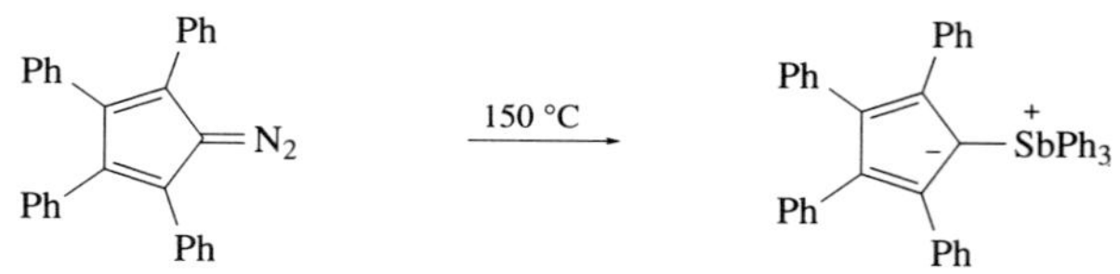

(**27**)

(43)

Further work using copper catalysts has resulted in the isolation of several more species ⟨86T3887⟩. In protic solvents these ylides are slowly cleaved to give triphenylantimony oxide and the corresponding methylene compound (Equation (44)).

(44)

Antimony compounds containing the $Sb(CF_3)_2$ group can loose this moiety on photolysis over 100 h ⟨88ZAAC(560)141⟩. As shown in Equation (45), two compounds are formed resulting from the simple replacement of $Sb(CF_3)_2$ by hydrogen and transformation of the CF_3 group into a C_3F_7 unit.

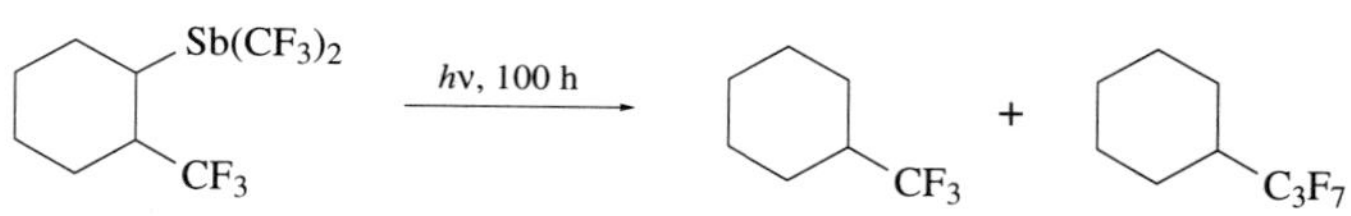

(45)

1.02.2.4 Reduction of Carbon–Bismuth Bonds

There appears to be only one reaction described where there is cleavage of a carbon–bismuth bond to give the corresponding hydrocarbon. Bis(diphenylbismutino)methane was synthesized in 1985 ⟨85CB1039⟩. It is possible to remove one of the diphenylbismuto groups using phenyllithium. This leaves an anion as shown in Scheme 40 which can, on addition of methanol, give the methyldiphenylbismutine.

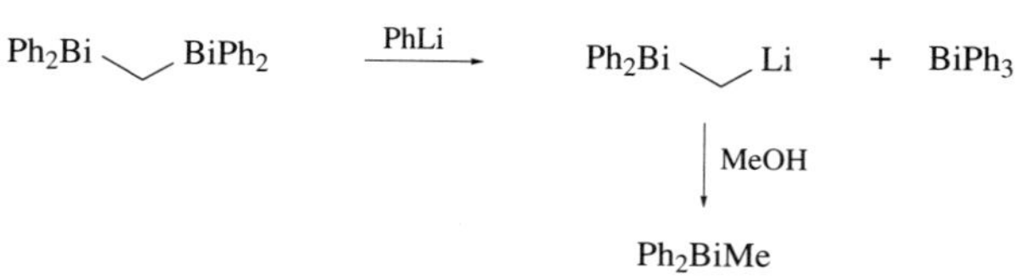

Scheme 40

1.02.3 REDUCTIVE SCISSION OF A CARBON–CARBON BOND TO TWO CARBON–HYDROGEN BONDS

There are two types of reaction which dominate this section. The first is the reductive cleavage where one carbon atom is that of a nitrile group. Here a carbon–nitrile bond is cleaved to furnish a carbon-hydrogen bond. The second reaction is the direct scission of an alkyl or aryl carbon–carbon bond to give two new hydrocarbons.

1.02.3.1 Reductive Cleavage when one Carbon is Part of a Nitrile Group

There are two methods of cleaving a nitrile to give the corresponding hydrocarbon and, formally, hydrogen cyanide, namely hydride reagents, or alkali metals in liquid ammonia.

1.02.3.1.1 Use of hydride reagents

The conversion can be accomplished using either lithium aluminum hydride or sodium borohydride. The 2-cyano-1,2-dimethylpiperidine (**28**) was transformed into the corresponding dimethylpiperidine using lithium aluminum hydride (Equation (46)) ⟨57JA5279⟩. Previous reports ⟨54MI 102-01, 54MI 102-02, 63BSF1428⟩ on the reduction of carbon–nitrile bonds using lithium aluminum hydride only gave satisfactory results in special cases.

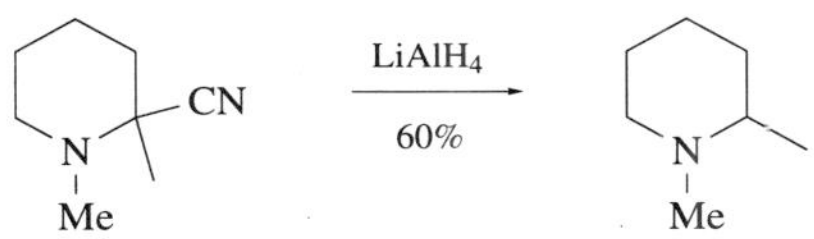

(28)

(46)

In studies relating to the synthesis of isoquinoline- and indole-alkaloids, the removal of nitrile groups using sodium borohydride was investigated because both reduction of the nitrile group to an amine unit *and* cleavage to the hydrocarbon was observed when lithium aluminum hydride was used, and a milder cleavage reagent was sought. The authors succeeded in removing the nitrile unit from various types of α-amino acids (Scheme 41) ⟨69TL3105⟩.

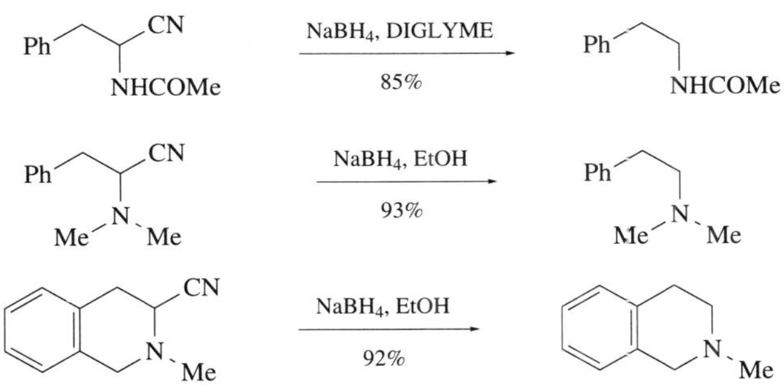

Scheme 41

1.02.3.1.2 Use of alkali metals in liquid ammonia

Sodium in liquid ammonia was investigated as a reducing reagent for the cleavage of nitriles in connection with the synthesis of (+)-maritidine (**29**) (Equation (47)) ⟨76TL57⟩. The nitrile was stirred for 15 minutes in THF at $-78\,^{\circ}C$ with sodium in liquid ammonia and a modest yield of (**29**) (58%) was obtained. As a prelude to this reaction several model studies were carried out ⟨76TL61⟩. Applying the same conditions to some simpler model systems gave higher yields (Scheme 42).

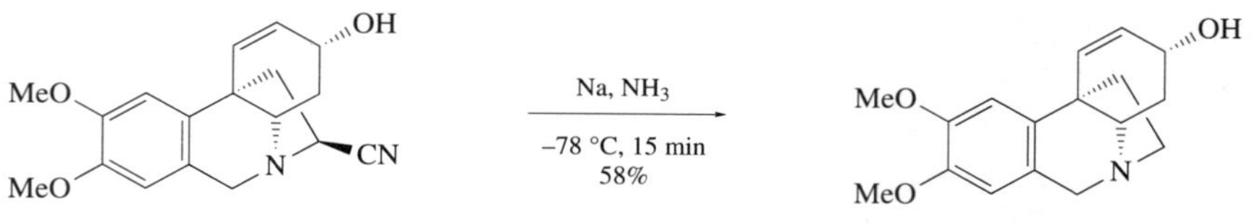

(29)

(47)

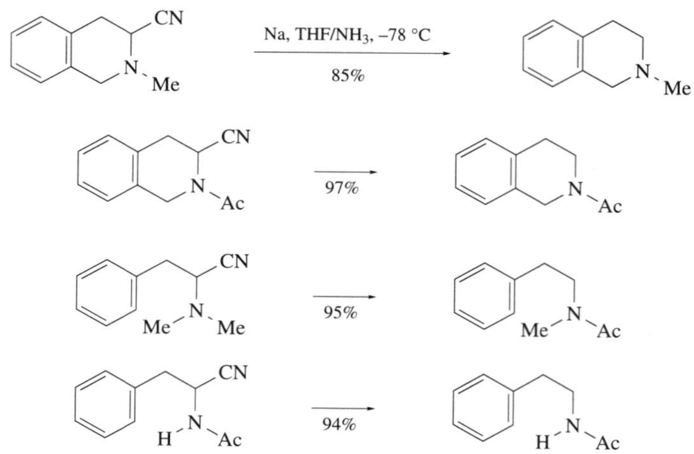

Scheme 42

1.02.3.2 Cleavage of Carbon–Carbon Bonds where both Products are Hydrocarbons

In the majority of cases these cleavage reactions result in the formation of at least one new aryl derivative. On the whole, the reaction conditions are harsh and the molecules involved are simple.

1.02.3.2.1 Cleavage using aluminum or aluminum/metal catalysts

It is possible to dealkylate alkylbenzenes over alumina/metal catalysts. The metals that have been used are rhodium and nickel. At 430 °C the dealkylation of toluene to methane and benzene is extremely effective. The dealkylation of xylene with a steam covered rhodium/aluminum oxide catalyst gives toluene in 93% yield and is much faster than the dealkylation of toluene itself (Scheme 43) ⟨71MI 102-01⟩.

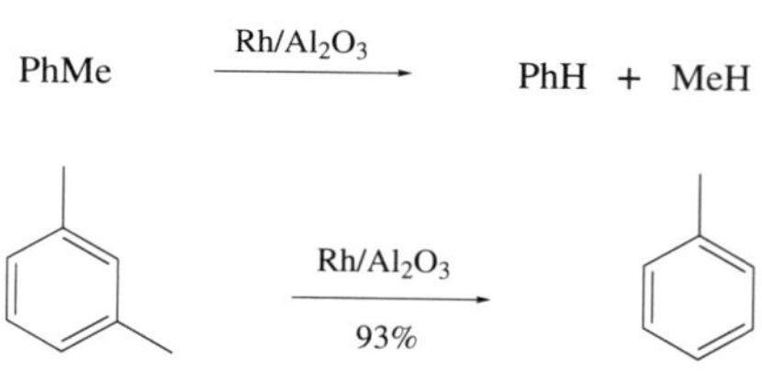

Scheme 43

In a rather harsh reaction (Equation (48)), toluene was converted into benzene in 95% yield using a cobalt/molybdate/alumina catalyst in the presence of sodium hydroxide and steam at 560–600 °C and 70 atm ⟨58IEC1677⟩. A similar catalytic system using nickel rather than rhodium has been studied. Various isomers of picoline, cresol and xylene were demethylated in the presence of water vapour and a nickel/aluminum oxide catalyst at temperatures between 400 °C and 440 °C. The results showed that the greater the deviation an isomer had from aromaticity the higher reactivity it possessed in the catalytic reaction ⟨65MI 102-01⟩.

$$\mathrm{PhMe} \xrightarrow[\text{70 atm, 560 °C; 95\%}]{\mathrm{Co/Mo/Al_2O_3}} \mathrm{PhH} + \mathrm{MeH} \qquad (48)$$

Lewis acids have been used in cleavage reactions. In particular aluminum chloride has been used to cleave the molecule shown in Equation (49). Aluminum chloride in benzene was shaken for five hours with the tetrachloride whereupon, 2,5-dichlorophenol was produced in 86% yield along with propane ⟨75OPP189⟩.

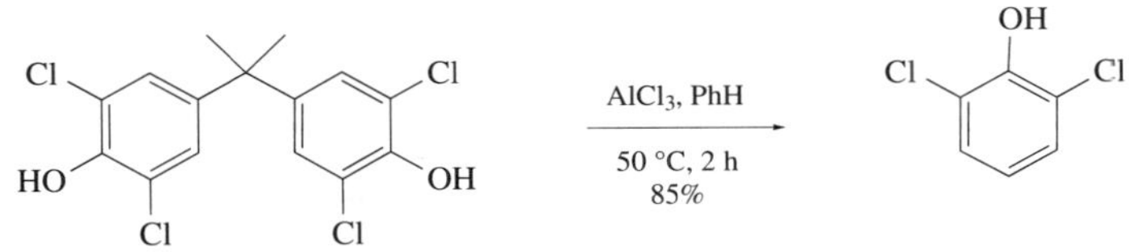

(49)

In an attempt to perform an intramolecular Friedel–Crafts reaction using compound (**30**), with aluminum chloride as the catalyst, an isopropyl group was cleaved (Equation (50)). The reaction is dependent on the nature, position and branching of substituents ⟨44BSF349⟩.

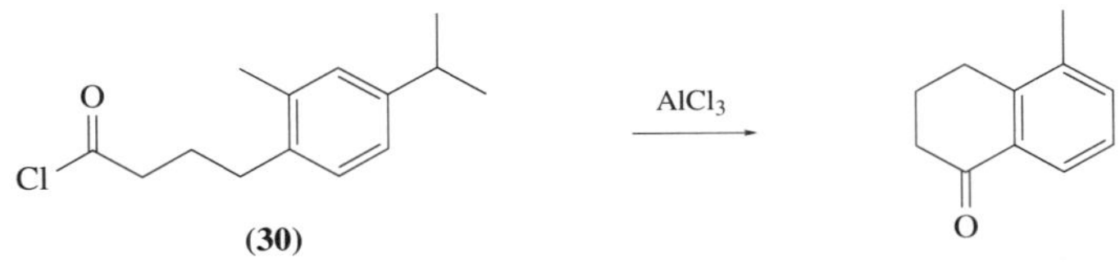

(**30**)

(50)

Lithium aluminum hydride has been used successfully to reduce the nitro compound featured in Equation (51) to give the corresponding benzylamine (90%) and nitro ester ⟨55LA177⟩. Dinitro compounds can also be cleaved using Raney nickel ⟨57CB1774⟩. A similar reaction was observed when the dinitro compound shown in Equation (52) was reacted with concentrated hydrochloric acid in ethanol. This reaction is claimed to have given a 100% yield ⟨58CB750⟩. The biphenyl (**31**) can be reduced to give toluene (Equation (53)) ⟨47CB202⟩.

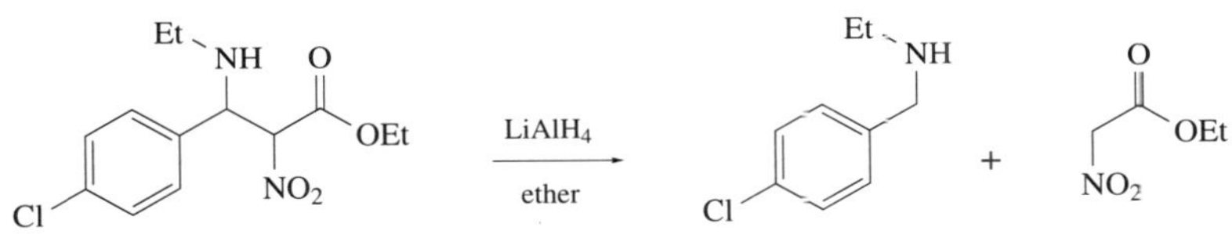

(51)

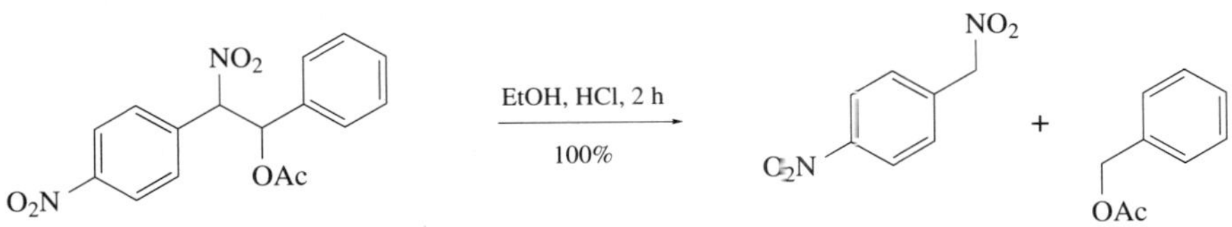

(52)

(**31**)

(53)

1.02.4 REDUCTION OF CARBON–BORON, –SILICON AND –GERMANIUM BONDS TO CARBON–HYDROGEN BONDS

1.02.4.1 Reduction of Carbon–Boron Bonds to Hydrocarbons

The cleavage of carbon–boron bonds can be effected under acid, neutral or basic conditions. The common factor in all the conditions employed is that the cleaving reagent contains an oxygen atom.

1.02.4.1.1 Cleavage of carbon–boron bonds under acidic conditions

Trialkyl boranes are relatively stable towards hydrolysis by water ⟨57CB2733⟩ and strong mineral acids ⟨86T5497⟩ although they can also be cleaved by anhydrous hydrogen fluoride (Equation (54)) ⟨72JA7859⟩. Compared to hydrogen fluoride, hydrogen bromide reacts slowly and removes only one alkyl group (Equation (55)) ⟨38JA115⟩.

$$R_3B + 3\,HF \xrightarrow{RT} 3\,RH + BF_3 \qquad (54)$$

$$R = Me, Et, Pr^i$$

$$Bu^n{}_3B + HBr \xrightarrow{55\text{–}60\,°C} Bu^nH + Bu^n{}_2BBr \qquad (55)$$

It has been known for several decades that one alkyl group is rapidly removed using acetic acid ⟨36JPR251, 55AG710⟩. A second alkyl group is cleaved at a much slower rate (Equation (56)) ⟨61JOC631⟩. This type of reaction has since been studied in some depth. A long chain fatty acid has been used to cleave a variety of boranes to the corresponding hydrocarbons. The results and yields of two of these reactions are detailed in Scheme 44 ⟨84JOM(260)17, 84JOM(270)9⟩.

$$(n\text{-}C_6H_{13})_3B + 2\,MeCO_2H \longrightarrow 2\,n\text{-}C_6H_{14} + n\text{-}C_6H_{13}B(O_2CMe)_2 \qquad (56)$$

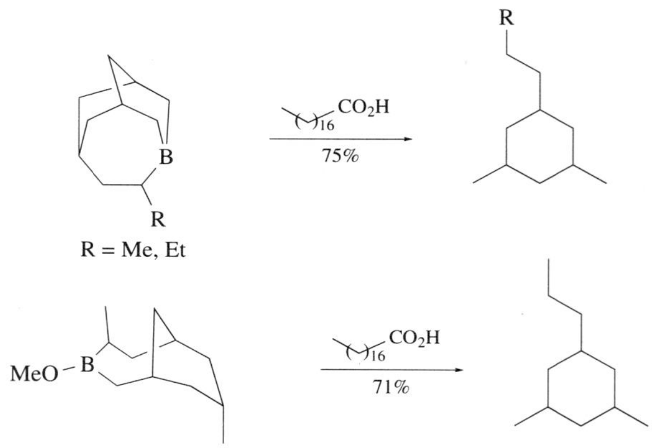

Scheme 44

The use of acids in carbon–boron bond cleavage was systematically studied in 1986, ⟨83JOM(255)135, 86T5497⟩ when ten different acids were used to investigate the cleavage of tri-*n*-hexylborane (Table 18 and Equation (57)). In the same publication several organoboranes were treated with acetic acid to yield the alkane according to the general reaction given in Equation (58). The full results are listed in Table 19. Further investigations involved changing the acid and using polyethers to aid the reaction (Equation (59) and Table 20). The last two entries in Table 20 are the results when isomerisation takes place before protonolysis.

Table 18 Reaction of tri-*n*-hexylborane with various acids.

Acid	*Yield of hexane* (%)
Acetic	60
Acetic + acetic anhydride	84
Propionic	80
Isobutyric	81
Butyric	80
Glycolic	55
Methanesulfonic	41
Hydrochloric	36
Phosphoric	12

(57)

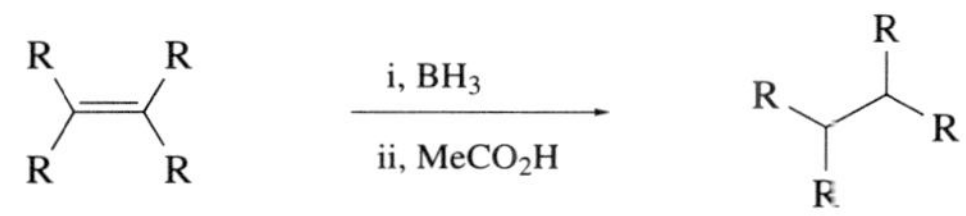

(58)

Table 19 Reaction of ethanoic acid with organoboranes derived from alkenes.

Alkene	*Yield of alkane* (%)
1-Pentene	50
1-Hexene	46
2-Methyl-1-butene	11
3-Methyl-1-butene	44
3,3-Dimethyl-1-butene	36
2-Methyl-2-butene	0.4
Styrene	50
α-Methylstyrene	32
trans-β-Methylstyrene	3
cis-2-Pentane	3
Cyclopentene	46
Cyclohexene	31
Norbornene	27

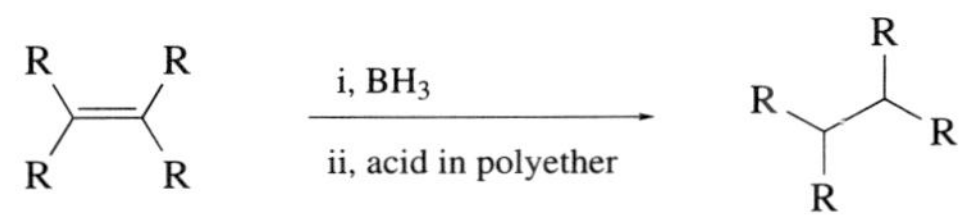

(59)

Table 20 The use of polyethers in cleavage reactions.

Alkene	*Yield of alkane* (%)	
	Proprionic acid in DIGLYME	*Caprylic acid in TRIGLYME*
1-Hexene	90	91
1-Octene	95	90
2-Hexene	50	
Cyclohexene	76	84
2,4,4-Trimethyl-1-pentene	82	
Styrene	88	84
2-Heptene	51	79
2,4,4-Trimethyl-2-pentene	5	
2-Heptene	85	90
2,4,4-Trimethyl-2-pentene	77	85

The conclusions drawn from this study on protonolysis were that the steric requirements of the alkyl groups attached to the boron play an important role in the rates of hydrolysis. Secondary alkyl groups protonolyze less readily than primary alkyl groups and more hindered alkyl groups are even more resistant. The stereochemistry of the protonolysis was established via deuterioboration of norbornene and dueterolysis of the product. The protonolysis was seen to proceed by retention of configuration at the migrating carbon atom.

1.02.4.1.2 Cleavage of the carbon–boron bond under neutral conditions

It is possible to cleave the carbon–boron bond in molecules such as (**32**) under neutral conditions using ethers such as THF and polyethers such as DIGLYME or TRIGLYME (Scheme 45) ⟨86T5497⟩.

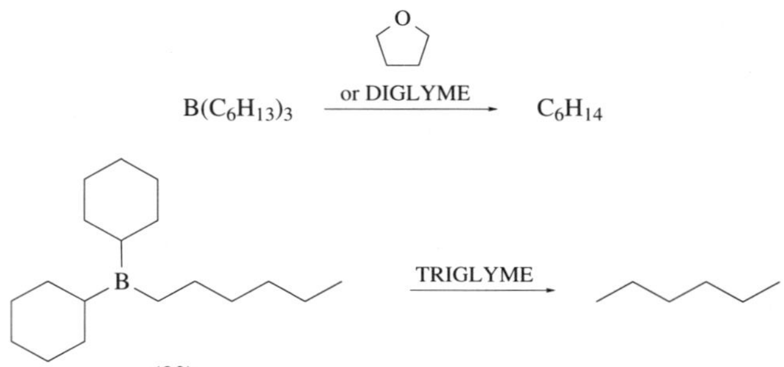

(32)

Scheme 45

Several organoboranes have been cleaved by alcohols. An interesting example of this is the use of the cyclic polyol octahydroxycyclobutane ⟨83CB1336⟩. A suspension of the polyol in mesitylene is reacted with triethylborane (activated with diethylboryl pivalate) to release eight equivalents of ethane and form the product (**33**) (Equation (60)).

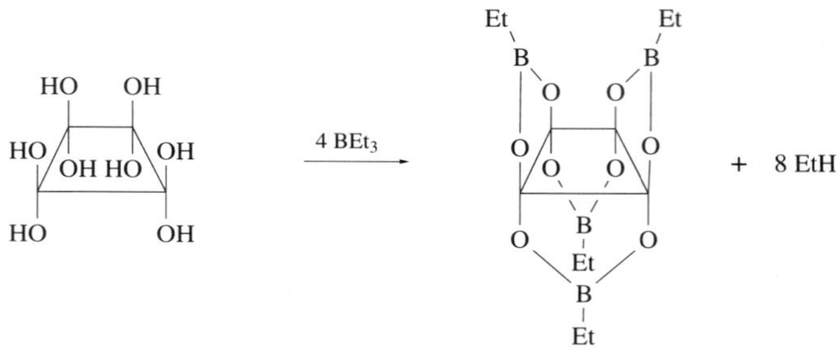

(60)

(33)

In a reaction of five-membered ring enamines with boranedimethyl sulphide with subsequent heating in methanol at 65 °C, some protonolyzed product was formed (Equation (61)). Other cases gave only the borate dimethyl ester ⟨87JOC4014⟩.

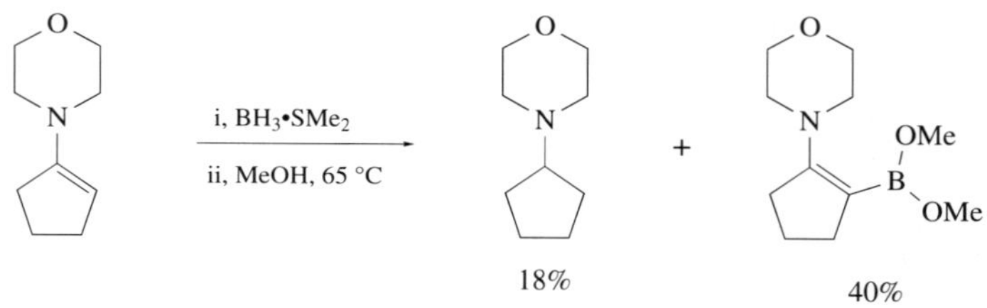

(61)

1.02.4.1.3 Cleavage of the carbon–boron bond with base

It has been reported that certain organoboranes can be cleaved using potassium hydroxide; an example is given below (Equation (62)) ⟨82JOM(226)115⟩.

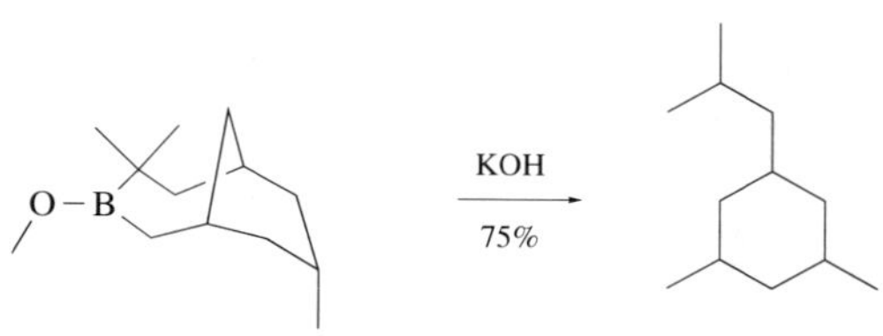

(62)

1.02.4.2 Reduction of Carbon–Silicon Bonds

There is a plethora of methods for the cleavage of carbon–silicon bonds to give carbon–hydrogen bonds, ranging from the use of acid or base to thermolysis.

1.02.4.2.1 Cleavage by acid

Acids have been used on several occasions to desilylate compounds and produce the corresponding hydrocarbon. Cleavage of an alkyl–silicon bond using concentrated sulphuric acid was first reported in about 1910 ⟨80JPR503⟩ and has since been used as a route to functionalize the silicon centre of organosilicon compounds (Equation (63)). In the case of TMS-R (where R is any of the alkyl groups shown in Equation (64)) one mole of methane is evolved per mole of tetraorganosilane reacted ⟨80JPR503⟩.

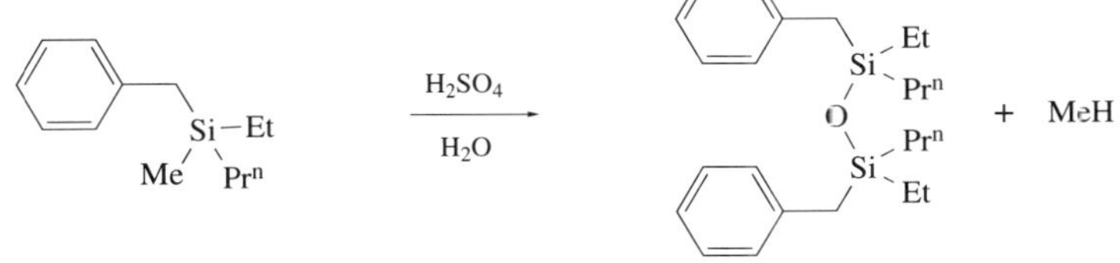

(63)

$$\text{TMS-R} \xrightarrow{H_2SO_4} Me_2Si(HSO_4)R \; + \; MeH \qquad (64)$$

$$R = -CH_2NMe_2,\ -CH_2CH_2NH_2,\ -CH_2CH_2CO_2H,\ -CH_2CCOMe$$

The use of the mineral acids, such as sulfuric acid, has been exploited in conjunction with ammonium fluoride to produce coumarins containing the $SiMe_2F$ group. The trimethyl silyl compound (**34**) was prepared and reacted with the mineral acid. This results in the cleavage of one methyl group from the silicon atom, to produce methane (Equation (65)). With only the acid present the methyl group was replaced by a hydroxyl group, whereas when ammonium fluoride was present, it was replaced by a fluorine atom ⟨80JPR503⟩. Not only can the silicon–methyl bond be cleaved, but cleavage of the coumarin–silicon bond can also occur (Equation (66)).

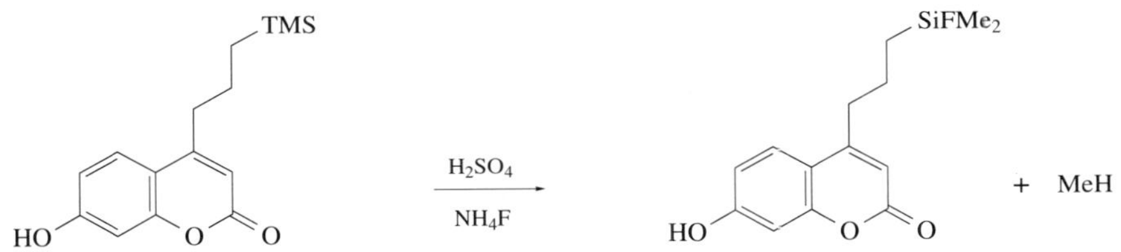

(65)

(**34**)

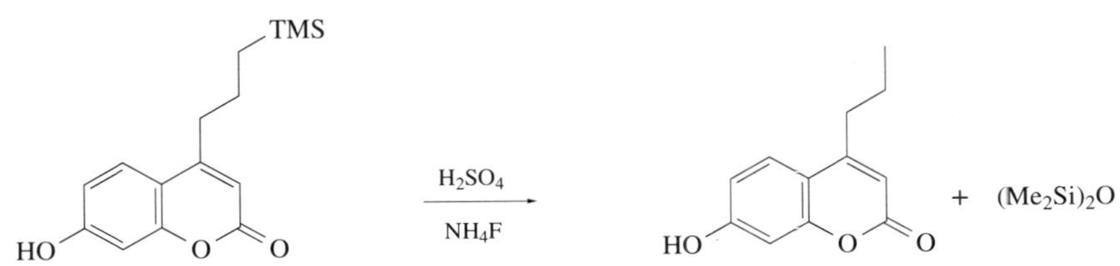

(66)

Trifluoroacetic acid has also been used to cleave carbon–silicon bonds. With the 1,4-bis(trimethylsilyl) compound shown in Equation (67), the acid cleaves the alkyl–silicon bond in preference to the vinyl–silicon bond, but the yield is very poor ⟨81ZOB420⟩.

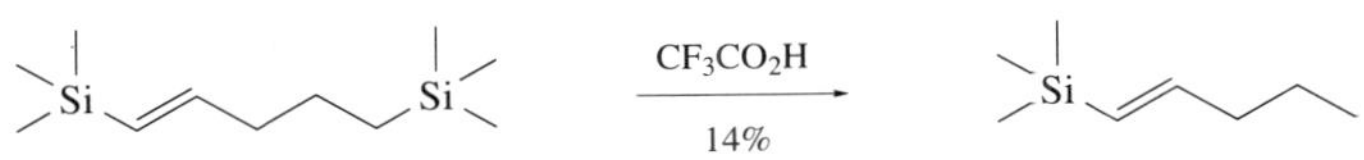

(67)

The cleavage of the carbon–silicon bond by electrophilic reagents such as Lewis acids is well established in organosilicon chemistry. In some cases the Lewis acid is used in conjunction with a hydrogen halide as in Equation (68) ⟨80JPR503⟩. However in others, such as the reaction of triethylsilane with aluminum tribromide, it is the only reagent used (Equation (69)) ⟨83ZOB806⟩.

$$Me_4Si + HCl \xrightarrow{AlCl_3} TMS\text{-}Cl + MeH \qquad (68)$$

$$Et_3SiH \xrightarrow{AlBr_3} EtH \qquad (69)$$

Simple *β*-trimethylsilyl ketones like (**35**) can be desilylated in ethanol to give the corresponding ketone (Scheme 46) ⟨70MI 102-02⟩. The ester analogue (**36**) requires hydrogen chloride for cleavage to produce the simple ester (Scheme 46).

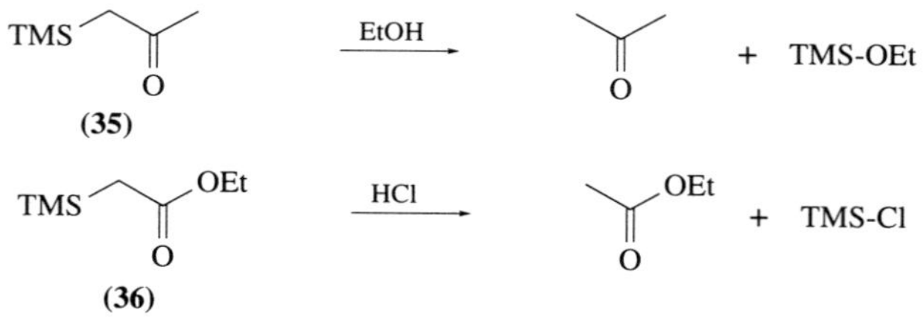

Scheme 46

1.02.4.2.2 Cleavage by base

There have been several reports that show the cleavage of a carbon–silicon bond to give a carbon–hydrogen bond using a base. One of the earliest of these reports showed that tetramethylsilane could be cleaved by potassium *t*-butoxide in DMSO (Equation (70)) ⟨67JOC4126⟩. This reaction was later modified to cleave a series of trimethylsilyl-compounds to the corresponding hydrocarbons. The new conditions employed were a 5% solution of potassium *t*-butoxide in 19 : 1 DMSO/water mixture at room temperature for 16 hours (Equation (71)). Some compounds needed 48 hours to desilylate completely ⟨82JA6809⟩.

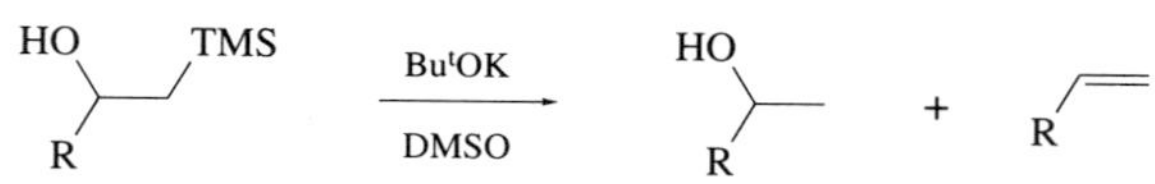

(70)

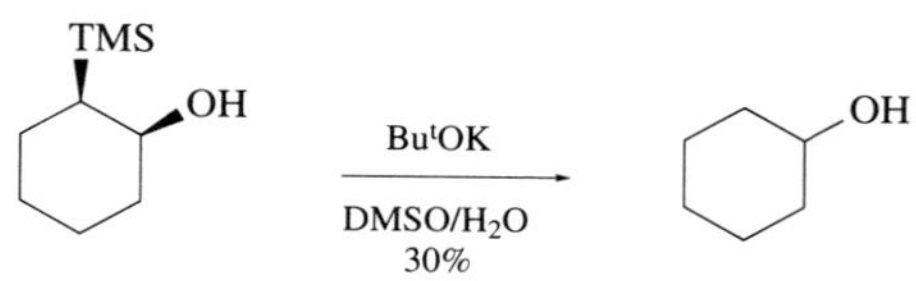

(71)

This cleavage is part of a reaction sequence resulting in stereospecific reductive methylation via a radical cyclization–desilylation mechanism ⟨86JA6826⟩. In the reaction sequence shown in Equation (72), it is possible to convert the tricyclic compound (**37**) to the methyl analogue (**38**) using five equivalents of potassium *t*-butoxide in DMSO at room temperature. The same research group has applied the reaction to steroids and to the production of acyclic systems (Equation (73)).

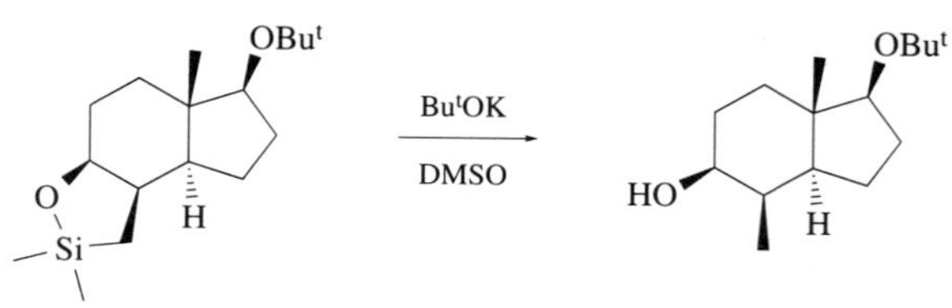

(72)

(**37**) (**38**)

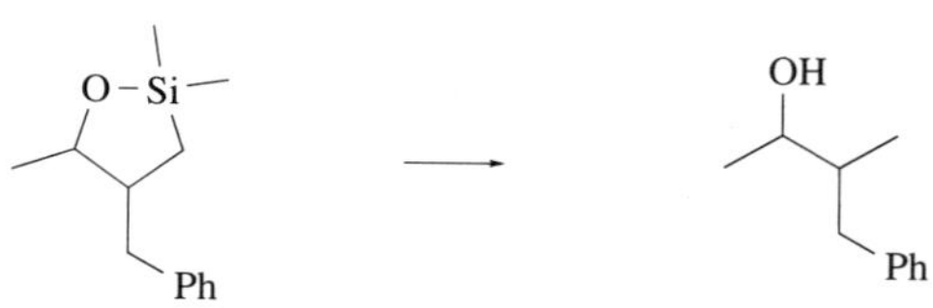

(73)

If fluoride ions in DMSO are used to desilylate vinyl silanes, then the resulting product is the parent alkene (Equation (74)) ⟨74TL3479⟩.

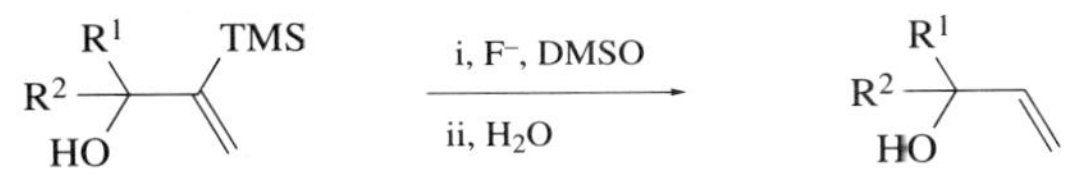

(74)

Potassium hydride was used to form alkenes of type **(39)** shown in Equation (75). The methyl substituted compound **(40)** was formed as a side product of the reaction ⟨87CB1049⟩. The use of potassium butoxide or caesium fluoride gave the same result.

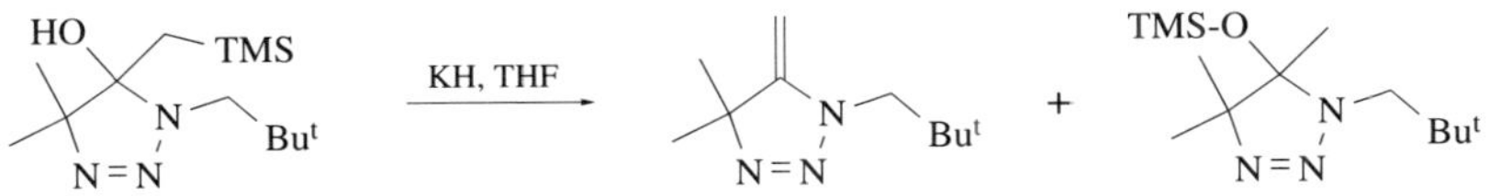

(75)

(39) **(40)**

1.02.4.2.3 Metal mediated desilylations

Several transition metal-mediated reactions of vinylsilanes have been described. The reaction of (*E*)-vinyl-1,2-dimethyl-1-trimethylsilane with the dinuclear platinum complex yields (*Z*)-2-butene as shown in Scheme 47. Reaction of the (*Z*)-vinyl-1,2-dimethyl-1-trimethyl silane yields primarily the (*E*)-2-butene ⟨76JOM(110)139⟩.

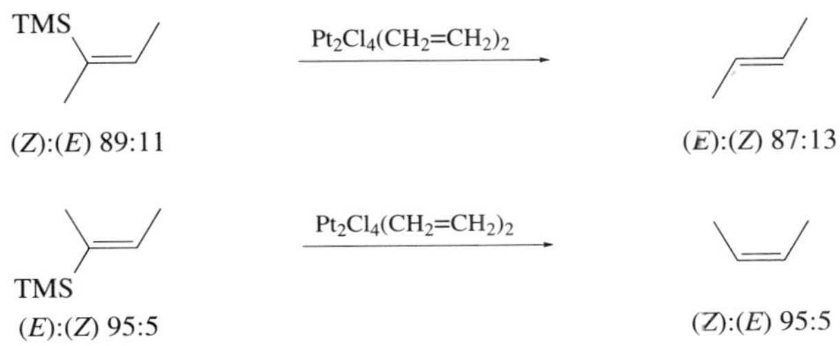

Scheme 47

In a side reaction of cyclopalladation as shown in Equation (76), the palladium complex **(41)** is broken down to give the butan-2-oxime complex. The authors suggest that this decomposition could be attributable to hydrogen chloride, formed by reaction of the *N*-hydroxyl moiety with the trimethylchlorosilane that is produced slowly in the process ⟨85OM1911⟩.

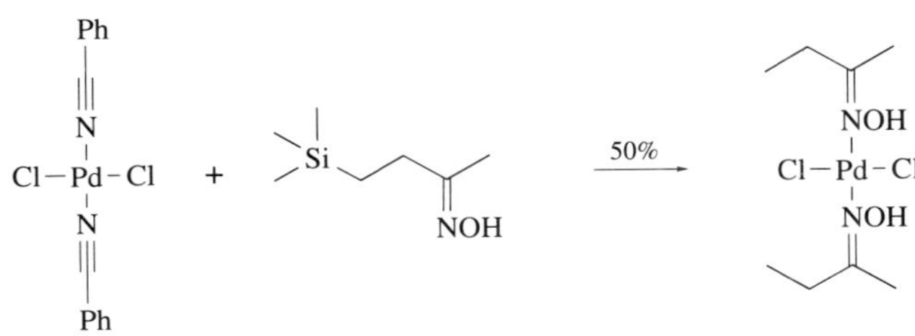

(76)

(41)

A combination of copper sulfate and methanol can cleave a cyclopentadiene ring from a silafulvene (Equation (77)) ⟨85OM584⟩.

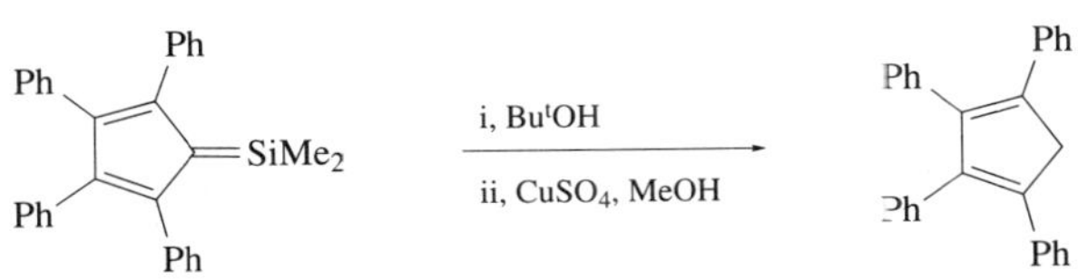

(77)

1.02.4.2.4 Desilylation by thermal decomposition and miscellaneous methods

The four-membered heterocycle shown in Equation (78) can be decomposed to ethane and diethylsulfide upon heating ⟨81JOM(214)145⟩.

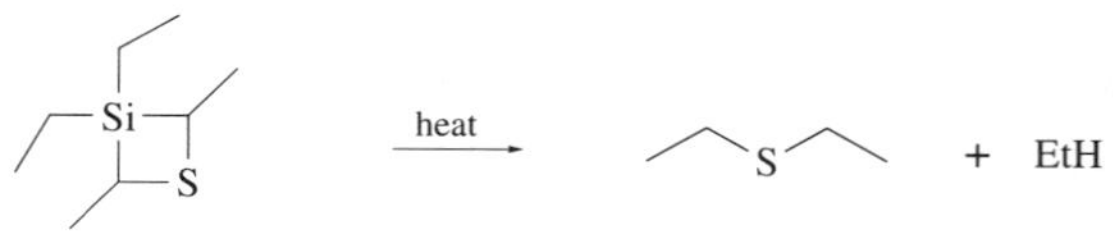

(78)

Aromatic *ipso*-desilylation (Equation (79)) is well documented in the literature ⟨74JCS(P2)596⟩. If there is a source of protons available then the substitution results in the formation of carbon–hydrogen bonds. The reaction has been used in the synthesis of oestrone (Equation (80)) ⟨80JA5253⟩.

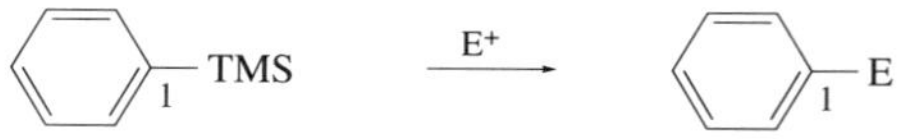

(79)

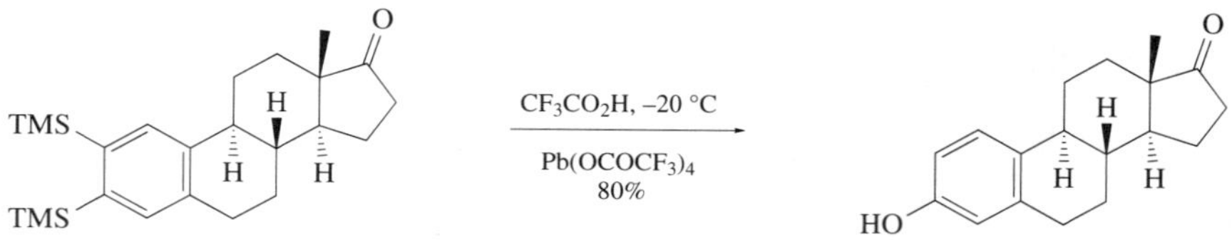

(80)

The desilylations so far have all concentrated on the desilylated products. A method for the preparation of trimethylsilyl esters of strong acids focuses attention on the silicon containing fragment (Equation (81)) ⟨77AG(E)399⟩. The method produces hydrocarbons as side products.

$$\text{Ar}-\text{TMS} \xrightarrow{\text{HX}} \text{TMS}-\text{X} + \text{ArH} \qquad (81)$$

$X = C_3F_5SO_3, C_3F_5CO_2, F_2P(O)O$

1.02.4.3 Cleavage of Carbon–Germanium Bonds to Carbon–Hydrogen Bonds

There are very few examples of the reduction of carbon–germanium bonds to carbon–hydrogen bonds. However there is a report in which a germanium bridged cyclohexadiene (**42**) was heated to give the corresponding benzene derivative and a germanium-containing species which polymerized (Equation (82)) ⟨73JA3078⟩.

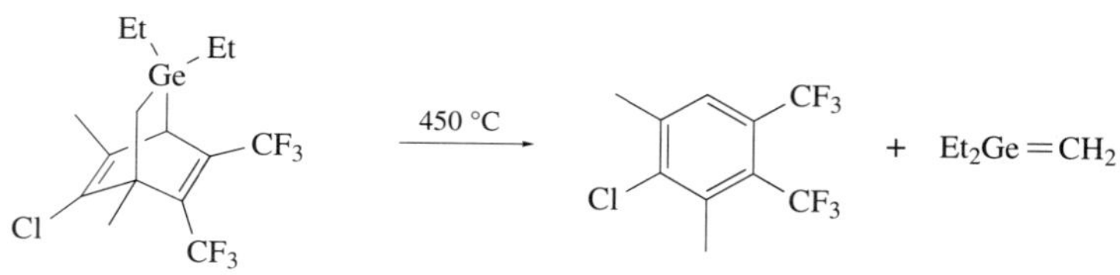

(**42**)

(82)

There have also been cleavages of carbon–germanium bonds to carbon–hydrogen bonds through photochemical means. For instance, the triethylaryl germanes shown in Equation (83) lost the aryl group on exposure to light of wavelength 250 nm ⟨86BCJ2807⟩ to furnish the hydrocarbon. The cleavage is thought to proceed by a radical mechanism (Scheme 48) ⟨86BCJ2807⟩.

$$Et_3GeAr \xrightarrow[\text{hexane}]{250\ \text{nm}} Et_3GeR + (Et_3Ge)_2 + ArH + C_{12}H_{26} \qquad (83)$$

Ar = Ph or *p*-toluene
R = *n*-C_6H_{13}, CH(Me)Bu, CH(Et)Pr

$$Et_3GeAr \longrightarrow Et_3Ge\bullet + Ar\bullet$$

$$RH + Ar\bullet \longrightarrow R\bullet + ArH$$

$$Et_3Ge\bullet + R\bullet \longrightarrow Et_3GeR$$

$$2\ Et_3Ge\bullet \longrightarrow (Et_3Ge)_2$$

RH = hexane

Scheme 48

Some work has been carried out to investigate the addition reactions of organo-metal alkoxides to carbon suboxide ⟨88G577⟩. When germanium alkoxides are exposed to carbon suboxide a malonate diester bonded to germanium is formed. This breaks down on addition of water to form dimethyl malonate (Scheme 49).

$$C_3O_2 + 2\ Bu^n{}_3GeOMe \longrightarrow (Bu^n{}_3Ge)_2C(CO_2Me)_2$$

$$(Bu^n{}_3Ge)_2C(CO_2Me)_2 + 2\ H_2O \longrightarrow H_2C(CO_2Me)_2 + 2\ Bu^nGeOH$$

Scheme 49

1.02.5 REDUCTION OF CARBON–METAL BONDS TO CARBON–HYDROGEN BONDS

The largest contributor to this section is, without question, the reduction of carbon–mercury bonds. Apart from this group, barring a few miscellaneous reactions, protonolysis reactions are involved, in general, removing unreacted organometallic reagent or producing deuteriated molecules. In these reactions, a carbon–metal bond (where the metal is an alkali or alkali earth metal) is reduced to a carbon–hydrogen bond by reaction with a proton donor, such as water or acid.

1.02.5.1 Reduction of Carbon–Mercury Bonds

The reduction of carbon–mercury bonds falls mainly into two categories, protonolysis and metal hydride demercuration. In addition there are a few miscellaneous reactions which will be discussed.

1.02.5.1.1 Protonolysis

There has been a great deal of attention focused on the protonolysis of organomercurials. It is possible to prepare organomercurials in high purity; in addition there are few problems with manipulation and they are monomeric in solution. These properties make them ideal candidates for mechanistic studies on the protonolysis of carbon–metal σ bonds. Several reviews ⟨68PAC79, 69MI 102-01, 69PAC1, B-72MI 102-02, 78T2827, 91COS(8)850⟩ and a book ⟨84JA3703⟩ have been written dealing with these studies.

As would be expected there are several mechanisms for the protonolysis reaction. The cleavage of diphenylmercury in acetic acid containing perchloric acid and the reaction of dialkylmercury with mercury nitrate exhibit only an S_E2 mechanism ⟨68PAC79⟩. Other protonolysis reactions, such as the reaction of an alkylmercury halide with hydrochloric acid, proceed through a four-centre transition state according to Scheme 50.

Scheme 50

The intramolecular nucleophilic participation of the chloride conjugate base is arguable. There is evidence, based on studies of protonolysis of unsymmetrical alkylmercurials, that suggests there is a three-centre transition state ⟨69JCS(B)1071⟩. Whatever the transition state, further reaction occurs by front side attack on the carbon centre, forming a transition state containing a pentacoordinate carbon atom ⟨84JA3703⟩. There are some reports of unimolecular S_E1 reactions ⟨68PAC79, 69JCS(B)1071⟩.

The carbon–mercury bonds are generally stable to water and alcohols, thus the protonolysis of these bonds requires stronger acids such as hydrochloric acid or sulfuric acid. Carboxylic acids are much less effective. In general the acid cleavage of dialkylmercurials is much more facile than that of alkylmercuric salts. The aryl–mercury bonds are cleaved more readily than alkyl–mercury bonds ⟨B-80MI 102-01⟩.

The protonolysis of solvomercurated products derived from the alkene leads back to the alkene ⟨77JOM(135)1⟩ and so other than the introduction of deuterium or tritium, this process has little synthetic utility. However the protonolysis of those alkenylmercurials derived from the solvomercuration of alkynes allows the preparation of stereoisomerically pure alkenes. Thus a 3 : 1 mixture of the vinylmercury compounds featured in Equation (84) were refluxed for two hours in acetic acid. Protonolysis occurred almost quantitatively, to give a 3 : 1 mixture of the corresponding esters ⟨80JCS(P1)1098⟩.

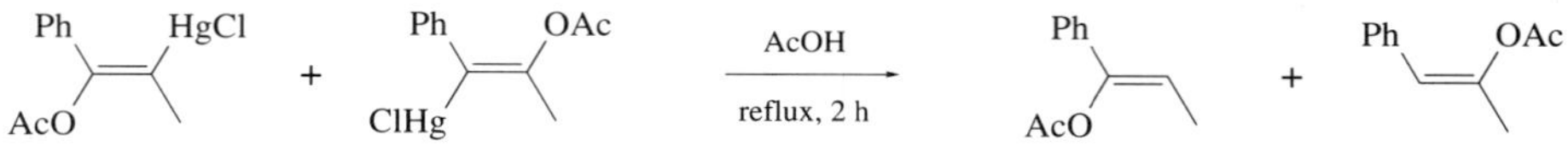

(84)

The alkenylmercury acetate in Scheme 51 undergoes protonolysis to give a product alkene with entirely inverted stereochemistry ⟨82JOC3707⟩. A mercurinium ion intermediate has been proposed and the steric interaction of the two phenyl groups forces the inversion of stereochemistry.

Protonolysis as the second step of intramolecular *C*-vinylation induced by mercury(II) salts has been described ⟨84JA4218, 85JA1726⟩ and some examples are given in Scheme 52.

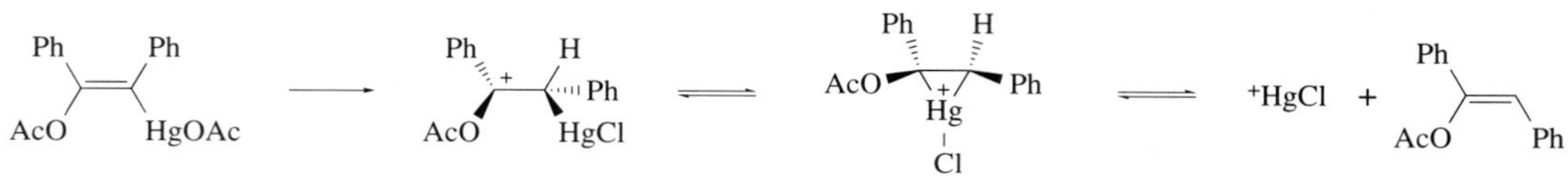

Scheme 51

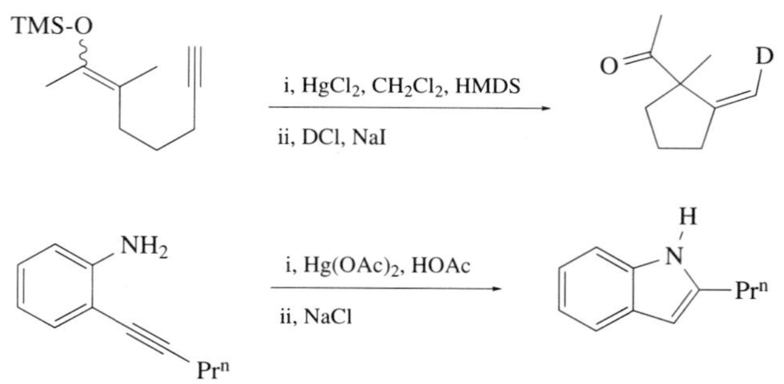

HMDS = hexamethyldisilazide

Scheme 52

1.02.5.1.2 Metal hydride demercuration

Both sodium borohydride and aluminum hydrides have been used to reduce organomercurials to produce the corresponding hydrocarbon. The mechanism through which sodium borohydride reduces the organomercurials is believed to be a free radical mechanism as shown in Scheme 53 ⟨76JA5973, 84TL5239, B-85MI 102-02⟩.

$$RHgX \longrightarrow RHgH + NaBH_3X$$

$$R\bullet + RHgH \longrightarrow RH + RHg\bullet$$

$$RHg\bullet \longrightarrow R\bullet + Hg$$

Scheme 53

There is a disadvantage with using sodium borohydride in the reduction of organomercurials, in that rearrangements of the initially formed radical can occur as illustrated by Equation (85) ⟨74JA870, 76JA5973⟩.

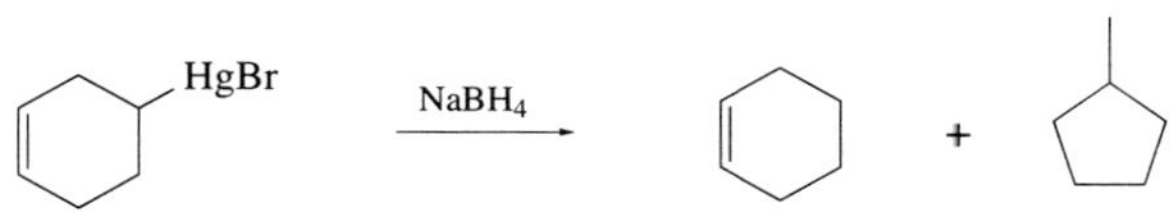

(85)

Reduction of *exo*- and *endo*-norbornylmercury(II) bromide with sodium borodeuteride provides *exo*-[2-$^2H_1$]norbornane as the major product (Equation (86)) ⟨70JA6611⟩. Other examples of this phenomenon can be found in the literature ⟨81JOC563⟩.

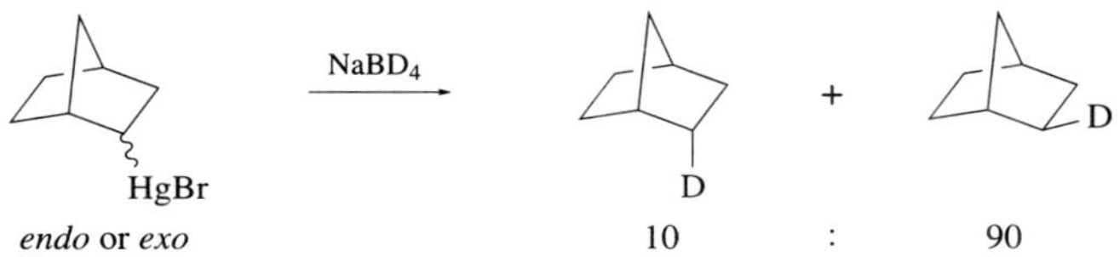

(86)

The stereoselectivities observed on cleavage of the carbon–mercury bond with sodium borohydride depend upon the nature of the radical intermediates and the reaction conditions. Attempts to find optimal conditions for this stereoselective cleavage have been made ⟨83JA7407⟩. The diastereoselectivity in metal hydride demercuration of α-mercury(II) carbonyl compounds is dependent on the nature of the solvent, the amount of hydride used, the mode of addition, the nature of the hydride source and the ligand on mercury, so any rationalization of these factors will be difficult ⟨86JOC2024⟩. Reductive elimination appears to be the major problem. This is often minimized by the use of alkaline borohydride (Equation (87)) ⟨84T2317⟩.

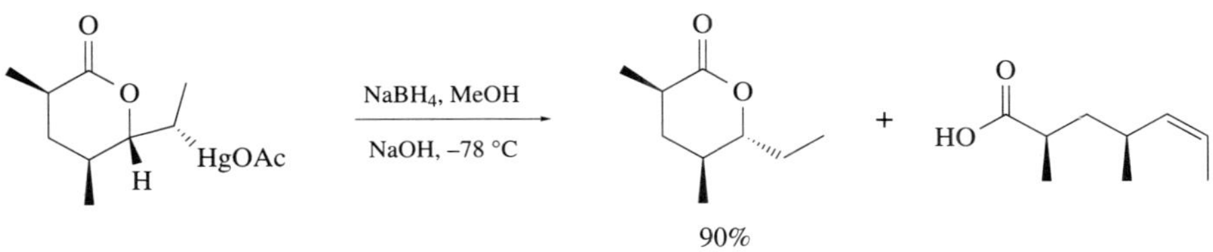

(87)

However, alkaline sodium borohydride must be avoided in the demercuration of organomercurials obtained by alkoxymercuration of vinyl ethers bearing conjugated electron-withdrawing groups, because of the danger of substantial reductive elimination. For simple vinyl ethers (Equation (88)) however, no problems are encountered ⟨83TL4923⟩.

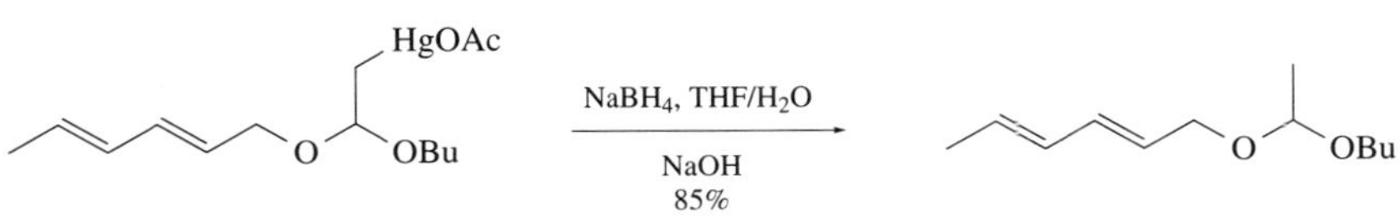

(88)

The products obtained from the oxymercuration-demercuration of alkenylacetates under standard conditions (sodium borohydride with 3 M sodium hydroxide) are diols. With methoxy- and hydroxyalkenes the yields are much lower because of competing deoxymercuration ⟨81JOC531⟩. Increasing the concentration of base increases the yield of hydrated products. Phase transfer reagents are sometimes used to avoid deoxymercuration and other side reactions (Scheme 54) ⟨79S891, 83T2863, 83TL3997, 84JOC2838, 86CC855⟩.

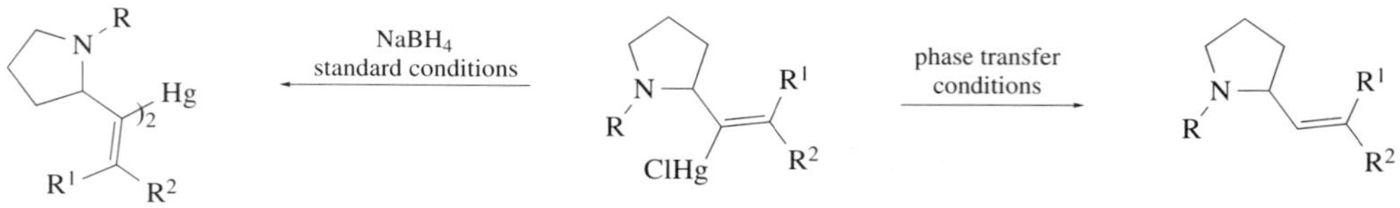

Scheme 54

There have been numerous publications on the use of sodium borohydride to cleave carbon–mercury bonds and, although there are other methods available, this is the method of choice ⟨91COS(8)850⟩.

A number of demercurations of RHgX or R^1HgOR^2 species use *n*-tributyltin hydride ⟨83JA6882, 84T2317⟩, or triphenyltin hydride ⟨84JA8313⟩, but complete removal of tin residues can prove difficult. As with sodium borohydride, there is also a problem with competitive deoxymercuration when using tin hydrides ⟨84T2317⟩. The presence of sodium acetate prevents this problem if triphenyltin hydride is used ⟨84JA8313⟩. The use of tri-*n*-butyltin hydride instead of alkaline sodium borohydride in the demercuration of peroxymercurials has led to improved yields of peroxides (Table 21) ⟨82JCS(P1)1797⟩.

Table 21 Comparison of yields for formation of peroxymercurials using Bu^n_3SnH and $NaBH_4$.

Peroxymercurial (X = HgBr)	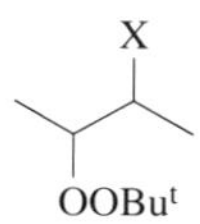	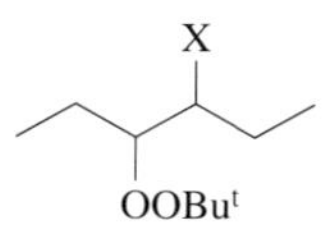	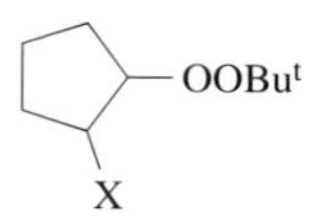	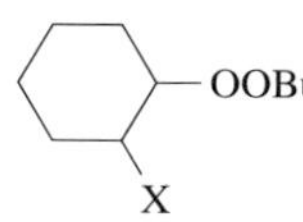
Yield of peroxide (%) (X = H)				
Bu^n_3SnH	63	24	59	61
$NaBH_4$/NaOH	0	0	2	13

Lithium aluminum hydride has been little used as the reducing agent in demercuration ⟨83JA6882, 86JA2094⟩. However the reduction of organomercury(II) halides with lithium aluminum hydride has been investigated and the findings suggest an electron transfer mechanism, involving attack of the alkyl radical on the metal hydride (Scheme 55) ⟨83TL1411⟩.

$$R\bullet + AlH_4^- \longrightarrow RH + AlH_3\bullet$$

$$AlH_3\bullet^- + RHgX \longrightarrow R\bullet + AlH_3 + Hg + X$$

Scheme 55

1.02.5.1.3 Miscellaneous methods

Although sodium borohydride is most often used for the reduction of carbon–mercury bonds, there has been a significant increase in the number of other reagents tried and utilized, for example thiols and sodium amalgam (*vide intra*). Reduction with hydrogen sulfide ⟨84JCS(P1)1689⟩, sodium dithionite ⟨79JOC228⟩, metals ⟨80JA337, 84AG(E)224⟩, alcohols ⟨88S144⟩, alkaline aqueous bases, ⟨84CC862⟩, aromatic amines ⟨80JCS(P1)1420⟩, Wilkinson's catalyst ⟨80JA337⟩ and electrochemical reductions ⟨85JOC673⟩ have also all been used.

Organomercurials react with thiols by a free radical substitution mechanism (Scheme 56). The reaction between the hexenylmercury chloride and benzene thiol is initiated by light or 2,2′-azobisisobutyronitrile (AIBN). A mixture of 1-hexene and methylcyclopentane is produced ⟨83JA1398⟩. A limitation in the use of certain thiol reagents is, as for most reagents used for demercuration, their ability to undergo elimination with reversion to the starting material ⟨91COS(8)850⟩.

$$PhS\bullet + RHgCl \longrightarrow R\dot{H}g(PhS)Cl$$

$$R\dot{H}g(PhS)Cl \longrightarrow R\bullet + PhSHgCl$$

$$R\bullet + PhSH \longrightarrow PhS\bullet + RH$$

Scheme 56

The sodium amalgam cleavage of alkylmercurials involves an ionic mechanism ⟨72JOC4341⟩. The reaction is also stereospecific with retention of configuration at the carbon centre. No rearrangement was observed in the rearrangement-prone nortricyclyl-norbornenyl system shown in Equation (89) ⟨81JOC563⟩. Sodium amalgam has also been used for reducing the amidomercurial derived from methyl undecalene ⟨85BSF75⟩.

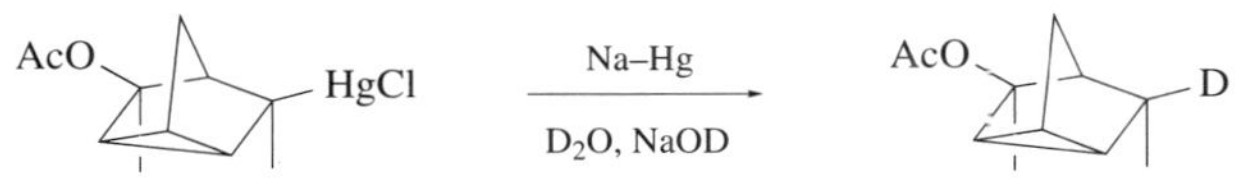

(89)

The replacement of mercury by hydrogen can be achieved using *trans*-metallation followed by hydrolysis (Scheme 57) ⟨82JOC3707, 84AG(E)224⟩.

Scheme 57

The latest addition to the numerous reagents for reduction of organomercurials is *N*-benzyl-1,4-dihydronicotinamide; it is thought to reduce the C—Hg bond via an electron transfer chain substitution mechanism ⟨81TL4495⟩.

1.02.5.2 Cleavage of Other Metals from Carbon to give a Carbon–Hydrogen Bond

1.02.5.2.1 Protonolysis

There are many carbon–metal bonds which can be reduced on addition of a proton donor to the reaction. The carbon–aluminum bond can be readily cleaved on addition of HX (where X is a hydroxyl or alkoxyl group, etc) to give the corresponding hydrocarbon as given by Equation (90) ⟨B-72MI 102-03⟩.

$$R{-}Al\langle \xrightarrow{H_2O \text{ or } ROH} RH \qquad (90)$$

In general, all organometallic compounds of the type RM (where M is lithium, sodium or potassium) and of the type RMgX (where X is a halogen) will undergo protonolysis readily, if not violently, when they come into contact with a proton donor (Scheme 58) ⟨B-80MI 102-01⟩.

$$RM \xrightarrow{H_2O} RH + MOH$$

$$RMgX \xrightarrow{H_2O} RH + Mg(OH)_2 + HX$$

M = Li, Na, K
X = Cl, Br, I

Scheme 58

The carbon–tin bond is very susceptible to attack by electrophiles. Water and aliphatic alcohols are generally inert, but phenols, mercaptans and carboxylic acids readily cleave the carbon–tin bond to give a carbon–hydrogen bond (Equation (91)) ⟨B-80MI 102-01⟩. The tin moiety can also undergo transmetallation to form alkyllithiums and so, indirectly, can be replaced by hydrogen (Scheme 59) ⟨76JOM(107)139⟩.

$$R^1{-}SnR^2{}_3 \xrightarrow{HX} R^1H + XSnR^2{}_3 \qquad (91)$$

Scheme 59

The reaction shown in Equation (92) is one of the very few examples of reduction of transition metal–carbon bonds which has any synthetic utility ⟨78AG(E)690⟩.

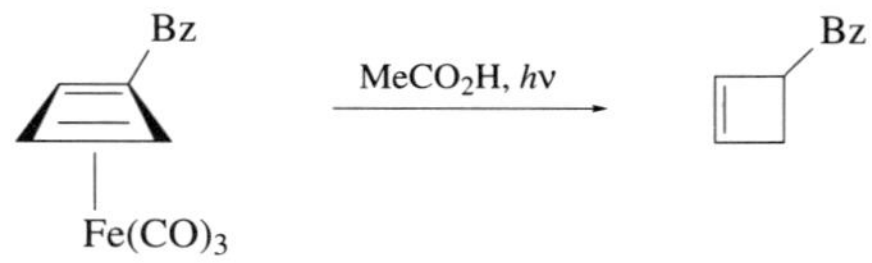

(92)

1.03
Two or More CH Bond(s) Formed by Addition to CC Multiple Bonds

KEITH JONES
King's College London, UK

1.03.1 REDUCTION OF ALKENES

1.03.1.1 General Methods for Alkene Reduction

The most common and effective method for the reduction of carbon–carbon double bonds is catalytic hydrogenation using molecular hydrogen over a heterogeneous catalyst. The heterogeneous nature of the catalyst allows for simple removal at the end of the reaction by filtration. The reduction of ethene was first reported in 1874 ⟨1874CB352⟩ but it was Sabatier and his group ⟨1897MI 103-01⟩ who explored this reaction extensively. Heterogeneous hydrogenation using a number of different metals as catalyst has been reviewed ⟨B-65MI 103-01, B-67MI 103-01, B-78MI 103-01, 79MI 103-01, B-85MI 103-01⟩. The mechanism of the reaction is extremely complicated and not well understood. However, the mechanism suggested in 1934 by Polanyi explains many of the features of importance to the organic chemist ⟨34MI 103-01⟩ (Scheme 1). Hydrogen is dissociatively adsorbed onto two catalytic sites on the metal surface while the alkene is adsorbed on two adjacent sites. A hydrogen atom is transfered to one carbon giving a monoadsorbed alkyl species which undergoes an irreversible transfer of the second hydrogen to give an alkane which is desorbed from the metal. Isomerisation and double bond migration can be understood in this mechanism as involving the monoadsorbed species undergoing a configurational change or returning to a doubly-adsorbed species which is different to the original. This mechanism also suggests that both these processes will be retarded at high hydrogen concentration. The type of catalyst used is the major factor influencing the pathway of the reaction (i.e., reduction versus isomerisation or double bond migration). The classic example of this is shown in Scheme 2, involving the hydrogenation of car-3-ene ⟨66JCS(C)41⟩. Hydrogenation over platinum yields *cis*-carane in high yield but the same reaction using palladium gives a mixture of products of which trimethylcycloheptane is the major one. This is caused by the propensity for palladium to give double bond migration followed by hydrogenolysis of the allylic cyclopropane and finally reduction of the C=C double bond. This is a property of the metal and is not affected much by the support. A decreasing order of activity in double bond migration is Pd > Ni ≫ Rh ≫ Ru ≈ Os > Ir ≈ Pt ⟨B-85MI 103-01⟩. For isomerisation, the following order of decreasing activity has been suggested, Pd ≫ Rh = Ru = Pt > Os > Ir. Another side reaction is the hydrogenolysis of allylic, benzylic and vinylic C—heteroatom bonds. The order of decreasing activity in hydrogenolysis of allylic and benzylic functions is Pd > Pt ≫ Rh = Ru ⟨B-85MI 103-01⟩ whereas the order of decreasing activity in hydrogenolysis of vinylic functions is Pt ≫ Ir > Rh > Os ≫ Ru = Pd ⟨71BCJ166⟩. The nature of the catalyst support can play a role in the reaction. In general if the initial product can be further hydrogenated a support such as calcium carbonate or barium sulfate is best as these seem to adsorb the intermediates less strongly than catalysts supported on carbon or alumina. Recently, a rhodium catalyst has been prepared by heating rhodium carbonyl with titanium dioxide. The resulting catalyst displays a strong metal-support interaction and is about twice as active as rhodium on alumina ⟨90RTC87⟩. The extent of double bond migration can be affected by the solvent. For example benzene has been found to inhibit migrations when used as a cosolvent ⟨B-85MI 103-01⟩.

* represents a catalytic site on the surface of the metal

Scheme 1

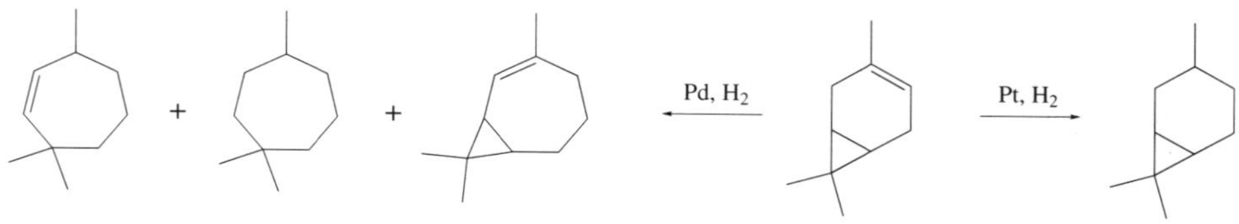

Scheme 2

Transition metal boride and aluminide reagents can be prepared by reaction of sodium borohydride and lithium aluminum hydride respectively with various transition metal salts. Their use as heterogeneous catalysts in the reduction of a variety of functional groups including alkenes has been reviewed ⟨86CRV763⟩.

The stereochemistry of hydrogenation of alkenes generally involves *cis* addition of hydrogen to the least hindered face of the alkene ⟨66MI 103-01⟩. The problem comes about in deciding which is the least hindered face of an alkene and avoiding isomerisation ⟨75JOC3073⟩. Norbornenes usually undergo hydrogenation from the *exo*-face unless the bridging carbon carries a large group ⟨72JOC1182⟩ although many factors such as reaction conditions and alkene purity can affect the outcome. A further complication arises when the alkene carries groups capable of binding to the surface of the catalyst and presenting the alkene in a particular orientation ⟨73JA6379⟩. This phenomenon is called haptophilicity. An example of this is shown in Equation (1) ⟨80JOC805⟩.

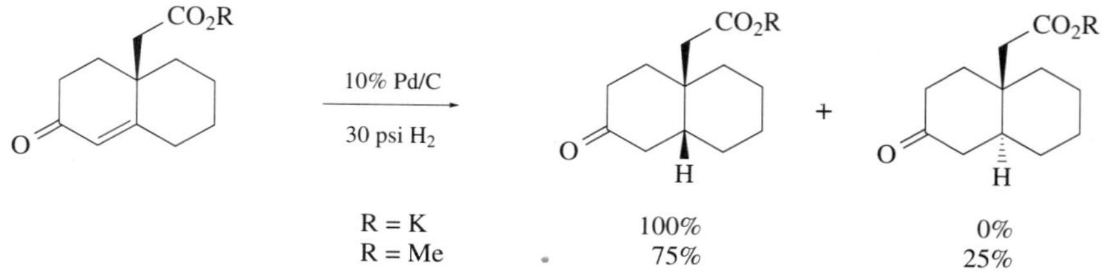

(1)

Although heterogeneous catalysts offer an excellent method for the reduction of alkenes, the use of transition metal complexes which are soluble in the reaction medium (homogeneous catalysts) is a viable and useful alternative. The pioneering work of Wilkinson and co-workers on rhodium tristriphenylphosphine monochloride ⟨65CC131, 65CI(L)560⟩ demonstrated the utility of such systems and this area has been reviewed ⟨76OR(24)1⟩. Homogeneous catalysts have advantages in that they are usually more selective than heterogeneous catalysts: they are not as susceptible to poisoning; they lead to cleaner labelling owing to negligible scrambling and they have been shown to be useful in asymmetric hydrogenations. A number of homogeneous catalysts have been used to reduce alkenes. In addition to Vaska's catalyst, $Ir(CO)(PPh_3)_2Cl$ ⟨62JA679⟩ the reactivity of which resembles Wilkinson's catalyst, there is the rhodium monohydride, $RhH(CO)(PPh_3)_3$ ⟨68JCS(A)2665⟩ and pentacyanocobalt(III) which is particularly useful for the reduction of electron-poor alkenes ⟨72BCJ2723⟩. A soluble catalyst can also be prepared by treatment of rhodium trichloride trihydrate with sodium borohydride in DMF ⟨71CC840⟩. The crystalline complex formed can be used for the reduction of alkenes.

Wilkinson's catalyst is selective for the reduction of mono- and disubstituted double bonds. In particular, terminal double bonds are very readily reduced ⟨67JCS(A)1574⟩. As can be seen from Table 1, simple cyclic alkenes and terminal alkenes are very rapidly reduced while as soon as steric hindrance is introduced either near the double bond or actually on the double bond, the rate of reduction slows dramatically. In addition, *trans* double bonds are less reactive than *cis* double bonds. The stereoselectivity of reduction is sensitive to both temperature and solvent as illustrated by the reaction in Equation (2). In benzene at 0 °C the product shown with an axial methyl is formed selectively but the use of ethanol as solvent or an increase in temperature gives a mixture of axial and equatorial methyl. Although specific deuteriation can be achieved by reduction of alkenes using Wilkinson's catalyst, the solvent can affect the course of this reaction (Scheme 3). Reaction of styrene with Wilkinson's catalyst and deuterium in benzene leads to clean addition of deuterium to give 1,2-dideuteriophenylethane but in chlorinated solvents or with ethanol/benzene a mixture of partially deuterated products is formed ⟨70CC495, 70CC497, 70CC571⟩. Most other functional groups are tolerated by Wilkinson's catalyst with the exception of aldehydes. Decarbonylation can be a serious problem but can be suppressed using high catalyst and hydrogen concentrations ⟨66TL1605⟩. Vaska's catalyst has been prepared by an improved route ⟨68IS101⟩ and reduces alkenes with similar steric preferences to Wilkinson's catalyst. However, it shows a greater propensity for not only *cis*/*trans* isomerisation but also regioisomerism of the double bond ⟨72JOM(42)163⟩.

Table 1 Relative rates of reduction of alkenes with Wilkinson's catalyst ⟨67JCS(A)1574⟩.

Alkene	*Relative rate*
Cyclohexene	31.6
Cyclopentene	34.3
1-Methylcyclohexene	0.6
1-Hexene	29.1
2-Methylpent-1-ene	26.6
cis-2-Pentene	23.2
cis-4-Methylpent-2-ene	9.9
trans-4-Methylpent-2-ene	1.8

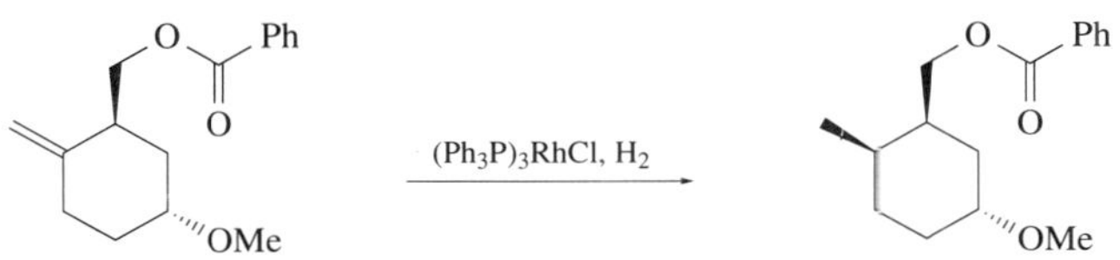

(2)

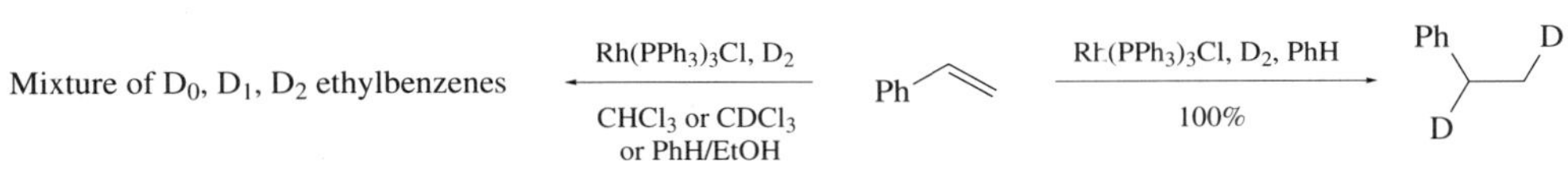

Scheme 3

$RhH(CO)(PPh_3)_3$ is a very selective catalyst, reducing 1-hexene to hexane while not reducing cyclohexene ⟨68JCS(A)2665⟩. Most other functional groups are tolerated and there is little or no isomerisation. The pentacyanocobalt(III) ion reacts with hydrogen to give a cobalt hydride species which is valuable for the reduction of conjugated dienes, styrenes and electron-poor alkenes ⟨72BCJ2723⟩. α,β-Unsaturated carboxylic acids are reduced in high yield (Equation (3)). However, this catalyst also promotes hydrogenolysis of allylic functions such as alcohols, acetates and halides. The catalyst formed from rhodium trichloride and sodium borohydride readily catalyses the reduction of simple alkenes ⟨71CC840⟩ but also reduces C=N, N=N and nitro groups ⟨73JCS(P1)2509⟩.

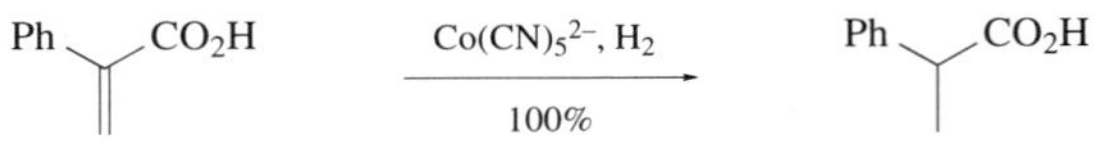

(3)

The reduction of alkenes using catalytic transfer hydrogenation involves the use of a hydrogen donor rather than molecular hydrogen. This approach has been reviewed ⟨85CRV129⟩. Both homogeneous catalysts and heterogeneous catalysts have been used in conjunction with a wide range of hydrogen donors to achieve the successful reduction of alkenes. The most common donors are trialkylammonium formates, cyclohexene, indoline and sodium phosphinate. The rate of reduction and indeed the regiospecificity of reduction can vary with different hydrogen donors ⟨83JOC1638⟩. Thus, along the series cyclohexene, 1-methylcyclohexene, 1-methyl-4-isopropylcyclohexene and 1-methyl-4-*t*-butylcyclohexene the rate of reduction increases by a factor of ten ⟨83JOC1638⟩. Among the heterogeneous catalysts, the most commonly used is palladium ⟨85CRV129⟩ whilst homogeneous catalysts based on palladium, platinum, ruthenium, rhodium, iridium, iron and nickel have been reported ⟨85CRV129⟩. The reaction presumably proceeds in a similar manner to the reaction using the same catalyst with molecular hydrogen except that the hydrogen donor is oxidised, for example in the case of cyclohexene to benzene. Benzylic hydrogenolysis can occur; indeed transfer hydrogenation has been reported to provide the best method for removal of the carbobenzyloxy group ⟨92TL2299⟩. Partial hydrogenolysis of the C—S bond in benzylic systems has been reported ⟨79JOC3442⟩. Carbon—halogen bonds can be cleaved by various catalytic transfer conditions although the C—F bond is stable. The order of ease of cleavage of C—X bonds is as expected, namely C—I > C—Br > C—Cl. The C—Cl bond can be cleaved using sodium phosphinate as the hydrogen donor ⟨77JCS(P1)443⟩. The advantages of this method are that special equipment such as hydrogenators and pressure equipment are not required as the source of hydrogen is usually a simple solid or liquid.

Reduction of alkenes using diimide was first reported in 1905 and has been reviewed ⟨91OR(40)91⟩. In all reductions using diimide, the diimide is generated *in situ*. There are a large number of methods for achieving this ⟨91OR(40)91⟩ but the most common are: (1) acid-promoted decomposition of sodium or potassium azodicarboxylate ⟨62CI(L)484⟩, (2) thermal or base-catalysed decomposition of arenesulfonyl hydrazines, for example 2,4,6-triisopropylbenzene sulfonyl hydrazine ⟨76T2157⟩ and (3) the reaction of hydrazine with an oxidising agent, for example hydrogen peroxide and copper(II) ⟨61TL353⟩, iodosobenzenediacetate ⟨87SC703⟩ or nickel salts ⟨84JCS(D)1477⟩. The reaction has been shown to involve a stereospecific *syn* addition of hydrogen (or deuterium) to one face (usually the least hindered) of the alkene ⟨73JA4346⟩ (Equation (4)) and has been suggested to involve a concerted type of mechanism ⟨61JA2957⟩. A comparison of the rate of reaction in different solvents has been carried out and acetonitrile has been shown to lead to the highest rates of reduction ⟨69MI 103-01⟩. A wide variety of alkenes are reduced by diimide although the rate of reaction varies with a number of factors. Table 2 shows the relative rates of reduction of some representative alkenes. The higher strain in cyclopentene leads to a faster reaction than with cyclohexene while increasing substitution slows the reaction down considerably. Conjugated dienes are more reactive than simple alkenes ⟨75JOC3599⟩ and conjugation with a carboxylic acid also speeds up the reaction. There is evidence that, usually, *trans*-alkenes are reduced more rapidly. The considerable rate differences between differently substituted alkenes makes this method one of the best for achieving selective reductions (Equation (5)) ⟨78T2981⟩. In addition to this selective reduction, a large number of functional groups survive diimide reduction. Alkynes are reduced more readily than alkenes but allylic and benzylic C—heteroatom bonds survive. Thus, allylic halides, esters, alcohols, amines and disulfides are reduced without reduction of the allylic function. Vinyl halides and vinyl ethers are reduced slowly again without reduction of the halide or ether function while alkenes conjugated to carbonyl groups undergo ready reduction without reaction at the carbonyl group. However, the reduction of the aldehyde group in aromatic aldehydes has been reported ⟨67JCS(C)1120⟩. Bonds which are usually very easily cleaved reductively such as N—N, N—O and O—O survive diimide reduction. The advantages of diimide reduction over catalytic hydrogenation include practical simplicity (no special equipment is required) and high selectivity.

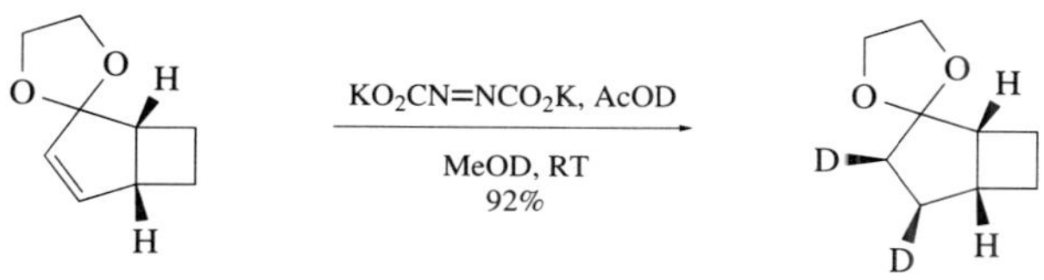

(4)

Table 2 Relative rates of reduction of alkenes with diimide.

Alkene	*Relative rate*	*Reference*
Cyclohexene	1	65JA2932
Cyclopentene	15.5	65JA2932
1-Pentene	20.2	65JA2932
trans-2-Pentene	2.59	65JA2932
cis-2-Pentene	2.65	65JA2932
2-Methylbut-2-ene	0.28	65JA2932
Fumaric acid	100	62AG(E)213
Maleic acid	10	62AG(E)213

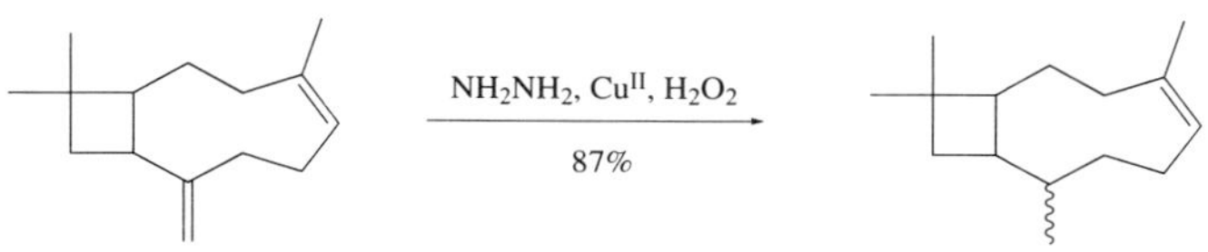

(5)

A two-stage procedure for the reduction of carbon–carbon double bonds involves the hydroboration of the alkene using borane to give an alkylborane followed by protonolysis to give an alkane ⟨59JA4108⟩. The area of borane reagents has been reviewed ⟨B-88MI 103-01⟩. Hydroboration of alkenes occurs readily on reacting borane with a wide variety of alkenes although the number of hydrogens of the borane replaced by alkyl groups depends strongly on the steric hindrance of the alkene. Thus unhindered alkenes react to give trialkylboranes while tetrasubstituted alkenes react to give monoalkylboranes. Addition to the double bond occurs in an anti-Markovnikov manner (the boron atom becomes attached to the least substituted carbon of the alkene) and the boron and hydrogen add in a *cis*-manner (Equation (6)). Although the resulting trialkylboranes are stable to water, they are cleaved by heating in carboxylic acids such as propionic, isobutyric and pivalic acids ⟨59JA4108⟩. Higher carboxylic acids can be used if higher temperatures are needed. Primary alkylboranes react most readily followed closely by secondary alkylboranes. Tertiary alkylboranes are far less reactive towards carboxylic acids ⟨71JINC1157⟩. The protonolysis reaction occurs with retention of configuration at the alkyl group ⟨68JCS(C)1474⟩.

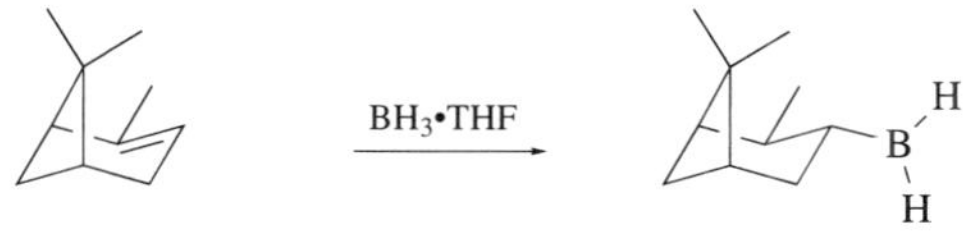

(6)

Although the methods described above constitute the most usual alkene reductions, a wide variety of other reagents have been reported to reduce carbon–carbon double bonds, particularly in α,β-unsaturated systems (see Section 1.03.1.5).

The use of sodium borohydride with cobalt(II) chloride or nickel(II) chloride for the direct reduction of alkenes has been reported ⟨79JOC1014⟩. Reduction of limonene using this system led to selective reduction of the less-substituted double bond (Equation (7)). It has been suggested that this reaction actually involves the formation of molecular hydrogen and a cobalt boride catalyst ⟨86CRV763⟩. Similar reductions using lithium aluminum hydride along with catalytic amounts of cobalt, nickel or titanium chlorides have been studied ⟨78JOC2567⟩.

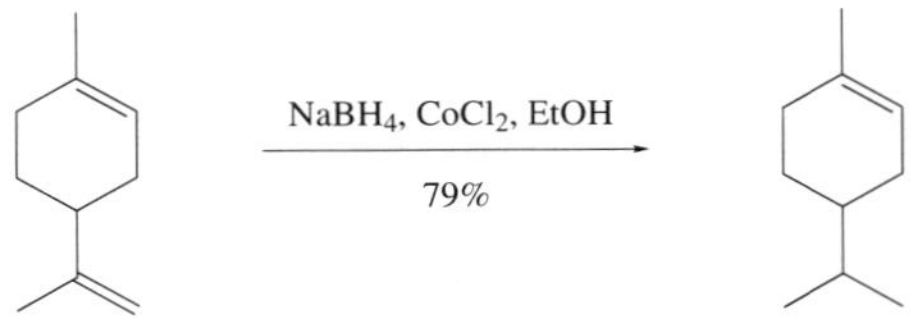

(7)

Some simple alkenes can be reduced using triethylsilane and excess trifluoroacetic acid ⟨67T2235⟩. Methylcyclohexene is reduced to methylcyclohexane (Equation (8)) although under the same conditions 1-pentene is not reduced.

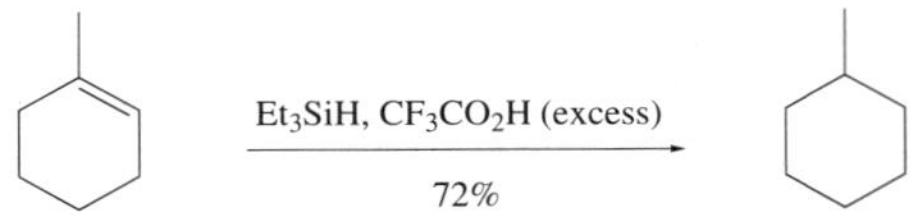

(8)

The enantioselective formation of alkanes by reduction of alkenes has become an extremely valuable process in organic chemistry. This will be treated in this section, although the substrates are inevitably heavily substituted alkenes, in order to make the presentation coherent. Undoubtedly the most impressive results have been obtained using homogeneous catalysts but the earliest results were achieved using heterogeneous catalysts. In 1956, Akabori *et al.* reported the reduction of alkenes using a catalyst consisting of palladium and silk fibroin, the water insoluble part of the silk protein from wild silkworms ⟨56MI 103-01⟩. Reduction of an oxazolone with this system was claimed to give amino acids in 30–70% enantiomeric excess (Equation (9)). It has been reported that this result is difficult to reproduce ⟨91TA843⟩, perhaps not surprisingly given the nature of the catalyst. More success has been achieved in reducing *N*-acetylcinnamic acid derivatives using a polymer-bound cationic rhodium catalyst ⟨84AG(E)435⟩. In this way, *N*-acetylphenylalanine has been prepared in high chemical and stereochemical yield (Equation (10)).

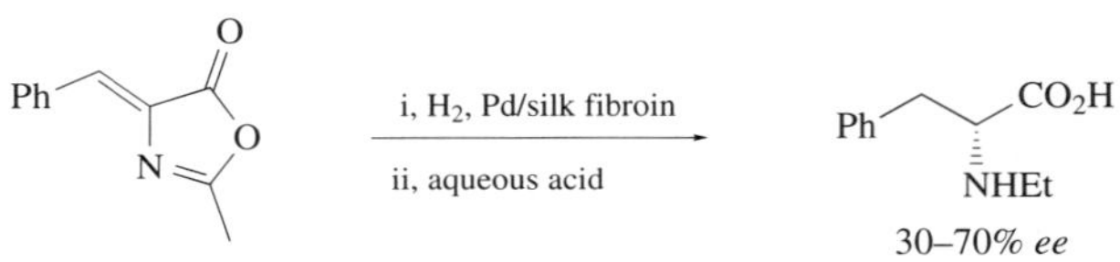

(9)

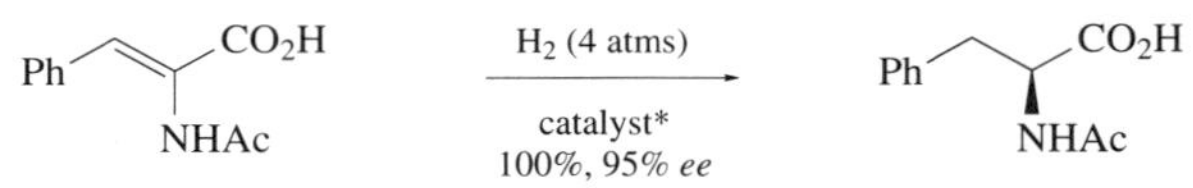

(10)

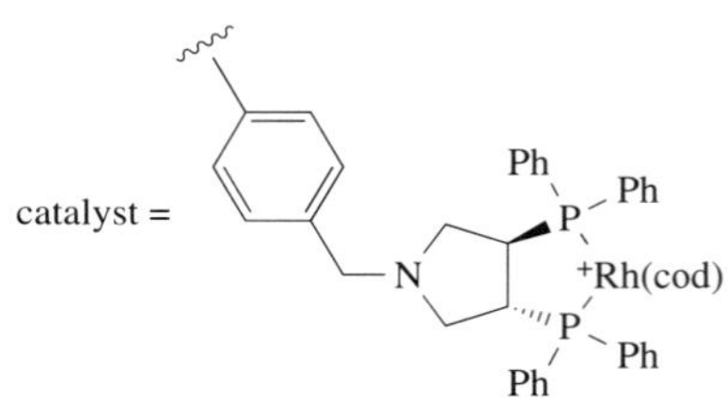

Homogeneous catalysts have provided a much more fruitful area for asymmetric reduction of alkenes. The first examples involved the modification of Wilkinson's catalyst by the addition of chiral tertiary phosphines. Reduction of simple alkenes proceeded in 3–15% *ee* ⟨68AG(E)942, 68CC1445⟩. The advent and use of chelating C2-chiral bisphosphines as ligands has revolutionised this area particularly with the *N*-acetylcinnamic acid substrates. For example, the preparation of *N*-acetylphenylalanine can be achieved in high chemical and stereochemical yield using the DIPAMP ligand (Equation (11)) ⟨77JA5946⟩. This system is used industrially for the synthesis of (*S*)-DOPA ((*S*)-3-(3,4-dihydroxyphenyl)alanine) which is used in the treatment of Parkinson's disease ⟨88BSF846⟩. A cationic rhodium complex with a chiral bisphosphinamide ligand is used for the synthesis of (*S*)-phenylalanine which is used in the nonnutritive sweetner aspartame ⟨89T6901⟩. Axially-chiral ligands such as BINAP have also been used with considerable success in such reductions ⟨84T1245⟩. The stereochemical results of these hydrogenations are sensitive to the initial hydrogen pressure and the optical yields obtained fall off dramatically at high initial hydrogen pressures. The mechanism of this hydrogenation has been the subject of extensive studies by Halpern ⟨87JA1746⟩ and Brown ⟨80JA3040⟩ and shown to involve reaction through the minor, less stable alkene/metal complex. Unfortunately, these rhodium-catalysed hydrogenations are successful with only a limited range of substrates. Although the carboxyl group of the *N*-acylcinnamic acids can be replaced with other electron-withdrawing groups and the phenyl group can be replaced with other aryl groups, alkyl groups or hydrogen, the amide function and the *β*-hydrogen seem to be essential requirements. In the light of this limitation, Noyori and Takaya have developed the use of ruthenium(II)-BINAP complexes for the hydrogenation of a wide range of prochiral alkenes ⟨90ACR345⟩. This catalyst is particularly useful for the reduction of unsaturated carboxylic acids ⟨87JOC3174⟩, for example in the synthesis of naproxen (Equation (12)) although high pressure is required. The double bond reduced does not have to be conjugated, although it does have to be near to the carboxyl group which complexes to the ruthenium (Equation (13)). Other functionalities capable of complexing with the ruthenium can also be reduced with good stereocontrol. Enamides can be reduced to give 1-benzylisoquinolines such as tetrahydropapaverine (Equation (14)) with excellent enantiocontrol ⟨86JA7117⟩. Allylic alcohols are also good substrates for the ruthenium(II)-catalysed hydrogenations ⟨87JA1596, 87JA4129⟩. Thus geraniol is reduced both chemo- and stereoselectively to (*R*)-citronellol using ruthenium(II)-(*S*)-BINAP (Equation (15)).

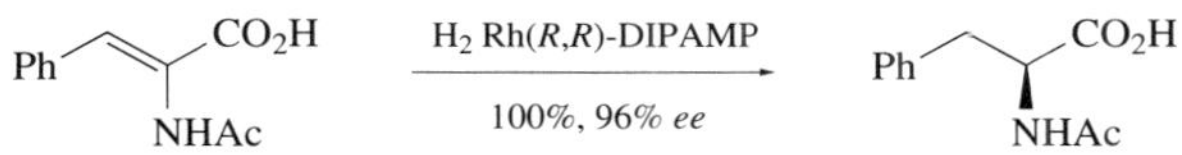

(11)

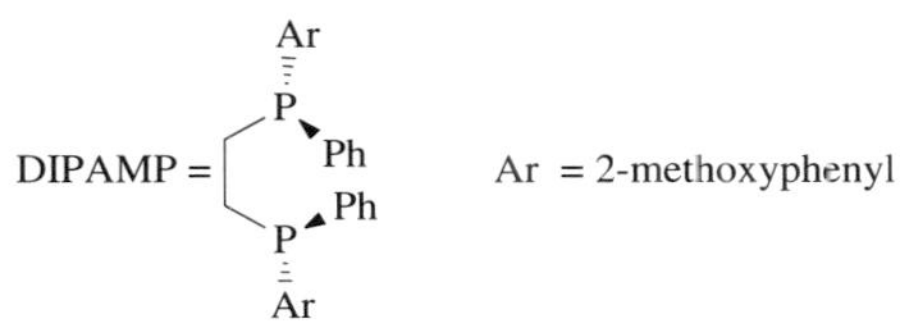

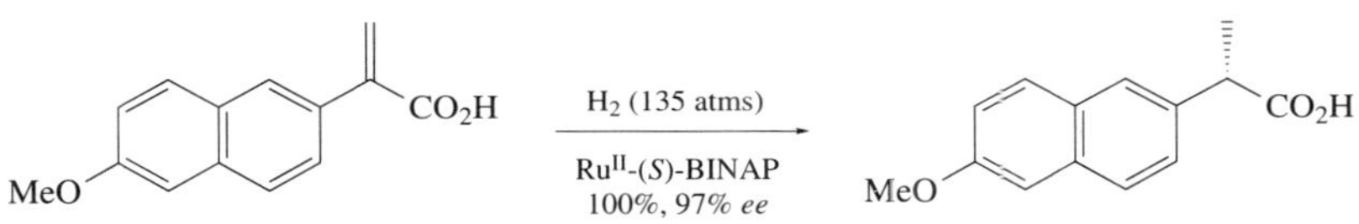

(12)

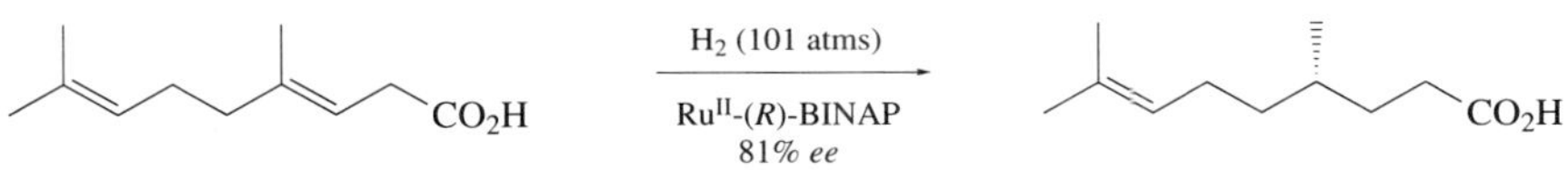

(13)

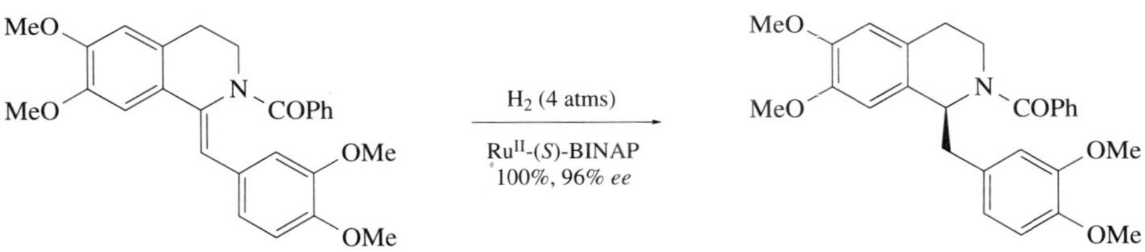

(14)

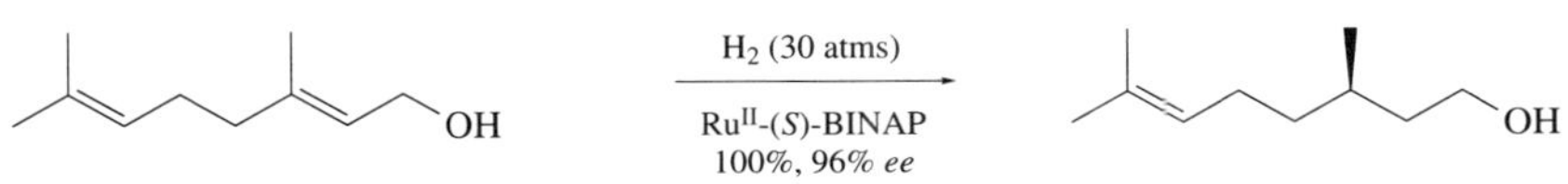

(15)

The use of baker's yeast for the enantioselective reduction of alkenes has also been reviewed ⟨90S1⟩. The most common substrates for this biotransformation are systems in which the double bond is conjugated with an electron-withdrawing group such as a carbonyl group. Allylic alcohols can be reduced with good enantioselectivity ⟨88T1299⟩. For example a phenylthio-substituted alkene has been reduced with high *ee* although the yield is moderate and the reaction time long (Equation (16)). Trisubstituted allylic alcohols can also be reduced with excellent stereocontrol as demonstrated by the reduction of geraniol to optically-pure citronellol ⟨87T4481⟩. Unsaturated esters can be reduced in moderate to good yields and with excellent enantioselectivity (Equation (17)) ⟨76HCA1832⟩. The reduction of the masked aldehyde in this reaction is a common feature of the baker's yeast reduction of unsaturated aldehydes. Cinnamaldehyde derivatives are readily reduced to optically-active alcohols. Thus a furylacrolein has been reduced using baker's yeast to the saturated alcohol (Equation (18)) in high chemical yield and with complete control of the stereochemistry ⟨88JCS(P1)3061⟩.

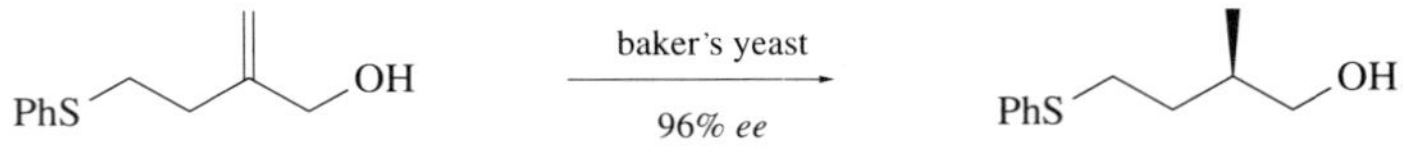

(16)

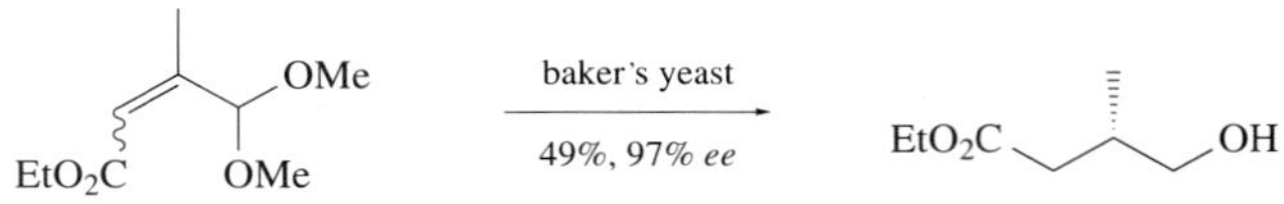

(17)

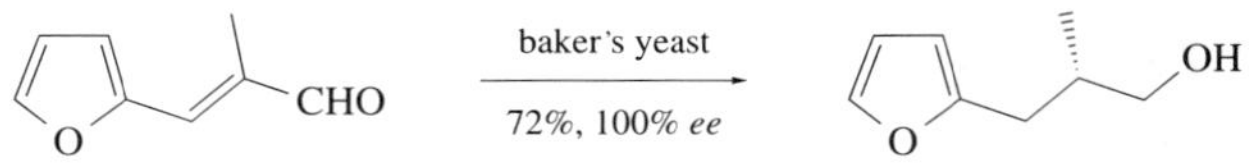

(18)

1.03.1.2 Reduction of Alkyl-substituted Alkenes

Alkyl-substituted alkenes are reduced by a wide range of heterogeneous catalyst systems although the reduction becomes more difficult as the number of alkyl groups on the double bond increases presumably due to the difficulty of adsorption of the alkene. An example illustrating this is the selective reduction of limonene (Equation (19)) which can be achieved to cleanly reduce the least-substituted double bond providing hydrogen uptake is stopped after one equivalent has been used ⟨83JOC3162⟩. However, care needs to be exercised in determining the most hindered double bond as shown by the examples in Equations (20) and (21). Reduction of the octalin (Equation (20)) proceeds readily at room temperature and atmospheric pressure to give a mixture of *cis*- and *trans*-decalins ⟨37JCS1136⟩ whereas addition of an extra ring means that drastic conditions are required (Equation (21)) ⟨78T1261⟩. Strain also increases the susceptibility of a double bond to reduction (Equation (22)). The reaction can be readily halted after reduction of the strained cyclobutane double bond ⟨52JA4867⟩. Simple cyclic alkenes are readily hydrogenated (Equation (23)). This reduction can be achieved using platinum oxide (in which case a prereduction of the catalyst is required) or palladium on carbon ⟨63OSC(4)304⟩.

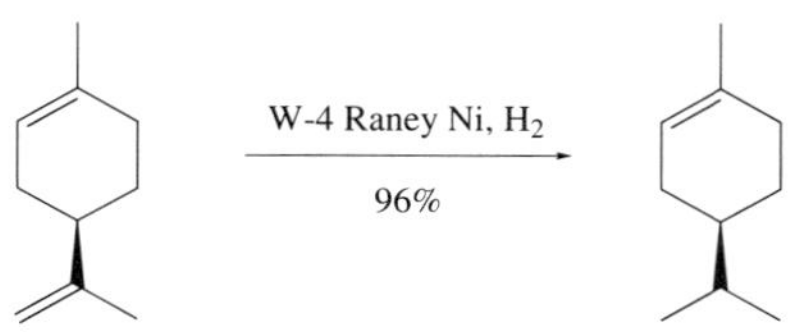

(19)

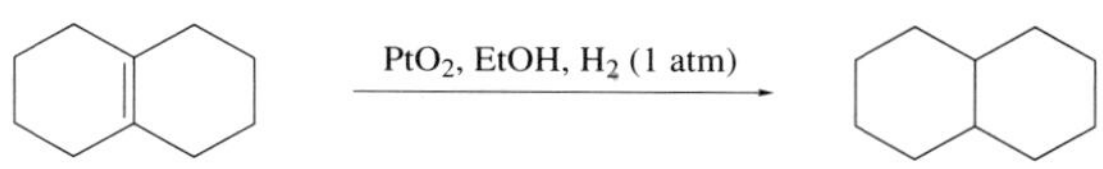

(20)

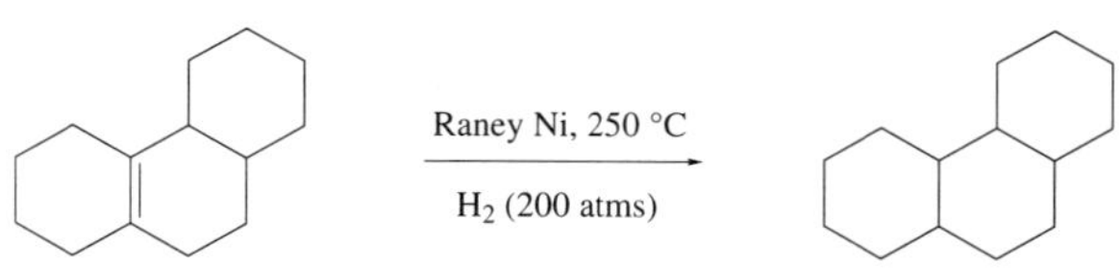

(21)

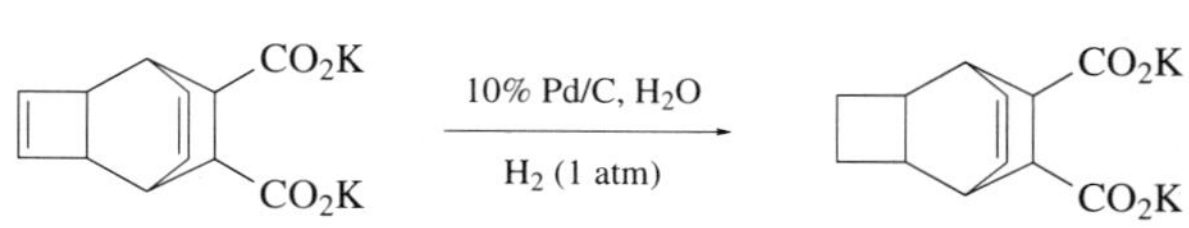

(22)

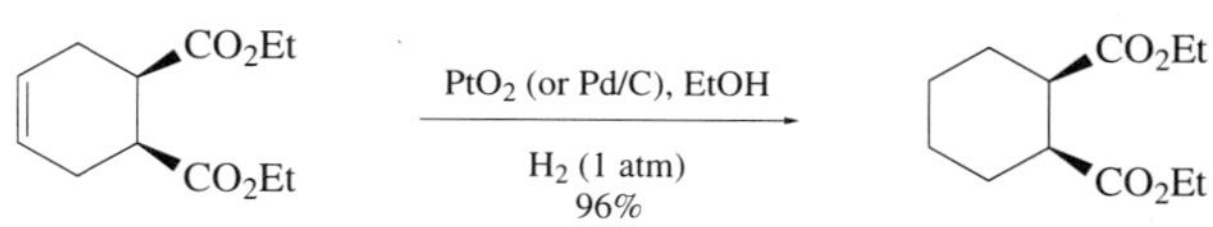

(23)

The reduction of simple alkyl substituted terminal alkenes (Equation (24)) has been studied extensively using Wilkinson's catalyst often as kinetic experiments ⟨67JCS(A)1574⟩. As can be seen from Table 1, the reduction of terminal double bonds is kinetically favoured. Although double bonds conjugated to carbonyl groups can be reduced using Wilkinson's catalyst (see below), the selective reduction of the terminal double bond in carvone (Equation (25)) is well known ⟨76OR(24)1⟩.

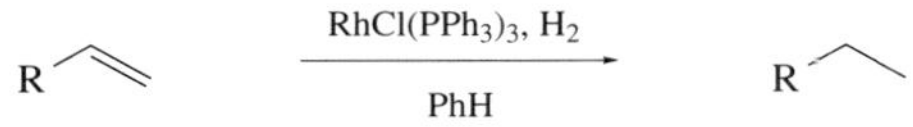

(24)

R = *n*-alkyl

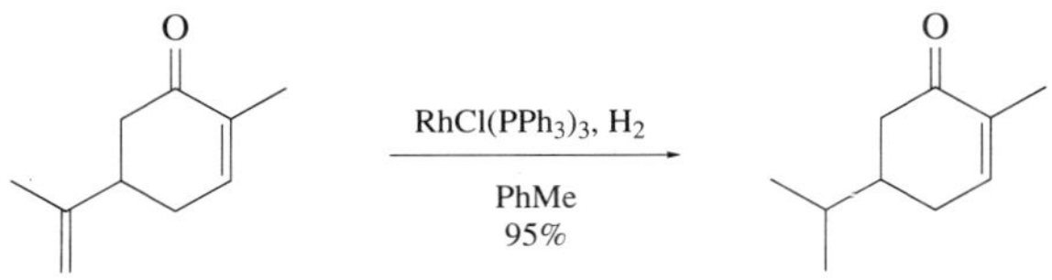

(25)

Simple alkenes have been reduced using palladium on carbon and cyclohexene ⟨75S448⟩ or triethylammonium formate (Equation (26)) ⟨78JOC3985⟩. Cyclic alkenes are also readily reduced, for example cyclohexene (Equation (27)) using indoline as the hydrogen donor ⟨75S448⟩.

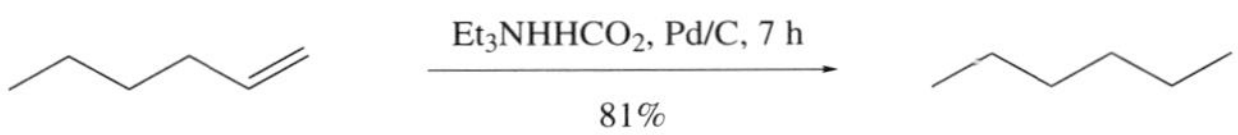

(26)

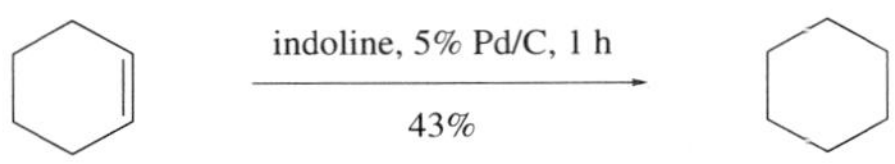

(27)

1.03.1.3 Reduction of Alkenyl-, Aryl-, Heteroaryl- and Alkynyl-substituted Alkenes

Conjugated dienes can undergo three types of reduction: selective 1,2-reduction of one of the double bonds, 1,4-reduction or complete reduction. An example of the latter two reactions is shown in Scheme 4 ⟨56ZOB1175⟩. A similar effect of the catalyst chosen is found in the reduction of mevinolin (Scheme 5). Palladium leads to 1,4-reduction whereas platinum leads to complete hydrogenation ⟨83JOC1991⟩. Conjugated dienes such as 1,3-pentadiene are reduced slowly by Wilkinson's catalyst although a simple rhodium hydride has been used to successfully reduce 1,3-butadiene (Equation (28)) to 1-butene although a higher temperature and pressure was required ⟨74JOM(70)89⟩. Reduction of 1,3-butadiene with pentacyanocobalt(III) occurs under milder conditions but gives a mixture of products (Equation (29)) ⟨72BCJ2723⟩. Dienes are also reduced under catalytic transfer conditions but the reaction is slower than with simple alkenes and can lead to a mixture of products (Equation (30)) ⟨78JOC3985⟩.

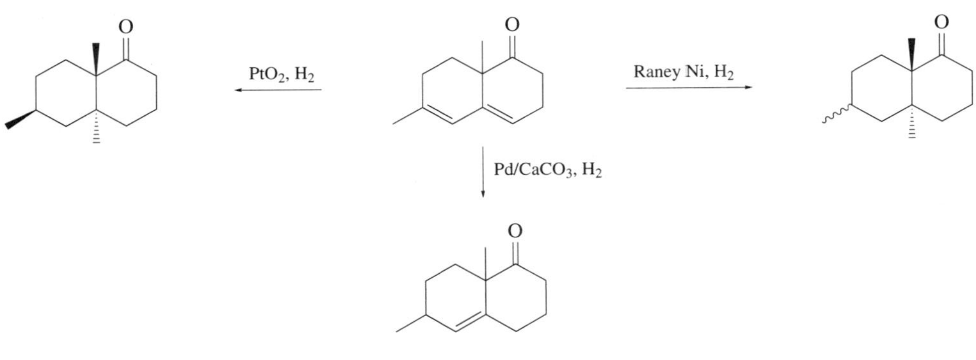

Scheme 4

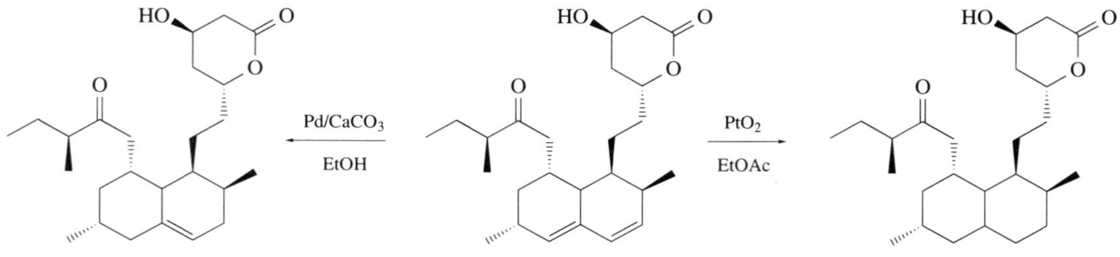

3:1 *trans*:*cis*

Scheme 5

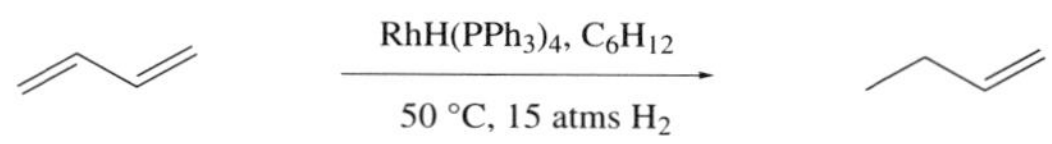

(28)

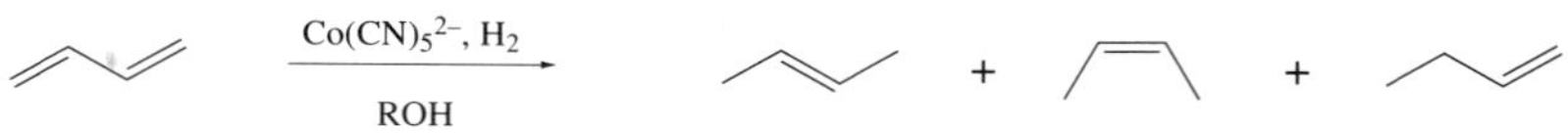

(29)

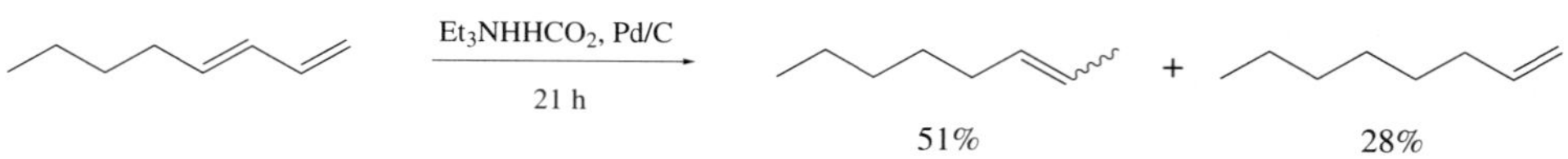

(30)

Allenes are generally reduced in a stepwise manner to alkenes and then alkanes with palladium being more selective than platinum ⟨64MI 103-01⟩. Allenes with a terminal double bond are selectively reduced at the terminal double bond. This is illustrated by the reduction 2,3-butadienoic acid over palladium to give crotonic acid (Equation (31)) ⟨54JCS3197⟩. A similar reaction for a phosphorus-substituted allene has been reported (Equation (32)) ⟨70JOC3510⟩ but yielding the opposite stereochemistry in the product.

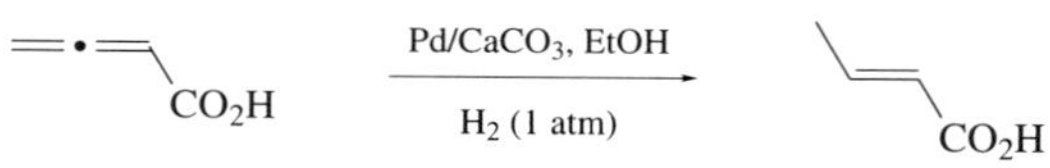

(31)

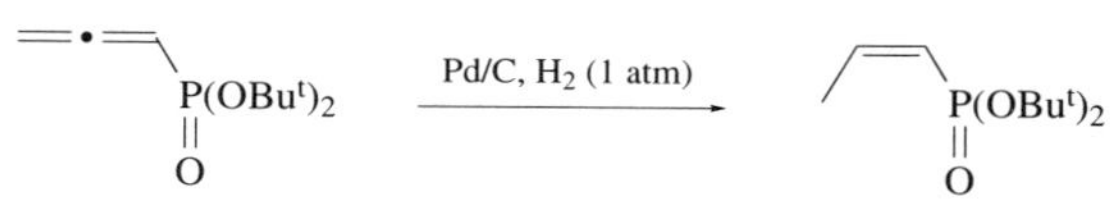

(32)

There are many examples of the reduction of aryl-substituted double bonds mostly of cinnamic acid derivatives using heterogeneous catalysts (Equation (33)) ⟨63OSC(4)136⟩. Phenylalanine is readily prepared from acetamidocinnamic acid by reduction using platinum under a slight pressure of hydrogen (Equation (34)) ⟨43OSC(2)491⟩. As discussed above (see Section 1.03.1.1), acetamidocinnamic acids are good substrates for homogeneous catalytic reduction (Equation (35)) ⟨69CC1365⟩. Styrenes can be reduced by many of the homogeneous catalysts including pentacyanocobalt(III) ⟨72BCJ2723⟩. Wilkinson's catalyst reduces styrene itself smoothly at room temperature ⟨67JCS(A)1574⟩ while Vaska's catalyst requires a higher temperature (Scheme 6) ⟨72RTC1057⟩. Conjugation of the alkene with an aromatic ring is no hinderance to reduction using catalytic transfer hydrogenation (Equation (36)) ⟨75S448⟩ although the reduction of stilbene (Equation (37)) is slow and requires a modified palladium catalyst ⟨78S5397⟩. There are two rather more unusual methods that have been reported to reduce the double bond in styrene systems. The reaction of 1,1-diarylethenes with lithium aluminum hydride leads to the alkane in good yield (Equation (38)). Electron-donating substituents on the aryl rings slow down the reaction and the solvent is important for a clean reaction ⟨76JOC3682⟩. A phase-transfer reduction of 1,1-diarylethenes and related systems has been reported using a low oxidation state cobalt catalyst and tetrafluoroboric acid ⟨83CC365⟩. Such activated double bonds are reduced in high yield (Equation (39)).

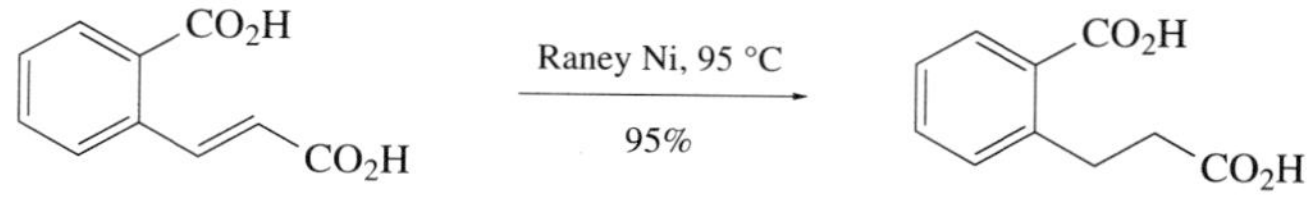

(33)

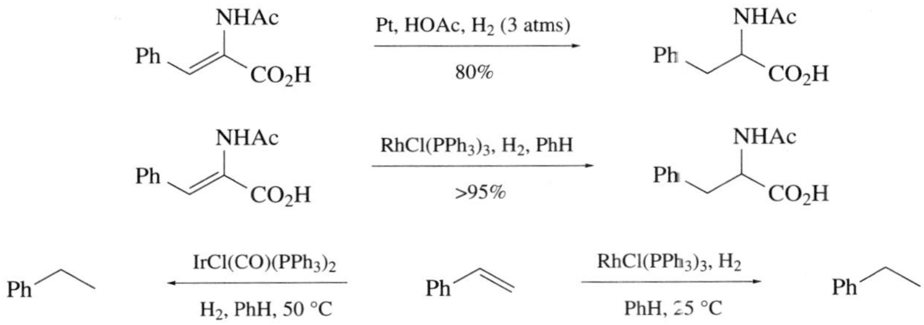

(34)

(35)

Scheme 6

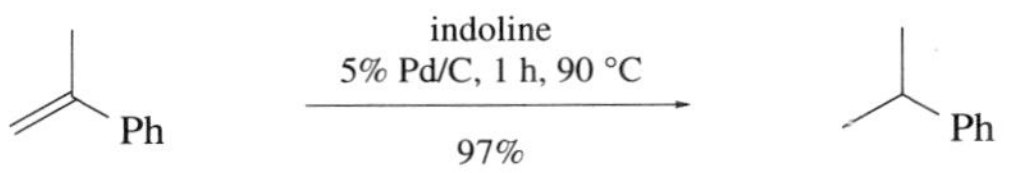

(36)

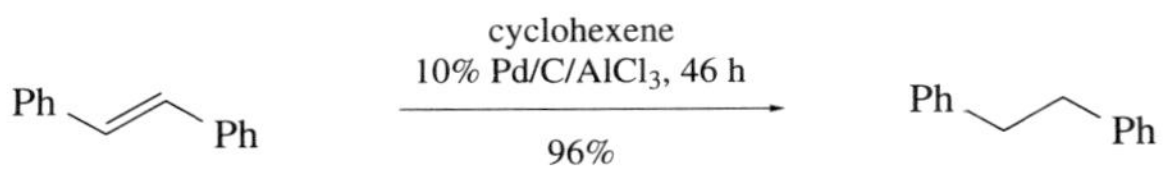

(37)

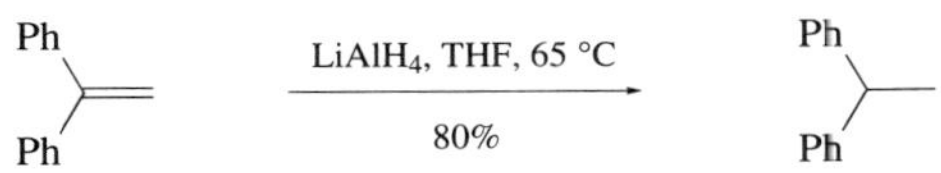

(38)

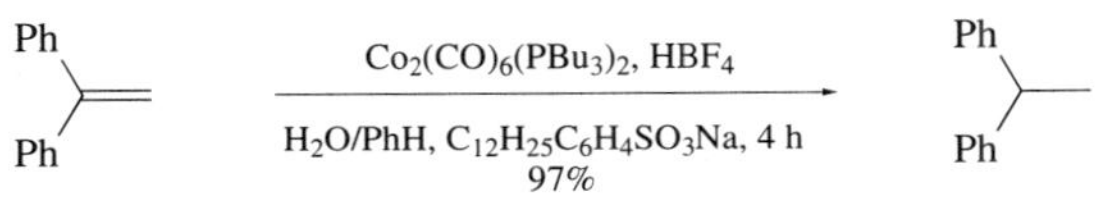

(39)

Heteroaryl-substituted double bonds are similarly straightforward to reduce. A wide variety of heteroaryl-substituted acrylic acids have been studied ⟨50JA3539⟩ and found to undergo clean reduction to the substituted propionic acid using Raney nickel, even if the heteroaromatic system is easily reduced (Equation (40)). Palladium on calcium carbonate has also been successfully used for the reduction of the quinoxaline-substituted system (Equation (41)) but the use of Raney nickel in this case has been reported to lead to a lactam by-product resulting from reduction of the heteroaromatic ring (Equation (42)) ⟨65JA1984⟩.

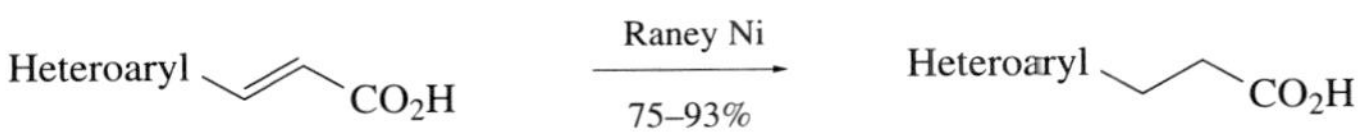

(40)

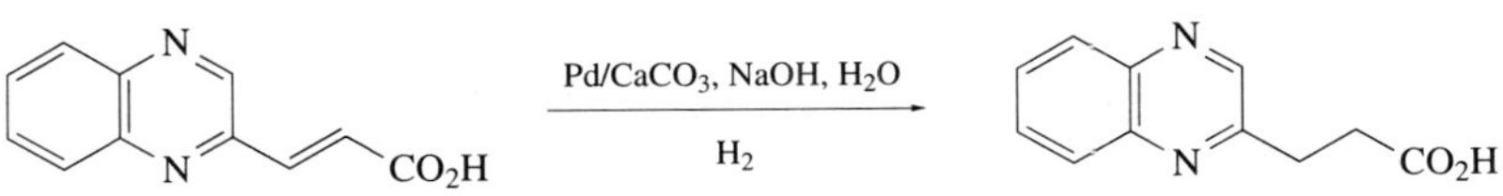

(41)

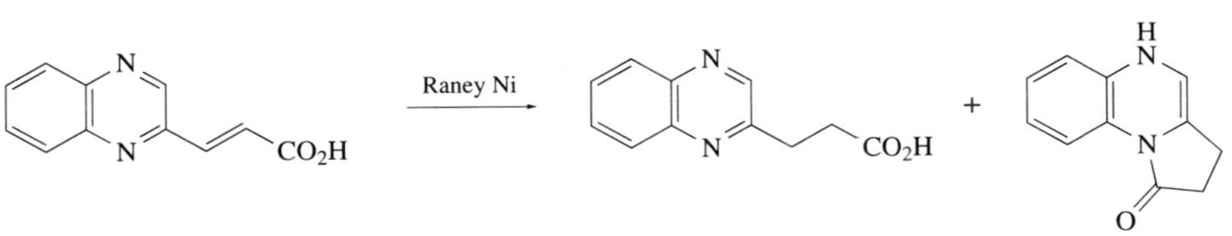

(42)

1.03.1.4 Reduction of Heteroalkyl-substituted Alkenes

Heteroalkyl-substituted alkenes can present problems for catalytic hydrogenation as a major side reaction is the hydrogenolysis of the C—heteroatom bond. However, vinyl halides are reduced to alkyl halides albeit rather poorly using diimide (Equation (43)) ⟨67JOC893⟩. A thorough study on the hydrogenation of an enol ether using a variety of metal catalysts has been carried out ⟨71BCJ166⟩. Some results of this study are shown in Equation (44). Palladium leads to the cleanest reduction while platinum gives the largest amount of hydrogenolysis. Oxygen-substituted alkenes such as vinyl acetate can be reduced using Wilkinson's catalyst (Equation (45)) ⟨68MI 103-01⟩, although it should be noted that it may be possible to reduce other double bonds in the presence of an enol ether (Equation (46)) ⟨71AJC513⟩. In this case, it is possible the major influence on the selectivity of the reaction is the steric hindrance of the enol ether double bond compared to the other double bond. Enol ethers are also somewhat resistant to diimide reduction, for example dihydropyran does not undergo reduction ⟨67JCS(C)1120⟩. However, a bisvinyl ether has been reported to undergo reduction (Equation (47)) ⟨75JOC3599⟩. Enol ethers are hydroborated to give exclusively 2-alkoxyalkylboranes regardless of the substitution pattern on the double bond. However, these organoboranes are susceptible to elimination ⟨68IJ691⟩.

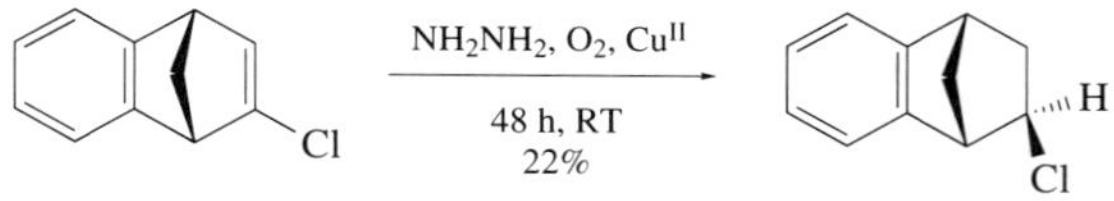

(43)

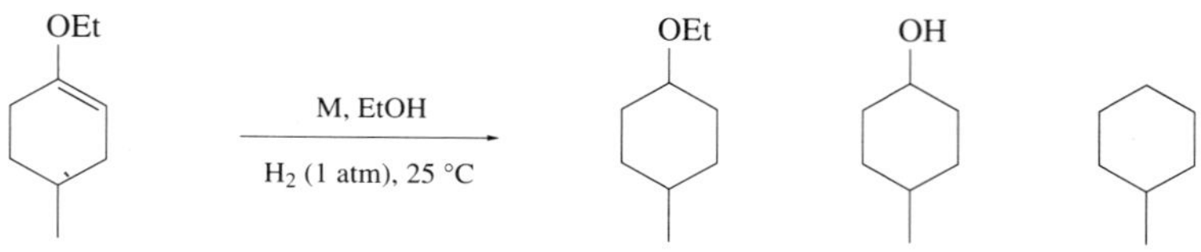

(44)

M = Ru	78%		
M = Rh	55%	30%	
M = Pd	98%		
M = Pt	45%		40%

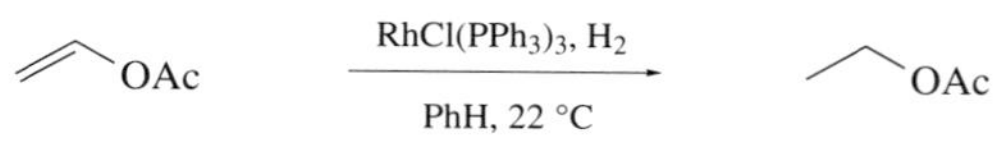

(45)

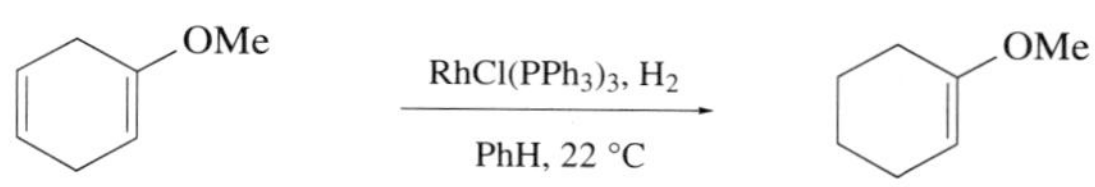

(46)

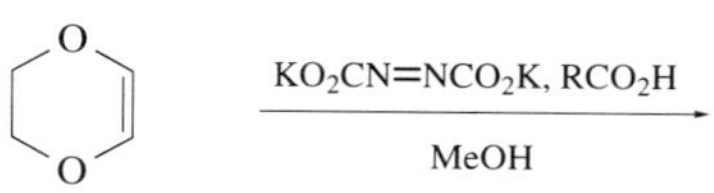

(47)

Enamines can be reduced using platinum oxide as the catalyst (Equation (48)) ⟨36CB2106⟩ although they are more usually reduced in other ways. Thus sodium borohydride in acetic acid reduces enamines to amines ⟨63BSF798⟩. For example the enamine derived from cholestenone undergoes reduction to give the homoallylic amine (Equation (49)) ⟨63JOC421⟩. Enamines also undergo reduction on heating with formic acid (Equation (50)) ⟨85ACS(B)187⟩. This reaction also works with enamines derived from morpholine (Equation (51)). An elegant deuterium labelling experiment has been carried out to shed light on the mechanism of this reaction and prepare specifically labelled amines (Scheme 7) ⟨57JA6210⟩. Nitrogen-substituted alkenes with the nitrogen in a variety of oxidation states are reduced to alkanes using Wilkinson's catalyst. Thus β-nitrostyrene is reduced to a nitroalkane ⟨66JCS(C)1894⟩ with no reduction of the nitro group (Equation (52)). The reduction of acetamidocinnamic acid derivatives has already been discussed. In common with simple vinyl ethers, simple enamines undergo no reduction of the double bond with diimide ⟨65CC71⟩.

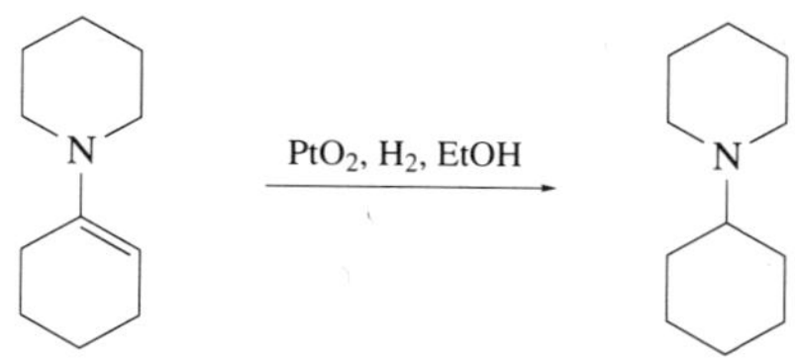

(48)

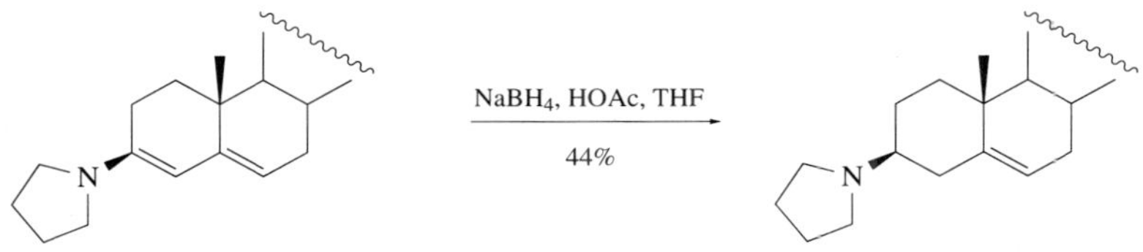

(49)

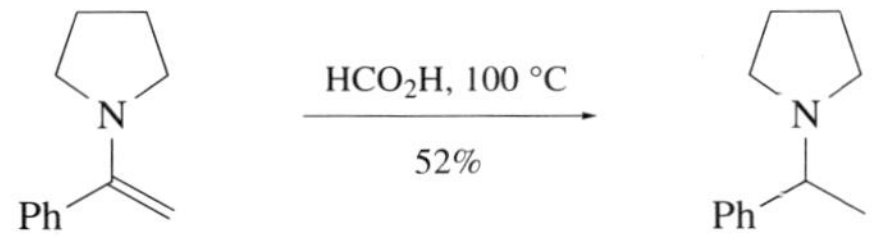

(50)

(51)

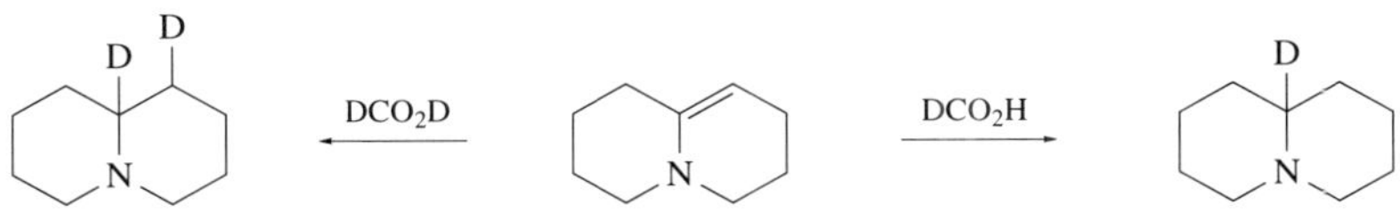

Scheme 7

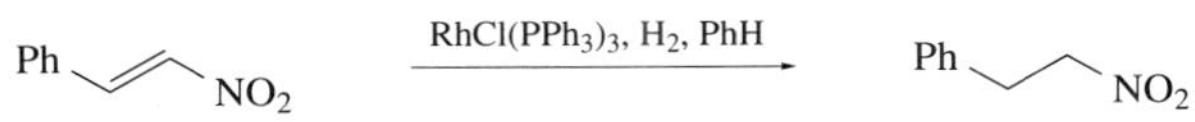

(52)

Vinylic sulfides are a particular problem since sulfur is an excellent catalyst poison and high catalyst ratios are required with long reaction times. Under such conditions hydrogenolysis of the C—S bonds is the most common reaction (Equation (53)). However, reduction of the same compound using a sulfur-treated palladium catalyst gave clean reduction of the double bond although the reaction was slow (Equation (54)) ⟨80JA1165⟩. Although there are no reports on the reduction of vinylsilanes using diimide, the clean reduction of the double bond in a vinylstannane (Equation (55)) has been reported by Kuivila and co-workers ⟨73JOM(61)167⟩

(53)

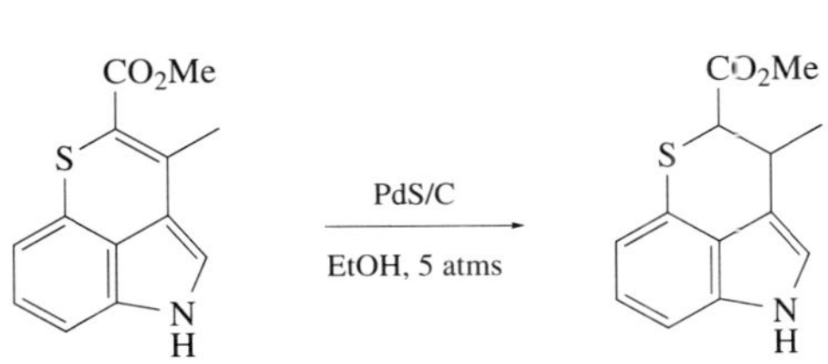

(54)

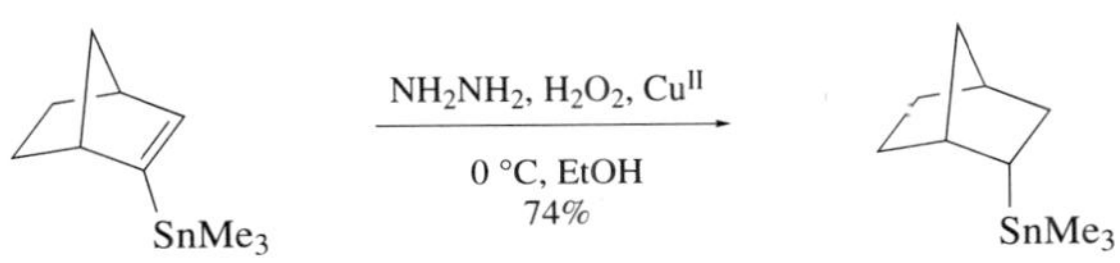

(55)

1.03.1.5 Reduction of Remotely Substituted Alkenes

Allylic substituents can also undergo hydrogenolysis although the activity of catalysts towards hydrogenolysis is rather different to that observed for vinylic systems. Palladium causes most hydrogenolysis followed by platinum and rhodium. Allylic chlorides (Equation (56)) are reduced in high yield by diimide ⟨76BCJ1437⟩ as are allylic bromides (Equation (57)) ⟨67JCS(C)1120⟩. Hydroboration of alkenes containing a range of functional groups is possible although for functionalities on or near the double bond some effect on the regioselectivity of addition is found. Allylic halides direct the boron unit to the proximal carbon of the double bond (Equation (58)). However, the resulting 2-haloboranes are unstable with respect to elimination to the alkene under either acid or base catalysis and hence such substrates cannot be used in the protonolysis reaction to give alkanes ⟨68MI 103-01⟩.

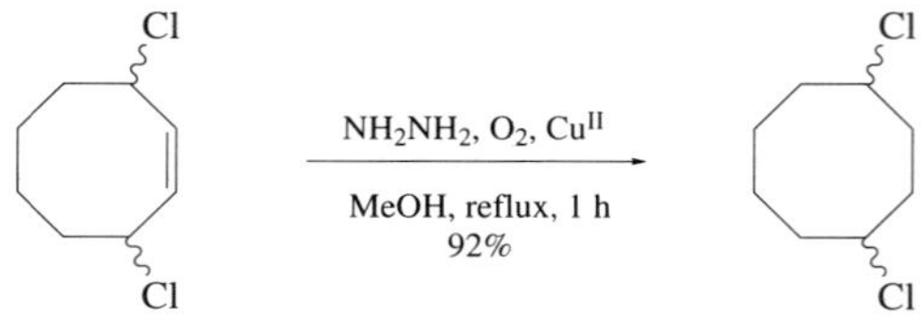

(56)

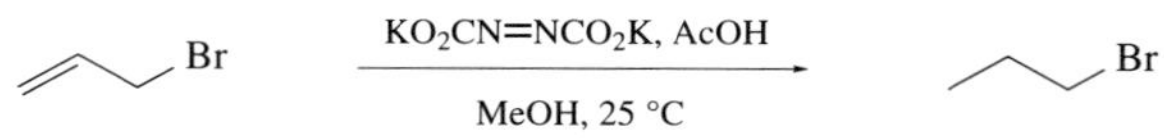

(57)

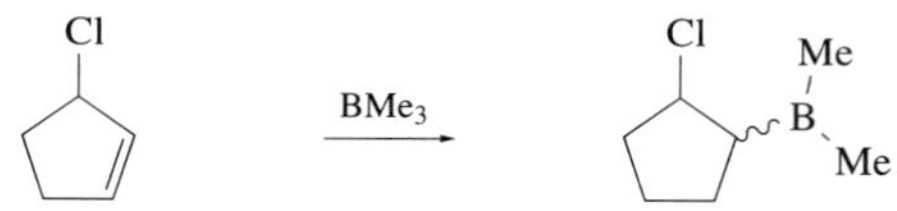

(58)

Rhodium on carbon gives clean reduction of an allylic acetate (Equation (59)) whereas both platinum oxide and palladium lead to extensive cleavage of allylic acetates ⟨79TL1641⟩. Allylic oxygen functions survive reduction of the double bond using diimide and cause no difficulties. For example, an allylic ozonide has been reported to undergo smooth reduction of the double bond (Equation (60)) ⟨87JMC1505⟩. The reduction of the double bond in allylic alcohols using lithium aluminum hydride has been reported and the mechanism of the reaction studied by NMR ⟨87BSF462⟩. Homogeneous catalysts have been used to control the facial selectivity of double bond reduction. Although the reduction of alkenes can occur with some diastereoselectivity, it is very hard to predict the result. Initially, Stork and Kahne ⟨83JA1072⟩ and subsequently Evans and Morrissey ⟨84JA3866⟩ realised that soluble catalysts combined with directing groups in the substrate could overcome this problem. Stork showed that cyclic allylic and homoallylic alcohols undergo diastereoselective reduction using a cationic iridium complex but perhaps the most striking example is the reduction of a hydroxyindenone (Equation (61)). Evans used both a cationic iridium complex and a cationic rhodium complex with a range of cycloalkenols but also demonstrated that acyclic systems undergo diastereoselective reduction using the rhodium catalyst and a higher pressure of hydrogen (Equation (62)).

(59)

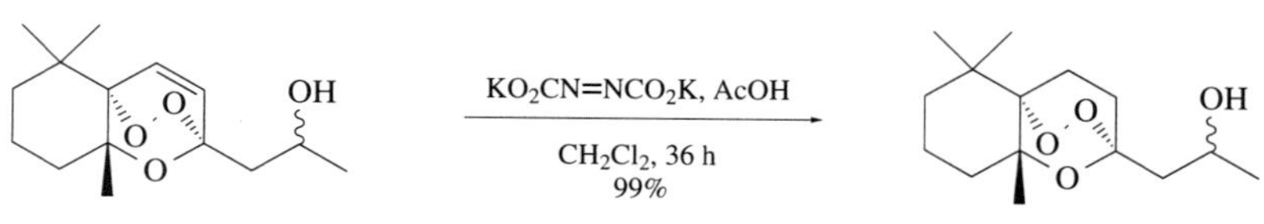

(60)

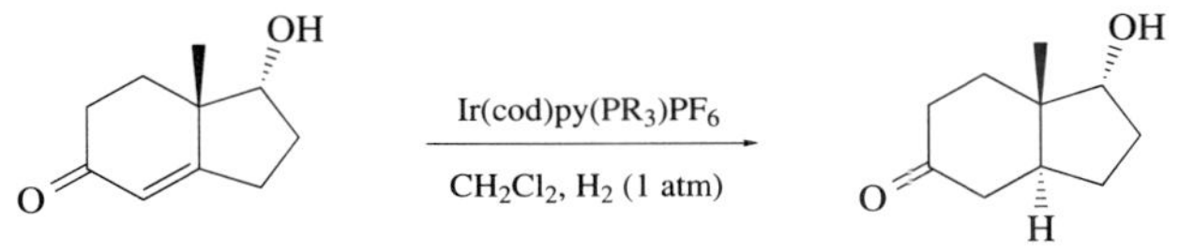

96% *trans*-indanone (61)

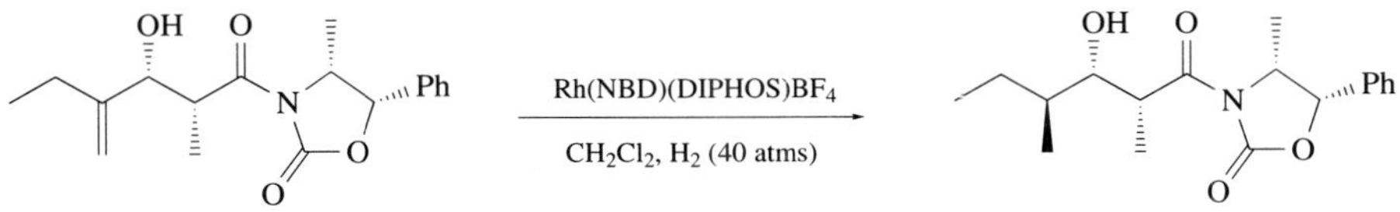

86% *de* (62)

Sulfur functionalities are not affected by diimide reduction (Equation (63)) ⟨74JOC3195⟩. One particularly useful alkene reduction which is often best achieved via hydroboration is the reduction of allylsulfides to sulfides. Thus allylmethylsulfide is reduced to methylpropylsulfide without causing problems of catalyst poisoning or hydrogenolysis (Equation (64)) ⟨59JA4108⟩.

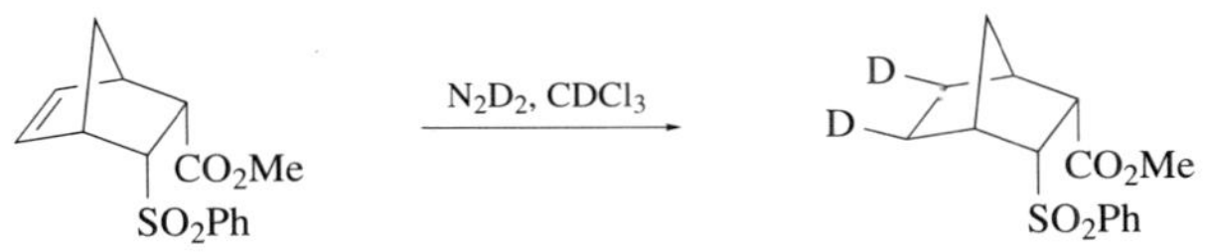

(63)

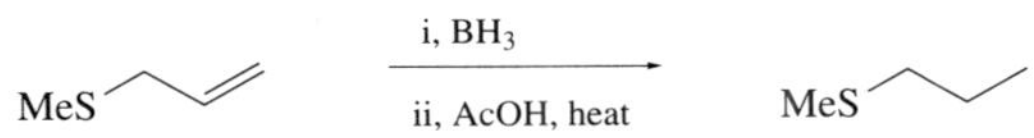

(64)

Using cyclohexanol as the source of hydrogen, catalytic transfer reduction of the double bond in a tetrahydropyridine has been reported to take place in high yield (Equation (65)) ⟨79MIP15154⟩.

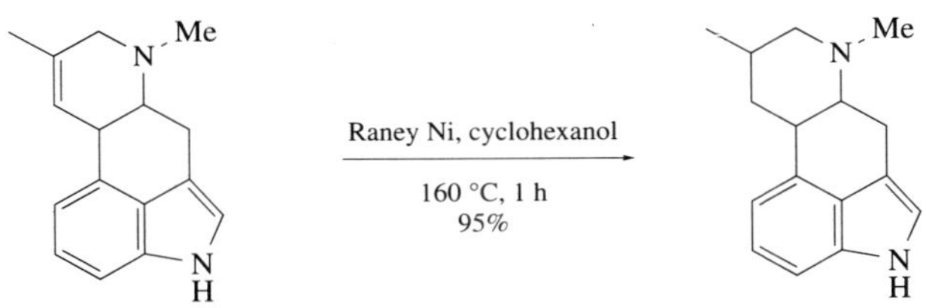

(65)

A particularly useful alkene reduction is the selective reduction of the carbon=carbon double bond of α,β-unsaturated carbonyl compounds. A wide variety of reagents have been reported to achieve this transformation. Palladium is probably the most suitable heterogeneous catalyst for this as reduction of the carbonyl group is rare (Equation (66)) ⟨83JOC3370⟩. In the reduction of eremorphilone using palladium, the conjugated double bond is reduced in preference to the rather unhindered double bond in the isopropenyl group (Equation (67)) ⟨67JOC2013⟩. The presence of a carbonyl group conjugated to the double bond can cause problems with homogeneous catalysts particularly where the carbonyl group is an aldehyde, when decarbonylation can be a side reaction. Reduction of acrolein using Wilkinson's catalyst gives propanal in yields varying between 61% and 82% depending on the concentration of catalyst and hydrogen pressure (Equation (68)) ⟨67JCS(C)270⟩. The reduction of the conjugated double bond in juglone has been reported (Equation (69)) ⟨67TL3457⟩. Reduction of α,β-unsaturated carbonyl compounds, nitriles, etc., proceeds smoothly using catalytic transfer technology and there are many examples ⟨74CRV567⟩. The reduction of crotonaldehyde has been reported in 81% yield (Equation (70)) ⟨78JOC3985⟩.

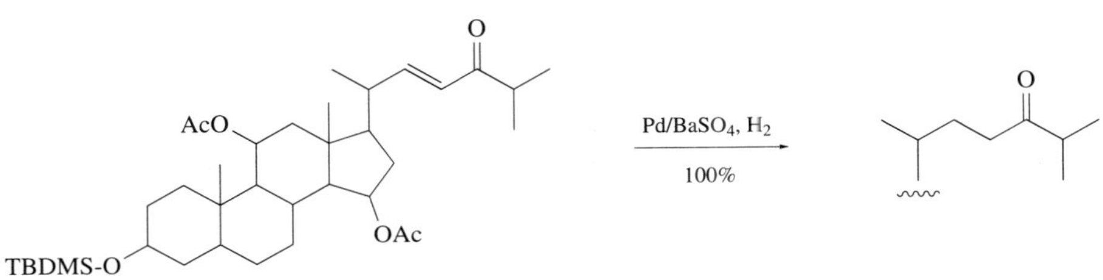

(66)

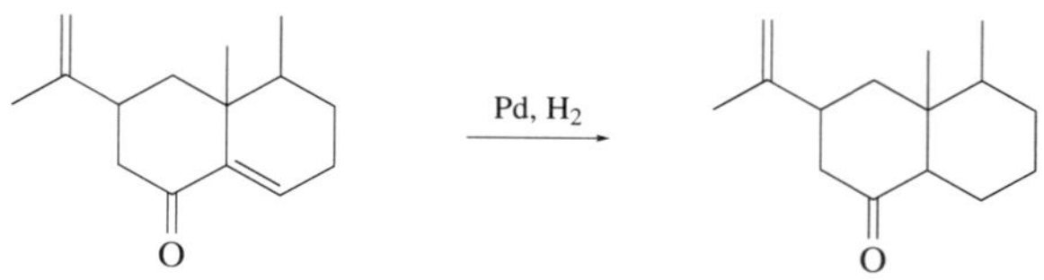

(67)

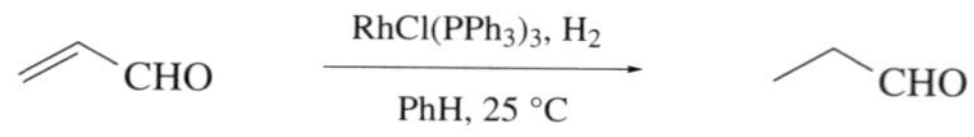

(68)

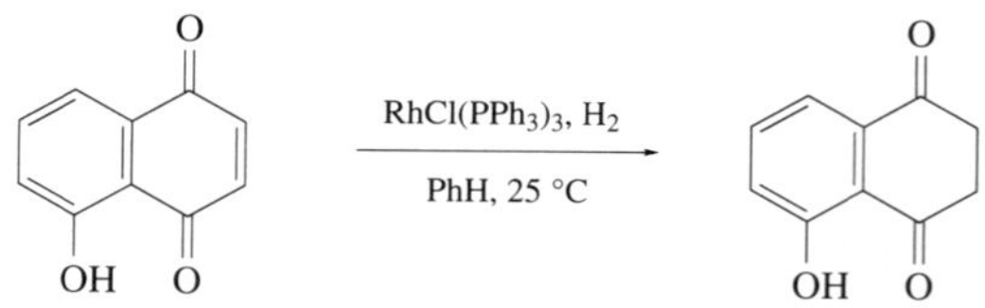

(69)

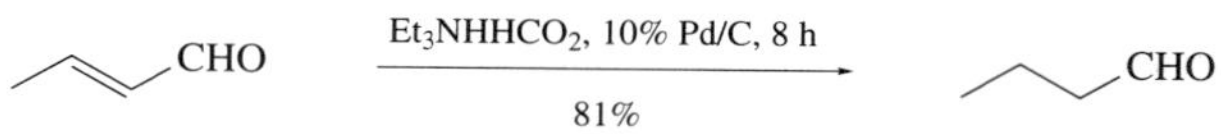

(70)

The reaction of tributyltin hydride with catalytic palladium(0) reduces cinnamaldehydes unless they have an electron-donating group in the *para* position. Some hints concerning the mechanism can be gleaned from the reactions shown in Scheme 8. Lipshutz *et al.* have reported the reduction of unsaturated ketones to saturated ketones using tributyltin hydride/copper(I) iodide ⟨89SL64⟩ which is suggested as involving hydridocuprates. The copper hydride reagent of Stryker and co-workers achieves these reactions cleanly ⟨88JA291⟩. Even cyclic ketones substituted at the β-position are reduced in high yield (Equation 71) along with the double bond in unsaturated esters. Isolated double bonds do not react.

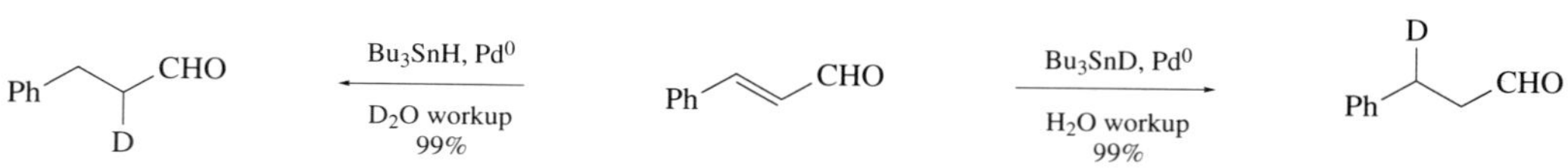

Scheme 8

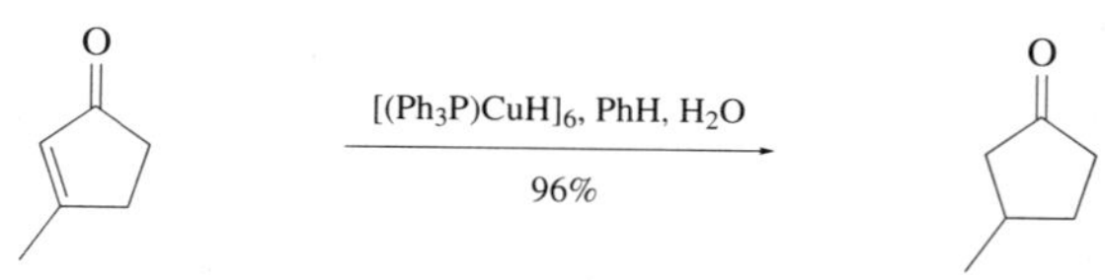

(71)

Cyclohexenones have also been reduced to cyclohexanones using methylcopper and diisobutylaluminum hydride (dibal-H) in hexamethylphosphoramide (HMPA), a system that reduces unsaturated esters to saturated esters ⟨86JOC537⟩. Silanes in the presence of a transition metal catalyst will reduce the carbon═carbon double bond of enones. Phenylsilane and catalytic molybdenum hexacarbonyl effect reduction of benzylidene acetone (Equation (72)) ⟨87JOC2576⟩ although the reaction with cyclohexenone gives cyclohexanone in poor yield. Rhodium(I) also catalyses the reaction of silanes and enones ⟨89JA6257⟩. The reaction presumably proceeds via hydrosilylation to give the silylenol ether followed by hydrolysis during workup (Scheme 9). Potassium triphenylborohydride has been reported to be a useful reagent for the conjugate reduction of both cyclic and acyclic enones (Equation (73)) ⟨88SC89⟩ although if an aromatic ring is also conjugated to the double bond, preferential 1,2-reduction occurs. An unusual system consisting of carbon monoxide, water and selenium reduces enones to ketones in the presence of a strong base such as 1,5-diazabicyclo[5,4,0]undec-5-ene (dbu) (Equation (74)) ⟨89BCJ1682⟩. Unsaturated esters, acids and amides also react but need higher temperatures while enals undergo reduction at the aldehyde. Magnesium in methanol will reduce unsaturated esters to saturated esters ⟨86TL2409⟩. Isolated double bonds and epoxides are unreactive and the use of deuteriated methanol leads to the overall addition of D_2 across the double bond (Equation (75)) ⟨87TL5287⟩.

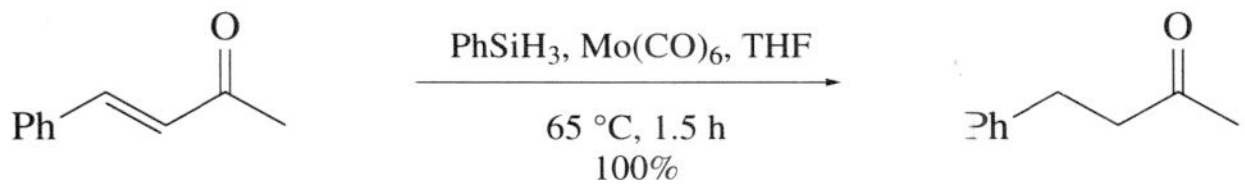

(72)

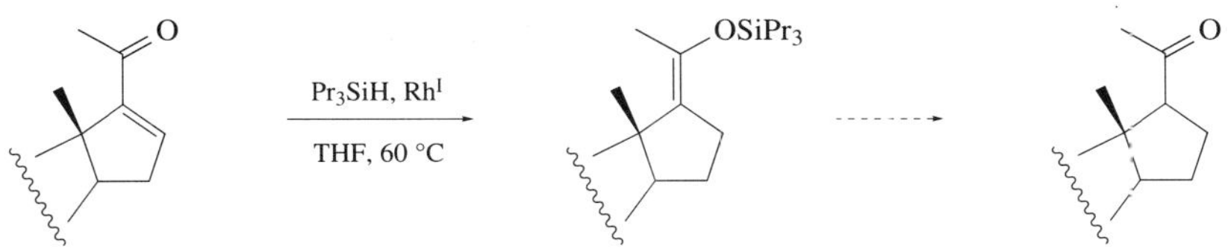

Scheme 9

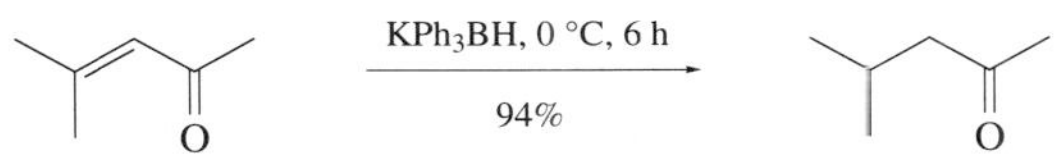

(73)

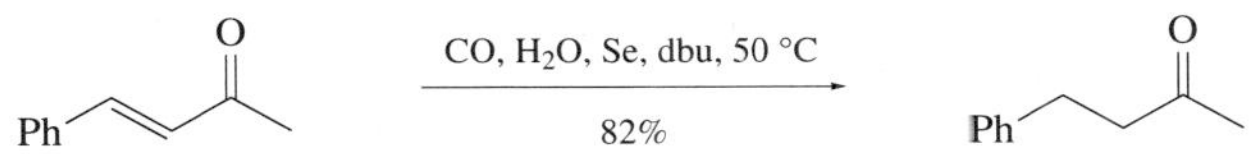

(74)

dbu = 1,5-diazabicyclo[5.4.0]undec-5-ene

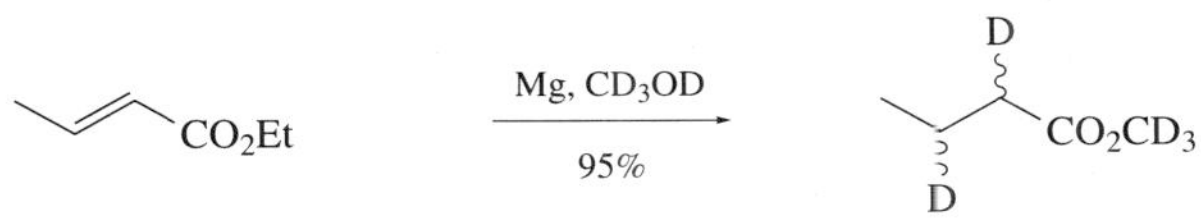

(75)

1.03.2 REDUCTION OF ARENES AND HETEROARENES

1.03.2.1 Types of Reaction

The reduction of arenes and heteroarenes can be accomplished by a variety of methods. Heterogeneous hydrogenation using a number of different metals as the catalyst has been reviewed ⟨B-65MI 103-01, B-67MI 103-01, B-78MI 103-01, 79MI 103-01, B-85MI 103-01⟩. This approach generally requires moderate temperatures and pressures to achieve the reduction of arenes, while some heteroarenes can be reduced under milder conditions. Such reductions are commonly used in industrial processes. More recently, homogeneous hydrogenation of arenes has been explored and reviewed ⟨79ACR324⟩. Dissolving metal reduction of arenes leads to partially reduced aromatic systems ⟨92OR(42)1⟩ and has found considerable use in synthesis ⟨86NPR35⟩. Reduction of arenes by hydride reducing agents is generally very difficult but the corresponding reaction of electron-deficient heteroarenes such as pyridine is much easier ⟨86AHC(39)1⟩. Electrochemical reduction particularly of polycyclic arenes ⟨B-72MI 103-01⟩ and heteroarenes ⟨84AHC(37)167⟩ has also found some applications. The following sections will consider each of these methods separately.

1.03.2.2 Heterogeneous Hydrogenation

The first reported reductions of aromatic rings involved the reduction of phenol and aniline ⟨04BSF101⟩ and benzene ⟨14BSF228⟩ using a nickel metal catalyst. Rhodium, ruthenium, iridium, palladium and platinum on a solid support will catalyse the reduction of benzene to cyclohexane under pressures of about 6.9 MPa of hydrogen and at temperatures of about 100 °C. However, rhodium and ruthenium are the best catalysts with rates approaching 100 times those of the other metals (Table 3). Generally, the metal is supported on either carbon or alumina. The choice of support is not usually important although the reduction of acetophenone to cyclohexylethanol using rhodium occurs in higher yield when alumina is used as the catalyst support ⟨79MI 103-01⟩. In common with many catalytic reactions, trace impurities in the substrate can dramatically affect the rate of reduction. Metal ions such as sodium can inhibit reduction in platinum oxide catalysed reactions. The traditional poisons for noble metal catalysts such as lead and mercury do not seem to have much effect on the rate of reduction ⟨B-67MI 103-01⟩. Acids can act as either inhibitors or promotors; HCl inhibits the reduction of toluene or benzoic acid over 5% rhodium on carbon or alumina ⟨61JOC1835⟩ whereas reduction over rhodium on alumina is promoted by acetic acid. ⟨55JA4262⟩.

Table 3 Relative rates of reduction of benzene.

Catalyst	*Relative rate*
Ir	1
Pd	1
Pt	1
Rh	77
Ru	100

Rhodium on alumina is probably the catalyst of choice for the hydrogenation of aromatic rings under the mildest conditions without benzylic hydrogenolysis. For example mandelic acid is reduced to hexahydromandelic acid using 5% rhodium on alumina without racemisation (Equation (76)) ⟨62JOC2288⟩. Hydrogenation of arenes using ruthenium catalysts is generally performed at 100°C and 6.9 MPa. Under these conditions, benzylic hydrogenation can occur as shown in Equation (77) ⟨58JA5270⟩. However, the hydrogenation of ephidrene (Equation (78)) using ruthenium has been reported to occur without racemisation ⟨58JA5270⟩. Platinum oxide in the presence of a trace of acetic acid can be used to hydrogenate benzyl alcohol without hydrogenolysis but as the amount of acetic acid is increased so does the amount of benzylic hydrogenolysis ⟨B-67MI 103-01⟩. Palladium is the preferred catalyst for the hydrogenolysis of benzylic functionalities and hence is least widely used for reduction of arenes.

(76)

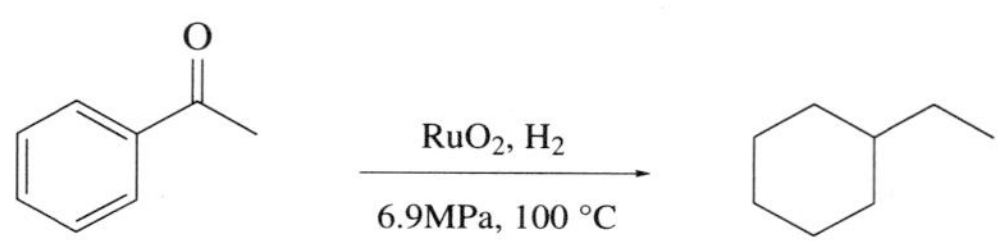

(77)

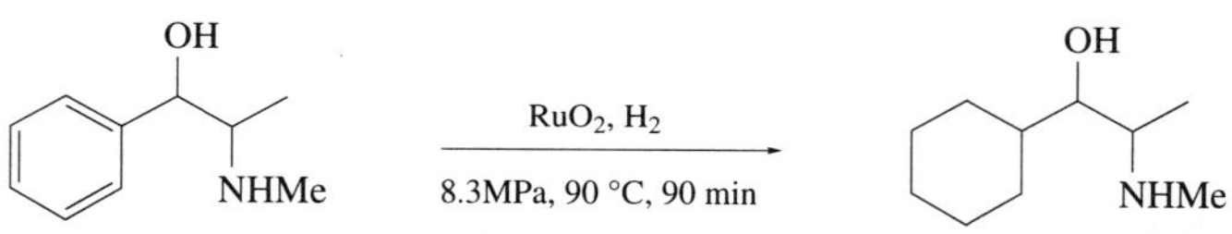

(78)

Although the effect of the support is usually small for alumina and carbon, a strong metal-support interaction has been described for Group 9 metals with transition metal oxides such as titanium dioxide ⟨87ACR389⟩. This has led to the development of a highly active rhodium catalyst formed by deposition of rhodium on titanium dioxide via decarbonylation of $Rh_4(CO)_{12}$ followed by activation by exposure to air ⟨90RTC87⟩. In the reduction of toluene at 1 atmosphere of hydrogen and 50°C, this catalyst was found to be about twice as active as the commercial rhodium/alumina.

Partial hydrogenation of arenes using heterogeneous catalysis is difficult to achieve as the intermediates are reduced faster than the arene and low selectivity is observed ⟨B-67MI 103-01⟩. The fusion of a less easily reduced aromatic ring can also allow partial reduction as shown in Equation (79) ⟨61JA3160⟩. The use of a modified ruthenium surface has been reported to enable the partial reduction of benzene ⟨63MI 103-01⟩.

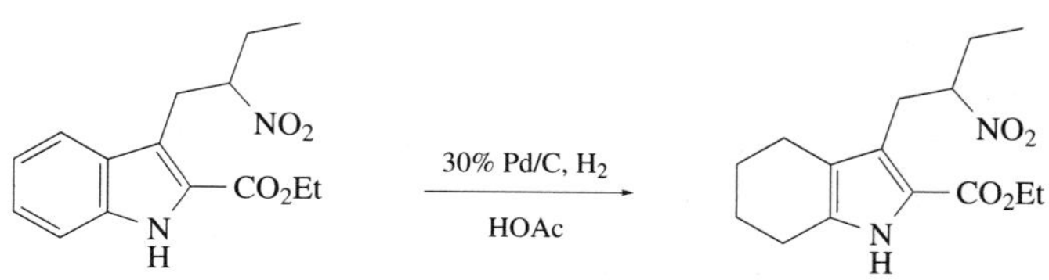

(79)

Alkyl-substituted benzenes can be reduced to alkyl-substituted cyclohexanes using a variety of catalysts although the reaction becomes slower with increasing number of substituents. With xylenes the product is predominantly the *cis* isomer indicating addition of hydrogen from one face ⟨57CRV895⟩ although this varies with catalyst and isomer (Table 4) ⟨B-67MI 103-01⟩.

Table 4 Reduction of dimethylbenzenes.

Isomer	*Catalyst*	*Ratio (cis : trans)*
1,2	Rh	89 : 11
1,2	Ru	93 : 7
1,3	Rh	74 : 26
1,3	Ru	85 : 15
1,3	PtO_2	81 : 19
1,4	Rh	64 : 36
1,4	Ru	70 : 30

Halogen-substituted aromatics suffer ready loss of the halogen. Generally, the C—F bond is stable ⟨55JA4262⟩ while the C—Cl and the C—Br bonds are cleaved with increasing ease ⟨53CB167⟩ and the C—I bond is cleaved rapidly. The use of nonpolar solvents has been reported to retard halogen loss ⟨49JA4154⟩.

Phenols can undergo ring reduction with three possible outcomes, namely reductive cleavage of the C—O bond, formation of a cyclohexanone or formation of a cyclohexanol. Reduction with cleavage of the C—O bond is prevalent with platinum and iridium catalysts ⟨B-67MI 103-01⟩. Formation of cyclohexanones is relatively rare but has been observed with palladium catalysts (Equation (80)) ⟨64JA118⟩. Formation of the cyclohexanol product is favoured using rhodium and ruthenium catalysts. The reduction of 1-naphthol to decanol proceeds in high yield (Equation (81)) using a rhodium catalyst ⟨71OS(51)103⟩. Ruthenium requires higher pressure but is equally successful in reducing phenols to cyclohexanols (Equation (82)) ⟨63JOC6⟩.

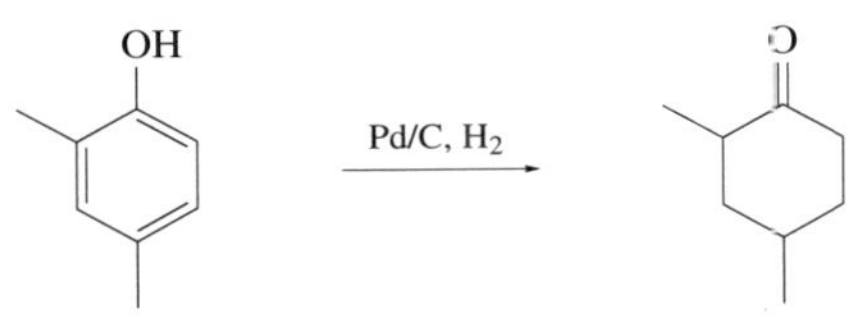

(80)

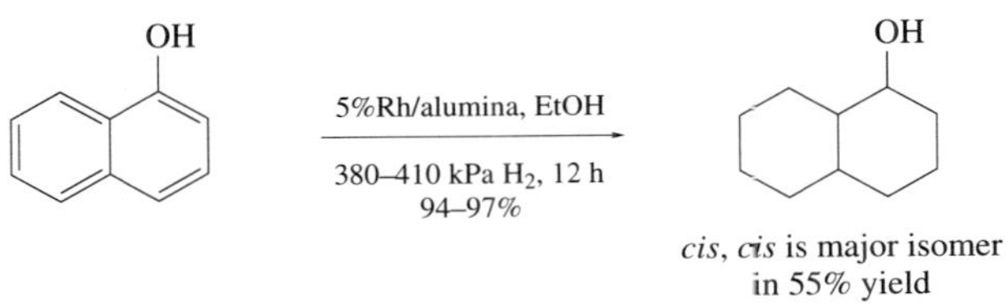

(81)

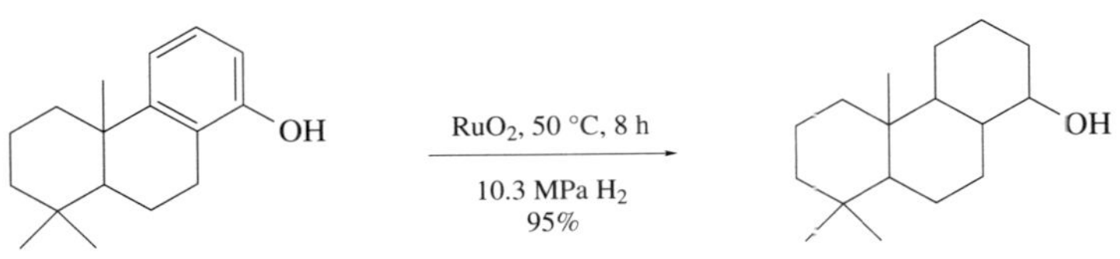

(82)

Anilines can be reduced to cyclohexylamines but suffer from the concommitant formation of dicyclohexylamines by reductive coupling ⟨B-67MI 103-01⟩. Again rhodium and ruthenium are the most active catalysts. Nitrobenzenes are more readily reduced at the nitro-group using catalysts such as platinum or palladium at low pressures and temperatures.

Heteroaromatic systems vary in their susceptibility to hydrogenation and with the nitrogen-containing rings, the presence of acid is often helpful to cause protonation and prevent poisoning of the catalyst by the more basic reduced product ⟨B-71MI 103-01⟩. Pyridine has been reduced to piperidine using a variety of catalysts with rhodium being the most active under mild conditions, platinum having been widely used, palladium and ruthenium being used industrially and Raney nickel requiring higher temperatures and pressures ⟨B-85MI 103-01⟩. Under acid conditions, the pyridinium salt is formed which is reduced more readily than the free base and quaternary pyridinium salts are reduced even more readily probably because the products are poorer catalyst poisons. In quinoline, selectivity of reduction is possible. Thus reduction using platinum in TFA causes reduction of the carbocyclic ring while use of nickel catalyst at high temperature causes reduction of the pyridine ring. Using nickel and high temperature and pressure, leads to the decahydroquinoline as a mixture of *cis* and *trans* isomers (Scheme 10) ⟨B-71MI 103-01⟩. Isoquinoline also undergoes selective reduction depending on the conditions (Scheme 11). In methanol with a platinum catalyst, reduction of the pyridine ring occurs whereas in strong acid, the carbocyclic ring is reduced ⟨75JOC1191⟩. Pyrimidine is reduced to the tetrahydropyrimidine using palladium or platinum catalysts under acidic conditions (Equation (83)) ⟨62CHE430⟩. Pyrrole is less readily reduced than benzene but can nonetheless be reduced to pyrrolidine using platinum or rhodium as the catalyst. Ruthenium at 100–200 °C and 10.34 MPa has been successful ⟨B-71MI 103-01⟩. The solvent can play an important role as shown in Scheme 12. In ether only the alkene double bond is reduced while in ethanol the pyrrole ring is reduced to give the pyrrolizidine ⟨66JOC870⟩. Indole resembles pyrrole in being difficult to reduce and the most successful catalysts are Raney nickel, ruthenium and copper chromite ⟨41JA1563⟩. Thus hydrogenation of indole over ruthenium at 100 °C and 85 atmospheres pressure gave octahydroindole whereas treatment with Raney nickel in ethanol at the same temperature and pressure gave indoline in 82% yield (Scheme 13).

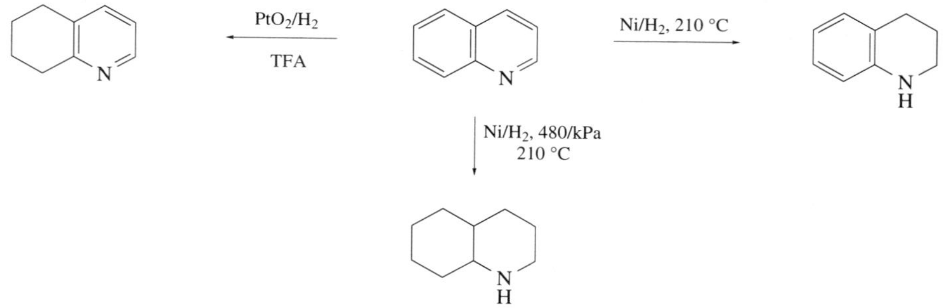

Scheme 10

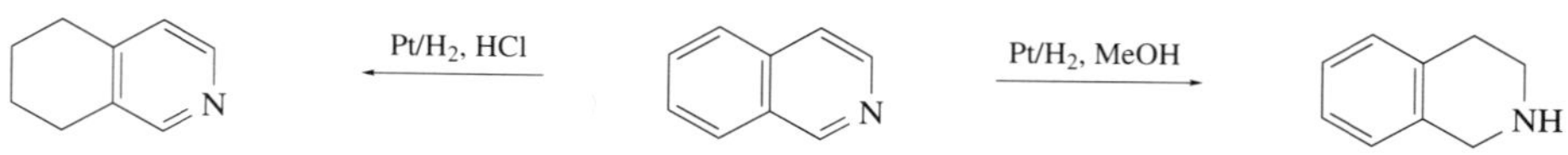

Scheme 11

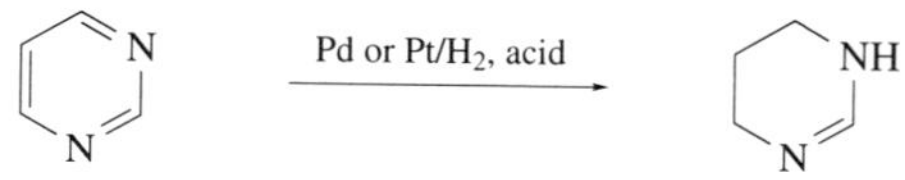

(83)

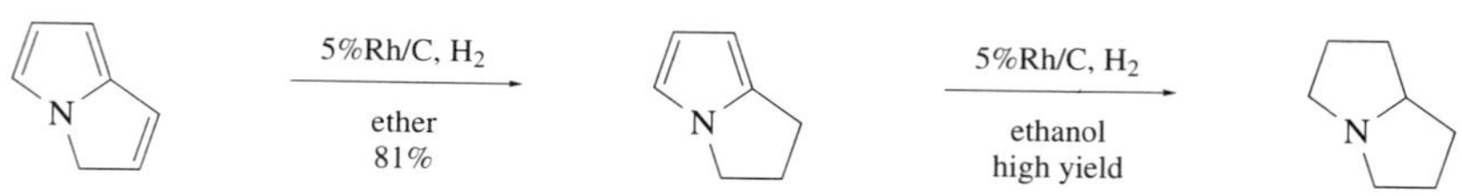

Scheme 12

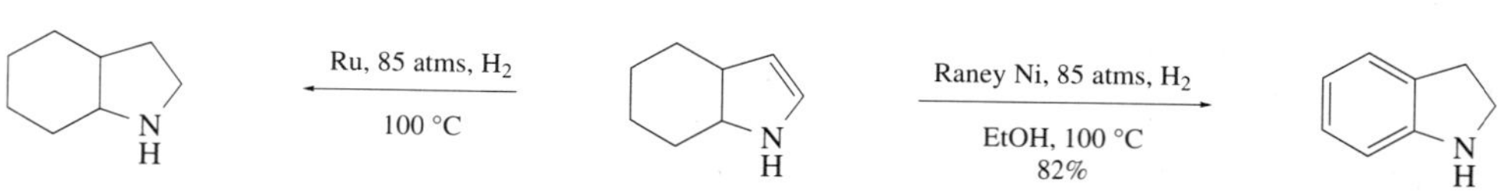

Scheme 13

Furans can undergo simple reduction to tetrahydrofurans particularly using palladium, ruthenium or rhodium catalysts. Furan has been reduced quantitatively to tetrahydrofuran using palladium and hydrogen at 690 kPa (Equation (84)) ⟨43OSC(2)566⟩. Under harsher conditions or with platinum catalyst, hydrogenolysis takes place ⟨67T215⟩. Benzofurans have been hydrogenated in the furan ring to give 2,3-dihydrobenzofurans using palladium at 100 °C and moderate pressure (Equation (85)) ⟨76CB2628⟩. There are very few reports concerning the hydrogenation of thiophenes although reduction over palladium to a tetrahydrothiophene is the key step in a biotin synthesis (Equation (86)) ⟨77JOC135⟩.

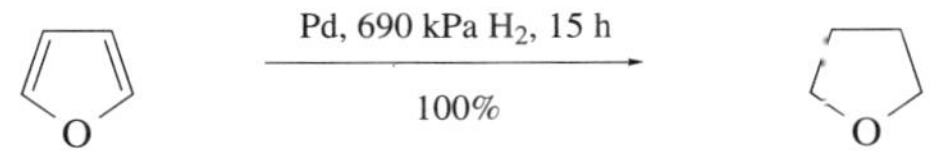

(84)

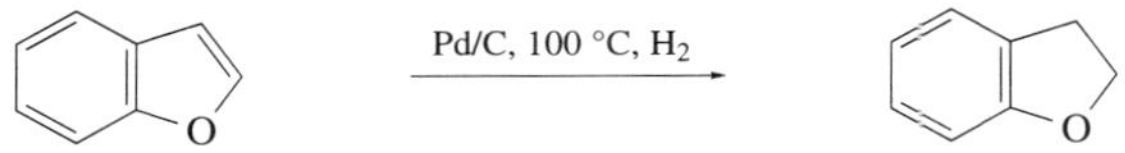

(85)

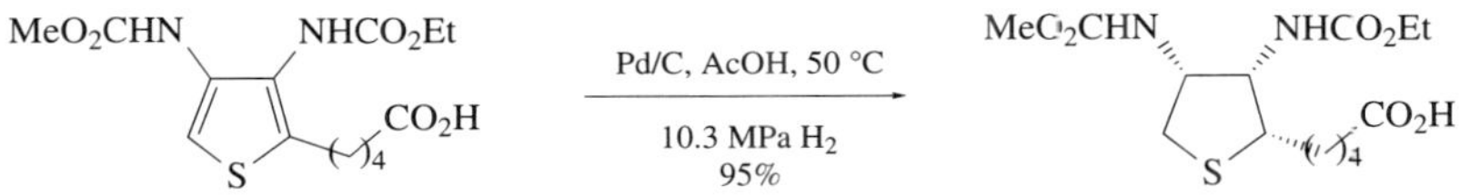

(86)

Hydrogenation of hydroxypyridines (pyridones) can result in hydrogenolysis in addition to reduction. 3-Hydroxypyridine is susceptible to reduction of the carbon—oxygen bond and gives a mixture of piperidine and 3-hydroxypiperidine (Equation (87)) whereas the 2- and 4-pyridones retain the oxygen, although in the case of 4-pyridone the product is 4-hydroxypiperidine (Equations (88) and (89)) ⟨57JA5279⟩ and ⟨58JA6412⟩. Amino-substituted pyridines also undergo hydrogenolysis to give piperidine (Equation (90)) ⟨61JOC818⟩. Carboxy-substituted pyridines can undergo decarboxylation on catalytic hydrogenation ⟨62JOC4046⟩. Nicotinic acid (Equation (91)) gives a mixture of piperidine-3-carboxylic acid and piperidine when reduced over rhodium, ruthenium or platinum although it has been reported that the use of rhodium on alumina in aqueous ammonia only gives reduction with no decarboxylation ⟨62JOC1406⟩. The presence of a carbonyl-substituent at the 3-position has also been reported to allow the isolation of the tetrahydro product. Thus the reduction of 3-acetylpyridine over palladium leads to the isolation of a tetrahydropyridine ⟨68JOC747⟩. Pyridines carrying substituents with remote functional groups generally present no complications in terms of reduction of the pyridine ring unless the functional group is particularly susceptible to reduction. Hydrogenation of the pyridine-3-acetic acid as the pyridinium salt was achieved over a mixed catalyst in 60–70% yield (Equation (92)) ⟨79CJC1977⟩. Benzylic hydrogenolysis is less of a problem in pyridines as shown by the reduction of the ketal of 2-acetylpyridine over platinum to give the piperidine ketal in high yield (Equation (93)) without hydrogenolysis of the benzylic carbon—oxygen bonds ⟨81T3615⟩.

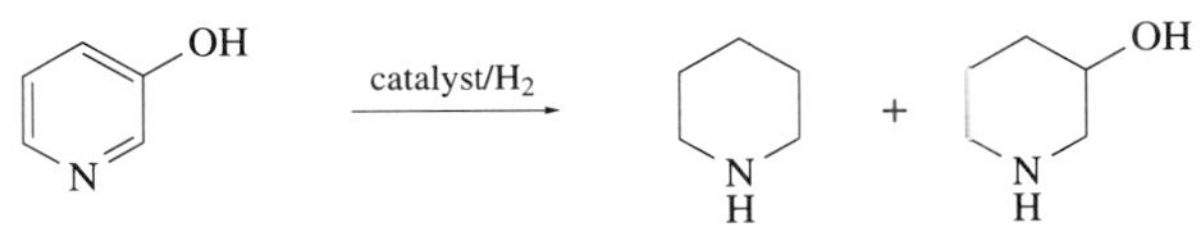

(87)

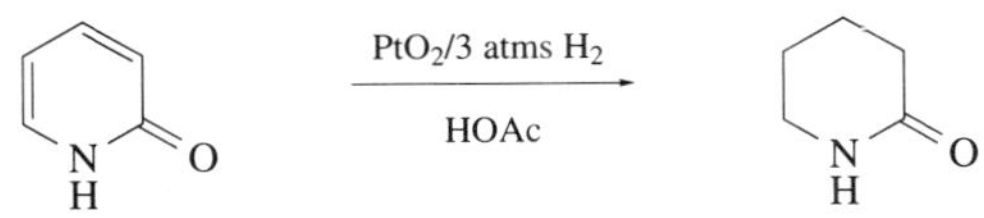

(88)

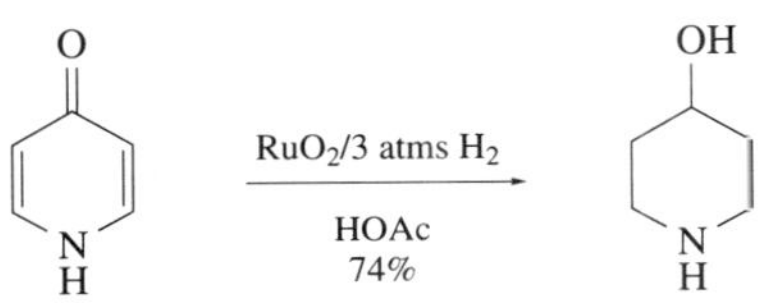

(89)

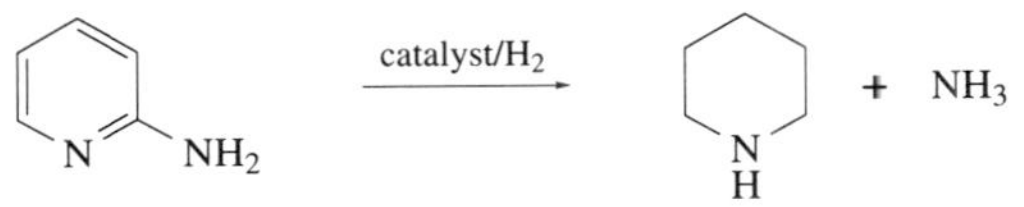

(90)

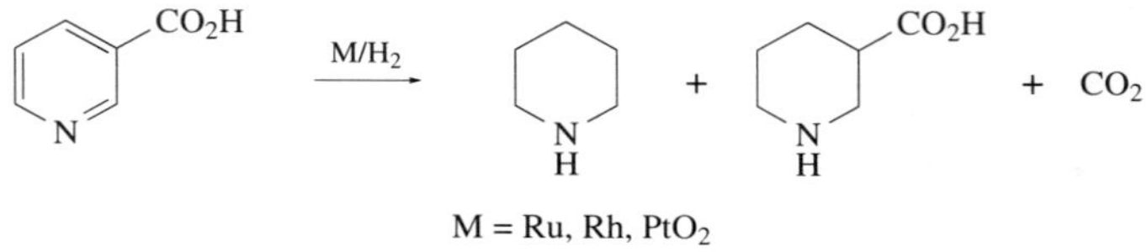

(91)

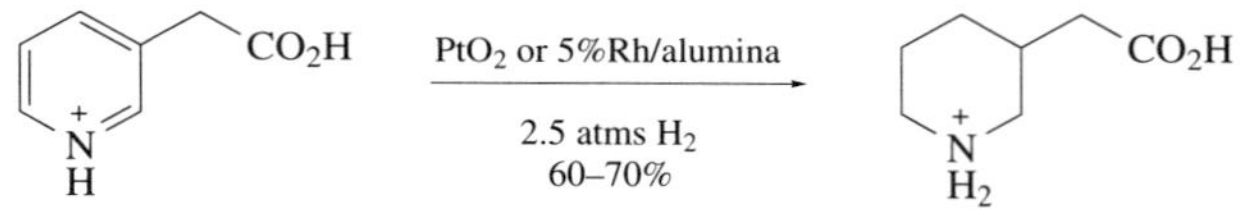

(92)

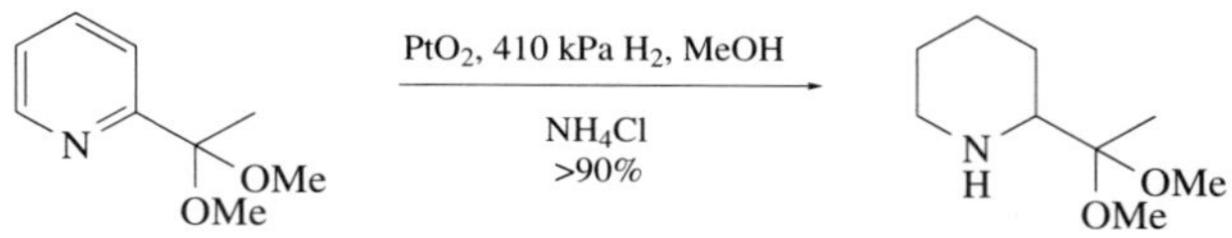

(93)

The introduction of electron-withdrawing substituents into pyrrole seems to increase the rate of hydrogenation. Thus *N*-phenylpyrrole is reduced with difficulty using Raney nickel at high temperature and pressure (Equation (94)) ⟨41JA1563⟩. However, the introduction of electron-withdrawing substituents on the nitrogen enable the smooth reduction of the pyrrole ring ⟨84JOC4203⟩. For example *N*-(*t*-butoxycarbonyl)pyrroles are reduced to the corresponding pyrrolidines using platinum on carbon at room temperature and atmospheric pressure (Equation (95)). The large amount of the *trans* isomer obtained in the reduction of the 2,5-disubstituted pyrrole is unusual (Equation (96)). The reduction of 2-substituted pyrroles carrying a carboxylic acid group in the side chain has been achieved using 5% rhodium on alumina to yield the amino acid after ester hydrolysis (Equation (97)) ⟨79CJC1977⟩. Given the difficulty of hydrogenation of indoles, there are few reports of such reactions of substituted indoles. Electron-withdrawing substituents on nitrogen (e.g., acetyl) lead to more selective reduction of the nitrogen-containing ring compared to unsubstituted indoles while *N*-alkyl indoles are more difficult to reduce ⟨B-85MI 103-01⟩. The reduction of 2-phenylindole to 2-phenylindoline using copper chromite at high temperature has been reported (Equation (98)) ⟨41JA1563⟩.

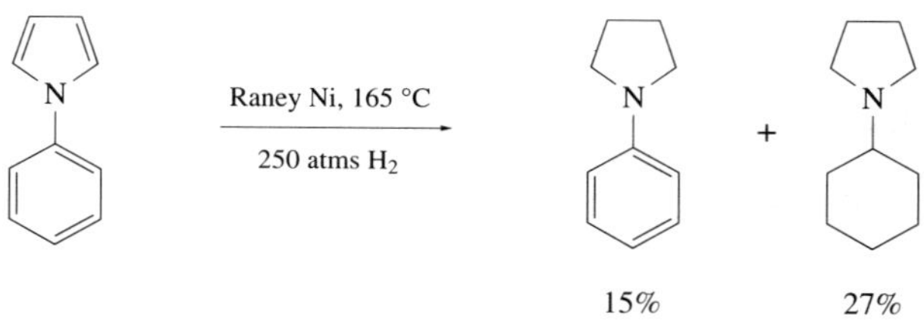

(94)

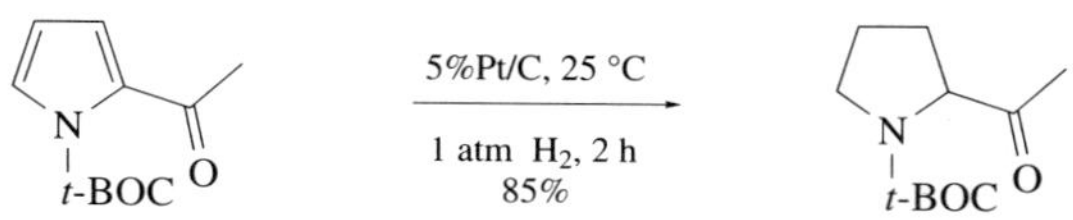

(95)

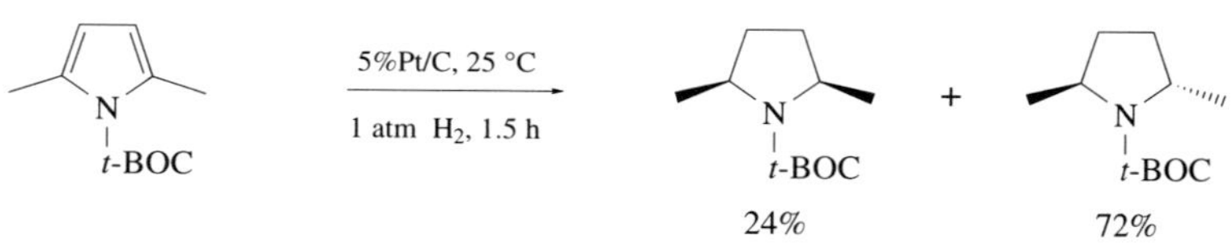

(96)

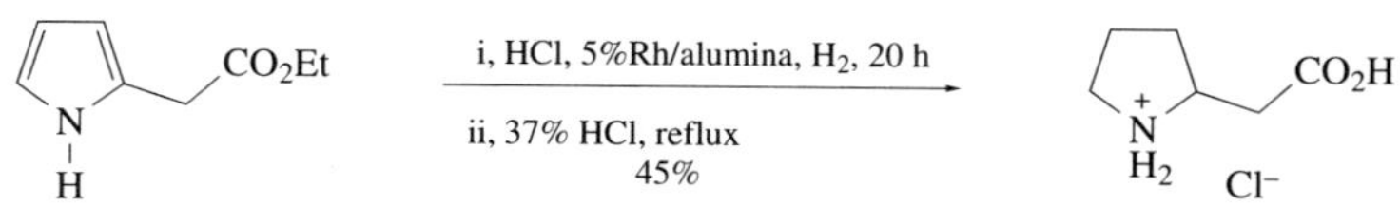

(97)

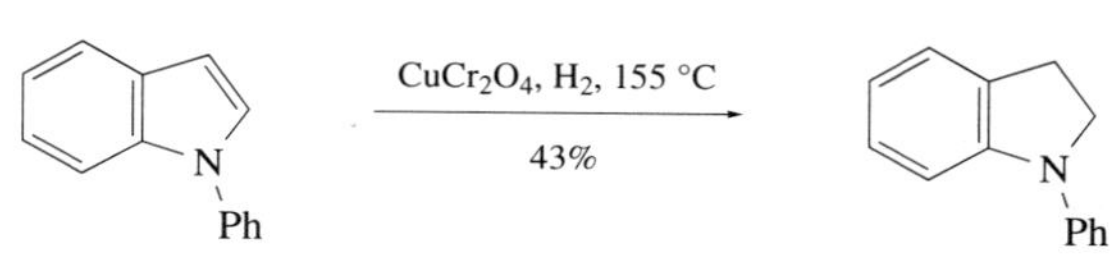

(98)

1.03.2.3 Homogeneous Hydrogenation of Arenes

Homogeneous hydrogenation of arenes can be achieved with a number of complexes. In 1974, Muetterties and Hirsekorn reported that π-allylCoL$_3$ complexes where L = phosphine or phosphite would reduce benzene to cyclohexane ⟨74JA4063⟩. No cyclohexenes are detected and the reaction is accomplished at 20°C and 1 atmosphere pressure of hydrogen with 0.01 molar equivalents of the cobalt complex ⟨76IC2379⟩. Hydrogenations using this catalyst are stereospecific giving the products of *cis* reduction. Thus, naphthalene gave *cis*-decalin as the sole product (Equation (99)) ⟨78JA2405⟩ and reduction of benzene with deuterium gave >95% of the all *cis* isomer of hexadeuteriocyclohexane ⟨75JA237⟩. Alkyl-substituents slow the reaction down as do electron-withdrawing substituents such as F or NO_2. Substituents which are capable of protonating the catalyst (OH and CO_2H) also inhibit the reaction. The mechanism of the reaction has been extensively studied ⟨79ACR324⟩ and the rate of reaction with alkenes has been shown to be broadly the same as the rate of reaction with arenes ⟨75JA237⟩. The allyl-rhodium complex, π-allylRh[P(OR)$_3$]$_2$ has been reported to reduce benzene under conditions of room temperature and atmospheric pressure of hydrogen ⟨79ACR324⟩. This catalyst does not display the stereospecificity of the cobalt complex and reduces alkenes some 10^5 times faster than arenes. A bis(rhodium) complex, [Rh(C_5Me_5)Cl_2]$_2$, has been reported by Maitlis and co-workers ⟨77CC427⟩ for the reduction of benzene. The presence of a base and a mildly coordinating solvent is required ⟨78JCS(D)617⟩. Reduction of electron-rich benzenes such as methoxybenzene is problematic (Equation (100)). Nitro-substituents are reduced (Equation (101)) but other electron-withdrawing substituents such as acetyl survive (Equation (102)). A ruthenium-based catalyst, (C_6Me_6)$_2$Ru has been reported to reduce 1,4-dimethylbenzene to a mixture of cyclohexane and cyclohexene derivatives under elevated temperature and pressure (Equation (103)) ⟨78JA7425⟩.

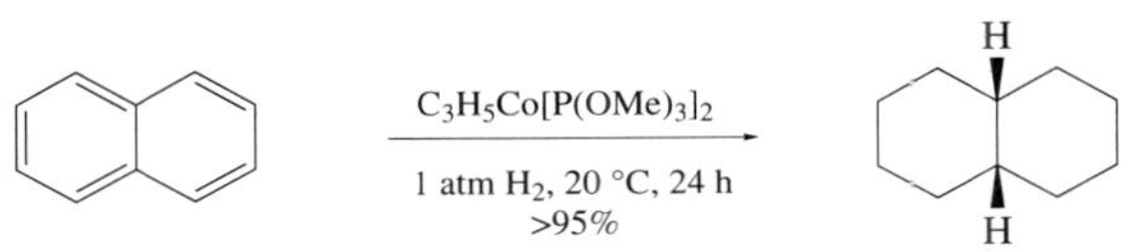

(99)

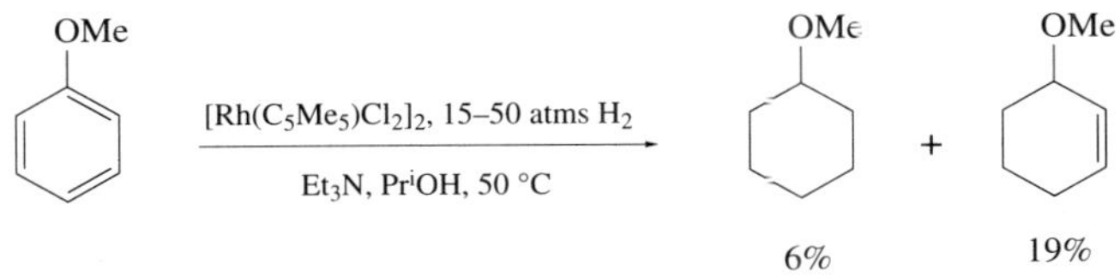

(100)

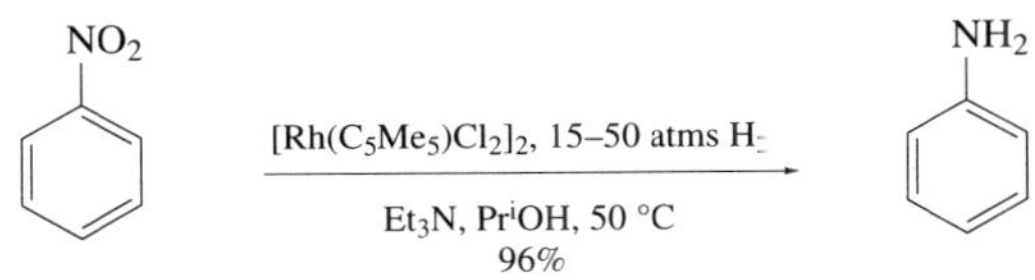

(101)

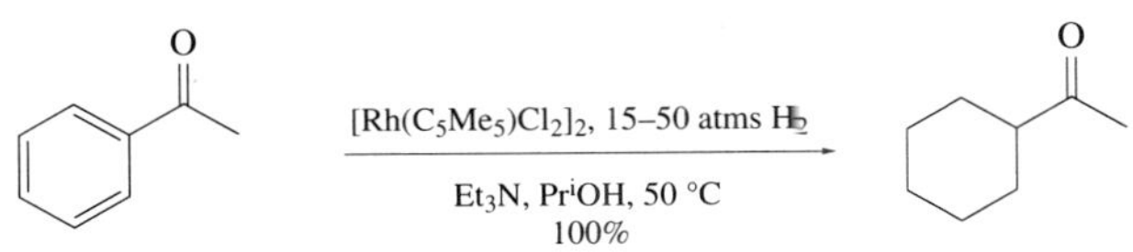

(102)

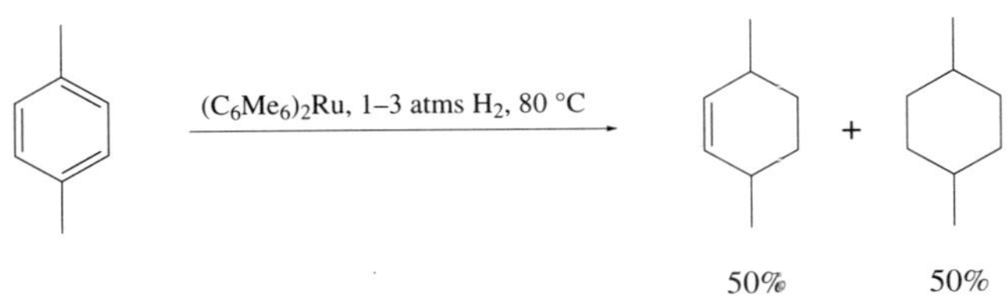

(103)

Several groups have explored the use of immobilised rhodium complexes as hydrogenation catalysts for arenes. The advantage of these systems lies in the ease of separation of the product. The first such system to be reported involved the reaction of $RhCl_3 \cdot 3H_2O$ with an anthranilic acid polymer to give the catalyst ⟨79JOC239⟩. Reduction of benzene to cyclohexane was achieved at room temperature in excellent yield (Equation (104)) but substituted benzenes are poor substrates for this system. For example nitrobenzene is reduced quantitatively to aniline. Stille and Sekiya reported the preparation of both rhodium- and cobalt-carbonylcyclopentadienyl complexes anchored to polystyrene via the cyclopentadienyl group ⟨81JA5096⟩. The rhodium complex acted as a hydrogenation catalyst reducing benzene at elevated temperature and pressure (100 °C and 4.5 MPa). This catalyst will reduce methoxybenzene (Equation (105)). In 1981, Schwartz and Ward reported the synthesis of a silica-bound allylrhodium hydride complex ⟨81JMOC397⟩. Reduction of benzene to cyclohexane was achieved at moderate pressures of hydrogen but naphthalene (Equation (106)) gave a mixture of tetralin and decalin consisting mainly of tetralin ⟨81JA5253⟩. This catalyst reduces methoxybenzene and aniline to the corresponding cyclohexanes in good yield although the reactions are a factor of 2 and 40 times slower respectively. A phosphinated polyacetylene polymer has been used to immobilise a bis(rhodiumnorbornadiene)chloride complex ⟨82CL603⟩. After activation of the catalyst and support by stirring in heptane under nitrogen, the quantitative reduction of benzene to cyclohexane occurred at 30 °C and 80 atmospheres pressure of hydrogen in 1 hour. Although methoxybenzene is reduced to methoxycyclohexane under these conditions, nitrobenzene is simply reduced to aniline while chlorobenzene and acetophenone are poor substrates.

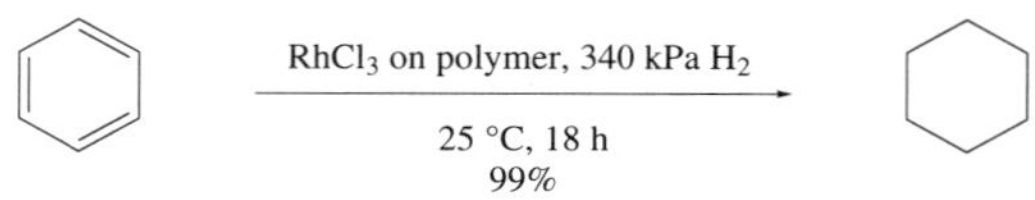

(104)

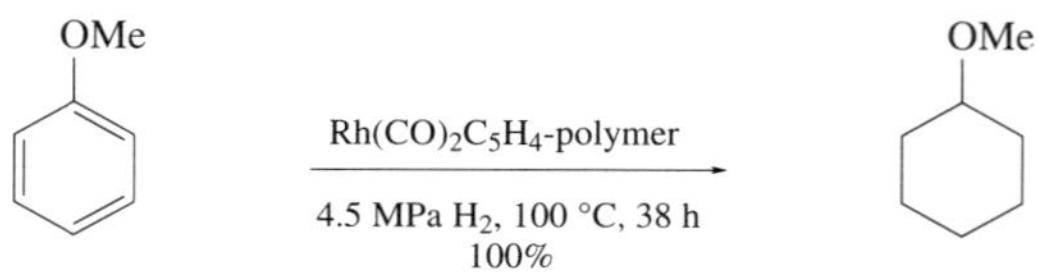

(105)

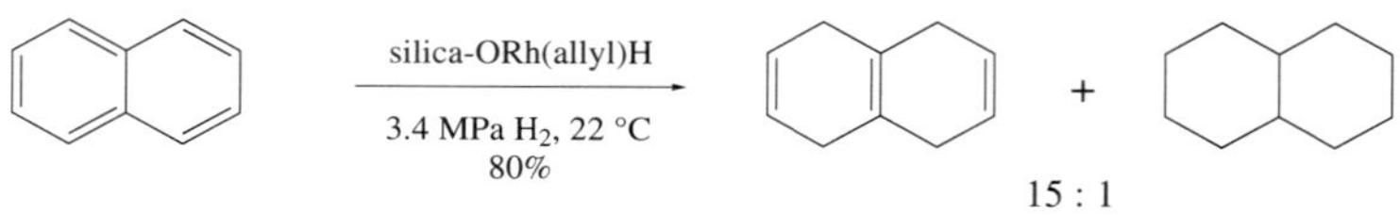

(106)

The partial reduction of arenes to cyclohexenes can be achieved using homogeneous transition metal complexes. Complexation of benzene to hexamethylbenzene ruthenium, cyclopentamethyl rhodium or cyclopentamethyliridium followed by hydride reduction gave the corresponding cyclopentadienyl complex (Scheme 14) which could be further reduced under more forcing conditions to the cyclohexadiene complex. Reaction of this with acid liberated cyclohexene in high yield and complete selectivity. The reaction has been made catalytic by the use of tetrafluoroboric acid ⟨84JCS(D)1747⟩. In a similar way, hydrogenation of a benzene–osmium complex using Pd/C as catalyst gave the cyclohexene complex in quantitative yield from which cyclohexene could be liberated with a mild oxidant ⟨88JA7906⟩. An attempt to make the process catalytic using the ruthenium complex failed.

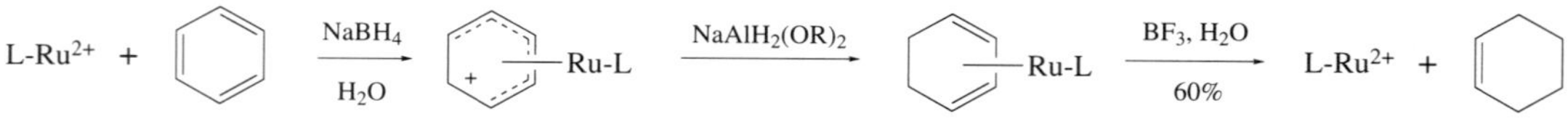

Scheme 14

A variation on the theme of homogeneous catalysis involves the use of phase-transfer systems. A system consisting of bis[rhodium-(1,5-hexadiene)chloride] with a hydrocarbon solvent, an aqueous buffer solution and a quaternary ammonium salt quantitatively reduced benzene to cyclohexane at room temperature and atmospheric pressure of hydrogen ⟨83OM1055⟩. Naphthalene is reduced to a mixture of *cis*-decalin and tetralin (Equation (107)). Methoxybenzenes are reduced to the corresponding cyclohexanes while phenols can give either cyclohexanones or cyclohexanols depending on the conditions. Acetophenone gives a mixture of products. Simple pyridines and furans are reduced in moderate yields. Equally mild conditions for the reduction of benzene have been reported for the two-phase system involving aqueous RhCl$_3$ and Aliquat 336 (methyltrioctylammonium chloride) with 1,2-dichloroethane as the organic phase ⟨83TL4139⟩. After 5 hours under 1 atmosphere pressure of hydrogen at 30 °C, a 51% conversion of benzene to cyclohexane was obtained with no other products. Further work on the mechanism of this reagent mixture has been reported ⟨87JOC2804⟩.

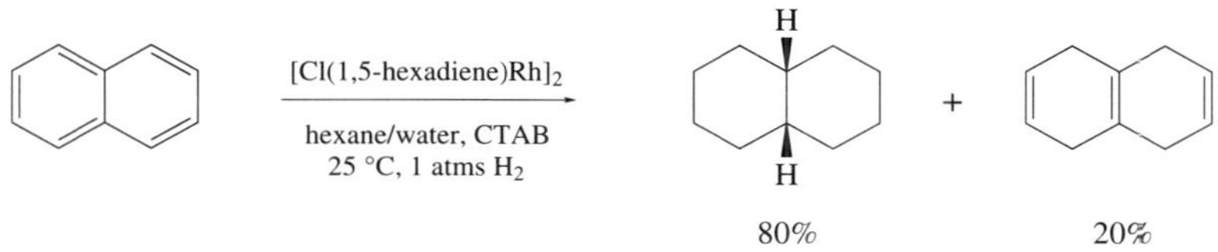

(107)

1.03.2.4 Dissolving Metal Reductions

This type of reduction has proven extremely useful in synthesis as the products are generally partially-reduced arenes which are ripe for further synthetic transformations. The first report of the dissolving metal reduction of an arene was in 1937 when toluene was reduced by sodium in ammonia ⟨37JA596⟩. However, the partial reduction of aromatic systems using an alkali metal in liquid ammonia was pioneered by Birch ⟨44JCS430⟩ and the reaction now bears his name ⟨88MI 103-02⟩. The reaction proceeds via a complicated mechanism involving radicals, radical anions and anions ⟨89T1579⟩ and the exact pathway followed depends on the substituents on the aromatic ring. This means that different types of aromatic molecule react in subtly different ways. Thus benzene and benzene rings substituted with electron-donating groups (alkyl, amino, alkoxy) need an alcohol to be present in the reaction mixture as ammonia is not a strong enough acid to protonate the intermediates. In addition, it is not possible to reductively alkylate such aromatics to give substituted cyclohexadienes. Activated benzenes, carrying for example carbonyl substituents, biphenyls and smaller polycyclic aromatics, do not require the presence of an alcohol for reduction and can be reductively alkylated. Finally, there are some larger polycyclic aromatics in which the final intermediate dianion is resistant to protonation owing to its stability. These systems are difficult to reduce. The metals generally employed are lithium, sodium and (occasionally) potassium, with lithium being the most reactive and potassium the least. The proton source in the reaction is also critical and the best yields are obtained for the less acidic proton donors ($Bu^tOH > EtOH >$ water) ⟨59JA3658⟩.

Although the reduction of benzene to 1,4-cyclohexadiene is often quoted as the archetypical Birch reduction, it was not until 1990 that a reliable method for this reaction (Equation (108)) was published ⟨90SC2165⟩. The differing results previously reported apparently arose from iron impurities present in liquid ammonia. The stereochemistry of the reduction has been studied using hexadeuteriobenzene and shown to give equal amounts of the *cis* and *trans* isomers ⟨68JA3590⟩. Polynuclear aromatics are more reactive than monocyclic systems ⟨70S161⟩. For example the reduction of naphthalene has been carried out under a variety of conditions and different products are isolated (Scheme 15) ⟨78JOC4283⟩.

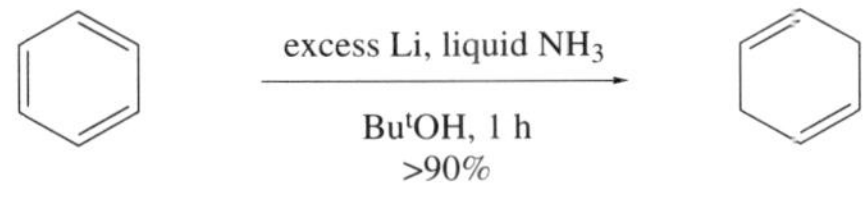

(108)

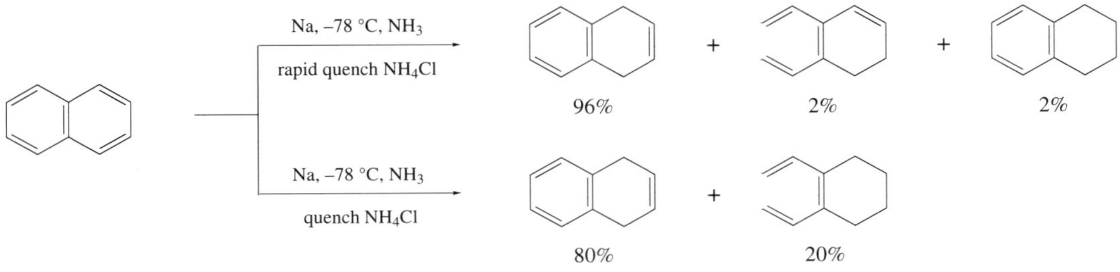

Scheme 15

The reduction of substituted benzenes is generally governed by whether the substituent is electron-donating or electron-withdrawing (Scheme 16). This feature has been extensively discussed by Birch *et al.* ⟨80JA3370, 80JA4074, 80JA6430, 81JA284⟩ and later by Zimmerman and Wang ⟨93JA2205⟩. Alkyl-substituted aromatics are reduced much as expected with the alkyl group(s) attached to alkenic carbons in the 1,4-cyclohexadiene product. Toluene is reduced in 88% yield to give 1-methyl-1,4-cyclohexadiene (Equation (109)) ⟨82AG(E)820⟩. The xylenes are reduced in a similar manner as shown in Equations (110)–(112) ⟨59JA3658, 73OSC(5)467⟩. The reduction of alkyl-substituted naphthalenes gave different results depending on the site of substitution. Thus 1-methylnaphthalene gave exclusively reduction in the unsubstituted ring on treatment with lithium in ammonia and ether followed by an aqueous quench (Equation (113)) while 2-methylnaphthalene gave a mixture of products with predominant reduction in the substituted ring (Equation (114)) ⟨72RTC1069, 81JOC1593⟩. However, it should be noted that the conditions for these two reactions are rather different.

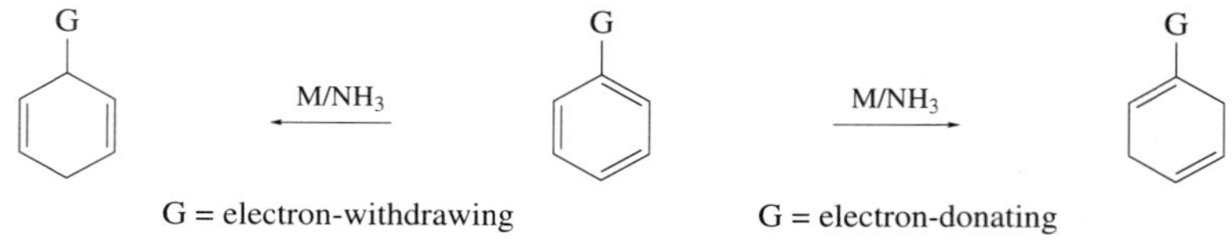

Scheme 16

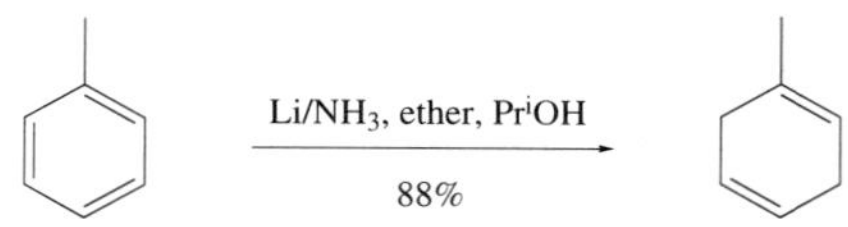

(109)

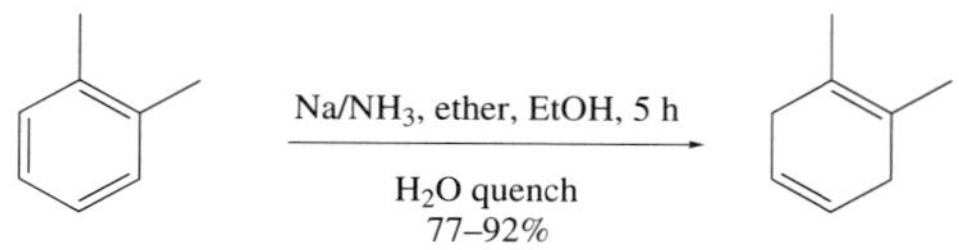

(110)

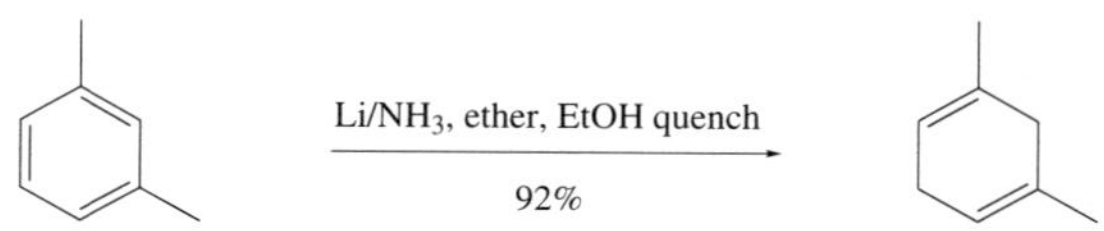

(111)

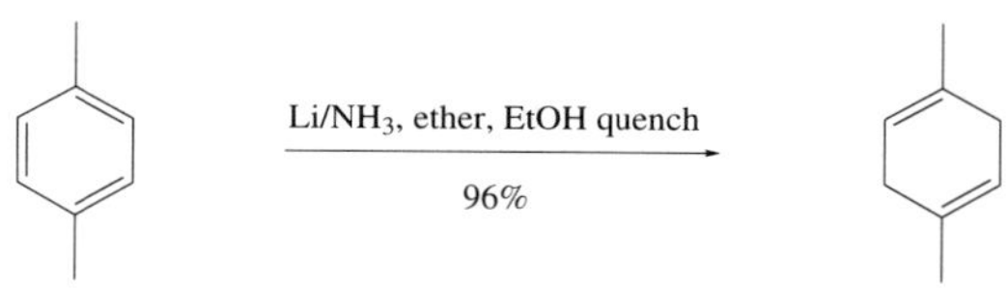

(112)

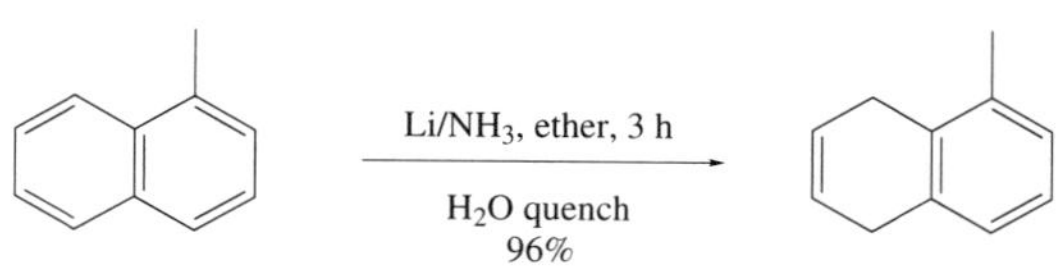

(113)

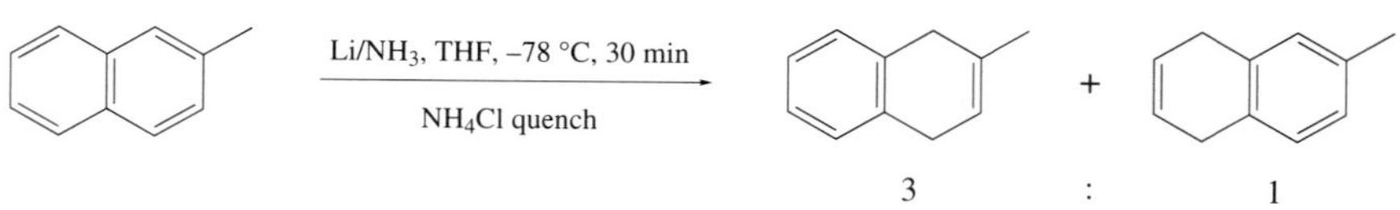

(114)

Carbonyl-substituted aromatic rings can be successfully reduced using Birch reduction conditions; indeed the presence of the electron-withdrawing group increases the rate of the reaction. However, preferential reduction of the carbonyl group can occur particularly with ketones. Reduction of acetophenone (Equation (115)) using potassium gave ethylbenzene in 78% yield ⟨54JCS113⟩. However, reduction of 1-acetylnaphthalene using sodium gave a dihydronaphthalene, with retention of the carbonyl group, in 81% yield (Equation (116)) ⟨82JCS(P1)875⟩ although the quenching agent used is important ⟨71JOC2588⟩. Reduction of indanone with lithium in the presence of a catalytic amount of cobalt metal gave clean reduction of the ketone with no ring reduction. Probably the most important substituent after the methoxy group in the Birch reduction is the carboxy group; this survives the reduction intact and there are many examples. Benzoic acid is reduced by sodium in ammonia in 89–95% yield ⟨73OSC(5)400⟩ to give the unconjugated isomer (Equation (117)). Similarly, the naphthalene-1-carboxylic acid is reduced quantitatively to the nonconjugated isomer (Equation (118)) with no reduction in the unsubstituted ring ⟨80SC627⟩. Esters and amides generally survive the Birch reduction conditions to give the expected products (Equations (119) and (120)) although benzaldehydes can be formed ⟨78JOC1007, 84JOC1544⟩.

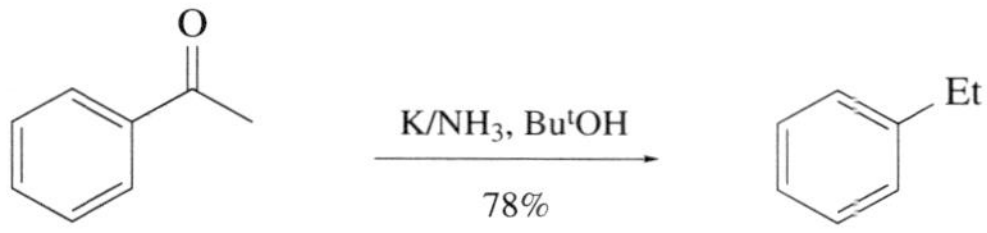

(115)

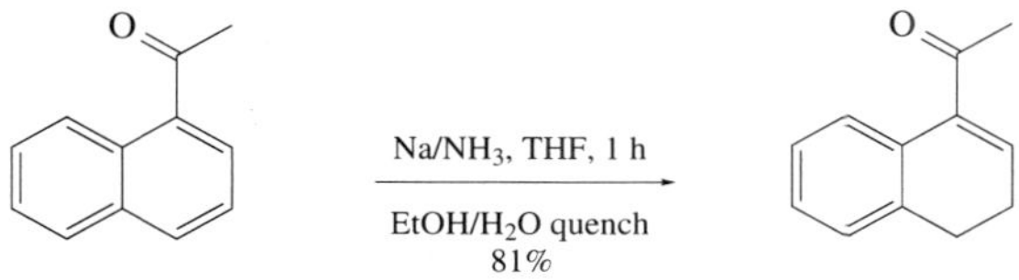

(116)

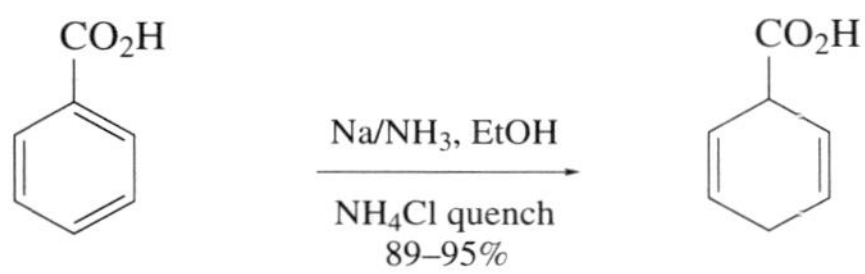

(117)

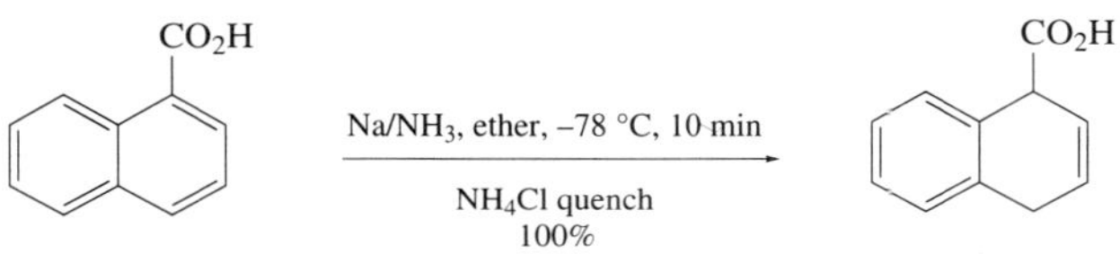

(118)

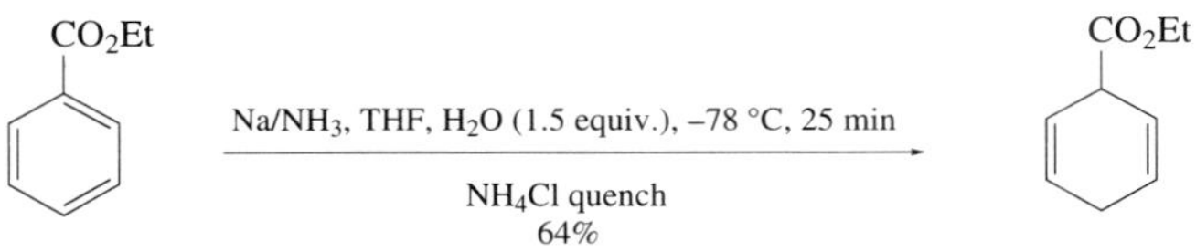

(119)

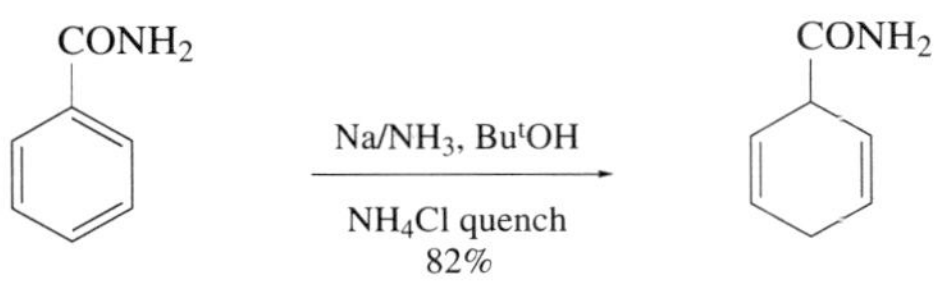

(120)

Reduction of halobenzenes is problematical as the carbon—halogen bond is usually cleaved rapidly. Reduction of a chloroalkane with lithium gave rise to a mixture consisting mainly of propylbenzene (Equation (121)) with a small amount of the dihydropropylbenzene ⟨78JCS(P1)1479⟩. The chloroindanone (Equation (122)) underwent reduction of both the carbonyl group and the chlorine atom on treatment with lithium in ammonia ⟨71JOC2588⟩.

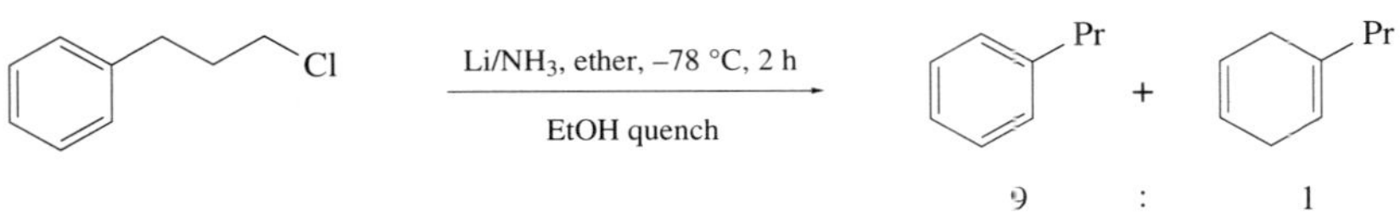

(121)

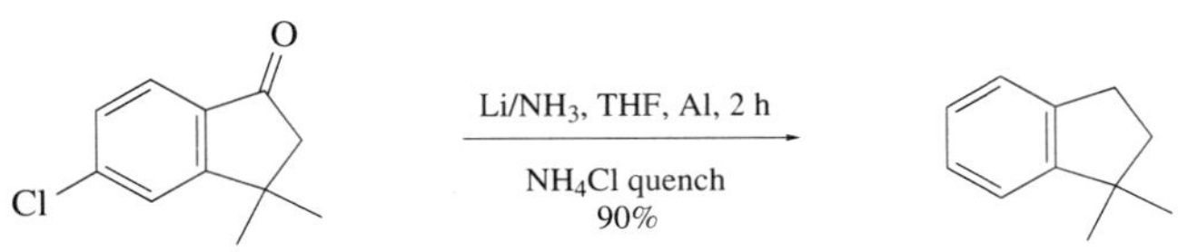

(122)

Aromatic ethers, in particular the methoxy-substituted benzenes, probably represent the most useful class of substrates for this reduction. The cyclohexadiene products have been widely used in organometallic chemistry ⟨B-85MI 103-02⟩ and in synthesis ⟨91JOC741⟩. Thus, the reduction of anisole using lithium in ammonia gave exclusively the 1-methoxy isomer (Equation (123)) in 75% yield ⟨88OSC(6)936⟩. This product can be rearranged to a variety of products by treatment with acids or base. This ability to use a methoxy-substituted benzene as a masked version of a cyclohexenone has been widely used in synthesis ⟨82JCS(P1)1381⟩. Alkyl-substituted aromatic ethers undergo reduction although the conditions required may vary. For example 1-methoxytetralin required the use of lithium for its reduction (Equation (124)) while the 2-methoxytetralin was reduced by sodium (Equation (125)) ⟨44JCS430⟩. However, there can be problems with the reduction of aryl ethers involving the cleavage of the alkoxy group. This is particularly likely to happen if the alkoxy group is situated *para* to an activating group as in the reduction of 4-methoxyacetophenone (Equation (126)). A study of this process using a variety of metals and conditions has been reported. The methoxynaphthalenes present a more complicated picture. 1-Methoxynaphthalene has been reduced to a mixture of naphthalene and 1,4-dihydronaphthalene (Equation (127)) on treatment with sodium in ammonia in the absence of an alcohol ⟨56CB2105⟩. This appears to be caused by the benzylic nature of the methoxy group since 2-methoxynaphthalene is reduced under the same conditions (Equation (128)) to give mainly the dihydronaphthalene (retaining the methoxy group) ⟨56CB2105⟩. On the other hand, in the presence of a proton source (ethanol) and with a water quench, the same compound gave the tetrahydronaphthalene (retaining the methoxy group) in 81% yield (Equation (129)) ⟨51JCS1945⟩.

(123)

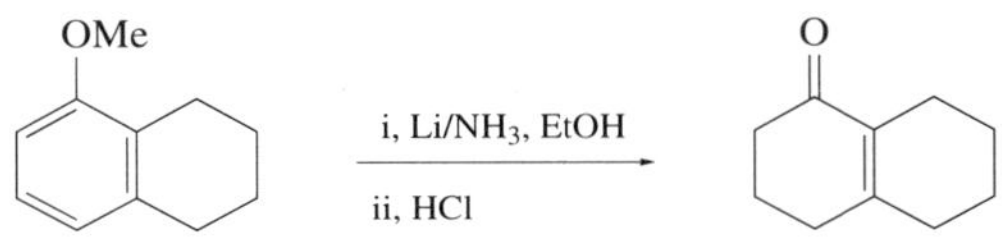

(124)

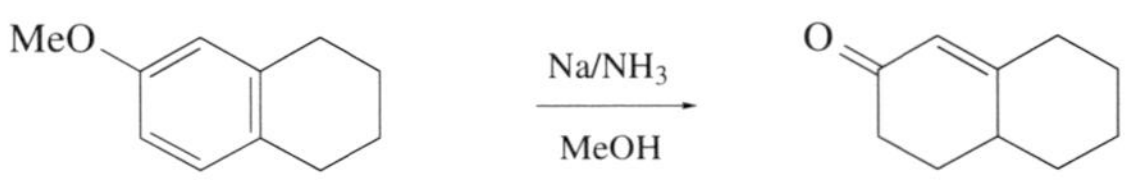

(125)

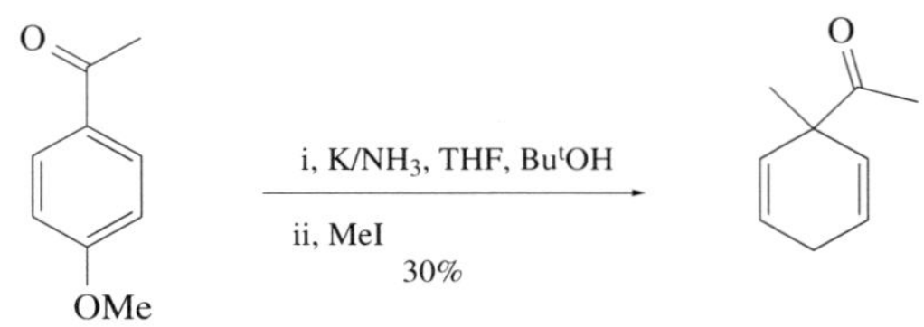

(126)

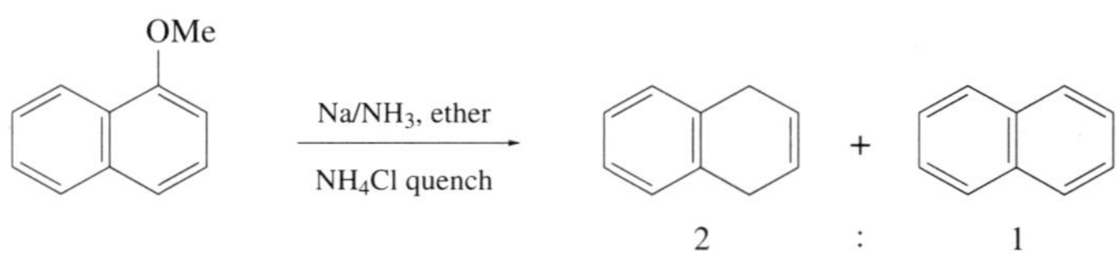

(127)

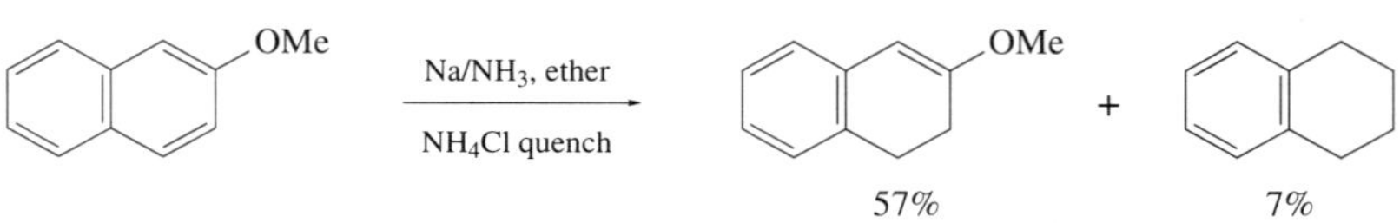

(128)

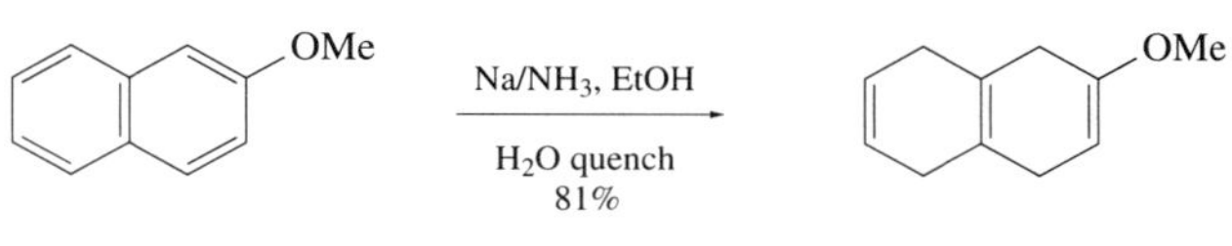

(129)

Phenols are generally rather poor substrates for Birch reduction owing to the formation of the radical dianion. This feature has been utilised to enable selective reduction. Reduction of a phenanthrene derivative using lithium in ammonia in the presence of *t*-butanol gave a product in which the phenolic ring was untouched (and the benzylic ketone reduced) while the methoxy-substituted ring was reduced (Equation (130)) ⟨67TL857⟩. Under forcing conditions involving a high concentration of lithium, Fried and co-workers reported the reduction of hydroxytetralin to the hydroxyoctalin (Equation (131)) in 76% yield ⟨67JA1044⟩.

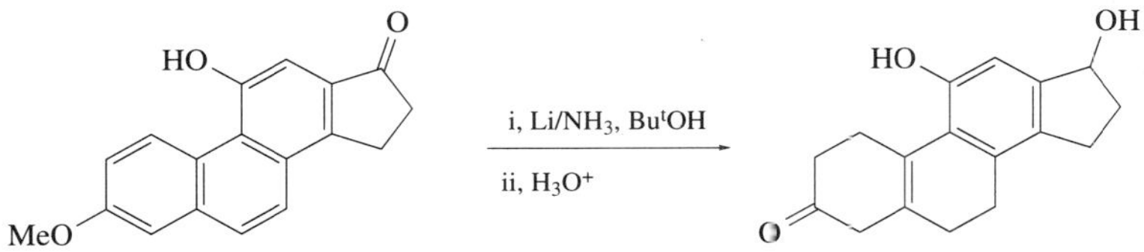

(130)

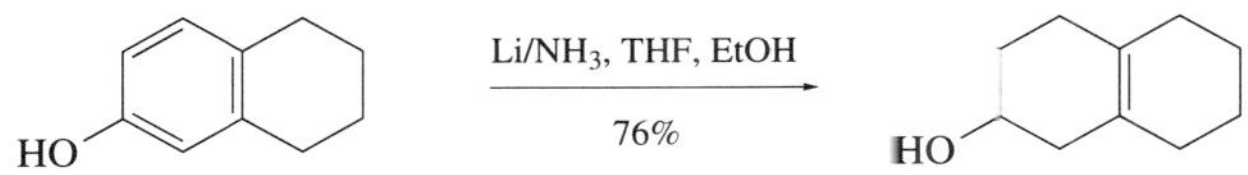

(131)

A study on the reduction of aromatic amines has been reported ⟨56JA4604⟩. The reduction of a range of 2-alkyl-substituted anilines to 2-substituted cyclohexenones was achieved although significant amounts of saturated ketones were isolated (Equation (132)).

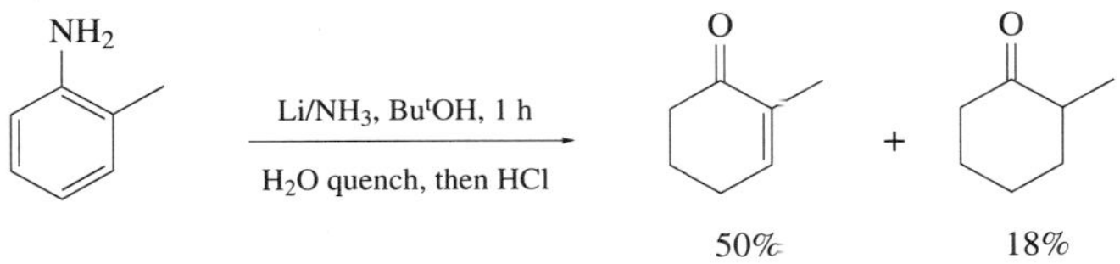

(132)

The reduction of arylsilanes using lithium in ammonia has been reported ⟨75JCS(P1)470⟩. The product of the reduction of a 4-methyl-substituted arylsilane was the isomer with the trimethylsilyl group and the methyl group attached to the tetrahedral carbons in the 1,4-cyclohexadiene product (Equation (133)). The removal of the trimethylsilyl group enables overall control of the regiochemistry of reduction to give isomers which are otherwise difficult to obtain.

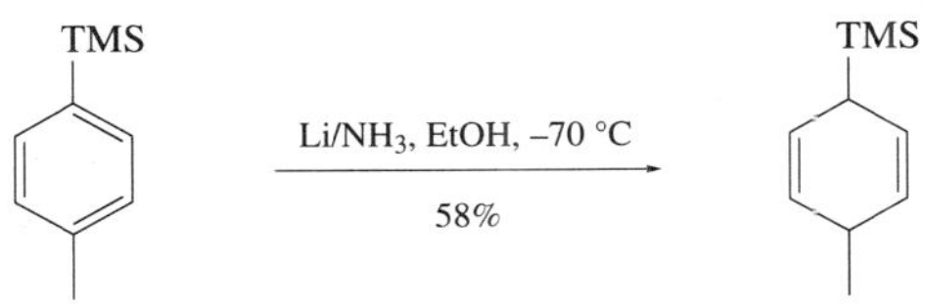

(133)

As might be expected, it is possible to reduce heteroaromatic systems under Birch-type conditions. The reduction of pyridine and simple 4-alkylpyridines proceeds to give 1,4-dihydropyridines (Equation (134)) in good yield ⟨75CC480⟩. 2-Substituted pyridines have been reduced and the products treated with acid and/or base to give ring opening followed by aldol cyclisation and dehydration ⟨75JA380⟩. Thus 2-methylpyridine was reduced with sodium in ammonia and, after an aqueous quench, the reaction mixture was subjected to treatment with sulfuric acid at 100 °C ⟨47JCS1270⟩. The product was cyclohexenone in 12% yield (Equation (135)). 2-Vinylpyridine after reduction with lithium in ammonia followed by acid treatment gave 2-methylcyclohexenone in 63% yield (Equation (136)) ⟨75JOC3606⟩. Quinoline and isoquinoline also undergo reduction in high yield. Quinoline was reduced (Equation (137)) to 1,4-dihydroquinoline using lithium in ammonia ⟨74TL2395⟩ whereas isoquinoline was reduced (Equation (138)) to 1,2,3,4-tetrahydroisoquinoline using sodium in ammonia in the presence of ethanol ⟨59T148⟩.

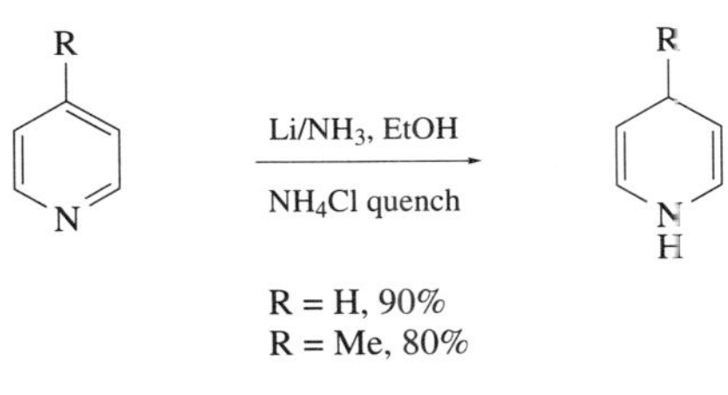

(134)

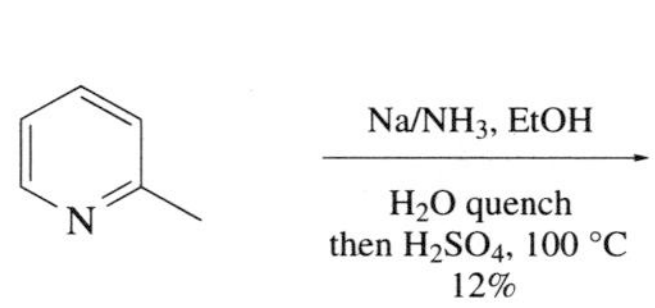

(135)

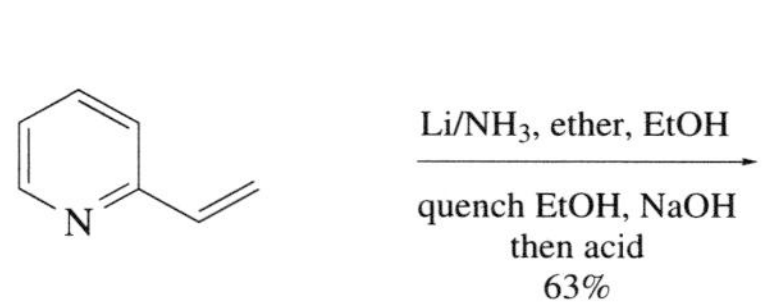

(136)

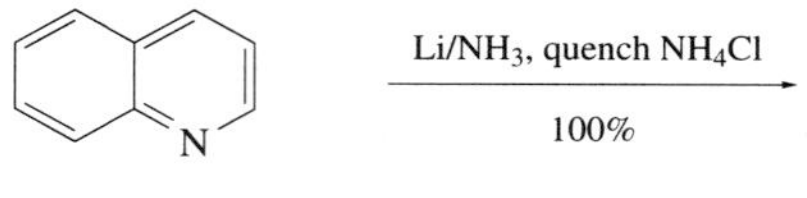

(137)

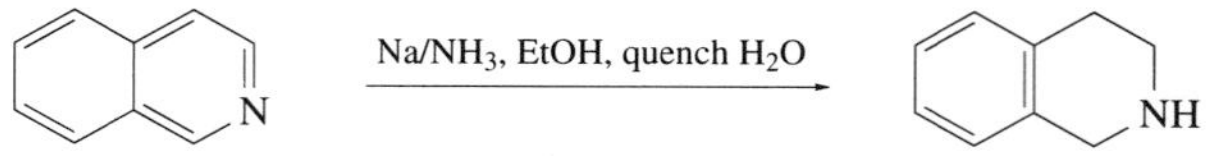

(138)

The π-excessive heterocycles are not so easily reduced and there are fewer examples of such reactions in the literature. Pyrrole is resistant to reduction as shown by the reaction of *N*-phenyl-substituted pyrrole with sodium in ammonia which resulted in reduction of the phenyl ring ⟨69AJC1321⟩. Furans carrying activating groups will undergo reduction. Furan-2-carboxylic acid is reduced to give the 2,5-dihydro product (Equation (139)) in 80% yield ⟨63MI 103-02⟩. Interestingly, the 3-isomer is reduced by lithium in ammonia in the presence of propan-2-ol to give the expected 2,3-dihydro product (Equation (140)) but in the presence of methanol or ethanol, the alcohol is incorporated and the ester is formed ⟨75BCJ1865⟩. In a similar fashion to furan, thiophene is only usefully reduced when carrying an activating group. The reduction of thiophene has been reported to give a complex mixture ⟨51JCS2556⟩ whereas thiophene-2-carboxylic acid can be reduced by lithium in ammonia followed by diazomethane treatment (Equation (141)) to give the 2,3-dihydrothiophene as its methyl ester ⟨79TL4985⟩. The benzannelated systems, as expected, usually undergo reduction in the benzene ring. Indole is reduced by lithium in the presence of methanol to the 4,7-dihydroindole (Equation (142)) in 51% yield ⟨71JOC279⟩. The *N*-methyl derivative behaves in the same manner but the *N*-phenyl derivative is reduced in the heterocyclic ring (Equation (143)) using sodium in the absence of alcohol to give the 2,3-dihydroindole ⟨69CI(L)1595⟩. In the oxygen series, benzofuran is reduced to the 4,7-dihydrobenzofuran (Equation (144)) even with the electron-donating methoxy group ⟨67JOC2794⟩.

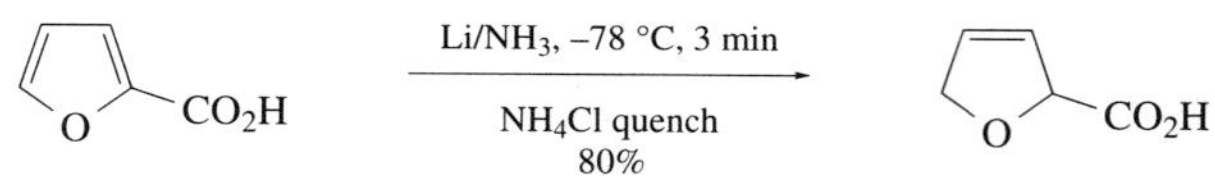

(139)

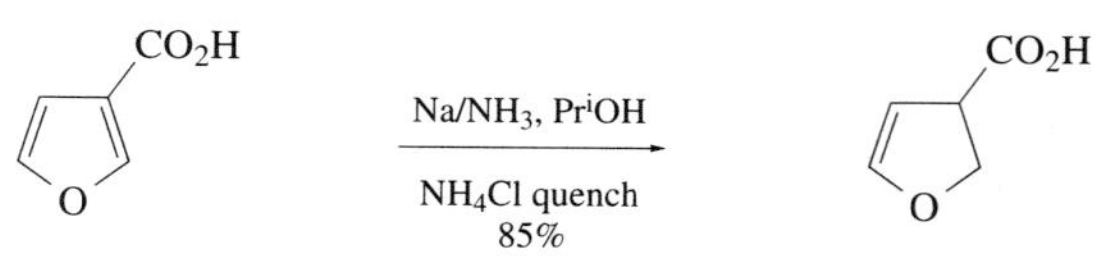

(140)

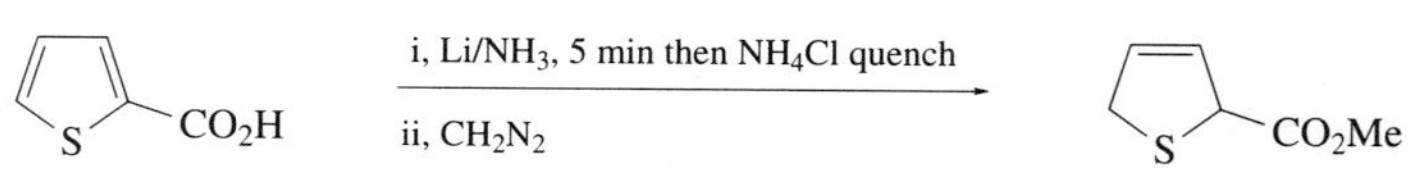

(141)

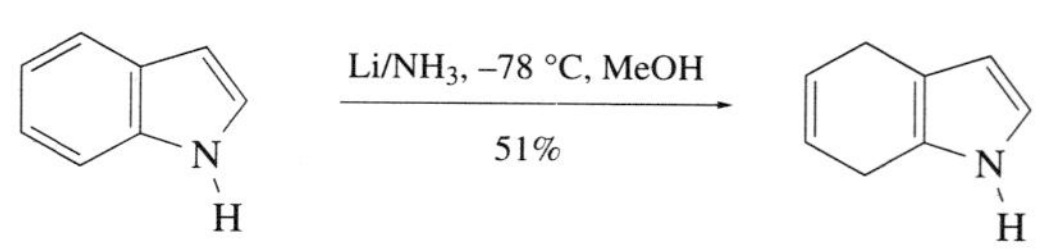

(142)

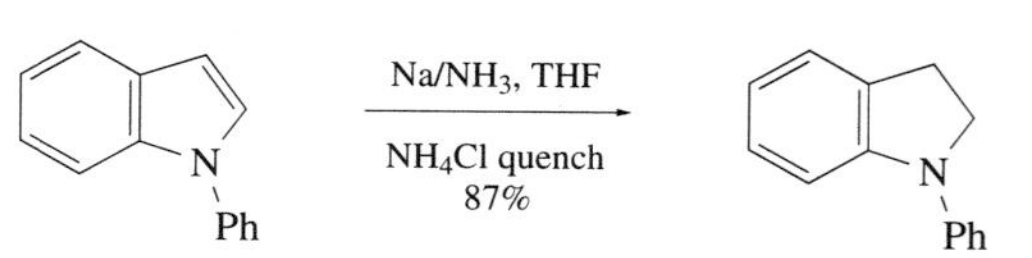

(143)

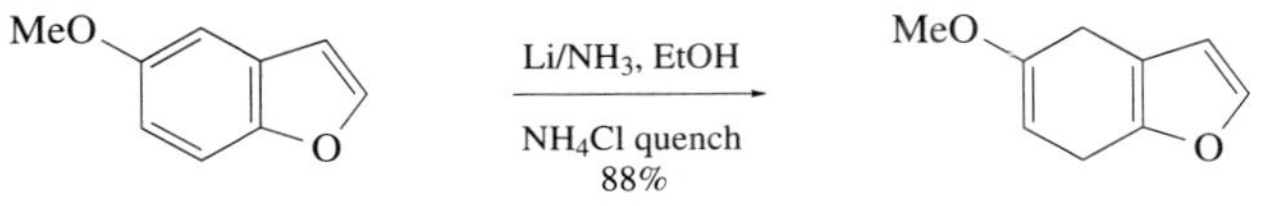

(144)

The use of alkali metals in other solvents has been explored and can be of some synthetic use. Reductions of aromatic rings utilising alkali metals in THF/HMPA have been reported ⟨87AG(E)204⟩. Reactions follow a similar pathway to the Birch reduction but are slower and offer no advantage for the reduction process, but are useful for reductive alkylation as there is no proton source. Crown ethers in THF have been used for reductions with alkali metals ⟨79AG(E)587⟩ and the reduction of benzoic acid by this methodology has been achieved. The well-known Bouveault–Blanc reduction conditions involving sodium in alcohol are not only applicable to carbonyl reduction and have been applied to the reduction of aromatic systems ⟨73OSC(5)398⟩. The reduction of a pyridine to a piperidine in the presence of a furan (Equation (145)) has been reported.

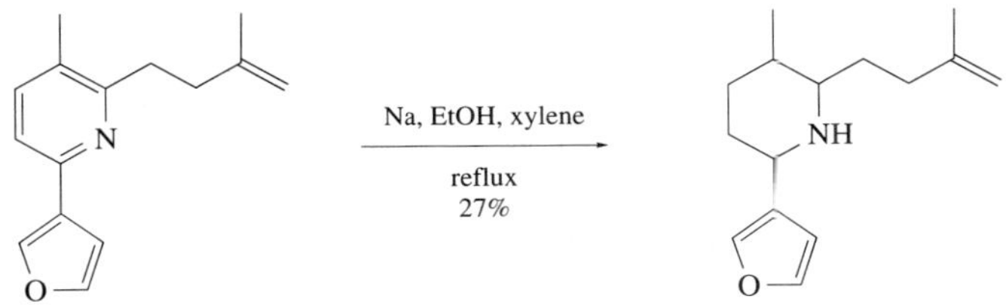

(145)

Probably the most widely explored alternative solvents for dissolving metal reductions are simple aliphatic amines. Under such conditions, the reduction generally proceeds further than the Birch reduction and gives fully-reduced products ⟨72S391⟩. Benkeser and co-workers were the first to report the reduction of benzene using lithium in ethylene diamine ⟨55JA3230⟩. This led to a mixture of cyclohexene and cyclohexane with no cyclohexadiene products formed (Equation (146)). The exact nature of the solvent is crucial to the course of these reactions as illustrated by the reduction of naphthalene (Scheme 17). Using ethylene diamine as solvent, the sole product is the fully-reduced decalin ⟨55JA3230⟩ whereas using a mixture of methylamine and ethylamine as solvent led to the formation of a mixture of isomeric octalins ⟨88OSC(6)852⟩. Replacement of lithium with the less reactive calcium (Equation (147)) enabled the reduction of aromatic rings to monoene products ⟨79JOC3737, 83JOC2796⟩. In synthetic terms the most useful conditions were found by using calcium in an amine containing an alcohol (Equation (148)) and this useful alternative to the Birch reduction has become known as the Benkeser reduction ⟨84TL2089⟩. These conditions offer significant advantages over the Birch conditions on larger scales. Reduction of anthracene using calcium in ethylene diamine (with butylamine added to prevent freezing) and *t*-butanol gave 76% of the hexahydroanthracene product along with a small amount of overreduction product (Equation (149)). Alkyl-substituted benzenes behave in a similar manner. Thus 1,2,4,5-tetramethylbenzene was reduced to a cyclohexa-1,4-diene (Equation (150)) using calcium in a mixture of methylamine and ethylene diamine ⟨83SC1103⟩.

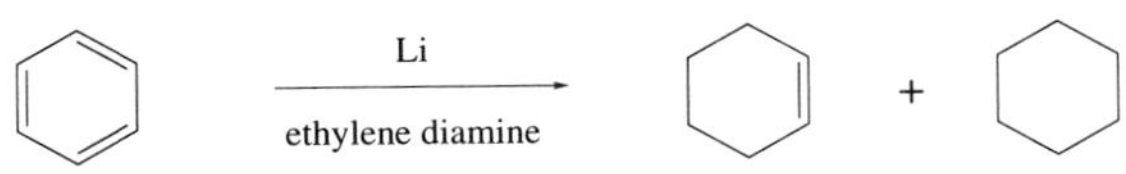

(146)

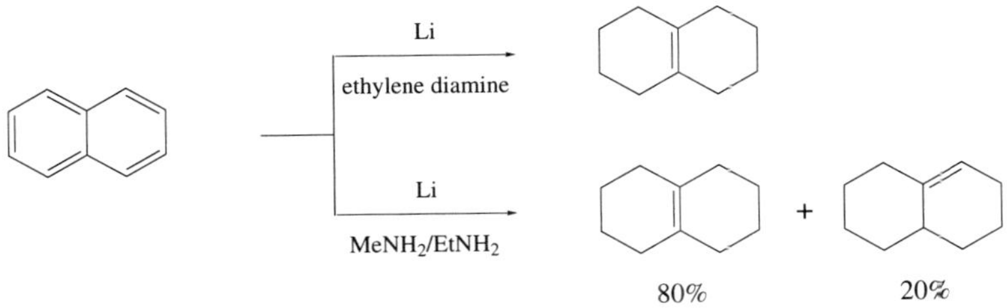

Scheme 17

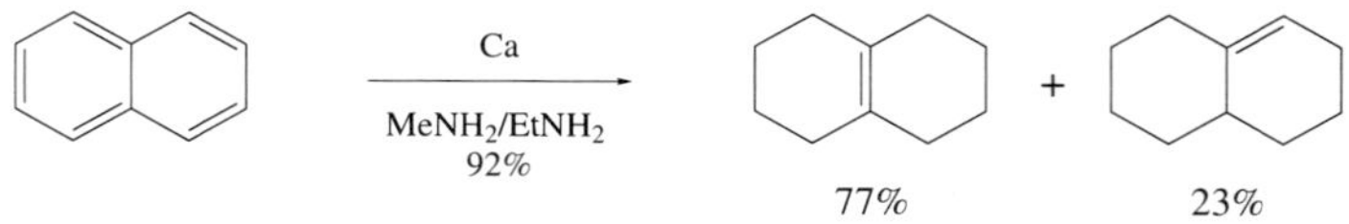

(147)

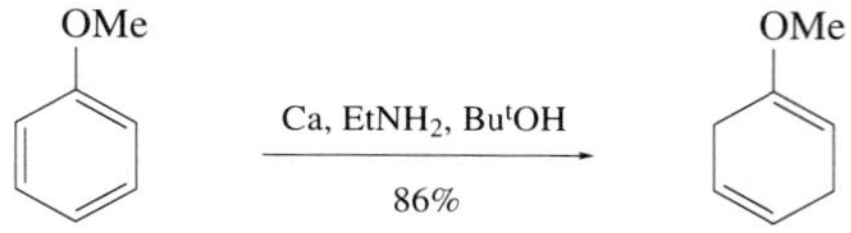

(148)

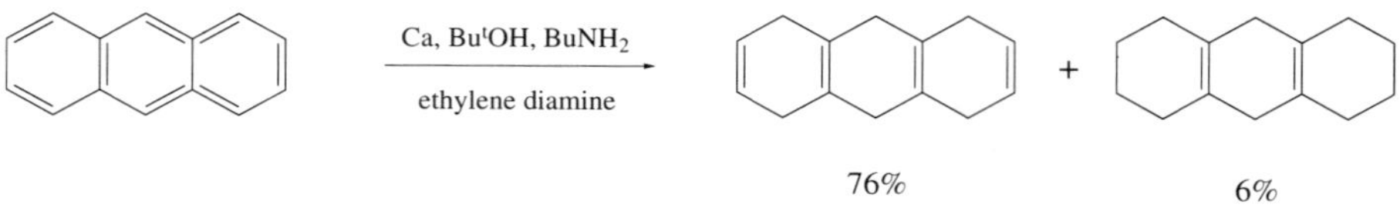

(149)

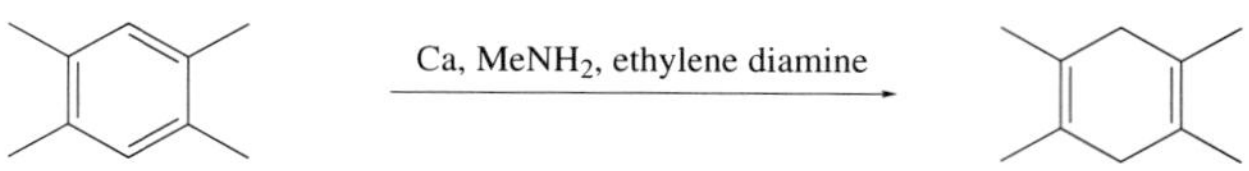

(150)

1.03.2.5 Hydride Reductions

Pyridine itself is not reduced by sodium borohydride but on treatment with lithium aluminum hydride a complex containing a mixture of 1,2- and 1,4-dihydropyridines is obtained ⟨63JA2236⟩. Although this mixture can act as a reducing agent, no dihydropyridines can be isolated from it. Pyridines bearing electron-withdrawing substituents and pyridinium salts can be reduced by sodium borohydride. Thus reduction of 3-cyanopyridine with sodium borohydride can be controlled to give either dihydro or tetrahydro product (Scheme 18) ⟨69TL3101⟩. The reduction of an *N*-methylpyridinium derivative leads to a mixture of products (Equation (151)) ⟨76JCS(P1)45⟩. The reduction of pyridine by sodium borohydride in the presence of chloroformates can be controlled to give 1,2-dihydropyridines with high selectivity (Equation (152)) ⟨72JOC1321⟩. Although pyrrole is inert to hydride reducing agents, indoles are reduced to indolines by tetrabutylammonium borohydride ⟨80H(14)1441⟩ and sodium cyanoborohydride (Equation (153)) under acidic conditions ⟨77S859⟩.

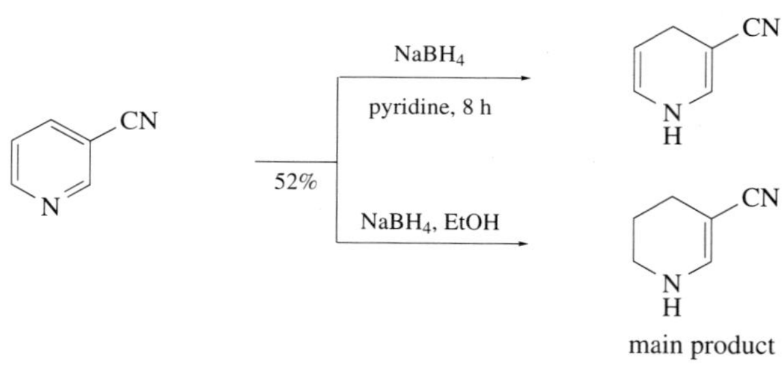

Scheme 18

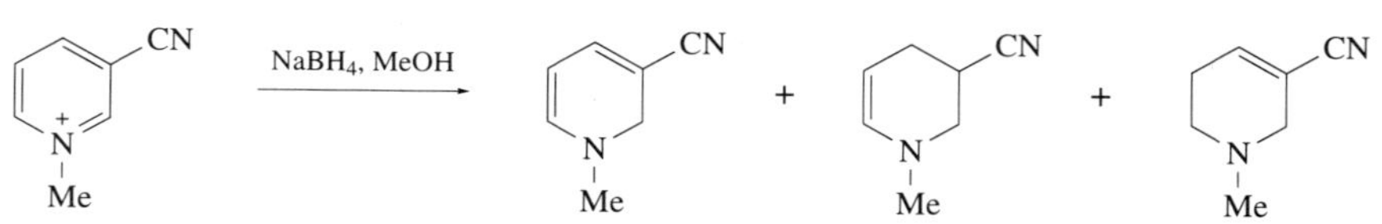

(151)

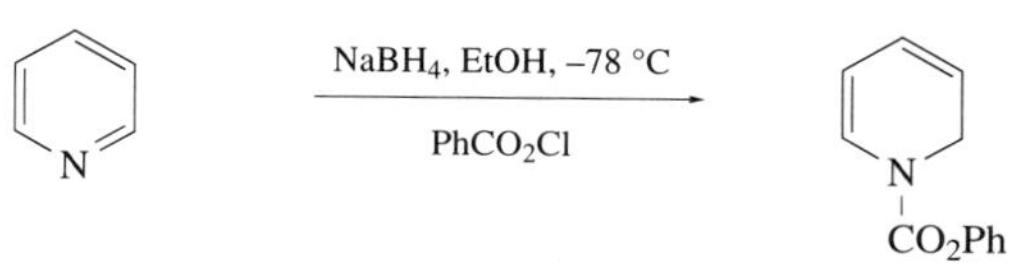

(152)

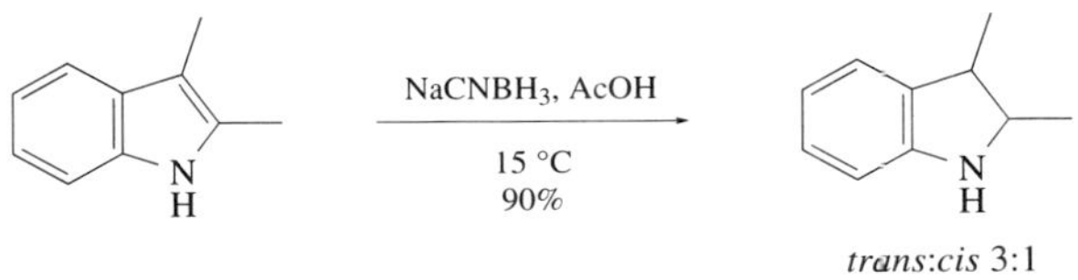

(153)

1.03.2.6 Electrochemical Reductions

Although electrochemical methods have been widely explored by industry, synthetic organic chemists have been reluctant to utilise this approach. The reduction of benzene to 1,4-cyclohexadiene has been reported ⟨81MI 103-01⟩ and Monsanto has also explored this process extensively. Polycyclic aromatics are easier to reduce ⟨42JA2365⟩ and naphthalene has been reduced to the 1,4-dihydro product on a preparative scale ⟨68BCJ2921⟩. This trend has been quantified by measurement of the redox potential of aromatic hydrocarbons ⟨89JA2325⟩.

Substituted benzenes are more amenable to reduction electrochemically. Phthalic acid has been reduced to a cyclohexene biscarboxylic acid (Equation (154)) in alkali at a mercury cathode ⟨49HCA1414⟩. Aromatic rings carrying electron-donating substituents can also be reduced. For example 2-methoxynaphthalene has been reduced to the dihydronaphthalene (Equation (155)) in 96% yield ⟨52JA4126⟩. The effect of substituents on the redox potentials of aromatic compounds has been reported ⟨91ZOR1137⟩.

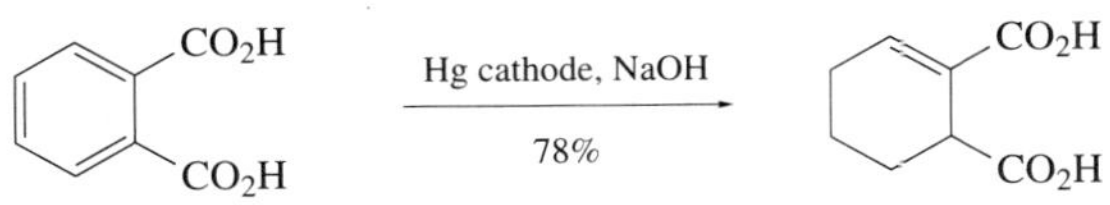

(154)

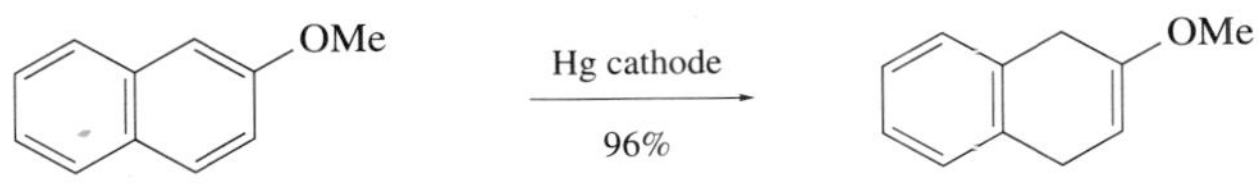

(155)

The reduction of heteroaromatics by electrochemical means has been explored ⟨65MI 103-02⟩. For example the reduction of pyridine to piperidine was reported in 1895 ⟨1895MI 103-01⟩ and the reduction of the quaternary pyridinium derivative of nicotinamide to the 1,4-dihydro derivative in quantitative yield has been reported ⟨65JOC1154⟩. This area has been reviewed ⟨77CJC3392⟩ and electron-withdrawing substituents make reduction easier. The reduction of thiophene-2-carboxylic acid (Equation (156)) to the dihydrothiophene in 70% yield has been reported ⟨75IZV888⟩.

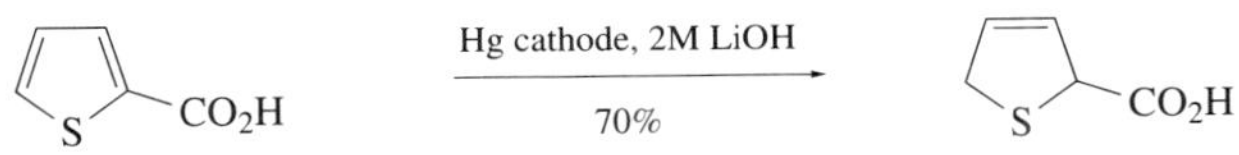

(156)

1.04
One or More CC Bond(s) Formed by Substitution: Substitution of Halogen

GAVIN L. EDWARDS
The University of New South Wales, Sydney, NSW, Australia

1.04.1 INTRODUCTION AND SCOPE

1.04.1.1 General Comments and Methods

If carbon–carbon bond formation is central to synthetic organic chemistry, then transformations involving the carbon–halogen bond must surely be the most important. The commercial availability of a wide range of halocarbons, as well as the ready accessibility of these compounds by simple functional group interconversions, helps consolidate their position as the most valuable functions in organic chemistry. This usefulness is still apparent despite the alternate strategy, much used in contemporary synthesis, of activation of alcohols via sulfonate ester formation. Numerous fundamentally different reactions have an overall outcome of replacement or substitution of a halogen function by a carbon group; representative examples are depicted in Scheme 1.

Nucleophilic substitution

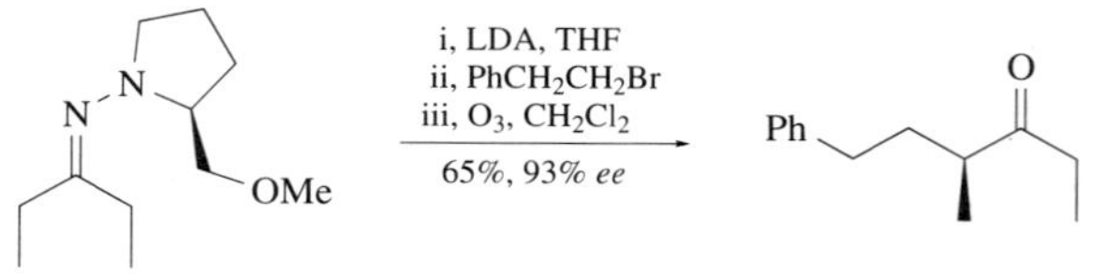

(1)

"Umpolung" by conversion to an organocuprate

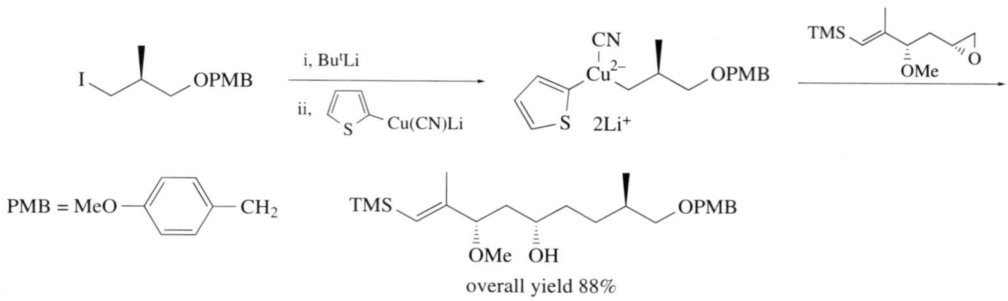

(2)

Free radical approach

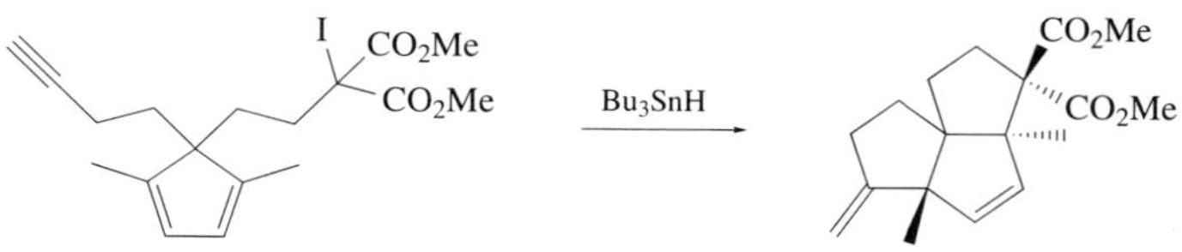

(3)

Electrophilic (Friedel–Crafts) approach

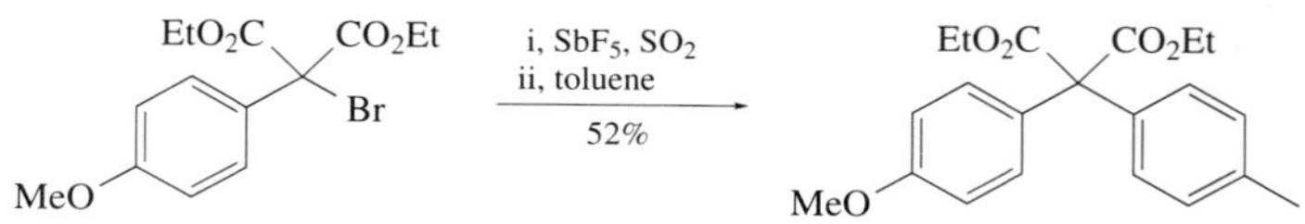

(4)

Organometallic-based syntheses

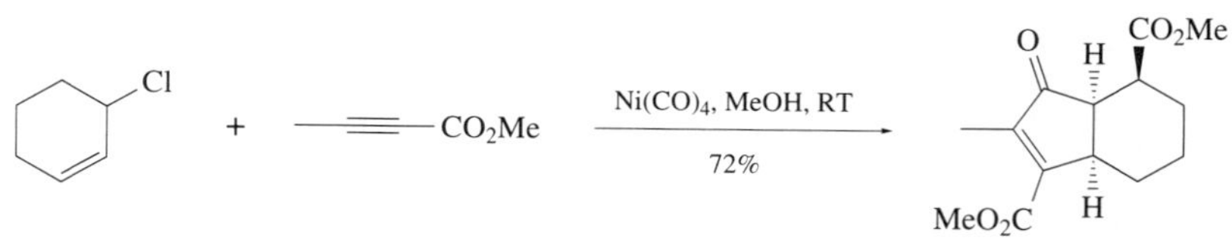

(5)

Scheme 1

Perhaps the most important class of reactions involves the use of halogen as a leaving group in nucleophilic substitution (Equation (1)) ⟨93CC424⟩, either directly or in a single electron transfer process. Alternatively, the carbon–halogen bond can serve as a precursor to an organometallic reagent which can itself act as a nucleophile, thus reversing the polarity of the original carbon–halogen bond (Equation (2)) ⟨93CC617⟩. Abstraction of the halogen atom (or homolysis of the C—X bond) can provide a carbon-centered radical which, following addition to a multiple bond, can give a product where a halogen atom has been replaced (in a formal sense) by carbon (Equation (3)) ⟨93PAC1153⟩. Activation of an alkyl halide by reaction with a Lewis acid can give electrophilic species which can result in carbon–carbon bond formation (Equation (4)) ⟨86TL4853⟩: the Friedel–Crafts alkylation is perhaps the best known reaction of this type although other electrophilic substitutions are possible. Conversion of an allyl halide into, for example, a π-allyl species can precede additions to multiple bonds or carbonylations where the final product has the original halogen atom apparently replaced by a carbon-based group (Equation (5)) ⟨93JOM(451)1⟩.

Coverage of all these areas is beyond the scope of this chapter, and discussion will be limited to displacement of halogens from alkyl halides by carbon nucleophiles, usually as carbanionic or organometallic species. The extent of the subject and the great variation possible in the carbon nucleophiles, means that the discussion will be structured according to the nature of the carbon nucleophile rather than the particular alkyl halide. Following some brief comments on a comparison of the different halogens, the first class to be mentioned will be carbanions with no stabilizing group, i.e. "simple" organolithium, -magnesium, -copper, etc., nucleophiles. Allyl and benzyl groups will be treated as normal carbon substituents, with special mention of enhanced or differing reactivity and ambident behavior being made where appropriate. (A separate review on allyl and benzyl carbanions is given in *Organic Reactions* ⟨82OR(27)1⟩.) This review will then consider carbanions based on a single stabilizing group, concluding with a brief section on the effects of combining two or three stabilizing groups. Some specific comments on reactions of polyhaloalkanes, as opposed to monohaloalkanes, will be made where appropriate. Many topics within this chapter have already been reviewed extensively, and relevant references will be given under individual section headings. Comparisons of metal ions in the organometallic component, as well as questions of diastereo- and enantioselectivity, will also be covered where relevant. Other aspects of carbanion and organometallic chemistry are covered throughout this work.

1.04.1.2 The Leaving Group—Comparison of F, Cl, Br, and I Derivatives

The effects of substrate structure and variation of reaction conditions on substitution reactions have been studied thoroughly; extensive discussions can usually be found in most good textbooks ⟨B-87MI 104-01, B-92MI 104-01⟩, and details will not be repeated here. However, one aspect which deserves some comment is the effect of changes of halogen upon reaction rate. (Astatine, having no isotopes with half-lives beyond a few hours, will not be considered in this review.) Both electronegativities and carbon–halogen bond strengths (Table 1) increase in the order I < Br < Cl < F; in fact, fluorine forms the strongest single bond to carbon of any of the elements ⟨92JA9576⟩, and conventional wisdom states that iodide and bromide are the best leaving groups, alkyl chlorides react at much slower rates and alkyl fluorides are very poor substrates indeed.

Table 1 Bond dissociation energies for methyl halides D(Me-X) (kcal mol^{-1}).

X	*Experimental value*	*Theoretical value (MP3)*[a]
F	108.1	105.1
Cl	83.5	79.4
Br	70.0	68.9
I	56.3	58.4

[a] MP3 = Third order Møller–Plesset perturbation calculations ⟨92JA9576⟩.

Some of the factors that could possibly affect the rate of halide displacement have been discussed ⟨79COC(1)493⟩. Many studies have reported on the reactivity of alkyl halides with (usually noncarbon) nucleophiles ⟨26MI 104-01, 37JCS1177, 37JCS1183, 53JA246⟩, and the general trend is that reactivity increases in the order F < Cl < Br < I. Measurements in the gas phase have shown that, in the reaction of MeX (X = F, Cl, Br) with various carbon and noncarbon nucleophiles, the leaving group ability follows the established order with F^- being the poorest, and Br^- the best leaving group ⟨74JA4027, 76CJC1643⟩; displacement of fluoride occurs at least 100 times slower than displacement of chloride. A more reactive substrate can be prepared by the Finkelstein reaction giving the alkyl iodide ⟨10CB1528⟩, or alternatively a catalytic amount of lithium iodide can be added to the reaction mixture if one of the lower halides (Cl, Br) is used. It is believed that this procedure interposes a Finkelstein-like step into the reaction sequence, where the more reactive alkyl iodide is prepared *in situ*, thus increasing the overall rate of substitution ⟨90JA3554⟩. If a polyhalo substrate is treated with a carbon nucleophile, then selectivity towards cleavage of the more labile C—X bond will be observed ⟨56JA3484, 57JA2311⟩. Displacement of fluoride has been reported (although the yield is not given in the text) in a double alkylation of ethyl acetoacetate with 1-chloro-2-fluoroethane (**1**) (Equation (6)), although there can be little doubt that the first step would involve displacement of Cl^- and kinetic factors would then favor the second carbon–carbon bond formation ⟨B-92MI 104-02⟩.

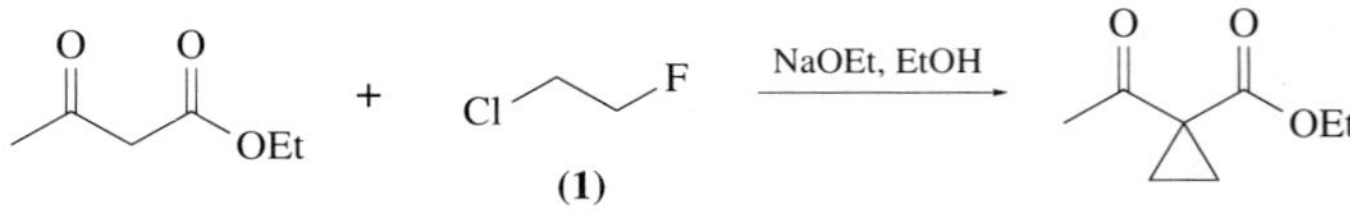

(6)

Although C—F bonds are generally considered to be inert towards nucleophiles, substitutions at saturated carbon atoms can occur ⟨63AFC63⟩ and several examples of C—C bond formation have been reported: as these reactions are relatively rare, and as fluorocarbons would not otherwise feature largely in this chapter, several reactions will be considered here. Benzylic organolithium reagents will couple with benzyl fluorides to afford phenyl-substituted ethanes in high yields, typically >90% ⟨76JA2580⟩. Displacement of fluoride by Grignard reagents has been observed for α-fluoroketones ⟨54JA4137, 61JCS3452⟩ and for α-fluorophenylacetic esters ⟨70BSF991⟩. In the latter case substitution occurs, presumably via the intermediacy of a magnesium alkoxide which then undergoes a 1,2-shift with concomitant magnesium-induced loss of fluoride (Scheme 2). (Similar reactions of other α-halocarbonyl compounds, where rearranged ketones are isolated in addition to simple 1,2-addition products, are known ⟨59JOC843, 74TL1451, 86CPB3599⟩.) These reactions do not appear to be generally applicable, as addition of Grignard reagents to α-fluoro aliphatic esters give normal products of 1,2-addition to the carbonyl group ⟨70BSF991⟩.

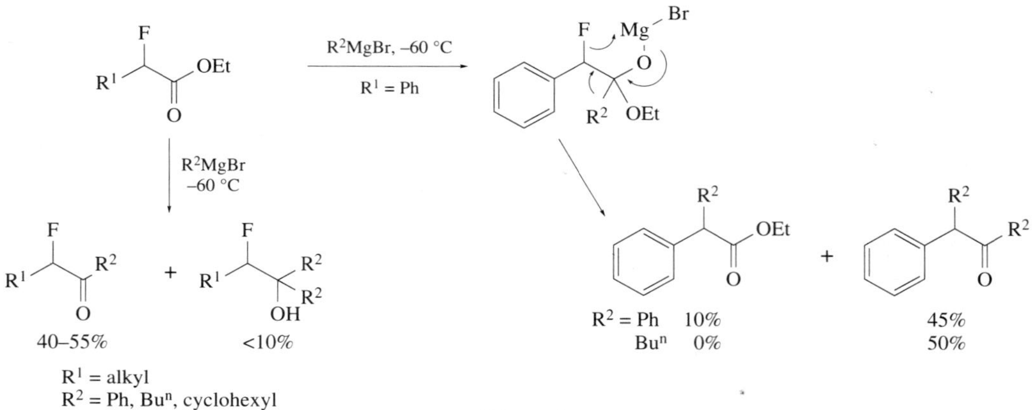

Scheme 2

1.04.1.3 Higher Valent Halogen Derivatives

The existence of hypervalent halogen compounds (where the halogen atom has more than one ligand) opens up, at least in theory, new possibilities for carbon–carbon bond formation. This class of compounds has been reviewed extensively ⟨B-83MI 104-01⟩, and the most commonly investigated species are trivalent iodine compounds. Although the chemistry of aryl, alkynyl, and alkenyl hypervalent iodine compounds is now well established ⟨92AG(E)274⟩, the preparations and reactions of alkyl derivatives are relatively rare. Simple alkyl iodinanes do not appear to be stable; while the electron-withdrawing nature of the iodine(III) center makes the attached carbon more prone to nucleophilic attack, it renders the compound liable to elimination and rearrangement in addition to the desired displacement of iodine by nucleophiles. When optically active 2-iodooctane (**2**) is treated with Cl_2 or Br_2, nonracemic 2-octyl chloride or bromide is isolated (Scheme 3). Oxidation of the iodine substituent to an unstable iodonium species is believed to precede displacement of IX_2; although the substitution occurs with predominant inversion of configuration, some racemization is observed ⟨54JA6040, 55JA5533⟩. Similar results are obtained when alkyl iodides are treated with peracids: primary iodides give displacement products, whereas secondary iodides give mixtures resulting from either displacement, elimination, or oxidation of the halogen-bearing carbon atom ⟨69JOC3974, 69JOC3978, 78JA4888, 80JA7760, 82JOC1904⟩. Useful yields of products can sometimes be obtained, as in the case of oxidation of a norbornyl iodide with peracetic acid ⟨77CC321⟩.

Electronegative fluorocarbon ligands are capable of stabilizing hypervalent iodine species, and polyfluorinated derivatives ($(R_fCH_2I(OTf))^+$, where R_f = perfluoroalkyl) can be isolated; treatment with noncarbon nucleophiles results in cleavage of the CH_2—I bond ⟨87BCJ3307, 91BCJ2008⟩. Displacement of the iodine-based unit by a carbon nucleophile is achieved if the iodonium triflate is treated with a trimethylsilyl ether in the presence of KF (Equation (7)), although the yields are

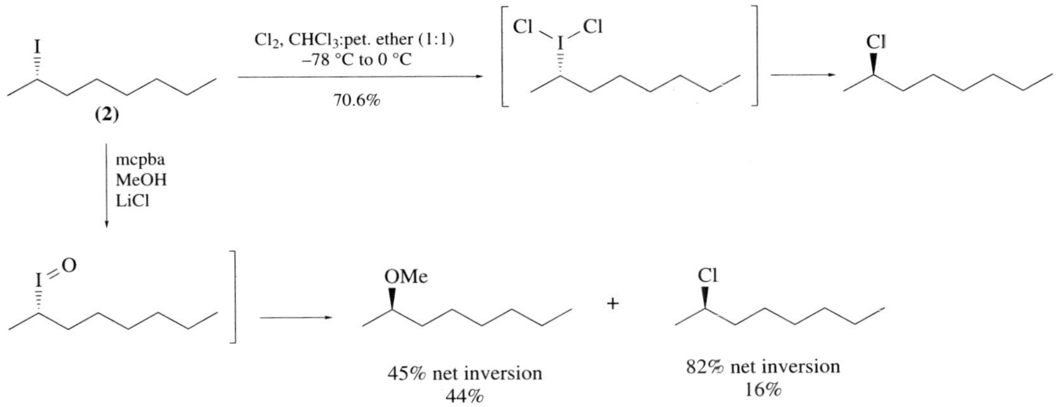

Scheme 3

variable ⟨87BCJ3823⟩. Substitutions in similar hypervalent iodine species are often accompanied by other reactions: Yagupol'skii and co-workers showed that the fluoroborate salts, on reaction with *N*,*N*-dimethylaniline, gave mainly the product of *N*-alkylation (70%) accompanied by some *para*-substituted dimethylaniline (19%) (Equation (8)) ⟨88JOU530⟩. Aniline gave only *N*-alkylated products.

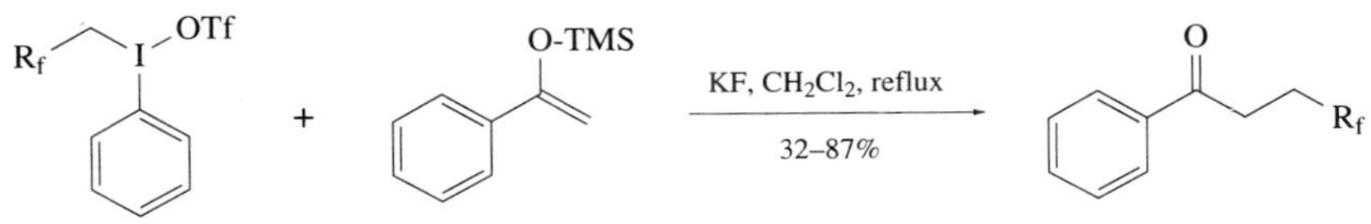

(7)

$R_f = CF_3, C_2F_5, i\text{-}C_3F_7, n\text{-}C_7H_{15}$

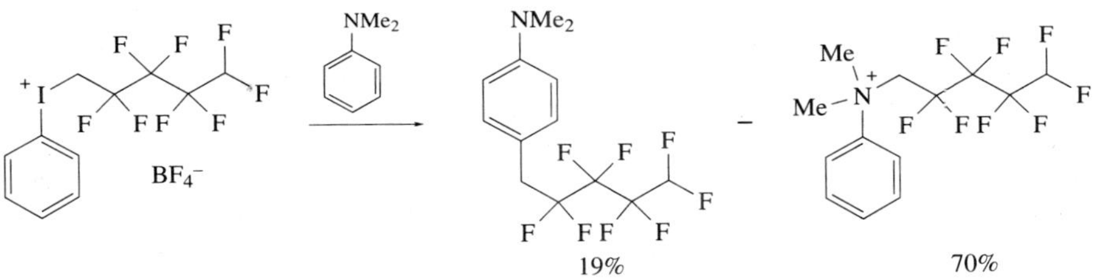

(8)

The decomposition of iodonium ylides, usually in the presence of copper(I) salts, gives compounds which formally have new carbon–carbon bonds in place of iodine, and which appear to arise from an initial loss of iodobenzene and the formation of a carbene or copper carbenoid ⟨70BCJ2506⟩. Reactions are often complicated by the formation of sideproducts. Work by Moriarty *et al.*, where intramolecular "displacement" of iodobenzene by an alkene leads to cyclopropane formation, casts doubt on the carbene proposal ⟨89JA6443, 90JA1297⟩. An alternative mechanism, involving initial addition of the alkene to the electrophilic iodine, transannular cyclization, and final reductive elimination of PhI to give the cyclic product, is suggested (Scheme 4). Moriarty *et al.* have used this intramolecular cyclopropanation reaction in an approach towards the 3,5-cyclovitamin D ring A synthon ⟨93TL4129⟩. Although enantiomerically pure *β*-keto esters were used for preparation of the iodonium ylides, diastereoselectivity in the ring closure step was poor.

1.04.2 CARBANIONS WITH NO STABILIZING GROUP

1.04.2.1 General Comments

The formation of a carbon–carbon bond with no heteroatom attached to either carbon atom can be accomplished readily by the reaction of a suitable organometallic reagent with a haloalkane, with acceptable yields being obtained with methyl, primary, and secondary haloalkanes. In most

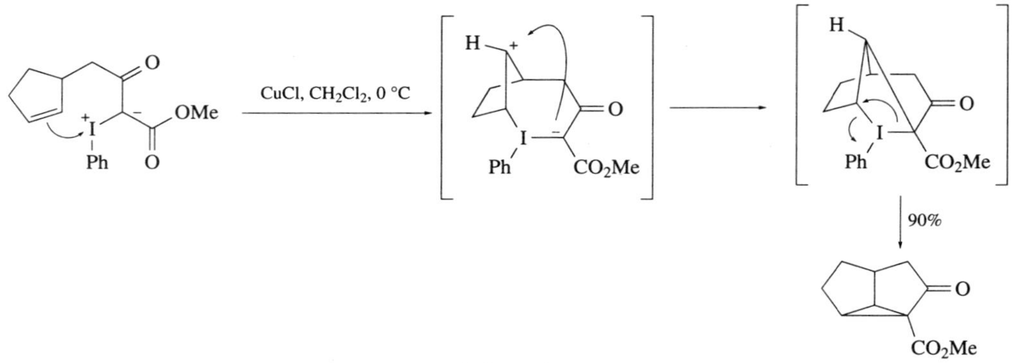

Scheme 4

cases, tertiary haloalkanes give poor yields as elimination tends to become the major pathway. The selection of the correct organometallic partner is vital for the success of the coupling—injudicious choice will allow a range of side reactions to compete or even to dominate. In general, organocopper reagents provide the most reliable route although acceptable yields can sometimes be obtained with other metals. The methods for preparation of the organometallic species will occasionally be considered, as both the metal itself as well as other attendant ions can markedly affect a reaction's outcome. If one (or both) of the carbon centers in the new carbon–carbon bond is chiral, then problems arise in that a degree of racemization can accompany coupling. Often, alternative routes need to be sought if complete stereocontrol is essential.

1.04.2.2 Group I Metals

Although organometallic compounds of Group I metals are usually represented simply as RM, it is now well known that these compounds are generally oligomeric both in solution and in the solid state. The x-ray crystal structures of organolithium compounds ⟨85AOC(24)353⟩, and also of organic derivatives of sodium, potassium, caesium, and rubidium ⟨87AOC(27)169⟩ have been reviewed extensively. Conceptually, the reaction of Group I organometallic compounds with alkyl halides is one of the simplest procedures for carbon–carbon bond formation; however, it is little used in synthesis. While stabilized carbanions with Li, Na, or K as counterions frequently can be alkylated smoothly and in high yield, the analogous reactions of alkyl halides with nonstabilized carbanions can be plagued by a range of competing reactions. These include metal–halogen exchange, α-metallation (giving a carbenoid), α- or β-elimination (giving a carbene or alkene, respectively), Wurtz-like processes (leading to mixtures of coupled products), and radical reactions arising from electron transfer processes in the course of the reaction ⟨B-74MI 104-01, 82JOC3521, B-87MI 104-02⟩. For example, in an extreme case where steric factors are likely to interfere with "normal" substitution reactions, the reaction of α-phenyl neopentyl chloride (**3**) with diphenylmethyllithium (**4**) gave a mixture of products (Equation (9)).

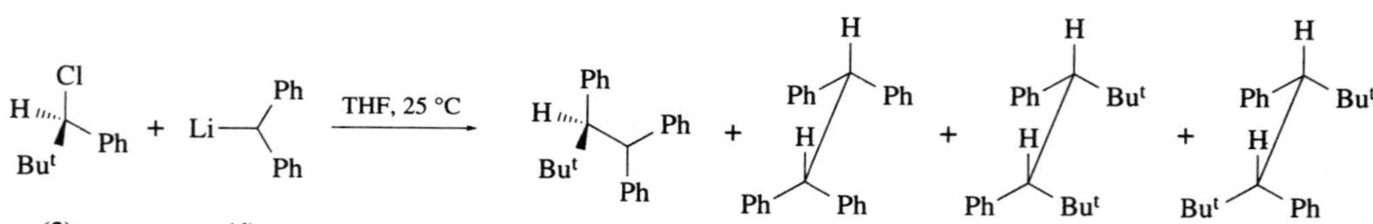

(3) (4) 50.5% 18.2% 10.4% 7.2% (9)

In a more straightforward example, a simple cycloalkyllithium (**5**) reacted with iodomethane to give, as the predominant product, an iodo compound resulting from lithium–iodine exchange (Scheme 5). Reaction of the cycloalkyllithium (**5**) with 1,2-dibromoethane resulted in bromination of the alkyllithium, with ethene being liberated as the by-product ⟨69JOC641⟩. Halogen transfer to the alkyllithium occurred with retention of configuration in both cases. These metal–halogen exchange reactions can also lead to mixtures of products ⟨44JA1515⟩, for example, the attempted alkylation of benzyl chloride with *n*-butyllithium gave *n*-pentylbenzene (21%) in addition to 1,2-diphenylethane (31%) and *n*-octane (a small quantity); the unpredictable nature of these reactions was demonstrated by the fact that no 1,2-diphenylethane was isolated from the reaction of benzyl chloride with 1-naphthyllithium. Where couplings are successful, the substrates of choice are usually bromo- or iodoalkanes, as the reactions of analogous chlorides will lead predominantly to dehydrohalogenation and other side reactions.

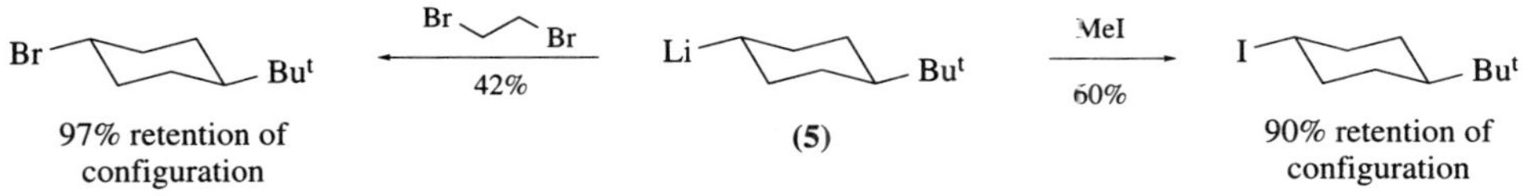

Scheme 5

Although organolithium reagents are well-known as strong bases, the simple alkyl derivatives are relatively weak nucleophiles and a high degree of aggregation in hydrocarbon solvents further inhibits their reactivity. Use of coordinating solvents, or the addition of ligands such as TMEDA, can reduce the degree of aggregation and enhance the reactivity of the organolithium species. Addition of even a small amount of ether will break up aggregates and enhance the rate of coupling markedly ⟨63JA2171⟩. Nevertheless the reactions are often unpredictable and better alternatives exist. In contrast to simple alkyllithiums, haloalkanes will readily alkylate vinyl, allyl, benzyl, aryl, allenyl, and alkynyl lithium reagents ⟨B-87MI 104-02⟩. Several examples of these reactions are discussed below.

The acidity of terminal alkynes and their ready conversion to lithium alkynides on reaction with *n*-butyllithium provides, following their subsequent alkylation with haloalkanes, a convenient and adaptable route to substituted alkynes ⟨91COS(3)271⟩. Although the basic nature of the lithium alkynide can lead to some dehydrohalogenation (especially with secondary and tertiary haloalkanes), substitutions are usually successful if primary haloalkanes are employed. This method has found widespread use and has been exploited as a stereoselective route to alkenes, for example, by Corey and Cheng in their prostaglandin and leukotriene syntheses (Equation (10)) (⟨B-89MI 104-01⟩ and references therein). Double deprotonation of a diyne (**6**) and treatment with α,ω-diiodoalkanes (**7**) (Equation (11)) has been developed by Semmelhack and Gallagher as an entry to enediyne structures ⟨93TL4121⟩. The reaction of vinylic lithium compounds with primary haloalkanes in THF at room temperature gives high yields of substituted alkenes (Equation (12)); elimination is the main reaction with secondary haloalkanes. The geometry of the double bond is retained during the alkylation reaction; as expected, the reactivity of the halides decreases in the order I > Br > Cl ⟨74TL3809⟩. The ready availability of configurationally pure vinylic lithium reagents makes this reaction a useful method for stereocontrolled alkene synthesis (⟨76TL4839⟩ and references therein). Deprotonation of allenes by *n*-butyllithium (or metal–halogen exchange with a bromoallene) gives a lithioallene that can be alkylated smoothly and in high yield with haloalkanes to give the substituted allene with only minimal contamination by the isomeric substituted alkyne (Equation (13)); an advantage of this procedure is that further lithiation and substitution can be carried out with the allene always being the major product ⟨75CC561⟩. However, choice of base, reaction conditions, and alkylating agent are important: Creary has shown that use of lithium diisopropylamide (LDA) to generate the allenyl anion and alkylation with benzyl chlorides leads to predominant formation of substituted alkynes with isolation of methylenecyclopropanes as by-products (Equation (14)) ⟨77JA7632⟩. Substituted allenes are formed in negligible amounts. The cyclopropanes presumably arise by proton transfer from the benzyl chloride to the allenyl lithium (giving a benzyl carbene or carbenoid), followed by addition of the carbene to the allene. Use of the more reactive benzyl bromides effectively suppresses the formation of carbene. Alternatively, addition of base to a mixture of benzyl halide and allene results in almost exclusive formation of methylenecyclopropanes.

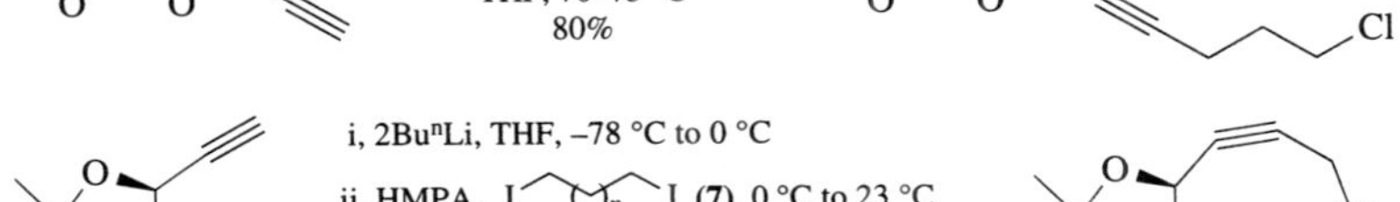

(10)

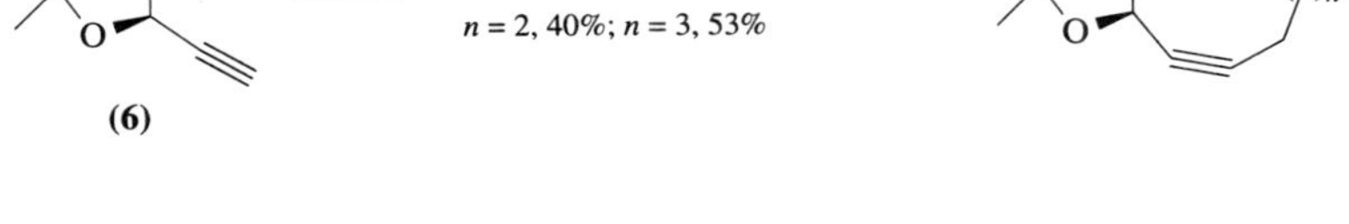

(11)

(**6**)

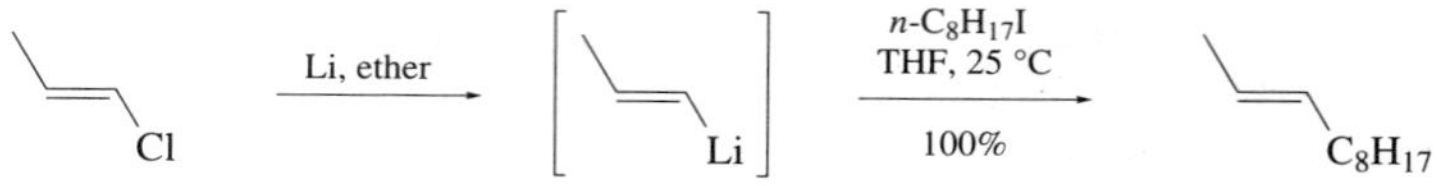

(12)

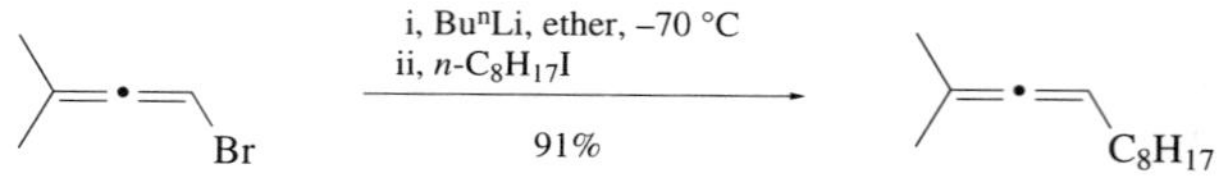

(13)

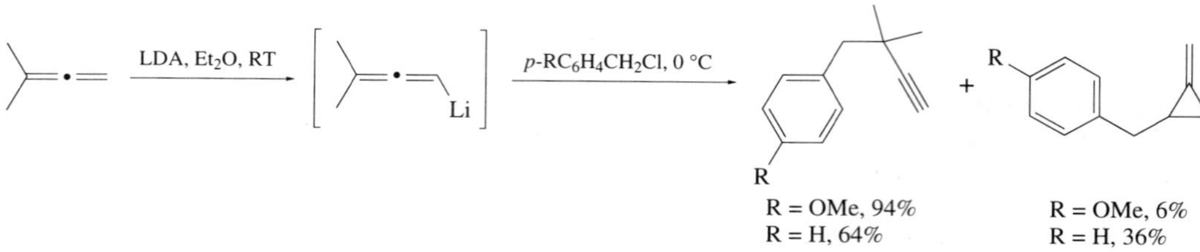

(14)

Resonance-stabilized organolithium species, such as allyllithium and benzyllithium reagents, are weaker bases and stronger nucleophiles compared with non-stabilized organolithium compounds, and will therefore react readily with alkyl halides to give coupled products. In a study of substitutions of cyclohexyl and 2-(phenyl)ethyl halides, yields were superior for bromo- and iodo- compared with chloro-substrates ⟨74JOC1168⟩. Where unsymmetrical allylic lithium reagents were used, $\alpha:\gamma$ selectivity was usually modest; however, alkylation was controlled by steric factors and predominated at the less-substituted (α) end (Equation (15)). Alkene geometry in the products was not reported; however, under these conditions the isolation of a mixture of *E* and *Z* isomers might be expected. The temperature dependency of alkene geometry vs. metal counterion in allylmetal systems has been studied ⟨91JA5893⟩.

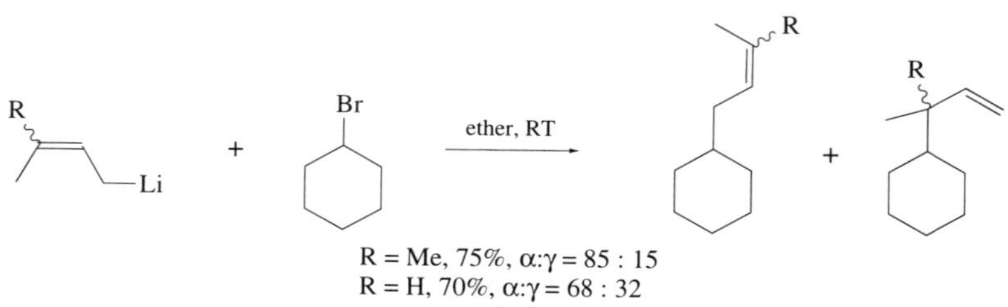

(15)

Another problem with the substitution of halogen by organolithium reagents concerns the stereochemistry of coupling. While the substitution reactions often follow second-order rate laws and exhibit halogen effects characteristic of an S_N2 process ⟨79JA2207⟩, simple alkyllithium reagents react with optically active secondary alkyl halides to give low yields of coupled product, with significant racemization of the stereocenter ⟨53JA3975⟩: in the reaction between n-butyllithium and (+)-2-bromobutane in hydrocarbon solvent, a 37% yield of (−)-3-methylheptane was recovered along with butene and butane, and the coupled product was almost completely racemized (98%). A study of benzyl, benzhydryl, and allyllithium species with optically active secondary alkyl halides revealed that the stereochemistry of the coupling process (in diethyl ether as solvent) is predominantly by inversion, although significant racemization is observed with optically active phenethyl chloride ⟨70JOC22⟩. However, the yields are still variable, and are often disappointing (40–60%).

The organic derivatives of heavier alkali metals display increased reactivity when compared with organolithium compounds. However, they suffer from similar synthetic drawbacks including a liability to thermal decomposition, poor solubility in hydrocarbon solvents, and a tendency to react with solvents (such as THF) used to enhance their solubilities. Although couplings occur with some net inversion, significant racemization is frequently observed. The substitutions often give mixtures of products, for example, the reaction of ethylsodium with (+)-2-chlorooctane gave octene (37%, isomer not specified), octane (8.5%), (+)-3-methylheptane (6.4%, approximately 20% racemization), and (+)-7,8-dimethyltetradecane (3.7%); extensive racemization is observed if the more reactive bromooctane is used ⟨50JA5127⟩. Allylsodium reacts with optically active 2-bromooctane in hexane to give 4-methyl-1-decene (resulting from inversion) in 83% yield with approximately 20% racemization ⟨50JA849⟩.

The major exception is the use of sodium alkynides for the preparation of substituted alkynes; couplings occur smoothly and in good yields (Equation (16)). Some relatively clean substitutions have been observed with simple organosodium (Equations (17) and (18)) and organopotassium compounds ⟨64AG(E)362⟩; however, these couplings in general only proceed satisfactorily with the more reactive halides (methyl, allyl, benzyl) and therefore the method has found little use ⟨64AG(E)287⟩. Side reactions can interfere or dominate even when a more reactive halide is employed; treatment of camphene (**8**) with a mixture of *n*-pentylsodium and potassium *t*-butoxide led to a vinylic metal species (believed to be a vinylsodium compound) which gave, upon reaction with iodomethane, a 1 : 1 mixture of the desired methylated compound (**9**) and the iodoalkene (**10**) resulting from metal–halogen exchange (Equation (19)) ⟨76HCA453⟩.

$$H{-}C{\equiv}C{:}\;Na^+ \xrightarrow[64\%]{\text{i, }CH_3CH_2CH_2I,\ NH_3;\ \text{ii, }NaNH_2;\ \text{iii, }Cl(CH_2)_3Br} ClCH_2CH_2CH_2{-}{\equiv}{-}CH_2CH_2CH_3 \qquad (16)$$

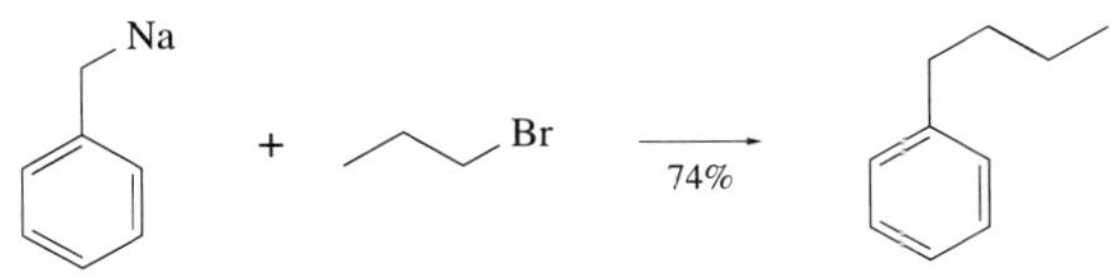

(17)

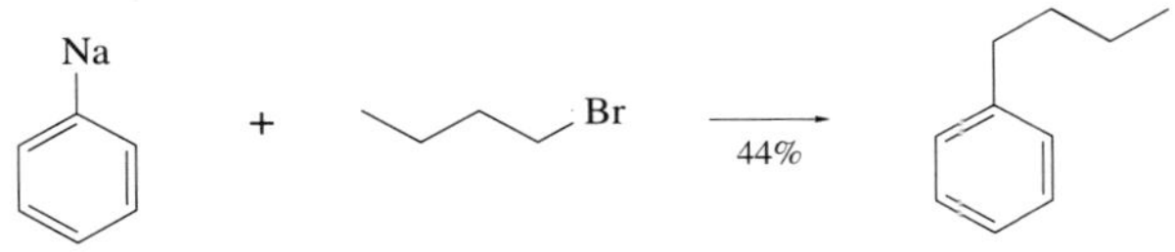

(18)

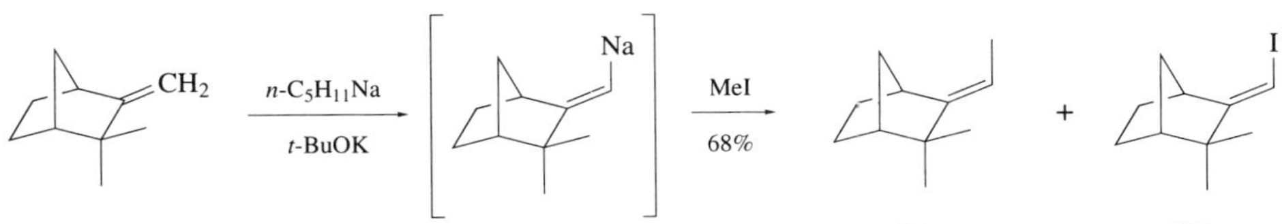

(8) (9) (10) (19)

One significant area where positive results have been achieved involves the alkylation of allylpotassium species which can be generated by treatment of alkenes with one of Schlosser's "superbases". The preparation and configurations of allyl organometallic species (M = Li, Na, K, Cs, MgX) have been reviewed by Schlosser *et al.* ⟨93T10175⟩. Allylpotassium reagents have been alkylated successfully with haloalkanes and the couplings have been applied to the introduction of unsaturated sidechains into several natural products. Djerassi and co-workers have shown that a range of steroids with unsaturated sidechains can be prepared in good yields (up to 95%) by this method if symmetrical allylpotassium reagents are used ⟨89JOC2117⟩; although steric bulk in substituents gave decreased yields, in general they are still acceptable (Scheme 6). Unsymmetrical allylpotassium units (such as prenylpotassium (**11**)) gave mixtures of substitution products, where the major product arose from attack at the least substituted terminus of the allyl group. In contrast, reaction of the allyl reagent derived from 2,3-dimethyl-2-butene gave a complex mixture of 13 products dominated by Wurtz coupling, and reaction of the steroidal iodide with metallated THF. Substitution with the allylpotassium reagent in this case was a minor pathway. Schlosser and Zhong have used the allylpotassium route to develop a short synthesis of α-santalol ⟨93TL5441⟩; interestingly the prenylpotassium reagent (**11**) reacted with the hindered bromoalkane (**12**) to give α-santalene (**13**) in 64% yield (Equation (20)). Nevertheless, despite the success of alkynide and allylmetal reagents, in general the heavier metals provide little benefit in carbon–carbon bond formation when compared with organolithium reagents, and more predictable results are obtained when the organometallic reagent is coupled with the alkyl halide under the influence of copper(I), either catalytically or in stoichiometric quantities.

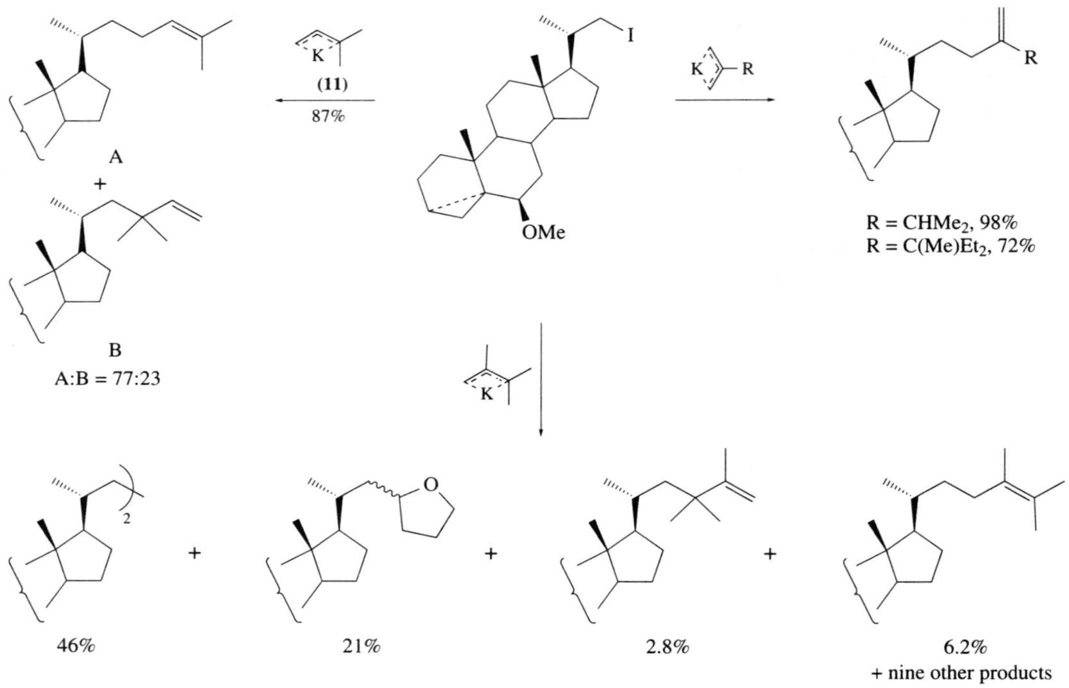

Scheme 6

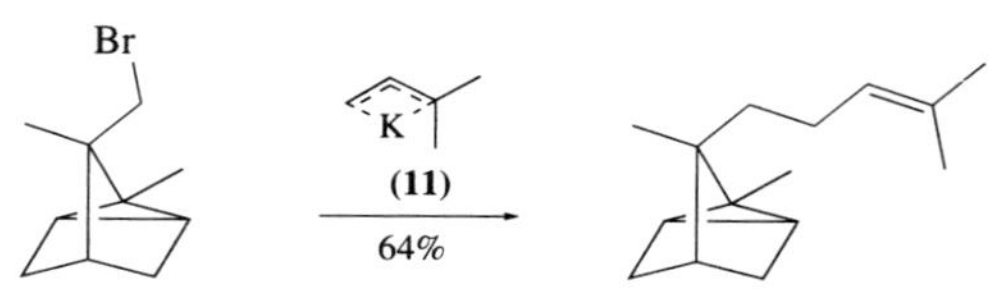

(12) (13) (20)

1.04.2.3 Group II Metals

Although the reaction of a Grignard reagent—which is generally the most readily formed reactive organometallic species—with an alkyl halide may conceptually seem to provide the simplest route to new carbon–carbon bonds, in practice it has little overall synthetic usefulness. The reactions of Grignard reagents with a variety of haloalkanes, and limitations of the process, have been discussed ⟨B-54MI 104-01, B-67MI 104-01, B-87MI 104-03⟩. This procedure has been used for the preparation of hindered alkanes: treatment of 2,3,3-trimethyl-2-chlorobutane with methylmagnesium chloride in ether at 20 °C gave a 38% yield of 2,2,3,3-tetramethylbutane ⟨44JA1389⟩. However, organomagnesium halides are normally insufficiently nucleophilic towards haloalkanes for couplings to occur at an acceptable rate, and similar side reactions as befall Group I organometallic-based substitutions can compete. Typical side reactions include α- and β-deprotonation leading to carbenes and alkenes respectively, metal–halogen exchanges, and single electron transfer processes providing free-radical derived products. In a similar manner to Group I organometallic species, the stereochemistry of reaction of optically active haloalkanes with organomagnesium reagents occurs with inversion, although the degree of racemization can be unpredictable. For example, pyrrylmagnesium bromide reacts with 2-bromooctane to give a mixture of 2- and 3-substituted pyrroles where displacement of bromide has occurred with complete inversion ⟨62JA4660⟩, and coupling between allylmagnesium bromide and 2-bromooctane gives 4-methyl-1-decene in 78% yield and with only 20% racemization ⟨50JA849⟩. In contrast to this, benzylmagnesium bromide reacts with 2-bromobutane to give 2-methyl-1-phenylbutane in only 17% yield, and substitution takes place with 91% racemization.

Benzyl Grignard reagents are capable of acting as ambident nucleophiles; while benzylmagnesium chloride reacts with iodomethane to give ethylbenzene (49%) as the sole product ⟨68JOC2203⟩, reaction with *t*-butyl halides gives some *ortho*- and *para-t*-butyltoluene in addition to neopentyl benzene ⟨70T2683⟩. Use of nonpolar solvents such as *n*-heptane, in which the Grignard reagent is insoluble, increased the proportion of ring-substituted products. In a similar manner to reactions of α-fluorocarbonyl compounds (as already discussed in Section 1.04.1.2), addition of RMgX to other α-haloketones can result in displacement of the halogen by addition to the carbonyl and subsequent Lewis-acid (Mg) induced rearrangement ⟨74TL1451, B-83MI 104-02, 86CPB3599⟩. Where two possible pathways for rearrangement exist (i.e. where two different alkyl groups in the intermediate alkoxide could migrate), then both products could be formed. That the pinacol-type rearrangement is initiated by an electrophilic interaction between Mg and the α-halogen was demonstrated by addition of alkyllithium reagents to α-halo ketones; only alcohols were isolated, which could be induced to rearrange by transmetallation with magnesium(II) ⟨72JOC4090⟩. The rearrangements appear to be under stereoelectronic control, as demonstrated in reactions of rigid decalone precursors (Scheme 7) ⟨72JOC4090⟩. The most successful substitutions with Grignard reagents involve the more reactive haloalkanes such as allylic and benzylic halides; however, even in these cases the yields are often only moderate. The reaction of Grignard reagents with a 1,3-dibromo-1-propene **(14)** offers some selectivity in substitution reactions ⟨93SL189⟩, whereas an organocopper reagent will replace both halogens, the analogous organomagnesium reagent will only substitute the more reactive (allylic) center (Scheme 8).

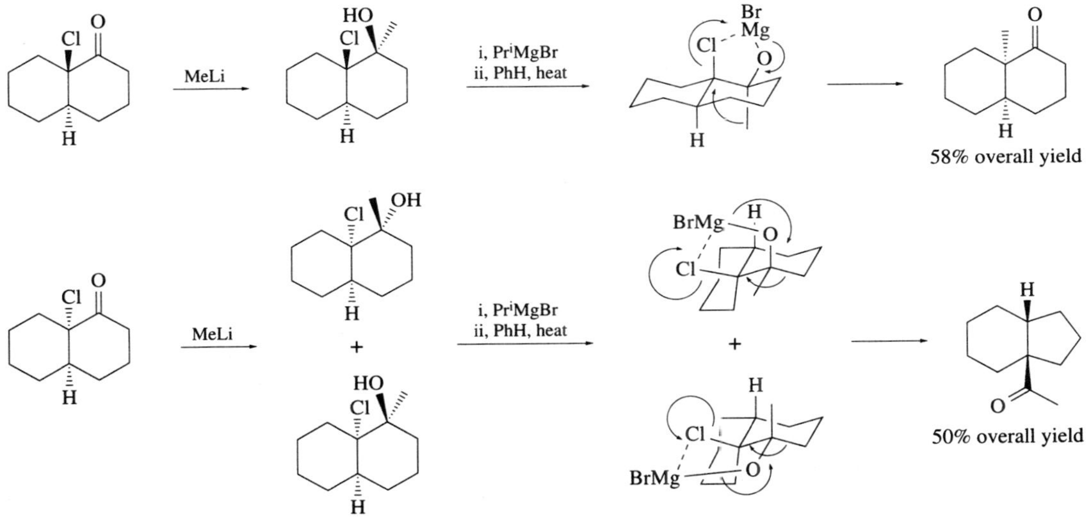

Scheme 7

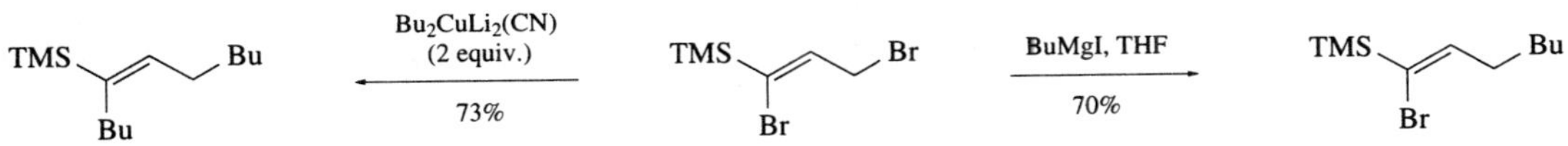

(14)

Scheme 8

The chemistry of organocalcium derivatives has produced erratic results. The outcome of reactions depended on the nature of the calcium used, as commercially available samples could contain various impurities such as sodium or magnesium that were believed to affect the preparation and reactivity of the organometallic species ⟨63JCS577⟩. While primary (and, to some extent, secondary) organocalcium reagents could be prepared by the direct reaction of alkyl iodides or bromides with calcium metal, preparation of the tertiary derivatives was often plagued by disproportionation reactions giving appreciable yields of alkene and alkane ⟨73JCC4268, 73T1069⟩. Preparation of the benzylcalcium compounds was complicated by significant Wurtz-type coupling ⟨73T1069⟩. Direct access to organocalcium reagents has been facilitated by the development of a highly reactive form of calcium by Rieke and co-workers ⟨90JOC5045⟩. Reduction of calcium halides by lithium biphenylide in THF at room temperature gave a colored, soluble calcium species which reacted with organic halides at temperatures as low as $-78\,°C$ for primary, secondary, and tertiary haloalkanes (including 1-bromoadamantane) or $-20\,°C$ for aryl halides. Fluorobenzene was converted to the organocalcium reagent at room temperature. The formation of Wurtz-coupled sideproducts was not reported, with the exception of rapid homocoupling of allylcalcium species with starting allyl halides. A novel carbocyclic ring synthesis using organocalcium reagents emerged from this work (Scheme 9). Reaction of highly active calcium with 1,3-dienes gave intermediate metallocycles (**15**) which reacted with α,ω-dihaloalkanes to give carbocycles in a highly regio- and stereospecific manner. In most cases the 1,2-addition products were observed, with the exception of reaction with 1,2-dihaloalkanes where 1,4-addition products dominated. 1,2-Elimination from the ethylene dihalides, giving ethene as a sideproduct, apparently did not compete. This procedure is an improvement on the magnesium-mediated method described earlier by Xiong and Rieke ⟨89JOC3247⟩, as the organocalcium metallocycles regularly gave dialkylation products (carbocycles) in higher yields while in some cases the magnesium metallocycle gave the monoalkylated product, requiring higher temperatures to effect cyclization. The regiochemistry (1,2-addition) was the same for both Ca and Mg species with the exception of addition to 1,2-dihaloethanes—only calcium gave an addition product, producing the 1,4-regioisomer.

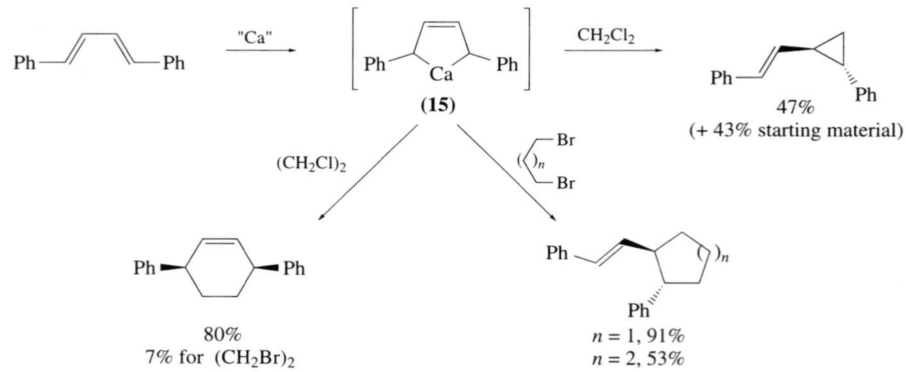

Scheme 9

Organobarium compounds have found relatively little application in organic synthesis; however, Yamamoto and co-workers have recently developed a straightforward method for the preparation of allylbarium reagents ⟨91JA8955⟩. Reaction of barium(II) iodide with lithium biphenylide in THF (the "Rieke method" used for preparation of active calcium and other metals) gave highly reactive barium which readily converted allylic chlorides to allylbarium reagents at $-78\,°C$. Although the structure of these allylbarium compounds is uncertain, Corey and Shieh have suggested that they may have η^3, π-complex structures which do not rapidly interconvert with isomeric σ-allylbarium forms ⟨92TL6435⟩. The remarkable α-selectivity in their reaction with carbonyl groups, as well as the excellent retention of double bond geometry compared with other allylmetal species (magnesium and calcium lead predominantly to γ-substitution, while lithium and cerium are less selective ⟨91JA8955⟩), led Corey and co-workers to investigate the alkylation of allylbariums with allyl halides ⟨92TL6435, 93TL5995⟩. Alkylations proceeded in good yields (55–66%) under mild conditions ($-78\,°C$ to room temperature), with complete α-selectivity and retention of double bond geometries in *both* allyl units. In cases where displacement of either bromide or chloride could have occurred, selective alkylation by displacement of bromide was observed. This remarkable selectivity of organobarium reagents has led to the development of a remarkably short, enantioselective synthesis of 2,3(*S*)-oxidosqualene (Scheme 10) ⟨93TL5995⟩.

1.04.2.4 Organocopper Derivatives and Copper-Catalyzed Couplings

Whereas carbon–carbon bond formation by substitution of a halogen with an organometallic reagent of a Group I or Group II metal often gives poor results with a range of competing reactions intervening, the use of copper to mediate the process—either in stoichiometric amounts or catalytically—has proven a great success and now forms one of the cornerstones of modern organic synthesis. The volume of literature on the topic is immense, and the various aspects of organocopper chemistry have been reviewed ⟨74AOC215, 75OR(22)253, 78PAC709, 84T641, 87S325, 91COS(3)207, 92OR(41)135⟩. Improvements and variations since the original work by Gilman *et al.*

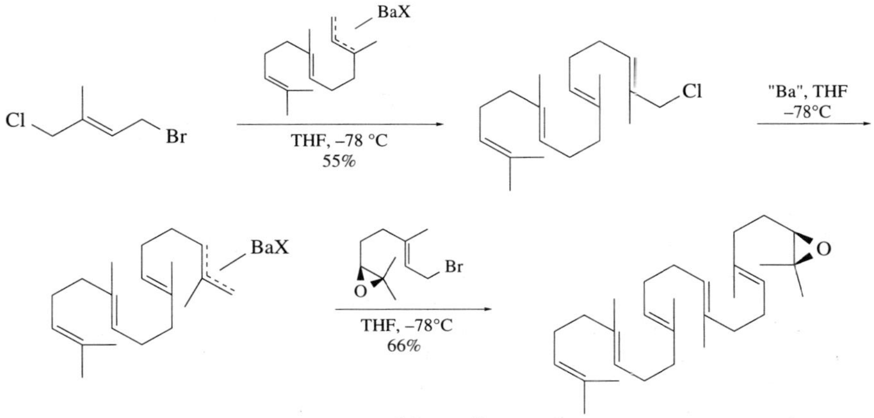

Scheme 10

⟨52JOC1630⟩ have meant that organocopper-mediated bond formation provides one of the most reliable reactions in synthesis.

One of the disappointments of early organometallic approaches to organic synthesis was that, although Grignard reagents were readily formed from a wide variety of substrates and many organolithium compounds were also accessible, coupling reactions with haloalkanes were generally unreliable. The early history of the effectiveness of transition metal salts as additives was not encouraging; for instance, Kharasch and Fields demonstrated that arylmagnesium bromides, in the presence of cobalt(II) chloride or silver(I) bromide, gave only biaryls on reaction with haloalkanes such as bromoethane ⟨41JA2316⟩. While a silver catalyst was shown by Tamura and Kochi to be effective for couplings where both alkyl groups were the same (e.g. Bu^nMgBr and Bu^nBr gave *n*-octane in 79% yield) ⟨71S303⟩, reaction of Grignard reagents with haloalkanes where the organic residues differed led to a mixture of all three possible coupling products (e.g. Bu^nMgBr and *n*-$C_6H_{13}Br$ gave *n*-octane, *n*-decane, and *n*-dodecane, in the ratio 1.9 : 1.4 : 1.0). Free radicals were implicated in the process ⟨71JA1483⟩. Ferric chloride has been used to catalyze the reaction between a bromocyclobutane (**16**) and vinylmagnesium bromide (Equation (21)) ⟨83CB882⟩.

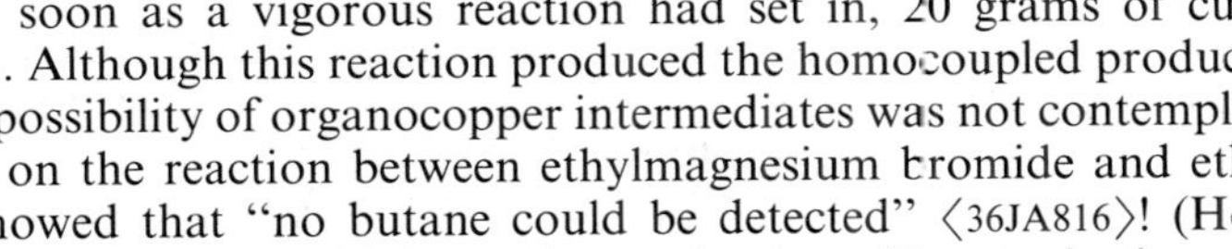

(16) (21)

The effect of copper on the substitution reactions has had a chequered history: an early report by Turner is of interest—he reported that, following addition of ethyl α-bromobutyrate (**17**) to magnesium in ether "as soon as a vigorous reaction had set in, 20 grams of cupric chloride were added" ⟨20MI 104-01⟩. Although this reaction produced the homocoupled product (**18**) in 30% yield (Equation (22)), the possibility of organocopper intermediates was not contemplated. Initial studies by Linn and Noller on the reaction between ethylmagnesium bromide and ethyl bromide in the presence of CuCl showed that "no butane could be detected" ⟨36JA816⟩! (However, a later re-investigation by Parker and Noller did detect the production of butane in the reaction ⟨64JA1112⟩.)

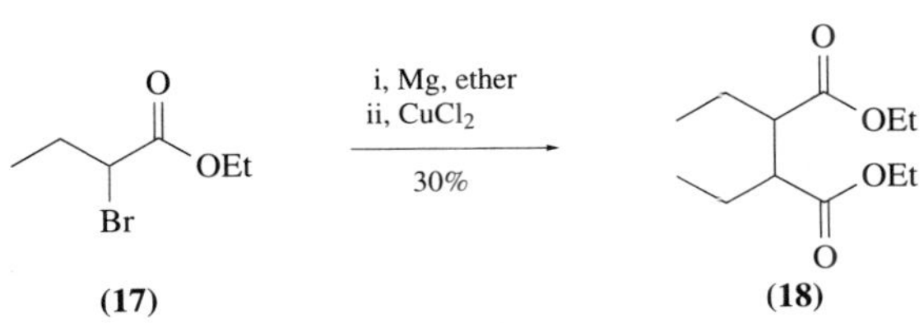

(17) **(18)** (22)

The early work on organocopper chemistry by Gilman *et al.* ⟨52JOC1630⟩ was followed by several other groups, including Corey and Posner ⟨67JA3911, 68JA5615⟩ who showed that lithium dialkylcuprates were excellent reagents for the displacement of bromide or iodide from haloalkanes by an alkyl ligand from copper. Although side reactions (reductions, eliminations, and homocouplings) were sometimes noted, yields were usually good to excellent. Reaction of 1-bromo-4-*t*-butylcyclohexane (**19**) with lithium dimethylcuprate (Equation (23)) occurred largely (but not exclusively) by inversion; some elimination and reduction products were also detected ⟨67JA3911, 75OR(22)253⟩. The reaction also proved useful for the displacement of both halogen atoms in a geminal dibromo substrate (**20**), although the reduction pathway was a significant contributor (Equation (24)) ⟨68JA5615⟩. The mechanisms involved are obviously complex; one of the more popular proposals involves initial, rate-determining oxidative addition of the haloalkane to the cuprate with inversion of configuration at the carbon atom of the haloalkane component, followed by reductive elimination ⟨91COS(3)207⟩. However, Bertz *et al.* have recently described the reactions of $Bu_2CuLi{\cdot}LiX$ with iodocyclohexane (for example, Equation (25)) and explained the product distribution as resulting from electron transfer from the cuprate to the alkyl iodide and subsequent cyclohexyl radical formation ⟨91JA631⟩. Yields of butylcyclohexane depended on the counterion X in the cuprate complex, decreasing in the order $Bu_2CuLi{\cdot}LiCN$ (70–80%) > $Bu_2CuLi{\cdot}LiBr$ (47–53%) ≈ $Bu_2CuLi{\cdot}LiOTf$ (44–50%) > $Bu_2CuLi{\cdot}LiI$ (29–32%). Formation of the radical intermediate apparently depends upon the reduction potential of the haloalkane; if the potentials are not sufficiently positive (as is likely for primary haloalkanes and secondary bromoalkanes) then transmetallations can account for side reactions.

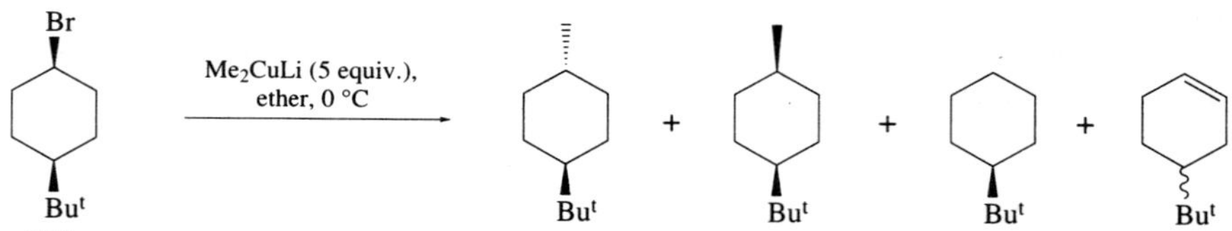

(**19**)

(23)

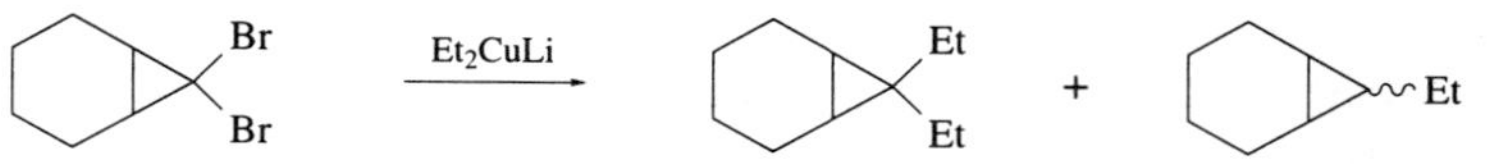

(**20**)

(24)

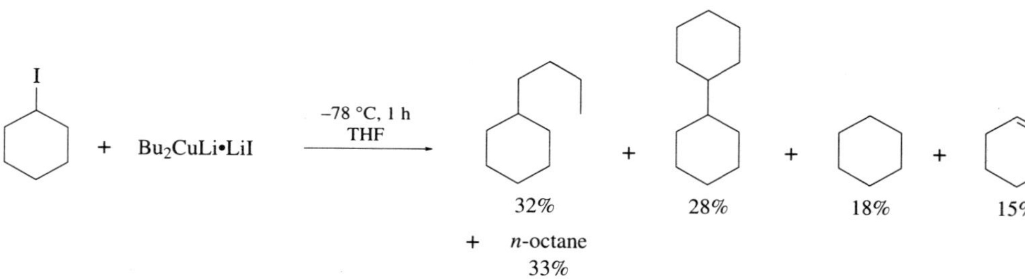

(25)

Whitesides *et al.* have investigated the reaction of diorganocuprates with a variety of alkyl halides, and high yields of coupled products were obtained ⟨69JA4871⟩. Branching in the cuprate was tolerated in that primary, secondary, and tertiary alkyl cuprates all coupled with primary alkyl halides in good to excellent yields (with yields being highest for alkyl iodides and bromides, and lower for the corresponding alkyl chlorides). Use of THF as solvent led to increased reactivity, even for the usually sluggish alkyl chlorides, and acceptable yields were then obtained. Couplings are most successful with primary alkyl halides; alkyl cuprate couplings with secondary and tertiary alkyl halides generally give poorer yields or fail completely. (It is noted that coupling between Ph_2CuLi and a secondary alkyl halide can be achieved in good overall yield.) A brief study on the relative effectiveness of catalytic versus stoichiometric copper showed that yields are somewhat lower in the catalytic case (e.g. *n*-iodooctane couples with Me_2CuLi in ether to give *n*-nonane in 97% yield, whereas MeLi in the presence of 5 mol.% CuI gives *n*-nonane in 64% yield) ⟨69JA4871⟩; nevertheless the couplings are still significantly more successful than in the absence of copper.

Lipshutz *et al.* have investigated couplings between secondary haloalkanes and the higher order cuprates derived from CuCN, which can be formulated as $Li_2[R_2Cu(CN)]$ ⟨81JA7672⟩. In general, the couplings proceed in excellent yields with both bromides and iodides. Substitution of halogen in a secondary haloalkane introduces the question of stereoselectivity. Whitesides *et al.* demonstrated that lithium diphenylcuprate reacted with a secondary alkyl bromide (**21**) with predominant (84–92%) inversion (Equation (26)) ⟨69JA4871⟩, and in a more extensive study by Lipshutz and Wilhelm, it was shown that secondary iodoalkanes gave essentially racemic mixtures on coupling with organocuprates whereas the corresponding bromides reacted with almost complete inversion ⟨82JA4696⟩. Given the mechanistic discussion of Bertz *et al.*, it is likely that radical intermediates intervene in the case of the iodides while the bromides undergo an S_N2-like displacement ⟨91JA631⟩. The reaction of organometallic reagents with unsymmetrical allyl halides opens up the possibility of regiocontrol by either S_N2 or S_N2' displacement of the halogen by the carbanionic alkyl group. In an extensive study, Yamamoto *et al.* have shown that the regioselectivity can be controlled with judicious choice of reagent ⟨80JA2318⟩. While Grignard reagents gave products of α-substitution (the S_N2 alternative), organocopper reagents favored γ-regioselectivity (S_N2' attack). The greatest control was observed when Lewis acid modified alkylcopper reagents were used in THF, with the best results being obtained consistently with $RCu{\cdot}BF_3$. Phenylcopper reagents were less selective. Use of lithium dialkylcuprates, or organocopper species derived from Grignard reagents (whether complexed with Lewis acids or not) gave poorer results.

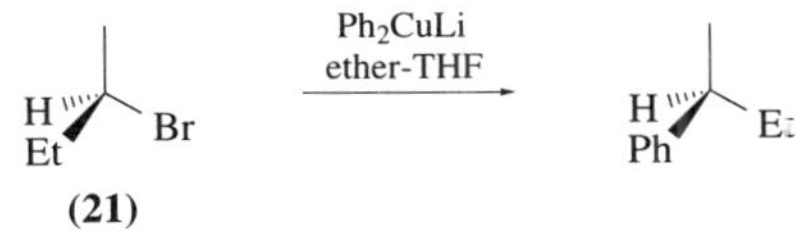

(21) (26)

Whereas stoichiometric organocopper reactions usually proceed best when the cuprates are derived from organolithium reagents, catalytic reactions normally involve Grignard reagents (which also have the added benefit of being easier to prepare). A major contribution to the area of organocuprate chemistry was made by Tamura and Kochi with the introduction of the THF-soluble cupric salt Li_2CuCl_4; here, the copper(II) species is reduced to an active copper(I) catalyst *in situ* ⟨71S303, 72JOM(42)205⟩. The catalyst has been shown to be superior to cuprous iodide for these cross-coupling reactions in a number of ways ⟨83JOC1912⟩: its use allows for lower reaction temperatures and provides higher yields of more easily purified products. Catalyst ratios as low as 1 mol.% still give excellent yields of product. A typical halogen dependency is observed with reactivity decreasing in the order $I > Br \gg Cl$. In contrast with the Ag-catalyzed couplings ⟨71S303⟩, copper-catalyzed reactions give predominantly the desired coupling products with only negligible amounts of unwanted homocouplings. A useful synthesis of ω-functionalized aldehydes has been developed by Volkmann *et al.* ⟨83JOC1767⟩. The Grignard reagent derived from 2-(2-bromoethyl)-1,3-dioxolane was treated with a variety of functionalized haloalkanes in THF containing Li_2CuCl_4 (5 mol.%) to give, following hydrolysis, ω-functionalized aldehydes in good to excellent yields (typically 70–85%). Functional groups in the haloalkane, such as methoxy, chloroalkyl, cyano, and ester, did not interfere with the magnesio-cuprate addition. The catalyst has proven useful for homologation reactions where copper-catalyzed coupling of a Grignard reagent with an α,ω-dibromoalkane, using as little as 1 mol.% catalyst, could be controlled to give a monosubstituted product in 60–96% yields (Equation (27)) ⟨74JA7101⟩. However, two-carbon homologations using this procedure ($RMgX/Li_2CuCl_4 + BrCH_2CH_2Br$) fail and metal–halogen exchange occurs giving ethene and a new bromoalkane R—X which undergoes coupling with remaining Grignard reagent to produce moderate yields of dimeric alkanes. Conversion of a vicinal dibromo compound to the corresponding alkene also occurs with preformed organocuprates ⟨73SC281⟩.

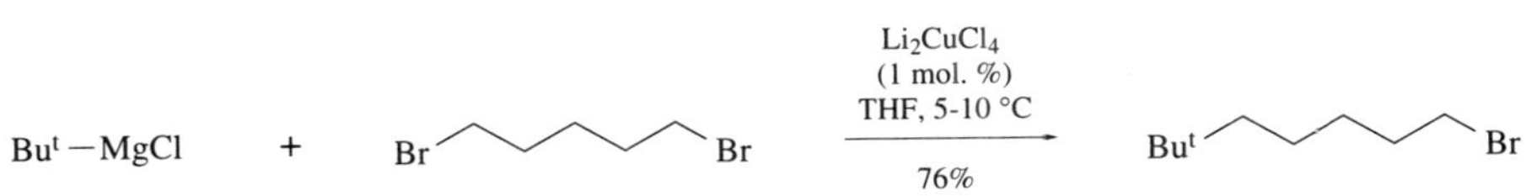

(27)

A disadvantage of stoichiometric lithium dialkylcuprate substitutions is that one of the alkyl ligands on copper is essentially "lost" and, while this may be of little concern if the precursor is readily available, loss of alkyl ligands that are expensive or not readily prepared can be distressing. Thus, the use of "dummy" ligands which are not transferred is a preferable procedure. Mandeville and Whitesides have studied a variety of ligands and conclude that ligands that produce relatively stable organocopper compounds, such as alkynyl and alkoxy ligands, are most successful (Equation (28)) ⟨74JOC400⟩. Corey *et al.* developed an alkynyl ligand derived from 3-methoxy-3-methyl-1-butyne which provides cuprates with appreciable solubility in THF, as well as acting as a "dummy" ligand that is not readily transferred ⟨78JOC3418⟩. Another useful higher order cuprate appears to be the mixed complex incorporating both 2-thienyl and cyano ligands as non-transferable groups; good yields of coupled products are obtained (Equation (29)) ⟨87S325⟩. It is assumed that $d\pi^*$ backbonding from the Cu(I) to the alkynyl or cyano ligands accounts for their limited transfer rates from cuprate reagents ⟨88JA2641⟩. Johnson and Dhanoa have reported that the dimsyl anion $MeS(O)CH_2^-$ acts as a nontransferable ligand ⟨87JOC1885⟩; the cuprates were prepared by treatment of DMSO with Bu^nLi followed by CuI and then 1 equivalent of the desired alkyllithium. Couplings with primary iodoalkanes proceed in high yields and the DMSO produced on hydrolysis of the reaction mixture is easily separated from the product. The alkylation reactions succeed where the transferable organic residue is derived from either an organolithium or a Grignard reagent ⟨86T2873⟩.

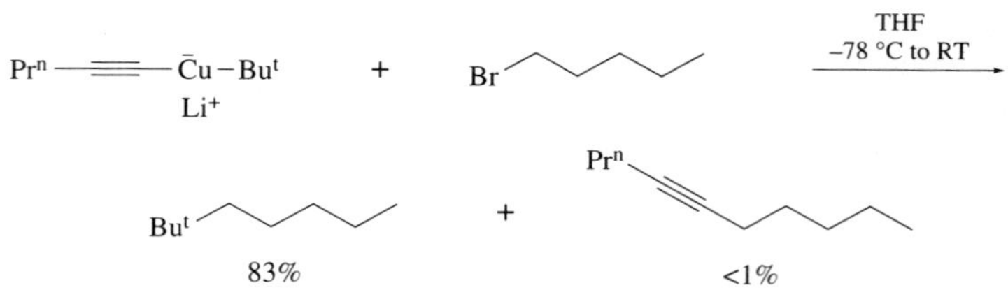

(28)

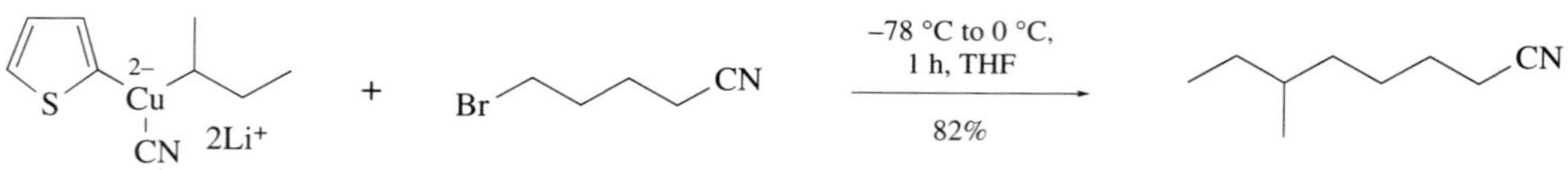

(29)

Another nontransferable group that has shown considerable success is the thiophenoxide ligand, which can be readily incorporated by treatment of an alkyllithium with CuSPh ⟨73JA7788⟩. The reactions of organobis(cuprates) based on this "dummy" ligand have been described by Wender and White ⟨88JA2218⟩; the addition of these novel reagents (**22**) to 3-halocycloalk-2-enones (**23**) provides a convenient method for the generation of spirocyclic compounds (Scheme 11). The observed halogen dependency is the reverse of that usually encountered, i.e. yields are similar for bromo- and chloroalkenones (typically 70–96%), and poorer for the iodo analogues (50%). Use of the thiophenoxide gave the highest yields, and preparation of the dicuprate from a Grignard, instead of organolithium, reagent also led to significantly reduced yields. It is interesting to note that the second addition resulting in quaternization only occurs readily in these intramolecular cases; addition of ordinary cuprates RCu(SPh)Li to the alkenones (**23**) only gave products of single addition unless much higher reagent concentrations were used (and yields were only modest even at saturation).

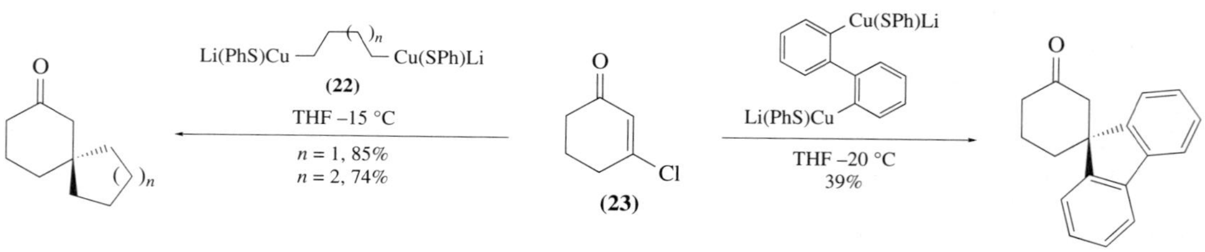

Scheme 11

One of the main bonuses of cuprate-based reactions is that a wide range of functional groups can be incorporated into the haloalkane component without the need for protection; alkene, ester, amide, ketone, nitrile, sulfonamide, as well as most popular protecting groups, are all tolerated remarkably well. While high yields are the norm, and can generally be relied upon in these reactions, very occasionally (and inexplicably) a single cuprate will fail to react with a haloalkane where others succeed. In a series of papers directed towards novel β-amino acids and naturally occurring alkaloids, Jefford *et al.* demonstrated that primary iodoalkanes (**24**) underwent nucleophilic substitution when treated with a range of lithium dialkylcuprates (Equation (30)) ⟨93TL1111, 93TL2911, 93TL7557⟩. Various primary, secondary, and tertiary alkyl groups were introduced in yields ranging from 42 to 96%, with most conversions occurring in over 70% yield. However, attempted introduction of a phenyl group by reaction with lithium diphenylcuprate was unsuccessful, with the reaction leading to either recovery of starting material or β-elimination ⟨93TL1111⟩. This result is intriguing as a similar β-(N-tosyl)amino alkyl iodide (**25**) underwent smooth substitution with Ph_2CuLi (Scheme 12) ⟨93TL5127⟩.

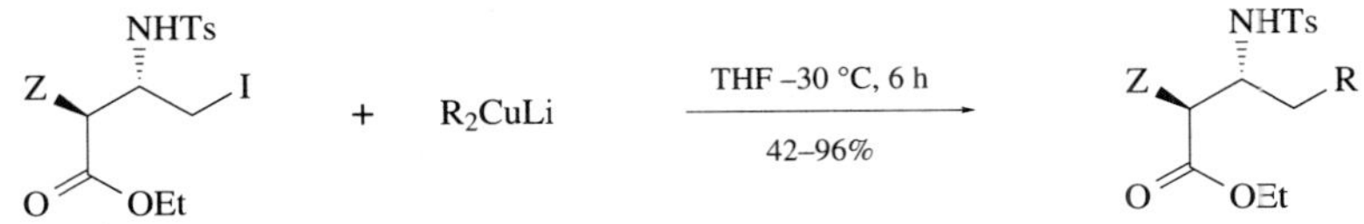

(30)

Z = H, OH
(24)

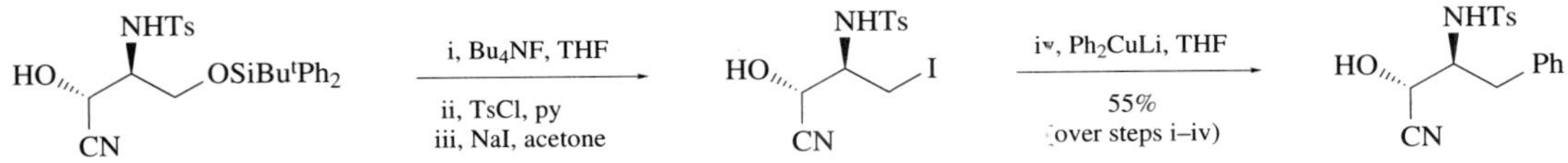

(25)

Scheme 12

Jeffery *et al.* have developed a novel synthesis of skipped enynes and diynes which avoids the necessity for preparation of carbanionic intermediates prior to formation of cuprates (Equation (31)). Coupling can be achieved in excellent yields by the reaction of a terminal alkyne with an allyl ⟨89TL2225⟩ or propargyl ⟨92TL5757⟩ halide in the presence of a catalytic amount of a copper(I) salt. The catalyst system containing CuI/Bu^n_4NCl (catalytic quantity of each)/Na_2CO_3 or K_2CO_3 in DMF gives 1,4-enynes or diynes in (typically) better than 70% yield. The reactions are highly chemo- and regioselective, giving products of exclusive α-attack in all cases except the reaction of 3-chlorobut-1-ene where substitution on the terminal (γ) carbon dominates (≈10 : 1). A variety of functional groups (alcohol, ester, chloroalkyl, trimethylsilyl) are tolerated without the need for protection.

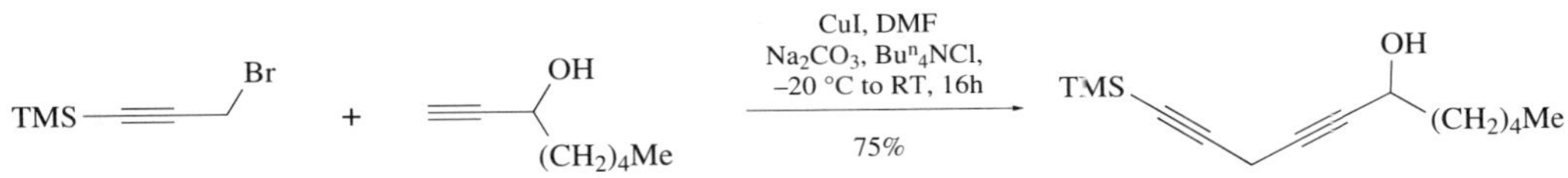

(31)

While organocopper reagents can tolerate a wide range of functionalities in the haloalkane component with which they couple, one aspect that has limited their usefulness is their preparation from organolithium or magnesium reagents that will react with these functional groups—any synthetic design must avoid leaving these functionalities in the cuprate. Therefore two strategies have usually been adopted: protection of sensitive functional groups, or strategic bond disconnection to place problematic functionalities in the haloalkane component. Work by Rieke *et al.* has gone some way towards solving this problem in that they have demonstrated that certain organocopper compounds (of uncertain overall composition) can be prepared directly by reaction between a halide substrate and "zerovalent copper", an active species usually produced by reduction of copper(I) complexes such as $Cu(I)I{\cdot}PR_3$ by lithium naphthalenide ⟨93JOC2483⟩. A range of functional groups such as ester, nitrile, less reactive halides (Cl, F), epoxides, and ketones are tolerated ⟨87JOC5056⟩. In an extensive series of papers (see ⟨93JOC2492⟩ for leading references), Rieke has shown that these organocopper derivatives undergo many of the reactions typical of other, more classical, copper-based reagents, including couplings with haloalkanes. Arylcopper reagents coupled with alkyl halides in variable yields (Equation (32)), however reactions were occasionally plagued with homocouplings of the alkyl residues of both the organocopper and the alkyl halide components, and the yields were sometimes disappointing. Alkylcopper reagents prepared by reaction of the zerovalent copper with primary haloalkanes couple with other haloalkanes to give unsymmetrically coupled products in moderate yields at best (Equation (33)); rapid homocoupling of the organic residues in the organocopper reagent competes and often dominates ⟨84JOC5280, 88JOC4482⟩. However, more promising results are outlined in another report ⟨92JA5110⟩. Reduction of $CuCN{\cdot}nLiCl$ ($n = 1,2$) at $-100\,°C$ by lithium naphthalenide in THF produced an active copper(0) species which underwent oxidative addition on treatment with allyl chlorides to give allylcopper reagents; at this low temperature homocoupling was negligible. The allylcopper reagents contained a wide range of extra functional groups (ketone, enone, epoxide, nitrile, ester, alkyl chloride, and carbamate) which were not affected. Treatment of the allylcopper reagent with allyl bromide gave a mixture of 1,5-dienes in excellent yield, indicating that coupling had occurred at both α and γ ends of the copper reagent; the major products corresponded to γ-attack. A new approach to thienyl-based organocopper reagents has been reported by Rieke *et al.* ⟨93JOC2492⟩, providing a route to stable alkylcopper reagents where homocouplings or eliminations do not compete.

However, their reactions with haloalkanes were unsatisfactory; for example, addition of the *n*-octylcopper reagent to iodomethane gave nonane in only 32% yield, and reaction with benzyl bromide gave only bibenzyl (48%) and toluene (20%).

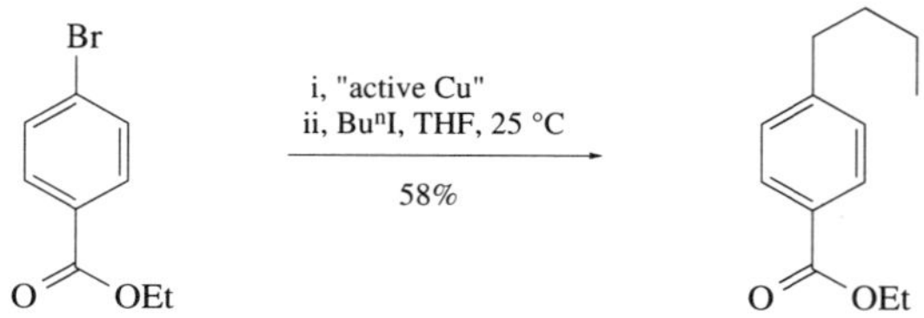

(32)

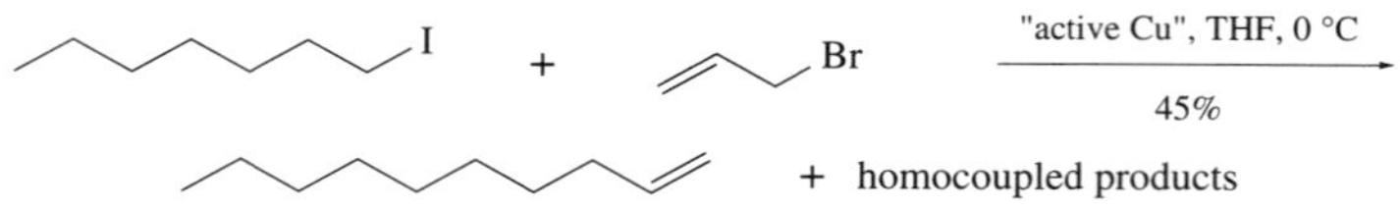

(33)

An alternative method for the preparation of organocopper reagents involves transmetallation of other organometallic species such as zinc- and titanium-based compounds (see Section 1.04.2.5 for the copper-catalyzed transfer of ligands from boron). Organozinc reagents are generally much less reactive than organomagnesium and organolithium compounds ⟨74AOC83⟩, and they do not generally react with alkyl halides unless the alkyl halide is in some way made more reactive (as in the case of an α-alkoxy haloalkane) or transition metal catalysis is employed. The copper(I)-catalyzed allylation of organozinc reagents proceeds in good to excellent yields ⟨87JOC4418⟩; the highly reactive nature of allyl halides means that chlorides and bromides show similar reactivity. Unsymmetrical allylating reagents gave mixtures of products, although γ-alkylation predominated. Excellent γ or S_N2' selectivity was observed by Nakamura *et al.* when a Gilman reagent R_2CuLi was treated with anhydrous $ZnCl_2$ (1 equiv.) prior to addition of the allyl halide; almost complete diastereofacial selectivity was also noted when chiral substrates were used ⟨89JA3091⟩. Virtually identical results were obtained when a dialkylzinc reagent was treated with catalytic $CuBr{\cdot}SMe_2$ (Equation (34)). A model involving interaction between a copper *d*-orbital and the antisymmetric π*-orbital of the alkene has been proposed by Nakamura to account for the observed 1,2-asymmetric induction ⟨91SL539⟩. Nakamura *et al.* have developed a route to zinc homoenolates (**26**) using anhydrous zinc chloride to regioselectively cleave one carbon–carbon bond in a cyclopropanone ketal (**27**) ⟨87JA8056, 87TL337⟩. Transmetallation with CuBr gave a zinc–copper reagent which, in the presence of hexamethylphosphoramide (HMPA), reacted with various carbon electrophiles including allyl bromides (Equation (35)).

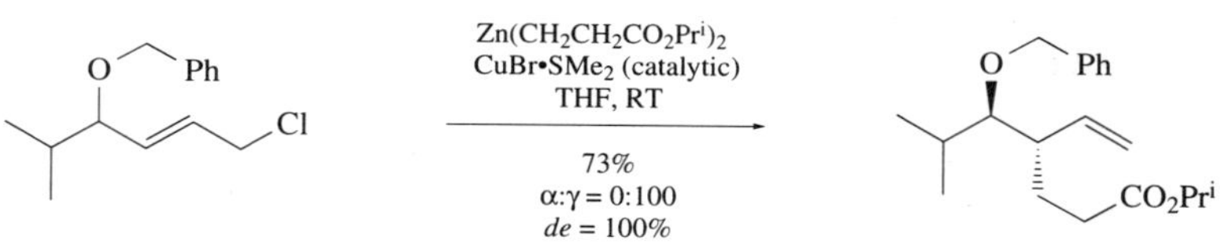

(34)

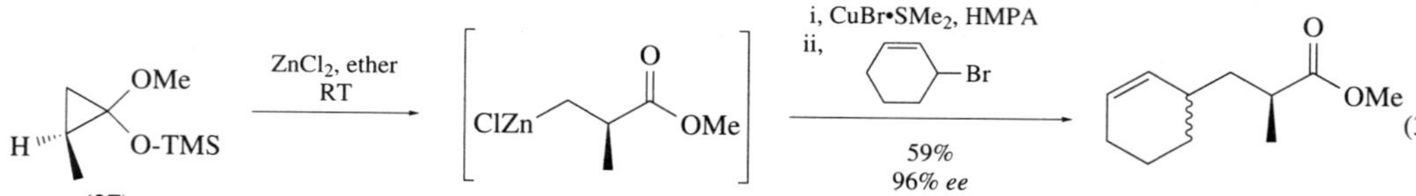

(35)

(27) **(26)**

In the preparation of organozinc species from oxidative addition reactions between haloalkanes and zinc metal, commercial samples usually display limited reactivity and require activation. A method used by Knochel and co-workers in an extensive series of reports on organozinc chemistry involves treatment of zinc dust in THF successively with 1,2-dibromoethane and then chlorotrimethylsilane. This activated zinc will then readily insert into carbon–halogen bonds. For example, 3-iodopropionitrile reacts with activated zinc to produce 2-cyanoethylzinc iodide which, following transmetallation with the soluble copper salt $CuCN{\cdot}2LiCl$ gave a new copper organometallic species that reacted regiospecifically with allyl halides giving γ-alkylated products in 83–99% yields

⟨88TL2395⟩. Alternatively, the α-substituted product could be obtained by treating the organozinc reagent with the allyl halide in the presence of 1 mol.% of $Pd(PPh_3)_4$ (Scheme 13). This procedure even allows the preparation and alkylation of zinc–copper organometallics containing acidic hydrogen atoms; amines (including indoles), amides, terminal alkynes, and (to some extent) alcohols were tolerated ⟨91JOC5974⟩. For example, 3-(2-iodoethyl)indole (**28**) could be converted to the zinc reagent, transmetallated with CuCN·2LiCl, and alkylated with an allyl bromide (limiting reagent) in 95% yield (Equation (36)).

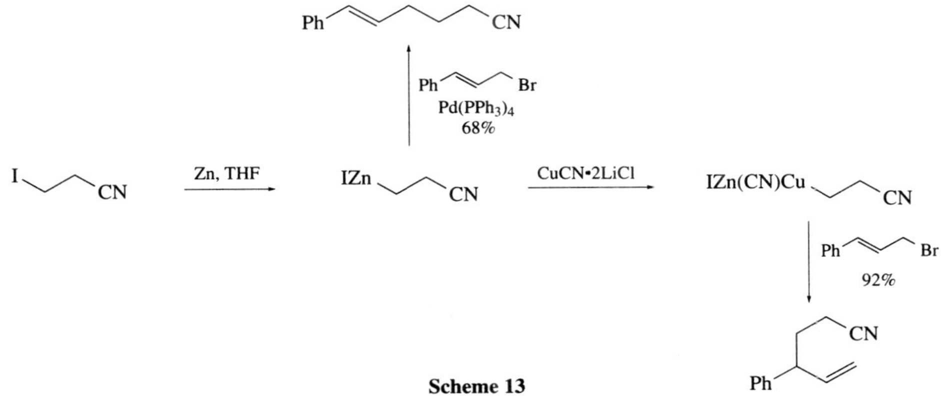

Scheme 13

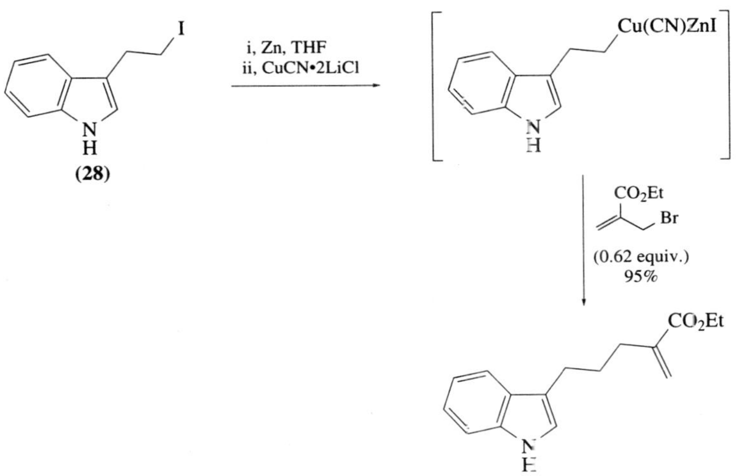

(28)

(36)

In an elegant extension of this work, Rao and Knochel combined carbocupration of alkynes using highly functionalized zinc–copper organometallics, with alkylation (or reaction with other electrophiles) as a new, stereospecific synthesis of polyfunctional alkenes ⟨91JA5735⟩. Addition of the cuprate $Me_2Cu(CN)Li_2$ to a functionalized organozinc allowed formation of the organocuprate. These reagents added efficiently and stereospecifically to alkynes (terminal or 1-methylthioalkynes in intermolecular reactions) giving vinylcuprates which could then be alkylated with allyl halides to give 1,4-dienes. An intramolecular variant (Scheme 14) allowed rapid construction of densely functionalized carbocycles. These intramolecular carbometallations proceeded more readily than similar reactions using lithium or magnesium copper reagents; yields ranged from 57 to 76%.

Rieke and co-workers have developed a procedure for the preparation of active zinc by reduction of zinc halides with lithium naphthalenide ⟨91JOC1445⟩; alkyl bromides containing a range of functional groups can be added to this zinc species to produce organozinc reagents which, following transmetallation with CuCN·2LiBr, gave organocuprates [RCu(CN)ZnX]. Reaction with a variety of allyl halides gave alkylation products where the γ-regioselectivity was, in general, superior to that obtained by Yoshida and co-workers ⟨87JOC4418⟩. Even relatively hindered allyl halides such as dimethylallyl bromide gave at least 95% γ-regioselectivity (Scheme 15). One interesting observation was that the regioselectivity showed little temperature dependence; reactions run at −78°C and allowed to warm gradually to room temperature gave almost the same results as reactions carried out at 0 °C.

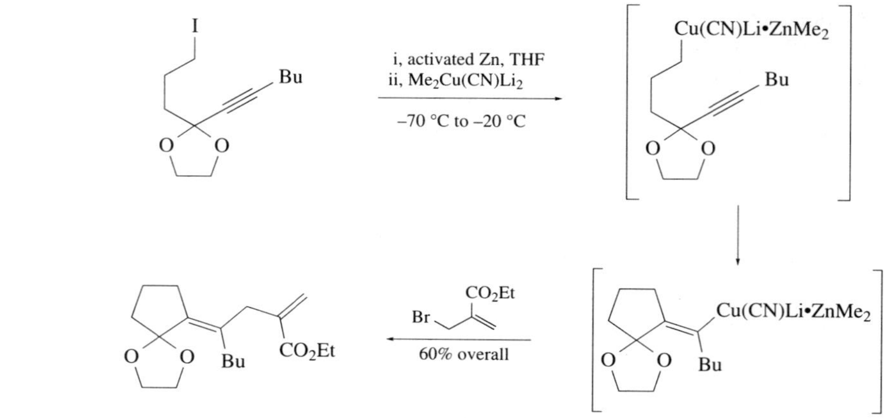

Scheme 14

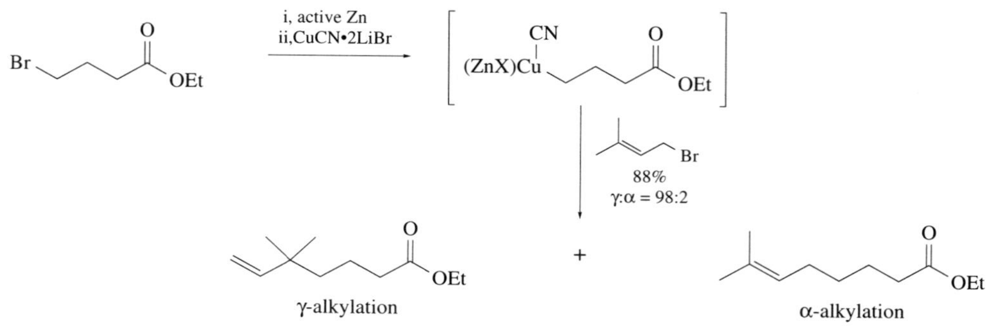

Scheme 15

While organotitanium reagents will add to carbonyl groups, they show little reactivity towards haloalkanes, including activated systems such as allyl halides ⟨92T5709⟩. However, addition of a catalytic amount of a copper(I) salt (6–7.5 mol.%) allows allylation of the organometallic reagent, giving a substitution product with excellent S_N2' regioselectivity; the selectivity is very high with allyl chlorides (usually better than 99 : 1), but poor with the corresponding bromide. It appears as though both copper and titanium are required to obtain high selectivity as the titanium reagent $R^1Ti(OR^2)_3$ is inert to allyl chlorides, and simple Gilman reagents R_2CuLi give more modest results. Superior S_N2' selectivity is obtained for "long chain" alkyltitanates (butyl, hexyl, etc.), but methyl-, benzyl-, allyl-, and phenyltitanium reagents were observed to give only moderate selectivity. The titanium reagents $R^1_{n+1}Ti(OR^2)_3Li_n$, ($n = 0,1$), show both excellent chemoselectivity (Equation (37)) towards allylation rather than conjugate addition, and diastereoselectivity in addition to allyl chlorides (e.g. Structure (**29**)) bearing a chiral center (Equation (38)). High yields and S_N2' regioselectivity were also reported for a reagent derived from Bu_2Zn and CuI·2LiCl ⟨92T5709⟩.

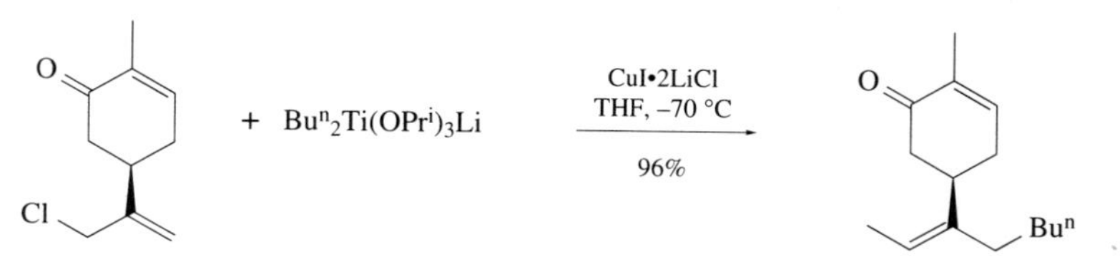

(37)

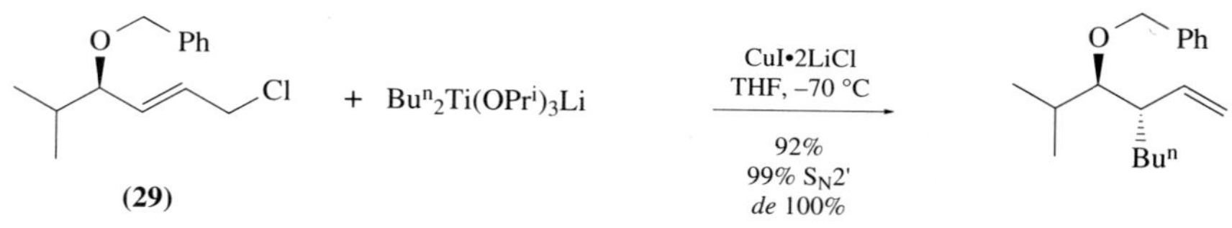

(**29**)

(38)

Organomanganese(II) chloride reagents ordinarily will not react with haloalkanes, and lithium trialkylmanganates such as $LiMnMe_3$ give only moderate yields of cross-coupled products, even with primary haloalkanes, whereas manganates with larger alkyl ligands give very poor yields and side reactions (including reduction, elimination, and homocouplings) intervene ⟨70TL315⟩. However, the addition of a catalytic quantity of Li_2CuCl_4 (3 mol.%) and also a polar solvent (*N*-methylpyrrolidone is superior to DMSO and DMF) to THF solutions of RMnCl gives a reagent that will effectively alkylate haloalkanes (Br, I) in very good to excellent yields (Scheme 16); both the copper catalyst and the polar solvent are essential for efficient alkylation. Primary, secondary, tertiary, vinyl, and aryl manganese chlorides give high yields, but allylmanganese chloride gives only a 22% yield of 1-tetradecene when treated with 1-bromoundecane. Phenylmanganese chloride reacted better in the absence of the polar solvent. Alkylations are only successful with primary haloalkanes, with little or no reaction occurring with secondary and tertiary substrates. The additions are highly chemoselective, and a wide variety of functional groups is tolerated in the bromoalkane, including chloroalkyl, sulfonate ester, carboxylate ester, carbonate, ketone, alcohol, and carboxylic acid (2 mol. equiv. RMnCl required for ROH and RCO_2H). Leaving groups on the β-carbon atom are also tolerated remarkably well. Yields were shown to be 10–15% higher for catalyzed RMnCl substitutions when compared with catalyzed Grignard substitutions ⟨93SL45⟩.

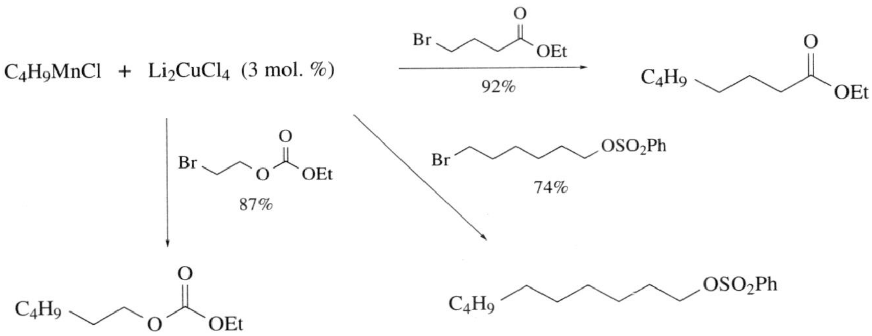

Scheme 16

1.04.2.5 Organoboron Compounds

While organoboron compounds are less frequently used for carbon–carbon bond formation by reaction with haloalkanes, nevertheless there are some cases where transfer of an alkyl group from boron (usually as a four-coordinate boron-*ate* complex) will occur. Lithium methyl(trialkyl)borates will selectively transfer an alkyl group to an allyl or propargyl halide in good to excellent yields (typically 53–96%) in the presence of a catalytic quantity of a copper(I) salt; in most cases the methyl group is not transferred. Substitution of allyl halides gives substituted alkenes (in the case of cinnamyl chloride, the product of γ-substitution predominates), whereas propargyl substrates give allenes (Scheme 17) ⟨77BCJ2199⟩.

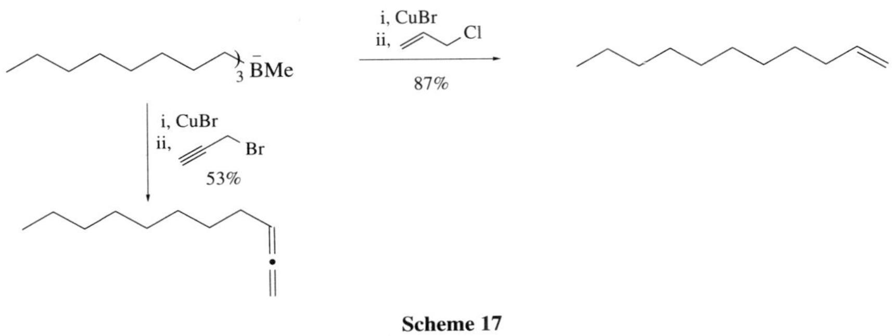

Scheme 17

Anions derived from α-halo esters ⟨68JA818⟩, ketones ⟨69JA2147⟩, and nitriles ⟨69JA6854⟩ (prepared using hindered bases) react with trialkylboranes to give alkylated products via an interesting rearrangement. Initial attack of the stabilized carbanion on the electrophilic boron center gives a boron-*ate* complex which then rearranges to a new trialkylborane where an alkyl group on boron has displaced a halogen atom. Synthesis of the precursor organoborane via a hydroboration reaction allows control of the stereochemistry of the final product (Scheme 18) ⟨69JA2150⟩. Where the borane is symmetrical, only one alkyl group will migrate and the other two groups are "wasted", however introduction of the 9-BBN ligand (which is not transferred) overcomes this problem. In the case of α-halo esters and ketones, Matteson has proposed that the intermediate borane (**30**) is unstable and rearranges to the boron enolate; the alkylated product is then liberated from boron by reaction with a proton source in the reaction mixture ⟨B-87MI 104-04⟩. Reactions in THF containing some Bu^tOH generally give very good to excellent yields of alkylated products. Complexation of the anion derived from dichloroacetonitrile provides a route to effect a double alkylation on carbon by sequential displacement of both halogen atoms ⟨70JA5790⟩.

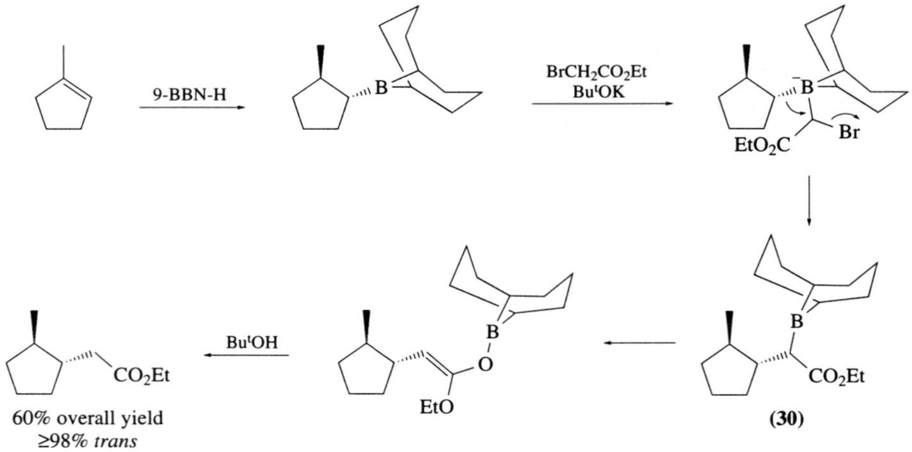

(**30**)

Scheme 18

Pelter *et al.* have reported a novel method (Scheme 19) for the synthesis of substituted ketones ⟨73CC544⟩ and 1,4-dicarbonyl compounds ⟨73TL4491⟩ using alkynylborates. Alkylation with reactive haloalkanes gives intermediate vinylboranes; both primary and secondary alkyl groups (R^1) will readily migrate from boron to carbon and the alkylation appears to be regiospecific, with the ligand from boron (R^1) migrating to the α carbon atom and the alkyl group (R^3) from the haloalkane being introduced to the β carbon center of the alkynyl residue. Yields of ketones (following oxidation) are excellent (78–88%). Protonation of the intermediate vinylboranes also can provide alkenes in good yields although the *E*/*Z* selectivity is relatively poor; use of the sterically demanding thexyl ligand on boron leads to greater selectivity ⟨75TL1633⟩.

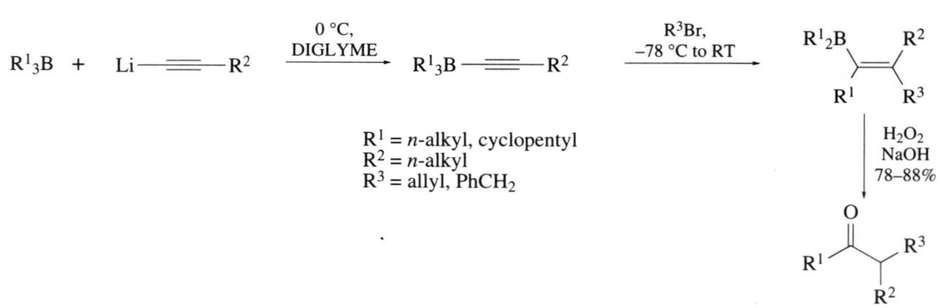

Scheme 19

1.04.2.6 Miscellaneous Species with No Stabilizing Group

A profusion of other organometallic reagents will react with haloalkanes to give coupled products, often in quite good yields. However, the applications of these alkylations are generally quite limited as similar or better results can usually be obtained using methodologies outlined in preceding sections. Nevertheless the chemistry is often interesting and some representative examples are given below.

π-Allylnickel reagents can be used to allylate haloalkanes; they are weakly basic and can tolerate other functionalities such as ketones, esters, and chloroalkyl groups ⟨72OR(19)115⟩, and yields range from good to excellent. Corey and Semmelhack used the coupling in a brief synthesis of α-santalene (**13**) ⟨67JA2755⟩; coupling occurred at the less hindered end of the allyl unit (Equation (39)). A cycloplatinated complex has been reported to react with iodomethane giving an adduct reminiscent of Wheland intermediates ⟨82JA6609⟩; this complex can be isolated as the fluoroborate salt, or decomposed on addition of cyanide (Scheme 20) to give the methylated arene ⟨82OM1366⟩. While cyclometallation reactions provide an effective method for regiospecific functionalization of organic substrates, the requirement for stoichiometric quantities of expensive transition metal precursors restricts their usefulness to cases where chelation-directed lithiation reactions are inappropriate.

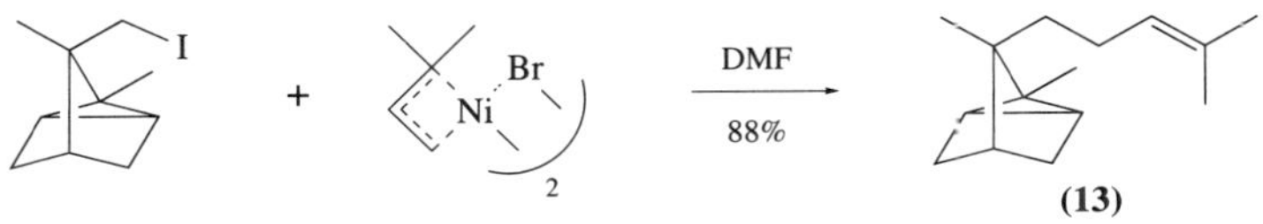

(**13**)

(39)

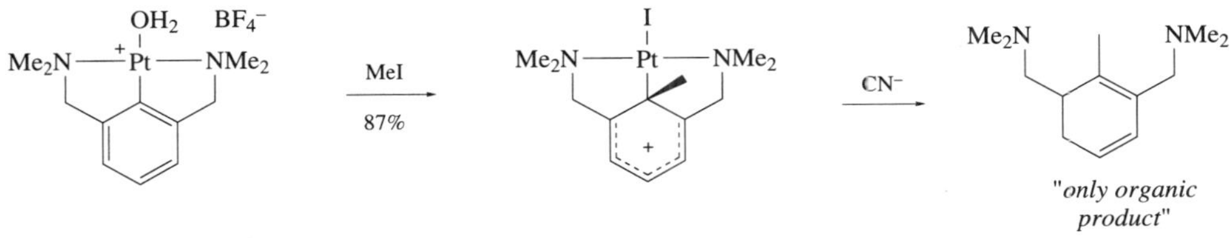

Scheme 20

Zinc alkyl reagents are generally inert to haloalkanes; in an early investigation Noller showed that dialkylzinc compounds would not react with primary, secondary, or allyl haloalkanes ⟨29JA594⟩. Tertiary halides gave coupling products, but in moderate yields. For example, diethylzinc reacted with *t*-butyl chloride in xylene to give 2,2-dimethylbutane in 45% yield. More recently, Reetz *et al.* have developed this reaction into a useful synthesis of compounds containing quaternary carbon centers where the presence of zinc chloride facilitates substitution of the halogen atom by an alkyl group ⟨80CC1202⟩. Although the reaction may have some carbocationic nature, the mild Lewis acidity of the zinc reagent is believed to reduce the tendency to rearrangement (Equation (40)). In contrast to saturated zinc alkyls, allylzinc halides are reactive and couple readily with allyl halides to give 1,5-dienes in good yields. Excellent regioselectivity is often achieved, where allylzinc bromide reagents couple γ to the zinc substituent, and attack the allyl halide to give products of S_N2-type reactions ⟨64BSF2485, 68CR(C)1614⟩. However, reactions with cinnamyl bromide and allylzinc, or with cinnamylzinc bromide and allyl halides, give mixtures of α- and γ-substituted products.

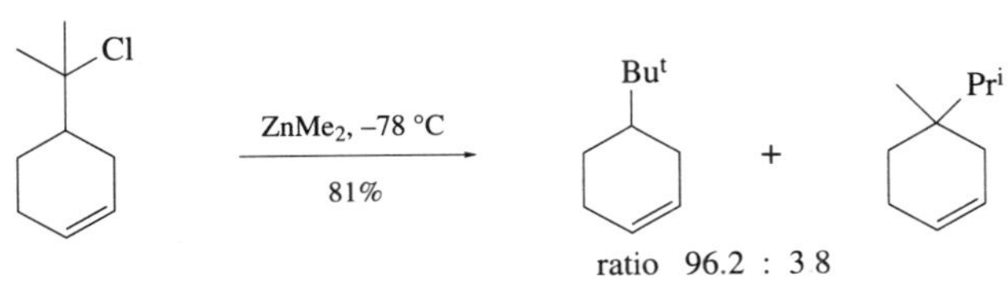

(40)

Organocadmium reagents are generally insufficiently nucleophilic to react with primary and secondary haloalkanes by displacement ⟨57JOC1326, 73JOC3189⟩, and reaction with tertiary halides leads to elimination of HX ⟨36RTC518, 57JOC1326⟩. However, Kollonitsch has contended that the reaction with tertiary haloalkanes results from the presence of magnesium salts in organocadmium reagents that are prepared by transmetallation of organomagnesium compounds and then used *in situ* ⟨66JCS(A)453⟩. For example, distilled diethylcadmium reacted with *t*-butyl chloride to the extent of only 4%, whereas addition of magnesium halides increased the conversion to 45%. Jones and Costanzo have reported that activated haloalkanes (allyl, benzyl) react with organocadmium reagents to give alkylated materials in variable yields ⟨73JOC3189⟩; this is in contrast to Cason and Fessenden who were unable to obtain any significant yields of coupled product with allyl bromide

⟨57JOC1326⟩. Alkylation of α-bromoketones and esters has been reported to proceed in modest to poor yield ⟨68JOC1675, 73JOC3189⟩; however, attempted alkylation of α-chloroesters with alkylcadmium chlorides failed and side reactions intervened. Where the α-bromoester is tertiary, condensation reactions dominate (Equation (41)), and racemization occurs in reactions involving secondary substrates (Equation (42)).

(41)

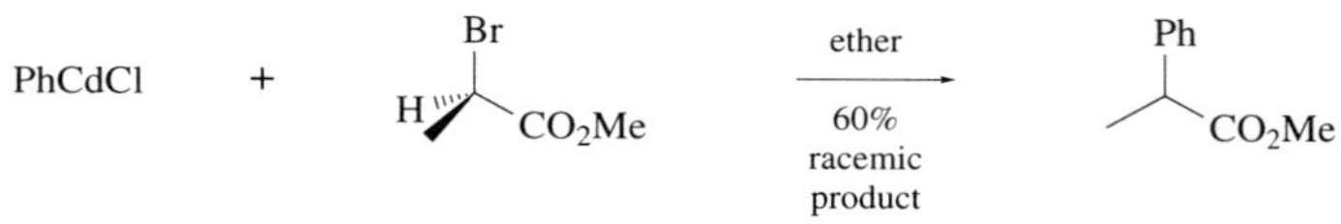

(42)

Organomercury compounds are usually quite inert to haloalkanes, with reactions only taking place under forcing conditions ⟨B-67MI 104-02, 78AG(E)27, 82T1713⟩. In an extensive study, Whitmore and Thurman showed that a range of haloalkanes gave no reaction with bis(*p*-tolyl)mercury on heating in toluene for over 300 h ⟨29JA1491⟩; reactions of both activated (allyl) and non-activated halides were generally unsuccessful with numerous side reactions intervening. A rare success was obtained with bromodiphenylmethane, where a 90% yield of triphenylmethane was recorded when the electrophilic bromoalkane was reacted in toluene with diphenylmercury for 170 h. A similar reaction with dibutylmercury gave much poorer conversion. Alkylations have been reported to occur at lower temperatures (although in modest yield) when Lewis acids such as aluminum bromide are employed as a catalyst ⟨82T1713⟩. Activated compounds such as α-mercuriocarbonyl reagents are reported to react readily and in the absence of a catalyst (Equation (43)) ⟨B-67MI 104-02⟩, however, Curtin and Hurwitz have shown that while trityl chloride reacts with these organomercurials readily, catalysis with stannic chloride is required for reaction with benzhydryl chloride ⟨52JA5381⟩. Carbocation intermediates are implicated. Organomercury compounds will react with allyl halides to give substitution products in the presence of palladium (II) catalysts; however, it is likely that the reaction proceeds by initial transmetallation of the aryl ⟨68JA5531⟩ or vinyl ⟨78JOM(156)45⟩ mercurial to give an organopalladium species which adds to the double bond of the allyl chloride. Elimination of $PdCl_4^{2-}$ from the alkylpalladium intermediate leads to a product of formal S_N2' substitution. The reactions are catalytic in palladium.

(43)

The reactions of alkylaluminum reagents with haloalkanes have been reported ⟨66JOC908⟩; whereas triethylaluminum reacted with benzyl halides and 2-phenylethylhalides to give coupled products in acceptable yields accompanied by some reduced material, reactions with primary, secondary, and tertiary haloalkanes were generally unreliable with reduction, elimination, and rearrangement products contributing to the total yield. Kennedy has reported that alkylation of tertiary chloroalkanes occurred in excellent yields in chloromethane as a solvent; substitution of cyclopentane as a solvent led to much slower reactions ⟨70JOC532⟩. Nevertheless, the carbocationic nature of the reaction suggests that its applications to synthesis are limited. One successful and useful procedure involving organoaluminum reagents centers on the hydroalumination of alkynes. Stereospecific *syn* addition of an aluminum hydride reagent gives a vinylalane; treatment with MeLi gives the *ate* complex which will alkylate reactive haloalkanes ⟨76JOC2214⟩. The reaction proved most successful with symmetrical and terminal alkynes; for example, hydroalumination of 1-octyne and subsequent allylation gave isomerically pure 1,4(*E*)-undecadiene in 68% yield. Allyltin reagents will react with haloalkanes to give substitution products; however, the reactions usually involve radical intermediates, or they are catalyzed by Lewis acids or transition metals ⟨B-87MI 104-05⟩. For example, in the reaction of bromoalkanes, homolysis of the carbon–bromine bond occurs either thermally or under irradiation to give an alkyl radical which adds to the allyl stannane. Subsequent cleavage of the carbon–tin bond gives products of S_H2' substitution (Equation (44)) ⟨75JOM(96)225⟩.

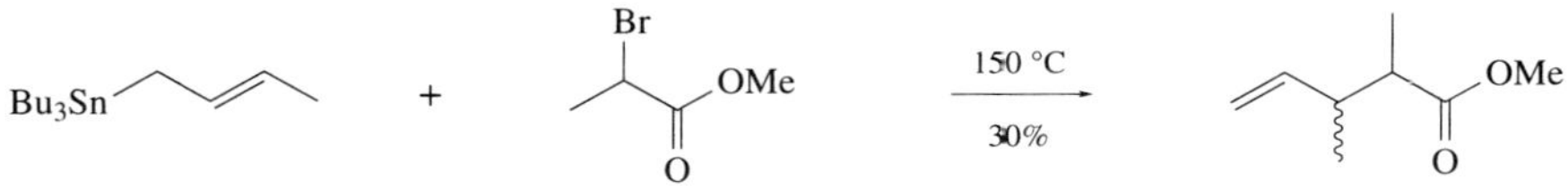

(44)

1.04.3 CARBANIONS WITH ONE STABILIZING GROUP

1.04.3.1 General Comments

The introduction of a substituent into an alkyl chain has the potential to lower the pK_a values of α-hydrogen atoms by providing stabilization of α-carbanions and therefore, at least in theory, a route to direct metallation. Electronegative elements such as oxygen can stabilize negative charges by induction, whereas boron with its vacant *p* orbital, and unsaturated functionalities such as carbonyl groups, can act as π-acceptors. However, the degree of influence varies: ethers and alkylamines have limited effects, and α-protons are removed only with strong bases (and then, only with some difficulty and with little practical value), while the nitro group has such a dramatic effect that hydroxide will readily convert many nitroalkanes to their corresponding nitronate salts. One advantage of this increased acidity lies in the reduced basicity of the conjugate bases; side reactions such as deprotonation of the haloalkane substrates by the carbanions are reported less frequently. In general, the nucleophilicities of the anions are enhanced and the choice of cation causes less trouble, with Group I metals proving quite acceptable in most cases. The proliferation of lithium amide bases has provided virtually limitless opportunities for the generation of lithium carbanions, and their alkylations proceed, in general, in good to excellent yields. None the less, there are no universal guidelines for this class of reactions; whereas many reactions proceed in high yield, others such as nitronates react in poor yields or give unchanged starting material. Two new factors are introduced for consideration with this group of carbanions. Asymmetric syntheses become possible by either the use of configurationally stable carbanions or by the addition of chiral auxiliaries to the molecule. The stabilizing group can provide a useful synthetic handle for further manipulation.

1.04.3.2 Enolates and Related Carbanions

Carbonyl groups and their reactions are fundamental to modern organic synthesis, and the development of reliable methods for α-alkylations has been one of the most keenly researched areas throughout the history of organic chemistry. At first sight, the process appears straightforward: the reaction of the neutral molecule with a base generates an anion α to the carbonyl group. The negative charge is readily stabilized, as the carbon–oxygen double bond has a relatively low-lying vacant π* orbital capable of delocalizing the electron density ⟨81JOC1693⟩. Subsequent addition of a haloalkane to the reaction mixture should give the product of nucleophilic substitution. However, the reaction is not always that simple, and several complicating factors need to be considered. An injudicious choice of base could result in competing nucleophilic addition to the carbonyl group. Unsymmetrical ketones can suffer competing α- and α′-deprotonation, giving regioisomeric enolates. While the geometry of the enolate is of little importance in achiral systems or where the stereochemistry of the product is of no consequence, in asymmetric synthesis the selective formation of either the (*E*)- or (*Z*)-enolate is vital. The enolate can act as an ambident nucleophile and therefore alkylation on oxygen and carbon can compete. In the alkylation of enolates that contain asymmetric elements the π-faces of the enolate are diastereotopic and therefore diastereomeric products can result. Simple ester enolates may not be stable with respect to ketene formation. Finally, if the rate of alkylation is slow relative to intermolecular proton transfer, then the residual enolate can act as a base and deprotonate alkylated material giving a new enolate; the ultimate outcome of this process would be a mixture of starting material and mono- and polyalkylated products.

In this section, the alkylation of enolates derived from ketones and aldehydes, as well as carboxylic acids and their derivatives, will be considered. (Various nitrogen-based enolate equivalents, such as oximes and hydrazones, are discussed briefly in Section 1.04.3.5.) However, the alkylations of enolates and related species have been reviewed extensively and frequently, and fuller discussions of regiospecific enolate generation ⟨76T2979⟩, the structure and reactivity of alkali metal enolates ⟨77T2737⟩, enamines ⟨82T1975⟩, stereoselective alkylations ⟨B-84MI 104-01⟩, tandem conjugate addition–alkylation procedures ⟨85S364⟩ and stereoelectronic influences ⟨89T4913⟩ can be found in these articles. Reviews by Caine ⟨91COS(3)1⟩ and Heathcock ⟨B-92MI 104-03⟩ provide detailed and excellent coverage.

Many solutions to the above mentioned problems have been provided, and nucleophilic attack by a base can be readily avoided by selection of strong, nonnucleophilic bases (such as the hindered alkali metal dialkylamides). In most cases organolithium or Grignard reagents can be considered inappropriate. Alternative methods for enolate generation that also avoid this problem include conjugate reduction of α,β-unsaturated compounds using metal/ammonia techniques ⟨65JA275, 75PAC553⟩, 1,4-addition of cuprates to enones and similar substrates ⟨85S364⟩, and preparation of enolates via enol acetates ⟨65JOC2502⟩ and silyl enol ethers ⟨68JA4462, 68JA4464, 80C265, 83TL1345⟩. Likewise the potentially serious problem of ester enolate instability finds an easy and acceptable solution with the application of lithium amide bases: whereas sodium ester enolates are readily unstable ⟨70JA3222⟩, the stronger lithium–oxygen interaction allows the preparation and even isolation of lithium ester enolates ⟨73JA3050, 73JOM(50)9⟩.

The prospect of competing *O*- and *C*-alkylation is a dilemma that deserves closer scrutiny. Theoretical studies on the alkylation of acetaldehyde enolate with methyl fluoride have indicated that, while the *C*-alkylated product is favored thermodynamically, the path to enol ether formation has a lower energy of activation and the *O*-alkylated compound is therefore kinetically favored ⟨86JA2659, 87JCPB765⟩. However, methyl fluoride is a "relatively 'hard' and unreactive electrophile" ⟨86JA2659⟩, and both studies examined "naked" enolates without consideration of cations, solvents, aggregation effects, and other factors. In reality *O*-alkylation does not pose a major problem especially when lithium enolates are employed; reactions in moderately polar solvents such as diethyl ether and THF also favor *C*-alkylation while highly polar solvents such as DMSO, which solvate cations strongly, lead to increased proportions of *O*-alkylated material (Equation (45)) ⟨68JOC2222, 71JOC1112, 77T2737⟩. The haloalkane approaches the enolate at an angle of 106° ⟨86JA2659⟩, and this stereoelectronic constraint dictates the outcome of intramolecular alkylations ⟨77CC233, 78JOC700⟩. If the ω-halo enolate will lead to a five-membered ring, *O*-alkylation dominates and the *exo*-methylene ether is formed (Scheme 21). In contrast, exclusive *C*-alkylation occurs for six-membered ring formation. These trends have been borne out in a recent theoretical study where it was shown that, while cycloalkanone formation is generally favored, cyclization to give an enol ether dominates for five-membered ring synthesis ⟨91JPO449⟩.

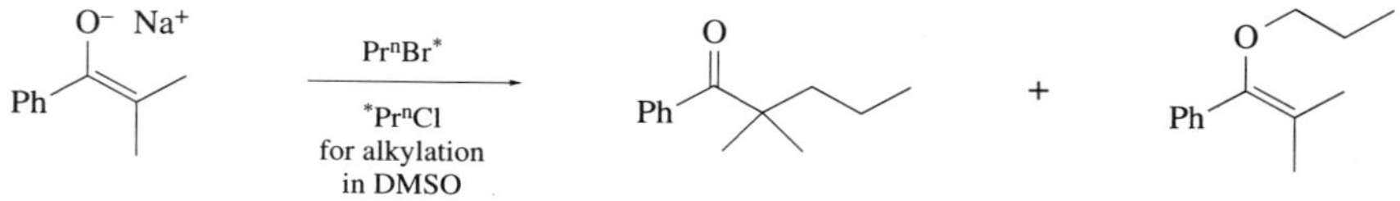

(45)

Solvent	*O:C ratio*
ether	0.00
DME	0.25
DIGLYME	0.27
DMSO	1.34

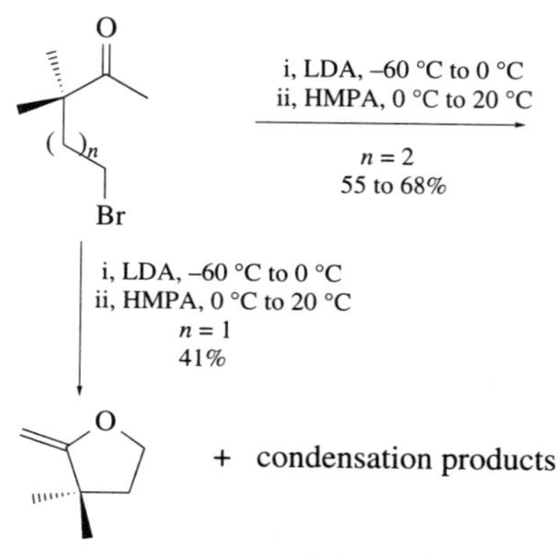

Scheme 21

Development of the hindered alkali metal amide bases has introduced a new and important control element into enolate chemistry. In addition to their ease of preparation and good solubility in organic solvents, the lithium amides allow regioselective enolate formation to be effected in unsymmetrical ketones. Where the two α-sites of a ketone impose different steric constraints on an approaching base, then the hindered amides will preferentially abstract a proton from the less substituted or less hindered α-position. At low temperatures and in the absence of acids (such as neutral ketones), these "kinetic" enolates can be formed with moderate to good selectivities. Alternatively, formation of an enolate under equilibrating conditions (in the presence of acids such as alcohol or an excess of the ketone) results in the formation of the most substituted (thermodynamic) enolate. These two extremes are illustrated in the intramolecular alkylation of an acetyl cyclopentane (**31**) (Scheme 22) ⟨78JOC2153⟩ and an α-phenyl ketone (**32**) (Scheme 23).

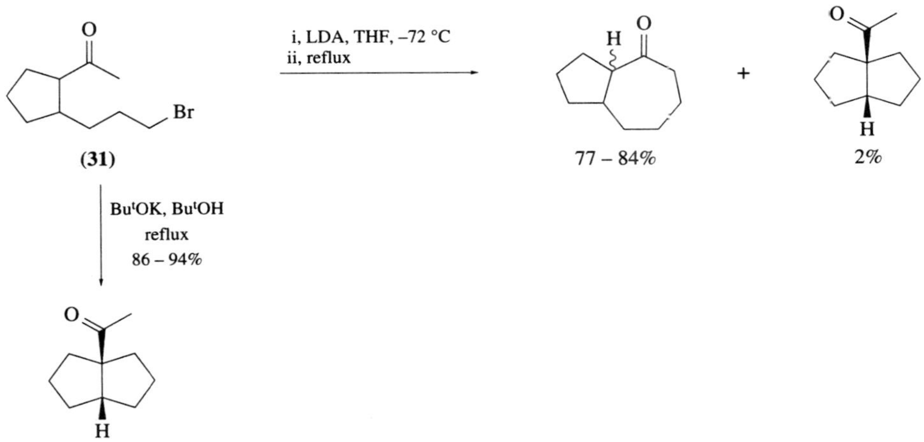

Scheme 22

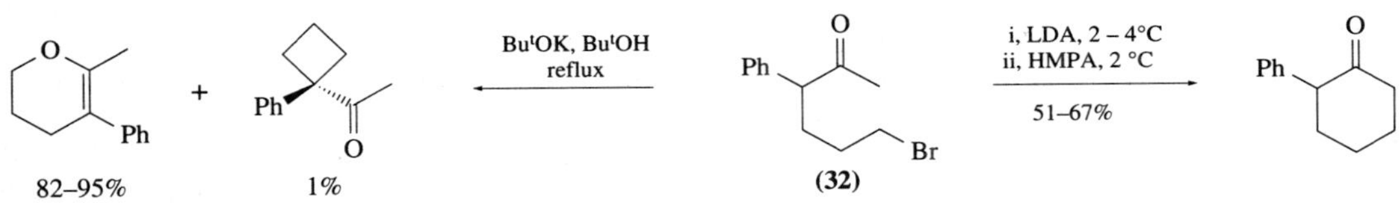

Scheme 23

Various hindered dialkylamides have been developed and employed for enolate formation and an excellent discussion of their relative merits can be found in Heathcock's review ⟨B-92MI 104-03⟩. A recently introduced, extremely hindered base is lithium bis(2-adamantyl)amide (LBAA) which has been developed by Collum and co-workers and may find future applications as a result of its simple synthesis: the parent amine can be prepared by reductive amination of adamantanone with NH_4Cl ⟨93TL5213⟩. The amide is monomeric under all conditions, and it readily forms mixed aggregates with alkyllithium reagents and lithium enolates, but not apparently with lithium halides. Enolization of 3-pentanone was studied under a variety of conditions, and LBAA frequently gave (*E*) : (*Z*) ratios of at least 50 : 1 (when the enolate was trapped as the silyl ether, *in situ*). Nevertheless, LDA remains one of the most widely used sterically encumbered bases as it gives, generally, satisfactory (*E*) : (*Z*) selectivity. Either isomer can be prepared selectively depending on the solvent employed, and a model developed by Ireland *et al.* ⟨76JA2868⟩ provides an explanation for the differing behavior: deprotonation in THF occurs via a closed, chair-like transition state giving the (*E*)-enolate, whereas the addition of a strongly coordinating solvent such as HMPA allows for a more open transition state leading to the (*Z*)-enolate ⟨B-92MI 104-03⟩. However, recent studies by Collum and co-workers have indicated that the process is significantly more complex, with added factors such as enolate aggregation and salt effects playing important roles ⟨91JA5751, 91JA9571, 93TL5213⟩.

Despite the potential pitfalls that could plague the formation of stereo- and regiodefined enolates, and the possibility of ambident reactivity, alkylations with activated and primary haloalkanes generally proceed satisfactorily although results with less reactive alkyl halides are somewhat more modest. One problem that occasionally complicates product purification arises when the rate of substitution is slow and proton transfer competes, resulting in enolate scrambling and polyalkylation. Classical solutions to these problems often involve several steps, such as the preparation of β-ketoesters, alkylation of the derived monoanion, and hydrolysis and decarboxylation to remove the activating ester group. More direct approaches center upon transmetallation of enolates giving species that can be selectively monoalkylated. Several groups have reported the usefulness of tin (IV) enolates (which can be mixtures of enolates and α-stannylated carbonyl compounds); Pereyre and co-workers ⟨67CR(C)1204, 69TL505, 70CR(C)100, 73JOM(55)273⟩ and Tardella ⟨69TL1117⟩ showed that reactive alkyl halides will alkylate tin enolates either in the presence of HMPA or in the absence of solvent. The reactions generally gave monoalkylated products with little or no polyalkylation being recorded. Jung and Blum demonstrated that, while the lithium enolate of acetaldehyde gives polymeric material on treatment with haloalkanes, transmetallation and subsequent alkylation of the tributyltin derivative leads to α-monoalkylated products in moderate (40–68%) yields (Scheme 24) ⟨77TL3791⟩.

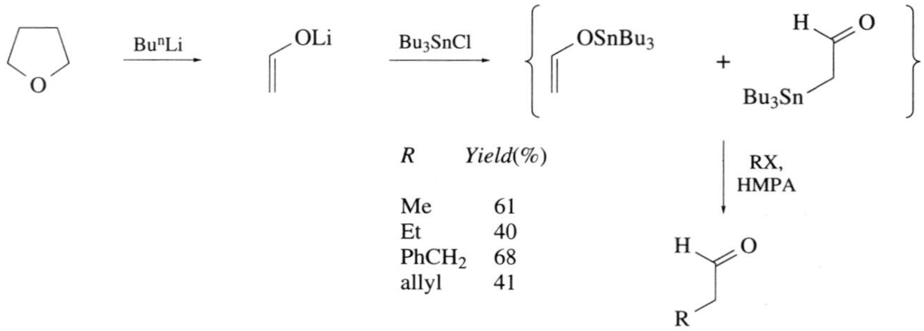

R	*Yield(%)*
Me	61
Et	40
$PhCH_2$	68
allyl	41

Scheme 24

Transmetallation with tin (IV) reagents provides an effective solution for the regioselective α-alkylation of cyclopentanone enolates derived from conjugate addition; in general, enolate equilibration would be expected to provide isomeric and polyalkylated by-products especially if unactivated haloalkanes are employed. However, sequential conjugate addition, formation of the tin (IV) enolate, and addition of HMPA prior to alkylation, suppresses the equilibration and regiospecific alkylation is effected ⟨84TL223, 84TL2487⟩; this approach has been used by Noyori and co-workers in a highly successful three-component coupling as a key step in a prostaglandin synthesis (Scheme 25) ⟨85JA3348, 88JA4718⟩. It is likely that the reactive species is actually a penta- or a hexacoordinate stannate rather than a neutral four-coordinate tin enolate. The Noyori reaction requires a large excess of the alkylating agent for acceptable yields to be obtained; when the enolate and allyl iodide are present in equimolar quantities the desired product is isolated in 55% yield (reduced from 81%) along with 35% of non-alkylated 1,4-adduct. The separation of the alkylated material from tin residues can be tedious, and therefore Noyori and co-workers have developed a zinc-based technique to overcome this shortcoming ⟨89JOC1785⟩. Addition of HMPA and dimethylzinc to the lithium enolates prior to alkylation suppresses proton exchange, enolate equilibration, and polyalkylation to less than 1%. However, a large excess of the alkylating agent is still required for optimum yields.

One solution to the problem of polyalkylation and enolate scrambling has been provided by

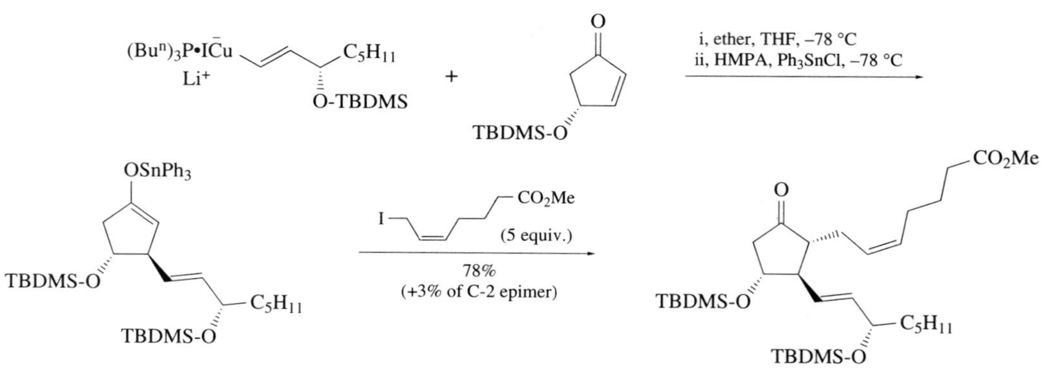

Scheme 25

Negishi *et al.*, via the use of triethyl boron-*ate* complexes of potassium enolates (Scheme 26) ⟨79TL845⟩. The reaction routinely gave high yields of monoalkylated products, whereas omission of BEt_3 gave the expected mixture of products. Use of BEt_3 and correct choice of metal cation (K) is critical: more hindered trialkylboranes do not allow complete complexation to the enolate oxygen atom and mixtures of products inevitably result. Similarly, use of lithium enolates also gives mixtures, as the stronger Li—O interaction apparently inhibits reaction of the enolate with the boron electrophile. The method is versatile because the choice of base can determine whether kinetic (using $KN(SiMe_3)_2$) or thermodynamic (using KH) enolates are generated and then trapped with BEt_3, as $\geq$85% regioisomerically pure species ⟨83TL1341⟩. The boron enolate methodology has been incorporated by Motherwell and co-workers into a conceptually unique approach to enolate chemistry ⟨91CC1399, 92TL6187⟩. The strategy centers on transition metal-mediated isomerization of allylic alkoxides and their derivatives as a direct route to stable enolates. (Although simple enols can be produced by a similar procedure, reaction of the enols with reactive halides was not reported ⟨91JA958⟩.) The lithium alkoxide or triethylboron-*ate* complex of an allylic alcohol was isomerized using Rh(I) catalysts, and the resultant enolate was monoalkylated cleanly with reactive alkyl halides (Equation (46)). Yields were variable (30–80%), depending on substrate structure. Where further isomerization of the enolate is possible, use of a Rh(I) catalyst gave an undesirable mixture of α- and α'-alkylated ketones. Using $(Cy_3P)_2NiCl_2$ effectively suppressed the second isomerization ⟨92TL6187⟩.

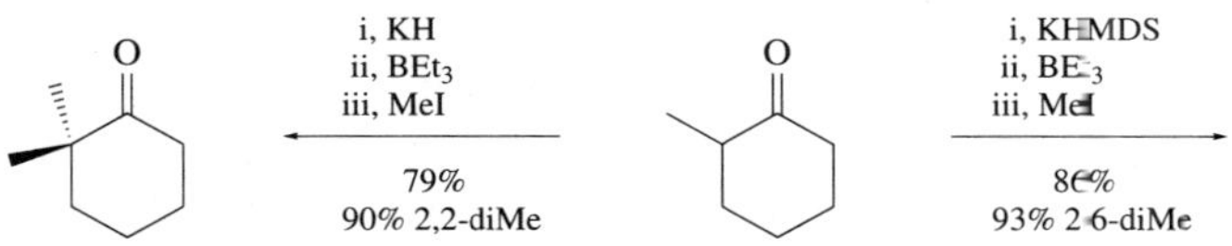

KHMDS = potassium hexamethyldisilazide

Scheme 26

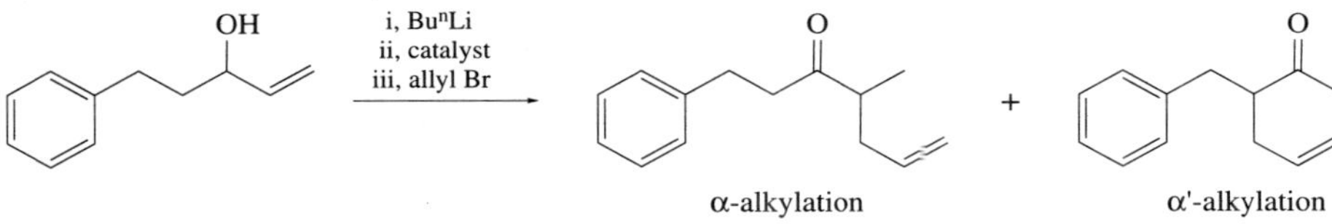

(46)

Catalyst	*Yield* (%)	$\alpha : \alpha'$
[Rh(dppe)]ClO_4	79	1 : 1.2
$(Cy_3P)NiCl_2$	78	15 : 1

Manganese (II) enolates have been used by Reetz and Haning to effect α-methylation of ketones in excellent yields ⟨93TL7395⟩. Regioselectivity is high, and polymethylated products are generally less than 3% of the total yield. The base of choice is $LiN(SiMe_3)_2$; substitution of LDA gives lower yields and significantly more polyalkylation. In addition, the reaction cannot be extended readily to other primary alkyl iodides; alkylations with Bu^nI "do not work as well".

Alkylation of silyl enol ethers can be readily effected by a number of methods. Addition of a mixture of enol ether and benzyltrimethylammonium fluoride in THF gives monoalkylated products in moderate to good yields, although some nonalkylated material often accompanies the desired product ⟨82JA1025, 85ACR181⟩. Yields are modest with nonactivated haloalkanes, and cyclopentanone silyl enol ether gave significant quantities of dialkylated material. However, silyl enol ethers can provide routes to materials that are not otherwise readily available. While enolates generally react with *t*-alkyl halides by an E_1 mechanism, silyl enol ethers readily alkylate these substrates in the presence of a Lewis acid such as $TiCl_4$ or $ZnBr_2$ ⟨79TL1519, 82AG(E)96⟩. This method also finds uses for alkylation of otherwise unreactive bridgehead halides; however, simple primary haloalkanes have failed to react until recently ⟨79BCJ1241, 79TL1519⟩. Benzylation of silyl enol ethers can be effected with silver (I) catalysis ⟨79BCJ1241⟩, and Jefford *et al.* have recently shown that primary iodoalkanes can be induced to react with the enol ethers in the presence of silver trifluoroacetate in yields that range from 27 to 83% but generally exceed 50% (Equation (47)) ⟨92TL1855⟩. Alkyl bromides and secondary iodoalkanes fail to react.

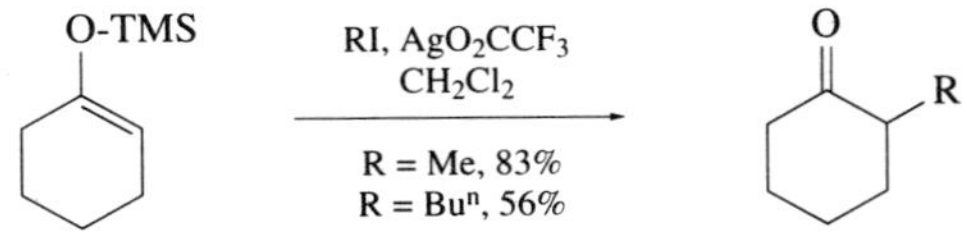

(47)

Most alkylations considered thus far in this section have centered on reactions that lead to racemic material. However, introducing a chiral element into an enolate renders the π-faces diastereotopic, and therefore facial discrimination in the alkylation step becomes crucial to avoid the formation of diastereomeric products. Other substituents already present in the enolate can affect the facial selectivity, as demonstrated in Noyori's prostaglandin synthesis which results in the 2,3-*trans* isomer as the dominant product (Scheme 25). Achiral carbonyl compounds can be modified by covalent inclusion of a chiral auxiliary to direct subsequent alkylation; this is one of the most intensely studied fields in carbonyl chemistry and a comprehensive coverage of all aspects is beyond the scope of this chapter. Nevertheless, several examples of asymmetric alkylations are presented in Equations (48)–(55). Most auxiliaries can be readily identified as being derived from the chiral pool; popular sources are terpenes, monosaccharides, and amino acids. For example, camphor-derived auxiliaries (Equations (48) and (49)) ⟨83TL3213, 87T1969⟩ and oxazolinones originating from amino acids such as valine ⟨82JA1737⟩ often provide products in high enantiomeric excess. An interesting double alkylation of a dianion has recently been reported which uses Evans' oxazolinone auxiliary to control the alkylation (Equation (50)) ⟨93T4147⟩. Each alkylation appears to be controlled locally with no long-range effects being detected. Another useful amino acid-derived auxiliary is the bicyclic lactam pioneered by Romo and Meyers ⟨91T9503⟩, which has found a variety of applications including the synthesis of chiral, nonracemic quaternary centers (Equation (51)) ⟨93JOC7507, 93TL7701⟩. While the source of all chirality must ultimately be natural, there are many auxiliaries that are prepared in nonracemic form by resolution and their relationship to natural products is not obvious. For example, highly stereoselective α-alkylations can be achieved using acyl iron reagents ⟨85CC209, 85TL3075, 86CC495⟩. While an acetyl group can be alkylated twice, the resulting α,α'-disubstituted acyl iron reagents are not enolizable. Therefore a conjugate addition–alkylation sequence is required (Equation (52)) ⟨86CC495⟩. Products are generally formed as single diastereoisomers. 2,2′-Binaphthol has been used as an auxiliary for the asymmetric alkylation of the derived crotonic ester ⟨91TL7281⟩. Deprotonation with LDA and alkylation gives the α-substituted esters exclusively in yields ranging from 32 to 83%, *de* $\geq$80% (Equation (53)). A new polycyclic oxazolinone has been employed for asymmetric alkylations of amide derivatives ⟨94TL721⟩. Yields are generally excellent and products are consistently isolated with diastereomeric ratios >140 : 1; in some cases only a single diastereoisomer is detectable (Equation (54)).

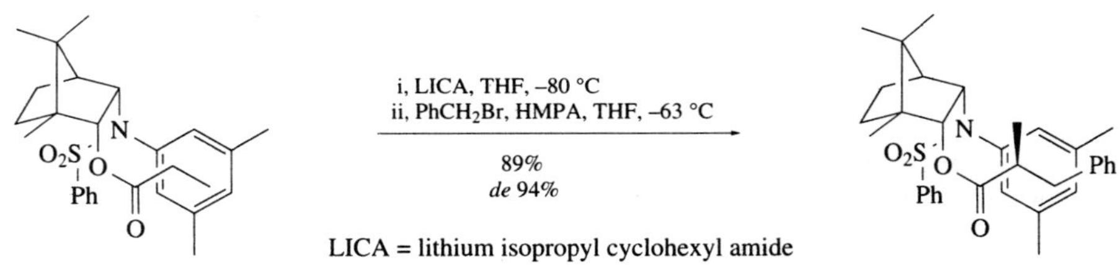

(48)

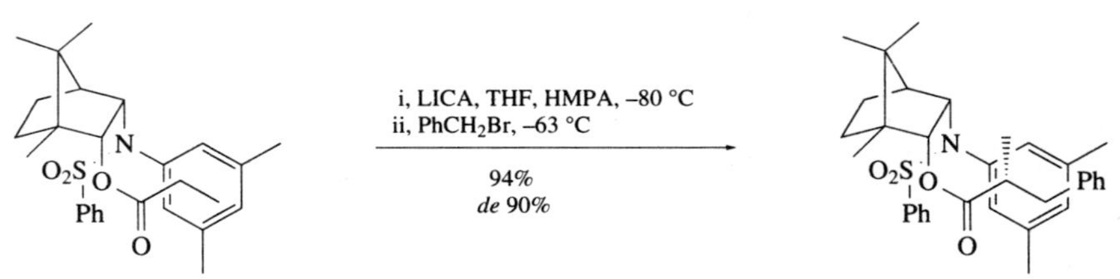

(49)

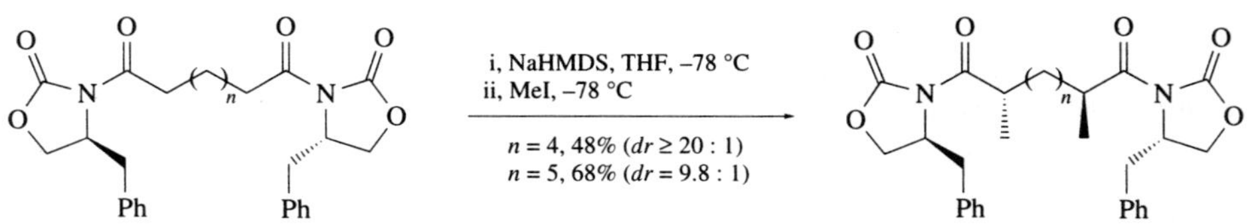

(50)

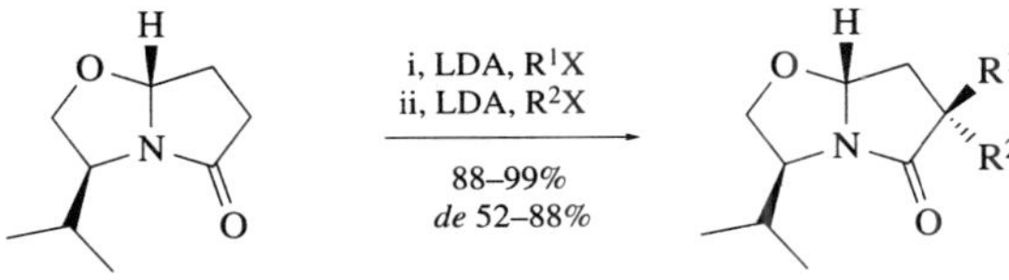

(51)

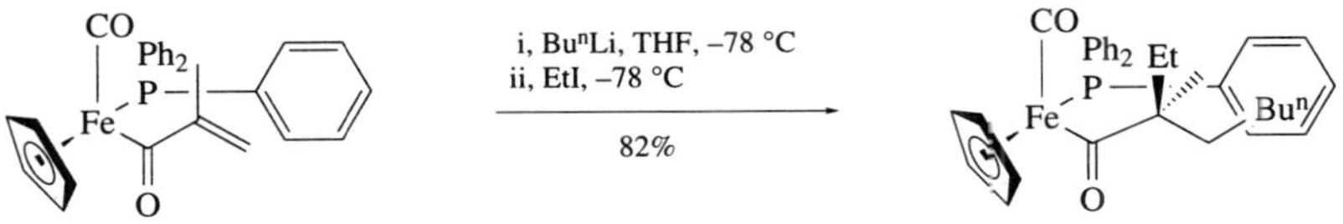

(52)

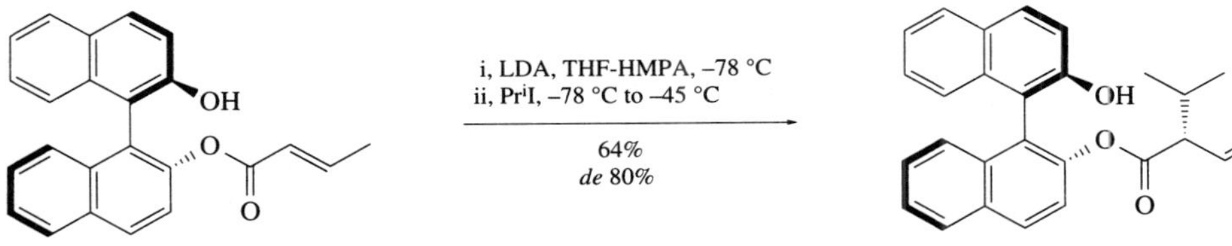

(53)

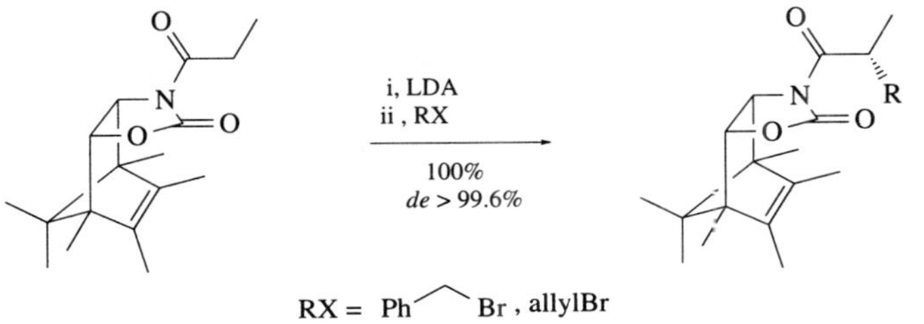

(54)

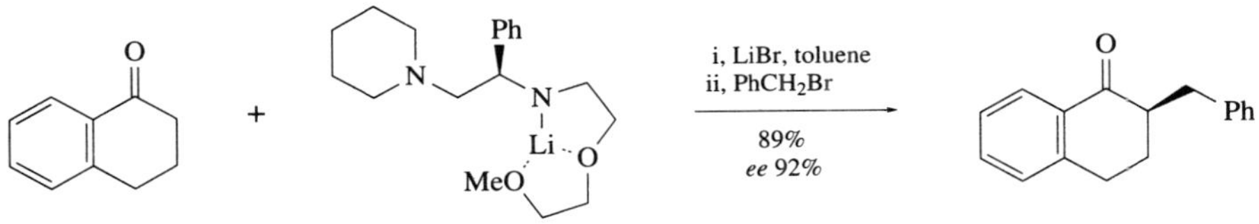

(55)

Alternatively, asymmetry can be introduced in α-alkylations with the chiral auxiliary present but bonded in a noncovalent fashion; however, results thus far have been variable. Chiral lithium amide bases can be used to effect enantioselective deprotonations and subsequent alkylations ⟨91TA1, 93TA1957⟩; for instance, Koga and co-workers have used a phenylglycine-derived lithium amide to give enolates that can be alkylated in up to 92% enantiomeric excess with reactive haloalkanes (Equation (55)) ⟨90CC1657⟩. Lithium bromide is essential for high optical yields, and the enantioselectivity is due to an enolate : LiBr : chiral amine complex. Asymmeteric α-alkylations have also been mediated by secondary ⟨93TL1963⟩ and tertiary amines ⟨89CPB1120⟩, and by ammonium salts under phase transfer conditions ⟨87JOC4745⟩; however, in most cases the optical yields vary over a wide range.

1.04.3.3 Oxygen Stabilized Carbanions

The chemistry described in this section is concerned solely with α-oxygenated carbanions, with a fuller discussion of enolates being found in Section 1.04.3.2. While oxygen is a π-donor and might be expected to destabilize a negative charge on atoms directly bonded to it, its relatively high electronegativity means that oxygen substituents can act to stabilize carbanions by inductive or σ-effects. Theoretical studies by Schleyer *et al.* suggest that bridged geometries may contribute to the stabilizing influence ⟨84JA6467⟩. The methods for preparation of carbanions with α-oxy substituents have been reviewed by Cheshire ⟨91COS(3)193⟩. Direct deprotonation may seem the most attractive method, and reaction of tetrahydrofuran with Bu^{n}Li as a route to acetaldehyde enolate is well known to all organic chemists ⟨77TL3791⟩. However, α-metallated ethers and acetals generally cannot be prepared by deprotonation at low temperature unless further activation (such as aryl, alkenyl, or alkynyl substituents) is provided. An exception was demonstrated by Corey and Eckrich, where *t*-butyl methyl ether (used as solvent in the reaction) was treated with Schlosser's base Bu^sLi/Bu^tOK at $-78\,^\circ C$ giving the lithio species which was alkylated with benzyl bromide in 83% yield (Equation (56)) ⟨83TL3165⟩. Facile removal of the *t*-butyl group results in the ether acting as an overall hydroxymethyl anion equivalent.

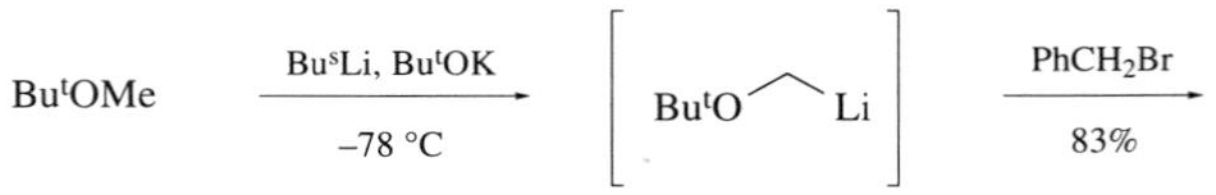

(56)

Both the groups of Evans ⟨74JA5560⟩ and Still ⟨74JA5561⟩ have examined the deprotonation and subsequent alkylation reactions of allyl ethers. Treatment of alkyl or silyl allyl ether with Bu^sLi at low temperature (below $-65\,^\circ C$) gave resonance-stabilized carbanions which could be quenched with haloalkanes to give a mixture of α- and γ-alkylated ethers; in almost all cases, the γ-regioisomer dominated. Evans *et al.* found that the proportion of γ-alkylation (Equation (57)) depended on the nature of the alkyl moiety of the ether ($\alpha:\gamma$ ratios varied from $\approx 1:1$ to $\approx 10:1$), but that other factors such as solvent or temperature had little effect. Still and MacDonald showed that allyl silyl ethers gave excellent yields of alkylated material where γ-alkylation again dominated in most cases, although the proportion depended on the structure of the haloalkane (Equation (58)) ⟨74JA5561⟩. The proportion of α-substituted allyl ether was lowest for small primary alkyl halides and significantly greater for secondary haloalkanes; when cyclohexyl iodide was used as alkylating agent, the α-regioisomer was the major product. Substitutions displayed the usual halogen dependency, the highest yields being obtained for iodides and bromides (with yields typically >95%), with chlorides giving poor conversions.

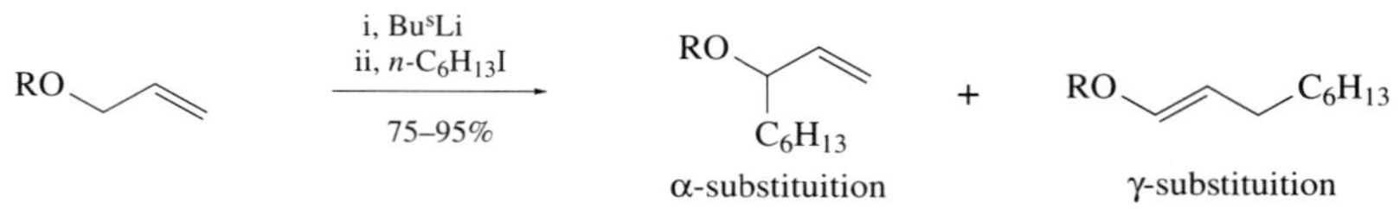

(57)

R	THP	Et	Bu^t
$\alpha:\gamma$	46 : 54	25 : 75	11 : 89

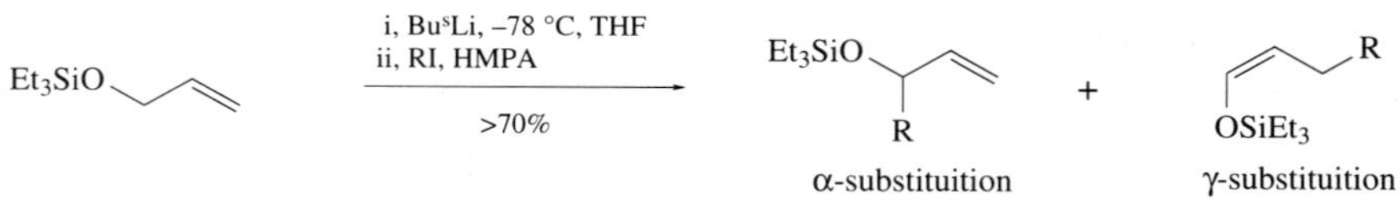

(58)

R	Me	Pr^i	cyclohexyl
$\alpha:\gamma$	3 : 97	40 : 60	61 : 39

The introduction of extra stabilization as two aryl groups allows formation and alkylation of *C,O*-dilithiated compounds. Yus and co-workers have shown that benzophenones may undergo two successive single electron transfer (SET) reduction steps on treatment with Li/naphthalene for 10 min at $-30\,^\circ C$ to give dilithio species which can be alkylated selectively on carbon, generally in good to excellent yields (Equation (59)) ⟨93T1327⟩. However, the reaction occasionally proved to be unpredictable; whereas methylation of the dianion derived from benzophenone proceeded smoothly giving methyl diphenyl carbinol in 92% yield, similar treatment of *ortho*-tolyl phenyl ketone gave only 32% of the desired material along with 51% of a ring-methylated, deoxygenated compound (Equation (60)). The temperature and timing of the reaction are vital for successful alkylation; Fry and co-workers showed that benzophenone and other diaryl ketones, on sonication with Li/4,4′-di-*t*-butylbiphenyl at room temperature for 24 h, and subsequent alkylation, gave *C*-alkylated, deoxygenated products exclusively ⟨90TL6155⟩. Substitution on carbon most likely precedes deoxygenation, and treatment of the original *C,O*-dilithio species with two different electrophiles provides a novel method for dialkylation (Scheme 27).

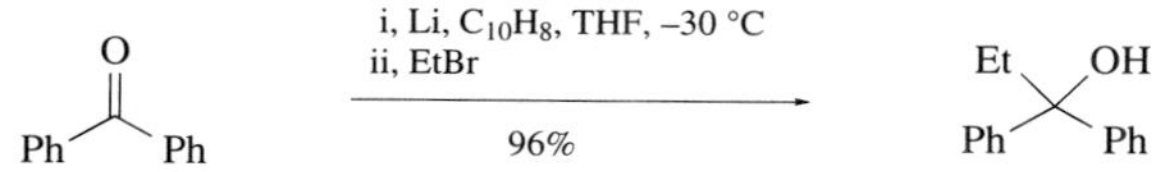

(59)

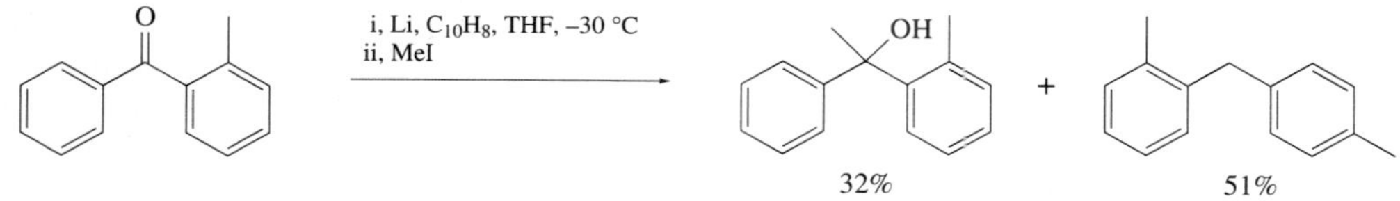

(60)

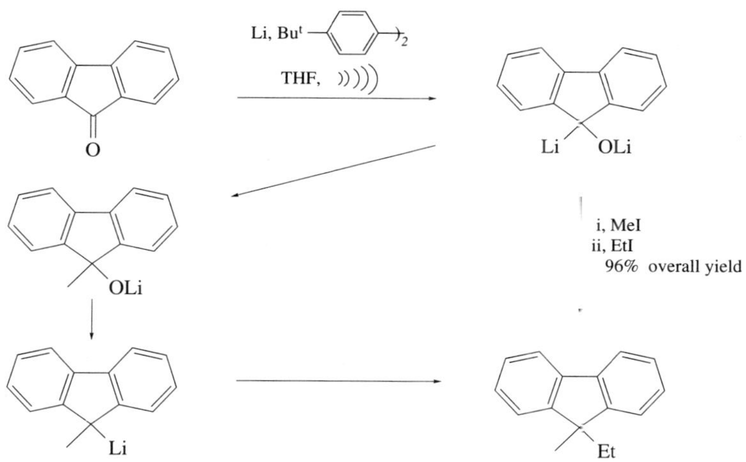

Scheme 27

While nucleophilic alkyllithium reagents might ordinarily be expected to react with esters (especially benzoates, where α-deprotonation is precluded) by addition to the carbonyl group, the introduction of steric hindrance can serve to block the approach of the organometallic component and therefore other pathways become important. Beak and Carter ⟨81JOC2363⟩ and Seebach and co-workers ⟨78HCA512⟩ have shown that benzoates with bulky groups (such as isopropyl and *t*-butyl) in the 2- and 6-positions are not attacked at the carbonyl group by bases such as Bu^{s}Li; instead, interaction with the ester carbonyl oxygen atom serves to direct lithiation to the α-position providing α-oxygenated carbanions which could be alkylated in moderate to good yields (Equation (61)). The reactions were successful with a number of haloalkanes; however, use of benzyl halides provided an exception since the isolated products were dimers of the organolithium reagent ⟨81JOC2363⟩. The degree of steric shielding at the carbonyl group is illustrated by the resistance of these esters to simple hydrolysis; cleavage could only be achieved by reductive methods ($LiAlH_4$).

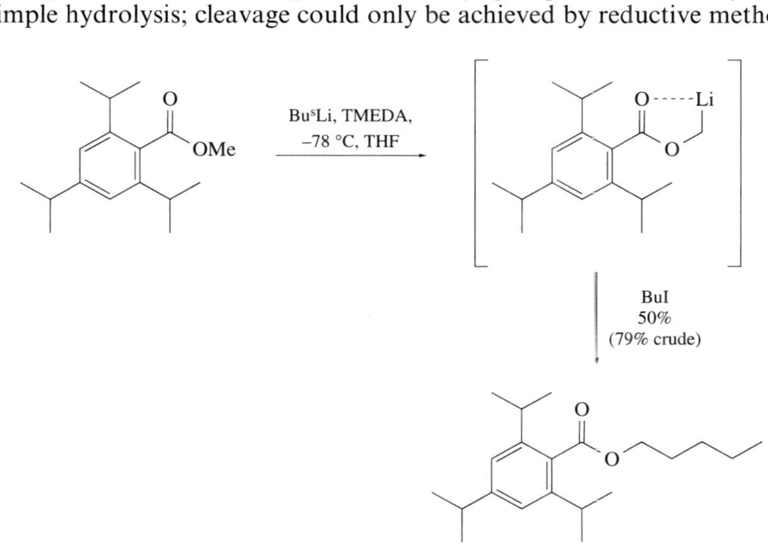

(61)

Hoppe *et al.* have demonstrated that the principle of deprotonation can be extended to a range of carbamates ⟨90AG(E)1424⟩. The carbamate derived from nonracemic (*ee* 97%) 1-phenylethanol was deprotonated with *s*-butyllithium and TMEDA at −78 °C to give a benzyl carbanion; at this low temperature, the organolithium species is essentially configurationally stable. Addition of an electrophile (*n*-propyl bromide) gave a tertiary carbamate in good optical and chemical yield (Equation (62)). Chiral α-oxygenated organolithium reagents are also accessible by means of a novel asymmetric deprotonation ⟨92AG(E)1459⟩. Treatment of carbamates with *s*-butyllithium in ether at −78 °C in the presence of the chiral diamine (−)-sparteine gave the chiral α-oxy carbanion in high diastereomeric excess ⟨92TL5323⟩. Addition of iodomethane gave the secondary carbamate in 81% yield, *ee* 96% ⟨90AG(E)1422⟩. Asymmetry is introduced at the deprotonation stage, where the *s*-butyllithium·(−)-sparteine complex discriminates between the enantiotopic α-protons in the carbamate (Equation (63)). In a related example, Schwerdtfeger and Hoppe have demonstrated that a neighboring chiral center can influence the course of metallation and subsequent alkylation steps: deprotonation of a carbamate derived from proline gave a lithio species that was methylated in 72% yield with a diastereomeric ratio of >95:5 (Equation (64)) ⟨92AG(E)1505⟩. Replacement of TMEDA by (−)-sparteine gave a lower yield of product (56%) with similar diastereoselectivity.

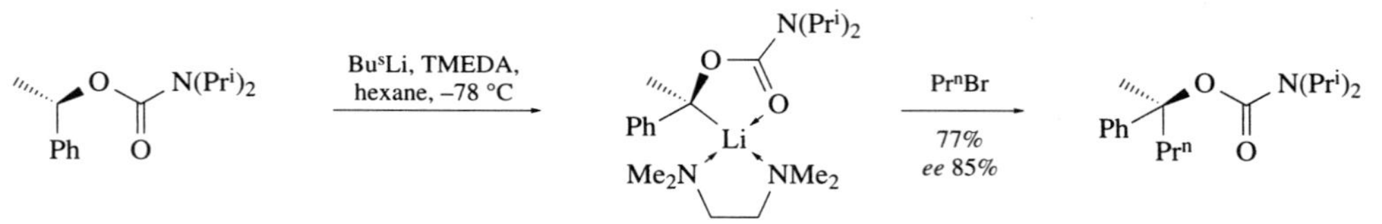

(62)

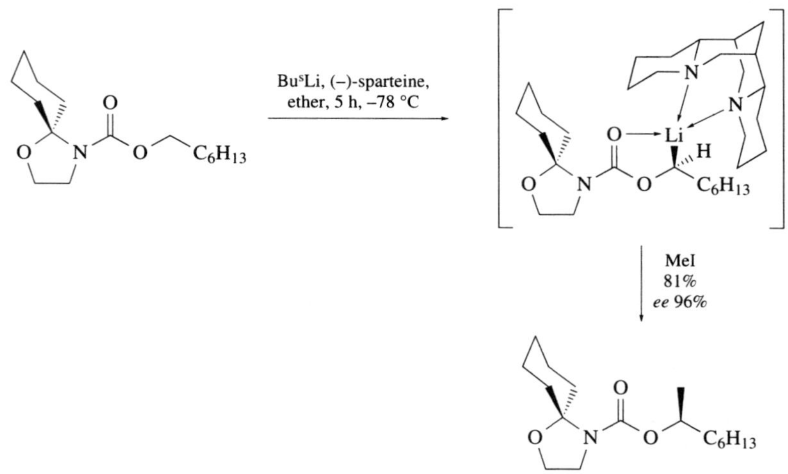

(63)

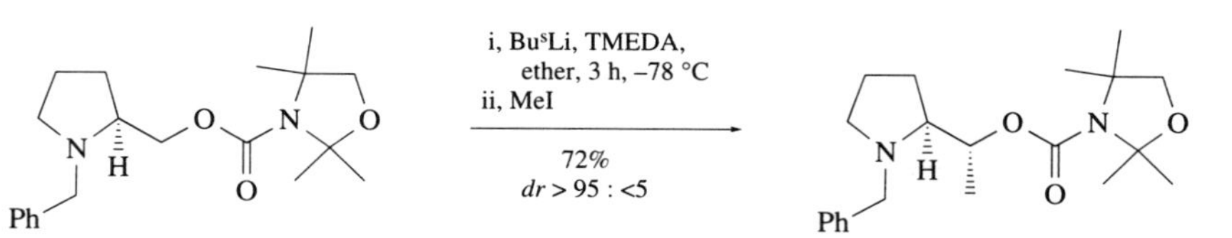

(64)

α-Alkoxystannanes are versatile reagents that provide another route to α-alkoxy carbanions, and can be prepared by a number of methods including addition of R_3Sn^- to aldehydes and subsequent alkylation of the α-stannyl alcohol ⟨78JA1481⟩. These compounds can also be obtained in high enantiomeric purity by asymmetric reduction of acyl stannanes with binaphthol-modified $LiAlH_4$ ⟨88JOC5584⟩ or from enantiomerically pure boronic esters ⟨89JA4399⟩. The carbanion can be generated by tin–lithium exchange or reaction of the α-alkoxystannane with alkyllithium reagents. This method provides a mild and regiospecific procedure for generation of these important carbanionic intermediates; lithio species which are not otherwise accessible by direct deprotonation can be prepared and alkylated in high yields (Scheme 28) ⟨78JA1481⟩. The tin–lithium exchanges are reported by McGarvey and co-workers to proceed most satisfactorily in dimethoxyethane as a solvent ⟨84JA3376⟩. An important feature of the α-alkoxyorganolithium species is their configurational stability at low temperatures; the additional chelation which is available in acetal groups further increases the barrier to inversion (and hence racemization) of the organolithium species ⟨80JA1201⟩. The α-alkoxystannane is converted to the lithio reagent with retention of configuration, and subsequent reaction with haloalkane electrophiles also occurs with retention of configuration ⟨84JA3376⟩. However, while simple α-alkoxylithio reagents (especially where added benzylic substitution is present) undergo efficient carbon–carbon bond formations when treated with iodoalkanes (Equation (65)), the reaction of the epimeric 4-*t*-butylcyclohexyl analogues gave very different results. Attempted alkylation with haloalkanes (I, Br, Cl) gave mixtures of products corresponding to protonation of an intermediate RLi with both retention and inversion of configuration, and also a significant proportion of enol ether (Equation (66)). In general, no alkylated material was recovered. No trends were apparent, although the intervention of radical intermediates was possible ⟨84JA3376, 88JA842⟩.

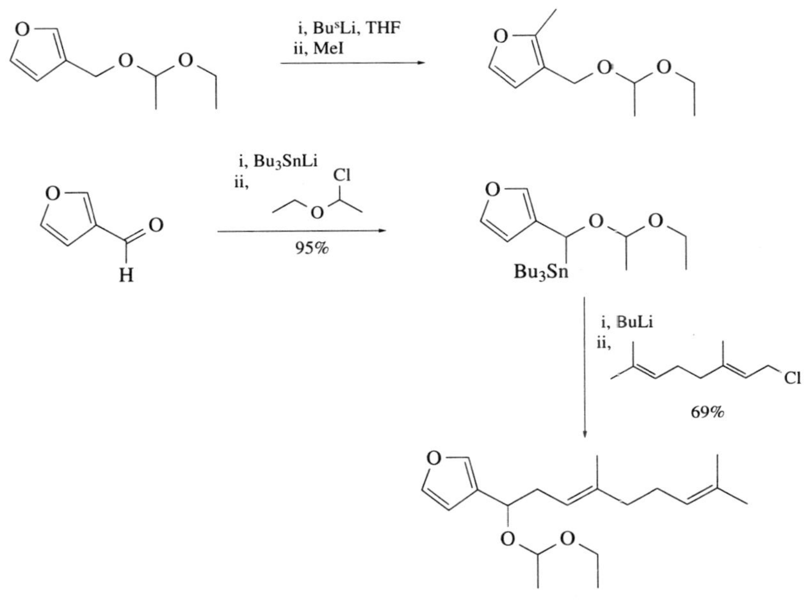

Scheme 28

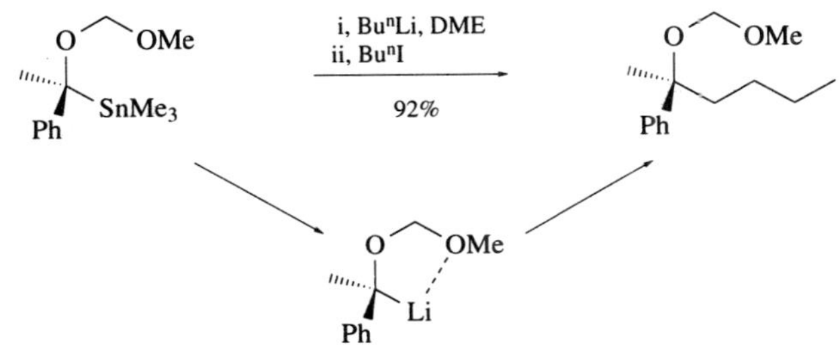

(65)

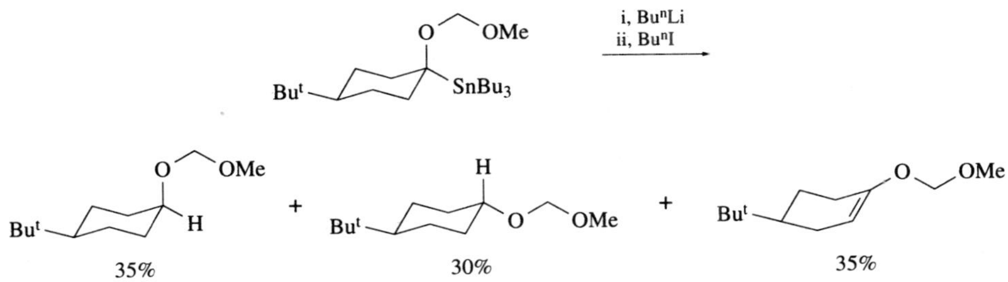

(66)

1.04.3.4 Sulfur Stabilized Carbanions

An assortment of different sulfur-containing functional groups can be employed to stabilize carbanions, and allow subsequent alkylations, with varying degrees of success. Several aspects of the alkylations of α-metallated sulfur-containing compounds will be described in this section; an excellent review by Krief of the alkylation of many classes of sulfur-stabilized carbanions has appeared ⟨91COS(3)85⟩, and more particularly the chemistry of sulfones ⟨77T2019, B-88MI 104-01, B-93MI 104-01⟩ and sulfoxides ⟨74ACR147, 81S185, B-88MI 104-01, 92TA961⟩ has also been reviewed on several occasions. The structural features of lithio derivatives of various sulfur-containing carbanions (such as sulfones, sulfoxides, and sulfoximes) have been reviewed by Boche ⟨89AG(E)277⟩.

Sulfur is highly nucleophilic and therefore competing alkylations on sulfur can occur. Anions derived from β-oxo dithioic acids are alkylated exclusively on sulfur to give the corresponding ketenedithioacetals ⟨86T3029⟩. Likewise, α-carbalkoxyenethiols, which can be regarded as tautomers of α-thioketoesters (thiopyruvates), can be deprotonated with sodium hydride and alkylated on sulfur (Scheme 29) ⟨85JOC3676⟩. Deprotonation of thiols with 1 equiv. of base leads to the corresponding thiolate; additional stabilization provided by allyl or benzyl substituents allows a second deprotonation to take place α to the sulfur atom when allyl or benzyl thiolates are treated with a second equivalent of a strong base such as Bu^nLi·TMEDA. The *C*,*S*-dianion derived from phenylmethanethiol (benzyl mercaptan) can be dialkylated with either an excess of a haloalkane, or with 1,3-dibromopropane giving 2-phenyltetrahydrothiophen (Equation (67)) ⟨74AG(E)202⟩. 2-Propenethiol can likewise be deprotonated twice under similar conditions (Bu^nLi, TMEDA, 0 °C) and the resulting dianion can be alkylated successively with two electrophiles. The reactivity of the two anions differs sufficiently to permit specific alkylation on carbon to occur first, although the allyl anion is alkylated with modest γ-regioselectivity ranging from 2 : 1 to 4 : 1 ⟨74AG(E)479⟩. The γ-alkylation is stereoselective, giving the (*Z*)-alkene with >95% control (Equation (68)). While addition of HMPA can reverse the regioselectivity of alkylation on carbon, product ratios are generally no better than 4 : 1 ⟨77CB1833⟩.

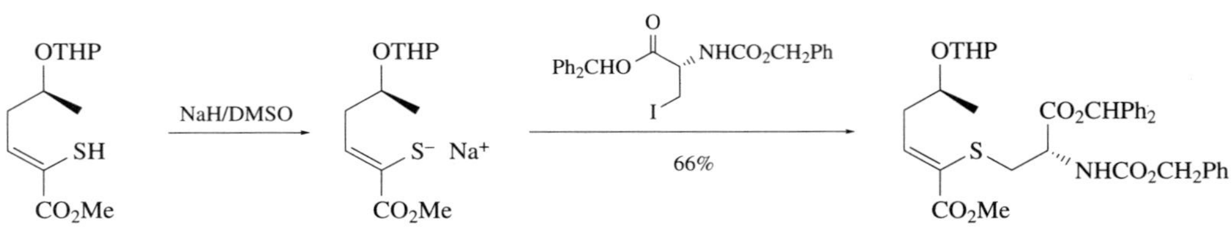

Scheme 29

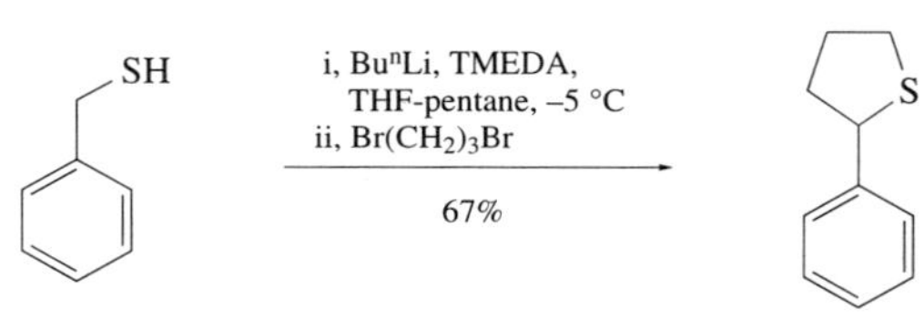

(67)

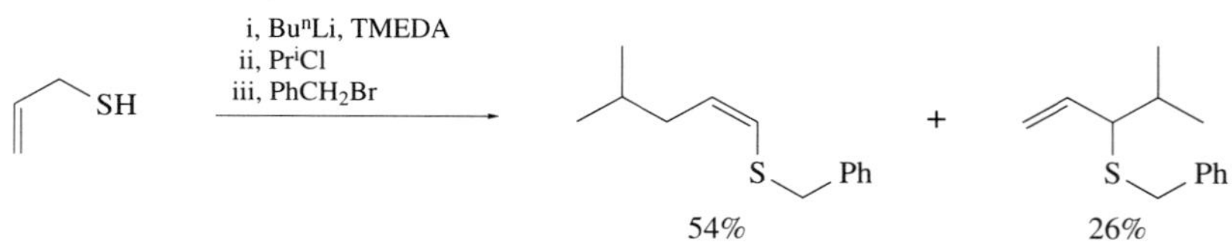

(68)

While a range of different types of reactions can be considered for formation of α-thio organometallic compounds from thioethers (dialkyl sulfides), the simplest and most direct approach uses the acidifying effect of the sulfur substituent to facilitate deprotonation. Simple thioethers can be deprotonated α to sulfur using strong bases such as Bu^nLi, either alone or in the presence of complexing reagents such as TMEDA, 1,4-diazabicyclo[2.2.2]octane, (dabco) and HMPA. Anions derived from simple dialkyl sulfides are quite basic, and alkylation reactions are often plagued by competing elimination ⟨67JOC1717⟩. Peterson has shown that dimethyl sulfide can be deprotonated and monoalkylated with 1-bromodecane in 32% yield; however, metallation of higher alkyl sulfides led to decomposition, presumably via β-elimination (Equation (69)). Reasonable yields of alkylated materials can be obtained using anions derived from alkyl phenyl sulfides, for example, Corey and Seebach have shown that thioanisole can be metallated at 0 °C by Bu^nLi in the presence of dabco, and the resulting phenylthiomethyllithium reagent can be alkylated with 2-iodopropane to give substituted material in 55% yield ⟨66JOC4097⟩. Dolak and Bryson have similarly reported successful alkylations; however, attempted benzylation gave only stilbene (78%) and unchanged sulfide (Equation (70)) ⟨77TL1961⟩.

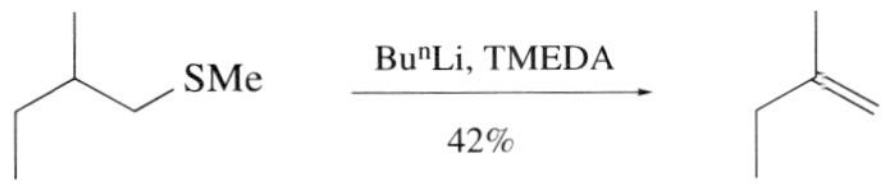

(69)

(70)

R	*Yield(%)*
Et	79
Allyl	82
$PhCH_2$	78 (stilbene)

Coordination-directed metallation of an SMe group has been used by Block and Aslam in an efficient approach to a synthetic equivalent for the methanethiol carbanion $HSCH_2^-$ ⟨85JA6729⟩. Protection of the sulfur with the 2-(tetrahydropyranyl) and 2-(tetrahydrofuranyl) groups provides a reagent that is cleanly metallated on the *S*-methyl group; there was no evidence of deprotonation at the tertiary position between oxygen and sulfur. This result is in contrast to the exclusive metallation of 1,3-oxathianes on C-2 ⟨85JOC657⟩; Block and Aslam have attributed the facile deprotonation of the *S*-methyl group to a favorable oxygen–lithium interaction directing the base, as well as steric hindrance and decreased acidity of the tertiary hydrogen ⟨85JA6729⟩. The stabilized carbanion can be alkylated with bromo- and iodoalkanes in 49–98% yields (Equation (71)).

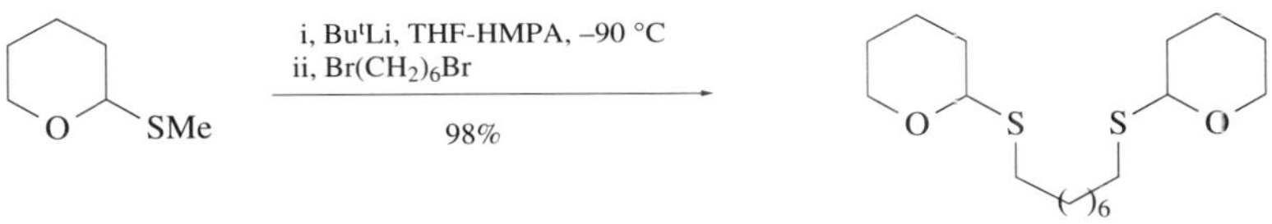

(71)

Alkyl vinyl sulfides are deprotonated selectively on the sp^2 carbon atom giving a vinyllithium species which can be alkylated in good to excellent yields (52–90%) and which, ultimately, can be viewed as an acyl anion equivalent. Yamamoto and co-workers extended the procedure to the dialkylation of α,ω-dihaloalkanes giving, ultimately, diketones in good yield ⟨73JA2694⟩. However, while 1,4-dibromobutane gave a 60% yield of 2,7-octanedione, a similar reaction with 1,3-dibromopropane gave the aldol product 3-methylcyclohex-2-enone (Scheme 30). Allyl and propargyl sulfides are metallated α to sulfur in the unsaturated group giving resonance-stabilized carbanions where subsequent alkylations generally occur selectively α to sulfur. An intramolecular variant has been used by Vedejs *et al.* to prepare 2-vinyltetrahydrothiophen (Equation (72)) ⟨78JOC1185⟩. Deprotonation of 2*H*-thiopyran using alkali metal amides leads to a resonance-stabilized anion which can be alkylated with *t*-butyl bromide in up to 54% yield exclusively α to sulfur (Equation (73)) ⟨78CC596⟩; this reaction illustrated one of the few successful alkylations with a tertiary haloalkane. Alkylation of the stabilized carbanion with bromocyclohexane gives a mixture of 2- and 4-substituted products in an 85 : 15 ratio. The lithiation and alkylation of allyl vinyl sulfide occurs exclusively on the allyl moiety; however, the regioselectivity of the alkylation depends upon the metallation conditions. While alkylations of carbanions derived from metallation with $Bu^nLi{\cdot}$dabco proceed with modest $\alpha:\gamma$ selectivity ($\approx 7:3$), deprotonations using Bu^sLi in the absence of added complexing agents leads to an allyl anion that can be alkylated with >95% α-selectivity ⟨73JA2693⟩. (This regiocontrol is in contrast with results obtained with 2-propenethiol dianions ⟨74AG(E)479⟩.) As expected, the highest yields are obtained with bromides and iodides, and chloroalkanes are considerably less reactive. Yamamoto and co-workers have demonstrated that the alkylated products will undergo a thio-Claisen rearrangement and subsequent hydrolysis when treated with $CaCO_3$ in hot aqueous DME, providing a convenient synthesis of γ,δ-unsaturated aldehydes (Scheme 31) ⟨73JA2693⟩. The rearrangement gives the (*E*)-alkene selectively, contaminated with less than 1% of the (*Z*)-isomer. Negishi *et al.* have studied the alkylation of 3-(phenylthio)prop-1-yne; 2 equiv. *n*-butyllithium generate an α,γ-dianion which is selectively alkylated by allyl bromides α to the sulfur substituent in excellent yields (Equation (74)) ⟨81JOC5041⟩. The corresponding sulfoxides and sulfones do not give coupled products in significant yields under similar conditions.

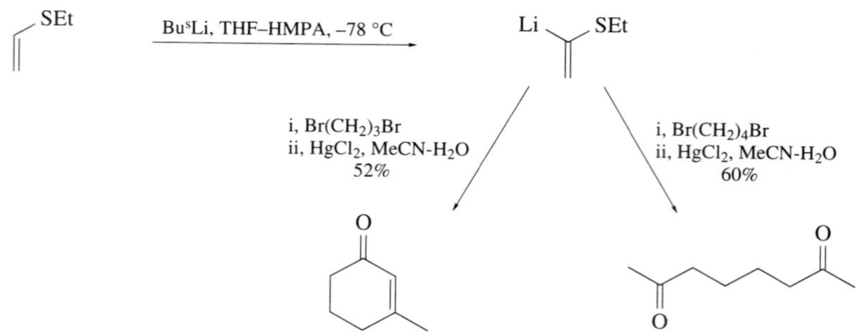

Scheme 30

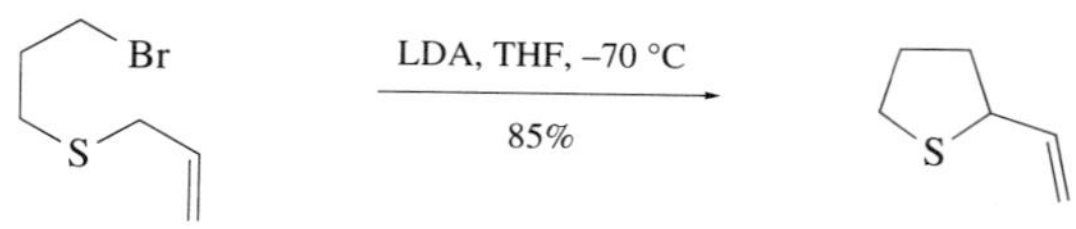

(72)

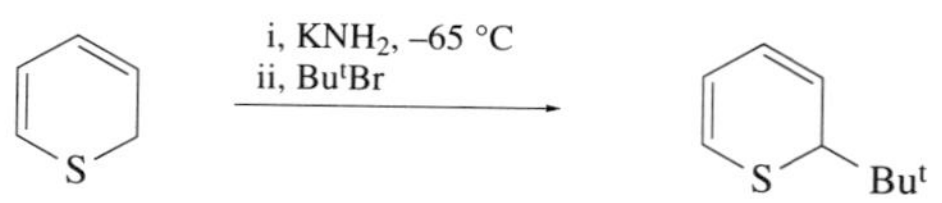

(73)

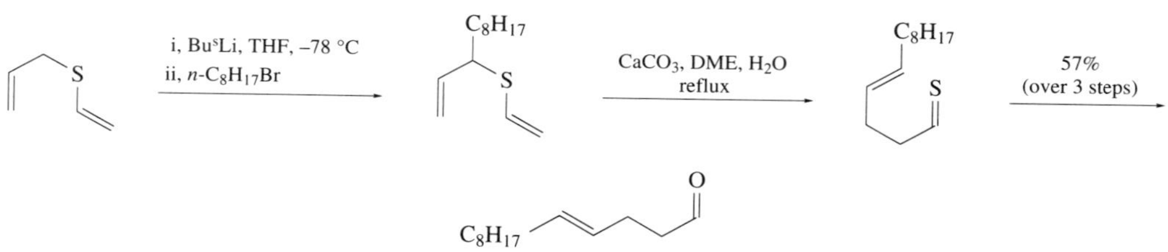

Scheme 31

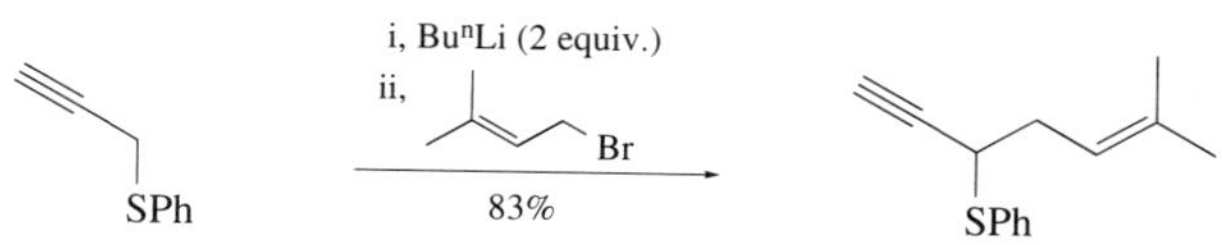

(74)

The effect of varying the oxidation state of sulfur on the acidity of neighboring hydrogen atoms can be judged by considering the pK_a values as determined by Bordwell and co-workers: methyl phenyl sulfone (29.0) and dimethyl sulfone (31.1) are more acidic than dimethyl sulfoxide (35.1), and methyl phenyl sulfide (thioanisole; extrapolated value ≈ 40) ⟨75JA7006, 78JOC3101⟩. Thus, it is apparent that the higher valence sulfur-containing functional groups have more marked acidifying effects and will lead to more stabilized carbanions.

The anion derived by deprotonation of dimethyl sulfoxide with strong bases such as sodium hydride, which is often referred to as the dimsyl anion, can be alkylated with primary alkyl bromides to give coupled products in good yields ⟨65CC29⟩. However, attempted benzylation fails ⟨65JA1345⟩; the highly basic nature of the carbanion leads to the formation of *trans*-stilbene (42%) plus "an oily mixture of sulfoxides." A similar reaction between the dimsyl anion and benzhydryl chloride gives tetraphenylethene in 98% yield. In general, sulfoxides can be deprotonated with lithium amides or methyllithium at low temperatures; use of *n*- or *t*-butyllithium promotes carbon–sulfur bond cleavage in addition to α-metallation ⟨74CJC761⟩, and *t*-butyllithium also racemizes aryl *t*-butyl sulfoxides within minutes at −78 °C. Trost and Bridges have demonstrated that aryl benzyl sulfoxides are readily deprotonated by lithium *N*-isopropylcyclohexylamide and the resulting anions are alkylated with allyl bromides or primary alkyl iodides ⟨75JOC2014⟩; thermolysis of the substituted products gives alkenes in 41–87% yields (Scheme 32). Allyl sulfoxides can likewise be deprotonated, and alkylations proceed with predominant α-regiocontrol ⟨74ACR147⟩.

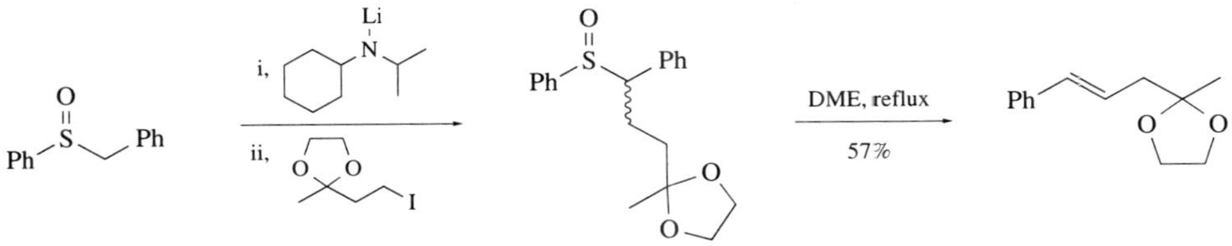

Scheme 32

Unsymmetrical sulfoxides are chiral at sulfur and therefore α-sulfinyl carbanions are diastereomeric and the possibility of diastereoselective alkylation exists. Durst *et al.* have shown that unsymmetrical alkyl benzyl sulfoxides can be deprotonated selectively at the activated (benzylic) position, and the resulting α-lithiosulfoxides can be methylated to provide substituted sulfoxides with varying diastereoselectivity ⟨71JA3077⟩: methylation of benzyl phenyl sulfoxide gave a 78% yield of monomethylated product as a 1.5 : 1 mixture of isomers, whereas benzyl *t*-butyl sulfoxide gave a 95% yield as a single diastereoisomer (Equation (75)). The intermediate α-lithio sulfoxide is deuteriated or carboxylated with retention of configuration; methylation occurs with inversion. The configuration of the sulfoxide is unaffected throughout the reaction. The mechanism of the reaction has been discussed extensively ⟨71JA3077, 86BCJ2801, 87JOC1414, 91COS(3)85⟩.

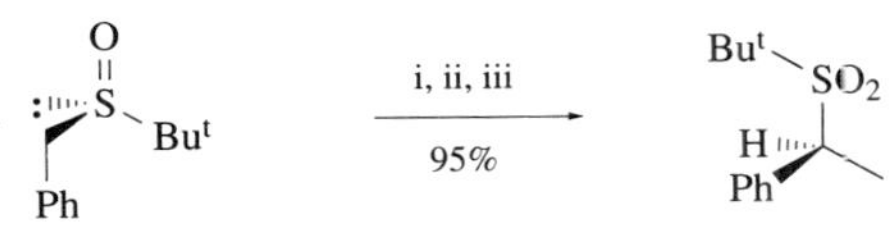

(75)

i, MeLi, THF, –60 °C; ii, MeI (product as a single diastereoisomer); iii, mcpba

Diastereoselective alkylations using sulfoxide-stabilized carbanions have been reviewed ⟨92TA961⟩; an example is the use of the sulfoxide to direct alkylation in the synthesis of biotin and biotin analogues ⟨78JA1558⟩. Deprotonation was best effected with methyllithium; use of *n*-butyllithium gave 20–25% carbon–sulfur bond cleavage in addition to low yields of the desired product. In all cases the alkylation product was accompanied by some recovered starting material, which arose from reprotonation of the anion (presumably by enolization of the ester) as carbanion formation was essentially complete. Addition of HMPA led to higher yields of alkylated material. A second alkylation could be carried out to give α,α′-disubstituted products with the two alkyl substituents in a *cis*-relationship (Scheme 33). Bravo *et al.* have shown that alkyl *p*-tolyl sulfoxides can be deprotonated with LDA or lithium tetramethylpiperidide (LITMP) at −75°C in THF and alkylated with lithium α-bromomethyl acrylate to give diastereomeric products in 55–88% yields with variable diastereoselectivities (*de* 22–90%) (Equation (76)) ⟨85TL2913⟩. Marginally better selectivities, but slightly lower yields, were obtained using a more hindered base (LITMP).

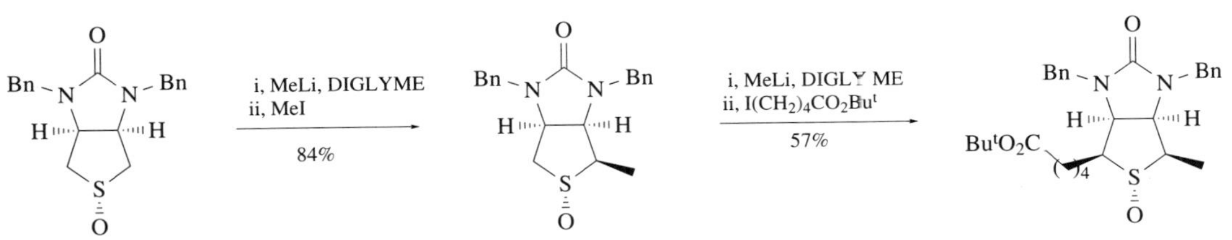

Scheme 33

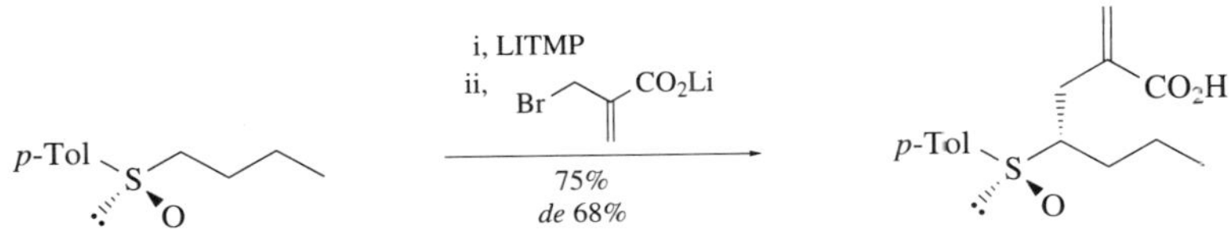

(76)

Cyclic sulfoxides—especially those with bulky substituents that can act as conformational "anchors"—are alkylated with high diastereoselectivities. Simpkins and co-workers have studied the alkylations of a 4-silyloxy thiopyranyl derivative, where deprotonation with LDA and alkylation with reactive haloalkanes gives excellent yields of products as single diastereoisomers. In a similar manner to the biotin synthesis described above, a second alkylation can be effected in the α'-position with excellent diastereoselectivity; the 2,6-*cis* product is formed, contaminated by only minor amounts of the 2,6-*trans* isomer ⟨92SL197⟩. Attempted asymmetric deprotonation with an enantiomerically pure lithium amide base gave, following subsequent methylation, the 2-methyl derivative in excellent yield and diastereoselectivity, but with only modest enantioselectivity (Equation (77)) ⟨93JCS(P1)3099⟩.

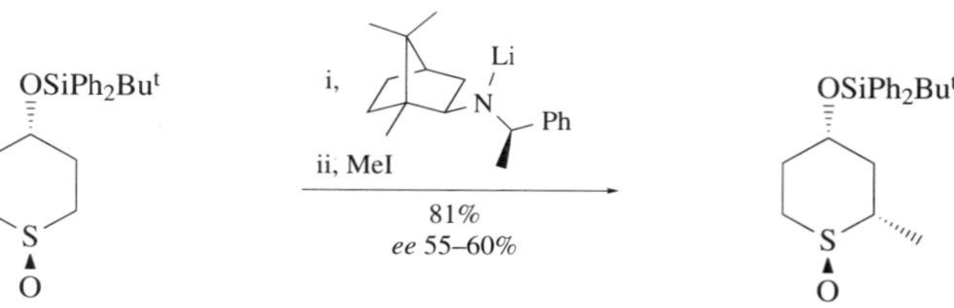

(77)

α-Sulfonyl carbanions have become the most widely used of the sulfur-stabilized carbanions as a result of their synthetic versatility and, ultimately, their facile removal. A sulfonyl group will act as a temporary activating substituent to direct alkylation, and desulfonylation is possible under reductive conditions ⟨B-88MI 104-02, 91TL1949⟩; substitution of the sulfonyl moiety using organometallic reagents is possible with vinyl ⟨90T6951⟩ and allyl ⟨83T3283⟩ sulfones; sulfones are increasingly useful in alkene synthesis, especially utilizing the Ramberg-Bäcklund ⟨68ACR209, 77OR(25)1⟩ and the Julia ⟨73TL4833, 85PS(24)97⟩ reactions. The required α-sulfonyl carbanions can be prepared by various routes including conjugate nucleophilic addition to vinyl sulfones ⟨78JA346⟩; however, the simplest and most widely used process simply involves deprotonation with strong bases such as alkali metal amides (including the dialkylamides such as LDA), sodium hydride, alkyllithiums, and even potassium alkoxides. Dimethyl sulfone can be deprotonated with 2 equiv. sodium amide and alkylated with 1-bromobutane giving di(n-pentyl) sulfone in 55% yield ⟨73JOM(59)53⟩. The greater acidity of α-protons (compared with sulfoxides) means that bases such as n-butyllithium can be used without competing carbon–sulfur bond cleavage intervening; alkylations with primary haloalkanes can be effected in excellent yields (Equation (78)) (for example, ⟨75TL1007⟩). In contrast, while organomagnesium compounds are also effective bases for α-deprotonation, the resulting α-sulfonyl Grignard reagents give only modest yields of coupled products on treatment with haloalkanes ⟨59JA2572⟩. However, Eisch and Galle have demonstrated that an α-sulfonyl Grignard reagent, prepared by conjugate addition of certain organomagnesium halides to 3-bromo-1-(phenylsulfonyl) propene, can undergo an intramolecular alkylation giving cyclopropanes in 40–76% yields as single stereoisomers (Scheme 34) ⟨79JOC3277⟩.

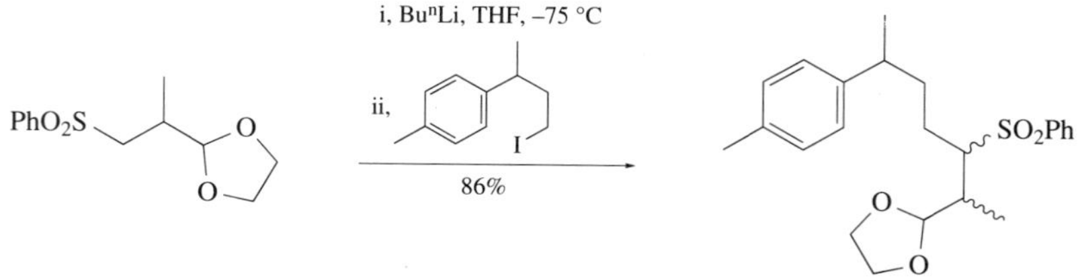

(78)

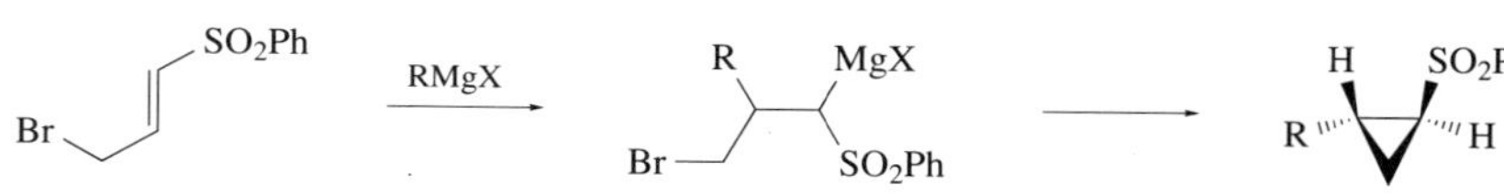

R = allyl, 76%; propargyl, 50%; Ph, 40%; $PhCH_2$, 54%; Me, 0%, Et, 0%; Bu^t, 0%

Scheme 34

In general, the alkylation of sulfones proceeds with good mono:dialkylation control. However, Pine *et al.* have shown that deprotonation and subsequent alkylation of methyl phenyl sulfone at 0°C gives significant levels of dialkylated material, with the ratio being relatively unaffected by reversing the order of addition ⟨90JOC2234⟩. Primary haloalkanes give products with a mono : dialkylation ratio of ≈3.6 : 1, and reactive (allyl, benzyl) haloalkanes give significantly more dialkylated material. Dialkylation is a stepwise process; despite the lower acidity of the α-protons in the monoalkylated material compared with the starting sulfone, proton transfer occurs to give a low concentration of a more reactive, substituted α-sulfonyl carbanion which is then alkylated at a significantly faster rate ⟨90JOC2234⟩. Addition of TMEDA leads to only a marginal increase in the proportion of monoalkylated material, however, changing the counterion to potassium, which will reduce ion pairing, gives a much higher mono : dialkylation ratio (Scheme 35).

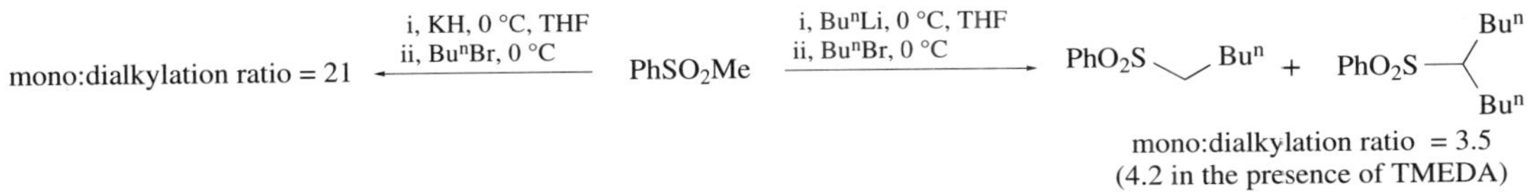

Scheme 35

Sulfonyl alkenes are readily metallated by strong bases, giving α-sulfonyl vinyllithium reagents which can be alkylated in good to excellent yields ⟨79JOC3279, 87TL989, 89JOC1491⟩. The metallating reagent must be chosen carefully as conjugate addition of the base can compete; for instance, while *s*- and *t*-butyllithium satisfactorily deprotonate a γ-morpholino vinyl sulfone, *n*-butyllithium gives a 1 : 1 mixture of α-vinyllithio and β-butyl α-lithioalkyl species (Scheme 36) ⟨93TL2003⟩. However, the problem is apparently substrate dependent, as others have reported no problems with *n*-butyl ⟨87TL989⟩ and methyllithium ⟨79JOC3279, 89JOC1491⟩. Eisch and Galle have developed an alternative route to circumvent this problem: conjugate addition of dimethyl amine gives the β-amino sulfone which can then be deprotonated with *n*-butyllithium and alkylated with iodoalkanes α to the sulfone ⟨80JOC4534⟩. Quaternization of the nitrogen and elimination with NaOH generates the α-substituted vinyl sulfone. Allyl sulfones are readily deprotonated to give resonance-stabilized carbanions which can be selectively alkylated α to sulfur ⟨91COS(3)85⟩. An intramolecular variant leading to methylenecyclopropanes has been studied by Jeffery and Stirling ⟨93JCS(P2)1617, 93JCS(P2)2163⟩. 2,5-Dihydrothiophen-*S*,*S*-dioxides (⟨92T8963⟩ and references therein) and 2,7-dihydrothiepin-*S*,*S*-dioxides ⟨93SL829⟩ can be alkylated selectively α to the sulfonyl group in satisfactory yields and therefore provide synthetic routes to substituted butadienes and hexatrienes respectively. Deprotonation and alkylation of the thiophen dioxide can be carried out satisfactorily at −78°C, whereas the thiepin is somewhat more sensitive and reactions must be effected below −100°C (Equation (79)).

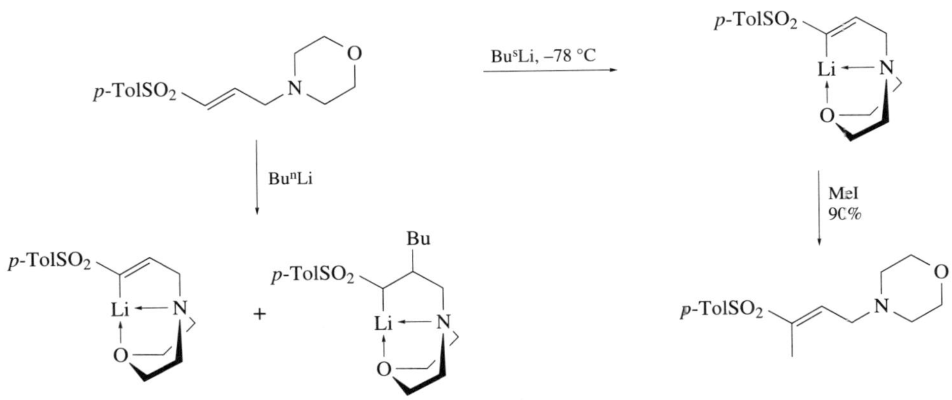

Scheme 36

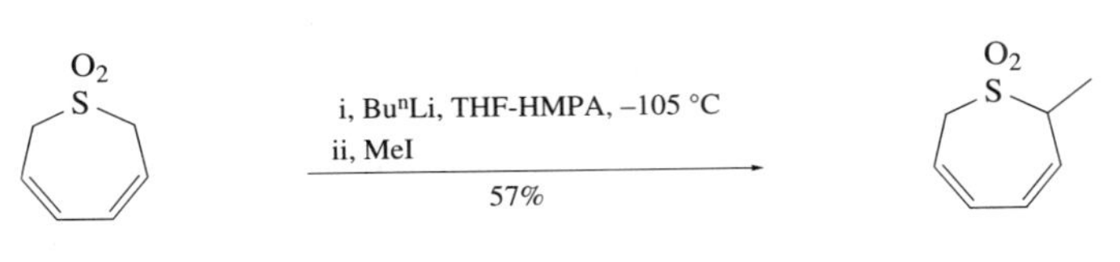

(79)

Treatment of alkyl sulfones with an excess of strong base can lead to the formation of dimetallated species ⟨56JA92, 77T2019, 89AG(E)277⟩. Where α-protons are available on both alkyl residues, α,α′-dimetallated intermediates are formed which lead to α,α′-dialkylated products on reaction with haloalkanes (Equation (80)) ⟨77OS(57)53⟩; where only one α-site is available for deprotonation, then α,α-dianions can be formed and dialkylated. Eisch *et al.* have shown that the dianion of methyl phenyl sulfone reacts with α,ω-dihalides to give cyclic products ⟨85JOC3674⟩. Studies by Gais and Vollhardt ⟨88TL1529⟩ suggest that, with alkyl phenyl sulfones, the *o*,α-dilithio species is formed initially which subsequently rearranges at higher temperatures to the α,α-dilithio compound ⟨91COS(3)85⟩.

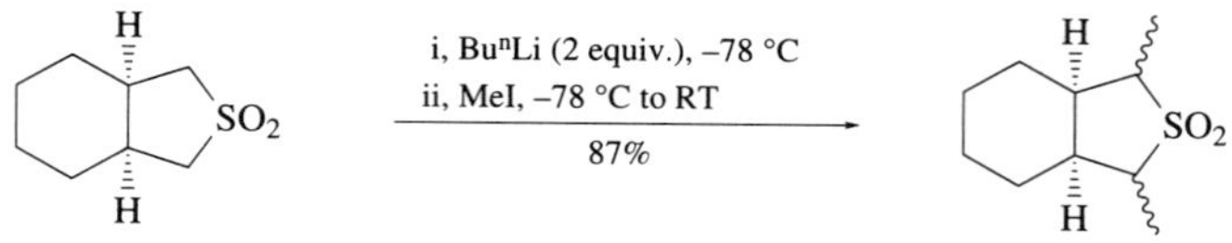

(80)

A final point of interest concerns the configurational stability of the α-sulfonyl carbanions. While the intermediate carbanions are chiral (⟨B-93MI 104-01⟩ and references therein), in general they are not configurationally stable and racemization is fast even at low temperatures. Exceptions are trifluoromethyl ⟨89AG(E)1025⟩ and *t*-butyl sulfones ⟨92JA4439⟩, where chiral nonracemic α-lithio species can be generated ($CF_3SO_2C(Li)(Me)CH_2Ph$ has an extrapolated half-life of 30 days at -78°C, whereas the *t*-butyl analogue has $t_{1/2} = 3$ h at -105°C) and subsequently alkylated with good enantioselectivity. However, racemic (or substantially racemized) products can usually be expected: Williams *et al.* have studied the deprotonation and alkylation of epimeric 5-(phenylsulfonyl)bicyclo[2.2.2]oct-2-enes and reported that while the alkylations are stereoselective, they never give just one product; the stereoselectivity results from other structural features in the molecule, not from the geometry of the parent sulfones (Equations (81) and (82)) ⟨90JOC3840⟩.

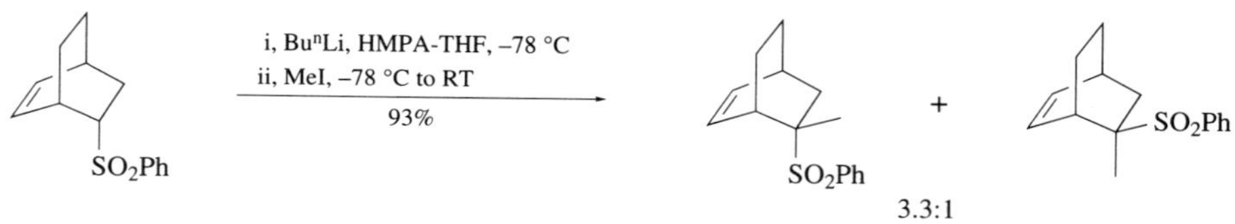

(81)

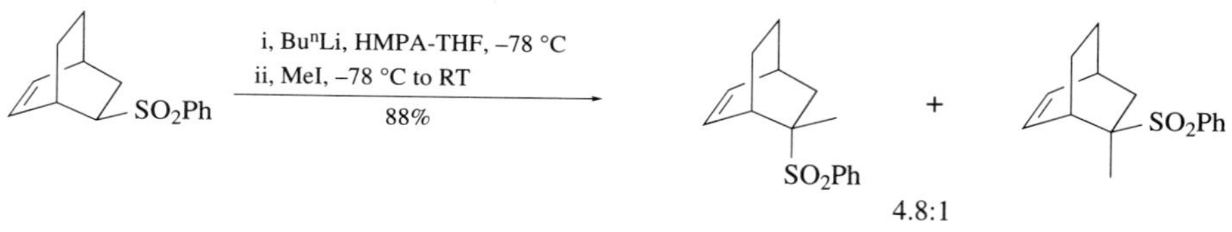

(82)

The sulfoximine is a sulfur-containing functional group at the same oxidation level as a sulfone which possesses the additional benefit of chirality at sulfur, and therefore diastereoselective alkylations become possible. However, reports thus far have been sporadic and results less than encouraging (⟨86AG(E)935⟩, for example). *S*-Allyl sulfoximines are deprotonated readily with *n*-butyllithium, and the resulting anions can be alkylated in 54–81% yield with primary haloalkanes ⟨91TL6497⟩. While the alkylations occur with complete regioselectivity α to sulfur, the diastereoselectivity is poor (1 : 1 to 2 : 1) (Equation (83)). More positive results have been reported recently by Trost and Matsuoka using *N*-nitrosulfoximines ⟨92SL27⟩. It was anticipated that the strongly electron-withdrawing *N*-nitro group would increase the acidity of the α-hydrogens and also improve the configurational stability of the carbanion through a shortening of the α carbon–sulfur bond. Deprotonation of the *S*-methyl group could be effected with sodium hydride (preferred for the first alkylation) or lithium hexamethyldisilazide (LiHMDS). The latter reagent is preferred for the second step as the stronger ion-pairing between Li^+ and the sulfoximinyl carbanion leads to greater diastereoselectivity compared with Na^+ or K^+. An attempted metallation with *n*-butyllithium led to decomposition of the *N*-nitrosulfoximine. The first alkylation of the *S*-methyl group was carried out in 48–69% yield, and the α,α-dialkylated material was then obtained in 50–96%

yield (Scheme 37). When the second alkylation is carried out at $-78\,^{\circ}C$, diastereomeric ratios approaching 10 : 1 are attainable.

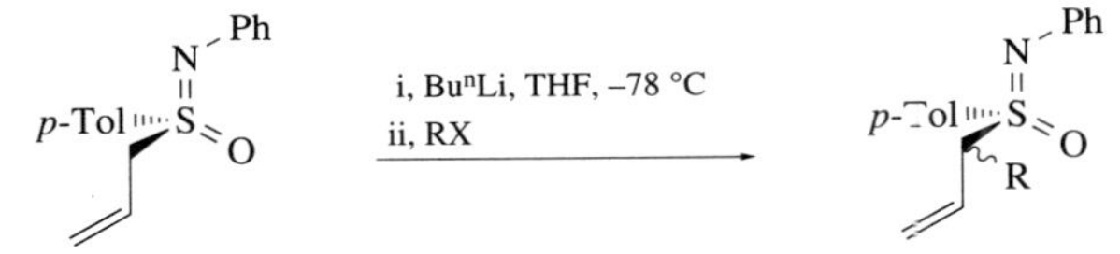

(83)

RX	*Yield(%)*	*d.r.*
MeI	69	1:1
EtI	81	1:1
Pr^{n}Br	54	1.3:1

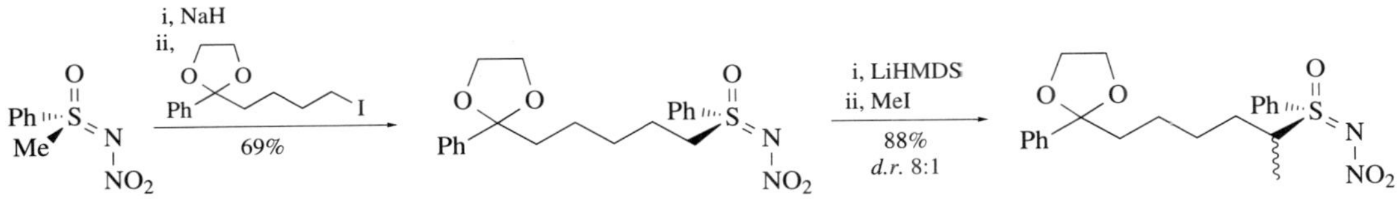

Scheme 37

Sulfonamides can likewise be expected to activate α-hydrogen atoms towards direct deprotonation by strong bases, and simple sulfonamides can be double deprotonated by treatment with strong bases such as LDA to give α,*N*-sulfonamide dianions. However, addition of iodomethane to the species generated from methanesulfonamide gave, in addition to unchanged starting material, *N*,*N*-dimethylmethanesulfonamide as the major product with only a trace of *C*-alkylated material ⟨84JOC1700⟩. Davis *et al.* have similarly reported a failure to alkylate the dianion derived from benzylsulfonamide ⟨90TL1653⟩. Introduction of a bulky group on nitrogen hindered alkylation at that site and C_α-alkylated material was isolated in good yields (Equation (84)); *N*-alkylation was effectively suppressed ⟨84JOC1700⟩.

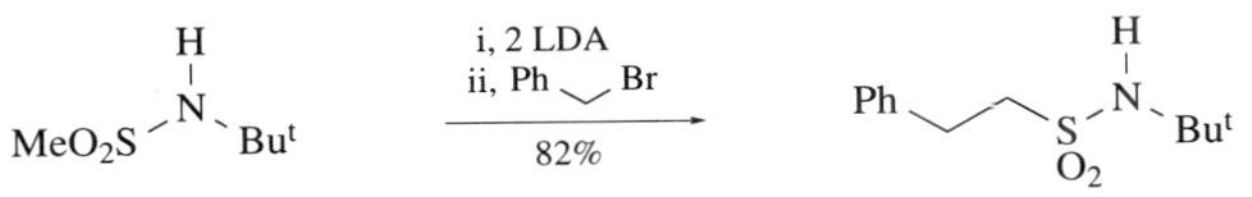

(84)

Davis *et al.* have developed an efficient route to nonracemic α-substituted sulfonamides using camphor-based chiral auxiliaries. Deprotonation of a camphor *N*-sulfonylimine with 1 equiv. LDA gives an azaenolate which is alkylated exclusively on nitrogen ⟨89TL1613⟩. A second equiv. of a strong base allows generation of a dianion where the second site of metallation is α to sulfur. Benzylation occurs exclusively α to sulfur in excellent yields; however, the products are isolated as a ≈1 : 1 mixture of diastereoisomers ⟨89TL1613, 91T173⟩. In general, camphor *N*-sulfonylimines give dianions on treatment with either 2 equiv. *n*-butyllithium or 3 equiv. LDA; while better results are obtained with LDA, the enantiomeric purities of the α-substituted sulfonamides (which can be liberated from the auxiliary by treatment with ethanolic HCl) are unacceptable. Significantly, alkylation α to sulfur can only be effected on the dianion: deprotonation of the fenchone *N*-sulfonylimine (which is incapable of forming an azaenolate) gives a sulfur-stabilized carbanion which cannot be alkylated ⟨90TL1653⟩. Improved diastereoselectivities are obtained on alkylation of dianions derived from the *N*-sulfonylimines of camphorsulfonamides (Scheme 38); in these cases α-alkylated benzylsulfonamides are obtained with enantiomeric excesses in the range 66–95%, with lower optical yields resulting from the alkylation of ethanesulfonamides ⟨93JOC4890⟩.

1.04.3.5 Nitrogen Stabilized Carbanions

Although the lone pair on nitrogen in an amine would be expected to destabilize a carbanion on a neighboring carbon atom by π-donation, the higher electronegativity of nitrogen relative to carbon means that, in a similar manner to α-oxygenated species, a nitrogen substituent is able to stabilize a negative charge by an inductive mechanism. Consequently hydrogen atoms on α-carbon atoms in

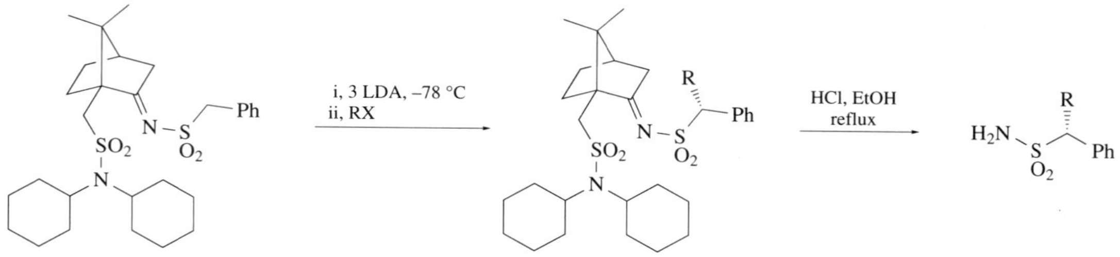

RX	*Yield(%)*	*ee(%)*
MeI	82	75
$PhCH_2Br$	79	93
Me_2CHI	58	90

Scheme 38

amines are rendered slightly acidic, and tertiary amines such as triethylamine can be metallated by *n*-butyllithium. The resulting σ-stabilized carbanions can be alkylated in low (14–25%) yields with *n*-iodobutane ⟨66JOC2061⟩. *N*-Methylpiperidine and other *N*-methyldialkylamines can be selectively deprotonated on the *N*-methyl carbon atom using one of Schlosser's mixed base systems, Bu^sLi/Bu^tOK, at 0°C; the resultant anions can be alkylated in good yields ⟨84TL1353⟩. However, the metallation of alkyl groups attached to nitrogen using this method is by no means universal, for example, *N*,*N*-dimethylaniline is only metallated in the phenyl ring, predominantly in the *ortho* position ⟨66JOC2047⟩. In addition, deprotonation of unmodified secondary and primary amines on α-carbon atoms is not feasible owing to the higher acidity of the amino hydrogen atoms. None the less, these classes of amines and other nitrogen-containing molecules can be deprotonated and efficiently alkylated via a wide variety of derivatives: conjugating groups can be employed to delocalize the negative charge and stabilize the anion (as in benzylamines and *N*-allylamines), electron-withdrawing substituents on nitrogen can establish dipoles (as in *N*-nitrosoamines), and heteroatom-containing groups can be used to direct deprotonation and stabilize the resulting organometallic reagent by chelation (such as amides). An extensive and systematic review of α-metalloamine synthetic equivalents has been published ⟨84CRV471⟩. Nitrogen substituents can also participate in the stabilization of negative charges, usually by conjugation, in a number of classes of compounds including oximes, nitriles, and hydrazones. Many examples in this latter group of stabilized carbanions have been developed as solutions to the problem of ketone alkylation.

One effective route to activation of α-hydrogen atoms involves the preparation and use of *N*-nitrosoamines ⟨75AG(E)15⟩. Deprotonation α to nitrogen is effected using bases such as Bu^nLi or LDA, and the resulting dipole-stabilized anions can be alkylated with a range of primary and other activated haloalkanes in typically 60–95% yields. However, the potential risks associated with this class of compounds would suggest that, except in rare cases where other methods fail, alternative approaches are preferable. One such approach, similar to the coordination-directed deprotonation of sterically encumbered benzoate esters, centers on deprotonation of amides. If no hydrogen atoms are available on the carbon atom α to the carbonyl group, and if nucleophilic addition to the carbonyl group is blocked, then coordination of the organolithium base to the amide oxygen atom directs deprotonation α to nitrogen ⟨83JA2080, 88JA8145⟩. However, while the organometallic intermediate can usually be alkylated in acceptable yields, regeneration of the free amine is often difficult and can require forcing conditions. Beak and Lee have studied the reactions of α'-lithioallylamides; lithiation could be readily effected with Bu^tLi at −78°C and subsequent alkylations were usually γ'-selective, although some exceptions were noted ⟨89JOC458⟩. The use of α,ω-dihaloalkanes and an excess of base allowed for the formation of spirocyclic compounds with the quaternary center α to nitrogen, although the yields were modest (Equation (85)).

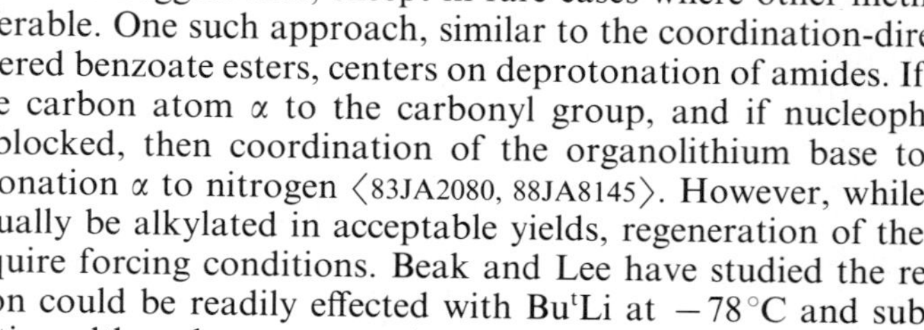

(85)

Vastly superior results have been obtained by employing either the formamidine (developed largely by Meyers *et al.* since the early 1980s) or the *t*-butoxycarbonyl functions to direct deprotonation and subsequently stabilize the organometallic intermediate. The formation of a 1 : 1 adduct prior to metallation of formamidines has been demonstrated by Meyers *et al.* using IR measurements ⟨83JA2082⟩. The resulting α-lithioformamidines can be alkylated in good to excellent yields; the reaction occasionally gives oxidation products (enamines) which are characteristic results of electron transfer processes (Scheme 39); however, these can be suppressed either by addition of HMPA (which presumably increases the rate of substitution reaction) or by transmetallation with 1-pentynylcopper. α,α′-Dialkylation of saturated heterocycles can be effected; however, there is little or no diastereoselectivity evident in the second alkylation ⟨84JA3270⟩. Removal of the formamidine, providing the free amine, can be accomplished using KOH, N_2H_4, or $LiAlH_4$. One advantage of the Meyers technology is that chiral auxiliaries can be attached directly to the imino nitrogen atom of the formamidine, allowing for the asymmetric alkylation of α-amino carbanions. Correct choice of the chiral auxiliary is vital: deprotonation and subsequent alkylation of tetrahydroisoquinolines via an (*R*)-phenethylamine-derived formamidine proceeded in excellent yields but with poor to moderate enantioselectivity, whereas the (bis)silyl ether (**33**) gave excellent (*S*)-selectivity (Scheme 40) ⟨83JA117⟩. The presence of an additional oxygen atom in the auxiliary, either as an alkyl or silyl ether, is essential for high enantioselectivity, presumably acting as an extra donor to chelate the lithium atom ⟨84T1361⟩. While the low basicity of silyl ethers would normally suggest that chelation would be less important for these auxiliaries ⟨90JA697⟩, it is likely that an oxygen–lithium interaction is still involved.

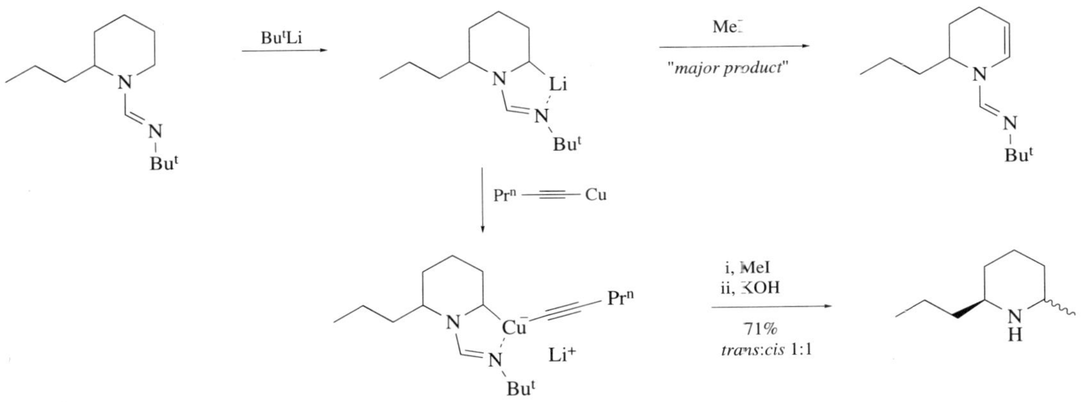

Scheme 39

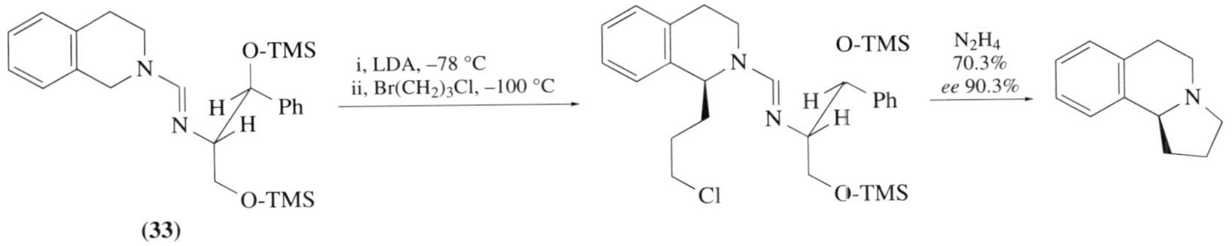

(33)

Scheme 40

More recently, formamidines derived from *O*-*t*-butyl valinol have been used to good effect in the synthesis of a range of natural products; some examples from the Meyers group include salsolidine ⟨87T5095, 89OS(67)60⟩, emetine (a formal total synthesis, ⟨91JOC6873⟩), corynantheine alkaloids ⟨91JOC2091⟩, homolaudanosine ⟨87T5095⟩, reticuline ⟨89H(28)295⟩, deplancheine ⟨86JOC3108⟩, yohimbone ⟨88JA4778⟩, and cryptaustoline ⟨92JA8483⟩. Several isoquinoline alkaloids, which can be prepared from similar starting materials using the formamidine directing group, are depicted in Scheme 41. The formamidine-based procedure has been extended recently to the asymmetric synthesis of alkylated tetrahydrobenzazepines ⟨93T1807⟩; while the optical yields are not always as high as for the corresponding isoquinolinyl compounds, enantiomeric excesses between 84 and 96% are generally observed. However, this method is not directly applicable to the α-alkylation of pyrrolidines or piperidines as the fully saturated heterocycles are only metallated as the achiral *N*-*t*-

butyl formamidines; introduction of oxygenated chiral auxiliaries prevents deprotonation from taking place. As a solution to this problem Meyers *et al.* discovered that the introduction of β,γ-unsaturation into the heterocycles allows deprotonation to occur at $-100\,°C$, and alkylation occurs to give the 2-substituted products in good yields with >90% enantiomeric excess ⟨85JA7974⟩. While some 2,3-enamine is formed as a result of γ-alkylation (8–30%) it is of little concern as the hydrazinolysis step decomposes the unwanted regioisomer (Scheme 42).

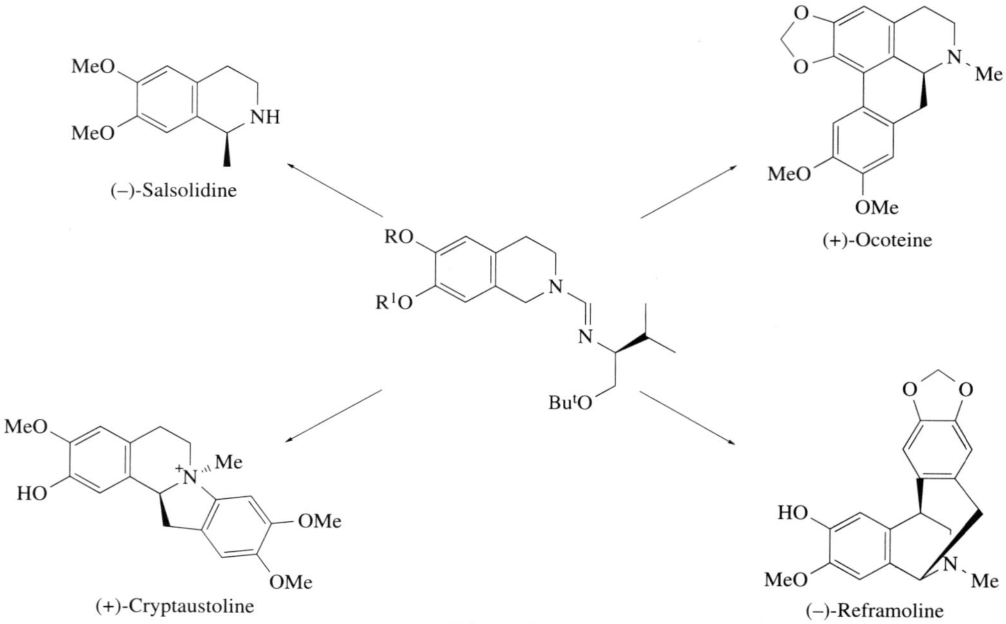

Scheme 41

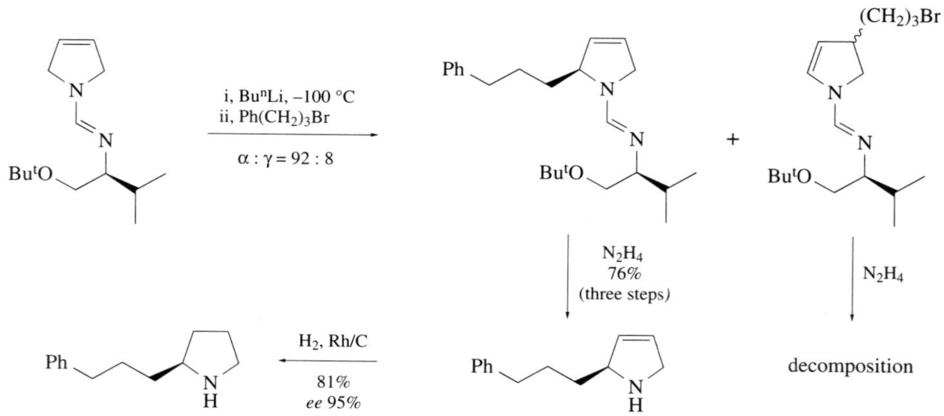

Scheme 42

Modification of the chiral auxiliary by replacement of the *O-t*-butyl group with the smaller *O*-methyl substituent allows metallation and alkylation to be carried out at a tertiary carbon atom in tetrahydroisoquinolines ⟨91TL5501⟩; the second alkyl group is introduced in good chemical and optical yields (Equation (86)), although the stereochemical outcomes are less satisfactory than for alkylation of the corresponding unsubstituted heterocycles. The facial selectivity for the second alkylation is the opposite of the first: excellent β-face selectivity is observed in the first alkylation leading to the tertiary center, whereas the second alkylation occurs with more moderate (*ee* 74–86%) α-face selectivity giving a quaternary carbon atom. The configurational stability of the intermediate lithio formamidine ⟨91TL5505⟩, and factors influencing this remarkable switch in facial selectivity ⟨91TL5509, 92JOC3819⟩, have been analyzed.

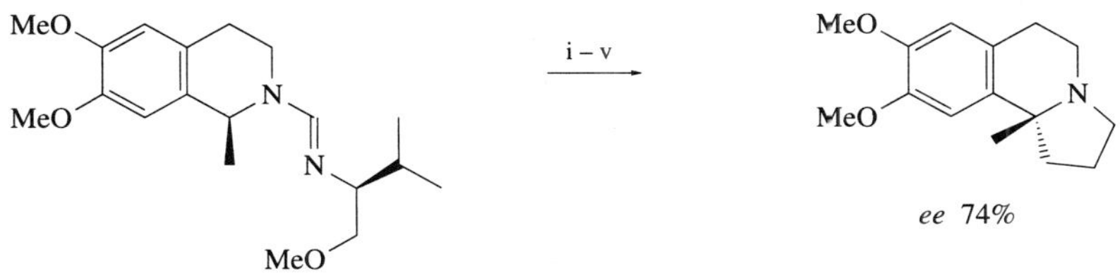

(86)

i, Bu^sLi; ii, $I(CH_2)_3OTBDMS$; iii, F^-; iv, PBr_3; v, N_2H_4

The *t*-butoxycarbonyl (*t*-BOC) group provides another versatile approach to α-aminocarbanions. Acyclic amines such as dimethylamine can be alkylated in good yields (Equation (87)) ⟨89TL1197⟩, and the *t*-BOC auxiliary can be removed under mild conditions. Oxazines can be alkylated using the *t*-BOC directing group, providing a useful 1-lithio-3-hydroxy-1-propylamine synthetic equivalent (Equation (88)) ⟨93JOC823⟩. While alkylation gives the oxazines as mixtures of diastereoisomers, this is of little consequence as liberation of the γ-hydroxyamine by treatment with acid removes the chiral center at C-2. Beak and Lee have shown that *t*-BOC protected, saturated nitrogen heterocycles can be alkylated on α-carbon atoms in yields that are variable, but generally greater than 60% ⟨93JOC1109⟩. The products can be lithiated a second time on the α′-carbon atom, and subsequent alkylation in piperidines and perhydroazepines occurs *trans* to the first alkyl group (Scheme 43) ⟨90JOC2578, 93JOC1109⟩.

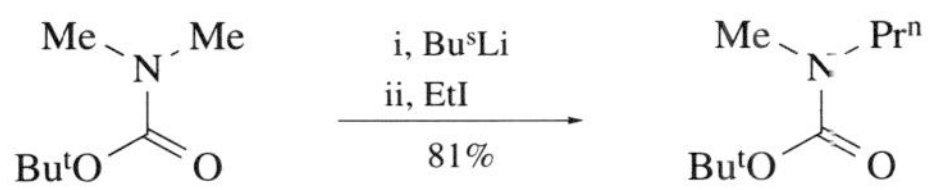

(87)

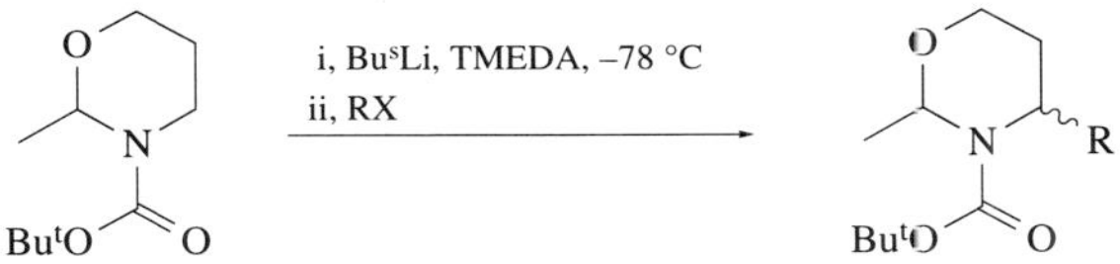

(88)

R	Me	$PhCH_2$	Bu^n
Yield (%)	80	55	52

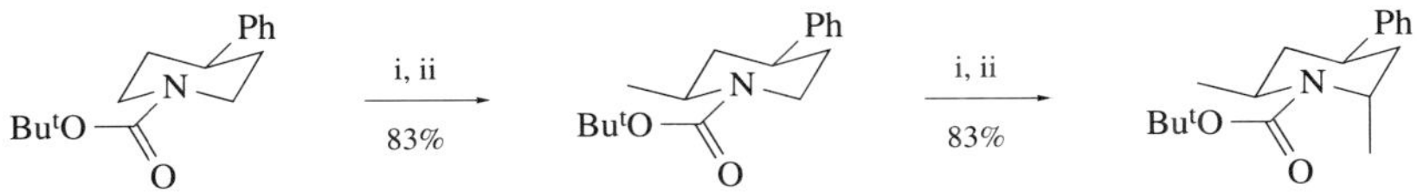

i, Bu^sLi, TMEDA, –78 to –20 °C; ii, MeI

Scheme 43

Meyers and Milot have demonstrated that the formamidine and *t*-BOC directing groups can provide complementary outcomes and are therefore equally valuable in the armoury of the synthetic chemist. Indolines and 1,2,3,4-tetrahydroquinolines can be deprotonated at C-2 α to nitrogen using the formamidine auxiliary to direct the alkyllithium base, whereas the *t*-BOC group directs the metallating reagent to the *ortho* site on the aromatic ring (Equations (89) and (90)) ⟨93JOC6538⟩. Application of this chemistry to decahydroquinolines likewise produces different—and equally valuable—results. While both groups direct the alkyllithium to effect deprotonation at C-2 in the *trans*-isomer (deprotonation at C-8a, giving a tertiary organolithium intermediate, is not observed), methylation of the formamidine-based anion leads to the equatorial product predominantly (in contrast, transmetallation with 1-pentynylcopper leads to reversal in stereoselectivity) whereas the *t*-BOC stabilized anion gives largely axial alkylation. The *cis*-perhydroquinoline gives essentially pure equatorial alkylation regardless of the auxiliary (in this case, transmetallation with the alkynyl-copper reagent results in loss of stereocontrol). Unfortunately, alkylation with *n*-iodopropane failed and electron transfer processes intervened giving the 2,3-enamine; "nucleophilic displacement on alkyl groups larger than methyl (except allylic or benzylic) appeared to be futile" ⟨93JA6652⟩.

Introduction of the *n*-propyl sidechain was effected instead by allylation and subsequent catalytic reduction.

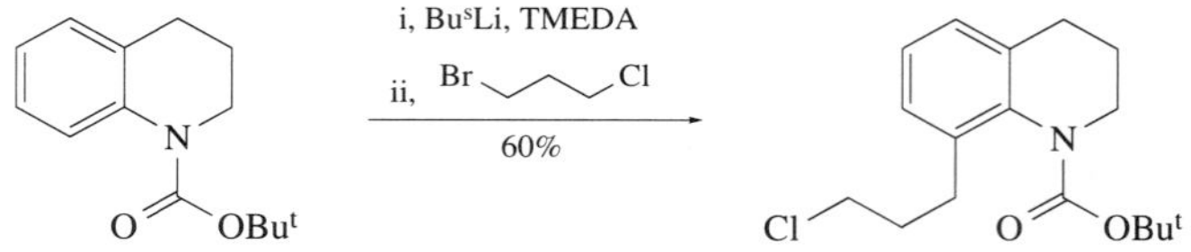

(89)

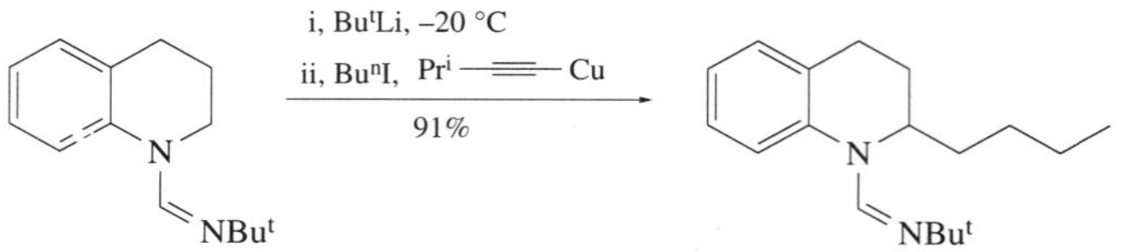

(90)

Dipole-stabilized carbanions can also be prepared by metallation of 2-(1-piperidinyl)oxazolines ⟨89JOC175⟩. Deprotonation is effected smoothly at $-23\,^{\circ}$C with Bu^{s}Li in the presence of TMEDA; however, the alkylations are disappointing. Methylation occurs in only 30% yield (giving a single diastereoisomer), and coupling either does not occur with other haloalkanes or occurs in negligible yields and with virtually no stereoselectivity. Transmetallation of the intermediate lithio species with cuprous cyanide gives a better yield of methylated material (53%), however, the previously observed stereoselectivity is lost. The mechanism of coupling with nonmethyl substrates almost certainly involves single electron transfer processes, as alkylation with a radical probe (5-hexenyl bromide) gives a mixture of hexenyl and cyclopentylmethyl-substituted compounds as well as disproportionation products (Scheme 44).

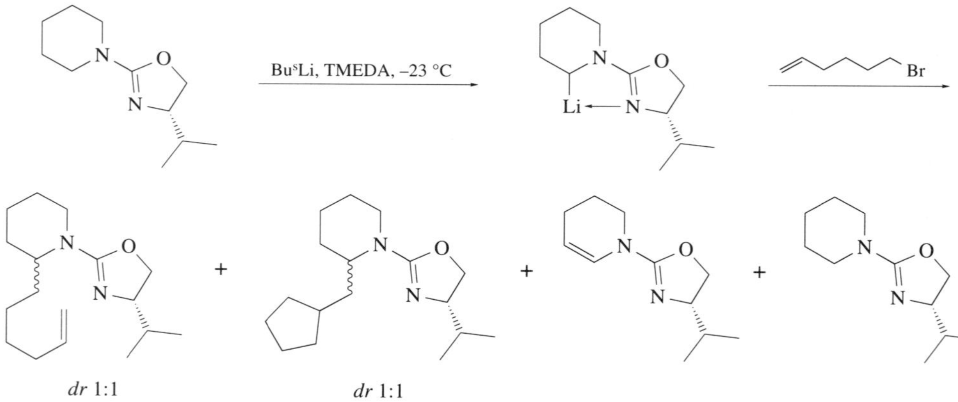

Scheme 44

Nitrogen substituents can participate in the stabilization of negative charges not only by inductive mechanisms as outlined above, but also by delocalization of the electron density into unsaturated functional groups such as imines, hydrazones, oximes, and nitroalkanes. Formation of monoanions or dianions provides a number of procedures for alkylation either α or β to the nitrogen substituent. Perhaps the simplest example involves deprotonation of *N*-alkyl imines, which is readily achieved using bases such as LDA. Subsequent alkylations occur in high yields giving masked ketones or aldehydes, as exploited by De Kimpe *et al.* in a synthesis of cyclic imines (Scheme 45) ⟨93TL4693⟩.

Chiral auxiliaries can be readily incorporated into the *N*-alkyl moiety, leading to facial discrimination in the alkylation of the intermediate imine anion. A number of groups, including Meyers *et al.*, have shown that these imines give chiral, nonracemic lithioenamines on deprotonation which can be alkylated in high chemical and optical yields (see ⟨81JA3081⟩ for leading references). For example, Meyers *et al.* demonstrated that cyclohexanone and cycloheptanone imines could be alkylated in 62–87% yield and with 80–99% enantiomeric excess ⟨81JA3081⟩. Unfortunately, cyclooctanone and indanone imines racemized extensively on hydrolysis, and larger rings (Equation (91)) and acyclic imines gave variable results (*ee* 18–97%) ⟨81JA3088⟩.

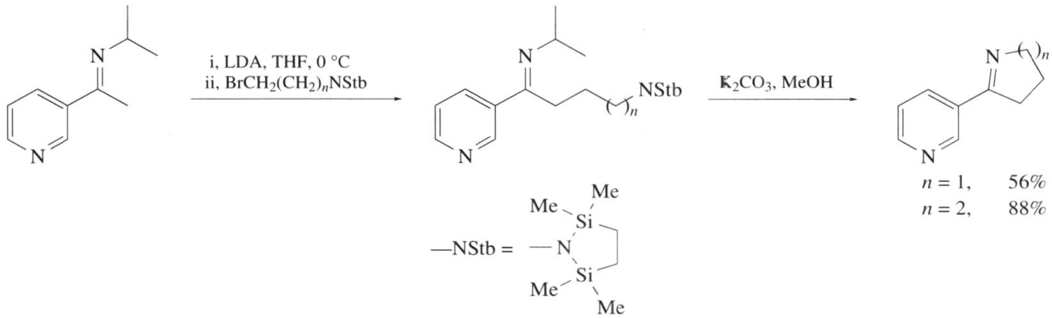

Scheme 45

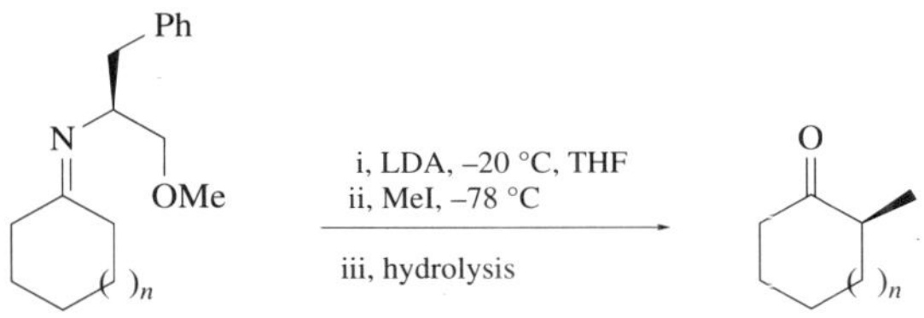

(91)

$n = 1$,	65%	*ee* 87%
$n = 2$,	90%	*ee* 85%
$n = 5$,	71%	*ee* 31%
$n = 10$,	75%	*ee* 37%

In a similar manner to oximes *N*,*N*-dialkylhydrazones act as carbonyl equivalents and provide an effective route for α-monoalkylation. Deprotonation can be effected with strong bases such as LDA, and alkylation of conformationally rigid hydrazone monoanions occurs with excellent axial selectivity (Equations (92) and (93)) ⟨85JA2078⟩. This result is in dramatic contrast with the alkylation of analogous ketone enolates where only modest selectivity is observed (see ⟨85JA2078⟩ for leading references). The improved selectivity with hydrazone anions is believed to be due to a combination of stereoelectronic effects and stereoselectivity in lithium coordination; aggregation of the intermediate organolithium species is not believed to play a major role in the outcome of the reaction. Metallation of unsymmetrical hydrazones usually occurs on the least substituted carbon atom, unless additional stabilization of the incipient negative charge is present; for example, the hydrazone of phenylacetone is metallated on the benzylic carbon atom ⟨76TL3⟩.

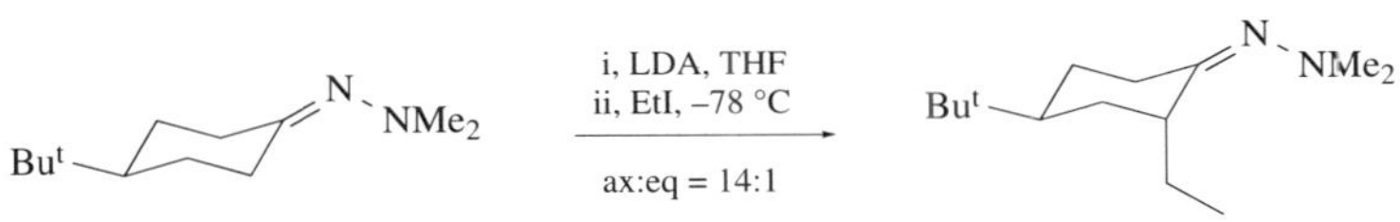

(92)

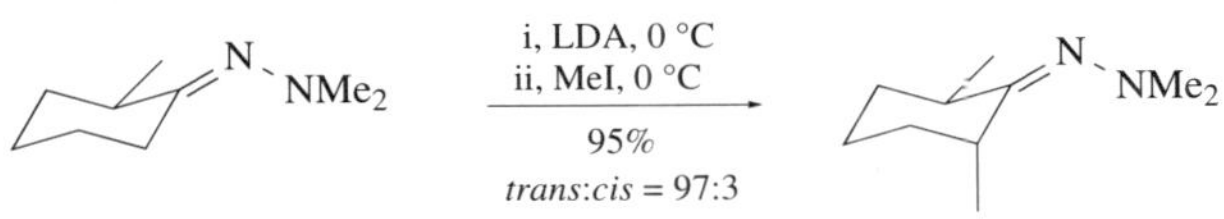

(93)

The development of hydrazones into useful chiral auxiliaries has been pioneered largely by Enders *et al.* ⟨B-84MI 104-02, 84T1345⟩. Ready access to both (*S*)- and (*R*)-enantiomers of 1-amino-2-methoxymethylpyrrolidine (*S*AMP and *R*AMP) from (*S*)-proline and (*R*)-glutamic acid respectively has provided a simple and successful route to either epimer of a range of chiral compounds. Deprotonation of the hydrazones can generally be effected in high yield with LDA, and alkylations at low temperature give the α-substituted hydrazones in excellent yields and diastereoselectivity. Cleavage of the hydrazones, usually by ozonolysis, regenerated the optically active substituted ketone or aldehyde (Equation (1), for example).

In contrast to *N*,*N*-dialkylhydrazones, Baldwin and co-workers have shown that *N*-monoalkylhydrazones are deprotonated readily on treatment with butyllithium, giving resonance-stabilized anions that can be alkylated selectively (but not exclusively) on the sp^2 hybridized carbon atom. Aldehyde *t*-butylhydrazones act as acyl anion equivalents, giving ketones in variable yields (15–

83%) on hydrolysis (Equation (94)) ⟨83CC1040⟩. Alkylation of the anion with iodomethane gives predominantly *N*-alkylated material; however, use of bulkier haloalkanes results in *C*-substituted products. Ketone hydrazones have been developed as α-amino carbanion equivalents ⟨86CC176⟩; alkylation of the intermediate lithio species gives a stable azo compound which can be reduced homolytically to give a tertiary alkane (Scheme 46). Alternatively, treatment of the azo compound with TFA results in loss of isobutene and formation of a benzophenone hydrazone which can either be hydrolyzed to give a hydrazine or hydrogenolyzed to provide a substituted amine.

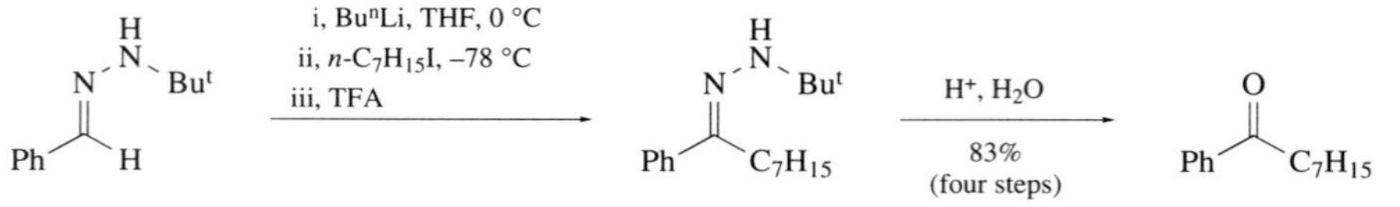

(94)

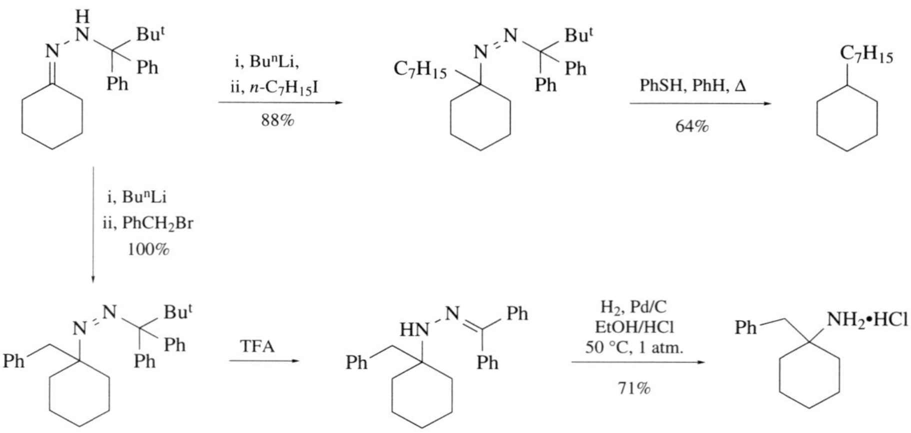

Scheme 46

While the hydroxyl proton is the most acidic hydrogen in simple oximes and addition of bases will readily give the oxy anion, this does not preclude a second deprotonation α to the sp^2 hybridized carbon atom on reaction with strong bases such as Bu^nLi ⟨69JA676, 76JOC439⟩. Anion formation with 2 moles of an organolithium base gives a *C,O*-dilithio species; the site of the second deprotonation (and hence the subsequent alkylation) is dependent upon the configuration of the starting oxime, occurring *syn* to the oxime oxygen indicating that the metallation is directed by chelation. These dianions can be alkylated on carbon with primary and secondary haloalkanes ⟨76JOC439, 76TL1439⟩ in good to excellent yields. An important and useful aspect is that the intermediate dilithio species are configurationally stable, and therefore a second metallation and alkylation can be carried out leading to a geminally dialkylated product (Equation (95)). However, this procedure cannot be extended to the alkylation of a tertiary center, as the lithiation step fails. Where the hydroxyl group is *syn* to a tertiary carbon atom, the *anti* position is metallated slowly in preference and the resulting anion can be alkylated in poor yields ⟨76TL1439⟩. Aldehydes can also be alkylated smoothly via the oxime dianions ⟨84TL263⟩;acetaldehyde oxime dianions can be alkylated in excellent yields. Metallation is effected at −78 °C and the dianion can be alkylated efficiently. In a similar manner to the ketone oximes, repetition of the procedure results in α,α-dialkylated aldehyde oximes (Scheme 47).

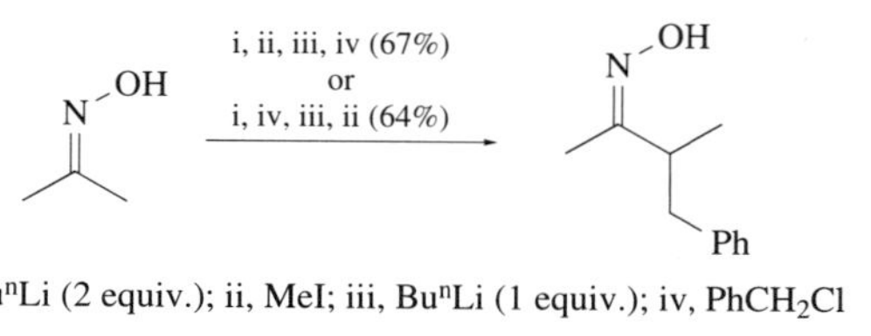

(95)

i, Bu^nLi (2 equiv.); ii, MeI; iii, Bu^nLi (1 equiv.); iv, $PhCH_2Cl$

Both cyano and isocyano groups are capable of stabilizing a negative charge; whereas the cyanomethyl anion is essentially planar indicating substantial delocalization with much of the negative charge on nitrogen ⟨87JA5996⟩, the isocyanomethyl anion has a pyramidal carbon with the

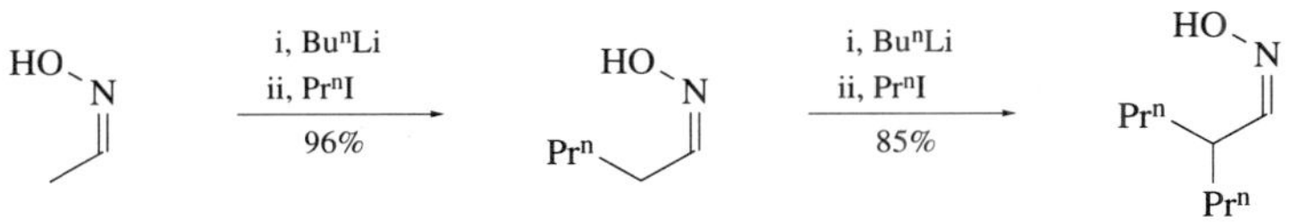

Scheme 47

charge localized on the CH_2 group ⟨87JA6004⟩. The various reactions of α-metallated isocyanides have been reviewed by Schöllkopf ⟨77AG(E)339⟩. Methyl isocyanide is deprotonated with *n*-butyllithium and alkylated in variable (27–70%) yields; however, the introduction of added stabilization for the carbanion in, for example, benzyl isocyanide, allows deprotonation with potassium *t*-butoxide and subsequent alkylations to occur in consistently higher yields ⟨77LA40⟩. Optically pure cyclopropyl isocyanides can be deprotonated at low temperature (LDA, −78 °C) and alkylated with iodomethane in high chemical and optical yields (Equation (96)) ⟨74JA3711⟩. While the anion appears to be configurationally stable below −50 °C, epimerization is observed at higher temperatures. Significantly, the configurational stability is independent of alkali metal (Li, Na, K) and is unaffected by added complexing agents such as crown ethers, TMEDA or HMPA. However, this behavior is not extended to acyclic isocyanides which are racemized to a large extent at temperatures where the cyclopropyl anion is resistant to epimerization ⟨77JA2631⟩. Nevertheless, the high barrier to inversion observed for the isocyanocyclopropyl anion is remarkable: analogous cyanocyclopropyl carbanions are racemized completely at −78 °C in aprotic solvents ⟨55JA6026⟩.

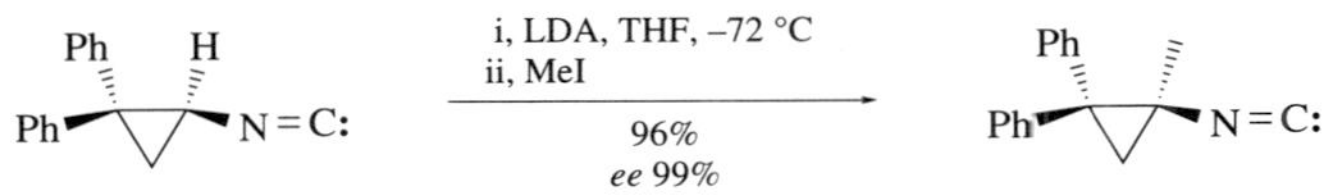

(96)

The cyano group has a modest acidifying effect on α-hydrogen atoms (the pK_a of acetonitrile has been measured as 31.3 in DMSO ⟨75JA7006⟩) and therefore relatively strong bases are required to effect deprotonation unless additional stabilization is available (for example, cyanofluorene is considerably more acidic, with a measured pK_a of 8.3 in DMSO ⟨75JA7006⟩). Methods for the preparation of nitrile-stabilized carbanions, and their subsequent reactions, have been reviewed extensively ⟨84OR(31)1⟩, and many examples of alkylations can be found in the tables accompanying that review. Strong bases such as LDA or organolithium reagents are frequently employed for anion generation, although there are occasional reports of the use of Grignard reagents. For example, Fauvarque *et al.* showed that the monoanions prepared with Grignard reagents can be monoalkylated in good yields, although some dialkylated material (<20%) was also recovered ⟨68CR(C)1162⟩. However, nucleophilic attack by the organolithium, or magnesium, reagent on the nitrile group is an alternative reaction that often takes place ⟨90S401⟩. In general, it is desirable to generate the carbanion rapidly and with high conversion to avoid side reactions such as condensations with neutral nitrile molecules. Addition of a second equiv. of strong base such as Bu^nLi or Bu^tLi can provide a dianion, often described as a ‘geminal’ dilithio species but more correctly represented as a *C*,*N*-dilithiated ketenimine. While dialkylation via this species is possible (Equation (97)) ⟨73JOM(47)417⟩, an alternative pathway involves formation and alkylation of a monoanion, deprotonation of the monoalkylated product and final alkylation of this second monoanion. This pathway can be especially relevant if the reaction between the base and the haloalkane is relatively slow ⟨74TL1377, 84OR(31)1⟩.

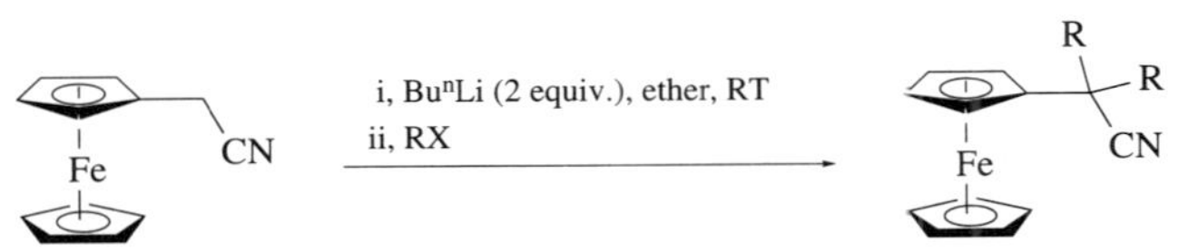

(97)

R	MeI	BuBr	$PhCH_2Cl$	$(CH_2Cl)_2$
Yield (%)	60	56	70	53 (cyclopropyl)

The nitro group has a remarkable acidifying effect on α-hydrogen atoms (for example, compare the value for CH_3NO_2, $pK_a = 11$, with acetone, $pK_a = 20$, ⟨53JA2439, 78JOC3095⟩), and the corresponding nitronates can be prepared quantitatively using relatively weak bases. These anions are important synthetic intermediates as the nitro group can be readily reduced to the amine ⟨88TL5733⟩ or hydrolyzed to a carbonyl function using either Nef or McMurry conditions ⟨73JOC4367⟩, and therefore the alkylation of nitronates would prove to be an important synthetic method. However, while some alkylations proceed in acceptable yields ⟨54JA3972⟩, the anions can act as ambident nucleophiles and reactions with haloalkanes are frequently dominated by *O*-alkylated products. Allyl halides react with sodium or potassium nitronates by this route giving carbonyl compounds following decomposition of the first-formed *aci*-nitro structures ⟨57HCA1250, B-69MI 104-01, 85SC1123⟩. Palladium (0) catalysis allows this reaction to become a useful procedure for oxidation of allyl chlorides to the corresponding α,β-unsaturated carbonyl compounds (Equation (98)) as *C*-alkylated products are produced in less than 10% yield ⟨85SC1123⟩. Successful *C*-alkylations were obtained by Kitajima and co-workers in a synthesis of polysubstituted 1-indanones ⟨93T537⟩; however, the desired carbocycles were isolated only with relatively hindered substrates where the aromatic nucleus was fully substituted. Substrates that were unsubstituted *ortho* to the nitromethyl group gave seven-membered heterocycles exclusively, resulting from *O*-alkylation (Scheme 48).

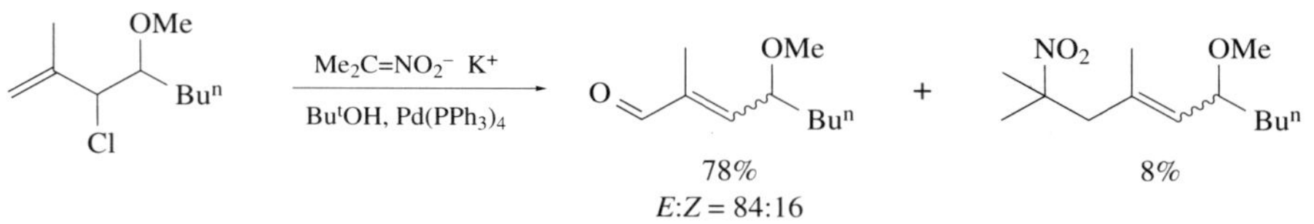

(98)

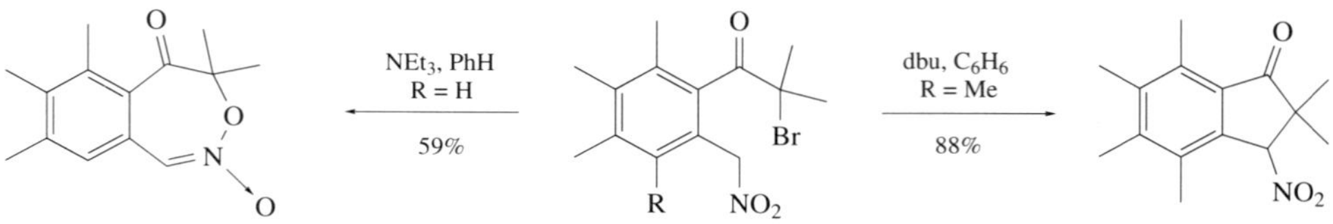

dbu = 1,5-diazabicyclo [5.4.0] undec-5-ene

Scheme 48

Relatively hindered nitrobenzyl halides are known to react with the 2-nitropropane anion to give *C*-alkylated products; however, the reactions do not proceed via conventional S_N1 or S_N2 mechanisms. Instead, a single electron transfer process intervenes where an intermediate benzyl halide radical anion loses a halide ion giving a benzyl radical which couples with the nitronate ⟨66JA5663⟩. A range of substrates have since been shown to give *C*-alkylated material by this process (Equation (99)), although the yields and product distributions are sensitive to reaction conditions (for example, see ⟨91TL4125⟩). Dickens and Luche have shown that sonication can affect the course of SET reactions ⟨91TL4709⟩; while *O*-alkylation can predominate in the absence of irradiation, sonication can increase the proportion of *C*-alkylated material dramatically.

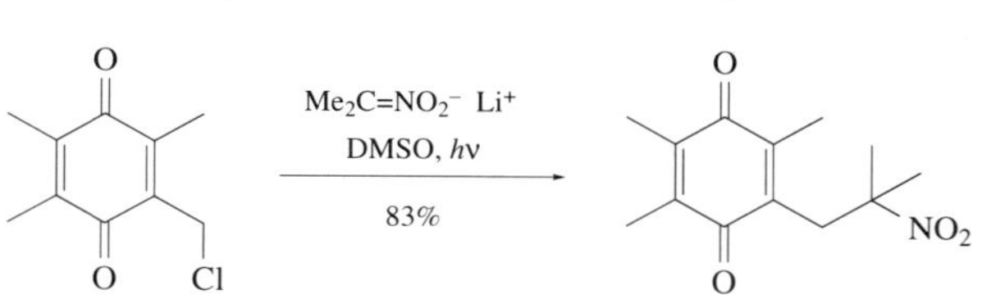

(99)

The nitro group is so effective at stabilizing negative charges that dianions can be formed readily on treatment with a second equiv. of a strong base such as Bu^{n}Li or Bu^{t}Li. 2-Aryl-1-nitroethanes can be deprotonated at the α- and β-positions (relative to the nitro group) giving a dianion that can be considered as a "super enamine." Alkylation occurs in good yields (56–76%) β to the nitro substituent ⟨76HCA2213, 79CB234⟩. Open chain and cyclic nitroalkanes can likewise be doubly deprotonated using one equivalent each of Bu^{n}Li and Bu^{t}Li in THF/HMPA or DMPU (*N*,*N*′-dimethylpropyleneurea); unsymmetrical nitroalkanes are deprotonated exclusively on the less substituted β-carbon atom (a kinetic "enolate"). However, the reactivity of these dianions with haloalkanes is limited: benzyl and cinnamyl bromide are alkylated in only 30–40% yields (Equation (100)) and allyl bromide and *n*-hexyl iodide failed to react ⟨77TL1161, 86CB575⟩.

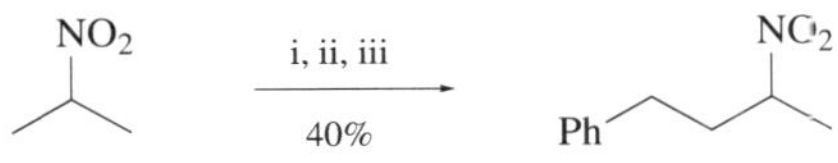

(100)

i, Bu^nLi; ii, Bu^tLi; iii, $PhCH_2Br$

(Phenylthio)nitromethane can be doubly deprotonated on the α-carbon atom and alkylated with haloalkanes in good yields ⟨76AG(E)505⟩; however, the poor nucleophilicity of the corresponding monoanion is demonstrated by the reaction with 1,5-dibromopentane (Scheme 49), where no cyclic material is isolated ⟨77TL1161⟩.

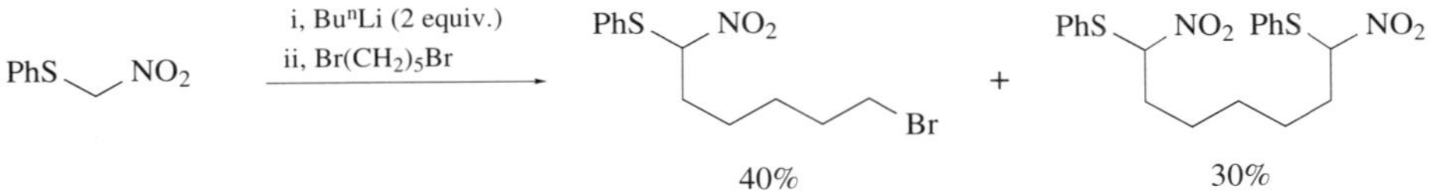

Scheme 49

1.04.3.6 Phosphorus Stabilized Carbanions

Phosphorus in both its tri- and pentavalent states should be capable of stabilizing a negative charge on an α carbon atom, and carbanions with a variety of different phosphorus-containing substituents are known and their alkylation reactions have been reported. Although the overwhelming majority of publications detailing these carbanions are in the area of Wittig and related alkene syntheses, organophosphorus reagents can be alkylated, and this is a legitimate route to higher alkyl phosphines and related species. For instance, in an early report Wittig and Rieber demonstrated that a tetramethylphosphonium salt could be deprotonated with phenyllithium to give an ylide which could then be alkylated with iodomethane (the yield was not reported) ⟨49LA(562)177⟩. A number of intramolecular examples have been described by Bestmann and Häberlein ⟨62ZN(B)787⟩ and by Märkl ⟨63AG(E)153, 63ZN(B)84⟩ (Schemes 50 and 51). An efficient synthesis of 1,5-dienes using phosphorane reagents has been devised by van Tamelen and co-workers ⟨70JA2139⟩. Deprotonation of the allylphosphonium salt (**34**) gave a resonance-stabilized phosphorane (**35**) which reacted cleanly with an allyl bromide to give the alkylated phosphonium salt. Only α-substitution was observed and the alkene geometry was retained in both components (Scheme 52).

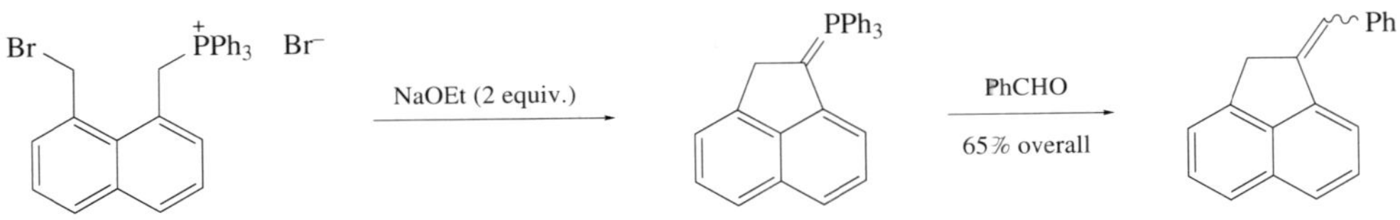

Scheme 50

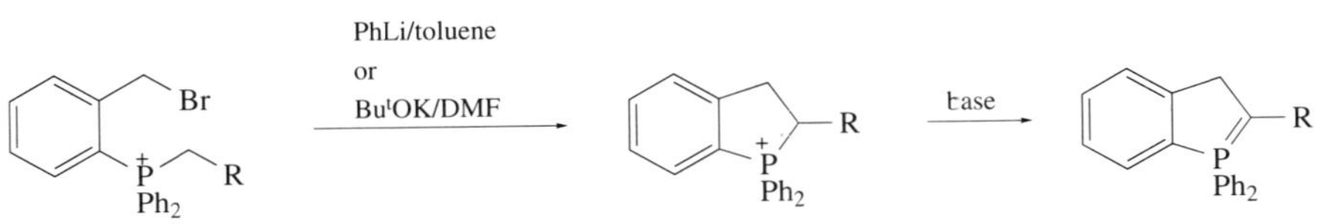

R = H, Ph

Scheme 51

Direct deprotonation of the α carbon in a phosphine using alkyllithium bases does not appear to be a useful procedure, as yields of the corresponding carbanions are variable ⟨65JOC1939, 66JOC2373⟩; indeed, the displacement of alkyl substituents such as benzyl from phosphorus has been reported ⟨76JA4805⟩. An exception is the use of Bu^nLi/TMEDA which is reported to deprotonate methyldiphenyl phosphine satisfactorily ⟨85IC3680⟩. However, complexation of borane to the phosphorus atom provides an excellent method for activating α hydrogen atoms in a variety of phosphines, and also for protection of the phosphorus against oxidation and alkylation. Although phosphine–boranes can be prepared by the reaction of a phosphine with borane, a more convenient procedure

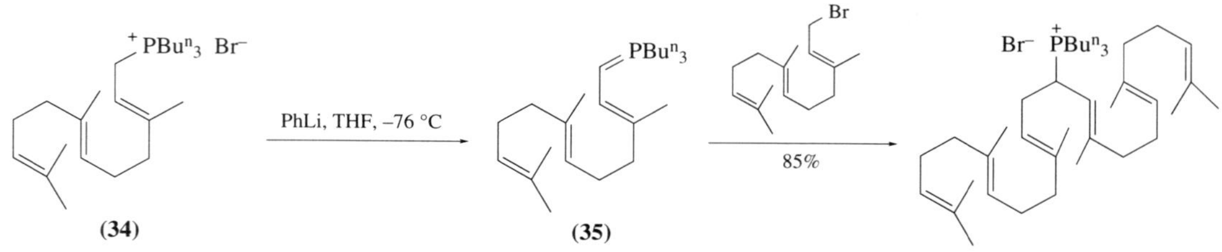

(34) **(35)**

Scheme 52

has been developed by Imamoto *et al.*: secondary and tertiary phosphine oxides undergo deoxygenation and *in situ* conversion to the phosphine–borane adducts on treatment with a mixture of $LiAlH_4$, $NaBH_4$, and $CeCl_3$ (approximate ratio 1.2 : 3 : 3 with respect to phosphine oxide), with yields generally falling in the range 62–96% ⟨85JA5301, 90JA5244⟩. The cerium chloride plays a dual role of activating the phosphine oxides towards reduction by $LiAlH_4$, and activating the $NaBH_4$ for formation of the phosphine–borane adduct ⟨85JA5301⟩. This method has the advantage of employing relatively stable phosphine oxides as starting materials, and allowing their direct conversion to the adduct without requiring isolation of (relatively) air-sensitive phosphines. While Schmidbaur *et al.* originally demonstrated that diphosphinomethane compounds, on complexation to 2 mol. BH_3, could be deprotonated and alkylated ⟨79AG(E)782⟩, Imamoto *et al.* have extended this early work to provide a general method for alkylation α to phosphorus. Deprotonation of $Ph_2P(BH_3)Me$ was readily effected using alkyllithium reagents or LDA, and addition of haloalkanes (allyl bromide, *n*-butyl bromide) gave alkylated products in good yields (≈60–70%). Alternatively, addition of $CuCl_2$ gave a dimeric product as a result of oxidative coupling (Scheme 53). The borane protecting group could be removed by addition of diethylamine or morpholine; the phosphine is regenerated with complete retention of configuration at phosphorus. The analogous reactions did not work with secondary phosphine–borane complexes, as the borane activated the P—H proton and it was abstracted with relatively weak bases such as KOH (Equation (101)) ⟨90JA5244⟩. Le Corre and coworkers have used this method for the preparation of phosphine–borane complexes with more substituted carbon centers attached to phosphorus (Scheme 54) ⟨93TL1011⟩. Yields were in the range 57–80%.

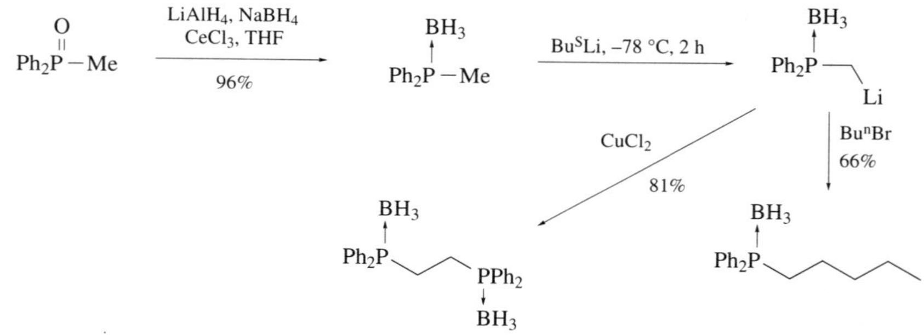

Scheme 53

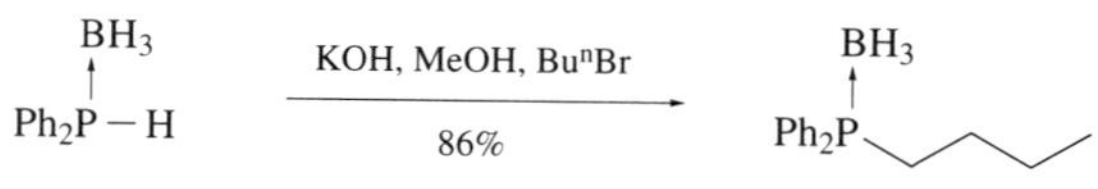

(101)

Phosphine oxides and phosphonates also exert an acidifying effect on neighboring hydrogen atoms. Phosphine oxides can be deprotonated on an α carbon using LDA ⟨87JOC2170⟩, and the methyl substituent on ethyl methyl phenyl phosphine oxide (**36**) was deprotonated selectively by *n*-butyllithium at −70 °C; quenching with benzyl bromide gave a single alkylated product ⟨89TL477⟩. However, introduction of a dimethylamino group into the ethyl side chain directs deprotonation (and therefore subsequent alkylation) into the C-2 group. Unfortunately, benzylation gave the monoalkylated phosphine oxide as a mixture of diastereoisomers (56%; 1.2 : 1); apparently the chiral phosphorus atom exerted little influence on the course of the alkylation step. In this case, the product was contaminated with some dialkylated material (24%) (Scheme 55).

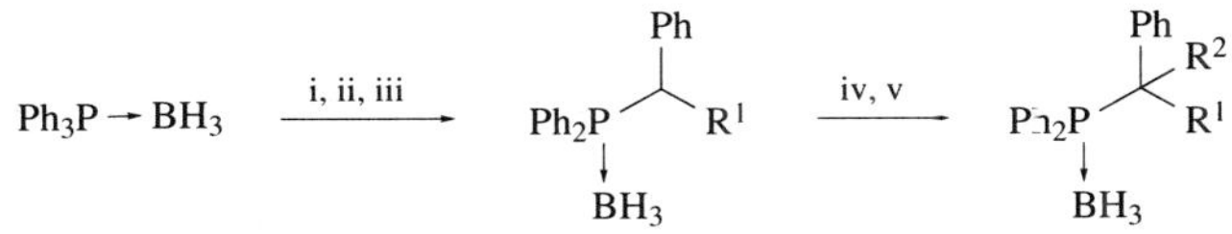

i, Li; ii, Me_3CCl; iii, $PhCH(R^1)X$; iv, Bu^nLi, 0 °C, THF; v, R^2X, –78 °C

R^1	R^2	*Yield* (%)
H	Me	74
H	Bu^n	80
CH_2Ph	Me	76

Scheme 54

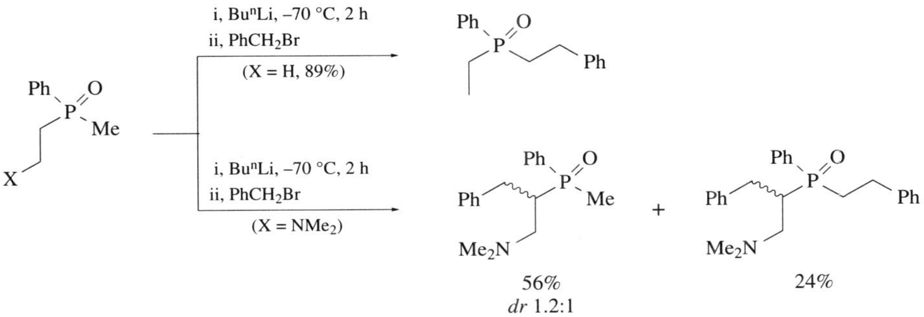

Scheme 55

Deprotonation of allyl phosphonates can be effected with *n*-butyllithium at $-60\,^\circ C$, and the resulting resonance-stabilized carbanions can be alkylated with a range of bromoalkanes; typically products are isolated in better than 75% yield ⟨74AG(E)407⟩. Coupling occurs exclusively on the α carbon atom. A novel approach to phosphorus-stabilized carbanions involves conjugate addition of organocopper reagents to vinylphosphorus compounds ⟨84JOC1522⟩. Reaction of lithium dibutylcuprate with the phospholene system (**37**) gave the carbanionic species (presumably a cuprate) which was alkylated cleanly with allyl bromide to give the disubstituted heterocycle (**38**) as a single diastereoisomer (Equation (102)); the chiral phosphorus controlled both the conjugate addition and the alkylation step. However, conjugate addition to, and alkylation of, the vinylphosphine oxide (**39**) gave the substituted phosphine oxide (**40**) as a 7 : 5 mixture of diastereoisomers (Equation (103)), implying limited stereocontrol in the alkylation of α-phosphinoyl anions in acyclic systems ⟨84JOC1522⟩.

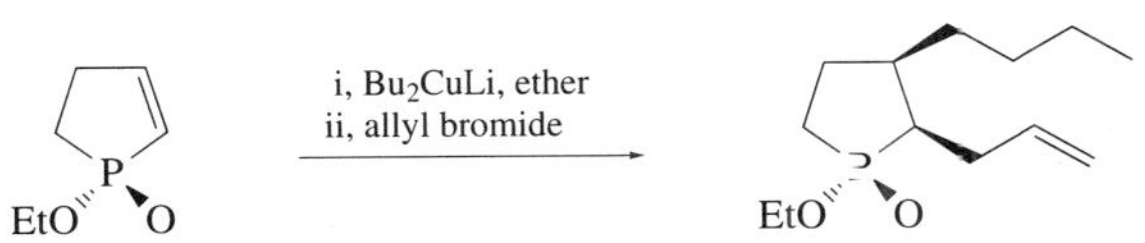

(**37**) (**38**)

single diastereoisomer (102)

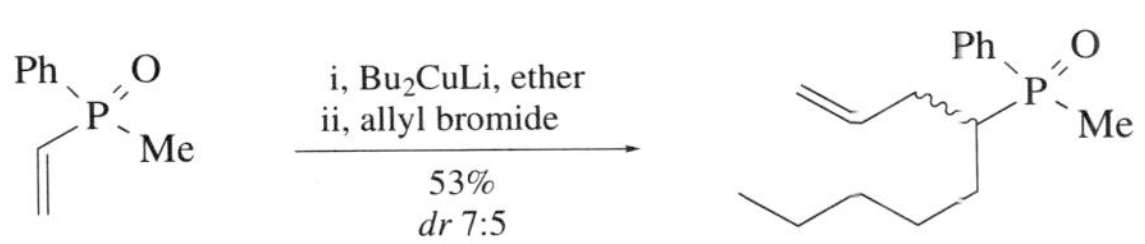

(**39**) (**40**) (103)

1.04.3.7 Silicon Stabilized Carbanions

Silicon is considerably less electronegative than carbon, and its highly polarizable nature means that a silyl substituent is capable of stabilizing a negative charge on an α carbon atom ⟨84JA6467⟩.

The predominant use of α-silyl carbanions features the addition to carbonyl compounds as the first step in the Peterson alkenation ⟨84S384, 90OR(38)1⟩. The review by Ager includes an extensive discussion of methods for the preparation of α-silyl carbanions, of which direct deprotonation with alkyllithium bases is the most direct but least successful approach, at least for simple alkyl silanes ⟨90OR(38)1⟩. Other methods include metal–halogen exchange (for example, Grignard reagent formation), transmetallation (such as Sn–Li interchange), single desilylation of geminal disilyl alkanes, and conjugate additions to alkenyl and alkynyl silanes. While direct deprotonation of alkylsilanes is an unreliable method for generation of carbanions, deprotonation and subsequent alkylation can be achieved in good yields in oxygenated systems where chelation to an oxygen substituent directs the base to an activated hydrogen atom (Equation (104)). The resulting α-silyl carbanion can be alkylated with iodomethane or with benzyl bromide in 74% and 60% yield, respectively ⟨92TL543⟩.

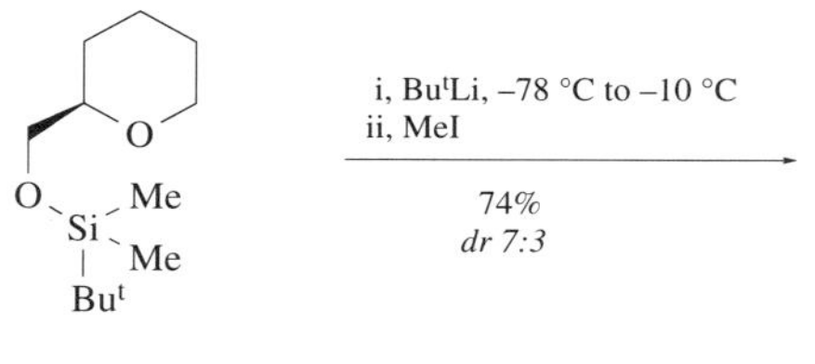

(104)

A range of silyl reagents of general constitution Z_3SiCH_2MgX has been developed as efficient alkylating and hydroxymethylating agents; the latter application is especially useful as treatment of the resulting alkylated silane with an oxidant (usually H_2O_2 or mcpba) in the presence of fluoride causes replacement of the silyl substituent with a hydroxyl with retention of configuration at the carbon center. Trialkoxysilylmethyl chlorides can be converted to the corresponding Grignard reagents and alkylated in good yields (62–70%) ⟨93JOM(451)C1, 93TL2111⟩; although the tris(ethoxy) derivatives slowly undergo self-condensation at room temperature, they are stable at −20 °C for extended periods and can even be manipulated successfully at 0 °C. The diisopropoxy(methyl) silylmethyl Grignard reagent was developed by Tamao *et al.* as a nucleophilic hydroxymethylation agent ⟨83JOC2120⟩. Coupling with *n*-alkyl and allyl halides occurred in excellent yields (>90%) in the presence of 10 mol.% of CuI catalyst in THF, and the subsequent oxidative desilylation was carried out with H_2O_2/KF in DMF. This process was extended to α-silyl carbanions derived from conjugate additions to vinyl silanes ⟨84TL1905, 84TL1909, 84TL1913⟩. A range of ligands on silicon was examined, with various amino and alkoxy groups being tested for their stability to organolithium and Grignard reagents ⟨84TL1905⟩. Yields of alkylated materials were generally good (60–70%) ⟨84TL1909⟩. The ability to introduce different ligands on silicon resulted in the development of an asymmetric variant on the reaction; however, the levels of optical induction obtained using ephedrine-derived systems were variable (*ee* 2–60%) and never outstanding (Scheme 56) ⟨84TL1913⟩.

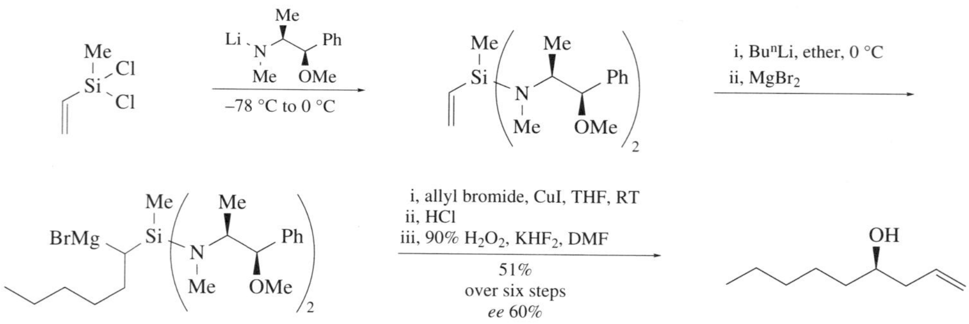

Scheme 56

The addition of Grignard reagents to 1-trialkylsilyl-1-alkynes is a regio- and stereoselective reaction which results in an α-silylalkenyl Grignard reagent which can be alkylated. Alkylcopper compounds, generated from Grignard reagents and CuBr, add in a conjugate manner to ethynylsilanes in the presence of triethyl phosphite; the resulting α-silyl alkenylcopper species can be alkylated with *n*-iodoalkanes in moderate to good yields (39–71%). The isomeric purity of the products is excellent (>99%), with resultant *syn* addition of the alkyl groups ⟨77TL1805⟩. Utimoto and co-workers have reported an intramolecular version of the reaction, which proceeds in excellent yields for the synthesis of five- and six-membered rings (Scheme 57) ⟨84TL1999⟩. Allylation of the vinyl Grignard reagent proceeded without need for copper catalysis. Attempted extension of the method to smaller ring sizes was less successful; the *exo*-alkylidenecyclobutane was isolated in only 9% yield, accompanied by alkylated open chain product (72%).

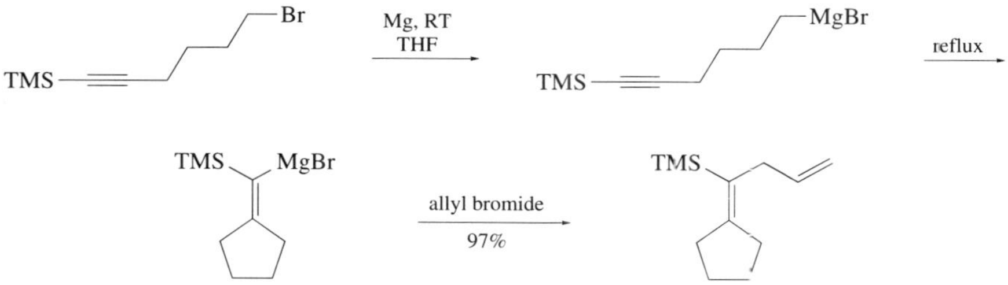

Scheme 57

In addition to the carbometallation of vinylsilanes, the analogous hydrometallation reactions are equally regio- and stereoselective. Hydromagnesiation of 1-trimethylsilyl-1-alkynes occurs on treatment with isobutylmagnesium bromide in the presence of a catalytic amount of Cp_2TiCl_2 (5 mol.%) in ether ⟨83TL1041⟩. Hydride is added regiospecifically β to the silicon center resulting in an α-silyl Grignard reagent which can be alkylated directly with reactive (methyl, allyl) halides, or with *n*-alkyl iodides under copper(I) catalysis giving substituted vinylsilanes with excellent E/Z selectivity (Equation (105)). Alternatively, hydroboration of alkynylsilanes with dicyclohexylborane gives vinylboranes which can be converted to the alkenyl borates by treatment with MeLi ⟨76JOC2941⟩. These borates can be alkylated directly in better than 70% yield with reactive halides, or in variable yield with *n*-alkyl halides following transmetallation with CuI (Equation (106)). In most cases excellent *syn* additions are observed. Hydroalumination reactions follow a similar course ⟨76JOC2214, 76JOC2215⟩. For example, a stereoselective synthesis of *Z*-1,2-dialkylalkenylsilanes may be achieved in good overall yields (72–84%) by the following sequence: hydroalumination with diisobutylaluminum hydride (dibal—H), formation of the *ate* complex on addition of MeLi, transmetallation with CuI·P(OEt)$_3$, and finally addition of the alkyl bromide or iodide (methyl, ethyl, butyl) ⟨84TL131⟩. In all cases, the *Z*-stereoselectivity was ≥94% (i.e. *syn* addition of H and alkyl) (Equation (107)).

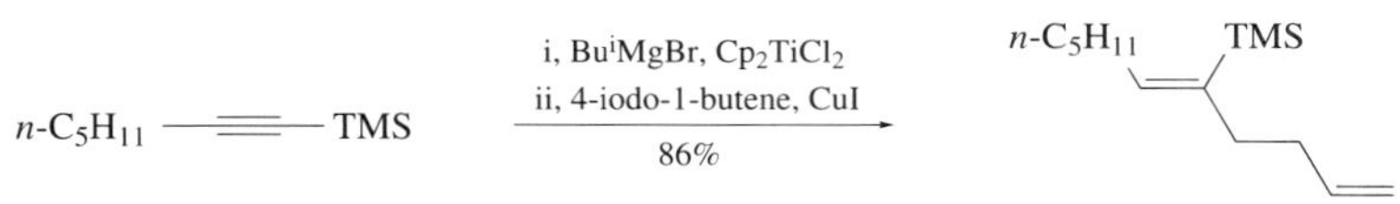

(105)

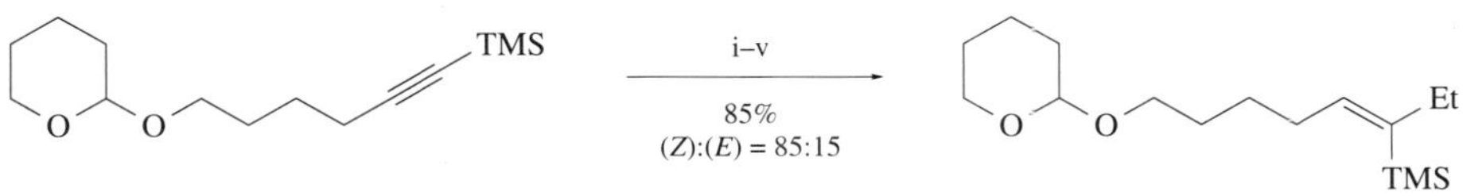

(106)

i, (*c*-C$_6$H$_{11}$)$_2$BH; ii, MeLi; iii, CuI; iv, P(OEt)$_3$, HMPA; v, EtI

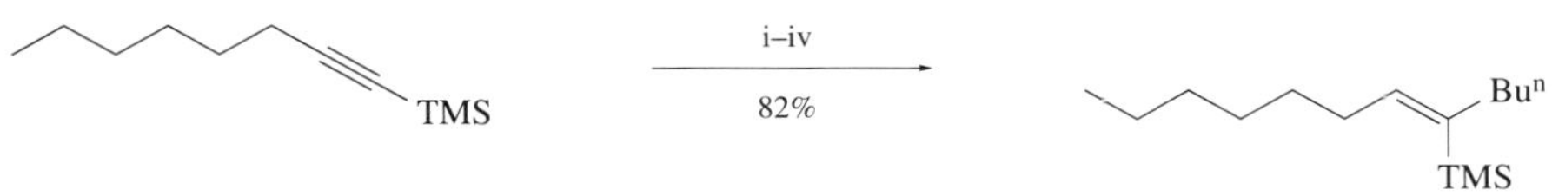

(107)

i, Dibal-H, ether; ii, MeLi; iii, CuI•P(OEt)$_3$, THF. iv, Bu^{n}I

Direct deprotonation α to silicon, and alkylation in high yields, can be achieved for allyl and benzyl groups attached to silicon. Bases such as Schlosser's base (Bu^{n}Li/KOBu^t), *s*-butyllithium, and LDA give delocalized carbanions which can be alkylated regioselectively depending on the nature of the groups attached to silicon and on the size of the haloalkane employed. Mixtures of products usually resulted, however, smaller groups on silicon and those with metal-complexing capabilities often gave higher proportions of α-alkylated products while bulkier silicon ligands favored γ-substituted products ⟨89JOC317, 91TL2879⟩. By replacing one of the groups on silicon with a chiral chelating ligand, highly enantioselective syntheses of benzyl ⟨89JA8737⟩, allyl ⟨89JOC317⟩, and propargyl ⟨93TL1449⟩ alcohols can be realized (Scheme 58). The introduction of a chelating ligand also heavily favored α-substitution.

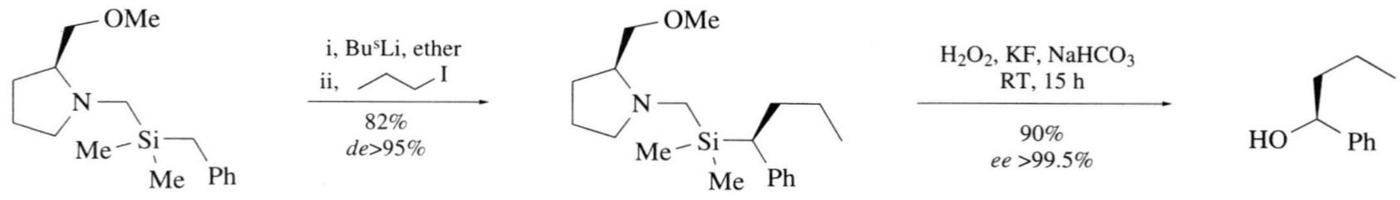

Scheme 58

1.04.3.8 Boron Stabilized Carbanions

Although the main uses of boron reagents in organic synthesis focus on reduction, hydroboration, and boron enolate chemistry, the formation and reaction of anions α to a boron atom have also played an important role. An *ab initio* study by Pross *et al.* has shown that the strong π-accepting nature of the boron leads to stabilization of a negative charge on an adjacent carbon center by overlap of the filled *p*-orbital on carbon with the vacant *p*-orbital on boron ⟨81JOC1693⟩. While the inherent stability of α-boryl carbanions suggests that deprotonation may provide a suitable route to these species, in practice a serious drawback is the reaction of the base at boron to give an *ate* complex rather than a carbanion. One solution to this problem that works well is the use of sterically hindered, nonnucleophilic bases which cannot form stable complexes with the organoborane. Thus Rathke and Kow have shown that amide bases such as LITMP will effectively deprotonate α to boron: *B*-methyl-9-borabicyclononane (*B*-methyl-9-BBN-H) can be alkylated on carbon by deprotonation and subsequent treatment with 1-bromobutane ⟨72JA6854⟩. Oxidation with alkaline H_2O_2 gave 1-pentanol in 42% yield (Scheme 59). One reason for the moderate yield is that incomplete deprotonation occurred; 2 mol. equiv. of base and 24 h exposure gave only 75% anion formation (as indicated by deuteriation). Other *B*-alkyl-9-BBN substrates could be deprotonated by LITMP, although the yields of carbanions were much lower ⟨72JA6854⟩. In a subsequent report, Kow and Rathke reported that vinylboranes could be deprotonated to give allyl anions which could be quenched with a variety of electrophiles; the products were examined following oxidation with alkaline H_2O_2 ⟨73JA2715⟩. The regioselectivity of substitution depends on the nature of the electrophile; carbanions reacted with silyl and carbonyl reagents γ to boron, however, protonation and alkylation (with MeI) occurred predominantly α to boron (Scheme 60).

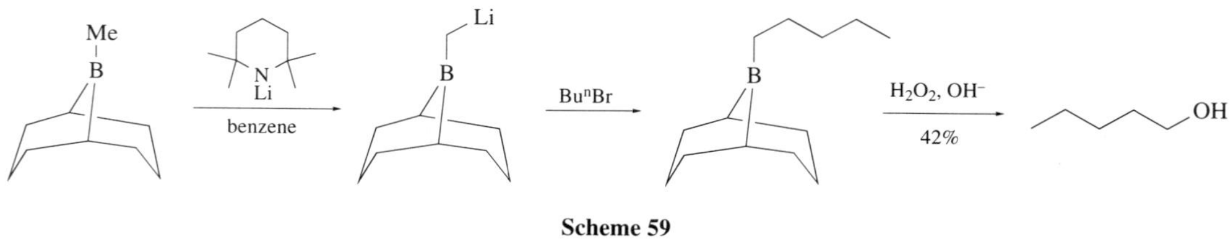

Scheme 59

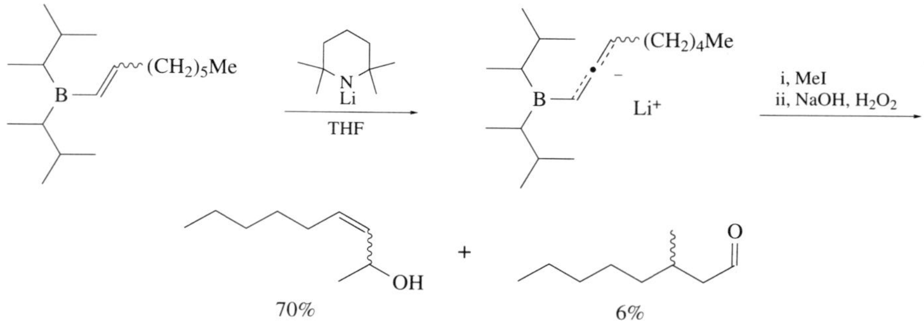

Scheme 60

An alternative approach, investigated initially by Wilson ⟨80JOM(186)297⟩ and subsequently by Pelter *et al.* ⟨83TL623, 83TL627, 83TL631⟩, uses sterically hindered ligands on boron to inhibit *ate* complex formation on reaction with moderately hindered bases; an added benefit is a decrease in the rate of aerial oxidation of the trialkylborane substrates. Dimesityl methyl borane (mesityl = 2,4,6-trimethylphenyl) and benzyl dimesityl borane could be readily deprotonated using hindered bases such as lithium dicyclohexylamide or mesityllithium; bases such as *n*-butyllithium and NaH gave *ate* complexes ⟨83TL623⟩. The boron-stabilized carbanions could be alkylated smoothly with primary haloalkanes; bromides and iodides gave the most satisfactory results (with yields being typically >85%), and chlorides were somewhat less reactive. The basic nature of the carbanion led to elimination rather than substitution when secondary haloalkanes were employed. The resultant alkylated boranes could either be oxidized to give primary alcohols (providing an efficient homologation procedure), or deprotonated again and alkylated to give dialkylated material ⟨83TL627⟩. Oxidation provided secondary alcohols in good to excellent yields (66–96%) (Scheme 61). In the analogous reaction with allyl dimesityl borane, deprotonation and subsequent alkylation gave the γ-alkylated product in a regio- and stereospecific manner for all haloalkanes except for iodomethane where some α-substitution was observed ⟨83TL631⟩. Oxidation converted the *E*-vinylboranes to aldehydes in excellent yields, and this procedure provides a useful three-carbon homologation sequence (Scheme 62). The γ-selectivity is in marked contrast to the earlier work of Kow and Rathke, where exclusive α-alkylation of the less sterically hindered *B*-allyl-9-BBN was reported ⟨73JA2715⟩.

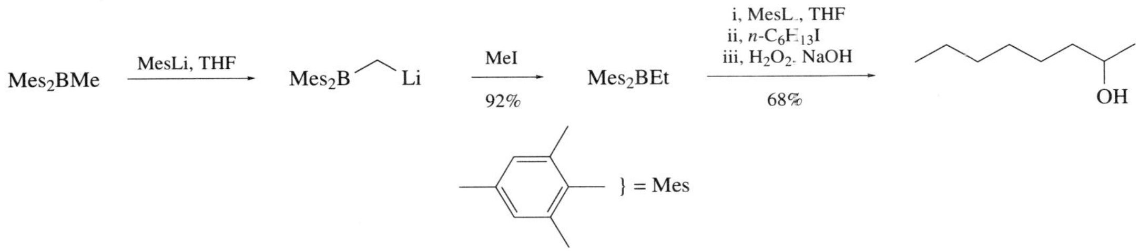

Scheme 61

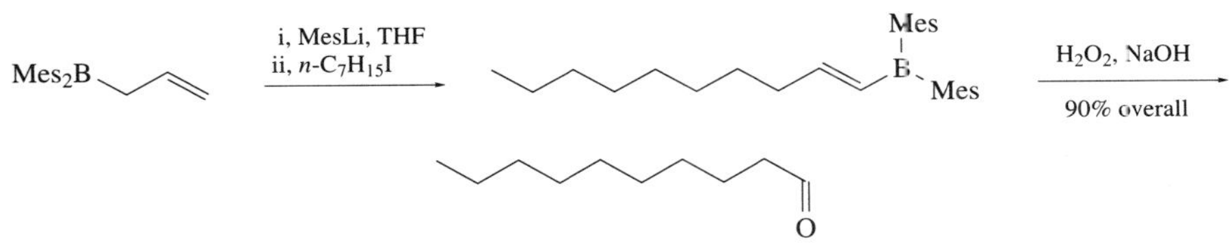

Scheme 62

Instead of blocking formation of *ate* complexes with bulky groups, a different approach to the problem uses heteroatom ligands (such as alkoxy) on boron that are capable of π-back donation, thus discouraging complexation of bases at boron. A simple and synthetically accessible example is the boronic ester, alkyl $B(OR)_2$. Unfortunately, carbanions stabilized by a single boronic ester cannot be prepared by deprotonation, a result in contrast to the facile deprotonation of diboronic esters by hindered bases such as LITMP ⟨77JA3196, 82OM20⟩. Similarly, while abstraction of a boryl group from geminal tri- and tetraborate esters leads to anionic species which are readily alkylated (for example, see ⟨70JOM(24)263, 76JOM(114)1⟩), attempts to cleave carbon–boron bonds in diboronic esters have been unsuccessful ⟨85OM1690⟩. However, destannylation of an α-trimethylstannyl boronic ester by treatment with MeLi at $-100\,^{\circ}C$ gave the α-lithio species; similarly treatment of an α-iodoboronic ester with Bu^{t}Li at $-100\,^{\circ}C$ also gave the organometallic reagent. Unfortunately, it failed to react with iodomethane ⟨85OM1690⟩. The successful alkylation of these α-boryl organometallic reagents has been reported by Knochel ⟨90JA7431⟩. Reaction of an α-haloboronic ester with zinc and subsequent transmetallation with CuCN·2LiCl gave a zinc–copper compound which reacted readily with a range of electrophiles including allyl bromides; yields were typically better than 80% (Scheme 63).

While geminal diboronic esters are unsatisfactory starting materials for the generation of stabilized carbanions, Zweifel *et al.* have developed a procedure using geminal diboranes as stable precursors to α-boryl anions ⟨73S37⟩. Dihydroboration of terminal alkynes provides a convenient and reliable

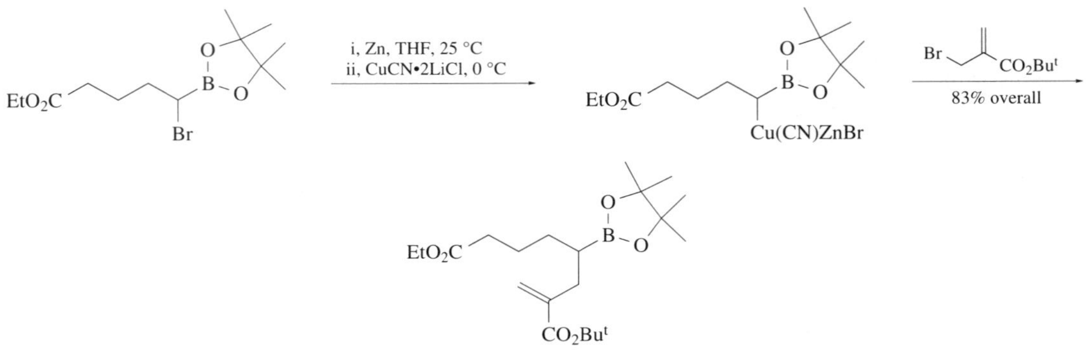

Scheme 63

entry to the sequence, and addition of 1 mole of methyllithium gives an *ate* complex which decomposes to yield a trialkylborane and intermediate α-borylcarbanion. Addition of a range of haloalkanes leads to alkylated products which can be oxidized under standard conditions to give secondary alcohols in 69–84% yields (Scheme 64).

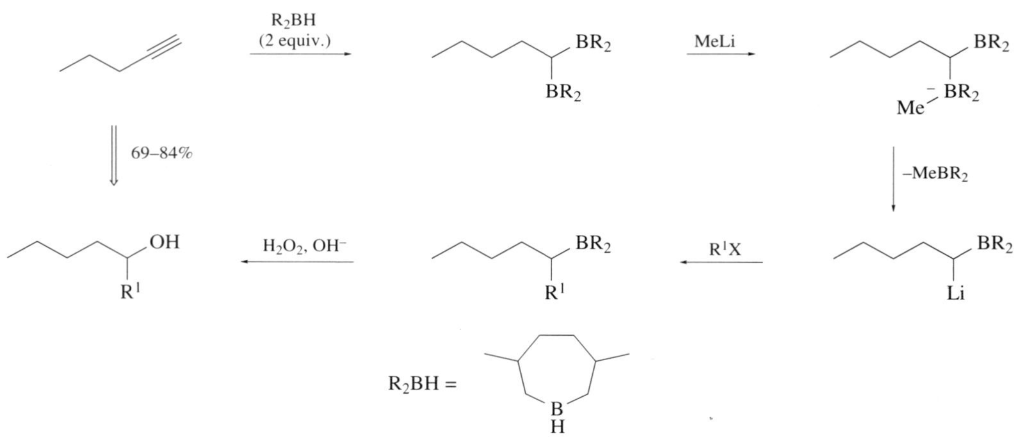

Scheme 64

1.04.3.9 Miscellaneous Stabilizing Groups

While functional groups based on other elements can be used to either facilitate the formation of, or stabilize the negative charge in, carbanions, most of the alternatives do not enjoy widespread use. α-Halocarbanions are frequently referred to as carbenoids ⟨72AG(E)473⟩; these highly reactive species must be prepared at low temperatures to prevent their decomposition. Geminal dihaloorganolithium reagents can be prepared in almost quantitative yields by deprotonation of the corresponding 1,1-dihaloalkane at −90 °C using Bu^{n}Li·TMEDA; alkylations proceed in 68–88% yields with primary haloalkanes ⟨77BSF765⟩. Halogen–metal exchange can occur in α,α-dihalo compounds on treatment with alkyllithium reagents when deprotonation is not possible; 1,1-dihalocyclopropanes give relatively stable carbenoids which can be alkylated in good yields ⟨77BCJ3288⟩. Selenium-stabilized carbanions can be readily prepared, generally by selenium–lithium exchange between selenoacetals and alkyllithium reagents, and the lithio species that result can generally be alkylated in good yields; these reagents have been reviewed in detail by Krief ⟨91COS(3)85⟩.

One of the most fascinating classes of activating/stabilizing groups involves the use of transition metals to stabilize the negative charge. α-Stannylcuprates can be prepared by transmetallation of an α-iodoalkyltin compound; the resulting reagent reacts with allyl halides, however the regio- and diastereoselectivities are generally poor ⟨92SL891⟩. Astruc and co-workers have shown that hexamethylbenzene can be hexaalkylated via the sandwich compound $[FeCp(C_6Me_6)]^+PF_6^-$ ⟨89CC614, 92SL57⟩; while Bu^{t}OK is a satisfactory base for deprotonation and alkylation with methyl, allyl, and benzyl halides ⟨82JA7549⟩, its use with 'normal' primary alkyl halides is precluded as dehydrohalogenation occurs at a faster rate than deprotonation/alkylation of the methyl groups, and the parent organometallic compound is recovered unchanged. Remarkably, KOH in DME proves to be sufficiently basic to allow the reaction to proceed; yields greater than 55% are reported (Equation (108)). Chromium carbenes can also be metallated α to the chromium(0) substituent, which renders the hydrogen atoms on a neighboring methyl group highly acidic ⟨74JA1230⟩. Deprotonation of chromium carbenes (bearing an additional oxygen or nitrogen stabilizing group) gives a stabilized anion that can be alkylated in good yields (Equation (109)) (⟨90JA2264, 93JA87⟩, and references therein).

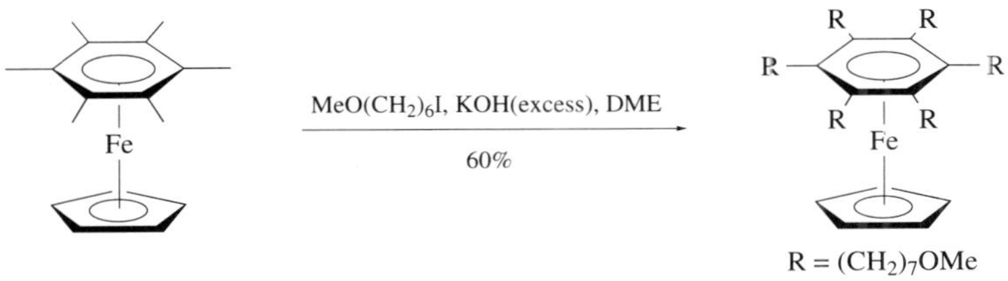

(108)

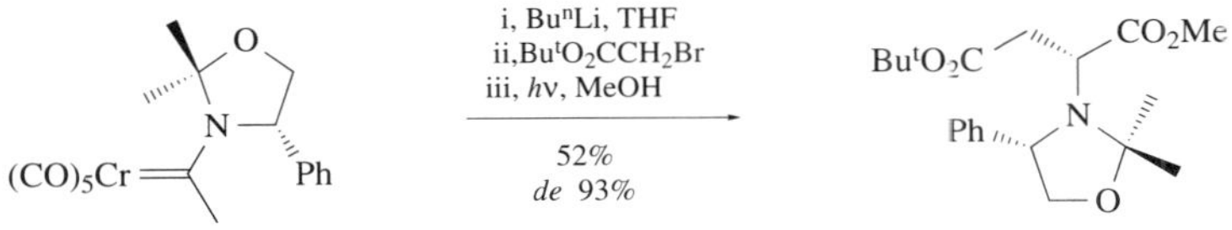

(109)

1.04.4 CARBANIONS WITH TWO AND THREE STABILIZING GROUPS

The union of the stabilizing influences of any two or three substituents described in the previous section provides a wealth of new chemistry centered upon an abundance of useful reagents, however, the range of possibilities makes the field so expansive that a truly comprehensive survey becomes impossible within the space limitations of this chapter. Nevertheless, some general comments can be made and specific examples provided. Introduction of a second or even a third stabilizing group will both increase the acidity of an α-hydrogen atom and also enhance the stabilization of the resulting carbanionic or organometallic species. Several important advantages of these multiply stabilized anions are apparent: deprotonation becomes the favored preparative route and it can generally be effected under milder conditions (compare the following pK_a values measured in DMSO: $MeNO_2$ (17.20), $PhSCH_2NO_2$ (11.93), $EtO_2CCH_2NO_2$ (9.25), $PhSO_2CH_2NO_2$ (7.20) ⟨78JOC3101⟩). The selective enhancement of acidity at a single site can improve the regioselectivity of deprotonation and subsequent alkylation (for example, the use of β-ketoesters as directing agents), and the di- and trifunctional reagents can often be readily transformed into a variety of functional groups. This last point is borne out in the frequent use of these species as synthetic equivalents for other unstable, or frequently unattainable, synthons. Alternatively, they are used for the procurement of certain products that are not readily available in high yields or purities by more direct methods. For example, β-ketoesters are well known synthetic equivalents that have been used classically for enolate monoalkylation: another possibility involves the addition of a second equiv. of a (strong) base to the β-ketoester enolate generating a dianion in the α'-position (for a recent review of dianion chemistry, see ⟨91T4223⟩). Alkylation will occur selectively at the more nucleophilic (basic) site; both concepts have been combined by Mori and Watanabe in a synthesis of the spiroketal pheromone isolated from *Andrena wilkella*, where the dianion of ethyl acetoacetate is alkylated first to introduce one chiral element, and then in a second step the monoanion is generated with K_2CO_3 for the addition of the second asymmetric center (Scheme 65) ⟨86T295⟩.

A fertile area for investigation has centered upon the development of new reagents that act as acyl anion equivalents; whereas nitroalkanes and hydrazones can be used in this manner (see Section 1.04.3.5), many other species also enjoy widespread use. Perhaps the most widely used of these are the reagents based upon dithianes ⟨89T7643⟩, an example of which can be found in the stepwise alkylation of a benzodithiepin used as a central part of Mori's syntheses of spiroketal pheromones (Scheme 66) ⟨85T2751, 85T3663⟩. Another approach uses cyanohydrin derivatives and α-dialkylaminonitriles ⟨83T3207⟩, as applied by Larchevêque and Lalande to the synthesis of *exo*-brevicomin (Scheme 67) ⟨85CC83⟩. Several other examples of acyl anion equivalents are depicted in Scheme 68

⟨79TL3375, 83JCS(PI)1131, 94SL121⟩, and many more examples are referred to in Wardell's review ⟨B-87MI 104-02⟩.

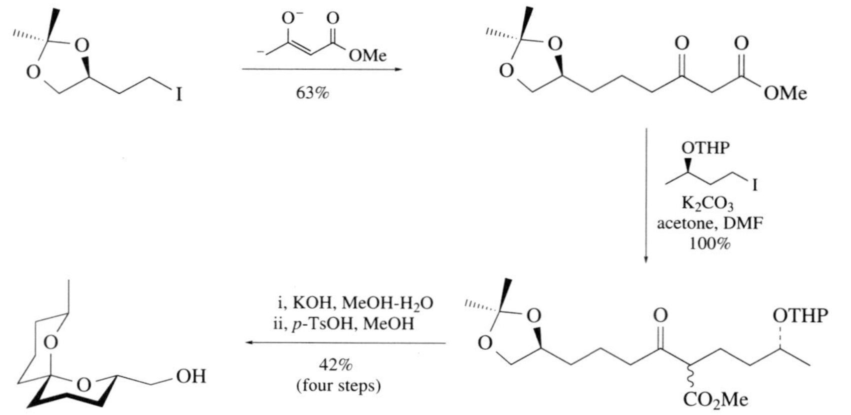

Scheme 65

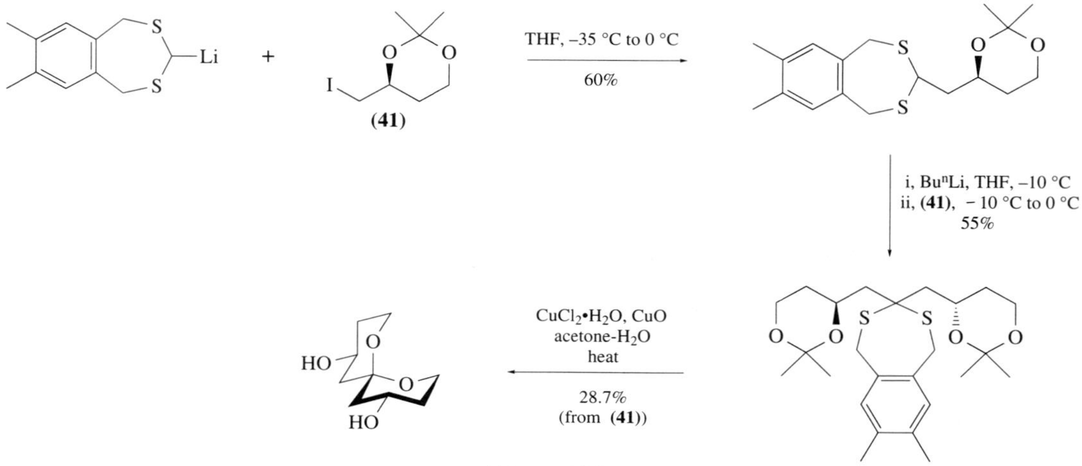

Scheme 66

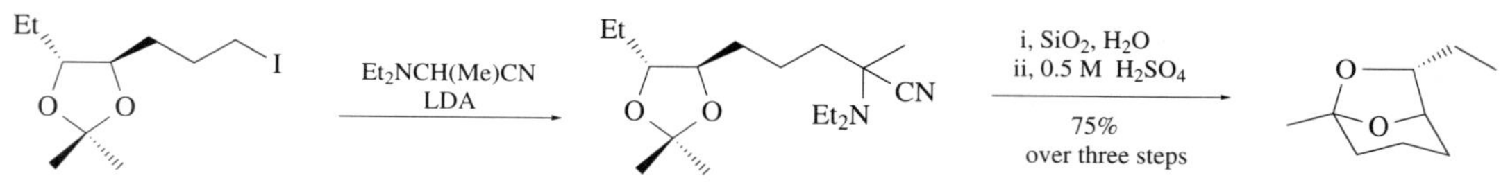

Scheme 67

A simple approach to (racemic) α-amino acids has been reported by Luche and co-workers where the trianion of *N*-*t*-BOC-protected glycine is generated by treatment with 3 equiv. LDA and then subsequently alkylated with a variety of primary and secondary haloalkanes in good to excellent yields (Equation (110)) (40–80%) ⟨92TL6461⟩. No dialkylation is reported. Some starting material is recovered; the basic nature of the trianion is believed to cause partial dehydrohalogenation of the haloalkane. Alternatively, masked glycine equivalents can be employed (Equation (111)) ⟨93TL1917⟩. Of greater importance, however, is the development of new methods for the synthesis of optically active amino acids which can be extended to the preparation of a range of structures of non-proteinogenic nature. One recent example is shown in Scheme 69 ⟨93TL63⟩. Many other synthetic approaches have been reviewed by Williams ⟨B-89MI 104-02⟩ and more recently by Duthaler ⟨94T1539⟩.

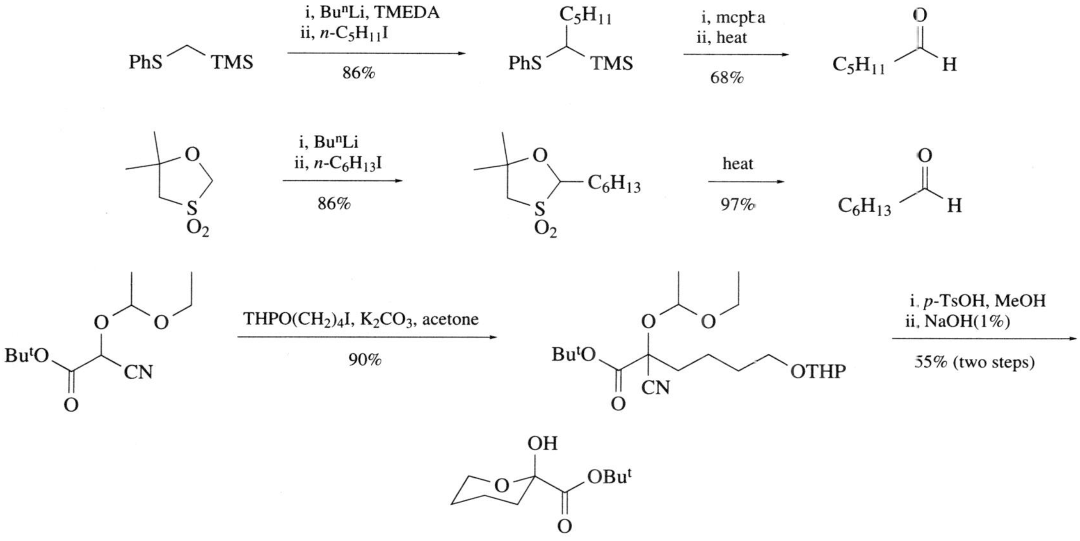

Scheme 68

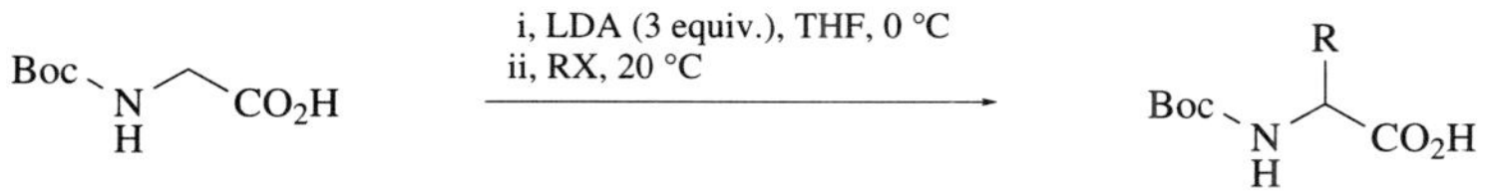

(110)

RX	*Yield(%)*
MeI	80
Bu^nBr	47
Me_2CHCH_2I	48
Pr^iI	41

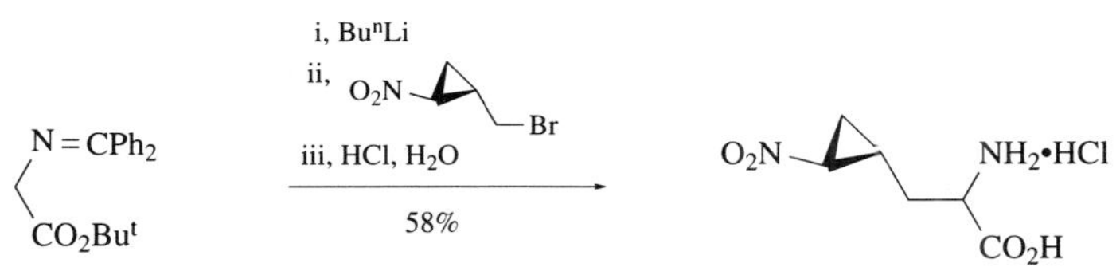

(111)

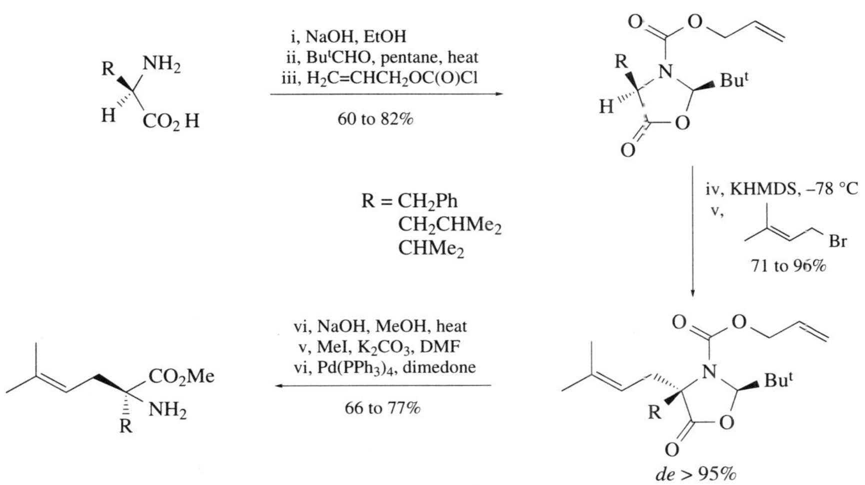

Scheme 69

An interesting example of a carbanion stabilized by boron is the *o*-carborane system, where deprotonation can be effected using *n*-butyllithium. While monolithiation would, in theory, provide a route to monoalkylated carboranes, facile proton transfer leads to the isolation of di-*C*-substituted products ⟨63IC1111, 92CRV209⟩. Development of an efficient alkylation procedure is becoming more important with the search for new neutron capture agents. Use of the *t*-butyldimethylsilyl (TBDMS) group to protect one carbon allows the monoalkylation to be carried out in good yields (70–86%); the silicon group can be removed almost quantitatively with tetra-*n*-butylammonium fluoride (tbaf) (Scheme 70) ⟨91JA5915, 92JOC1384⟩. Kabalka *et al.* have shown that *C*-methyl *o*-carborane can be lithiated with *n*-butyllithium and alkylated with bromoalkanes in high yields (76–96%) to give 1,2-di-*C*-alkylated-*o*-carboranes; catalytic lithium iodide assists the alkylation (Scheme 71) ⟨92TL7687⟩. However, a 4-bromo-1-butyne failed to alkylate in the presence of catalytic LiI; the lithiocarborane is relatively basic and elimination is the major pathway unless an excess of LiI is added. In contrast, Gomez and Hawthorne have shown that the silyl-protected carborane alkylates 4-bromo-1-butene in 81% yield ⟨92JOC1384⟩.

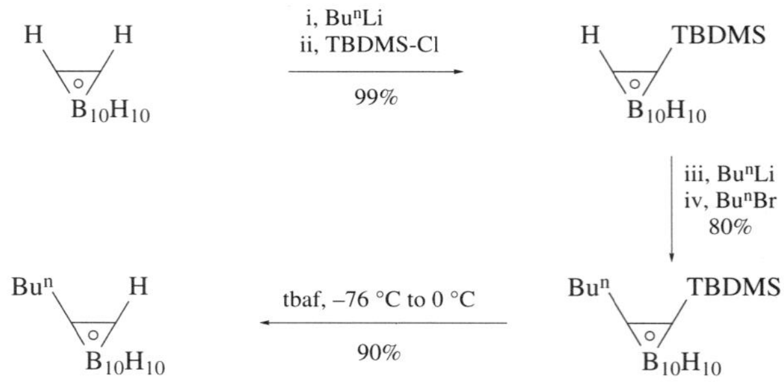

Scheme 70

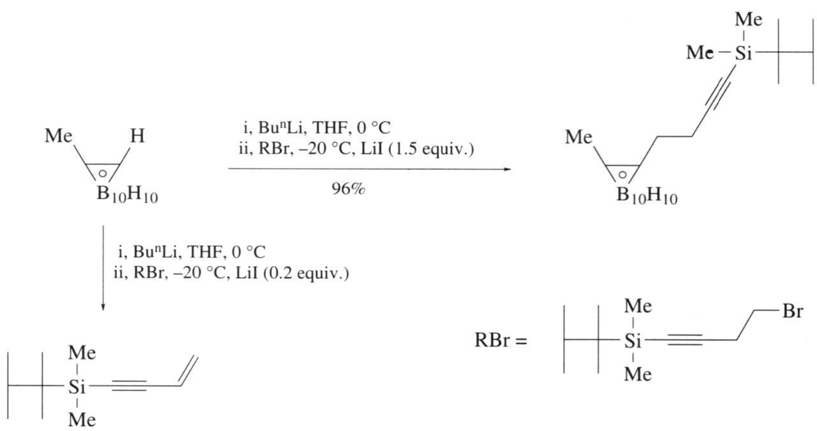

Scheme 71

1.04.5 POLYHALOALKANES

The reactions of dihaloalkanes as electrophiles have already been embraced within this chapter, and in general the reactions follow the basic guidelines set down in Section 1.04.1.2. Where two different halogens are present in an α,ω-dihaloalkane, then the reactivity generally follows the order I > Br > Cl > F and selective replacement can often be effected (Equations (6) and (10); Scheme 10); 3-bromo-1-chloropropane is a commercially available, useful reagent for the introduction of the 3-chloropropyl side chain (Equation (89); Scheme 40). Substrates such as 1,3-dibromopropane or butane frequently will give mixtures of mono- and disubstituted material as well as unchanged starting dihalide (Equations (11), (71) and (85); Schemes 9, 30 and 49), although clean mono-alkylation is sometimes attainable (Equation (27)). While disubstitutions of 1,2-dihalo compounds are known (Schemes 9 and 97), problems can arise with 1,2-dihaloalkanes, where some nucleophiles react at halogen leading to elimination and generation of the alkene (Scheme 5). Disubstitutions of 1,1-dihalo substrates are also achievable (Schemes 9 and 24).

1.04.6 ACKNOWLEDGMENTS

I would like to thank Professor David Black for reading sections of this chapter, and especially my wife Alison for her patient assistance in the preparation and checking of this manuscript.

1.05
One or More CC Bond(s) Formed by Substitution: Substitution of Chalcogen

TIMOTHY N. BIRKINSHAW
Astra Charnwood, Loughborough, UK

1.05.1 SUBSTITUTION OF OXYGEN FUNCTIONS

Radical replacements of oxygen functionality stand alone as a separate area. Epoxides are also treated separately due to their distinctive reactivity.

1.05.1.1 Radical Reactions

The use of radicals in organic synthesis has been reviewed by Giese ⟨B-86MI 105-01⟩, Neumann ⟨87S665⟩ and Curran ⟨88S417, 88S489⟩ and includes the addition to carbon multiple bonds of radicals generated from alcohol derivatives. These reactions will be covered in broader terms in Chapters 1.07 and 1.08. Many of these reactions can also be carried out with radicals generated from halogens (Chapter 1.04), which in turn are often prepared from oxygen compounds and so represent a formal C—O to C—C transformation.

1.05.1.1.1 Intermolecular reactions

Barton and Crich ⟨85TL757⟩ substituted tertiary alcohols, including adamantanol via thermolysis of a mixed oxalate e.g. (**1**) in the presence of alkenes (Equation (1)). The scope is limited by the need to use very reactive alkenes and by the side reaction which gives the thio ether (**2**).

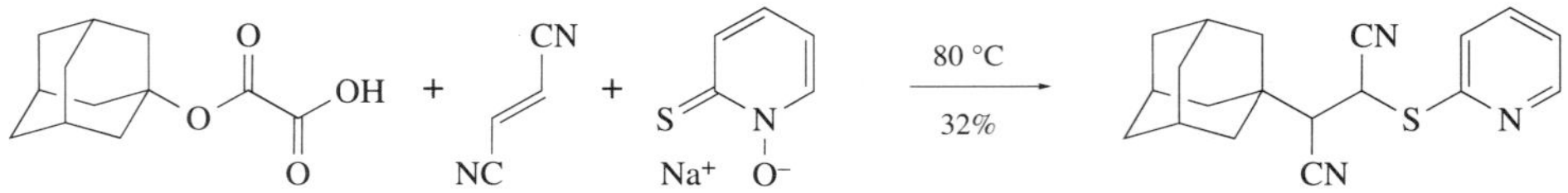

(**1**)

(1)

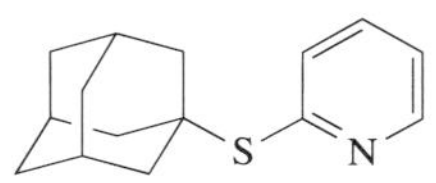

(**2**)

Moderate to good yields could be obtained from radical addition to a variety of electron deficient alkenes using Bu_3SnH (generated *in situ* from the chloride and $NaBH_4$) and 2,2′-azo-bisisobutyronitrile(AIBN) ⟨84AG(E)69⟩ (Scheme 1). This method can be extended to more complex systems such as the galactofuranose (Equation (2)) where competing hydrogen abstraction from the reagent gives a 10% yield of deoxygenated compound (**4**) along with 35% of the required alkylated product (**3**) as a single stereoisomer, with approach of the alkene from the less hindered face. A mixture of isomers is obtained when the steric demands are not quite so clear cut.

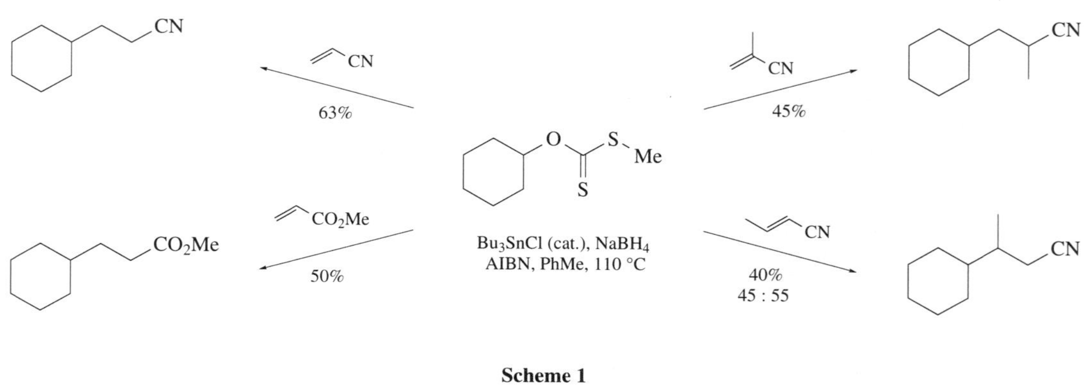

Scheme 1

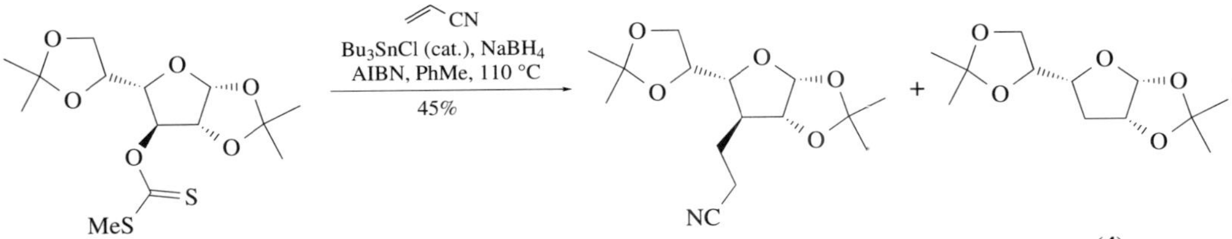

(3) 75 : 25 (4) (2)

Keck *et al.* ⟨85JOC4317⟩ used a radical allylation as the key step in the synthesis of pseudomonic acid C from L-lyxose (Equation (3)). Thermolytic generation of a radical from the thiocarbonate (**5**) gave the product (**6**) derived from the approach of allyl tin to the less sterically congested face of the sugar. The reaction is analogous to the radical reduction method using Bu_3SnH; the most reactive group is abstracted, generating Bu_3Sn radical which continues the chain. Initiation with AIBN proved less satisfactory as did the use of xanthate (**7**) or imidazole thiocarbonate (**8**) as radical precursors ⟨82JA5829⟩. Radical attack on allyltributyltin generates tributyltin radicals which then attack the sulfur and continue the radical chain. The reaction is limited to allyl and methallyl stannane; with crotylstannane (**9**) the alkane is too unreactive and allylic H-abstraction occurs to give butadiene ⟨85T4079⟩. Mori and Takechi ⟨90H(31)1189⟩ used similar conditions to effect radical allylation in their synthesis of a prostacyclin analogue (Equation (4)). Allyl attack was again from the less hindered face.

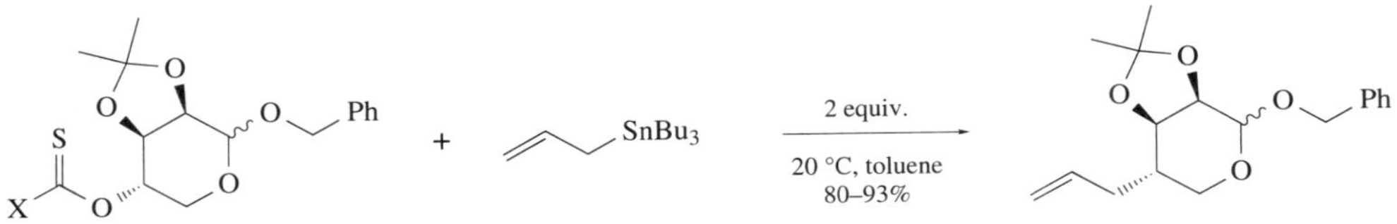

(3)

(**6**)

(**5**) X = OPh
(**7**) X = SMe
(**8**) X = Imidazole

(**9**)

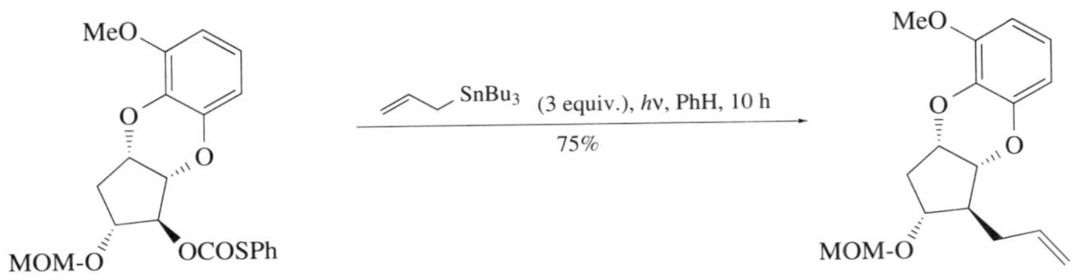

(4)

The isomeric hexoses (**10**)–(**13**) were allylated to give the axially substituted compounds (**14**) and (**15**) ⟨91CC690⟩. The latter readily isomerized to the equatorial isomer (**16**), showing that the α-keto radical reacts under kinetic control (Scheme 2). Roberts and co-workers have effected radical allylation in a monocyclic system (Equation (5)) ⟨90JCS(P1)2352⟩. Here the facial discrimination of the radical intermediate was not quite so great and an 87 : 13 (eq : ax) mixture of isomers was obtained.

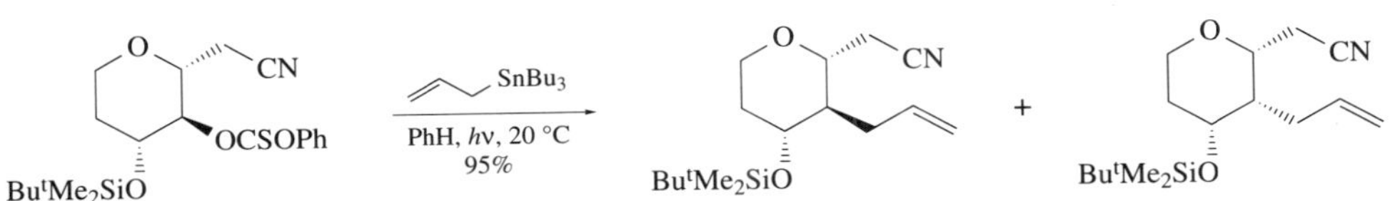

87 : 13 (5)

Allylic and benzylic alcohols can be reductively coupled by low-valent titanium, a reaction which proceeds via a radical intermediate. McMurray *et al.* ⟨78JOC3249⟩ have investigated the reaction,

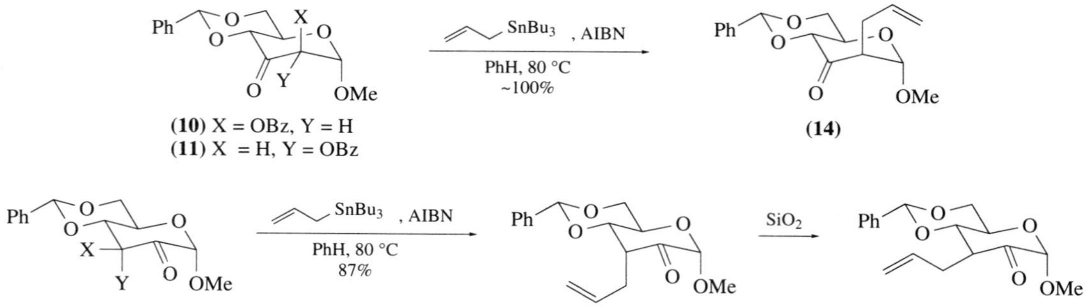

(10) X = OBz, Y = H
(11) X = H, Y = OBz

(14)

(12) X = OBz, Y = H
(13) X = H, Y = OBz

(15) **(16)**

Scheme 2

first discovered by van Tamelen and Schwartz ⟨65JA3277⟩, using $TiCl_3/LiAlH_4$ to generate tin(II) *in situ* and found that the reaction works well (68–95%) with benzylic and simple allylic systems (Equations (6) and (7)) but gives mixtures of products with unsymmetrical alcohols and does not work at all with nonallylic alcohols. Attempts to couple allylic diols intramolecularly gave only reduction products.

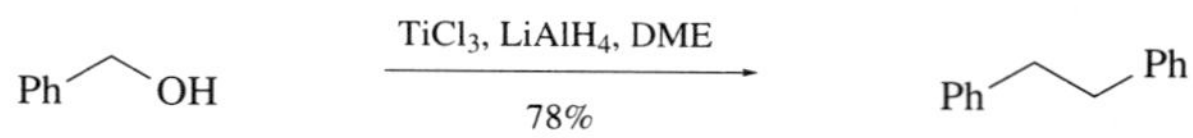

(6)

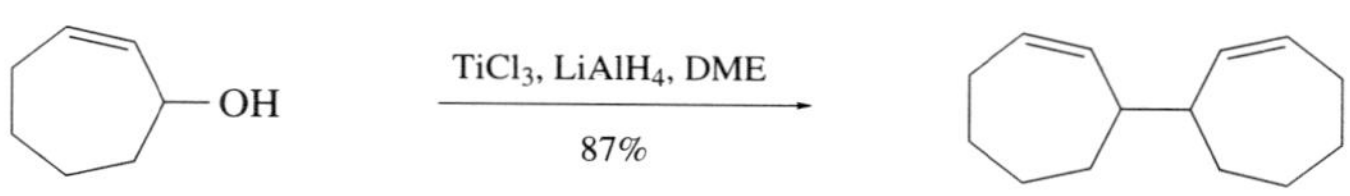

(7)

1.05.1.1.2 Intramolecular reactions

Intramolecular reactions have allowed a greater variety of functional groups to be used as radical traps as the problem of hydrogen abstraction is reduced by a more favourable entropy factor. Much work has been done in this area on both oxygen and bromine derived radicals ⟨85JOC546, 86JA3102, 91ACR139⟩. Sharma *et al.* ⟨90TL4931, 91T519⟩ have used intramolecular radical reactions in the synthesis of fungal metabolites from sugars (Equations (8) and (9)). In both cases only one stereoisomer was formed.

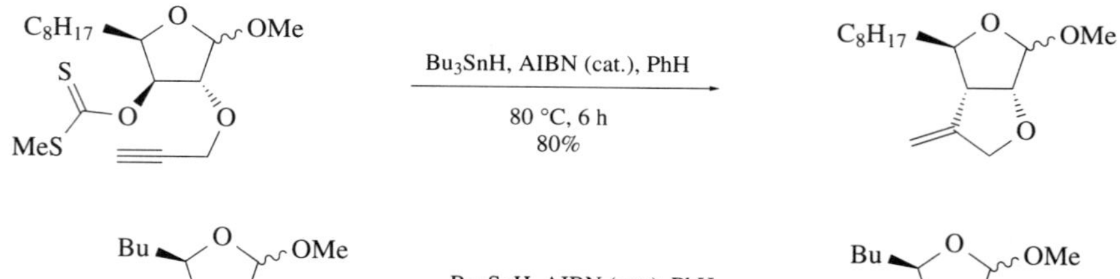

(8)

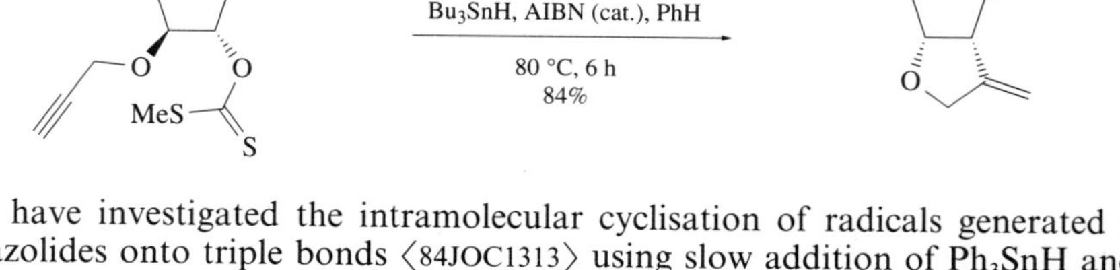

(9)

Clive *et al.* have investigated the intramolecular cyclisation of radicals generated from thiocarbonylimidazolides onto triple bonds ⟨84JOC1313⟩ using slow addition of Ph_3SnH and AIBN in refluxing benzene. The addition to terminal alkyne was low yielding (Equation (10)) but those to substituted alkynes proceeded in good yield (Equation (11)). Of the additions to nitriles (Equations (12), (13) and (14)) only that forming the [5,5] ring system gave an unsatisfactory yield (Equation (12)). Competing reduction of the oxygen function occurred in low yield in some cases.

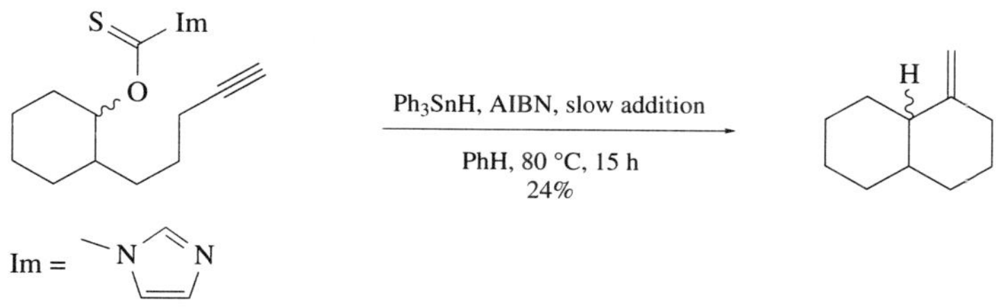

(10)

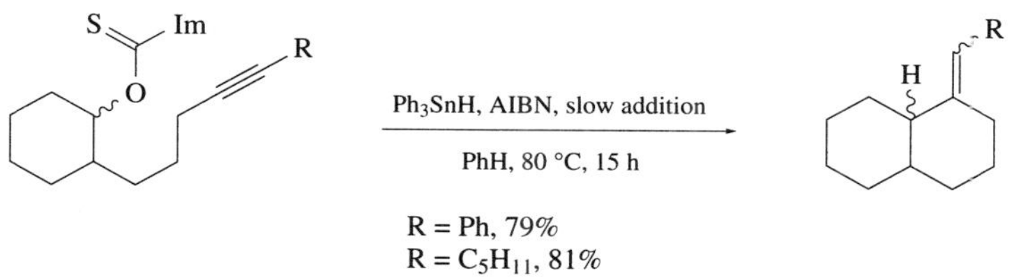

(11)

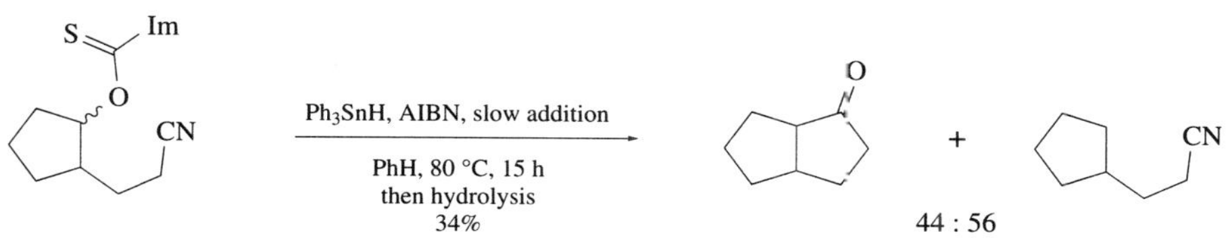

(12)

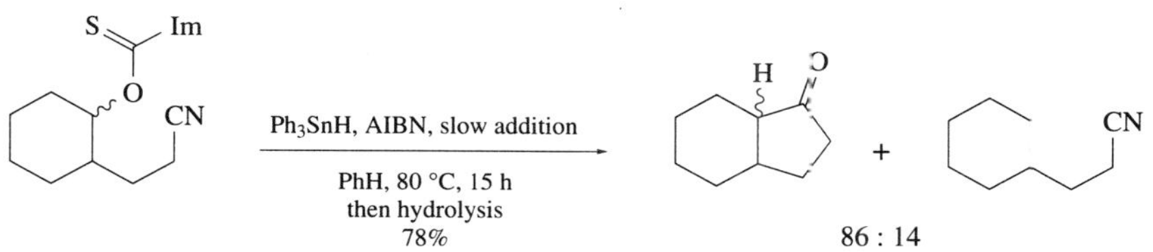

(13)

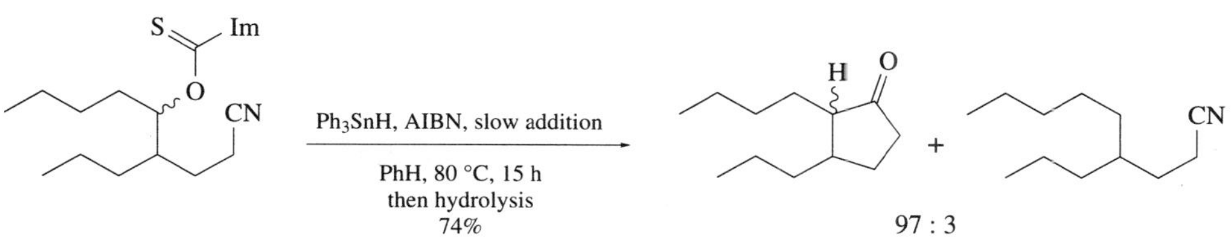

(14)

A cyclopentane ring can be constructed in good yield by intramolecular cyclisation of the tetrahydrofuran xanthate (**17**) (Equation (15)) ⟨92TL5261, 93JOC2894⟩. A small amount of the thio compound (**18**), derived from a competing radical chain propagation reaction, was also isolated. This can be converted to the required compound by radical desulfurisation. Bartlett et al. have shown that oxime ethers, in contrast to carbonyl groups, act as effective radical traps in intramolecular cyclisations to give cyclopentanes and cyclohexanes (Equation (16)) ⟨88JA1633⟩. Competing reduction is a problem in some cases and mixtures of *cis* and *trans* isomers are formed.

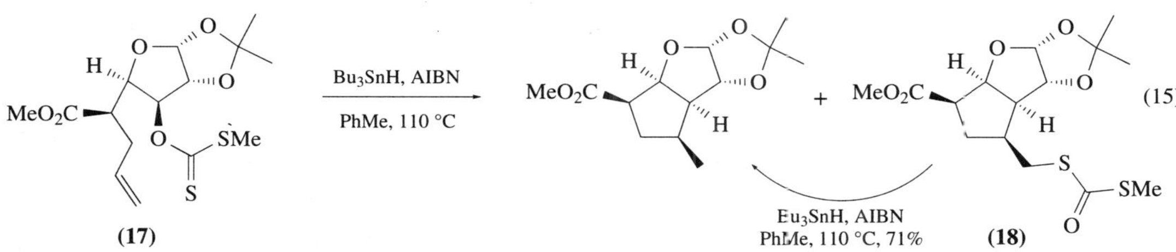

(**17**) (**18**) 10%

(15)

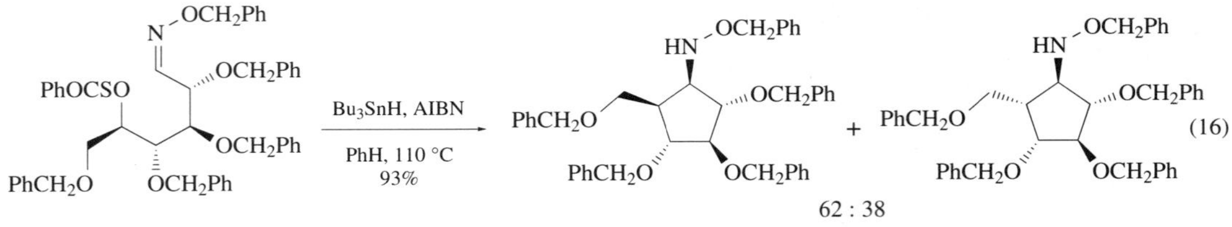

Ziegler *et al.* ⟨93JA2581⟩ have used the selectivity of a thiocarbonate derived radical ⟨90JOC1416⟩ cyclising (in a 5-*exo*-dig fashion) onto an alkyne to prepare a substituted cyclopentene (Equation (17)). The starting material is a 1.6 : 1 mixture of diastereoisomers but the isomer ratio of the product is not quoted. RajanBabu has used an enol ether as the radical trap in a synthesis of a Corey lactone (Equation (18)) ⟨88JOC4522⟩. The unusual 1,5 *trans* stereochemistry is rationalized by postulating a boat-like transition state (**19**) which removes the unfavourable steric interaction between the benzyloxy group and the cyclic acetal that would be found in the more usual chair transition state (**20**) which would give rise to the *cis* stereochemistry. With the opposite chirality at the benzyloxy centre the expected 1,5-*cis* stereochemistry is obtained, while the desoxy compound gives a 77 : 23 mixture of *cis* : *trans* isomers.

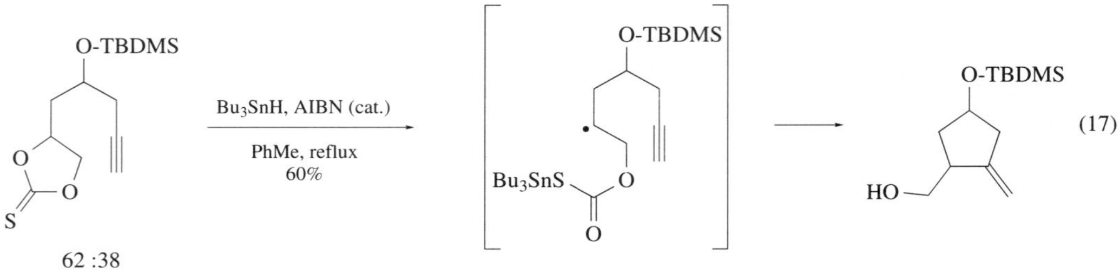

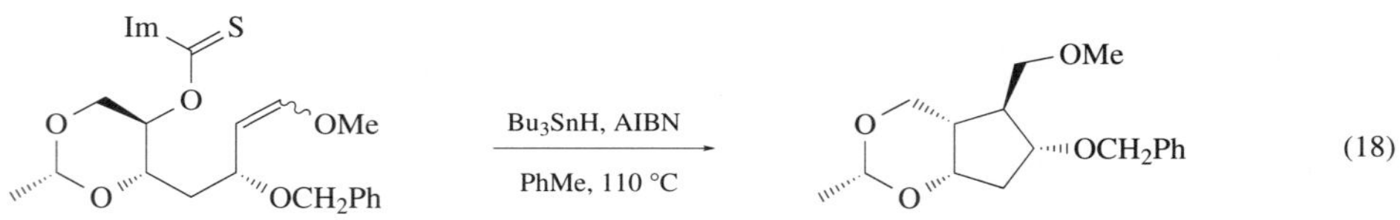

(19) **(20)**

The tricyclic skeleton of the sesquiterpene silphinene was prepared by radical cyclisation of the thiocarbonate (**21**) (Equation (19)) creating the quaternary carbon centre and the correct stereochemistry at the adjacent stereogenic centre ⟨88TL107⟩.

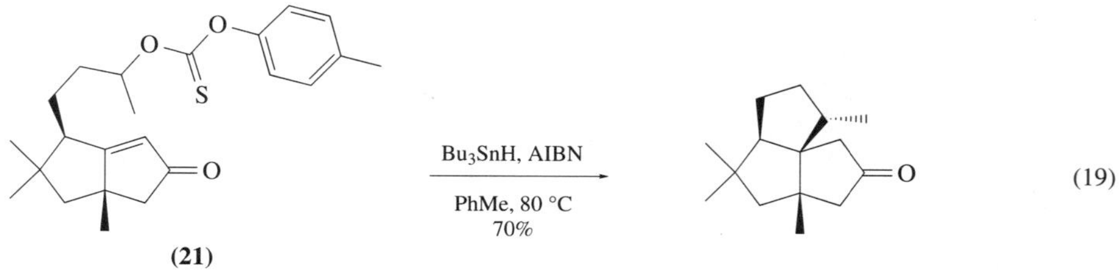

A limitation of radical cyclisations can be seen in Snider and Kulkarni's syntheses of tricyclic sesquiterpenes (Scheme 3) ⟨85TL5675⟩. The two required compounds (**25**) and (**26**) were each isolated in only 8% yield from the radical cyclisation of compound (**22**); the major products (**23**) and (**24**) being derived from tributyltin cleavage of the thiocarbonylimidazole. The reduction product (**27**), derived from competing hydride abstraction from tributyltin hydride, was not observed.

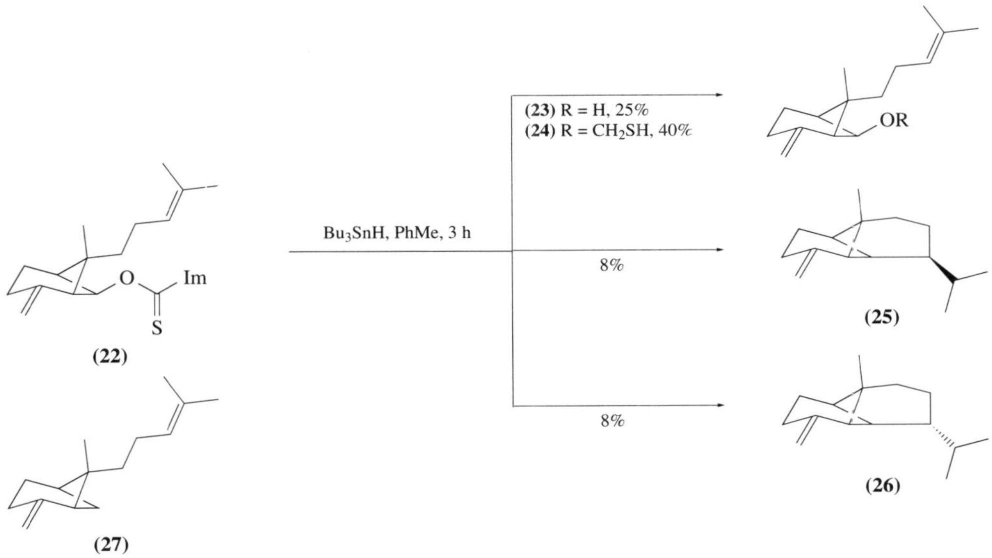

Scheme 3

1.05.1.2 Displacement of Alcohol Derivatives

Alcohol derivatives which are good leaving groups can be displaced by nucleophiles under neutral or basic conditions. The hybridization of the carbon atom β to the alcohol carbon atom has a substantial effect on the reactivity at the α carbon atom and so this section is divided according to the nature of the adjacent group. Some reactions are specific to that section while others are more general in their scope.

1.05.1.2.1 Alkyl alcohol derivatives

(i) Sulfates, sulfonates and phosphorus derivatives

Alkyl sulfonates and sulfates react readily with nucleophiles. A systematic study of comparative leaving group abilities of the different sulfonate (and other) leaving groups does not appear to have been done (and would be nucleophile dependent) but the relative ease of displacement parallels the acid strengths of the leaving groups and their rate of solvolysis and is shown in Figure 1 ⟨82S85⟩. For most purposes tosylates suffice (mesylates can be deprotonated from the methyl group) but the more reactive triflates as well as dialkyl sulfates and halides also react in a similar fashion (see Chapter 1.04). Cuprates and copper-catalysed reagents are the best carriers for simple alkenyl and alkyl nucleophiles (Equations (20) and (21)) ⟨87H(26)1291, 89JOC5831⟩. In some cases halides may be preferable ⟨83JOC3334⟩. Displacement occurs with inversion of configuration via an S_N2 mechanism ⟨89JOC5831⟩. It is possible to displace two different sulfonate groups sequentially (Equation (22)) ⟨89TL3999, 90JOC4417⟩; since dialkylcuprates are more reactive than the copper-catalysed Grignard reagents, in that they tend to displace both sulfonates, the latter reagents are preferred for the first step. It is claimed that secondary and tertiary tosylates give low yields or do not react at all ⟨74AG50⟩ but this is contradicted for secondary tosylates by the example given above. Recently, displacement of a bridgehead mesylate in a gibberellin has been reported ⟨93CC1111⟩. A hetero-atom β to the nucleofuge usually causes rate retardation ⟨92JCS(P1)95⟩ but rate enhancement can also occur

⟨89JOC5831⟩. The displacing nucleophile can come from an aromatic Grignard reagent ⟨86BCJ3097⟩ or from an anion generated reductively (Equation (23)) ⟨86JOM(310)135⟩.

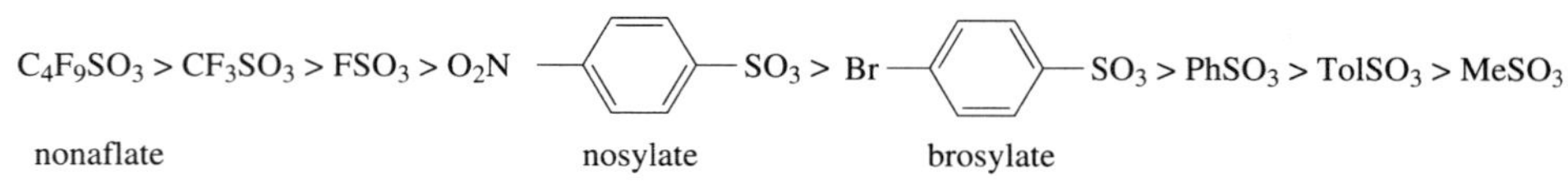

Figure 1

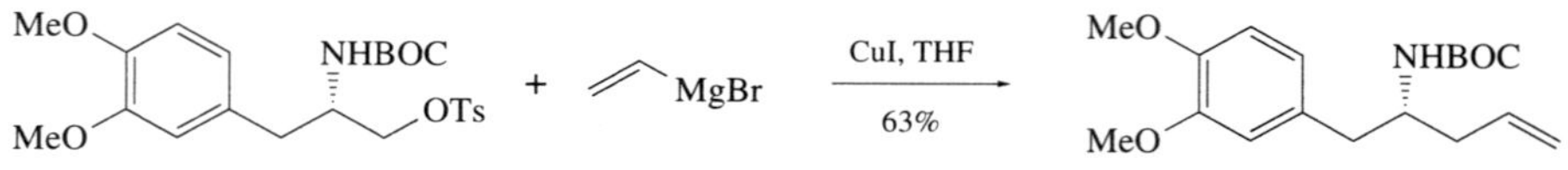

(20)

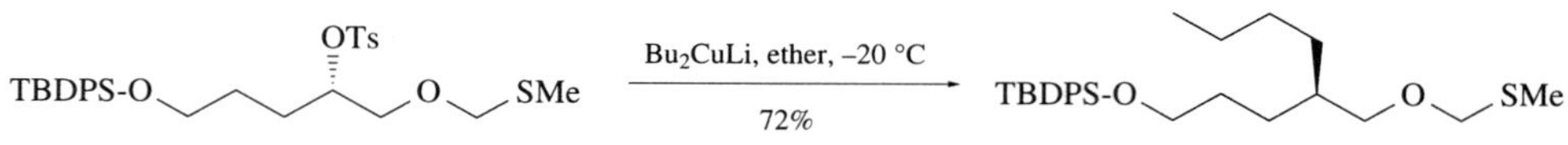

(21)

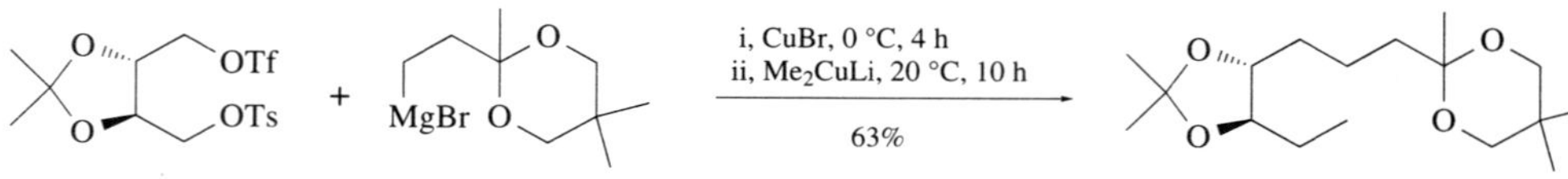

(22)

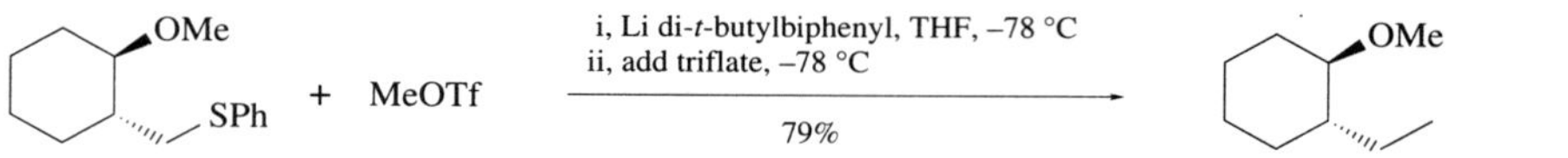

(23)

Sulfonates and sulfates alkylate stabilized anions such as alkynes ⟨92JCS(P1)95⟩ (Equation (24)) ⟨63HCA1792⟩, cyanide (Equation (25)) (in this case under mild phase-transfer conditions) ⟨88HCA47⟩, phosphonate anion ⟨93JOC6174⟩, sulfur-based anions ⟨66AJC521, 80CJC716⟩, boron-stabilized anions ⟨87JOC5484⟩ and those derived from carbonyl compounds including dianions of β-ketoesters ⟨86H(24)3149⟩ (Equation (26)). Sulfates give better yields than bromides and iodides, with little *O*-alkylation or di-alkylation occurring (Equation (27)) ⟨90SC125⟩. Regio- and diastereoselective alkylations are possible ⟨74LA369, 88BCJ571⟩. The abilities of several alkylating agents to alkylate ketone enolates have been compared ⟨93JOC4469⟩. Cyclohexanone and triflates ⟨77CRV69⟩ gave by far the best results with no dialkyl products formed and the highest yields obtained (Equation (28)) ⟨93JOC4469⟩. Significant amounts of dialkylation occurs with methyl iodide and benzyl bromide under similar reaction conditions. Triflates are more difficult to work with as they cause polymerisation of THF; exchange of alkyl groups can occur when using diethyl ether as the solvent. Hexane is a better solvent.

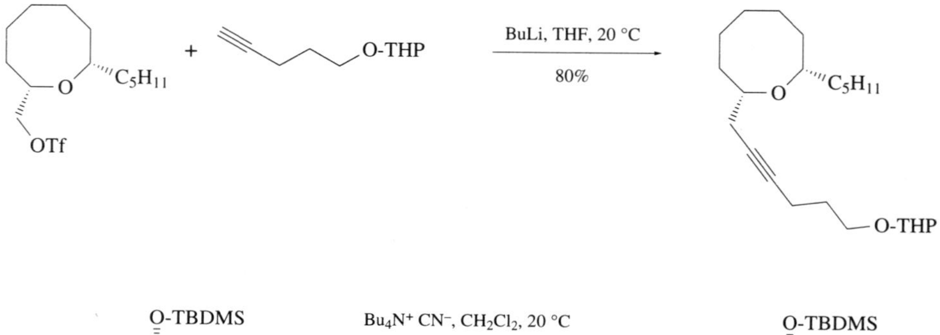

(24)

(25)

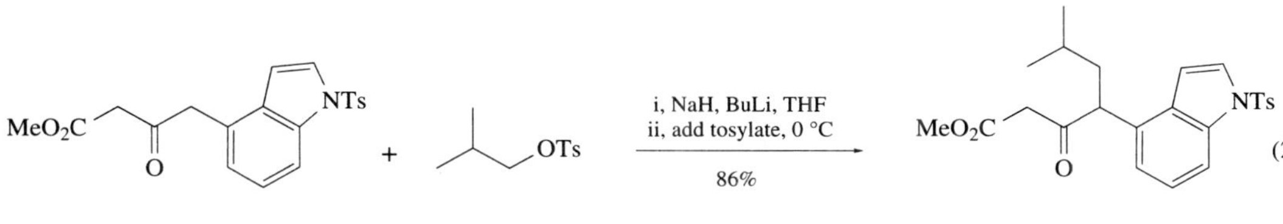

(26)

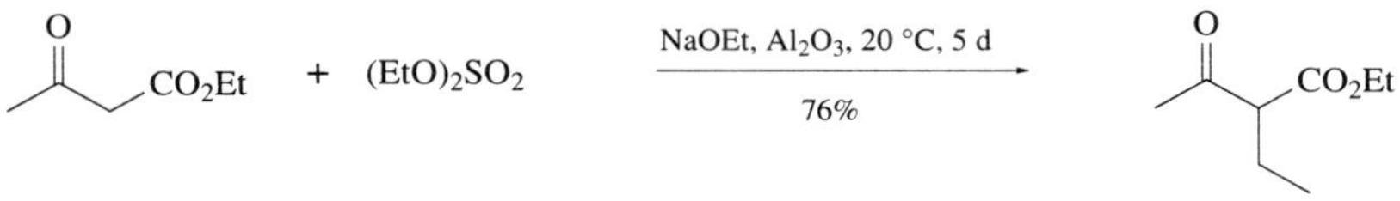

(27)

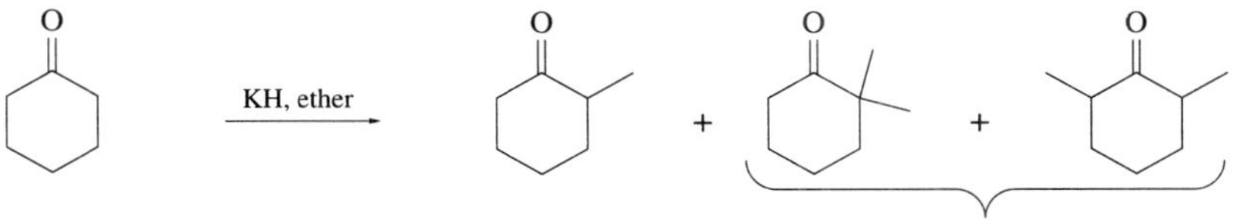

(28)

MeI, ether	0 °C to 25 °C	70%	15% + 15% SM
MeOTf, hexane	−30 °C to 25 °C	91%	0% + 9% SM

Alcohols can be displaced by cyanoacetate anion under Mitsunobu conditions (Equation (29)) ⟨72TL1279, 76TL2455, 81S1⟩, whereas 1,3-dicarbonyl compounds give largely *O*-alkylation. Bissulfones react in high yield ⟨93JOC5892⟩ under Mitsunobu conditions and cyanide has also been used as the nucleophile ⟨76HCA2100, 85SC663⟩.

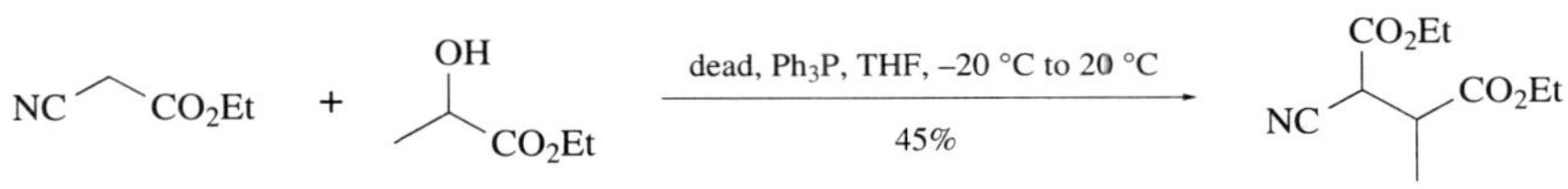

(29)

dead = diethyl azodicarboxylate

Replacement of hydroxyl by cyanide is achievable under a wide range of conditions. Trifluoroacetate esters can be replaced by cyanide in refluxing THF–hexamethylphosphoramide (HMPA) ⟨88SC445⟩. This is only useful for primary alcohols and can be used to chemoselectively replace one hydroxyl group of a diol. Trimethylsilyl chloride and sodium cyanide convert a range of primary, secondary, tertiary (including adamantanol) and benzylic alcohols ⟨81JOC2985⟩ (Equation (30)). The presence of a catalytic amount of sodium iodide is necessary.

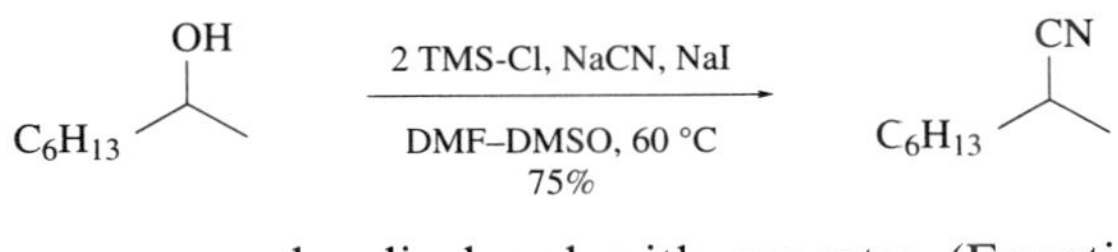

(30)

Triflates α to an ester group can be displaced with cuprates (Equation (31)) ⟨90TL2149⟩ or β-ketoesters (Equation (32)) ⟨93TL2051⟩ in high optical purity (≥94% *ee*). The choices of cuprate and solvent appear to be critical to the success of the reaction, reduction of the triflate being the main side reaction. The corresponding tosylates readily eliminate tosic acid under similar conditions. Displacement of a benzoate group α to a carbonyl moiety has been accomplished using samarium diiodide ⟨92H(34)2247⟩. The reaction proceeds via reductive generation of an enolate and subsequent aldol reaction (Equation (33)).

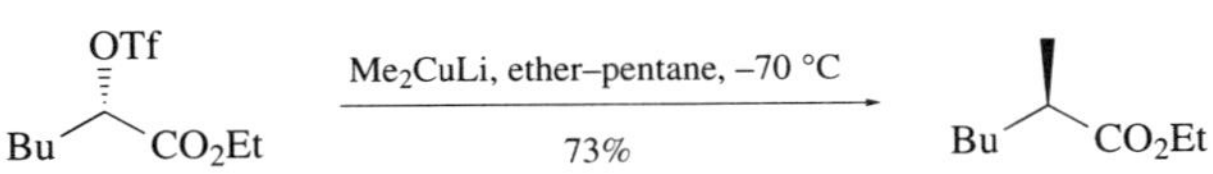

(31)

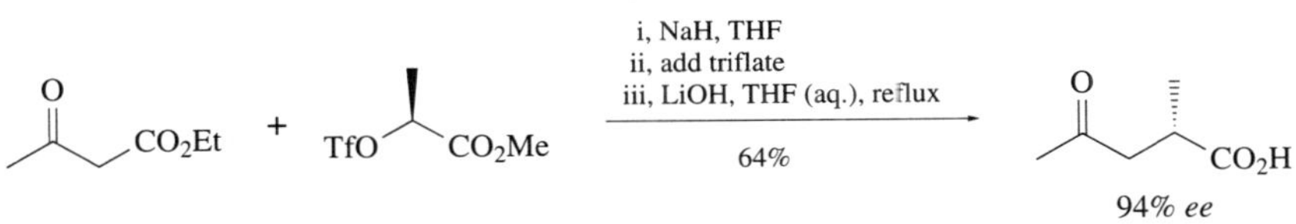

(32)

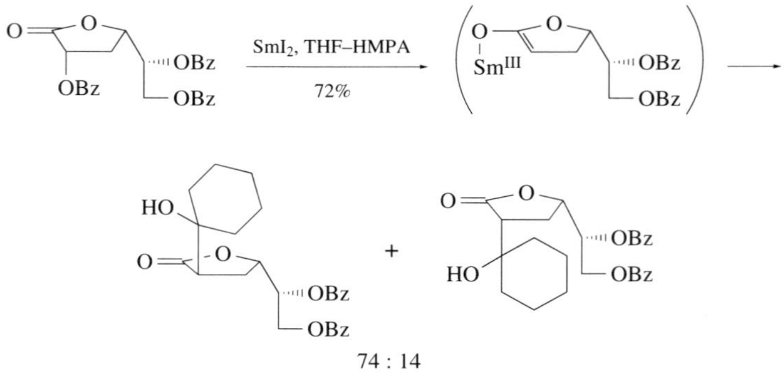

(33)

Chiral alkylations have been performed with sulfonates. Alkylation of the 'Enders hydrazone' (**28**) with methyl tosylate gives better enantiomeric excesses (*ees*) than with methyl iodide (Equation (34)) ⟨87CC358⟩. Triflates also react with the chiral amino acid synthon (**29**) in much higher yields than the corresponding iodides (Equation (35)) ⟨86TL3403⟩. One possibility for sulfates and sulfonates not open to halides is to have a chiral leaving group. Duhamel *et al.* have studied a number of chiral sulfate and sulfonate methylating agents with glycinate anions and obtained *ees* up to 61%, although most *ees* were in the range 0–20% (Equation (36)) ⟨88T5495⟩.

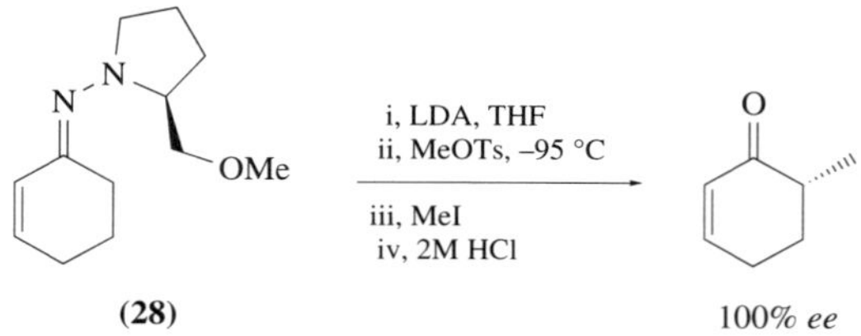

(**28**)

(34)

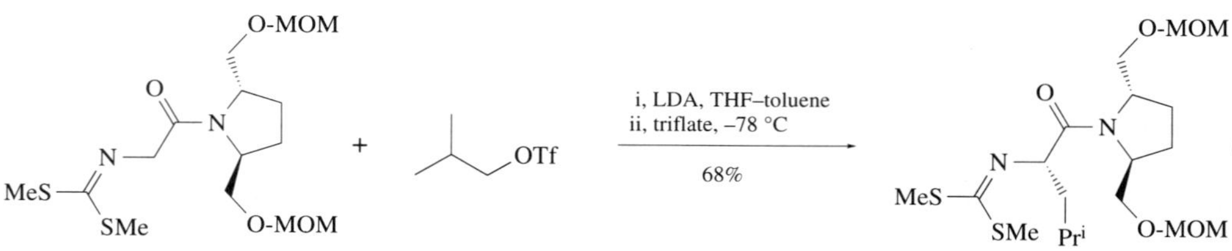

(**29**)

(35)

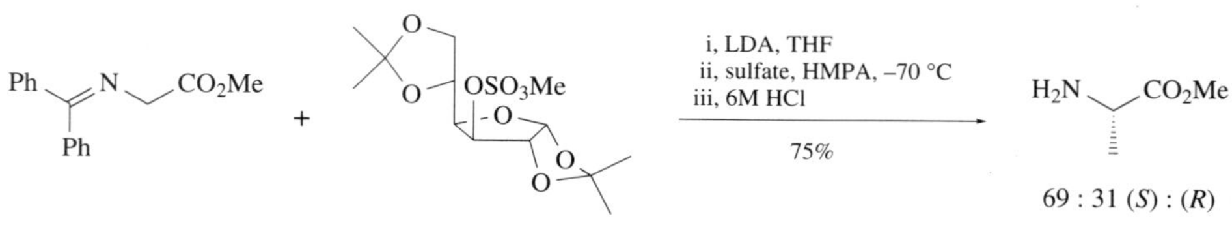

(36)

(ii) Friedel-Crafts reactions

Friedel-Crafts alkylations have been reviewed by Olah *et al.* ⟨91COS(3)293⟩ (which includes a section on alkylation with oxygen functions) and in a monograph ⟨B-84MI 105-01⟩.

The relative ease of alkylation of alcohols decreases in the order benzylic ≈ allylic ≈ tertiary > secondary > primary > methyl, so the Friedel-Crafts reaction of alcohols is most useful for benzylic and allylic alcohols. A wide variety of catalysts can be used including $SnCl_4$, $TiCl_4$, $ZnCl_2$, $BF_3 \cdot Et_2O$, $AlCl_3$–$MeNO_2$, P_2O_5, H_2SO_4, $TeCl_4$ (but not $SeCl_4$) ⟨86BCJ3617⟩, various sulfonic and carboxylic acids (including polymer-bound super acids such as Nafion-H) as well as the more usual $AlCl_3$. K10-Montmorillonite has been used as a catalyst for alkylation with primary and secondary alcohols; the former gave substantial amounts of rearrangement products but good yields were obtained with the latter ⟨93TL1197⟩.

Shen *et al.* have achieved a Friedel-Crafts alkylation of the normally unreactive nitrobenzene (Equation (37)) ⟨90JOC3961⟩. $AlCl_3$ coordinates the nitro group and further deactivates the benzene ring. Sulfuric acid does not do this and so the reaction proceeds in moderate yield. The *ortho*:*meta*:*para* ratio is 19:70:11. The selectivity of this Friedel-Crafts alkylation is poorer than that of nitration reactions suggesting that the former procedes via an early transition state.

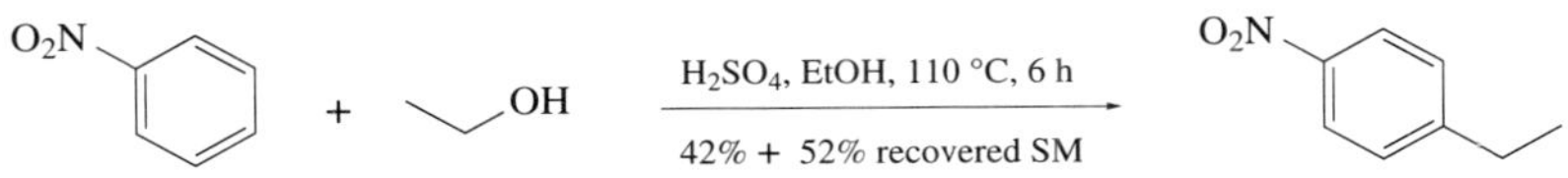

(37)

Phosphorus acid derivatives are also substrates for Friedel-Crafts alkylations (Equation (38)) ⟨71S203, 72S142⟩. $(MeO)_2POCl$, $(Pr^iO)_3P$ and $(Pr^iO)_3O$ are also substrates.

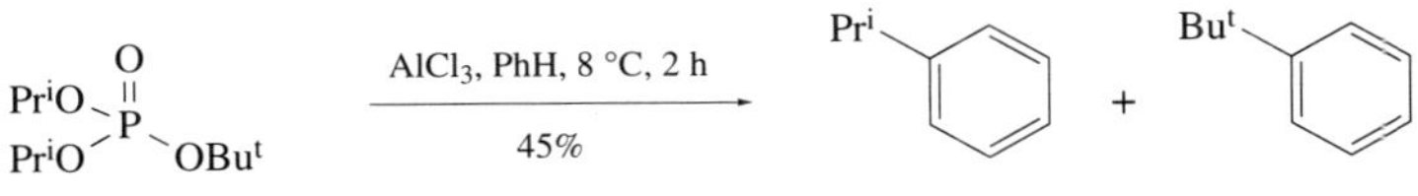

(38)

85 : 15

Lactones undergo Friedel-Crafts alkylation to give, initially, arylalkanoic acids which can be further cyclized. Co-ordination of Lewis acid (LA) to the carbonyl group weakens the C—O bond with attack of the aromatic nucleus occurring at the carbon atom (Scheme 4). The lactone (**30**) reacts readily with benzene at low temperature to give the acid (**31**) (Equation (39)) ⟨88T1567⟩. Friedel-Crafts acylation of the acid requires much more forcing conditions (polyphosphoric acid (ppa) at 120 °C). The lactone (**32**) alkylates and then acylates dimethyl catechol in one step under more forcing conditions (Equation (40)) ⟨75CPB1917⟩. Lactone (**33**) reacts selectively at the ester oxygen atom and not the α-hydroxyl group (Equation (41)) ⟨62BSF1882⟩. Esters are also substrates (Equations (42) and (43)) ⟨63JCS4522, 69AG496⟩. In the latter the more reactive benzylic ester reacts preferentially.

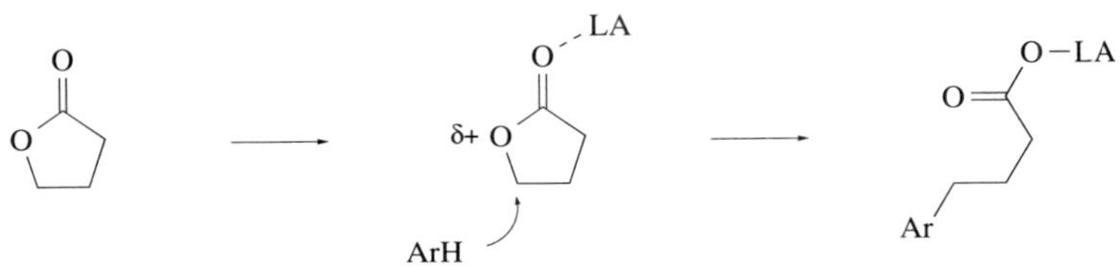

Scheme 4

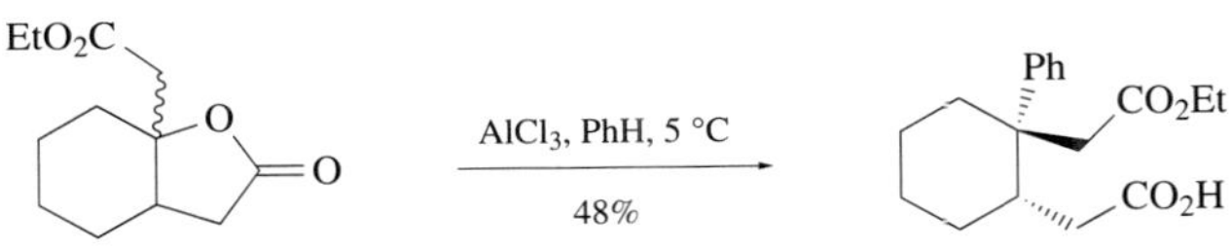

(39)

(30) **(31)**

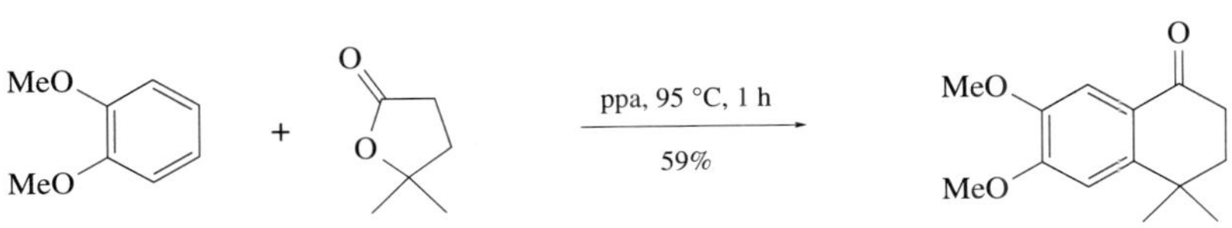

(40)

(32)

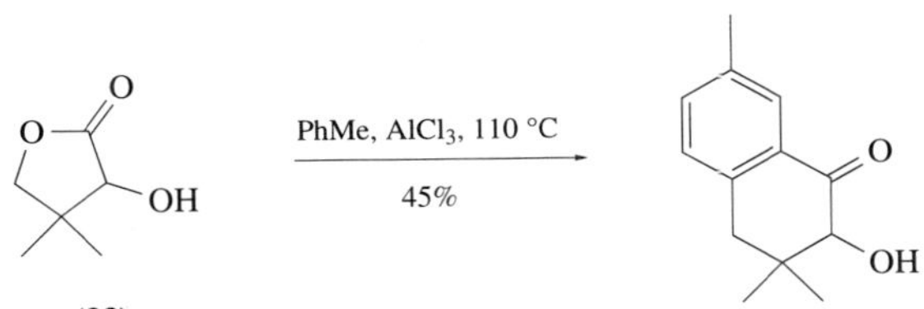

(41)

(33)

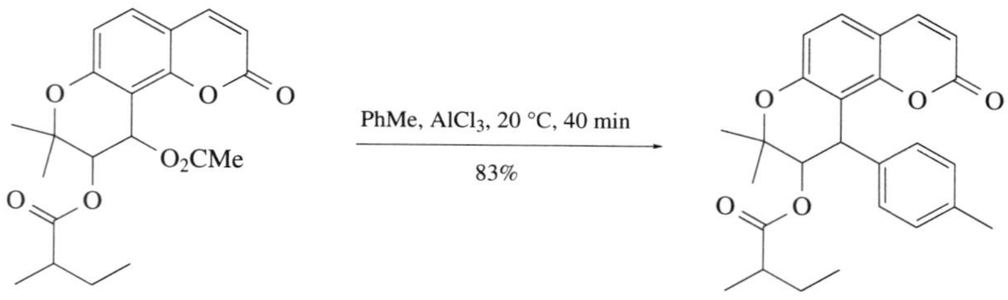

(42)

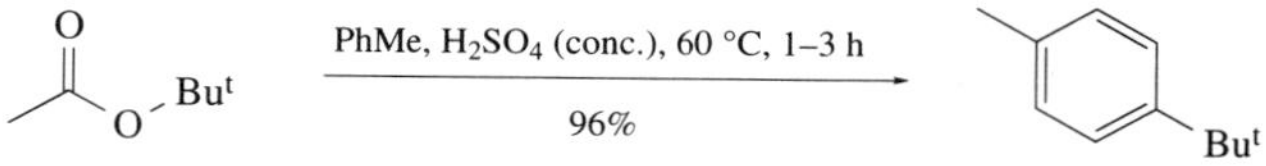

(43)

Piccolo and co-workers have investigated the reaction of a number of optically active 2-sulfonyloxypropanoates and 3-sulfonyloxybutanoates with aromatic systems, catalysed by $AlCl_3$ ⟨85JOC3945, 91JOC183⟩. These reactions proceed in good yield and high stereospecificity with inversion of configuration. The *ee*s of the products need to be treated with a certain amount of caution as most were calculated from comparison of $[\alpha]_D$ values. Mesylates are preferred over tosylates, triflates and chlorosulfonates as they give the best *ee*s and yields and are the easiest to prepare from the corresponding alcohols. Representative examples are shown in Scheme 5 and Equations (44) and (45). The reaction is thought to proceed via a cyclic transition state (**34**), as proposed by Suga and co-workers ⟨79BCJ2377⟩. The aluminum coordinates the sulfonate group and the arene approaches from the opposite side to give inversion of stereochemistry. When no arene is present chloride ion forms the nucleophile and 2-chloropropanoate and 3-chlorobutanoate are the products obtained in high yields and optical purities. The sulfonates also react with other aromatic systems apart from benzene, including toluene, chlorobenzene and naphthalene. Optical yields, where determined, are lower than with benzene and poor regioselectivities were observed.

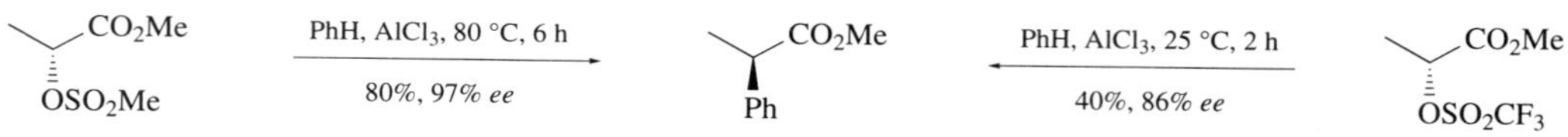

Scheme 5

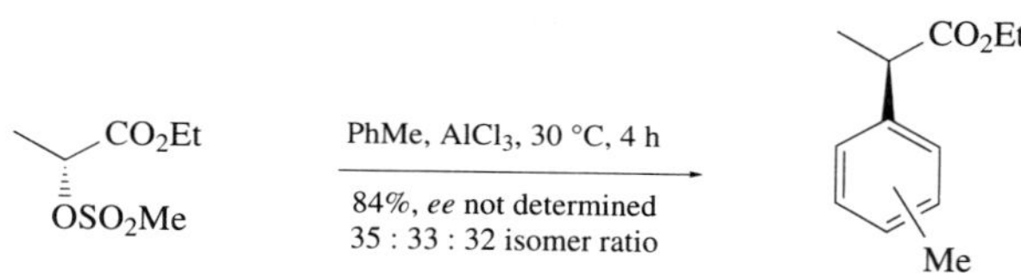

(44)

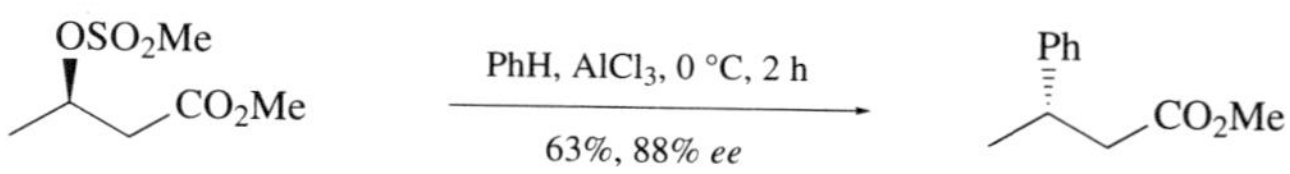

(45)

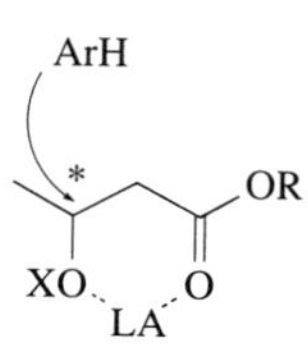

(34)

Effenberger and Weber ⟨87AG(E)142, 88CB421⟩ found that the stereochemical outcome of Friedel-Crafts alkylation using threonine triflates depended on the relative stereochemistries of the two chiral centres (Scheme 6). The different behaviours of (*RR*) and (*RS*) diastereomers is explained in terms of the different conformations of intermediate carbonium ions. The yields are low as elimination to the α,β and β,γ enones (**35**) and (**36**) occurs in 25–30% and 35–40% yields, respectively.

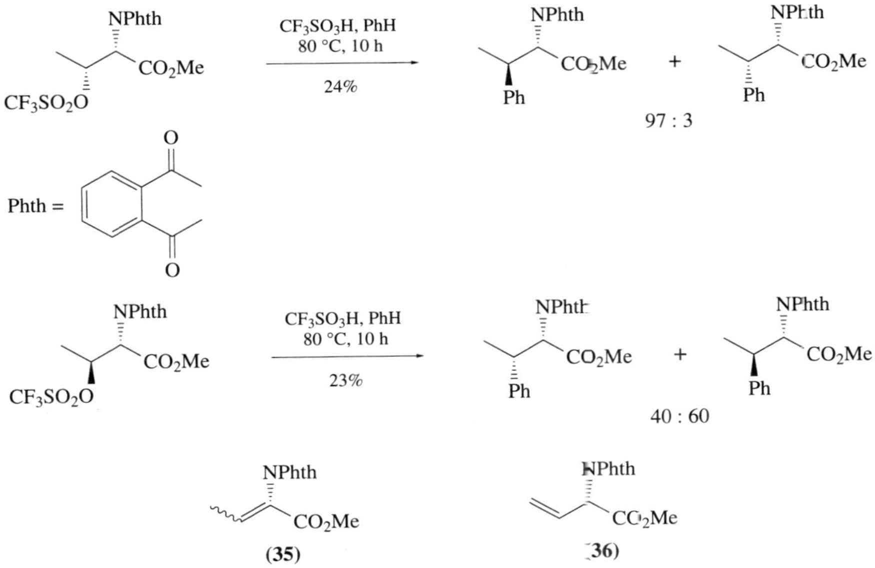

Scheme 6

(iii) Cyclisations

Intramolecular displacements of oxygen functionalities are well documented in the literature, especially for the formation of three-, five- and six-membered rings, both carbo- and heterocyclic. Lithium diisopropylamide (LDA) treatment of the sulfonate (**37**) gives the cyclopentane (**38**) in moderate yield with good diastereoselectivity (Equation (46)) ⟨88CC760⟩. Potassium hexamethyldisilazide (KHMDS) as base gave a lower *de* (88 : 12). Six-membered rings ⟨86TL943⟩ and heterocyclic systems can also be prepared (Equation (47)) ⟨90CL593, 92BCJ3487⟩. In the absence of HMPA a 21% yield (81 : 19 isomer mix) was obtained. The corresponding iodide gave eliminated product. Cyclopropanes may also be prepared (Equation (48)) ⟨86AG(E)754, 93TL1917⟩. Intramolecular Mitsunobu coupling of hydroxy-bis-sulfones and hydroxy-nitrosulfones proceeds in high yield to give cyclized products with ring sizes of three to seven (Equation (49)) ⟨92JOC3757, 92TL6723, 93JOC5892⟩. Annulation reactions with diols are also high yielding (Equation (50)). This is in marked contrast to the Mistsunobu reactions mentioned earlier.

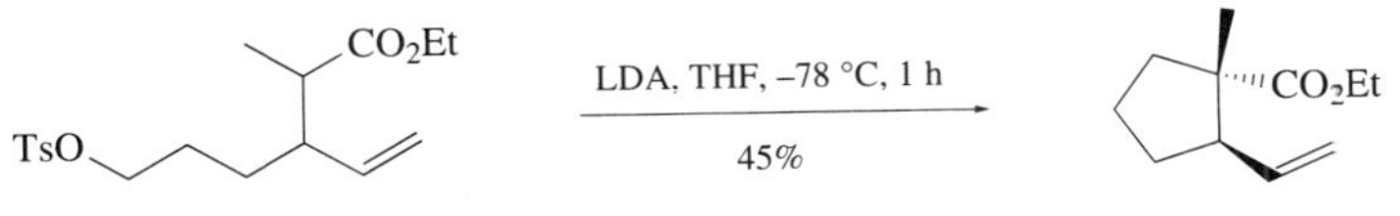

(**37**) (**38**) 92 : 8 (46)

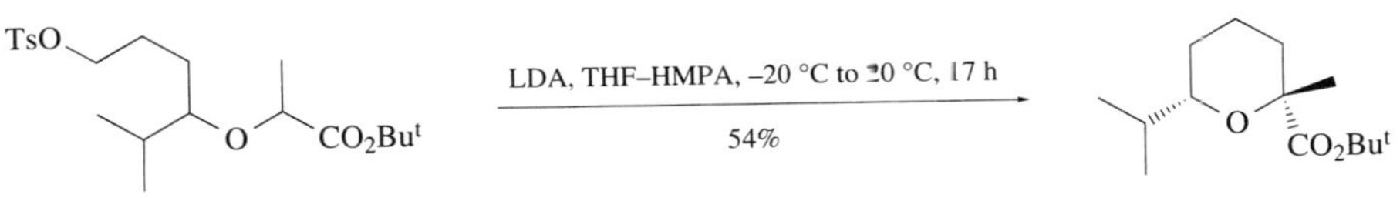

(47)

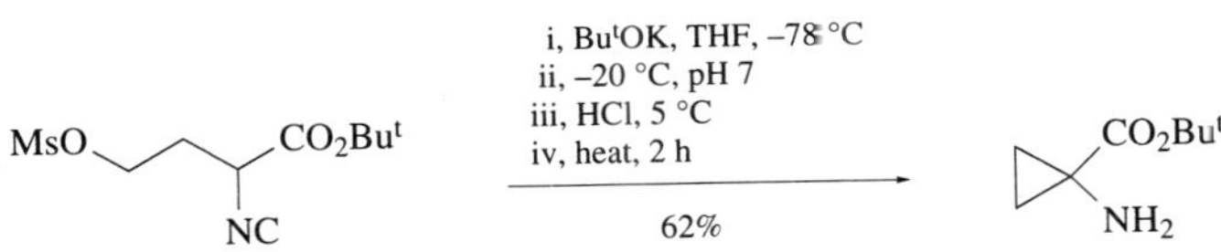

(48)

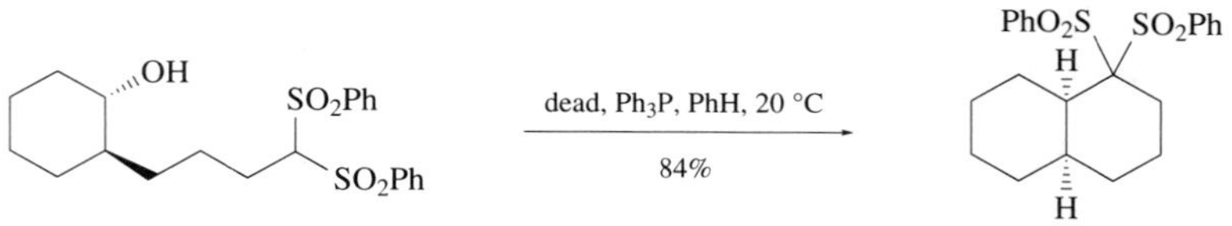

(49)

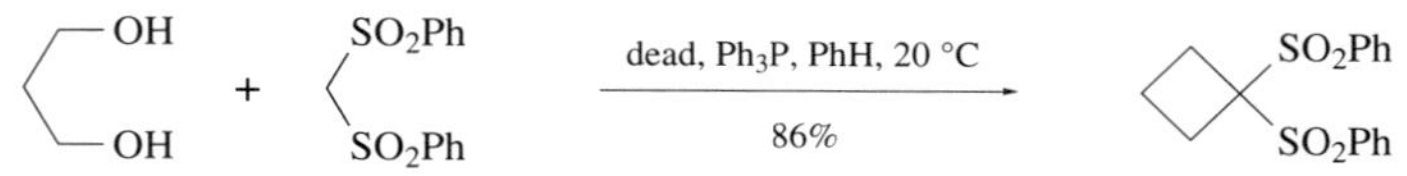

(50)

Polycyclic molecules may also be prepared by intramolecular displacement of oxygen leaving groups, including the [1.1.0]bicyclobutane (**39**) (Equation (51)) ⟨82JOC2564⟩ and spirocycle (**40**) (Equation (52)) ⟨87JCS(P2)1221⟩. Cyclisation of the anion of the sulfone (**41**) requires more forcing conditions (110 °C) (Equation (53)) ⟨88S854⟩. Dialkyl phosphate can be used as the leaving group (Equation (54)) ⟨88H1575⟩; (in this case the anion is generated by reductive cleavage of an amino-nitrile). Annulation can be effected when a bissulfonate is reacted with a molecule with two nucleophilic centres such as a diester (Equation (55)) ⟨80S389⟩. This can be used to prepare five- and six-membered rings in 34–71% yields. Moderate to good diastereoisomeric excesses can be obtained using dimenthyl succinate as the diester with dihalides and bistosylates (Equation (56)) ⟨85JA3343⟩. Five- and six-membered rings can be obtained from tosylates giving ratios of isomers >95 : <5 *trans* : *cis*. The chirality is dependent on the enolate geometry.

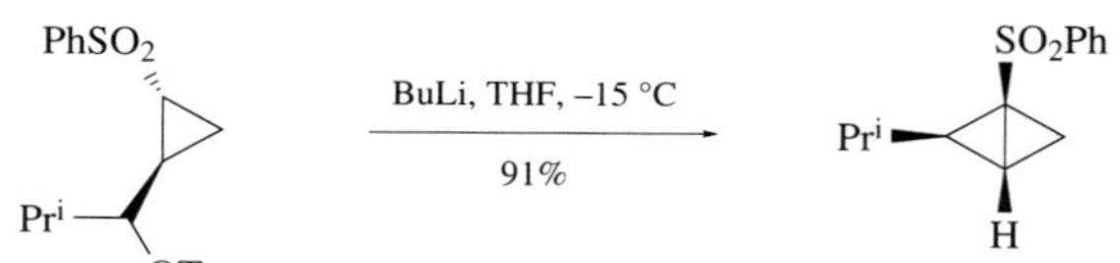

(**39**)

(51)

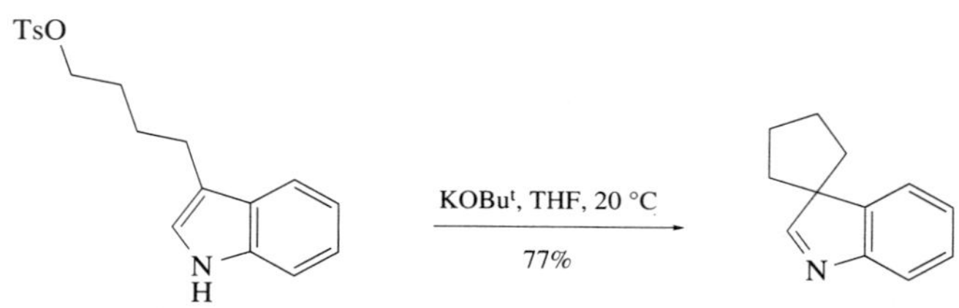

(**40**)

(52)

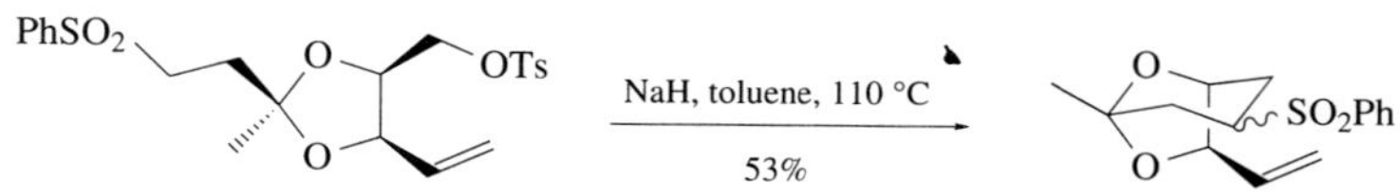

(**41**)

(53)

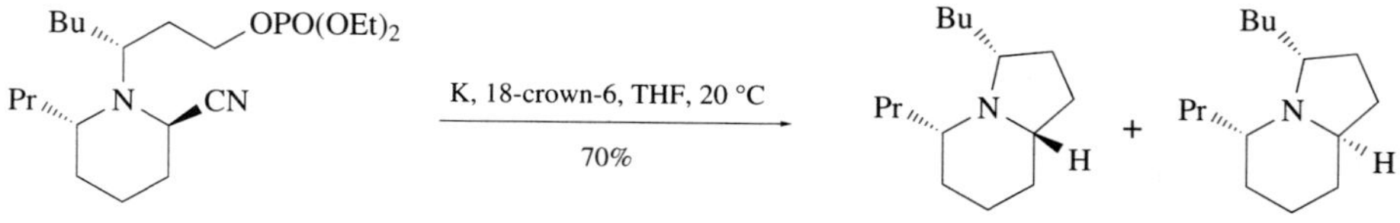

90 : 10

(54)

(55)

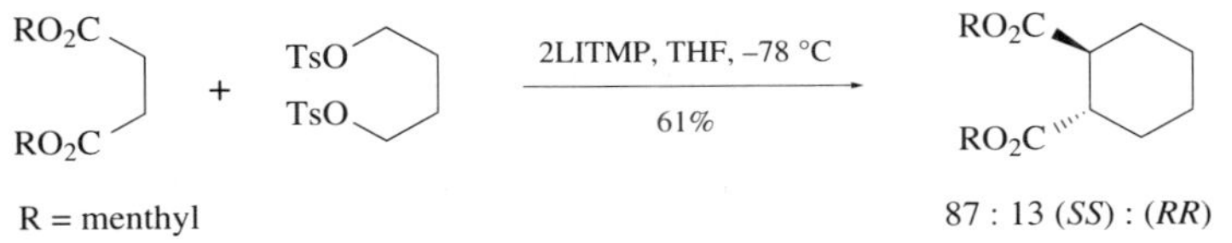

(56)

Intramolecular Friedel-Crafts alkylations are possible (Equations (57) and (58)) ⟨89HCA1537, 90JOC1338⟩. In the former case, rearrangement of the secondary carbonium ion to tertiary ion occurs.

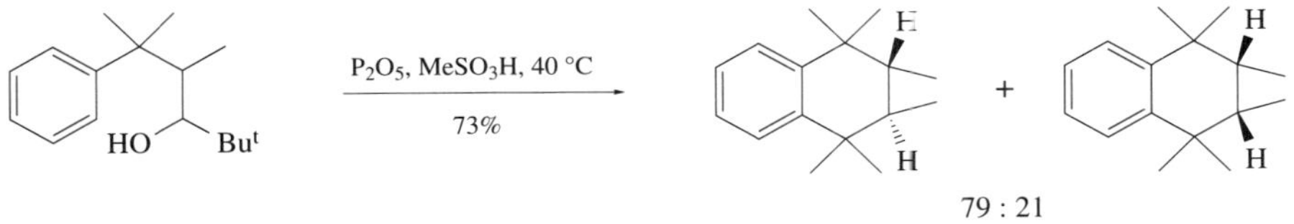

(57)

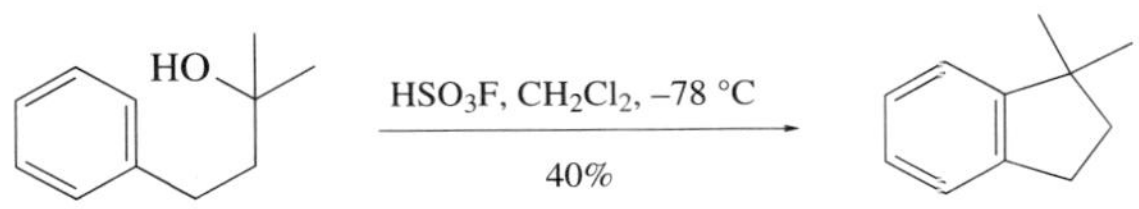

(58)

(iv) Cyclic sulfates

Cyclic sulfates (see review ⟨92S1035⟩, popularized by Gao and Sharpless ⟨88JA7538⟩, are a special case as they behave very much like epoxides (*vide infra*); they are more reactive (the much better leaving group more than compensates for the lower ring-strain) and in some cases both oxygen functions can be displaced. Copper-catalysed Grignard reagents ⟨88JA7538⟩ (Equation (59)) and dithiane (with the cyclic sulfate of a 1,4 diol) (Equation (60)) ⟨91S347⟩ have been used as nucleophiles and annulation occurs with malonate (Equation (61)) ⟨93JOC3767⟩.

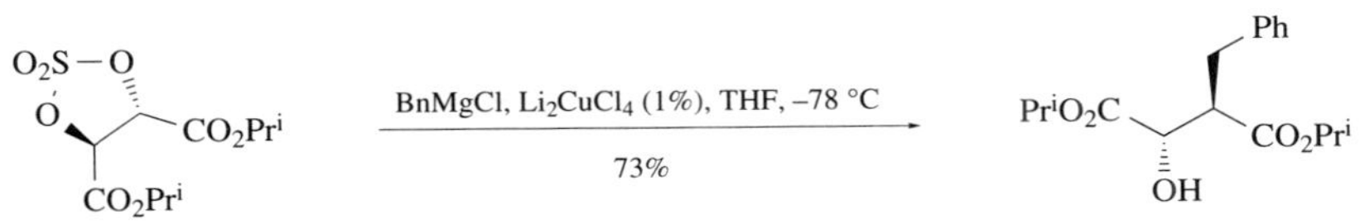

(59)

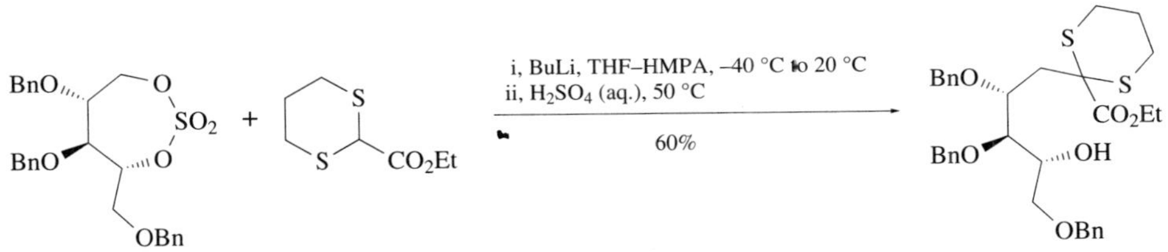

(60)

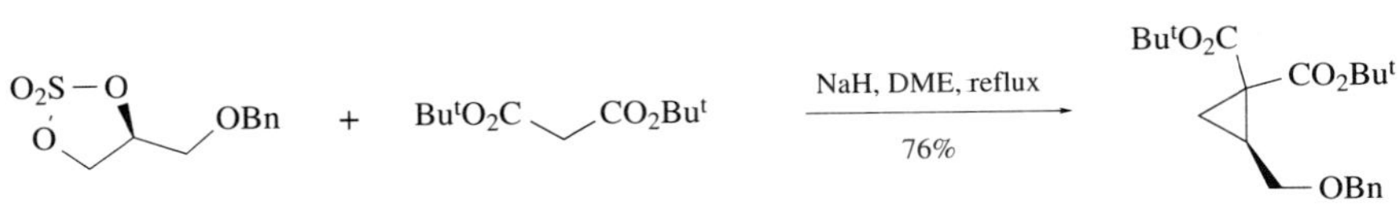

(61)

(v) Miscellaneous reactions

Hydroxyl groups can be displaced by cyanide ions under conditions which generate the chloride *in situ* ⟨67JOC855, 80S1007⟩. DMSO is by far the best solvent, giving yields in the range 45–70% for primary and secondary alcohols.

Phenoxide is the leaving group in the reaction of the phenoxy acetic acid dianion with the 9-borabicyclo[3.3.1]nonyl (9-BBN) derivative (Equation (62)) ⟨90JOC6356⟩. The reaction goes via an 'ate' complex, with simultaneous migration of a group from boron and elimination of phenoxide. Yields are in the range 42–90%. Three- to seven-membered ring ethers can be opened with carbon monoxide and trialkylsilane under cobalt catalysis to give one-carbon chain-extended diols (Equation (63)) ⟨89JA7938⟩ but the reaction has only been thoroughly investigated for oxiranes.

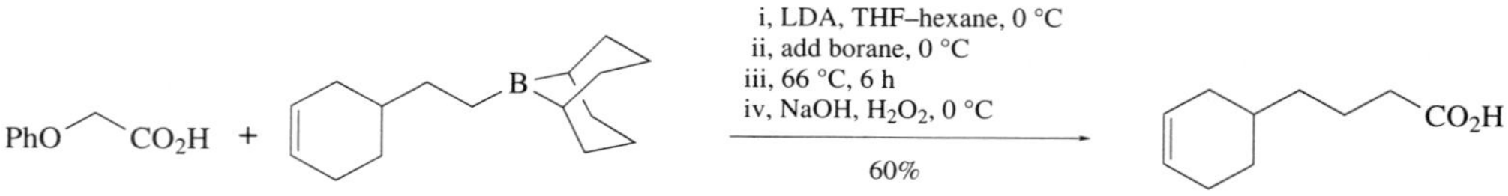

(62)

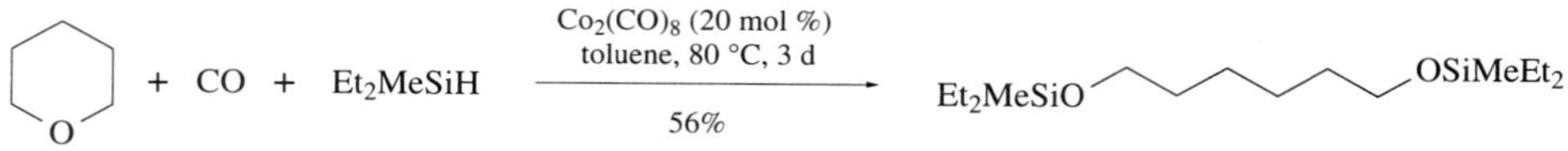

(63)

Tetrahydrofuran is ring-opened by (*R*)- and (*S*)-(−)-1-amino-2-methoxymethylpyrollidine hydrazones in the presence of a silyl triflate in high *de* (Equation (64)) ⟨93S1092⟩. The silane acts as a Lewis acid, activating the cyclic ether. *β*-Quaternary acids and esters can be prepared from tertiary alcohols and 1,1-dichloroethene in concentrated sulfuric acid in moderate to good yield (Equation (65)) ⟨65AG967, 85S493⟩.

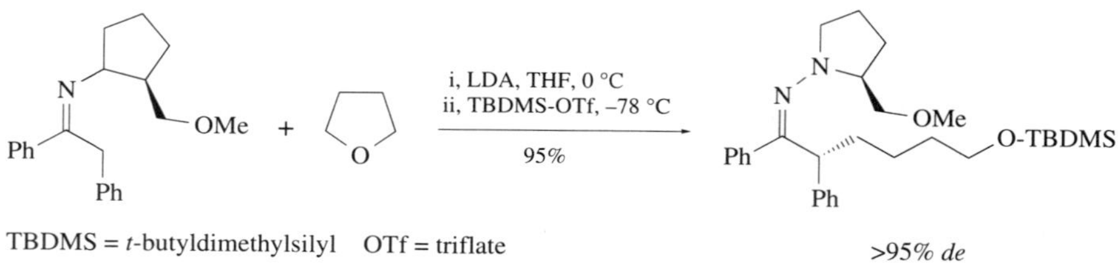

(64)

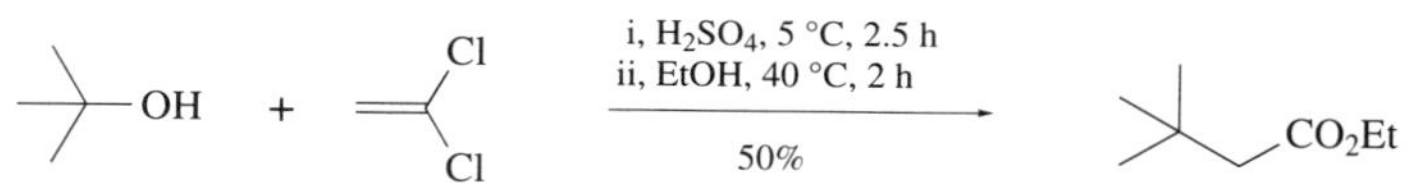

(65)

Simple tertiary alcohols can be methylated to give quaternary compounds using Me_2TiCl_2 ⟨81CC237⟩ or Me_3Al at 150 °C ⟨72CC595⟩.

1.05.1.2.2 Allylic alcohol derivatives

Allylic alcohols and their derivatives have proved to be highly versatile molecules due to their high reactivity and utility. The adjacent double bond has a great effect on the hydroxyl group, stabilising a positive charge at the allylic position and complexing to transition metals (especially palladium). Attack of a nucleophile can be at the oxygen centre (S_N2) or at the distal carbon atom (S_N2') (with either *syn* or *anti* attack) with movement of the double bond (Scheme 7). Alternatively the stabilisation of the positive charge over the allyl system makes an S_N1 mechanism possible, especially under acidic conditions. The stereochemistry of the reaction greatly depends on the reagent used as well as on the substrate itself. Allylic substitution has been reviewed ⟨80T1901⟩ as have intramolecular reactions ⟨92T7383⟩.

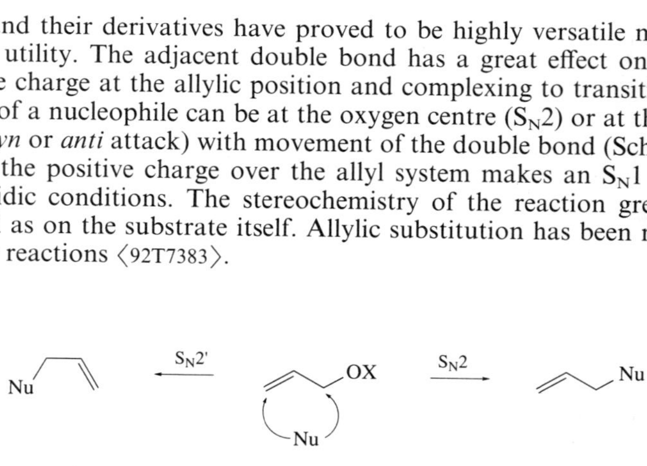

Scheme 7

(*i*) *Main group metals and copper*

A great range of leaving groups is available on derivatisation of allylic alcohols. As well as sulfonate esters, carboxylic esters, carbonates, phosphates and ethers have been used. Allylic tosylates readily alkylate enolates ⟨84JA1079⟩, zinc reagents (Equation (66)) ⟨87JOC4418⟩ and 1,3-dicarbonyl compounds (Scheme 8) ⟨88CB1461⟩. In the last case attack occurs at the less hindered end of the allylic system. Under copper catalysis organometallic reagents readily displace acetate and similar leaving groups. The regiochemistry of attack at allylic acetates has been studied by a number of groups including Bäckvall *et al.* (Scheme 9) ⟨90JA6615⟩ (and references cited there). They found that the regiochemistry of copper-catalysed Grignard additions to primary allylic acetates was dependent on a number of factors. α-Addition was favoured by rapid addition of the Grignard reagent, THF and lower temperatures. γ-Attack is favoured by slow addition of Grignard reagent, higher concentrations of catalyst, higher temperatures and ether as solvent. Heterosubstituents at the δ-position also had a γ-directing effect, as did addition of $BF_3 \cdot OEt_2$. The solvent effect is most marked when CuCN is used as catalyst. The remarkable change is believed to be due to competition between two different catalytic cycles involving different copper species. Other groups have found high γ-selectivity using CuCN catalyst with allylic pivalates (copper halides tend to give mixtures of α and γ substitution products) ⟨86JOC2884⟩ and solvent effects ⟨73CC1085⟩. However each system appears to have its own idiosyncrasies of selectivity (Equation (67)) ⟨83S804⟩.

In cyclic systems the nucleophile often attacks *anti* to the leaving group, that is, the nucleophile and leaving group are on opposite sides of the allylic moiety. *syn* Attack occurs under some conditions for example allylic carbonates and carbamates where the nucleophile is delivered to the γ-carbon atom from the same face as the leaving group (Equations (68) and (69)) ⟨87JOC897, 88TL5661, 89JOC3239⟩. Allyloxybenzthiazoles show a similar effect (γ, *syn*) in 6-membered rings in reactions involving copper nucleophiles ⟨90JOC2294⟩. Denmark and Marble have made use of an asymmetric S_N2' reaction where the leaving group blocks one face of the allyl system and directs the nucleophilic attack (Equation (70)) ⟨90JOC1984⟩. In this system *ees* of up to 95% were obtained. Displacement of the dialkyl phosphate unit in compound (**42**) occurs in high *ee* with inversion of configuration (*anti* attack at the γ-carbon atom) (Equation (71)) ⟨91SL251⟩. In this case phosphate proved superior to other leaving groups in respect of both chirality transfer and regioselectivity. Ethylidene-1,3-dithiane attacks from the γ-position to displace phosphate ⟨79JOC3428⟩.

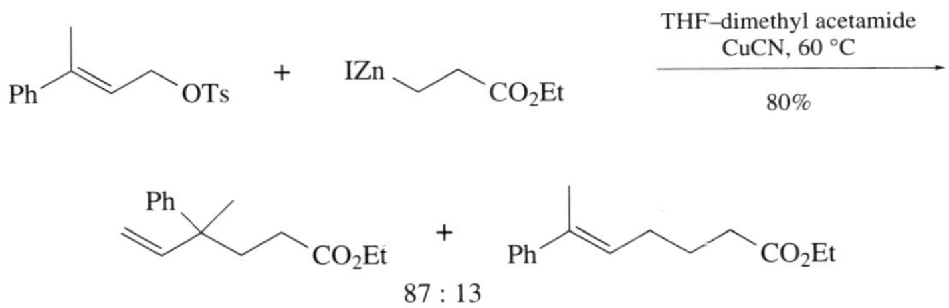

(66)

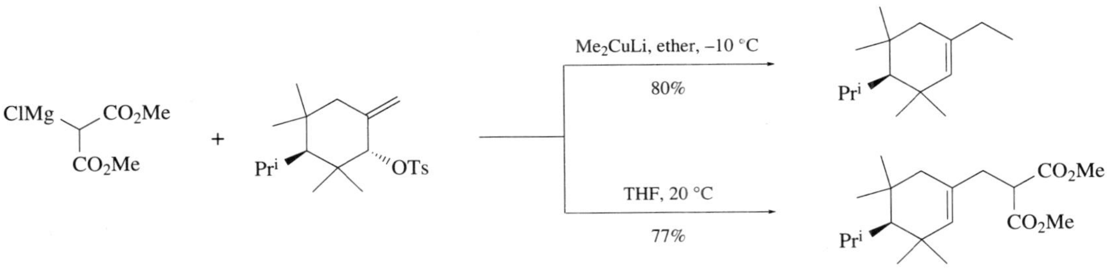

Scheme 8

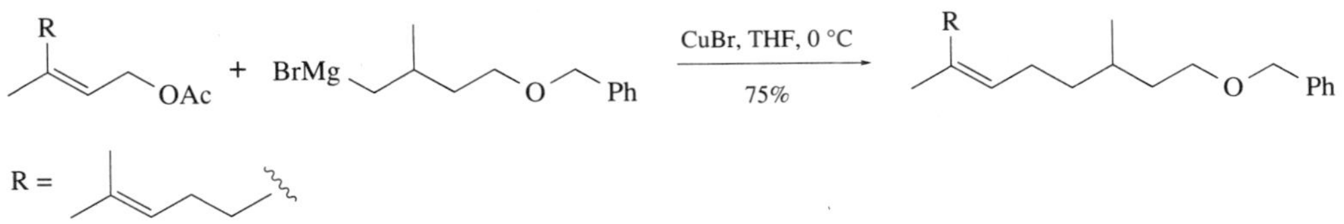

(67)

Solvent	*Addition time*	*Temp.* (° C)	*cat.*	(%)	
THF	1.5 min	−30	Li_2CuCl_4	2	96 : 4
THF	2.5 h	0	Li_2CuCl_4	5	14 : 86
THF	2 min	−78	CuCN	10	100 : 0
ether	3 min	20	CuCN	10	8 : 92

Solvent	*Addition time*	*Temp.* (° C)	*cat.*	(%)	
THF	4 min	−70	Li_2CuCl_4	2	99 : 1
THF	30 h	0	Li_2CuCl_4	10	15 : 85
THF	25 min	0	CuCN	10	99 : 1
ether	3.5 h	0	CuCN	10	0 : 100

Scheme 9

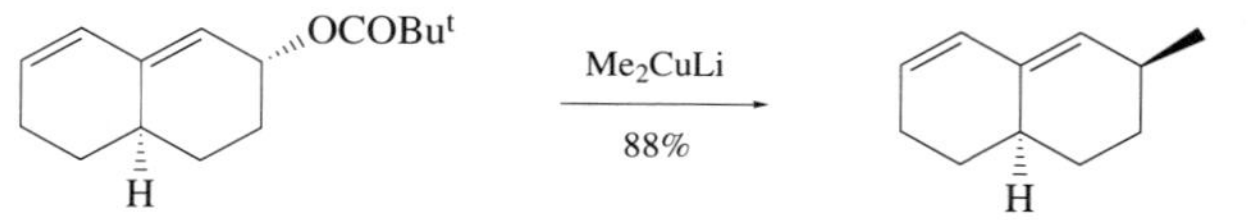

96 : 4 *cis* : *trans* 80 : 20 *trans* : *cis*

(68)

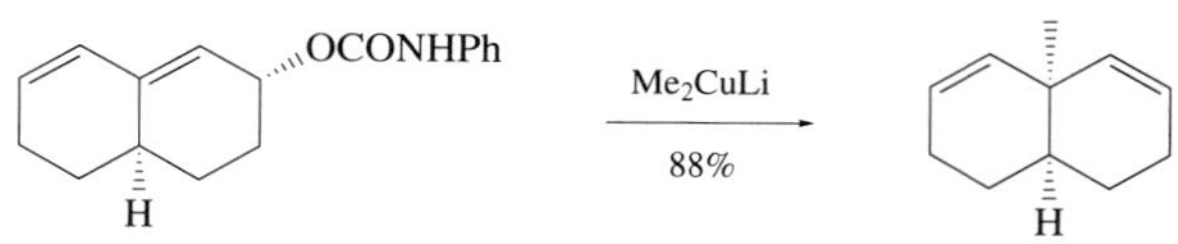

(69)

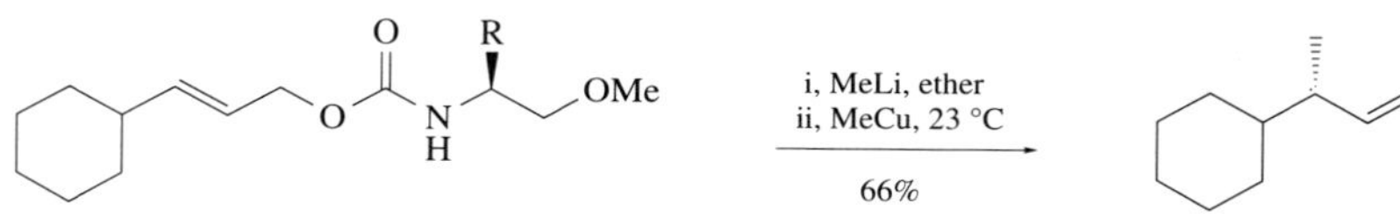

R = naphthyl 91% *ee*

(70)

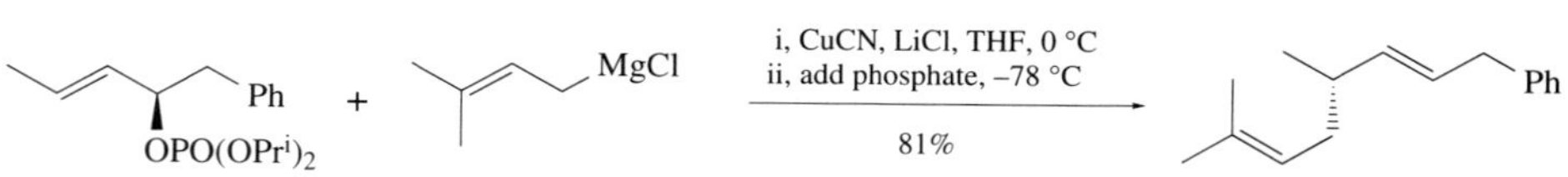

(42)

(71)

The reaction of copper reagents with γ-mesyloxy-α,β-enoates has been extensively studied, especially systems with a chiral centre at the δ-position ⟨92SL769⟩. Unlike many allylic compounds, the acetate group in γ-acetoxy-α,β-enoates is not displaced by organocopper reagents, reduction being the main reaction ⟨85CC1452, 88CC449⟩. Mesylate is displaced in an *anti* S_N2' fashion to give an α-alkylated γ,δ-enoate ⟨89JA4864⟩. Alkyl or dialkylcyanocuprate–boron trifluoride complex is the best reagent, giving >99% α-alkylation in 78–98% yield. The chirality of the new chiral centre is determined by both mesylate chirality and double bond geometry; the new double bond is always (*E*). It is thus possible to prepare all four isomers (i.e. two enantiomeric pairs) of α,δ dichiral compounds in two ways (Scheme 10) ⟨89JA4864⟩. Alkyl groups at the α-C are tolerated (Equation (72)) ⟨89JOC4055⟩ as are protected δ-amino groups (Equation (73)) ⟨90AG(E)801, 91JOC4370⟩. Zinc cuprates obviate the need for a Lewis acid in the reaction, so reducing problems with acid-sensitive groups ⟨92JOC1024⟩. Reduction of the mesyloxy group occurs with vinyl copper reagents but use of vinyl zinc cuprates overcomes this problem ⟨93T9479⟩. In a few cases a *syn*-S_N2' reaction is observed ⟨93JOC1207⟩. For some *syn*-mesylates, for example those with a protected ring nitrogen atom attached at the δ-carbon centre, *syn*-S_N2' is the predominant mode of reaction (Equation (74)). Only with bulky groups on copper (e.g., Bu^t,Pr^i) are significant amounts of *syn*-S_N2' products seen with other substrates.

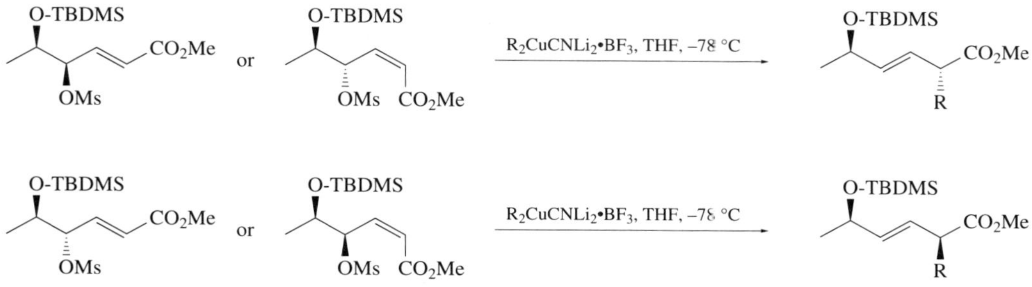

R = Me, Et, Pr, Bu

Scheme 10

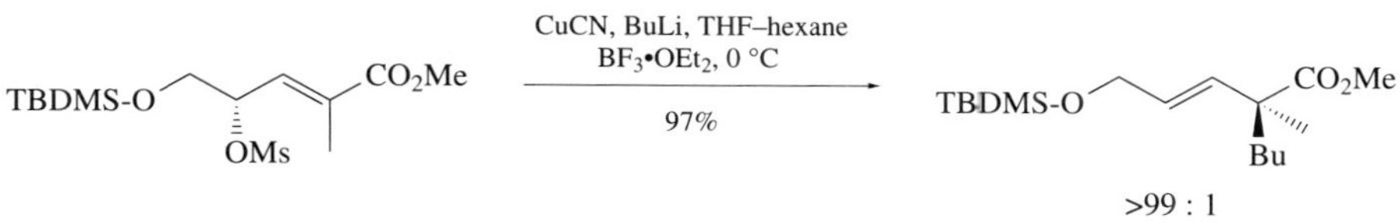

(72)

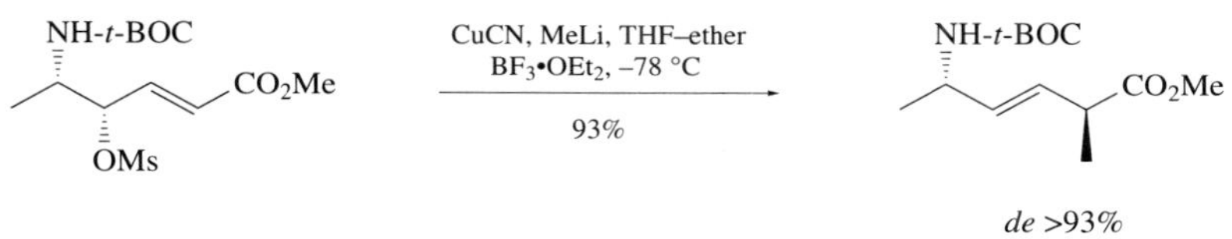

(73)

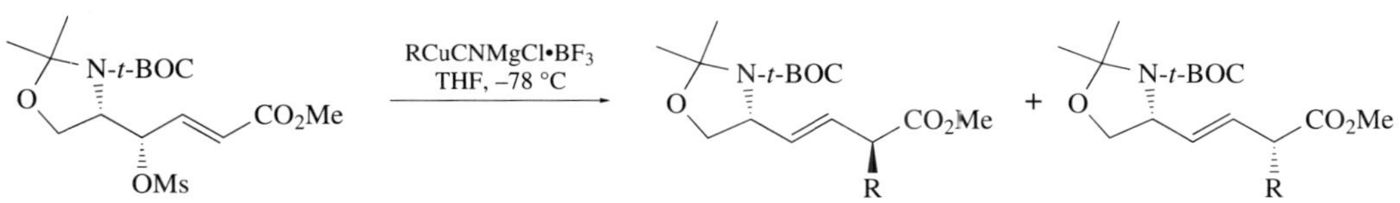

(74)

R	*Yield* (%)	
Et	96	36 : 64
Bu^i	98	25 : 75
Bu^t	83	2 : 98

Lewis acid catalysed reactions involving silyl enol ethers give good yields of the γ,δ unsaturated ketones (Equation (75)) ⟨84CB322⟩. An adjacent thioether acts as a directing group in the TMS triflate catalysed reaction of the unsymmetrical acetate (**43**) (Equation (76)) ⟨91TL4311, 93BCJ848⟩. Isomer ratios were in the range 89 : 11 to 100 : 0. The reaction probably proceeds via an episulfonium ion. Mukaiyama *et al.*, found that the trityl cation catalyses the reaction of allyl ethers (secondary and tertiary) with silyl enol ethers (Equations (77) and (78)) ⟨79S751, 86CL1009⟩. Tertiary ethers react at the γ position while secondary ones give poor α,γ selectivity.

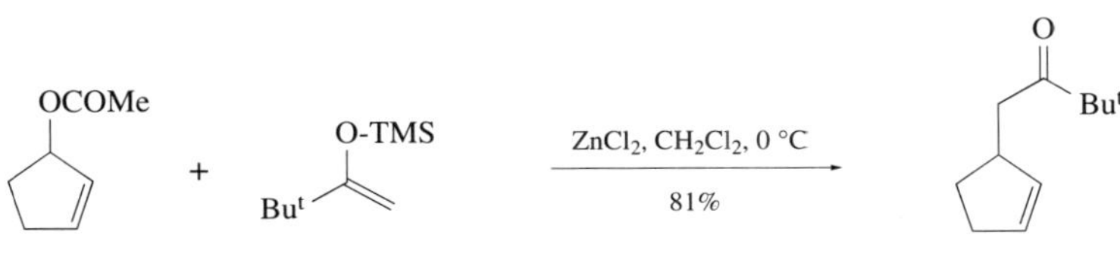

(75)

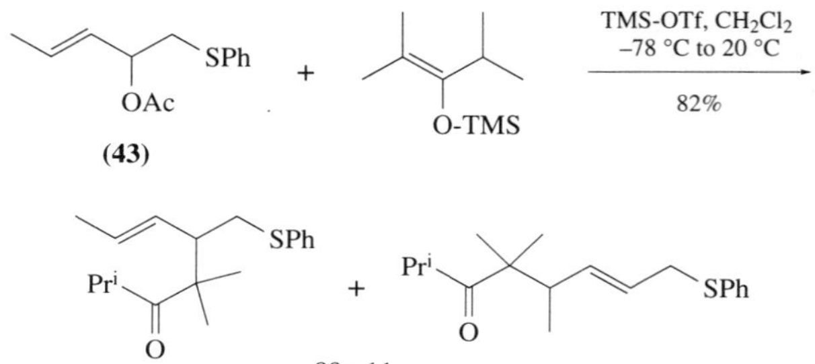

(43)

89 : 11 (76)

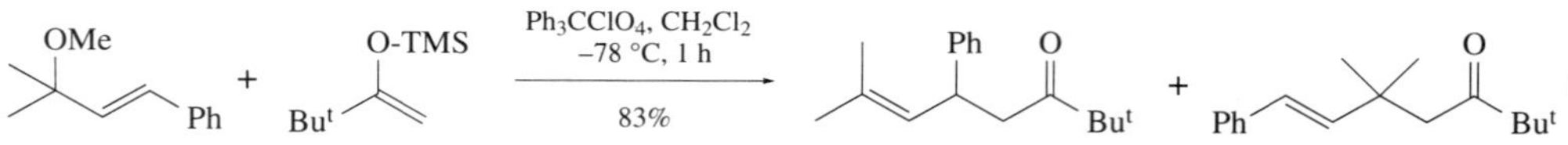

92 : 8 (77)

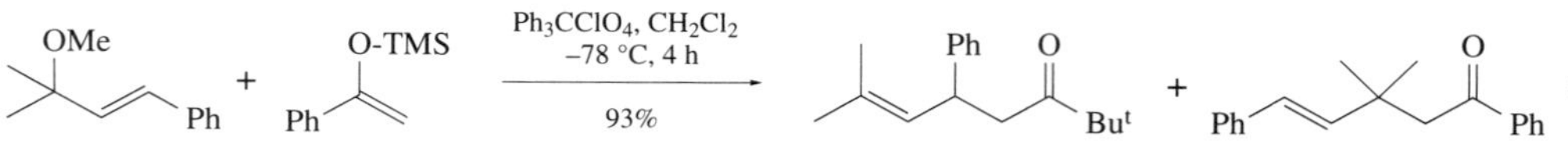

48 : 52 (78)

Allyl vinyl ethers usually undergo a [3,3] sigmatropic rearrangement (the Claisen rearrangement) (Chapter 1.17). However, on treatment with 3 M $LiClO_4$ in Et_2O, allyl vinyl ethers rearrange in a [1,3] sigmatropic fashion to give γ,δ enones (Equations (79)–(81)) ⟨91JA5488⟩. The reaction is not concerted, as shown by the different stereochemical features of the reactions displayed in Equations (80) and (81) (the former gives largely retention; the latter exclusively inversion) and by a crossover experiment with the two ethers (**44**) and (**45**) which gives all four possible products. Pearson's and Grieco's groups have independently discovered that, under the same conditions, allylic acetates and alcohols react with silyl enol ethers and other nucleophiles to give γ,δ enones ⟨92JOC2986, 92TL4735, 93CC510⟩. Representative examples are shown in Equations (82) and (83) and Scheme 11. The concentration of $LiClO_4$ is critical; concentrations lower than 3 M give substantial amounts of by-products. The products obtained are those with the most highly substituted double bond.

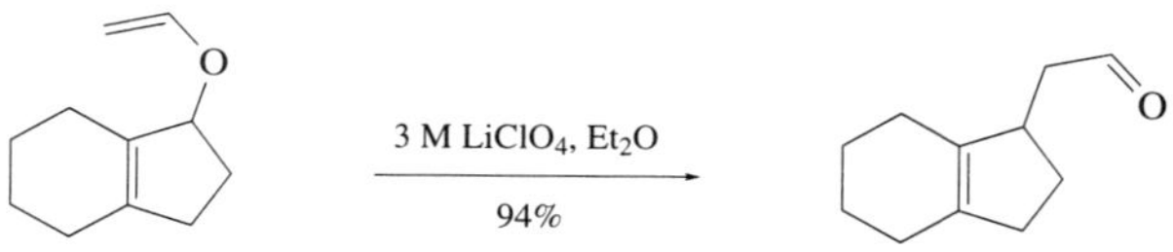

(79)

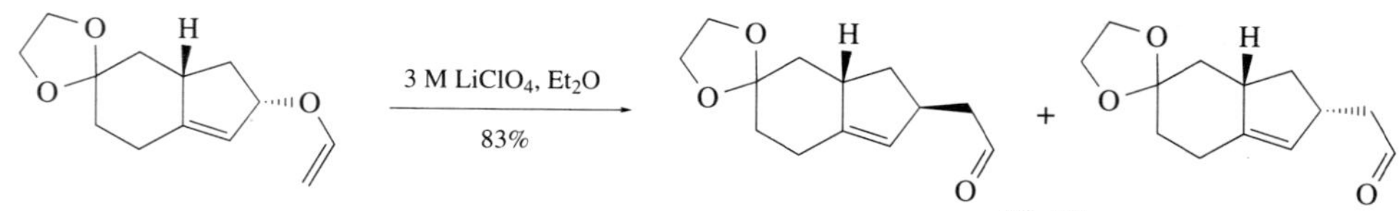

16 : 84 (80)

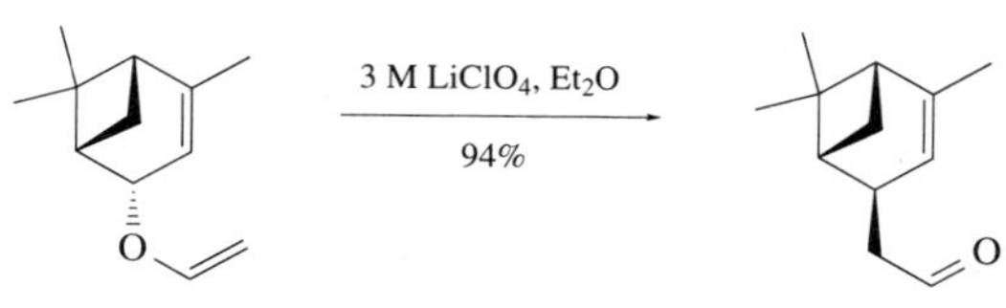

(81)

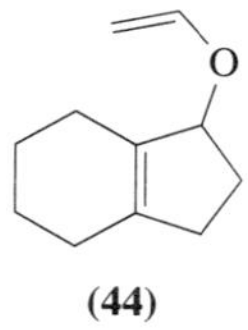

(**44**)

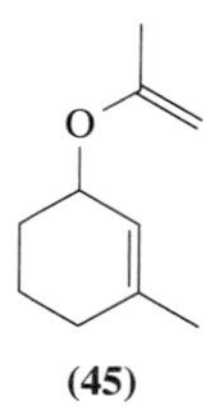

(**45**)

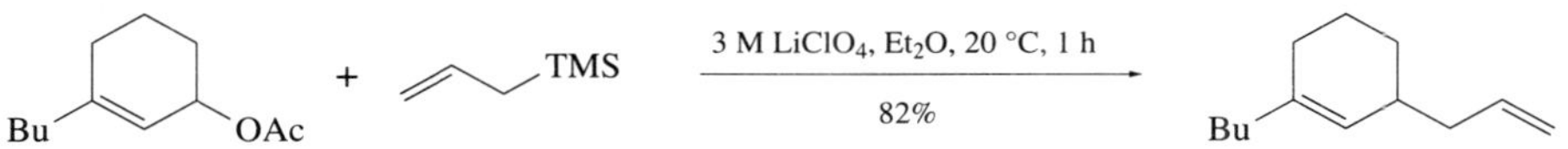

(82)

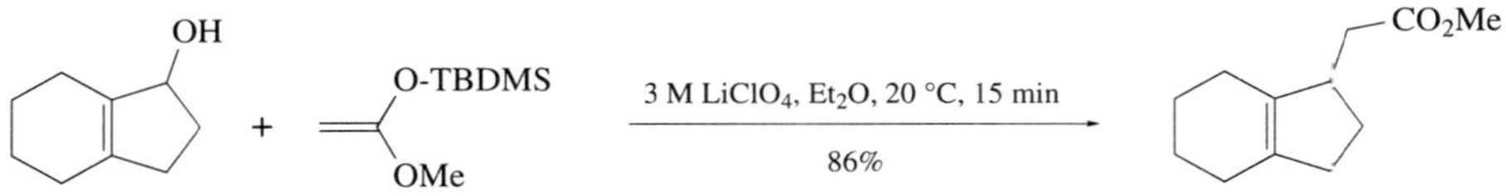

(83)

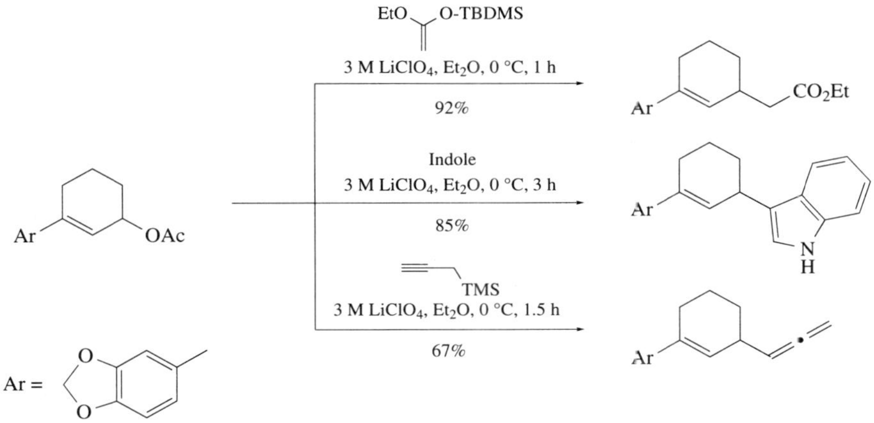

Scheme 11

β,γ-Unsaturated nitriles are the products obtained on treating allylic methoxymethyl (MOM) ethers with TMS cyanide (Equation (84)) ⟨87CL1167⟩. The corresponding α,β unsaturated nitriles are not detected. For the cyclic systems (5- and 6-membered rings) shown in Equation (85) ⟨87TL5521⟩ the mechanism of the reaction is probably via Michael attack followed by elimination of acetate (or silyl ether) rather than by a simple S_N2 or S_N2' pathway. Similar reactions on a 2-nitroallyl acetate system have been carried out ⟨84HCA261⟩. Ethers may also be displaced ⟨84CL1505⟩. The displacement reaction for the cyclic ether (**46**) and similar compounds is solvent dependent. No reaction occurs in ether, elimination to a diene occurs in DME ⟨93TL4591⟩ while in TMEDA S_N2' ring opening takes place (Equation (86)) ⟨92TL2617⟩.

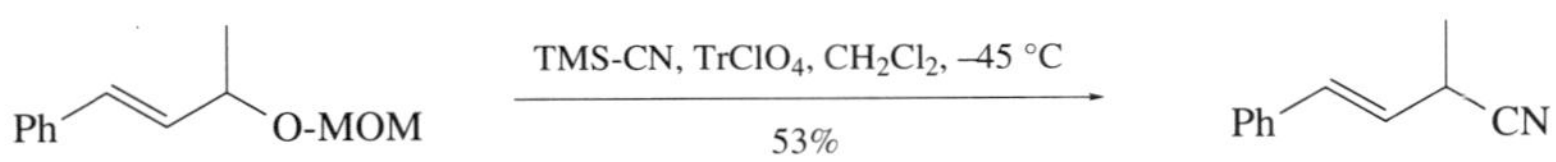

(84)

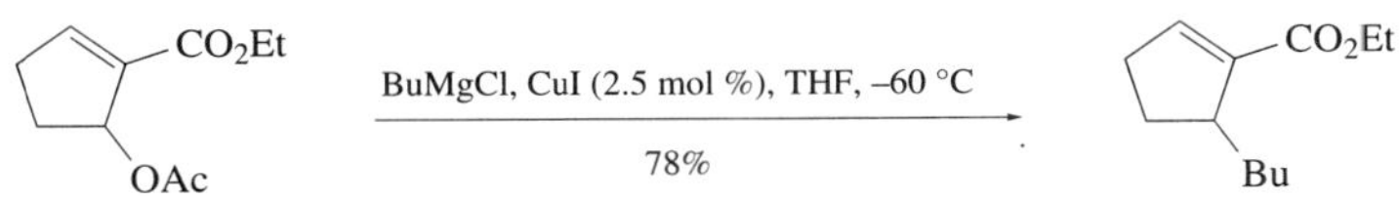

(85)

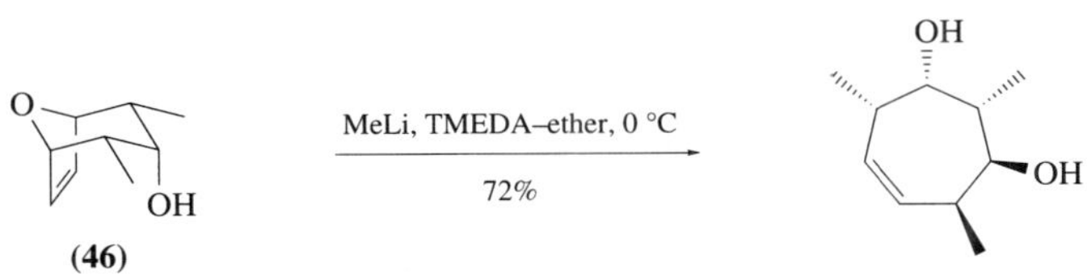

(**46**)

(86)

(ii) π-Allylic intermediates

Loss of the leaving group from an allylic system leaves an allyl cation which can readily co-ordinate to certain transition metals, especially palladium and nickel but also iron, molybdenum and tungsten. These allyl transition metal complexes can then react with a variety of nucleophiles including carbon-based systems.

(a) Palladium and nickel. Allylic substrates with a good leaving group (e.g., halide, ester, carbonate) readily react with palladium(0) species (usually in the presence of a phosphine ligand) to give a π-allylpalladium complex (**47**)(Scheme 12). Nucleophiles can then attack at either of the terminal carbon atoms, in most cases from the opposite face to the palladium (i.e., the reaction proceeds with overall retention of configuration). However some 'hard' nucleophiles react with inversion. (*Z*) Double bonds are often at least partly isomerized (via π–σ–π complexes) to the more stable (*E*)-configuration. Nucleophilic attack is usually directed at the less substituted end of the allylic system. The regioselectivity of malonate attack on dienyl acetates has been studied but there is disagreement concerning the details of the selectivity and mechanism ⟨93JA6609, 93TL7513⟩. The area has been extensively reviewed ⟨80ACR385, B-80MI 105-01, 82COMC-I(8)799, 82PAC197, B-85MI 105-01, 86T4361, 87JOM(334)243, 92TA1089⟩. The simplest nucleophile is carbon monoxide, which can readily act as a ligand for many transition metals including palladium. Carbonylation of allyl phosphates and acetates can be accomplished in the presence of a hindered base to give β,γ enolates (Equations (87) and (88)) ⟨93JOC1538⟩. The reactions proceed best under pressure but acceptable yields can be obtained at atmospheric pressure. The carbonylation of allylic acetates requires the presence of sodium bromide to displace acetate from the coordination sphere of palladium, otherwise acetate rejoins to the allyl unit. Carbonylation occurs at the least substituted end of the system, with isomerisation to the (*E*)-alkene taking place in the disubstituted examples. The reaction proceeds with inversion of configuration (Equation (89)) as CO migrates intramolecularly from palladium to the allyl unit, rather than attacking from the opposite face. Other examples involving carbamates ⟨92SL323⟩ and cyclic carbonates ⟨87TL3497⟩ have been reported.

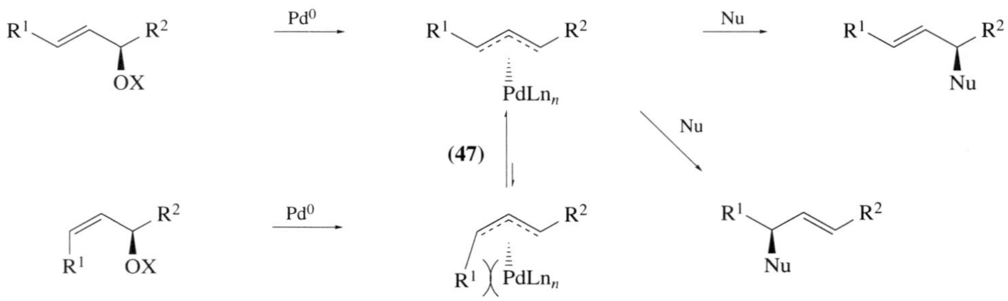

Scheme 12

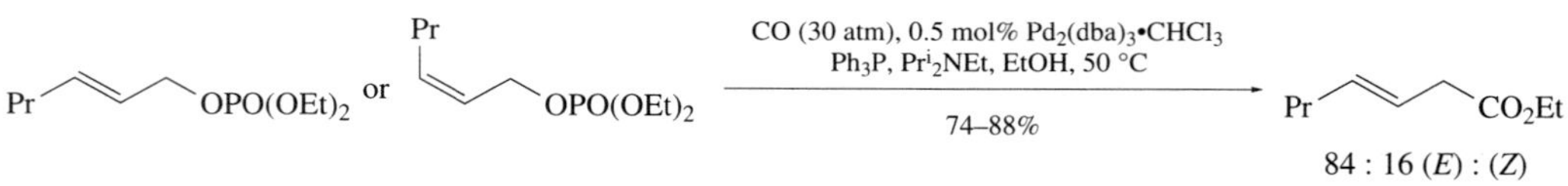

(87)

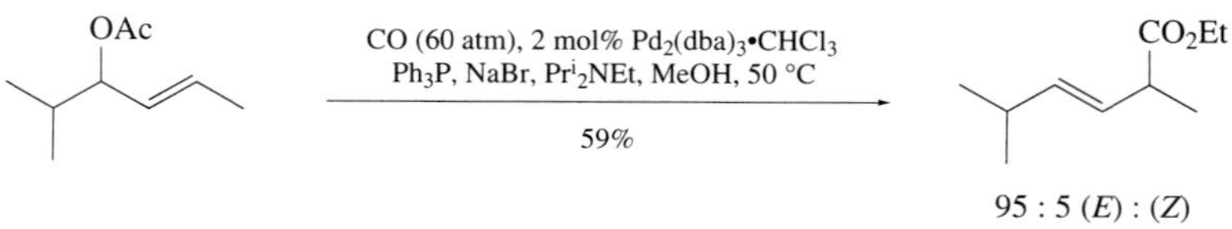

(88)

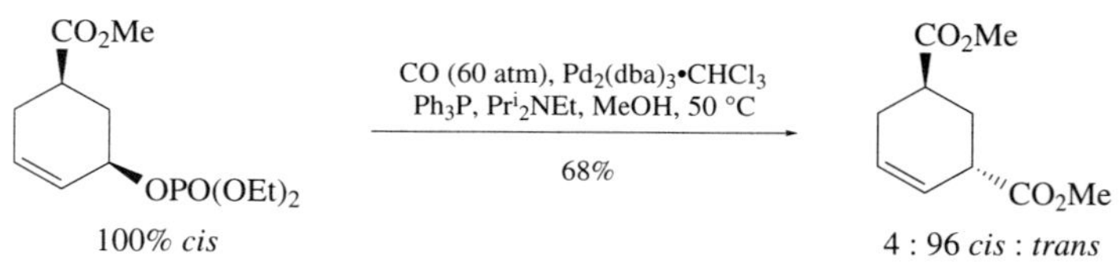

(89)

Cyanation of allylic acetates and carbonates can be carried out using TMS-CN ⟨93JOC16⟩ (Equation (90)). Alkene isomerisation was observed; the (*E*) : (*Z*) ratios depend on the bulk of the substituents.

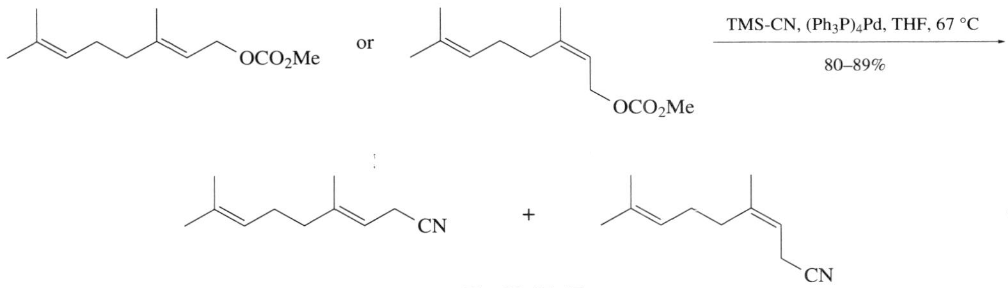

71 : 29 (*E*):(*Z*)

(90)

Compounds with an acidic hydrogen atom attached to a carbon centre will act as nucleophiles in the palladium-catalysed allylation reaction. Malonate anion displaces phosphate, acetate and other leaving groups (Equations (91) and (92)) ⟨82TL5549, 89JOC2759⟩. The overall stereochemistry of replacement of the leaving group is retention of configuration, with isomerisation of acyclic (*Z*) double bonds. When carbonates are used as leaving groups, alkoxide is generate *in situ* and so it is not necessary to provide a base to deprotonate the nucleophile (Equation (93)) ⟨85JOC1523⟩. Isocyanocarboxylates also react in good yields ⟨87TL4849⟩. Reactions can be carried out intramolecularly (Equation (94)) ⟨84TL5921, 89TL617⟩. In the last case there is a very high degree of chirality transfer. Most compounds with two electron withdrawing groups attached to a carbon centre (i.e. —CH(EWG)$_2$) will act as nucleophiles. Less stabilized nucleophiles such as nitro compounds, nitriles, sulfones and ketones can be allylated using diallyl carbonate ⟨85JOC1523⟩. Allenyl phosphates also react with stabilized nucleophiles (Equation (95)) ⟨88S983⟩.

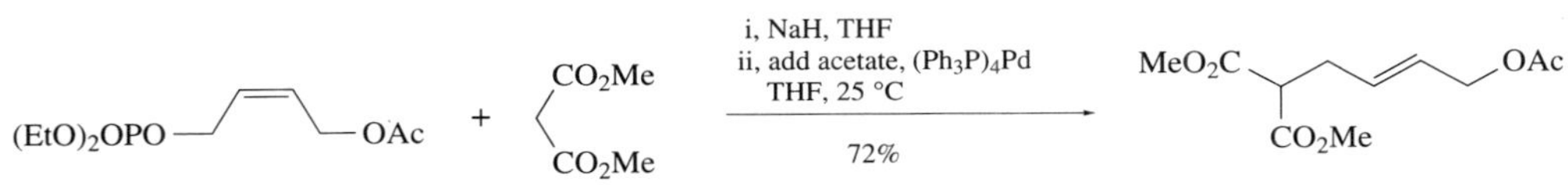

(91)

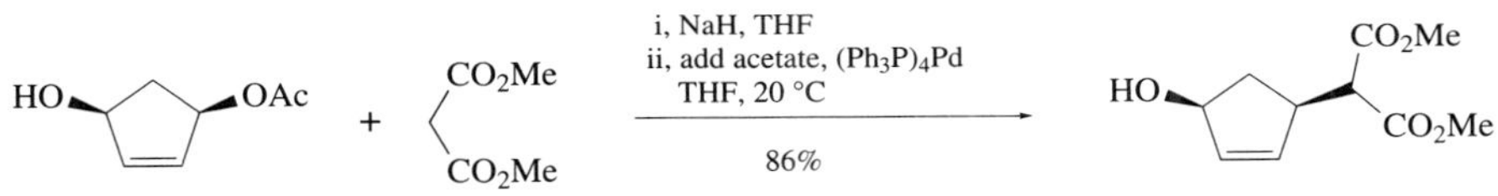

(92)

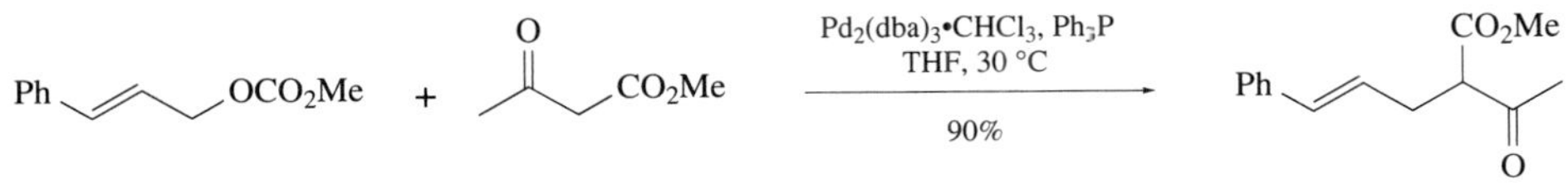

(93)

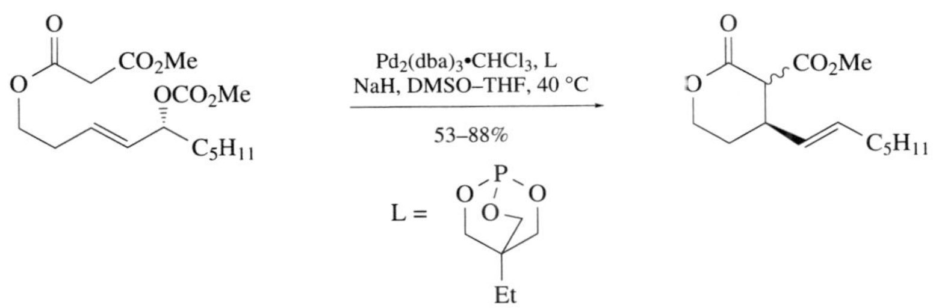

(94)

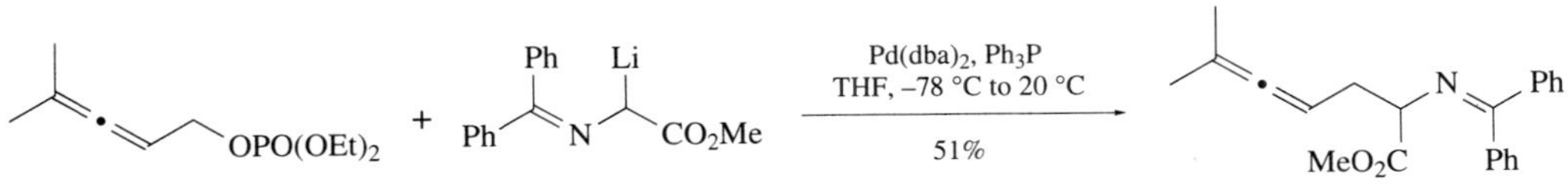

(95)

When the allylic unit contains an allylic silyl or stannyl group a palladium-coordinated trimethylenemethane (TMM) synthon is formed which can be used to annelate a five-membered ring onto activated alkenes (Equation (96)) ⟨86AG(E)1, 89JA7487⟩. Spirosystems can be prepared with moderate diastereoselectivity ⟨89TL1495⟩. Trost *et al.* have obtained some diastereoselectivity in the reaction using a cyclic chiral auxiliary (Equation (97)) ⟨89JA6482⟩. This type of reaction also encompasses additions to vinyl sulfoxides ⟨89TL1803⟩, imines ⟨86CC1201⟩, aldehydes ⟨90JA408⟩ and dienes ⟨87JA3483⟩ (generating cycloheptenes). A related reaction is the palladium-catalysed addition of 2-bromozincmethyl allylic ethers to ketones and imines ⟨88TL3579, 93TL3309⟩.

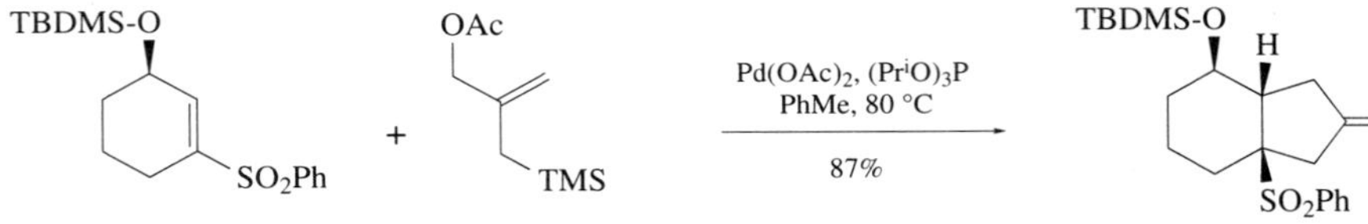

(96)

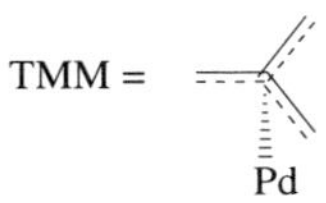

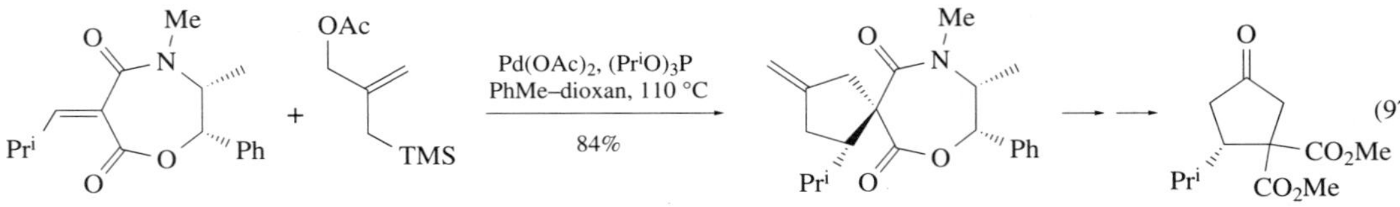

(97)

A number of groups have investigated enantioselective allylic substitutions using chiral ligands to coordinate the palladium with varying success (reviews:- ⟨91COS(4)585, 92TA1089, 93AG(E)547⟩. Most of the systems have symmetrically substituted allyl units with termini rendered enantiotopic by the chiral ligands on palladium. Most ligands used only give high *ee*s in a few cases. Representative examples are shown in Scheme 13 ⟨88PAC7, 91HCA232, 93AG(E)566, 93TL1769, 93TL3149⟩. Asymmetric [3 + 2] cycloadditions (up to 73% *ee*) ⟨89TL375⟩, enolate alkylations (57% *ee*) ⟨86TL4573⟩ and alkylation with chiral enolate (and achiral ligands) (40–82% *ee*) ⟨91CC372⟩ have been reported.

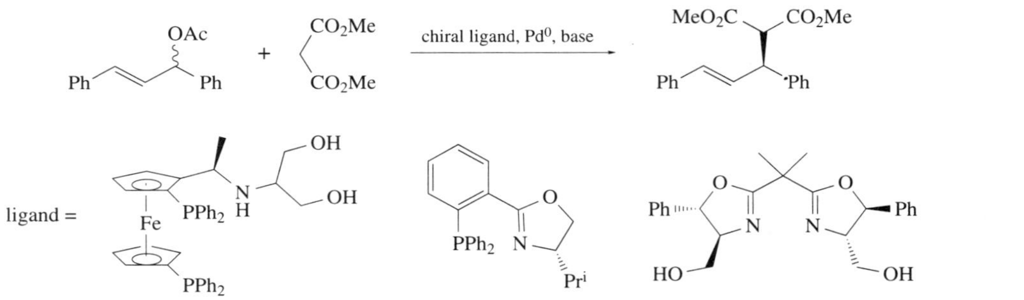

Scheme 13

Vinyl tin ethers have been used as nucleophiles and give clean monoalkylation of the resultant ketones ⟨80TL2591, 87CL1237⟩ (Equation (98)). The overall stereochemistry is retention of configuration; the tin does not co-ordinate to the palladium before adding, but approaches from the opposite face. Examples of additions of Grignard reagents are rare due to competing reactions; advantages are gained on using copper-based chemistry, and zinc and tin reagents have been used ⟨83JOC5302, 89JA4981⟩. Sato and co-workers have reacted a Grignard reagent with chiral silylallyl phosphates with high regio- and stereo-selectivities (Equation (99)) ⟨93CC1595⟩. Palladium gave a much better regiocontrol than nickel and copper and better enantiopurities than iron (<93% *ee*). The silyl unit is only partially responsible for the stereocontrol as the *t*-butylallyl compound gave poor regiocontrol.

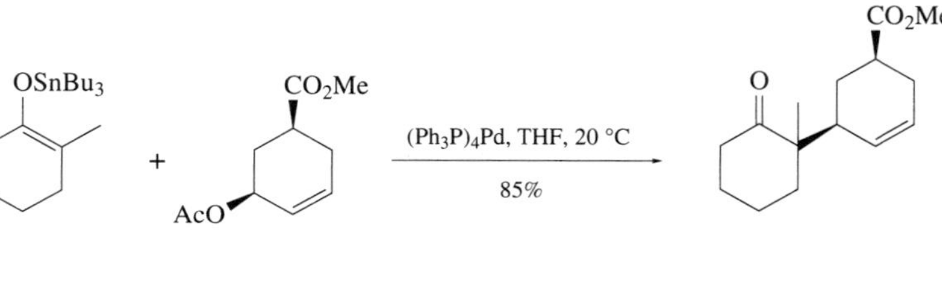

(98)

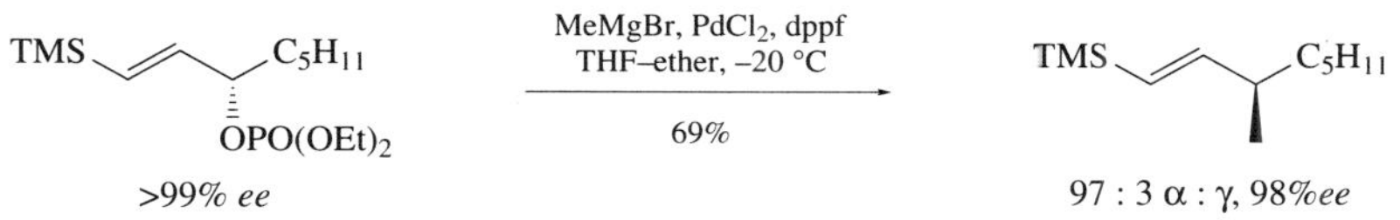

(99)

Allylic acetates and carbonates will react intramolecularly with alkenes and alkynes under palladium (or nickel) catalysis in a formal 'metallo-ene' type reaction ⟨90PAC1941⟩. The reaction proceeds via a π-allyl palladium species with the olefin inserting into the palladium–allyl bond with overall inversion of configuration (Scheme 14) ⟨88TL4705⟩ except in a few bicyclic systems where stereochemical constraints do not allow the formation of *trans* ring junctions ⟨87HCA1477⟩. The reaction can be combined with carbonylation ⟨89TL5883, 94T415⟩. Stereochemical features have been investigated ⟨88TL4709, 90TL1265⟩ and a very high degree of chirality transfer can be obtained with optically-active allylic alcohols (Scheme 14) ⟨90TL6995⟩. The (*Z*)-allylic acetate isomerizes *in situ* to the (*E*)-alkene of opposite chirality, probably via a π–σ–π palladium complex (*vide supra*). A similar phenomenon has been reported with malonate attack on allyl systems ⟨89JOC469⟩. No chirality transfer is seen with the terminally non-substituted allylic acetates. The chirality transfer is catalyst and substrate concentration dependent.

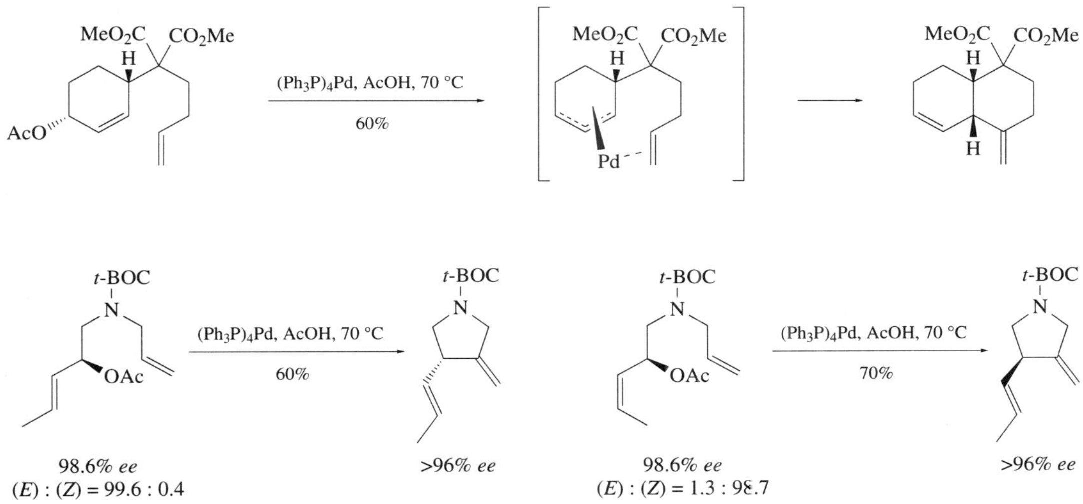

Scheme 14

The reactivity of allylic acetates and alcohols can be reversed in the presence of palladium catalyst and certain reducing metals. Allylic acetates and two equivalents of samarium(II) iodide ⟨86TL1195⟩ or one equivalent of tin(II) chloride ⟨88CC44⟩ in the presence of a palladium complex add to ketones and aldehydes (Equation (100)). Allyl alcohols also react ⟨88JA4473⟩. The reaction gives an allyl metal species which then adds to the carbonyl group. Regio- and stereo-selectivities with these samarium mediated reactions are not very high. Diethylzinc may also be used as the reducing agent ⟨93TL7619⟩.

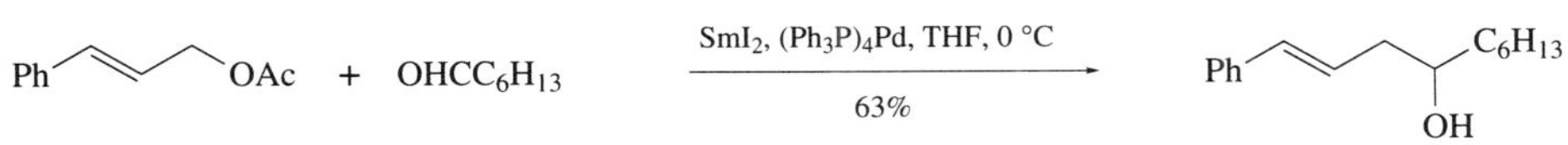

(100)

Nickel salts ⟨82COMC-I(8)713, 82COMC-I(8)773⟩ have also been used in reactions involving allylic alcohol derivatives and they catalyse a similar variety of reactions ⟨86JOM(317)383, 88JOM(339)199⟩ including Grignard additions ⟨83S701, 87TL4547⟩. Some of these reactions utilize chiral phosphine ligands and generate *ee*s of 2–90% (Equation (101)) ⟨85TL3259, 86T2043⟩. These reactions, which do not work with palladium based reagents, are very substrate dependent. Intramolecular insertions into alkenes have been performed ⟨88TL6433⟩. These reactions proceed with a high degree of chirality transfer (T. N. Birkinshaw and W. Oppolzer, unpublished results) and high stereoselectivity ⟨90TL1265⟩.

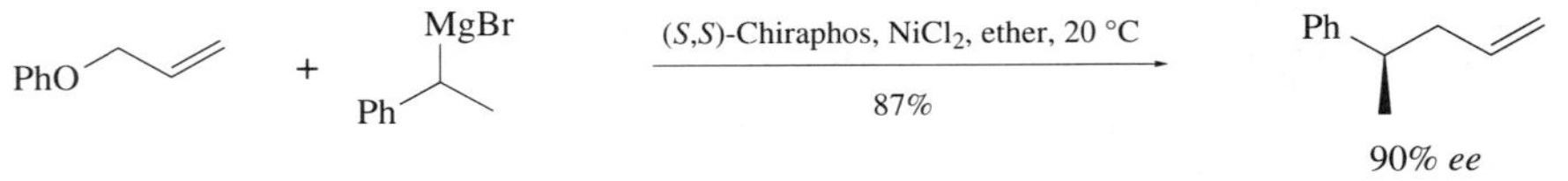

(101)

(b) Iron. Iron carbonyl complexes ⟨82COMC-I(4)377, 82COMC-I(8)939⟩ of allylic ethers and carbonates can be prepared from diiron nonacarbonyl or $Bu_4N[Fe(CO)_3NO]$ ⟨87JOC974⟩. Iron, which can form η^3, η^4 as well as η^5 complexes, tends to react with nucleophiles at the site at which the leaving group was attached; being configurationally more stable than the corresponding palladium complexes (Scheme 15) ⟨85JOM(285)C13⟩. Regioselectivities range from 64:36 to 100:0. Chirality transfer is high (Equation (102)) ⟨88JOC4419⟩. The stoichiometric iron complexes of γ-alkoxyenoates react with silylenol ethers in the presence of a Lewis acid (Equation (103)) ⟨93TL4497⟩. (*Z*) Double bond geometry is retained if $ZrCl_4$ is used as Lewis acid. The intermediate iron complexes can be prepared from vinylic epoxides and cyclic sulfonates ⟨90SL224⟩. The iron complexes also allylate electron-rich aromatic systems ⟨87TL5415⟩.

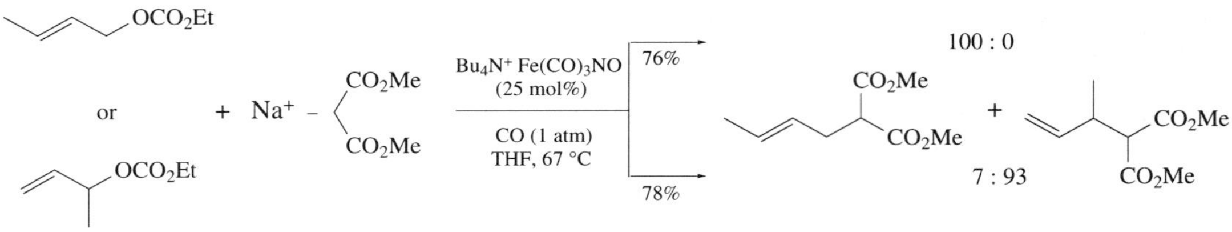

Scheme 15

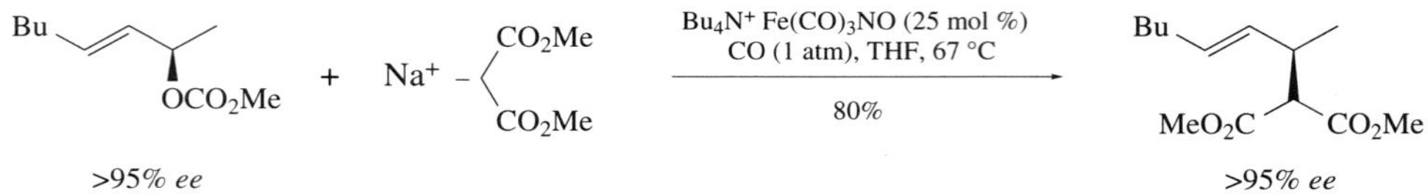

(102)

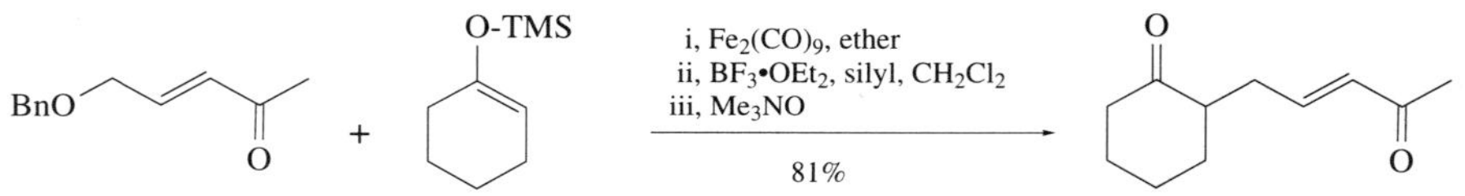

(103)

(c) Other metals. A number of other metals catalyse the reaction of allylic alcohol derivatives with nucleophiles. Molybdenum catalyses the reaction of allyl acetates ⟨87T4817⟩ and carbonates ⟨87BCJ1525⟩ with a different regiochemistry to that displayed in the corresponding palladium catalysed reactions; attack at a secondary centre is preferred over attack at a primary site with some anions but requires much higher temperatures ($\sim 100\,^\circ C$). Tungsten also gives a different regiochemistry, tending to react at the more substituted end regardless of the nature of the nucleophile ⟨84JA6837, 87JA2176⟩. Ruthenium complexes catalyse allylation of aldehydes ⟨89JOM(369)C51⟩ while rhodium has been used for allylation under neutral conditions ⟨84TL5157⟩ as well as for carbonylations ⟨85CL1477⟩. A number of different metals has been used and these results have been surveyed ⟨85JOM(296)269⟩).

(iii) Rearrangements

Rearrangements such as the [2,3]-Wittig rearrangement will be covered elsewhere (see reviews ⟨70AG(E)763, 86CRV885, 91COS(3)975, 93TL8047⟩).

(iv) Miscellaneous reactions

Carbenes, generated from diazo-compounds, insert into the C—O bond of allylic ethers under rhodium or copper catalysis (Scheme 16) ⟨86JA6060, 93TL4385⟩. In this case rhodium gives mostly the product from C—H insertion. Reduced zirconocene ('Cp_2Z-') inserts into allyl ethers to give allylzirconium species which can react with electrophiles such as aldehydes (Equation (104)) ⟨92TL1295⟩. Propargylic ethers give mostly acetylenic compounds with some allenic by-products ⟨92TL3769⟩. C—C Bonds may be formed intramolecularly (Equation (105)) ⟨93JA8835⟩. Stereoselectivities range from 75:25 up to 100:0 for the formation of four- and five-membered rings. Zirconocene catalyses selective γ-displacement of alkyl ethers probably via a zirconocycle intermediate ⟨93JA8485⟩.

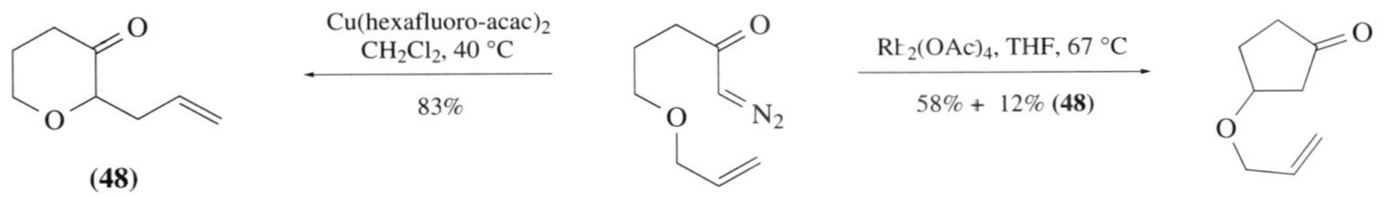

Scheme 16

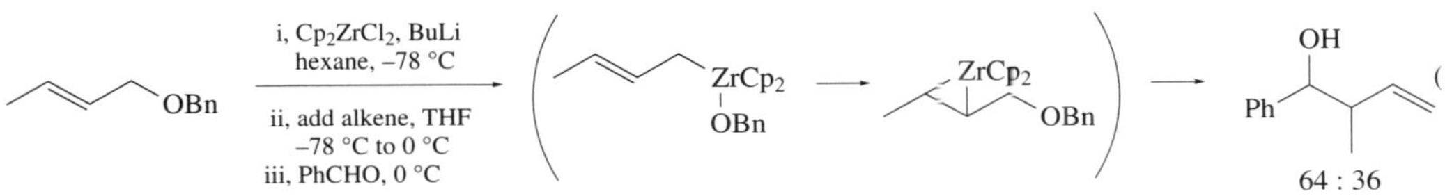

(104)

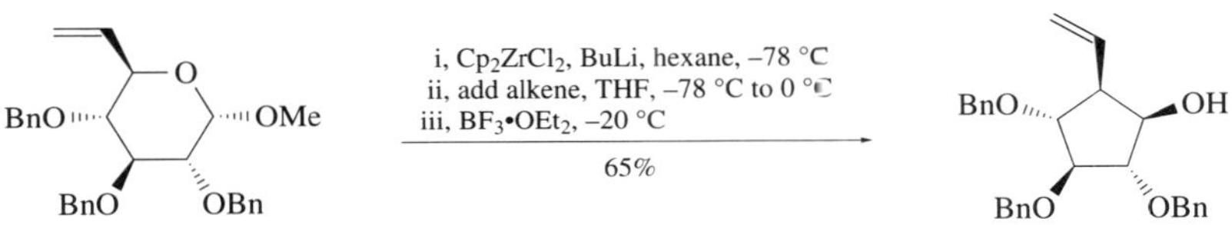

(105)

1.05.1.2.3 *Propargylic alcohol derivatives*

Propargylic alcohol functions (and halides) are readily isomerized to allene derivatives when treated with nucleophiles. The reaction of the former compounds with copper-based reagents has been studied by Macdonald *et al.* (Scheme 17) ⟨80JOC4740⟩. Their results suggest that the extent of direct displacement of the leaving group (acetate, methyl carbonate or tosylate) is dependent on the nature of the copper species and steric bulk on the alkyne, but much less so on solvent, temperature and leaving group (Equation (106)). Bulkier groups on copper give no allenic products ⟨78CC876⟩. Cleavage of the leaving group can be a problem with some reagents. Badouy and Gore ⟨81JCR(S)278⟩ have shown that the steric bulk of the cuprate reagent (in their case mixed ones) favours alkyne over allene formation (Equation (107)). Direct displacement of a silyloxy group with Grignard reagent has been reported (Equation (108)) ⟨76JCS(P1)1384⟩.

Scheme 17

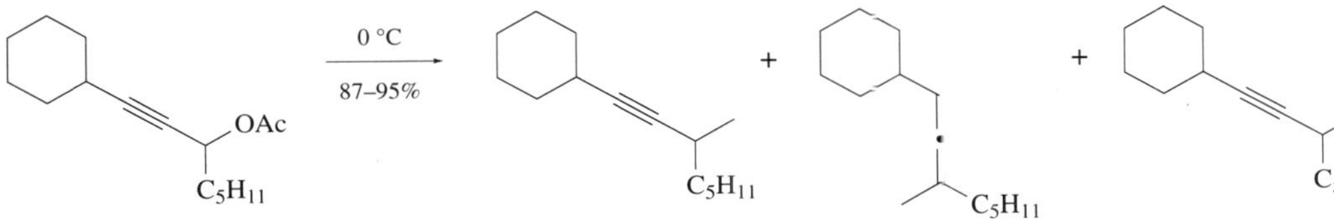

(106)

Me_2CuLi	ether	26	:	21	:	53
$MeCu{\bullet}MgBrI$	THF	72	:	17	:	11
$MeCu{\bullet}LiBr{\bullet}MgBrI$	THF	0	:	100	:	0

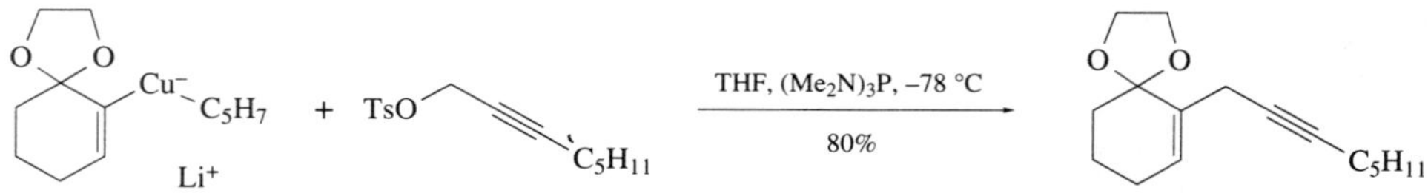

(107)

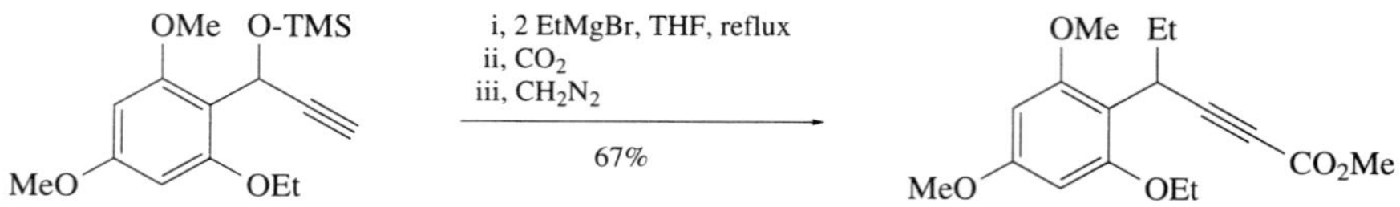

(108)

Displacement of the mesylate derived from the alcohol (**49**) appears to be direct but actually proceeds through the intermediacy of a chiral allenic bromide, which can be isolated (Scheme 18) ⟨88G857⟩. Terminal propargylic mesylates can be displaced by reaction with zincates, giving allenylzinc species, which can subsequently react with electrophiles ⟨93JOC6166⟩. Both regio- and stereoselectivities are good (Equations (109) and (110)).

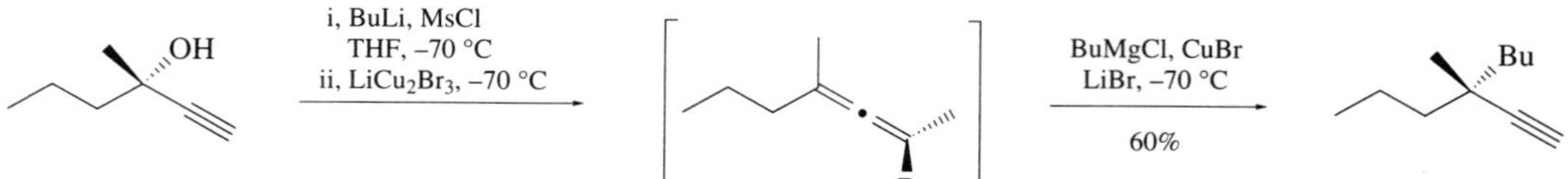

(**49**)

Scheme 18

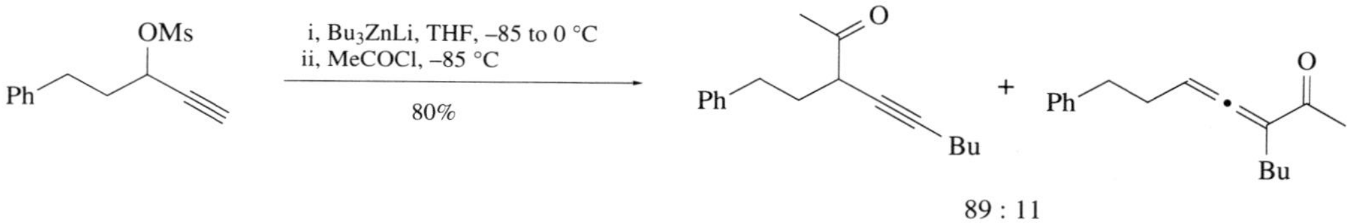

(109)

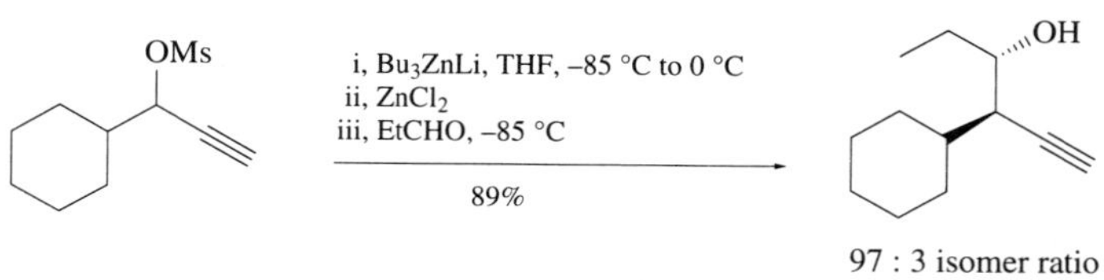

(110)

The anions of α-phenoxy alkynes react with boranes in a similar fashion to α-phenoxy acid derivatives (see Section 1.05.1.2.1(v)) (Equation (111)) ⟨79SC809, 82ACR178⟩. In the absence of HMPA allenes are produced.

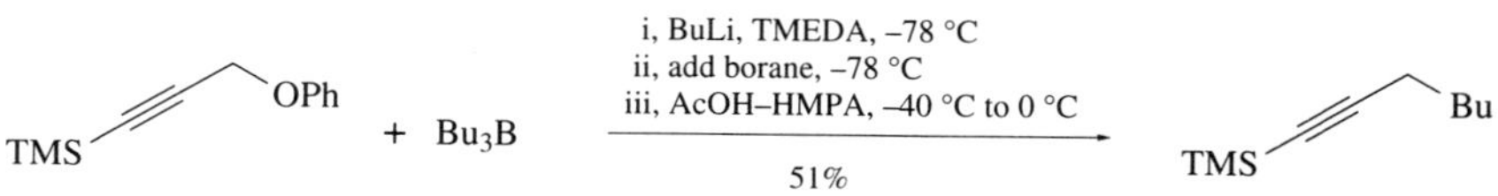

(111)

The Nicholas reaction ⟨87ACR207⟩ occurs between the dicobalt hexacarbonyl complexes of propargylic alcohols or ethers and nucleophiles and it is a simple way of avoiding the problems of isomerisation to allenes. Alkynes readily react with dicobalt octacarbonyl to give dicobalt hexacarbonyl complexes. Treatment of these compounds with either protic acid (usually HBF_4) or Lewis acid gives the stabilized carbonium ion (**50**) which can then react with a variety of nucleophiles (Scheme 19). The carbonium ion will also undergo Friedel-Crafts type alkylations. Addition of aluminum alkyne to the cobalt complex of the propargylic ester (**51**) with subsequent oxidative removal of the cobalt gives the skipped diyne (**52**) (Equation (112)) ⟨83TL2239⟩. 1,3-Dicarbonyl compounds can be used as the nucleophile ⟨78TL4349⟩. Reaction of Evans-type boron enolates with cobalt alkynes gives good yields and selectivities (Equation (113)) ⟨87JA5749, 93TL2581⟩. Intramolecular aldol condensations have been reported in a ten-membered ring (calichemycin—esperamycin ring system) ⟨89TL851⟩ or with the alkyne unit as an exocyclic feature ⟨93SL769⟩. Allylsilanes also react inter- ⟨80TL1595⟩ and intramolecularly (Equation (114)) ⟨86JA3128⟩. Six- to eight-membered rings were prepared with the intermediate cobalt complexes being further reacted. Radical dimerisations of the cobalt complex have been reported ⟨94TL363⟩.

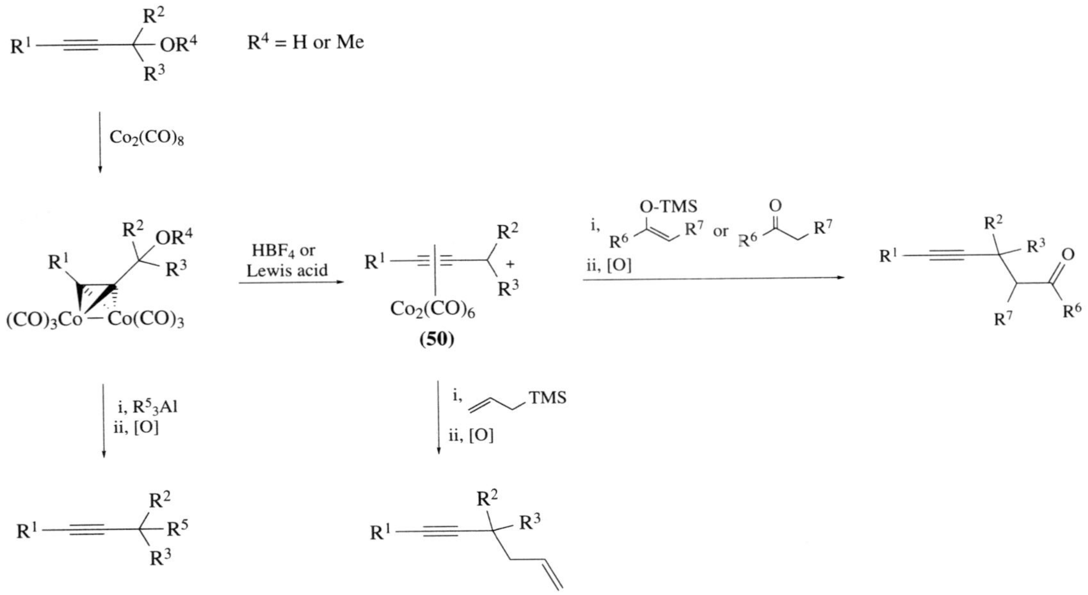

Scheme 19

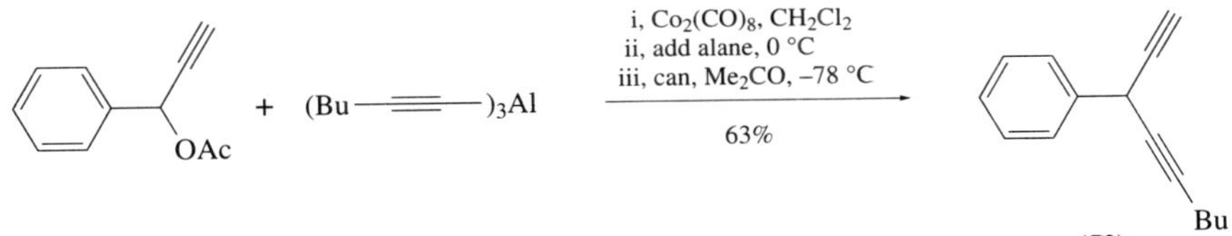

(51) (52) (112)

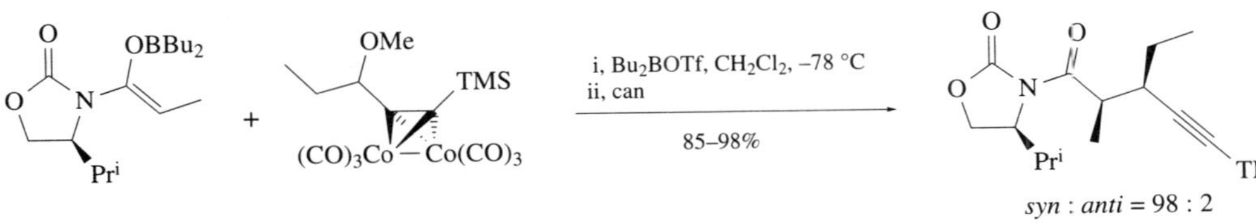

(113)

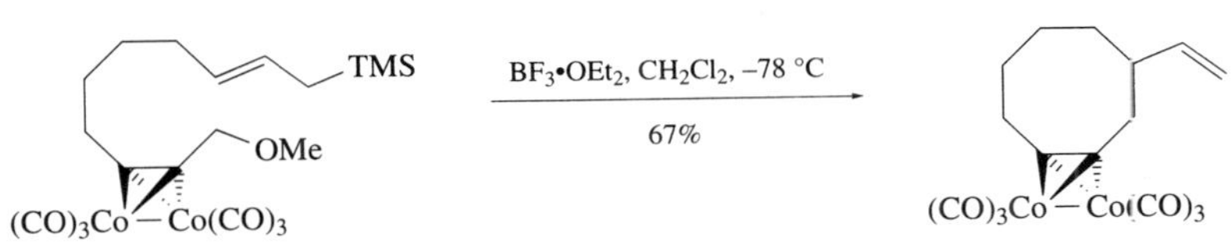

(114)

The intermediate carbonium ions are, in principle, chiral but undergo rapid epimerisation at the carbonium centre ⟨87JA5749⟩ so that, although they will react with high *de* with optically active substrates only racemic products are obtained from reaction with optically inactive nucleophiles. Caffyn and Nicholas have prepared the complex **(53)** with a phosphine ligand which is stable enough for the two diastereomers to be separated ⟨93JA6438⟩ (the alkyne–cobalt complex is a chiral centre) (Equation (115)). These compounds reacted with silylenol ethers with modest to good diastereoselectivity (1.1 : 1–10.5 : 1).

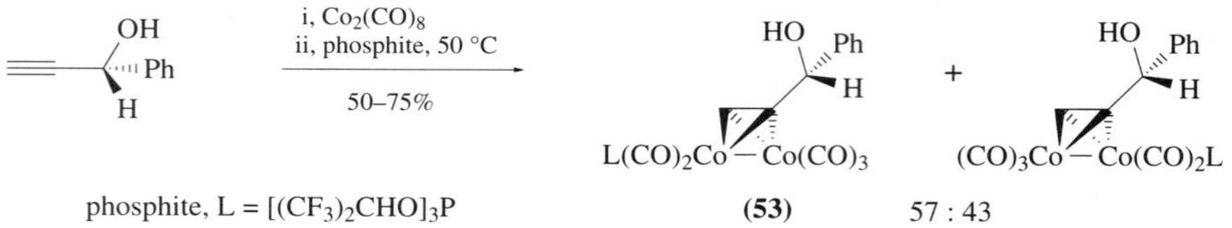

(115)

1.05.1.2.4 Benzylic alcohol derivatives

The increased reactivity of benzylic compounds compared to the corresponding alkyl derivatives means that a greater choice of leaving groups is available; reactions proceed under mild conditions. Halides (often chlorides) (see Chapter 1.04) are used but many oxygen functions can be readily displaced due to the stabilising effect of the adjacent aromatic ring. Hydroxyl groups can be displaced by a cyanide ion, using trialkylphosphine/carbon tetrachloride to generate the chloride *in situ* (Equation (116)) ⟨67JOC855, 80S1007⟩. Benzyl alcohol reacts with *o*-nitroarylacetonitrile **(54)** under Mitsunobu conditions (Equation (117)) ⟨91TL7195, 93H(35)349⟩. Both of these methods work for allylic and primary alcohols, while the latter procedure has been used on secondary alcohols. Triflates may be displaced by alkynes, even in the absence of base (Equation (118)) ⟨82JOC4577⟩. The reaction probably proceeds via an S_N1 mechanism. Benzylic hydroxyl groups may be substituted by *in situ* preparation of the acetate followed by displacement with cyanide or alkyne under slightly acidic conditions (Scheme 20) ⟨86S689⟩. This largely gets round the problem of a competitive elimination process observed with some secondary benzylic compounds. Similar conditions allow the alkylation of silylenol ethers with benzylic compounds (Equations (119) and (120)) ⟨84CB322, 86AJC2095, 90H(31)1359⟩. Stereochemical details are not given for most examples cited in the literature. Benzylic alcohols can be allylated with allyltrimethylsilane/BF_3 in variable yield (Equation (121)) ⟨82JOC2125⟩. The reaction is limited to bisbenzylic and tertiary benzylic alcohols; other alcohols give low yields. Neighbouring groups may effect the course of a reaction. The *o*-methoxy group is necessary for displacement of the carbonate of the alcohol **(55)**, for in its absence only the dimethylamino group is displaced (Equation (122)) ⟨90SC3347⟩. Some *p*-nitro-*t*-benzylic esters and ethers can be displaced by carbon nucleophiles via an electron-transfer mechanism (Equation (123)) ⟨88JOC1475⟩.

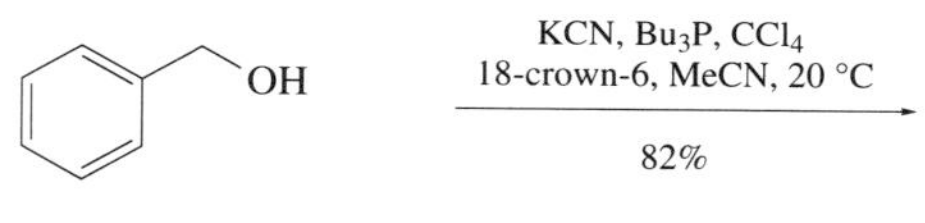

(116)

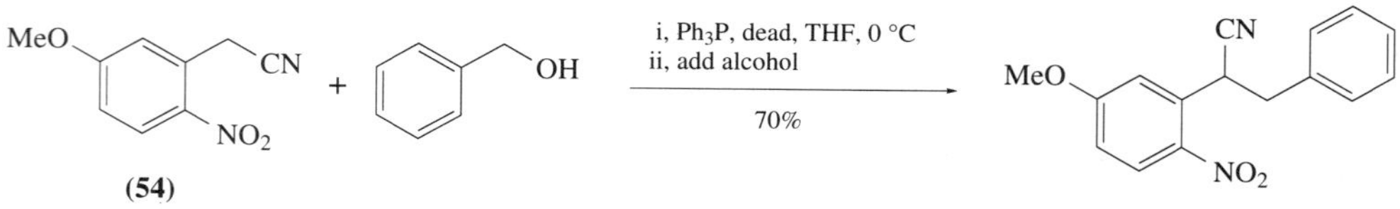

(117)

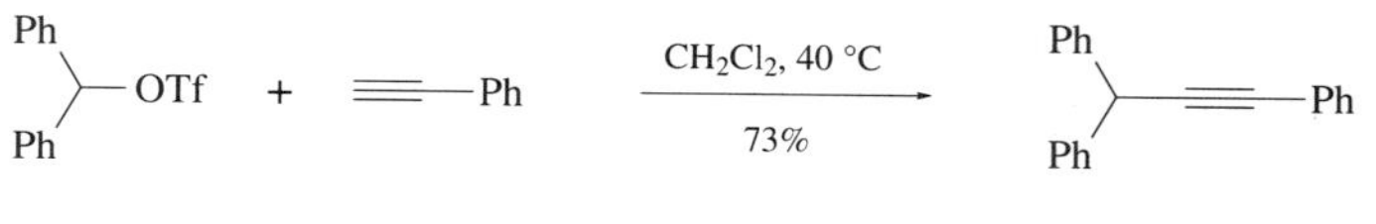

(118)

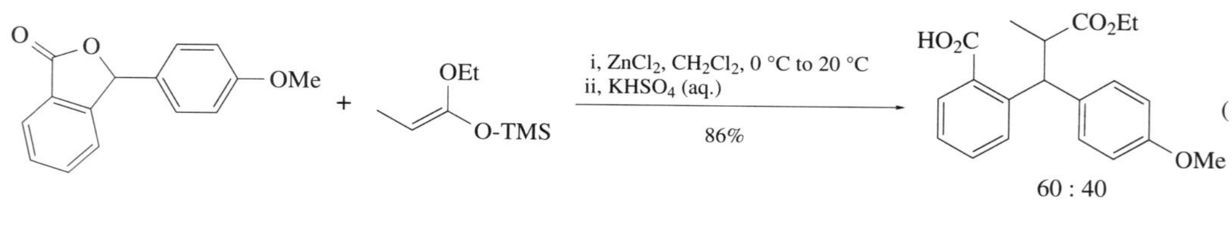

(119)

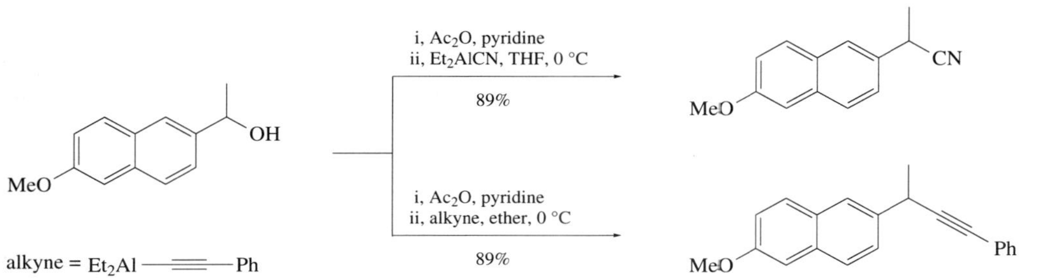

Scheme 20

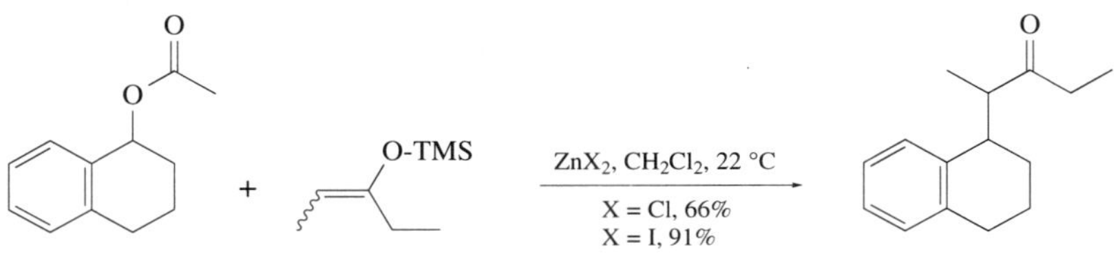

(120)

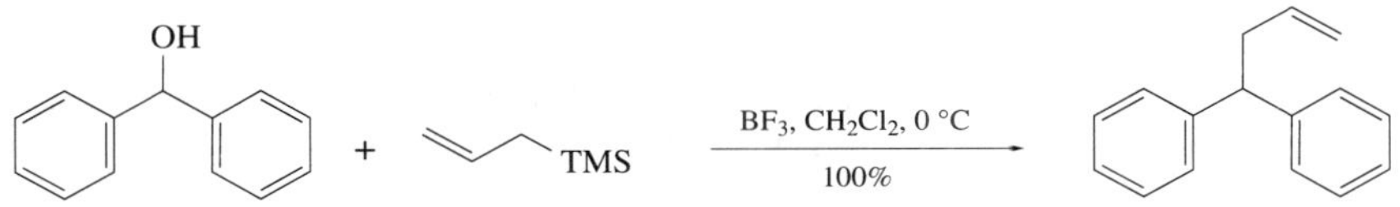

(121)

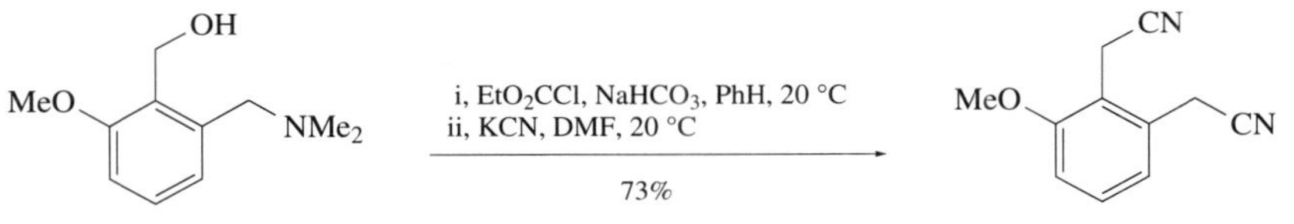

(55)

(122)

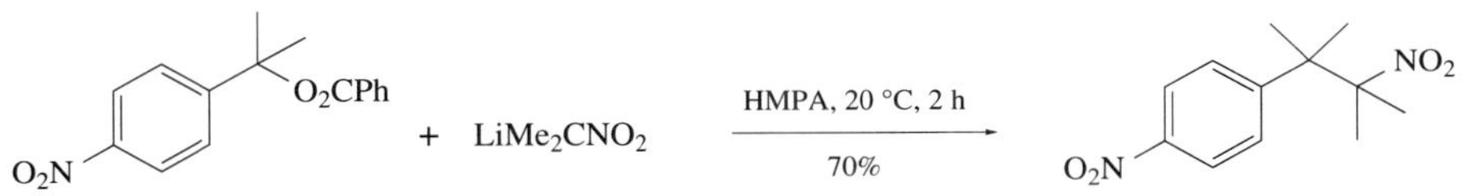

(123)

2-*t*-Butylphenol is alkylated in moderate yield to give only one isomer (Equation (124)) ⟨81S125⟩; similarly the anthraquinone featured in Equation (125) is alkylated at its most reactive position ⟨75BCJ2939⟩.

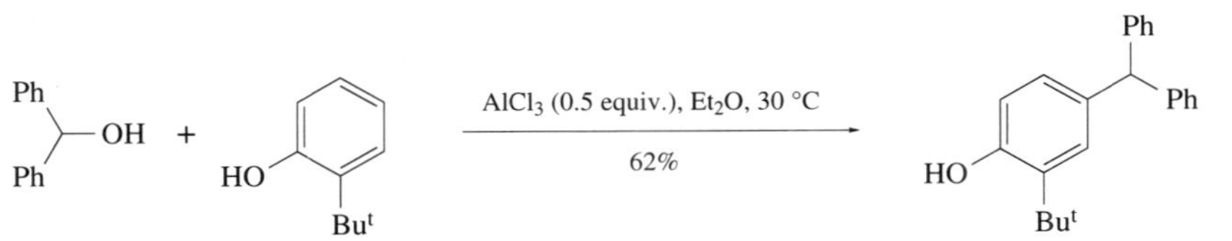

(124)

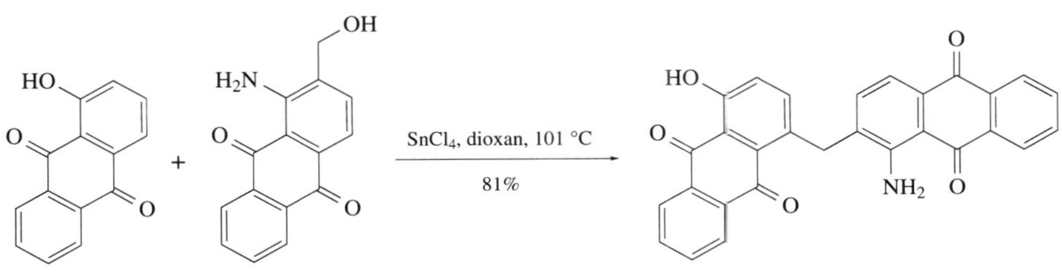

(125)

The perfluorinated sulfonic acid resin Nafion-H has been used to catalyse benzylation of benzene and substituted benzenes ⟨91JOC2089⟩. The advantages of this catalyst are the relatively mild conditions employed and the ease of work-up (filtration). Small amounts of the bisbenzyl ethers are found (Equation (126)) but yields are generally quite good; *ortho*/*para* selectivities are similar to those obtained with more conventional systems (Equation (127)). Bisbenzylation is possible. Bisbenzyl compounds can also be prepared from the Friedel-Crafts alkylation of chromium tricarbonyl complexes of benzyl alcohols (see below) ⟨93SL323⟩.

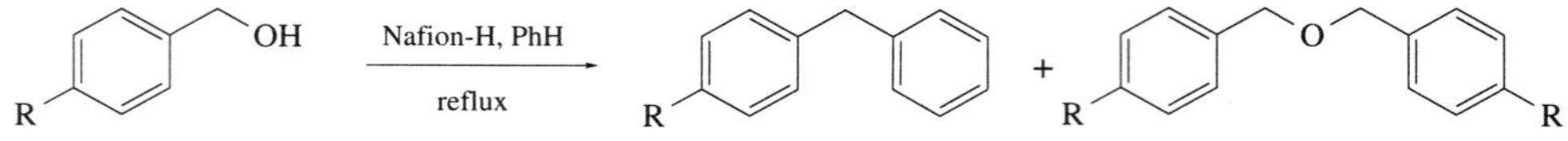

(126)

R	*Yield* (%)	*Yield* (%)
H	54	16
4-Me	81	3
4-Bu^t	85	2
4-NO_2	52	19

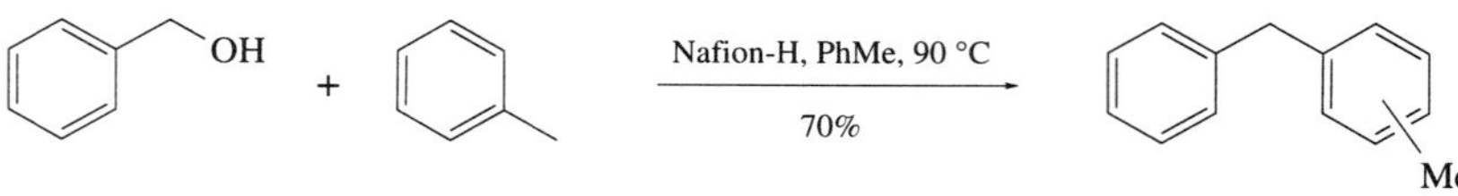

(127)

o : *m* : *p* 42 : 3 : 55

Simple benzyl alcohols are carbonylated by carbon monoxide in the presence of dicobalt octacarbonyl catalyst, often at high temperature and pressure ⟨B-80MI 105-02⟩. Chain extended esters and amides can be prepared in variable yields (20–82%) under much milder conditions ⟨82BCJ643, 88JOC3306⟩ (Equations (128) and (129)). Reductive carbonylation can be performed under similarly mild conditions (25 °C, 1 atm) using trimethylsilane as the reducing agent (Equation (130)) ⟨90JOC5923⟩. The one-carbon extended silyl ethers are the products.

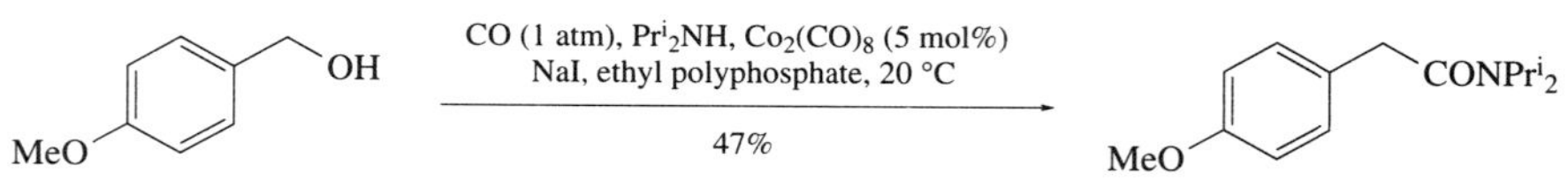

(128)

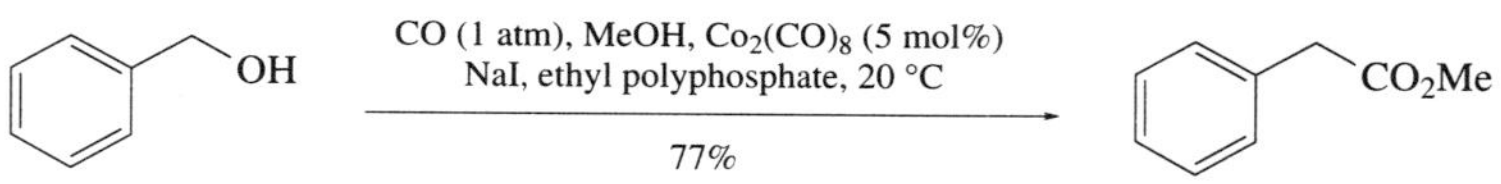

(129)

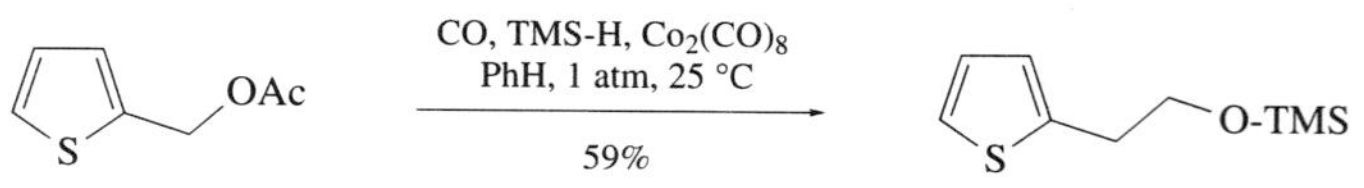

(130)

Arene chromium tricarbonyl complexes are readily prepared from the corresponding arenes. They can display quite different chemical properties from the parent arenes ⟨B-91MI 105-01, 93SL323⟩, for example they have the ability to stabilize a carbonium ion at a benzylic position. The metal can be readily removed under mild oxidative conditions such as atmospheric oxygen and sunlight. In some respects arene chromium tricarbonyl complexes are the aromatic equivalents of the cobalt complexes of alkynes but unlike such cobalt complexes the chromium complexes are configurationally stable, with implications for the synthesis of chiral materials. The area has recently been reviewed ⟨93SL323⟩.

The chromium complexes of benzyl alcohols (**56**) readily form the carbonium ions (**57**) which react with Grignard reagents or undergo an intramolecular reaction (Scheme 21 and Equation (131)) ⟨87JOM(336)143, 89TL3581⟩. The latter case demonstrates the stereochemical course of the reaction which involves a double inversion at the benzylic centre. Benzylic acetates react with silyl enol ethers ⟨83TL2837⟩ and trimethylaluminum (Equations (132) and (133)) ⟨86JOC2859, 86TL967, 86TL2479⟩ and the corresponding alcohols react with allyl silanes and β-dicarbonyl compounds ⟨86JOM(299)119⟩. Bicyclic (6,5 and 6,6) systems have also been used. In these cases inversion of configuration at the benzylic centre occurs (Equation (134)) ⟨85TL767, 86S386, 90JOM(382)121⟩.

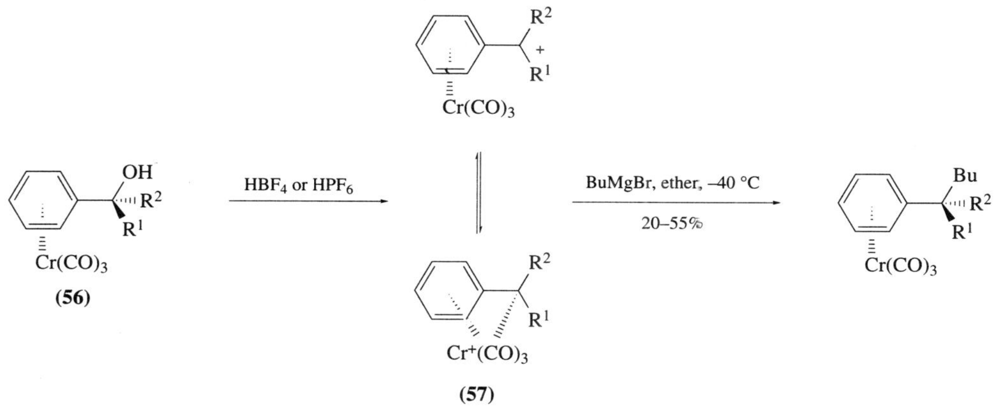

Scheme 21

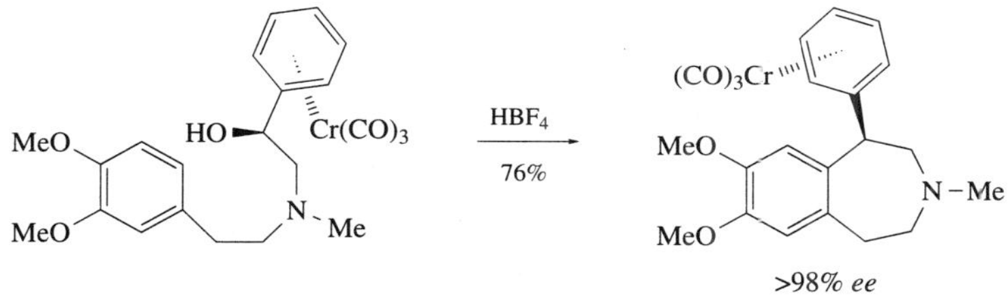

(131)

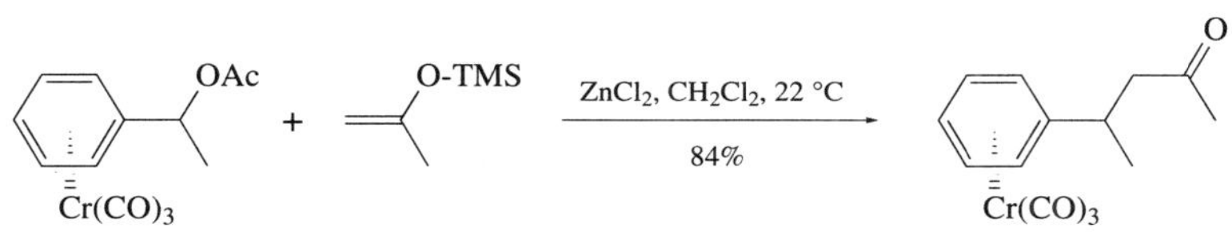

(132)

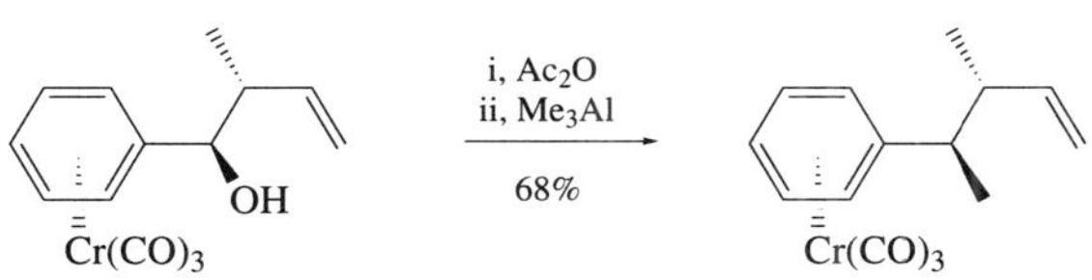

(133)

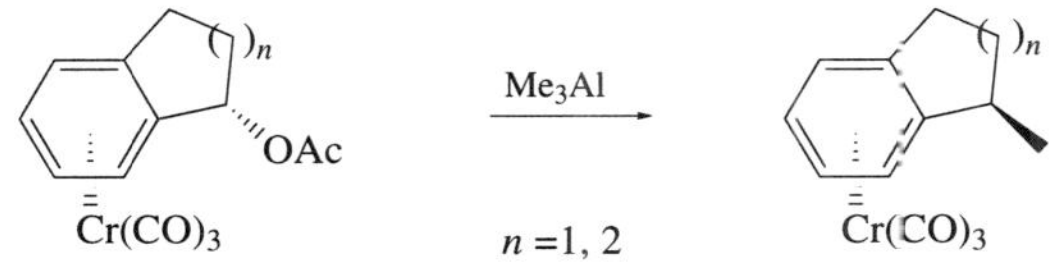

(134)

ortho Substituents can affect the stereochemical course of a reaction as the *syn* conformer is much less stable than the *anti* conformer (Equation (135)); loss of stereochemical integrity has been seen in some systems ⟨93SL323⟩. However the two diastereomerically pure acetates **(58)** and **(59)** give the corresponding products stereospecifically (Scheme 22) ⟨86JOC2859⟩.

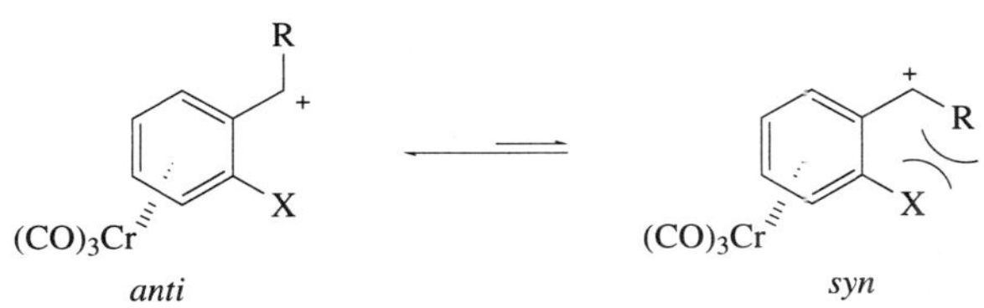

(135)

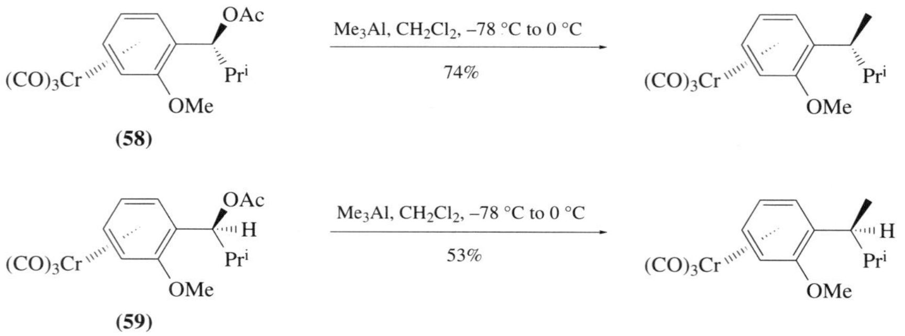

Scheme 22

1.05.1.3 Opening of Epoxides

Epoxides (oxiranes) are special cases of oxygen-functionalised groups in that two adjacent carbon atoms are involved. Displacement of oxygen at one centre still leaves the other C—O bond in place, and so the process constitutes a synthesis of alcohols. Although this topic will be covered in Chapter 2.02.1 (for related rearrangements see Chapter 1.13.4) the following text gives an overview of the area from the point of view of carbon nucleophiles displacing oxygen leaving groups to give a carbon atom with no attached heteroatoms. Vinyl epoxides (ethenyloxirane derivatives) which can react via S_N2 or S_N2' fashions, glycidic epoxides and oxetane derivatives are also discussed.

Synthetically useful reactions of epoxides have been reviewed ⟨84S629⟩ and sections 2.3 and 2.4 of that review (which deal with the opening of epoxides with carbon nucleophiles and intramolecular reactions) are particularly relevant to this discussion.

1.05.1.3.1 Simple epoxides

Carbon–carbon bond formation by the opening of epoxides with carbon nucleophiles, both stabilized and non-stabilized, is very well documented in the literature and has been widely used in synthesis. A good overview of the area is provided in the review ⟨84S629⟩.

(i) Unstabilized nucleophiles

The carbon–oxygen bonds of epoxides are strained. However, since alkoxides are normally poor leaving groups, coordination of the epoxide oxygen atom (to metal or Lewis acid) greatly facilitates displacement of the oxygen atom. Organolithium, organomagnesium, organoaluminum and organocopper are the most common reagents used in this process and each has its own advantages; in particular alkyl, alkenyl and aryl copper reagents have found widespread use as, in general, they have proved superior to lithium and magnesium reagents.

The reaction of styrene oxide (**60**) with carbon nucleophiles illustrates the effect of different metals. Three different products can result depending on the reagent used. Methyl Grignard gives a 50:50 mixture of the rearrangement product (**63**) (common for this type of reagent due to the Lewis acidity of magnesium halides) and the product (**61**), the less Lewis acidic dialkyl magnesium gives only the product (**61**) formed by attack at the more substituted carbon atom while normal cuprates and higher order cuprates give a reversal of selectivity (Scheme 23) ⟨77TL3407, 78TL2399, 82JA2305⟩. Halohydrin formation can also be observed.

Simple lithium and magnesium reagents are now little used to open epoxides due to the superiority of cuprates and other reagents. Very polar solvents such as HMPA and DMSO do allow acceptable yields to be obtained with alkenyl and alkynyl lithium reagents (Equations (136), (137) and (138)) ⟨75TL51, 78S26, 81S490, 81S875⟩. Several groups have reported that $BF_3 \cdot Et_2O$ catalyses the addition of lithium reagents to epoxides in good yields ⟨82TL5229, 83JA5946, 83TL5121⟩. The method is especially useful for alkyne nucleophiles where copper reagents are not useful (Equation (139)) ⟨83TL391⟩. Ganem and co-workers have studied the mechanism and synthetic utility of this process (Equation (140)) ⟨84JA3693⟩. Alkylboron fluorides are not involved and the reaction is believed to go via $BF_3 \cdot Et_2O$ coordination to the oxygen atom followed by nucleophilic attack. In a few cases traces

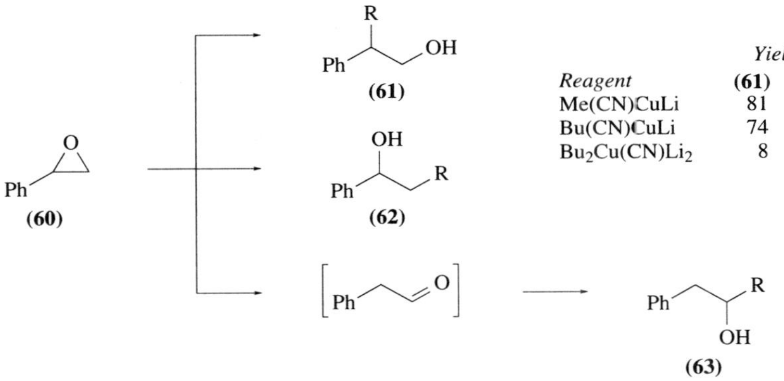

(60) (61) (62) (63)

Reagent	Yield (%) (61)	(62)
Me(CN)CuLi	81	18
Bu(CN)CuLi	74	21
$Bu_2Cu(CN)Li_2$	8	85

Scheme 23

of product arising from attack on the THF solvent were observed. Nonetheless this is now the method of choice for opening epoxides with alkynes.

(136)

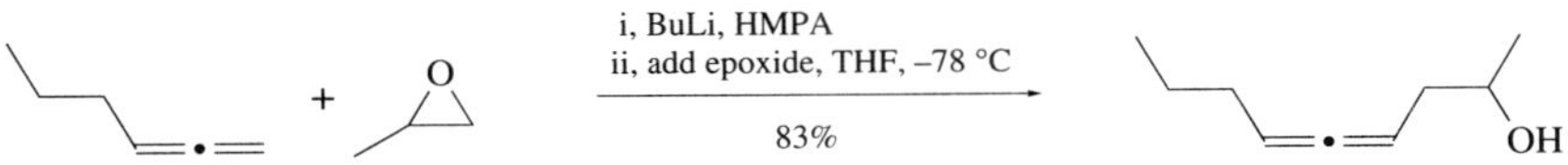

(137)

(138)

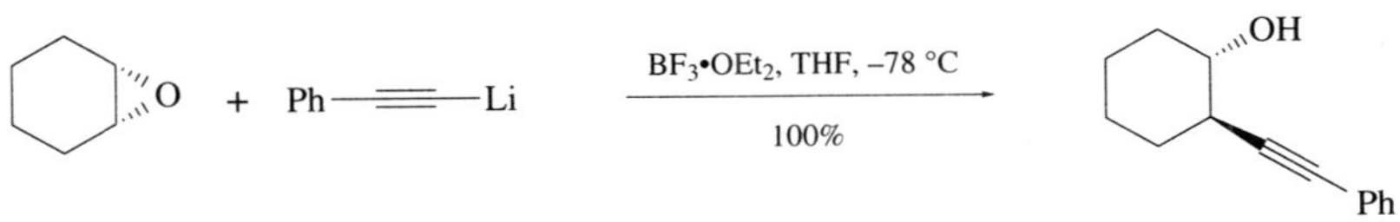

(139)

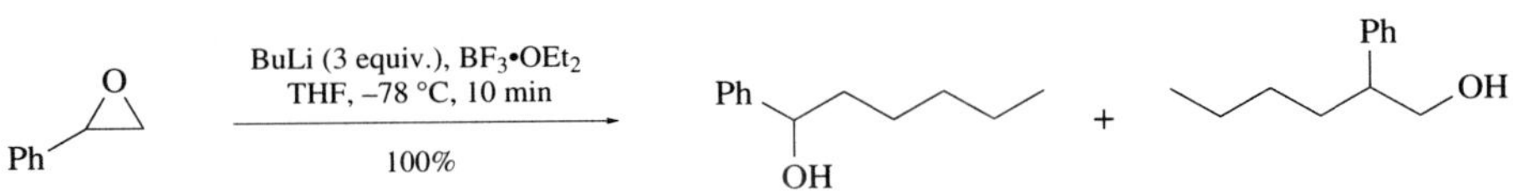

(140)

64 : 36

The dianion of the allylic alcohol (**64**) (deprotonated with Schlosser's base) opens cyclohexene oxide to give, after oxidation, the lactone (**65**) (Equation (141)) ⟨83SC237⟩.

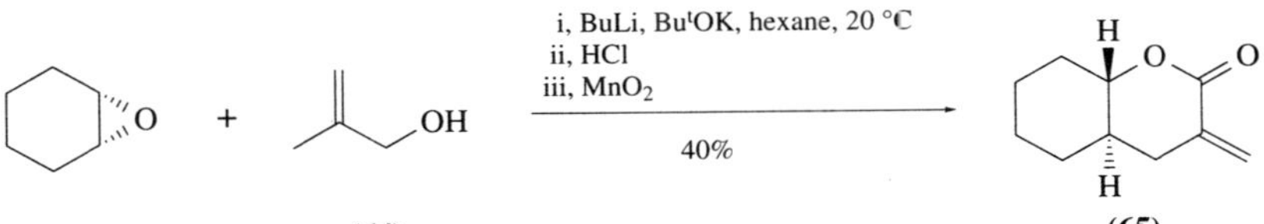

(141)

(**64**) (**65**)

Lithium reagents have been converted into samarium and other lanthanide reagents in order to open epoxides in high yield (Equation (142)) ⟨86AG(E)760⟩. Similarly gallium reagents (prepared with a catalytic amount of trimethylgallium) can be used (Equation (143)) ⟨84TL5423⟩.

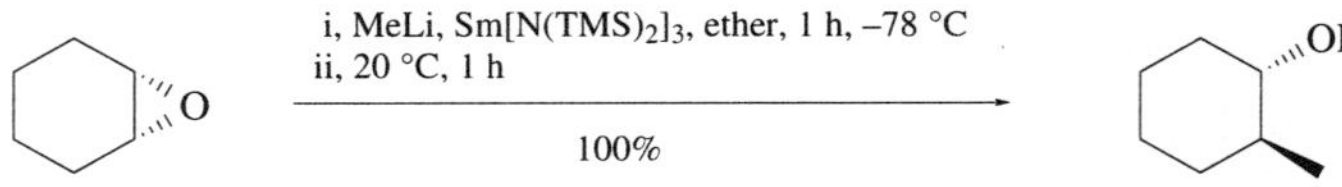

(142)

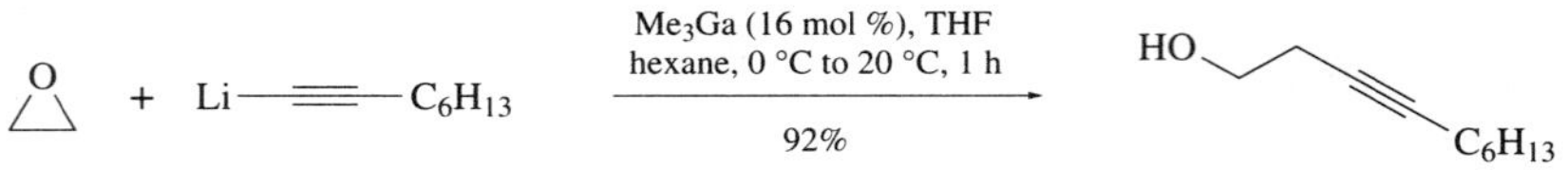

(143)

Organocuprates have proved to be very useful reagents for opening epoxides and have found wide application in organic synthesis (see reviews ⟨87S325, 90SL119, 92OR(41)135, 75OR(22)253⟩). Copper reagents can be generated from Grignard reagents and a catalytic amount of copper salt, (usually copper iodide) and they react with epoxides (Equations 144–146) ⟨78CL773, 82TL1267, 86TL5791⟩. In the last case (Equation (146)) the hydroxyl group must be protected before Grignard addition. In the first case (Equation (144)) absence of catalyst gives a mixture of bromohydrin and rearrangement product. Yields are in the range 41–77%. In Equation (147) attack from the γ-position of the allylic Grignard reagent (**5**) occurs in the absence of catalyst while, with 10% CuI present, attack occurs from the α-position ⟨78TL4069⟩. Preformed 'lower-order' organocuprates (Gilman reagents) add to epoxides (Equation (148)) ⟨74JA5254⟩, (Equation (149)) ⟨80TL4365⟩ but unlike mixed cuprates (Equation (150)) ⟨77TL3407⟩, suffer from the disadvantage that only one of the alkyl groups is used.

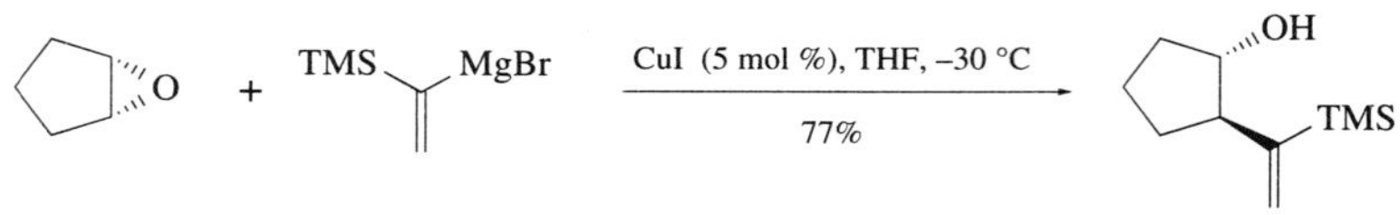

(144)

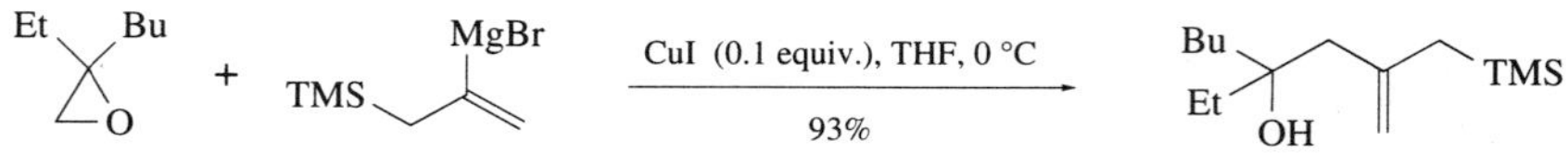

(145)

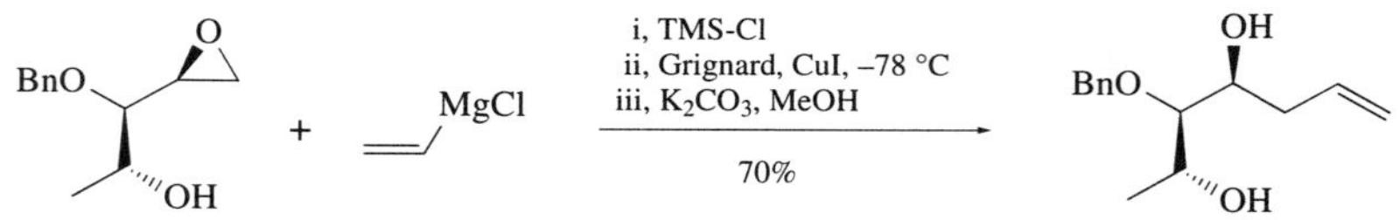

(146)

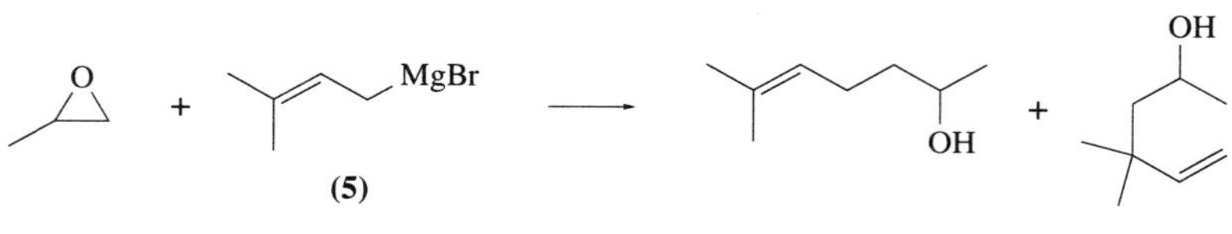

(147)

(**5**)

No CuI	9	91
10% CuI	99	1

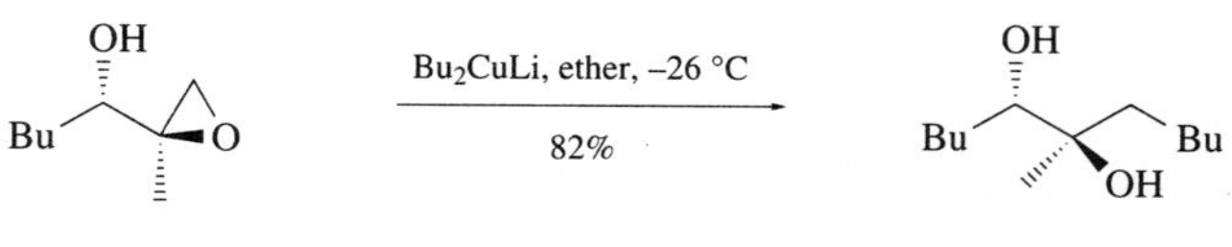

(148)

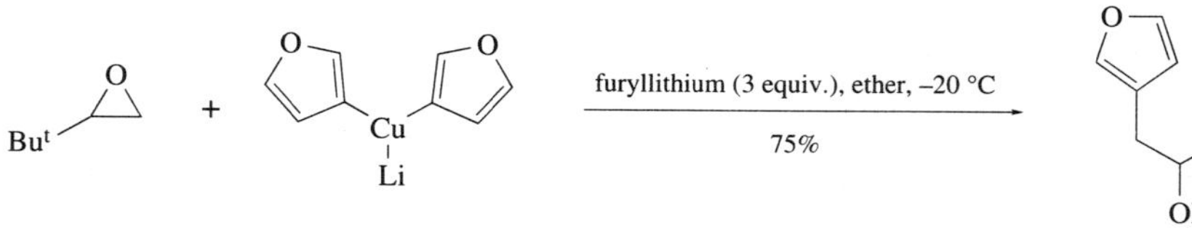

(149)

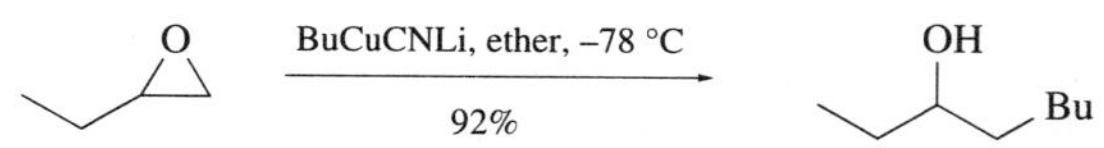

(150)

'Higher-order' cuprates have been developed which readily open epoxides in good yield. (Equations (151) and (152)) ⟨82JA2305, 84JOC1147⟩. Higher order cuprates which have dummy or non-transferable ligands such as thiophene and cyanide also open epoxides (Equation (153)) ⟨85JOM(285)437, 88TL893⟩. The greater difficulty of preparation of these latter reagents must be balanced against the value of half of the ligand being discarded in a reaction involving a simple dialkylcuprate. Allyl groups can be used to open epoxides by preparing the cuprates from allyltin reagents ⟨90JA4063⟩ or by using highly activated copper prepared by the reduction of copper salts ⟨93TL3063⟩ (Scheme 24). In the latter case a methyl group is used as a dummy ligand. Both of these methods avoid Wurtz coupling of the allyl moieties.

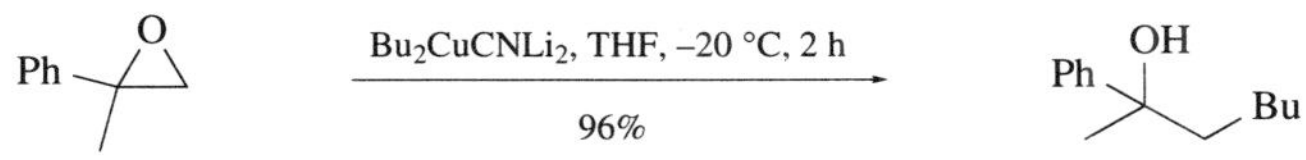

(151)

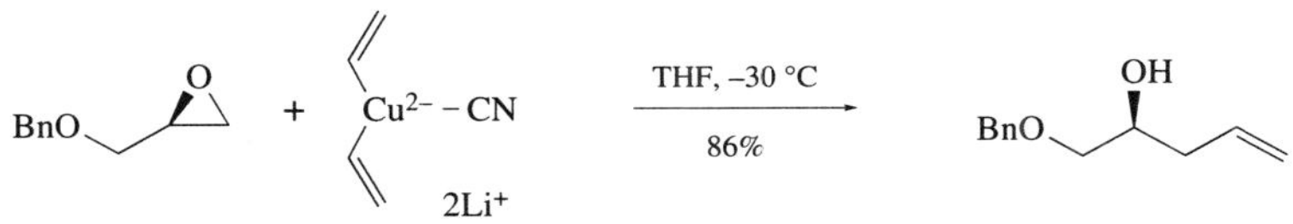

(152)

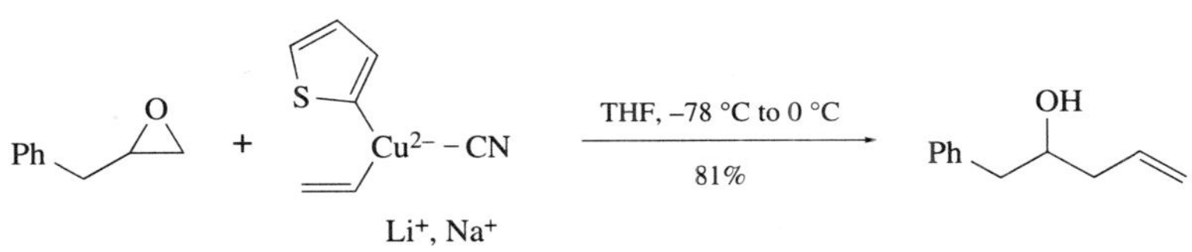

(153)

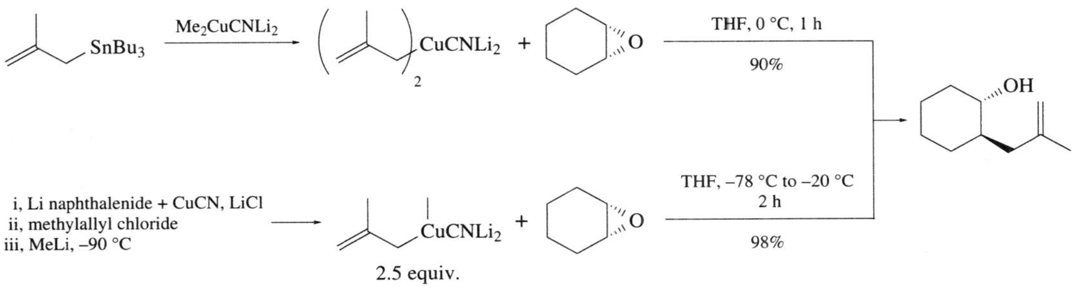

Scheme 24

Lewis acid catalysis, specifically $BF_3 \cdot OEt_2$ catalysis, has proved beneficial to some cuprate additions to epoxides ⟨84TL5959⟩ (Equation (154)). Other Lewis acids gave low yields.

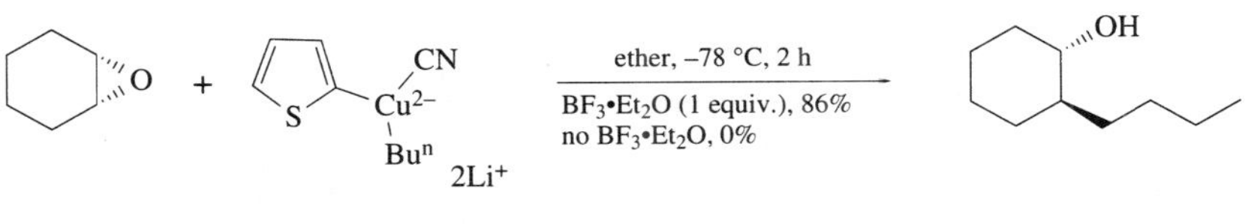

(154)

Organoaluminum reagents open epoxides readily but have found greatest utility for the transfer of alkenyl and alkynyl groups. Alkenylaluminum reagents are readily prepared from alkynes via carbo-alumination ⟨84OR(32)375, 87ACR65, 91COS(4)865⟩ (see also Chapter 2.19) and give products of well-defined stereochemistry (Equations (155) and (156)) ⟨85JCS(P1)1589, 93AG(E)1068⟩. The former example uses an aluminate species while the latter uses a simple alane. Boron reagents have also been used to open epoxides ⟨74T3037, 77T1945⟩. Alkenyl or alkynyl boronates open epoxides by transfer of the alkenyl or alkynyl group to give the boronate complexes (**66**) and (**67**) which can be further manipulated, especially by oxidative cleavage of boron (Schemes 25 and 26).

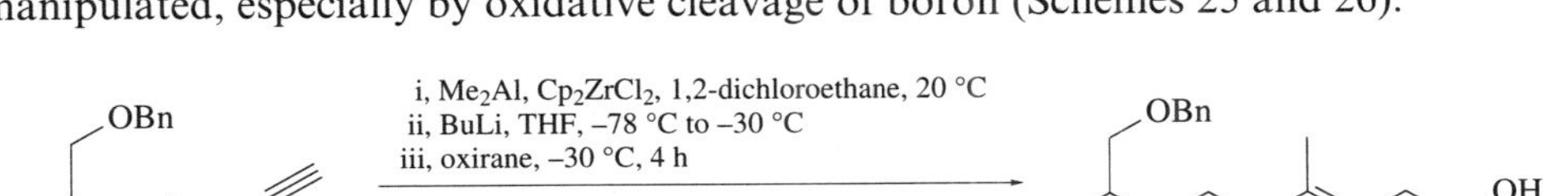

(155)

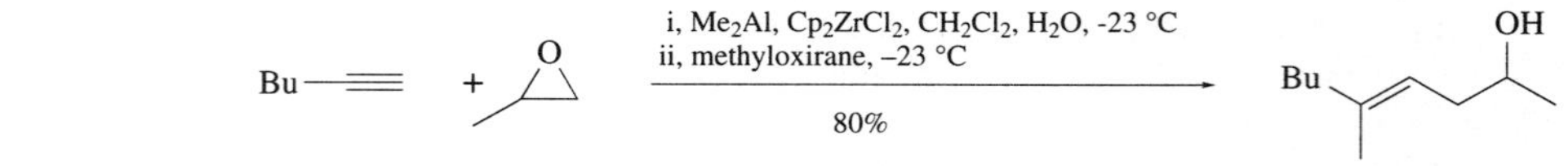

(156)

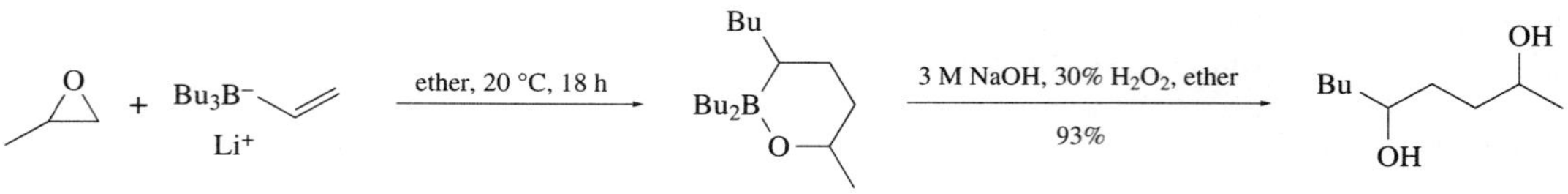

(66)

Scheme 25

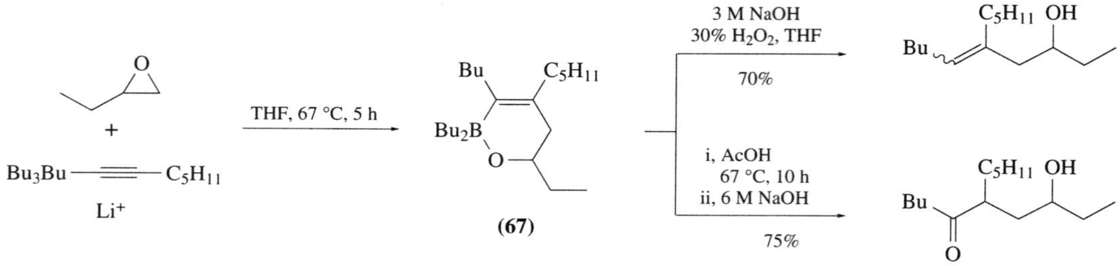

(67)

Scheme 26

Epoxides can be opened under high pressure (5 kbar) by a manganese pentacarbonyl species with subsequent migration of carbon monoxide and attack of the acyl complex on a Michael acceptor (Equation (157)) ⟨88JOC4892⟩. Yields are in the range 40–99%.

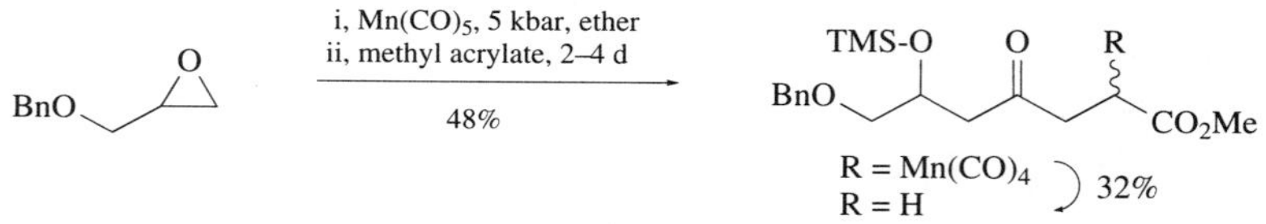

(157)

(*ii*) *Stabilized nucleophiles*

The simplest carbon nucleophile is cyanide which in its various guises will open epoxides. Potassium cyanide in aqueous DMF can be used (Equation (158)) ⟨86BCJ1363⟩. Silylcyanides react in the presence of acidic catalysts such as aluminum-based compounds ⟨86S1032, 87JOC1013⟩, lanthanides ⟨87TL5513, 90TL6209⟩, calcium and other metal oxides (Equations (159) and (160)) ⟨90CL481⟩ (the high regioselectivity in the latter case is believed to derive from the formation of a calcium chelate) and basic catalysts such as potassium cyanide (Equation (161)) ⟨90JOC2016⟩ and aluminum cyanides (Equations (162) and (163)) (note the *trans* diaxial opening in the former case) ⟨70JCS2365, 88S943⟩.

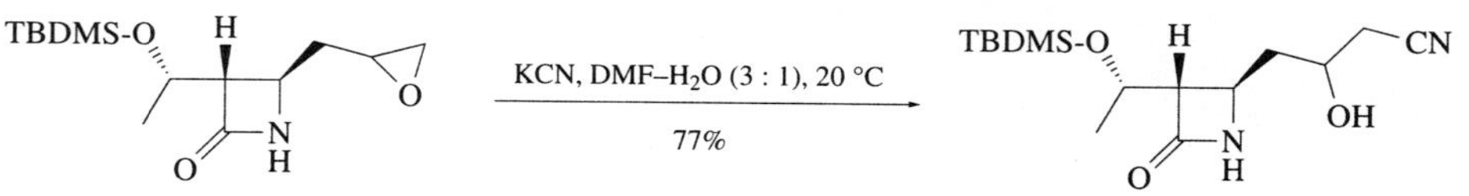

(158)

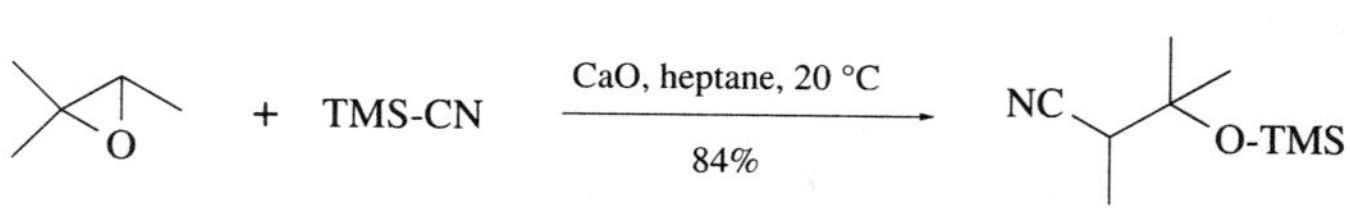

(159)

94 : 6 regiochemistry

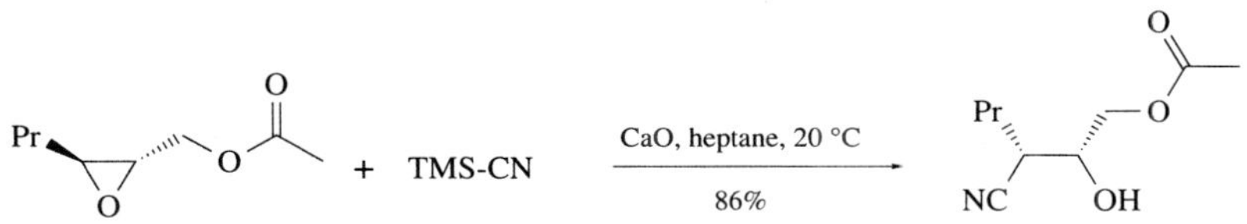

(160)

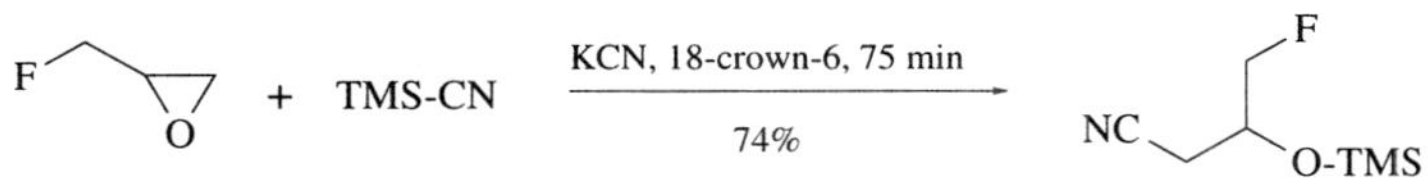

(161)

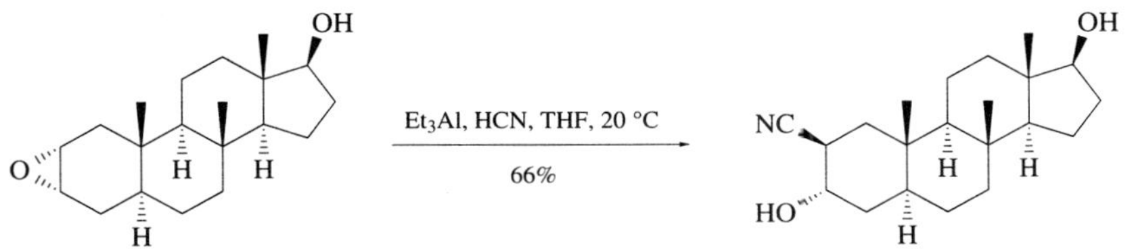

(162)

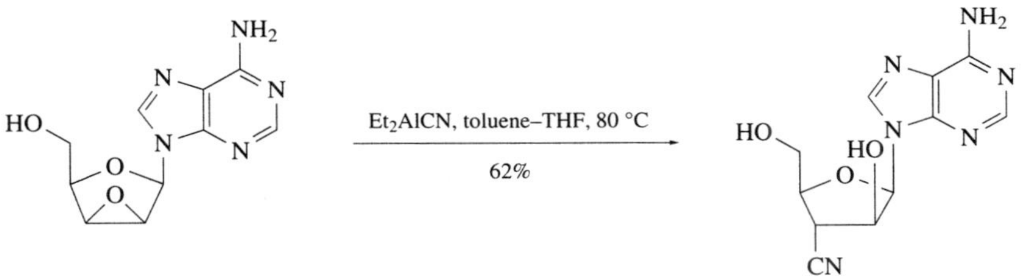

(163)

Epoxides can be opened by a variety of enolate anions including dialkyl malonate anions and the anion of acetate esters. The dianions of carboxylic acids are particularly useful as the products can be readily lactonized (Equation (164)) ⟨83JOC1455⟩. In the latter reaction the absence of HMPA causes the yield to fall to 55%. Anions of tertiary amides (Equation (165)) ⟨81JOC2831⟩ and dianions of secondary amides (Equation (166)) also react ⟨88SC1159⟩. The former reaction is moderately stereoselective in ether when the substituents on nitrogen are large. The presence of a Lewis acid in the latter reaction improves yields and reduces reaction times. Dianions of acetoacetates react at the terminal carbon atom as normal ⟨78JOC788⟩. The aluminum enolates of *t*-butyl esters open epoxides in poor to moderate yields (12–71%) and with *syn*:*anti* selectivities of 62:38 to 95:5 ⟨93JOC7304⟩. A variety of α-substituted acetate anions open epoxides and the incorporated substituent allows further functionalisation of the molecule. Representative examples are shown in Equations (167) ⟨74CL385⟩, (168) ⟨87TL1881⟩, (169) ⟨75CC537⟩ and (170) ⟨86AG(E)754⟩. Anions α to nitriles also open epoxides in good yield ⟨78S221⟩. The product obtained depends on the work up conditions employed (Equations (171) and (172)) ⟨80SC49⟩. When α-silyl acetonitriles are used, silyl transfer accompanies carbon–carbon bond formation ⟨79JCS(P1)26⟩. The intermediate anion can react further with epoxides or other nucleophiles ⟨79BCJ2389⟩. Epoxides can also be opened by isocyanide anions in good yield to give oxazines or isocyanoalcohols depending on work-up conditions (Scheme 27) ⟨73AG(E)323⟩.

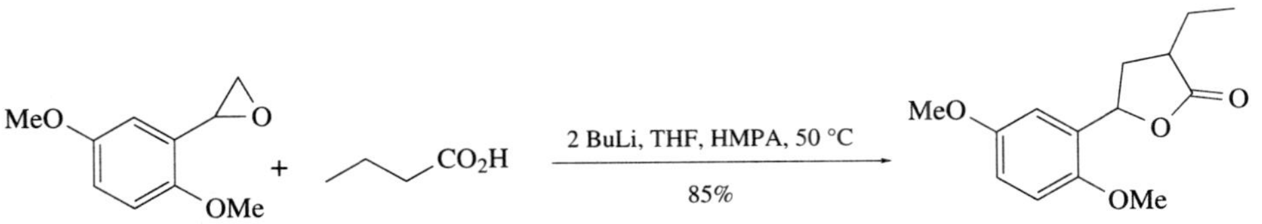

(164)

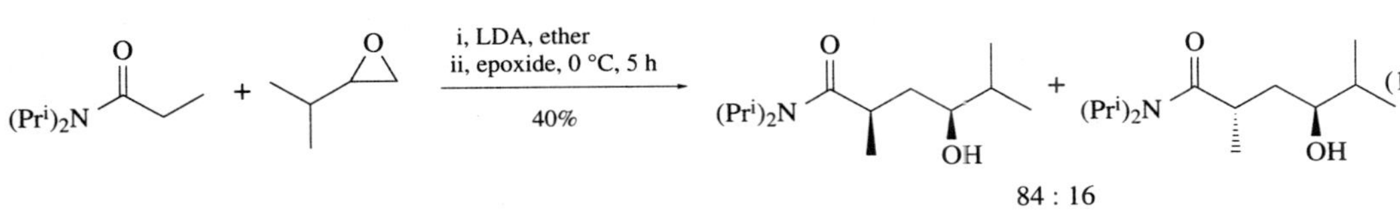

(165)

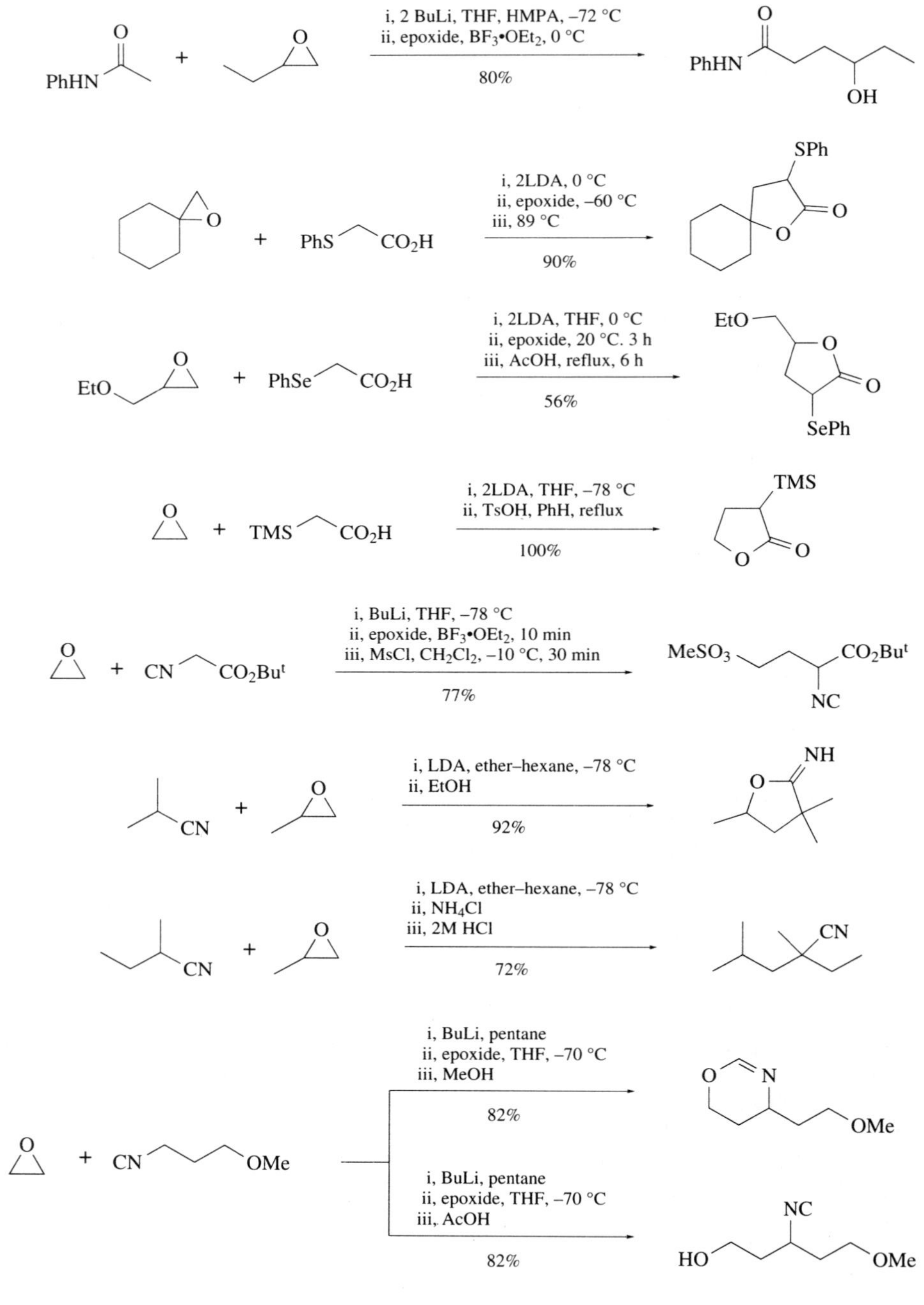

(166)

(167)

(168)

(169)

(170)

(171)

(172)

Scheme 27

Sulfur-stabilized anions can be used to open epoxides (Equation (173)) ⟨82TL2379⟩. Allylic sulfur-stabilized anions can attack from the α-carbon atom (Equation (174)) ⟨88CL645⟩ or predominantly from the γ-carbon atom (Equation (175)) ⟨75TL4027⟩. The regioselectivity of the reaction is ascribed to the intramolecular chelation of lithium. Dithianes are also good nucleophiles (Equation (176)) ⟨65AG(E)1075, 77AG(E)264, 93JOC2181⟩. Other sulfur-stabilized anions that can be used to open epoxides include those derived from sulphoxides ⟨72TL4605⟩, β-keto sulfoxides (dianions) (Equation (177)) ⟨74JOC732⟩, tris(methylthio)methane ⟨93SC811⟩ (which is a high-yielding carboxylate anion synthon) and sulfones (Equations (178)–(180)) ⟨93TA399, 88TL2059, 85TL4747⟩. Ylides generated from sulfoxonium salts give oxetanes from epoxides (Equation (181)) ⟨83JOC5133⟩.

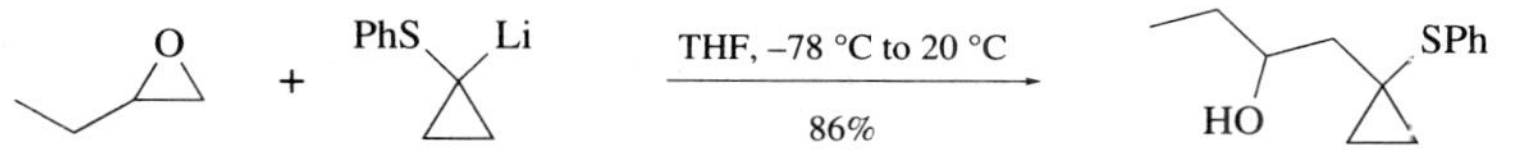

(173)

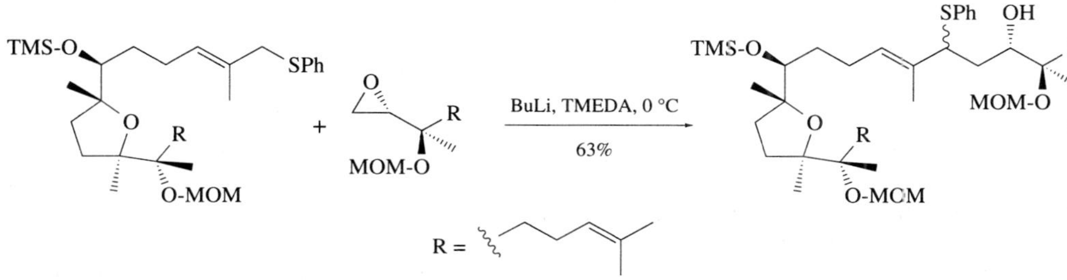

(174)

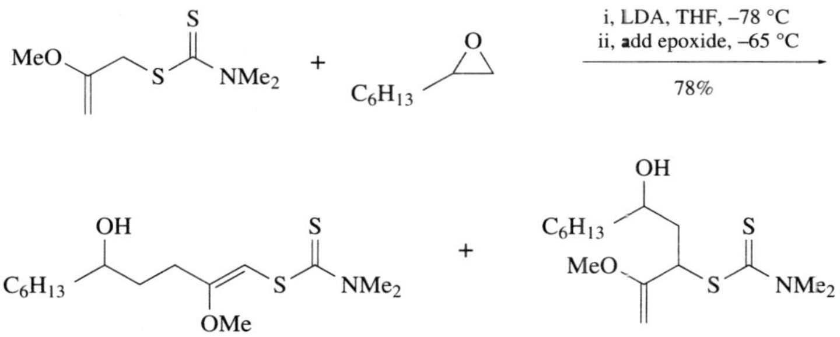

81 : 19

(175)

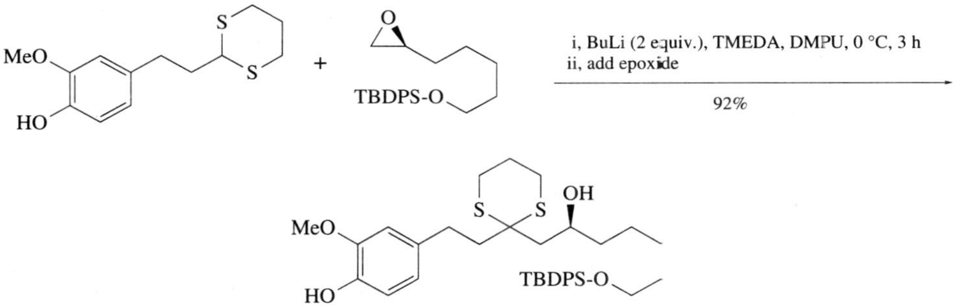

(176)

DMPU = 1,3-dimethyl-3,4,5,6-tetrahydro-2(1*H*)-pyrimidinone

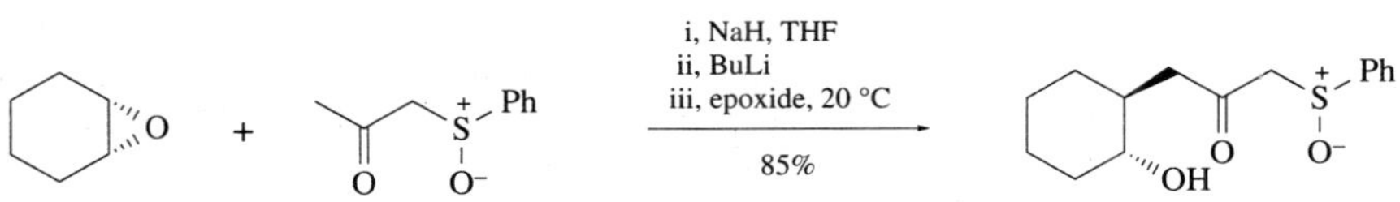

(177)

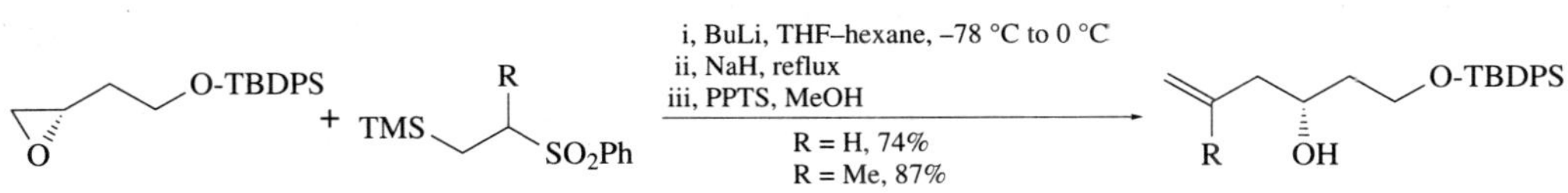

(178)

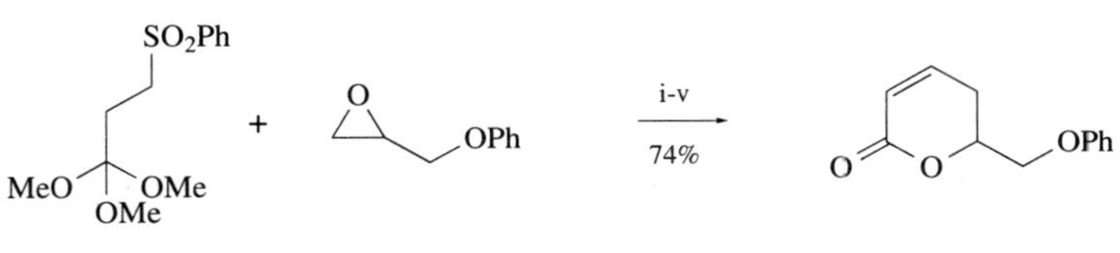

(179)

i, BuLi, THF, HMPA; ii, epoxide, 20 °C; iii, H_3O^+; iv, TsOH; v, Et_3N or dbu

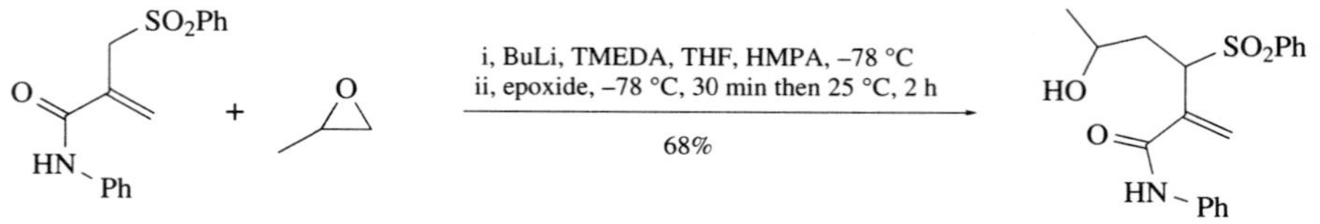

(180)

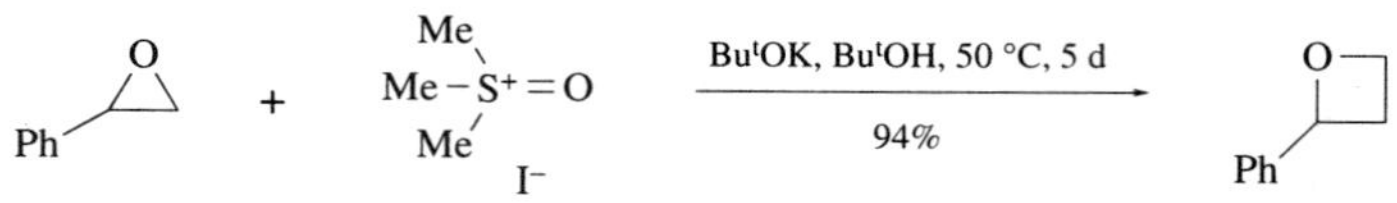

(181)

Other species that have been used to open epoxides include anions derived from methyl oxazoles (Equation (182)) ⟨81JOC1410⟩, imines ⟨75S256⟩, ketones ⟨80JA6163⟩, hydrazones ⟨76TL11⟩ as well as Wittig and phosphonate reagents ⟨67JA5850, 93TL3539⟩. Pelter *et al.* have recently published a study of the opening of epoxides with dimesitylboron stabilized anions which, on oxidative work-up, gave 1,3 diols ⟨93T3007⟩. Transition metals catalyse epoxide opening with carbon monoxide ⟨82COMC-I(8)799, 84JA6093, B-91MI 105-02⟩ but often require excess epoxide in order to reduce further reaction of the carbonyl compounds that are produced. A recent method using rhodium catalysis appears to have overcome this problem (Equation (183)) ⟨93JOC4187⟩.

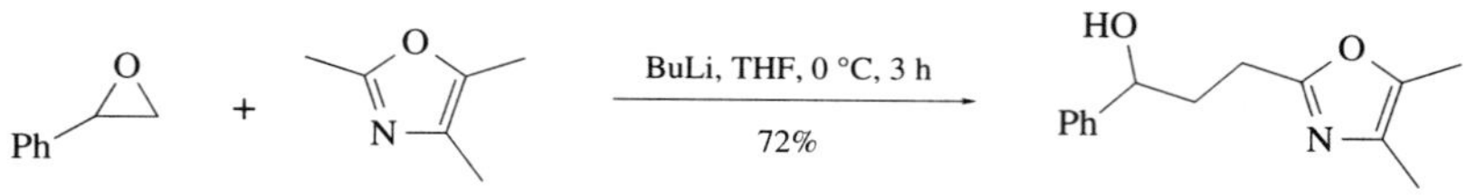

(182)

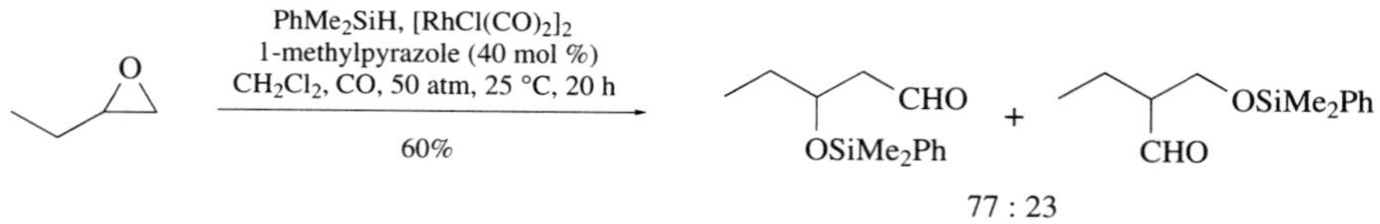

(183)

(iii) Friedel-Crafts and acid-catalysis

Epoxides and other cyclic ethers can be used in Friedel-Crafts alkylations ⟨48BSF818, 48BSF822, 48BSF825, 48BSF827⟩; reactivity decreases with increasing ring size 3 > 4 > 5 > 6 and with less substitution tertiary > secondary > primary ⟨60MI 105-01⟩. Friedel-Crafts alkylations have been reviewed by Olah *et al.* ⟨91COS(3)293⟩ (this review includes a section on alkylation with oxygen functions) and in a monograph ⟨B-84MI 105-01⟩.

Isomerisation of epoxides to aldehydes can compete with alkylation; the extent of isomerization is dependent on the Lewis acid employed. With mild acid catalysts such as $BF_3 \cdot Et_2O$, $ZnCl_2$ and magnesium halides isomerisation is faster than alkylation and the products isolated are aldehydes or aldehyde-derived products (Scheme 28) ⟨78BCJ174⟩. Strong Lewis acid catalysts such as $AlCl_3$ or $SnCl_4$ give the expected oxirane ring-opened products (Equation (184)) ⟨68JHC337, 75BCJ960⟩. The ring-opening proceeds to give the more stable carbonium ion, although this intermediate must be strongly coordinated by the catalyst as Suga and co-workers have shown that almost complete stereochemical integrity can be retained in this reaction ⟨75BCJ960⟩ (Equation (185)).

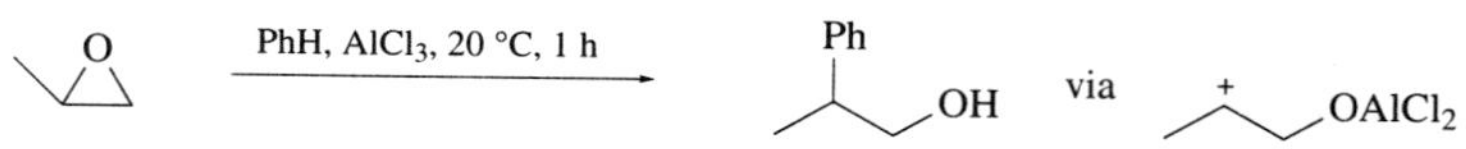

(184)

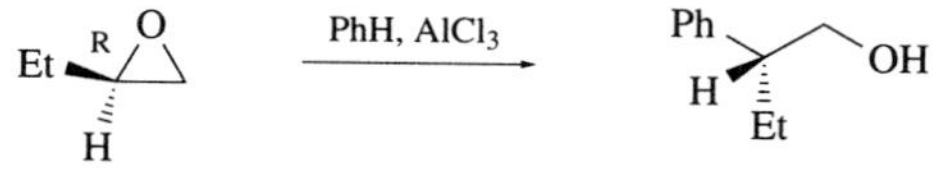

(185)

Epoxides can also react with nonbenzenoid aromatic systems as exemplified by the reaction of oxirane with substituted azulenes (Equation (186)) ⟨72JOC2957⟩, and reaction of the epoxide (**68**) with indole (Equation (187)) ⟨83LA226⟩. Vinylepoxides also undergo stereoselective Friedel-Crafts reactions ⟨93CPB55⟩.

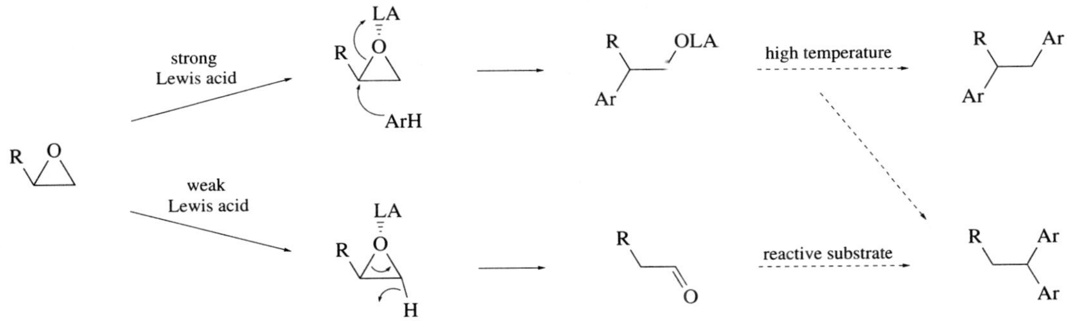

Scheme 28

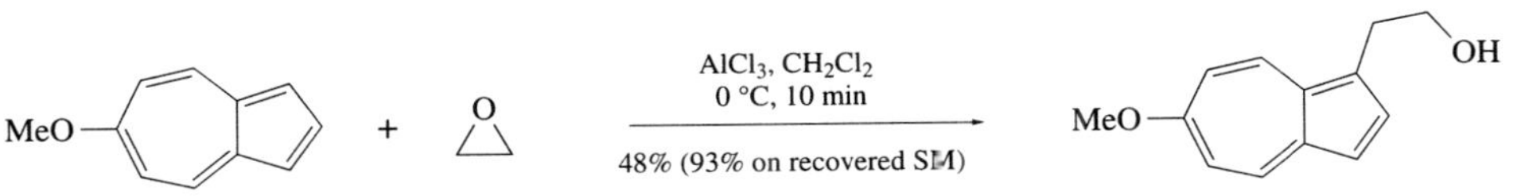
(186)

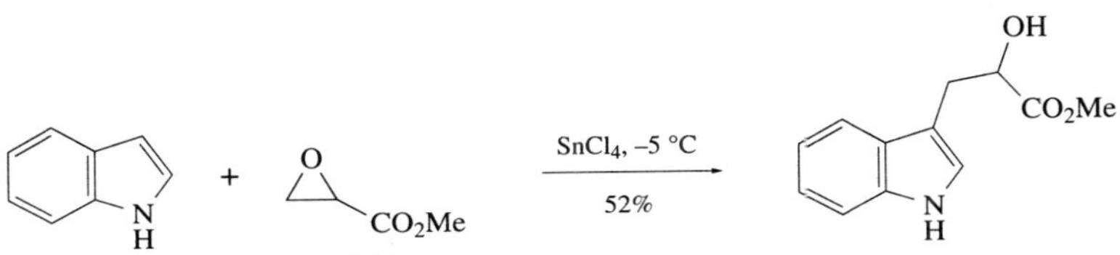
(68)
(187)

Allylsilanes and stannanes will alkylate oxiranes ⟨79S446⟩ under Lewis acid catalysis (Equation (188)). This method is particularly useful in some intramolecular reactions (see below).

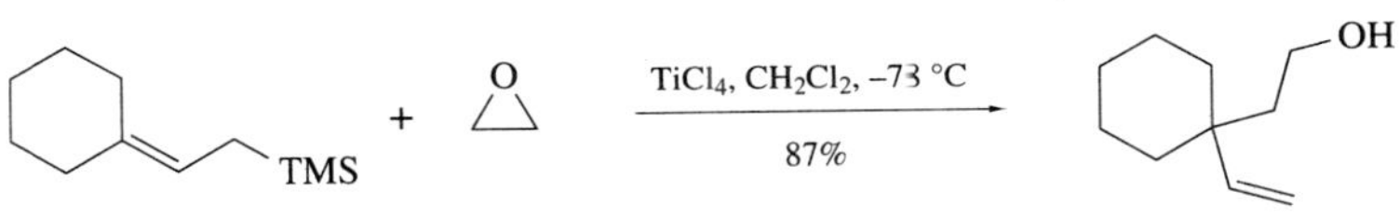
(188)

(iv) Intramolecular reactions

Intramolecular cyclisations of anions onto epoxides rely, for a large part, on the rapid generation of the nucleophile and a slow rate for the intermolecular attack.

Direct *ortho* lithiation of the phenyl ether **(69)** at $-100\,^{\circ}C$ followed by warming gave the corresponding bicyclic compound in 64% yield (Equation (189)) ⟨78JOC3800⟩. The observed stereochemical outcome follows from Baldwin's Rules (5-*exo* tet versus 6-*endo* tet) ⟨76CC734⟩. Transmetallation can also be used to generate the nucleophile (Scheme 29) ⟨80JOC922⟩. The presence of magnesium bromide alters the regiochemistry, giving the product of epoxide opening at the terminal carbon atom. Formation of an intermediate bromohydrin by attack of Br^- on the terminal carbon atom, with subsequent ring formation is the rationale for this difference in behaviour.

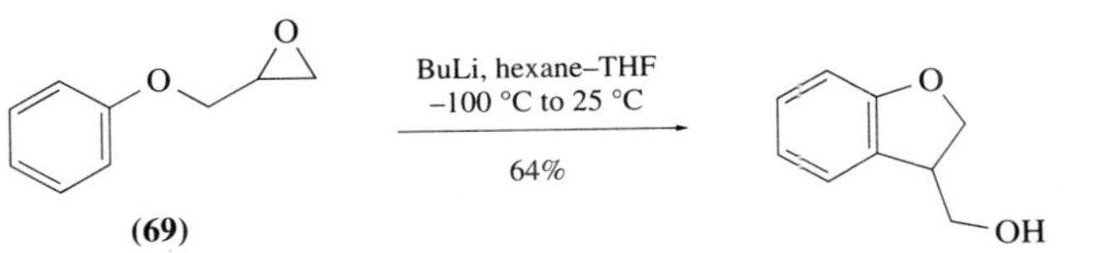
(69)
(189)

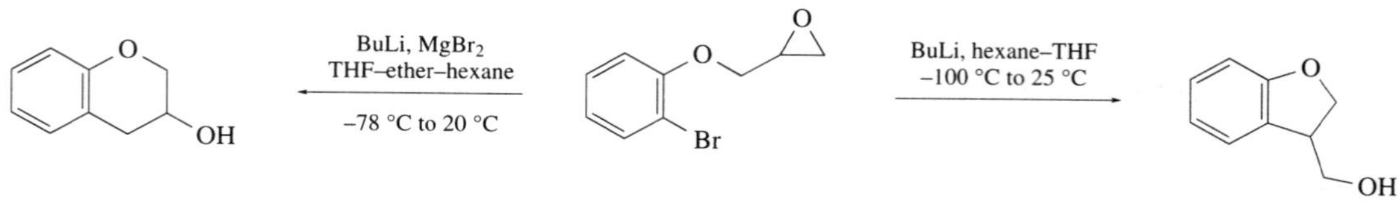

Scheme 29

The effect of ring size on the intramolecular attack of epoxides has been studied by a number of groups. Cooke and Houpis have studied the cyclisations of a series of ω-iodoepoxides (Scheme 30) ⟨85TL3643⟩. The course of the reaction is influenced by the substitution pattern and the presence of Lewis acids. Copper bromide reverses the selectivity of cyclisation of 4-iodobutyloxirane. These results suggest that the observed selectivity is a fine balance between several factors. Similar reactions can be used to prepare strained tricyclic systems ⟨66JA2257, 82JOC3211⟩.

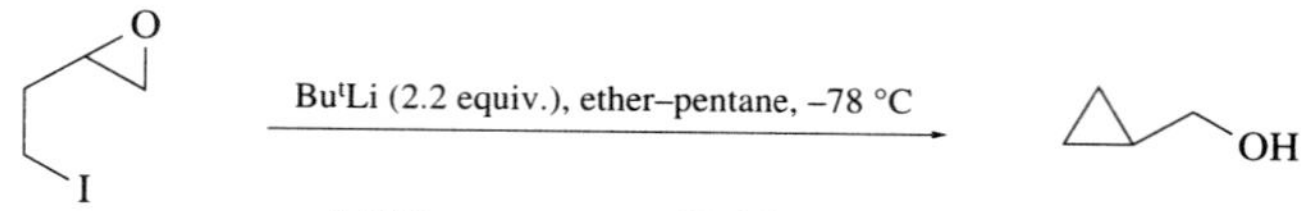

Additive	*Yield (%)*
none	77
$MgBr_2$	77

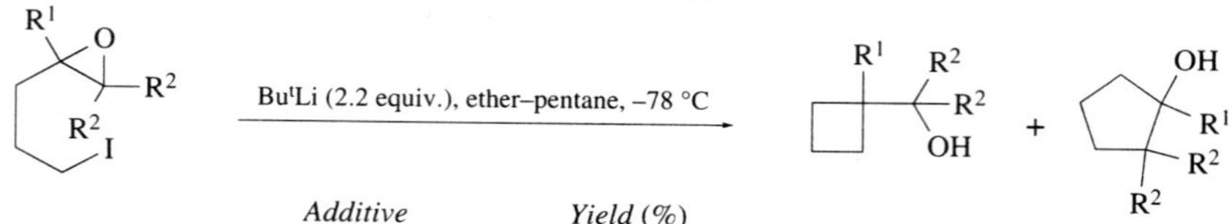

	Additive	*Yield (%)*	cyclobutane		cyclopentane
$R^1 = R^2 = H$	none	82	88	:	12
$R^1 = Me, R^2 = H$	none	45	5	:	95
$R^1 = H, R^2 = Me$	none	66	95	:	5
$R^1 = R^2 = H$	$MgBr_2$	60	15	:	85
$R^1 = Me, R^2 = H$	$MgBr_2$	0			
$R^1 = H, R^2 = Me$	$MgBr_2$	49	25	:	75
$R^1 = R^2 = H$	$CuBr{\bullet}Me_2S$	30	<1	:	>99

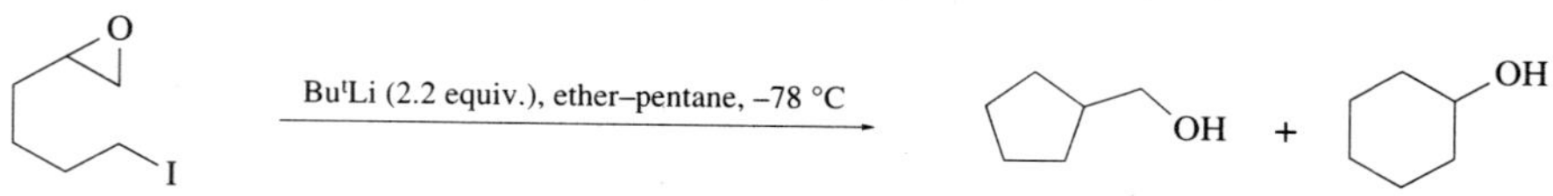

Additive	*Yield (%)*	cyclopentane		cyclohexane
none	78	91	:	9
MeMgBr	56	94	:	6
$BF_3{\bullet}OEt_2$	40	>99	:	<1
$CuBr{\bullet}Me_2S$	59	19	:	81

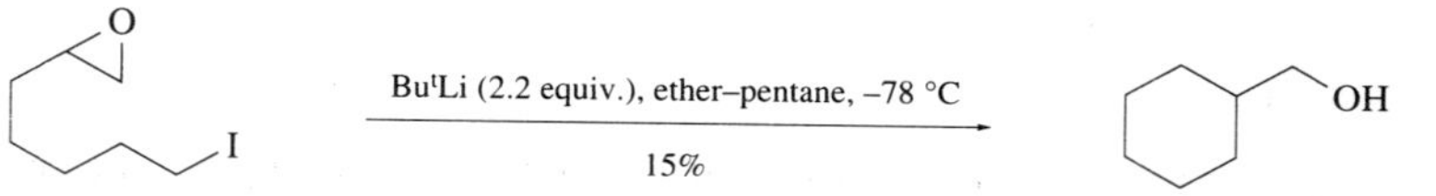

Scheme 30

Stork *et al.* have studied the intramolecular reactions of α-cyanocarbanions and epoxides ⟨74JA5268, 74JA5270⟩. They found that with equal substitution of the two ends of the epoxide cyclisation occurs to give the smaller possible ring. Attack at the distal end occurs only when it is the less substituted; cyclobutane ring systems are not easily made by this method (Scheme 31). The effect of epoxide geometry on the outcome of the reaction has been studied ⟨75TL585⟩. Sulfone anions also react to give products of various ring sizes ⟨76JOC3648, 81CJC1415⟩ but the outcomes are dependent on the base used. LDA deprotonation gives cyclopropanes from γ-epoxysulfones but fails to give cyclobutanes from δ-epoxysulfones. Deprotonation with methylmagnesium iodide gives cyclobutanes from γ-epoxysulfones and cyclopentanes from δ-epoxysulfones via the formation of iodohydrins and subsequent displacement of the distal iodine atom (Scheme 32). The same effect is seen with nitrile substrates. Further substitution around the epoxide unit can also alter the regiochemistry. Vinyl sulfones (Equation (190)) ⟨85TL6301⟩ and other stabilized anions are also suitable nucleophiles, particularly for the formation of small rings (Equations (191) and (192)) ⟨81TL1751, 82JOC2564, 82TL1343⟩. The cyclisation of the epoxyamide (**70**) sets up the correct *trans* β-lactam stereochemistry for the synthesis of thienamycin-type antibiotics (Equation (193)) ⟨88TL815⟩. Macrocyclisations have also been accomplished (Equation (194)) ⟨84TL5781⟩.

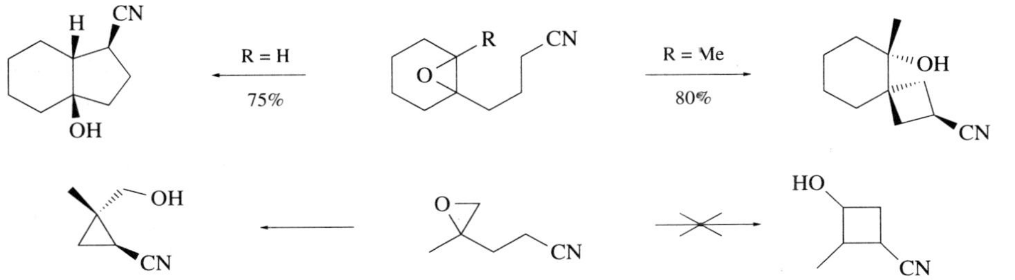

Scheme 31

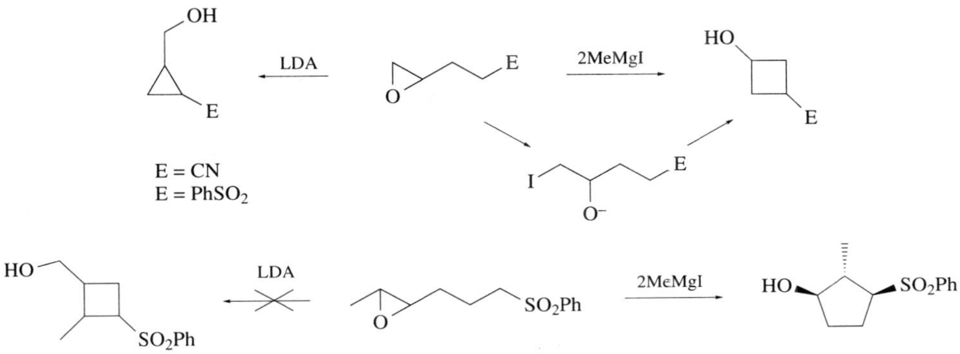

Scheme 32

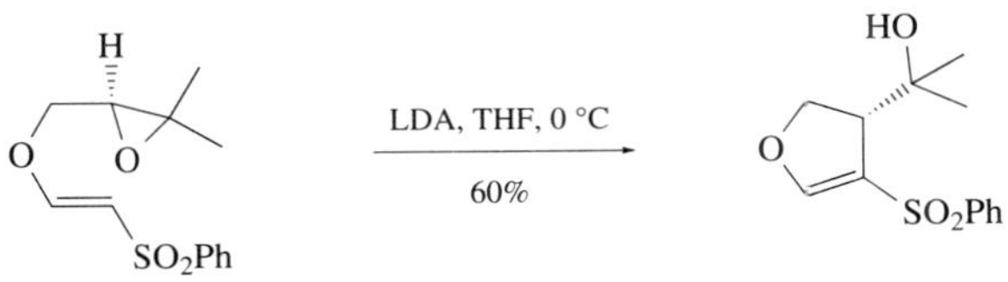

(190)

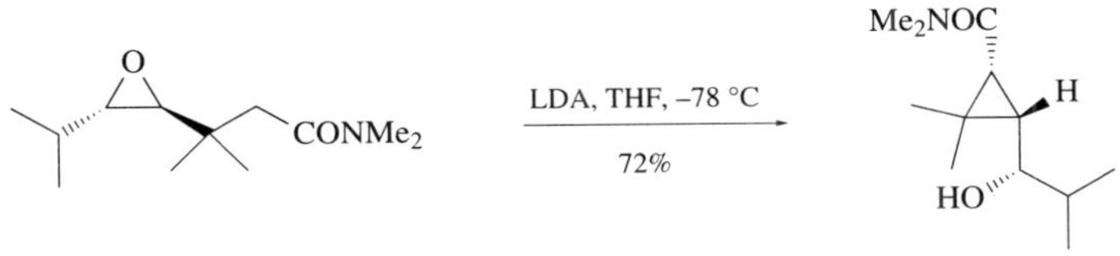

(191)

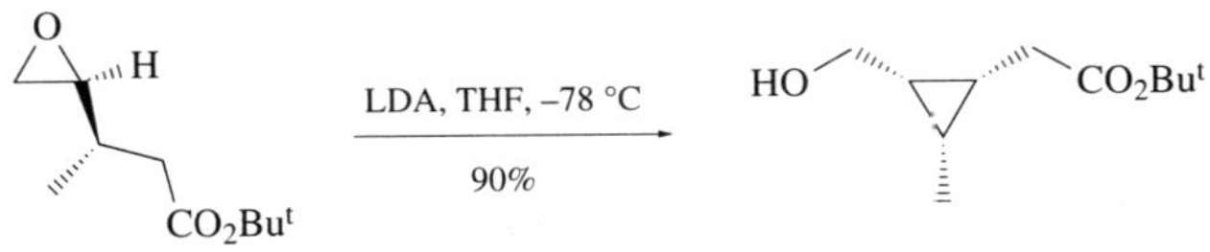

(192)

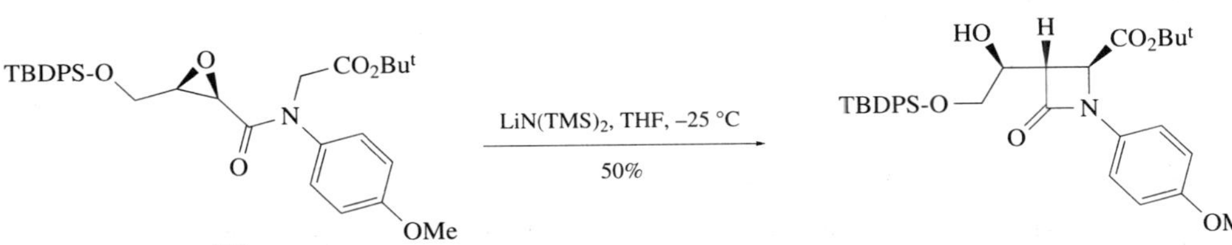

(193)

(**70**)

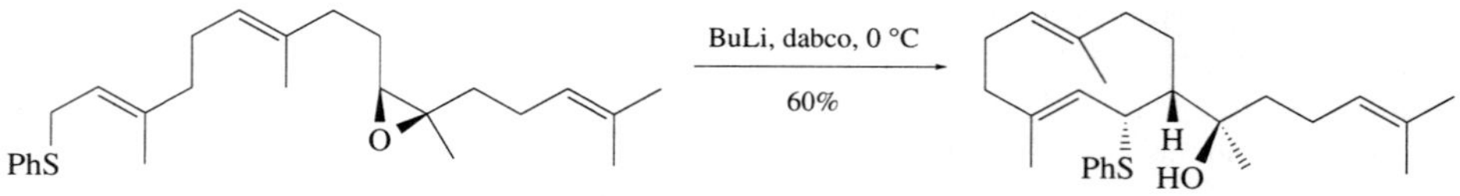

(194)

dabco = 1,4-diazabicyclo[2.2.2]octane

The intramolecular Friedel-Crafts cyclisation (Equation (195)) ⟨83JOC2499⟩ gives mainly the expected seven-membered ring compound as well as the six-membered ring isomer; the reaction also works with heteroaromatics such as the furan **(71)** (Equation (196)) ⟨83JOC4572⟩. A combined intramolecular epoxide opening followed by a Friedel-Crafts alkylation was discovered serendipitously in the process described in Equation (197). The epoxides **(72)** and **(73)** undergo Lewis acid catalysed intramolecular epoxide opening with subsequent lactonisation to give the same 60 : 40 mixture of cyclopropanes (Scheme 33) ⟨93JA2970⟩. The $SnCl_4$–$MeNO_2$ combination is the same as that used by van Tamelen and Anderson in their work on the cyclisation of squalene oxide ⟨72JA8225⟩. Similar cationic cyclisations can give pentacyclic structures ⟨93TL7849⟩. Cyclisations of allylsilanes and stannanes take place readily with good regio- and stereo-selectivity (Equations (198) and (199)) ⟨89JOC3114, 87TL2753⟩. In the latter case six-membered ring formation is prefered. A number of groups have carried out related work ⟨B-81MI 105-01, 83CJC214, 84CC585, 84CC1273, 86CJC584, 89TL4845, 90JCS(P1)1226⟩ concerning the regio- and stereoselectivities of such cyclisations.

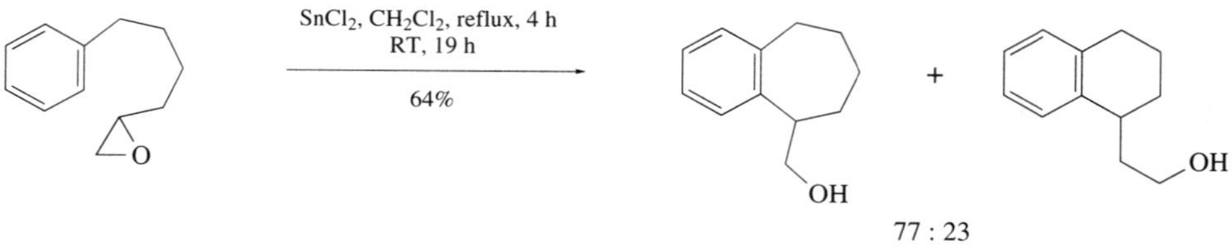

(195)

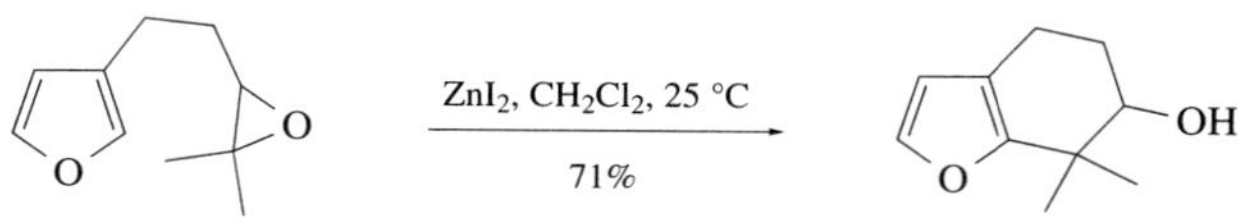

(196)

(71)

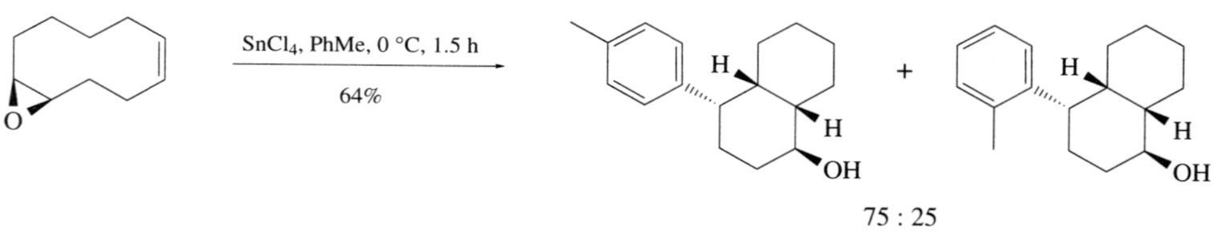

(197)

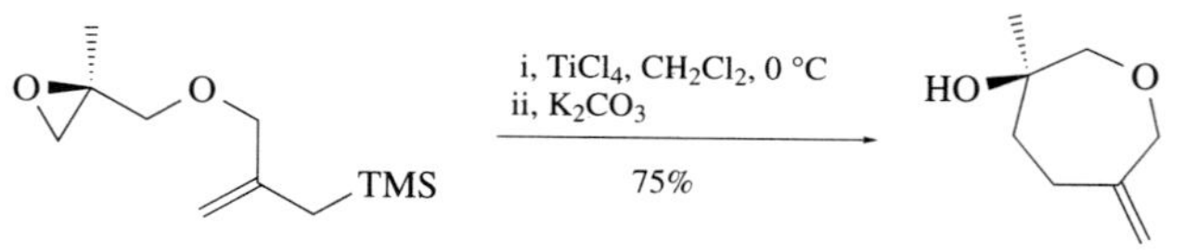

(198)

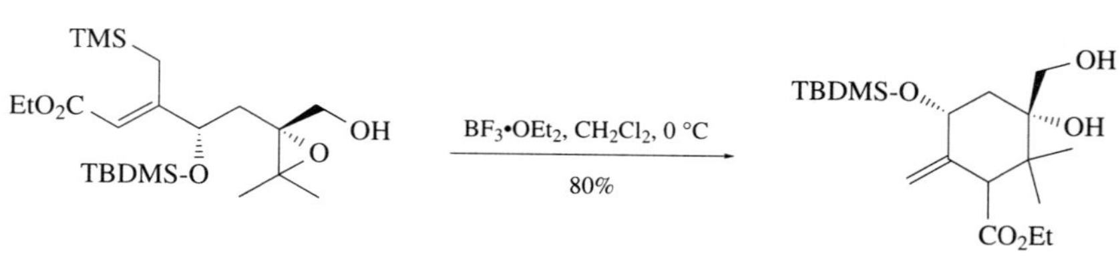

(199)

(v) Stereo- and regiochemistry

With unsymmetrically substituted epoxides there is the possibility of attack at either end of the epoxide unit. In the majority of cases attack occurs at the least substituted end of the epoxide with

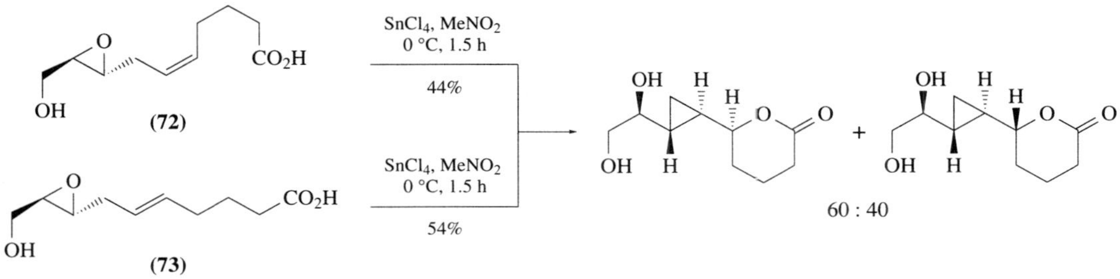

Scheme 33

inversion of configuration (S_N2) (e.g., Equation (200)) ⟨84JOC3928⟩. There are exceptions to this generalization, mostly involving aromatic epoxides (Equation (201)) ⟨90H(31)1997⟩ (similar reactions occur with MeLi and MeMgI), (Equation (202)) ⟨86TL2015⟩ which are attacked at the more reactive benzylic position, (see also Scheme 23). Alkylaluminum reagents can open aromatic epoxides with reversed regiochemistry and retention of configuration (Scheme 34) ⟨91TL1059⟩ (and see ⟨89H(29)249⟩ in Section 1.05.1.3.2). MO calculations provide an explanation for this phenomenon in that they suggest that there is considerable carbonium ion character at the carbon atom next to the phenyl group; migration of the methyl group takes place from the aluminum species co-ordinated to the oxygen atom. Silyl epoxides are also exceptions to the normal rule. In this case selectivity is determined by the bulk of the silyl group and the presence or absence of Lewis acid ($BF_3 \cdot Et_2O$) (Scheme 35) ⟨93TL3695⟩. Lewis acid and bulky silyl groups favour 'conventional' regioselectivity. The reactions of symmetrically substituted epoxides (i.e. both ends of the epoxide are mono- or disubstituted) are much harder to predict but electronic, steric and coordination factors come into play. Attack on silyloxyepoxides is probably affected by the steric bulk of the silyl group (Equation (203)) ⟨89TL6963⟩. Several groups have studied the regioselectivities of cuprate and alane addition to epoxyacids ⟨81JOC5242, 85TL4683⟩ and epoxyesters ⟨73JOC4263, 73JOC4346⟩. The regioselectivities of additions to glycidic systems are covered in the following section. A theoretical study of the addition of nucleophiles to oxiranes conjugated with unsaturated groups such as alkenes (see next section), ketones and aldehydes, esters and enol ethers has been undertaken ⟨88JOC139⟩. With ketones and aldehydes the prefered site of attack for both hard and soft nucleophiles is the carbonyl group while for esters it is the epoxide carbon atom which lies next to the carbonyl group.

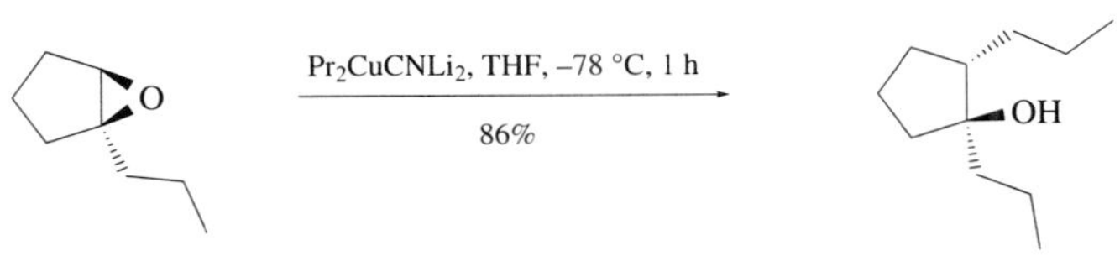

(200)

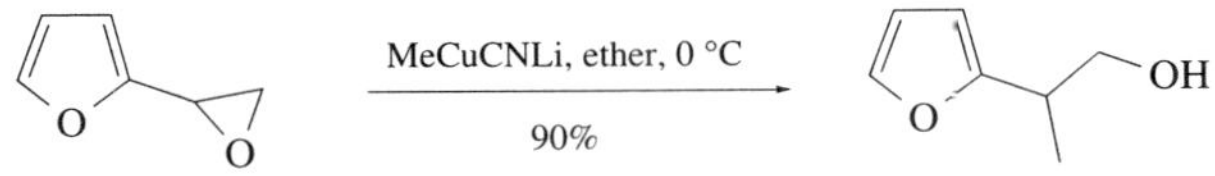

(201)

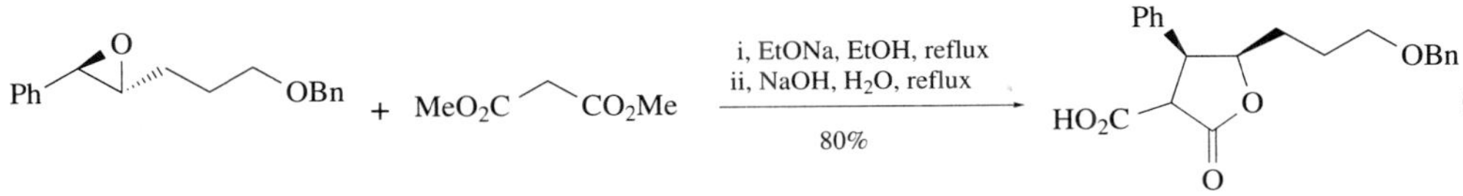

(202)

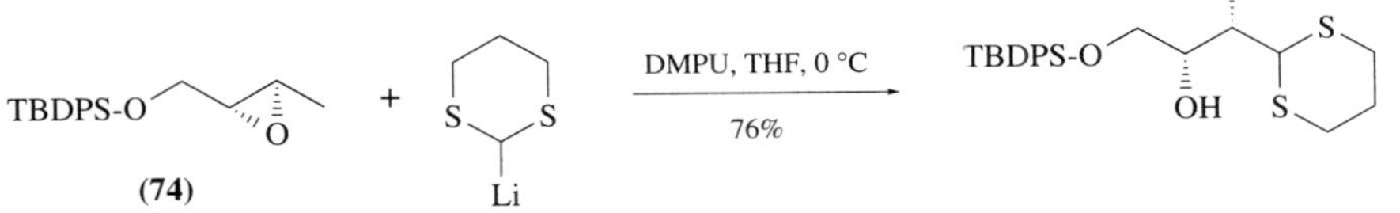

(203)

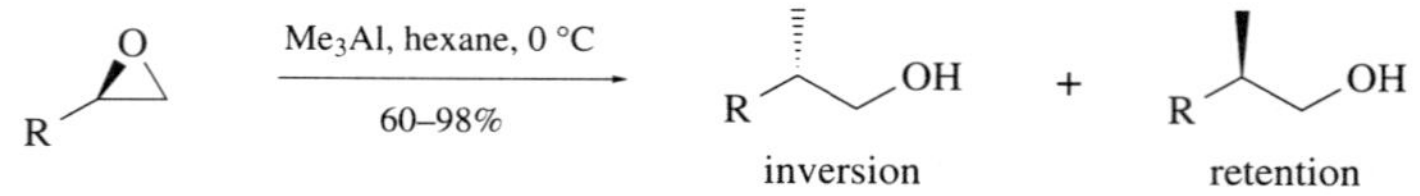

R	inversion		retention
hexyl	98	:	2
decyl	97	:	3
phenyl	15	:	85
pentafluorphenyl	35	:	65

Scheme 34

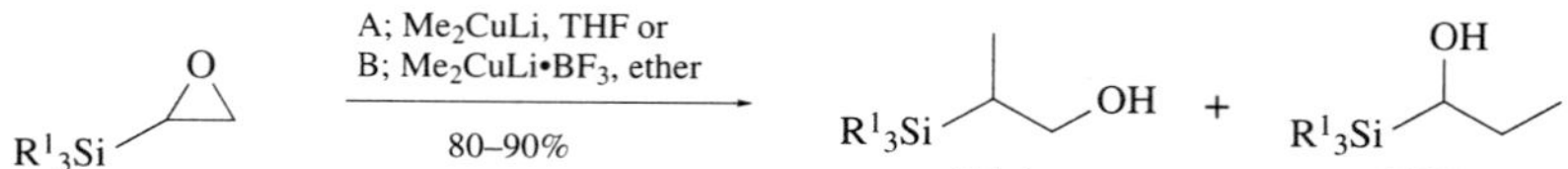

(74a) **(74b)**

R$^1{}_3$Si	**(74a)** : **(74b)** *A*	*B*
Pr^{i}Me$_2$Si	100: 0	94: 6
Bu^{t}Me$_2$Si	98: 2	34: 66
Pr$^i{}_3$Si	62:38	0:100

Scheme 35

Attack on epoxide units appended to six-membered rings (and other conformationally restricted systems) usually occurs to give the *trans*-diaxial product. Scheme 36 shows sequential *trans* diaxial openings in an anhydrosugar ⟨85TL4111⟩. Equations (204) and (205) ⟨88SC151, 90S285⟩ contrast the reaction profiles of β and α epoxysugars. Equatorial attack can occur if steric factors or neighbouring groups exert directing effects. Attack of nucleophiles on the epoxycyclohexanols **(75)** **(76)** and silyl ethers **(77)** **(78)** appears to be controlled by these factors (Scheme 37) ⟨77JOC394⟩; (see also ⟨87JOC34, 89TL685⟩).

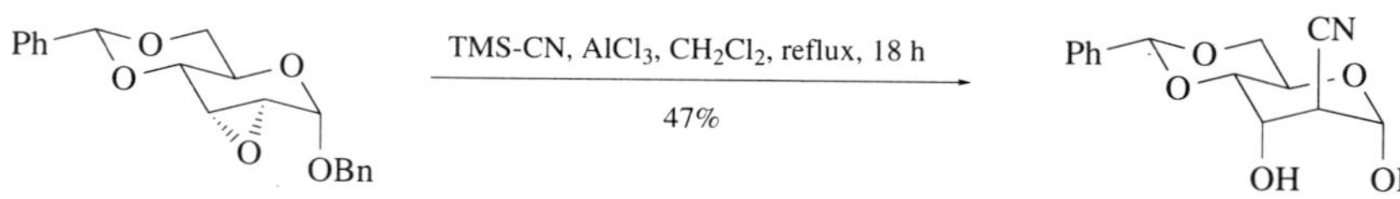

(204)

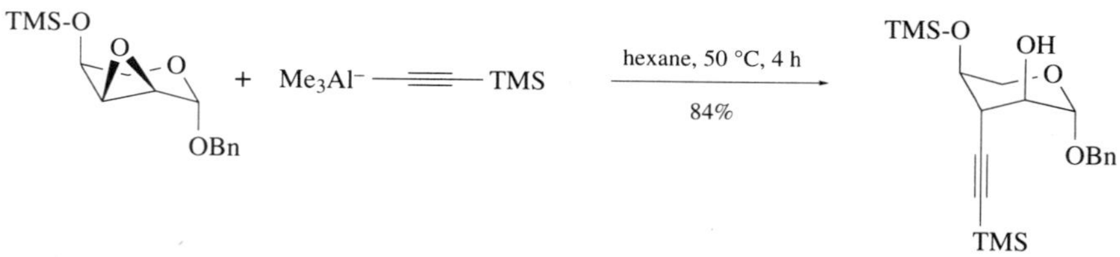

(205)

In other bicyclic ring systems the regiochemistry of attack is largely determined by stereochemical considerations. For 6-oxabicyclo[3.1.0]hexanes the factors governing the regiochemistry of nucleophilic attack are not always very clear; steric hindrance and coordination of the nucleophile to other groups in the vicinity of the oxirane can play a role. The regioselectivities observed for the nucleophilic opening of the epoxides **(79)** and **(80)** are different; the selectivity can be reversed by using an alanate nucleophile instead of an alane (Scheme 38) ⟨85JOC3923⟩. Cuprate and dithiane attack the epoxide **(81)** at different sites. The cuprate is probably directed to the more hindered site by the methoxymethyl group (Scheme 39) ⟨83JA1988⟩. Selectivity is seen in the opening of the nucleoside epoxide **(82)** (Equation (206)) ⟨85TL97⟩. This is the usual site of attack in this type of structure. Opening of the epoxide **(83)** in a prostaglandin synthesis was less successful; only 30% of the required isomer was obtained (Equation 207) ⟨70TL311⟩.

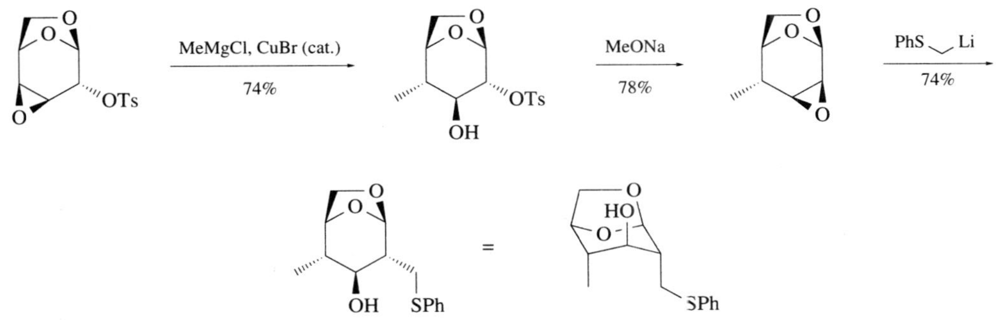

Scheme 36

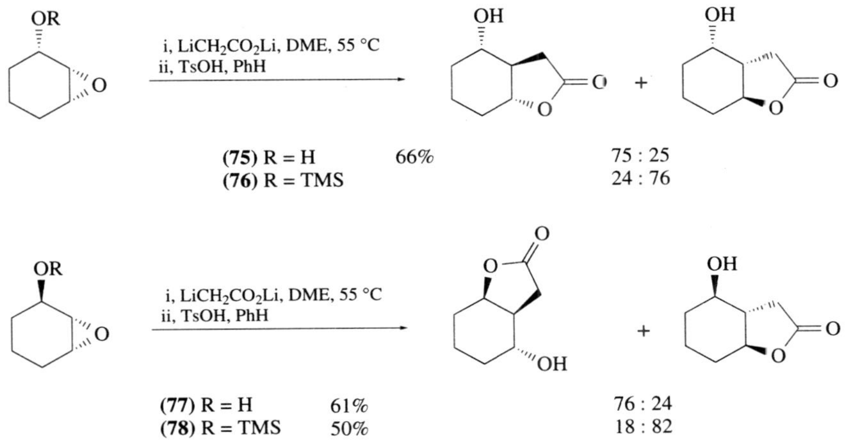

Scheme 37

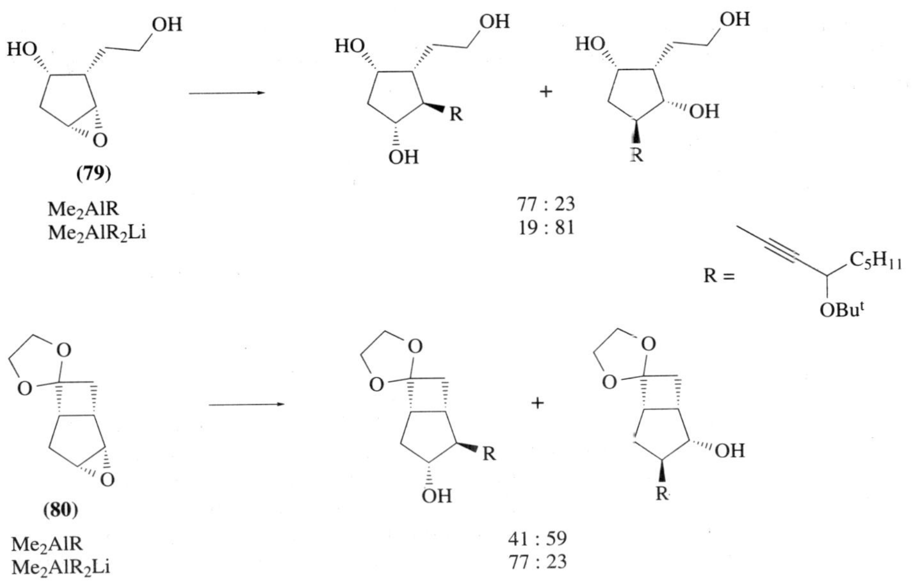

Scheme 38

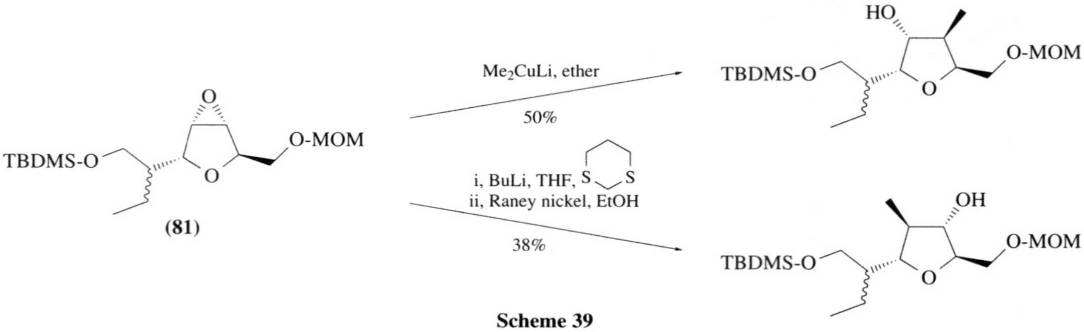

(81)

Scheme 39

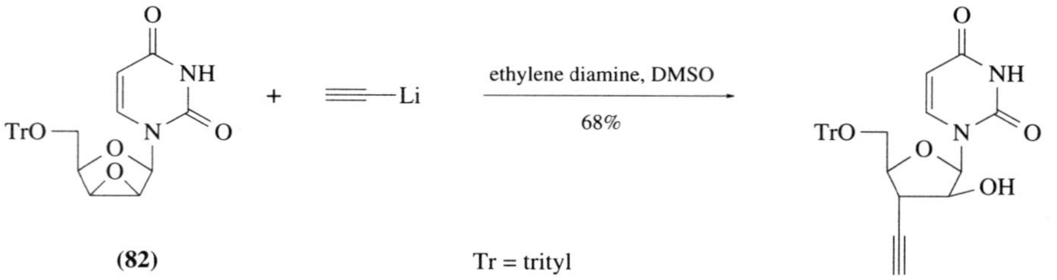

(82) (206)

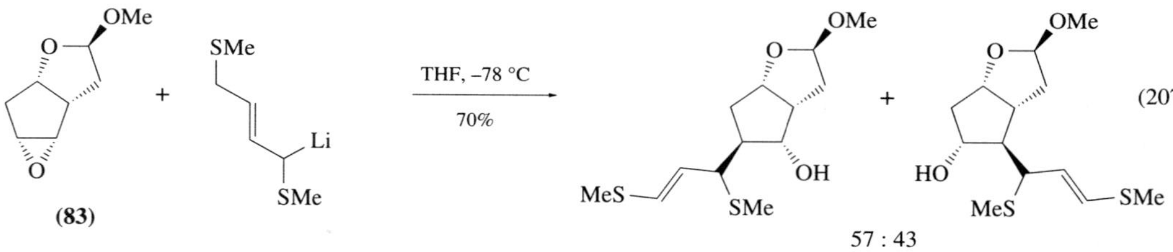

(83) (207)

Selective opening of one epoxide in the presence of a second can be achieved if there is sufficient difference in the reactivities of the two epoxides; in the case of the example shown in Equation (208) nucleophile attack occurs on the terminal epoxide ⟨93CC614⟩.

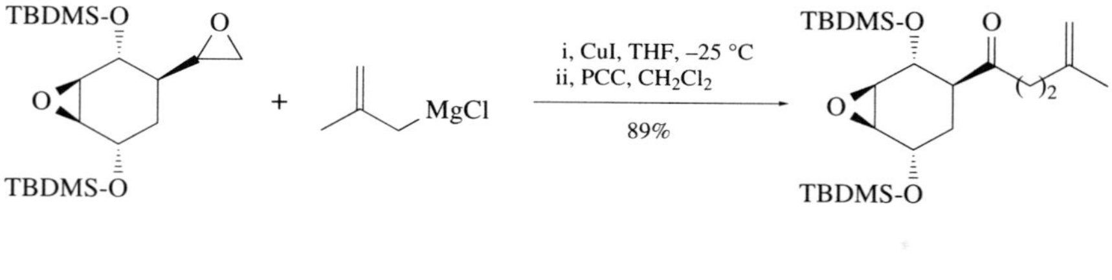

(208)

1.05.1.3.2 Glycidic epoxides

Glycidols (epoxyalcohols) and their derivatives ⟨91CRV437⟩ provide interesting examples to illustrate the subtleties of epoxide ring-openings as the neighbouring group often influences the course of the reaction. This is especially true when the neighbouring group is another leaving group. The adjacent oxygen functionality can have a directing effect on the regiochemistry of attack on the epoxide ring. Pfaltz and Mattenberger have studied the regioselectivity of attack of a number of different carbon nucleophiles on benzyl protected allylic and homoallylic epoxyalcohols (Schemes 40 and 41) ⟨82AG(E)71, 82AG(S)161⟩. Trimethylaluminum in the presence of a catalytic amount of butyllithium opens these epoxides to give 1,2 diols (C-3 attack). In contrast cuprate, lithium and Grignard reagents gave mixtures of 1,2 and 1,3 diols (C-2 attack). A similar dichotomy is seen in cyclic systems ⟨83JOC409⟩. The related homoallylic cases show the same regiochemistry, giving 1,3 diols with aluminum or mixtures of 1,3- and 1,4-diols with other metals (Scheme 42). Roush *et al.* ⟨83TL1377⟩ and Oshima and co-workers ⟨82TL3597⟩ found similar high selectivities for attack at C-3 of the epoxyalcohols with aluminum species giving 1,2-diols (Equations (209)–(211)). More reactive substrates and longer reaction times by-pass the need for the formation of the ate complex. The selectivity is greatly reduced when the C-3 substituent is oxygenated. This may be due to inductive deactivation of C-3, or it may be due to the incorporation of an alternative site for aluminum coordination. Copper(I)-catalysed Grignard addition to similar substrates ⟨83JOC4131⟩ gives the expected 1,3-diols (addition at C-2) with regioselectivities of 69 : 31 to 87 : 13 (Equations (212) and (213)). Selectivities were not improved when the benzyloxy group was one carbon atom further away from the reactive unit. See ⟨77JOC394, 87JOC34, 89TL685⟩ for examples in cyclic systems (Section 1.05.1.3.1(iv)).

	Yield (%)	*ratio*
(*Z*) MeLi, ether, 20 °C, 22 h	82	59 : 41
(*Z*) Me_2CuLi, ether, –20 °C, 30 min	43	47 : 53
(*Z*) Me_3Al, BuLi (0.3 equiv.), toluene, –20 °C, 3 h	76	>99 : 1
(*E*) Me_2CuLi, ether, –40 °C, 1 h	83	78 : 22
(*E*) Me_3Al, BuLi (0.3 equiv.), toluene, –20 °C, 1 h	76	>99 : 1

Scheme 40

	Yield (%)	*ratio*
(*Z*) Me_2CuLi, ether, –20 °C, 2 h	83	59 : 41
(*Z*) Me_3Al, MeOLi (0.3 equiv.), toluene, 0 °C, 6 h	87	95 : 5
(*E*) Me_2CuLi, ether, –40 °C, 15 min	90	68 : 32
(*E*) Me_3Al, MeOLi (0.3 equiv.), toluene, 0 °C, 3 h	85	99 : 1

Scheme 41

(209)

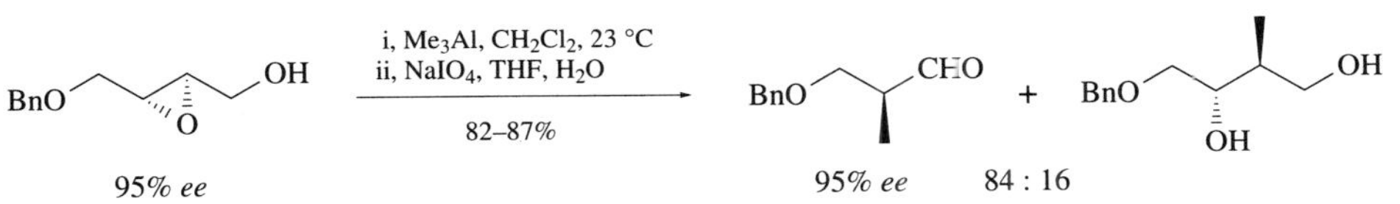

(210)

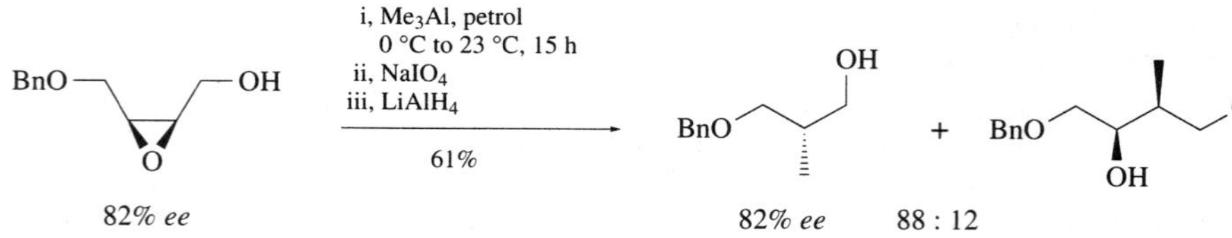

(211)

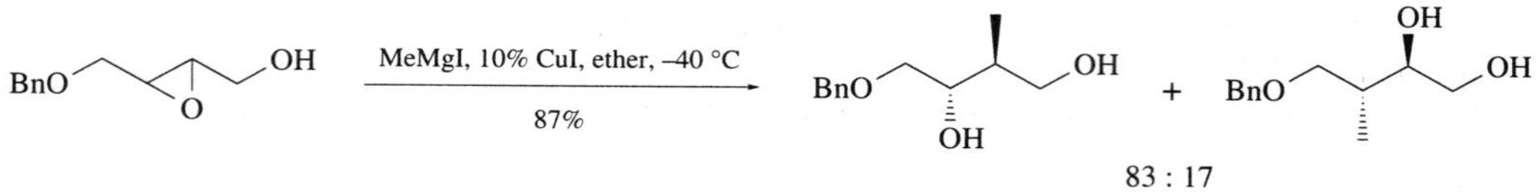

(212)

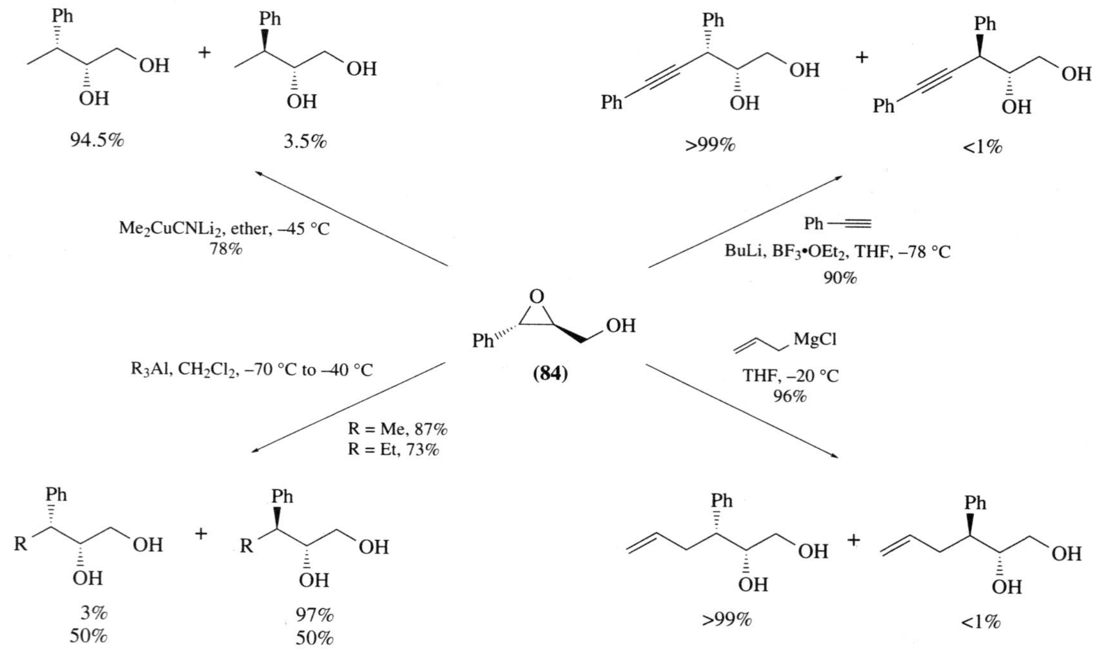

Scheme 42

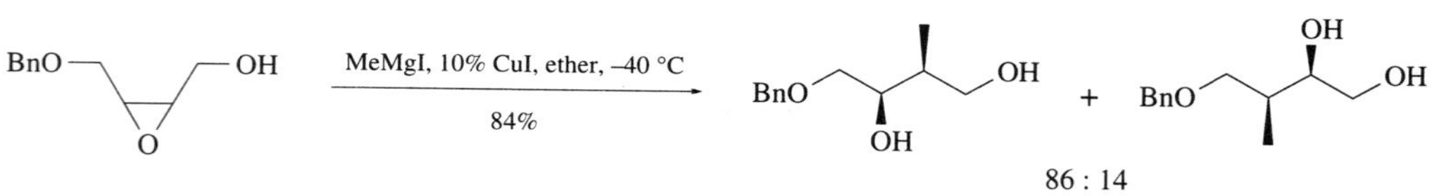

(213)

Epoxide opening of the aromatic substrate (**84**) ⟨89H(29)249⟩ occurs at the benzylic position (C-3) (with inversion of configuration) with Grignard and cuprate nucleophiles (Scheme 42); retention of configuration is observed when trimethylaluminum is the nucleophile. Triethylaluminum gives much poorer stereocontrol. Aluminum reagents generally give much poorer selectivities suggesting that an S_N1 pathway is competing with the normal S_N2 route when an adjacent group is capable of stabilizing a carbocation; in contrast to the result obtained for the aliphatic glycidol (**85**) (Equation (214)) ⟨84LA1165⟩.

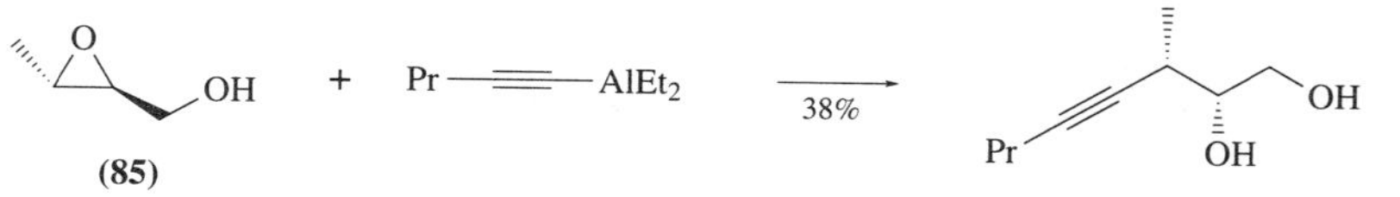

(214)

Treatment of the epoxyalcohol (**86**) with lithium acetylide (**87**) and $BF_3 \cdot Et_2O$ gave mixtures of the alcohols (**88**) and (**89**) (Scheme 43) ⟨84CC202, 87T2701⟩. The latter compound arises from direct attack on the epoxide, formation of the former compound involves a Payne rearrangement ⟨62JOC3819⟩. (The rearrangement does not occur if the $BF_3 \cdot OEt_2$ is added to the alkyne before the epoxide ⟨89H(29)249⟩). Under the same conditions the isomeric epoxyalcohol (**90**) gave only the expected product (Equation (215)) arising from attack at the less hindered carbon atom.

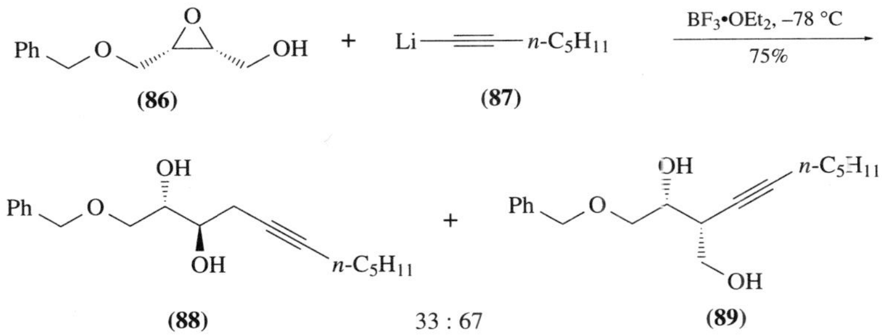

Scheme 43

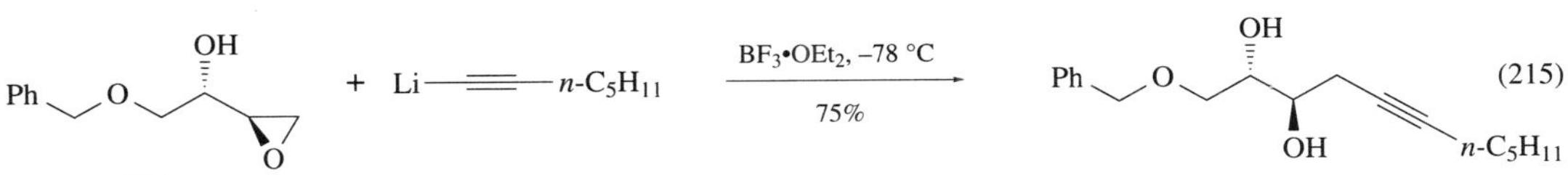

(**90**)

Sharpless and co-workers found that higher order cuprates open the epoxide (**91**) selectively, in moderate yield, with complete retention of chirality (Equation (216)) ⟨89JOC1295⟩. Opening of glycidyl tosylate (**91**) with simple Grignard and lithium reagents gave unsatisfactory results. The bisaddition product (**92**) is formed as a by-product in varying amounts. Copper-catalysed Grignard addition to the antipode gave the hydroxytosylate (**93**), again with complete retention of chirality and only a small amount of epoxide (Equation (217)). The optical rotation of this epoxide indicates that it is derived from ring closure of the major product and not from direct displacement of tosylate. A major side reaction is epoxide opening by the halide ion associated with the Grignard reagent. Halohydrin formation is significantly reduced, but not eliminated, when Li_2CuCl_4 is used as the catalyst. Phenyllithium-boron trifluoride etherate also affords the hydroxytosylate in modest yield (66%), with no sign of products resulting from tosyl displacement. Starting material (38%) was also recovered. A similar result was obtained by Rotstein and Walker who opened both enantiomers of glycidal tosylate (**94**) with Grignard reagent (**95**) under copper catalysis to give the corresponding optically active tosylate (**96**) in >99.5% *ee* after crystallisation (Equation (218)) ⟨93TA1521⟩. Selective epoxide opening of glycidal tosylate using diethylaluminum cyanide is also possible (Equation (219)) ⟨89JOC1295⟩. Attack of the anion of the thionitrile (**97**) on the homoallylic epoxide-mesylate gives epoxide opening then displacement of the mesylate group by the reformed anion to give the cyclopentane (**98**) (Equation (220)) ⟨85TL3121⟩.

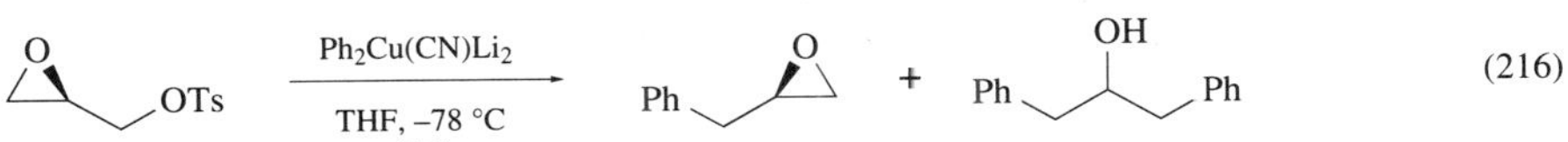

(**91**) 88% *ee* 88% *ee* (**92**)

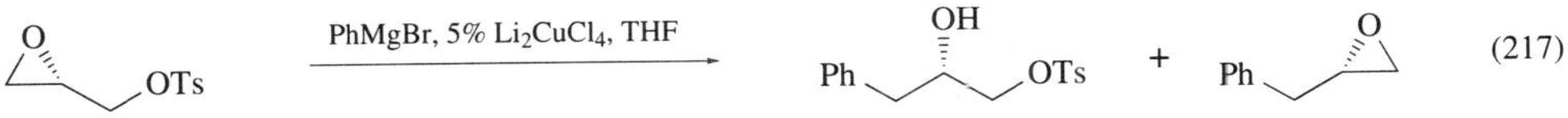

(**93**) 94% *ee* 79%, 94% *ee* 5%, 88% *ee*

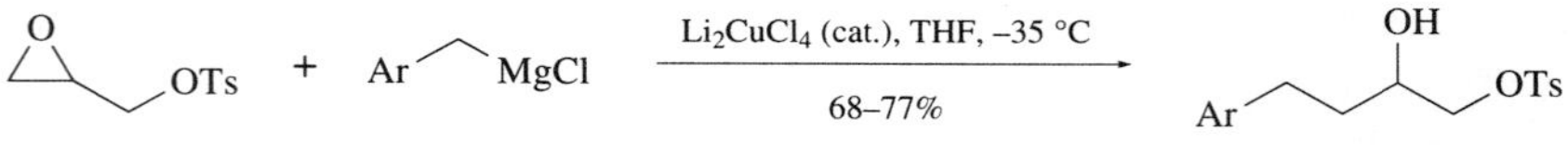

(94) (95) (96) (218)

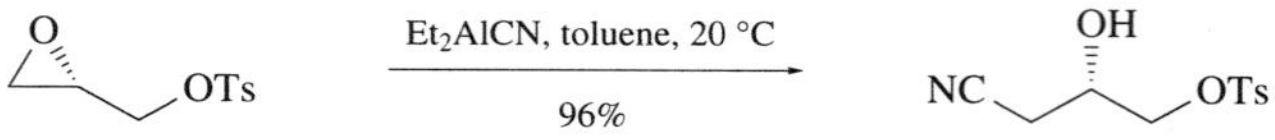

(219)

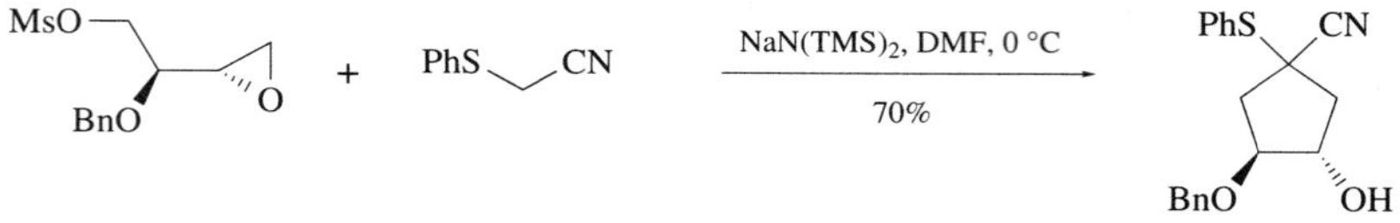

(97) (98) (220)

In contrast French workers ⟨87T2701⟩ found that the glycidyl tosylate (**99**) is attacked by both alkyl copper-catalysed Grignard reagents and cuprate reagents with displacement of the tosyl group (Equation (221)). Alkenyl and alkynyl reagents did not react; starting material was recovered unchanged. The corresponding tosylate (**100**) is inert to both ring opening and tosyl group displacement. This suggests that there is a fine balance between attack at epoxide and displacement of the sulfonate group.

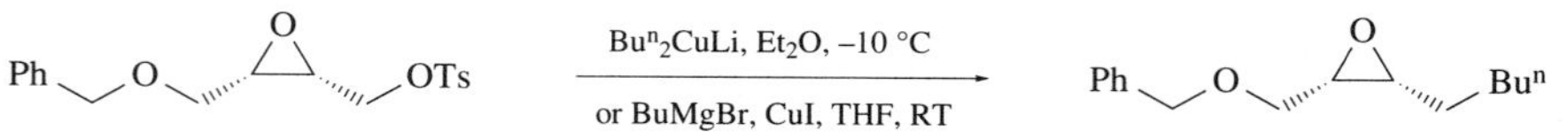

(99) (221)

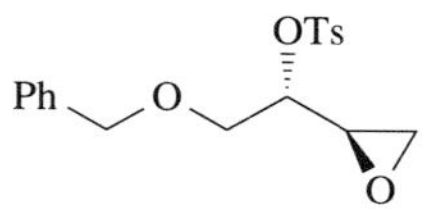

(100)

Aitken and co-workers have studied the stereochemical course of the reaction of various glycidol derivatives using the anion of the aminonitrile (**101**) (Scheme 44) ⟨93SL599⟩. The reaction can take two routes, either via direct displacement of the leaving group or via epoxide opening and subsequent closing to give product of opposite chirality. It was found that with triflate as the leaving group the reaction proceeded via route A; with bromide, chloride and benzenesulfonate route B was more prominent. The corresponding tosylate, mesylate and nosylate did not react. The regioselectivity of the benzenesulfonate is unexpectedly poor considering that sulfonates are generally considered to be better leaving groups that bromides ⟨82S85⟩. The lack of reactivity of the mesylate is also surprising.

Reaction of optically pure epichlorohydrin with the aluminate (**102**) gave only the product from epoxide opening (Equation (222)) ⟨90T4503⟩. Similar selectivity is seen in the reaction of dimethyl malonate with epichlorohydrin (Equation (223)) ⟨89HCA1301⟩; initial attack occurs at the epoxide unit with formation of the chlorohydrin as an intermediate.

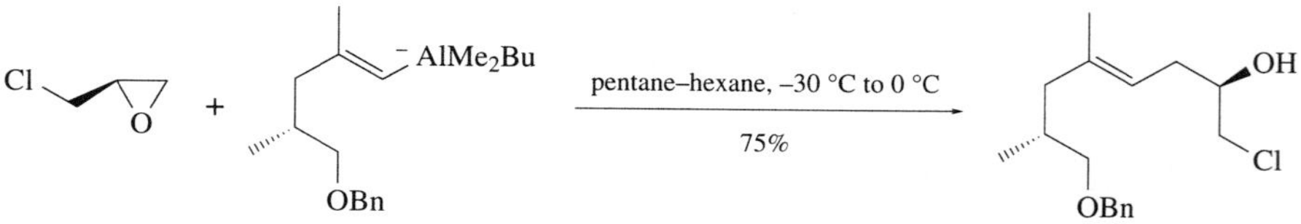

(102) (222)

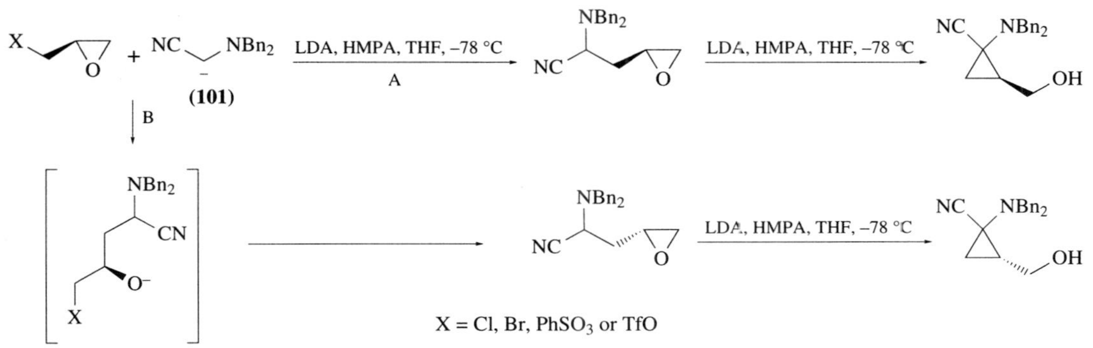

Scheme 44

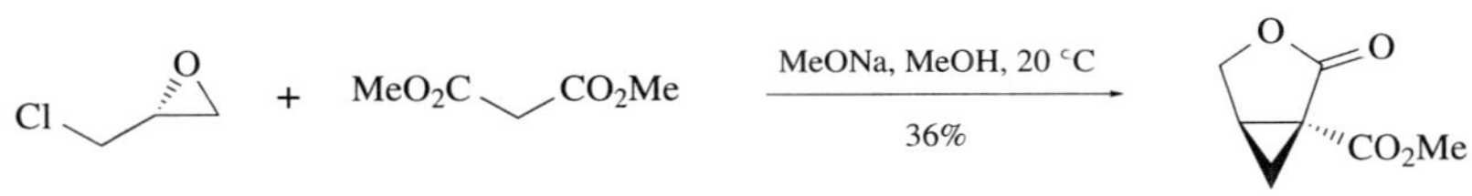

(223)

1.05.1.3.3 Vinylogous epoxides

Vinylepoxides (monoepoxydienes or ethenyloxiranes) can be opened by nucleophiles at any of three positions; namely either end of the epoxide (one of which is vinylic) or in an S_N2' fashion by attack at the distal end of the alkene. The modified neglect of diatomic overlap (MNDO) study of oxiranes by Font and co-workers mentioned above ⟨88JOC139⟩ suggested that hard nucleophiles should attack the epoxide (the two ends being of roughly equal reactivity) and soft nucleophiles should target the alkene (Figure 2). The same is true where the alkene is oxygenated at C-3 (an enol ether). These theoretical results are largely borne out in practice, where copper reagents are 'soft' and most other metals are 'hard', except that attack at the unsubstituted end, C-1, is fairly rare; there appears to be little study or rationale of the C-1/C-2 selectivity. In contrast propargyl epoxides are attacked at C-1 by cyanide and cuprate reagents ⟨72JOU2530, 91TA943, 93TL8263⟩.

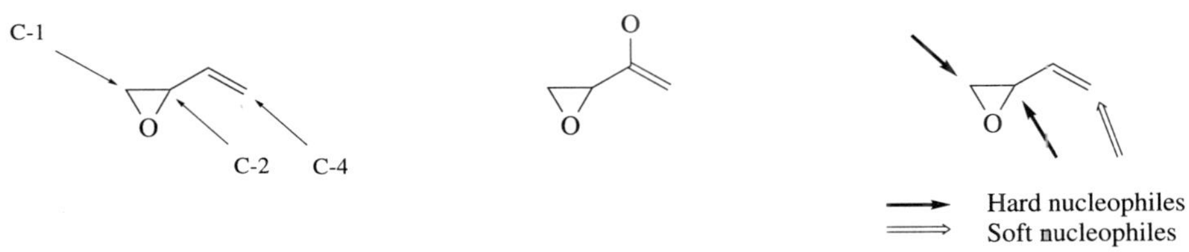

Figure 2

Sulfone anion (**103**) opens epoxide (**104**) at C-1, with subsequent lactonisation, in moderate yield ⟨88TL2059⟩ (Equation 224) (see also ⟨81SC723⟩). The proportion of HMPA was critical in the reaction of allene anion with vinylepoxide, concentrating the attack of the anion at C-1 (Scheme 45) ⟨79S508⟩. In the absence of HMPA a mixture of products is seen. A Wittig reagent and cobalt octacarbonyl also react through attack at C-1 (Scheme 46) ⟨88TL1763, 89JA7938, 90JOC324⟩.

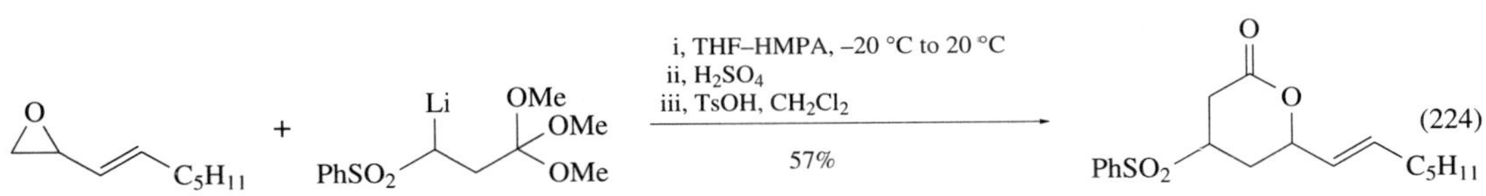

(224)

(**103**) (**104**)

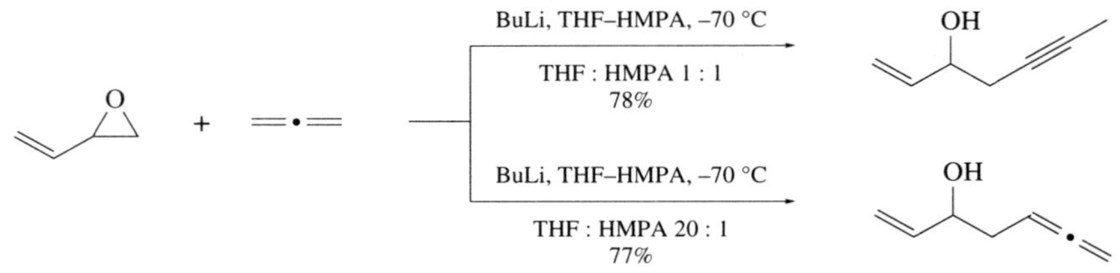

Scheme 45

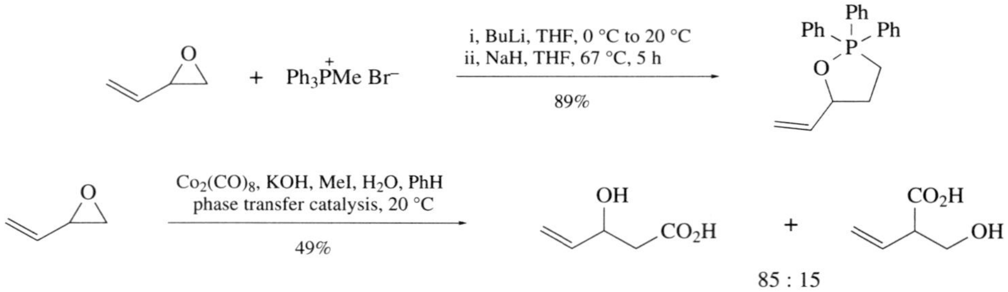

Scheme 46

Opening of vinylepoxides by attack at C-2 is favoured by Grignard, dialkylmagnesium and lithium reagents ⟨70JA4978, 70JA4979, 71JA3047, 73JOC4263, 81JOC5379⟩. Aluminum reagents open vinyl epoxides by direct displacement at C-2 (Equation (225)) ⟨83S284⟩. An intramolecular reaction gives a six-membered ring product from C-2 opening (Equation (226)) ⟨90JA1661⟩. Iron carbonyl inserts into the C-2 position of cyclohexadiene monoepoxide (Equation (227)) ⟨81JCS(P1)270⟩. Vinyl epoxides react with allylsilanes in the presence of $BF_3 \cdot OEt_2$ (Equation (228)) ⟨87CL963⟩. 1,2-Addition occurs unless the vinyl epoxide has a phenyl group at C-4 or is dialkylated at C-4, whereupon the addition goes 1,4.

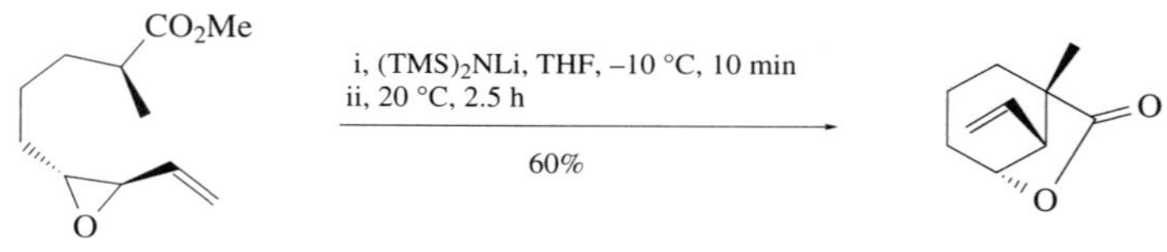

(225)

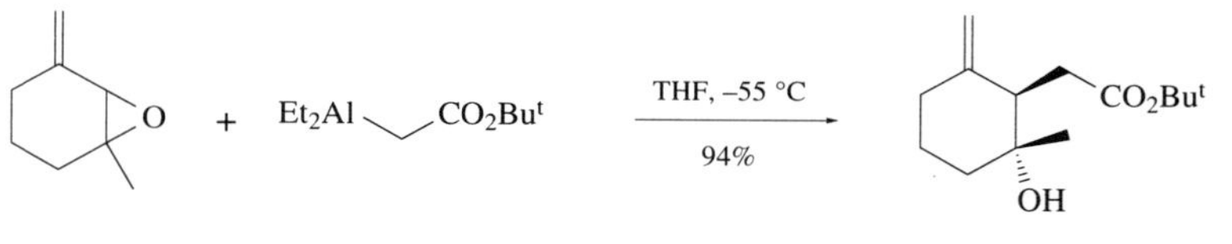

(226)

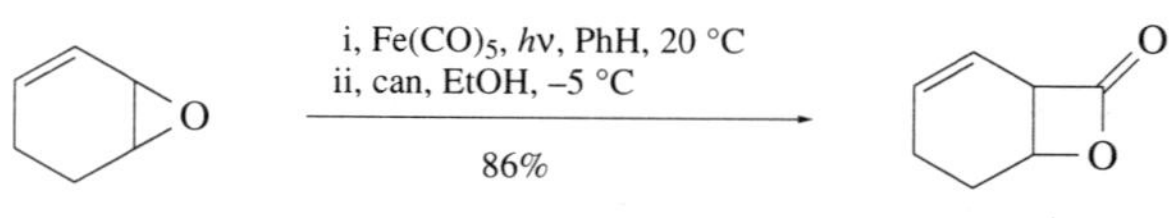

(227)

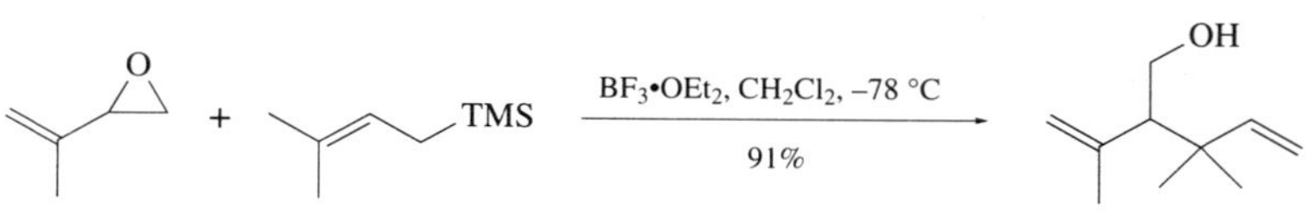

(228)

Copper reagents tend to favour C-4 attack (S_N2' or 1,4 addition) but not exclusively so. The regio- and stereo-chemistry of these reactions in both cyclic ⟨87JOC4898⟩ and acyclic systems ⟨88JOC4274, 89T391⟩ have been studied and reviewed ⟨89CRV1503⟩. In acyclic cases the highest $S_N2' : S_N2$ ratio is obtained with mixed cuprates ($RCuCNLi_2$) in ether; THF and the 'harder' dialkylcopper reagents give poorer S_N2' : S_N2 ratios ⟨88JOC4274, 89T391⟩. Ratios of *anti* : *syn* attack are generally >9 : 1. Scheme 47 shows typical results indicating that the product ratios are dependent on both reagent type and substrate stereochemistry. The (*Z*) analogue of the benzyloxy substrate in Scheme 47 gives (*E*) : (*Z*) ratios of ~15 : 85. In cyclic systems addition of cuprates is largely, if not exclusively *anti* and 1,4 (S_N2'), especially with cyanocuprates ⟨87JOC4898⟩. Significant amounts of products of 1,2 addition are only observed with alkenyl and phenyl cuprates. Sequential 1,4 additions can be performed (Scheme 48). The effect of solvent can be significant; in THF, cuprate reagents add 1,2 to the vinylepoxide while in ether 1,4 addition occurs (Scheme 49) ⟨82CC1331⟩. Organozinc reagents also add in a 1,4 fashion ⟨93SL277, 93SL279⟩.

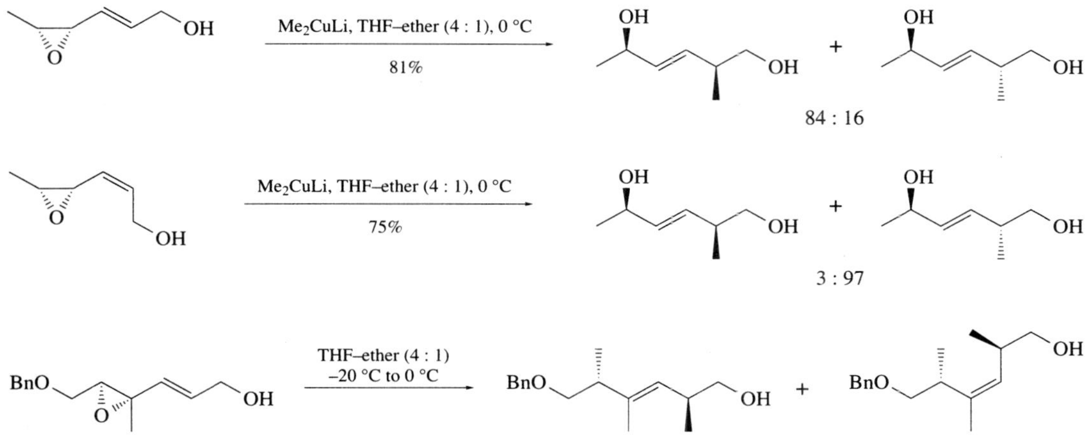

Reagent	*Yield* (%)	*anti* : *syn*	(*E*):(*Z*)
Me_2CuLi	89	99 : 1	55 : 45
MeCuCNLi	76	91 : 9	75 : 25

Scheme 47

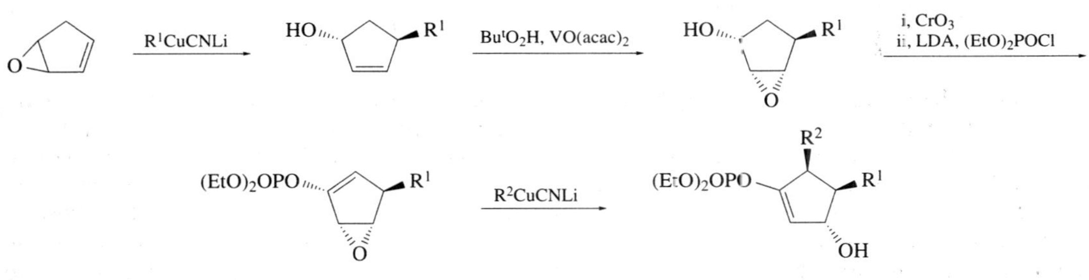

Scheme 48

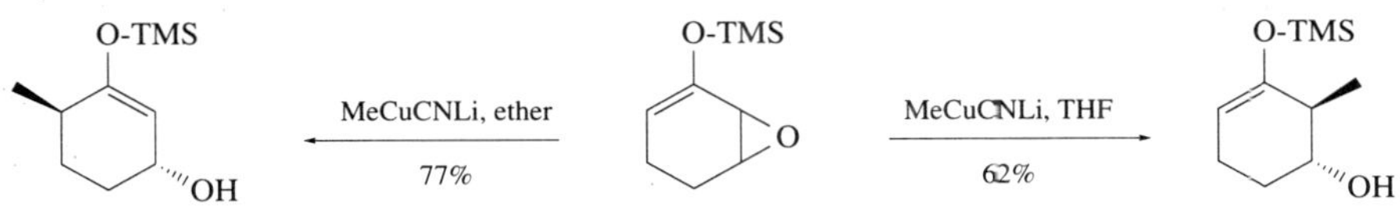

Scheme 49

The special case of the γ,δ-epoxides of $\alpha,\beta,\gamma,\delta$-dienoates has been studied with both copper and aluminum nucleophiles. 'Harder' copper reagents such as methyl copper and dimethyl cuprate favour S_N2 attack while 'softer' cyanocuprates favour S_N2' reactions; the presence of BF_3 does not alter the ratios greatly, but does increase the proportion of *syn* product (Scheme 50) ⟨89T435, 92SL769⟩. Alanes and aluminates react in an S_N2 *anti*-mode. In the presence of traces of water trimethylaluminum adds only to the epoxide unit, with predominant *anti* attack at the δ position and only a small amount of product resulting from attack at the γ position (Equation (229))

⟨91JOC6483⟩. When the epoxide is oxygenated at ε (epsilon), ζ (zeta) or η (eta) positions, only the product of γ *anti* attack is seen (Equation (230)).

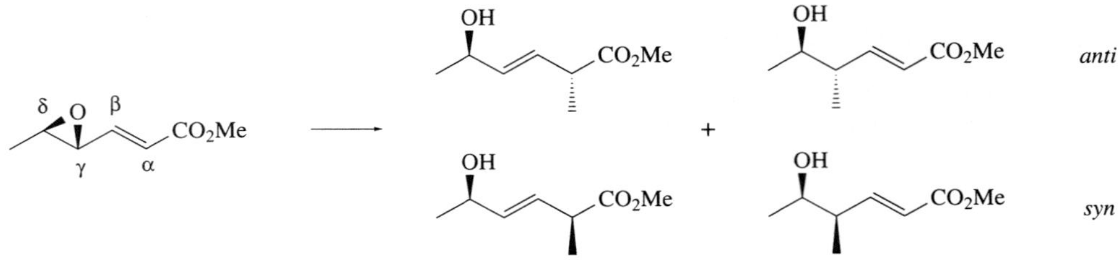

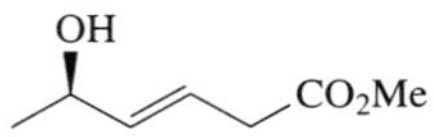

(105)

Reagent	Yield (%)	anti : syn	S_N2' : S_N2	anti : syn
MeCu	65	87 : 13	25 : 75	97 : 3
MeCuCNLi	59 + 12% **(105)**	99 : 1	68 : 32	97 : 3
$Me_2CuCNLi_2 \cdot 5BF_3$	84	30 : 70	79 : 21	54 : 46
Me_3Al	47	45 : 55	11 : 89	83 : 17

Scheme 50

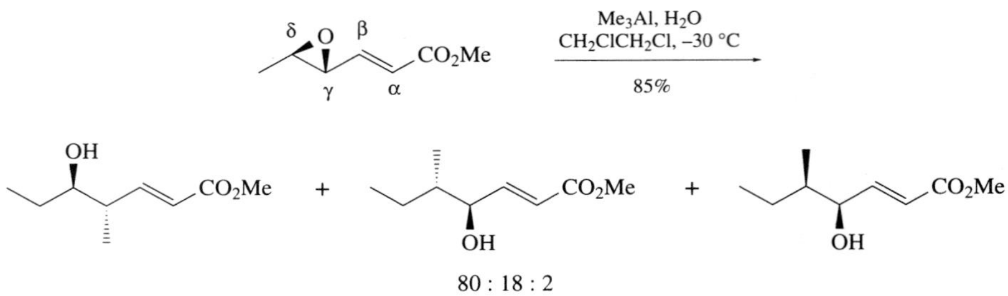

(229)

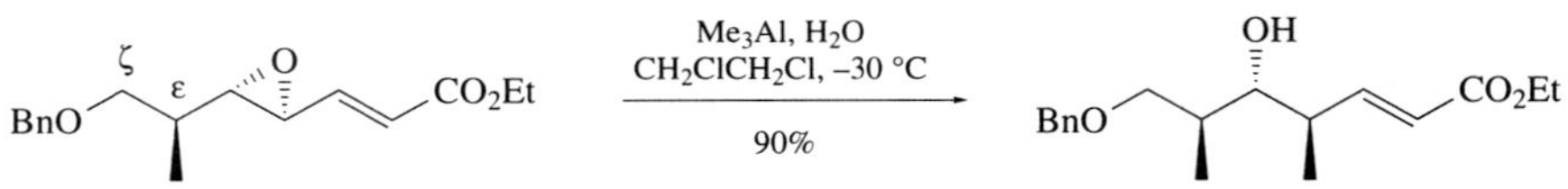

(230)

Vinyl epoxides are readily opened by transition metal catalysts to give π-allyl species which readily react with nucleophiles in a similar fashion to the π-allyl species prepared from allyl acetates etc. ⟨82COMC-I(8)101⟩. The regiochemistry of attack of the incoming nucleophile (which approaches from the same face as the alkoxide group) is influenced by stereochemical factors. For example, the presence of a bulky silyl group determines the position of attack on the vinyl epoxides (**106**) and (**107**) (Scheme 51) ⟨88TL343⟩. Aryl and alkenyl tin reagents can act as carbon nucleophiles ⟨89T979⟩; the products are mostly those resulting from 1,4 attack (Equation (231)). Palladium-catalysed carbonylations are also possible, albeit in low yield ⟨93TL2135⟩. Nickel catalyses the addition of vinyl boron species (Equation (232)) ⟨82JOM(233)13⟩. Vinyl epoxides can be used as allylating agents, again via a π-allyl species, in high yield and selectivity using chromium(II) (Equation (233)) ⟨90JOC1705⟩.

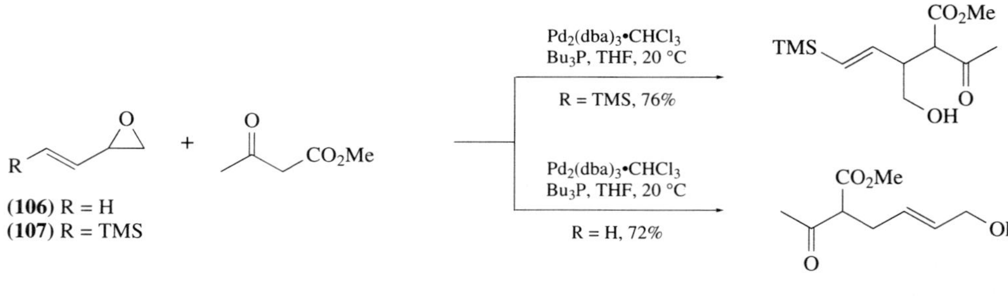

(**106**) R = H
(**107**) R = TMS

Scheme 51

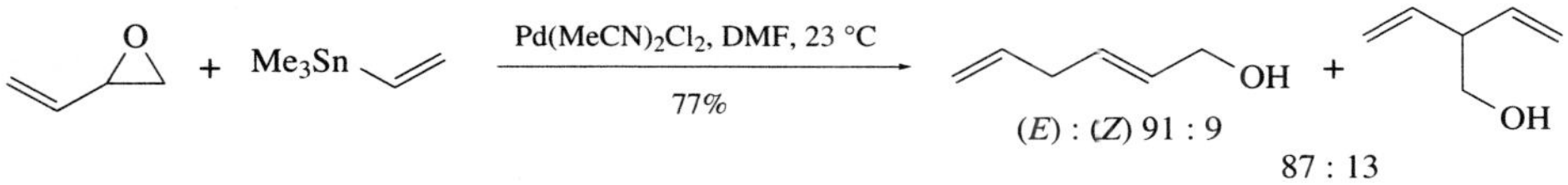

(231)

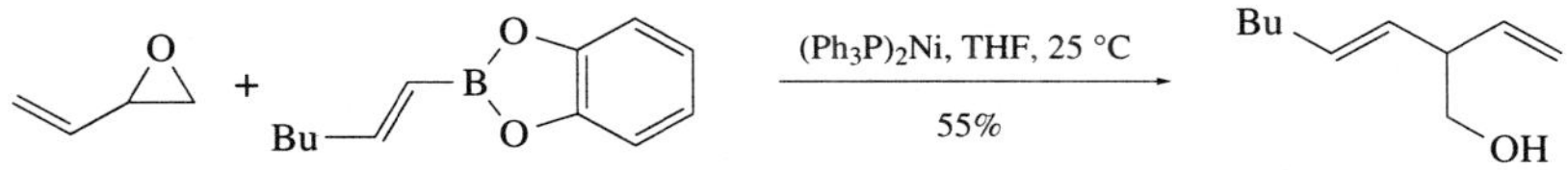

(232)

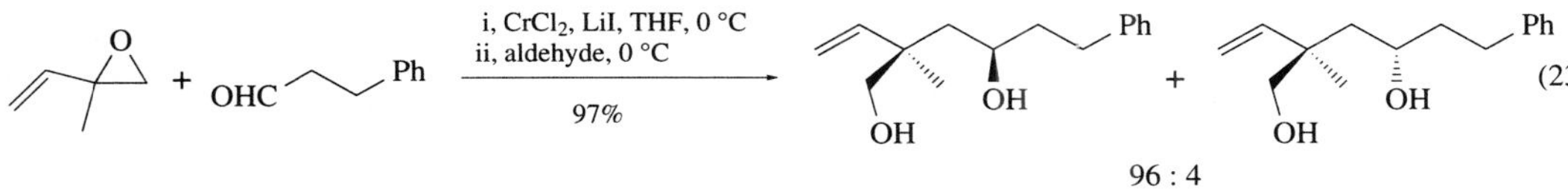

(233)

1.05.1.3.4 *Oxetanes and β-lactones*

(i) Oxetanes

Oxetanes can be opened by carbon nucleophiles in a similar fashion to epoxides, but not quite so readily. Grignard reagents ⟨76CB3375⟩, cuprates ⟨79TL1503⟩, lithium reagents ⟨87CB1695⟩ and stabilized nucleophiles ⟨75JOC2963, 84TL1159⟩ have all been used as nucleophiles (Equations (234)–(237)). With trialkylsilane and cobalt carbonyl catalysis oxetane gives a chain extended diol ⟨89JA7938⟩. Intramolecular cyclisations can be effected in a similar fashion to epoxides (Equation (238)) ⟨84TL4549⟩. Cyclisation to cyclohexanes and cyclopropanes are also possible.

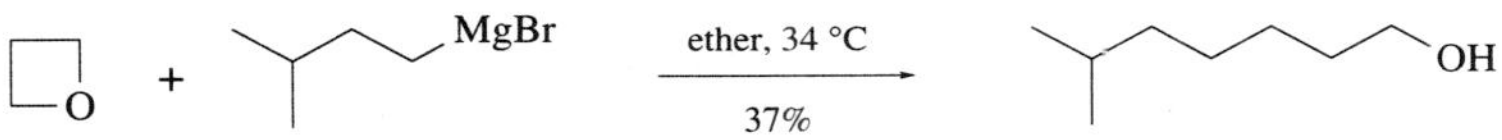

(234)

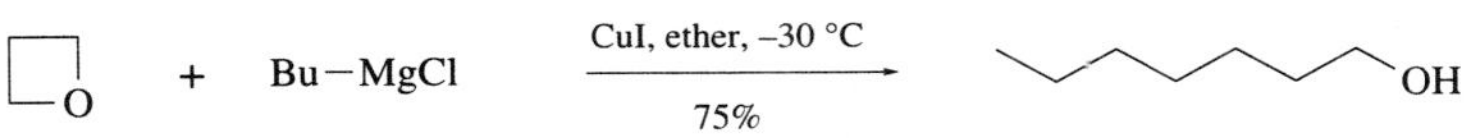

(235)

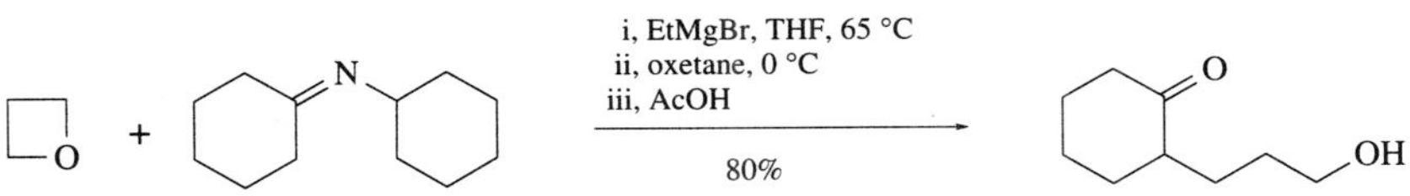

(236)

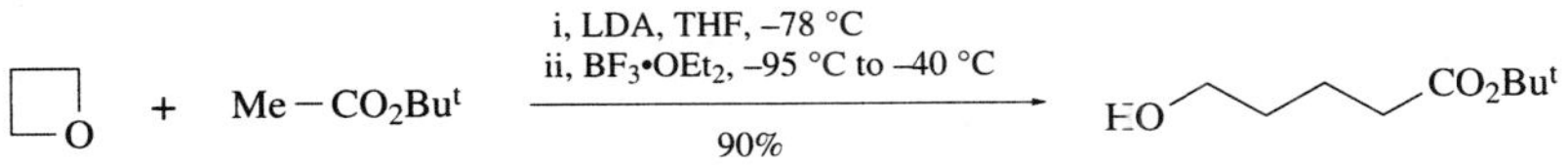

(237)

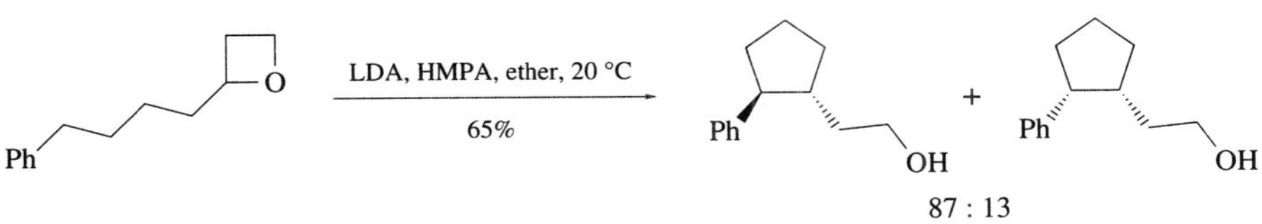

(238)

Allylsilanes and stannanes will also alkylate oxetanes under Lewis acid catalysis (Equation (239)) ⟨85JOC2782⟩. 2-Phenyloxetane undergoes Friedel-Crafts alkylation, in this case with opening at the more reactive benzylic position (Equation (240)) ⟨73ACS3944⟩.

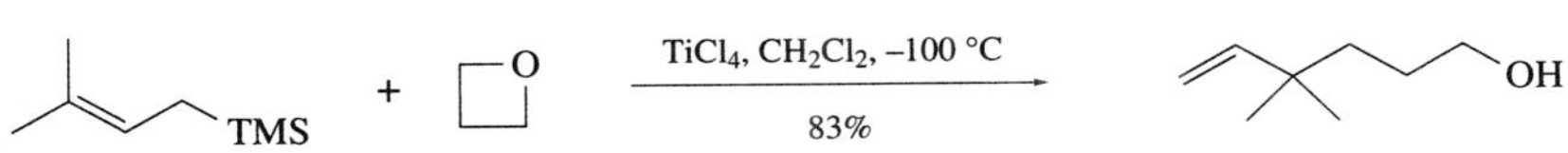

(239)

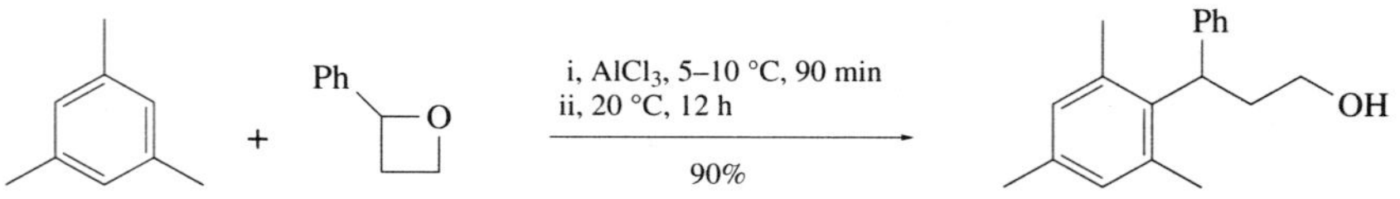

(240)

Vinyl oxetanes are opened in a 1,3 fashion (S_N2') by a variety of nucleophiles to give homoallylic alcohols in yields ranging from 22–84% (Equation (241)) ⟨90SL341⟩. Palladium catalysis can be applied to poorer nucleophiles such as organomercurials and 1,3-dicarbonyl anions ⟨88TL5069, 89TL3487⟩ (Equations (242) and (243)).

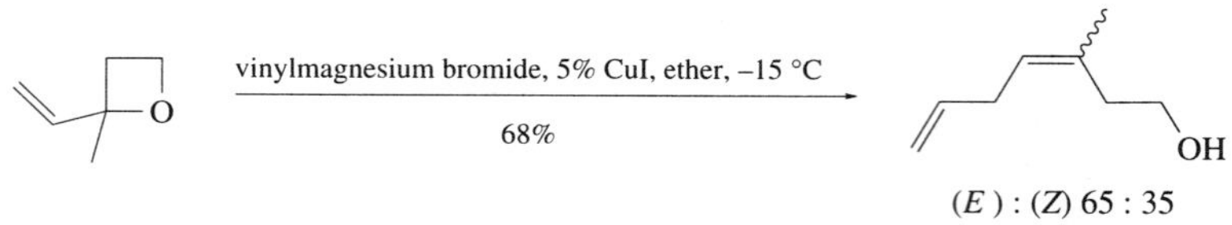

(241)

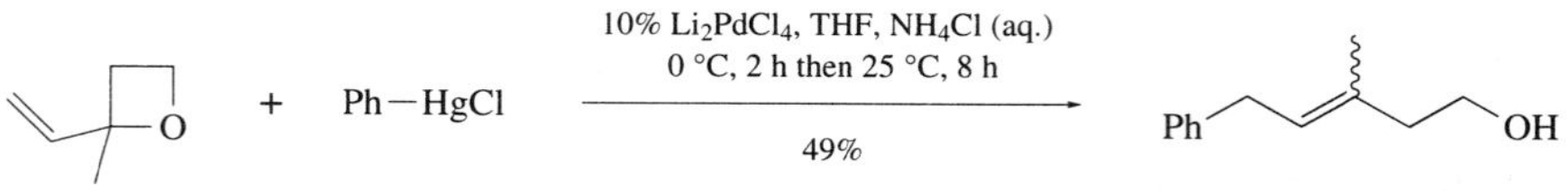

(242)

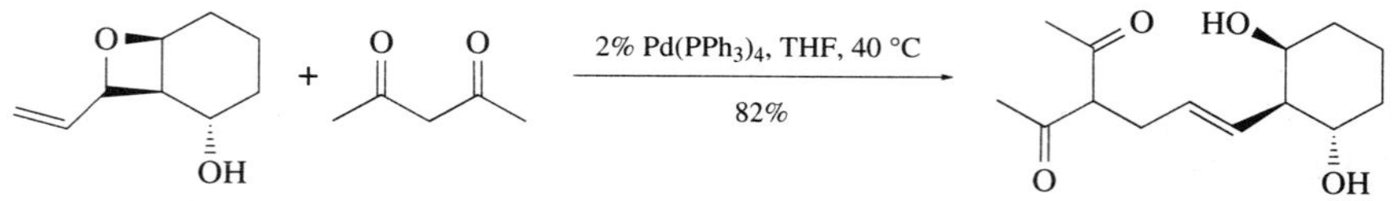

(243)

(ii) β-Lactones

β-Lactones (2-oxetanones) unlike most other lactones and esters, can be opened through attack by soft nucleophiles at the alkyl group bonded to the ring oxygen atom. Hard nucleophiles (e.g., alkyllithiums) attack at the carbonyl carbon atom ⟨87HCA1326⟩. Copper-catalysed Grignards and alkyllithiums add to give products of 3-carbon extension (Equation (244)) ⟨80TL935, 80TL2181⟩. With the oxetanone (**108**) inversion of configuration at the stereogenic centre occurs as expected ⟨82TL3587⟩. Mixed cuprates have also been used as nucleophiles ⟨87JA4649⟩.

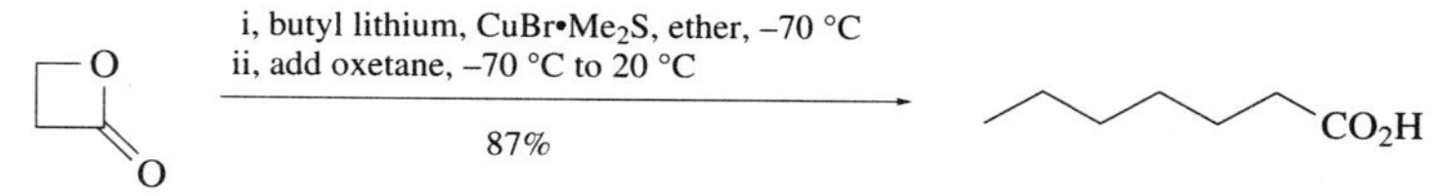

(244)

1.05.2 SUBSTITUTION OF SULFUR FUNCTIONS

A useful review of sulfone chemistry has been published ⟨B-93MI 105-01⟩.

1.05.2.1 Radical Reactions

Reactions of radicals have been reviewed by Giese ⟨B-86MI 105-01⟩, Neumann ⟨87S665⟩ and Curran ⟨88S417, 88S489⟩ and include the addition to carbon multiple bonds of radicals generated from sulfur

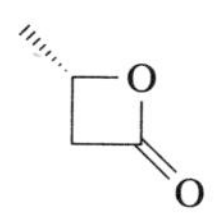

(108)

derivatives. Radicals from sulfur compounds are not as readily formed as those from selenides. Allylic sulfur compounds can act as radical traps (S_H2').

Radicals can be generated from xanthates with visible light (and an acylxanthate as a 'trigger') or with UV light (Equation (245)) ⟨89H(28)171⟩. Alkyl radical generation from acid chloride and thiopyridyl salt in the presence of an allylic sulfide gives the substituted enones via radical addition to the alkene and loss of a sulfur radical, which continues the radical chain process (Equation (246)) ⟨86JCS(P1)1613⟩. Keto-esters can be allylated by generation of the manganese radicals and trapping with an allylic sulfide ⟨90TL357⟩. Poor yields and bisallylated products are obtained with some substrates. Cyclisations in 6-*exo*-trig ⟨90TL823⟩ and 6-*endo*-trig ⟨91TL843⟩ fashions have been carried out (Equation (247)). Radicals generated thermally or photochemically from hydroxamate esters can cyclize, with loss of sulfur, to produce cyclopropanes ⟨90TL6085⟩; the intervention of activated alkenes generates cyclopentanes (Equations (248) and (249)) ⟨90TL4203⟩. The phenylsulfinyl radical (generated *in situ* from allylic sulfones) can act as the propagator ⟨85CC897⟩. In the example shown in Equation (250), the phenylsulfinyl radical is trapped by the intermediate primary radical.

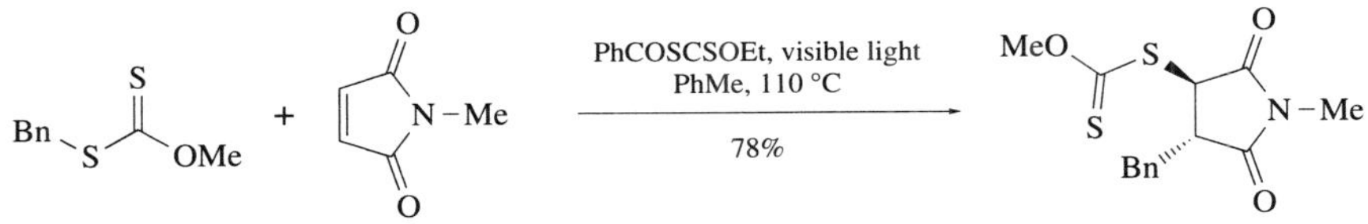

(245)

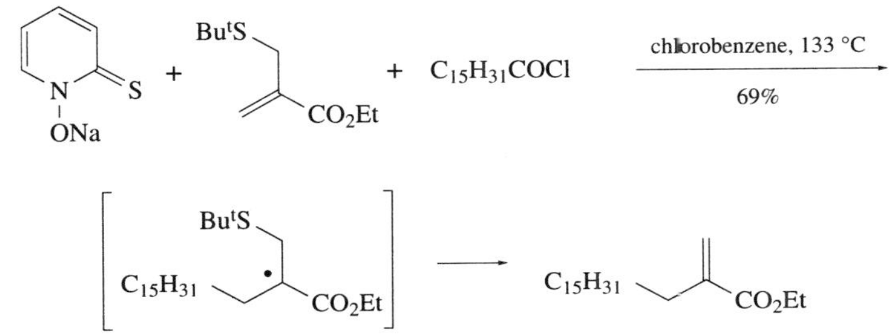

(246)

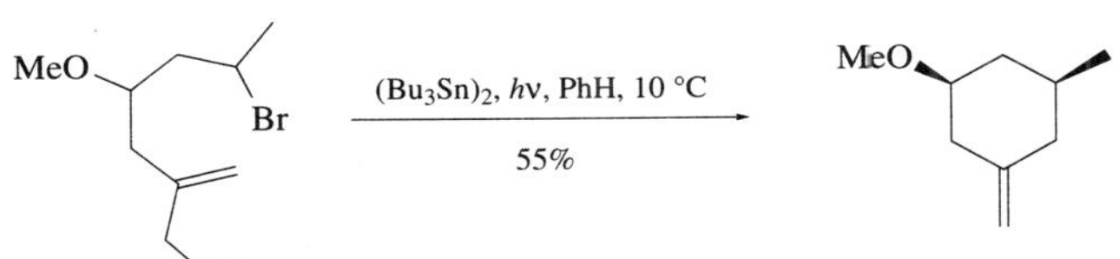

52 : 48

(247)

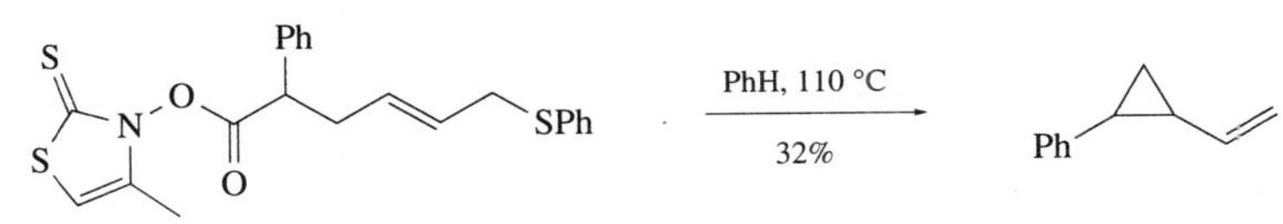

(248)

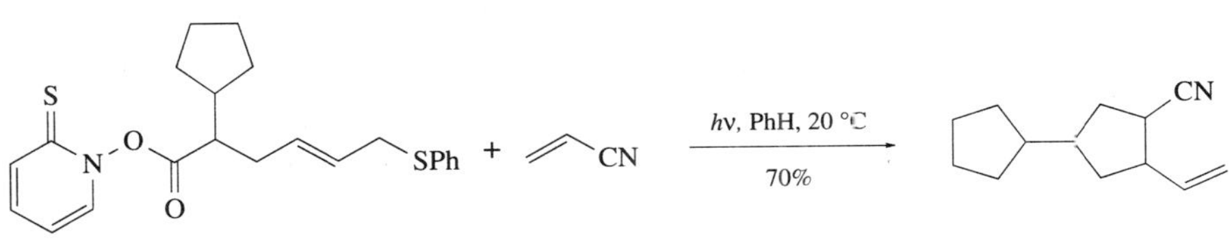

(249)

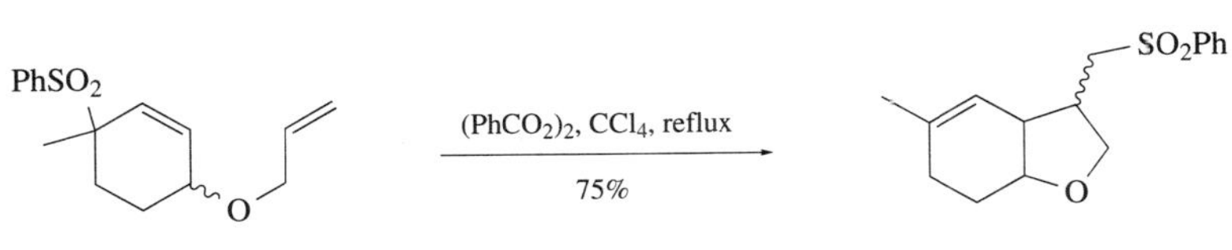

(250)

1.05.2.2 Thermal and Photochemical Reactions

Synthetically useful extrusion reactions of sulfur compounds that result in the formation of a C—C bond have been reviewed ⟨88T6241⟩.

Photochemical desulfurization of bisbenzylsulfides in the presence of phosphite has been much used to prepare [2.2]-cyclophanes (Equations (251) and (252)) ⟨75S807, 89BCJ164⟩. Yields are very variable and seem to depend on the degree of strain in the system formed ⟨89CB347, 89CB1113⟩. The reaction, which also works for alkyl and allyl sulfides ⟨69JOC1233, 73JA1547⟩, proceeds via a radical mechanism. Extrusions can be selective for one sulfur atom ⟨73C308⟩ (Equation (253)). Note that benzylic sulfides are normally moderately thermally stable.

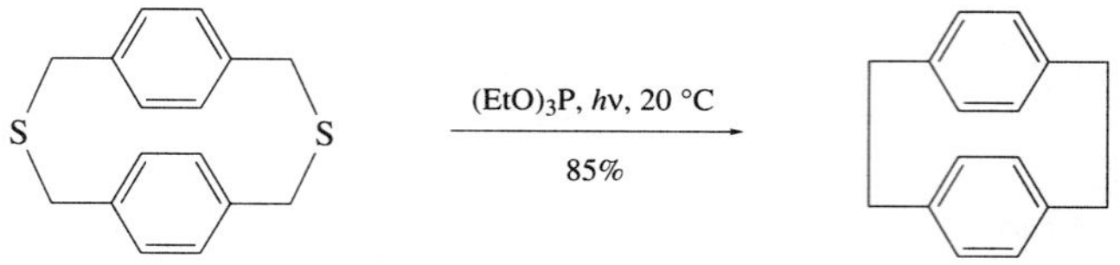

(251)

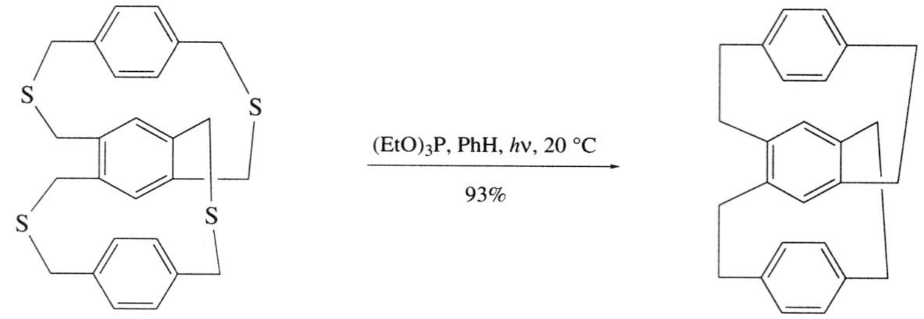

(252)

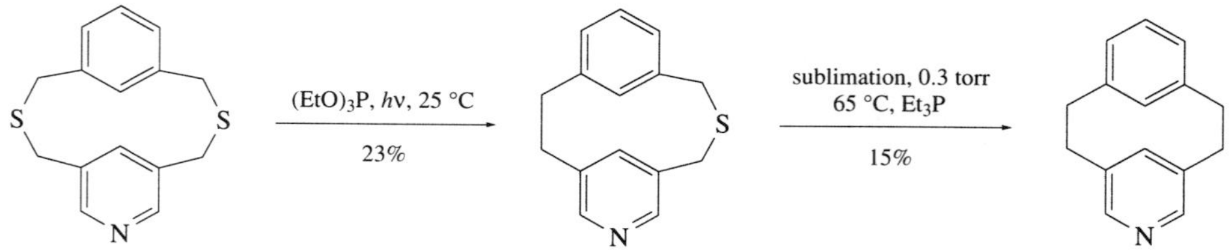

(253)

Sulfur dioxide can be extruded from cyclic sulfites both thermally and photochemically in high yield (Equations (254) and (255)) ⟨78CJC512, 83CJC757⟩. In the photochemical reactions a phenyl group is needed to act as a chromophore. Sulfur dioxide is also extruded from cyclic sulfones under similar conditions. This has been used to prepare cyclophanes ⟨75AG(E)559, 78SC285, 91JHC289⟩. Bridges of four to eight carbon atoms can be built in this way. In some cases (e.g., sulfolenes) SO_2 can be extruded via an electrocyclic reaction to give a diene, which can then recyclize to form a cyclobutene. Extrusion of SO_2 from the benzo[*c*]thiophenedioxide (**109**) involves a disrotatory process and this is followed by a conrotatory ring closure (Equation (256)) ⟨72CC541⟩. The driving force for cyclisation is rearomatisation. The photochemical reaction has reversed topicity (N.B. the Woodward–Hoffmann rules). Hydride abstraction can be a competing process as shown by the different outcomes of the thermal and photochemical reactions in Equation (257) ⟨81JCS(P1)1846⟩. Photochemical extrusion of SO_2 from four-membered rings gives cyclopropanes in high yields (Equation (258)) ⟨78S579⟩. Thermal reactions often give significant amounts of alkenes, especially at higher temperatures ⟨71JA676⟩. *β*-Lactams can be prepared via photochemical extrusion of SO_2 but not in very high yield or stereochemical purity (Equation (259)) ⟨83JOC494⟩.

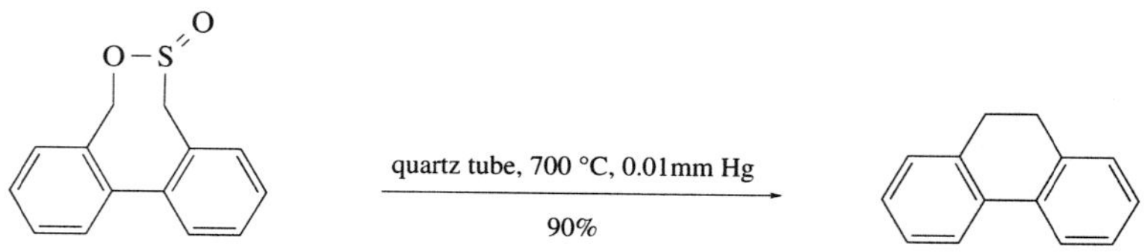

(254)

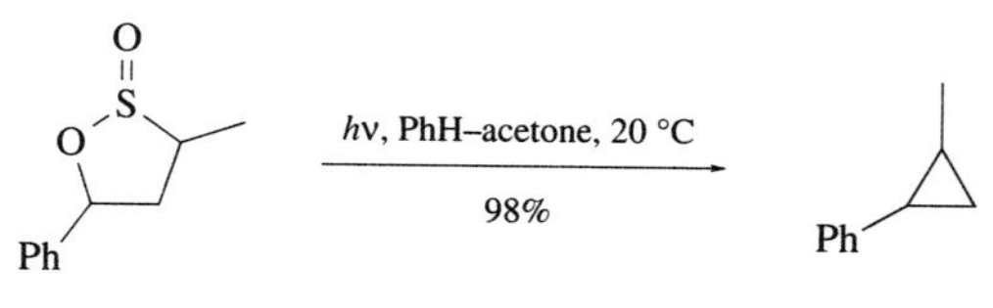

(255)

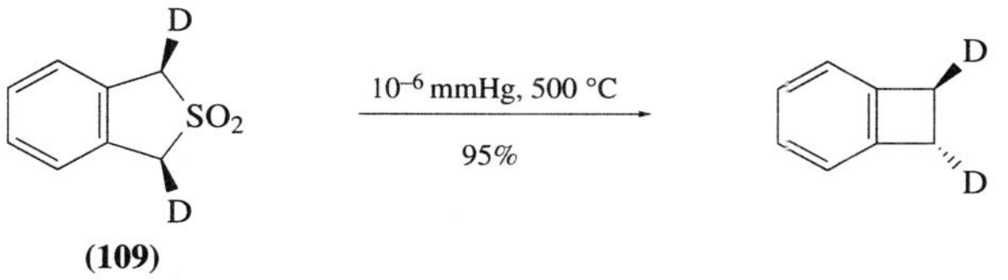

(109)

(256)

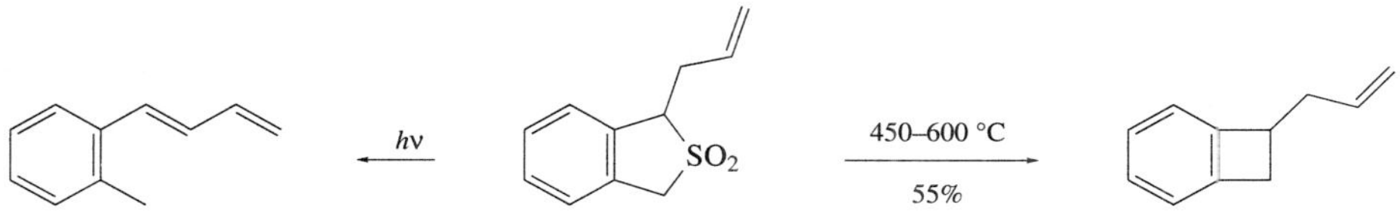

(257)

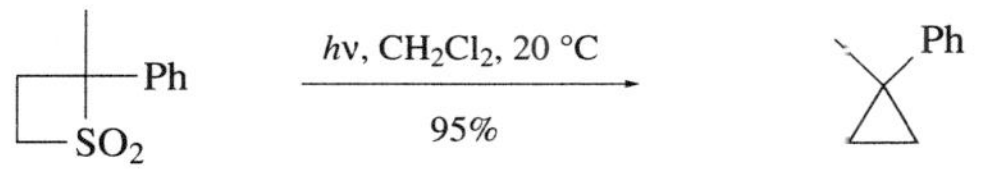

(258)

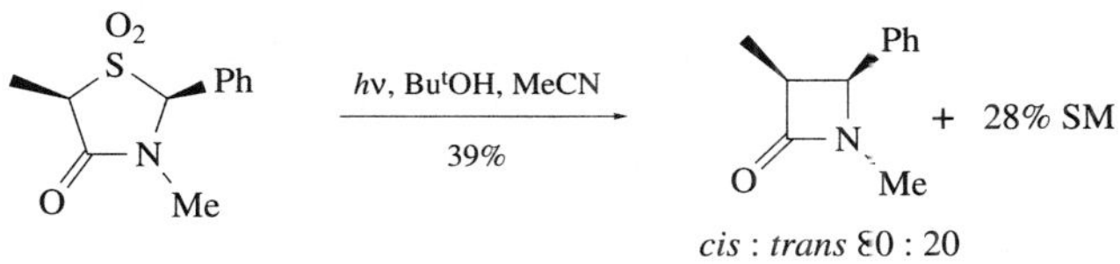

(259)

1.05.2.3 Ring Contractions and Rearrangements

1.05.2.3.1 [1,2]-Shifts

[1,2]-Shifts of sulfur compounds are of two types, for example Wittig rearrangements ⟨B-75MI 105-01, 91COS(3)975⟩ of α anions and Stevens rearrangements ⟨91COS(3)913⟩ of ylides. Both reactions are 'forbidden' and proceed via radical pair intermediates ⟨70AG(E)763⟩. The two methods are commonly used to prepare cyclophanes. For example, the ylide of a bipyridyl (Equation (260)), prepared *in situ*, spontaneously rearranges to the carbocyclic compound ⟨81JOC1747⟩. Treatment of thietane with diazomalonate gives the ring expanded product resulting from rearrangement of the intermediate ylide (Equation (261)) ⟨72JOC1721⟩. An example of intramolecular ylide formation gives cyclic products (Equation (262)) ⟨88TL6005⟩. The reaction is limited to migrating groups without a β-hydrogen atom which otherwise eliminates in competition to the migration process. The Wittig-type rearrangement generally gives better yields and is easier to perform than the Stevens rearrangement (Equation (263)) ⟨75TL219⟩ and has been much used in cyclophane chemistry ⟨82JA2560, 85CB1230, 88JOC4472⟩.

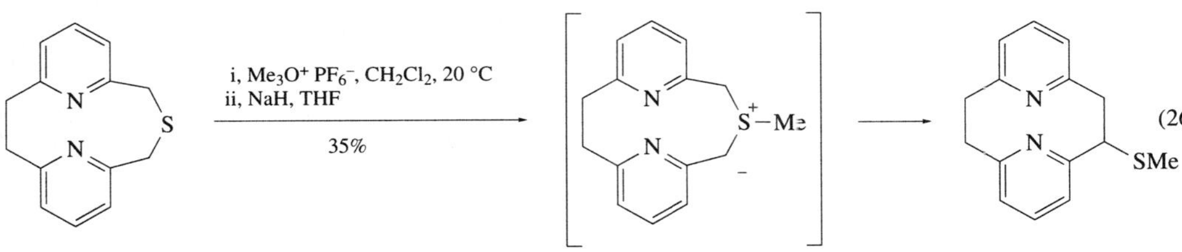

(260)

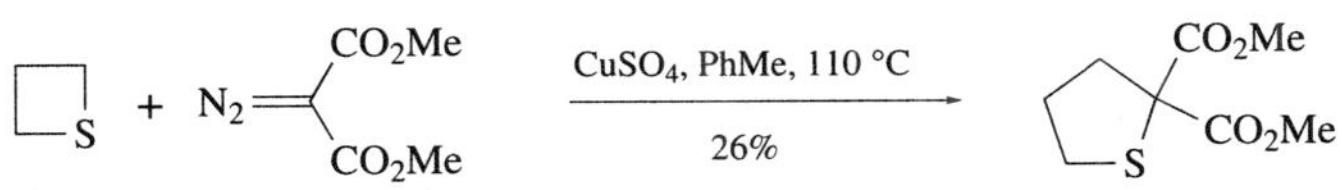

(261)

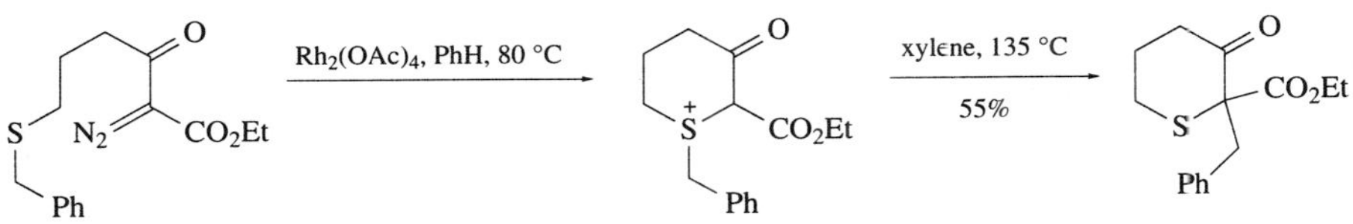

(262)

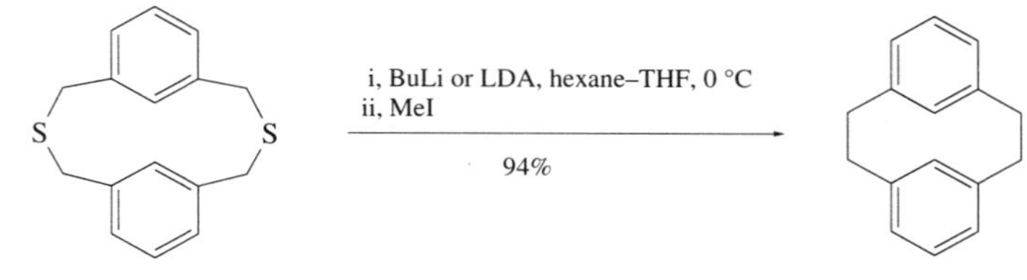

(263)

Deprotonation of the cyclic sulfide (**110**) gives a cyclopentane derivative. The reaction does not go through a [1,2]-shift but probably involves elimination of malonate followed by intramolecular addition of that to the intermediate moiety thioaldehyde unit (Equation (264)) ⟨88JA5932⟩.

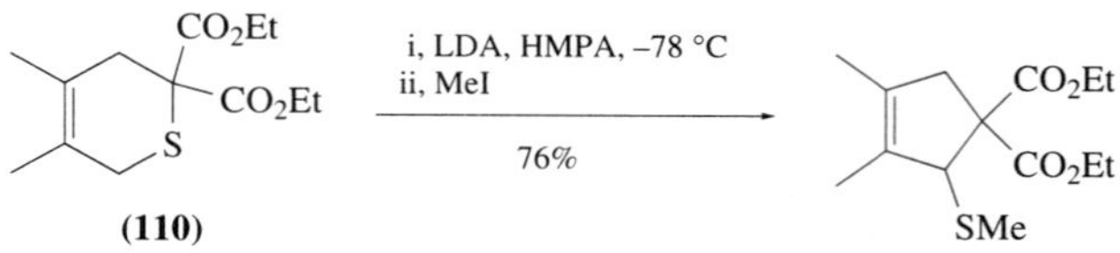

(264)

Sulfur-containing four-membered ring compounds undergo an anionic [1,2]-shift that results in ring contraction. Thus thietanonium salts, when treated with butyl lithium, give cyclopropanes stereoselectively via a sulfur(IV) intermediate (Equation (265)) ⟨71JA676⟩. The stereochemistry of the rearrangement of cyclic sulfoxides has been studied ⟨82CC589⟩ and shown to be stereospecific in most cases (Scheme 52), with retention of configuration at the migrating carbon atom and inversion at the migration terminus. Although the stereochemistry can be accounted for by a concerted mechanism it is believed that the reaction involves a biradical intermediate. Cyclic sulfones also rearrange in a similar manner ⟨63CI(L)1203⟩.

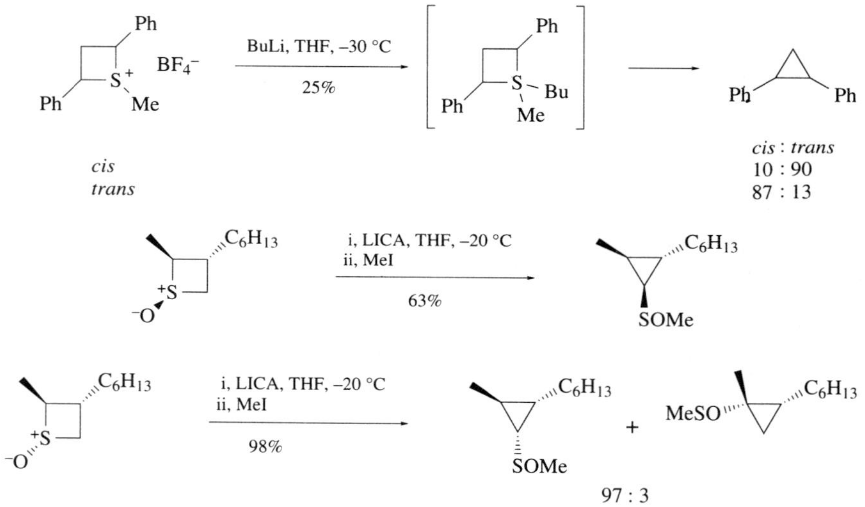

(265)

Scheme 52

1.05.2.3.2 [2,3]-Wittig rearrangements

Allylic sulfur compounds readily form ylides which rearrange via a [2,3]-Wittig reaction forming a new C—C bond (Equation (266)) (see review ⟨84ACR358⟩). Treatment of the appropriate allylic sulfide with an alkylating agent and then a strong base generates the ylide (**111**) which rearranges to a ring expanded product (Equation (267)) ⟨78JOC1185⟩. Selectivity between two allylic groups can be achieved (Equation (268)) ⟨78TL519⟩ and ring sizes of up to 17 atoms have been prepared ⟨77HCA1361⟩. Another way of generating ylides which then undergo rearrangement is by rhodium-catalysed insertion of a carbene (generated from a diazo compound) ⟨86CRV919⟩. The intermediate ylides, many of which are stable, rearrange on heating (Equation (269)) ⟨87CL1569⟩. Copper compounds and $BF_3 \cdot OEt_2$ can also be used to catalyse the reaction, with the copper system giving fewer by-products (Equation (270)) ⟨80TL395⟩. Intramolecular insertions give six-and seven-ring sulfur compounds although the yields for some of the seven-ring compounds are poor due to competing β-elimination (Equation (271)) ⟨88TL6005⟩. Simmon–Smith conditions generate sulfur ylides which rearrange *in situ* to give rearranged products incorporating one extra carbon atom (Equation (272)) ⟨82JOC4005, 82TL3019⟩. Chiral [2,3]-rearrangement of an ylide with a chiral auxiliary gave low yields and *ees* (6–47% and 4–56% respectively) ⟨82CL1871⟩. Anions of allyl sulfides (i.e., non-ylides) also rearrange via a [2,3] reaction, but at much lower temperatures (Equation (273)) ⟨81CL315, 82CL351⟩.

(266)

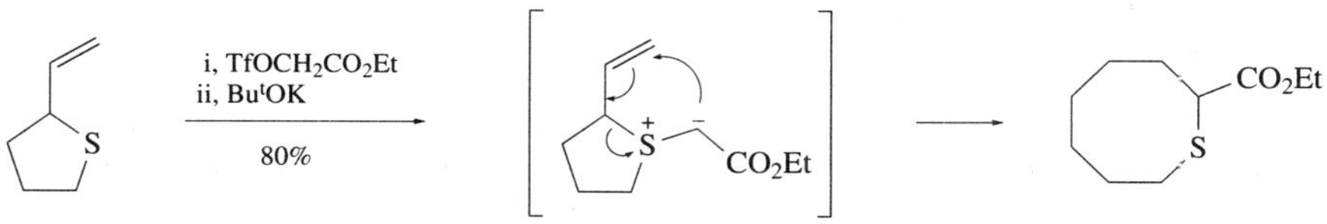

(111)

(267)

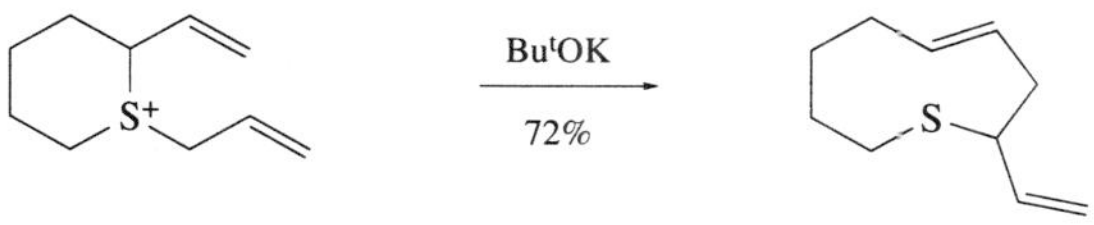

(268)

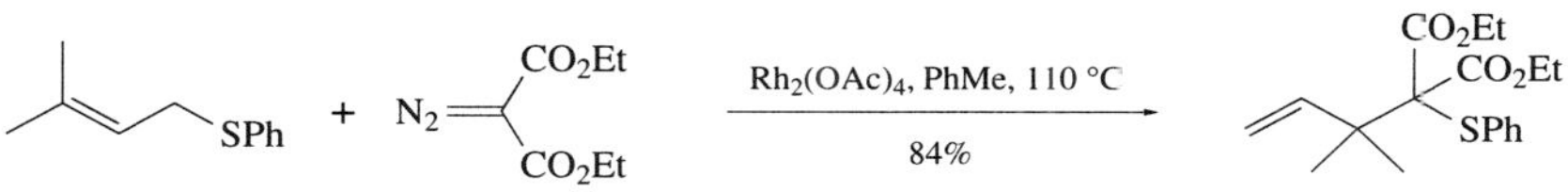

(269)

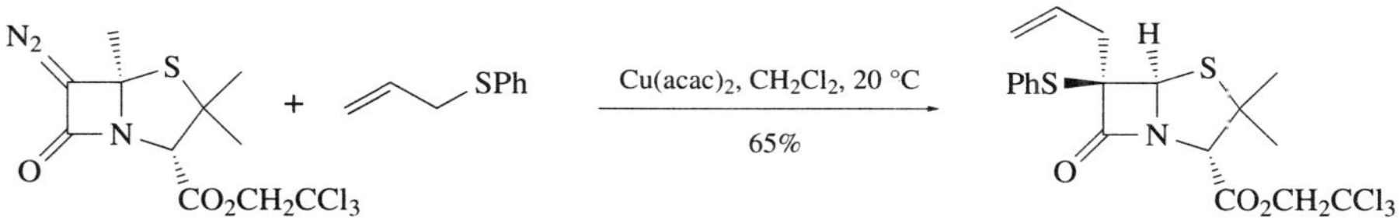

(270)

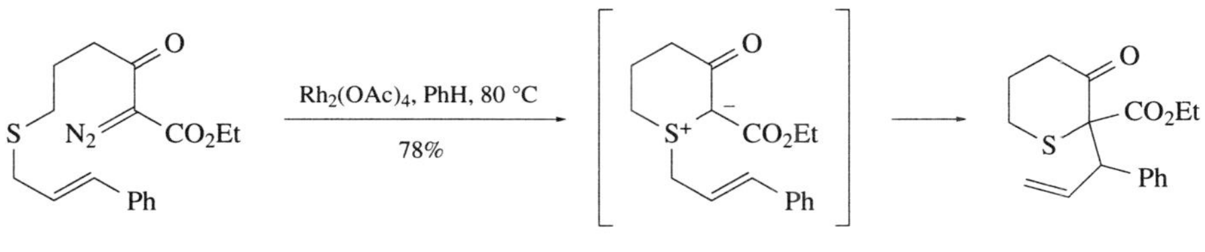

(271)

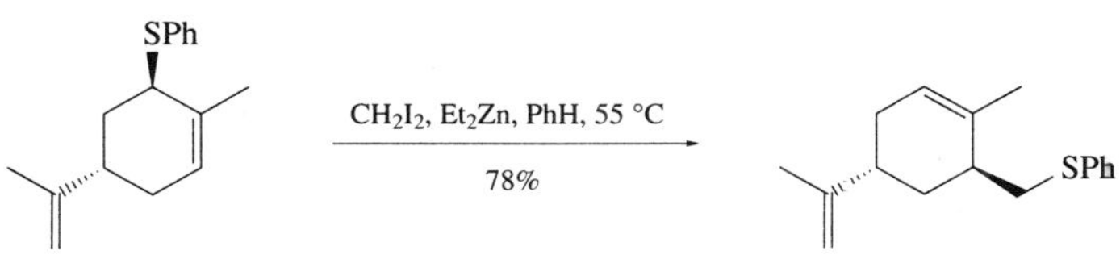

(272)

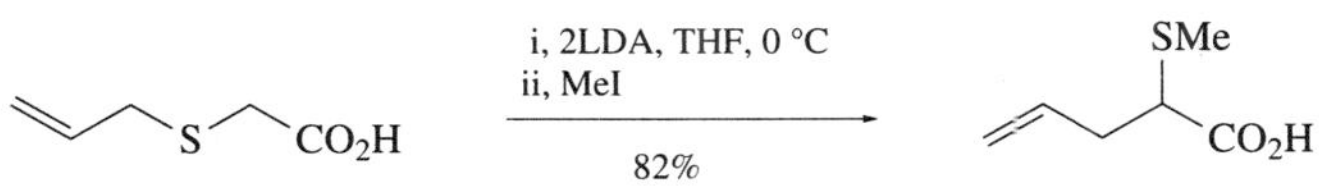

(273)

1.05.2.3.3 Other rearrangements

Allylic sulfides add to activated alkynes under Lewis acid catalysis (Equation (274)) ⟨84JOC1985⟩. The stereochemical outcome depends on the Lewis acid employed ($AlCl_3$ gives largely (*E*)-alkene while $ZnCl_2$ gives mostly (*Z*)-alkene). Suitably substituted allyl groups show that the reaction proceeds with allylic rearrangement via a sulfonium species. Sulfur compounds can also undergo a thio-Claisen rearrangement.

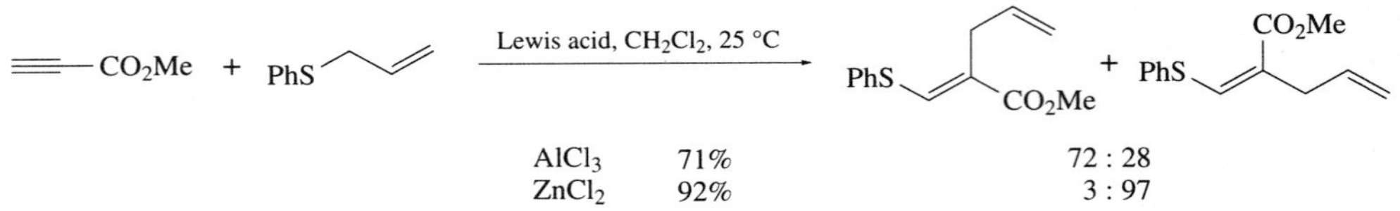

(274)

1.05.2.4 Metal Exchange Reactions

Sulfur can be readily exchanged for lithium and then alkylated to form a new C—C bond ⟨89ACR152⟩. Reductive lithiation can be achieved with a number of reagents including lithium in ammonia ⟨74TL3955⟩, lithium naphthalenide ⟨89CB703⟩, lithium 4,4′-di-*t*-butylbiphenylide (LDBB) ⟨93JA3855⟩ and lithium 1-(dimethylamino)naphthalenide (LDMAN) ⟨86T2803⟩. In all these cases the sp^{3}C—S bond is cleaved selectively. These sulfur-derived anions undergo reaction with electrophiles such as CO_2 and aldehydes (Equations (275) and (276)) ⟨84TL4349, 93JA3855⟩. At low temperatures a (*Z*)-alkene is formed exclusively from an allyl sulphide (Equation (277)) ⟨87JA4710⟩. Attack is at the least substituted end of the system, in contrast to the similar reaction in the presence of titanium isopropoxide ⟨86T2803⟩. Stabilized anions can be generated by metal exchange using a tin reagent on thiocyanohydrins (Equation (278)) ⟨85CL1149⟩.

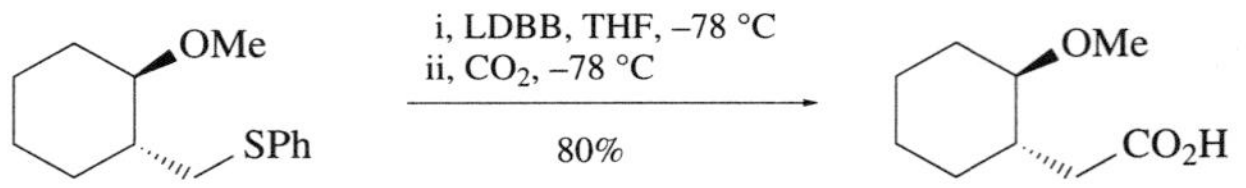

(275)

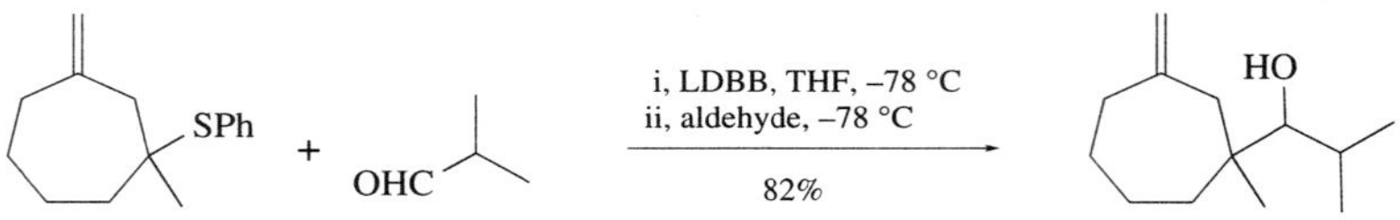

(276)

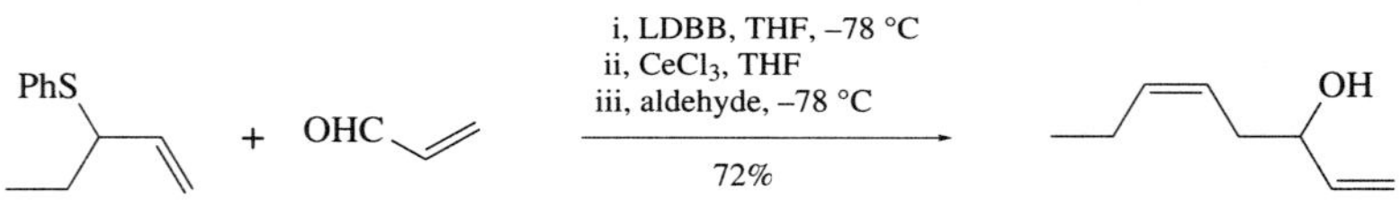

(277)

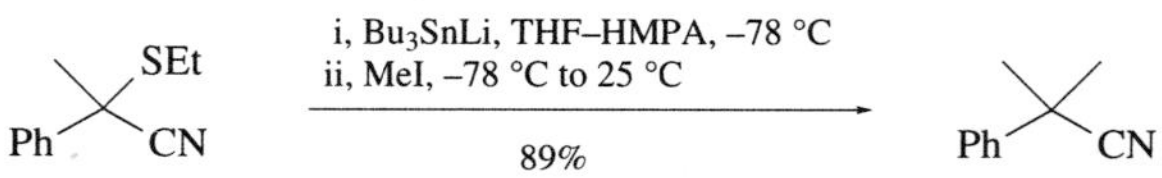

(278)

Dialkylcuprate cleaves an α-ketosulfoxide to generate an enolate which can then be alkylated (Equation (279)) ⟨81JA2886⟩. The mechanism is not known. α-Sulfonylketones are reductively cleaved to give enolates which, once again, can be alkylated (Equation (280)) ⟨85JOC3846⟩.

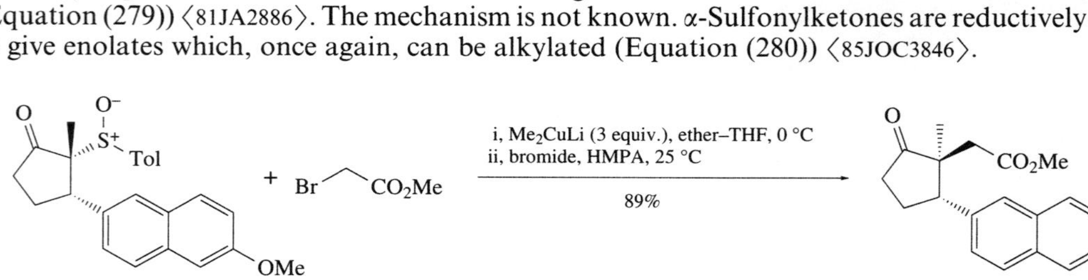

(279)

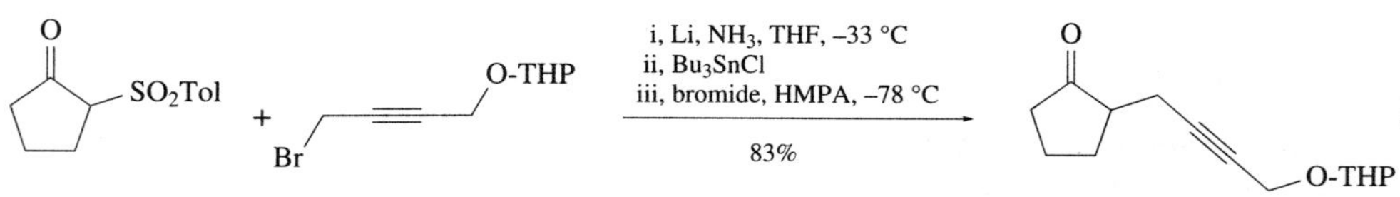

(280)

1.05.2.5 Sulfur Leaving Groups

There are many reactions which involve alkylation α to a sulfur group with subsequent reductive cleavage of the C—S bond giving overall, a formal C—S to C—C conversion. This section covers only those reactions where sulfur is displaced directly. Direct displacements of sulfur groups at alkyl carbon atoms are rare; mostly allylic and benzylic sulfur systems are involved except for some intramolecular reactions.

1.05.2.5.1 Allylic and benzylic sulfur systems

Allylic and benzylic sulfur leaving groups, especially sulfones, are readily displaced by nucleophiles in a similar fashion to oxygen-based leaving groups. However sulfones require more forcing reaction conditions than, for example, the corresponding acetates (see Section 1.05.1.2.2).

(i) Main group metals and copper

Copper-based nucleophiles displace sulphone as sulfinate in both S_N2 and S_N2' fashion (Equations (281) and (282)) ⟨83T3283⟩. Attack is usually at the less substituted end of the allylic system but a number of factors including solvent, the attendant halogen atom and alkene geometry affect the outcome. As with oxygen leaving groups using ether as solvent favours γ attack. Displacement is in the *anti* mode (Equation (283)) ⟨87JA6396, 88JA5216⟩. Sulfoxide groups can also be displaced with similar yields and selectivities (Equation (284)) ⟨85JCS(P1)1171⟩. Copper catalysis may not always be necessary ⟨89TL3837⟩. Aluminum alkenes and alkynes displace allylic sulfones under Lewis acid catalysis at the less substituted carbon centre (Equations (285) and (286)) ⟨86JA1098⟩.

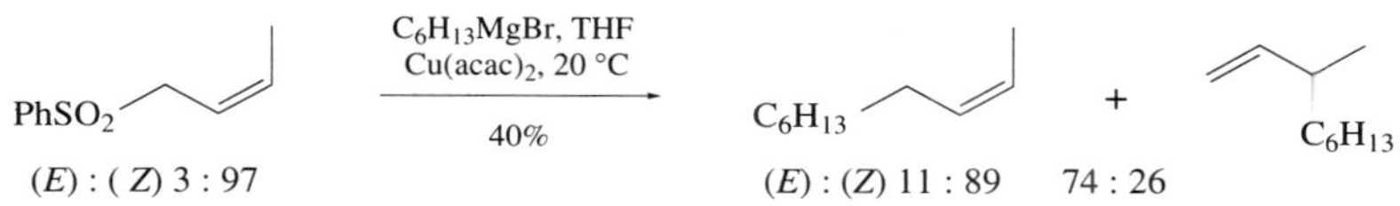

(281)

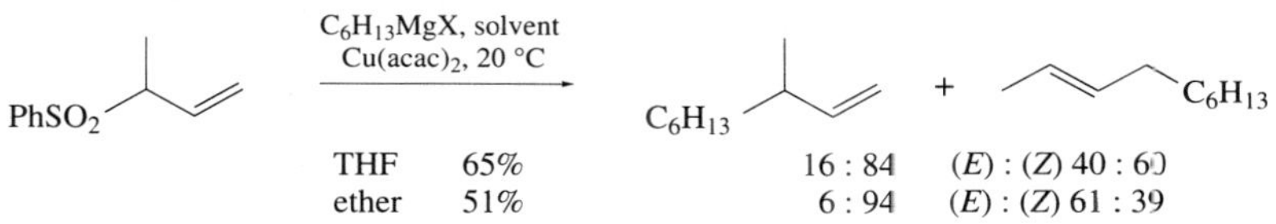

(282)

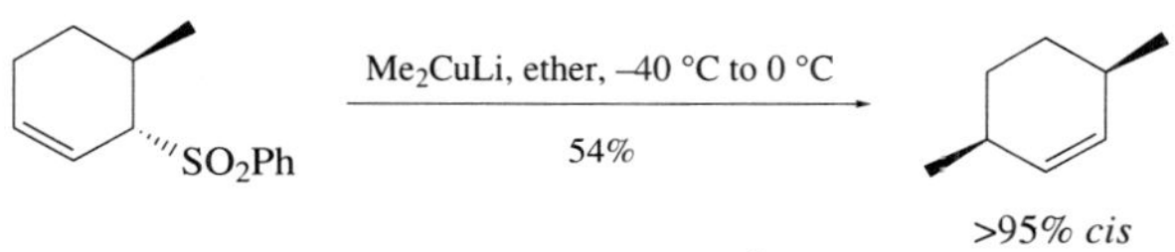

(283)

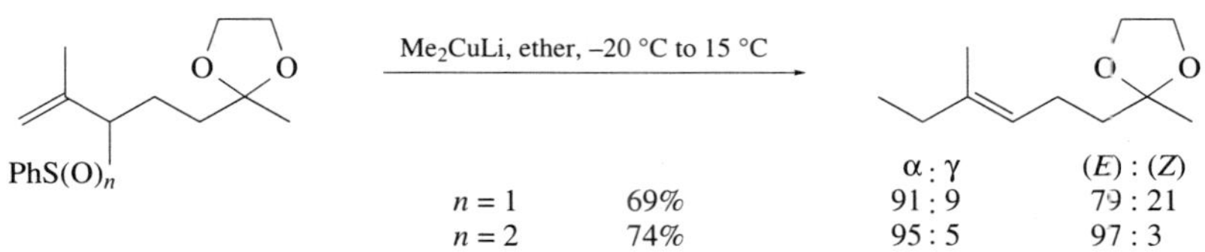

(284)

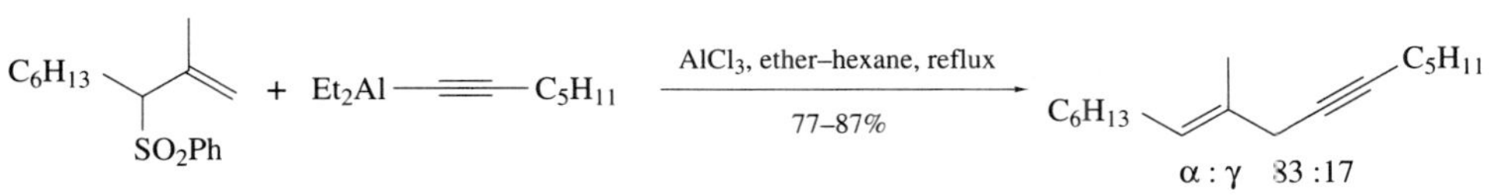

(285)

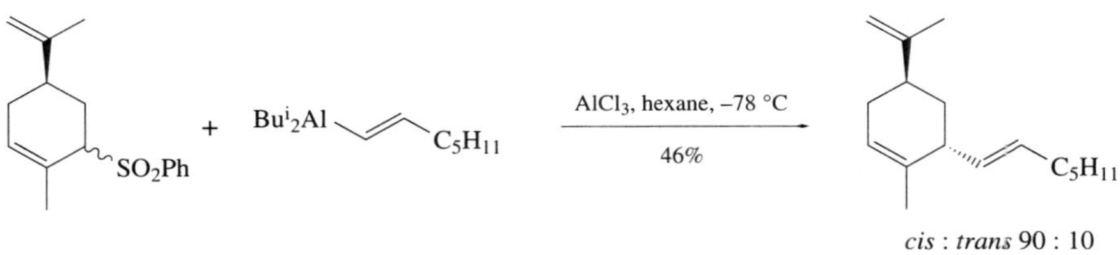

(286)

A number of other sulfur leaving groups have been used including the simple phenylthio-moiety ⟨77JOM(142)1, 80JA1064⟩. Thiobenzothiazoles are readily displaced by copper-catalysed Grignard reagents in either S_N2 and S_N2' fashion depending on the solvent (Scheme 53) ⟨83JCS(P1)2953⟩. Dialkylcopper reagents are less selective. Isothiouronium salts are displaced under very mild, non-basic conditions using copper catalysis in aqueous solution (Equation (287)) ⟨84JOC1460⟩. Sulfonamidsulfinyl ⟨84TL2135⟩ and 1-phenyltetrazoyl-5-thio ⟨88TL4105⟩ have also been used as leaving groups.

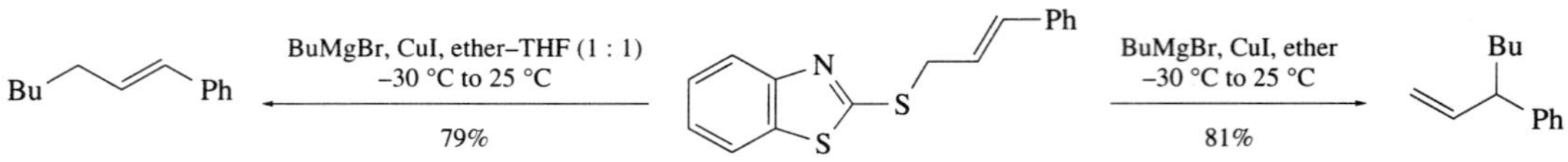

Scheme 53

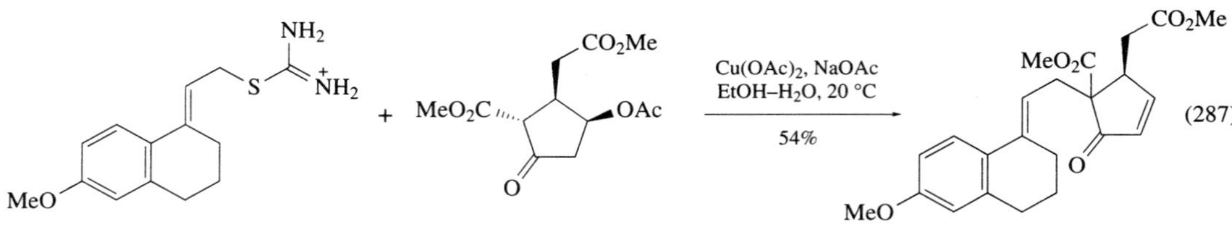

(287)

Electron-deficient tertiary benzylic sulfones can be displaced by nucleophiles such as malonate and nitroalkanes under conditions that favour electron transfer processes ⟨88JOC1475⟩.

(ii) Transition metal catalysis

Palladium and other transition metals catalyse the replacement of sulfur by carbon nucleophiles. Transition-metal mediated C—S bond cleavage reactions have been reviewed ⟨90S89⟩. Under standard conditions sulfone is displaced by malonates and other stabilized anions to give, for α,β-unsaturated sulfones, mixtures of regioisomers (Equation (288)) ⟨80JA5979⟩. The regioselectivity depends on the substitution pattern and the steric bulk of the nucleophile and proceeds with net retention of configuration (i.e., a double inversion) (Equation (289)) ⟨87JA6396⟩. Nickel ⟨86JOM(317)383⟩ and molybdenum complexes ⟨89BCJ2913, 90JOC1127⟩ also catalyse this type of reaction but with different regiochemical consequences (Equation (290)). Molybdenum is selective for alkylation at the more substituted end of the allylic system, but the reaction products indicate a fine balance between steric and electronic factors; in some cases selectivities up to 100 : 0 have been achieved ⟨89BCJ2913⟩. Other leaving groups have been used with palladium catalysis including phenyltetrazoylthio (**112**) ⟨88TL4105⟩ and thiophosphonate (**113**) ⟨83JOC1293⟩. Molybdenum hexacarbonyl catalysed the displacement of phenylthio but the reaction is slower than that for the sulfones ⟨89BCJ2913⟩. Molybdenum also catalyses benzocyclobutane formation from acylsilanes (Equation (291)) ⟨90TL2713⟩. Chromium, tungsten and palladium can also be used as catalysts. Nickel is the best catalyst for cross-coupling of allyl sulfides with Grignard reagents ⟨79TL3425, 84CL1505⟩. Modest (up to 57% *ee*) enantioselection has been obtained using chiral ligands ⟨86T2043⟩.

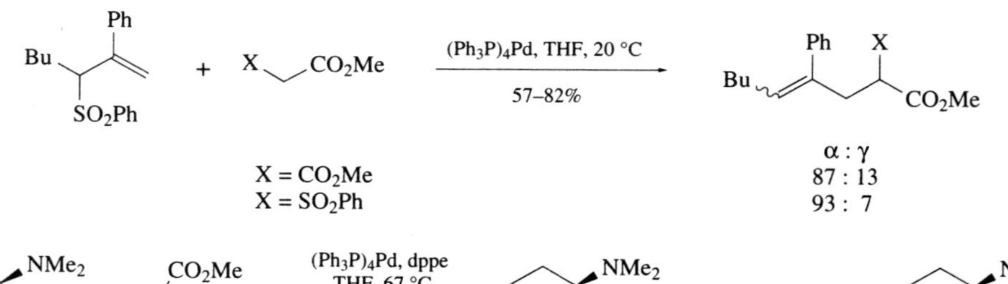

(288)

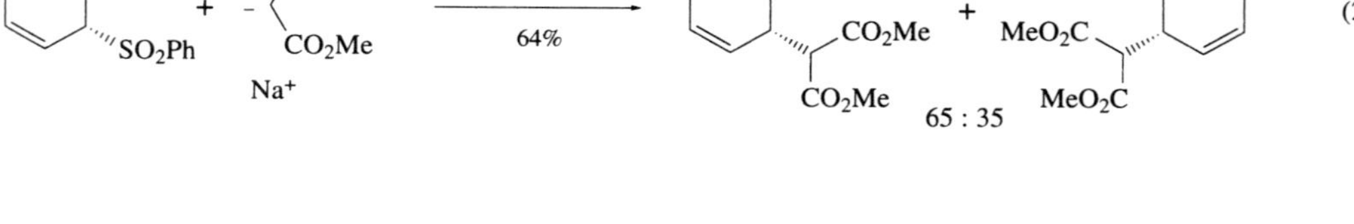

(289)

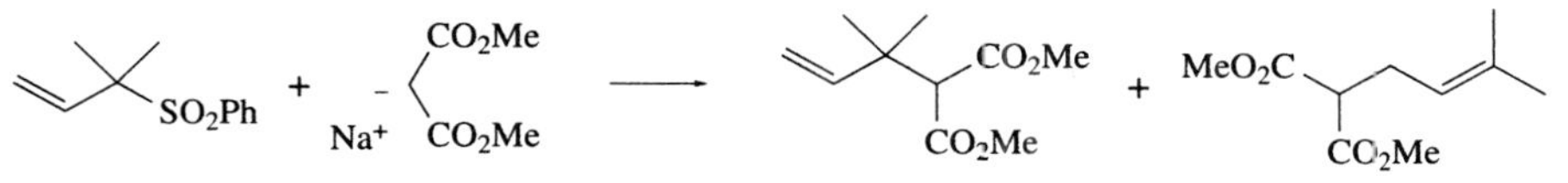

(290)

$Mo(CO)_6$, PhMe	61%	90 : 10
$(Ph_3P)_4Pd$, THF	67%	68 : 32
$Ni(dppe)_2$, THF	92%	39 : 61

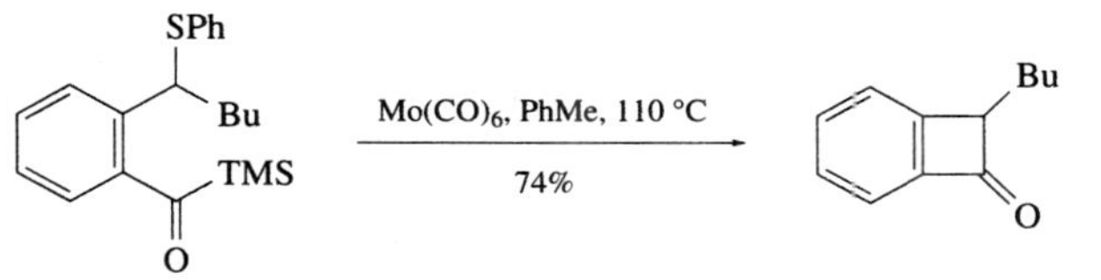

(291)

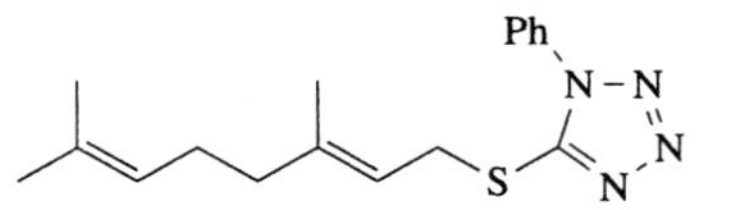

(112)

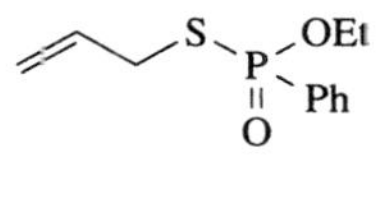

(113)

Benzylthiols can be carbonylated under high pressure (60 atm) using a cobalt catalyst (Equation (292)) ⟨85JOC147⟩. Dibenzyldisulphides can be carbonylated to give thioesters ⟨85TL2609⟩.

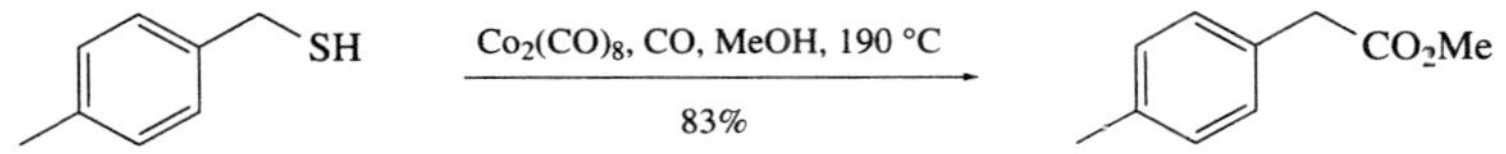

(292)

(iii) Lewis acid catalysed reactions

Silylenol ethers and allylstannanes (but not allylsilanes) can be alkylated with allylic and benzylic sulfur compounds in the presence of Lewis acids (Equations (293) and (294)) ⟨87CL1691, 87CL1695⟩. A silver salt is used to activate the thiotetrazole leaving group in this alkylation reaction (Equation (295)) ⟨90TL265⟩. Yields are in the range 50–98% for allylic and benzylic substrates. Aluminum trichloride catalyses Friedel-Crafts alkylation of allylic and benzylic sulfones (Equations (296) and (297)) ⟨84JA7260, 90IJC(B)390⟩.

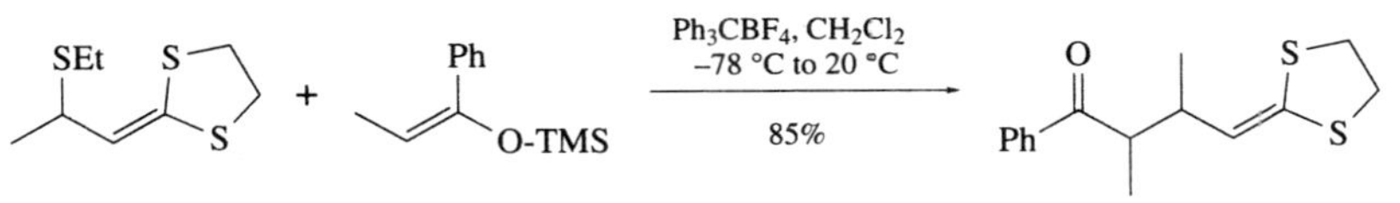

(293)

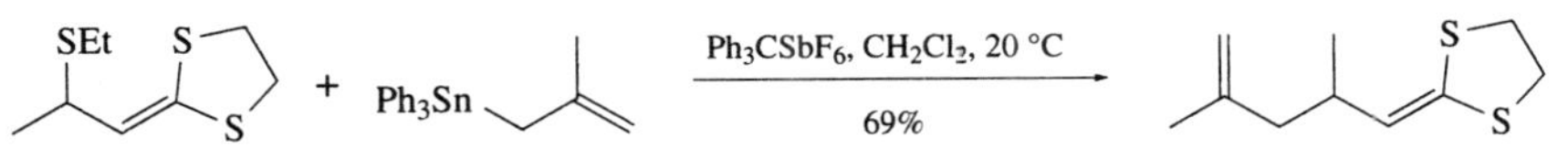

(294)

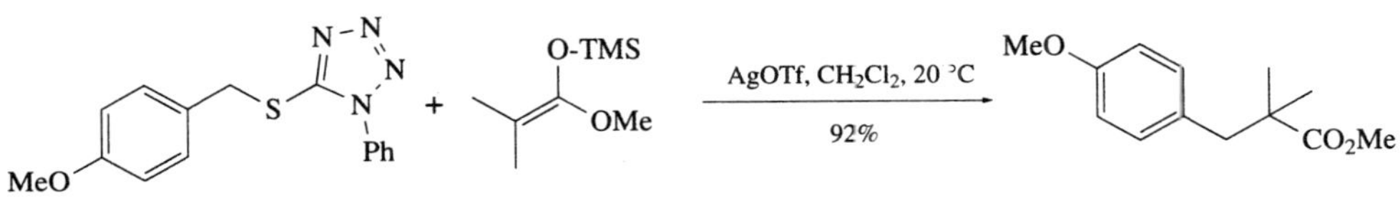

(295)

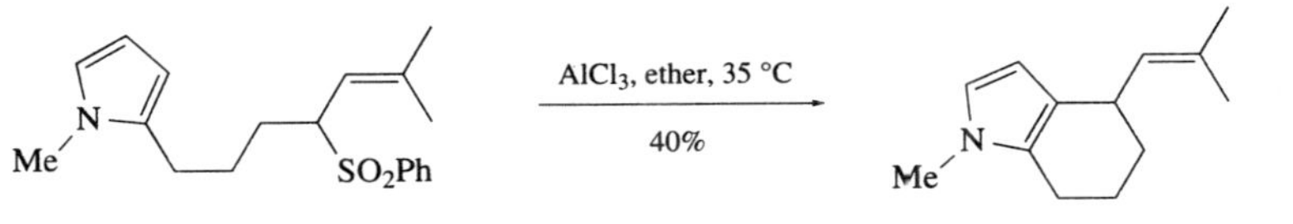

(296)

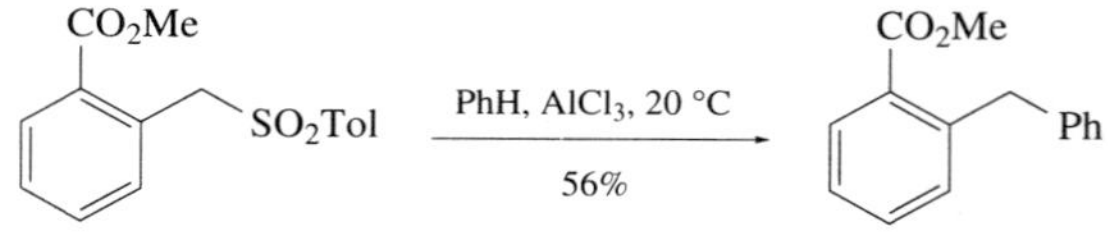
(297)

1.05.2.5.2 *Intramolecular reactions*

When symmetrical 1,3-diphenylthiopropane derivatives are deprotonated α to the sulfur atom they readily cyclize with loss of thiophenol to give cyclopropanes exhibiting *trans* ring geometry (Equation (298)) ⟨82BCJ2965, 87CL809⟩. Moderate diastereoselectivities, which depend on the bulk of the silyl unit, are obtained when a chiral auxiliary is attached (Equation (299)) ⟨88JA7185⟩. Other internal nucleophiles will displace sulfur (providing it is not possible to deprotonate α to sulfur) (Equation (300)) ⟨82T693⟩ if the nucleophile is of higher or comparable acidity to the sulfur leaving group (Equations (301) and (302)) ⟨66CC757, 85TL6115⟩. A sulfur group may also be displaced by an internal nucleophile generated from an addition process such as a Michael reaction (Equations (303) and (304)) ⟨87JOC218, 81JOC5027⟩.

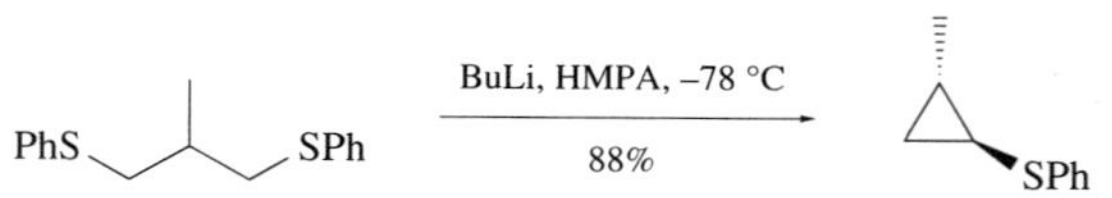
(298)

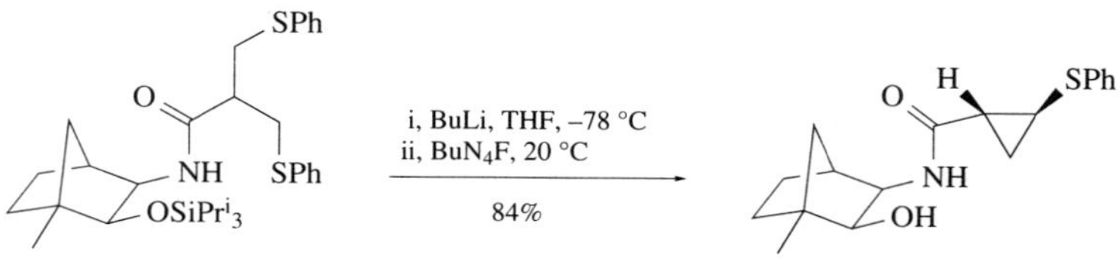
(299)

83% *de*

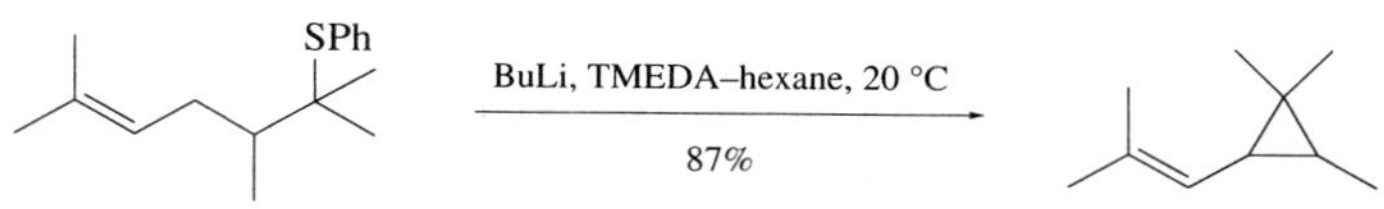
(300)

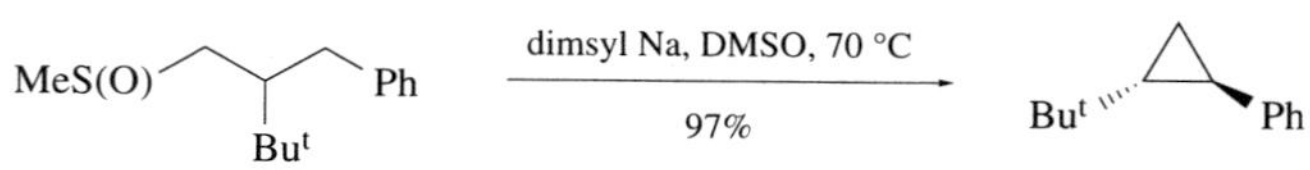
(301)

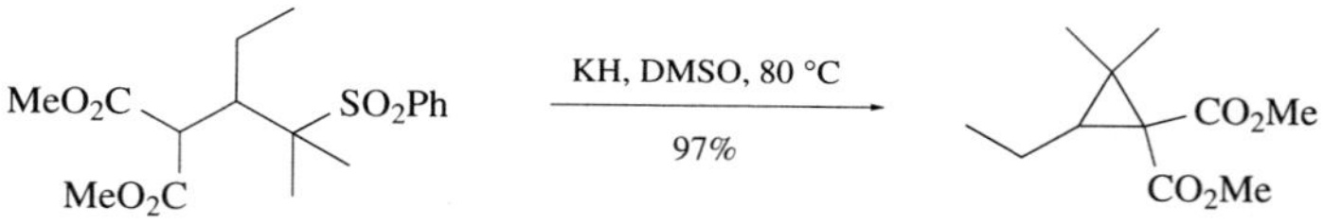
(302)

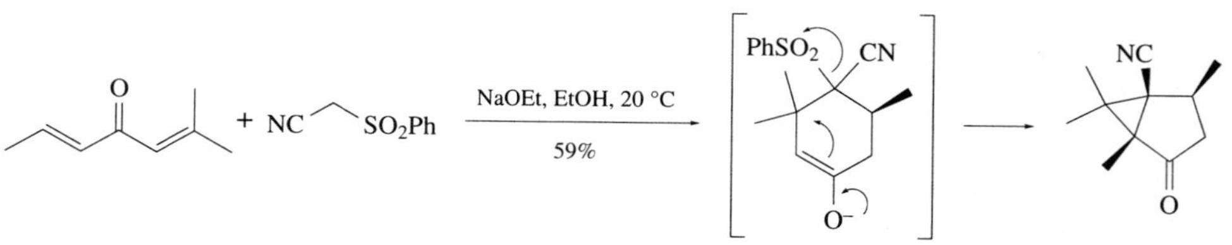
(303)

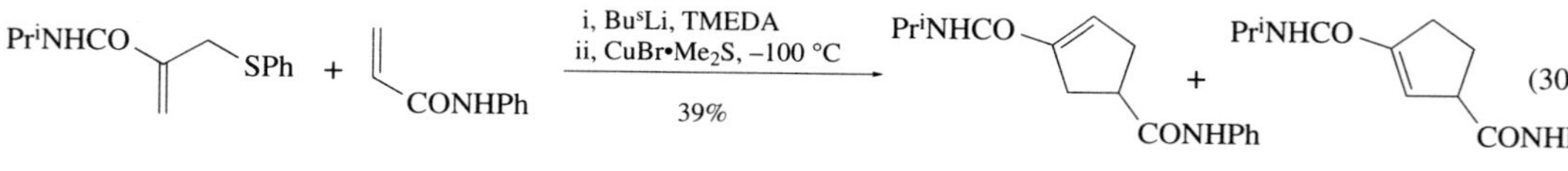
(304)

75 : 25

1.05.2.5.3 Sulfonium salts

Sulfonium salts can act as alkylating agents with dialkylsulfide acting as the leaving group. When the groups on sulfur are not the same problems of chemoselectivity can arise. The important factors are that the nucleophile must be insufficiently basic to generate an ylide from the sulfonium salt and that the counterion must be nonnucleophilic. The soft leaving group favours carbon vs. oxygen alkylation of 1,3-dicarbonyl compounds ⟨80TL4811, 83JOC1362⟩. Five-membered ring thioethers are ring-cleaved while six-membered ring thioethers react at the exocyclic carbon atom except when the sulfur centre is allylic (Scheme 54). This endocyclic vs. exocyclic attack has been studied with other nucleophiles ⟨76JOC1052⟩. A thallium salt of a β-ketoester opens a tricyclic compound selectively (Equation (305)) ⟨71TL3791⟩. Intramolecular alkylation is possible (Equation (306)) ⟨68AG(E)456⟩ as is Friedel-Crafts alkylation ⟨78AG(E)274⟩. Sulfonium salts with three different substituents are optically active and have been used in asymmetric alkylations ⟨89JOC2374⟩. Unfortunately the reaction is neither chemo- nor enantioselective (Equation (307)).

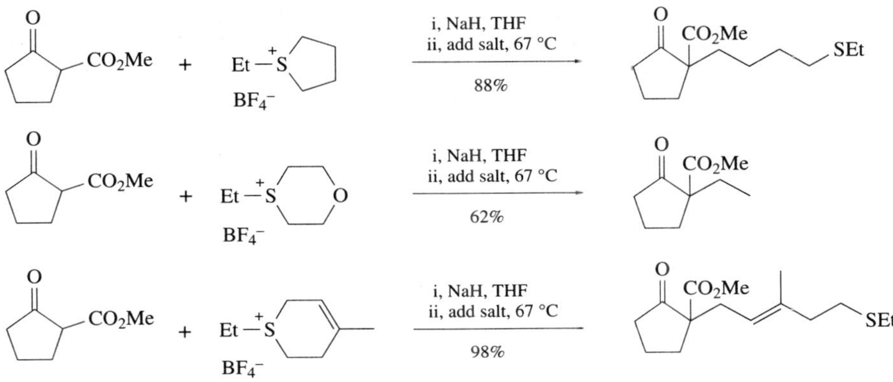

Scheme 54

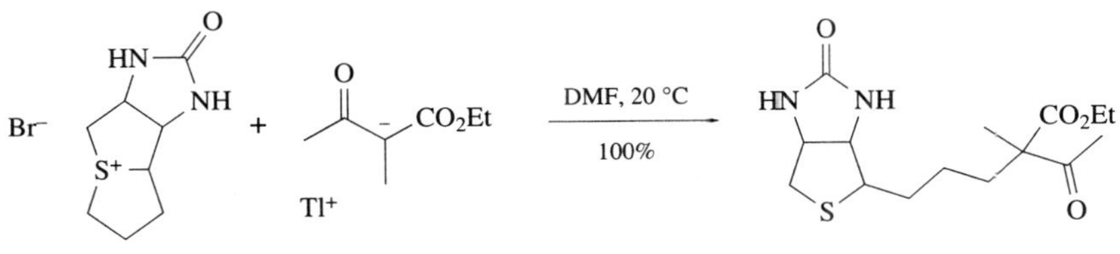

(305)

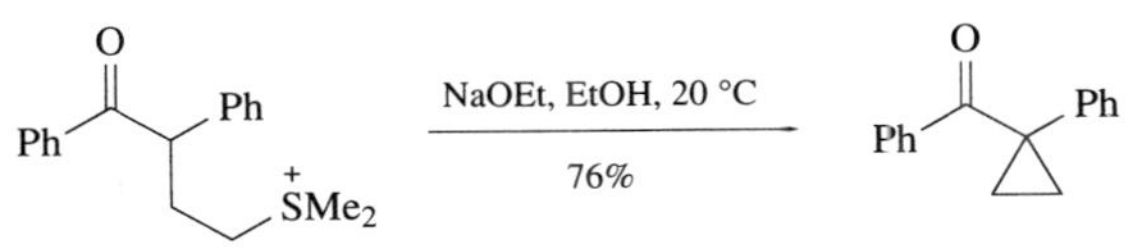

(306)

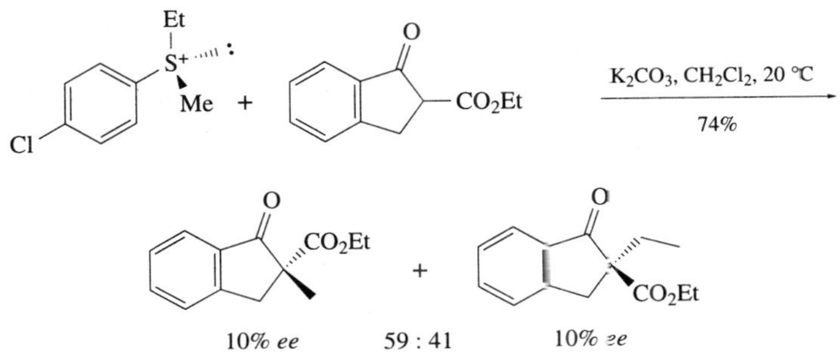

(307)

1.05.2.5.4 Miscellaneous reactions

Acylation of the oxathianedioxide (**114**) (two equivalents) and then *in situ* hydrolysis gives γ-hydroxyketones in 53–89% yield (Equation (308)) ⟨87CC449⟩.

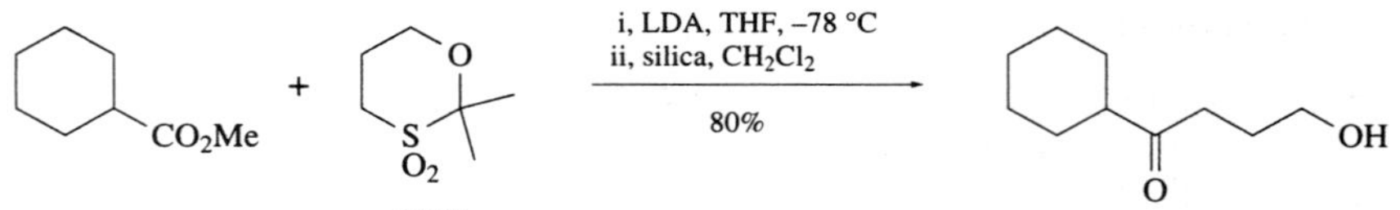

(114)

(308)

1.05.2.6 Opening of Thiiranes

Like epoxides thiiranes (thioepoxides) ⟨84CHEC(7)131⟩ are readily ring-opened by a similar range of nucleophiles. Grignard reagents attack, via an S_N2 process, at the less hindered carbon centre (Equation (309)) ⟨78M609⟩. The allyl lithium (**115**) attacks from the γ-position and the resultant thio anion displaces chloride ion to give a thietane (Equation (310)) ⟨85S1069⟩. Cyanide opens the ring under Lewis acid catalysis, again at the less hindered position (Equation (311)) ⟨83TL2311⟩. Malonate derivatives open thiiranes and the resultant thiol anion cyclizes onto esters and nitriles to give thiolactones and thioimidates (Equations (312) and (313)) ⟨86BCJ2321⟩. Phenyl thiirane is an exception to the norm, giving a 70 : 30 selectivity of attack at the more substituted carbon centre.

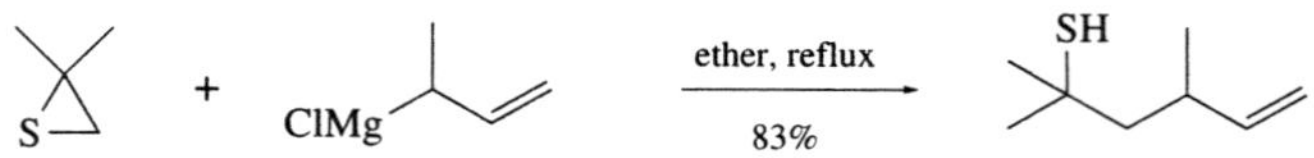

(309)

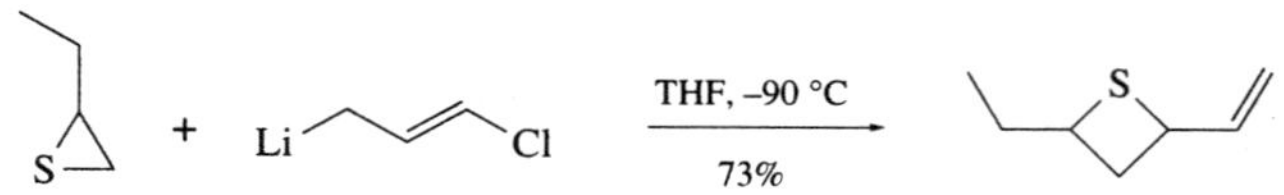

(115)

(310)

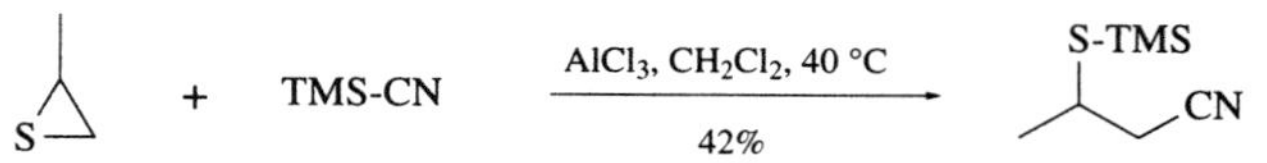

(311)

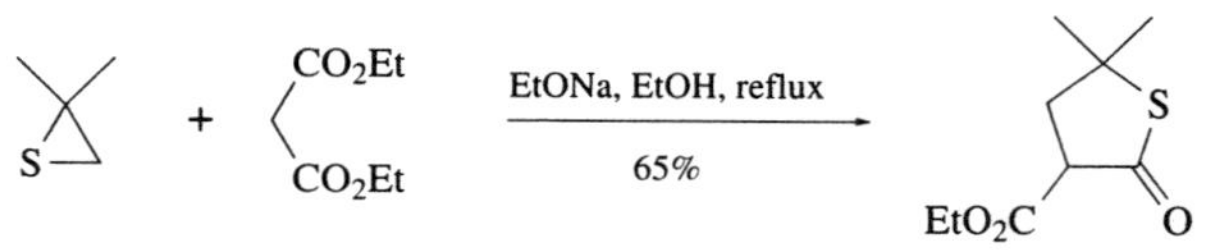

(312)

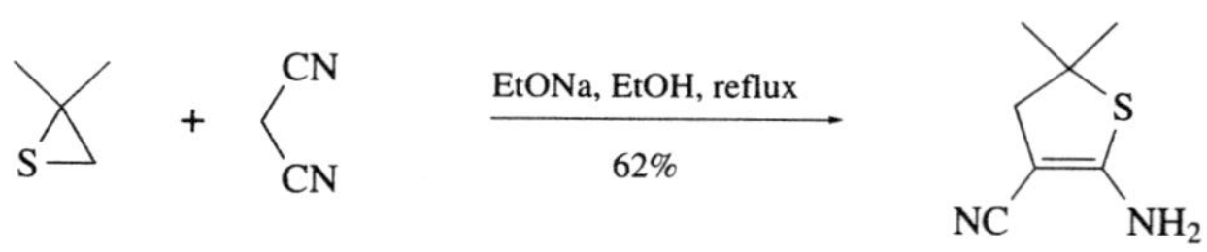

(313)

Anions α to isocyanide open thiiranes to give, after cyclisation, the six-membered ring heterocycles (Equation (314)) ⟨79LA451⟩. Cobalt-catalyzed carbonylation gives β-thiol acids (Equation (315)) ⟨87OM1625⟩. The thiirane is opened at the more substituted position. A similar reaction can be performed on thietanes ⟨89JOC20⟩. Under Lewis acid catalysis methylenethiiranes (**116**) add to enones to give tetrahydrothiophenes in variable yields (7–86%) (Equation (316)) ⟨84TL4011⟩. Intramolecular cyclisation onto an alkene has also been reported ⟨89JOC4660⟩.

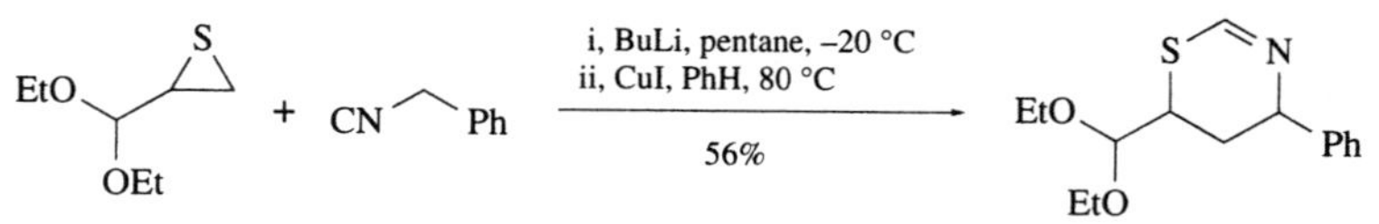

(314)

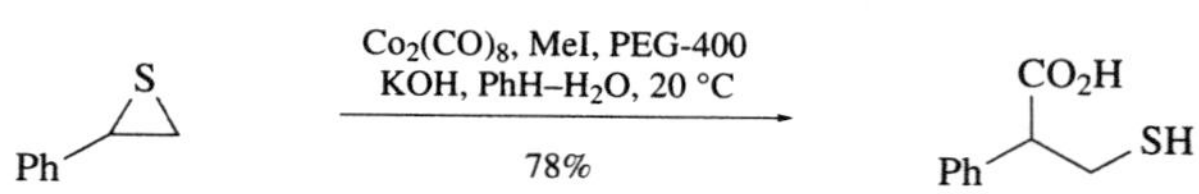

(315)

PEG = polyethylene glycol

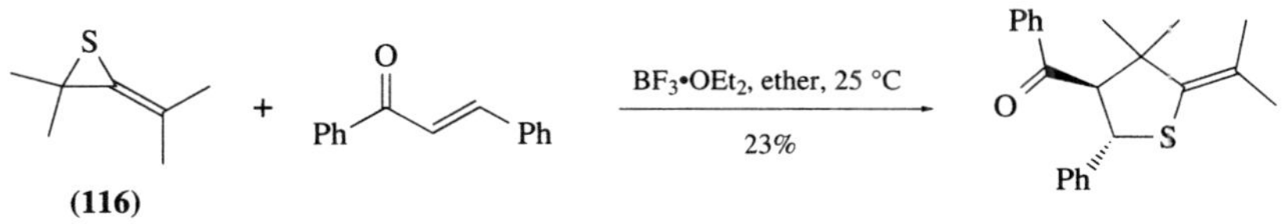

(116)

(316)

1.05.3 SUBSTITUTION OF SELENIUM AND TELLURIUM

The chemistries of selenium ⟨B-86MI 105-02⟩ and tellurium ⟨91S793, 91S897⟩ have been reviewed.

1.05.3.1 Thermal and Radical Reactions

Vacuum pyrolysis of selenides and tellurides results in extrusion of the free element and formation of a C—C bond (see review of extrusion reactions ⟨88T6241⟩); intramolecular reactions have been reported for tellurium (Equation (317)) ⟨75TL1893⟩ and both inter- and intramolecular reactions for selenium (Equation (318)) ⟨80SC595, 81CL627, 82BCJ182⟩; the last ploy has been used to prepare paracyclophanes. The method is limited to benzylic compounds. [2,2]-Paracyclophanes can also be prepared photochemically in good yield ⟨87BCJ4027⟩. Detelluration can be achieved with bisallylic substrates in those cases where deselenation fails ⟨82JOC1641⟩.

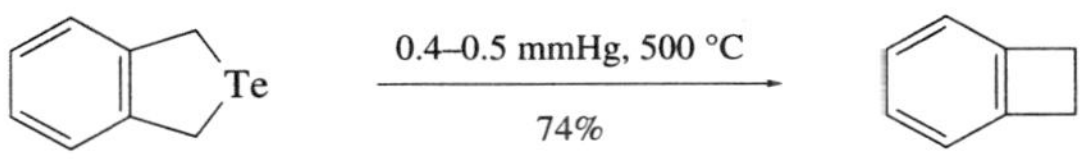

(317)

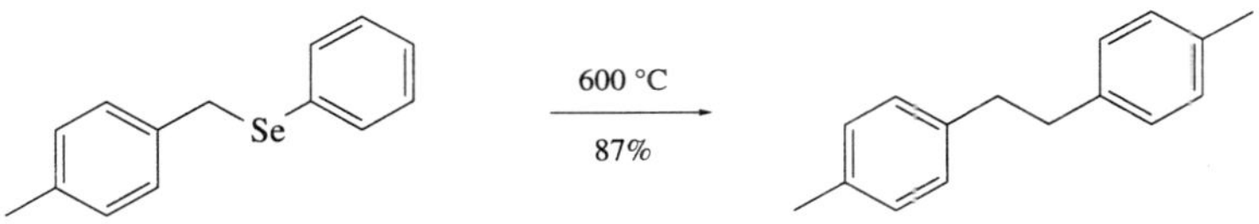

(318)

Cleavage of carbon—selenium and carbon—tellurium bonds takes place homolytically to form carbon-centred radicals which will add to the normal range of radical acceptors ⟨B-86MI 105-01⟩. Phenyl selenium compounds have several advantages over other radical precursors (of which a pleasant odour is not one) including ease of synthesis from a variety of sources (see Chapter 2.14). Radicals derived from selenium and tellurium add to acrylates in moderate to good yields ⟨82JOC3348, 89TL1209⟩ (Equations (319) and (320)). Phenylselenium was much better than either iodide or mercury acetate as a radical precursor. Reduced compounds are the main by-products and stereochemical integrity is only retained in the [5,5] bicyclic system. Selenium-derived radicals α to carbonyls also add to alkenes in 35–89% yields (Equation (321)) ⟨90TL227, 90TL5697⟩. In the absence of a hydride source the intermediate radical is trapped by a phenylselenyl moiety. Mixtures of isomers are obtained with some of the nucleoside radical acceptors ⟨90TL227⟩. Tin compounds are useful radical traps; reacting with selenium derived radicals to give 1,4-dicarbonyl compounds (Equation (322)) ⟨90TL6669⟩ and γ,δ-enoates (Equation (323)) ⟨88JA4815, 91SL412⟩ and alkenes ⟨83TL1357⟩.

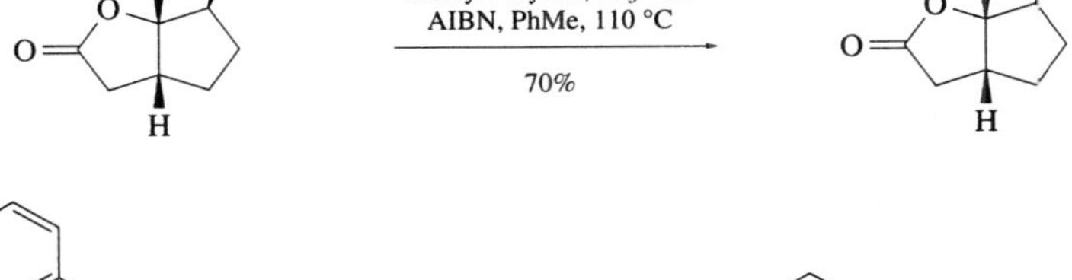

(319)

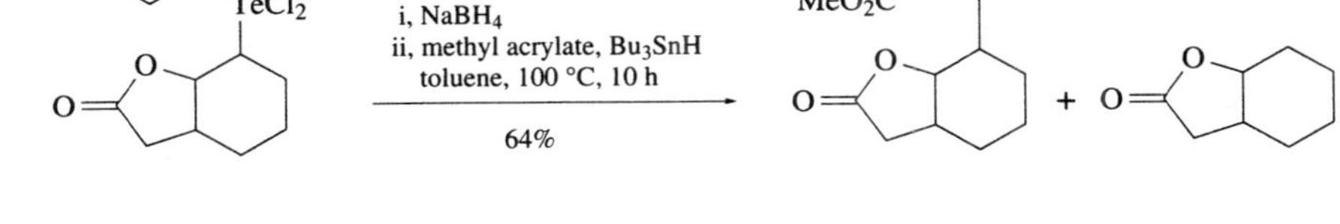

(320)

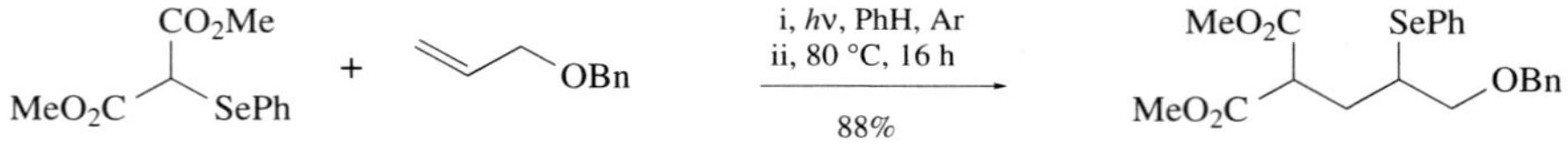

(321)

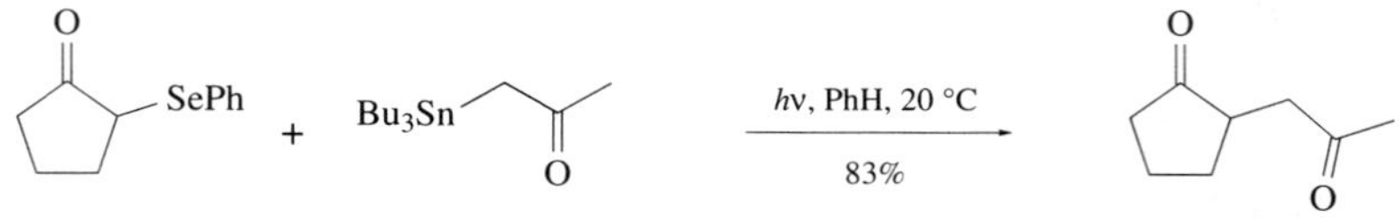

(322)

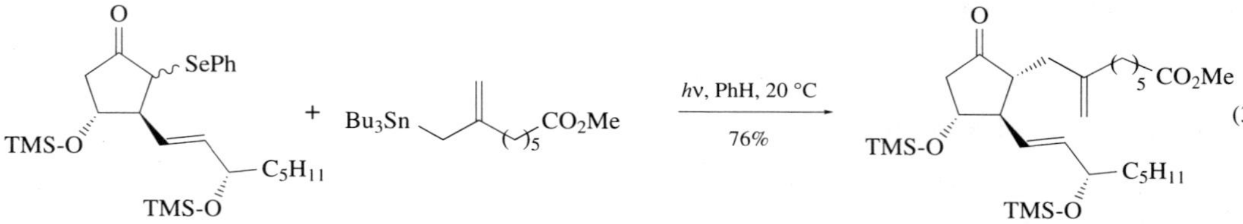

(323)

Radicals derived from selenium compounds have been much used in intramolecular C—C bond formation. The α-cyano-compounds (**117**) cyclize in 75–91% yields as mixtures of isomers (Equation (324)) ⟨87JOC4943⟩. The photochemically initiated cyclisations shown in Equation (325) ⟨91CC354⟩ give the *trans* compounds with selenium transfer. Annellations are sensitive to substitution and stereochemistry. The nucleoside analogue (**118**) gives a single isomer at the new chiral centre (Equation (326)) while its C-3 epimer gives a mixture of compounds ⟨91T2237⟩. Cyclisation of the compound (**119**) gives a 50 : 50 mixture of products, while its analogue (**120**) gives a single isomer (Equation (327)) ⟨85TL6431⟩. This method does not work for compounds involved in the formation of six-membered rings. Bicyclic systems can be produced by similar cyclisation of reactions (Equation 328) ⟨90JOC1786⟩; spiro, heterospiro and bridged systems are also available; mixtures of isomers are obtained in some cases ⟨88TL1315, 89TL3865⟩. Selenium compounds can be cyclized onto aldehydes at ambient temperature (Equation (329)) ⟨93CC429⟩.

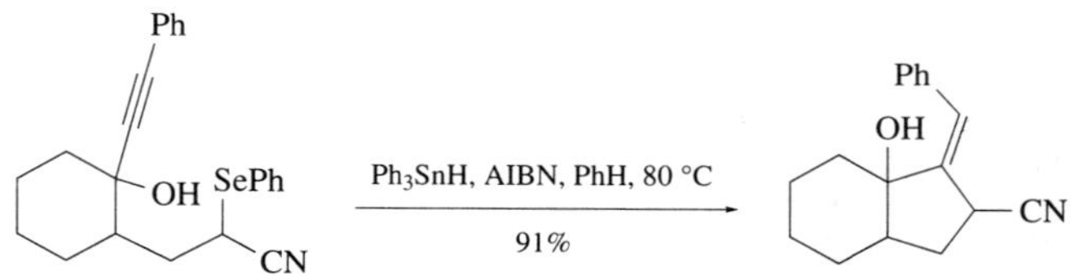

(117)

(324)

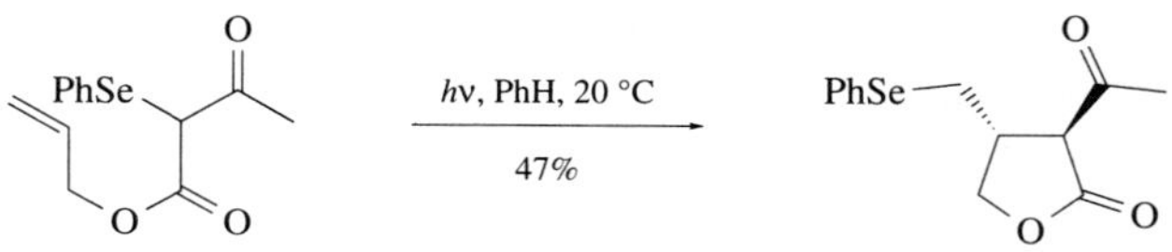

(325)

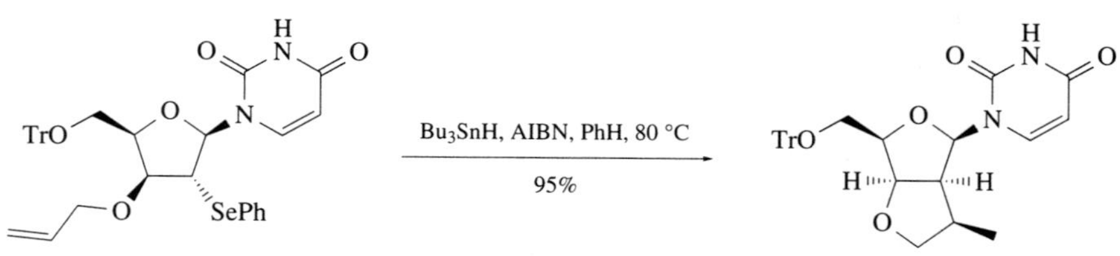

(118)

(326)

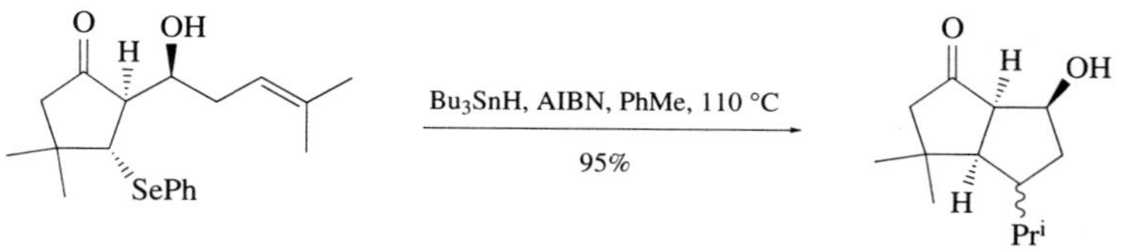

(119)

50 : 50

(327)

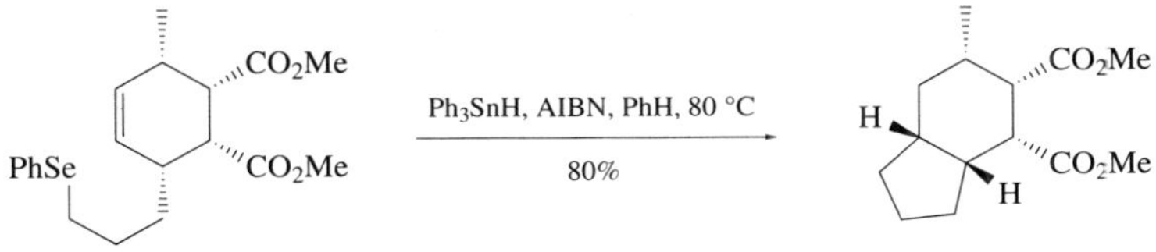

(328)

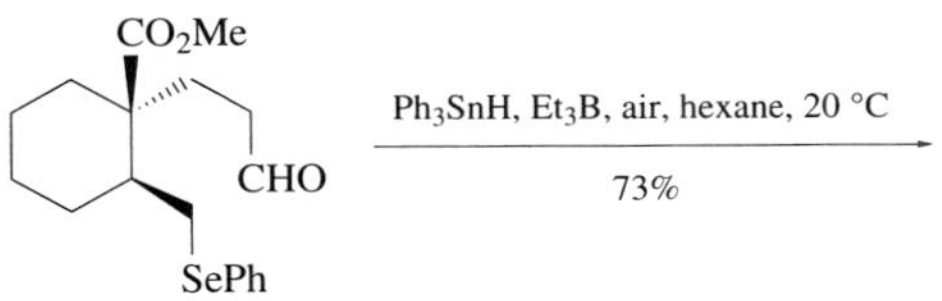

(329)

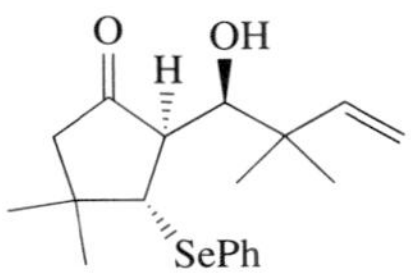

(120)

Free radical ring expansion takes place regioselectively in 1,3-dicarbonyl systems to give six- and seven-membered 1,4-dicarbonyl products (Scheme 55) ⟨88JA2565, 89TL6129⟩. Inter- and intramolecular hydride extraction are competing processes especially with *N*-benzyl piperidones, giving rise to reduced products.

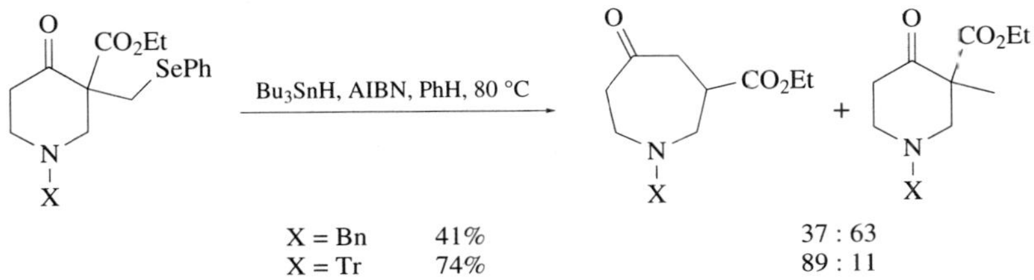

Scheme 55

1.05.3.2 Metal–Exchange Reactions

Organoselenium and organotellurium compounds are readily transmetallated with reagents such as butyllithium. Alkyl lithiums have a high affinity for selenium and attack the selenium centre preferentially, in contrast to sulfur-based systems where α-deprotonation is the normal mode of attack. That bond is cleaved which gives the most stable lithium species. Thus benzylic and allylic selenides form the benzyl and allyl lithium species, (Equations (330) and (331)) ⟨85TL1093, 86CC457, 81TL4341⟩ which readily couple with primary halides or add to carbonyl compounds and enones (up to 100% 1,4 addition). In some cases potassium diisopropylamide was found to be a better reagent than LDA or butyl lithium (Equation (332)) ⟨86TL1723⟩. A few examples of displacement of tellurium by carbon via tellurium–lithium exchange are known but these are probably of limited synthetic value except for allyl and benzyl tellurides where the corresponding lithium (and magnesium) species are difficult to obtain (Scheme 56) ⟨87AG(E)1187, 89TL7353⟩.

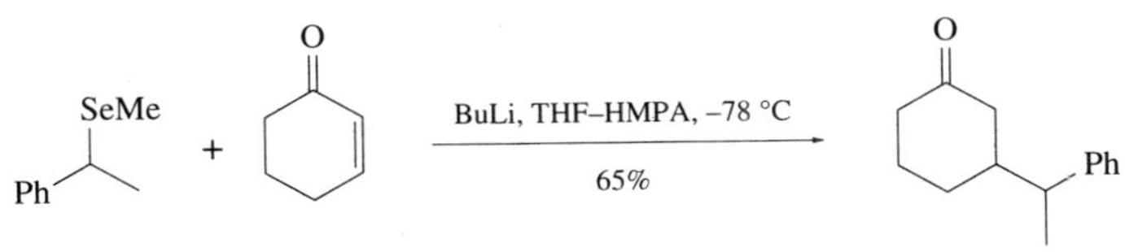

(330)

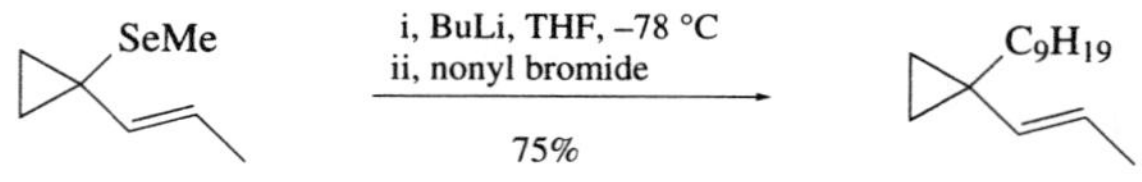

(331)

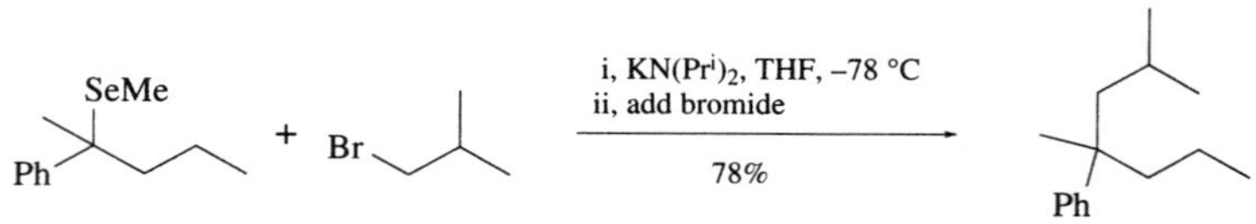

(332)

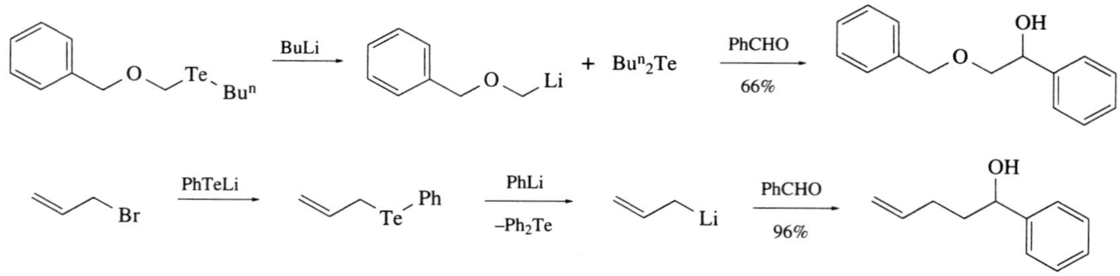

Scheme 56

The benzyl lithiums derived from selenides will add to unactivated double bonds to form five-membered rings, regenerating alkyl lithiums ⟨87CC1214, 91TL417⟩.

1.05.3.3 α-Carbanion Formation

Treatment of the selenoester (**121**) with an alkoxide results in the extrusion of metallic selenium, giving 1,3-dicarbonyl compounds in 46–86% yields (Equation (333)) ⟨78CL1007⟩.

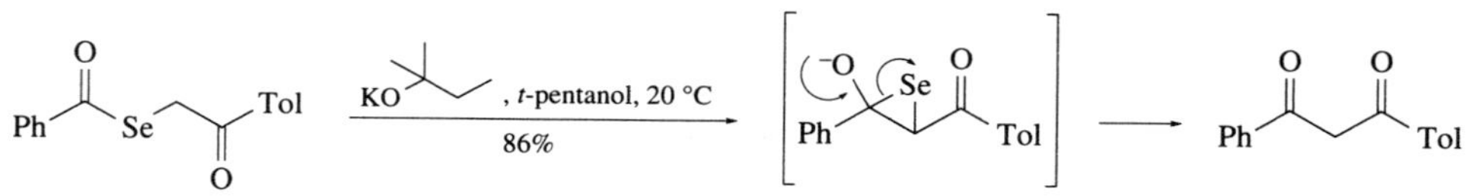

(333)

(**121**)

1.05.3.4 Migrations of Selenides and Tellurides

α-Oxygen substituted selenides and tellurides undergo oxidative or base-catalysed rearrangement of the pinacol type which leads to migration of an alkyl group and loss of selenium or tellurium ⟨85JOC5200, 86JCS(P1)1983⟩. These reactions will be covered elsewhere. Under some conditions, for example using thallium hydroxide and a haloform to generate carbenes, replacement of the selenium and hydroxyl moieties by C(hal)$_2$ is observed yielding cyclopropanes ⟨87TL1549⟩.

1.05.3.5 Miscellaneous Reactions

Copper-catalysed Grignard reagents displace phenylselenyl groups, with S_N2 inversion (Equation (334)) ⟨91TL5709⟩. Selenophenoxide can be used as a leaving group, in a similar fashion to thiophenoxide, in the molybdenum-catalysed allylation (Equation (335)) ⟨89BCJ2913⟩. Yields and selectivities are little different from the sulfur analogues. Selenium can also be used in place of sulfur in the copper-catalysed insertion of diazo compounds into carbon—chalcogen bonds ⟨80TL395⟩.

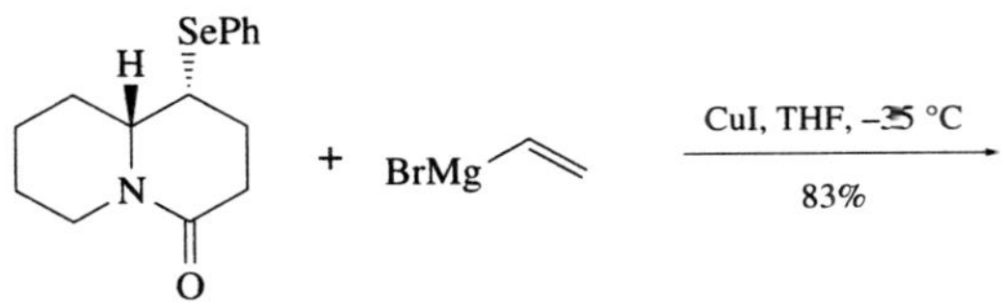

(334)

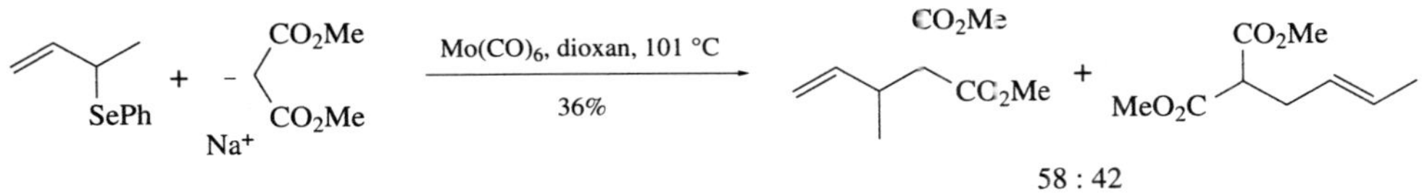

(335)

Tellurium is eliminated from tellurides in high yield when treated with palladium and a base ⟨88TL3533⟩ (Equation (336)). The reaction is intramolecular as no scrambling occurs when unsymmetrical tellurides are demetallated (Equation (337)). It is postulated that the reaction procedes via a tellurium–palladium complex (**122**) which undergoes a double alkyl migration, elimination of tellurium and coupling of the alkyl (or alkyl and aryl groups) (Scheme 57). The reaction is stoichiometric in palladium.

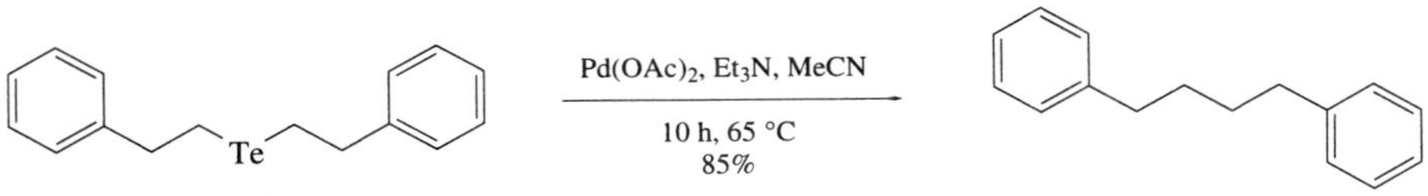

(336)

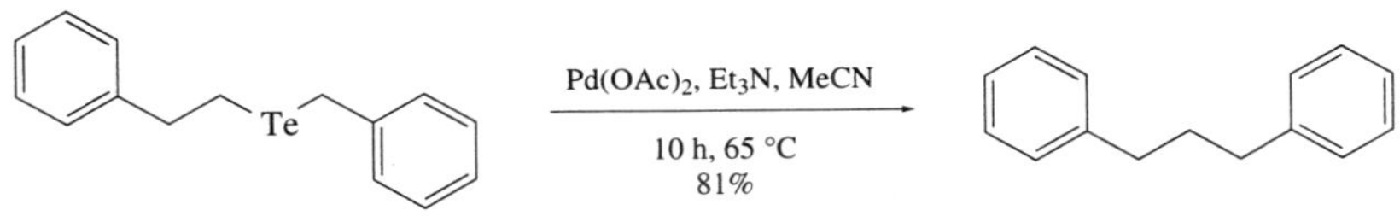

(337)

(122)

Scheme 57

Tetraalkyltellurium reagents add to alkynes in 16–98% yields (GC; based on tellurium) to give *cis* alkenes, with dialkyltellurium as by-product (Equation (338)) ⟨93OM473⟩. The reaction is believed to proceed via a radical mechanism.

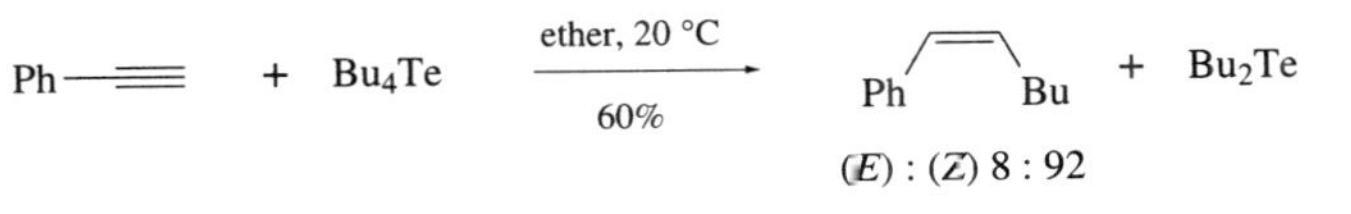

(338)

Trimethylselenium and trimethyltellurium ions in super acid media ($HSbF_6$) methylate toluene ⟨88JOM(348)199⟩. While the selenium reagent gives an *ortho*:*meta* ratio of ca. 2:1 (similar to the sulfur and oxygen analogues) tellerium gives a ratio *ortho*, *meta*, *para* of 93:6:1.

1.06
One or More CC Bond(s) Formed by Substitution: Substitution of Carbon-Nitrogen, -Phosphorus, -Arsenic, -Antimony, -Boron, -Silicon, -Germanium and -Metal Functions

PHILIP C. BULMAN PAGE, HEATHER L. McFARLAND and ANDREA P. MILLAR
University of Liverpool, UK

1.06.1 SUBSTITUTION OF NITROGEN FUNCTIONS

1.06.1.1 Nitro Compounds

Aliphatic nitro compounds are versatile synthetic intermediates. The highly electron-withdrawing nitro group facilitates α-carbanion formation (and subsequent carbon–carbon bond forming processes) and is easily converted into other functional groups ⟨91S423⟩.

Kornblum discovered that substitution reactions of the nitro group in various types of tertiary nitro compounds can be induced by relatively soft nucleophiles through single electron transfer chain processes ($S_{RN}1$) (Scheme 1) ⟨75AG(E)734⟩. These reactions proceed by a chain sequence in which radical anions and free radicals are intermediates. They occur readily under mild conditions and in excellent yields. In contrast to S_N2 displacements they are rather insensitive to steric hindrance and are useful for the synthesis of highly branched structures that are otherwise difficult to prepare.

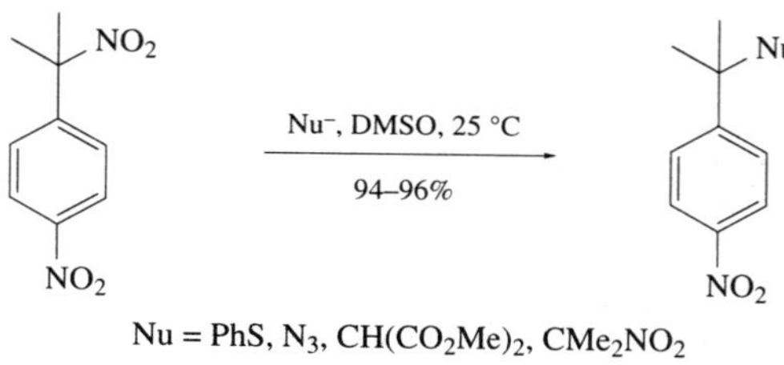

Scheme 1

A more general substitution was discovered in the early 1980s whereby primary, secondary and tertiary allylic nitro compounds undergo displacement of the nitro group by various nucleophiles with the aid of metal catalysts such as palladium phosphine complexes ⟨82CC821, 82JA3727⟩ and Lewis acids ⟨86CC1285⟩. For example, the reaction of 1-(nitromethyl)-3,4-dihydronaphthalene with sodio-2-carboethoxycyclopentanone in DMF at 20°C using 1–5 mol% of both tetrakis-(triphenylphosphine) palladium(0) and 1,2-bis(diphenylphosphino)ethane (dppe) gave the nucleophilic displacement product in 64% yield (Equation (1)) ⟨86JOC4375⟩.

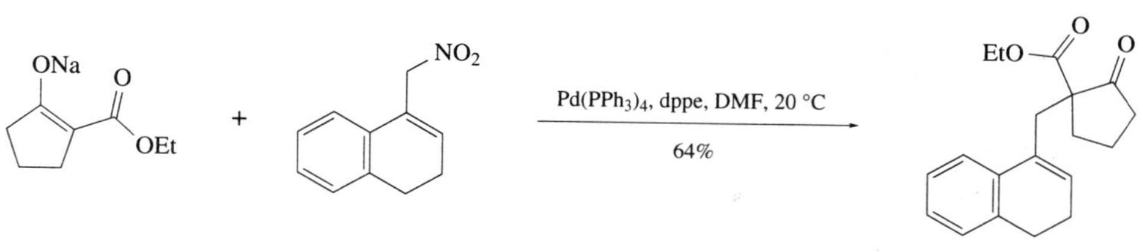 (1)

Acyclic allylic nitro compounds also serve as reactive substrates for tetrakis(triphenylphosphine)palladium(0)-catalysed substitution reactions (Scheme 2) ⟨86JCS(P1)1439⟩. The regiochemistry of such substitution of tertiary allylic nitro compounds is governed primarily by the steric factors of the nucleophiles. Generally, bulky nucleophiles attack selectively at the less hindered terminus (Scheme 2, entry 1), whereas less bulky groups tend to give direct substitution (entry 2). The regiochemistry is also controlled by the steric and electronic effects of the α-substituents. Again, bulky alkyl groups assist substitution at the less hindered carbon atom (entry 3), while less bulky groups result in direct substitution (entry 4). Groups which are electron withdrawing, such as esters, cause selective attack at the terminal carbon atom with allylic displacement (entry 5).

A B

Entry	*R*	*Nu*⁻	*Yield* (%)	*Ratio* **A**:**B**
1	Me	*p*-$MeC_6H_4SO_2C^-(Me)CO_2Et$	90	0:100
2	Me	$^-CH(CN)CO_2Me$	75	99:1
3	*n*-C_6H_{13}	$^-CH(CO_2Me)_2$	78	42:58
4	Me	$^-CH(CO_2Me)_2$	60	73:27
5	CO_2Et	$^-CH(CO_2Me)_2$	75	0:100

Scheme 2

In contrast to the palladium-catalysed substitution of allylic nitro groups by various nucleophiles, a novel uncatalysed reaction, specific for cyclic α-nitroalkyl enones, was reported in 1989 ⟨89TL2413⟩. For example, the reaction of 2-(nitromethyl)cyclohexenone with ethyl sodio-2-cyanobutanoate in DMF at room temperature gave the nucleophilic displacement product in 72% yield (Equation (2)) ⟨89TL2413⟩. Mechanistic studies concluded that the formation of the S_N2 product occurs by a double addition–elimination sequence to give the thermodynamically more stable product. The high reactivity of cyclic 2-(nitroalkyl)enones is however due to the leaving ability of the NO_2 group. The β-functionalized cyclic enone products were used for further synthetic manipulation.

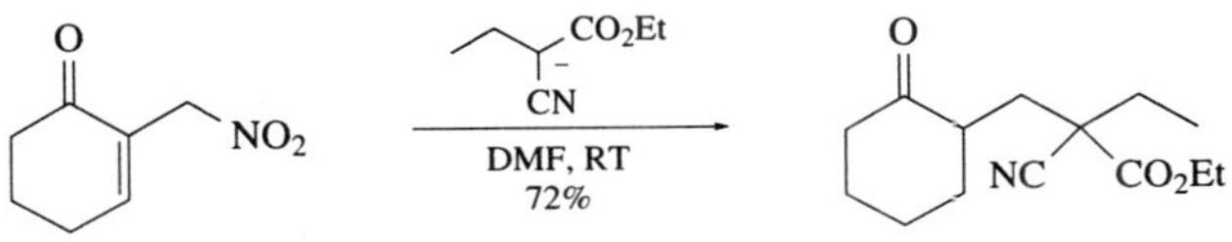

(2)

Lewis acid-catalysed nucleophilic substitution reactions of tertiary, benzylic and allylic nitroalkanes normally proceed through an S_N1 mechanism involving a carbocation intermediate. Inter- and intramolecular Friedel–Crafts reactions of these compounds can best be achieved by using two equivalents of tin tetrachloride (Scheme 3). Other Lewis acids such as aluminum trichloride, titanium tetrachloride and zinc chloride are less effective ⟨86CC1285⟩.

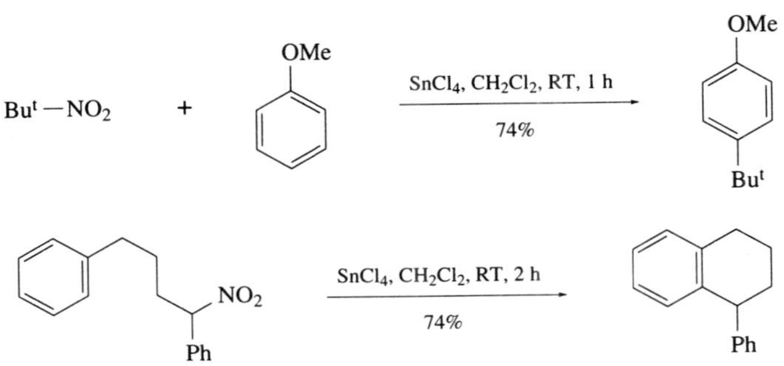

Scheme 3

Benzylic and tertiary allylic nitro compounds react with allytrimethylsilane rapidly at −10 °C in the presence of one equivalent of tin tetrachloride (Equation (3)) ⟨86CC1285⟩.

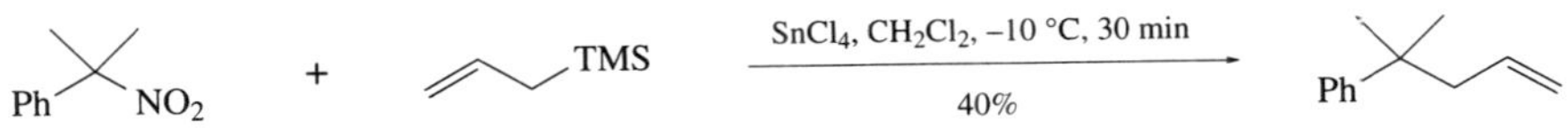

(3)

Secondary and tertiary allylic nitro compounds can undergo Lewis acid-catalysed substitution by silyl enol ethers. For example, 1-trimethylsilyloxycyclohexene reacts at ca. −20 °C in the presence of one equivalent of tin tetrachloride to produce allylated products (Equation (4)) ⟨86CC1285⟩.

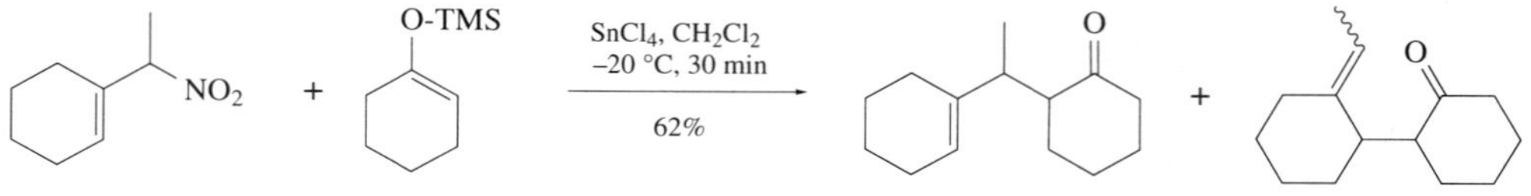

(4)

35 : 65

Secondary and tertiary allylic nitro compounds also react with cyanotrimethylsilane in the presence of 0.2 equivalents of tin tetrachloride to give β,γ-unsaturated nitriles as a mixture of isomers (Equation (5)) ⟨86TL3025⟩. Titanium tetrachloride is also an effective catalyst for these transformations, but zinc chloride, aluminum chloride and boron trifluoride etherate mediate slow, poor-yielding reactions. Ester and nitrile functional groups were tolerated in the substrate when tertiary allylic nitro groups were replaced by nitrile at room temperature. In general, secondary allylic nitro compounds were poorer substrates than tertiary species.

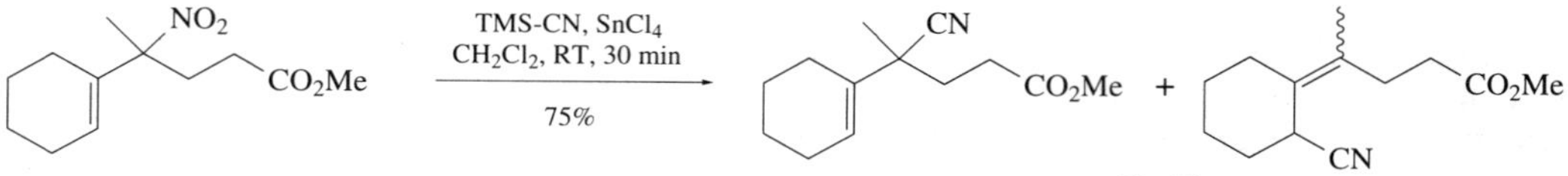

(5)

30 : 70

Lewis acids are also required for the displacement of the nitro groups in primary, secondary, tertiary and benzylic β-nitro sulfides ⟨87JOC4133⟩. These sulfides can be prepared in a one-pot procedure by reaction of aldehydes or ketones with nitro compounds and thiols in the presence of amines. Their treatment with cyanotrimethylsilane or allyl trimethylsilane in the presence of a Lewis acid ($AlCl_3$, $SnCl_4$ or $TiCl_4$) gave the substitution products as shown (Scheme 4). Similar treatment of 2-nitrobutane gave no substitution reaction, showing the importance of the β-phenylthio function. The reactions are thought to proceed via episulfonium ions, and the ratio of products depends on the electron donating or withdrawing nature of the substituents. The predominant product is formed by attack of the nucleophile at the incipiently more electropositive terminus of the episulfonium ion intermediate. A 1,2-migration of the thiophenyl group occurs when the nitro group is primary and the thiophenyl group is located on a tertiary or benzylic position (Scheme 4, entry 1) ⟨89JOC4998⟩.

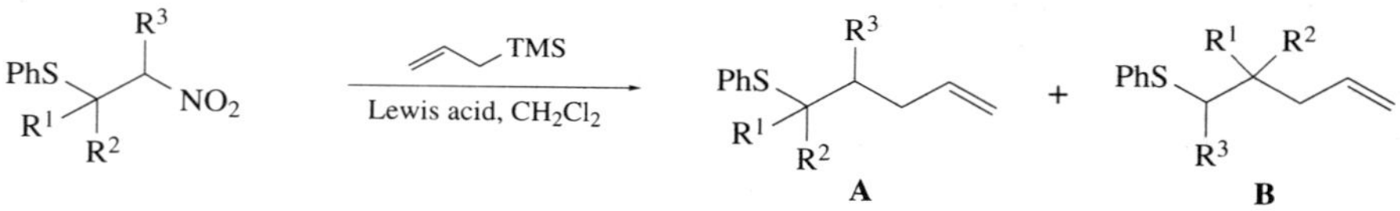

Entry	*R¹*	*R²*	*R³*	*Lewis acid*	*Yield (%)*	*Ratio* **A : B**
1	Me	Me	H	$AlCl_3$	72	7 : 93
2	Et	H	Me	$TiCl_4$	55	75 : 25

Scheme 4

Stereoselectivity in the formation of carbon–carbon bonds by nitro displacement using pure *anti* and pure *syn* β-nitrosulfides was examined. In each case the *anti* isomer reacted more rapidly with nucleophiles than the corresponding *syn* isomer. Cyclic compounds could also be formed through this process by intramolecular nitro group displacement in β-nitrosulfides with retention of configuration (Equation (6)).

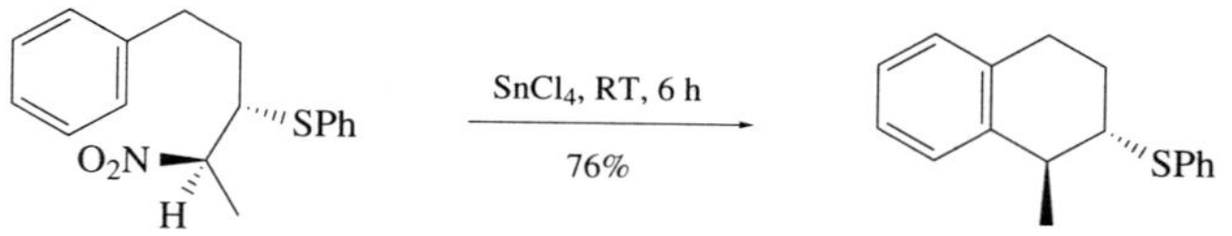

(6)

1.06.1.2 Quaternary Ammonium Salts

Many examples of carbon–carbon bond formation with cleavage of carbon–nitrogen bonds in quaternary ammonium salts can be found in the literature. A representative selection is described below.

Reactions of various quaternary ammonium salts with enolate anions derived from activated methylene compounds provide a synthetic route for 2-aryl-prop-1-ene systems (Scheme 5) ⟨89SC2415⟩. Sodium hydride as base and acetonitrile as solvent were found to give the best results. The enolates derived from ethyl malonate, ethyl acetoacetate and ethyl cyanoacetate were treated with quaternary ammonium salts derived from 2-aryl-3-(*N*,*N*-dimethylamino)-prop-1-enes. The reactions were generally clean and efficient, giving 2-arylalk-1-ene systems in high yields.

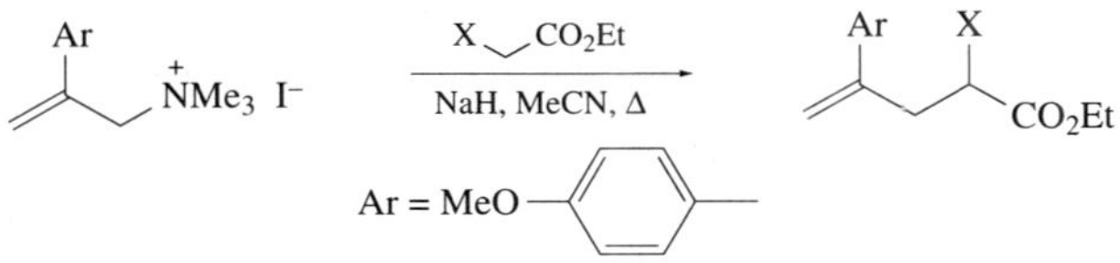

X	*Yield* (%)
CO_2Et	93
MeCO	69
CN	24

Scheme 5

3-Substituted isoprenylsilanes are useful, not only as nucleophilic reagents for the introduction of a 1,3-diene skeleton, but also as highly reactive 1,3-dienes. They can be prepared by the displacement reaction between an ammonium halide salt of 2-[(dimethylamino)methyl]-3-[(trimethylsilyl)methyl]-1,3-butadiene and a Grignard reagent in the presence of dilithium tetrachlorocuprate as a catalyst (Scheme 6) ⟨87JOC2947⟩. Oddly, the use of other catalysts such as cuprous chloride, cuprous bromide and cuprous iodide did not give satisfactory results. Furthermore, efficient reaction often required a large excess of Grignard reagent.

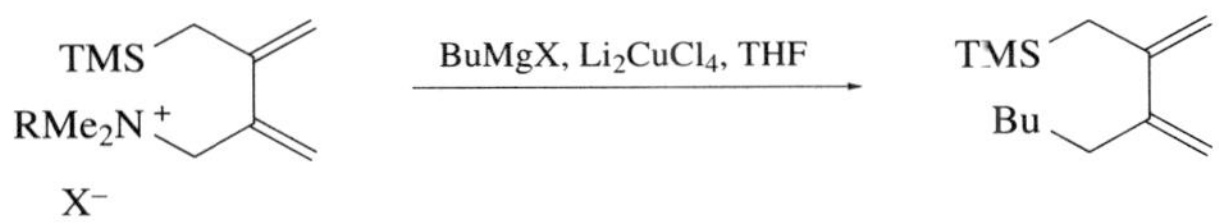

R	*X*	*Yield* (%)
Me	I	88
Bu	I	52
$PhCH_2$	Cl	92

Scheme 6

(*E*),(*Z*)-Diene-containing pheromones can be prepared by nucleophilic substitution of an (*E*),(*Z*)-dienyl ammonium salt, as illustrated by the synthesis of 7(*E*),9(*Z*)-dodecadien-1-yl acetate, the sex pheromone of *Lobesia botrana*, a major pest of vineyards (Scheme 7) ⟨80TL67⟩. For these reactions, use of lithium diorganocuprate reagents as nucleophiles was found to give better yields of the (*E*),(*Z*)-dienes with greater selectivity than did use of copper-catalysed Grignard reagents.

RM	*Yield* (%)	*Ratio* (E),(Z) : (E),(E)
$MeMgBr/Li_2CuCl_4$	53	90 : 10
Me_2CuLi	68	92 : 8

Scheme 7

The transformation of 2-(1-(dimethylamino)ethyl)-3-(2-hydroxyethyl)-1*H*-indole into the corresponding nitriloalcohol has been used in alkaloid synthesis. The reaction involves conversion of an amine into a quaternary ammonium iodide followed by displacement with KCN/18-crown-6 in acetonitrile to give the desired nitriloalcohol in 82% yield (Equation (7)) ⟨81JOC3293⟩.

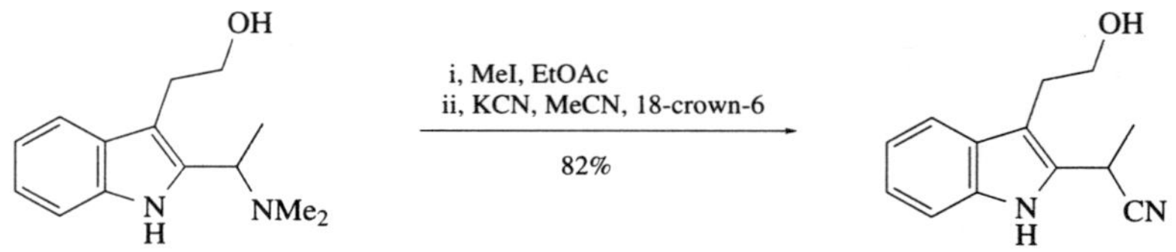

(7)

Another example of displacement of an ammonium salt, involving formation of a new carbon–carbon bond, is used in a novel route to 6-*t*-butyl-2-arylpyridines (Scheme 8) ⟨92SC351⟩. The starting material, 4,4-dimethyl-1-piperidino-pentan-3-one, was quaternised with methyl iodide. Reaction with various aryl β-ketoesters in the presence of base provided the crude substituted products which were hydrolysed and decarboxylated to give the 1,5-diketones in fair yields. Cyclization provided the 6-*t*-butyl-2-arylpyridines.

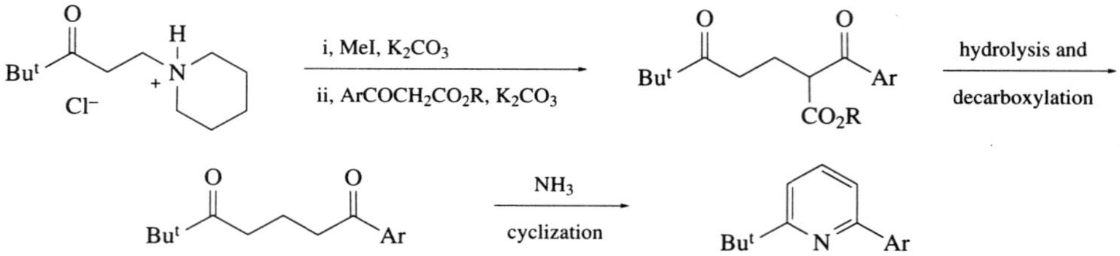

Scheme 8

Benzyl- and allyltriethylammonium halides can undergo catalytic carbonylation to give the corresponding acids in the presence of dicobalt octacarbonyl ⟨83JOC1919⟩. These reactions were carried out at 65 °C under a slow stream of carbon monoxide. Irradiation using a sunlamp was necessary (Scheme 9).

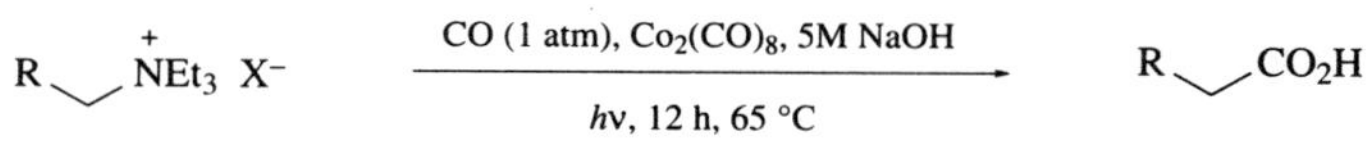

R	*X*	*Yield* (%)
C_6H_5	Br	85
2-MeC_6H_4	Cl	85

Scheme 9

An allylic quaternary ammonium salt can be prepared from diisopropylethylamine by addition to methallyl phenyl sulfoxide and trimethylsilyl triflate in dry dichloromethane at 0 °C. This salt has been shown to alkylate sodium dimethyl malonate regioselectively in THF at room temperature, using tetrakis(triphenylphosphine) palladium(0) as a catalyst to give the product in 60% yield (Scheme 10) ⟨88TL2257⟩.

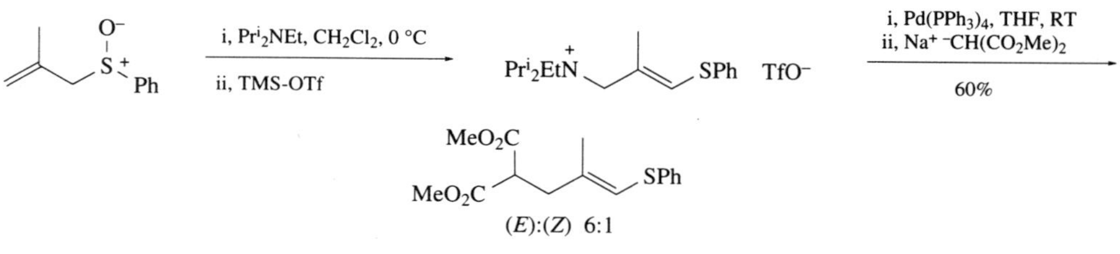

Scheme 10

2-Azanorbornene-2-spiro-1′-aziridinium triflate can be used for the aminoethylation or *N*-methyl-aminoethylation of carbonyl compounds. This spiroaziridinium salt can be prepared by an aqueous Diels–Alder reaction between cyclopentadiene and the immonium ion derived from formaldehyde and 2-bromoethylamine hydrobromide; subsequent treatment with silver triflate provides the spiro-aziridinium triflate which is extremely stable at ambient temperature in the absence of moisture ⟨92TL4401⟩. The ability of 2-azonianorbornene-2-spiro-1′-aziridinium triflate to function as an alkylating agent is well illustrated by its reaction with the enolate derived from a bicyclic lactone. This reaction provided the alkylated lactone in 74% yield (Scheme 11). This spiroaziridinium triflate has been used to alkylate a variety of enolates in good yields.

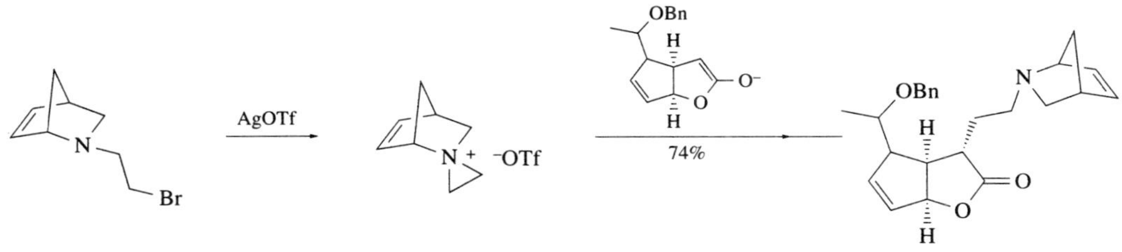

Scheme 11

1,2-bis-(Cyanomethyl)arenes are useful synthetic building blocks for a variety of benzazepine and indene derivatives which have important biological activities. They can also be used in the preparation of benzobiphenylenes and cyclo-octatetraenes. The preparation of 3-methoxy-1,2-bis-(cyanomethyl)benzene from 3-methoxy-1-(dimethylaminomethyl)benzene involves a cyanide displacement of the dimethylamino unit of 3-methoxy-2-hydroxymethyl-1-(dimethylaminomethyl)benzene, via the benzyl chloride, itself produced *in situ* by displacement of a quaternary ammonium salt (Scheme 12) ⟨90SC3347⟩. Concurrent displacement of both the dimethylamino and hydroxy groups occurred when a hydroxymethyl substituent was sited *ortho* to the alkoxy group.

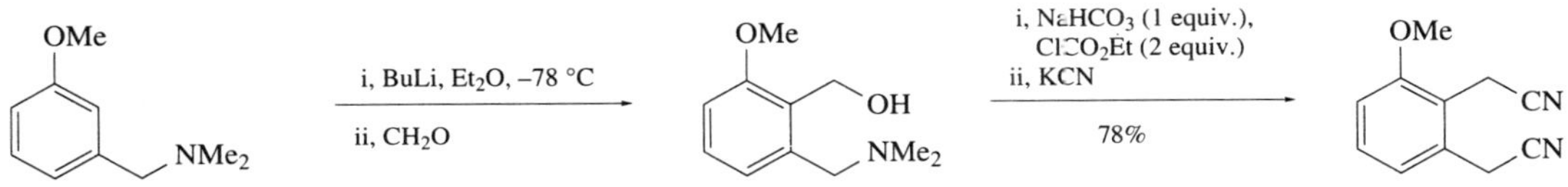

Scheme 12

The condensation of a quaternary ammonium salt derived from 2-thiolo-3-diethylaminomethylimidazole with ^{14}C-labelled ethyl acetamido cyanoacetate in the presence of sodium ethoxide was used to produce labelled 2-methylthiohistidine after hydrolysis and decarboxylation in 83% yield ⟨58JA4405⟩. Greater yields were obtained by the use of ion-exchange resins for purification rather than by isolation via the mercury complex as had previously been carried out. Cleavage with sodium in liquid ammonia provided labelled 2-thiolohistidine, a key intermediate for the synthesis of ergothionene (Equation (8)).

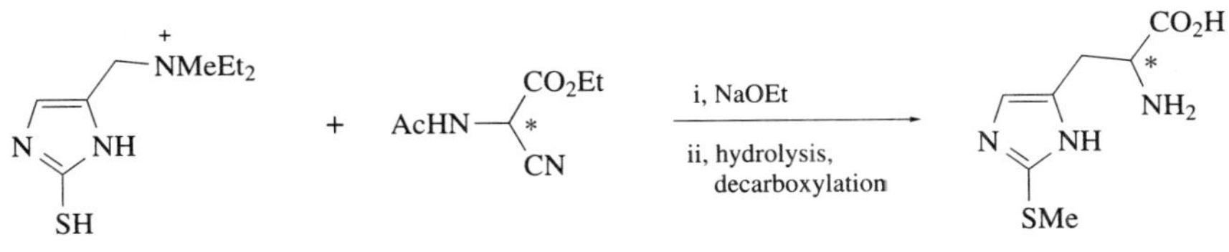

(8)

1.06.1.3 Tertiary Amines

Classically, Mannich reactions of enolisable carbonyl compounds to yield *β*-aminoketones have involved treatment of an acidic mixture of formaldehyde and a simple secondary amine such as dimethylamine with the ketone in ethanol under reflux for several days. Alternatively, the reaction of enolates or silyl enol ethers with iminium ions such as *N*,*N*-dimethylmethyleneiminium ions may be used. *β*-Aminoketones have also been synthesized from aminals; for example, reactions between boron enolates and *N*,*N*,*N*′,*N*′-tetramethyldiaminomethane (TMDM) are often used for the generation of *N*,*N*-dimethylmethyleneiminium ions (Equation (9)) ⟨90CC1532⟩. The mechanism involves a Mannich process, in which the boron atom activates the aminal by functioning as a Lewis acid. The reactions do not proceed with the less Lewis acidic lithium enolate. Reduced yields were however obtained for dimethylaminomethylation at sterically hindered positions of ketones using this method.

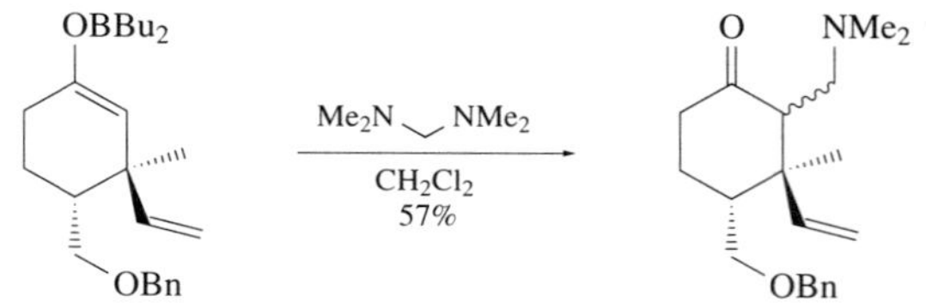

(9)

Tertiary amines have been used in a number of alkylation reactions. For example, a β-aminoketone formed by reaction of an aryl methyl ketone with paraformaldehyde and dimethylamine hydrochloride was used to alkylate potassium cyanide with displacement of dimethylamine to give a β-aroylpropionitrile (Scheme 13) ⟨58JOC1281⟩. Only amines possessing an aryl group adjacent to the carbonyl group undergo this displacement. The reaction has been used to prepare pyrrolidines which are potential antihypertensive agents ⟨47JCS1190⟩.

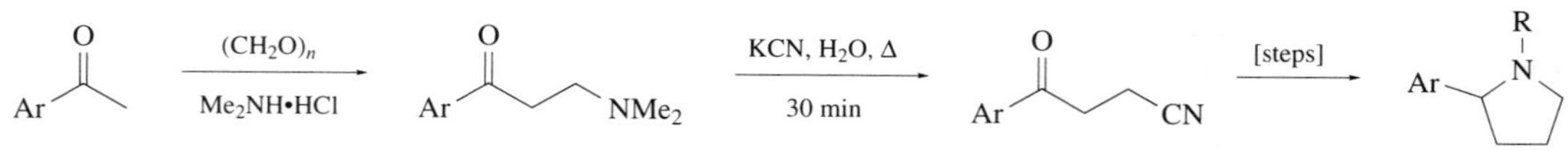

Scheme 13

A further example is the alkylation of indole with diethylaminoacetonitrile, which provides a convenient method for the preparation of indole-3-acetonitrile on a small scale (Equation (10)) ⟨53JA3589⟩. Diminished yields were obtained upon attempted scale up. The authors suggest that this is due to the great sensitivity of the cyanomethylation reaction to temperature fluctuations.

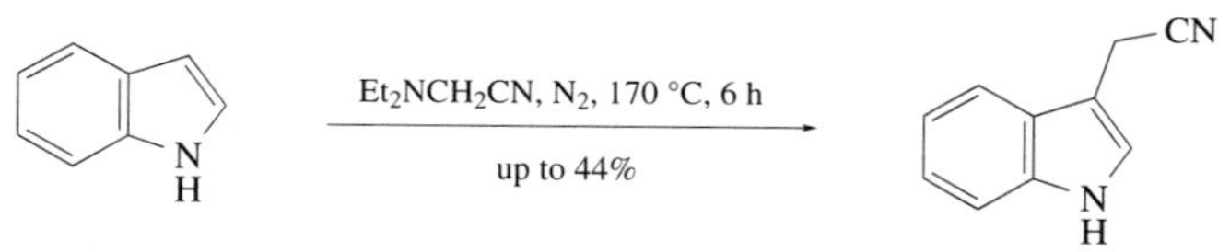

(10)

N-Unsubstituted α-alkylated pyrrole-α-acetonitriles are a difficult class of compounds to prepare, but are potentially useful in indole synthesis. One viable route involves treatment of α-branched pyrrole tertiary amines with *p*-toluenesulfonic acid in the presence of excess sodium cyanide in acetonitrile–water solution at reflux (Scheme 14). These nitriles are thought to arise through an elimination–addition mechanism, via the azafulvenes. The α-branched tertiary amines were produced by reaction of 6-dimethylamino-1-azafulvene with an excess of various alkyl lithium reagents ⟨87SC863⟩.

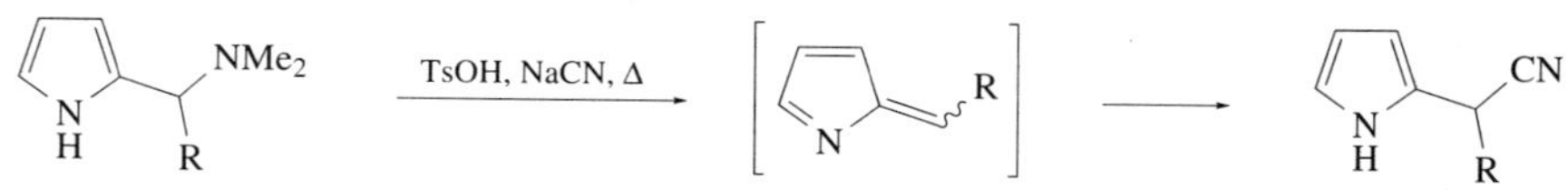

R	*Yield* (%)
Bu^t	78
Bu^s	96
H	70

Scheme 14

An unusual example of alkylation involving tertiary amines is that of 3-(dimethylaminomethyl)indole (gramine) with triethoxycarbonylmethane, again through an elimination–

addition process (Scheme 15) ⟨49JA663⟩. Further treatment of gramine with boiling 95% aqueous ethanolic sodium cyanide for eighty hours gave indole-3-acetamide, which could be smoothly converted into indole-3-acetic acid in high yield by hydrolysis with sodium hydroxide (Scheme 15) ⟨48JA3770⟩.

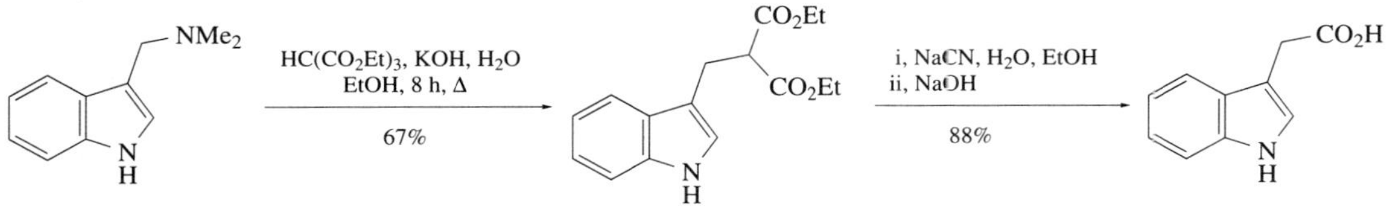

Scheme 15

Similar base catalysed alkylation of a Reissert compound with gramine in the presence of sodium occurs to give 1-cyano-1-skatyl-2-benzoyl-1,2-dihydroisoquinoline in 46% yield ⟨50JA2134⟩. Further basic hydrolysis provides 1-substituted isoquinolines in quantitative yields (Equation (11)).

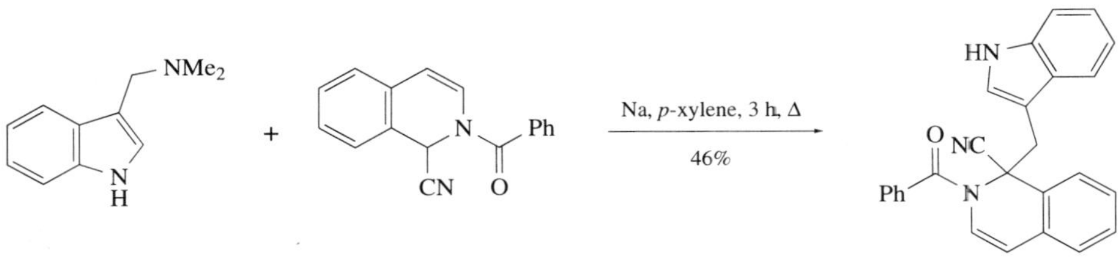

(11)

3-Substituted 2-*exo*-methylene-cycloalkanones can be prepared from 2-(nitro-methyl) cyclohexenones by conjugate addition of organocuprates followed by elimination of the nitro group. This methodology has been extended to asymmetric synthesis using chiral amines in place of the nitro group. The required (*S*)-2-(2-(methoxymethyl)-1-(pyrrolidinyl)methyl)-2-cyclohexen-1-one can be prepared in 89–94% yield from 2-(nitromethyl)-2-cyclohexen-1-one and (*S*)-2-(methoxymethyl)-pyrrolidine. Optically active 3-substituted 2-*exo*-methylene-cyclohexanones were obtained in 90% enantiomeric excess by zinc bromide assisted asymmetric conjugate addition of lithium diorganocuprates, followed by elimination of the pyrrolidine in aqueous work-up (31–72% yields) (Scheme 16). The absence of zinc bromide did not affect the *ee* but was found greatly to decrease the yields ⟨90JOC408⟩.

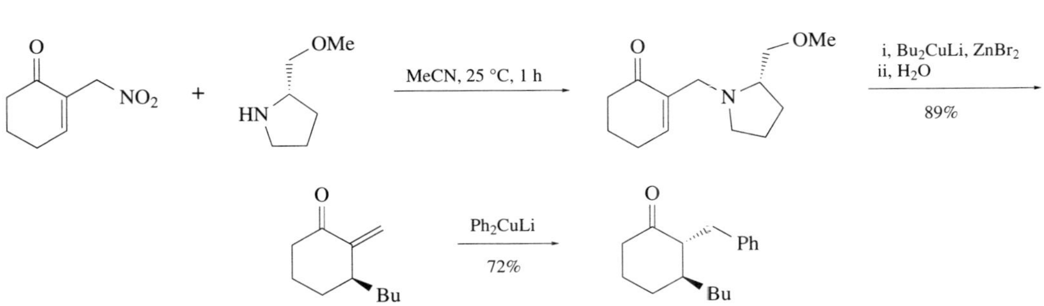

Scheme 16

Triflates have also been used to convert primary amines into reactive leaving groups for nucleophilic substitution. Lithium dimethyl and diphenyl cuprates have been found to react with *N*,*N*-bistrifluoromethanesulfonimides of benzylic or allylic primary amines to give the substitution products (Scheme 17) ⟨78TL4727⟩. The reactions are specific for benzylic and allylic bistrifluoromethanesulfonimide groups and are carried out with five equivalents of cuprate in ether or ether/THF as solvent. Investigation into the mechanism of these substitution reactions concluded that direct displacement at the allylic position seemed more probable than displacement at the terminus followed by rearrangement of the double bond.

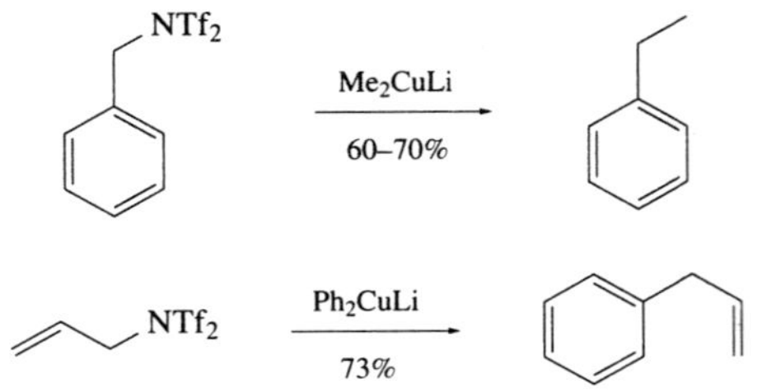

Scheme 17

1.06.1.4 Secondary Amines

Investigation into the alkylating properties of the secondary amines *N*-(α-(3-indolyl)ethyl)-*N*-isopropylamine, *N*-(α-(3-indolyl)methyl)-*N*-*t*-butylamine, and *N*-(α-(3-indolyl)ethyl)-*N*-*t*-butylamine with dialkyl and diaryl acetamidomalonates showed most rapid alkylation using *N*-(α-(3-indolyl)ethyl)-*N*-*t*-butylamine (Equation (12)) ⟨57JA2217⟩. The reactions could be followed semi-quantitatively by titration of the volatile amines evolved. All three indoles gave excellent yields and required lower temperatures or shorter reaction times than those reported for gramine, perhaps because of the bulky groups attached to the acyclic nitrogen atom.

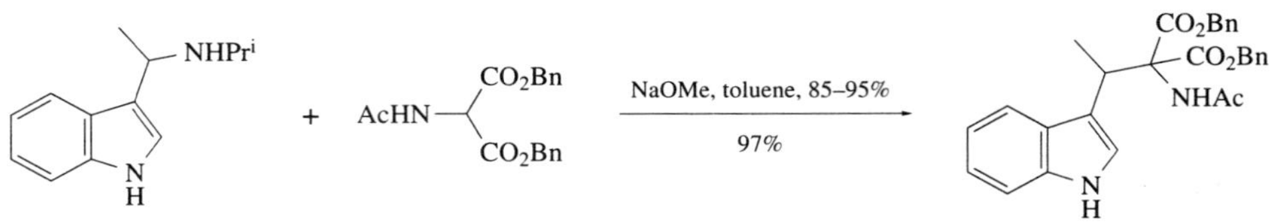

(12)

In a different type of reaction, the secondary amine *N*,*N*-diphenyl trimethylenediamine breaks down smoothly at 230–250 °C, in the presence of hydrobromic acid, to give aniline and 1,2,3,4-tetrahydroquinoline ⟨60JOC463⟩. Decomposition of the monohydrobromide was found to be fastest using 0.1 M hydrobromic acid. The proposed mechanism involves intramolecular nucleophilic displacement of the ammonium salt (Scheme 18).

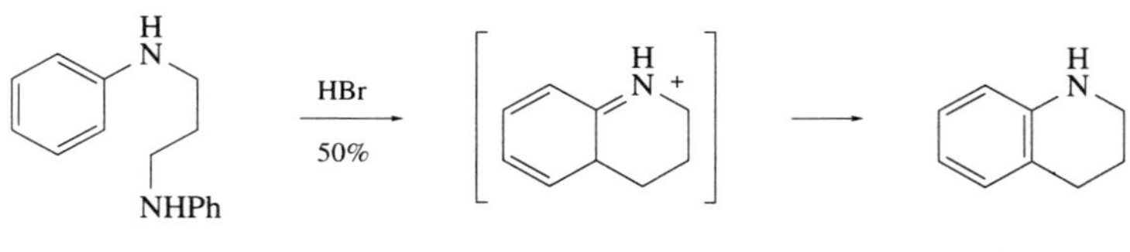

Scheme 18

1.06.1.5 Diazotization of Primary Amines

Diazotization with formation of a new carbon–carbon bond involves the generation of a diazonium ion, commonly from a primary amine by the action of nitrous acid followed by nucleophilic displacement. The diazonium ion intermediate can decompose with loss of nitrogen; the incipient carbocation may be trapped by reaction with an external nucleophile, the classical reaction for aromatic amines, or may rearrange or decompose, a common fate for aliphatic amines ⟨76AG(E)251, 87CL1163⟩. The rate-determining step in the case of aliphatic primary amines is thought to be the formation of a nitrosamine which isomerises to a diazonium salt in the presence of acid. This salt then rapidly loses nitrogen with attack by an external nucleophile to form the carbocation. The process has been thoroughly reviewed ⟨B-68MI 106-01, B-78MI 106-01, B-86MI 106-01⟩.

A representative diazotization followed by attack by an external nucleophile is provided by the reaction of the hydrochloride salt of a protected 2-aminomethyl pyrrole with nitrous acid and a porphobilinogen lactam ester (Scheme 19) ⟨71JA2738⟩.

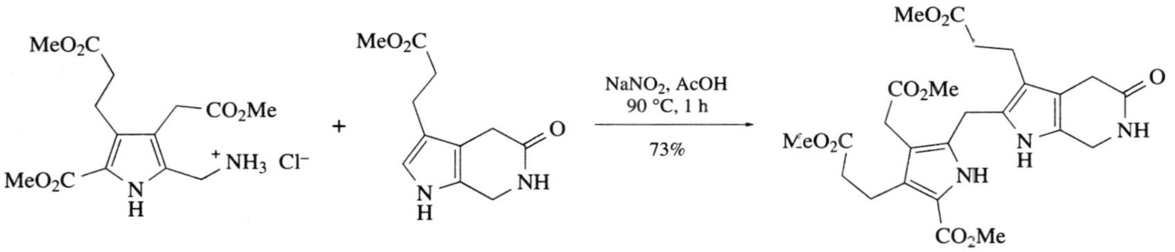

Scheme 19

An unusual and interesting example with intramolecular carbon–carbon bond formation is seen in the diazotization of 9-aminomethyl-1,4-dimethyltriptycene, carried out by treatment of the hydrochloride salt with ten equivalents of sodium nitrite in acetic acid at room temperature for fifteen hours. The production of the carbocation at the α-position of the bridgehead and the presence of the methyl group at one of the *peri*-positions of the triptycene skeleton are found to favour intramolecular insertion as opposed to skeletal rearrangement. The transformation is shown in Equation (13) ⟨87CL1163⟩. In this case the intermediate carbocation inserts into a C—H bond of the 1-methyl group as opposed to suffering attack by a nucleophile. The reaction is thought to take place via a pentacoordinated species which undergoes proton loss, favoured by the close proximity of the participating carbon atoms.

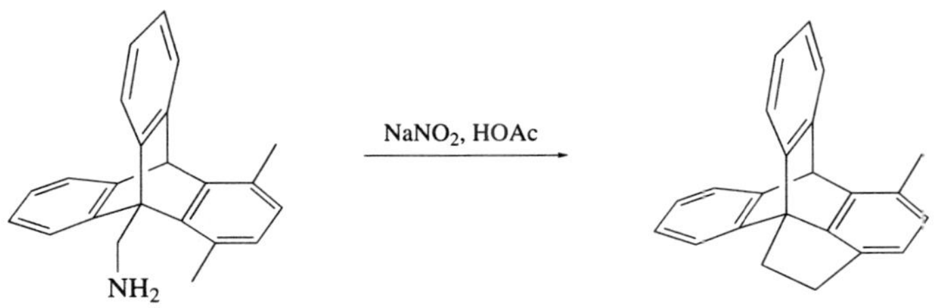

(13)

A further intramolecular example thought to involve a Wagner–Meerwein rearrangement is observed when 18-amino-5α-pregnane-3β,20-diol is treated with nitrous acid in acetic acid to give 14β,18-cyclo-5α-pregnene-3β,20-diol (Scheme 20) ⟨64T1259⟩. Wagner–Meerwein rearrangement to give the tertiary carbenium ion is followed by attack on the carbocation by the other bridgehead carbon atom with loss of a proton.

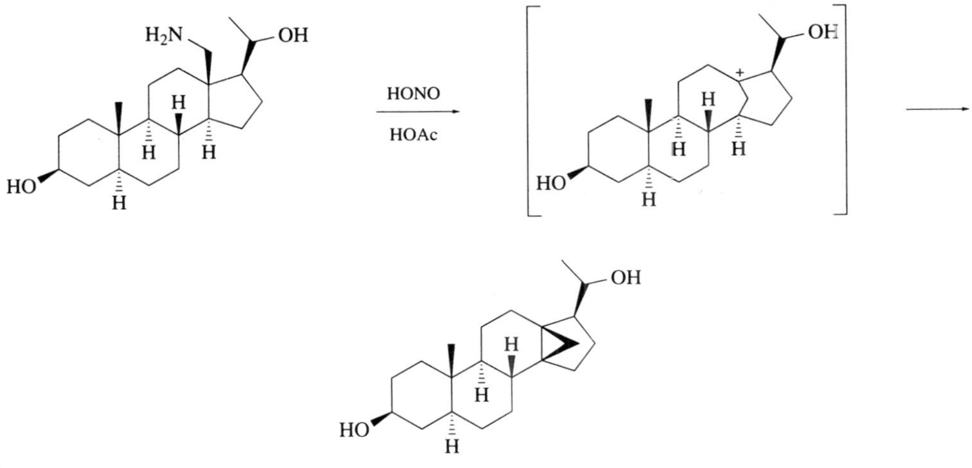

Scheme 20

1.06.1.6 Functionalized Benzotriazoles

Functionalized benzotriazole moieties are well known as versatile synthons for the introduction of heteroalkyl groups in reactions with carbon nucleophiles (Equation (14)), and there are therefore many examples of carbon–carbon bond formation through carbon–nitrogen bond cleavage with displacement of benzotriazole units ⟨91JOC4439⟩. The methodology, developed by Katritzky, has proved popular, no doubt partly due to the ease of preparation of the substituted benzotriazole reagents from readily available starting materials. Reactions with carbon nucleophiles generally proceed under mild conditions, and the benzotriazole unit may be recovered at the end of the reaction ⟨89JCS(P1)225⟩. Amines ⟨87JCS(P1)805⟩, hydrazines ⟨89JCS(P1)2297⟩, aminoesters ⟨89S323, 89S747⟩, aminoketones ⟨90T987⟩, ethers ⟨89JOC6022⟩, sulfides ⟨91HCA1924⟩, amides and thioamides ⟨88JOC5854⟩ have all been prepared using this chemistry.

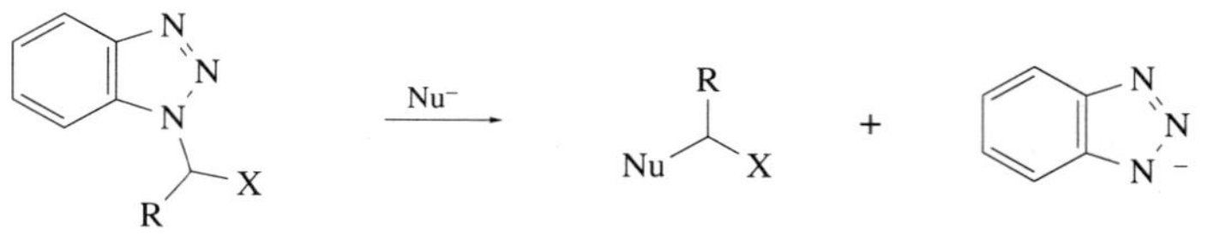

(14)

In the simplest cases, aromatic secondary amines may be prepared from 1-(1-(arylamino) alkyl)benzotriazoles using Grignard reagents (Scheme 21) ⟨87JCS(P1)805⟩.

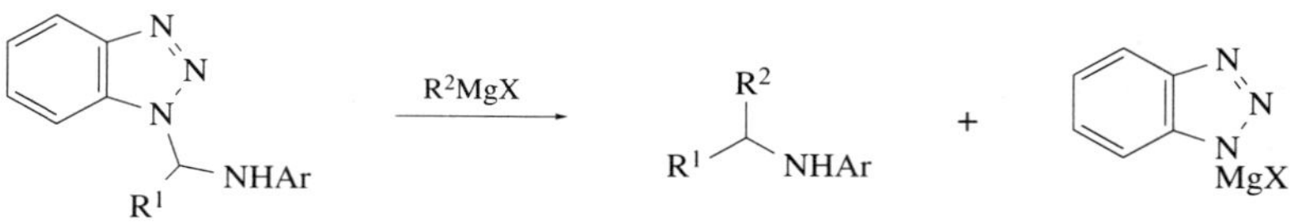

R¹	*R²MgX*	*Yield (%)*
Prⁱ	MeMgI	80
Me	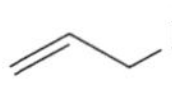 MgBr	62
H	BnMgCl	78

Scheme 21

This methodology utilising Grignard reagents has been extended to the syntheses of many types of compound, including 1,1-disubstituted hydrazines, 2-acyl-1,1-disubstituted hydrazines ⟨89JCS(P1)2297⟩, amides and thioamides ⟨88JOC5854⟩. Similarly, symmetrical secondary amines can be prepared by reaction of bis(benzotriazol-1-ylmethyl)amine with Grignard reagents at room temperature (Equation (15)) ⟨89JCS(P1)225⟩.

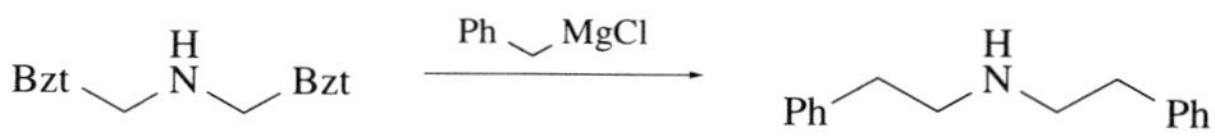

(15)

Bzt = 1-benzotriazolyl

A similar reaction takes place between bis(benzotriazol-1-ylmethyl)hydroxylamine and Grignard reagents or lithium phenylacetylide, giving *N*,*N*-disubstituted hydroxylamines in high yields (Scheme 22). It is thought that a nitrone is the intermediate species.

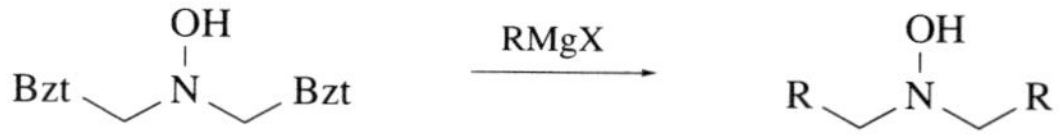

RMgX	*Yield (%)*
PhMgBr	94
Ph—≡—Li	87

Scheme 22

N,*N*-Dialkyl-2-alkynamines are of considerable pharmaceutical interest and may be prepared by reaction of 1-(dialkylaminoalkyl)benzotriazoles with alkynyl lithiums at 20 °C (Scheme 23) ⟨89S31⟩.

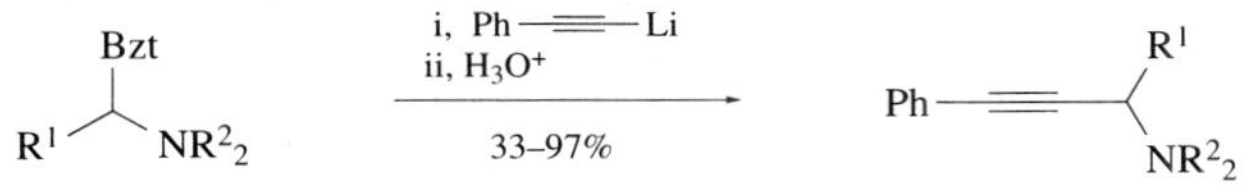

R¹	*R²*	*Yield (%)*
H	Me	79
Pr^i	Bn	95

Scheme 23

Nucleophilic displacement of dialkylamino (1-benzotriazolyl) acetic esters by a carbanion such as an organozinc derivative gives 2-dialkylaminoalkanoic esters in good yield (Equation (16)) ⟨89S323⟩. In this case, Grignard reagents give poor results, perhaps due to deprotonation and/or reaction of the Grignard reagent at the ester group. However, organozinc reagents react smoothly at 20–25 °C to give the amino esters in good yields. A general method for the preparation of β-aminoesters has also been reported using zinc powder and chloromethylsilane. Reaction with α-1-dialkylamino-1-(1-benzotriazolyl)alkanes and bromoesters gives the desired products in good yields ⟨89S747⟩.

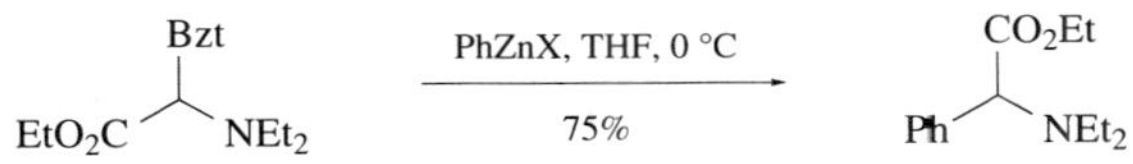

(16)

N-Substituted 1-(benzotriazolyl)-2,2-diethoxyethylamines react with Grignard reagents to give the corresponding 2-aminoalkanol diethyl acetals. Hydrolysis provides the corresponding α-amino-aldehydes (Scheme 24) ⟨90S1173⟩.

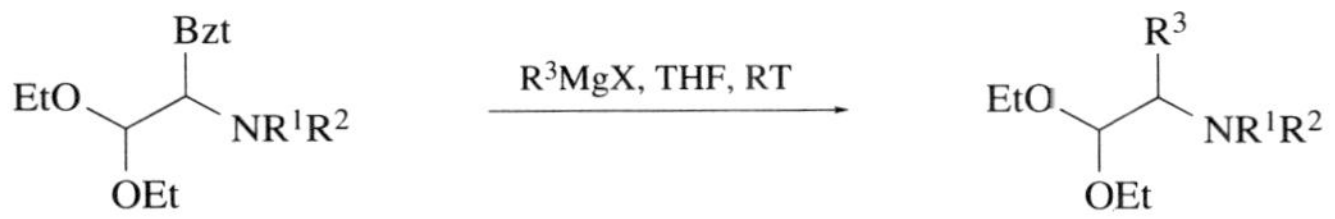

R¹	*R²*	*R³*	*Yield (%)*
Ph	H	Ph	71
Bn	Bn	Et	96
Ph	Me	Et	91

Scheme 24

Symmetrical vicinal tertiary and secondary diamines are important in polyazamacrocycle synthesis and in metal chelation, and are useful intermediates for the construction of nitrogen heterocycles. A simple two step procedure for their preparation using benzotriazole chemistry has been reported ⟨90JOC3209⟩. Double condensation of an amine with glyoxal and benzotriazole provides the bis(benzotriazolyl) products in high yields. Treatment with two equivalents of Grignard reagent induces displacement of both benzotriazole units, giving the vicinal diamines (Scheme 25).

Grignard or organolithium reagents can be used in the synthesis of primary amines by reaction

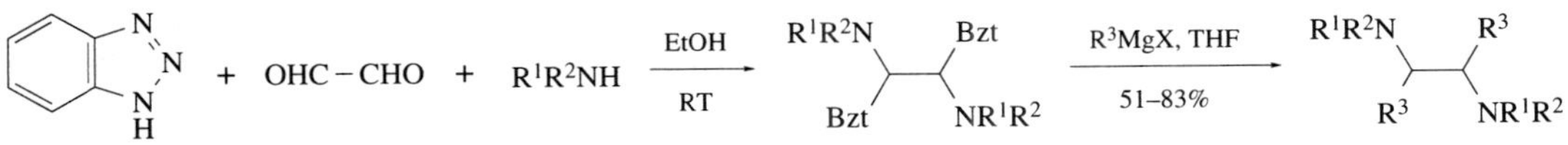

R¹R²NH	*R³MgX*	*Yield (%)*
BnNH	PhMgBr	81
BnNH	MeMgBr	83
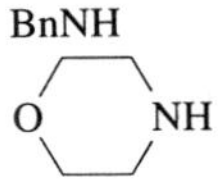	Bu^{n}MgCl	68

Scheme 25

with *N*-triphenylphosphorylidene-1-(benzotriazol-1-yl) methylamine (Scheme 26), which can be prepared from the readily available 1-azidomethylbenzotriazole by Staudinger phosphorylation ⟨89TL3303⟩. The resulting phosphazenes can be hydrolysed during basic workup to give primary amines, which may be converted into hydrochlorides for isolation or subjected *in situ* to various phosphazene transformations. Alkyl, cycloalkyl, aralkyl, aryl and heteroaryl Grignard reagents and lithium phenylacetylide give equally good yields.

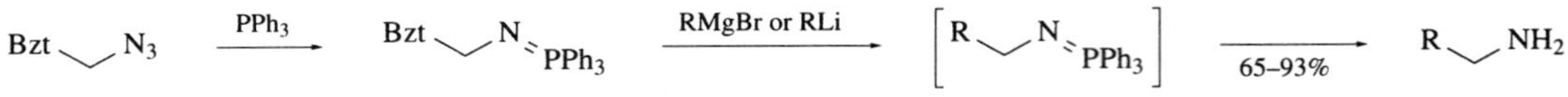

Organometallic	*Yield* (%)
n-$C_{12}H_{25}MgBr$	87
Ph—≡—Li	86
c-$C_6H_{11}MgBr$	77

Scheme 26

Symmetrical and unsymmetrical bis-heterocyclic alkanes, which are important in the food industry, in dyestuff chemistry and as intermediates in the synthesis of various heterocyclic macromolecules, have been prepared using benzotriazole chemistry. These compounds have proved difficult to synthesize, but the use of benzotriazole intermediates has provided a general and convenient method for their preparation ⟨93JOC4376⟩. Symmetrical bis(heteroaryl)alkanes can be prepared in excellent yields by reaction of methyl *N*-(α-benzotriazol-1-ylalkyl) carbamates with an excess of 2-methylthiophene or 2-methylfuran in dichloromethane at room temperature using $ZnCl_2$ as the Lewis acid catalyst (Scheme 27). Unsymmetrical bis-heterocycles can be formed by isolation of the monosubstituted intermediate and further reaction with a different heterocyclic molecule. This methodology provides a route to a wide range of bis(heteroaryl)alkanes in excellent yields (Scheme 28).

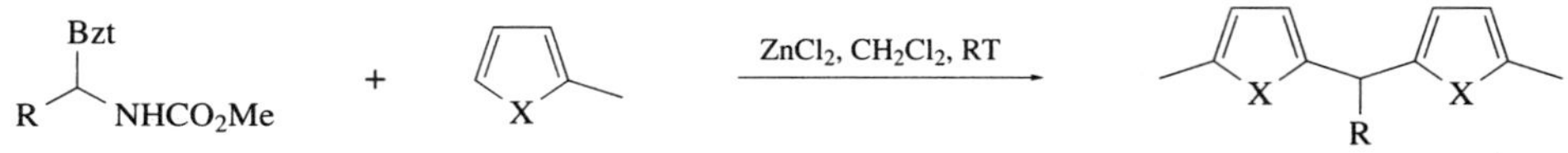

X	*R*	*Yield* (%)
O	Pr^i	88
O	Ph	90
S	Pr^i	90
S	Ph	94

Scheme 27

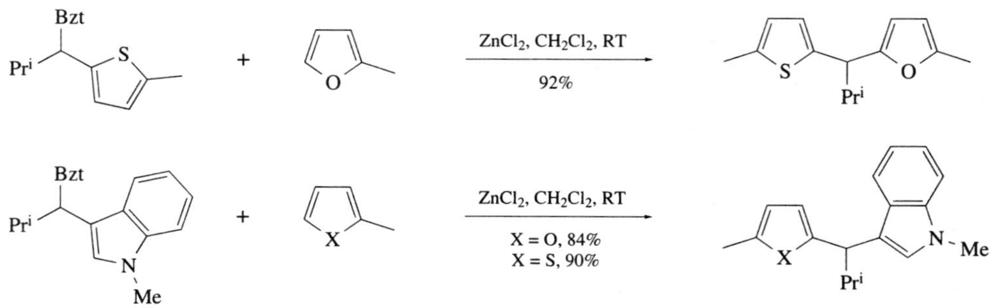

Scheme 28

Grignard reagents can also displace the benzotriazole group of various α-benzotriazoyl methyl ethers to give the corresponding methyl ethers (Scheme 29) ⟨91JCS(P1)3295⟩. It has been found that α-(*N*-benzotriazolyl) undecyl, benzyl and octyl methyl ethers react smoothly with butylmagnesium iodide to give the expected products in 57–68% yield. Octylmagnesium iodide was found to react in a similar way with α-(*N*-benzotriazoyl) butyl methyl ether. Ethers with the general formula

R^1R^2CHOMe can thus be formed by nucleophilic attack onto the carbocations formed on loss of the benzotriazole group. The heteroatom, which is separated from the leaving group by one carbon atom, assists in the departure of the benzotriazole anion.

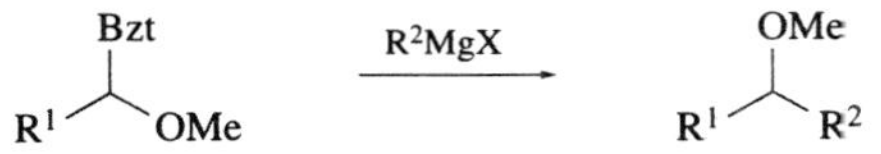

R^1	R^2MgX	*Yield* (%)
Bn	BuMgI	68
Bu	n-$C_8H_{17}MgI$	60
$C_{10}H_{21}$	BnMgBr	50

Scheme 29

Other examples in which the benzotriazole group is displaced include the preparation of *t*-alkyl sulfides by reaction of α-1-benzotriazolylsulfides with Grignard reagents (Equation (17)) ⟨91HCA1924⟩.

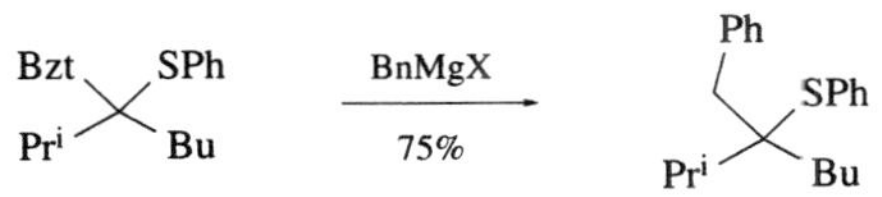

(17)

1.06.1.7 Azides

Carbon–carbon bond forming reactions via azides seem to be relatively uncommon. β-Azido substitution can be achieved in cyclic enol ethers in excellent yields by treatment of several tips enol ethers with trimethylsilyl azide and iodosobenzene in dichloromethane solution at −15 to −18°C. The products were isolated in place of the expected α-azidocyclohexanone and are thought to arise via enonium ions ⟨92JA3993⟩. The β-azido enol ether reacts with carbon nucleophiles in the presence of Lewis acids with azide displacement to form the nucleophilic substitution products without loss of the enol ether moiety, again via the enonium ions (Scheme 30).

1.06.1.8 Ring Opening of Aziridines

Aziridines have been used on a number of occasions as electrophiles in reactions with carbon nucleophiles. It has been shown that boron trifluoride etherate promotes nucleophilic ring opening of *N*-substituted aziridines by diorganocopper lithium reagents, providing a route to primary and secondary amines (Scheme 31) ⟨85TL1153⟩.

A cuprate reagent has also been used in the absence of Lewis acid for such a carbon–carbon bond formation, in a stereocontrolled synthesis of (±)-perhydrohistrionicotoxin. The key step involved reaction of an acylaziridine with lithium dibutylcopper to afford the lactam (Equation (18)) ⟨75JOC2009⟩.

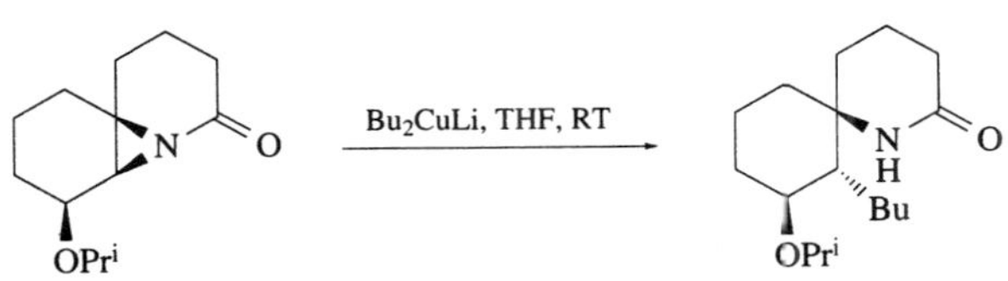

(18)

Aziridines are activated towards cleavage by nucleophiles by the presence of an electron-withdrawing group (such as tosyl or *t*-butoxycarbonyl) bonded to the nitrogen atom. For example, reaction of *t*-butyl 2(*S*)-*N*-tosylaziridine carboxylate with copper catalysed Grignard reagents gave the ring opened products in greater yields than those observed for the corresponding unactivated aziridines. A reduction product, perhaps arising through hydride transfer, was also observed as a by-product in most reactions with the activated substrates. The C3—N cleavage was favoured over the C2—N cleavage for the bulky butyl, ethyl and isopropyl reagents, while the reverse proved true

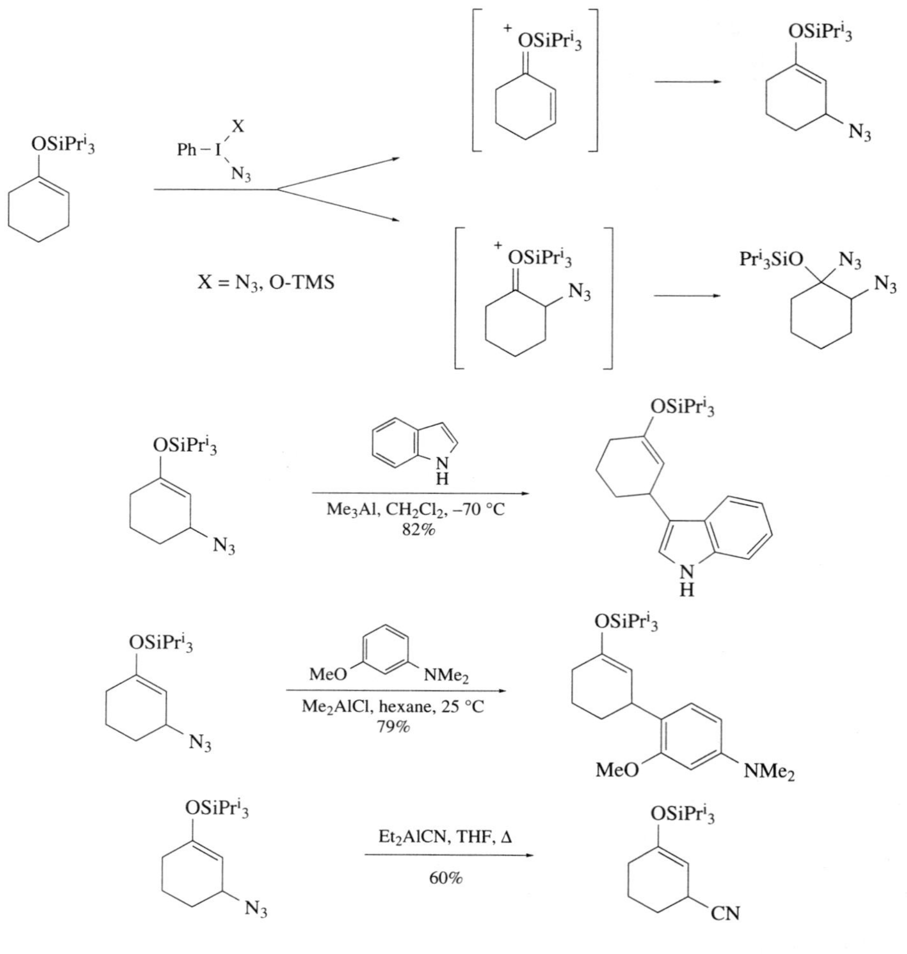

Scheme 30

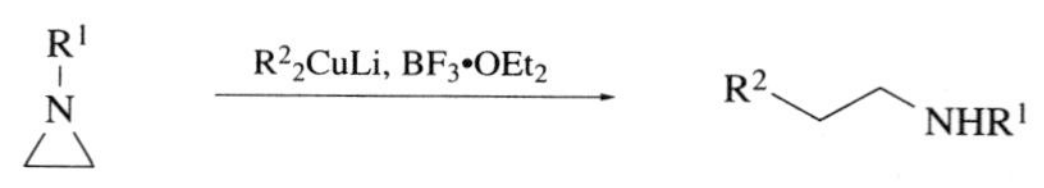

R^1	*R^2*	*Yield (%)*
Bn	Me	80
Me	Bu	94
DMB	Ph	80

DMB = 4,4'-dimethoxybenzhydryl

Scheme 31

for the methyl cuprate reagent. The C3—N cleavage products were converted into the parent amino acids in high optical yields (>95% *ee*) (Scheme 32) ⟨89CC1852⟩.

Highly selective nucleophilic opening of oxiranes by cyanotrimethylsilane under lanthanide catalysis has been reported. The reaction was also applied to aziridines derived from easily accessible α-amino acids. Optically pure 2-substituted aziridines react with cyanotrimethylsilane in the presence of ytterbium trinitrile to give *β*-amino nitriles in optically pure form by selective attack of cyanide at the less substituted carbon atom of the ring (Scheme 33) ⟨90TL6379⟩. The authors suggest that this reaction provides a new route for the synthesis of optically active *β*-lactams, which could be formed from the *β*-amino acids derived, in turn, from the optically active *β*-amino nitriles.

Almost quantitative cleavage of *N*-tosylaziridine rings can be achieved with dianions of glyoxylic

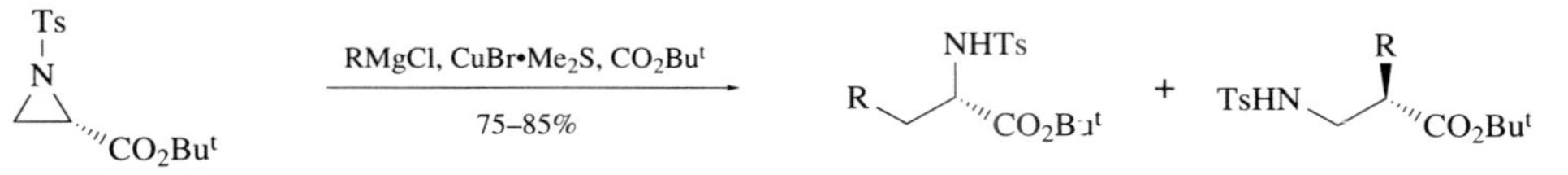

R	*Ratio* **A**:**B**
Pr^i	47:28
Me	30:55

Scheme 32

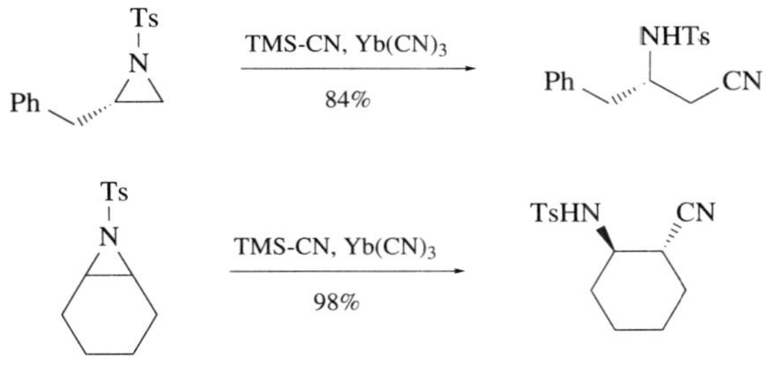

Scheme 33

acid derivatives, such as bis(ethylthio)acetic acid, using potassium as the counterion (Equation (19)) ⟨80CJC716⟩. The 2,2-bis(ethylthio)carboxylic acids so formed were easily converted into the corresponding α-ketoacids using *N*-bromosuccinimide.

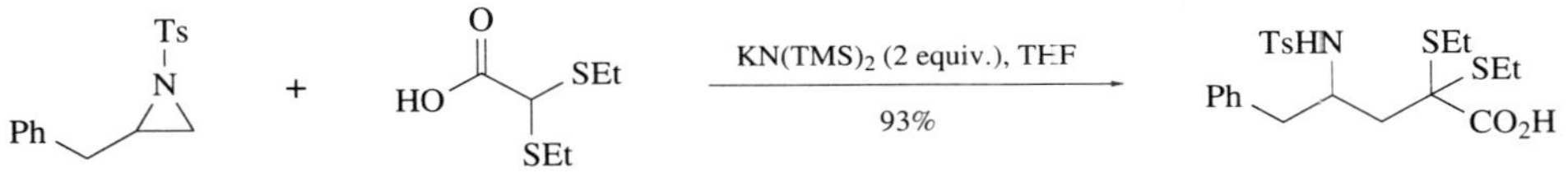

(19)

1.06.1.9 Radical Mediated Processes

Radical mediated cleavage of carbon–nitrogen bonds with formation of carbon–carbon bonds is unusual.

One example involves the use of 2-cyanopropyl radicals to introduce a tertiary carbon atom bearing an electronegative group into the C9 position of an anthracene derivative. 9-*t*-Alkyl anthracenes are usually prepared by the Grignard reaction of *t*-alkyl-magnesium halides with anthrone followed by dehydration. Cyano- and alkoxycarbonyl groups are however too reactive to be introduced by this process. Oki and co-workers have found that the action of 2-cyanopropyl radicals upon anthrone, especially under basic conditions, provides the substitution product in 88% yield. The 2-cyanopropyl radicals were generated by thermal decomposition of azobisisobutyronitrile (AIBN) (Scheme 34) ⟨77TL2441⟩. Reduction with sodium borohydride affords a mixture of geometrical isomers of the alcohol; reaction with phosphorus pentoxide then yields the desired 9-substituted anthracene as the major product.

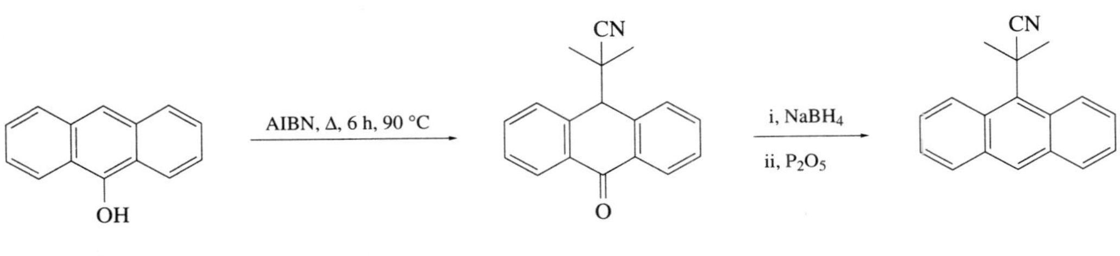

Scheme 34

In an alternative reaction, treatment of anthrone with dimethyl azobisisobutyrate using the conditions described above was used to produce the substitution product (**1**) in 62% yield (Equation (20)).

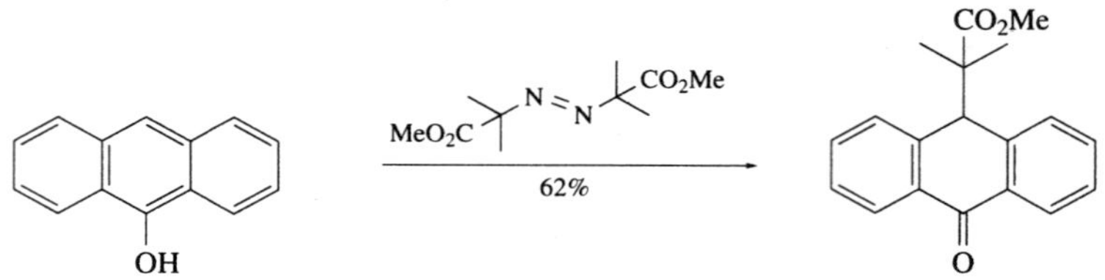

(**1**)

(20)

Kinetic studies into the isonitrile-nitrile rearrangement by flash-pyrolysis using inhibitors has led to the proposal of a free radical chain mechanism ⟨84TL3441⟩. Flash-pyrolysis gives almost quantitative yields of nitriles with complete retention of stereochemistry (Scheme 35).

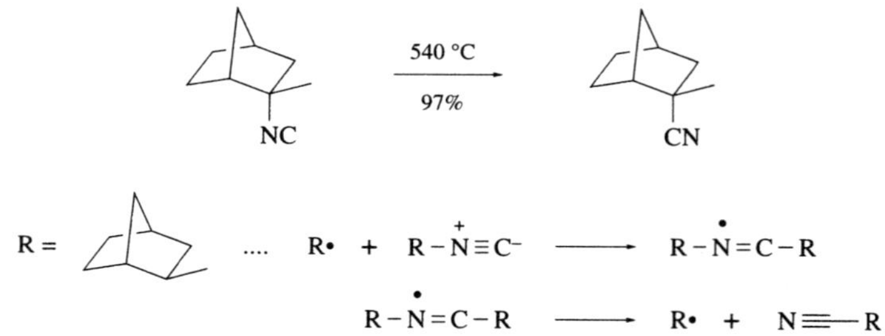

Scheme 35

1.06.1.10 Rearrangements

The following reactions illustrate rearrangements in which new carbon–carbon bonds are formed by substitution of carbon–nitrogen bonds.

Iodide ion-mediated rearrangement of *N*-vinylaziridines possessing a carbonyl group at the β-position occurs to give pyrrolines (Scheme 36) ⟨67JA3600⟩. The mechanism of this reaction is presumably similar to that proposed by Heine for iodide ion catalysed rearrangement of *N*-acylaziridines ⟨63JA2743⟩, involving initial nucleophilic attack by halide ion at an aziridine carbon atom with ring cleavage. Internal alkylation of the resulting negative charge with expulsion of iodide then occurs.

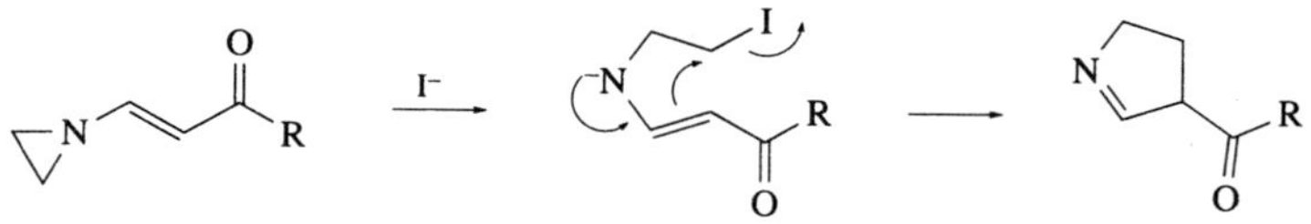

Scheme 36

The key step in a total synthesis of the alkaloid (±)-crinine involves just such a rearrangement of a highly functionalised *N*-vinylaziridine, induced by treatment with anhydrous sodium iodide in diethylene glycol dimethyl ether at 145 °C (Equation (21)) ⟨67JA3600⟩.

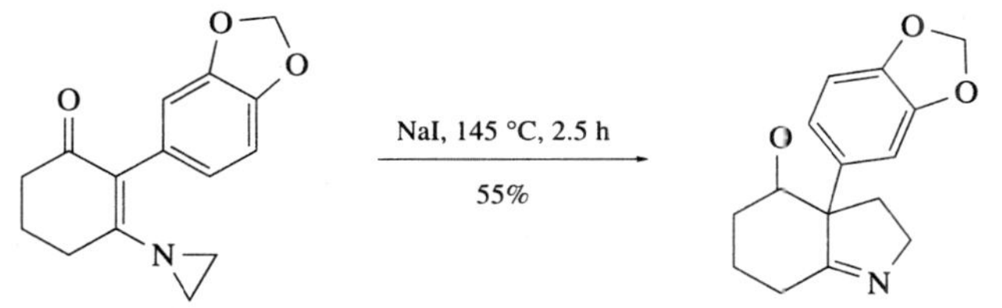

(21)

Phosphorus-substituted heterocycles can be obtained by rearrangement of phosphonium substituted vinyl aziridines, prepared by addition of aziridines to prop-2-ynyl triphenylphosphonium bromide or 2-phenylethynyl triphenylphosphonium bromide in methylene chloride (Scheme 37) ⟨78JOC4207⟩. Heating these activated aziridines under reflux in acetonitrile induces rearrangement to give the phosphonium substituted 2-pyrrolines in good yields, in a similar process to the iodide-mediated rearrangement discussed above. (2-Methyl-2-pyrrolin-3-yl) triphenylphosphonium bromide has been found to undergo both *N*-alkylation and Wittig reaction with aldehydes in the normal manner ⟨78JOC4207⟩.

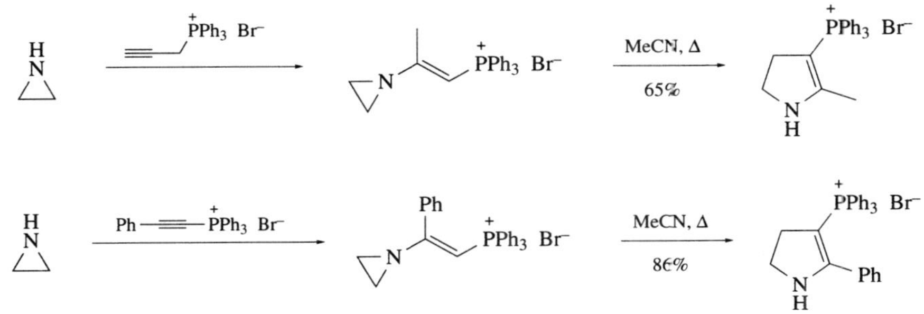

Scheme 37

1.06.1.11 Electrophilic Substitution

As discussed above, there are many examples of nucleophilic substitution of nitrogen containing groups with attack of a carbon nucleophile, thus introducing a new carbon–carbon bond. Electrophilic substitution involving loss of electrofugal nitrogen is however much less common. In one rare example, methyl dialkylcyanoazocarboxylates can lose nitrogen to form mixtures of cyano esters and methylated nitriles upon treatment with lithium methoxide and methyl iodide, the chemoselectivity depending upon the reaction conditions (Scheme 38) ⟨71JA4318⟩. The formation of methylated nitriles required at least stoicheiometric quantities of base. The *in situ* methylation of cyclohexyl azocarboxylates provided predominantly the products of equatorial methylation.

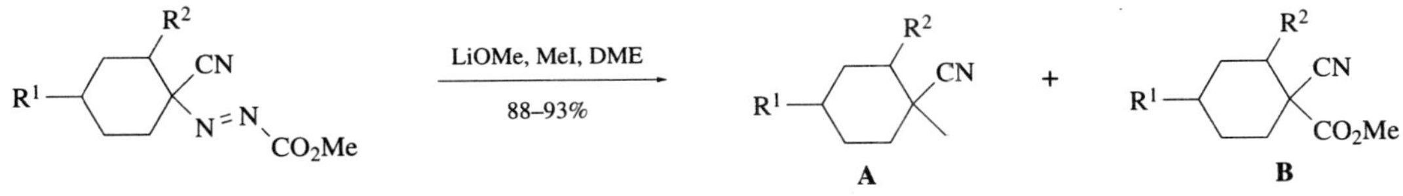

R¹	*R²*	*Ratio* **A:B**
H	H	77:13
H	Me	63:25
Bu^t	H	70:23

Scheme 38

1.06.2 SUBSTITUTION OF PHOSPHORUS, ARSENIC AND ANTIMONY FUNCTIONS

1.06.2.1 Substitution of Phosphorus Functions

Reactions involving simultaneous sp^3 carbon–phosphorus bond cleavage and sp^3 carbon–carbon bond formation are very rare in the literature. The vast majority of carbon–phosphorus bond cleavage reactions where carbon–carbon bonds are formed involve Wittig or related alkenation processes, and all involve phosphonium ylides.

The use of phosphorus ylide reactions to prepare cyclopropane compounds involves carbon–carbon single bond formation, either through conjugate addition of the ylide carbanion to activated double bonds with subsequent expulsion of the tertiary phosphine (Scheme 39) ⟨58JA4386, 62AG(E)116⟩, or through epoxide cleavage by nucleophilic attack of the ylide carbanion, followed by thermal decomposition of the oxaphospholane formed (Scheme 40) ⟨59JA6330, 62JA676, 62JA677, 62JA3944, 64T1695⟩. The suggested mechanism involves cleavage of the oxaphospholane carbon–phosphorus bond to produce a carbanion which cyclises with concurrent expulsion of phosphine oxide ⟨64QR1964⟩. Good yields of products are only achieved when the oxaphospholane intermediate bears a carbanion stabilising group in an appropriate position.

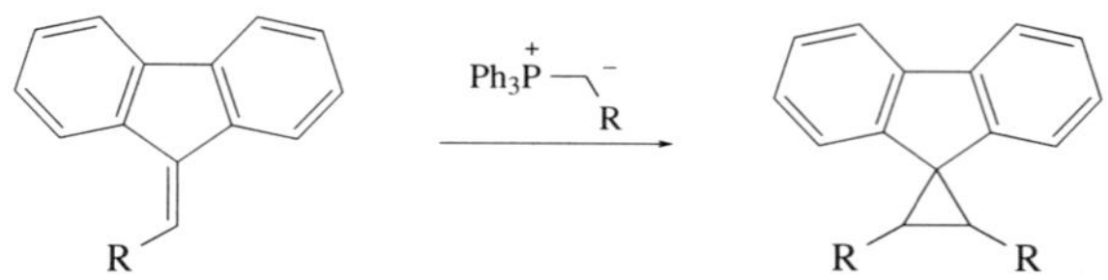

R	*Yield* (%)
C_3H_7	50
H	23

Scheme 39

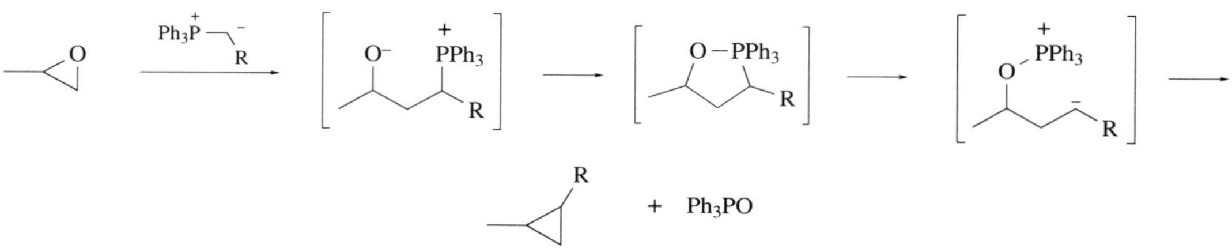

R	*Yield* (%)
Ph	21
C_6H_{13}	26

Scheme 40

In a similar process, cyclopropylketones may be prepared by treatment of esters of 3-hydroxypropylphosphonium salts with potassium *t*-butoxide in refluxing *t*-butanol ⟨71JOC2379⟩. The corresponding cyclopropylketones were produced in 42–59% yields (Scheme 41).

1.06.2.2 Substitution of Arsenic Functions

All the reactions in which the breaking of an sp^3 carbon–arsenic bond and the formation of a new sp^3 carbon–carbon bond take place involve an arsonium ylide as the reactive intermediate. In a typical example, treatment of triphenylarsine with 2-bromomethylnaphthalene at elevated temperatures afforded (2-naphthyl)methyl triphenyl arsonium bromide in good yield ⟨76JOM(108)175⟩. Deprotonation using sodium hydride generated 2-naphthylmethylene triphenylarsenane. The reactions of this ylide with a range of substituted benzaldehydes at room temperature were explored. The corresponding *trans* epoxides were produced in good yields in a process mechanistically similar to the more well known sulfur-based epoxide formation (Scheme 42).

An example by Still involves the stereoselective synthesis of *trans* epoxides from aliphatic

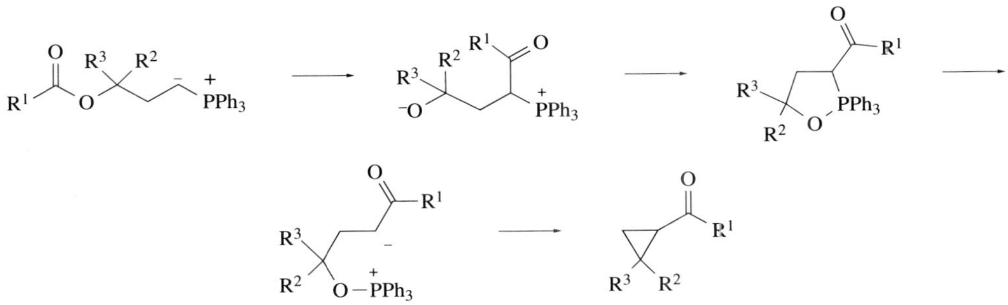

R^1	R^2	R^3	Yield (%)
Ph	H	H	59
Me	Ph	PhCO	54

Scheme 41

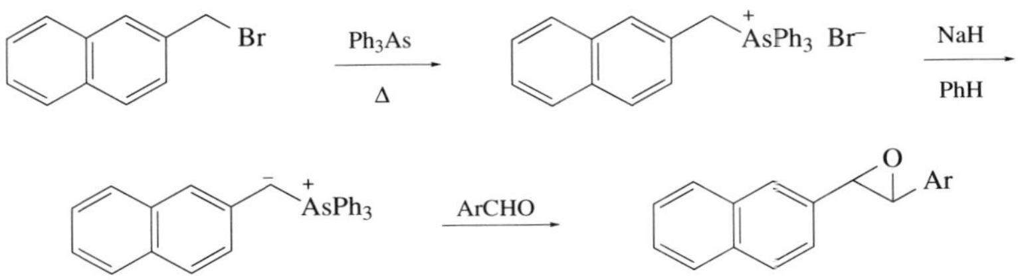

Ar	Yield (%)
Ph	90
4-ClC_6H_4	92

Scheme 42

aldehydes ⟨81JA1283⟩. The route again proceeds in a highly stereoselective manner. For example, the reaction of triphenylarsonium ethylide with octanal to produce largely (*E*)-dec-2-ene oxide proceeds in 76% yield (Equation (22)). Interestingly, optimum selection for the *trans* product is observed under the same salt-free conditions used to give *cis* stereochemistry in the Wittig reaction. The orange ylide was prepared by deprotonation of ethyltriphenylarsonium fluoroborate with potassium hexamethyldisilazide (KHMDS) in 10% HMPA–THF at -40 °C. The solution is then chilled to -78 °C before addition of the aldehyde.

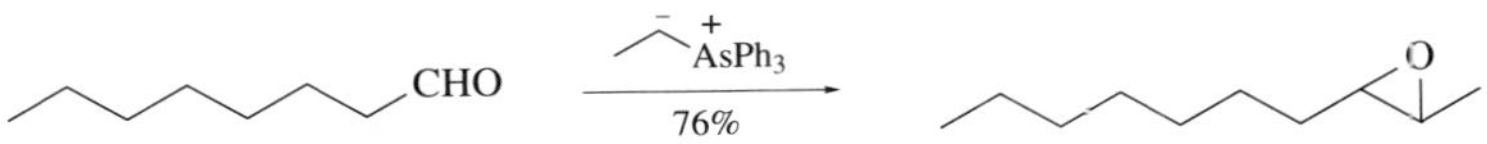

98% (*E*) (22)

In a further example by Mioskowski, vinylic epoxides were synthesized in high yields from allylic arsonium ylides and carbonyl compounds in very high yields (55–95%) (Scheme 43) ⟨83TL4419⟩. This is a particularly useful reaction, since vinylic epoxides cannot easily be prepared from allylic dimethylsulfonium ylides, which undergo [2,3]-sigmatropic rearrangements, nor from allylic diphenylsulfonium ylides, which are difficult to prepare. Since triphenylarsine is more nucleophilic than diphenylsulfide, it reacts well with allylic bromides to produce the corresponding allylic arsonium salts, and hence the corresponding allylic arsonium ylides.

An unusual variation involves the reaction of aromatic aldehydes with *β*-hydroxyethyltriphenylarsonium bromide in the presence of solid potassium hydroxide at room temperature. These reactions produce 2,3-epoxy-3-aryl-propanols in good yields under solid–liquid phase transfer conditions (Scheme 44) ⟨88TL5295⟩. It is of note that bases such as triethylamine, piperidine and potassium carbonate do not induce this reaction at room temperature, and furthermore, standard conditions such as treating the arsonium salt with strong bases at -78 °C followed by addition of

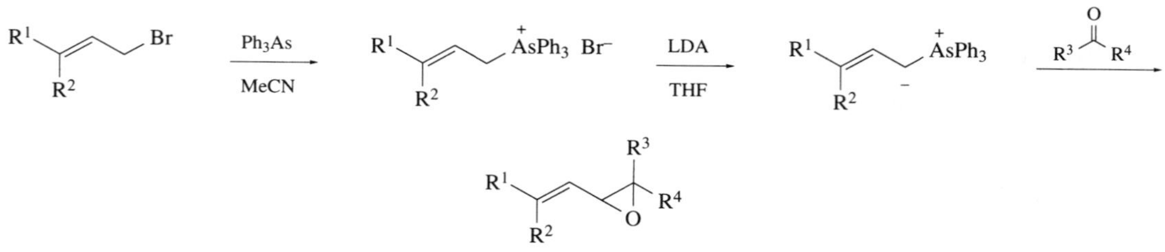

R^1	R^2	Aldehyde/ketone	Yield (%)
Ph	H	PhCHO	83
H	H	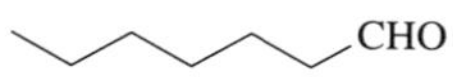	65
H	H	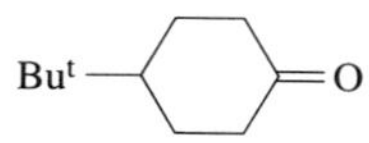	70
Ph	H	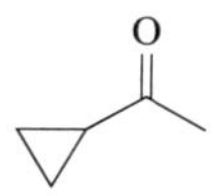	81
Ph	H	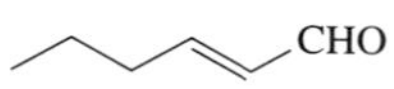	75

Scheme 43

the aldehydes did not produce the derived alcohols in reasonable yields. Only by using a solid–liquid phase transfer process could the desired products be obtained in good yields (40–84%).

Ar	Yield (%)
Ph	83
py	77

Scheme 44

A vinylogous version of this reaction is also known, where 5-aryl-4,5-*trans*-epoxypent-2-en-1-ols were conveniently synthesized in a one-step reaction of aldehydes with (*E*)-4-hydroxybut-2-enyl triphenyl arsonium salts under the above conditions ⟨91TL3999⟩. During the reaction of the (*Z*)-salt, the original double bond is partially isomerized (Scheme 45). When potassium fluoride–alumina, which is less basic than potassium hydroxide, was used as the base and acetonitrile used as the solvent instead of THF, the yield increased, but the ratio of isomers did not change. Use of the (*E*)-4-hydroxybut-2-enyl triphenyl arsonium salt produced the 5-aryl-4,5-*trans*-epoxy-(*E*)-pent-2-en-1-ols in good yields (53–78%) with no isomerization of the double bond.

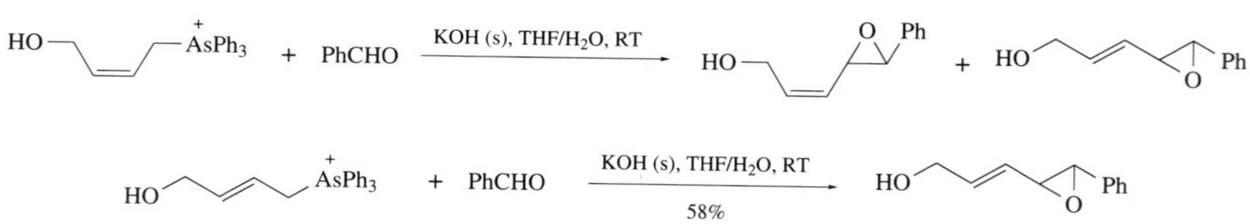

Scheme 45

A further interesting variation of this type of reaction is found in the synthesis of (trimethylsilylethynyl) cyclopropanes via an arsonium ylide ⟨88S321⟩. This involves the reaction of 3-(trimethylsilyl)-2-propynylidenetriphenylarsorane, generated *in situ* from the corresponding arsonium salt and butyllithium, with α,β-unsaturated ketones to produce cyclopropanes in 96–100% yield with high stereoselectivity (Equation (23)).

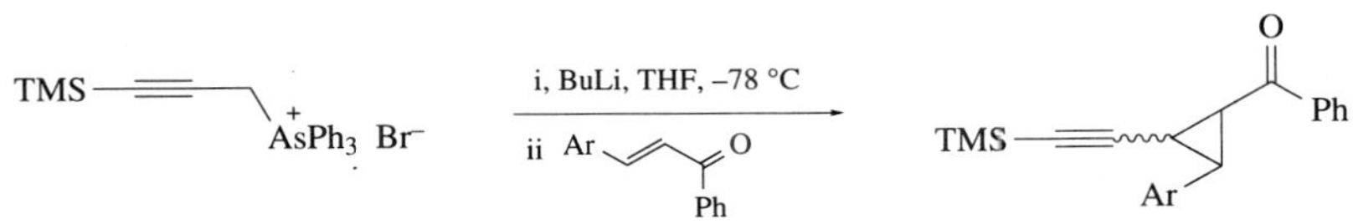

(23)

1.06.2.3 Substitution of Antimony Functions

Although several reports have appeared dealing with organoantimony chemistry ⟨88JOM(25)351⟩, few have dealt with the use of organoantimony reagents in sp^3 carbon–carbon bond forming reactions. One example by Henry *et al.* in 1960 involves the reaction of triphenylmethylstibonium salts with benzophenone (Scheme 46) ⟨60JA563⟩. Triphenylmethylstibonium iodide can be prepared by first treating triphenylstibine with trimethyloxonium tetrafluoroborate to produce the quaternary boron tetrafluoride salt of triphenylstibine, which is easily converted to the corresponding iodide.

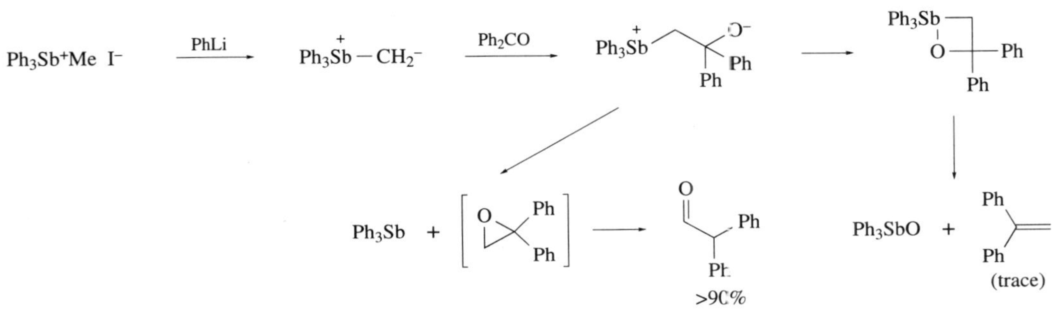

Scheme 46

The authors envisaged two possible pathways for the subsequent reaction with benzophenone, leading either to diphenylethylene or to the corresponding epoxide (Scheme 46). The latter pathway is supported by the isolation in almost quantitative yield of diphenylacetaldehyde, a rearrangement product of the unstable epoxide, with only a trace of diphenylethylene observed.

Another reaction using organoantimony reagents for the formation of sp^3 carbon–carbon bonds is found in the use of bis(diphenylstibinyl)methane as a precursor to the reactive species diphenylstibinylmethyllithium, which reacts with ketones, haloalkanes and other organometallic reagents (Scheme 47) ⟨83CB473⟩. bis(Diphenylstibinyl)methane is treated with phenyllithium to give diphenylstibinylmethyllithium, with cleavage of one carbon–antimony bond and generation of triphenylstibine.

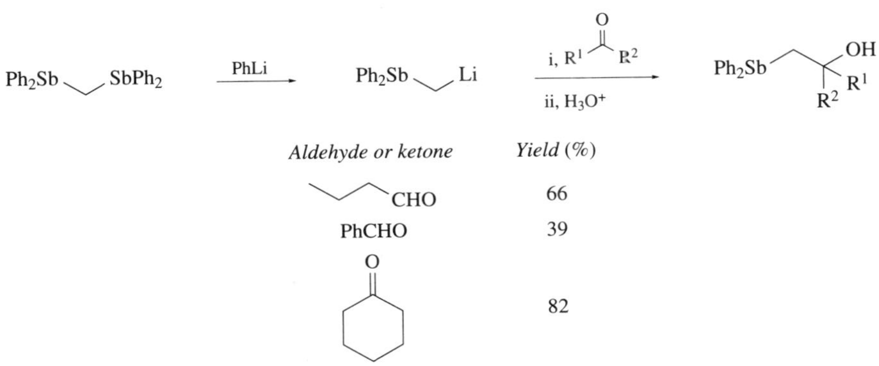

Scheme 47

A paper in 1990 by the same author also concentrates on the reactions of carbonyl compounds with organoantimony reagents mediated by a range of strong bases ⟨90CC85⟩. While treatment of quaternary stibonium salts with weakly nucleophilic strong bases such as lithium diisopropylamide (LDA) or potassium *t*-butoxide affords stibonium ylides, strong nucleophiles such as organolithiums or Grignard reagents produce pentaorganostiboranes. The quaternary stibonium salt used in this case was benzyltributylstibonium bromide which, upon treatment with alkyl- or phenyllithiums, gave pentaorganostiboranes which reacted with aromatic aldehydes with transfer of the benzyl group to produce alcohols (Scheme 48). Benzyltrialkylstibonium bromides are easily prepared by mixing a trialkylstibine with benzylbromide at room temperature.

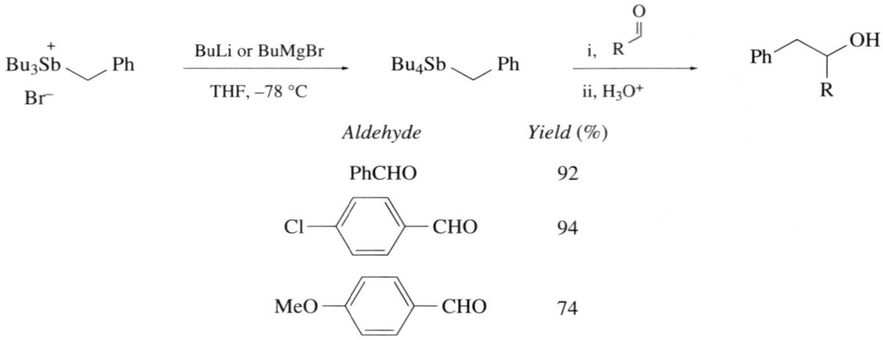

Scheme 48

The reactions of pentaorganostiboranes with ketones and acyl chlorides have also been examined (Scheme 49). In the absence of any additional catalyst, benzyltetrabutylstiborane reacts with acyl chlorides to give the corresponding benzyl ketones in good yields (70–90%). No butyl ketones, resulting from transfer of a butyl group from antimony, were observed. Alkyl groups may however be transferred selectively from pentaalkylstiboranes; for example, a methyl group is transferred from tributyldimethylstiborane. Both ethyl and butyl groups are transferred from tributyldiethylstiborane ⟨92JOC774⟩.

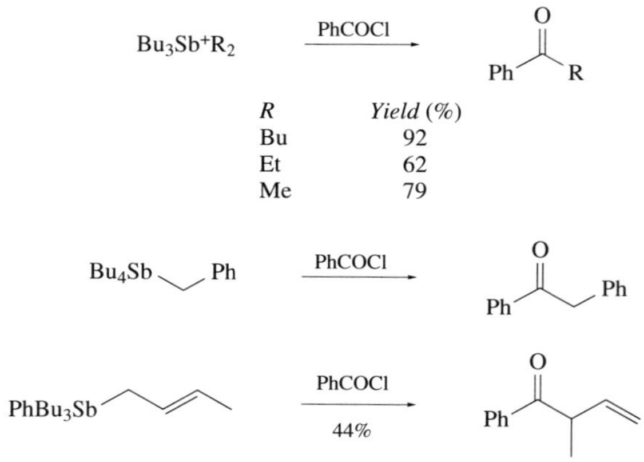

Scheme 49

1.06.3 SUBSTITUTION OF BORON, SILICON AND GERMANIUM FUNCTIONS

1.06.3.1 Substitution of Boron Functions

Trialkylboranes are important reagents for the formation of sp^3 carbon–carbon bonds, and are readily available through hydroboration processes. They may be used in various types of addition reactions, particularly conjugate additions induced by oxygen radicals, bases and electrochemical processes (Scheme 50). Another important area of these carbon–carbon bond forming reactions involves substitution by alkyl groups at sp^2 centres in various alkenes and aromatic heterocycles in a Friedel–Crafts type alkylation.

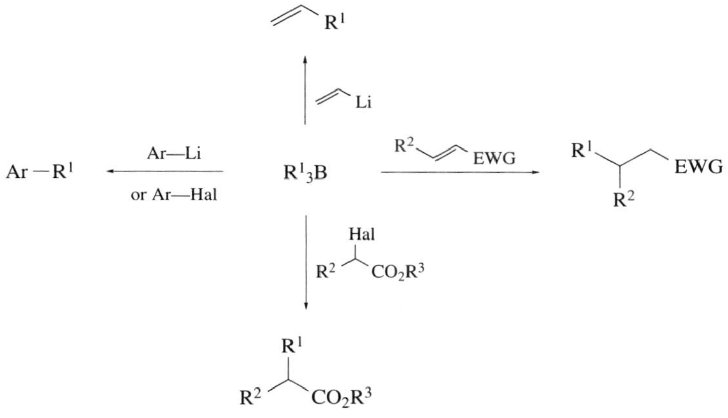

Scheme 50

1.06.3.1.1 Conjugate addition to α,β-unsaturated ketones and aldehydes

One of the first examples of conjugate addition reactions with trialkylboranes, published by Hawthorne in 1965, is a reductive alkylation of quinones with trialkylboranes in the preparation of alkylhydroquinones (Scheme 51) ⟨65JA4585⟩. This reaction nicely illustrated the potential use of trialkylboranes in organic synthesis. The reactions proceed in nearly quantitative yield, and the only limitation on the reaction is the availability of the trialkylborane being used. Previously, preparation of alkylhydroquinones had involved much less efficient, often multistep, processes, such as electrophilic alkylation of hydroquinones or acylation by Friedel–Crafts reactions followed by reduction.

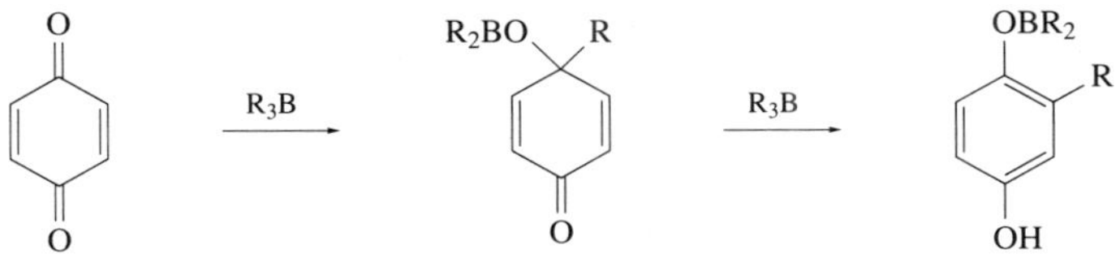

R	*Yield (%)*
C_6H_{13}	99
Bu	86
Bu^s	94
Bn	90

Scheme 51

Much of the work involving organoboranes as reagents for the formation of new sp^3 carbon–carbon bonds was carried out by Suzuki and Brown in the 1970s. Results of a collaboration between the two groups on the reactions of organoboranes as a means of synthesizing aldehydes was published in 1967 and describes the conjugate addition reactions of trialkylboranes to acrolein ⟨67JA5709⟩. The reaction first produces an intermediate enolborinate which undergoes hydrolysis to yield the corresponding aldehyde (Scheme 52). This reaction is particularly convenient since it can be carried out in one pot, by adding water to a THF solution of the organoborane, and then introducing acrolein into the reaction mixture. This procedure results in concomitant conjugate addition and hydrolysis to produce the aldehyde. The reaction takes place over about one hour at room temperature (in some cases it is complete within ca. five minutes) and is very general.

A similar reaction of organoboranes with methyl vinyl ketone was also examined by Suzuki and Brown ⟨67JA5708⟩, giving the corresponding methyl ketones in almost quantitative yields. Again, both the addition and hydrolysis may occur within a few minutes at 25°C. Since methyl ketones can be readily converted into the corresponding acids using the haloform reaction ⟨67JA5710⟩, this

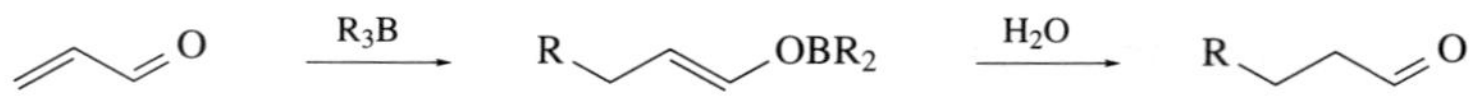

R	*Yield* (%)
Bu	55
Bu^s	96
c-C_6H_{11}	77

Scheme 52

process could also provide a simple synthesis of carboxylic acids. Several illustrative examples are shown in Scheme 53.

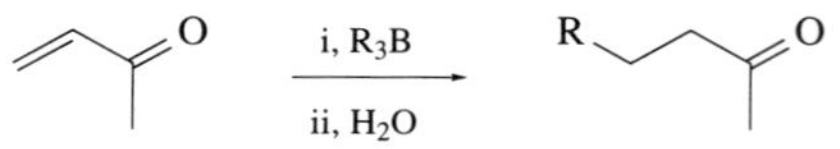

R	*Yield* (%)
Me_2CHCH_2	70
	80
Bu	86

Scheme 53

The corresponding addition to β-substituted α,β-unsaturated ketones and α,β-alkynyl ketones does not occur spontaneously, requiring some form of radical initiator. Suitable radical initiators include ultraviolet light and/or traces of oxygen; for example, acetylethyne readily undergoes conjugate addition in the presence of a catalytic amount of oxygen (Scheme 54) ⟨70JA3503⟩. The initially formed allenic intermediates undergo hydrolysis to produce the corresponding α,β-unsaturated methyl ketones in good yield. This reaction illustrates the versatility of conjugate addition reactions with organoboranes, even with unreactive substrates. *In situ* hydrolysis in this case minimizes the attack of the allenic intermediate by the free-radicals.

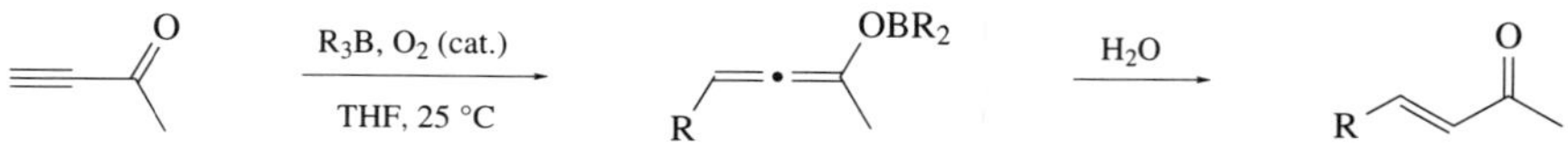

R	*Yield* (%)
Et	77
c-C_6H_{11}	65

Scheme 54

The principal disadvantage of all of these reactions is that only one of the three substituent groups on the boron atom can normally participate under the mild conditions used, but the simplicity and rapidity of the reactions perhaps provides adequate compensation. The methodology can be extended to allow conjugate addition to many α,β-unsaturated carbonyl compounds including α-methylacrolein ⟨68JA4165⟩, α-bromoacrolein ⟨68JA4165⟩ and α-methylenecycloalkanones (Scheme 55) ⟨68JA4166⟩. β-Substituted α,β-unsaturated carbonyl compounds do not however react spontaneously with organoboranes, but require an initiator.

Conjugate addition of trialkylboranes to crotonaldehydes or other α,β-unsaturated carbonyl compounds may proceed spontaneously as discussed above, under irradiation or in the presence of radical initiators ⟨71JA1506, 71JA1508⟩. A further method of inducing a reaction between trialkylboranes and crotonaldehyde has been developed using excess pyridine to promote reaction at room temperature over eight days (Equation (24)) ⟨73TL787⟩. Use of only equimolar quantities of pyridine (three equivalents) results in much lower yields of the aldehydic product. Tertiary amines do not, in general, promote the reaction to such an extent, although 4-picoline afforded a yield of

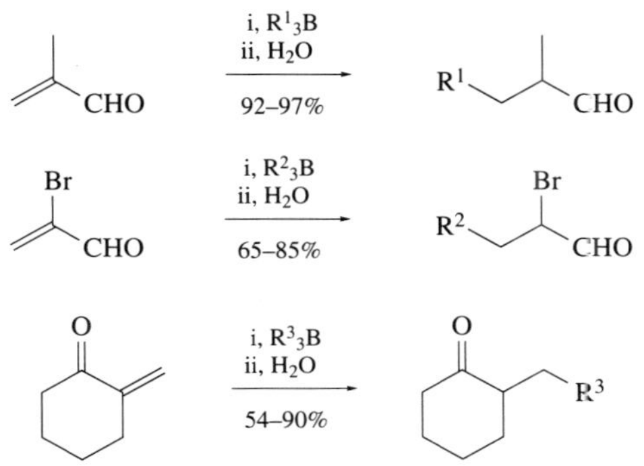

R¹	*Yield* (%)	*R²*	*Yield* (%)	*R³*	*Yield* (%)
Bu	95	Bu	85	*c*-C_5H_9	85
c-C_6H_{11}	92	*c*-C_6H_{11}	65		

Scheme 55

96% after eight days. The most effective solvent for the reaction was toluene, and the best yield (ca. 99%) was obtained using tricyclopentylborane as the substrate. It has been suggested that the promotion of the reaction by pyridine involves a radical intermediate.

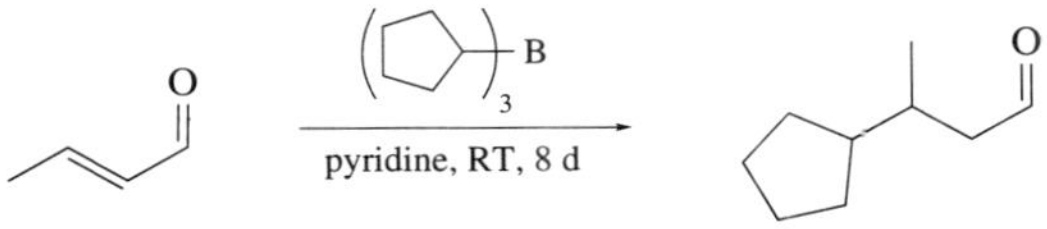

(24)

A particularly interesting variation is the reaction with 1-acyl-2-vinylcyclopropanes, where the products after hydrolysis are the corresponding γ,δ-unsaturated ketones, isolated in good yields, predominantly with the (*E*)-configuration (Scheme 56). The authors suggest a mechanism involving electron transfer, as shown in Scheme 56, for the reaction of cuprous methyltrialkylboranes with 1-benzoyl-2-vinylcyclopropane ⟨72JA5495⟩.

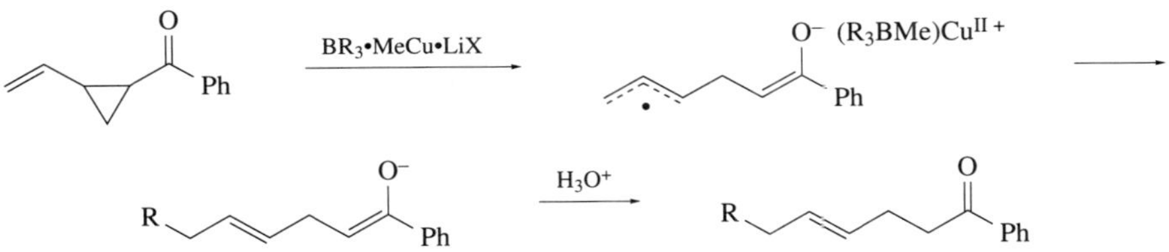

R	*Yield* (%)
Pr	84
Bu	81

Scheme 56

An example of such substitution in quinone systems using organoboranes is the preparation of 2-alkyl-1,4-naphthaquinones from 1,4-naphthaquinones. Although the preparation of 2-alkylquinones may be carried out in up to quantitative yields by the reaction of quinones with alkyl radicals produced by treatment of trialkyl boranes with catalytic amounts of oxygen (Scheme 57), quinones other than *p*-benzoquinone are usually relatively unreactive ⟨71JOM(33)25⟩. A modification developed in 1992, in which trialkylborane is added to the quinone and water in the presence of oxygen, followed by oxidation with silver oxide, affords 2-alkyl-1,4-naphthaquinones in good yields (Scheme 58) ⟨92S461⟩. 2,3-Dialkyl-1,4-naphthaquinones are produced in the presence of magnesium bromide.

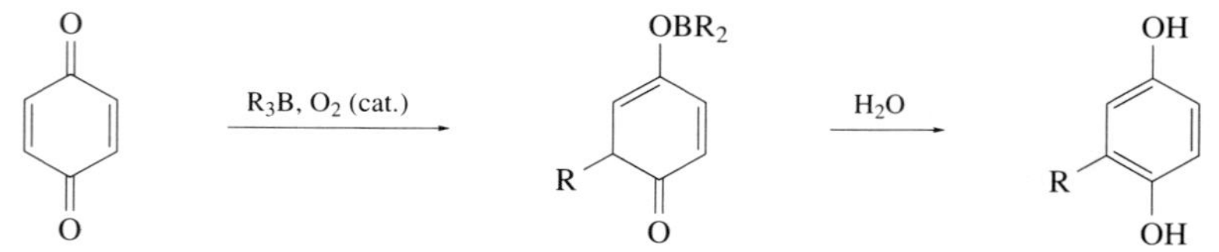

R	*Yield* (%)
Et	95
Bu	85

Scheme 57

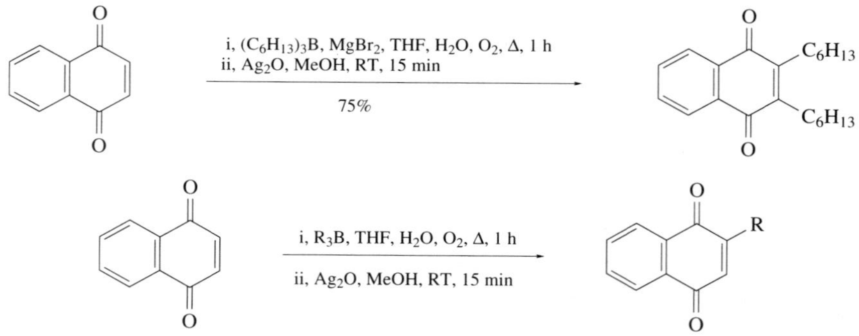

R	*Yield* (%)
c-C_5H_9	94
$(CH_2)_3Ph$	79

Scheme 58

1.06.3.1.2 Conjugate addition to α,β-unsaturated carboxylic acid derivatives

Simple carboxylate derivatives such as acrylonitrile and ethyl acrylate do not undergo addition of organoboranes through any of the processes described above, perhaps because of ready polymerisation or because the unstable intermediates undergo other reactions. These difficulties are overcome when cuprous methyltrialkylborates, which are readily obtained from trialkylboranes by addition of alkyllithium followed by a copper(I) salt, are used as reagents ⟨76TL255⟩. For example, alkyl nitriles are produced in good yields from the reaction with acrylonitriles in 19–93% yields (Scheme 59). The choice of cuprous halides used in the generation of the cuprous methyltrialkylborate is crucial: in the preparation of hexanenitrile, use of cuprous iodide gave only a 19% yield, while cuprous bromide gave an 84% yield. While many other methods for producing alkylnitriles may be available, including electrochemical methods ⟨75CL523⟩, this reaction does provide a novel and general synthetic procedure. A similar reaction with ethyl acrylate proceeds readily under the same conditions (Equation (25)).

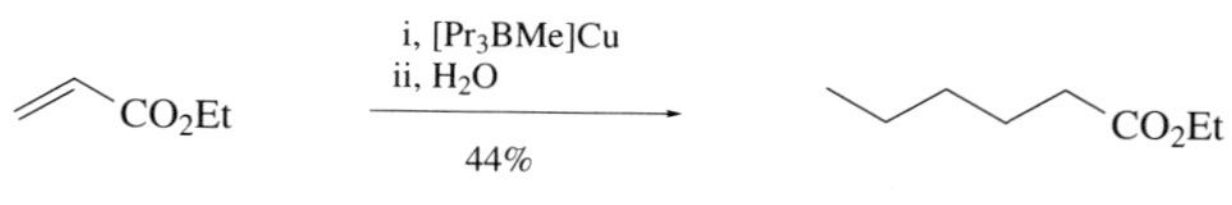

(25)

Organoboranes also undergo clean electrochemically mediated reactions with acrylates in good yields (51–94%), providing a convenient preparative method for substituted carboxylic esters (Equation (26)) ⟨78BCJ339⟩. The electrolyte of choice for the reaction was tetrabutylammonium iodide. The suggested mechanism for the reaction involves independent reactions at the anode and cathode.

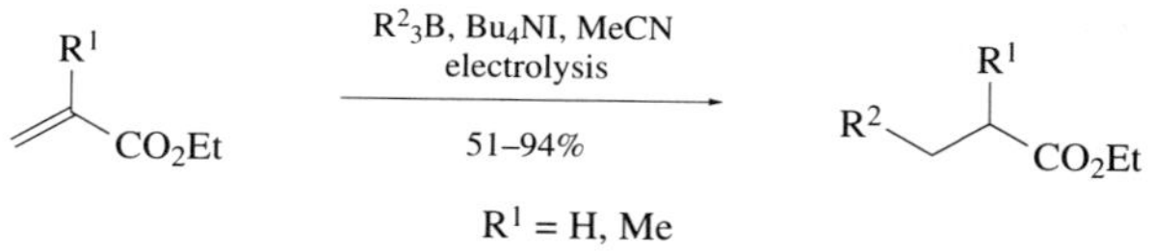

(26)

R^1 = H, Me

$$R_3B \xrightarrow{MeLi} [R_3BMe]Li \xrightarrow{CuX} [R_3BMe]Cu{\cdot}LiX$$

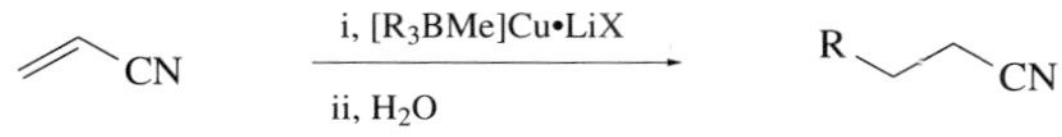

R	*X*	*Yield (%)*
Pr	I	19
Pr	Br	84
Pr	Cl	65
Bu	Br	88

Scheme 59

1.06.3.1.3 Conjugate addition to vinyl and alkynyl epoxides

Suzuki and Brown have also examined conjugate addition reactions of organoboranes to various α,β-unsaturated epoxides. Like β-substituted α,β-unsaturated ketones, these species fail to react with organoboranes spontaneously, requiring some form of radical initiator. Again, suitable radical initiators include ultraviolet light and/or traces of oxygen.

In the presence of catalytic oxygen or other free radical initiators, 1,3-butadiene monoepoxide undergoes conjugate addition with trialkylborane, providing the corresponding 4-alkylbut-2-en-1-ols in high stereochemical purity and reasonable yields (44–75%) (Scheme 60) ⟨71JA2792⟩. One interesting aspect of this particular reaction is that THF, although it is the normal solvent for such reactions, could not be used, as considerable quantitatives of 4-(2-tetrahydrofuryl)but-2-en-1-ol were formed in the reaction mixture, presumably resulting from free radical mediated conjugate addition of THF to the epoxide as a competing side reaction (Equation (27)). This difficulty was circumvented by using benzene as the solvent.

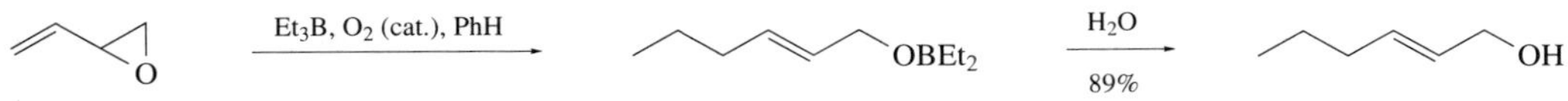

Scheme 60

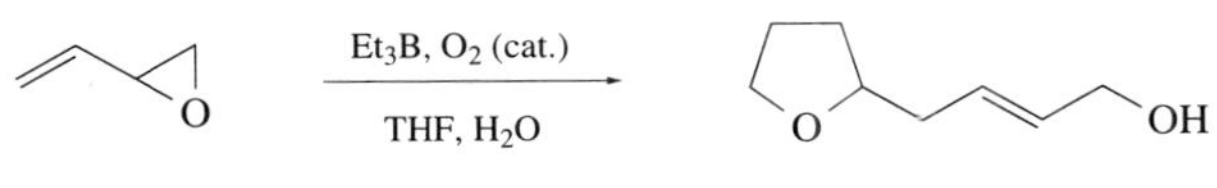 (27)

The analogous reaction with 3,4-epoxy-1-butyne is also known ⟨73S305⟩. Conjugate addition produces allenic derivatives which were hydrolysed to give the corresponding allenic carbinols (Scheme 61). The reaction was successful using both triethyl and tricyclopentylborane; the reactions were carried out in both benzene and tetrahydrofuran, benzene producing the higher yields in most cases. 3,4-Epoxy-3-methyl-1-butyne was also used as a substrate, although reduced yields were obtained.

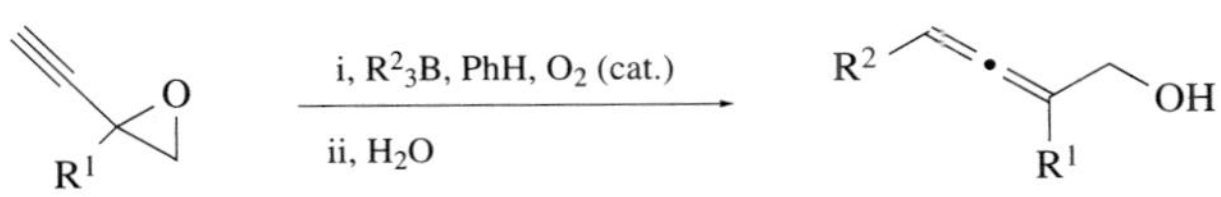

R¹	*R²*	*Yield (%)*
H	Et	53
Me	Et	47

Scheme 61

1.06.3.1.4 Aromatic and alkenic substitution

Most of the organoboron reactions discussed so far involve some kind of conjugate addition. A different type of reaction is found in the substitution of proton or halide substituents in aromatic rings or alkenes in a reaction mimicking electrophilic substitution. The reactions are successful with aliphatic trialkyl boranes.

In 1967 Zweifel devised an elegant synthesis of both (*Z*)- and (*E*)-1,2-disubstituted alkenes involving boron chemistry ⟨67JA3652, 68JA6243, 71JA6309, 72JA6560⟩. (*E*)-Vinylboranes, prepared by addition of dialkylboranes to alkynes, were shown to undergo transfer of one ligand from boron to the alkenic carbon atom followed by a highly stereoselective elimination process upon treatment with electrophiles. It is interesting that, although in the absence of base a thermal *syn* elimination takes place to give the (*E*)-alkenes, an *anti* elimination occurs in the presence of base to give the (*Z*)-alkenes (Scheme 62). The process was later modified by Evans through the use of boronic esters to prevent competition between ligands in the transfer of the alkyl group ⟨76JOC3947⟩.

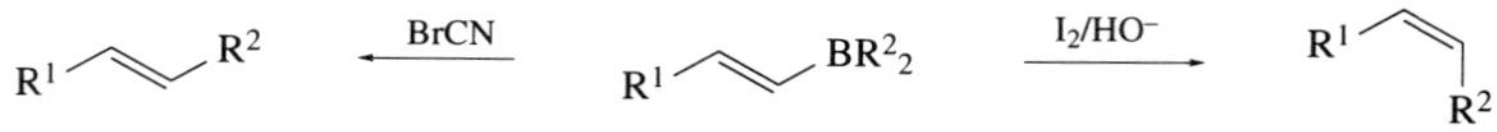

Scheme 62

Subsequently, a modification allowing the stereoselective preparation of trisubstituted alkenes was reported by Levy. For example, reaction of ethylmagnesium bromide with 1-octyne in the presence of cuprous bromide dimethyl sulfide complex leads, after treatment with iodine, to a 63% yield of the (*E*)-2-ethyl-1-iodo-1-octene. Subsequent sequential treatment of this vinyl iodide with butyllithium, triethylborane and elemental iodine at low temperature leads to a 75% yield of (*E*)-4-ethyl-3-decene (Equation (28)) ⟨78JOC1279⟩.

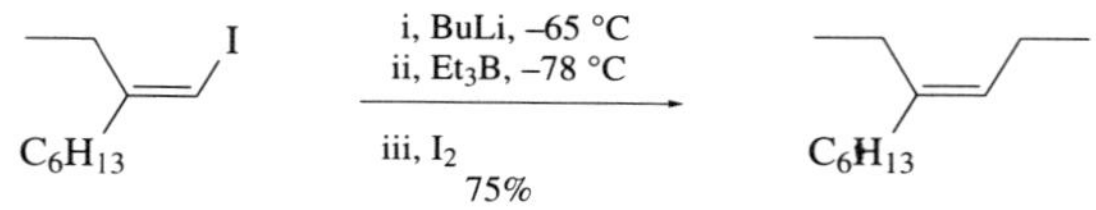

(28)

Treatment of the vinyl iodide intermediate with butyllithium leads to halogen–metal exchange. Subsequent transfer of the alkyl group from boron occurs with retention of configuration at the vinyl carbon atom. The mechanism of the subsequent rearrangement is similar to the halogenation reactions of vinylboranes. Reaction of the vinyllithium with the trialkylborane gives an 'ate' complex. Addition of iodine leads to migration of an alkyl group from boron to carbon; *syn* elimination of dialkyliodoborane then gives the product alkene (Scheme 63).

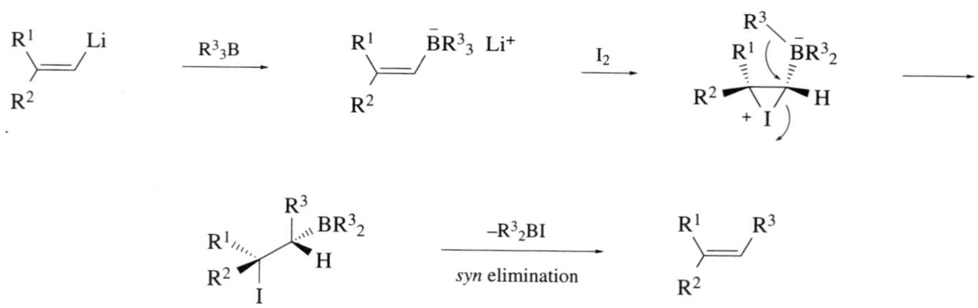

Scheme 63

Furans and pyrroles respond poorly to normal methods of aromatic substitution such as the Friedel–Crafts reaction. Ring destruction and polymerization are commonly observed rather than substitution. However, organoboranes can be used to introduce alkyl groups using metallated species generated via halides or by deprotonation of furans and pyrroles, usually at the C2 position (Scheme 64) ⟨78JOC4684, 79BCJ1865, 79TL2313⟩.

The 2-alkylated heterocycles are produced in variable yields. The mechanism for the reaction is similar to that of trisubstituted alkene preparation (Equation (28)), and is thought to proceed in three stages: (i) reaction of the trialkylborane with the carbanion derived from the heterocyclic

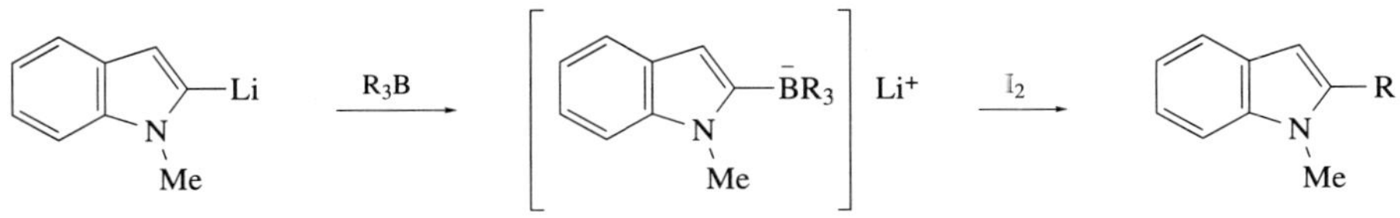

R	*Yield* (%)
Et	79
c-C_5H_9	79

R	*Yield* (%)
C_6H_{13}	82
c-C_5H_9	73

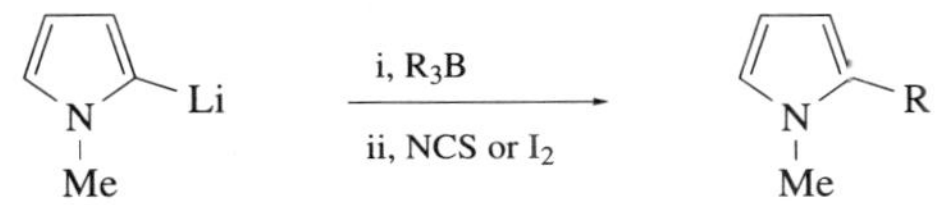

R	*Yield* (%)
Pr	92
C_6H_{13}	70

Scheme 64

compound to give a borate species; (ii) electrophilic attack by iodine at the 5-position in the heterocycle, with migration of an alkyl group from boron to the C2-position and (iii) collapse to the product with loss of dialkylhaloborane and halide (Scheme 64) 〈73JA3080〉.

Coupling of vinyl boranes and vinyl boronates with aryl and vinyl halides is a well known and reliable reaction (Scheme 65) 〈81TL127, 87CL25, 87S373, 91COS(3)241〉. Direct substitution, in which transition metal-catalysed cross coupling of trialkylboranes with aryl and 1-alkenyl halides takes place with formation of a carbon–carbon single bond, is also a mild and straightforward reaction 〈80PAC669〉. Use of a palladium catalyst containing bis(diphenylphosphine)ferrocene (dppf) ligand enables the coupling of 1-alkenyl and aryl halides with trialkylboranes such as 9-alkyl-9-borabicyclo-[3.3.1] nonanes to take place in the presence of bases 〈86TL6369〉 (Scheme 66). Use of the dppf ligand helps to suppress competing β-hydride elimination in the transferred alkyl groups 〈84JA158〉. The base used is commonly sodium methoxide, and yields are good to excellent (60–99% in most cases). An unusual application of this reaction is seen in the synthesis of functionalized organotin compounds using 9-(ω-stannylalkyl)-9-BBN as the organoboron reagent 〈91SL687〉. Use of ($PdCl_2$(dppf)) or $Pd(PPh_3)_4$ as the palladium catalyst and potassium phosphate as the base in dioxane solution was essential for successful coupling, in the case shown presumably through an addition–elimination process. The products may be cyclodestannylated in the presence of acid catalysts to provide carbocyclization products (Scheme 67).

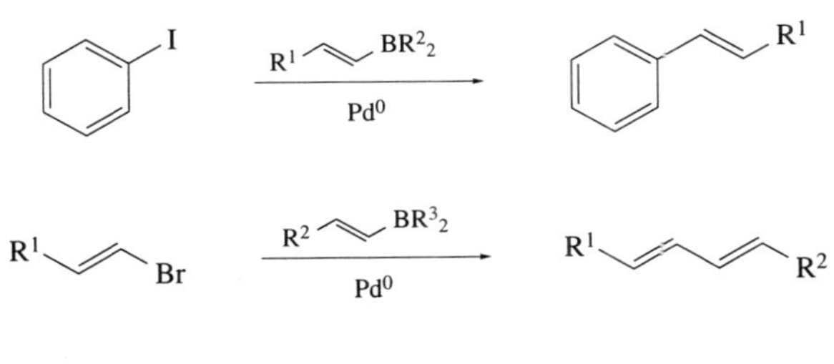

Scheme 65

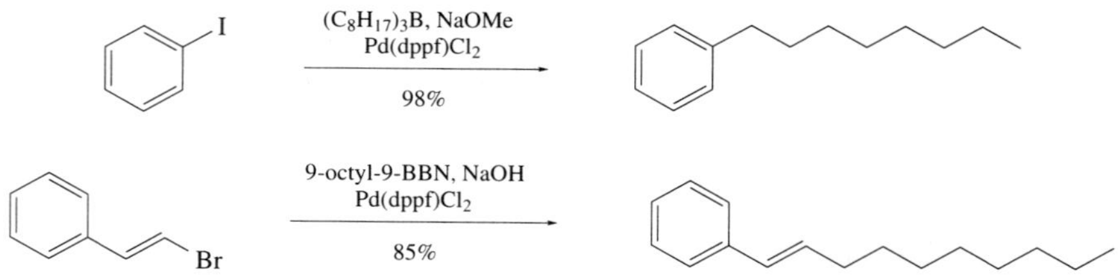

Scheme 66

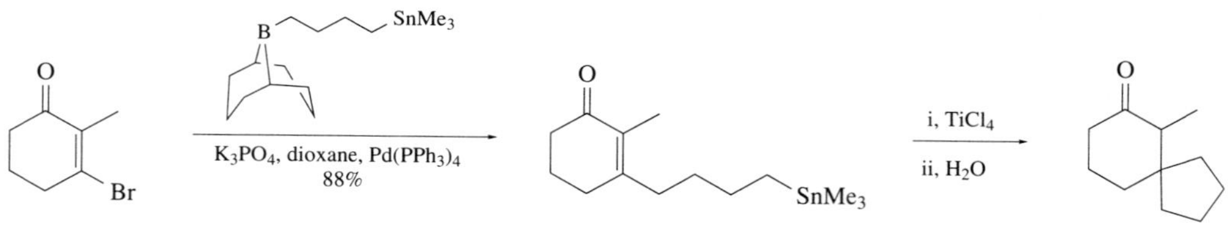

Scheme 67

1.06.3.1.5 *Aliphatic substitution*

A useful example of substitution of a halide by an alkyl group from an organoborane at an sp^3 carbon atom involves α-haloesters as substrates, providing a simple synthesis of esters (Scheme 68) ⟨68JA818, 68JA1911⟩. The reaction may also be carried out using dihaloesters to provide either the mono- or disubstituted products in yields of 47–98%.

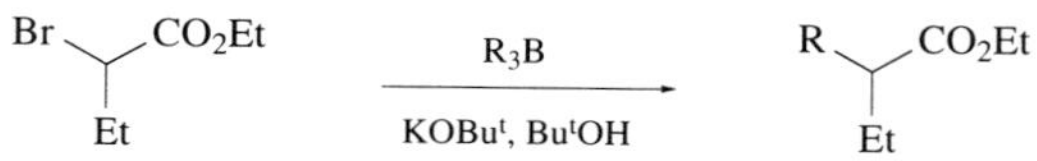

R	*Yield (%)*
Et	87
C_6H_{13}	63

Scheme 68

In another carbon–carbon bond forming reaction using organoboranes, a ligand is transferred from a trialkylborane to displace the acetoxy group in α-imino-α-acetoxy carboxylates, producing the corresponding α-substituted α-iminocarboxylates, which may be hydrolysed to give α-amino acids in good yields (Scheme 69) ⟨85CC1168⟩.

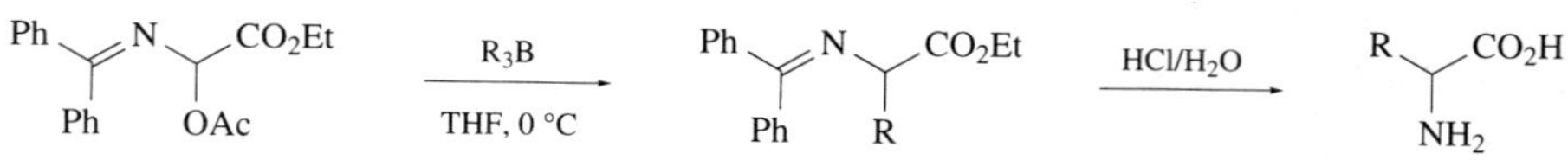

R	*Yield (%)*
c-C_6H_{11}	90
Bu^t	59
Ph	57
C_8H_{17}	92

Scheme 69

A rather unusual and interesting reaction reported by Hunter involves an acetal as the substrate, the leaving group being one of the alkoxy moities (Equation (29)) ⟨89TL2013⟩. The methodology may be extended to the ring opening of cyclic acetals. Products were isolated in 64–75% yield. The lithium butyltriallylborate used in this case was produced by treating triallylborane with one equivalent of butyllithium in THF at 0 °C.

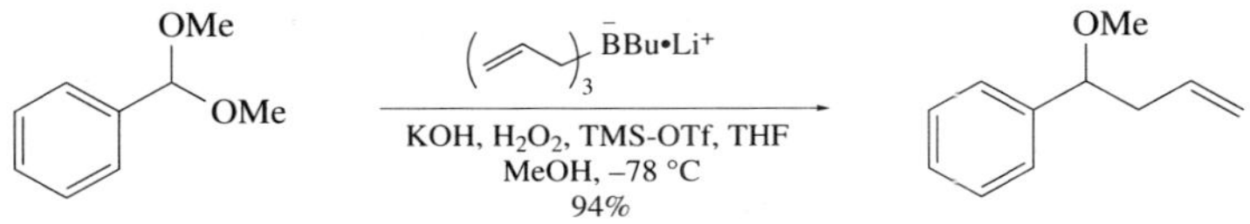

(29)

Allylboranes can undergo a typical organometallic addition to the carbonyl group of a wide range of aldehydes and ketones in dichloromethane solution (Scheme 70) ⟨91TL6749⟩.

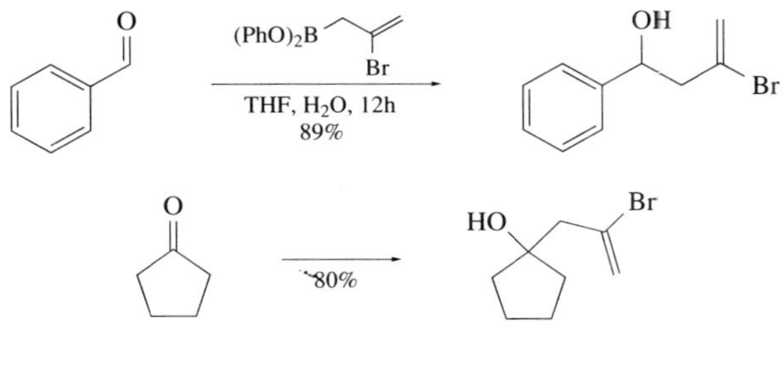

Scheme 70

In contrast to the reactivity of allylboranes, allylboronates normally react with ketones with transfer of one alkyl group from boron only under high pressure conditions. Allylboronates will however react with α-ketocarboxylic acids in the presence of triethylamine to give α-hydroxy-carboxylic acids (Scheme 71) ⟨91TL4619⟩. The reaction may proceed through a bicyclic transition state.

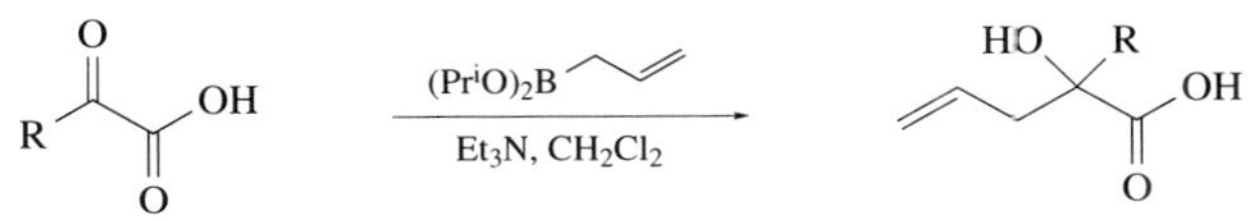

R	*Yield* (%)
Ph	94
Me	86
Et	90

Scheme 71

1.06.3.2 Substitution of Silicon Functions

1.06.3.2.1 Alkylation

Simple substitution reactions of sp^3 carbon–silicon bonds which form new sp^3 carbon–carbon bonds are not common. The vast majority of such reactions involve desilylation, effected by any one of a number of fluoride ion reagents such as caesium fluoride or *t*-butylammonium fluoride (tbaf), followed by attack of the anionic desilylated species, or other intermediate, at a carbon electrophile such as an alkyl halide, aldehyde, ketone, α,β-unsaturated ketone or carbon dioxide ⟨75JA3257⟩.

Silanes in trifluoroacetic acid have been used to effect the ionic hydrogenation of alkenes by protonation followed by hydride transfer from the silanes ⟨74S633⟩. Although tetrasubstituted silanes do not react with alkenes in the absence of catalyst, in the presence of trifluoroacetic acid and Lewis acids a methyl group may be transferred from tetramethylsilane to an alkene double bond (Scheme 72). The reaction is believed to take place through initial protonation, and is successful with tri- and tetra-substituted ethylenes which can form tertiary carbenium ions upon protonation, to produce saturated hydrocarbons containing quaternary carbon atoms ⟨80CC748⟩. One interesting aspect of this reaction is that, although the reaction proceeds with other tetramethyl group 14 derivatives including those of tin and germanium, it does not proceed with TMS-Br, TMS-Cl or

TMS-O-TMS. Even so, these reactions are useful for the synthesis of hydrocarbons with quaternary carbon atoms.

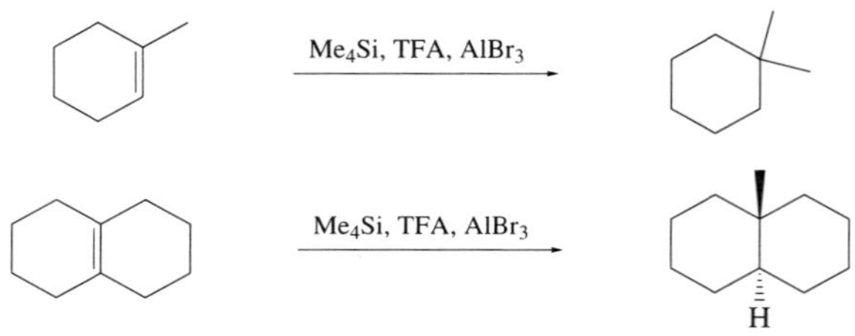

Scheme 72

In an interesting application of the fluoride induced nucleophilic addition process, α-halosilanes can be converted by treatment with fluoride ion into reactive species, formally the α-halocarbanions, which can be used in the synthesis of stilbenes, epoxides, cyclopropanes, benzazepines and phthalidylisoquinolines (Scheme 73) ⟨91JOC3908⟩. For example, α-halosilanes are desilylated using caesium fluoride in DMF in the presence of α,β-unsaturated ketones to give cyclopropanes through a conjugate addition/halide displacement process.

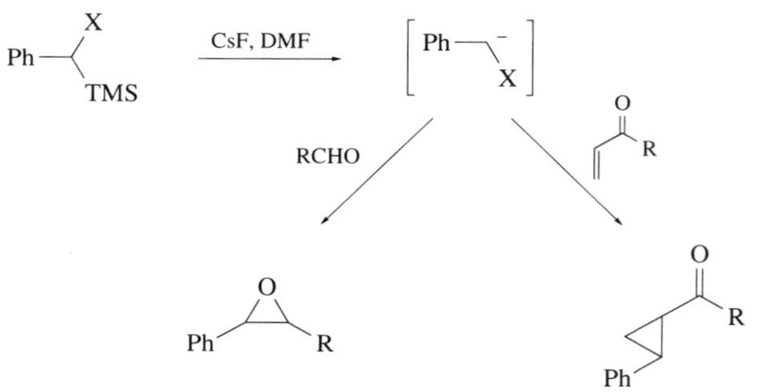

R	*Yield* (%)
Ph	87
4-MeOC$_6$H$_4$	58

R	*Yield* (%)
Me	50
OMe	59

Scheme 73

Alkylation is achieved in moderate yield upon cleavage of the silyl group of 3-trimethylsilyloxetan-2-ones, readily available by [2 + 2]-cycloaddition reactions, in the presence of methyl iodide. The reaction presumably proceeds via an enolate intermediate, generated upon addition of the 2 + 2 adduct to a rapidly stirred suspension of tris((dimethylamino)sulphur trimethylsilyl)difluoride (TASF) and methyl iodide (five molar equivalents) in tetrahydrofuran at 0 °C (Equation (30)). Other fluoride sources, including caesium fluoride and tetrabutylammonium fluoride, were examined, but with little success ⟨89TL6829⟩.

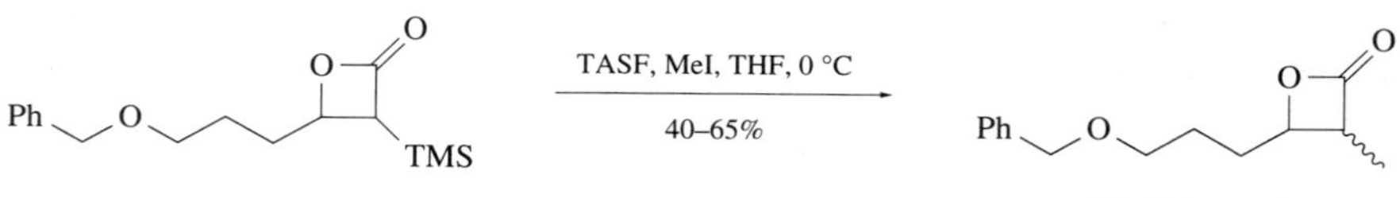

(30)

85:15 *trans*:*cis*

The ready loss of the trimethylsilyl unit from 2-trimethylsilyl-1,3-dithiane has been used in a tandem bisalkylation with epoxides ⟨94SL511⟩. 2-Lithio-2-trimethylsilyl-1,3-dithiane is alkylated with epoxides in the presence of crown ethers to give the alkoxide, which undergoes rapid migration of the silyl moiety from carbon to oxygen, generating a new 2-lithio-1,3-dithiane species which reacts with a second molecule of epoxide. The process was used in a synthesis of enantiomerically pure C2-symmetric 1,5-diols (Scheme 74).

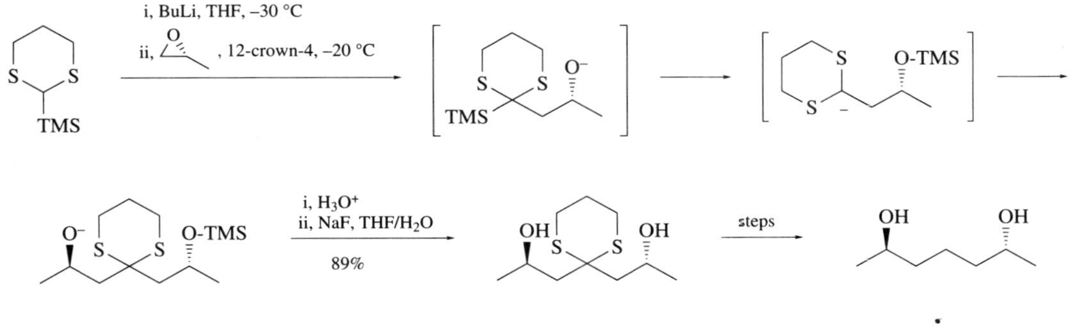

Scheme 74

1.06.3.2.2 Hydroxylation/aldol reactions

A simple example of a hydroxyalkylation reaction involving carbon–silicon bond cleavage, reported by Kuwajima, involves the treatment of α-trimethylsilyl esters with a carbonyl compound in the presence of a catalytic amount of tbaf in an aprotic solvent such as THF ⟨76TL1699⟩. β-Silyloxyesters are produced by migration of the silyl group, and can subsequently be hydrolysed using acetic acid and water to produce the β-hydroxy esters (Scheme 75). Only small amounts of isolable by-products were detected except where aliphatic aldehydes were used; these tended to undergo self-condensation, with only small amounts of the crossed aldol adduct being produced.

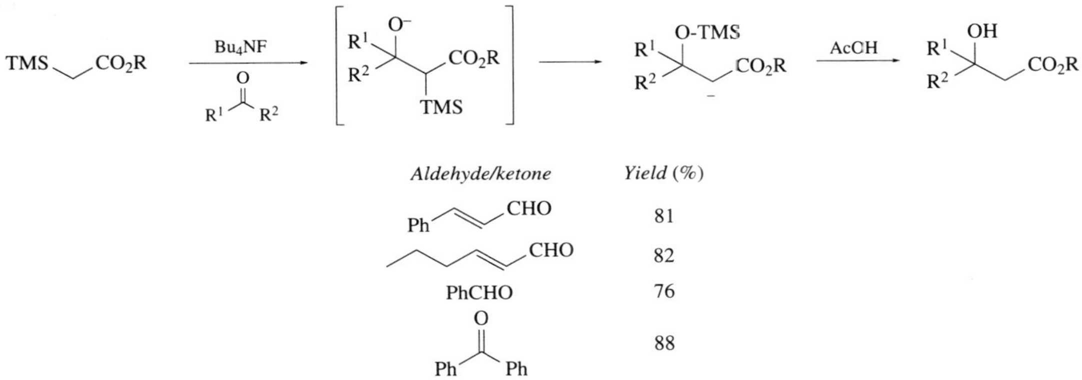

Scheme 75

In a related reaction, *N*,*N*-dimethyl-2-trimethylsilylacetamide may be deprotonated by LDA to form the lithium enolate, which has been shown to add to epoxides with ring cleavage and similar silyl group migration (Scheme 76) ⟨78JOC1947⟩.

Acyl silanes undergo a rearrangement induced by fluoride ion in which alkyl group migration from

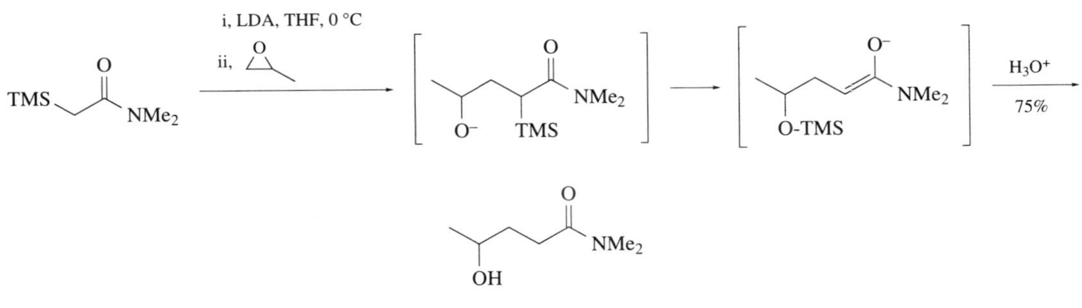

Scheme 76

silicon to carbon occurs, followed by Brook-type rearrangement, ultimately producing secondary alcohols (Scheme 77) ⟨87TL4455⟩.

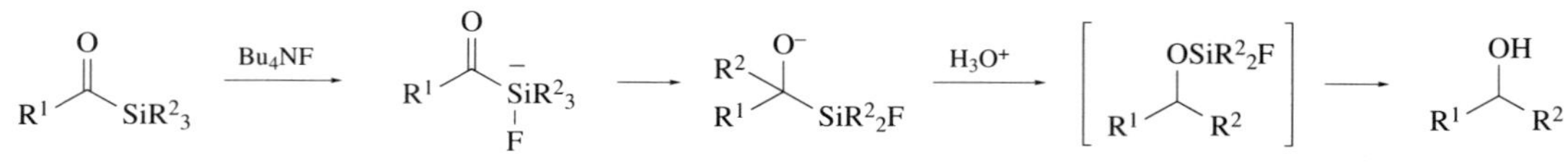

R^1	R^2	Yield (%)
$Me(CH_2)_5$	Me	68
Ph	Ph	54

Scheme 77

An interesting example by Paquette involves the reaction of cyclopropyl anions with carbonyl species ⟨84JA6442⟩. Attempts to deprotonate the cyclopropanes led only to the formation of self-condensation products. Although protons attached to three-membered rings exhibit greater acidity than normal, those cyclopropanes carrying carbonyl, nitro and sulfonyl groups are markedly less acidic than their acyclic counterparts. This has been explained by the formation of an exocyclic π-bond upon proton abstraction, maximising charge delocalization and increasing *p*-character. As a result, ring strain and chemical reactivity are enhanced. Cyclopropanes bearing electron-withdrawing substituents which also contain an α-trimethylsilyl group can however be efficiently desilylated in the presence of carbonyl group electrophiles to form the alkylated products (Scheme 78). The electron-withdrawing substituent may be a nitro, cyano or carboxyl group. Use of caesium fluoride produced substantial amounts of the silyl enol ether derivatives along with the expected aldol products. The best catalyst for desilylation in this case was tbaf in THF as a solvent.

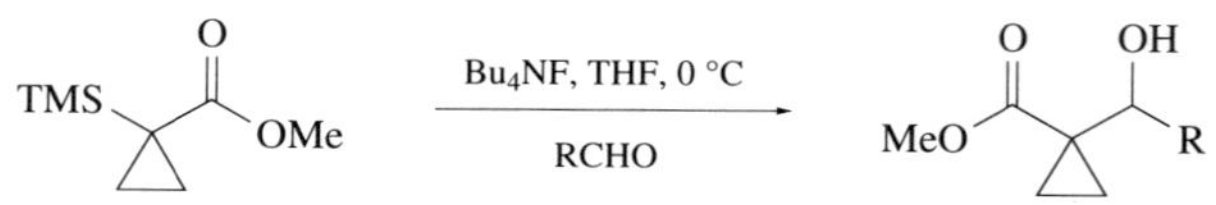

R	Yield (%)
Me	90
Bu^t	51

Scheme 78

1.06.3.2.3 Acylation

Desilylation of a silylcyclopropane derivative has also been carried out using carbon dioxide as the electrophilic species in an efficient conversion into the corresponding cyclopropyl carboxylic acid after aqueous workup (Equation (31)) ⟨91SL919⟩. Interestingly, the reaction proceeds in a very facile manner using one atmosphere of carbon dioxide. In this case the most successful fluoride ion reagent was caesium fluoride, used in ten-fold excess. In THF poor solubility proved a problem and silylcyclopropane was recovered. In solvents such as DMF and pyridine, however, the reactions proceed smoothly in quantitative yields.

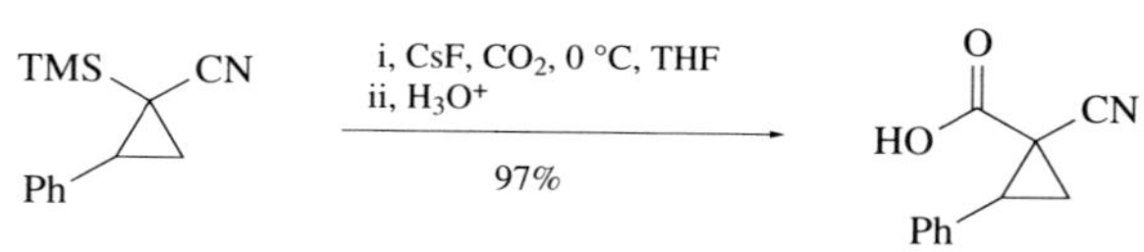

(31)

Desilylation–acylation of cyclopropylsilanes mediated by Lewis acid is also known. Lewis acid mediated synthesis of cyclopropyl ketones from cyclopropyltrimethylsilane and simple acid chlorides in the presence of aluminum trichloride was reported by Calas (Scheme 79) ⟨76S737⟩. Previously, cyclopropyl ketones had generally been prepared from acyclic carbonyl compounds, cyclization being the last step.

R	*Yield* (%)
Me	60
Ph	50

Scheme 79

Under the Calas reaction conditions, only very small amounts (<1%) of ring cleaved products were observed. This is remarkable, since the action of acyl chlorides on cyclopropane or methylcyclopropane normally results exclusively in the formation of chloroketones by ring cleavage, suggesting either a change in reaction mechanism or modification of the reactivity of the cyclopropyl ring by the trimethylsilyl group. α,β-Unsaturated acyl chlorides react similarly, giving rise to α,β-unsaturated cyclopropylketones in good yields.

1.06.3.2.4 1,3-Dipole formation

A further type of reaction which leads formally to the generation of sp^3 carbon–carbon bonds from sp^3 carbon–silicon bonds is found in 1,3-dipolar cycloaddition reactions. A silicon-containing 1,3-dipole precursor is treated with either fluoride ion or Lewis acid to generate the 1,3-dipole by desilylation. Cycloaddition to a dipolarophile leads to the formation of cyclic systems with the formation of a new carbon–carbon bond.

(i) Azomethine ylides

The 1,3-dipolar cycloaddition of an azomethine ylide with an alkene leads to a simple synthesis of a pyrrolidine ⟨92TL6607⟩. Desilylation of appropriately substituted *N*-(silylmethyl)amines has proved a very useful method for preparing nonstabilized azomethine ylides, which readily react with electron-deficient alkene dipolarophiles including α,β-unsaturated esters, ketones, nitriles and amides (Scheme 80), and is usually achieved by treatment with Lewis acid or fluoride ion. The reactions all proceed in good to excellent yields and can be carried out in one pot. The reactivity of the alkene depends upon the nature of the alkenic substituent, which in all the cases studied was an aromatic ring.

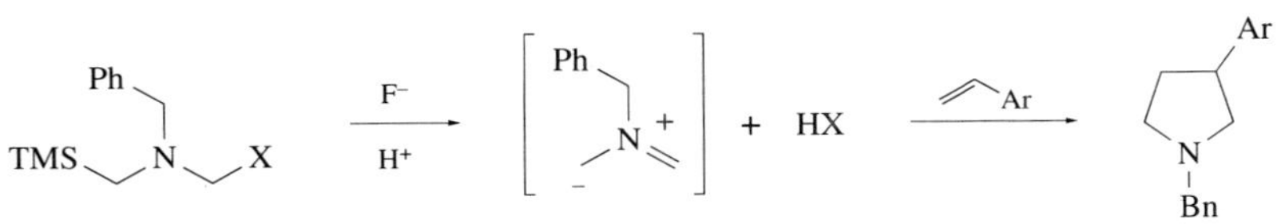

X = CN, OMe

Ar	*Yield* (%)
3-nitrophenyl	85
4-cyanophenyl	76
4-methoxyphenyl	20
2-pyridyl	79

Scheme 80

In one rather unusual example, the azomethine ylide is generated using photo-induced electron transfer (PET) to initiate desilylation, instead of fluoride ion ⟨92CC1313⟩. The ylide precursor, bis(trimethylsilylmethyl)benzylamine, is easily prepared and the two trimethylsilyl groups act as activating groups. The irradiation process is carried out using a 450-W Hanovia medium pressure lamp with a pyrex filter of >280 nm, in the presence of 9,10-dicyanoanthracene (DCN). Irradiation is carried out in a methanol–water solution (9 : 1). The process results in the formation of an amine cation radical which easily loses an electron and silyl cation to form the trimethylsilylallyliminium salt which undergoes further desilylation to generate the nonstabilized azomethine ylide capable of reaction with dipolarophiles including benzophenone, methyl cinnamate and phenylvinylsulfones (Scheme 81) ⟨87TL2649⟩.

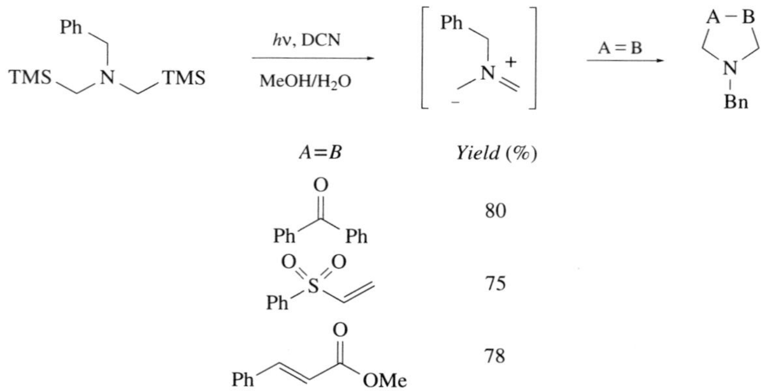

Scheme 81

(ii) Thiocarbonyl ylides

Treatment of alkoxymethylthiomethyl silanes with fluoride ion leads formally to the generation of thiocarbonyl ylides capable of 1,3-dipolar cycloaddition to aldehydes to form 1,3-oxathiolanes (Scheme 82) ⟨92CL1073⟩. The reaction is said to take place stepwise, by initial nucleophilic addition followed by ring closure under acidic conditions.

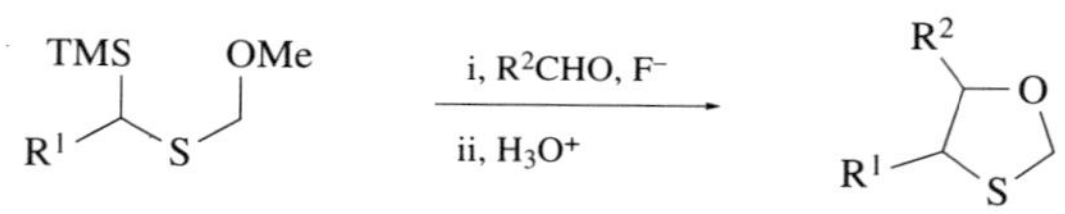

R^1	R^2	*Yield* (%)
H	Ph	84
H	Bu^t	87
Ph	4-MeOC$_6$H$_4$	82

Scheme 82

Alternatively, treatment of the alkoxymethylthiomethylsilanes with a silyl enol ether in the presence of Lewis acid leads to the formation of 3-hydroxythiolanes, again purportedly through a two-step process, in this case involving the alkoxymethylthiomethylsilane as the electrophile attacking the silyl enol ether. The product then undergoes desilylation, using a fluoride ion, leading to ring closure (Equation (32)).

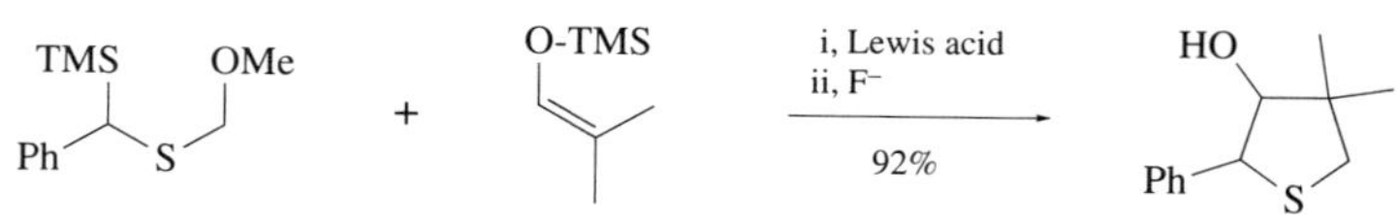

(32)

In both types of reaction, yields were good to excellent. *t*-Butylammonium fluoride, used in catalytic quantity, was found to be the best source of fluoride.

(iii) Trimethylenemethane

Of all the 1,3-dipoles generated through carbon–silicon bond cleavage, the best known is certainly the trimethylenemethane (TMM) system invented by Trost. In this system, reaction of 2-acetoxymethyl-3-allyltrimethylsilane with dipolarophiles under conditions of palladium(0) catalysis leads to efficient annelation to give five-membered carbocycles or heterocycles bearing an *exo*-methylene unit through a formal 1,3-dipolar cycloaddition process. The reactive intermediate may be formulated as a delocalized trimethylenemethane species (Scheme 83) ⟨79JA6429⟩. The reagent is prepared as shown in Scheme 84 by silylation of α-methylallyl alcohol followed by acetylation.

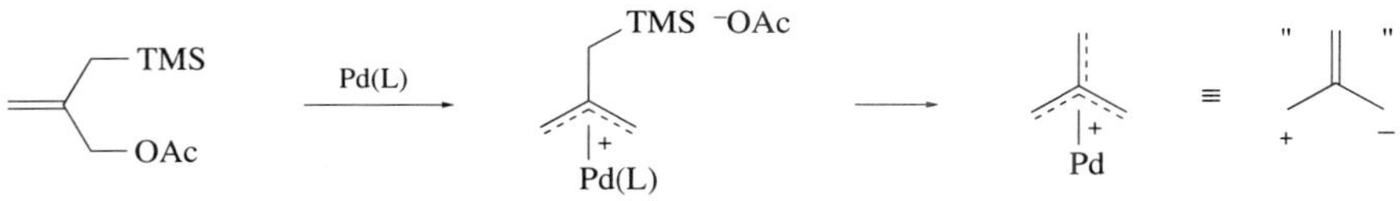

Scheme 83

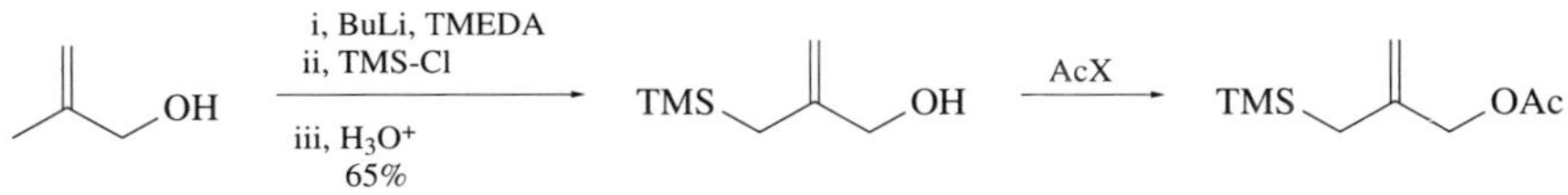

Scheme 84

Palladium-catalysed 1,3-dipolar addition of TMM systems to a range of dipolarophiles has been thoroughly reviewed, including the chemoselectivity, regioselectivity and diastereoselectivity of the process, in both intermolecular and intramolecular reactions ⟨86AG(E)1⟩. Palladium catalysts which may be used typically include tetrakis(triphenylphosphine) palladium(0) and palladium acetate/triisopropylphosphite, which Trost has shown to increase the efficiency of the reaction ⟨86TL1445⟩. Scheme 85 illustrates a range of intermolecular reactions with dipolarophiles containing electron withdrawing groups including ester, nitrile and sulfone ⟨79JA6429⟩. Many similar examples are known ⟨88JA1602, 89AG215, 90T1859, 91JA9007⟩.

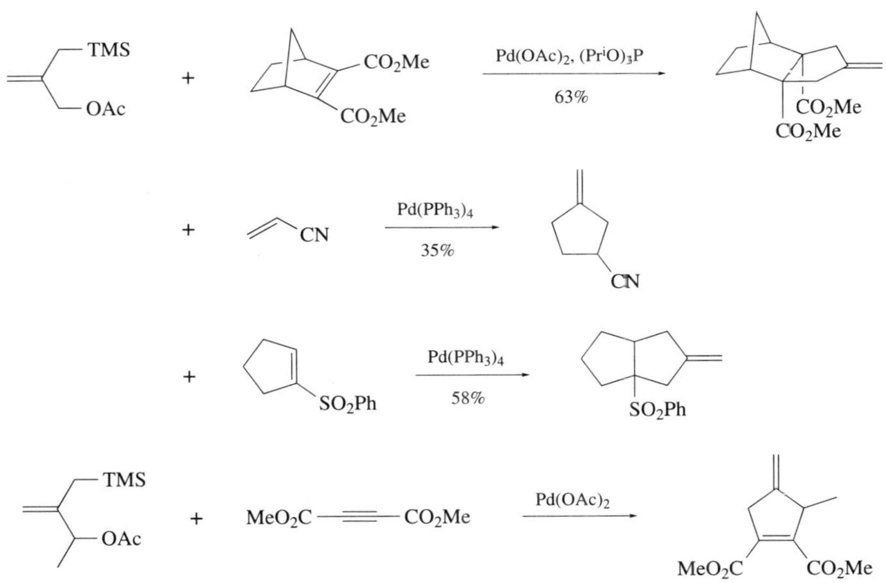

Scheme 85

TMM dipolar addition reactions are known to be diastereoselective. For example, dipolar addition to bicyclo[2.2.1]non-1-ene 1,2-dicarboxylates are highly *exo* selective ⟨86TL1445⟩, a reaction used in an approach to (±)-albene ⟨82JA6668⟩ (Schemes 85 and 86), and high diastereoselectivity is seen in the addition to simple α,β-unsaturated esters ⟨88JA3687⟩ and other derivatives (Equation (33)) ⟨86JA6051, 89JA6482⟩.

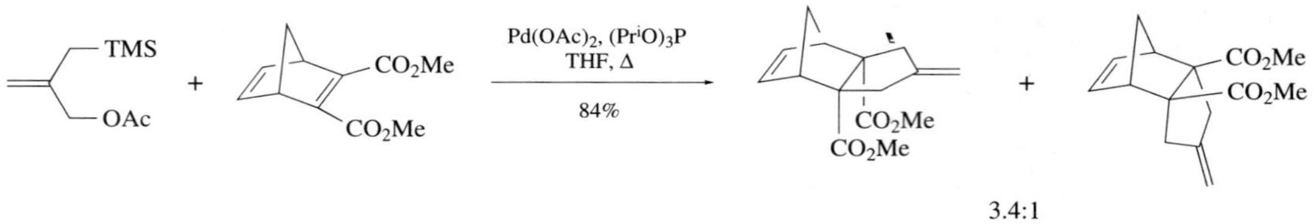

Scheme 86

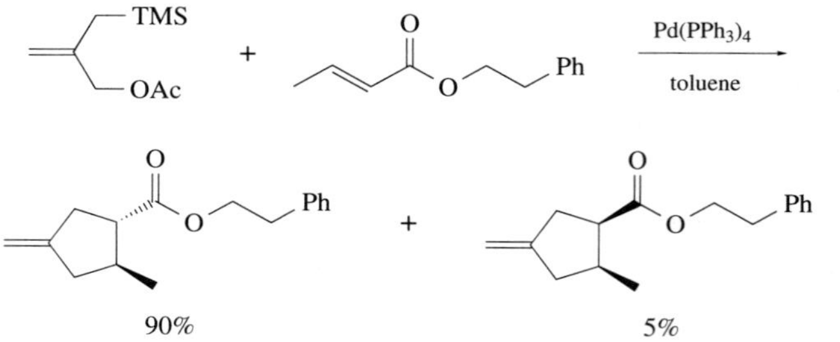

(33)

An optically pure vinyl sulfoxide has also been shown to undergo highly diastereoselective dipolar addition with TMM (Equation (34)) ⟨89TL1803⟩, as have cyclic γ-alkoxy vinyl sulfones (Equation (35)) ⟨89JA7487, 89TL1495⟩.

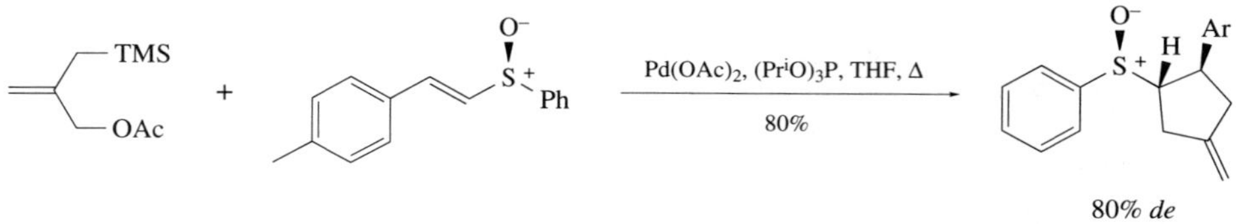

(34)

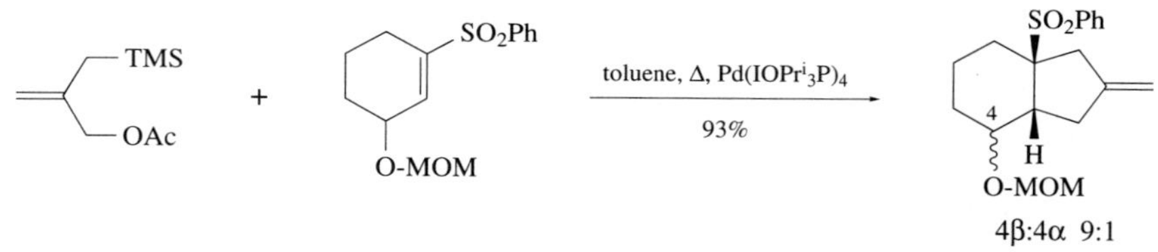

(35)

TMM dipolar cycloaddition reactions are successful even with tropone (a [6 + 3]-cyclization) ⟨87JA615⟩ and some dienoate systems (Scheme 87) ⟨87JA3483⟩.

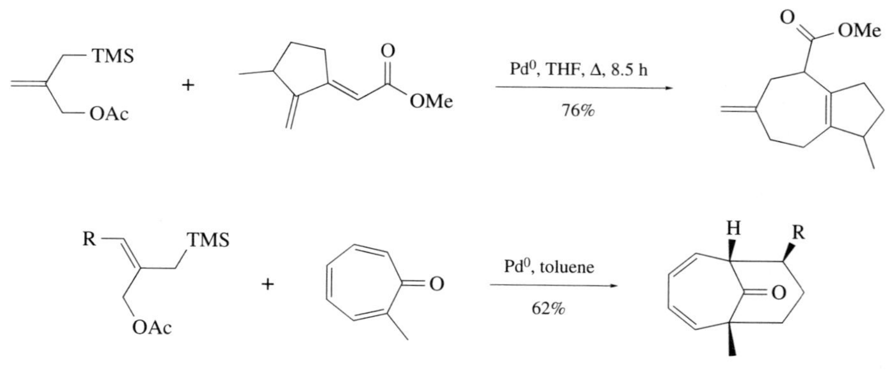

Scheme 87

Carbon-heteroatom double bonds can also be successful dipolarophiles. The carbonyl groups of α,β-unsaturated and α,β-acetylenic ketones, in cases where the double bond is sterically hindered, provide an interesting illustration of functionalized tetrahydrofuran synthesis using this procedure ⟨89JA5902, 89JA6482, 90JA408, 93TL7183, 93TL7903⟩. In the example illustrated, indium tris(acetylacetate) is used as cocatalyst (Equation (36)).

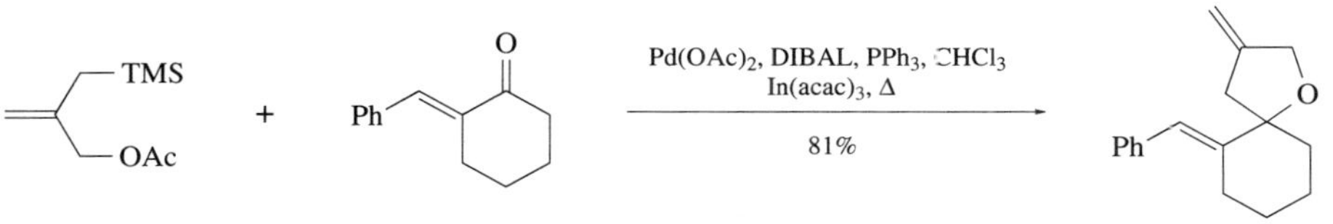

(36)

Imines also undergo dipolar addition to give the corresponding substituted pyrrolidines ⟨86CC1201⟩. Equation (37) shows such a case in which, unusually, a nickel catalyst is used.

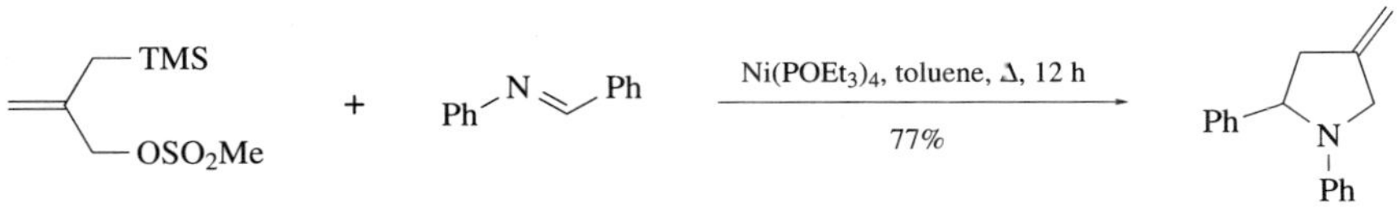

(37)

A number of instances of TMM cyclizations in intramolecular situations have appeared in the literature, with the formation of bicyclic (Scheme 88) or even tricyclic (Equation (38)) carbocyclic products ⟨82JA3733, 82JA6879, 82TL1267, 91JA7350, 91JA7363, 92JOC686⟩, as well as heterocycle formation by addition to aldehydes, seen in an approach to the phyllanthocin ring system (Equation (39)) ⟨89H(28)321⟩.

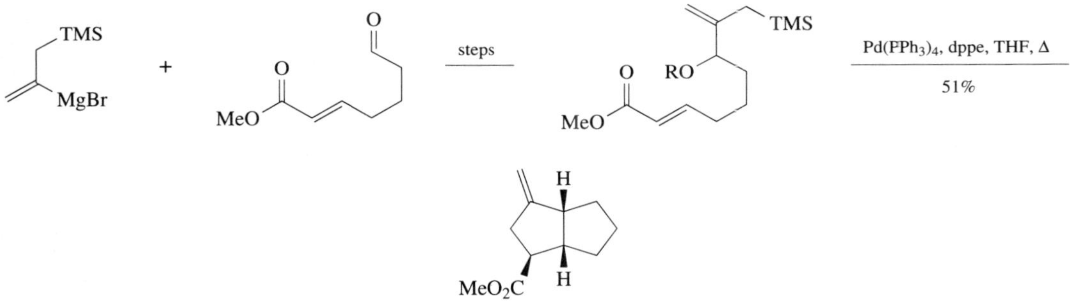

Scheme 88

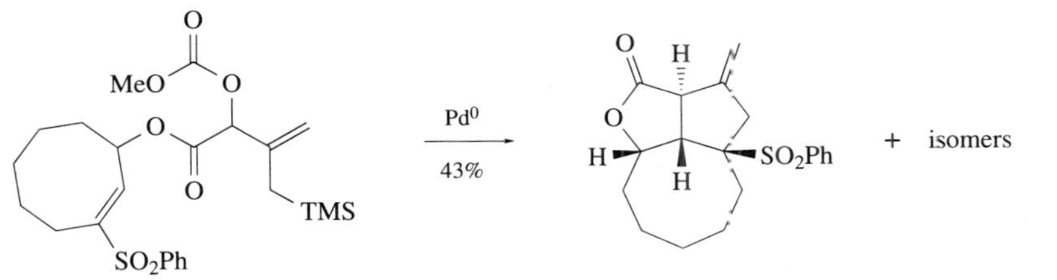

(38)

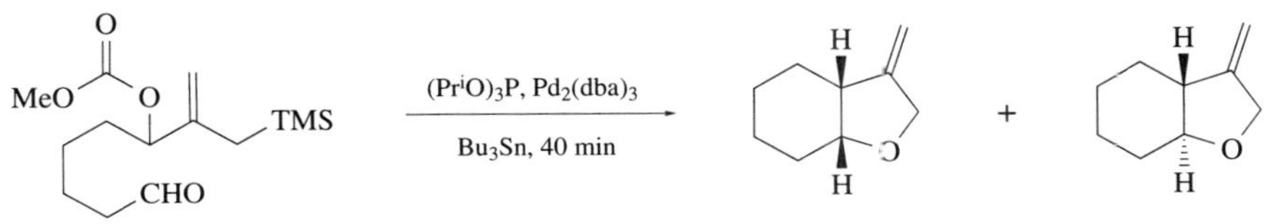

(39)

1.06.3.2.5 Free radical mediated reactions

Titanium dioxide has been used as a catalyst for photochemical oxidation, in the presence of silver sulphate, for the generation of free radical species from a range of benzyltrimethylsilanes, leading to the formation of diarylethanes in good yields with loss of the silyl moieties (Scheme 89) ⟨92CC59⟩. The reaction is said to take place by initial cation radical formation followed by a very fast carbon–silicon bond cleavage, leading to the formation of benzyl radicals.

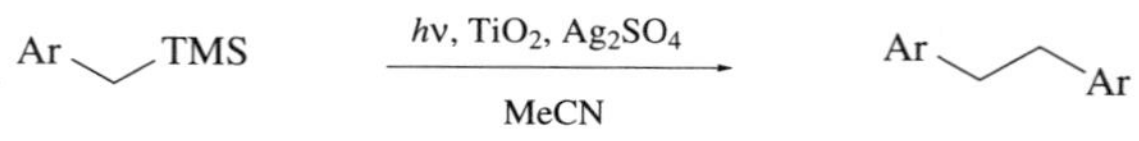

Ar	*Yield (%)*
Ph	74
4-MeC_6H_4	71
3-$CF_3C_6H_4$	50

Scheme 89

Silver sulfate is used as the reducible species in this case and the whole process is carried out using a 500 W high-pressure mercury lamp in acetonitrile as solvent. Under these conditions the reactions proceed efficiently (50–74% yield). The rate of the reaction is highly dependent upon the nature of any ring substituent; it is depressed in the presence of electron-withdrawing substituents and increased by electron-donating substituents. For example, the irradiation time for the reaction when the substrate contained an electron-donating methyl group was two hours, while, when an electron-withdrawing trifluoromethyl group was present, the irradiation time increased to twenty hours.

Some allylic and benzylic silanes undergo similar oxidative coupling induced by oxovanadium species ⟨93TL5601⟩. Treatment of cinnamyltrimethylsilane with $VO(OEt)Cl_2$ in dichloromethane leads to regioselective oxidative coupling to form the 1,5-hexadiene (Scheme 90). The reaction does not occur with less reactive silanes such as allyltrimethylsilane or methallyltrimethylsilane. Interestingly, however, cross coupling of each of these species with the more reactive cinnamyltrimethylsilane is successful in producing the corresponding unsymmetrical dienes in moderate yields (40–50%) together with a small amount of the self-coupling product of cinnamyltrimethylsilane (6–12%).

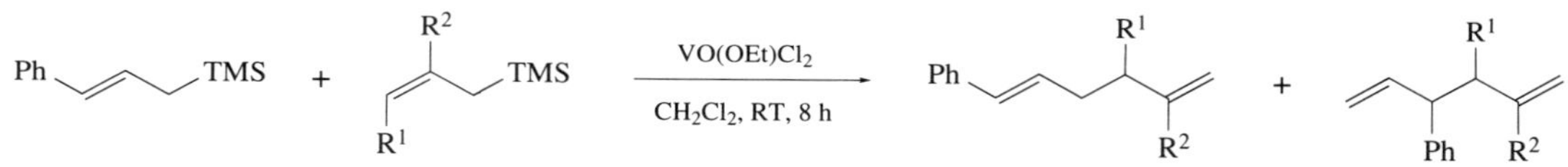

A **B**

R¹	*R²*	*Yield (%)*	*Ratio* **A : B**
H	H	40	2.5 : 1
H	Me	51	1.5 : 1
Me	H	56	2.0 : 1

Scheme 90

1.06.3.3 Substitution of Germanium Functions

There are very few reactions of alkyl organogermanium species in which a new carbon–carbon bond is formed at the expense of the sp^3 carbon–germanium bond ⟨86S776⟩. Reactions of species containing sp^2 carbon-germanium bonds are much more common ⟨88JOM(25)348, 88OM739⟩.

One example of such a reaction of an alkyl organogermanium species involves the preparation of the germanium reagent, triphenylgermylacetyl triphenylgermane, which can act as an acetate equivalent in a Lewis acid-mediated reaction with aldehydes to form the corresponding aldol products (Scheme 91) ⟨87CL495⟩. The germanium reagent was prepared using known methodology for the synthesis of acyl germanes ⟨85CL1829⟩ by reacting the corresponding ester with three equivalents of triphenylgermyllithium at room temperature for ten minutes.

The reagent also reacts with α,β-unsaturated aldehydes to produce the corresponding aldol products, and, perhaps surprisingly, does not undergo any conjugate addition (Equation (40)). The

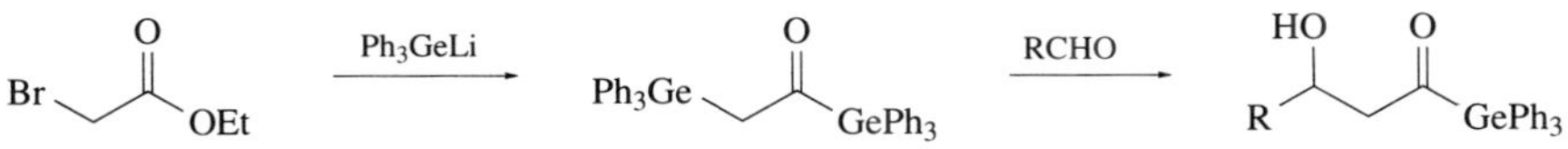

R	*Yield* (%)
Ph	81
PrCHO	75

Scheme 91

yields of the reaction range from 68–81% and no by-products are produced. The reagent does not react with ketones or acetals under similar conditions.

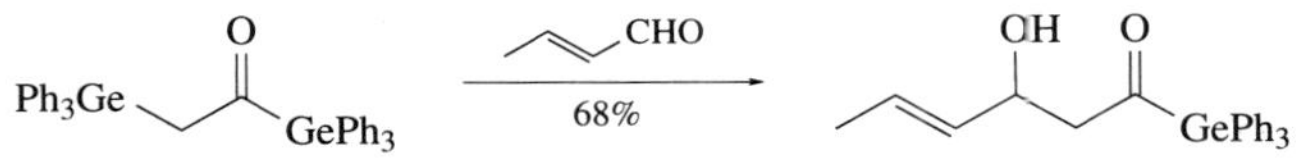

(40)

Ethyl (trimethylgermyl)acetate reacts with carboxylic acid derivatives such as acyl chlorides, acyl imidazoles, thioesters and esters via its lithium enolate ⟨90OM1325⟩. The authors' studies of the reaction suggest that the most likely of the possible reaction pathways is that which involves initial formation of the 3-oxoalkanoic ester enolate through the loss of chlorotrimethylgermane (Scheme 92).

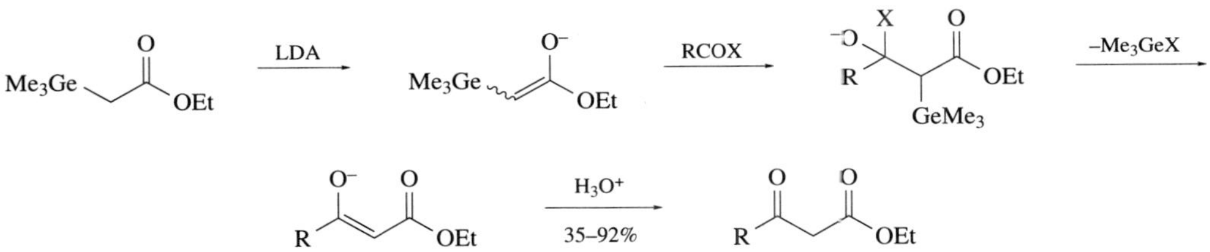

R = Me, Pr, Ph; X = Cl, imidazolide, CN, OEt, SCH_2Ph

Scheme 92

A rather more unusual reaction, investigated by Seyferth ⟨69JOM(16)503⟩, involved the ring expansion of 1,1-diethyl-1-germacyclobutane by dichlorocarbene insertion into a germanium–carbon bond. When 1,1-diethyl-1-germacyclobutane is allowed to react with phenyl(bromodichloromethyl) mercury (in slight excess) in benzene solution at reflux, the major product obtained, in 35% yield, was shown to be 1,1-diethyl-2,2-dichloro-1-germacyclopentane (Equation (41)). Ring expansion has therefore occurred, in the first case of carbene insertion into a carbon–germanium bond.

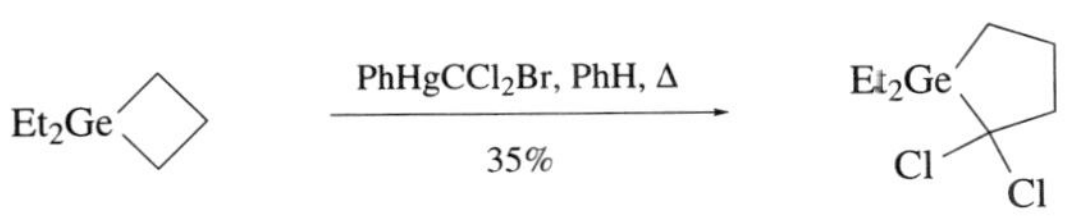

(41)

1.06.4 SUBSTITUTION OF METAL FUNCTIONS

In the literature, reactions are very rare in which two organometallic species containing different metals, each having the metal atom bonded to an sp^3 carbon atom, undergo exchange to give the product of coupling of the alkyl groups together with a new bimetallic species.

In one such reaction, conjugated dienes can be activated towards nucleophilic attack by complexation with tricarbonyliron followed by treatment with fluoroboric acid under an atmosphere of carbon monoxide. The resulting cationic salts are known to undergo nucleophilic addition by a wide variety of soft nucleophiles. These include triphenylphosphine, pyridine, malonates, secondary amines, electron-rich aromatic compounds and organocadmium reagents ⟨92CC13⟩. For example, dialkylcadmium reagents were shown to react with an acyclic allyliron complex to provide mainly

(*Z*)-alkenes in reasonable yields. The reactions proceed regioselectively and with retention of alkene configuration (Equation (42)) ⟨75TL3617⟩.

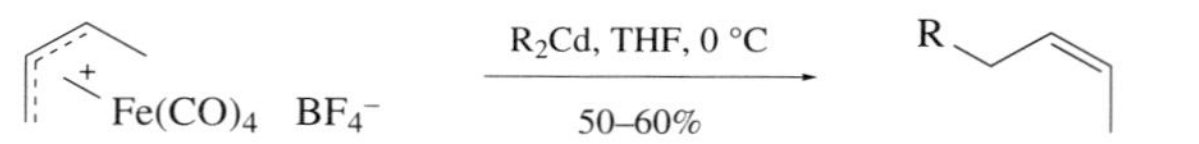

(42)

Highly functionalized copper zinc reagents can be prepared by transmetallation from the corresponding functionalized zinc compounds with cuprous cyanide in THF solution. These react effectively with the above cationic salts to give a variety of polyfunctionalized isoprenyl derivatives including isoprenyl compounds and β,γ-unsaturated ketones (Scheme 93).

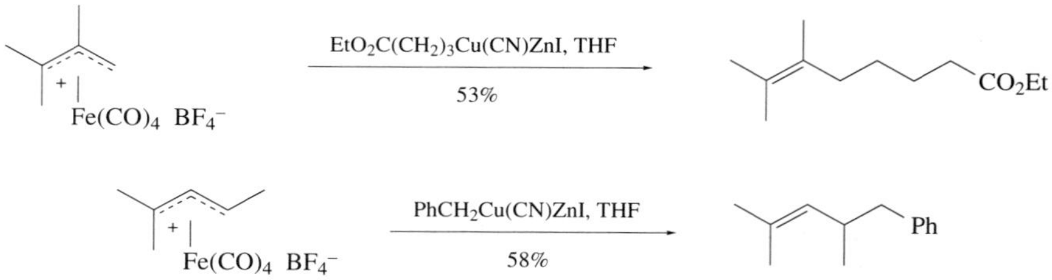

Scheme 93

The cations undergo preferential nucleophilic attack at the less substituted allyl terminus with complete regioselectivity, yielding principally alkenic products with a trace of ketones (Scheme 94). Ketones are thought to arise by initial addition to a coordinated carbon monoxide ligand, producing a metal–acyl intermediate. This is followed by migration of the acyl group to the less substituted terminus of the allyl unit and detachment of the iron carbonyl moiety. One example shows approximately equal amounts of alkene and ketone products, possibly due to steric factors at the allyl terminus of the cation. However, this is not reflected upon reaction with benzylic copper reagents.

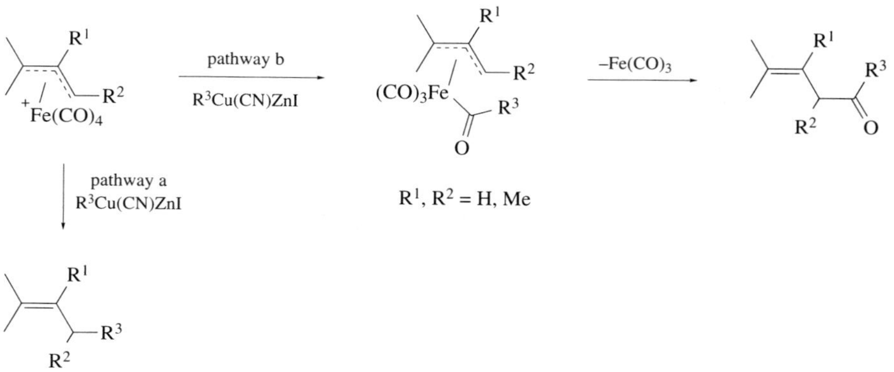

Scheme 94

1.07
One or More CC Bond(s) Formed by Addition: Addition of Carbon Electrophiles and Nucleophiles to CC Multiple Bonds

ALAN ARMSTRONG
University of Bath, UK

1.07.1 INTRODUCTION

Addition of carbon electrophiles and nucleophiles to CC multiple bonds encompasses a wide range of important bond forming processes. The emphasis of this chapter will be on important principles and on reactions displaying particular synthetic utility. Since it is the formation of tetracoordinate products that is of interest, virtually all the addition processes discussed will involve addition to alkenes rather than to alkynes.

1.07.2 ADDITION OF CARBON ELECTROPHILES TO CC MULTIPLE BONDS TO GIVE TETRACOORDINATE PRODUCTS

The initial product of formal electrophilic addition to an alkene is a carbocation (Scheme 1); a general discussion of this process appears elsewhere in this volume (see Chapter 1.07.2.3). The carbocation can undergo any of the subsequent transformations typical of these species (reaction with a nucleophile, loss of proton to form an alkene, rearrangement, etc.). In many cases the final product contains functionality (e.g. an alkene) that is the subject of later chapters, so only limited coverage is presented here. Also included in this section are several important methods for carbon–carbon bond formation (such as hydroformylation) that do not involve obvious electrophilic or nucleophilic reactants.

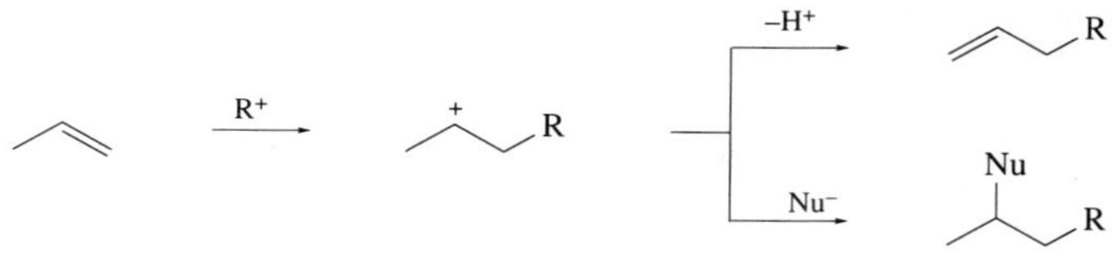

Scheme 1

1.07.2.1 Addition of Alkanes to Alkenes

The addition of alkanes to alkenes can be carried out under thermal conditions, where a radical mechanism is almost certainly operating, or under acid catalysis, where carbocation intermediates are involved ⟨B-64MI 107-01, 90AG(E)1371⟩. The first step under acidic conditions is addition of a proton to the alkene to give the carbocation (**1**), which abstracts hydride from the alkane to give the new carbocation (**2**), capable of undergoing addition to the alkene with formation of a new C—C bond (Scheme 2). Given the obvious potential problems of regioselectivity, carbocation rearrangement and polymerisation, this process is rarely used in the laboratory, but is of industrial importance.

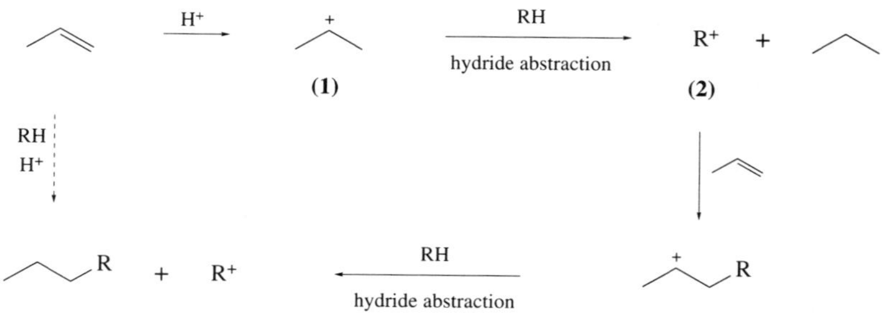

Scheme 2

1.07.2.2 Addition of Alkenes to Alkenes

Reaction of alkenes with alkenes can be accomplished under acidic conditions, again via a carbocation intermediate (Scheme 3). Intermolecular examples are largely limited to dimerization in view of the obvious likelihood of product mixtures in any attempted reaction between two different alkenes ⟨68RCR789⟩. Even the dimerization reactions suffer from problems of polymer formation, and may give poor regioselectivity with respect to both initial carbocation formation and in the position of the final double bond. Dimerization can also be accomplished using transition metal catalysts ⟨86CRV353⟩ or bases ⟨74S309⟩. Intramolecular examples are of greater synthetic utility, and much attention has been focused on cascade processes where several rings are formed in the same reaction. A spectacular precedent is, of course, provided by nature in the enzymatic cyclization of squalene oxide to dammaradienol during the course of steroid biosynthesis (Equation (1)).

This type of cyclization has been successfully mimicked in the laboratory using various electrophilic activating groups ⟨76AG(E)9, B-84MI 107-01, 91Txi⟩. An impressive example, where six new stereocentres are created, is shown in Equation (2) ⟨82JOC161⟩.

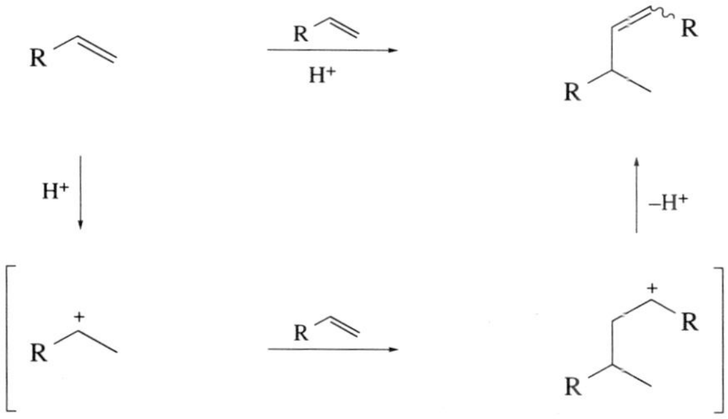

Scheme 3

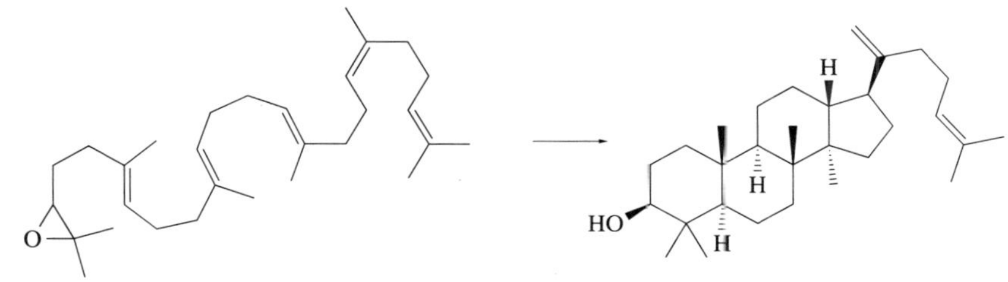

(1)

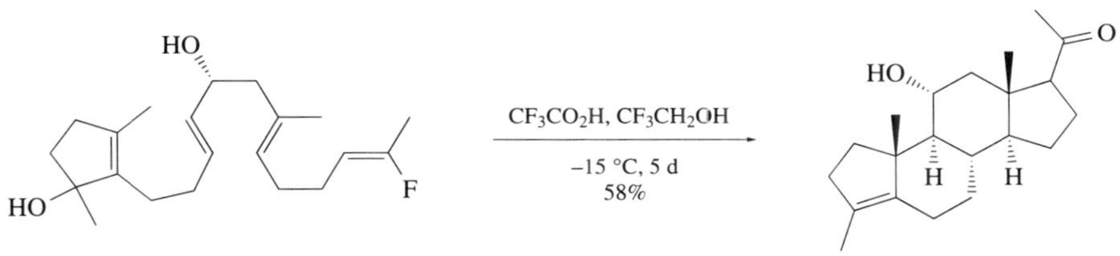

(2)

1.07.2.3 Addition of Electrophiles to Alkenes bearing Cation Stabilizing Groups

A common obstacle to the use of electrophilic addition processes such as those described above is the possibility of the formation of more than one intermediate carbocation, either because of nonregioselective addition of the electrophile to the alkene or because of subsequent rearrangement. In either case, a mixture of products may result. Alkenes bearing heteroatomic functionality capable of cation stabilization, on the other hand, can display high regioselectivity in reacting with electrophiles. Allyl silanes, for example, undergo electrophilic addition in the presence of a Lewis acid resulting in the formation of positive charge β- to silicon (Scheme 4) ⟨B-81MI 107-01⟩. Final loss of the silyl group generates a new alkene. Typical electrophiles include acid chlorides, alkyl halides, α,β-unsaturated carbonyl systems (see Section 1.07.3.2.3(iv)) and acetals.

Another class of compounds that react readily with electrophiles with high or exclusive regio-

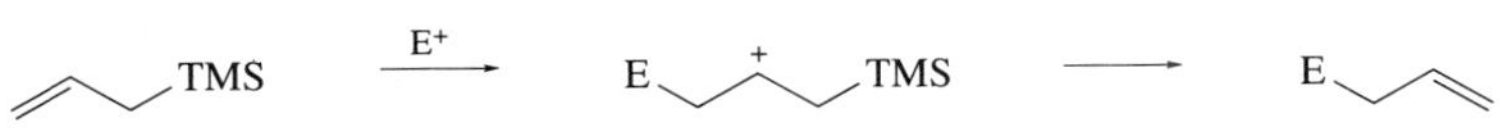

Scheme 4

selectivity are those with heteroatomic functionality bonded directly to an alkene unit, for example, enol ethers and enamines (Scheme 5). However, the final product in this case is usually a ketone, so transformations of this type will not be covered in detail in this section.

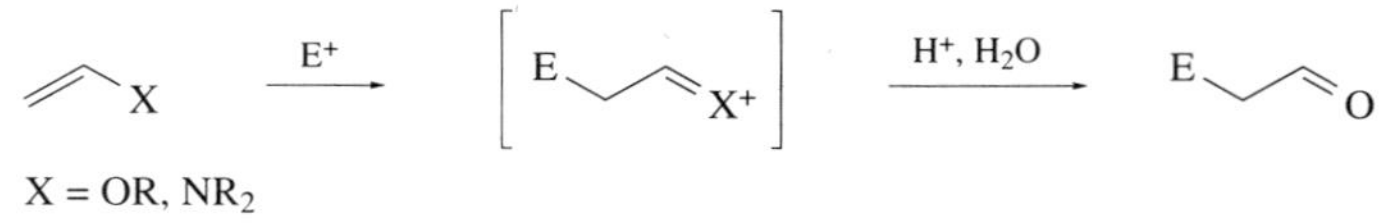

Scheme 5

1.07.2.4 Hydroformylation and Hydrocarboxylation

Hydroformylation and hydrocarboxylation (Equation (3)) are extremely important industrial processes for the preparation of aldehydes and carboxylic acid derivatives, respectively. With appropriate choice of catalyst and conditions, many reactions can also be effected readily in the laboratory. These processes and their possible mechanisms have been reviewed ⟨B-80MI 107-01, 82COMC-I(8)101, 91COS(4)913⟩. Although exact mechanistic details are uncertain, two distinct general pathways are often written. One possibility (Scheme 6) begins with the insertion of the alkene into a metal hydride bond, followed by carbonyl insertion and reaction of the metal–acyl complex with H—X. In the alternative (Scheme 7), a metal–carboxyl complex (**3**) is formed initially and it is this into which the alkene inserts. Again, reaction with H—X completes the transformation. Both mechanisms are possible depending on the particular reaction, the metal catalyst and the ligands employed.

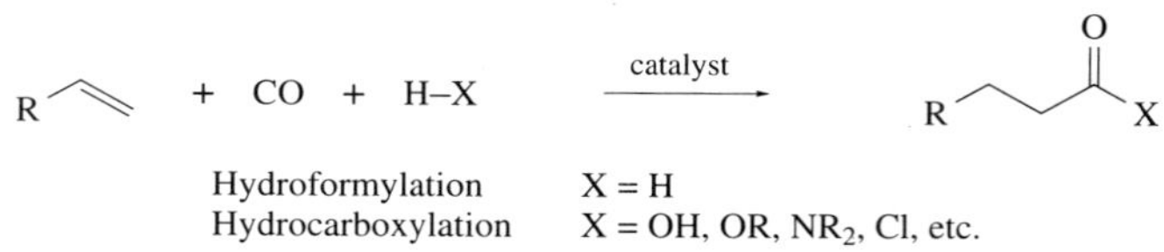

(3)

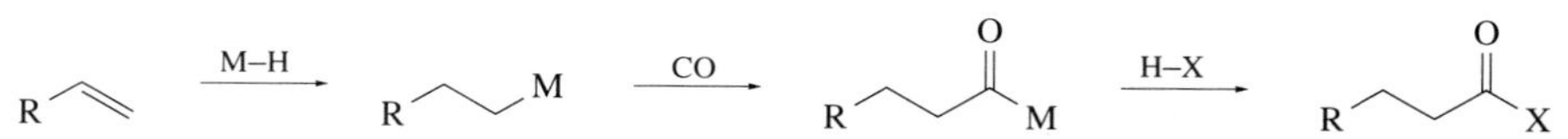

Scheme 6

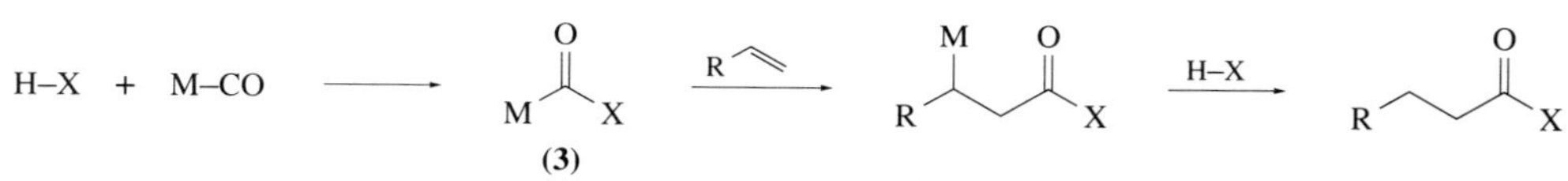

(3)

Scheme 7

1.07.2.4.1 Hydroformylation

While many transition metal complexes can act as catalysts for hydroformylation, those derived from cobalt (e.g., $Co_2(CO)_8$) are most widely used industrially ⟨B-80MI 107-01⟩. Rhodium-based catalysts are more reactive and so allow lower pressures and temperatures to be used. Many common functional groups, including acetals, alcohols and ethers, are tolerant of the hydroformylation conditions. For unsymmetrical alkenes, ratios of regioisomeric products are dependent upon the substrate structure and the exact reaction conditions, including the nature of the ligands on the transition metal. The reaction appears to be sensitive to steric effects, and the order of reactivity is generally straight chain terminal alkene > straight chain internal alkene > branched chain alkene. The aldehyde group usually becomes attached to the less hindered end of the alkene. However, electronic effects are also influential. For example, addition to allyl ether (**4**) gives selectively the regioisomer (**5**) (Equation (4)) ⟨74CA(81)120222t, 80AG(E)178⟩.

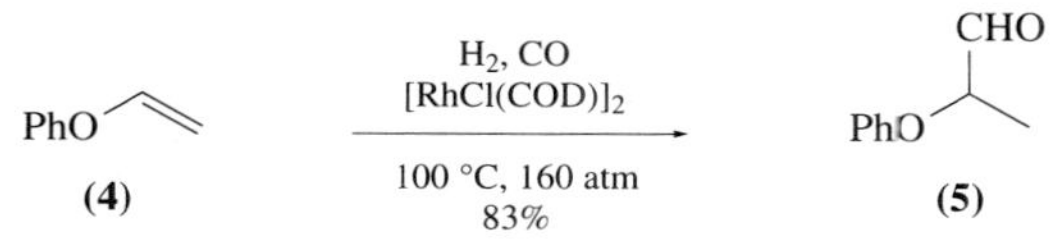

(4)

Hydroformylation proceeds with *syn*-stereochemistry, as shown by the stereoselective conversion of 1-methylcyclohexene to *trans*-2-methylcyclohexanecarbaldehyde (Equation (5)) ⟨77JA1058⟩. The asymmetric variant has received attention, with variable results ⟨93MI 107-01⟩; the need to control chemoselectivity (hydrogenation often competes) and regioselectivity complicate the problem. Success using chiral phosphinephosphite rhodium complexes is illustrated in Scheme 8 ⟨93JA7033⟩. Use of a vinyl acetate as substrate here assures good regioselectivity; the simple terminal alkene (**6**) gave poorer results in this respect.

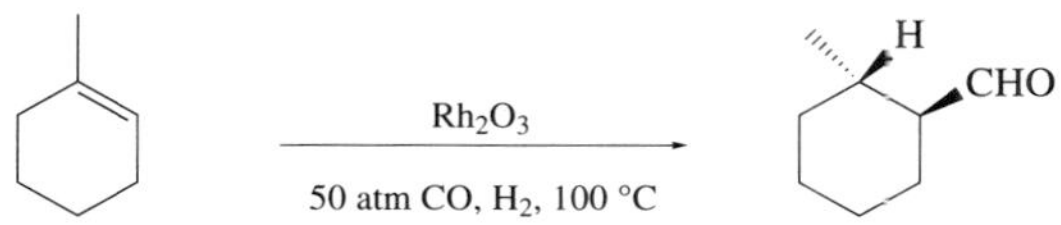

(5)

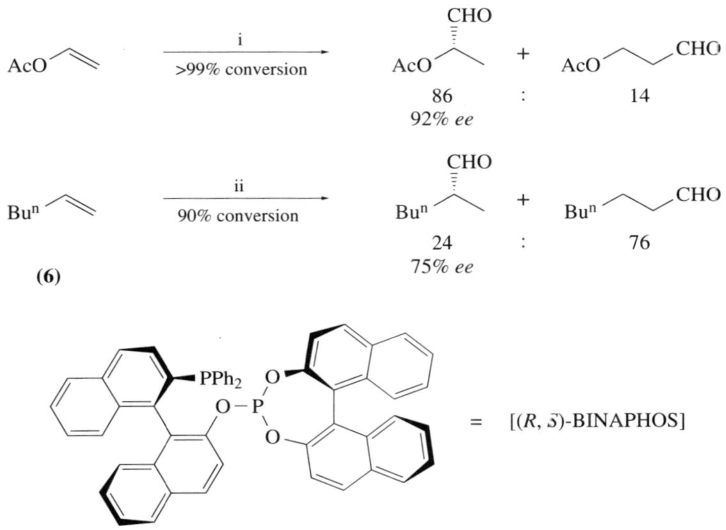

i, 0.25 mol% $Rh(acac)(CO)_2$, 0.54 mol% [(*R*, *S*)-BINAPHOS], 1:1 H_2:CC, 100 atm, benzene, 60 °C, 36 h;
ii, 0.1 mol% $Rh(acac)(CO)_2$, 0.22 mol% [(*R*, *S*)-BINAPHOS], 1:1 H_2:CO, 100 atm, benzene, 30 °C, 93 h

Scheme 8

1.07.2.4.2 Hydrocarboxylation

As shown in Equation (3), hydrocarboxylation processes are carried out using carbon monoxide and a catalyst in a protic solvent, H—X ⟨B-80MI 107-01, 82COMC-I(8)101⟩. Since there are several possibilities for H—X (water, alcohols, acids, amines, etc.), many different carboxylic acid derivatives are accessible. While mineral acids alone can be used, catalysts based on transition metals are usually employed. Nickel-based catalysts are common as they give high chemoselectivity. As is the case for hydroformylation, the product ratios with unsymmetrical alkenes are strongly dependent on the reaction conditions, but terminal alkenes generally give predominantly the α-methyl branched acid with nickel catalysts. Alper *et al.* have reported a system for regioselective hydrocarboxylation under mild conditions (Equation (6)) ⟨83CC1270⟩.

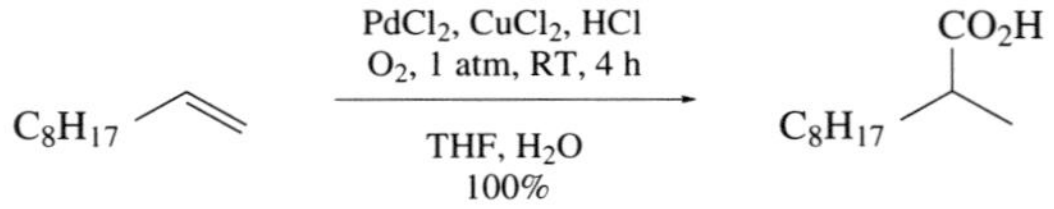

(6)

The asymmetric variant of this process was used in a synthesis of the anti-inflammatory drug ibuprofen (**7**) (Equation (7)) ⟨90JA2803⟩.

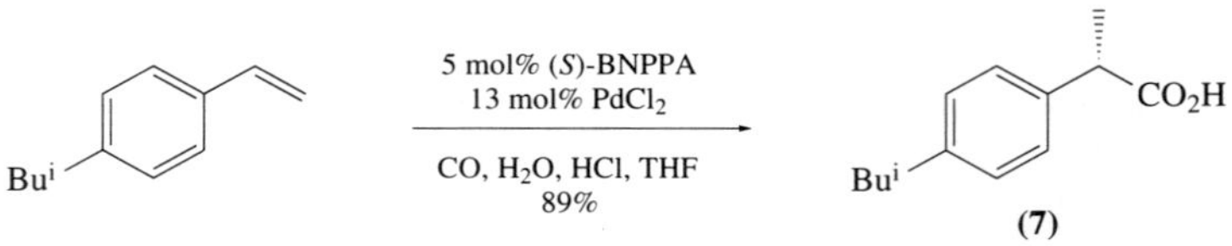

(7)

(7)
83% *ee*

BNPPA = 1,1'-binaphthyl-2,2'diyl hydrogen phosphate

1.07.2.4.3 Dicarboxylation

Dicarboxylation, the 1,2-addition of two ester groups to an alkene, can be achieved under mild conditions (1–3 atm, 25 °C) in an alcohol solvent (Equation (8)) ⟨76JA1810⟩. The process is carried out catalytically in palladium(II) with copper(II) chloride as the stoichiometric co-oxidant. In the absence of sodium acetate, acyclic alkenes give β-methoxy esters as the major product. The addition proceeds predominantly with *syn*-stereochemistry ⟨76JA1806⟩.

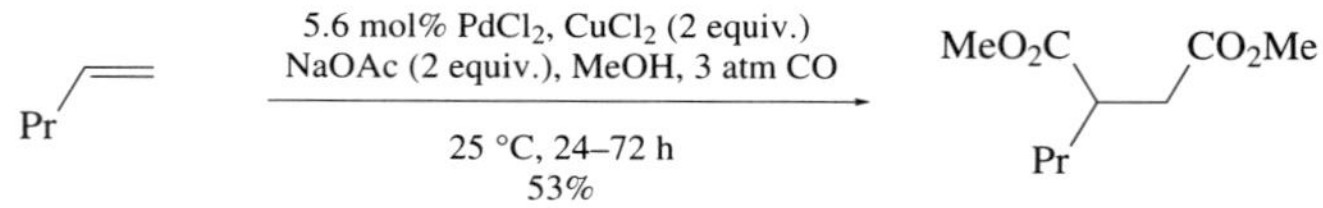

(8)

1.07.2.5 Cyclopropanation

One of the most common methods for cyclopropane formation is the formal addition of a carbene or carbene equivalent to an alkene (Equation (9)) ⟨91COS(4)951⟩. Since most carbenes are electrophilic, these reactions are in general fastest with electron-rich alkenes. Diazo compounds are commonly used as the alkylidene source, and the process can be carried out thermally, photochemically or in the presence of a metal catalyst. Under thermal conditions, the reaction proceeds via a 1,3-dipolar cycloaddition followed by loss of nitrogen (Scheme 9); the intermediate pyrazoline (**8**) is often not observed as it is unstable. The reaction is stereospecific, i.e. the *cis* : *trans* ratio of the starting alkene is retained in the product cyclopropanes, but the reaction is often slow for simple, unstrained alkenes.

(9)

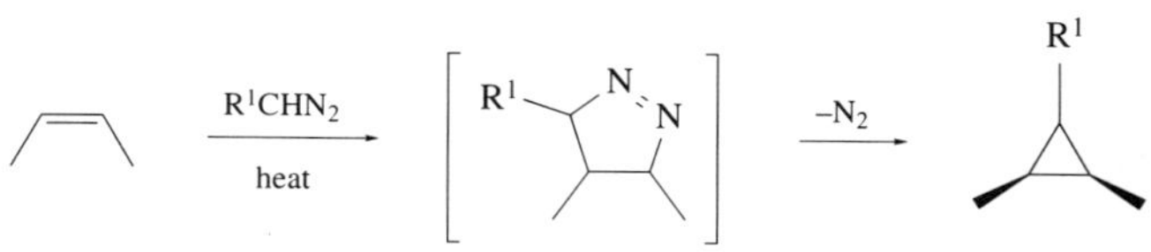

(8)

Scheme 9

When photochemical conditions are employed, the reaction proceeds via a carbene intermediate. Stereospecificity depends on the spin state of the carbene; singlet carbenes, generally obtained when a diazo compound is photolysed directly, add stereospecifically, whereas triplet carbenes, obtained in the presence of a sensitizer, display non-stereospecific addition (Equation (10)) ⟨56JA5447, 59JA3383, 62JA1015⟩.

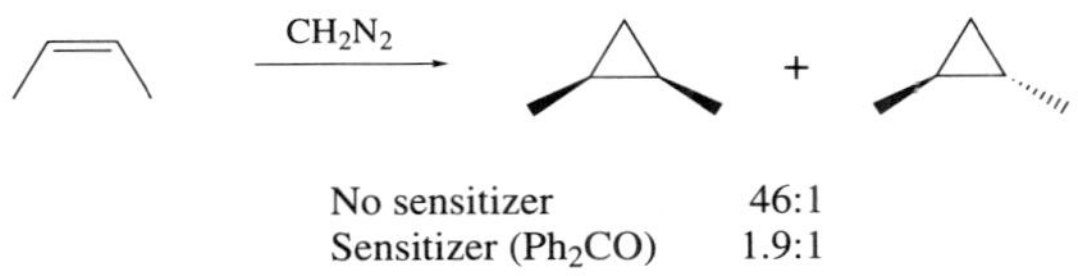

(10)

In the mid-1980s the use of metal catalysts largely superseded the thermal and photochemical methods of alkene cyclopropanation using diazoalkanes ⟨86CRV919⟩. Soluble copper complexes and rhodium(II) carboxylates are particularly popular. The metal carbenoid species involved are less prone to rearrangement than are free carbenes, and react stereospecifically with respect to alkene geometry. Diazocarbonyl compounds are commonly employed ⟨91COS(4)1031⟩. Another issue of diastereoselectivity arises here, that of the relative stereochemistry between the substituents on the alkene and those on the carbenoid. Poor results are usually obtained, however (Equation (11)) ⟨84OM44⟩. Intramolecular examples are more successful in this respect (Equation (12)) ⟨77TL109⟩.

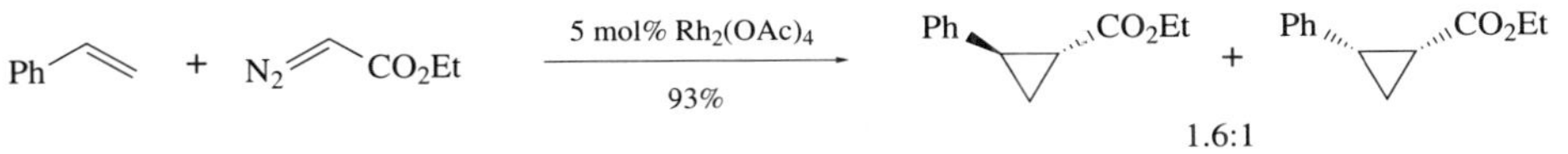

(11)

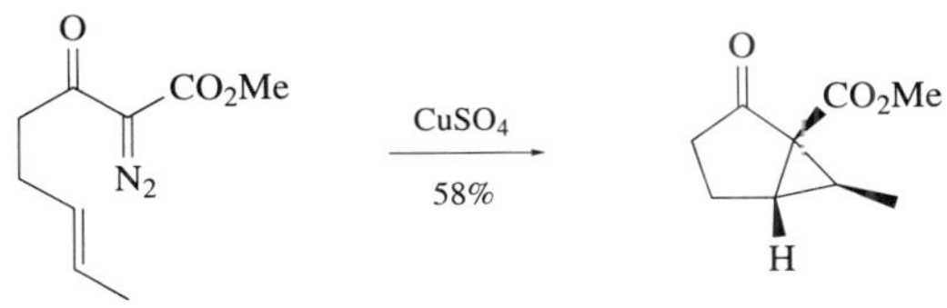

(12)

In 1994, a great deal of effort was directed towards developing an asymmetric variant, although much remains to be achieved in terms of efficiency (an excess of alkene is often employed) as well as diastereo- and enantioselectivity. An example is shown in Equation (13) ⟨94JA2223⟩.

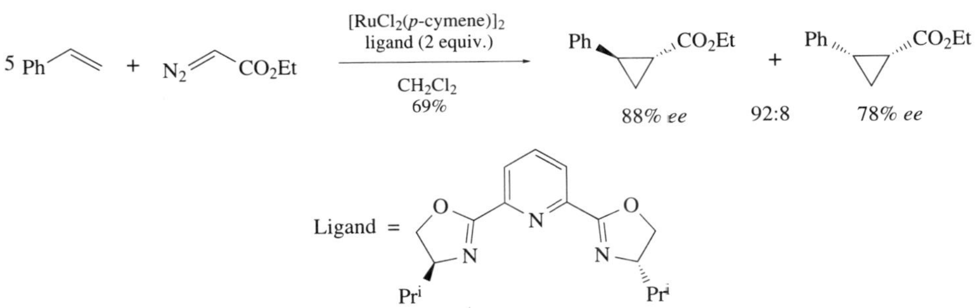

(13)

Geminal dihaloalkanes are also widely used as precursors to carbene equivalents in cyclopropanation reactions, the best known example of this being the Simmons–Smith reaction (Equation (14)) ⟨73OR(20)1⟩. This procedure generally suffers less from side reactions than do procedures involving the use of diazo compounds, and is compatible with many common functional groups (halides, alcohols, amines, esters, amides, ketones, ketals). While the exact mechanism remains uncertain, a species such as ICH_2ZnI is believed to be involved. The transformation is stereospecific, proceeding with retention of the alkene geometry. Electronic effects have a weak influence, with a small preference for electron-rich alkenes being observed, but steric effects often play a dominant role. Importantly, heteroatomic functionality can exert powerful directing effects (Equation (15)) ⟨93CRV1307⟩, allowing control of regio- and stereochemistry.

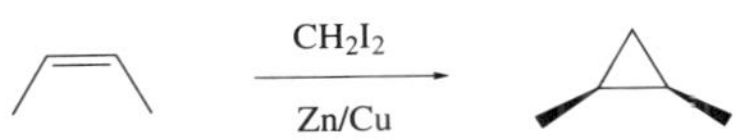

(14)

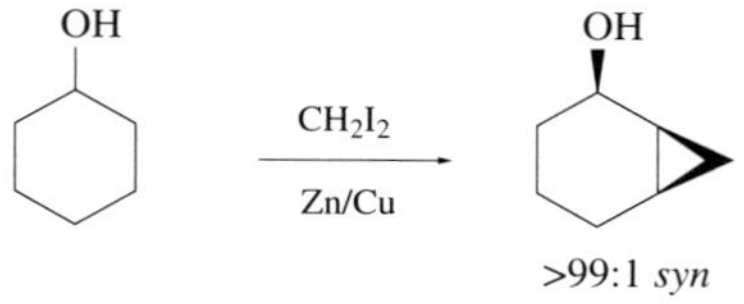

(15)

Many alternative experimental procedures have been developed for the generation of the Simmons–Smith cyclopropanation reagent. For transfer of substituted alkylidenes, use of the substituted diiodomethane $RCHI_2$ and diethyl zinc has proved successful (Equation (16)) ⟨69T2647⟩, but the disubstituted variants $R_2CI(ZnI)$ are apparently too unstable for use in cyclopropanation.

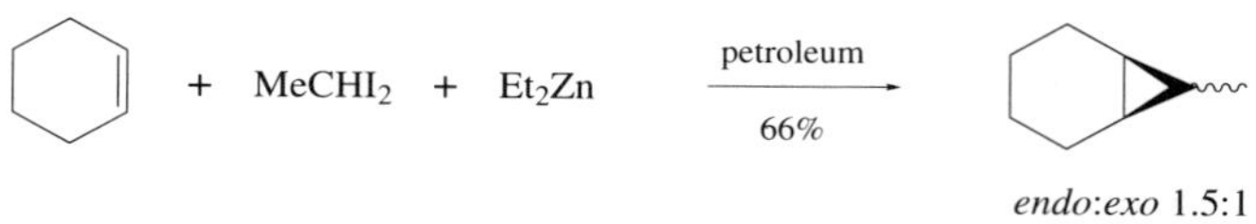

(16)

Methods for cyclopropane formation involving initial nucleophilic attack on the alkene are covered elsewhere in this chapter (see Section 1.07.3.2.1(iii)).

1.07.3 ADDITION OF CARBON NUCLEOPHILES TO CC MULTIPLE BONDS TO GIVE TETRACOORDINATE PRODUCTS

1.07.3.1 Introduction

This section considers addition of carbon nucleophiles to alkenes to form new carbon–carbon single bonds (Equation (17)). Nucleophilic attack on the electron-rich alkene π-cloud is greatly facilitated by the presence of electron-withdrawing groups on the alkene, and the vast majority of reactions fall into this category. Addition reactions of this type are often known as conjugate addition, or, in the case where a carbonyl group activates the alkene, 1,4-addition. When stabilized nucleophiles are used, the process is sometimes called the Michael reaction after one of the earliest contributors to the field (see Section 1.07.3.2.2(i)).

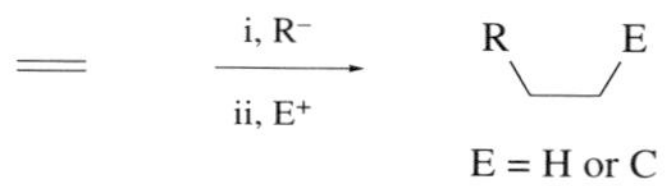

(17)

Developments in alkene activation by transition metal complexes have allowed the reaction of unactivated alkenes with organometallic reagents to become a viable synthetic transformation and these processes are also discussed.

1.07.3.2 Addition to Electron-deficient Alkenes

1.07.3.2.1 Introduction and general aspects

Nucleophilic addition to activated carbon–carbon multiple bonds to give tetracoordinate products encompasses a wide range of important bond-forming processes. Since this chapter can survey only a small fraction of the enormous range of nucleophiles, substrates and reaction conditions that have been used, the emphasis will be placed upon the general principles that have emerged for the prediction of reaction products, and upon examples that illustrate particular synthetic utility or selectivity. For more detailed coverage of nucleophilic addition to alkenes, the reader is referred to relevant reviews ⟨91COS(4)1, 91COS(4)69, B-92MI 107-01⟩.

(i) Reactivity and regioselectivity

The range of electron-withdrawing groups that activate alkenes towards nucleophiles is large, and includes carbonyl compounds (aldehydes, ketones, esters, amides, acids), nitriles and heteroatomic functionality (sulfoxides, sulfones, nitro groups, phosphonates, etc.). To a first approximation, it is reasonable to assume that the ability of these groups to activate alkenes will correlate with their relative ability to stabilize an adjacent carbanion, so the order of reactivity with some common activating groups is likely to be:

$$NO_2 > COR > CO_2R > SO_2R > CN \sim CONR_2$$

In a survey of the reactivity of α-heterocarbanions in conjugate addition reactions, Krief gave the order of reactivity shown in Figure 1, and pointed out that this followed both their electrolytic reduction half-wave potential (important for additions proceeding via an electron transfer mechanism; *vide infra*) and their LUMO energy ⟨80T2531⟩.

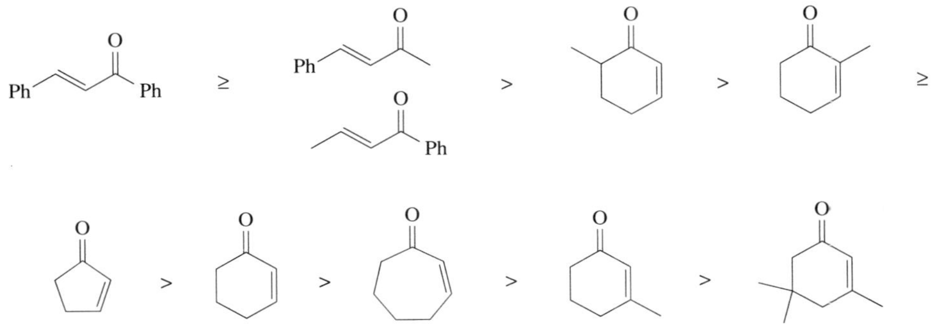

Figure 1 Order of reactivity of some conjugate acceptors towards α-heteroatoms.

For α,β-unsaturated carbonyl compounds, it is of course necessary to consider the possibility of 1,2-addition competing with 1,4-addition (Scheme 10). Several factors influence the relative importance of the two pathways; in general, anything that facilitates reaction reversibility will tend to favour formation of the 1,4-adduct over the 1,2-adduct. This is because, for a reversible process or one involving a late transition state, the position of attack is governed by the relative stabilities of the intermediates formed; the oxyanion (**9**) is clearly less stable than the enolate ion (**10**). For reactions having an early transition state (likely to be irreversible processes), frontier molecular orbital analysis can be applied ⟨B-76MI 107-01⟩, and used as a theoretical basis for the widely used hard/soft acids and bases (HSAB) concept. Overlap of the LUMO of the conjugate acceptor with the HOMO of the nucleophile must be considered as this leads to maximum stabilization. For 'soft' nucleophiles (those with a high energy HOMO), attack will be at the position of the conjugate acceptor which has the highest LUMO coefficient. Conversely, 'hard' nucleophiles (low energy HOMO) will attack at the position of highest partial charge. For acrolein, the simplest conjugate acceptor, calculations indicate that the greatest partial positive charge resides on the carbonyl carbon atom (Figure 2). However, since the largest LUMO coefficient is on the β-carbon, 'soft' nucleophiles should undergo 1,4-addition in preference to direct attack on the carbonyl group. These ideas explain the effect of the nature of the (organo)metal associated with the carbanion, one of the most important aspects in determining the regioselectivity of addition. Since the selectivity displayed by various organometallic reagents with specific substrates will be discussed in the proceeding sections, only general comments will be made at this point. Organolithium reagents ('hard') tend to afford products of 1,2-addition, whereas organocopper reagents ('soft') effect 1,4-addition. Organomagnesiums (Grignard reagents) show intermediate properties, but 1,2-addition predominates for most substrates. Steric factors are also important, bulky anions usually attacking 1,4. Thus, addition of *t*-butylmagnesium chloride to cyclohexenone gives a good yield of the 1,4 addition product (Equation (18)) ⟨41JA758⟩.

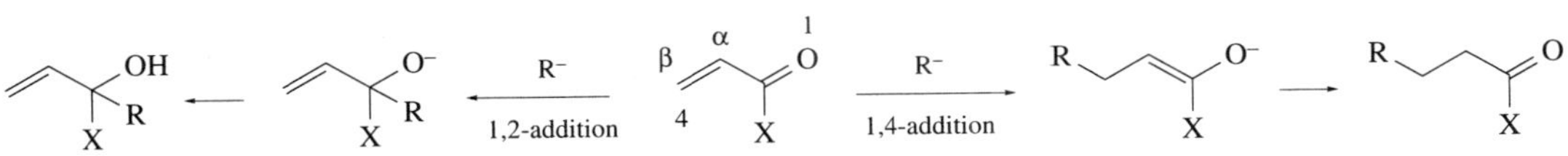

(9) **(10)**

Scheme 10

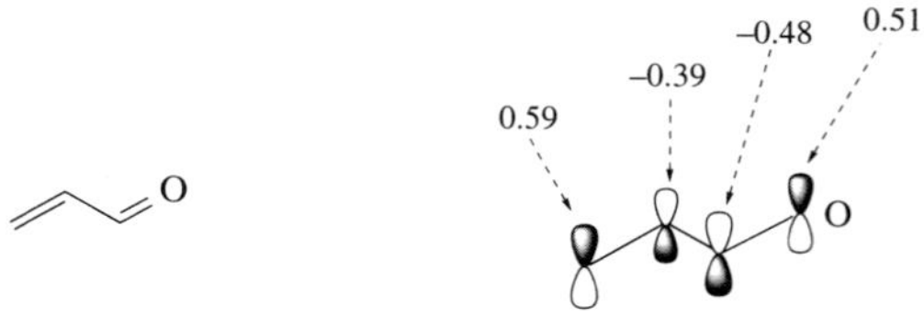

Figure 2 LUMO coefficients for acrolein.

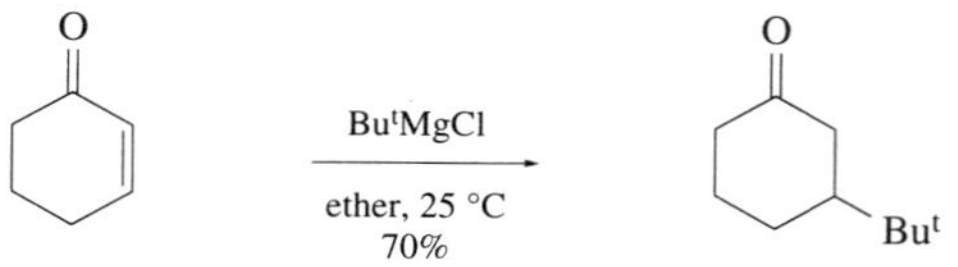

(18)

Generalizations concerning 1,2- vs. 1,4-addition may also be made according to the class of carbonyl acceptor. In general, α,β-unsaturated aldehydes display a strong tendency for 1,2-addition. For α,β-unsaturated ketones, the product ratio is subject to the largely predictable factors outlined throughout this section. Steric hindrance in the substrate can have a great effect on the ratio of addition products, with bulk in the region of either the carbonyl group or the β-carbon atom reducing the amount of attack at that position. β,β-Disubstituted α,β-unsaturated carbonyl systems are generally poor conjugate acceptors for this reason.

Appropriate choice of solvent can also control the addition product ratio. It has been found empirically that polar solvents such as hexamethylphosphoramide (HMPA) or DMF promote 1,4-addition, whereas solvents of low polarity favour 1,2-addition. A model that explains this trend has been put forward by Cohen *et al.* ⟨87JA7923⟩. An equilibrium is considered to exist between contact ion pairs (CIP) and solvent separated ion pairs (SSIP) (Scheme 11). To a first approximation, it is assumed that contact ion pairs undergo only 1,2-addition, via a four-centre transition state (**11**). Solvent separated ion pairs, on the other hand, are assumed to react only by 1,4-addition. This mode of attack is preferred whether the transition state is early (attack at position with highest LUMO coefficient) or late (giving the more stable product, the enolate). The effect of polar solvents such as HMPA is thus readily explained by their tendency to promote SSIP formation. The model also provides an explanation for the increased proportion of 1,4-addition observed for bulky nucleophiles.

Another prediction arising from this model is that lower temperatures, which strongly favour

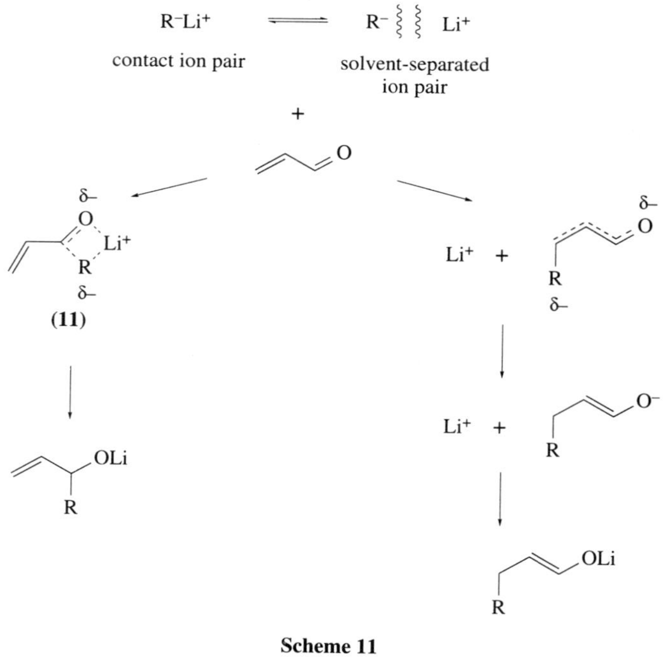

Scheme 11

SSIP, should promote conjugate addition for reactions under kinetic control. This prediction was confirmed for a study of the addition of bis(phenylthio)methane to cyclohexenone (Scheme 12) ⟨87JA7923⟩. For reversible processes, of course, higher temperatures (and longer reaction times) would be expected to favour formation of the thermodynamically more stable, conjugate addition product. This is clearly seen in the addition of α-alkoxy ester enolates to cyclohexenone (Scheme 13) ⟨76JOC4044⟩; evidently the addition of these relatively stable carbanions is reversible.

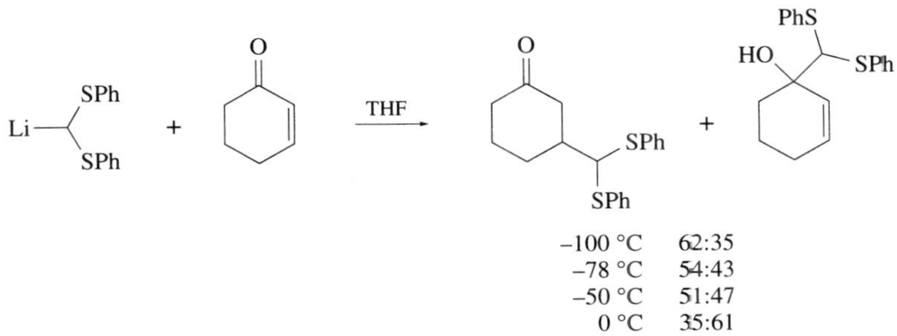

Scheme 12

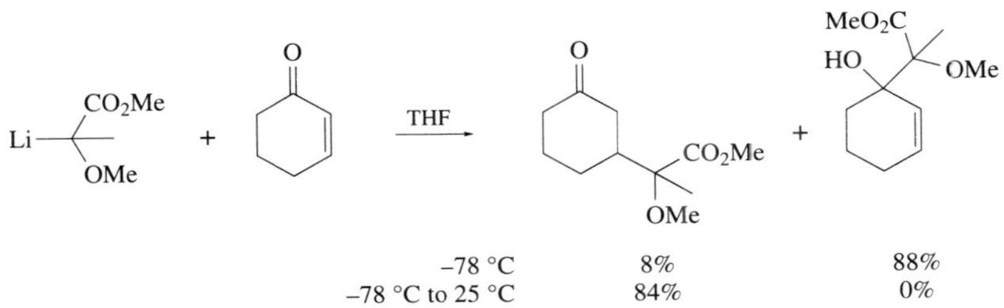

Scheme 13

A general strategy to provide the conjugate addition product using a nucleophile which might otherwise add in a 1,2-fashion is to place an anionic group α- to the carbonyl group, thus disfavouring direct (1,2-) attack. An example of this approach, known as charge-directed conjugate addition, is shown in Scheme 14 ⟨77JA642⟩.

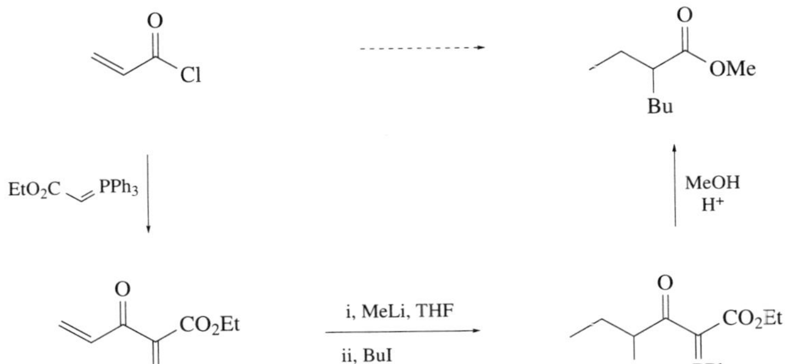

Scheme 14

Another strategy for promoting 1,4-addition over 1,2-addition in the addition of unstabilized nucleophiles to α,β-unsaturated aldehydes was reported by Yamamoto and co-workers ⟨94JA4131⟩. The key to this approach involves complexation of the carbonyl group with the extremely bulky Lewis acid aluminum tris(2,6-triphenylphenoxide) (ATPH) (**12**) (Equation (19)). Attack in a 1,2-fashion on the complexed substrate is presumably strongly disfavoured for steric reasons.

Conjugate addition of lithium enolates of ketones and esters, silyl enol ethers and silyl ketene acetals was also shown to be successful.

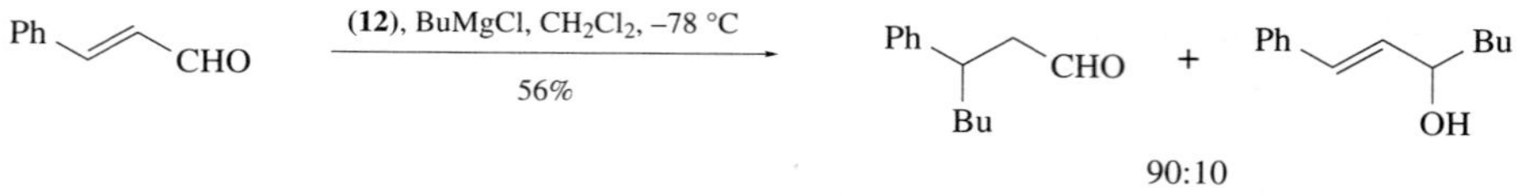

(19)

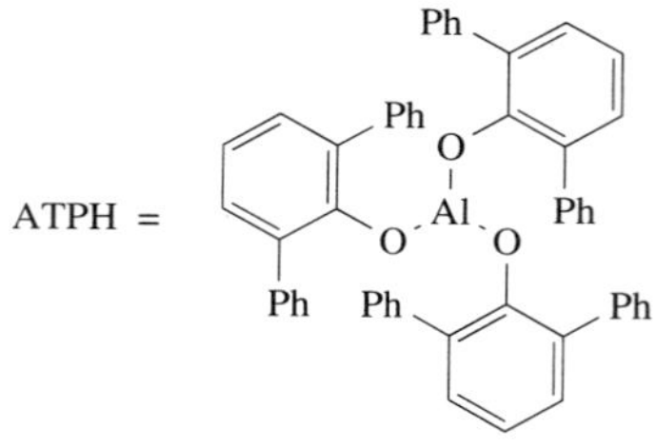

(12)

(ii) Intramolecular processes

While most of the examples of addition of carbon nucleophiles to alkenes presented in this chapter are intermolecular, the corresponding intramolecular processes are generally feasible provided the nucleophile can be generated in the presence of other functionality in the molecule. It is also important that any stereoelectronic requirements for the addition reaction are met; these are summarized in Baldwin's rules for ring closure ⟨76CC734⟩, a special set of which was presented for ring closure reactions involving enolate nucleophiles ⟨77CC233, 82T2939⟩. Examples of intramolecular reactions will be given in later sections.

(iii) Trapping processes

An outstandingly useful aspect of many of the conjugate addition reactions described here is the possibility of trapping the initially formed carbanion with an electrophile, thus allowing the formation of two new carbon–carbon bonds from an alkene (Equation (20)) ⟨77TL3215⟩. Such processes have been reviewed ⟨85S364, 90OR(38)225, 91COS(4)237⟩.

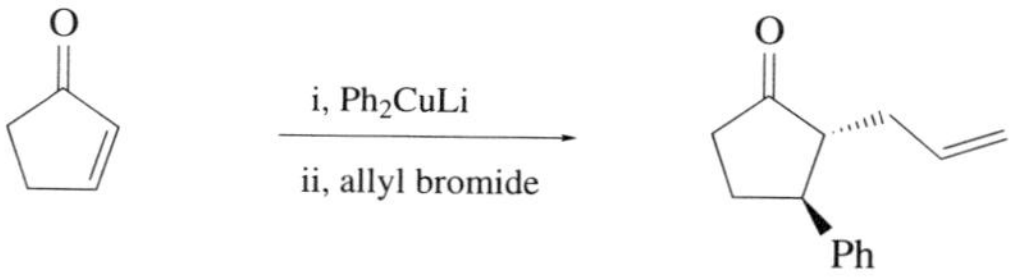

(20)

Several methods for cyclopropanation of alkenes rely on nucleophilic addition where the anion stabilizing group of the nucleophile subsequently acts as a leaving group (Scheme 15). Rotation about the central C—C single bond before closure of the cyclopropane ring is possible, and so the configuration of the alkene may be lost in the product cyclopropane. Sulfur ylides are the most widely used reagents for cyclopropanations of this type. Dimethylsulfoxonium methylide (**13**), introduced by Corey, is commonly used (Equation (21)) ⟨65JA1353, 87T2609⟩. The simpler sulfur ylide, dimethylsulfonium methylide (Me_2SCH_2), generally affords epoxides from enones via reaction at the carbonyl group. Optically active anions such as structure (**14**), derived from sulfoximines, have been used to provide cyclopropanes with moderate enantiomeric excesses (Equation (22)) ⟨73JA7418⟩.

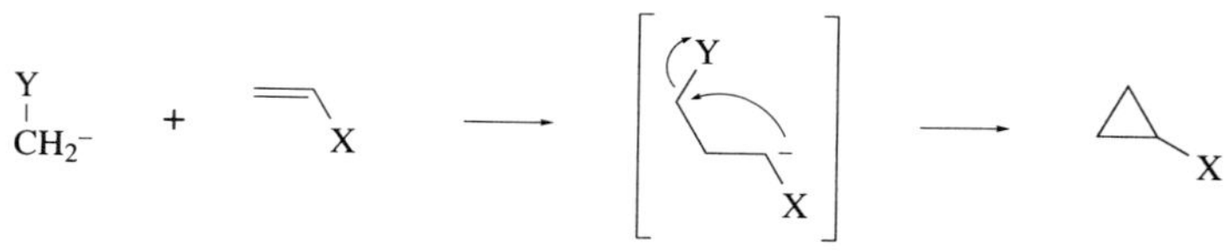

Scheme 15

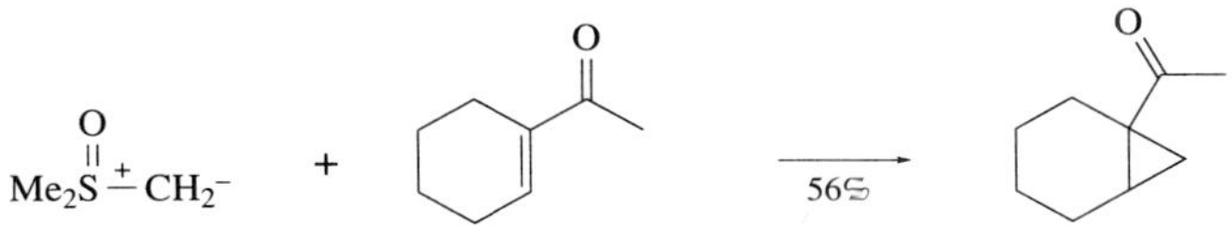

(13)

(21)

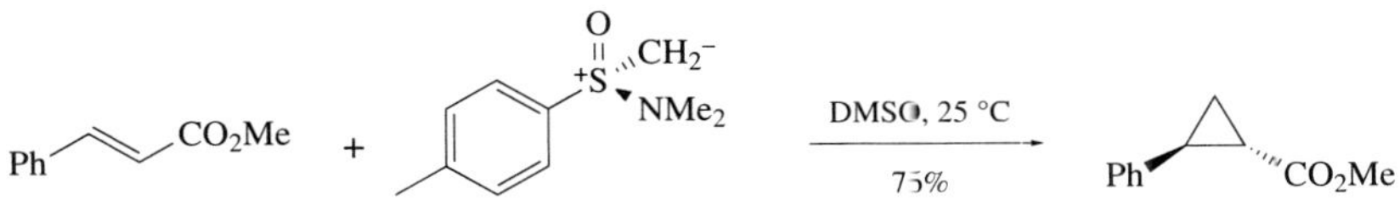

(14)

30% *ee*

(22)

(iv) Stereoselectivity

Control of relative stereochemistry in the reaction of achiral nucleophiles with chiral substrates is often an important issue in conjugate addition chemistry. These reactions can be divided into two main categories—those where chirality is present in the main body of the conjugate acceptor (Equation (23)), and those where a chiral electron-withdrawing group activates the alkene (Equation (24)). For reactions of the type in Equation (23), the stereochemistry observed in many examples can be rationalized by simple ideas involving steric approach control, especially for cyclic conjugate acceptors. Rationalization of stereoselectivity with acyclic substrates with chirality at the γ-position has largely been based on extensions of the Felkin–Anh model ⟨83TL2231, 87CC1572, 92TL7521, 94MI 107-01⟩.

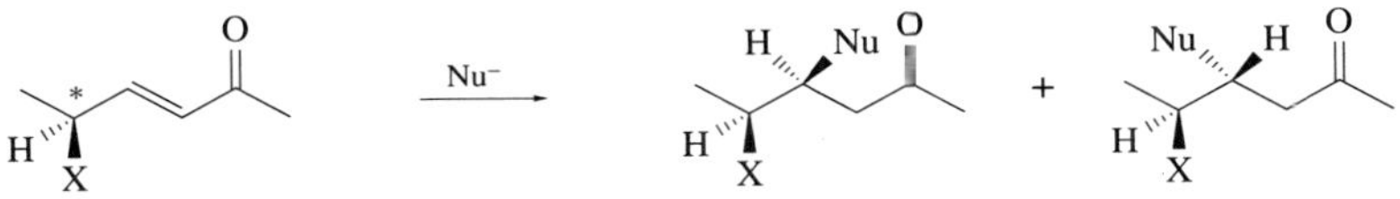

(23)

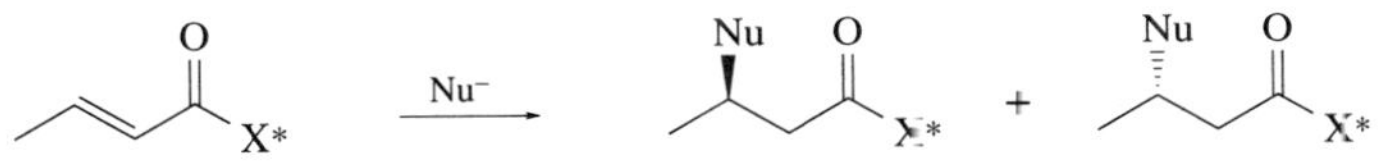

(24)

Use of the electron-withdrawing group as a chiral auxiliary (as in Equation (24)) has been a widely used strategy in asymmetric synthesis ⟨91COS(4)199, 92CRV771, 94MI 107-01⟩. Chiral carboxylic acid derivatives such as esters and amides are particularly suitable as their introduction is simple and they are easily removed following the conjugate addition reaction. The sultam derivatives of Oppolzer provide a good example of this approach (Equation (25)) ⟨87HCA2201, 87T1969⟩. The sulfoxide is another unit that has seen widespread use as a chiral auxiliary, particularly in the α-position of an α,β-unsaturated carbonyl compound. It is noteworthy that the product stereochemistry can sometimes be controlled by choice of reaction conditions. Using dialkyl magnesiums in THF, reaction occurs through the conformation where the sulfur–oxygen dipole opposes the carbonyl dipole (Equation (26)). Nucleophilic attack on the less hindered face (opposite to the bulky *p*-tolyl group) followed by reductive removal of the sulfoxide moiety affords the product of (*S*)-configuration for the sulfoxide enantiomer shown in Equation (26). Addition of zinc bromide promotes chelation between the sulfoxide and carbonyl oxygens, and attack on the less hindered face now affords the (*R*)-product (Equation (27)) ⟨87ACR72⟩.

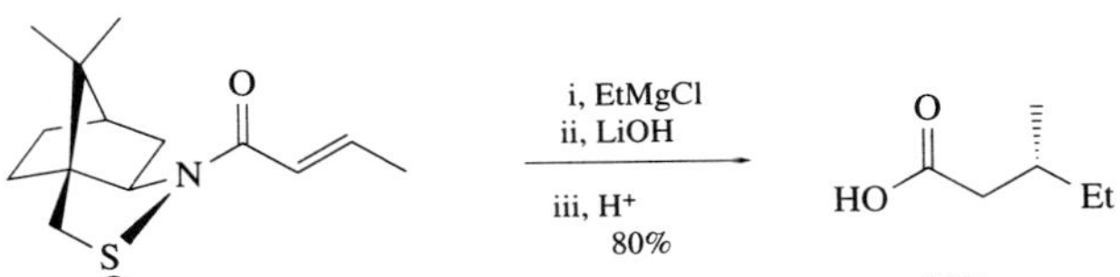

89% *ee*

(25)

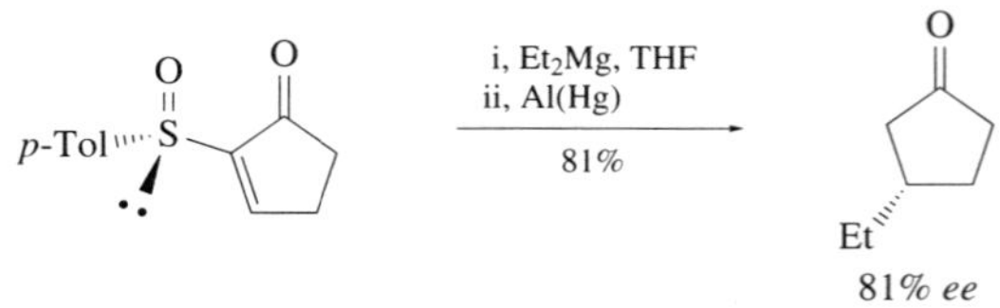

(26)

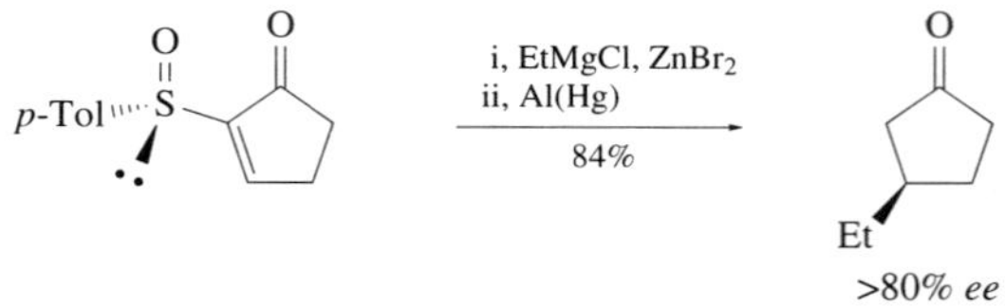

(27)

The development of practical chiral catalysts for asymmetric conjugate addition of unstabilized nucleophiles to prochiral electron-deficient alkenes is one of the most important current goals in asymmetric synthesis ⟨91COS(4)199, 92CRV771, 94MI 107-01⟩. The ideal situation of being able to add catalytic quantities of a chiral ligand to effect high enantioselectivity in the addition of common organometals (organolithiums, Grignards, cuprates) to a range of conjugate acceptors remains to be achieved.

1.07.3.2.2 *Addition of stabilized nucleophiles*

(i) Addition of enolates and enolate equivalents

The use of stabilized nucleophiles in conjugate addition processes has its origins in the nineteenth century with the study by Komnenos ⟨1883LA(218)145⟩ and then by Michael ⟨1887MI 107-01, 1887JPR349⟩ on the addition of malonate salts to α,β-unsaturated esters (Scheme 16). Since this early work, a huge range of stabilized nucleophiles and acceptors have been shown to react in this way, and the Michael reaction, as it is commonly known, has been the subject of several reviews ⟨91COS(4)1⟩. The reversible nature of addition reactions involving relatively stable carbanions means that these nucleophiles generally undergo selective 1,4-addition to α,β-unsaturated carbonyl compounds. Relatively weak bases (e.g., amines) are required. Since the initial addition product is the enolate **(15)**, which can itself effect deprotonation of the starting material, the reaction can often be performed with only catalytic amounts of added base. If stoichiometric base is employed, then **(15)** can potentially be alkylated, allowing formation of a further carbon–carbon bond.

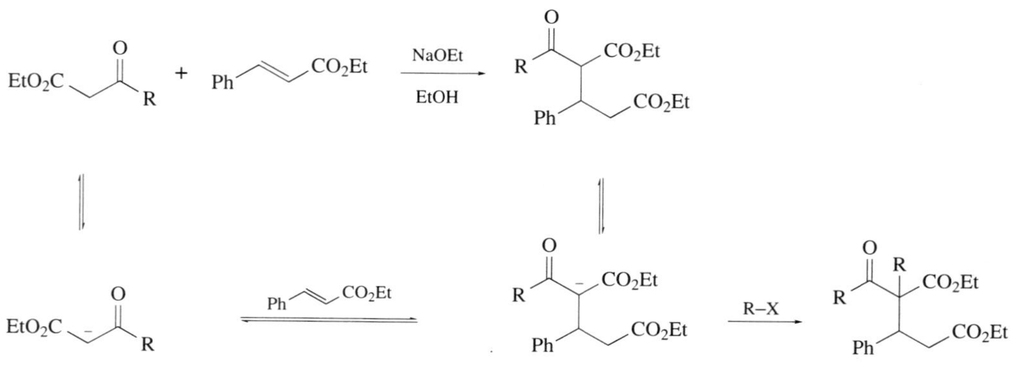

Scheme 16

Michael addition of ketone enolates is an important process, and forms part of the outstandingly useful Robinson annelation (Equation (28)). This transformation has been reviewed many times ⟨76S777, 76T3, 79COC(1)1043⟩ and will not be discussed in detail here, since the emphasis of this review is on the Michael reaction itself. When unsymmetrical ketones are used, reaction usually takes place at the more substituted α-position (Equation (29)) via formation of the thermodynamically more stable enolate ⟨66JOC646⟩. This preference may be reversed by using enamines in place of ketone enolates (Equation (30)) ⟨63JA207⟩.

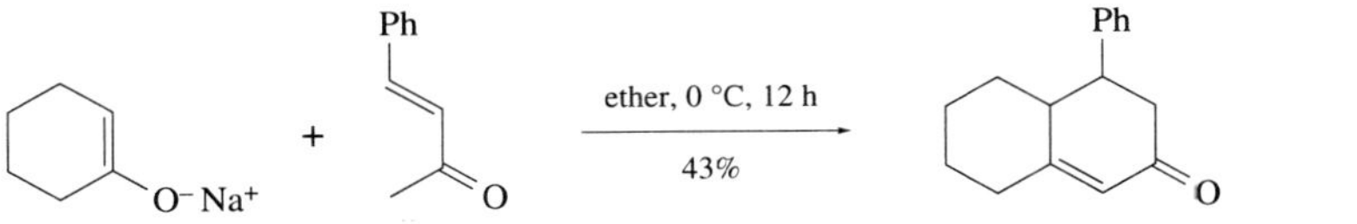

(28)

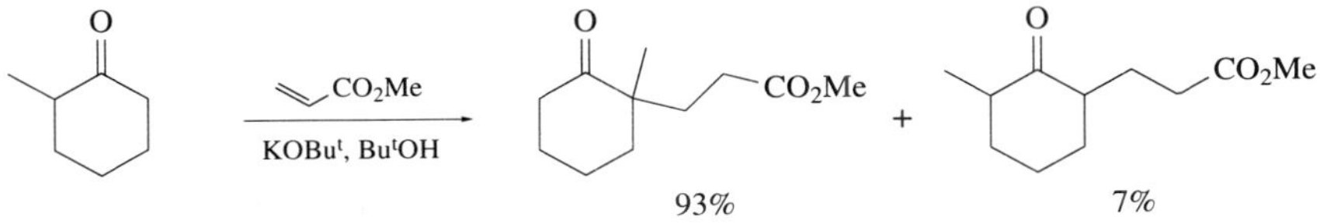

(29)

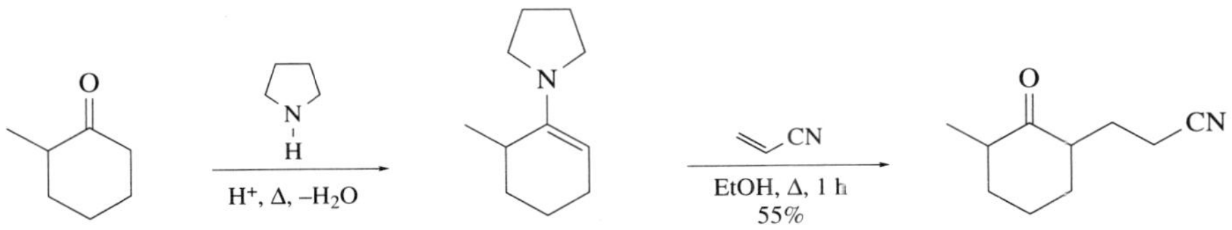

(30)

The Michael reaction was traditionally performed under protic conditions, but work in the early 1970s, primarily from the groups of Boeckman and Stork, pioneered the successful use of aprotic conditions. Previous attempts at quenching kinetic enolates, generated in aprotic solvents, with simple enones had led to scrambling of the enolate (using the enone as proton source) and to polymerization of the enone. Stork demonstrated that introduction of a trialkylsilyl group in the α-position of the enone prevented polymerization (Scheme 17) ⟨73JA6152⟩. This methodology was originally believed to be unsuitable for the trapping of regio-unstable kinetic enolates due to the rapid rate of enolate equilibration relative to conjugate addition. It was then discovered that this was not the case provided that water and protic solvents were rigorously excluded (Scheme 18) ⟨74JA6179, 74JA6181⟩.

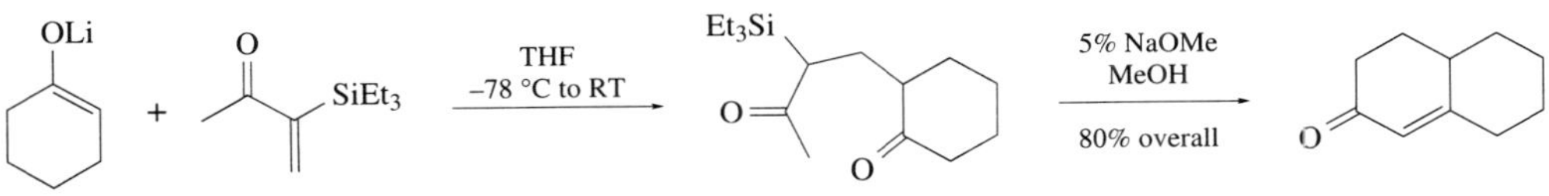

Scheme 17

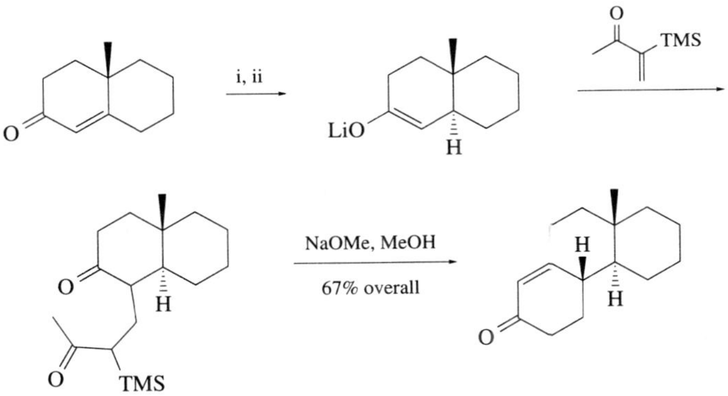

i, Li, NH_3, Bu^tOH, THF, TMS-Cl, Et_3N; ii, MeLi, GLYME

Scheme 18

Intramolecular Michael addition of enolates can be performed successfully to provide carbocycles (Equation (31)) ⟨78TL2445⟩ or heterocycles (Equation (32)) ⟨88TL2195⟩.

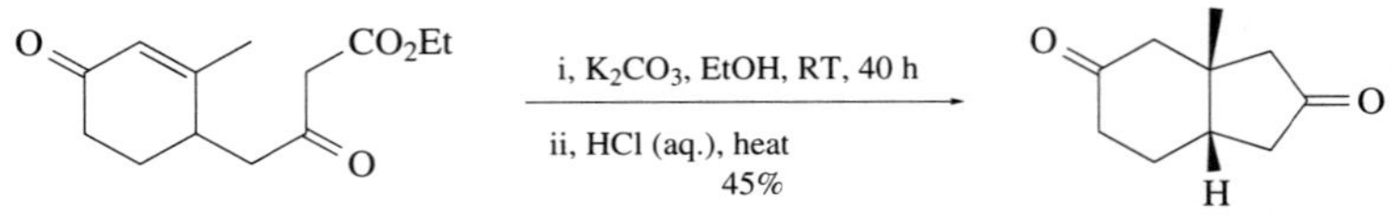

(31)

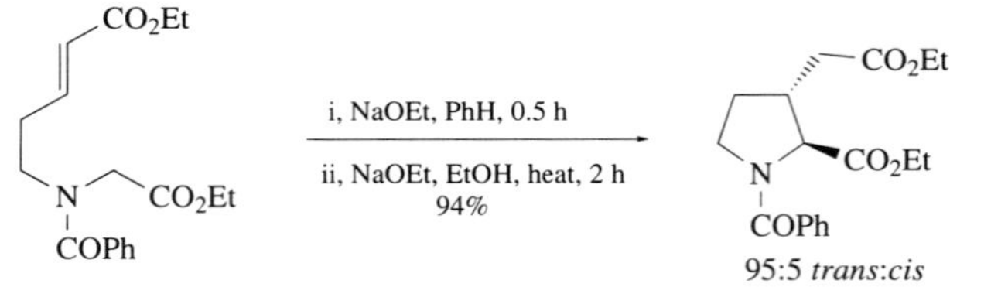

(32)

An extremely useful alternative to the base-mediated addition of enolate derivatives mentioned above is the use of silyl enol ethers (Equation (33)) ⟨76BCJ779⟩ and silyl ketene acetals (Equation (34)) ⟨76CL163⟩ in the presence of a Lewis acid. It is noteworthy that this procedure, known as the Mukaiyama–Michael reaction, can be used for the preparation of quaternary carbon centres. 1,2-Addition does not compete, and regiodefined kinetic or thermodynamic silyl enol ethers ⟨83S1⟩ do not undergo scrambling.

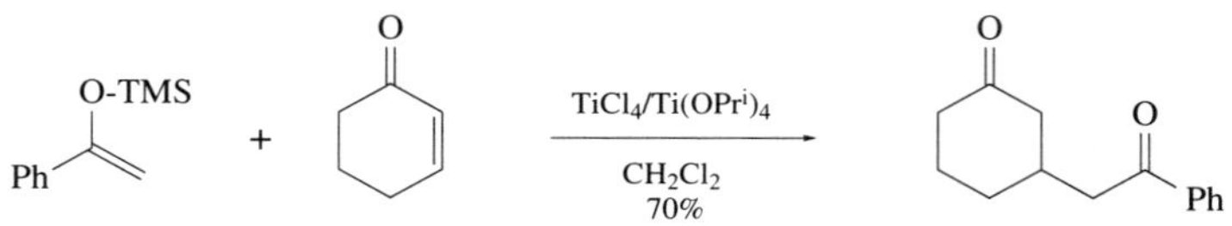

(33)

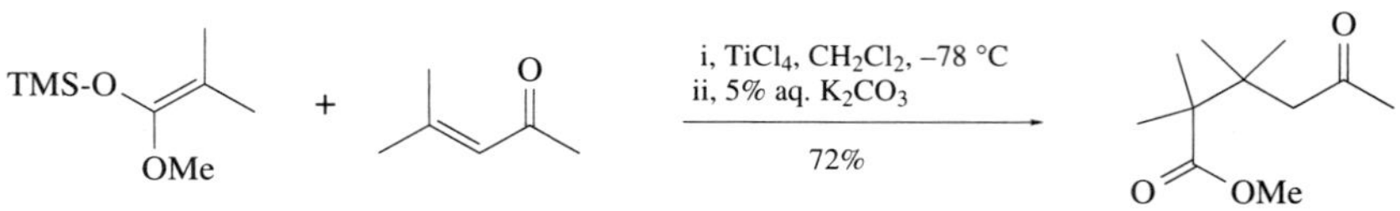

(34)

(ii) Addition of acyl anion equivalents

Conjugate addition of acyl anion equivalents is a powerful method for the formation of 1,4-dicarbonyl compounds (Scheme 19). Simple 1,3-dithianes generally undergo 1,2- rather than 1,4-addition with $\alpha\beta$-unsaturated carbonyl derivatives, although the actual product ratios are strongly dependent on the reaction conditions. Study of this process provided information about many of the general effects of temperature and solvent on 1,2- vs. 1,4-addition (see Section 1.07.3.2.1(i)). A key observation was the discovery by Brown that addition of HMPA resulted in high selectivity for the 1,4-product (Equation (35)) ⟨79CC100, 82H(18)187⟩. This is in accord with the SSIP/CIP concept presented in Section 1.07.3.2.1(i).

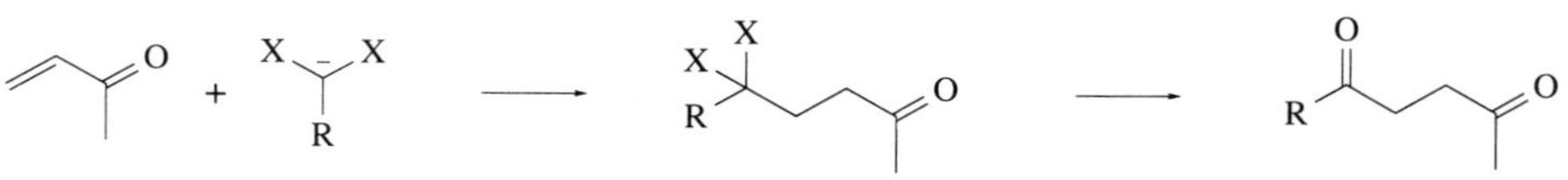

Scheme 19

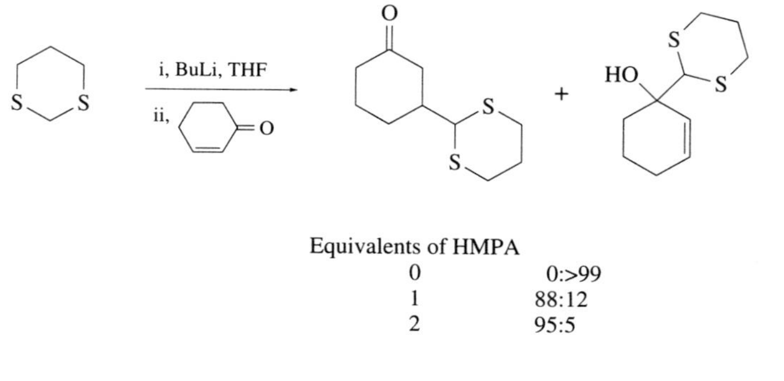

(35)

Equivalents of HMPA	
0	0:>99
1	88:12
2	95:5

Several useful alternatives to 1,3-dithianes have been developed which undergo addition with good 1,4-selectivity. Schlessinger and co-workers introduced the glyoxalate-derived dithiane (**16**) for this purpose and demonstrated the possibility of conversion of the dithiane to a carbonyl group or to a ketal following conjugate addition (Scheme 20) ⟨73TL2595⟩. Stork and Maldonado found that cyanohydrin derivatives (**17**) obtained from simple aldehydes underwent a mixture of 1,2- and 1,4-addition (Equation (36)) ⟨74JA5272⟩. A general strategy to favour the desired 1,4-pathway, however, was to make the addition reaction reversible, by use of a more stabilized anion. Thus, protected cyanohydrins such as structure (**18**), derived from α,β-unsaturated aldehydes, provided predominantly the conjugate addition product (Equation (37)) ⟨74JA5272⟩.

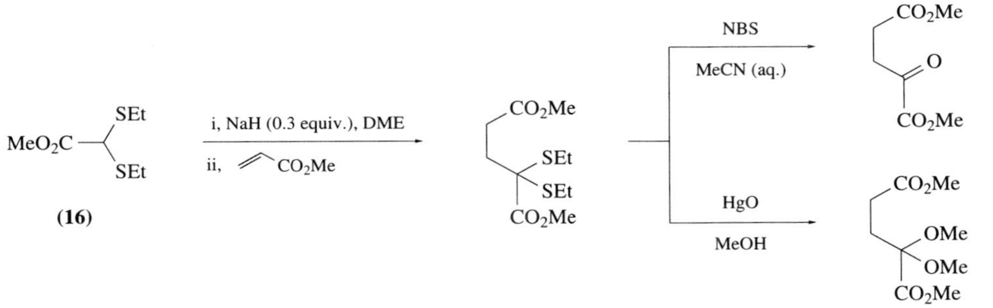

Scheme 20

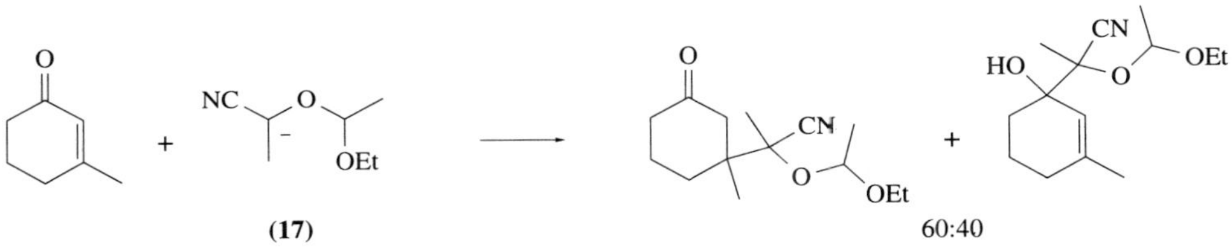

(36)

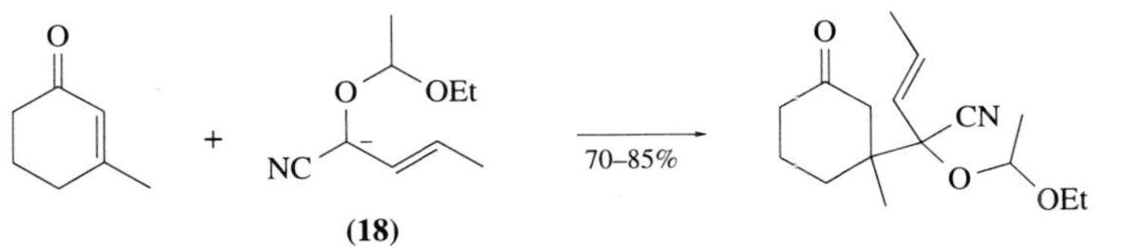

(37)

Since the nitro group can be converted into a carbonyl group, Michael addition of nitro anions also represents a useful method for the preparation of 1,4-dicarbonyl compounds (Scheme 21) ⟨73JOC4367⟩.

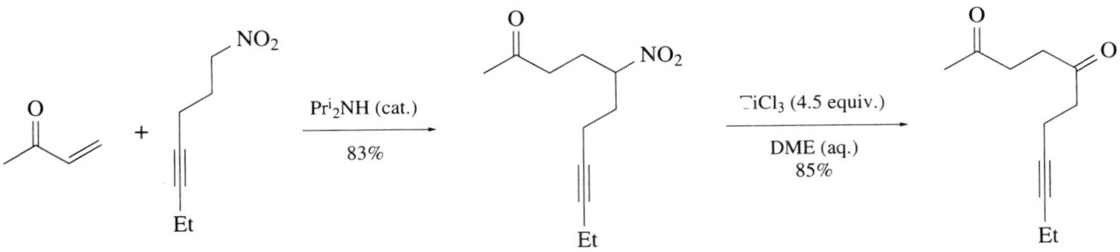

Scheme 21

1.07.3.2.3 *Addition of unstabilized nucleophiles: alkyl, alkenyl, alkynyl and aryl anion equivalents*

The conjugate addition of alkyl, alkenyl, alkynyl and aryl anion equivalents to activated alkenes represents an extremely important class of synthetic transformations. As mentioned previously, the nature of the organometal is one of the major factors in determining the extent to which conjugate addition to alkenes occurs relative to alternative processes (e.g. 1,2-addition), and illustrative examples of the reactivity of the common organometals are presented here.

(i) Organolithiums

The tendency of organolithium reagents to add in a 1,2-fashion to α,β-unsaturated carbonyl compounds was mentioned earlier (see Section 1.07.3.2.1(i)). Few exceptions exist, with the charge directed strategy of Scheme 14 being one of the few general ways of promoting conjugate addition of organolithiums. α,β-Unsaturated aldehydes are of course particularly susceptible to 1,2-attack, but their imine derivatives undergo conjugate addition with organolithiums and the aldehyde can be regenerated (Scheme 22) ⟨87TL5279⟩.

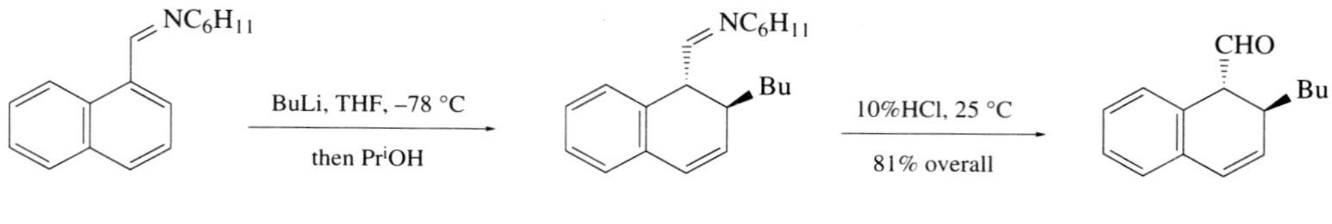

Scheme 22

Alkenes activated by groups other than carbonyls can react by addition of organolithiums to the β-carbon atom. Where a heteroatomic activating group is bonded directly to the alkene, the initial product of addition is an α-heterosubstituted organometallic, the synthesis of which is also covered elsewhere in this series (see Chapter 1.07). Some examples of addition to this type of alkene are given in Equations (38)–(40). Addition to nitroalkenes has proved useful in natural product synthesis (Equation (38)) ⟨82TL3171⟩. While reaction of organolithiums with α-unsubstituted vinylic sulfones leads to exclusive α-deprotonation ⟨79JOC3279⟩, Isobe *et al.* has shown that α-(trimethylsilyl)vinyl phenyl sulfones (**19**) undergo clean conjugate addition (Equation (39)) ⟨80CL331⟩. Simple vinyl silanes themselves also undergo clean addition of organolithiums (Equation (40)) ⟨52JA4582⟩.

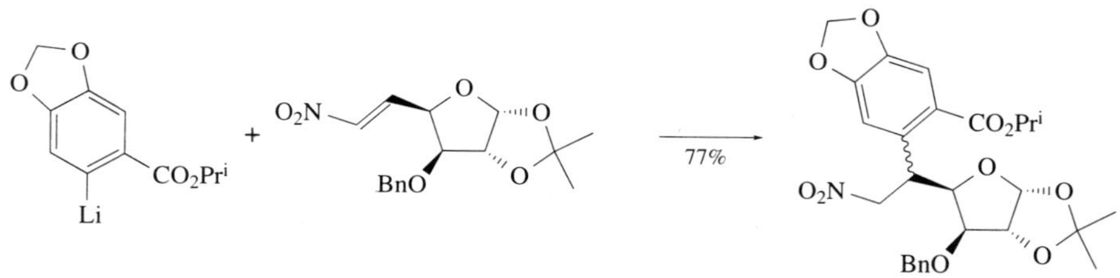

(38)

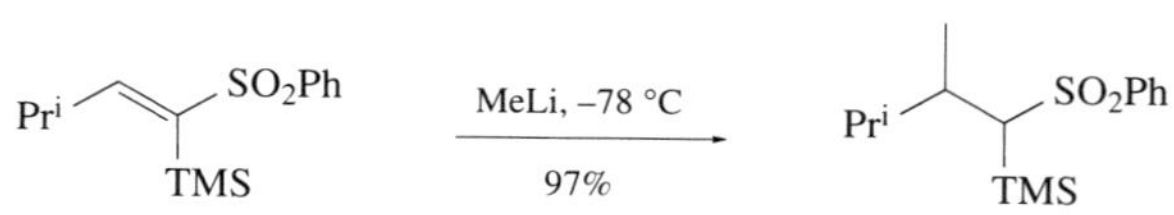

(**19**)

(39)

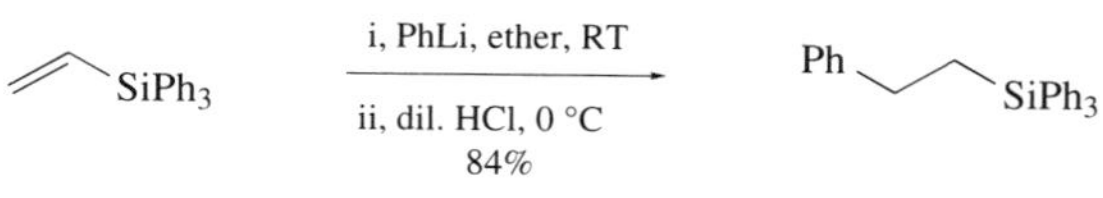

(40)

(ii) Organomagnesiums

Grignard reagents usually undergo 1,2-addition to α,β-unsaturated carbonyl compounds, but there are examples where particular structural features of the substrate, particularly steric hindrance in the region of the carbonyl group, cause exclusive 1,4-addition. *N,N*-Disubstituted cinnamides were reported as long ago as 1905 to react with Grignards in this way ⟨05MI 107-01⟩. An interesting example of this, highlighting also the differing reactivity of Grignard reagents compared with organocuprates, is seen in the addition to amide **(20)** (Scheme 23) ⟨83TL5089⟩.

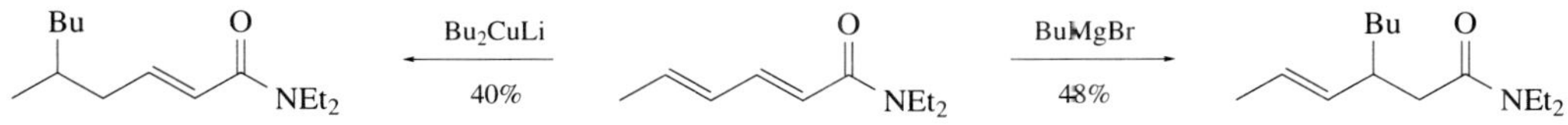

(20)

Scheme 23

A major advance in the history of conjugate addition was the report by Kharasch and Tawney that the addition of catalytic amounts of copper(I) salts caused Grignard reagents to undergo 1,4-addition to isophorone (Equation (41)) ⟨41JA2308⟩. While this process was widely used for many years, it was often unreliable, and has largely been superseded by the use of stoichiometric organocopper reagents.

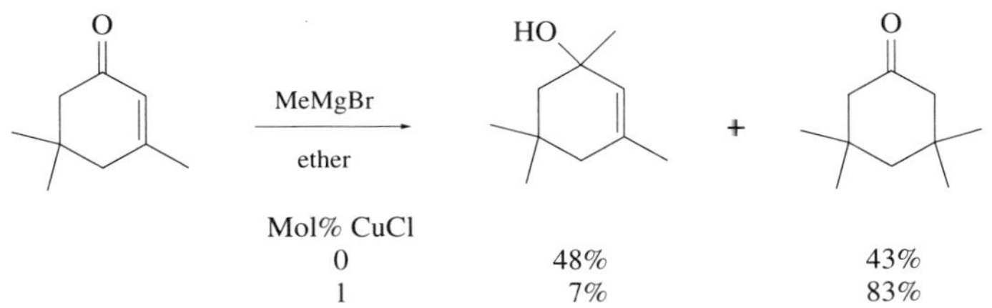

(41)

(iii) Copper-based reagents

Organocopper-based reagents are immensely useful for conjugate addition processes, and several comprehensive reviews have appeared ⟨72OR(19)1, 91COS(4)169, 92OR(41)135⟩. Following the early demonstration by Kharasch and Tawney ⟨41JA2308⟩ that the presence of catalytic amounts of copper(I) salts caused Grignard reagents to undergo 1,4-addition, House *et al.* studied the addition of various organometallics to (*E*)-3-penten-2-one, and concluded that organocuprates were responsible for the selective addition in the earlier work ⟨66JOC3128⟩. Several different types of organocopper reagent have been employed since then, differing in the stoichiometry and ratios of copper to lithium. Monoorganocopper species, RCu, prepared according to Equation (42), are less widely used than organocuprates because they are generally less stable and less reactive. Lower order organocuprates, R_2CuLi, are usually prepared *in situ* by reaction between two equivalents of RLi and copper(I) iodide in ether (Equation (43)). There are many examples of their use in conjugate addition reactions, delivering alkyl (Equation (44)) ⟨87TL3791⟩, alkenyl (Equation (45)) ⟨82TL2627⟩ or aryl groups (Equation (46)) ⟨77ACS(B)667⟩. It is not generally possible to transfer alkynyl groups from organocopper reagents ⟨69JOC3615⟩, but a combination of monopentynylcopper and iodotrimethylsilane has proved successful ⟨93JOC7238⟩; organozincs and organoaluminums have also been used for conjugate addition of alkynyl groups (see Section 1.07.3.2.3(iv)).

$$CuX + RLi \longrightarrow RCu + LiX \qquad (42)$$

$$CuI + 2RLi \longrightarrow R_2CuLi + LiI \qquad (43)$$

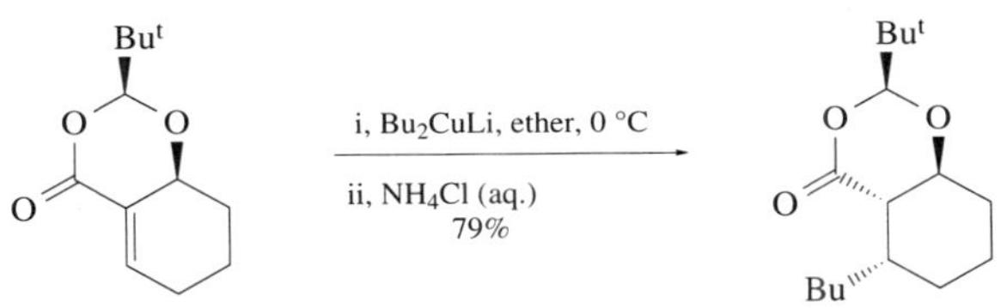

(44)

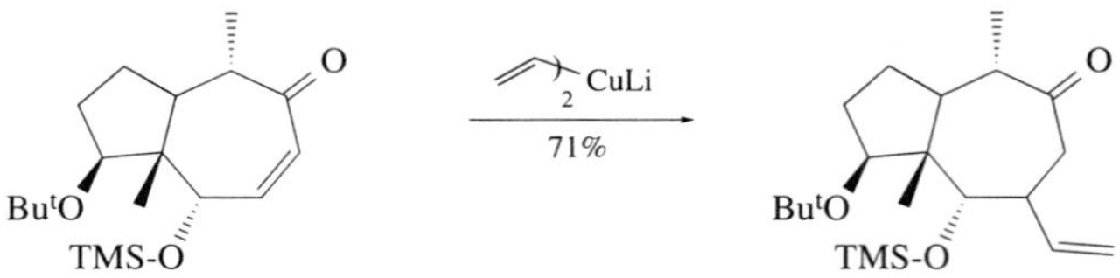

(45)

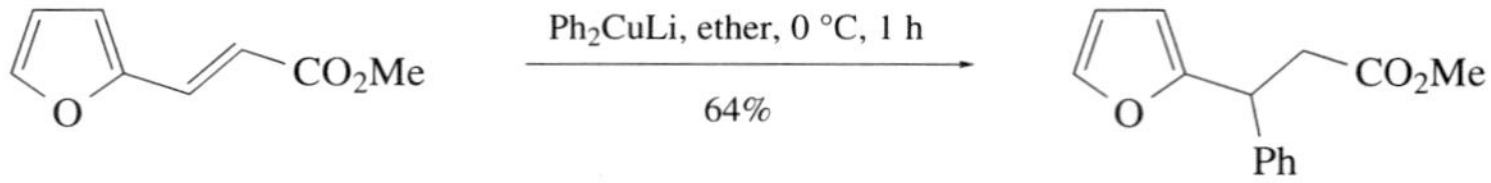

(46)

In terms of the conjugate acceptor, the scope of the organocuprate conjugate addition reaction is wide; even α,β-unsaturated aldehydes, prone to undergo 1,2-addition with most organometallic reagents, can participate cleanly in conjugate addition with organocuprates (Equation (47)) ⟨80T2305⟩, but the regioselectivity for these acceptors is highly dependent on the particular substrate and the reaction conditions ⟨84PAC91⟩.

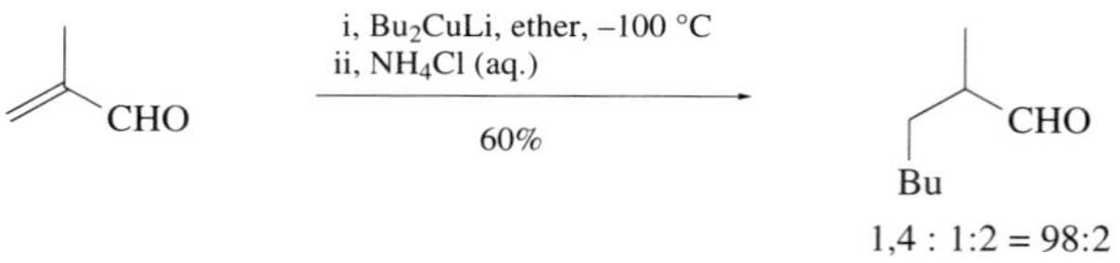

(47)

During conjugate addition of lower order organocuprates R_2CuLi, only one of the two R groups is transferred, with the other being lost as RH on workup. While this is not a problem where simple, commercially available organolithium compounds are involved, the process is clearly unsatisfactory for more valuable organolithiums. Much effort has therefore been expended in developing mixed cuprates, R_TR_DCuLi, where R_T is the transferable ligand and R_D is a nontransferable, 'dummy' ligand. Alkynes have been used for this purpose ⟨74JOC400⟩, as have cyanocuprates $R_TCu(CN)Li$ prepared from one equivalent R_TLi and CuCN; however, these species are relatively unreactive and so have not seen widespread use. The rate of addition is dramatically increased by using the higher order organocuprates $R_TR_DCu(CN)Li_2$, with the 2-thienyl group as the nontransferable ligand. The preparation of this higher order mixed cuprate is shown in Scheme 24, and an example of its use in Equation (48) ⟨85JOM(285)437⟩. The reagent (2-thienyl)Cu(CN)Li (**21**) is relatively stable and is commercially available (a 'cuprate in a bottle' precursor to higher order organocuprates) ⟨87TL945⟩.

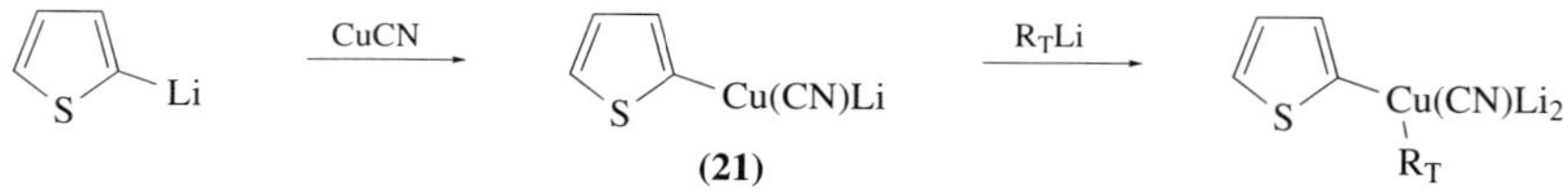

Scheme 24

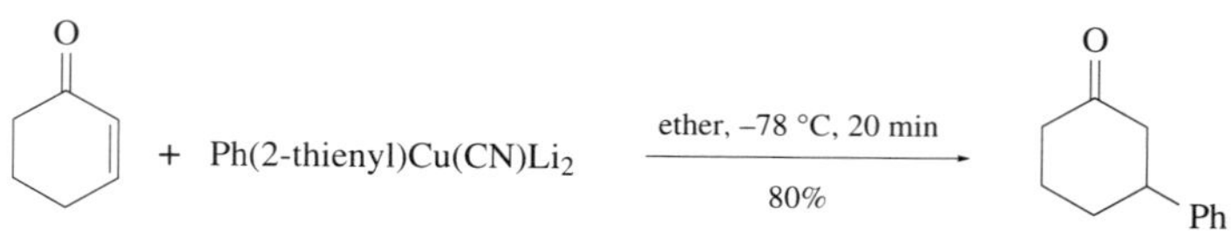

(48)

Thus, changing from lower to higher order species is one method for increasing the reactivity of organocuprates. Another approach that can greatly increase organocuprate reactivity as well as improving 1,4- vs. 1,2-selectivity is the addition of Lewis acids ⟨86AG(E)947, 91COS(4)139⟩. Yamamoto showed that addition of boron trifluoride etherate to monoorganocoppers produced a new complex, $RCu \cdot BF_3$, capable of effecting reaction where lower order cuprates failed. For example, this new reagent system was used for conjugate addition to α,β-unsaturated acids (Equation (49)) ⟨82JOC119⟩. Boron trifluoride etherate has also been shown to increase the reactivity of higher order cuprates, allowing conjugate addition even to very hindered substrates (Equation (50)) ⟨84TL5959⟩.

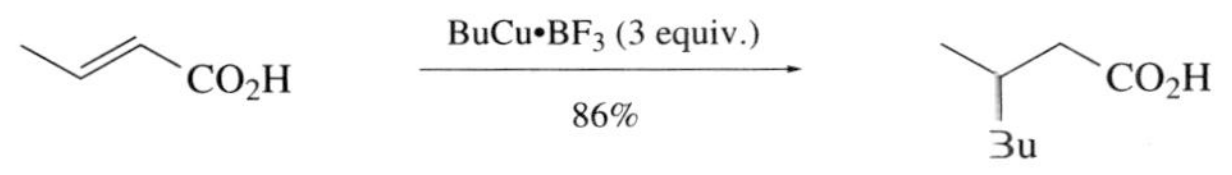

(49)

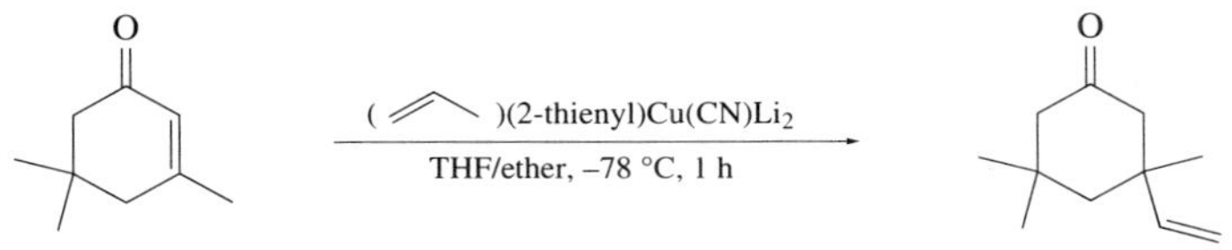

(50)

Equiv. $BF_3 \cdot OEt_2$	Yield
0	<5%
1	98%

Trimethylsilyl chloride is another additive that has proved extremely effective at increasing the rate of conjugate addition processes. Lower order organocuprates by themselves undergo little addition to α,β-unsaturated amides, but this transformation is cleanly achieved in high yield in the presence of trimethylsilyl chloride (Equation (51)) ⟨86TL1047⟩. Use of trimethylsilyl chloride together with HMPA has also been recommended, particularly for additions to α,β-unsaturated aldehydes ⟨86TL4029⟩.

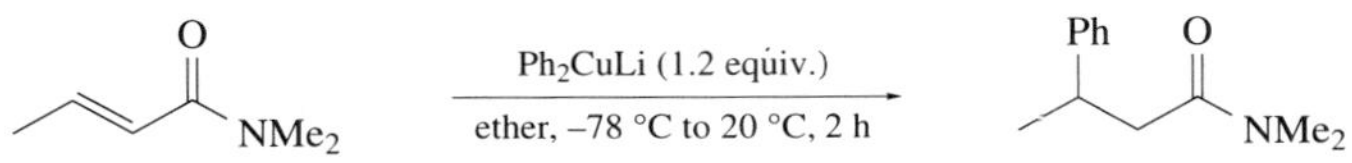

(51)

Equivs. TMS-Cl	Yield
0	0%
1	86%

The mechanism of addition of organocuprates to enones has been intensely studied for some time, but the precise details are still uncertain ⟨94JA2902⟩. Although never directly observed, it is generally considered that an intermediate organocopper(III) species (**22**) is formed, which undergoes reductive elimination to give the β-alkyl enolate (**23**) (Scheme 25). The mechanism of formation of structure (**22**) is more controversial, however, with both one- and two-electron processes postulated. House noted a correlation between the success of organocopper conjugate addition reactions and the redox potential of the substrates, suggesting that a single electron transfer mechanism was operating ⟨76ACR59⟩. While the ability of organocuprates to undergo electron transfer processes has indeed been demonstrated, it seems that long-lived radical anion intermediates are not involved ⟨79JA4236⟩. Spectroscopic studies have shown that mixing of an organocuprate with an enone at low temperature results in reversible formation of a complex which is presumed to result from a copper *d*–alkene π^* interaction ⟨94JA2902⟩. If such species actually lie on the reaction pathway (this is still not certain) then it is not currently known how they are transformed into the β-alkylated enolate (**23**).

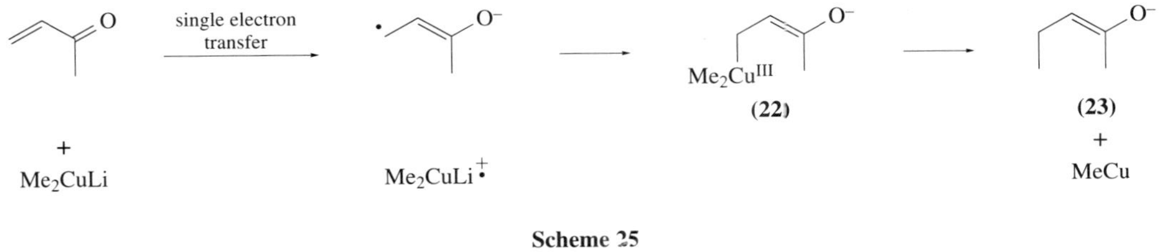

Scheme 25

(iv) Other organometallic reagents

Many other organometallics have been shown to be capable of conjugate addition processes, but their use has been far less widespread than that of the common organometallics detailed above. This section will concentrate on methods that have proved successful where the usual reagent of choice (organocuprates) fail.

Historically, organozincs were among the first organometallics to be synthesized, but they have not been widely used as they are less reactive and less easy to handle than are, for example, Grignard reagents. One useful process is the reaction of dialkylzincs in the presence of a catalytic amount of nickel (acetylacetonate) (Equation (52)) ⟨84JOC931⟩. Under these conditions, addition can even be effected to *β*,*β*-disubstituted enones, which are often poor substrates for organocuprates due to steric hindrance.

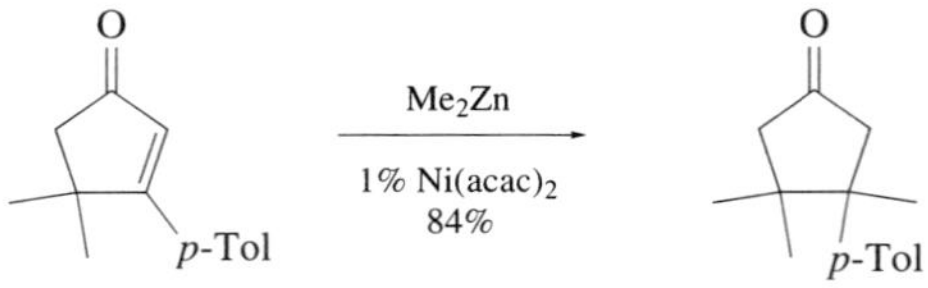

(52)

As mentioned earlier (see Section 1.07.3.2.3(iii)), conjugate addition of alkynyl groups is not usually possible with organocuprates. Alkynyl aluminums, generated by treatment of the corresponding alkynyl lithium with diethyl aluminum chloride, provide a solution to this problem for conjugate addition to *α*,*β*-unsaturated ketones (Equation (53)) ⟨71JA7320⟩. However, the reaction is restricted to those enones that can adopt the s-*cis* conformation; thus, cyclic enones like 2-cyclohexenone do not participate. A more general method appears to be the trialkylsilyl triflate-mediated addition of alkynyl zincs (Equation (54)), successful with both cyclic and acyclic enones ⟨90TL7627⟩.

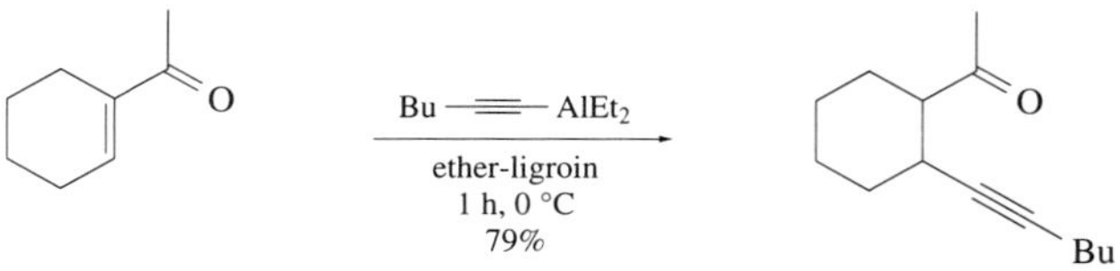

(53)

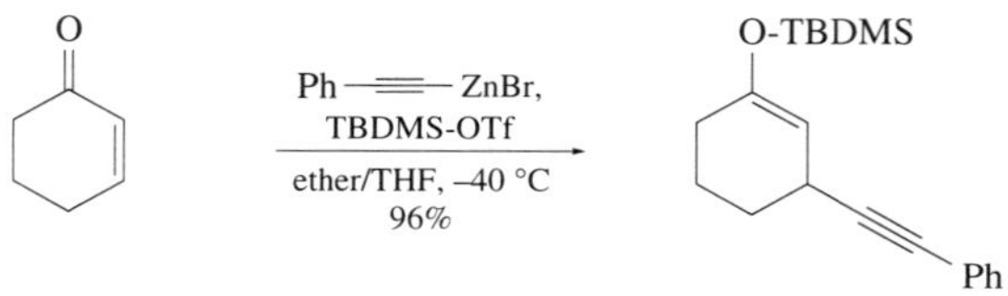

(54)

Delivery of allyl groups is another area where organocuprate methodology is sometimes unreliable. Trialkyl allylsilanes react cleanly with *α*,*β*-unsaturated ketones in the presence of a Lewis acid (Equation (55)) ⟨77JA1673, 82PAC1⟩. The comparison to Grignard and cuprate methodology in Scheme 26 highlights the utility of the method ⟨82PAC1⟩. Substituted allylsilanes can be used, and react cleanly at the position *γ*- to silicon. While *α*,*β*-unsaturated esters and nitriles are poor acceptors under these Lewis acid-mediated conditions, they do undergo allylation in the presence of fluoride anion (Equation (56)) ⟨83TL1909⟩. This process is believed to proceed via a pentacoordinate allylic silicate species ⟨89SL1⟩. Emphasizing the complementarity of the two sets of reaction conditions, and in contrast to their behaviour with Lewis acids, *α*,*β*-unsaturated ketones give mixtures of 1,2- and 1,4-addition in the fluoride-promoted process ⟨78TL3043⟩.

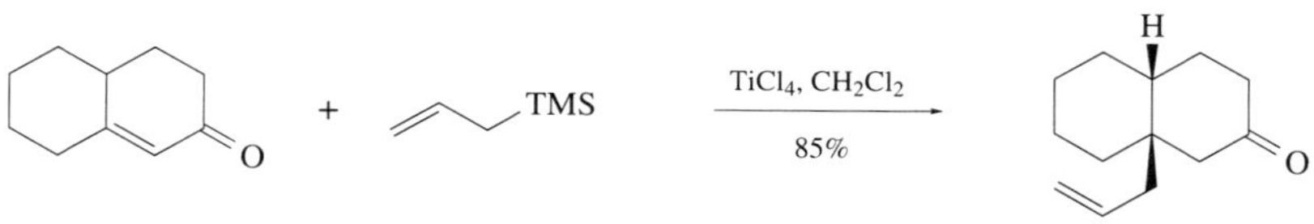

(55)

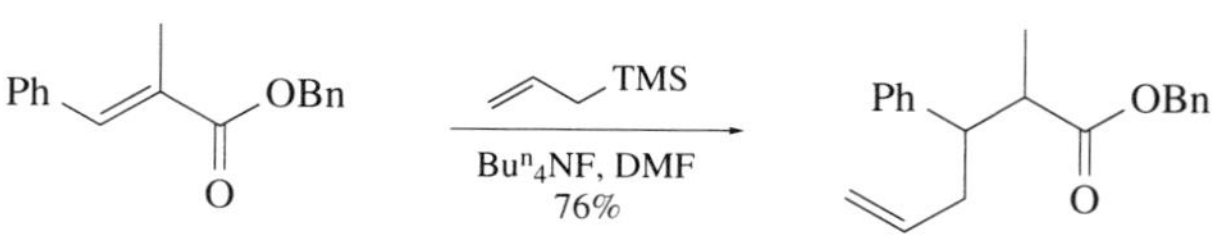

(56)

The allylsilane conjugate addition reaction can be carried out intramolecularly, the example in Equation (57) also demonstrating that 1,6-addition is possible ⟨88JOC50⟩. Interestingly, this same substrate (**27**) provided a mixture of intramolecular 1,2- and 1,4-addition products when fluoride-initiation was employed ⟨88JOC50⟩.

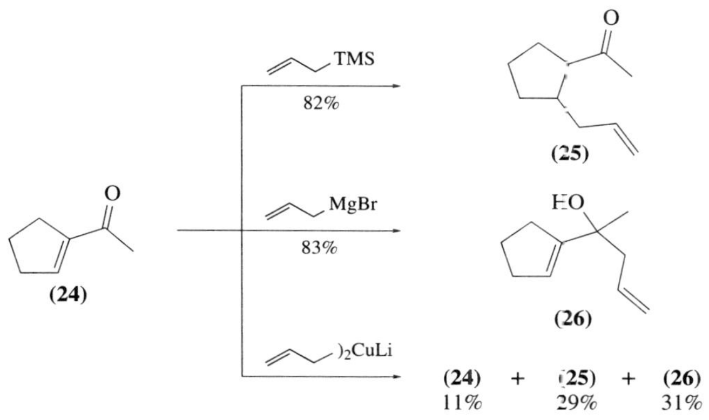

Scheme 26

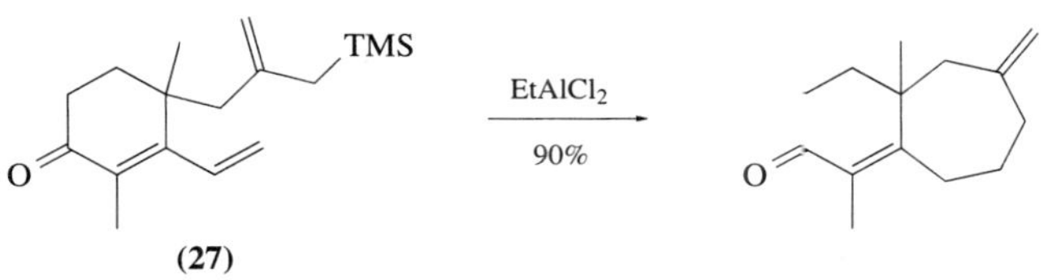

(57)

1.07.3.3 Addition to Unactivated Alkenes

1.07.3.3.1 Addition of stabilized nucleophiles

Hegedus *et al.* have shown that terminal alkenes can be alkylated with stabilized nucleophiles (p*K*a ~ 10–17) mediated by stoichiometric palladium chloride (Equation (58)) ⟨80JA4973⟩. The presence of two equivalents of triethylamine is essential if reduction to palladium(0) is to be avoided. The method was extended to internal disubstituted alkenes and to the use of ketone enolates, and even phenyllithium, as nucleophiles by addition of HMPA to the reaction mixture. The intermediate π-alkyl palladium species that is presumed to be involved may be carbonylated, giving the overall product of difunctionalization of the alkene (Equation (59)) ⟨80JA4980⟩.

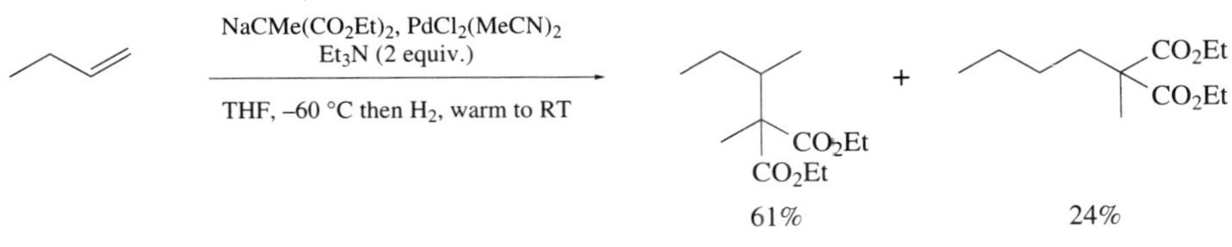

(58)

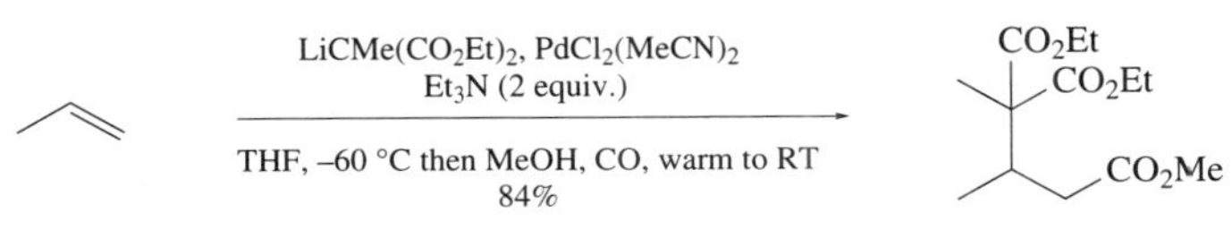

(59)

1.07.3.3.2 Addition of unstabilized nucleophiles

Addition of carbon nucleophiles to unactivated alkenes to form a new carbon–carbon bond and a new organometal (Equation (60)), a process known as carbametallation, was reviewed in the early 1990s ⟨91COS(4)865⟩. This chapter will concentrate on developments and reactions of particular synthetic utility, with emphasis on the more common organometals. Since the initial addition product is a carbanion, Chapter 1.07.1.4 should also be consulted.

(60)

In general, rather forcing reaction conditions are required for direct nucleophilic addition to unactivated alkenes, and since most organometals will react with other functional groups before adding to nonpolar alkenes, the range of other functionality tolerated in the substrate is limited. The carbanion product may itself react with starting alkene, so polymerisation is another potential problem. For both these reasons, the intermolecular addition of organolithiums to isolated alkenes is of little synthetic value. The intramolecular variant, however, can be performed successfully since the initial cyclization is entropically more favourable than possible subsequent intermolecular reactions. Bailey *et al.* have shown that generation of the organolithium may be effected cleanly in the presence of the alkene by halogen–metal exchange using the iodide (Scheme 27) ⟨87JA2442, 94T5957⟩. Non-terminal alkenes generally react sluggishly in this 5-exo-trig process, although cyclization onto vinyl silanes and 1-phenyl alkenes has been performed successfully ⟨94T5957⟩. Tandem cyclizations are also possible, an impressive example involving the formation of the thermodynamically unfavourable *trans*-isomer of the bicyclo[3.3.0]octane ring system (Equation (61)) ⟨92JA8053⟩. Vinyl lithiums have also been used as nucleophiles ⟨88JA4788⟩.

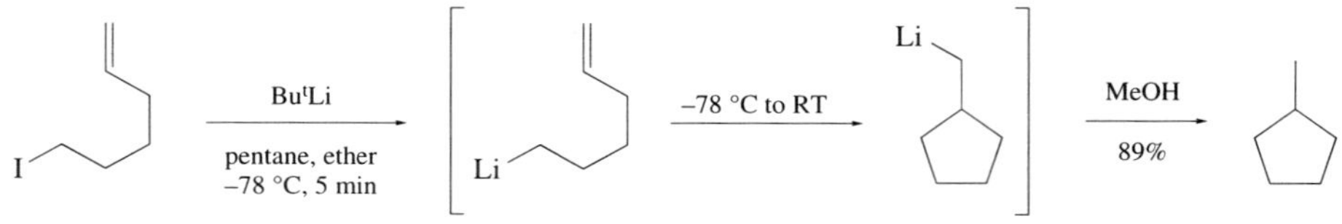

Scheme 27

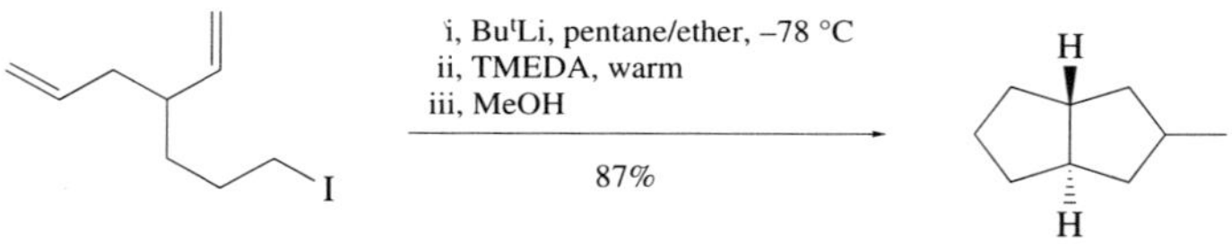

(61)

The presence of adjacent heteroatom functionality facilitates addition of organolithiums to alkenes. For example, allylic alcohols are good substrates in the presence of TMEDA (Equation (62)) ⟨69TL707, 72JOC4236⟩. A common side-reaction is formal S_N2' displacement of the hydroxyl group, and this is the major pathway for 2-cyclopentenol and 2-cyclohexenol.

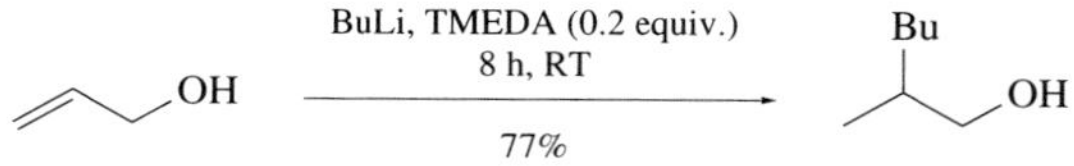

(62)

Reaction of unactivated alkenes with Grignard reagents similarly requires severe reaction conditions, although again the intramolecular version has been performed successfully ⟨77AOC(16)131⟩. A major advance in terms of regio- and stereoselectivity has been the discovery of the zirconium-catalysed process by Dzhemilev and Vostrikova ⟨85JOM(285)43⟩ and its development by Hoveyda and co-workers ⟨91JA5079, 92JA6692, 93JA6614⟩. Addition to simple terminal alkenes occurs with high regioselectivity (Scheme 28). The intermediate may be quenched with aldehydes, halogen sources (I_2 or nbs), or converted to the primary alcohol via the organoborane. Disubstituted alkenes generally react sluggishly in the catalysed carbomagnesation process, unless there is ring strain or a neighbouring heteroatom. As is the case using organolithiums, the presence of heteroatomic functionality is beneficial to the reaction rate and so allylic alcohols react faster than simple alkenes. Of great importance is the high stereoselectivity that can be obtained with these substrates; as shown in Scheme 29, the sense of stereoselectivity is reversed if the hydroxyl group is protected. If the hydroxyl functionality occupies the homoallylic position, the same stereoisomer is obtained whether or not it is protected; the stereochemical outcome is sensitive to the chirality at the allylic carbon atom ⟨91JA8950⟩. The effect of chiral zirconium catalysts on the diastereoselectivity ⟨93JOC4237⟩ and enantioselectivity ⟨93JA6997, 94JA3123⟩ of the reaction has been examined.

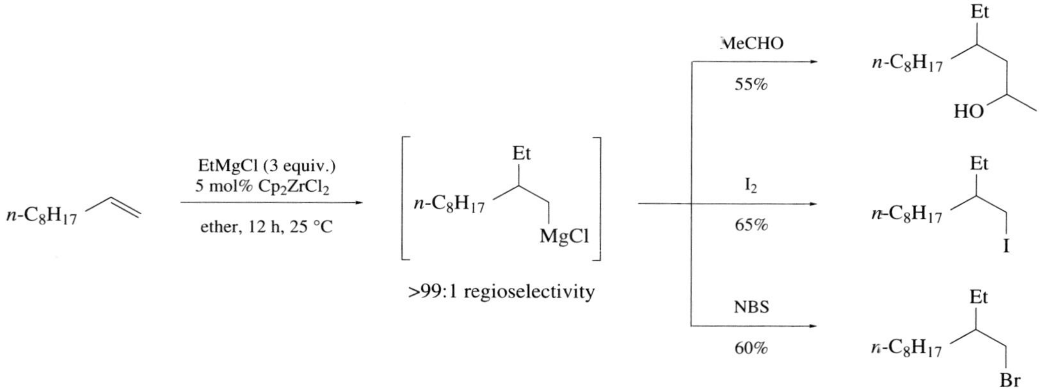

Scheme 28

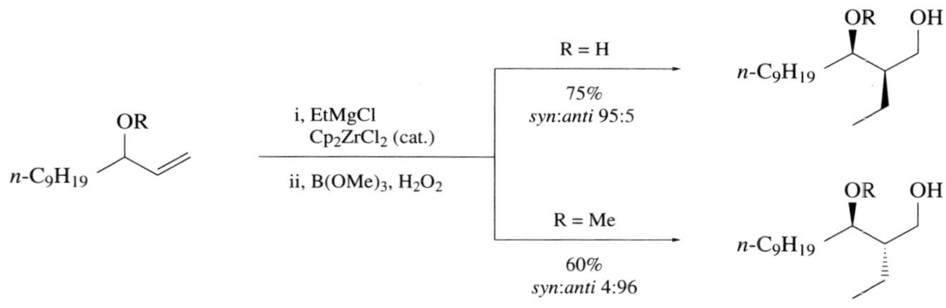

Scheme 29

1.08
One or More CC Bond(s) Formed by Addition: Addition of Carbon Radicals and Electrocyclic Additions to CC Multiple Bonds

ANDREW J. CLARK and PAUL C. TAYLOR
University of Warwick, Coventry, UK

1.08.1 ADDITION OF CARBON RADICALS TO CARBON–CARBON MULTIPLE BONDS TO GIVE TETRACOORDINATE PRODUCTS

1.08.1.1 Introduction

The ability to form carbon–carbon bonds is one of the primary tasks in the construction of organic molecules. In the 1980s and 1990s this has been increasingly achieved by the addition of carbon-centred radicals to carbon–carbon multiple bonds. This review will focus on the reactivity, chemoselectivity, regioselectivity and stereoselectivity of carbon radical additions. Other reviews on intermolecular and intramolecular radical addition reactions have been published by Curran ⟨88S417, 88S489, 91COS(4)715, 91COS(4)779, 91CRV1237⟩. These contain examples of carbon, oxygen, nitrogen and other heteroatom-centred radical additions and cyclisations and are recommended reading. Curran himself recommends as 'the bible of radical chemistry, new testament' a review by Giese ⟨89HOU(E19a)⟩. The reader is also directed to two books which cover the important concepts and principles ⟨B-86MI 108-01, B-92MI 108-01⟩. Acyclic diastereofacial selection in intermolecular radical reactions covering the literature up to October 1993 has been reviewed ⟨94SL1⟩. Other reviews on the stereochemical outcome of intermolecular additions ⟨89AG(E)969, 91ACR296⟩ and intramolecular additions ⟨91ACR139⟩ have also appeared. Retrosynthetic planning ⟨91SL63⟩ and computer-assisted retrosynthetic analysis ⟨90JCI458, 90JOC9⟩ have also been reviewed. In addition there are a number of earlier reviews which cover the basic principles ⟨83AG(E)753, 84SCI883, 85AG(E)553, 87T3541, 88CSR361⟩. A review of transition metal mediated radical reactions has been published ⟨94CRV519⟩.

This chapter is organised according to the classification of the reaction (e.g., intermolecular, intramolecular or tandem radical additions) and the type of method by which it is conducted. For a more detailed account of the preparation of carbon-centred radicals from radical precursors see Chapters 1.19 and 1.21.

1.08.1.2 Basic Principles

1.08.1.2.1 Synthetic advantages of carbon-centred radical reactions

Carbon-centred radicals are highly reactive intermediates which often show great chemoselectivities, regioselectivities and stereoselectivities upon addition to carbon–carbon multiple bonds. They can be prepared under essentially neutral conditions (see Chapters 1.19 and 1.22), and thus many of the undesired side reactions that are often associated with basic anionic reagents (such as epimerisation) are generally avoided. Neutral carbon-centred radicals are less sensitive to solvent effects than ionic reagents because they do not have associated with them a degree of aggregation or ion pairing. Consequently, they are particularly suited for the synthesis of crowded bonds (such as quaternary centres) and for carrying out transformations in highly hindered frameworks ⟨82JA2321⟩. There are some limiting factors however, when choosing a solvent for a given radical reaction. Mainly these concern the rate of radical/solvent reactions (which are particularly important when dealing with relatively long-lived radicals) and the solubility of reagents. There are two further differences between ionic and radical reactions that should be mentioned. First, carbon-centred radicals can only attack functional groups if the bimolecular rate constant k exceeds 10^2 mol^{-1} s^{-1} (see Section 1.08.1.4.1). This means that carbon-centred radicals are generally inert to such functionalities as OH, NH_2 or CO_2H groups (unlike basic carbanions, or electrophilic carbocations) ⟨B-86MI 108-01⟩. This is primarily due to the endothermic nature of homolysis of the strong O—H, or N—H bonds. Consequently, it is not necessary to protect such functionalities, and this can cut down on the number of steps in a synthesis. An example is the cyclisation of (**1**) via the vinyl radical (**2**); the facile formation of the quaternary centre should also be noted (Scheme 1) ⟨82JA2321⟩. The other consequence is that radical reactions do not need to be conducted under strictly anhydrous conditions. On the other hand, most radicals will react with triplet oxygen and so reactions should generally be carried out under an inert atmosphere. Another major difference between ionic and radical reactions is that carbon radicals do not normally undergo facile β-elimination of OR or NR_2 groups; this is because a less stable π C—C bond would be formed at the expense of the stronger σ C—O or σ C—N bond. Weaker bonds such as β C—Br ⟨68QR549⟩, C—SR ⟨76JA951⟩, C—SnR_3 ⟨85T4079⟩ and C—Co ⟨83ACR343⟩ do undergo β-elimination of Br, SR, SnR_3 and Co radicals, respectively, fast enough to be problematic or, conversely, synthetically useful (Scheme 2).

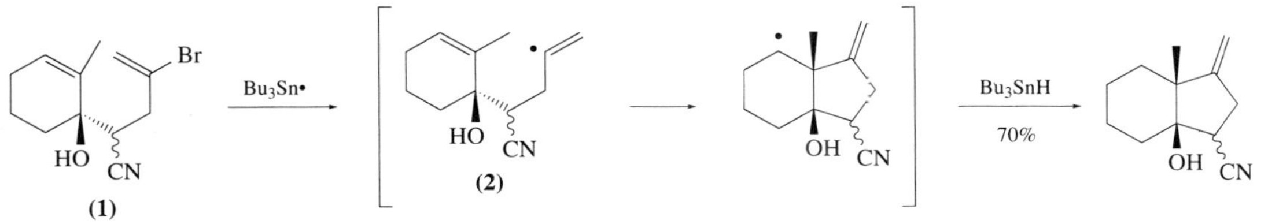

Scheme 1

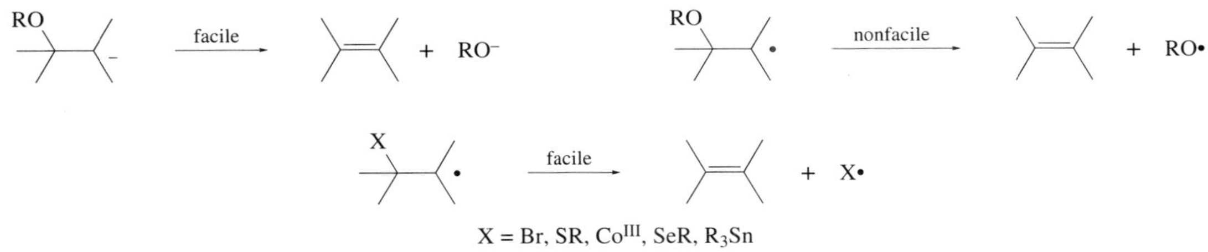

Scheme 2

1.08.1.2.2 Stability and structure of alkyl and vinyl radicals

The relative stability of substituted carbon-centred radicals can be approximated by comparing the C—H bond dissociation energies *D*(C—H) in a number of compounds ⟨86TCC1⟩. The lower the *D*(C—H) the more stable the radical (Table 1).

Table 1 C—H bond dissociation energies.

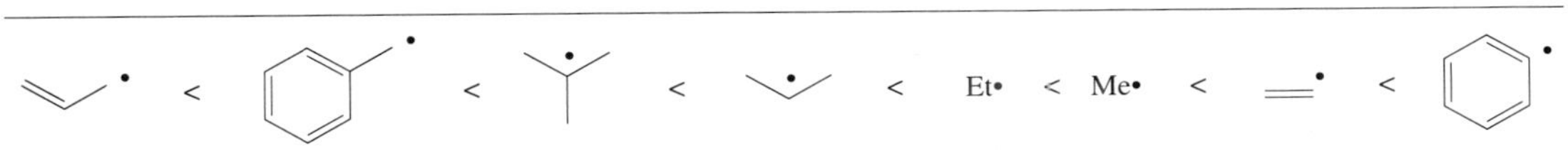

Allyl	Benzyl	3°	2°	1°	Me	Vinyl	Aryl
D_{298}(R–H) (kcal mol^{-1})[a] (kJ mol^{-1} in parentheses)							
86 (360)	88 (368)	95.8[b] (401)	96 (402)	100 (418)	105 (439)	105 (439)	111 (464)

[a] Except where noted, these values are from ⟨B-88MI 108-01⟩. [b] From ⟨90ACR375⟩.

Carbon-centred radicals can adopt either a planar or pyramidal structure (Figure 1) ⟨B-73MI 108-01⟩. Conjugating substituents favour the planar structure while alkyl and heteroatom substituents can cause pyramidalisation ⟨74JA2201⟩. The barrier to inversion in the pyramidal structures is very low and as a consequence stereochemical information at the proradical centre is lost ⟨83HCA2210⟩.

Vinyl radicals also have a low barrier to inversion ⟨64JCP(40)213⟩ which enables either stereoisomeric precursor to be used. Hence, both the (*E*) and (*Z*) isomers (**3**) are converted to the carbocycle (**4**) via the interconverting (*E*)- and (*Z*)-vinyl radicals (Scheme 3) ⟨82JA2321⟩. Vinyl radicals can either exist in the bent form or the linear form (Figure 2). The latter is preferred with conjugating substituents ⟨75JA4269⟩.

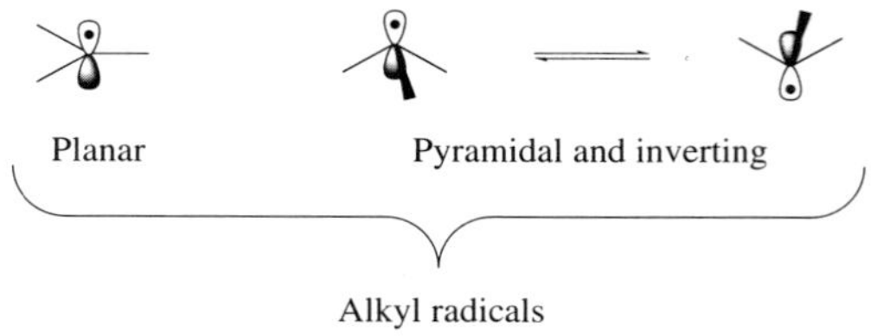

Figure 1 Structure of carbon-centred radicals.

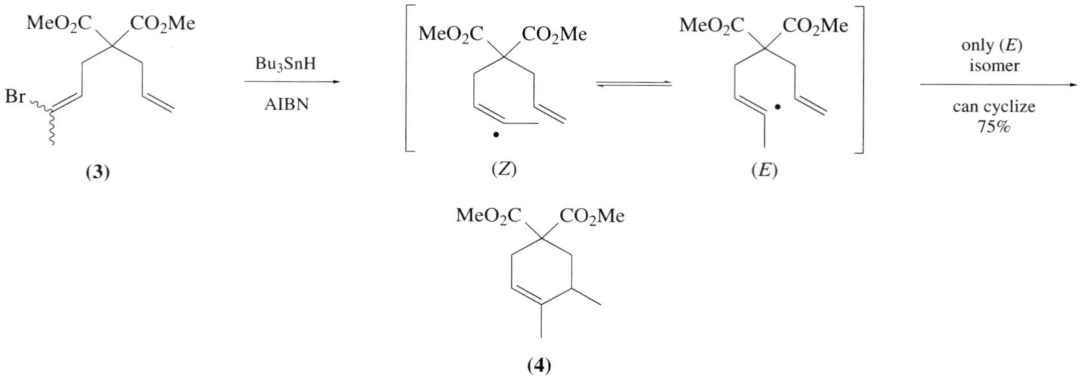

AIBN = 2,2'-azobisisobutyronitrile

Scheme 3

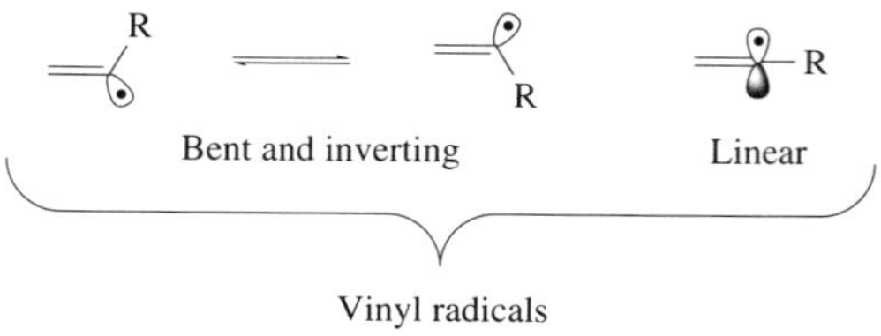

Figure 2 Structure of vinyl radicals.

1.08.1.3 Intermolecular Radical Additions

1.08.1.3.1 Introduction

Carbon-centred radicals can undergo intermolecular addition reactions with carbon–carbon double and triple bonds. The formation of a new σ C—C bond (88 kcal mol^{-1}, 368 kJ mol^{-1}) is at the expense of a π C═C bond (54–59 kcal mol^{-1}, 226–247 kJ mol^{-1}) and is a highly favourable exothermic process (Equation (1)). This process is normally not reversible unless the starting radical is particularly stable relative to the product radical, or the σ C—C bond is weak or the π C═C bond is strong. Theoretical studies for radical additions to alkenes indicate that the transition state is early and that the angle of attack of the radical on to the alkene is tetrahedral. All the carbons are slightly pyramidalised and there is little charge separation (Figure 3) ⟨86JOC2874⟩.

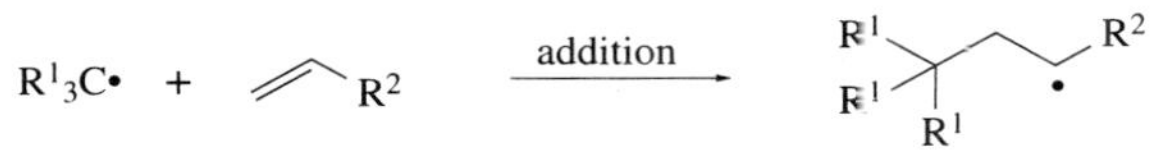

(1)

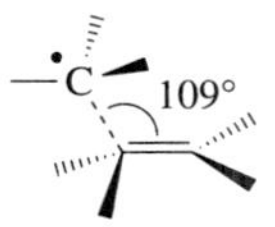

Figure 3 Radical attack on an alkene.

1.08.1.3.2 Electronic nature of radicals

All carbon-centred radicals can be classified as nucleophilic, electrophilic or ambiphilic depending upon the substituents attached to the carbon radical. The character of the radical will determine the types of addition reactions that it can undergo.

(i) Nucleophilic radicals

Simple alkyl radicals can be considered nucleophilic and react preferentially with electron-poor alkenes in high yields ⟨B-86MI 108-01⟩. The early transition state for radical additions to alkenes allows reactions to be rationalised using frontier molecular orbital (FMO) theory ⟨B-76MI 108-01⟩. The SOMO of a nucleophilic radical will combine with the LUMO of an alkene (Figure 4). Electron-withdrawing substituents attached to the alkene lower the LUMO and hence increase the rate of reaction. This can be illustrated by the following examples of cyclohexyl radical additions (Table 2). The addition to acrolein is some 8500 times faster than to hex-1-ene ⟨83AG(E)753⟩.

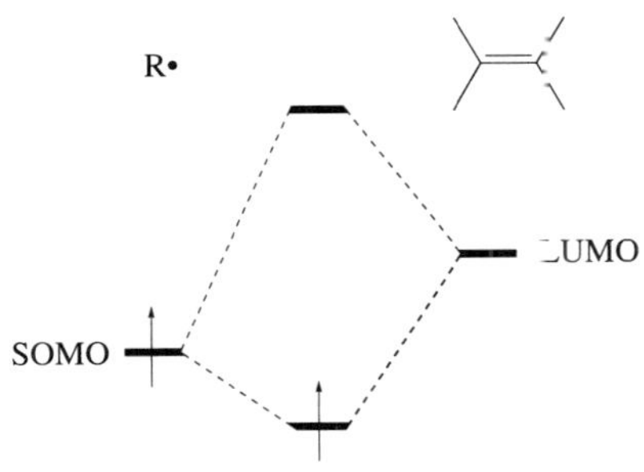

Figure 4 Orbital interaction for nucleophilic radicals.

Generally, intermolecular addition of nucleophilic radicals to unactivated alkenes are too slow as to render them synthetically useful. Addition of alkyl substituents at the radical centre often gives rise to a small acceleration in the rate of addition to alkenes despite an increased steric demand ⟨83TL703⟩. This is because the addition of electron-donating substituents serves to raise the energy of the SOMO causing a corresponding rate acceleration. Consequently, tertiary radicals are more reactive than secondary radicals, which in turn are more reactive than primary radicals ⟨83TL703⟩. The most synthetically useful reaction of nucleophilic radicals is addition to the β position of activated alkenes and alkynes. Substitution of a further electron-withdrawing group at the α position of the alkene can increase the rate of addition even further, while substitution at the β position generally retards the rate of addition due to steric effects (Tables 3 and 4). Substitution at the β position also affects the regioselectivity of the addition. As the β group becomes larger, the tendency for attack at the α position increases (Figure 5) ⟨83AG(E)753⟩. For a more complete treatment of

Table 2 Relative rates of addition of cyclohexyl radicals to alkenes.

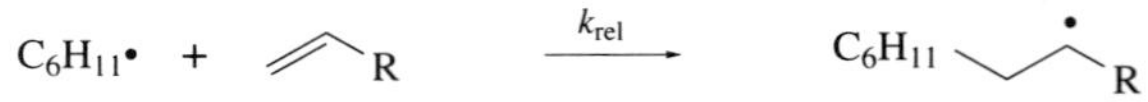

k_{rel} is the relative rate compared to that of addition to styrene (k_a ~4 x 10^4 M^{-1} s^{-1})

R	k_{rel}	*R*	k_{rel}
CHO	34	COMe	13
CN	24	CO_2Et	6.7
Cl	0.12	OAc	0.016
Bu	0.004	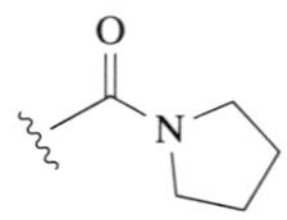	4.5[a]
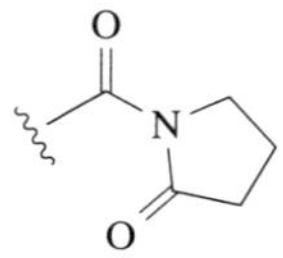	28[a]	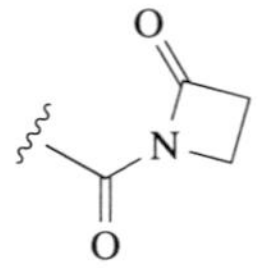	200[a]

Except where noted, these values are from ⟨83AG(E)753⟩. [a] From ⟨93TL4489⟩.

radical additions by FMO theory and more information on the rate of addition of carbon-centred radicals to alkenes, see ⟨83AG(E)753⟩. Nucleophilic radicals can also add to internal alkenes and to alkynes.

Table 3 Influence of the substituents Z on the rate of addition of C_6H_{11}• to acrylates at 20°C.

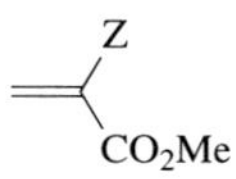

Z	k_{rel}	*Z*	k_{rel}
CN	310	CO_2Me	150
Cl	12	C_6H_5	6.4
H	1	Me	0.75

Table 4 Influence of the substituents Y on the rate of addition of C_6H_{11}• to acrylates at 20°C.

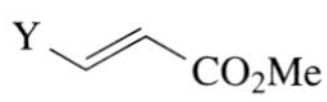

Y	k_{rel}	*Y*	k_{rel}
CN	6.0	CO_2Me	5.0
Cl	0.67	C_6H_5	0.009
H	1	Me	0.011

Values from ⟨83AG(E)753⟩.

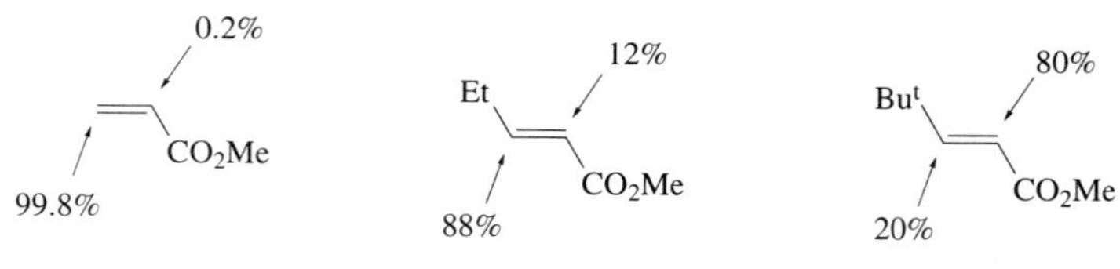

Figure 5 Regioselectivity of addition of cyclohexyl radicals to substituted acrylates

(ii) Electrophilic radicals

Carbon-centred radicals that contain two electron-withdrawing groups such as esters or nitriles have relatively low-lying SOMOs and react preferentially with electron-rich alkenes that contain high energy HOMOs. These are called electrophilic radicals and they have been far less studied than their nucleophilic counterparts (Table 5). For more information, see ⟨86CB444, 86JOC5007, 87JOC205, 88CB2063⟩.

Table 5 Relative rates for the addition of the malonyl radical to styrenes at 110°C.

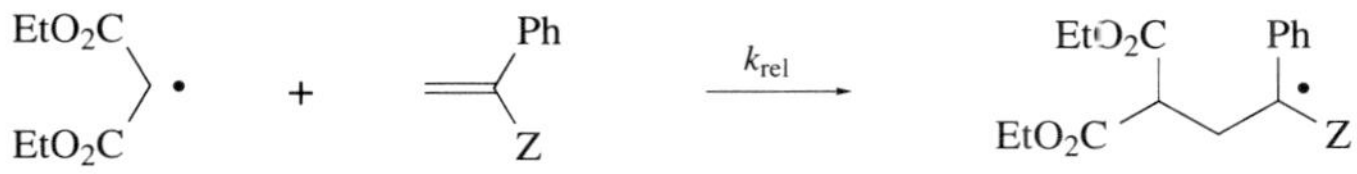

Z	k_{rel}	*Z*	k_{rel}
$N(Et)_2$	6.1	OMe	2.1
Me	1.0	C_6H_5	1.2
CO_2Et	0.27	COC_6H_5	0.18

Values from ⟨88CB2063⟩.

(iii) Ambiphilic radicals

There are another class of carbon-centred radicals called ambiphilic radicals that have SOMO energies intermediate between that of nucleophilic and electrophilic radicals. These in theory can combine with either electron-rich or electron-deficient alkenes with close-lying HOMOs and LUMOs, respectively. Accurate rate studies for the addition of such radicals have been carried out and show that the rate can be accelerated by addition to both electron-withdrawing alkenes and electron-rich alkenes ⟨88CB2063, B-89MI 108-01⟩.

1.08.1.3.3 Stereoselectivity

(i) Cyclic radical additions

Cyclic radicals that contain β substituents can exhibit a high degree of diastereoselection in the addition to activated alkenes ⟨89AG(E)969⟩. In a series of cycloalkene radicals, high *trans* diastereoselectivities were achieved (Table 6). When R = OEt stereoselectivities were found to increase upon changing the solvent from hexane to acetonitrile. This was explained by the increase in solvation of the ether oxygen atom and hence its greater steric bulk as a directing group. The effect of substituents on the alkene has also been investigated and found to have an influence on the selectivity ⟨83TL5343, 84CB2132⟩. Thus, styrene gave a higher diastereoselectivity than chloroacrylonitrile, and the selectivity was found to increase as the rate of addition to the alkene decreased ⟨85CB1345⟩. If more than one directing group is attached to the ring then the direction of addition is determined by the overall geometry of the ring ⟨84AG(E)69, 85JA4332⟩.

Table 6 Stereoselectivity of addition of cyclic radicals to activated alkenes at 20°C.

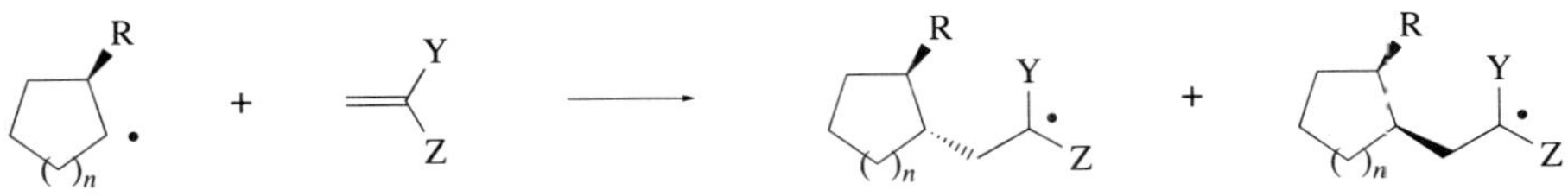

n	*Y*	*Z*	*R*	*Solvent*	anti/syn	n	*Y*	*Z*	*R*	*Solvent*	anti/syn
1	H	CN	OEt	hexane	68:32	1	H	CN	OEt	CH_2Cl_2	77:23
1	H	CN	OEt	CH_3CN	81:19	1	H	CN	Me	CH_2Cl_2	92:8
1	H	Ph	OEt	CH_2Cl_2	90:10	1	Cl	CN	OEt	CH_2Cl_2	72:28
2	H	CN	OMe	CH_2Cl_2	65:35	2	H	NHAc	OMe	CH_2Cl_2	>95:<5

Values from ⟨89AG(E)969⟩.

In six-membered rings, β substituents also direct addition via *anti* attack, although the stereoselectivity is generally not as high as in the analogous five-membered ring case (Table 6). Both axial and equatorial substituents direct *anti*, leading to high selectivity ⟨83TL5343, 84CB2132⟩. The *anti* selectivity can be increased if the alkene component has a substituent at the reacting radical addition site. Further substituents on the cyclic radical can either increase or decrease the selectivity depending upon the overall geometry of the molecule; however, with two axial or equatorial substituents adjacent to the radical centre, stereoselectivity is increased ⟨84TL2743⟩.

(ii) Acyclic radical additions

In the 1990s stereocontrol in acyclic radical addition reactions has attained prominence. High stereoselectivities can now be achieved by either chiral auxiliary or substrate control. For a thorough treatment of this field the reader is directed to two reviews ⟨91ACR296, 94SL1⟩. Most studies have involved proton, deuteron or halogen abstraction by chiral radicals and so do not come under the remit of this chapter. However, as all radical addition reactions to alkenes eventually lead to other radicals which are then trapped by various reagents, specific examples of 1,2-induction will be found in the text where they illustrate a method.

Radical additions to various alkenes containing chiral auxiliaries lead to good diastereoselection. 1,4-Asymmetric induction has been achieved with a number of auxiliaries (compounds (**5**)–(**6**)) ⟨91ACR296, 94SL1⟩. Porter and co-workers have shown that good stereoselectivity (80/1) can be accomplished in the addition of *t*-butyl radicals to the acrylamide (**5**) ⟨89JA8311, 91JA1791⟩. Addition of cyclohexyl radicals to the related trisubstituted acrylamide (**6**) is regioselective (α position) and highly diastereoselective (125/1) ⟨90TL1679, 91TL707⟩. The main drawback with this method is the difficulty of removing the auxiliary. The oxazolidine auxiliary does not suffer from this problem as it can be removed with acid hydrolysis without loss of chirality. Addition of the cyclohexyl radical to oxazolidine (**7**) gave an 80/1 mixture of stereoisomers ⟨91JA7788, 92JA7664⟩. Giese has shown that high diastereoselectivities can be achieved in radical additions to methylacrylamides (**8**) followed by trapping of the resultant radical (Scheme 4) ⟨93TL2445⟩. This high 1,4-induction was rationalised via attack opposite from the shielding pyrrolidine group in the conformation shown in Figure 6. As expected this model accurately predicts that the 1,4-induction will increase with the increasing size of the initial radical (R·). Radical addition to activated alkenes containing other chiral auxiliaries such as the imide (**9**) ⟨91JA5918, 92JA7007⟩, the camphor sultam (**10**), ⟨90JA6738, 92JA7664⟩ and the bis(methoxymethyl)pyrrolidine (**11**) ⟨93TL2445⟩ have also been explored.

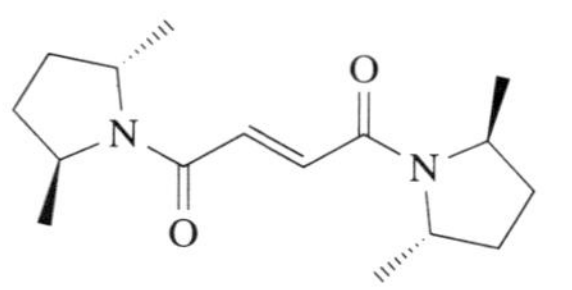

(5)

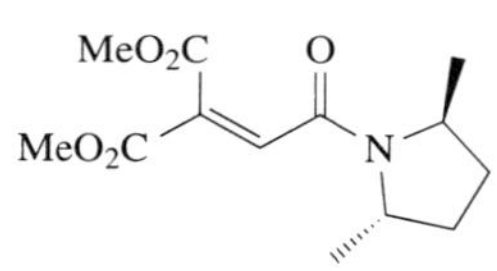

(6)

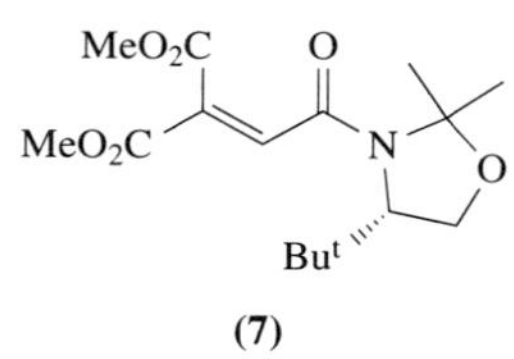

(7)

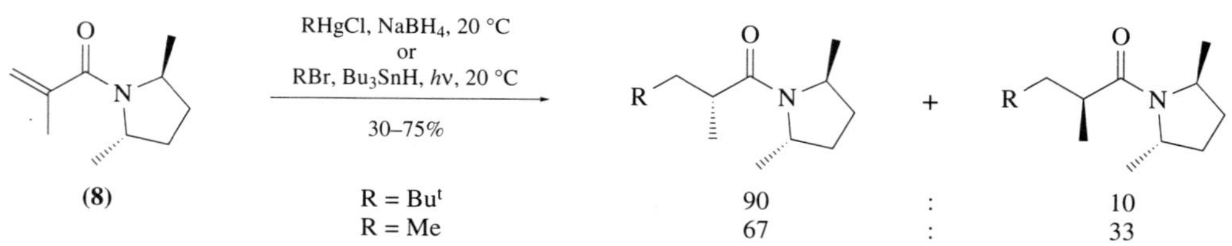

Scheme 4

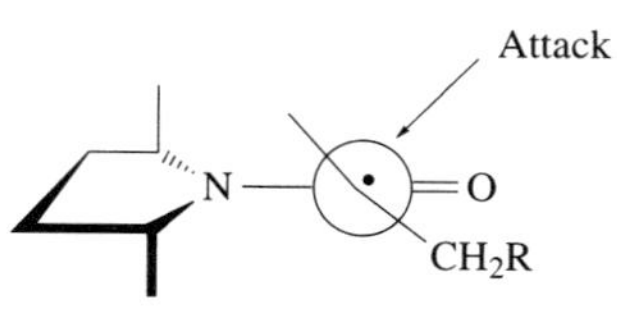

Figure 6 Radical attack on methylacrylamides.

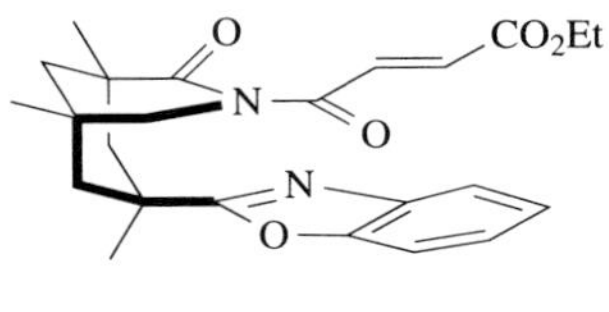

(9)

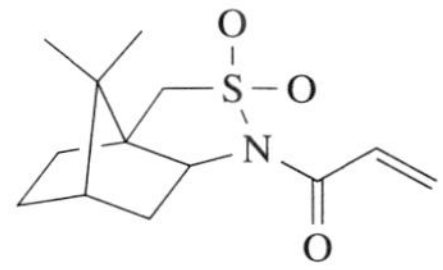

(10)

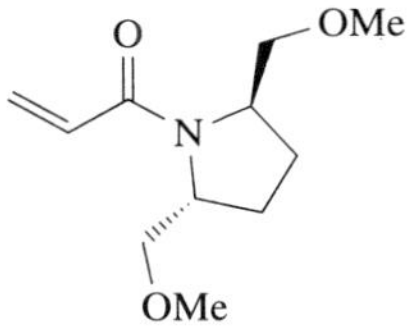

(11)

The vast majority of substrate-controlled 1,2-asymmetric inductions that have been studied involve reactions of conformationally restricted conjugated radicals ⟨91ACR296, 94SL1⟩. Addition of *t*-butyl radicals to diethyl citraconate (**12**) followed by reduction of the intermediate radical by Bu^tHgH furnishes essentially one diastereoisomer (*threo*/*erythro* ratio, 40/1). This is in marked contrast to the similar reaction with the cyclic alkene citraconic anhydride (**13**) which furnishes the opposite diastereoisomer (18/1) after transesterification (Scheme 5) ⟨91SL423⟩. The 1,2-induction for (**12**) has been explained by postulating that the conformation which leads to the lowest energy transition state is that where $A^{1,3}$ strain is minimised (Figure 7) ⟨91SL425⟩. Attack of the radical trapping agent from the less hindered face then leads to the observed product. This explanation has been used to rationalise the preferential formation of *syn* isomers from trapping of conformationally restricted α-ester, α-amide and α-benzyl radicals ⟨94SL1⟩.

(12) *threo* *erythro* **(13)**

i, Bu^tHgCl, $NaBH_4$, H_2O, CH_2Cl_2; ii, EtOH, reflux, evaporation, K_2CO_3, EtI, acetone, reflux

Scheme 5

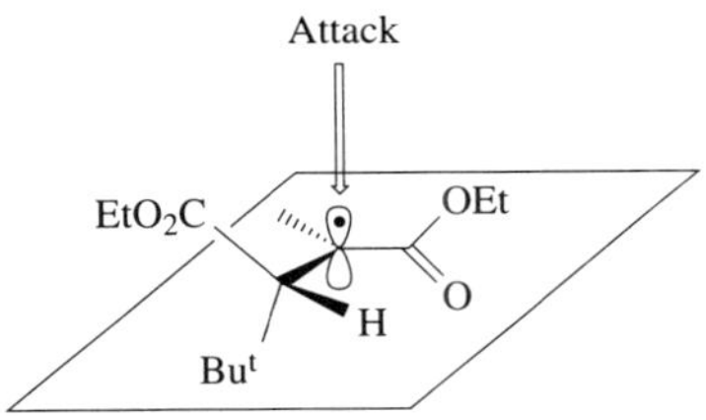

Figure 7 Radical attack on diethyl citraconate

The effect of the groups X (alkyl, aryl, OR, OH, F, CO_2Et), R^1 (alkyl, aryl) and R^2 (alkyl, H) on the diastereoselectivity of additions to alkenes and other radical traps has been extensively investigated (Table 7) by, for example, Hart ⟨91SL412, 92JOC4457⟩, Guindon ⟨91JA9701, 92JA4912, 93TL5885⟩, Giese ⟨92JOC3994, 92SL442, 92TL2673, 93HCA441, 93JA8585, 93TL2445, 93TL5885⟩, Curran ⟨91TL6307, 92JA4436, 92JA7007, 93JA8585, 93T4821, 93T4841⟩ and Porter ⟨91JA7002, 91JA7788, 91TL707, 92JA7664⟩. This has allowed an assessment of the steric and stereoelectronic factors which affect the diastereoselectivity to be probed ⟨91ACR296, 94SL1⟩. In general *syn* selectivity has been observed in the addition of chiral secondary α-ester and chiral α-amide radicals to allyltributylstannane whereas *anti* selectivity has been observed for reactions of conformationally restricted β-hydroxylated α-esters and amides. This *anti* selectivity has been shown to be solvent dependent (Table 7) ⟨92JOC4457⟩. Asymmetric induction in diester substrates can be increased by the addition of lanthanides such as $Eu(fod)_3$ ⟨94CC987⟩. The preference for *syn* products has been rationalised by considering that most radical reactions have early transition states and so ground state conformations should be taken into account. ESR and theoretical models have indicated that the lowest energy conformation of ester-substituted radicals is that which minimises allylic $A^{1,3}$ strain (Figure 7) ⟨93TL5885⟩. Attack then occurs at the least hindered side of the molecule to produce *syn* products preferentially. Work by Curran on the addition to secondary and tertiary α-amide radicals has highlighted that models based upon ground state conformations may be inadequate ⟨93T4821, 93T4841⟩. He emphasises the need to take into account the steric energy cost to approaching reagents, and that such factors could outweigh ground state energy-based requirements.

Table 7 Relative 1,2-asymmetric inductions in allylation of conformationally restricted radicals.

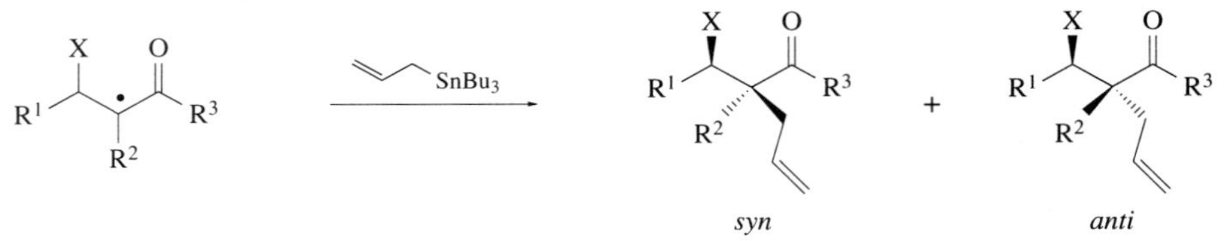

R^1	R^2	R^3	*X*	*Solvent*	*Yield* (%)	*Temperature* (°C)	syn/anti	*Ref.*
Me	H	OEt	OMe	THF	75	−78	4.9/1	92JOC4457
Me	H	OEt	OH	PhMe	75	−78	1/1.5	92JOC4457
Me	H	OEt	OH	THF	75	−78	3.3/1	92JOC4457
Bu^t	H	NMePh	OAc	PhH	64	25	9/1	91JOC4335
Bu^t	H	NMePh	OH	THF	64	25	1/13	91JOC4335

Asymmetric 1,4-induction in the addition of chiral radicals to alkenes has also been observed ⟨91ACR296, 94SL1⟩. The reactions of conformationally restricted α-ester and α-amide radicals containing chiral auxiliaries such as the (−)-8-phenylmenthol **(14)** ⟨93T6419⟩, 2,5-dimethylpyrrolidine **(15)** ⟨90JA6740⟩, 2,5-dimethoxymethylpyrrolidine **(16)** ⟨92JA7007⟩, camphor sultam **(17)** ⟨90JA6738⟩ and imide **(18)** ⟨92JA7007⟩ have been investigated. Addition of the imide radical **(18)** to allyl stannane followed by hydrolysis gave the acid **(19)** as essentially one diastereomer (96/4) (Equation (2)) ⟨92JA7007⟩. Similar selectivities were obtained for the other auxiliaries ⟨91ACR296, 94SL1⟩. Radical allylation of the camphor sultam radical derivative **(17)** at −20 °C was more selective than allylation of the corresponding lithium enolate.

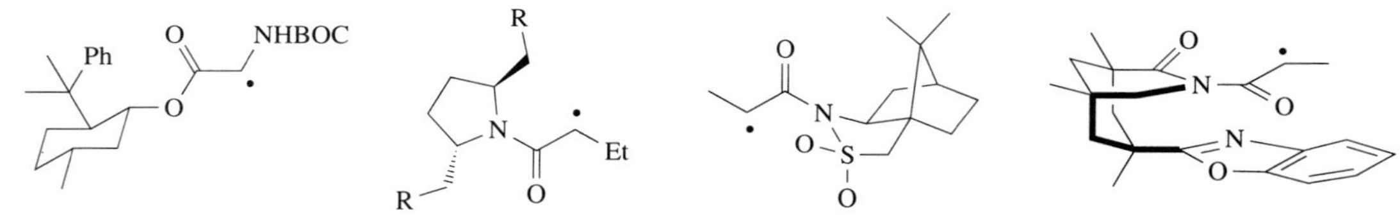

(14) **(15)** R = H **(16)** R = OMe **(17)** **(18)**

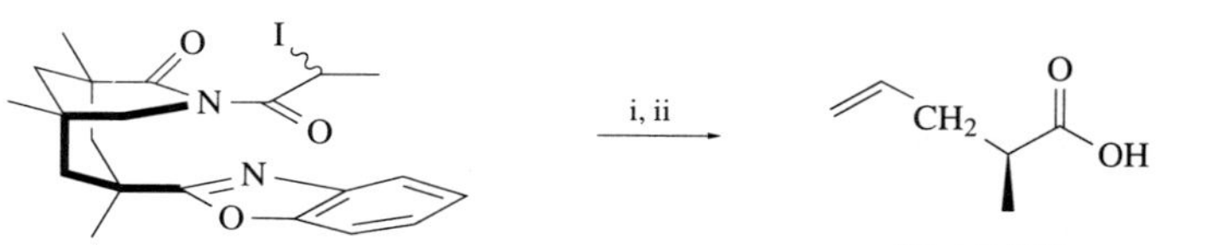

(S)-(+)-**(19)** (2)

i, (0.35 M) allyltributylstannane (2.0 equiv.), AIBN, 5 mol%, C_6H_6, 80 °C, 4 h, 38%;
ii, LiOH•H_2O, 30% H_2O_2, THF, 18 h, RT, 55%

Addition of the chiral radical **(15)** to ethyl acrylate furnished the desired product **(20)** in 84% *de* (Equation (3)). The diastereoselectivity could be increased if the radical was allowed to react with the chiral acrylamide also derived from the same auxiliary, to give the product **(21)** (Equation (4)) ⟨90JA6740, 91JA7002⟩. Presumably the increased selectivity is due to a 'matched' double asymmetric induction.

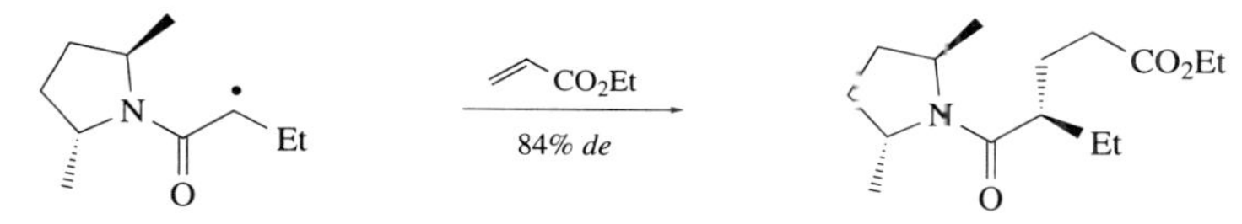

(15) (20) (3)

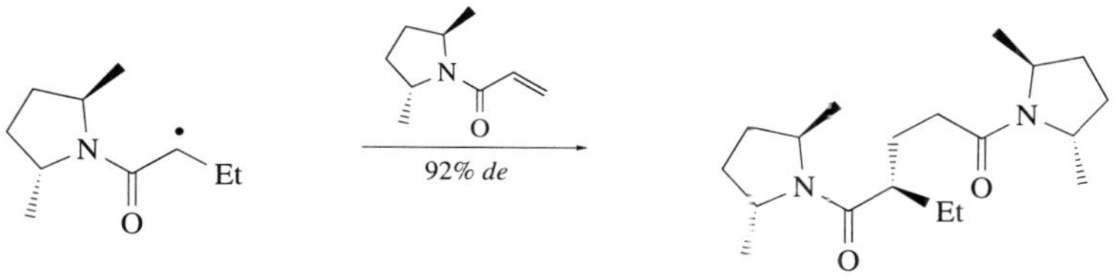

(15) (21) (4)

1.08.1.3.4 *Methods for conducting radical reactions*

(i) *Introduction*

For a given radical addition to an alkene to be successful certain reactivity and selectivity requirements have to be met ⟨B-86MI 108-01⟩. First, the rate of addition k_a of the radical (**22**) to the alkene (**23**) must be faster than any destructive reactions of the radical such as radical–radical recombination or disproportionation (Scheme 6). These latter unwanted reactions normally take place at rates approximating to the diffusion-controlled limit, and for additions to be synthetically useful, chain reactions are normally required. In chain reactions the concentration of the reacting radical is maintained at a lower concentration (normally 10^{-7}–10^{-8} M) than that of the alkene or alkyne partner (the concentration of the latter is variable but is typically of the order of 1 M) ⟨B-73MI 108-02⟩. This concentration difference enables radical additions to carbon–carbon multiple bonds to be synthetically useful providing the rate of addition k_a exceeds 10^2 mol^{-1} s^{-1}. Further addition of this newly formed radical (**24**) to another molecule of alkene or alkyne (polymerisation) is normally avoided by trapping the radical. Trapping may be achieved with hydrogen donors (Equation (5)), heteroatom donors (Equation (6)) or electron donors (Equation (7)) (see Scheme 7), or intramolecular homolytic bond cleavage of a suitably substituted β leaving group (Equation (8)) (see Section 1.08.1.2.1). The rate of trapping must be faster than that for polymerisation but less than that for trapping of the original radical (**22**) otherwise no addition of this radical will take place. Despite these limiting factors, a number of successful chain reaction methods can be used. For further discussion see ⟨B-86MI 108-01, 91COS(4)715⟩.

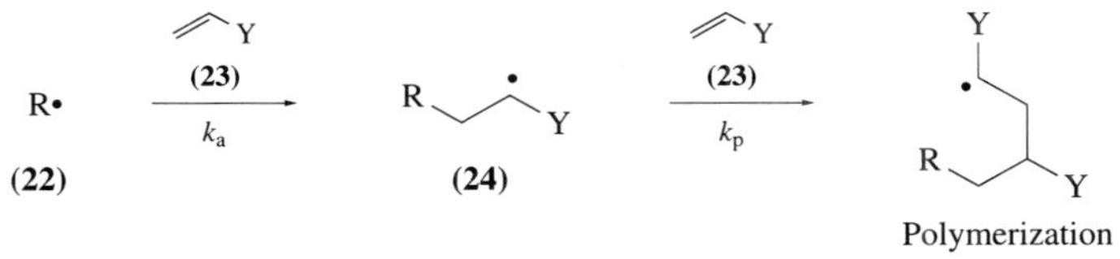

Scheme 6

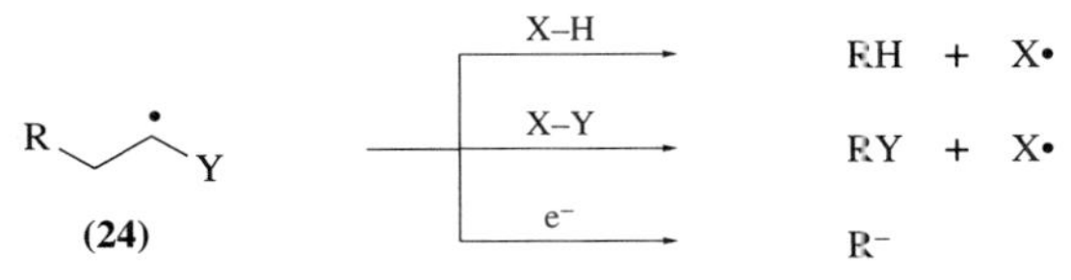

(5)
(6)
(7)

Scheme 7

(8)

(ii) Initiation of chain reactions

Radical chain reactions are normally initiated by either photolytic or thermal homolytic cleavage of a chemical initiator such as a peroxide ⟨79COC(1)909⟩ or an azo compound ⟨80CRV99⟩. The most common initiator is 2,2′-azobisisobutyronitrile (AIBN), which decomposes on heating (60–120 °C) or photolysis to two isobutyronitrile radicals. The amount of initiator required in any given reaction depends upon the efficiency of the chain transfer step. Chains with a slow chain transfer step will require more initiator than faster ones. For more information see ⟨91COS(4)715, B-92MI 108-01⟩. Triethylborane has also been used as an initiator to form $Bu_3Sn\cdot$ from Bu_3SnH ⟨89BCJ143⟩.

(iii) Chain reactions

(a) Tin hydrides. This is one of the most common methods for facilitating radical reactions ⟨87S665⟩. Typically trialkyl- or triaryltin hydrides are reacted with alkyl iodides in the presence of an activated alkene and AIBN ⟨85AG(E)553⟩. The chain carrier is the $Bu_3Sn\cdot$ **(25)** radical which reacts with the alkyl iodide to form the desired nucleophilic alkyl radical **(22)**. This radical then adds to the activated alkene partner **(23)** to form the new electrophilic radical **(24)**, which is trapped by Bu_3SnH to form the desired product **(26)** and regenerate the chain transfer reagent $Bu_3Sn\cdot$ **(25)** (Scheme 8). A full treatment of the factors which effect the outcome of tin-mediated cyclisations has been published by Giese ⟨B-86MI 108-01⟩ and Curran ⟨91COS(4)715⟩.

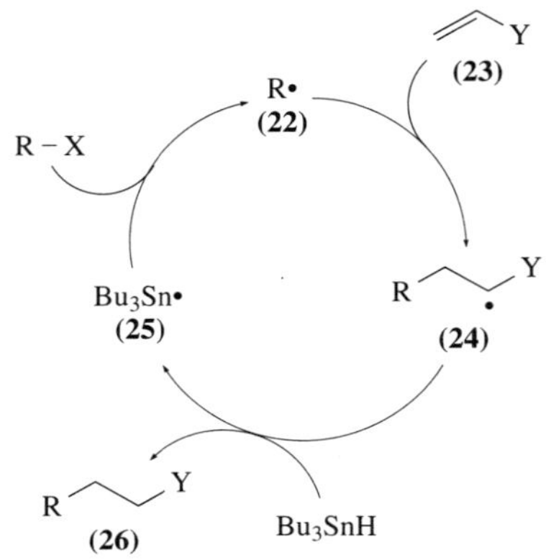

Scheme 8

Accurate kinetic rate measurements for the reactions of a number of radicals with alkenes and hydrogen and halogen abstraction reactions are known. For example, competitive trapping of the cyclohexyl radical **(27)** by Bu_3SnH ($k_H 3 \times 10^5\ M^{-1}\ s^{-1}$) (instead of addition to methyl acrylate ($2 \times 10^6\ M^{-1}\ s^{-1}$)) is generally avoided by employing a low concentration of the hydride donor with respect to the alkene (Scheme 9) ⟨91COS(4)715⟩. Either a huge excess of the alkene can be used, or a slow dropwise addition of the reducing agent can be employed via a syringe pump. Alternatively, a low concentration of Bu_3SnH can be generated by the reduction of a catalytic amount of Bu_3SnCl by $NaBH_4$ or $NaCNBH_3$ *in situ* ⟨63JOC2165, 86JA303, 86JOC3726⟩. For the success of the next step the reverse is required, that is, radical **(28)** must be trapped before competitive addition of another molecule of methyl acrylate can occur. This can be accomplished because the electronic pairing between the electrophilic/ambiphilic radical **(28)** and methyl acrylate is now mismatched (see Sections 1.08.1.3.2.2 and 1.08.1.3.2.3) thus slowing the rate of the polymerisation reaction. The yields of intermolecular addition reactions of nucleophilic radicals to electron-poor alkenes can be increased if the alkene becomes more activated. This is because the rate of addition k_a is increased while the rate of polymerisation k_p is decreased.

Other radical precursors such as bromides ⟨88JOC1432⟩, selenides ⟨89T935⟩, xanthates ⟨84AG(E)69⟩ and tertiary nitroalkanes ⟨85T4013⟩ can be used; however, chlorides are not generally applicable as precursors because abstraction by the $Bu_3Sn\cdot$ radical is slow enough for competitive hydrostannylation of the partner alkene or alkyne to be observed. Iodides are the precursors of choice since abstraction by $Bu_3Sn\cdot$ radicals occurs at rates approaching the diffusion-controlled limit ⟨84JA343⟩. Primary, secondary and tertiary alkyl, vinyl, allyl, allenyl and aryl radicals have all been

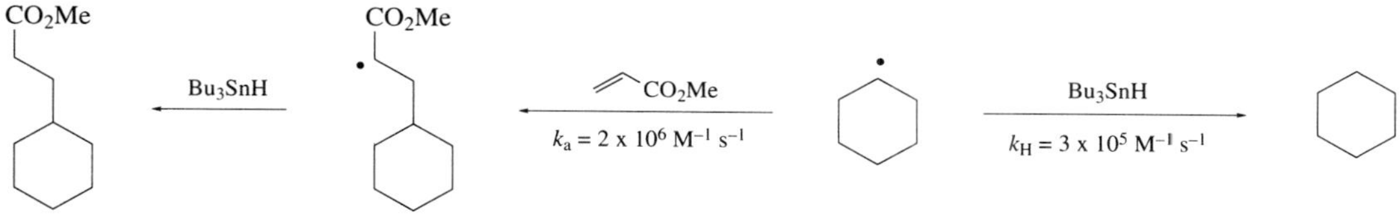

(28) **(27)**

Scheme 9

prepared using this method ⟨85AG(E)553, B-86MI 108-01, 91COS(4)715⟩. The tin method has been used widely in the formation of carbon–carbon bonds ⟨85AG(E)553⟩, as in the synthesis of natural products ⟨91CRV1237⟩ and sugar derivatives ⟨89AG(E)969⟩. The reaction of electrophilic radicals with electron-rich alkenes can also be accomplished by the tin method ⟨86CB444⟩.

(b) Germanium hydrides. Like tin hydrides, germanium hydrides can mediate chain reactions between radicals prepared from alkyl iodides and bromides with alkenes. Whereas trialkylgermanium radicals are as reactive as trialkyltin radicals in halogen abstractions the corresponding hydride is a much poorer hydrogen donor ⟨85JA4594, 87JOC3509, 88T6295⟩. This characteristic can be useful for minimising the amount of reduction product obtained in radical addition reactions ⟨91COS(4)715⟩.

(c) Silicon hydrides. Organotin hydrides are extremely toxic ⟨B-85MI 108-01⟩ and often difficult to remove from the reaction products ⟨90S448⟩. Silicon hydrides would be a preferable alternative; however, the Si—H bond is much stronger than the Ge—H or Sn—H bond, and hydrogen donation from simple alkyl silanes is too slow to maintain chains ⟨86JA4132⟩. Tris(trimethylsilyl)silane (TTMSS) has a relatively weak Si—H bond (79 kcal mol^{-1}, 330 kJ mol^{-1}), and this enables it to take part in radical chain reactions as a hydrogen donor ⟨87JA5267, 88JOC3641, 89TL681⟩. Despite the relative weakness of the Si—H bond in TTMSS, it is still stronger than in germanium or tin hydrides, and consequently it is a poorer hydrogen atom donor ⟨92ACR188⟩. This often enables increased yields of addition products to be formed at the expense of the undesired reduction products. Addition of the cyclohexyl radical (**27**) to methyl acylate in the presence of TTMSS gave the desired product (**29**) in 85% yield (Equation (9)) ⟨89TL681⟩. Roberts has shown that simple trialkylsilanes (Si—H bond energy 90 kcal mol^{-1}, 377 kJ mol^{-1}) can be used as hydrogen donors in radical chain reactions, providing a suitable polarity reversal catalyst is present (Scheme 10) ⟨89CC1387, 90TL5093, 91JCS(P1)103⟩. Other silicon hydrides have been used with varying success ⟨92ACR188⟩.

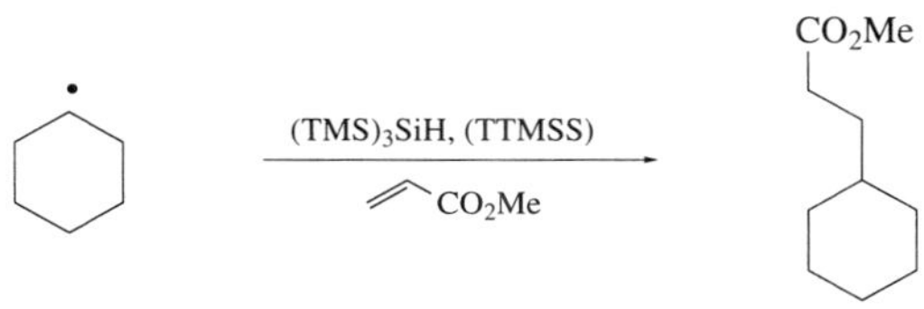

(9)

(27) **(29)**

$$R\bullet + XSH \longrightarrow RH + XS\bullet$$

$$XS\bullet + R_3SiH \longrightarrow XSH + R_3Si\bullet$$

Scheme 10

(d) Carbon hydrides. The C—H bond strength is even greater than the Si—H bond strength, and so only activated C—H bonds such as those adjacent to carbonyl groups are capable of sustaining a radical chain process. Unlike the examples already mentioned, an excess of donor reagent does not cause problems of premature reduction of the initial radical prior to the addition to the alkene. This is because reduction of the initial radical regenerates the trapping agent and another starting radical. Only suitably activated C—H bonds can be used in synthetically useful procedures. Hence, the electrophilic radical (**30**) undergoes addition to the electron-rich alkene (**31**) in the presence of di-*t*-butylbenzoyl peroxide (DTBP) as a radical initiator (Equation (10)) ⟨61IZV1099⟩. Other electrophilic radicals adjacent to esters and lactones have also been employed ⟨B-86MI 108-01⟩.

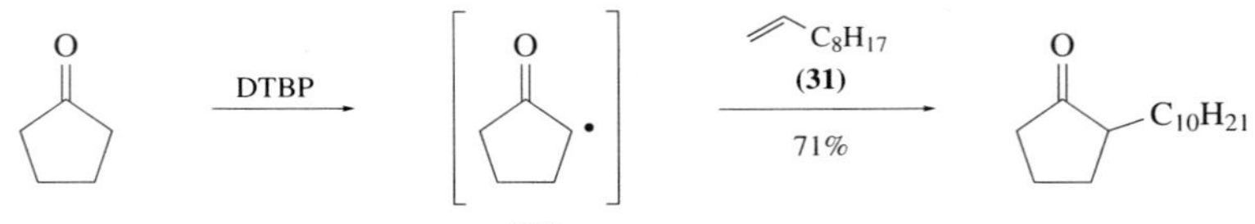

(30)

(10)

(e) Mercury hydrides. The use of mercury hydrides as donor reagents was pioneered by Giese and co-workers, who have shown their wide applicability in intermolecular radical addition reactions ⟨85AG(E)553, B-86MI 108-01, 88CRV487⟩. The Hg—H bond is weaker than the corresponding Sn—H bond, and mercury hydrides are therefore better hydrogen atom donors ⟨84CB3160⟩. Consequently, the rate of reduction k_H of the initial radical **(22)** is often faster than the rate of addition k_a to the desired alkene **(23)**. Radical additions using this method are therefore limited to very reactive alkenes only (Scheme 11). However, there are a number of advantages to using the mercury method. First, the reagents are prepared *in situ*, and reactions can be accomplished under mild conditions. Second, no thermal or photochemical initiation is necessary. This is due to the spontaneous decomposition of the reactive mercury hydrides. Third, mercury hydrides act as both the radical precursor and the radical trap; finally, reactions carried out using this method are easier to work up and purify than those utilising the corresponding tin or germanium hydrides ⟨91COS(4)715⟩.

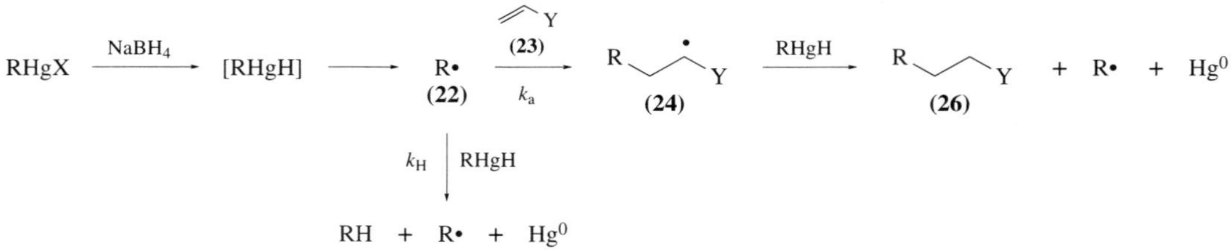

X = halide or acetate

Scheme 11

The mercury hydride reagents are normally generated *in situ* by the reduction of alkyl mercury halides or alkyl mercury acetates. Alternatively, alkyl mercury hydrides can be prepared by hydroboration of alkenes, followed by transmetallation of the resulting alkyl boranes with mercuric acetate followed by reduction. This enables alkenes to couple with other alkenes ⟨81AG(E)965⟩. Use can be made of the regioselectivity and stereoselectivity of the hydroboration reactions of alkenes to couple the terminal alkene unit in **(32)** to methyl vinyl ketone (MVK) selectively (Equation (11)) ⟨81AG(E)965⟩. For a comprehensive account of this methodology see ⟨88CRV487⟩.

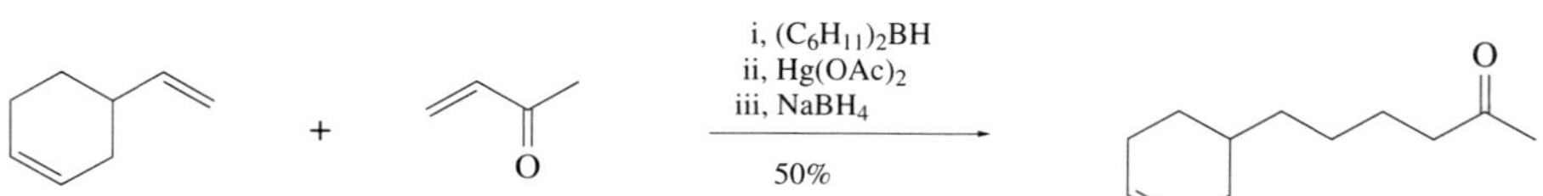

(32)

(11)

(f) The Barton method. One disadvantage with the hydrogen atom donor reagents is that the radical chain is terminated by reduction. Trapping with a heteroatom enables a functional handle to be introduced into the final product. The best known method for conducting this type of group transfer reaction was introduced by Barton and his co-workers ⟨87MI 108-01, B-92MI 108-01⟩. Homolysis of thiohydroxamate esters **(33)** either photolytically with a tungsten lamp or thermally generates carboxy radicals which undergo loss of CO_2 to form nucleophilic radicals **(22)** ⟨85TL5943⟩. Addition to activated alkenes or alkynes followed by trapping of the new radical as the pyridyl sulfide **(34)** furnishes the functionalised products (Scheme 12). Primary, secondary and tertiary radicals can be prepared in high yield using this method ⟨87MI 108-01⟩. With unreactive alkenes trapping of the initial radical **(22)** before addition can become competitive. To maximise formation of the addition products an excess of alkene is normally required. Good yields are only obtained for activated alkenes, alkynes and doubly activated internal alkenes ⟨86JCS(P1)39⟩. Other classes of thiohydroxamate esters have been evaluated for their efficiency in carrying out these transformations, with varying success ⟨86JCS(P1)39, 92T7091⟩. The thiohydroxamate esters can be prepared from the

corresponding carboxylic acids by a number of methods: a procedure utilising dipyridyl disulfide and tributylphosphine has been recommended ⟨92T7083⟩. The resulting pyridyl sulfides are useful synthetic precursors for a number of functional groups. Barton has developed a two-carbon (Equation (12)) ⟨93TL6505⟩ and a three-carbon (Equation (13)) ⟨92TL5017⟩ homologation procedure giving rise to acids, amides, esters, aldehydes and ketones.

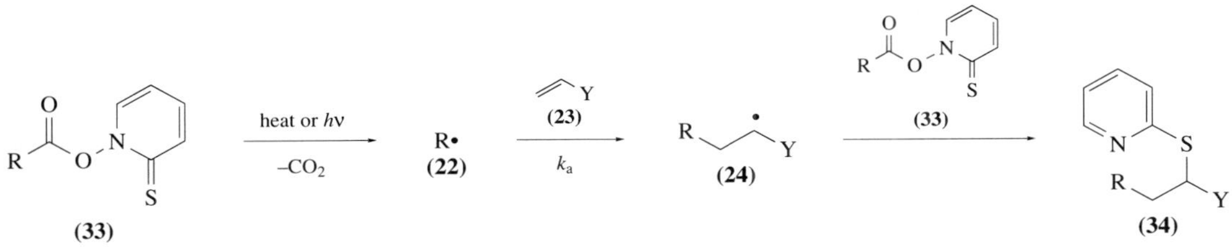

Scheme 12

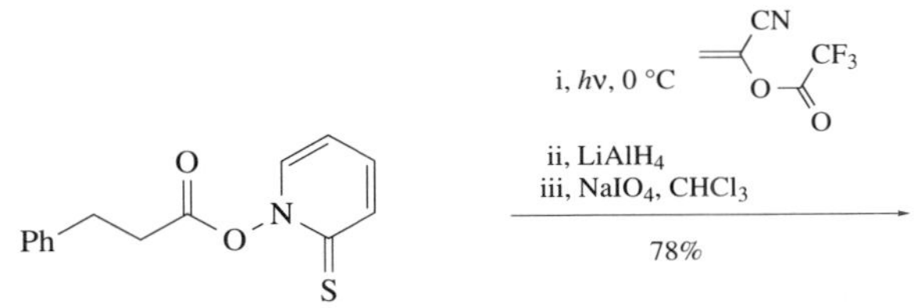

(12)

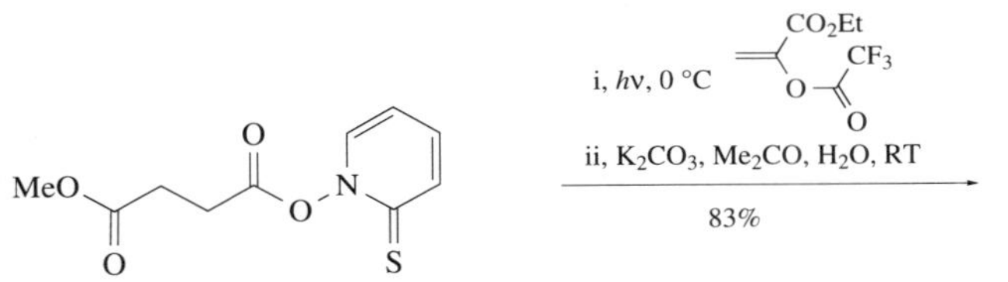

(13)

(g) The borane method. Alkyl radicals (**22**) can be generated from trialkyl boranes (**35**) by reaction with oxygen ⟨72AG(E)692⟩. The adduct radicals are trapped as boron enolates (**36**) and can undergo further reactions such as aldol reactions (Scheme 13) ⟨73JA967⟩. Addition products (e.g., **37**) are limited to those from reactions with activated alkenes such as α,β-unsaturated aldehydes and ketones. Polymerisation is observed for additions with nitrile- and ester-activated alkenes, presumably because the adduct radicals polymerise far quicker than they add to the borane ⟨72AG(E)692⟩. A disadvantage of this method is that only one alkyl group is transferred from boron; however, modification of the ligands on boron can circumvent this problem ⟨71JA3777⟩. Atom transfer additions have also been initiated by triethylborane either in the presence of Ph_3SnH or in its absence. Hence, addition of three equivalents of *t*-butyl iodide to MVK in the presence of triethylborane gave the desired adduct (**38**) and the by-product (**39**) arising from addition of an ethyl radical to the alkene after quenching with methanol (Equation (14)) ⟨88TL1041⟩. In 1989, Nozaki extended this methodology to additions of alkyl radicals on to alkynes ⟨89TL3155⟩. The radicals are generated from iodides and undergo smooth atom transfer additions to give iodoalkenes in good yield (Equation (15)). This method has also been used to initiate the addition of *t*-butyl radicals to chiral unsaturated keto-sulfoxides in an asymmetric radical addition ⟨93JA464⟩.

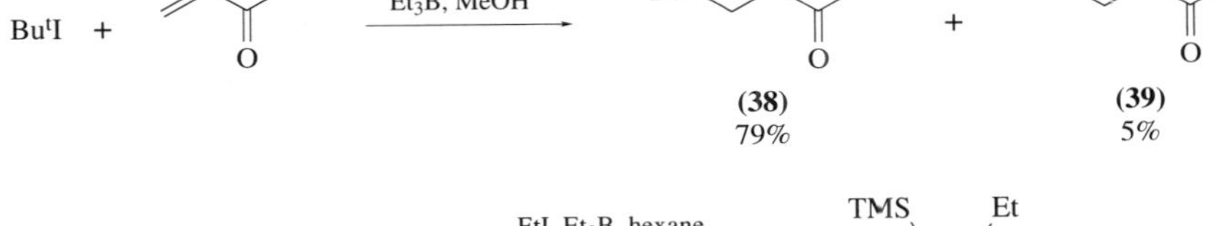

(14)

(15)

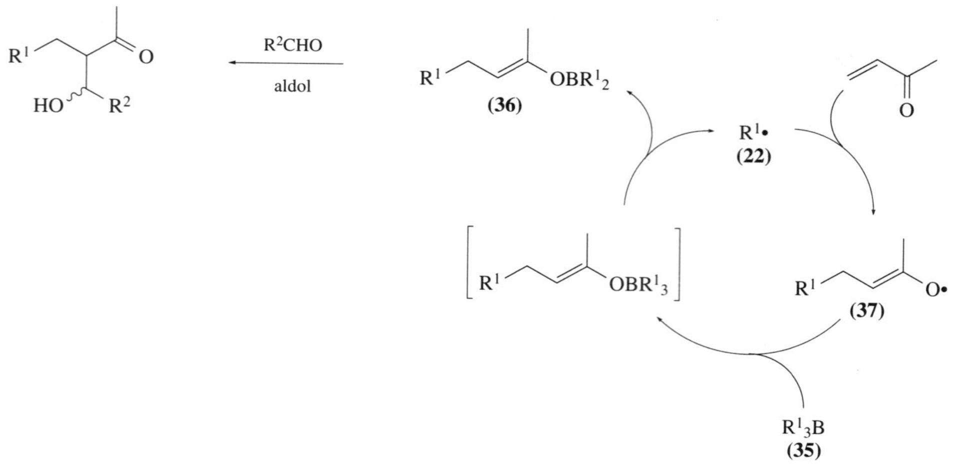

Scheme 13

(h) The fragmentation method. As already mentioned in Section 1.08.1.2.1, relatively weak bonds such as C—Br, C—SnR_3 and C—SR, can fragment if they are located β to a radical. This unimolecular process is generally fast and liberates a radical as a leaving group which can propagate the chain. This method for conducting radical chain reactions has the advantage over the metallic hydride donors in that the problem of reduction of the initial radical (**22**) no longer exists and that the process is not reductive. Consequently, this method is useful for conducting radical additions to unactivated alkenes and for retaining a functionality handle in the molecule after addition ⟨88S489⟩. The use of allyl and vinyl stannanes has become popular for generating the $Bu_3Sn\cdot$ as the chain carrier (Equation (16)) ⟨82JA5829⟩. Although the procedure is very versatile, crotyl stannanes are not normally tolerated ⟨85T4079⟩. Johnson has shown the applicability of allyl cobalt complexes for carrying out similar reactions, and demonstrated that methyl substitution at the allylic sites is tolerated (Equation (17)) ⟨83ACR343⟩.

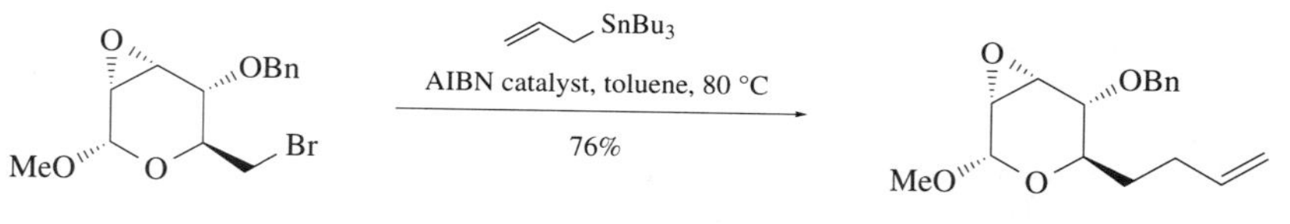

(16)

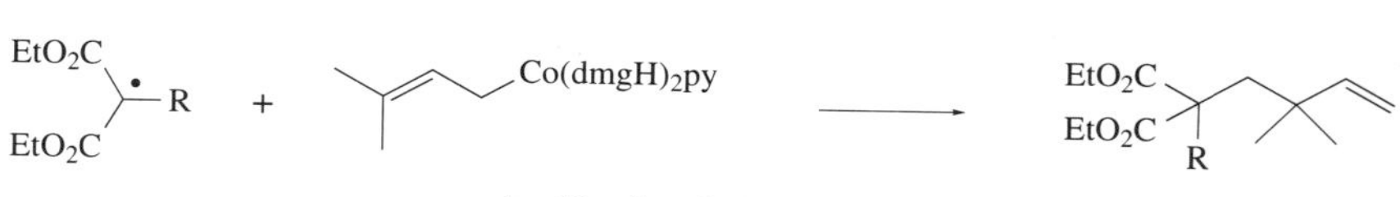

(17)

dmgH = dimethylglyoximate

(i) The atom transfer method. Photolysis of the alkyl iodide (**40**) in the presence of the alkene (**41**) and a catalytic amount of $(Bu_3Sn)_2$ furnishes the radical addition product (**42**), where the product radical has been trapped by iodine atom transfer from another molecule of the starting material (Equation (18)) ⟨89JA6265, 91COS(4)715⟩. The initial radical is formed either by direct photolytic cleavage of the iodide or from abstraction by the $Bu_3Sn\cdot$. This method is particularly suited for the formation of adducts where the rate of addition of the radical to the alkene is slow ⟨89JOC3140⟩. This is because, unlike the metal hydride or Barton methods, atom transfer from the starting iodide to the initial radical reproduces both entities, that is, the initial radical is not trapped before addition. The limiting factor is the rate of atom transfer ⟨89JOC1826⟩. If this is too slow then polymerisation can occur. One way to facilitate fast atom transfer steps is to use reactive iodides as radical precursors and to ensure that the adduct radical is less stable than the initial radical (i.e., the initial radical is stabilised). These criteria generally restrict this type of reaction to the addition of electrophilic radicals to electron-rich alkenes.

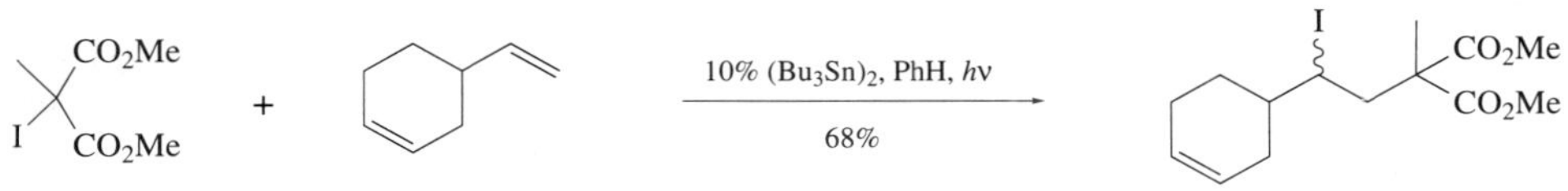

(40) **(41)** **(42)** (18)

(iv) Nonchain reactions

(a) Organocobalt group transfer. Photolysis or thermolysis of the relatively weak Co^{III}—C bond (18–25 kcal mol^{-1}, 75–105 kJ mol^{-1}) in RCo(salen), RCo(salophen) or $RCo(dmgH)_2py$ complexes (salen (*N,N′*-bis(salicyclidene)ethylenediamino); salophen (*N,N′*-bis(salicyclidene)*o*-phenylenediamino); dmgH (dimethylglyoximato); py (pyridine)) occurs homolytically to generate an alkyl radical **(22)** and a Co^{II} complex ⟨88CSR361⟩. Addition of this radical **(22)** to an alkene **(23)** followed by retrapping with the Co^{II} complex furnishes a new alkyl Co^{III} complex **(43)** that can undergo elimination of $Co^{III}H$ to form alkenes **(44)** (Scheme 14). The precursor alkyl cobalt complexes can be prepared by reaction of alkyl halides or tosylates with supernucleophilic Co^{I} complexes prepared *in situ* ⟨76AG(E)417⟩. Branchaud and co-workers have used this method to produce the sugar derivative **(45)** from the organocobaloxime **(46)** (Equation (19)) ⟨88TL167⟩. In this case the unsaturated product is formed by *β*-elimination of $HCo^{III}(dmgH)_2py$ by 'dehydrocobaltation', from the intermediate cobaloxime. The more activated the radical alkene acceptor, the more facile is this elimination process. The sp^2–sp^2 coupling of two alkenes via a 'hydrocobaltation–dehydrocobaltation' procedure has been demonstrated by Pattenden and co-workers (Scheme 15) ⟨90JCS(P1)2709⟩. The cobaloxime radical precursor **(47)** is formed by the addition of cobalt and hydrogen across the activated alkene **(48)** ⟨88CC1110⟩. Photolysis in the presence of styrene followed by dehydrocobaltation gave the addition product.

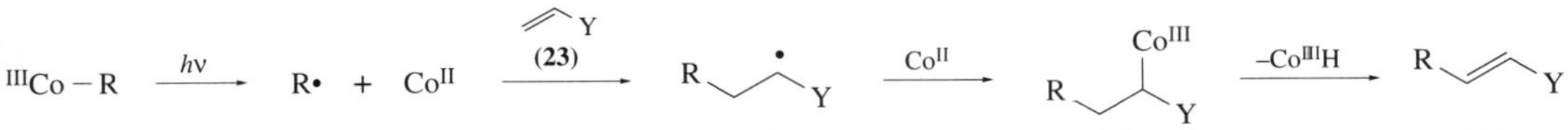

(22) **(24)** **(43)** **(44)**

Scheme 14

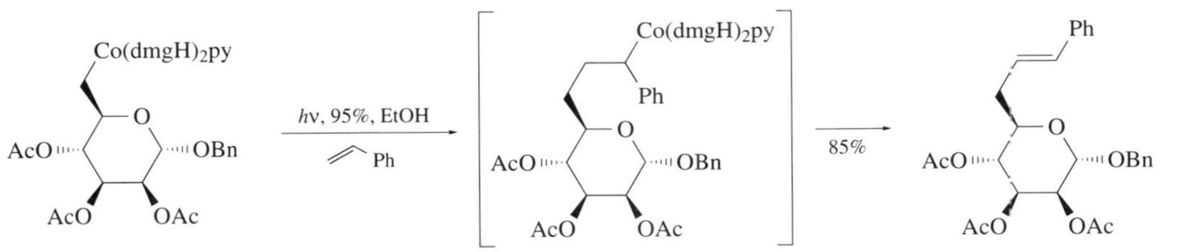

(46) **(45)** (19)

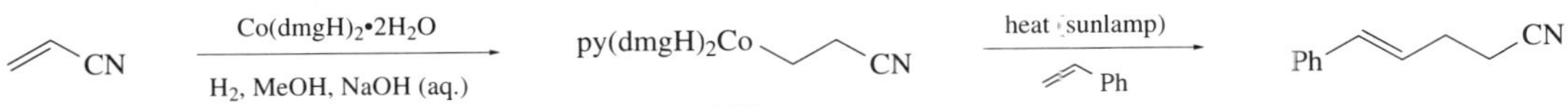

(48) **(47)**

Scheme 15

Reductive cobalt-mediated additions have been achieved with a catalytic amount of reduced vitamin B_{12} in the presence of an electrochemical or chemical reductant (such as zinc) ⟨83JA7200⟩. As in the oxidative cases the radical precursors are normally alkyl halides, and the initial radicals are formed by homolytic cleavage of Co^{III} alkyl complexes. Radical addition followed by reduction and reformation of the Co^{I} complex completes the cycle.

(b) Electron transfer processes. Nucleophilic alkyl radicals can be oxidised to cations by a number of metallic species. For a review on the use of titanium, vanadium, chromium, niobium, manganese, iron, cobalt, copper and ruthenium complexes in oxidative and reductive radical reactions, see ⟨94CRV519⟩. The most widely studied metal-mediated method for conducting radical additions to alkenes involves the use of two equivalents of $Mn^{III}(OAc)_3$ in conjunction with the desired alkene, and an activated radical precursor (Equation (20)) ⟨74JA7977, 74JOC3456⟩. The radical precursor (normally an enol derivative of a β-keto ester, β-keto acid, β-diketone, ester, ketone or an aldehyde) is oxidised to the enol radical, which can undergo addition to an electron-rich alkene ⟨88JOC2137⟩. Further oxidation of this newly formed nucleophilic radical by a second equivalent of $Mn^{III}(OAc)_3$ yields the corresponding carbocation, which can undergo reaction with a nucleophile to furnish the lactone (**49**) ⟨87TL175⟩. Alternatively, if no suitable nucleophile is present then elimination of a proton may result ⟨83JOC1373⟩. $Mn^{III}(OAc)_3$ is slow to promote the oxidation of 2° or 1° nucleophilic radicals to carbocations and so $Cu^{II}(OAc)_2$ is often added to perform this transformation.

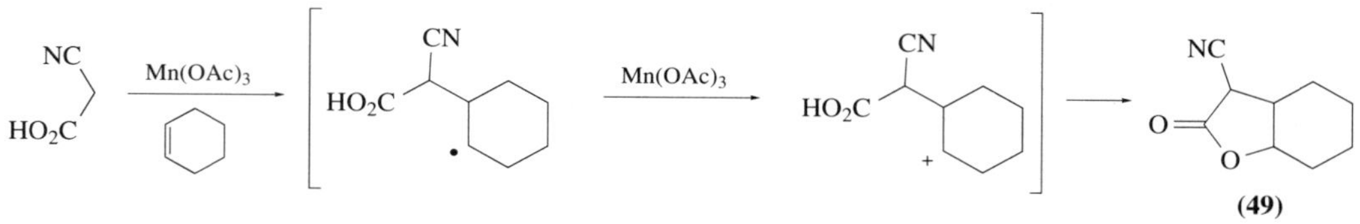

(20)

(49)

1.08.1.4 Intramolecular Radical Additions

1.08.1.4.1 Introduction

This section is organised similarly to that on intermolecular additions. The reader is strongly recommended to read the previous section for a more detailed account of the mechanisms and principles involved in carrying out radical reactions and also to consult the reviews by Curran ⟨88S417, 88S489, 91COS(4)715, 91COS(4)779, 91CRV1237⟩, Giese ⟨B86-MI 108-01⟩ and Ramaiah ⟨87T3541⟩. Radical cyclisation reactions are normally easier to carry out than intermolecular radical additions. This is because most cyclisations proceed at a much quicker rate ($k_{c\ \text{5-hexenyl cyclisation}}$, $2 \times 10^6\ s^{-1}$) ⟨87JOC3509⟩ than in intermolecular additions. Like intermolecular reactions, the rate of cyclisation of the initial radical and the rate of trapping of the final radical must be greater than termination of the chain by unwanted side reactions. One requirement is that the initial radical must not be trapped before cyclisation has occurred. This is not normally a problem even though the initial radical and the cyclised radical often have the same nucleophilicity. In practice, low concentrations of trapping agent can be employed but are not normally required unless the cyclisation is particularly slow.

1.08.1.4.2 The 5-hexenyl radical: regioselectivity

The factors which effect radical cyclisation reactions are best illustrated by ring closure of the 5-hexenyl radical (**50**) (Equation (21)). The 5-hexenyl radical cyclises at 25°C to furnish both the 5-*exo trig* product (**51**) and the 6-*endo trig* product (**52**) in a ratio of 98 : 2. The rate of cyclisation has been measured and found to be of the order of $2 \times 10^6\ s^{-1}$ for the 5-*exo* cyclisation and $4 \times 10^5\ s^{-1}$ for the 6-*endo* cyclisation ⟨87JOC3509⟩. The reaction itself is highly exothermic, irreversible and under kinetic control. In contrast to the intermolecular additions already described, the radical acceptor does not need to be activated, although cyclisation is faster for activated alkenes. The factors which determine the regioselectivity of the cyclisation have been the focus of considerable theoretical attention. The reader is directed to reviews of Beckwith ⟨85T3925⟩ and Houk ⟨87JOC959⟩ or the synopsis by Curran ⟨91COS(4)779⟩. In brief, MM2 and STO-3G calculations seem to indicate that the less strained transition state is that which leads to the smaller ring, and that cyclisation occurs via a strain-free, chair-like transition state (Figure 8). The same conclusion has been reached by Baldwin using a vector analysis approach ⟨76CC734⟩.

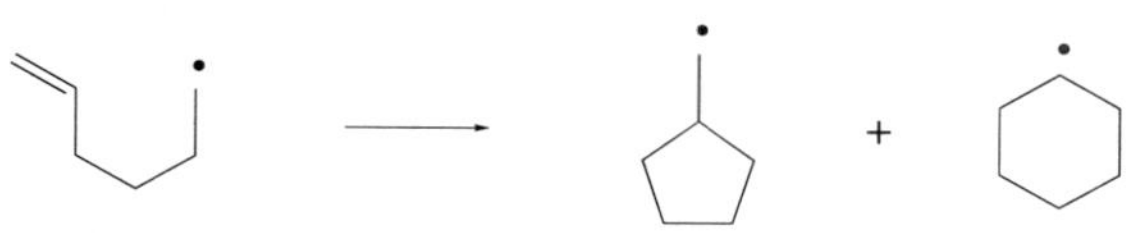
(21)

(50) (51) (52)
98 : 2

Figure 8 The chair-like 1,5-transition state.

The effects of substituents on the outcome of cyclisation of the 5-hexenyl radical have been thoroughly investigated. A methyl group at the reacting alkene centre will reduce the rate of cyclisation on to that carbon atom primarily due to steric effects ⟨85T3925⟩. Consequently, substitution at the 5 position furnishes mainly the 6-*endo* product (Equation (22)). Substitution at the 2, 3 and 4 positions increases the rate of 5-*exo* cyclisation, probably by raising the energy of the ground state relative to the transition state. Substitution of alkyl groups at the reacting radical centre has little effect on the outcome of the cyclisation (Equation (23)). The incorporation of heteroatoms (such as oxygen or nitrogen) in the 3 position of the chain causes a marked increase in the rate of cyclisation, with the 5-*exo* product predominating (Equation (24)). This effect is due to better overlapping of the radical with the alkene partner in the 5-*exo* transition state primarily due to the smaller bond angles and bond lengths associated with the carbon–heteroatom bonds than with the carbon–carbon bonds ⟨85T3925⟩. Conversely, incorporation of a silicon atom into the chain favours the 6-*endo* product (Equation (25)) ⟨85T3979⟩.

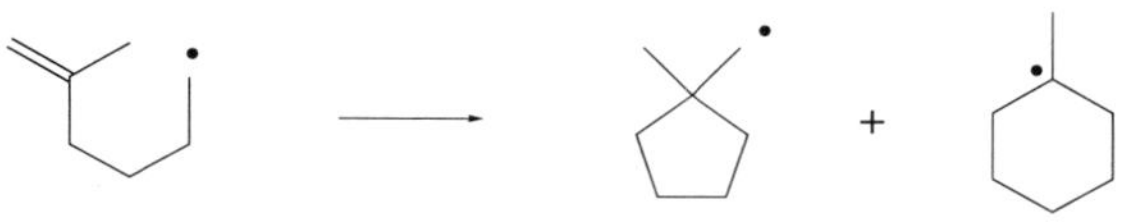
(22)

33 : 66

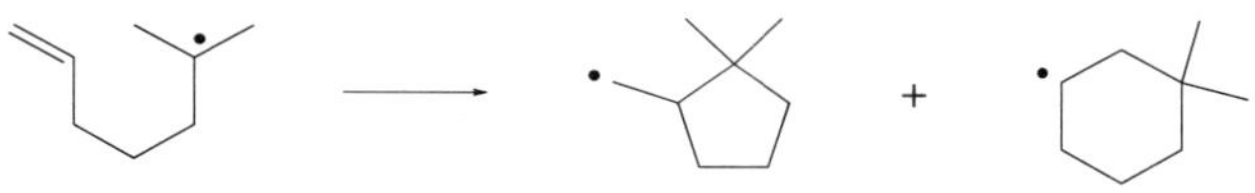
(23)

99 : 1

(24)

(25)

Electrophilic radicals with two radical stabilising groups often undergo reversible cyclisations to furnish 6-*endo* products preferentially ⟨71ACR386⟩. 6-*endo* Products have also been observed to form preferentially in radical cyclisations of keto-substituted radicals ⟨89JOC3140⟩.

1.08.1.4.3 Stereoselectivity in substituted 5-hexenyl radical cyclisations

The stereochemical outcome of the cyclisation of 1-, 2-, 3- and 4-substituted 5-hexenyl radicals has been studied and has been rationalised by the model proposed by Beckwith ⟨83AJC545⟩. The cyclisation occurs via a chair-like transition state with the substituents occupying pseudoequatorial positions (Figure 8). Hence, cyclisation of 2- and 4-substituted 5-hexenyl radicals gives *trans* products while cyclisation of 3-substituted 5-hexenyl radicals gives *cis* products predominantly (Equation (26)) ⟨85T3925⟩. Small alkyl substituents at C-1 generally give *cis* products while larger alkyl groups and ether substituents at C-1 predominantly give *trans* products. Other studies indicate that the minor diastereomeric products are produced from less stable chair and boat transition states ⟨89JA1759⟩. Rajanabubu and co-workers have shown that the directing effect of a substituent at the 4 position is important in determining the stereochemical outcome of cyclisations which lead to fused rings ⟨89JA1759⟩. Cyclisation of the radical (**53**) gives the *trans*-1,2 product (**54**), presumably via a low-energy boat conformation (Equation (27)).

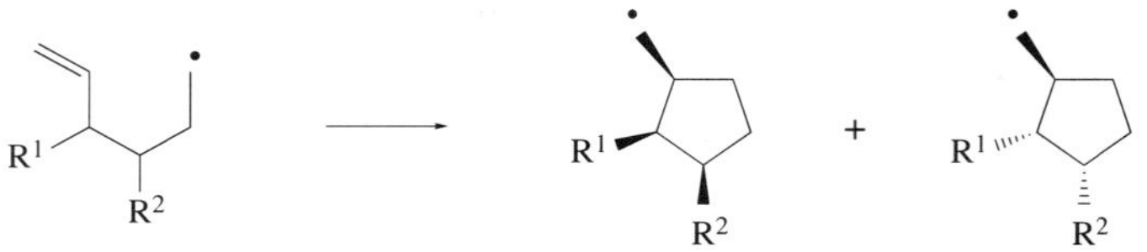

(26)

R^1 = H, R^2 = Me	75	:	25
R^1 = Me, R^2 = H	17	:	83

(**53**) (**54**) (27)

1.08.1.4.4 Allyl, vinyl and aryl cyclisations

The cyclisation of aryl and vinyl radicals is much more rapid than the corresponding alkyl radicals ($k_c \approx 3 \times 10^8$ s^{-1}) and consequently higher yields are often obtained in simple 5-*exo* cyclisations ⟨85T3925⟩. A vinyl radical cyclisation is a useful transformation because the product contains a double bond that can be further functionalised if needed (Equation (28)) ⟨88S417⟩. Vinyl radicals can be prepared from vinyl halides but are harder to form than alkyl radicals due to the increase in strength of the sp^2 carbon–halogen bond. Bromo- and iodo-substituted alkenes give the best results, with less reactive precursors giving lower yields ⟨83TL1871⟩. The geometry of the initial vinyl halide precursors is irrelevant as the barrier to inversion of the vinyl radical is low (Scheme 3) ⟨92TL3543⟩. The *exo*:*endo* ratio is normally dependent upon the concentration of any added radical trap. For example, cyclisation of (**55**) with a 0.02 M concentration of Bu_3SnH gave (**56**) and (**57**) in a ratio of 75:25, while cyclisation with a higher concentration (1.7 M) gave the 5-*exo* product exclusively ⟨86TL4529⟩. This is because at low Bu_3SnH concentrations the cyclised radical (**58**) can undergo a rearrangement to give the ring-expanded product (**59**) (Scheme 16).

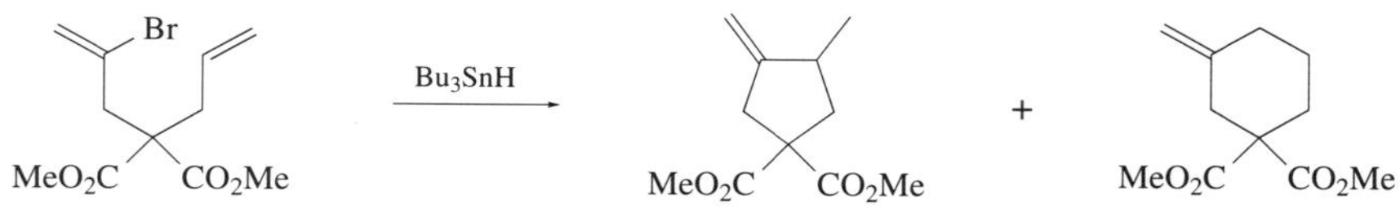

(**55**) (**56**) (**57**) (28)

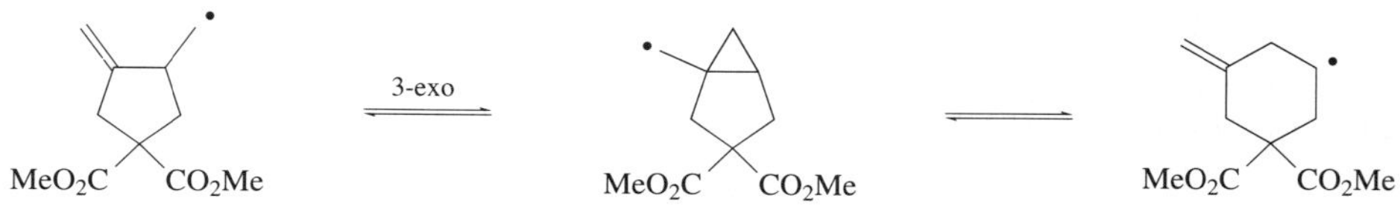

(**58**) (**59**)

Scheme 16

Allyl radicals can undergo radical cyclisations, although they are not as reactive as their vinyl radical counterparts (delocalisation energy 14.0–14.5 kcal mol^{-1}, 58.6–60.7 kJ mol^{-1}) ⟨88JA6911⟩.

Allenyl radicals have been used in carbocyclisations to form vinylidene-substituted cyclopentanes ⟨93TL5251⟩.

Aryl radical cyclisation has become a popular method for the formation of benzo fused five- and six-membered rings (Equation (29)) ⟨91COS(4)779⟩. Aryl radical cyclisations can been directed 6-*endo* by steric crowding on the 5-*exo* carbon ⟨90TL2315⟩ or by cyclisation on to suitably activated alkenes. Hence, Glover and co-workers have shown that 6-*endo* cyclisation of **(60)** is faster ($k_c = 4.6 \times 10^8$ s^{-1}) than 5-*exo* cyclisation ($k_c = 1.5 \times 10^8$ s^{-1}) at 80 °C ⟨93JOC2115⟩. The ratio of products was found to be dependent upon the nitrogen substituent. Radical cyclisation of precursors that contain suitably activated rings can undergo a ring expansion in a process related to that observed with vinyl radicals. This process is similarly dependent upon the concentration of the radical trapping agent employed (Equation (30)) ⟨86TL2833⟩.

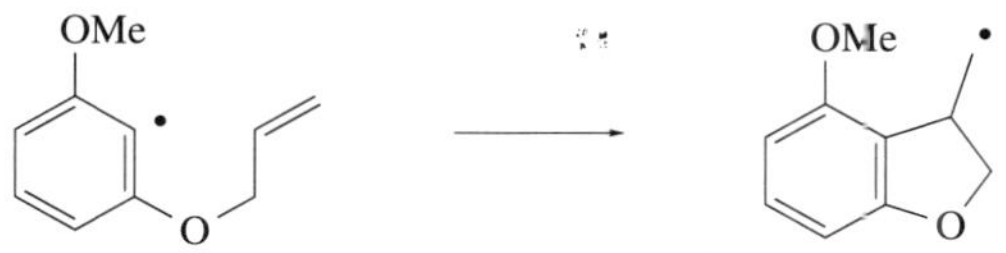

(29)

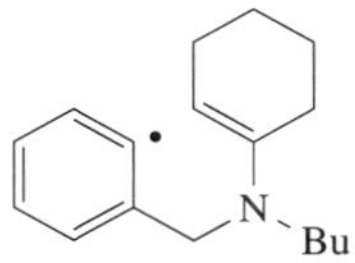

(60)

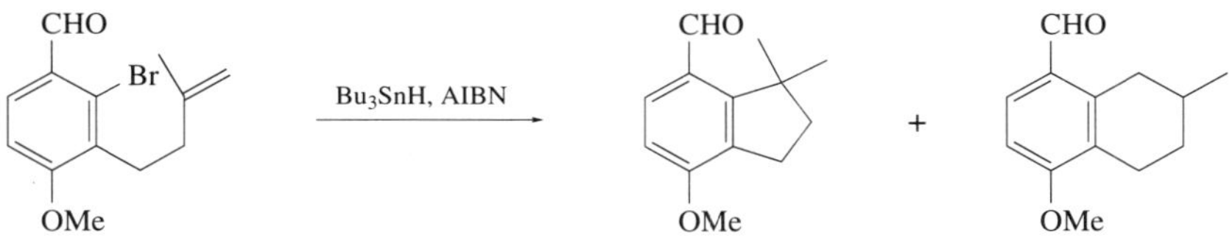

(30)

0.15M 91 : 9
0.02M 43 : 57

1.08.1.4.5 Formation of six-membered rings

Cyclisation of the 6-heptenyl radical **(61)** is possible although it is less regioselective than cyclisation of the 5-hexenyl radical (Equation (31)). The rates for 6-*exo* cyclisation at 25 °C ($k_c = 5 \times 10^3$ s^{-1}) and 7-*endo* cyclisation ($k_c = 7 \times 10^2$ s^{-1}) are much slower than for the corresponding 5-*exo* and 6-*endo* cyclisations ⟨74CC472⟩. Cyclisations can be accomplished if low concentrations of trapping agents are used or atom transfer cyclisations are utilised. One competing reaction is often reduction of the initial radical by intramolecular 1,5-hydrogen transfer to form the allyl radical (Equation (32)) ⟨88JA5900⟩. This problem can be addressed if suitable substituents are located on the chain to block the hydrogen transfer reaction or if the alkene component is activated. Aryl and vinyl radical 1,5-hydrogen transfer processes are more exothermic than for simple alkyl radicals, and Curran and his co-workers have developed this facile rearrangement for the preparation and cyclisation of many heterosubstituted radicals ⟨88JA5900, 93JA6051⟩.

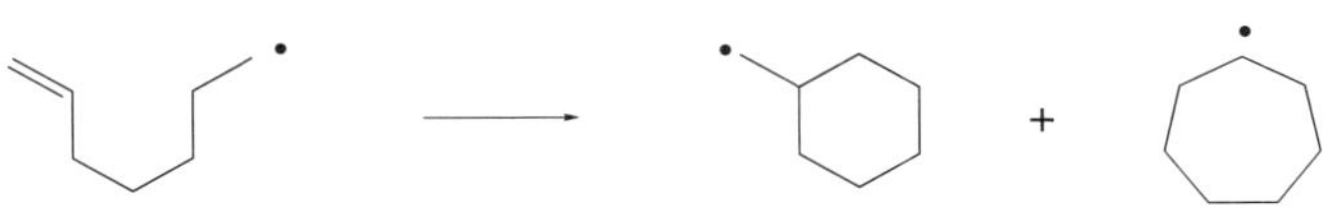

(31)

(61) 85 : 15

(32)

1.08.1.4.6 Formation of other ring sizes

7-*exo* and 8-*exo* Alkyl radical cyclisations are disfavoured processes and are not normally synthetically useful. The best method for the formation of medium-sized rings utilising a radical strategy is via ring expansion of smaller rings by a process analogous to that observed in vinyl radical cyclisations (see Section 1.08.1.4.4). For a description of these procedures the reader is directed to the excellent review by Dowd ⟨93CRV2091⟩. 7-*endo* and 8-*endo* Aryl radical cyclisations have been reported using $Mn(OAc)_3$ to form the cycloheptane (**62a**) and the cyclooctane (**62b**) (Equation (33)) ⟨91T8663⟩. Larger rings can be prepared by a radical macrocyclisation approach. This has been shown to be successful for 11–16-membered rings, although an activated radical acceptor is required ⟨86JA2787⟩. An *endo* attack is the natural mode for macrocyclisation, and a number of natural products have been successfully synthesised utilising this approach (Equation (34)) ⟨92JCS(P1)1313⟩. Cyclisation on to both activated alkenes and alkynes is possible ⟨92T3413⟩. Electrophilic radicals prepared from iodomalononitriles have been used in macrocyclisations on to unactivated alkenes using atom transfer conditions ⟨90JA9401⟩. Chiral induction in a radical macrocyclisation process has been reported ⟨89JA8309⟩.

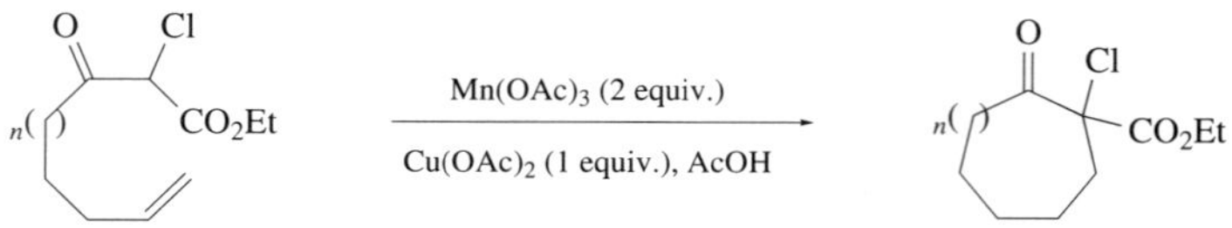

(33)

(**62**) **a**; n = 1, 58%
b; n = 2, 47%

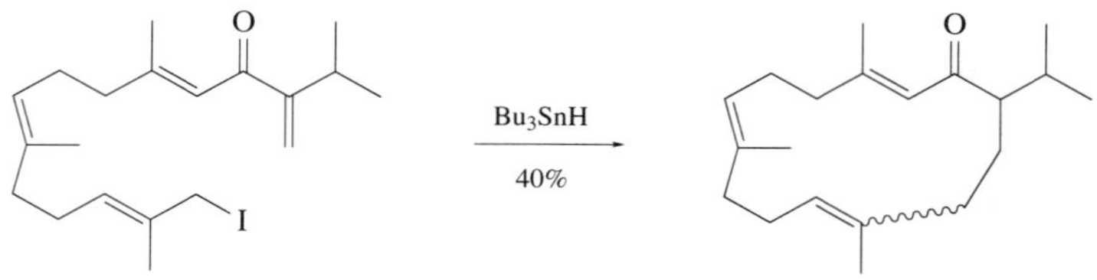

(34)

(*E*):(*Z*) = 4:1

Cyclisation of 3-butenyl and 4-pentenyl radicals to produce cyclopropanes and cyclobutanes, respectively, is possible in theory, although both these reactions are reversible with the equilibrium generally favouring the open-chain radicals (Scheme 17) ⟨91COS(4)779⟩. Cyclisation and trapping of 3-*exo* and 4-*exo* cyclisation products is possible if the cyclised radical is particularly stable with respect to the starting open-chain radical. Cyclisation and trapping with Bu_3SnH to form cyclopropanes is possible when the cyclised radical is stabilised by virtue of it being allylic ⟨94JOC718⟩, benzylic ⟨92JOC4284⟩ or α to a keto group ⟨92TL7307⟩. Vinyl cyclopropanes have been formed by 3-*exo* cyclisation followed by fragmentation of a PhS· (Equation (35)) ⟨94TL4531⟩. Cyclopropane formation has also been reported in alkenes having intrinsically large amounts of ring strain with favourable geometric alignment of the homoallylic radical and the alkene (Equation (36)) ⟨93TL2091⟩.

Scheme 17

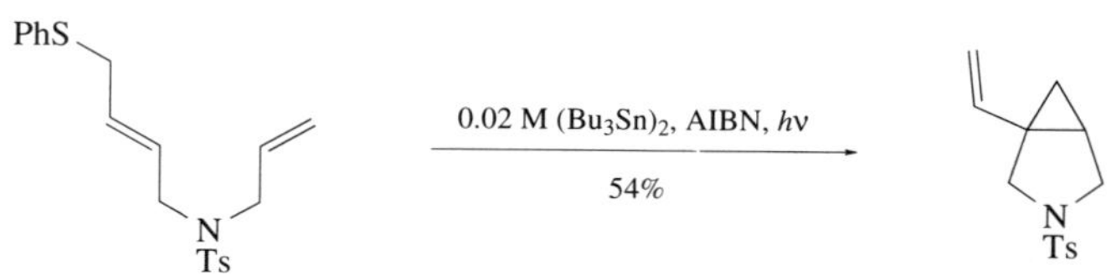

(35)

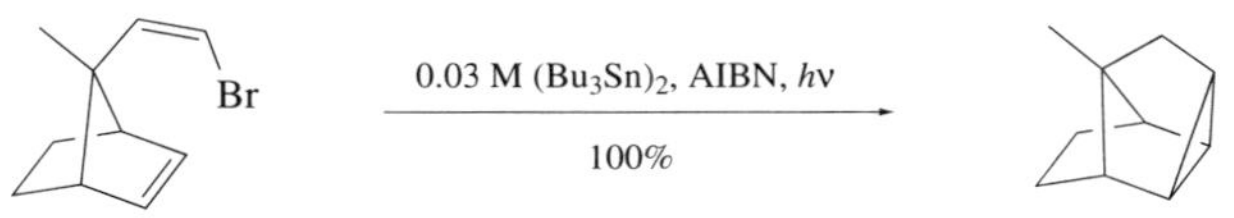

(36)

1.08.1.4.7 Methods for facilitating radical cyclisation reactions

(i) Introduction

The following sections indicate the different methods available to facilitate radical cyclisation reactions. The reader is strongly advised to first read Sections 1.08.1.4.1 and 1.08.1.4.2 to become familiar with the factors which affect the success of radical chain reactions.

(ii) Radical chain methods

(a) Tin hydrides and germanium hydrides. The most commonly employed method to facilitate radical cyclisation reactions is the tributyltin hydride method ⟨B-86MI 108-01, 87S665, 91COS(4)779, 91CRV1237⟩. The mechanism for the chain reaction has already been mentioned (see Section 1.08.1.3.4(ii)(a)). Typically, trialkyl- or triaryltin hydrides and AIBN are reacted with alkyl, vinyl or aryl halides that contain a suitably positioned radical acceptor (Scheme 18). The chain carrier is the $Bu_3Sn\cdot$ radical which reacts with alkyl, vinyl or aryl halides (**63**) to form carbon-centred radicals (**64**) ⟨85AG(E)553⟩. Due to the increased rate of radical addition in intramolecular reactions compared to intermolecular addition, cyclisation of nucleophilic and electrophilic radicals can be achieved without the need to activate the radical acceptor. Trapping of the initial radical (**64**) with Bu_3SnH before cyclisation can be avoided if low concentrations of the reagent are employed (typically 0.02 M concentrations are required for efficient cyclisation of alkyl radicals). For particularly slow cyclisations very low concentrations of Bu_3SnH are often required ⟨86JA2787⟩. These conditions can be accomplished by utilising either a syringe pump addition of Bu_3SnH or by using one of the catalytic procedures ⟨63JOC2165, 86JA303, 86JOC3726⟩.

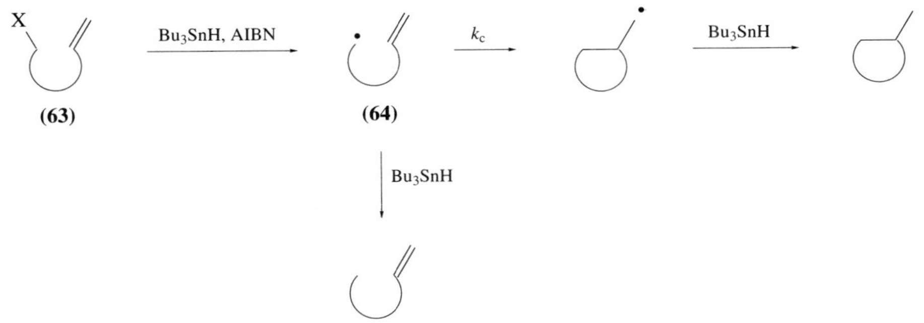

Scheme 18

The rate of 5-*exo* or 6-*exo* cyclisation of vinyl and aryl radicals is much faster than the corresponding alkyl radicals, and these cyclisations can be accomplished in high yields with the minimum of reduction products ⟨88S417⟩. Allenyl radicals have also been generated and cyclised successfully using this method ⟨93TL5251⟩. Various radical precursor groups can be employed, with the order of abstraction by $Bu_3Sn\cdot$ being I > Br > SePh ≈ OC(S)Me > Cl > SPh ⟨91CRV1237⟩. Iodides are the precursors of choice since abstraction by $Bu_3Sn\cdot$ radicals occurs at rates approaching the diffusion-controlled limit ⟨84JA343⟩. Thiophenyl precursors have been used where halo precursors were inconvenient and difficult to prepare or were unstable. For example, acylamino radicals have been generated in this manner ⟨84JA8209⟩. The mild and neutral nature of radical cyclisations mediated by Bu_3SnH is particularly suited for transformation of sugars and sugar derivatives. Various five- and six-membered ring functionalised carbocycles have been prepared by 5-*exo* or 6-*exo* radical cyclisation of sugar derivatives (Equation (37)) ⟨86JA3102⟩. This is a particularly powerful strategy for the formation of highly functionalised frameworks ⟨93CRV2779⟩. The tin method has been used frequently to prepare heterocyclic units ⟨91COS(4)779, 91CRV1237⟩. γ-Lactones are easily prepared by the reductive cyclisation of unsaturated bromoacetals followed by Jones oxidation (Equation (38)) ⟨83JA3741⟩. The increased rate of 5-*exo* cyclisation observed for heteroatom-substituted chains leads to particularly high regioselectivity. This accelerating effect can even overcome factors that might otherwise lead to competitive 6-*endo* cyclisation (Equation (39)) ⟨85JOC5620⟩. Cyclisation of chains with oxygen in the 1, 2, 3 and 4 positions have all been reported. Numerous cyclisations to form five- and six-membered nitrogen heterocycles have also been achieved ⟨B-86MI 108-01, 91COS(4)779, 91CRV1237, B-92MI 108-01⟩. Stannylation of alkynes can be used to generate vinyl radicals that may undergo cyclisation on to suitably positioned acceptors (Equation (40)) ⟨89TL225⟩. Hydrostannylation normally occurs at the less hindered terminal end of alkynes. Vinyl radicals can also be generated from vinyl halides utilising the tin method. Yields were shown to increase when moving from chloro, via bromo, to iodo precursor groups ⟨83TL1871⟩. Aryl radical cyclisations have also been successfully carried out using the Bu_3SnH method, as have radical macrocyclisations and radical transannular reactions ⟨91COS(4)779⟩.

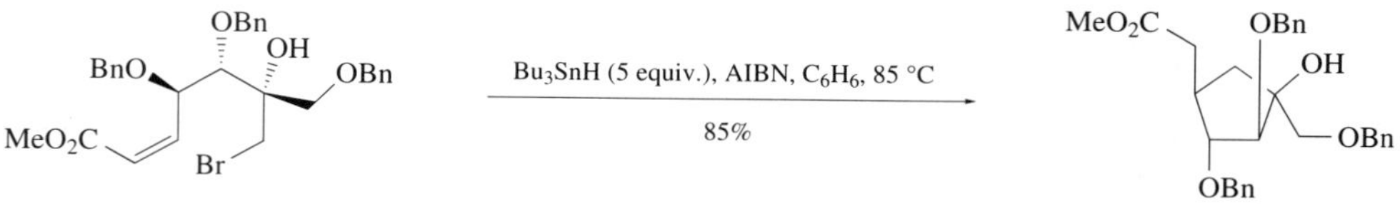

(37)

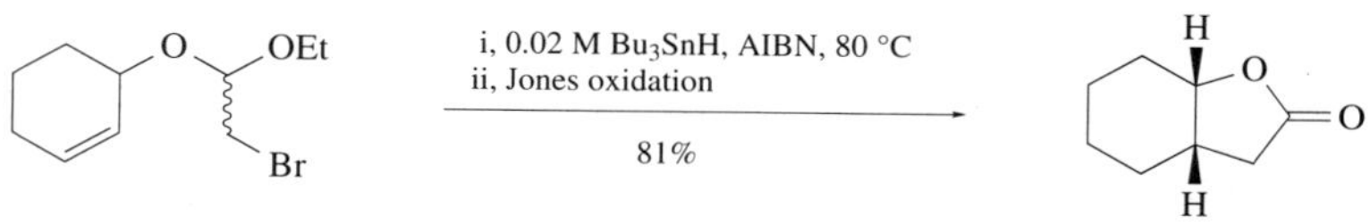

(38)

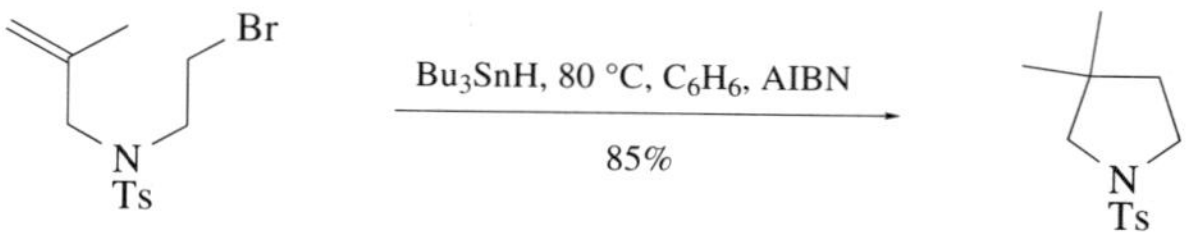

(39)

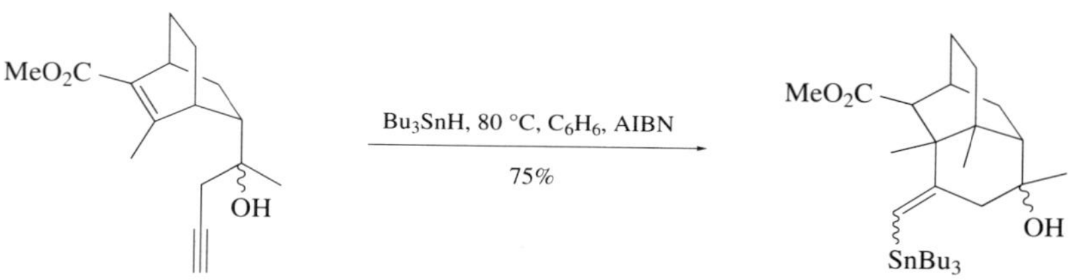

(40)

Trialkylgermanium hydrides have been less widely employed in radical cyclisation reactions. The mechanism of the radical chain reaction is identical to that of cyclisation with Bu_3SnH, the major difference being that Bu_3GeH is a far less reactive hydrogen atom donor ⟨87JOC3509⟩. This reagent is useful for mediating relatively slow cyclisations.

(b) Silicon hydrides. Trialkyltin hydrides react with many functional groups via a radical chain mechanism and play an important role in synthesis. However, among the disadvantages suffered by organotin hydrides are: (i) they are extremely toxic ⟨B-85MI 108-01⟩, (ii) they are often difficult to remove from the reaction products ⟨90S448⟩ and (iii) they are expensive. Organosilicon hydrides are a desirable alternative; however, the Si—H bond is much stronger than the Sn—H bond and hydrogen donation from simple alkyl silanes is too slow to maintain chain reactions ⟨86JA4132⟩. Tristrimethylsilylsilane (TTMSS) can be used in radical chain reactions because its Si—H bond energy is low enough to sustain the chain ⟨92ACR188⟩. Despite the relative weakness of the Si—H bond it is still stronger than in germanium hydrides or tin hydrides and so it is a poorer hydrogen atom donor. Giese and co-workers have demonstrated that intramolecular cyclisation reactions can be mediated by the use of TTMSS (Equation (41)) ⟨89TL681⟩. Cyclisation of 6-bromohexene (**65**) with TTMSS led to a 24:1 ratio of cyclised products to uncyclised products, while with Bu_3SnH the ratio was 6:1 under the same conditions. This observation suggests that the TTMSS reagent may be useful for mediating slow cyclisations. TTMSS has been used to mediate radical macrocyclisation of the precursor (**66**) to give the 14-*endo-trig* product (**67**), which was a key intermediate

in the sythesis of (−)-zearalenone (Equation (42)). Use of the silane led to the same products and yields as the Bu_3SnH-mediated reaction but was easier to work up ⟨90TL3641⟩. The TTMSS radical can be successfully added to alkynes to form a vinyl silane radical which can undergo a reductive vinyl radical cyclisation ⟨93TL6787⟩.

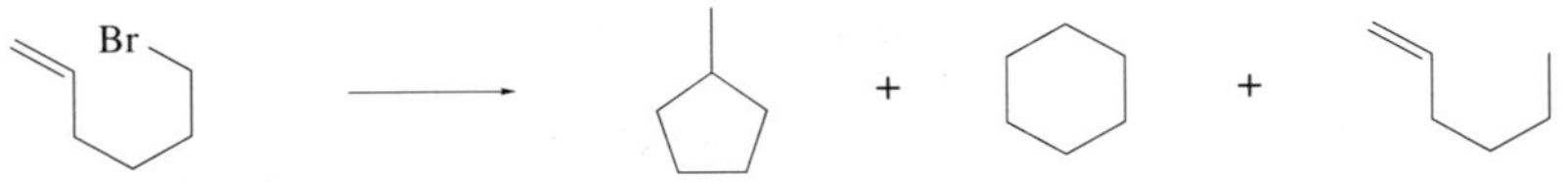

(41)

(65)						
$(TMS)_3SiH$	93	:	2	:	4.1	
Bu_3SnH	83	:	1.2	:	15	

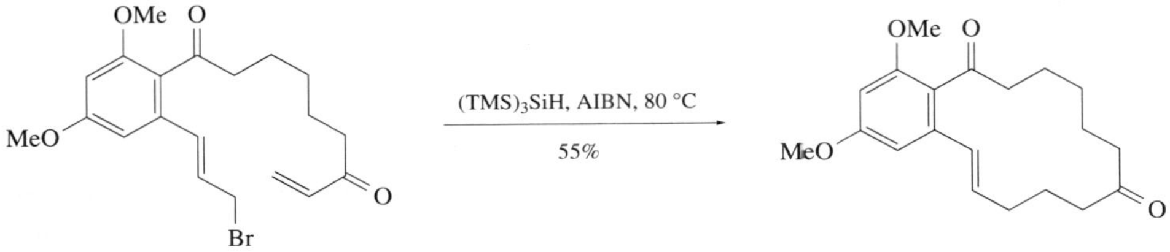

(42)

(66) **(67)**

(c) Mercury hydrides. Most reductive radical cyclisation reactions are carried out using the tin method; however, mercury hydrides can also be used to generate radicals suitable for cyclisation ⟨85AG(E)553, 88CRV487⟩. The Hg—H bond is much weaker than the Sn—H bond and consequently mercury hydrides are better hydrogen donors. This limits the methodology to particularly rapid cyclisations. Precursors can be prepared by oxymercuration of alkenes to give alkylmercury salts, which can be reduced to the mercury hydrides *in situ* (Equation (43)). The procedure is simple to carry out and purification is less problematic than with the tin method ⟨82JOC2231⟩.

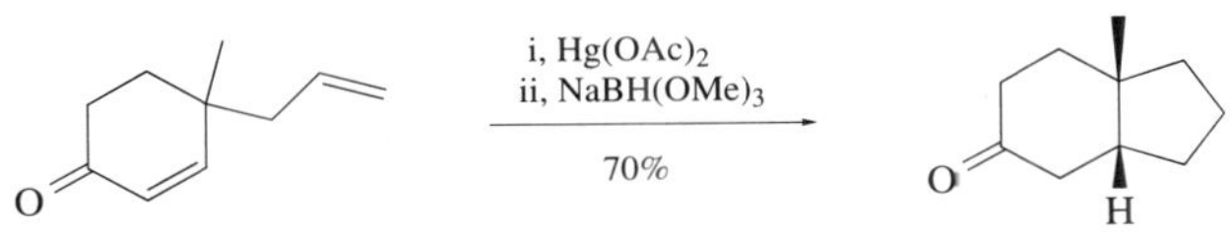

(43)

(d) Hydrogen atom transfer reactions. Pioneering work in the area of hydrogen atom transfer cyclisations was carried out by Julia ⟨71ACR389, 74PAC553⟩. He showed that stabilised radicals **(68)** prepared from benzoyl peroxide and cyanomalonates **(69)** cyclised to give mixtures of *exo* and *endo* products, and that the reaction was reversible due to the stability of the initial radical **(68)** (Scheme 19) ⟨67BSF2641, 69BSF2427⟩. In fact this method is probably the most useful for carrying out reversible cyclisations to give thermodynamic mixtures of products. Cyclisation of **(69)** at high temperatures gave the 6-*endo* product predominantly while at lower temperatures the 5-*exo* product predominated. Hydrogen abstraction from C—H bonds is very slow relative to abstraction from metal hydrides, consequently ring opening to give back the stable electrophilic radical **(68)** is favoured over hydrogen atom abstraction. This method is generally limited to simple molecules.

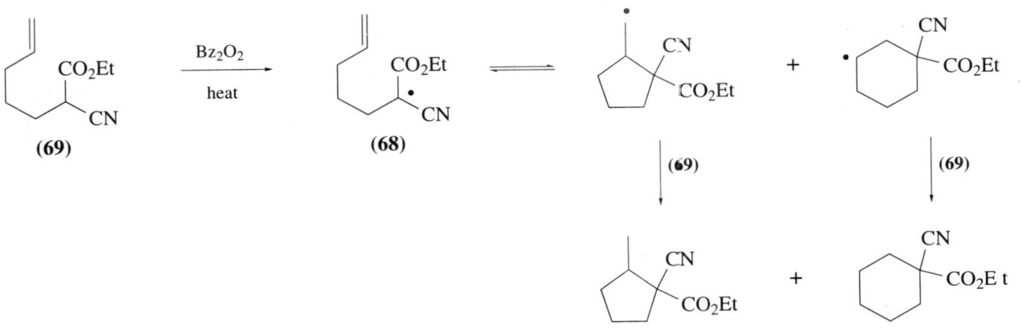

Scheme 19

(e) The Barton method. The Barton method (see Section 1.08.1.3.4(iii)(f)) has been used to facilitate radical cyclisation reactions. Homolysis of thiohydroxamate esters initially gives radicals which rearrange to cyclised radicals and are trapped as the pyridyl sulfide to furnish functionalised products (Equation (44)) ⟨87MI 108-01, 91COS(4)779, B-92MI 108-01⟩. Although not as commonly employed as the tin method it has the advantage of being nonreductive in nature. The radical precursors are prepared from acids via the acid chlorides and the initial radicals are formed by decarboxylation. This is often a good method for forming radicals where other carbon–heteroatom bond precursors are difficult or impossible to prepare ⟨87TL1413⟩. Minimisation of the thiohydroxamate concentration in theory should allow relatively slow cyclisations to be conducted.

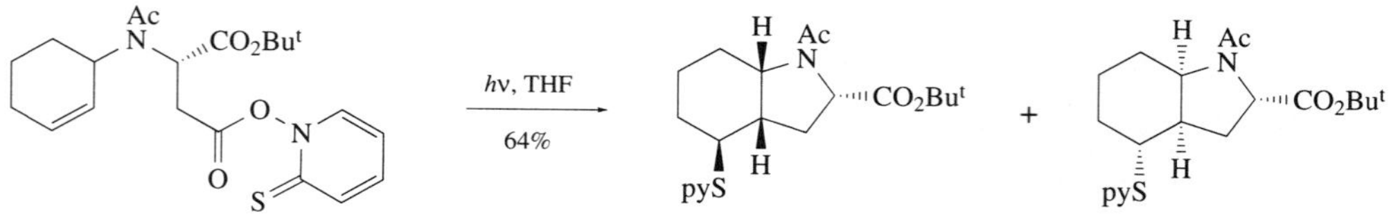

(44)

(f) The fragmentation method. As already mentioned in Section 1.08.1.3.4(iii)(h), relatively weak bonds such as C—Br, C—SnR_3 and C—SR can fragment if they are substituted β to a radical. This unimolecular process is generally fast and liberates a radical as a leaving group. This radical may then react with a radical precursor to propagate the chain. Radical cyclisation reactions carried out utilising this method have the advantage over the metallic hydride donors in that the problem of reduction of the initial radical no longer exists and that the process is not reductive. Consequently, this method should be useful for conducting particularly slow radical cyclisations and for introducing functionality into the molecule after addition (Equation (45)) ⟨85TL3311⟩. The use of allyl and vinyl stannanes ⟨90JA5601⟩ have become very popular methods of generating the chain carrier ⟨91CRV1237⟩.

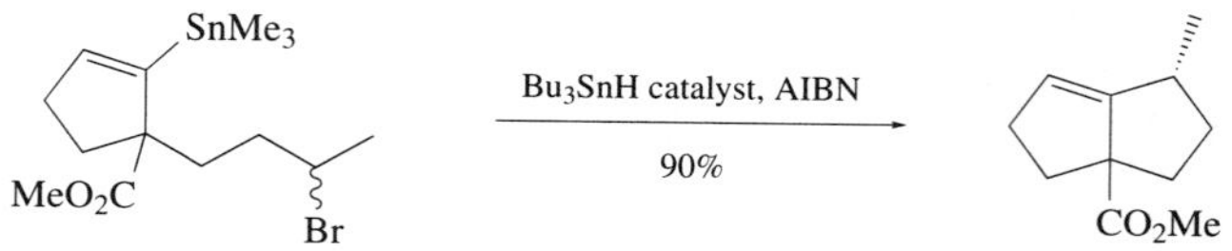

(45)

(g) Atom transfer reactions. Atom transfer cyclisations (see Section 1.08.1.3.4(iii)(i)) offer versatile alternatives to metal hydride mediated radical cyclisation reactions ⟨88S489, 91COS(4)779, 91CRV1237⟩. Both heteroatoms (e.g., halogens) and hydrogen atoms (see Section 1.08.1.4.7(ii)(d)) can undergo atom transfer reactions although the halo atom transfer cyclisations are the more versatile because the products contain a functionality handle (i.e., a halogen). Atom transfer cyclisations are particularly suited for mediating slow cyclisation reactions. Irradiation of **(70)** with a catalytic amount of $(Bu_3Sn)_2$ formed the bicyclic iodide **(71)**, whereas normal reductive radical cyclisation with Bu_3SnH furnished only the reductive product **(72)** even with low concentrations (0.02 M) of Bu_3SnH (Scheme 20) ⟨89JOC3140⟩. The Bu_3SnH-mediated cyclisation fails because the initial radical is reduced prior to cyclisation, while the atom transfer method succeeds because the initial radical has sufficient lifetime to undergo cyclisation. One way to facilitate fast atom transfer steps is to use the reactive iodides as radical precursors and to ensure that the final radical is less stable than the initial radical ⟨89JOC3140⟩. Consequently, α-iodo esters, and nitriles, are excellent precursors for atom transfer reactions. Even relatively unstable unconjugated alkyl radicals can undergo efficient atom transfer cyclisation as long as the adduct radical is even less stable (Equation (46)) ⟨89JA6265⟩. Thermoneutral iodide atom transfer reactions can also be employed; however, the rate of transfer is relatively low $1 \times 10^5\ M^{-1}\ s^{-1}$, and other side reactions can occur ⟨86TL5821⟩. Due to the absence of hydrogen atom donors in atom transfer cyclisations these reactions are particularly suited for forming six- and seven-membered rings and for facilitating tandem processes where the cyclisations are particularly slow ⟨89JA6265⟩.

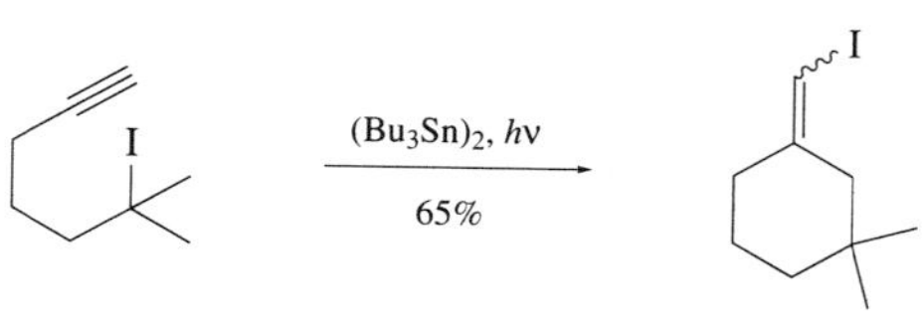

(46)

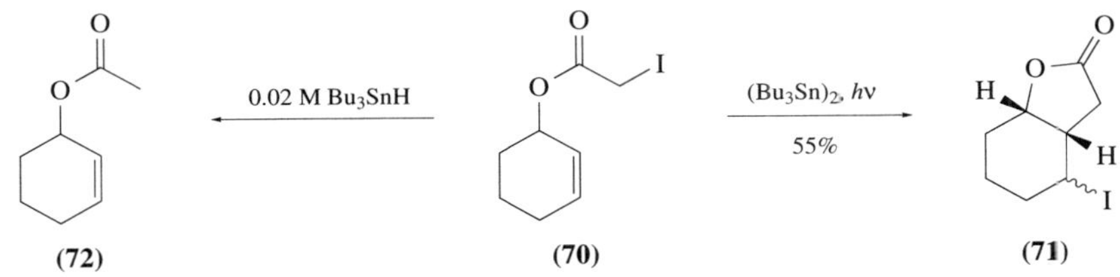

Scheme 20

(iii) Nonchain reactions

(a) Organocobalt group transfer. Photolysis or thermolysis of the relatively weak Co^{III}—C bond (18–25 kcal mol^{-1}, 75–105 kJ mol^{-1}) in RCo(salen), RCo(salophen) or $RCo(dmgH)_2py$ complexes occurs homolytically to generate a carbon-centred radical and a Co^{II} complex (see Section 1.08.1.3.4(iv)(a)) ⟨88CSR361, 94CRV519⟩. These initial radicals can undergo cyclisation followed by trapping *in situ* by Co^{II} to generate a new RCo^{III} complex (Scheme 21). These RCo^{III} complexes can then undergo a variety of reactions such as elimination or oxidation to furnish functionalised products via a formal oxidative radical cyclisation (Scheme 22) ⟨84TL4317⟩. The most common organocobalt complexes used are those with Schiff-base ligands such as dimethylglyoximato ($(dmgH)_2$), salen or salophen ⟨88CSR361, 94CRV519⟩. The initial RCo^{III} complexes can be formed by reacting supernucleophilic Co^{I} complexes with a variety of substrates including alkyl and aryl halides and tosylates ⟨76AG(E)417⟩. Hence, the vinyl ether bromoacetal halide **(73)** reacts with $Co^{I}(dmgH)_2$ to form the bicyclic cobaloxime **(74)** via a 5-*exo* alkyl radical cyclisation followed by trapping with $Co^{II}(dmgH)_2$. Elimination of $HCo^{III}(dmgH)_2$ by photolysis or insertion of oxygen followed by reduction leads to the functionalised products **(75)** and **(76)**, respectively (Scheme 22) ⟨90JCS(P1)2691⟩. This oxidative method has been used in approaches towards kainic acid ⟨87CC166⟩ and the alkaloid physovenine ⟨92T6875⟩. Radical cyclisations have also been initiated by the addition of $Co^{I}(dmgH)_2$ to oxiranes ⟨91TL243⟩.

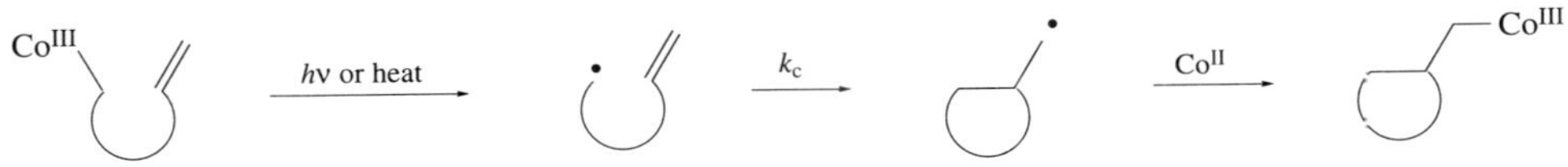

Scheme 21

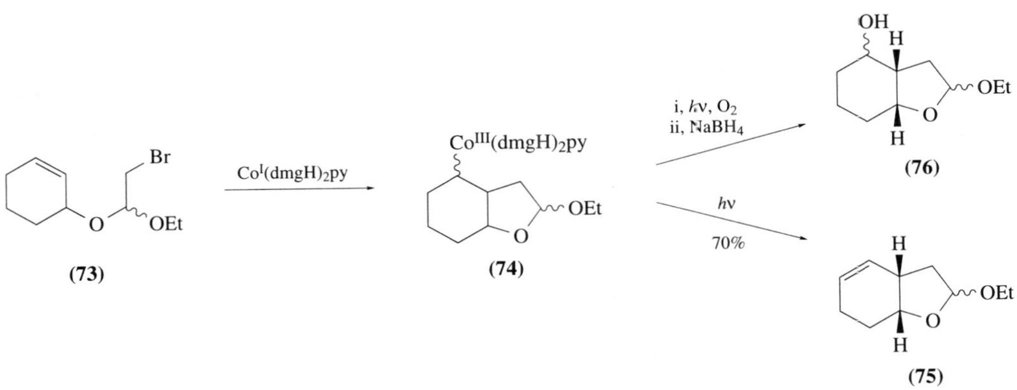

Scheme 22

Scheffold has used the reduced form of vitamin B_{12} in catalytic amounts to mediate the cyclisation of the alkyl halide **(77)** (Equation (47)). The reduced vitamin B_{12} can be produced electrochemically or by reduction with zinc metal ⟨80JA3642⟩. A novel reductive procedure utilising a catalytic amount of $Co^{II}Cl_2$ and two equivalents of a Grignard reagent has been used to produce and cyclise aryl radicals ⟨94CC41⟩. This offers a high-yielding, catalytic, cheap, 'industrial friendly' alternative to the utilisation of toxic trialkyltin hydrides in mediating reductive aryl radical cyclisation reactions.

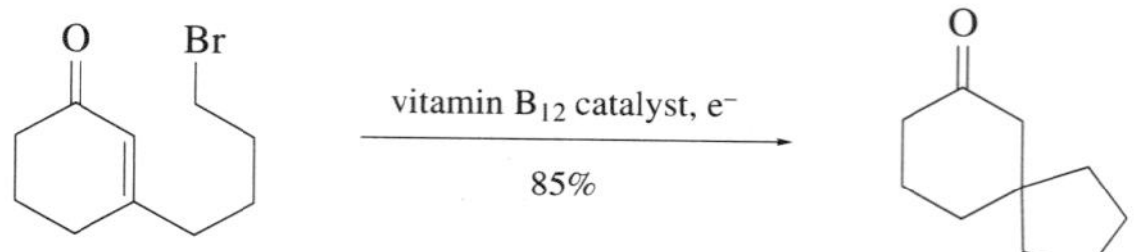

(77)

(47)

Cyclisation of aryl radicals can be accomplished by the addition of Co^{I}salen to aryl iodides **(78)** (Equation (48)) ⟨86TL2303⟩. Cyclisation followed by trapping with Co^{II}salen furnishes RCo^{III}salen complexes **(79)**, which can be further functionalised to halides, sulfides, selenides, nitriles, alcohols, amines and sulfonic acids ⟨90JCS(P1)2709⟩. Minor amounts of chiral induction in cobalt-mediated aryl radical cyclisations has been observed ⟨89TL5485⟩.

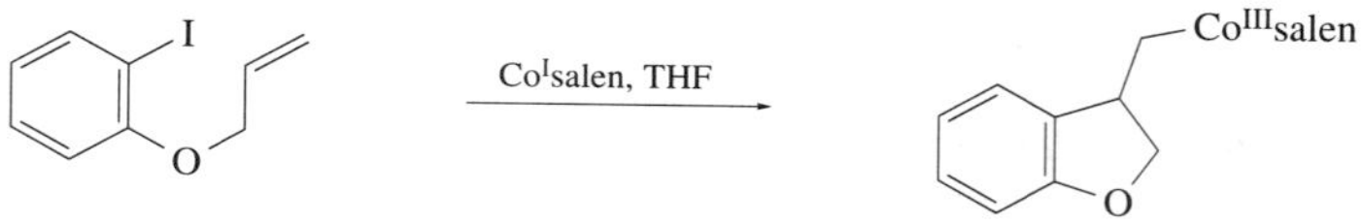

(78) (79)

(48)

(b) Manganese-mediated cyclisations. $Mn^{III}(OAc)_3$ has been used to oxidise *β*-ketoesters, *β*-diketones and *β*-diesters to radicals which can undergo oxidative radical cyclisations ⟨94CRV519⟩ (see also Section 1.08.1.3.4(iv)(b)) The reactions are terminated by oxidation of the final radical to a carbocation with either another equivalent of $Mn^{III}(OAc)_3$ (for 3° radicals) or by an equivalent of $Cu^{II}(OAc)_2$ (for 1° or 2° radicals) (Scheme 23). Elimination of a proton ⟨83JOC1373⟩, or reaction of the carbocation with an appropriately positioned nucleophile then terminates the reaction ⟨87TL175⟩. The slow step in the overall transformation is normally the formation of the Mn^{III}-enolate which then rapidly loses Mn^{III} to give the free radical ⟨88JOC2137⟩. This method has been used to mediate 5-*exo*, 6-*exo*, 6-*endo*, 7-*endo* and 8-*endo* radical cyclisations ⟨88TL5209⟩.

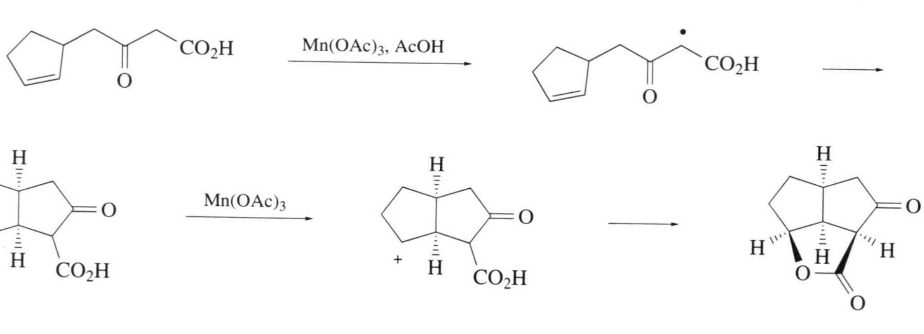

Scheme 23

Overoxidation is sometimes a problem with products that contain an enolisable hydrogen atom. This problem can be overcome by blocking the oxidation site with a chlorine atom ⟨88JOC2137⟩. The solvent of choice for $Mn^{III}(OAc)_3$-mediated cyclisations is normally AcOH; however, EtOH and DMSO have been used for acid-sensitive substrates ⟨93JOC6217⟩. Snider has investigated the effect of the ligand on the outcome of the cyclisation ⟨93JOC6217⟩. Asymmetric induction in an $Mn^{III}(OAc)_3$ cyclisation has also been reported (23–92% *de*) ⟨93JOC7640⟩.

(c) Other metal-mediated radical cyclisation reactions. For a review on metal-mediated radical cyclisation reactions which include reactions mediated by manganese, iron, ruthenium, copper, chromium and cobalt, see ⟨94CRV519⟩. Many of these metals mediate the chlorine atom transfer cyclisation of perchlorocarbonyl compounds and so do not fall into the remit of this chapter. However, the use of two equivalents of $Cr^{II}(OAc)_2$ has been shown to mediate the cyclisation of *α*-bromoesters (Equation (49)) ⟨92SL743⟩. The second equivalent reduces the final radical to an organochromium compound that can undergo *β*-elimination of a suitably orientated leaving group (OH) or addition to aldehydes and ketones (Equation (49)).

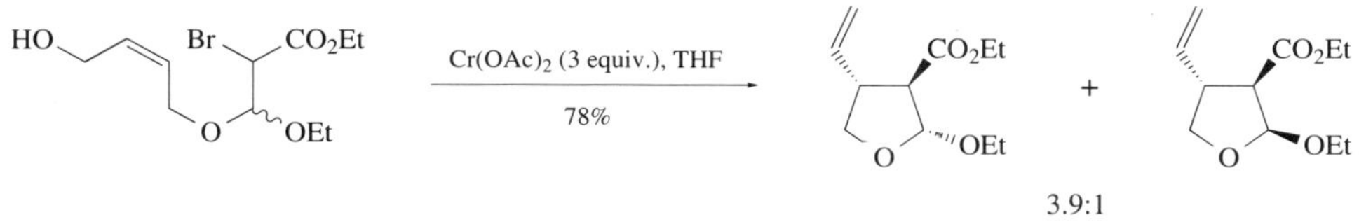

(49)

Samarium diiodide is becoming a popular reagent for conducting radical cyclisations, particularly of aryl and vinyl radicals ⟨92SL943⟩. Both aryl and vinyl radicals can be prepared from their corresponding bromides or iodides by reaction with SmI_2 in hexamethylphosphoramide (HMPA). Alkyl radicals can also be generated by reduction of alkyl halides with SmI_2; however, further reduction to the alkyl samarium prior to cyclisation is possible. Curran has measured the rate constants for reduction of alkyl radicals to alkyl samariums by SmI_2 in HMPA ⟨93TL1717⟩. Whereas 1° and 2° radicals are reduced easily by SmI_2 in HMPA ($1 \times 10^6\ M^{-1}\ s^{-1}$), 3° alkyl, vinyl and aryl radicals do not undergo this reaction. Consequently, cyclisation of aryl or vinyl radicals followed by reduction of the newly formed alkyl radical to the alkyl samarium is a facile process. Trapping of the intermediate alkyl samarium by an electrophile then gives functionalised products (Equation (50)) ⟨92JA6050⟩. The effects of additives upon the reduction reaction with SmI_2 has been investigated ⟨90SL773⟩.

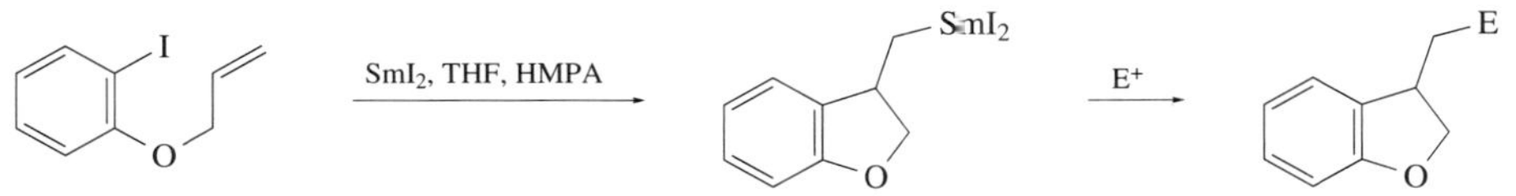

(50)

1.08.1.5 Tandem Processes

1.08.1.5.1 Intramolecular/intramolecular sequences

The process of two or more sequential radical addition reactions is known as a tandem or 'cascade' process. These transformations often enable complex structural frameworks to be produced from relatively simple radical precursors. For reviews on tandem radical reactions, see ⟨91COS(4)779, B-92MI 108-02⟩. The most common variety of tandem processes involve two or more sequential radical cyclisation reactions (normally of the 5-hexenyl type). For example, Curran and co-workers have synthesised the linear triquinane natural product hirsutene via a Bu_3SnH-mediated cyclisation (Equation (51)) ⟨85JA1448⟩. During such reactions the same general rules governing the regioselectivity and stereoselectivity of normal radical cyclisation reactions apply. Care must be taken when designing a tandem process to ensure that the rate of addition of all the intermediate radicals is faster than any termination step (such as reduction with tributyltin hydride). Consequently, if using a reductive method it is important to keep the concentration of the hydrogen donor low, or use a catalytic sequence. The most common variants of these reactions involve the initial Bu_3SnH-mediated cyclisation of an alkyl radical on to an alkene or alkyne ⟨91COS(4)779⟩. Other tandem processes have been conducted utilising the fragmentation method ⟨90CC1543⟩, the halogen transfer method ⟨89JA6265⟩, the organocobalt method ⟨92TL2851⟩ and the manganese(III) method ⟨90JA2759⟩. An approach towards complex, fused skeletons is the tandem macrocyclisation/transannular cyclisation method pioneered by Pattenden and co-workers. Hence, the taxane skeleton was formed via a novel 14-*endo* macrocyclisation followed by a 6-*exo* transannular cyclisation (Equation (52)) ⟨92TL4843⟩. Other studies have shown that 6,6-fused, 5,7-fused, 5,6-fused and 5,7,5-ring-fused cyclic systems can be prepared by this methodology ⟨94TL2413, 94TL2417⟩.

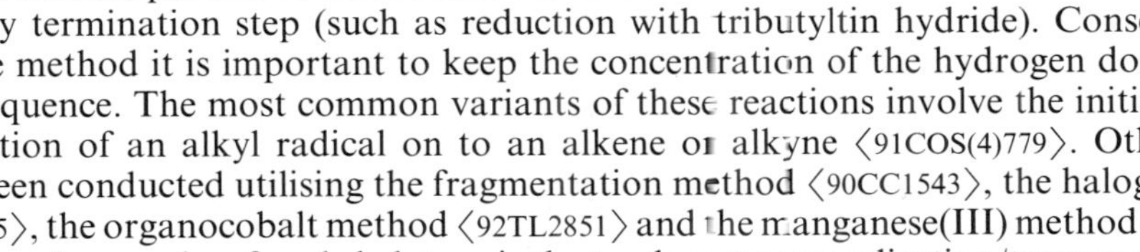

(51)

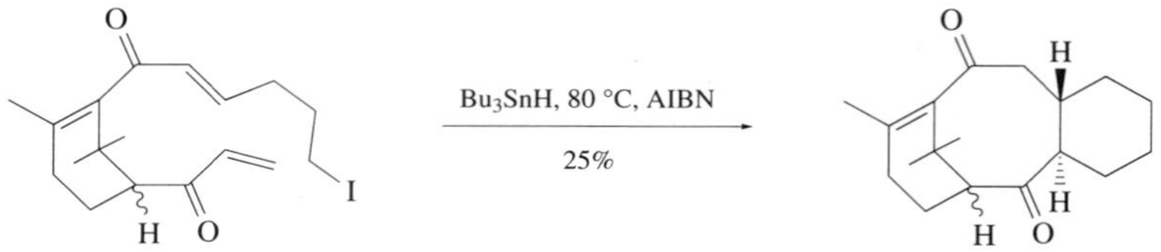

(52)

1.08.1.5.2 Intramolecular/intermolecular processes

The combination of an intramolecular cyclisation followed by an intermolecular addition is only possible if the rate of the cyclisation reaction is much faster than that of addition. Slow cyclisations can be utilised if the initial radical and the acceptor are electronically unmatched. This type of sequence has been conducted by most of the general radical methods ⟨91COS(4)779, 91CRV1237⟩; an example utilising the fragmentation method ⟨87JOC2958⟩ is shown in Equation (53).

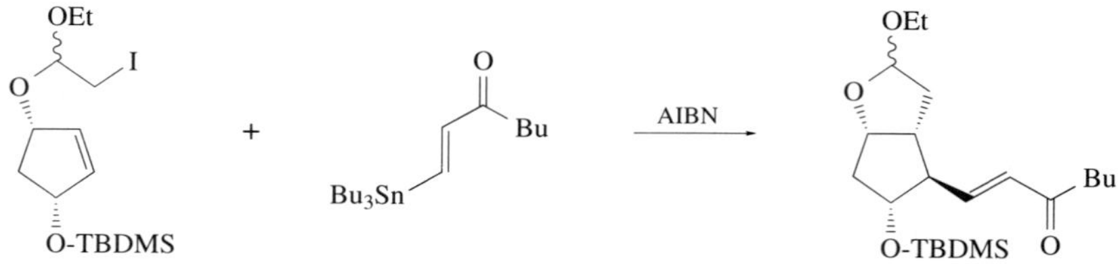

(53)

1.08.1.5.3 Intermolecular/intramolecular additions

This type of tandem process is much less common than those already described. One major problem is that the initial radical and the final radical are often similar electronically and that a further unwanted intermolecular addition can take place. This is often observed for the tin hydride ⟨86TL5893⟩ or the Barton method ⟨88CC285⟩. An elegant solution to this problem is to use the fragmentation method to terminate the sequence at the desired point (Equation (54)) ⟨89TL2501⟩. Alternatively, Curran and co-workers have shown that the atom transfer method can be employed by careful planning of a sequence in which the final radical is the least resonance stabilised and so abstracts iodine from the initial precursor before further addition can take place ⟨87JA6558⟩.

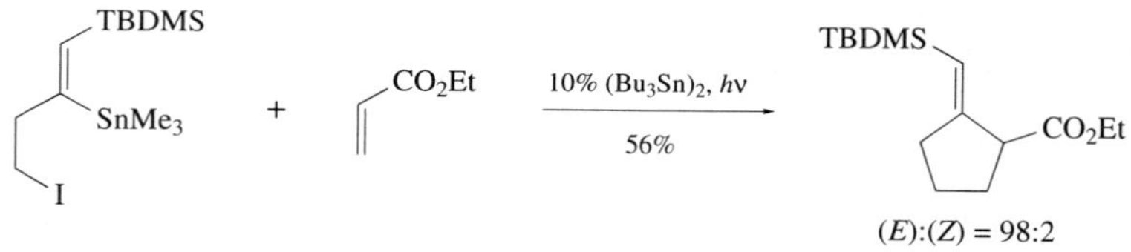

(54)

1.08.1.5.4 Intermolecular/intermolecular additions

Sequential intermolecular/intermolecular additions are particularly difficult to perform without polymerisation occurring. A solution to this problem involves the suitable electronic pairing of the initial nucleophilic radical derived from (**80**) with the activated alkene (**81**) ⟨88JA1288⟩. Upon addition the new electrophilic radical produced adds to the allyl stannane (**82**) at a greater rate than the electronically mismatched alkene (**81**) (Equation (55)). Elimination of the stannane radical terminates the sequence and acts as a chain carrier.

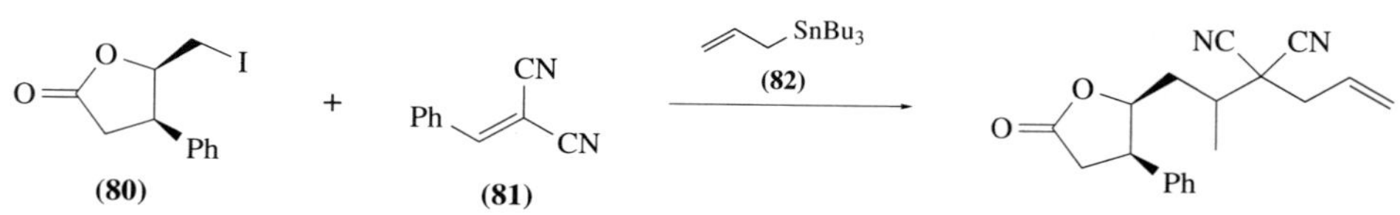

(55)

1.08.2 ELECTROCYCLIC ADDITIONS TO CARBON–CARBON MULTIPLE BONDS TO GIVE TETRACOORDINATE PRODUCTS

1.08.2.1 Introduction

1.08.2.1.1 Scope

The term 'electrocyclic addition' usually applies to reactions which proceed with a concerted cyclic movement of electrons. As it is often difficult to establish that a particular addition is truly concerted this criterion has been interpreted somewhat loosely. The result is that stepwise cycloaddition processes have been included if the intermediates appear to be short-lived species.

The remaining restrictions are as follows. Only reactions of C═C bonds are considered. Furthermore, in the cyclic products the two carbon atoms derived from the C═C bond must each be bonded to four carbon atoms. Most of these reactions are summarised by Equation (56). The one class of reaction not covered by Equation (56), but within the scope of the chapter, is the addition of carbenes (Equation (57)).

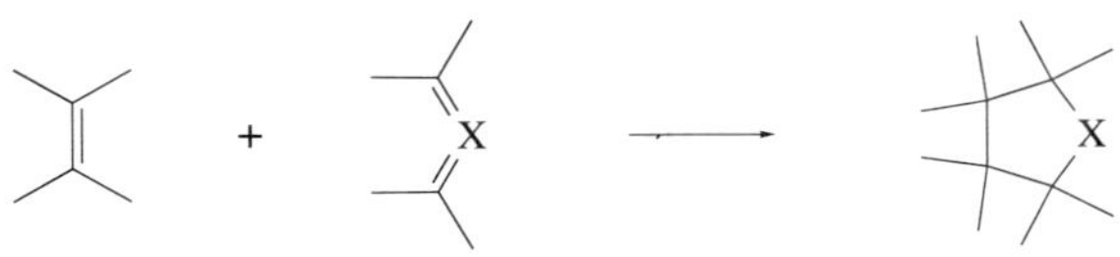

(56)

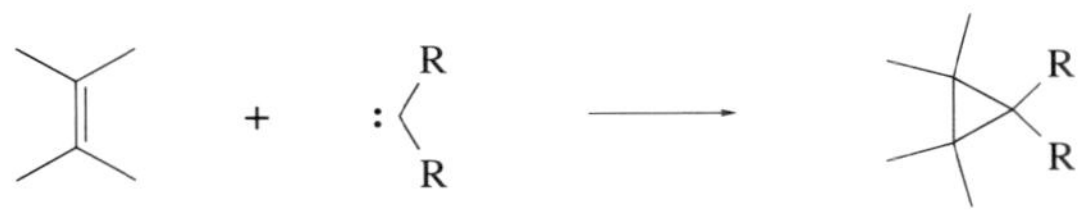

(57)

Finally, it should be remarked that, if an electrocyclisation produces a new C═C bond as well as a new C—C bond (e.g., the conventional Diels–Alder reaction), it most properly belongs in Chapter 1.17. However, a very brief account of the Diels–Alder reaction is included for the sake of completion.

1.08.2.1.2 General mechanistic aspects

Since the universally acclaimed work of Woodward and Hoffmann ⟨69AG(E)781⟩, chemists have been able to benefit from a deep understanding of the mechanisms of cycloaddition reactions. It is not intended to discuss these ideas at length here, but a reminder of some of the points pertinent to this chapter is not out of place. All the reactions described herein involve an alkene, which has two π electrons, as one partner. The full range of reactions to be discussed can be categorised as either $2\pi + 2\pi$ or $2\pi + 4\pi$. The 2π components covered are alkenes, carbenes and allyl cations. The 4π components are dienes (the Diels–Alder reaction) and heterodienes; carbonyl, thiocarbonyl and azomethine ylides; allyl and azaallyl anions and trimethylene methane.

The first point to consider is that for small π systems such as these, the antarafacial mode of cycloaddition is geometrically very unlikely and suprafacial addition is therefore the norm, as illustrated for the Diels–Alder reaction in Figure 9. Suprafacial $2\pi + 2\pi$ reactions are photochemically allowed and thermally forbidden by the Woodward–Hoffmann rules; suprafacial $2\pi + 4\pi$ additions are, conversely, thermally allowed and photochemically forbidden.

However, there are many cycloaddition reactions which break these rules. This is because not all

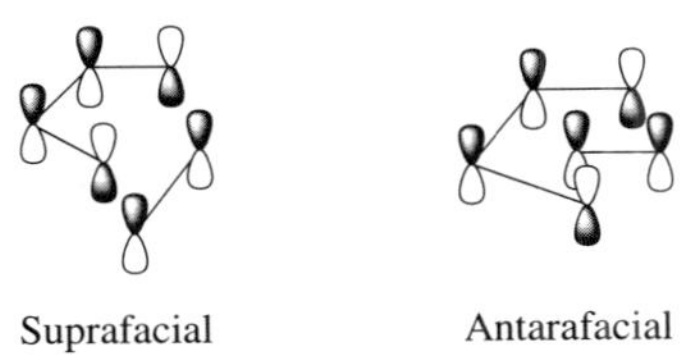

Figure 9 Suprafacial and antarafacial addition for small π systems.

cycloadditions proceed by a truly concerted pathway. The extreme opposite of the concerted reaction is the two-step reaction (Scheme 24). In practice many reactions fall somewhere between these two limits. Knowledge of which type of mechanism applies in a particular reaction is not merely of academic importance. The regiochemical and stereochemical outcomes of reactions may be profoundly affected by the preferred pathway, as discussed in the subsequent sections.

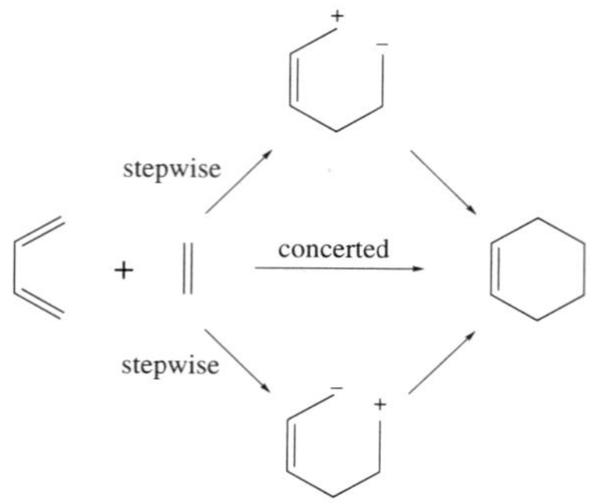

Scheme 24

The FMO approach to cycloadditions allows the outcome to be predicted by considering the overlap of either the HOMO of the 2π or 4π component with the LUMO of the alkene, or the LUMO of the 2π or 4π component with the HOMO of the alkene. Clearly it is important to know which case applies. This is done by estimation of the two HOMO–LUMO energy gaps. The smaller gap represents the more important contribution, and it is this one which should be considered. In practice, this means that if the alkene bears electron-withdrawing substituents its LUMO will be important and its interaction with the HOMO of the 4π component should be considered. In the case of an electron-rich alkene the alkene HOMO 4π component LUMO should be considered. Similar arguments apply to $2\pi + 2\pi$ additions.

These fundamental observations are necessary for the discussions of regiochemical and stereochemical considerations which follow.

1.08.2.1.3 General regiochemical aspects

If both partners in a cycloaddition are unsymmetrically substituted there are two possible regioisomeric products. If one regioisomer predominates the reaction is said to be regioselective. Clearly it is desirable for the reaction to be highly regioselective, leading to only one isomer. In addition, it is important to be able to predict which of the regioisomers will be preferred.

For truly concerted cycloadditions the origins of any regioselectivity lie in the polarisation of the HOMO and LUMO by their substituents. As mentioned in Section 1.08.2.1.2, it is important to know which HOMO–LUMO interaction is important. Once this has been decided, the effect of the substituents on the coefficients of the termini of the reacting π systems should be estimated. Both these factors can be easily assessed by reference to standard texts ⟨B-76MI 108-01⟩. The preferred regioisomer can then be predicted by matching the large coefficient terminus of one partner with the large coefficient terminus of the other.

The effect of substituents on the polarisation of the π system is often relatively small. For this reason, the regioselectivity of concerted cycloadditions is often unspectacular. The use of, in particular, Lewis acid catalysts to enhance the polarisation, and hence regioselectivity, of Diels–Alder cycloadditions is widespread and is discussed later. However, this principle has not been widely applied to other categories of cycloadditions.

Cycloadditions which proceed by 'more stepwise' pathways often have higher regioselectivities. The origin of such regioselectivity may result from minimisation of steric interactions but, more often than not, it is due to the ability of a substituent to stabilise developing charge in the intermediate. For instance, the reaction of azaallyl anion (**83**) with styrene proceeds via the intermediate in which the developing negative charge is stabilised by a phenyl group to yield the more sterically congested regioisomer (**84**) (Equation (58)) (see Section 1.08.2.4.2). A basic understanding of the mechanism of a particular cycloaddition is very important in prediction of its regiochemical outcome.

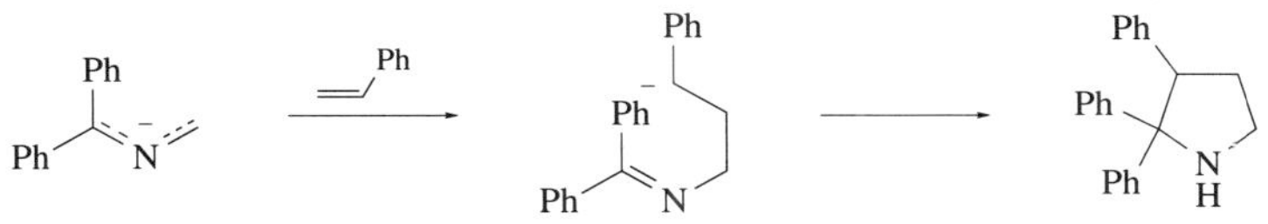

(83) (84) (58)

1.08.2.1.4 General stereochemical aspects

If all four reacting carbon centres in a cycloaddition are prochiral, a total of 16 stereoisomers can be formed, corresponding to eight enantiomeric pairs. The immense power of cycloaddition chemistry in synthesis is due to the fact that a great deal of selectivity is observed, and the goal of producing just one of these 16 isomers in a wide range of cycloaddition reactions now appears possible.

In a perfectly concerted cycloaddition reaction there will be absolute correspondence between both the stereochemistry of the starting materials and the stereochemistry of the product. In the case of a thermal $2\pi + 4\pi$ addition, retention of the stereochemistry of both components is predicted from FMO considerations. Similarly, the outcome of other classes of cycloadditions can be predicted. Partial loss of stereochemistry of the starting materials in a cycloaddition is strongly indicative of a nonconcerted mechanism. However, complete conservation of stereochemistry should not be taken as proof of a concerted mechanism as many stepwise cycloadditions proceed via strongly coordinated intermediates in which the stereointegrity is retained (e.g. see azaallyl anion addition in Section 1.08.2.4.2).

If it is accepted that antarafacial addition will not occur, a concerted $(2\pi + 4\pi)$ cycloaddition has only four possibilities left for isomerism. For example, the cycloaddition in Equation (59) can give a mixture of the four isomers **(85)**–**(88)**. Product **(85)** and its enantiomer **(87)** result from so-called *endo* addition whereas **(86)** and **(88)** result from *exo* addition. In fact, the *endo* product is usually favoured, perhaps due to secondary orbital effects (overlap of π orbitals of substituents) ⟨B-76MI 108-01⟩, but this effect is not universal. In practice, *endo*/*exo* product mixtures usually result. One of the major challenges in cycloaddition chemistry is to achieve the highest possible *endo*/*exo* selectivity.

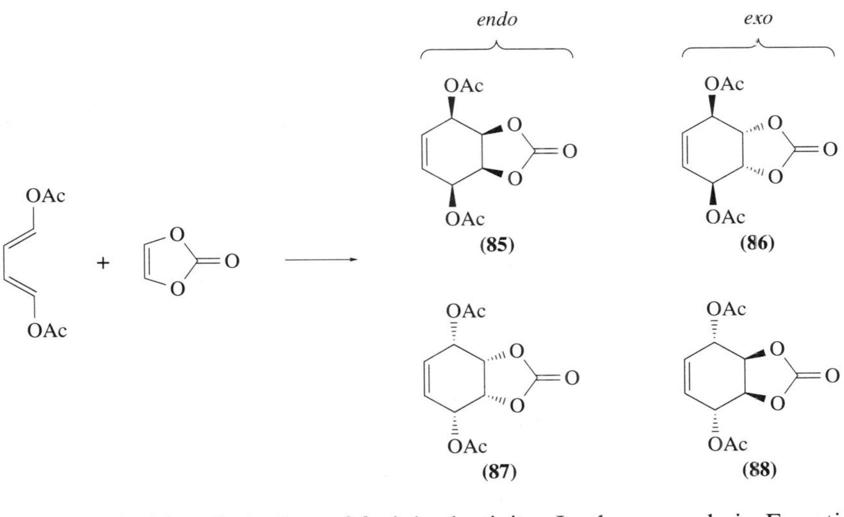

(85) (86) (87) (88) (59)

The final stereochemical hurdle is that of facial selectivity. In the example in Equation (59) the two possible enantiomeric *endo* products result from the addition of the 4π component to both prochiral faces of the alkene. If, due to a chiral environment, addition to one of these prochiral faces is favoured, the reaction will be enantioselective. In practice, this task is made easier by attaching the chiral environment to one of the reacting partners in the form of a chiral auxiliary. In this case the products are diastereomers, not enantiomers, and the success of the reaction will be judged by its diastereoselectivity, not enantioselectivity.

Three stereochemical features will be highlighted in the subsequent discussion: degree of retention of stereochemistry of starting materials, *endo/exo* selectivity and enantioselectivity (or diastereofacial selectivity if a chiral auxiliary is involved).

1.08.2.2 Formation of Three-membered Rings

1.08.2.2.1 Additions of free carbenes

The addition of carbenes to alkenes to produce cyclopropanes is a very important process (Equation (60)). Free carbenes are in general much too reactive to be useful (with some exceptions ⟨B-92MI 108-03⟩) due to competing reactions, in particular C—H bond insertion. However, there are other methods of 'carbene transfer' which achieve the same result in a more controlled fashion ⟨B-92MI 108-03⟩. Of these, alkylidene ylides (e.g., sulfonium ylides) are prominent. However, the mechanism of the 'carbene transfer' from these ylides is clearly stepwise and excludes them from the scope of this chapter.

(60)

1.08.2.2.2 Additions of metal carbenoids

Two main classes of 'carbene transfer' remain to be discussed. March provides an excellent introduction to the first protocol, the Simmons–Smith procedure ⟨B-92MI 108-03⟩. This involves an organozinc reagent derived from a diiodoalkane and zinc (usually activated as a copper couple). Bond formation proceeds as if concerted, with retention of stereochemistry of the alkene and some diastereofacial selectivity (e.g., Equation (61)) ⟨82JOC1615⟩.

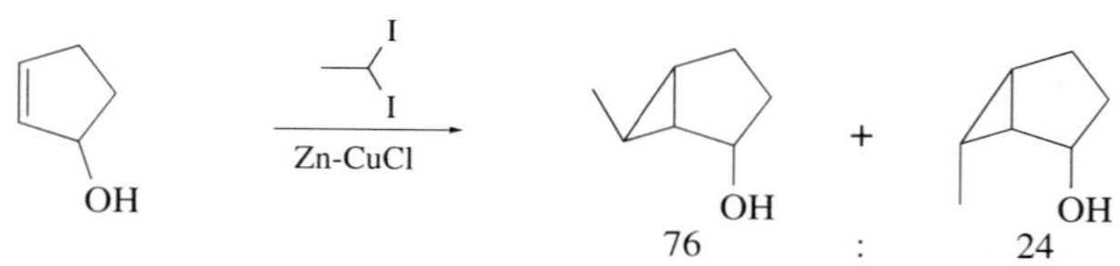

(61)

Another development in the Simmons–Smith methodology is the discovery of an enantioselective version ⟨92CL61⟩. In this case the substrates are allylic alcohols, the organozinc reagent is diethylzinc and the asymmetric induction is due to a chiral diol (Equation (62)). Although *ee* values are modest, these initial results are most promising and invite further investigation.

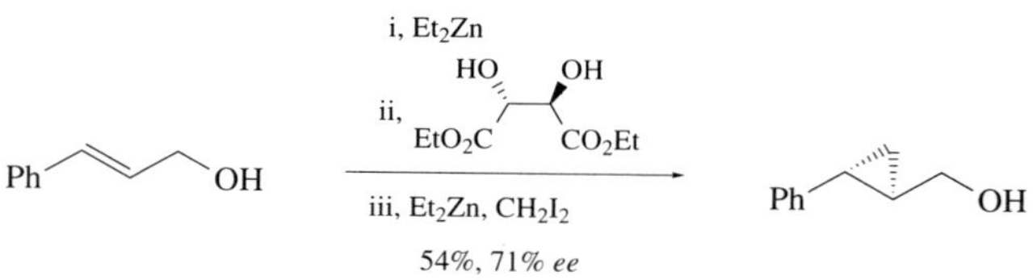

(62)

Some significant developments in carbene chemistry have involved metal-stabilised keto-carbenoids (**89**) (the term 'carbenoid' is usually employed when free carbenes are not involved). The most usual precursors of (**89**) are diazo compounds (**90**), but dibromides, sulfur ylides and iodonium ylides have also been used ⟨86ACR348, 86CRV919, 87TCC75, 91COS(4)1031⟩. The two prevalent metals are copper(I) and rhodium(II), but palladium(II) has also proved useful.

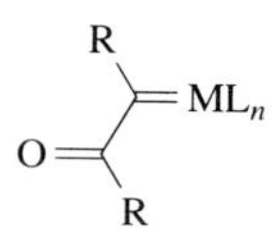

(89)

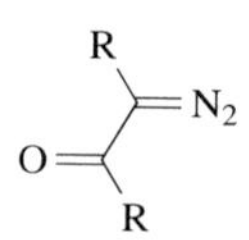

(90)

It is worth noting that, although the metal carbenoid species are much more selective than the corresponding free carbene systems, mixtures will usually result if more than one alkene is present in a molecule ⟨91COS(4)1031⟩. Carbene cycloadditions proceed with retention of stereochemistry of the alkene (Equation (63)) ⟨84OM44⟩.

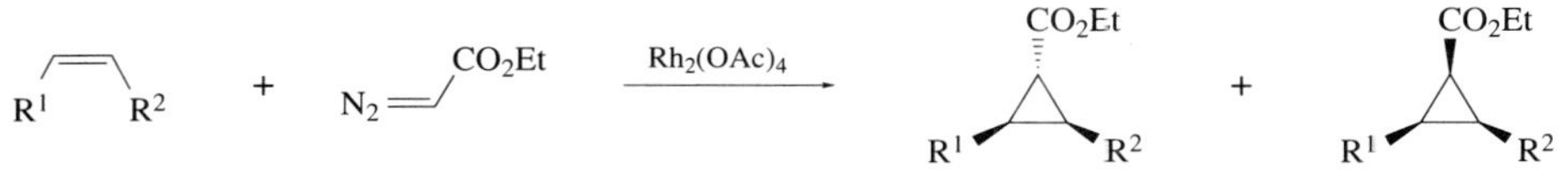

(91) (92) (63)

The reaction illustrated in Equation (63) also highlights the need to consider diastereofacial selectivity. Unfortunately, this selectivity is often rather poor. It is possible to increase the proportion of the *trans* isomer (**91**) significantly ⟨82TL2261⟩, but the *cis* isomer (**92**) remains largely inaccessible.

The asymmetric cyclopropanation of alkenes by ketocarbenoids is a topic of interest. Both copper and rhodium catalysts have been modified with chiral ligands for this purpose and the results have been most impressive ⟨91RTC305⟩. The early leaders in the field were the salicylaldehyde-derived imines (**93**) developed by Aratani as ligands for copper ⟨85PAC1839⟩. A typical example is the preparation of a chrysanthemic acid precursor (**94**) (Equation (64)). The *ee* values were found to be useful, but the problem of *cis*/*trans* mixtures again proved to be a problem.

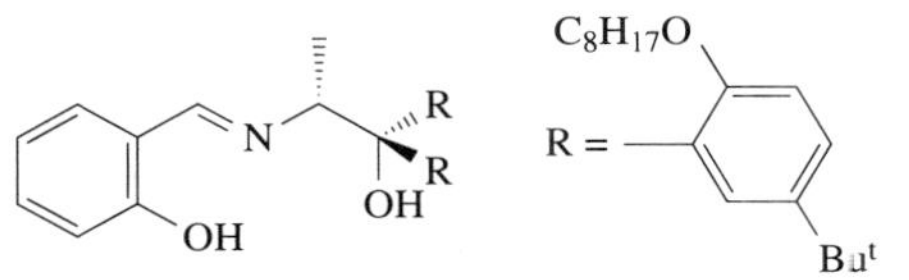

(**93**)

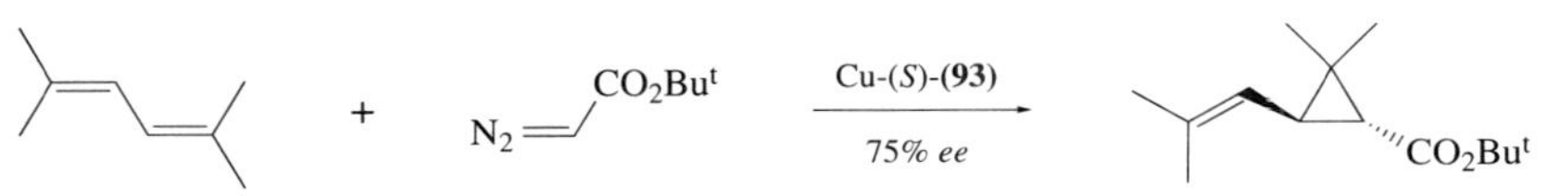

+ *cis* isomer (64)

(**94**)

Semicorrin and bisoxazoline ligands with the general structures (**95**) and (**96**) have been shown to be excellent chiral ligands for copper catalysts ⟨91HCA232, 91JA726, 91RTC305, 91TL7373⟩. One example where the enantioselectivity is outstanding and the *cis*/*trans* ratio is very good is given in Equation (65) ⟨91JA726⟩. However, catalysts with ligands of the type (**95**) and (**96**) have not been shown to be effective with more substituted alkenes; in such cases the Aratani catalysts (utilizing ligand (**93**)) remain the most useful ⟨94T4519⟩.

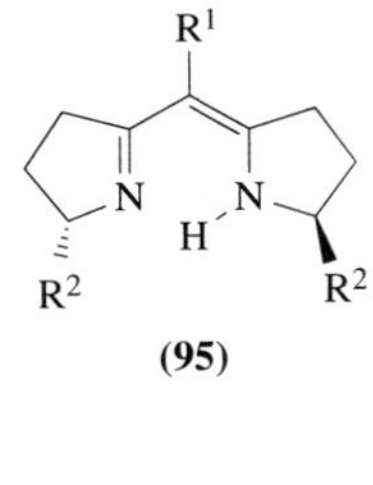

(**95**)

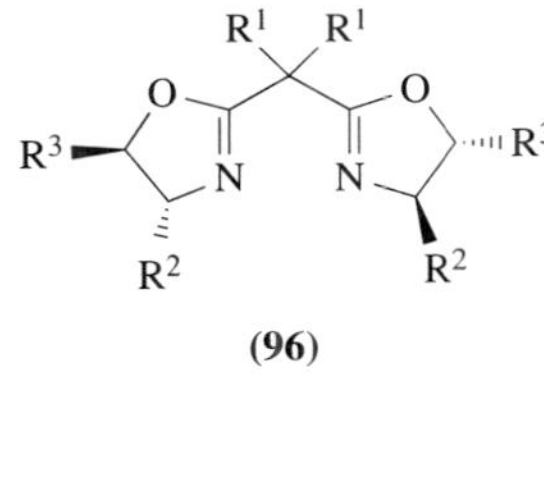

(**96**)

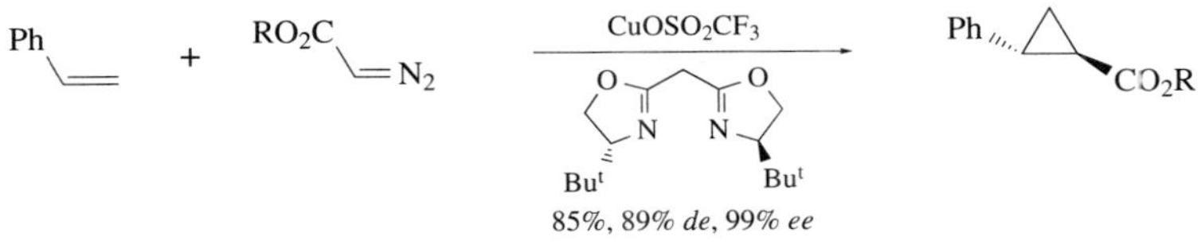

85%, 89% *de*, 99% *ee* (65)

R = 2,6-di-*t*-butyl-4-methylphenol

Chiral ligands have also been used to modify rhodium catalysts. The most effective ligands have been of the 5*S*-MEPY type (**97**), the rhodium complexes of which catalyse the asymmetric cyclopropanation of mono- and disubstituted alkenes with respectable *ee* values but poor *cis*/*trans* ratios ⟨91RTC305⟩.

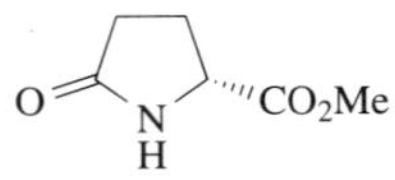

(97)

As so often with cycloaddition reactions, the problems of stereoisomerism can be avoided by way of intramolecular reactions. Indeed, the intramolecular cyclopropanation of alkenes by carbenes has proved a useful reaction ⟨91COS(4)1031⟩. Bicyclic skeletons with cyclopropane rings fused with cyclobutanes and cyclopentanes are accessible by this route. The reactions shown in Scheme 25 are typical and highlight the stereochemical control possible in these processes ⟨77TL109⟩.

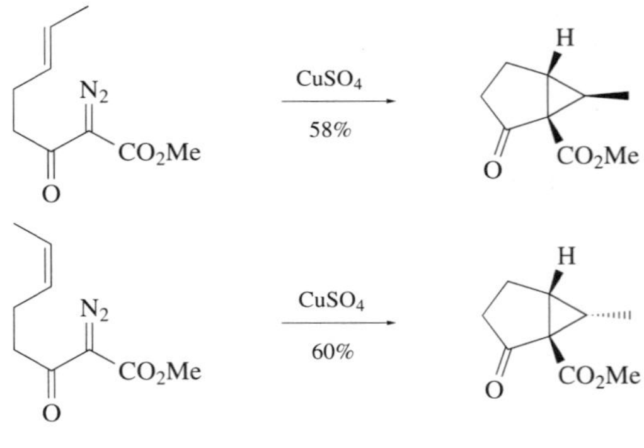

Scheme 25

In view of the promising results obtained from intramolecular cyclopropanation reactions, it is surprising that little progress has been made with the asymmetric version. However, the rhodium-5*S*-MEPY complexes mentioned above hold promise ⟨94T4519⟩. The example shown in Equation (66) illustrates the potential of these reactions.

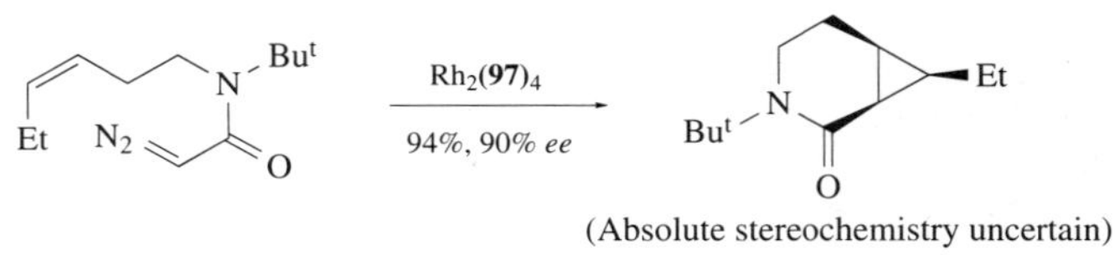

(66)

1.08.2.3 Formation of Four-membered Rings

The addition of one alkene to another is an obvious route to cyclobutanes. This kind of reaction can be achieved in three ways, as described below: (i) thermally, (ii) photochemically or (iii) by metal catalysis.

1.08.2.3.1 Thermal [2+2] additions

The thermal $2\pi + 2\pi$ addition of alkenes is forbidden by the Woodward–Hoffmann rules. However, such additions do occur and have been reviewed ⟨91COS(5)63⟩. The mechanism is therefore expected to be stepwise, and experimental evidence supports this postulate. For the purpose of understanding the results of thermal $[2+2]$ additions, a mechanism via a diradical intermediate is sufficient and will be used here.

The thermal dimerisation of acrylonitrile (**98**) is illustrative (Equation (67)). The reaction is highly regioselective, giving the so-called head-to-head product. However, the stereoselectivity is very poor, a mixture of *cis* and *trans* isomers being observed. These phenomena are readily explained by the intermediacy of diradical (**99**). This is the most thermodynamically favourable diradical, as both radical centres are stabilised by substituents. The possibility for free rotation around carbon–carbon bonds in (**99**) is evident. Little stereoselectivity would therefore be expected. Similar results are observed with other alkene dimerisations ⟨91COS(5)63⟩.

(67)

(**98**) (**99**)

Alkenes bearing electron-withdrawing substituents are found to react readily with electron-rich alkenes. Tetracyanoethylene has been widely used as a substrate ⟨91COS(5)63⟩. The example in Equation (68) is typical. In this case both the regioselectivity and stereoselectivity are good ⟨64JOC940⟩. The latter selectivity is probably due to formation of the thermodynamically favoured product (**100**), in which steric interactions are minimised.

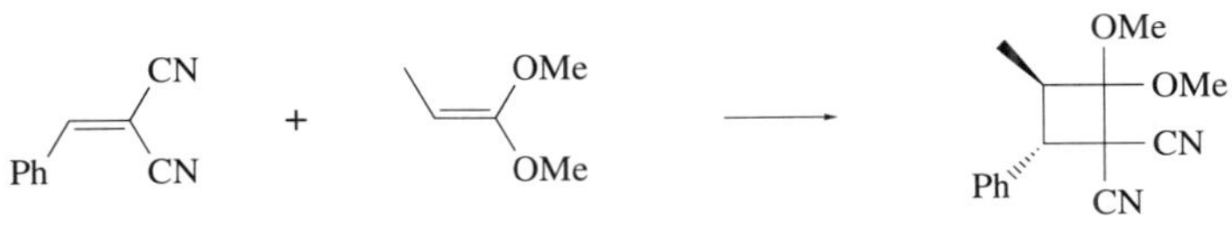

(68)

(**100**)

Diastereoselectivity in thermal [2 + 2] additions can be good. For example, reaction of tetracyanoethylene with a chirally substituted alkene was found to proceed with very high *de* values, giving essentially one product (**101**) in 94% yield (Equation (69)).

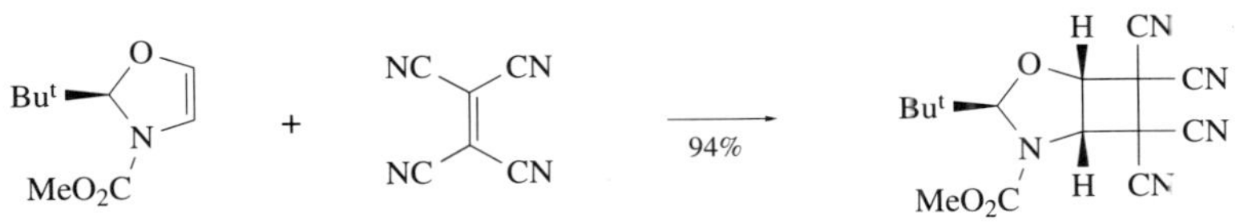

(69)

(**101**)

Intramolecular thermal alkene cycloadditions have also been investigated. This can be a useful method of producing polycyclic skeletons. The reaction in Equation (70) provides a simple example, apparently proceeding with good stereoselectivity ⟨77JA4417⟩. A more exotic variation is the construction of the hydrocarbon (**102**) (Equation (71)) ⟨84JOC1353⟩.

(70)

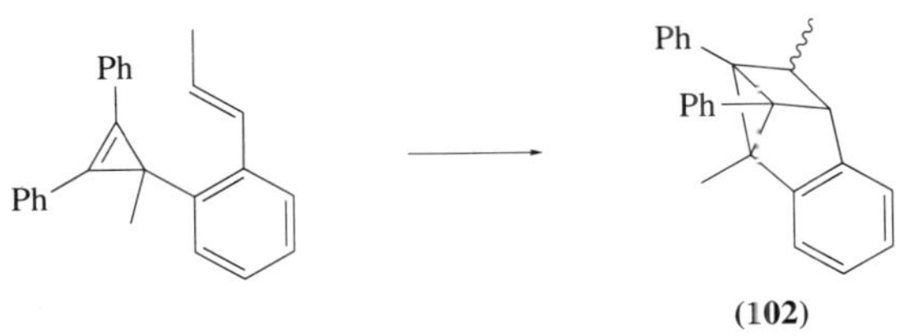

(71)

(**102**)

The high regioselectivity and diastereofacial selectivity observed in thermal [2 + 2] additions, coupled with the fact that the stereoselectivity can also be high in favourable cases, makes these reactions potentially very useful synthetically.

1.08.2.3.2 Photochemical [2 + 2] additions

In contrast to the thermal [2 + 2] additions discussed above, the photochemical addition of one alkene to another is allowed by Woodward–Hoffmann theory. However, such cycloadditions are considered not to be truly concerted processes, but stepwise, proceeding via a diradical intermediate ⟨91COS(5)123⟩.

The regiochemistry of photochemical alkene cycloadditions is rather unpredictable. For example, replacement of H by CH_3 in the enone **(103)** has a huge effect on the regiochemical outcome (Equation (72)). The regiochemistry varies under the influence of a number of factors. The polarity of the alkenes, steric interactions and relative stabilities of intermediate diradicals have all been used to attempt to rationalise the direction of regioselectivity. However, rationalisation remains difficult. The most important point is that regioselectivity can be very good.

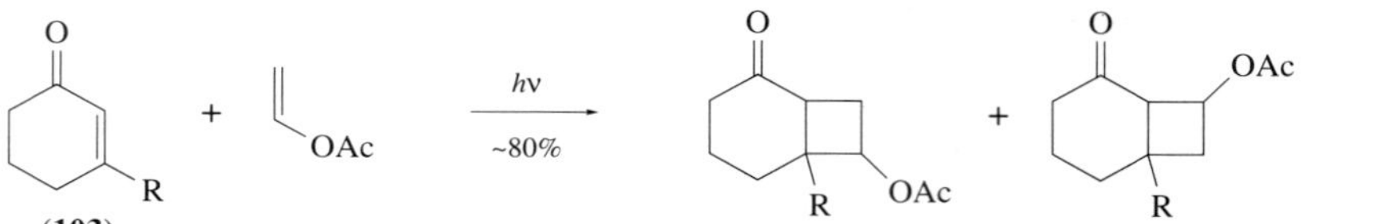

(103) (72)

R = H	100	:	0
R = Me	20	:	80

The reaction medium can also have a profound effect on regioselectivity. The reaction in Equation (73) illustrates the result of changing solvent polarity ⟨68CC982⟩. The preorganisation of the substrates is also significant. Addition of potassium dodecyl sulfate to a cyclohexane solution of **(104)** and **(105)** reverses the selectivity ⟨82CJC425⟩ (Equation (74)). Variation of regiochemistry with temperature has also been noted ⟨78TL2057⟩.

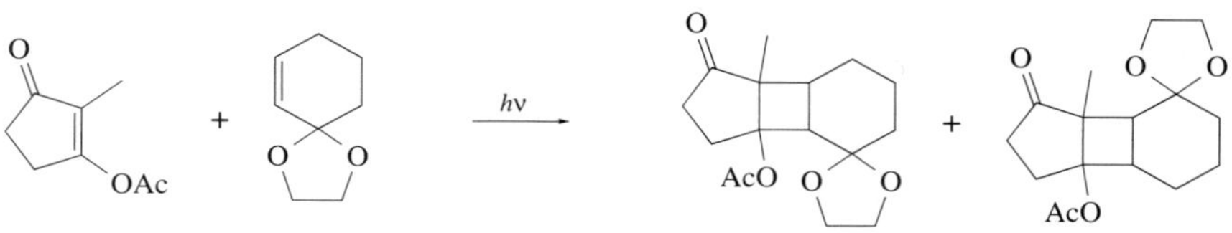

(73)

Cyclohexane	2	:	98
Methanol	55	:	45

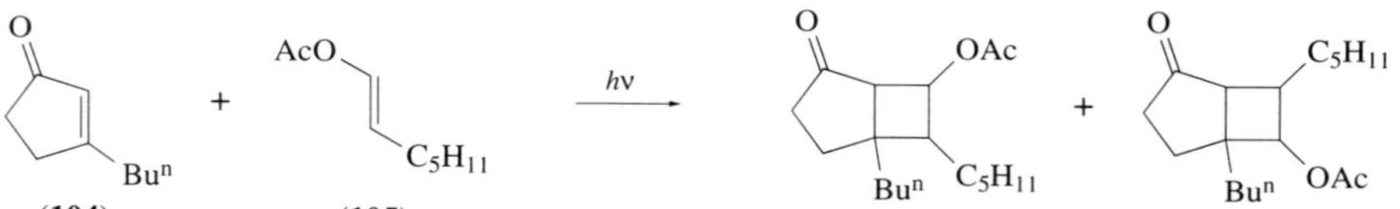

(104) **(105)** (74)

Without potassium dodecyl sulfate	0	:	100
With potassium dodecyl sulfate	70	:	30

The stereoselectivity of photochemical cyclobutane formation is rather easier to predict. If it is geometrically feasible for the stereochemistry of the alkene to be lost, it will be. This point is demonstrated by the reaction of a cyclopentene and a cyclohexene (Equation (75)) ⟨82CJC425⟩. The six-membered ring is flexible enough for *cis/trans* isomerism to occur, whereas the stereochemistry of the geometrically constrained five-membered ring is retained. Loss of stereochemistry is almost always observed with acyclic alkenes.

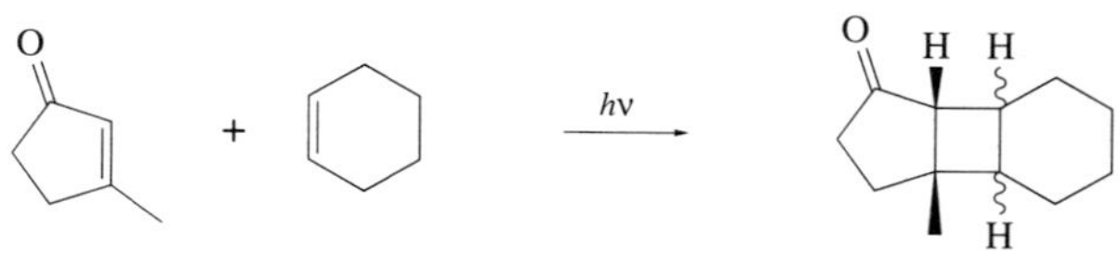

(75)

The second point to note is that, if stereochemistry is retained, the *anti* arrangement around the new carbon–carbon bonds will nearly always result, as illustrated in Equation (76).

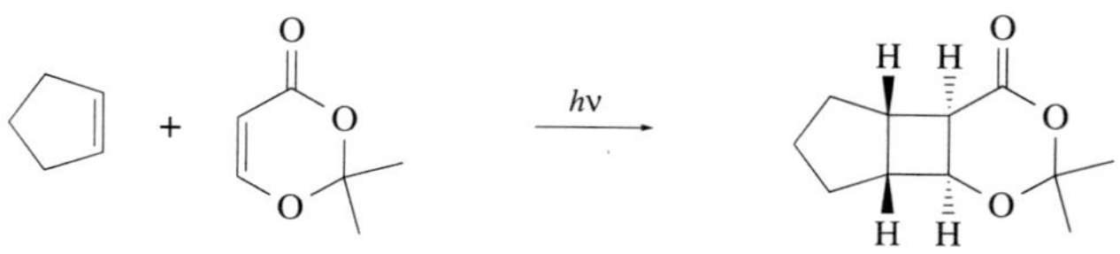

(76)

Diastereofacial selectivity is often rather poor in photochemical alkene cycloadditions. However, highly diastereofacial addition of a chirally substituted alkene to cyclopentenone did permit a synthesis of bourbonene, but unfortunately there was no regioselectivity in the photochemical reaction (Scheme 26) ⟨68JA6171⟩.

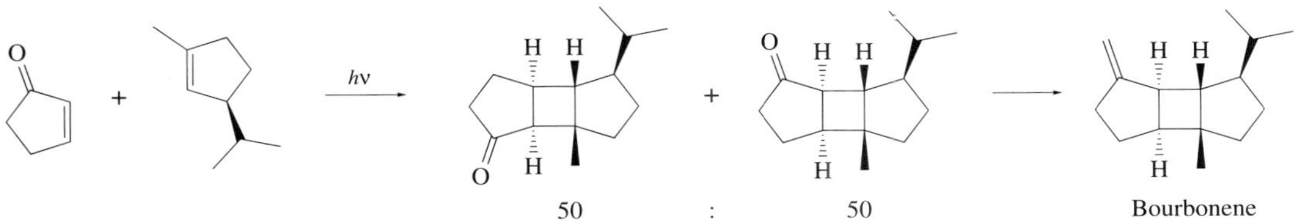

Scheme 26

The attachment of a chiral auxiliary to one of the alkenes is an obvious strategy for preparation of optically pure cyclobutane derivatives ⟨92CRV741⟩. For example, Meyers's group successfully used the substituted alkene (**106**) to prepare (−)-grandisol (Scheme 27) ⟨86JA306⟩. The diastereofacial selectivity in the crucial [2 + 2] addition step was good (>90% *de*)

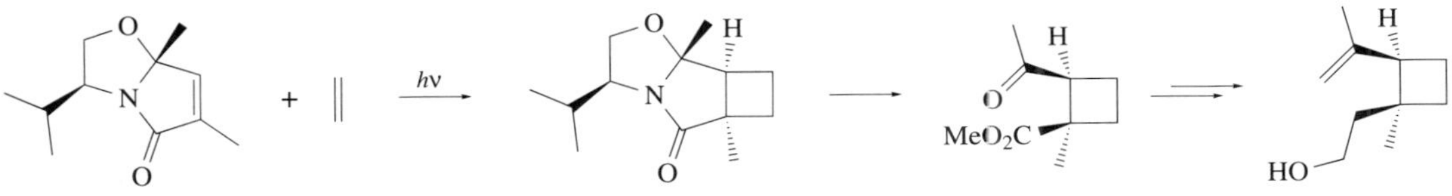

(106) Grandisol

Scheme 27

It would be appropriate here to correct a misconception arising from original literature ⟨82JA1742, 85TL4707, 85TL6163, 87TL2833, 88TL2613⟩, and persisting in review articles ⟨91COS(5)123, 92CRV741⟩, regarding the effects of chiral ester groups on [2 + 2] additions. The figures which should have been highlighted from these papers were the *de* values, which are generally low (4–84%). The *ee* values quoted, after removal of the chiral auxiliary, simply reflect the extent to which the authors were successful in separating, or otherwise, the diastereomeric products. The use of chiral ester groups as auxiliaries for these reactions should not, therefore, be considered particularly successful.

Circularly polarised light has been used to promote asymmetric photochemical additions of alkenes ⟨92CRV741⟩, but so far *ee* values have been mainly in single figures.

As pointed out by Crimmins ⟨91COS(5)123⟩, the main problem with intermolecular [2 + 2] photochemical cycloadditions as synthetic tools is that the regio- and stereoselectivities, though sometimes very good, are rather unpredictable. It is no surprise, therefore, that intramolecular versions of these reactions have become more popular. Many such reactions have been reported, often using a chiral building block to produce optically pure products ⟨91COS(5)123⟩.

The most important generalisations regarding these intramolecular reactions are that regioselectivity is usually very high, that the stereoselectivity is not so good (alkene stereochemistry can still be lost) and that diastereofacial selectivity can be extremely high. Two examples (Scheme 28) have been chosen to illustrate the utility of ring-fused cyclobutanes as precursors to larger rings ⟨80CC1195, 80HCA1198⟩.

1.08.2.3.3 Metal-catalysed [2 + 2] additions

In their review, Kricka and Ledwith complained of a paucity of examples of metal-catalysed cycloadditions of alkenes ⟨74S539⟩. In fact, although only dimerisation reactions had been studied, quite a number of metals had been used. However, there has been little further development of this area. Work has concentrated on copper triflate catalysts (e.g., ⟨80TL2477, 83T485⟩) and titanium catalysts (e.g., ⟨72JA6867⟩), as illustrated in Equation (77).

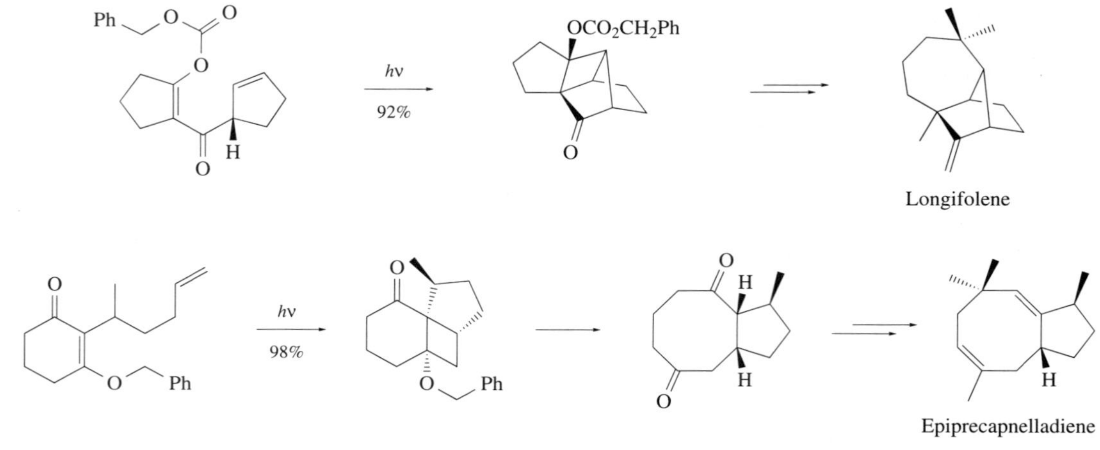

Scheme 28

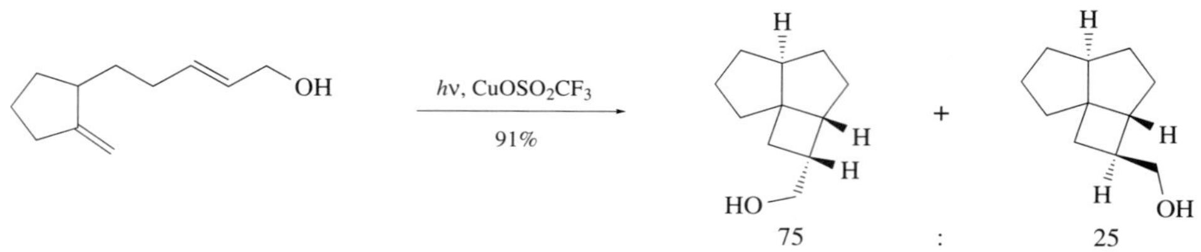

(77)

A breakthrough in [2 + 2] cycloaddition methodology was the discovery that a chiral titanium complex catalysed the reaction of a ketene dithioacetal (**107**) with an electron-deficient alkene (**108**). The cyclobutane derivative (**109**) was produced in 96% yield and 98% *ee* (Equation (78)) ⟨92JA8869⟩. In this case the titanium centre is presumably acting as a Lewis acid.

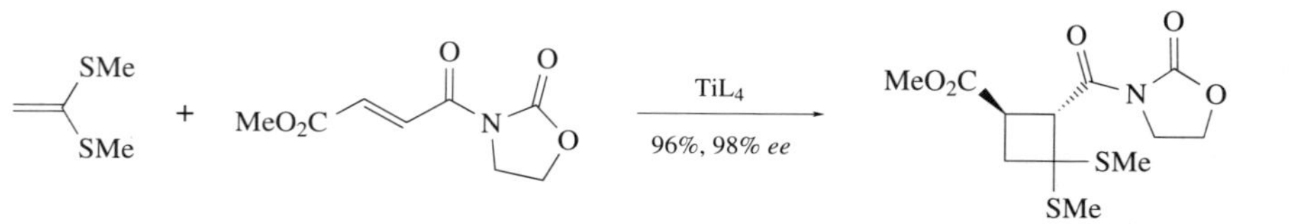

(78)

(107) **(108)** **(109)**

1.08.2.4 Formation of Five-membered Rings

1.08.2.4.1 [3 + 2] Additions of three-carbon fragments

(i) Additions of trimethylenemethane

Trimethylenemethane (TMM) (**110**) is an intriguing molecule; whether it is a dipole or a diradical (and, if the latter, whether it is in the singlet or triplet state) has been the subject of numerous studies ⟨78ACR446⟩. Whatever its structure, TMM and its derivatives represent an accessible and versatile class of 4π three-carbon fragments which, on addition to alkenes, yield cyclopentanes. Free TMM (**110**) is too reactive to be synthetically useful ⟨86AG(E)1⟩ but the successful use of derivatives is described below.

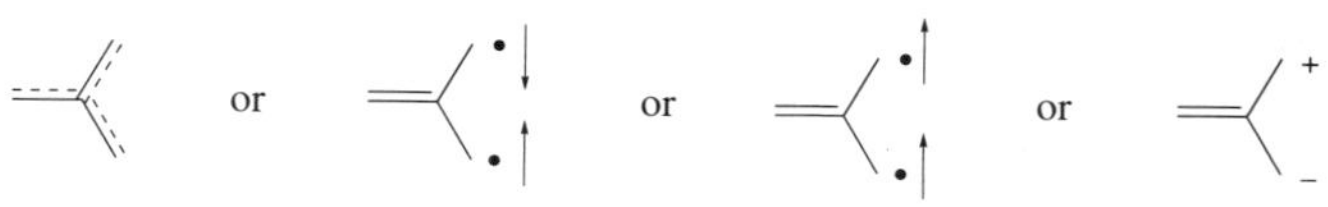

(110)

(a) 2-Alkylidenecyclopentane-1,3-diyls. One of the reactions of TMM which must be avoided, if trapping is to be successful, is ring closure to the corresponding cyclopropane (Equation (79)) ⟨B-90MI 108-01⟩. For cyclic derivatives such as (**111**) the product of ring closure would be highly strained, and these 2-alkylidenecyclopentane-1,3-diyls (henceforth called 'diyls') therefore have a significant lifetime ⟨78ACR446⟩.

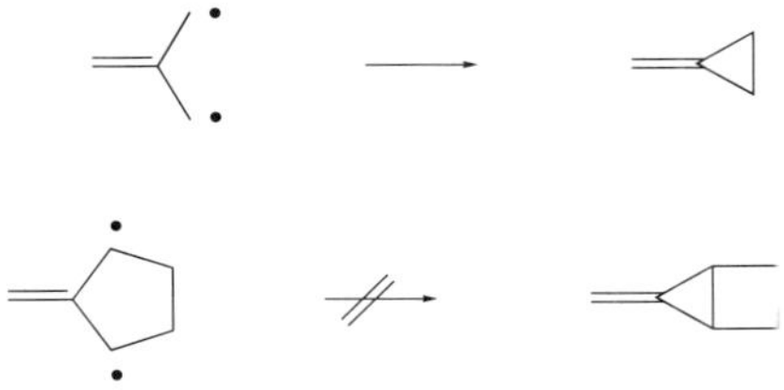

(79)

(**111**)

Little has described how functionalised bicyclic diazenes (**112**) can be used to generate useful diyl species ⟨86CRV875, 91COS(5)239⟩, either photochemically or thermally. The diyl typically equilibrates before cycloaddition (Scheme 29). Intermolecular reaction with an electron-deficient alkene, though successful in terms of trapping of the diyl, led to a mixture of regio- and stereoisomers. This poor selectivity, which was presumably due to the highly reactive nature of the diyls, indicated the need for further steric constraints.

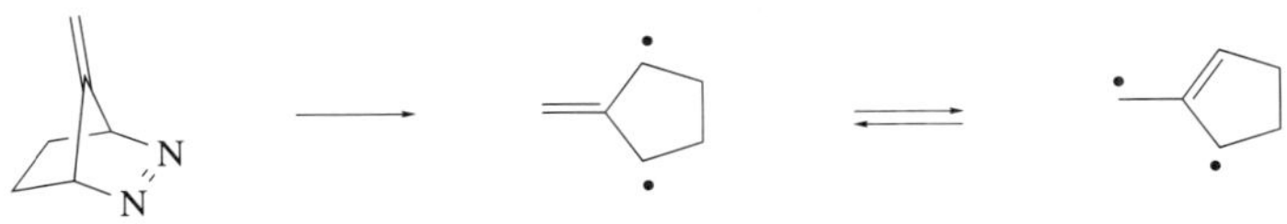

(**112**)

Scheme 29

Two examples serve to demonstrate the excellent selectivities which can be attained in intramolecular diyl cycloadditions. The first shows how the stereochemistry of the alkene component is conserved (H^1 relative to CO_2Me), how the *cis* ring junction (H^1 relative to H^2) is always preferred and how the *anti* configuration (H^1 relative to H^3) is preferred to the *syn* (Equation (80)) ⟨81JA2744⟩.

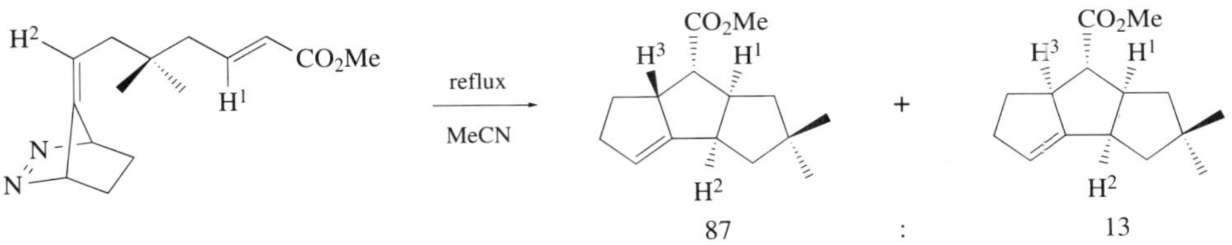

(80)

The second example illustrates the effect of introducing a chiral centre into the tether. The conformational preference of the substituent ($OSiR_3$ in this case) should lead to a marked preference for one of the possible cyclic transition states. Indeed, the tricyclic product (**113**) was isolated almost pure (Equation (81)) ⟨85JA2495⟩. The close analogy between structure (**113**) and a range of natural products has been noted ⟨91COS(5)239⟩.

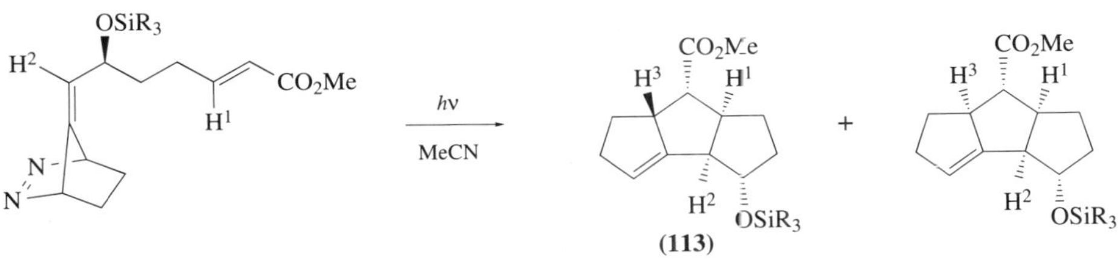

(**113**)

(81)

Little has remarked that the intermolecular version of the reaction may return to prominence in the long term due to the more ready availability of the starting materials ⟨86CRV875⟩. The intermolecular reaction also appears to hold promise for asymmetric induction.

(b) Trimethylenemethane acetals. The presence of an acetal substituent on TMM appears to aid stability. Nakamura and co-workers showed that the intermediate **(114)** was generated by the thermolysis of a methylenecyclopropanone acetal and could be trapped with electron-deficient alkenes (Scheme 30) ⟨89JA7285, 90JOC5553⟩. The reaction was found to be stereospecific, the stereochemistry of the alkene being retained.

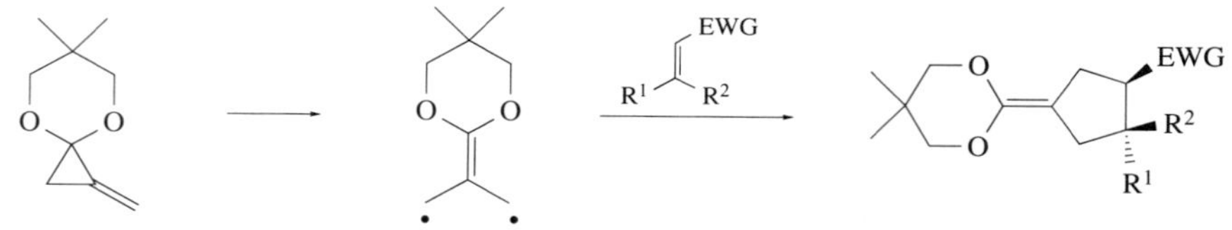

(114)

Scheme 30

Further investigation by the same group established that substituted TMM acetals react highly stereoselectively with alkenes, the preferred stereoisomer resulting from the *endo* transition state (e.g., Scheme 31, Equation (82)) ⟨92JA8707⟩. However, the regioselectivities of the cyclisations were rather poor, except when a *trans*-alkene and a hindered TMM were used (Scheme 31, Equation (83)).

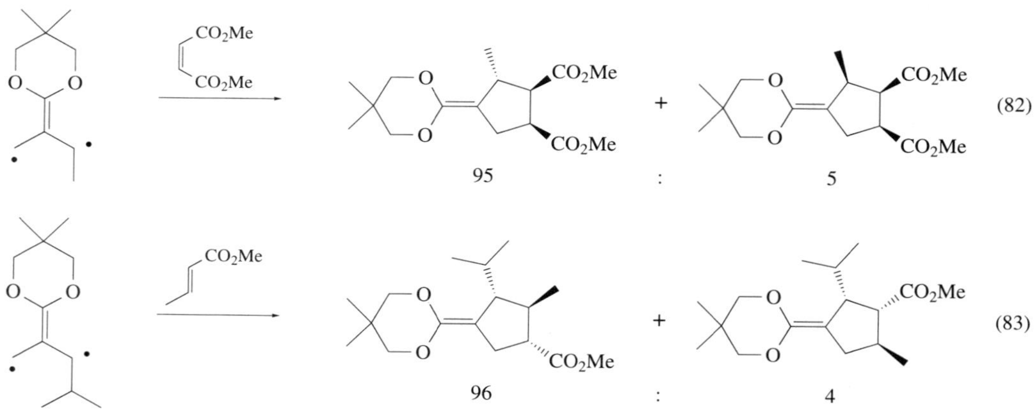

Scheme 31

The simplicity of the cycloadditions of TMM acetals, which can be prepared in a few steps, makes this method of cyclopentane synthesis potentially very important, especially in view of the possible use of catalysis to improve the regioselectivity and to induce asymmetry.

(c) Transition metal–trimethylenemethane complexes. A number of synthetic equivalents of dipolar TMM have been proposed, including the isobutene derivative **(115)** ⟨86JA4683, 88PAC107, 93JOC6716⟩; however, cyclisations using most of these reagents proceed in two distinct steps and are thus beyond the scope of this chapter. In contrast, certain metal complexes of TMM react with alkenes in what can appear a 'concerted' fashion, as no intermediates have been observed. Although the mechanism is almost certainly stepwise ⟨86AG(E)1⟩, these reactions are usually classified as cycloadditions and are therefore discussed here.

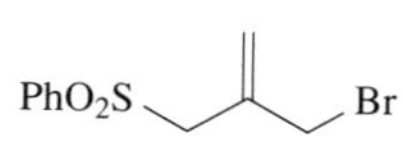

(115)

As Trost remarked, palladium(0) complexes of TMM **(116)** are ideal TMM equivalents as they bind TMM with sufficient strength to stabilise it, but not so tightly that it becomes unreactive ⟨86AG(E)1⟩. In addition, the ability of palladium(0) compounds to ionise bonds from carbon to poor leaving groups permits a straightforward preparation of the complexes from the readily available TMM synthetic equivalent (**117**). Derivatives of (**117**) are also accessible by a range of methods ⟨86AG(E)1⟩.

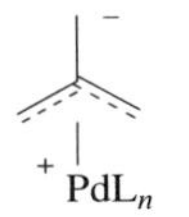

(116)

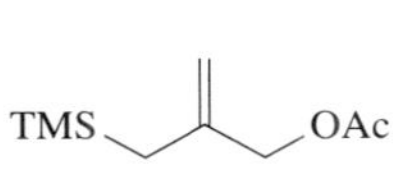

(117)

Unsubstituted TMM complexes (**116**) were used to determine the fundamental characteristics of the intermolecular cycloadditions of these compounds to alkenes. The two typical examples in Scheme 32 demonstrate (i) a marked preference for reaction with electron deficient alkenes, on the less hindered face and (ii) that the stereochemistry of a *trans*-alkene is retained. However, the stereochemistry of *cis*-alkenes can be partly lost. The latter observation is just one indication of the fact that the reactions of TMM complexes are not truly concerted.

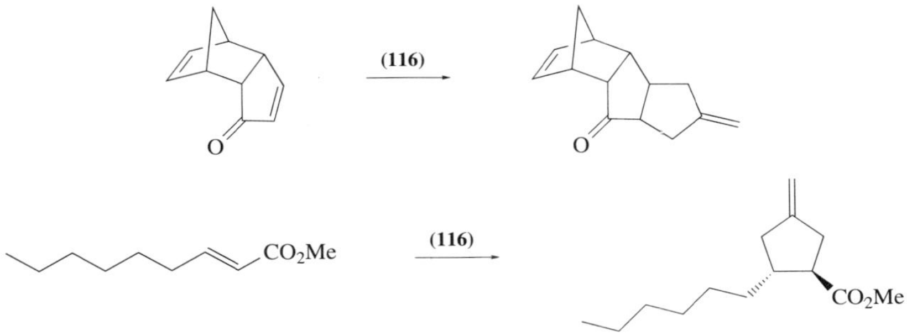

Scheme 32

Reactions of substituted TMMs introduce two new stereochemical problems (Equation (84)) ⟨86AG(E)1⟩. Cycloadditions of monosubstituted TMMs are most unusual in having *three* possible regiochemical outcomes. The two possible relative stereochemistries (between R and Z) for each regioisomer mean that there are six possible products. However, the reactions are highly regioselective, which means that four of the possibilities are not observed ⟨92S151⟩. Unfortunately the diastereoselectivity is generally poor, resulting in mixtures of *cis* and *trans* products. Trost and co-workers have shown that the stereochemical problems can be rectified by epimerisation ⟨86AG(E)1⟩, or circumvented by incorporation of a phenylthio group ⟨91JA9007⟩. The reactions of symmetrically disubstituted TMMs follow the same rules, yielding only one regioisomer of a possible three (rather than six if the substituents were different), but all four possible diastereoisomers are obtained (Equation (85)) ⟨92S151⟩.

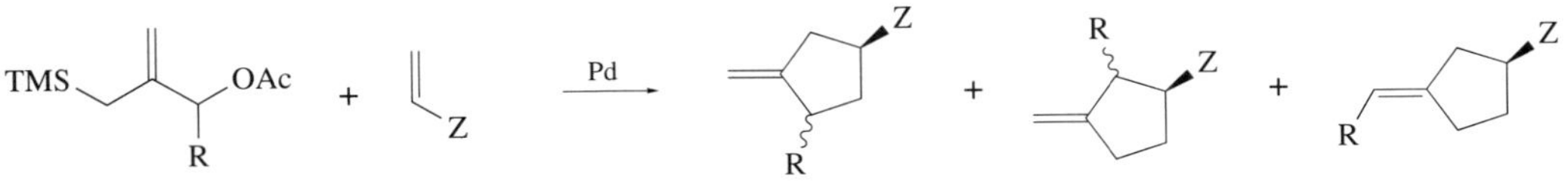

(84)

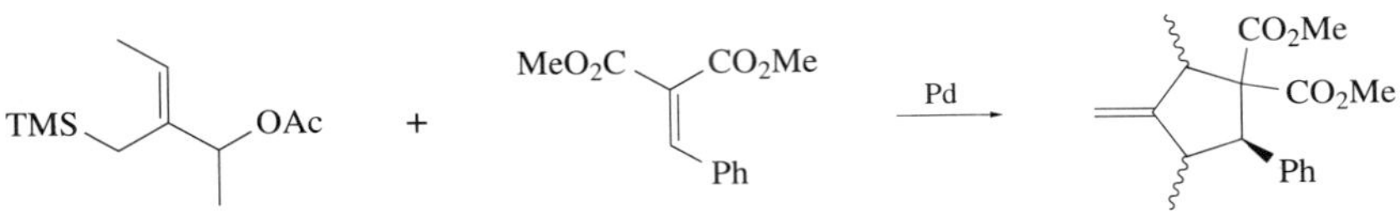

(85)

One other important aspect of the regiochemistry of cycloadditions of TMMs is that it is independent of the TMM precursor. Hence, for example, (**118**) and (**119**) both lead to the same product mixture.

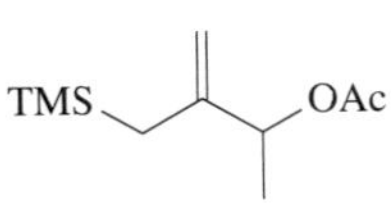

(118)

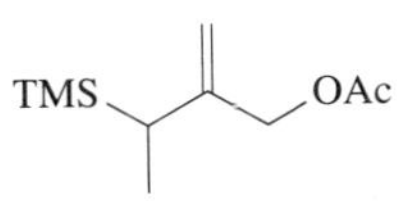

(119)

Intramolecular cycloadditions of palladium–TMM complexes have also been investigated ⟨91JA7350, 92JOC686⟩. The cyclisations are highly regioselective and usually proceed with retention of the configuration of the *trans*-alkene and, although most examples do give rise to mixtures of diastereoisomers, one product usually predominates. For example, the products **(120)** and **(121)** (R = Me), which bear a bridgehead alkyl substituent, were formed in a ratio of 2.4 : 1. Furthermore, the stereochemical problem is obviated by the facile isomerisation of the products **(120)** and **(121)**.

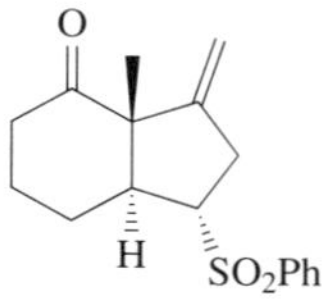

(120)

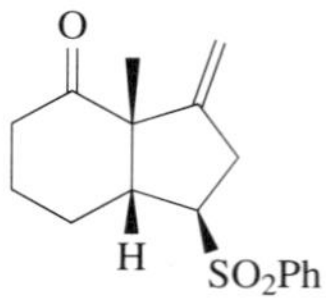

(121)

The synthetic application of palladium–TMM complexes has largely been restricted to unsubstituted TMM. The first two examples show how palladium–TMM can be instrumental in constructing cyclopentane building blocks (Schemes 33 and 34) ⟨90CC1436, 90T1859⟩. The excellent tolerance of palladium–TMM for remote functional groups are illustrated by Equation (86) and Scheme 35 ⟨92JA7375, 93JOC4298⟩.

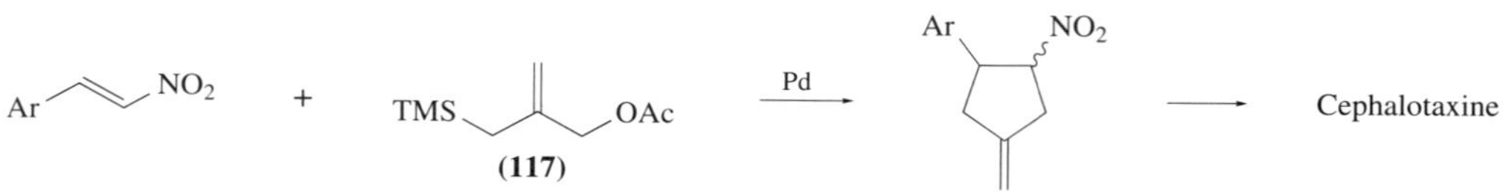

Scheme 33

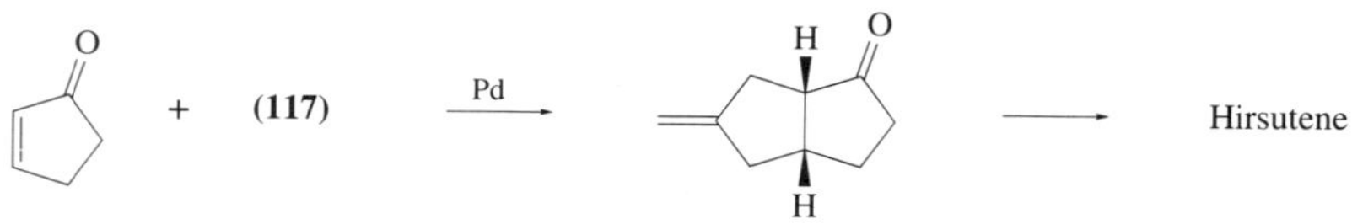

Scheme 34

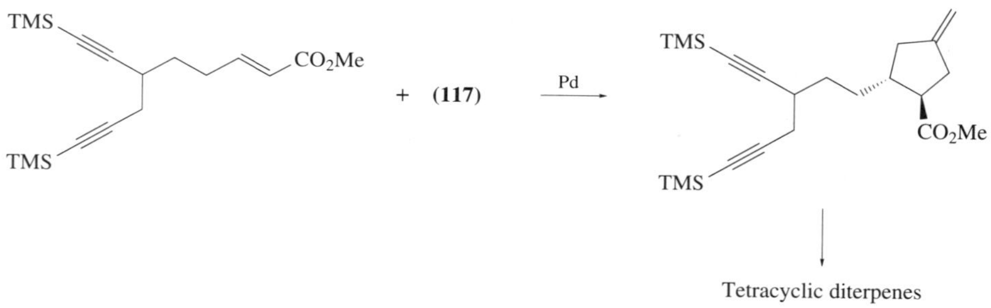

(86)

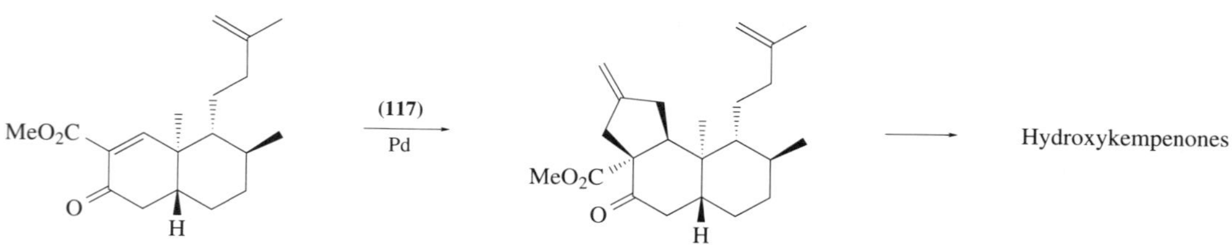

Scheme 35

The related nickel(0)- and palladium(0)- catalysed additions of methylenecyclopropanes to electron-poor alkenes provide another synthetic route to methylenecyclopentanes ⟨86AG(E)1⟩.

Two significant developments in methylenecyclopropane addition chemistry illustrate the potential of the process. The first example shows how a trimethylsilyl derivative (**122**) can be used as a synthetic equivalent of methylenecyclopropane (Scheme 36) ⟨87CB1933⟩. Its use avoids the difficult reaction conditions required by methylenecyclopropane itself. Equation (87) highlights the excellent stereoselectivity observed when a chiral auxiliary is employed ⟨89LA739⟩.

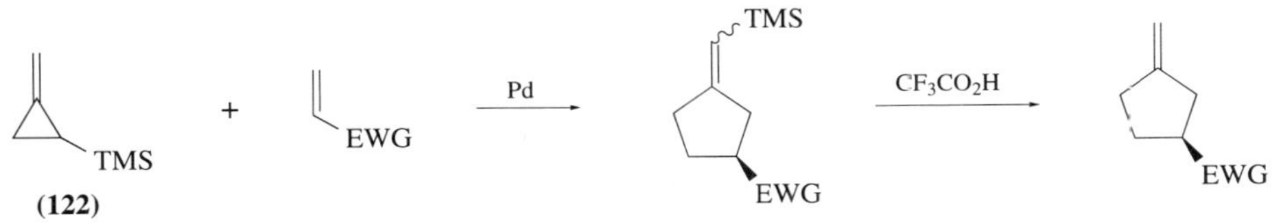

(122)

Scheme 36

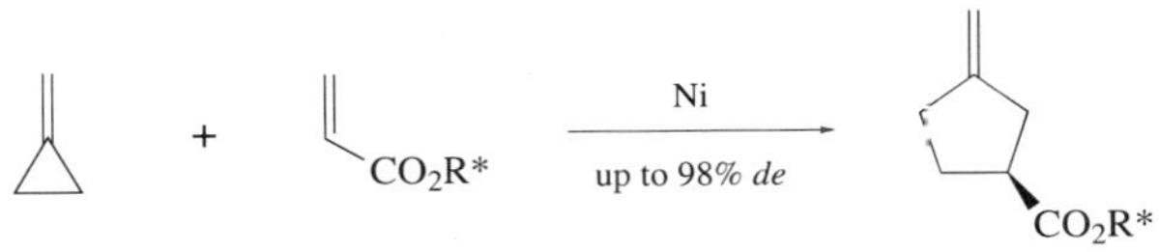

(87)

R^* = chiral auxiliary

(ii) Additions of allyl anions

The addition of an allyl anion, as a 4π component, to a 2π alkene is a potentially useful method for synthesis of cyclopentanes. An early example of such a reaction was published by the group of Kauffmann ⟨74AG(E)468⟩. A carbamoyl-stabilised lithium allyl compound (**123**) was found to react with its parent alkene (**124**) to produce the cyclopentane (**125**) (Equation (88)).

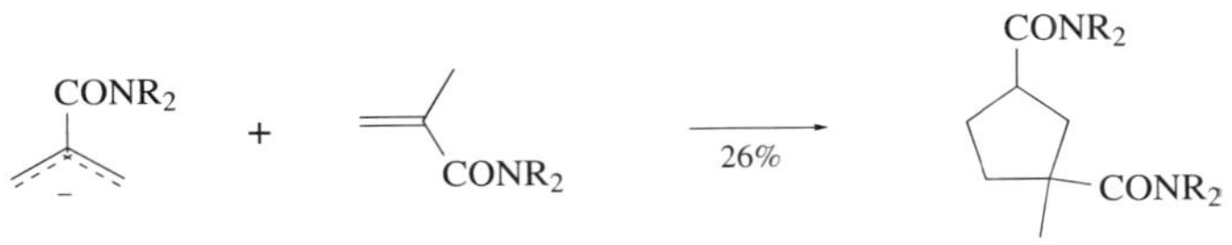

(88)

(123) **(124)** **(125)**

The chemistry described in this paper was later developed into a useful synthetic process by Beak *et al.* ⟨86JOC4627⟩. Addition of a similar anion to a variety of carbamoyl-substituted alkenes gave reasonable yields of the cyclopentanes (**126**) (Equation (89)). The regioselectivity and stereoselectivity were found to be good, with the exception of the configuration at C-4; two epimers were often observed. The stereochemical outcome is consistent with a stepwise mechanism proceeding via anion (**127**). The apparent need for carbamoyl substituents is presumably a reflection of the ability of the nitrogen atom to coordinate to lithium in the intermediates.

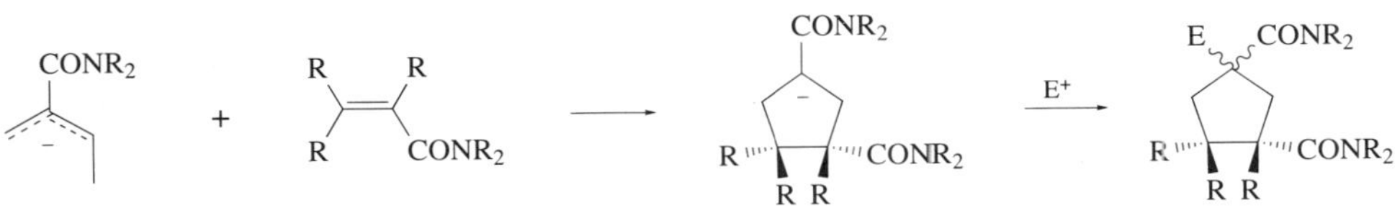

(89)

(127) **(126)**

(iii) Additions of allyl cations

Allyl cations are 2π fragments which are expected to react thermally with 4π partners, that is, they should rightly be categorised as dienophiles. Indeed, this strategy has been used in synthesis of seven-membered rings ⟨86T4611⟩. Noyori argued that allyl cations could be induced to react, in a symmetry-forbidden manner, with alkenes if a nonconcerted mechanism was encouraged ⟨78JA1791⟩.

He postulated that this could be achieved by stabilising the cationic intermediate with an electron-releasing group as discussed below.

When α,α'-dibromoketones (**128**) were reacted with $Fe_2(CO)_9$ (**129**), intermediates (**130**) were formed which added successfully to a range of aryl-substituted alkenes with excellent yields in some cases (Scheme 37) ⟨78JA1791⟩. Despite the stepwise nature of the reactions, they proceeded with retention of alkene stereochemistry and are therefore synthetically useful.

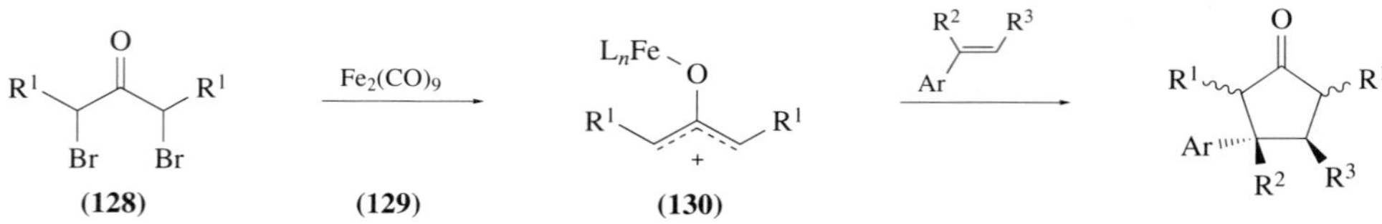

Scheme 37

The main disadvantage of the methodology is the limited range of α,α'-dibromoketones which can be used ⟨86T4611⟩, the most severe limitation being the failure of the simplest example, α,α'-dibromoacetone, to react. This means that the products will inevitably bear alkyl substituents, which may be unwanted. Unfortunately this problem was not resolved by the introduction of a mixed lanthanum–tin chloride reagent ⟨89BCJ2348⟩.

An alternative approach to the generation of allyl cations has been developed by the group of Hoffmann ⟨88T3899⟩, who investigated the use of alkoxy and trialkylsilyloxy groups for the stabilisation of the cationic intermediates. These variations permitted the introduction of allylic alcohols as starting materials. The allylic alcohols (**131**) react with Lewis acids such as titanium tetrachloride to form the allyl cations (**132**) (Equation (90)). Despite the observation of one [3 + 2] adduct in a reaction with a diene, the reactions of these cations with simple alkenes do not appear to have been studied ⟨88T3899⟩.

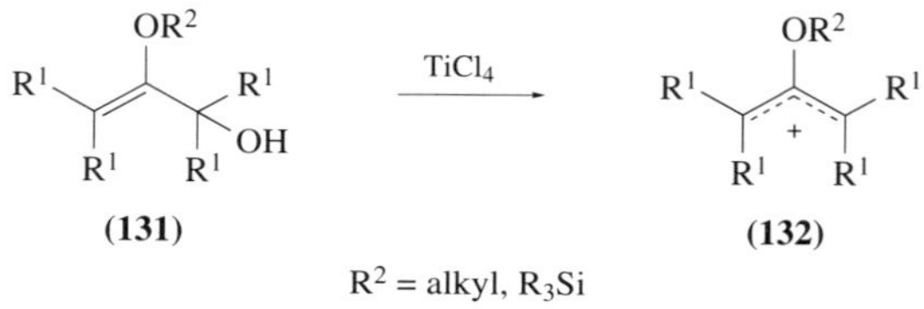

(90)

(*iv*) Meta *photocycloaddition of arenes*

The photocycloaddition of arenes to alkenes has been intensely studied from a theoretical and mechanistic point of view ⟨93CRV615⟩. The simplest example, the reaction of benzene with ethene, is shown in Equation (91). Variation by substitution both on the benzene nucleus and on the alkene is possible ⟨93CRV615⟩. The intramolecular version of the *meta* photocycloaddition is illustrated by the reaction of phenylpentene (**133**) (Equation (92)). It is clear that such reactions are uniquely powerful for the rapid construction of polycyclic skeletons. The production of cyclopentane units is of particular interest.

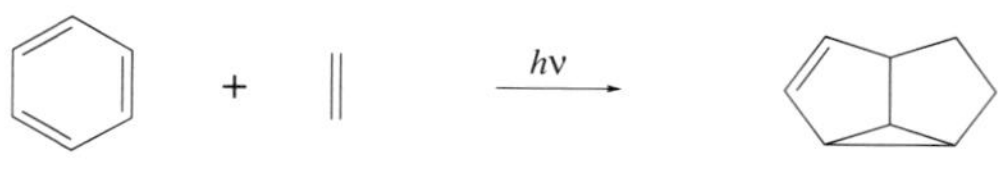

(91)

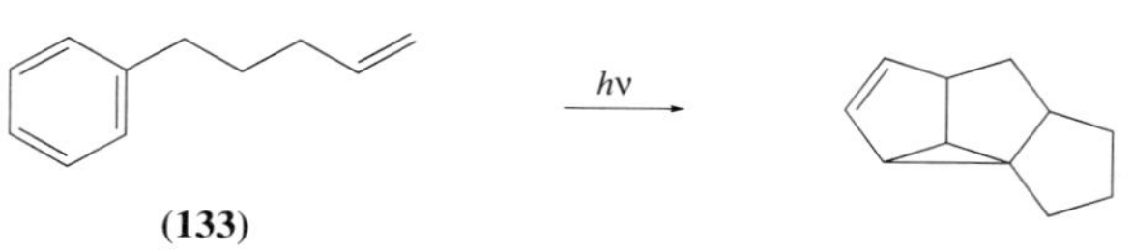

(92)

The intermolecular *meta* photocycloaddition reaction has not found much synthetic application ⟨93CRV615⟩. However, Wender and co-workers have shown that the intramolecular version is a valuable synthetic tool ⟨90PAC1597⟩. For example, their synthesis of silphinene (**134**) is remarkably short (Scheme 38). The production of two isomers in the first step is typical of these cycloadditions, but, if the number of possible isomers is considered, this selectivity is excellent. The reviews by Cornelisse ⟨93CRV615⟩ and Wender ⟨90PAC1597⟩ are recommended for further reading.

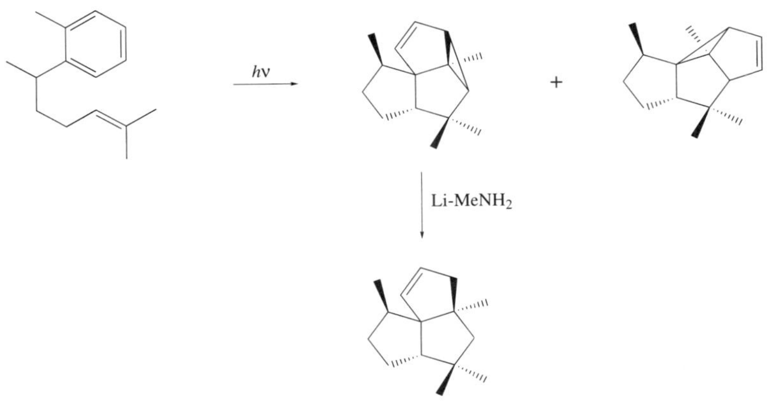

(134) (±)-Silphinene

Scheme 38

The enantiomerically pure arylpentene (**135**) can be rapidly converted to (−)-retigeranic acid (**136**) (Scheme 39). The development of a catalytic asymmetric version of the process remains a significant challenge.

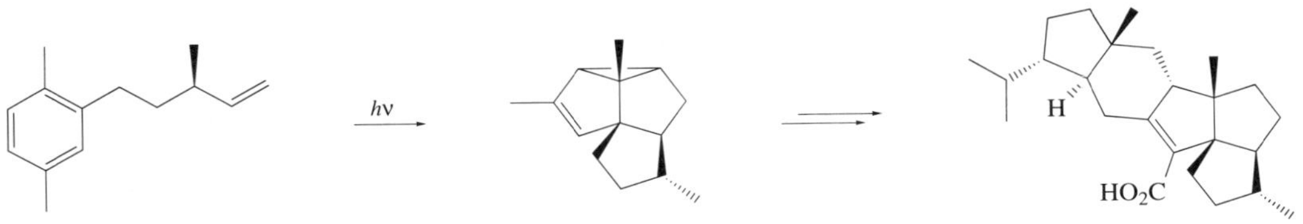

(135)

(136) (−)-Retigeranic acid

Scheme 39

1.08.2.4.2 *[3 + 2] Additions of CXC fragments*

(i) Additions of carbonyl ylides

Disconnection of a tetrahydrofuran gives, among other possibilities, an alkene and a carbonyl ylide (Scheme 40). Some simple methods of preparing carbonyl ylides and representative examples of their cycloaddition reactions are discussed below. The problems posed by carbonyl ylides are largely due to their highly reactive and transient nature.

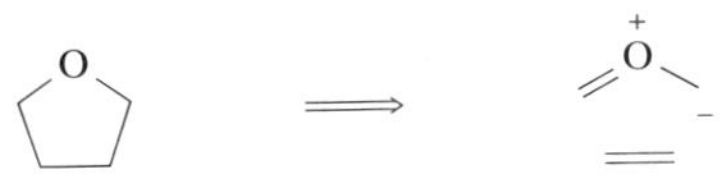

Scheme 40

The earliest approaches to carbonyl ylides involved thermolysis or photolysis of certain epoxides (Scheme 41) ⟨B-76MI 108-02⟩. A logical extension of this chemistry was to facilitate the thermolysis by incorporating a group which would undergo extrusion, nitrogen from oxadiazolines (**137**) or carbon dioxide from dioxolanones (**138**) being appropriate ⟨91COS(4)1069⟩. The simple cycloaddition chemistry of these compounds is summarised in Padwa's review ⟨B-84MI 108-01⟩. The regiochemistry of the reactions is predictable from FMO theory if one uses the carbonyl ylide HOMO in reactions with electron-poor alkenes and the ylide LUMO in reactions with electron-rich alkenes ⟨91COS(4)1069⟩.

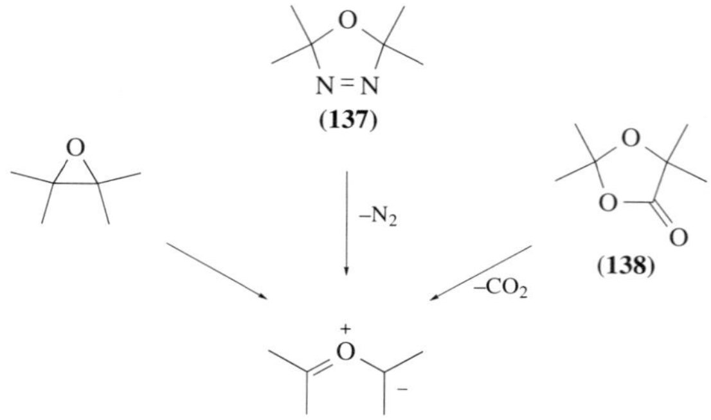

Scheme 41

FMO theory is also useful for prediction of the stereochemical outcome of the reaction. Hence, for example, the epoxide (**139**) opens in a symmetry-allowed fashion on thermolysis, that is, with conrotation, to produce the carbonyl ylide (**140**), which, in turn, reacts with the alkene trap with retention of alkene stereochemistry, as expected (Scheme 42). The *endo*/*exo* selectivity cannot, of course, be predicted from FMO theory. In fact, mixtures of stereoisomers usually result.

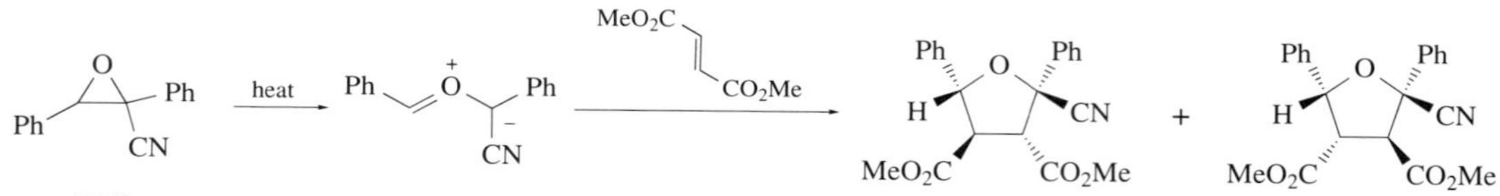

Scheme 42

The thermal and photochemical ring-opening reactions described above do have synthetic limitations as these conditions can accelerate other, unwanted reactions. A method which avoids use of these more forcing conditions is the reaction of a carbene or carbenoid with a carbonyl compound (Equation (93)). This chemistry has been extensively studied by the group of Padwa, and reported, along with leading references to related work, in a review ⟨91ACR22⟩; the combination is particularly effective when the carbene–carbonyl reaction is intramolecular, as in Scheme 43.

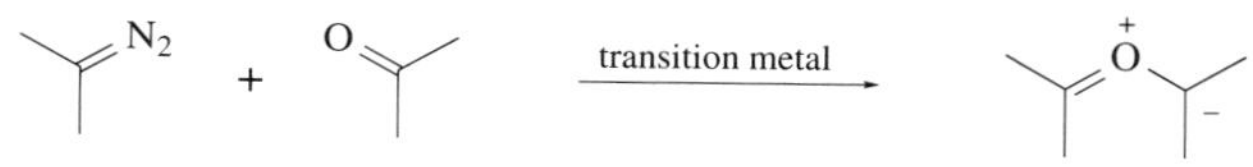

(93)

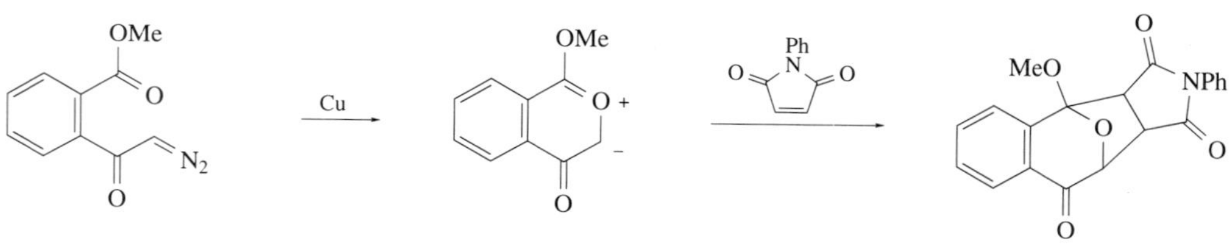

Scheme 43

Padwa's group have also investigated intramolecular trapping of carbonyl ylides ⟨91ACR22⟩. The example shown in Scheme 44 is typical and demonstrates how this 'tandem cyclisation' procedure can quickly assemble a polycyclic skeleton. The regiochemistry and relative stereochemistry become excellent when the alkene and ylide are tethered in this manner.

The major disadvantage of the rhodium carbenoid approach is that an α-keto group is a prerequisite. Alternative methodology, based on halosilane elimination, offers an answer to this problem ⟨93TL5943⟩, although only arylcarbonyl ylides have been prepared in this way. The reaction of (**141**)

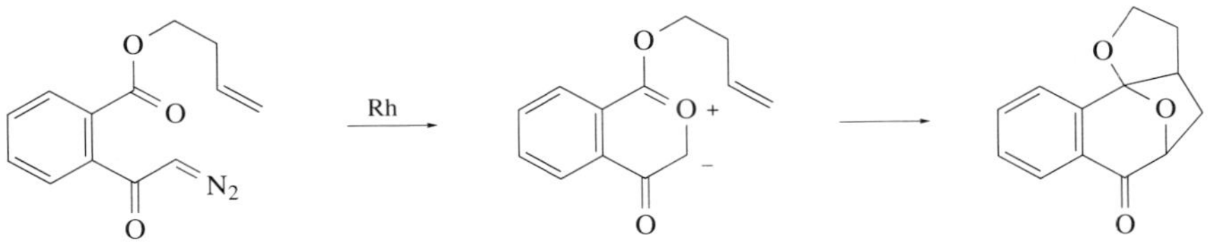

Scheme 44

with caesium fluoride promotes the elimination of the elements of trimethylsilylchloride to generate the carbonyl ylides (**142**) (Equation (94)). Intermolecular trapping of these ylides proceeds with retention of stereochemistry of the alkene. Regiochemistries and relative stereochemistries are not outstanding, however.

(**141**) (**142**) (94)

The rhodium carbenoid and halosilane elimination routes to carbonyl ylides are complementary. The use of these routes in intramolecular cycloadditions to alkenes provides a powerful method for the construction of tetrahydrofuran-based polycyclic compounds.

(ii) Additions of thiocarbonyl ylides

The methods of generation of the sulfur compounds (**143**) are similar in many respects to those used for their oxygen counterparts. Hence, extrusion of nitrogen from heterocyclic compounds (**144**) ⟨76T2165⟩, addition of carbenes to thiocarbonyl groups ⟨83TL4181, 84BSB511, 85TL3011⟩ and elimination from silyl precursors (**145**) ⟨86TL4039⟩ all provide access to the reactive intermediates (**143**) (Scheme 45).

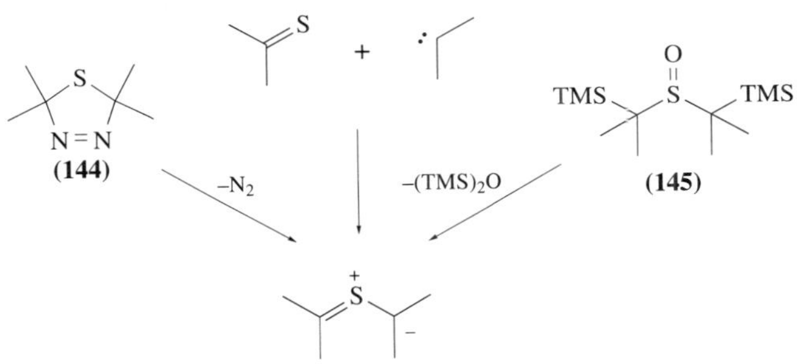

Scheme 45

The cycloadditions of thiocarbonyl ylides to some, mainly electron-deficient, alkenes have been studied (e.g., Scheme 46). Prediction of the regiochemistry of the reaction appears difficult ⟨76T2165⟩ but, as expected, the reactions proceed with retention of stereochemistry of the alkene and with rather good *endo*/*exo* selectivity.

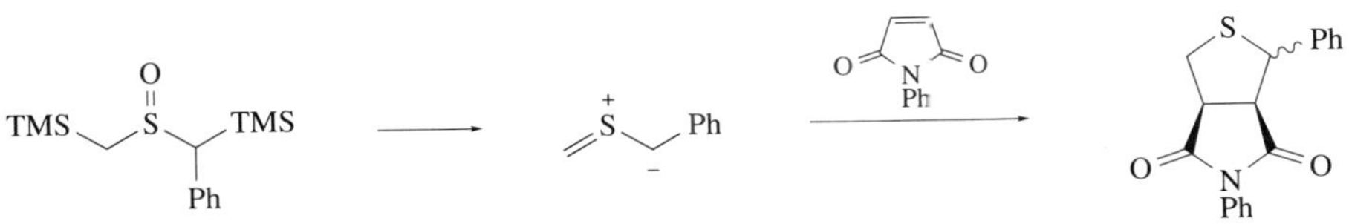

Scheme 46

There have been few synthetic applications of thiocarbonyl ylides ⟨91COS(4)1069⟩. Only a handful of intramolecular examples are known ⟨91COS(4)1111⟩ and there appears to have been no report of an asymmetric version of the reaction. This may be due to the tendency of the thiocarbonyl ylides (**143**) to isomerise to the corresponding thiirane, which as an intramolecular reaction can compete favourably with an intermolecular cycloaddition.

(iii) Additions of azomethine ylides

The nitrogen analogues of carbonyl and thiocarbonyl ylides are called azomethine ylides, and these ylides have become widely used in organic synthesis. The methods of generation of these reactive intermediates are similar to those described above for carbonyl and thiocarbonyl ylides, and the reactivity of the ylides towards alkenes is also comparable. As there are already a number of reviews dealing with generation and trapping of stabilised azomethine ylides ⟨88H(27)981, B-88MI 108-02, 89AHC(45)231, B-90MI 108-01, 91COS(4)1069, 91COS(4)1111⟩, this chemistry will be briefly summarised. In this review a more detailed discussion focuses on work involving nonstabilised azomethine ylides and asymmetric azomethine ylide cycloadditions.

The principal routes to stabilised azomethine ylides (**146**) are from aziridines such as (**147**), by thermolysis or photolysis, from readily accessible oxazolidines such as (**148**), and from imines such as (**149**), by thermal isomerisation (which can be acid-catalysed). These three methods are illustrated in Scheme 47. The intermolecular cycloaddition of a stabilised azomethine ylide to an alkene is efficient when the alkene possesses electron-withdrawing groups. In this case the regioselectivity can be predicted by considering the dipole HOMO and the alkene LUMO. The 2,4-disubstituted pyrrolidine product thus predominates if the alkene is monosubstituted (e.g., Equation (95)).

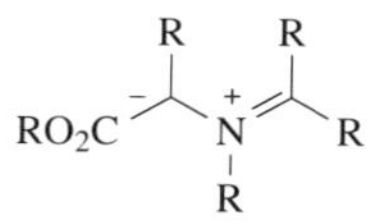

(146)

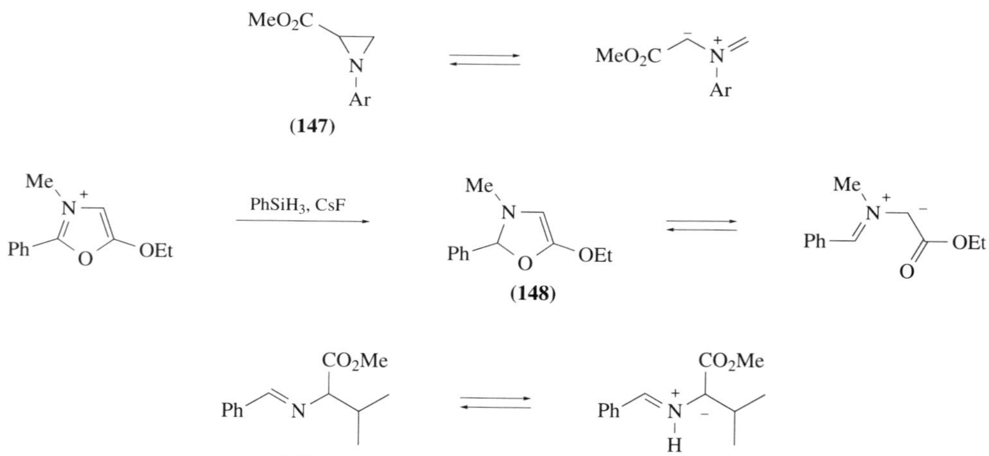

Scheme 47

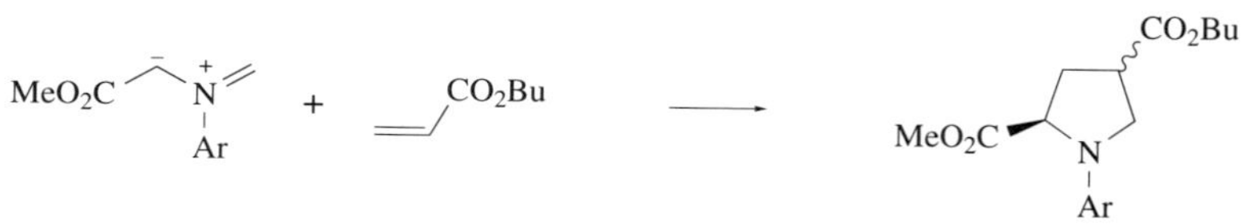

(95)

The stereoselectivity of azomethine ylide cycloadditions is also typical of a concerted reaction. Hence, the stereochemistry of the alkene is retained. Although stereoisomers are usually observed, *endo*/*exo* selectivity can be very high. The synthesis of *allo*-kainic acid (**150**) illustrates all these points (Scheme 48). The electron-withdrawing groups are found in the 2 and 4 positions in the cycloadduct. The *trans*-alkene stereochemistry is observed as an *anti* relationship in the product, and the *endo*/*exo* selectivity is excellent.

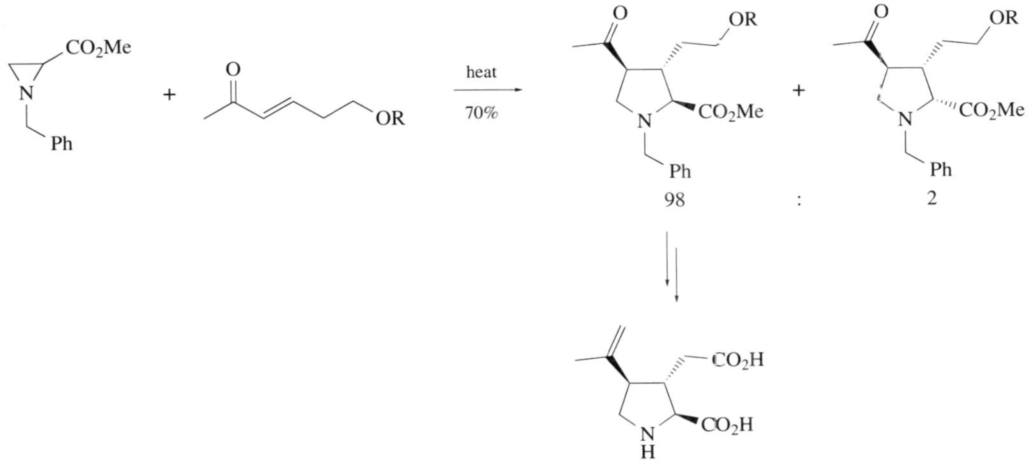

(**150**) *allo*-kainic acid

Scheme 48

The attempted parallel synthesis of kainic acid from the *cis*-enone (**151**) (Scheme 49) highlights one of the main problems of the aziridine, or any other thermal, approach to azomethine ylides, that is, the relatively high temperatures required for the thermolysis. Under these conditions the thermal isomerisation of the enone from *cis* to *trans* is fast enough to pose a problem. Another potential pitfall is that, although the aziridine thermolysis is stereospecific, the stereochemistry of the product ylide can be lost under the reaction conditions.

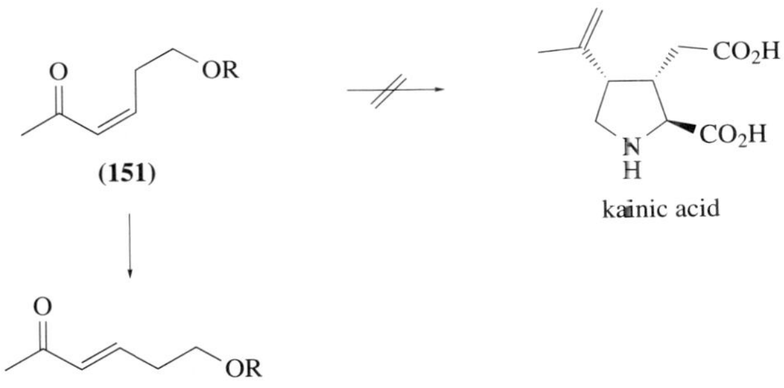

Scheme 49

Although electron-deficient alkenes are required for intermolecular cycloaddition to stabilised azomethine ylides, intramolecular reactions with unactivated alkenes can proceed smoothly ⟨85T3547⟩. This type of reaction was recognised as appropriate for the construction of the skeleton of the marine alkaloid sarain A (Equation (96)) ⟨91JOC3210, 92JOC7056⟩. The latter paper also described a general strategy for the preparation of diazabicycloalkanes (**152**) using intramolecular azomethine ylide cycloadditions.

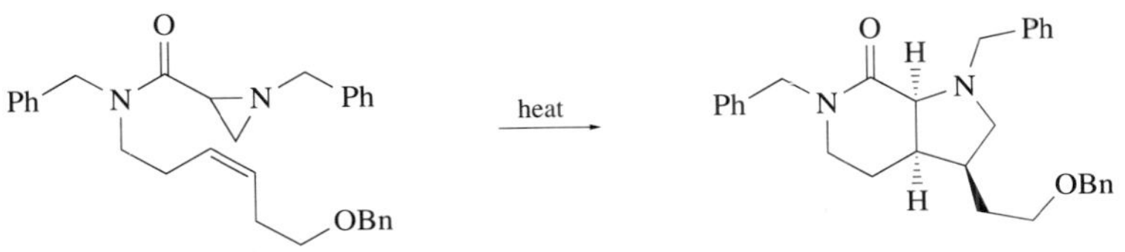

(96)

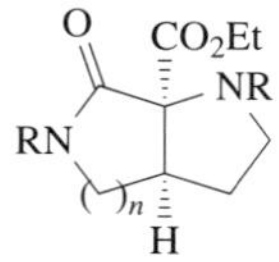

(152) n = 1–3

All of the reactions outlined so far have involved stabilised azomethine ylides, which are more accessible than their nonstabilised counterparts. Silyl precursors to carbonyl and thiocarbonyl ylides have already been described, and this methodology has also been applied to the formation of nonstabilised azomethine ylides.

The disadvantages of the earlier desilylation routes to azomethine ylides are highlighted in a paper which describes an apparently superior method ⟨93TL4861⟩. Bis(trimethylsilyl) derivatives of certain trialkylamines were reacted with silver fluoride. The azomethine ylide intermediates were trapped by alkenes (and other dipolarophiles) in quantitative yield to produce pyrrolidines. The utility of the process was illustrated by the synthesis of pyrrolizidine and indolizidine skeletons (Scheme 50). Although the regioselectivity and stereoselectivity are modest, the combined yield of 98% is very good for the trapping of a reactive intermediate.

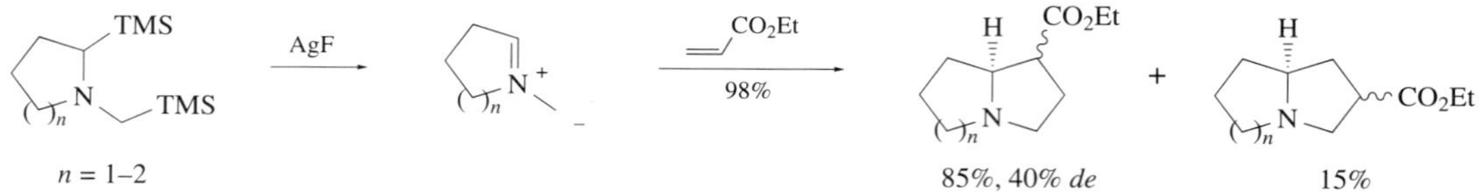

Scheme 50

Azomethine ylide cycloaddition to alkenes is one of the few categories of five-membered ring formation where significant asymmetric induction has been realised; this has been achieved through the use of chiral auxiliaries. In principle, the auxiliary could be placed on either the ylide or the alkene partner in the cycloaddition. Indeed, successful applications of both these strategies have been reported ⟨94AG(E)683, 94JOC4⟩, and high levels of stereoselection are observed with the use of well-known, recoverable auxiliaries.

The attachment of Oppolzer's chiral sultam to either an aziridine or imine precursor has been demonstrated to lead to azomethine ylides which react with alkenes with, in some cases, high *endo* selectivity and with excellent diastereofacial selectivity (Scheme 51) ⟨94JOC4⟩. Interestingly, the stereochemistry of the product depends on which method is used to generate the ylide. This can be explained by assuming that the aziridine-derived ylide is *cis* whereas a *trans*-ylide results from the imine tautomerisation.

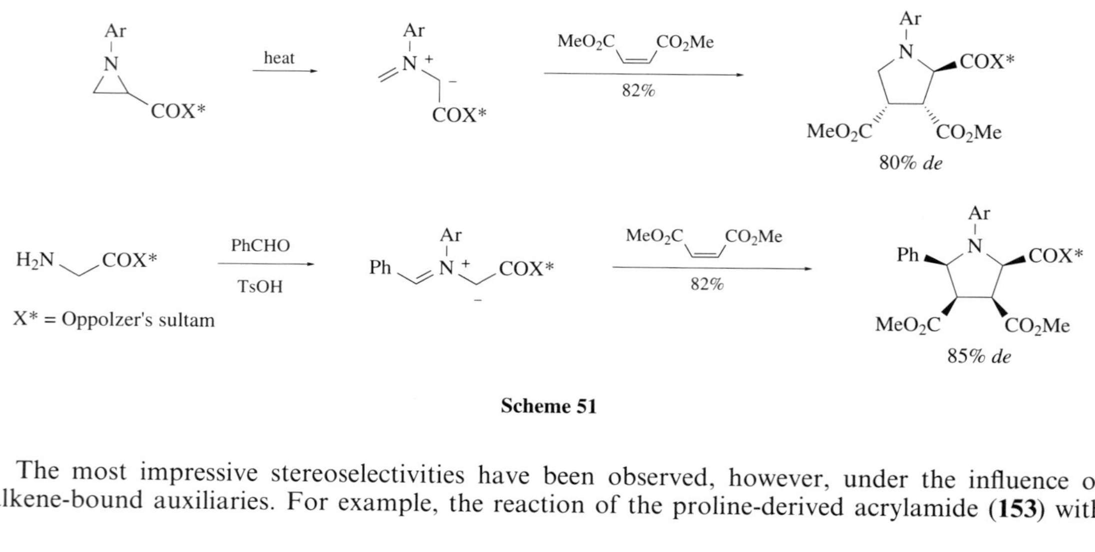

Scheme 51

The most impressive stereoselectivities have been observed, however, under the influence of alkene-bound auxiliaries. For example, the reaction of the proline-derived acrylamide **(153)** with various imine-derived azomethine ylides proceeded, in several cases, with essentially complete control of stereochemistry to yield only one isomer (Scheme 52) ⟨94AG(E)683⟩. It was also demonstrated that the auxiliary could be removed by acid hydrolysis.

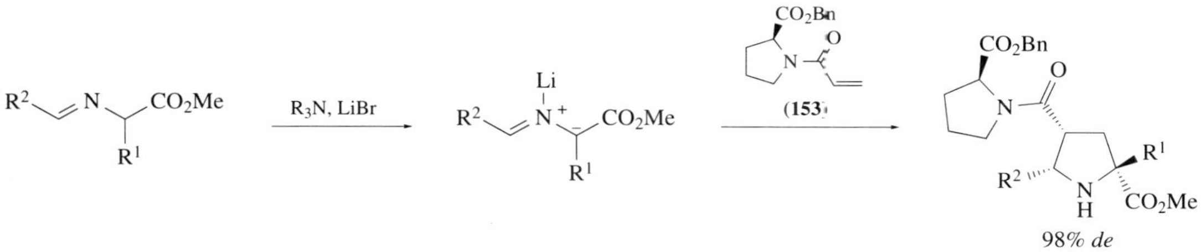

Scheme 52

(iv) Additions of azaallyl anions

2-Azaallyl anions (**154**) are the nitrogen analogues of allyl anions (see Section 1.08.2.4.1); cycloaddition of these entities to alkenes, when followed by protonation, is equivalent to the addition of the corresponding azomethine ylide (see Section 1.08.2.4.2.iii). Indeed, the azaallyl addition is complementary to the azomethine ylide addition, since the latter species generally require electron-poor alkenes ⟨94TL2641⟩.

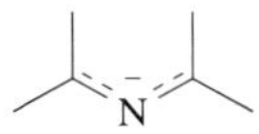

(154)

The early work on the cycloaddition reactions of azaallyl anions was carried out by the group of Kauffmann ⟨74AG(E)627⟩. Two methods of preparing the anions were described, the lithiation of an imine (**155**) by a dialkylamide and the ring-opening reaction of a lithiated aziridine (**156**). The reactions with alkenes are typified by high regioselectivity, for example as in Scheme 53. Although this was, perhaps, indicative of a stepwise mechanism, the stereochemistry of the reaction suggested otherwise, the alkene stereochemistry being retained. A stepwise mechanism via a strongly lithium-bound intermediate, as proposed by Kauffmann, would explain both these phenomena, and appears most plausible.

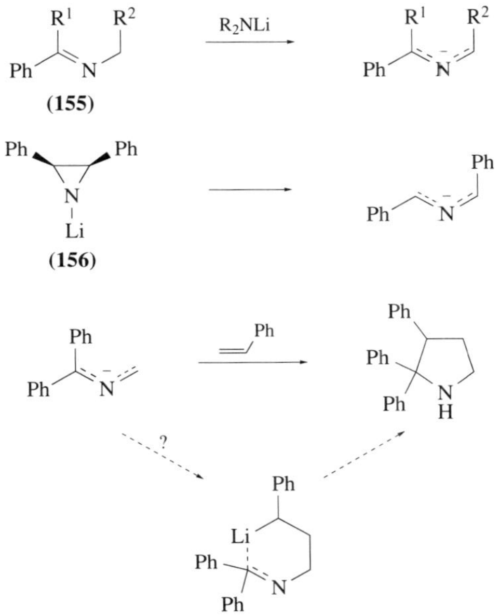

Scheme 53

Both the methods described by Kauffmann for the preparation of azaallyl anions are restricted to cases where the anions are stabilised enough for their formation to be thermodynamically possible but not so much that they do not react ⟨B-90MI 108-01⟩. These restrictions are perhaps responsible for the fact that Carruthers's 1990 review of the area revealed few developments.

Although a desilylation strategy was successful in producing azaallyl anions, this route was limited to the production of stabilised anions ⟨87JOC2523⟩. Nonstabilised azaallyl anions became more readily accessible when the reaction of trialkylstannylmethanimines (**157**) with alkyllithiums was investigated ⟨92JA1329⟩. Precursors (**157**) were prepared by an aza-Wittig reaction (Equation (97)). Reaction with either methyl- or butyllithium and an alkene trap produced pyrrolidine products in high yields. Although the stereochemistry of the alkene was found to be retained, mixtures of regioisomers and diastereoisomers were usually observed (Scheme 54). The preference for the more sterically crowded regioisomer is consistent with the results from the Kauffmann group ⟨74AG(E)627⟩.

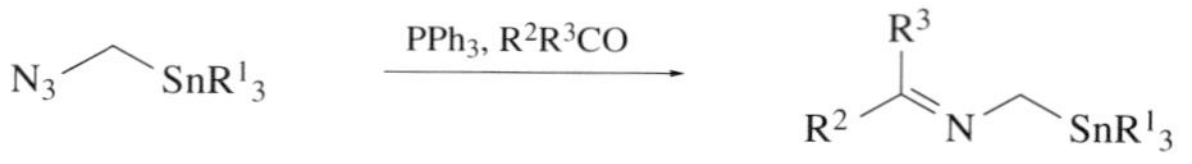

(**157**)

(97)

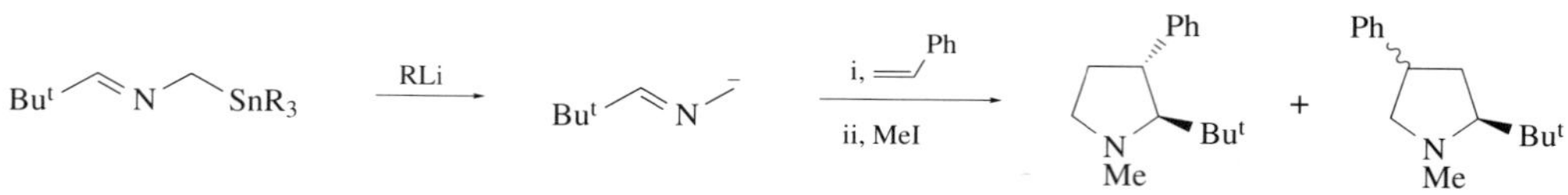

Scheme 54

One of the remaining problems with the azaallyl anion methodology was the need for the activating (usually aryl) group on the alkene, which might well be an unwanted substituent. This restriction can be removed by use of vinyl sulfides, selenides and silanes. The heterosubstituent on the product pyrrolidine can thus be removed readily after it has fulfilled its function as an activator ⟨92JA1329⟩.

As expected, the intramolecular addition of azaallyl anions to alkenes has proved less problematical than the intermolecular reaction. Hence, stabilised anions have been shown to add successfully to unactivated alkenes with a reasonable degree of stereoselectivity (Equation (98)) ⟨86JA2769⟩. Nonstabilised azaallyl anions have also been shown to take part in intramolecular cycloadditions ⟨92JA1329⟩.

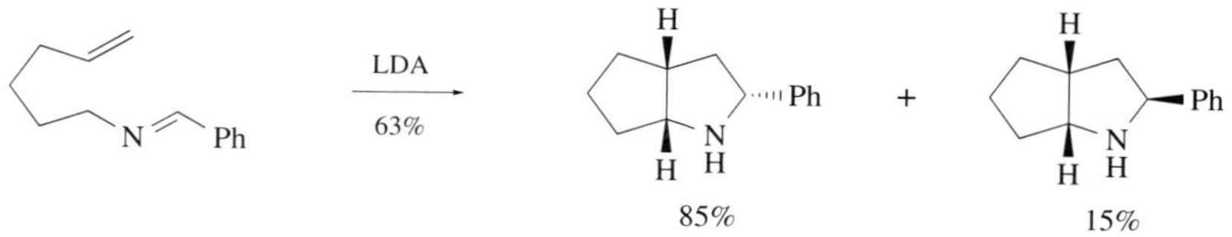

(98)

1.08.2.5 Formation of Six-membered Rings

The reaction of a diene, or a heterodiene, with a dienophile is usually known as the Diels–Alder reaction. This process has won recognition as one of the most powerful synthetic methods. As a new C=C bond is formed in the reaction, it is classified in Chapter 1.17. However, to permit comparison with the related cycloadditions in this chapter a very brief account is provided here. The other restrictions on scope have been applied. Thus, Diels–Alder reactions of alkenes with dienes, or heterodienes with two carbon termini, qualify.

1.08.2.5.1 [4 + 2] Diels–Alder additions of dienes

(i) *Intermolecular additions*

Intermolecular Diels–Alder reactions ⟨91COS(5)315⟩ generally proceed with retention of both alkene and diene stereochemistries, but the regiochemistry and *endo*/*exo* selectivity is often poor. The reaction in Equation (99) is typical. Without doubt the most significant advance in Diels–Alder chemistry was the discovery that Lewis acids catalyse the process. Thus, aluminum trichloride controls the reaction in Equation (99), making it highly regioselective and *endo* selective while reducing the optimum temperature from 120 °C to 20 °C ⟨67JOC869⟩.

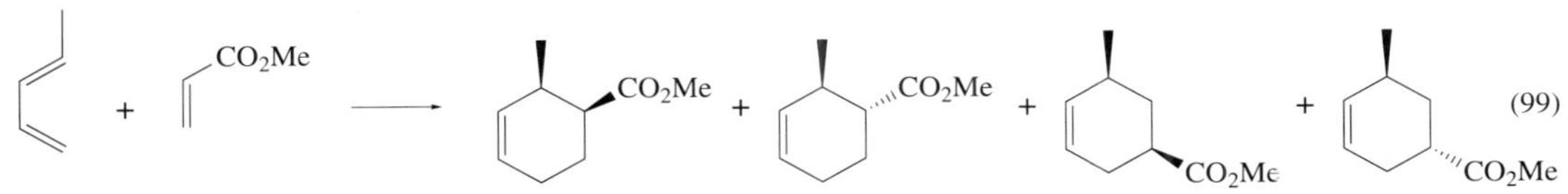

(99)

No catalyst	45	:	39	:	11	:	5
0.15 equiv. $AlCl_3$	93	:	5	:	2	:	0

In 1983, other methods of improving on the reactivity and selectivity of Diels–Alder reactions were investigated. The reduction in reaction time, increase in yield and improved selectivity observed when using the sodium carboxylate (**159**) in water, rather than the traditional ester (**158**) in benzene, are most impressive results (Equation (100)) ⟨83TL1897⟩. High pressure and ultrasound have been used to promote Diels–Alder reactions (e.g., Equation (101)). Although less attractive than Lewis acids they may prove useful in reactions where the dienophile lacks a group which is able to coordinate to a Lewis acid ⟨91COS(5)315⟩.

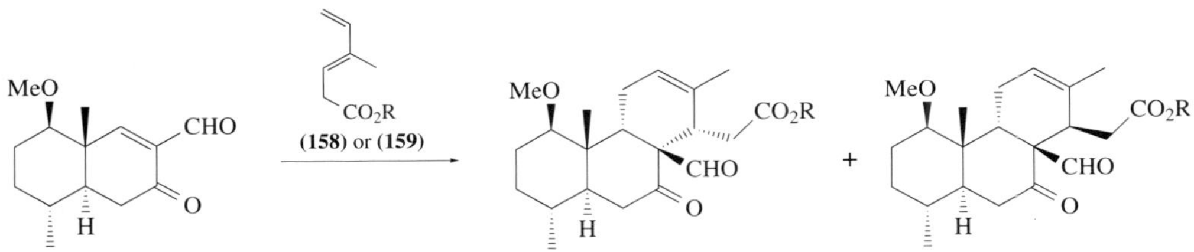

(100)

(**158**) R = Et, benzene, 288 h, 52%	46	:	54
(**159**) R = H, water, 5 h, 100%	75	:	25

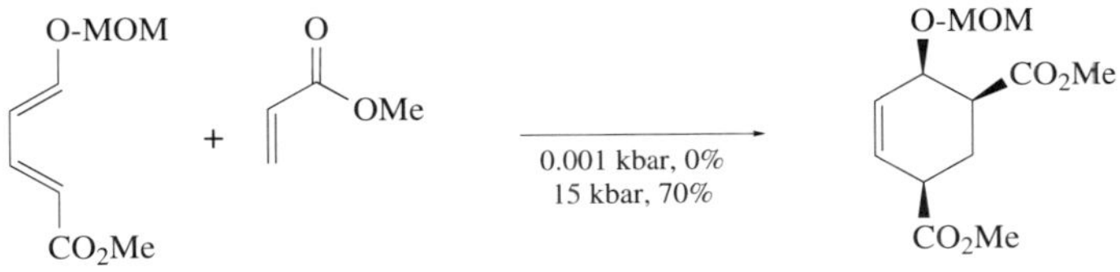

(101)

High diastereofacial selectivity has also been achieved in the intermolecular Diels–Alder reaction. The chiral auxiliary approach has proved extremely effective ⟨91COS(5)315⟩. The auxiliary can be attached to either the alkene (Equation (102)) or, less commonly, the diene (Equation (103)). Once again, the use of Lewis acid catalysts is central to the success of the reactions.

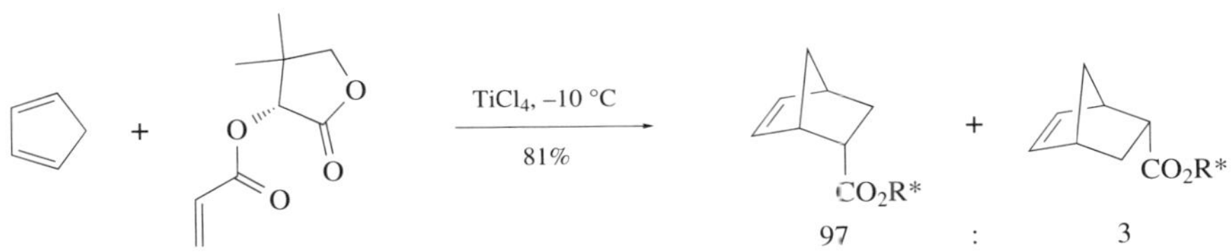

(102)

97 : 3

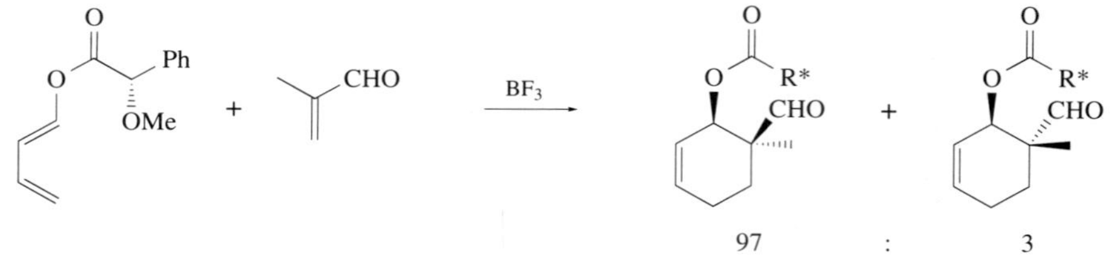

(103)

Chiral Lewis acids have been used to catalytically control the absolute stereochemistry of Diels–Alder reactions ⟨92CRV1007⟩. Impressive results have been obtained in this area, with chiral boron, aluminum and titanium Lewis acids being particularly effective (e.g., Scheme 55) ⟨89JA5493, 91JA7794, 91TA643⟩.

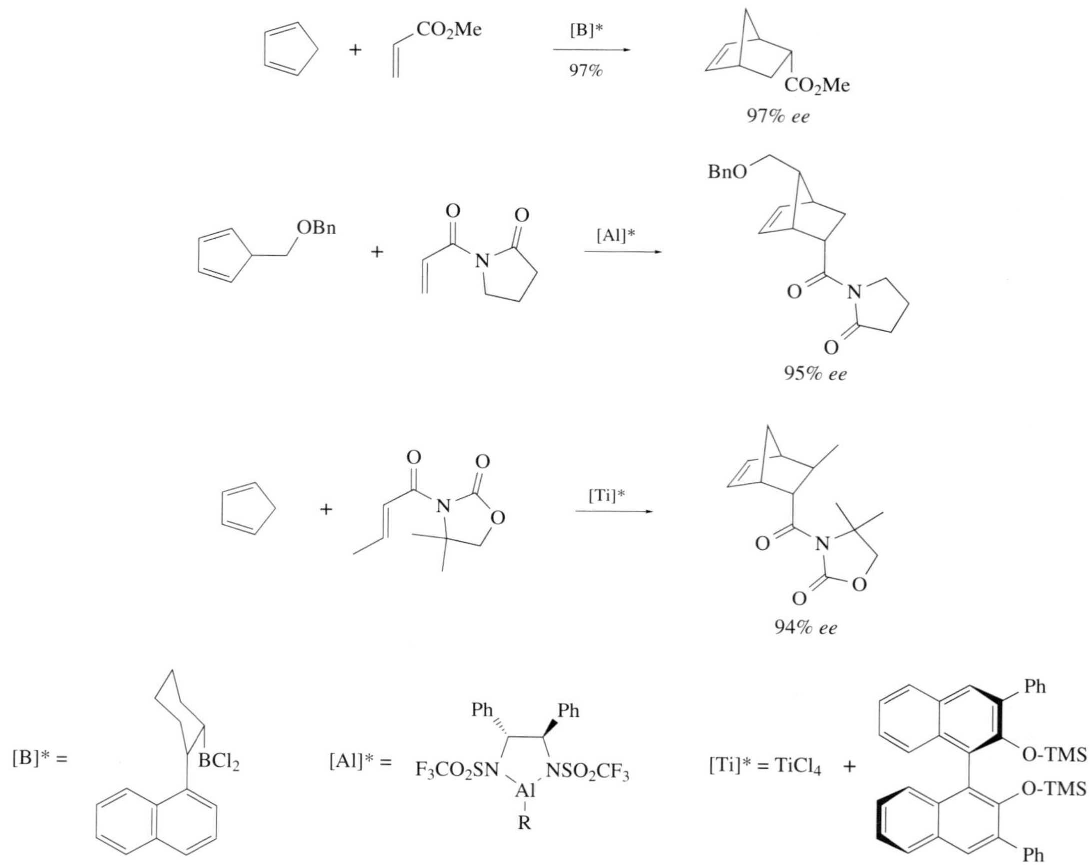

Scheme 55

(ii) Intramolecular additions

The intramolecular Diels–Alder reaction has also found widespread use in synthesis ⟨91COS(5)513⟩. The steric constraints of the tether will usually obviate the problem of regiochemistry. However, mixtures of stereoisomers do result. The example in Equation (104) is typical. As there is a pre-existing chiral centre, *endo/exo* and diastereofacial selection are both involved. The importance of Lewis acid catalysis is demonstrated once again. With dimethylaluminum chloride the reaction proceeds at room temperature and with much higher stereoselectivity ⟨84JOC5277⟩.

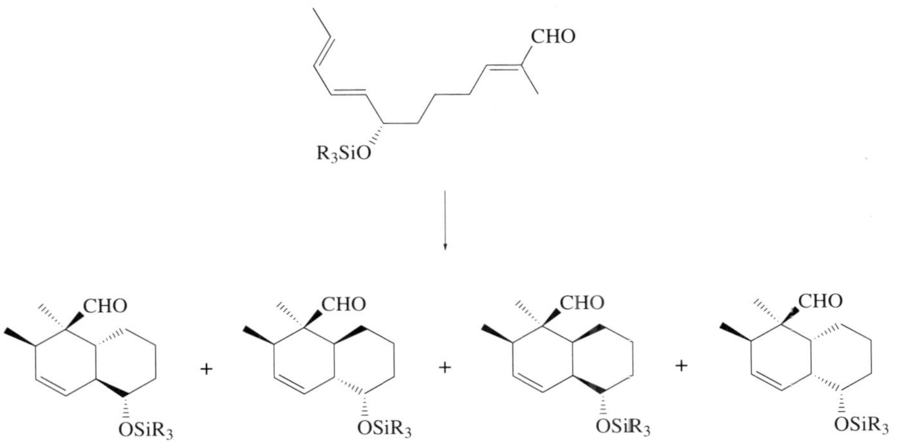

(104)

Thermal	27	:	27	:	33	:	13
$EtAlCl_2$	15	:	75	:	10	:	0

Asymmetry has been induced in intramolecular Diels–Alder reactions by chiral auxiliaries ⟨91COS(5)513⟩, in a similar manner to the intermolecular counterparts, and also by the use of chiral Lewis acid catalysts ⟨92CRV1007⟩. The intramolecular Diels–Alder reaction in Equation (105) illustrates how two new rings and four contiguous new chiral centres can be formed in one reaction with almost complete control of stereochemistry ⟨91TA1305⟩.

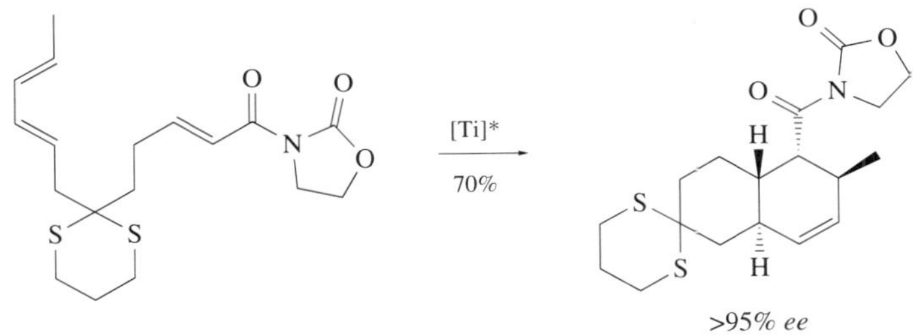

(105)

1.08.2.5.2 [4 + 2] Diels–Alder additions of heterodienes

Diels–Alder reactions of alkenes and heterodienes with two carbon termini (**160**) are not common ⟨91COS(5)451⟩. 2-Azabutadienes have been employed in this connection: Scheme 56 illustrates an elegant method for the preparation of the heterodiene by desilylation of an imide. Addition to maleic anhydride proceeded with high *endo* selectivity to give one major product ⟨82JA1428⟩.

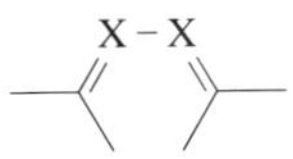

(160)

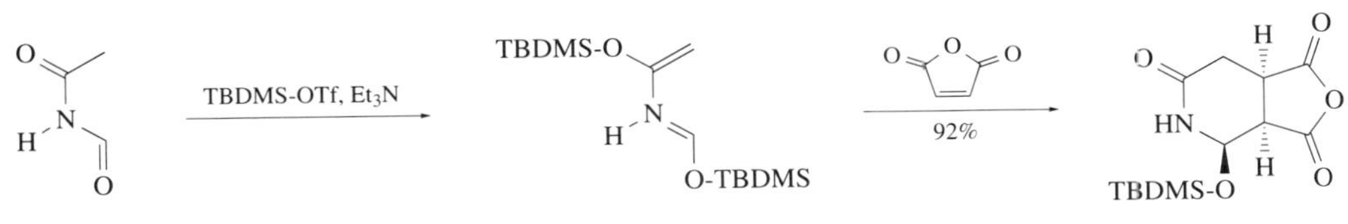

Scheme 56

1.09
One or More CH and/or CC Bond(s) Formed by Rearrangement

IAIN COLDHAM
University of Exeter, UK

1.09.1 TYPES OF REACTION

This chapter will address the preparation of saturated carbon atoms with no attached heteroatoms by rearrangement reactions. Reviews in this area have normally focused on a specific rearrangement reaction, although a number of more general reviews are available ⟨B-63MI 109-01, B-73MI 109-01, B-92MI 109-01⟩. The following sections are arranged according to the type of migrating group. In most cases it is easy to distinguish which is the migrating group and which the backbone and this review will use the conventional terminology in which both the breaking and forming bonds are to the same migrating carbon or hydrogen atom. When the group migrates without its bonding electrons the rearrangement is said to involve the migration of the substituent as a cation; with its bonding electrons the substituent migrates as an anion and with a single bonding electron as a radical. The use of sigmatropic rearrangements and electrocyclic reactions will also be discussed for the preparation of new C—C or C—H bonds.

1.09.1.1 Substituent Migrates as a Cation

The migration of a substituent as a cation (electrophilic rearrangement) is possible by a stepwise mechanism, normally involving initial anion formation, interaction of the anion with the migrating group R and rearrangement to give a new anion (Scheme 1). This type of rearrangement is, however, rare in comparison with the migration of a substituent as an anion. The migrating group R must be one which is able to support an additional pair of electrons such as an aromatic ring.

Scheme 1

If the migrating group R is a saturated carbon atom then a concerted mechanism would necessitate inversion at the carbon-centred R group. The forming C—C and breaking C—C bonds would then have to become colinear in the three-membered ring transition state and this situation would clearly not be favoured. As a result, there are no known examples of such a concerted 1,2-electrophilic rearrangement. The related 1,2-Wittig and 1,2-Stevens rearrangements have been found to involve radical pair intermediates and these rearrangements are discussed under migration of a substituent as a radical in Section 1.09.1.3 and in Chapter 1.18.

1.09.1.1.1 1,2-Electrophilic migration

A new C—C bond can be formed on metallation of 1,1-diaryl-2-haloalkanes or 1,1,1-triaryl-2-haloalkanes, by migration of an aromatic ring. The best method appears to be the use of sodium metal to form the carbanion (**1**) at the migration terminus ⟨57JA4985, 57JA5455⟩. An aromatic group then migrates to give the more stable carbanion (**2**) in which the negative charge is delocalised around the remaining aromatic substituents (Scheme 2). If the reaction is quenched with an alcohol then a new C—C bond (by rearrangement) and a new C—H bond (by quenching) are formed. Quenching with CO_2 is also possible and this sets up two new C—C bonds.

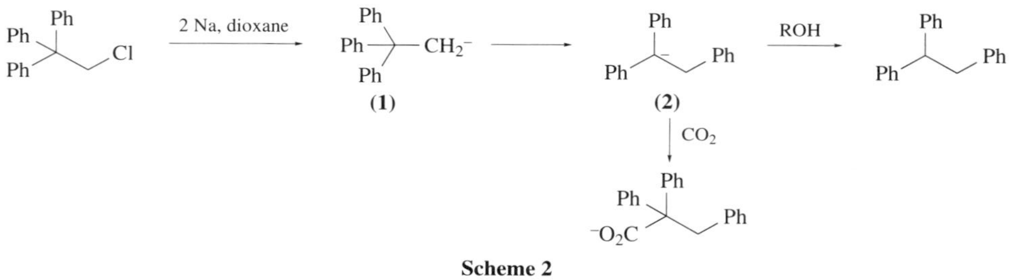

Scheme 2

In addition to the sodium salt it has been shown that this rearrangement can be effected with the lithium or potassium salt of (**1**). The potassium derivative can be prepared from the chloride with potassium metal in DME ⟨61JA412⟩. Rearrangement of the potassium salt of (**1**) occurs at $-50\,^{\circ}C$ whereas the corresponding lithium salt must be warmed to $0\,^{\circ}C$ or higher for any rearrangement to occur. These facts suggest that the reaction is indeed anionic rather than radical as the propensity for rearrangement increases with the increasing ionic character of the carbon–metal bond. Further confirmation of the anionic nature of the rearrangement has been obtained from the chloride (**3**) (Scheme 3). The potassium salt rearranges instantaneously and can be prepared from the dialkylmercury derivative, which in turn can be prepared from the magnesium Grignard of the starting halide (**3**) ⟨61JA1196⟩. Neither the magnesium nor the mercury derivatives undergo rearrangement and the lithium derivative (**4**) must be heated before rearrangement occurs. The new C—C bond forms to the phenyl group rather than the methyl group as the phenyl group is better able to stabilise the negative charge during migration. Rearrangement of 2-phenyl-2-(*p*-tolyl)propyllithium occurs with preferential migration of the phenyl group which lends further support to the carbanion rather than free radical mechanism.

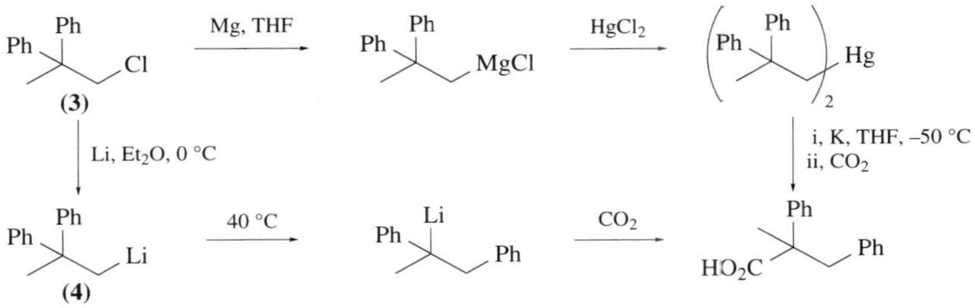

Scheme 3

1.09.1.1.2 *Truce–Smiles rearrangement*

The Smiles and related rearrangements have been reviewed ⟨70OR(18)99⟩. Like the 1,2-electrophilic migrations discussed in Section 1.09.1.1.1, the Truce–Smiles rearrangement allows the formation of a new C—C bond to an aromatic ring ⟨58JA3625⟩. Yields are best with 2,6-disubstituted aryl sulfones such as (**5**) ⟨59JA481⟩. Deprotonation with Bu^{n}Li in ether has been used most commonly and gives a red anion which, after rearrangement is quenched to give the *o*-benzylarenesulfinic acid (Scheme 4).

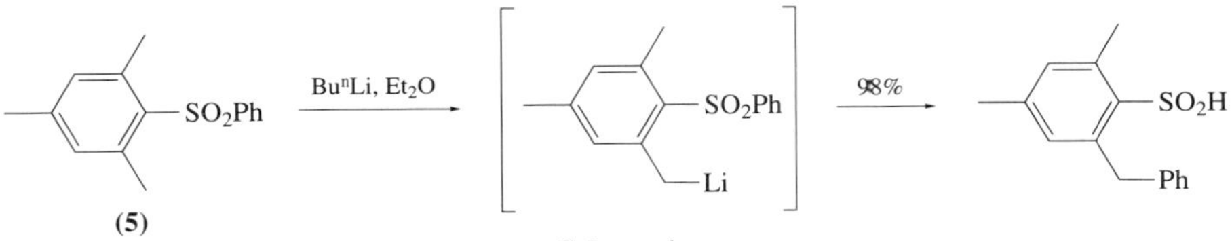

(**5**)

Scheme 4

The migration of a *p*-tolyl group maintains the *p*-tolyl configuration in the product, so demonstrating that attack of the carbanion takes place on the aromatic ring at the carbon atom which is bonded to sulfur. However, the migration of a 1-naphthyl group gives the 2-naphthyl derivative ⟨70JOC1828⟩. Migration of a tertiary alkyl group can be effected under similar conditions ⟨86JA3466⟩ although in this case the mechanism involves radical intermediates.

1.09.1.2 Substituent Migrates as an Anion

Rearrangement reactions which involve formation of a new C—C or C—H bond by migration of a substituent as an anion are relatively common in organic chemistry (nucleophilic rearrangements) ⟨B-60MI 109-01⟩. They can be represented as shown in Scheme 5 with initial cation formation by loss of X$^-$, followed by migration of the R group with its bonding electrons to give a new cation (**6**) at the migration origin. The extent to which the R group migrates before complete ionisation at the migration terminus depends on a number of factors, not least the leaving group ability of X and the stability of the initial cation. Inversion of configuration at the migration terminus is often observed (when this centre is chiral), suggesting some interaction of the R group with the developing cation rather than complete ionisation. When the R group contains a chiral centre at the migrating carbon, then the configuration at this centre is retained completely. Nucleophilic rearrangements give rise to a new C—C or C—H bond at the migration terminus, leaving the more stable, usually more substituted cation (**6**), with the positive charge at the migration origin. This cation can then react with a nucleophile or lose a proton (or other positively-charged species) to give the final rearranged product. If one of the substituents which remains at the migration origin is an alcohol or alkoxide then the new carbocation will be quenched to give a carbonyl group. This is the most common type of substrate for nucleophilic rearrangements and sets up a new C—H or C—C bond α to a carbonyl group.

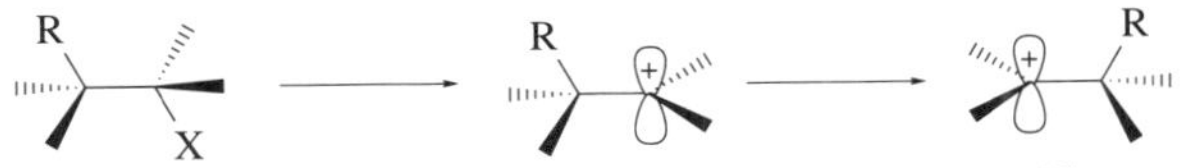

(**6**)

Scheme 5

A discussion of which group migrates in the rearrangement and further stereochemical implications are outlined in the following sections and in Chapter 1.18. A discussion of Wagner–Meerwein rearrangements, pinacol rearrangements, rearrangements of epoxides, semipinacol rearrangements, Favorskii rearrangements and Wolff rearrangements are given, as these illustrate the types of functional group transformations which are possible by nucleophilic rearrangement.

1.09.1.2.1 Wagner–Meerwein rearrangement

In this section attention will focus on the new C—C bond formed and the type of bicyclic ring system generated by a Wagner–Meerwein rearrangement. There are a number of general reviews in this area ⟨B-60MI 109-01, B-63MI 109-01, 85MI 109-01, 91COS(3)705⟩. This section will also cover formation of a new C—H or C—C bond when the migrating bond is not part of a ring system.

The Wagner–Meerwein rearrangement is a nucleophilic rearrangement involving 1,2-migration of a ring carbon atom in a bicyclic or polycyclic molecule. As such it gives rise to a new bicyclic or polycyclic ring system, and so can be a powerful technique for the generation of complicated structures, with C—C bonds in positions difficult to access by conventional chemistry. Original work by Wagner and Brickner centred on the rearrangement of α-pinene (**7**) to bornyl chloride (**8**) (Equation (1)) ⟨1899CB2307⟩ and Meerwein and Van Emster proposed the involvement of cationic intermediates ⟨22CB2500⟩. There have been a number of mechanistic studies (see general reviews and references therein), the most widely accepted involving a two electron, three-centre bond, termed a carbonium ion or non-classical carbonium ion.

(**7**) (**8**) (1)

Wagner–Meerwein rearrangements are best known in terpenoid systems where the presence of branched bicyclic rings promote cationic rearrangements. The initially formed carbocation is normally generated by, either loss of a good leaving group or, protonation of an alkene. Rearrangement then generates a new C—C bond with retention of absolute configuration at the migrating carbon and a new, usually more stable, bicyclic carbocation. This tends to limit the new C—C bond to be part of a five- or six-membered ring. In concerted rearrangements the best alignment would occur when the leaving and migrating groups are antiperiplanar. Such an arrangement is rarely the case in constrained bicyclic systems, however this stereochemical preference allows the prediction that the best aligned bond will be the one that migrates. For example, the *exo*-brosylate (**9**) rearranges exclusively to the *endo*-brosylate (**10**), whereas the *endo*-brosylate (**11**) rearranges quantitatively to the brosylate (**12**) (Scheme 6) ⟨64JA2722⟩. In both cases the four-membered ring is opened, with exclusive migration of the carbon atom of the four-membered ring which is *anti* to the brosylate leaving group. Of the six σ-bonds adjacent to the carbocation (**13**), only two are roughly coplanar with the empty *p*-orbital. Wagner–Meerwein rearrangement gives the bicyclic system (**14**), (Equation (2)) whereas methyl group migration (Nametkin rearrangement) gives the carbocation (**15**) (Equation (3)). Both these pathways generate a new C—C bond and a new carbocation which can undergo further rearrangement or be quenched by attack of a nucleophile or by proton loss.

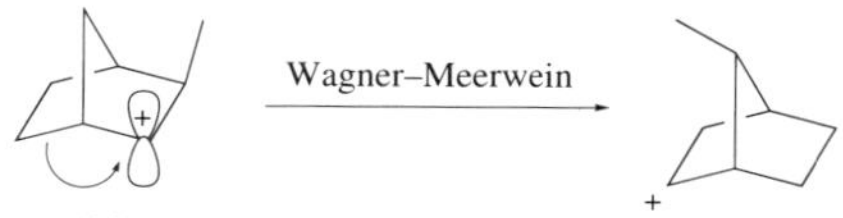

(**13**) (**14**) (2)

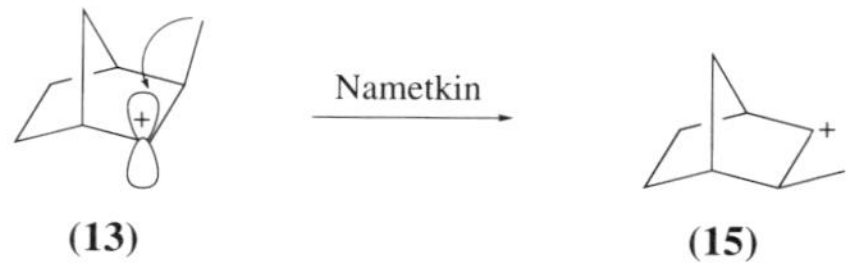

(**13**) (**15**) (3)

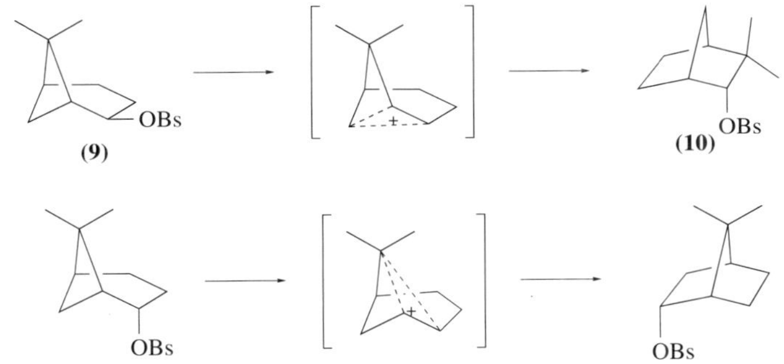

(9) (10) (11) (12)

Scheme 6

The possibility of a number of competing rearrangement pathways may limit the synthetic usefulness of a strategy based on the Wagner–Meerwein rearrangement. Protonation of fenchyl alcohol (**16**) gives rise to a number of rearranged products of which the alkene (**17**) (the product of a Wagner–Meerwein rearrangement) and the alkene (**18**) (the product of a Nametkin rearrangement) predominate (Equation (4)) ⟨76ACR257⟩. In addition, racemisation of optically-pure starting material is possible, particularly if a symmetrical carbonium ion is formed. Treatment of optically-pure norbornyl brosylate (**19**) with acetic acid gives racemic norbornyl acetate (Equation (5)) ⟨52JA1154, 49JA2953⟩. Racemisation can sometimes be avoided by incorporation of an appropriate substituent ⟨76JCS(P1)359⟩.

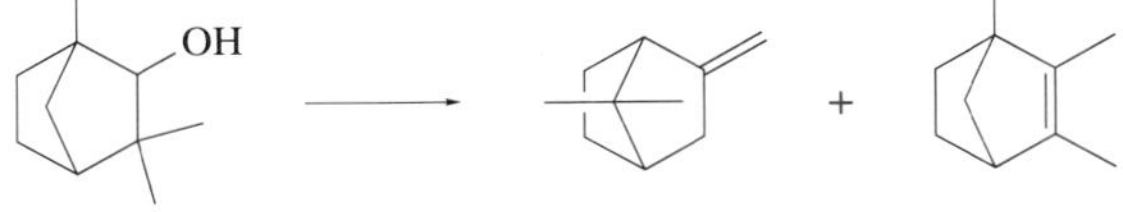

(4)

(16) (17) (18)

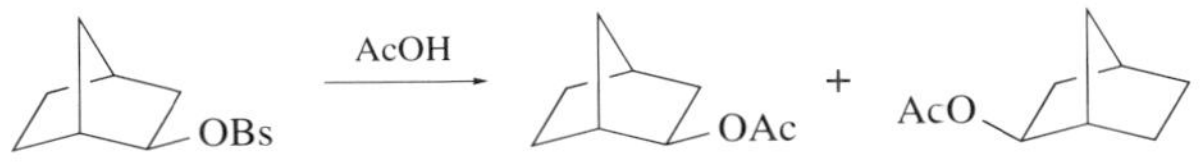

(5)

(19)

By careful consideration of the possible rearrangement pathways, the Wagner–Meerwein rearrangement can be used successfully in syntheses involving the construction of new C—C bonds and complex bicyclic structures. So far in this section, only rearrangements which give the bornane skeleton (a bicyclo[2.2.1]heptane ring system) have been outlined. Wagner–Meerwein rearrangements have been used for the synthesis of other ring systems such as the bicyclooctanes based on the [3.2.1], [2.2.2] and [3.3.0] arrangements. Examples of each of these structural classes will be described.

The bicyclo[2.2.1]heptane ring system has received the most study and has been the subject of mechanistic work. This ring system can be prepared by rearrangement of the bicyclo[3.1.1], [2.2.1] or [3.2.0] carbocations as outlined in Scheme 7. New C—C bond formation by rearrangement of the bicyclo[3.1.1] carbocation has been discussed. The bicyclo[2.2.1] skeleton can also form a useful approach to the bornane system by Wagner–Meerwein rearrangement. For example, rearrangement of the secondary carbocation, generated by tosylation of the alcohol (**20**), takes place to give the more stable tertiary carbocation which loses a proton to give (−)-β-santalene (**21**) (Equation (6)) ⟨74JCS(P1)1938⟩. Formation of a new C—C bond in the bicyclo[2.2.1]heptane system by rearrangement of a bicyclo[3.2.0] carbocation is less common ⟨65JOC4230⟩.

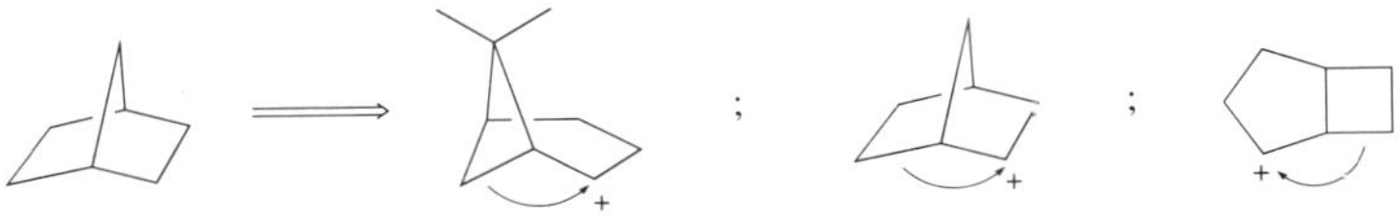

[2.2.1]

Scheme 7

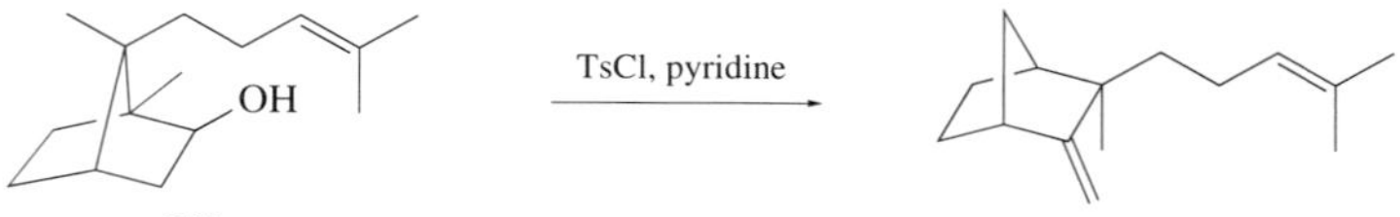

(20) **(21)** (6)

The bicyclo[3.2.1]octane ring system has been prepared from a number of different bicyclic rings as illustrated in Scheme 8. Common to many examples of Wagner–Meerwein rearrangements is the presence of a bridgehead methyl or hydroxyl group to stabilise the new carbocation. For example, rearrangement of the bicyclooctane (**22**) with $BF_3 \cdot OEt_2$ generates the carbocation (**23**), which is stabilised by the methyl substituent ⟨67T793⟩. Suitably activated bicyclo[4.2.0]octane rings (often conveniently accessed by [2 + 2] cycloaddition) can be used to prepare the bicyclo[3.2.1]octane ring system. This approach has been used towards the synthesis of α-caryophyllene, as illustrated in Scheme 9 ⟨64JA1652⟩ and towards the trichothecenes ⟨81JOC3376⟩. New C—C bond formation by rearrangement of the bicyclo[2.2.2] carbocation has also provided a useful route to the bicyclo[3.2.1]octanes ⟨B-63MI 109-01, 82HCA371⟩. Conversely, the bicyclo[3.2.1] carbocation may be used for the construction of the bicyclo[2.2.2]octane, although this is much rarer ⟨72JCS(P1)1274⟩.

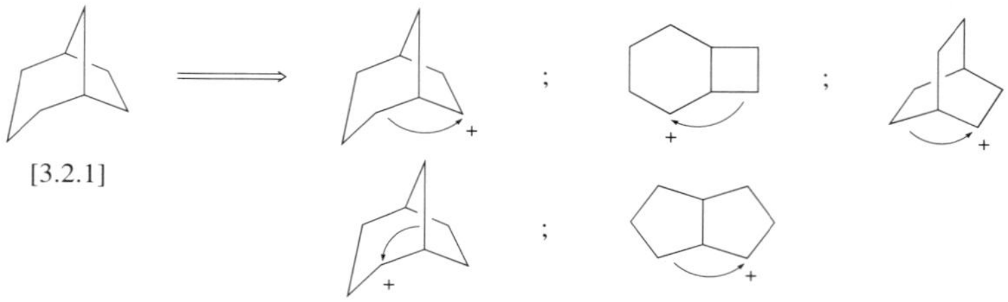

Scheme 8

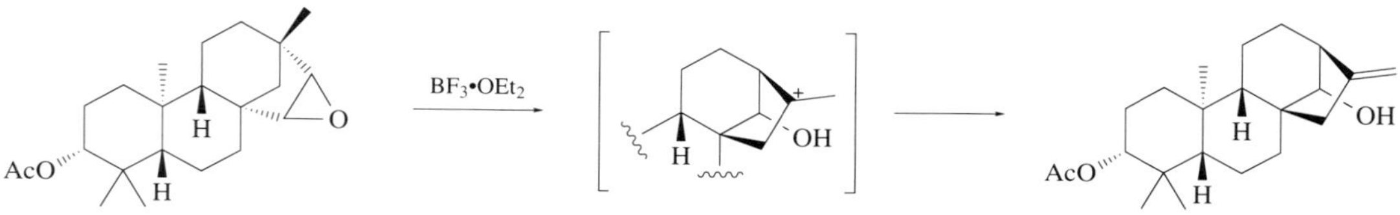

(22) **(23)**

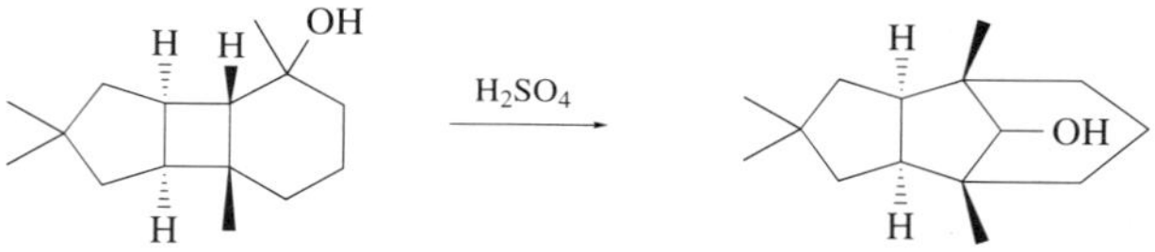

Scheme 9

The formation of a new C—C bond as part of a bicyclo[3.3.0]octane ring can be accomplished from the bicyclo[3.2.1]octane or bicyclo[4.2.0]octane ring systems (Scheme 10). From a bicyclo[3.2.1]octane, ring contraction must take place ⟨81TL4313, 84JOC2937⟩, whereas a combined ring expansion–contraction of a bicyclo[4.2.0]octane gives rise to the [3.3.0] ring system. This type of Wagner–Meerwein rearrangement of the [4.2.0] ring system has been used in the synthesis of isocomene (**24**) (Equation (7)) ⟨81JA82⟩ and quadrone ⟨84JOC4094⟩.

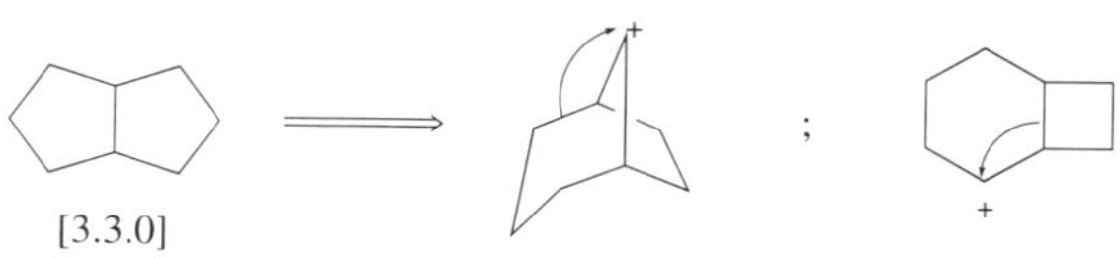

Scheme 10

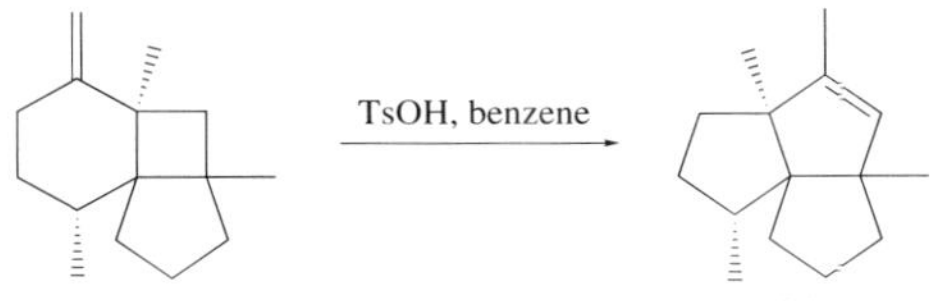

(24) (7)

Other bicyclic ring systems may undergo Wagner–Meerwein rearrangement, as may systems in which the carbocation is not part of the bicyclic framework. These latter cases are best described as simple ring expansions, exemplified by the Demjanov rearrangement shown in Equation (8). Like the Wagner–Meerwein rearrangement, the ring expansion of the cyclopropylmethyl cation is believed to involve nonclassical carbonium ions ⟨59JA4390⟩. Its usefulness in terms of setting up a new C—C bond by rearrangement is limited by the mixture of products normally formed and because scrambling of the carbon atoms occurs. In the related four- to five-membered ring expansion reactions there is no evidence for nonclassical ions ⟨07CB4959⟩.

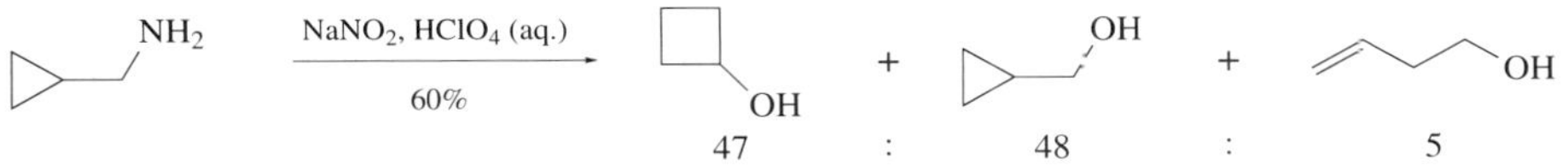

(8)

Related to the Wagner–Meerwein rearrangement is the 1,2-alkyl shift known as the Nametkin rearrangement. This is most commonly observed in ring systems where the original carbocation is generated adjacent to a geminal dialkylated or other tetrasubstituted carbon atom. For example, the alkene (**18**) is formed by dehydration of the alcohol (**16**) (*vide supra*). A combination of Nametkin and hydride shifts occur in steroidal systems, in which an axially-oriented methyl group or hydrogen atom can undergo 1,2-migration to an adjacent carbocation. This can generate a number of new C—C and C—H bonds, exemplified by the rearrangement in acid of the alkene (**25**) to the alkene (**26**) (Equation (9)) ⟨56JCS2419⟩.

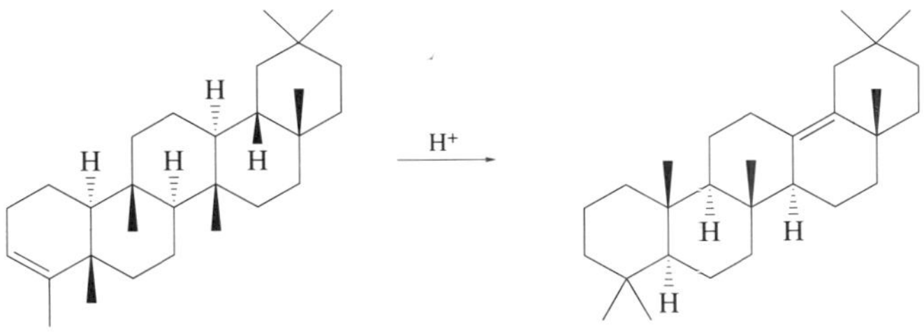

(25) (26) (9)

Finally, in acyclic systems 1,2-migration of an alkyl group to an adjacent carbocation allows a new C—C bond to be formed. The absolute configuration of the migrating group is retained in the rearrangement. This is illustrated in the conversion of the amine (**27**) to the alcohol (**28**) (Equation (10)) ⟨73CB1376⟩. These rearrangements are normally restricted to neopentylic systems. A mixture of products is often generated, because the new tertiary carbocation can be quenched by reaction with a nucleophile or by loss of a proton. These drawbacks can be avoided by using the related pinacol rearrangement, a more controlled and reliable method for C—C bond formation in acyclic systems.

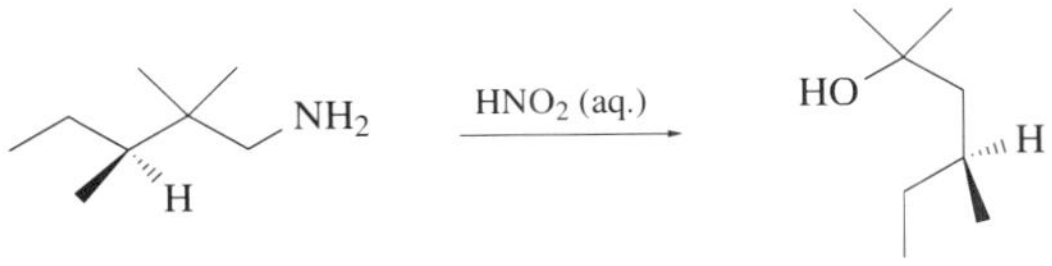

(27) (28) (10)

1.09.1.2.2 Pinacol rearrangement

The pinacol rearrangement of vicinal diols involves a 1,2-shift with formation of a new C—H or C—C bond α to a carbonyl group. Original work by Fittig ⟨1860LA(114)54⟩ used tetramethylethylene glycol (pinacol) which rearranges in sulfuric acid to methyl *t*-butyl ketone (pinacolone) (Equation (11)). The pinacol rearrangement is not restricted to tetrasubstituted diols, but may also occur in a variety of substituted 1,2-dioxygenated systems. A number of reviews are available ⟨60QR357, B-66MI 109-01, 91COS(3)721⟩. Sulfuric acid has been most commonly employed as the catalyst although many other acids (e.g., perchloric acid, phosphoric acid, trifluoroacetic acid, tosic acid) and Lewis acids (e.g., $BF_3 \cdot OEt_2$, TMSOTf (trimethylsilyl triflate)) can effect the rearrangement. The use of cold H_2SO_4 is often desirable as this minimises dehydration.

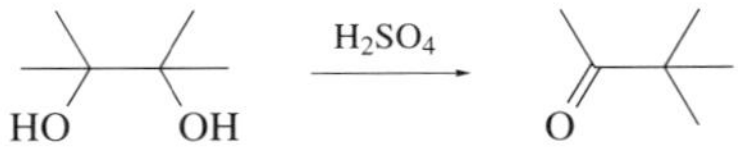

(11)

When the pinacol rearrangement is performed on vicinal diols in which all four substituents are identical there is no ambiguity as to the identity of the migrating group. This type of substrate is therefore particularly popular and often used in model studies. Vicinal diols with two different substituents may give a mixture of two isomeric products on 1,2-migration. Similarly, with three different substituents up to three carbonyl compounds may result and four different substituents may allow up to four new products each with a new C—C or C—H bond α to the carbonyl group. Clearly this mixture of isomeric products may limit the synthetic value of the pinacol rearrangement, unless there is a clear preference for one regioisomeric product. In such cases, when there is more than one type of functional group which may migrate, an understanding of the different migratory aptitudes is important in determining which new C—C or C—H bond is formed. This, however, is not the only factor which must be borne in mind, as the relative stabilities of the two possible carbocations from an unsymmetrical diol may dictate from which carbon centre migration occurs. Indeed this is normally of paramount importance in determining the regiochemistry as $[^{18}O]H_2O$ studies have demonstrated the likelihood of a carbocation intermediate in the pinacol–pinacolone rearrangement ⟨58JCS403⟩.

Rearrangement of tetrasubstituted vicinal diols provides a useful route to α-trisubstituted and spirocyclic ketones. For example, the diol (**29**) prepared by reductive coupling of cyclopentanone, rearranges in sulfuric acid to the spirol[4.5]decane ring system (**30**) (Equation (12)) ⟨54JA2753⟩. The tetrasubstituted vicinal diol (**31**) with two different substituents also gives rise to only one carbocation intermediate regardless of which hydroxyl group is lost. This type of diol can rearrange to two isomeric ketones (**32**) and (**33**) and therefore acts as a direct comparison between the migratory aptitudes of the two substituents R^1 and R^2 (Equation (13)). In general, the group which migrates to form the new C—C or C—H bond is the one which is better able to stabilise a positive charge. Hence aryl groups normally migrate more readily than alkyl groups and at a similar rate to the migration of a hydrogen atom ⟨59JA4733⟩. The presence of electron-donating substituents on the aromatic ring or increased substitution on the alkyl group will enhance the propensity for migration, although conformational and neighbouring group effects may also play a role in determining which group migrates. For example, the ethyl group migrates in preference to the methyl group in the rearrangement of the diol (**34**), although the selectivity is not high (Equation (14)) ⟨22CB1960⟩. In contrast, rearrangement of the diol (**35**) proceeds with exclusive phenyl group migration to give the ketone (**36**), because the phenyl group is better able to support a positive charge via a bridged phenonium ion (Equation (15)) ⟨63JA1548⟩.

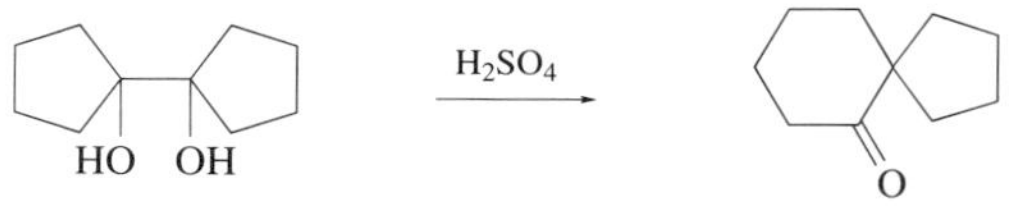

(29) **(30)**

(12)

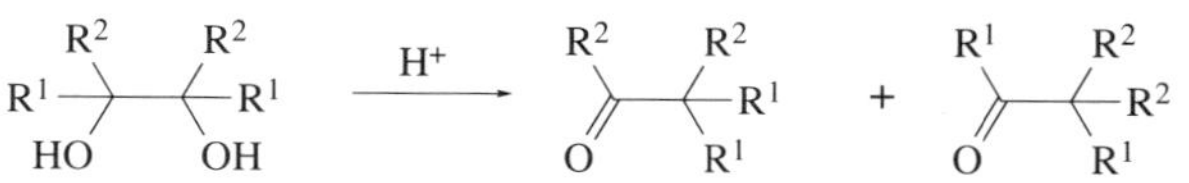

(31) **(32)** **(33)**

(13)

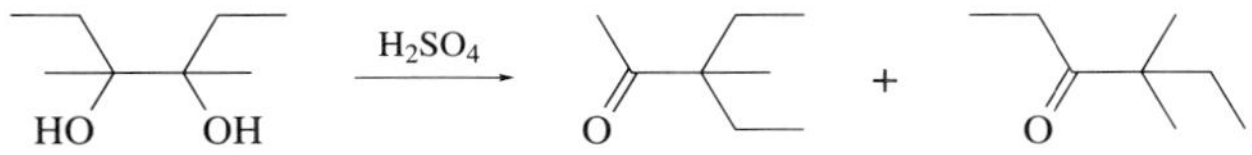

(34) 4 : 1 (14)

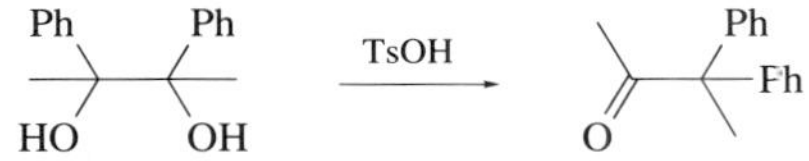

(35) (36) (15)

Although a comparison of migratory aptitudes can be used to predict which of two groups migrate in a symmetrical vicinal diol, in an unsymmetrical diol a consideration of the stabilities of the two intermediate carbocations must also be taken into account. Hence the diol (**37**) rearranges under kinetic conditions (cold H_2SO_4) with methyl group migration as this proceeds through the more stable intermediate carbocation. Under thermodynamic conditions (acetic acid, cat. H_2SO_4) the product of phenyl group migration (**38**) is formed (Scheme 11).

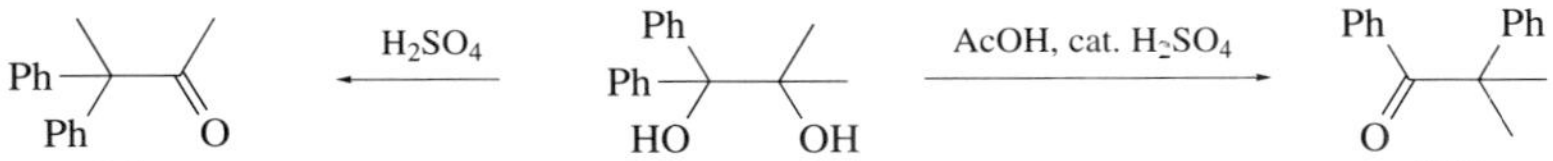

(36) (37) (38)

Scheme 11

Rearrangement of trisubstituted diols provides a route to α-trisubstituted aldehydes or to ketones. It may be possible to control, to some extent, the outcome of such rearrangements by use of appropriate conditions. Under kinetic control with mild, dilute acid the aldehyde may be preferred, whereas harsher conditions often favour the ketone. For example, rearrangement of the diol (**39**) with aqueous oxalic acid gives a significant amount of the aldehyde (**40**), whereas cold, concentrated H_2SO_4 allows essentially exclusive formation of the ketone (**41**) (Scheme 12) ⟨55JA5517⟩.

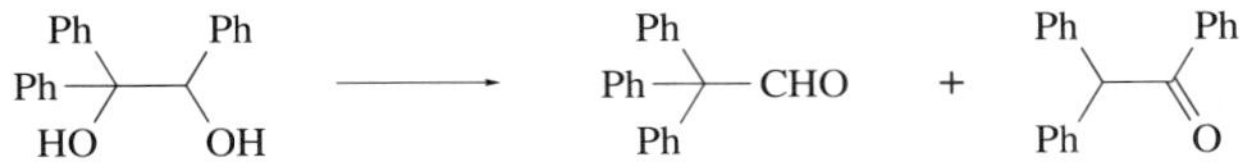

(39) (40) (41)

	(40)		(41)
oxalic acid	30	:	70
conc. H_2SO_4	0	:	100

Scheme 12

The combination of a more stable tertiary carbocation intermediate and the greater preference for hydrogen atom migration over alkyl group migration often leads to a new C—H bond α to a ketone as the major product. Conformational effects may also play a role in determining the outcome of the rearrangement as illustrated in Scheme 13 ⟨59JA1497⟩. Both the *cis*- and *trans*-diols rearrange via the more stable tertiary carbocation but give different ratios of hydrogen atom or carbon atom migration. In addition to the regiochemical control required to effect a successful and useful pinacol rearrangement, the stereochemistry at the migration terminus may also need to be controlled. If the rearrangement proceeds via a planar carbocation then a mixture of products is possible, particularly in acyclic systems. This may further limit the synthetic usefulness of the rearrangement in setting up a new C—C or C—H bond.

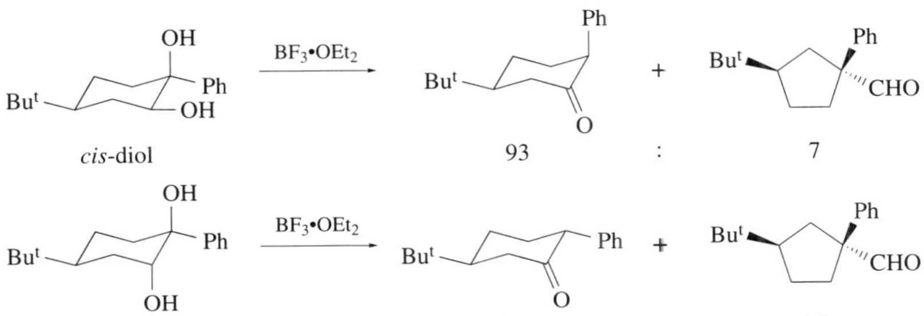

cis-diol 93 : 7

trans-diol 61 : 39

Scheme 13

Rearrangement of vicinal diols in which one centre is a primary alcohol normally gives rise to aldehydic products. This is to be expected based on the formation of the more stable secondary or tertiary carbocation followed by hydrogen atom migration. The product aldehydes may, however, be unstable under the acidic reaction conditions. Conversion of the primary or secondary hydroxyl group of a vicinal diol to a leaving group (e.g., with MsCl or TsCl) allows migration from the more substituted carbon centre. This reversal of regiochemistry (based on carbocation stability) is often a very controlled and useful modification of the pinacol rearrangement and is discussed in Section 1.09.1.2.4 under the semipinacol rearrangement.

1.09.1.2.3 Rearrangement of epoxides

An epoxide may rearrange when activated by protonation or Lewis acid complexation of the oxygen atom of the epoxide group. Ring opening of the epoxide can generate the same carbocation intermediate as that proposed for the pinacol rearrangement (Section 1.09.1.2.2). This carbocation may undergo a 1,2-shift to give a new C—H or C—C bond α to the new carbonyl group. Alternatively, the epoxide may open to give a vicinal disubstituted compound (e.g., a vicinal diol or halohydrin) which may itself rearrange under the reaction conditions. The type of reaction which occurs depends chiefly on the acid used but may also be affected by the substitution pattern, the solvent and the temperature. As in the pinacol rearrangement, the regiochemical and stereochemical outcome must be controlled for successful use in a synthesis. As would be expected, the more stable carbocation will be preferred and the substituent with the higher migratory aptitude will preferentially form the new C—H or C—C bond. The rearrangement of epoxides has been comprehensively reviewed by Rickborn ⟨91COS(3)733⟩. The factors that affect carbocation formation or ring opening to (normally) a vicinal halohydrin and aspects of control in the rearrangement are discussed in this section.

The choice of protic or Lewis acid for the rearrangement of a given epoxide may be crucial in controlling the regiochemical outcome and hence which new C—H or C—C bond is formed. The use of a proton as the acid source may not promote rearrangement at all, as the conjugate base is often nucleophilic enough to allow direct epoxide ring-opening and the formation of a vicinal disubstituted compound. Although this does not preclude the use of an acid catalyst such as H_2SO_4 or TFA, the rearrangement is normally better promoted by a Lewis acid. The choice of Lewis acid is important as the presence of bromide or iodide counterions allow initial halohydrin formation before rearrangement. Lithium perchlorate ($LiClO_4$) or the more commonly used boron trifluoride (BF_3) are both thought to allow initial carbocation formation followed by rearrangement. These reactions are normally carried out by heating in nonpolar solvents.

Like the pinacol rearrangement the regiochemistry is determined by which C—O bond breaks and which substituent (one of four) migrates. This can be predicted on the basis of carbocation stability and migratory aptitude. For example, the epoxide (**42**), when treated with $LiClO_4$, gives the aldehyde (**43**) in which a new C—H bond has been formed by migration of a hydrogen atom to the tertiary carbocation (Equation (16)) ⟨71JA1693⟩. Anhydrous $LiClO_4$ should be used in order to aid solubility. The presence of only small amounts of catalyst and a weakly nucleophilic counterion provide relatively mild conditions.

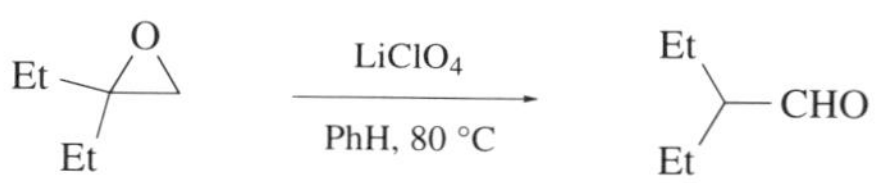

(**42**) (**43**) (16)

The importance of the choice of Lewis acid is illustrated in Scheme 14 which depicts the rearrangement of 1-methylcyclohexene oxide. With $LiClO_4$ the tertiary carbocation intermediate is formed and hydride migration then takes preference over alkyl migration. With LiBr (which requires the solubiliser hexamethylphosphoramide (HMPA)) 2-methylcyclohexanone is not produced, therefore precluding carbocation intermediates. In this case the *trans*-bromohydrins are formed, which can only rearrange by alkyl migration to give the ring contracted products. Hence lithium halide catalysis may reverse the regiochemical outcome, although in acyclic epoxides a mixture of regioisomers is often obtained. Lithium halide catalysis has been shown to promote the rearrangement of spirocyclic epoxides. This is most useful as a route to cyclopentanones as outlined in Equation

(17) ⟨75CR(C)791⟩. In substituted cyclobutanes the new C—C bond is formed to the more substituted centre.

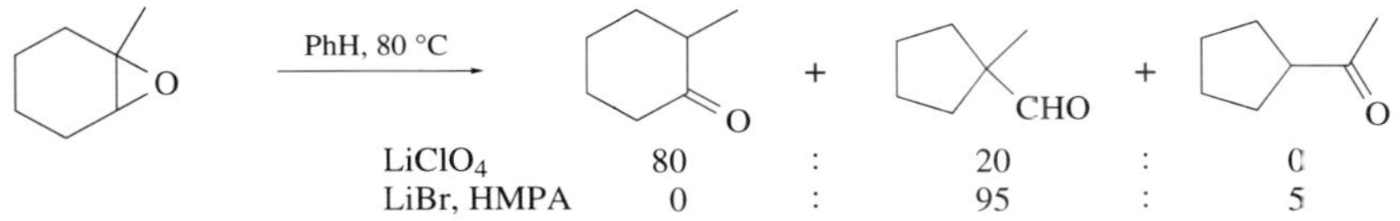

Scheme 14

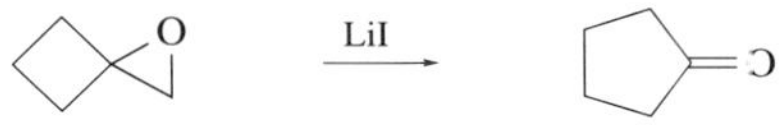

(17)

The relative migratory aptitudes of the two substituents determines which group migrates in the rearrangement. The stereochemistry may also influence the outcome of the rearrangement. The two diastereoisomeric epoxides of 2-butene rearrange to give a different ratio of products. In each case 2-butanone is the major product as expected on the basis of the higher migratory aptitude of a hydrogen atom over a methyl group. However, some isobutyraldehyde, the product of methyl group migration, is observed from the *trans*-isomer (Scheme 15) ⟨69T4445⟩. This can be rationalised by invoking rapid migration to the carbocation intermediate from the first-formed conformer which is correctly set up for 1,2-migration. Rotation of the OBF_3^- group away from the bulkier substituent on the adjacent carbocation allows the prediction of which new C—H or C—C bond will be formed. From the example in Scheme 15, the OBF_3^- group will prefer to rotate away from the methyl group on the adjacent carbocation, so setting up hydrogen atom migration from the *cis*-epoxide but methyl group migration from the *trans*-epoxide. Due to the lower migratory aptitude of a methyl group, further rotation must compete with methyl group migration.

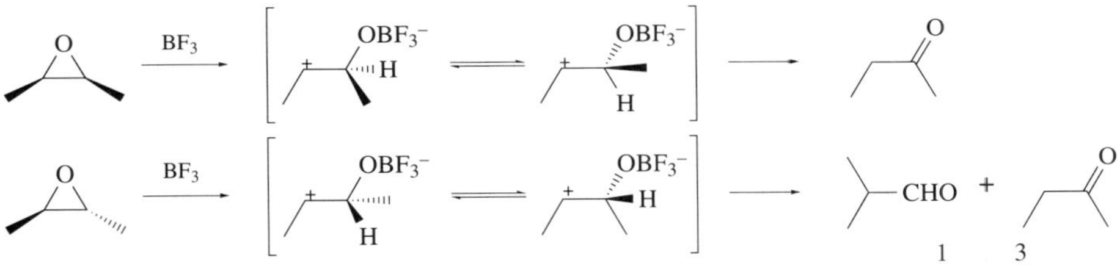

Scheme 15

In cases where epoxide opening is relatively slow and migration is fast, then inversion at the migration terminus would be expected. In practice a mixture of products is often observed with some preference for inversion, suggesting the intermediacy of a carbocation but with some stereochemical memory. Rearrangement of isomeric 1,1-disubstituted epoxides (derived from *exo*-methylene steroids with $BF_3 \cdot OEt_2$ in benzene) results in similar product ratios of both epimeric aldehydes ⟨69T1479⟩. However, work with bulky aluminum catalysts has allowed the synthesis of aldehydes with nearly complete inversion at the migration terminus ⟨92T3749⟩. For example, the epoxide (**44**) rearranged to the aldehyde (**45**) with excellent diastereoselectivity (Equation (18)). The choice of triphenylsilyl as the protecting group for the adjacent alcohol gave the highest diastereomeric excess. Inversion at the migration terminus has also been observed in the rearrangement of acyl epoxides. Acyl group migration in the rearrangement of epoxides (**46**) and (**47**) gave the 1,3-dicarbonyl compounds (**48**) (Equation (19)) and (**49**) (Equation (20)) respectively ⟨78JA1605⟩.

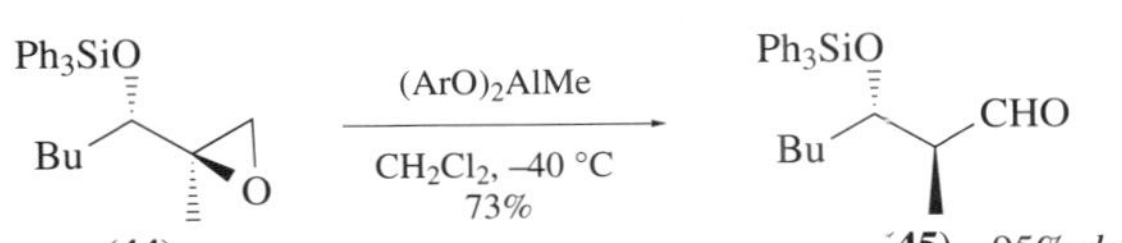

(18)

(**44**) (**45**) 95% *de*

Ar = 4-bromo-2,6-di-*t*-butylphenyl

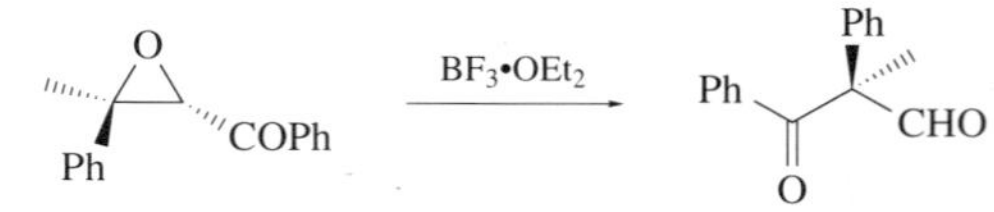

(46) (48)

(19)

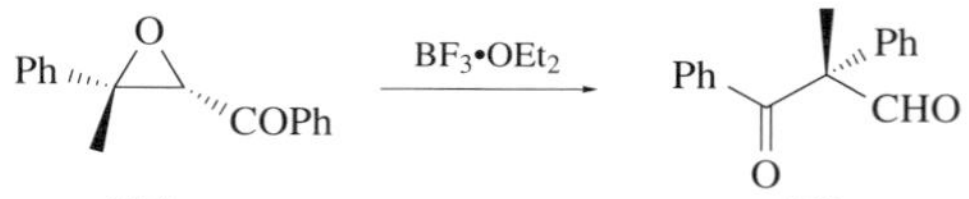

(47) (49)

(20)

Rearrangement of acyl epoxides derived from cyclohexenones provides a convenient access to cyclopentanones by ring contraction (Equation (21)) ⟨85TL985⟩. This can be compared with the ring expansion outlined in Equation (17).

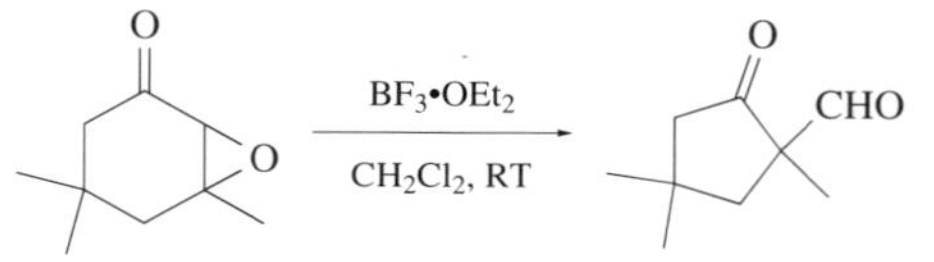

(21)

Lewis acid treatment of an epoxide containing an adjacent quaternary carbon may allow new C—C bond formation by a 1,2-shift from the quaternary centre. This gives rise to a homoallylic alcohol after loss of a proton from the new tertiary carbocation. Epoxide rearrangement promoted by a Grignard reagent usually results in the formation of alcohol products. This occurs via ring opening of the epoxide with the halide, to give the magnesium salt of the halohydrin which then rearranges on heating in a nonpolar solvent. The product aldehyde and/or ketone is then captured by the Grignard reagent to give a secondary and/or tertiary alcohol. For example, rearrangement of cyclohexene oxide with MeMgBr gives the alcohol **(50)** as the major product (Equation (22)) ⟨69CR(C)567⟩. With MeMgI a mixture of the alcohol **(50)** together with substantial amounts of the product derived from hydrogen atom migration is obtained. A very similar ratio of C—C : C—H formation is observed on rearrangement with $MgBr_2$ or MgI_2. The presence of such mixtures in this and other epoxide rearrangements and the absence of a standard set of conditions to promote rearrangement to a single, predictable regio- and stereoisomer often detracts from what is a useful method for the formation of a new C—H or C—C bond α to a carbonyl group.

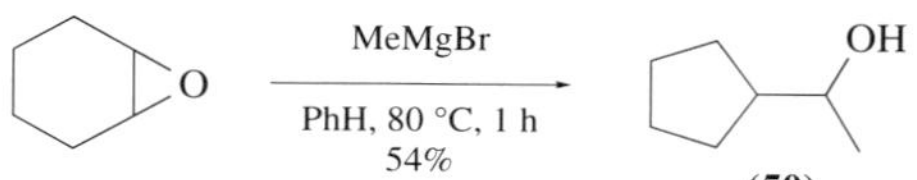

(50)

(22)

1.09.1.2.4 Semipinacol rearrangement

The semipinacol rearrangement has many similarities to the pinacol and epoxide rearrangements and, like these, provides a route to carbonyl compounds with a new C—H or C—C bond α to the carbonyl group ⟨91COS(3)777⟩. The semipinacol rearrangement has come to represent all 1,2-migrations of 2-heterosubstituted alcohols related to the pinacol rearrangement. The conditions for the rearrangement vary with the heteroatom (X in Equation (23)). Most commonly the semipinacol rearrangement refers to the case when X is a nitrogen atom, although many such rearrangements with X being an oxygen, sulfur, selenium or halogen atom are known. The transformation has advantages over the related pinacol and epoxide rearrangements in that it is generally more controlled and proceeds under milder conditions. The regiochemical outcome is more defined because migration normally occurs from the carbon atom bearing the hydroxyl group. The criteria for which of the two groups on this carbon atom migrate can be judged on the basis of their relative migratory aptitudes (see Section 1.09.1.2.2) ⟨59JA4733, 81TL4069⟩, although this is commonly overriden by stereochemical effects or the relief of ring strain. Inversion of configuration at the migration terminus is normally observed and this has obvious value in chiral synthesis. The reaction works best when migration occurs from a tertiary origin to generate a ketone (or ester). Competing epoxide formation can be minimised by avoiding basic conditions or certain leaving groups (e.g., $X = SR_2$).

(23)

The semipinacol rearrangement of a 2-amino alcohol with HNO_2 is known as the Tiffeneau–Demjanov rearrangement ⟨60OR(11)157⟩. The preferred rearrangement conditions involve dissolving the amino alcohol in dilute aqueous acid and adding aqueous sodium nitrite at 0 °C, maintaining a pH 4–6. This generates HNO_2 which reacts with the amine to give an intermediate diazo compound which then loses nitrogen on rearrangement. Addition of acid and nitrite is continued until gas evolution ceases. The reaction is quite effective for ring expansion, as shown in Scheme 16. The diazo intermediate may alternatively be generated by the reaction of a ketone with a diazoalkane. In such cases the rearrangement works best with cyclic ketones and with ether as the solvent, because polar aprotic solvents promote epoxide formation. Ethyl diazoacetate with Lewis acid catalysis may also be employed and provides the homologated β-keto ester.

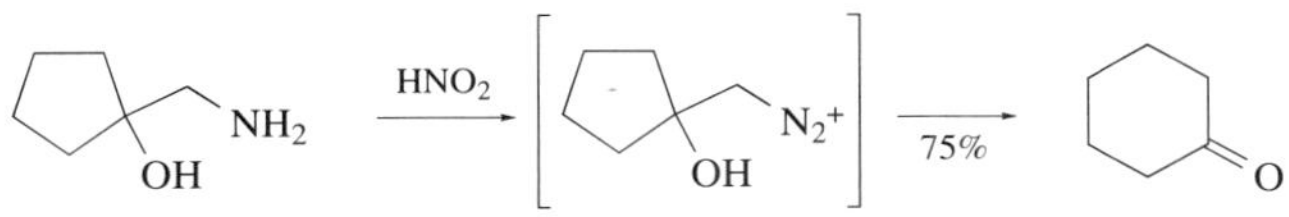

Scheme 16

The more substituted group normally migrates to form the new C—C bond, although conformational effects may take precedence, particularly in cyclic systems where the group which is antiperiplanar to the leaving group becomes involved. This is clearly illustrated in Scheme 17 in which the isomeric 2-amino alcohols give different products on treatment with $HClO_{4(aq)}/NaNO_{2(aq)}$ ⟨65JCS2513⟩. The equatorially-oriented amino group allows ring contraction (Equations (24) and (25)) whereas the axially-oriented amino group forces hydrogen atom migration (Equation (26)) or epoxide formation (Equation (27)), depending upon which group is antiperiplanar to the breaking C—N bond. In each case inversion at the migration terminus has occurred, so discounting carbocation intermediates.

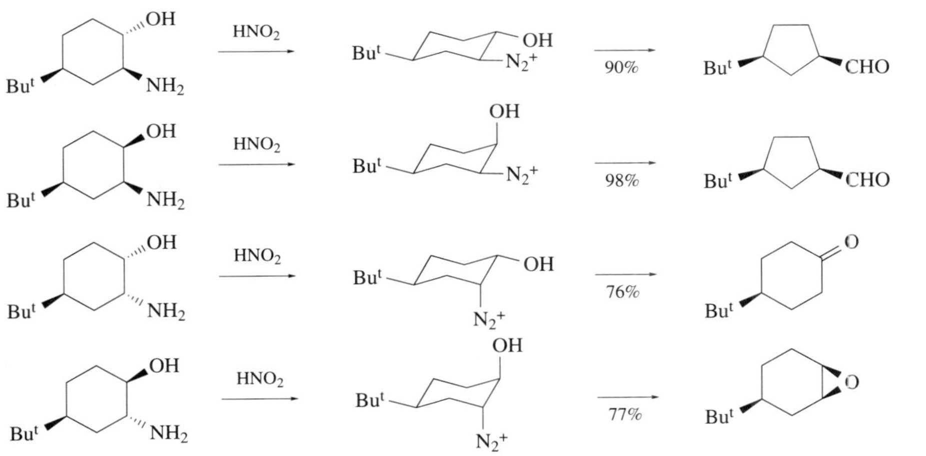

(24)

(25)

(26)

(27)

Scheme 17

A convenient and important modification of the pinacol rearrangement (X = OH, Equation (23)) (see Section 1.09.1.2.2 for a discussion of the pinacol rearrangement) involves the conversion of one of the alcohol groups to, normally, a mesylate ($X = OSO_2Me$) or tosylate ($X = OSO_2Tol$). This provides access, on rearrangement, to complimentary regioisomeric products in which migration takes place from the tertiary to the secondary or primary centre. In the pinacol rearrangement migration takes place to the tertiary centre. A Lewis acid is used to promote the reaction, typically Et_3Al or Et_2AlCl in cold CH_2Cl_2, or $LiClO_4$ in THF. In the latter case the conditions are particularly mild, especially if $CaCO_3$ is added to neutralise the acid formed in the reaction. An example of the use of this transformation by Corey *et al.* is shown in Equation (28) as part of an approach to the sesquiterpene longifolene ⟨64JA478⟩. The complete inversion of configuration at the migration terminus ⟨83TL4997⟩ has been made use of in a number of syntheses, for example to (*S*)-naproxen (Scheme 18), which uses Et_3Al in the rearrangement step ⟨92TA1551⟩.

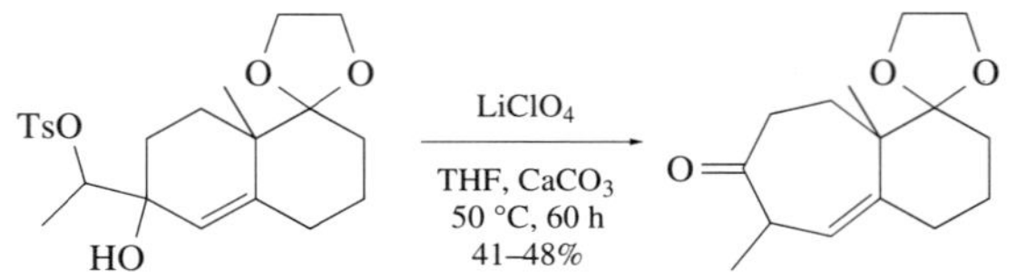

(28)

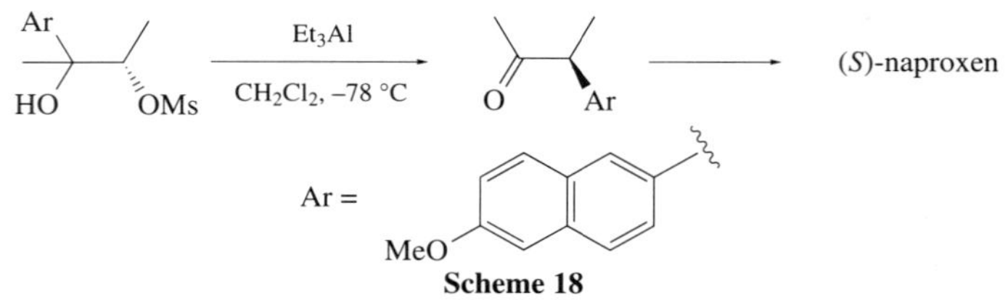

Scheme 18

2-Hydroxy sulfides (X = SR, Equation (23)) and selenides (X = SeR) are known to undergo semipinacol rearrangement, the outcome being determined, at least in part, by the reaction conditions. Protic acids such as TsOH or HBF_4 favour loss of the hydroxyl group as water and migration from the carbon atom bearing the sulfur (or selenium) group. Hydrolysis then gives the ketone with the opposite regiochemistry from the normal semipinacol rearrangement. The potential for episulfonium ion formation and hence products derived from it, restrict C—C bond formation to the expansion of small rings, particularly cyclopropanes, as developed by Trost *et al.* ⟨73JA3068⟩. For example, the ring expansion of the 2-hydroxy sulfide (**51**) occurs with aqueous HBF_4 in ether at room temperature to give the cyclobutanone (**52**) (Equation (29)).

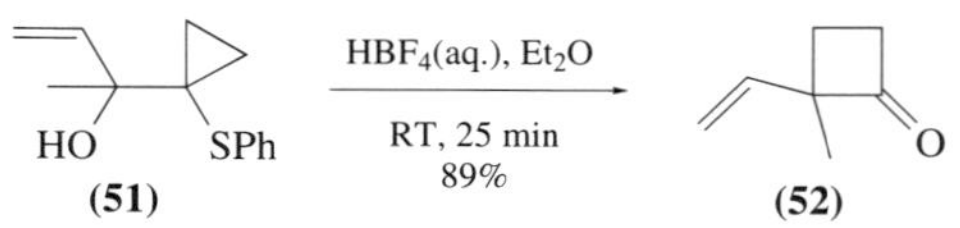

(29)

The use of Lewis acid conditions instead of protic acids favours loss of the sulfur or selenium group, followed by rearrangement. The preferred conditions make use of silver(I) or copper(I) as their BF_4^- or triflate salts, or the use of dichlorocarbene generated *in situ* from TlOEt in $CHCl_3$ ⟨84TL2713⟩. Substituents (but not small rings) at the migration terminus (the carbon bearing the sulfur or selenium group) promote rearrangement. Semipinacol rearrangement with 2-hydroxy sulfones can also be promoted by Lewis acids, in particular Et_2AlCl. This is exemplified in Equation (30) ⟨92JA5432⟩.

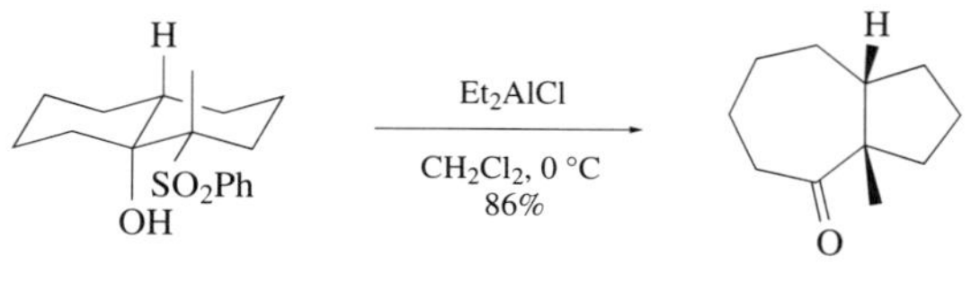

(30)

Semipinacol rearrangement of halohydrins provides another route to α-substituted carbonyl compounds. The hydroxyl proton can be removed with a strong base such as Bu^nLi or RMgX, then rearrangement occurs on heating, normally in nonpolar solvents such as benzene at 80 °C. Epoxides may be formed and this has been implicated in a number of such rearrangements. The halohydrin (**53**) undergoes hydrogen atom migration (as expected) to give a new C—H bond α to the ketone (**54**). The minor product (**55**) presumably arises from the epoxide (Equation (31)) ⟨55JA5083⟩. Halohydrins may be prepared from epoxides using a lithium halide as discussed in Section 1.09.1.2.3.

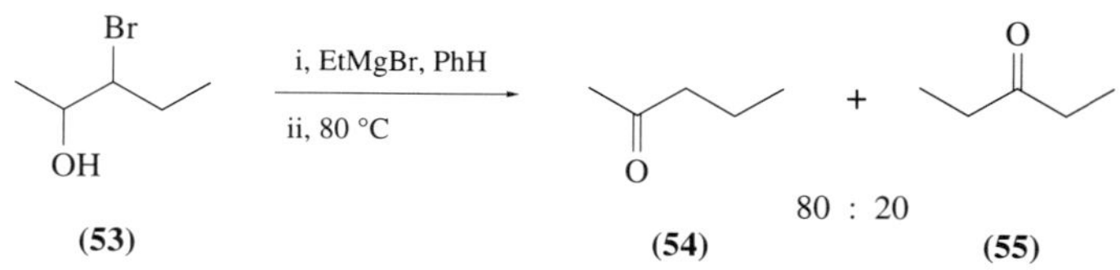

(31)

Rearrangements of halohydrins with a second halogen atom at the migration terminus are known, as are those with a second oxygen-bound substituent at the migration origin. The latter (2-halo acetals) provide access to α-substituted esters under particularly mild conditions, especially when the migrating group is electron-rich. Acetals such as (**56**) rearrange on heating in alcohol solvent (e.g., 1,2-ethanediol) containing a mild base (e.g., sodium acetate) to avoid acetal hydrolysis (Equation (32)) ⟨85S505⟩. Silver tetrafluoroborate in methanol or other Lewis acids may be used to promote the rearrangement ⟨84AG(E)413⟩.

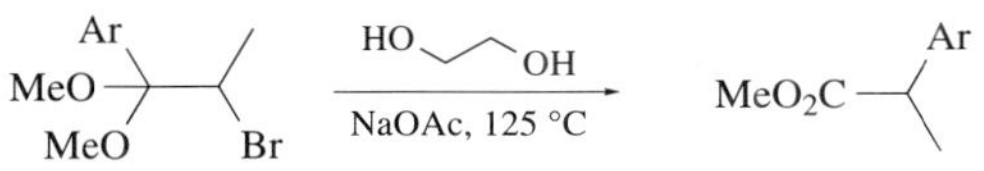

(**56**)

(32)

The presence of a good leaving group vicinal to an alcohol is implicated in the semipinacol rearrangement. Inversion at the migration terminus normally occurs unless a carbocation intermediate is formed in which rotation is faster than migration. A carbocation may be formed from an alkene, as in the cyclisation onto an acetal (Scheme 19) ⟨89JA1514⟩. Cyclisation occurs on treatment with a Lewis acid, particularly $SnCl_4$ in CH_2Cl_2, to give the carbocation which rearranges to the ketone.

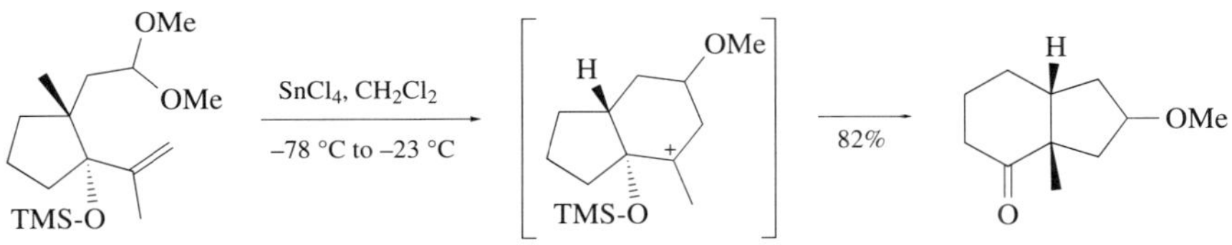

Scheme 19

1.09.1.2.5 Favorskii rearrangement

The Favorskii rearrangement is related to the rearrangements described in this section as the overall transformation involves the migration of a substituent as an anion to give a new C—C bond α to a carbonyl group. In the Favorskii rearrangement the starting material is an α-halo ketone which is transformed, via carbanion intermediates, into a carboxylic acid or its derivative. A simplified representation of the reaction is outlined in Equation (33), although the situation is more complicated than depicted as up to two regioisomers of the product may result. The scope and limitations of the rearrangement, named after early work by Favorskii ⟨1894MI 109-01⟩, are described in this section and in a number of reviews ⟨50BSF35, 60OR(11)261, 70RCR732, 91COS(3)839⟩.

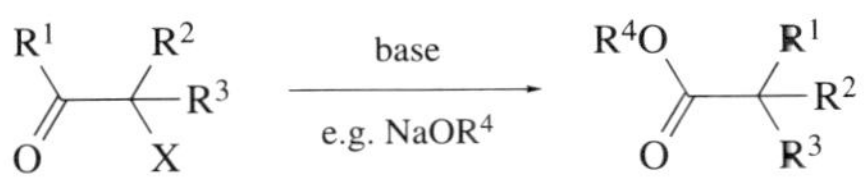

(33)

The Favorskii rearrangement is most useful for the preparation of branched-chain aliphatic carboxylic acids and their derivatives and for ring contraction of cyclic ketones. In acyclic systems the reaction works best when the carbon atom bearing the halide is substituted. The conditions for the rearrangement may vary depending on the structure of the α-halo ketone. Most commonly sodium methoxide in methanol or in diethyl ether is used. Other bases such as hydroxides or carbonates of Group 1 metals may be used, although this may lead to hydroxy ketone formation. There are examples however where these bases are particularly appropriate, such as the rearrangement of 2-chlorocycloheptanone to cyclohexanecarboxylic acid which occurs in 69% yield with hot aqueous K_2CO_3, or the rearrangement of some 17-bromo-20-ketosteroids. The use of ammonia or other amines has been reported but is of limited scope. The choice of alkoxide may affect the yield and the use of NaOEt or $NaOPr^i$ (or NaOBn) may have some benefits over NaOMe (Equation (34)) ⟨40JA2590⟩.

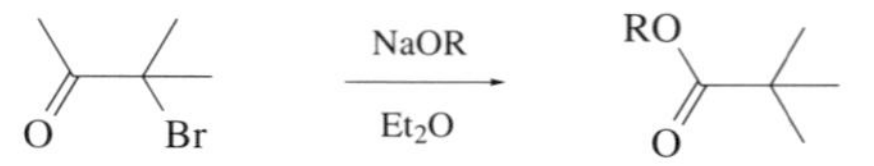

(34)

R = Me, 39%
R = Et, 61%
R = Pr^i, 64%

Usually the α-halo ketone is added to a fairly concentrated solution or suspension of the alkoxide at 0 °C or room temperature in an alcohol or ethereal solvent. Heterogeneous reactions (e.g., NaOMe in Et_2O) require longer reaction times and may be accelerated by heating. Excess alkoxide may improve the yield of the rearranged product. α-Bromo ketones are more prone to reduction or disproportionation in the presence of strong bases than α-chloro ketones, but both have been used extensively for Favorskii rearrangements.

The two isomeric α-chloro ketones (**57**) and (**58**) rearrange to the same ester on treatment with NaOMe in Et_2O (Scheme 20) ⟨51JA3900⟩. Such examples, together with the work of Loftfield using ^{14}C labelled α-chloro ketones ⟨51JA4707⟩ demonstrate the involvement of a common intermediate, generally accepted to be the cyclopropanone.

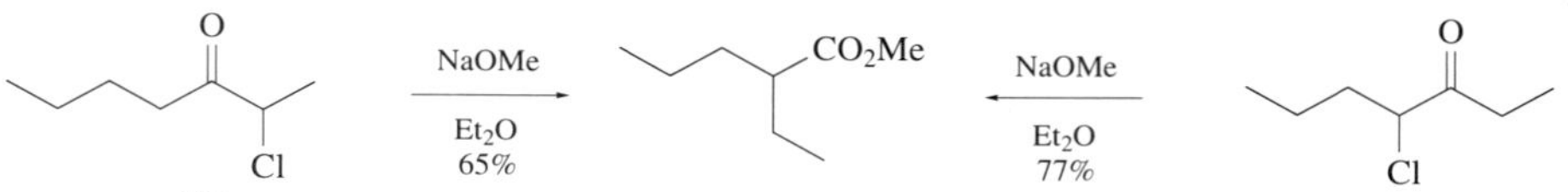

(**57**) (**58**)

Scheme 20

Enolate formation is thought to be followed by cleavage of the carbon–halogen bond to give either the oxy-allyl cation, or directly the cyclopropanone (Scheme 21). The three-membered ring breaks down to give the more stable carbanion (primary > secondary > tertiary) and this determines the regiochemical outcome.

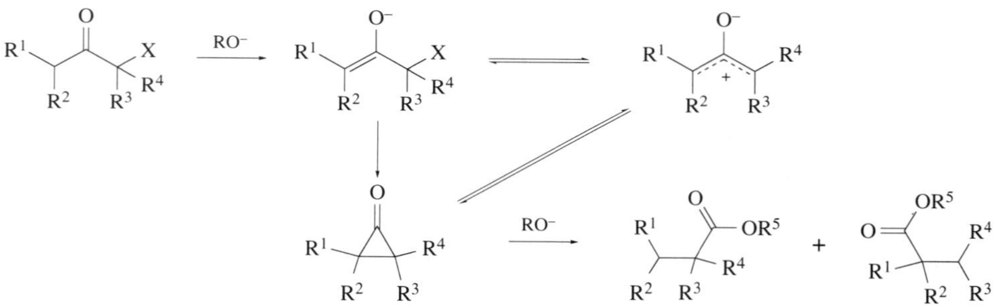

Scheme 21

The stereochemical outcome at the new carbon centre α to the carbonyl group must also be taken into account. If the rearrangement proceeds via the oxy-allyl cation then a mixture of stereoisomers should result. If direct displacement of the halogen occurs then inversion at this centre should be observed. The pathway taken seems to be affected by the structure of the α-halo ketone and by the solvent. The diastereomeric α-chloro ketones (**59**) and (**60**) rearrange stereospecifically in Et_2O to the esters (**61**) (Equation (35)) and (**62**) (Equation (36)) respectively ⟨60JA4307⟩. However, in methanol the reaction is nonstereospecific, suggesting that the more polar, protic solvent favours the intermediate oxy-allyl cation. Steric factors may also affect the stereochemistry if direct ring closure to the cyclopropanone is hindered, thereby favouring oxy-allyl cation formation.

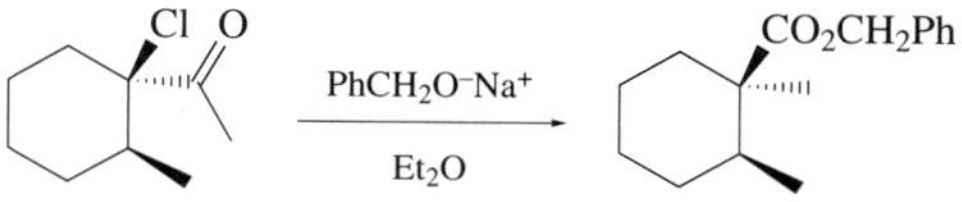

(**59**) (**61**)

(35)

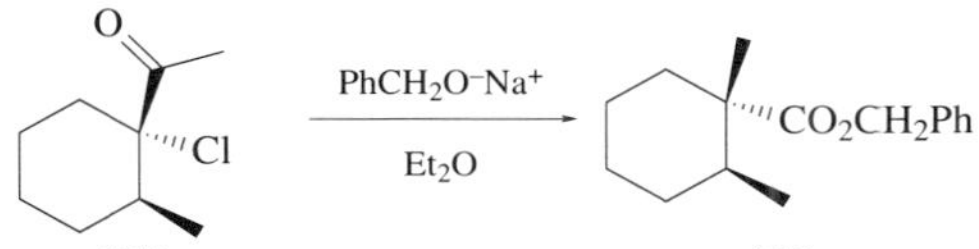

(60) **(62)** (36)

The sensitivity of α-halo ketones makes them prone to the formation of a number of side products. These may include α-alkoxy ketones, unsaturated ketones and epoxyethers (especially from aryl substituted α-halo ketones) and products derived from them. Epoxyether formation is the main pathway in the attempted rearrangement of α-bromo ketone (**63**), although the corresponding chloride (**64**) can be induced to rearrange on careful exclusion of water ⟨54JA35⟩. The isomeric α-bromo ketone (**65**) rearranges easily and ether is the preferred solvent (Scheme 22) ⟨50JA2884⟩.

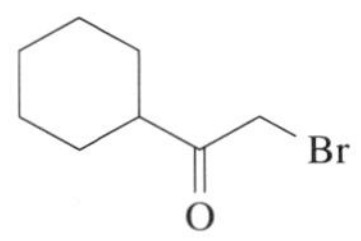

(63)

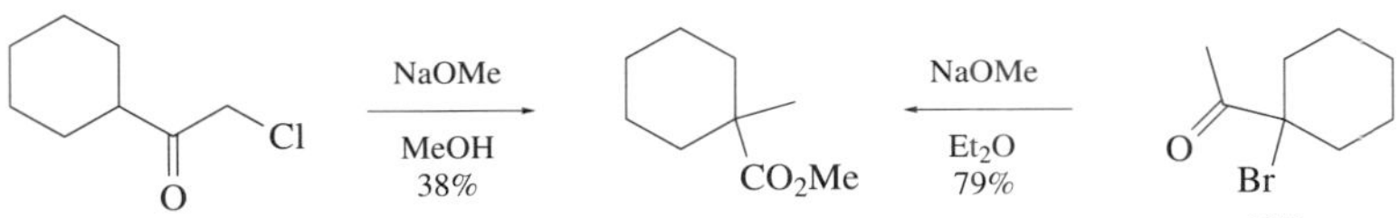

(64) **(65)**

Scheme 22

The application of the Favorskii rearrangement to ring contraction has been particularly successful for the preparation of three– or five–ten-membered rings. The preparation of four-membered rings from 2-halocyclopentanones usually fails except in certain constrained systems such as those leading to the cubanes. Ring contraction of 2-chlorocyclohexanone can be accomplished by heating with NaOMe in Et_2O as outlined in Equation (37) ⟨59OS(39)37⟩. The preparation of cyclopropanecarboxylic acids and their derivatives can be achieved from 2-halocyclobutanones using NaOH, KOH or NH_3, and proceeds with inversion at the carbon bearing the halogen atom. This is believed to occur via a semipinacol-type (or semibenzilic) pathway rather than via enolate and cyclopropanone formation. This mechanism is also thought to occur in the rearrangement of α-halo ketones which have no α′-hydrogen atom.

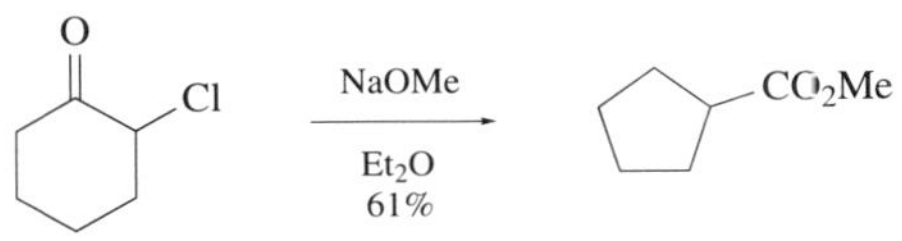

(37)

While α-halo ketones can be transformed into α-substituted carboxylic acids, esters or amides, dihalo or polyhalo ketones normally give, on rearrangement and elimination, unsaturated carboxylic acids and their derivatives.

1.09.1.2.6 Wolff rearrangement

Like the Favorskii rearrangement, the Wolff rearrangement leads to the formation of a new C—C bond α to a carboxylic acid or its derivative. The starting material for the rearrangement is an α-diazo ketone and the discovery of its transformation into carboxylic acids and derivatives was made by Wolff ⟨12LA(394)23⟩. Many α-diazo ketones have been shown to undergo the rearrangement and several reviews are available ⟨42OR(1)38, 60AG535, 67RCR260, 75AG(E)32, 91COS(3)887⟩. The overall process is outlined in Scheme 23. In most cases R^1 is a hydrogen atom and the α-diazo ketone can be prepared by the reaction of an acid chloride (RCOCl) with diazomethane. Other methods for the preparation of α-diazo ketones, such as the transfer of a diazo group from tosyl azide to an enolate, are given in Chapter 3.09

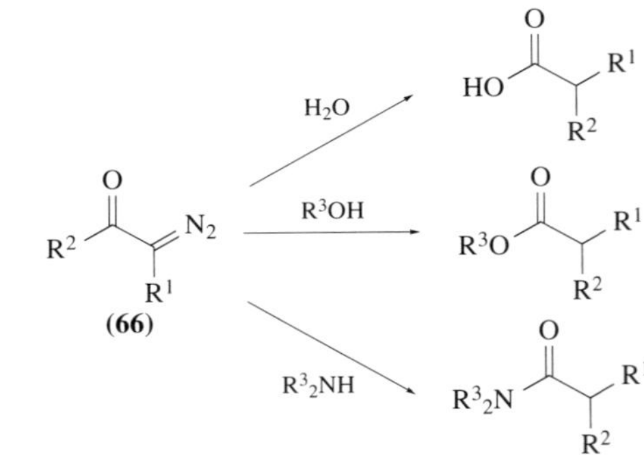

Scheme 23

When the rearrangement occurs with the α-diazo ketone (**66**; $R^1 = H$), the reaction allows the one-carbon homologation of the carboxylic acid (RCO_2H) in a three-step process called the Arndt–Eistert synthesis ⟨35CB200⟩. This involves the conversion of the carboxylic acid (RCO_2H) to its acid chloride (see Chapter 5.01), which in turn is converted to its α-diazomethyl ketone using diazomethane. Rearrangement then gives the homologated acid (RCH_2CO_2H) or its derivative. The conditions for the rearrangement normally involve heating the diazo ketone in an appropriate solvent (depending upon which acid derivative is required), often in the presence of a transition metal catalyst, the most effective and popular being silver oxide or occasionally silver benzoate or silver nitrate. Photolysis of the α-diazo ketone using a medium-pressure mercury arc is an alternative and effective method, permitting the use of lower reaction temperatures to promote rearrangement.

The mechanism of the Wolff rearrangement may depend on the substrate and the conditions of the reaction. Carbene intermediates have been postulated although these will no doubt be associated with any metal present in the reaction medium. Migration of the R group in the α-diazo ketone (**66**) is followed by ketene formation. The ketene is trapped by water, an alcohol, a thiol or an amine to give the product carboxylic acid, ester, thioester or amide. The R group is known to migrate with retention of configuration, as illustrated in Equation (38) ⟨41JA1674⟩. Note however that racemisation may occur at high temperatures or in the presence of a catalyst, if the chiral centre contains an enolisable hydrogen atom. Racemisation can be avoided by using photochemical conditions ⟨83JCS(P1)2387⟩. Aqueous dioxane is commonly employed as the solvent when a carboxylic acid is required. A mild base such as sodium carbonate, sodium thiosulfate or a tertiary amine is often used in silver catalysed rearrangements. If the yield of the carboxylic acid is only mediocre then this may be improved by resorting to ester or amide formation.

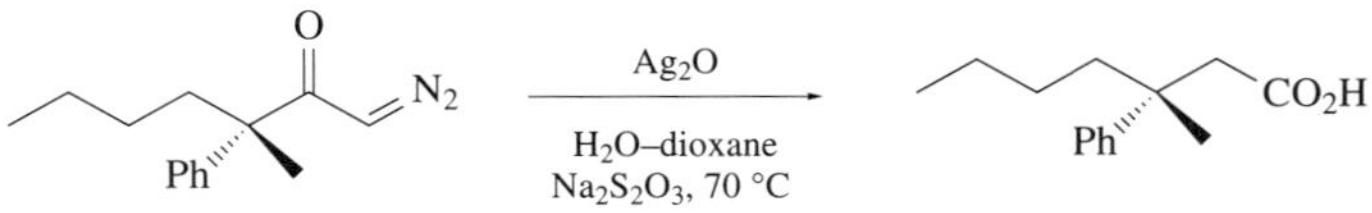

(38)

For the preparation of esters, the α-diazo ketone may be heated with silver oxide in an alcoholic solvent such as methanol. Alternatively, photochemical activation in an alcoholic solvent is a useful method for rearrangement to the ester. The use of metal catalysis is not always necessary and indeed detrimental for α-substituted diazo ketones. For example, the diazo ketone (**67**) did not rearrange with silver oxide in methanol. However, excellent yields of rearranged product (**68**) can be obtained by heating in benzyl alcohol (Equation (39)). High temperatures (170–190 °C) and the presence of a tertiary amine such as isoquinoline gave the highest yields ⟨48JOC763⟩.

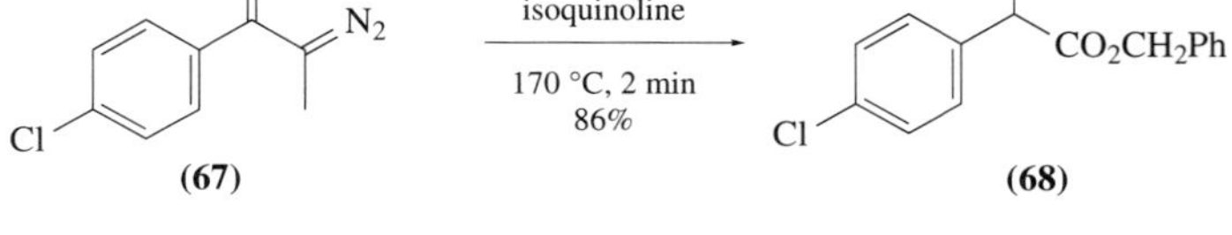

(39)

The formation of a new C—C bond at a saturated carbon atom α to a carboxylic amide may be accomplished by Wolff rearrangement in the presence of ammonia or an amine. As with the formation of carboxylic acids or esters, one of three general methods may be employed. Addition of aqueous ammonia or an amine and aqueous silver nitrate to the diazo ketone in dioxane followed by heating gives the amide. Without silver catalysis higher temperatures are needed. In these cases rearrangement may be effected by using the amine (e.g., aniline) as the solvent. Alternatively, photolysis of the α-diazo ketone in the presence of the amine in ether or benzene leads to the carboxylic amide. For example, the simple α-diazo ketones **(69)** rearrange on photochemical activation to give the amides **(70)** (Equation (40)) ⟨59CB528⟩.

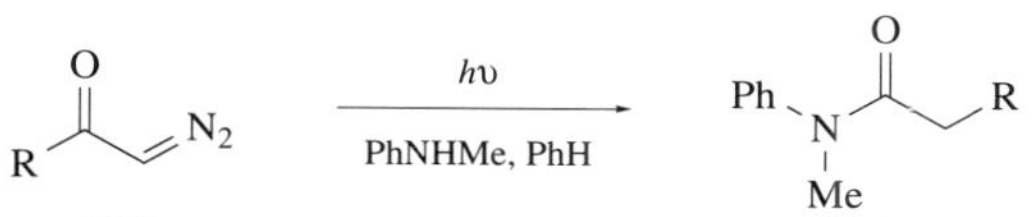

(69) **(70)** (40)

R = Me, 45%
R = Ph, 77%
R = CH_2Ph, 80%

Ring contraction of cyclic α-diazo ketones is possible, particularly under photochemical activation ⟨55CB934⟩. Most examples involve the formation of cyclobutane or cyclopentane carboxylic acids or their derivatives. Contraction of larger rings is possible although side-products may predominate, depending upon the conditions used. The cyclic α-diazo ketones (**71**; $n = 1$–6) rearrange smoothly under photochemical conditions (Equation (41)) ⟨69CB3877⟩. Using silver oxide catalysis, the major product is the unrearranged α,β-unsaturated ketone formed by 1,2-shift of a hydrogen atom.

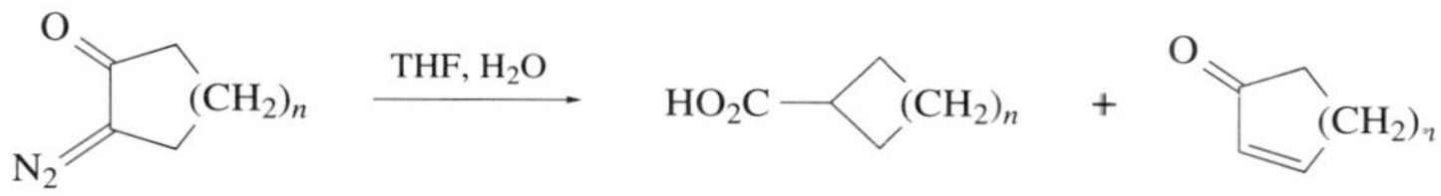

(71) (41)

hν	(n = 1–6)	62–95%	–
Ag_2O	(n = 3–6)	2–15%	54–70%

The reactivity of α-diazo ketones makes them prone to competing reaction pathways. These may include 1,3-dipolar cycloaddition, diazo-coupling, cyclopropanation, hydrogen atom abstraction (giving products of insertion) or reaction with a soft nucleophile (giving ylide formation). Careful choice of reaction conditions may avoid these side products and provide a useful method for C—C bond formation by the Wolff rearrangement.

1.09.1.3 Substituent Migrates as a Radical

The formation of a new C—C or C—H bond by rearrangement with the migration of a neutral species is less common than when the substituent migrates as an anion. Migration of a substituent as an anion normally occurs from a carbocation or other electron-deficient species, while a substituent may migrate as a radical from a carbanion or a radical species. When carbanions are involved, such as the 1,2-Wittig and 1,2-Stevens (including sulfur ylide) rearrangements (see Sections 1.09.1.3.1 and 1.09.1.3.2), dissociation into the radical and radical anion takes place (Scheme 24). Rapid migration of the radical and recombination with the radical anion gives the rearranged product.

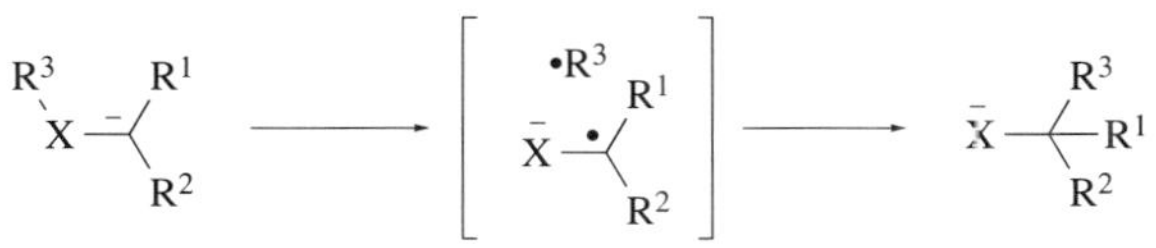

Scheme 24

Alternatively, the formation of a free radical may allow the migration of a substituent as a radical. Such rearrangements can be represented by the 1,2-shift of the R^3 group of the radical **(72)** (Equation (42)) ⟨B-63MI 109-01, B-65MI 109-01⟩. Free radicals are rapidly destroyed by bimolecular reactions leading to either coupling or disproportionation. The 1,2-shift generates a new radical which is in turn capable of further reaction. Free radicals can be generated in a number of ways but show little tendency to rearrange, except when a vicinal aromatic group is present which may undergo a 1,2-shift as illustrated in Scheme 25 ⟨47JA2916, 52JA5381⟩. Dilution to a 1 M solution in chlorobenzene generally gives improved yields of the rearranged product. By comparison, no rearrangement of Me_3CCH_2CHO in *t*-butyl hydroperoxide (tbhp) occurs and in general 1,2-shifts of an alkyl group or a hydrogen atom do not occur. This has made possible the controlled chemistry of carbon radicals for C—C and C—H bond formation as described in this volume, (e.g., chapter 1.08.1).

(72)

(42)

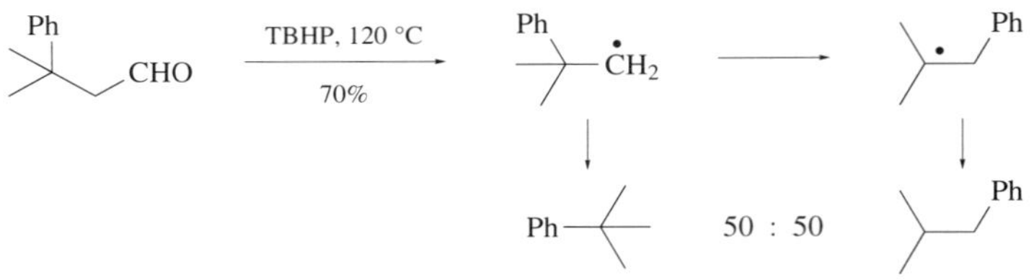

Scheme 25

1.09.1.3.1 1,2-Wittig rearrangement

The migration of a substituent as a radical and the formation of a new C—C bond by rearrangement of α-alkoxy carbanions was developed by Wittig ⟨42LA(550)260, 47LA(557)205⟩ after early work by Schorigin ⟨24CB1634, 25CB2028⟩. The transformation is outlined in Scheme 26 and involves the formation of a carbanion, normally by direct deprotonation with strong base, α to the oxygen atom of the ether. Rearrangement gives the alcohol product ⟨70AG(E)763, 91COS(3)975⟩.

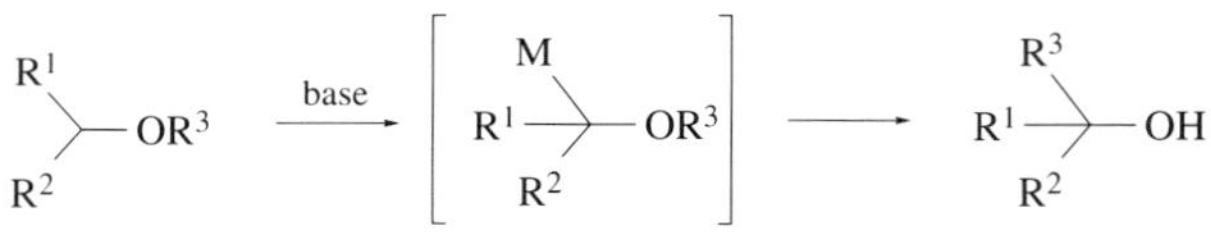

Scheme 26

Initial studies on benzyl ethers **(73)** used either phenyllithium in THF ⟨42LA(550)260⟩ or potassium amide in diethyl ether ⟨51JA1437⟩ to effect deprotonation (Equation (43)). The rate of rearrangement depends on the metal counterion and generally follows the order Li > Na > K, with no reaction being observed with zinc or magnesium compounds. As illustrated in Table 1 the relative migratory aptitudes of the R group are in the order benzyl ≈ allyl > alkyl > aryl. When the alkyl group R contains a β-hydrogen atom, elimination to benzyl alcohol is possible and yields are often poor, particularly with straight-chain alkyl groups. Benzyl alcohol, the alkene and other minor by-products such as benzaldehyde may often be observed. When the migrating group R is allylic the rearrangement proceeds through a 2,3-shift (see Section 1.09.1.4.4) with some competing 1,2-shift depending upon the substitution pattern ⟨70CC4⟩.

The new C—C bond is formed by rearrangement of the R group as a radical ⟨66JA78, 68TL2809⟩ although a substantial degree of retention at the migrating carbon is normally observed. This is ascribed to a rapid recombination of the radical and radical anion which are held in a solvent cage. The optically-active benzyl ethers (**74**; R = H, Ph) rearrange in THF on addition of BuLi with partial inversion of the migrating group (Equation (44)) ⟨61LA(642)1, 63LA(663)22⟩.

Table 1 Rearrangement of benzyl ethers.

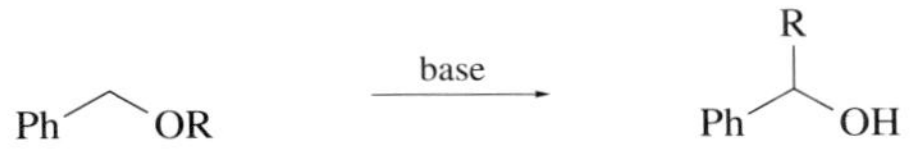

(73) (43)

R	*Conditions*	*Yield (%)*	*Ref.*
Ph	Excess Na, 100 °C, 20 h	30–35	a
	2 KNH_2, Et_2O, reflux, 18 h	0	b
Me	PhLi, Et_2O, RT, 40 h	35	c
	1.1 Bu^tLi, THF, TMEDA, –60 °C, 1 h	52	d
Et	2 KNH_2, Et_2O, reflux, 19 h	0 (40% $PhCH_2OH$)	b
	PhLi, Et_2O, RT, 40 h	0 (44% $PhCH_2OH$)	c
	1.1 Bu^tLi, THF, TMEDA, –60 °C, 10 min	34 (33% $PhCH_2OH$)	d
Bu^s	2 KNH_2, Et_2O, reflux, 140 h	27	b
Pr^i	1.1 Bu^tLi, THF, TMEDA, –60 °C, 10 min	62	d
Bu^t	1.1 Bu^tLi, THF, TMEDA, –74 °C, 30 min	45	d
	5 MeLi, THF, RT, 48 h	51	e
1-Adamantyl	2 MeLi, THF, RT, 48 h	54	e
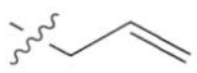	2 KNH_2, Et_2O, reflux, 15 h	65	b
CH_2Ph	2 KNH_2, Et_2O, reflux, 15 h	61–67	b
	PhLi, Et_2O, RT, 40 h	30	c
	5 MeLi, THF, RT, 48 h	65	e

[a] ⟨24CB1634⟩. [b] ⟨51JA1437⟩. [c] ⟨42LA(550)260⟩. [d] ⟨68TL2809⟩. [e] ⟨62JOC1933, 66JA78⟩.

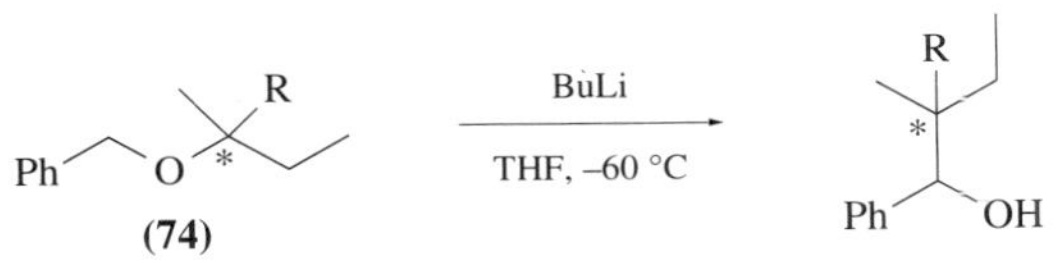

(74) (44)

R	*Retention at C* (%)*
H	60
Ph	90

The use of allyl rather than benzyl ethers gives a mixture of the 1,2- and 1,4-rearrangement products (Equation (45)) ⟨69TL821, 77TL3485, 89T2649⟩. In such cases Bu^nLi has been used to deprotonate α to the oxygen atom and yields of the 1,2-product are normally in the range 20–30%. With β-alkoxyalkyl allyl ethers in which chelation to the alkoxy group can occur, the 1,2-products are formed with high diastereoselectivity ⟨87TL1043⟩.

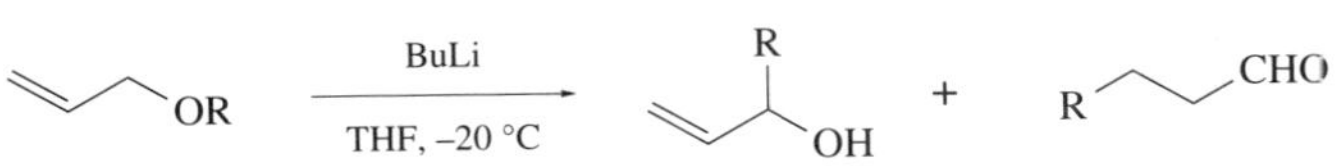

(45)

The problem of controlling the regioselectivity in the deprotonation has been addressed using anion-stabilising groups such as ketones or nitriles ⟨54JA494, 60JCS3521⟩ in which KOH in EtOH or $NaOBu^n$ in Bu^nOH were used as bases, although the range of migrating groups appears to be limited. More recent work has focused on methods which allow complete regioselectivity in the anion formation using exchange reactions. For example, a benzyl group migrates on heating a benzyl stannylmethyl ether after tin–lithium and lithium–zirconium exchange ⟨89OM1593⟩. Treatment of an acetal with lithium metal in THF exchanges one of the alkoxy groups for lithium to give the required α-alkoxy lithium which can undergo rearrangement by a 1,2-shift (Scheme 27, Table 2)

⟨90JOC5532⟩. As with direct deprotonation methods, substantial amounts of benzyl alcohol are formed when the migrating group is a primary alkyl group.

Table 2 Rearrangement from acetals.

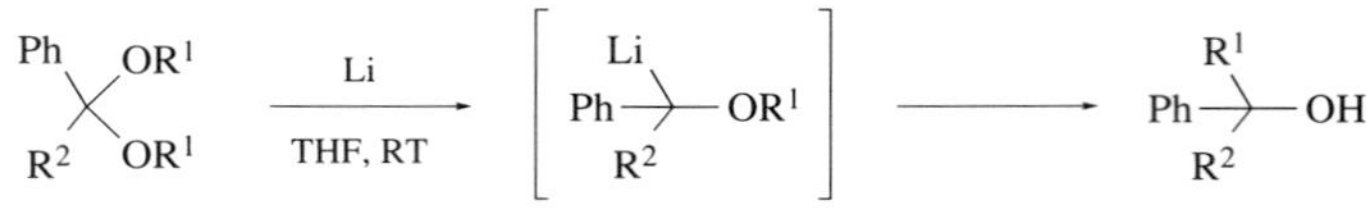

Scheme 27

R¹	*R²*	*Yield (%)*
Me	H	70
Et	H	41 (33% $PhCH_2OH$)
Bu^n	H	42 (29% $PhCH_2OH$)[a]
Me	Me	73
Me	Bu^n	75

[a]K, THF, RT.

Carbanion formation with sulfur to lithium ⟨92JA375⟩ and selenium to lithium ⟨92CB1957, 93TL297⟩ exchange have allowed an investigation into the stereocontrol at the carbanion centre. Either diastereoisomer of the selenide **(75)** gives rise to the organolithium **(76)** via the radical intermediate. The 1,2-Wittig rearrangement of the organolithium **(76)** creates the new C—C bond which gives the alcohol (77), in which complete inversion of configuration has occurred at the migration terminus (Scheme 28). The 1,2-Wittig rearrangement, although lower-yielding and less frequently used than the 2,3-Wittig rearrangement (see Section 1.09.1.4.4 and Chapter 1.18), can set up a new C—C bond with a high degree of stereocontrol.

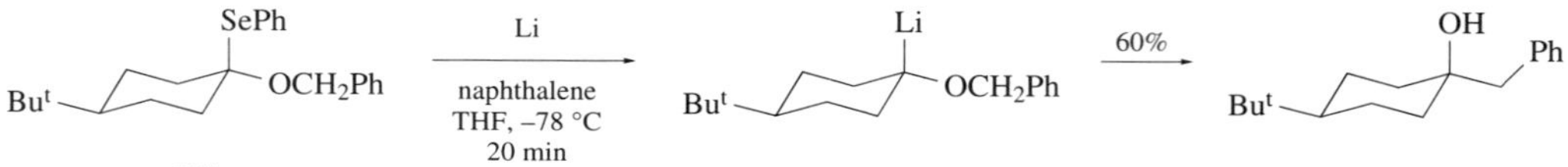

(75) **(76)** **(77)**

Scheme 28

1.09.1.3.2 1,2-Stevens rearrangement

This section will cover the formation of a new C—C bond by 1,2-rearrangement of ylides ⟨91COS(3)913⟩. An alkyl group (especially benzyl) of an ammonium ylide can migrate from the quaternary nitrogen atom to the carbanion α to the nitrogen atom (Stevens rearrangement) ⟨70OR(18)403, B-76MI 109-01⟩. Similarly, an alkyl group of a sulfonium ylide can migrate from the sulfur atom to the carbanion α to the sulfur atom ⟨B-75MI 109-01⟩. Initial studies, performed by Stevens *et al.*, were conducted with acyl-stabilised ammonium ylides (Equation (46)). Aqueous NaOH (or sodium alkoxide in alcohol) can be used to effect deprotonation and rearrangement of the ammonium salt ⟨28JCS3193, 30JCS2107⟩. The new C—C bond is formed with retention of stereochemistry at the migrating carbon centre ⟨47JCS93, 52JA5179, 62AG(E)155, 63JCS3397, 69TL3415⟩ as studied for an α-methylbenzyl migrating group (Equation (47)), particularly if the reaction is kept at 0 °C ⟨83JCS(P1)1009⟩. In this case both the stereoselectivity (99% retention at the migrating carbon) and intramolecularity (99.9%) are very high. The use of NaOMe or a higher temperature reduces the stereoselectivity, as may the presence of substituents on the migrating carbon centre or around the aromatic ring of the migrating α-methylbenzyl group. Electron-withdrawing substituents on the aromatic ring of the migrating benzyl group increase slightly the rate of rearrangement whereas electron-donating substituents tend to decrease the rate ⟨32JCS55, 32JCS1926⟩.

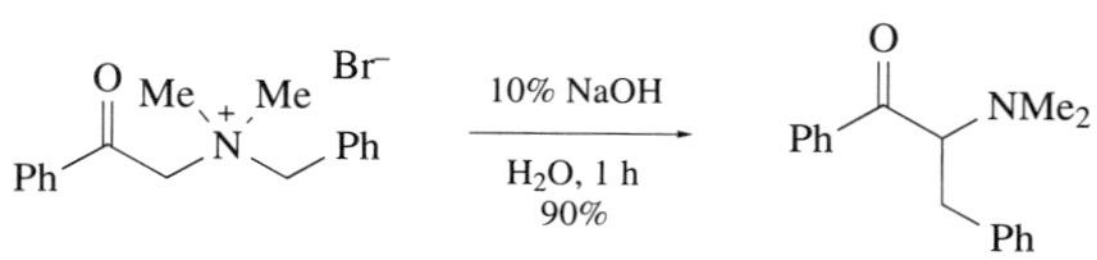

(46)

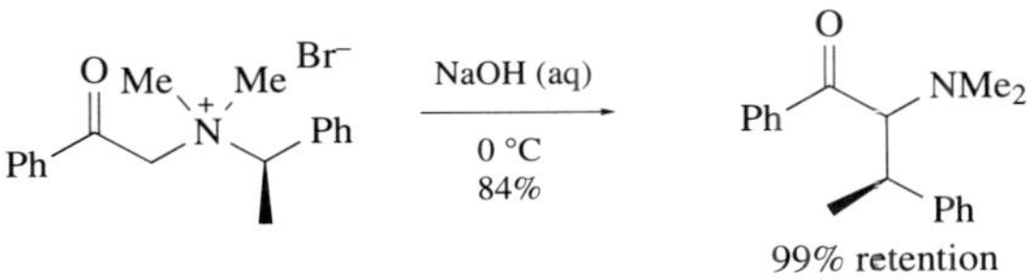

(47)

As with the 1,2-Wittig rearrangement, the alkyl group is thought to migrate as a radical in a solvent cage ⟨69TL3415, 70JA1769, 83JCS(P1)1009⟩. Small amounts of radical coupled products can sometimes be isolated. The corresponding acyl-stabilised sulfonium ylide (**78**) has been described as undergoing an analogous rearrangement ⟨32JCS69⟩ although it has been shown that the choice of solvent is crucial ⟨68JOC70, 68TL5381, 69TL2619, 70LA(737)158⟩. With polar, protic solvents such as MeOH a mixture of the enol ether (**79**) and the Sommelet–Hauser product (see Section 1.09.1.4.4) (**80**) is formed. With aprotic solvents such as THF, $CHCl_3$ or benzene the 1,2 product (**81**) is preferred (Scheme 29). Studies on the 1,2-rearrangement of sulfonium ylides also suggest the intermediacy of radicals ⟨69TL2619, 69JA1237, 70CC576⟩.

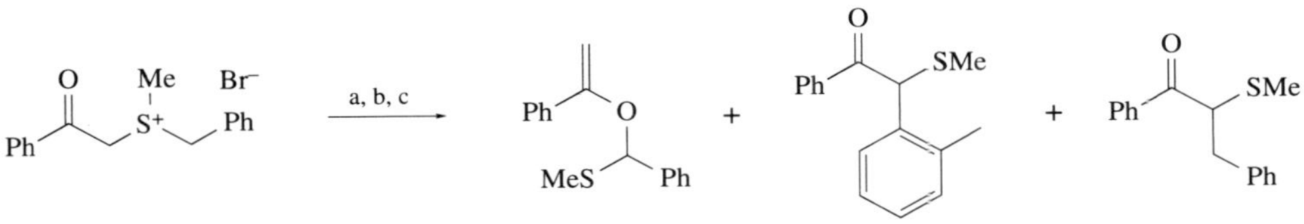

	(78)	(79)	(80)	(81)
a, 5% Na, MeOH, 60 °C, 2 h		78%	3%	–
b, 15% Na, MeOH, 60 °C, 2 h		trace	69%	–
c, i, NaOMe, MeOH 0 °C; ii, THF, 43 h		–	–	73%

Scheme 29

Although NaOH and NaOMe can be effective bases for acyl-stabilised ylide formation, stronger bases such as $NaNH_2$ or Bu^nLi are generally required. Dibenzyldimethylammonium bromide rearranges on treatment with Bu^nLi in Et_2O ⟨67JOC1155⟩ or KNH_2 in refluxing toluene ⟨55JOC1119⟩ (Equation (48)). The choice of solvent and temperature are important as the more polar KNH_2 in liquid ammonia at lower temperature gives almost exclusive Sommelet–Hauser product. The necessity of using strong bases may promote side reactions such as direct displacement. This can sometimes be minimised using lower temperatures although this can favour the Sommelet–Hauser rather than the 1,2-Stevens product. Methyl group migration is less favoured than migration of a benzyl group and benzyltrimethylammonium iodide (**82**) gives predominantly the Sommelet–Hauser product with $NaNH_2$ in liquid ammonia ⟨63OSC(4)585⟩. The use of PhLi ⟨48LA(560)116⟩ or Bu^nLi ⟨65JOC3888, 67JOC1155⟩ in Et_2O, however, allows competing 1,2-Stevens rearrangement to the amine (**83**) (Equation (49)).

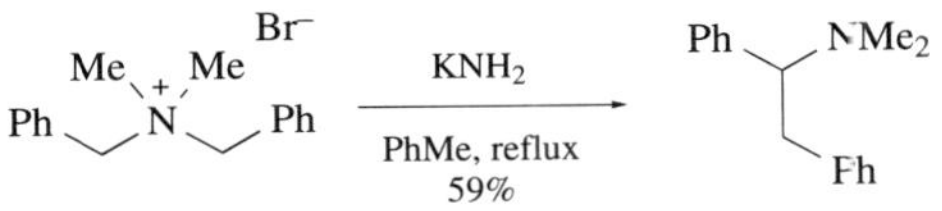

(48)

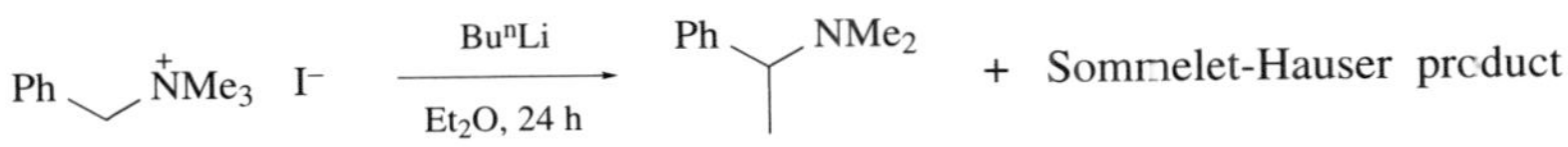

(49)

(**82**) (**83**)

Conditions	*Yield* (%)
RT	25
68 °C	32

Aryl groups do not normally migrate and the presence of a β-hydrogen atom often allows Hofmann elimination. As a result of the lack of regioselectivity in the deprotonation, a mixture of products may be formed ⟨79JOC2348⟩. A solution to this may lie in regiospecific anion formation by desilylation (or other exchange reactions). With allyl-substituted quaternary amines 1,2-, 1,4- and 2,3-rearrangement pathways are possible and may occur concurrently. These facts tend to limit the scope and usefulness of the rearrangement.

When the heteroatom forms part of a ring 1,2-rearrangement may give rise to ring contracted or ring expanded products, particularly for four- or five-membered rings. For example, the transformation of the dihydropyrrole (**84**) into the tetrahydropyridine (**85**) can be accomplished by refluxing the ylide in benzene (Equation (50)) ⟨81JCS(P1)1953⟩.

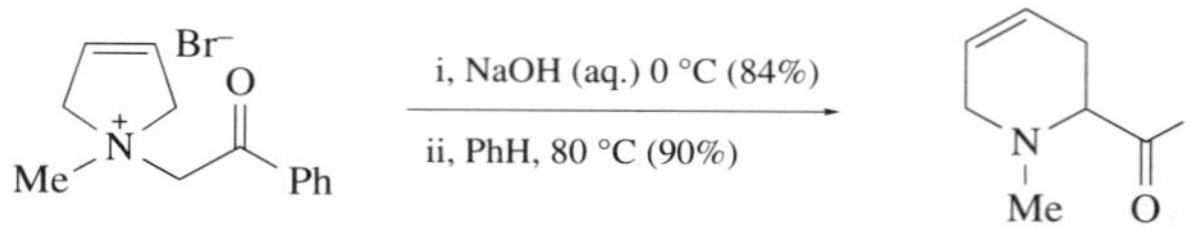

(**84**) (**85**) (50)

An alternative method for ylide formation which has been receiving more recent attention is the reaction of the heteroatom with a carbene ⟨77ACR179, 86ACR348, 91CRV263⟩. Sulfides in particular ⟨72CC860, 72JOC1721, 84JOC1917, 90T6501⟩, but also amines ⟨93JA1177⟩, ethers ⟨86JA6062, 92JOC3479⟩ and halides ⟨81JOC5094⟩, react to give an ylide which may undergo 1,2-rearrangement. Intramolecular capture of a carbene, normally with Cu(II) or Rh(II) catalysis leads to cyclic products, as represented by the examples in Scheme 30. As expected, a benzyl or allyl group migrates in preference to an alkyl group. Sommelet–Hauser (from benzyl-substituted ylides) and 2,3-sigmatropic rearrangements (from allyl-substituted ylides) which often compete with the 1,2-rearrangement will be discussed in Section 1.09.1.4.4.

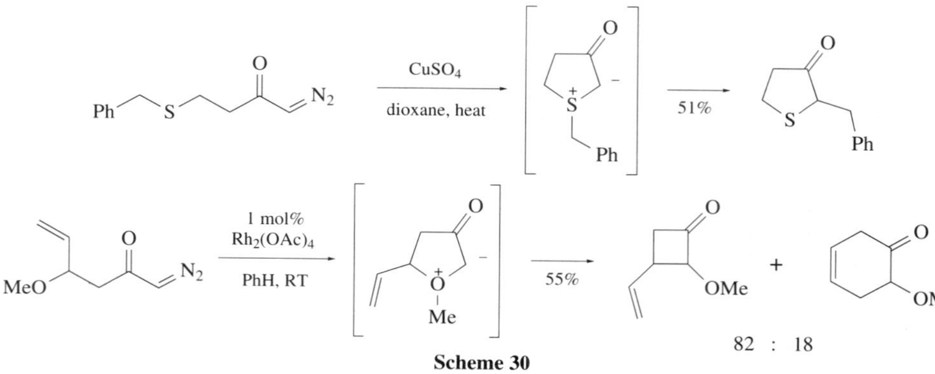

Scheme 30

1.09.1.4 Sigmatropic Rearrangements

This section deals with the transformation of an unsaturated to a saturated carbon centre with the formation of a new C—C or C—H bond at this centre by a sigmatropic rearrangement. As the σ-bond shifts across one or more π-systems the double bonds become reorganised in the process. Such transformations have been rationalised by Woodward and Hoffmann in terms of the application of orbital symmetry concepts ⟨65JA2511, 69AG(E)781⟩. This section will cover migrations across π-systems involving 1,3-, 1,5-, 1,7-, 2,3- and 3,3-sigmatropic rearrangements which lead to the formation of a new C—C or C—H bond at a saturated carbon centre. The related 1,2-sigmatropic rearrangements have been covered in the previous sections.

1.09.1.4.1 1,3-Sigmatropic rearrangements

The migration of a carbon atom by a 1,3-sigmatropic rearrangement occurs suprafacially across the π-system and many examples are known ⟨68ACR152, 68QR338, 76CRV187, 91COS(5)999, 93OR93⟩. High temperatures are commonly required unless use is made of anion-assisted systems. Thermal 1,3-sigmatropic rearrangements, in accord with the Woodward–Hoffmann rules have been shown to proceed stereospecifically with inversion of configuration at the migrating carbon atom ⟨67JA5503, 69TL2607, 69JA4322⟩. For example, both the bicyclo[3.2.0]heptene (**86**) and the bicyclo[2.1.1]hexene (**87**) rearrange on heating, with inversion at the migrating carbon centre (Equations (51) and (52) respectively).

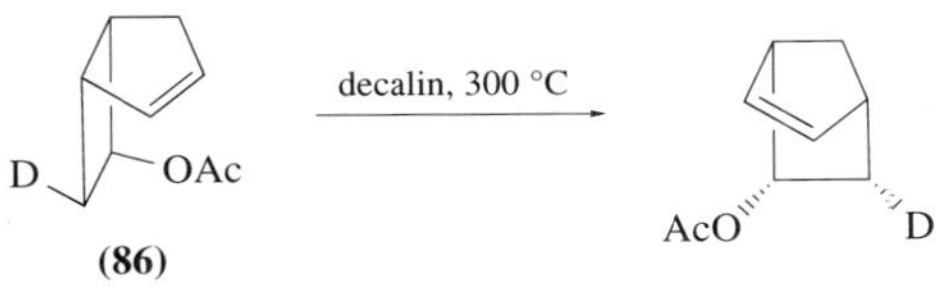

(86)

(51)

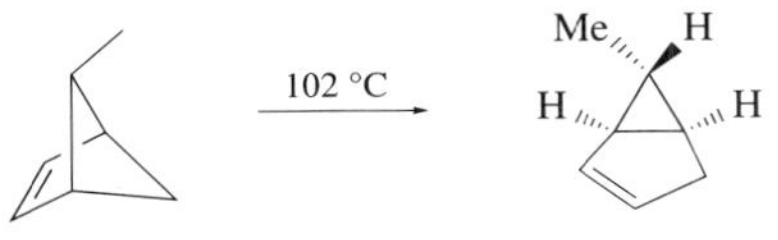

(87)

(52)

Rearrangements with retention of configuration at the migrating carbon centre are known, particularly for charge-accelerated rearrangements ⟨81JA7661⟩. The presence of an alkoxy group increases the rate of rearrangement allowing the use of lower temperatures. Magnesium and zinc salts can be used although lithium and especially potassium alkoxides react most rapidly, often at room temperature. The allylic or homoallylic alcohol is normally treated with Bu^nLi or KH in a polar solvent such as THF, DIGLYME or HMPA. Complexing agents such as 18-crown-6 can accelerate the rearrangement. Representative 1,3-sigmatropic rearrangements of acyclic homoallylic alcohols are shown in Equation (53) and Table 3. The rearrangement works best when the homoallylic alcohol is substituted with bulky or aromatic groups to give predominantly the *cis*-alkene product. Some fragmentation to the ketone is often observed as a by-product. Homoallylic amines are also known to undergo 1,3-sigmatropic rearrangement on heating ⟨80JCS(P1)1458⟩.

Table 3 Rearrangement of homoallylic alkoxides.

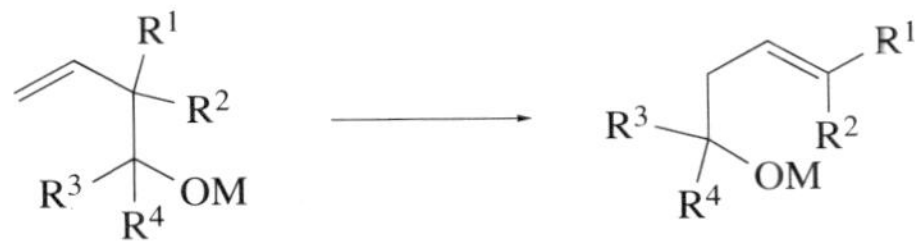

(53)

M	*R¹*	*R²*	*R³*	*R⁴*	*Solvent*	*Temp* (°C)	*Time* (h)	*Yield* (%)	(E) : (Z)	*Ref.*
Li	H	Me	Et	Et	DIGLYME	162	158	7	28 : 72	a
Li	H	Pr^i	Et	Et	DIGLYME	162	48	27	30 : 70	a
Li	H	Ph	Et	Et	THF	65	6	76	25 : 75	a
Li	H	Me	Et	Ph	DIGLYME	162	144	77	b	a
K	Me	Me	H	Ph	HMPA	75	12	85	b	c
Li	H	Me	Pr^i	Bu^t	THF	25	12	98	19 : 81	a
MgBr	H	Me	Pr^i	Bu^t	THF	25	96	84	29 : 71	d
Li	H	Me	Bu^t	Bu^t	THF	25	1.2	91	14 : 86	a
Li	H	Pr^i	Pr^i	Pr^i	THF	65	48	86	50 : 50	a
Li	H	Me	C_6H_{11}	Bu^t	THF	25	12	81	19 : 81	a
Li	H	Ph	Pr^i	Pr^i	THF	25	0.7	78	b	a
K	$CH_2{=}CH_2$	Me	H	C_6H_{13}	THF	−7	0.1	60–80	100 : 0	e

[a]⟨78JA2134⟩. [b]*(E)* : *(Z)* ratio undetermined. [c]⟨79TL2779⟩. [d]⟨69JA5162⟩. [e]⟨77TL2559⟩.

Treatment of an allylic alcohol with base, such as KH in HMPA, allows 1,3-sigmatropic rearrangement by migration of an allyl, benzyl or other anion-stabilising group. This has allowed the transformation of 1-substituted cycloalkenols to 3-substituted cycloalkanones (Equation (54)) ⟨81JA7661, 93OR93⟩ or a two-carbon ring expansion ⟨76CC846, 78JOC1050, 82JOC798⟩.

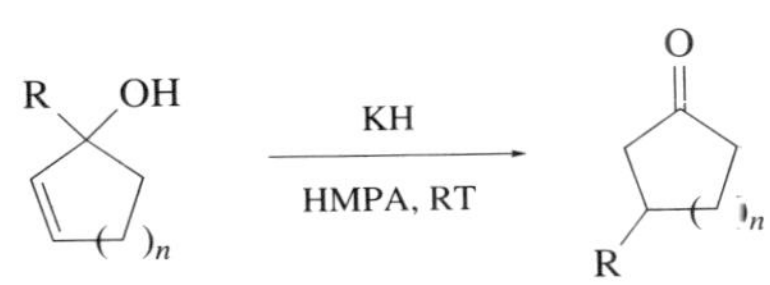

(54)

1,3-Sigmatropic rearrangements when one of the atoms of the breaking σ-bond is a heteroatom are known. Allyl vinyl ethers rearrange with migration of the allyl group using $LiClO_4$ activation ⟨91JA5488, 92TL4735⟩. This allows the conversion of a C—O bond to a C—C bond as exemplified in Equation (55). The 3,3-sigmatropic rearrangement may compete with the 1,3 process when the migrating group is allylic.

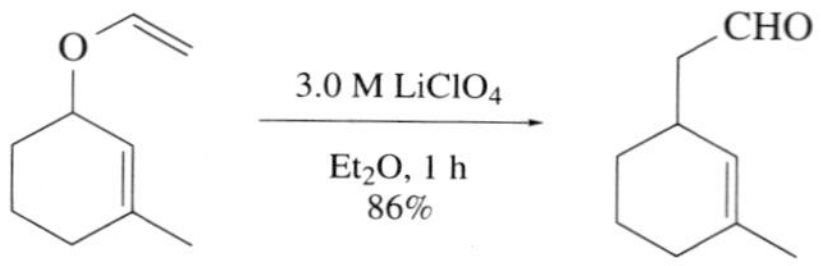

(55)

The ring expansion of vinylcyclopropanes (Equation (56)) and vinylcyclobutanes can be classed as a 1,3-sigmatropic rearrangement ⟨85OR(33)247, 91COS(5)899⟩. These rearrangements can be promoted thermally or photochemically and are accelerated when the ring is substituted by an alkoxy group.

(56)

1.09.1.4.2 1,5-Sigmatropic rearrangements

The migration of both a hydrogen atom and a carbon-centred group by a 1,5-sigmatropic shift is known and has been reviewed ⟨68QR338, 69CRV103, B-71MI 109-01, 76CRV187⟩. The rearrangement is promoted thermally and is characteristic of *cis*-1,3-pentadienes and *cis*-1-alkyl-2-vinylcyclopropanes (Scheme 31).

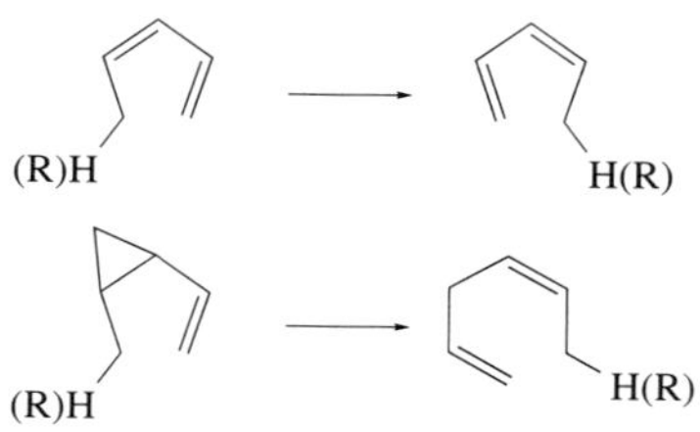

Scheme 31

The suprafacial, concerted nature of the 1,5-sigmatropic rearrangement has been demonstrated by Roth *et al.* who heated the diene (**88**) to 250 °C to effect the 1,5-hydrogen shift ⟨70CB426⟩. A mixture of the dienes (**89**) and (**90**) was produced, both of which arise by a suprafacial migration of the hydrogen atom (Equation (57)).

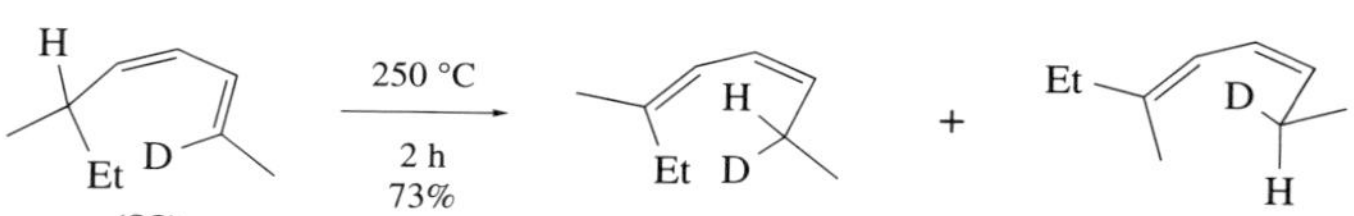

(**88**) (**89**) 40 : 60 (**90**)

(57)

The 1,5-sigmatropic rearrangement, which takes place by a suprafacial pathway, is a thermally-allowed process and so can occur in cyclic systems ⟨66C229⟩. These rearrangements occur at lower temperatures than in the acyclic series and many such examples have been reported. 1,3-Cycloheptadienes and cycloheptatrienes rearrange at 60–140 °C ⟨69RTC266⟩, whereas cyclopentadienes rearrănge at room temperature ⟨65T2329⟩. Further acceleration of rearrangement occurs with anion-assisted systems. As with other sigmatropic rearrangements, potassium alkoxides are particularly effective. The potassium salt of the alcohol (**91**) rearranges by a 1,5-hydrogen shift 10^5 times faster than the alcohol itself (Equation (58)) ⟨80JA3972⟩.

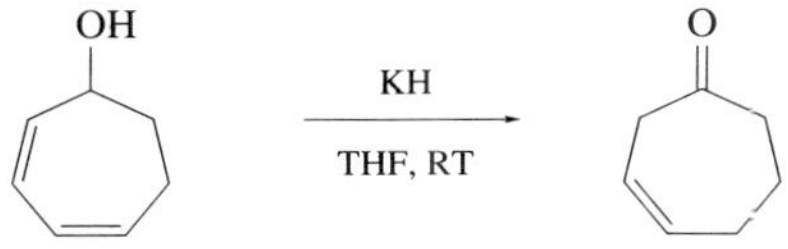

(91)

(58)

The formation of a new C—C bond by a 1,5-sigmatropic shift of an alkyl group is considerably slower than the migration of a hydrogen atom. The rearrangement is known in cyclic systems and follows a suprafacial pathway with retention of configuration at the migrating group ⟨70CC1168⟩. The 1,5-alkyl shift is normally followed by more rapid 1,5-hydrogen shifts leading to the most stable diene. Representative is the rearrangement of the cyclopentadiene (**92**) which requires heating to 320 °C and leads to the cyclopentadiene (**93**) (Scheme 32) ⟨68RTC298⟩.

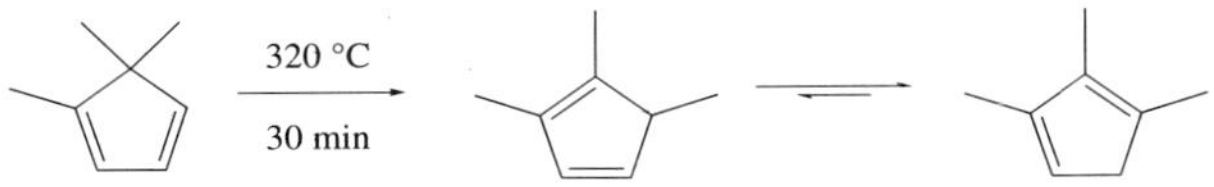

(92) (93)

Scheme 32

1,5-Sigmatropic rearrangements in substituted indenes have allowed a comparison of the relative migratory aptitudes of a number of substituents ⟨71JA650, 80JCS(P1)714, 93JCS(P1)681⟩. The rate of migration has been found to follow the order CHO > COMe > H > vinyl > CO_2Me > CN ≈ C≡CH > aryl > alkyl. Such rearrangements commonly employ diphenyl ether as the solvent and temperatures in the region of 200–300 °C. Accelerated migrations of alkyl, vinyl, aryl and cyclopropyl groups in cyclopentadiene derivatives have been observed when activated by an alkoxy substituent ⟨84CC990⟩.

1.09.1.4.3 1,7-Sigmatropic rearrangements

Thermal 1,7-shifts are thought to proceed by an antarafacial process and therefore require open-chain systems, namely 1,3,5-heptatrienes ⟨68QR338, 76CRV187⟩. In cyclic trienes the 1,5-sigmatropic shift is preferred as this may proceed thermally through a symmetry-allowed suprafacial pathway. For an antarafacial 1,7-hydrogen shift a helical-like transition state has been postulated. This is only possible if the central double bond of the triene and at least one of the other two double bonds is *cisoid*. If this is not the case, then a 1,5 shift, an electrocyclic ring closure or a 4 + 2 cycloaddition usually takes place. For example, the (*Z,Z*)-1,3,5-undecatriene (**94**) undergoes a thermal 1,7-hydrogen shift above room temperature to give the triene (**95**) (Equation (59)), whereas even above 100 °C the corresponding (*Z,E*)-, (*E,Z*)- and (*E,E*)-isomers do not undergo 1,7-hydrogen shifts but follow alternative reaction pathways ⟨75HCA1016⟩.

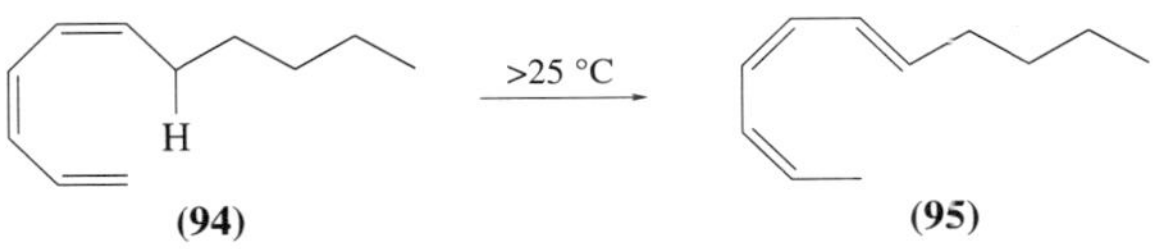

(94) (95)

(59)

An early and well-known example of a 1,7-sigmatropic rearrangement is the interconversion of vitamin D and precalciferol (Equation (60)) ⟨49BSF501, 57RTC689, 64RTC1173⟩. The 1,7-hydrogen shift has been shown to be reversible ⟨70HCA173, 87JA8051⟩ and the position of equilibrium depends on the substitution pattern of the triene.

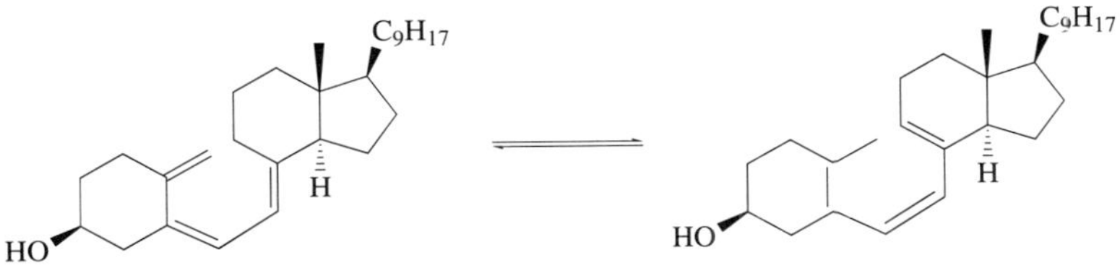

(60)

Photochemically-induced 1,7-sigmatropic rearrangements proceed by a suprafacial pathway and have been observed in cycloheptatrienes. A 1,7-hydrogen shift ⟨63AG(E)688, 65RTC241⟩ or a 1,7-alkyl shift ⟨62JOC2291, 67JA1880⟩ occurs on irradiation of the cycloheptatriene in, for example, benzene, ether or methanol with a high-pressure mercury lamp. The rearrangement competes with formation of the bicyclo[3.2.0]heptadiene.

1.09.1.4.4 2,3-Sigmatropic rearrangements

The 2,3-sigmatropic rearrangement reaction is a well-known and widely-used method for C—C bond formation. The reaction occurs by an irreversible, suprafacial process which is symmetry-allowed, involving carbanion formation α to a heteroatom and rearrangement through a five-membered ring envelope-shaped transition state. Most commonly an allylic framework is used, as generalised in Equation (61), although examples with benzylic, propargylic and allenylic substrates are known. The stereoselectivity at the new double bond and at the two new chiral centres must be high for a useful transformation. These factors, together with general synthetic methods, will be discussed in the following two subsections which are classified according to the type of heteroatom (X in Equation (61)).

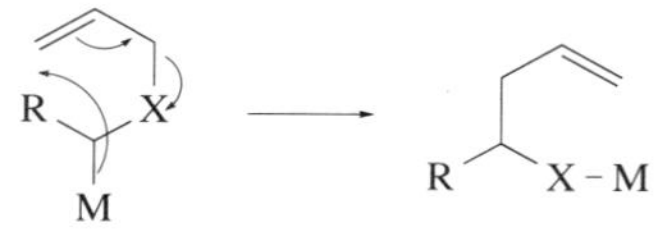

(61)

X = heteroatom
M = metal

(i) 2,3-Rearrangement of ethers

The 2,3-rearrangement of ethers (X = O) is known as the 2,3-Wittig rearrangement and has been the subject of a number of reviews ⟨79AG(E)563, 86CRV885, 91COS(3)975, 91COS(6)873, 91S594⟩. The carbanion is normally generated by direct deprotonation α to the oxygen atom using Bu^nLi or lithium diisopropylamide (LDA). This deprotonation must be regioselective except when symmetrical ethers are employed, such as in the early work by Hauser and Kantor using diallyl ether and KNH_2, Et_2O ⟨51JA1437⟩. The regioselectivity will be dependent on the substituent R (Equation (61)) which can be classified into one of three categories: hydrocarbon groups (vinyl, alkynyl or aryl), acyl groups and unstabilised systems in which R = H, alkyl. This last category requires regiospecific anion formation, for example, by tin–lithium exchange or reductive lithiation of O,S-acetals. The conditions used to effect the rearrangement and the selectivities achieved tend to depend upon which category of R group is present and so each substrate will be discussed in turn.

To generate a carbanion stabilised by a vinyl, alkynyl or aryl group Bu^nLi in THF is normally used. The Bu^nLi is best added at low temperature (often −78 °C) to minimise side reactions such as the 1,2-Wittig rearrangement. When bisallylic ethers are employed as substrates, deprotonation takes place α to the oxygen atom on the less substituted allylic group as this generates the more stable carbanion ⟨81JA6492, 83JOC279⟩. In most cases rearrangement gives very good stereoselection in favour of the (*E*)-isomer of the C=C bond. The stereoselection across the new C—C bond, pertinent to the review in this section (see Chapter 1.18 for C=C formation), is normally high in favour of the *syn* diastereoisomer from the (*Z*)-alkene (Equation (62), Table 4). This diastereoselectivity is often >90 : 10 and therefore provides a useful route to homoallylic alcohols of defined stereochemistry. As would be expected, the use of a γ-substituted allylic ether with (*E*) geometry favours the alternative *anti* diastereoisomer (Table 4) although in these cases the degree of stereoselection is often lower.

With an electron-withdrawing R group (Equation (61)) such as an acyl group the carbanion is normally generated using LDA. Representative examples of Equation (63) are shown in Table 5. With a carboxylic acid, two equivalents of LDA are required and low temperatures are normally employed ⟨81TL69⟩. The use of ester enolates has been documented (Table 5) and selectivities can be high in favour of the *syn* diastereoisomer, especially in the presence of Cp_2ZrCl_2 or Cp_2TiCl_2 ⟨86TL4581, 88TL4763, 92BCJ1841⟩. Lewis acids such as trimethylsilyl triflate (TMSOTf) or Bu^n_2BOTf with a tertiary amine such as Et_3N or Pr^i_2NEt in CH_2Cl_2 provide alternative reaction conditions to promote the 2,3-Wittig rearrangement ⟨86TL4511, 91TL4647⟩.

Table 4 2,3-Wittig rearrangement with hydrocarbon stabilising groups.

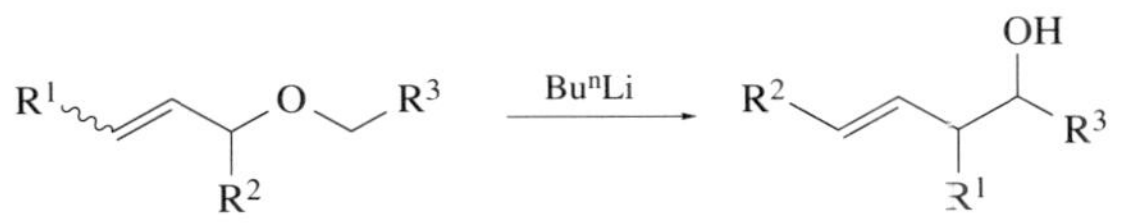

(62)

R^1 (E or Z)	R^2	R^3	*Conditions*	*Yield* (%)	(E) : (Z)	syn : anti	*Ref.*
Me (Z)	H	Ph	Et_2O–hexane, TMEDA, –80 °C			100 : 0	a
Me (Z)	Pr^i	Ph	THF	89	100 : 0	93 : 7	b
Me (Z)	H	$CH=CH_2$	THF, –85 °C	88		92 : 8	c
Me (Z)	H	$CMe=CH_2$	THF, –85 °C	71		95 : 5	c
Me (Z)	Pr^i	$CH=CH_2$	THF	84	100 : 0	92 : 8	b
Me (Z)	H	C≡CH	THF, –85 °C	56		90 : 10	d
Me (Z)	H	C≡CMe	THF, –85 °C	55		100 : 0	d
Me (Z)	H	C≡CTMS	THF, –85 °C	74		100 : 0	d
R^e (Z)	H	C≡CTMS	THF, –78 °C	77		98 : 2	f
Me (Z)	Pr^i	C≡CH	THF	89	100 : 0	91 : 9	b
Me (Z)	Pr^i	C≡CTMS	THF	62	100 : 0	>98 : 2	b
Me (Z)	Me	C≡CTMS	THF, –85 °C	64	>99: 1	>99 : 1	g
Bu^n (Z)	Me	C≡CTMS	THF, –85 °C	96	100 : 0	99 : 1	h
Et (Z)	Me	C≡CMe	THF, –78 °C	75	100 : 0	>99 : 1	i
Me (E)	H	Ph	Et_2O–hexane, TMEDA, –80 °C			50 : 50	a
Me (E)	Pr^i	Ph	THF	95	82 : 18	24 : 76	b
Me (E)	Me	Ph	THF, 0 °C		83 : 17		j
Me (E)	H	$CH=CH_2$	THF, –85 °C	81		16 : 84	c
Me (E)	H	$CMe=CH_2$	THF, –85 °C	70		28 : 72	c
Me (E)	Pr^i	$CH=CH_2$	THF	75	100 : 0	40 : 60	b
Me (E)	H	C≡CH	THF, –85 °C	72		1 : 99	d
Me (E)	H	C≡CMe	THF, –85 °C	65		1 : 99	d
Me (E)	H	C≡CTMS	THF, –85 °C	72		73 : 27	d
R^e (E)	H	C≡CTMS	THF, –78 °C	93		60 : 40	f
Me (E)	Pr^i	C≡CH	THF	63	92 : 8	11 : 89	b
Me (E)	Me	C≡CH	THF, –85 °C	62	100 : 0	9 : 91	d

[a]⟨70CC4⟩. [b]⟨84JOC1842⟩. [c]⟨81JA6492, 83JOC279⟩. [d]⟨84T2303⟩, R^e = (*S*)-CH(Me)O-TBDMS. [f]⟨88TL4587⟩. [g]⟨84TL565⟩. [h]⟨85CL1723⟩. [i]⟨91JOC5973⟩. [j]⟨71JA3556⟩.

High yields and *syn* diastereoselectivities have been achieved in the 2,3-sigmatropic rearrangement of α-allyloxyacetamides using LDA or lithium hexamethyldisilazide (LHMDS) in THF at −78 to −85°C ⟨85CL1729⟩. Rearrangement when the carbanion is stabilised by an oxazoline or dihydrooxazine is also known to occur in THF at −78°C, deprotonation being effected with LDA or Bu^nLi ⟨83TL513⟩.

Regioselective carbanion formation α to the oxygen atom by tin–lithium exchange, followed by 2,3-Wittig rearrangement, was introduced by Still and Mitra ⟨78JA1927⟩. The trialkyltin group can be exchanged for lithium using Bu^nLi in THF at −78°C. This generates a relatively unstabilised carbanion which undergoes rearrangement at low temperature to give the homoallylic alcohol (Equation (64), Table 6). The alternative sulfur to lithium exchange has been effected using lithium naphthalenide in THF ⟨89JA2981⟩.

The use of exchange reactions with an alkyl group at the carbanion centre has allowed the discovery that rearrangement proceeds with inversion of stereochemistry at this centre ⟨92AG(E)647, 92JA375, 92TL5795⟩. For example, transmetallation of the stannane (**96**) with Bu^nLi gives the homoallylic alcohol (**97**) with essentially complete *anti* diastereoselectivity and with essentially no loss of enantiomeric purity through an inversion pathway (Equation (65)).

Table 5 2,3-Wittig rearrangement with acyl stabilising groups.

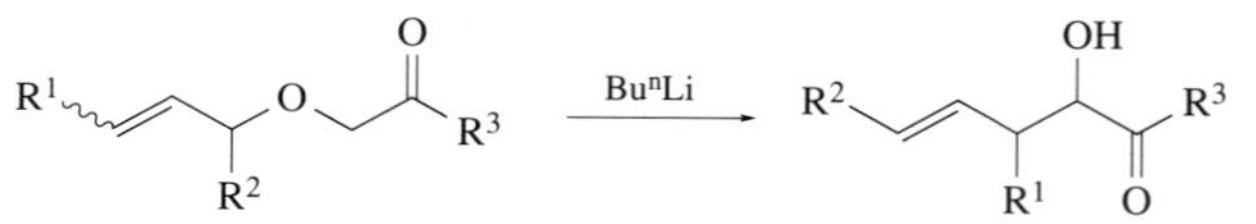

(63)

R^1 (E or Z)	R^2	R^3	Conditions	Yield (%)	(E) : (Z)	syn : anti	Ref.
H	Me	OH	2 LDA, THF, −78 °C	87[a]	>95 : 5		b
H	C_7H_{15}	OH	2 LDA, THF, −78 °C	74	>95 : 5		b
Me (*Z*)	H	OH	2 LDA, THF, −78 °C	73[a]		75 : 25	b
Me (*E*)	H	OH	2 LDA, THF, −78 °C	60[a]		35 : 65	b
Me (*Z*)	H	OPr^i	LDA, THF, −100 to −20 °C	20		67 : 33	c
Me (*E*)	H	OPr^i	LDA, THF, −100 to −20 °C	14		82 : 18	c
Me (*E*)	Me	OPr^i	LDA, THF, −100 to −20 °C	72	60 : 40	80 : 20	c
Me (*Z*)	H	OPr^i	LDA, THF, Cp_2ZrCl_2, −100 to −20 °C	15		88 : 12	c
Me (*E*)	H	OPr^i	LDA, THF, Cp_2ZrCl_2, −100 to −20 °C	47		98 : 2	c
Me (*E*)	Me	OPr^i	LDA, THF, Cp_2ZrCl_2, −100 to −20 °C	91	0 : 100	99 : 1	c
Me (*E*)	C_8H_{17}	OPr^i	LDA, THF, Cp_2ZrCl_2, −100 to −20 °C	70	0 : 100	99 : 1	c
Me (*E*)	CH_2OBn	OPr^i	LDA, THF, Cp_2ZrCl_2, −100 to −20 °C	72	60 : 40	100 : 0	d
Me (*E*)	CH_2OBn	OPr^i	LDA, THF, Cp_2TiCl_2, −100 to −20 °C	53	98 : 2	98 : 2	d
Me (*Z*)	H	OMe	TMSOTf, Et_3N, CH_2Cl_2	86		53 : 47	e
Me (*E*)	H	OMe	TMSOTf, Et_3N, CH_2Cl_2	83		92 : 8	e
Me (*Z*)	H	OMe	Bu_2BOTf, Pr^i_2NEt, CH_2Cl_2, −78 °C to RT	54		30 : 70	f
Me (*E*)	H	OMe	Bu_2BOTf, Pr^i_2NEt, CH_2Cl_2, −78 °C to RT	55		92 : 8	f

[a]Quench with TsCl. [b]⟨81TL69⟩. [c]⟨86TL4581⟩. [d]⟨88TL4763, 92BCJ1841⟩. [e]⟨86TL4511⟩. [f]⟨91TL4647⟩.

Table 6 2,3-Wittig–Still rearrangement.

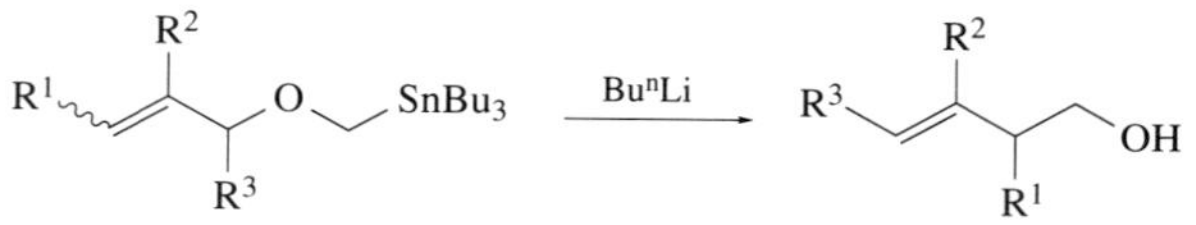

(64)

R^1 (E or Z)	R^2	R^3	Conditions	Yield (%)	(E) : (Z)	Ref.
H	Me	Bu^n	THF, −78 °C, 30 min	95	3 : 97	a
H	H	$C_7H_{15}{}^n$	THF, −78 °C, 30 min	95	40 : 60	a
Me (*Z*)	H	$C_7H_{15}{}^n$	THF, −78 °C, 30 min	91	100 : 0	a
Me (*Z*)	H	Pr^i	THF, −78 to 0 °C	72	100 : 0	b
Me (*E*)	H	$C_7H_{15}{}^n$	THF, −78 °C, 30 min	96	35 : 65	a
Me (*E*)	H	Pr^i	THF, −78 to 0 °C	72	53 : 47	b
Me (*E*)	H	CH_2NHTr	hexane, −110 °C	80	60 : 40	c

[a]⟨78JA1927⟩. [b]⟨85TL5013⟩. [c]⟨92T6425⟩.

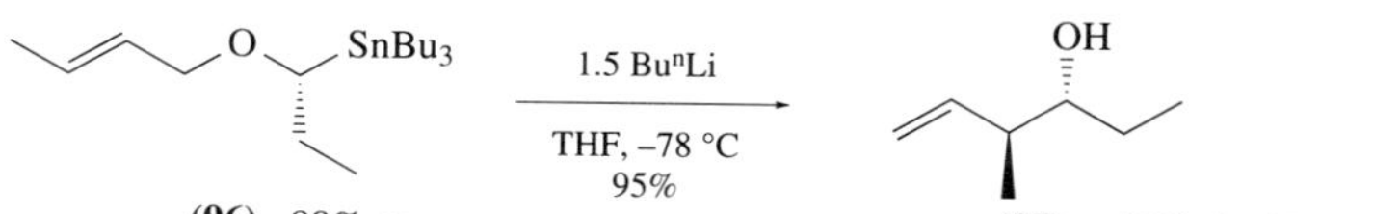

(96) 88% *ee* **(97)** >99% *de*, 87% *ee*

(65)

The related 2,3-thia-Wittig–Still rearrangement in which the ether oxygen atom has been replaced by a sulfur atom proceeds efficiently using transmetallation with Bu^nLi ⟨92TL7323, 93CB1227⟩, although rearrangement of *N*-allyl aminomethylstannanes to the homoallylic amine has been shown to occur by a 1,2-rearrangement ⟨93JCS(P1)1275⟩. Sulfur and nitrogen ylides rearrange by a 2,3-sigmatropic process as discussed below.

Many examples of asymmetric induction in the 2,3-Wittig rearrangement have been investigated using the three general methods for the rearrangement described in this section. As the rearrangement is suprafacial, transfer of chirality across the allyl system is observed. Examples of diastereoselection using a chiral auxiliary (especially on the acyl group) or with an adjacent allylic chiral centre, or using a chiral base on cyclic allylic ethers are known and have been discussed in the given reviews, for example ⟨91S594⟩.

(ii) 2,3-*Rearrangement of ylides*

A number of reviews ⟨B-71MI 109-02, 79AG(E)563, 91COS(3)913, 91COS(6)873⟩ cover the 2,3-sigmatropic rearrangement of allyl-substituted ylides ($X = NR_2$, SR, OR, halogen, Equation (61)) often referred to as the 2,3-Stevens rearrangement after early work by Stevens ⟨32JCS1932⟩. Ylides can be generated by deprotonation (or desilylation) of ammonium or sulfonium salts (prepared by alkylation of amines or sulfides) or by direct reaction of the amine, sulfide, selenide, ether or halide with a carbene. The ylide so formed rearranges (often at room temperature) through a five-membered ring transition state with the formation of a new C—C bond in an acyclic or a cyclic substrate, the latter potentially providing a useful ring expansion or contraction method. Each of these aspects will be dealt with in turn.

Deprotonation of an ammonium or sulfonium salt is a common method for ylide formation and can be effected with a variety of bases. Mild conditions such as K_2CO_3 in EtOH or H_2O can be used for generation of sulfonium ylides. Slightly harsher conditions such as NaOH, NaOMe or $NaNH_2$ are normally needed for ammonium ylide formation. Table 7 gives a number of examples of the 2,3-Stevens rearrangement when the carbanion is stabilised by a hydrocarbon (aryl, alkenyl or alkynyl) group (Equation (66)). If one of the substituents on the heteroatom is a benzyl group then deprotonation and rearrangement may give rise to accompanying Sommelet–Hauser rearrangement. The Sommelet–Hauser rearrangement has been reviewed ⟨70OR(18)403, 91COS(3)913⟩ and may be favoured with protic solvents.

Table 7 2,3-Ylidic rearrangement with hydrocarbon stabilising grops.

(66)

R^1	R^2	XR^3	R^4	*Conditions*	*Yield* (%)	syn : anti	*Ref.*
H	H	SEt	Ph	NaOEt	a		b
Me	Me	SEt	Ph	NaOEt	a		b
Ph	H	SEt	Ph	K_2CO_3	a	60 : 40	b
Ph	H	SEt	CH=CHPh	K_2CO_3, EtOH	100	c	d
Me	Me	NMe_2	Ph	$NaNH_2$, NH_3, –33 °C	44[e]		f
Me	Me	NMe_2	CH=CMe$_2$	$NaNH_2$, NH_3, –33 °C	65		f
H	H	NMe_2	CH=CH$_2$	NaOMe, DMSO, RT	50		d
Ph	H	NMe_2	CH=CHPh	NaOMe, DMSO, RT	65[g]	91 : 9	d
H	H	NMe_2	C≡CPh	NaOMe, DMSO, RT	85		d
Ph	H	NMe_2	C≡CPh	NaOMe, DMSO, RT	91	86 : 14	d
H	Ph	NMe_2	C≡CPh	NaOMe, DMSO, RT	85	10 : 90	d
Ph	H	NMe_2	C≡CH	NaOMe, DMSO, RT	99	100 : 0	d

[a]High yield. [b]⟨69CC293⟩. [c]Ratio not determined. [d]⟨80JCS(P1)1436⟩. [e] Together with 42% Sommelet–Hauser product. [f]⟨72HCA2233⟩. [g]Together with 28% 1,2-product.

Deprotonation adjacent to an acyl group in a sulfonium or ammonium salt followed by rearrangement allows transformation to a homoallylic sulfide or amine (Equation (67), Table 8). In these examples and those in Table 7 (and with the 2,3-Wittig rearrangement), the diastereoselectivity is dependent upon whether the substituent is located preferentially *exo* or *endo* to the five-membered ring transition state. A mixture of diastereoisomers is normally observed, especially from acyclic ylides, although selectivities can be good. The double bond stereoselection is normally in favour of the *E* isomer ⟨86T2979, 92JOC428⟩.

Table 8 2,3-Ylidic rearrangement with acyl stabilising groups.

(67)

R^1	R^2	XR^3	R^4	Conditions	Yield (%)	syn : anti	Ref.
H	H	SMe	Ph	2M K_2CO_3 (aq.)	69		a
H	H	SMe	Me	2M K_2CO_3 (aq.)	63		a
Me	H	SMe	Ph	2M K_2CO_3 (aq.)	94	b	a
Me	Me	SMe	Ph	2M K_2CO_3 (aq.)	96		a
Me	Me	SMe	Me	2M K_2CO_3 (aq.)	90		a
Me	Me	SMe	OEt	2M K_2CO_3 (aq.)	91		a
H	H	SEt	Ph	NaOEt	c		d
Me	Me	SEt	Ph	NaOEt	c		d
Ph	H	SEt	Ph	K_2CO_3	c	b	d
Me	Me	NMe_2	Ph	2M NaOH (aq.), RT, 1 h	98		e
Me	Me	NMe_2	Me	10M NaOH (aq.), RT, 1 h	60		e
Me	Me	NMe_2	OMe	NaOMe, DMSO, RT	93		e
Ph	H	NMe_2	Ph	2M NaOH (aq.), 60 °C	98	27 : 73	e
Ph	H	NMe_2	Ph	NaOH (aq.), $CHCl_3$, 0 °C		17 : 83	f
H	Ph	NMe_2	Ph	NaOH (aq.), $CHCl_3$, 0 °C		67 : 33	f
Ph	H	NMe_2	Bu^t	NaOH (aq.), Et_2O, RT		10 : 90	f

[a]⟨80JA2125⟩. [b]⟨Ratio not determined⟩. [c]High yield. [d]⟨69CC293⟩. [e]⟨80JCS(P1)1450⟩. [f]⟨81JCS(P1)1953⟩.

Analogous to the 2,3-Wittig–Still rearrangement is the rearrangement of relatively unstabilised ylides, normally generated by desilylation using CsF in diglyme or MeCN ⟨79JA6452, 85JOC2965⟩ or in hexamethylphosphoramide (HMPA) ⟨90JA1999⟩. For example, the ammonium salt (**98**) can be desilylated with CsF in HMPA and the resulting ylide rearranges to give the (*E*)-homoallylic amine (**99**) (Equation (68)). Quite remarkable is the complete (*Z*) selectivity in the rearrangement of the ylide generated from the ammonium salt (**100**) (Equation (69)) ⟨90JA1999⟩. This rearrangement is thought to proceed by initial deprotonation to give the allyl anion. Proton transfer then sets up the required ylide for 2,3-sigmatropic rearrangement to the (*Z*)-homoallylic amine (**101**).

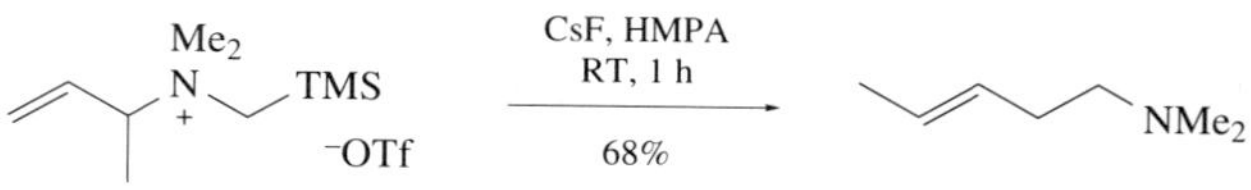

(98) **(99)** (68)

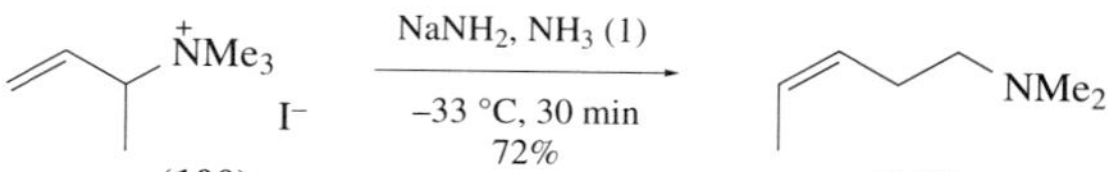

(100) **(101)** (69)

The above methods for ylide generation require the prior formation of a sulfonium or ammonium salt, followed by deprotonation (or desilylation) of the salt. An alternative method, which also encompasses oxonium and halonium ylide formation, involves the use of a carbene. The carbene can be generated from a diiodoalkane and Et_2Zn ⟨82TL3019⟩ or more commonly from a diazoalkane and copper(II), rhodium(II) or light ⟨72JA3870, 81JOC5094⟩. Table 9 gives examples of the conversion of an allylic sulfide (or selenide) into a homoallylic sulfide (or selenide) (Equation (70)) and Table 10 gives further acyclic examples from allylic amines, ethers and halides (Equation (71)). Slow addition of the diazoacetate or diazomalonate to a catalytic amount of anhydrous $CuSO_4$ or

$Rh_2(OAc)_4$ and the allylic substrate generally gives the best results. In most cases the 2,3-sigmatropic rearrangement occurs with very little or no competing cyclopropanation or 1,2-Stevens rearrangement.

Table 9 2,3-Sulfonium and selenonium ylide rearrangement from carbenes.

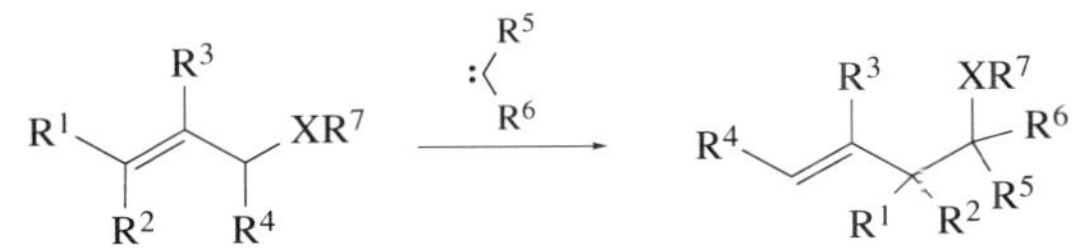

(70)

XR^7	R^1	R^2	R^3	R^4	R^5	R^6	Conditions	Yield (%)	(E) : (Z)	Ref.
SPh	H	H	H	H	H	H	Et_2Zn, CH_2I_2, PhH, 55 °C	55		a
SPh	Me	Me	H	H	H	H	Et_2Zn, CH_2I_2, PhH, 55 °C	80		a
SePh	Me	Me	H	H	H	H	Et_2Zn, CH_2I_2, PhH, 55 °C	75		a
SPh	Me	Me	H	Me	H	H	Et_2Zn, CH_2I_2, PhH, 55 °C	75	100 : 0	a
SPh	Me	Me	H	H	Me	H	Et_2Zn, $MeCHI_2$, PhH, 55 °C	70		a
SPh	H	H	H	H	CO_2Me	CO_2Me	$N_2C(CO_2Me)_2$, $h\nu$	32		b
SPh	H	H	H	H	CO_2Me	CO_2Me	$N_2C(CO_2Me)_2$, $CuSO_4$, 100 °C	90		b
SBu^n	H	H	H	H	CO_2Me	CO_2Me	$N_2C(CO_2Me)_2$, $CuSO_4$, 100 °C	93		b
SPh	Me	H	H	H	CO_2Me	CO_2Me	$N_2C(CO_2Me)_2$, $CuSO_4$, 100 °C	92		b
SPh	H	H	Me	Bu^n	CO_2Me	CO_2Me	$N_2C(CO_2Me)_2$, $CuSO_4$, 100 °C	70	90 : 10	c
SMe	H	H	H	H	CO_2Et	H	N_2CHCO_2Et, $Rh_2(OAc)_4$, 60 °C	91		d
SEt	H	H	H	H	Ph	H	N_2CHPh, $CuSO_4$, 110 °C	49		e
SEt	H	H	H	H	Ph	Ph	N_2CPh_2, $CuSO_4$, 110 °C	38		e

[a]⟨82TL3019⟩. [b]⟨72JA3870⟩. [c]⟨73JOC2572⟩. [d]⟨81JOC5094⟩. [e]⟨72JOC3791⟩.

Table 10 2,3-Ammonium, oxonium and halonium ylide rearrangement from carbenes.

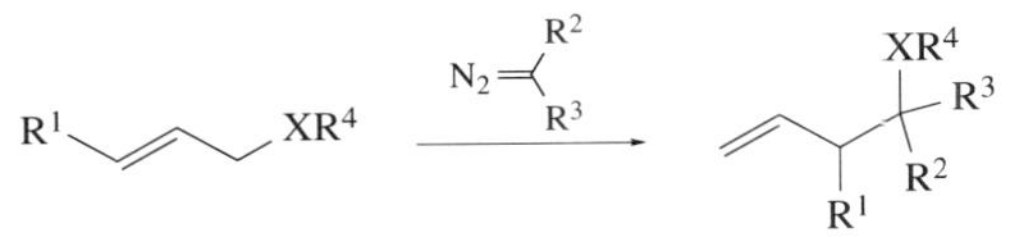

(71)

XR^4	R^1	R^2	R^3	Conditions	Yield (%)	syn : anti	Ref.
NEt_2	H	CO_2Me	CO_2Me	$CuSO_4$, 100 °C	40		a
NEt_2	Me	CO_2Me	CO_2Me	$CuSO_4$, 100 °C	27		a
NMe_2	H	CO_2Et	H	$Rh_2(OAc)_4$, 60 °C	60		b
NMe_2	Me	CO_2Et	H	$Rh_2(OAc)_4$, 60 °C	79	75 : 25	b
NMe_2	Ph	CO_2Et	H	$Rh_6(CO)_{16}$, 60 °C	78	72 : 28	b
OMe	H	CO_2Me	CO_2Me	$CuSO_4$, 100 °C	60		a
OMe	Me	CO_2Me	CO_2Me	$CuSO_4$, 100 °C	75		a
OEt	Me	CO_2Me	CO_2Me	$CuSO_4$, 100 °C	15		a
Cl	H	CO_2Me	CO_2Me	$h\nu$	53		a
Cl	H	CO_2Me	CO_2Me	CuCl, 110 °C	32		a
Cl	H	CO_2Et	H	$Rh_2(OAc)_4$	5[c]		b
Br	H	CO_2Me	CO_2Me	$h\nu$	38		a
Br	H	CO_2Et	H	$Rh_2(OAc)_4$	21[d]		b
Br	H	CO_2Et	CO_2Me	$Rh_2(OAc)_4$	86		b
I	H	CO_2Et	H	$Rh_2(OAc)_4$	98		b

[a]⟨72JA3870⟩. [b]⟨81JOC5094⟩. [c]Together with 90% cyclopropanation. [d]Together with 55% cyclopropanation.

Other copper salts have been shown to be effective catalysts for the reaction. For example, the carbene generated from the compound (**102**) with $Cu(acac)_2$ reacts with allyl phenyl sulfide or allyl methyl sulfide to give a mixture of the diastereoisomeric β-lactams (**103**) 〈82JCS(P1)2757〉. A related ammonium ylide gave the 2,3-rearranged product in 75% yield as a single diastereoisomer 〈71JA2342〉. Ylides have been formed through intramolecular addition of an ether or amine to a carbene, followed by a 2,3-sigmatropic rearrangement (Scheme 33) 〈86JA6060, 86JA6062, 92TL6193, 93TL4385〉. This gives rise to 2-allyl-3-ketotetrahydrofurans and other saturated heterocycles with both good yields and good diastereoselectivities.

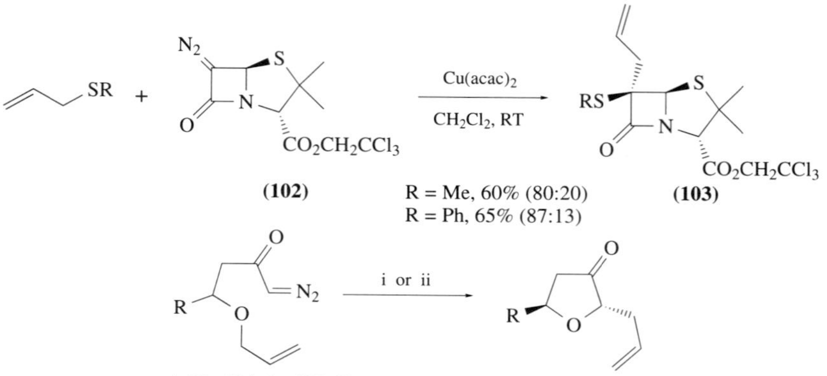

i, $Rh_2(OAc)_4$, CH_2Cl_2, R = H, 70%; R = Me, 65% (93 : 7)

ii, $Cu(acac)_2$, THF, reflux, R = Pr^i or Pr^n, 83% (>97 : 3)

Scheme 33

When the 2,3-sigmatropic rearrangement forms a new C—C bond within a ring then a useful ring expansion or contraction is possible. In ring expansions 〈84ACR358〉 which give eight-membered rings the (*Z*)-alkene is preferred (Equation (72)) 〈75JA6878〉, whereas for nine-membered or larger rings the (*E*)-alkene is favoured (Equation (73)) 〈78JOC4831〉 (except when the carbanion is axially-oriented 〈78TL523〉). Rearrangement with ring contraction (Scheme 34) may be slower than in the acyclic series. The sulfonium ylide (**104**) can be isolated and requires heating to 180 °C to effect rearrangement 〈72JA3870〉. However, the ylides formed by deprotonation of the salts (**105**) and (**106**) rearrange at 0 °C or room temperature to give the *cis*-disubstituted five-membered rings 〈81JCS(P1)1953〉.

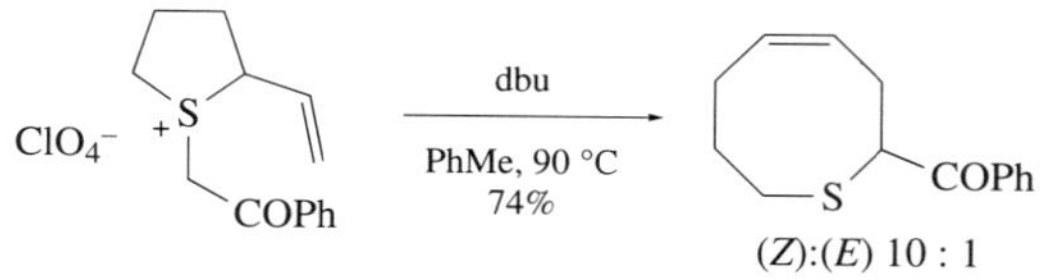

(72)

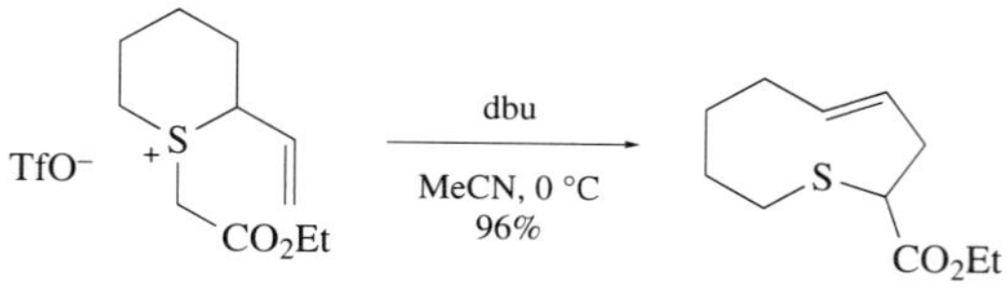

(73)

1.09.1.4.5 3,3-Sigmatropic rearrangements

A great many examples of the 3,3-sigmatropic rearrangement have been documented and the reaction has found particular use for the stereocontrolled preparation of C—C bonds. This is because the reaction proceeds through a symmetry-allowed, highly-ordered six-membered ring transition state, which (unless constrained conformationally) prefers the chair-shape and therefore allows a prediction of the stereochemical outcome across the new C—C bond and at the new C=C bond(s). The basic framework of the reaction is shown in Equation (74) and involves the interconversion of 1,5-dienes. With the all-carbon system the reaction is known as the Cope rearrangement, whereas when X is a heteroatom the reaction is termed a Claisen rearrangement.

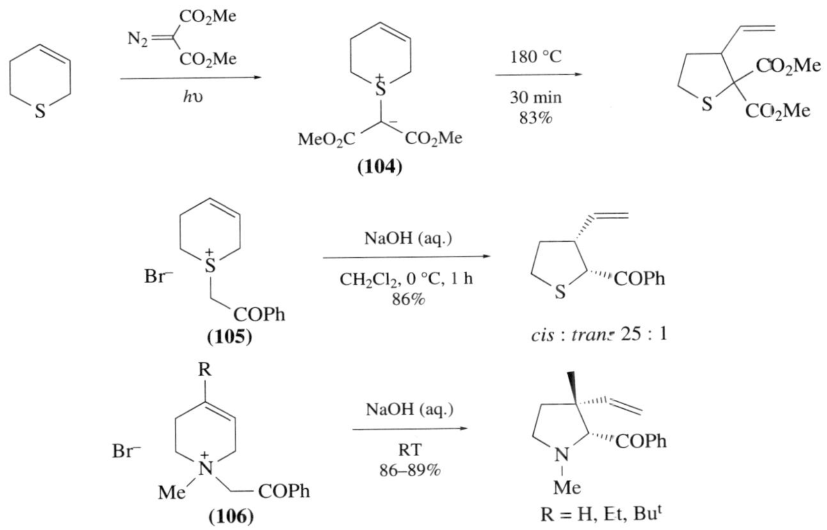

Scheme 34

As the general synthetic methods for each classification (and within each classification) differ to some extent, the reaction will be discussed in the following two subsections.

(74)

$X = CH_2$: Cope
$X = O$: Claisen

(i) Cope rearrangement

The Cope rearrangement is so named after work in the early 1940s by Cope ⟨40JA441, 41JA1843, 41JA1852⟩. The rearrangement has found many synthetic applications, particularly when modified as the oxy-Cope and anionic oxy-Cope variants. As a result, many accounts of the Cope rearrangement have been published and readers should refer to the following reviews and references therein ⟨63AG(E)115, 75OR(22)1, 84CRV205, 91COS(5)785⟩. The parent Cope rearrangement is often carried out by heating the 1,5-diene neat or in a high-boiling solvent such as xylene or decalin. The choice of solvent is taken on the basis of the boiling-point required to effect the rearrangement as the polarity seems to have little influence on the rate of the reaction. For volatile dienes gas-phase reactions in static or flow systems may be used ⟨75OR(22)1⟩.

The 3,3-sigmatropic rearrangement of 1,5-hexadienes is a reversible process and the position of equilibrium depends on the substitution pattern and/or strain of the dienes. It is well known that the thermodynamic stability of an alkene increases with increasing substitution and it is therefore not surprising that the equilibrium lies in favour of the more substituted product. On heating 3-methyl-1,5-hexadiene (Equation (75)) a mixture of the two isomeric dienes results, with the equilibrium lying in favour of the isomer 1,5-heptadiene ⟨65JPC1278⟩. Using less vigorous conditions, more complete conversion than with alkyl substituted dienes may be observed when the diene is substituted by a group capable of *p*-orbital overlap with the newly formed C═C bond. Some of the early work by Cope used dienes such as those in Scheme 35 in which a phenyl group (Equation (76)) ⟨44JA1684⟩ or an acyl group (Equation (77)) ⟨41JA1843, 41JA1852⟩ comes into conjugation with one of the C═C bonds on rearrangement. This type of 3,3-sigmatropic rearrangement can, in general, be effected at temperatures in the range 150–200 °C. As mentioned above, the difference in strain between the two 1,5-dienes can influence the position of equilibrium. Some *cis*-1,2-divinylcyclopropanes such as the parent diene (**107**) are only stable at low temperature (e.g., −20 °C) and rearrange rapidly at room temperature ⟨73CC319⟩, while others are stable enough to be isolable and require heating to promote the Cope rearrangement. The rearrangement of divinylcyclopropanes gives rise to a new C—C bond in a cycloheptadiene ring, for example (**108**) in

Equation (78). Rearrangement of *cis*-1,2-divinylcyclobutanes normally requires heat and gives access to the cyclooctadiene ring.

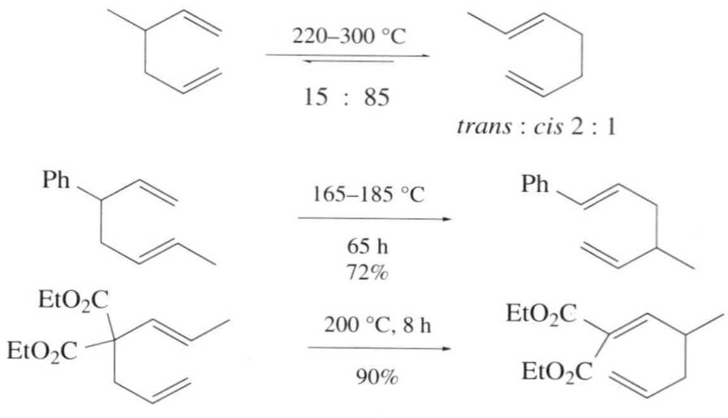

(75)

(76)

(77)

Scheme 35

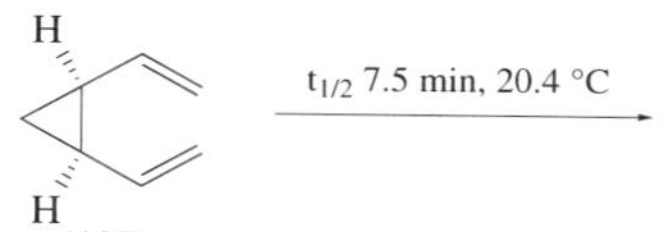

(78)

(107) **(108)**

When there are one or more substituents at C-3 and/or C-4 of the 1,5-diene, rearrangement can give a mixture of (*E*)- and (*Z*)-isomers of the product diene. The (*E*):(*Z*) ratio depends on the orientation of the substituent in the transition state. As this is normally chair-shaped, the substituent generally prefers the less-hindered equatorial position and this leads to the (*E*)-isomer of the product ⟨62T67⟩. The degree of stereocontrol across the new C—C bond is also very high and the preferred diastereoisomer can be related to the alkene geometry of the starting 1,5-diene. This is a result of the chair-shaped transition state which imparts stereospecificity to the rearrangement such that the (*E,E*)- and (*Z,Z*)-isomers of the starting diene give predominantly the *syn*-diastereoisomer of the product, whereas the (*E,Z*)-isomers give the *anti*-diastereoisomer ⟨87JOC5198⟩. In addition, as a consequence of the ordered transition state, chirality transfer across the allylic system is possible ⟨84MI 109-01⟩.

When the 1,5-diene is substituted by a hydroxy or alkoxy group at C-3 and/or C-4, the 3,3-sigmatropic shift is known as the oxy-Cope rearrangement. This gives rise, initially, to an enol or enol ether which is converted to the corresponding carbonyl compound. Early work on this rearrangement ⟨64JA5017, 64JA5019, 65JA1150, 66TL6115⟩ demonstrated that the reaction provided a convenient method for the preparation of δ,ε-unsaturated carbonyl compounds, although temperatures in excess of 200°C are normally required. Despite this the oxy-Cope rearrangement has found a number of applications as its irreversibility has allowed the synthesis of large rings which are disfavoured by the parent Cope rearrangement. For example, the diene (**109**) rearranges on heating to the ten-membered ring (**110**) (Equation (79)) ⟨70TL509⟩, whereas the equilibrium for the Cope rearrangement lies in favour of the six-membered ring.

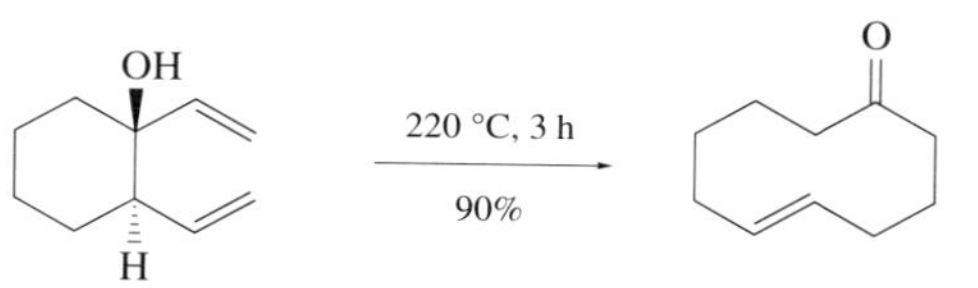

(79)

(109) **(110)**

The simple diene (**111**) undergoes an oxy-Cope rearrangement to give the unsaturated aldehyde (**112**) (Equation (80)) when distilled at 380°C ⟨65JA1150⟩ and the effect of substitution on each of the six carbon atoms has been investigated ⟨67JA3462⟩. A major side-reaction in the oxy-Cope rearrangement is a fragmentation by a competing retro-ene process. This was found to reduce the yield of the Cope rearrangement product particularly when the 3-hydroxy-1,5-diene is substituted at C-1, C-5 and/or C-6. Competing fragmentation in the oxy-Cope rearrangement may be minimised by the use of an alkoxy or acyloxy group in place of the hydroxy group, or by the use of *N*-methyl-2-pyrrolidinone as the solvent ⟨78CC972⟩.

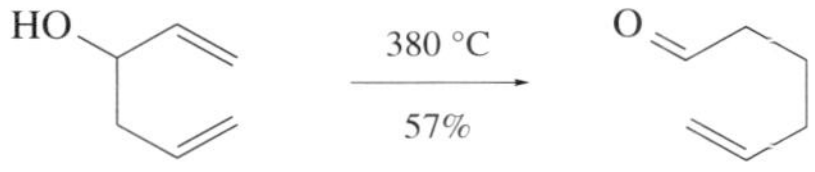

(111) (112) (80)

In addition to the thermal activation of the Cope and oxy-Cope rearrangements, it is known that these rearrangements can be promoted by the addition of an acid or metal catalyst ⟨84AG(E)579, 84CRV205⟩. A few examples of acceleration of the Cope rearrangement with acids such as H_2SO_4 or HCl, or with Lewis acids such as $AlCl_3$, $BF_3 \cdot OEt_2$ or Al_2O_3 have been reported ⟨84CRV205⟩. More examples have been documented with the use of a metal catalyst, especially a palladium(II) complex, but also nickel(0) and mercury(II) salts ⟨84CRV205, 84AG(E)579, 90JA3945⟩. The preferred complex seems to be $PdCl_2(PhCN)_2$ which is used normally in benzene or THF at room temperature. Some examples are given in Scheme 36 illustrating the relative ease of these processes (the diene **(113)** requires heating to a temperature of 150 °C for 5 h to effect the thermal rearrangement) ⟨66JOM(6)412, 73JOM(49)477⟩. The rearrangement of acyclic dienes can be promoted using palladium(II) catalysis and this can be effective with less than one equivalent of $PdCl_2$ ⟨80JA865⟩. The oxy-Cope rearrangement of tertiary alcohols, for example the diene (**114**), proceeds at room temperature in the presence of 10 mol% $PdCl_2(PhCN)_2$ ⟨83TL1157⟩. A major drawback, however of these metal-catalysed 3,3-sigmatropic rearrangements in acyclic dienes is the necessity for the diene to be substituted at either C-2 or C-5.

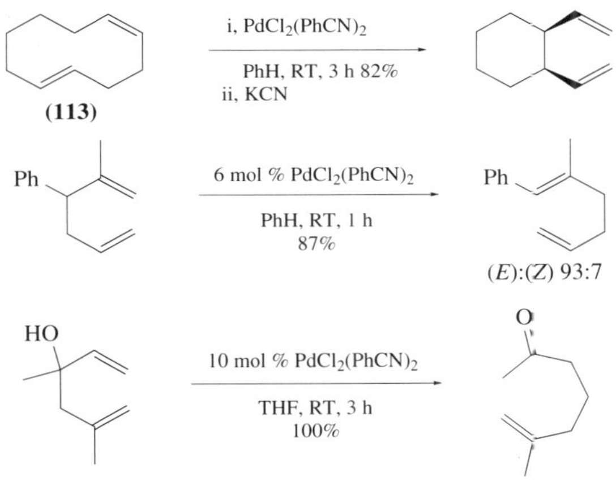

Scheme 36

An alternative and now common method for catalysing the oxy-Cope rearrangement is the addition of a base to deprotonate the alcohol. This was popularised by Evans and Golob ⟨75JA4765⟩ who showed that enormous rate enhancements are possible. The anionic oxy-Cope rearrangement (Scheme 37) is often best accelerated as the potassium salt (using KH or potassium hexamethyldisilazide (KHMDS) as the base, normally in THF), although sodium alkoxides may work well (deprotonation with NaH). Whereas the oxy-Cope rearrangement requires high temperatures, deprotonation of the alcohol at room temperature may be followed directly by rearrangement. In cases which are reluctant to proceed at reflux in THF, the use of DIGLYME as the solvent or the addition of 18-crown-6 or TMEDA may promote the reaction. An improvement in the yield of the rearranged product (the δ,ε-unsaturated carbonyl compound) may result from cleaning the KH with iodine ⟨86JOC1124⟩.

The 1,3-sigmatropic rearrangement may compete with the anionic oxy-Cope rearrangement,

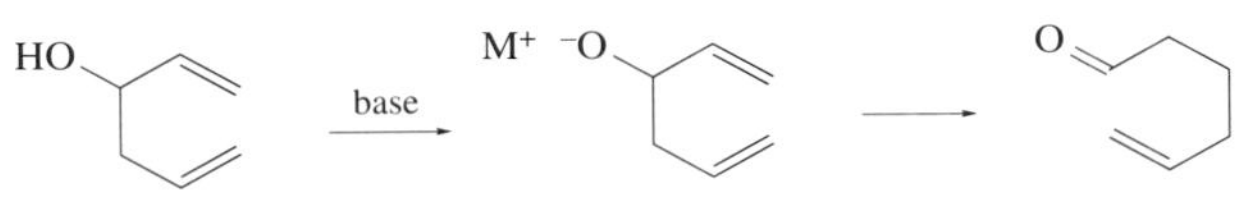

Scheme 37

although the 3,3-sigmatropic process is usually favoured ⟨77TL2559⟩. Like the parent Cope rearrangement the anionic oxy-Cope rearrangement has been demonstrated to prefer the chair-shaped transition state. This is illustrated in the conversion of the potassium salt of the 3-hydroxy-diene (**115**) into the ketone (**116**) on heating (Equation (81)) ⟨80JA774⟩. The diastereoselectivity across the new C—C bond of the ketone (**116**) reflects the chair-like transition state with the alkoxy group in the pseudoequatorial position. In acyclic systems the preference for the pseudoequatorial alkoxy group is diminished, resulting in poor chirality transfer from the carbon bearing the hydroxy group across the allylic system ⟨92JA5018⟩. However, the rearrangement of dienes in which one of the C=C bonds is constrained to a ring often proceeds with a high degree of stereocontrol and is a powerful technique for the construction of complex bicyclic and polycyclic ring systems.

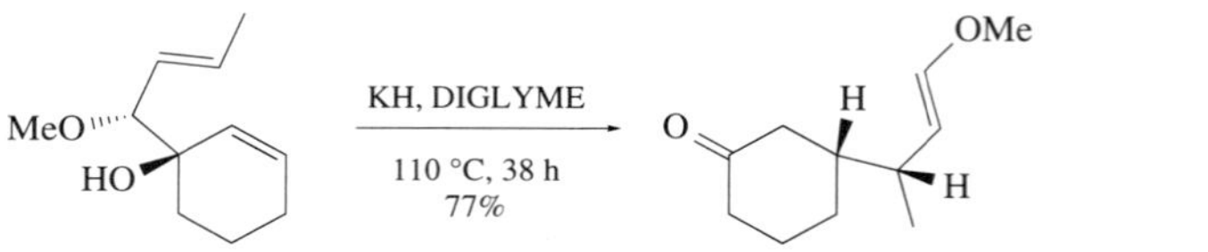

(**115**) (**116**) *trans*:*cis* 96:4 (81)

(ii) Claisen rearrangement

Since the first report by Claisen ⟨12CB3157⟩, the rearrangement of allyl vinyl ethers and modifications of this 3,3-sigmatropic process have been the subject of a great many mechanistic and stereochemical studies, and have formed the key step in many synthetic strategies ⟨75OR(22)1, 80T3, 84CRV205, 88CRV1423, 89S71, 91COS(5)827, 93MI 107-01⟩. The rearrangement proceeds with the formation of a new C—C bond at the expense of a C—O (or other C–heteroatom) bond through a highly ordered six-membered ring transition state. As a result, the rearrangement shows a high degree of stereospecificity with the stereochemical outcome at the new C—C bond determined by the geometry of the double bonds in the starting material.

The Claisen rearrangement has been successful with a wide variety of substrates including allyl aryl ethers, allyl vinyl ethers (in open-chain or cyclic systems) and substrates in which the allyl group has been replaced by a propargyl or an allenyl group. In addition, aza-Claisen, thia-Claisen and metalla-Claisen rearrangements have been investigated. A number of methods for the preparation of allyl vinyl ethers are known ⟨91COS(5)827⟩. These commonly involve the reaction of an allylic alcohol with a simple alkyl vinyl ether using a catalytic amount of a protic or Lewis acid (particularly $Hg(OAc)_2$) ⟨57JA2828⟩. The various modifications of the Claisen rearrangement, especially those of Eschenmoser ⟨64HCA2425⟩, Johnson ⟨70JA741⟩ and Ireland ⟨72JA5897⟩ provide simpler methods for accessing the required allyl vinyl ether arrangement and each will be discussed in turn (see Chapter 1.18).

Aromatic Claisen rearrangements involve the rearrangement of allyl aryl ethers with migration of the allyl group through an allylic transposition to the *ortho* position of the aromatic ring (Scheme 38) ⟨B-63MI 109-01⟩. Further 3,3-sigmatropic rearrangements to the *para* position may take place to give a mixture of the *ortho* and *para* substituted products. The rearrangement can be promoted by heating the allyl aryl ether neat or in an appropriate solvent (e.g., dimethyl aniline) in the temperature range 150–200 °C ⟨75OR1⟩. A lower temperature may be employed when the reaction is conducted in the presence of a Lewis acid catalyst such as BF_3, BCl_3, Et_2AlCl or a transition metal derivative ⟨84CRV205⟩.

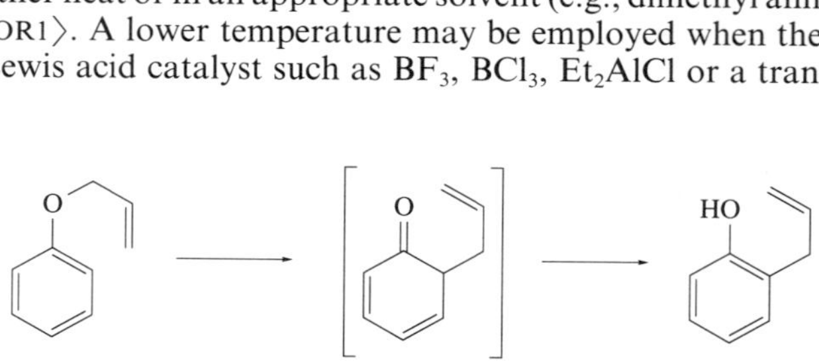

Scheme 38

Aliphatic Claisen rearrangements of allyl vinyl ethers give rise to γ,δ-unsaturated carbonyl compounds. The rearrangement is normally effected by heating the substrate neat or in a solvent such as decalin and generally requires temperatures in the range 100–200 °C. Allyl vinyl ether itself rearranges to 4-pentenal at 255 °C (Equation (82)) ⟨38JA1905⟩ although lower temperatures can be used at the expense of longer reaction times and reduced yields ⟨50BSF121⟩.

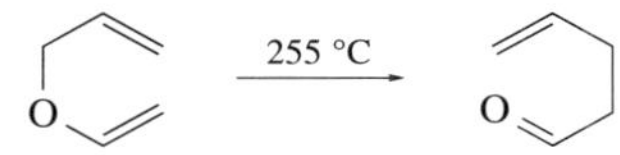

(82)

The presence of a substituent α to the oxygen atom in the allyl unit of the allyl vinyl ether gives rise, on rearrangement, to predominantly, if not entirely, the (*E*)-alkene product ⟨60JA4681⟩. This arises from the preferential orientation of the substituent in the less-hindered pseudo-equatorial position of the chair-shaped transition state. The preference for the chair form in the rearrangement of acyclic allyl vinyl ethers is of the order of 95–96% ⟨74T1959⟩ and this accounts for the high degree of stereocontrol in the Claisen rearrangement. *Ab initio* quantum mechanical calculations also support the chair-shaped transition structure ⟨88JA2314⟩. As a result, the stereochemical outcome at the new C—C bond can be predicted on the basis of the geometry of the C=C bonds in the allyl vinyl ether. This is illustrated in Scheme 39 in which the (*E,E*)- and (*Z,Z*)-isomers give predominantly (≥95%) the *threo* diastereoisomer whereas the (*E,Z*)- isomers give predominantly (≥95%) the alternative *erythro* diastereoisomer ⟨68HCA1457, 75HCA1293⟩. When the allyl unit of the allyl vinyl ether forms part of a ring system then conformational constraints may favour the boat-shaped transition structure with the consequence that a reduced or even reversed stereoselection may result.

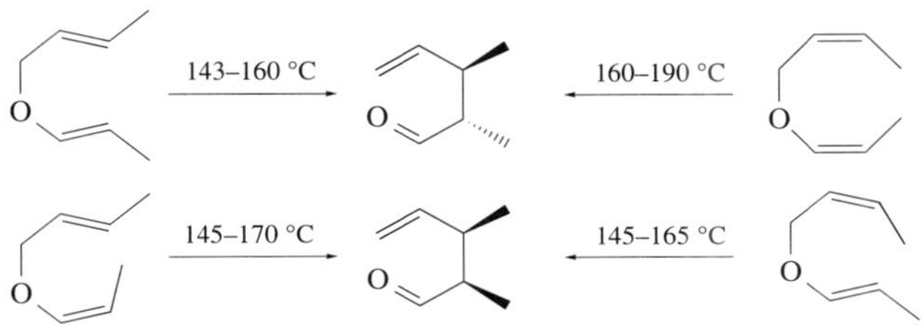

Scheme 39

A reduction in the temperature required to effect the Claisen rearrangement may be made in the presence of a catalyst ⟨84CRV205⟩. Most successful have been the use of alkylaluminium or palladium based catalysts, rearrangement taking place often at room temperature. Of note is the use of Et_2AlSPh in 1,2-dichloroethane which allowed the rearrangement of the allyl vinyl ether (**117**) to the γ,δ-unsaturated aldehyde (**118**) at room temperature (Equation (83)) ⟨81TL3985⟩. The corresponding thermal rearrangement required heating the allyl vinyl ether (**117**) in a sealed tube to 180 °C, but gave a high stereoselection at the new C=C bond (>95% (*E*)). The catalysis of the Claisen rearrangement with bulky aluminium reagents has been successful ⟨88JA7922, 90JA316⟩. A reduction or reversal of the stereoselection normally observed in the thermal Claisen rearrangement is also observed with palladium catalysts. A catalytic amount of the palladium(II) salt $PdCl_2(MeCN)_2$ in THF or benzene is effective in promoting the Claisen rearrangement of allyl vinyl ethers at room temperature ⟨86TL6267, 87TL5879, 89SC2109⟩. Competing cleavage of the allyl vinyl ether is, however, observed especially if the vinyl group is unsubstituted.

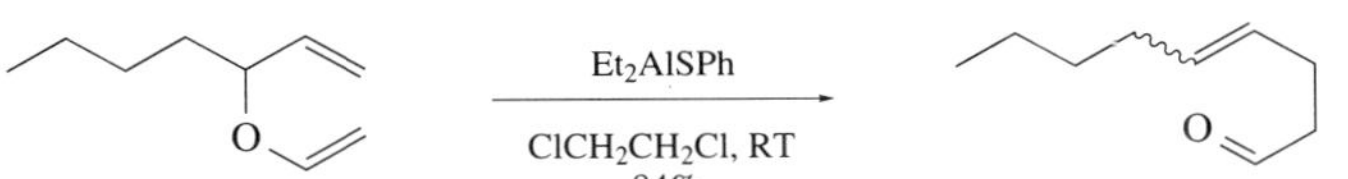

(**117**) (**118**) (*E*) : (*Z*) 39 : 61

(83)

In 1961 Meerwein *et al.* reported the use of the Claisen rearrangement for the preparation of γ,δ-unsaturated carboxylic amides (Scheme 40, conditions a) ⟨61LA(641)1⟩. This discovery was later developed by Eschenmoser and co-workers ⟨64HCA2425, 69HCA1030⟩. This method obviated the need for the prior formation of an allyl vinyl ether using an acid catalyst. The general procedure involves heating a dialkyl acetal of an *N,N*-dialkylamide with an allylic alcohol in the temperature range 100–180 °C. Alternatively, treatment of the lithium salt of the allylic alcohol with the methylated amide gives the Claisen product at room temperature ⟨85JOC5909⟩ (Scheme 40, conditions b).

As expected, the Meerwein–Eschenmoser modification of the Claisen rearrangement occurs through a chair-shaped transition state and this allows control over the stereochemistry at the new C—C bond, as depicted in Scheme 41 ⟨71CB3679⟩. The use of the dimethyl acetal of a propionamide (or higher amide) results in the preferential formation of the (*Z*) geometry of the ketene *N,O*-acetal. Therefore, the choice of (*E*)- or (*Z*)-allylic alcohol determines the stereochemical outcome of the rearrangement; the (*E*)-allylic alcohol giving the *erythro* product, the (*Z*)-allylic alcohol giving the

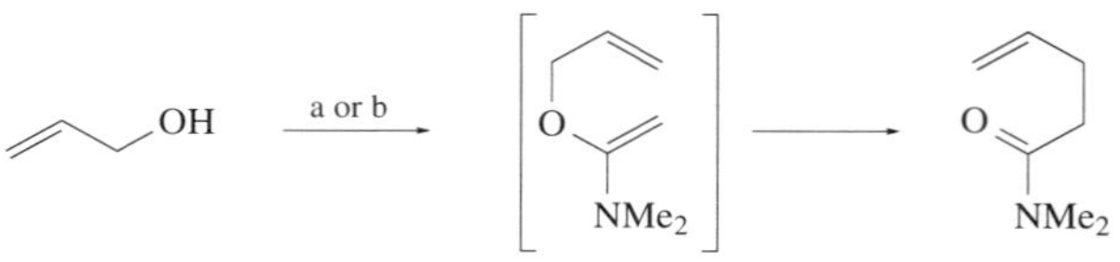

a: $MeC(OEt)_2NMe_2$, 130 °C, b: i, Bu^nLi, THF; ii, $MeCONR_2$, $MeSO_2CF_3$, THF, RT

Scheme 40

threo product. With a substituent in the α position of the allyl unit the (*E*) geometry of the γ,δ-unsaturated amide predominates, although competitive elimination to a diene may be observed ⟨66TL6425⟩.

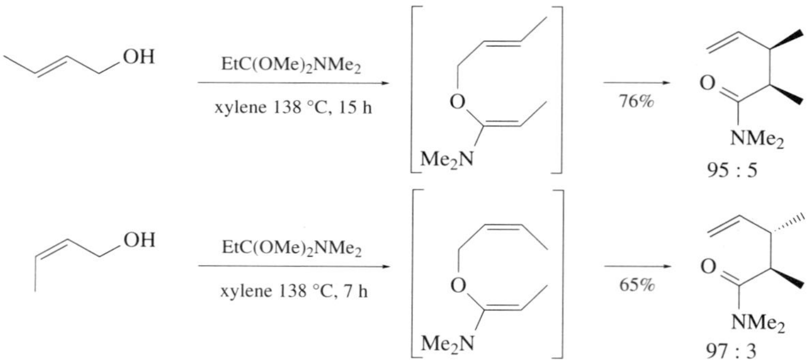

Scheme 41

Related to the Meerwein–Eschenmoser method is that of Johnson *et al.* ⟨70JA741⟩ in which the allylic alcohol is reacted with an *ortho*-ester in the presence of a catalytic amount of a weak acid. Like the Meerwein–Eschenmoser method, a temperature in the range 100–180°C over a period of 1–48 h is generally required for ketene acetal formation and rearrangement, although microwave irradiation is known to accelerate this process ⟨92JCS(P1)311⟩. The Johnson modification of the Claisen rearrangement gives access to γ,δ-unsaturated esters in one step from allylic alcohols. For example, heating the allylic alcohol (**119**) with excess triethyl orthoacetate and a catalytic amount of propionic acid gives the γ,δ-unsaturated ester (**120**). The corresponding allyl vinyl ether rearrangement via ether (**121**) requires a two-step operation giving the γ,δ-unsaturated aldehyde (**122**) in lower overall yield and with a lower stereoselectivity (Scheme 42).

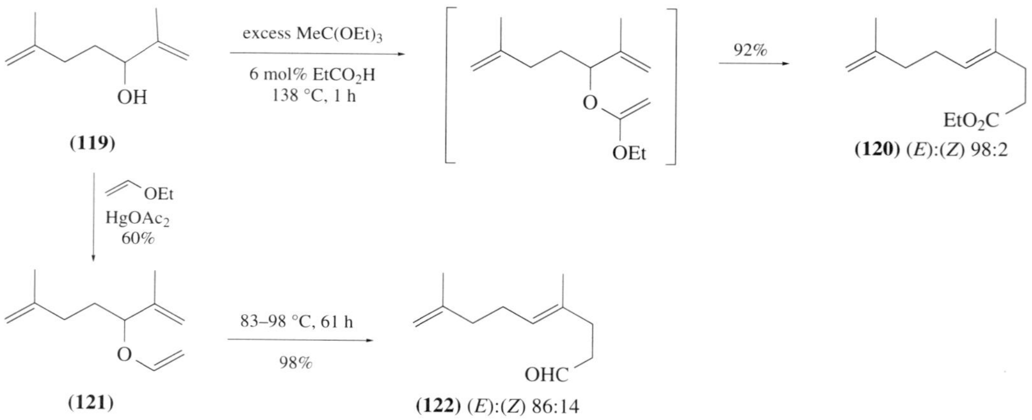

Scheme 42

A more controlled and now popular method for effecting the Claisen rearrangement was developed by Ireland *et al.* ⟨72JA5897, 75TL3975, 76JA2868⟩ who used acylated allylic alcohols (Scheme 43). The method requires only one equivalent of an acylating agent and lithium dialkylamide to deprotonate to the corresponding enolate. Control over the enolate geometry is possible, unlike earlier related methods using a β-keto ester (Carroll modification) ⟨40JCS704, 40JCS1266, 41JCS507, 43JA1992⟩ or using other bases ⟨49JA1150, 49JA2439, 60JOC1815, 72SC27⟩. The Ireland–Claisen rearrangement proceeds at low temperatures to give γ,δ-unsaturated carboxylic acids ⟨93MI 109-01⟩. The problem of aldol side products is avoided by formation of the silyl ketene acetal from the enolate prior to rearrangement. This is accomplished normally by the addition of trimethylsilyl chloride or *t*-butyldimethylsilyl chloride (TBDMS-Cl) (which gives less C-silylation) to the enolate in THF at $-78\,^\circ$C, or by an *in situ* quench with the silyl chloride. On warming these silyl ketene acetals the 3,3-sigmatropic rearrangement takes place. This generates the silyl ester of the rearranged product which can be hydrolysed in acid or base to the corresponding carboxylic acid.

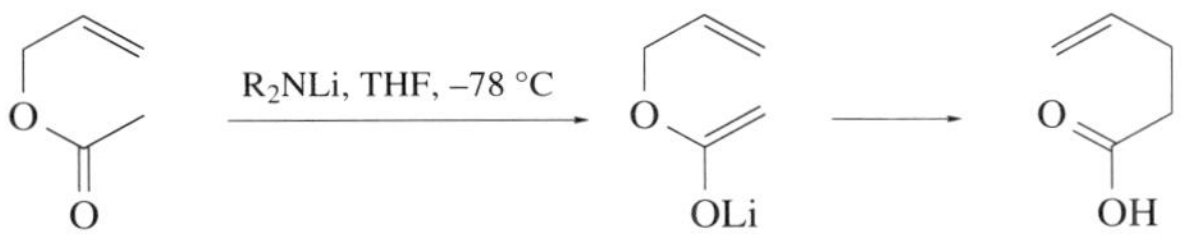

Scheme 43

As the acyclic Ireland–Claisen rearrangement proceeds through a chair-shaped transition state, the control over the enolate geometry allows a control over the diastereoselectivity at the new C—C bond. Enolisation at $-78\,^\circ$C in THF followed by trapping with the silyl chloride favours the (*E*)-silyl ketene acetal, whereas THF/23% hexamethylphosphoramide (HMPA) as the solvent favours the (*Z*)-silyl ketene acetal (Scheme 44) ⟨75TL3975⟩. The use of THF/DMPU (*N*-*N*′-dimethylpropyleneurea) may improve the selectivity further ⟨91JOC650⟩. On warming, the (*E*)-isomer (with the (*E*)-allylic alcohol) rearranges to predominantly the *erythro* diastereoisomer and the (*Z*)-isomer to the *threo* diastereoisomer (Scheme 44). The diastereoselectivity is reversed from the (*Z*)-allylic alcohol. When the allylic group is constrained to a ring then the boat-shaped transition state may be favoured ⟨91JOC3572⟩.

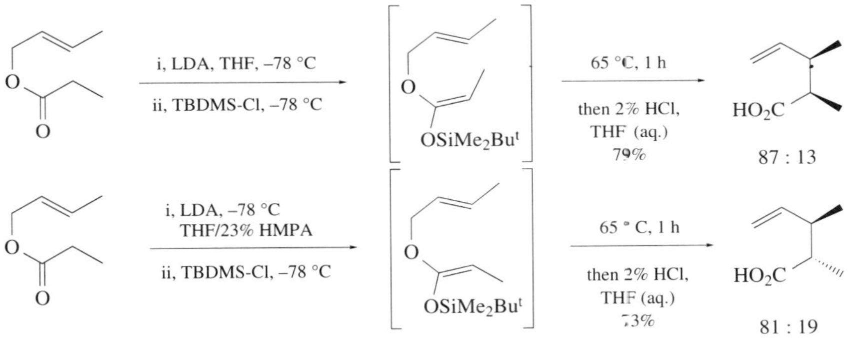

Scheme 44

The preference for a defined transition state in the Claisen rearrangement makes possible the transfer of chirality across the allyl unit. In addition, a chiral centre at a position other than α to the oxygen atom in the allyl unit may influence the stereochemical outcome of the rearrangement. Noteworthy in the Ireland–Claisen rearrangement is the use of a chiral boron reagent to generate a boron enolate which rearranges with high diastereoselectivity and enantioselectivity ⟨91JA4026⟩. Future developments and applications of the Claisen rearrangement will enhance further the status of this useful method for the stereocontrolled formation of C—C bonds.

1.09.1.5 Electrocyclic Reactions

This section of the chapter will give a brief outline of electrocyclic processes that create a new C—C σ bond. This new σ bond is formed between the terminal ends of a polyene, with concomitant reorganisation of the π system. The C—C bond becomes part of a ring (containing one less π bond than the polyene) and the rearrangement occurs with almost total stereospecificity. A description of the orbital symmetry involved in electrocyclic reactions has been discussed by Woodward and Hoffmann ⟨65JA395, 69AG(E)781, B-70MI 109-01⟩ and by Fukui ⟨71ACR57⟩. The most important of these electrocyclisations for C—C σ bond formation are the 1,3-diene-cyclobutene interconversion, the 1,3,5-triene-cyclohexadiene interconversion and the Nazarov cyclisation. A general review of these and other electrocyclisations can be found in the monograph by Marvell ⟨B-80MI 109-01⟩. A discussion of aspects of electrocyclisations with particular reference to the formation of the new C=C bond(s) is given in Chapter 1.17.

1.09.1.5.1 Cyclobutene formation

The interconversion of a cyclobutene and a butadiene can be promoted thermally or photochemically ⟨B-80MI 109-01, 91COS(5)675⟩. As cyclobutenes have a high degree of ring strain they are, in general, thermodynamically less stable than butadienes. Four electron electrocyclisations have therefore been used most widely for the controlled preparation of a diene, rather than for cyclobutene formation. Particularly noteworthy in this regard is the formation of *o*-quinodimethanes from benzocyclobutenes and their subsequent use for the preparation of bicyclic and polycyclic ring systems. Benzocyclobutenes are actually thermodynamically more stable than their open-chain dienes, although the diene can be trapped, normally by a dienophile in a Diels–Alder reaction. For example, the *o*-quinodimethane (**123**), formed by extrusion of SO_2 at 110 °C can be trapped by a dienophile (Scheme 45). In the absence of a dienophile, the benzocyclobutene (**124**) can be isolated in good yield ⟨86CJC793⟩. Similarly, the parent benzocyclobutene can be isolated by heating the corresponding sulfone neat at 280–300 °C (or in the vapour phase at ≥460 °C) ⟨59JA4266⟩. Other cyclobutenes which tend to be more stable than their ring-opened dienes include difluorinated, polyfluorinated or other polysubstituted cyclobutenes. In addition, bicyclic butenes may be more stable than their monocyclic dienes if considerable ring strain is present in the diene. This is the case with medium ring dienes containing a double bond of (*E*) geometry. On warming (*E*,*Z*)-1,3-cyclooctadiene (**125**) to 80 °C, bicyclo[4.2.0]octene (**126**) is formed essentially quantitatively (Equation (84)) ⟨65JA3996⟩. In contrast, the reduced ring strain in (*E*,*Z*)-1,3-cyclodecadiene, the two-carbon homologue of (**125**), favours the diene over the bicyclo[6.2.0]decene ⟨67TL4901⟩.

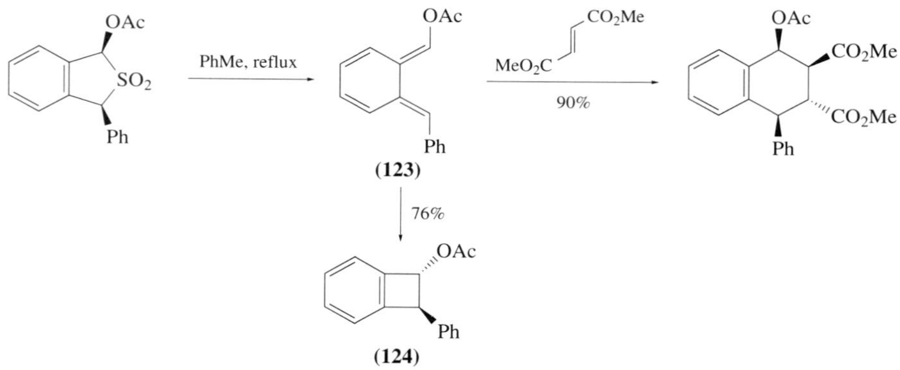

Scheme 45

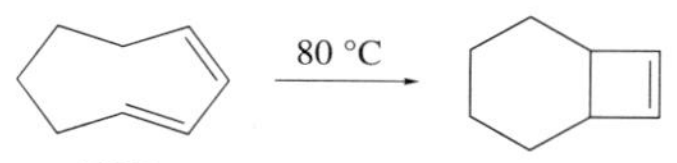

(84)

Despite the lower thermodynamic stability of simple cyclobutenes, they can be prepared from 1,3-dienes under photochemical conditions ⟨62JA4141, 62PCS334, 65T1001, 78AG(E)150⟩. Irradiation of butadiene itself gives rise to cyclobutene in moderate yield, together with side products which include bicyclobutane ⟨63JA4045⟩. Electrocyclisation takes place from the excited singlet state of the 1,3-diene. Normal conditions involve irradiation of a solution of the 1,3-diene in Et_2O with a medium-pressure mercury lamp. The stereochemical outcome at the new C—C bond can be interpreted on the basis of the Woodward–Hoffmann rules ⟨65JA395, 69AG(E)781⟩, which state that a four electron electrocyclisation occurs by a thermal conrotatory or photochemical disrotatory pathway. Irradiation of 2,4-hexadiene causes rapid (*E*)–(*Z*) isomerisation to give a mixture of the (*E*,*E*)- (35.7%), (*Z*,*E*)- (30.5%) and (*Z*,*Z*)- (33.7%) isomers (Scheme 46). Further irradiation leads to *cis*-3,4-dimethylcyclobutene (**127**) which, from rate studies, was shown to arise from the (*E*,*E*) diene by a disrotatory process ⟨68JA4498⟩.

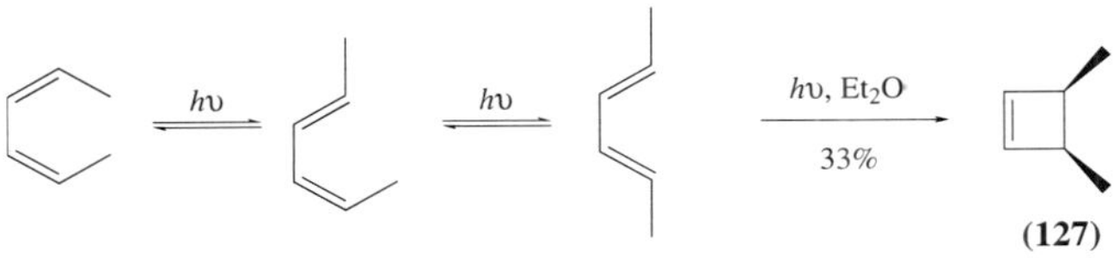

(127)

Scheme 46

Photochemical electrocyclisation of cyclic 1,3-dienes gives rise to bicyclo[*n*.2.0]alkenes by a disrotatory process ⟨62JA4141, 62JOC1910, 66JA1073⟩. (*Z*,*Z*)-1,3-Cyclooctadiene (**128**) can be transformed into bicyclo[4.2.0]octene (**126**) using a mercury arc lamp. Addition of a sensitiser (e.g., acetophenone) causes prior isomerisation to the (*Z*,*E*)-isomer, which undergoes electrocyclisation on warming to 80 °C by a conrotatory closure (Scheme 47) ⟨67JA112⟩.

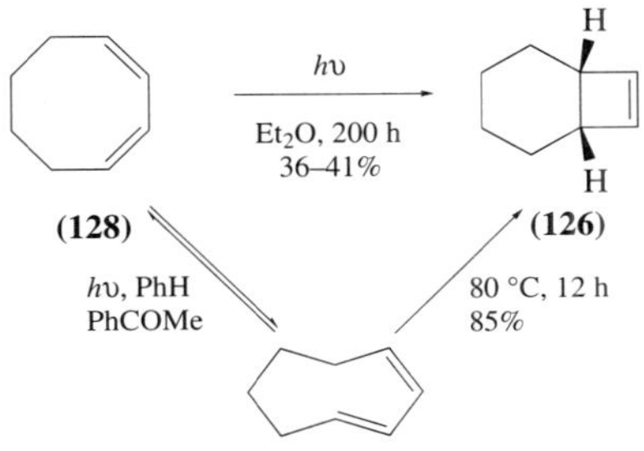

(125)

Scheme 47

1.09.1.5.2 Cyclohexadiene formation

Unlike most butadiene-cyclobutene electrocyclisations, the thermal hexatriene-cyclohexadiene electrocyclisation lies in favour of the ring-closed product. The parent 1,3,5-hexatriene (Equation (85)) in which the central C=C bond has (*Z*) geometry can be converted into 1,3-cyclohexadiene on heating to ≥110 °C ⟨64JCS3080⟩. The rearrangement of hexatrienes is normally performed by heating the triene in a hydrocarbon solvent such as pentane, hexane, cyclohexane or xylene. Although photochemical conditions have also been documented to effect this electrocyclisation, these conditions are used mostly for the reverse process, ring-opening of cyclohexadienes. The electrocyclisation of hexatrienes has been the subject of a number of reviews ⟨63AG(E)1, B-80MI 109-01, B-81MI 109-01, 83RCR61, 91COS(5)699⟩.

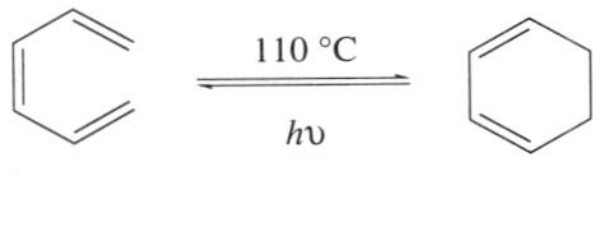

(85)

An important and early example of the six-electron electrocyclisation came from studies on the rearrangement of precalciferol (previtamin D) to pyrocalciferol and isopyrocalciferol ⟨55BSF1341, 61T146⟩. In this system thermal and photochemical induced cyclisations give complementary stereochemical results. This led to the development of the Woodward–Hoffmann rules for electrocyclic reactions ⟨65JA395, 69AG(E)781⟩. A consideration of orbital symmetry allows the prediction that a thermal six-electron electrocyclisation should occur by a disrotatory mechanism, whereas the photochemical reaction would prefer the conrotatory pathway. This has been verified in the rearrangement of both acyclic and cyclic trienes. The electrocyclisation of (*E*,*Z*,*E*)-2,4,6-octatriene (**129**) in pentane at 132 °C (in a sealed tube) gives the *cis* isomer of dimethylcyclohexadiene (**130**) by a disrotatory closure (Equation (86)) ⟨65TL385, 65TL391, 73T3781⟩. The cyclisation is stereospecific, giving the *cis* product with a diastereoisomeric purity of >99.5%. The corresponding (*E*,*Z*,*Z*)-triene (**131**) interconverts readily with the (*Z*,*Z*,*Z*)-isomer at 110 °C via consecutive 1,7-hydrogen shifts. On further heating to 178 °C in hexane this mixture cyclises to give the *trans*-dimethylcyclohexadiene (**132**) and the cyclohexadiene (**133**) derived from the diene (**132**) by a 1,5-hydrogen shift (Equation (87)) ⟨65TL385, 73T3781⟩.

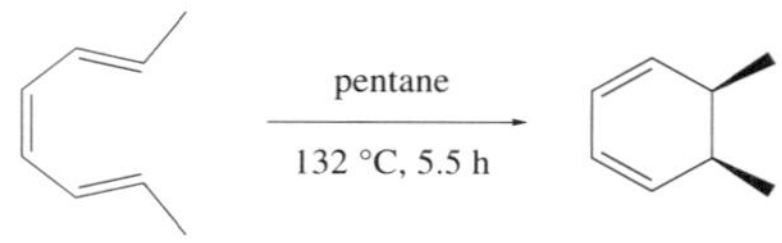

(86)

(129) **(130)**

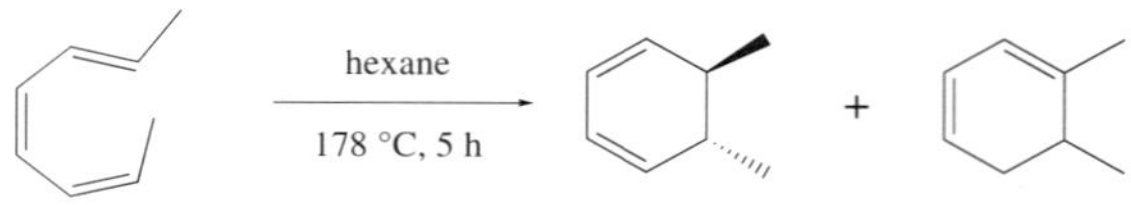

(87)

(131) **(132)** **(133)**

The 1,7-hydrogen shift requires that the central C=C bond and at least one of the other C=C bonds of the triene are of (*Z*) geometry. In these cases, the 1,7-hydrogen shift competes favourably with the electrocyclisation process. This gives rise to a new triene which may undergo electrocyclisation, or a further 1,7-hydrogen shift to give a different geometrical isomer of the original triene. On further heating, the same diastereoisomer of the cyclohexadiene may, therefore, result from both the (*Z*,*Z*,*Z*)- and (*E*,*Z*,*Z*)-trienes. Applications of the electrocyclisation reaction for C—C σ bond formation in cyclohexadienes are, therefore, limited by competing 1,5- and 1,7-hydrogen shifts ⟨75TL3555, 78T2783, 79JCS(P2)810, 87JA8051⟩, and by the difficulty of preparing the required triene geometrically pure.

The photochemical electrocyclisation of trienes should occur by a conrotatory pathway and indeed, irradiation of (*E*,*Z*,*E*)-2,4,6-octatriene (**129**) in ether with a high pressure mercury arc lamp, gives some *trans*-dimethylcyclohexadiene (**132**) ⟨62TL549⟩. However, a considerable amount of starting material is reisolated together with polymeric products. Substitution at C-2 of the acyclic triene improves the yield and increases the rate of formation of the cyclohexadiene to some extent ⟨68BSF3401, 69BSF3665, 73T1393, 75TL3741⟩. Irradiation of trienes is known to cause (*E*)–(*Z*) isomerisation ⟨68JOC3679, 71JA1532⟩ and the formation of bicyclic alkenes ⟨73PAC197, 78AG(E)150, B-80MI 109-02, 87RTC613⟩. The mixture and complexity of products that arise from the photochemical electrocyclisation tend to limit this method to the reverse ring-opening process.

Ring-fused trienes give rise, on electrocyclisation, to bicyclic dienes. When the central C=C bond of the triene is part of a ring then this double bond is locked in the (*Z*) configuration and the reaction is facilitated. For example, the triene (**134**) rearranges on heating at 125 °C to give the bicyclic diene (**135**) (Equation (88)) ⟨73T3797, 83JOC4272⟩. Electrocyclisation may occur at low temperatures when cyclisation leads to the formation of an aromatic ring ⟨72CC1034⟩. The triene (**136**), formed by photochemical ring-opening of *trans*-1,2-dimethyl-1,2-dihydronaphthalene, undergoes thermal disrotatory electrocyclisation at room temperature (Equation (89)) ⟨72HCA3005⟩.

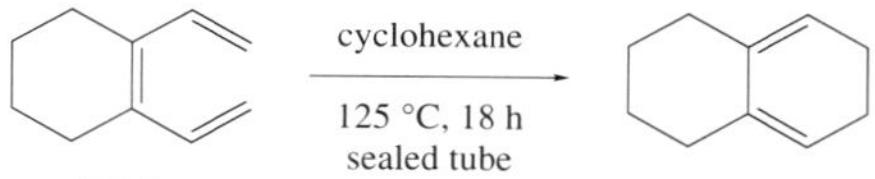

(88)

(134) **(135)**

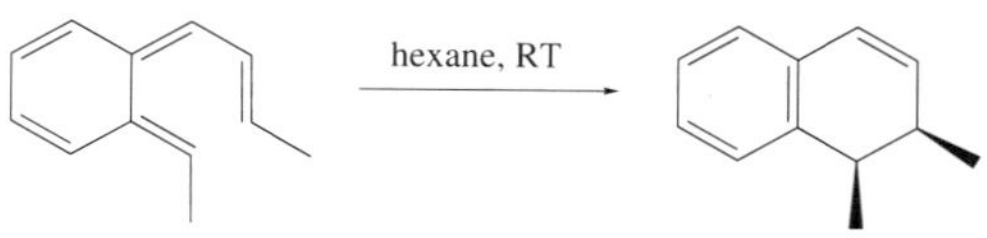

(89)

(136)

The formation of a new C—C bond across the ring junction of a bicyclic diene is possible by electrocyclisation of a medium ring triene. The general transformation (for the (*Z*,*Z*,*Z*)-triene) is outlined in Equation (90)). The proximity of the terminal ends of the triene in such cyclic systems can cause large rate effects. Electrocyclisation of cycloheptatrienes to norcaradienes ($n = 1$, Equation (90)) occurs at low temperatures, although the equilibrium lies in favour of the triene in the unsubstituted case ⟨67AG(E)402, B-80MI 109-01⟩. Electron-withdrawing groups, especially at C-7 (or suitably bridged or ring-fused derivatives), stabilise the bicyclic form. Likewise, cyclooctatriene is favoured over bicyclo[4.2.0]octadiene ($n = 2$, Equation (90)) unless substituted at C-7 and/or C-8 ⟨52JA4867, 68TL5215⟩.

(90)

Larger ring trienes ($n = 3,4$, Equation (90)) can be converted cleanly to the bicyclic form on heating. The stereochemical outcome across the new C—C bond depends on the geometry of the triene, the rearrangement following a thermal disrotatory pathway ⟨81JOC687⟩. For example, (*Z,Z,Z*)-cyclononatriene (**137**) cyclises to the *cis*-fused bicyclo[4.3.0]nonadiene (**138**) on heating at 100 °C in cyclohexane (Equation (91)) ⟨65TL377⟩, whereas (*E,Z,Z*)-cyclononatriene (**139**), which rearranges at a slightly faster rate than the all-(*Z*) isomer (**137**), gives the *trans*-fused diene (**140**) on electrocyclisation (Equation (92)) ⟨65TL391⟩. However, attempted photochemical electrocyclisation gave none of the diene product ⟨73PAC197, 81T3263⟩. Using cyclodecatrienes it has been shown that increased substitution at C-1 and C-6 (the terminal ends of the triene) slows the rate of electrocyclisation ⟨63JA4033, 73JA3932, 82JA6874⟩.

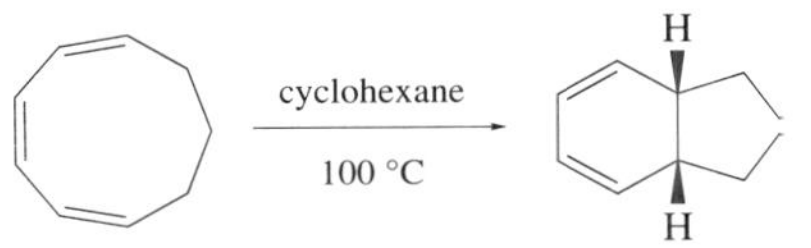

(137) (138)

(91)

(139) (140)

(92)

1.09.1.5.3 Nazarov cyclisation

The formation of a new C—C bond by electrocyclisation of a divinyl ketone is named after studies by Nazarov ⟨42IZV392, 43IZV389, 44IZV65⟩. A series of papers by Nazarov (see ⟨57JGU693, 60JGU765⟩ and references therein) triggered the development of this reaction for the synthesis of cyclopentenones. Reviews of the cyclisation have been fairly comprehensive ⟨83S429, 84S529, 91COS(5)751⟩. The original work involved the hydrolysis of a dienyne to an allyl vinyl ketone. This allyl vinyl ketone isomerises to a divinyl ketone and it is this species which, on acid catalysis, undergoes an electrocyclisation to the cyclopentenone. Both protic and Lewis acid conditions effect the cyclisation, the most commonly used acids being H_3PO_4, H_2SO_4, HCl, $BF_3 \cdot OEt_2$, $AlCl_3$, $SnCl_4$ or $FeCl_3$. The reaction normally proceeds at reasonable rates at room temperature or on slight warming. Protonation or Lewis acid complexation of the oxygen atom of the carbonyl group of a divinyl ketone generates a pentadienyl cation (Scheme 48). This cation undergoes a four-electron electrocyclisation to generate a new C—C bond in a five-membered ring containing an allyl cation. Substituents capable of stabilising the intermediate cation (e.g., aromatic groups α to the carbonyl group) increase the rate and yield of the reaction. The intermediate allyl cation may be trapped by a nucleophile (e.g., the acid counteranion or the solvent) ⟨72JCS(P1)2271⟩ or lose a proton. Loss of a proton and enol-keto tautomerism leads to the cyclopentenone product(s) (Scheme 48).

Divinyl ketones are, in general, unstable compounds and the Nazarov cyclisation is therefore often effected from a precursor which can be transformed into the divinyl ketone. Subsequent electrocyclisation of the divinyl ketone to the cyclopentenone then takes place under the reaction conditions used for its formation. Common methods for the hydration–electrocyclisation of dienynes employ phosphoric acid or concentrated HCl at 50–90 °C. For example, the dienyne (**141**), on warming with concentrated HCl, is transformed into the divinyl ketone (**142**), which cyclises to the cyclopentenone (**143**) (Scheme 49) ⟨43IZV389⟩. With unsymmetrical dienynes there exists the problem of regioselectivity during the hydration of the alkyne. This generates up to two divinyl ketones.

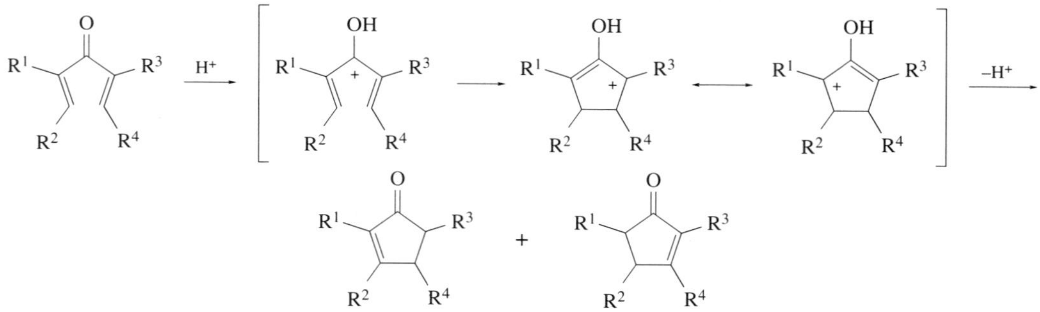

Scheme 48

Unsymmetrical divinyl ketones will cyclise to the cyclopentenone with the most highly substituted C═C bond.

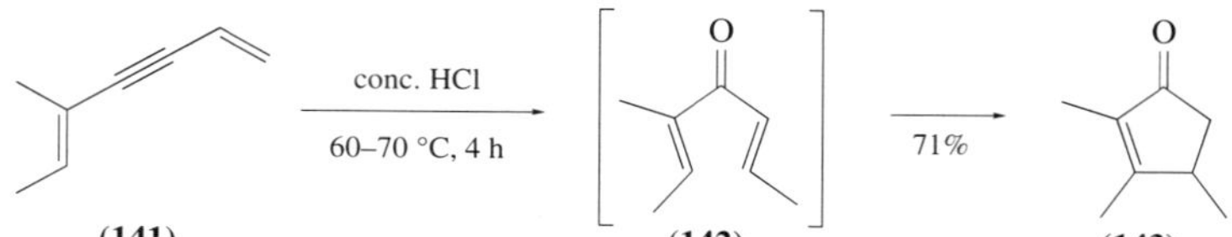

(141) **(142)** **(143)**

Scheme 49

The dienyne may be prepared by dehydration of an acetylenic alcohol. If the acidic conditions for this step promote hydration of the alkyne then the cyclopentenone may be formed directly. In fact, the acid-catalysed transformation of acetylenic diols to cyclopentenones is well documented. The acid H_2SO_4 in methanol is often used for this multistep transformation ⟨76JCS(P1)410, 76TL3585, 79JA1599, 81BCJ2747⟩.

A number of allyl vinyl ketones and divinyl ketones can be prepared and have been shown to undergo Nazarov cyclisation. Common conditions for the cyclisation use phosphoric acid in formic or acetic acid at 50–90 °C to give the cyclopentenone in 50–80% yield ⟨44IZV65, 46IZV633, 47IZV205, 52JCS1430⟩. Other methods for the formation and electrocyclisation of a divinyl ketone include the acylation of an alkene with a carboxylic acid (using polyphosphoric acid at 55–100 °C) ⟨57JIC169, 68T545, 68TL2101, 70BSF2981⟩ or with an acid halide (using a Lewis acid such as $AlCl_3$ in CH_2Cl_2) ⟨53JCS124, 59JCS4017, 79TL4553, 84JOC3870⟩. Alternatively, the acylation of a vinyl silane with an acid halide (using a Lewis acid such as $AlCl_3$ and/or $SnCl_4$ in CH_2Cl_2) ⟨79TL1995, 79TL1999, 80JOC1046, 80JOC3017⟩ or (less commonly) the acylation of a vinyl cuprate with an acid halide ⟨88HCA168⟩ give access to the required divinyl ketone. Elimination of a β' heteroatom from an α,β-unsaturated ketone ⟨53JCS2247, 56BSF1653, 59JCS4017, 85CL1137⟩ or oxidation of a 1,1-divinyl carbinol ⟨52JCS1430⟩ provide further alternative methods for the preparation of a divinyl ketone.

The Nazarov cyclisation of a divinyl ketone generates the thermodynamically most stable cyclopentenone with the C═C bond in the most substituted state. Control over the position of the C═C bond is possible with a suitably positioned trialkylsilyl group ⟨82JA2642, 83HCA2377, 83HCA2397, 86T2821, 88HCA168⟩. Desilylation is generally preferred over deprotonation, and the known β-cation stabilising effect of a silyl group helps to reduce side reactions resulting from the intermediate carbocations. For example, the divinyl ketone **(144)** undergoes electrocyclisation via the allyl cation **(145)** on treatment with $FeCl_3$ in CH_2Cl_2. Loss of the silyl group directs the position of the C═C bond to the less substituted side of the cyclopentenone **(146)** (Scheme 50) ⟨82JA2642, 83HCA2377⟩. The stereochemical outcome α to the carbonyl group is dependent upon which side the enolate is protonated. From acyclic divinyl ketones this commonly results in a mixture of diastereoisomers. In cyclic systems, such as the divinyl ketone **(144)**, the diastereoselectivity can be high. If the whole of the divinyl ketone is incorporated into a ring, then the Nazarov cyclisation will lead to a bicyclic product ⟨68TL3635, 75JA928, 80BCJ169, B-80MI 109-03⟩.

The stereochemical outcome across the new C—C bond is often obscured by the loss of a proton at one of these centres during the cyclopentenone formation. Observation of the expected thermal conrotatory nature of this electrocyclisation ⟨65JA395, 69AG(E)781⟩ is possible if the intermediate allyl cation is quenched by a nucleophile or if the proton lost is *exo* to the five-membered ring. The latter situation has been described by Woodward ⟨B-70MI 109-01⟩ for the electrocyclisation of the divinyl ketone **(147)** (Equation (93)). The resultant tricyclic ketone **(148)** has *trans* stereochemistry, as

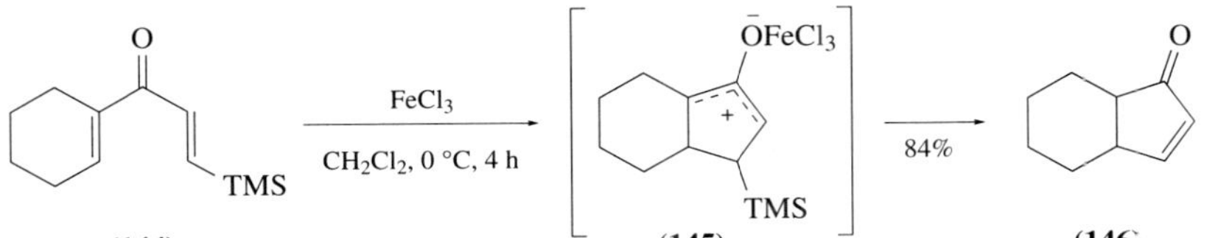

(144) (145) (146)

Scheme 50

expected for a conrotatory electrocyclisation. The *cis* stereoisomer of the same ketone was reported to be produced by a disrotatory photochemically activated electrocyclisation.

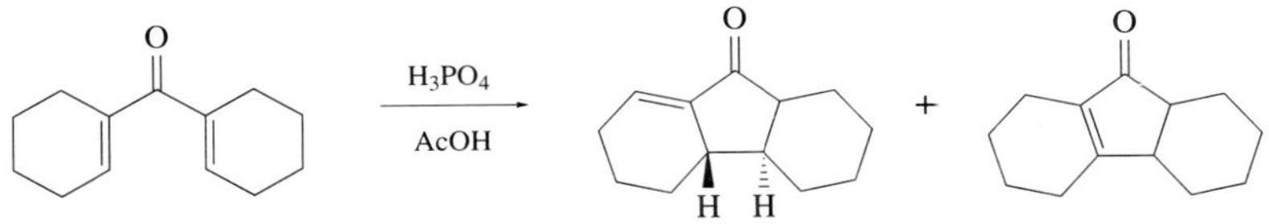

(147) (148) (93)

The Nazarov cyclisation, being an electrocyclic ring closure of a pentadienyl cation, is not restricted to the cyclisation of divinyl ketones. Divinyl chlorides, vinyl allene oxides and other suitably substituted unsaturated systems may be converted to the required pentadienyl cation necessary for a four-electron electrocyclisation. Methods for pentadienyl cation formation and electrocyclisation have been reviewed by Denmark ⟨91COS(5)751⟩.

1.10
One or More =CH Bond(s) Formed by Substitution or Addition

MARTIN A. HAYES
Xenova Group, Slough, UK

1.10.1 ONE OR MORE =CH BOND(S) BY SUBSTITUTION

1.10.1.1 Reduction of =C–Halogen Bonds

1.10.1.1.1 Reduction of aryl halides

The reduction of aryl halides has been reviewed ⟨80S425, B-84MI 110-03, 91COS(8)895⟩. A wide variety of synthetic methods have been used to achieve this transformation. These methods include catalytic hydrogenation and reduction by: (i) tin hydrides, (ii) complex hydrides, (iii) metals, and (iv) electrochemical means. The ease of hydrogenolysis of an aryl halide bond is observed to follow the order I > Br > Cl > F; this difference in cleavage rate allows the selective reduction of mixed halogenoarenes.

(i) Reduction of aryl fluorides

The chemistry of aryl fluorides is featured in review articles ⟨68FCR315, B-76MI 110-01⟩. Generally rather severe conditions are required to hydrogenolyse the aryl C—F bond. The majority of successful syntheses fall into three categories: (a) reduction with LAH in combination with other reagents, (b) hydrogenolysis via catalytic hydrogenation, and (c) miscellaneous methods. Beckwith and Goh ⟨83CC907⟩ found that a mixture of LAH and di-*t*-butylperoxide under irradiation effectively reduced a range of fluorinated aromatics in good yield (Equation (1)). Imamoto *et al.* reported that LAH in combination with $CeCl_3$ in DME or THF to be effective for the reduction of aryl halides, including fluoronaphthalene (Equation (2)) ⟨85CL1491⟩. Alsop *et al.* noted that the LAH reduction of perfluorotoluene proceeded regioselectively yielding 46.9% of the product defluorinated *para* to the fluoromethane group (Equation (3)) ⟨62JCS1801⟩. LAH in refluxing Et_2O was reported to reduce pentafluorobenzene to 1,2,4,5-tetrafluorobenzene mainly ⟨62JCS3253⟩.

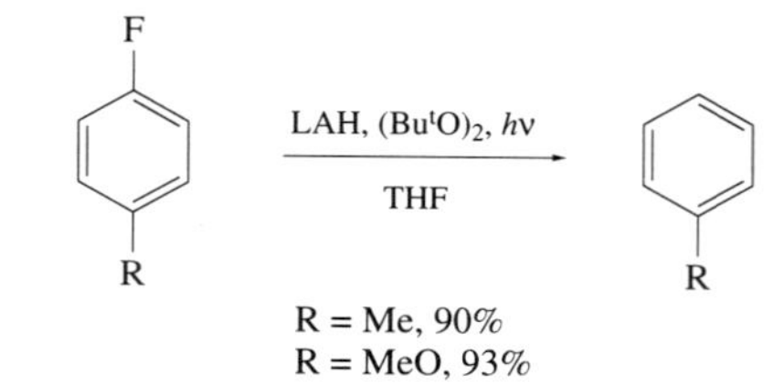

(1)

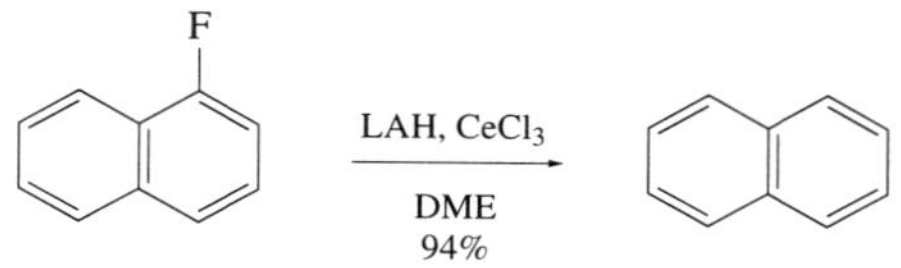

(2)

(3)

The reduction of *p*-fluorobenzoic acid under catalytic hydrogenation conditions has been described ⟨20MI 110-01⟩. In 1963 Oelschlager and Schmersahl ⟨63AP(296)324⟩ reported the reduction of fluoroaniline. Only a poor yield (20%) of aniline was obtained.

Many miscellaneous reductions of aryl fluorides are reported in the literature. These include reduction with Raney nickel or copper alloys ⟨87OPP442⟩. This method gives excellent yields of defluorinated arene but lacks selectivity as noted by the simultaneous reduction of the C—F bond and the methyl ketone moiety on reaction of *p*-fluoroacetophenone. Respess and Tamborski reported that Grignard reagents in combination with $CoCl_2$ reduced hexafluorobenzene to pentafluorobenzene in 91% yield (Equation (4)) ⟨69JOM(18)263⟩. The reduction of α-fluoronaphthalene to naphthalene with EtSH–$AlCl_3$ has been achieved (Equation (5)) ⟨82TL689⟩. The regioselective defluorination of perfluoroaryl diazonium salts on treatment with base has been reported (Equation (6)) ⟨64JOC3042⟩.

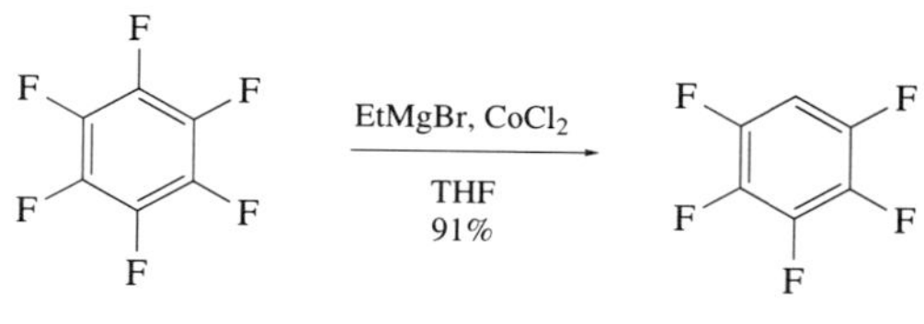

(4)

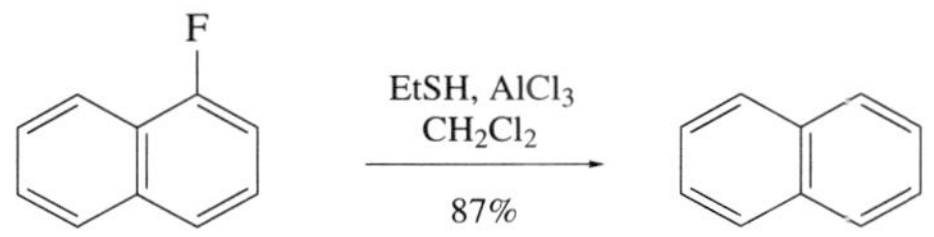

(5)

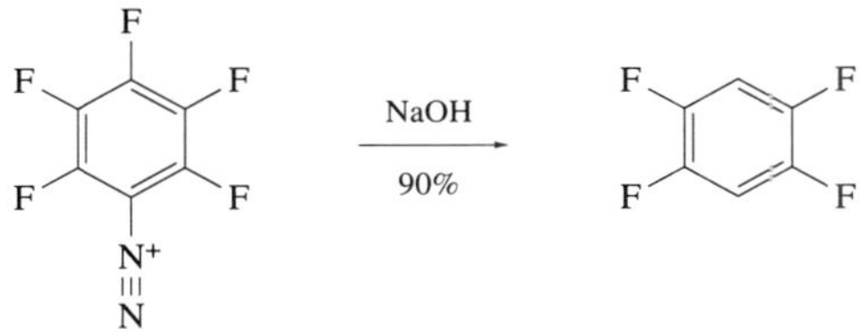

(6)

(ii) Reduction of aryl chlorides

The replacement of chloride by hydrogen in an aryl chloride is significantly easier than the reduction of the C—F bond in aryl fluorides. The synthetic methods of choice are catalytic hydrogenation and reduction with Raney nickel.

Catalytic hydrogenation has been extensively studied as a method of dechlorinating aryl halides ⟨91COS(8)895⟩. Many catalysts are known to be effective, including Adam's catalyst (PtO_2), Pd/C ⟨46JA261⟩ and Pd/Si ⟨83ZN(B)1697⟩. Hydrogenolyses using the above catalysts exhibit a degree of chemoselectivity, the reaction tolerating functionality such as phenols and aromatic nitro groups. A paper by Grushin and Alper ⟨91OM1620⟩ has extended the scope of this reaction. Using a rhodium complex of the type $L_2Rh(H)Cl_2$ (where L=(Cyp)$_3$P or ($Pr^i{}_3P)_2$) under phase transfer conditions, chloroarenes were reduced to the corresponding arene in excellent yields. Carboxylic acid, amine ether and ketone functionality were tolerated as was a trifluoromethyl group (Equation (7)). Catalytic hydrogen transfer reactions with hydrogen donors such as formate have been used successfully (Equation (8)) ⟨85TL1381⟩. Reduction of chloroarenes with complex hydrides such as LAH and $NaBH_4$ are notoriously difficult ⟨91JOC4718⟩. Complex hydrides in combination with Lewis acids reduce chloroarenes to arenes, a good example being the reduction of chloronaphthalene with a mixture of LAH and $TiCl_4$ ⟨73CL291⟩. $NaBH_2(OCH_2CH_2OMe)_2$ with $NiCl_2$ in THF is an extremely potent dechlorinating system ⟨93TL3263⟩; pentachlorophenol was completely dechlorinated to phenol (and a little cyclohexanol) with the above reagent in two hours at 68 °C (Equation (9)). A combination of LAH and $CeCl_3$ has also been described as a good reagent for the dechlorination of aryl chlorides ⟨85CL1491⟩. Aromatic halides have been dehalogenated by electrochemical means in a highly efficient manner ⟨B-83MI 110-01⟩. A development in this methodology ⟨91SC2377⟩ has been to effect the electrochemical reduction in the presence of a catalytic amount of $SmCl_3$. Under the reaction conditions ether, nitrite and ester functionalities remained untouched (Equation (10)).

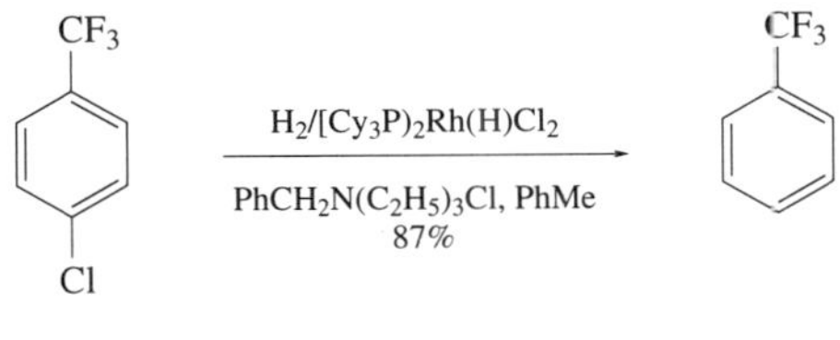

(7)

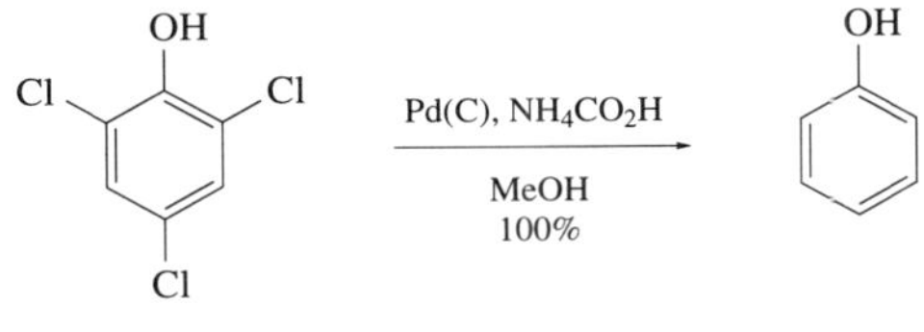

(8)

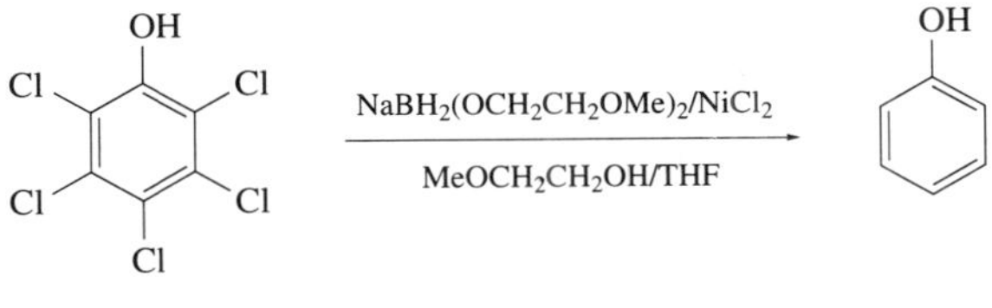

(9)

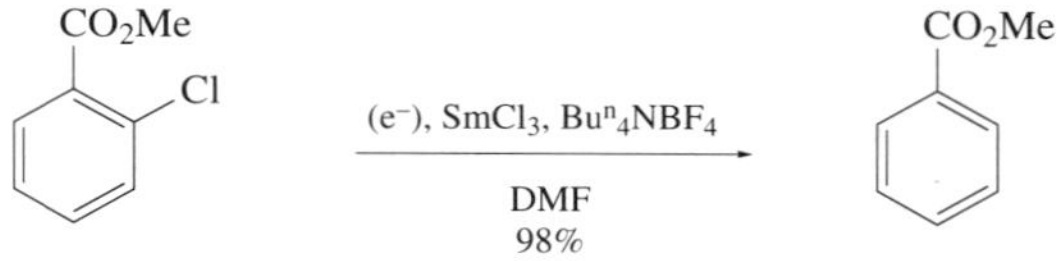

(10)

Reduction of aryl chlorides with Raney nickel and related alloys has been widely used but the method suffers from a lack of chemoselectivity ⟨87OPP442⟩ although both catalytic hydrogenation ⟨65JCS5045⟩ and LAH reductions ⟨69JOC3918⟩ have been carried out to selectively reduce out chlorine in mixed chlorofluoro systems. Miscellaneous methods of reduction include the use of a combination of zinc, $NiCl_2$, PPh_3 and NaI, which reduces *p*-chloroanisole almost quantitatively ⟨82JOC2622⟩ and the reduction of chlorobenzene with triethyl silane in the presence of Pd/C and light ⟨69JOC638⟩.

(iii) Reduction of aryl bromides

The conversion of aryl bromides to their corresponding arenes is probably the most widely studied of all the aromatic halide to arene reductions. The synthetic methods which have been utilised to carry out this reduction include catalytic hydrogenation ⟨73OSC(5)346⟩ and reduction with complex hydrides ⟨69JOC3918⟩ and tin hydrides ⟨87S665⟩.

Palladium and platinum based catalysts have been extensively used for catalytic hydrogenation of aromatic bromides. The use of hydrogen donors (e.g., HCO_2H) rather than molecular hydrogen is a development that avoids the hazards of using hydrogen gas (Equation (11)) ⟨82S876⟩. Reduction with LAH and other complex hydrides is a general method for reduction of bromoaromatics, in contrast to chloro- and fluoroaromatics. LAH reduction frequently occurs at, or just above, room temperature ⟨69JOC3918⟩. LAH combined with $TiCl_4$ is a powerful reagent for the reduction of aromatic bromides ⟨73CL291⟩.

Narisada *et al.* have reported the selective reduction of aryl halides (specifically bromides and iodides) with $NaBH_4$–Cu_2Cl_2 in MeOH. This system reduced aryl bromides to the corresponding arenes, working particularly well for α-methoxy- and α-carbomethoxy-bromobenzenes (Equation (12)) ⟨89JOC5308⟩. It has been shown that both triphenyl ⟨60JOC2203⟩ and tributyltin hydrides ⟨87S665⟩ reduce bromoarenes with high yield and selectivity.

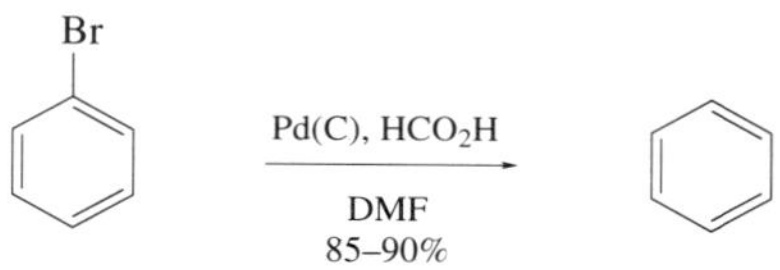

(11)

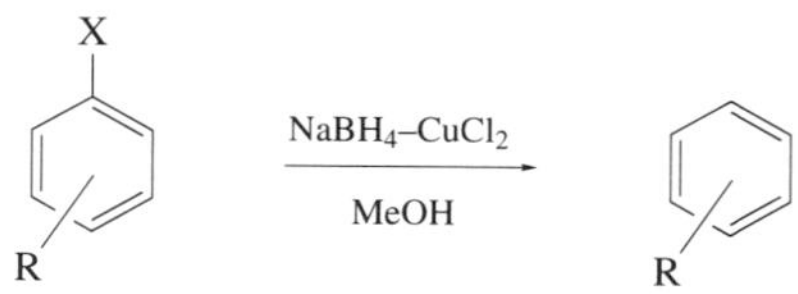

(12)

X = Br, I
R = Me, MeO

Electrochemical reductions provide an elegant way to obtain unusually substituted thiophene derivatives. Dapperheld *et al.* found that simply by varying the metal salts and counter ions in solution it was possible to change the substitution pattern of the product on reduction of tetrabromothiophene and a range of derivatives ⟨90S403⟩. 2,3,5-Tribromothiophene is reduced by zinc in AcOH to 3-bromothiophene ⟨73OSC(5)149⟩. *p*-Bromofluorobenzene was reduced to fluorobenzene in quantitative yield by electrochemical reduction in DMF in the presence of $SmCl_3$ ⟨91SC2377⟩.

Other important methods for reduction of bromarenes are Caubère's complex reducing reagents, i.e. mixtures of $NiCl_2$, NaH and *t*-pentyl alcohol ⟨83AG(E)599⟩, and also the combination of EtSH and $AlCl_3$ ⟨84JOC3641⟩.

(iv) Reduction of aryl iodides

Aryl iodides are reduced to arenes on reaction with hydrides such as $NaBH_4$ ⟨69JOC3923⟩ and by hydrides in combination with metal salts (e.g., $NaBH_4$–$CeCl_3$–MeOH ⟨89JOC5308⟩). In the latter example all iodo derivatives investigated were smoothly converted to the corresponding arene in quantitative yield at 0°C. $NaBH_4$ in combination with Cp_2TiCl_2 ⟨81JOM(204)345⟩ reduced *p*-bromoiodobenzene to bromobenzene, though a small amount (ca. 12%) of debrominated product was also observed (Equation (13)). Catalytic hydrogenation and Raney nickel have been much used for the reduction of iodobenzenes and the methodology is amply covered in review articles ⟨91COS(8)895⟩ and ⟨80S425⟩.

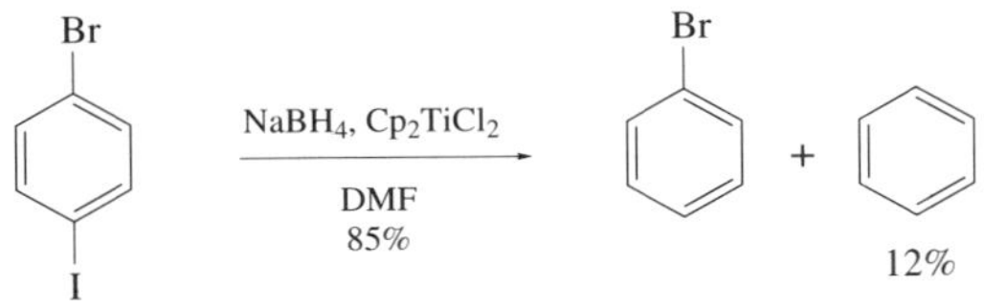

(13)

1.10.1.1.2 Reduction of vinyl halides

The reduction of a vinyl halide to the corresponding alkene has been successfully achieved in many ways ⟨91COS(8)895⟩. These include: (i) dissolving metal reductions, (ii) reductions with complex hydrides or tin hydrides, (iii) electrochemical reductions, and (iv) hydrogenolysis via catalytic hydrogen ⟨B-84MI 110-03, 80S425⟩.

(i) Reduction of vinyl fluorides

The ease of reaction generally follows the order of bond strengths for the carbon–halogen bond, C—F being the most difficult bond to cleave. Reduction of a vinyl fluoride to the corresponding alkene can be achieved by reaction with either LAH or $NaBH_4$. Normant and co-workers found that treatment of some trifluorinated allylic alcohols with LAH in refluxing diethyl ether led to the replacement of one terminal fluorine atom (Equation (14)) ⟨78S128⟩. The product was obtained as a 3 : 1 mixture of (*E*)- and (*Z*)-isomers in 82% yield. Dmowski reported stereo- and regiospecific reduction of (*E*)- and (*Z*)-1-phenylfluoropropenes ⟨85JFC(29)273⟩. The substrates were treated with LAH in GLYME at either −15°C to 10°C or between 0–30°C. The higher temperature reaction led to replacement of two vinyl fluorine atoms, whereas the lower temperature reaction gave a monoreduced product, the fluorine in the benzylic position being replaced. Snieckus and co-workers reduced difluoroalkenes to monofluoroalkenes in excellent yield and stereoselectivity using $LiBEt_3$/THF (Equation (15)) ⟨93TL415⟩. Elegant syntheses of (*E*)- and (*Z*)-1,2,3,3,3-pentafluoropropene have been reported ⟨89JFC(44)1637⟩. The (*E*)-isomer was formed by reduction of hexafluoropropene with Bu_3P, followed by hydrolysis (Scheme 1). Isomerisation of the (*E*)-product with SbF_5 yielded the corresponding (*Z*)-isomer (Equation (16)). A stereoselective reduction of 2,2-difluorovinylphosphine oxides to monofluorovinylphosphine oxides with $LiAlH(OBu^t)_3$ has been published ⟨92SL833⟩. Fluorovinylsilanes were reduced with LAH to vinyl silanes ⟨88T4135⟩. Burton and Johnston published a vinyl fluoride reduction that featured an interesting reversal in the chemoselectivity. Treatment of 1-chloroperfluorocyclopentene with $NaBH_4$ in DIGLYME unexpectedly led to a replacement of a fluorine atom ⟨64JA5361⟩.

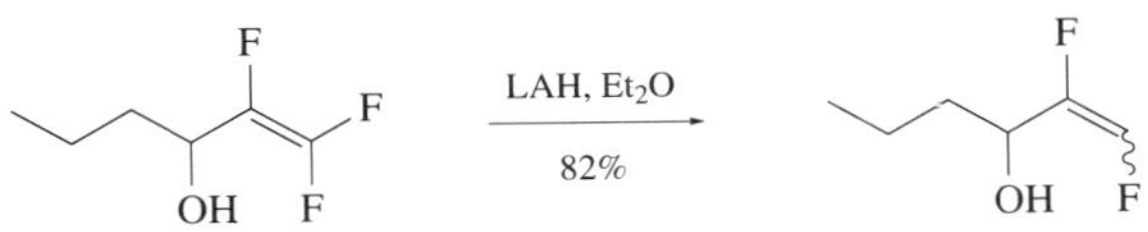

(14)

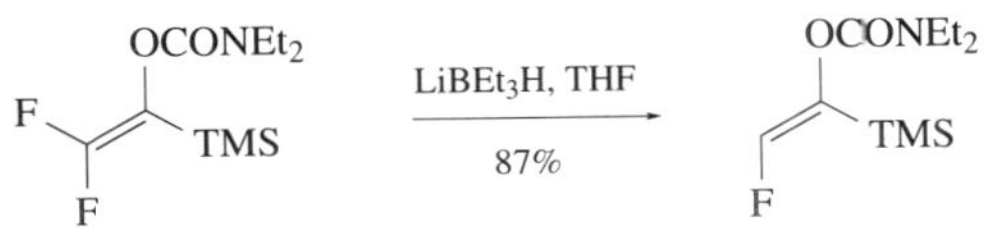

(15)

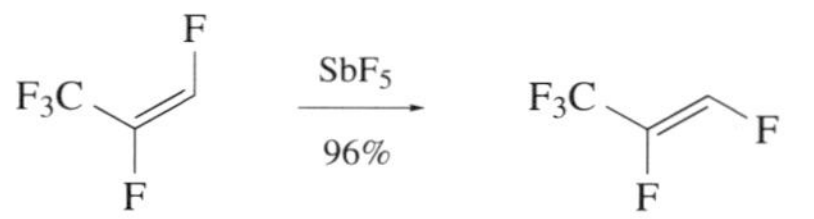

(16)

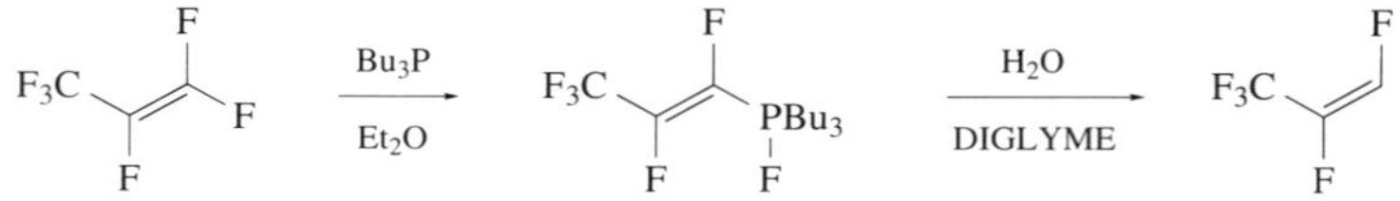

Scheme 1

The catalytic hydrogenation of vinyl fluorides generally leads to mixtures of products via competing hydrogenolysis and saturation reactions ⟨69BSF842⟩; a straightforward method for C—F bond cleavage without concomitant double bond saturation has not yet been reported ⟨91COS(8)895⟩.

(ii) Reduction of vinyl chlorides

Vinyl chlorides are reduced by complex hydrides without concomitant double bond reduction while, in contrast, reduction under catalytic hydrogenation conditions is plagued by problems of overreduction and hydrogenolysis leads to complex mixtures of products. The combination of LAH with $TiCl_4$ proved to be an excellent system for reducing vinyl chlorides to alkenes ⟨73CL291⟩. Complex reducing agents (CRAs) are a variation of this hydride based reduction methodology ⟨81JOC1270⟩. 1-Chlorocyclohexene and also 4-methyl-1-chlorocyclohexene were reduced by a mixture of $NaH/Ni(OAc)_2/t\text{-}C_5H_{11}OH$ in high yield (≥90%) (Equation (17)). Only small quantities of saturated products were observed.

Dissolving metal reductions are also an important route to alkenes from the corresponding vinyl chlorides. Gassman and Marshall published an efficient procedure for the synthesis of 7,7-dimethoxybicyclo[2.2.1]heptene from 7,7-dimethoxy-1,2,3,4-tetrachlorobicyclo[2.2.1]hept-2-ene ⟨68OS(48)68⟩ by reduction of the former compound with Na in *t*-butyl alcohol; both vinylic and allylic chlorine atoms were removed (Equation (18)). Zinc in AcOH is a highly chemoselective reducing system, reduction of a vinyl chlorosulfone proceeding in excellent yield (86%) with selective loss of the chlorine atom (Equation (19)) ⟨85S676⟩. Iron pentacarbonyl has also been used with some success for the reduction of vinyl chlorides ⟨73TL447⟩.

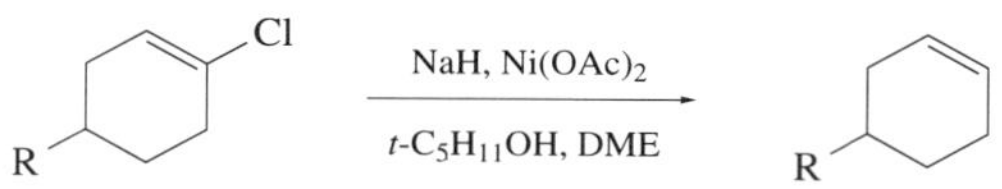

(17)

R = H, 90%
R = Me, 97%

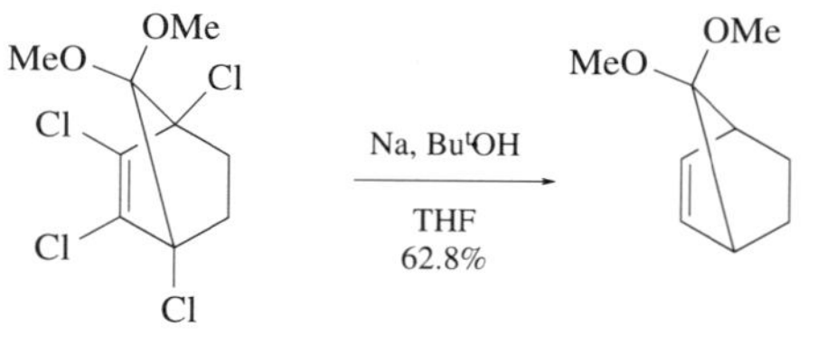

(18)

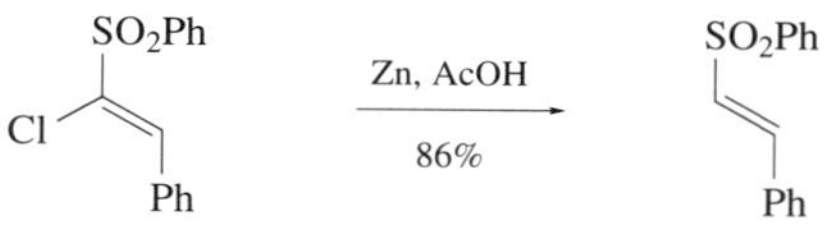

(19)

(iii) Reduction of vinyl bromides

The reduction of vinyl bromides to the corresponding alkenes is more straightforward than the reduction of vinyl fluorides and chlorides. Indeed additional synthetic methods, such as catalytic hydrogenation, are suitable for the reduction of vinyl bromides.

The LAH reduction of vinyl bromides is a much used procedure for alkene synthesis ⟨80JOC3513⟩, the reaction proceeding in some cases via a free radical mechanism. Sodium hydride, together with $Ni(OAc)_2$, also exhibits good selectivity for vinyl bromide reduction over saturation of the double bond. 1-Bromocyclooctene was reduced to cyclooctene at room temperature, with only 5% of the fully saturated product being formed ⟨81JOC1270⟩.

Metal–halogen exchange with subsequent hydrolysis of the intermediate vinylmetal species has been used to effect, overall, a reduction ⟨92JOC5805, 93JOC4897⟩. Thus 1,1-dibromo and 1-bromo-1-chloroalkenes were treated with lithium trialkyl zincates or BuLi at low temperatures to give the corresponding 1-bromo- and 1-chloro-alkenes respectively, after hydrolysis of the vinylmetal intermediate. Yields and selectivities depend on the reactivity of the substrate and on the method by which the vinylmetal intermediate is generated. An illustrative example is the reaction of chlorobromoalkene (**1**) with $(Bu)_3ZnLi$, which gave, on hydrolysis, a 92% yield of the (*E*)-alkene (Equation (20)). It was suggested that a chelating effect involving the alkoxy group led to the high selectivity observed. A chelation effect was also proposed by Grandjean and Pale ⟨93TL1155⟩ to explain the high stereoselectivity observed when dibromoalkene epoxide (**2**) was treated with MeLi–THF at −105 °C. No (*Z*)-isomer was observed (Equation (21)).

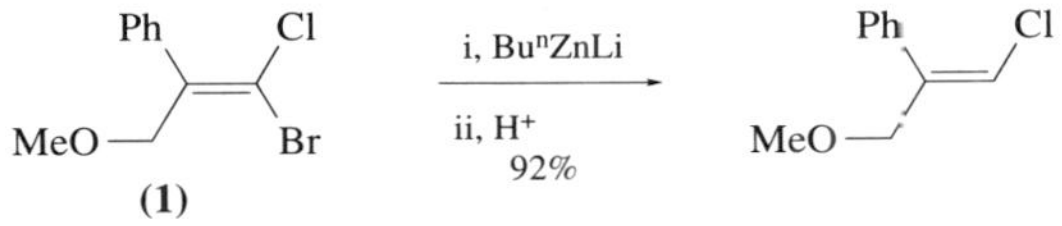

(20)

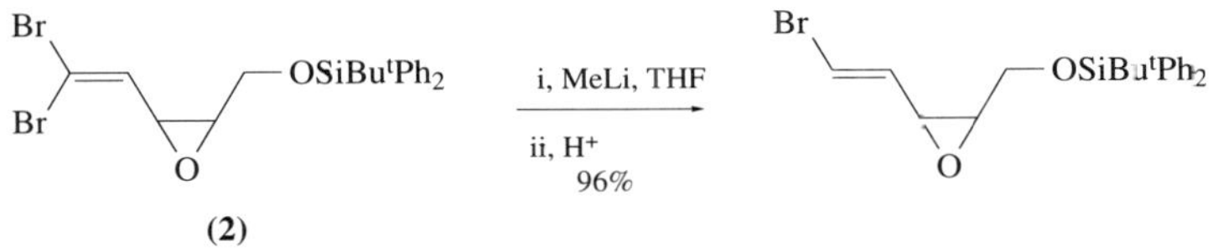

(21)

Vinyl bromides were studied as substrates for a model $NAD(P)^+$–NAD(P)H reduction system ⟨85JOC3283⟩. Reduction of (*Z*)- and (*E*)-ethyl 2-bromocinnamates with *N*-benzyl-1,4-dihydronicotinamide in the presence of catalytic $ClRh(Ph_3)_3$ was achieved with good yield. Interestingly, the (*E*)-vinyl bromide was reduced with concomitant interconversion to the (*E*)-alkene, indicating the intermediacy of a vinyl radical (Equation (22)).

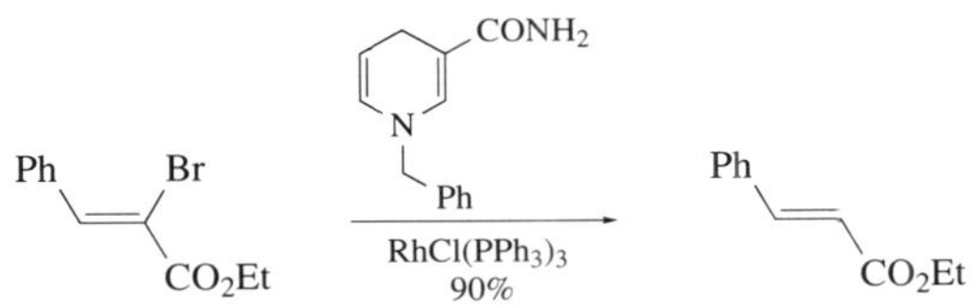

(22)

Catalytic hydrogenation is a popular route to alkenes from vinyl bromides. Kupchan and Afonso prepared ethyl 2-methylcrotonate from ethyl 2-methyl-3-bromocrotonate in 71% yield via catalytic hydrogenation over $Pd/BaSO_4$ ⟨60JOC2217⟩.

(iv) Reduction of vinyl iodides

It is well documented that the ease of reduction of vinyl halides follows the strength of the C—X bond and therefore one would assume that the C—I bond is the easiest to cleave, however this is a reaction that is little studied. Reduction with complex hydrides, for example $NaBH_4$ ⟨71JFC(1)121⟩ or $NaAlH_4$ ⟨82JFC(20)157⟩ works well, as does reduction by Zn/HCl ⟨53JCS922⟩. 3-Iodocyclohexenone was reduced by *N*-benzyl-1,4-dihydronicotinamide (BNAH) in the presence of $ClRh(Ph_3)_3$ ⟨85JOC3283⟩. Ochiai and co-workers have advocated the use of vinyl iodonium tetrafluoroborates as vinyl cation equivalents ⟨88T4095⟩. These vinyl iodonium salts were shown to react with a wide variety of nucleophiles, a representative example being the reaction with diphenylcopperlithium, $(Ph)_2CuLi$ (Equation (23)). Vinyl iodium tetrafluoroborates, on treatment with base, yield carbene intermediates, which undergo insertion reactions ⟨88JA6565⟩ to form cyclopentenes, and/or rearrangement to form alkynes (Equation (24)). It would appear reasonable to assume that cleavage of the sp^2-C—I bond might be achieved readily with hydride reagents such as $NaBH_4$ or LAH.

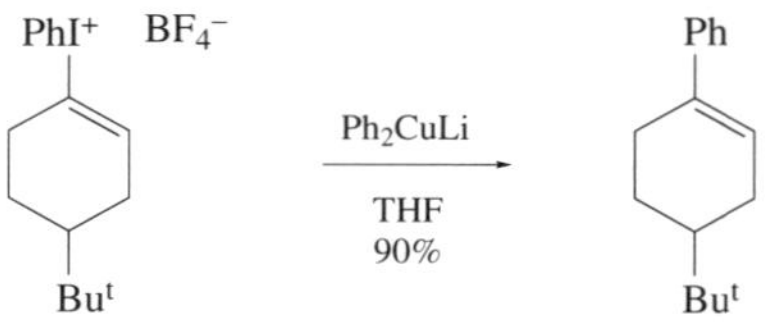

(23)

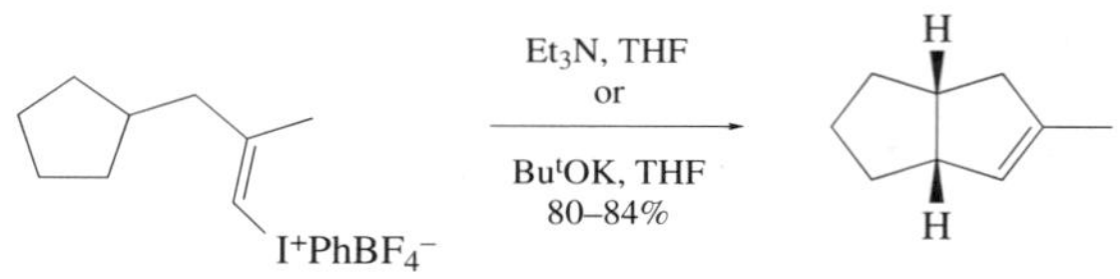

(24)

1.10.1.2 Reduction of =C—O Bonds

1.10.1.2.1 Reduction of phenols and derivatives

Deoxygenation of phenol has been carried out by heating phenol in the presence of dry LAH at 350 °C. The harsh conditions of this reaction obviously makes this transformation unrealistic for most synthetic chemists. This type of direct displacement is considerably more successful when applied to polycyclic aromatic hydrocarbons containing phenolic groups. Treatment of 1-hydroxynapthalene with $AlCl_3$ in EtSH at reflux for five hours yields the corresponding hydrocarbon in 76% yield (Equation (25)) ⟨82TL689⟩. Hydroiodic acid in acetic acid at reflux has also been used for similar transformations ⟨79JOC4813⟩. Phenolic ethers, particularly methyl ethers are similarly reduced using the aforementioned reagents.

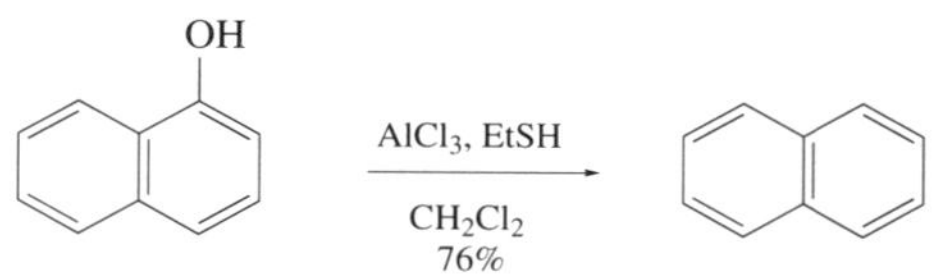

(25)

Phenolic ethers have been displaced by refluxing with sodium in EtOH ⟨11CB2134⟩. Regiospecific reductive demethoxylation of 1,2,3-trimethoxybenzene was reported by Azzenza *et al.* ⟨89S28⟩; the electron-transfer type reduction was carried out using finely divided alkali metal in THF. The intermediate anions were found to react with a range of electrophiles (Equation (26)) ⟨90JOC5386⟩.

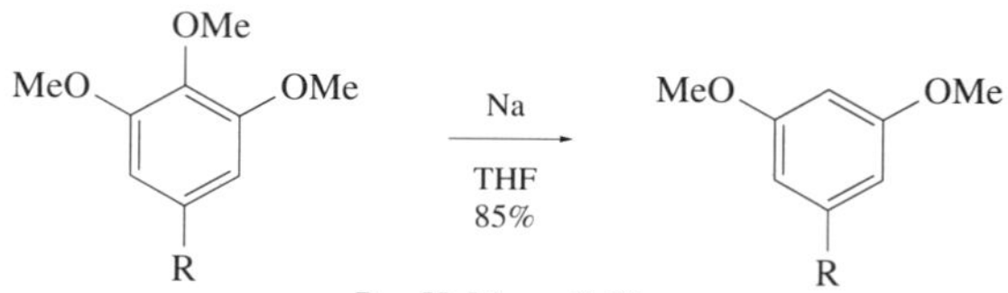

R = H, Me, n-C_5H_{11}

(26)

Phenols, when converted to the corresponding phosphate ester by reaction with diethylchlorophosphidate and base, are reduced to hydrocarbons by dissolving metal reduction. Treatment of the phosphate ester with either Li/NH_3(l) or Na/NH_3(l) provides good yields of aromatic hydrocarbons ⟨55JCS522⟩. The reaction has also been carried out using activated titanium metal in THF ⟨78JOC4797⟩. Shono and co-workers reduced phenolic phosphate esters electrochemically ⟨79JOC4508⟩: this methodology has been used for a synthesis of the polyketide olivetol **(3)** (Scheme 2).

A number of related phenolic derivatives have been found to be susceptible to hydrogenolysis, the overall reaction being deoxygenation. Most common are the phenolic ethers formed by reaction of phenol with 2-chlorobenzoxazole ⟨66JA4271⟩ or 1-phenyl-5-chlorotetrazole (Equation (27)) ⟨84JMC1575⟩. Both ethers are hydrogenolysed under standard conditions (H_2, 5% Pd/C) giving reasonable yields of hydrocarbon.

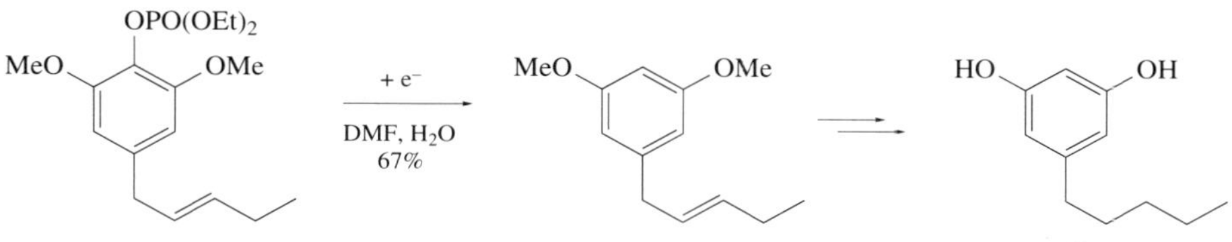

(3)

Scheme 2

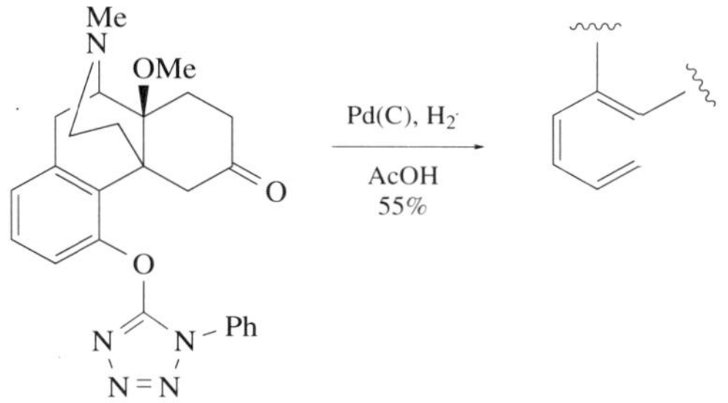

(27)

A methodology developed in the 1980s and 1990s has begun to dominate phenolic deoxygenation. Conversion of phenol to the corresponding trifluoroalkanesulfonate ester, with subsequent displacement by either radical or catalytic means, has led to a deoxygenation reaction that is high yielding and chemoselective. The preparation of perfluoroalkanesulfonic esters has been reviewed ⟨82S85⟩ as have the synthetic transformations of both aryl and vinyl triflates ⟨93S735⟩.

One of the earliest examples of the displacement of an aryl trifluoroalkanesulfonate by hydrogen was reported by Subramanian ⟨84S481⟩. A range of aryl nonaflates were prepared by reaction with n-$C_4F_9SO_2F$ and base. These intermediates were then subjected to palladium-catalysed hydrogenolysis leading to good yields (84–100%) of reduced products (Scheme 3). The main problem encountered was a lack of chemoselectivity, that is the alkenic double bonds and C—halogen bonds were reduced; quinoline derived nonaflates were overreduced. Reduction of phenol triflates by homogenous palladium(0) catalysis has overcome this problem. Reduction with trialkylammonium formate as hydrogen donor, combined with either PPh_3 or dppf as a chelating ligand, gave good yields (60–98%) of deoxygenated products and tolerated ketone, alkene, nitro, α,β-unsaturated ester and amide functionalities (Equation (28)) ⟨86TL5541⟩. Aryl triflates have also been reduced by a nickel(0) based catalyst using MeOH as the hydrogen donor ⟨91CL2017⟩. A binary ligand system based on nickel combining both PPh_3 and 1,1′-bis(diphenylphosphino)nitrate (dppn) was found to be most successful (Equation (29)). This reduction was also found to be tolerant of a range of functionalities and gave good yields. Use of MeOD as a deuterium source yielded a route to regiospecifically labelled arenes. Chen and He reported a one-pot reduction of arylperfluoroalkenesulfonates which was closely related to Cacchi's original procedure ⟨88S896⟩. Good yields were reported for this highly chemoselective methodology. The selectivity of this method was further exemplified during a synthesis of (+)-4-demethoxydaunomycinone ⟨90JCS(P1)428⟩. The 4-OH phenol was triflated selectively, due to the greater acidity of the 4-OH over the 6- and 11-OH groups. Palladium-catalysed reduction with HCO_2H and dppf as the coordinating ligand led to an 82% yield of the desired product. Deprotection with TFA yielded (+)-4-demethoxydaunomycinone (**4**) (Scheme 4). Similar results were reported by Chen and co-workers using Pd $(PPh_3)_2Cl_2$ or $Pd(PPh_3)_4$ in combination with perfluoroalkanesulfonates of the type $ArSO_2(CF_2)_2O(CF_2)_2H$ ⟨86CC1452⟩. Highly hindered, electron-rich phenols were deoxygenated via the corresponding triflates under both homogenous and heterogenous conditions ⟨90JOC991⟩. Trioxygenated aromatics were efficiently deoxygenated via the corresponding triflate to give moderate to good yields of resorcinol dimethyl ether (**5**) (Equation (30)). Mechanistic studies indicated a number of hydrogen donors were present in the reaction medium ⟨86TL5541⟩.

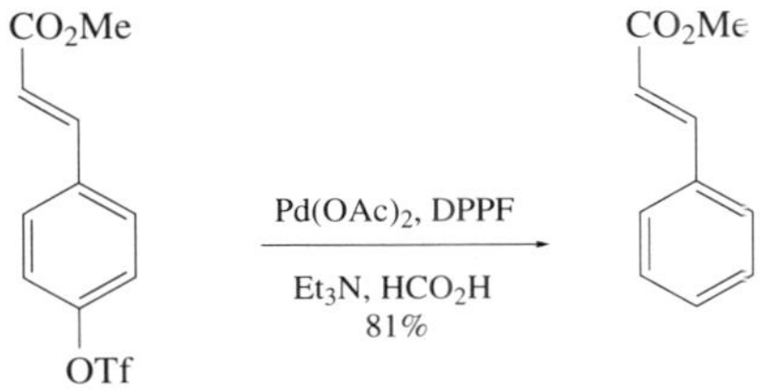

(28)

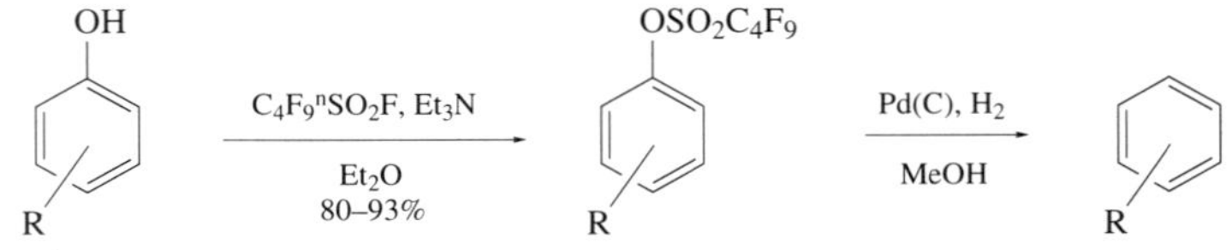

R = H, *o*, *p*-Me, MeO, Cl

Scheme 3

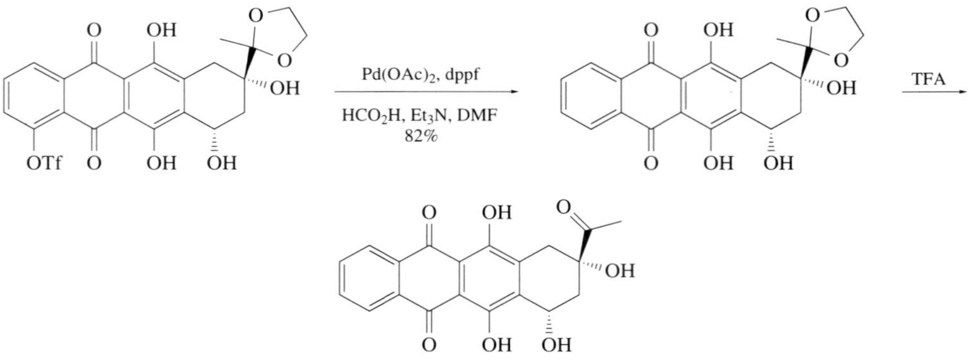

(4)

Scheme 4

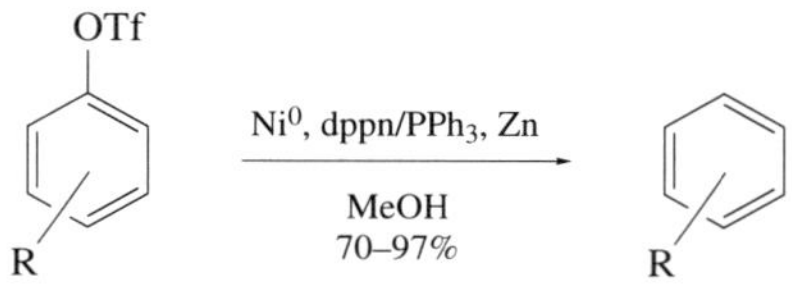

(29)

R = Me, MeO, CN, MeCONH, MeCO

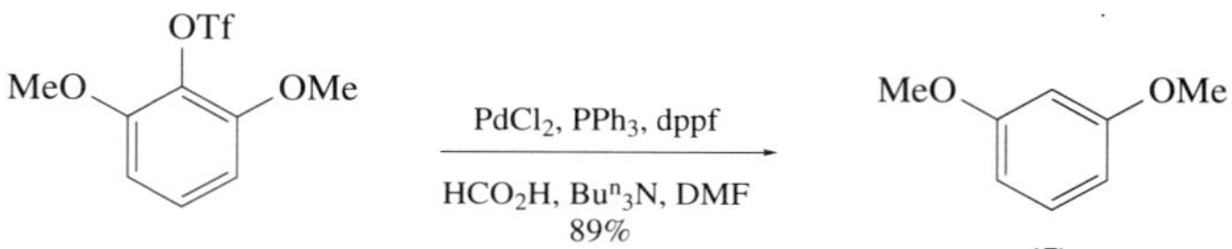

(30)

(5)

1.10.1.2.2 Reduction of enol ethers and derivatives

The reduction of an enol derivative to the corresponding alkene is a difficult transformation and has been tackled by a variety of strategies. Enol ethers have been reduced to alkenes by dibal-H ⟨66JOC321⟩ and also by sodium in a Birch-type reduction ⟨70AJC1641⟩. Treatment of silyl enol ethers with borane followed by subsequent acid-induced elimination give good yields of alkene ⟨75TL4005⟩.

The reduction of the enol phosphate moiety is generally carried out by dissolving metal reduction using lithium metal in Bu^{t}OH together with ammonia or amines (RNH_2(R = H, Me, Et)). The method was originally described by Ireland ⟨69TL2145⟩ as a way of transforming ketones to alkenes. Many natural products have been synthesised utilising this methodology including a quassinoid ⟨84JA3353⟩, (+/−)-seychellene ⟨85JOC2668⟩ and the potent antitumor antibiotic (−)-ptilocaulin (Equation (31)) ⟨84JA721⟩. Ishihara and co-workers described a chemoselective reduction of an enolphosphate in the presence of both phosphonate and halide functionality with lithium dibutylcuprate in TMEDA/THF ⟨87JOC300⟩. The closely related enol phosphoramidate is also reduced by lithium in amines ⟨72JA5098⟩. An important olefin unit was introduced into the indacene subunit of ikarugamycin using this reduction method (Equation (32)) ⟨87TL31⟩.

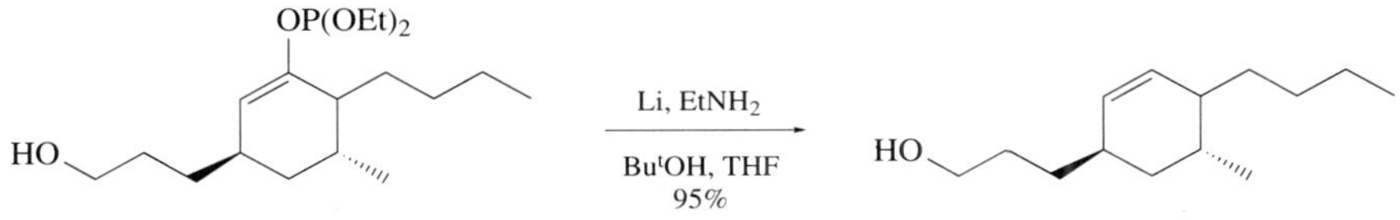

(31)

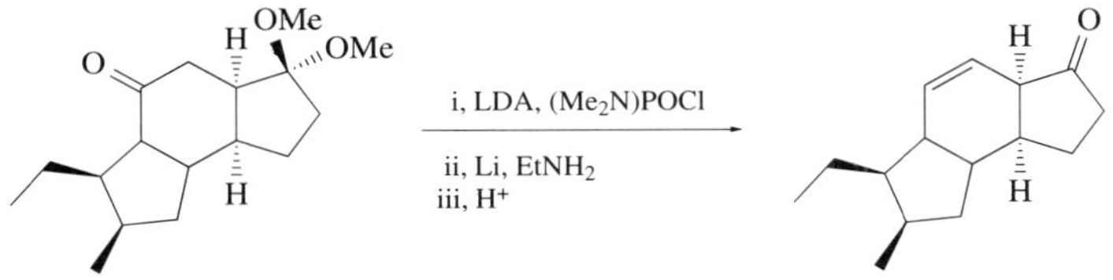

(32)

Vinyl triflates are reduced to alkenes by palladium catalysis with either tin hydrides, silicon hydrides ⟨86JA3033⟩ or formate ⟨84TL4821⟩ acting as the hydrogen source. Vinyl triflates were efficiently reduced to alkenes in high yield by trialkylammonium formate under palladium(0) catalysis ⟨84TL4821⟩. A range of steroidal dienoyl triflates were reduced to dienes in a highly chemoselective manner, the reaction tolerating a range of functionality (Equation (33)) ⟨90OS(68)138⟩. Donnelly has used this methodology to synthesise a series of rare neoflavonoids (Equation (34)) ⟨90JCS(P1)2851⟩. Similarly Ortar and co-workers prepared a series of 6,7-didehydroöestrogens using this methodology ⟨90SC1293⟩.

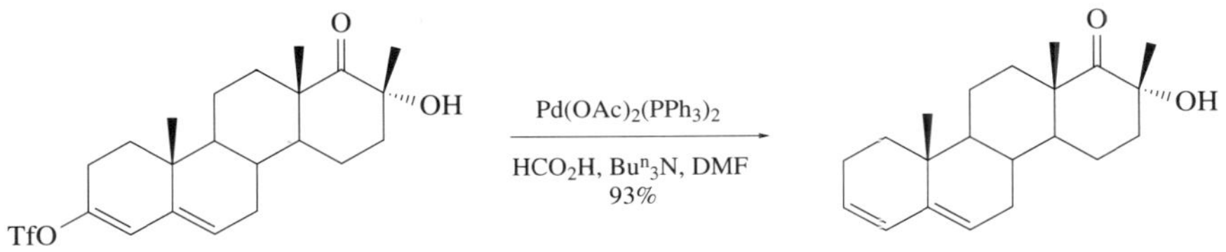

(33)

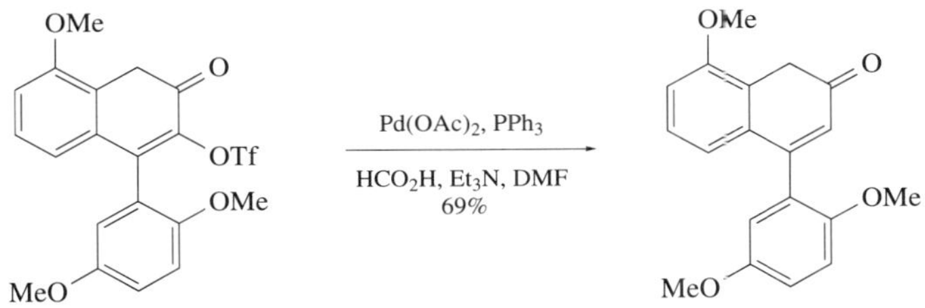

(34)

Elegant mechanistic studies were carried out by Scott and Stille on the reduction of 4-*t*-butylcyclohexenyl triflate ⟨86JA3033⟩. A wide range of reducing agents were investigated and it was found that both Bu_3SnH and Et_3SiH effectively reduced vinyl triflates in the presence of $Pd(Ph_3)_4$ and LiCl (Equation (35)). Other more nucleophilic reducing agents were not effective giving rise to problems of S—O cleavage and catalyst decomposition. Interestingly, an attempted reduction with tributylammonium formate yielded no alkene.

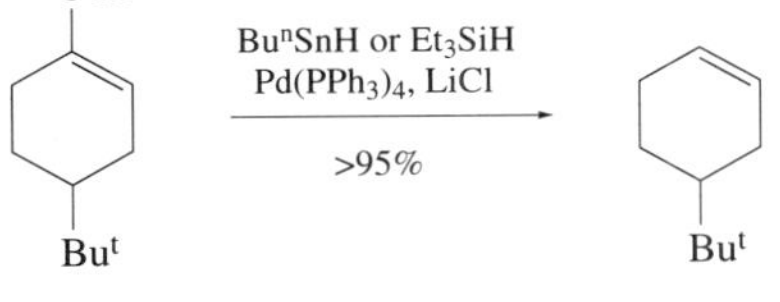

(35)

1.10.1.3 Reduction of =C–Sulfur, Selenium and Tellurium Bonds

1.10.1.3.1 Reduction of aryl sulfides, selenides and tellurides

(i) Reduction of aryl sulfides

An enormous number of aryl thiols and sulfides have been desulfurised by reaction with Raney nickel ⟨62CRV347, 62OR(12)356⟩ and this method remains a powerful tool for the reduction of aryl sulfides (for an example see ⟨83JOC4419⟩). The extensive use of this reagent has also highlighted its drawbacks. The reagent is pyrophoric, it is difficult to weigh out accurately, it loses activity on storage and displays such high chemical reactivity that over-reduction of substrates is a common occurrence.

Nickel boride, formed from the *in situ* reduction of nickel(II) salts with $NaBH_4$ displays a number of advantages over Raney nickel. The reagent is more chemoselective and the amount of reagent used can be accurately determined. A number of benzothiophene derivatives have been desulfurised with nickel boride in high yield with short reaction times (15–60 min) ⟨90CC819⟩. Ethyl, methyl and phenyl thioethers were also reduced in moderate to excellent yields with nickel boride (Equation (36)) ⟨86SC779⟩.

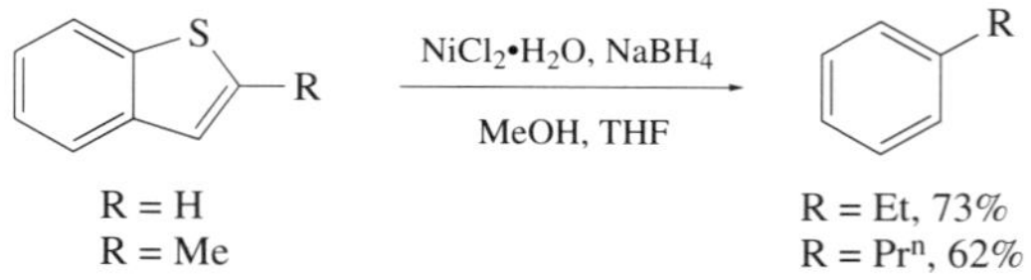

(36)

Dissolving metal reductions, particularly $Li/NH_3(l)$, are known to effect rapid desulfurisation of aromatic thiols and thioethers. This reaction has been reviewed ⟨77T2019, 78CRV363⟩. The role of alkali metals as desulfurising agents has also been extensively studied. The reduction is known to proceed via a single electron transfer mechanism and this limits the practical use of this method to simple thiol and sulfides due to the over-reduction of neighbouring functional groups.

Advances in transition metal-mediated C—S bond cleavage have been reviewed ⟨90S89⟩. Homogenous nickel(0) species have been shown to be efficient desulfurising reagents, especially when used in combination with a complex hydride (Equation (37)) ⟨86JA7763⟩. A novel nickel complex, $LiAlH_2(THF)_n \cdot (C_{10}H_8N_2)Ni$ (formed by reaction of $LiAlH_4$ with 2,2′-bipy-(1,5-cyclooctadiene) nickel) desulfurised phenoxathiin to diphenyl ether. Without the presence of hydride the complex remains a desulfurisation reagent but yields different products. Both reactions are thought to proceed via single electron transfer mechanisms, the differing products being rationalised in terms of the greater anionic character of nickel in the Ni–$LiAlH_2$ complex. A similar organonickel complex, which has been used for the reduction of aromatic thiols and thioethers, is nickelocene–$LiAlH_4$ (Equation (38)) ⟨85CC1610⟩. 2-Napthalene thiol was reduced to napthalene and the reaction was found to work for a number of aromatic thiols. The reagent was highly chemoselective, tolerating esters and alkenic double bonds, though reduction of aromatic nitro groups was observed. Further study on this reagent has revealed the facile desulfurisation of thiols including dibenzothiophene ⟨88JOC4466⟩. The same research group has employed $NiBr_2 \cdot DME \cdot PPh_3$ and $LiAlH_4$ in a similar series of reductions ⟨89JOC4474⟩.

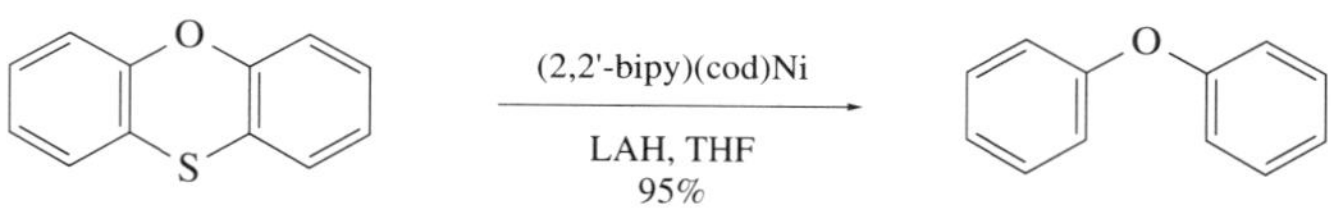

(37)

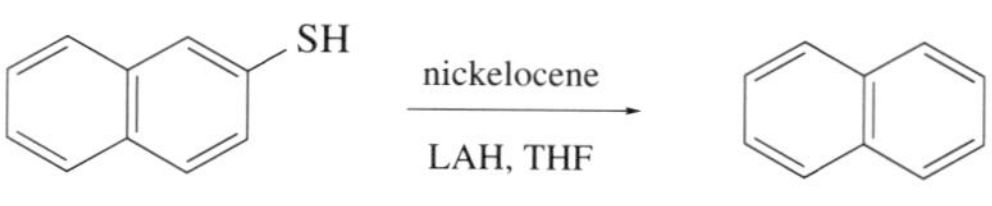

(38)

An older but no less effective desulfurisation method was described by Wenkert and Ferreira ⟨82CC840⟩. The reaction of secondary Grignard reagents with alkylthioarenes, in the presence of a 1 : 1 mixture of $NiCl_2$ and PPh_3, led to the substitution of thioalkyl groups by hydrogen (Equation (39)). It was observed that manipulation of the ligands on nickel led to alkylation rather than substitution. In a related study an indole thioether was reduced to the indole in 50% yield ⟨85JOC1125⟩.

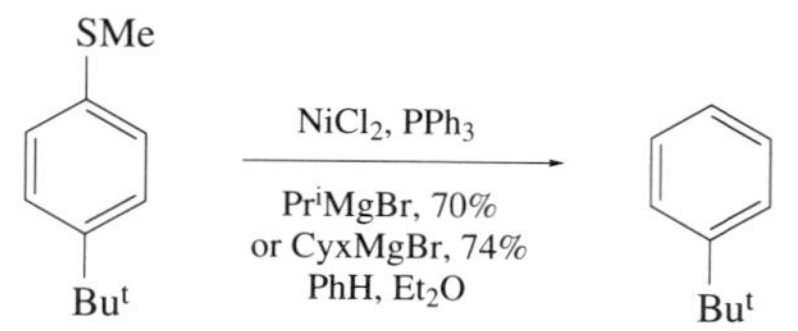

(39)

Metal carbonyls such as $Co_2(CO)_8$, ⟨83TL5329⟩, $Mo(CO)_6$ and $Mn_2(CO)_{10}$ ⟨74JOM(73)359⟩ have also come to the fore as effective desulfurisation reagents. The most studied complex in this group is $Mo(CO)_6$ ⟨80CC169⟩. The actual species carrying out the reduction is probably a molybdenum(II) species, tetrakis(acetodimolybdenum), generated *in situ* from reaction of $Mo(CO)_6$ and HOAc. Good yields of hydrocarbon were obtained at ca. 120 °C (Equation (40)). It was observed that if the Mo reagent was preadsorbed onto silica the same reductions could be accomplished at room temperature. Desulfurisations with $Mo(CO)_6$ occur via a free radical mechanism and with good chemoselectivity compared with other metal carbonyl reagents ⟨85JOC5413⟩. Nickel complex reducing agents (NiCRAs), mixtures of NaH, $Ni(OAc)_2$ and t-$C_5H_{11}OH$, with or without a coordinating ligand are known to reduce aryl thiols and thioethers ⟨88TL2963⟩. Aryl sulfoxides and sulfones are also reduced to hydrocarbons with this reagent system (Equation (41)) ⟨90JOC6194⟩.

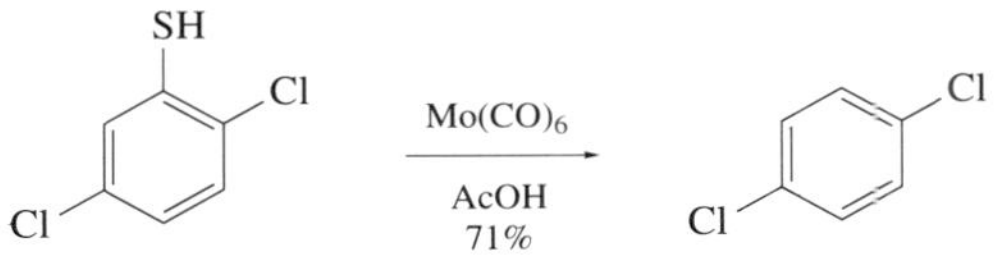

(40)

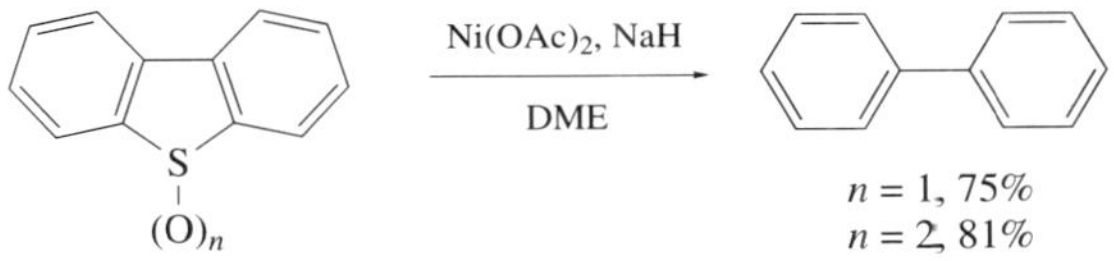

(41)

(ii) Reduction of aryl selenides

The reduction of the carbon–selenium bond has been the focus of much study due to the extensive use of organoselenium reagents in synthetic organic chemistry. The majority of methods published up to 1995 concentrate on the cleavage of alkyl-C—Se bonds rather than aryl-C—Se bonds.

Wiseman and Gould reduced dibenzoselenophenoxide to biphenyl with Raney nickel (Equation (42)) ⟨54JA1706⟩. Phenyl alkyl selenides undergo cleavage of the alkyl C—Se bond on exposure to Ph_3SnH ⟨91COS(8)835⟩. Clive and co-workers also described the use of Ph_3SnH at elevated temperatures (ca. 120 °C) for reducing the R—SePh bond (Equation (43)) ⟨80JA4438⟩. The reduction proceeded equally successfully with diselenoacetals.

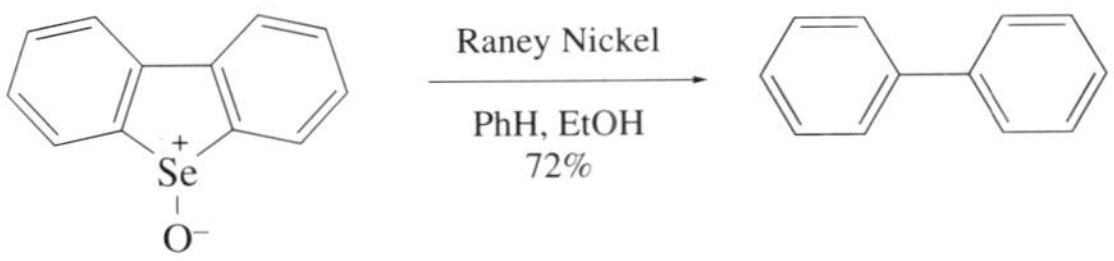

(42)

$$\text{R–SePh} \xrightarrow{Ph_3SnH} \text{R–H} \quad (43)$$

Kreif and co-workers utilised Raney nickel for the reduction of alkylphenyl selenide ⟨76TL2643⟩. Dissolving metal reductions with $Li/NH_3(l)$ proceed in a similar fashion ⟨91COS(8)835⟩.

(iii) Reduction of aryl tellurides

Despite considerable interest in organotellurium chemistry ⟨92PS(67)183⟩, little work has been done on the selective reduction of the carbon–tellurium bond ⟨89TL1209⟩. Transmetallation of alkyl and aryl telluriums with BuLi was described by Sonoda and co-workers ⟨87AG(E)1187⟩. Reaction of the aryl-Li species with electrophiles, (e.g., H_2O or PhCHO) yielded substituted arenes (Scheme 5). Phenyl tellurol (PhTeH) is formed by the reaction of diphenylditelluride with $NaBH_4$ in EtOH ⟨82TL1181⟩. Diaryl telluride (Ar_2Te), when treated with Grignard reagents (RMgX) under nickel(II) or cobalt(II) catalysis yielded cross- and homo-coupled products of the type Ar—R, Ar—Ar and

R—R. A similar reaction has been achieved using Li in THF ⟨75JOM(88)175⟩. Reductions with nickel or copper powder have been attempted, but lack chemoselectivity ⟨92PS(67)183⟩. Reduction of the aryl-C—Te bond in R-TePh with Ph_3SnH was reported by Clive and co-workers ⟨80JA4438⟩.

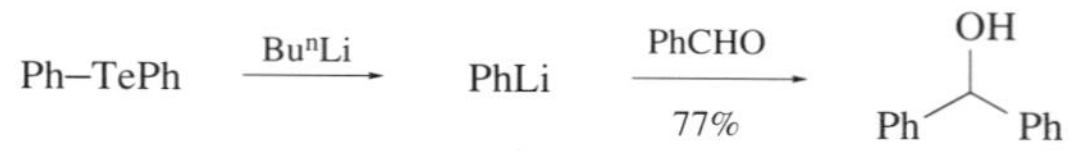

Scheme 5

1.10.1.3.2 Reduction of vinyl sulfides, selenides and tellurides

(i) Reduction of vinyl sulfides, sulfoxides and sulfones

The use of Raney nickel for the transformation of vinyl sulfides to alkenes is well documented ⟨62CRV347, 66OR(12)356⟩. The problems arising from the use of this reagent are well known, in the case of vinyl sulfide concomitant reduction of the olefinic double bond is most troublesome. Nonetheless, Corey and Schmidt utilised a vinyl sulfide reduction with deactivated Raney nickel in a synthesis of the C-15–C-20 segment of rifamycin (Equation (44)) ⟨79TL2317⟩. Nickel boride has also been used to reduce vinyl sulfides to alkenes ⟨73JCS(P1)654⟩.

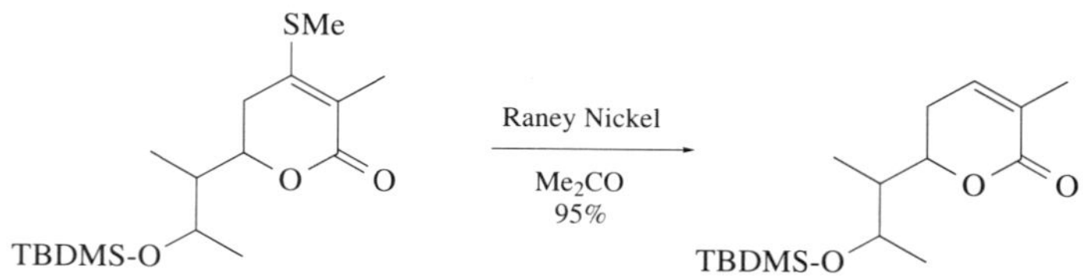

(44)

Trost and Ornstein developed an elegant desulfurisation procedure using a combination of secondary Grignard reagents and $(PPh_3)_2NiCl_2$ in Et_2O/THF ⟨81TL3463⟩. Moderate to good yields of alkene were realised with high stereospecificity and no overreduction observed (Equation (45)). The methodology was used in a synthesis of the douglas fir tussock moth pheromone (**6**). In a related procedure bisthioalkenes were desulfurised to vinyl sulfides by treatment with Pr^i–MgBr and $NiCl_2PPh_3$ ⟨82CC840⟩. Again manipulation of the ligands on nickel led to alkylation rather than reduction. Treatment of dithioacetal with the higher order cuprate, $Me_2CuCNLi_2$ also led to vinyl sulfide formation (Equation (46)) ⟨90CC1284⟩. Caubère's nickel complex reducing agent, NiCRA(bipy) was found to reduce vinyl sulfides in good yield with moderate to excellent stereoselectivity ⟨90JOC6194⟩. Ketene dithioacetals were stereospecifically reduced to vinyl sulfides by SmI_2 (Equation (47)) ⟨93JOC6541⟩.

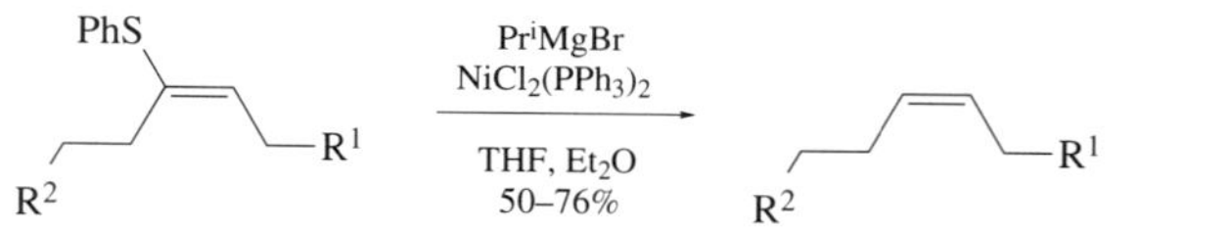

(45)

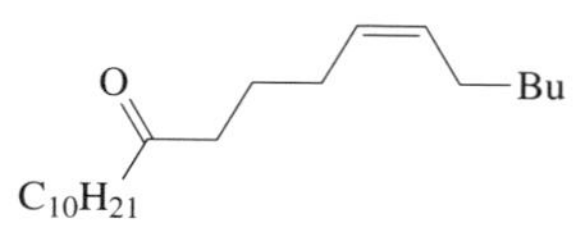

(6)

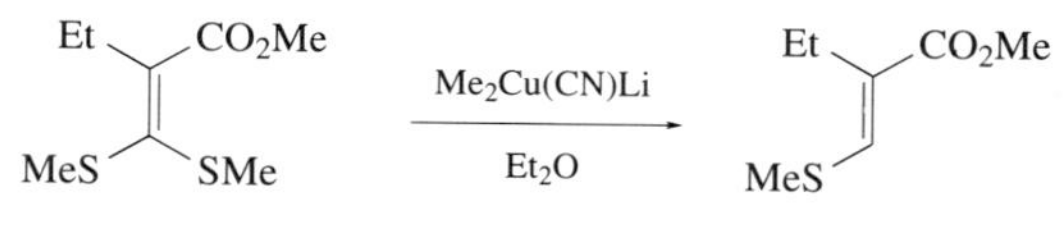

(46)

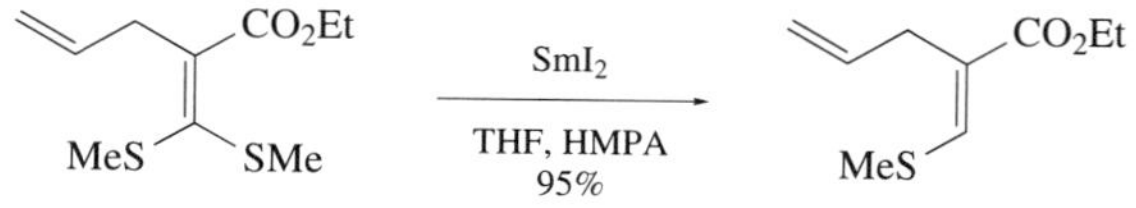

(47)

The direct conversion of a vinyl sulfoxide to an alkene is a rare transformation but the nickel complex reducing agents reported in 1990 are known to effect this reaction ⟨90JOC6194⟩. Sodium hydride poor NiCRAs gave the corresponding alkene with modest stereoselectivity. Trost *et al.* reported an interesting reaction of a vinyl sulfoxide, which, upon refluxing in ethanolic NaOEt, rearranges and loses sulfenate giving an allylic alcohol (Equation (48)) ⟨83JA5075⟩. Theobald and Okamura reported a stereospecific reduction of vinyl sulfoxides using Bu^{t}Li and MeOH as the proton source (Equation (49)) ⟨87TL6565⟩. The use of MeOD provided a route to stereospecifically labelled alkenes.

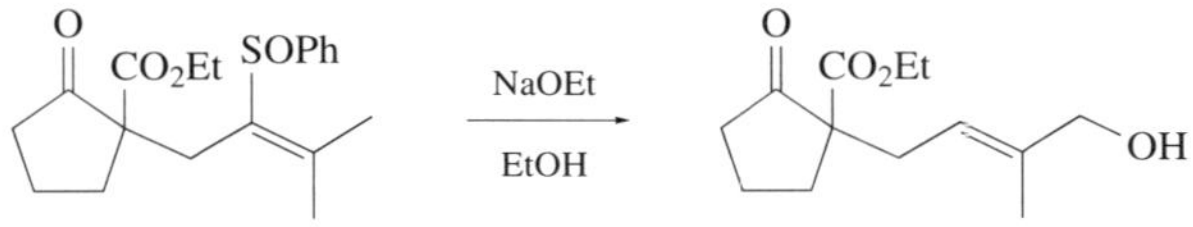

(48)

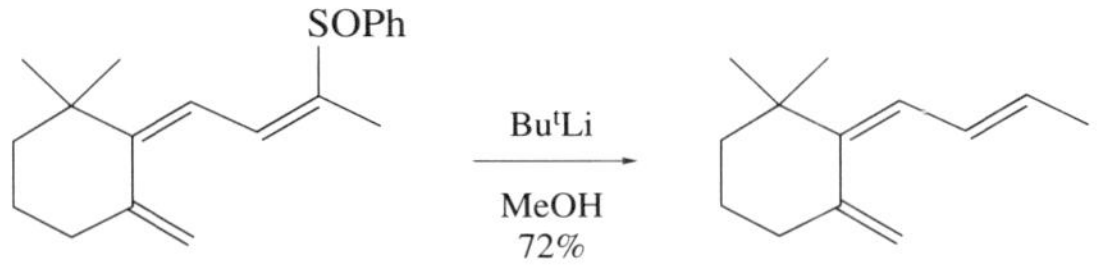

(49)

An excellent review by Simpkins highlights the importance of vinyl sulfones in organic synthesis ⟨90T6951⟩. Many different methodologies have been used for the reduction of vinyl sulfones to alkenes. An early paper by Umani-Ronchi described the use of Al-(Hg) amalgam for the stereospecific reduction of α,β-unsaturated sulfones ⟨73CC351⟩ (see also ⟨82JOC4713⟩). A modification of this methodology using Na-(Hg) amalgam in phosphate buffer as solvent was used by Djerassi and co-workers in a synthesis of the marine sterol xestosterol (Equation (50)) ⟨86TL2187⟩. A synthesis of vinyl fluorides also features an amalgam based reduction of a vinyl sulfone ⟨90TL5449⟩. Potassium graphite, C_8K, is another reagent that has been studied in this connection. High yields of desulfonated products were observed but extensive (*E*)–(*Z*)-isomerisation occurred ⟨77JCS(P1)123⟩. Nickel complex reducing agents (NiCRAs) have also been examined and, although they do cleave the C—S bond, a considerable amount of sulfone is recovered and again (*E*)–(*Z*)-isomerisation is evident ⟨90JOC6194⟩.

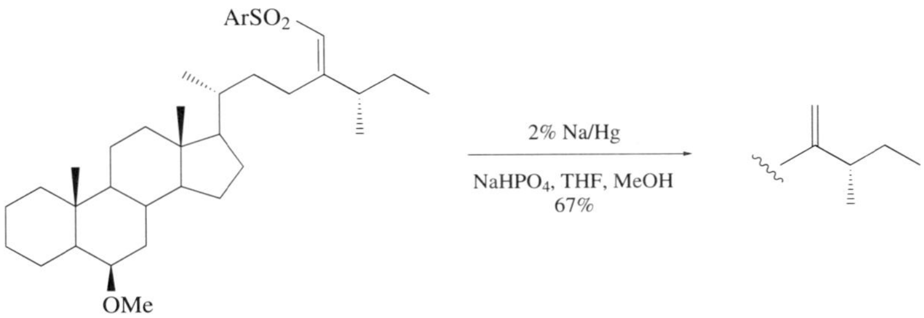

(50)

Julia and co-workers described a stereospecific reduction of vinyl sulfones using Grignard reagents (e.g., Bu^{n}MgCl) in the presence of the transition metal catalysts $Pd(acac)_2$ or $Ni(acac)_2$ ⟨83TL4311, 87T859⟩. Palladium catalysis gave greatest stereoselectivity (>98.5%) depending on the choice of ligand (Equation (51)). Desulfonation using sodium dithionite has been described by the same group ⟨82TL3265⟩; sulfones treated with dithionite and bicarbonate in aqueous DMF are reduced in a highly stereospecific manner (up to 100%) to the corresponding alkenes (Equation (52)). The reduction involves a Michael-type addition of HSO_2^-, followed by loss of SO_2 and expulsion of sulfinate in an *anti* fashion ⟨86T2475⟩. Pak and co-workers reported an efficient desulfonylation of a vinyl sulfone using Mg in EtOH, catalysed by $HgCl_2$ ⟨93TL4541⟩. Unsaturated cyclic sulfones are ring opened at the sp^2C—S bond by ultrasonically dispersed potassium metal ⟨87JOC2224⟩. Sodium hydrogen telluride was used to desulfonate α-methylthiosulfones to give vinyl sulfides as mixtures of (*E*)- and (*Z*)-isomers (Equation (53)) ⟨89S42⟩.

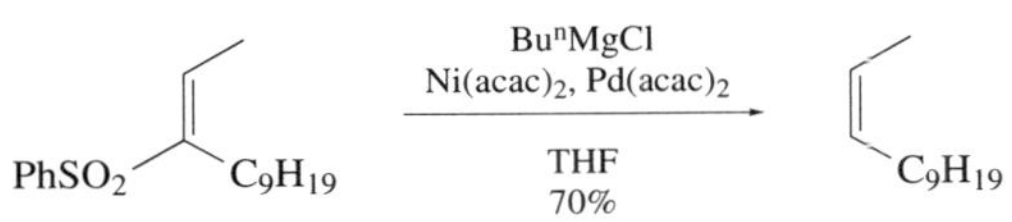

(51)

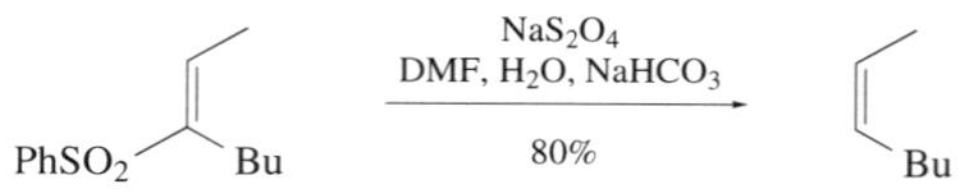

(52)

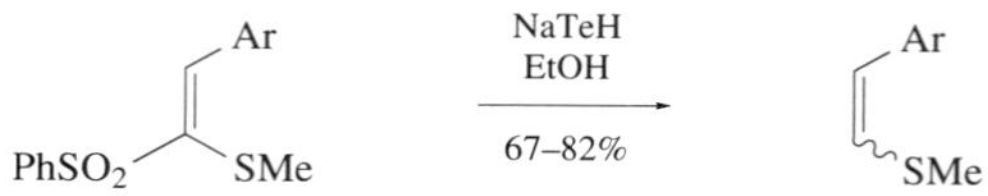

(53)

A useful synthesis of vinyl fluorides ⟨91JA7439⟩ relies on the stereospecific displacement of a vinyl sulfone by Bu_3SnH. The resulting vinyl stannanes were stereospecifically destannylated by treatment with methanolic NH_3 (Scheme 6) ⟨93OS(72)216⟩. The method was also used for the stereospecific introduction of deuterium label simply by destannylating in MeOD containing NaOMe.

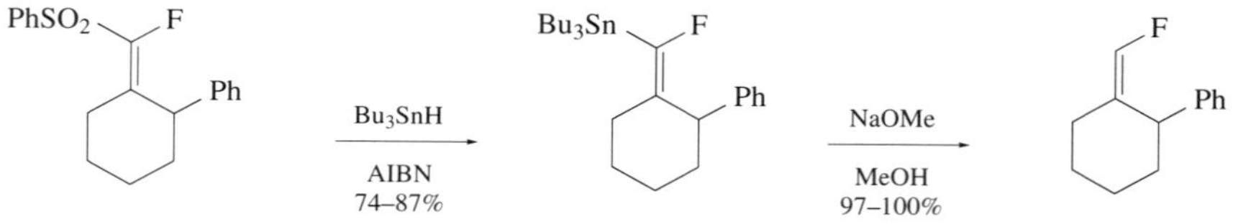

Scheme 6

Vinyl sulfones have been desulfonated by an addition/elimination mechanism on treatment with Bu_3SnLi ⟨83CC619⟩. Addition of Bu_3SnLi to an unsaturated sulfone yields a tributylstannyl sulfone which undergoes β-elimination on contact with silica gel to form the corresponding alkene (Scheme 7). It was observed that this β-elimination also took place on heating the intermediate in xylene. In a related study vinyl silanes were prepared by trapping the intermediate anionic species formed on the addition of Bu_3SnLi to an unsaturated sulfone with TMS-Cl. The β-elimination of this silylstannyl sulfone was studied and found to be stereoselective ⟨83CL1457⟩.

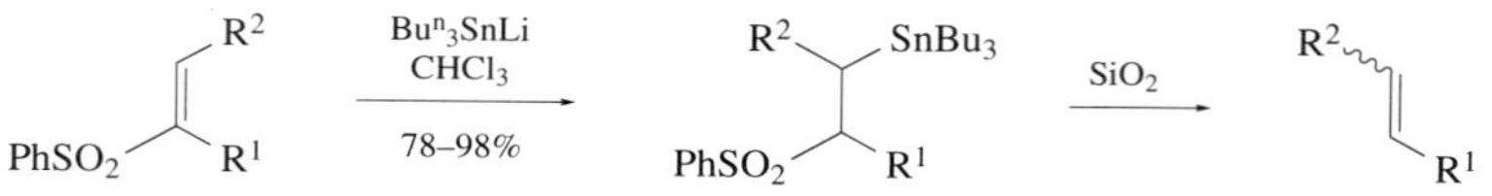

Scheme 7

(ii) Reduction of vinyl selenides

The chemistry of vinyl selenides has been reviewed ⟨83JOM(253)131⟩, as has the reduction of selenium–carbon bonds ⟨91COS(8)835⟩. There exist few general methods for the reduction of a vinyl selenide to an alkene.

It is possible to reduce the sp^2C—Se bond in a vinyl selenide with nickel boride ⟨84CC1417, 88JOC3815⟩. A variety of selenosulfones were reduced in up to 95% yield, though considerable (*E*)–(*Z*)-isomerisation was noted (Equation (54)). The *in situ* derived reagent spared isolated double bonds, ketones and aromatic nitro groups. It was also observed that hydrogen was incorporated from MeOH (and not THF), suggesting that the reaction does not proceed via a radical mechanism. Deselenisations of vinyl selenides to alkenes tend not to be carried out with Raney nickel, probably due to overreduction of alkene double bonds by this reagent. Selenium is more sensitive to nucleophilic attack than sulfur and as such reacts with strong bases/nucleophiles. In the case of vinylic selenides, reaction with Bu^nLi followed by protonolysis leads to alkene formation. Krief has shown that a 1,1-bis(methyl)selenoalkene, upon treatment with Bu^nLi in THF, forms the corresponding 1-lithio-1-selenoalkene ⟨82TL3411⟩. Upon electrophilic quench with D_2O/H_2O the 1-selenoalkene is formed in a ca. 40 : 60 mixture of (*E*)- and (*Z*)-stereoisomers (Equation (55)). Deselenisation of vinyl selenides via hydride radicals is an efficient method of synthesising alkenes ⟨82TL3411⟩. Treatment of vinylic selenide with Bu_3SnH–AIBN at 90 °C leads to the formation of alkene (no stereochemical details were reported). Allylic alcohols were also reduced under similar conditions forming alkenes. The authors also reported the use of P_2I_4 as an efficient reagent for deselenisation. Allenes were formed by reaction of β-hydroxyselenides with P_2I_4. The oxidation of vinyl selenides with mcpba leads to allene formation ⟨92PS(67)203⟩.

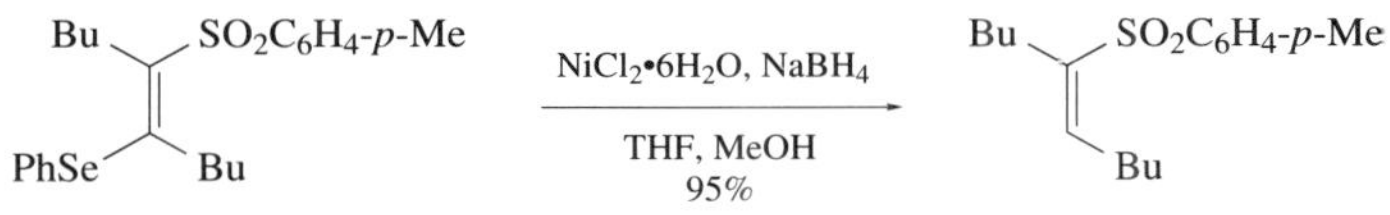

(54)

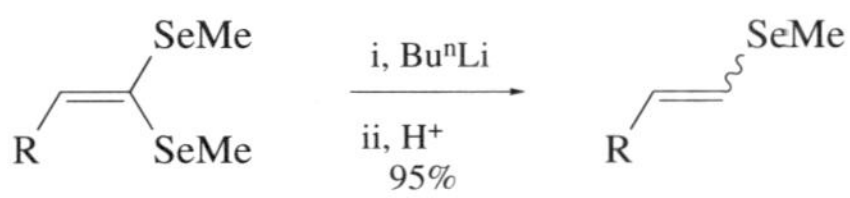

(55)

(iii) Reduction of vinyl tellurides

The chemistry of vinyl tellurides has been reviewed ⟨91SR1⟩. An important method for breaking the sp^2-C—Te bond is via transmetallation/hydrolysis. The exchange reaction with lithium is well established ⟨75CB314⟩ as exemplified by Brandsma and co-workers ⟨92JOM(441)1⟩. They found that thiatellurin, when treated with two equivalents of BuLi, yielded a dilithiated adduct which was stable at −105 °C. This compound was subsequently treated with a variety of electrophiles, including H_2O which led to the formation of divinyl sulfide (Equation (56)).

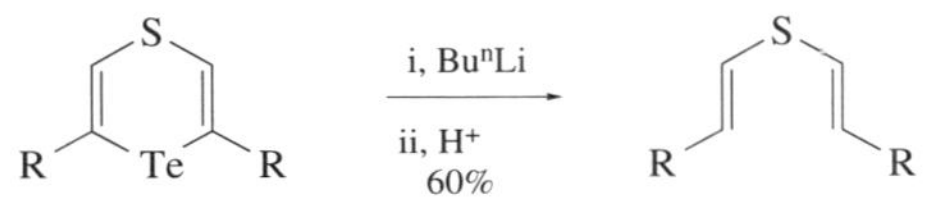

(56)

R = H, Me, Bu^t

Similarly a vinyllithium was formed from the reaction of vinyl telluride and BuLi ⟨89TL7353⟩. The vinyllithium was quenched with electrophiles including PhCHO and $C_8H_{17}Br$ (though not H_3O^+) with strict retention of double bond geometry (Equation (57)). Similarly vinyl tellurides have been transformed into cuprates, which also react with electrophiles (e.g. α,β-unsaturated ketones) again with retention of double bond geometry ⟨90SC1681⟩.

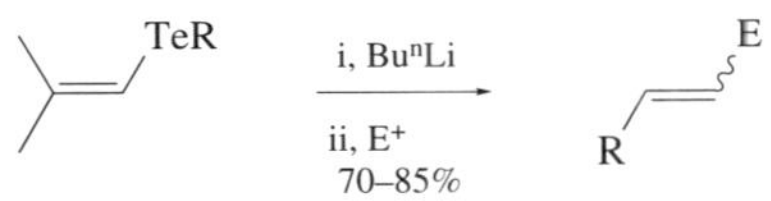

(57)

Raney nickel acts as a catalyst for the reduction of vinyl tellurides ⟨92PS(67)183⟩. The problems with using Raney nickel for chemoselective reactions have been emphasised previously.

1.10.1.4 Reduction of =C—N Bonds

1.10.1.4.1 Reduction of arylcarbon–nitrogen bonds

Occasionally aromatic nitro groups have been directly displaced from arenes by reaction with $NaBH_4$ ⟨68HCA371, 83JA4017⟩ (Equation (58)). The reaction is by no means general and cleavage of the aryl-C—N bond is more usually carried out via the corresponding aryl amine.

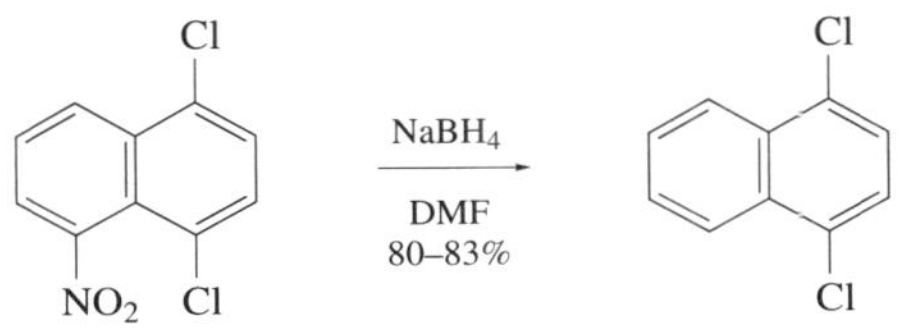

(58)

Many synthetic methodologies exist for the reduction of an aryl nitro group to aryl amines. These include reduction by (i) metals/H^+, (ii) catalytic hydrogenation ⟨B-76MI 110-01⟩ (iii) catalytic hydrogen transfer with HCO_2H/Pd ⟨77JCS(P1)443⟩, and (iv) complex hydride/transition metal catalyst combinations such as $NaBH_4/NiCl_2(PPh_3)_2$ ⟨79JCS(P1)2409⟩. A detailed examination of this reaction is beyond the scope of this chapter.

The reductive removal of a primary aromatic amine is commonly carried out by reduction of the corresponding diazonium salt ⟨B-78MI 110-01⟩. The diazonium salt is formed by reaction of arylamine with sodium or potassium nitrite at low temperature, most diazonium salts being unstable above 5°C. Tetrafluoroborate ⟨86JCS(P1)873⟩ and hexafluorophosphate salts ⟨63JOC568⟩ are notable exceptions. The diazonium salt can then be reduced with a variety of reagents including hypophosphorous acid, H_3PO_2 ⟨49JA2137⟩, EtOH ⟨42JCS748⟩, formaldehyde ⟨39JA2418⟩ and by metals such as zinc ⟨40JCS207⟩. Sn/HCl ⟨87JCS(P1)1951⟩, $NaBH_4$ ⟨61JA1251⟩, $Bu^n{}_3SnH$ and Et_3SiH ⟨70T4609⟩ have also been utilised.

Hypophosphorous acid remains the most selective reducing agent for this transformation ⟨49JA2137⟩. It exhibits a high degree of selectivity for the protonolysis of diazonium salts over a range of functionalities present on the arene ring. Indeed even aliphatic amine groups remain inert to diazotisation/reduction with H_3PO_2 (Equation (59)).

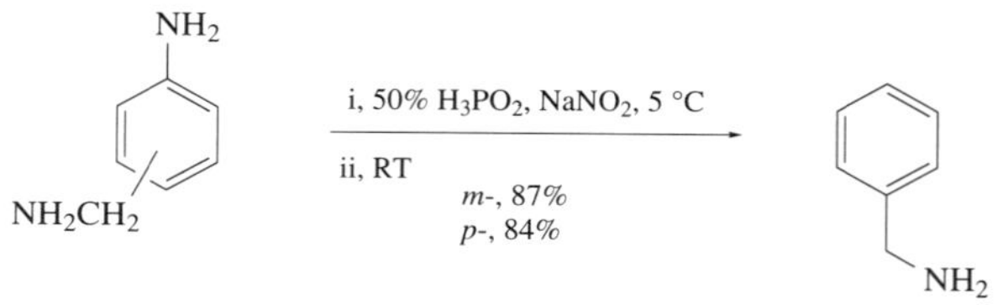

(59)

Sodium borohydride in MeOH or DMF is an effective reagent/solvent combination for the reduction of diazonium salts (Equation (60)) ⟨61JA1251⟩. Treatment with *t*-butyl nitrite in DMF also has been described as a useful method ⟨77JOC3494, 86JCS(P1)873⟩. Cadogan and Molina reported a simple 'one-pot' deamination procedure using pentyl nitrite in refluxing THF (Equation (61)) ⟨73JCS(P1)541⟩. Other examples include the use of $Bu^n{}_3SnH$ in polar solvents for the reduction of diazonium tetrafluoroborate salts, although this reagent lacks chemoselectivity, reducing aromatic nitro groups also. Triethylsilane exhibits improved chemoselectivity compared with $Bu^n{}_3SnH$ (Equation (62)) ⟨70T4609⟩.

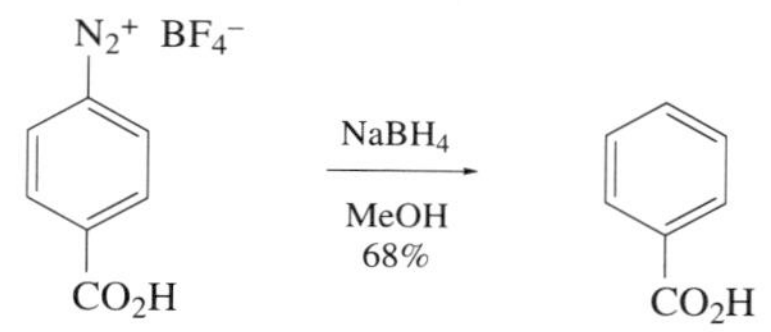

(60)

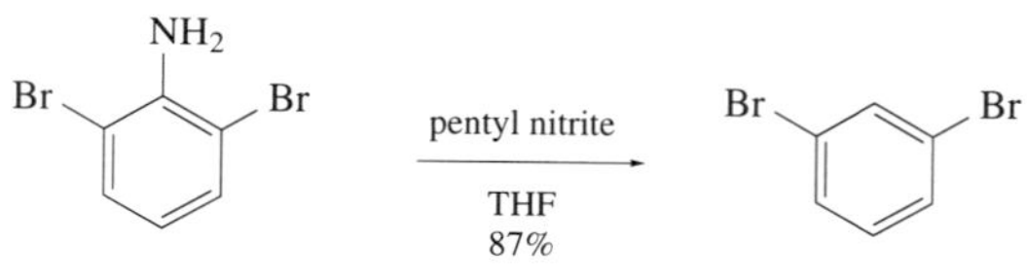

(61)

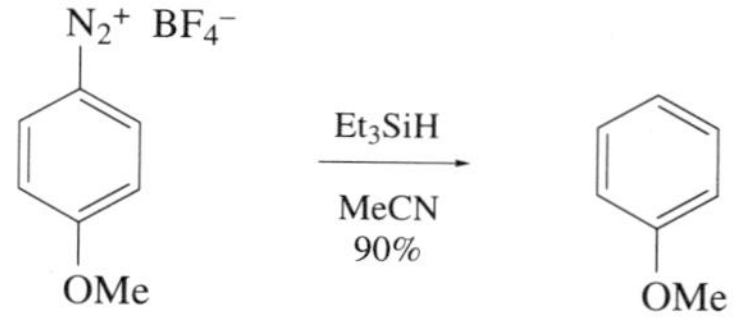

(62)

The mechanism of the reduction depends on the reaction conditions, the type of reducing agent and the nature of the diazonium salt. Diazonium salts decompose to yield aryl cations and N_2 in polar solvents. Subsequently arenes are formed by hydrogen transfer from either the solvent or the reducing agent. A free radical mechanism has been proposed for this type of reduction ⟨43JCS86⟩. The reduction of diazonium salts with $Bu^n{}_3SnH$ proceeds via two pathways, one involving hydrogen atom abstraction from $Bu^n{}_3SnH$, the other involving reaction of an arene radical with solvent ⟨70T4609⟩. Evidence that the reduction of diazonium salts with formamide proceeds intramolecularly has come from labelling studies ⟨77JOC3494⟩.

Hydrazines of the type ArNH—NH_2 are oxidised by a variety of reagents including HgO and

$K_3Fe(CN)_6$ to give azo compounds ArN=NH ⟨77S339⟩. These species readily decompose to ArH and N_2 ⟨73TL3497⟩.

1.10.1.4.2 Reduction of vinylcarbon–nitrogen bonds

Lewis and co-workers reported the unusual replacement of an amine for hydrogen during studies on the reaction of AlH_2Cl with enamines ⟨63PCS19⟩. Later studies found that enamines were efficiently cleaved to alkenes with either chloroalane or alane (AlH_3) itself (Equation (63)) ⟨68T4489⟩. Mechanistically the process is an addition/elimination, for example the addition of alane to the double bond followed by elimination yielding alkene and an amine. The methodology was applied to a series of pyrrolidine-derived enamines. The same group reported enamine cleavage was also observed on treatment with diborane, B_2H_6 (Scheme 8) ⟨64TL2039⟩. A report by Singaram and Brown *et al.* extended this boron based reduction methodology ⟨89JA384, 91JOC1543⟩: a series of aldehyde and ketone derived enamines were subjected to reduction by 9-BBN forming high yields of the corresponding alkenes (Scheme 9). Terminal alkenes and dienes, cycloalkenes, heterocyclic and acyclic (*Z*)- and (*E*)-alkenes were synthesised using this methodology. Interestingly, the hydroborating reagent and workup conditions were found to play a major role in determining the stereochemistry of the alkene formed from reaction of enamines derived from acyclic ketones. For example, hydroboration of (*E*)-(1)-morpholino-1-phenylprop-1-ene by 9-BBN followed by methanolysis gave the (*Z*)-alkene in isomerically pure form in 80% yield. On the other hand hydroboration of the same substrate by borane–dimethyl sulfide complex (BMS), followed by methanolysis and oxidation with H_2O_2 yielded (*E*)-1-phenylprop-1-ene as the pure geometric isomer.

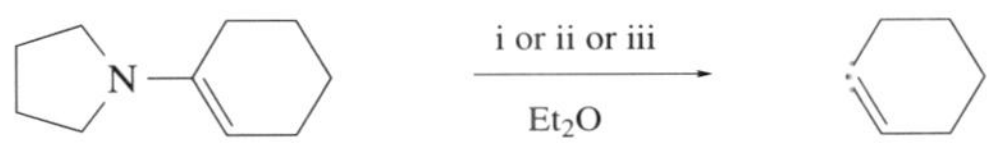

(63)

i, $AlHCl_2$, 75%; ii, AlH_2Cl, 75%; iii, AlH_3, 80%

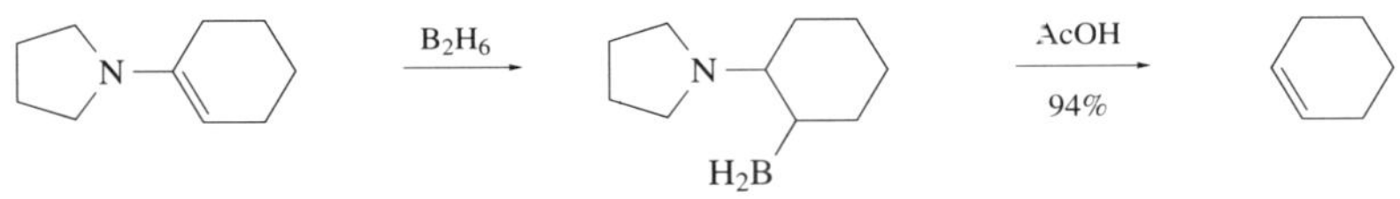

Scheme 8

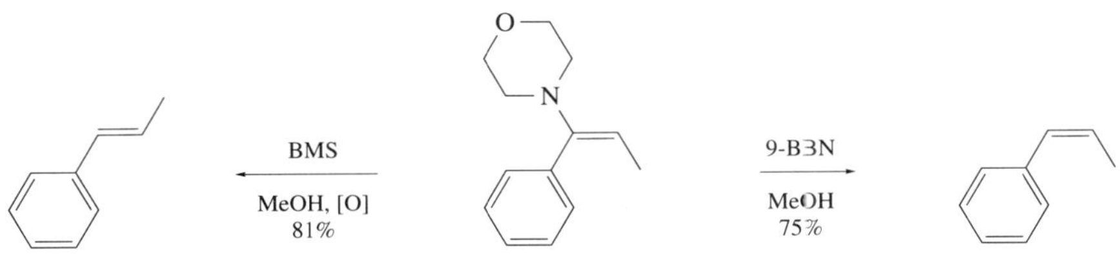

Scheme 9

The mechanistic rationale given for these transformations is as follows: hydroboration by borane–dimethyl sulfide, followed by methanolysis and oxidation proceeds via an intermediate dimethylboronate ester, which on oxidation forms the *N*-oxide. This species is thought to undergo a *syn* elimination forming the (*E*)-alkene (Scheme 10). Hydroboration with 9-BBN, followed by methanolysis involves the coordination of MeOH to the boron atom followed by an *anti* elimination (Scheme 11).

Barton and co-workers published an elegant alkene synthesis based on the radical induced elimination of a nitro group from a nitroalkene ⟨93TL8051⟩. Nitroalkenes were reacted with ethyl trithiocarbonate to yield *β*-nitrotrithiocarbonates (Scheme 12). Adducts of this kind underwent a radical-induced elimination on exposure to a combination of Barton esters, visible light and heat to form alkenes as mixtures of (*E*)- and (*Z*)-stereoisomers. Ono and co-workers have investigated the use of tin hydride-based methodologies to effect the transformation of nitroalkenes to alkenes ⟨85T4013⟩.

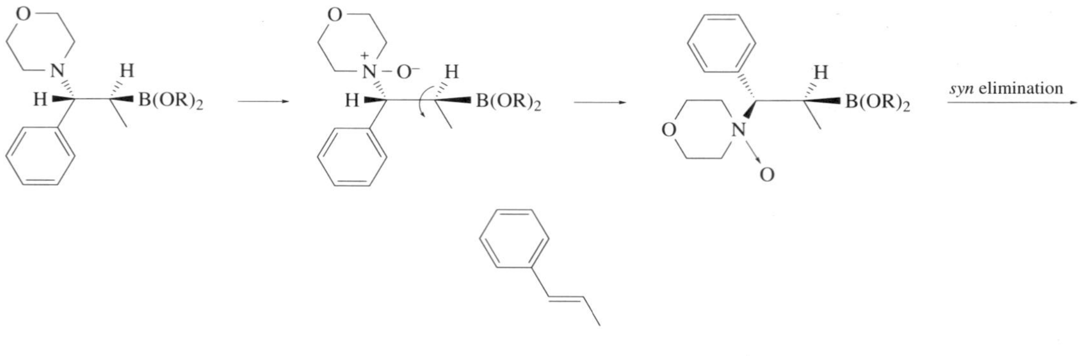

Scheme 10

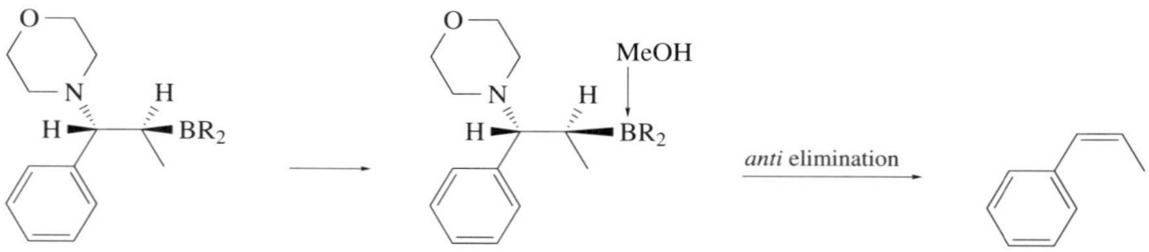

Scheme 11

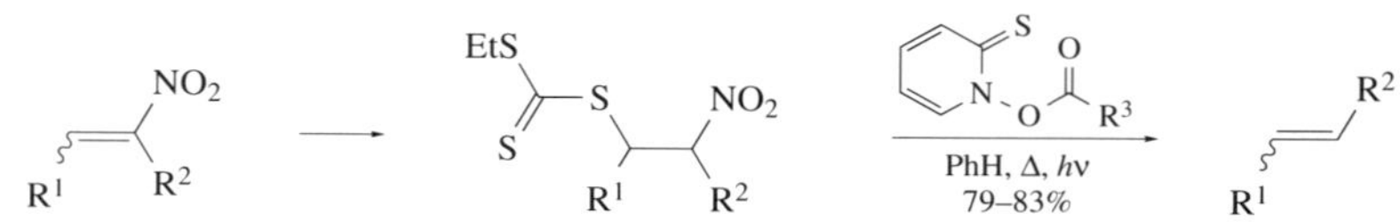

Scheme 12

1.10.1.5 Reduction of =C—P, =C—As, =C—Sb, =C—Bi Bonds

1.10.1.5.1 Reduction of arylcarbon–phosphorus, –arsenic, –antimony and –bismuth bonds

(i) The reduction of arylcarbon–phosphorus bonds

A variety of methods exist for reducing aryl-C—P^{III} bonds, the commonest being the reduction with alkali metals. The reduction yields the corresponding phosphide anion, which can be quenched with water or alcohols yielding secondary phosphines ⟨B-72MI 110-01⟩. The alkali metal reduction of tertiary phosphines has been enhanced by ultrasound ⟨85JCR(S)18⟩. It has been reported that metal–PPh_3 complexes undergo C—P bond cleavage under specific conditions ⟨85CRV171⟩. Almost quantitative amounts of benzene were formed when complexes of the type $RuH_2(PPh_3)_4$ were heated in pyrrolidine ⟨80JOM(193)37⟩.

Two general methods exist for the cleavage of aryl phosphonium salts, these being hydrolysis or reduction with LAH ⟨83PS(15)359⟩. Generally the alkaline hydrolysis of a phosphonium salt is used to prepare the corresponding phosphine oxide either with or without fracture of the aryl-C—P bond ⟨91COS(8)835⟩.

Posner and Lu described the hydrolysis of a phosphonium bromide with KOH (Equation (64)) ⟨85JA1424⟩. Similarly Minami *et al.* hydrolysed cyclobutene phosphonium perchlorate to cyclobutenephosphine oxide and benzene (Equation (65)) ⟨83JOC2569⟩.

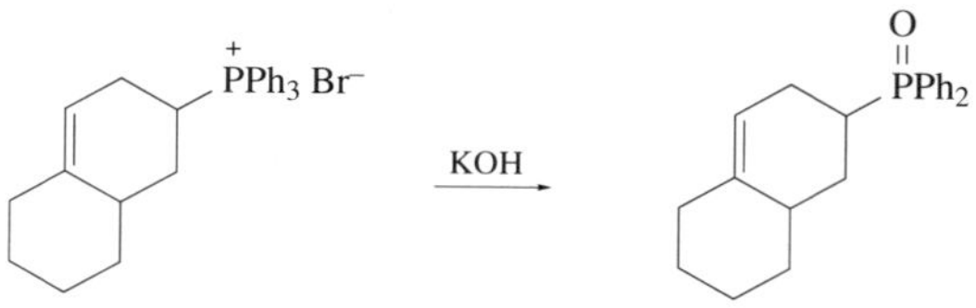

(64)

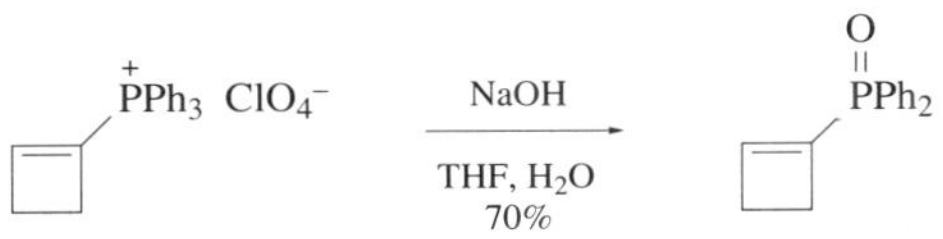

(65)

Extensive studies were carried out on the hydrolysis of σ-chlorophenyltriphenylphosphonium iodide ⟨67JOC2701⟩ with methoxide. Sodium naphthelenide has been used for the selective cleavage of one aryl-C—P bond in tertiary diphosphines ⟨86JOM(312)53⟩. Bickelhaupt and co-workers reduced 1-phenylphosphoniumadamantane bromide with elemental sodium to yield 1-phosphadamantane (Equation (66)) ⟨83T4225⟩.

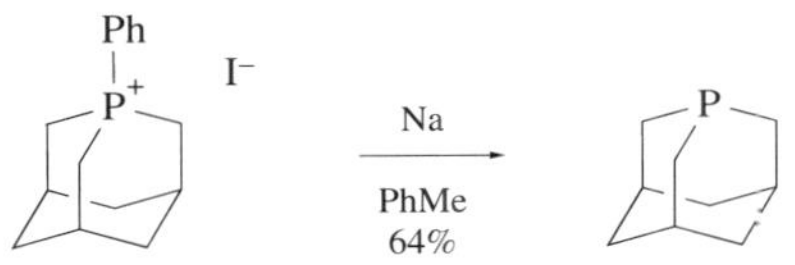

(66)

There are few general methods available for the reduction or hydrolysis of aryl-C—P^V bonds ⟨91COS(8)835⟩. Hydrolysis of aryl phosphonates and phosphates generally occurs at the P═O bond rather than at the P—C bond. Methods which have given P—C bond cleavage include heating with solid KOH. In certain cases reduction with hydride has produced P—C bond cleavage. Dibenzophosphepin was treated with NaH to yield, initially, the ring opened product (Equation (67)) ⟨80PS(8)243⟩.

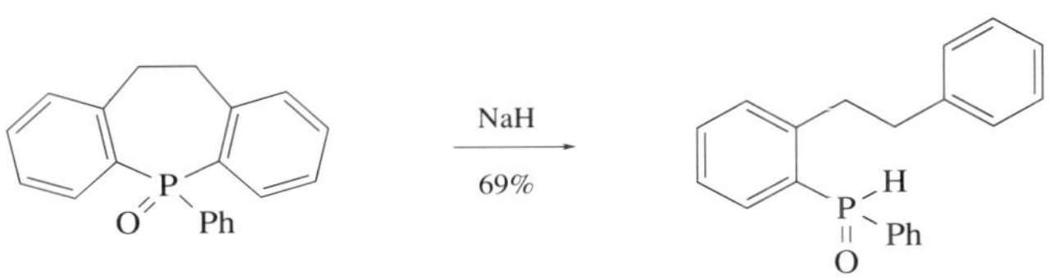

(67)

(ii) Reduction of arylcarbon–arsenic, –antimony and –bismuth bonds

The chemistry of organic arsenic, antimony and bismuth compounds has been reviewed ⟨B-94MI 110-01⟩. Triorganoarsenic compounds, diarsines and diorganoarsenic halides are reduced by alkali metals to afford the corresponding diorganoarsenic species (Scheme 13) ⟨67JOC2627⟩. Selectivity in the reduction of unsymmetrical tertiary arsines is thermodynamically controlled. In the case of diphenyl ethyl arsine this leads to formation of phenyl ethyl arsine anion ⟨74S328⟩. Arsenide anions react with protic acids (e.g., H_2O) to give hydrides ⟨65JCS4120⟩. Protonolysis of triorganoarsenic(III) compounds with HI leads to formation of organoarsenic(III) diiodides (Equation (68)) ⟨86JCS(D)997⟩. Triphenylarsine was reduced electrochemically to Ph_2AsH ⟨74JA5032⟩. Triorganoarsenic(IV) dihalides, under thermolytic conditions, form diorganoarsenic(III) halides ⟨57ZAAC348⟩. Wittig and Hellwinkel reported that pentaorganoarsenic(V) compounds when treated with electrophiles such as EtOD formed triorganoarsenic(III) compounds (Scheme 14) ⟨64CB769⟩.

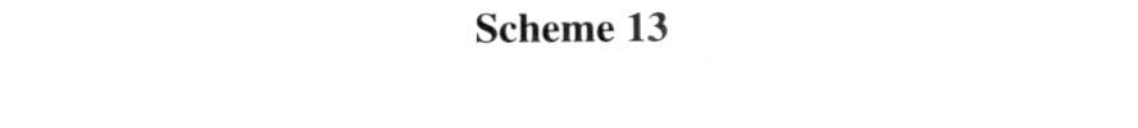

Scheme 13

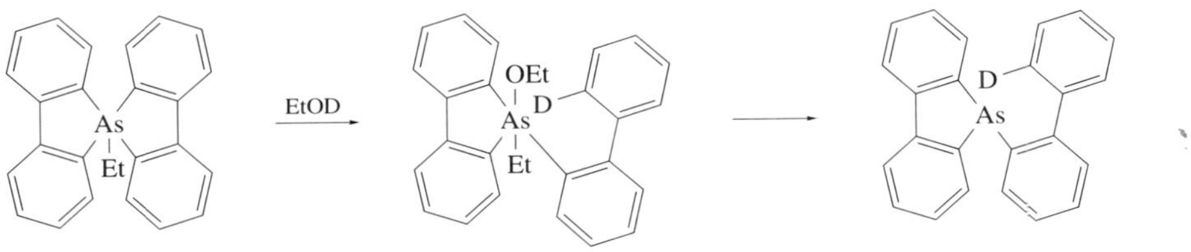

Scheme 14

(68)

The cleavage of the aryl-C—As bond has also been achieved using sodium hydrogen sulfite ⟨28JCS3046⟩ and by aqueous alkali at elevated temperatures ⟨40CA(34)6407⟩. Arsenous acids react with HgO breaking the As—C bond ⟨34CB317⟩. Organoantimony(III) and bismuth(III) hydrides are thermally unstable. Antimony hydrides can be formed by hydride reduction of organoantimony halides at low temperatures ⟨75ZAAC(418)158⟩. Only $MeBiH_2$ and Me_2BiH are known ⟨61CB1447⟩. Protonolysis of triorganoantimony and -bismuth compounds with mineral acids gives diorgano-antimony and diorganobismuth(III) halides ⟨85JOM(256)C1⟩ and ⟨66ZAAC(343)196⟩. Reaction of Ph_3Bi with benzoic acid leads to the formation of $PhBi(OCOPh)_2$ ⟨51JA2880⟩.

Pentaorganoantimony and -bismuth compounds react with electrophiles such as HF, HCl and Br_2 leading to the formation of tetraorganoantimony and -bismuth halides ⟨52LA(577)26, 52LA(578)136⟩. Pentaphenyl antimony (Ph_5Sb) disproportionates to Ph_3Sb and Ph—Ph at 160–200°C, as does Ph_5As ⟨B-94MI 110-02⟩. Treatment of pentaphenyl antimony with Cl_2 led to the formation of Ph_4SbCl (Equation (69)). Quaternary halides (Ph_4SbBr_3) thermolise at 180°C to give Ph_3SbBr_2 and PhBr. Quaternary bismuth halides (e.g., Ph_4BiCl) are unstable at room temperature and disproportionate to give Ph_3Bi (Scheme 15) ⟨52LA(578)136⟩.

$$Ph_5Sb \xrightarrow[76\%]{Cl_2} Ph_4SbCl \qquad (69)$$

$$Ph_5Bi \xrightarrow[54\%]{HCl} Ph_4BiCl \longrightarrow Ph_3Bi$$

Scheme 15

1.10.1.5.2 Reduction of vinylcarbon–phosphorus bonds

An efficient route to substituted alkenes and allenes from vinyl phosphines occurs with metallation of an intermediate phosphine with MeLi, followed by hydrolysis of the resultant vinyllithium. This approach was taken by Schmidbaur and co-workers during the synthesis of 1-(diphenylphosphino)-1,3-diphenylallene (Equation (70)) ⟨87CB1403, 89CB259⟩. Vinyl phosphonium salts are generally hydrolysed with either base or acid to give the corresponding alkenes.

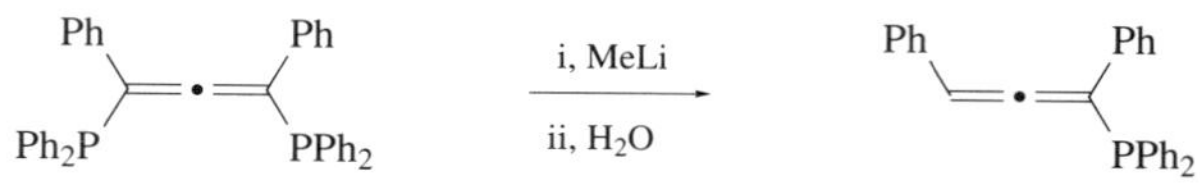

(70)

Larpent and Meigian hydrolysed a series of water-soluble vinyl phosphonium salts stereo-specifically with NaOH ⟨93TL4331⟩. The same group reported the preparation of regiospecifically labelled fumaric and cinnamic acids from D_2O/HCl hydrolysis of a sulfonated vinyl phosphonium salt ⟨88TL4577⟩. Fluorinated vinyl phosphonium salts were hydrolysed with NaOH ⟨83JA650, 85JA2811⟩: the fluoroalkenes were formed in good yields with complete retention of alkene geometry (Equation (71)). Enol ethers ⟨90TL2901⟩ and ketenethioacetals ⟨79CB28⟩ have also been synthesised via hydrolysis of the corresponding vinyl phosphonium salts.

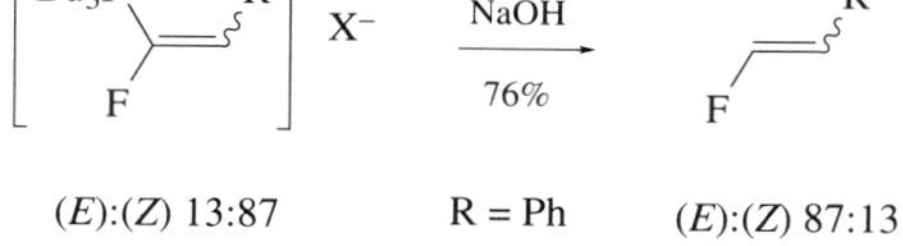

(71)

(*E*):(*Z*) 13:87 R = Ph (*E*):(*Z*) 87:13

1.10.1.6 Reduction of =C—B, =C—Si, =C—Ge Bonds

1.10.1.6.1 Reduction of aryl boranes, silanes and germanes

(i) Reduction of aryl boranes

The aryl-C—B bond appears to have similar reactivity to alkenylcarbon–boron bonds, though little data exist which allow direct comparison. The reduction of aryl boranes is treated briefly in a review ⟨B-87MI 110-01⟩ and a book ⟨B-61MI 110-01⟩ on organoboron chemistry. Aryl boronic acids undergo protodeboronation on treatment with a carboxylic acid, for example HCO_2H ⟨61JA2159⟩. The reaction proceeds via the formation of a six-membered cyclic transition state (Scheme 16).

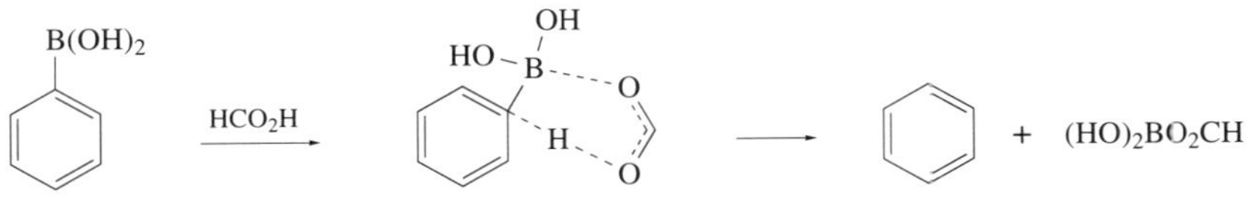

Scheme 16

The cleavage of the arylcarbon–boron bond was studied in the early 1930s. Hydrolytic cleavage was effected by heating $PhB(OH)_2$ with water, acid or base ⟨30JCS2171⟩. Dimethylamino- and 4-methoxy-α-napthyl-boronic acids were readily deboronated under mild conditions ⟨48JA232⟩. The presence of metal salts (e.g. $CuCl_2$) has also been found to promote aryl-C—B bond cleavage ⟨30JCS2171⟩. Ammonical $AgNO_3$ proved efficient at deboronating phenyl boronic acid ⟨38JA121⟩ Dearylation of borinic acids is known, and has been effected by prolonged exposure to heat. Treatment with aqueous ethanolic dimethylaminoethanol has also proved effective ⟨58JCS1451⟩. Tri α-napthyl boron gave napthalene and trimethyl borate on treatment with excess methanol ⟨55JOC9⟩.

(ii) Reduction of aryl silanes

The chemistry of aryl silanes has been extensively covered in a large number of books and review articles, for example ⟨B-81MI 110-01, B-89MI 110-01, 93SL171⟩.

The *ipso* substitution of silicon for hydrogen in an aryl silane is a well-studied reaction. Eaborn reported in 1975 that the C—Si cleavage occurred via a mechanism analogous to electrophilic aromatic substitution via a Wheland-type intermediate ⟨75JOM(100)43⟩. This *ipso* substitution of silicon is favoured over replacement of hydrogen due to the stabilisation of the incipient carbocation β to silicon in one of the canonical forms which can be drawn for the intermediate. The β-effect has also been the subject of intensive study and is now thought to arise due to a combination of four effects, namely inductive, field, $(p–d)\pi$ bonding and hyperconjugation ⟨93SL171⟩.

The competing electrophilic substitution of hydrogen over silicon is known to occur but, to date, appears to be limited to reaction with nitrating agents (e.g., NO_2^+) ⟨72JCS(P2)172⟩ though, even with nitrating agents, *ipso* substitution for silicon has been observed ⟨57JCS498⟩.

A number of factors need to be taken into account when discussing the protodesilylation reaction. The type of substitution on silicon can have a profound effect on the rates of C—Si bond cleavage. Silicon with electron-rich groups attached (e.g., SiR_3 where R = $MeOC_6H_4$) has a greatly increased C—Si cleavage rate, this effect being due to the greater stabilisation of the β-carbocation ⟨B-89MI 110-01⟩.

It should also be noted that both steric and electronic factors play significant roles in determining aryl silane reactivity. Silanes of the type $ArSiCl_3$ are cleaved more slowly than Ar-TMS, reflecting the weak electron-withdrawing effect of halogen, whereas those of the type $ArSiPh_3$ are also cleaved more slowly than Ar-TMS, this effect being a combination of steric bulk and electron donation. Ring substituents, especially those exerting powerful directing effects (e.g. R_2N ⟨61CA(55)14340⟩ and OMe ⟨60JCS179⟩) are known to promote reactions that dominate over the *ipso* substitution of silicon and therefore ring substitution occurs without loss of Si. The nature of the electrophile may also exert some influence on the mode of substitution in aryl silanes. With two SiR_3 groups present on an aryl nucleus it is known that the first C—Si bond cleaves some 40 times faster than the second, thereby allowing for substitution of silicon by different nucleophiles ⟨84AG(E)539⟩.

Most commonly, the method of cleavage used synthetically is treatment of the aryl silane with either mineral acid (e.g., HCl ⟨60JCS1566⟩) or carboxylic acid (e.g., CF_3CO_2H ⟨77JA5483⟩).

An efficient regioselective desilylation of a disilylated steroid was achieved with TFA, subsequent oxidation of the remaining C—Si bond leading to a concise synthesis of (+/−)oestrone (Equation (72)) ⟨84AG(E)539⟩. Effenberger also described the use of a number of carboxylic and sulfonic acids as efficient desilylating reagents ⟨78S755⟩. Cleavage of C—Si bonds in aryl and other aromatic systems initiated by nucleophilic attack at silicon is a less favourable process than C—Si cleavage initiated by electrophilic attack at carbon. However, in recent years this type of desilylation has come to the fore from a synthetic standpoint.

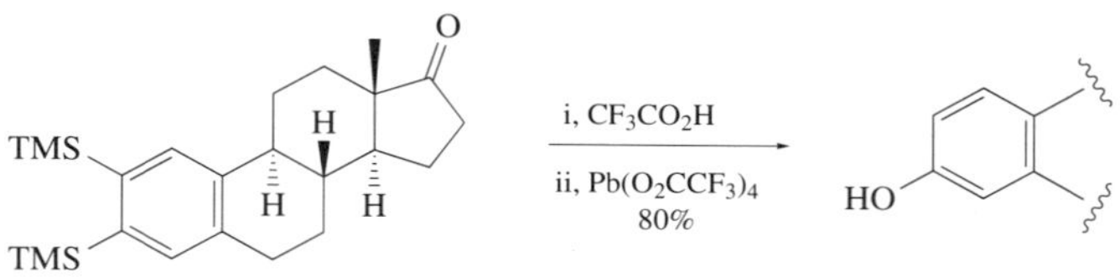

(72)

The most commonly used reagents for desilylation via nucleophilic catalysis are butoxide and F^-. Butoxide was used by Vollhardt *et al.* in a synthesis of terphenylene ⟨83CC502⟩. An interesting one-pot protodesilylation and carbodesilylation was described by Dunogues *et al.* ⟨90TL6179⟩: a bis-silylated fluoroarene was treated under Friedel–Crafts conditions with acetyl chloride in the presence of KF. A 79% yield of the desilylated fluorobenzophenone was isolated (Equation (73)). Caesium fluoride in DMF/H_2O has also proved a popular method for desilylating aryl silanes (Equation (74)) ⟨89CPB36, 89JOC4372⟩. Interestingly a tris-silylated arene was desilylated to the bissilylated product simply on standing in CH_2Cl_2 ⟨85JA1379⟩. The reason for this reaction occurring spontaneously is probably the relief of the massive steric congestion built into the starting material. Selective protodesilylations of bis-silylated arenes with t-butyl ammonium fluoride (tbaf) have been reported ⟨87AG(E)1035⟩. The bis-silylated cobalt complex shown in Equation (75) was treated with tbaf at room temperature for two hours to afford 91% of the monosilylated product.

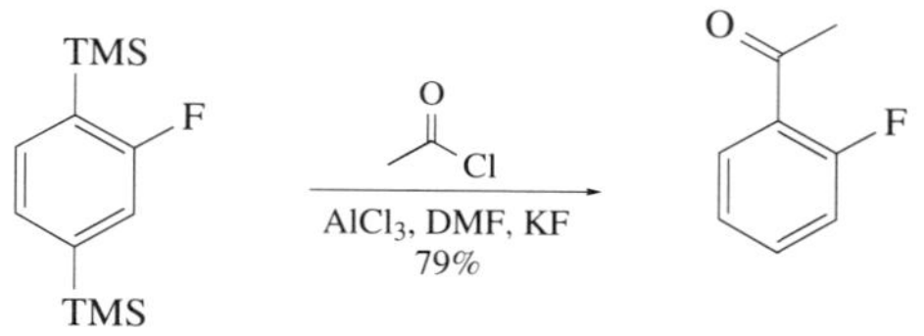

(73)

(74)

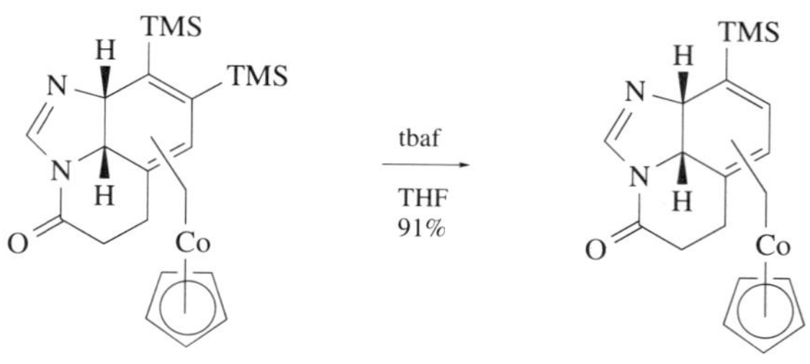

(75)

(iii) *Reduction of aryl germanes*

Aryl germanes have a lower carbon–metal bond energy than aryl silanes and therefore are expected to be more susceptible to reaction with electrophiles. $ArGeMe_3$ was readily degermylated with HCl in aqueous dioxane ⟨64JCS4804⟩. The reaction was observed to be significantly slower when D_2O was used in place of H_2O, indicating that the attachment of a proton to the aryl ring was the rate determining step in the protiodegermylation. Methanolic perchloric acid has also been utilised for the cleavage of $ArGeEt_3$ ⟨60JCS1566⟩. Trifluoroacetic acid in $CFCl_3$ was used to degermylate mixtures of trimethylgermylfluoroarenes (Equation (76)) ⟨87JFC(36)63⟩. The C—Ge bond in an aryl germane is also broken by methanolic sodium methoxide ⟨76JCS(P2)925⟩. Tetraphenylgermane formed triphenylgermane on treatment with Br_2 ⟨50JA5566⟩. Kobayashi *et al.* investigated the photochemical behaviour of several aryl alkyl germanes ⟨86BCJ2807⟩. It was found that the aryl-C—Ge bond cleaved readily under photolytic conditions, generating germyl radicals. Aryl radicals reacted with solvent forming benzene or its derivatives. The $Et_3Ge\cdot$ species subsequently reacted with solvent derived radicals to form $Et\cdot GeR$ or dimerised to form $Et_3GeGeEt_3$ (Equation (77)). Aryl dialkyl digermanes of the type $(ArMe_2Ge)_2$ undergo Ge—Ge bond cleavage on photolysis ⟨88OM1869⟩.

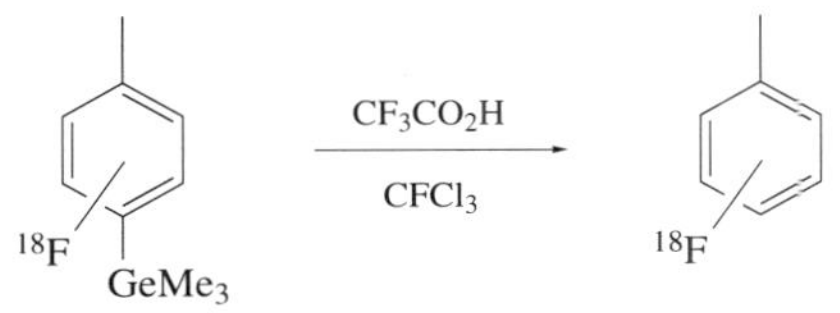

(76)

$$Et_3GeAr \xrightarrow[\text{hexanes}]{h\nu,\ 250\ \text{nm}} Et_3GeR + (Et_3Ge)_2 + ArH + C_{12}H_{26} \quad (77)$$

The halogenodegermylation of aryltrimethylgermanes has been used in the synthesis of ^{77}Br and ^{131}I radiolabelled arenes ⟨87JOC664⟩. The short reaction times (ca. one minute for completion) makes this an ideal synthetic route to radiolabelled compounds containing radionuclei with short half lives (Equation (78)).

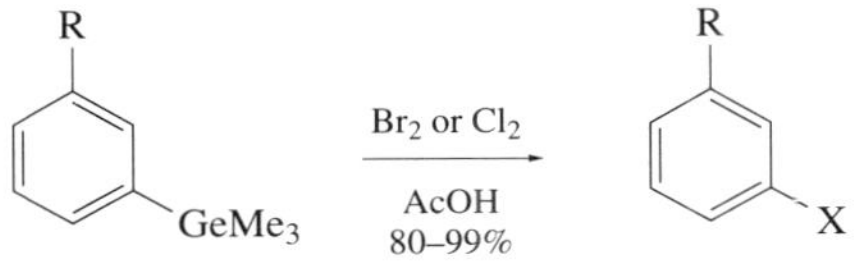

(78)

1.10.1.6.2 Reduction of vinyl boranes, silanes and germanes

(i) Reduction of vinyl boranes

The hydroboration of alkynes is an important synthetic transformation which has been featured in many reviews ⟨B-75MI 110-01, B-88MI 110-01⟩. Depending on the conditions of the subsequent cleavage of the intermediate vinylboron species a range of functionality can be introduced. In this section the replacement of boron by hydrogen (most commonly carried out by protonolysis) is of prime importance. Protonolysis can be carried out under such mild conditions that many functional groups remain unreactive.

All three alkenyl groups are cleaved from a trialkenyl borane on treatment with carboxylic acid at 0 °C ⟨59JA1512⟩. The protonolysis is known to proceed with retention of configuration thus leading to a stereospecific synthesis of (*Z*)-alkenes ⟨71JA3395⟩. Acetic acid has been used to cleave the sp^2-C—B bond of an allenylborane ⟨74JA5620⟩. Brown has reported a synthesis of (*Z*)-1-haloalk-1-enes, via hydroboration with $(cyclohexyl)_2BH$ or 9-BBN, the intermediate organoboranes being cleaved with AcOH (Equation (79)) ⟨89JOC6064⟩. This method has also been used for a (*Z*)-enyne synthesis ⟨89TL6487⟩ and also for the synthesis of (*Z*)-vinyl silanes (Equation (80)) ⟨90TL5113⟩. Alkenyl dialkyl boranes are preferentially cleaved (e.g. by methanolysis) at the alkene–boron bond, ⟨86JOC4512⟩. A variety of 9-BBN derivatives were treated under such conditions giving high yields (>86%) of the corresponding alkene (Equation (81)). Acid catalysis (pivalic acid) was found to be required for sterically hindered derivatives. Treatment with HCl leads to protonolysis ⟨74S339, 75JOC2845⟩.

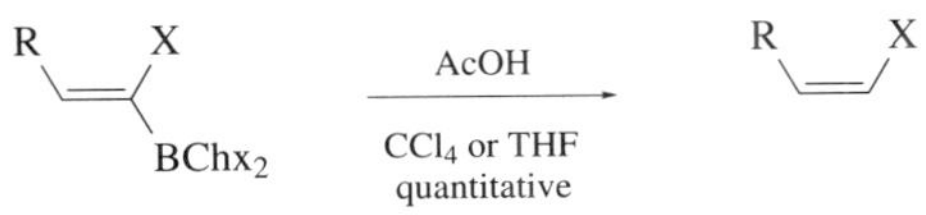

(79)

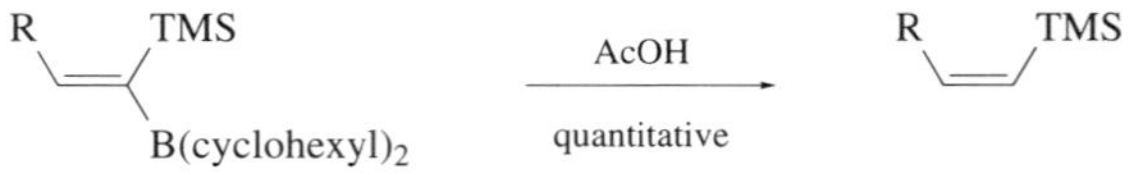

(80)

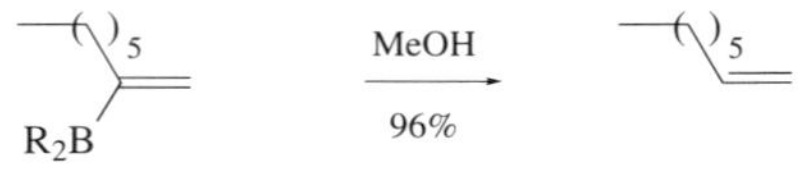

(81)

R_2 = 9-BBN

Negishi described an elegant method for cleaving the boron–carbon bond in alkenyltrialkylborates ⟨76JOC3484⟩. Representative alkenyldialkylboranes were treated with Bu^{n}Li. The borate intermediate so formed was subsequently hydrolysed stereospecifically with NaOH forming the (*Z*)-alkene (Scheme 17). Corey and Ravindranathan utilised $Ag(NH_3)_2NO_2$ to break a vinyl–boron bond in the presence of an acetal group (Equation (82)) ⟨72JA4013⟩. Catalytic cleavage of a vinyl borane by $Pd(OAc)_2$ under essentially neutral conditions was observed by Yamamato *et al.* (Equation (83)) ⟨78CC702⟩.

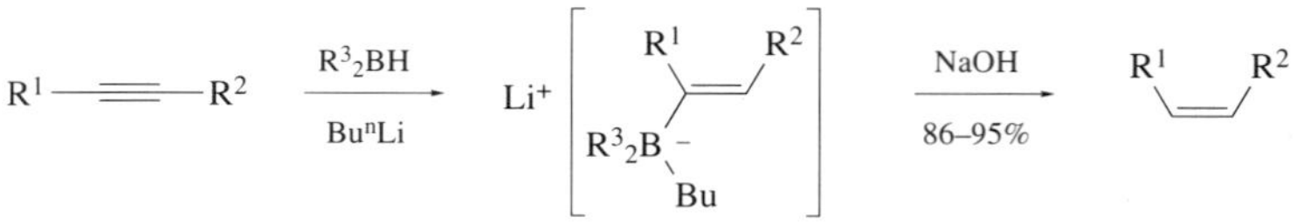

Scheme 17

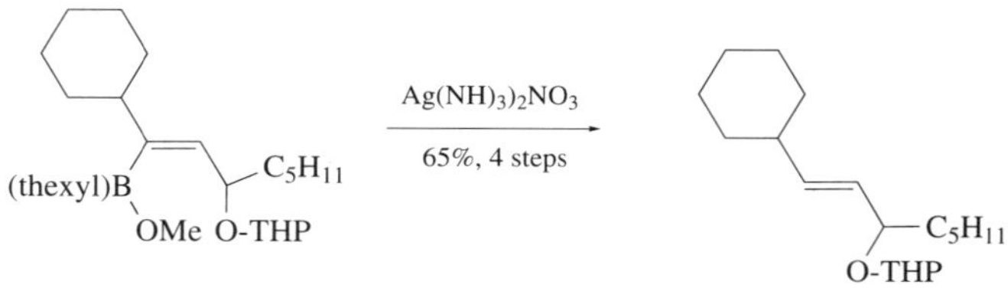

(82)

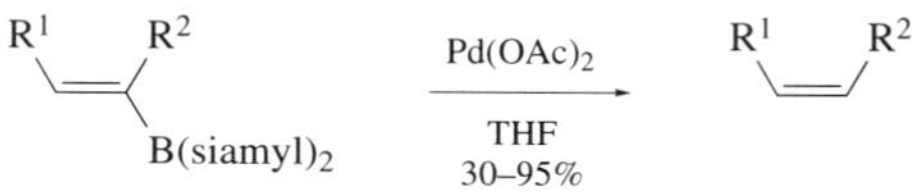

(83)

R^1 = Bu^n, Hex^n, Ph, Oct^n
R^2 = Bu^n, Et, Ph, H, $(CH_2)_7CO_2Me$

(ii) Reduction of vinyl silanes

The chemistry of vinyl silanes has been extensively reviewed ⟨79S761, B-81MI 110-01, B-89MI 110-01, 89OR(37)57⟩. Sommer and co-workers reported that *β*-trimethylsilylstyrene was cleaved by mineral acid to yield styrene ⟨54JA1613⟩. The stereospecificity of the reaction was subsequently investigated ⟨73JA3416⟩, Koenig and Weber finding that trimethylsilylstyrene on treatment of DCl was converted to *β*-deuteriostyrene with complete retention of alkene geometry. The mechanistic details of this substitution have been investigated and are described in detail in a review ⟨89OR(37)57⟩. This substitution of silicon for hydrogen with preservation of the double bond and retention of configuration appears to be general and a number of efficient reagent–solvent combinations have been reported to carry out this transformation. Thus a variety of acids have been utilised. These include hydroiodic acid (HI) ⟨76JOC2215⟩ or iodine in H_2O/benzene (Equation (84)) ⟨83TL1041⟩, HF ⟨83T935⟩, HBr ⟨71JOM(33)321⟩, AcOH ⟨82JOC4713⟩ and particularly HCl in either MeOH ⟨85JCS(P1)2399⟩ or acetone ⟨86JA3516⟩ (used in a synthesis of protomycinolide(IV)), or diethyl ether ⟨81JOM(217)43⟩. Fluoroboric acid is also an efficient desilylating agent for vinyl silanes ⟨84JA3668⟩. Isomerisation of the product alkene can be a problem. A mild reagent for this transformation is *p*-toluene sulfinic acid (TsH), a much quoted example of its use being the cleavage of the acid sensitive sesquiterpene **(7)** (Equation (85)) ⟨77TL4305⟩. Desilylations with TsH are known to lack stereospecificity compared with mineral acid based transformations. Sensitive systems have been successfully desilylated with

TsOH in mixed aqueous-organic solvent systems (Equation (86)) ⟨84TL1999⟩. This methodology has been extended to the synthesis of stereospecifically deuteriated alkenes. Useful reagent combinations are I_2–D_2O ⟨75TL2825⟩, DCl-organic solvent ⟨73JA3416⟩, or AcOD–D_2O ⟨79JOC155⟩. Vollhardt and co-workers used CF_3CO_2D to probe the differing reactivities between C-1 and C-2 bissilylated alkenes on protodesilylation ⟨86JA1359⟩.

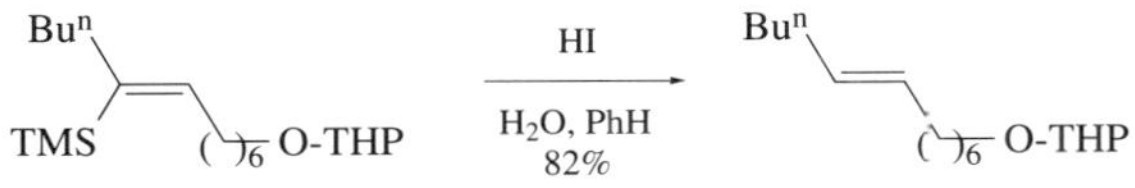

(84)

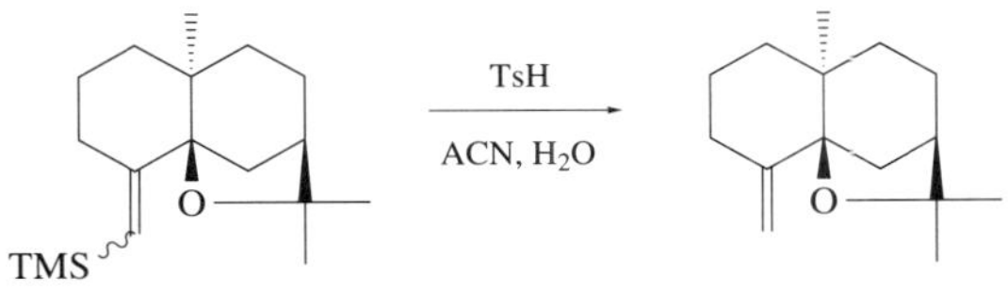

(7)

(85)

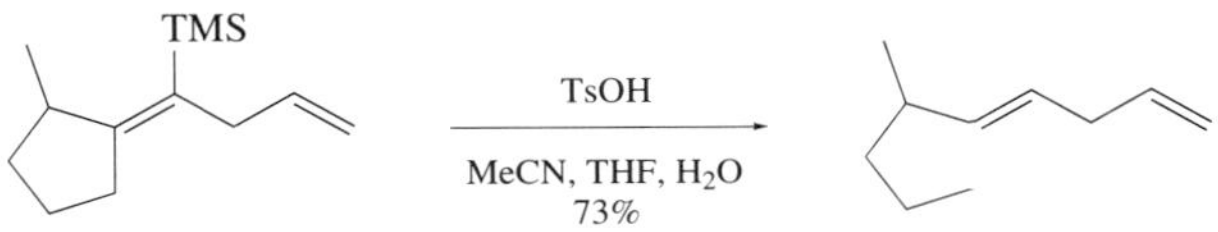

(86)

Vinyl silanes are known to be desilylated under conditions of nucleophilic catalysis ⟨89OR(37)57⟩ specifically where good nucleophiles for silicon (i.e., F^-, or alkoxide) are employed. Conditions tend to be harsh or prolonged for efficient desilylation (e.g., ⟨75JA5603⟩) unless other factors come into play, for example the stabilisation of the putative vinyl anion intermediate. Effective stabilising groups include hydroxy ⟨74TL3479⟩, halogen ⟨81S999⟩ and cyano ⟨82CC56⟩ moieties. Protodesilylation then occurs under considerably milder conditions. *t*-Butyl ammonium fluoride (tbaf) has been used by many groups for efficient desilylation of 'stabilised' vinyl silanes (Equation (87)) ⟨79JA4420, 83JOC3761⟩.

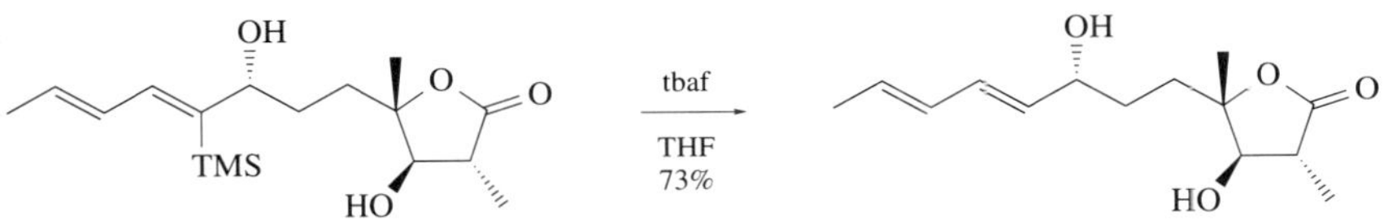

(87)

Examples of alkoxide catalysis include those of Tamao *et al.* ⟨75JA5603⟩ who used NaOEt in DMSO at 130 °C to cleave bis-silylated alkenes. The vigorous conditions led to loss of stereointegrity of the product alkene (Equation (88)). Kusakabe and Sato showed that a vinyl silane containing an acid sensitive acetonide could be smoothly converted to the corresponding alkene using NaH in HMPA (Equation (89)) ⟨86CL1473⟩.

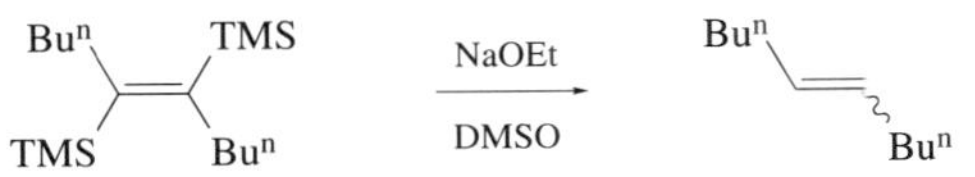

(*E*) (*E*), 90%; (*Z*), 6%

(88)

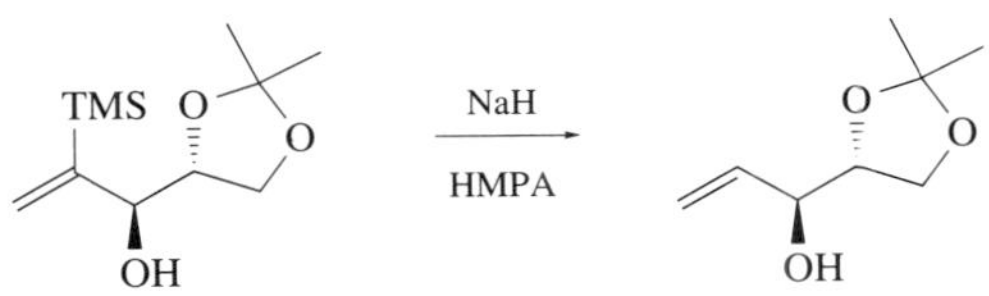

(89)

(iii) Reduction of vinyl germanes

Vinyl germanes, have not been exploited extensively in synthetic chemistry. For example, very few reports exist which describe their reaction with electrophiles.

Oda and co-workers described a straightforward synthesis of vinyl germanes from the platinum-catalysed hydrogermylation of terminal alkynes ⟨84TL3221⟩. The regiochemical and stereochemical outcomes were dependent on the catalyst used. The vinyl germanes were treated with bromine or iodine to form vinyl halides in good yield, with excellent stereoselectivity in the case of iodine. Cyclic alkenyl germanes give excellent yields of vinyl iodides when treated with iodine ⟨85CL53⟩.

Piers has extended this methodology in an elegant synthesis of the tetraquinane diterpenoid (+)-crinipellin B ⟨93JOC11⟩. Ikenaga *et al.* exploited the reactivity of vinyl germanes ($PhCH{=}CHGeMe_3$) with arene diazonium tetrafluoroborates (ArN_2BF_4) under palladium catalysis to yield products of the type PhCH=CHAr ⟨90CL185⟩.

1.10.1.7 Reduction of =C—Metal Bonds

1.10.1.7.1 Reduction of arylcarbon–metal bonds

The substitution of a metal for a proton in an arylmetal species is most commonly carried out by simple protonation by electrophiles such as H_2O or mineral acids. Less commonly, reduction by complex hydrides is used ⟨B-87MI 110-02⟩.

PhLi reacts with a variety of proton donors including H_2O, ROH and RCO_2H to yield arenes ⟨B-87MI 110-02⟩. The protonolysis of an aryl-Li intermediate was reported by Parham and co-workers (Equation (90)) ⟨76JOC1184⟩.

$$\text{PhLi} \xrightarrow[\text{quantitative}]{\text{ROH}} \text{PhH} \qquad (90)$$

Aryl Grignard reagents, ArMgX, are readily hydrolysed to arenes with H_2O or dilute mineral acid (Equation (91)) ⟨B-87MI 110-03⟩. The organometallic chemistry of calcium, strontium and barium has been less thoroughly investigated. Organo calcium, strontium and barium 'Grignard' reagents have been prepared at low temperature and appear to show similar reactivity towards electrophiles ⟨73T1069, 78JCS(D)657⟩. Diaryl beryllium reagents, such as Ph_2Be, are readily synthesised but their chemistry remains poorly investigated ⟨82COMC-I(1)121⟩. The tantalum moiety of aryl tantalum(III) compounds undergoes *ipso* substitution with a variety of atoms or groups including F, CN, SeCN and NO_2 ⟨B-87MI 110-04⟩. Reduction with $NaBD_4$ in EtOD is an efficient method for the deuteriodethallation of aryl tantalum(III) species (Equation (92)) ⟨73TL1375⟩.

$$\text{ArMgX} \xrightarrow[\text{quantitative}]{\text{H}_2\text{O or H}^+} \text{ArH} \qquad (91)$$

$$\text{ArTa}^{\text{III}}\text{Z}_2 \xrightarrow[\text{or NaBD}_4\text{, EtOH}]{\text{NaBH}_4\text{, EtOD}} \text{ArD} \qquad (92)$$

Aryl stannanes, on protonolysis with carboxylic and mineral acids, yield arenes ⟨87MI 110-05⟩. Eaborn *et al.* prepared a variety of deuteriobenzenes from the corresponding aryltin derivatives ⟨73JCS(P1)137⟩. Aryl trimethylstannanes on treatment with AcOD, containing 10% D_2O, yielded labelled arenes in good yield and high (>98%) isotopic purity (Equation (93)). Seitz and co-workers reported a related method for the synthesis of ^{3}H-tamoxifen ⟨81SC281⟩. σ-Stannylated tamoxifen was tritiodestannylated on treatment with excess $(CF_3CO_2)_2O$ and T_2O. The destannylation procedure was equally applicable to other starting materials (Equation (94)).

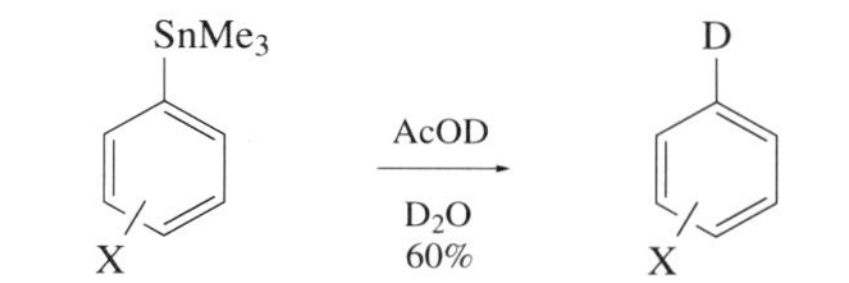

(93)

X = *o*-Me, *m*-Me, *p*-Me, *m*-Cl, *p*-Cl, *m*-OMe, *m*-CF_3, *p*-F

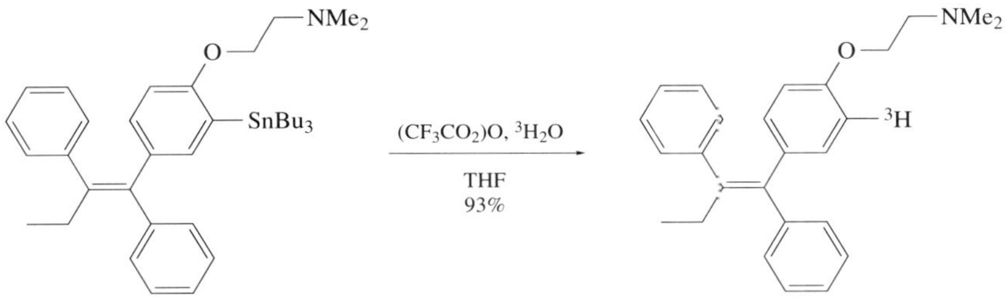

(94)

Aryl mercurials are cleaved by AcOH/perchloric acid giving, in the case of diphenylmercury, benzene and mercury(II) salts. The bond-breaking process is known to be a simple S_E2-type electrophilic substitution.

1.10.1.7.2 Reduction of vinylcarbon–metal bonds

Organoalkali metal compounds react with a range of proton sources including H_2O, ROH, RCO_2H, RNH_2 and mineral acids. Alkenyllithiums react with H_2O or H^+ to form alkenes with retention of configuration (Equation (95)) ⟨77JA8356, B-87MI 110-02⟩. Vinyl Grignard reagents are also readily hydrolysed in H_2O forming alkenes (Equation (96)). Carbomagnesiation of alkynes via Normant reagents $RCu \cdot MgX_2$ leads to the formation of intermediate vinyl cuprates which also break down on protonolysis to form alkenes (Equation (97)) ⟨78FAC709⟩. Vinyl derivatives of the remaining Group 2 metals (calcium, strontium, barium and beryllium) are less well known though the similar reactivities of Ca, Sr, and Ba 'Grignard' reagents has been reported ⟨B-87MI 110-03⟩.

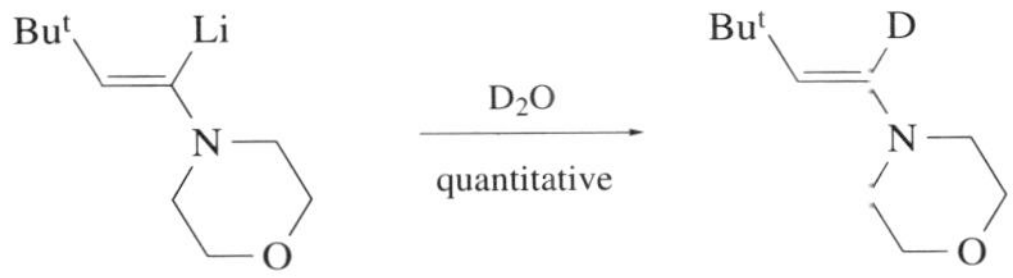

(95)

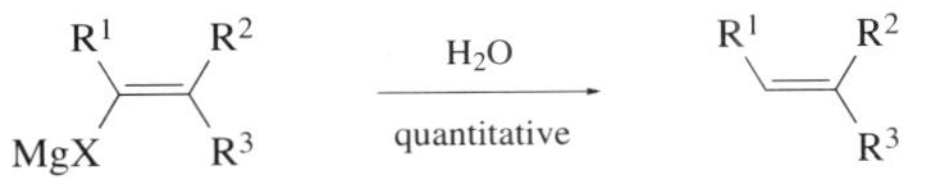

(96)

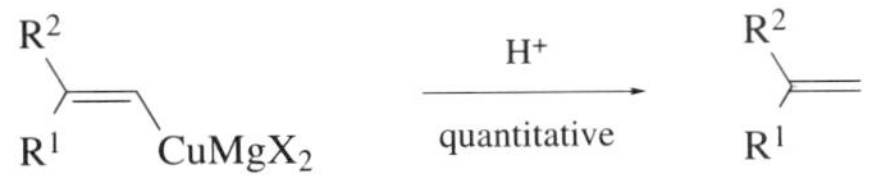

(97)

Vinylaluminum species are commonly formed by the hydro- or carboalumination of alkynes. The hydroalumination reaction has been reviewed ⟨84OR(32)1, B-87MI 110-02⟩. Vinyl alanes are generally cleaved by reaction with proton donors such as MeOH or H_2O (Equation (98)) ⟨63JOC1254, 65CB824⟩. The scope of the carboalumination reaction has been greatly broadened by the observation that catalysis with zirconium leads to a 100% *cis* carbometalation ⟨81JA4985, 81PAC2333⟩.

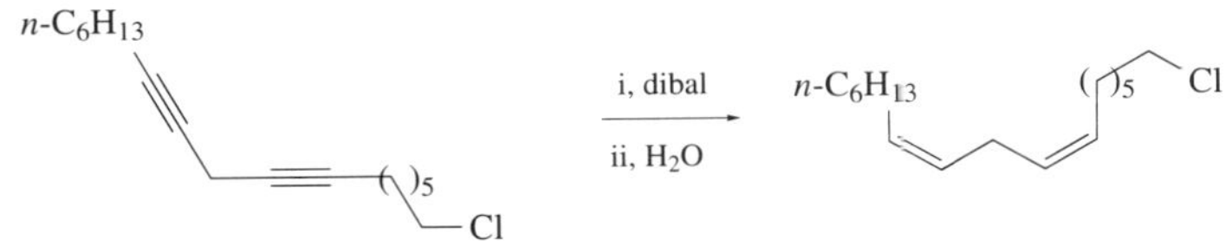

(98)

Lipshultz and co-workers used a *in situ* generated Schwartz reagent, $Cp_2Zr(H)Cl$, for the hydrozirconation of alkynes ⟨90TL7257⟩. Reaction of the intermediate vinyl-Zr species with electrophiles such as H_2O, D_2O and N-bromosuccinimide (NBS) led to the formation of alkenes in good yield and without loss of acid-sensitive functionality (Scheme 18).

Vinyl-Ni complexes of the type $Ni(acac)(PPh_3)(C(Ph){=}C(R_2)CH_3)$ are cleaved with both electrophilic and nucleophilic reagents (e.g., H_2O, TsOH, LAH) to form alkenes in good to excellent yields ⟨81JA3002⟩. The protiodethallation of monovinylthallium(III) compounds proceeds with

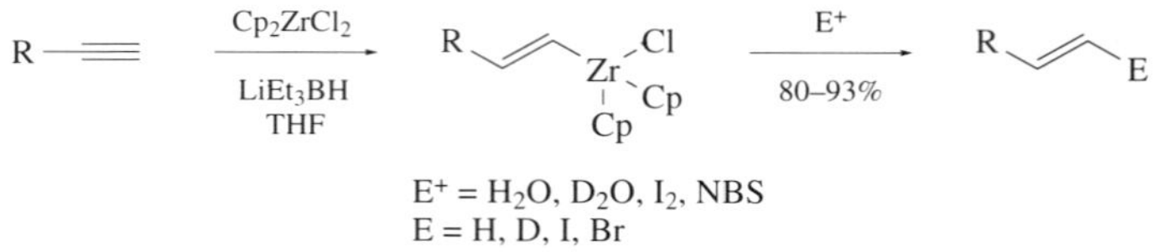

Scheme 18

retention of configuration of the alkene upon heating in AcOH ⟨74BCJ2663⟩ or by reduction with $NaBH_4$ in protic solvents (Equation (99)) ⟨81JCS(P1)991⟩.

Vinyl mercurials are cleaved to the corresponding alkene by either protonolysis or reduction, both methods being dealt with in extensive reviews ⟨78T2827, B-85MI 110-02, 91COS(8)835⟩. The acetoxymercuration of alkynes, with subsequent hydrolysis of the intermediate vinyl mercurial, is an important route to stereochemically pure alkenes ⟨80JCS(P1)1098⟩. Protodemercuration of a range of alkylphenyl alkynes with AcOH takes place with strict retention of configuration at the site of substitution. A combination of HCl/NaI has been used for the demercuration of vinyl mercurials (Scheme 19) ⟨85JA1726⟩.

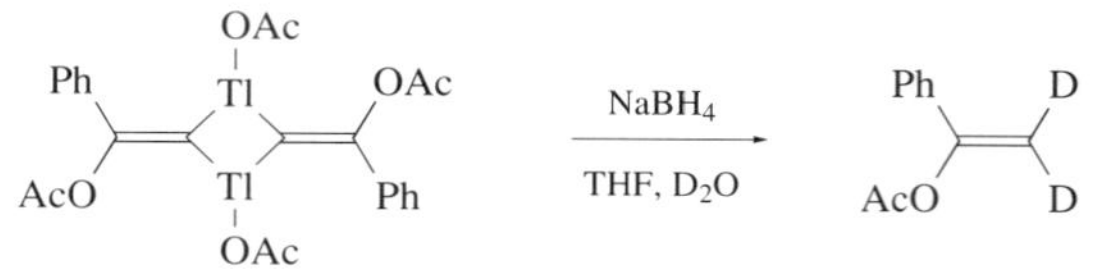

(99)

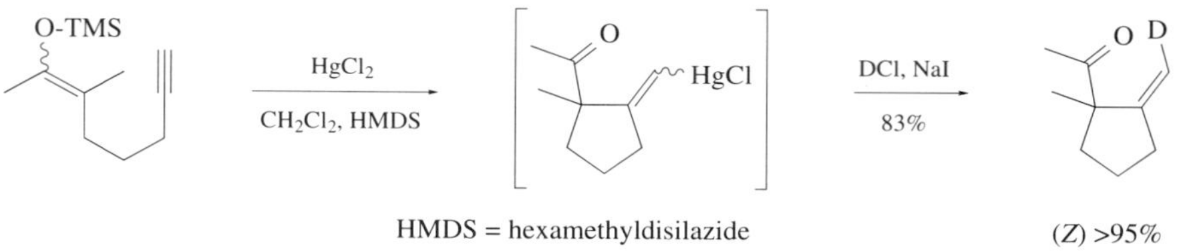

Scheme 19

A non-cage free radical mechanism is thought to operate during the $NaBH_4$ mediated reduction of organomercurials ⟨B-85MI 110-01⟩. Interestingly LAH appears to have been little used to effect reductions of this type. The main problem encountered on reduction of organomercurials is reverse deoxymercuration; this difficulty is usually overcome by using a basic homogenous reaction medium. Use has been made of $NaBH_4$ in NaOH for effectively cleaving vinylmercurials ⟨84JOC3411⟩.

The use of phase transfer catalysis is a powerful methodology for cleavage of vinyl mercurials where competing side reactions have proved troublesome. The reduction of vinyl mercurial with $NaBH_4$ in CH_2Cl_2 has been carried out with benzyltriethylammonium chloride as the phase transfer catalyst (Equation (100)) ⟨79S891, 83TL3997⟩.

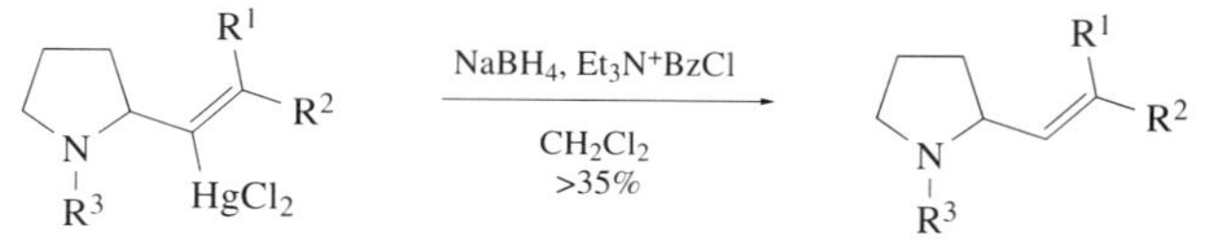

(100)

The mercury atom in a vinyl mercurial is susceptible to Hg–metal exchange. This exchange process, followed by protonolysis, leads to efficient alkene formation. Bis(halogenomercurio)methylenecyclohexanes were transmetallated with either lithium metal or Bu^tLi. After one hour the reaction was quenched with D_2O leading to the formation of exocyclic methylene compounds in ca. 50% yield ⟨84AG(E)224⟩. Zinc dust in AcOH was utilised to reduce vinyl mercurials where the more standard $NaBH_4$/NaOH conditions gave divinyl mercurials ⟨82JOC3707⟩. Miscellaneous methods for reducing organomercurials include Na–Hg amalgam ⟨82JOC3707⟩ and reaction with thiols ⟨83JA1398⟩.

Eaborn and co-workers studied the destannylation of a variety of β-trimethylstannylstyrenes under both acidic (AcOH–MeOH) and basic (NaOH–MeOH) conditions. A number of significant factors including double bond geometry, substituents and reaction conditions were observed to play a major role in determining the rate of sp^2-C—Sn bond cleavage ⟨80JOM(201)233⟩. Kinetic data are available for the protonolysis of vinyl trialkylstannanes on treatment with HCl–MeOH ⟨82OM586⟩.

Stereoselective syntheses of three insect sex pheromones including (6(*E*), 11(*Z*))-hexadeca-6,11-dien-1-yl acetate (from the wild silkworm *Antheraea polyphemus*) have been reported ⟨84JOC5175⟩. In the final step of the synthesis the (*E*)-double bond was generated by stereospecific protonolysis of both vinyl stannane and vinyl borane moieties with AcOH (Equation (101)). Comins and co-workers reported a synthesis of dihydrocannivonine, the intermediate vinyl stannane being destannylated using oxalic acid ⟨84TL4867⟩. A series of α-methylene-γ-valerolactones was synthesised using an intramolecular radical cyclisation of acetylenic ketones onto propiolates ⟨90TL5039⟩. The resultant vinyl stannanes were protonolysed by treatment with HCl in diethyl ether.

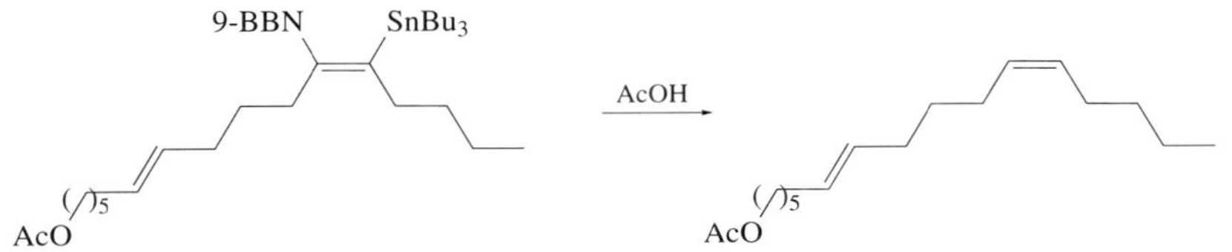

(101)

Casson and Kocienski transmetallated a series of α-alkoxyvinyl stannanes with BuLi leading to formation of the corresponding α-alkoxyvinyllithiums. Protonolysis with D_2O–H_2O led to formation of enol ethers (Scheme 20) ⟨93SL1133⟩. The cleavage of polyfluorovinylstannanes has been investigated ⟨61JOC2934⟩. The fluorostannyl alkenes were treated with reagents such as iodine, HBr, AcOH and EtONa. It was found that =C—Sn bond cleavage took place a great deal more readily in the perfluorovinylstannyl series than with the corresponding protiovinylstannanes.

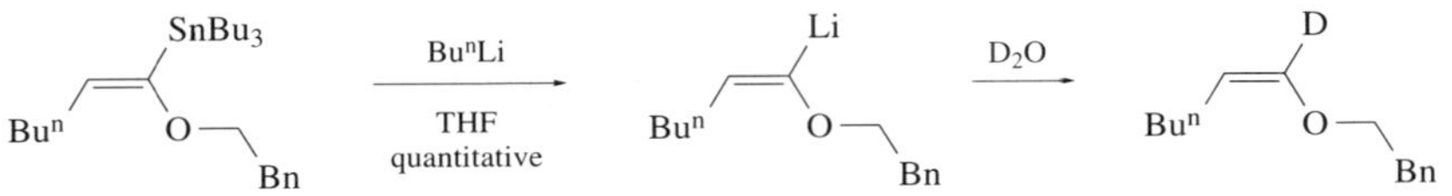

Scheme 20

1.10.2 ONE OR MORE =C—H BONDS BY ADDITION

1.10.2.1 Addition to Alkynes

1.10.2.1.1 Addition of hydrogen to alkynes

A vast array of synthetic methodologies exist for reducing an alkyne to an alkene. These include: (i) hydrogenation under the influence of both heterogenous and homogenous catalysis, (ii) dissolving metal reductions, (iii) transition metal/complex hydride reductions, (iv) hydrometallations, with subsequent hydrolysis, and (v) catalytic hydrogen transfer under both heterogenous and homogenous catalysis. Each method will be described in more detail later in the article with comment on the stereospecificity, mechanism and efficiency where appropriate.

(i) Addition of hydrogen to alkynes by heterogeneous catalysts

The hydrogenation of alkynes under heterogenous catalysis is featured in many reviews ⟨B-84MI 110-01, 85CRV129, B-85MI 110-01, 91COS(8)417⟩. Generally palladium-, platinum- and nickel-based catalysts are used, the metals being dispersed on inert supports such as carbon or silica. Polymer bound $PdCl_2$-based catalysts have been introduced more recently with some success ⟨79MI 110-01⟩. Reactions are generally carried out under an H_2 atmosphere, typically 1–30 atm. Hydrogenations of this kind must be carefully monitored to ensure semihydrogenation occurs, significant yields of over-reduced products being widely reported ⟨91COS(8)417⟩. Addition of amines, particularly pyridine and quinoline, are used to halt the hydrogenation at the alkene level. The most frequently used heterogenous catalyst is the Lindlar catalyst, Pd–$CaCO_3$, $Pd(OAc)_2$ poisoned with quinoline ⟨73OS880⟩. Alkynes reduced with this catalyst generally give (*Z*)-alkenes, the reaction being most selective at low temperature ⟨77T457⟩.

The vast majority of palladium- and nickel-based heterogenous hydrogenation catalysts give higher yields of the less thermodynamically stable (*Z*)-isomer, though the actual mechanism is not well understood ⟨91COS(8)417⟩. Analogies have been made with the *syn*-addition of hydrogen observed by mononuclear homogenous hydrogenation catalysts as studied by Muetterties *et al.* ⟨81ICA1⟩.

Good chemoselectivity is observed with catalysts of this type. Chemoselective reduction of triple bonds in the presence of conjugated and non-conjugated double bonds is known ⟨73S457⟩. The reduction of diynes leads to the formation of ((*Z*),(*Z*))-alkenes ⟨77T1845⟩.

Developments in heterogenous hydrogenation catalysis include the use of finely divided palladium metal dispersed in a siloxane polymer matrix ⟨90JOC3452⟩. The catalyst is reported to be highly stereoselective, reducing dec-5-yne to dec-5-ene in a 20:1 ratio of ((*E*):(*Z*)) stereoisomers, in quantitative yield (Equation (102)). Kuno and co-workers reported the use of a Pd-zeolite catalyst, modified with diphenyldiethoxysilane, for the stereospecific reduction of alkynes to alkenes ⟨92CL1725⟩.

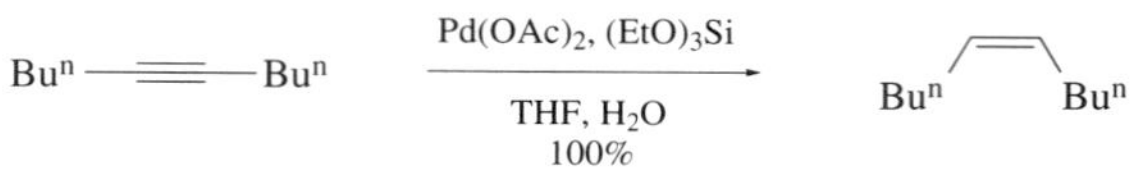

(102)

(ii) Addition of hydrogen to alkynes by homogeneous catalysis

The reduction of alkynes by hydrogenation under homogenous catalysis is the subject of many reviews ⟨76OR(24)1, 82COMC-I(1)121, 91COS(8)443⟩. Alkynes were reduced to alkenes using Wilkinson's catalyst [RhCl(PPh$_3$)$_3$] ⟨81MI 110-01⟩ in benzene containing either phenol or 2,2,2-trifluoroethanol ⟨68MI 110-01⟩. Similarly hydrogenations using D_2 and [Rh(COD)(PPh$_3$)(Py)]PF$_6$ as the catalyst converted phenylacetylene into (*Z*)-PhD=CHD ⟨77JOM(141)113⟩. In the presence of [RhH(PMe$_2$Ph)$_5$]PF$_5$ internal and terminal alkynes are reduced to alkenes ⟨85JMOC213⟩. Dinuclear rhodium and iridium complexes of the type Rh$_2$(μ-X)(CO)$_2$(Ph$_2$ECH$_2$EPh$_2$)$_2^+$ and [Ir$_2$Cl(CO)$_3$Ph$_2$PCH$_2$PPh$_2$)$_2$] were reported to be potent alkyne hydrogenation catalysts ⟨82CJC1363⟩.

Novel arene–Cr(CO)$_3$ complexes are known to catalyse a variety of synthetic transformations ⟨93S643⟩. Catalysts of the form MBz·Cr(CO)$_3$ and NP·Cr(CO)$_3$, (MBz = methyl benzoate and NP = napthalene) are known to reduce disubstituted alkynes with high stereospecificity and yield (87–100%) to form the (*Z*)-alkenes (Equations (103) and (104)) ⟨85JOC1147⟩.

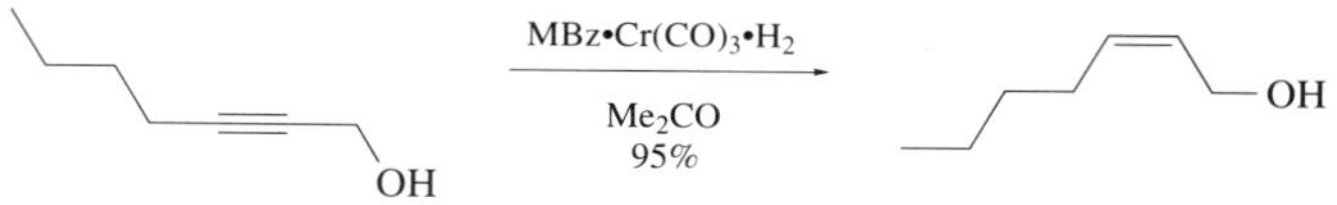

(103)

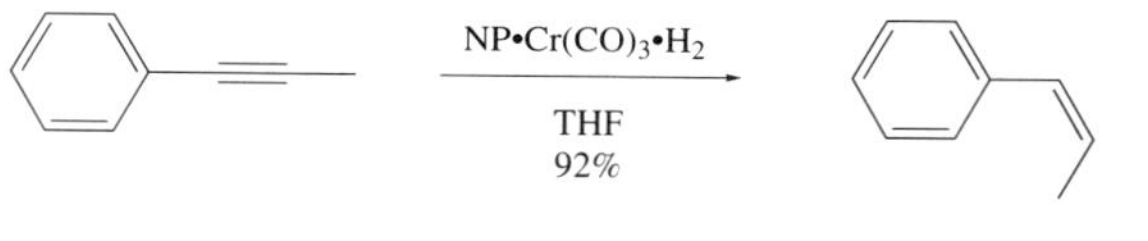

(104)

This methodology is reinforced by many other organometallic catalysts which have exhibited good stereoselectivity for the synthesis of (*Z*)-alkenes. These include Ti(Cp)$_2$(CO)$_2$ ⟨66BCJ1178⟩, Ti(Cp)$_2$Cl$_2$ ⟨86CC1317⟩, and the iron cluster μ_5-C$_5$H$_5$Fe-μ_3-(CO)$_4$ ⟨79JOM(178)43⟩.

The formation of (*E*)-double bonds by catalytic hydrogenation via homogenous catalysis is an uncommon transformation. Acetylene dicarboxylic acid dimethylester is reduced to dimethyl fumarate by hydrogen in the presence of RhH$_2$(η^2-O$_2$COH)(P—Pr$^i_3$)$_2$. The rhodium complex was reported to catalyse the isomerisation of (*Z*)- to (*E*)-alkene bonds (Equation (105)) ⟨83JA6273⟩.

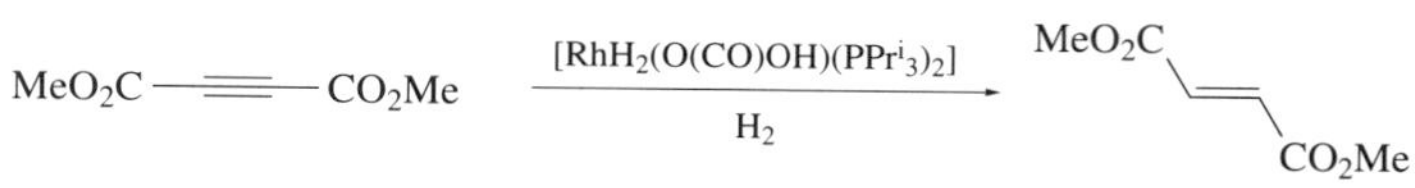

(105)

(iii) Addition of hydrogen to alkynes by other methods

(a) Dissolving metal reductions. The discovery that alkynes were stereospecifically reduced to (*E*)-alkenes by Na/NH$_3$(l) was a major breakthrough in organic synthesis ⟨41JA216⟩. This method

then complemented the known methods for synthesising (*Z*)-alkenes and it remains powerful methodology. Variations on this theme include reductions with Li/RNH_2 ⟨55JA3378⟩, Na/HMPA ⟨74JOC747⟩ and Yb/NH_3 ⟨78JOC4555⟩. Calcium/RNH_2 mixed solvent systems has also been used to perform alkyne reductions but problems are encountered with extensive double bond migration ⟨84JOC1662⟩. Alkyne reduction has also been carried out with zinc either in combination with $BrCH_2CH_2Br$ or as a Zn–Cu couple (Equation (106)) ⟨80TL1069⟩ This reduction tends to give (*Z*)-products. Little work appears to have been done to determine the mechanism of reduction with zinc.

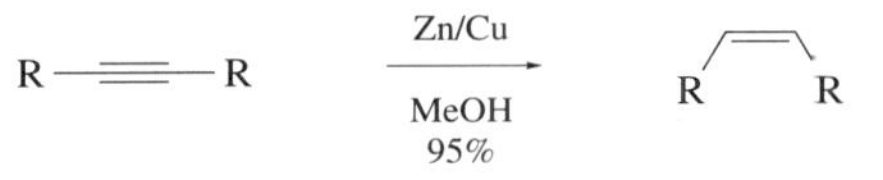

R = Ph, CH_2OH (106)

Dissolving metal reductions exhibit reasonable chemoselectivity, tolerating many functionalities including acetals, ethers and acetates. The reduction of enynes leads to preferential reduction of the triple bond ⟨84CC735⟩.

(b) Catalytic transfer hydrogenation. The catalytic transfer hydrogenation of alkynes to alkenes has been reviewed ⟨85CRV129, 74CRV567⟩. The reaction can be carried out under both heterogenous ⟨80JOC4926⟩ and homogenous ⟨89TL4657⟩ palladium catalysis, though both methods have been plagued by over-reduction and the formation of ((*E*):(*Z*))-isomer mixtures. A report highlights the use of homogeneous palladium(0) catalysis for the preparation of (*Z*)-alkenes in high stereochemical purity ⟨93CC386⟩. Disubstituted alkynes were reduced in 77–96% yield using $Pd_2(dba)_3$–PBu_3 as catalyst and HCO_2H–NEt_3 as the hydrogen source (Equation (107)). A high degree of stereoselectivity (>98%) was observed.

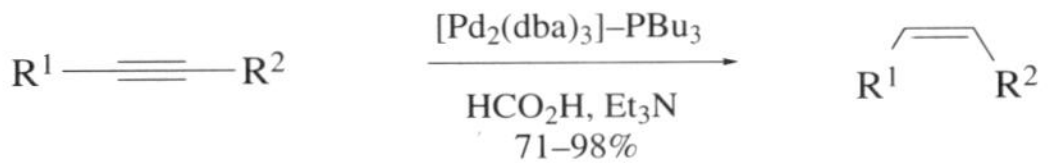

R^1 = *n*-C_8H_{17}, Ph, Et, *n*-C_5H_{11}
R^2 = H, Me, CO_2Me, CH_2OH (107)

Catalytic transfer hydrogenation reactions exhibit a number of favourable characteristics over classical hydrogenation reactions, in that no special apparatus is required, the reactions are more economical and considerably less hazard is involved. Potentially these characteristics will lead to this type of reaction being exploited more fully.

(c) Low valent metal reductions. The reduction of alkynes by low valent chromium(II) species has been reported ⟨64JA4358⟩. A number of alkynes were reduced stereospecifically to (*E*)-alkenes in excellent yield by reaction with $CrSO_4$ in DMF/H_2O. The mechanism leading to the observed stereochemistry was rationalised in terms of a bis-metallated alkyne intermediate, with subsequent hydrogen transfer (Equation (108)).

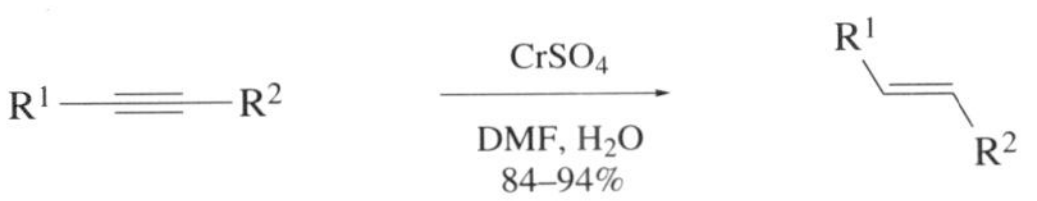

R^1 = H, CO_2H, Ph, Me, CH_2OH
R^2 = CH_2OH, Ph, CO_2H (108)

A stereospecific reduction of both terminal and disubstituted alkynes with low valent niobium and tantalum species has been described ⟨90TL365⟩. Dodec-1-yne was reduced with $NbCl_5$ and zinc in DME/C_6H_6. The reaction was quenched with NaOD/D_2O to afford dideuteriated 1,2-[$^2H_2$]-dodec-1-enes. Similarly, low valent tantalum, formed from the reaction of $TaCl_5$ and Zn gave good yields of (*Z*)-alkenes at considerably faster rates than those observed for the low valent niobium reductions (Scheme 21) ⟨92JOC1615⟩.

(d) Transition metal hydride reductions. A wide range of transition metal salts in combination with LAH, $NaBH_4$ and other hydride reducing agents are known to effect the reduction of alkynes to alkenes. Dialkylalkynes are cleanly reduced to (*Z*)-alkenes by LAH at low temperature in the presence of $NiCl_2$ ⟨78JOC2567⟩. A similar reaction is observed using LAH and stoichiometric quantities of $FeCl_2$ ⟨78JOC2567⟩. Sodium borohydride in the presence of cobalt(II) is also an effective

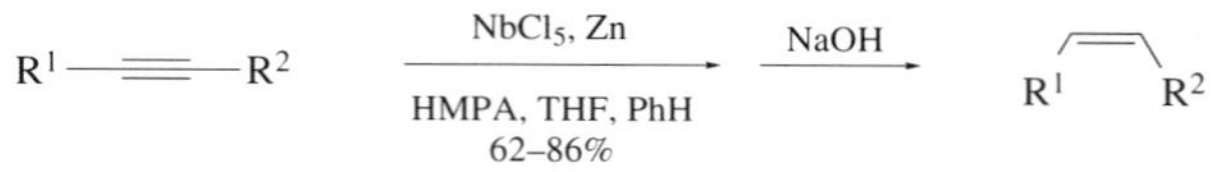

$R^1 = n\text{-}C_{10}H_{21}$, Ph, $n\text{-}C_5H_{11}$, Bu^t, TMS
$R^2 = n\text{-}C_{10}H_{21}$, $n\text{-}C_5H_{11}$, $n\text{-}C_6H_{13}$, $n\text{-}C_7H_{15}$

Scheme 21

combination ⟨79JOC1014⟩. Related reagent mixtures include MgH_2–CuI ⟨78JOC757⟩ and NaH–$Ni(OAc)_2$ ⟨77TL1069⟩. Synthetically this reaction is restricted to compounds lacking functionality sensitive to hydride reducing agents.

Alkynes have also been reduced with transition metal salts such as $CoCl_2 \cdot 4PPh_3$ and SmI_2, with ROH as the proton source: (*Z*)-alkenes were formed in good yield (Equation (109)) ⟨91TL5559⟩. Interestingly, addition of HMPA led to a reversal of stereoselectivity. The use of the copper(I) hydride reagent $[(Ph_3P)CuH]_6$ for selectively reducing terminal alkynes to alkenes and disubstituted alkynes to (*Z*)-alkenes has been documented (Equation (110)). The reaction is highly chemoselective tolerating propargylic alcohols and acetates. Related copper(I) hydride 'ate' complexes also gave the corresponding alkene ⟨90TL2397⟩.

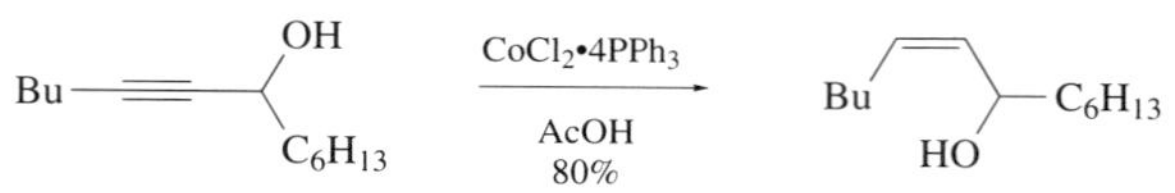

(109)

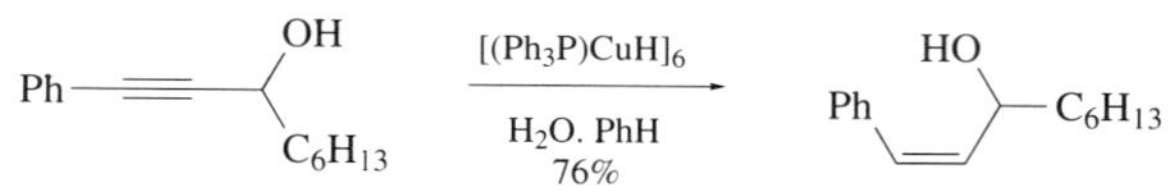

(110)

(e) Reduction with diimide. The reduction of alkynes with diimide has been reviewed ⟨91COS(8)471⟩. This reduction has been known for many years and remains a most simple and elegant way of reducing a triple bond to a (*Z*)-alkene with high stereospecificity and chemoselectivity. The reduction with diimide is extremely chemoselective tolerating ester, amide, and allylic halide functionality. Iodinated alkenes have been prepared by diimide reduction of 1-iodoalkynes ⟨72JA9256⟩. The main drawback is that dialkylalkenes are as reactive as the starting alkynes, which can lead to the production of fully saturated compounds ⟨91COS(8)471⟩.

The reagent can be generated in many ways, the most popular method being the oxidation of hydrazine with molecular O_2 or H_2O_2 in the presence of copper(II) and/or a carboxylic acid ⟨61TL347⟩. The transfer of hydrogen occurs in a *syn* fashion and generally the reaction proceeds in high yield, as in the reduction of diphenylacetylene to (*Z*)-stilbene ⟨61TL347⟩. The high degree of stereoselectivity was fully exploited during Cornforth's synthesis of chiral acetate ⟨69NAT1212⟩.

(f) Hydrometallation of alkynes to alkenes. The hydrometallation of alkynes with boron, aluminum, zirconium and other metals with subsequent hydrolysis of the metal–alkene bond is covered in other sections of this article.

1.10.2.1.2 Addition of C—H to alkynes

The addition of C—H to alkynes is reviewed in Chapter 1.12.

1.10.2.2 Addition to Allenes

1.10.2.2.1 Addition of hydrogen to allenes

Allenes are important synthetic intermediates due to their ability to react with nucleophiles, electrophiles and radical species and their chemistry has been extensively reviewed ⟨B-80MI 110-01, B-82MI 110-01, B-84MI 110-02, 93S165⟩. The semihydrogenation of allenes to alkenes has been successfully achieved by hydrogenation under both heterogeneous and homogeneous catalysis. Allenes have also been reduced to alkenes by: (i) dissolving metal reductions, (ii) diimide and (iii) by complex hydrides.

Heterogeneous catalytic hydrogenation over both palladium and platinum have been described. On addition of D_2 to buta-1,2-diene using palladium on alumina as catalyst three products were formed, (*Z*)-but-2-ene, (*E*)-but-2-ene and but-1-ene in the ratio 40 : 53 : 7 respectively ⟨63JA2881⟩. Reduction of allenes over platinum catalysts tends to lead to complete saturation of both allene double bonds ⟨B-84MI 110-02⟩.

Wilkinson's catalyst, $RhCl(PPh_3)_3$, was investigated for the reduction of allenes to alkenes ⟨72TL1391⟩. The reduction proved to be regiospecific and highly stereoselective for acyclic allenes such nona-1,2-diene and 3-ethylpenta-1,2-diene (Equation (111)). However, a low rate of conversion was observed with nona-1,2,6-triene. It was thought that this may be due to the ability of the triene and the product dienes to coordinate to Rh.

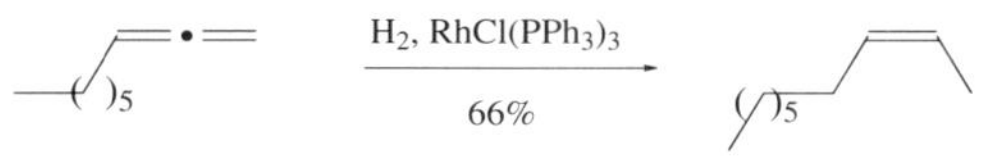

(111)

The diisobutylaluminum hydride (dibal-H) reduction of allenes has been investigated ⟨80TL51⟩. Addition of dibal-H at room temperature or at 60 °C led to the formation of an intermediate vinylalane, which, on protonolysis, yielded an alkene. It was found that the more substituted double bond of an allene was selectively reduced, leading to the formation of an alkene product with the opposite regioselectivity observed for reduction by catalytic hydrogenation (Equation (112)). Reduction of an allenyl silane with dibal-H gave a mixture of two products in the ratio 79 : 21. The products were identified as the vinyl silane and 1-trimethylsilyldecane (Equation (113)).

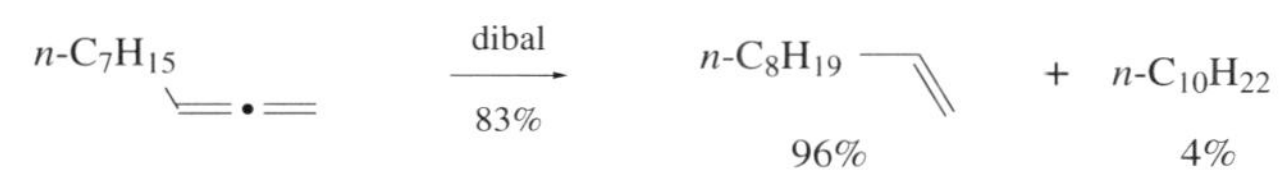

(112)

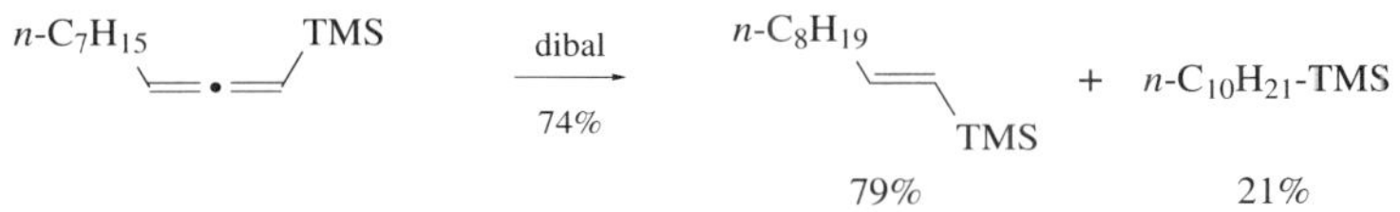

(113)

Diimide preferentially reduces allenes to (*Z*)-alkenes. The reagent reduces the less substituted end of the allene and does so via approach from the less hindered face. Reductions of this type can be complicated by overreduction to alkanes because allenes and alkenes have similar reactivities towards diimide. However, nona-1,2-diene is reduced to (*E*)-nona-2-ene in 32% yield; only a small amount (ca. 1%) of the (*Z*)-isomer was observed ⟨80JOC1604⟩. Cyclic allenes for example cyclonona-1,2-diene and cyclonona-1,2,6-triene were efficiently converted to (*Z*)-cyclononene and ((*Z*),(*Z*))-cyclonona-1,5-diene, respectively ⟨70TL4243⟩. Cyclopropylallene was regio- and stereospecifically reduced to the corresponding vinylcyclopropane. (Equation (114)) ⟨73TL2517⟩. Dissolving metal reductions, particularly Na–NH_3(l) have been used extensively for reducing allenes to alkenes. Acyclic alkyl allenes (*e.g.*, octa-1,2-diene and nona-1,2-diene) give excellent yields of the corresponding 2-(*E*)-alkenes when treated with Na–NH_3(l) ⟨70JOC347⟩. Interestingly, nona-2,3-diene yielded an equimolar mixture of (*E*)-non-2-ene and (*E*)-non-3-ene in 85% yield. The dissolving-metal reduction of phenyl substituted allenes gave primarily alkyl benzenes.

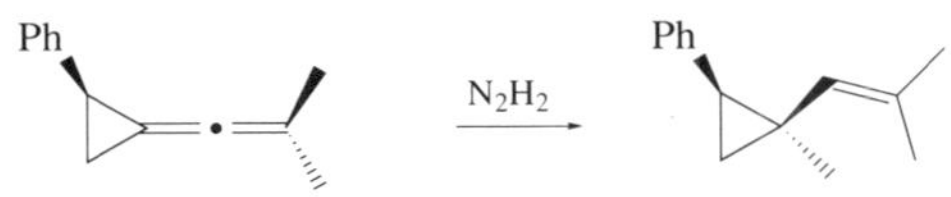

(114)

The reduction of a series of 1,2-cyclic allenes (from cyclonona-1,2-diene to cyclotetradeca-1,2-diene) with Na–NH_3(l) was investigated in the presence and absence of a proton source (EtOH) (Scheme 22). Profound differences in product ratios were observed ⟨71TL2075⟩. More (*E*)-alkene was formed in the presence of a proton donor and it was suggested that this reflected the intermediacy of a more stable (*E*)-allylic radical species. ((*Z*),(*Z*))-Cyclonona-1,5-diene ⟨67JOC4143⟩ and ((*Z*),(*Z*))-cyclodeca-1,6-diene ⟨68CJC84⟩ were prepared by dissolving-metal reduction of cyclonona-1,2,6-triene and cyclodeca-1,2,6-triene respectively (Scheme 22). Reduction of trisubstituted cyclic allenes under these conditions is both regiospecific and stereoselective ⟨75S194⟩. 1-Methylcyclonona-1,2-

diene gave predominately (*Z*)-1-methylcyclonon-1-ene. Cyclopropyl allenes on reduction with Na–NH_3(l) led to 1 : 1 mixtures of (*E*) : (*Z*)-vinylcyclopropanes in ca. 70% yield (Equation (115)) ⟨78TL3483⟩.

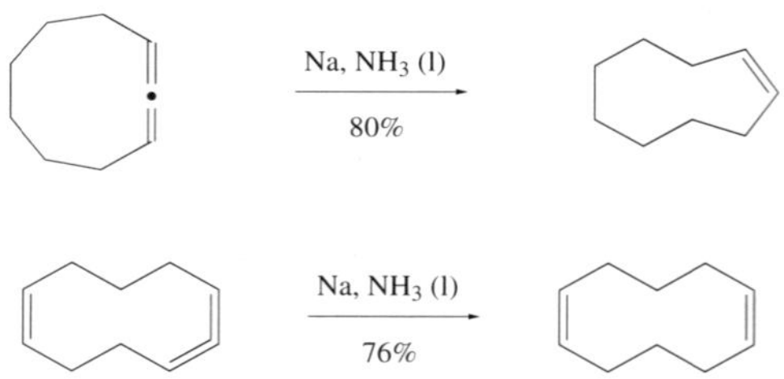

Scheme 22

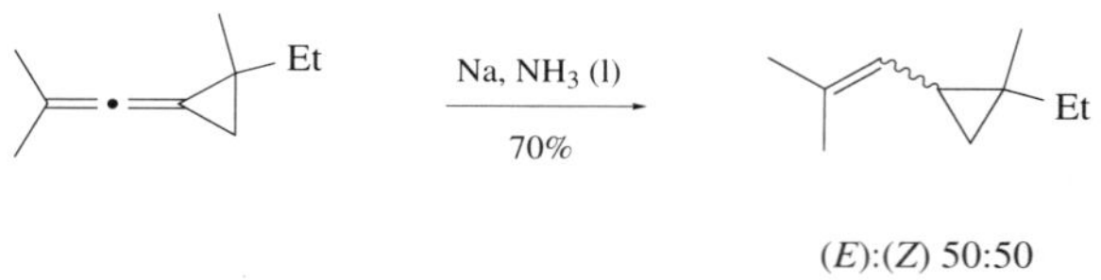

(115)

1.10.2.2.2 Addition of C—H to allenes

The addition of C—H to allenes is reviewed in Chapter 1.12.

1.11
One or More =CC Bond(s) Formed by Substitution or Addition

PATRICK G. STEEL
University of Durham, UK

1.11.1 INTRODUCTION

Although this chapter is concerned with the formation of C—C= bonds by substitution or addition, a liberal view has been taken of the mechanistic interpretation of these classifications. Consequently, many processes that are only formally substitution processes have been included, for example, addition/elimination and cross-coupling reactions. In this respect a significant number of these procedures can be considered as substitutions in either direction. However, in order to minimise repetition most have only been listed once and the interested reader is advised to consider both possibilities.

1.11.2 BY SUBSTITUTION

1.11.2.1 Substitution of Halogen

1.11.2.1.1 Substitution of alkyl halide

(i) Friedel–Crafts alkylation

Alkylation of aromatic and unsaturated aliphatic hydrocarbons with alkyl halides is dominated by the Friedel–Crafts reaction ⟨1884AC(P)449⟩. This particular method has been the subject of a number of reviews and monographs ⟨91COS(3)293⟩. Although both classes of hydrocarbons undergo the reaction under similar conditions, the synthetic utility of alkene alkylation is limited since it tends to produce complex product mixtures through competing alkene oligomerisation ⟨82MI 111-01⟩. Although much of the work on this reaction has been carried out using alkyl halides these are not the only species to undergo the reaction; alkanes, alkenes, alkynes, alcohols, ethers, epoxides and certain carboxylic acid derivatives are also viable electrophiles. The use of these species will be discussed where appropriate elsewhere in this chapter.

The reactivity of the alkyl halides decreases down the group F > Cl > Br > I. Although the use of alkyl fluorides can lead to selective reactions (Equation (1)) ⟨64JOC2317⟩, the low cost and ready availability of alkyl chlorides and bromides make these compounds the most commonly used reagents. The use of alkyl iodides is limited owing to problems of side reactions and/or decomposition. A similar trend in reactivity is found in the nature of the alkyl moiety, with benzyl and tertiary halides being the most reactive; followed, in order, by secondary, primary and methyl halides. However, steric factors can affect this order (see below).

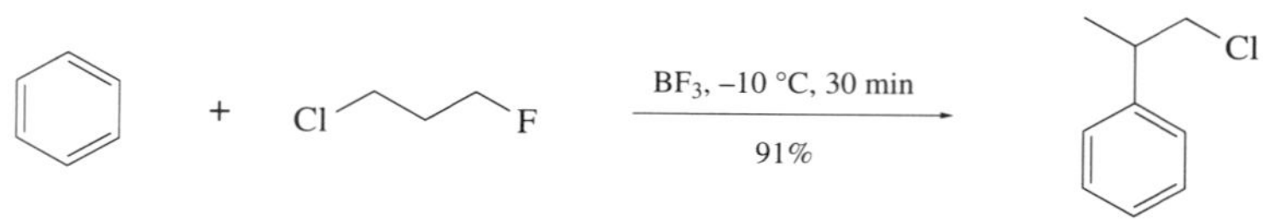

(1)

The general class of catalysts and co-catalysts is large and includes acidic halides and chalcogenides, metal alkyls, acidic cation-exchange resins, protic acids, and Brønsted–Lewis superacid combinations. There have been a number of attempts to order these, particularly the Lewis acids, according to their reactivity ⟨72JA7448, 85MI 111-01⟩. With the exception of the solid supported reagents most of these species are moisture sensitive and cannot be recycled. However, lanthanide trihalide compounds are efficient catalysts for the alkylation of a wide range of aromatic substrates and can be recovered, after aqueous workup of the reaction mixture, and reused with no significant loss of catalytic activity ⟨86CL357⟩.

Alkylation of arenes with primary alkylhalides frequently affords products derived from isomerisation of the initially formed carbocation. For example, alkylation of toluene with *n*-propylchloride affords only a mixture of isopropyltoluenes ⟨79NJC269⟩. Similar side reactions including isomerisation and hydride transfer products are also observed with higher halides (Scheme 1) ⟨58JA5867, 79CL1003⟩. In the latter example, alkylation by the tertiary pentyl carbocation is retarded due to unfavourable steric factors, thus permitting the contrathermodynamic carbocation rearrangement to the corresponding *sec*-pentyl species. A similar result was obtained commencing with methylbutene. This tendency for isomerisation to precede alkylation can be suppressed by the incorporation of electron-accepting groups such as —CO_2H, —CO_2Me and —CN into the alkyl chain (Equation (2)) ⟨87BAU327⟩. An alternative pathway for the alkylation of arenes, which avoids the problems of isomerisation, is via the reduction of the corresponding aryl ketone prepared by a Friedel–Crafts acylation reaction.

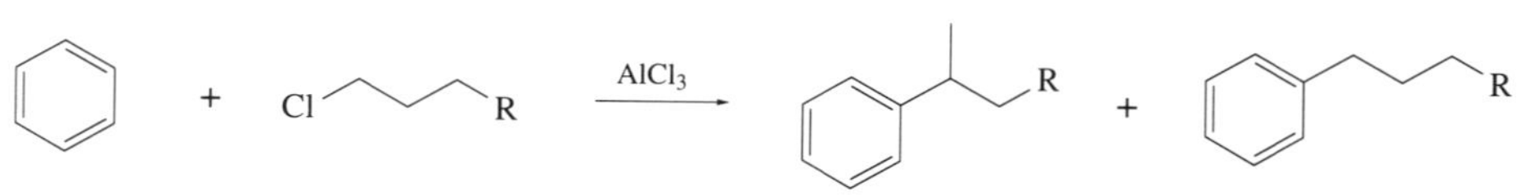

(2)

R =	Me	76%	24%
R =	CO_2H	2%	98%
R =	COMe	0%	100%
R =	CO_2Me	0%	100%

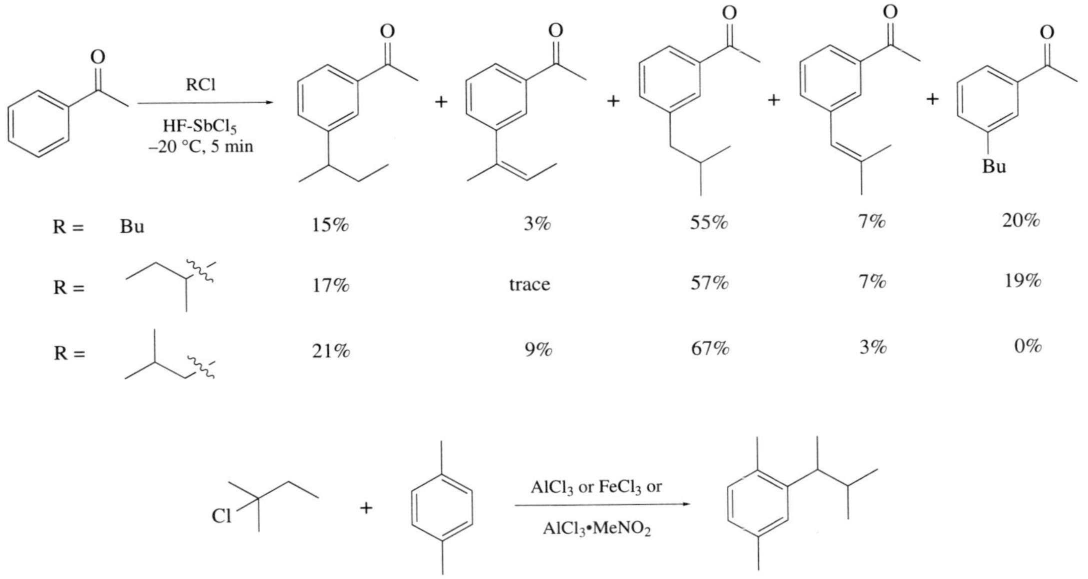

Scheme 1

Asymmetric induction may be observed in the Friedel–Crafts reaction. For example, both (−)-2-chloro-1-phenylpropane (**1**) and (+)-1-chloro-2-phenylpropane (**2**) react with benzene in the presence of aluminum trichloride to afford (−)-1,2-diphenylpropane (**4**) ⟨83BCJ1089⟩. To account for the retention of configuration, this reaction is postulated to proceed via a double inversion mechanism involving the phenonium intermediate (**5**) (Scheme 2). Similar intermediates are suggested to occur in the reaction of 1,2-dihaloethanes ⟨50JGU450⟩ and reactions commencing with optically active 1,*n*-dihaloalkanes proceed with moderate stereoselectivity ⟨80CC86⟩. However, in many of these reactions with polyhaloalkanes, complex product mixtures result due to competing isomerisation, polysubstitution and dehydrogenation pathways ⟨63CA(59)2732, 56JOC3595⟩. Other examples of asymmetric arene alkylation exist. However, many of these are heteroatom assisted ⟨87AG(E)683⟩ and, as such, are beyond the scope of this chapter.

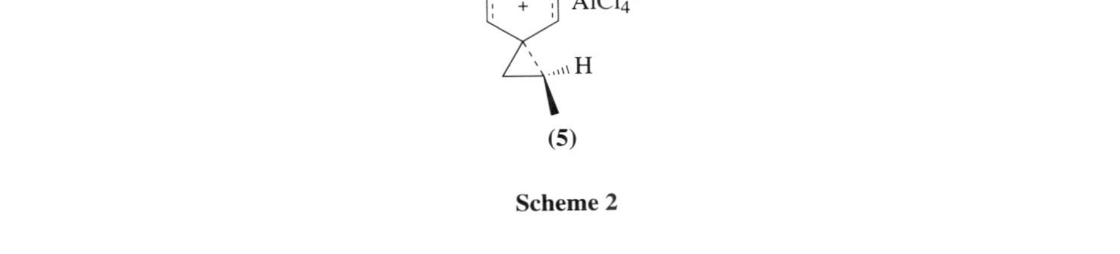

Scheme 2

(ii) *Substitution by organometallic* sp² *carbanions*

Carbon–carbon bond-forming reactions between vinylic organometallic species and a range of electrophiles has been comprehensively reviewed ⟨91COS(3)241⟩. The range of alkylating agents involved is large and varied and includes alkyl halides, alkyl tosylates, epoxides, esters and related derivatives. Each of these will be considered where relevant in this chapter.

Nominally, the most direct method for the formation of this unit is the Wurtz–Fittig reaction in which an alkylated aromatic is generated through the sodium metal mediated coupling of an aryl bromide and an alkyl iodide ⟨69T5771⟩. However, the yields are not good and the same result is best achieved in two discrete steps. In this respect the S_N2 alkylation of a vinyl- or aryl Grignard or lithium species with an alkyl halide provides the simplest entry to this structural class ⟨75S434⟩. Vinyllithiums combine efficiently with primary alkyl bromides and iodides, but alkyl chlorides afford much lower yields. The reaction is not efficient in solvents other than THF, nor does it proceed with secondary alkyl halides. Similar coupling reactions of organomagnesium reagents in THF or hexamethylphosphoramide (HMPA) have also been reported ⟨66BSF1888⟩. Trisubstituted alkenes can be prepared in this fashion, indicating that steric bulk in the vinyl anion moiety is not a problem. Since alkene isomerisation is not normally observed, then, provided clean generation of the vinyllithium is possible, either *cis* or *trans* alkenes can be prepared. Under certain conditions some chemoselectivity can be observed. For example, in the presence of TMEDA vinyllithiums undergo selective monoalkylations with α,ω-dibromoalkanes ⟨74JA7101⟩ (however, see ⟨90SC2349⟩). Cycloalkylation is also possible, although apart from lithium, the use of other (ω-haloalkenyl) organometallics (Al, Zn, Zr, Si) requires the presence of an additional silyl or second metal group on the metal bearing carbon ⟨88JA5383⟩.

This chemistry can easily be extended to allenyllithiums which are readily generated by either direct deprotonation or lithium–halogen exchange. These couple efficiently and rapidly with primary alkyl bromides or iodides to produce substituted allenes ⟨75CC561⟩. Competing formation of the corresponding alkyne can be suppressed by careful control of the reaction conditions and the presence of additives such as HMPA ⟨79TL937⟩.

Due to transmetallation and other competing pathways many vinylic and aryllithium species do not combine efficiently with allylic halides. In these cases the use of the corresponding Grignard reagent, either prepared directly or via transmetallation with magnesium bromide, can circumvent this problem ⟨84JOC737⟩ with the reaction proceeding in ethereal solvents at ambient temperatures. The use of other transition metals such as zinc and copper as transmetallating agents also proves satisfactory (see below). However, even with these modifications the use of nonpolar solvents or less reactive alkyl halides requires elevated temperatures for the efficient coupling with aryl and vinyl Grignard reagents. α-Haloketones combine with aryl Grignard reagents in a two-step process, involving initial carbonyl attack followed by a (1,2) aryl migration of the intermediate magnesiohalohydrin (Scheme 3) ⟨59JOC843⟩. Vinyl Grignard reagents behave similarly ⟨73CL667, 81TL2243⟩ and in this respect it has been noted that, although both α-chloro and α-bromo ketones are substrates, the former are more efficient. Tertiary chlorides also undergo this rearrangement and this provides an efficient entry into quaternary vinyl-substituted carbon centres (Equation (3)). The conceptually similar reaction of α-haloketones with phenylcopper reagents fails due to competing reduction. However, α-arylketones may be obtained from the reaction of α-bromoketones with arylboranes ⟨69JA6852⟩, or the coupling of phenylcopper reagents with *in situ* generated α-bromoenamines ⟨87JOC3697⟩ or α-halotosylhydrazones ⟨75JA7372⟩.

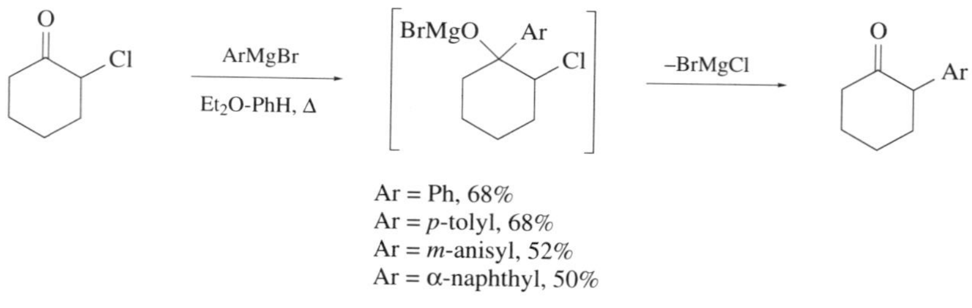

Scheme 3

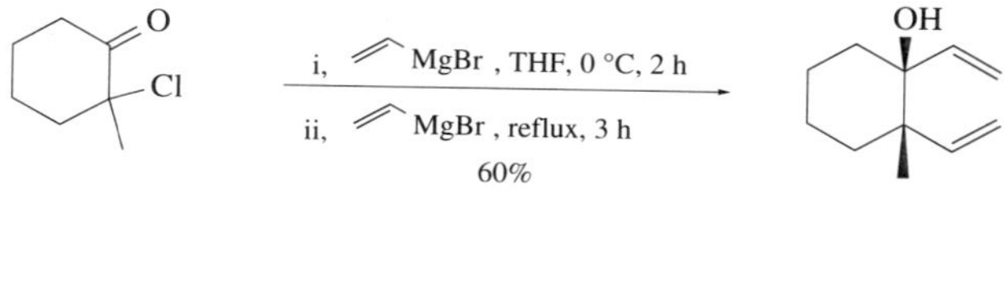

(3)

The scope of these reactions between aryl and vinyl Grignard and lithium reagents can be greatly extended through the use of copper salts. Primary iodides and tosylates (see below) react in the presence of copper(I) iodide under much milder conditions than those required for direct alkylation ⟨92OR135⟩. In this respect a common strategy is to generate the lithium divinyl- or diarylcuprate, R_2CuLi. These react efficiently with both primary alkyl iodides and bromides, and activated secondary halides. In many cases additional additives, for example, phosphines, sulfides, etc. are required, whilst the choice of solvent can also prove to be crucial ⟨76OS(55)103⟩. One problem with the use of this methodology is the loss of 1 equiv. of a potentially valuable aryl or alkenyl unit. In certain cases, such as those involving reactive alkylating agents (e.g., allylic halides) the use of HMPA can promote efficient transfer of both vinyl groups. With less reactive electrophiles the addition of 3 equiv. of triethylphosphite has been shown to promote the second alkylation ⟨79S826⟩. This problem is avoided through the use of either heterocuprates containing nontransferable ligands ⟨87JOC1885⟩ or higher order cuprates ⟨87S325⟩. These latter species frequently provide for greater reactivity when compared with their homocuprate analogues. For example, the higher order vinylcuprate (**9**) reacts smoothly with the secondary alkyl iodide (**6**) whereas the corresponding homocuprate (**8**) does not (Equation (4)) ⟨81JA7672⟩. In contrast, whilst the analogous aryl species, $Ph_2Cu(CN)Li_2$, reacts with primary chlorides but does not couple with secondary iodides, the corresponding homocuprate, Ph_2CuLi, combines efficiently with both primary and secondary bromides. These reactions are predominantly S_N2 processes as evidenced by the almost complete inversion of stereochemistry observed with enantiomerically pure substrates. In all of these transformations there is almost complete retention of alkene geometry ⟨71HCA1939⟩.

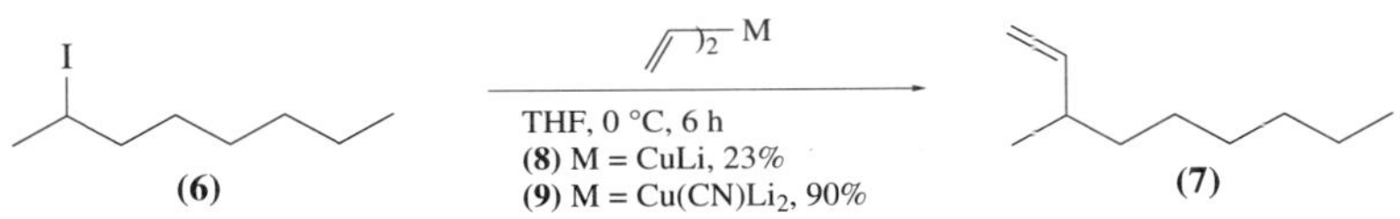

(4)

With Grignard reagents a substoichiometric amount of the copper species can be used. As with the organolithium-derived reagents these species show considerable differences in reactivity when compared with the parent species. For example, the direct coupling with haloacetates and haloalcohols is possible ⟨77TL3263⟩. A further degree of chemoselectivity is observed with copper-catalysed organomanganese chloride reagents which, in the presence of *N*-methylpyrrolidine (NMP), selectively and efficiently alkylate primary bromides in the presence of chlorides, sulfonates, acids and esters ⟨93SL45⟩. A CuX–TMEDA catalyst system has been demonstrated to be a useful combination for the alkylation of aryl Grignard reagents with primary iodides and benzylic halides ⟨72BCJ2582⟩. However, as with the stoichiometric reactions these catalytic methods tend to give much lower yields with less reactive electrophiles. Some improvements in reactivity can be realised through the use of dilithium tetrachlorocuprate ⟨71S303⟩ although in some cases excess Grignard reagent is required for good yields to be obtained. Iodides ⟨71S303⟩ and tosylates ⟨74AG(E)82⟩ (see below) provide the best substrates.

Similar catalysis of vinyl carbanion alkylation can be achieved through the use of other transition metal catalysts. Besides organolithium and Grignard reagents a variety of other carbanionic species have been used including those based on B, Al, Sn, Zr, Cd and Hg ⟨91COS(3)435⟩. In this respect complexes of nickel and palladium are particularly prominent. In addition to allylic halides these catalysts have found applications to alkylations with a variety of other allylic electrophiles with a particular emphasis on those containing oxygenated leaving groups (see below). The use of unactivated C sp^3 electrophiles is limited by competing reduction and β-hydride elimination. The latter problem can be overcome through the use of B-alkyl-9-BBN (9-BBN = 9 borabicyclo[3.3.1.]nonyl) derivatives ⟨91PAC419⟩ (below), whilst the use of diorganozincs derived from arylmagnesium chlorides (not bromides) permits efficient coupling with relatively unreactive neopentyl iodides ⟨93JOC4866⟩. Of the other vinylmetallic species employed, those containing aluminum and zirconium are of particular importance since they can readily be accessed through the hydrometallation of the corresponding alkyne. Although simple alkylation is only efficiently observed with reactive alkyl halides, for example, MOM-Cl (MOM = methoxymethyl) ⟨76S625⟩, enhanced reactivity is realised through the formation of the corresponding 'ate' complex ⟨84OR(32)375⟩. Alternatively, productive alkylation of these and other related vinyl organometallics is possible in the presence of an additional transition metal catalyst (above).

α-Aryl carbonyl compounds can be obtained through the nickel phosphine catalysed coupling of arylzinc reagents with α-haloesters ⟨83OM598⟩ and α-halonitriles ⟨87S40⟩. The corresponding cadmium reagents can be alkylated with primary and secondary α-bromoesters and secondary allylic bromides but fail to combine efficiently with simple alkyl bromides and a variety of related electrophiles ⟨73JOC3189⟩. These latter reactions proceed via free radical intermediates, as evidenced by ESR spectroscopy and the complete loss of optical activity when R-(+)-bromopropionate was used as the electrophile.

1.11.2.1.2 Substitution involving vinyl halides

(i) Alkyl nucleophiles

Although the direct reaction of C sp^3 organometallic reagents with sp^2 halides is known, most of these couplings proceed by initial metal–halogen exchange followed by nucleophilic substitution in the reverse direction ⟨93JOC5976⟩ (for alkenyl organometallics and alkyl halides, see Section 1.11.2.1.1(i)). However, with some fluoroalkenes a S_NV (nucleophilic vinylic substitution) process is observed with both organolithium and Grignard reagents (Scheme 4). However, the initial product can readily undergo further reaction to form the corresponding acetylide ⟨88BCJ1625⟩.

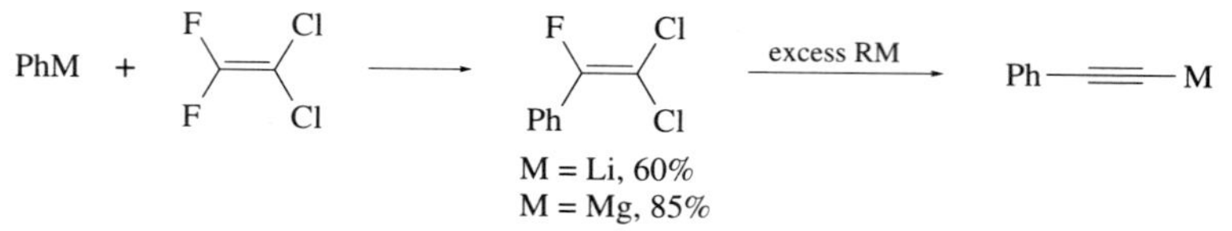

Scheme 4

Alkyliron, -cobalt and -copper reagents may be employed for the substitution of both alkenyl and aryl bromides, chlorides and fluorides in good yields. The use of Me_4FeLi_2 provides useful levels of discrimination between different halogens. The coupling may also be achieved using a substoichiometric quantity of $FeCl_2$ together with the appropriate Grignard reagent ⟨93CB1447, 68JA5615⟩. In general, the use of a transition metal catalyst is necessary for the efficient substitution of alkenyl halide. These processes are of much synthetic use since they proceed with retention of alkene stereochemistry, providing good routes to di-, tri- and tetrasubstituted alkenes. Alkenyl halides may be coupled with alkyl-, vinyl- or aryllithiums in palladium, ruthenium or rhodium catalysed processes ⟨79JOC2408⟩. However, the use of organolithium reagents in these reactions is not normally as efficient as that of the corresponding Grignard or organozinc species. The range of catalysts used is large, with those based on nickel being the most widely used. A great variety of nucleophiles (alkyl, aryl and alkenyl Grignard) and alkenyl halides (iodide, bromide, chloride and even fluorides ⟨73JOM(50)C12⟩) can be coupled with this reagent combination ⟨76BCJ1958, 78OS(58)127⟩. The reactions are strongly exothermic and care must be taken to add the Grignard reagent slowly to the catalyst and alkenyl halide. Alkenyl Grignards are somewhat less reactive and, whilst cyclohexyl Grignards couple smoothly under nickel catalysis, the use of other secondary Grignard reagents, with either nickel or palladium catalysts, frequently results in alkene isomerisation. This isomerisation is varied and the extent depends considerably on the nature of all three reaction partners ⟨73JOM(50)C12, 84JA158, 86JOM(312)343⟩. In this respect alkenyl bromides tend to be the halides of choice, since depending on the particular catalyst, both the presence of β-hydrogen atoms and also some degree of branching are acceptable within the Grignard component. Chiral alkenyl ⟨84TL5769⟩ and allenyl bromides ⟨91JOM(417)289⟩ can be coupled efficiently with a variety of organometallic reagents under transition metal catalysis with little deterioration of optical integrity. However, the stereochemical outcome can depend dramatically on the specific conditions. Whereas alkylation of the more reactive allenyl iodides occurs with retention of configuration, chloro- and bromoallenes react predominantly by an invertive pathway ⟨85JOC3042⟩. This can lead to the fully substituted, enantiomerically pure, allene products of opposite configuration to the starting halide ⟨91JOM(417)289⟩. The nature of the cuprate species is also critical; homocuprates giving substituted allenes with inversion of configuration whilst the use of heterocuprates (cyano- and bromo-) favour retention. The steric bulk of the organocopper species is also significant, tertiary groups leading to allenic products whereas straight-chain primary groups frequently result in the production of alkynyl by-products. Alkenyl- and allenyliodides may be directly displaced by Grignard reagents in the presence of the soluble copper catalyst, dilithium tetrachlorocuprate ⟨77JOM(128)1⟩ and with these substrates even tertiary alkyl Grignards react with little isomerisation.

Dihaloalkenes are also acceptable substrates. 1,1-Dichloroalkenes undergo a double displacement reaction on treatment with an alkyl Grignard/nickel catalyst combination. Using an alkylzinc chloride as the nucleophile and palladium catalysis the *trans* chlorine atom can be replaced more rapidly and, consequently, it is possible to achieve a stepwise dialkylation with different alkyl groups ⟨87JA1257⟩. There is a similar difference in reactivity between (*Z*) and (*E*) bromoalkenes although the extent and nature of this difference is very dependent on the catalyst used. For example, (*E*)-bromoalkenes are found to be more reactive in palladium-catalysed coupling processes ⟨86TL2529⟩ whilst the reverse is true for the combination of a primary alkyl Grignard reagent (not MeMgX) and an iron catalyst, $Fe(DBM)_3$ (DBM = dibenzoylmethido-). However, with this iron catalyst and the more sterically demanding secondary and tertiary alkyl Grignard reagents, preferential reactions with (*E*)-bromoalkenes occur with no complications due to rearrangement of the Grignard reagent being observed ⟨76JOC502⟩.

Selective monoalkylation of 1,2-dichloroalkenes with Grignard reagents can be realised through the use of excess alkene and a nickel catalyst ⟨81TL315⟩. Unsymmetrical 1,2-dihaloalkenes may be selectively monoalkylated depending on the lability of the leaving group ⟨76CL1237⟩. Leaving ability is found to decrease in the order Br > Cl > SR > OR. Within this series it is possible to achieve two sequential alkylations with different Grignard reagents ⟨85JCS(P1)1115⟩.

There are several alternatives to the use of alkyl Grignard reagents. Foremost amongst these are the organozinc reagents which couple smoothly under both nickel and palladium-based catalysis ⟨92T9577⟩. Both primary and secondary alkyl zincs are normally satisfactory although in certain cases alkyl isomerisation can occur with the latter species ⟨80JA3298⟩. In contrast to the corresponding Grignard reagents, a wide variety of functionality can be tolerated in both components ⟨91JOC1445, 93CRV2117⟩. For example, Grignard-labile functionality, such as unmasked carbonyl groups, can be present in the vinyl halide moiety as evidenced by the use of both α- and β-halocarbonyl compounds ⟨91TL4453⟩. The latter provide an entry to β-alkylated-α,β-unsaturated carbonyl compounds (see below) (Scheme 5) while β,γ-unsaturated esters may be accessed through the use of Reformatsky-type reagents ⟨81JOM(209)109⟩. Although with these latter reagents, (*E*)-alkenyl bromides react with predominant retention of stereochemistry, significant scrambling of alkene geometry has been observed with the corresponding (*Z*) isomers (Equation (5)).

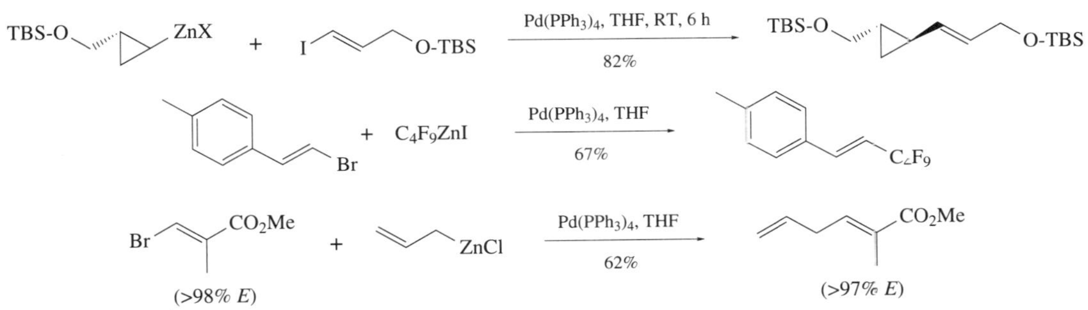

Scheme 5

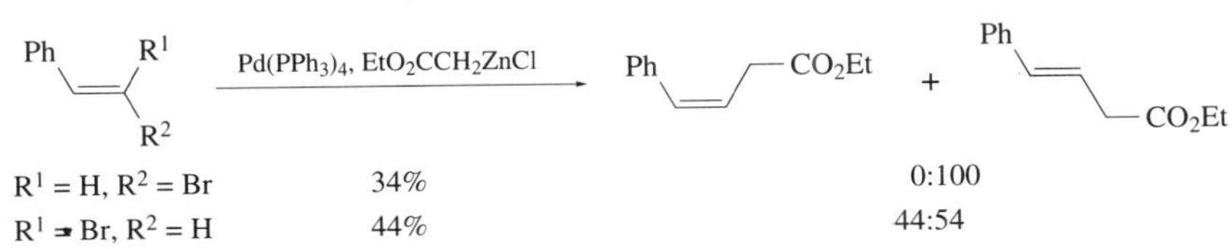

(5)

The analogous lithium enolates may also be coupled with alkenyl bromides. These reactions require the use of a stoichiometric amount of a nickel promoter ⟨77JA4833⟩. In a number of cases the same product can be obtained from the equivalent stannylated ester albeit with moderate yields.

(ii) Alkenyl nucleophiles

The direct coupling (or dimerisation) of alkenyl halides can simply be achieved by heating the reactants in the presence of an activated copper species. Since the temperature requirement frequently precludes the use of many sensitive substrates, milder conditions have been sought. These have included the development of more active copper catalysts, for example the species obtained through treatment of Cu(OTf) with aqueous ammonia in acetone ⟨75JOC3649⟩. The nature of these catalytic species remains ill-defined and more recent progress has been obtained through the use of specific transition metal complexes. A variety of metal centres have been exploited for this purpose, notably those involving nickel ⟨81JA6460, 93BCJ2446⟩ and palladium ⟨85CC971⟩.

The use of such transition metal complexes provides the principal methods for the synthesis of dienes through the coupling of dissimilar alkenyl halides. In general, provided that functionality can be tolerated, it is better to use the more sterically hindered Grignard reagent. In cases in which there is little difference between the two alkenyl halides, the direction of the addition can significantly affect the yield of the reaction (Scheme 6) ⟨86TL2881⟩.

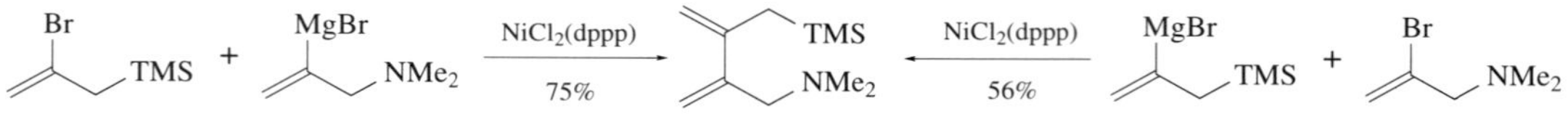

Scheme 6

Nickel and palladium catalysts are generally employed with the latter providing the greatest selectivity and efficiency. Vinyl iodides are the most reactive halides and offer advantages over the corresponding acyclic triflates in that they can readily be obtained stereochemically pure with either the (*Z*) or (*E*) geometry ⟨87JA813⟩. A large variety of alkenyl organometallic species may be employed including those based on Li, Mg, Cu, Cd, Hg, B (see below), Si, Ti, Ce and more importantly Sn ⟨86AG(E)508, 92S803⟩, Al, Zr and Zn ⟨87JA2393⟩. The reactions proceed rapidly under mild conditions with retention of configuration in both components ⟨78TL191⟩ whilst substitution in either component does not adversely affect the rate or stereochemical outcome ⟨83BSF332⟩. The reaction can also be extended to include the use of a variety of alkynylorganometallic reagents ⟨87JA2138⟩.

The use of lithiodivinylcuprates require the addition of $ZnCl_2$ as a co-catalyst ⟨81TL959⟩. These not only accelerate the reaction but, in the case of divinylmagnesiocuprates, permit the transfer of both vinyl ligands. This is believed to occur through the dissociation of the cuprate into an alkenylzinc and a mixed alkenylzinc cuprate ⟨84T2741⟩. In general, the direct use of alkenylzinc reagents, which can be easily accessed from the corresponding organolithium by transmetallation, provides both higher yields and greater stereocontrol ⟨91TL6683⟩. When prepared using activated zinc, a wide range of functionality can be incorporated into the organometallic unit ⟨93T29⟩. These reactions are most efficiently catalysed by a palladium(0) species, usually $Pd(PPh_3)_4$ ⟨82ACR340⟩. This methodology is also amenable for the alkylation of both aryl and allenyl halides (see below). Although the use of zinc-based reagents affords significant advantages over the other organometallic species, these reactions frequently require an excess of the nucleophile to achieve complete substitution of the vinyl halide. Polyhaloalkenes are also acceptable substrates. With both 1,1- and 1,2-dichloroethylenes, selective monosubstitution has been reported ⟨84TL6001, 87JA1257, 87TL1649⟩.

Similar palladium or nickel-catalysed couplings can be achieved using vinyl alanes ⟨88OS(66)60⟩. The presence of a co-catalyst, either $ZnCl_2$ or $CdCl_2$, is found to be essential for efficient coupling ⟨78JA2254⟩. With the exception of some vinyl iodides the reactions are usually stereospecific, proceeding with retention of configuration. In this respect, being somewhat more selective, palladium phosphine complexes are generally preferred as catalysts to those based on nickel ⟨76JA6729⟩. The vinyl alanes can be readily prepared by hydroalumination of alkynes which results in the formation of the (*E*)-alkenyl alane. This is in contrast to, and complementary with, alkyne carbocupration which affords (*Z*)-alkenyl cuprates. Consequently di- and trisubstituted alkenes of either geometry can be easily obtained from the same alkyne. Good yields are normally observed, with a range of functionality being tolerated in the vinyl halide component ⟨77JA3168⟩, although the presence of excessive steric hindrance in the alane moiety can hinder the coupling ⟨76JA6729⟩. Zirconium reagents formed in a similar fashion may be used as alternatives to alkenyl alanes with a similar degree of success with the added advantage that hydrozirconation is more tolerant of a wider range of functionality than is hydroalumination ⟨78JA2252, 78TL1027⟩. In general, the efficiency of the cross-coupling is enhanced by transmetallation of either alane or zirconane to the corresponding alkenylzinc (rel. rate Al : Zr : Zn = 2 : 3 : 2000) ⟨87JA2393⟩. Similar palladium catalysis can be employed in the synthesis of enyne and allenynes. Milder, nonbasic conditions are found in the presence of copper salts ⟨83S32⟩ or in aqueous conditions ⟨92SL715⟩. Copper salts may be avoided through the use of the corresponding aryl or vinyl triflate in an amine base solvent system (see below) ⟨93TL6403⟩; whilst neutral coupling conditions, that are relatively insensitive to oxygen, moisture or other catalyst poisons, can be achieved without palladium through the use of stoichiometric copper(I) salts. However, this method does require relatively high reaction temperatures and the use of HMPA as the solvent ⟨89SC2199⟩.

(iii) Aryl nucleophiles

In an analogous fashion, vinyl halides may be coupled with a variety of aryl nucleophiles. Many of the features of this process are similar to the coupling of an alkenyl organometallic and an alkenyl halide (see Section 1.11.2.1.2(ii)) and will not be repeated here. Although this coupling can be achieved in either sense, the union of an aryl Grignard and vinyl halide is generally found to the more efficient strategy. The principal exception to this is found when there are problems of steric crowding in the halo component.

Although arylzinc reagents and nickel or palladium catalysts are the favoured combination, other transition metals have found use as the catalyst centre, for example, Fe(III) ⟨86BCJ1656⟩. Similarly, other organometallic components can be used, notably RLi (see above). Alternatively, with aryl nucleophiles it is possible to avoid the requirement for the catalyst through the use of a diarylcuprate ⟨69JA4871⟩ or arylmanganese-ate complexes. However, both methods have limitations; only one aryl ligand is transferred from the cuprates whilst the manganese reagents are restricted to alkenyl iodides and result in scrambling of the alkene geometry ⟨89IZV2623⟩.

(iv) Addition–elimination displacements of vinyl halides

Substitution of the chlorine atom in β-chloroenones with carbon nucleophiles can readily be achieved through their reaction with a variety of active methylene compounds ⟨66CRV161⟩. However, acceptable yields of the β-alkylated vinyl ketone are only obtained with those compounds possessing a single acidic hydrogen atom. Reagents with two acidic hydrogen atoms undergo a very rapid further reaction, leading to aromatic side-products. Hindered β-haloenones, such as chloromethylene camphor ⟨16CB25⟩ undergo facile displacement of the halide on treatment with Grignard reagents, while most β-haloenones undergo preferential 1,2 addition. However, the resultant carbinols undergo rapid rearrangement on mild acid hydrolysis to afford alkylated α,β-unsaturated ketones ⟨51BSFC13⟩. Direct substitution of β-chloroenones can be achieved using diethyl or dialkenyl cadmium reagents, but the yields are poor ⟨59AC(P)541⟩.

Direct displacement of halogen occurs readily on treatment of β-haloketones with a variety of copper-based reagents. With simple dialkyl cuprates, cyclic β-haloenones produce the products of halogen substitution in good yield, but acyclic or exocyclic β-haloenones frequently afford the saturated ketone product (Equation (6)) ⟨76JOC636⟩. These problems of conjugate addition can be circumvented by the use of lithium(phenylthio)alkylcuprates ⟨82CJC1256⟩. In general, the bromides are the substrate of choice since, although the reactivity of the haloenone increases in the order Cl < Br < I, iodoenones are considerably less easy to handle.

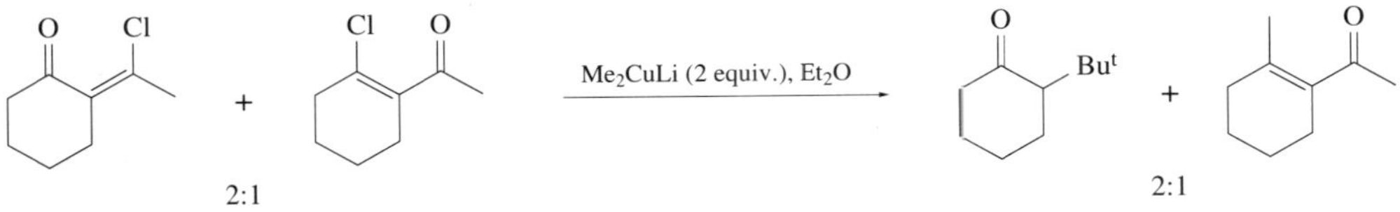

(6)

β-Haloenoates are also suitable substrates in these substitution reactions. For example, ethyl β-bromoacrylate undergoes efficient alkylation on treatment with the methyl triethylborate copper complex ([Et_3BMe]Cu). As with many of these addition–elimination reactions this is a stereospecific reaction with retention being the favoured pathway ⟨77TL3369⟩. The stereoselectivity is, however, substrate/reagent dependent. (*Z*)-2-β-Chlorocinnamates react with Grignard reagents in the presence of a copper(I) catalyst with retention of configuration whereas reaction of the corresponding (*E*)-isomer shows no stereochemical preference ⟨83ACS(B)173⟩.

A number of other carbanion reagents may also be employed. For example simple ketone enolates combine efficiently with β-chloroacrylate ⟨78CJC419⟩, and allyl phosphorus ylides react with β-chloroenones to provide a general polyene synthesis (Scheme 7) ⟨75TL1359⟩. Radical-based strategies are also possible although the stereochemical outcome is somewhat variable ⟨93TL3041⟩.

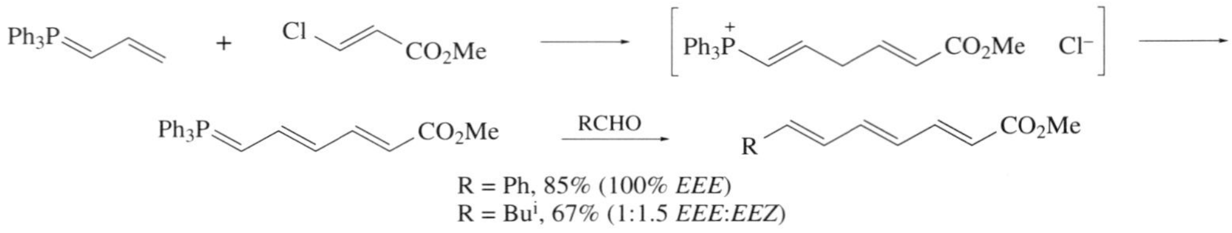

Scheme 7

Alternative coupling partners exist, notably β-halovinylsulphones. These are particularly good substrates for displacement of either fluorine, chlorine or bromine with lithium dialkylcuprates. Good selectivity is observed, although only at low temperatures with chlorides and bromides. Iodides similarly can be displaced, with high stereoselectivity, on treatment with *n*-butyl copper-bis(diisopropylsulphide) complex. The corresponding dialkyl cuprate is not stereospecific (Equation (7)) ⟨74JOC3174⟩. Substitution of halides from sulphone-activated systems may also be achieved with active methylene compounds ⟨92JOC1161⟩. Reactions using the analogous sulphoxides are also possible ⟨92JOC1718⟩.

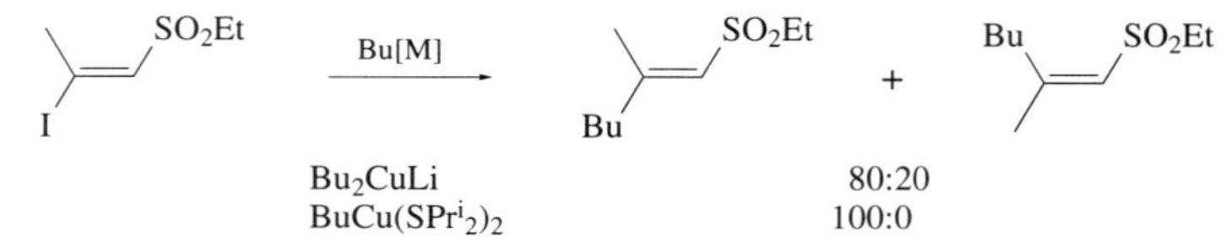

(7)

1.11.2.1.3 *Substitution of aryl halides*

(i) Alkyl nucleophiles

These coupling reactions bear considerable resemblance to those involving alkenyl halides. Primary Grignard reagents are the most reactive species, whilst secondary and tertiary Grignard reagents frequently undergo isomerisation prior to reaction. Reactivity can also be controlled by the nature of the catalyst. In general, with primary Grignard reagents and nickel catalysts, the reactivity of the halides decrease as the group is ascended (ArI > ArBr > ArCl ≫ ArF) while for the same reagents and palladium catalysts both aryl chlorides and fluorides are relatively inert (ArI > ArBr ≫ ArCl, ArF). Thus, with nickel catalysis the use of aryl chlorides often affords the best balance between rate and product selectivity. However, under more forcing conditions all the aryl halides may be substituted in this fashion. Likewise with palladium catalysis the use of bromoarenes is favoured. With the exception of alkoxy substituents which accelerate the reaction, aryl ring substituents influence the reaction in much the same way as they do for nucleophilic aromatic substitution (see Section 1.11.2.1.3(iv)). However, these reactions are subject to some steric demands with hindered, notably *ortho*-substituted, aryl halides failing to react with Grignard reagents bearing β-hydrogen atoms. In these cases reductive elimination of the corresponding alkene occurs preferentially ⟨76BCJ1958⟩. These competing pathways can, to a certain extent, be regulated by the choice of catalyst. For example, with $NiCl_2$(dppp) isopropylmagnesium chloride couples with chlorobenzene to afford isopropylbenzene in 96% yield; while using $NiCl_2(PEt_3)_2$, reduction is the principal pathway. In general, for most coupling reactions, switching from a monodentate to a bidentate phosphine ligand favours coupling with respect to reduction. Similarly, the chiral pentyl Grignard reagent couples with chlorobenzene, in the presence of $NiCl_2$(dppp), with little degradation of optical purity whilst with $NiCl_2$(dmpe) significant stereochemical scrambling is observed (Equation (8)) ⟨72JA9268⟩.

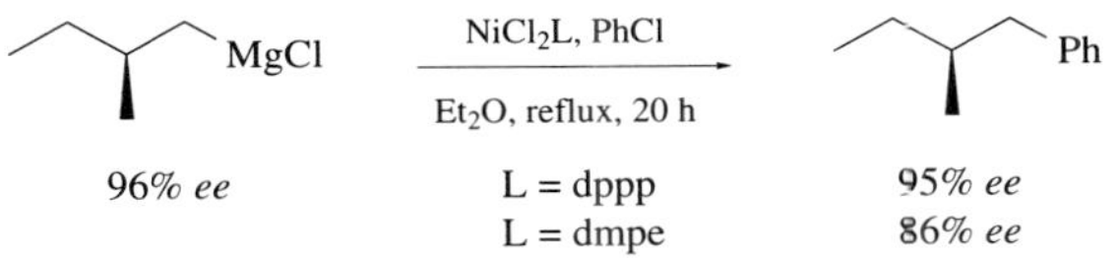

(8)

Further enhancements can be achieved through the use of the equivalent zinc and tin reagents. Both are more tolerant of a wider range of functionality than the corresponding Grignard reagent. Coupling reactions of tetraorganostannanes require the presence of HMPA as a co-solvent to ensure efficient ligand transfer. The use of such additives ⟨79JA4992, 84CL1225⟩, or the presence of internal coordination sites on the tin ligand ⟨92JA6556⟩ is frequently essential for efficient alkyl transfer from tin reagents. These palladium-catalysed couplings of alkyl stannanes usually proceed most smoothly with the aryl iodide. However, with allyltributyltin higher reactivity is observed with the aryl bromide ($ArI < ArBr > ArCl$) ⟨77CL301⟩. Although most aryl chlorides are unreactive to these conditions, alkylations with tetrabutyltin can be achieved on arene activation through complexation with chromium hexacarbonyl ⟨87CC1755⟩.

Reformatsky reagents can also be used to provide an entry into a variety of α-arylated carbonyl compounds. The catalysts of choice are either $Pd(PPh_3)_4$ or $Ni(PPh_3)_4$. The latter compound provides greater versatility, allowing coupling with iodides, bromides and chlorides, whilst the palladium catalyst is only suitable for use with aryl iodides. Efficient reaction is only achieved in the presence of a strongly coordinating co-solvent such as DMF, HMPA or NMP ⟨79JOM(177)273⟩.

Under palladium catalysis and in the presence of tributyltin fluoride, aryl bromides are alkylated by silylenol ethers ⟨82JA6831⟩ or silylketene acetals ⟨91JOC261⟩ and it is hypothesised that the active nucleophile is the corresponding α-stannylketone.

Difunctional reagents have also been reported for both reaction partners. *bis* Grignard reagents produce the dicoupled product in low yield ⟨75JA4405⟩. Although complete substitution is normally observed with di- and polyhaloarenes, selectivity can be achieved either through the use of excess electrophile, or through the application of different leaving groups ⟨91CL2073⟩. A number of substituent groups, including methyl, methoxy and oxazolinyl, demonstrate useful directing effects ⟨82JOC4319, 84OM630⟩. This is also true of heteroaryl halides.

(ii) Alkenyl nucleophiles

The more efficient way of achieving this connection is normally the reverse process, namely a coupling reaction between an aryl organometallic and alkenyl halide (see Section 1.11.2.1.2(iii)). However, in cases in which there is severe steric hindrance in the arene moiety the use of vinyl nucleophiles and aryl halides can offer some advantages. In these cases the use of a copper salts allows the inclusion of a variety of functional groups into the arene unit ⟨83JOC1912⟩. The most efficient coupling reaction is observed with aryl iodides. In certain cases, in which difficulties arise in the direct formation of the cuprate, the corresponding zinc species (accessible either through transmetallation ⟨81TL3851⟩ or directly via reaction with activated zinc ⟨93T29⟩) together with a palladium catalyst can provide a convenient alternative strategy. As with the copper-catalysed displacements, aryl iodides are usually the substrates of choice; the reactivity reflects the ease of oxidative addition of the aryl halide to palladium(0) ⟨77ACR434⟩, that is $PhI > p\text{-}ClC_6H_4Br > PhBr > o\text{-}MeC_6H_4Br > o\text{-}MeOC_6H_4Br$. Vinyl stannanes can similarly be combined with aryl halides in the presence of palladium catalysis. With the exception of a free amino group a broad spectrum of functional groups are compatible with the reaction conditions, although electron-withdrawing groups result in lower reactivity ⟨87JOC422⟩.

Alkenyl alanes or the corresponding zirconium species, readily accessible by hydrometallation of an alkyne, also undergo nickel or palladium-catalysed coupling with aryl halide (Scheme 8) ⟨76CC596, 78JA2254⟩. This strategy can provide access to tetrasubstituted arylalkenes ⟨87S816⟩ although the geometry of the alane component is restricted to the (*E*) isomer. With sterically demanding substrates the addition of $ZnCl_2$ can aid the conversion ⟨78JA2254⟩ although in extreme cases it is frequently better to convert the alkenyl alane into the corresponding iodide and then couple with an aryl Grignard reagent.

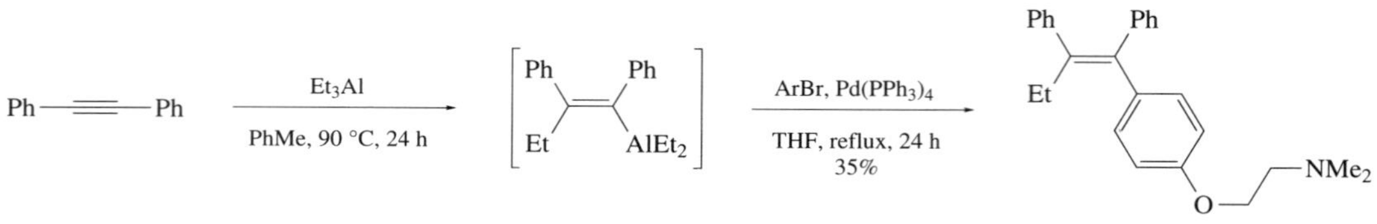

Scheme 8

This connection can also be attained through the use of vinyl boranes ⟨79CC866⟩ and vinyl boronates. The latter have been found to be more effective with 2-alkenyl boron reagents. Both classes of reagent require the involvement of a palladium(0) catalyst and also the presence of a base. The strength of this base can, particularly with vinyl boronates, control the regiochemistry of the processes. Stronger bases such as ethoxide favour a head-to-head coupling whilst triethylamine and other weaker bases afford the products of a head-to-tail union (Equation (9)) ⟨81JOM(213)C53⟩.

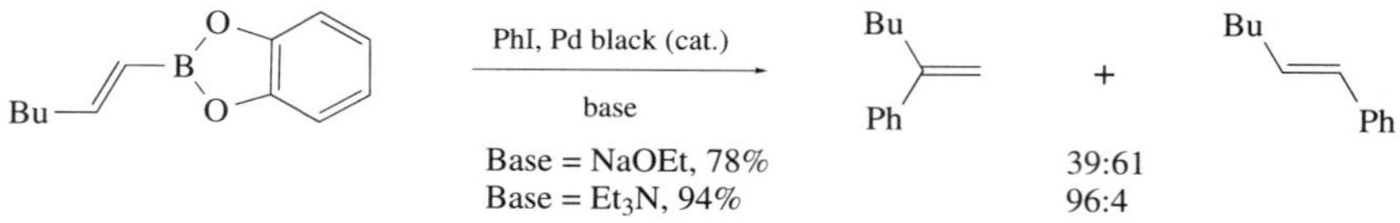

(9)

Aryl alkynes are obtained from aryl halides on treatment with a cuprous acetylide ⟨72OS(52)128⟩. This, the Stephens–Castro coupling reaction can be carried out more simply using palladium catalysis ⟨75TL4467⟩. A milder, more efficient, variant is obtained under phase transfer conditions ⟨82T631⟩. Similar methodology has been applied to the synthesis of enynes and allenynes ⟨83S32⟩.

(iii) Aryl nucleophiles

The recent developments in the synthesis of biaryls has been reviewed ⟨90AG(E)977⟩. Symmetrical biaryls may be accessed through the dimerisation of aryl halides on heating in the presence of a copper salt. This, the Ullmann reaction, works for all but aryl fluorides with reactivity increasing as the following series is descended (ArCl < ArBr < ArI). The reaction requires both a highly polar solvent (pyridine, quinoline, DMF and tetramethylurea are common choices) and high temperatures ($\leqslant$260 °C). In general, provided it is compatible with the temperature requirement, a broad spectrum of functionality is tolerated. Alkyl, alkoxy and *ortho*-alkoxycarbonyl groups accelerate the reaction whilst free protic groups, OH, NH_2, CO_2H, etc., inhibit the reaction. The Ullmann reaction is sensitive to steric effects and for particularly crowded aryl halides it is advisable to form the corresponding aryl organometallic and couple using transition metal catalysis (see below). In order to circumvent the high temperatures, a variety of more activated copper species can be prepared; for example by potassium naphthalenide reduction of cuprous iodide ⟨79JOC3445⟩, or by combination of Cu(OTf) and aqueous ammonia in acetone ⟨82JCS(P1)2299⟩. The latter combination allows the reaction to proceed efficiently at room temperature.

Nickel salts can also be used to replace the copper species in the Ullmann reaction ⟨81JA6460, 89JOC4840⟩. Although similar to the classical Ullmann reaction in terms of substituent effects, this modification is somewhat more sensitive to the steric demands of the reaction partners. Likewise, protic functional groups tend to afford products of simple reduction whilst the presence of a nitro group completely inhibits the coupling reaction. Although less reactive than the corresponding nickel-catalysed coupling of aryl Grignard and lithium reagents, this process is more compatible with sensitive functional groups. A further attraction of this modification is that by using zinc together with triphenylphosphine as a stoichiometric reducing agent, the coupling process can be rendered catalytic in nickel ⟨77TL4089, 86JOC2627⟩. With this system it is most profitable to use the aryl chloride, which affords a greater ratio of coupling to reduction, than the corresponding bromide and iodide.

The Ullmann reaction can be used for coupling of different aryl halides provided there is sufficient difference in reactivity. Apart from the halide itself, the electronic effects of the other arene substituents are important. In general, the best results are achieved through the coupling of the iodide of the electron-rich arene, which undergoes rapid oxidative addition to form the corresponding aryl copper species, and the bromide of an electron-deficient arene. Alternatively, some success can be achieved through the use of a large excess of the active copper species and running the reaction at a temperature below that required for dimerisation of the less reactive component ⟨77TL2015⟩.

Greater specificity in heterocoupling, together with a much reduced temperature requirement, are obtained using preformed arylcopper species. Simple diarylcuprates couple with unfunctionalised iodoarenes. However, for efficient transfer, the presence of various additives, for example, lithium bromide or triphenyl phosphine together with the 'correct' solvent is necessary ⟨69JA4871⟩. One attractive option which allows the coupling to be achieved at ambient temperatures is to use a preformed arylcopper reagent which contains a ligating *ortho* substituent (Equation (10)) ⟨80JA790⟩. Steric hindrance is not normally a problem with this reaction and the only restriction with its use is the requirement for the activating *ortho* groups in both components. An alternative and more versatile strategy involves the coupling of two arene units through the low temperature oxidative coupling of higher order mixed diarylcuprates ⟨93JA9276⟩.

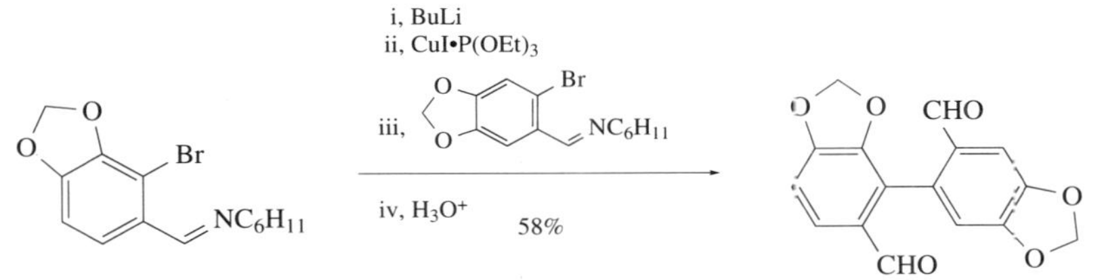

(10)

The best alternative to these high temperature processes are the nickel or palladium-catalysed coupling of aryl halides and aryl Grignard reagents ⟨78OS(58)127⟩. Aryllithium reagents generally produce much poorer yields. As with alkenyl coupling reactions it is more efficient to use the Grignard reagent derived from the more hindered arene. In general, there is no apparent restriction on the degree of substitution provided that not more than two *ortho* substituents are present, preferably both on the Grignard component ⟨86T2111⟩. A wide range of catalytic species have been employed using both nickel and palladium complexes. Whilst the palladium-catalysed reaction shows greater chemoselectivity, tending to be restricted to activated aryl bromides ⟨86T2111⟩, nickel catalysts are more tolerant of sterically hindered aryl halides including a broad range of aryl bromides and iodides. Asymmetry may be incorporated into this process through the use of chiral ferrocenyl phosphine nickel complexes ⟨88JA8153⟩. Finally, selective mono- or polyarylation is possible with certain nickel catalysts ⟨80TL845, 82JOC4319⟩. In most of these reactions the use of ultrasound is found to significantly accelerate the reaction ⟨86BCJ2044⟩.

Arylzinc reagents also couple efficiently under similar conditions. These reagents are additionally attractive being not only more tolerant of functionality in the electrophilic component, but are also readily accessible through transmetallation of the corresponding aryllithium ⟨88OS(66)67⟩. Similarly, arylstannanes have been employed with the transfer of a single aryl unit from tetraphenyltin being possible ⟨79JA4992⟩.

Another good source of aryl nucleophiles are the corresponding boronic acids, although these do not couple with aryl chlorides ⟨91PAC419⟩. Arylboranes have also been employed in this regard and both tolerate a wide spectrum of functionality. Steric effects appear to be relatively unimportant and efficient ligand transfer can be realised with very small amounts of catalyst ⟨81SC513⟩. Again, many of these species are readily available through transmetallation of the corresponding aryllithium ⟨87TL5093⟩. This sequence may be used in an iterative fashion for the regiospecific construction of polyphenols (Scheme 9) ⟨87TL5097⟩.

(iv) Nucleophilic aromatic substitution

S_NAr displacement of aryl halogen requires the presence of an appropriately sited activating group or heteroatom. For a given compound the ease of halide displacement is in the order $F > Cl \sim Br \sim I$. Displacement by alkyllithium reagents normally fails due to competing lithium–halogen exchange reactions. However aryllithium and Grignard reagents will displace the halogen from 2-fluoro- and 2-bromo-2-aryloxazolines ⟨85T837⟩, and 2-aryloxazoles ⟨87CL19⟩. Chiral biaryls can be prepared using this strategy, either through the use of chiral auxiliaries ⟨93S250⟩ or chiral catalysts ⟨92JA8732⟩.

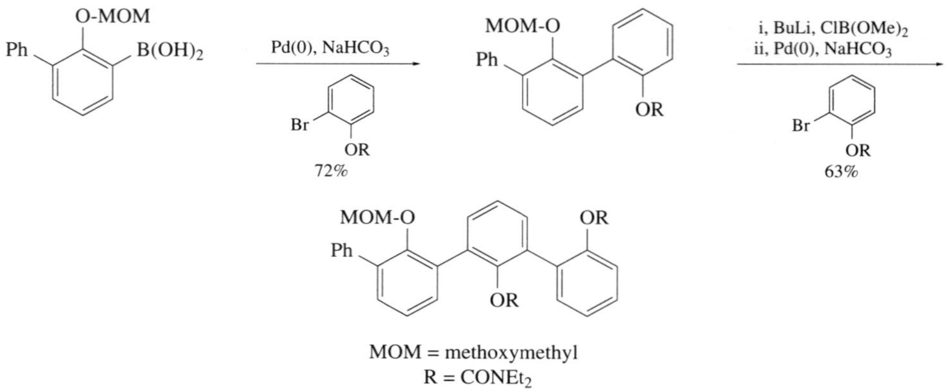

Scheme 9

Active methylene compounds are also suitable nucleophiles with diarylation rarely being a significant problem ⟨16CB2222, 51CRV273⟩. These stabilised nucleophiles also provide for a very efficient substitution process on reaction with haloarene transition metal complexes. The use of more basic, reactive carbanions tend to afford products of hydrogen as opposed to halogen substitution (Scheme 10) ⟨81T3957, 90BSF401⟩.

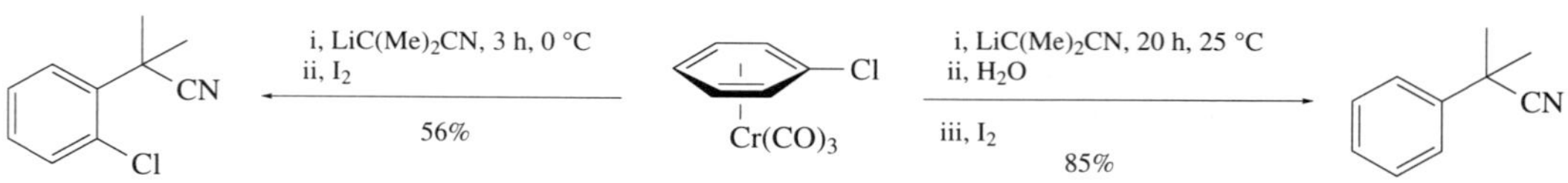

Scheme 10

Displacement of aryl halogen may also take place via the $S_{RN}1$ pathway and in this case the order of reactivity is found to be I > Br > Cl ≫ F. Aryl fluorides are sufficiently inert such that exclusive monosubstitution of halofluorobenzenes is observed. Both hydrocarbon and heteroaromatic halides are acceptable substrates. Although unsubstituted arylhalides do react, in many cases the substitution fails without the presence of an activating group (see below). With the exception of the nitro group and protic functionality, most substituents are tolerated provided that they do not react with the nucleophile or create excessive steric hindrance. In the latter case reduction can be a competing process ⟨75CPB2620⟩.

The $S_{RN}1$ reaction provides an efficient route to α-arylketones through the reactions of ketone enolates ⟨91COS(4)451⟩. Whilst nitrile-stabilised carbanions behave similarly, the use of ester enolates is not nearly so effective. β-Dicarbonyl nucleophiles can be used although the success of the reaction is very dependent upon the particular reagent/substrate combination. In a related process unsymmetrical biaryls can be prepared using phenoxide anions through an electrochemical-mediated reaction (Equation (11)) ⟨88CC203⟩. In this reaction synthetically useful yields are only observed in the presence of specific substituents ⟨88TL1289⟩. A similar coupling can be achieved using arylamide anions ⟨87TL4653⟩.

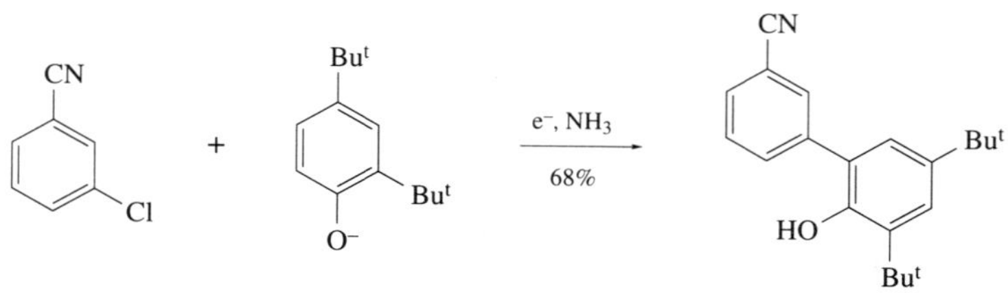

(11)

Most unstablised hydrocarbon nucleophiles tend to be too basic for the solvents used in these reactions although some limited success can be obtained with conjugated hydrocarbons. However, in many cases these processes are complicated by competing reduction of the nucleophile ⟨73JOC3020⟩.

1.11.2.2 Substitution of Oxygen

1.11.2.2.1 Substitution of alkyl–oxygen leaving groups

(i) Friedel–Crafts alkylations

As with alkyl halides the reactivity of alcohols to Friedel–Crafts substitution decreases along the series benzyl, allyl, tertiary > secondary > primary > methanol. Whilst reactions with secondary and the more reactive alcohols proceed smoothly, primary alcohols require an excess of the Lewis acid and invariably afford products resulting from rearrangement of the initial formed carbocation. Enhanced reactivity is observed with the corresponding mesylates and tosylates. A highly stereoselective alkylation of benzene occurs on reaction with sulfonate esters of lactic acid and related analogues. The reaction proceeds with almost complete inversion of configuration at the reacting centre reflecting the S_N2 nature of this pathway ⟨91JOC183⟩. The reaction is postulated to proceed by way of the chelated intermediate (**10**) (Scheme 11). If the acid catalyst employed cannot form this complex, for example CF_3SO_3H, then complete racemisation is observed ⟨87AG(E)142⟩. In an unusual process trifluoroacetic acid reacts with unhindered arenes in the presence of sodium borohydride to produce 1,1-diaryltrifluoroethane ⟨85S756⟩. The reaction presumably proceeds by way of the corresponding trifluoroacetaldehyde since more hindered arenes yield the corresponding 1-aryltrifluoroethanol and similar conversions have been reported for chloral and other reactive aldehydes ⟨64MI 111-01⟩.

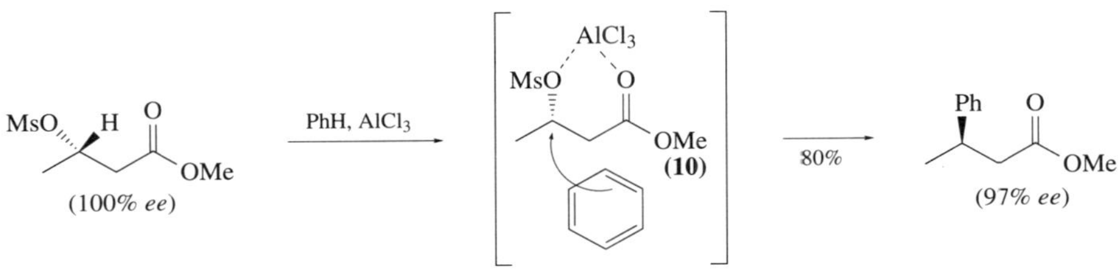

Scheme 11

Whilst acyclic ethers are generally unreactive, alkylations can be achieved with their cyclic analogues. The reactivity of these compounds is proportional to ring strain and consequently decreases with increasing ring size (Scheme 12) ⟨48BSF811⟩. Similarly, alkyl substitution in the α-position enhances reactivity consistent with the cationic nature of the reaction (Scheme 13) ⟨60DOK(131)27⟩. With higher homologues reaction is frequently accompanied by competing side reactions including rearrangements, condensations, intramolecular reactions, etc. ⟨58DOK31⟩. Epoxides undergo regioselective ring opening at either carbon atom depending upon the nature of the Lewis acid used. Mild species, such as $MgBr_2$ and $ZnBr_2$, lead to diarylmethane products via ring opening to the aldehyde whilst stronger Lewis acids promote direct S_N2 ring opening to afford β-hydroxyarylalkanes which can react further to produce diarylalkanes ⟨78BCJ174⟩. Epoxides are opened with almost complete stereocontrol to produce, initially, the product with inverted configuration at the reacting centre ⟨75BCJ960⟩. Exceptional selectivity is observed in the Friedel–Crafts reaction of 5,6-epoxycyclodecene (**11**) (Equation (12)), which on treatment with an arene and $SnCl_4$ affords the decalinol (**12**) as a single isomer ⟨81JOC2709⟩. The invertive pathway is also observed with oxetane substrates, although again the nature of the Lewis acid can critically affect the optical purity of the product (Scheme 14) ⟨82BCJ167⟩. Preferential reaction occurs at the most substituted carbon atom indicating the involvement of considerable carbocationic character in the transition state. As the ring size is increased the selectivity decreases, although inversion through an S_N2 type process is still the favoured pathway.

(12)

(11) **(12)**

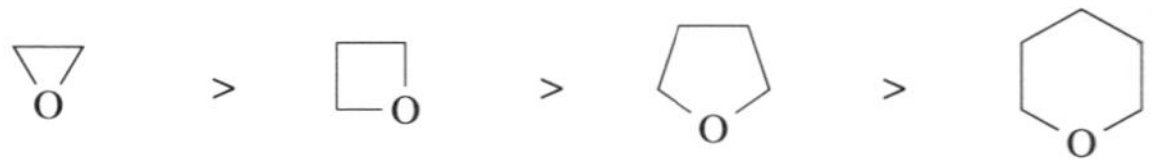

Scheme 12

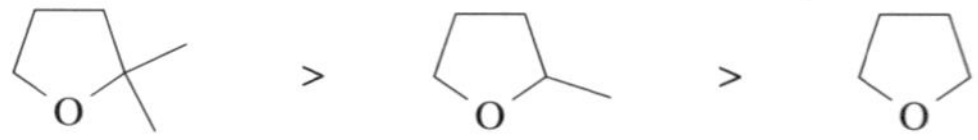

Scheme 13

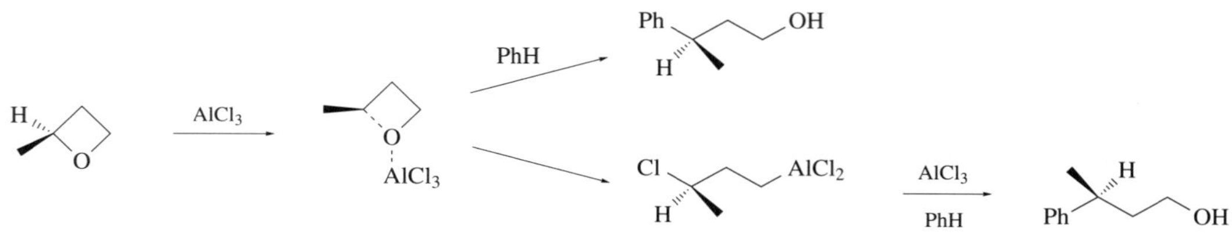

Scheme 14

Similar cyclic ion pair intermediates (**13**) are postulated to account for the high degree of asymmetric induction observed in alkylations using lactones as substrates (Scheme 15) ⟨67JA5421⟩. These reactions are often complicated by the propensity of the first formed arylalkanoic acids to cyclise affording the corresponding tetralone. As with ethers, simple acyclic esters are not sufficiently reactive to be effective substrates.

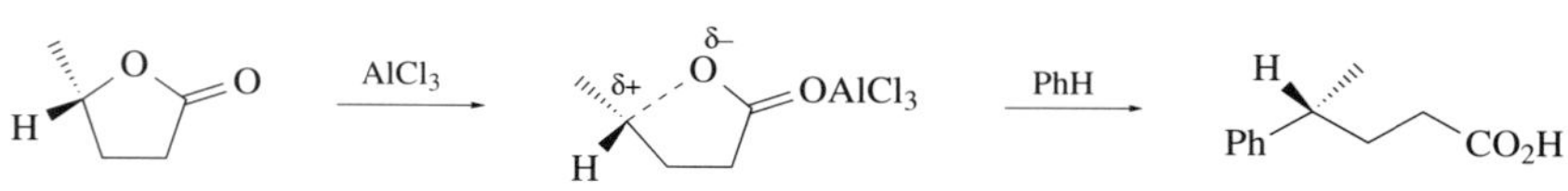

(100% *ee*) (**13**) (40% *ee*)

Scheme 15

(ii) Substitution by organometallic sp[2] *carbanions*

The most direct method for achieving this coupling is through the reaction of an alkyl tosylate with a vinyl- or arylcuprate or copper-modified Grignard reagent ⟨92OR135⟩. This reaction is very similar to the corresponding process using alkyl halide electrophiles. However, both primary and secondary tosylates react more efficiently and readily than do the analogous halides ⟨73JA7777⟩. Other alcohol activating groups can be used with equal efficacy, notably allylic acetates and β-lactones. The latter substrates provide a useful three-carbon homologation sequence (Equation (13)) ⟨80TL935⟩. ω-Ethenyl lactones undergo an S_N2' displacement and this can provide easy access to dienoic acids ⟨81CL1307, 82TL3583⟩. This last method is also applicable to larger lactones in which there is no competing strain-assisted direct displacement (Equation (14)). These last S_N2' displacements are examples of the transition metal-catalysed alkylation of Grignard reagents that can be achieved with a variety of allylic electrophiles. Alcohols, esters, sulfonates, phosphates and ethers have all been employed with some success. The nature of the leaving group can affect the rate of substitution. For example, for the palladium-catalysed reactions of vinyl alanes the order has been found to increase along the series $OSiR_3 < OPO(OEt)_2 < OAlMe_2 < OAc$ ($\approx$Cl) and within this series some selectivity may be observed (Equation (15)) ⟨81TL3737⟩. Although allyl alcohols may be coupled directly ⟨76JOC266⟩ (see below) it is frequently more efficient to carry out the reaction in two discrete steps, that is activation and substitution. In one particularly attractive strategy, allyl alcohols are coupled efficiently via the *in situ* conversion to the corresponding iminium salt (**14**) (Scheme 16) ⟨83TL5745⟩. Phenyl and vinyl Grignard reagents react regioselectively at the α-carbon atom whilst the addition of CuX/HMPA favours the formation of products of allylic substitution.

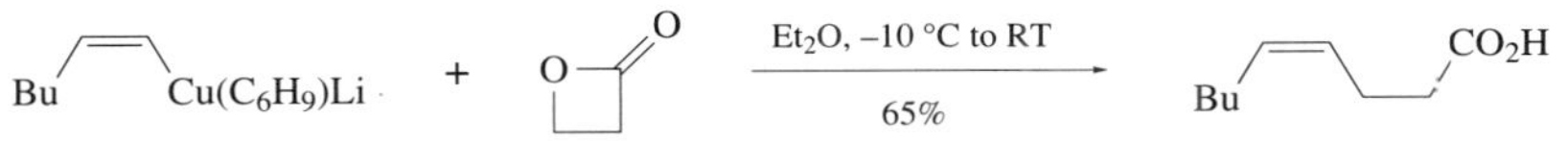

(13)

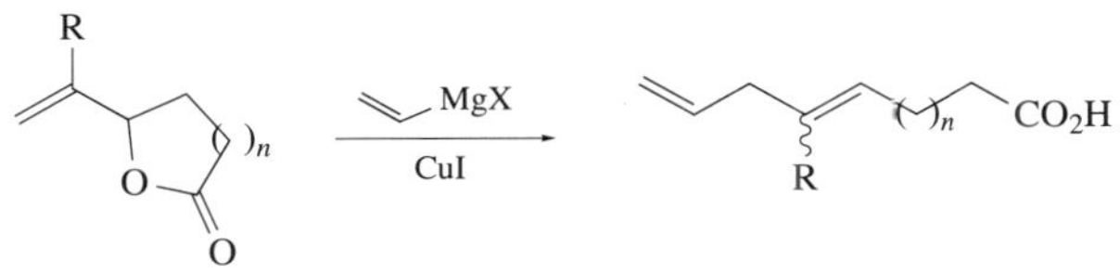

(14)

$n = 0, 1, 2$; R = H, Me

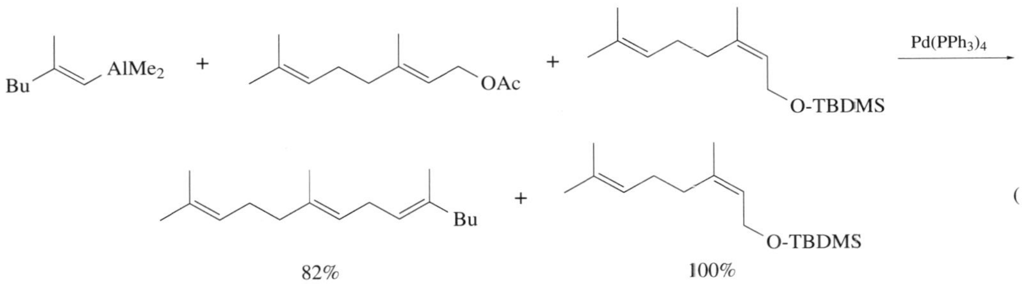

(15)

TBDMS = *t*-butyldimethylsilyl

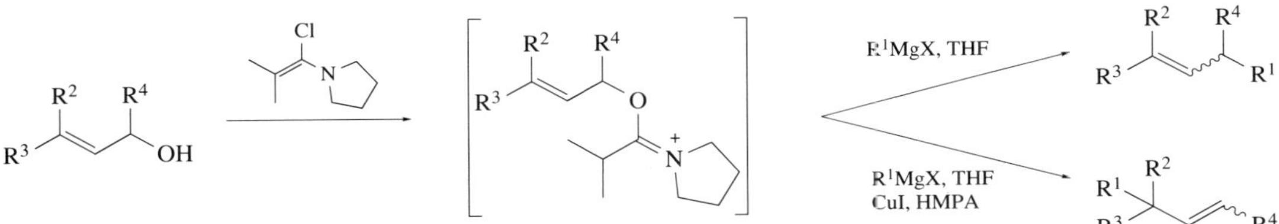

(14)

$R^2 = R^3 = R^4$ = H, alkyl; R^1 = Ph, vinyl

Scheme 16

In general, as with the corresponding displacement of allylic halides, the regiochemistry of this reaction is very dependent upon the nature of the nucleophile, electrophile and catalyst and it is difficult to make any general predictions ⟨91COS(3)468⟩. For example, copper-catalysed addition of vinyl Grignards to allylic phosphates affords products of direct S_N2 displacement ⟨77JCR(M)2101⟩ whereas aryl Grignards react with with allylic acetates in the presence of a nickel catalyst in an S_N2' fashion. Similarly, in the treatment of allylsilyl ethers with aryl Grignard reagents, Pd catalysis affords predominantly the disubstituted alkene and nickel(II) catalysis generates the terminal alkene (Equation (16)) ⟨81CC313⟩. In these transition metal-catalysed processes, the harder nickel centre tends to favour attack at the more substituted carbon atom, reflecting the greater degree of carbocation character in the η^3allyl ligand, whilst the softer palladium forms an essentially symmetrically bound η^3allylpalladium species which is attacked at the position of least steric hindrance. In an exception to this, allylic alcohols are substrates in nickel(II)-catalysed displacement reactions, albeit requiring a large excess of the Grignard reagent, in which exclusive terminal substitution is observed ⟨77JOM(127)371⟩. When these transition metal catalysts contain chiral ligands then significant asymmetric induction can be observed (Equation (17)) ⟨85TL3259⟩.

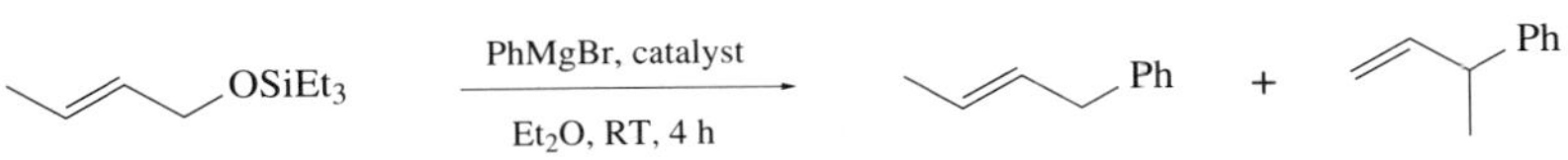

(16)

catalyst	$NiCl_2$(dppf)	75%	12	88
	$PdCl_2$(dppf)	85%	96	4

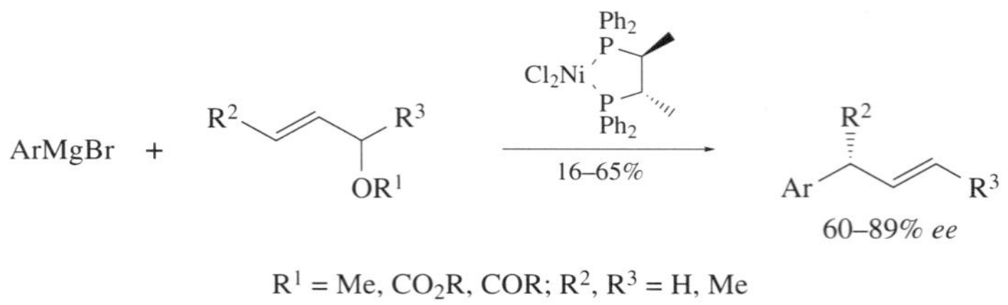

(17)

The other major class of oxygen-based electrophiles are those containing epoxides. Although most organometallic epoxide opening reactions require the use of copper-based reagents to obtain good yields, the direct coupling of heteroaromatic and allenyllithium reagents with unhindered epoxides occurs efficiently. With the latter reagents, the rate and efficiency of the coupling can be significantly enhanced through the use of HMPA as a co-solvent. With 1,2-disubstituted epoxides regioselectivity can be problematical and the results obtained are very substrate/nucleophile dependent. In certain cases selective alkylation of unsymmetrical epoxides can be obtained, for example, predominant attack is observed at C-2 of β,γ-epoxyalcohols ⟨83JOC4131⟩. The efficiency of these direct couplings can be enhanced by the addition of Lewis acids, notably $BF_3 \cdot OEt_2$, to the epoxide ⟨84JA3693⟩. This modification is sufficiently active to permit the ring-opening alkylation of oxetanes whilst the reduction in ring strain renders THF sufficiently inert to be used as the solvent for this reaction (Equation (18)). The direct intramolecular alkylation of epoxides with organolithiums is also possible ⟨78JOC3800⟩. Again, these reactions can be enhanced through the addition of an appropriate Lewis acid, although the nature of this latter species can be critical. In both these cases the use of such catalysts may alter the regioselectivity of the process. For example, the presence of magnesium bromide can promote a variety of side reactions through the formation of halohydrin and aldehydic intermediates (Scheme 17) ⟨80JOC922⟩.

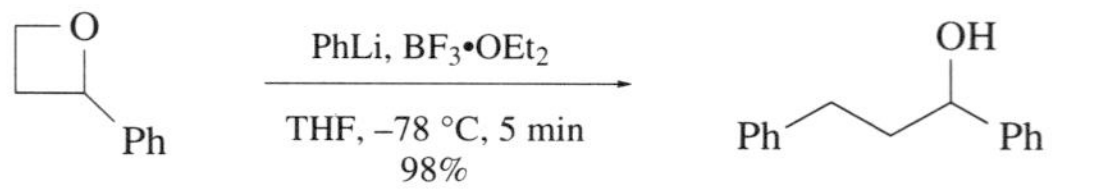

(18)

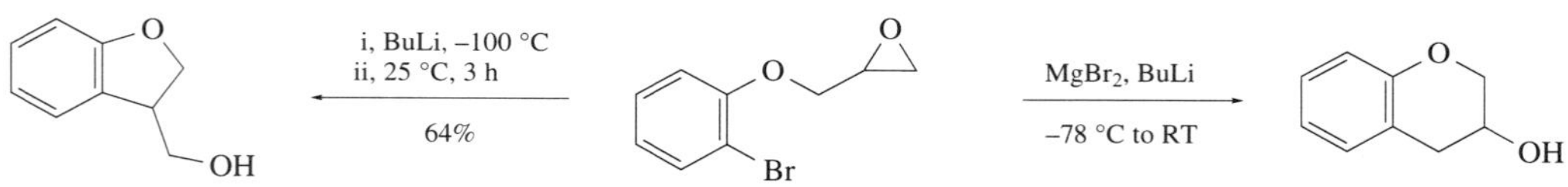

Scheme 17

With unhindered epoxides, aryl and vinyl Grignard reagents afford comparable yields with or without copper catalysis, whilst with more hindered substrates the requirement for copper-based reagents, in either stoichiometric or catalytic quantities, becomes essential ⟨78TL4069, 79TL1503⟩. In this respect both copper(I) halides and cyanides are suitable sources of copper. Lower order lithiocuprates regiospecifically open terminal epoxides in good yield, whilst more hindered trisubstituted epoxides are best cleaved using higher order cuprates ⟨84JOC3928⟩. As above, 1,2-disubstituted epoxides often give mixtures of regioisomers with the major product being controlled by steric factors within the substrate ⟨79TL4343⟩, although electronic factors can influence the outcome ⟨85TL4683⟩. As with other displacements using vinylic cuprates the major problem is the lack of efficient transfer of the second vinyl ligand. The use of mixed lower order cuprates, RCu(CN)Li ⟨77TL3407⟩ or R_TR_sCuLi ⟨80T1961⟩ offers an economical route to homoallylic alcohols. The latter species can be formed by the *in situ* addition of lithium acetylide to the vinyl copper species formed after transfer of the first vinyl unit thus permitting efficient use of both vinyl groups of a divinylcuprate. In this respect, unlike their aliphatic analogues, many heteroaryl copper species are essentially inert, frequently finding a role as a nontransferable ligand in mixed and higher order cuprates. One particularly attractive method that circumvents this problem is to use an active elemental form of copper generated *in situ* by lithium naphthalide reduction of a CuI–PBu_3 complex. In addition to providing increased reactivity the resulting reagent, which can be generated from either vinyl or aryl halides, provides access to functionalised species not available through standard organolithium or Grignard technology ⟨87JOC5057⟩. Alternatively, the activity of the copper species towards epoxides can be enhanced by the addition of Lewis acids ⟨86AG(E)947⟩. Under these conditions those cuprates which generate lithium halide salts in their preparation frequently produce halohydrin by-products. This is not observed with the lower- or higher-order cyanocuprates. The latter, in combination with Lewis acids, afford an extremely powerful nucleophile with respect to epoxides (Equation (19)) ⟨86T5607⟩. In addition to epoxides a number of other functional groups including allylic ethers and acetals also undergo substitution ⟨84TL3075⟩.

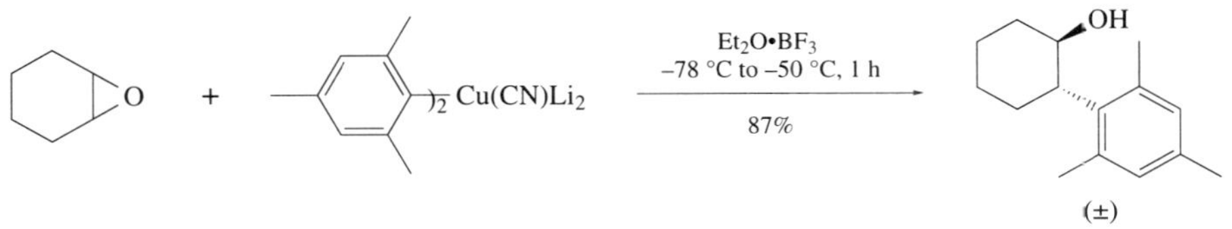

(19)

Magnesiocuprates obtained from the addition of Grignard reagents to alkynes combine efficiently with terminal epoxides but only sluggishly with more substituted species. Higher yields may be obtained through the formation of the corresponding 'ate' complex by addition of lithium acetylide ⟨78TL2465⟩. Similar behaviour is observed with the alkenylaluminate obtained from the addition of butyllithium to an alkenyl alane ⟨76CC17⟩. The corresponding lithiodialkenylcuprate prepared from acetylene can be prepared stereospecifically as the (*Z*) isomer and as such provides a convenient entry to (*Z*) homoallylic alcohols ⟨80T1961⟩. In common with most copper reagents these compounds do not react with esters. Consequently, this functionality may be located in either reaction partner (Scheme 18) ⟨87TL1781⟩. However, in some cases in which the ester is contained on the cuprate moiety products of alkene isomerisation have been recorded ⟨83JOC4621⟩.

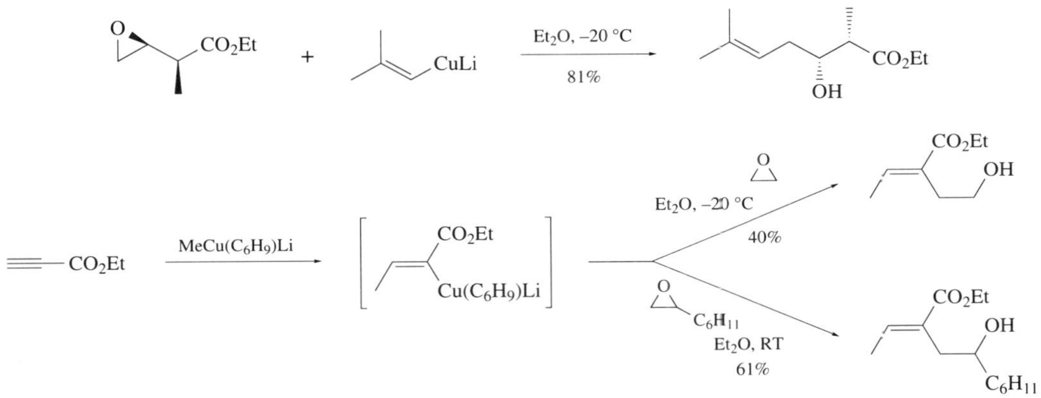

Scheme 18

Vinylic epoxides are ambident electrophiles and as such can alkylate carbanions to produce up to three isomeric products. Grignard reagents tend to provide a mixture of products favouring 1,2 addition, with the eventual outcome dependent on the nature of the reacting partners ⟨70JA4978⟩. Copper-based reagents favour 1,4 addition through an S_N2' pathway ⟨78S528, 78TL2027⟩ as does the use of aryl and vinyl stannanes in conjunction with a palladium catalyst ⟨89T979⟩. Similar selectivity is observed in the reactions of cyclic vinyl epoxides. Vinyl epoxides derived from α,β-epoxyketone derivatives (enolates, enolethers and enol phosphates) can undergo selective allylic displacement to provide a convenient route to functionalised α-aryl- and α-vinylketones. However the regiochemistry does depend on the substitution pattern (Scheme 19) ⟨82JA3165⟩.

1.11.2.2.2 Substitution of alkenyl–oxygen leaving groups

(i) Substitution by organometallic carbanions

Enol phosphates may be simply converted to the corresponding alkenes on treatment with reactive cuprates such as lithium dibutylcuprate. However, less reactive reagents such as lithium dimethylcuprate give poor yields whilst more hindered enol phosphates are unreactive ⟨76TL4405⟩. Enol phosphates may also be displaced by a variety of Grignard reagents using nickel catalysis ⟨81S1001⟩. The only palladium species to be effective in this coupling is tetrakis(triphenyl-

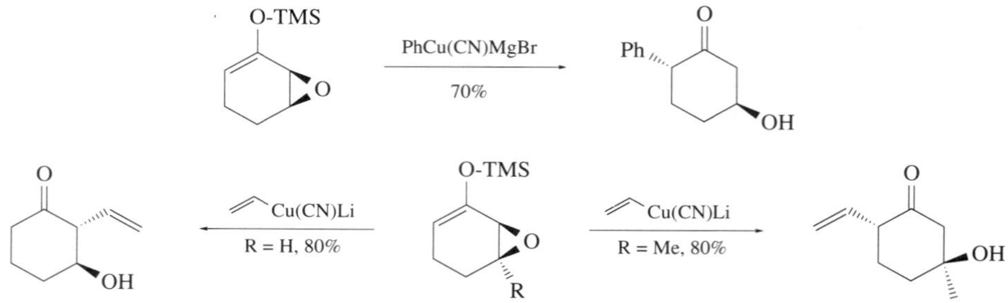

Scheme 19

phosphine)palladium although, in this case, much longer reaction times are required. The presence of β-hydrogen atoms in the Grignard component can be problematical although the use of bidentate phosphine ligands, for example, $NiCl_2$(dppp), can circumvent this difficulty ⟨83TL5137⟩. Although palladium(0) catalysis is not suitable for coupling of enol phosphates with either organolithium or Grignard reagents, the less nucleophilic organoaluminum compounds are effective, providing the desired alkene in good to excellent yield ⟨84BCJ108⟩. These reactions are highly stereospecific and a range of organoaluminum compounds can be used with alkenyl- and alkynylaluminum ligands showing greater migratory aptitude than alkyl groups. With triethylaluminum it is more effective to use diphenylphosphates, although diethylphosphate can be efficiently displaced in refluxing 1,2-dichloroethane. Unlike the corresponding nickel-catalysed Grignard displacements, a variety of other vinyl leaving groups are stable, notably vinyl sulphides, and this can lead to the stereocontrolled synthesis of tri- and tetrasubstituted alkenes. Palladium catalysis is also required for the displacement of enol phosphates with trialkylmanganates, R_3MnM ($M = Li$, MgBr), prepared *in situ* from Li_2MnCl_4 and the organometallic species ⟨87CL2203⟩. Reagents derived from organolithiums are superior to those from the corresponding Grignard reagent whilst higher alkyl reagents tend to give side-products from reduction and vinyl coupling. These manganese reagents can also efficiently displace the triflate group from an enol triflate without the need for a co-catalyst. With phenyl, benzyl and alkylmagnesiumbromide the amount of Li_2MnCl_4 can be reduced to a catalytic quantity.

Enol (or vinyl) triflates are, in general, more reactive than the corresponding enol phosphate and can readily be replaced by a variety of organometallic reagents and are a particularly attractive source of cycloalkyl electrophiles, being readily and stereospecifically obtained from the corresponding cyclic ketone ⟨88ACR47⟩. Aryl-, vinyl- and alkylcuprates, including Me_2CuLi, afford the corresponding alkene in high yield and with complete stereospecificity ⟨80TL4313⟩. The reaction is relatively independent of the steric environment in either parameter with the only restriction being that acyclic vinyl triflates can be difficult to prepare regiospecifically. The parent organolithium reagents are not suitable for these reactions since they lead to S—O bond cleavage. The major C—C bond-forming use of vinyl triflates is in the palladium-catalysed coupling reactions with organostannanes ⟨90OS(68)116⟩ or organoboron compounds ⟨93JOC2201⟩. In the presence of $\leqslant 2$ mol% of tetrakis(triphenylphosphine)palladium and at least one equivalent of LiCl, a variety of organostannanes (alkyl, vinyl, acetylenic and aryl) couple efficiently ⟨93S735⟩. The presence of the lithium salt is essential and aryl groups may be coupled if the equivalent arylzinc chloride is employed. In this and the corresponding displacement of aryl or vinyl halides other additives may be equally effective, including copper iodide ⟨90JOC5359⟩ and copper oxide ⟨93JOM(460)127⟩. Vinyl groups are transferred more readily than alkyl groups whilst for the latter group β-hydrogen atom elimination is not a problem. In general, both trimethyl and tributyl stannanes can be used, both offering different advantages. The former provides water soluble by-products and shorter reaction times, but is considerably more toxic and expensive, and can be less easy to handle. The reaction proceeds in most polar solvents, for example, DMF, DMSO, HMPA and *N,N′*-dimethylpropyleneurea (DMPU), which not only solubilise LiCl but also act as ligands for the intermediate complexes in the catalytic cycle ⟨86JA3033⟩.

Various vinyl ethers may also be converted to substituted alkenes on treatment with Grignard reagents in the presence of nickel catalysts. Both alkyl enol ethers ⟨84JOC4894⟩, and silyl enol ethers ⟨80TL3915⟩ undergo the reaction with aryl and alkyl Grignard reagents. The coupling is most facile using nickel acetylacetonate as the catalyst when it proceeds in refluxing ether. However, under these conditions there is scrambling of alkene geometry. Greater selectivity can be achieved through the use of $NiCl_2$(dppp) or related chelating phosphines. With higher alkyl Grignard reagents, competing reduction through β-hydrogen elimination can be suppressed through the use of the ferrocenyl ligand (dppf). Cyclic vinyl ethers, dihydrofuran (DHF) and dihydropyran (DHP) are also substrates producing di- and trisubstituted homo- and bishomoallylic alcohols respectively in good yield, with predominant retention of alkene stereochemistry (see below). Although vinyl Grignard reagents are not suitable, a range of other alkyl and aryl Grignards are effective and provided the reaction is quenched efficiently almost complete stereoselectivity can be observed ⟨92JCS(P1)3419⟩. Grignard reagents possessing β-hydrogen atoms are considerably less reactive, giving product mixtures containing reduction and isomerised by-products. In general, pyran substrates are less reactive requiring higher catalyst loadings, longer reaction times and give more modest yields (50–60%), failing completely with Grignard reagents possessing β-hydrogen atoms ⟨92JCS(P1)3431⟩.

The opposite alkene stereochemistry can be obtained from the corresponding lithiated vinyl ether through ring-opening reactions with alkyllithiums in the presence of CuCN (Scheme 20). The intermediate cyanocuprate can be trapped with a range of electrophiles to provide the desired substituted homoallylic alcohol of high isomeric purity ⟨89JA2363⟩. The identical transformation has also been reported to occur without the addition of CuCN ⟨91TL5903⟩.

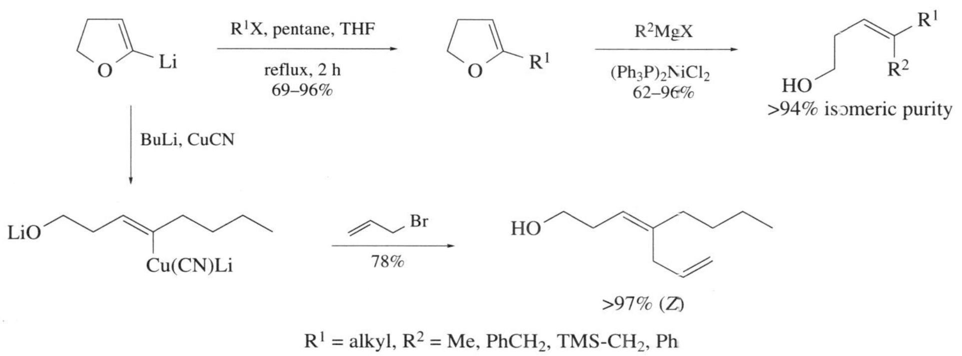

Scheme 20

Treatment of the vinyl ether–iron complex (**15**) with lithium enolates followed by subsequent acid-promoted elimination affords the α-alkenylketone (**16**) ⟨88OS(66)95⟩. Similarly, treatment of the (1,2-dialkoxyethylene) iron complexes (**17**) with cuprates, Grignard reagents or enolates affords moderate yields of either the *cis* or *trans* alkenes (Scheme 21) ⟨84JA7264⟩. Finally diketene combines with primary alkyl Grignard reagents in the presence of cobalt ⟨82BCJ3555⟩ or nickel catalysts ⟨77CL103, 83CPB4346⟩ to afford 3-methylene carboxylic acid in good yields.

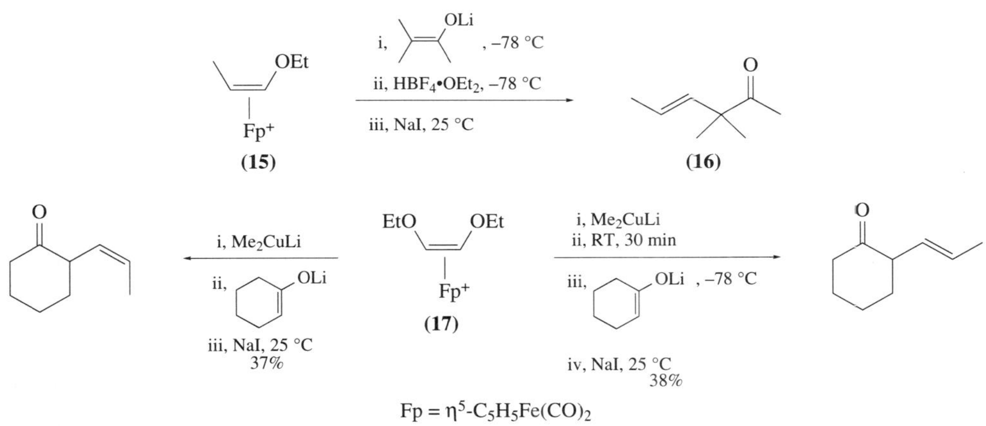

Scheme 21

(ii) Displacement of vinyl–oxygen by addition–elimination

Regiospecific displacement of a vinyl alkoxy group from β-alkoxy enones is not possible with either organolithium or Grignard reagents since preferential 1,2 attack occurs. However, on acidic workup, hydrolysis of the vinyl ether occurs with concomitant elimination of water to afford the isomeric β-alkylated enone (Equation (20)) ⟨73JOC1775, 89S477⟩. Similar transformations are also possible commencing with the corresponding silylenol ether ⟨82SC795⟩. The acidic workup step can be avoided through the use of cerium-catalysed Grignard reagents ⟨92SC1953⟩.

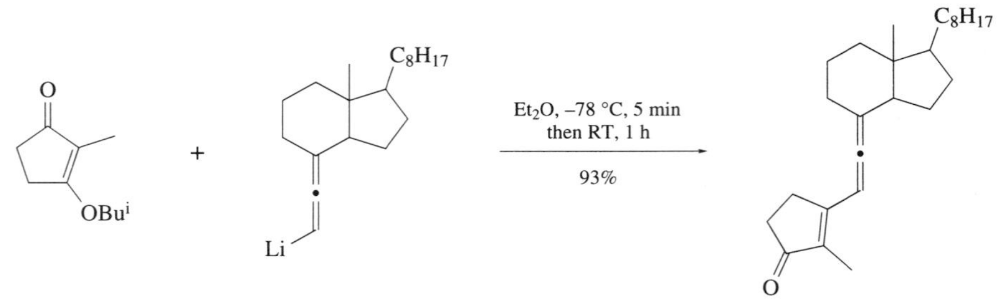

(20)

ipso Substitution via addition–elimination of an oxygen-centred leaving group is possible through the use of lithium dialkylcuprates or related reagents. There is considerable variety in the nature of the leaving group; β-alkoxy ⟨73CC907, 88SC343⟩, β-acetoxy- ⟨74TL925⟩ and β-phosphoryloxy-αβ-unsaturated carbonyl compounds ⟨84OS(62)14⟩ all afford good yields of the β-alkylated enone (enoate). The corresponding sulphonates can also be used ⟨90ACS842⟩, although elimination and a second conjugate addition to afford the activated carbonyl compound can compete ⟨81JOC197⟩.

As with the analogous displacement of β-haloenones the reactions proceed with predominant retention of stereochemistry. With acyclic β-acetoxy substrates the use of phosphine-stabilised cuprates and very low temperatures (~ −110°C) are required for good stereoselectivity. Similar temperature requirements for high selectivity are also found for the readily accessible enol phosphates of β-ketoesters. For these substrates the process is further limited to primary dialkylcuprates. With the more reactive β-phosphoryloxy ketones even lithium di(*t*-butyl)cuprate reacts efficiently and can provide an efficient entry to tetrasubstituted alkenes, (Equation (21)) ⟨79CJC1431⟩. Higher order cyanocuprates are required for efficient substitution of a pivolate from β-pivaloyloxy sulphones. The stereochemical outcome, although good, is found to vary according to the bulk of the transferred alkylcuprate ⟨87CC207⟩.

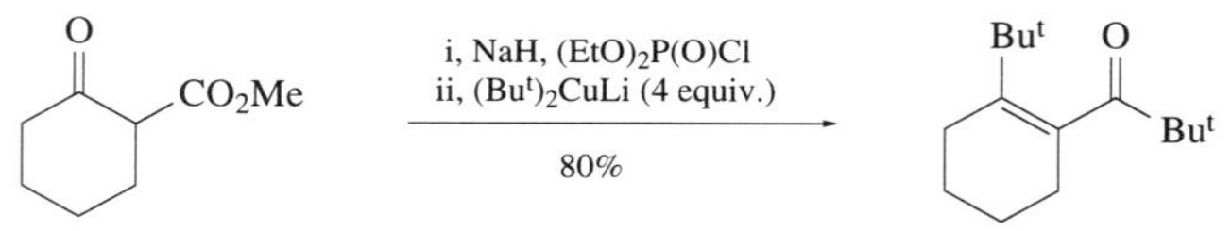

(21)

1.11.2.2.3 *Displacement of aryl–oxygen leaving groups*

(i) Substitution by organometallic carbanions

As with enol ethers, aryl methyl ethers can undergo displacement by Grignard reagents under conditions of nickel catalysis. However, the reaction requires relatively vigorous conditions and is limited to aryl Grignard reagents ⟨79JA2246⟩. Much better yields are obtained using aryl phosphates although a large excess of the Grignard reagent is required ⟨81TL4449⟩. Similarly, good yields are obtained much more rapidly under mild conditions using the corresponding tetrazolyl ether ⟨88TL5553⟩. Aryl triflates, being less reactive than their alkenyl counterparts, are not displaced by Gilman type cuprates, undergoing competing S—O bond scission instead. However, clean coupling is obtained on reaction with higher order cyanocuprates and with these latter reagents a wide range of functionality is compatible ⟨83TL2723⟩. Aryl triflates are also excellent substrates for palladium-mediated coupling with a wide range of organostannanes ⟨87JA5478⟩, organoboranes ⟨93JOC2201⟩ and trialkylaluminums ⟨89JCS(P1)2513⟩.

Arylstannanes are particularly suitable coupling partners for electron-rich aryl triflates providing a convenient route to unsymmetrical biaryls ⟨92JOC678⟩. Enhanced yields in this process may be obtained through the use of the equivalent arylzinc species. Symmetrical biaryls are obtained when the aryl triflate is treated with only 0.5 equivalents of hexamethyldistannane. These products may also be accessed through modified Ullmann reactions involving either ultrasound activation ⟨86CL407⟩ or transition metal catalysis ⟨93SL568⟩.

Aryl triflates may also be substituted, in a palladium-catalysed process, by silylketene acetals ⟨91JOC261⟩ or siloxycyclopropanes ⟨88JA3296⟩ to afford, in moderate to good yields, α- or β-arylated carbonyl compounds respectively (Scheme 22). In this reaction the presence of halide salts are deleterious to the process. Similar products may also be obtained in the palladium-catalysed reaction of zinc Reformatsky reagents with phenyl fluoroalkylsulphonates. The analogous alkynyl and benzyl zinc reagents also react efficiently ⟨87TL2387⟩. The resultant phenylalkynes can also be accessed directly from the parent alkyne through a palladium-mediated Heck reaction ⟨86TL1171⟩, (see also Section 1.11.1.8.2).

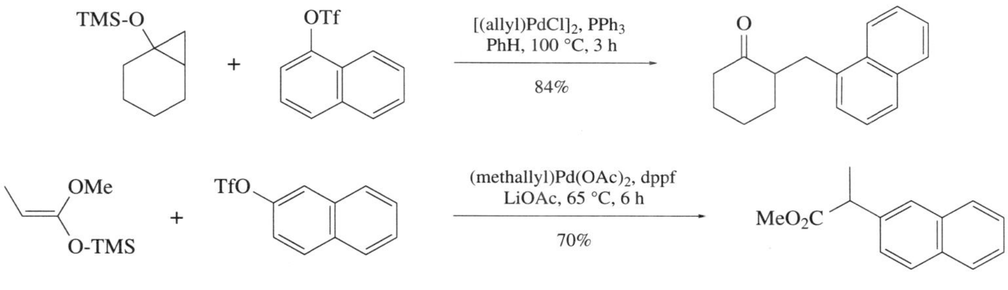

Tf = trifyl (trifluoromethanesulfonyl)

Scheme 22

(ii) Nucleophilic aromatic substitution

Oxygen-based leaving groups may be displaced from activated arenes through an S_NAr process. In this respect the displacement of methoxy groups from 1-methoxy-2-oxazolinyl arenes provides an attractive route into enantiomerically pure biaryls ⟨85JA682⟩. Other activating groups may be employed provided that they are inert to the reaction conditions. In this respect certain highly hindered arylketones and nitriles have proved successful ⟨51CRV273⟩. An alternative possibility to achieve arene activation is via complexation with a transition metal. However, in this case the process is normally one of non *ipso* substitution with the alkoxide being lost on protonation. Suitably substituted heterocycles also undergo *ipso* substitution through an addition–elimination pathway ⟨86H(24)3337⟩. For example, sulphonates may be selectively displaced from the 2-position of pyridines on treatment with a range of Grignard reagents.

$S_{RN}1$ processes are possible with oxygen nucleofuges. However, only aryldialkoxy-phosphates show any synthetic potential ⟨74JA112⟩.

1.11.2.3 Substitution of Other Chalcogens

1.11.2.3.1 Substitution of alkyl–chalcogen leaving groups

Displacement of alkyl sulfur leaving groups with vinyl organometallic nucleophiles is relatively rare and is restricted to allylic substrates. As with many allylic displacements regioisomeric products frequently result. Allylic sulfur ylides ⟨77JOM(142)1⟩, sulfones ⟨83T3283⟩ and sulfoximines ⟨91JA1442⟩ may be displaced by Grignard reagents in the presence of copper salts. Other cuprates are less effective. Tertiary allylic sulfones, which cannot be deprotonated, undergo direct displacement without the need for a transition metal catalyst ⟨87BSF487⟩. Nickel catalysts are required for the displacement of allylic seleno- ⟨80TL87⟩ and thioethers ⟨79TL3425⟩ whilst alkyl sulfones combine with vinyl alanes in the presence of palladium catalysts ⟨86JA1098⟩. Tertiary alkylsulfones have also found a role in Friedel–Crafts alkylation reactions (Equation (22)) ⟨84JA7260⟩.

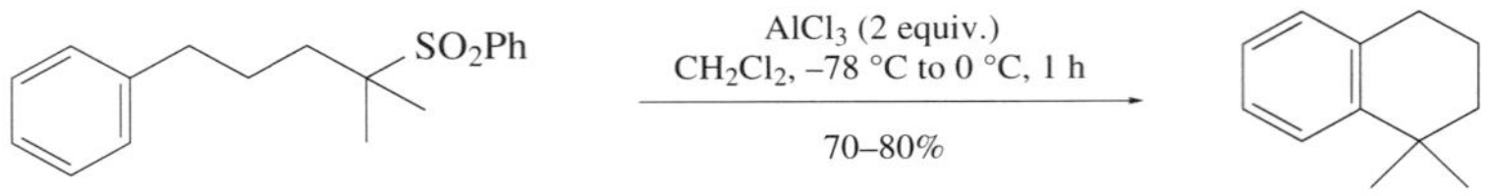

(22)

1.11.2.3.2 Substitution of alkenyl- or aryl-chalcogen leaving groups

Although β-styrylmethylthioether can be efficiently combined with lithium di-*n*-butylcuprate all other cuprate/substrate combinations are ineffective in this addition ⟨73JOC2747⟩. This is probably not a direct displacement but rather follows an addition elimination pathway (see below).

Grignard reagents, but not the corresponding organolithiums, may be utilised in the presence of a nickel catalyst ⟨79CC637, 79TL43⟩. In general, alkenyl thioethers are more labile than aryl thioethers, whilst all these sulfur containing compounds are more reactive than the corresponding oxygen species and this trend continues as the group is descended. In most of these reactions, there is predominant retention of alkene stereochemistry although in some cases this is found to depend on the exact nature of the catalytic species. A similar observation can be made for the corresponding selenides. Displacement of vinyl sulfide can provide access to dienes and through the use of dienylsulfides, trienes and polyconjugated aromatic species ⟨88PAC79, 89TL243⟩. Cyclic vinyl thioethers are cleaved by primary alkyl Grignard reagents although an excess of the Grignard reagent is required if high yields are to be obtained ⟨85JOC3828⟩. Thiophene, selenophene and tellurophene undergo a double substitution reaction to give (*Z*,*Z*) dienes stereospecifically with the exception that phenyl magnesium bromide affords exclusively the (*E*,*E*) isomer (Equation (23)). The yields are normally good, although the less reactive aliphatic Grignard reagents result in significant ($\sim$10%) amounts of arylated dienes formed by transfer from the phosphine ligands of the nickel catalyst ⟨84CC617⟩. Benzothiophene undergoes the analogous reaction and in this case, consistent with the reactivity order alkenyl > aryl, the alkenyl–sulfur bond is cleaved first. The difference is not large and in many cases the cleavage of both bonds is competitive. However, selectivity can be observed on the basis of steric size (Scheme 23) ⟨83T2289⟩. Aryl thiols, thioethers, sulfoxides and sulfones are all cleaved under similar conditions ⟨79CC637, 93CC1682⟩. Aryl selenides behave similarly with methyl Grignards and are generally found to be more reactive (ArSeMe $\gg$ ArCl > ArSMe). Within this series, some selectivity can be observed ⟨80TL87, 91G59⟩. The coupling is less efficient with Grignard reagents possessing β hydrogen atoms. Selectivity is also observed with ketene dithioacetals which can undergo stepwise substitution, with the *trans* alkylthio group being replaced more rapidly, to afford trisubstituted alkenes in a stereospecific fashion (Scheme 24) ⟨82CC840⟩.

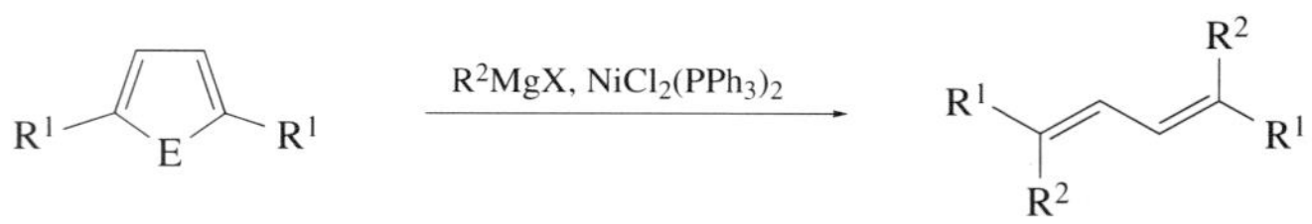

(23)

E = S, Se, Te; R^1 = alkyl *not* aryl

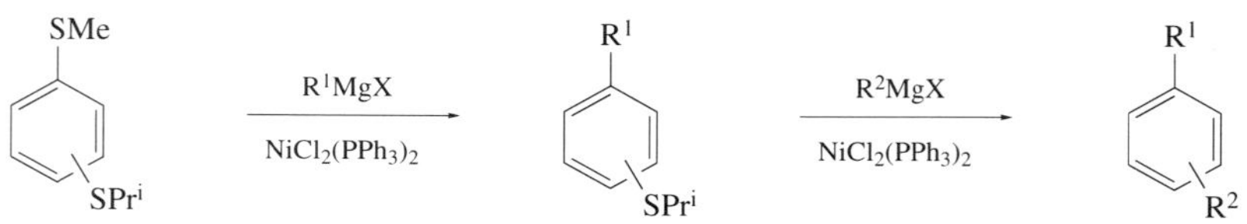

Scheme 23

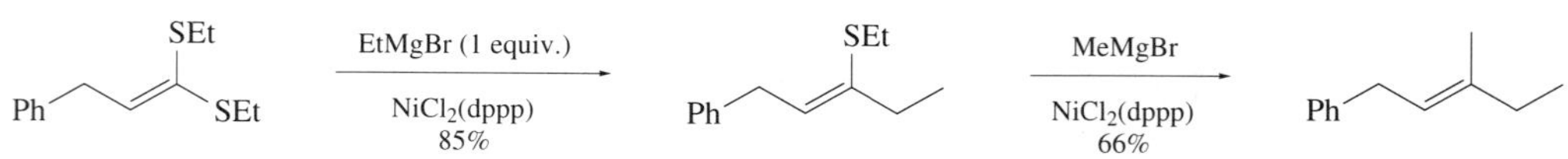

Scheme 24

Diarylselenides are converted to biaryls on treatment with Raney nickel in the absence of solvent ⟨68ACS1883⟩. Diaryl sulfides and sulfones behave similarly on a variety of solid supports ⟨65JOC3897⟩.

Vinyl sulfones are good substrates for transition metal-catalysed displacement by Grignard reagents ⟨85BSF762⟩. The coupling is highly dependent on the Grignard/catalyst combination. Aryl Grignard reagents couple under either nickel or iron catalysis whereas the use of methyl magnesium bromide is only possible with nickel catalysts. In the latter case, the geometrical purity of the product alkene is found to decrease in the order Cl > Br > I ⟨82TL2469⟩. Higher alkyl Grignard reagents, notably secondary and tertiary species which with other catalysts afford significant amounts of competing reduction and isomerisation, couple more efficiently using $Fe(acac)_3$ catalysis ⟨85BSF772, 88T111⟩.

Provided there are not propargylic hydrogen atoms present, alkynyl sulfones are displaced by phenyllithium or phenylmagnesium bromide, in a single electron transfer process, to afford the corresponding phenylalkyne ⟨79JOC3444⟩. Similar displacements are also possible from vinyl and aryl sulfones ⟨84TL4851⟩. Vinyl selenones undergo direct substitution on treatment with dimethylcopperlithium ⟨84JOC1230⟩. (*E*)-alkenes are produced on treatment of vinyl sulfones with trialkylboranes ⟨74BCJ503⟩. Symmetrical diaryl sulfones are cleaved on reaction with 2 equiv. of lithium triethylborohydride to produce one equivalent of the alkylated arene. This transformation is much less effective with arylethyl sulfones ⟨83OM779⟩.

Vinyl sulfoximines are displaced, with complete retention of stereochemical integrity, on treatment with an organozinc reagent in the presence of a nickel catalyst and a magnesium, lithium or zinc salt as a co-catalyst ⟨92TL461⟩. The use of the analogous Grignard or organolithium reagent produces the α-metallated sulfoximine and, although substitution does occur on addition of a second equivalent of the Grignard reagent, the alkene is produced nonselectively. Styryl sulfonylchlorides may be coupled with alkenes in a ruthenium(II) mediated reaction to afford either (*E*,*E*)-1,4-diaryl-1,3-butadienes or the haloalkylated alkene depending on the nature of the alkene substituent (Equation (24)) ⟨85JOC5045, 88BCJ1231⟩.

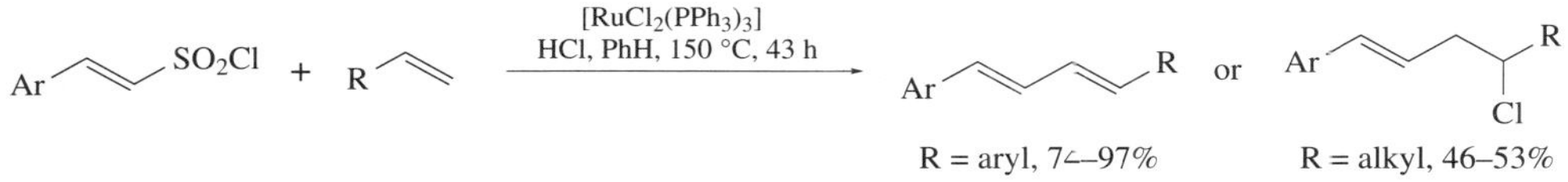

(24)

β-Thioalkyl-α,β-unsubstituted carbonyl compounds undergo ready substitution on treatment with a range of copper-based reagents. With the less reactive esters the reaction occurs with retention of configuration ⟨73CL1097⟩. At low temperatures ketones are also suitable substrates although more reactive cuprates can lead to double conjugate addition ⟨73CC907⟩. With moderately reactive, unhindered (*E*)-vinylogous thioesters (β-alkylthioenones) high selectivity with inversion is found in Et_2O, whilst in THF, retention occurs. With one exception, the equivalent (*Z*)-vinylogous thioesters uniformly afford products with retention of alkene stereochemistry (Equation (25)) ⟨86JOC4687⟩.

These vinylogous thioesters can be accessed by the addition of organometallic reagents to α-oxoketenedithioacetals ⟨86T3029⟩. The two alkylthio groups may be sequentially substituted to afford tri- and tetra-substituted enones with high stereoselectivity. Invariably, the initial substitution results in alkylation *cis* to the ketone (Equation (26)) ⟨85JA4679⟩. The corresponding β-alkylthioenoates do not undergo substitution but are instead reduced to the parent enoate. β-Methyl-α,β-unsaturated esters are obtained following rearrangement on hydrolysis of the initial 1,2 adduct formed by addition of methylmagnesium bromide. More sterically demanding Grignard reagents result in initial 1,4 displacement of an alkylthio group (Scheme 25) ⟨85JCS(P1)1289⟩.

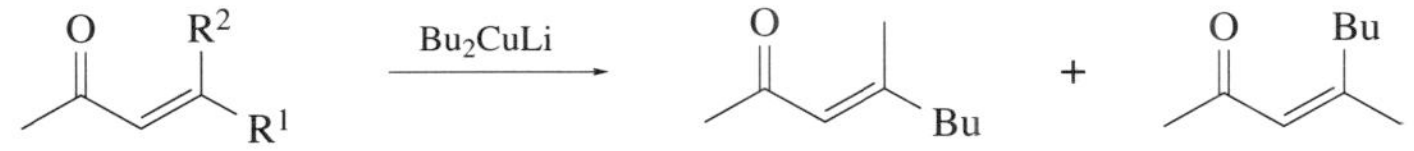

(25)

R^1 = SMe, R^2 = Me	$THF{\bullet}Me_2S$, –78 °C, 93%	97:3
R^1 = SMe, R^2 = Me	Et_2O, –65 °C, 95%	5:95
R^1 = Me, R^2 = SMe	Et_2O, –78 °C, 61%	4:96
R^1 = Me, R^2 = SMe	THF, –78 °C, 64%	8:92

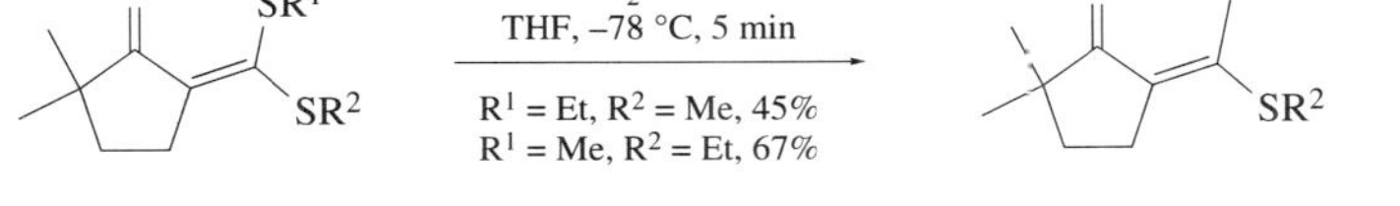

(26)

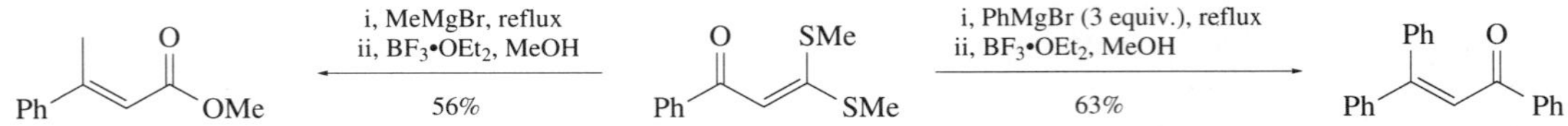

Scheme 25

Aryl sulfones may be similarly displaced from a variety of activated alkenes ⟨89TL173, 90JOC5446⟩. Similarly, ketone enolates may be used to displace a β-sulfonyl or β-sulfinyl group from α,β-unsaturated carbonyl compounds ⟨90AJC1375⟩. The phenylseleno group of β-selenosulfones can be replaced by a range of active methylene compounds and alkylselenocuprates. The use of a selenocuprate is necessary to eliminate problems of double conjugate addition and transfer of the other copper ligands. Complete retention of stereochemistry is observed and consequently trisubstituted sulfones may be formed stereospecifically. The corresponding Cu(SePh)-catalysed Grignard addition may be used with unhindered substrates albeit offering a less predictable stereochemical outcome ⟨87JOC4258⟩.

1.11.2.4 Substitution of Nitrogen

The displacement of amino nitrogen from an sp^3 carbon atom by a vinyl or aryl organometallic reagent normally requires some additional activation such as the release of strain in the ring opening of *N*-tosylaziridines ⟨89JOC5324⟩. As with epoxide opening reactions, cuprates are the most effective reagents, with coupling normally proceeding with enhanced efficiency in the presence of Lewis acids. On quaternisation, allylic amines may be displaced by vinylcuprates ⟨85JA6137⟩ and aryl Grignard reagents in the presence of Li_2CuCl_4 ⟨87JOC2947⟩. The use of butyl and benzylammonium salts affords higher yields, presumably due to enhanced solubility.

Cuprates may be employed for the displacement of allylic nitro groups ⟨84CC274⟩ whilst tertiary and benzylic nitro compounds may be employed in the Friedel–Crafts alkylation of activated arenes ⟨86CC1285⟩. Alkylated arenes are also accessible through the reaction of α-diazoacetates with phenyllithium although the same coupling can be achieved far more efficiently using the corresponding aryl ⟨73JOC2574⟩. In a similar fashion β-γ-unsaturated esters are obtained from the related coupling of α-diazoacetates with vinyl chloroboranes ⟨91SL684⟩.

β-Vinyl nitrogen groups may be displaced from a variety of activated alkenes in an addition–elimination process. Enaminones undergo substitution on reaction with both Grignard ⟨31CB2543, 93TL4667⟩ and organolithium reagents ⟨78JOC4248⟩, although the former frequently required doubly activated enaminones for efficient coupling. Similar substitution reactions have been reported for noncarbonyl activated substrates, for example, nitroenamines ⟨69CB2966⟩. Other nucleophiles, notably ketone and ester enolates, may be employed with some degree of success ⟨87CL1775⟩. Arylnitro groups are readily displaced in a variety of S_NAr reactions ⟨78T2057⟩ whilst aryl diazonium salts can be efficiently replaced in an $S_{RN}1$ process ⟨93T235⟩.

Early work on the copper-catalysed coupling of an aryl diazonium salts with activated alkenes, the Meerwein arylation, has been comprehensively reviewed ⟨76OR225⟩. The reaction has been refined such that the requirement for preparation and isolation of the diazonium salt can be circumvented through *in situ* diazotisation by treatment of the free amine with an alkyl nitrite ⟨77JOC2431⟩. Further simplification can be achieved on replacing the copper salt with a low valent palladium complex. This modification extends the process to include alkenes substituted with both electron-donating and -withdrawing groups. Only acrylonitrile and enamines are not efficiently arylated. Problems are observed with hindered alkenes whilst 1,2 disubstituted alkenes undergo a competitive isomerisation ⟨81T31⟩. *In situ* generation of the diazonium salt is also possible in this palladium-catalysed reaction, although only in strongly acidic conditions ⟨81JOC4885⟩. Since diazonium salts are amongst the most reactive of leaving groups some selectivity can be observed with a wide variety of other substituents (Equation (27)). Similar products may be obtained under neutral conditions in the 'Stille' cross coupling of vinylstannanes with arylazo compounds ⟨83JOC1333⟩. Recent developments have enabled these and the related arene alkylations with trialkylstannanes to be carried out in aqueous media ⟨90IZV2665, 93JCS(P1)1943⟩. Finally, aryl and allyl trialkylammonium iodides may be substituted by a range of Grignard reagents in a nickel phosphine-catalysed reaction. As with many of these coupling reactions alkyl Grignard reagents possessing β-sp^3 hydrogen atoms are not suitable, functioning instead as reducing agents ⟨88CC975⟩.

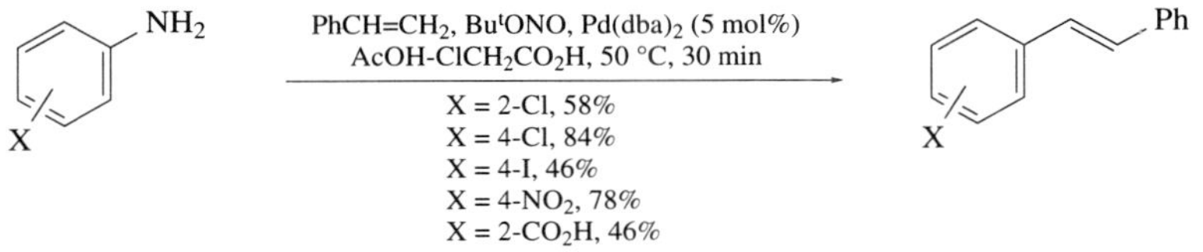

(27)

1.11.2.5 Substitution of Boron

Dialkylalkenylboranes, readily accessible via hydroboration of alkynes, can be converted into a range of substituted alkenes (Scheme 26) ⟨71JA6309⟩. Enantiomerically pure (>99%) alkenes of defined geometry are readily accessible through the use of homochiral hydroborating agents prepared from monoisopinocamphenylorganoboranes ⟨91TA277⟩. Treatment of 1,1-((*Z*)-1-bromoalkenyl)dialkylboranes (**19**) with base, followed by protonolysis of the resulting alkylated alkenylborane affords, the disubstituted (*E*)-alkene (**20**) ⟨67JA5086⟩. The isomeric (*Z*)-alkenes (**22**) are obtained on treatment of (*E*)-alkenyldialkylboranes (**21**) with sodium hydroxide and iodine ⟨67JA3652⟩. Trisubstituted alkenes (**24**) can also be prepared in a similar fashion commencing from disubstituted alkynes ⟨75S376, 88JOC6009⟩. Despite the high stereoselectivity of these conversions, all suffer from the need to generate the starting alkenylborane through hydroboration of an alkene with the dialkylborane. In order to obtain chemoselectivity in this process it is necessary to use large, sterically demanding alkyl groups which obviously restricts the possible alkenes that can be prepared. Furthermore, synthetic efficiency is low as only one of the alkyl groups may be transferred.

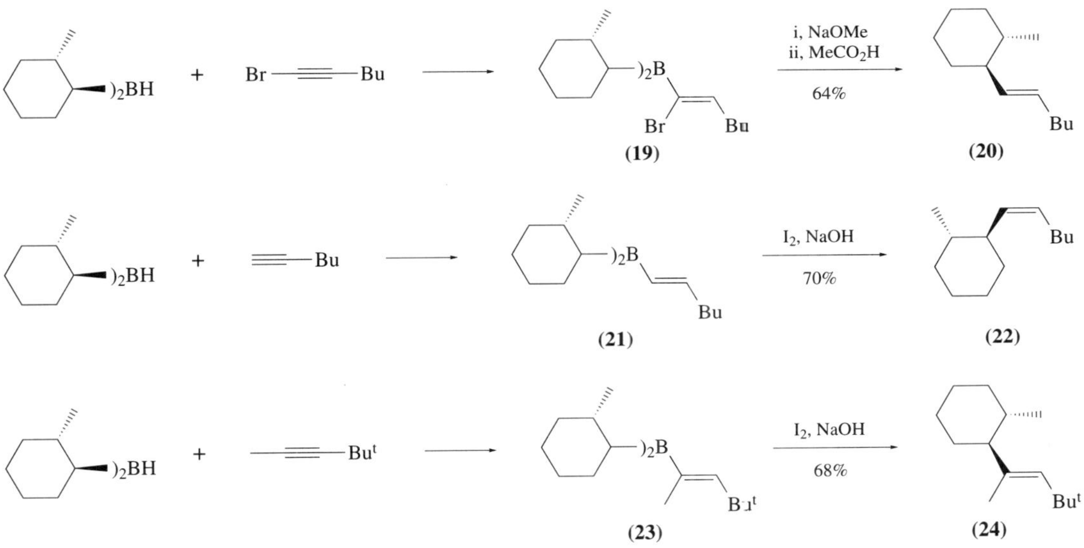

Scheme 26

Most of these difficulties can be circumvented through the use of alkyldihaloboranes followed by controlled reduction ('hydridation') (Scheme 27) ⟨88JOC239⟩. However, the formation of (*Z*)-alkenes via transfer of a primary alkyl group suffers from competing (*E*)-vinyl iodide formation although this can be minimised by carrying out the reaction at low temperatures. Similar side products complicate the synthesis of trisubstituted alkenes through the addition of vinyllithiums to B-alkyl-9-BBN derivatives ⟨80S945⟩. Alternatively, hydrolysis of the alkenyldibromoborane followed by treatment of the resultant boronic acid derivative (**25a**) with a stereodefined vinyllithium yields the 'ate' complex (**26**) which, on reaction with iodine and base, then affords the substituted alkene of inverted geometry (Scheme 28) ⟨76JOC3947⟩. The order of addition to the boron centre may be reversed. This variation, the combination of a vinyldialkoxyborane (**25b**) with an organolithium or

Grignard reagent ⟨88JOC239, 88JOC6009⟩, allows for the incorporation of groups not accessible by hydroboration notably aryl and heteroaryl, whilst the addition of α-halocarbanions can provide an entry to α-alkenylketones and related species ⟨86JOC3398⟩. Both (*Z*) and (*E*) isomers can be formed selectively although the reaction leading to trisubstituted β,γ-unsaturated carbonyl compounds is nonselective. For these targets, high selectivity and moderate overall yields can be obtained through the use of (2-(tributyltin)alkenyl)boranes ⟨93JOC1949⟩.

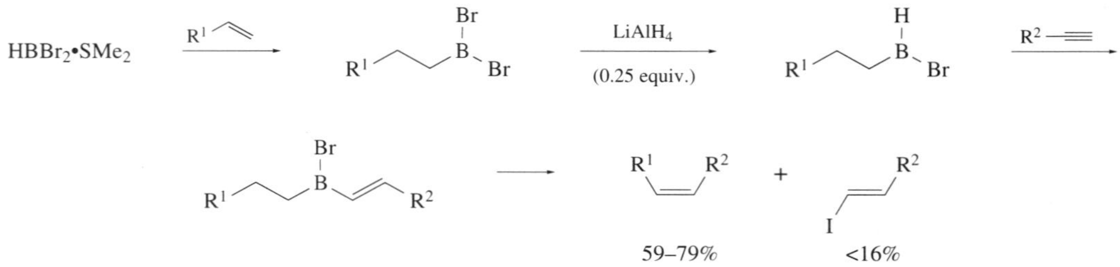

Scheme 27

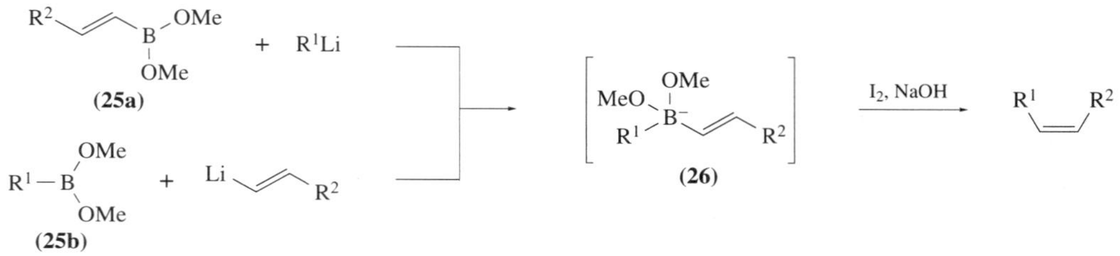

Scheme 28

In a similar fashion selective diene and enyne synthesis is possible (Scheme 29) ⟨86MI 111-01⟩. Symmetrical (*Z,E*)-dienes can be prepared from the corresponding dialkenylborane although good yields are only obtained with boranes containing a nonmigrating group ⟨73JOC1617, 76JOC1785⟩. The use of alkyldialkenylboranes, including those derived from thexylborane, is not satisfactory as competitive alkyl group migration can occur; this problem can be circumvented by prior oxidation with trimethylamine-*N*-oxide, to yield the corresponding dialkenylborinate ⟨68JA6243⟩. Unsymmetrical dienes can also be prepared using similar strategies to those illustrated for simple alkenes. (*Z,E*) Dienes are synthesised through the addition of lithium acetylide to dialkylalkenylboranes followed by treatment with boron trifluoride and *in situ* protonolysis ⟨78JOM(156)159⟩ and (*E,E*) dienes from (1-haloalk-1-enyl)alkenylboranes ⟨91T343⟩. The equivalent symmetrical (*E,E*) and (*Z,Z*) dienes can be efficiently obtained from the appropriate alkenylborane in a copper bromide–methylsulphide catalysed procedure ⟨80JOC549⟩. This is postulated to proceed via the dimerisation of the corresponding alkenylcopper compound, as are the related processes involving treatment of either alkenyldibromoboranes ⟨80JOC389⟩ or dialkenylchloroboranes ⟨77JA5652⟩ with three equivalents of methyl copper.

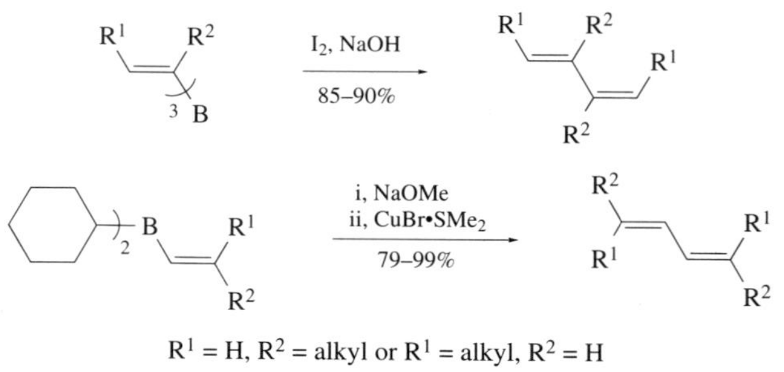

Scheme 29

Many of these methods, particularly those directed towards conjugated (*E,E*) dienes, have been superseded by developments in the palladium-catalysed coupling of organoborane derivatives with a wide variety of haloalkenes, -alkynes and -arenes as well as the corresponding triflates. The reactivity varies in the order I > Br > OTf ≫ Cl ⟨91PAC419, 93JOC2201⟩. Chlorides may be employed, albeit with some difficulty, their successful replacement requiring the use of bidentate phosphine ligands ⟨92T8117⟩. The reaction proceeds with complete retention of alkene stereochemistry in both components. Both boranes and boronates may be employed with the latter providing greater efficiency ⟨93ACS221⟩, particularly when a (*Z*)-alkenyl boron group is to be coupled ⟨86TL3745⟩.

The reactions require the presence of a base (alkoxide, acetate, hydroxide) to facilitate the transmetallation between the organopalladium intermediate and organoborane. However, as noticed previously (see Section 1.11.2.1.3(ii)) the nature of the base can alter the regiochemistry of the addition ⟨81JOM(213)C53⟩. Although both Heck- and Suzuki-type couplings can occur with the same substrate, reasonable selectivity for either pathway can be achieved although the precise outcome can depend upon the reaction conditions (Equation (28)) ⟨93TL3599⟩.

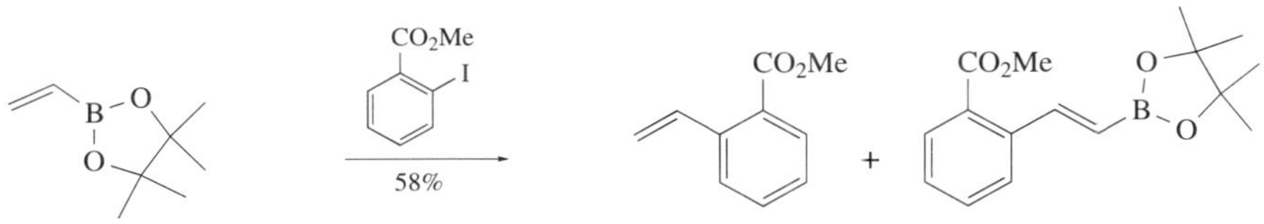

(28)

Pd(OAc)$_2$ (5 mol%), phenanthroline, Bu^{t}OK, MeCN, 45 °C, 66 h — 100% — 0%
Pd(OAc)$_2$, Ph$_3$P, Bu$_3$N, PhMe, 120 °C, 18 h — 13% — 87%

The organoboron component may be alkenyl ⟨90OS(68)130⟩, aryl ⟨81SC513, 92SL207⟩ or alkyl ⟨89JA314⟩. In alkylation reactions the nature of the phosphine ligand can be crucial since complications due to phosphine quaternisation or competitive phenyl transfer from the phosphine can occur. These problems can be minimised through the use of substituted phosphine ligands, for example, tris(2-methoxyphenyl)phosphine. Further improvements can be achieved through the addition of excess phosphine although this results in a slower reaction ⟨92TL6679⟩. Both alkyl borane and boronic acid derivatives may be employed. The latter couple more efficiently as the free acid ⟨90TL6509⟩ whilst further enhancements of both the rate and the yield are obtained through the use of thallium hydroxide as the base (rel. rate KOH = 1, TlOEt = 5, Ag_2O = 30, TlOH = 1000) ⟨87JA4756⟩. Methyl, primary and secondary alkyl groups may be employed. Although secondary alkyl groups only couple slowly, *β*-hydrogen elimination is not normally a problem.

Aryl boronates also offer advantages over the corresponding boranes in that they are more accessible, air stable reagents (being prepared by transmetallation from the corresponding aryllithium reagent), that only require weak bases to facilitate the cross-coupling reaction. The use of tetraaryl borates ($NaAr_4B$) with vinyl and aryl triflates can eliminate the need for a base altogether, although an equimolar quantity of the borate is required ⟨92TL4815⟩.

Polysubstituted alkenylboranes may be employed with halide groups undergoing preferential substitution, and as a result, selective synthesis of trisubstituted alkenes ⟨86CL1329, 93CL1429⟩ and unsymmetrical polyenes is possible ⟨88CL809⟩. Multiple functionalisation is also tolerated in the 'acceptor' component and this can provide an alternative entry to *α*-vinyl- and arylketones, etc. ⟨92BCJ2863⟩. Finally, alkenylation and alkylation may be achieved in both an inter- and intramolecular fashion providing a convenient synthesis of cycloalkenones (Equation (29)).

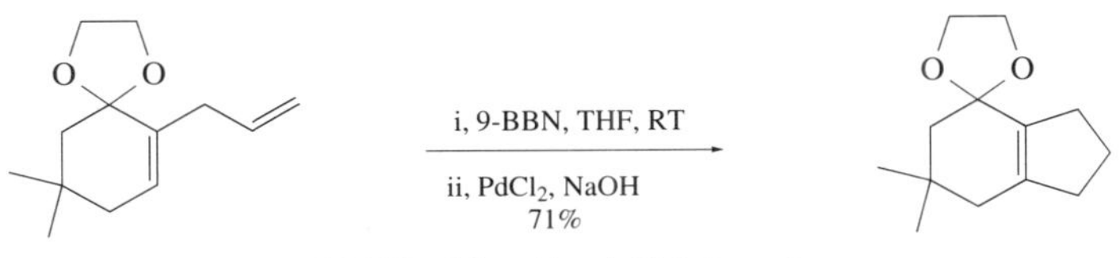

(29)

9-BBN = 9-borabicyclo[3.3.1]nonyl

1.11.2.6 Substitution of Silicon and Germanium

Although aryl and vinyl silanes undergo a vast number of electrophilic *ipso* substitution reactions, relatively few lead to simple alkene/arene coupling/alkylation. Biaryls may be prepared from aryl silanes whilst vinyl silanes may be dimerised, although geometrical mixtures of isomers result. Stereospecific alkene synthesis is possible from vinyl silanes in a indirect fashion via epoxidation, cuprate opening and Peterson elimination ⟨89JOC868⟩. Arenes may be efficiently allylated on reaction with either allyl silanes or allyl germanes in the presence of thallium trifluoroacetate ⟨81TL4491⟩. The reaction probably proceeds via the corresponding organothallium salt as does the analogous thallium(III) promoted arylation with aryl silanes ⟨74TL3391⟩. Although the C—Si (and C—Ge) bond is less polar than the corresponding C—B bond and consequently activation by transition metal catalysts more difficult, vinyl and styryl silanes may be employed in the Heck reaction ⟨71TL4701⟩. Vinyl silanes combine with aryl halides under palladium catalysis to afford styrene derivatives. These arise through re-addition of the initially formed palladium hydride and subsequent antielimination of the trimethylsilyl and palladium moieties. Addition of silver salts suppress the re-addition and styryl silanes are isolated ⟨86JOC5286⟩. Similar results have been observed in the coupling of aryl diazonium salts with vinyl silanes under palladium catalysis. Although α-substituted alkenyl silanes react with high selectivity ⟨88CL873⟩, substitution of the β-isomers proceeds with loss of regio- and stereochemical integrity (Scheme 30) ⟨86JCS(P1)1959⟩. Similar observations have been recorded for the analogous germyl derivatives ⟨90CL185⟩.

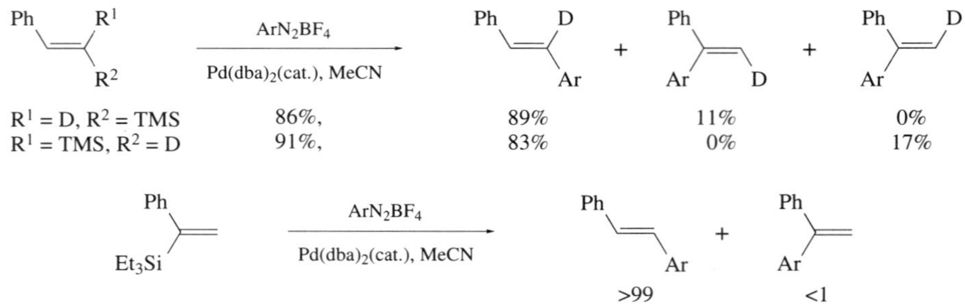

Scheme 30

Efficient substitution of silicon frequently requires the formation of a pentacoordinate silicon species. Commonly generated by addition of fluoride, these organofluorosilicates are effective alkenyl and aryl transfer reagents affording, in the presence of palladium salts, routes to 1,3- and 1,4-dienes as well as substituted arenes ⟨82OM542, 91SL845, 93CRV1371⟩. In the presence of various late transition metal salts, homocoupling to produce symmetrical (*E*,*E*) dienes is possible ⟨70JPR150, 79TL1137, 79TL1141⟩.

Vinyl, alkynyl and allyl silanes couple with aryl, alkenyl and allyl iodides and bromides in the presence of (trisdiethylamino)sulfoniumdifluoro trimethylsilicate (TASF) and allylpalladium chloride dimer. The coupling is stereoselective, clean, high yielding and tolerant of a wider range of functionality (including unprotected alcohol and carbonyl groups) than most related transition metal coupling processes ⟨88JOC918⟩. However, under these conditions the reactions of aryl halides is limited to the iodides and even these require the presence of HMPA as a co-solvent. Likewise, the addition of methylphosphite is beneficial for couplings with alkenyl halides. Similarly the nature of the silyl groups can significantly affect the efficiency of the coupling ⟨89JOC268⟩.

TASF can also act as an arene methylating reagent ⟨88TL97⟩ and consequently alternative sources of fluoride, such as tetra-*n*-butylammonium fluoride (tbaf), can prove advantageous with aryl halides. Similarly, competitive transfer of the methyl group from aryl methyl difluorosilanes can occur. In this case the use of the ethyl analogue suppresses this pathway and allows for the efficient synthesis of biaryls in the presence of either tbaf or potassium fluoride ⟨89CL1711⟩. Alkoxydialkylsilylalkenes behave analogously and have advantages over halosilanes in that they are stable and can be easily regiospecifically prepared through intramolecular hydrosilylation (Scheme 31) ⟨89TL6051⟩.

Finally, vinyl triflates couple efficiently with alkenyl, aryl, alkynyl and alkyl organofluorosilicates. Although migration of an alkyl group is slow, those containing β-hydrogen atoms prove to be acceptable substrates ⟨90TL2719⟩.

1.11.2.7 Substitution of a Metal

Direct substitution of a metal by a carbon nucleophile is relatively rare. Soft nucleophiles (β-dicarbonyl compounds, etc.) displace lead from aryl- ⟨80TL965⟩ and vinylleadtriacetates ⟨88JCS(P1)2847⟩. Otherwise, formation of carbon–carbon bonds through substitution of a metal occurs principally in the dimerisation of organometallic units. In order not to repeat material

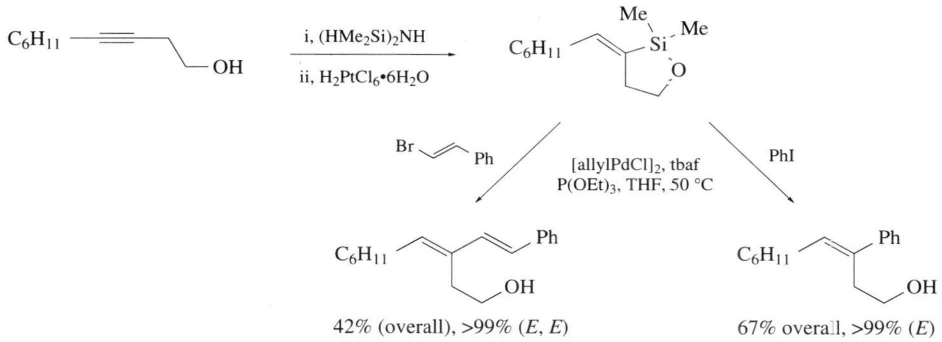

tbaf = tetra-*n*-butylammonium fluoride

Scheme 31

covered elsewhere, this section will not refer to those methods involving transiently generated organometallic species.

The simplest method to obtain symmetrical dienes through this connection is the thermal dimerisation of a vinylcopper or silver salt. The reaction proceeds in good to excellent yields with almost complete retention of monomer stereochemistry ⟨71JA1379⟩. The same conversion may be achieved on addition of a variety of transition metal salts, or, more efficiently, by bubbling oxygen through a cooled solution of the vinylcopper or cuprate ⟨74JOM(77)269, 76JA3732⟩. Other oxidising species may also be employed. In this respect various nitrogen-containing oxidising agents, for example, NO_2, $MeNO_3$ and $LiNO_3$, are particularly effective for the silver-promoted coupling of vinyl and aryl Grignard reagents ⟨72BCJ1120⟩. The latter can frequently be coupled more efficiently on combination of the aryl Grignard reagent with a thallium salt ⟨76OS(55)48⟩. A wide range of aryl and vinyl Grignard and lithium reagents may be dimerised on treatment with a variety of transition metal complexes including Li_2CuCl_4 ⟨86T1399⟩, $Pd(PPh_3)_2Cl_2$ ⟨87JOM(334)181⟩ and Cr, Co^{II} and Mn^{II} salts ⟨76JOM(113)99⟩.

Other organometallic species also couple on addition (transmetallation) with copper salts, notably alkenyl alanes ⟨70JA6678⟩, zirconanes ⟨77TL1303⟩ and stannanes ⟨87JOC4296⟩. Alkenyl- and arylstannanes are also dimerised on treatment with palladium chloride in HMPA ⟨89S633⟩. A similar dimerisation is obtained on reaction with palladium acetate and *tert*-butylhydroperoxide and with this catalyst combination selective cross-coupling with allylstannanes occurs (Equation (30)) ⟨87CL5⟩.

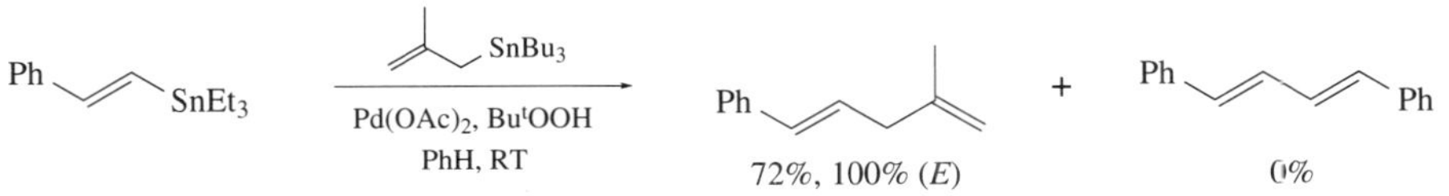 (30)

Provided that the toxicity is not considered a problem, symmetrical biaryls can be efficiently prepared from the corresponding arylthallium bis(trifluoroacetate) on treatment with a catalytic amount of Li_2PdCl_4. Given that aryl thallation is relatively simple to control, a variety of differently substituted products can be easily accessed by this method ⟨83JOC1924⟩.

Vinylmercury compounds provide an attractive entry to symmetrical dienes ⟨86MI 111-02⟩. Vinyl and aryl mercuric halides dimerise in high yield on treatment with $LiPd_2Cl_4$ ⟨76JOC2241⟩. A wide range of functionality may be present whilst steric hindrance presents no problems. However, the reaction does require stoichiometric palladium chloride and the use of HMPA as a solvent. These problems may, in part, be circumvented through the use of either divinylmercury reagents which dimerise on treatment with catalytic quantities of palladium(0) phosphine complexes ⟨74TL3207⟩. However such divinyl mercurials are considerably more difficult to handle than the corresponding halide salts and more importantly the reactions proceed with a small but significant ($<15\%$) loss of stereochemical integrity. Similar loss of stereochemical information occurs on addition of stoichiometric cuprous chloride to recycle the palladium catalyst. Interestingly, in this case inversion of alkene geometry is the favoured pathway (Equation (31)) ⟨74CB1654⟩. The use of the rhodium carbonyl complex $[Rh_2(CO)_4Cl_2]$.LiCl avoids these difficulties and cleanly and efficiently provides the head-to-head coupled dimer ⟨77JOC1680⟩. Replacing LiCl by the more soluble tetraalkylammonium halides can provide for a faster reaction ⟨84IZV2580⟩. The head-to-tail diene is obtained selectively using palladium catalysis and switching to a non-polar solvent, for example, benzene, in the absence of lithium chloride (Equation (32)). The addition of a base, for example, triethylamine, to neutralise the HCl generated affords enhanced yields, whilst the process can be rendered catalytic in palladium on addition of 2 equiv. of cuprous chloride ⟨78JOC1468⟩.

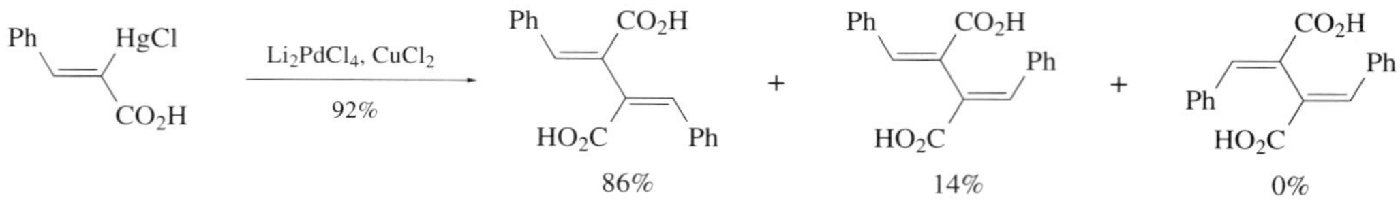

(31)

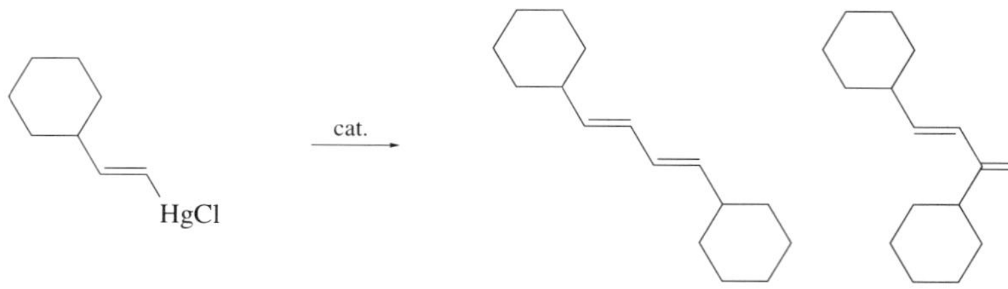

(32)

cat.		
$PdCl_2$ (10 mol%), 2 CuCl, 2 Et_3N, PhH, RT	0%	98%
$[ClRh(CO)_2]_2$ (1 mol%), 2 LiCl, HMPA, RT	90%	0%

Finally, the dimerisation of aryl mercurials can provide a particularly useful, and complementary, modification to the classical Ullmann reaction. Aryl mercurials containing an amino group, which completely inhibits the normal copper-catalysed reaction, are converted to biaryls on treatment with the palladium dichloride/copper(I) chloride catalyst system ⟨76JOC2661⟩.

1.11.2.8 Substitution of Hydrogen

1.11.2.8.1 Friedels–Craft alkylations using alkenes and alkanes

The intermediate carbocation in the Friedel–Craft alkylation of arenes can also be generated from the corresponding alkene and this strategy is of considerable industrial importance, notably in the manufacture of ethyl- and isopropylbenzene. Through careful choice of catalyst and reaction conditions it is possible to observe some degree of control over product formation. For example *ortho*-alkylation of phenol can be realised on treatment with 1-hexene in the presence of aluminum phenolate (Equation (33)) ⟨86UKZ1090⟩. Similarly, provided the catalyst and/or the conditions are mild, then alkylation without rearrangement may be observed ⟨72ZOR1484⟩. Control in this class of Friedel–Crafts alkylation may also be achieved through the use of functionalised alkenes ⟨91JA1364⟩. For example, the use of protonated nitroalkenes as the electrophilic component provides an efficient entry to α-arylated ketones which can be difficult to prepare through classical Friedel–Crafts alkylation (Equation (34)) ⟨89JOC733⟩.

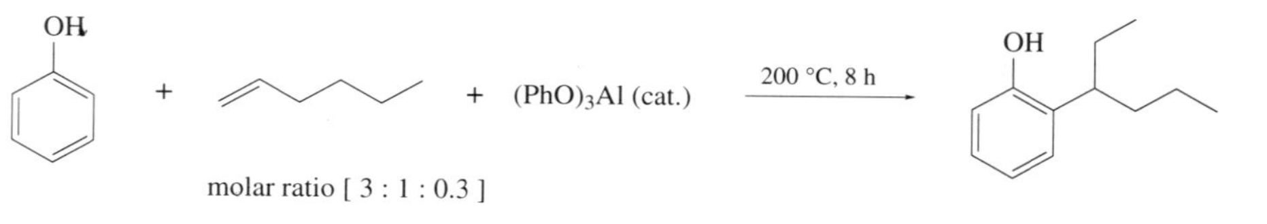

(33)

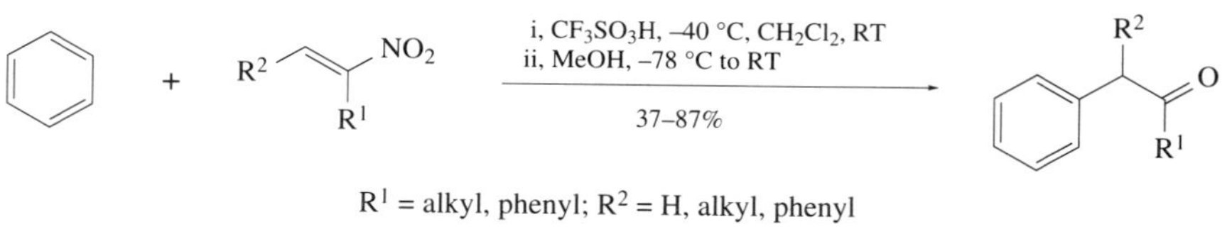

(34)

R^1 = alkyl, phenyl; R^2 = H, alkyl, phenyl

Alkylation of arenes with alkanes can also be achieved. However, this process is accompanied by competing isomerisation and disproportionation reactions and is consequently not of great synthetic viability.

1.11.2.8.2 Organopalladium-mediated substitution of alkenes

Organopalladium derivatives lacking β hydrogen atoms combine with unhindered alkenes in a process that results in effective substitution of vinylhydrogen. This, the Heck reaction, proceeds via a *syn*-addition–*syn*-elimination pathway and, where possible, leads preferentially to the (*E*) alkene isomer ⟨91COS(4)833⟩. The initial organopalladium species is normally generated *in situ* by either transmetallation of a palladium(II) salt with an organometallic reagent, direct palladation (of arenes), or by oxidative addition of an alkyl, alkenyl or aryl halide, triflate, diazonium salt, etc. to a low-valent palladium complex ⟨85MI 111-02⟩. The reaction may be rendered both catalytic in palladium and enantioselective (see below). Ruthenium complexes may also be employed and in some cases show considerably greater activity than the corresponding palladium species ⟨92JOM(423)405⟩.

Electrophilic aromatic substitution with palladium acetate occurs readily although, as with many aromatic substitution reactions, mixtures of isomeric products arise. However, some selectivity can be observed in cases in which the arene contains a good ligating group (Scheme 32) ⟨81T173⟩. A second drawback to this approach is the requirement for a stoichiometric quantity of the initial palladium salt although a number of catalytic modifications have been reported ⟨84T2699⟩.

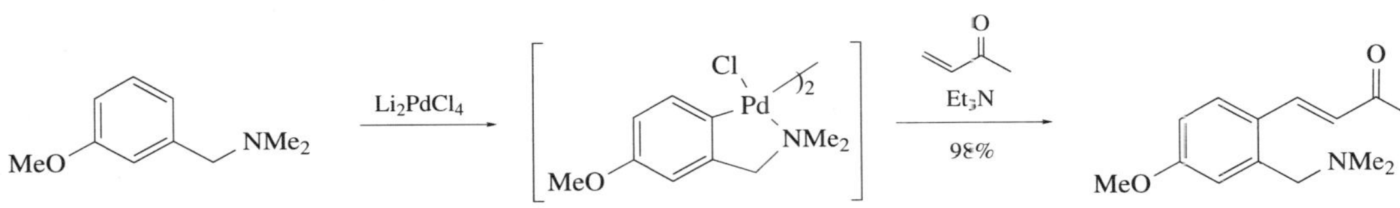

Scheme 32

Transmetallation of hard organometallic species is complicated by competing redox processes. Although good synthetic yields can be realised with the appropriate choice of palladium ligands, particularly with stabilised anions, the fact that the process cannot be rendered catalytic limits its use ⟨84T2415⟩. As a result, generation of organopalladium species by a transmetallation process is only viable using soft organometallic species such as organoboranes ⟨75JOC1083⟩, organomercurials, organosilanes ⟨74JOM(72)163⟩ and organothallium reagents ⟨75JOM(99)C8, 80BCJ553⟩.

Transmetallation of alkyl, aryl and alkenyl mercurials (RHgX and R_2Hg), is fast and selective although the use of alkyl groups is restricted to those lacking a β hydrogen atom. The preparation of the corresponding arylmercurials is relatively facile and tolerant of a wide range of functionality. The coupling reactions are normally highly regioselective, favouring substitution at the least hindered terminus of the alkene and at the β-position of Michael acceptors ⟨68JA5518⟩. (*E*/*Z*) alkene isomerisation can be problematical although in some cases use of the corresponding thallium salt can be advantageous ⟨84JA5274⟩. Diene synthesis is possible although the yields are frequently low due to readdition of the palladium hydride species to afford π-allyl palladium complexes. The addition of tertiary amine bases alleviates this problem and synthetically useful yields of 1,3-and 1,4-dienes may be realised ⟨81TL5231⟩. Although such an approach can be rendered catalytic in palladium through the addition of an oxidant, for example, $CuCl_2$ ⟨69JA6707⟩, it is normally more efficient to generate the organopalladium species via oxidative addition of RX to a palladium(II) salt. These are all catalytic with the only requirements being the presence of a base, to neutralise the acid formed in the reaction, and in most cases a triarylphosphine to solubilise the various catalytic intermediates. In this respect temperature control is critical to achieve a balance between reaction rate and palladium solubility (the latter decreases at higher temperatures). A variety of electrophiles may be employed with the reactivity decreasing down the series: $—N_2^+ > —OTf > —I > —Br > —Cl > —SO_2Cl > —COCl$.

Whilst intermolecular alkene alkylation is limited to benzylic halides ⟨74JA1133⟩, intramolecular processes are known for a range of substrates including those containing β-sp^3 hydrogen atoms.

However, it should be noted that many of these processes afford halide products ⟨85TL1519⟩, presumably via a competing, radical-mediated atom transfer pathway. Multiple C—C bond-forming processes are possible in situations in which the intermediate alkylpalladium hydride can be intercepted by a second alkene prior to elimination (Scheme 33) ⟨93JA9421, 94T359⟩.

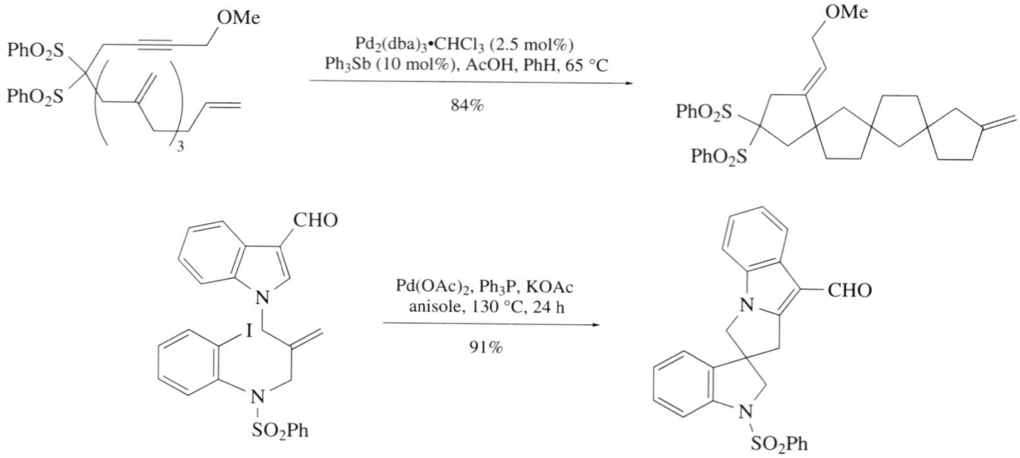

Scheme 33

Alkene arylation is most effectively achieved with aryl iodides and in many cases these couplings occur at ambient temperatures in the absence of any phosphine ⟨79JOC4078⟩. In contrast, aryl bromides require excess phosphine ligands to maintain catalyst solubility and to minimise the effect of competing phosphine quaternisation. This, and problems associated with competing phenyl transfer from the phosphine, can be alleviated by the use of more bulky substituted phosphines, for example, tri-*ortho* tolylphosphine ⟨78JOC2941⟩. However, even relatively unreactive aryl chlorides may be utilised in the presence of a NaI–$NiBr_2$ mixture which probably results in the *in situ* generation of the corresponding iodide ⟨88JA2655⟩. A number of other aryl groups may similarly be employed including stannanes ⟨87JOC5064⟩, acid chlorides ⟨82JOM(233)267, 84JOM(265)323⟩, sulfonyl chlorides ⟨90JCS(P1)2207⟩, sulfinates ⟨70JOC3273⟩ and sulfonates ⟨86TL1171⟩ and diazonium salts (see above). A wide range of other substituents is tolerated including, when diazonium salts are employed, the *ortho*-carboxy group. In general, electron-withdrawing groups accelerate and electron-donating groups retard the reaction. Consequently, coupling is frequently complicated by competitive reduction for these less reactive electron-rich arenes. Dienes behave similarly although synthetically useful yields are only obtained with dienes additionally conjugated to an electron-withdrawing group ⟨78JOC5018⟩. As with simple alkenes, preferential attack occurs at the least substituted diene terminus. Butadiene combines with reactive arenes to afford products of double arylation. With unsymmetrical alkenes some regioselectivity can be obtained by varying the nature of the phosphine ligands. Using aryl triflates, bidentate phosphines favour α-arylation whilst β-aryl alkenes are produced with monodentate ligands. However, bidentate palladium phosphine complexes are only suitable for the coupling of aryl halides in the presence of a halide-sequestering agent ⟨92JOC3558⟩.

The Heck reaction between vinyl iodides and activated alkenes provides a very efficient route to conjugated dienes ⟨78JOC3898⟩. Similar results are obtained through the use of vinyl triflates ⟨85JOC2302⟩. The coupling reaction is sensitive to steric restrictions, in either component. As with alkene arylation, acceptable yields of dienes are only obtained when the acceptor alkene is activated by an electron-withdrawing group. In these cases predominant formation of the (*E*,*E*) isomer results due to isomerisation through reversible addition–elimination of the palladium–hydride species generated in the reaction. In situations in which there are different hydrogen atoms available for elimination this isomerisation can result in the formation of alkene regioisomers. This complication can be suppressed through the use of milder reaction conditions accessible through a combination of DMF as the solvent and the presence of an alkali metal acetate as a co-promoter ⟨88TL905⟩. Various other additives including silver ⟨87JOC4130⟩ and thallium salts ⟨91TL687⟩ have also been utilised successfully. Recent developments using phase transfer catalysis ⟨84CC1287⟩, high pressures ⟨94SL189⟩ or aqueous conditions ⟨92SL715⟩ have enabled the reaction to be carried out at much lower temperatures.

Finally through the use of chiral phosphines the Heck reaction may be rendered enantioselective. The highest enantioselectivities are obtained with vinyl triflates ⟨92TL2589⟩ although similarly good results may also be obtained from vinyl iodides, provided a silver salt is present (Equation (35)) ⟨94T371⟩.

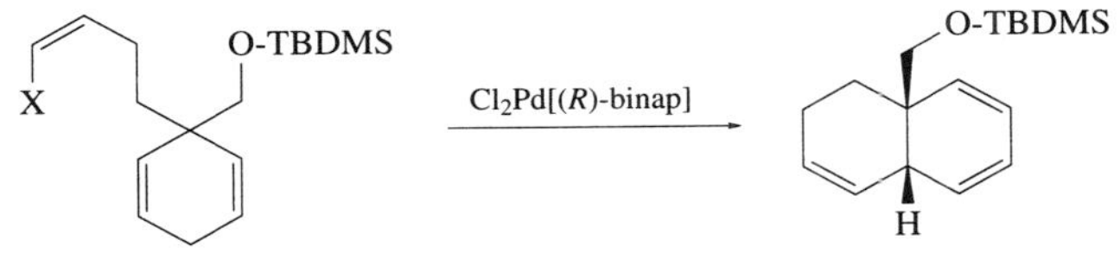

(35)

X = I, NMP, Ag_3PO_4, 60 °C, 84 h, 67% 80% *ee*
X = OTf, PhMe, K_2CO_3, 60 °C, 55 h, 35% 92% *ee*

1.11.2.8.3 Miscellaneous methods

Through cobalt-mediated radical coupling procedures it is possible to effectively substitute a vinylic hydrogen atom ⟨91TL6273⟩. This is effectively an alkyl equivalent of the Heck reaction, a process that is normally severely restricted due to β-H elimination. This 'substitution' process may be combined with a series of radical cyclisations enabling the incorporation of structurally complex alkyl groups (Equation (36)) ⟨90JCS(P1)2709⟩. The reverse connection, effective substitution of alkyl hydrogen by an alkene unit, can be achieved through the use of *in situ*-generated alkylidene carbenes (Equation (37)) ⟨83JOC5251⟩. Similar ketocarbene insertion reactions can provide access to alkyl substituted arenes and related cyclic polyenes ⟨91COS(4)1031⟩. Finally, in the presence of an oxidising agent, direct substitution of electron deficient arenes is possible with a variety of organolithium and Grignard reagents ⟨78HCA449⟩.

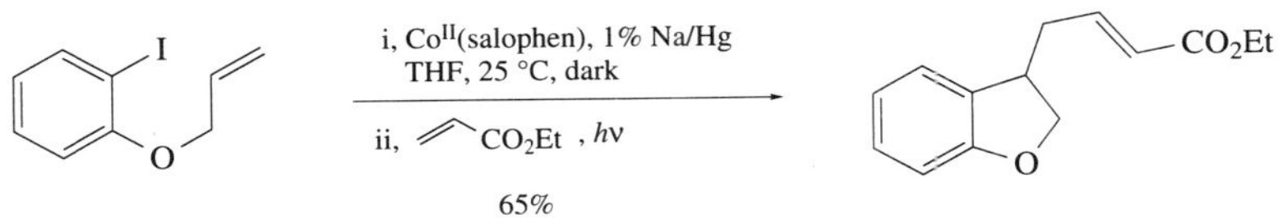

(36)

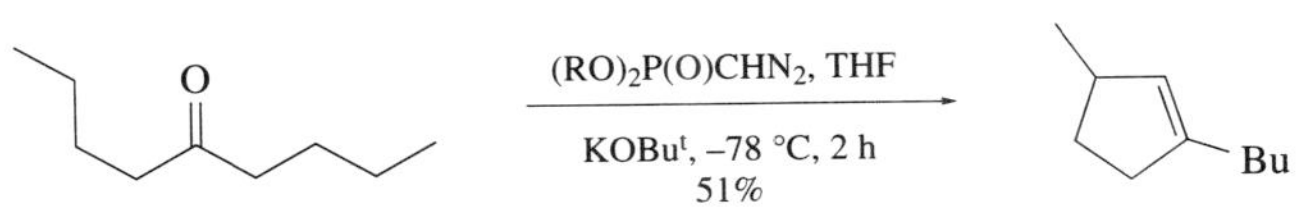

(37)

1.11.3 BY ADDITION

1.11.3.1 Nucleophilic Addition to Allenes

Direct addition of organometallic units (carbometallation) to unactivated allenes is possible, although limited to the more reactive species such as organolithium and Grignard reagents ⟨72JOM(34)1⟩. Intermolecular reactions are possible and tend to occur at the central carbon atom ⟨71TL3785⟩. However, low reactivity is a common problem, for example, organocuprates do not add ⟨89T413⟩, and in general the addition is more facile in an intramolecular fashion ⟨69TL2313⟩. In these cases the length of the tether controls the regiochemistry of the resultant cycloalkene (Equation (38)) ⟨92TL5311⟩. In many cases these reactions probably proceed by a radical pathway since the product ratios are similar to those obtained under free radical conditions ⟨78JCR(S)416, 82BSF269⟩. Softer nucleophiles, such as active methylene compounds, may be employed in the presence of a palladium catalyst to afford moderate yields of unsymmetrical 2,3-dialkyl-1,3-butadienes (Equation (39)) ⟨73JOC1483⟩.

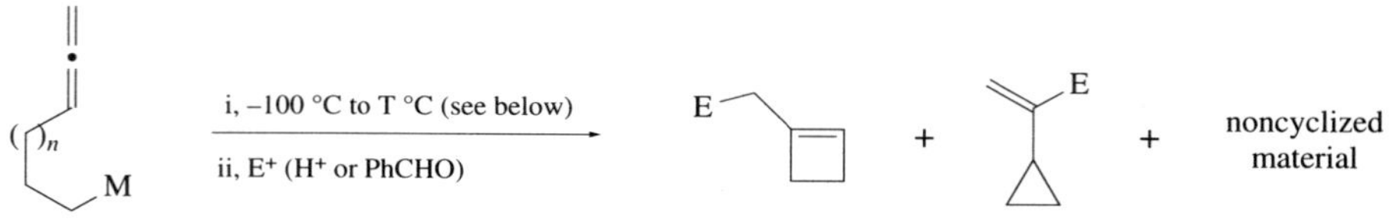

(38)

$n = 1$	M = MgBr	RT		0%	100%	0%
$n = 1$	M = Li	–78 °C	100%	100*%	0%	0%
$n = 2$	M = Li	–78 °C	86%	3%	50%	47%
$n = 2$	M = Cu(CN)Li	0 °C	86%	0%	80%	20%

(* 7:3 mixture of exo / endocyclic alkene isomers)

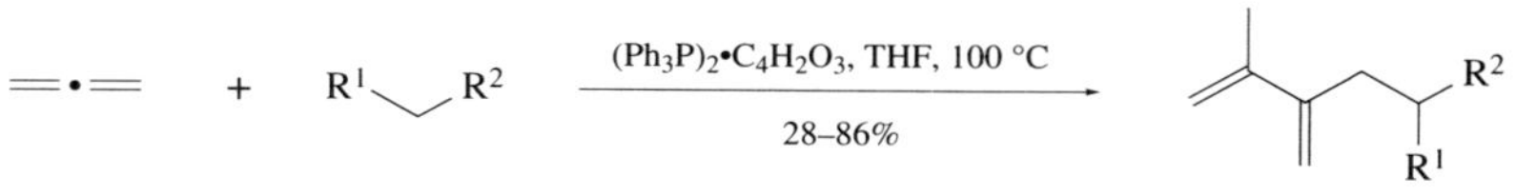

(39)

R^1, R^2 = MeCO-, CO_2Et, CN

Simpler, more efficient addition is obtained with allenes containing an activating group, either alkoxy- or electron-withdrawing, and with these substrates the use of copper-based reagents is particularly effective. 1-Alkoxyallenes combine with both lithiumorganocopper and cuprates to produce the (*Z*) and (*E*) alkenic products, respectively (Scheme 34). Exclusive addition occurs at C-3 with high stereoselectivity being observed in both cases ⟨84IJ113, 92BSF171⟩. The corresponding magnesio species are not selective ⟨76TL947⟩. In contrast, allylmetal reagents (Zn, Cu) preferentially attack at C-1. Allenes bearing large substituents undergo elimination of zinc alkoxide leading to effective substitution of the alkoxy group. Similar products are obtained if the alkoxyallene is metallated prior to addition of the allylzinc reagent. A variety of carbonyl groups may be employed, in which case the reaction is essentially a Michael addition. The use of vinyl nucleophiles provides an easy entry to 2-substituted butadienes ⟨92TL7029, 92TL3563⟩. These products can also be obtained through S_N2' displacement of an allylic leaving group ⟨91CB2075⟩.

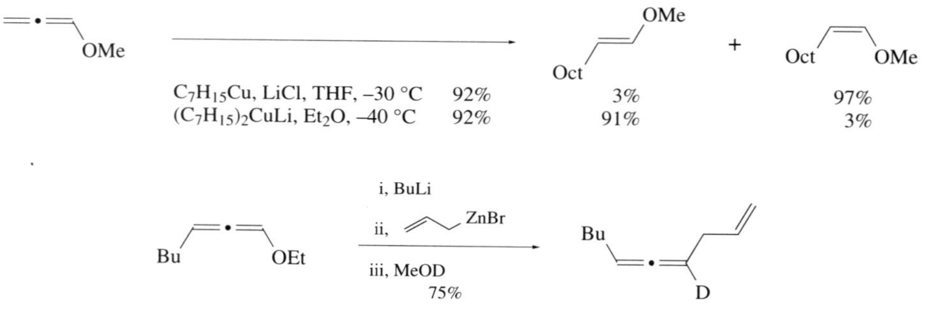

Scheme 34

Enynes are obtained through the palladium-catalysed coupling of terminal alkynes with allenes. The presence of a strong electron-withdrawing group on the allene is essential for high yields and regioselectivity, although these are still very catalyst/substrate dependent ⟨90JA2816⟩. On treatment with a polymer supported bimetallic nickel–chromium catalyst, α,ω-alkenylallenes undergo cyclisation to form cyclic 1,4-pentadiene compounds (Equation (40)) ⟨88JA5231⟩. A similar coupling can be achieved in an intermolecular fashion on treatment with Wilkinson's catalyst, although with somewhat variable allene regioselectivity ⟨86JOM(317)373⟩. Allenes may also participate in Heck-type reactions with haloalkenes and arenes. The intermediate palladium π-allyl species can be trapped by an appropriate nucleophile to produce functionalised butadienes of well defined stereochemistry. The (*E*) isomer is generally favoured with substituted alkenyl halides and through the use of hindered phosphines, polar aprotic solvents and good nucleophiles. With vinyl bromide lower selectivity is observed, although the use of bulky less nucleophilic anions provides for preferential formation of corresponding (*Z*) isomer (Scheme 35) ⟨90PAC1867, 92BSF273⟩.

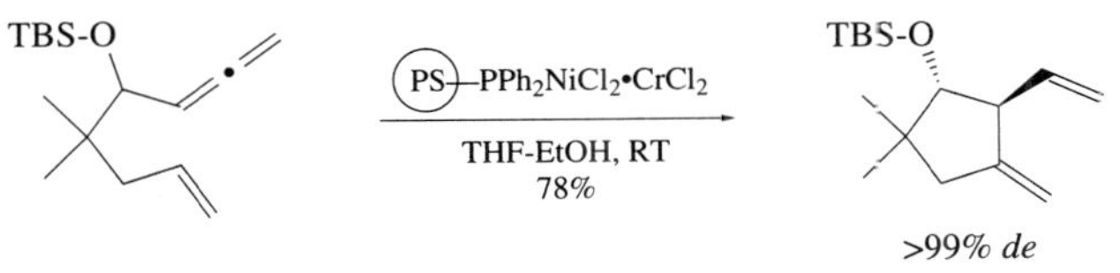

(40)

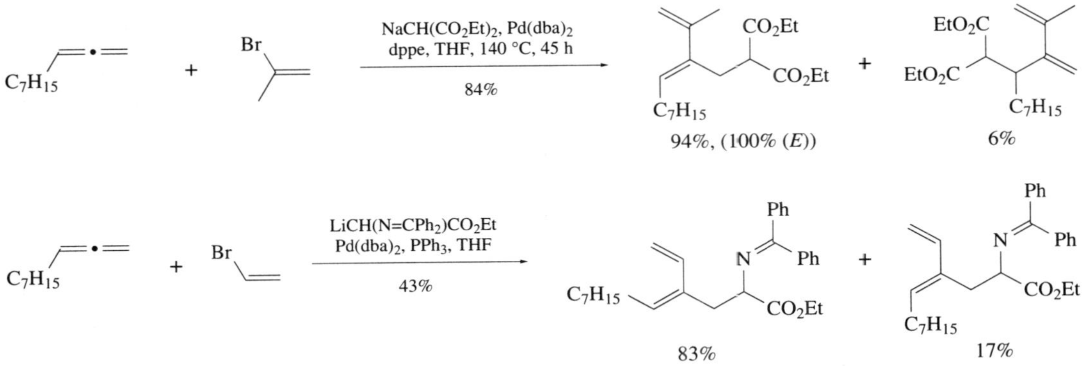

Scheme 35

1.11.3.2 Free Radical Additions to Allenes

The addition of simple methyl radicals ($CH_3^{\bullet}$, $CX_3^{\bullet}$) to allenes occurs preferentially at a terminal carbon atom ⟨72JOC3881⟩. However, this appears to be an anomalous result and most other radical species attack at the central carbon atom to produce an allylic radical intermediate. Incorporation of an acceptor group onto the allene greatly facilitates the addition and can enhance the selectivity ⟨91TL3659⟩. As with alkene substrates, intramolecular radical cyclisations involving allenes generally follow an *exo*-mode of ring closure ⟨92SL721, 93TL607⟩. However, reduction of the allyl radical is frequently nonselective producing mixtures of the *endo*- and *exo*-cyclic alkene ⟨84JOC2125, 88T7037⟩, although in certain cases the nature of the tether can provide some degree of control ⟨85T4001⟩. Finally, allene–enynes may be employed in Bergmann-cycloaromatisation sequences which proceeds via radical intermediates (Scheme 36) ⟨93JOC1651⟩.

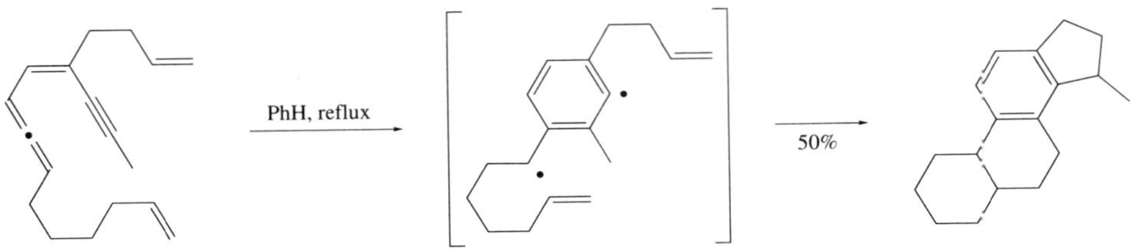

Scheme 36

1.11.3.3 Cycloaddition Reactions Involving Allenes

Allenes undergo a range of cycloaddition processes including [2 + 2], [4 + 2] and [1,3] dipolar cycloadditions. The first group can occur with alkenes, alkynes and other allenes and can be carried out under both thermal and photochemical initiation.

Thermal allene dimerisation leads to 1,2-diethylenecyclobutane derivatives but is frequently accompanied by the formation of higher oligomers. Only when the temperature and conditions are carefully regulated are synthetically useful yields obtained. The combination of two different allenic species requires the use of a large excess of the less reactive allene to prevent interference from homodimerisation ⟨72JOC950⟩. As with most allene cycloadditions the regiochemistry of these dimerisations is very substrate dependent.

Ketenes react with allenes to produce methylenecyclobutanones. Good yields may be obtained but the reaction generally shows relatively poor stereoselectivity ⟨85T2759⟩ although in certain cases, notably small cyclic allenes, very high selectivity can be observed together with a degree of asymmetric induction ⟨72TL1189⟩. The initial adducts frequently undergo thermal rearrangement to form bicyclo-[4.n.0]-alkenones in moderate yields ⟨80JCR(S)98⟩.

Provided that the alkene is activated by suitable electron-donating substituents allene–alkene cycloadditions can similarly provide synthetically viable yields. Simple terminal alkenes and cyclohexene do not react; neither do sterically encumbered alkenes ⟨75AG(E)709⟩. The use of unsubstituted allenes requires the use of excess acceptor alkenes, otherwise further reaction occurs to produce octahydronaphthalene products ⟨58JA751, 59JA2723⟩. In general, the reactivity of the allene component is enhanced by substitution. However, in these cases problems of regioselectivity can arise although the reaction preferentially occurs at the less substituted allene double bond. Although not normally the case, the reaction can be exceptionally stereoselective ⟨68JA4187⟩. In this respect, the stereochemistry of the alkene is usually maintained ⟨82JA3676⟩ and some degree of asymmetric induction from a chiral allene can be observed ⟨69CC1225, 92JOC1146⟩. Both the regio- and stereoselectivity of the thermal allene–alkene cycloaddition can be affected by a number of external variables. For example, carrying out the reaction under elevated pressure results in a faster reaction with greater conservation of allene and alkene geometry, albeit with a corresponding decrease in the regioselectivity (Equation (41)) ⟨90JOC711⟩.

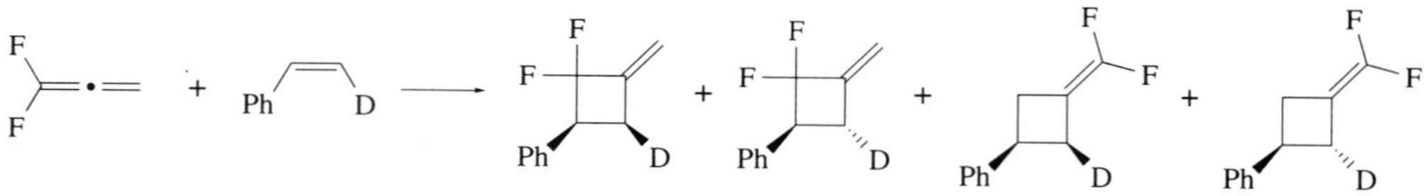

(41)

Δ, 70 °C	83%	(59:41)	17%	(79:21)
13 kbar, RT	70%	(86:14)	30%	(95:5)

A highly stereoselective [2 + 2] cycloaddition between alkenes and allenic esters is obtained on catalysis by Lewis acids ⟨86JOC3643⟩. The reaction is limited to those activated allenes which can easily form a vinyl cation intermediate; simple alkyl allenes do not react. Reflecting the formation of this intermediate, very high regioselectivity is observed, with addition only occurring at the remote allene double bond (Equation (42)). Lewis acid catalysed [2 + 2] cycloaddition is also possible in the opposite electronic sense with 1-heterosubstituted allenes (notably alkylthio ⟨92JA8869⟩) combining with electron-deficient alkenes to form good yields of the methylene cyclobutene. In this particular case the use of chiral Lewis acids has been shown to impart high asymmetric induction to the cycloaddition. Alkyl allenes do react with alkynes to produce alkylidene cyclobutenes ⟨88JOC4807⟩ although in certain cases ene-type processes are found to compete, affording cross-conjugated trienes ⟨74JCS(P1)1209⟩.

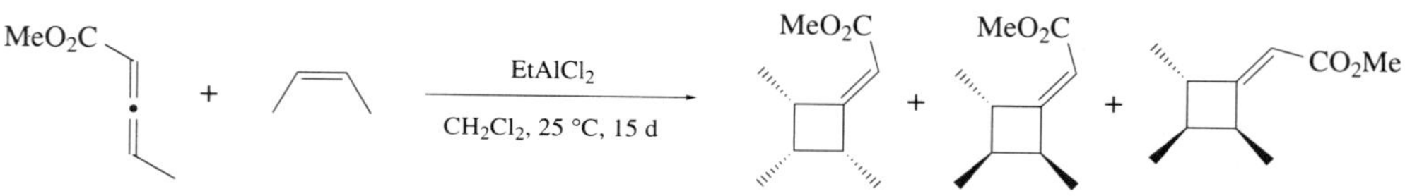

(42)

70%	3%	1%

In contrast to these thermal [2 + 2] cycloadditions which favour production of the head-to-tail regioisomers, the equivalent photochemical activated process leads predominantly to the head-to-head coupled adduct ⟨64JA5570, 64TL3695⟩. Within these isomers a variety of stereochemical outcomes are possible and the precise ratio is very dependent on the exact reaction conditions ⟨88ACS(B)259⟩. Synthetic efficiency is best achieved with electron-deficient alkenes, particularly enones, and in this respect the product stereochemistry may be predicted from a consideration of a set of empirical rules ⟨80T719⟩. The reaction may be carried out in both an inter- and intramolecular fashion. In the latter case the shape of the tether controls the regiochemistry of the addition although with larger tethers both regioisomers can be formed ⟨82JOC3297⟩. Similar variations in the regiochemistry have also been observed on varying the nature of the enone component ⟨85TL989, 85TL1429⟩.

Allenes also undergo a variety of Diels–Alder cycloadditions participating as both diene and dienophile components, the former requiring the presence of additional conjugation. These can occur both intra- and intermolecularly and normally exhibit high stereoselectivity ⟨89JA3717⟩. In the intramolecular case the nature of the tether can be crucial. For example, cyclisation of a vinyl allene system to afford hydrindane is more facile than that of the homologous substrate leading to a decalin ring system (Scheme 37) ⟨83JOC4370⟩. However, alkyl allenes and other unactivated allenes show low dienophile activity requiring forcing conditions for efficient coupling and only display poor regio- and stereoselectivity. Increased reactivity is observed in strained allenes ⟨90JA6416⟩ or by generation of an α-carbocation (Equation (43)) ⟨91TL6473⟩. Enhanced efficiency is also observed on incorporation of electron-withdrawing groups to the allene ⟨90TL4859⟩ and this can be additionally accentuated through the use of Lewis acid catalysis ⟨90JCS(P1)2185⟩. Consequently, providing that the correct diene–allene combination is obtained, a highly selective and efficient cycloaddition can be obtained. Which allene double bond participates depends on the balance of electronic properties of both components (Scheme 38). For example, electron-poor allenes and electron-rich dienes combine at the allene 1,2 double bond ⟨82JCR(S)300⟩, as do electron-rich allenes with electron-deficient dienes ⟨80TL4787⟩. Electron-rich allenes similarly add to electron-rich dienes at the more electron-rich allene bond, although this is not strictly a cycloaddition, proceeding by an electron transfer process ⟨93TL8431⟩.

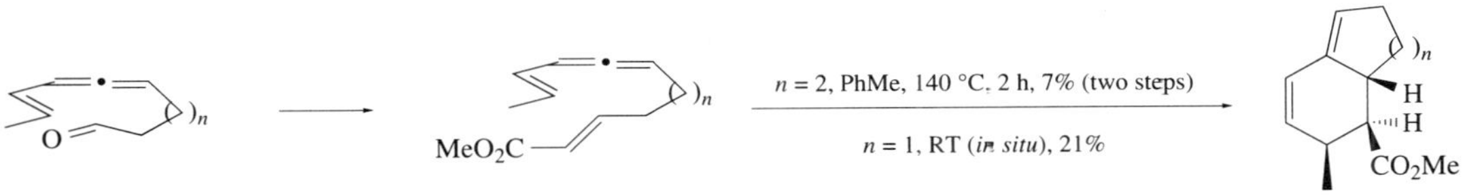

Scheme 37

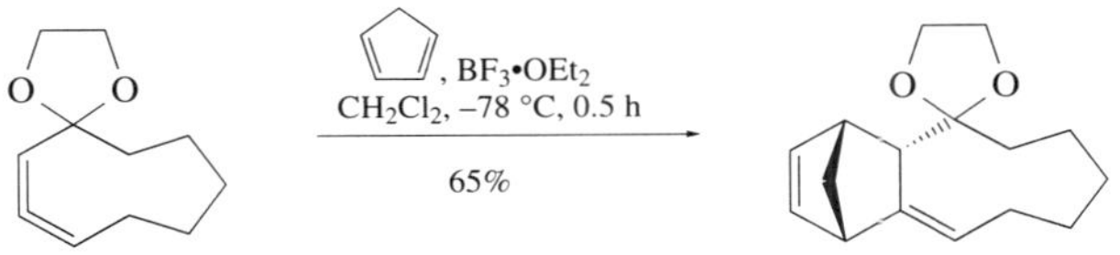

(43)

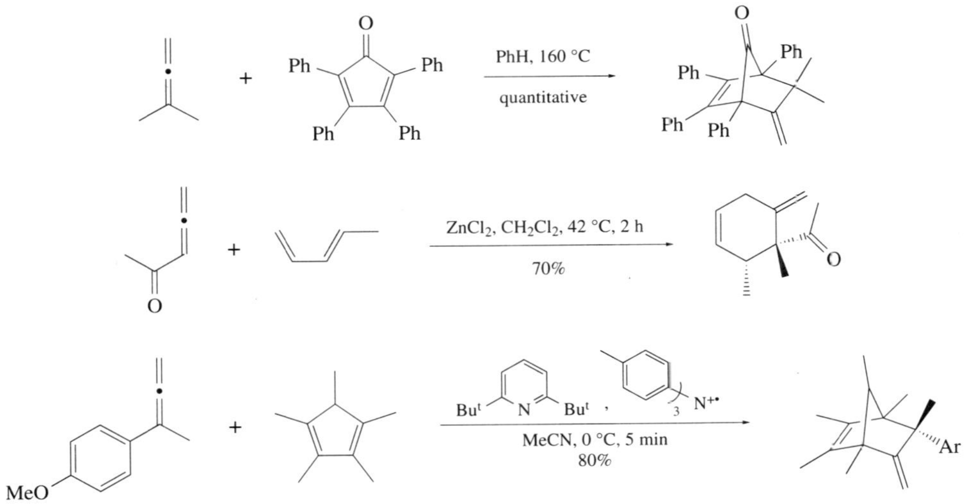

Scheme 38

1.11.3.4 Carbene Additions to Allenes

Although allenes react readily with electrophilic carbenes the addition is not an efficient method for the synthesis of alkenic compounds, as double addition to form spirocyclopentanes frequently occurs. Monoaddition is possible, particularly in the reaction of substituted allenes with dihalocarbenes. In this case preferential reaction occurs at the more substituted allene double bond

⟨91ZOR488⟩. Depending on the mode of carbene generation some regiocontrol may be obtained since singlet carbenes add to the 1,2-bond of dimethylallene and triplet carbenes to the 2,3-bond ⟨80JA1611⟩. Similarly, whilst generation of the carbene from diazoalkanes under either photolytic or copper-catalysed initiation affords considerable amounts of spiropentane products, the use of palladium initiators allows the isolation of methylene cyclopropanes in good yields ⟨87ZOR2545⟩. Both steric and electronic factors are significant, as demonstrated by the fact that terminal cyclopropanes are formed preferentially, although aryl allenes react nonselectively.

Transition metal carbene complexes, $L_nM{=}C(R)Ph$ ($L_nM = (CO)_4Fe$, $(CO)_5Cr$, $(CO)_5Mo$ and $(CO)_5W$; R = OEt) react with allenes to form the corresponding trimethylenemethane complexes. These may subsequently be trapped by alkynes to produce 4-methylenecyclopentenes ⟨91CB2335⟩. The carbene complex, $(CO)_5W{=}(H)Ph$, behaves differently, combining with 1- and 1,2-disubstituted allenes to produce methylenecyclopropanes in a highly stereoselective fashion ⟨90CC858⟩. Finally in a related process, photolysis of a monosubstituted allene in the presence of chromium hexacarbonyl affords moderate yields of the cyclopentadienone (**27**) formed from 2 equiv. of the allene and one molecule of carbon monoxide (Equation (44)) ⟨87AG(E)357⟩.

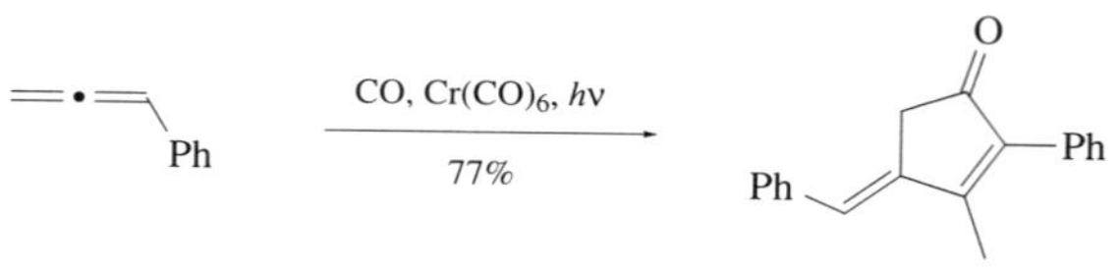

(**27**)

(44)

1.12
One or More C=C Bond(s) Formed by Addition

ANDREW C. REGAN
University of Manchester, UK

1.12.1 REDUCTION OF ALKYNES TO ALKENES

1.12.1.1 Heterogeneous Catalytic Hydrogenation

The heterogeneous catalytic hydrogenation of alkynes to form alkenes has been reviewed in several monographs, usually as part of a more general review of hydrogenation ⟨B-65MI 112-01, B-71MI 112-01, B-78MI 112-01, B-79MI 112-01, B-85MI 112-01, B-85MI 112-02⟩. There have also been some reviews specifically of this area ⟨73S457, 77T1845⟩. Both terminal and internal alkynes are readily hydrogenated over metal catalysts, with internal alkynes generally being reduced to give (*Z*)-alkenes, often with high stereoselectivity (Equation (1)). This is one of the most important methods for stereoselective synthesis of (*Z*)-disubstituted alkenes.

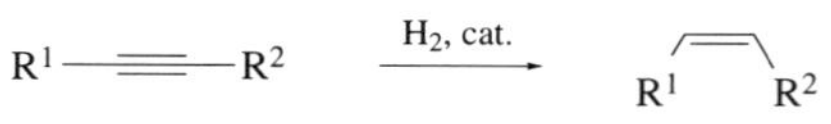

(1)

1.12.1.1.1 Catalysts

Alkynes are usually more strongly adsorbed to the catalyst than alkenes, allowing partial hydrogenation to the alkene to be achieved selectively. The most common catalysts to be employed are palladium and nickel. Platinum is not usually selective, with the product alkene being reduced at a similar rate to the alkyne ⟨52QR101⟩, but has been used successfully on some occasions (e.g., Equation (2)) ⟨44HCA1297⟩. The method of preparation of the catalyst, the nature of the support, solvent and any additives can all have important effects on the reaction.

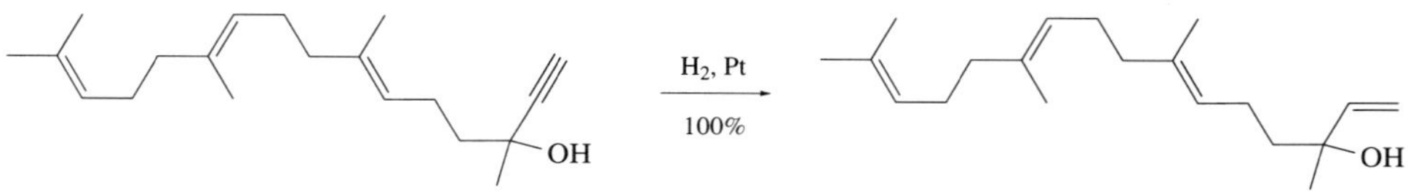

(2)

Palladium catalysts generally afford the greatest selectivity. Although colloidal palladium was used in early work, this has now been replaced by supported catalysts for synthetic applications. Pd—C is not always selective, and often requires the reaction to be interrupted after uptake of one equivalent of hydrogen (e.g., Equation (3)) ⟨70HCA1713⟩. The selectivity of Pd—C can be greatly improved by addition of quinoline (e.g., Equation (4)) ⟨47HCA1911⟩.

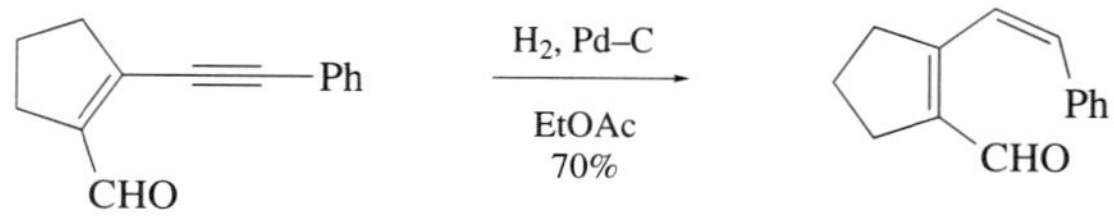

(3)

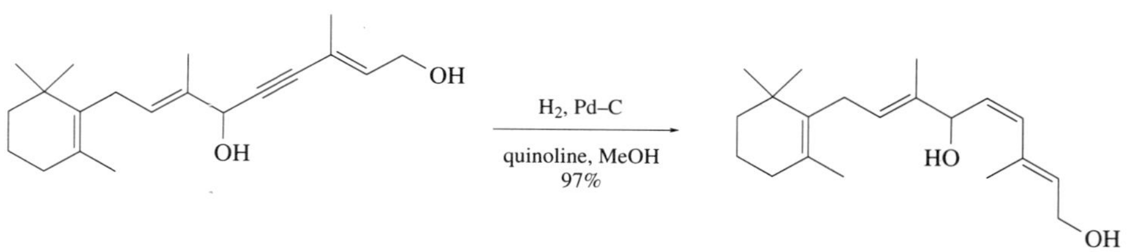

(4)

The most commonly used supports are calcium carbonate, barium sulfate, and barium carbonate. Pd—$CaCO_3$ has been used both without additives, and with the addition of quinoline ⟨42CRV77⟩. The best known modification of Pd—$CaCO_3$ is Lindlar's catalyst ⟨66OS(46)89, 73OSC(5)880⟩, obtained by treatment of Pd—$CaCO_3$ with lead acetate, which lowers its activity, apparently by affecting the morphology of the metal ⟨87JOC3126⟩. This has been used many times, both with and without modifiers, to achieve partial reduction of alkynes. Hydrogenations over Lindlar's catalyst often slow down or stop spontaneously after the uptake of one equivalent of hydrogen. For example, the reduction of the alkynoic acid (**1**) stopped after 95% of the theoretical amount had been adsorbed

(Equation (5)) ⟨82JOC1221⟩. However, the structurally similar amide (**2**) did not stop or slow down after the uptake of one equivalent of hydrogen (Equation (6)).

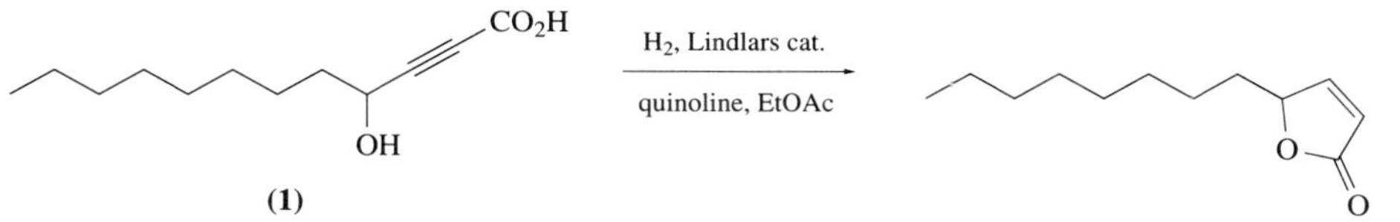

(5)

(**1**)

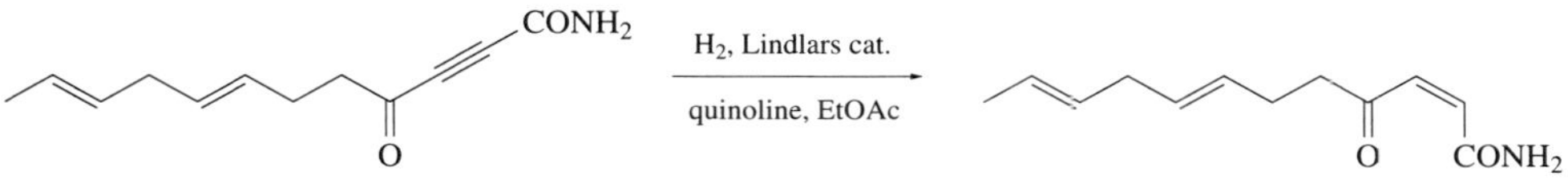

(6)

(**2**)

Lindlar recommended that quinoline should be used as a modifier with his catalyst for high selectivity ⟨66OS(46)89⟩, and there are many examples of this strategy (e.g., ⟨83TL1739, 83BCJ2539, 83JCS(P1)387⟩). However, its use is often omitted, and this may be responsible for the lower selectivities sometimes reported. Lindlar also suggested 2,2′-(ethylenedithio)diethanol as a more effective additive than quinoline ⟨73OSC(5)880⟩.

Other modifiers have also been used with Lindlar's catalyst, particularly other amines, for example pyridine, piperidine, aniline, and diethylaniline ⟨51JA5073, 54JA4446, 65JOC1752, 83JCS(P1)387, 83JOC4430⟩. Other types of modifier include chlorobenzene, hydroxides ⟨62JOC2398, 62JOC4323⟩, and sulfur compounds.

Representative examples of hydrogenations using Lindlars catalyst are shown in Table 1.

Palladium on barium sulfate has also often been used, and in some cases has been found to be more selective than the Lindlar catalyst. For example, the former catalyst was successful for the reduction shown in Equation (7), whereas Lindlar's catalyst was not ⟨73JOC3652⟩.

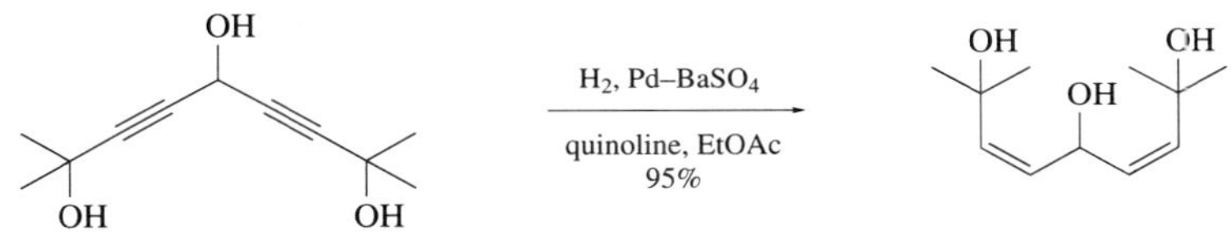

(7)

Again, quinoline has often been used as an additive, and this system has been claimed to be more reproducible than the Lindlar procedure ⟨56JA2518⟩. The use of neat pyridine as a solvent ⟨B-67MI 112-01⟩ has been claimed to be even better than quinoline "for all cases examined," with uptake of hydrogen stopping sharply at the alkene stage, with no over-reduction ⟨82JOC4254⟩. Some examples of the use of Pd—$BaSO_4$ are shown in Table 2. This catalyst has also been used for the formation of annulenes by partial reduction of cyclic alkynes ⟨70TL5139⟩.

Raney nickel has also been used as the hydrogenation catalyst, although it has now generally been replaced by Lindlar's catalyst. It may be useful in certain cases, and its selectivity can be achieved by interruption of the hydrogenation, or by the use of piperidine or zinc additives ⟨52JA295, 53JCS3294⟩. Freshly prepared Raney nickel is not selective, but is improved by aging in ethanol ⟨53JCS3156⟩. Raney nickel has been found useful for hydrogenation of 2-alkyne-1,4-diols, where palladium catalysts are often unsuitable (Equation (8)) ⟨51JA1066, 53CR(236)1044⟩; however, palladium on charcoal with small amounts of NaOH, KOH, or Et_3N has since been found to be better for these substrates ⟨62JOC2398⟩. A later nickel catalyst is the P-2 form, which is prepared from $Ni(OAc)_2$ and $NaBH_4$ in EtOH ⟨73JOC2226⟩, and is particularly stereoselective when used with ethylenediamine as an additive (Equation (9)) ⟨75T1846, 83JOC2266, 83JOC2274⟩.

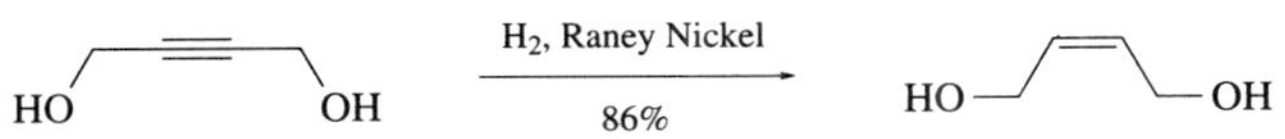

(8)

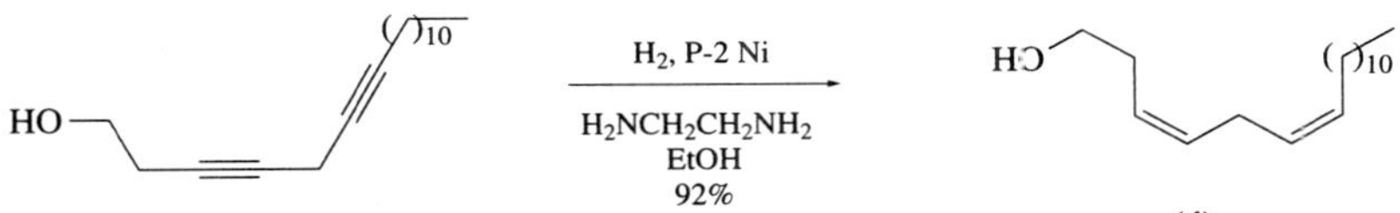

(9)

(**3**) (**4**)

Table 1 Hydrogenations of alkynes over Lindlar's catalyst.

Entry	*Substrate*	*Conditions*	*Yield (%)*	*Ref.*
1	Ph—≡	quinoline, hexane	~100 (not isolated)	73OSC(5)880
2		PhH	99	90JOC1670
3		hexane	92	90TL5749
4	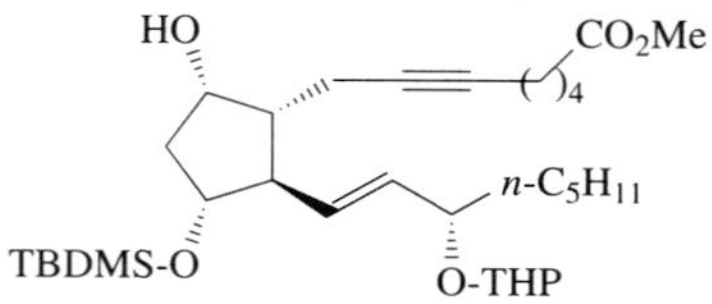	PhH, cyclohexane	96	84AG(E)847
5	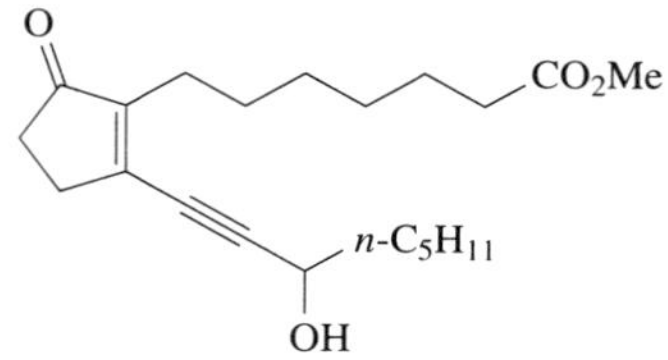	EtOH	89	88BCJ2859
6	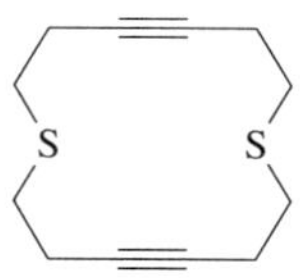	EtOAc	98	88CB2239
7		EtOAc	79	83JOC5221
8		Et_3N, THF	90	83TL4883

The stereoselectivity of the reduction for a given catalyst may vary markedly with even small changes in the structure of the substrate, and similarly, for a given substrate, it may be necessary to try several different catalysts before satisfactory results are obtained. For example, the diyne alcohol (**3**) is reduced to the (*Z,Z*)-dienol (**4**) with high stereoselectivity using P-2 nickel (Equation (9)) ⟨83JOC2266⟩, whereas the triyne alcohol (**5**) gives a mixture of products both using P-2 nickel and also with Pd—$BaSO_4$ and Pd—$CaCO_3$ with Et_3N. Lindlar's catalyst with quinoline, on the other hand, gave good stereoselectivity for the all-(*Z*)-trienol (**6**) (Equation (10)) ⟨83JOC2270⟩.

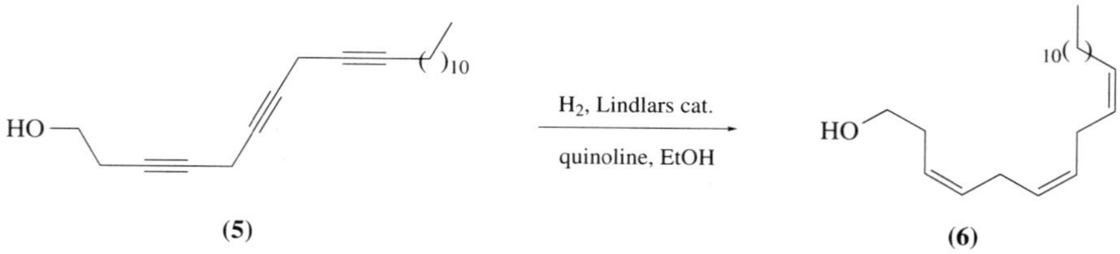

(**5**) (**6**) (10)

In order to obtain good selectivity, the alkene product must be desorbed rapidly from the catalyst, relative to further reduction or isomerization. Usually the alkyne is more strongly bound than the alkene, and competes effectively for the catalytic sites, blocking re-adsorption of the alkene, or displacing it. The order of reactivity with a palladium catalyst is usually as follows: terminal alkyne > internal alkyne > alkene (terminal or internal), and this is clearly seen in the sequence of hydrogenations shown in Scheme 1 ⟨61T16⟩. If both the triple bonds of a diyne are internal, then hydrogenation gives the diene.

Enynes can be selectively hydrogenated to dienes; this is highly selective for nonconjugated systems, but less so for conjugated ones ⟨73S457, 77T1845⟩.

Table 2 Hydrogenations of alkynes over Pd—$BaSO_4$.

Entry	*Substrate*	*Conditions*	*Yield (%)*	*Ref.*
1	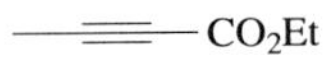	quinoline, Et_2O	93–98	90OSC(7)226
2	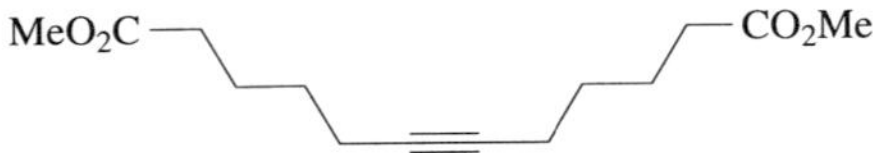	quinoline, MeOH	97	56JA2518
3	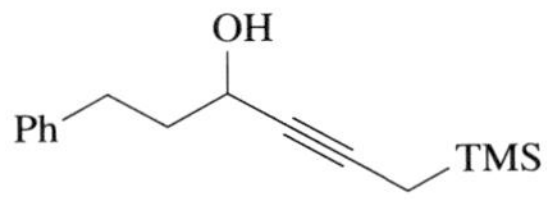	hexane	96	84CC534
4	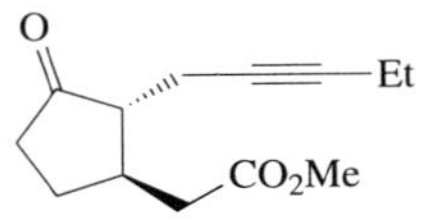	pyridine	~100	82JOC4254
5		quinoline, EtOH	>70	80TL1739
6	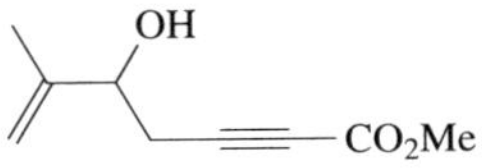	pyridine	71	84JOC747
7	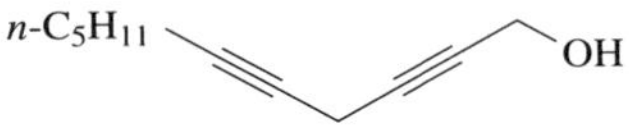	pyridine	70	83TL409
8	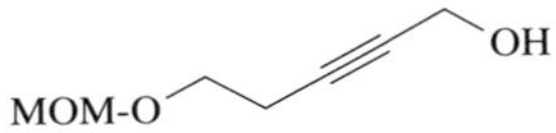	pyridine	~100	83JA5373

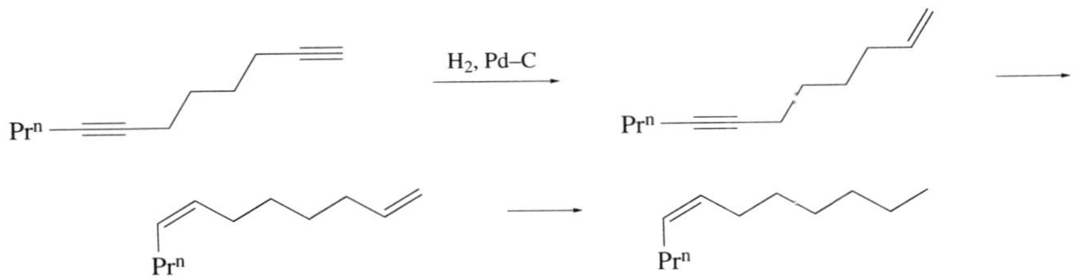

Scheme 1

Towards the end of the reaction, when the alkyne concentration becomes low, the alkene may start to react. The function of amine additives appears to compete effectively with the alkene, but not with the alkyne, for catalytic sites, and to be unaffected by the reaction conditions, and hence remain at constant concentration throughout the reaction ⟨71RTC961⟩.

In some cases, the alkene product may undergo double-bond migration or *cis–trans* isomerization on the catalyst faster than further reduction, and in such circumstances it may be beneficial to stop the reaction before one equivalent of hydrogen is absorbed ⟨83BCJ353, 83JA2923⟩.

The effect of temperature on selectivity has been studied. Lowering the temperature to $-10\,°C$ to $-30\,°C$ has been found to improve the selectivity using Lindlar's catalyst ⟨77T1845⟩. With other catalysts, however, the opposite effect has been observed ⟨64MI 112-01⟩.

1.12.1.1.2 Stereoselectivity

Catalytic hydrogenation of internal alkynes usually affords the (*Z*)-alkene often with high stereoselectivity. However, varying amounts of the (*E*)-alkene are also always formed, which is generally believed to result from isomerization of (*Z*)-alkene, although some may also be formed as an initial product ⟨73S457, B-85MI 112-02⟩.

Palladium is the most stereoselective of the group 10 catalysts, with platinum being less so. Lindlar's catalyst is generally the most stereoselective of the supported palladium catalysts, being relatively inactive towards alkene isomerization and hydrogenation. The highest stereoselectivity is achieved in nonpolar solvents (ranging from pentane and hexane to THF and ethyl acetate), and it is often lower in alcohols ⟨B-85MI 112-02⟩. The temperature may be an important factor in determining stereoselectivity, and subambient temperatures (0–20 °C) are often better than room temperature.

Pd—$BaSO_4$ has been widely used for formation of (*Z*)-alkenes, and in some cases has been found to be more satisfactory than the Lindlar catalyst. Examples of stereoselective reductions using both Lindlar and Pd—$BaSO_4$ catalysts are shown in Tables 1 and 2.

1.12.1.1.3 *Catalytic hydrogenation of functionalized alkynes*

It is often possible to reduce an alkyne in the presence of other functional groups, because of the selective absorption of the alkyne on the catalyst. However, functional groups adjacent to the alkyne may cause problems, especially if hydrogenolysis of the product is possible ⟨57JCS1622⟩.

Propargylic alcohols can usually be hydrogenated to (*Z*)-alkenols. Glycols are more sensitive than alcohols to the catalysts and conditions; they undergo hydrogenolysis more readily ⟨62JOC4323⟩ and may also cyclize to tetrahydrofurans in acid conditions ⟨62JOC2398⟩. Palladium is usually the preferred metal, with both Lindlar's catalyst and Pd—$BaSO_4$ finding widespread use. Palladium with small amounts of KOH also works well ⟨62JOC4323, 67ANY(145)91⟩, although large amounts of alkali lead to *cis–trans* isomerization ⟨59JCS2902⟩. Reduction of alkynols and glycols is often less stereoselective than with unfunctionalized alkynes ⟨75JOC1588⟩. Use of subambient temperatures can be helpful in minimizing hydrogenolysis and isomerization (Equation (11)) ⟨69JA5901, 77T1845, 79T2039⟩. Reduction of alkynols may sometimes be accompanied by isomerization to the ketone.

Alkyne epoxides are reduced readily to the epoxy alkene, provided that the product is not an allylic epoxide (Equation (12)) ⟨82JCS(P1)545⟩.

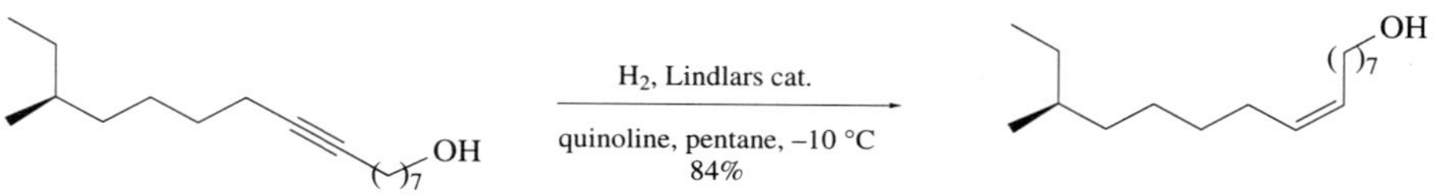

(11)

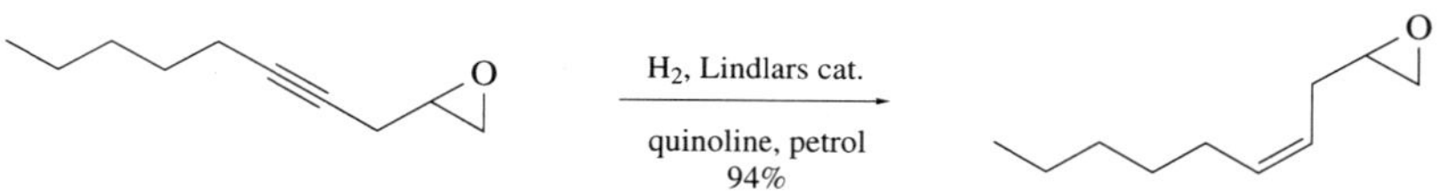

(12)

With amino alkynes, hydrogenolysis may be important, and this is sensitive to the structure of the substrate ⟨61JOC1073⟩. Hydrogenolysis has been reduced by using platinum rather than palladium, and by reducing the amine salt rather than the free amine ⟨49CR(228)1501, 50BSF1305, 61JOC3772⟩.

Alkynyl aldehydes and ketones are reduced readily to the corresponding alkenes, if the carbonyl group is not conjugated with the alkyne. Internal conjugated alkynes are often unsatisfactory substrates (Equations (13) and (14)) ⟨49JCS737, 49JCS1827, 49JCS2028, 72JOC2989⟩ and better results are obtained with terminal alkynes (Equation (15)) ⟨70HCA485, 72JOC2989⟩.

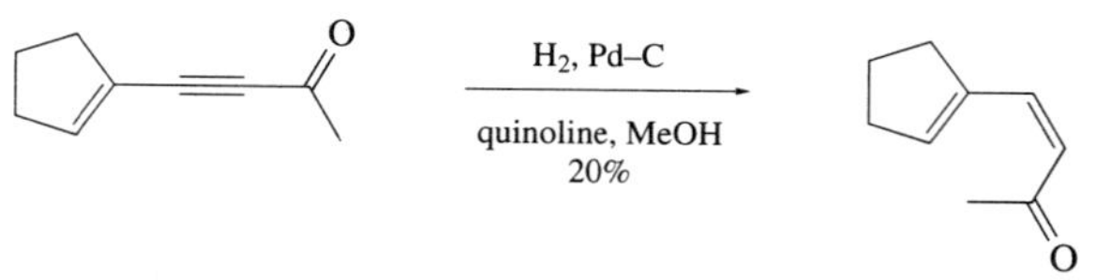

(13)

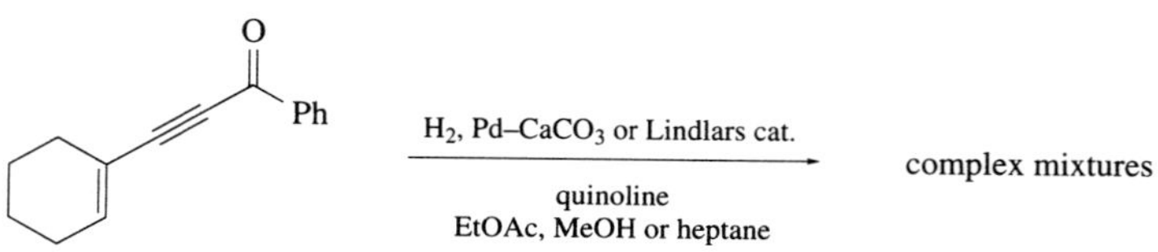

(14)

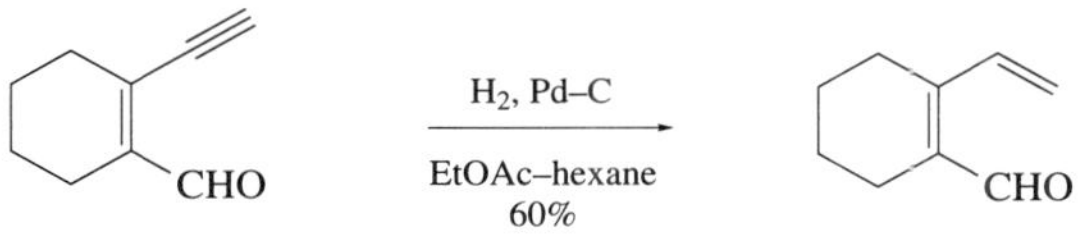

(15)

1.12.1.2 Homogeneous Catalytic Hydrogenation

Homogeneous catalysts have been rarely used for the hydrogenation of alkynes. Palladium chloride has been used for the reduction of diphenylethyne to *cis*-stilbene (Equation (16)), and was found to be more selective than 10% palladium on carbon ⟨83BCJ353⟩.

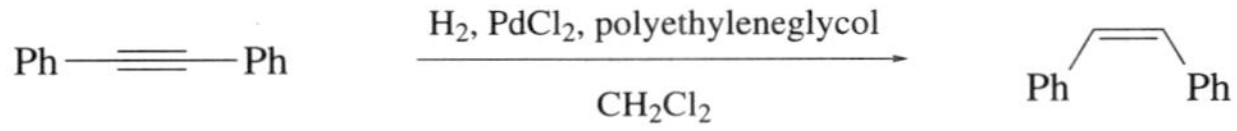

(16)

Wilkinsons catalyst (and also [Rh(OAc)(PPh_3)$_3$]) usually gives complete reduction of the triple bond ⟨71JCS(A)3224⟩. However, in benzene with acidic cosolvents (e.g., CF_3CO_2H or PhOH) selectivity is possible with terminal alkynes, because the reduction of the alkyne (but not the alkene) is accelerated ⟨68MI 112-01⟩. Another rhodium catalyst, [Rh(COD)PPh_3Py]PF_6 in the presence of benzoate ion, is also effective for the selective reduction of terminal alkynes (Equation (17)) ⟨77JOM(141)113⟩.

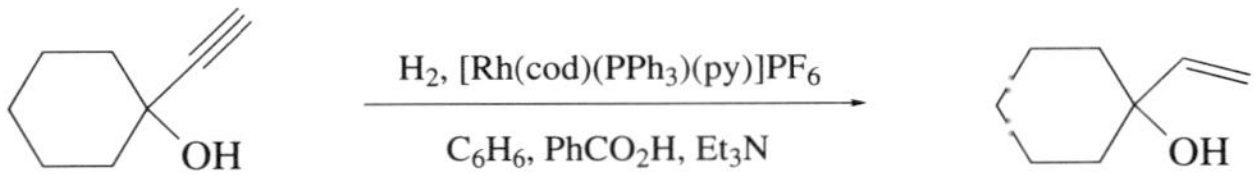

(17)

Chromium complexes of the type [(arene)Cr(CO)$_3$] catalyze hydrogenation of internal alkynes to (*Z*)-alkenes with high selectivity (Equation (18)) ⟨85JOC1147, 85JOC3246⟩. 1-Phenylpropyne and tetradec-7-yne are reduced to (*Z*)-alkenes using these catalysts.

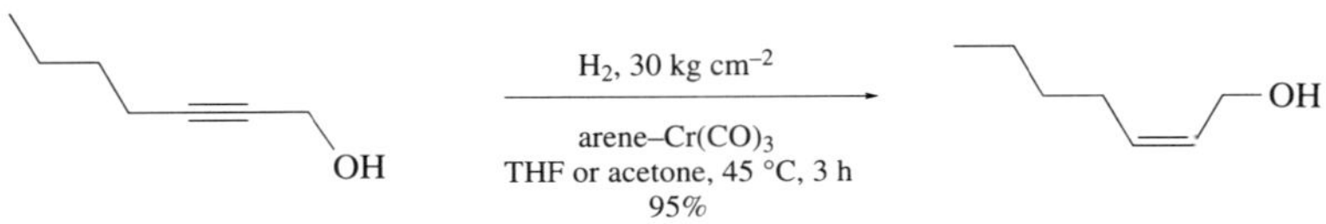

(18)

Opposite stereoselectivities with two different rhodium catalysts have been observed for the same substrate dimethyl acetylenedicarboxylate (DMAD); [(μ-H)$_2$$H_2$$Rh_2$(triphos)$_2$]$^{2+}$ gives dimethyl maleate, albeit very slowly (Equation (19)) ⟨88IC4429⟩, whereas [RhH_2(OC(=O)OH)(PPr^i_3)$_2$] gives dimethyl fumarate (Equation (20)) ⟨83JA6273⟩. The latter catalyst also reduces diphenylethyne to *trans*-stilbene, in contrast to $PdCl_2$ (see above). On the other hand, with [$RhCl_3$(Py)$_3$] and $NaBH_4$ in DMF, diphenylethyne is reduced to *trans*-stilbene, but DMAD is reduced with (*Z*)-selectivity to dimethyl maleate ⟨69CC1503⟩.

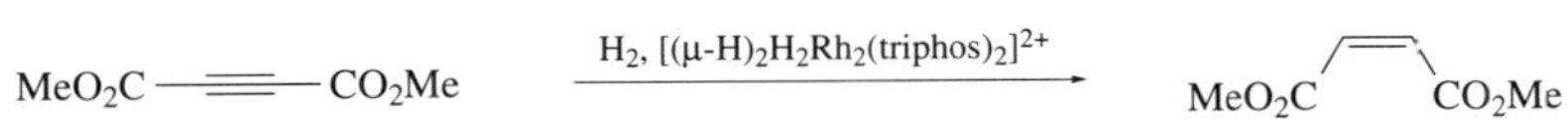

(19)

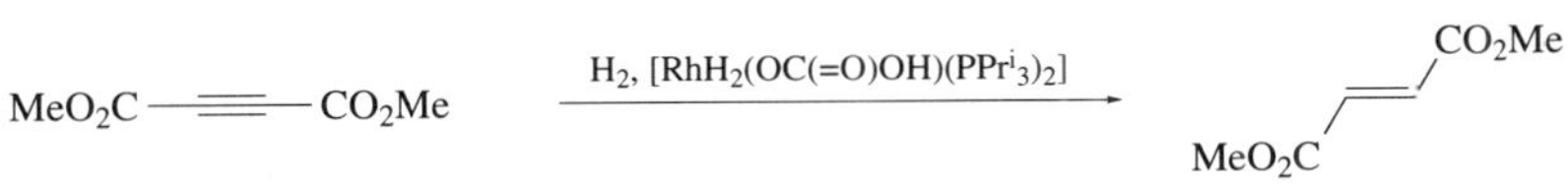

(20)

Catalytic transfer hydrogenation is also possible using formic acid and Pd—C catalyst (Equation (21)) ⟨80JOC4926⟩, or alternatively, using a homogeneous palladium catalyst (Equation (22)) ⟨93CC386⟩.

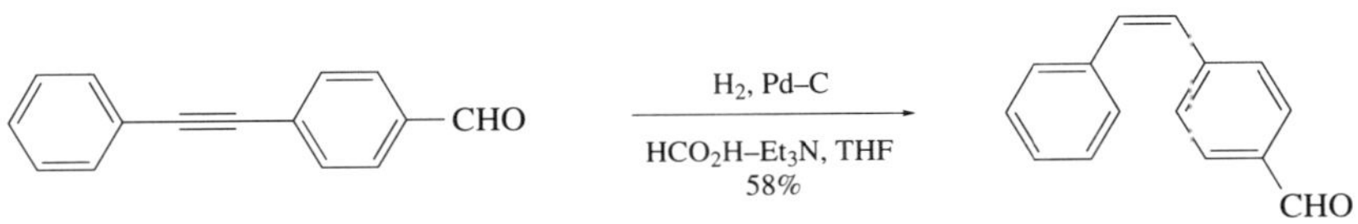

(21)

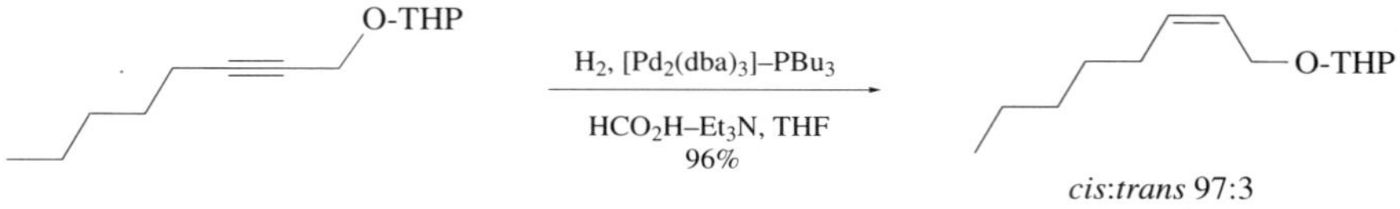

(22)

1.12.1.3 Dissolving Metal Reductions

The first dissolving metal reductions of dialkyl alkynes to give *trans*-alkenes using sodium in liquid ammonia were reported by Campbell and Eby ⟨41JA216⟩. Before this, the reduction of alkynes to (*Z*)-alkenes by catalytic hydrogenation was well known; however, there was no satisfactory method for a *trans*-selective reduction.

Shortly afterwards, terminal alkynes were successfully reduced to 1-alkenes, also using sodium in liquid ammonia, provided that an ammonium salt was added to the reaction mixture (Equation (23)) ⟨43JA2020⟩. Without the ammonium salt, substantial metallation of the acidic alkyne occurs.

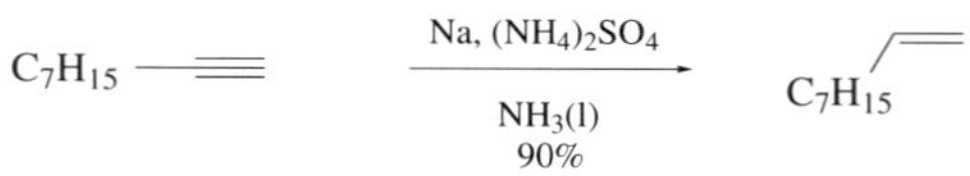

(23)

Other metal–amine combinations include lithium or calcium and an alkylamine. Lithium and ethylamine reduce internal alkynes to *trans*-disubstituted alkenes at $-78\,^{\circ}C$ (Equation (24)), but at $17\,^{\circ}C$ over-reduction to the alkene also occurs, especially if excess lithium is used ⟨55JA3378⟩. Calcium and an amine also gives this reaction in 70–88% yields, but some migration of the π-bond occurs (Equation (25)) ⟨84JOC1662⟩. Lithium in water reduces internal alkynes to *trans*-alkenes ⟨86AG(E)167⟩.

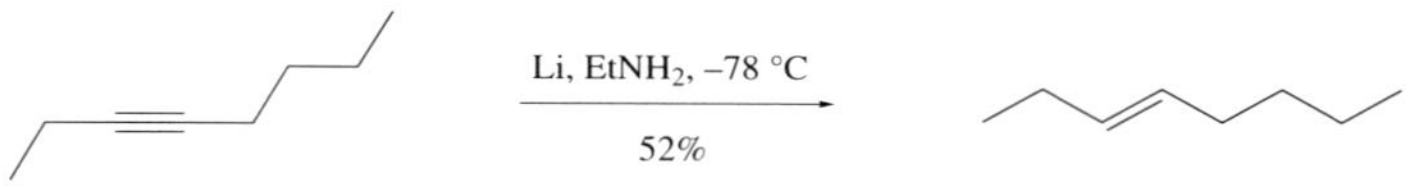

(24)

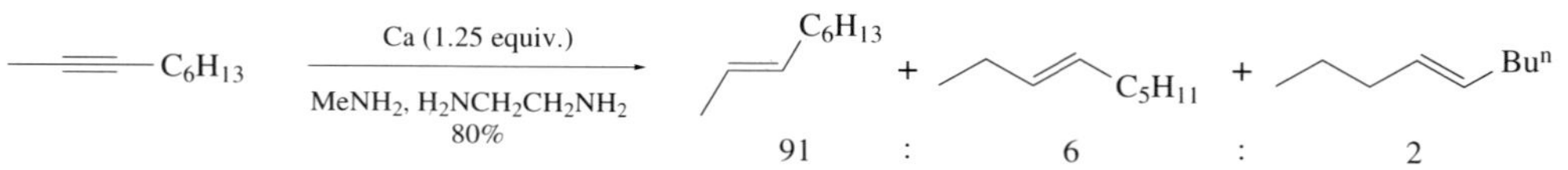

(25)

The combination of sodium in hexamethylphosphoramide (HMPA) gives different results depending on the conditions ⟨74JOC747⟩. In the presence of *t*-butanol as a proton donor *trans*-alkenes are obtained in good yields. However, without the *t*-butanol present, a mixture of *cis*- and *trans*-alkenes is formed, and some π-bond migration also occurs.

Ytterbium in liquid ammonia gives *trans*-alkenes in good yields; however, this reagent is very expensive ⟨78JOC4555⟩.

Zinc has been used for the reduction of alkynes. With a zinc–copper couple in an alcohol solvent, terminal alkynes are reduced to 1-alkenes in very good yield (Equation (26)), and disubstituted alkynes give the (*Z*)-alkenes (Equation (27)), in contrast to the stereoselectivity of the other dissolving metal reductions discussed above ⟨80TL1069, 87TL5395⟩. The combination of zinc and 1,2-dibromoethane is also effective. Conjugated diynes are reduced to (*Z*)-enynes by zinc and dibromoethane (Equation (28)) ⟨84CC735⟩. In this case the triple bond closer to the heteroatom is reduced, and the carbon–oxygen bond is not cleaved. Conjugated enynes can be reduced to dienes without affecting the alkene double bond (Equation (29)); 1,4-diethoxybut-2-yne is reduced by zinc, dibromoethane, and $LiCuBr_2$ without cleavage of the propargylic oxygen functionalities (Equation (30)) ⟨84CC735⟩.

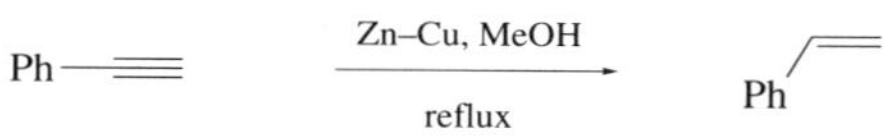

(26)

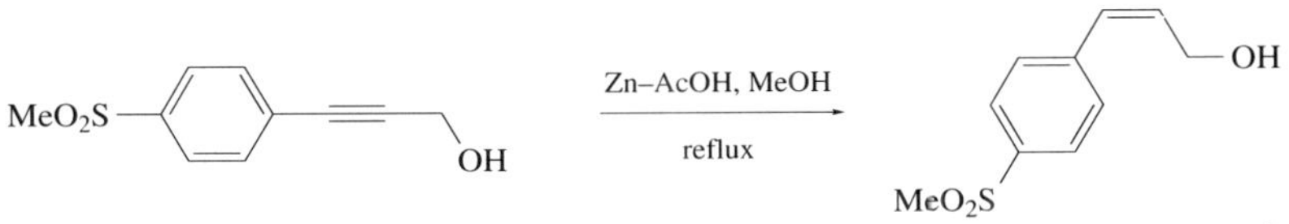

(27)

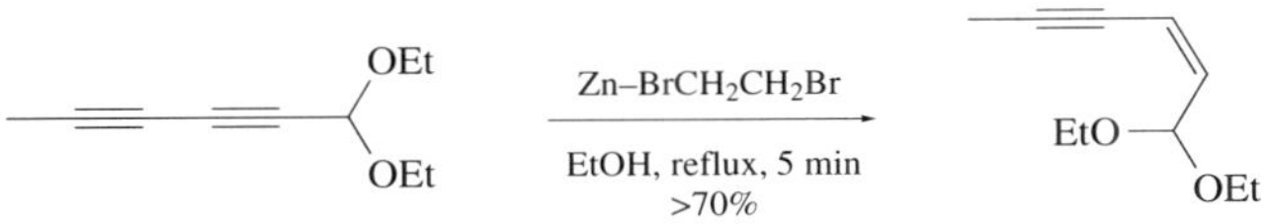

(28)

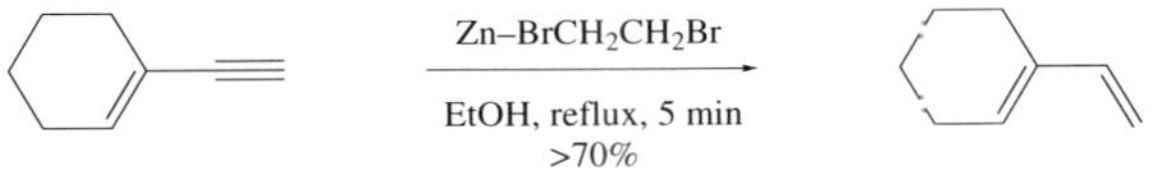

(29)

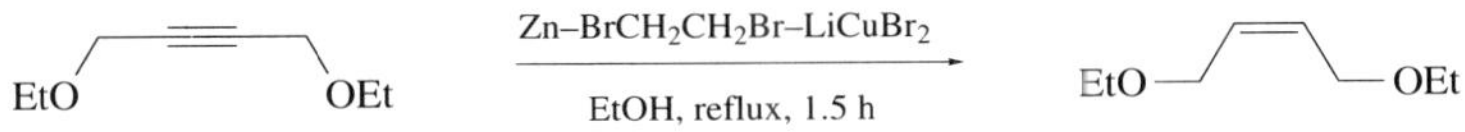

(30)

1.12.1.4 Hydride Reducing Agents

1.12.1.4.1 Hydroboration

Alkynes can be reduced to alkenes by hydroboration followed by protonolysis of the resulting alkenylborane, and this results in overall *cis*-addition of H_2 ⟨B-78MI 112-02, B-88MI 112-01, 91COS(8)703⟩. Borane–THF generally gives complex mixtures as a result of further reaction of the initially formed alkenylborane, especially with terminal alkynes, whereas internal alkynes can give reasonable yields of monoaddition products ⟨61JA3834, 67JA291, 71JA3395⟩. However, use of more sterically hindered dialkylboranes gives selective monoaddition, and disiamylborane is most generally used reagent, readily reacting with both terminal and internal alkynes (Scheme 2) ⟨59JA1512, 61JA3834, 67JA291, 71JA3395, 79JOC2328, 82JOM(225)63⟩. The resulting dialkylalkenylboranes undergo protonolysis preferentially at the alkenyl–boron bond to give the corresponding alkenes after treatment with a carboxylic acid at 0 °C; the stereoselectivity for the formation of (*Z*)-alkenes is generally high ⟨59JA1512, 61JA3834, 81T2617⟩.

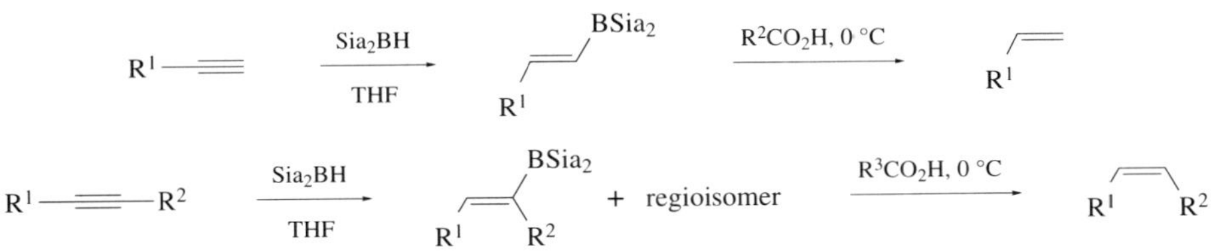

Scheme 2

Various reagents have been used for the hydroboration of alkynes. With terminal alkynes the boron atom becomes attached to the terminal carbon atom. With unsymmetrical internal alkynes the regioselectivity of the addition of the borane is determined by both steric and electronic factors, depending upon the particular reagent used, and representative examples are shown in Table 3 ⟨83TL1433⟩. This regioselectivity is, of course, irrelevant if protonolysis is to follow, but is important for subsequent carbon–carbon bond-forming reactions (see Section 1.12.3.1.1).

Dicyclohexylborane has also been used successfully, and has been recommended as having significant advantages over disiamylborane, especially in terms of storage of the reagent and thermal stability of the alkenylborane products ⟨B-88MI 112-02⟩. It may also show different selectivity to disiamylborane, for example disiamylborane has been used to achieve monohydroboration of a conjugated diyne, whereas dicyclohexylborane reacts with both triple bonds (Scheme 3) ⟨70JA4068⟩. Both reagents react selectively with the triple bond of an enyne ⟨70JA4068, 71JA1308⟩, whereas 9-borabicyclo[3.3.1]nonane (9-BBN-H) shows the opposite selectivity ⟨79JOC2328⟩.

Dicyclohexylborane has also been used to reduce a skipped triyne system to the all-(*Z*)-triene

Table 3 Regioselectivities of hydroboration of internal alkynes.

	Pr^n—≡—		Ph—≡—	
	↑	↑	↑	↑
BH_3•THF	40	60	74	26
Sia_2BH	39	61	19	81
Catecholborane	40	60	27	73
9-BBN-H	22	78	65	35
Br_2BH•SMe_2	25	75	64	36
Mes_2BH	10	90	2	98

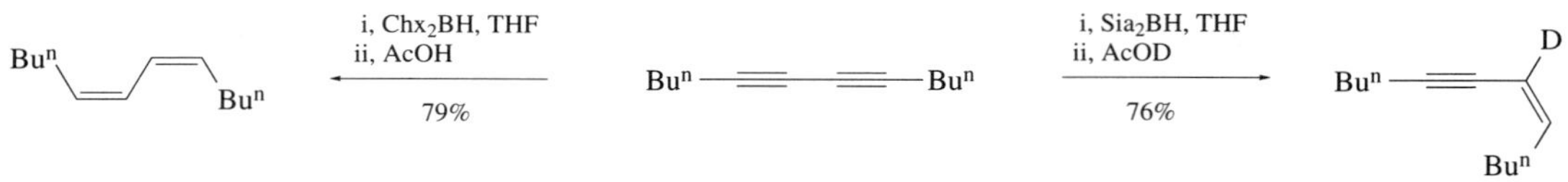

Scheme 3

⟨88AG(E)298⟩, and to selectively reduce a propargylic ether in the presence of a conjugate diene ⟨84TL1655⟩.

9-BBN—H is useful for the hydroboration of internal alkynes and the regioselectivity is a result of both steric and electronic effects ⟨79JA96⟩. With terminal alkynes, diaddition is mostly obtained, and disiamylborane is generally a better choice of reagent. The alkenylboranes formed with either 9-BBN—H or dicyclohexylborane can also be hydrolyzed to the alkenes with methanol alone (Scheme 4), whereas those formed from disiamylborane are more stable, and require the addition of a catalytic amount of a carboxylic acid ⟨86JOC4512⟩.

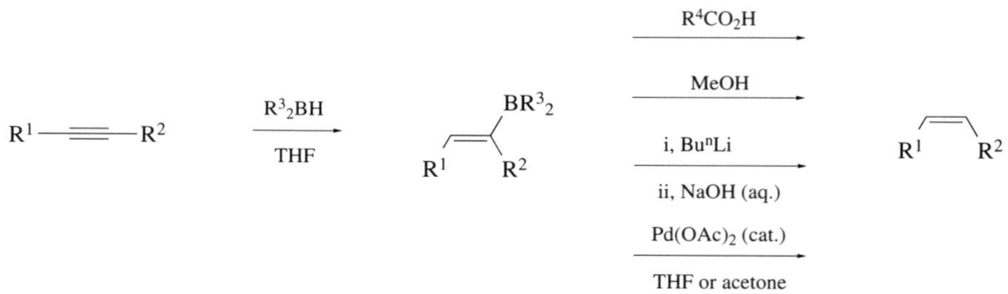

Scheme 4

Where the substrate contains acid-sensitive functional groups, the hydrolysis of the alkenylborane may be performed under basic conditions by first converting it into the boronate with Bu^nLi, followed by treatment with aqueous NaOH (Scheme 4) ⟨76JOC3484⟩. Alternatively, both acidic and basic conditions may be avoided by using a neutral procedure, involving simply the addition of a catalytic amount of $Pd(OAc)_2$; the required proton is derived from the solvent (Scheme 4) ⟨78CC702, 80BCJ1670⟩.

Dimesitylborane is a very hindered dialkylborane which reacts readily with terminal and internal alkynes, but very slowly with alkenes; and it is also more regioselective than any other borane reagent ⟨83TL1433⟩.

Catecholborane has also been used for hydroboration of alkynes; however, it is less reactive than dialkylboranes, and requires refluxing in THF ⟨72JA4370, 83JOC5080, 85JA3626⟩. The resulting alkenylboranes may be useful for other synthetic transformations ⟨76T981⟩.

Other reagents which have been used for hydroboration of alkynes include the haloboranes $ClBH_2$ ⟨73JOC1617⟩, and $Br_2BH\cdot SMe_2$ ⟨80JOC389⟩; the latter reagent can be used in dichloromethane, and is selective for an internal alkyne over a terminal alkene, unlike 9-BBN-H ⟨83JOC644⟩.

1.12.1.4.2 Hydroalumination

Hydroalumination of an alkyne, followed by hydrolysis of the resulting alkenyl aluminum compound, results in overall reduction of the alkyne to an alkene ⟨B-78MI 112-02, 84OR375, 91COS(8)733⟩. The reagent most often used for this process is diisobutylaluminum hydride (dibal-H) ⟨60LA(629)222, 78JOC2739⟩, and this will reduce both terminal and internal alkynes. The reagent is selective for alkynes over alkenes (Equation (31)) ⟨77TL3641⟩, and terminal alkynes react faster than internal alkynes. The reduction is stereoselective, and gives the (*Z*)-alkene from an internal alkyne ⟨68JOM(11)P7⟩. The stereoselectivity is usually very good, for example 1-phenylpropyne gives only the (*Z*)-alkene using dibal-H, whereas hydroalumination over a nickel catalyst gives 6% of the *trans*-isomer (Equation (32)) ⟨68JOM(12)P33⟩. A number of side reactions may interfere with the hydroalumination. Terminal alkynes undergo metallation of the terminal C—H to a small extent, although this is less serious when the alkyl group attached to the triple bond is secondary or tertiary, rather than primary ⟨84OR(32)375⟩. With conjugated terminal alkynes, metallation competes with addition to a much greater extent, for example phenylethyne gives 20% metallation (Equation (33)) ⟨84OR(32)375⟩. Other side reactions include over-addition, exchange of C—H and aluminum in the product, and dimerization of the intermediate alkenylaluminum compounds ⟨91COS(8)733⟩. The intermediate *syn*-alkenylaluminum compound may also isomerize to the *anti*-isomer, via a second addition of the aluminum reagent.

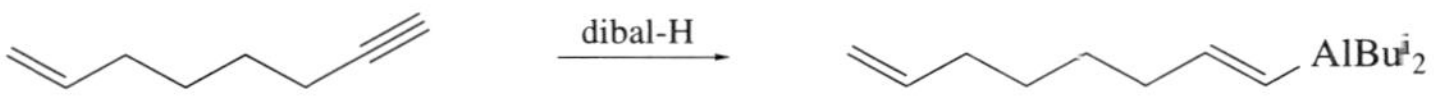

(31)

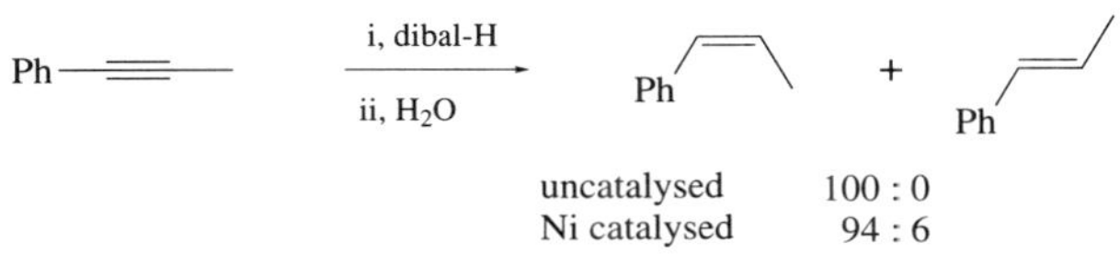

(32)

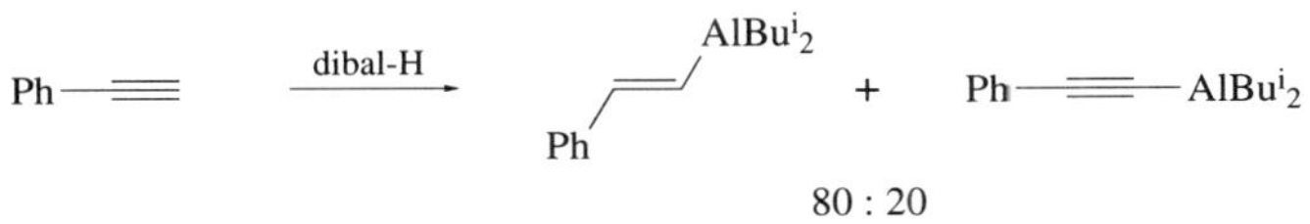

(33)

For the reaction of dibal-H with terminal alkynes, it is best to use a hydrocarbon solvent, since over-addition occurs in ether, while metallation takes place in tertiary amine solvents. Internal alkynes give negligible over-addition, especially in hydrocarbon solvents, but do give a few percent of dimerized product ⟨84OR(32)375⟩. However, the use of a twofold excess of dibal-H over alkyne reduces the dimerization. The complex $Ni(acac)_2$ catalyzes the addition but the regio- and stereoselectivity are lower ⟨68JOM(12)P33⟩.

With conjugated enynes, if the alkene is terminal (i.e. an alk-1-en-3-yne) then mainly polymeric products are obtained; however, if both unsaturated units are internal, then the alkyne undergoes hydroalumination preferentially (Equation (34)) ⟨77JOC3981⟩. A conjugated ynone has been reduced with dibal-H in HMPA, with the stereoselectivity being dependent upon the conditions (Equation (35)) ⟨87JOC1624⟩. In the absence of MeCu the (*E*)-isomer is favoured, whereas with MeCu present, a 1.5 : 1 ratio of (*Z*)- to (*E*)-isomers is formed. Terminal alkynes can also be reduced with dibal-H/MeCu to give terminal alkenes in good yield (Equation (36)) ⟨82JOM(226)C55⟩.

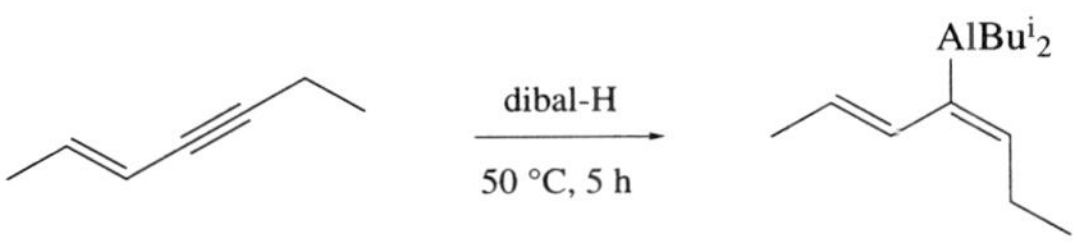

(34)

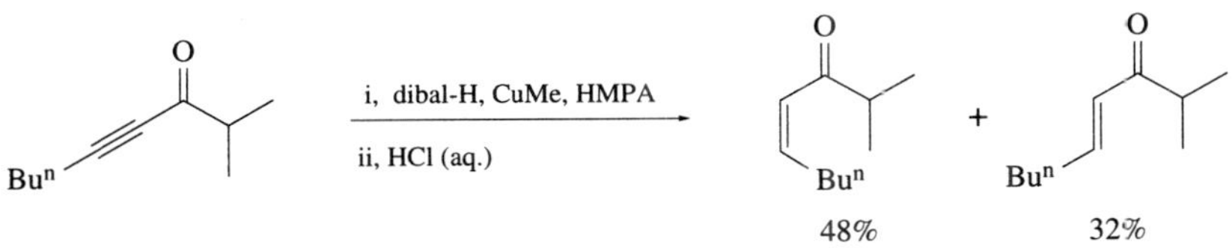

(35)

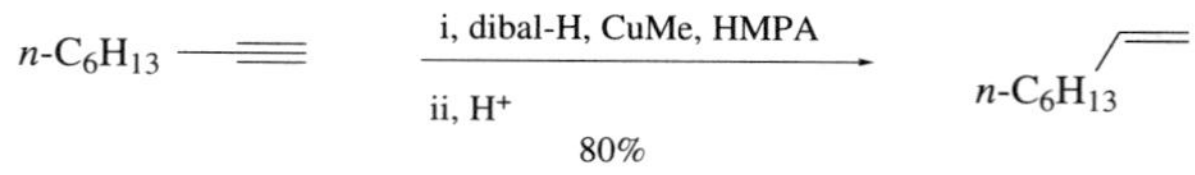

(36)

The reduction of a tertiary propargyl amine with dibal-H gives the opposite stereoselectivity to that normally observed, with formation of the (*E*)-isomer ⟨79TL3145, 86JMC112⟩. Similarly, a tertiary aminodiyne has been reduced selectively to give the (*E*)-enyne in 46% yield, accompanied by 19% of the (*Z*, *E*)-diene (Equation (37)) ⟨84JMC1539⟩.

Propargyl silanes can be selectively reduced to the (*Z*)-allylsilanes (Equation (38)) ⟨84S113⟩.

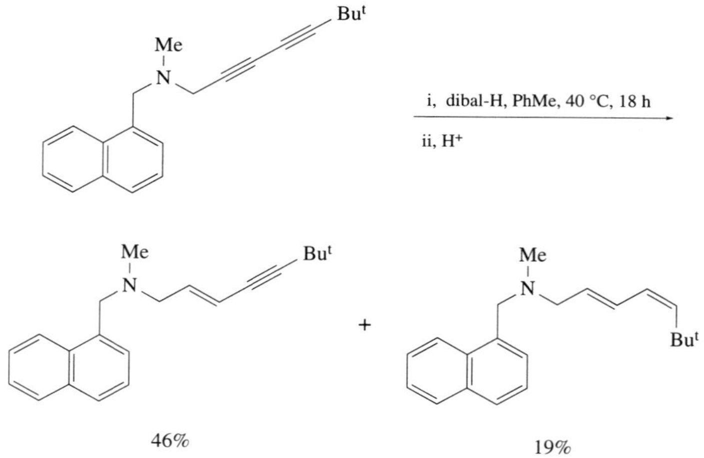

(37)

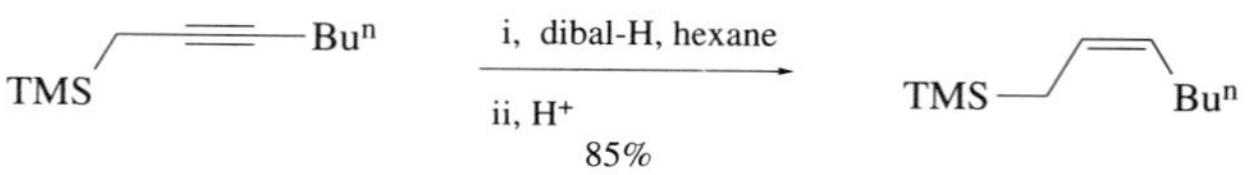

(38)

In an unusual reaction dibal-H reduced a long-chain skipped diyne to give the (*E*,*E*)-diene (Equation (39)) ⟨63JOC1254⟩. One alkyne bond of a diynol (Equation (40)) was selectively reduced to give the (*E*)-enynol using a combination of dibal-H and Bu^nLi in a procedure which is claimed to be milder and more selective than the use of LAH ⟨84S730⟩. The same combination of reagents reduces a 1-silyldiyne to either the (*E*)-enyne or the (*Z*, *E*)-diene, depending upon the conditions (Equation (41)) ⟨83JA1383⟩.

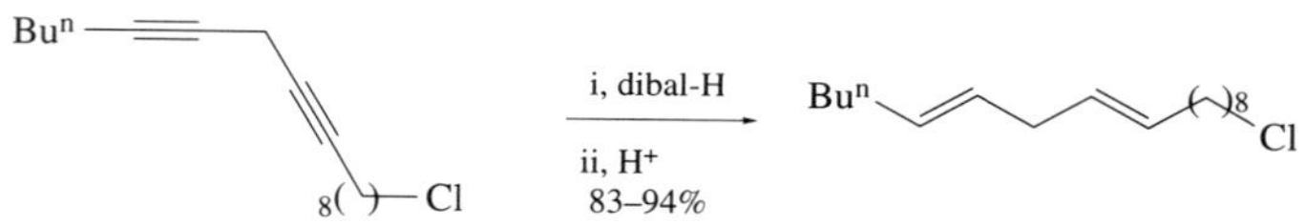

(39)

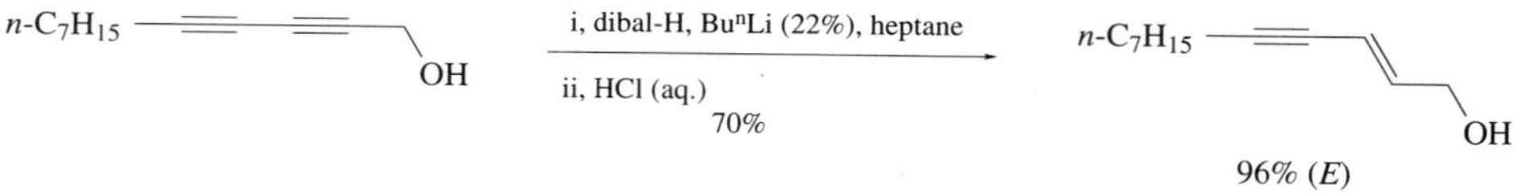

(40)

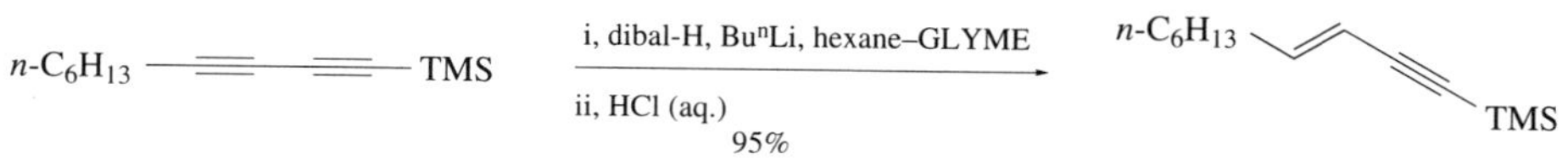

(41)

Lithium aluminum hydride does not generally reduce alkynes, except at elevated temperatures (120–150 °C/DIGLYME/THF) ⟨84OR(32)375⟩, unless there is another functional group present, particularly if this is an oxygen-containing group adjacent to the alkyne. For example, 1-phenyl-1-propyne reacts regioselectively at a lower temperature (Equation (42)).

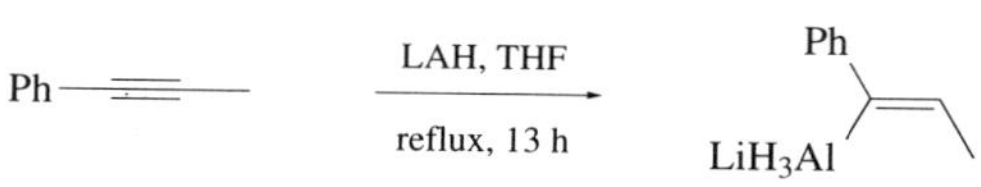

(42)

Reduction of propargyl alcohols by LAH to (*E*)-allylic alcohols is a standard method (Equation (43)) ⟨83TL4899⟩. The solvent is important for stereoselectivity, for example *trans*-addition normally occurs in THF, but *cis*-addition may predominate in diisopropyl ether ⟨70JA4898, 74JOC968⟩. The enynol (**7**) gives the dienol (**8**) (Equation (44)) ⟨54JCS1854⟩, and the aminodiol (**9**) is similarly reduced to the alkene (**10**) (Equation (45)) ⟨59HCA401⟩. In the reduction of the alcohol (**11**) the chiral centre adjacent to the alkyne is not affected (Equation (46)) ⟨78CC621⟩.

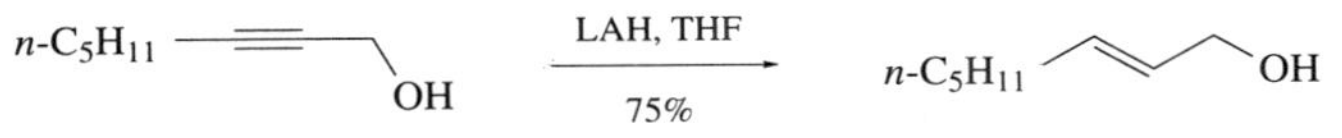

(43)

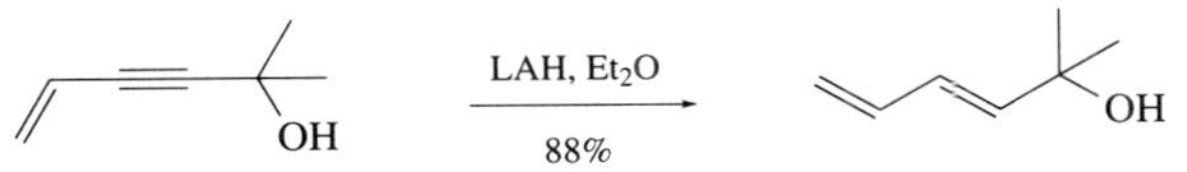

(44)

(**7**) (**8**)

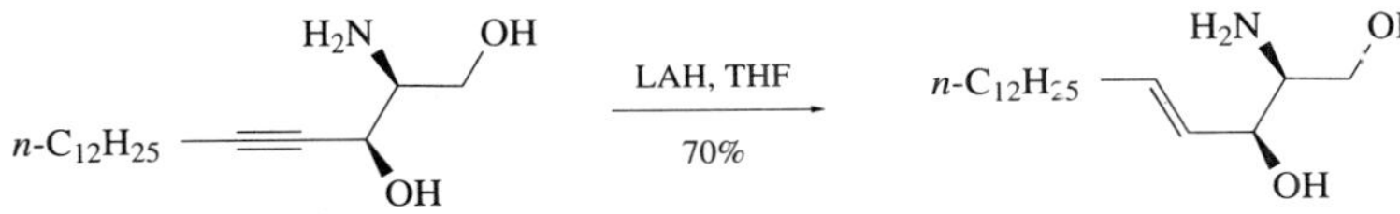

(45)

(**9**) (**10**)

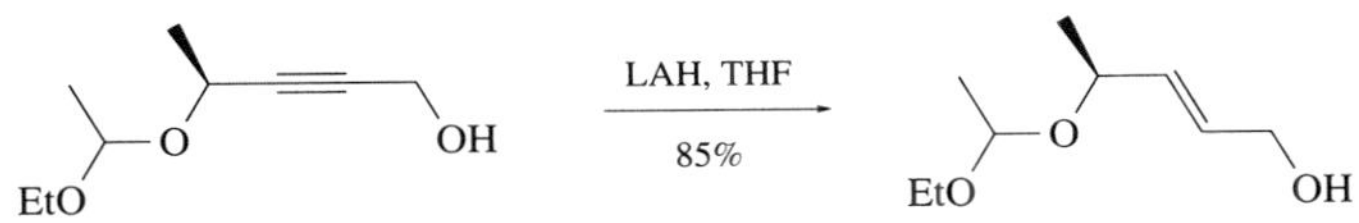

(46)

(**11**)

Where a compound contains several alkyne groups, selective reduction may be possible, for example LAH reduces only the alkyne adjacent to the hydroxyl group in (**12**) (Equation (47)) ⟨83TL4883⟩. When there is an ester or carboxylic acid group adjacent to an alkyne, then both groups are reduced to give an allylic alcohol. For example the ester (**13**) gives the (*E*)-alkenol (**14**) (Equation (48)) ⟨54JCS3217⟩, and ethyne dicarboxylic acid (**15**) gives (*E*)-butene-1,4-diol (**16**) (Equation (49)) ⟨51JA5444⟩. An ether–oxygen on the carbon adjacent to the alkyne may also facilitate reduction; for example the benzyl-protected C-glycoside (**17**) is reduced to the (*E*)-alkene (**18**); however, an elevated temperature (90 °C) in GLYME is required (Equation (50)) ⟨83TL4833⟩.

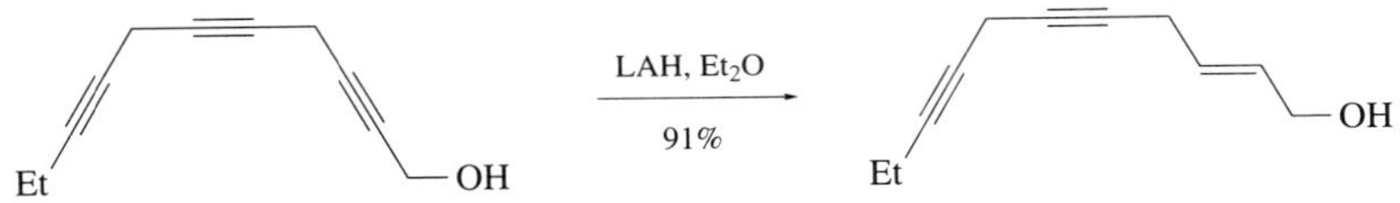

(47)

(**12**)

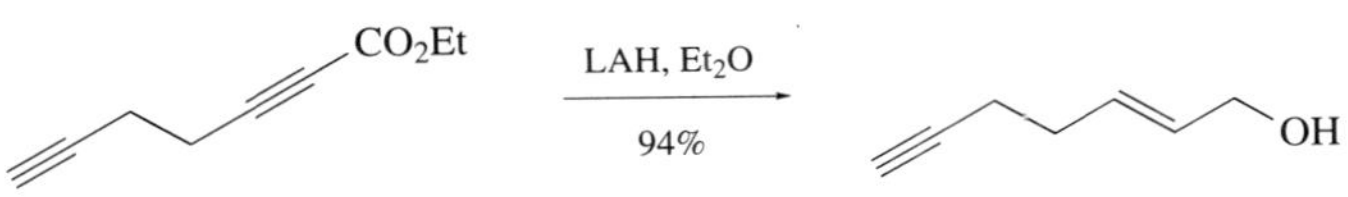

(48)

(**13**) (**14**)

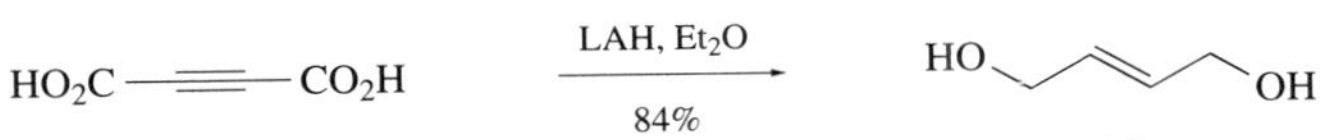

(49)

(**15**) (**16**)

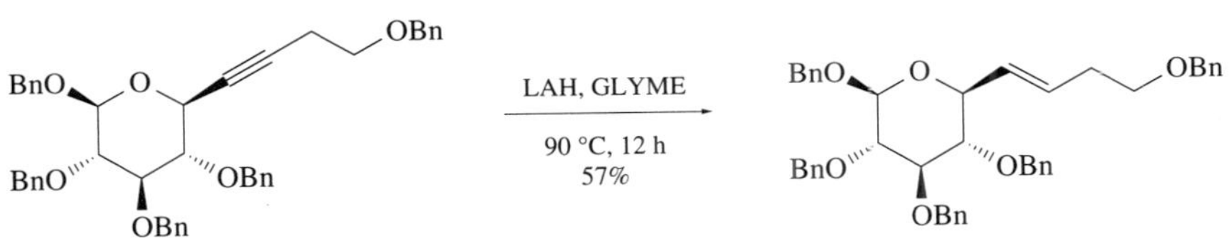

(50)

(**17**) (**18**)

It is not always necessary for the alcohol group to be adjacent to the alkyne in order for reduction with LAH to be successful; for example, the alkynol (**19**), where the hydroxyl group is remote from the alkene, is reduced to the (*E*)-alkenol (Equation (51)) ⟨77S561, 81CB292, 84TL1555⟩.

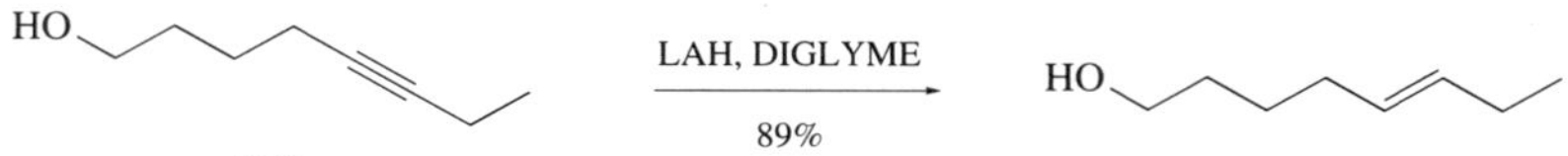

(**19**) (51)

Propargyl amines are reduced by LAH to the (*E*)-allylic amines ⟨87JOC5044⟩, and 2-alkynenitriles are reduced chemoselectively to the alkenenitriles ⟨79S430⟩.

Sodium bis(2-methoxyethoxy)aluminum hydride (Red-Al) also reduces propargyl alcohols to allylic alcohols (Equation (52)) ⟨76JOC3497, 86OS(62)182⟩.

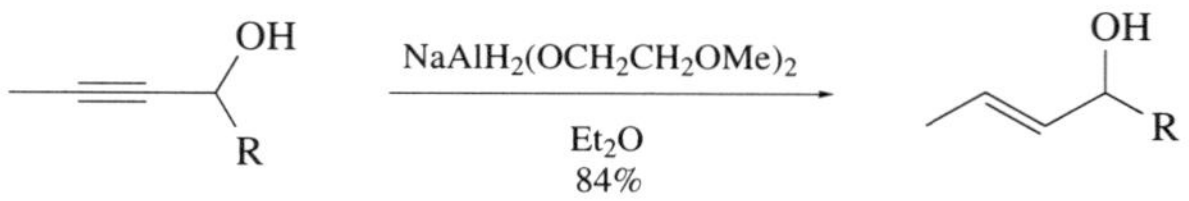

(52)

Unfunctionalized alkynes can be reduced by $LiHAlMe(Bu^i)_2$ ⟨67JA5085⟩, and this reagent also reduces conjugated internal alkynes, with no addition to the second triple bond (Equation (53)) ⟨77S52⟩. With unsymmetrical conjugated alkynes there is no selectivity.

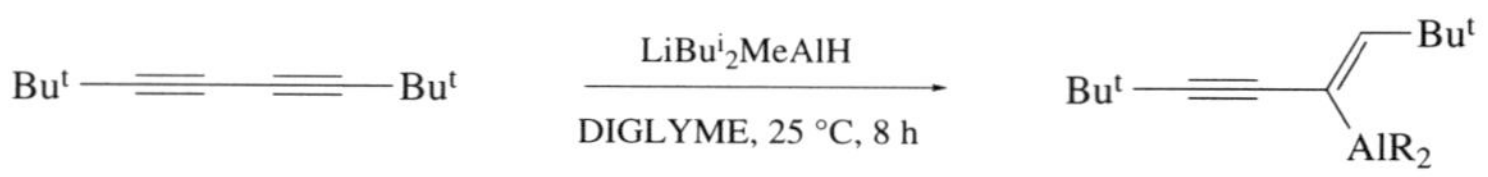

(53)

Simple alkynes can be reduced by LAH in the presence of various transition metal salts. The (*Z*)-alkenes are usually obtained, and the metal salt is generally a halide of one of the transition metals from titanium to nickel ⟨78JOC2567⟩. For example, 2-hexyne is reduced by LAH in the presence of $NiCl_2$ to (*Z*)-hex-2-ene in excellent yield (Equation (54)) ⟨77TL4481⟩, and terminal alkynes are also reduced successfully. Other halides which have been used include $TiCl_4$ ⟨76TL15⟩ and $NbCl_5$ ⟨82CL157⟩.

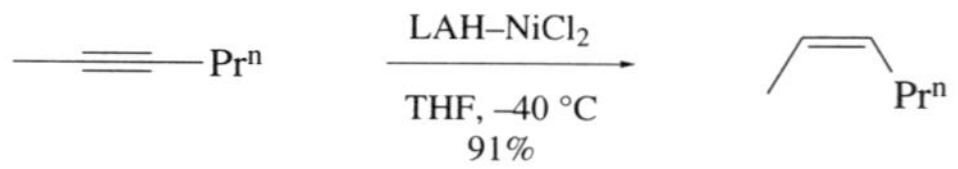

(54)

1.12.1.4.3 Other hydride reducing agents

$NaBH_4$ has also been used to generate a hydrogenation catalyst from $Ni(OAc)_2$ in order to reduce an internal alkyne to give the (*Z*)-alkene selectively (Equation (55)) ⟨84TL2179⟩. Similarly, the alkyne (**20**) was reduced using aminoantipyrine as an additive (Equation (56)) ⟨88CA23052⟩. Alkynes conjugated with amides can also be reduced by $NaBH_4$, in the presence of catalytic amounts of $FeCl_2$ and lipoamide, to give a mixture of geometrical isomers, with moderate selectivity for the (*E*)-isomer (Equation (57)) ⟨85CL1851⟩. Alkyl- or phenyl-substituted alkynes give mostly (*Z*)-alkenes under these conditions.

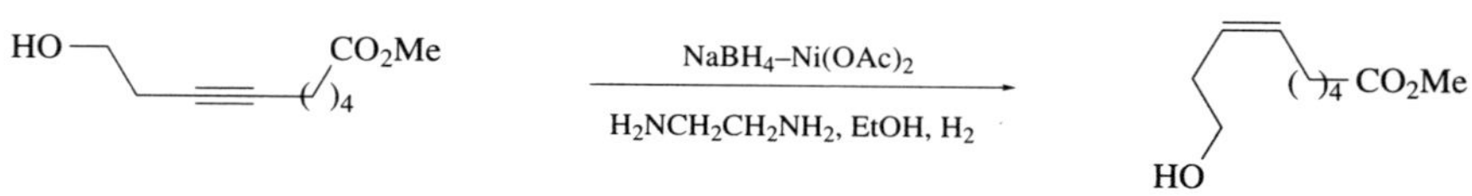

(55)

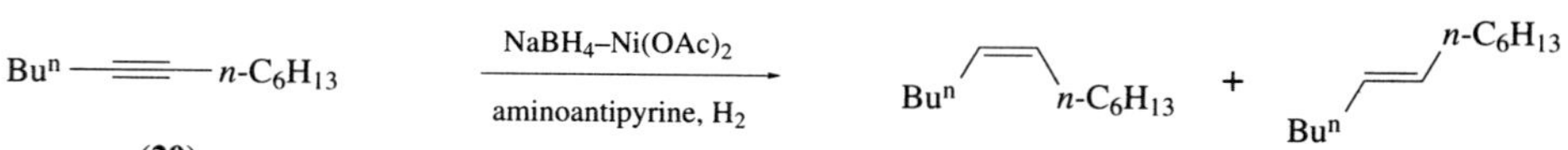

(**20**) 90% 6% (56)

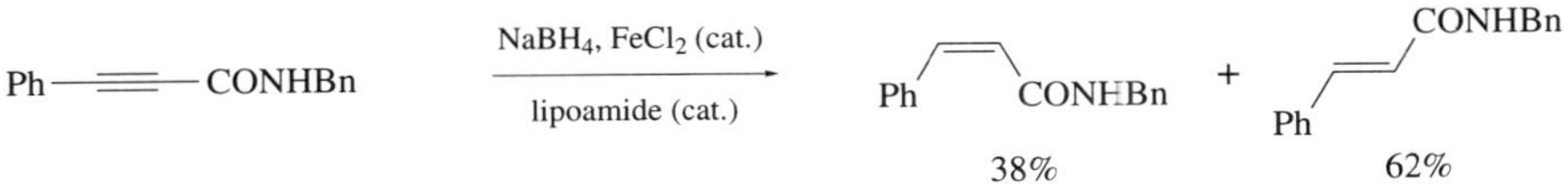

(57)

Diphenylethyne is reduced by $NaBH_4$—$PdCl_2$—PEG200 to give (*Z*)-stilbene (Equation (58)) ⟨85T2387⟩; in this and other examples the benzene ring and nitrile groups were unaffected.

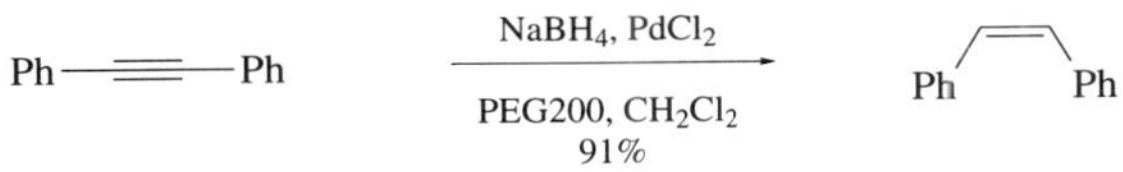

(58)

Only one alkyne in the sulfone diyne (**21**) is reduced by $NaBH_4$ in the presence of Et_3N, to give mostly the (*Z*)-enyne (**22**) (Equation (59)) ⟨90SC3477⟩; in the absence of the base no reduction occurs.

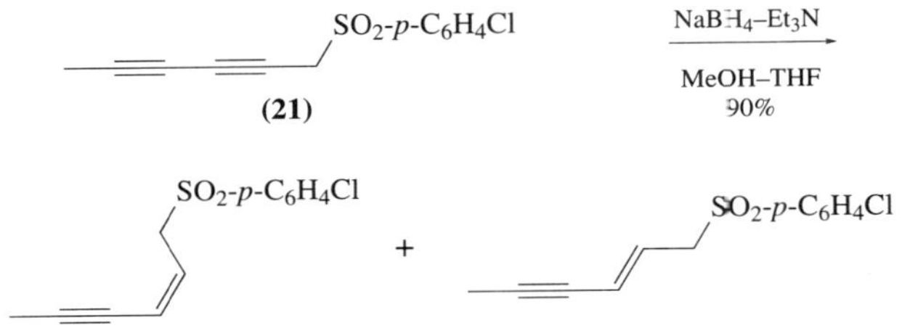

(59)

(**22**) 85 : 15

$NaBH_4$ has been used without any additives to reduce the propargylic alcohol (**23**), in which the alkyne is also conjugated to an allene, to the (*E*)-enol (Equation (60)) ⟨74S573⟩.

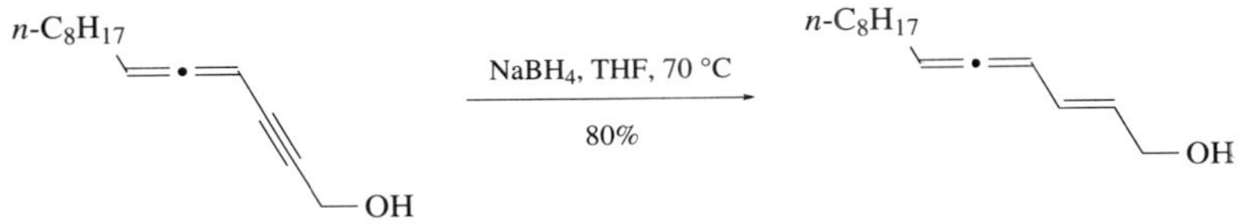

(60)

(**23**)

Other metal hydrides which have been used in the presence of transition metal salts include $NaH/Ni(OAc)_2$ ⟨77TL1069⟩, and MgH_2 in the presence of CuI ⟨78JOC757⟩ or Cp_2TiCl_2 ⟨78CC30⟩. A copper hydride (generated from CuBr and Redal) has been used to reduce conjugated alkynoic esters, giving a mixture of isomers with the (*Z*)-isomer predominating ⟨75JOC3619⟩.

Silanes have occasionally been employed in the presence of transition metals as hydride donors, to reduce internal alkynes to the corresponding (*Z*)-alkenes. The combination of $(EtO)_3SiH$ and a catalytic amount of $RhCl_3$ reduces an alkyne without affecting an ester group ⟨92JOC4786⟩. Terminal alkynes and propargyl alcohols are not suitable substrates, however. Trichlorosilane has also been used in the presence of KF and CuF_2 ⟨79TL1141⟩.

Hydrozirconation of alkynes, followed by protonolysis of the resulting alkenylzirconium complexes gives overall *cis*-reduction of the alkyne ⟨91COS(8)667⟩; however, because this procedure is stoicheometric in reagent, it is not usually the method of choice.

The pyrophoric intermetallic hydride $LaNi_5H_6$, prepared by adsorption of hydrogen onto $LaNI_5$ alloy, reduces 2-butyne-1,4-diol to the (*Z*)-alkene, but this method is not superior to hydrogenation over Lindlar's catalyst (Equation (61)) ⟨87JOC5695⟩.

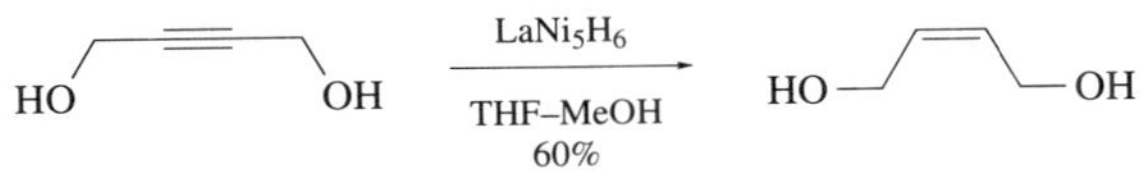

(61)

1.12.1.5 Miscellaneous Reducing Agents

Chromium(II) salts, for example, $CrSO_4$ in aqueous DMF, have been used to reduce alkynes to (*E*)-alkenes in excellent yields (Equation (62)) ⟨64JA4358⟩. Terminal alkynes are also reduced successfully. This reduction is selective for alkynes because simple alkenes do not undergo reduction by chromium(II) ⟨66JA4964, 86JOC253⟩. $Cr(ClO_4)_2$ has also been employed for the reduction of terminal alkynes (Equation (63)), and in contrast to $CrSO_4$, reduces internal alkynes to (*Z*)-alkenes ⟨79JOC3471⟩. 2-Butyn-1-ol is reduced with complete stereoselectivity to (*E*)-2-buten-1-ol using $Cr(ClO_4)_2$ in aqueous DMF, whereas in DMF—Et_3N the (*Z*)-isomer is formed preferentially, albeit with much lower stereoselectivity ⟨79JOC3471⟩. Conjugated ynones are reduced to the (*E*)-enones using either $CrSO_4$ or $CrCl_2$ ⟨86S184⟩.

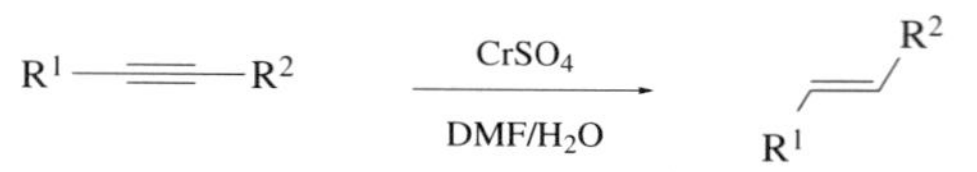

(62)

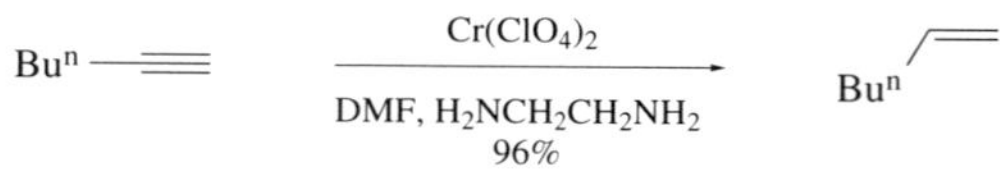

(63)

Internal alkynes are reduced stereoselectively to the (*Z*)-alkenes by the complex $(CuHPPh_3)_6$ (Equation (64)) ⟨90TL2397⟩, and also in very high yield and stereoselectivity with SmI_2 in the presence of catalytic $(PPh_3)_4CoCl_2$ ⟨91TL5559⟩. Butyl Grignard reagents in the presence of CuI are also effective for the *cis*-reduction of alkynes ⟨76JOC4089⟩.

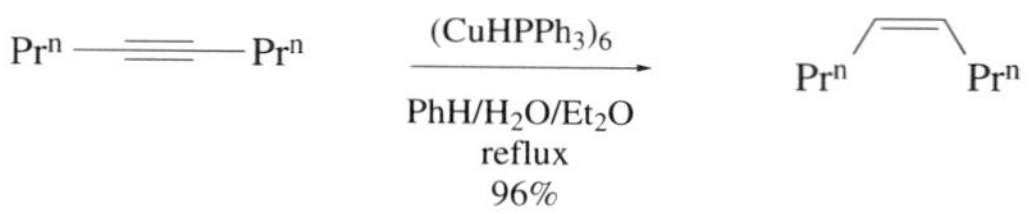

(64)

1.12.2 ELECTROCYCLIC ADDITIONS

This section includes formal cycloaddition reactions, which may be nonconcerted and involving radical or ionic processes, as well as concerted electrocyclic reactions. Cycloaddition reactions involving alkynes have been reviewed ⟨B-78MI 112-03, B-90MI 112-01⟩. Many cycloaddition reactions involving alkynes are catalyzed by transition metals, and this area has also been reviewed ⟨88CRV1081⟩.

1.12.2.1 [2 + 2]-Cycloaddition Reactions

[2 + 2]-Cycloaddition reactions between alkynes to alkenes may occur either photochemically or thermally, and result in the formation of cyclobutenes.

Photochemical cycloaddition gives products which may react further by photochemical electrocyclic ring opening or addition to a second molecule of alkene, and this reaction has been reviewed ⟨B-78MI 112-04⟩. Thus diphenylethyne gives a good yield of the [2 + 2] adduct (Equation (65)) ⟨70JOC1237⟩. With DMAD and ethene, it is difficult to isolate any cyclobutene, because the ring opening and double addition are fast ⟨71JA782⟩.

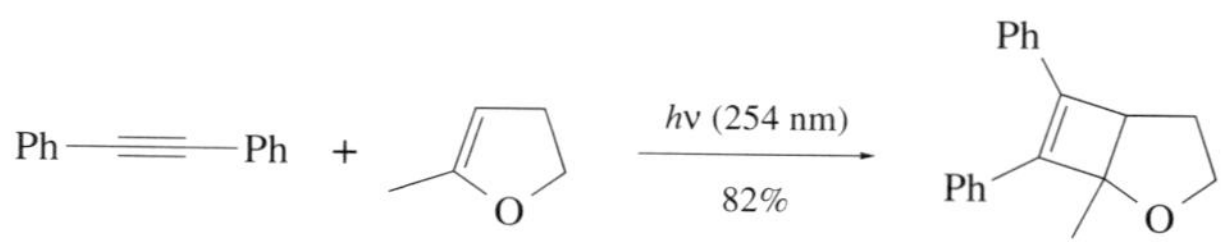

(65)

1-Hexyne undergoes photochemical [2 + 2]-cycloaddition to cyclohexenone to give the regioisomeric cyclobutenes with a 9 : 1 selectivity (Equation (66)) ⟨79T77⟩. In an intramolecular version of this reaction the regioselectivity is complete because of the geometric constraints imposed by the tether; however, the diastereoselectivity is poor (Equation (67)) ⟨82JA5568, 84JA2115⟩.

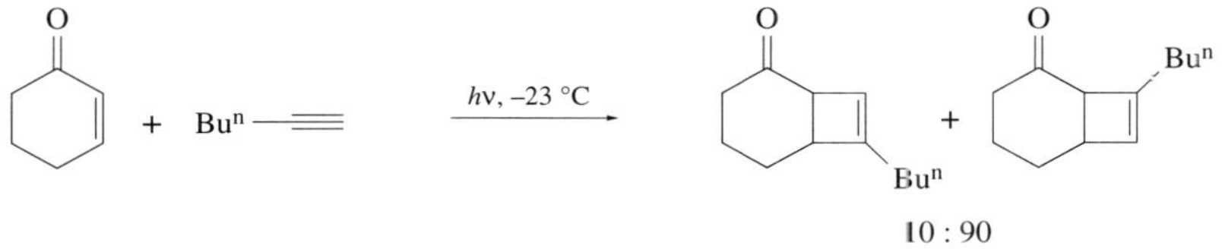

(66)

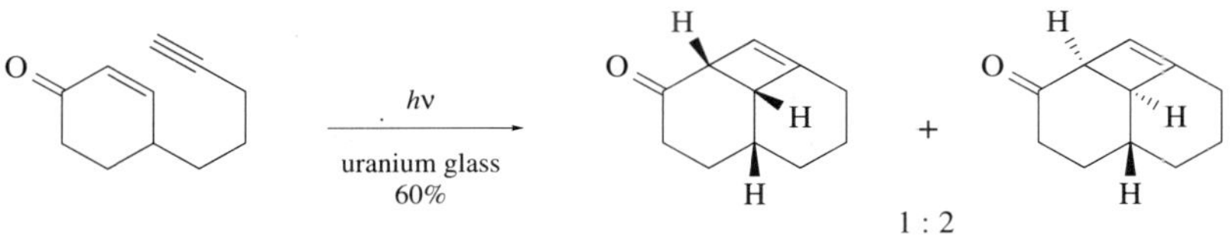

(67)

Alkynes undergo [2 + 2]-photochemical cycloaddition to aromatic rings. With hexafluorobenzene the cyclobutene can be isolated in high yield (Equation (68)) ⟨77JA3504⟩. With benzene the cyclobutene products usually undergo thermal electrocyclic ring opening to the cyclooctatetraene ⟨75T2855⟩ while with naphthalene and diphenylethyne the initially formed cyclobutene undergoes a further intramolecular cycloaddition; this second step can be reversed by heating, allowing the cyclobutene to be isolated (Equation (69)) ⟨71AJC2325⟩.

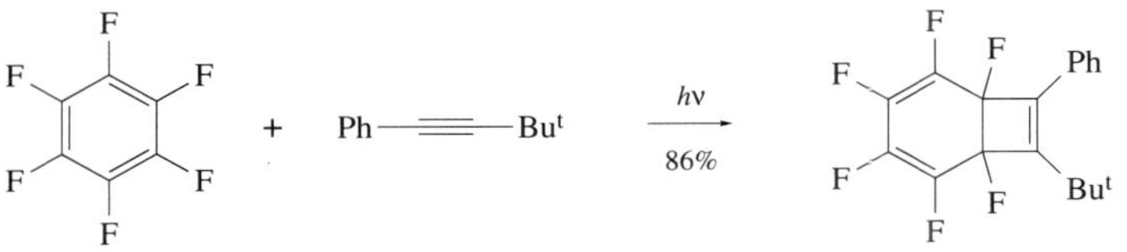

(68)

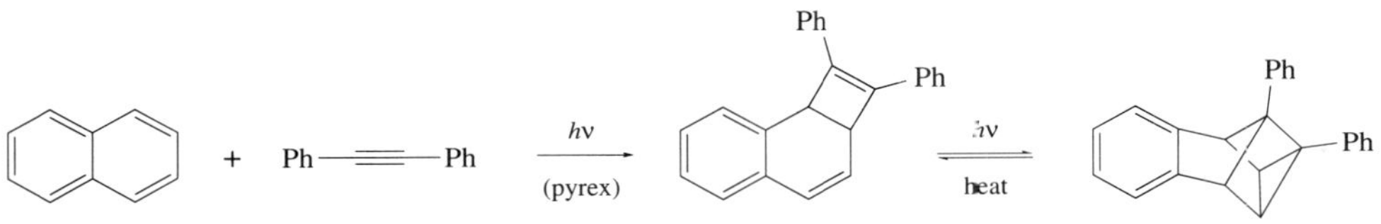

(69)

Quinones undergo [2 + 2]-cycloaddition to the C=C bond (with diphenylalkyne) to give cyclobutenes if the lowest excited triplet state is (π,π^*) (e.g., Equation (70)) ⟨69JOC520⟩, whereas addition to the C=O bond occurs if the (n,π^*) state is lower in energy ⟨67JCS(C)1625, B-74MI 112-01⟩.

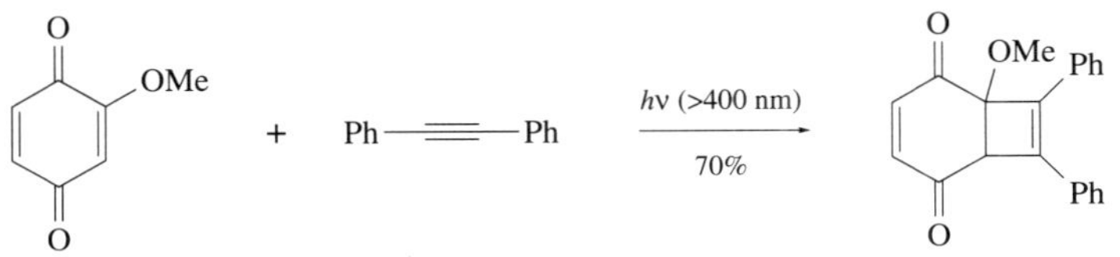

(70)

Cyclobutene formation from alkynes and alkenes can also be catalyzed by Lewis acids ⟨79JOC248, 79JOC253⟩. Transition metal catalysis is also possible ⟨88CRV1081⟩; however, this is limited to norbornene and similar alkenes (Equation (71)) ⟨64CB2451⟩, and to certain alkynes (e.g., DMAD).

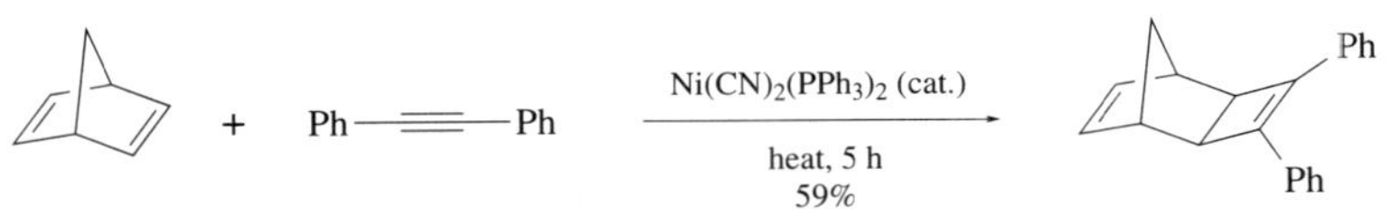

(71)

Transition metal catalyzed cycloaddition of two molecules of an alkyne to give cyclobutadienes is well studied; Fe(CO) complexes are almost universally employed, since the cyclobutadiene may be readily liberated from the product complex ⟨83JOC1262⟩. The chemistry of cyclobutadienes has been reviewed ⟨77CRV691, 80T343, B-85MI 112-03⟩.

Cycloaddition of two molecules of an alkyne using a rhenium catalyst gives a 3-methylene cyclobutene rather than a cyclobutadiene (Equation (72)) ⟨84CB3169⟩.

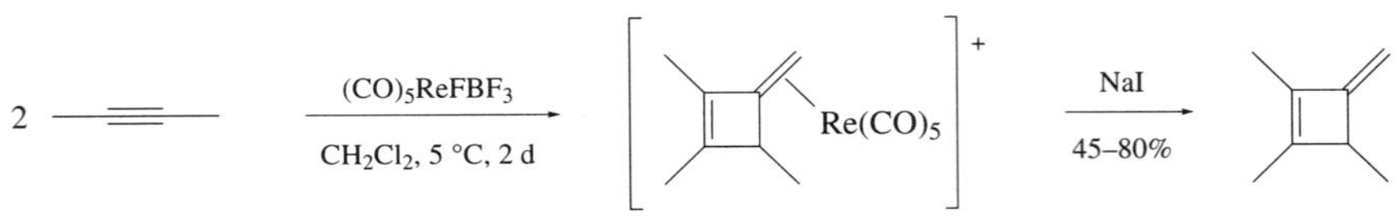

(72)

The enamine of cyclohexanone (**24**) adds to DMAD to give the cyclobutene (**25**) (Equation (73)) ⟨86JOC2004⟩. In similar reactions the cyclobutenes can often be isolated, particularly at low temperature, but usually undergo electrocyclic ring expansion upon heating.

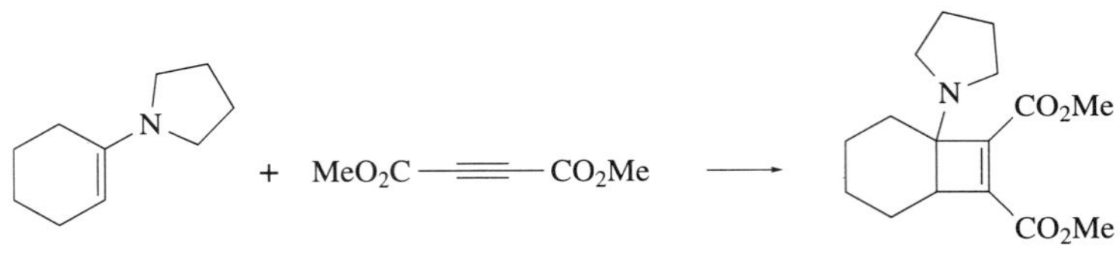

(24) **(25)** (73)

Two sequential [2 + 2]-cycloadditions to alkynes can be observed; thus DMAD first adds to one double bond of norbornadiene, and then diphenylethyne adds to the other, catalyzed by a ruthenium complex ⟨94AG(E)580⟩.

Cycloadditions of alkynes have also been performed with ketenes ⟨83JOC3382, 84TL5043, 90OS(68)32⟩, including cyanoketene ⟨73TL9⟩ this is a good route to the formation of cyclobutenones.

Heterocycles have been used as the alkene component in [2 + 2]-cycloadditions to alkynes; for example dihydropyridines add to DMAD to give fused cyclobutenes (Equation (74)) ⟨73JCS(P1)2818⟩, and the enamine (**26**) gives cyclobutene (**27**) (Equation (75)) ⟨72TL5269⟩. The dihydropyran (**28**) undergoes a similar addition to DMAD; however, the product then undergoes ring opening at the aminoacetal, followed by ring closure and electrocyclic ring expansion to give the highly substituted cyclohexadiene derivative (**29**) (Scheme 5) ⟨75TL3151⟩.

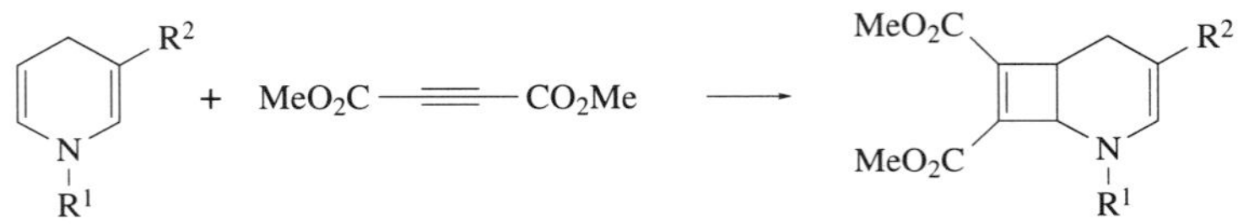

(74)

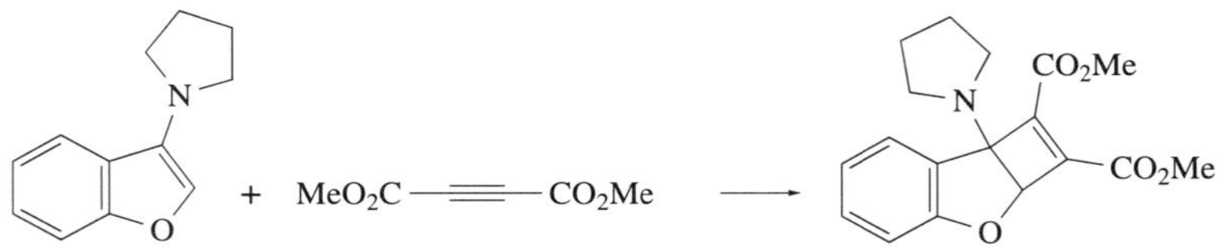

(26) **(27)** (75)

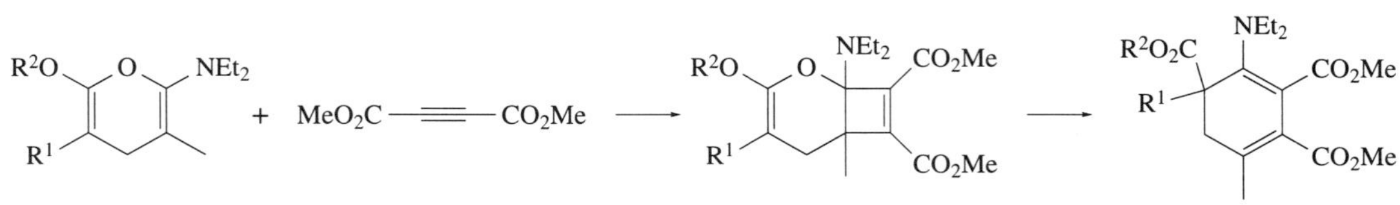

(28) **(29)**

Scheme 5

For the combination of allenes and alkynes there are two competing reactions: the [2 + 2]-cycloaddition and an "ene" reaction. 1,1-Dimethylallene and bis(trifluoromethyl)ethyne give a mixture of the two cyclobutenes (Equation (76)) ⟨71CC1144⟩, whereas tetramethylallene gives only the "ene" product ⟨74JCS(P1)1209⟩. Both the [2 + 2] and the "ene" reactions appear to occur via the same diradical intermediate ⟨74JCS(P1)1844⟩. Allenes can undergo reaction at either double bond; in the example above with 1,1-dimethylallene, reaction occurs mainly at the less-substituted double bond; however, with (**30**) and the hexafluorobutyne the more-substituted double bond reacts, presumably because this results in a relief of bond-angle strain in the cyclopropane ring (Equation (77)) ⟨75JA4413⟩.

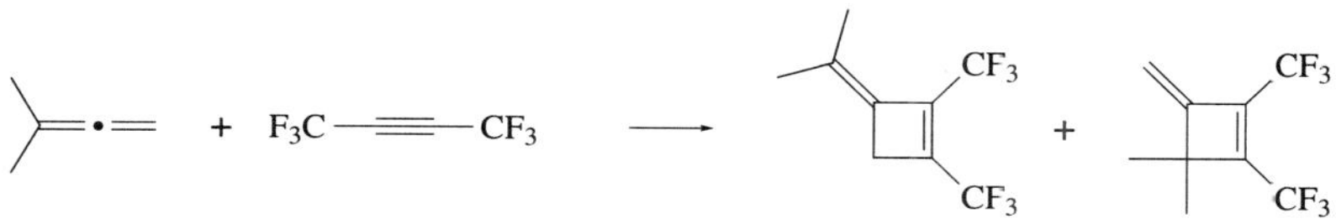

(76)

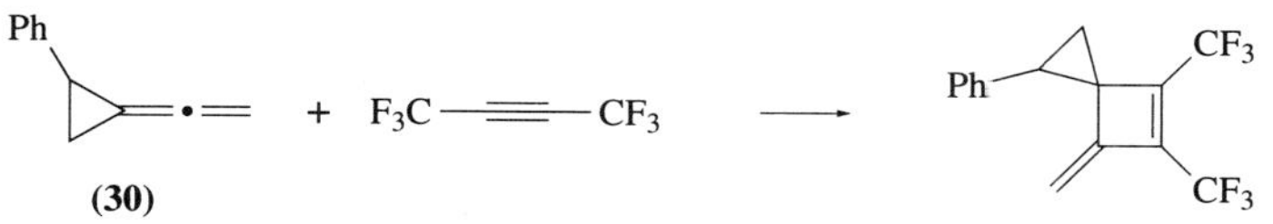

(30)

(77)

[2 + 2]-Cycloadditions can be catalyzed by Lewis acids, giving the cyclobutenes if the alkene is 1,2-disubstituted (Equation (78)) ⟨76JOC3061, 82OM397⟩, and the ene product if the alkene is monosubstituted (see below).

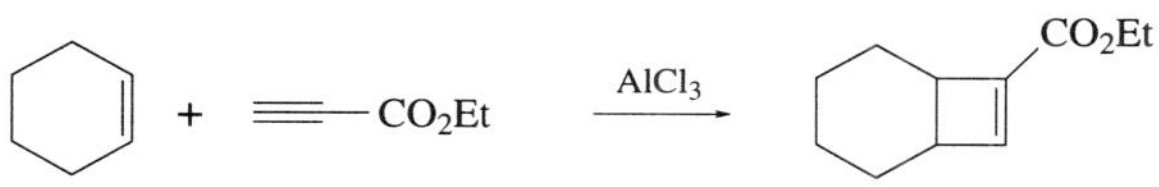

(78)

1.12.2.2 Diels–Alder Reactions

Simple isolated alkynes are not normally reactive as dienophiles in the Diels–Alder reaction ⟨B-90MI 112-01⟩. For thermal cycloadditions between simple dienes and alkynes, drastic conditions are required ⟨83AG(E)778⟩. However, the Diels–Alder reaction between butadiene and diphenylethyne proceeds in good yield at 20 °C in the presence of an iron catalyst (Equation (79)) ⟨68JOC3948, 83AG(E)778⟩. Some cyclic alkynes, for example cyclooctyne, react as dienophiles, as a result of the relief of ring strain ⟨89CB123⟩, and arynes can similarly serve as reactive dienophiles ⟨62AG(E)415, B-67MI 112-02⟩. An *ortho*-quinone dimethide is a very reactive diene which can be generated *in situ* (e.g., by a thermal electrocyclic ring opening of the benzcyclobutane (**31**)) and it reacts with a simple unactivated alkyne, for example propyne, to give the regioisomeric adducts in quantitative yield (Equation (80)) ⟨76TL881⟩.

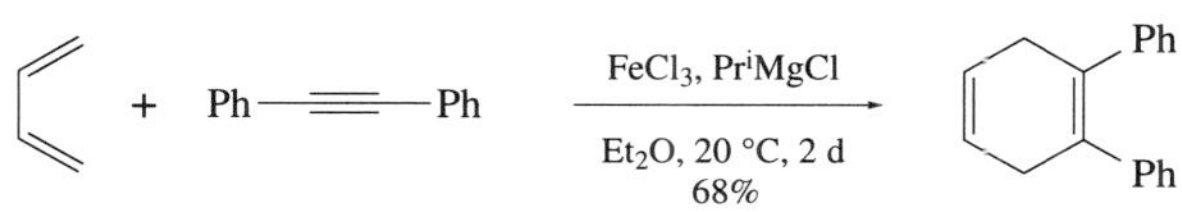

(79)

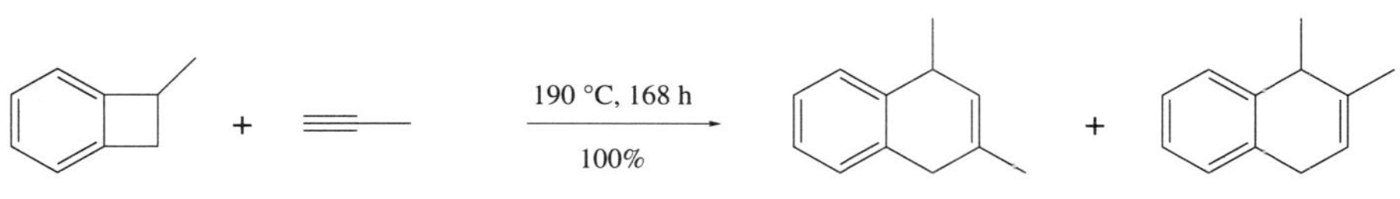

(31) (excess) 65 : 35

(80)

Addition of activating groups, which can be removed later, results in dienophiles which are synthetic equivalents for ethyne, and the use of these has been reviewed ⟨84T2585⟩.

Alkynes activated by electron-withdrawing groups react well as dienophiles, for example ethyne activated by a butanoyl group reacts thermally with butadiene (Equation (81)), and benzoylethyne reacts readily with cyclopentadiene (Equation (82)) ⟨46JCS52⟩. But-3-yn-2-one reacts with a steroid-derived diene to give the Diels–Alder adduct in 91% yield (Equation (83)) ⟨86TL5833⟩. The chiral heterobicyclic diene (**32**) also reacts with high *endo* face-selectivity with the same dienophile (Equation (84)) ⟨84JOC2510⟩.

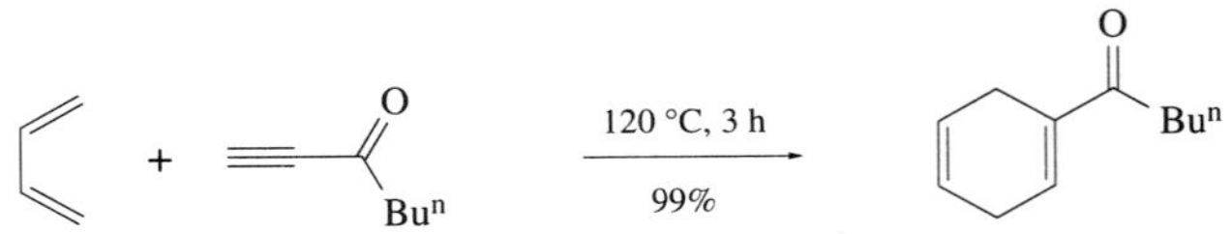
(81)

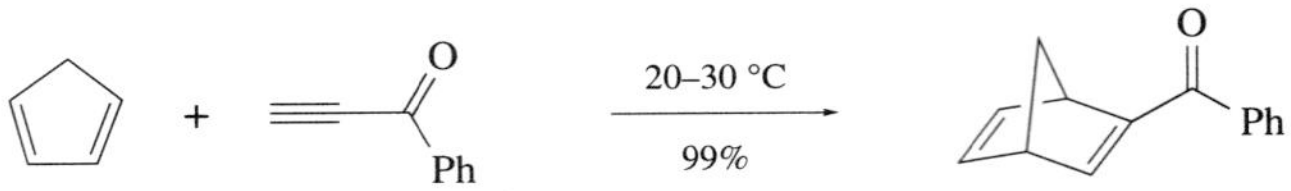
(82)

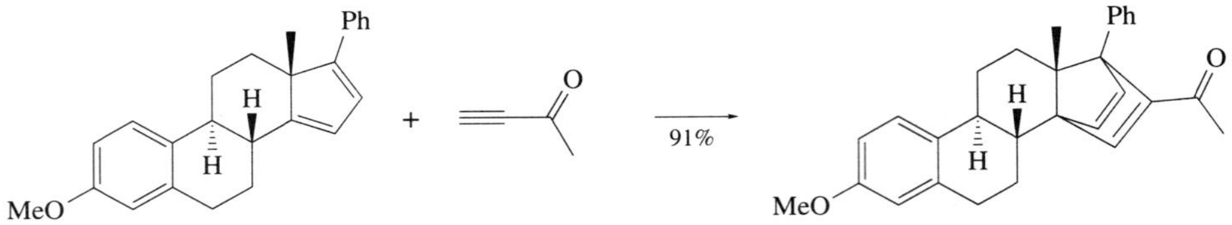
(83)

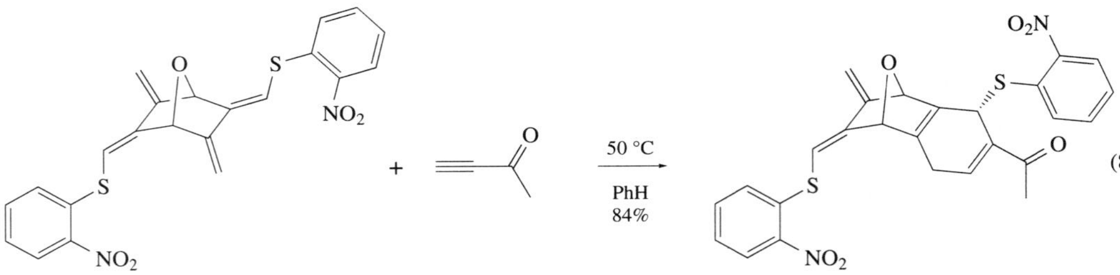
(84)

(32)

A formyl-activating group has been used in the reaction of alkynal (**33**) with the symmetrical bis(silyloxy)diene (**34**) at 115 °C in benzene (Equation (85)) ⟨84JOC1898⟩.

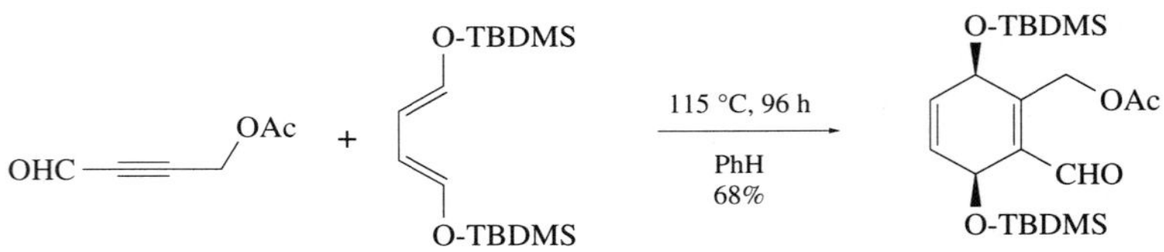
(85)

(33) (34)

Methyl propiolate has been widely used as a dienophile, and reacts with cyclopentadiene at only 50 °C in an uncatalyzed reaction (Equation (86)) ⟨87CB531⟩. With a protected 1-aminobutadiene the same dienophile gives only a 10% yield of the Diels–Alder adduct, with the normal regioselectivity ⟨84JA2443⟩; with a 2-silyloxybutadiene a 73% yield and good regioselectivity is obtained ⟨84TL4357⟩. However, with 1-trimethylsilylpenta-1,3-diene a mixture of regioisomers is obtained (Equation (87)) ⟨81S753⟩.

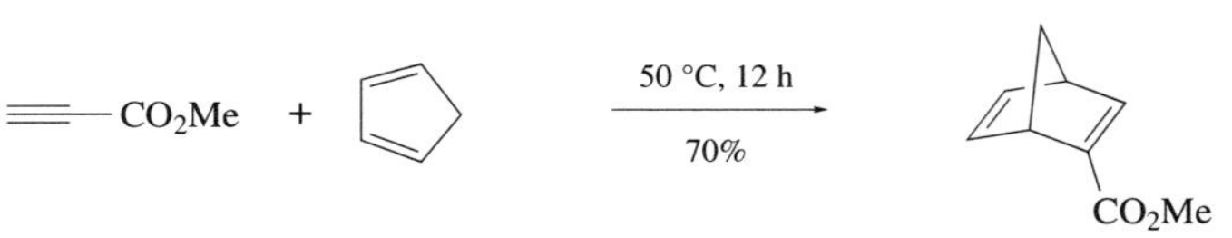
(86)

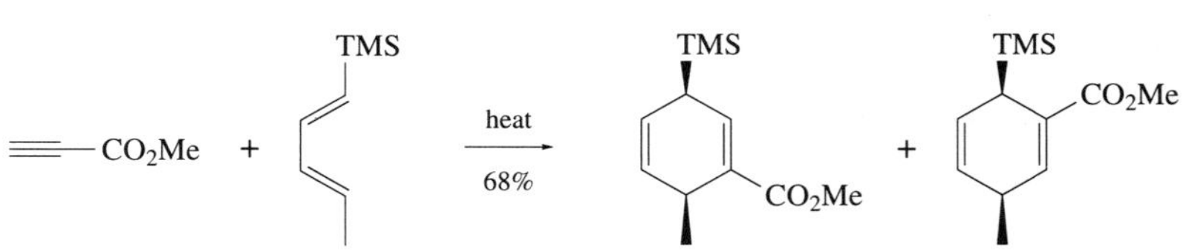
(87)

3 : 1

There have been a few examples of reaction of chiral dienes with methyl propiolate. The threonine-derived diene (**35**) gives a 3 : 1 ratio of diastereoisomers (**36**) and (**37**) when reacted in benzene at 110 °C (Equation (88)); this reverses to a 1 : 1.7 ratio at 220 °C, which in turn improves to 1 : 10 at high pressure and room temperature ⟨87JA5167⟩. An alkynic *ortho*-ester has been used as an equivalent for ethyl propiolate in a low-temperature addition to cyclohexa-1,3-diene, catalyzed by TMS-OTf (Equation (89)) ⟨88TL3407⟩.

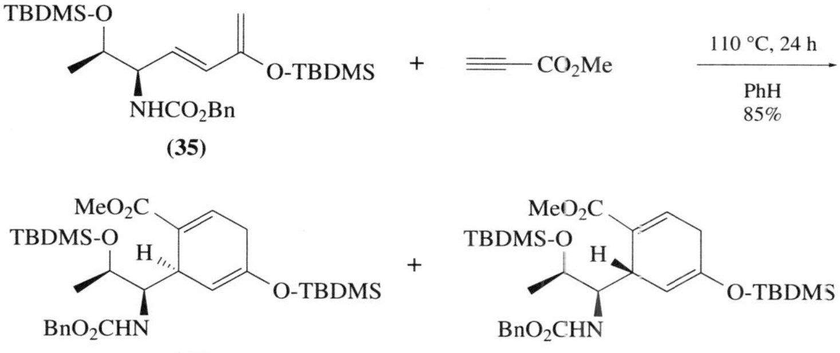

(**35**)

(**36**) 3 : 1 (**37**) (88)

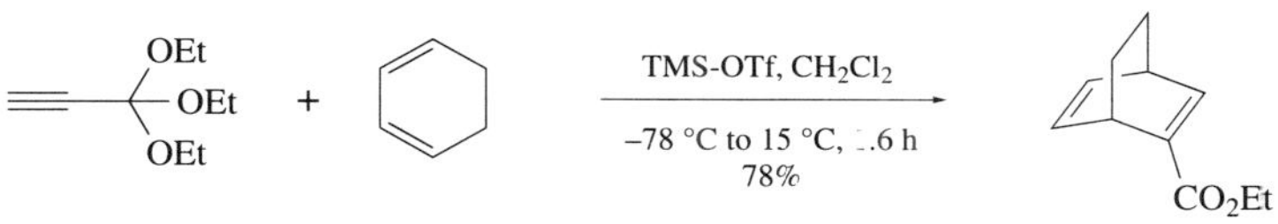

(89)

Among alkyne dienophiles with two activating groups, dimethyl and diethyl ethyne dicarboxylate have been extensively used with a wide range of dienes, and some representative examples are shown in Table 4. As well as acyclic and carbocyclic dienes, a variety of heterocyclic dienes have been reacted successfully with these two dieneophiles. Furans (including furan itself) react well (entry 5). *N*-protected pyrroles usually give poor yields in the thermal reaction; however, the addition of several equivalents of $AlCl_3$ gives good yields at lower temperatures (entry 6).

Photochemical [4 + 2]-cycloadditions are also possible, using pyrroles ⟨68AG(E)67⟩, furan ⟨68CC552⟩, and thiophene ⟨73CB674⟩; thermal [4 + 2]-cycloadditions have also been documented ⟨66AG(E)1039⟩.

Alkenylketene acetals have been used as dienophiles with either DMAD or methyl propiolate; this results in adducts which readily undergo aromatization via elimination to give oxygenated benzoate esters (Scheme 6) ⟨78CL649, 79JA7001⟩.

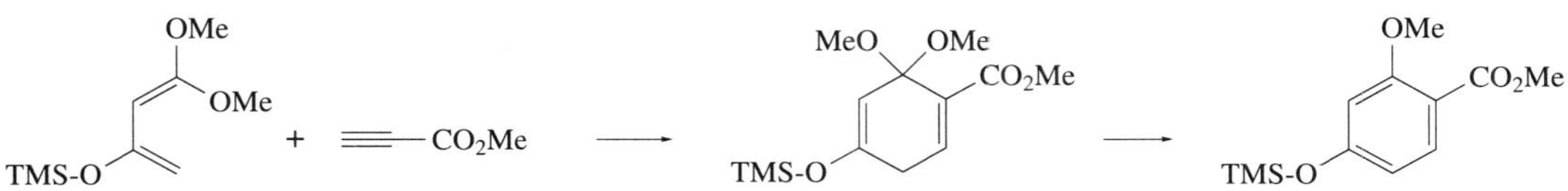

Scheme 6

Ethyne dicarboxylic acid itself has been used as a dienophile, for example with the diene (**38**) (Equation (90)) ⟨55JA73⟩. The diacid chloride has also been used in a cycloaddition with cyclopentadiene ⟨82CB804⟩. The reactions of butynedinitrile, DMAD and diethyl ethyne dicarboxylate as dienophiles with furan-3,4-dinitrile have been compared ⟨62JOC3520⟩.

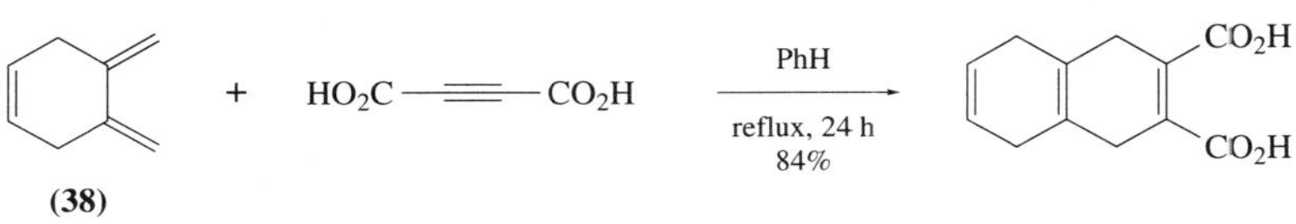

(**38**) (90)

Bis(trifluoromethyl)ethyne is another symmetrical and very reactive dienophile which has been quite often used (Equation (91)) ⟨80TL3323⟩. Its thermal reaction with *N*-benzoylpyrrole is unusual in that it gives a quantitative yield of the Diels–Alder adduct ⟨82JOC4779⟩, and with the unprotected

Table 4 Examples of Diels–Alder additions to dimethyl and diethyl ethynedicarboxylate.

Entry	*Substrate*	*Product*	*Conditions*	*Yield (%)*	*Ref.*
1	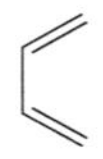	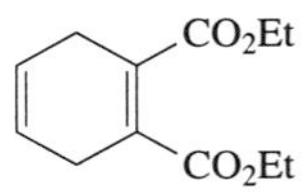	40–45 °C, 134 h pressure bottle	98	83JOC273
2	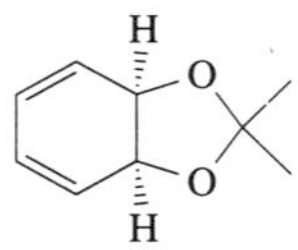	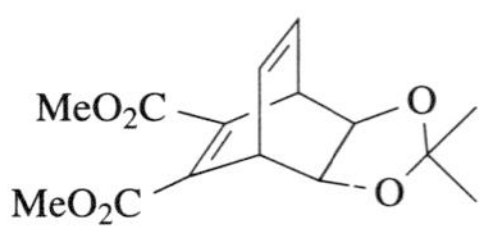	PhH, reflux or H_2O	85	91JCS(P1)1255
3	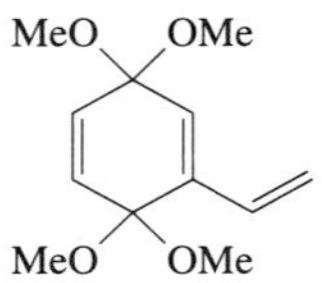	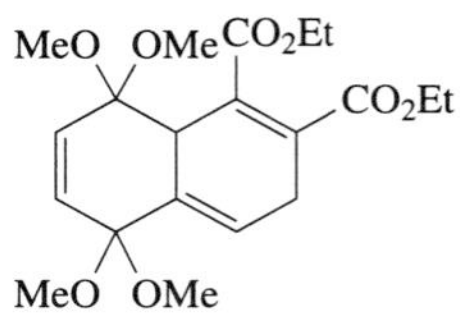	PhH, 160 °C, 72 h sealed tube	83	89JA5948
4	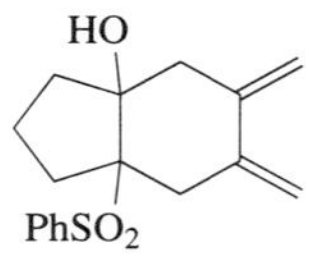	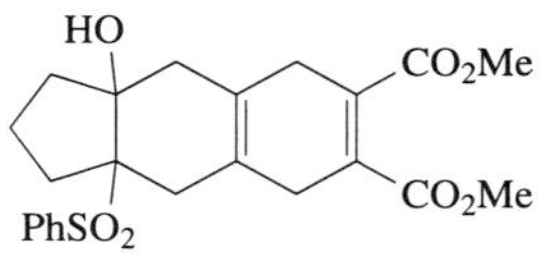	THF, reflux	95	83TL1129
5	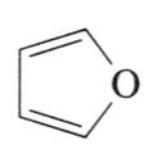	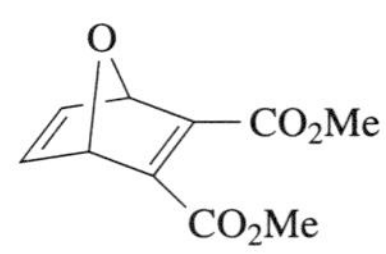	Et_2O, reflux, 130 h	75	66AG1057
6		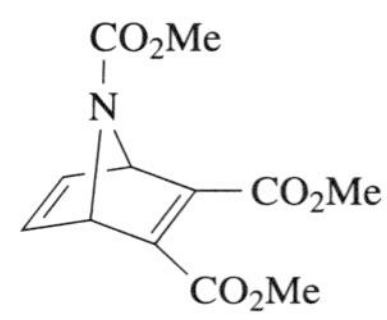	$AlCl_3$, CH_2Cl_2, 40 °C, 1 h	90	69CJC2391

pyrrole itself cycloaddition still occurs, with a second equivalent undergoing addition of the pyrrole N—H atoms ⟨75TL363⟩. It also reacts with silicon- and germanium-containing heterocyclic dienes ⟨72JA6229, 72JOM(42)C21, 73JA3078⟩.

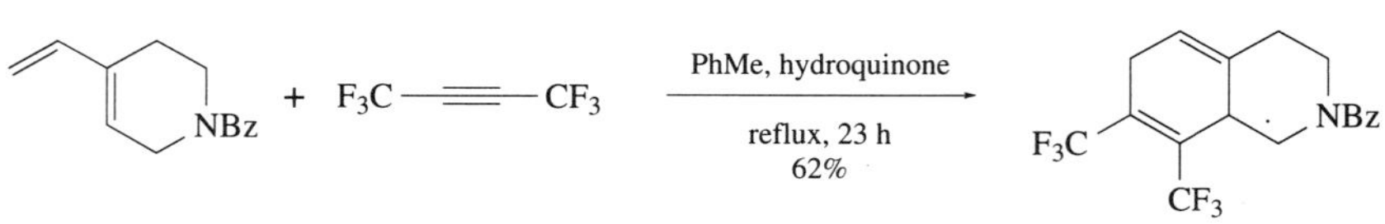

(91)

Dewar benzenes can be prepared from cyclobutadienes and alkynes; in Scheme 7 DMAD adds to a cyclobutadiene ⟨71S139, 86CB244⟩ and in another example a stabilized tricyclic diene undergoes a purely thermal reaction with methyl phenylpropiolate ⟨81TL1675⟩.

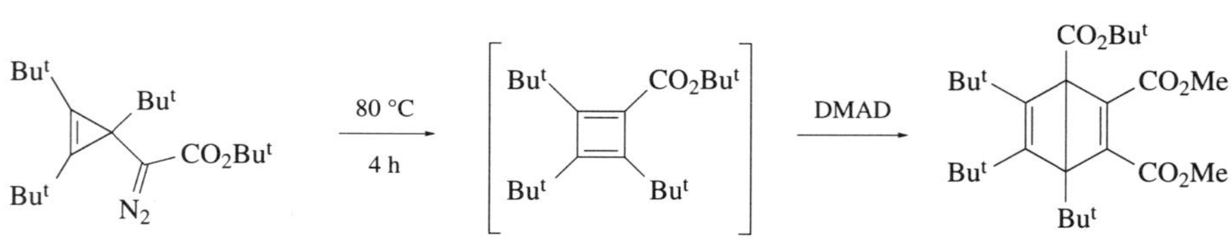

Scheme 7

Alkynes have also been used as dienophiles with heterodienes; for example electron-rich pyrimidines undergo addition to DMAD ⟨75LA1113⟩. 1-Amino-substituted 2-azabutadienes react with methyl propiolate at low temperature without a catalyst to give substituted pyridines (Equation (92)) ⟨75JA4409⟩. Oxazoles can react as heterodienes with unactivated alkynes; in the first example shown in Scheme 8 the adduct loses benzonitrile to give the furan ⟨83TL2473⟩. An intramolecular version of this reaction is also shown in Scheme 8 ⟨81JOC2065⟩.

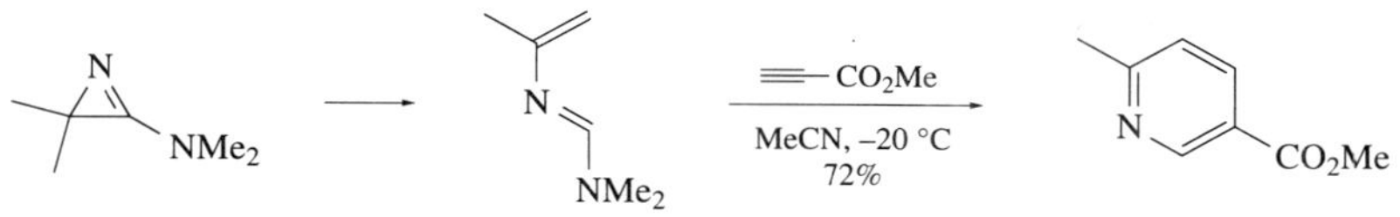

(92)

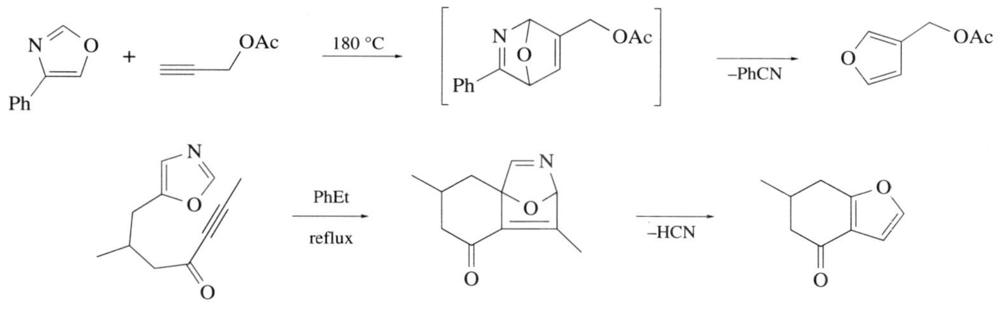

Scheme 8

Intramolecular Diels–Alder cycloadditions of alkynes are also possible ⟨84OR1⟩. The link between the alkyne and the diene generally has to have at least three atoms (e.g., ⟨81S753⟩), and two examples are shown in Equations (93) ⟨85TL5393⟩ and (94) ⟨84OR(32)1⟩. Intramolecular Diels–Alder reactions with an ester linkage have been employed in the synthesis of a naphthalene derivative ⟨66JOC2376⟩ and in a stereoselective synthesis of the A and B rings of gibberellic acid ⟨73TL4477⟩. Intramolecular Diels–Alder reactions of alkynes have also been catalyzed by, for example, compounds of nickel ⟨89JA6432⟩, aluminum ⟨89TL7157⟩, and rhodium ⟨90JA4965⟩.

The 1,2-diazine (**39**) undergoes an intramolecular Diels–Alder reaction, with loss of nitrogen from the adduct to give a dihydroindole (Equation (95)) ⟨87JA2717⟩.

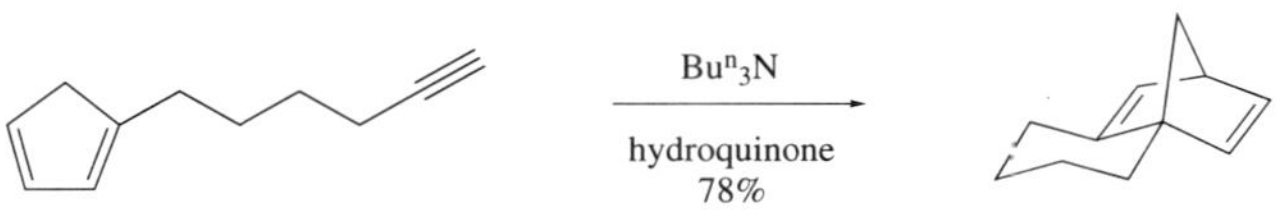

(93)

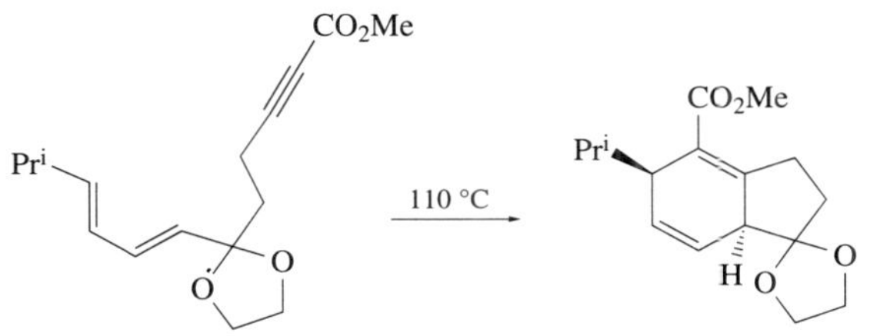

(94)

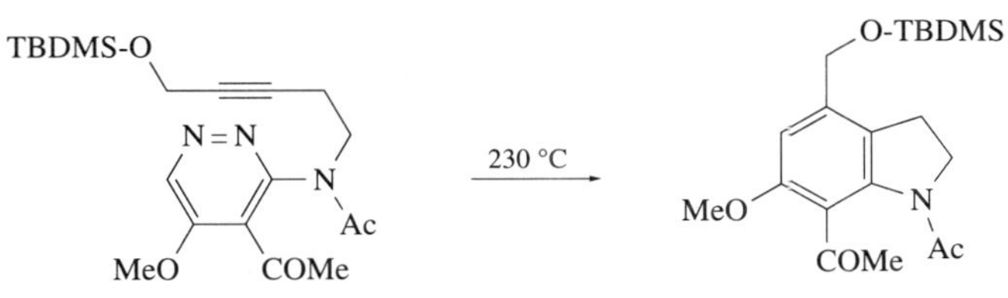

(95)

(**39**)

1.12.2.3 1,3-Dipolar Cycloaddition Reactions

1,3-Dipolar cycloaddition reactions have been reviewed ⟨B-84MI 112-01⟩. Within the scope of this chapter, the 1,3-dipole is limited in structure to those where atoms 1 and 3 are both carbon in order to obtain a nonheterosubstituted alkene product.

Stabilized azomethine ylides add to DMAD, and the ylide can be generated by a thermal conrotatory ring opening of an aziridine. The cycloaddition to DMAD then takes place under the same conditions, faster than isomerization of the ylide, resulting in good stereoselectivity in the formation of the cycloadduct (Scheme 9) ⟨67JA1753, 69AG(E)604⟩. Nonstabilized azomethine ylides have also been prepared *in situ*, by desilylation of silylmethyl imminium salts; addition to DMAD gives the pyrrolidine ⟨79JA6452, 84JOC3314⟩.

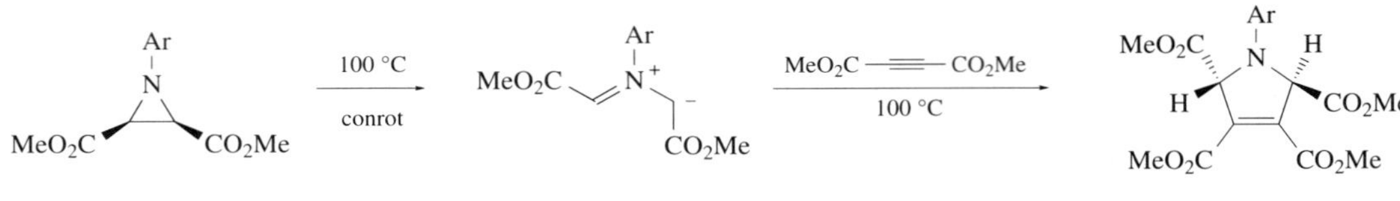

Scheme 9

Intramolecular addition of azomethine ylides generated from aziridines has also been accomplished with an ester linkage containing 3 or 4 atoms; here, terminal alkynes give better yields than internal ones, and oxidation of the adduct yields bicyclic pyrroles ⟨85JOC2309⟩. The azomethine ylides can also be generated from an oxazole precursor ⟨89TL2627⟩, and the linkage can also include an aryl ring ⟨85T3547⟩. An intramolecular addition where the azomethine ylide is part of a dihydroisoquinoline ring has been reported ⟨86JOC1159⟩.

Nitrile ylides have been used as 1,3-dipoles in cycloadditions to alkynes. An intramolecular example is shown in Equation (96) ⟨77JHC947⟩.

An intramolecular cycloaddition of a carbonyl ylide to a terminal alkyne gives the complex tricyclic system shown in Scheme 10 ⟨83TL1675⟩.

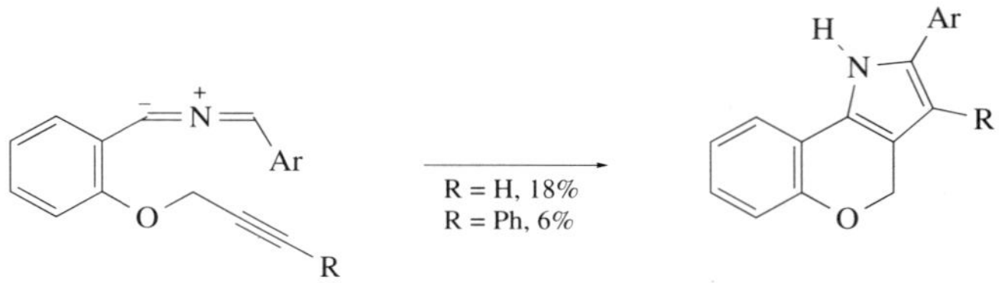

(96)

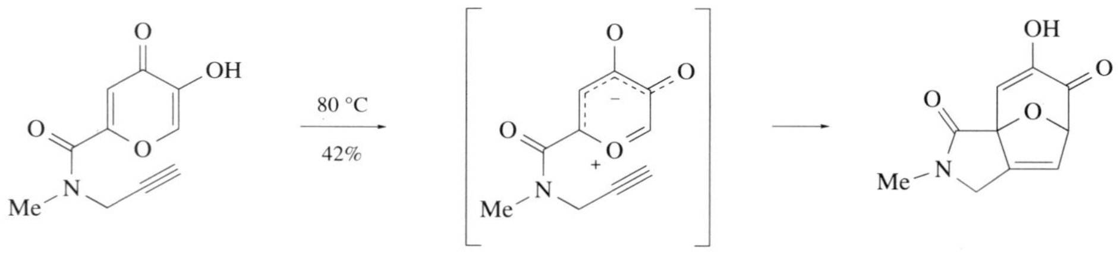

Scheme 10

1.12.2.4 Ene Reactions

Ene reactions have been reviewed ⟨69AG(E)556, 78AG(E)476, 80ACR426, 91COS(5)1⟩. Ene reactions between alkenes and alkynes result in the formation of 1,4-dienes. Alkynes are more reactive than alkenes as enophiles in both intermolecular and intramolecular ene reactions ⟨69AG(E)556, 78AG(E)476⟩. Intermolecular ene reactions occur thermally between simple unactivated ene components and electron-deficient alkynes (e.g., DMAD, alkyl propiolates, dicyanoethyne) at temperatures below 200 °C (Scheme 11) ⟨62LA(651)141, 69AG(E)556⟩. Disubstituted alkenes give both higher yields and regioselectivity than monosubstituted alkenes, and limonene gives 70–80% of the ene product with methyl propiolate ⟨79TL2625⟩.

Lewis acid catalysis allows ene reactions to occur at room temperature with activated alkynes. For example, methyl propiolate reacts with a variety of alkenes (1,1-di-, tri- and tetrasubstituted) in the presence of $AlCl_3$ or $EtAlCl_2$ (Equation (97)) ⟨76JOC3061, 79JA5283, 79JOC2802⟩. With 1,2-disubstituted alkenes [2 + 2]-cycloaddition occurs, and with monosubstituted alkenes a mixture of

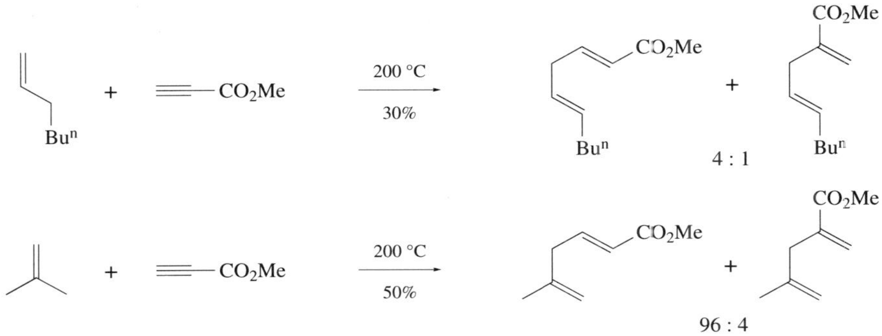

Scheme 11

[2 + 2] and ene products is formed. For alkenes with nonbasic substituents, one equivalent of Lewis acid usually gives optimal yields; however, if basic functional groups are also present, an extra equivalent of catalyst should be added for each basic site ⟨79JA5283⟩. Similar results are obtained using DMAD ⟨79TL1963, 81CJC288⟩.

With DMAD and trisubstituted alkenes, the regioselectivity of transfer of the hydrogen has been studied ⟨79JA1906, 80JOC2773⟩, as indicated in Equation (98).

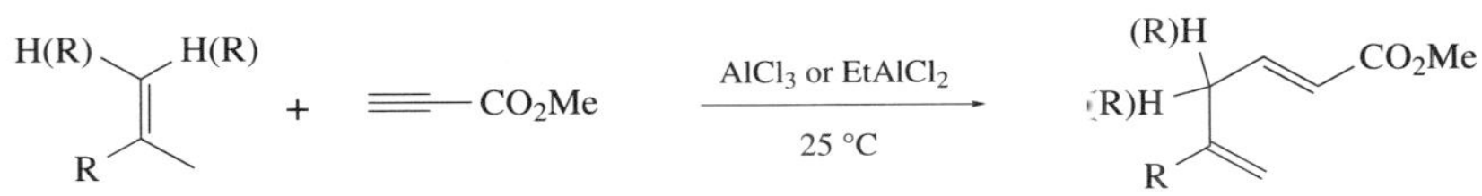

(97)

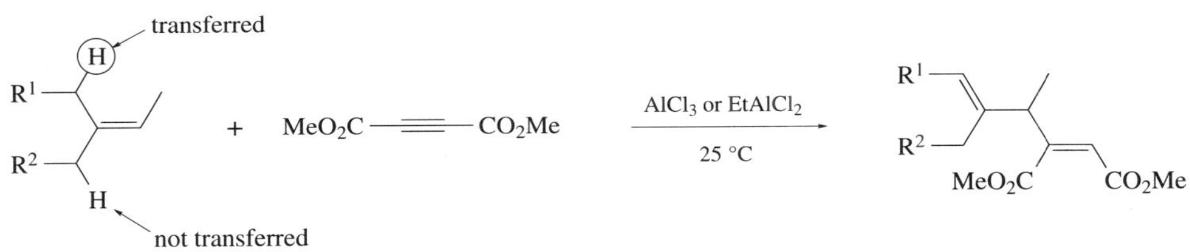

(98)

3-Butyn-2-one is also an effective enophile in Lewis acid-catalyzed reactions ⟨77TL2831⟩.

Intramolecular ene reactions result in the formation of a new ring. Dehydrolinloöl (**40**) cyclizes in quantitative yield upon heating to 200°C (Equation (99)) ⟨71LA(747)60, 72TL2013⟩. Similar results have been obtained by heating in a microwave oven for 15 minutes ⟨86TL4945⟩. Stork has used this type of intramolecular ene reaction to construct the five-membered ring in a prostaglandin synthesis ⟨76JA6747⟩.

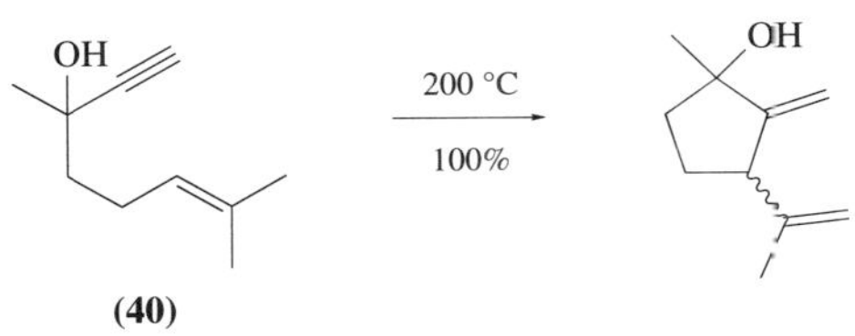

(99)

(40)

The effect of alkyne substitution upon reactivity in the formation of five-membered rings has been investigated; the order of reactivity is ester > terminal alkyne > internal alkyne, and the temperatures required vary between 135 and 225°C (Scheme 12) ⟨78JOC2161⟩. Higher temperatures are required for intramolecular ene reactions of alkynes which give the corresponding six-membered rings (Scheme 13) ⟨78JOC2161⟩, and there are few examples of this process.

Oxygen-containing heterocycles can be constructed using an ether linkage ⟨87CL2347⟩ and the same order of reactivity was found as above.

Intramolecular ene-type products can also be obtained using palladium catalysis, with a reversed regioselectivity for the transfer of the hydrogen atom compared with the thermal reaction (Scheme 14) ⟨85JA1781, 85TL4887, 87TL1611⟩.

Enols react as the ene component with alkynes in intramolecular ene reactions (the Conia

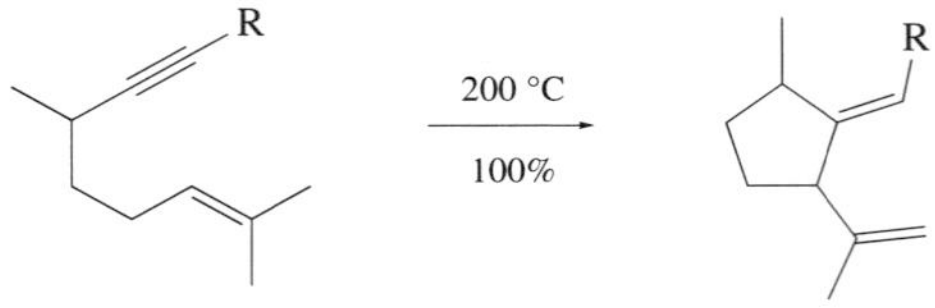

R	*Conditions*	*Yield*
H	210 °C, 62 h	>95%
CO_2Me	135 °C, 24 h	>95%
Me	225 °C, 48 h	15%

Scheme 12

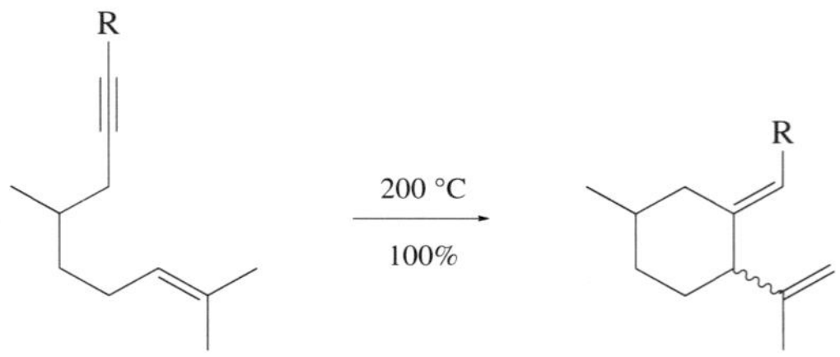

R	*Conditions*	*Yield*
H	225 °C, 48 h	17%
CO_2Me	225 °C, 60 h	85%

Scheme 13

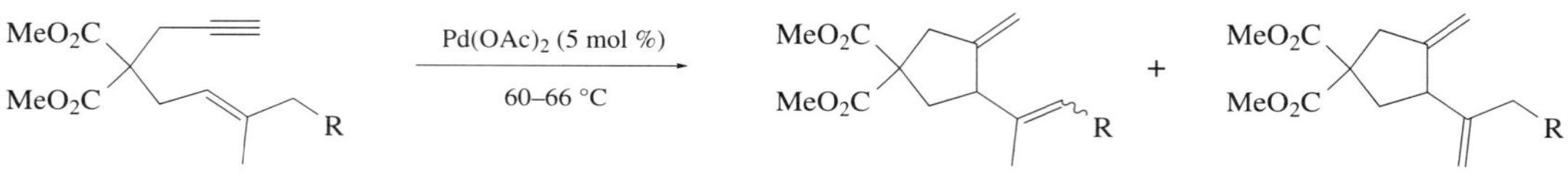

	80%	6 : 94
Thermal, 625 °C	83%	100 : 0

Scheme 14

reaction), either thermally ⟨75S1⟩, or at much lower temperatures using zinc or tin halides (Equation (100)) ⟨81JCS(P1)1516⟩.

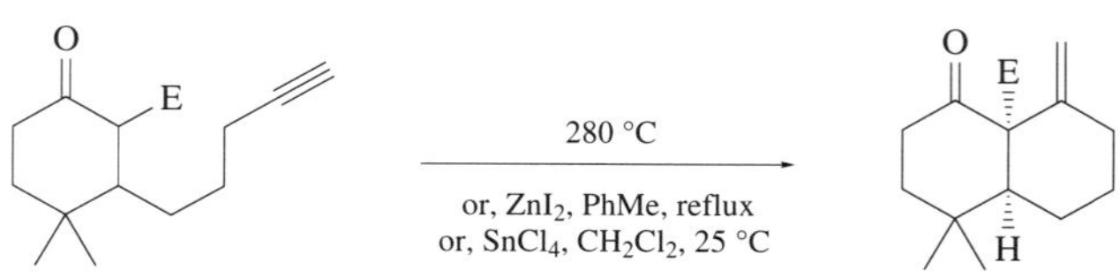

(100)

E = CO_2Et

Use of alkynyl silylenol ethers and mercuric chloride allows milder conditions to be employed; mercuration of the terminal alkyne occurs and allows introduction of electrophiles stereoselectively (Scheme 15) ⟨85JA1726, 87BSF1006, 87BSF1015, 87TL3923⟩.

1.12.2.4.1 Metallo-ene reactions

The metal used for metallo-ene reactions with alkynes is usually nickel (formed by reaction of $Ni(CO)_4$ with an allyl halide) and this strategy has been reviewed ⟨67AG(E)124⟩.

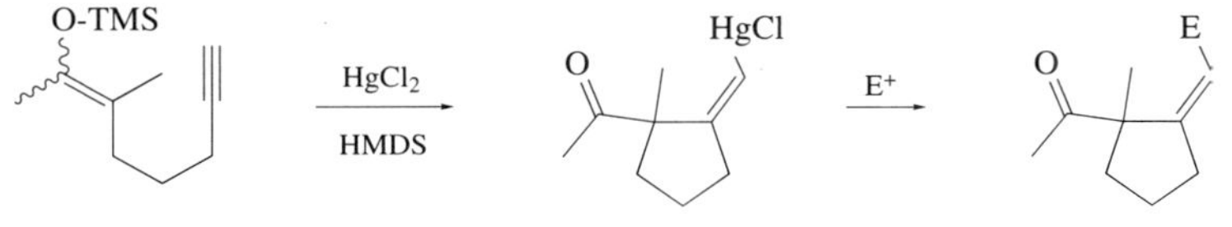

E = H, D, CO_2Me, COMe

Scheme 15

Ethyne itself can participate as the enophile, as in Scheme 16; insertion of CO into the metallo-ene product results in the unsaturated ester (**41**).

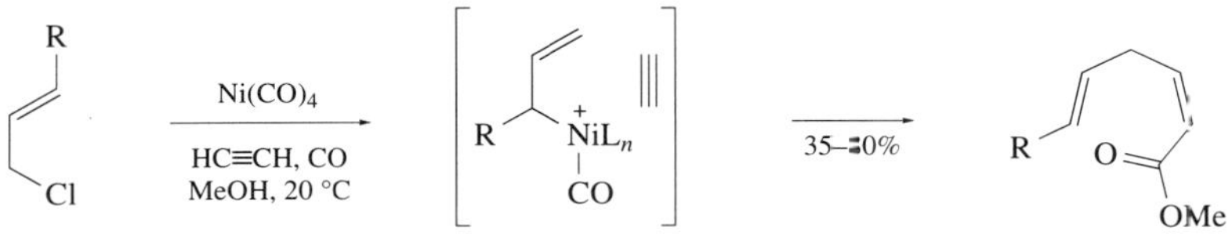

(**41**)

Scheme 16

With unsubstituted allyl chloride, the intermediate acyl nickel species undergoes cyclization to give a cyclopentenone and this process has been used in the synthesis of methylenomycin B ⟨85TL6397⟩.

An intramolecular version of this reaction gives a bicyclic cyclopentenone (Equation (101)) ⟨87TL4745⟩.

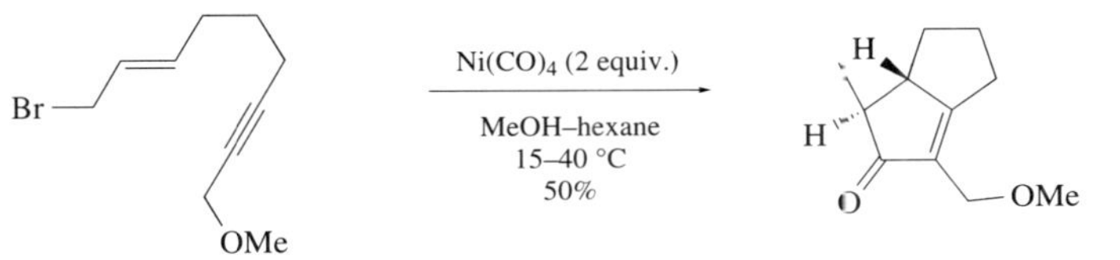

(101)

1.12.2.5 [3 + 2]-Cycloaddition Reactions

[3 + 2]-Cycloaddition reactions result in five-membered ring formation, with the alkyne contributing two of the atoms. For example, the silylallene (**42**) reacts with 3-butyn-2-one in a [3 + 2]-reaction catalyzed by $TiCl_4$ (Equation (102)) ⟨83T935⟩.

Trimethylene methane additions to alkynes have been achieved using alkenylcyclopropane precursors; in the diphenyl-substituted example shown in Equation (103) good regioselectivity is obtained ⟨85AG(E)316, 87TCC77⟩.

An intramolecular version of this reaction is shown in Equation (104); here a palladium catalyst is used instead of nickel ⟨89TL7107⟩.

The same type of product can be obtained by cycloaddition of a diradical, prepared by thermal extrusion of nitrogen, onto DMAD (Scheme 17) ⟨85JOC2400⟩; in this case, a bicyclic product is obtained from an intermolecular addition.

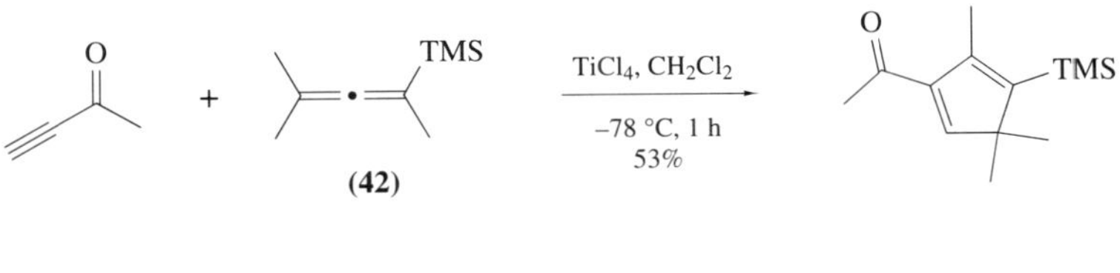

(**42**)

(102)

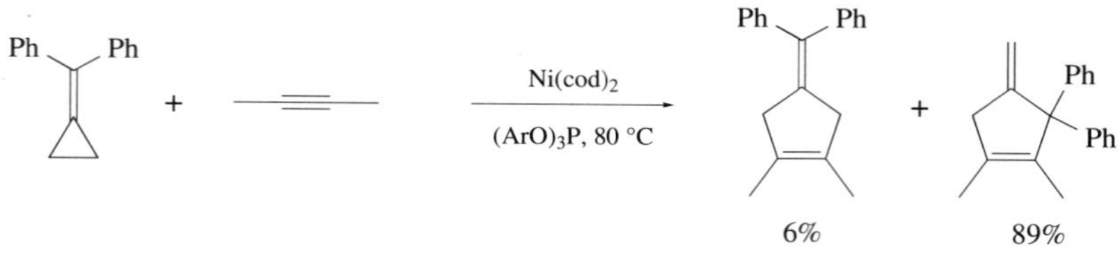

(103)

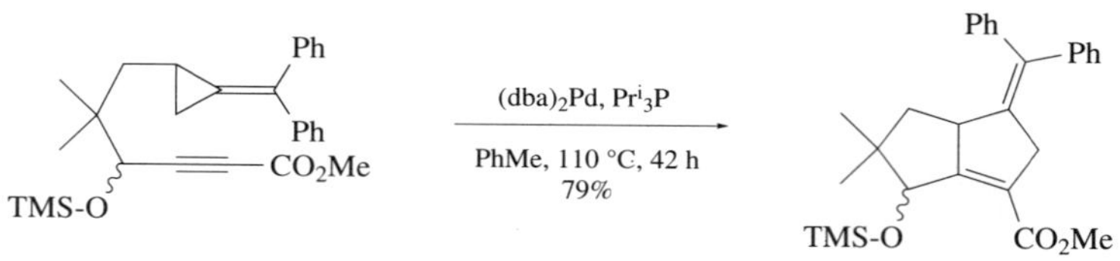

(104)

dba = dibenzylideneacetone

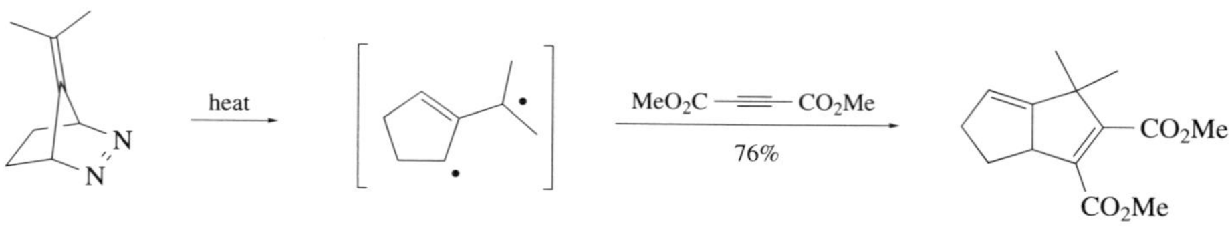

Scheme 17

1.12.2.6 The Pauson–Khand Reaction

The Pauson–Khand reaction involves the formation of a cyclopentenone from an alkyne, an alkene, and carbon monoxide, in the presence of $Co_2(CO)_8$, and is a formal [2 + 2 + 1]-cycloaddition (Equation (105)). It was first reported by Khand and Pauson ⟨73JCS(P1)975, 73JCS(P1)977⟩, and has been reviewed ⟨85T5855, 91OR(40)1⟩. The reagents are usually heated together in a hydrocarbon or ethereal solvent and yields are typically in the range of 40–60%. The mechanism involves formation of the complex $Co_2(CO)_6 \cdot R^1C{\equiv}CR^2$ as the first step ⟨91OR(40)1⟩. Ethyne and terminal alkynes work best, whereas internal alkynes give lower yields. The alkene structure is more limited in scope.

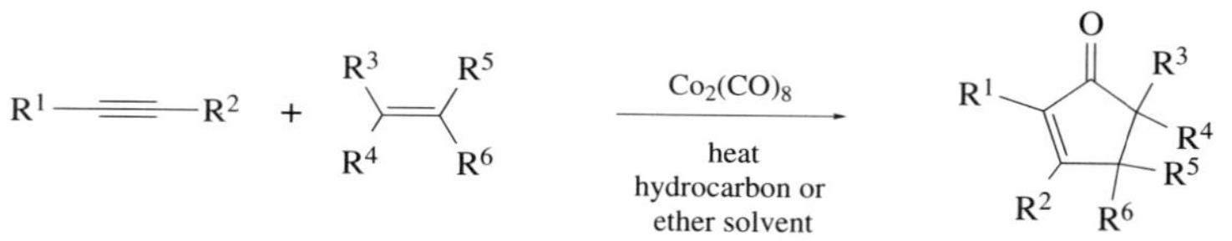

(105)

In the intermolecular reaction, simple alkenes suffer from a lack of both reactivity and regioselectivity. Ethene itself is an exception, and reacts well with many alkynes, for example with terminal alkynes (Equation (106)). There is usually high regioselectivity with respect to the alkyne, with the alkyl group becoming incorporated at the 2-position of the cyclopentenone. However, with terminal alkenes, there is usually little regioselectivity with respect to the alkene, although it remains good for the alkyne (Equation (107)) ⟨88TL999⟩.

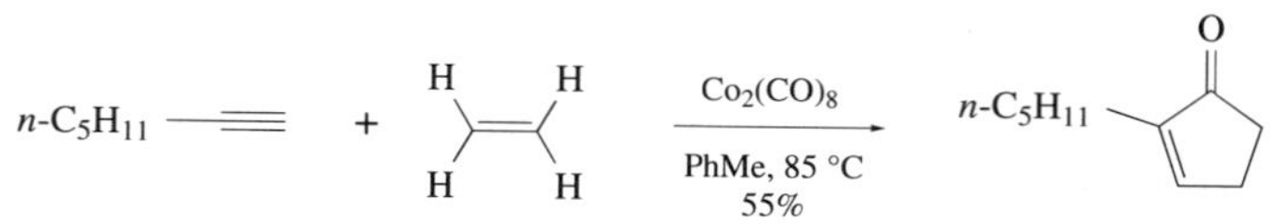

(106)

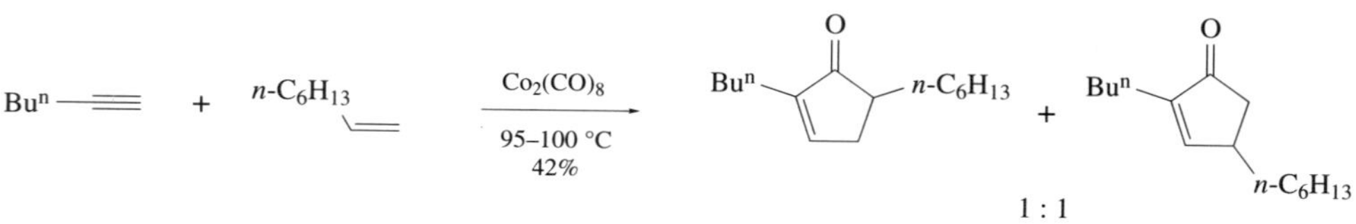

(107)

The reaction between internal alkynes and terminal alkenes gives reduced yields, but high regioselectivity for incorporation of the alkene (Equation (108)) ⟨88TL999⟩. Allyl ethers and homoallyl thioethers also give high regioselectivity as alkene components ⟨91OR(40)1⟩.

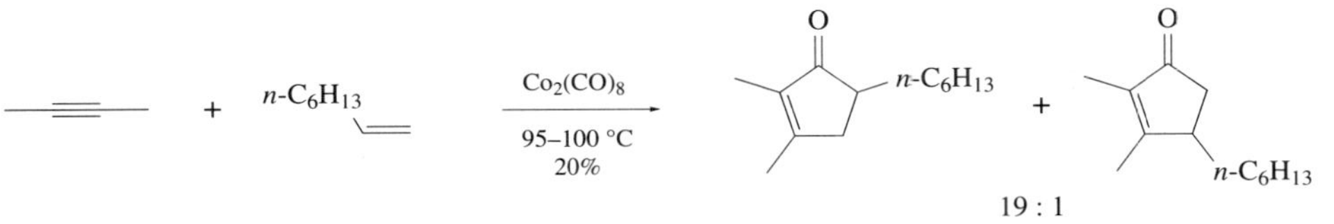

(108)

Alkynes substituted with electron-withdrawing groups do not generally undergo the Pauson–Khand reaction, although intramolecular reaction is possible ⟨89IZV495⟩. With electron-withdrawing groups on the alkene an anomalous reaction occurs, giving dienes with high regioselectivity.

Styrenes are intermediate in reactivity between terminal alkenes and alkenes substituted with an electron-withdrawing group, and so give a mixture of cyclopentenones and 1,3-dienes, often in good overall yield (Equation (109)) ⟨78JCR(M)4454⟩. More-substituted acyclic alkenes are not useful substrates, unless a homoallylic heteroatom is present.

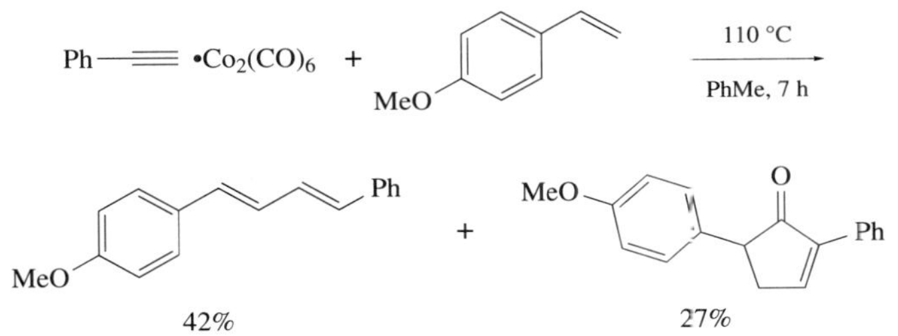

(109)

Monocyclic alkenes react with terminal alkynes in 30–70% yields, and the yield can sometimes be substantially improved by the addition of a trialkylphosphine oxide (Equation (110)) ⟨88JOM(354)233⟩. Cyclohexene gives poor yields, whereas cycloheptene and cyclooctene give moderate yields, but even then only with phenylethyne, and not terminal alkyl ethynes ⟨77JCR(M)168⟩. Cyclohexa-1,3-diene reacts twice with phenylethyne, first in a Diels–Alder reaction, and then in a Pauson–Khand reaction (Equation (111)) ⟨78JCR(M)4401⟩.

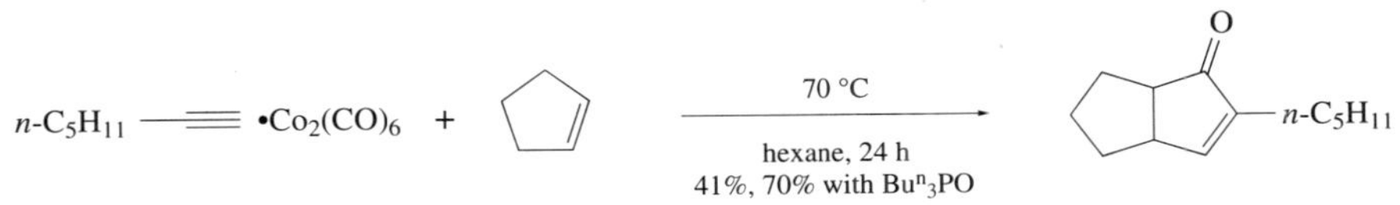

(110)

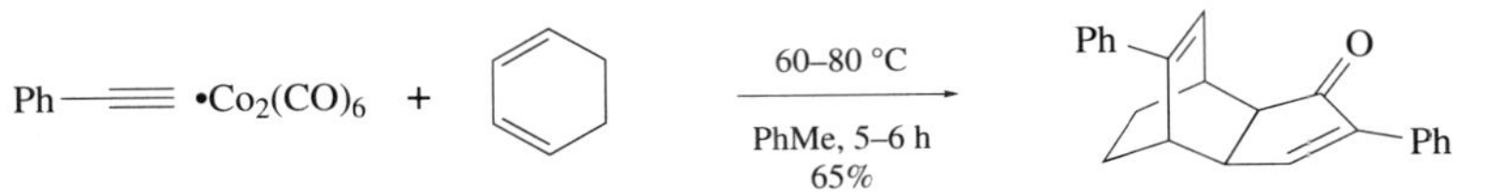

(111)

Ring-fused cyclobutenes give good yields (Equation (112)) and the cyclopentenone is formed on the *exo*-face with complete regioselectivity ⟨77JCR(M)153⟩. The ring-fused cyclopentene in Scheme 18 undergoes isomerization before Pauson–Khand reaction with *t*-butyldimethylsilyl (TBDMS)-protected propargyl alcohol giving an angular fused triquinane ⟨85T5995⟩.

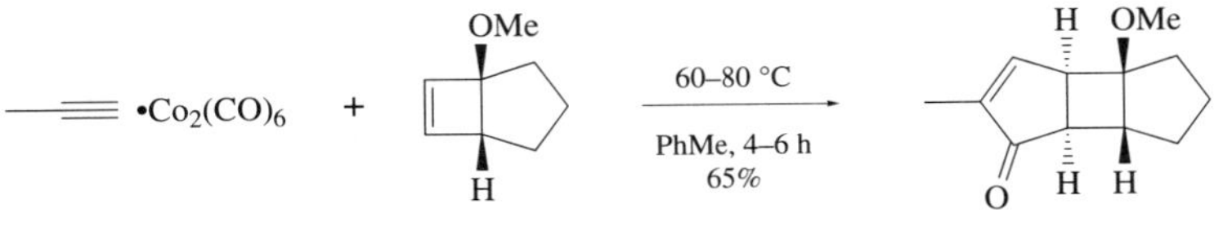

(112)

Norbornene reacts with internal alkynes in 20–40% yields, and with terminal alkynes in 30% to >90% yields (Equation (113)); 100% *exo* selectivity is observed ⟨89TL4021⟩. Dry-state absorption onto silica instead of traditional liquid phase reaction improves the yield in this and many other Pauson–Khand reactions. Addition of (*R*)-(+)-glyphos to the reaction in Equation (113) results in

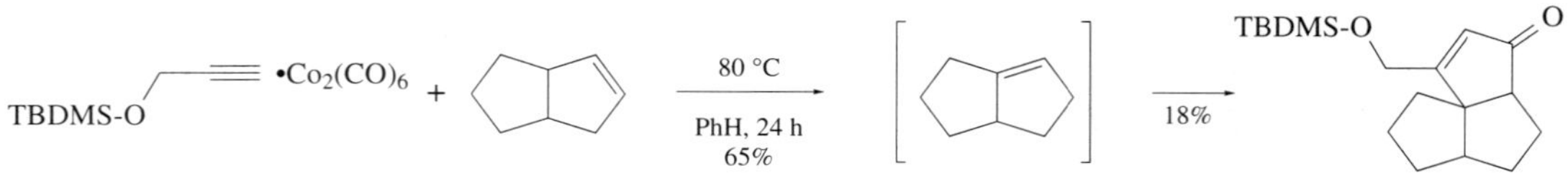

Scheme 18

an asymmetric coupling in 100% *ee* ⟨88JOM(355)449⟩. Norbornadiene reacts in moderate to low yields with alkynes at one or both double bonds, respectively. Heterocyclic bridged alkenes have also been used ⟨85JOC5215⟩.

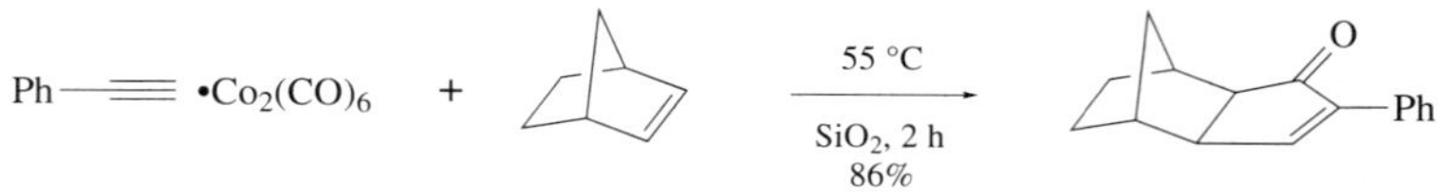

(113)

Intramolecular Pauson–Khand reactions are possible using enynes which result in five- or six-membered rings (Equation (114)) ⟨86JA3835⟩. Reactions which would give four-membered rings result in alkyne trimerization instead ⟨81JOC5436⟩. Terminal and internal alkenes react well, as do trisubstituted alkenes with terminal alkynes. Dry-state absorption conditions work best for (protected) hydroxyl-containing substrates ⟨86TL1241⟩. Heteroatom-linked enynes have also been used successfully ⟨89S472⟩. Asymmetric intramolecular cyclizations are possible using chiral auxiliaries attached to the enyne (Equation (115)) ⟨90JA9388⟩.

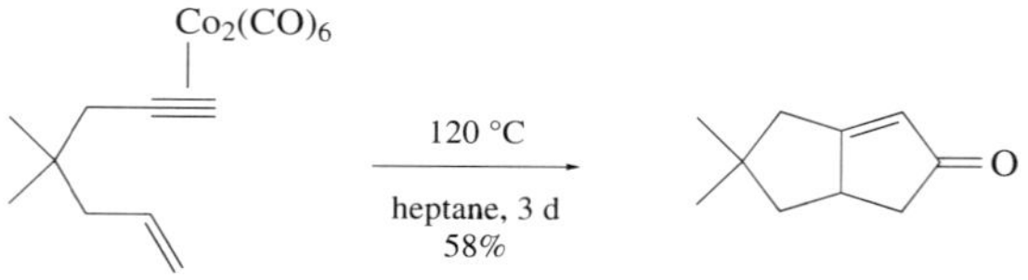

(114)

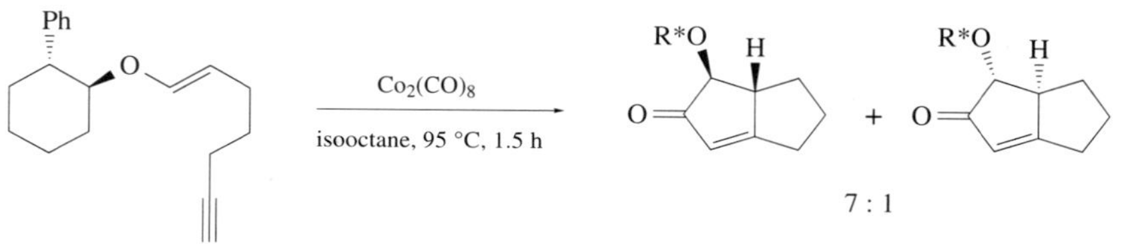

(115)

1.12.2.7 Cyclotrimerization and Cyclo-oligomerization of Alkynes

The cyclotrimerization of alkynes to benzenes has been an intensively studied reaction, and has been reviewed ⟨B-85MI 112-03⟩. Many catalysts are available, and the preparation of highly substituted benzenes is often made simple by this method. Equation (116) shows the synthesis of symmetrical hexaalkyl benzenes using a palladium catalyst which is particularly straightforward to prepare ⟨87JOC1161⟩. Reactions of this type between diynes and alkynes are possible (e.g., Equation (117)) using either cobalt or Wilkinson's catalysts; such reactions have been reviewed ⟨84AG(E)539⟩.

Cyclotetramerization of alkynes is also possible using nickel catalysts ⟨48LA(560)1⟩, but cyclotrimerization may compete (Equation (118)) ⟨61JA2944⟩.

Other multicomponent cycloadditions involving alkynes, catalyzed by transition metals, can be found in the review by Schore ⟨88CRV1081⟩.

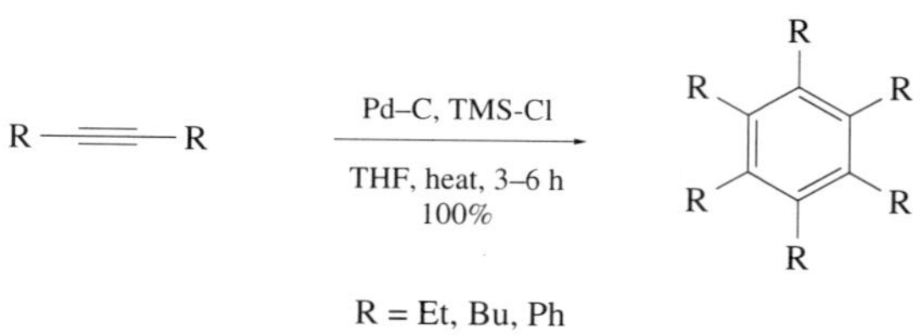

(116)

(117)

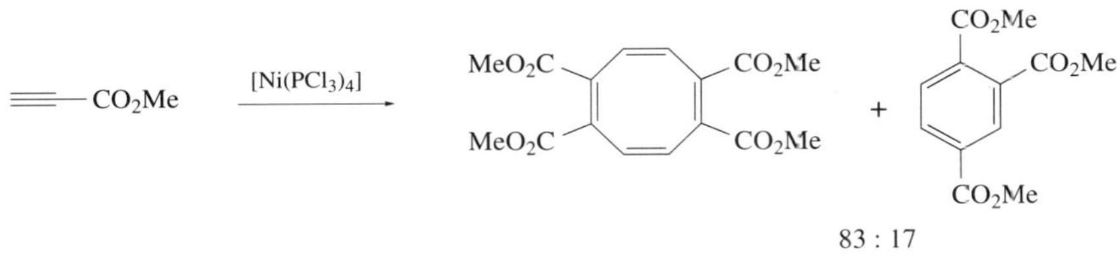
(118)

1.12.3 IONIC ADDITIONS

1.12.3.1 Hydrometallation of Alkynes followed by C—C Bond Formation

Hydrometallation of alkynes followed by replacement of the carbon–metal bond by a carbon–carbon bond results in alkene formation by overall addition of hydrogen and carbon. A wide range of hydride-containing reagents may be used for this process, and the hydrometallation steps are the same as those discussed in Section 1.12.1.4.

1.12.3.1.1 Hydroboration followed by C—C bond formation

Hydroboration of alkynes by dialkylboranes, as discussed above in Section 1.12.1.4.1 gives alkenylboranes (see also Chapter 2.18.1). The C—B bond can be replaced by a C—C bond, using a variety of reagents, resulting in overall addition of hydrogen and a carbon substituent to the alkyne. With internal alkynes, the regioselectivity of hydroboration becomes important, and is generally controlled by steric effects, with the boron atom becoming attached to the less sterically hindered carbon atom ⟨B-78MI 112-02⟩.

Hydroboration of a terminal alkyne with 9-BBN-H, followed by reaction with aldehydes gives allylic alcohols ⟨77JOC579⟩. Hydroboration of terminal alkynes with catecholborane, followed by palladium-catalyzed carbonylation in methanol gives α,β-unsaturated methyl esters ⟨81CL879, 86CL1329⟩. Alternatively, the intermediate borane can be coupled with aryl or alkenyl halides (Scheme 19) ⟨81JOM(213)C53⟩.

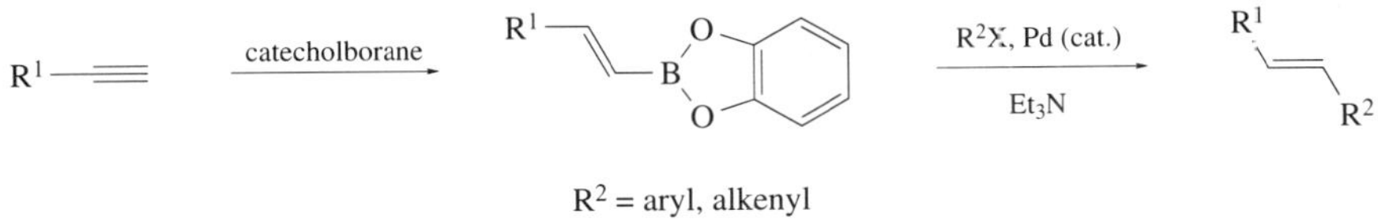

Scheme 19

Treatment of alkenylboranes derived from terminal alkynes with iodine and NaOH or NaOMe results in transfer of one of the alkyl groups on boron onto the alkene, with inversion of the alkene geometry, giving a (*Z*)-alkene (Scheme 20) ⟨82JOC1792, 82JOM(225)C1, 82T2355⟩. On the other hand, treatment with either BrCN ⟨72JA6560, 77CC852⟩ or $Pd(OAc)_2/Et_3N$ ⟨80BCJ1670⟩ results in retention of configuration and gives the (*E*)-alkene.

A variation on this reaction is the use of a bromoalkylborane dimethylsulfide complex for the hydroboration step; reaction with NaOMe and I_2 then gives the alkene (Scheme 21) ⟨82JOC171, 82JOC3806, 82JOC5407⟩. 1,3-Dienes can also be prepared by hydroboration of two equivalents of an internal alkyne by thexylborane, followed by oxidation and reaction with NaOMe and I_2 as above (Scheme 22) ⟨68JA6243⟩. Chloroborane reacts with two equivalents of an alkyne to give the dialkenylchloroborane; reaction with either iodine or MeCu then gives coupled dienes of different stereochemistry ⟨73JOC1617, 75JA5606⟩. 1,3-Dienes are also obtained by successive reaction of thexyl borane with a 1-haloalkyne followed by a terminal alkyne; treatment of the intermediate borane

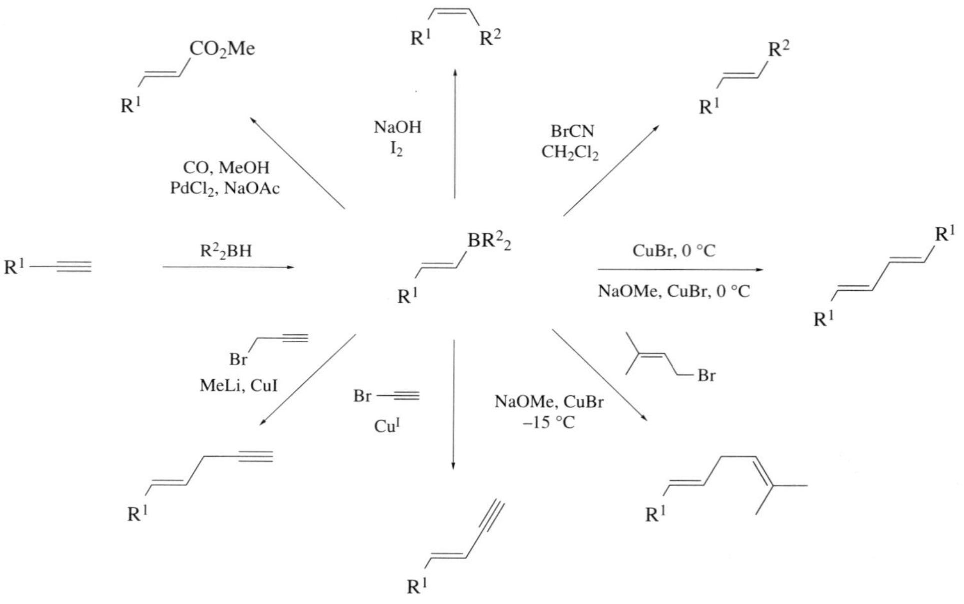

Scheme 20

with methoxide results in C—C bond formation by alkenyl migration; protonolysis liberates the diene ⟨73CC606⟩.

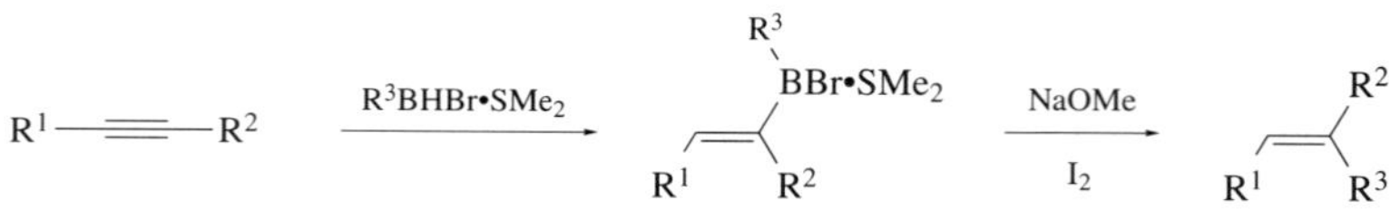

Scheme 21

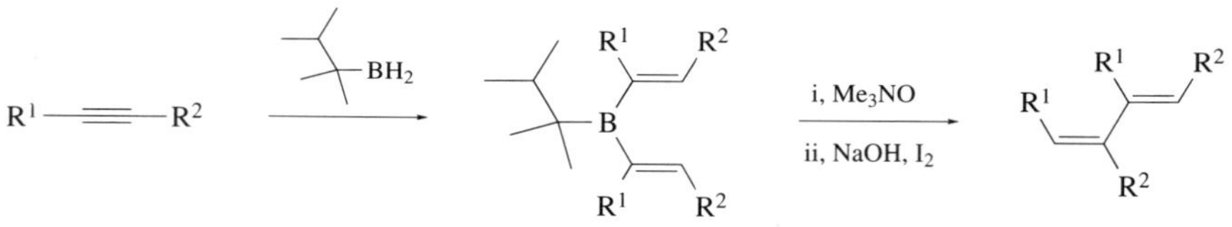

Scheme 22

The C—B bond in the alkenylborane can be replaced by an alkyl group from reactive alkylating agents such as allyl, propargyl, or benzyl bromide under a variety of conditions, often involving copper or palladium reagents (Scheme 20). Allyl bromide reacts using, for example, NaOMe/CuBr·SMe_2 ⟨80JOC550⟩, MeCu ⟨80JOC1640⟩, or CuI/MeLi ⟨80BCJ1471⟩; these last conditions can also be used for alkylation with propargyl bromide. Catalytic $Pd(OAc)_2$ is also successful for allyl bromide ⟨80BCJ1670⟩, and $Pd(PPh_3)_4$ has been used for allyl and benzyl bromides, and also aryl halides ⟨79CC866, 80TL2865, 82JOC2117⟩. 1-Bromoalkynes can also be used as alkylating agents, again using either a copper(I) salt ⟨84JA462⟩ or catalytic palladium(0) ⟨79TL3437, 81T2617, 81TL127, 85JA972, 85PAC1749⟩. Alternatively, the same products may be obtained using a 1-lithioalkyne in the presence of I_2 ⟨73CC874, 86TL539⟩.

By using alkylthexylborane as the reagent for hydroboration of ethyl propiolate, the alkenylborane (**43**) is formed. The alkyl group on boron rather than the thexyl group is selectively transferred upon treatment with bromine and, depending upon the conditions, either the (*E*)- or (*Z*)-isomer of the α,β-unsaturated ester may be obtained (Scheme 23) ⟨74JOC2321⟩.

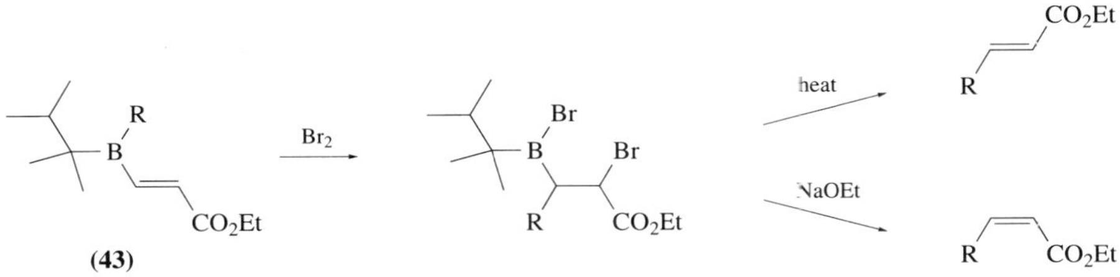

Scheme 23

1.12.3.1.2 Hydroalumination followed by carbon–carbon bond formation

Addition to alkynes of the aluminum hydride reagents described in Section 1.12.1.4.2 results in alkenylalane intermediates (see also Chapter 2.19.5). With suitable reagents, alkenes can be produced, with a new carbon–carbon bond being formed with retention of the stereochemistry of the alkenylalane ⟨84OR(32)375⟩. The alkenylalane intermediates react as alkenyl carbanion equivalents; however, their nucleophilicity is low because of the empty *p*-orbital on aluminum, and reaction with carbon electrophiles is often not very satisfactory. However, addition of an organolithium reagent generates an intermediate alanate, which then has a complete octet, is negatively charged, and is much more reactive towards carbon electrophiles ⟨84OR(32)375⟩.

Addition of dibal-H to unfunctionalized terminal alkynes results in *cis*-addition of hydrogen and aluminum to give the alkenylaluminum intermediate (**44**), which can then be alkylated with the reactive choromethylmethyl ether to give the alkene (Scheme 24) ⟨76S816⟩. While simple primary alkyl halides do not react with (**44**), they can undergo aminomethylation ⟨80TL3763⟩. Vinyl and aryl halides react with (**44**) to give conjugated (*E*)-alkenes, using either nickel(0) ⟨87JA2393⟩ or palladium(0) ⟨76CC596, 76JA6729, 87JA2393, 87TL1649, 87TL4875⟩ as catalysts (Scheme 24). This is a poor reaction starting from internal alkynes; however, addition of $ZnCl_2$ gives improved yields and shorter reaction times (Equation (119)) ⟨78JA2254⟩. 1-Haloalkynes can also be used as electrophiles with (**44**), giving conjugated *trans*-enynes from terminal alkynes (Scheme 24) ⟨78JA2254⟩.

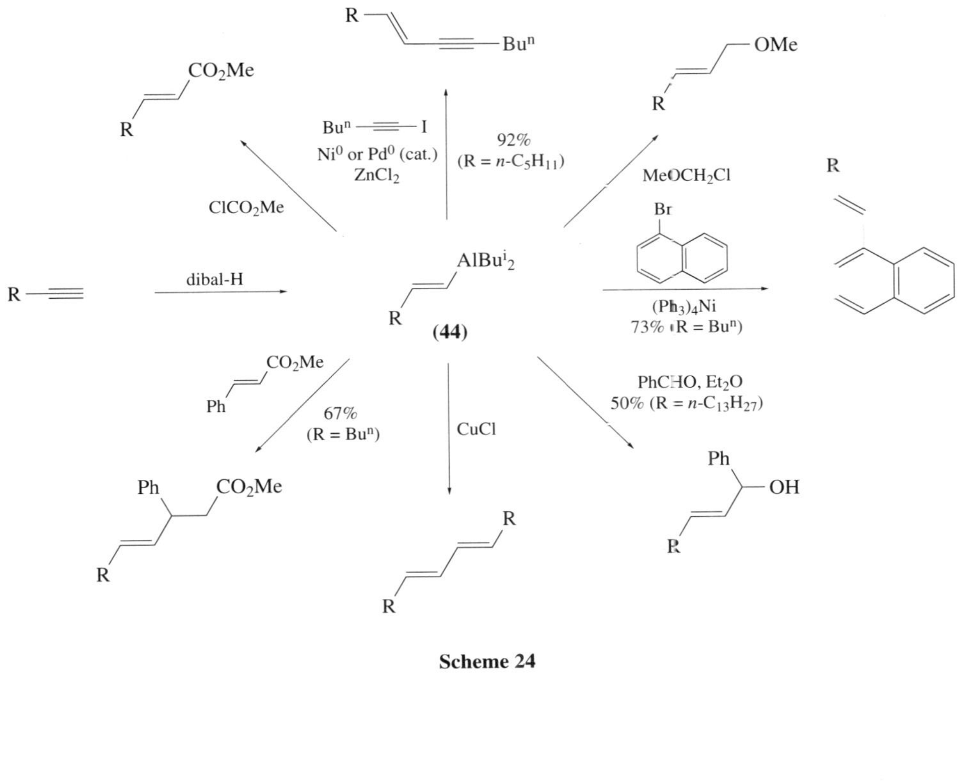

Scheme 24

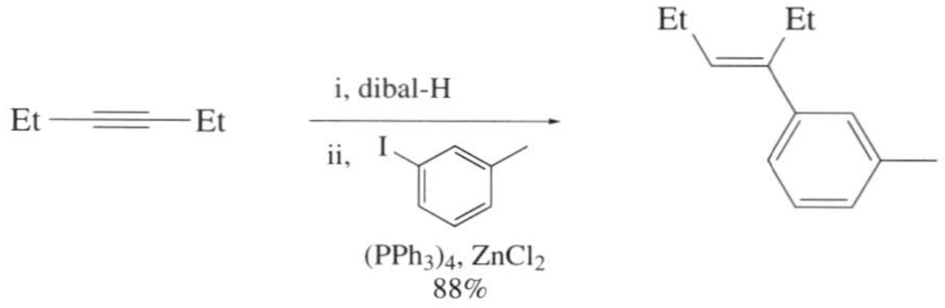

(119)

Addition of (**44**) to aldehydes results in a stereoselective synthesis of allylic alcohols ⟨67JA2754, 82TL2087, 87TL5129⟩. Ketones react similarly to give tertiary alcohols, although if they are hindered then competing reduction of the carbonyl group may occur ⟨84OR(32)375⟩. Michael addition is also possible with α,β-unsaturated ketones and esters (Scheme 24) ⟨72TL4083⟩, and reaction with chloroformates gives (*E*)-α,β-unsaturated esters ⟨78TL2357⟩.

Self-coupling of (**44**) with CuCl gives symmetrical, geometrically pure (*E*,*E*)-dienes (Scheme 24) ⟨70JA6678⟩ and cross-coupling with allylic halides using CuCl is also possible, to give 1,4-dienes ⟨74S658⟩.

Reaction of dibal-H with two equivalents of an internal alkyne generates the dienylalane with new C—C bond formation; protonolysis then yields the symmetrical (*E*,*E*)-diene with high stereoselectivity (Equation (120)) ⟨60LA(629)222⟩.

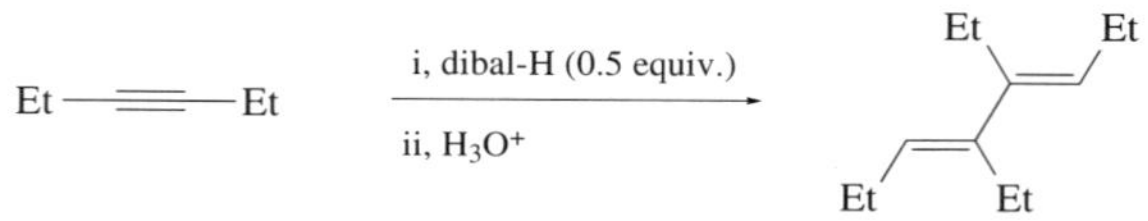

(120)

The intermediate aluminum enolate formed by addition of dibal-H to methyl propiolate can be alkylated by bromocyclohexene (Equation (121)) ⟨87JOC1624⟩. Other allylic bromides react similarly.

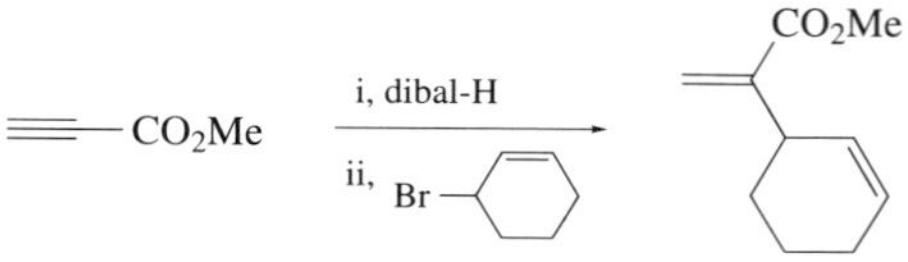

(121)

Reaction of the intermediate (**44**) with methyllithium results in formation of an alanate, which reacts with a wider range of electrophiles, as illustrated in Scheme 25. Alkylation with simple alkyl halides is possible ⟨76TL1927⟩. Epoxides can also be used if not too hindered, giving homoallylic alcohols ⟨75S632, 76CC17⟩. Carboxylation with CO_2 gives α,β-unsaturated carboxylic acids ⟨67JA2754⟩.

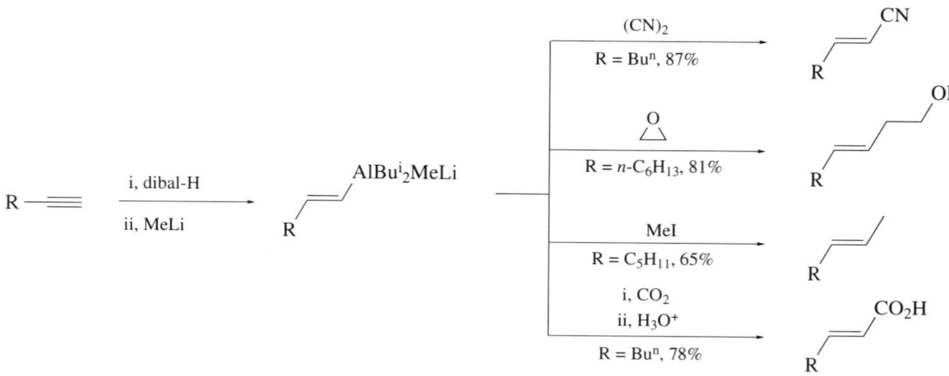

Scheme 25

Disubstituted alkynes undergo *trans*-addition of $LiBu^i_2MeAlH$, and the resulting alanate is thus the opposite geometrical isomer to that formed with dibal-H and MeLi; the alanate can be carboxylated, or trapped by formaldehyde (Equation (122)) ⟨67JA5085⟩. Cyanogen reacts with both the alanate and alane intermediates to give the alkenyl nitriles as geometrical isomers (Equations (123) and (124)) ⟨68JA7139⟩.

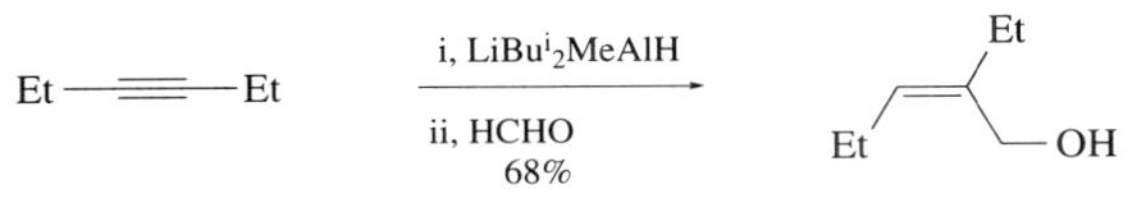

(122)

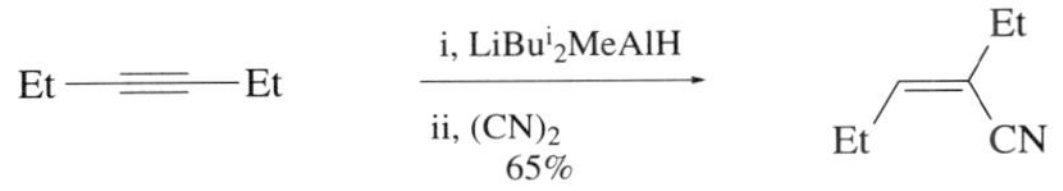

(123)

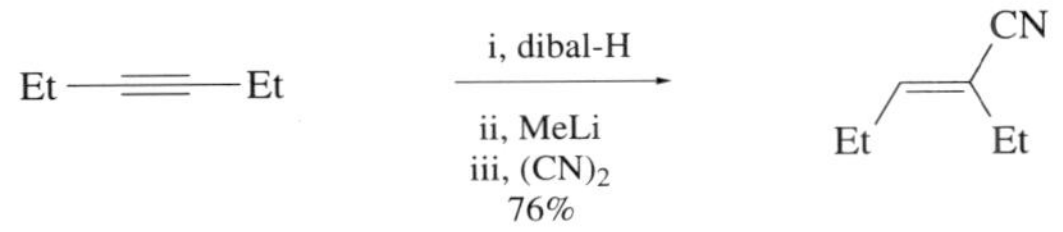

(124)

1.12.3.1.3 Hydrosilylation followed by carbon–carbon bond formation

Alkynes undergo platinum-catalyzed hydrosilylation to give isolable, stable alkenyl silanes (see Chapter 2.18.2), which can then undergo electrophilic substitution to give alkenes (Scheme 26) ⟨77TL3317, 79S761⟩ (see also Chapter 1.11.1.6).

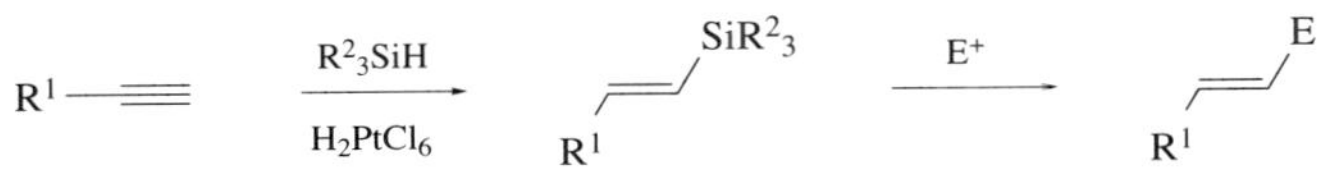

Scheme 26

Alkynes also undergo hydrosilylation with $HSiCl_3$; reaction with KF then gives an alkenylsilicate dianion intermediate which can be reacted with a variety of reagents catalyzed by palladium, resulting in new carbon–carbon bond formation ⟨82OM542⟩. Without additional reagents, self-coupling is catalyzed by $PdCl_2$, giving symmetrical (*E*,*E*)-1,3-dienes.

1.12.3.1.4 Hydrostannylation followed by carbon–carbon bond formation

Terminal alkynes undergo hydrostannylation by Bu^n_3SnH regioselectively to give (*E*)-alkenylstannanes with *cis*-addition of the hydrogen and tin atoms (Scheme 27) (see Chapter 2.19.6). Treatment with Bu^nLi then gives the corresponding alkenyllithium compound, which can be reacted with an appropriate electrophile, for example to effect alkylation ⟨75JOC2265, 77TL3847⟩; alternatively, cuprate formation allows Michael addition to an enone ⟨75JOC2265, 76TL4705, 78JOC3450⟩.

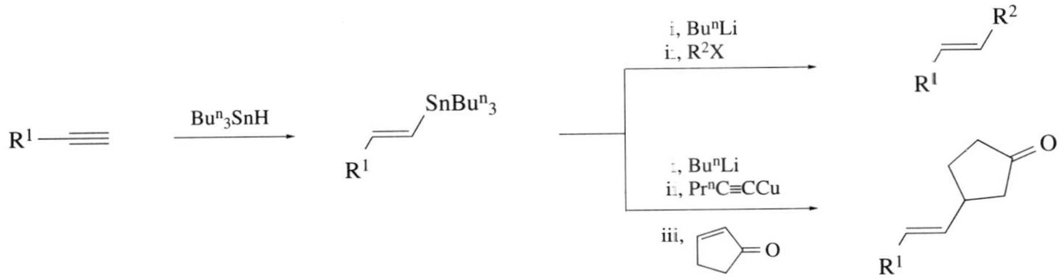

Scheme 27

Treatment of ethyne itself with Ph_3SnCu or $(Ph_3Sn)_2CuLi$ followed by an alkyl halide gives (*Z*)-alkenylstannanes; these species can be transmetallated with Bu^nLi and then reacted with an electrophile with retention of alkene geometry (Scheme 28) ⟨82TL2797⟩.

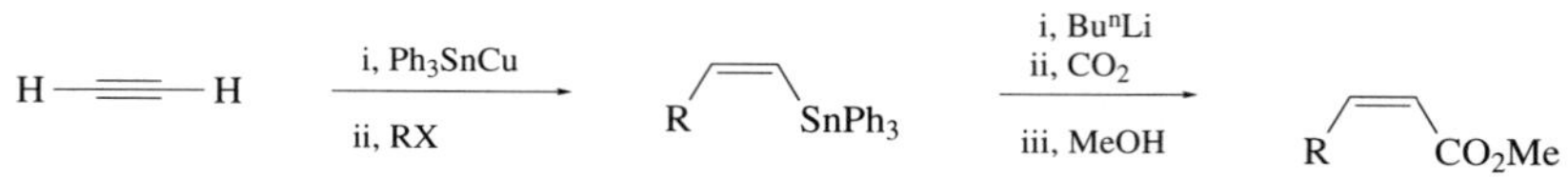

Scheme 28

1.12.3.1.5 Hydrozirconation followed by carbon–carbon bond formation

Both terminal and internal alkynes undergo hydrozirconation with Cp_2ZrHCl, giving (*E*)-alkenyl zirconium compounds (**45**), resulting from *cis*-addition at the H and Zr atoms (Scheme 29). In the case of terminal alkynes this process is also regioselective, the zirconium atom becoming attached to the terminal carbon atom. The C—Zr bond can be transformed into a C—C bond with a variety of reagents, resulting in stereoselective formation of di- or trisubstituted alkenes ⟨85MI 112-04⟩.

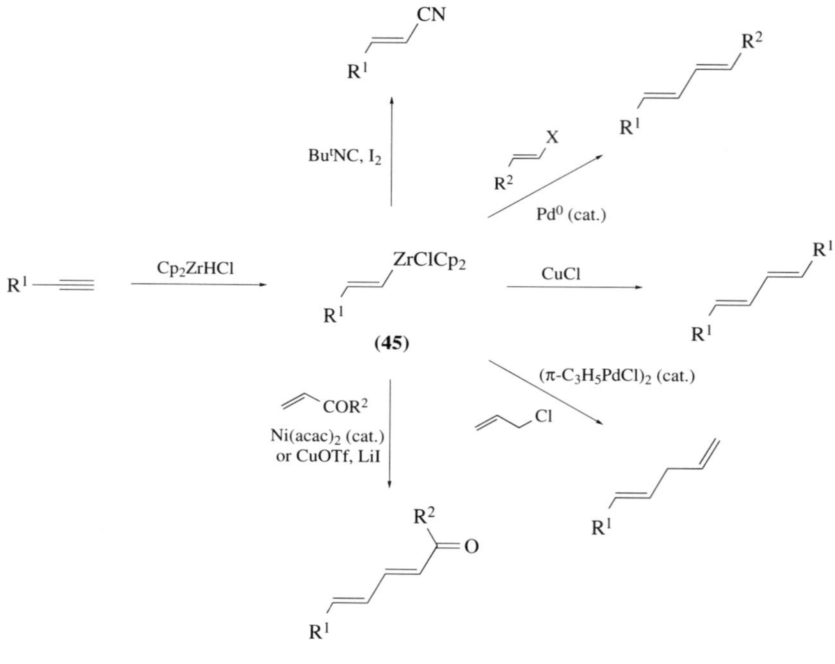

Scheme 29

Self-coupling occurs in the presence of $CuCl_2$, giving symmetrical conjugated dienes ⟨77TL1303⟩. Unsymmetrical conjugated dienes can be obtained by coupling with alkenyl halides, catalyzed by palladium(0) ⟨78TL1027, 82TL63, 87JA2393, 87TL4875, 87TL4879⟩. Allylation may be achieved using either $(\pi\text{-}C_3H_5PdCl)_2$ ⟨80JA7381, 81TL4655⟩, or allyl chloride with catalytic $(\pi\text{-}C_3H_5PdCl)_2$ ⟨81TL2629⟩.

Michael addition to α,β-unsaturated ketones is possible in the presence of either CuOTf and LiI ⟨77TL1303⟩, or catalytic $Ni(acac)_2$ ⟨77JA8045⟩. Replacement of zirconium by CN may be achieved using *t*-butyl- or trimethylsilylisonitrile and iodine, giving α,β-unsaturated nitriles ⟨87TL295⟩.

Treatment of enynes with $ZrCp_2$ results in bicyclic organozirconium compounds (**46**); protonation then gives alkenyl cycloalkanes; whereas treatment with carbon monoxide gives cyclopentenones (Scheme 30) ⟨85JA2568, 86JOC4080, 86TL2829, 87TL917⟩.

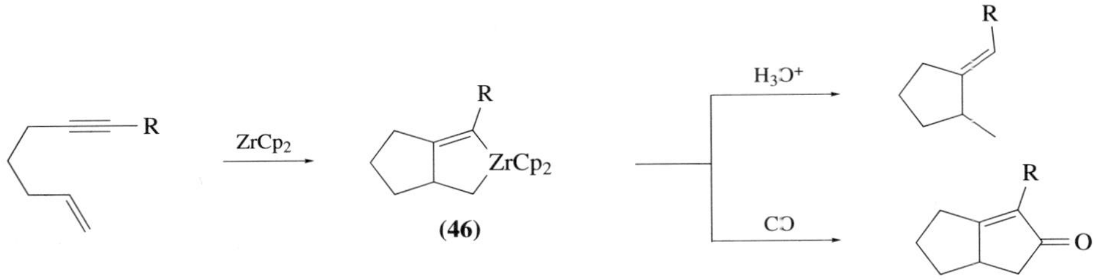

Scheme 30

1.12.3.2 Ionic Additions of Stabilized Carbanions to Activated Alkynes

The conjugate addition of stabilized carbanions and organometallics to alkynes activated by a conjugative electron-withdrawing group has been covered by several reviews ⟨B-69MI 112-01, B-78MI 112-05, 91COS(4)1, B-92MI 112-02⟩. Further reactions may ensue after the initial addition, such as cyclization, or addition of a second molecule of the nucleophile.

The anion of diphenylacetonitrile adds to ethyl propiolate in 92% yield using a quaternary ammonium hydroxide as the base (Equation (125)) ⟨53HCA1809⟩. Addition of an oxazolone to 1-phenylpropynone initially gives the (*Z*)-adduct; however, this can be isomerized to the (*E*)-isomer upon heating (Scheme 31) ⟨71AG(E)653⟩.

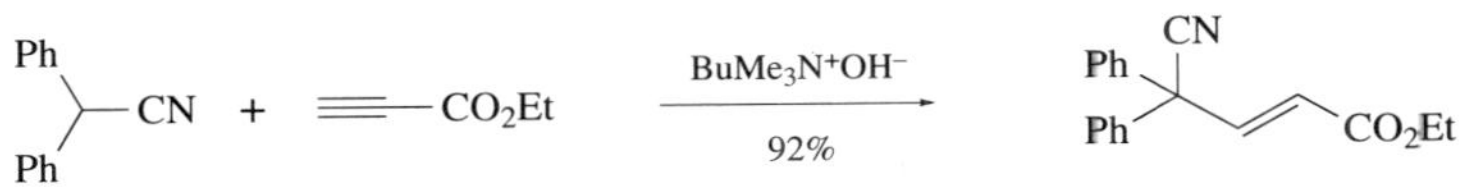

(125)

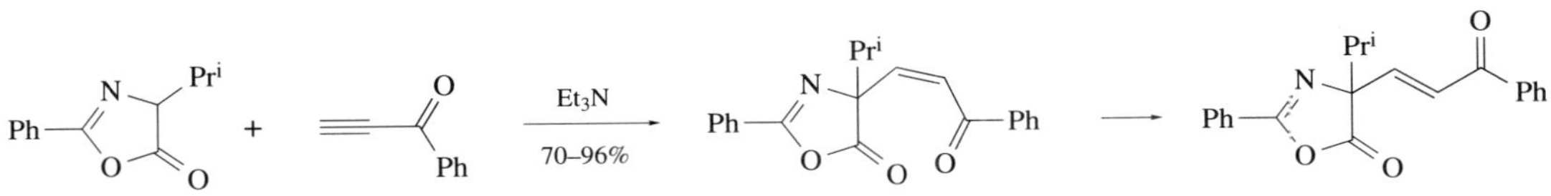

Scheme 31

The reaction of dimethyl malonate with DMAD results in two products (Equation (126)) ⟨65JCS4355, 67JCS(C)1986⟩, which are the result of sequential addition of the initial adduct to two further molecules of DMAD, followed by isomerization and Dieckmann cyclization ⟨64JOC423⟩. Replacement of the malonate with ethyl cyanoacetate in this reaction gives cyclic products resulting from one molecule of nucleophile and two molecules of DMAD (Equation (127)).

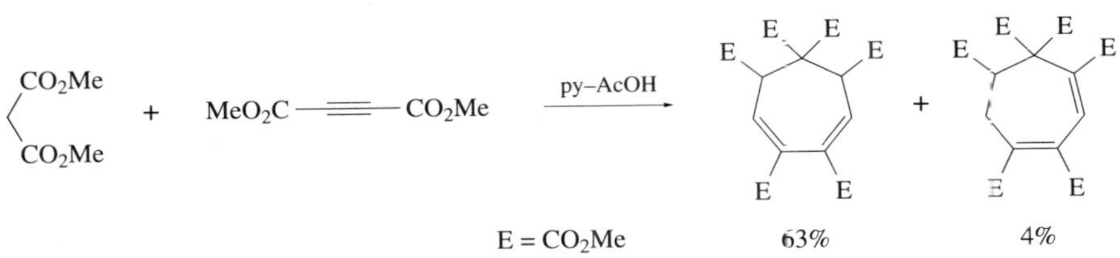

(126)

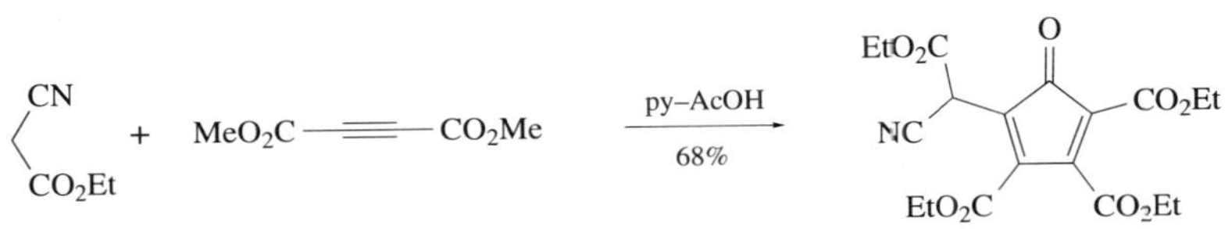

(127)

Conversely, the successive Michael additions of a ketone enolate to two molecules of ethyl propiolate, followed by ring closure, gives the benzene derivative featured in Equation (128) ⟨69JOC4134⟩.

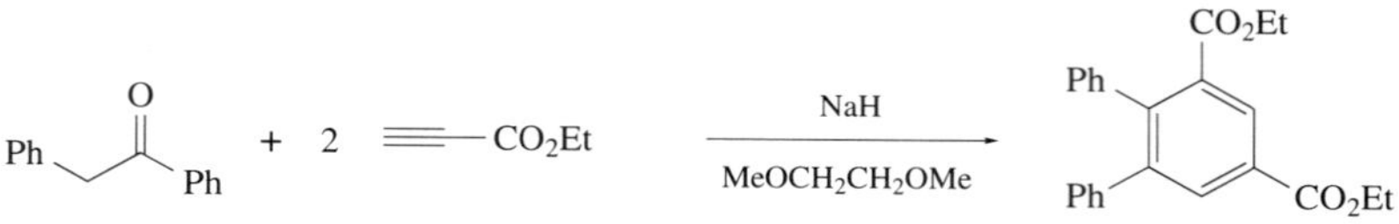

(128)

Products of the Robinson annelation can be obtained by the addition of ketone enolates to alkynyl ketones (Equation (129)), however the yields are usually low because the initial Michael adduct is formed preferentially as the (*E*)-isomer ⟨50JA494⟩, and better yields are obtained using alkoxy-substituted alkynes.

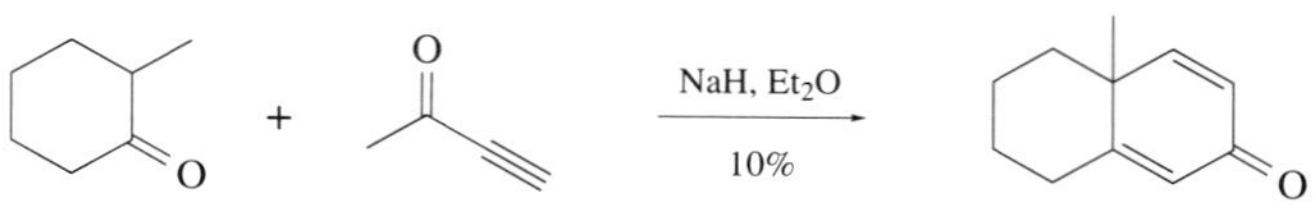

(129)

Intramolecular additions are also known, for example the *β*-ketoester (**47**) adds to its alkynyl ketone side chain in a 6-*endo*-dig process to give the bicyclic enone (**48**) (Equation (130)) ⟨86TL5455⟩. Corresponding five-membered ring compounds can also be synthesized; however, yields for the seven- and eight-membered ring compounds are poor. Intramolecular conjugate addition may also be followed by trapping of the intermediate enolate, for example by alkylation (Equation (131)) to give a tricyclic enone ⟨87TL3457⟩.

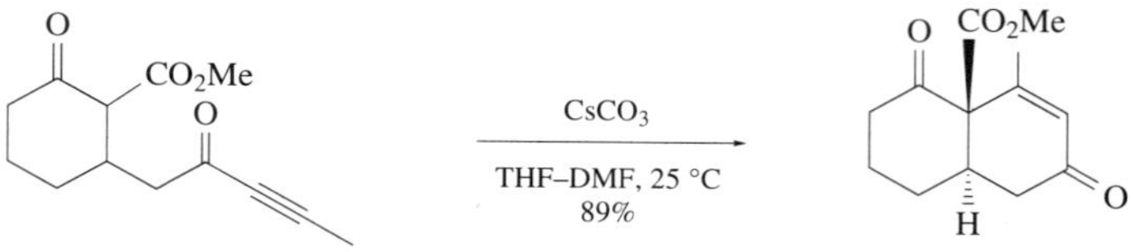

(**47**) (**48**)

(130)

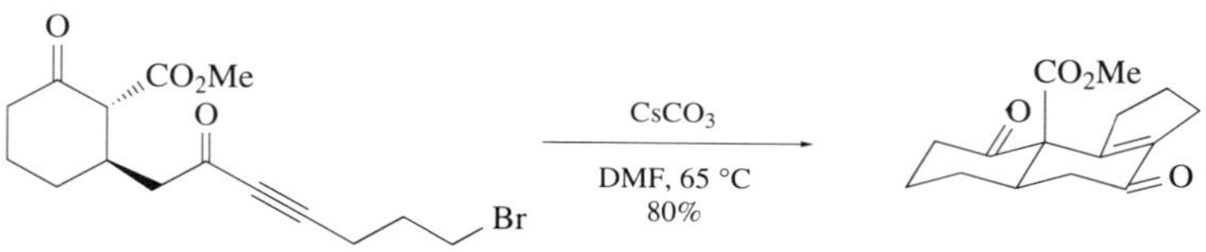

(131)

In the example shown in Equation (132) the second addition which forms the ring is not the 6-*endo*-dig Michael addition; instead, a 5-*exo*-dig cyclization occurs onto the second alkyne ⟨68T4285⟩. Another 5-*exo*-dig cyclization is shown in Equation (133); in this case a bridged bicyclic system is formed ⟨79JA1284⟩.

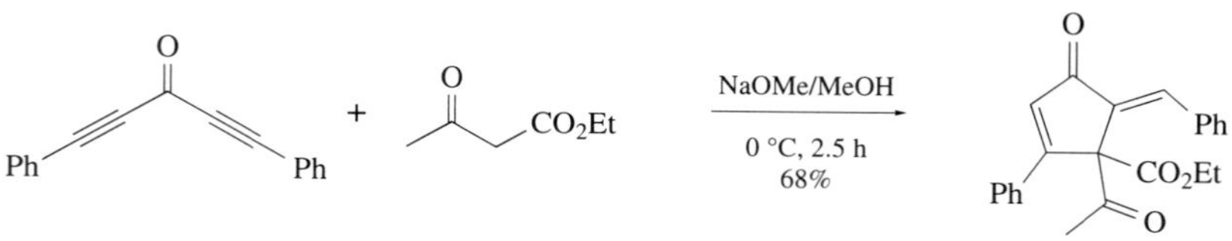

(132)

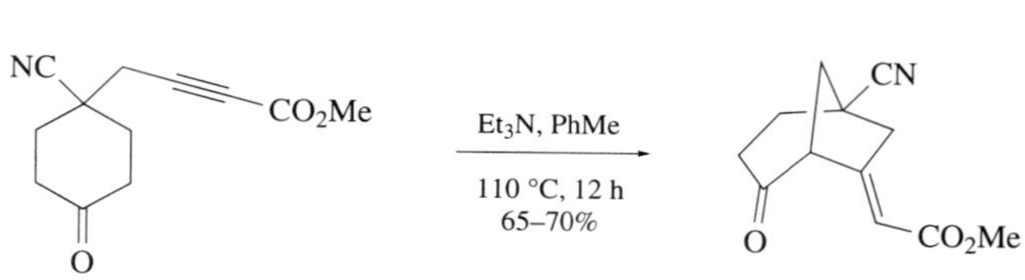

(133)

1.12.3.3 Carbometallation of Alkynes

Addition of an organometallic reagent to an alkyne generates an alkenylmetallic compound, which is often not isolated, but is reacted *in situ* with another electrophile. Both protonation and reaction with carbon-centred electrophiles give alkenes with no attached heteroatoms. Carbometallation of alkynes has been reviewed ⟨81S841, 91COS(4)865⟩.

1.12.3.3.1 Additions of organolithium and Grignard reagents to alkynes

The addition of noncuprate organometallics to activated alkynes does not usually give good results because multiple addition can easily occur ⟨72OR(19)1⟩.

The addition of organolithium reagents to simple alkynes is also not usually a synthetically useful reaction ⟨91COS(4)872⟩. Terminal alkynes are deprotonated, as are disubstituted alkynes (at the propargylic position). Grignard reagents also deprotonate terminal alkynes, but addition to alkynes occurs in moderate yields in the presence of certain transition metal complexes ⟨91COS(4)877⟩. Intramolecular addition gives better results. With disubstituted alkynes with no hydrogen atoms α to the triple bond, addition may occur, as illustrated by the addition of Bu^tLi to diphenylethyne (Equation (134)); however, some metallation of the aromatic rings also occurs ⟨69JOC1936⟩. Grignard reagents also add to diphenylethyne in high yield ⟨75JOM(96)C47⟩.

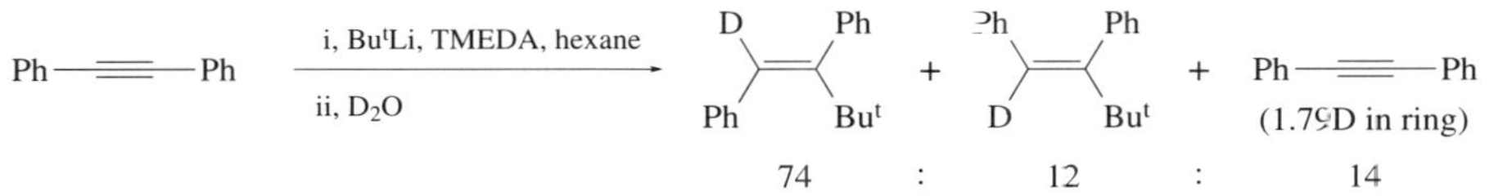

(134)

Intramolecular addition of organolithium and Grignard reagents to alkynes is possible, for example in a simple 5-alkynyl Grignard reagent (Equation (135)) ⟨58TL1457⟩.

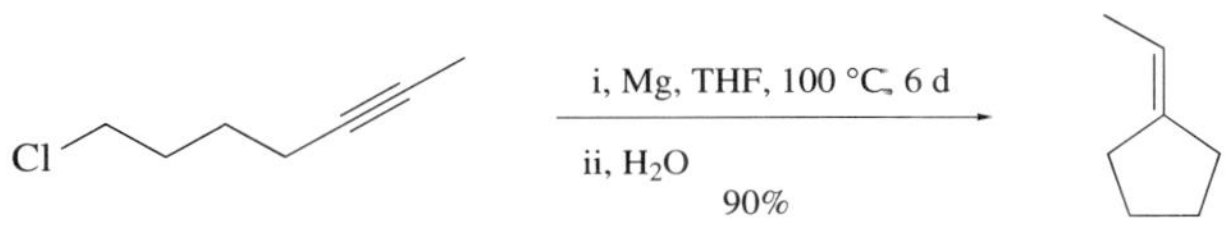

(135)

Biphenyl systems can also undergo intramolecular addition after activation as either the organolithium or organomagnesium compound ⟨66JA3027⟩.

With propargyl alcohols yields of addition products may be high, because the intermediate alkenyllithium compound can be stabilized by coordination of the alkoxide oxygen atom, and this also results in overall *trans*-addition to the alkyne (Scheme 32) ⟨74TL2161⟩. Trapping the intermediate with CO_2 gives the lactone.

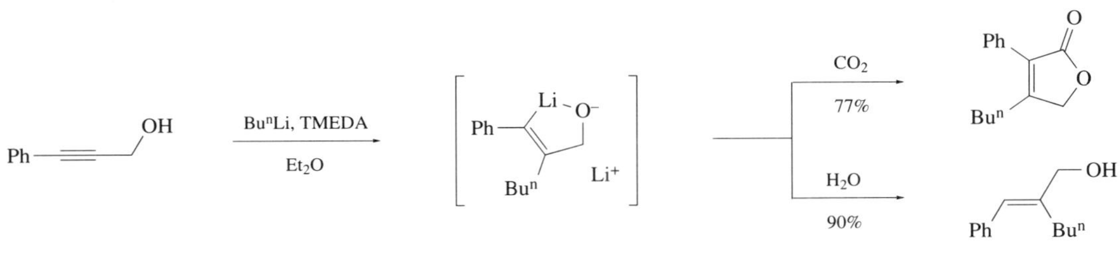

Scheme 32

Addition of Grignard reagents to propargyl alcohols gives overall *trans*-addition as for organolithium reagents ⟨77TL167⟩, however, if the hydroxyl group is moved one carbon atom further away, then the regioselectivity is much reduced ⟨71TL3777⟩.

Benzynes show unusual reactivity towards nucleophilic addition, and difunctionalization is also possible by trapping the intermediate arylmetal species with an electrophile ⟨85T837, 88JA7178, 88JCS(P1)839, 88T127⟩.

1.12.3.3.2 Addition of organocopper reagents to alkynes

The addition of organocopper reagents to alkynes is the most useful carbometallation reaction, and is included in several reviews ⟨78PAC709, 81S841, B-83MI 112-01, 84T641, 91COS(4)896⟩.

(i) Addition of organocopper reagents to activated alkynes

The conjugate addition of organocuprates to α,β-unsaturated carbonyl compounds, including alkynes, has been reviewed ⟨72OR(19)1, 75OR(22)253, 91COS(4)169⟩.

The addition of the divinylcuprate to the ester **(49)** at −78 °C occurs in a *cis*-fashion (Equation (136)) ⟨81S841⟩, and this reaction seems to be general for other organocopper reagents ⟨73TL1611, 81S841⟩. Addition of lithium dimethylcuprate to methyl phenylpropiolate at −80 °C gives *cis*-addition, resulting in the (*Z*)-product **(51)** after protonation, but warming the intermediate organocuprate **(50)** to 25 °C results in isomerization, and the (*E*)-isomer **(52)** is then obtained (Scheme 33) ⟨69JA1853, 73JCS(P2)1971⟩. Reaction of *n*-butylcopper·9-BBN-H with DMAD also results in *cis*-addition (Equation (137)); however, the use of other Lewis acid ligands on boron gives less selectivity, except for BEt_3 ⟨79JOC1744, 86AG(E)947⟩.

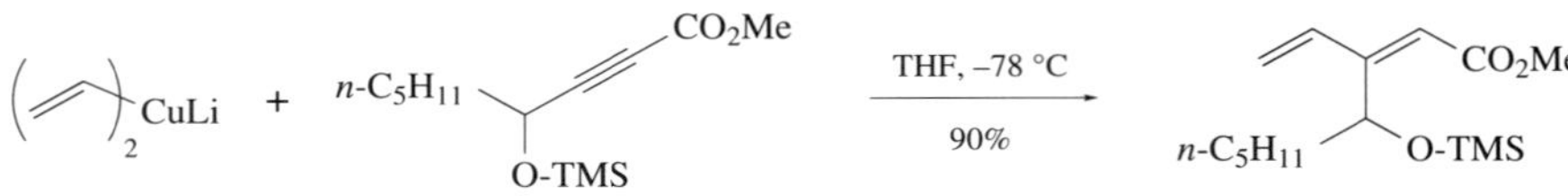

(49) (136)

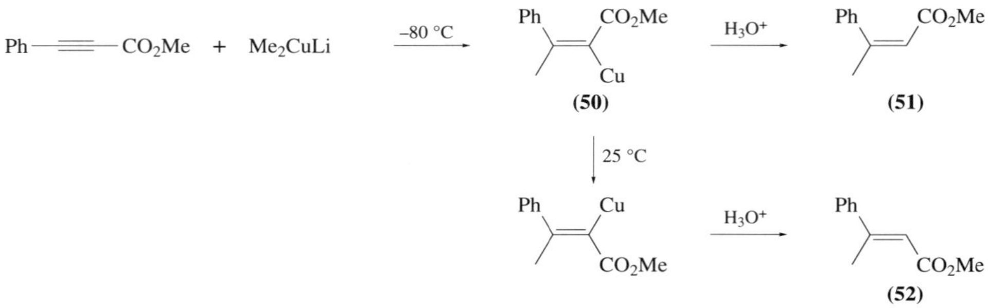

Scheme 33

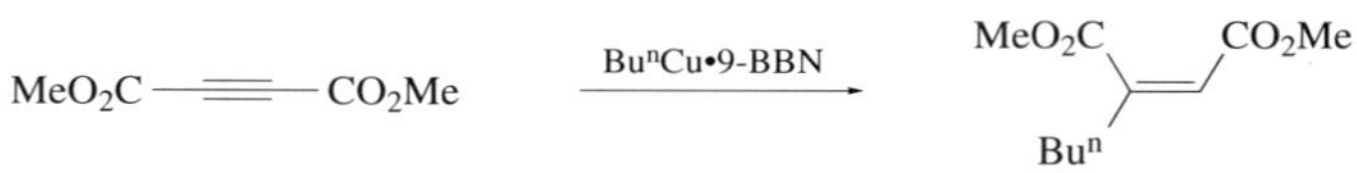

(137)

Conjugate addition of organocuprates may be followed by trapping of the intermediate, for example using an acid chloride as in Equation (138) ⟨81JOC3696⟩. Here again there is overall *cis*-addition to the alkyne, but with electrophiles which are less reactive than acid chlorides, equilibration of the intermediate enolate may occur.

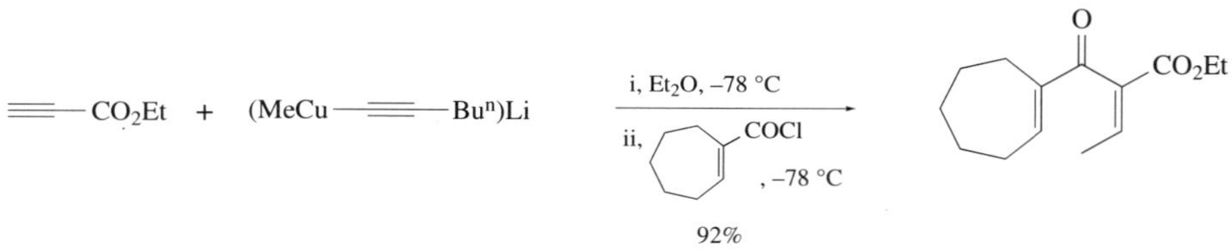

(138)

(ii) Additions of organocuprate reagents to unactivated alkynes

Organocuprates are the most important and useful of the organometallic reagents for addition to unactivated alkynes. Almost completely stereoselective *cis*-addition occurs, and high regioselectivity is obtained with terminal alkynes ⟨81S841⟩. A wide variety of types of organocuprate reagent may be used, for example $RCu \cdot MgX_2$, R_2CuMgX, R_2CuLi, and RCu(X)Li (where $X = OBu^t$, SPh, CN, etc.). The choice of reaction conditions (solvent, cosolvent, ligand) is critical for each reagent and the alkyne involved ⟨74BSF1656⟩; for example in the addition of MeCu to phenylethyne coupling reactions compete ⟨71TL2583, 74JOM(67)467⟩; however, Bu^nCu adds to 1-hexyne in 85% yield (Equation (139)), in the presence of $MgBr_2$. With reagents of the type $RCu \cdot MgX_2$ and terminal alkynes, high yields of the coupled products are obtained ⟨75JOM(96)471⟩. $Me_2S \cdot MgX_2$

complexes of organocuprates have been found to give improved and more reproducible results (Scheme 34) ⟨86OS(64)1⟩. Grignard reagents modified by the addition of copper(I) salts will similarly add to terminal alkynes only in the presence of excess $MgBr_2$ ⟨88TL4313⟩.

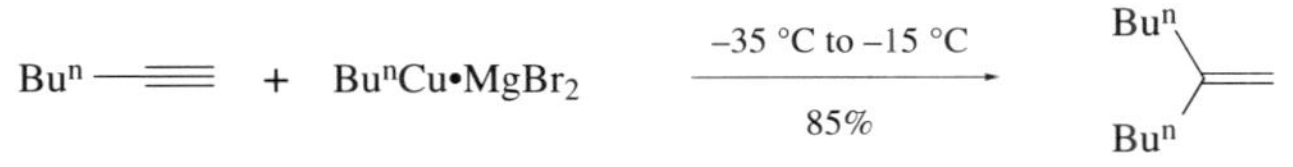

(139)

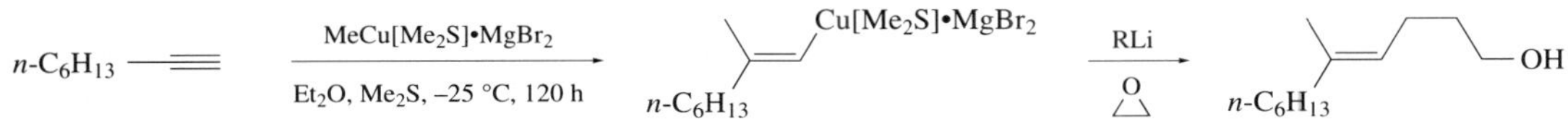

Scheme 34

Ethyne itself is the most reactive unfunctionalized alkyne towards the addition of organocuprates (Equation (140)) ⟨84OS(62)1⟩, and the lithium diorganocuprates, R_2CuLi, work best in this case. However, reagents of the type R_2CuLi do not add to ethyne or other alkynes when R = Ph, allyl, vinyl, or alkynyl. Ethyne can also undergo oligomerization initiated by the addition of the organocuprate, and the intermediate can then be trapped by a suitable electrophile (Scheme 35) ⟨81S841, 82TL5151, 85TL3285⟩.

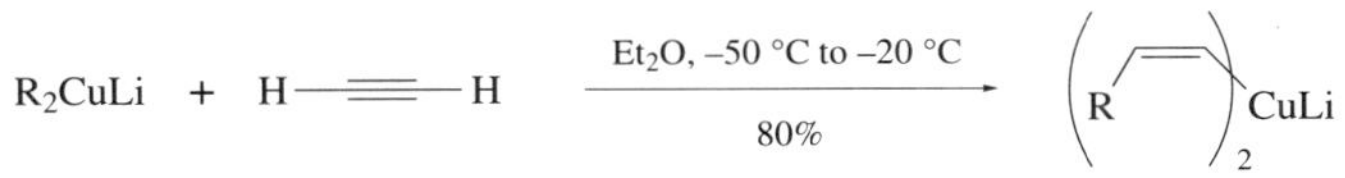

(140)

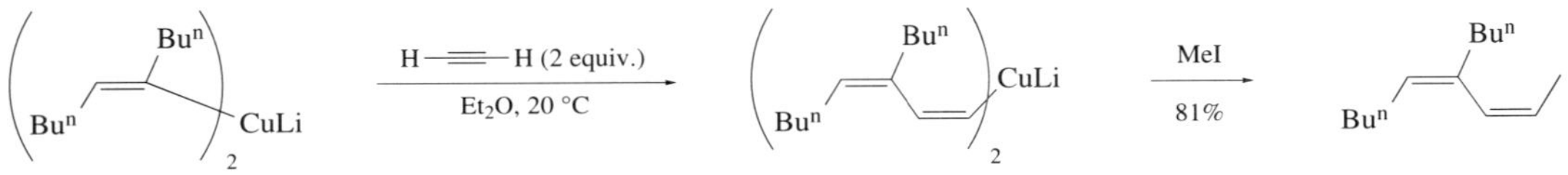

Scheme 35

Secondary alkyl cuprates often add to terminal alkynes with low regioselectivity (Equation (141)) ⟨77TL2023⟩, whereas additions of secondary and tertiary Grignard reagents using CuCl and excess $MgBr_2$ in THF are regioselective ⟨88TL4313⟩.

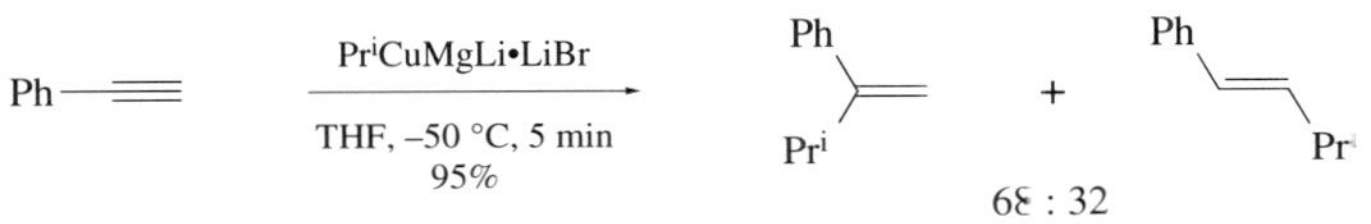

(141)

Organocuprates do not usually add to disubstituted alkynes which do not have extra functional groups; however, intramolecular reactions of this type are possible (Equation (142)) ⟨84JOC1574⟩.

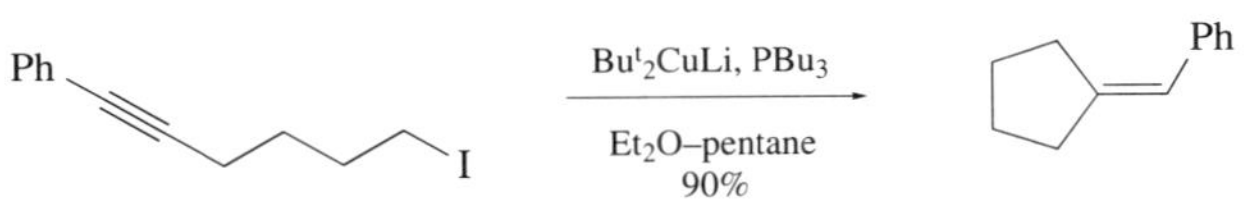

(142)

The presence of functional groups on the alkyne can alter the regioselectivity of the addition reaction, particularly when a heteroatom is present at a propargylic or homopropargylic position, which may give rise to products from chelation or steric control ⟨81S841⟩.

The intermediate alkenyl copper or cuprate species formed in the above reactions can be protonated, or alternatively trapped with a variety of carbon-centred electrophiles ⟨79JOC3888, 81S841⟩, and a good summary of possible transformations can be found in the review by Knochel ⟨91COS(4)896⟩ and the compilation of methods by Larock ⟨B-89MI 112-01⟩. Some selected examples are shown in Scheme 36. The dialkenyl cuprates are more reactive than the alkenyl copper reagents, and are required for reaction with certain electrophiles. The reactivity of an alkenyl copper can be enhanced by addition of HMPA, or an alkynyllithium, or magnesium salts (e.g., for reaction with epoxides ⟨79JOC3888⟩). The intermediate cuprate can also be coupled with an alkynyl phenyl iodosyl salt, giving a stereoselective synthesis of enynes ⟨87JA7561⟩.

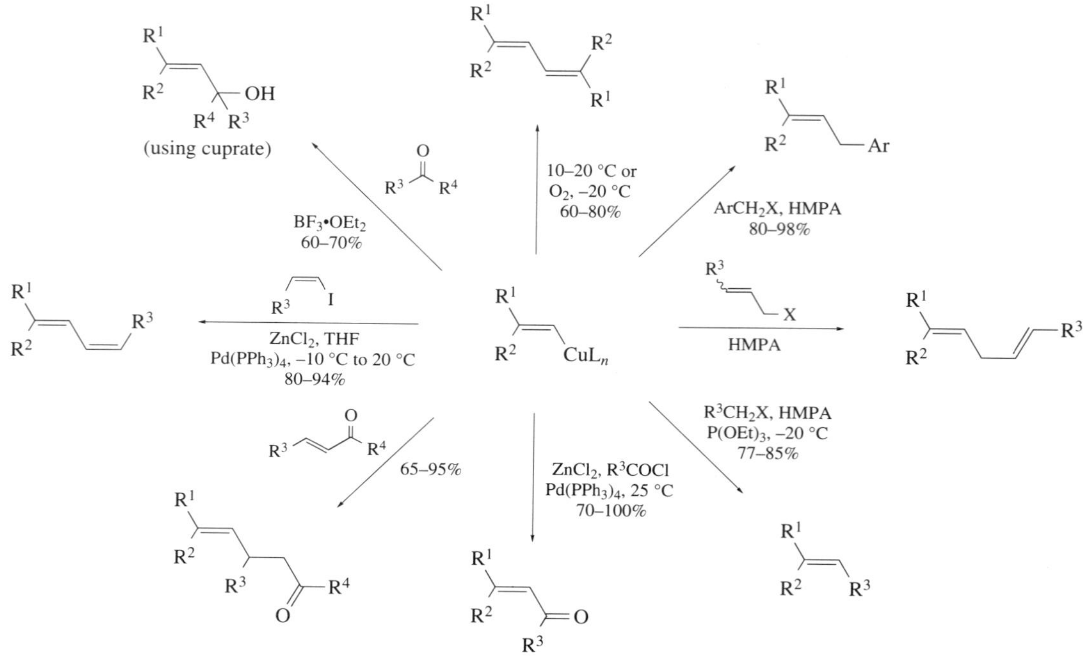

Scheme 36

1.12.3.3.3 Addition of organoboron reagents to alkynes

Although trialkylboranes will react with alkynes, this reaction is of limited synthetic utility, because severe reaction conditions are required, and further reaction is often possible ⟨91COS(4)886⟩.

Carboboration of terminal alkynes by BBr_3 and an organozinc reagent followed by palladium-catalyzed cross-coupling allows the synthesis of stereochemically defined trisubstituted alkenes (Equation (143)) ⟨88TL1811⟩.

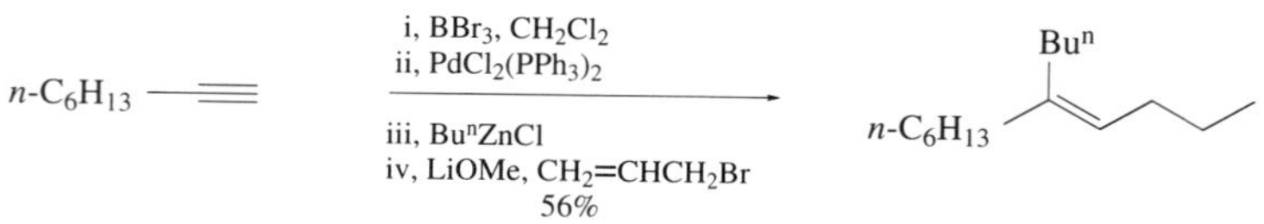

(143)

1.12.3.3.4 Addition of organoaluminum reagents to alkynes

Formation of alkenylaluminum compounds from alkynes and subsequent useful transformations have been reviewed specifically ⟨84OR(32)375, 91COS(4)888⟩, and as part of a more general review on carbometallation of alkynes ⟨81S841⟩.

Trialkylaluminum and chlorodialkylaluminum compounds add to terminal alkynes in the presence of Cp_2ZrCl_2, giving alkenylalane intermediates (**53**) resulting from *cis*-addition, which can then react with a variety of reagents to give further C—H or C—C bond formation (Scheme 37). This process has been reviewed ⟨81PAC2333, 85MI 112-04⟩. Various functional groups can be tolerated on the alkyne, including protected and free hydroxyl groups, sulfide, halogen and alkene moieties ⟨81JOC4093⟩.

Protonation of the alkenyl alanes (**53**) results in formation of 1,1-disubstituted alkenes (Scheme 37) ⟨60LA(629)222⟩. The intermediate (**53**) can be coupled with organic halides, catalyzed by palladium(0), nickel(0), or zinc(II) and this process has been reviewed ⟨82ACR340, 87TL2221⟩. The halides can be aryl, vinyl, alkynyl, allyl, or benzyl. Treatment with ethyl chloroformate results in stereoselective formation of α,β-unsaturated esters ⟨78TL2357, 81JOC4093, 82TL27⟩. Michael addition of (**53**) is also possible with α,β-unsaturated ketones, for example methyl vinyl ketone ⟨81PAC2333⟩.

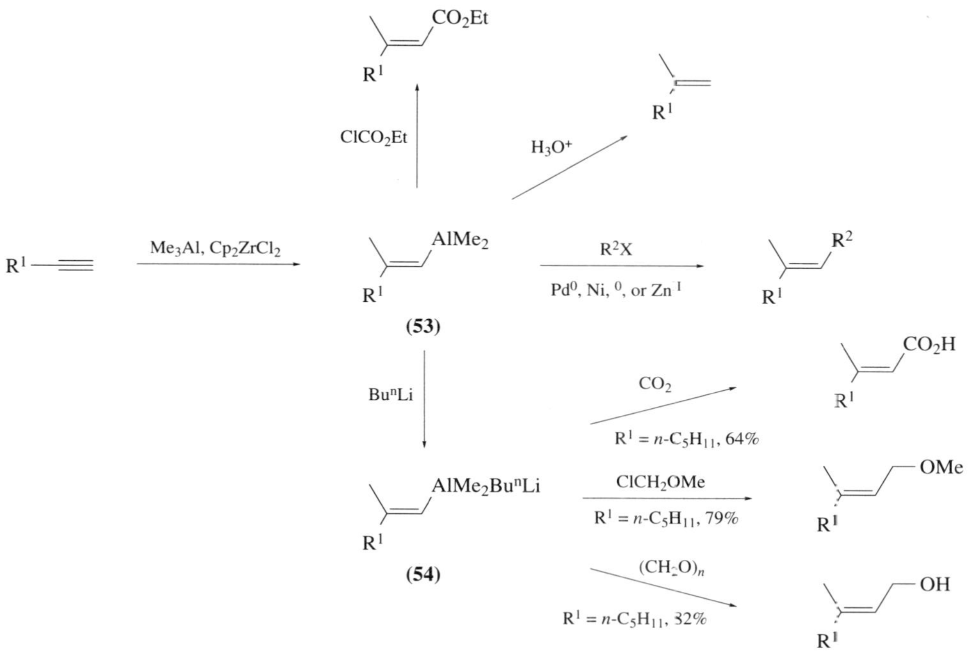

Scheme 37

Treatment of the intermediate (**53**) with Bu^nLi gives the more reactive alanate (**54**), which reacts with, for example, formaldehyde, ketones, chloromethyl methyl ether, CO_2, and epoxides (Scheme 37) ⟨78TL2357, 80JOC5223, 81JOC4093⟩.

Internal alkynes also add Me_3Al in the presence of Cp_2ZrCl_2, although the regioselectivity is usually poor, except in special cases, for example 1-phenylpropyne ⟨84OR(32)375⟩. Addition to symmetrical internal alkynes followed by protonation gives (*Z*)-trisubstituted alkenes (Equation (144)) ⟨78JA2252⟩. 3-Alkyn-1-ols undergo regioselective addition of trimethylaluminum in the presence of $TiCl_4$, resulting in stereoselective formation of trisubstituted homoallylic alcohols (Equation (145)) ⟨81JOC807, 85JOC2124⟩.

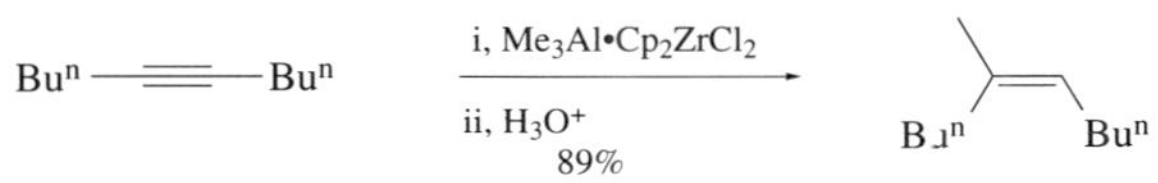

(144)

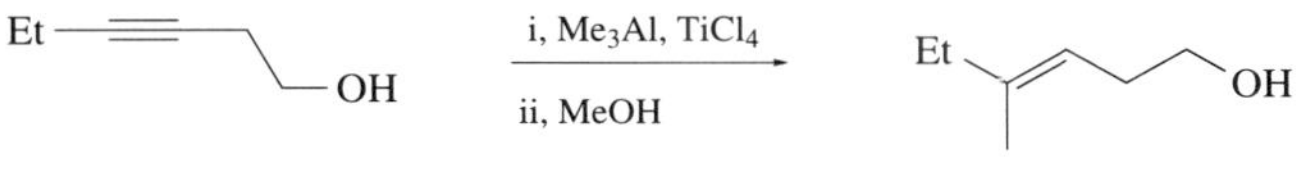

(145)

Allyl and benzyl alanes also undergo zirconium-promoted addition to terminal and internal alkynes, although the regioselectivity is moderate ⟨84TL5863⟩.

1.12.3.3.5 Addition of organozinc reagents to alkynes

Organozinc reagents add to alkynes more readily than to alkenes ⟨91COS(4)879⟩; for example allyl zinc bromide readily adds to terminal alkynes, and also to propargyl alcohol (Equation (146)) ⟨76TL3083⟩.

The zinc enolate of diethyl methylmalonate adds to terminal alkynes, but often gives a mixture of regioisomers (Equation (147)) ⟨75CR(C)999⟩.

(146)

(147)

The same portfolio of regioisomeric products are formed in the addition of allyl zinc bromide to the terminal and internal aminoalkynes (Equation (148)); the internal alkynes add the allyl zinc reagent to the carbon atom distal to the amino group with complete regioselectivity, whereas with the terminal alkynes the regioselectivity is reversed ⟨72JOM(44)69⟩. Propargyl zinc bromides also add to functionalized alkynes regioselectively to give enynes ⟨79JOM(166)1⟩. Nonfunctionalized internal alkynes do not react with allyl zinc reagents; however, intramolecular reaction is possible (Equation (149)), giving the *syn*-addition product ⟨78CR(C)265⟩. Dialkylzinc reagents add to both terminal and internal alkynes promoted by a zirconium complex ⟨88S1, 83OM563⟩. Addition of crotyl zinc bromide to the bromoalkyne (**55**) results in an intermediate which cyclizes to a cyclobutene (Equation (150)) ⟨84JA6105⟩.

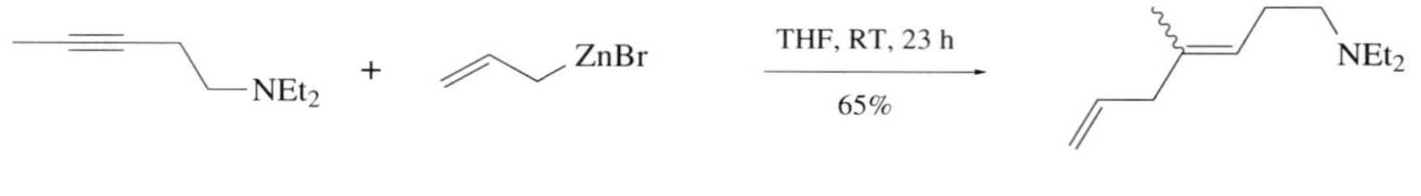

(148)

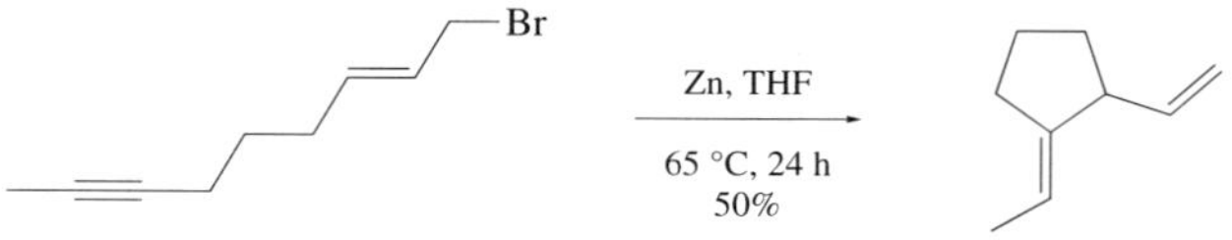

(149)

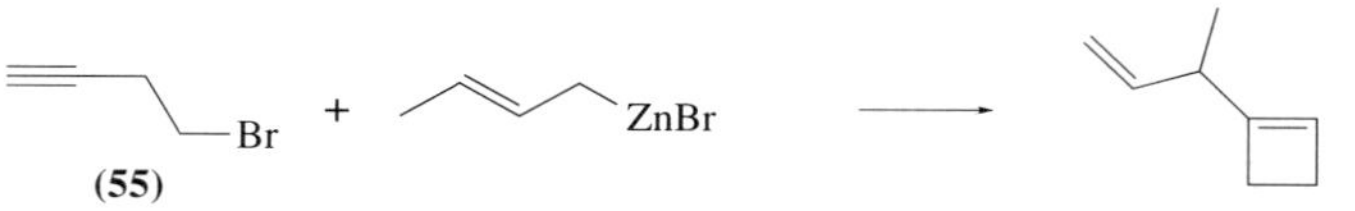

(**55**)

(150)

1.12.3.3.6 Other carbometallation reactions of alkynes

Organosilver reagents add to alkynylnitriles to give *cis*-addition of the alkyl group and hydrogen (Equation (151)) ⟨79TL3327, 81JOM(206)257⟩. Conjugated diynes also undergo addition of organosilver reagents, and the intermediates can be trapped to give further C—C bond formation, for example with allyl bromide or CO_2 (Equation (152)) ⟨81RTC337⟩.

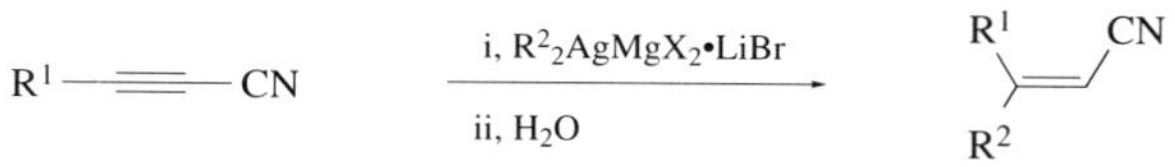

(151)

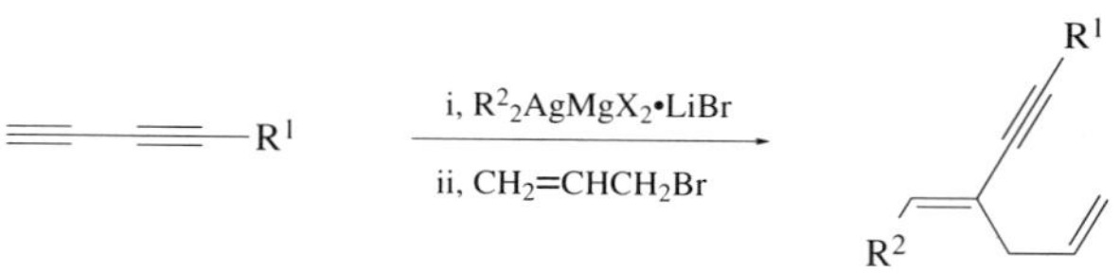

(152)

Intramolecular carbomercuration is possible with alkynyl silyl enol ethers, giving five- or six-membered rings ⟨83S801⟩; acylation of the intermediate results in a second carbon–carbon bond formation, and overall *cis*-addition with high stereoselectivity (Scheme 38).

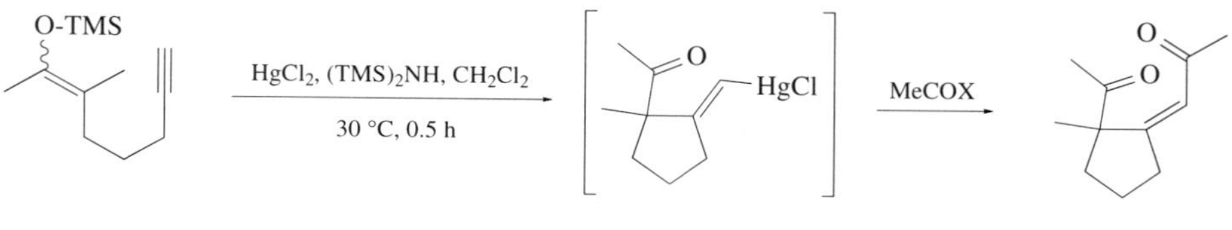

Scheme 38

The nickel complex $MeNi(acac)_2PPh_3$ adds to alkynes, and the C—Ni bond in the resulting alkenyl nickel complexes can be replaced by either C—H using TsOH or LAH, or by C—C using Me_3Al or MeLi (Equation (153)) ⟨81JA3002⟩. Reaction with CO and MeOH results in α,β-unsaturated esters.

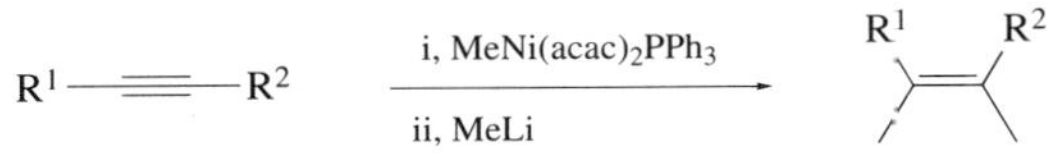

(153)

Carbon dioxide adds to alkynes in the presence of $Ni(cod)_2$, and the intermediate cyclic nickel complex can undergo further carbon–carbon bond formation with CO, alkyl halides, or even addition to another alkyne (Scheme 39) ⟨84JOM(266)203, 84JOM(270)C15⟩.

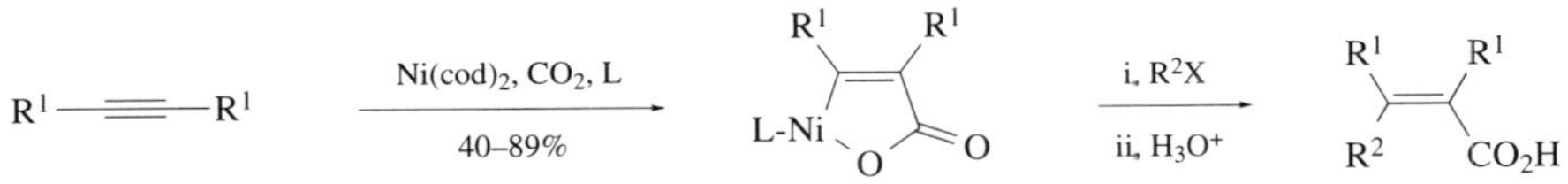

L = bidentate N or P ligand

Scheme 39

Hydrogen cyanide adds to internal alkynes, in the presence of a nickel or palladium catalyst, to give α,β-unsaturated nitriles (Equation (154)) ⟨82CC1231⟩.

(154)

1.12.3.4 Addition of Sulfur, Selenium, and Tellurium Reagents to Alkynes

Terminal alkynes undergo regioselective addition of $ArSO_2SePh$ to give alkenylsulfones (Scheme 40); treatment with an organocuprate then results in conjugate addition–elimination of the phenylselenide, and finally Na(Hg) amalgam reduction yields terminal alkenes ⟨86TL2187, 87JOC4258⟩.

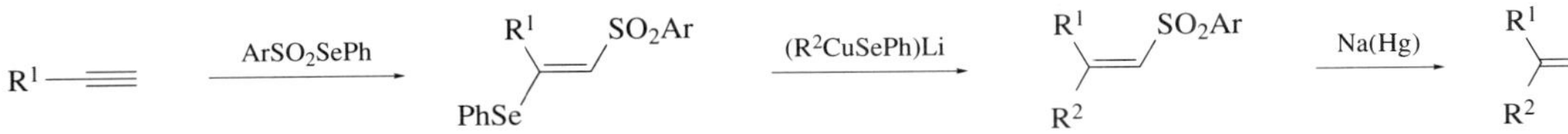

Scheme 40

Terminal alkynes also undergo *trans*-addition of PhTe—H to give (*Z*)-alkenyltellurides; the tellurium can then be replaced by a Grignard reagent ⟨82TL1181⟩, or by carbonylation ⟨87JOC4859⟩, both with retention of alkene geometry (Scheme 41).

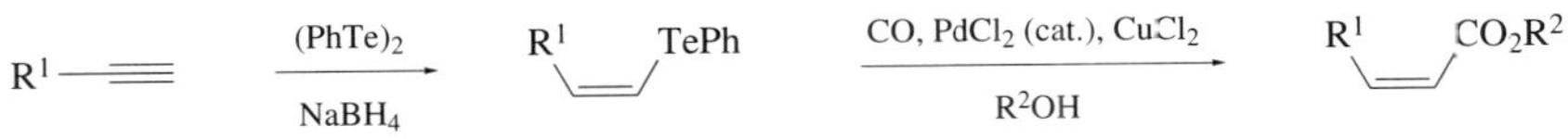

Scheme 41

1.12.3.5 Addition of Iron Reagents to Alkynes

Internal alkynes can be converted into the alkynyl iron complex (**56**). The triple bond is then activated towards the addition of nucleophiles, including carbon-centred nucleophiles such as Me^-, Ph^-, and malonate anions. Transfer of a CO ligand from the iron followed by reaction with alcohol results in stereoselective formation of α,β-unsaturated esters (Scheme 42) ⟨86JA1940, 88ACR229⟩.

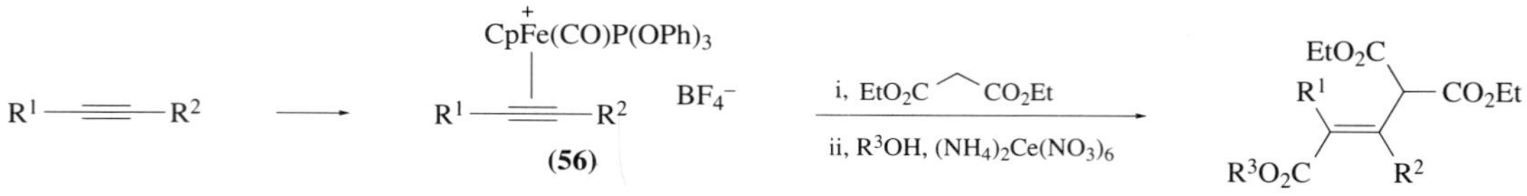

Scheme 42

1.12.3.6 Palladium-catalyzed Additions to Alkynes

A reductive Heck-type coupling reaction between an alkyne and an aryl iodide yields aromatic alkenes (**57**) (Equation (155)) ⟨84TL3137⟩. The reducing agent is formate, and the Ar and H are added stereoselectively *cis* to the triple bond. Propargyl alcohols can also be used ⟨85T5121⟩.

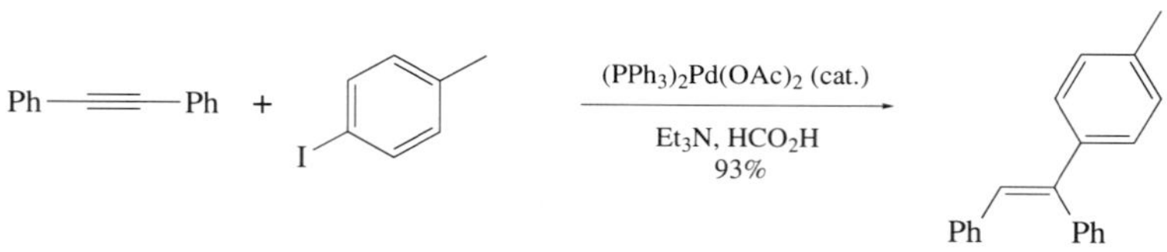

(155)

(**57**)

Terminal alkynes can be dimerized to give enynes in the presence of catalytic $Pd(OAc)_2$ and PAr_3 ⟨87JA3486⟩. Unsymmetrical coupling is also possible between a terminal alkyne and an alkyne bearing a carboxylate ester. Enynes can be cyclized to methylenecyclopentanes in the presence of a trialkylsilane, acetic acid, and a palladium catalyst (Equation (156)) ⟨87JA3161⟩. This process is also possible without the reductive step, giving 1,2-bis(alkylidene)cyclopentanes ⟨87JA3483⟩. Cascade cyclizations are also possible, where an organopalladium intermediate undergoes intermolecular addition to an alkyne, and this is followed by intramolecular cyclization onto an alkene ⟨93TL153, 93TL157⟩.

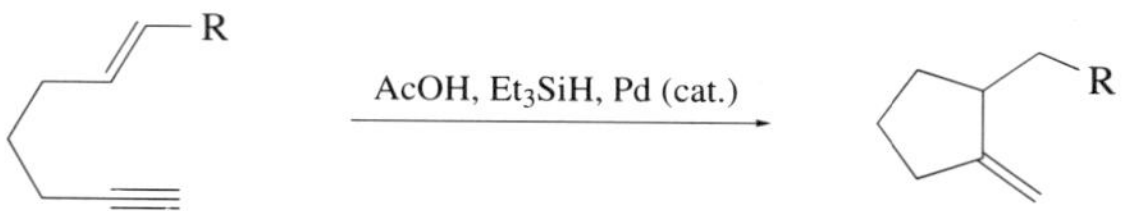

(156)

Palladium-catalyzed carbonylation of alkynols gives α-methylene lactones with five- or six-membered rings; fused bicyclic lactones can also be prepared in this way ⟨81JA7520⟩.

1.12.4 FREE RADICAL ADDITIONS

Free radical addition reactions have been the subject of many reviews, and additions to alkynes are included in several of these articles ⟨B-86MI 112-01, 88S417, 88S489, 91COS(4)715, 91COS(4)779, B-92MI 112-01⟩.

1.12.4.1 Intermolecular Free Radical Additions

Alkynes have been relatively little used in intermolecular additions of carbon-centred free radicals; alkenes are more reactive ⟨82AG(E)768⟩, and much more often used. The radical formed by hydrogen abstraction from propan-2-ol adds to propiolic acid with subsequent lactonization, in moderate overall yield ⟨62CR(254)1817⟩ and *t*-butyl radical adds to methyl propiolate to give mainly the (*Z*)-product ⟨82AG(E)768⟩; this stereoselectivity is rationalized by invoking steric hindrance of the *t*-butyl group during the hydrogen abstraction by the intermediate radical addition product. The radical generated from 7-bromoheptan-2-one by catalytic vitamin B_{12}, with a chemical or electrochemical reductant undergoes addition to ethyl propiolate, with a 7 : 1 (*E*) : (*Z*) selectivity ⟨83JA7200, 87PAC363⟩.

Free radicals generated from trialkylboranes react with conjugated ynones in an S_H2 process, yielding α,β-unsaturated ketones as a mixture of geometrical isomers ⟨70JA3503⟩.

1.12.4.2 Intramolecular Free Radical Additions

The simplest example of this type of reaction is the 5-hexynyl radical, which, like the 5-hexenyl system, cyclizes in the 5-*exo* mode rather than the 6-*endo* mode, to give methylenecyclopentane ⟨79JCS(P2)1535, 85T3925⟩. Many examples of this type of process can be found in the reviews listed above. Heteroatom linkages have also been widely used, for example the bromoacetal (**58**), which after radical cyclization is oxidized to the α-methylene lactone (Equation (157)) ⟨88TL6487⟩, and highly functionalized sugar-derived iodide (**59**) (Equation (158)) ⟨89TL57⟩. Aryl radicals have also been cyclized in 5-*exo* mode (Equation (159)) ⟨85TL6001⟩.

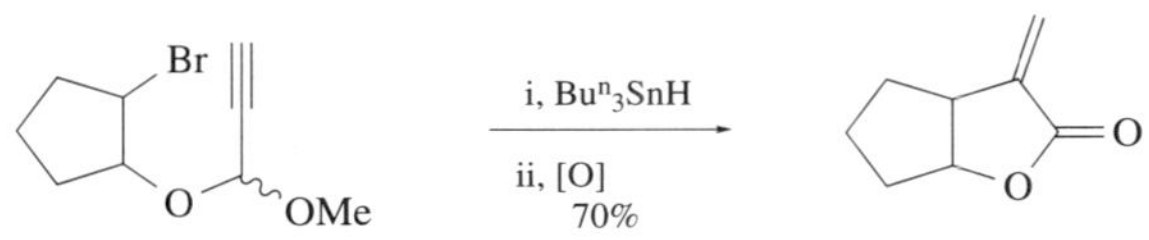

(157)

(58)

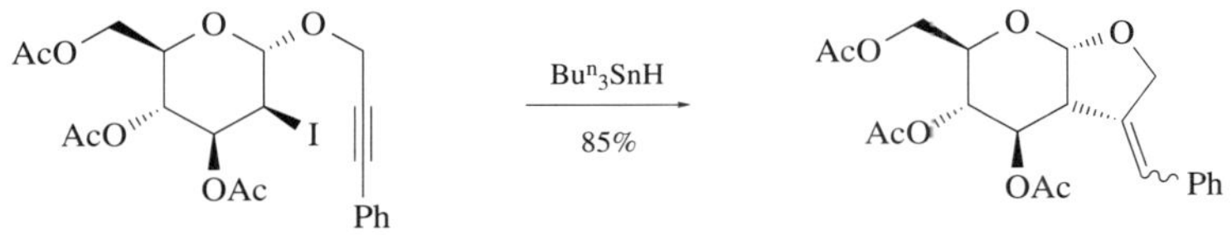

(158)

(59) (*E*):(*Z*) 1:3

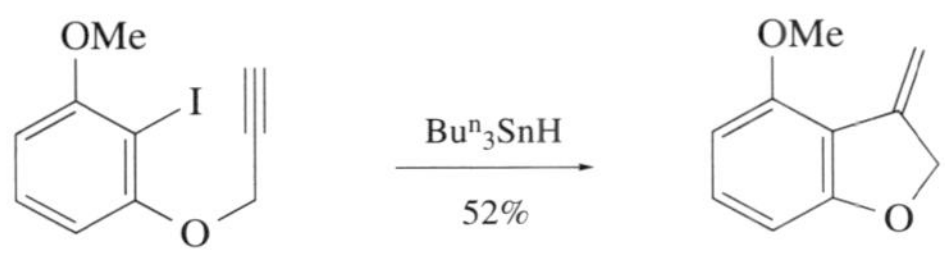

(159)

This preference for 5-*exo* cyclization is not universal, however, and although the pyrrolidinone (**60**) gives clean 5-*exo* cyclization (Equation (160)), the compound lacking the TMS group cyclizes to give a mixture of five- and six-membered ring products ⟨85T3959⟩. Intermolecular radical addition to an activated alkene can be followed by cyclization onto an alkyne, as in Equation (161) ⟨85CC980⟩.

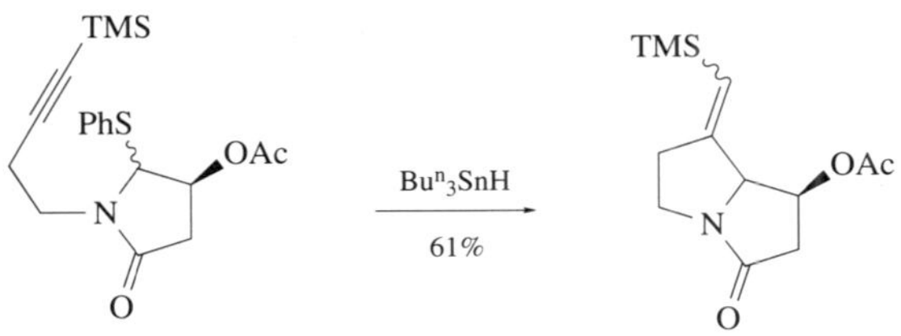

(160)

(60)

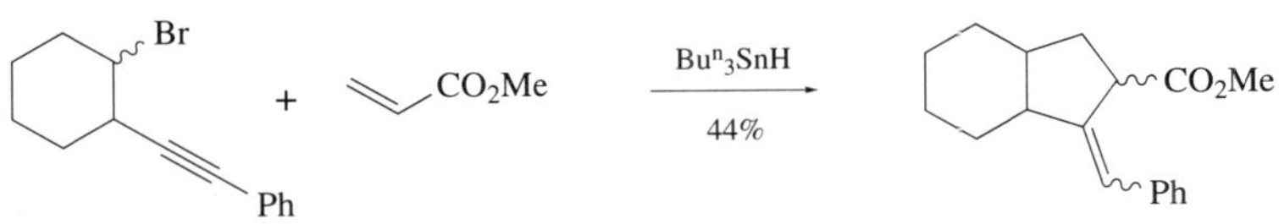

(161)

The 6-heptynyl system also cyclizes in the *exo* mode (e.g. Equation (162)) ⟨84JOC4244⟩. When the chain linking the radical to the alkyne is part of a β-lactam, then 7-*endo* cyclization is preferred ⟨82TL2505⟩. Again, addition of a substituent to the alkyne terminus reverses the selectivity; in this case phenyl substitution results in 6-*exo* cyclization. The cyclohexyl radical formed from (**61**) similarly cyclizes in 6-*exo* mode onto the phenyl-substituted alkyne in good yield (Equation (163)) ⟨84JOC1313⟩.

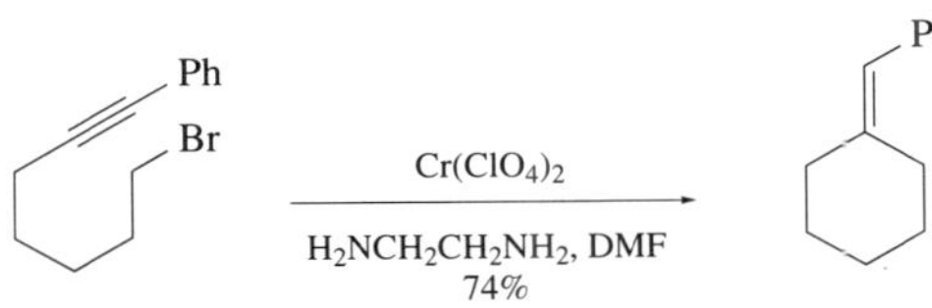

(162)

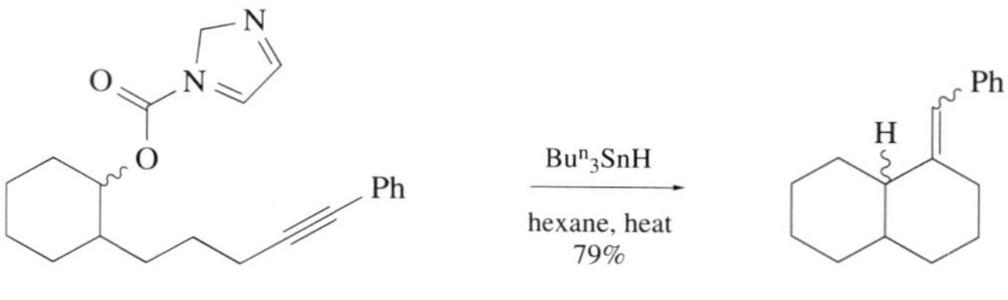

(61)

(163)

When the radical is stabilized, as in cyano ester **(62)**, the 7-*endo* mode is preferred (Equation (164)) ⟨71ACR386⟩. In contrast, the same type of radical in a hexynyl system undergoes 5-*exo* cyclization onto the alkyne, whereas the corresponding alkene undergoes 6-*endo* cyclization ⟨62CR(255)959⟩. Nitro groups can serve as precursors to radicals, and in Equation (165) cyclization onto the alkynyl ether gives a methylene tetrahydrofuran ⟨85T4013⟩.

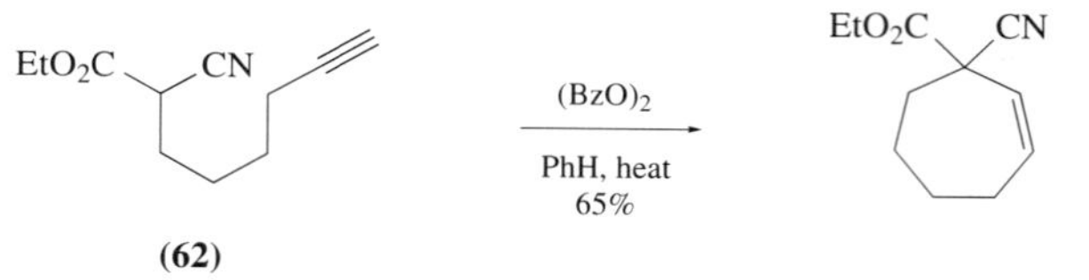

(62)

(164)

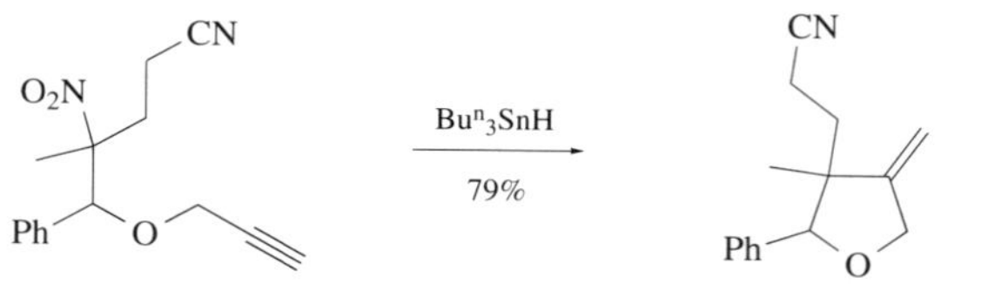

(165)

Acyl radicals undergo cyclization onto alkynes; in Equation (166) the radical is generated from a selenocarbonate ⟨86TL641⟩. Radical anions generated by reduction of carbonyl groups cyclize onto alkynes, resulting in unsaturated alcohols. The radical anions can be generated using a variety of reducing agents; for example Na/THF or sodium naphthalenide (Equation (167)) ⟨81JOC2622, 86TL2243⟩, Na/NH_3 ⟨79JA7107⟩, Zn/TMS-Cl ⟨83TL2821⟩, and cathodic reduction ⟨60JCS1045, 68CR(C)1834⟩. Ultraviolet light can also be used to generate the ketyl radical ⟨87TL4547⟩.

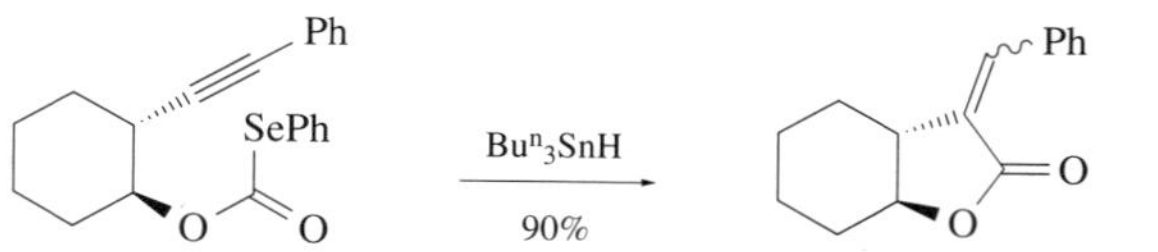

(166)

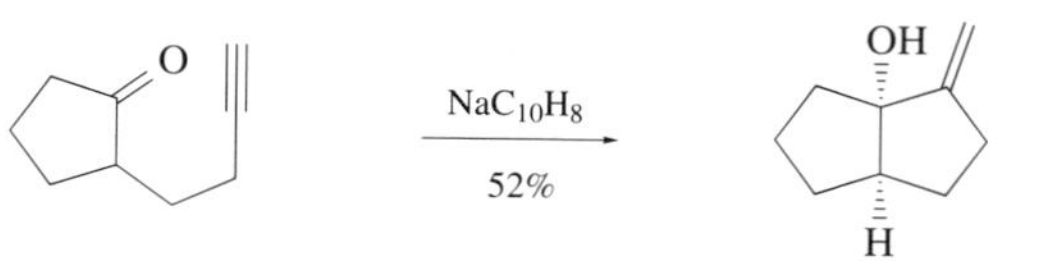

(167)

The cyclization onto the alkyne can be part of a tandem radical cyclization; for example cyclization onto an alkene (Equation (168)) can be followed by cyclization of the intermediate radical onto the terminal alkyne ⟨85TL4991⟩. This approach has been used as a key step in the synthesis of tricyclic sesquiterpenes. A ketyl radical generated by reduction by SmI_2 has also been used in a tandem cyclization (Equation (169)) ⟨88JA5064⟩. The order of cyclizations can be reversed (Equation (170)) such that the initial radical cyclization is onto the alkyne, and the resulting alkenyl radical cyclizes onto a cyclohexene unit ⟨83JA3720⟩. In Scheme 43 a different type of sequence takes place, with initial intermolecular addition of a stannyl radical to the alkyne, followed by 6-*exo* cyclization of the resulting alkenyl radical. Protodestannylation then results in the loss of the tin group and formation of the *exo*-methylene group ⟨87JA2829, 88OS(66)75⟩.

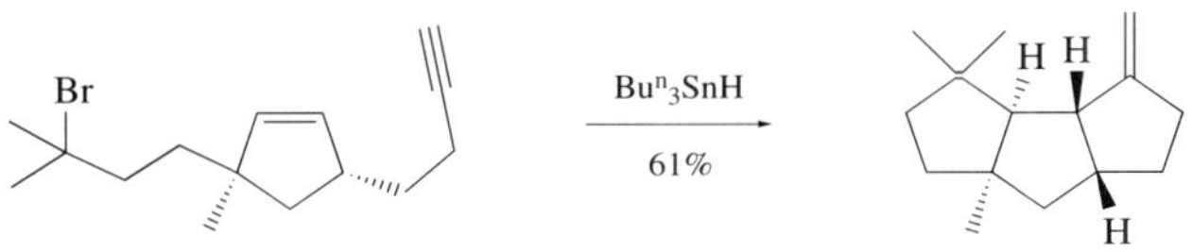

(168)

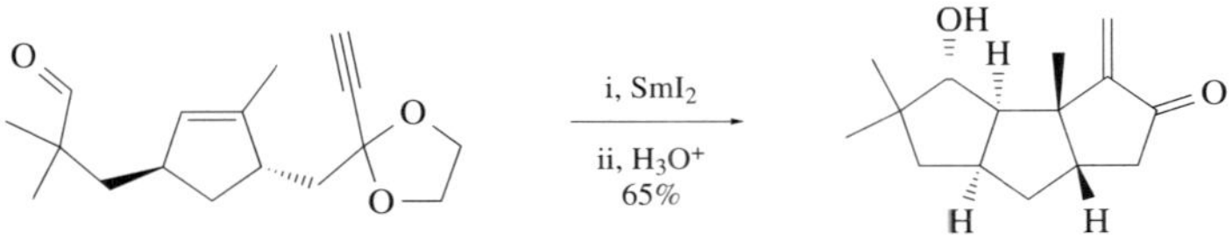

(169)

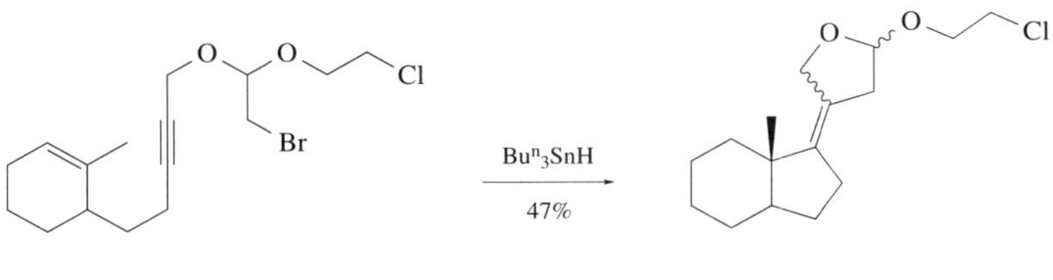

(170)

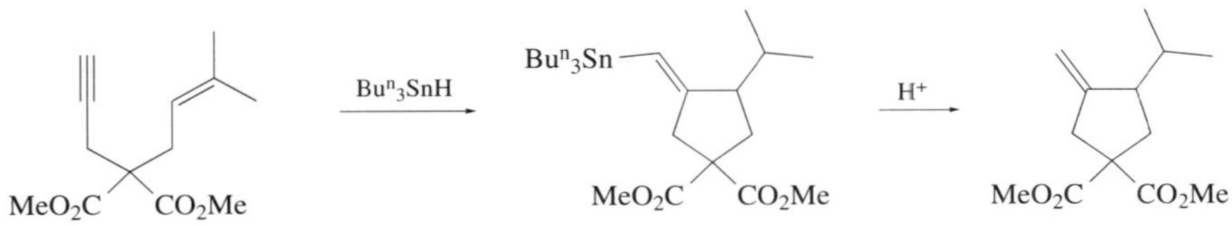

Scheme 43

1.12.5 CARBENE AND OTHER ADDITIONS

1.12.5.1 Addition of Simple Carbenes and Carbenoids to Alkynes

Addition of carbenes to alkynes is a good method for the preparation of cyclopropenes ⟨B-69MI 112-02, B-78MI 112-03⟩. Carbenes substituted with α-carbonyl groups add to internal alkynes ⟨B-87MI 112-01, 87TCC(137)75, 88TCC137⟩. Copper catalysis was generally used, and although this is successful for internal alkynes, terminal alkynes preferentially undergo C—H insertion. This limitation can be overcome by the use of bis(trimethylsilyl)ethyne followed by removal of the silyl groups after carbene addition (Scheme 44), and is equivalent to the overall addition to ethyne itself ⟨87JOC4875⟩. Enynes react selectively at the alkene double bond, giving alkynyl cyclopropanes. Diazoalkanes can also undergo 1,3-dipolar cycloaddition to alkynes; with alkynes substituted with electron-withdrawing groups, this is often faster than formation of the carbene. Another side reaction which can occur with phenylcarbenes is indene formation via a diradical intermediate; this is the major product in the reaction of diphenylcarbene with terminal alkynes, but the minor one with dimethylethyne ⟨73JA6286⟩.

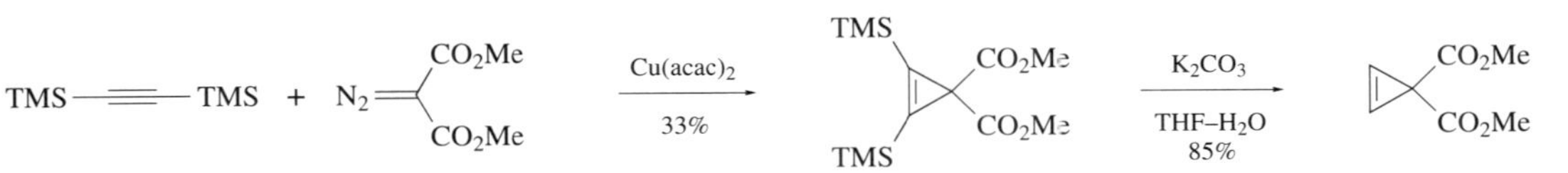

Scheme 44

Rhodium-catalyzed carbene formation from α-diazoesters is generally better than the use of copper catalysts, and addition of the carbene derived from ethyl diazoacetate to alkynes is the method of choice for the synthesis of cyclopropene-3-carboxylic acid esters (Equation (171)) ⟨78TL1239⟩. These cyclopropene carboxylic esters can also be made to undergo ring expansion to give cyclobutenes ⟨71JA3828⟩. Terminal alkynes usually react successfully; however, phenylethyne fails because the alkyne polymerizes. In some cases on using carbenes substituted with α-carbonyl groups furans are formed instead of cyclopropenones; which pathway is followed depends upon the particular carbenoid and alkene used ⟨91COS(4)1031⟩.

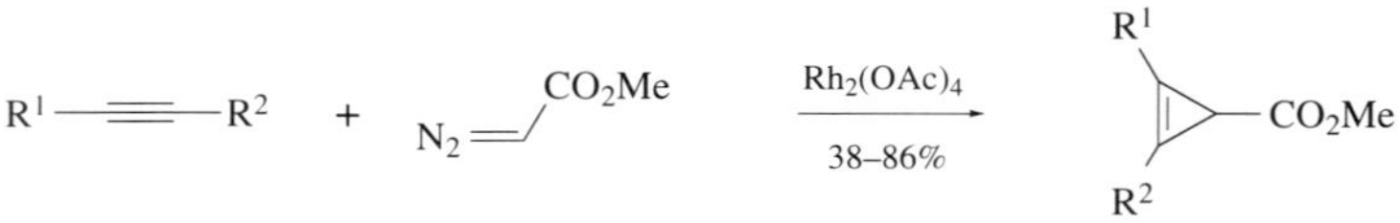

(171)

1,1-Dihalocarbenes, which are readily prepared, react sluggishly with alkynes compared to alkenes ⟨91COS(4)999⟩. The resulting dihalocyclopropenes are unstable; however, hydrolysis results in a good method for the preparation of cyclopropenones (Equation (172)) ⟨77CB154⟩. Again, internal alkynes add dihalocarbenes, whereas terminal alkynes do not, and enynes react preferentially at the alkene double bond. An alkylidene carbene can be generated by α-elimination of an enoltriflate; addition to an alkyne then gives the alkylidene cyclopropene ⟨75JA3854⟩.

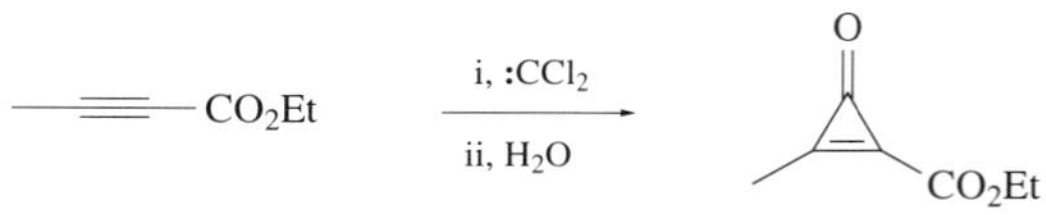

(172)

Intramolecular addition of carbenes to alkynes is also possible and, although it is much less studied than with alkenes, the subject has been reviewed ⟨92T5385⟩. The highly strained bicyclic cyclopropene (**63**) is thought to be initially produced; this species forms a vinyl carbene (**64**), which gives various products, often with only a single C—C bond derived from the alkyne (Scheme 45) ⟨90JOC414⟩.

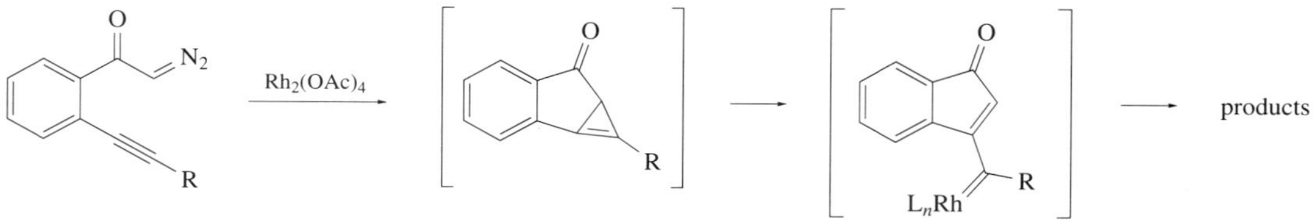

(63) **(64)**

Scheme 45

In Scheme 46 the vinyl carbene produced adds to a double bond of the diene; [3,3]-rearrangement of the divinylcyclopropane (**65**) finally yields the tetracyclic compound (**66**) ⟨91JOC2523⟩. An unusual example of a sequential double reaction with two alkynes is shown in Scheme 47. The initial intramolecular attack on the first alkyne gives a vinyl carbene as usual; this is then trapped by a second intermolecular addition of 2-butyne giving a cyclopropene ⟨91TL3755⟩. An intramolecular version of the second addition is known ⟨92T5385⟩.

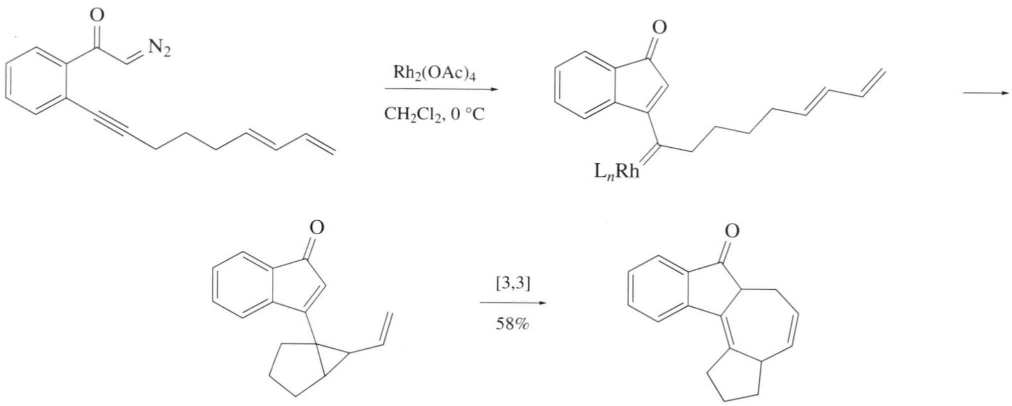

(65) **(66)**

Scheme 46

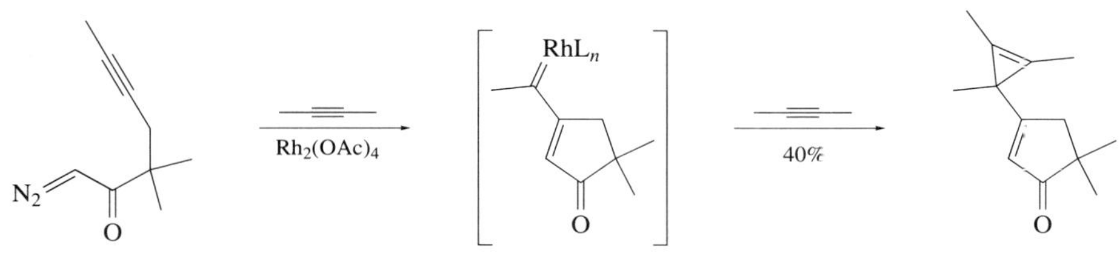

Scheme 47

1.12.5.2 Reaction of Fischer Carbene Complexes with Alkynes

Chromium carbene complexes react with alkynes to give cyclobutenones (Scheme 48) ⟨87JOM(334)9⟩. However, the most useful reaction of Fischer carbene complexes with alkynes is that of an arylalkoxy chromium carbene complex with carbon monoxide and an alkyne (Scheme 49) ⟨81JA7677⟩. This reaction has been reviewed ⟨88CRV1081⟩. The chromium can be easily removed from the product complex by CO to give a naphthol, or oxidatively to give a quinone. The reaction is general for both terminal and internal alkynes, and is highly regioselective with respect to the alkyne, being completely so for terminal alkynes. The π-system attached to the carbene may be aryl, heteroaryl, or alkenyl. Intramolecular reactions of this type are also recorded ⟨93JA10671⟩.

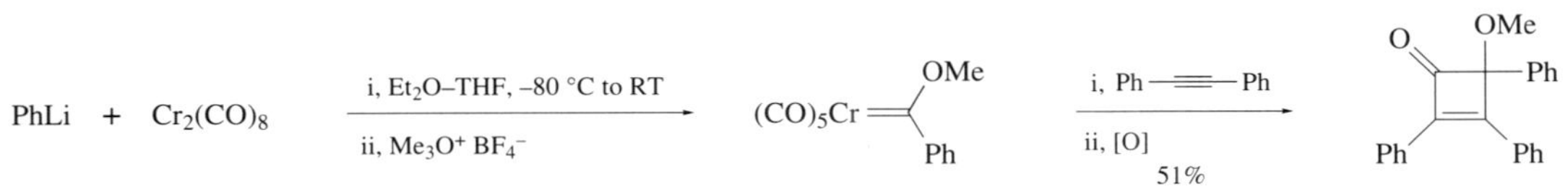

Scheme 48

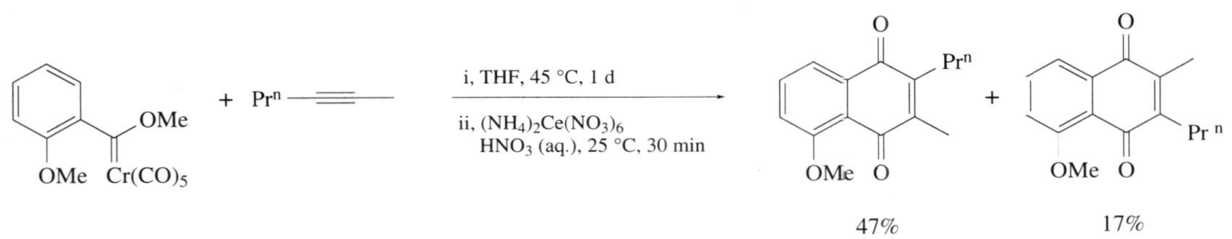

Scheme 49

1.13
One or More C=C Bond(s) by Elimination of Hydrogen, Carbon, Halogen or Oxygen Functions

JONATHAN M. PERCY
University of Birmingham, UK

1.13.1 BY ELIMINATION OF HYDROGEN

1.13.1.1 Dehydrogenation of Hydrocarbons

The most demanding dehydrogenation would involve the regioselective removal of two hydrogen atoms from an unactivated alkane. Active iridium and rhodium catalysts have been described ⟨84TL1279, 87JA8025⟩, which achieve the transformation, consuming an alkene oxidant during the catalytic cycle. Unsymmetrical alkanes have been oxidized with modest regio- and stereoselectivities ⟨92JA9492⟩. Though none of the current methods appear suitable for use in the synthesis of complex molecules, the area is developing rapidly and significant progress is being made.

The best known examples of reactions in this class lie outside the scope of this chapter and involve the aromatisation of polycyclic hydroaromatic compounds ⟨78CRV317⟩. A number of reagents have been employed to effect dehydrogenations, via hydride abstraction from benzylic or allylic positions. The reagent of choice 2,3-dichloro-5,6-dicyano-1,4-benzoquinone (ddq)(**1**), is commercially available and chemically reactive, although the cost of the reagent is relatively high. Chloranil (**2**) is a less expensive alternative. Reactions are normally performed in inert, high boiling solvents including toluene, chlorobenzene, xylene and 1,4-dioxan. The corresponding dihydroquinol is formed during the reaction, and can be removed by filtration or chromatography on alumina. Solvents must be dry to prevent nucleophilic attack by water on the quinone leading to decomposition. Nucleophilic functional groups should be protected.

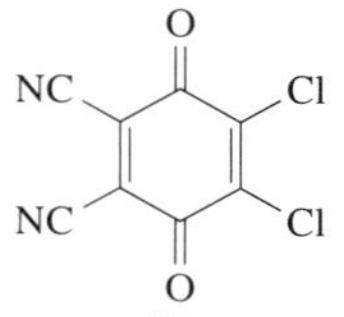

(1)

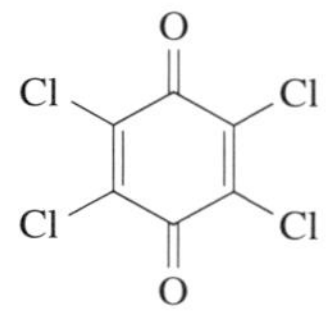

(2)

When hydride abstraction generates a highly stabilised carbenium ion, dehydrogenations occur smoothly under mild conditions ⟨91TL3679⟩. Oxidations of this type are unusual, but appealing, when further unwanted oxidation steps are impossible (Equation (1)).

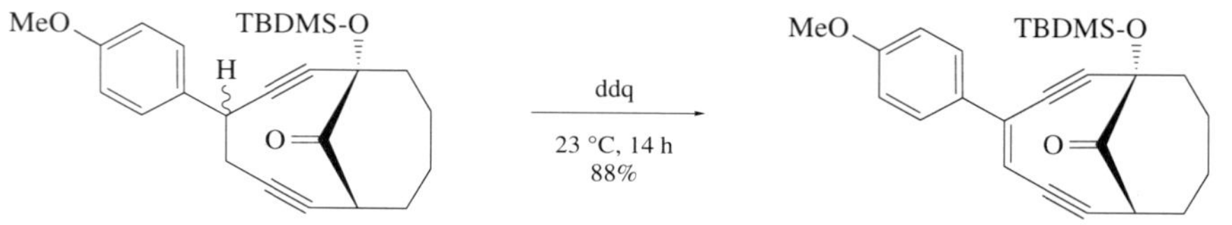

(1)

TBDMS = *t*-butyldimethylsilyl

1.13.1.2 Dehydrogenation of Ketones

The direct dehydrogenation of saturated ketones is a common and well-used reaction. The mechanism probably involves initial enolisation of the ketone, followed by nucleophilic attack on the quinone oxidant. Elimination of the hydroquinol follows to form the α,β-double bond (Scheme 1). Bulky substituents at the β-carbon reduce the rate of oxidation, presumably by impeding the nucleophilic addition step ⟨87BCJ4468⟩. A range of ketones have been oxidised in this way (Table 1, entries 1–7).

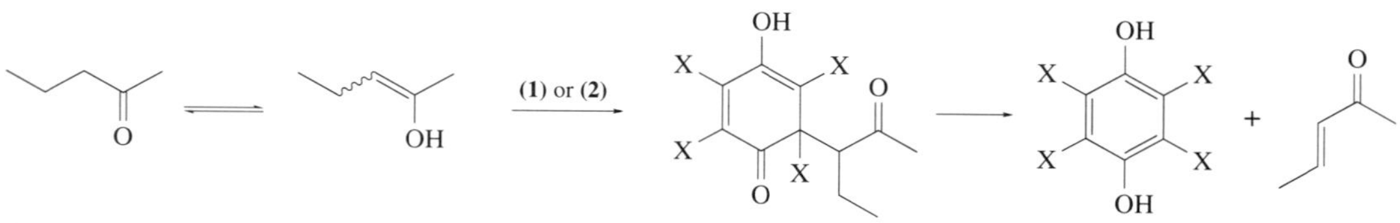

Scheme 1

A substituted cycloheptenone (entry 1) underwent efficient dehydrogenation when refluxed with ddq in benzene ⟨81JOC1⟩. Flavanone and chromanone substrates underwent smooth oxidation to

Table 1 Dehydrogenations with ddq.

Entry	*Substrate*	*Product*	*Yield (%)*
1	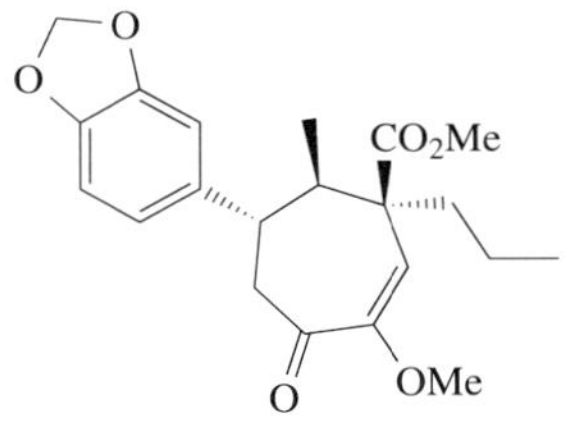	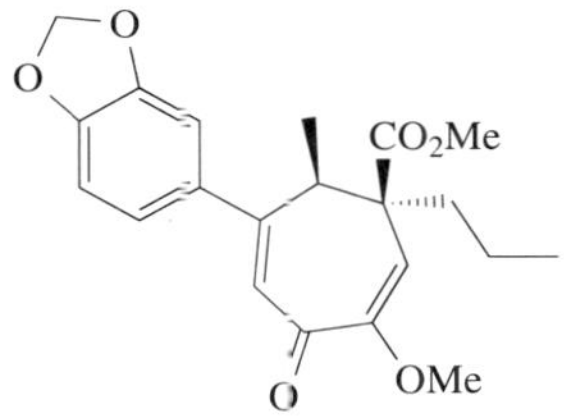	90
2	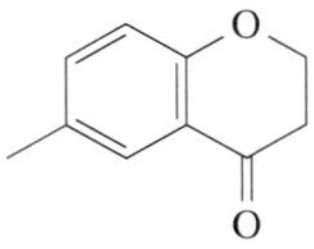	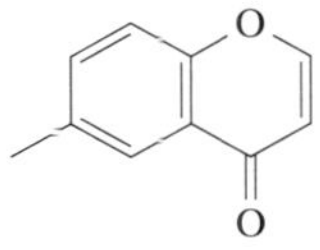	75
3	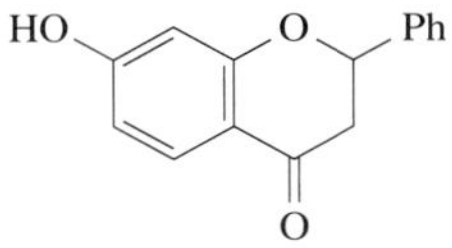	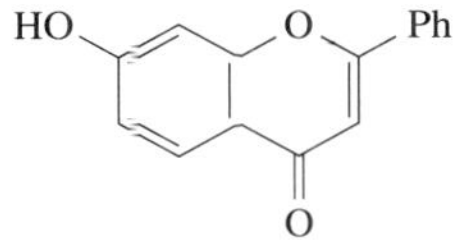	50
4	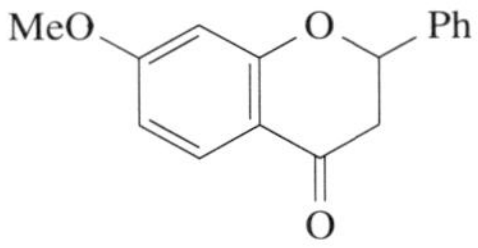	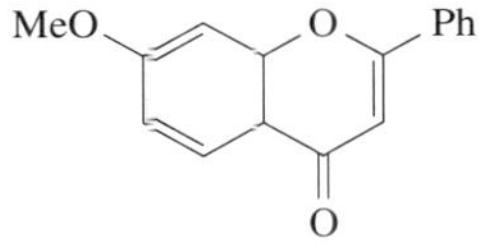	75
5	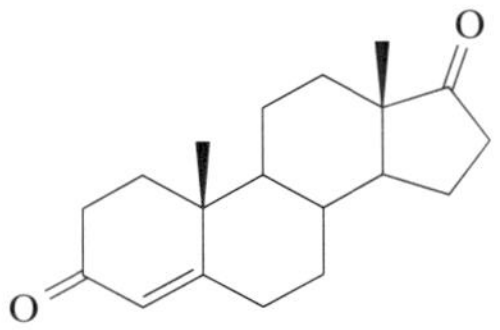	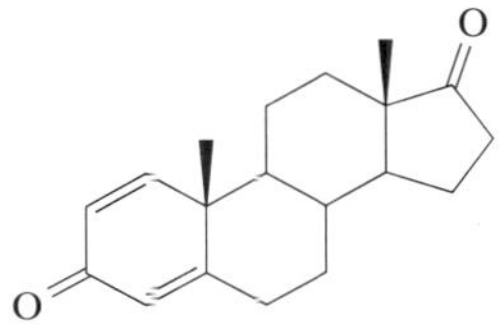	84
6	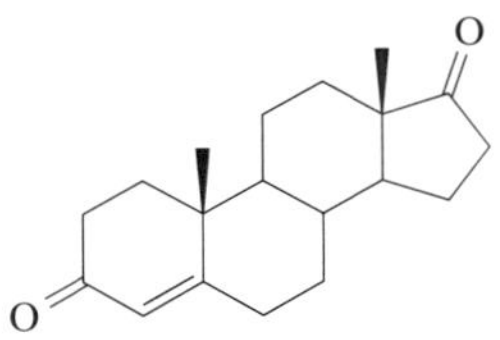	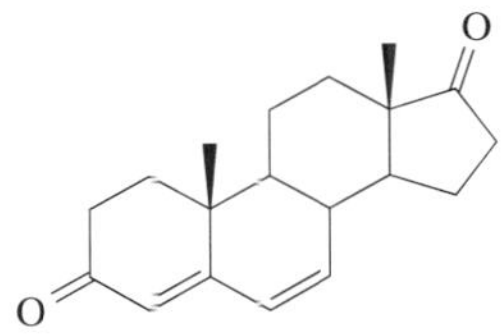	98
7	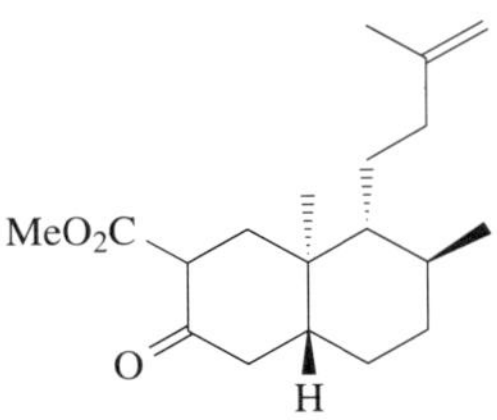	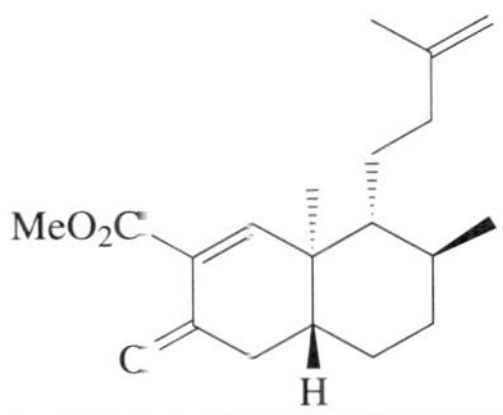	77

the corresponding flavones and chromones (entries 2–4) ⟨83S310⟩. The effect of the presence of a nucleophilic functional group was a reduction in yield, possibly through nucleophilic attack on the oxidant. The addition of an acid catalyst appeared to accelerate the reaction, presumably by increasing the rate of enolisation ⟨67JCS(C)1720⟩. Oxidation was complete within 5 hours, when *p*-toluene sulfonic acid was added to the reaction mixture (entry 6); in the absence of the catalyst, a reaction time of 15 hours was required (entry 5). The regiochemical outcome of the reaction also changed; oxidation occurred in the B-ring when the catalyst was present. The β-ketoester oxidation (entry 7) is unusual, but the observation is entirely consistent with the proposed mechanism via an enol ⟨92JA7375⟩.

Attempts to oxidise **(3)** and **(4)** using a ddq procedure were unsuccessful. The α-and β-faces of the steroid A-ring in **(3)** are both sterically hindered around the β-carbon atom, impairing attack on the oxidant ⟨89T6409⟩. In **(4)**, enolisation of the β,γ-unsaturated ketone may occur to generate a sterically hindered enol ⟨92JA5959⟩. Flavanone oxidation has been studied by a number of groups, and alternative methods use thallium salts ⟨92JCS(P1)2565⟩.

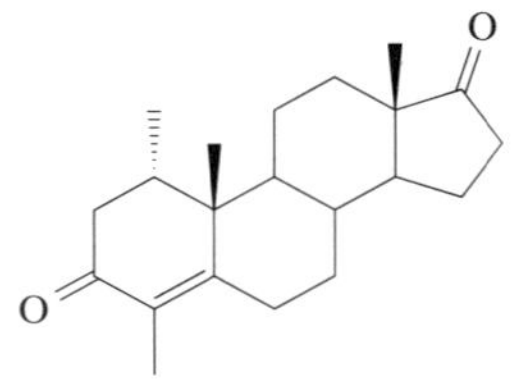

(3)

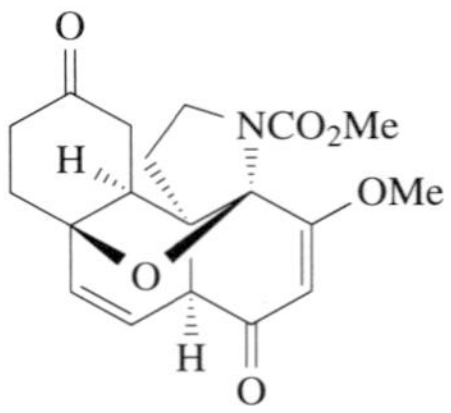

(4)

Ketone-to-enone conversions have been achieved by other methods. Scheme 2 shows a fortuitous procedure, applied to the synthesis of prostanoid intermediates ⟨83TL555⟩. The straightforward one-step oxidation used commercial (99%) copper(II) bromide, and offered significant advantages over more conventional multistep procedures ⟨80JOC4702⟩. However, the method does not appear to have found other applications.

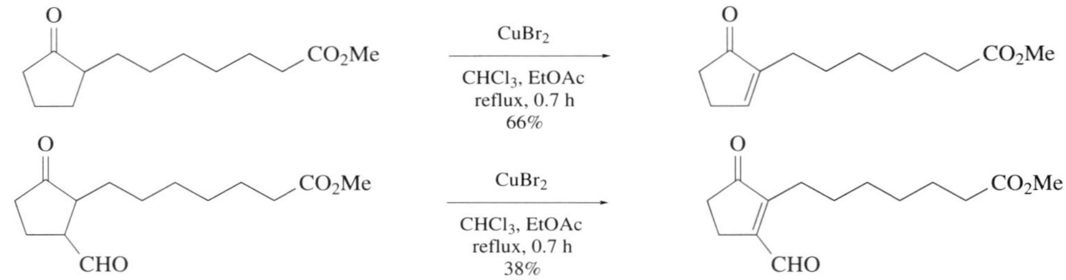

Scheme 2

Cation radical oxidants achieved a regioselective oxidation of substituted cyclopentanones to afford the more substituted cyclopentenone products in modest yield (Scheme 3) ⟨90T2371⟩. The salts could be preprepared or generated *in situ*. An excess of oxidant was required, and the reactions were performed in aqueous acetonitrile in the presence of collidine. Lead tetraacetate/copper(II) acetate has been used to dehydrogenate β-ketoesters and β-ketoamides (Scheme 4). The amides were easy to prepare and were oxidised in better yield than the corresponding esters. The procedure could be used on a multigramme scale ⟨93TL3021⟩.

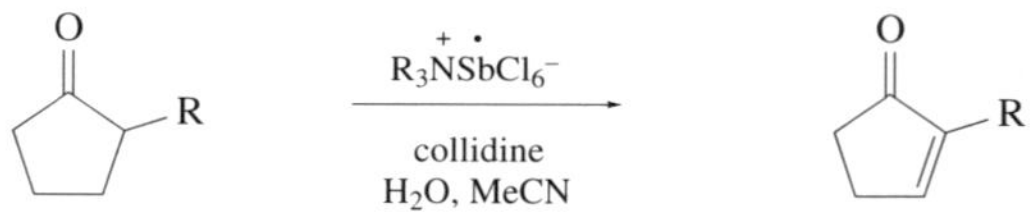

R = Me, 50% yield/65% conversion
R = Bu^t, 40% yield/65% conversion

Scheme 3

1.13.1.3 Dehydrogenation of Silyl Enol Ethers

The oxidation of a silyl enol ether is a valuable synthetic method, allowing the conversion of ketones to enones in a stepwise procedure. The method combines well with the developments in regiocontrolled silyl enol ether formation. Effective oxidants include ddq, palladium(II) compounds and tritylium salts.

Initial studies using ddq deployed a large excess of the oxidant to obtain enones in moderate yields ⟨78TL3455⟩. Tritylium salts, generated *in situ*, oxidised enol ethers to cyclic enones in moderate yield. The products were usually contaminated with ketone. Acyclic silyl enol ethers were oxidised in low yield, and ddq failed to yield oxidation products ⟨77JOC3961⟩. Both methods were refined by Fleming and Paterson in an elegant study combining conjugate addition methods with silyl enol ether oxidation ⟨79S736⟩. Table 2, entries 1–7 show the scope of the reaction.

The ease of oxidation varied with level of substitution at the carbon atom β- to the carbonyl

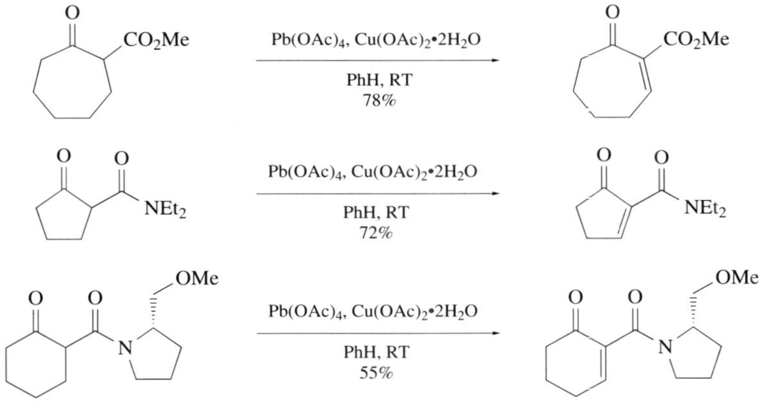

Scheme 4

Table 2 Dehydrogenations of silyl enol ethers with ddq.

Entry	*Substrate*	*Product*	*Yield (%)*
1	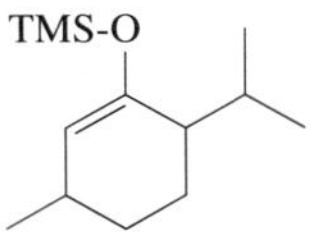	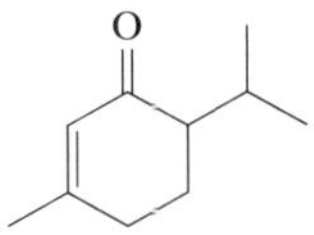	74
2	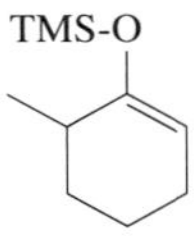	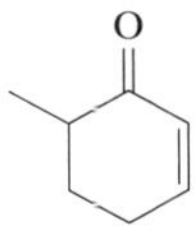	73
3	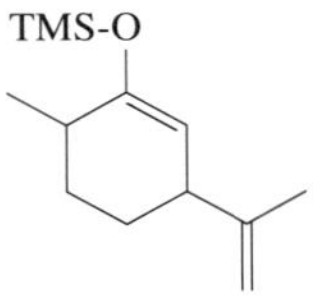	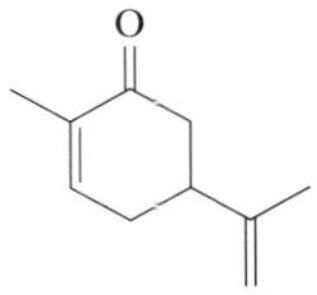	52
4	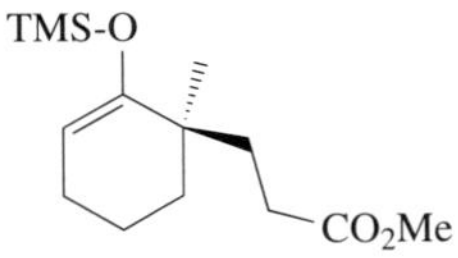	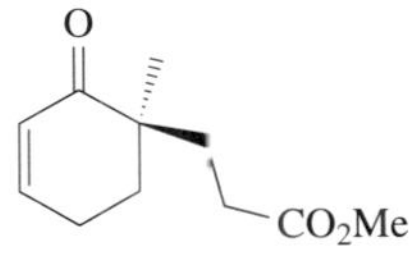	>80
5	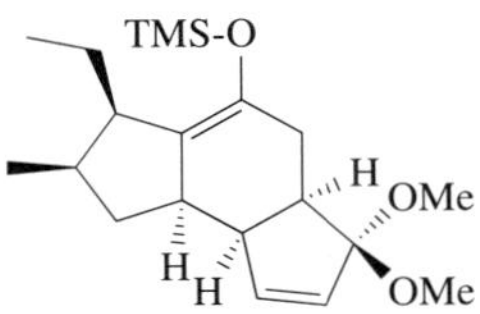	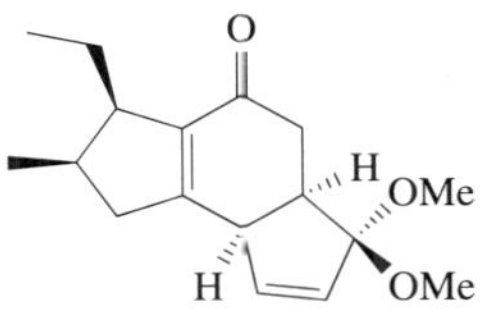	>45
6	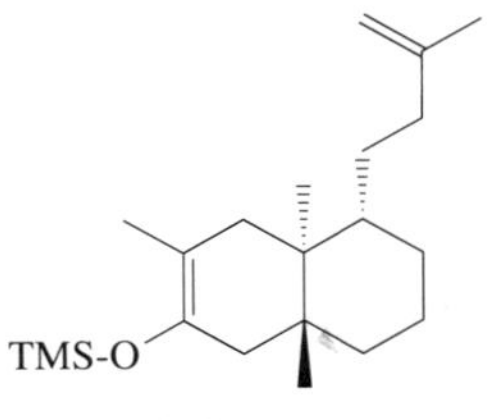	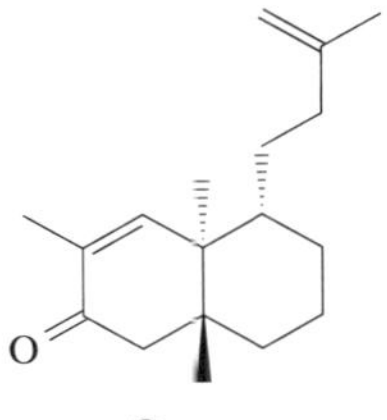	65
7	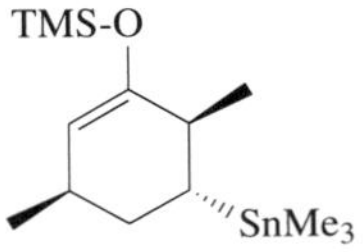	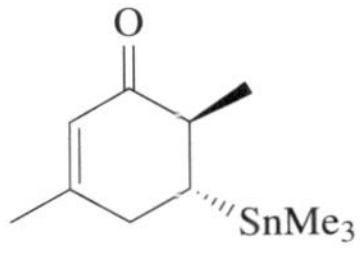	>57

group. When the position was trisubstituted (entry 1), one equivalent of ddq afforded high yields of enone. When the position was less substituted, ddq was used in excess, and collidine was required to remove the dihydroquinol formed in the reaction (entries 2 and 3). Unless the oxidation was rapid, the dihydroquinol acted as a general acid catalyst and protonated the silyl enol ether, leading to the formation of the saturated ketone. The oxidation has been applied in syntheses of complex molecules ⟨92JA7375⟩ and the conditions were compatible with the presence of a methyl ester ⟨90TL879⟩, an allylic ketal ⟨90JA9284⟩ and a trialkylstannyl group ⟨91TL5219⟩. A final example described an oxidation of an O-silyl imidate with applications in the steroid field ⟨88JA3318⟩. Equation (2) shows the reaction, which tolerated the presence of a range of functional groups in the D-ring of the steroid. A four-fold excess of silylating agent was used, so the nucleophilic functional groups were probably protected *in situ*.

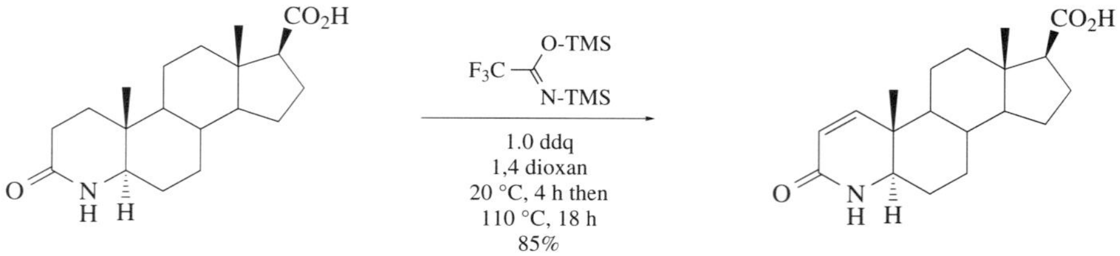

(2)

Oxidation with palladium acetate, usually in acetonitrile, has proved a general and valuable method. One equivalent of the oxidant was required, though 0.5 equivalents could be used when benzoquinone was present as a co-oxidant. The stoichiometric method is expensive to perform but valuable products have been obtained. The reaction conditions are very mild; high yields of enones were obtained by stirring the enol ether and oxidant in acetonitrile; workup was usually facile. Small amounts (generally $\ll 10\%$) of saturated ketone were also formed. Table 3 shows a range of examples including several from recent natural product syntheses. Acyclic silyl enol ethers were oxidised to *E*-enones or enals ⟨78JOC1011⟩. Exposed ⟨89LA2151⟩ and protected ⟨93CC619⟩ hydroxyl groups were tolerated. Other examples described steroid manipulations ⟨91TL3357⟩, homochiral cyclohexenone syntheses ⟨93TA21⟩ and a building block for an asymmetric synthesis of $(-)$-$\Delta^{9(12)}$-capnellene ⟨94T655⟩. Catalytic versions of the reaction were developed in Tsuji's laboratory ⟨83TL5635⟩. Silyl enol ethers, enol acetates and enol carbamates underwent efficient oxidation in the presence of palladium acetate, dppe and diallyl carbonate. However, relatively high temperatures were required to initiate the catalytic cycle; the reactions were performed in acetonitrile or benzonitrile at reflux (Equation (3)). Enol acetates underwent the catalytic reaction in high yield (Equation (4)). A range of conditions were described ⟨86T2971⟩, so a careful choice may be essential for optimum yields to be obtained. The tributyltin methoxide coreagent was required to convert the stable enol acetate to the reactive tin enolate *in situ*. Enol allyl carbamates underwent dehydrogenation in the presence of catalytic amounts of palladium acetate and triphenylphosphine (Equation (5)). The reactions were performed on a 0.1 mol scale and afforded high isolated yields ⟨87S992⟩.

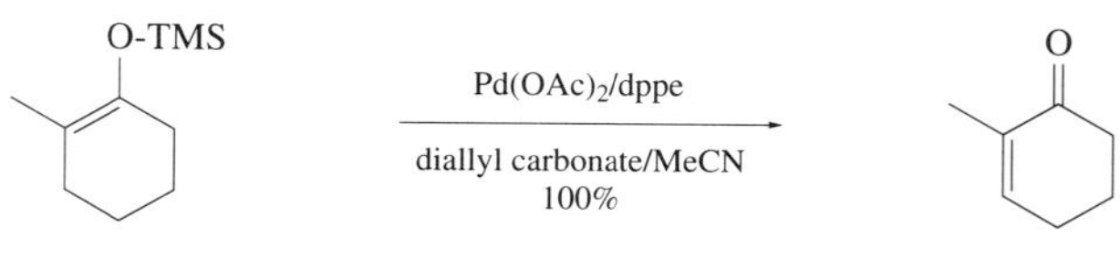

(3)

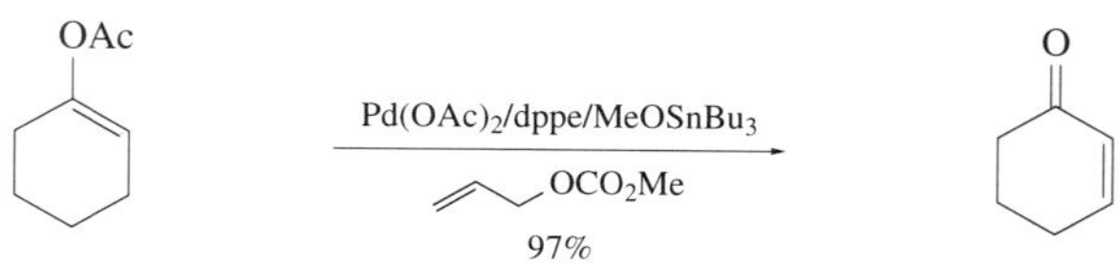

(4)

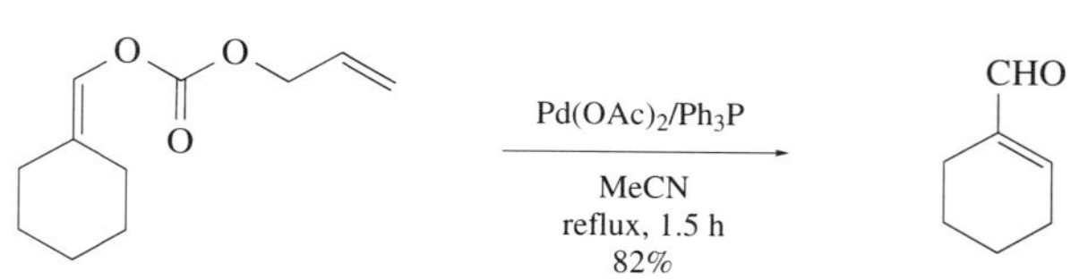

(5)

Ketene acetals and aminals also underwent oxidation with catalytic palladium acetate in acetonitrile or benzonitrile in the presence of two equivalents of allyl methyl carbamate (Equation (6)). The reactions in benzonitrile were performed at five-fold higher concentration.

Table 3 Dehydrogenations with palladium acetate.

Entry	*Substrate*	*Product*	*Yield (%)*
1	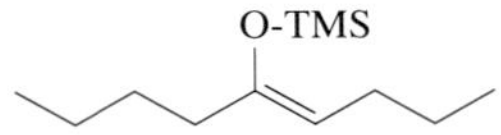	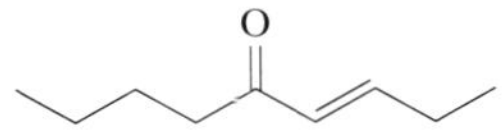	97
2	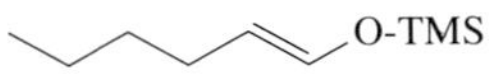	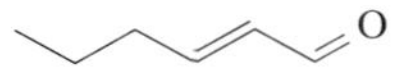	92
3	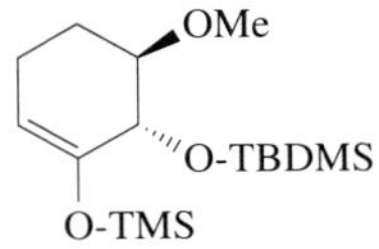	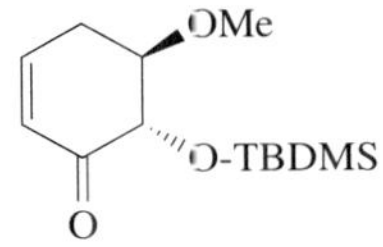	>88
4	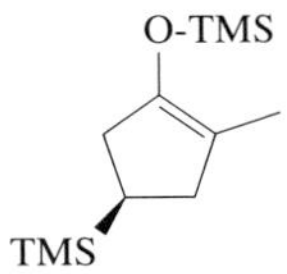	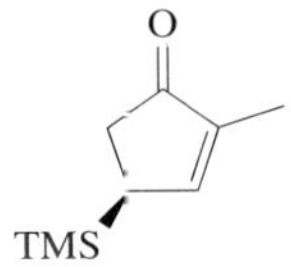	>74
5	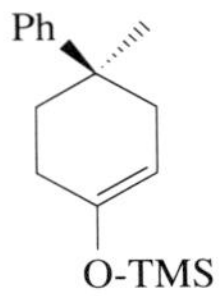	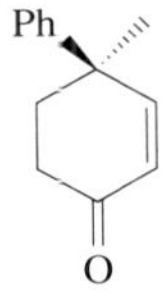	high
6	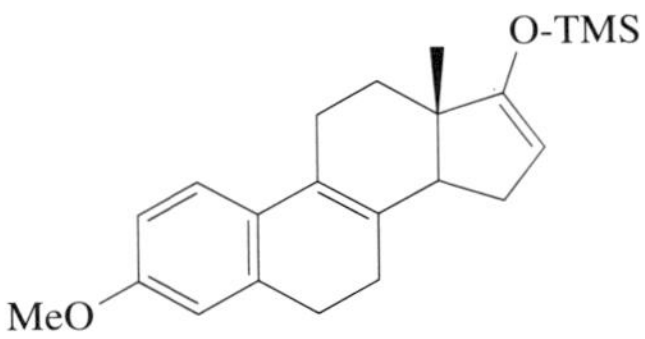	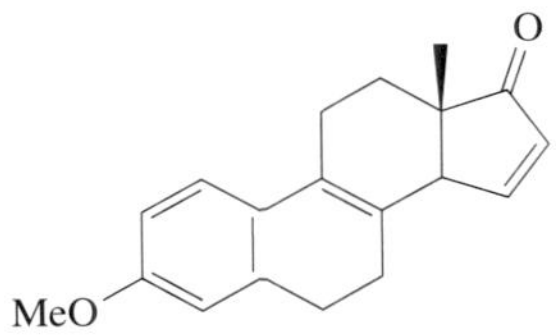	84
7	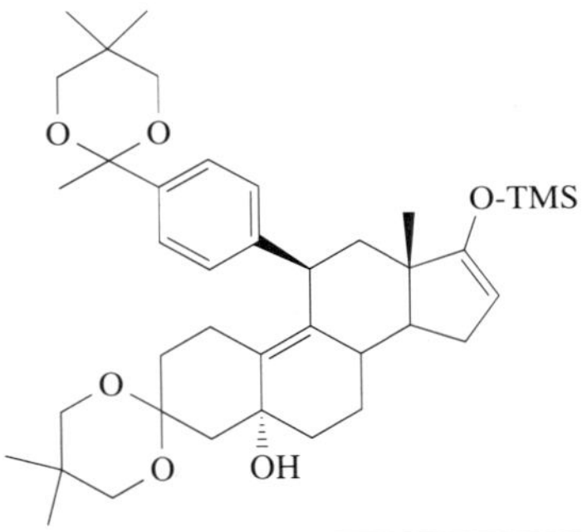	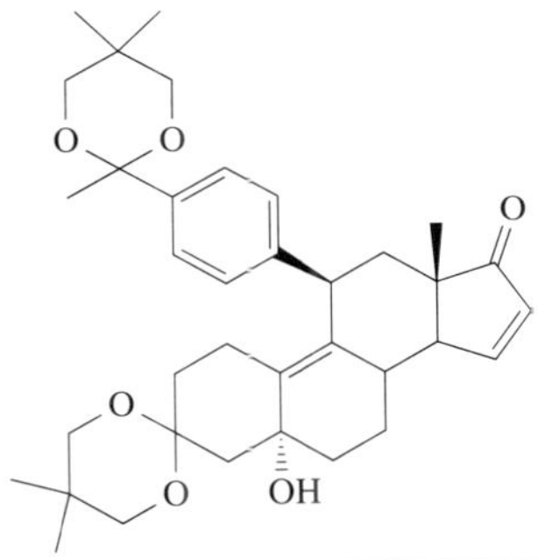	92

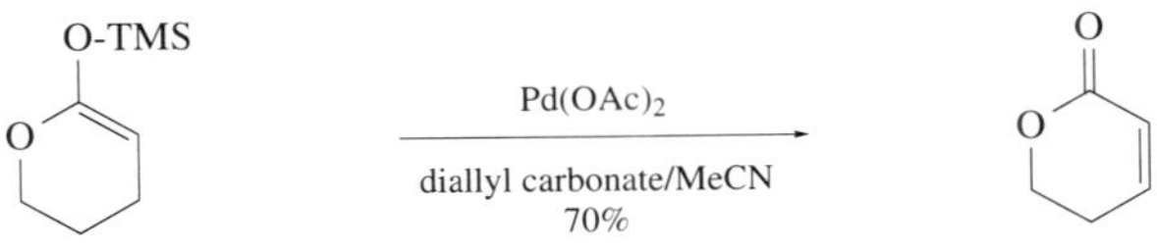

(6)

It is surprising that the catalytic chemistry developed by Tsuji has not found wider application in target syntheses. Despite the higher cost of the palladium reagents, the most popular method appears to be the stoichiometric oxidation.

1.13.2 ELIMINATION OF CARBON FUNCTIONS

1.13.2.1 Elimination of Hydrogen Cyanide

The range of examples in the literature is very limited, reflecting the modest nucleofugacity of the cyanide anion. Reaction mechanisms are most likely to be E1cb or E1cb-like. Enamines were prepared by elimination of hydrogen cyanide, following alkylation of α-aminonitriles, a useful early class of acyl anion equivalent. Equation (7) shows a typical result ⟨79S127⟩.

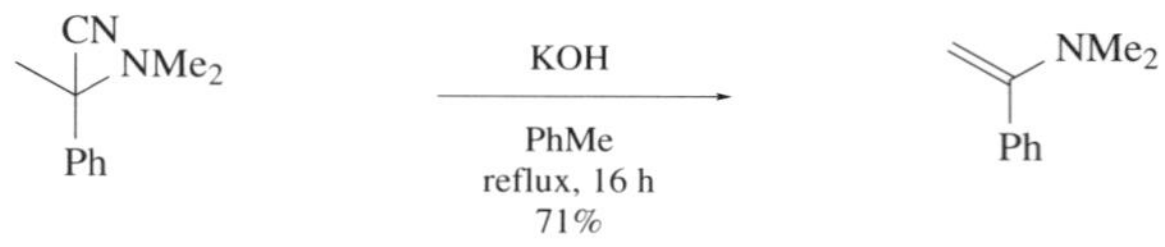

(7)

The experimental procedure appears straightforward. Shorter reaction times were possible when potassium *t*-butoxide replaced potassium hydroxide in the role of base. Other eliminations of hydrogen cyanide are shown in Equations (8)–(11). Forcing conditions were used in most cases, as exemplified in Equation (8) ⟨92M939⟩. In some cases, cyanide loss occurred directly from the enolate, though full experimental details were not reported ⟨77TL3187⟩. Equation (10) shows a typical strong base procedure ⟨56JA82⟩. Equation (11) is interesting because of the high reactivity of TCNE towards neutral aromatic nucleophiles. The very mild conditions for the elimination reflect the high degree of carbanion stabilisation in the conjugate base ⟨62ACS(B)523⟩. In other cases, direct elimination of hydrogen cyanide occurred in the reaction mixture ⟨59OS(39)68⟩.

(8)

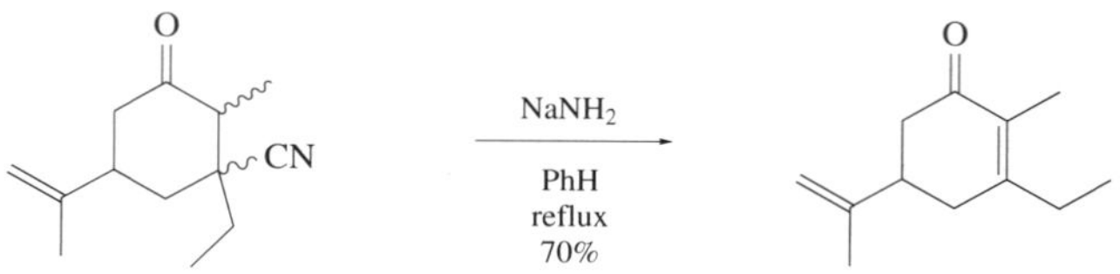

(9)

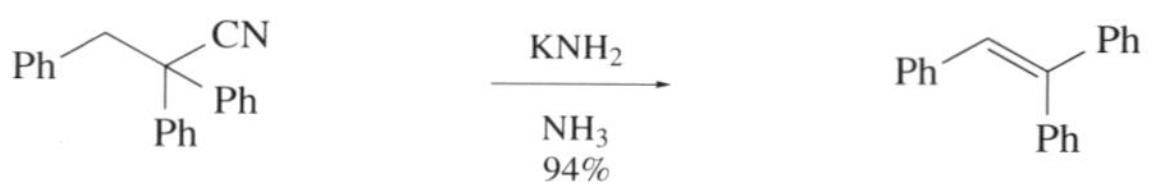

(10)

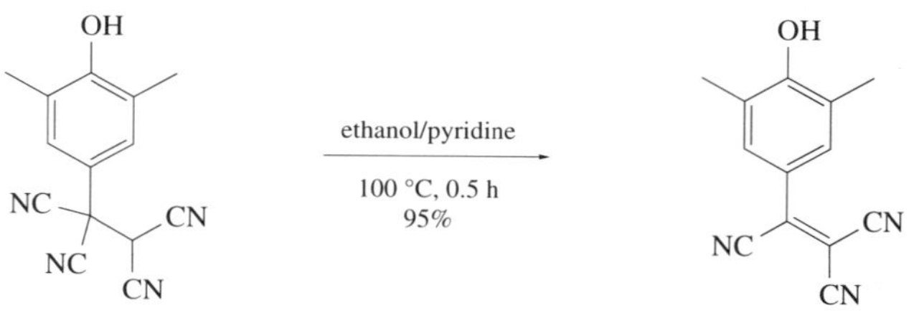

(11)

1.13.2.2 Elimination of Carbon Oxides

1.13.2.2.1 Decarboxylation

The decarboxylation of carboxylic acids to form alkenes has been achieved with lead tetraacetate. The reagent is commercially available and inexpensive, and its preparation is facile. Polar solvents (acetic acid, acetonitrile, DMF) were used in most cases. The literature contains many examples of successful decarboxylations using the reagent ⟨72OR(19)279⟩.

The ease of decarboxylation is related to the stability of the carbenium ion formed at the carboxyl-bearing carbon atom. Tertiary acids undergo decarboxylation at room temperature with photochemical initiation at 350 nm. Running the reaction at low temperature prevents oxidation of the nucleophilic trisubstituted alkene products. Side reactions have proved problematic in procedures using this reagent; for example, α,β-unsaturated ketones are oxidised at the α′-position ⟨92S235⟩. The decarboxylation of secondary and even primary acids has been achieved in the presence of a low concentration of copper(II) acetate monohydrate to promote the oxidation of the first-formed alkyl radical to the corresponding carbenium ion. Rearrangement and poor regioselectivity have attended the decarboxylation reaction, in common with other carbenium ion-dependent processes. Some conversions are shown in Table 4, entries 1–6.

Table 4 Decarboxylations with lead tetraacetate.

Entry	*Substrate*	*Product*	*Yield* (%)
1	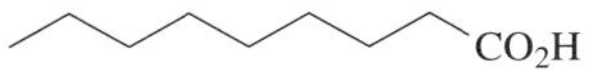	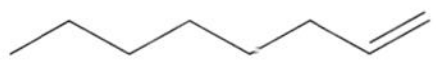	87
2	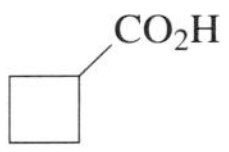		78
3	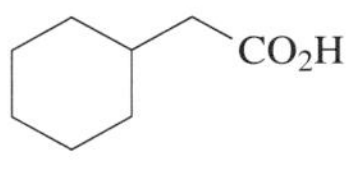	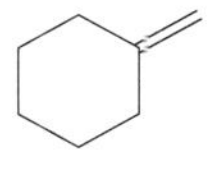	84
4	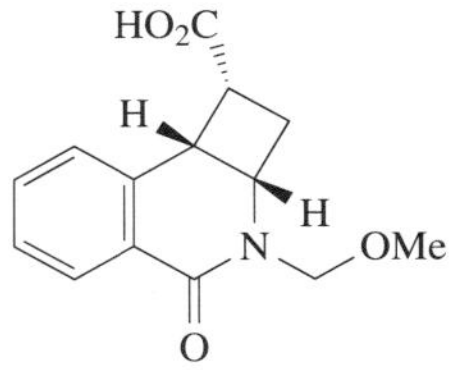	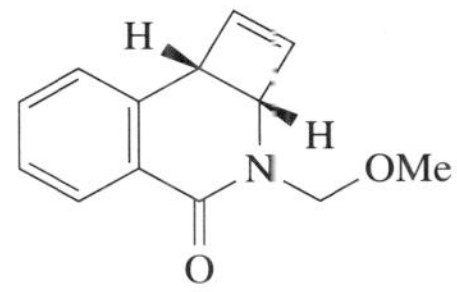	90
5	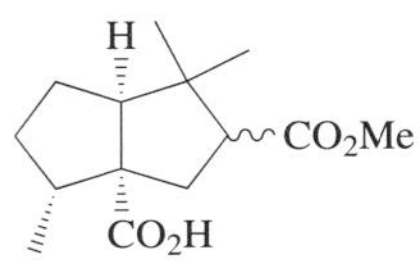	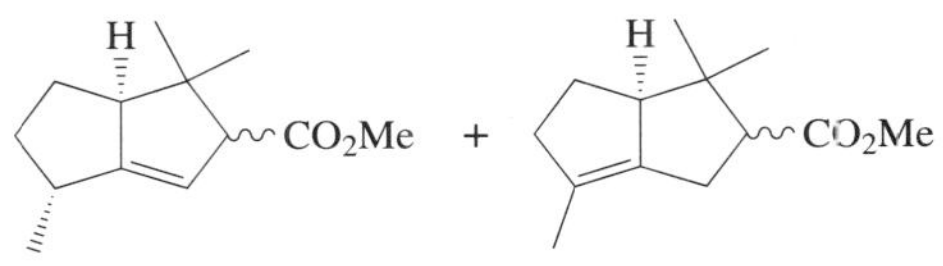	56
6	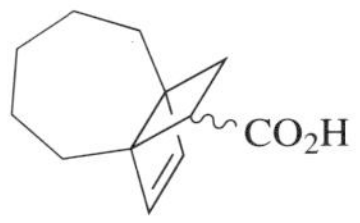	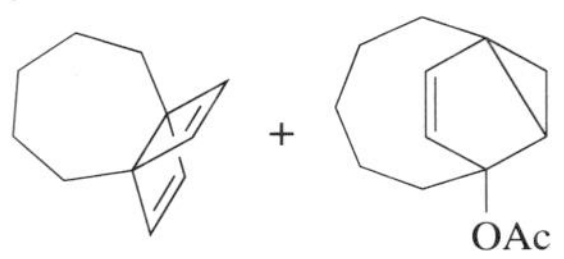	19 + 37

Entries 1 to 3 ⟨68T2215⟩ and entry 4 ⟨91H(32)2089⟩ represent typical results. Entry 5 displayed a low degree of regioselectivity ⟨61JA927⟩ while, in entry 6, the dienic propellane was not the major reaction product ⟨93S460⟩. In the absence of copper acetate, none of the desired alkene was obtained. Ring opening and rearrangement product predominated; considerable strain relief resulted upon opening the propellane bond.

Non-Kolbé electrolysis of carboxylic acids has proceeded with decarboxylation to form alkenes. Table 5 lists some examples. Entries 1 ⟨80CPB282⟩, 2 and 3 ⟨74JOC2486⟩ are examples of efficient syntheses of α,β-unsaturated lactones. Electrolyses were performed in aqueous pyridine solutions containing triethylamine with graphite electrodes. Entries 4 to 6 were performed in acetonitrile/ethanol containing potassium hydroxide ⟨88CB1151⟩. The non-Kolbé electrolysis proceeded through a carbenium ion intermediate and was assisted by the well-known stabilising effect of the β carbon–silicon bond. The influence of reaction conditions on the selectivity of the non-Kolbé process has been discussed, though prescriptive guidelines are not yet available ⟨93CB1623⟩.

When a carbanion stabilising group is present, mild catalytic decarboxylation methods become feasible. Allyl esters of α-cyano- and β-ketocarboxylic acids underwent decarboxylation to cyanoalkenes ⟨87S992⟩ and enones ⟨84S1009⟩, respectively, in the presence of palladium catalysts. Some examples are shown in Schemes 5 and 6. Direct access to an α-fluoroenone (Equation (12) is particularly appealing because the chemistry of such species is almost unexplored ⟨94T487⟩.

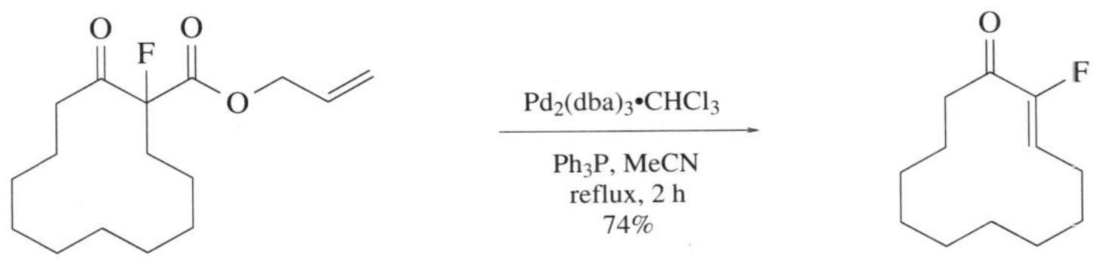

(12)

Table 5 Decarboxylation under non-Kolbé conditions.

Entry	*Substrate*	*Method*	*Product*	*Yield* (%)
1	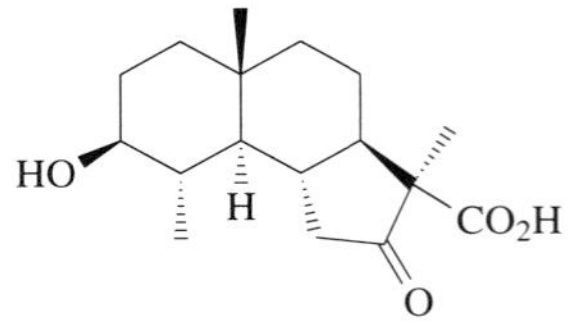	a	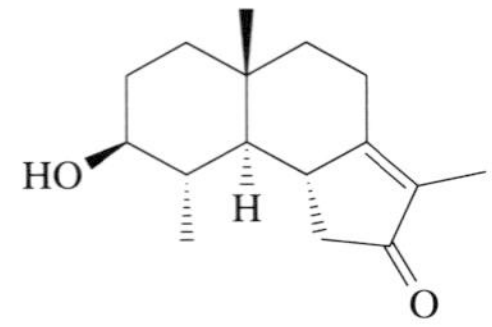	67
2	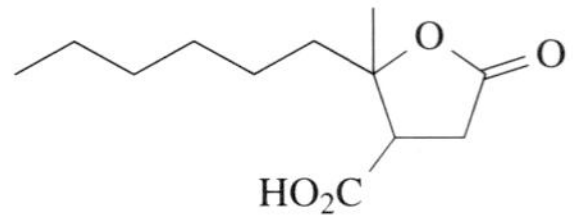	a	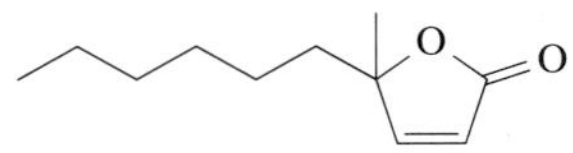	85
3	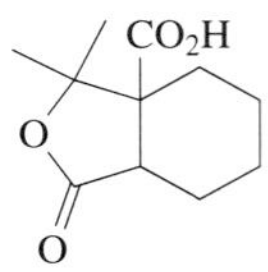	b	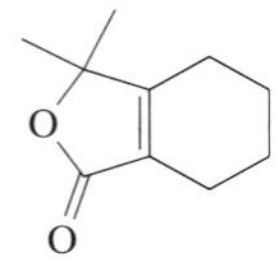	95
4	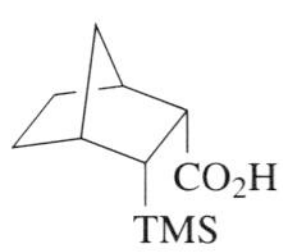	b	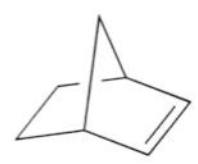	60
5	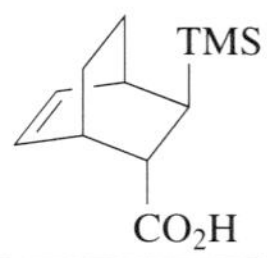	b	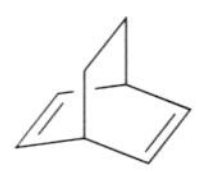	82

[a] Graphite electrode, aqueous pyridine, triethylamine. [b] Graphite electrode, ethanolic potassium hydroxide, acetonitrile.

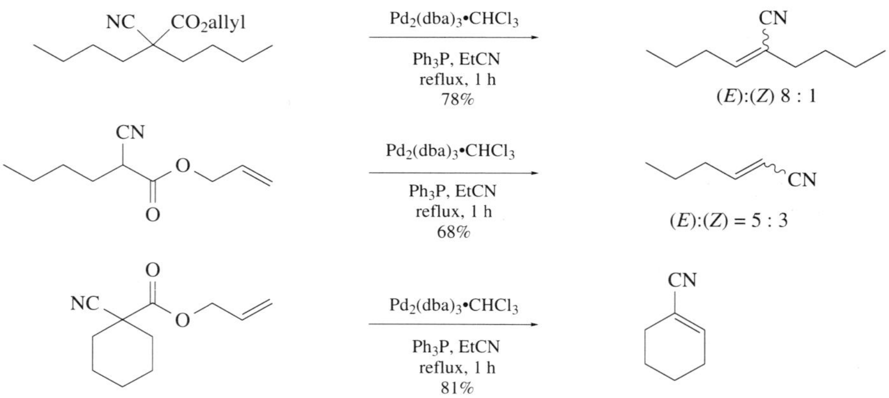

Scheme 5

1.13.2.2.2 Didecarboxylation

The strategic significance of the didecarboxylation reaction has diminished with the introduction of modern alkyne equivalents for cycloaddition chemistry. De Lucci and Modena provided an excellent survey of the area up to 1984 ⟨84T2585⟩. The main methods include lead tetraacetate oxidation, anodic oxidation and a transition metal complex-mediated procedure.

Some reactions with lead tetraacetate are shown in Table 6, entries 1–4. Both diacids and cyclic anhydrides underwent didecarboxylation with the lead reagent.

Cyclic anhydrides fused to four-membered rings were particularly reactive (entry 1) ⟨61JA1705⟩; *trans*-1,2-dicarboxylic acids were more reactive than the *cis*-isomers (entries 2 and 3) ⟨79S117⟩. The

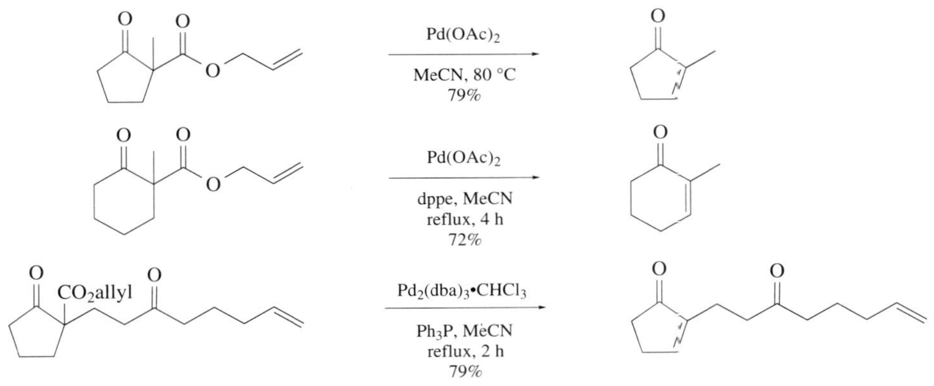

Scheme 6

Table 6 Didecarboxylations with lead tetraacetate.

Entry	*Substrate*	*Product*	*Yield (%)*
1	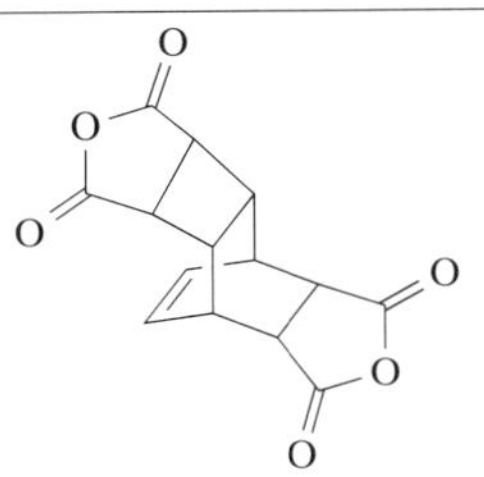	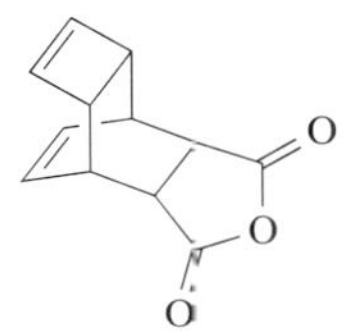	56
2	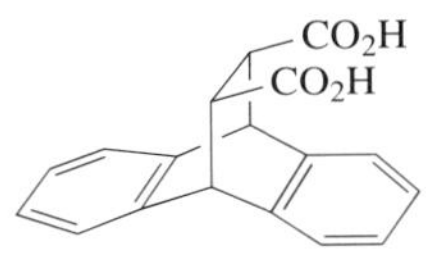	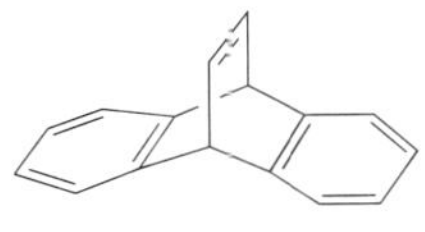	18
3	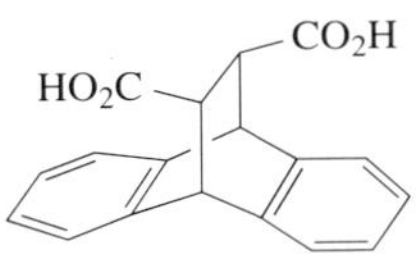	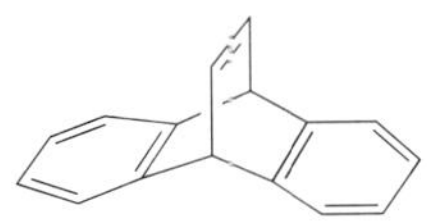	29
4	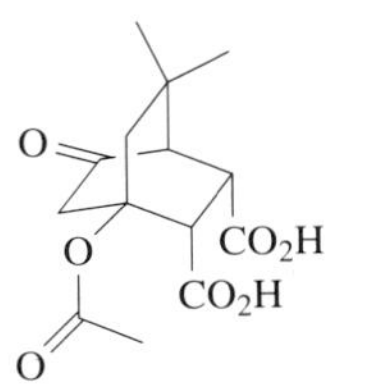	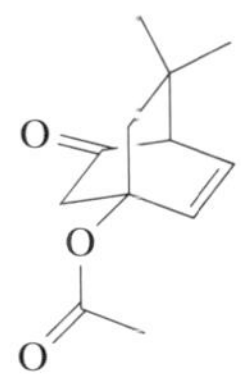	

most common experimental procedure used the diacid and the presence of oxygen was advantageous; yields were moderate at best. A range of other functional groups survives the conversion (entry 4) ⟨68JA113⟩. Electrochemical didecarboxylations have been performed on cyclic anhydrides and dicarboxylic acids. Moderate to good yields have been obtained, but scales were limited and special equipment and forcing conditions were required. Experimental details for these procedures are few and far between, though aqueous pyridine containing triethylamine appeared to be the usual reaction medium. Some examples are shown in Table 7, entries 1–6.

A range of other functionalities including amide groups (entry 1) ⟨80AG(E)463⟩, carbon–carbon double bonds (entry 2) ⟨80JOC5378⟩, and ketone carbonyl groups (entry 3) ⟨81JOC1874⟩ was compatible with the electrochemical reaction conditions. Entries 4 and 5 represent key steps in general approaches to linear and angular triquinanes ⟨85AG(E)973⟩. Entry 7 provided some experimental details; in this case, the reaction was performed in refluxing methanolic sodium methoxide with

Table 7 Didecarboxylations under electrochemical conditions.

Entry	*Substrate*	*Product*	*Yield* (%)
1	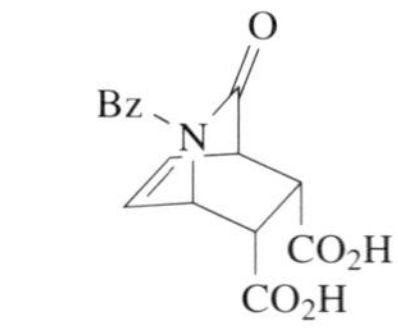	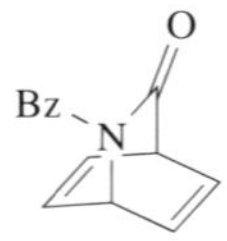	45
2	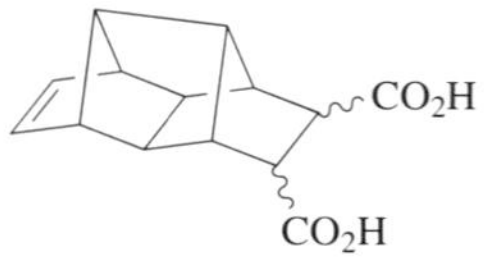	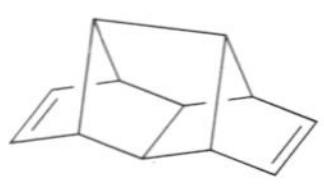	47
3	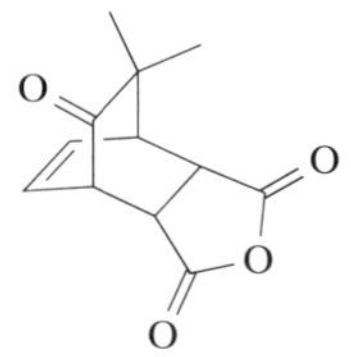	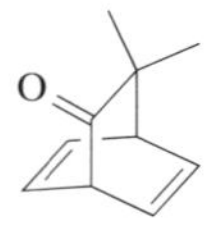	57
4	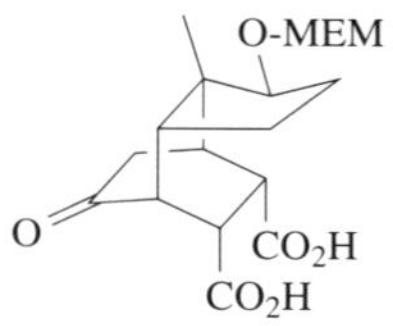	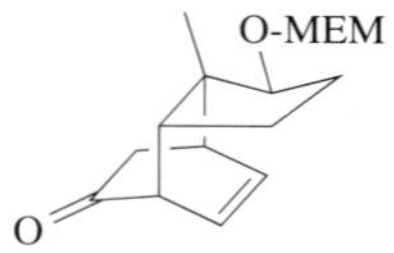	45
5	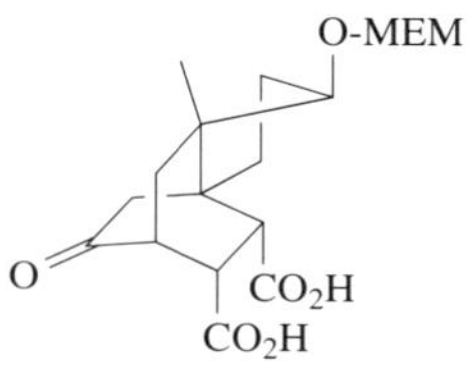	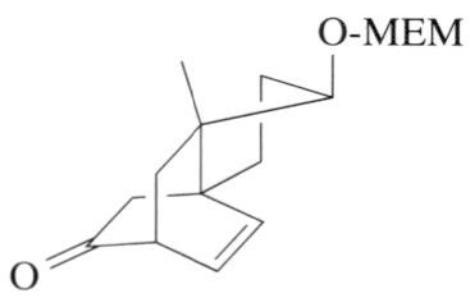	55
6	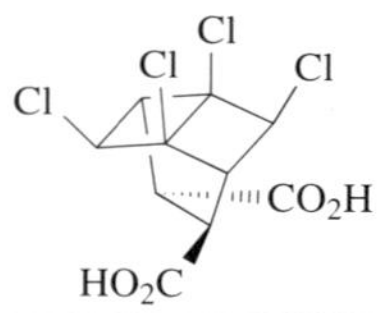	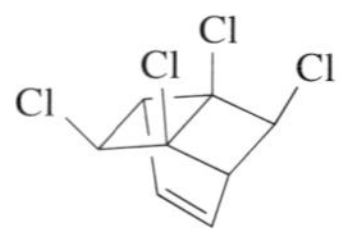	54

MEM = methoxyethoxymethyl

platinum electrodes ⟨94JCS(P1)793⟩. The alkene was obtained in moderate yield (54%); dehydrochlorination by elimination or reduction processes was not reported.

The complex bis(triphenylphosphine)nickel dicarbonyl (**5**) first described by Trost ⟨71TL2603⟩ has found application in the preparation of a range of bicyclic and polycyclic systems. The reagent is commercially available (though fairly expensive) and reacts with cyclic anhydrides in refluxing DIGLYME. In many cases, an excess of the reagent was required. Workup was usually complicated by the contamination of the product with two equivalents of triphenylphosphine which had to removed by chromatography followed by repeated recrystallisation ⟨84JA6702⟩. Monocyclic alkenes were not formed cleanly; reduction and aromatisation led to mixtures of products. Table 8, entries 1–5 show some successful and attempted conversions. Other functional groups including ketone (entry 4) ⟨86LA1578⟩ and ester (entries 1 and 2) ⟨71TL2603, 84JA6702⟩ carbonyl groups withstood the reaction conditions but some 1,4-reduction of an enone occurred (entry 3) ⟨91JOC997⟩. The reagent appeared to be highly sensitive to subtle changes in the molecular architecture of the substrate. Entries 4 and 5 differed only in the absence or presence of a bridgehead methyl group, yet this minor alteration reduced the yield to zero in the latter case ⟨86LA1578, 92M587⟩. Despite the limitations and variation in reaction yield, the nickel-mediated procedure has remained a useful method for the manipulation of architecturally complex polycyclic hydrocarbons ⟨87JA4626⟩.

Table 8 Nickel complex-mediated didecarboxylations.

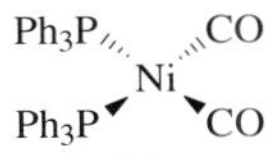

(5)

Entry	*Substrate*	*Equivalents of* **(5)**	*Product*	*Yield* (%)
1	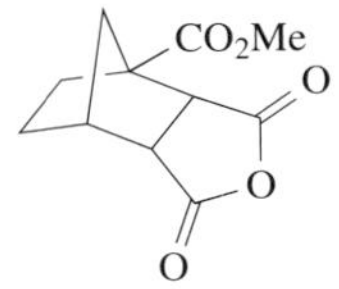		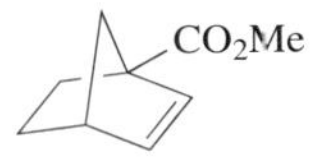	53
2	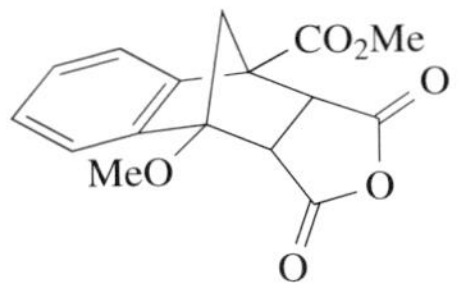	1.5	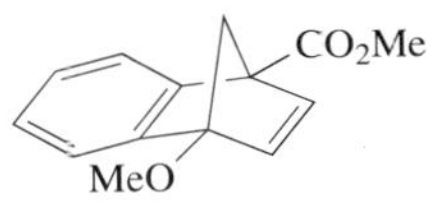	50
3	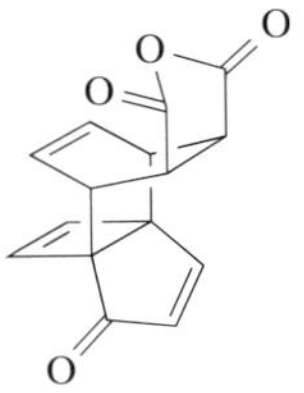	1.5	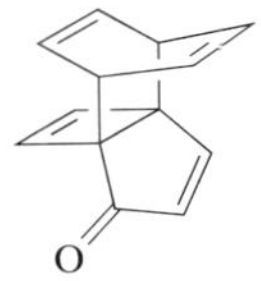	43
4	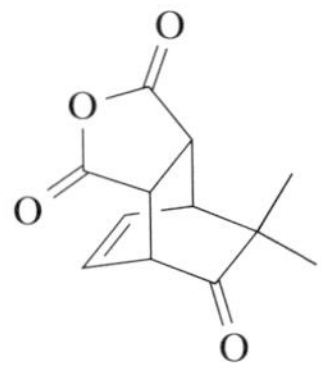	0.2	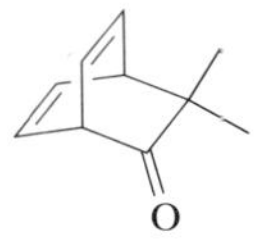	64
5	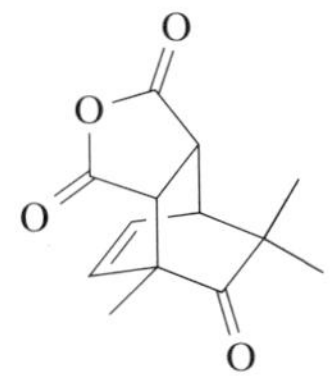		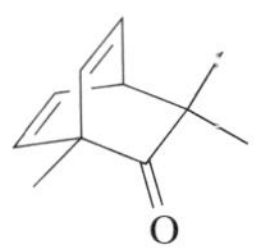	0

1.13.2.2.3 Decarboxylation/dehydration

Decarboxylation, coupled with loss of a hydroxyl group, has been achieved under mild conditions making β-hydroxyacids useful alkene precursors. Elimination can be performed directly, or in a stepwise procedure, via a β-lactone ⟨93S441⟩; both processes are stereospecific. In the more direct route, the hydroxyacid was refluxed with an excess (5 to 6 equivalents) of DMF dimethylacetal in dry chloroform. Reaction times ranged from 2 to 10 hours, and good yields of alkenes were obtained. In some cases, DMF neopentyl acetal was used. The cost of both reagents is fairly high, but valuable products have been obtained, and workup was usually straightforward. The reaction mechanism involves activation of the hydroxyl group followed by decarboxylation via an antiperiplanar (E2-like) transition state with significant partial positive charge development at the hydroxyl-bearing carbon atom. Substituents capable of stabilising the developing charge accelerated the reaction. Table 9, entries 1–5 show a range of examples. Highly substituted alkenes formed smoothly (entry 1) ⟨75TL1545⟩. Entry 2 described the preparation of a (*Z*)-alkene ⟨84T1269⟩ In entry 3, a pure diastereoisomer was obtained via a silylcuprate conjugate addition to an unsaturated ester, followed by enolate trapping with an aldehyde ⟨92JCS(P1)3351⟩; in the ensuing decarboxylation/dehydration the allylsilane was obtained as the pure (*E*)-stereoisomer. Similar methodology afforded (*E*)- or

(*Z*)-disubstituted allylstannanes ⟨87JCS(P1)2259⟩. Dienes have also been prepared using the dineopentyl acetal reagent (entries 4 and 5) ⟨75HCA153, 84JOC2079⟩.

Table 9 Dehydration/decarboxylations with amide acetals.

Entry	*Substrate*	*Reagent*	*Product*	*Yield (%)*
1	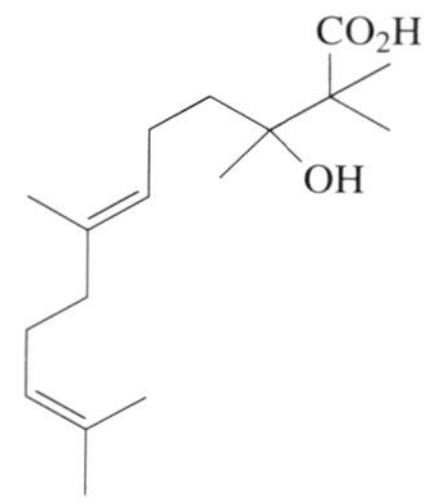	a	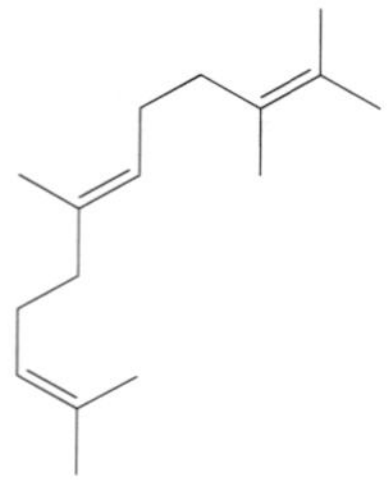	70
2	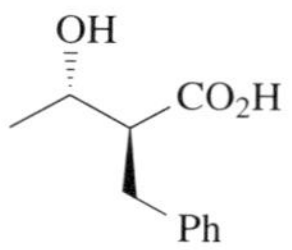	a	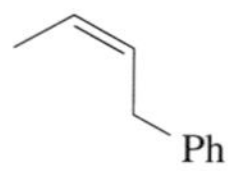	high
3	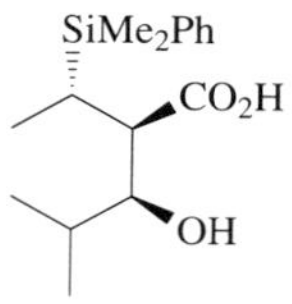	a	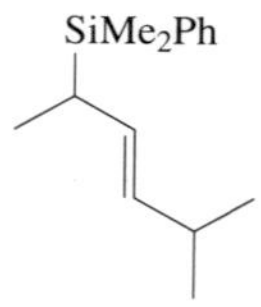	91
4	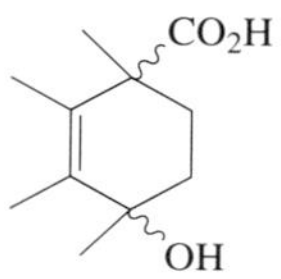	b	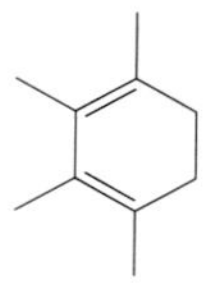	94
5	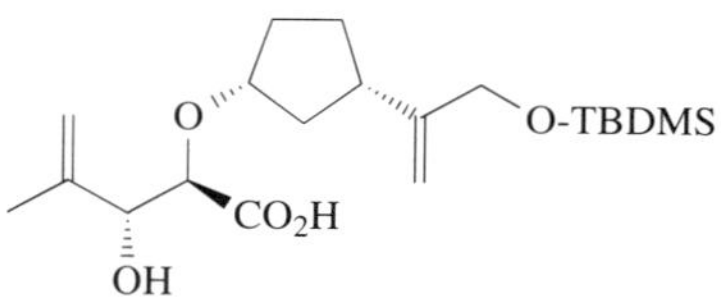	b	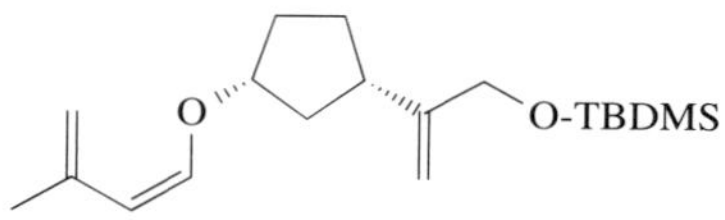	68

[a] DMF dimethyl acetal. [b] DMF dineopentyl acetal.

The reagent of choice for conversion of *β*-hydroxy acids to *β*-lactones appears to be benzene sulfonyl chloride. Dehydration usually occurred in dry pyridine with an excess of the halide. The rate of carbon dioxide elimination was strongly dependent on structure. *Trans*-disubstituted *β*-lactones were more reactive than the *cis*-isomers. Bulky substituents retarded the reaction, the effect on the *cis*-isomers being more pronounced. Substituents capable of stabilising a developing positive charge at C-2 accelerated the pyrolysis. Forcing conditions were required in some cases and sealed tube procedures have been used ⟨81JOC3359⟩. A particularly convenient procedure generated *β*-lactones directly in good yield by trapping thioester enolates with aldehydes and ketones (Scheme 7) ⟨91JOC1176⟩. Pyrolysis was performed by Kugelrohr distillation from 230–400 mesh silica gel for more volatile alkenes, or by refluxing with silica gel in benzene or cyclohexane for the less volatile species. The silica gel catalysis was effective only in highly substituted cases. Table 10 shows some transformations achieved using this method. Cost-effectiveness and ease of purification were combined in this appealing alternative to Wittig and related methods for alkene assembly. *α*-Methylene-*β*-lactones underwent rapid Diels–Alder cycloaddition with dienes (Scheme 8). Pyrolysis revealed alkylidene norbornenes, the formal products of cyclopentadiene/allene cycloaddition ⟨93CB1481⟩. Simple allenes are poor dienophiles, so the equivalence is useful even if the requirement for a pyrolysis procedure limited the scale to approximately 1 mmol. *β*-Lactone pyrolyses have also been performed in DMF ⟨84TL4881⟩ and collidine at reflux (Equation (13)). The developing positive charge receives some allylic stabilisation in this case.

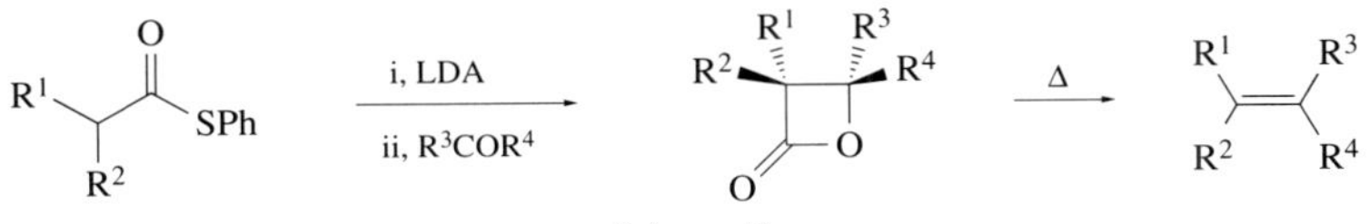

Scheme 7

Table 10 Silica-mediated β-lactone pyrolyses.

Entry	*Substrate*	*Product*	*Yield* (%)
1	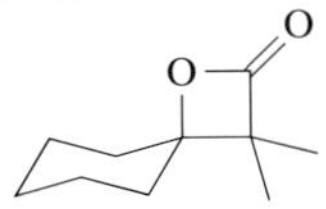	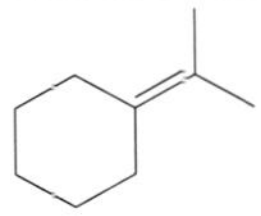	88
2	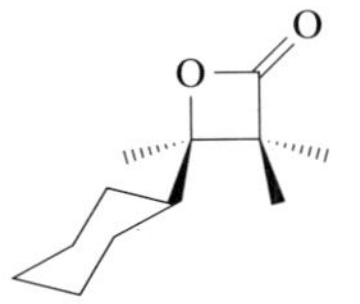	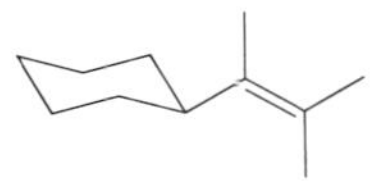	95
3	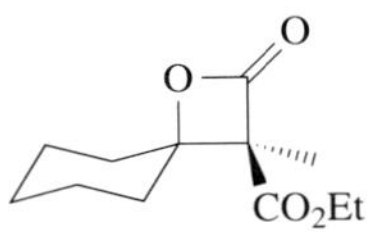	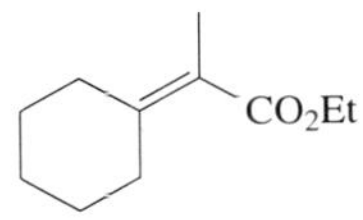	97
4	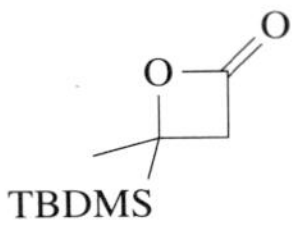		93
5	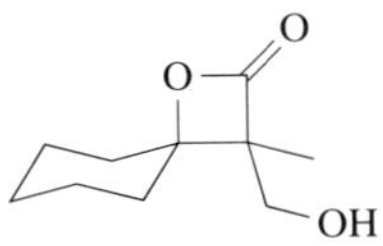	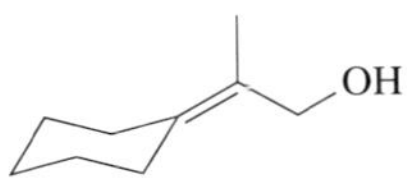	77

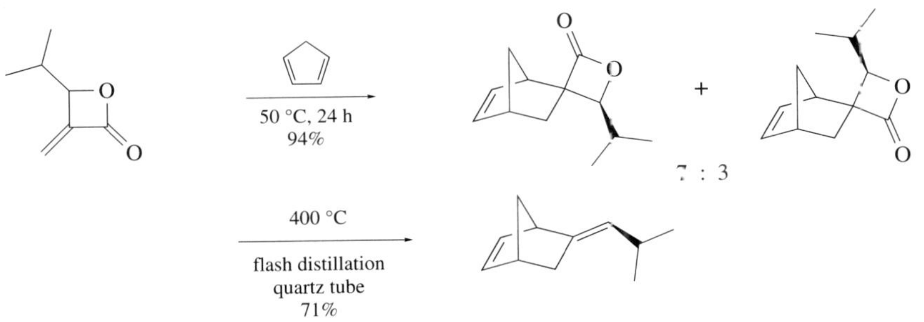

Scheme 8

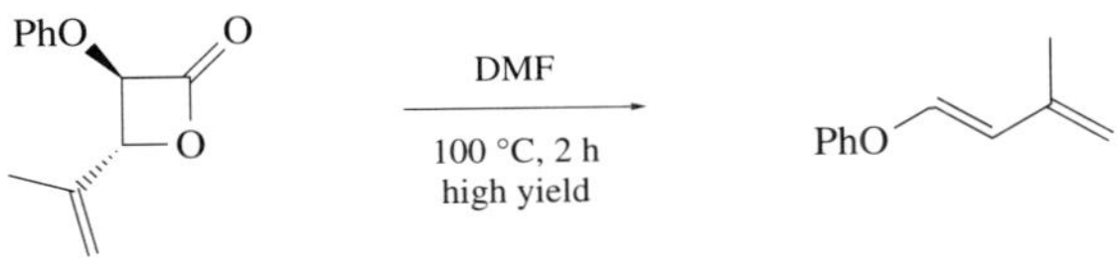

(13)

Alternatively, catalytic chemistry has been developed ⟨92TL6783⟩ allowing the direct conversion of β-hydroxy acids to alkenes and enol ethers (Equation (14)).

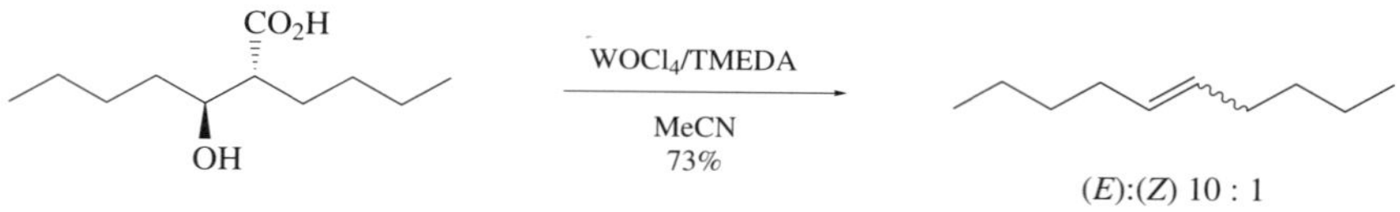

(14)

1.13.2.2.4 Decarboxylation/dehalogenation

Useful examples of the reaction are very limited. β-Bromoesters were converted to alkenes in moderate to excellent yield upon heating in hexamethylphosphoramide (HMPA) at 140 °C (Equation (15)). Synthesis of the bromoesters imposed the main limitation upon this methodology ⟨83TL1475⟩. The reaction was also performed successfully in the noncarcinogenic 1,3-dimethyl-2-imidazolidininone (**6**). A related procedure was used in a synthesis of (±)-acarenoic acid (Equation (16)). Heating the bromoester in pentan-3-one with sodium iodide led to the formation of furanone in excellent yield. The reaction proceeded only in the ketone solvent, and formed part of a versatile furanone synthesis ⟨87BCJ2917⟩.

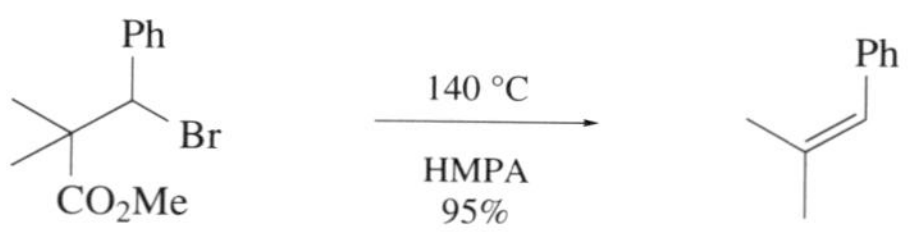

(15)

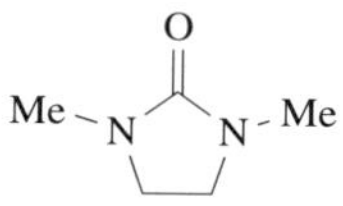

(6)

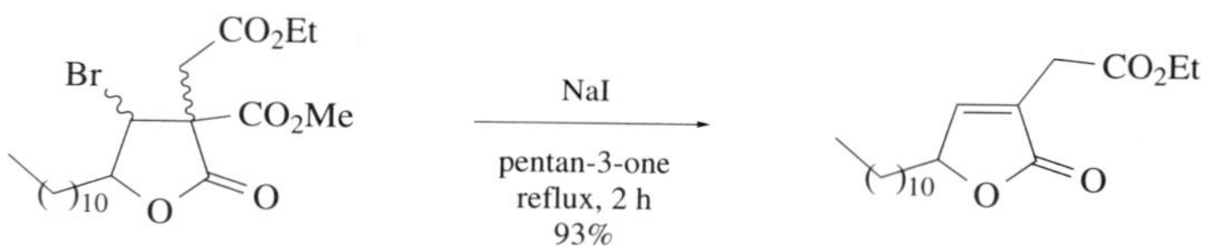

(16)

1.13.3 BY ELIMINATION OF HALOGEN (OR H-HAL)

1.13.3.1 Elimination of Dihalides

The elimination of a dihalide has only limited use in the synthetic chemist's repertoire. The most convenient starting material for a 1,2-dihalide is usually an alkene ⟨72HOU(5/1b)180⟩. As Kociénski pointed out ⟨91COS(6)975⟩, 'Unless there are very pressing reasons, one cannot advocate the synthesis of alkenes from alkenes: to travel in circles is the domain of astronomers not chemists!' Alkenes form 1,2-dibromides in high yield and under mild conditions. A significant limitation for alkene protection by dibromide formation has been the severity of the conditions required for, and the low efficiency of, debromination. Simple alkenes containing robust functional groups were obtained by heating the corresponding dibromide in DMF at 155–160 °C ⟨91S827⟩. A sodium selenite/cysteine reducing system was used to achieve debromination under mild aqueous conditions (Equation (17)). Excellent yields of alkenes were obtained and functional groups such as the ketonic carbonyl group were tolerated. The reaction system was odourless and exhibited a distinctive range of colour changes on progression to completion ⟨90CC730⟩. An alternative procedure used a tellurium reagent and tolerated some useful and delicate functional groups ⟨90TL6291⟩ (Scheme 9).

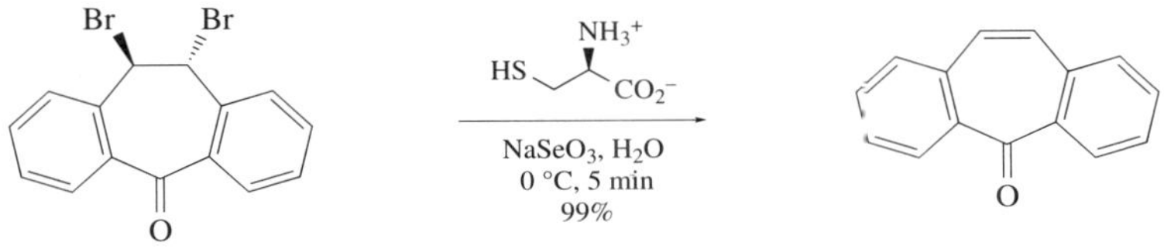

(17)

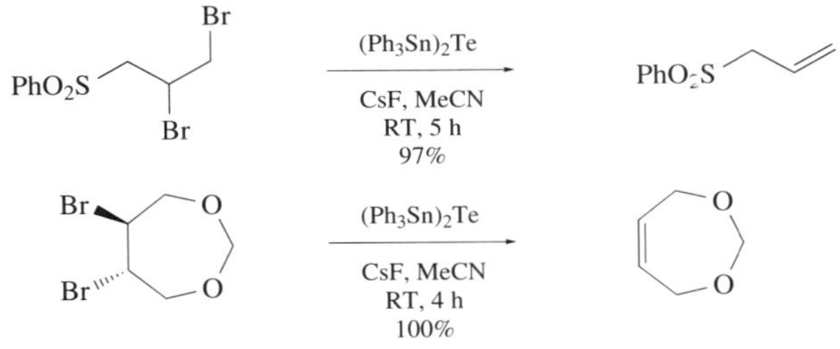

Scheme 9

Though the cost of reagents would be high, chemistry of such subtlety may find applications in total syntheses of complex molecules. Debromination of 1,4-dibromides to form dienes is a useful method, particularly when *ortho*-quinodimethanes are formed. A range of metals can be used to achieve the transformation. Trapping conditions are usually employed, so that the reaction mixture also contains a dienophile, to allow bi- and polycyclic systems to be constructed in an efficient manner. Cava and co-workers reported an example in which iodide anion was used to initiate didebromination in a 1,4-sense, forming an *ortho*-quinodimethane, which was trapped *in situ* (Equation (18)) ⟨82JOC409⟩.

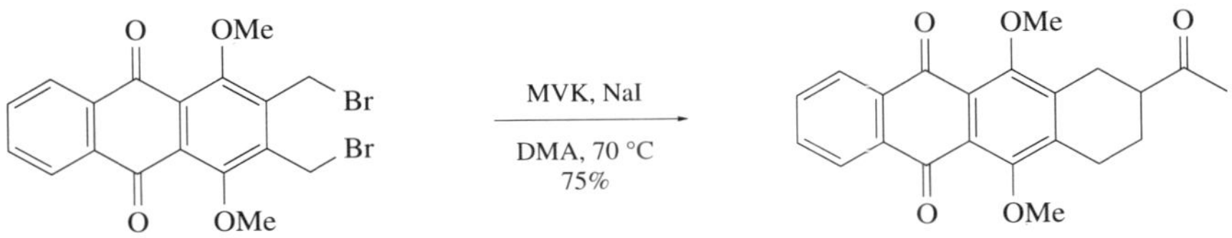

(18)

MVK = methyl vinyl ketone

More forcing conditions are required for the elimination of chlorine. However, the reaction is worthwhile because chloroalkenes undergo efficient photochemical [2 + 2] additions to enones to afford cyclobutane products. Reductive dechlorination reveals the carbon–carbon double bond for further transformation (Scheme 10) ⟨93S460⟩. High yielding free radical dechlorinations have been reported. Equation (19) shows an application of the tin method in a simple cyclohexane derivative.

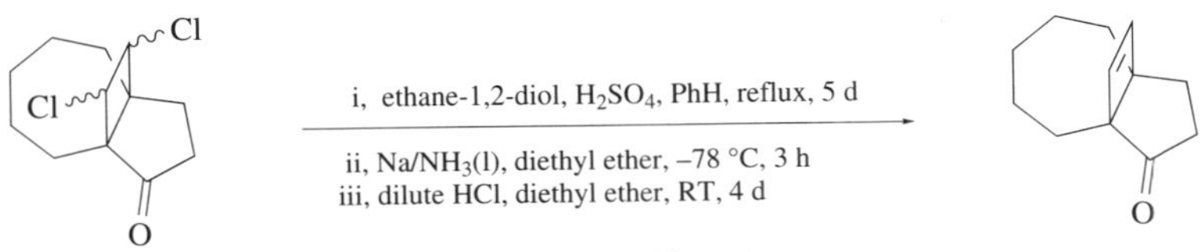

Scheme 10

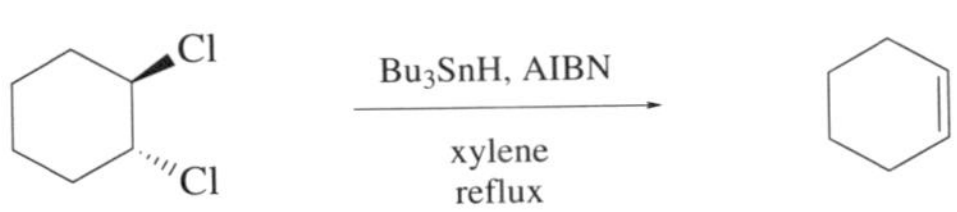

(19)

AIBN = 2,2′-azobisisobutyronitrile

Dechlorination occurred upon exposure of a dichlorodifluoroenone to zinc metal in an interesting synthesis of γ,γ-difluoro-β,γ-enones (Equation (20)) ⟨93JOC5163⟩. A range of zinc-based reaction conditions were investigated; sonication was essential for the success of the elimination.

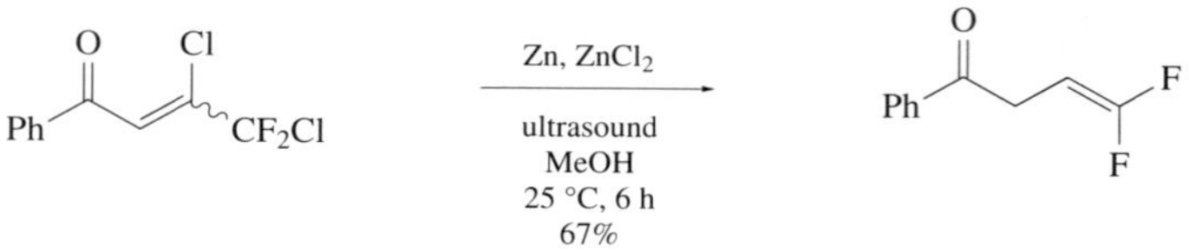

(20)

Despite the high strength of the carbon–fluorine bond, mild defluorinations have been described. Pez and co-workers ⟨92JOC2856⟩ have described conversions of perfluorocyclohexane and perfluorodecalin to perfluorobenzene and perfluoronaphthalene, respectively, with sodium/benzophenone ketal in THF. Chambers has described a mild defluorination procedure using either a sodium/mercury amalgam or tetrakis(dimethylamino)ethene (Equations (21) and (22)) ⟨94JCS(P1)3115⟩.

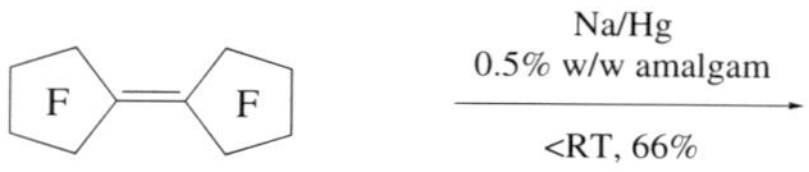

(21)

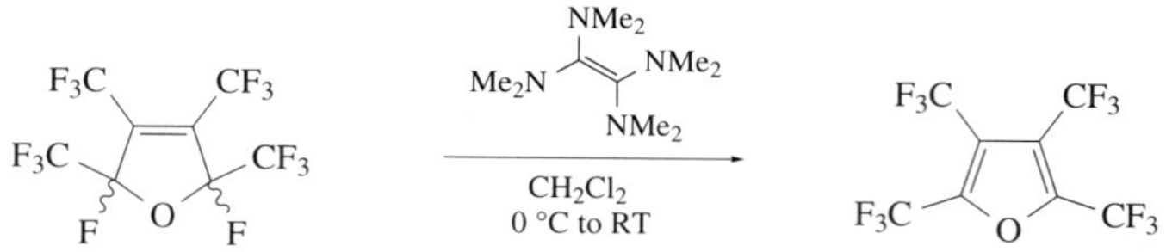

(22)

1.13.3.2 Elimination of Hydrogen Halides

The mechanistic aspects of these reactions were studied extensively during the halcyon days of physical organic chemistry. In simple unsymmetrical haloalkanes, the elimination may be regiochemically ambiguous; it is unlikely that any modern synthetic sequence would include such a step. The issues affecting such reactions ⟨91COS(6)949⟩ and detailed mechanistic aspects ⟨B-89MI 113-01⟩ have been reviewed.

1.13.3.2.1 Dehydrofluorination

Dehydrofluorination in simple unactivated systems required the use of strong bases. One recent systematic study revealed that lithium diisopropylamide (LDA) was required for clean dehydrofluorination (Equation (23)) and inferred that the solvent of choice was ether or ether/hexane ⟨88T2855⟩. The elimination of the fluoride anion received some electrophilic assistance from the lithium cation; the cation was only available for this role when weakly solvated in the less polar solvents.

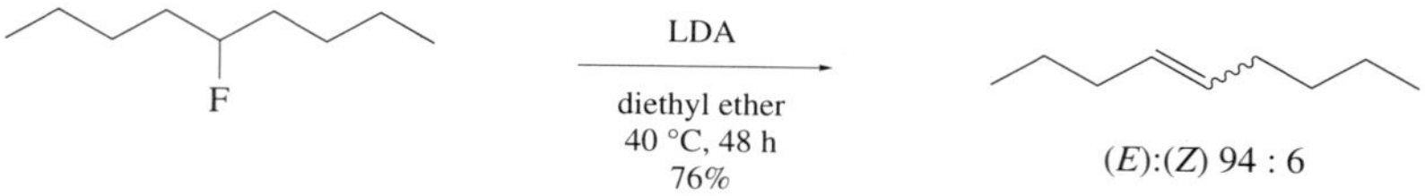

(23)

Poor yields of alkene were obtained when alkoxide bases were used in alcohol solvents. Vigorous conditions were required because fluoride ion elimination involves a transition state with a high degree of E1cb character.

Synthetically useful dehydrofluorinations have been performed on derivatives of trifluoroethanol. Strategically, these methods transform the relatively inert trifluoromethyl group into a manipulable difluorovinylic unit ⟨89TL1641, 92CC1477⟩. Dehydrofluorination occurred in good yield upon exposure of a trifluoromethyl dithioketal to methyllithium in ether (Equation (24)) affording a useful building block for the synthesis of tetrathiafulvene derivatives ⟨93JOC4685⟩.

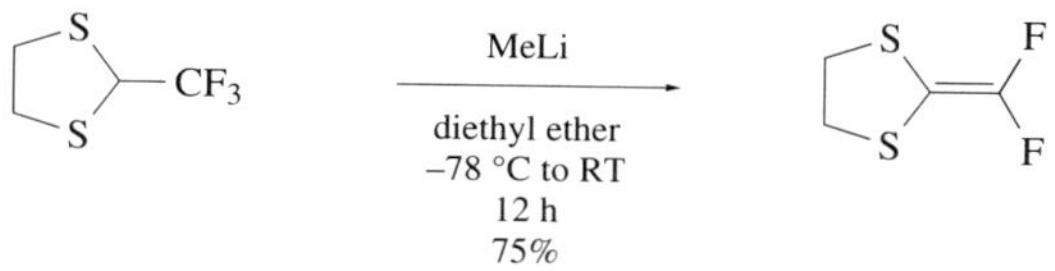

(24)

1.13.3.2.2 Dehydrochlorination

As the nucleofugacity of the halide anion (X^-) increases, the character of the transition state for HX elimination becomes more E2-like, and progressively weaker bases and milder conditions can be used.

A range of bases including sodium acetate and amines has been used to dehydrochlorinate β-chloro carbonyl compounds. Equations (25)–(27) show some examples which occurred in moderate to good yields. In Equation (25), the reaction required four days at room temperature with excess triethylamine in ether ⟨91JOC7177⟩. Sodium acetate was used to effect dehydrochlorination in a useful enone synthesis (Equation (26)) ⟨88OSC(6)327⟩. A range of cyclic enones was prepared in high yield by this route. In a reactive system derived from a quinone, dehydrochlorination proceeded instantaneously with diethylamine in ether at room temperature (Equation (27)) ⟨88OSC(6)210⟩. The dehydrochlorination of a cyclobutane dinitrile was also performed in good yield using triethylamine (Equation (28)) ⟨88OSC(6)427⟩. A cyclopropene was prepared in good yield from the corresponding chlorocyclopropane upon treatment with potassium *t*-butoxide in THF under mild conditions (Equation (29)) ⟨81OS(60)53⟩.

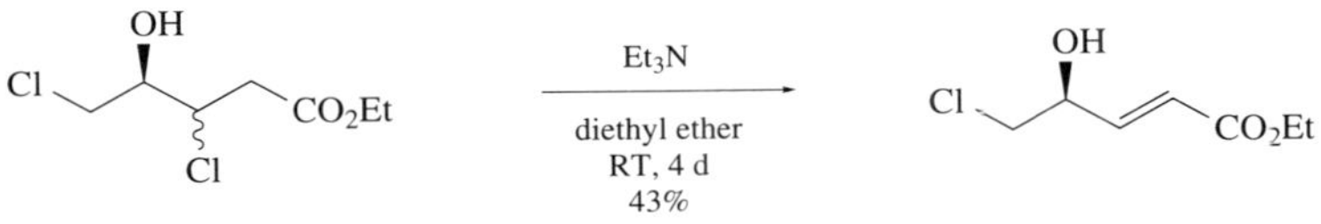

(25)

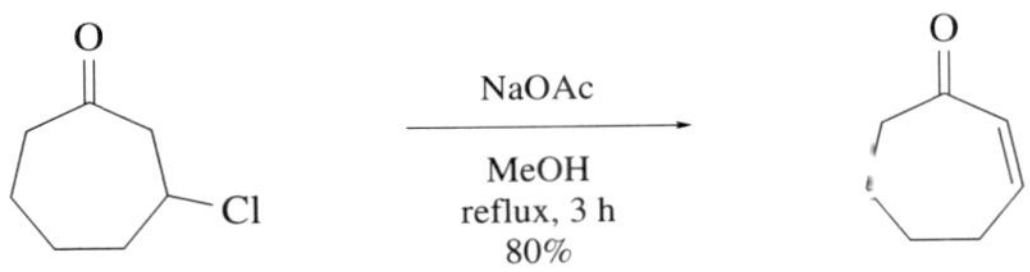

(26)

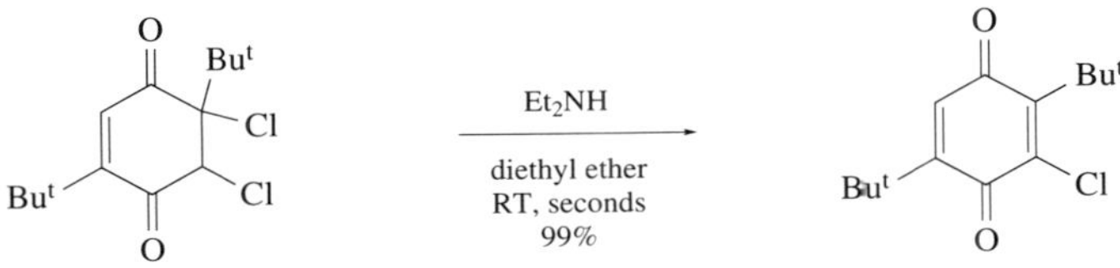

(27)

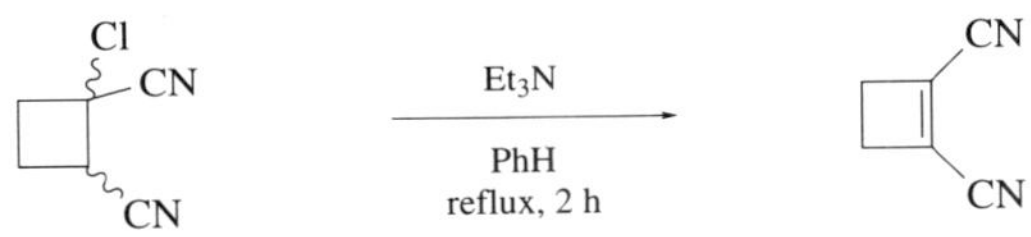

(28)

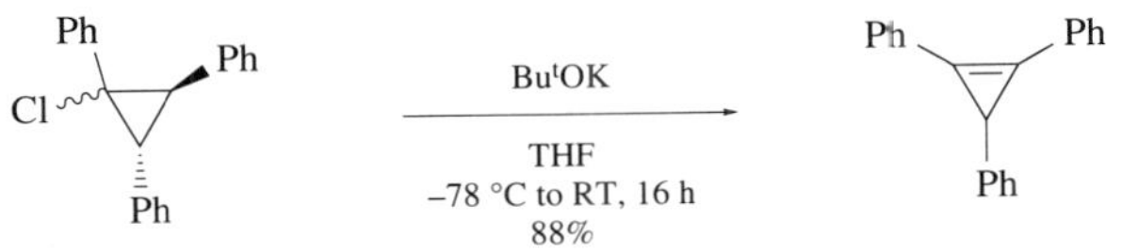

(29)

Unactivated substrates required more strongly basic conditions. The multiple dehydrochlorinations performed by Stoddart and co-workers ⟨92JA6330⟩ required a large (8.5-fold) excess of potassium *t*-butoxide at low temperature (Equation (30)). Lithium carbonate has also been used in dehydrochlorination reactions (Equation (31)) ⟨80JOC2036⟩. An alternative approach to dehydrochlorination was used by Viehe and co-workers ⟨85T3331⟩. Treatment of α,α-dichloroamides with aluminum chloride in dichloromethane, led to the isolation of chloroenamides in good yield (Equation (32)).

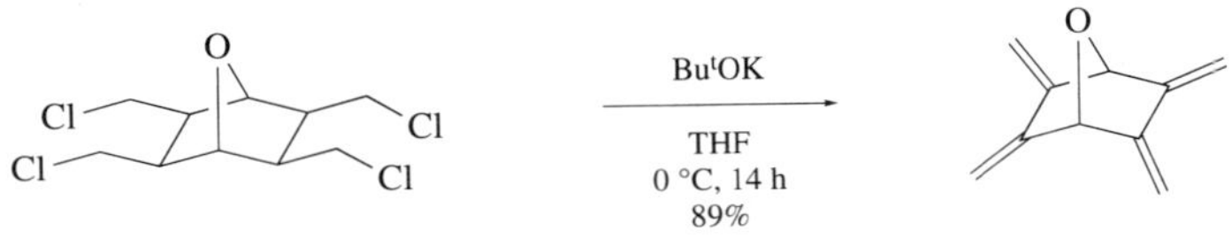

(30)

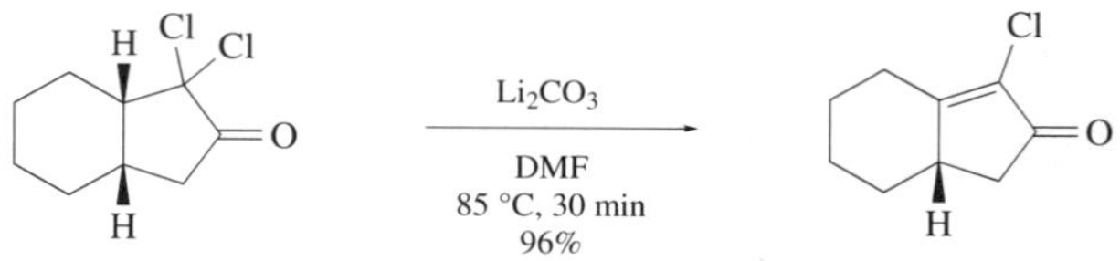

(31)

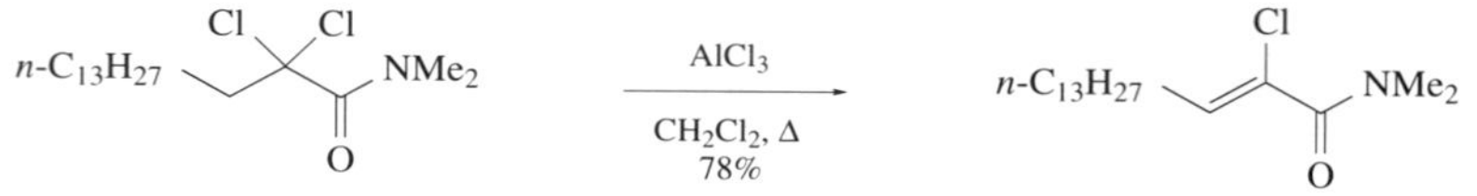

(32)

The procedure was also applied successfully to cyclic cases. Dehydrochlorination failed with bases such as lutidine and collidine, even at 178 °C. Bases such as 1,5-diazabicyclo[5.4.0] undec-5-ene (dbu) are of limited use for dehydrochlorination. The sharp contrast between activated and unactivated systems was emphasised by a paper which described a useful synthesis of vinyl sulfones ⟨78JOC1208⟩. Attempts to dehydrochlorinate at the sulfide level required prolonged reaction times at high temperature, and led to mixtures of regioisomeric vinyl sulfides. The β-chlorosulfones reacted rapidly at low temperature to afford quantitative yields of regioisomerically pure vinyl sulfones (Scheme 11).

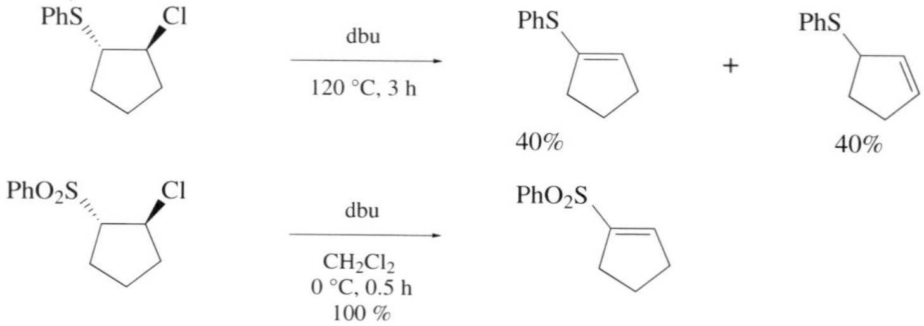

Scheme 11

1.13.3.2.3 Dehydrobromination

Bromides have proved slightly easier to eliminate than the corresponding chlorides. Useful bases for this purpose include alkoxides, amidines and phosphazenes. Potassium *t*-butoxide has proved popular, although the choice of solvent appears to be important. The base has been used in petroleum ether in the presence of a phase transfer catalyst (18-crown-6) at reflux temperature. Hindered secondary bromides underwent elimination in excellent yield (Equation (33)). However, less hindered bromoalkanes were converted to *t*-butyl ethers in a side reaction, accounting for 10–20% of the product mixture ⟨79S372⟩. The importance of the antiperiplanar conformation in the E2 elimination was revealed by a study of spiroketals (Equation (34)).

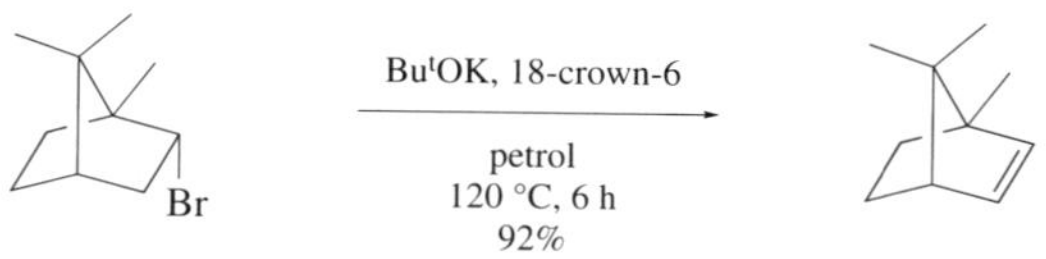

(33)

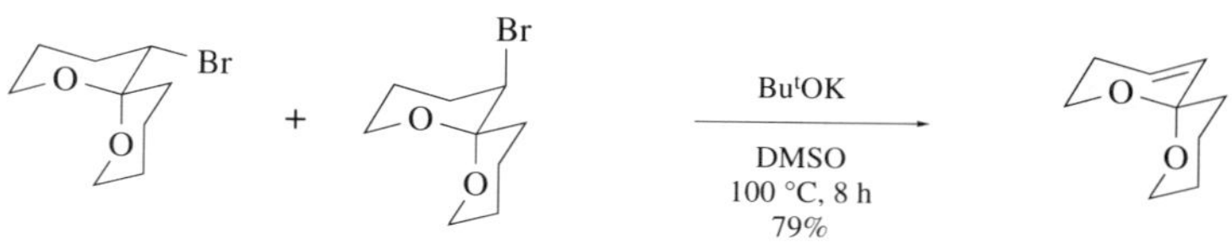

(34)

The axial bromide was consumed within 150 minutes in DMSO containing potassium *t*-butoxide at 100 °C. The equatorial epimer required a four-fold longer reaction time for dehydrobromination ⟨93JOC2501⟩. The product alkene underwent double bond isomerisation upon prolonged heating. When the reaction was performed in THF, proton abstraction occurred at the α-carbon atom, followed by spiroketal ring opening. The use of amide bases produced the same result. A simple workup procedure involved saturating the DMSO solution with salt and extracting the volatile spiroketal products into pentane. The combination of dipolar aprotic solvent and hindered alkoxide base appears to represent the procedure of choice ⟨92JA9419⟩. A useful desymmetrisation procedure used a potassium alkoxide base in THF. A significant degree of carbanion stabilisation was provided by the ionised carboxylate group, allowing the elimination to proceed at low temperature (Equation (35)) ⟨93CC116⟩. A bromoalkane survived exposure to potassium *t*-butoxide in pentane at lower temperature, allowing a useful isomerisation to be achieved (Scheme 12). Dehydrobromination required treatment with dbu at higher temperature ⟨93CC1038⟩. This amidine base ⟨72S591⟩ has allowed valuable alkenes to be isolated in good yield. Alkylation of the amidine is normally slow in comparison with the elimination reaction. In unreactive cases, dbu has been used as the reaction solvent (Equation (36)) ⟨93TA879⟩. More reactive bromides underwent dehydrobromination in nonpolar solvents containing dbu (Scheme 13) ⟨92JA9673, 92JOC5557⟩. An application involving a sensitive substrate was described by Ganem during the synthesis of senepoxide (Equation (37)) ⟨78JA6483⟩. Dehydrobromination occurred in benzene at room temperature; at higher temperature, decarboxylation accompanied the elimination.

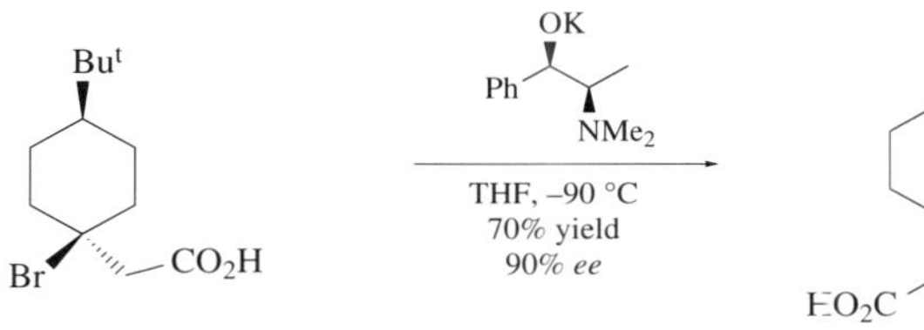

(35)

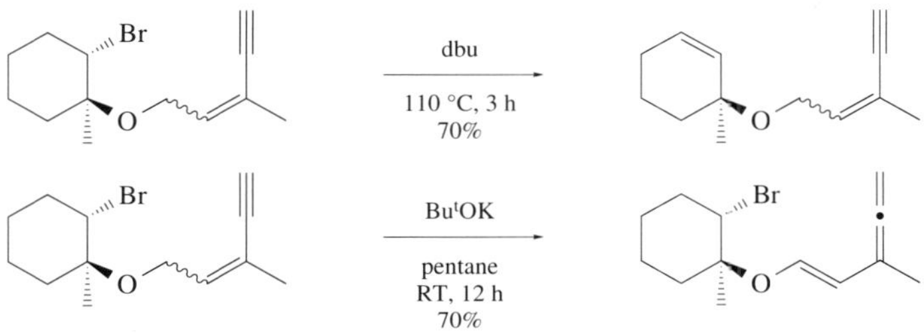

Scheme 12

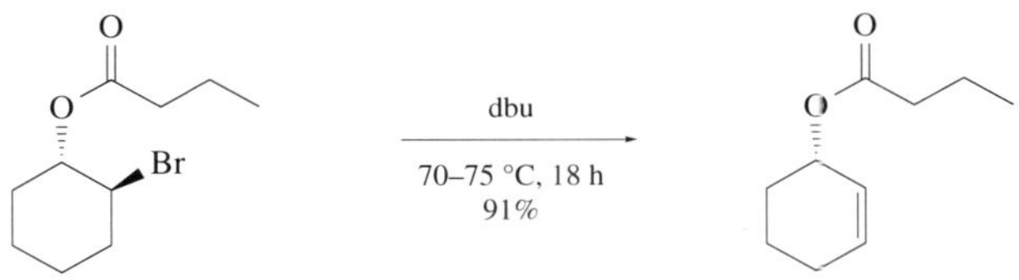

(36)

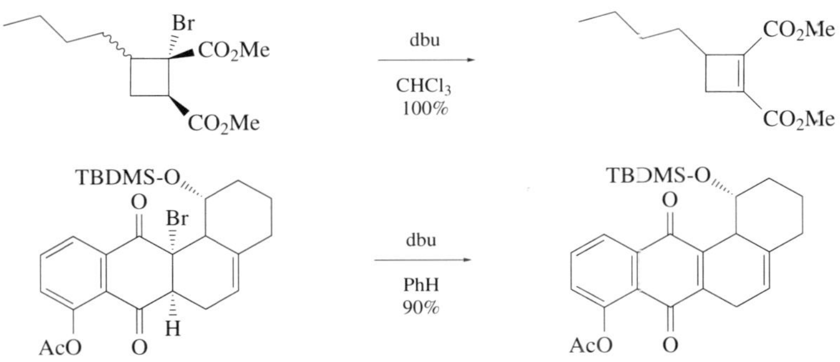

Scheme 13

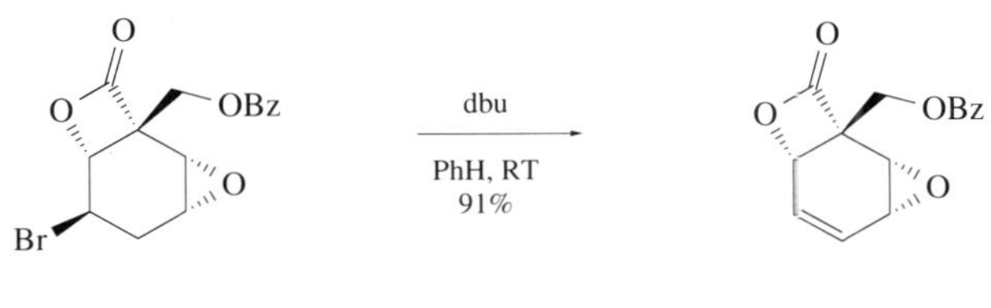

(37)

In some applications, even hindered amidine bases were alkylated. Schwesinger and co-workers have described extremely hindered and yet reactive phosphazene bases which may be useful in such circumstances ⟨87AG(E)1167⟩. A 96% yield of 1-octene was obtained upon treatment of 1-bromooctane with (**7**) after 6 hours at room temperature. Only 3% of alkylation product was obtained. In contrast, the phase transfer method of von Dehmlow afforded a 75% yield of alkene after 3 hours at 60 °C accompanied by a significant amount (18%) of *t*-butyl octyl ether.

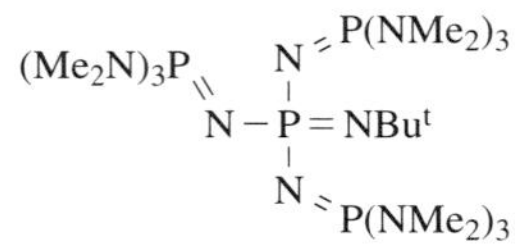

(**7**)

The most reactive base for E2 elimination is phosphazenium fluoride (**8**) ⟨91AG(E)1372⟩. This extraordinary 'naked fluoride' reagent is unstable but effected the dehydrobromination of a primary substrate at −78 °C! Fluoroalkane (10%) was also formed in the reaction. The reagent found use in a synthesis of the architecturally complex and thermally sensitive isopagodane skeleton ⟨94AG(E)117⟩.

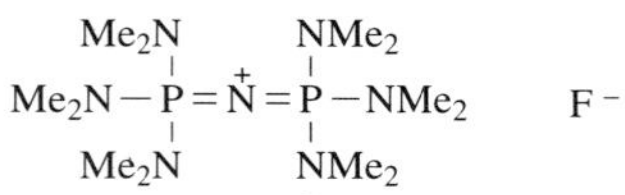

(**8**)

Amidine bases failed to provide good yields of bromoalkenes from 1,2-dibromides. Morpholine and DMSO were required to achieve the dehydrobromination in good yield (Equation (38)) ⟨90S843⟩.

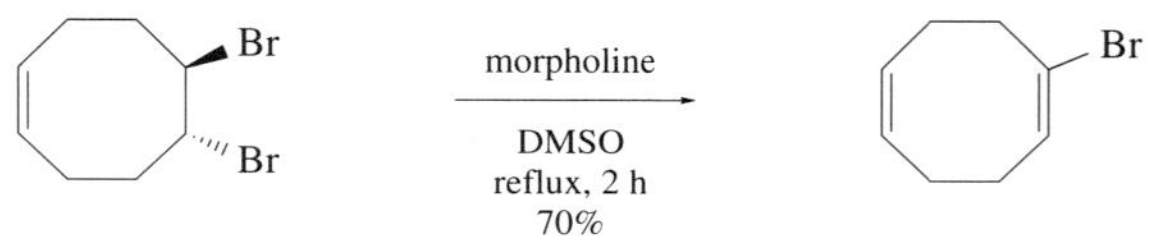

(38)

Bromoalkenes were obtained when α,ω-dibromides were heated in HMPA at high temperature. Amine and amidine bases were unsuccessful in this role ⟨84S885⟩.

1.13.3.2.4 Dehydroiodination

Examples of the elimination are relatively scarce. An efficient vinyl isocyanide synthesis was achieved using an excess of potassium *t*-butoxide at low temperature (Equation (39)) ⟨89TL3335⟩. Fluoroalkenes have been prepared via dehydroiodination using dbu in dichloromethane (Equation (40)) ⟨91TL1215⟩.

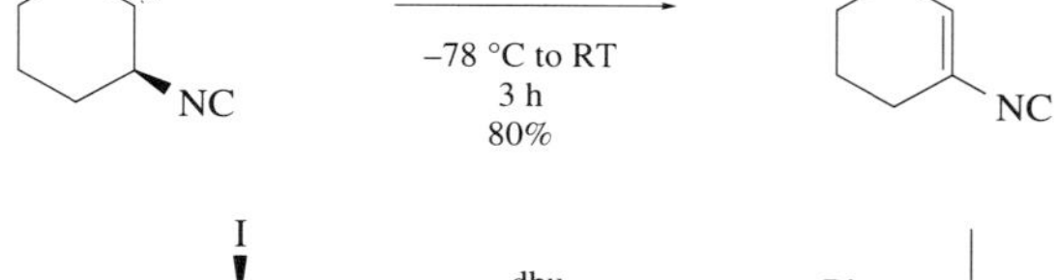

(39)

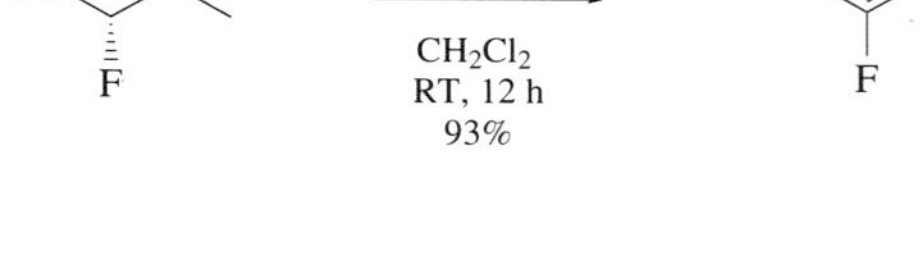

(40)

1.13.4 BY ELIMINATION OF OXYGEN FUNCTIONS

1.13.4.1 Dehydration

1.13.4.1.1 Using Burgess' reagent

Burgess' reagent ([*N*,*N*-diethyl-*N*-[(methoxycarbonyl)amino]sulphonyl] ethanaminium hydroxide, inner salt) (**9**) has been used to dehydrate a range of secondary and tertiary alcohols under mild conditions ⟨73JOC26⟩. The reagent is crystalline, commercially available and readily prepared in high yield on a 1.5 mol scale (Scheme 14) ⟨90JA8433⟩.

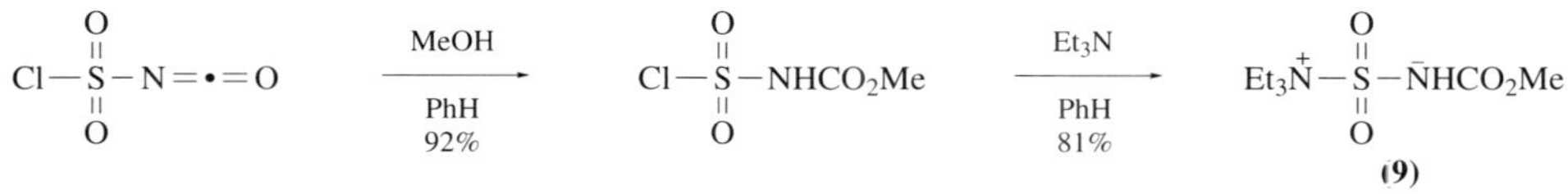

(9)

Scheme 14

Dehydration occurred smoothly in benzene or toluene at 50 °C, normally using a slight excess (1.1 to 1.5 equivalents) of the reagent. Workup is always simple, often involving hydrolysis and solvent extraction. However, when the product alkenes were volatile, the reactions were performed neat, allowing the alkenes to distil from the reaction mixture. Table 11, entries 1–5 show a range of alcohols dehydrated successfully using the reagent. The products include highly acid-sensitive species (entries 1 and 2) ⟨73JOC26⟩, a densely functionalised quinocarcinol methyl ester precursor (entry 3) ⟨85JA1421⟩ and a sensitive propellane (entry 4) ⟨84JA1518⟩. In entry 5, dehydration afforded only the nonconjugated diene in good yield ⟨90JA9284⟩.

Table 11 Dehydrations with Burgess' reagents.

Entry	*Substrate*	*Method*	*Product*	*Yield (%)*
1	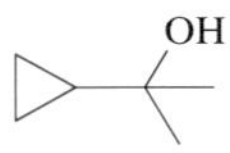	a	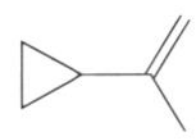	66
2	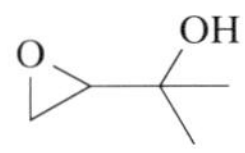	a	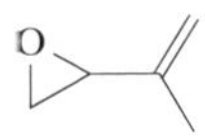	69
3	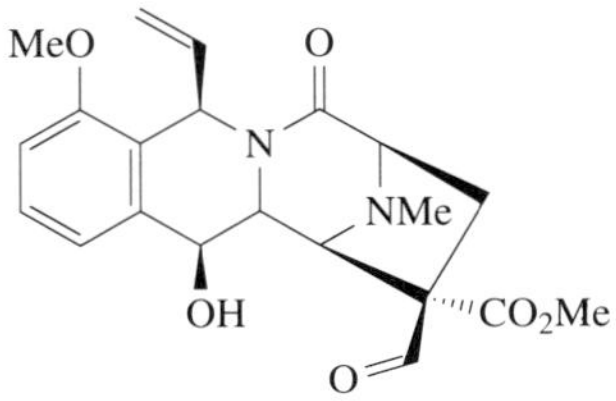	b	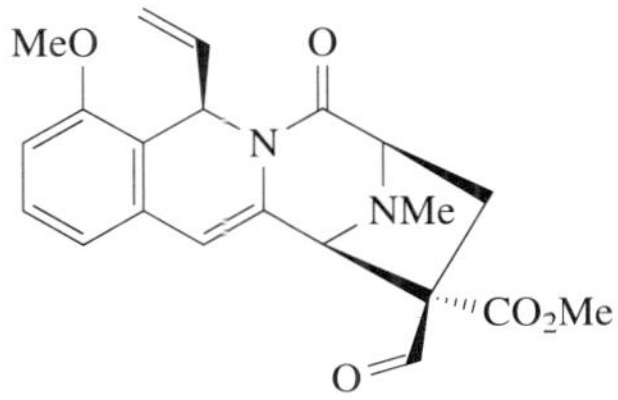	53
4	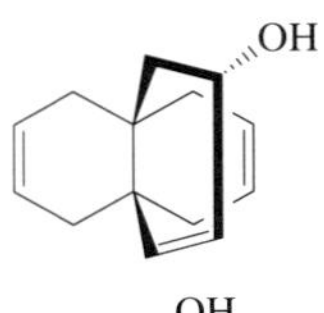	b		92
5	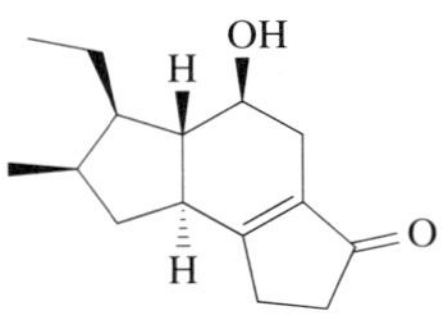	a	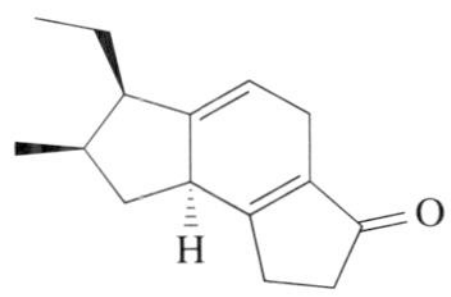	66

[a] Neat reagent. [b] Benzene solution, 50 °C.

A mechanism has been proposed for the reaction which involves a *syn*-elimination from an intermediate sulphamate ester. However, the ester is a good leaving group, and the products of E1 elimination have been observed where stabilised carbenium ions are accessible. McCague used Burgess' reagent to prepare a tamoxifen analogue, producing the same 2.8 : 1 mixture of (*Z*) and (*E*)-alkenes from both diastereoisomeric alcohols (Scheme 15) ⟨87JCS(P1)1011⟩. Mixtures of products have been obtained where elimination could occur across a number of possible C—C bonds. Thus *endo*-2-methyl-2-norborneol was converted to *endo*- and *exo*-cyclic alkenes (1 : 1) in 66% yield (Scheme 16) ⟨90JA8433⟩, while a mixture of unsaturated esters resulted upon dehydration of a pyrethroid precursor (Scheme 17) ⟨87T853⟩.

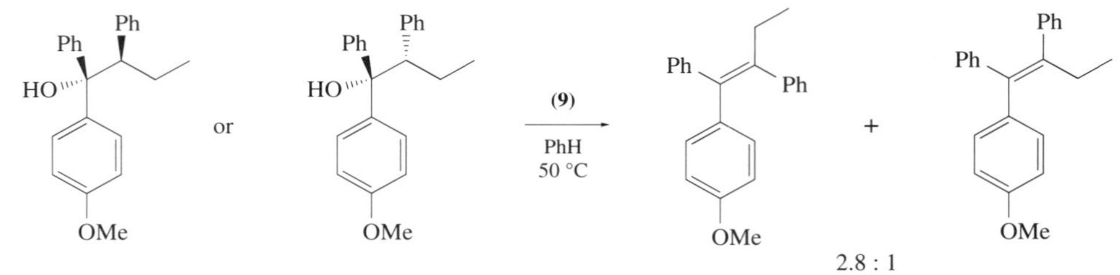

Scheme 15

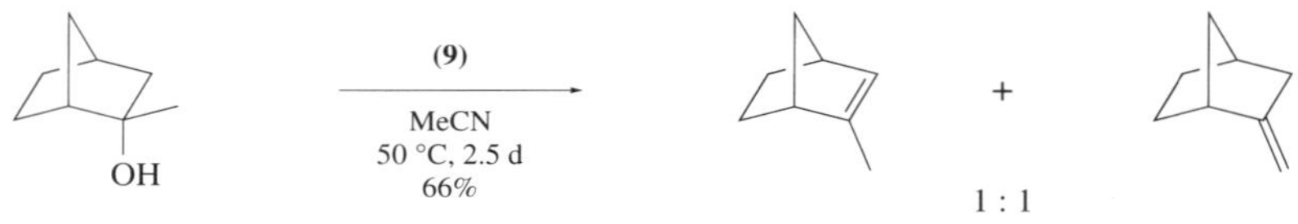

Scheme 16

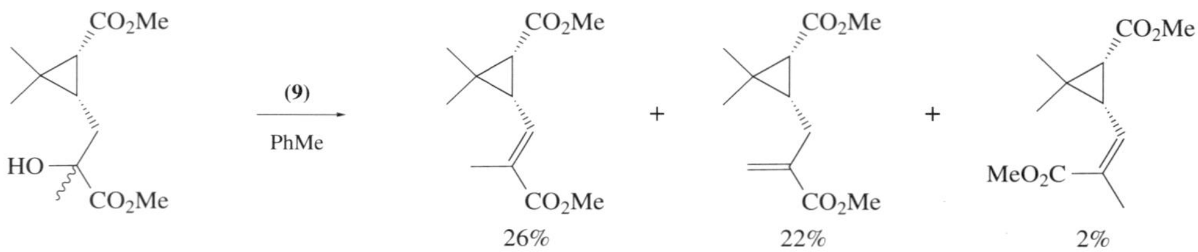

Scheme 17

1.13.4.1.2 Using Martin's sulfurane reagent

Martin's sulfurane (bis(α,α-bis[trifluoromethyl]benzyloxy)diphenyl sulfur) (**10**) is a moisture-sensitive and expensive commercial reagent. However, the material has been prepared on a 250 g scale, and stored indefinitely in the absence of moisture ⟨77OS(57)22⟩. The mechanism of dehydration has been investigated ⟨72JA5003⟩. Like Burgess' reagent, tertiary alcohols reacted via an E1-like mechanism; for secondary alcohols, an E2-like pathway was inferred while primary alcohols formed stable ethers.

The reagent appears to be particularly useful for the dehydration of alcohols prone to carbenium ion rearrangement. Even cyclopropyl alcohols underwent elimination. Table 12, entries 1–7 show the scope of the dehydration. Entry 4 underwent skeletal [3,3]-rearrangement following dehydration ⟨86JA3739⟩. In the extreme case of entry 1, the tricyclopropyl alcohol rearranged, resulting in a low (32%) yield of the dehydration product ⟨72JA5003⟩. The development of angle strain in the alkene retards the rate of elimination ⟨82TL1343, 85JA4964⟩. Entries 5 and 6 indicate that the more acidic proton was removed when a choice was available ⟨86JOC3098, 92JOC5208⟩. A sensitive diene was prepared in entry 7 ⟨88LA543⟩.

1.13.4.1.3 Dehydration by other methods

Treatment of alcohols with strong acids has been used to achieve dehydration. The method is most useful when highly stabilised carbenium ions are accessible, and proton loss is unambiguous

Table 12 Dehydrations with Martin's sulfurane reagent (**10**).

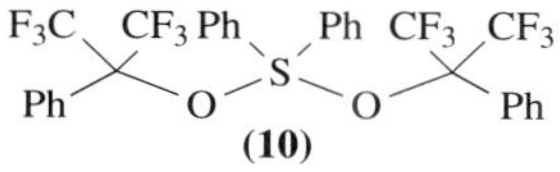

(10)

Entry	*Substrate*	*Temperature* (°C)	*Product*	*Yield* (%)
1	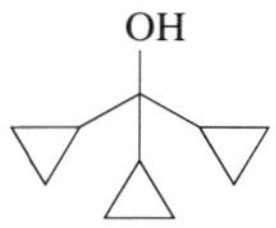	25	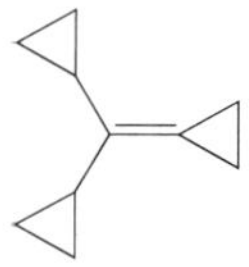	32
2	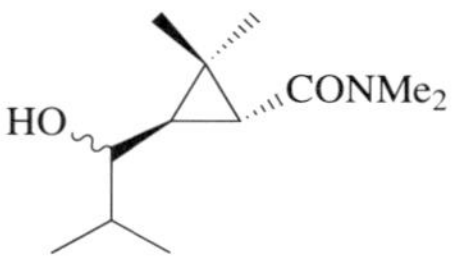	–78	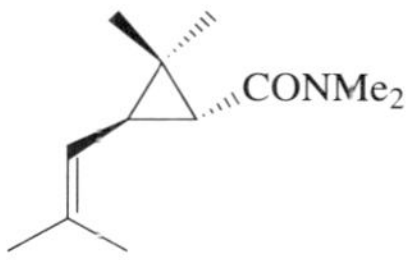	92
3	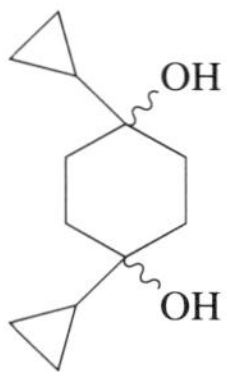	25		52
4	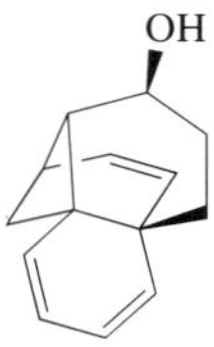	25		60
5	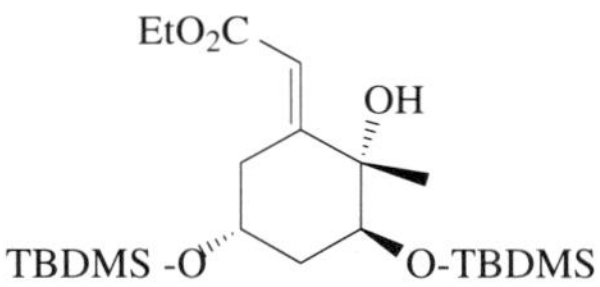	0	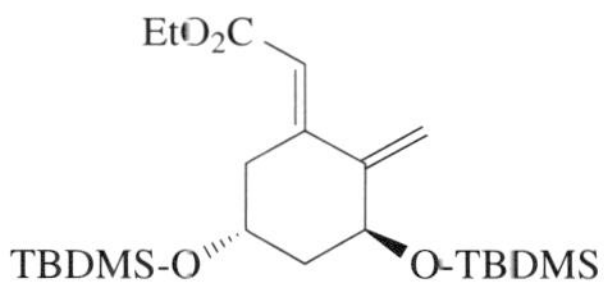	80
6	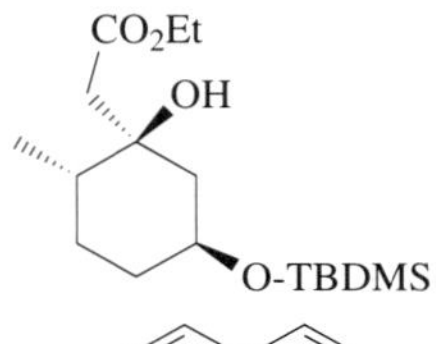	25	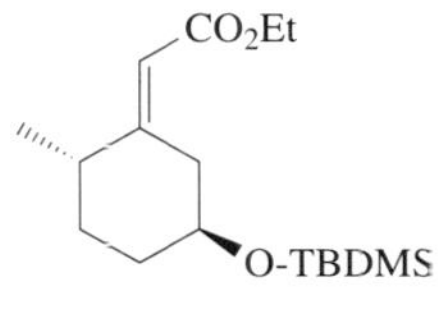	93
7	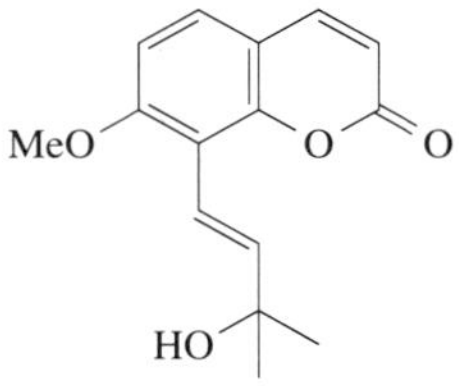	25	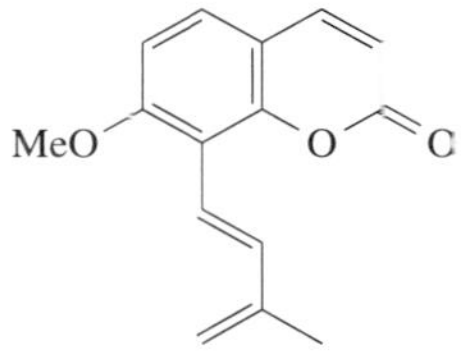	>50

or can be thermodynamically controlled. An efficient acid-catalysed dehydration will almost invariably be more economical than any other method. A recent example involved an acid-catalysed dehydration of a tertiary benzylic alcohol. The reaction was performed in benzene using Dean–Stark apparatus and a catalytic quantity of *p*-toluene sulfonic acid to afford the more stable endocyclic alkene in good yield (Equation (41)) ⟨94JOC406⟩.

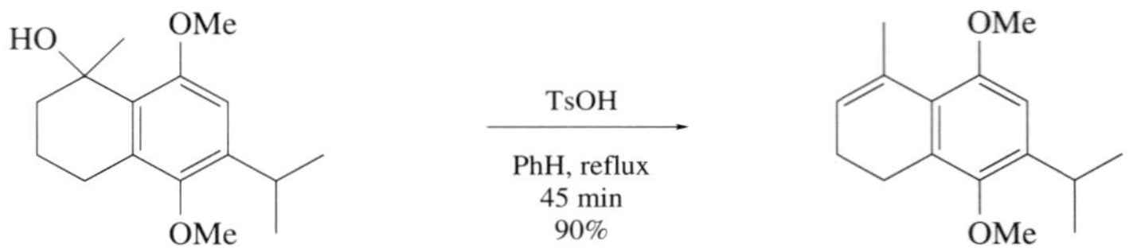

(41)

Supporting the acid catalyst on silica gel allowed the efficient dehydration of secondary and tertiary alcohols. Refluxing the alcohol with the supported reagent in benzene afforded good yields of Δ^2- and Δ^3-steroids (Equation (42)) ⟨85S1159⟩. A strong acid ion exchange resin was used to dehydrate sterols. Double bond migration accompanied dehydration when a sterol was refluxed with Nafion-H in chloroform (Scheme 18) ⟨87CC490⟩. Other dehydrations were performed with a sulfuric acid catalyst ⟨90OSC(7)363⟩ and in trifluoroacetic acid ⟨86TL1449⟩.

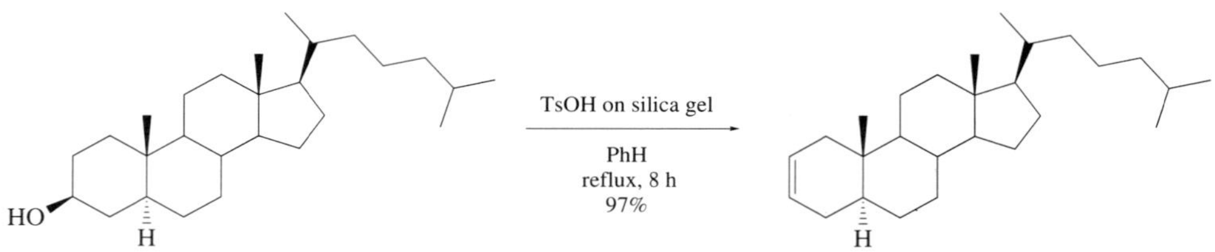

(42)

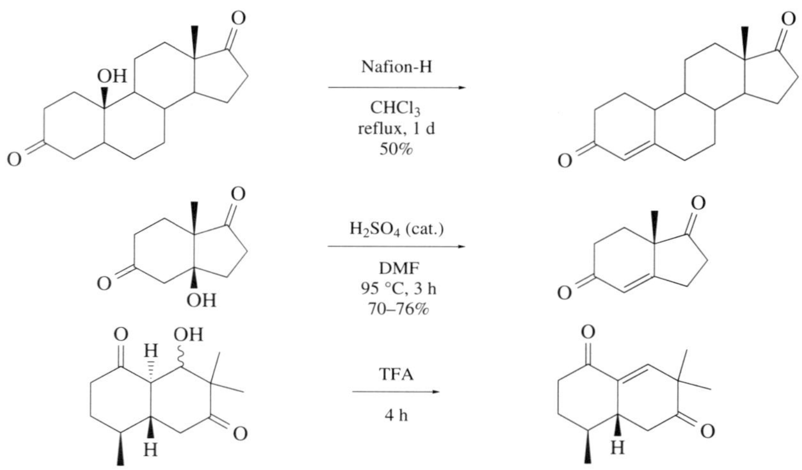

Scheme 18

A useful method involved generation of the expensive sulfonyl diimidazole reagent *in situ*. Selective reaction at a steroidal hydroxyl group was followed by elimination. This is an attractive method because the byproducts from the reaction are particularly easy to remove (Equation (43)) ⟨88JPR309⟩.

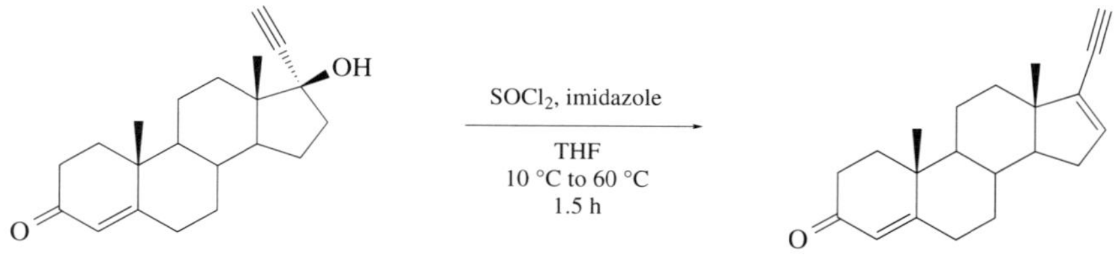

(43)

Synthetically useful nitroalkenes were generated by a dehydration reaction on alumina ⟨92JOC2160⟩. An unusual but general dehydration allowed the direct conversion of primary and secondary alcohols to alkenes. The reaction produced volatile byproducts and involved an easy workup (Equation (44)) ⟨88SC1795⟩.

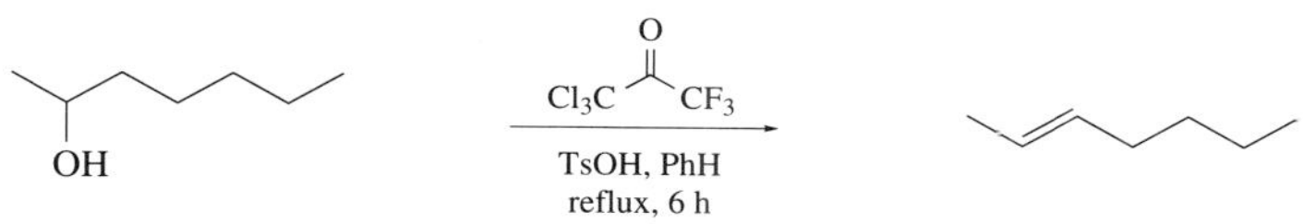

(44)

1.13.4.2 Elimination of Alcohols (H—OR)

The low nucleofugacities of simple alkoxide anions means that elimination usually occurs under E1cb conditions. Some useful reactions have been reported and are shown in Scheme 19. Potassium *t*-butoxide ⟨92JOC1613⟩, potassium hexamethyl disilazide (KHMDS) ⟨94JOC394⟩ and dbu ⟨93T6717⟩ have all found applications. Fragmentation of the leaving group to release acetone prevented the back reaction. In the final example, the amidine proved superior to a range of hydride and amide bases.

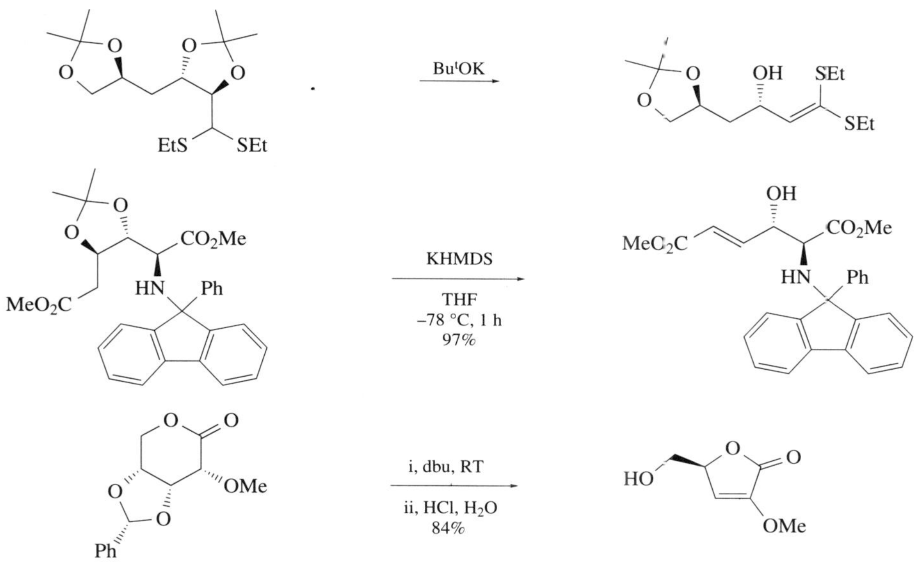

Scheme 19

Superbase-mediated eliminations from simple homoallyl ethers form a interesting class of reactions. A range of dienes and trienes were prepared by alkoxide elimination using LIDAKOR (LDA/Bu^tOK) in THF (Scheme 20) ⟨90T2411⟩.

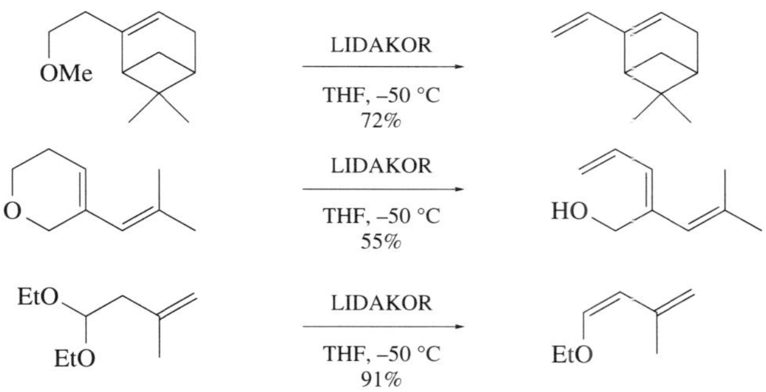

Scheme 20

Pivalate esters underwent elimination at slightly higher rates. Eliminations of alkoxy groups from acetals and ketals have afforded enol ethers. ⟨82TL323, 82TL631⟩. A method of choice which uses readily available reagents has been described ⟨88JOC5576⟩. The eliminations were highly regioselective, with removal of the least hindered proton. A range of substituents was tolerated in the ketal moiety (Equation (45)). Under acidic conditions, ethanol was eliminated from an acetal to form an enol ether. A significant amount of hemiacetal was formed in the reaction (Equation (46)) ⟨94JOC311⟩.

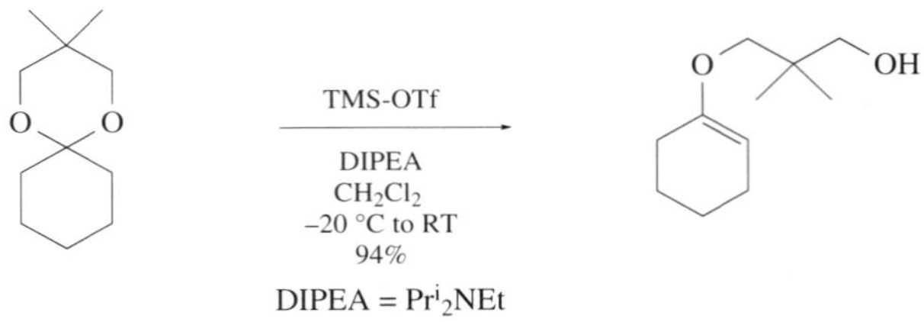

(45)

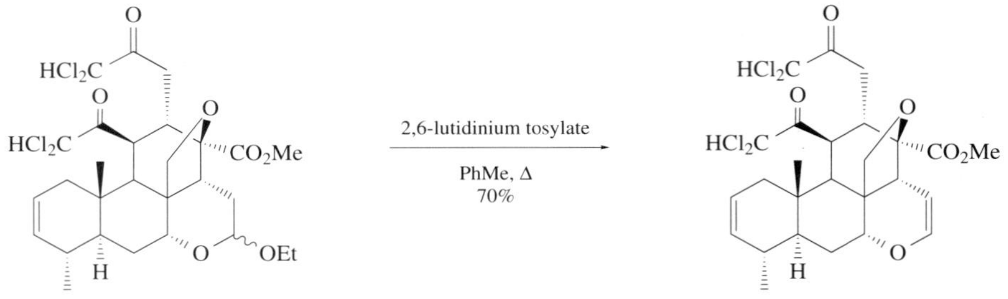

(46)

1.13.4.3 Eliminative Ring Opening of Epoxides

Epoxides open readily under Lewis acid or Brönsted base conditions ⟨84S629⟩. Chiral amide bases have been used in desymmetrisation procedures involving epoxide opening by elimination. Equation (47) shows an example ⟨93CC884⟩.

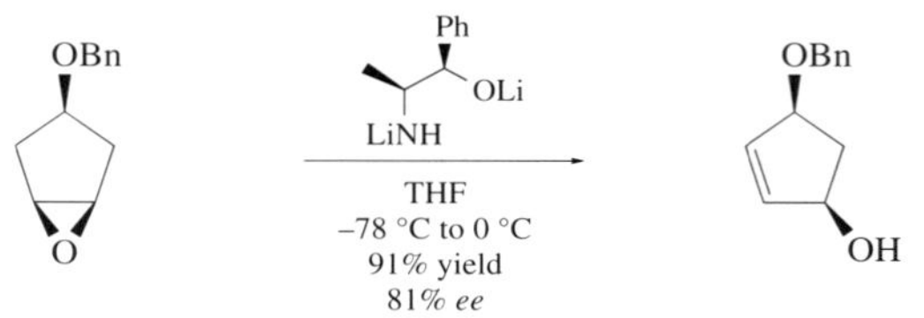

(47)

The products were obtained in moderate enantiomeric excess; changing the solvent from THF to a benzene–THF mixture allowed the other enantiomer to be isolated. Magnesium amides appear to combine basicity with Lewis acidity in a useful way, allowing epoxides to be opened in high yield ⟨86TL299⟩. The amides were prepared in toluene by reacting the lithium amide with Grignard reagent. An excess (3–4 fold) of the amide was required, reactions were run at 0 °C, and the choice of solvent was critical. Poor yields were obtained in THF, with and without additives such as HMPA. Proton abstraction usually occurred from methyl groups rather than from methylene positions. Table 13, entries 1–5 show some typical examples; the amide reagents tolerated other functional groups (entries 1–3) ⟨86TL299⟩. In entry 3, the less acidic proton was removed. The regioselectivity displayed in entry 5 was exceptional ⟨92T10265⟩ and was reversed when the elimination was performed in monoglyme. The procedure was deployed in an asymmetric synthesis of *R*-(+)-perrillaldehyde from limonene ⟨88SC1905⟩.

Epoxide opening can also occur under E1cb conditions using weak bases. A useful sequence used arsonium ylide chemistry to prepare β,γ-epoxy aldehydes; treatment with triethylamine in ether led to the formation of γ-hydroxy-α,β-unsaturated aldehydes (Equation (48)) ⟨89TL179⟩.

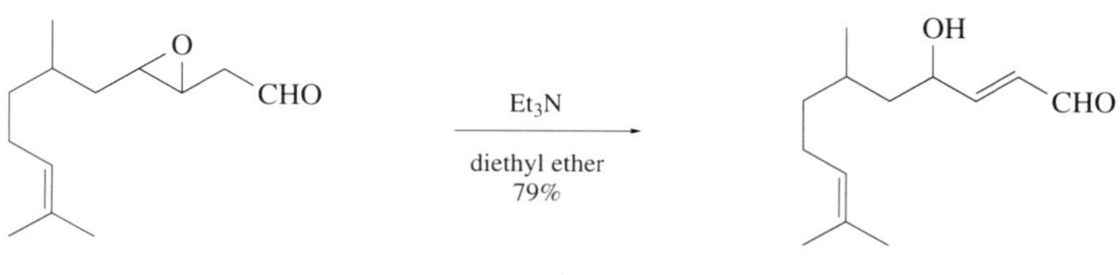

(48)

Table 13 Epoxide opening with magnesium amide bases.

Entry	*Substrate*	*Base*	*Product*	*Yield (%)*
1	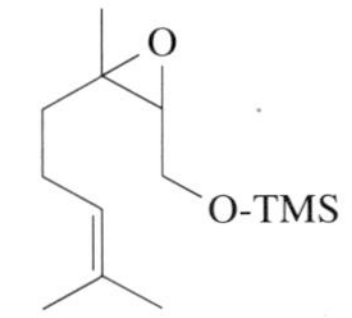	a	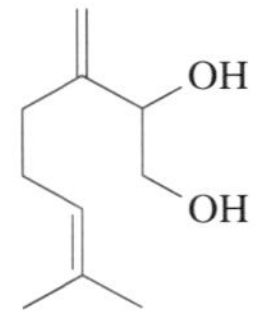	96
2	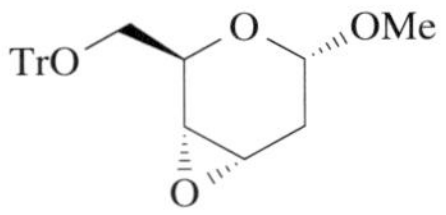	a	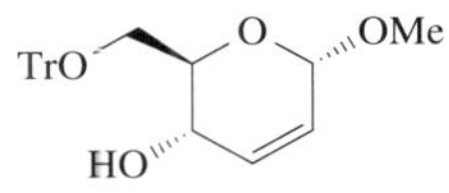	94
3	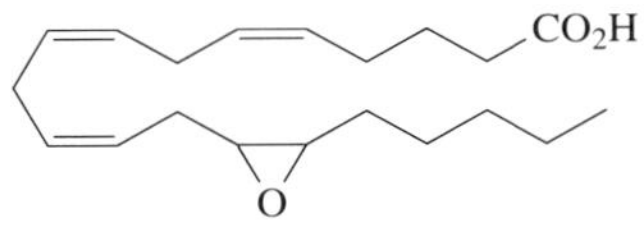	a	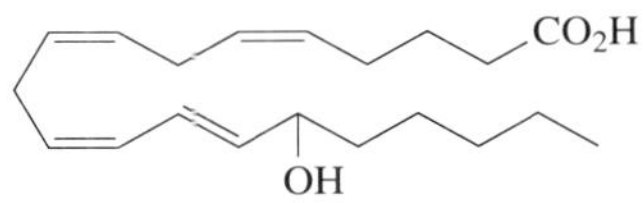	51
4	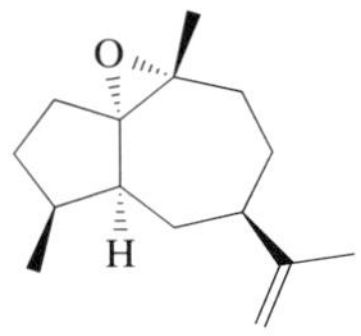	b	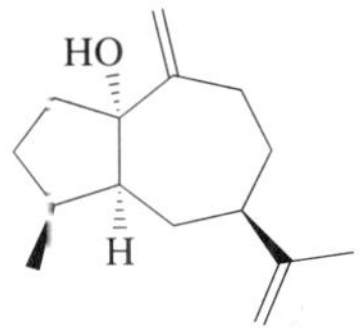	74
5	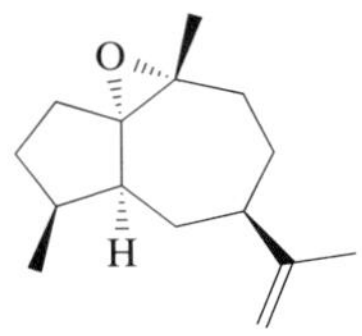	c	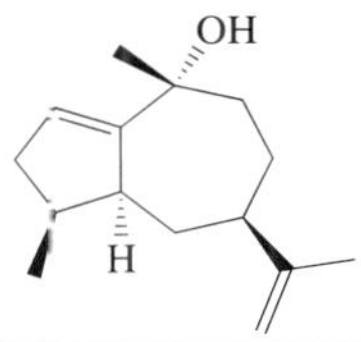	92

[a] MeMg(Pr^{i}NC$_6$H$_{11}$), PhMe, 0 °C, 1.5–7.5 h. [b] MeMg(Pr^{i}NC$_6$H$_{11}$), DME, 0 °C. [c] MeMgBr, LDA, PhMe, –5 °C, 6 h.

The product described in Equation (48) corresponds to the addition to aldehydes of a β-acyl vinyl anion equivalent. Similar eliminations were used to prepare an enone intermediate during a total synthesis of (±)-breynolide ⟨92JA9419⟩. Using an amidine base at low temperature ensured that the most acidic proton was removed (Equation (49)). Epoxide opening was used to trigger an anionic cyclisation during the total synthesis of an elfamycin natural product (Scheme 21) ⟨85JA1691⟩. Useful pyranone syntheses have also involved eliminative epoxide openings (Scheme 22) ⟨93TA1795⟩.

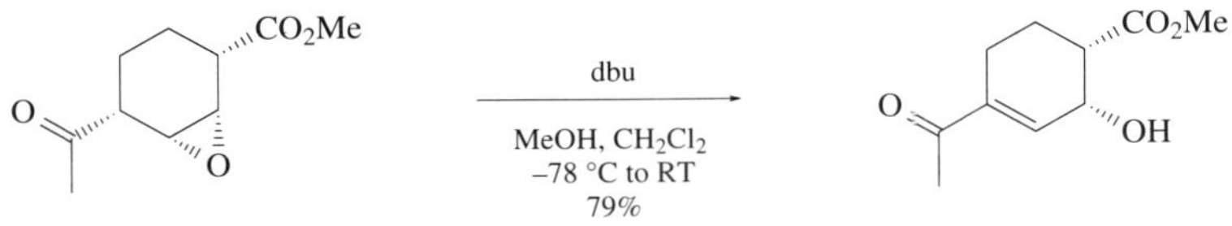

(49)

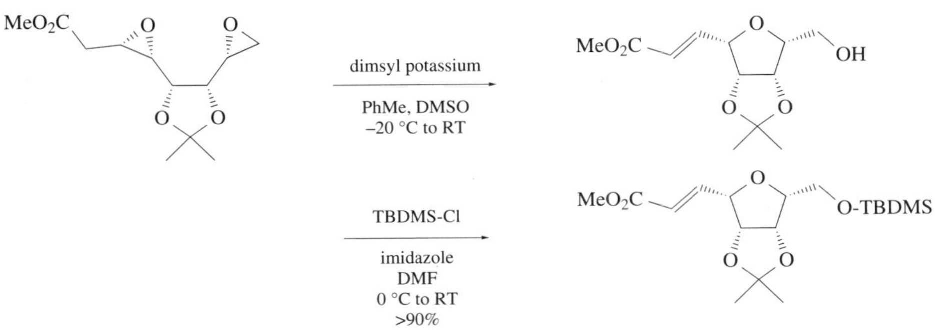

Scheme 21

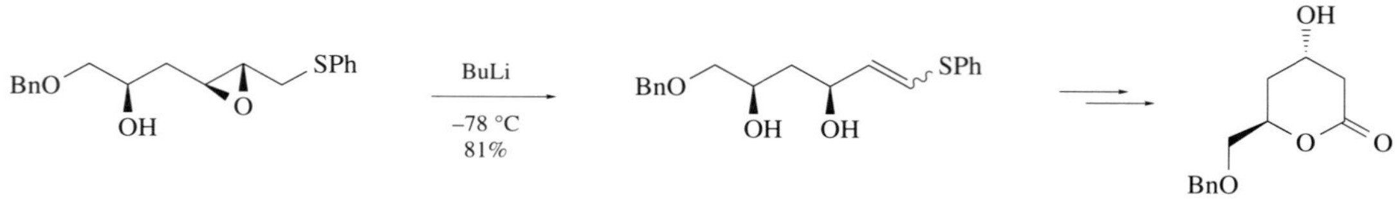

Scheme 22

1.13.4.4 Elimination of a Carboxylic Acid (H—OCOR)

Acetate anion has been eliminated in high yield from activated and unactivated substrates. An alkylidene furanone synthesis combined a furan metallation procedure with a high-yielding acetate elimination in acidic media (Equation (50)) ⟨92BCJ2366⟩, presumably via an E1-type mechanism.

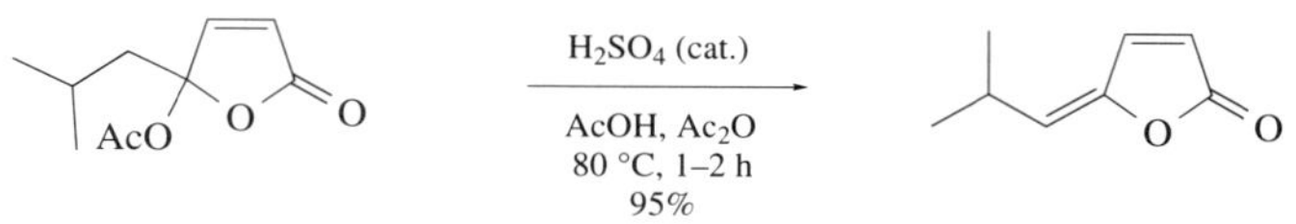

(50)

Acetate anion was eliminated from an unactivated substrate upon treatment with a transition metal catalyst (**11**) ⟨91T8155⟩. The regiochemical outcome of the reaction was unexpected. Deprotonation occurred from a methylene position in the ring rather than from the more accessible methyl group. Mild reaction conditions were employed to achieve this interesting transformation (Equation (51)). Trifluoroacetylation of a tertiary alcohol *in situ* led to an elimination reaction (Equation (52)) ⟨93OSC(8)210⟩. Conversion to the corresponding mesylate also led to a good yield of enone, but sulfur-containing byproducts were difficult to remove from the product. The preparation of a particularly reactive trifluoroacetylating agent was reported ⟨90OSC(7)506⟩.

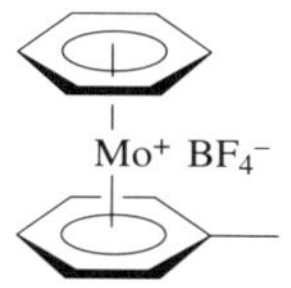

(11)

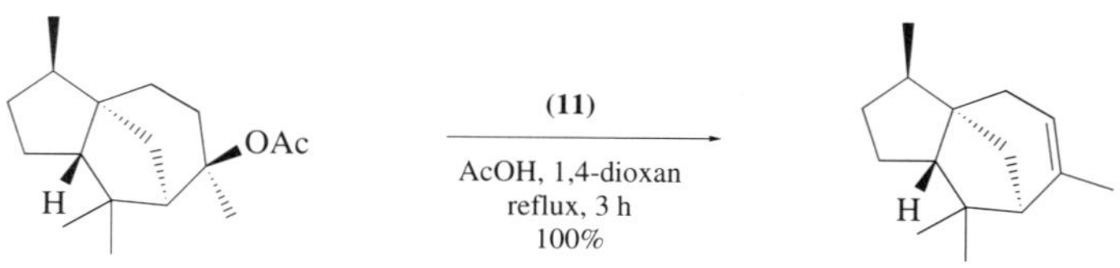

(51)

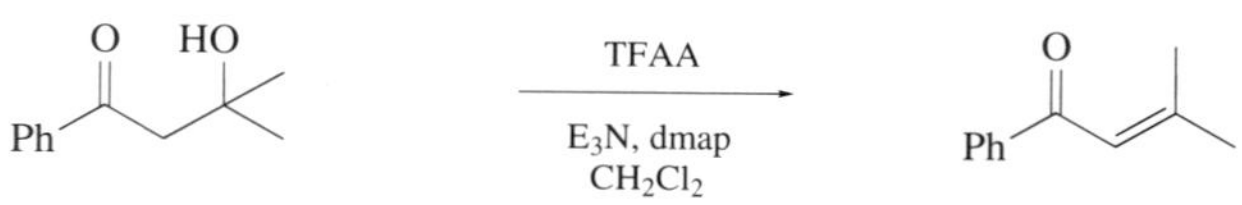

(52)

dmap = 4-dimethylaminopyridine; TFAA = trifluoro acetic anhydride

A wide range of E1cb eliminations has been described. The elimination of a pivalate unit occurred upon Kugelrohr distillation of the ester from anhydrous potassium carbonate; this was used in a synthesis of (*R*)-ionone on a gramme scale (Equation (53)) ⟨92HCA1023⟩. Other useful bases included aqueous potassium hydroxide ⟨91TL3739⟩, Hunig's base ⟨89CPB2882⟩ and dbu (Scheme 23) ⟨90CC84, 92JOC2270, 94JOC281⟩.

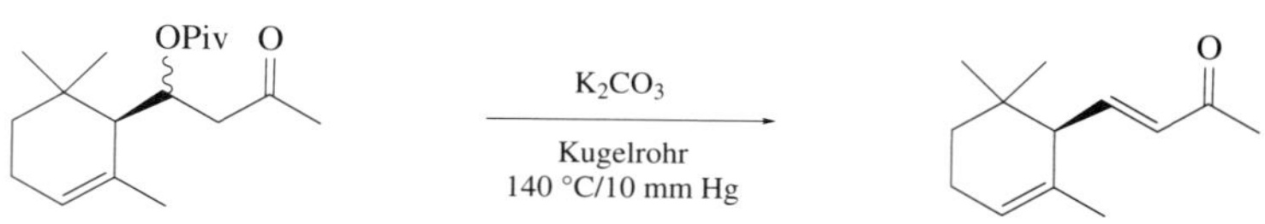

(53)

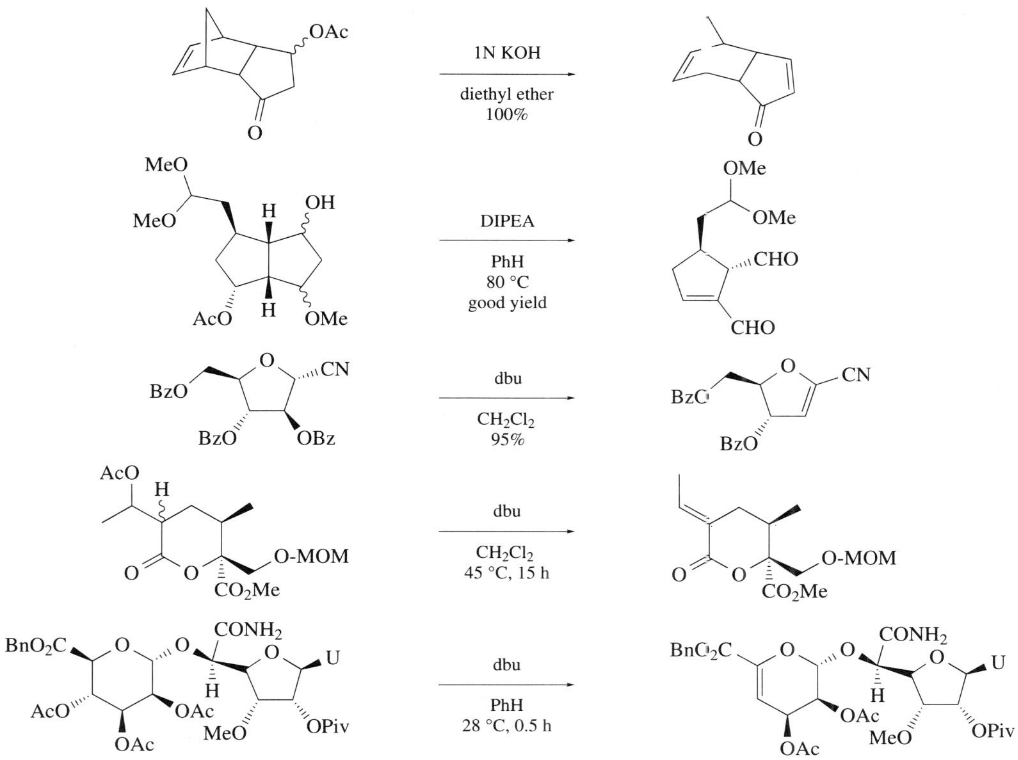

U = uridine; MOM = methoxymethyl

Scheme 23

1.13.4.5 Elimination of a Sulfonic Acid

A popular method of alkene formation involves formation of the methanesulfonate (mesylate) ester followed by treatment with base ⟨90SC1463⟩. In some cases, isolation of the ester was unnecessary. Activated substrates underwent elimination directly in high yield upon treatment with methanesulfonyl chloride and an amine base (Scheme 24) ⟨92JOC3615, 94JOC332⟩. In hindered cases, 4-dimethylaminopyridine (dmap) was used to facilitate the esterification ⟨94T1397⟩. Other mesylate

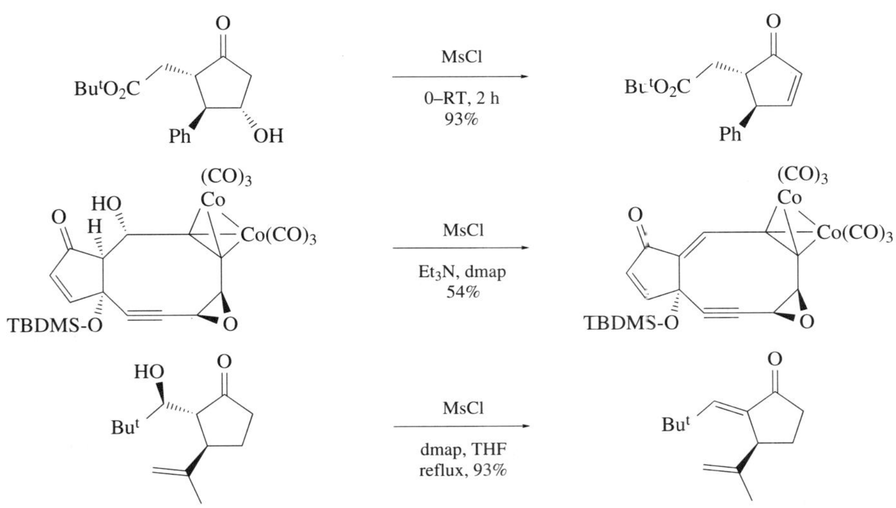

Scheme 24

eliminations have been performed with alkoxide bases. Table 14, entries 1–5 provide a selection of examples.

Table 14 Mesylate eliminations with alkoxide bases.

Entry	*Substrate*	*Conditions*	*Product*	*Yield (%)*
1	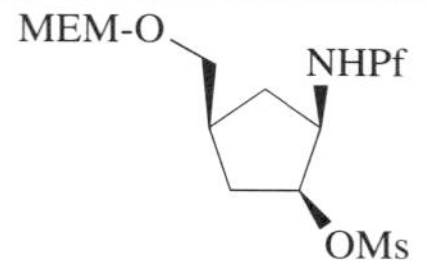	a	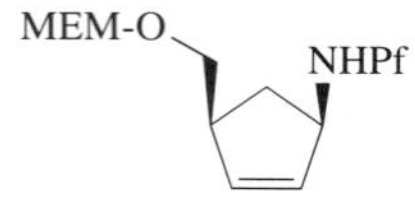	78
2	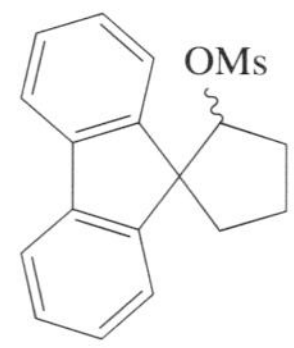	b	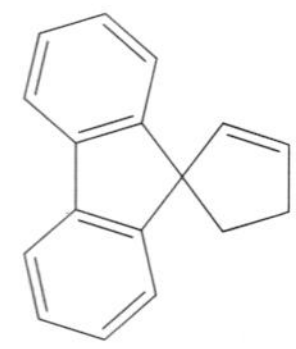	58
3	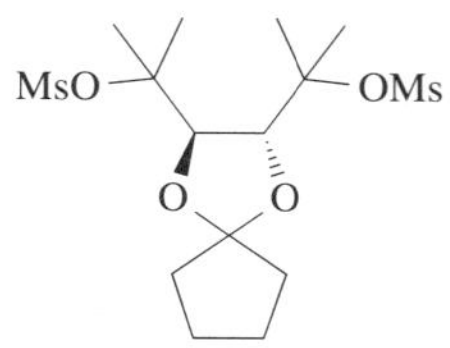	b	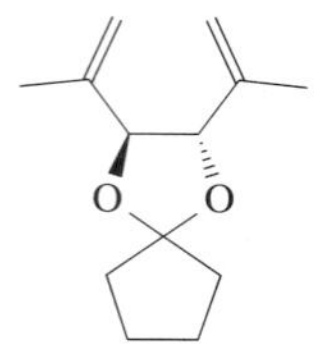	84
4	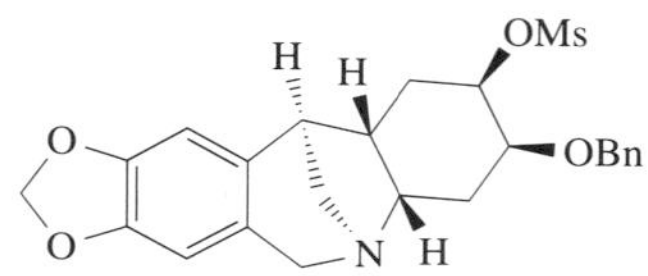	b	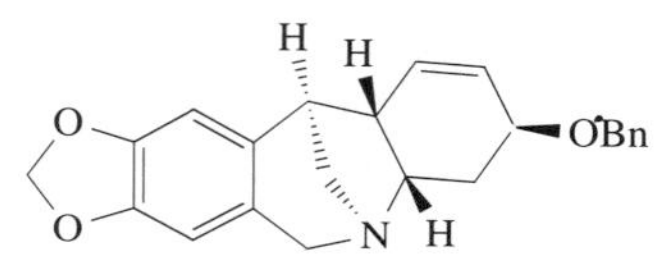	70
5	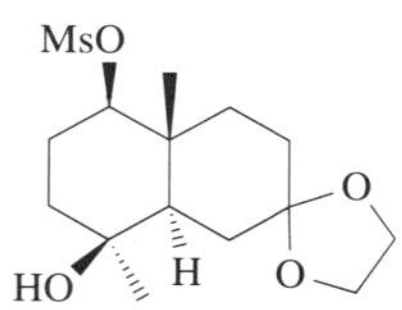	c	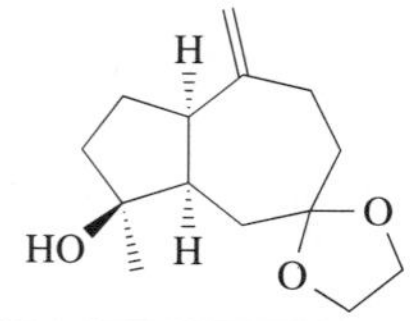	76

[a] $MeOCH_2CH_2ONa$, DMF, 0 °C. [b] Bu^tOK, DMSO. [c] $MeCH_2C(Me)_2ONa$, PhH, reflux.

Pf = 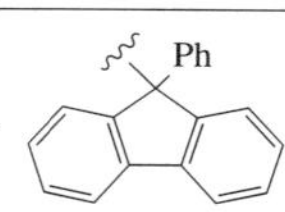

In entry 1, the 9-phenyl-9-fluorenyl (Pf) protecting group was used to prevent proton abstraction α- to the amino group ⟨94JOC394⟩. The regiochemical outcome was unambiguous in entries 2 ⟨92JA8818⟩ and 3 ⟨92JA4128⟩. In entry 4, the major regioisomer is shown ⟨92JOC7285⟩. In entry 5, the mesylate was locked in an equitorial environment on the *trans*-fused decalin system, and the antiperiplanar transition state required for the E2 elimination could not be reached. Instead, an interesting elimination reaction involving simultaneous ring expansion occurred ⟨94JOC374⟩. The phosphazenium naked fluoride base (**8**) converted a steroidal benzene sulfonate to the corresponding diene in excellent yield under very mild conditions (Equation (54)) ⟨91AG(E)1372⟩. In comparison, only a 75% yield of the alkene was obtained after 4 days in the presence of potassium *t*-butoxide in DMSO at 80 °C. Amidine bases have also been used in a number of cases. Equation (55) shows a high yielding example ⟨92JA9419⟩. In Equation (56), aromatisation followed elimination with 1,5 diazabicyclo[4.3.0]non-5-ene (dbn) in benzene solution ⟨90JA9284⟩. Palladium-catalysed mesylate elimination in the absence of a base led to the formation of a range of interesting glycals (Scheme 25) ⟨92JA1491⟩.

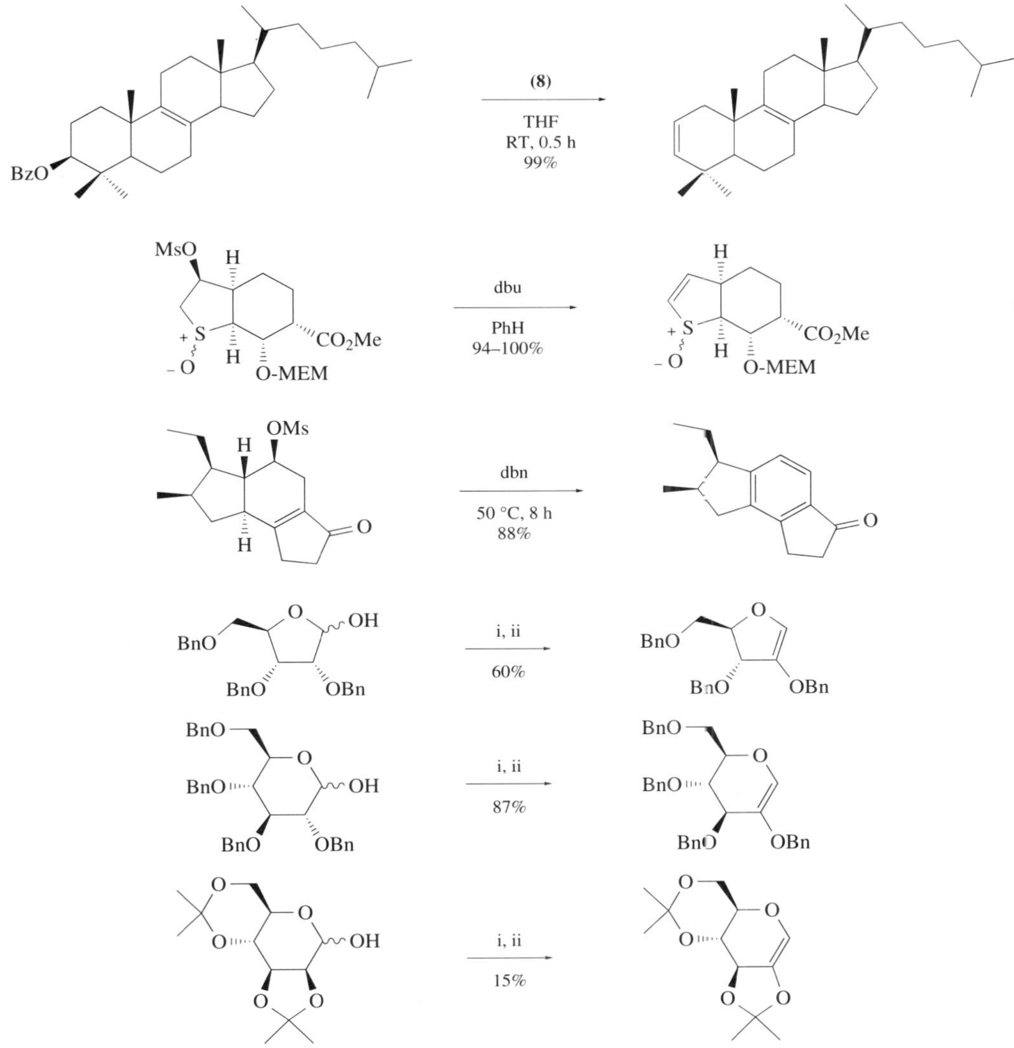

(54)

(55)

(56)

i, Ms_2O, collidine, CH_2Cl_2, RT, 1 h
ii, $Pd(PPh_3)_4$, 50 °C, 12 h

Scheme 25

1.13.4.6 Elimination of 1,2-Diols

Vicinal eliminations involving two hydroxyl groups (or their derivatives) have formed the subject of a review ⟨84OR(30)457⟩. The reactions have found useful applications in the areas of carbohydrate, nucleoside and enediyne chemistry. The Corey–Winter reaction was developed to deoxygenate 1,2-diols via thionocarbonate formation followed by treatment with a diaminophosphorane ⟨82TL1979⟩. Scheme 26 shows examples from the enediyne ⟨93TL4121⟩ and nucleoside ⟨89JOC2217⟩ fields. The mildness of the reaction conditions allowed a range of other functional groups to survive the dideoxygenation procedure. Organometallic procedures using bis(cyclooctadienyl)nickel afforded lower yields in the enediyne case, and involves using hydrogen sulfide gas in the workup.

Table 15, entries 1–5 highlight a range of vicinal eliminations performed under free radical conditions. Treatment of a ribonucleotide bis(xanthate)ester using the usual tin method conditions afforded a dideoxynucleoside in 93% yield (entry 1), offering a clear advantage over the Corey–Winter method which afforded the dideoxynucleoside in only 51% yield ⟨89JOC2217⟩. Halides also served as effective leaving groups in free radical elimination. Entries 2 ⟨90S411⟩ and 3 ⟨90TL3829⟩ illustrate high yielding eliminations.

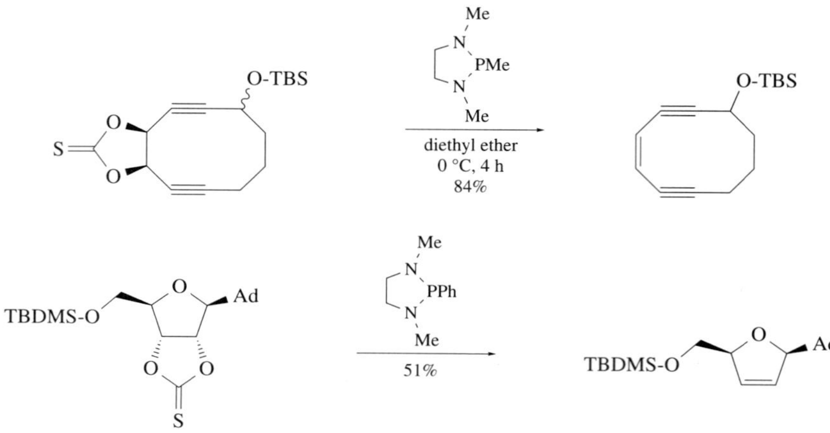

Scheme 26

Table 15 Eliminations via free radicals.

Entry	*Substrate*	*Conditions*	*Product*	*Yield (%)*
1	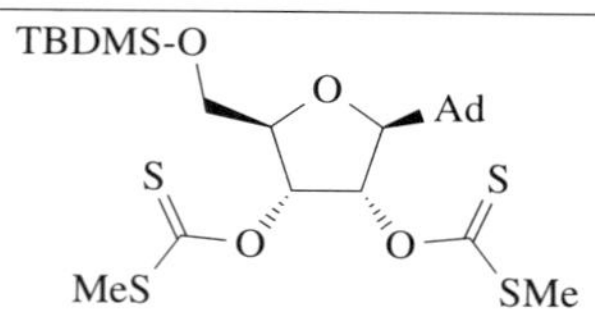	a	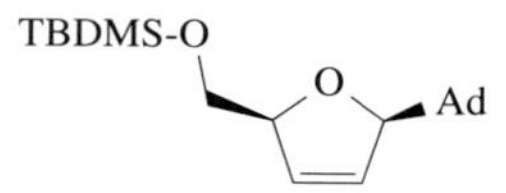	93
2	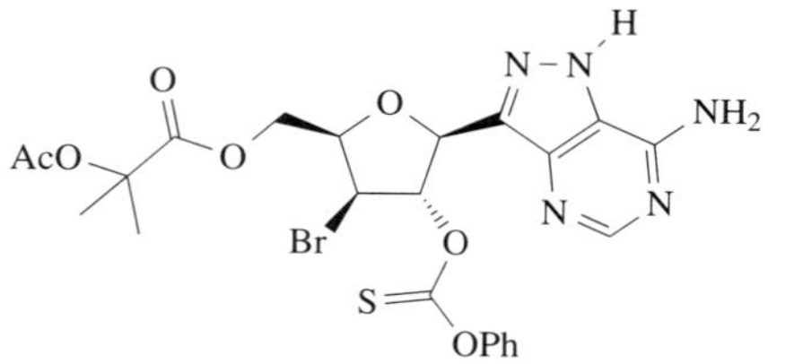	a	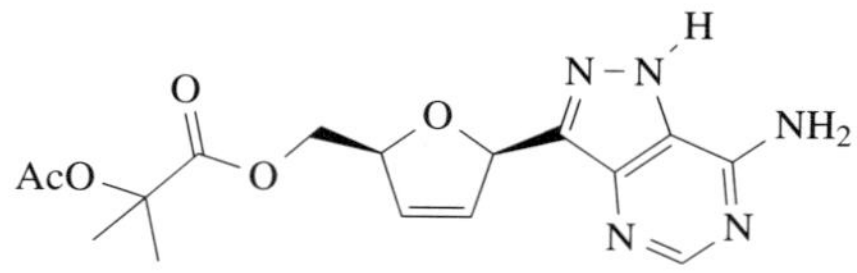	58
3	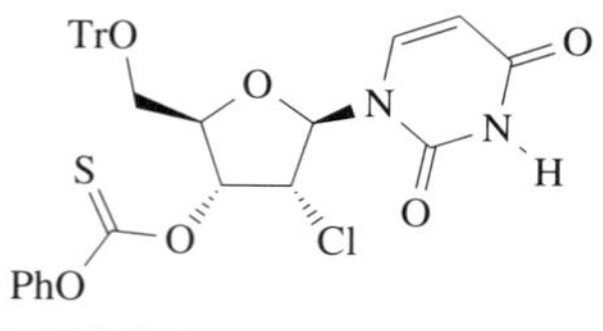	b	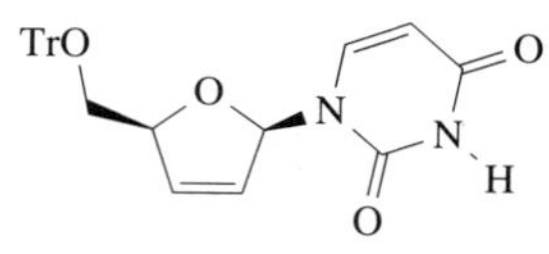	84
4	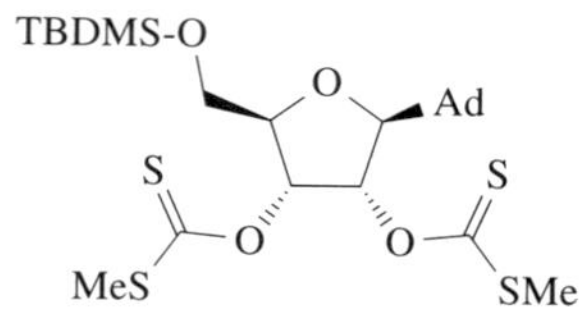	c	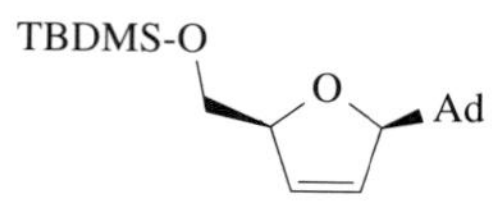	70
5	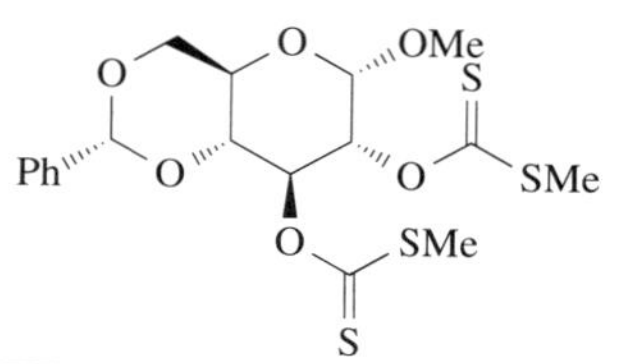	c	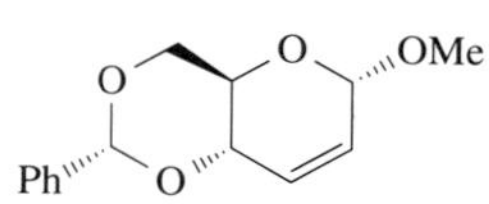	76

[a] Bu_3SnH, AIBN, PhMe, reflux. [b] Bu_3SnH, AIBN, PhMe, 60 °C. [c] Ph_2SiH_2, AIBN, PhMe, reflux.

Barton has developed chemistry that avoids the use of toxic tin compounds; instead, a silane serves as the chain carrier (entries 4 and 5) ⟨91TL2569⟩. The method developed by Samuelsson *et al.* (Equation (57)) is particularly suitable for larger scale operations because all the byproducts from the reaction are soluble in water ⟨90JOC4273⟩. Treatment of dibenzoates with samarium diiodide (Equation (58)) afforded elimination products in moderate to good yield ⟨92TL573⟩. The method is noteworthy because of the routine use of benzoate esters in carbohydrate chemistry. Short reaction times and facile workup also add to the appeal of the samarium-mediated reaction.

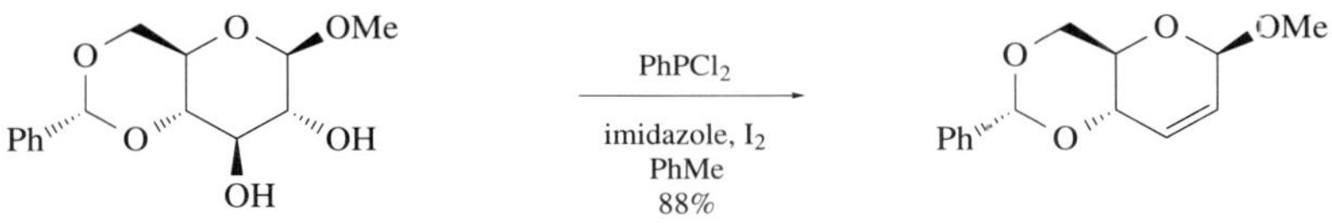

(57)

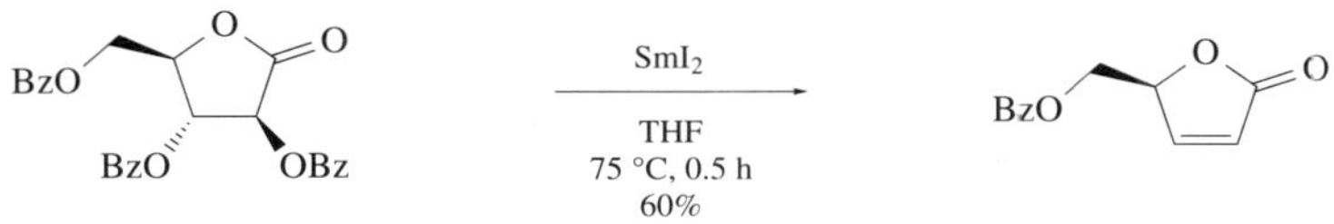

(58)

1.13.4.7 Deoxygenation of Epoxides

One-pot sequences are available involving nucleophilic ring opening, followed by elimination. An acid-sensitive glycal was prepared by a mild procedure (Equation (59)) ⟨77CB2127⟩. Epoxide opening was efficient, but the elimination step proceeded in only moderate yield. A simple and inexpensive combination of reagents (chlorotrimethyl silane/sodium iodide in acetonitrile) converted a *cis*-epoxide into a (*Z*)-alkene in good yield ⟨81TL3551⟩. A range of metals has been used in deoxygenation reactions, including low-valent titanium ⟨78JOC3249⟩, niobium ⟨82CL157⟩, and samarium ⟨80JA2693⟩. A procedure based on an organotungsten intermediate tolerates the presence of ester and ether groups ⟨81OS(60)29⟩, but the most sensitive method involves an organotitanium reagent (Equation (60)) ⟨90JA6408⟩.

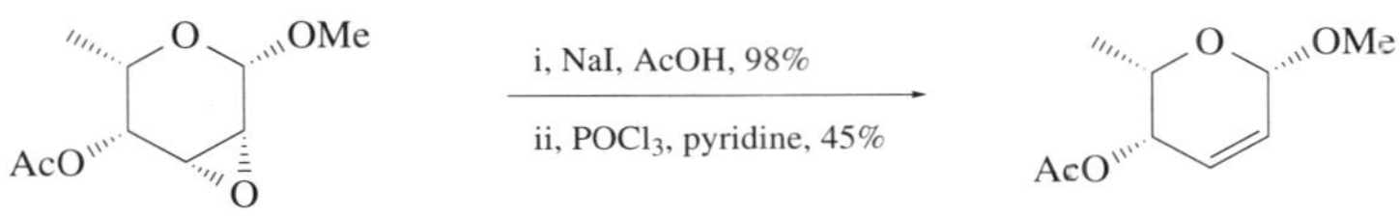

(59)

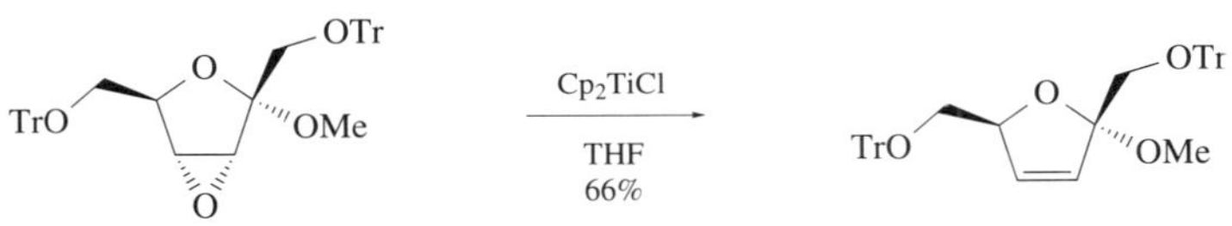

(60)

Attempts to deoxygenate the furanoid epoxide by other methods led exclusively to the formation of the furan, via elimination of the alkoxy group.

1.14
One or More C=C Bond(s) by Elimination of S, Se, Te, N, P, As, Sb, Bi, Si, Ge, B or Metal Functions

ANITA R. MAGUIRE
University College Cork, Republic of Ireland

1.14.1 INTRODUCTION

Elimination reactions involving heteroatomic groups often provide very useful synthetic routes to alkenes, e.g., alkenes are formed efficiently by elimination of sulfoxide or selenoxide groups. Furthermore, many alkene syntheses involve elimination of a heteroatomic moiety as a step in the overall process, e.g., the Wittig and the Peterson reactions. In the following discussion of these transformations, particular emphasis is placed on the synthetic utility of the elimination reactions.

1.14.2 BY ELIMINATION OF SULFUR, SELENIUM OR TELLURIUM FUNCTIONS

1.14.2.1 Elimination of Sulfide, Selenide or Telluride Groups

1.14.2.1.1 Elimination of sulfide groups

Elimination of sulfides to form alkenes can be effected on treatment with base ⟨88JA2066, 88JA5075⟩, e.g., BuLi ⟨79JA3283⟩ or KOH/HMPA ⟨65CI(L)1768⟩ (Equation (1)). Similarly on treatment with potassium 3-aminopropylamide, vinyl sulfides are transformed to terminal alkynes by elimination followed by isomerization ⟨78JOC3083⟩; vinyl ethers under the same conditions lead to diene formation. Elimination of the elements of a thiol from thioacetals and thioketals to form vinyl sulfides can be brought about by treatment with cuprous ions in the presence of an amine ⟨75JOC812, 76JOC3218, 80OS(59)202⟩. Extension of this methodology to the formation of dienes by elimination of two equivalents of thiophenol is illustrated in Equation (2).

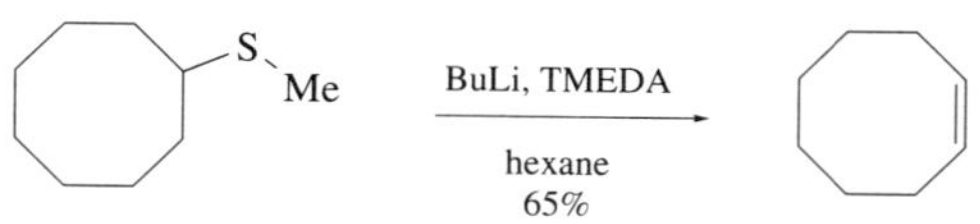

(1)

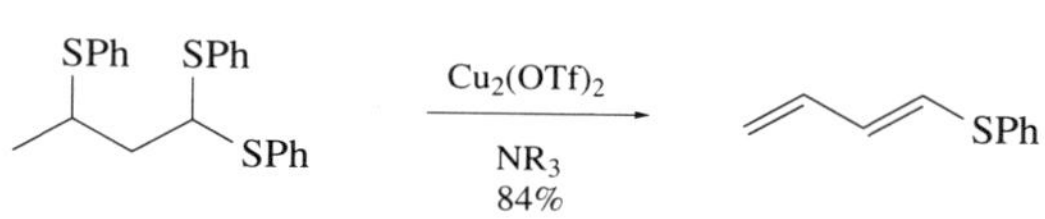

(2)

Elimination of β-substituted sulfides to form alkenes has been investigated, most commonly with phenylthio derivatives, but occasionally with alkylthio groups—thus β-hydroxy sulfides can be transformed to alkenes on treatment with methyllithium and *o*-phenylene phosphorochloridite ⟨72TL737⟩, thionyl chloride, phosphorus triiodide, or diphosphorus tetraiodide ⟨79TL4111, 81TL4009⟩, *N*-ethyl-2-fluoropyridinium tetrafluoroborate, triethylamine and lithium iodide ⟨78CL413, 80JOC3549, 88TL659⟩, electrochemically ⟨78TL2807⟩, using titanium tetrachloride and zinc or lithium aluminum hydride ⟨74CL1523, 75CL871⟩, by transformation to *S*-methylthiocarbonates followed by treatment with tributyltin hydride ⟨77TL4223⟩, or by transformation to the benzoate followed by reduction with sodium naphthalenide ⟨86TL1343⟩ or lithium in liquid ammonia ⟨72JA4758⟩. Equation (3), showing an elimination used in the synthesis of (−)-periplanone-B ⟨86TL1343⟩, illustrates the synthetic utility of this transformation. Application of this reductive elimination in the methylenation of ketones is illustrated in Scheme 1 ⟨72JA4758⟩; this method compares favourably with the Wittig reaction giving efficient reaction even with the highly hindered tricyclic ketone shown.

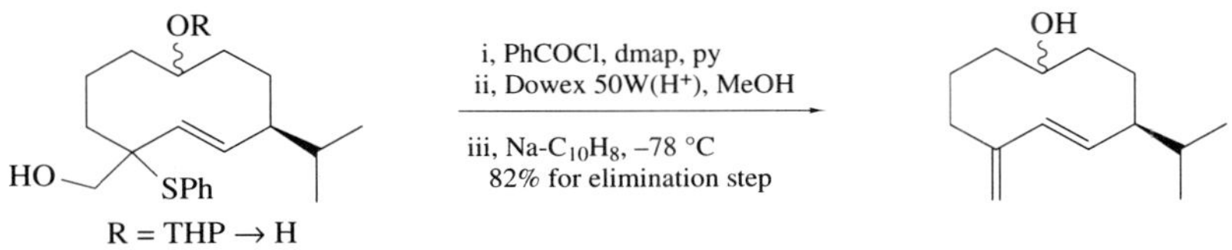

(3)

dmap = 4-dimethylaminopyridine

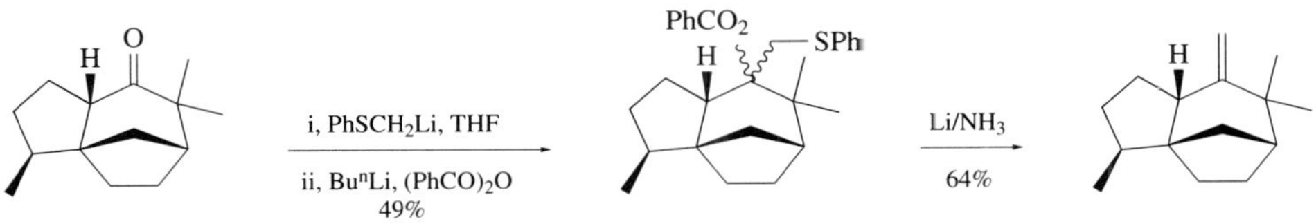

Scheme 1

Elimination of β-hydroxy sulfides using *N*-ethyl-2-fluoropyridinium tetrafluoroborate, triethylamine and lithium iodide ⟨78CL413, 80JOC3549, 88TL659⟩ proceeds stereoselectively to form the (*Z*)- and (*E*)-alkenes from *syn* and *anti* isomers of the β-hydroxy sulfides, respectively. This elimination has been employed in a stereoselective synthesis of (*Z*)-α,β-disubstituted acrylates (Equation (4)) ⟨88TL659⟩. Formation of appreciable amounts of the more stable (*E*)-isomer were observed only when extended conjugation was present (R = PhCH=CH).

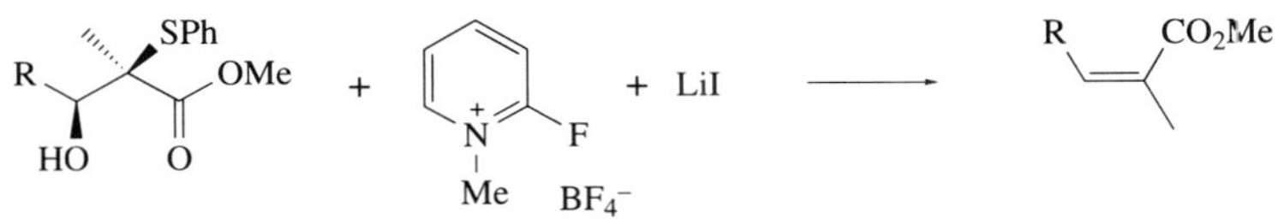

(4)

R = Ph, $Ph(CH_2)_2$, *n*-C_7H_{15}

β-Oxygenated sulfides can be eliminated to form alkenes on treatment with sodium in liquid ammonia ⟨79CC83⟩ or TMS-Cl and NaI (Equation (5)) ⟨81JOC231⟩ β-Hydroxy thiols can be transformed to the corresponding alkenes via the oxathiolanes on treatment with lithium diisopropylamide (LDA) (Scheme 2) ⟨74JCS(P1)433, 88JCS(P1)813, 88JCS(P1)817⟩.

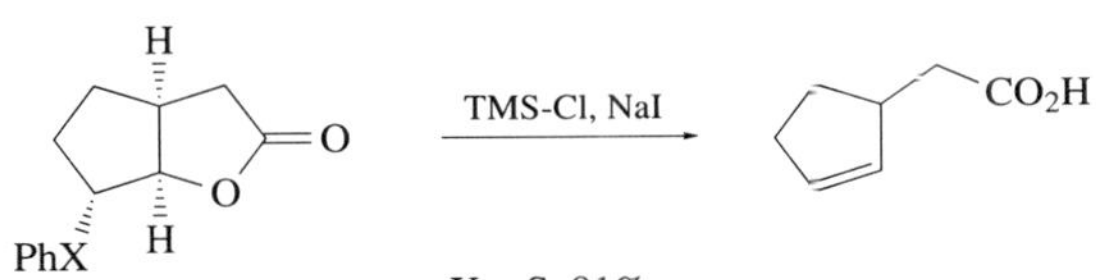

(5)

X = S, 91%
X = Se, 84%

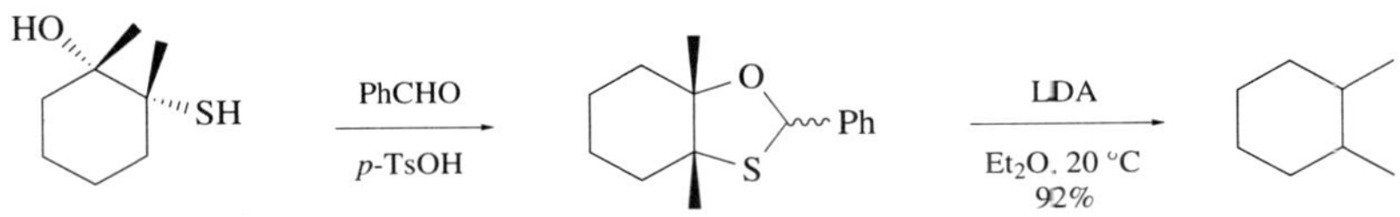

LDA = lithium diisopropylamide

Scheme 2

Similarly, β-chloro or bromo sulfides can be reduced to alkenes using tributyltin hydride (Equation (6)) ⟨76JA951, 77TL4223⟩, or by use of metals such as Mg, Zn or Na ⟨44JOC310⟩. Allyl 2-pyridyl sulfides, on alkylation with tributylstannylmethyl iodide, undergo elimination of the stannyl and sulfide groups to form 1,3-dienes directly (Scheme 3) ⟨82TL2205⟩ (see also Section 1.14.6.3). *β*-Nitro sulfides eliminate stereoselectively on treatment with tributyltin hydride ⟨84TL5319, 88CC1278⟩; this methodology has been employed in the synthesis of (*E*)-allyl alcohols (Equation (7)) (see Section 1.14.3.5).

(6)

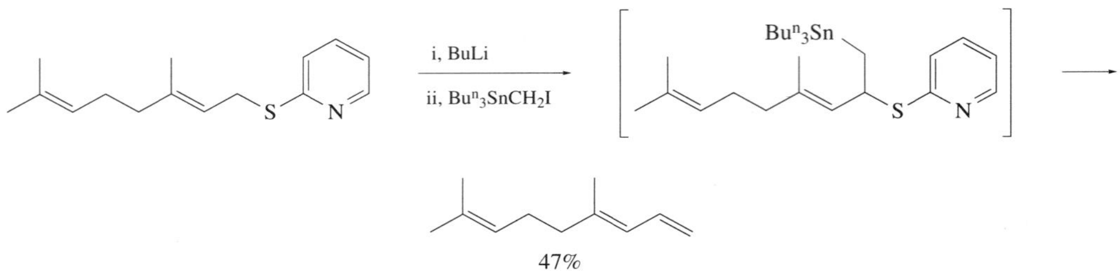

Scheme 3

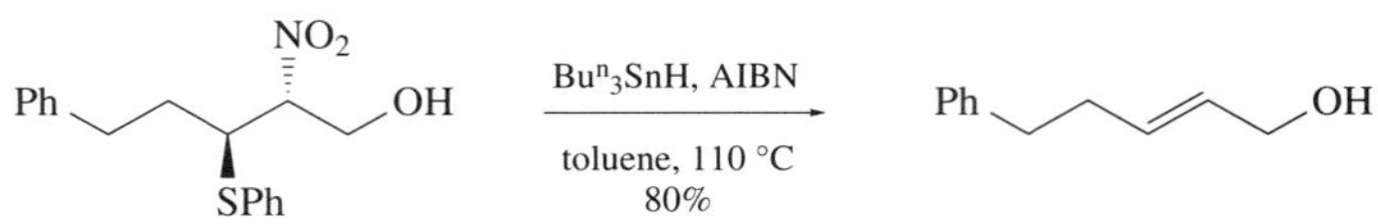

(7)

AIBN = 2,2'-azobisisobutyronitrile

Pyrolysis of α-halo sulfides produces alkenes (Scheme 4) ⟨81AG(E)585⟩; this transformation, which resembles the Ramberg–Bäcklund reaction of α-halo sulfones, may proceed via a thiirane which then loses a sulfur atom to form the alkene. Removal of sulfur from episulfides to produce alkenes has been discussed ⟨80T557⟩. On treatment with triethylphosphite episulfides undergo a highly stereoselective removal of sulfur to form the alkene in a *syn* fashion (Equation (8)) ⟨59JA578⟩ indicating that attack by phosphorus occurs at sulfur rather than at carbon, in contrast to the reactivity pattern observed with epoxides. Other reagents which transform episulfides to alkenes include phenyllithium ⟨59JA578⟩, methyl iodide ⟨49JCS282, 64JOC3258⟩, diphosphorus tetraiodide, tributyltin hydride or Raney nickel ⟨83TL1657⟩, lithium aluminum hydride ⟨62CI(L)1236⟩, 3-methyl-2-selenoxobenzothiazole ⟨76S200⟩, and ethyl diazoacetate and cupric salts ⟨75JA2553⟩.

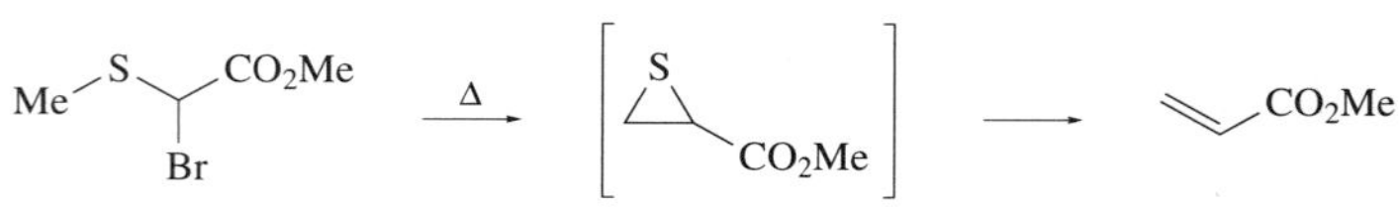

Scheme 4

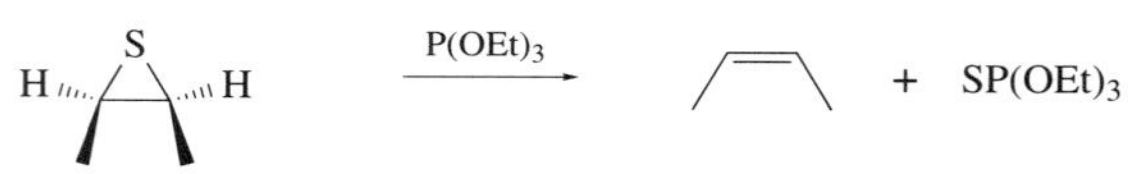

(8)

1.14.2.1.2 *Elimination of selenide groups*

Reductive elimination of *β*-hydroxyselenides provides a useful route to alkenes ⟨78T1049, 79ACR22⟩. Once it was observed that the adducts derived from benzeneselenenyl trifluoroacetate addition to alkenes could revert to alkenes under certain conditions (Equation (9)) ⟨74CC100, 74JOC428⟩, investigation of conditions which could be employed for the transformation of *β*-hydroxyselenides to alkenes was undertaken. Reich demonstrated that methanesulfonyl chloride and triethylamine (both reagents being used in excess) resulted in good yields of the alkenes ⟨75CC790⟩. The elimination occurs in an *anti* fashion and is believed to be the reverse of the electrophilic addition of selenium reagents to alkenes, proceeding via an episelenonium ion (Scheme 5) ⟨79JA6638⟩. Krief and co-workers ⟨76TL1385, 76TL3227, 76TL3743, 78TL2693, 82CC564⟩ subsequently demonstrated that this transformation can also be achieved using toluenesulfonic acid in pentane, perchloric acid in ether, trifluoroacetic acid anhydride and triethylamine, thionyl chloride and triethylamine, phosphochloridite and sodium hydride, phosphorus oxychloride and triethylamine, or phosphorus triiodide and triethylamine; both phenylseleno and methylseleno derivatives were investigated. *Anti* elimination was observed under these conditions also. Application of this stereoselective elimination to the synthesis of allyltrimethylsilanes is illustrated in Equation (10) ⟨87TL2061, 90T1885⟩. Clive and Kalè demonstrated that trimethylsilyl chloride and sodium iodide can similarly transform β-hydroxyselenides into alkenes ⟨81JOC231⟩.

(9)

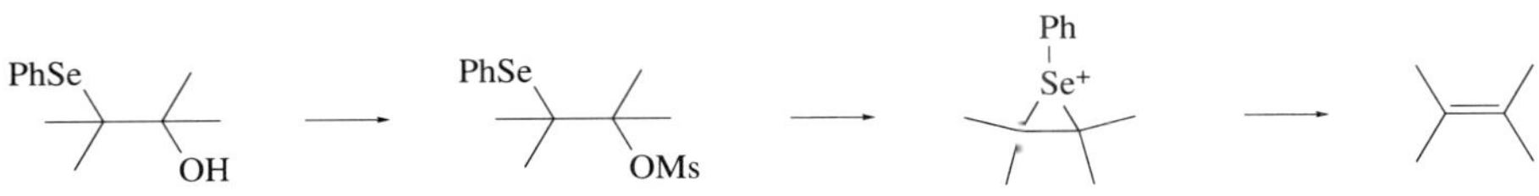

Scheme 5

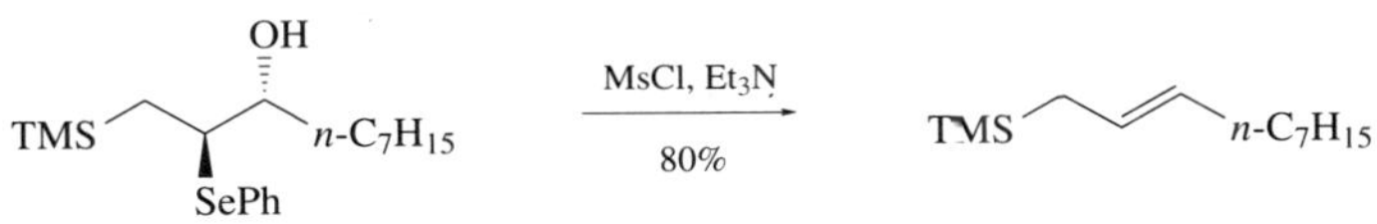

(10)

As β-hydroxyselenides can be derived by nucleophilic attack of selenide anions on an epoxide ⟨76TL1385⟩, or by reduction of α-seleno aldehydes or ketones ⟨76TL3227, 79ACR22⟩, or by attack of an α-selenocarbanion on an aldehyde or ketone ⟨76TL1385, 82CC564⟩, or by reduction of β-hydroxyselenoxides ⟨79JA6648⟩, elimination of these compounds can be employed for the transformation of epoxides to alkenes, of aldehydes and ketones to alkenes, or of α-seleno ketones to alkenes (Scheme 6). The elimination of β-hydroxyselenides is efficient, even for the formation of tetrasubstituted alkenes ⟨76TL3743, 82CC564⟩ and in some cases is preferable to the Wittig reaction for the preparation of such hindered alkenes. Equation (11) illustrates the use of the elimination in the synthesis of highly substituted dienes ⟨87CC1540⟩.

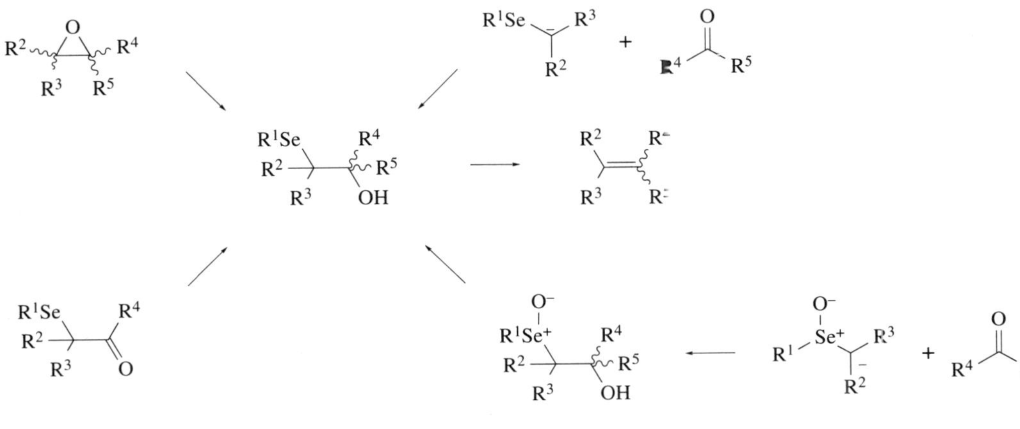

Scheme 6

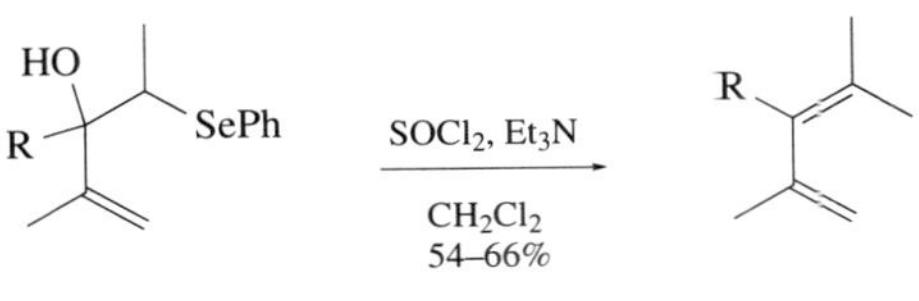

(11)

R = Ph, $(CH_2)_2Ph$, $(CH_2)_5Me$

The elimination of β-hydroxyselenides can be applied to the formation of α,β-unsaturated esters, lactones ⟨78TL2693⟩ and lactams ⟨86JOC3108⟩; use of this transformation was made in Meyers' synthesis of (*S*)-(−)-deplancheine (Scheme 7) ⟨86JOC3108⟩.

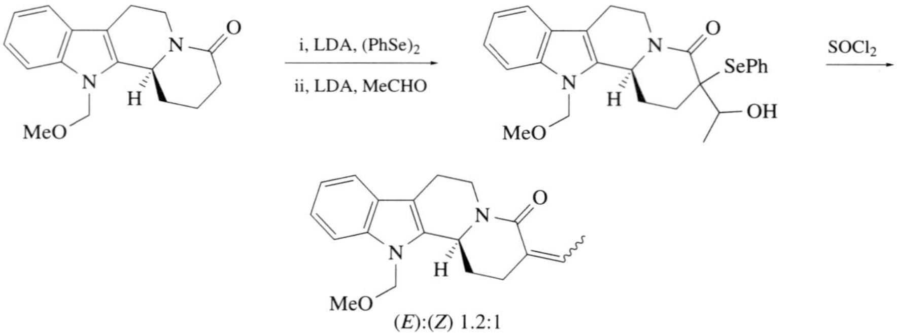

Scheme 7

Elimination of β-oxygenated selenides (in which the selenide bears an ether, silyloxy or lactone group) to form an alkene can be brought about using sodium in ammonia ⟨79CC83⟩ or chlorotrimethylsilane and sodium iodide (Equation (5)) ⟨81JOC231⟩. A similar transformation is possible with β-oxygenated sulfides albeit at higher temperatures ⟨81JOC231⟩.

β-Haloselenides can be eliminated to form alkenes on treatment with phenylselenolate ⟨80TL1877⟩. Destannylselenenylation ⟨92JOC3145⟩ (analogous to known destannylsulfurization processes ⟨85JOC3622⟩), to form alkenes can be brought about by fluoride ion, Lewis acid, silica gel, bases or a variety of other conditions (Equation (12)) (see Section 1.14.6.3).

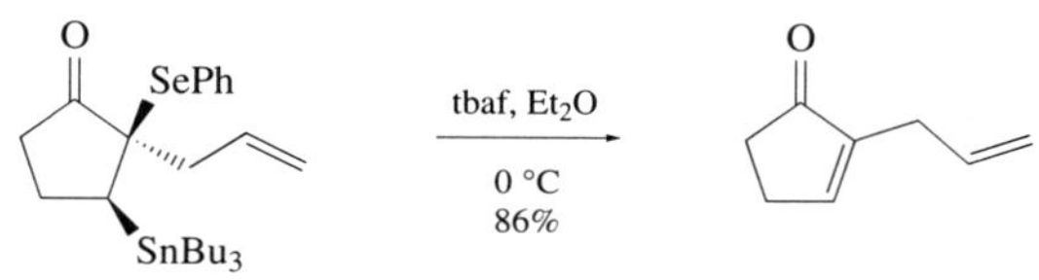

(12)

Treatment of a selenide, containing an appropriately positioned cyclopropane ring, with tributyltin hydride or triphenyltin hydride results in formation of a cyclopropylmethyl radical which undergoes ring opening to form an alkene (Scheme 8) ⟨89CC332⟩.

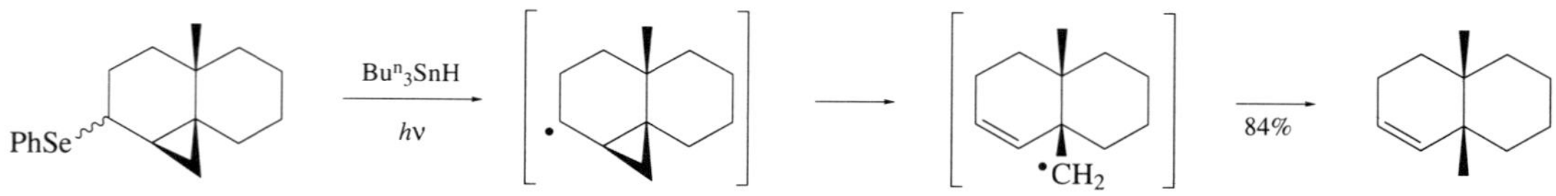

Scheme 8

1.14.2.1.3 *Elimination of telluride groups*

Primary tellurides form alkenes in high yields on treatment with chloroamine-T; the elimination is believed to proceed via the adduct shown in Scheme 9 ⟨81CL447, 89JOC4391⟩.

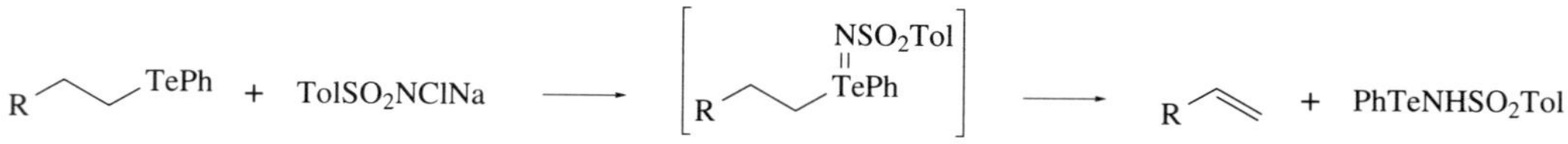

Scheme 9

Alkene inversion occurs on treatment of alkenes with tellurium(IV) chloride followed by sodium sulfide; this is believed to involve elimination via an epitelluride (Scheme 10) ⟨81TL1919, 83JOC3918, 86S1⟩. Further examples in which epitellurides are proposed as alkene precursors have been described ⟨91S793, 91S897⟩.

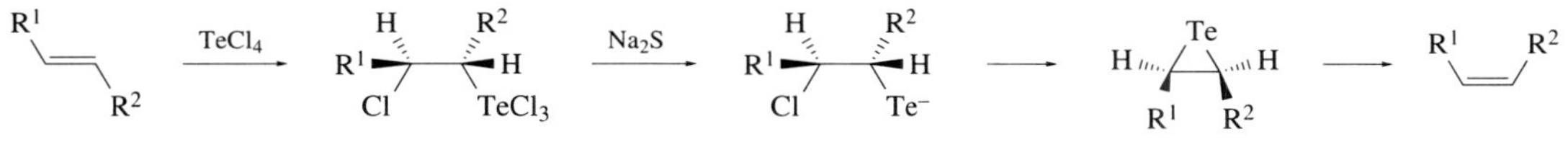

Scheme 10

1.14.2.2 Elimination of Sulfoxide, Selenoxide or Telluroxide Groups

1.14.2.2.1 Elimination of sulfoxide groups

Thermal elimination of alkyl aryl sulfoxides to form alkenes was first reported by Kingsbury and Cram in 1960 ⟨60JA1810⟩; since then this has become widely accepted as a useful synthetic route to alkenes ⟨91COS(6)1011⟩. Cram's proposed mechanism involved a stereospecific *cis* elimination as shown in Scheme 11.

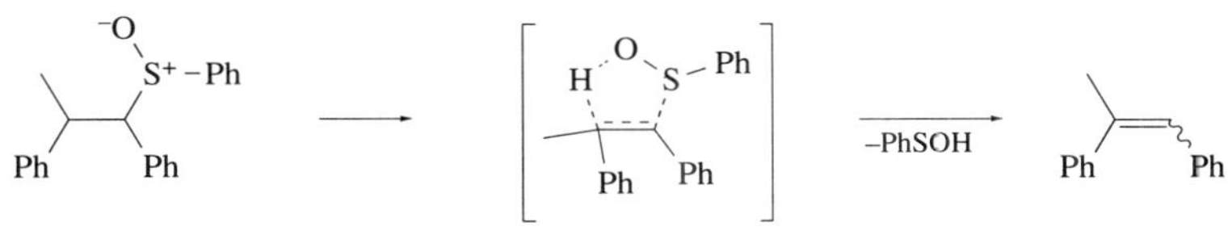

Scheme 11

Subsequently it was demonstrated ⟨64JOC2699, 65CC29⟩ that pyrolysis of dialkyl sulfoxides to form alkenes is also possible, although higher temperatures are required than for aryl sulfoxides. Eliminations involving pyridyl- ⟨78HCA998, 87JCR(S)353, 93JOC1579⟩ and pyrazinylsulfoxides ⟨88H(27)261⟩ have also been reported. In unsymmetrical dialkyl sulfoxides there is an enhanced tendency towards thermolytic cleavage of the bond connecting sulfur to the more highly substituted alkyl group ⟨67JOC102⟩. The sulfenic acid produced in the pyrolysis of sulfoxides can be trapped and identified (see for example ⟨70CC1683, 73IJS205, 77JCS(P1)1574⟩).

In the 1970s, following Cram's initial report, considerable investigation of this reaction was undertaken to determine more fully the conformational and electronic factors involved (⟨76JA4887⟩ and references therein). For example, the rate of elimination in benzenesulfinyl substituted cycloalkanes is dependent on ring size (Equation (13)), indicating energy differences in achieving the conformation required for the *cis* elimination ⟨67JOC1631⟩.

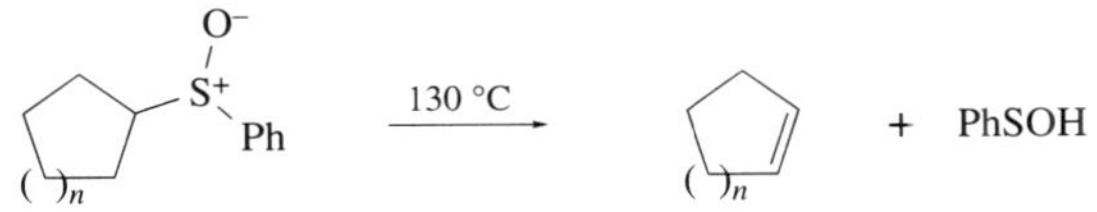

(13)

$n = 1,2,3$
Relative rates = 25:1:120

Jones *et al.* ⟨76JCS(P1)459⟩ demonstrated in a series of steroids that the stereochemistry at the sulfoxide could control the regiospecificity of the elimination process as shown in Equation (14). The regioselectivity of sulfoxide elimination in 1,3-difunctional sulfoxides has been studied ⟨89JCR(S)30, 91JCR(S)62⟩; elimination occurs selectively away from an alkoxy group and towards an alkylthio group (Equation (15)).

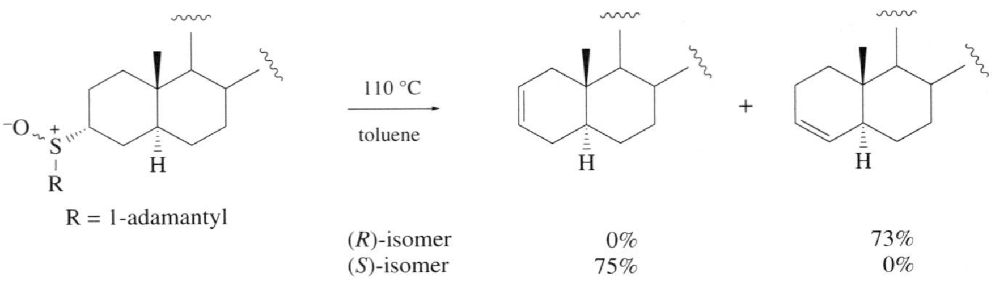

(14)

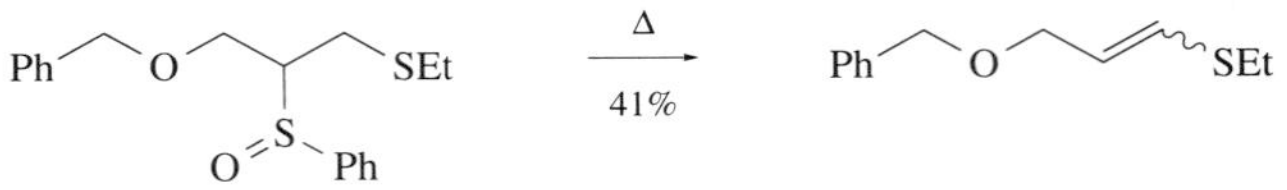

(15)

There are many examples of the application of sulfoxide thermolyses to alkene synthesis, e.g., Ley *et al.* reported the preparation of a dihydrofuran acetal derivative via sulfoxide elimination ⟨87TL221⟩, while Takano *et al.* employed the thermolysis of an aryl sulfoxide in the presence of calcium carbonate to form (+)-goniothalamin quantitatively as a single isomer having the (*E*) configuration (Equation (16)) ⟨92TA853⟩, illustrating the preference for formation of the thermodynamically more stable isomer whenever possible given the restraint that the elimination occurs in a *cis* fashion. Synthesis of novel avermectin derivatives containing terminal alkenyl groups was possible by methyl sulfoxide pyrolysis, again in the presence of calcium carbonate to neutralize the sulfenic acid produced ⟨91SL873⟩. In the last example, only one of the two diastereoisomeric sulfoxides eliminated to form the alkene in good yield; the other failed to eliminate at temperatures below which degradation of the molecule occurs.

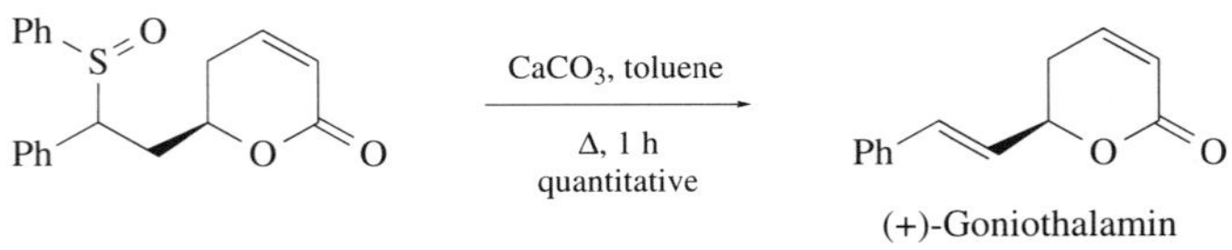

(16)

The principal synthetic application of the thermal elimination of sulfoxides has been in the transformation of esters, ketones and related compounds into the corresponding α,β-unsaturated compounds. Trost has described this process in detail ⟨76JA4887, 78ACR453, 78CRV363⟩. Introduction of a sulfenyl group α to a carbonyl compound is a facile process; subsequent oxidation with a wide variety of oxidizing agents, e.g., mcpba or $NaIO_4$, followed by thermolytic extrusion of a sulfenic acid constitutes an exceptionally mild and general route to the α,β-unsaturated derivatives (Scheme 12). Frequently the elimination process may be conducted without purification of the intermediate sulfoxide.

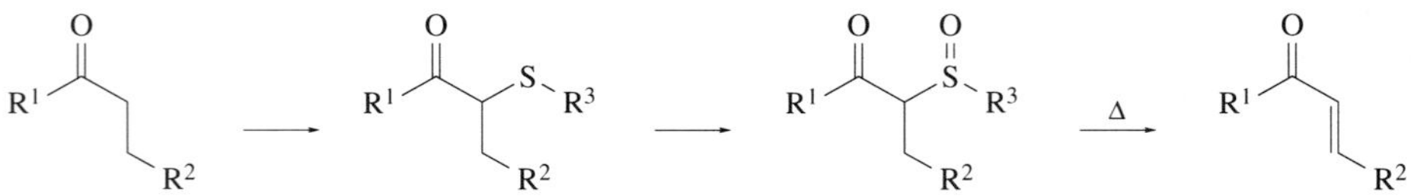

Scheme 12

The mildness of this method of introduction of unsaturation is evidenced by the wide variety of functional groups which can tolerate this sequence ⟨76JA4887⟩. For example, the tricyclic α-phenylsulfinyl ketone (**1**) eliminates readily to form the cyclopentenone (**2**) in over 90% yield (Equation (17)).

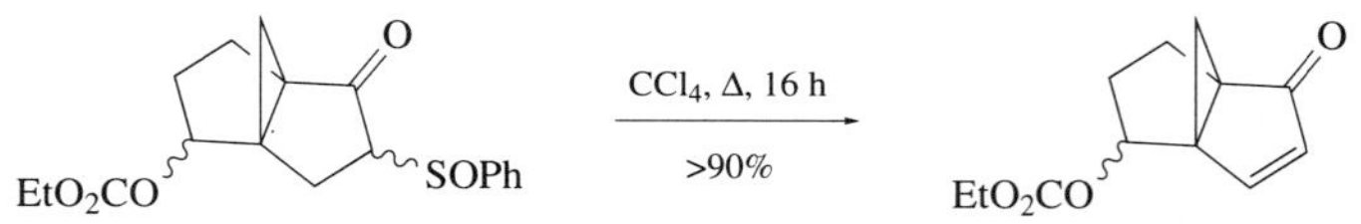

(**1**) (**2**)

(17)

α,β-Unsaturated ketones, esters, lactones and lactams have been synthesized by this approach ⟨76JA3022, 76JA4887, 78ACR453, 86JA3385⟩; some examples are shown in Table 1. The stereochemical preference in the elimination process is illustrated in entry 2—formation of the methyl ester of the queen substance (**4**) with exclusively (*E*) stereochemistry resulted from pyrolysis of the methyl sulfoxide (**3**) at 110 °C ⟨78ACR453⟩. Entries 5 and 6 illustrate that this methodology may also be extended to the synthesis of α,β-unsaturated tosylhydrazones ⟨87JOC1218⟩ and oxazolines ⟨87TL5509⟩. Use of the pyrazinylsulfinyl group as a leaving group for sulfoxide elimination has been applied to the synthesis of α,β-unsaturated nitriles (Equation (18)) ⟨88H(27)261⟩.

Table 1 Sulfoxide elimination to form α,β-unsaturated ketones, esters, lactones, lactams and related compounds.

Entry	*Sulfoxide*	*Alkene*	*Yield* (%)	*Ref.*
1	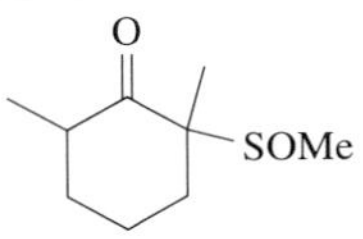	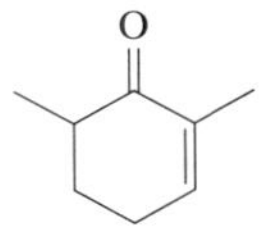	84	76JA4887
2	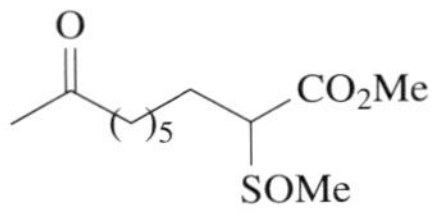 (**3**)	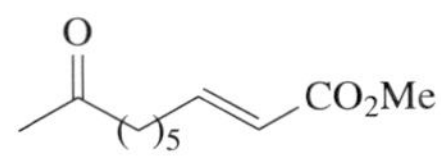 (**4**)	>86	78ACR453
3	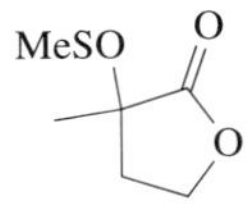	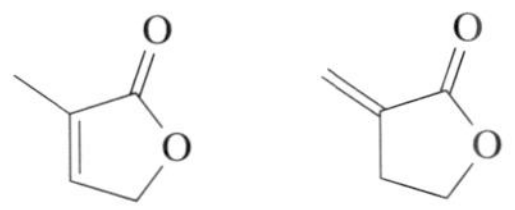 87 : 13	95	76JA4887
4	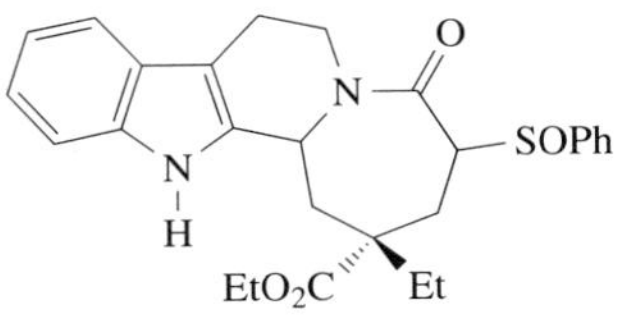	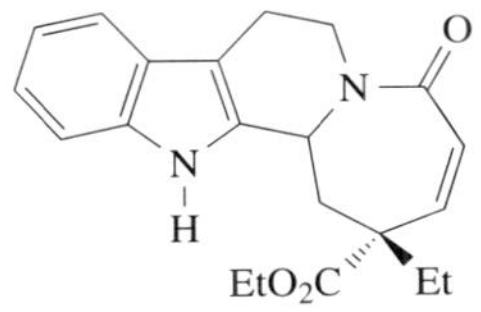	86 from sulfide	76JA3022
5	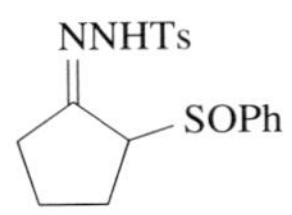	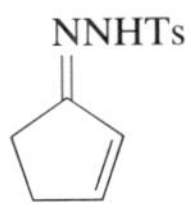	78	87JOC1218
6	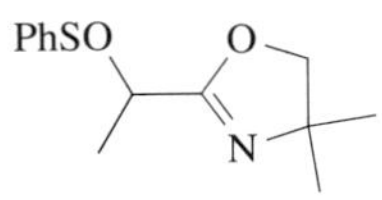	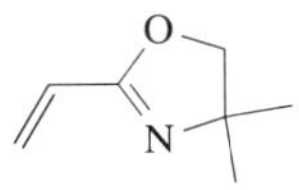	88	87TL5509

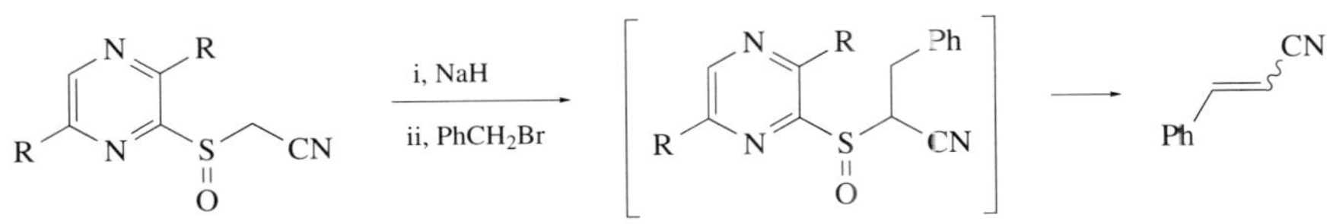

(18)

R = Et, Pr^i

The temperatures required for the elimination reactions are substantially lower for the compounds with adjacent carbonyl groups than for simple alkyl sulfoxides ⟨76JA4887⟩, and depend strongly on the nature of the second substituent on the sulfoxide—aryl sulfoxides may be eliminated at temperatures of 25–80 °C, whereas temperatures of 110–130 °C are necessary with alkyl sulfoxides. The relative facility of elimination in the carbonyl containing compounds can be explained by the effect of dipole–dipole interactions in the ground-state conformation of the sulfoxide and the conjugative stabilization of the newly introduced double bond ⟨76JA4887⟩. In cases where the higher temperature associated with the alkyl sulfoxide elimination is damaging, use of phenyl sulfoxide can overcome the problem. For example, when the methyl sulfoxide (**5a**) was employed, a retro-Diels–Alder reaction resulted in formation of ethyl benzoate, but the phenyl sulfoxide (**5b**) eliminated smoothly to form the α,β-unsaturated ester (**6**) (Scheme 13) ⟨73JA6840, 76JA4887⟩.

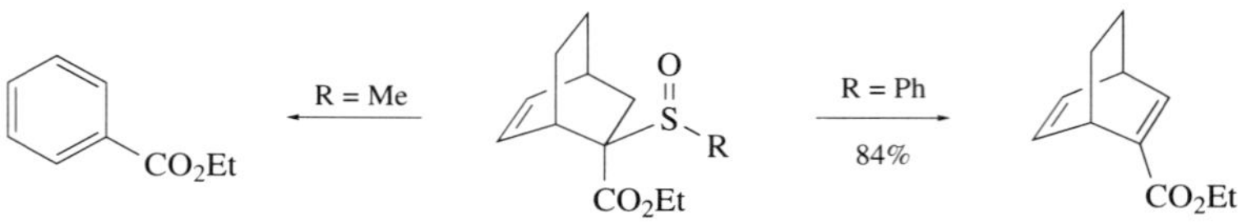

(**5**) **a**; R = Me
b; R = Ph

(**6**)

Scheme 13

Trost and Parquette introduced the use of methyl 2-pyridinesulfinate as a reagent for the sulfinylation–desulfinylation of ketones and esters to form their α,β-unsaturated derivatives (Scheme 14) ⟨93JOC1579⟩. Using potassium hydride as a base, sulfinylation of ketones proceeds efficiently at room temperature; thermolysis then leads directly to the enones. Ketones are selectively targeted in compounds containing both keto and ester groups. Addition of metal salts or dihydropyran increases the efficiency of the elimination process. This procedure clearly has the advantage of obviating the necessity for an oxidation step.

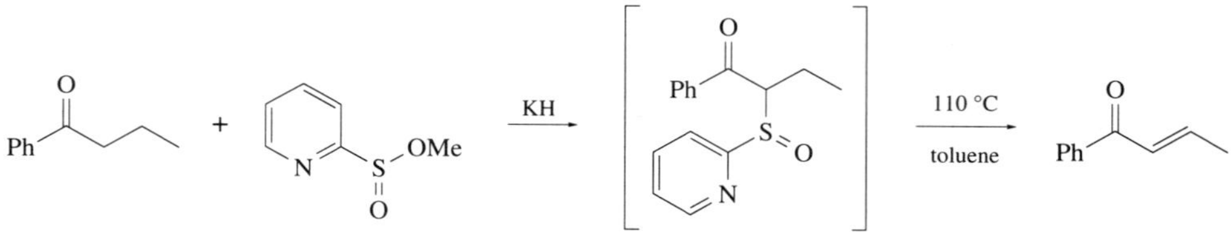

Scheme 14

In certain cases use of trapping agents (such as calcium carbonate, trimethyl phosphite and 2-thiolbenzothiazole) for the sulfenic acid produced in the thermolysis can be advantageous—either in simplifying the purification procedure or in improving the reaction efficiency. For example, the efficient conversion of oestrone methyl ether to its unsaturated derivative required the use of trimethyl phosphite as a sulfenic acid trap for efficient reaction ⟨78CRV363⟩.

The regiochemical preference for sulfoxide elimination in acyclic systems is generally $C{=}CCH_2 \sim C{\equiv}CCH_2 > ArCH_2 \sim CH_3 \gg C{-}CH_2 \ggg C{-}H$ ⟨78CRV363⟩. In the acyclic α-sulfinyl ester (**7**) exclusive elimination towards the methyl group was observed following the expected order of ease of H abstraction. Closure to form the *exo*-methylene lactone indicates the synthetic utility of this regiospecific elimination (Scheme 15) ⟨75TL4197⟩.

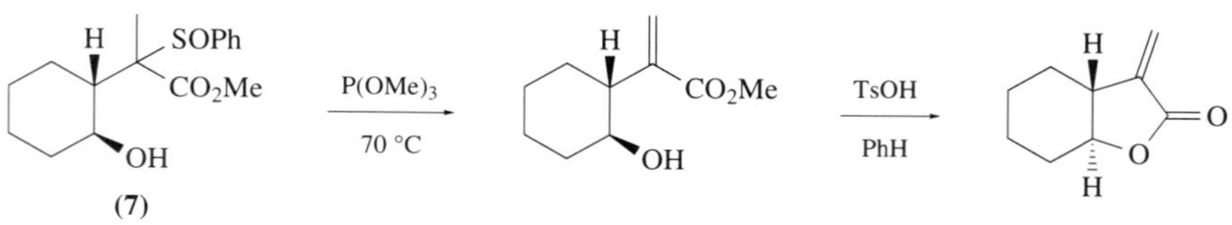

(**7**)

Scheme 15

In cyclic systems, provided the elimination occurs in a *cis–syn* fashion, *endo*-cyclic alkene formation is preferred. Thus internal elimination was observed in the keto sulfoxide (**8**) to form carvone (Equation (19)) ⟨76JA5017, 77JA8116⟩ (see also entries 1 and 3 in Table 1). However due to geometric constraints only *exo* cyclic elimination was observed for (**9**) (Equation (20)) while its C(3) epimer generated the *endo*-cyclic product exclusively ⟨74TL1097, 78CRV363⟩. Similarly, *exo* elimination is preferred in a cyclic β-keto sulphoxide (*exo* : *endo* = 5 : 1) while in the corresponding hydroxysulfoxide, *endo* elimination predominated ⟨92T33⟩. Therefore both regiochemical and stereochemical preferences are reversed on reducing the ketone to the corresponding alcohol, indicating the influence of conformational factors on the course of the thermolytic elimination process. In large-ring cyclic ketones, elimination to form the (*E*) alkene is observed ⟨78CRV363, 92TL7181⟩.

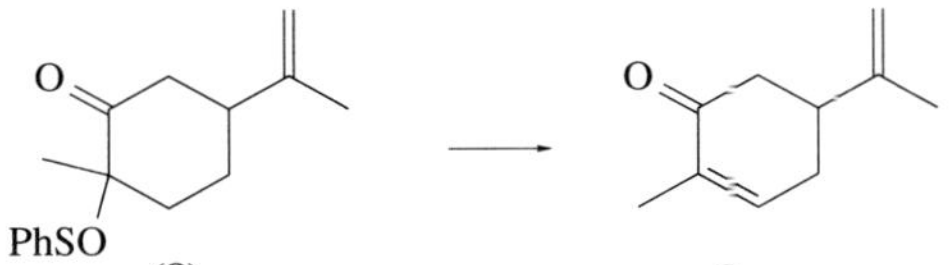

(8) Carvone (19)

(9) (20)

As the presence of a sulfur substituent α to a carbonyl group enhances considerably the acidity of the adjacent α-hydrogen atom, regiospecific alkylation of a carbonyl compound followed by sulfoxide elimination provides a useful route to alkylated α,β-unsaturated derivatives, an approach which has been employed widely ⟨74JA7165, 78CRV363⟩, and which complements the Wittig reaction and related processes. The preparation of an α,β-unsaturated ester (Equation (21)) illustrates this process. This transformation has also been conducted under phase-transfer conditions ⟨87SC1839⟩.

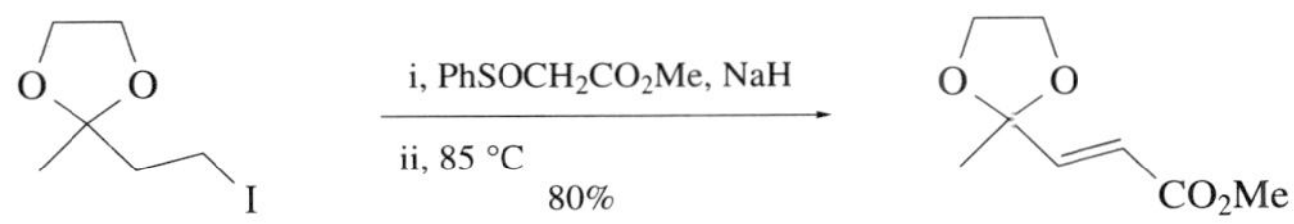

(21)

Following this alkylation–elimination procedure, disubstituted alkenes are usually obtained with the (*E*) configuration, while trisubstituted alkenes are formed as a mixture of (*E*) and (*Z*) isomers. This overall sequence was employed in Oppolzer's synthesis of acorenone (Equation (22)) ⟨77HCA2388⟩. An extension of this alkylative elimination sequence to sulfoxides bearing activating groups other than carbonyl groups, e.g., aryl, sulfenyl, sulfinyl and nitrile groups, was also reported ⟨75JOC2014⟩.

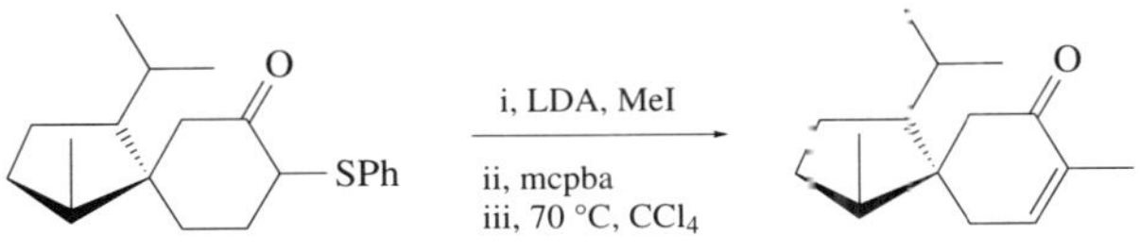

Acorenone (22)

Paterson and Fleming extended this route to α,β-unsaturated carbonyl compounds to include elimination of β-sulfinyl substituted compounds as illustrated in Equation (23). Application of this methodology to the synthesis of α-methylene lactones and esters ⟨79TL993⟩, ketones ⟨79TL995⟩ and α-alkylidene ketones, aldehydes, esters and lactones ⟨79TL2179⟩ was demonstrated. Similarly Crich and Lim have observed facile elimination in a β-pyridinesulfinyl ester ⟨87JCR(S)353⟩. An example of *syn*-elimination of a β-benzenesulfinyl group to form a spirocyclic butenolide (**10**) is illustrated in Equation (24) ⟨92JCS(P1)2303⟩, see also ⟨76SC357⟩.

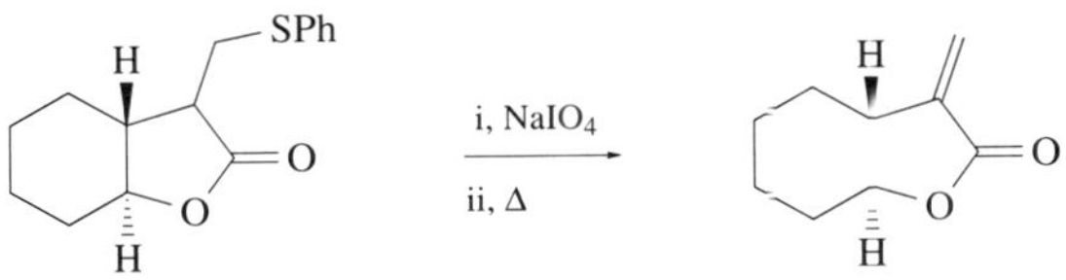

(23)

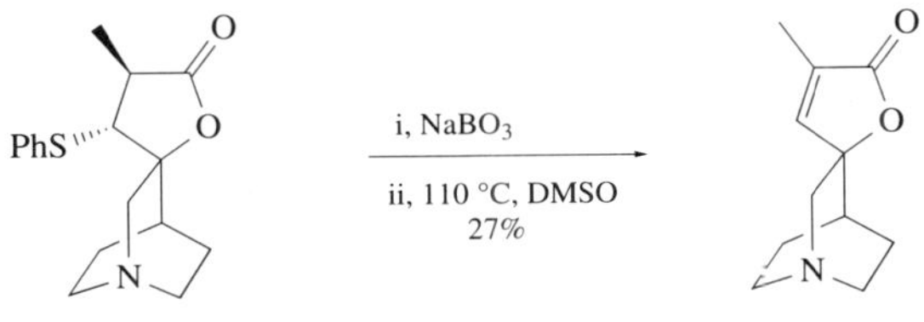

(**10**) (24)

While sulfoxide elimination is usually brought about thermally, the corresponding base-catalysed process, believed to occur via an E2 mechanism, has been observed when aliphatic sulfoxides ⟨64JA1561⟩ were treated with potassium *t*-butoxide in DMSO.

Formation of alkenes from *β*-hydroxysulfoxides on treatment with NBS, NCS, SO_2Cl_2 ⟨73JA3420⟩ or *o*-phenylene phosphorochloridite ⟨72TL649⟩ has also been reported. Durst and co-workers suggested that *β*-sultines are formed as intermediates in this process followed by facile loss of SO_2 (Scheme 16) ⟨73JA3420⟩. Stereoselective elimination of *β*-bromosulfoxides to form alkenes occurs on treatment with tributyltin hydride ⟨78JA3874⟩. Regiospecific alkene formation by silicon-directed sulfoxide elimination has been achieved by thermal elimination of sulfenic acid from *β*-trimethylsilyl sulfoxides followed by protodesilylation (Scheme 17) ⟨82CC281⟩ (see Section 1.14.5.1.3).

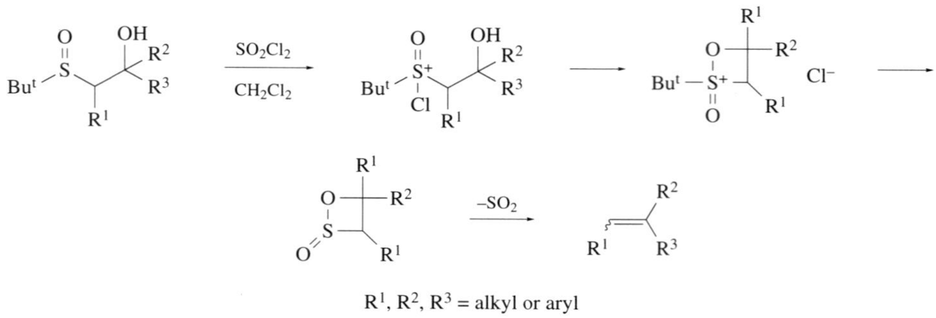

Scheme 16

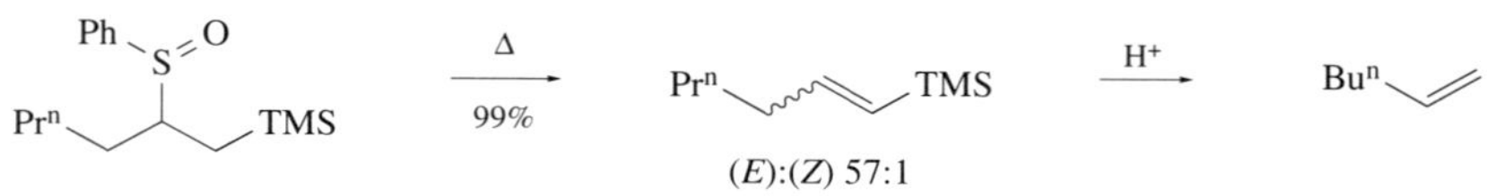

Scheme 17

Thermolysis of allylic sulfoxides to form dienes has been reported (Equation (25)) ⟨82JOC4801⟩. However allylic rearrangement of the sulfoxides prior to elimination may occur leading to isomeric dienes ⟨81TL4137⟩. Formation of 1-nitro-1,4-cyclohexadienes (which are readily converted to aromatic nitro compounds) by sulfoxide elimination has been reported ⟨88JOC251⟩. Sulfoxide elimination to form allenes, in good chemical yields but in low enantiomeric purities, is illustrated in Scheme 18 ⟨92TL4985⟩. This is believed to proceed via a stable *σ*-copper species. Very little allene formation is observed from the diastereomeric sulfoxide mesylate, indicating the strict steric and/or electronic requirements for this process.

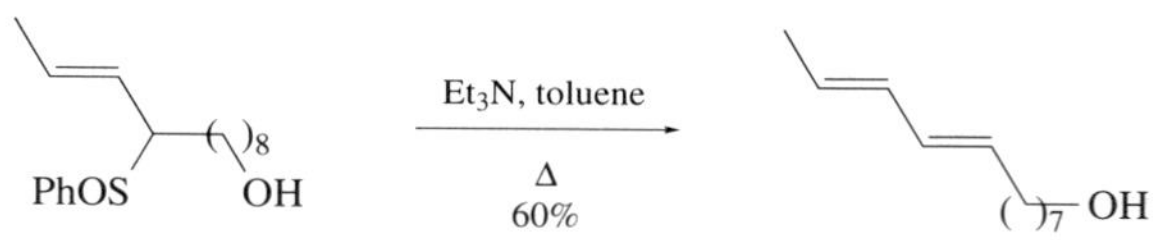

(25)

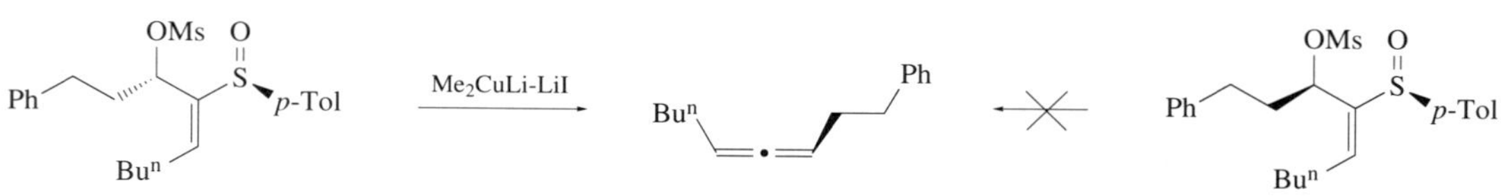

Scheme 18

1.14.2.2.2 *Elimination of selenoxide groups*

Selenoxide *syn*-elimination is one of the most useful methods for introducing a C=C bond, largely because of the extremely mild conditions required to effect elimination of selenenic acids from selenoxides bearing a hydrogen atom at the β-position, allowing preparation of sensitive alkenes such as α-methylene derivatives of lactones, even β-lactones ⟨78T1049, 93JOC322⟩. Following the publication of isolated examples of the reaction (Equation (26)) ⟨67MI 114-02, 70CC86, 71JOC2561⟩, the enormous synthetic potential of this method was recognized only in 1973 when Sharpless *et al.* ⟨73JA2697, 73JA6137, 73TL1979, 75CS9⟩ and Reich *et al.* ⟨73JA5813⟩ independently demonstrated its utility in the synthesis of alkenes (Equation (27)). Selenoxide eliminations have been widely used in synthesis since this time and have been discussed in many reviews and books ⟨78T1049, 79ACR22, B-84MI 114-01, B-86MI 114-01, B-86MI 114-02, B-86MI 114-03, B-87MI 114-01, B-87MI 114-02, B-87MI 114-03, 91COS(6)1011⟩.

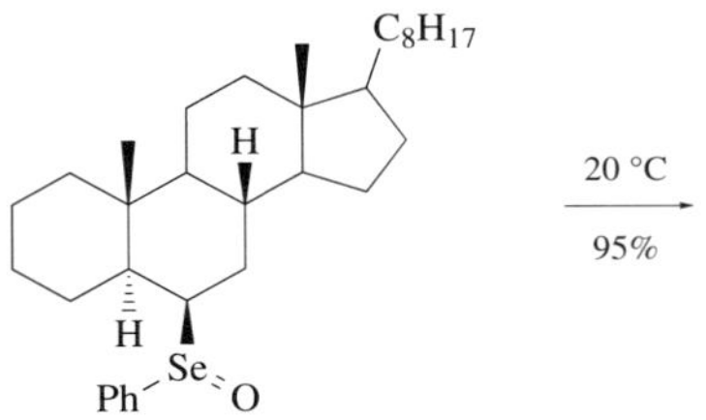

(26)

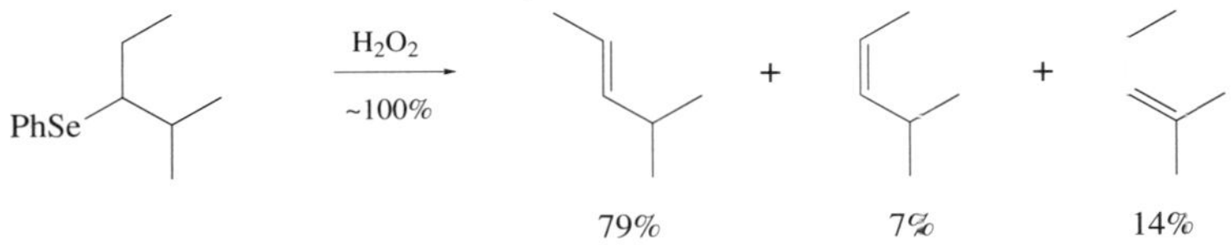

(27)

While the selenoxide *syn* elimination is very similar mechanistically to the analogous sulfoxide elimination process, requiring all five participating centres to be coplanar as shown in Scheme 19 ⟨73TL1979⟩, the rate of elimination from selenoxides is much faster than from sulfoxides; in fact these eliminations often occur below room temperature, approximately 100 °C lower than the corresponding sulfoxide elimination process ⟨74JOC120, 74TL1097, 76JA4887, 78JOC1697, 78T1049⟩.

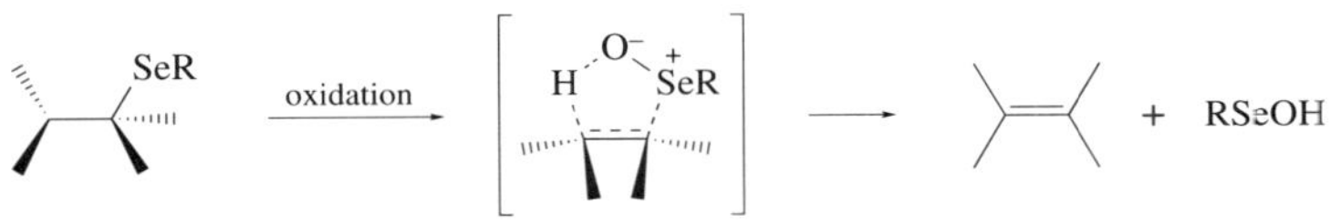

Scheme 19

Many aryl selenoxides containing a suitably positioned hydrogen atom are unstable and fragment spontaneously at room temperature ⟨78T1049, 79ACR22⟩. Frequently the selenide is oxidized to the selenoxide which is not isolated but eliminates directly *in situ* to form the alkene on warming to room temperature or in some cases on addition to a refluxing solvent such as carbon tetrachloride or hexane ⟨76JOC1396, 78JOC1697⟩ at which temperature the elimination is essentially spontaneous. The oxidation to the selenoxide can be achieved with a range of diverse reagents ⟨78T1049, B-87MI 114-02⟩, e.g., hydrogen peroxide ⟨73TL1979⟩, ozone ⟨75JA5434⟩, sodium periodate ⟨73CC695⟩, mcpba and other peracids ⟨75JA3250⟩, *t*-butyl hydroperoxide ⟨78JOC1689⟩ and sodium perborate ⟨84JCS(P1)1593⟩. Use of *t*-butyl hydroperoxide has been reported to minimize side reactions which may occur in the elimination process ⟨78JOC1689⟩ while the presence of alumina may be beneficial ⟨78TL1141⟩ also. Side reactions which may occur during the selenoxide elimination include epoxidation of the double bond ⟨78JOC1697⟩, addition of the selenenic acid or one of its disproportionation products to the newly formed double bond ⟨76JOC2503, 78JOC1689, 78JOC1697⟩, isomerization of double bonds already present in the compound ⟨81T263⟩ and reduction of the selenoxide to selenide by the selenium(II) by-products ⟨76JOC2503⟩. It is recommended that the elimination is conducted in the presence of an amine, e.g., triethylamine ⟨76JOC2503, 78JOC1697⟩ as this removes the selenenic acid as it is formed, thereby reducing the possibility of side products. Water decreases the rate of the elimination process, via hydrate formation; therefore it is recommended that anhydrous conditions are employed for efficient reaction ⟨73TL1979⟩.

The synthesis of colnelic acid (Equation (28)) ⟨87TL4917⟩ by Corey *et al.* employed elimination of a methyl selenoxide, formed by oxidation of the selenide with 2-benzenesulfonyl-3-phenyloxaziridine, to give an equimolar mixture of the (*E*) and (*Z*) isomers. Alkyl selenoxides eliminate more slowly than aryl selenoxides ⟨78TL1141⟩. The presence of electron-withdrawing substituents on the aryl ring further increases the rate of the reaction ⟨75JA1597, 75JOC947, 78JOC1697⟩; *o*-

nitrophenylseleno groups have been found to be particularly useful. The 2-pyridylseleno group has been demonstrated to result in extremely efficient elimination ⟨80TL5037, 85JCS(P1)373, 85T4347⟩. While the advantage of the phenylseleno group is the ready availability and ease of introduction of this group, in cases where the elimination proves difficult use of the more reactive groups can be advantageous (Scheme 20) ⟨75JA5434, 84JOC3796⟩.

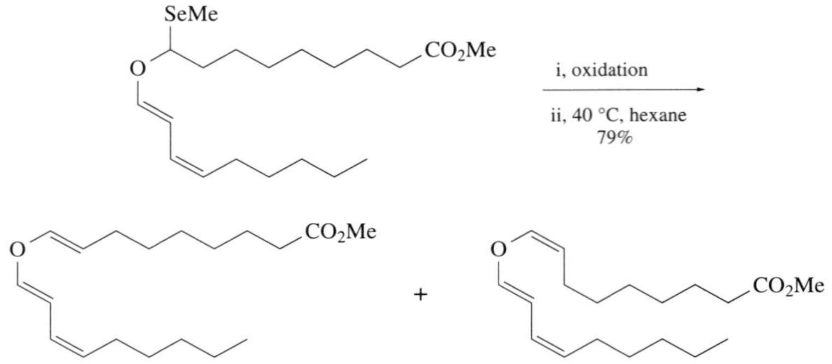

(28)

(*E*):(*Z*) 1:1

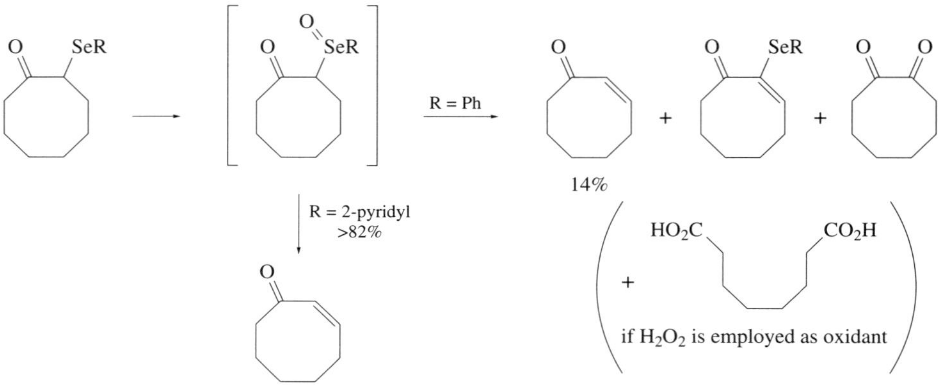

Scheme 20

In general, the same regiochemical preferences are observed as for sulfoxides, i.e. C=C—CH ~ C≡C—CH > $PhCH_2$ ~ CH_3 > CH_2 > CH ⟨73JA5813, 73TL1979, 75JA3250⟩. Therefore when elimination in more than one direction is possible, elimination towards the less-substituted position is generally observed, see for example Equation (27), i.e., β-alkyl substitution slows the elimination process, except when formation of a conjugated alkene is possible, e.g., β-phenyl substituents enhance the rate of the elimination ⟨78JOC1697⟩. In cyclic systems elimination to form an *endo* alkene is preferred over elimination to produce the *exo* product ⟨73JA5813, 75JA5434, 75JOC542⟩, provided the transition state required for the *syn* elimination is conformationally accessible. Thus the selenoxide in Equation (29) eliminates in an *endo* fashion to form the butenolide even though *exo* elimination would result in greater conjugation in the product. However in the selenide (**11a**) elimination occurs on oxidation to form selectively the more stable *endo* alkene while the epimeric compound (**11b**) produces the *exo* product only ⟨74JOC120⟩; presumably the transition state required for *endo* elimination is not readily attained (Equations (30a) and (30b)). Interestingly, in a synthesis of the cyclohexanediol fragment of nonaromatic β-milbemycins ⟨87CC880⟩, (Equation (31)), exocyclic elimination was preferred in a cyclohexanol derivative, while in the corresponding cyclohexanone, endocyclic elimination predominated.

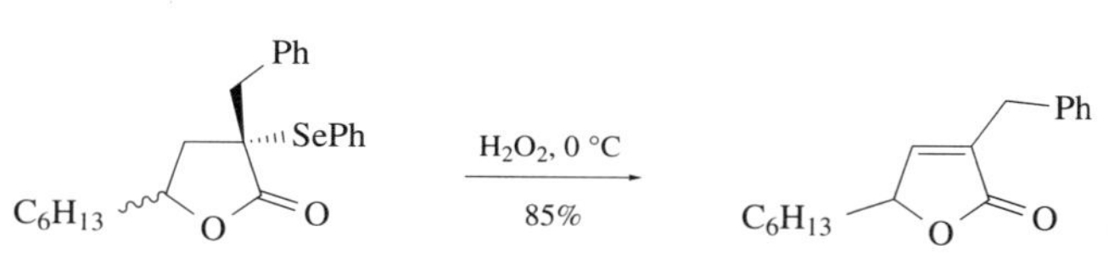

(29)

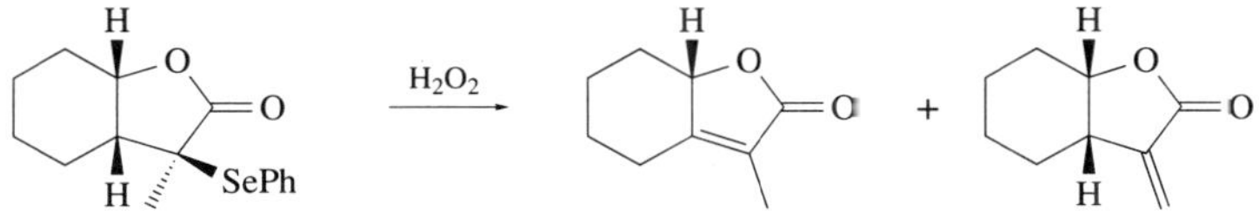

(30a)

(11a) *endo* 9:1 *exo*

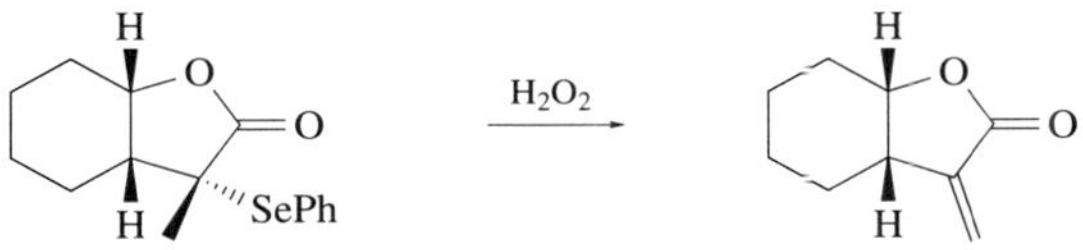

(30b)

(11b) *exo* only

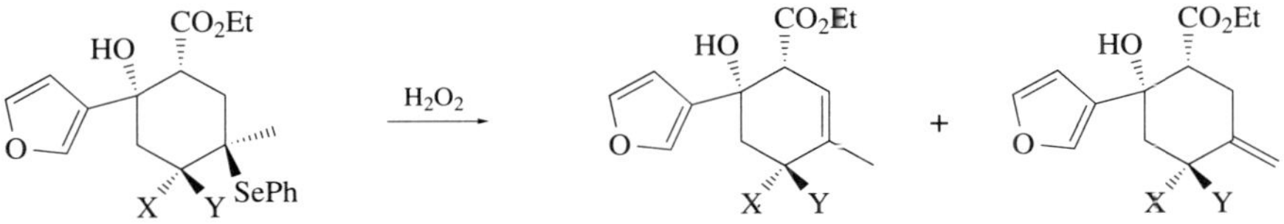

(31)

X, Y = O	85	:	15
X = H, Y = OH	10	:	90

When the selenoxide is substituted by a heteroatom at the β-position this may dramatically affect the regiochemical preference for the elimination process ⟨B-87MI 114-02⟩. In general hydroxy, alkoxy or acetoxy groups direct elimination away from the substituent (Equation (32)) ⟨74JOC429⟩. Nitrogen substitution in the form of amino ⟨75JOC3313, 85JA3891⟩, amido ⟨81JOC4727⟩ or azido ⟨79TL2697⟩ groups also directs elimination to the allylic position while by contrast, isothiocyanate directs elimination towards the formation of the vinylic product ⟨83JOC5246⟩ (Equation (33)). In contrast to the behaviour with oxygen substituents elimination towards a β-thiobenzoyloxy group to form the vinylthiobenzoate is favoured over formation of the allyl derivative (Equation (34)) ⟨88JCS(P1)575⟩. The effect of halogen substituents is less predictable and often both regioisomers are observed ⟨81JOC1846, 81JOC3721⟩. Groups which have an acidifying effect on an adjacent hydrogen atom (e.g., carbonyl ⟨73TL1979⟩, cyano ⟨82CC871⟩, sulfone ⟨83JOC4986⟩, sulfoxide ⟨81JA3486⟩, nitro ⟨82TL4733⟩) usually accelerate the rate of elimination towards the substituent, frequently regiospecifically. Elimination in secondary selenoxides bearing a tributylstannyl group in the 3-position has been investigated ⟨91JCS(P1)429⟩; when elimination towards a methyl group is possible then homoallylic stannanes are produced selectively, otherwise a mixture of allylic and homoallylic stannanes is formed (Equations (35a) and (35b)). Thus the tributyltin substituent has little effect on the regioselectivity of the selenoxide elimination.

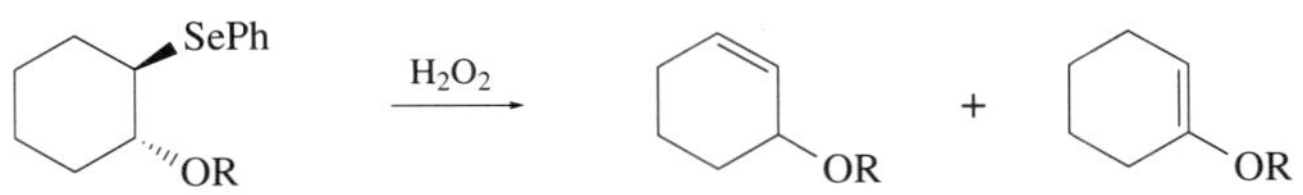

(32)

R = H, Ac, alkyl >97 : <3

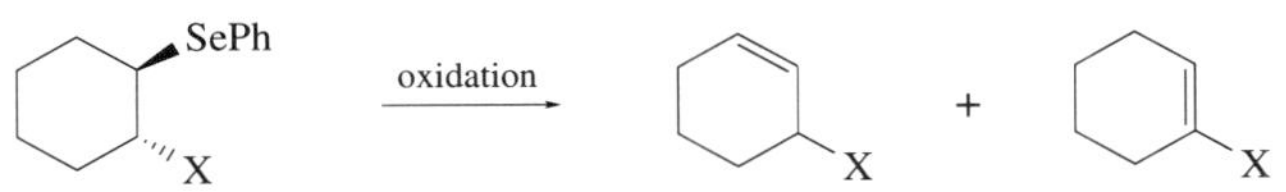

(33)

X = NHAc	100	:	0
$X = N_3$	60	:	40
X = NCS	17	:	83

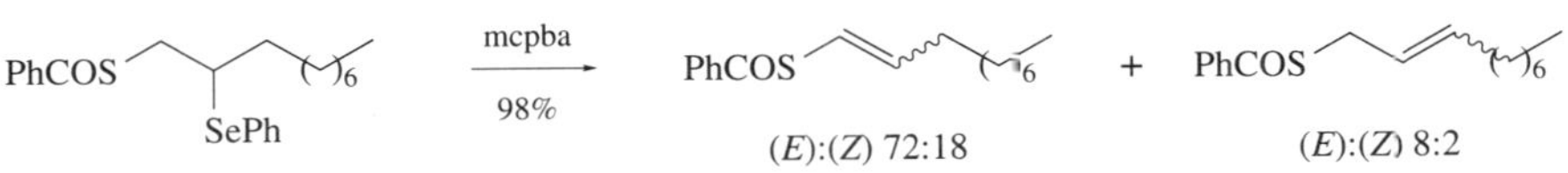

(34)

(*E*):(*Z*) 72:18 (*E*):(*Z*) 8:2

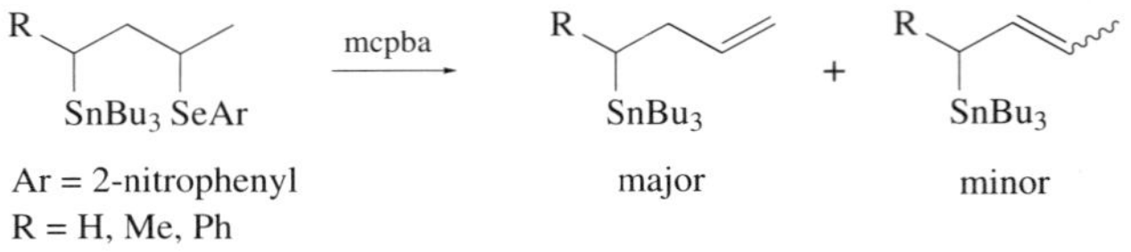

(35a)

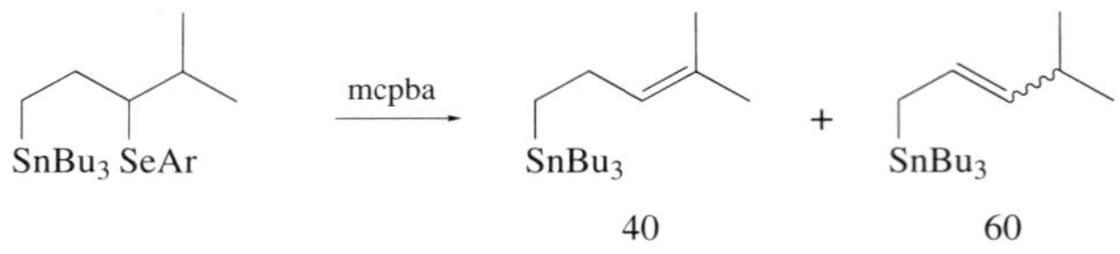

(35b)

The stereochemical outcome of the elimination process is again similar to that observed in sulfoxide eliminations: in acyclic systems disubstituted alkenes are generally formed preferentially with the (*E*) configuration, as for example, in the Sharpless synthesis of the methyl ester of a naturally occurring pheromone (Equation (36)) ⟨73JA6137⟩, while trisubstituted alkenes are formed as a mixture of (*E*) and (*Z*) isomers ⟨73JA5813, 75JA5434⟩.

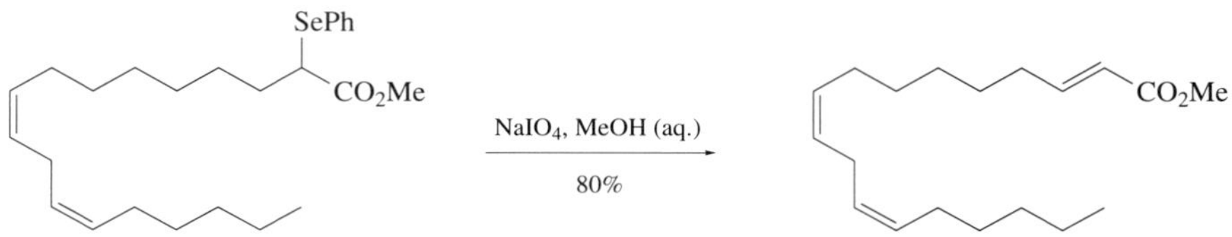

(36)

Synthesis of terminal alkenes by selenoxide elimination has been widely employed in synthesis ⟨B-84MI 114-01⟩, for example in a preparation of vernolepin (Equation (37)) ⟨76JA1612⟩. Elimination to form terminal alkenes is best achieved by using the *o*-$NO_2C_6H_4Se$ group ⟨75JOC947⟩ or the 2-pyridylseleno group ⟨82TL2105⟩ which eliminate more readily than phenylseleno on oxidation to the corresponding selenoxide ⟨75JA1597⟩. Addition of the eliminated selenenic acid to terminal double bonds can be a major side reaction leading to β-hydroxyselenides. However use of an amine to trap the selenenic acid as it is formed or an excess of the oxidizing agent to oxidize the Se(II) species to seleninic acid can reduce the extent of this side reaction considerably (Equation (38)) ⟨78JOC1697⟩. The synthesis of selenides as precursors of terminal alkenes has been conducted by nucleophilic displacement reactions using ArSeNa (Scheme 21) ⟨78CC518⟩. Grieco developed a particularly useful route for direct transformation of terminal alcohols to selenides by treatment of the former compounds with *o*-nitrophenylselenocyanate in the presence of tributylphosphine ⟨80JOC3537⟩. This route has the added advantage that selective reaction at primary alcohols occurs under these conditions; Equation (39) illustrates the application of this selectivity in Grieco's synthesis of estradiol ⟨80JOC2247⟩. Replacement of the *o*-nitrophenylselenocyanate in this sequence by *N*-phenylselenophthalimide (N-PSP) is possible ⟨79JA3704⟩.

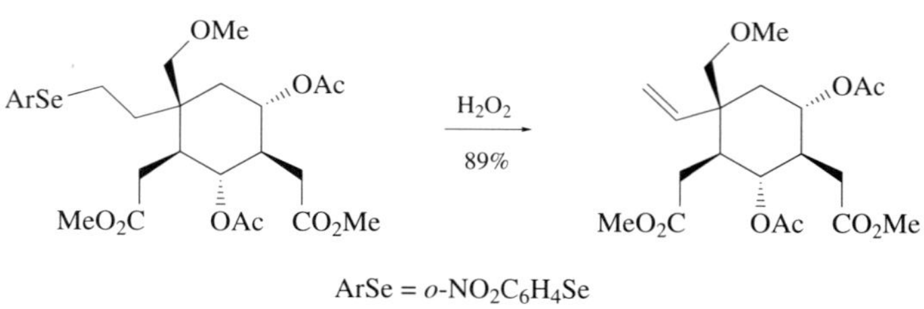

(37)

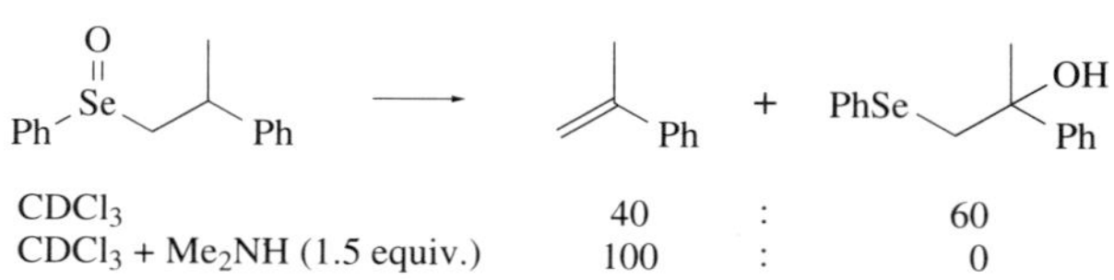

(38)

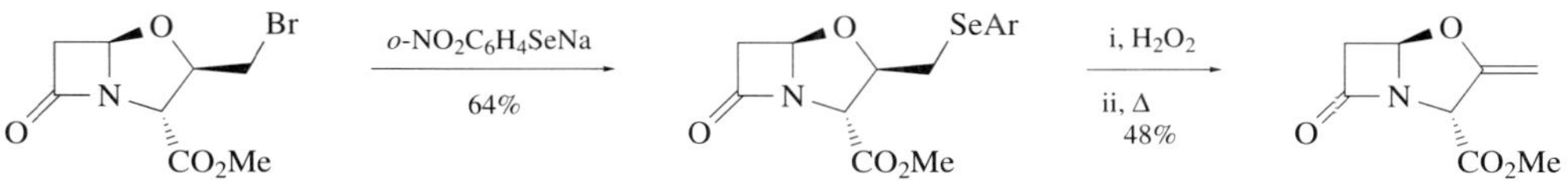

Scheme 21

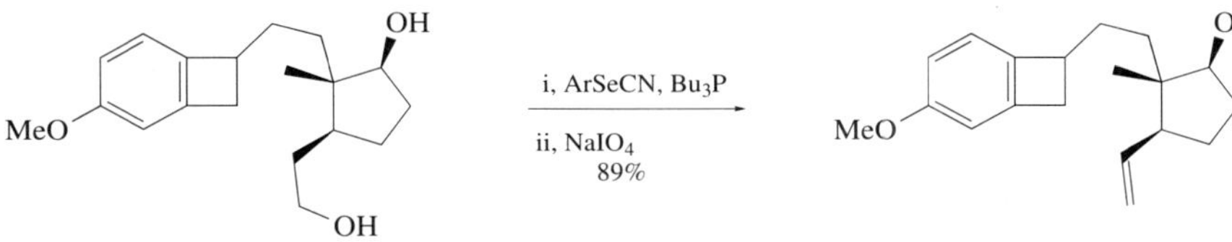

(39)

Elimination of vinylic selenoxides to form alkynes or allenes has been studied by Reich and Willis ⟨80JA5967⟩ and Back and co-workers ⟨89JOC4146⟩. As shown in Equation (40), when a *cis*-hydrogen atom is present elimination to form the alkyne is preferred, if not then elimination to form the allene predominates. Acetylenic ⟨81TL5111⟩ and allenic ⟨83JOC4776, 87TL1737⟩ sulfones have been prepared by this approach.

(40)

$R^1 = H$, $R^2 = Pr^i$	85%	15%
$R^1 = Pr^i$, $R^2 = H$	<3%	81%

Formation of α,β-unsaturated carbonyl compounds by selenoxide elimination, a powerful synthetic method first developed by Sharpless *et al.* ⟨73JA6137⟩ and Reich *et al.* ⟨73JA5813, 75JA5434⟩, is now the most widely used application of organoselenium chemistry. Scheme 22 illustrates this general method—the enol or enolate of a carbonyl compound is treated with an electrophilic selenium reagent to introduce the α-seleno substituent, which is then oxidized to the selenoxide which undergoes *syn* elimination. This mild, efficient and regioselective method of introducing α,β-unsaturation has been applied to aldehydes, ketones, esters, lactones, amides, lactams and β-dicarbonyl compounds. Equations (41)–(43) display examples of these transformations.

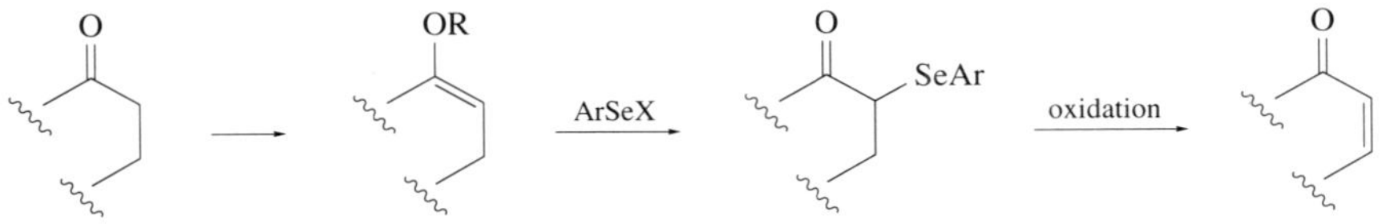

X = Br, Cl, SePh, etc.
R = H, metal, SiR_3, Me

Scheme 22

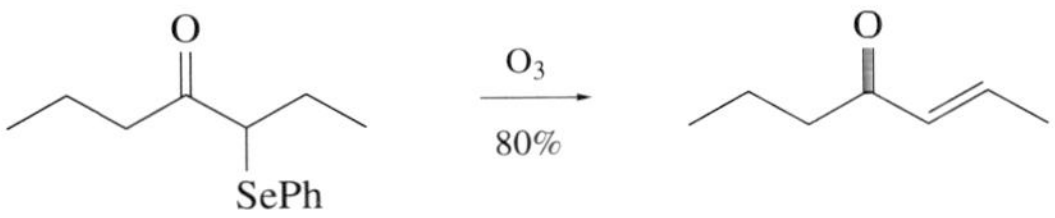

(41)

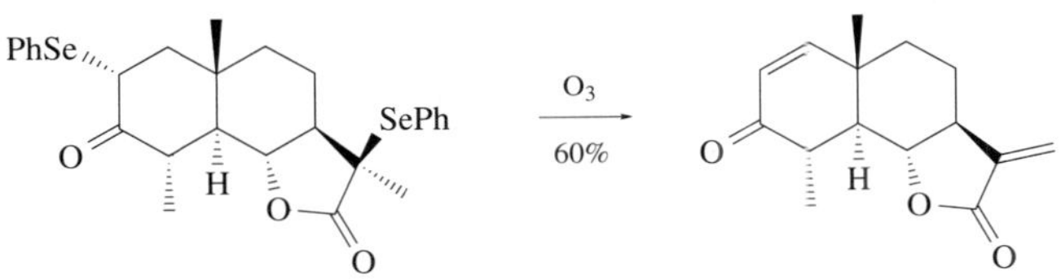

(42)

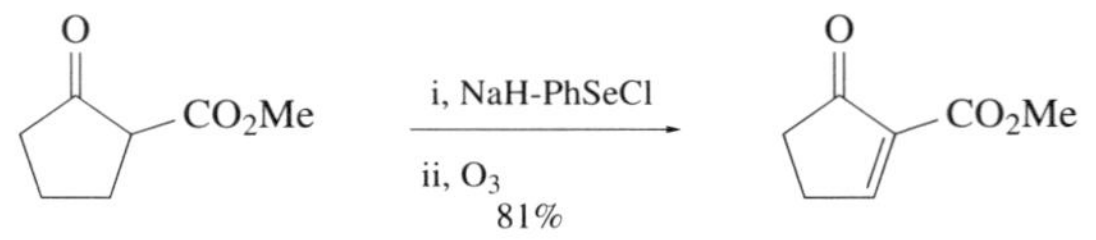

(43)

With aldehydes and ketones introduction of the α-seleno group can be achieved simply by treatment with phenylselenenyl chloride in ethyl acetate; oxidation and elimination may then be carried out directly to form the unsaturated products (Scheme 23) ⟨73JA6137⟩. Reich developed an alternative procedure ⟨75JA5434⟩ in which the enolate is formed by deprotonation, usually with LDA, to form the kinetic enolate followed by addition of PhSeCl or PhSeBr (Scheme 24). For acyclic aldehydes and ketones elimination occurs to form the (*E*)-enones preferentially (Equation (41)) ⟨73JA6137, 75JA5434⟩. In cyclic systems, the stereochemistry of the elimination is generally controlled by ring size and by conformational effects ⟨73JA6137⟩. Clearly the direction of the selenenylation of unsymmetrical ketones is determined by the regiochemistry of the enolization; therefore the alkene, resulting from formation of the kinetic enolate, is introduced on the less substituted side of the ketone ⟨75JA5434⟩. An application of this methodology in the synthesis of a naturally occurring cyclopentenone is shown in Equation (44) ⟨92JCS(P1)2683⟩.

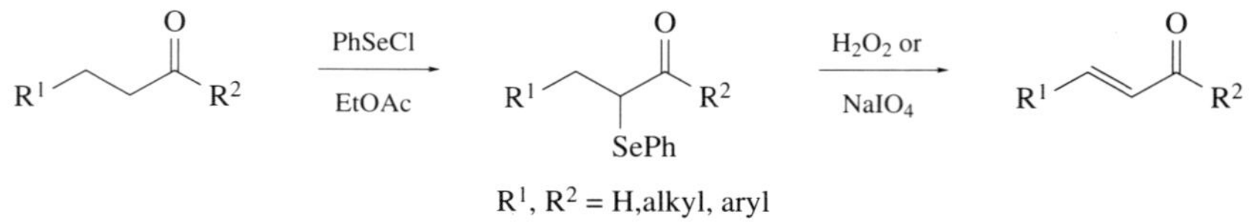

Scheme 23

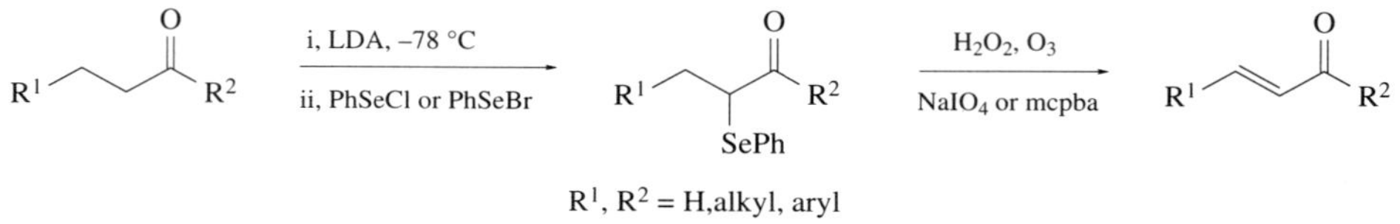

Scheme 24

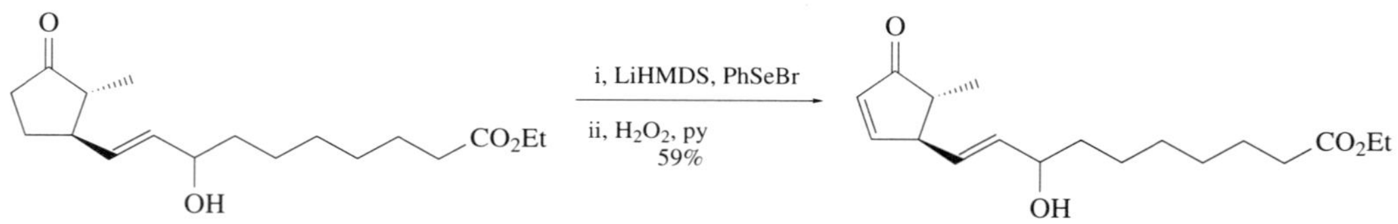

(44)

If selenenylation of unsymmetrical ketones with the opposite regiochemistry to that obtained from the kinetic enolate is required, an alternative possibility is to react the enol acetates or silyl enol ethers (which can be formed on the thermodynamically favoured side of the ketone) with electrophilic selenium reagents such as PhSeCl or PhSeBr or $PhSeOCOCF_3$ (Scheme 25 ⟨74JOC2133, 75JA5434⟩ and Scheme 26 ⟨84CC1345⟩). The methodology involving silyl enol ethers was employed in the synthesis of hydrindan derivatives ⟨92JOC3173⟩.

Scheme 25

Selenenylation of enones usually occurs at the α′-position producing α,β,α′,β′-unsaturated dienones, as shown in Equation (45) ⟨75JA5434⟩, or if this position is blocked substitution occurs at the γ-position producing the α,β,γ,δ-unsaturated ketone following oxidation and elimination (Equation (46)) ⟨81JA1501⟩.

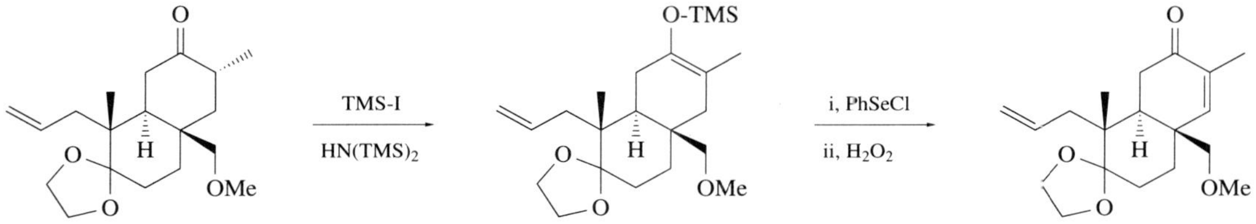

Scheme 26

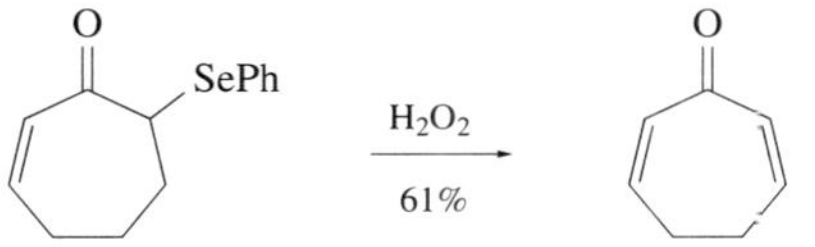

(45)

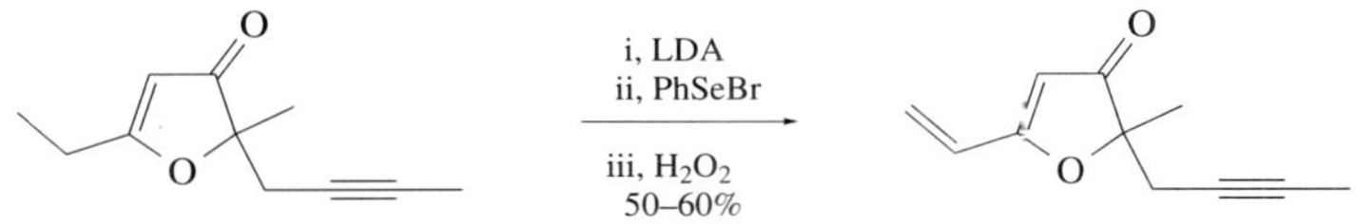

(46)

In some cases the yields of cyclic enones obtained may be low—presumably because the conformational preference of the cyclic ketone conflicts with the transition state required for the *syn* elimination thereby allowing alternative reaction pathways to be followed. For example when 2-phenylselenocyclooctane is treated with hydrogen peroxide a mixture of products is formed (Scheme 20) ⟨75JA5434⟩. In addition to the product of the *syn* elimination, Pummerer-type reaction or formation of vinyl selenides is observed. In such cases, the yield of the enone can be improved by using the more reactive 2-pyridylseleno group in place of the phenylseleno group.

With esters and lactones generation of the enolate under basic conditions is required prior to treatment with PhSeCl or PhSeBr or PhSeSePh followed by oxidation and elimination to form the α,β-unsaturated ester or lactone ⟨73JA6137⟩. Formation of the (E)-α,β-unsaturated ester is favoured as illustrated in Equation (36). With lactones, formation of the endocyclic alkene is usually preferred as shown in Equation (29) ⟨75JOC542⟩, as observed also with sulfoxides. However Grieco has used *exo* elimination of selenoxides to form α-methylene lactones extensively in synthesis by ensuring that the seleno group and the β-hydrogen atom are *trans* oriented thereby directing the *syn* elimination to occur in an *exo* fashion (Equations (30a) and (30b)) ⟨74JOC120⟩. Equation (42) demonstrates the use of this methodology in the synthesis of tuberiferine, where a double elimination is achieved to introduce both the α,β-unsaturated cyclohexenone and the α-methylene lactone functionalities in a single step ⟨76CC582⟩.

Relatively few examples of the formation of α,β-unsaturated amides and lactams are known. In the selenenylation of the lactam shown in Scheme 27 ⟨77JHC681⟩, two equivalents of the base were required to prevent formation of diselenated products. Similarly for efficient selenenylation of nitriles two equivalents of base are required ⟨75JOC2165⟩.

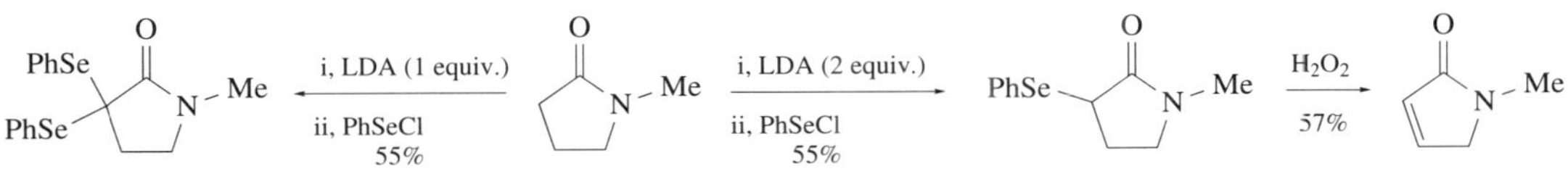

Scheme 27

Selenenylation of numerous β-dicarbonyl compounds followed by transformation to their unsaturated derivatives has been undertaken ⟨B-87MI 114-02⟩. One example of the synthesis of an unsaturated β-keto ester is shown in Equation (43) ⟨75JA5434⟩. As the acidity of β-dicarbonyl compounds is increased compared to the simpler carbonyl compounds, use of milder bases for the introduction of the seleno group is possible. However sodium hydride is commonly employed for this transformation.

Elimination of selenoxides derived from β-seleno ketones to form α,β-unsaturated ketones is also possible (Equation (47)) ⟨73TL1979, 81TL1809⟩. The observed regiospecificity of the elimination may be due to the formation of the conjugated alkene. Similarly selenoxide elimination to form $\alpha,\beta,\gamma,\delta$-

unsaturated ketones is possible as shown in Scheme 28; the β-seleno aldehyde is essentially acting as a crotonaldehyde equivalent ⟨86TL2949⟩.

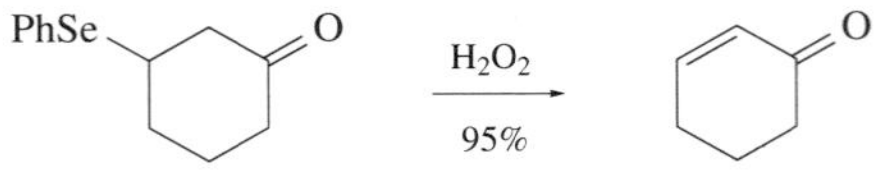

(47)

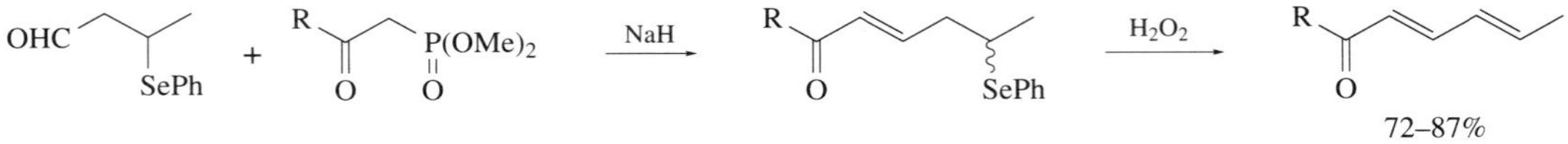

Scheme 28

Selenoxide elimination has also been employed in the synthesis of α,β-unsaturated nitro compounds ⟨81CC261⟩, nitriles ⟨74TL2279, 75JOC2165, 77JA5210⟩, sulfones ⟨80JOC1486⟩, sulfoxides ⟨75JA5434⟩, phosphonates ⟨81TL3097⟩ and phosphine oxides ⟨79CB355⟩ (Equations (48) and (49)).

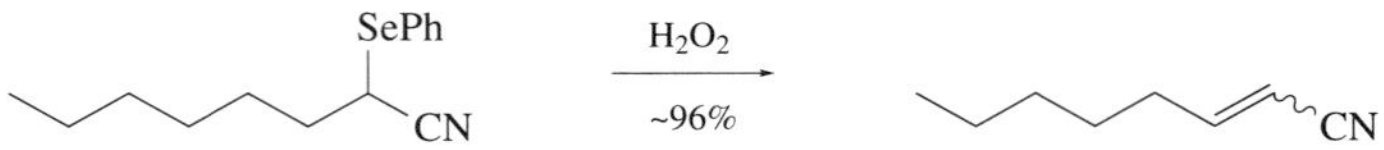

(E):(Z) 54:46

(48)

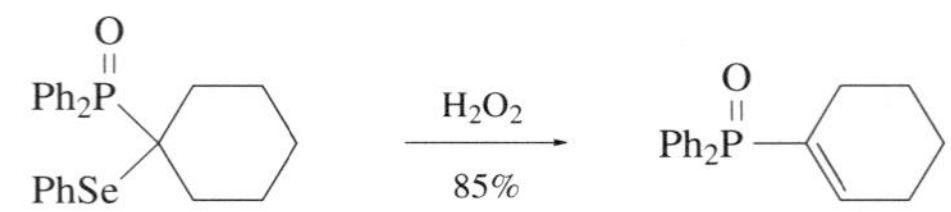

(49)

An interesting extension of the *syn* elimination process is in the functionalization of alkenes via selenium reagents; treatment of alkenes with electrophilic selenium reagents, e.g., PhSeBr, $PhSeO_2CCF_3$, gives the *trans*-substituted selenide ⟨74CC100, 74JOC428, 74JOC429, 78JOC1689⟩ which may then be oxidized and eliminated ⟨74JOC429⟩ to form an allylic substituted product or a vinylic substituted product depending on the nature of the substituent (Scheme 29; see also Equations (32) and (33)). Ring opening of epoxides to give β-hydroxyselenides, which usually proceeds with a high degree of stereoselectivity of selenide attack at the less-substituted carbon atom, followed by oxidation and elimination, furnishes allylic alcohols (Scheme 30) ⟨73JA2697⟩. This protocol has been widely applied in synthesis (Equation (50)).

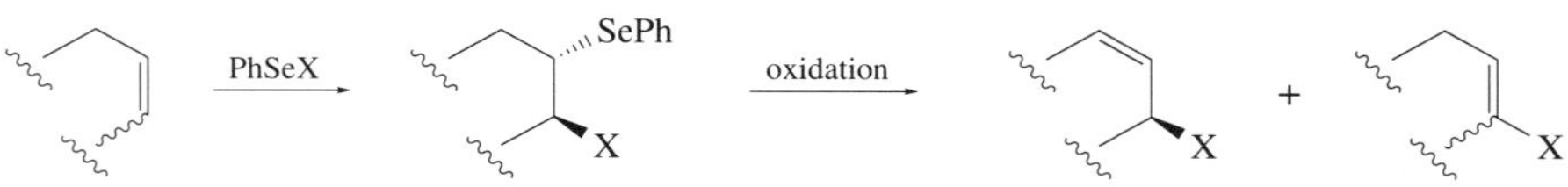

X = e.g. OH, OR, OAc, Cl, Br, NHCOMe

Scheme 29

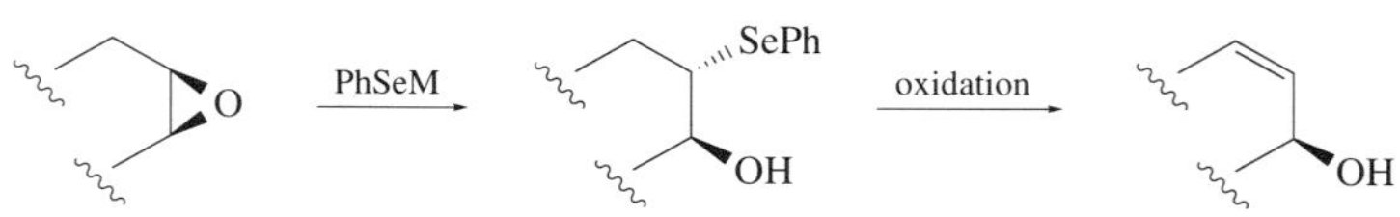

Scheme 30

(50)

R = H, Ac, alkyl

Allylic selenoxides rearrange readily to form allylic alcohols via a [2,3]-sigmatropic rearrangement ⟨72JA7154, 73JA2697, 73TL1979⟩. Therefore, while allylic sulfoxides eliminate to form 1,3-dienes

⟨74ACR147, 82JA7051⟩ in the case of selenoxides the equilibrium lies much more towards the selenenate ester which is hydrolysed to the allylic alcohol as illustrated in Scheme 31. This has been exploited by Reich for the preparation of allylic alcohols (Equation (51)) ⟨75JOC2570⟩ and by Clive *et al.* for the 1,3-transposition of primary allylic alcohols (Scheme 32) ⟨78CC770⟩. Back has reported [2,3] sigmatropic rearrangement of a dienyl selenoxide to produce an allenic alcohol (Scheme 33); use of an enol ether (in addition to an amine) as a scavenger for the selenenic acid formed was found to be necessary ⟨92TL4725⟩.

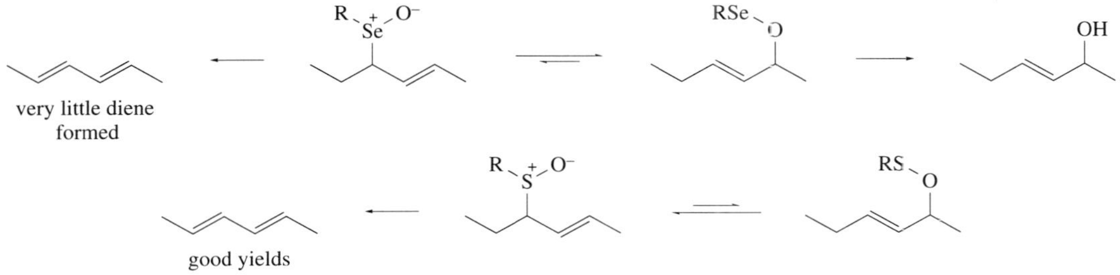

Scheme 31

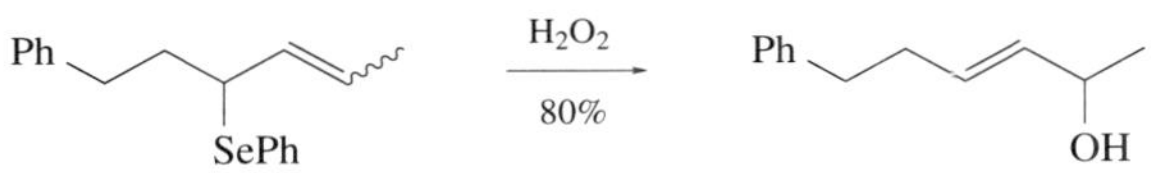

(51)

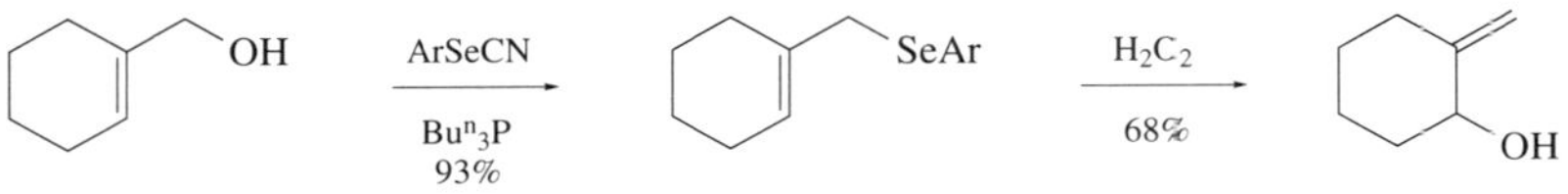

Scheme 32

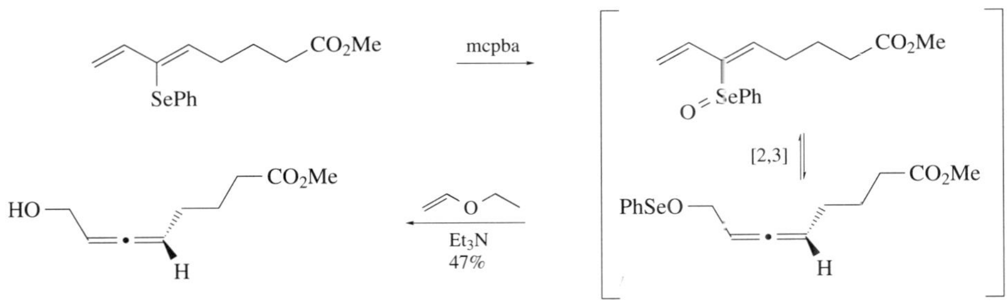

Scheme 33

Asymmetric selenoxide elimination to form chiral allenes has been reported, albeit with moderate enantioselectivity (Scheme 34) ⟨92CC46⟩.

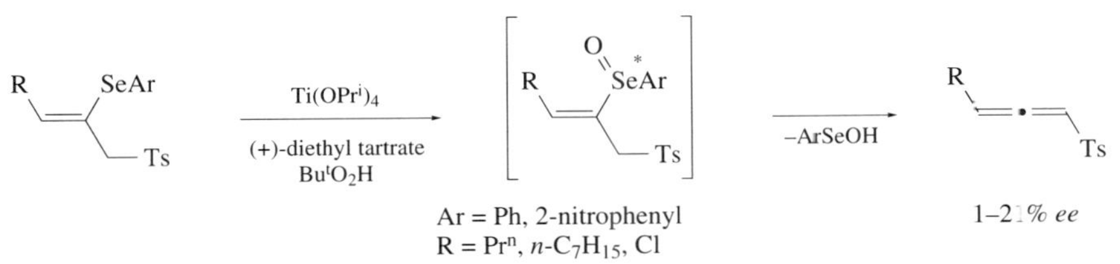

Scheme 34

Direct introduction of a selenoxide group by use of benzeneseleninyl chloride was briefly investigated as shown in Equation (52) ⟨75JA5434⟩ but proved less convenient than the two-step (selenenylation followed by oxidation) procedure. A variation of the selenoxide elimination reaction using

$PhSeCl_3$ for the introduction of the selenium substituent has been reported (Scheme 35) ⟨87CS515⟩. Formation of a similar intermediate (bearing a $PhSeCl_2$ group) by chlorination of *β*-chloro selenides with sulfuryl chloride followed by hydrolysis and selenoxide elimination has been applied to the synthesis of chloroalkenes ⟨87TL1463⟩.

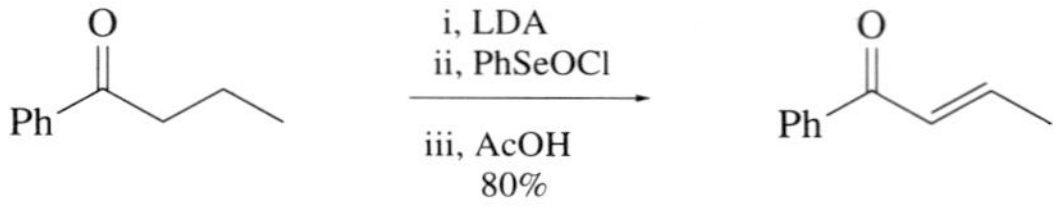

(52)

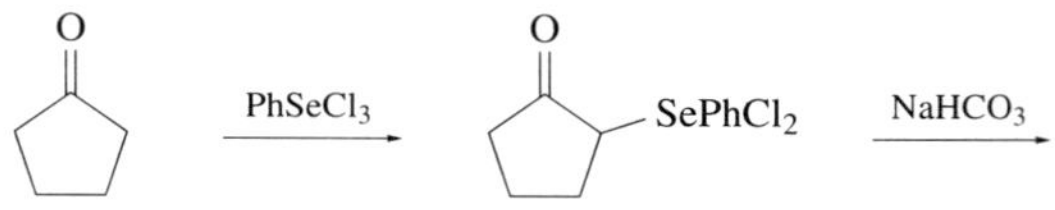

Scheme 35

Carbanions may be generated adjacent to selenoxides by treatment with a base such as LDA, provided the selenoxide cannot undergo *syn* elimination, e.g., methyl phenyl selenoxide, or the selenoxides are handled below 0 °C to prevent the selenoxide fragmentation ⟨75JA3250⟩. The resulting carbanions react with alkyl halides, aldehydes and ketones and may subsequently undergo elimination to form an alkene. Scheme 36 gives an illustrative example of this alkylative elimination process.

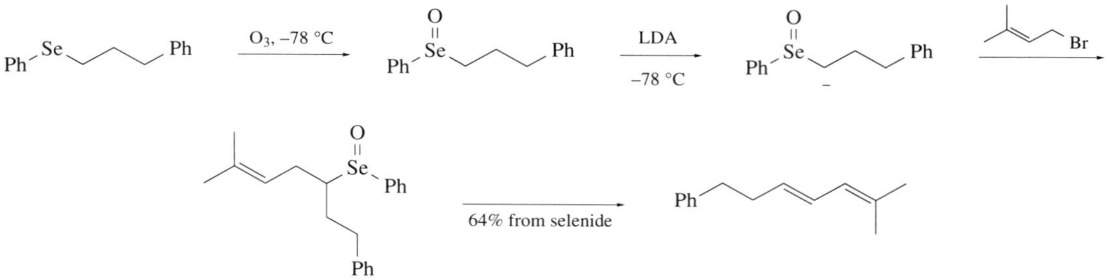

Scheme 36

Selenoxide elimination has also been employed for the conversion of enones into *β*-alkyl enones ⟨74JOC2133, 75JA5434⟩; the copper enolate resulting from a conjugate addition to an enone can be trapped with PhSeBr or PhSeSePh (frequently a mixture of the two is employed) to form the selenide which can subsequently be oxidized to effect elimination to form the *β*-substituted enone (Scheme 37).

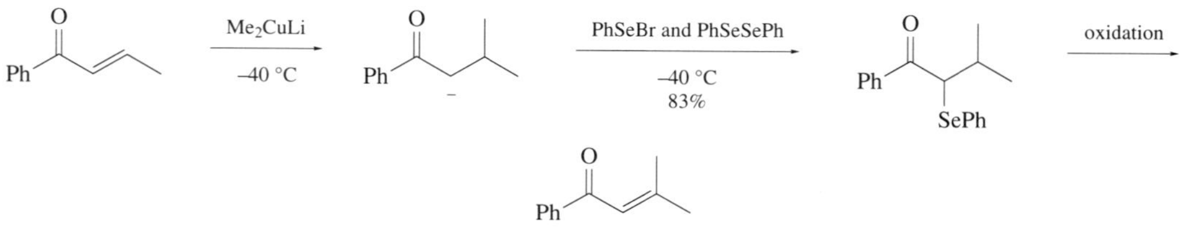

Scheme 37

1.14.2.2.3 Elimination of telluroxide groups

Syn elimination in telluroxides to form alkenes has also been observed ⟨75CS9⟩; however this process is much less useful than the corresponding selenoxide or sulfoxide elimination processes and therefore has attracted much less attention. Equation (53) shows that, as for selenoxides and sulfoxides, elimination towards the less-substituted position is preferred, with greater regioselectivity observed for the telluroxide elimination than that observed with selenoxides and sulfoxides ⟨83JA2748⟩. Also when a *β*-oxygen- ⟨82TL1177, 83JOM(250)203, 83TL4347, 85TL895⟩ or amido- ⟨89JCS(P1)1775⟩ substituent is present, elimination occurs away from the substituent to form the allylic product as illustrated in Equation (54). Telluroxide elimination also exhibits the same stereochemical preference as selenoxide elimination—formation of the (*E*) isomer of the alkene is preferred; however, unlike selenoxides, the stereoselection is sensitive to the amount of oxidant present and when an excess of oxidant is employed formation of the (*Z*) isomer is also observed ⟨85CC1037⟩. Interestingly, the cyclohexyl telluroxide derivative undergoes elimination only at elevated temperatures (200–240 °C); conformational factors may be responsible for decreasing the rate of the elimination ⟨83JA2748⟩. Similarly, the telluroxide shown in Equation (55) eliminated only on refluxing in aqueous THF containing sodium hydroxide ⟨89JOC4391⟩. Elimination to form terminal alkenes has been achieved in good yields from 4-methoxyphenyltelluroxides ⟨81CC277⟩, while good yields of alkenes are obtained on treatment of secondary alkylphenyltellurides with mcpba ⟨86S1⟩. Oxidative elimination of a *β*-phenyltelluro group to form an *α*,*β*-unsaturated ketone is also possible ⟨89CL607⟩.

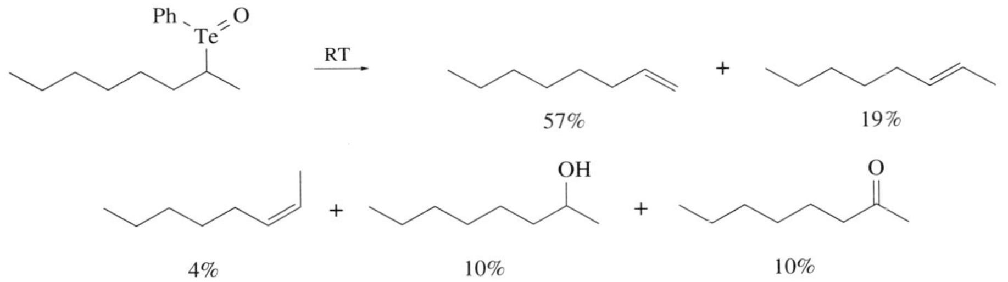

(53)

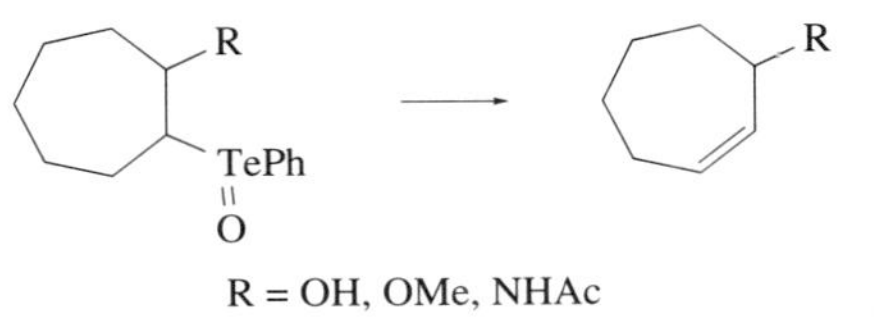

(54)

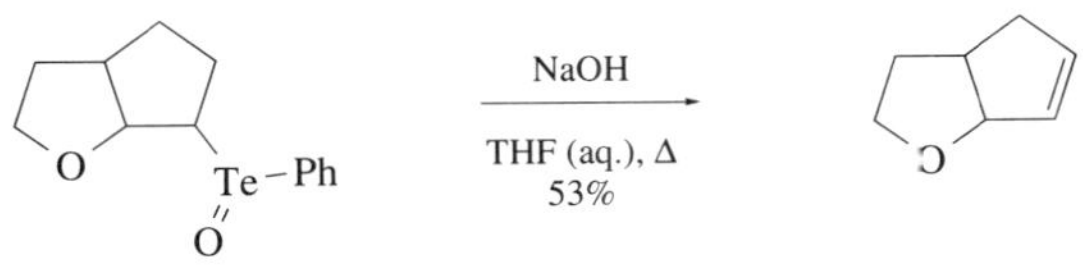

(55)

The telluroxides may be formed either by oxidation of the corresponding tellurides ⟨80JOC274⟩ with, for example, mcpba ⟨85TL895⟩ or *t*-butyl hydroperoxide ⟨75CS9⟩, or by treatment of the appropriate tellurium dibromide with aqueous base ⟨81CC277, 90CE261⟩. In a related process, tellurides may be oxidized with chloroamine T ⟨81CL447, 83S1006, 89JOC4391⟩ to form alkenes in good yields.

In comparison with selenoxide elimination, telluroxide fragmentation frequently requires much higher temperatures (ca. 200 °C) ⟨83JA2748⟩ making this a much less useful reaction synthetically. Possible reasons which have been suggested for this decreased reactivity of telluroxides include the longer bond lengths which may result in a greater distance between the oxygen atom and the *β*-hydrogen atom and the fact that telluroxides form stable hydrates. Furthermore, control of the level of oxidation of tellurium ⟨81CC277⟩ is much more difficult than for selenium—the tellurenic acid produced by the elimination readily reduces the telluroxide to telluride, lowering the reaction yield considerably, and also the telluroxides are easily overoxidized to tellurones, thereby limiting the oxidants which may be employed to form the telluroxide. Adduct formation with excess oxidant, e.g., mcpba, has been proposed ⟨85TL895⟩. Formation of alcohols via a 1,2-shift may compete with the telluroxide elimination process ⟨75CS9⟩.

The formation of an allylic alcohol in the oxidation of an allylic telluride (Scheme 38) is believed to indicate that a [2,3] sigmatropic rearrangement, similar to that observed in allylic selenoxides, occurs also in allylic telluroxides ⟨85TL921, 91S897⟩.

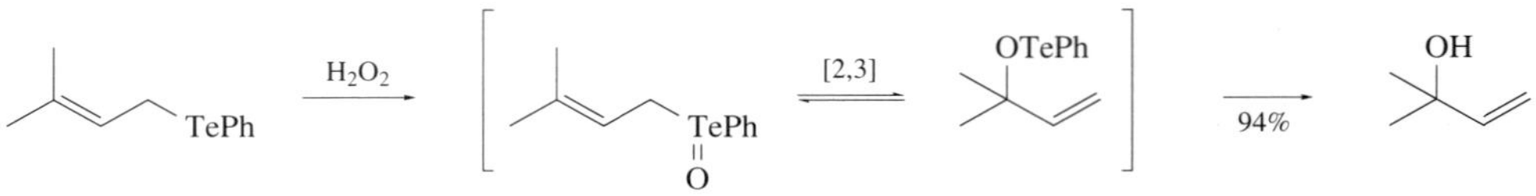

Scheme 38

1.14.2.3 By Elimination of Sulfone, Selenone or Tellurone Groups

1.14.2.3.1 Elimination of sulfone groups

Alkene formation by sulfone elimination has been discussed in detail by Simpkins ⟨B-93MI 114-01⟩. Three different elimination processes can be identified—base-catalysed elimination, elimination of β-substituted sulfones (including the Julia reaction) and the Ramberg–Bäcklund reaction. Of these, the reactions of β-substituted sulfones are by far the most widely used in synthesis.

(i) Elimination of sulfones

Fenton and Ingold first reported elimination of sulfinic acids from sulfones on treatment with sodium ethoxide ⟨30JCS705⟩; later work in this area demonstrated that potassium *t*-butoxide in DMSO is a particularly efficient method of effecting sulfone elimination ⟨63JA2739, 64JA1561, 71JOC1898⟩. In simple sulfones, regioselective Hofmann-type elimination occurs to produce the less-substituted alkene (Equation (56)). An E2 or E1cB mechanism is believed to operate (especially if the hydrogen atom is being removed from an activated position) ⟨75CC940, 87MI 114-01⟩. Benzenesulfonyl groups are widely employed but other groups have also been used, for example the dichloromethylsulfonyl group ⟨89ZOR654⟩.

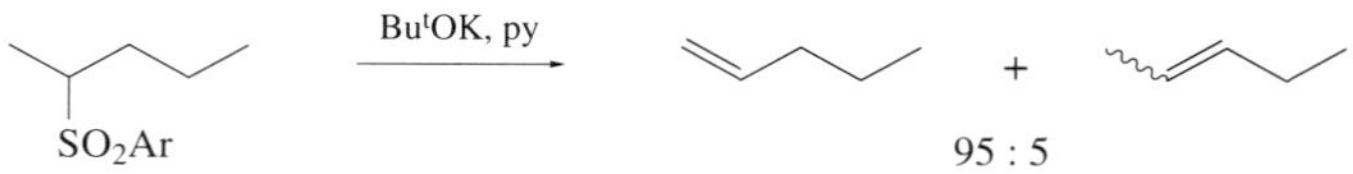

(56)

Eliminations of activated sulfones, e.g., allylic and homoallylic sulfones, have been employed in synthesis. For example, Julia's synthesis of retinoic acids utilizes Bu^{t}OK in THF to produce the (*E*)-alkene unit required in the heart of the polyene (Scheme 39) ⟨73BSF746⟩. Sulfones stabilize α-carbanions particularly efficiently; this enables facile alkylation of the sulfone prior to base-induced elimination which allows considerable versatility in the alkene synthesis. The elimination reaction displays selectivity for the (*E*)-alkene; thus when the (*E*)-isomer of the homoallylic sulfone shown in Equation (57) was treated with base (*E*,*E*)-2,4-undecadiene was produced as the major product ⟨86T4807⟩.

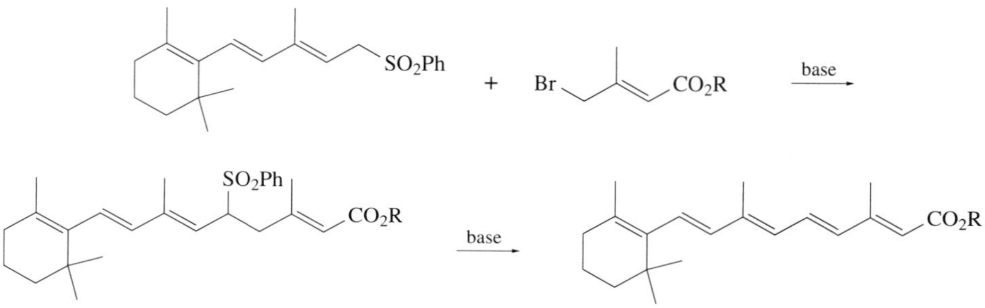

Scheme 39

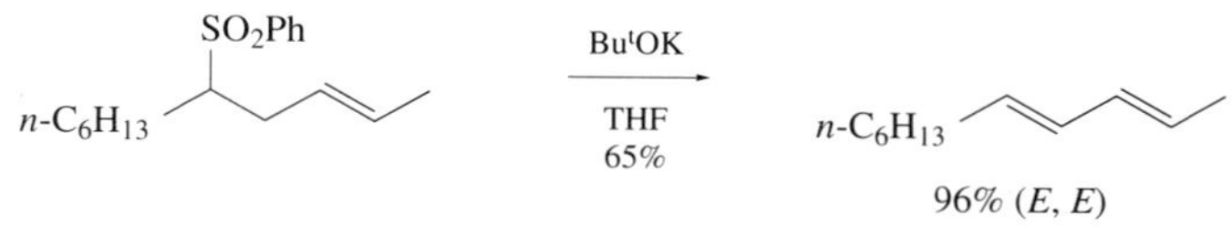

(57)

Elimination in β-sulfonyl carbonyl compounds to form α,β-unsaturated derivatives is relatively facile (Equation (58)) ⟨82CL165, 88TL2059⟩ and occurs under mild conditions, for example with 1,5-diazabicyclo[5.4.0]undec-5-ene (dbu), basic alumina ⟨86TL3733, 93JCS(P1)987⟩ or with sodium carbonate in DMF (Equation (59)) ⟨77JOC1349⟩. It is noteworthy that the sulfone in Equation (60) eliminates under acidic conditions. While elimination to form $\alpha,\beta,\gamma,\delta$-unsaturated derivatives is frequently quite efficient (Equation (61)) ⟨86TL3337⟩, in one example (Equation (62)) Trost *et al.* found that efficient elimination occurred only in the presence of a palladium catalyst ⟨80JA5979⟩.

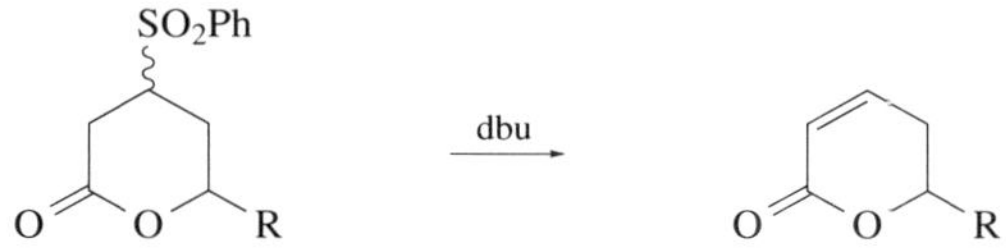

(58)

dbu = 1,5-diazabicyclo[5.4.0]undec-5-ene

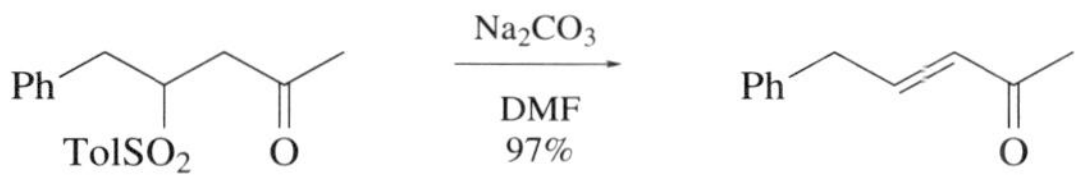

(59)

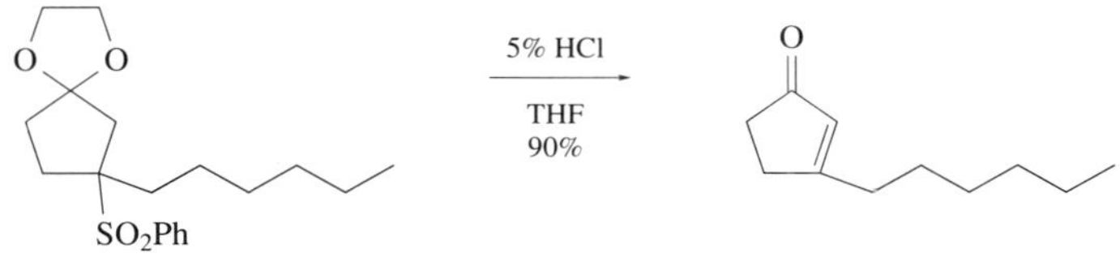

(60)

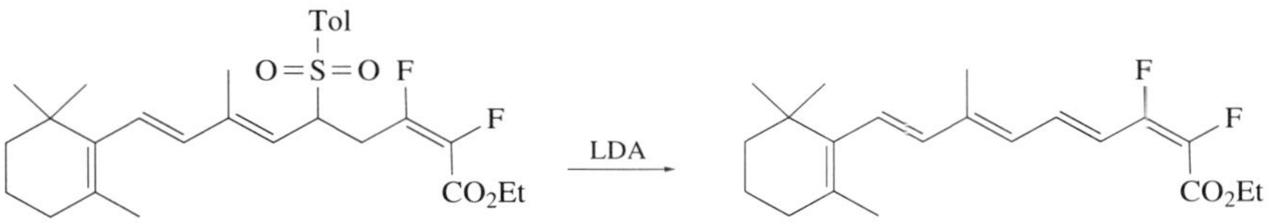

(61)

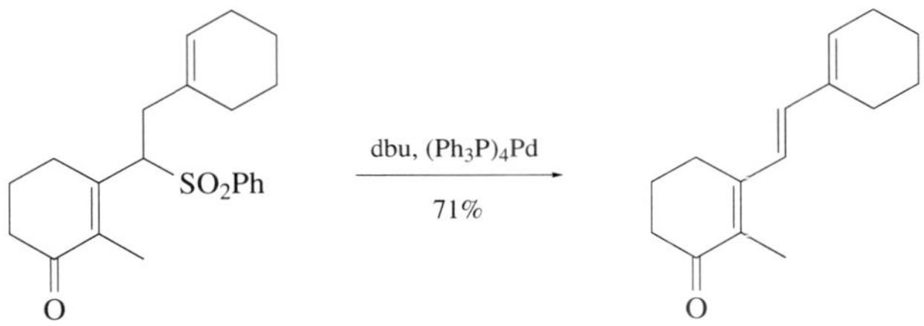

(62)

An interesting example of sulfone elimination to form cyclopent-1-enecarboxamide derivatives is illustrated in Scheme 40 ⟨89JOC1647⟩. When the lithium derivative of a β'-sulfonyl acrylamide is reacted with an activated alkene to form a cyclopentyl derivative as an anion, elimination of the benzenesulfonyl group occurs spontaneously to form the α,β-unsaturated amide. In addition to acting as a leaving group in this process, the sulfone group stabilizes the anion and controls the regioselectivity of the addition to the alkene.

Elimination of sulfones to form (nitroaryl)ethylene derivatives occurs under mild conditions ⟨87S1142⟩. Base-induced elimination of β-alkoxy or β-acetoxy-substituted sulfones produces dienes or alkynes depending on the nature of the sulfone (Equations (63) and (64)) ⟨84JA3670, 86JOC3830, 86JOC3834, 86JOC3896⟩. The presence of a geminal heteroatom, e.g., S, O or N, appears to facilitate base-induced sulfone elimination (Equation (65)) ⟨83TL4993⟩. Cyclic sulfones can be transformed to cycloalkenes on treatment with butyllithium followed by lithium aluminum hydride (Equation (66)) ⟨77OS(57)53, 87JA3730⟩.

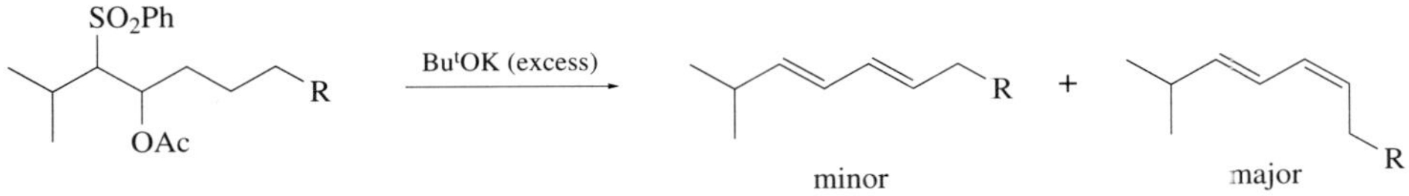

(63)

$R = C_5H_{11}$

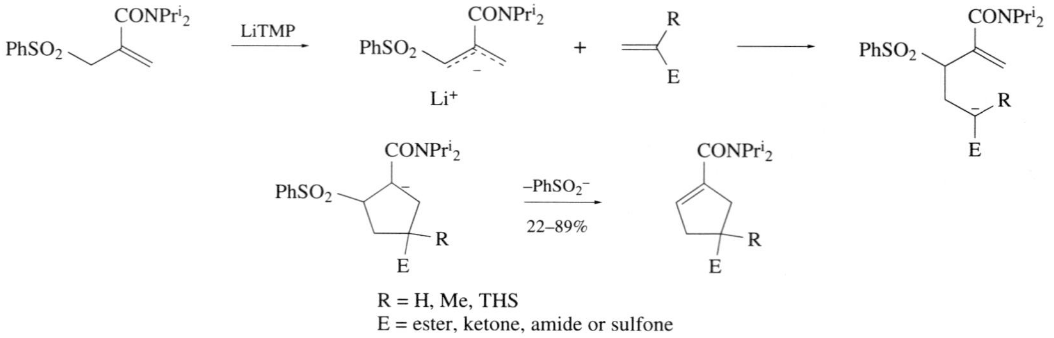

Scheme 40

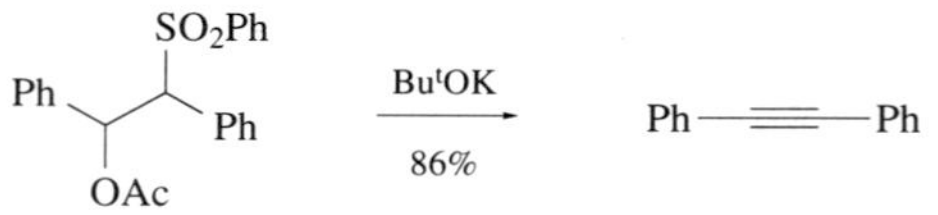

(64)

(65)

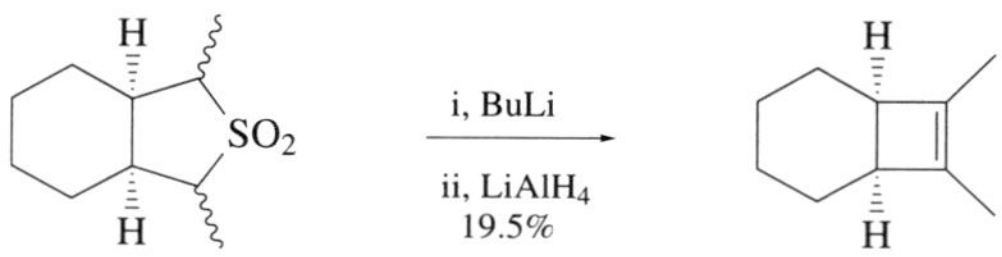

(66)

(ii) Elimination of β-substituted sulfones

(a) Julia reaction. Reductive elimination of sulfones with leaving groups at the β-position, especially hydroxyl groups or derivatives thereof, has been widely investigated. β-Hydroxysulfones are readily formed by reaction of a sulfonyl carbanion with an aldehyde or ketone; reductive elimination of these compounds directly, or following derivatization of the hydroxyl group, furnishes a useful route to alkenes in a process which has been named the Julia reaction (Scheme 41) ⟨73TL4833⟩. Two reviews of this process are available ⟨81CI(L)548, 85PS(24)477⟩. An example of the use of this alkenation procedure (Equation (67)) is a step in Ley's synthesis of a fragment of the ionophore tetronasin ⟨90TL5525⟩. Alternatively the β-hydroxysulfones may be obtained by reduction of β-keto sulfones formed by acylation of the sulfonyl carbanion ⟨88JCS(P1)53, 88TL6357⟩.

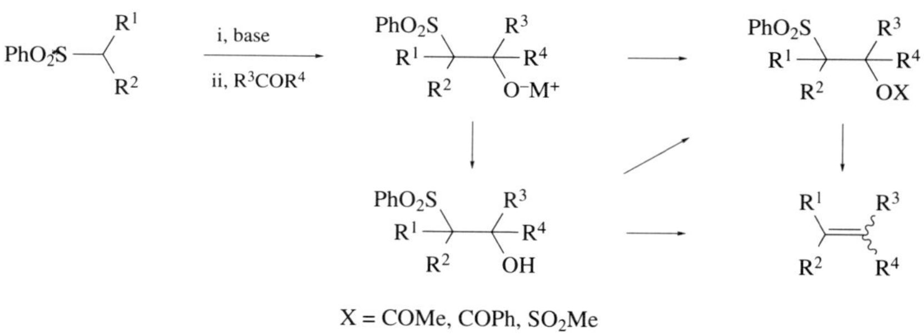

Scheme 41

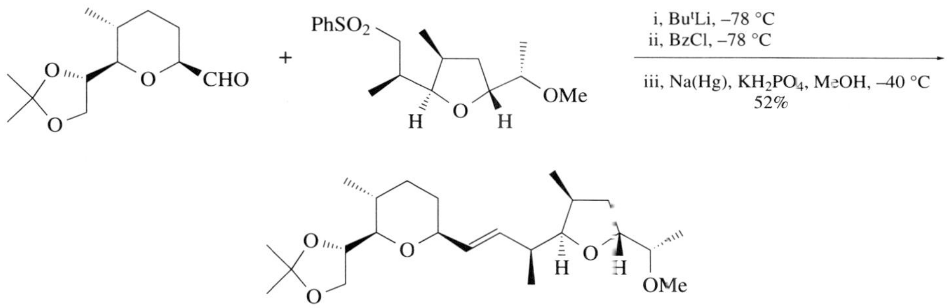

(67)

Sodium amalgam in methanol, sometimes with the addition of a cosolvent as recommended by Kocienski *et al.* ⟨78JCS(P1)829⟩, is frequently employed to effect the reductive elimination of β-hydroxysulfones, e.g., Equation (68) ⟨77JOC2036⟩. Alternatively, functionalization of the hydroxyl group prior to treatment with sodium amalgam, for example with MsCl ⟨73TL4833, 82TL1963⟩ or $SOCl_2$/pyridine ⟨86JA2776⟩, or an acyl group ⟨87CC1342⟩, or a silyl group (Equation (69)) ⟨79JCS(P1)1290⟩, may lead to more efficient alkene formation. However, radical reduction may also be employed provided the hydroxyl group is first transformed into an appropriate group (Equation (70)) ⟨77TL4223, 91TL2703⟩; in the case of the diene illustrated in Equation (71), elimination of the β-hydroxysulfone occurred directly on treatment with Bu^n_3SnH without prior functionalization of the hydroxyl group ⟨81TL2675⟩. Similarly, samarium iodide effects efficient elimination of β-hydroxy imidazolyl sulfones without prior functionalization of the hydroxyl group (Scheme 42) ⟨90TL7105⟩. Electrochemical reductive elimination of β-hydroxysulfones can also be used especially for the production of terminal alkenes ⟨78CL69, 78MI 114-01⟩.

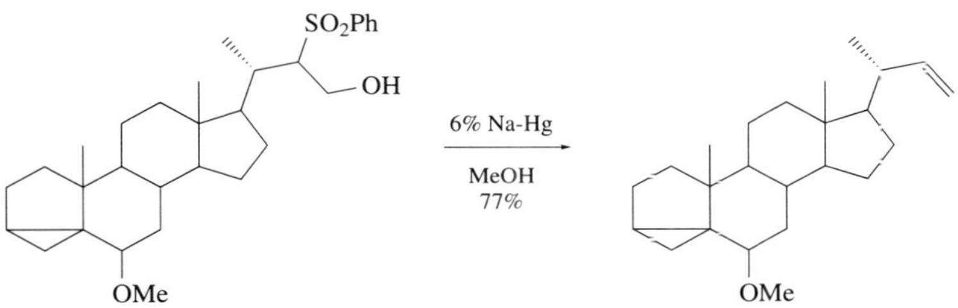

(68)

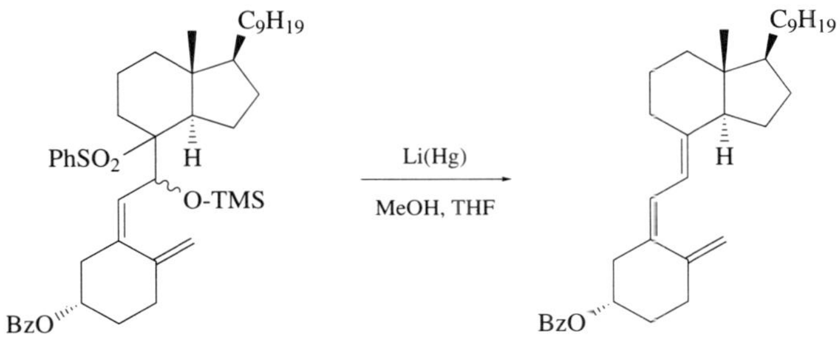

(69)

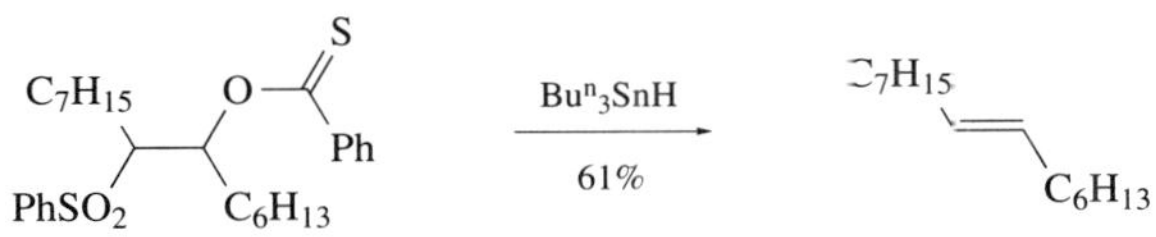

(70)

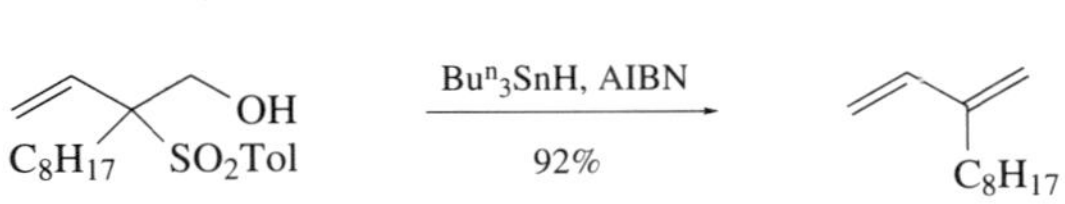

(71)

The stereochemical preference of the elimination for selective formation of (*E*)-alkenes is illustrated in Equation (72) ⟨78JCS(P1)829⟩. Furthermore, when conjugated double bonds are formed, the stereochemistry of the pre-existing alkene unit is unaffected. Mechanistically, it is believed that electron transfer to the sulfone group followed by expulsion of a benzene sulfinate anion or the corresponding radical produces the carbanion which subsequently undergoes elimination in an *anti*-

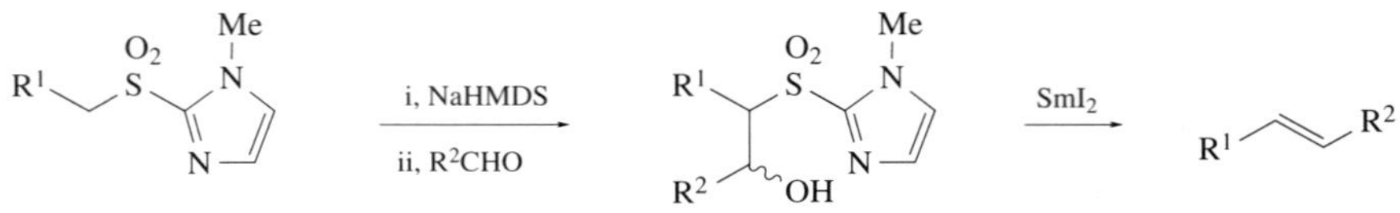

Scheme 42

periplanar fashion to produce the alkene. Thus the stereoselectivity of the process can be explained by rapid equilibration in the carbanionic intermediate so that the substituents are as far apart as possible before the elimination occurs (Scheme 43) ⟨80JCS(P1)1045⟩. As a consequence of the *anti*-periplanar requirement of the process, elimination is difficult in cyclic systems in which the leaving group cannot easily adopt an *anti*-periplanar arrangement to the carbanionic lone pair ⟨81CI(L)548⟩. Interestingly, while use of the silyloxy group, or alternatively an acetoxy group, produces the desired triene in Equation (69), use of a benzoyloxy leaving group results in a mixture of stereoisomers about the central double bond ⟨79JCS(P1)1290⟩.

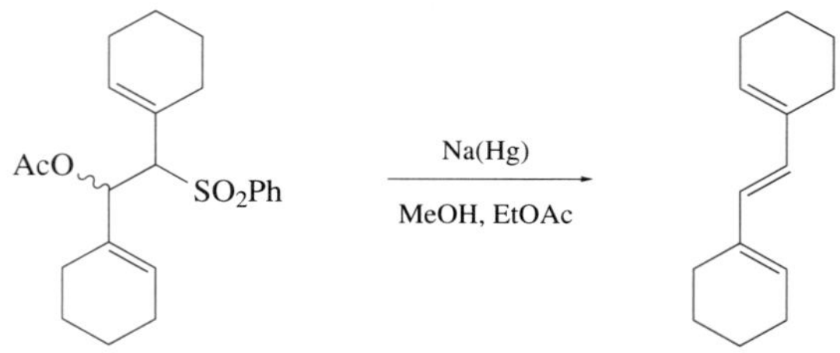

(72)

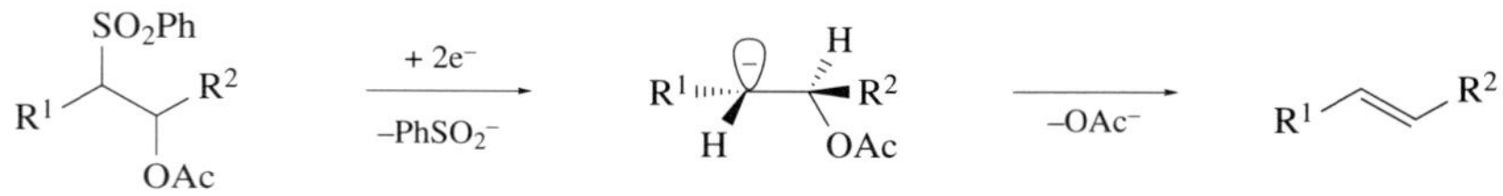

Scheme 43

(b) Elimination of β-silyl, nitro, stannyl and sulfonyl sulfones. Fluoride induced elimination of β-silyl sulfones is a very mild route to unsaturated systems (see Section 1.14.5.1.3) ⟨79TL2649, 88JOC2688⟩. Treatment with tetra-*n*-butylammonium fluoride (tbaf) in refluxing THF usually produces the alkene in high yields, e.g., Equation (73). Extension of this strategy to the production of vinylogous systems gives (*E*)-dienes in good yields (Equation (74)) ⟨84TL1219⟩.

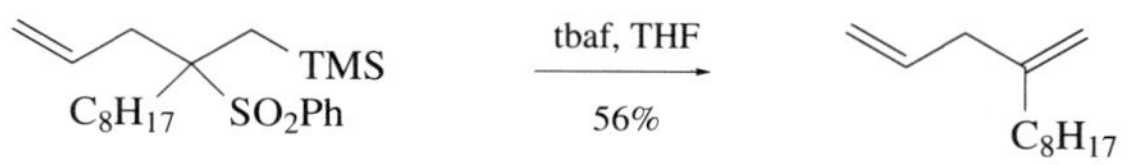

(73)

tbaf = tetra-*n*-butylammonium fluoride

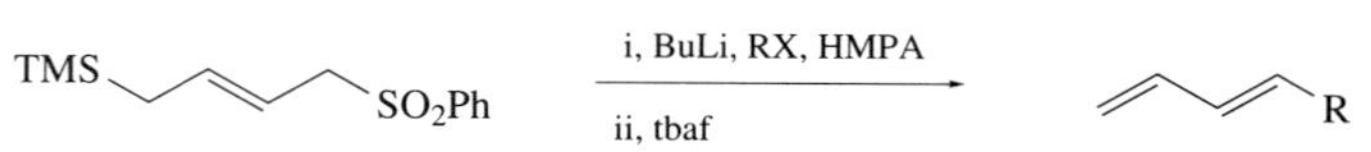

(74)

R = alkyl, benzyl

β-Stannyl sulfones are even more labile towards elimination on treatment with fluoride ion (see Section 1.14.6.3) (Equation (75)) ⟨82TL2205, 83CC619, 85JOC3622⟩ and frequently are not isolated but eliminate to form the alkene either directly or on contact with silica gel (Equation (76)), thereby furnishing a useful method for methylenation. The more stable 2-pyridyl sulfone derivative can be isolated prior to elimination.

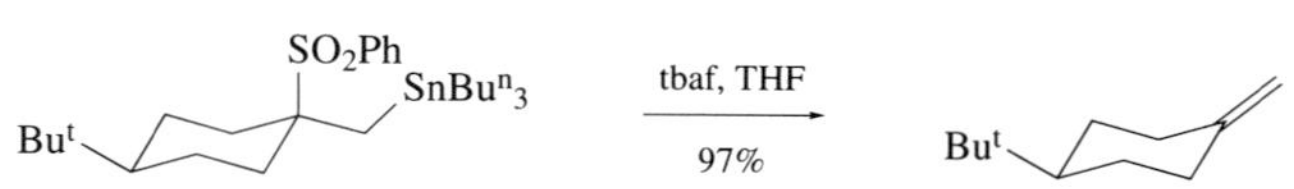

(75)

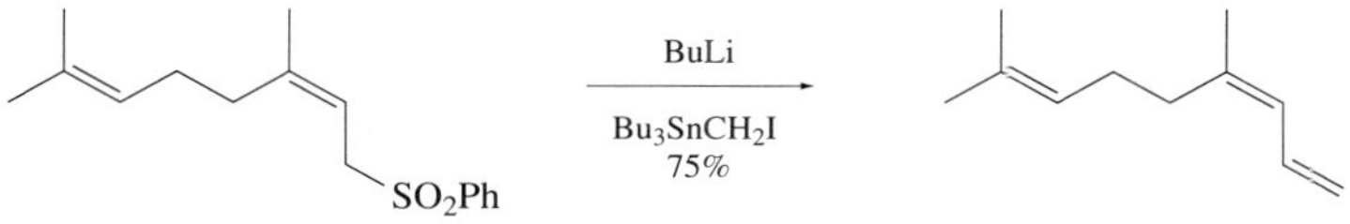

(76)

Elimination of *β*-nitrosulfones under reducing conditions to form alkenes has been investigated (see Section 1.14.3.5) ⟨78TL763, 81CL1139, 87JOC5111⟩. 1,4-Cyclohexadienes have been synthesized using this reductive elimination as illustrated in Scheme 44 ⟨88JOC251⟩. While Na_2S and NaTeH may be employed for this transformation, Bu^n_3SnH has the advantage of effecting a highly stereospecific *anti*-elimination (Equation (77)). Only Bu^n_3SnH gives satisfactory results in eliminations to form alkenes which are not conjugated.

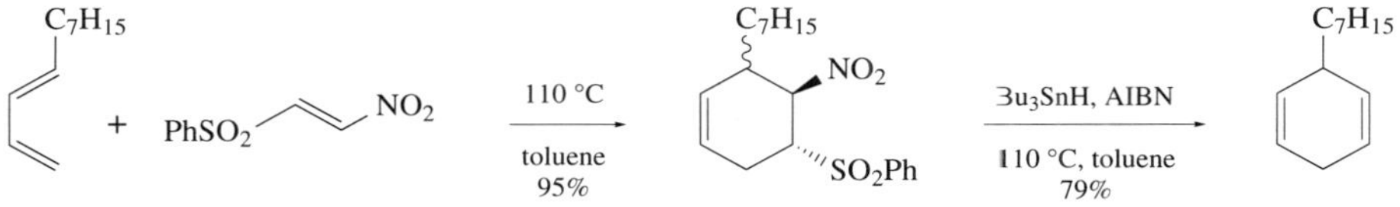

Scheme 44

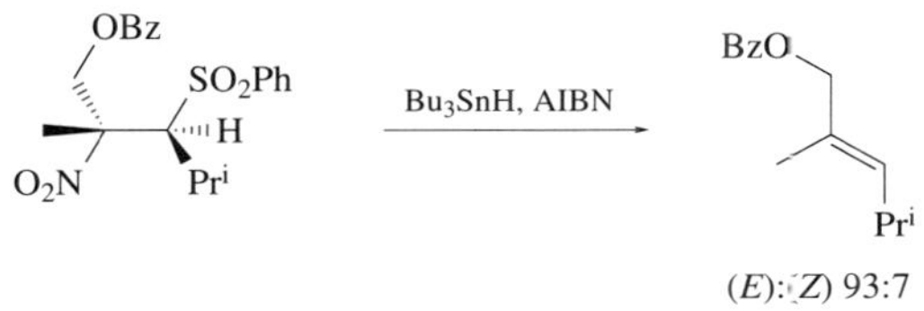

(77)

Reductive desulfonylation of 1,2-disulfones is achieved using either sodium amalgam in buffered methanol ⟨84JOC596⟩ or magnesium in methanol ⟨85JOC1749⟩ (Equation (78)).

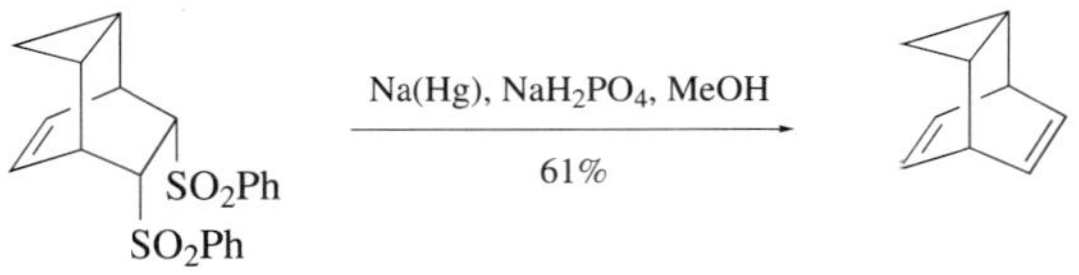

(78)

(iii) Ramberg–Bäcklund reaction

Base treatment of a sulfone bearing an *α*-halogen atom (or another leaving group) and an *α*′-hydrogen atom results in elimination of SO_2 and formation of an alkene (Equation (79)). This transformation was first reported in 1940 ⟨40MI 114-01⟩ and has attracted considerable attention both from a mechanistic point of view and due to its synthetic utility. Many reviews of this reaction have appeared ⟨B-67MI 114-01, 68ACR209, 68MI 114-04, 70ACR281, 77OR(25)1, B-78MI 114-01, 82T2857, 91COS(3)861, B-93MI 114-01⟩. Equation (80) illustrates application of this reaction to the synthesis of conjugated enediynes ⟨88JA4866⟩.

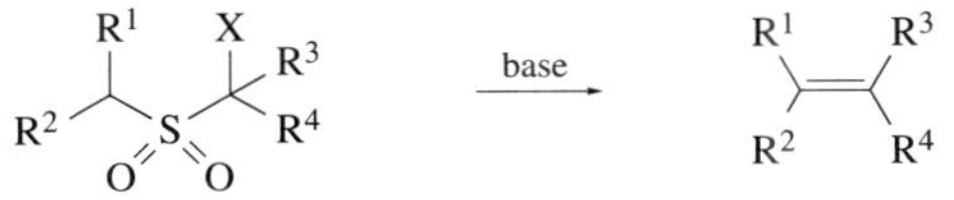

(79)

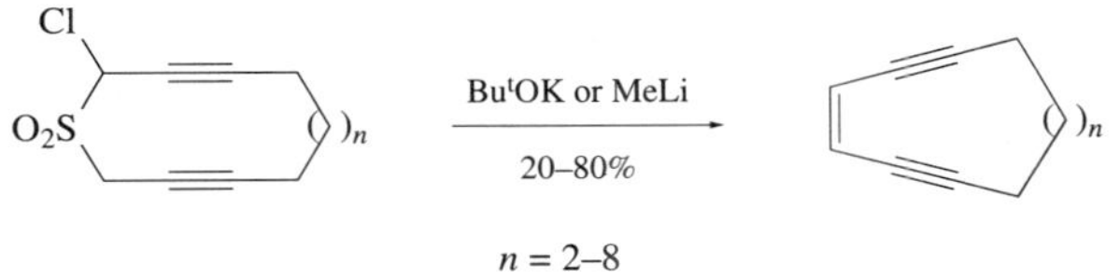

(80)

The mechanism of the reaction is believed to be that shown in Scheme 45, being established predominantly through Bordwell's research ⟨51JA5187⟩. Intramolecular displacement of the leaving group by the sulfonyl-stabilized carbanion produces an episulfone intermediate which is unstable; loss of SO_2 produces the alkene under the basic reaction conditions. The leaving group is most commonly chloride but other halides, sulfinate ⟨80JOC1719, 90JA8084⟩ or triflinate ⟨86JA2358⟩ may also be used. Isolation of an episulfone (Equation (81)) further established the intermediacy of this species in the Ramberg–Bäcklund reaction ⟨89TL3267⟩.

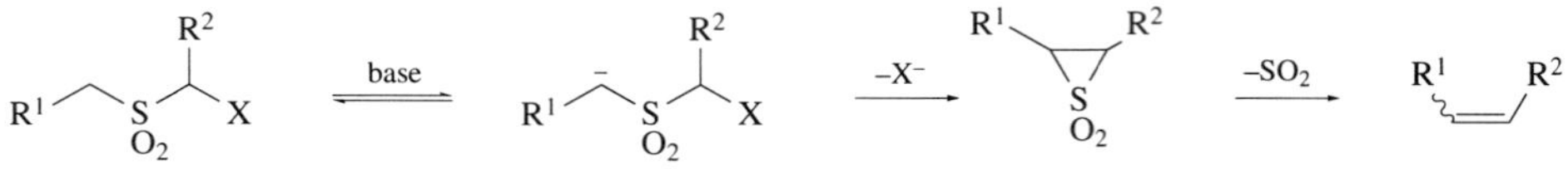

Scheme 45

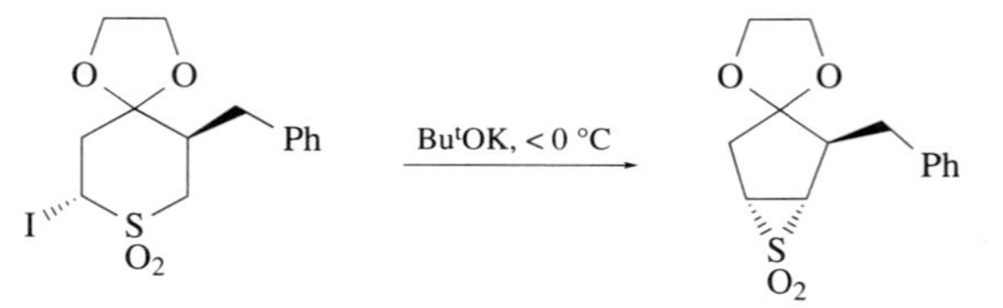

(81)

The bases employed for the reaction are generally NaOH, NaOMe or $KOBu^t$ in H_2O, MeOH, Bu^tOH, THF or DMSO. The use of phase-transfer conditions has also been reported ⟨82S504⟩. Meyers developed an alternative method in which *in situ* chlorination and Ramberg–Bäcklund reaction take place on treatment of the sulfone with KOH, Bu^tOH and CCl_4 (Equation (82)) ⟨69JA7510, 74TL1105⟩. These conditions efficiently transform diallylic sulfones to trienes (Equation (83)) ⟨82TL5043⟩; the new double bond is formed selectively with (*E*) geometry. However, these conditions are not suitable for use with simple primary dialkyl sulfones as dichlorination occurs more rapidly than episulfone formation resulting in formation of vinylsulphonic acid salts via a thiirene dioxide intermediate (Scheme 46) ⟨73TL1843⟩.

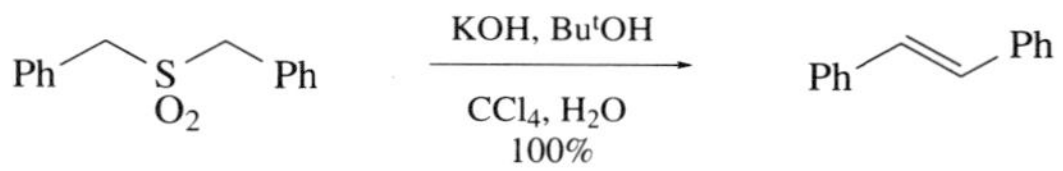

(82)

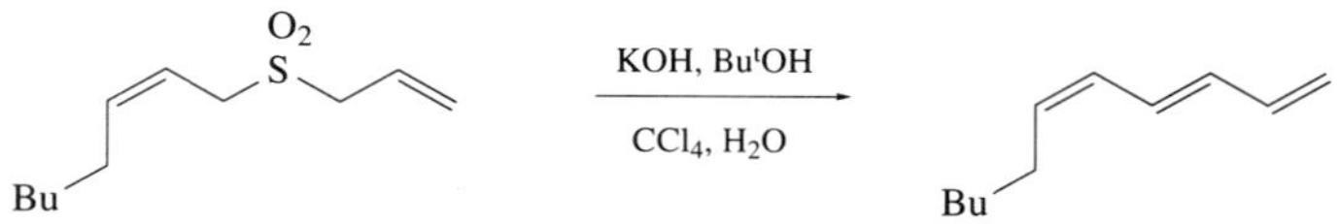

(83)

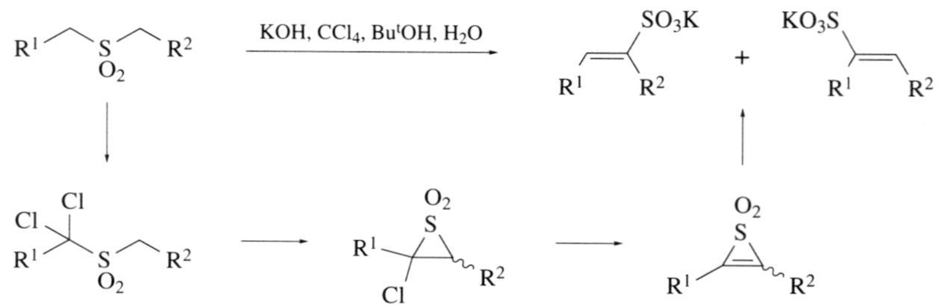

Scheme 46

Application of the Ramberg–Bäcklund reaction to the preparation of strained alkenes has proved successful, e.g., the *anti*-Bredt alkene shown in Equation (84) was obtained in this way ⟨83HCA1090⟩. However, in cyclic sulfones the Ramberg–Bäcklund reaction is successful only when certain geometric requirements are met ⟨69JA3870, 83HCA1090⟩, otherwise 1,2-elimination competes (Equations (85a) and (85b)). Thus if the H—C—S—C—X atoms are in a W-type arrangement, the Ramberg–Bäcklund reaction is most favourable ⟨83HCA1090⟩. The reaction has also been applied to the synthesis of medium-size rings, e.g., in a synthesis of (+)-eremantholide A ⟨91JA9682⟩.

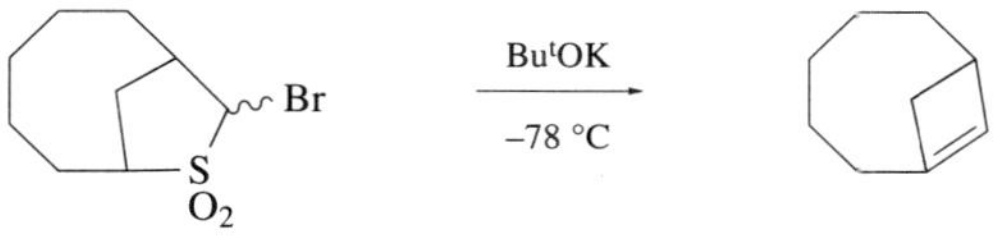

(84)

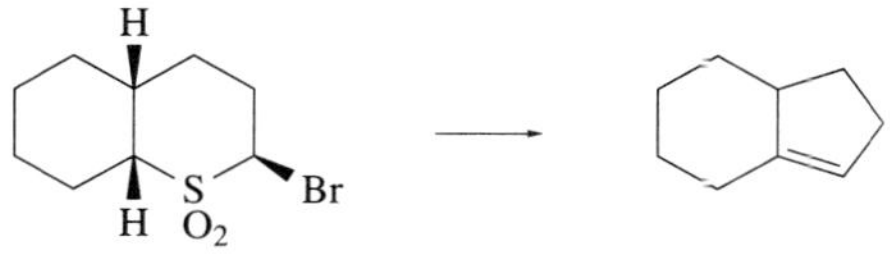

(85a)

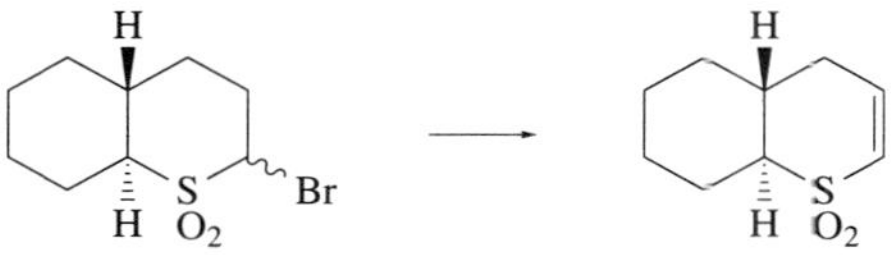

(85b)

Homologation of alkenes to dienes via the Ramberg–Bäcklund reaction has been achieved (Scheme 47) ⟨83JA6164, 86JA4568, 87OS(65)90⟩. Free-radical halosulfonylation with bromomethanesulfonyl bromide followed by base-induced elimination (for example, with 1,5-diazabicyclo-[4.3.0]non-5-ene (dbn)) produces the bromomethyl alkenyl sulfone, which can undergo a vinylogous Ramberg–Bäcklund reaction to give the diene, formed preferentially with (*Z*) stereochemistry ⟨88CL2009⟩.

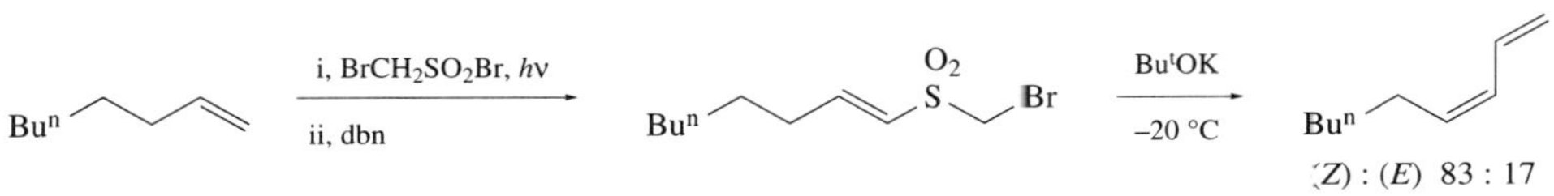

Scheme 47

Synthesis of α,β-unsaturated esters and acids can be accomplished (Equation (86)) ⟨78JOC4884⟩. The Ramberg–Bäcklund approach has been employed in many cyclopentenone syntheses (for example, ⟨89T455⟩). The formation of the cyclopentenone derivatives shown in Equation (87) illustrates a number of features ⟨85JOC2110⟩; firstly use of an additional sulphone group in place of the halogen allows regioselective alkylation prior to the elimination process thus increasing the versatility of this process; secondly use of triflone allows elimination to be conducted under relatively mild conditions presumably due to the leaving group ability of triflinate.

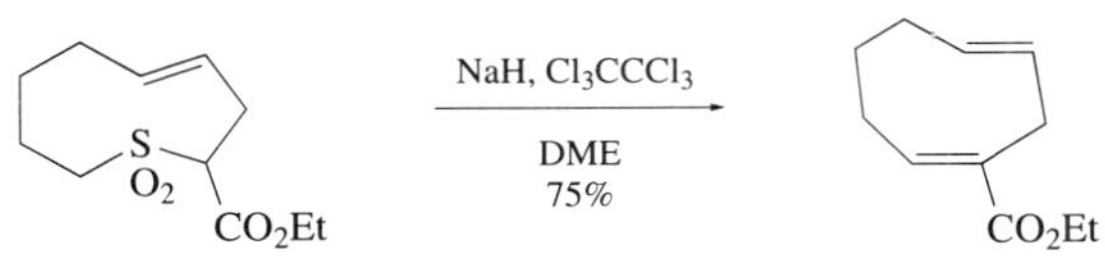

(86)

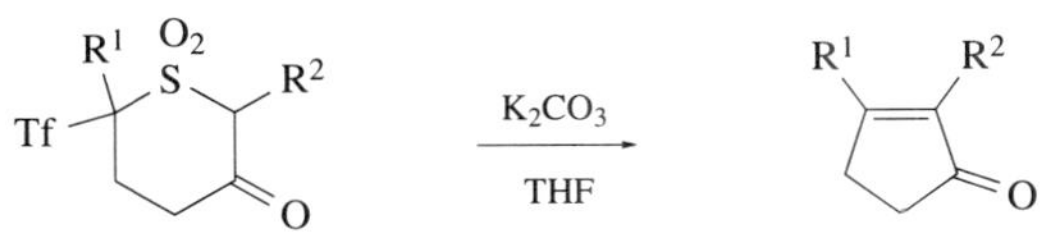

(87)

Michael addition of sulfinate to unsaturated sulfones bearing a suitably positioned leaving group can result in a Ramberg–Bäcklund reaction (Equation (88)) ⟨81T417⟩. As an alternative, episulfones, as a mixture of *cis* and *trans* isomers, may be formed by reaction of primary sulfonyl chlorides with triethylamine; thermolysis produces the corresponding symmetrically substituted (*E*)- and (*Z*)-alkenes through loss of SO_2 (Scheme 48) ⟨89TL3131⟩.

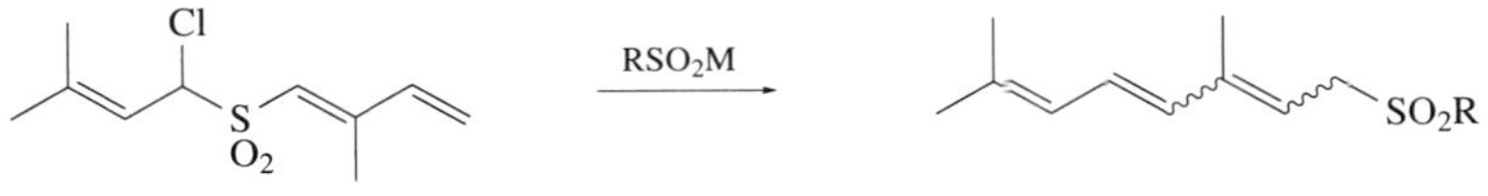

(88)

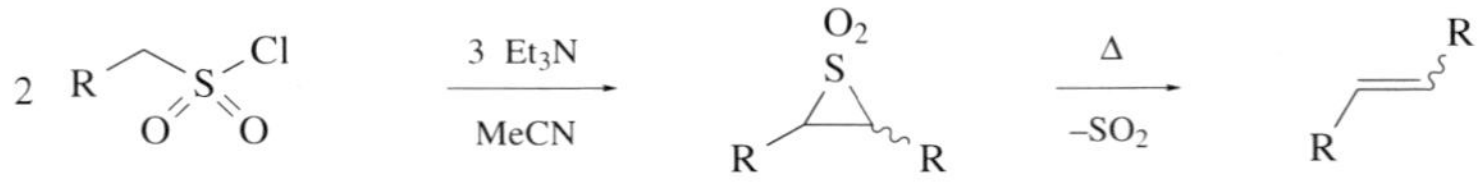

R = H, alkyl, aryl

Scheme 48

While episulfone intermediates in the Ramberg–Bäcklund reaction are usually insufficiently stable to isolate, the corresponding thiirene dioxides can be isolated and characterized (Scheme 49) ⟨71CC22, 71JA476⟩ prior to loss of SO_2. Extrusion of SO_2 from episulfones is believed to occur in a nonconcerted fashion ⟨68JA429, 68JA435, 74TL3987⟩, in contrast to the concerted chelotropic eliminations of SO_2 from 2,5-dihydrothiophene-1,1-dioxide and 2,7-dihydrothiepin-1,1-dioxide (Equation (89)) ⟨66JA2857, 69JA5682, 88CC1044, 89JCR(S)275⟩. Extrusion reactions in sulfur, selenium and tellurium compounds which are synthetically useful, including reactions which form alkenes, have been reviewed ⟨88T6241⟩.

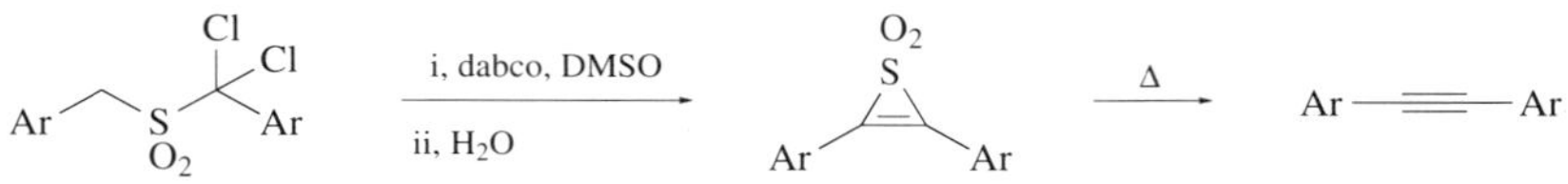

dabco = 1,4-diazabicyclo[2.2.2]octane

Ar = Ph, *p*-ClPh, *p*-MePh

Scheme 49

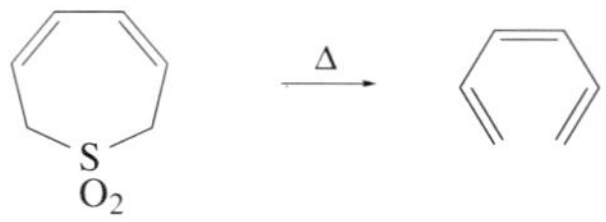

(89)

1.14.2.3.2 *Elimination of selenone and tellurone groups*

In contrast to elimination reactions of sulfones, much less is known about selenone and tellurone elimination. One of the reasons for this is that selenones are not readily available; furthermore, both selenones and tellurones are very susceptible to nucleophilic displacement of seleninate and tellurinate anions. However, selenones can be prepared by oxidation of selenoxides (for example, with mcpba), and elimination to form an alkene has been reported (Equation (90)) ⟨B-83MI 114-01⟩, while Cava *et al.* reported pyrolysis of an alkyl aryl tellurone, formed by sodium periodate oxidation of a telluroxide, to form an alkene *inter alia* (Equation (91)) ⟨81CC277, 82CC164⟩. Oxidation of selenides or tellurides in MeOH with mcpba leads to the corresponding methyl ethers, believed to be formed by initial oxidation to the selenone or tellurone followed by nucleophilic displacement by methanol ⟨83TL4347, 85JCS(P1)471⟩. The ability of the selenone moiety to act as a leaving group has been demonstrated in a fragmentation sequence leading to an enol ether ⟨81JOC5246⟩.

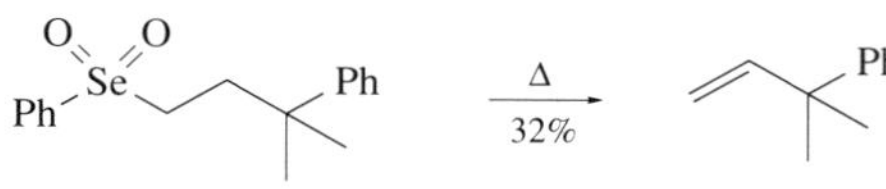

(90)

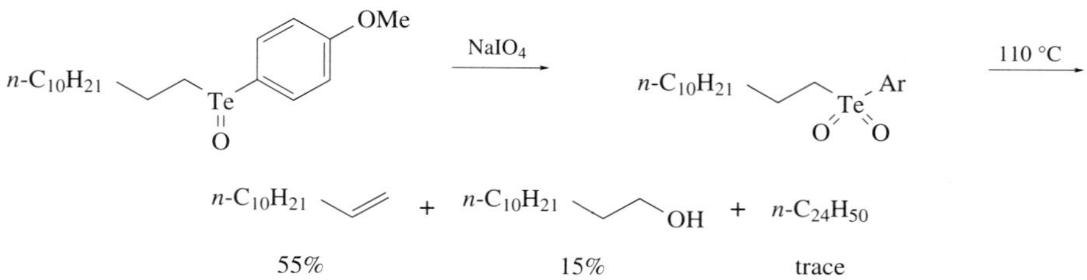

(91)

1.14.2.4 Elimination of Sulfilimine, Selenimide, Tellurium Imide, Sulfinamide, Sulfoximine, Sulfonamide and Sulfonate Groups

As sulfoxide and selenoxide elimination provides an extremely useful route to alkenes, investigation of the reactivity of the corresponding sulfilimines and selenimides was undertaken. Oae ⟨71T4921, 71T6101⟩ reported that the sulfonimido analogues eliminate much more rapidly than the sulfoxides in a concerted *cis* fashion (Equation (92)). Sharpless reported that oxidation of a selenide with chloroamine-T under phase-transfer conditions results in rapid formation of the alkene via a selenimide as shown in Scheme 50 ⟨75CS9⟩; the corresponding selenoxide elimination is much slower. Similarly on treatment with chloroamine-T, tellurides are transformed to the corresponding alkenes presumably via a tellurium imide ⟨81CL447, 83S1006, 89JOC4391⟩. The elimination of the telluride shown in Equation (93) on oxidation with chloroamine-T resulted in a higher yield of the alkene than the corresponding telluroxide elimination shown in Equation (55) ⟨89JOC4391⟩. However, as formation of the imide derivatives of sulfides and selenides is less attractive than simple oxidation, these eliminations offer little advantage over the sulfoxide ⟨76JA4887⟩ and selenoxide elimination processes.

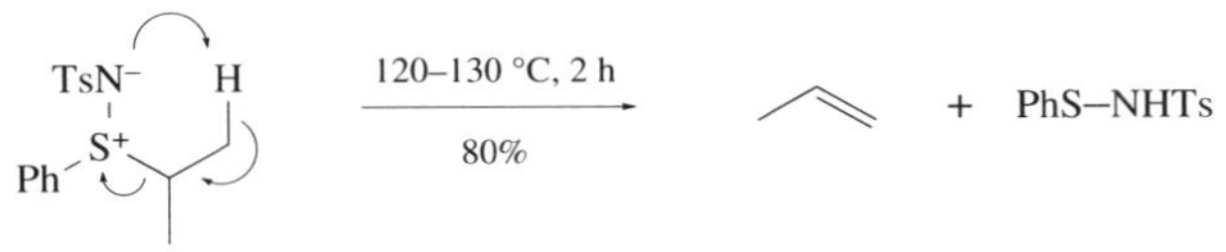

(92)

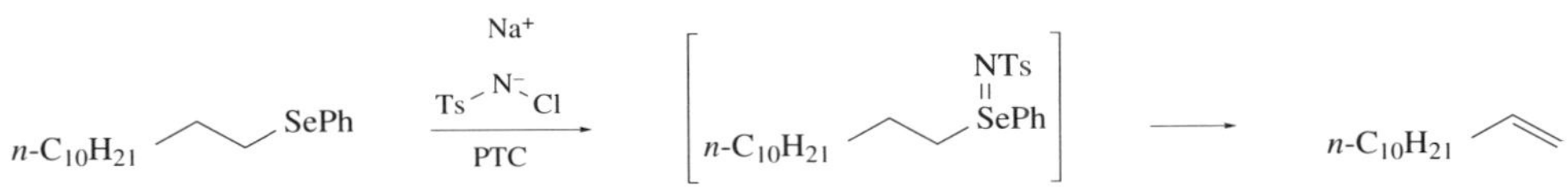

Scheme 50

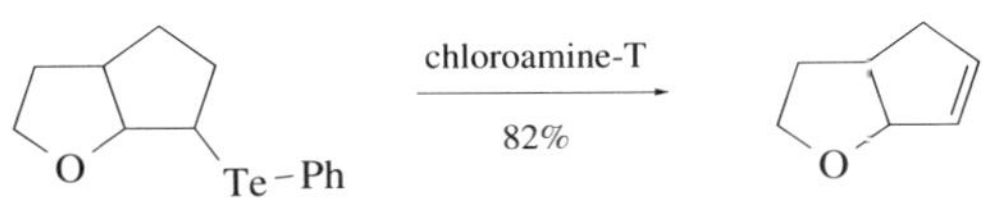

(93)

Corey and Durst have demonstrated that *cis*-elimination of *β*-hydroxy sulfinamides to produce alkenes, believed to proceed as illustrated in Scheme 51, may be brought about thermally ⟨66JA5656, 68JA5548, 68JA5553⟩. Scheme 52 illustrates how this elimination may be used to transform carbonyl compounds to alkenes. Fragmentation of unsaturated sulfinamides to form terminal alkenes is illustrated in Equation (94) ⟨88TL3255⟩.

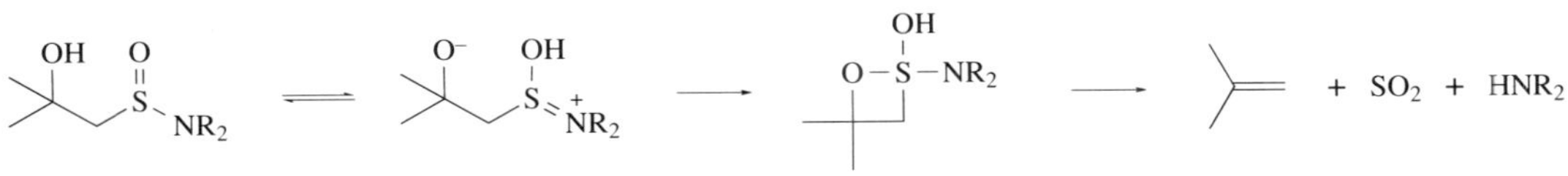

R = H, alkyl, aryl

Scheme 51

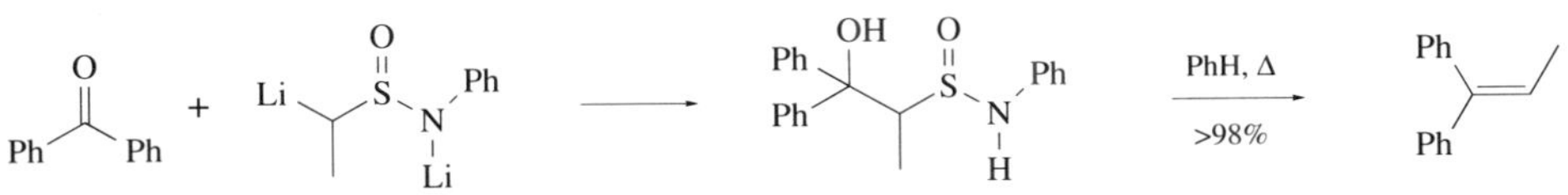

Scheme 52

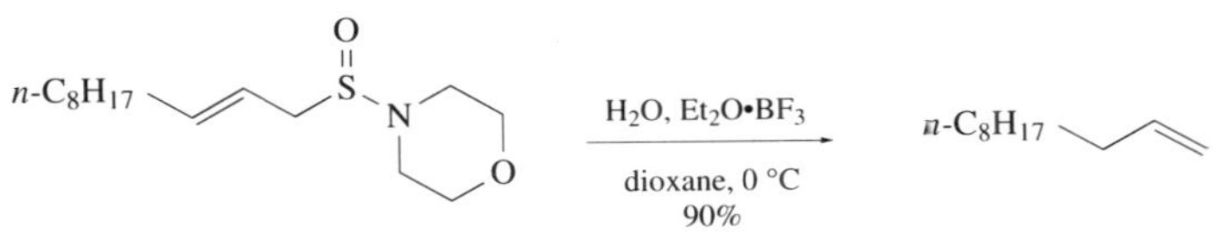

(94)

Reductive elimination of β-hydroxy sulfoximines with aluminum amalgam furnishes alkenes (Equation (95)) ⟨73JA6462, 79JA3602⟩. Mixtures of *cis* and *trans* isomers are formed in this elimination process. Base-induced elimination of sulfoximines is also possible; on treatment with potassium dimsylate in DMSO, benzothiazines (Equation (96)) undergo an elimination of the sulfoximine group to form an alkenyl sulfinamide ⟨91T8855⟩; this reaction is believed to occur via an Elcb mechanism. The stereoselectivity of the elimination is low. Nucleophilic attack of lithium cyanide or lithium dimethylphosphonate on vinylsulfoximines is followed by elimination of the sulfoxime group to form α,β-unsaturated nitriles or phosphonates (Equation (97)) ⟨91TL3119⟩.

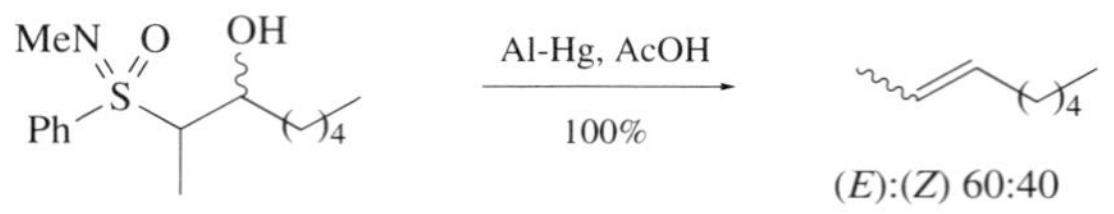

(95)

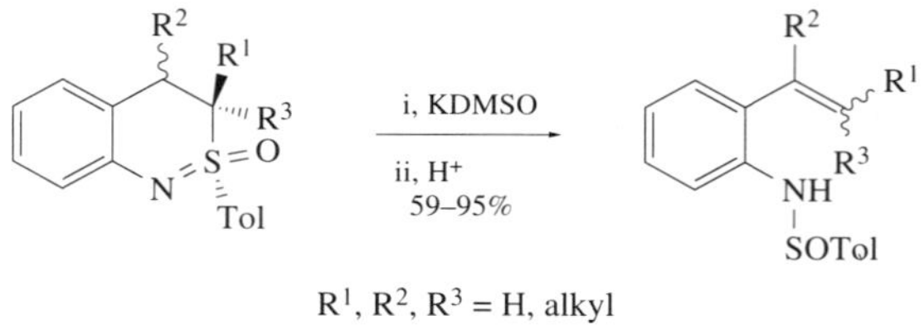

(96)

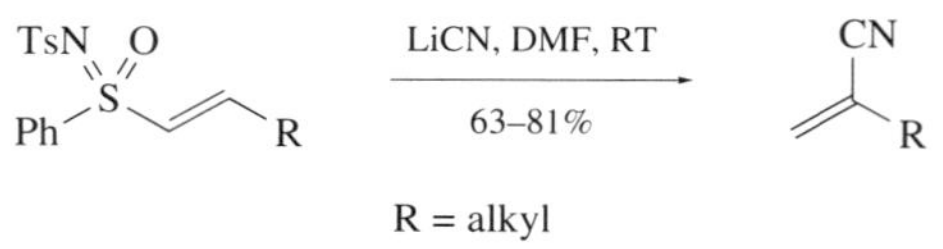

(97)

Elimination of the sulfonamide shown in Equation (98) to form a naphthalene derivative occurs readily ⟨92T2515⟩. Formation of (nitroaryl)ethylene derivatives by elimination of sulfonamide or sulfonate ester groups is illustrated in Scheme 53 ⟨87S1142⟩.

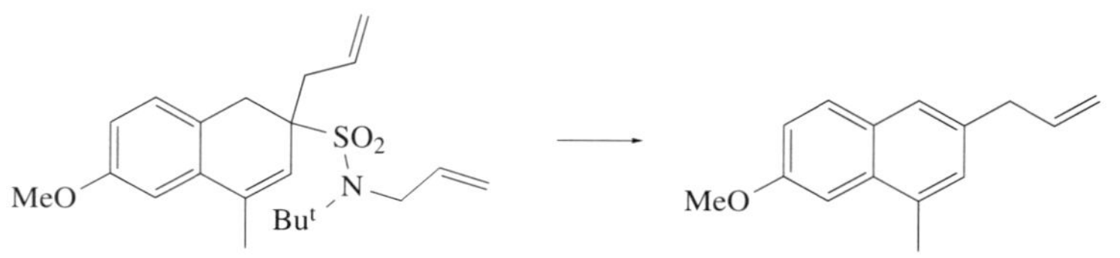

(98)

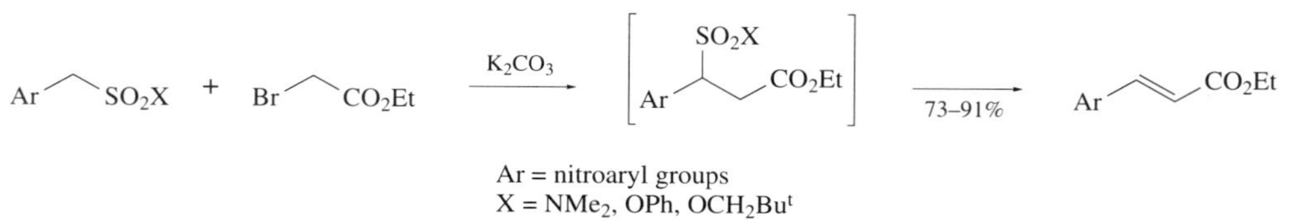

Scheme 53

Fragmentation of 1,2-dithietane-1,1-dioxides on treatment with lithium cyanide forms (Z)-alkenes (Equation (99)) ⟨93SL839⟩.

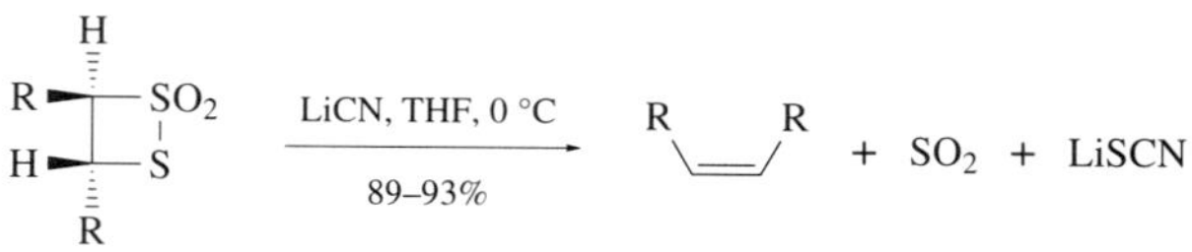

(99)

1.14.2.5 Elimination of Sulfonium, Selenonium and Telluronium Salts and Ylides

Sulfonium salts undergo elimination to form alkenes in a process which is similar to that observed with ammonium salts; thermal decomposition of sulfonium hydroxides and base-induced elimination of sulfonium salts are well established ⟨B-81MI 114-01⟩. Considerable investigation of the mechanistic aspects of this elimination has been undertaken ⟨B-81MI 114-01⟩; see, for example, ⟨90JA6950⟩. Elimination to form the less-substituted alkene is preferred ⟨88T2913⟩. On treatment with nucleophilic alkyllithiums such as methyl, phenyl or *n*-butyllithium, sulfonium *n*-alkylides lead to significant amounts of alkenes through *β*-elimination, in addition to ylide formation (Equation (100)) ⟨B-78MI 114-02⟩. An interesting example of this elimination process is illustrated in Equation (101) whereby a polymeric sulfonium salt was thermolysed to effect elimination and produce a highly conjugated polymer ⟨92JCS(P1)3225⟩. Elimination of sulfonium salts in medium rings on treatment with potassium *t*-butoxide has been observed (Equation (102)) ⟨88CB2239⟩.

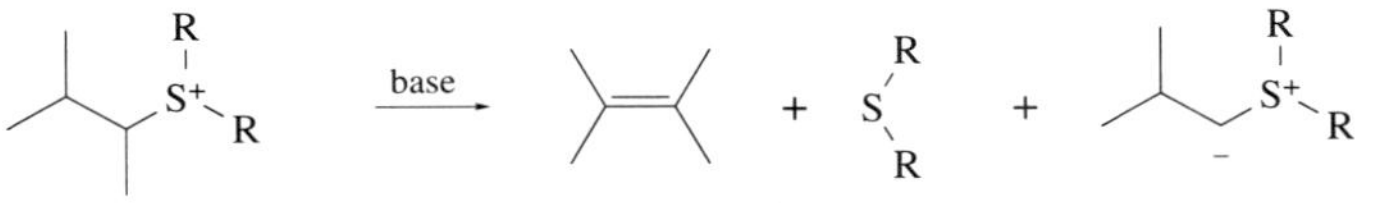

(100)

R = alkyl, aryl

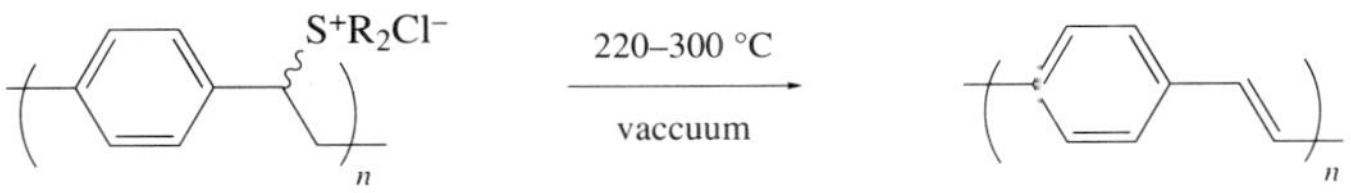

(101)

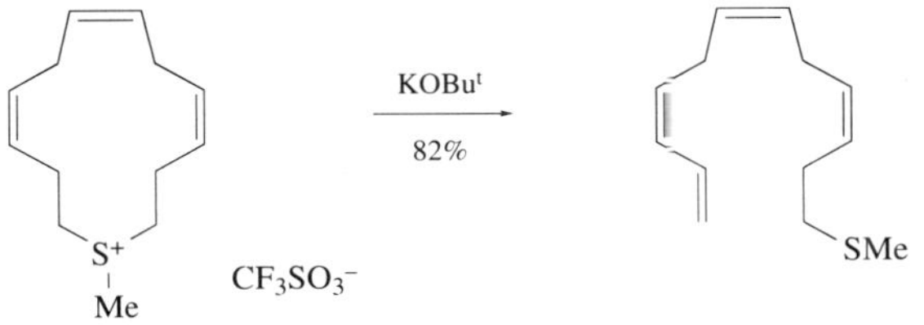

(102)

Sulfonium ylides bearing a *β′*-hydrogen can undergo an *α*,*β′*-elimination process on treatment with base (Scheme 54) ⟨60CB2819, 65TL2899, B-78MI 114-03⟩.

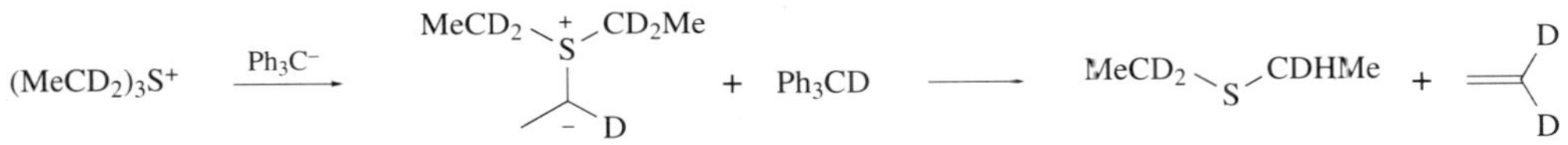

Scheme 54

Krief and co-workers ⟨78TL187, 79TL4233, 80TL1997⟩ have demonstrated that elimination of selenonium salts to form alkenes can be employed as an alternative to selenoxide elimination; alkylation of the selenide with reagents such as dimethyl sulfate, methyl fluorosulfate or methyl iodide followed by base-induced elimination (for example, using potassium *t*-butoxide in DMSO) produces alkenes in good yields (Equation (103)).

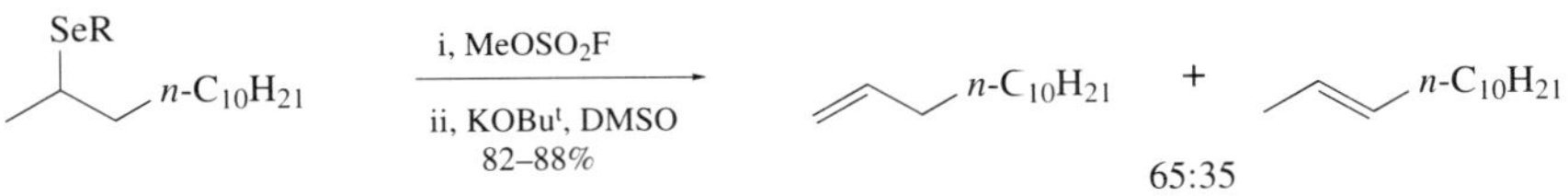

(103)

R = Me, Ph

While sulfonium and selenonium ylides condense with aldehydes and ketones to produce epoxides, the analogous telluronium salts and ylides react with aldehydes and ketones to produce alkenes (Equation (104)) ⟨83TL2599, 87TL801, 88JOC4862, 91S897⟩.

(104)

1.14.3 BY ELIMINATION OF NITROGEN FUNCTIONS

1.14.3.1 Elimination of Amine Oxides—the Cope Reaction

Although elimination of amine oxides to form alkenes and hydroxylamine derivatives had been reported earlier ⟨1898CB1553⟩, the synthetic utility of this transformation was first recognized by

Cope ⟨49JA3929⟩. Reviews of the Cope reaction have appeared ⟨60CRV431, 60OR(11)317, B-68MI 114-01, 91COS(6)1011⟩. Frequently the Cope reaction is carried out by oxidizing an amine with reagents such as hydrogen peroxide ⟨63OSC(4)612⟩ or mcpba ⟨86JA1039⟩; without isolation of the amine oxide, pyrolysis (usually at 100–150 °C) effects elimination to provide the alkene. Cram demonstrated that Cope elimination occurs in dry DMSO or THF at room temperature ⟨62JA1734⟩. The synthetic utility of the Cope reaction is that, due to the fact that it occurs under relatively mild conditions, little isomerization of the alkene is observed (Equation (105)) ⟨52JA2137, 57JA960⟩ compared with the Hofmann elimination, and there are few side reactions. Occasionally rearrangement of a benzyl or allyl group can occur to form an *O*-alkylated hydroxylamine ⟨49JA3929⟩.

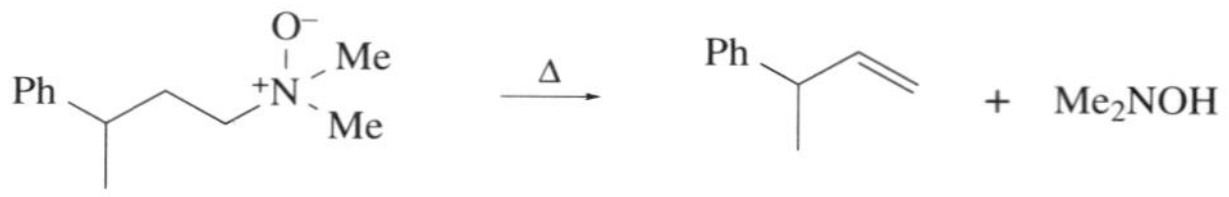

(105)

The elimination of amine oxides follows an Ei-type mechanism via a five-membered planar cyclic transition state as shown in Equation (106); thus the amine oxide derived from *threo*-2-amino-3-phenylbutane produces the *cis* alkene, while the *erythro* amine gives the *trans* isomer ⟨54JA5740⟩. Furthermore, *cis*- and *trans*-2-phenylcyclohexyldimethylamine oxides selectively form different alkenes as shown in Equations (107a) and (107b) ⟨57JA960⟩, due to the stereochemical requirements of the planar cyclic transition state. Cyclooctyldimethylamine oxide produces only *cis*-cyclooctene on pyrolysis ⟨53JA3212⟩; originally this was believed to indicate that the elimination must occur in a *syn* fashion. However Bach *et al.* later demonstrated that, in an analogous ylide elimination of a cyclooctyl ammonium derivative which proceeds via a similar cyclic mechanism, formation of *trans*-cyclooctene is possible ⟨71JA7118, 72JA8608⟩. Therefore the formation of only *cis*-cyclooctene in the Cope reaction could not be taken as definitive proof of a *syn* elimination. Instead Bach confirmed the *syn* nature of the amine oxide elimination using deuterium labelling studies as shown in Equation (108) ⟨73JOC1742⟩. Kwart *et al.* have used kinetic isotope effects to probe the nature of the transition state of the Cope reaction ⟨78JA3927, 81JA4650⟩.

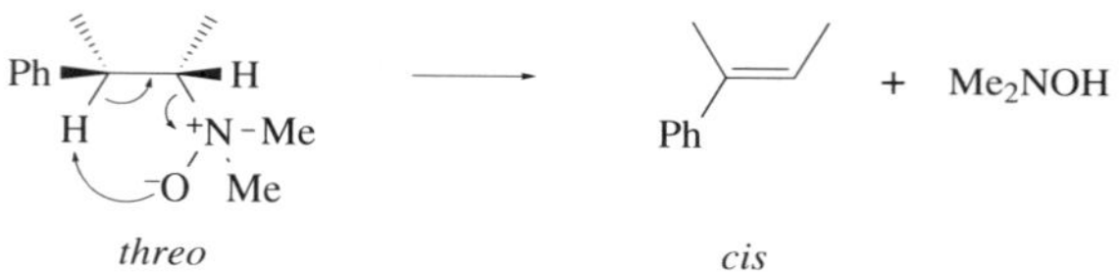

(106)

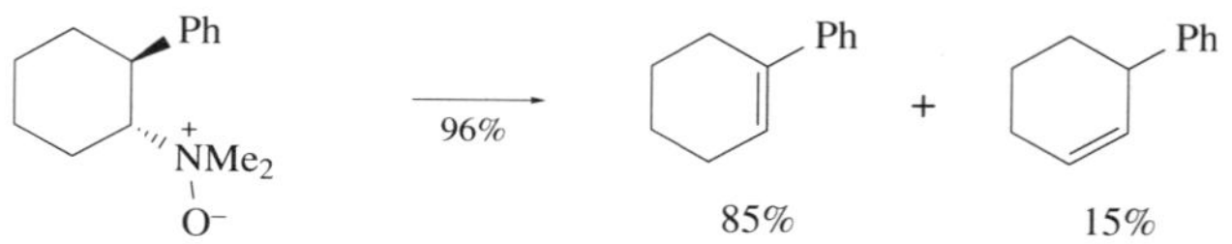

(107a)

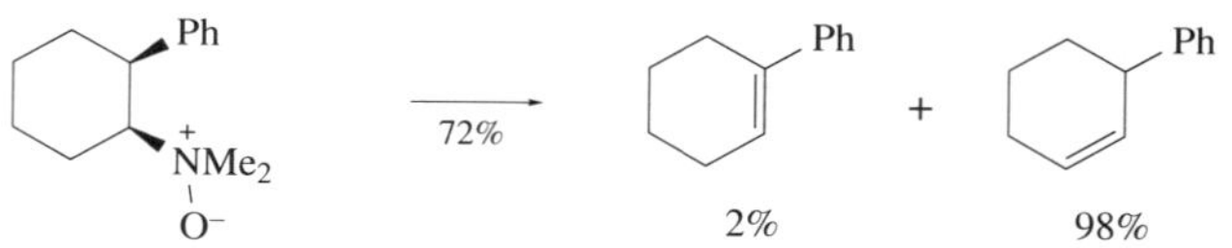

(107b)

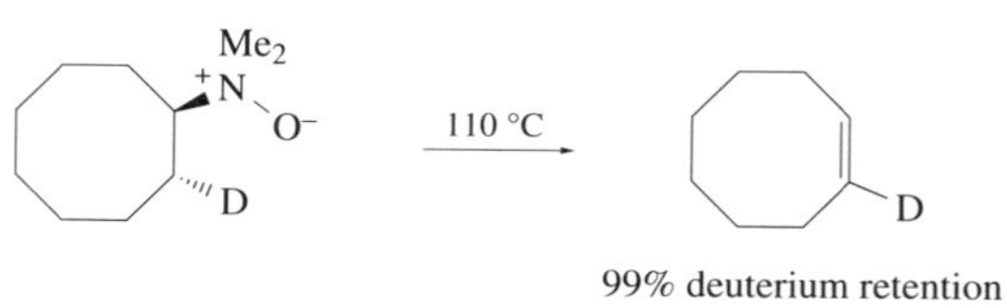

(108)

In contrast to the Hofmann elimination, alkyl substitution has little effect on the direction of elimination in the Cope reaction; in acyclic amines it is governed almost entirely by the number of hydrogen atoms at the β-position, except in the case of a bulky *t*-butyl group or if a conjugated alkene may be formed ⟨57JA4720⟩ (Equations (109) and (110)). Equation (109) also demonstrates the preference for formation of the *trans* alkene if both *cis* and *trans* alkenes may be formed.

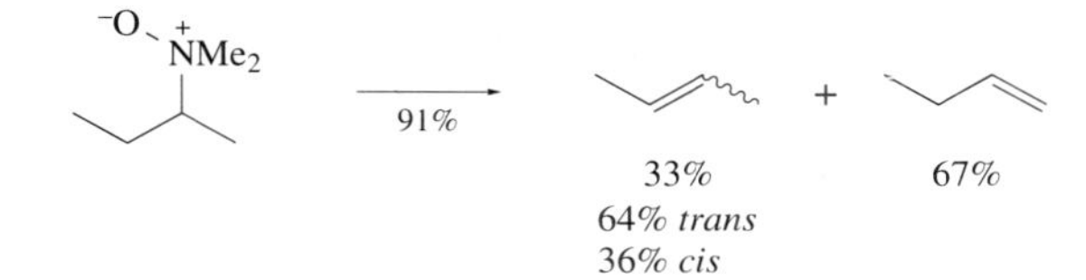

(109)

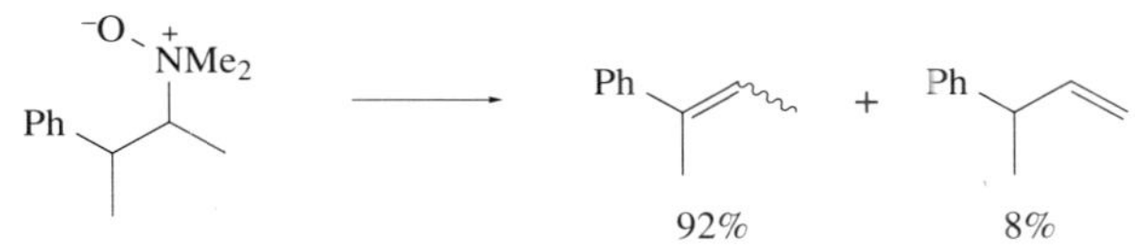

(110)

In cyclic systems the direction of elimination is controlled by the requirement of a planar cyclic transition state (Equations (107a) and (107b)) ⟨57JA960⟩. Similarly in the menthyl and neomenthyl derived amine oxides (Equations (111a) and (111b)) ⟨58JA355⟩, a *cis* β-hydrogen atom is available only in the 2-position in the neomenthyl derivative and therefore only 2-menthene is produced. In contrast, the menthyl derived amine oxide has *cis* β-hydrogens at both the 2- and 4-positions and therefore both menthene isomers are formed. When elimination may occur in an *endo* or *exo* cyclic fashion in cycloalkylamine oxides, elimination to form the *endo* alkene is generally preferred (Equation (112)) for ring sizes of 5 and 7–10 ⟨60JA4663⟩. However, in cyclohexyl derivatives, elimination is almost entirely *exo* cyclic ⟨57JA4729⟩. It is believed that the planar cyclic transition state which would be required for *endo* cyclic elimination in the cyclohexyl derivatives is not readily accessible. In medium-size rings (9 or 10) the Cope reaction forms the *trans* cycloalkenes ⟨55JA1628⟩ in a synthetically useful process. Cope investigated eliminations of amine oxides to form 2-norbornene and 2,5-norbornadiene ⟨59JA2799⟩; the *exo* amine oxide derivatives gave higher yields than the *endo* isomers. Lower yields were obtained in these Cope eliminations than in the Hofmann eliminations that form the same bicyclic alkenes.

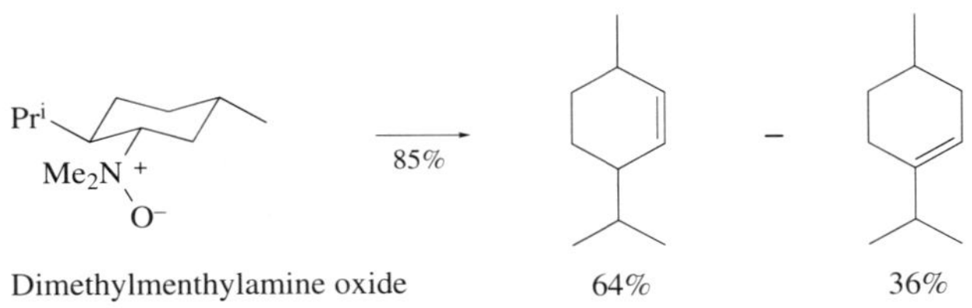

(111a)

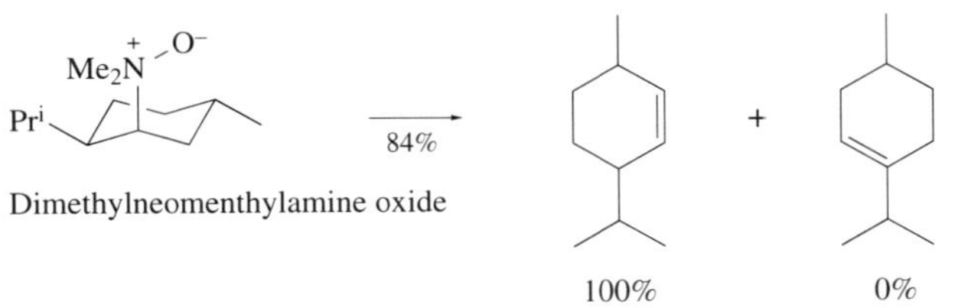

(111b)

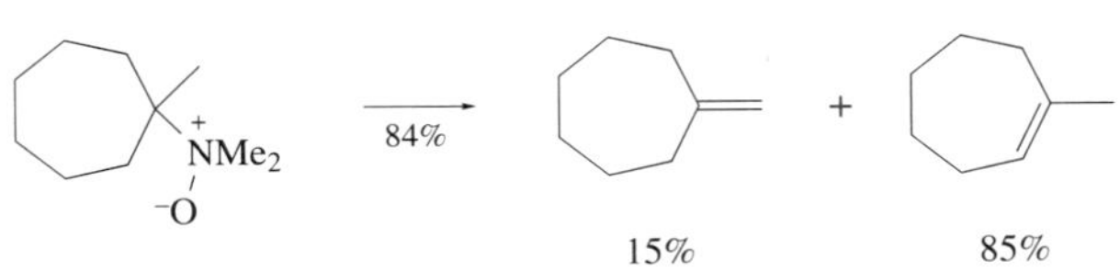

(112)

Heterocyclic tertiary amines may be cleaved on oxidizing to the corresponding amine oxides followed by elimination ⟨60JA4656⟩. Cleavage of *N*-methylpiperidine oxide does not occur on heating as the steric requirements of a cyclohexyl ring do not allow the preferred planar transition state to be attained. However, on increasing the ring size to 7 and 8, the cleavage to produce unsaturated hydroxylamine derivatives becomes increasingly easy (Equation (113)). *N*-Methyl-α-pipecoline oxide undergoes elimination (Equation (114)) ⟨60JA4656⟩ to form the unsaturated hydroxylamine in addition to the bicyclic side product believed to be formed by nitrone cycloaddition to the initially formed alkene. Only the *trans* isomer of the amine oxide underwent elimination.

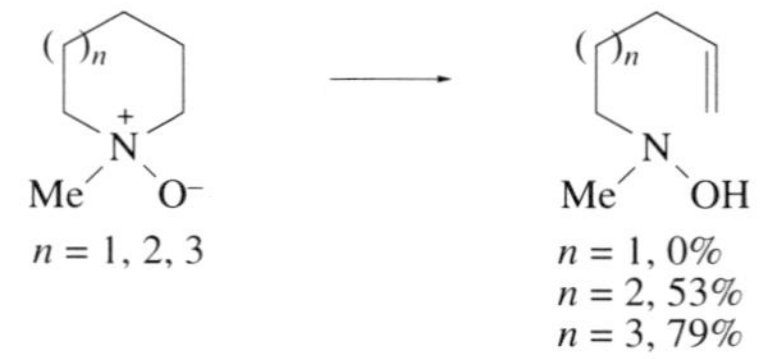

(113)

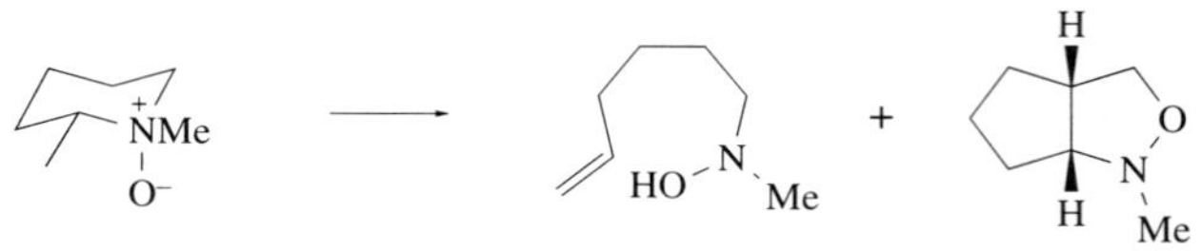

(114)

While the Cope reaction can provide a useful synthetic route to alkenes, it has also been used to prepare dialkylhydroxylamines, particularly unsymmetrical derivatives (Equation (115)) ⟨55JCS769⟩.

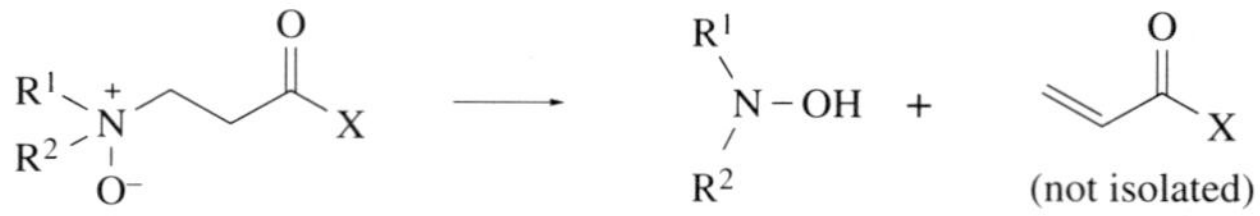

(115)

X = Ph, OAlkyl

Application of the Cope reaction in synthesis is illustrated by its use in the synthesis of Vitamin B_{12} (Equation (116)) ⟨86JA1039⟩, and in the synthesis of norsecularines ⟨88H(27)2367⟩. Elimination to form a conjugated diene is illustrated in Equation (117) ⟨88MI 114-01⟩. The reverse Cope elimination, in which unsaturated hydroxylamines undergo cyclization, is also possible; an example of this, in which pyrrolidine and piperidine *N*-oxides are formed, is illustrated in Scheme 55 ⟨90JOC3007⟩. The sila-Cope reaction ⟨82JA7666⟩, where a *β*-silyl substituent is eliminated in place of the hydrogen atom has been used in a pheromone synthesis (Equation (118)) ⟨90TL1865⟩ (see Section 1.14.5.1.3).

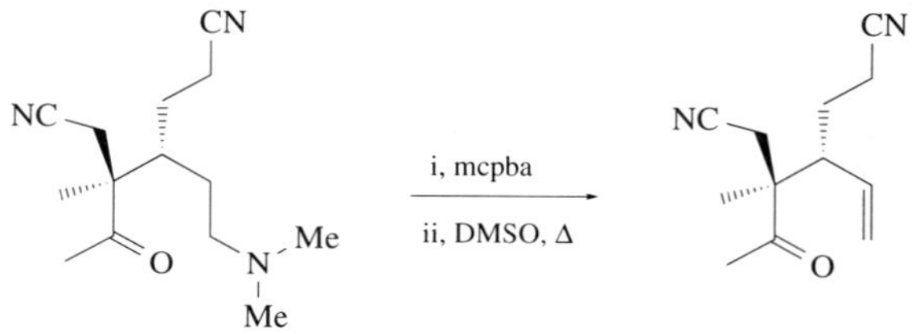

(116)

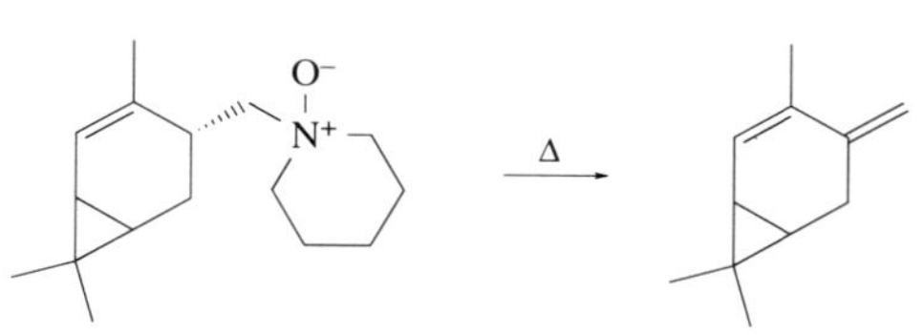

(117)

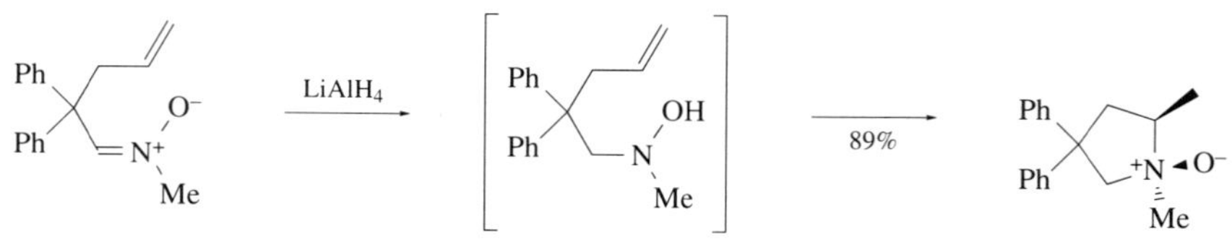

Scheme 55

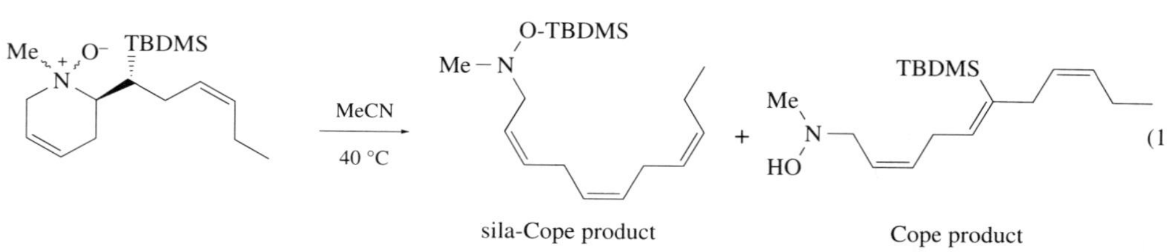

(118)

1.14.3.2 Elimination of Quaternary Ammonium Salts

1.14.3.2.1 Hofmann elimination of quaternary ammonium hydroxides

Thermal decomposition of a quaternary ammonium hydroxide to form an alkene, an amine and water (the final step of the Hofmann degradation process) was first reported in 1851 ⟨1851LA(78)253⟩. The Hofmann elimination has been the subject of a number of reviews ⟨60OR(11)317, B-68MI 114-02, B-73MI 114-01⟩. Hofmann degradation (Scheme 56) involves the treatment of an amine with sufficient methyl iodide to convert it to the corresponding quaternary ammonium iodide, treatment with silver oxide to convert it to the hydroxide salt, which is then thermally decomposed (100–200 °C) by distilling an aqueous or alcoholic solution of the hydroxide often at reduced pressure. This process has been used extensively for structural determination of amines, especially alkaloids. Nucleophilic attack of hydroxide to produce the alcohol as a side product can occur under the conditions employed for the Hofmann elimination ⟨60OR(11)317⟩. Improved yields in the elimination step have been reported by concentrating the reaction mixture and conducting the elimination under reduced pressure ⟨71JCS(C)1327⟩; lower temperatures are required to effect elimination under these conditions. Use of an ion-exchange resin for the preparation of quaternary ammonium hydroxides is possible ⟨76OS(55)3⟩. An alternative procedure for the elimination involves heating the ammonium salt with alkali hydroxides in polyethylene glycol monomethyl ether ⟨79LA1925⟩. Hofmann elimination proceeded on treatment of quaternary benzyltetrahydroisoquinoline related alkaloids with a large excess of diazomethane ⟨88H(27)751⟩.

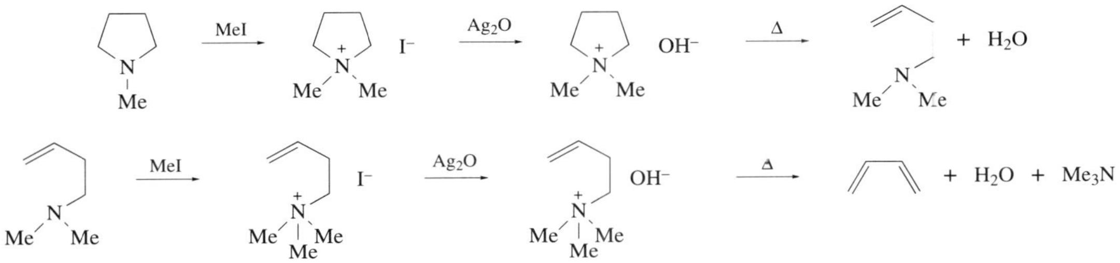

Scheme 56

Extensive investigation of the mechanism of the Hofmann elimination has been undertaken. The usual mechanism is E2 ⟨61JA3861⟩ leading to the product of *trans* elimination ⟨56JA790⟩; in the Hofmann degradations of *threo* and *erythro* 1,2-diphenylpropylamines ⟨56JA790⟩ the *erythro* isomer produces the (*Z*)-alkene while the *threo* isomer produces the (*E*)-alkene (Equations (119a) and (119b)). However on changing the base to *t*-butoxide in *t*-butanol both *threo* and *erythro* isomers produce the same (*E*)-alkene, indicating a change in the mechanism of the reaction with the stronger base. In certain cases *cis* elimination is observed (Equation (120)) ⟨67JA6701, 67JCS(B)509⟩. In the case of some highly hindered ammonium salts an alternative Ei mechanism is involved ⟨63JA1949⟩.

(119a)

(119b)

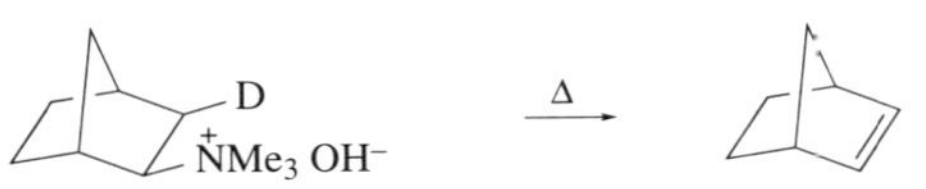

(120)

The direction of elimination of ammonium salts has been summarized in the Hofmann Rule: the β-hydrogen atom is removed most readily if it is located on a methyl group, next on a methylene group and least readily from a methine group. Therefore, elimination proceeds to give the least substituted alkene. This preference is illustrated in Equation (121). However, for cyclic compounds elimination within the ring is controlled largely by the requirement that the amino group and hydrogen atom are *trans* to each other (Scheme 57) ⟨50AP(283)36, 52JCS153, 58JA355, 60OR(11)317⟩. When both β-hydrogen atoms are *trans* to the ammonium group, elimination from the more substituted position occurs (pathway (a)), i.e., contrary to the Hofmann rule giving instead the Saytzeff product. Whenever a conjugated alkene may be formed the direction of elimination is that which produces the conjugated product; see for example Equation (122) where elimination of the *cis* hydrogen atom occurs to form the conjugated alkene ⟨60JA4692, 62JOC293⟩.

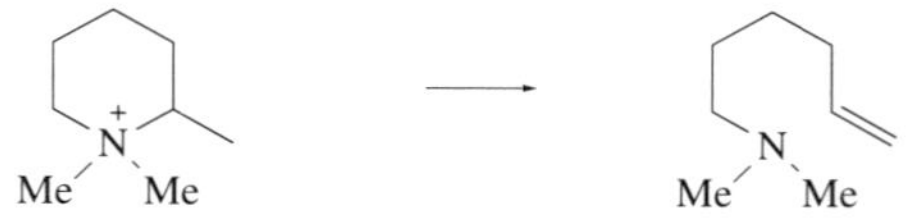

(121)

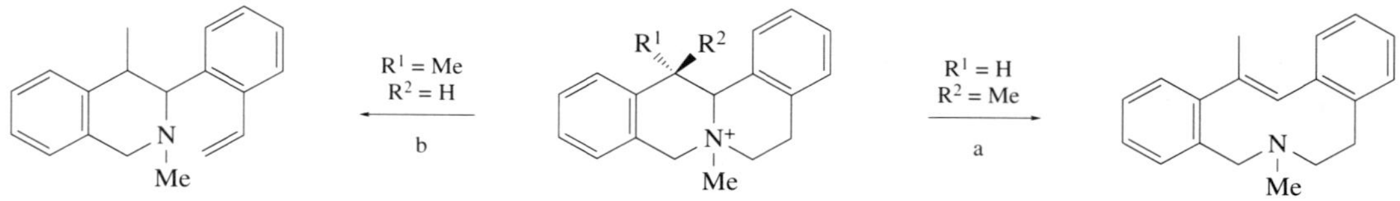

Scheme 57

(122)

The stereochemistry of the alkene formed in the elimination process depends on the substituents; aliphatic alkenes such as 2-pentene and cyclooctene are formed as mixtures of *cis* and *trans* isomers ⟨57JA4720, 68JA5561, 73OSC(5)315⟩, while stilbene is formed as the more stable *trans* isomer ⟨32JCS1932⟩. The stereochemical course of the reaction can be controlled by the *trans* coplanar stereochemical requirement for the elimination (Equations (119a) and (119b)) ⟨56JA790⟩.

Hofmann elimination has been used in the synthesis of dienes and trienes (Equation (123)) ⟨73OSC(5)608, 92JCR(S)309⟩. Equation (124) illustrates the use of the Hofmann elimination in the synthesis of 2-vinyltryptamines ⟨92S383⟩; higher yields were obtained by use of sodium hydroxide rather than with silver oxide.

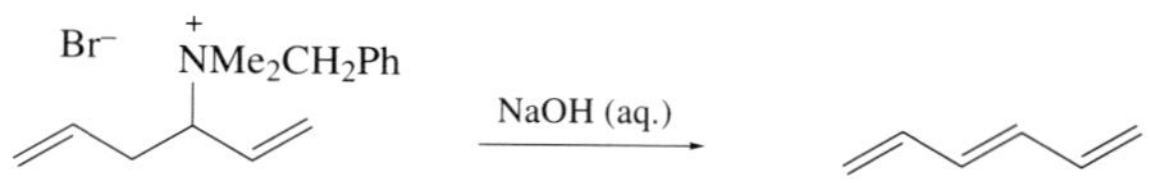

(123)

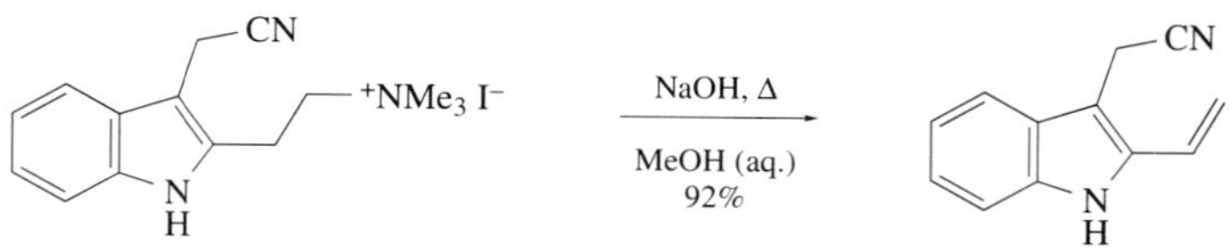

(124)

1.14.3.2.2 *Elimination of quaternary ammonium salts on treatment with base*

When quaternary ammonium halides are treated with a strong base, e.g., potassium amide or phenyllithium, elimination of the amine occurs to form an alkene ⟨57LA(612)102⟩. However this is mechanistically quite different to the Hofmann elimination as it is an α',β-elimination proceeding via an ylide (Scheme 58) ⟨71JA7118, 72JA8608, 74CC819, 74CC820, 91JA4712⟩. Products of *syn* elimination are usually formed in this elimination process; therefore the alkenes may be isomeric to those obtained in the Hofmann elimination which generally proceeds in an *anti* fashion. The mechanism of the elimination is dependent on the conditions employed; while treatment of cyclooctylammonium iodide with dimsyl anion in DMSO results in elimination via an ylide intermediate, use of potassium *t*-butoxide in DMSO leads to E2 elimination ⟨79TL3815, 91JA8821⟩. The regiochemistry in the elimination of trimethylammoniumalkyl ethers, as their iodide salts, on treatment with aqueous sodium hydroxide has been studied ⟨87ZOR1660⟩.

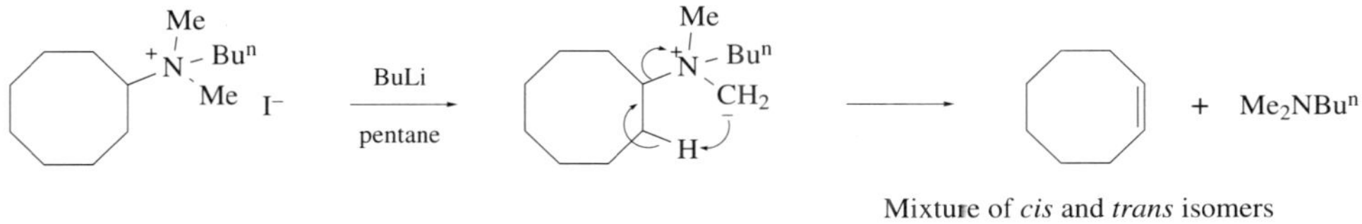

Scheme 58

The aziridinium ion shown in Equation (125) underwent eliminative cleavage in addition to substitution on treatment with base ⟨88JOC1733⟩; in less-substituted aziridinium ions only substitution is observed.

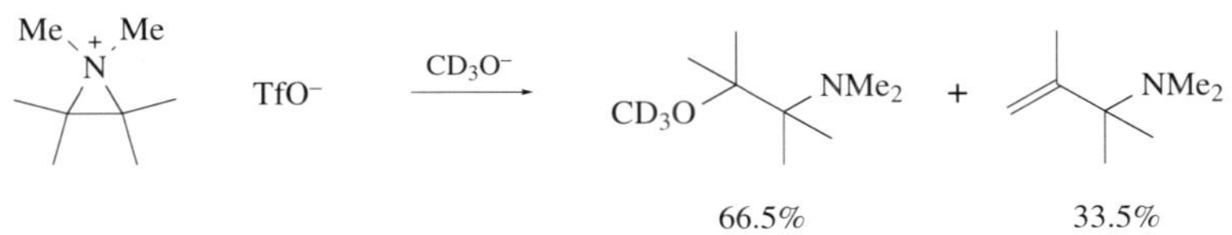

(125)

Elimination of pyridines from pyridinium cations to form alkenes has been investigated ⟨91JA6950⟩, for example in the synthesis of 2-vinylthiophene ⟨63OSC(4)980⟩. Base-induced elimination (dbu) of pyridine from a pyridinium salt has been used in a sequence for the α-alkoxyalkylation of α,β-enones via pyridiniosilylation ⟨91TL2043⟩. Base-induced elimination of related systems such as isoquinilinium cations has also been investigated ⟨89JA2258⟩. Katritzky *et al.* developed a mild alternative to the Hofmann elimination which involves conversion of primary amines, bearing primary or secondary alkyl groups, to a pyridinium salt (Scheme 59) which subsequently undergoes elimination to form the alkene ⟨81CC96⟩. With primary alkyl groups, elimination occurs on heating to 150–180°C ⟨82JOC3506⟩, but with secondary alkyl groups elimination occurs at 20°C via a carbonium ion intermediate ⟨82JCS(P1)2347⟩.

Elimination of β-silyltetraalkylammonium salts can be brought about on treatment with caesium

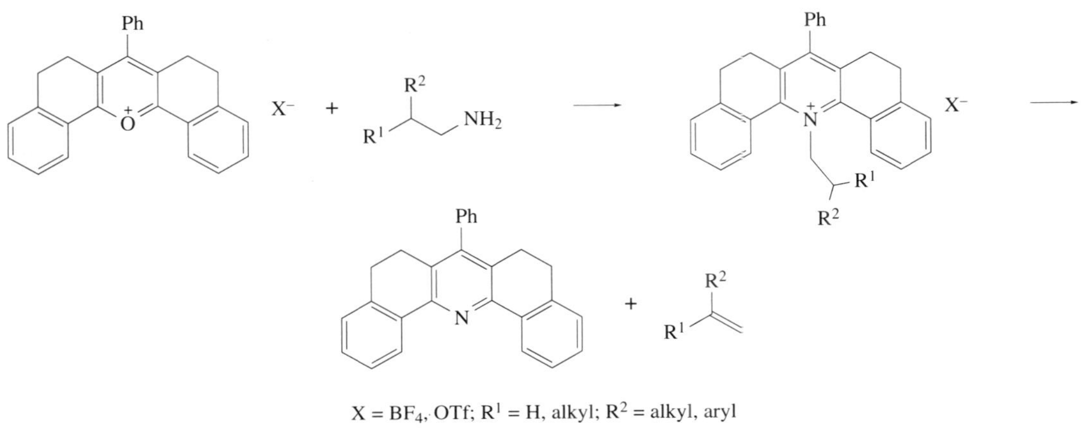

X = BF$_4$, OTf; R^1 = H, alkyl; R^2 = alkyl, aryl

Scheme 59

fluoride ⟨82JA7666⟩ in a silyl equivalent of the Hofmann elimination (Equation (126)) (see Section 1.14.5.1.3).

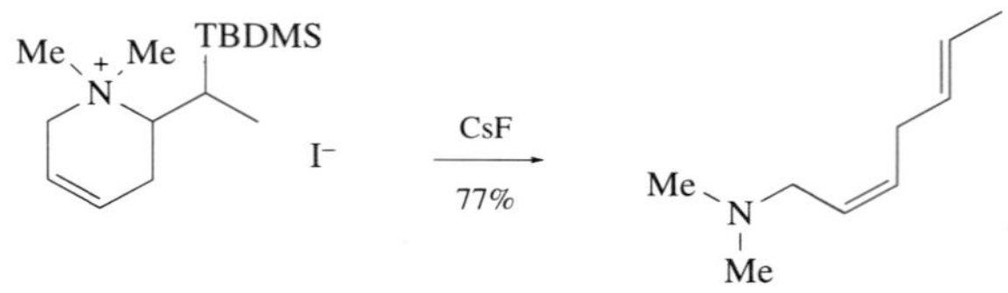

(126)

1.14.3.3 Alkenes from Arenesulfonylhydrazones

1.14.3.3.1 The Bamford–Stevens and Shapiro reactions

Decomposition of toluene-*p*-sulfonylhydrazones, bearing α-hydrogen atoms, with strong bases to form alkenes was first reported by Bamford and Stevens ⟨52JCS4735⟩. This important transformation of an aldehyde or ketone to an alkene via its hydrazone derivative (Scheme 60) has been thoroughly investigated and widely applied in alkene synthesis since that time. Excellent reviews of this transformation have appeared in the literature ⟨76OR(23)405, 83ACR55, 90OR(39)1⟩. The original conditions employed were sodium in ethylene glycol, i.e., protic conditions; an aprotic version of the Bamford–Stevens reaction is also possible using solvents such as DIGLYME. The most commonly employed base is sodium alkoxide but sodium hydride ⟨66LA(691)41⟩, lithium hydride ⟨67BSF675⟩, sodium amide ⟨66LA(691)41⟩ and lithium aluminum hydride ⟨62TL1261, 64G537, 65CB3236, 85JOC5460⟩ have also been used to decompose the hydrazones to form alkenes. Decomposition of dihydrazones by alkali has been reported ⟨67JCS(C)1964⟩.

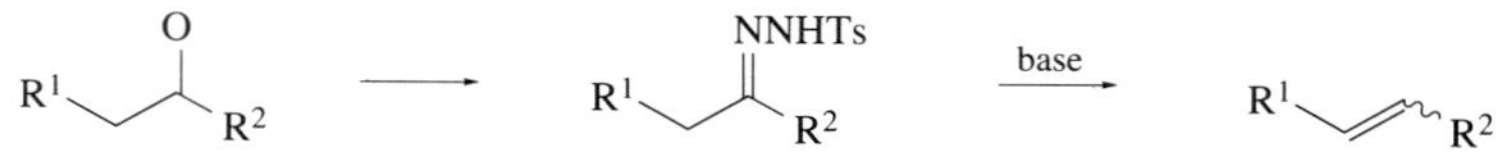

Scheme 60

Perhaps the most significant development in terms of the synthetic utility of the transformation was the discovery that alkyllithiums ⟨67JA5734, 67JA5736⟩ may be used to decompose the toluene-*p*-sulfonylhydrazones to form vinyllithiums which give alkenes on protonation; this modification became known as the Shapiro reaction. The synthetic utility of the Shapiro reaction was enhanced by extension to (2,4,6-triisopropylbenzene)sulfonylhydrazones (trisylhydrazones) which fragment more rapidly to form alkenes than tosylhydrazones and obviate the necessity for excess base ⟨78JOC147⟩. In 1992, use of *N,N*-diethylaminosulfonylhydrazones in the Shapiro reaction was reported ⟨92MI 114-01⟩. While the Shapiro reaction is in general extremely useful and successful, if a large excess of alkyllithium is employed, substitution can compete with elimination ⟨69CC1395⟩. This side reaction can also occur with highly α-substituted derivatives ⟨90OR(39)1⟩. With tosylhydrazones derived from aldehydes, addition of the alkyllithium to the azomethine bond occurs rather than elimination to form the alkene as illustrated in Scheme 61 ⟨77TL135⟩, representing a major limitation of the Shapiro methodology. Tosylhydrazones containing only tertiary α-hydrogen atoms do not undergo elimination on treatment with alkyllithiums leading instead to substitution products ⟨75TL1811, 76OR(23)405⟩; however by use of LDA in place of the alkyllithium, trisubstituted alkenes can be prepared (Equation (127)) ⟨78JOC1404⟩. LDA has also been employed in the synthesis of disubstituted alkenes (Equation (128)) ⟨87JOC3346⟩.

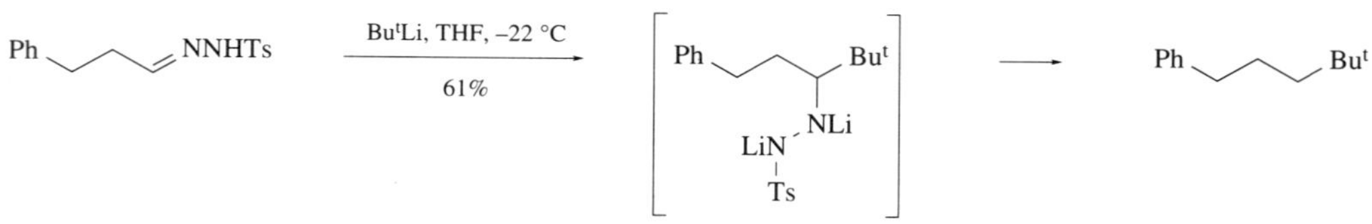

Scheme 61

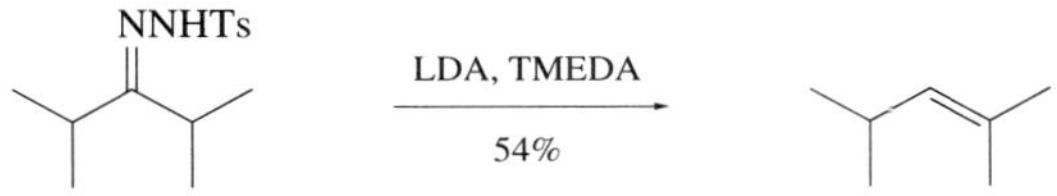

(127)

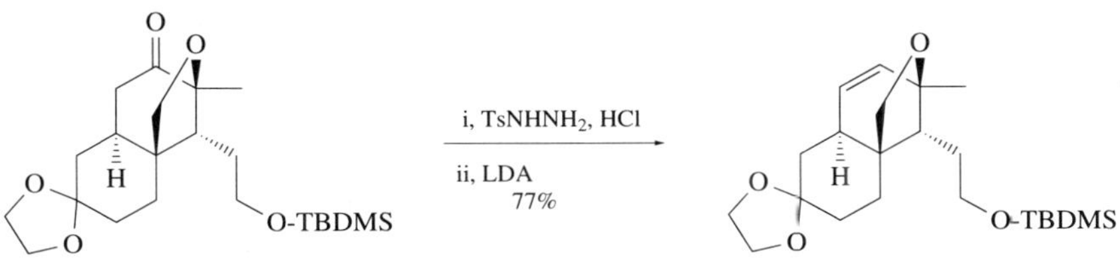

(128)

The products obtained from the Bamford–Stevens reaction frequently depend on the solvent employed ⟨68JA531⟩, the nature of the base ⟨67JA471, 67JA1442⟩ and the number of equivalents of base present ⟨59T305, 65JA659⟩. The reaction is believed to proceed initially to form a diazo compound then via either a carbene or carbocation intermediate ⟨72JA7081⟩ with subsequent formation of alkenes from these intermediates (Scheme 62) (discussed in detail in ⟨76OR(23)405⟩). The carbocation mechanism is frequently found under protic conditions while the carbene mechanism is found in aprotic conditions; however, there are exceptions to this generalization. With simple unbranched ketones (e.g., cyclohexanone) good yields of alkenes may be obtained (Equation (129)) ⟨69T3107⟩, but with branched ketones or strained cyclic ketones, rearranged products or insertion products are often present in the reaction mixtures thereby limiting the synthetic utility of the Bamford–Stevens reaction ⟨52JCS4735⟩. However, the reactions of cyclopropanecarboxaldehyde tosylhydrazones under aprotic conditions have proved useful in the synthesis of cyclobutene derivatives (Equation (130)) ⟨65CB4022⟩. In contrast to the Bamford–Stevens reaction where rearranged products are commonly obtained, the Shapiro reaction generally leads cleanly to unrearranged alkenes ⟨67JA5734⟩, even in neopentyl and highly strained systems ⟨68JA800⟩, and is therefore much more useful synthetically (Equation (131)) ⟨71JCS(C)3173⟩.

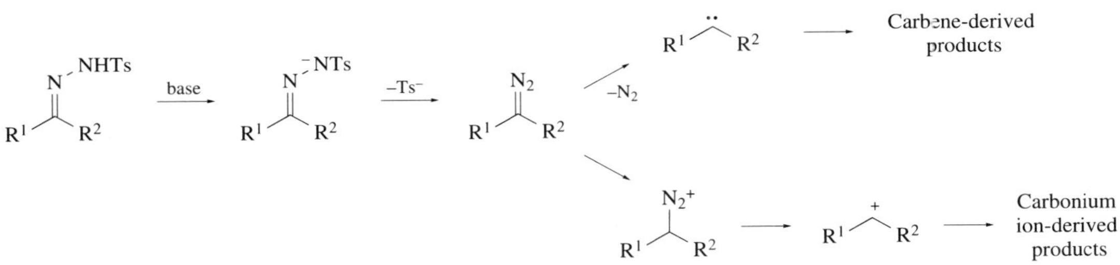

Scheme 62

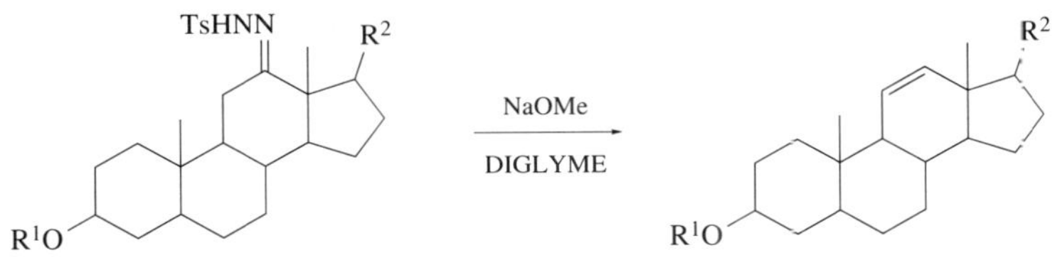

(129)

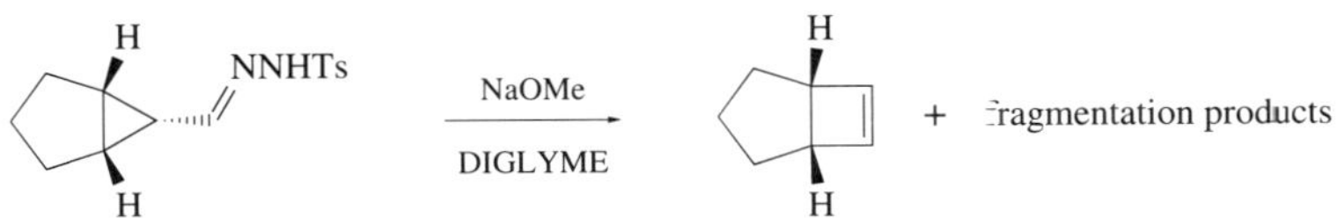

(130)

(131)

The regiochemical outcome of the tosylhydrazone decomposition also depends on the reaction conditions (Scheme 63). While the Bamford–Stevens reaction in general leads to the more-substituted alkene ⟨60JA634, 64JOC2788⟩, under Shapiro conditions the less-substituted alkene is almost always the exclusive product ⟨67JA5734⟩; in the methyllithium-induced decomposition of the hydrazone derived from phenylacetone, the terminal alkene is produced rather than the conjugated product ⟨76OR(23)405⟩. The regioselection of the Shapiro reaction was important in Paquette's synthesis of precapnelladiene (Equation (132)) ⟨85JA7352⟩. The mechanism which is believed to operate for the Shapiro reaction is shown in Scheme 64 ⟨74JOC2302, 83ACR55⟩. Thus the alkyllithium transforms the tosylhydrazone to the dianion, loss of tosylsulfinate and nitrogen produce the vinyllithium which gives the alkene on protonation. In ether or hydrocarbon solvents, the dianion formed is almost invariably the *syn* dianion ⟨75TL1811⟩ (except for α,β-unsaturated ketones ⟨77JA3414⟩) which therefore controls the regiochemistry of the product (Scheme 65); however, in strongly cation-coordinating solvents such as TMEDA ⟨78JOC147⟩, the directional influence of the nitrogen anion no longer operates and the least substituted vinyl anion is produced instead. However Trost and Nanninga have observed the formation of a mixture of regioisomers on treatment of the tosylhydrazone shown in Equation (133) with alkyllithiums ⟨85JA1293⟩; the ratio of regioisomers depended on the nature of the alkyllithium and the solvent.

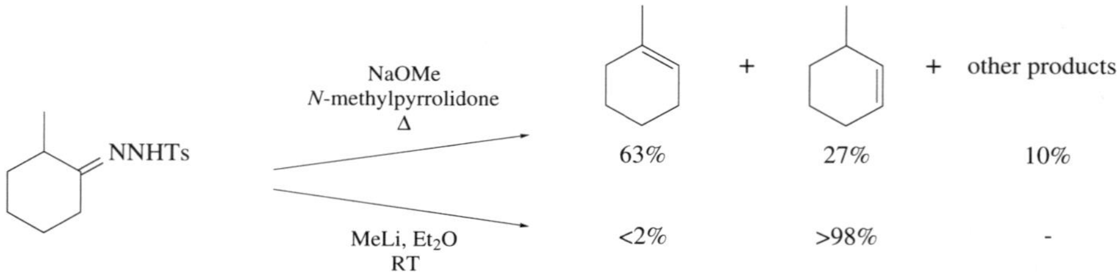

Scheme 63

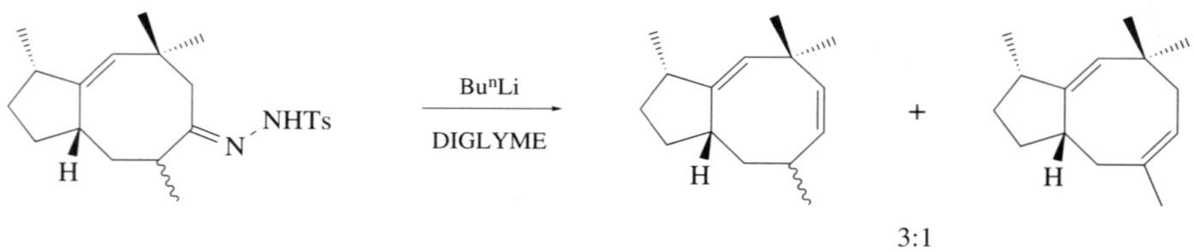

(132)

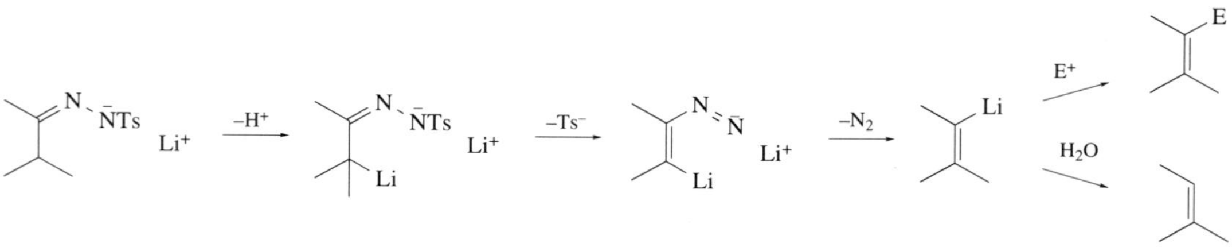

Scheme 64

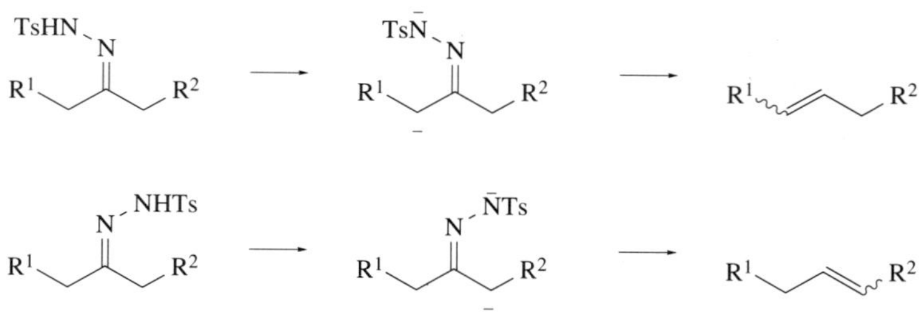

Scheme 65

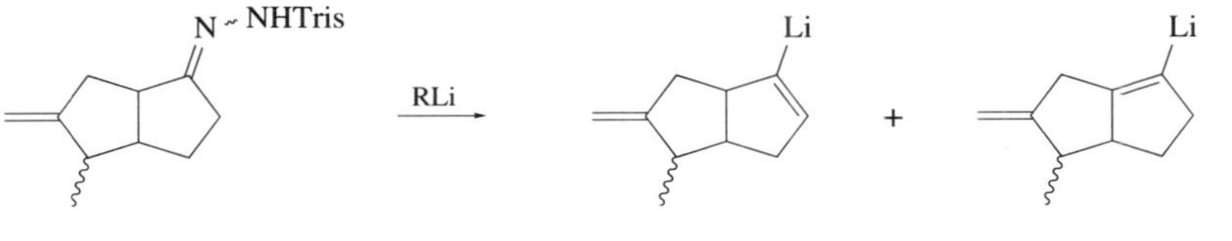

(133)

Tris = 2,4,6-triisopropylbenzenesulfonyl

The Shapiro reaction produces selectively the (*E*)-vinyllithium, thereby leading to the *cis* alkene ⟨68TL345⟩, as illustrated in Equations (134) ⟨78JOC147⟩ and (135) ⟨82OM1449⟩; however, if branching is present at the α'-position, the degree of stereoselection is reduced considerably. Investigation of the stereochemical preference in eliminations in medium-size rings (7–15 atoms) has demonstrated that the ratio of *cis*-to-*trans* isomers depends on the ring size ⟨75T1035⟩. In the LDA-induced formation of trisubstituted alkenes there is also a stereochemical preference for the formation of the *cis* isomer ⟨78JOC1404⟩. A low *trans*-to-*cis* product ratio was reported in the Bamford–Stevens reaction indicating a low degree of stereoselection in this process ⟨60JA634⟩. Marshall *et al.* demonstrated that in large rings, the Bamford–Stevens reaction also leads to a mixture of *cis* and *trans* isomers ⟨84JA6006⟩.

(134)

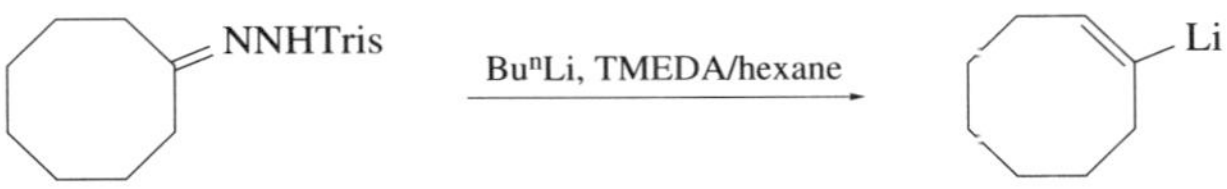

(135)

When applied to α,β-unsaturated ketones the Shapiro reaction leads to diene formation ⟨68JA4762, 82TL3733⟩. Application of this diene formation in a synthesis of compactin is illustrated in Equation (136) ⟨87JOC3541⟩.

(136)

Originally the Shapiro reaction was limited to the formation of simple alkenes by protonation of the alkenyllithium ⟨67JA5734⟩. Later it was discovered that trapping of the vinyllithium with electrophiles other than a proton can be achieved if the vinyllithium is generated using an excess of Bu^{n}Li in TMEDA ⟨75TL1815, 76TL2947⟩, or alternatively by use of (2,4,6-triisopropylbenzenesulfonyl)hydrazones ⟨78JOC147⟩ which fragment very rapidly (use of this derivative obviates the necessity for excess base). The alkenyllithiums thus formed have been trapped by a very wide range of electrophiles (e.g., alkyl halides, DMF, CO_2, aldehydes, silyl chlorides, etc.) ⟨90OR(39)1⟩ (Equation (137)) ⟨76TL2287⟩. The versatility of the Shapiro reaction is extended further by the ability to alkylate the dianion prior to fragmentation by keeping the temperature of the reaction below that required for the fragmentation (typically 0–20 °C); subsequent addition of a further equivalent of base brings about the elimination reaction (Scheme 66) ⟨78JOC1409, 80JOM(186)155, 81JCS(P1)2848, 81JOC1315⟩. Shapiro and Lipton have demonstrated the application of this methodology in the synthesis of homoallylic alcohols (Scheme 67) ⟨78JOC1409⟩. Notably the elimination from the trianion occurs preferentially in the direction away from the oxygen atom in the β-hydroxytosylhydrazones ⟨78JOC1409⟩.

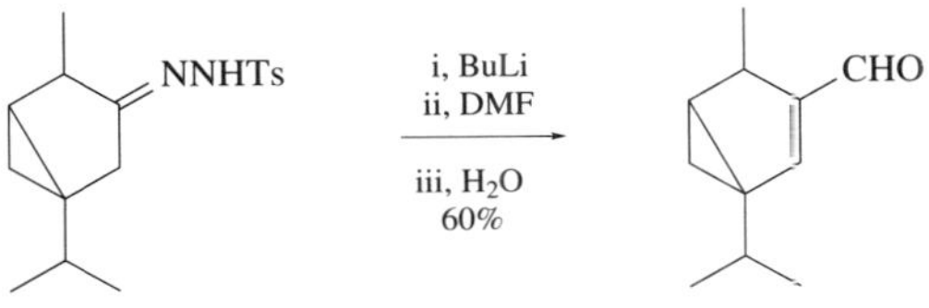

(137)

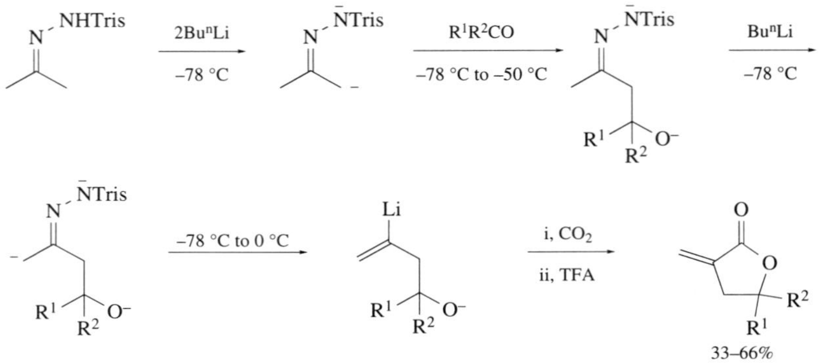

Scheme 66

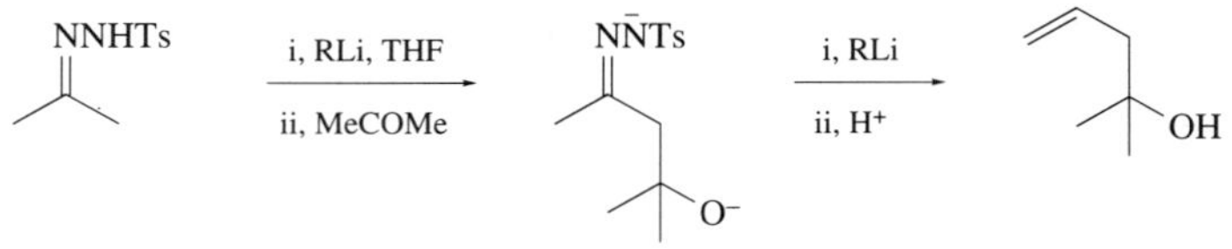

Scheme 67

When applied to α-keto amides (Scheme 68) the Shapiro reaction generates the dianion which protonates or alkylates at carbon, thereby acting as a synthon for the propenoic acid anion ⟨81CC65, 81T3935⟩. Baldwin and Bottaro used a Shapiro reaction on an amide derivative to produce an acyl anion equivalent (Equation (138)) ⟨81CC1121⟩.

E = H^+, alkyl halide, aldehyde, ketone, etc.

Scheme 68

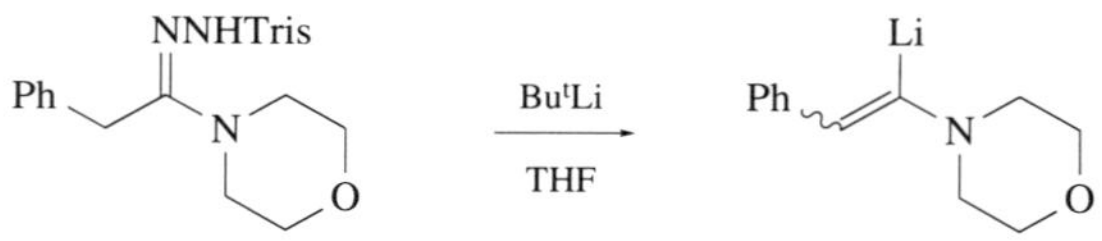

(138)

The Shapiro reaction has been employed for the synthesis of polycyclic alkenes (Equation (139)) ⟨87JOC2644, 87TL1985, 87TL6159⟩, while a triple Shapiro degradation was employed in the synthesis of [4.4.4]propellahexaene ⟨87JA3174⟩. When the hydrazone bears a leaving group in the α-position, elimination occurs at the monoanion stage to form a tosylazoalkene; this may be employed for the preparation of lithioalkadienes (Scheme 69) ⟨81JA5314⟩. With the trifluoromethyl derivative shown in Equation (140) elimination occurred to form the difluoroallene ⟨89JFC(44)161⟩.

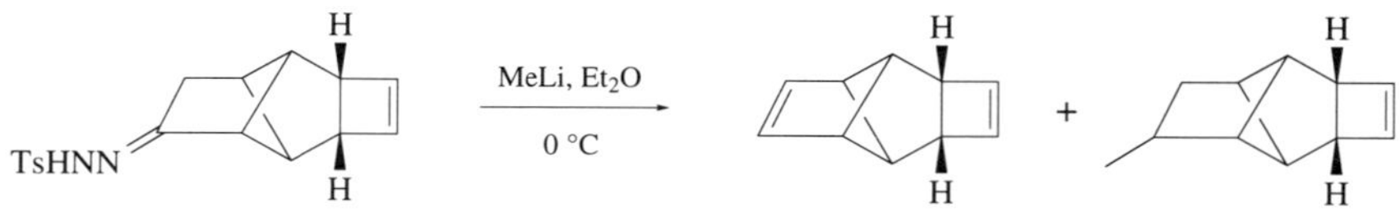

(139)

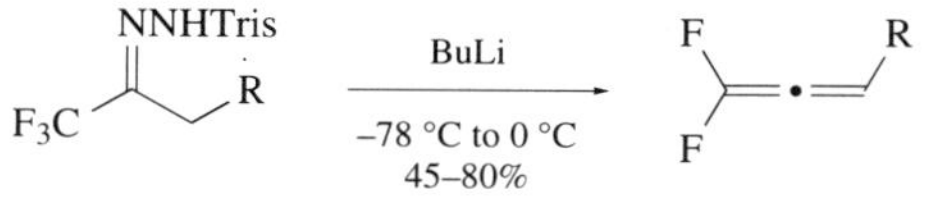

(140)

R = H, alkyl, Ph

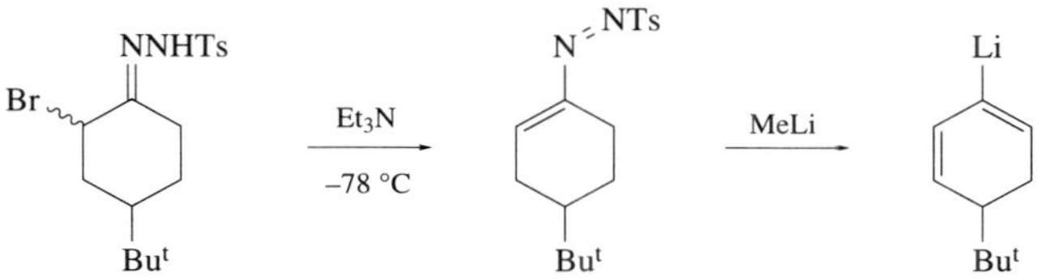

Scheme 69

1.14.3.3.2 Alkenes from hydrazones by alternative routes

Tosylhydrazones of α,β-unsaturated carbonyl compounds can be transformed to alkenes in which the double bond has migrated towards the carbon centre which was originally the carbonyl carbon atom, on treatment with sodium cyanoborohydride in acidic conditions ⟨75JOC923, 76JA2275, 78JOC2299, 79OPP201, 84JA3353, 85JOC2798⟩, or catecholborane ⟨76JOC574, 80OS(59)42, 86JA6276, 86JA7686⟩. The reaction is believed to proceed via a 1,5-sigmatropic rearrangement in an intermediate diazene as illustrated in Scheme 70. The migration occurs cleanly even if the double bond is moving out of conjugation and proceeds stereoselectively to furnish the (*E*) isomer as the major product. The transformation can prove useful in preparing less-substituted alkenes, in moving double bonds out of conjugation (see, for example, Equation (141)) ⟨75JOC923⟩ and in moving an endocyclic double bond to an exocyclic position.

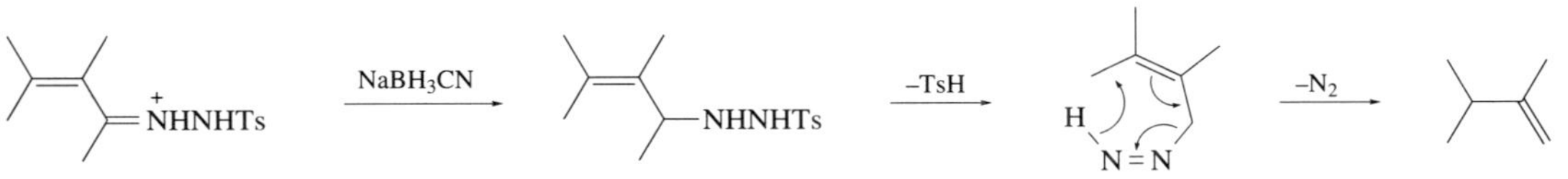

Scheme 70

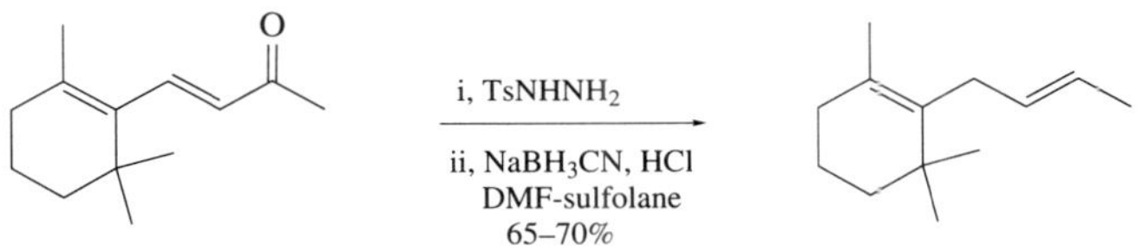

(141)

While 1,2-dihydrazones are converted to alkynes by treatment with reagents such as Ag_2O, HgO or $Cu(II)/O_2$/pyridine ⟨53JA2153, 70TL761, 83T3279⟩, formation of a small amount of an alkene has been observed ⟨70TL761⟩.

1.14.3.4 Elimination of Amine Derivatives

Elimination of amines under conditions other than those employed for the Hofmann and Cope reactions is also possible. Deamination in Mannich bases has been discussed in a review ⟨90T1791⟩. This is a particularly facile process in β-amino ketones and other carbonyl derivatives; for example, Equation (142) illustrates the elimination of azetidine to form an α,β-unsaturated carboxylic acid ⟨73OS(53)13⟩, while Equation (143) illustrates the elimination of dimethylamine to form functionalized decalin derivatives ⟨90HCA640⟩. The stereochemistry of the elimination of the secondary amine illustrated in Equation (144) is dependent on the conditions employed ⟨88BCJ3211⟩.

(142)

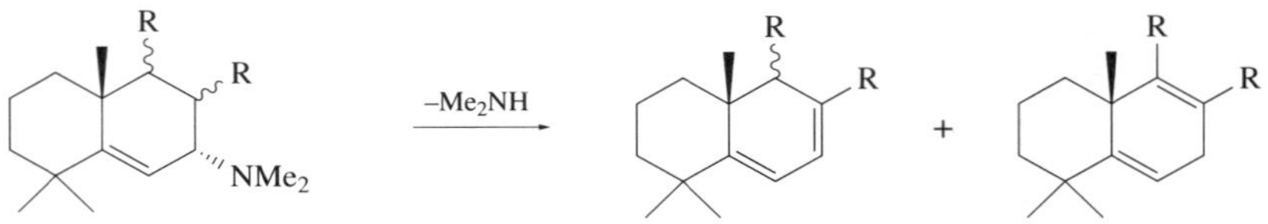

(143)

R = CO_2Me or CN

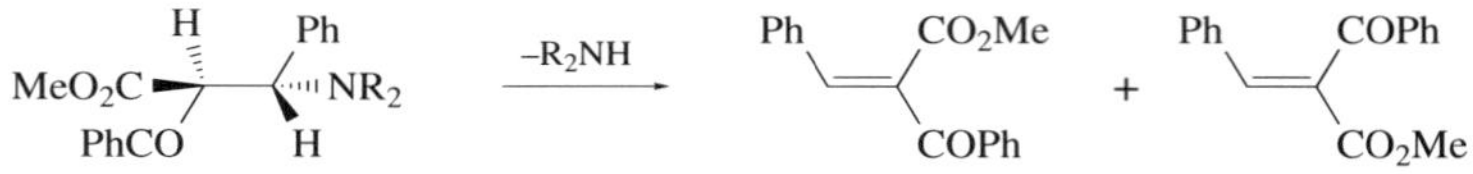

(144)

MeCN, 20–60 °C	100%	0%
AcOH	51–62%	38–49%

$R_2 = -(CH_2)_4-, -(CH_2)_5-, -(CH_2)_6-$

Secondary amines are readily transformed to a magnesium amide which decomposes on heating to 150–250 °C to form an alkene (Scheme 71) ⟨78JOC4750⟩. The elimination follows *syn* stereochemistry and is believed to proceed via a six-membered cyclic transition state as shown (Scheme 71). A similar reaction occurs with zinc and aluminum amides. An amine bearing a *β*-chloro substituent eliminates on treatment with magnesium ⟨63JOC878⟩.

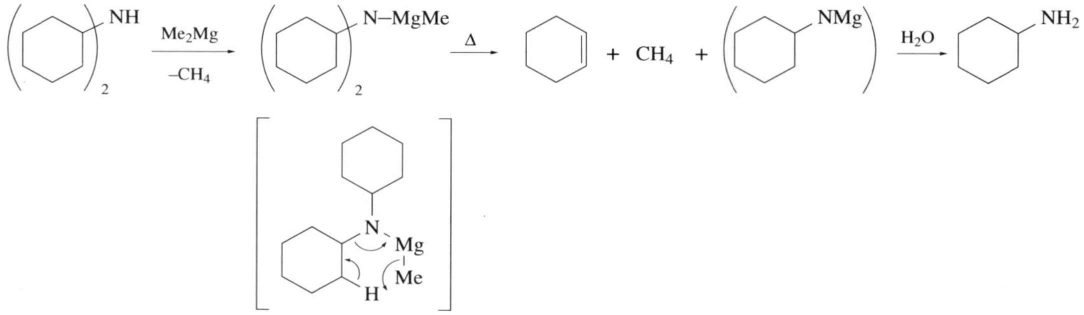

Scheme 71

Primary amines, bearing secondary alkyl groups, on transformation to *N*,*N*-disulfonimide derivatives followed by pyrolysis (160–200 °C) eliminate very cleanly to form alkenes with good stereocontrol (Scheme 72) ⟨81TL199⟩.

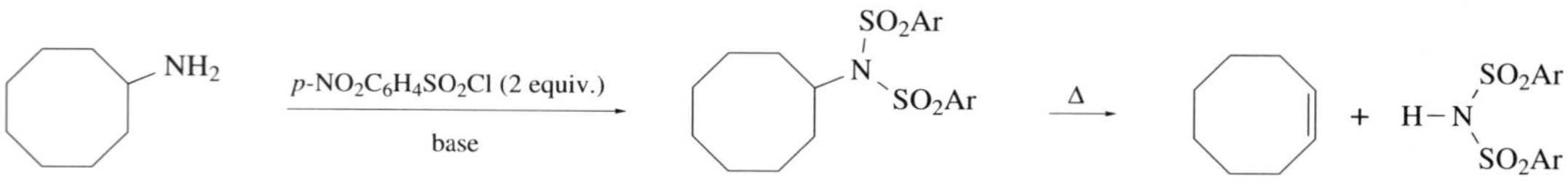

Scheme 72

Elimination of aziridines to form alkenes has been reviewed ⟨80T557⟩. On nitrozation with a variety of reagents, such as *n*-butyl nitrite or nitrosyl chloride, aziridines deaminate to form alkenes with retention of the stereochemistry of the aziridine in the product (Equation (145)) ⟨64JOC1316, 69TL4001⟩. Treatment of aziridines with difluoroamine leads to alkene formation with a similar retention of stereochemistry ⟨67JA1761⟩. Oxidation of *N*-aminoaziridines with manganese dioxide again results in alkene formation; the stereochemistry of the aziridine is not retained when phenyl substituents are present ⟨70JA1784⟩. *N*-Alkylaziridines can be deaminated by reaction with mcpba (Equation (146)) ⟨70AG(E)374⟩, ferrous iodide ⟨74CL165⟩, dichlorocarbene ⟨72TL3827⟩ or the copper carbenoid derived from ethyl diazoacetate ⟨72TL4659⟩. The reaction with mcpba presumably proceeds via an *N*-oxide intermediate. When conducted in benzene in the dark, this reaction proceeds stereospecifically and produces the alkene efficiently; in dichloromethane both the yield and stereospecificity are much lower.

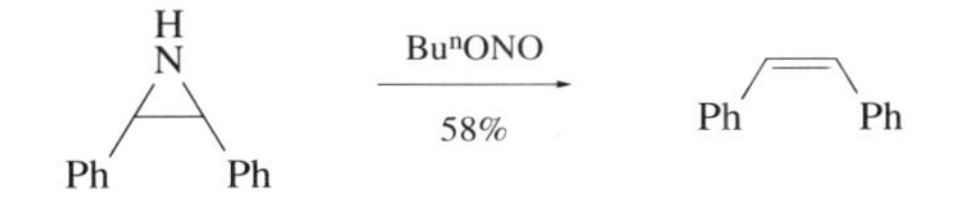

(145)

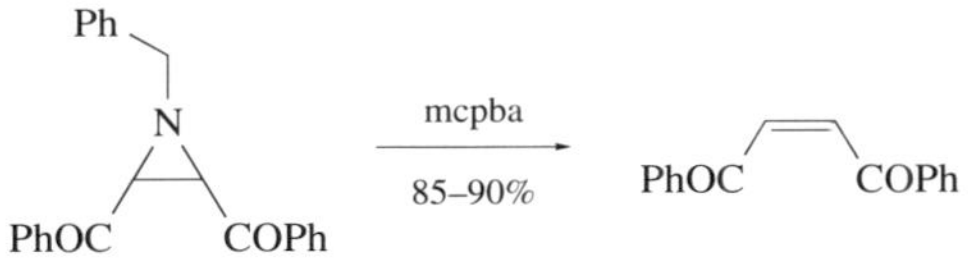

(146)

Elimination of amides to form alkenes ⟨60CRV431, 60OR(11)317⟩ can occur on pyrolysis with ⟨44JCS322, 46JA502, 49JCS1074⟩ or without ⟨58JOC996⟩ phosphorus pentoxide. An example of this elimination in a colchicine derivative is shown in Equation (147) ⟨44JCS322⟩. In the reaction with P_2O_5, formation of a nitrile from the amido group is likely, while in the absence of P_2O_5 the reaction appears to be analogous to acetate pyrolysis, proceeding with similar regiochemistry; however, higher temperatures are required for amide elimination ⟨58JA4588, 59JA651⟩. Acetamides in which the nitrogen atom bears a tertiary alkyl group eliminate on refluxing in 15% HCl ⟨48JA4045⟩.

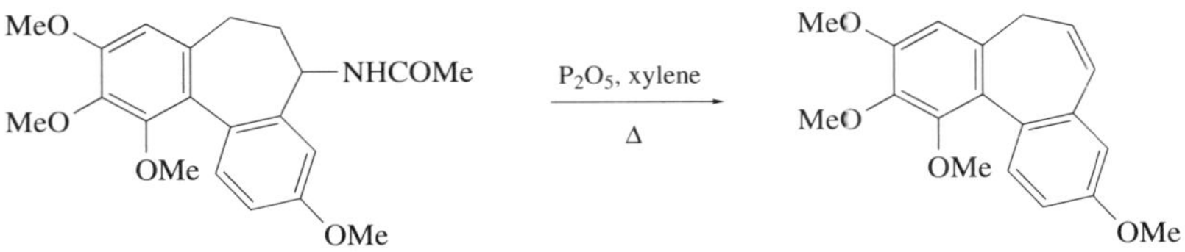

(147)

Deamination of amines with nitrous acid frequently results in the formation of alkenes (among other products), via elimination in intermediate diazonium or carbonium ions ⟨B-68MI 114-03⟩; an example is shown in Equation (148) ⟨89JOC5424⟩. However, rearrangements are common ⟨89CB389⟩ and this process is rarely synthetically useful. Elimination of nitrogen from diazoalkanes also leads to alkenes ⟨B-86MI 114-04, 87TCC75⟩; for example the α-diazoester shown in Equation (149) leads to the *cis*-enoate on catalytic decomposition with rhodium(II) acetate ⟨70CJC2110, 81TL4163⟩. Thermal decomposition of *N*-nitrosoacetamides (readily prepared from amines) leads to diazoalkanes; if this thermolysis is conducted in the presence of rhodium(II) acetate the diazoalkane is intercepted forming rhodium carbenoids; a 1,2-hydrogen migration then leads to alkenes (Scheme 73) ⟨90JA3717⟩. Palladium-catalysed decomposition of 3,3-pentamethylenediazirine affords cyclohexene (Equation (150)) ⟨90JMOCL1⟩, presumably via a carbene intermediate.

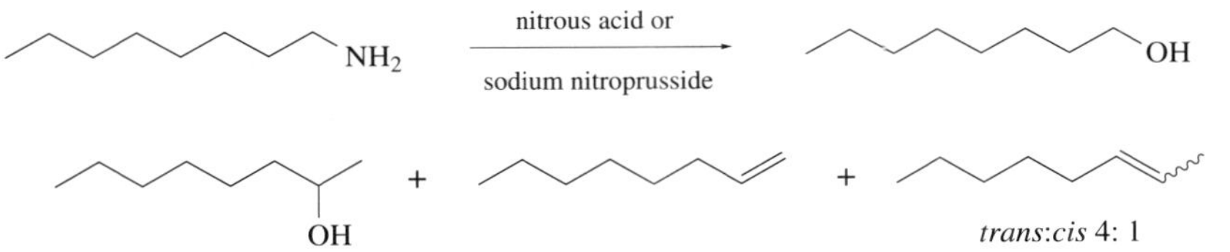

(148)

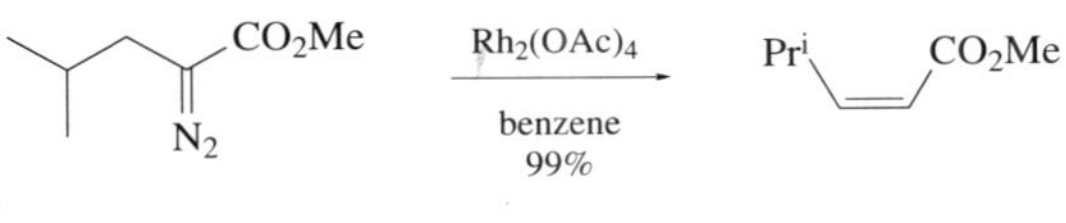

(149)

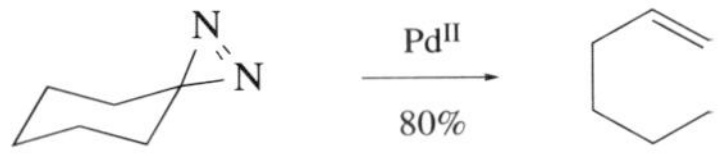

(150)

1.14.3.5 Elimination of Nitro Groups

Nitrous acid elimination may be effected thermally from quaternary nitroalkanes ⟨88TL2805, 89BSF409, 90JOC5801⟩ or from an angular tertiary nitro group on contact with silica gel or neutral alumina ⟨84JOC670⟩ or by treatment with a base, for example, sodium methoxide ⟨71JA4316⟩. Thermal elimination of nitrous acid from quaternary nitroalkanes proceeds selectively to give the

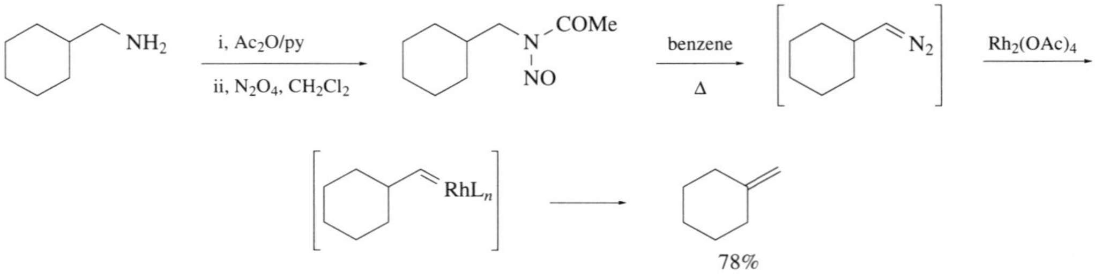

Scheme 73

less-substituted alkene (Equation (151)) ⟨88TL2805⟩. Notably, thermal nitrous acid elimination produces β,γ-unsaturated esters in good yields with no evidence of double-bond migration ⟨90JOC5801⟩.

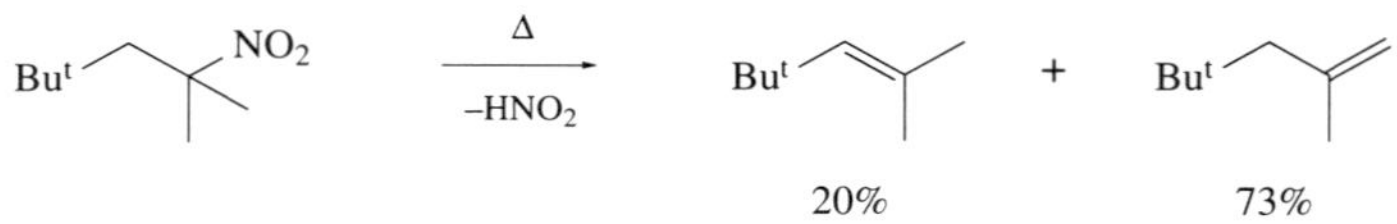

(151)

Treatment of β-substituted nitro compounds with tributyltin hydride produces alkenes via a radical elimination—this has been observed with β-nitrosulfides as illustrated in Equation (7) ⟨84TL5319⟩, β-nitrosulfones as shown in Scheme 44 and Equation (77) ⟨81CL1139, 87JOC5111, 88JOC251⟩ and β-dinitro compounds ⟨81CL1139⟩. Formation of (*E*)-allyl alcohol derivatives in a stereoselective fashion by elimination of β-nitrosulfides is shown in Equation (7) ⟨88CC1278⟩. The elimination of the nitro and sulfide/sulfone groups occurs stereospecifically in an *anti* conformation (see Sections 1.14.2.1.1 and 1.14.2.3.1) while the elimination of a dinitro compound is nonstereospecific as illustrated in Equation (152) ⟨81CL1139⟩. Elimination of β-dinitro compounds to form tetra-substituted alkenes can also be achieved by use of sodium sulfide ⟨71JA4316⟩ or calcium amalgam ⟨77JOC2944⟩. Tin(II) chloride effects elimination in α,β-diaryl vicinal dinitro compounds but not in the corresponding aliphatic compounds ⟨79BCJ1107⟩; the reaction is believed to proceed via a radical anion intermediate. Only the *trans* alkene is produced irrespective of the stereochemistry of the starting material. Best results are obtained by use of an excess of tin(II) chloride in ethanol, although DMF may also be employed for this transformation. In contrast to the eliminations under free-radical conditions, thermolysis of symmetrical dinitrotetraalkylethanes forms alkylated 1,3-butadienes by elimination of nitrous acid ⟨88CI06⟩.

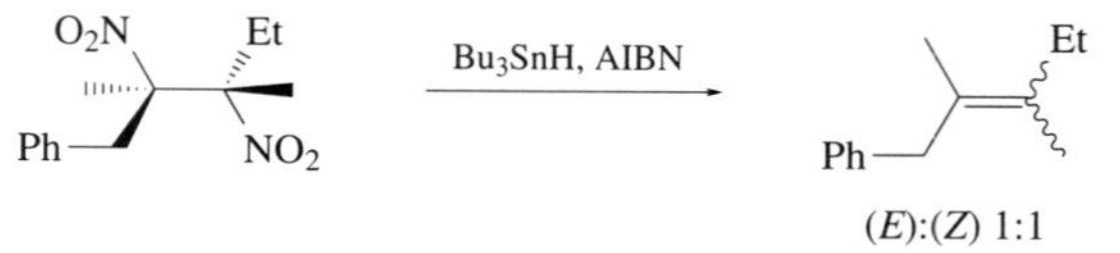

(152)

1.14.4 BY ELIMINATION OF PHOSPHORUS, ARSENIC, ANTIMONY OR BISMUTH FUNCTIONS

1.14.4.1 Elimination of Phosphorus Groups

Elimination of phosphorus groups almost always involves concurrent loss of β-oxygenated groups; the affinity of phosphorus for oxygen is largely responsible for this reactivity. The most widely employed alkene formation involving elimination of phosphorus groups is the Wittig reaction ⟨53LA(580)44⟩ and modifications thereof. As these transformations are discussed in detail in Chapter 1.16 only the elimination step in these processes will be described here. Many reviews of these extremely useful reactions have appeared (for example ⟨63QR406, 65OR(14)270, 74CRV87, 77OR(25)73, B-79MI 114-01, 89CRV863⟩).

1.14.4.1.1 Elimination of phosphine oxide—the Wittig reaction

Elimination of a phosphine oxide from a phosphonium betaine, formed by condensation of a phosphonium ylide with an aldehyde or ketone, is believed to proceed via an oxaphosphetane intermediate (Scheme 74). The relative importance of betaine versus oxaphosphetanes as intermediates has been studied in detail ⟨89CRV863⟩, and it appears that oxaphosphetanes are more likely as genuine intermediates even though some evidence for the existence of the betaine intermediate has been obtained, for example when benzaldehyde is treated with methylenetriphenylphosphorane the zwitterion could be isolated as its hydrobromide. Treatment with base regenerates the betaine which decomposes on heating to 65°C to give triphenylphosphine oxide and styrene ⟨54CB1318⟩. As the elimination occurs in a *syn* fashion the stereochemistry of the resulting alkene is dependent on the stereochemistry of the betaine. Thus the *erythro* diastereomer of the betaine leads to the *cis* alkene while the *threo* leads to the *trans* alkene (Scheme 75) ⟨68AG(E)650⟩. Therefore the stereochemical outcome of the Wittig reaction is dependent on the factors which control the stereochemistry and reversibility of the formation of the zwitterion (see for example ⟨64AG(E)250⟩). While (*Z*) isomers are preferred from reactions with 'nonstabilized' ylides, mixtures of (*Z*)- and (*E*)-isomers are commonly obtained.

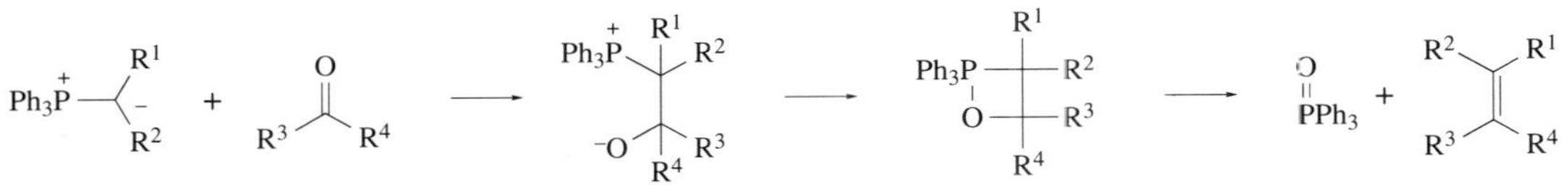

Scheme 74

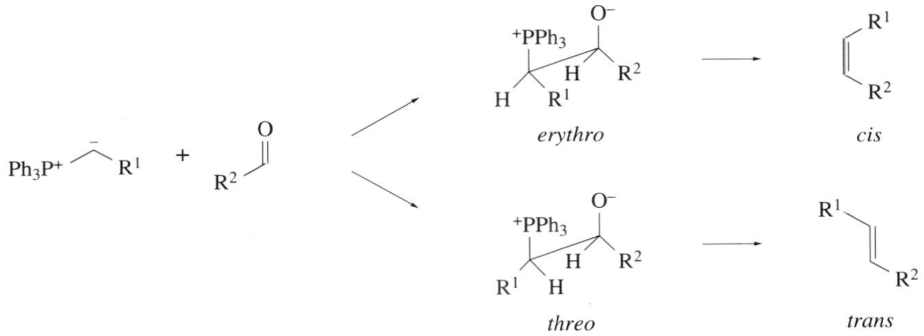

Scheme 75

1.14.4.1.2 Extension to phosphoryl derivatives—the Horner–Wadsworth–Emmons modifications

Carbanions obtained by deprotonation of phosphine oxides, $R_2P(O)CH_2R$, phosphinates, $RP(OR)(O)CH_2R$, or phosphonates, $(RO)_2P(O)CH_2R$, may be used in place of the phosphonium ylide in the Wittig reaction resulting in elimination of phosphinate, phosphonate or phosphate anions, respectively ⟨58CB61, 59CB2499, 74CRV87, 77OR(25)73⟩. The phosphinate modification has received little attention as the carbanions are more difficult to prepare than either of the other two. Use of phosphonates can offer considerable advantage over the Wittig reaction ⟨61JA1733⟩, for example, ease of workup, availability of reagents and convenient reaction conditions. Mechanistically these reactions are believed to be similar to the Wittig reaction again proceeding via a four-membered cyclic transition state possessing some double-bond character (Scheme 76) ⟨74CRV87⟩. There is strong evidence for the existence of the intermediate oxyanion; in some cases, the protonated forms, *β*-hydroxyphosphonates, are isolated from the reaction instead of the elimination products (Scheme 77) ⟨66JA5654⟩. Efficient elimination occurs only if the phosphonate bears an electron-withdrawing group *α* to the phosphorus atom, i.e, R^1 or R^2, to activate the intermediate (Equation (153)) ⟨68CC1699⟩. The stereochemistry of the alkene resulting from the Horner or Wadsworth–Emmons modifications is controlled by the stereochemistry in the intermediate oxyanion as the elimination occurs in a *syn* manner, similar to the Wittig reaction. However *trans* alkene formation predominates in the phosphoryl reactions; reversibility in the formation of the oxyanion eventually results in the formation of the thermodynamically more stable isomer ⟨77OR(25)73, 84T5153⟩. Use of barium hydroxide as the preferred base for a Horner–Wadsworth–Emmons reaction with a base-sensitive aldehyde has been reported ⟨93SL774⟩.

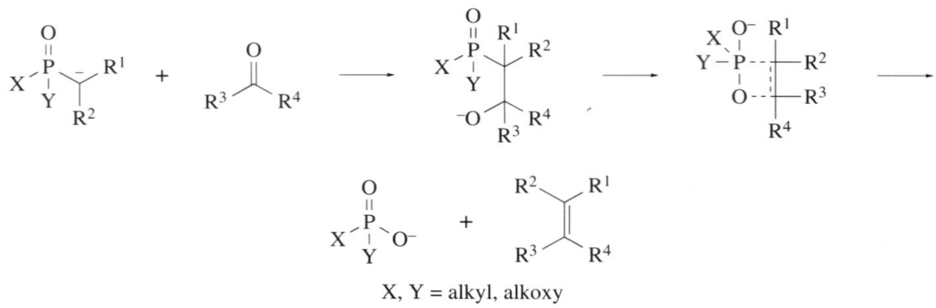

Scheme 76

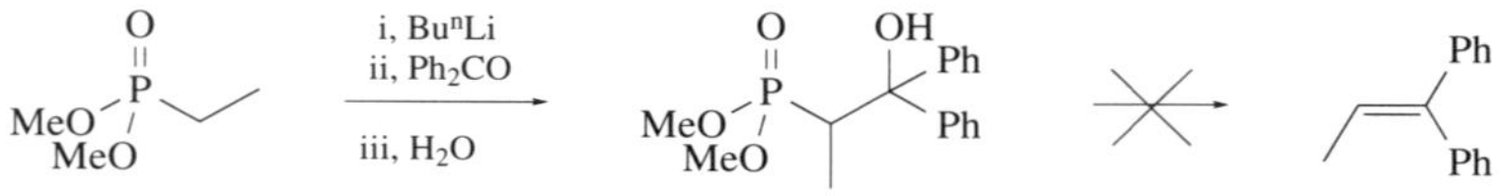

Scheme 77

 (153)

1.14.4.1.3 *Elimination of phosphonium groups*

Unlike ammonium salts, phosphonium salts decompose to give a hydrocarbon and phosphine oxide rather than an alkene ⟨29JCS2342, 60OR(11)317⟩. However betaines, like those formed as intermediates in the Wittig reaction, undergo elimination of phosphine oxide to form alkenes. These betaines may be formed from epoxides by treatment with triphenylphosphine ⟨55CB1654⟩ or diphenylphosphide followed by quaternization with methyl iodide ⟨71JA4070, 73JA822⟩. As the epoxide ring opening to form the betaine occurs with inversion of configuration followed by *syn* elimination of the phosphine oxide, this results in an overall change of stereochemistry of the alkene unit as illustrated in Scheme 78, and therefore may be used for isomerization of alkenes ⟨80T557⟩. (Alternatively the β-hydroxy phosphine derivatives formed by epoxide ring opening with diphenyl phosphide can be oxidized to the corresponding phosphine oxide prior to elimination with sodium hydride ⟨74CC142⟩ as discussed below.) Both di- and trisubstituted alkenes can be formed under these conditions; *trans*-cyclooctene and *cis,trans*-1,5-cyclooctadiene are produced with >99.5% stereospecificity from the corresponding *cis* epoxides ⟨71JA4070⟩. However, formation of *trans*-cycloheptene was not possible ⟨71JA4070⟩; whereas betaine decomposition usually occurs at 25°C, the betaine derived from cycloheptene oxide is stable at this temperature. On refluxing in THF elimination occurs to form the vinyl phosphonium salt (Scheme 79). If dbu is present, then *cis*-cycloheptene is also formed. Clearly the four-membered cyclic transition state required for *syn* elimination is not easily accessible in the seven-membered ring. While the diphenylphosphide reaction may be conducted with keto epoxides under carefully controlled conditions, the presence of an ester group is not compatible with the reaction conditions, as the anion attacks the ester group ⟨73JA822⟩.

An unusual elimination from a phosphonium ylide believed to proceed via an oxaphosphinane has been reported ⟨89CC1464⟩.

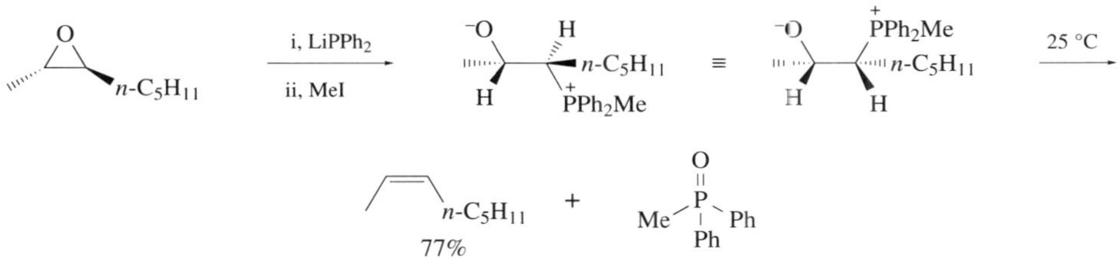

Scheme 78

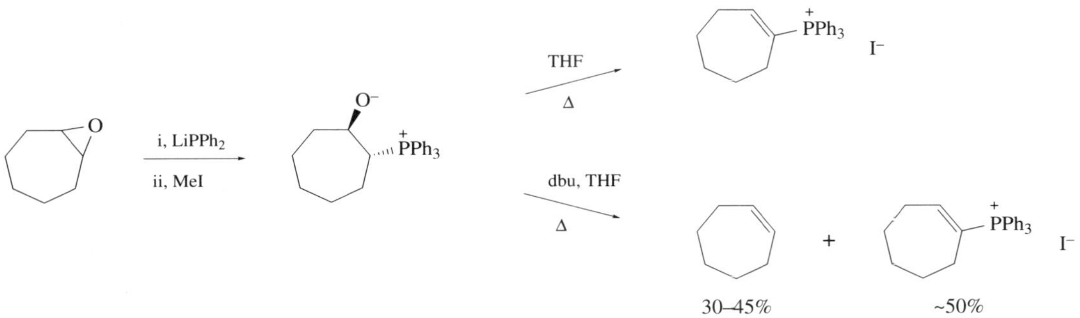

Scheme 79

1.14.4.1.4 Elimination of β-hydroxy phosphine oxides

Syn elimination of β-hydroxy phosphine oxides on treatment with sodium hydride may be employed for alkene isomerization as illustrated in Scheme 80 ⟨74CC142⟩. This modification of Vedejs' method ⟨71JA4070, 73JA822⟩ has the advantage that the diphenylphosphinate salt produced in the elimination is water soluble, thereby facilitating workup.

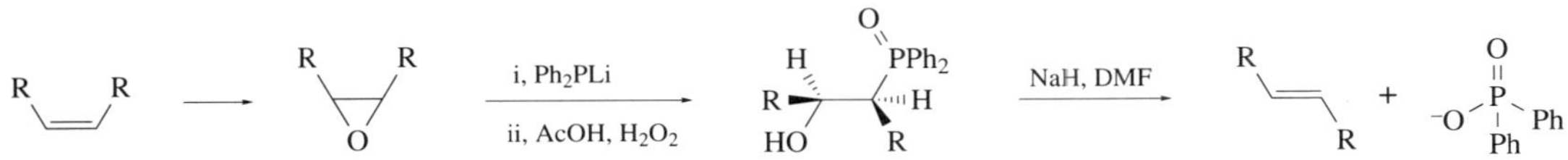

Scheme 80

Warren and co-workers have developed stereoselective syntheses of alkenes by elimination of phosphinate anions from β-hydroxy phosphine oxides ⟨81CC100, 83TL111, 83TL5293, 88JCS(P1)1799⟩. Addition of the lithium derivatives of alkyl diphenylphosphine oxides to aldehydes forms the phosphinoyl alcohols, predominantly as the *erythro* diastereomer, in good yields. Isolation of the alcohol as a single diastereomer is possible. *Syn* elimination occurs on treatment with sodium hydride in DMF at 50°C to form the (*Z*) alkene (Scheme 81) ⟨81CC100, 84JCS(P1)243⟩. When a secondary alkyl phosphine oxide is employed, a mixture of diastereomers is formed; however, these may be separated and eliminated separately to produce both the (*E*)- and (*Z*)-trisubstituted alkenes as single isomers ⟨83TL111⟩. Alternatively, trisubstituted alkenes can be formed by reaction of primary alkyl phosphine oxides with ketones, as in a synthesis of (*Z*)-α-bisabolene (Scheme 82) ⟨83TL111⟩. However, this approach can lead to difficulties in some cases; for example, one hydroxy phosphine oxide adduct reverted to starting materials on treatment with base while another decomposed rapidly even at −50°C preventing diastereomer separation and thereby producing alkene mixtures.

An alternative approach to β-hydroxy phosphine oxides involves the reduction of β-keto phosphine oxides with sodium or lithium borohydride; this produces the opposite stereoisomer to that described above ⟨85JCS(P1)2307⟩. β-Keto phosphine oxides (readily obtained by acylation of alkyl phosphine oxides with esters or lactones) give the *threo* alcohols selectively; these can be purified to a single diastereomer and eliminated to form the (*E*) alkene as shown in the synthesis of feniculin (Scheme 83) ⟨81CC100, 83TL5293⟩. With secondary alkyl phosphine oxides, acylation with esters is

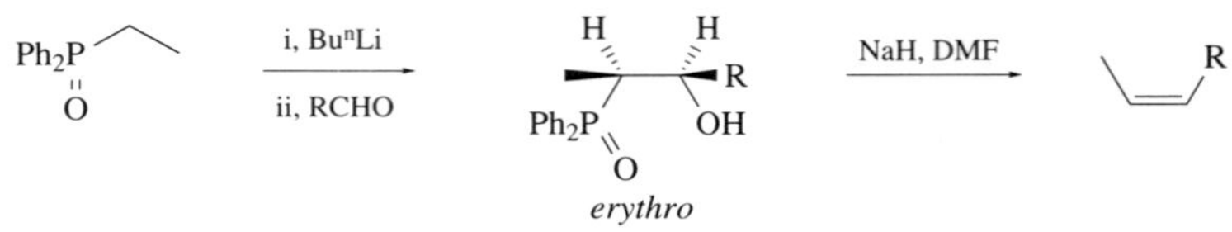

R = alkyl, aryl

Scheme 81

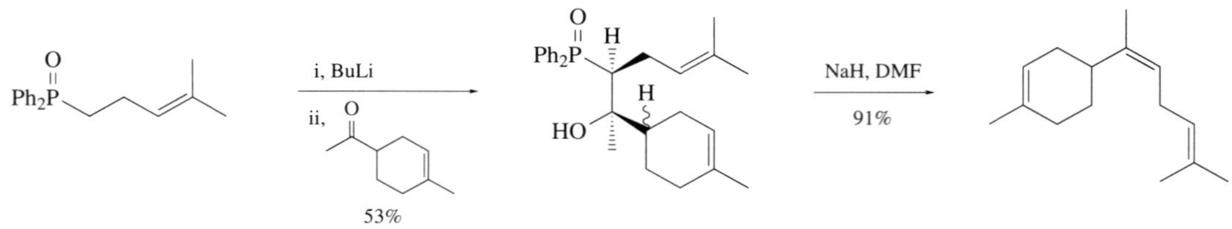

Scheme 82

inefficient but on reacting the appropriate copper derivative with an acid chloride the keto derivatives are readily obtained (Scheme 84) ⟨83TL111⟩. Reduction of these ketones leads to a mixture of diastereomers which are separable and therefore can be transformed to either the (*E*) or (*Z*) alkene. Alternatively, if a mixture of diastereomers of β-hydroxy phosphine oxides is formed by alkylation with an aldehyde, then oxidation to form the corresponding β-keto phosphine oxide followed by stereoselective reduction can be employed to form the (*E*) alkene (Scheme 85) ⟨83TL5293⟩.

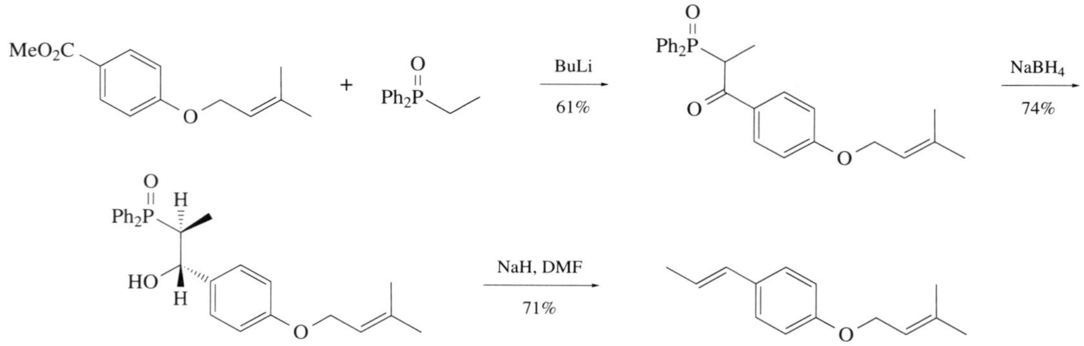

Scheme 83

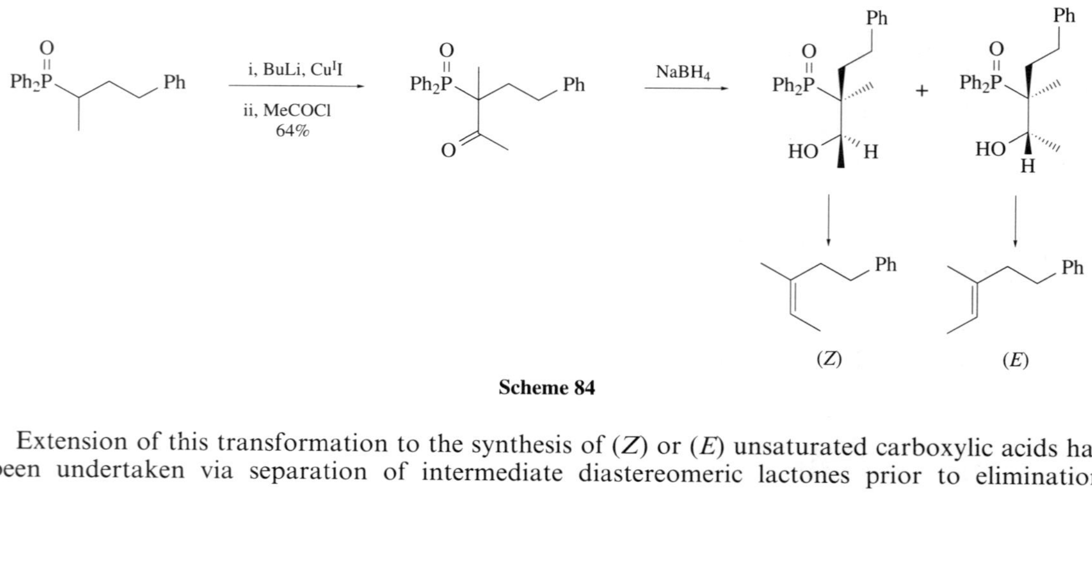

Scheme 84

Extension of this transformation to the synthesis of (*Z*) or (*E*) unsaturated carboxylic acids has been undertaken via separation of intermediate diastereomeric lactones prior to elimination

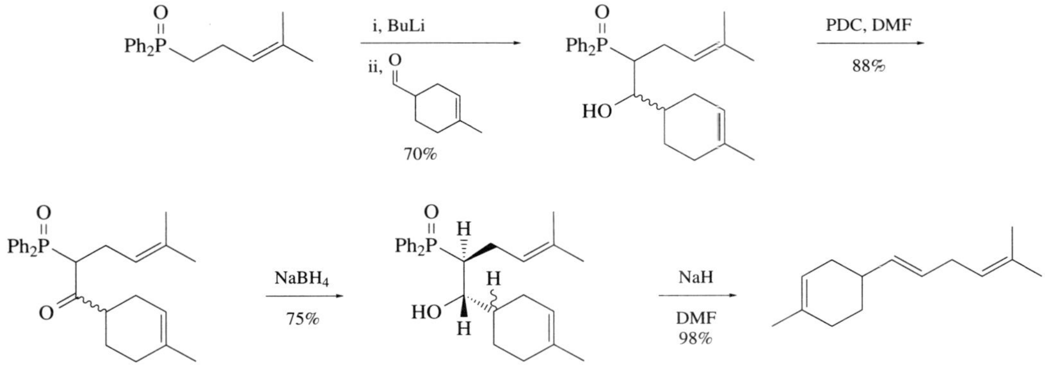

Scheme 85

⟨85TL505, 86TL2265, 88JCS(P1)1799, 92JCS(P1)2155⟩. Application of this methodology to the synthesis of unsaturated alcohols ⟨84TL357, 88JCS(P1)2971, 92JCS(P1)3407⟩ and dienols ⟨91JCS(P1)1485⟩ has also been reported. Formation of an allyl sulfide ⟨83JCS(P1)1169⟩ and conjugated dienes ⟨77JCS(P1)550, 77JCS(P1)1452⟩ have also been observed.

The synthetic utility of this methodology is demonstrated in Wittman and Kallmerten's synthesis of oudemansins ⟨87JOC4303⟩ whereby stereoselective formation of the desired alkene was achieved by use of the lithium borohydride reduction (Scheme 86); use of the Wittig or Wadsworth–Emmons reactions led to mixtures of alkene isomers. Interestingly, in this case the alkene formed directly on reduction with lithium borohydride.

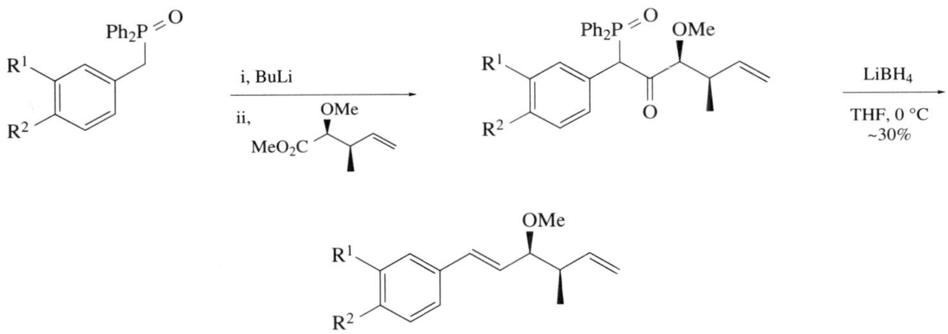

R^1 = H, OMe; R^2 = H, Cl

Scheme 86

1.14.4.1.5 Elimination of β-hydroxyphosphinothioic amides

Johnson and Elliott have demonstrated that β-hydroxyphosphinothioic amides, on treatment with methyl iodide in the presence of a base such as pyridine, undergo *syn* elimination to form alkenes ⟨82JA7041⟩. This method has been employed for the methylenation of ketones and for the preparation of di-, tri- and tetrasubstituted alkenes. As the diastereoisomeric β-hydroxyphosphinothioic amides can be separated prior to elimination, the alkenes can be formed in isomerically pure form (Scheme 87). Optical resolution in the alkene formation is possible by use of a homochiral phosphinothioic amide ⟨82TL5005⟩.

1.14.4.1.6 Elimination of β-hydroxyphosphonic acid bisamides

This methodology has been developed by Corey and co-workers ⟨66JA5652, 66JA5653, 68JA6816, 69JOC3053⟩. While β-hydroxyphosphonates, or their anions, do not eliminate to form alkenes unless

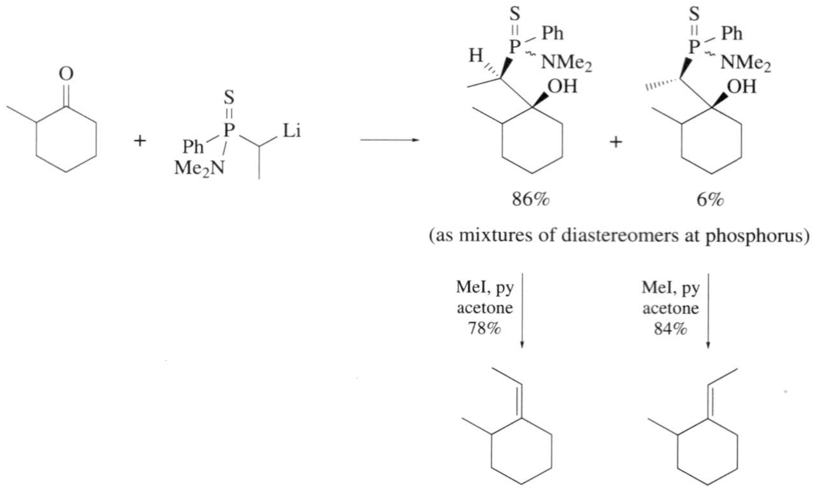

Scheme 87

they bear an adjacent activating group (Scheme 77 and Equation (153)) ⟨66JA5654, 68CC1699⟩, β-hydroxyphosphonic acid bisamides decompose to form alkenes on heating via a *syn* elimination ⟨66JA5652, 66JA5653, 68JA6816⟩. This reaction can be used for the synthesis of mono-, di-, tri- and tetrasubstituted alkenes; the ease with which tetrasubstituted alkenes are formed is notable. The β-hydroxyphosphonamides can be prepared either by reaction of simple α-lithio phosphonamides with aldehydes or ketones (the intermediate alkoxy anions do not eliminate) or by reduction of β-ketophosphonamides; as observed with the phosphine oxide derivatives opposite stereochemical outcomes are obtained by these complementary methods leading to stereoselective synthesis of both (*E*) and (*Z*) alkenes (Scheme 88). The phosphonamide route leads efficiently to *trans*-1-*t*-butyl-1-alkenes in contrast to the preference for formation of the *cis* isomer from the conventional Wittig reaction.

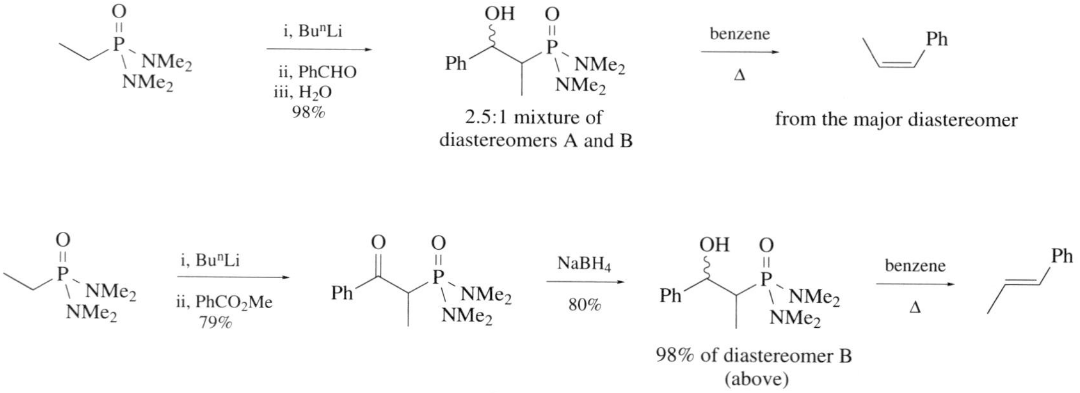

Scheme 88

Epoxide ring opening with the diamidophosphite anion also produces β-hydroxyphosphonamides which can subsequently be transformed to alkenes by thermolysis ⟨69JOC3053⟩. Evidence was obtained that the elimination is *syn*, proceeding via a betaine in a fashion similar to that observed with β-hydroxy sulfinamides (Scheme 89) ⟨68JA5553⟩; thus opening of the epoxides derived from *cis*- and *trans*-but-2-ene furnished the diastereoisomeric β-hydroxyphosphonamides which could be eliminated to *trans*- and *cis*-but-2-ene, respectively ⟨80T557⟩.

Application of the phosphonamide elimination to the synthesis of conjugated dienes has also been demonstrated (Scheme 90) ⟨69JOC3053⟩. Conditions had to be controlled to ensure α-alkylation occurred as the adduct formed by γ-alkylation did not eliminate. The thermolysis was conducted in mineral oil containing calcium carbonate to maintain neutrality in the reaction mixture.

Elimination of 1,2-diaryl-2-hydroxyethanephosphonamides under acidic conditions produced styrene derivatives stereoselectively ⟨88PS(40)69⟩.

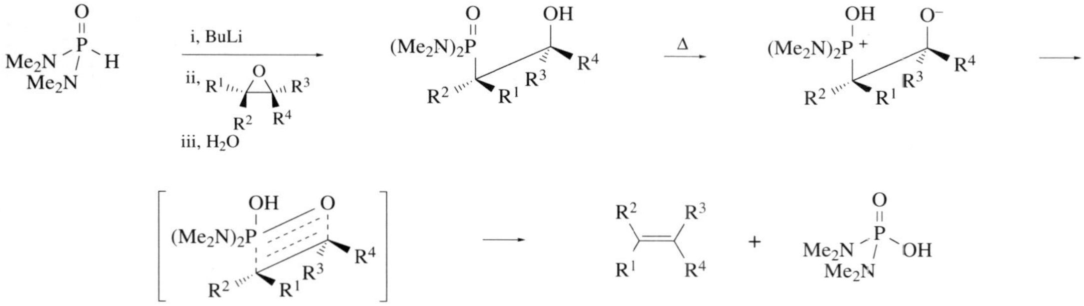

a; $R^1 = H, R^2 = Me, R^3 = H, R^4 = Me$
b; $R^1 = H, R^2 = Me, R^3 = Me, R^4 = H$

Scheme 89

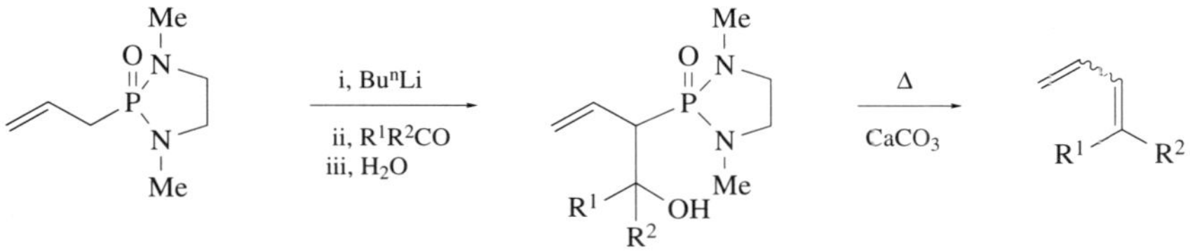

R^1, R^2 = H, alkyl, phenyl

Scheme 90

Chiral phosphonic diamides have been developed which exhibit high diastereoselectivity in their reactions with substituted cyclohexanones (Equations (154a) and (154b)) ⟨84JA5754⟩.

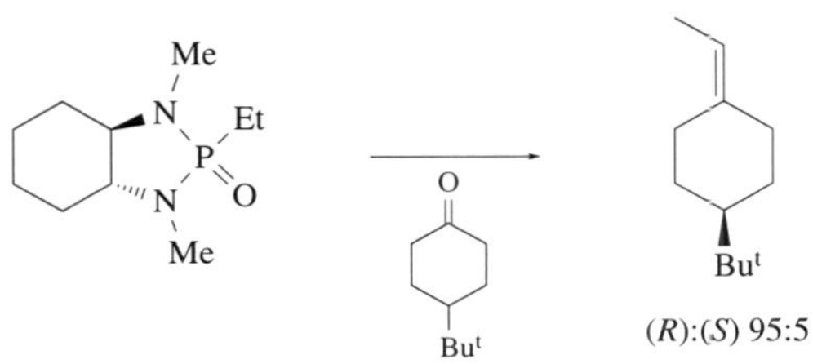

(*R*):(*S*) 95:5

(154a)

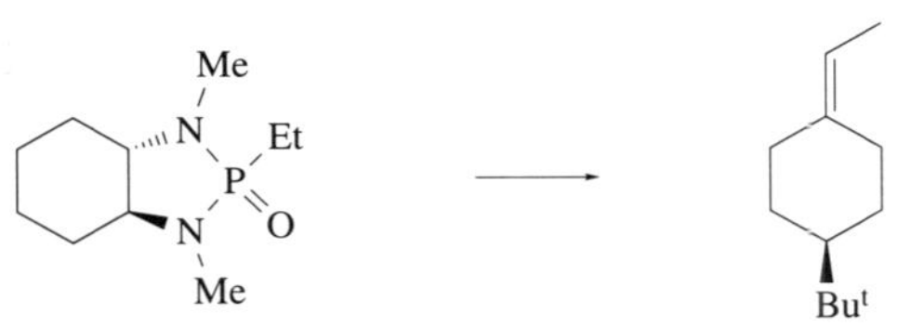

(*R*):(*S*) 5:95

(154b)

1.14.4.1.7 Elimination of phosphonothioate esters

While β-hydroxyphosphonates which do not bear an adjacent activating group, or their anions, do not eliminate to form alkenes ⟨66JA5654⟩ as shown in Scheme 77, the anions of β-hydroxyphosphonothioates eliminate under mild conditions (25 °C is sufficient when conjugated alkenes are formed) to form 1,1-disubstituted, tri- or tetra-substituted alkenes (Scheme 91) ⟨66JA5654⟩. The rate of elimination is increased by silver(I) catalysis. Only the anions of β-hydroxyphosphonothioates derived from the reaction of aldehydes with dimethyl *n*-alkylphosphonothioates failed to eliminate.

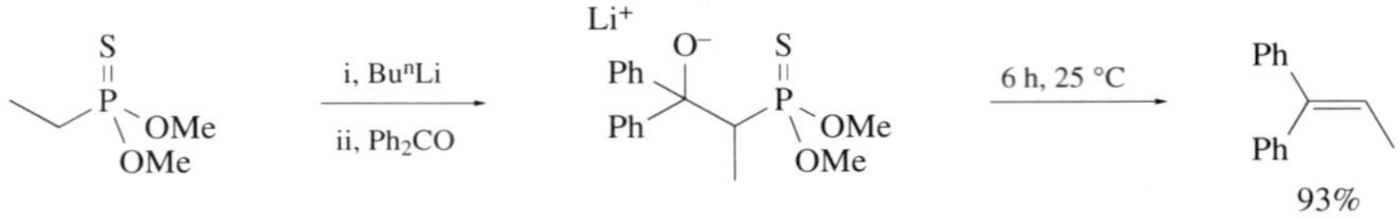

Scheme 91

1.14.4.1.8 *Elimination of triethyl phosphate*

Deoxygenation of epoxides can be effected with trialkyl phosphites ⟨57JOC1118⟩; this transformation is believed to proceed by epoxide ring opening with the phosphite followed by *syn* elimination of $(EtO)_3P{=}O$. In contrast desulfurization of episulfides with trialkyl phosphite occurs by attack of the phosphite on sulfur ⟨59JA578⟩.

1.14.4.2 Elimination of Arsenic, Antimony and Bismuth Functions

1.14.4.2.1 *Elimination of arsenic groups*

Elimination of methyldiphenylarsine oxide from the *trans*-β-hydroxy arsonium salt on treatment with sodium hydride leads to a mixture of alkene and epoxide (Equation (155)) ⟨75CS9⟩. Investigations of condensations of arsonium ylides with aldehydes have demonstrated that this behaviour is typical of β-oxy arsonium compounds as discussed below.

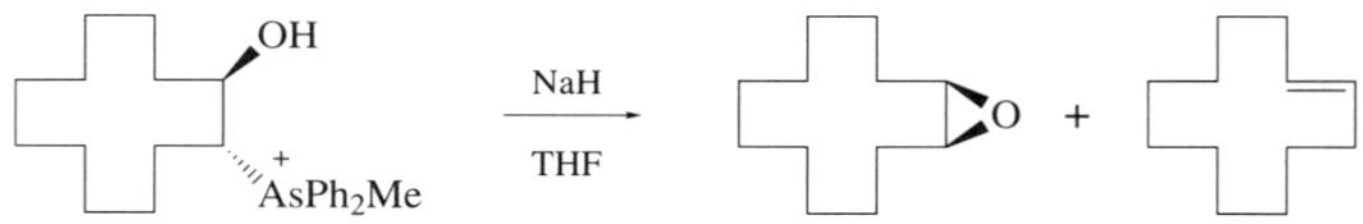

(155)

Alkene formation by reaction of arsonium ylides ⟨B-70MI 114-01⟩ with carbonyl compounds in a Wittig-type reaction is well documented, involving elimination of arsine oxide from a β-oxyarsonium intermediate. This area has been reviewed ⟨82AOC(20)115, 87CSR45, B-90MI 114-01⟩. Alkenes and/or epoxides are generated in reactions of arsonium ylides with carbonyl compounds—elimination of arsine oxide is in competition with arsine displacement to form epoxides; the product distribution and the diastereoselectivity can be affected by the nature of the ylide, the carbonyl compounds and the reaction conditions. Thus arsonium ylides display behaviour intermediate to that of sulfonium ylides, which lead to epoxides exclusively, and phosphonium ylides, which form alkenes only. Arsonium ylides are more reactive than phosphonium ylides, e.g., fluorenylidenetriphenylarsorane reacts with *p*-dimethylaminobenzaldehyde to give the alkene in excellent yield (Equation (156)), whereas the corresponding phosphorus reagent failed to react with this aldehyde ⟨60JOC183⟩. In the arsonium ylide-mediated methylenation of benzophenone to form 1,1-diphenylethene, diphenylacetaldehyde is also obtained which may result from rearrangement of 1,1-diphenyloxirane ⟨61JA417⟩.

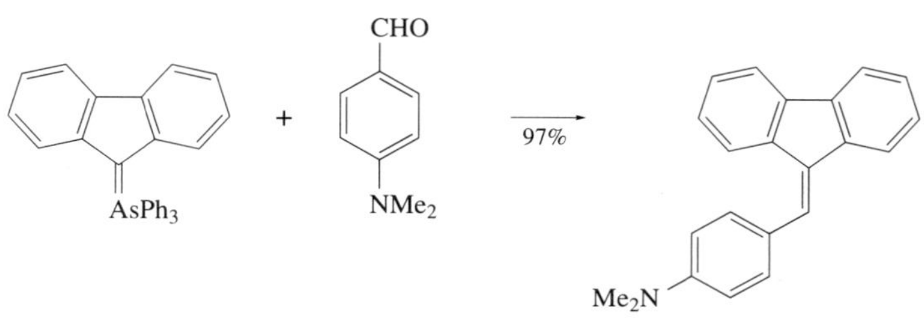

(156)

In general, if the arsonium ylide is stabilized (with a carbonyl group for example), elimination of arsine oxide to form alkenes is preferred; nonstabilized ylides lead to epoxides instead ⟨72T343, 89CRV863⟩. Formation of the (*E*)-alkene or *trans* epoxide usually predominates ⟨83TL4419, 86TL4583, 87JOC3558, 87TL2155, 87TL2159⟩. Semistabilized ylides may lead to products of either pathway: the products may vary depending on the reaction conditions. For example, in the reaction shown in Scheme 92, epoxide formation predominated in the presence of a lithium salt while formation of the (*E*)-alkene was preferred under salt-free conditions ⟨88BSB271⟩.

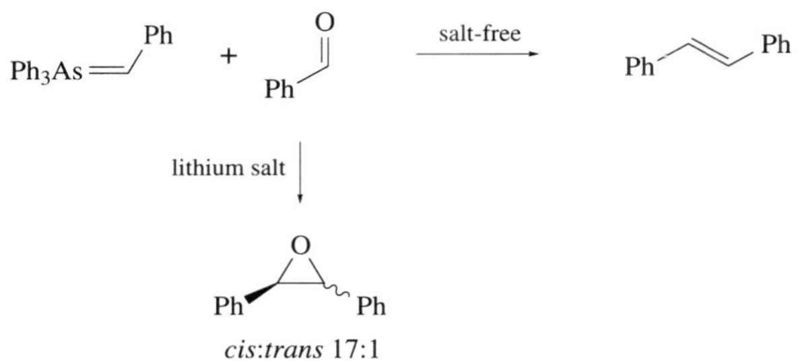

Scheme 92

Huang *et al.* ⟨B-90MI 114-01⟩ have developed the synthesis of (*E*)-α,β-unsaturated aldehydes ⟨85TL6447⟩, ketones, amides ⟨87TL2159, 88TL3949⟩ and polyenones ⟨87TL2155⟩ by the reaction of triphenylarsonium salts with aldehydes in the presence of a base. Application to the synthesis of a number of unsaturated natural products has been undertaken ⟨86TL4583, 87JOC3558, 88LA377, 88TL3949⟩; Scheme 93 illustrates the use of this methodology in the synthesis of trichonine ⟨88LA377⟩. Insect sex pheromones containing (*Z*,*E*)-conjugated dienes have been synthesized using a condensation with an arsonium salt to introduce the (*Z*) alkene followed by a Wittig reaction ⟨87JOC3558⟩. In addition, a catalytic Wittig-type reaction catalysed by tri-*n*-butylarsine has been developed which involves elimination of tributylarsine oxide in the catalytic cycle ⟨89JOC2027⟩.

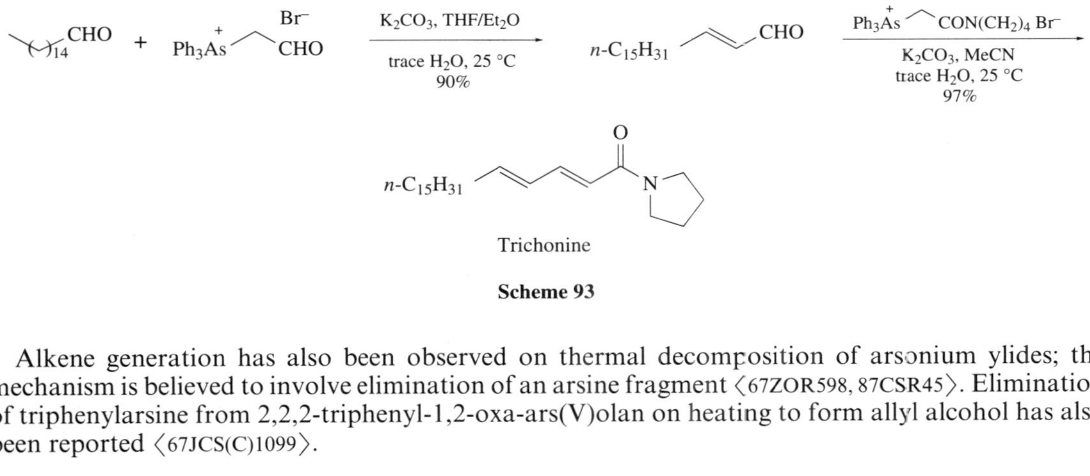

Scheme 93

Alkene generation has also been observed on thermal decomposition of arsonium ylides; the mechanism is believed to involve elimination of an arsine fragment ⟨67ZOR598, 87CSR45⟩. Elimination of triphenylarsine from 2,2,2-triphenyl-1,2-oxa-ars(V)olan on heating to form allyl alcohol has also been reported ⟨67JCS(C)1099⟩.

1.14.4.2.2 Elimination of antimony and bismuth groups

Alkene formation involving elimination of antimony groups has also been investigated (reviewed in ⟨B-70MI 114-02, 87CSR45, 92ACR182⟩). In 1960, Henry and Wittig reported that only a trace of alkene formation was observed in the reaction of a stibonium ylide with benzophenone ⟨60JA563⟩. Later, Lloyd demonstrated fulvene formation on reaction of triphenylstibonium tetraphenylcyclopentadienylide with aryl aldehydes ⟨72T343⟩. Reaction of tributylstibonium ylides with aldehydes and ketones leads to alkene formation ⟨78TL4399⟩ involving elimination of tributylstibonium oxide (which readily forms a hydrate) from a β-oxy stibonium intermediate (Scheme 94) ⟨90JCR(S)388⟩. Trialkylstibines mediate a condensation of α-bromocarboxylic acid derivatives, α-bromonitriles or α-bromoketones with aldehydes and ketones to form α,β-unsaturated esters, amides, nitriles and enones via the intermediacy of a stibonium salt (Equation (157)) ⟨86TL2903, 89JOM(366)87, 89SC83, 89SC501, 92ACR182⟩. Remarkably, this reaction proceeds in the absence of base.

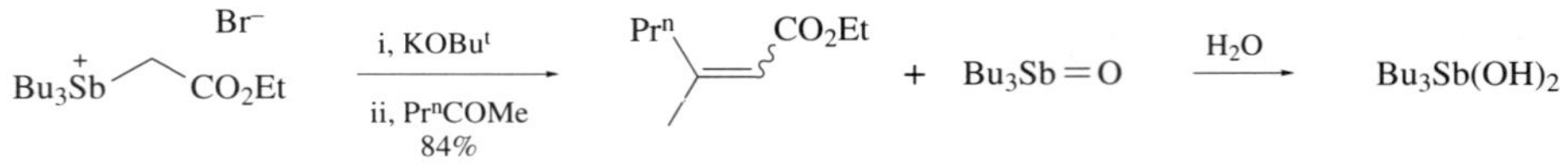

Scheme 94

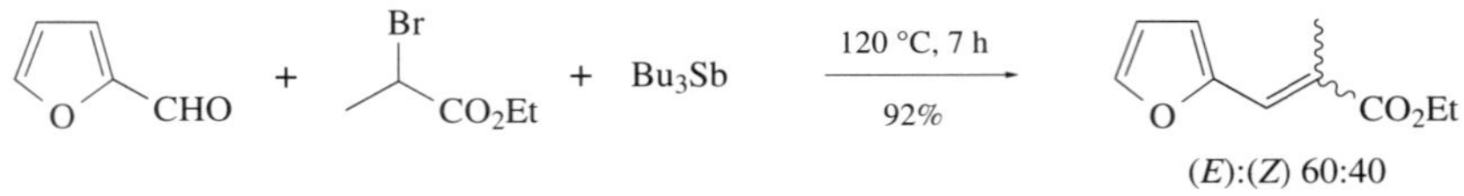

(157)

Relatively few reports of bismuthonium ylides have appeared ⟨67CC1042, 72T343, 87CSR45, 88S319, 89CC1749⟩. Structural properties and relative reactivity of stibonium and bismuthonium ylides bearing electron-withdrawing substituents have been discussed; these ylides do not appear to participate in alkene forming reactions ⟨88JCS(P2)1829⟩.

1.14.5 BY ELIMINATION OF SILICON, GERMANIUM OR BORON FUNCTIONS

1.14.5.1 Elimination of Silicon Groups

Alkene formation by elimination from organosilanes with a leaving group at the β-position, typically hydroxy or halo substituents, is very useful synthetically and has been the subject of many reviews (see for example ⟨78CSR15, 79COC(3)541, B-81MI 114-02, B-83MI 114-02, 84S384, 85S717, B-88MI 114-02, B-88MI 114-03, 89JOM(360)39, 91COS(1)729, 91COS(6)1000⟩). The Peterson alkenation involving elimination from β-hydroxysilanes in a process similar to the Wittig reaction is probably the most frequently employed example of an elimination reaction involving silicon ⟨91SL764⟩.

1.14.5.1.1 Elimination of β-hydroxysilanes

While earlier examples of elimination from β-hydroxysilanes had been reported ⟨47JA1551⟩ the synthetic utility of this transformation was recognized and demonstrated by Peterson in 1968 ⟨68JOC780⟩ and subsequently developed by Peterson, Hudrlik, Chan and co-workers ⟨74JOC3264, 74TL171, 75JOC2263, 78TL2383⟩. The trimethylsilyl group is commonly employed but phenyldimethylsilyl has also been used ⟨89TL229⟩. This silyl group has the advantage that the β-hydroxysilane is sufficiently stable to purify by chromatography over silica prior to elimination. Elimination can be effected under acidic (typically acetic acid, sulfuric acid or boron trifluoride etherate) or basic (typically potassium *t*-butoxide or potassium hydride or sodium hydride) conditions ⟨84S384⟩. Johnson and Tait have successfully employed aqueous hydrofluoric acid to effect elimination ⟨87JOC281⟩. Alternatively, the hydroxy group can be activated with thionyl chloride ⟨74TL171⟩, acetyl chloride ⟨74JOC3264, 78TL2383⟩, mesyl chloride ⟨74JOC3264, 75JOC2263⟩ or trifluoroacetic anhydride ⟨74TL171⟩ prior to elimination, which is frequently effected by fluoride ion (Equation (158)) ⟨85BCJ2812⟩. This process can be particularly advantageous in cases where isomerization of alkenes is a problem or with terminal alkenes. In a synthetic approach to mitomycins, Danishefsky used new conditions for the hydroxysilane elimination; the intermediate in this case was stable to the usual elimination conditions. However, on treatment with 2,3-dichloro-5-6-dicyano-1,4-benzoquinone (ddq), quinone formation and hydroxysilane elimination proceeded in quantitative yield (Equation (159)) ⟨88JOC3391⟩.

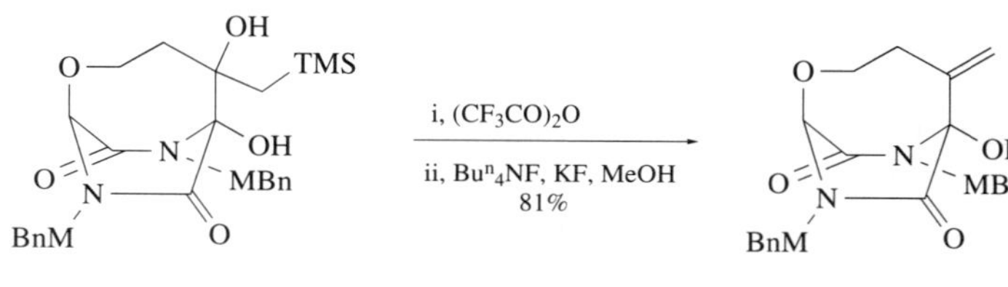

(158)

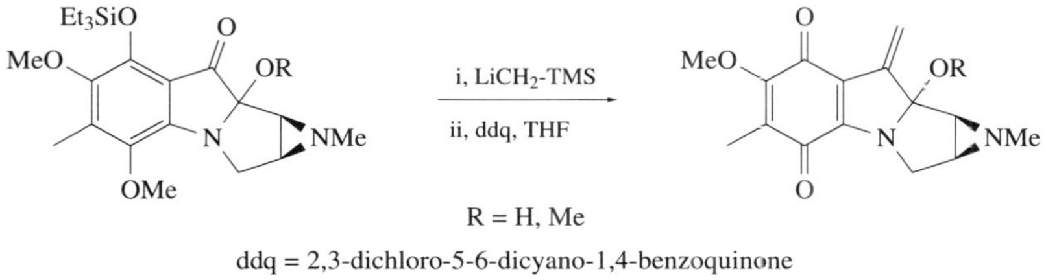

(159)

The stereospecificity of the elimination process is noteworthy: under acidic conditions elimination occurs in an *anti* fashion while *syn* elimination is observed under basic conditions (Scheme 95) ⟨75JA1464⟩. Thus, provided a single diastereomer of the β-hydroxysilane can be obtained, this can be converted into either isomer of the alkene. Evidence for the *anti*-periplanar geometry of the acid-induced elimination includes the fact that the β-hydroxysilane (**12**) is stable to acid ⟨76CC697⟩. Through use of β-hydroxysilanes which are chiral at silicon, Larson *et al.* have demonstrated that the acid-catalysed elimination process proceeds with inversion of configuration at silicon while retention at silicon occurs when the elimination is conducted with potassium hydride ⟨88T3781⟩.

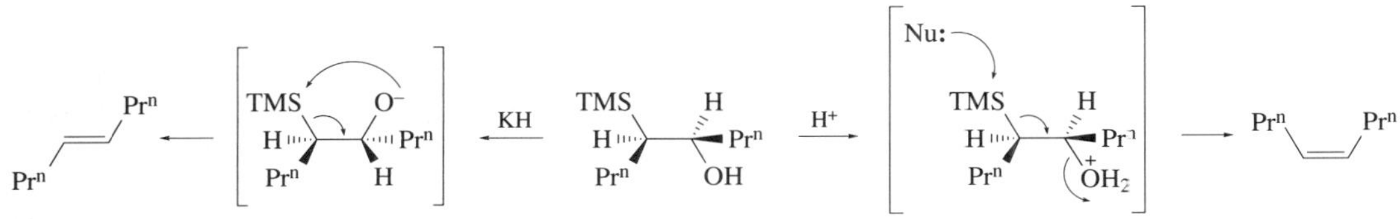

Scheme 95

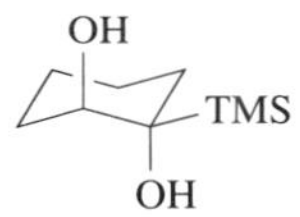

(12)

The β-hydroxysilanes used in the elimination reactions are frequently obtained by condensation of an α-silylcarbanion with a carbonyl compound (Scheme 96) ⟨68JOC780⟩. When magnesium is used as the countercation, the hydroxysilane can be isolated as elimination does not occur spontaneously; however, with alkali metals the alkene is frequently isolated directly from the reaction mixture ⟨79COC(3)541⟩. When lithium ions are involved and if the substituent X is capable of stabilizing an anion, then the elimination of the β-oxidosilane is usually rapid and the alkene is isolated directly; however, if X is not anion-stabilizing then the β-hydroxysilanes are generally isolable and eliminate on treatment with acid or base. While a four-centred Wittig-type intermediate can be envisaged for the elimination from the β-oxidosilane, it is also possible that transfer of the silyl group to the oxygen atom can occur to form a carbanion which then loses the siloxy anion to form the alkene ⟨73JOC2664⟩. In competitive situations, elimination of a silyl group occurs rather than a phosphorus moiety ⟨62JOC3647, 68JOC780⟩. In the case of a β-gorgonene precursor (Equation (160)) methylenation could be achieved using trimethylsilylmethylmagnesium chloride but was not possible using the corresponding Wittig reagent ⟨73TL3497⟩. Furthermore the by-product of the Peterson alkenation, hexamethyldisiloxane, is much more easily removed from alkene products than triphenylphosphine oxide due to its volatility ⟨68JOC780⟩. An example of the Peterson methodology in synthesis, in which good selectivity is observed, is illustrated in Equation (161) ⟨88T4339⟩. The chemoselectivity of the Peterson alkenation can be modified by use of titanium or cerium reagents; in the presence of titanium tetrachloride ⟨81TL5031⟩, aldehydes but not ketones are methylenated with trimethylsilylmethylmagnesium chloride; use of cerium trichloride ⟨87JOC231⟩ leads to better results with base-sensitive aldehydes and ketones.

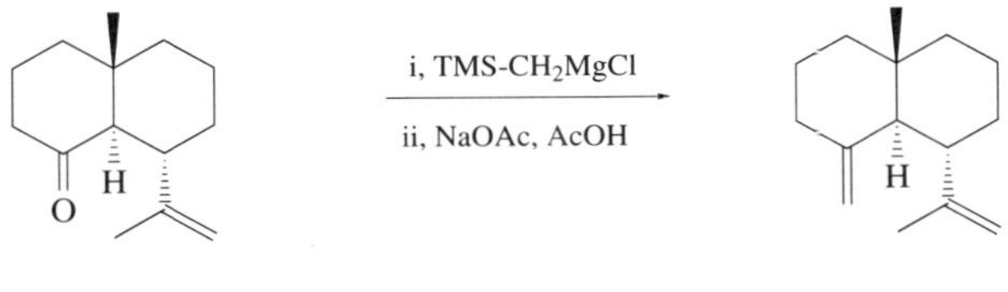

(160)

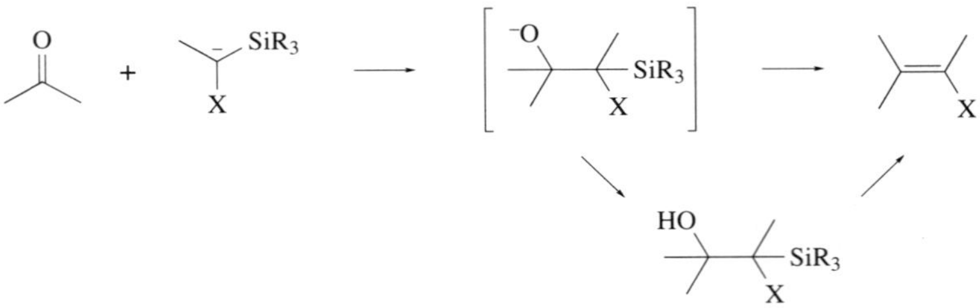

Scheme 96

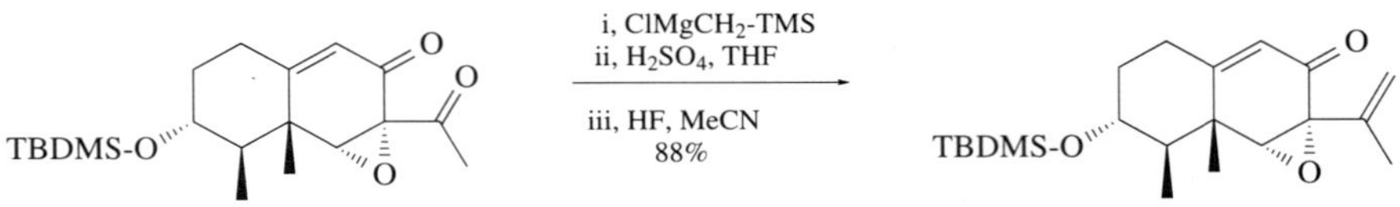

(161)

β-Hydroxysilanes can be formed by regiospecific ring opening of epoxysilanes (Scheme 97) ⟨75JOC2263, 76JOC2615⟩, or from β-ketosilanes (Scheme 98) ⟨74TL1133, 75SC15, 76JOC2940⟩ on treatment with organometallic reagents or, in the case of β-ketosilanes, hydride reducing agents ⟨75JA1464⟩. Ring opening of epoxysilanes with silicon tetrafluoride followed by elimination with potassium hexamethyldisilazide has been used to synthesize fluoroalkenes ⟨89TL967⟩. Organosilyl anions attack epoxides to form β-oxidosilanes which undergo spontaneous *syn* β-elimination to form alkenes (Scheme 99) ⟨76JA1265, 76S199⟩; as this step proceeds stereospecifically, it provides an excellent method for isomerization of alkenes ⟨80T557⟩. β-Hydroxysilane precursors for the elimination may also be formed by Baeyer–Villiger oxidation of β-silylketones and hydrolysis of the lactone thus formed ⟨80JA6894⟩, furnishing a useful route to unsaturated carboxylic acids. Treatment of ethyl trimethylsilylacetate with two equivalents of Grignard reagent followed by treatment with acid to effect elimination of the intermediate β-oxidosilane produces 1,1-disubstituted ethylenes as illustrated in Equation (162) ⟨82TL1035⟩.

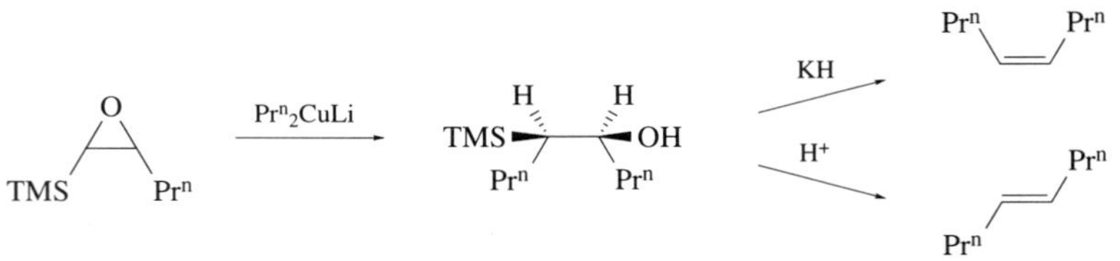

Scheme 97

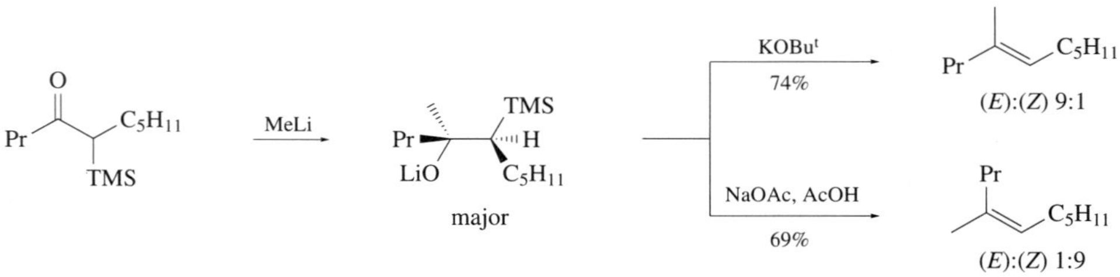

Scheme 98

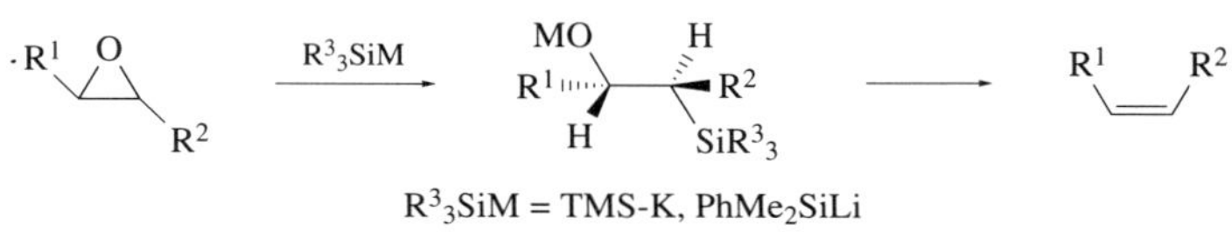

$R^3{}_3SiM$ = TMS-K, $PhMe_2SiLi$

Scheme 99

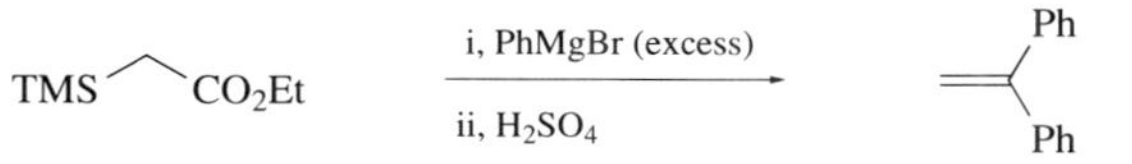

(162)

Elimination of β-hydroxysilanes has been employed to prepare many unsaturated systems such as allenes (Scheme 100) ⟨78JOC1526⟩, α,β-unsaturated esters ⟨74BCJ2529, 74JA1620⟩, acids ⟨75CC537⟩, amides ⟨75SC359, 78JA1548⟩, lactams (Equation (163)) ⟨78S746, 86JOC558⟩, nitriles ⟨74TL4005⟩ and enynes (Equation (164)) ⟨81JA5568, 82TL719⟩.

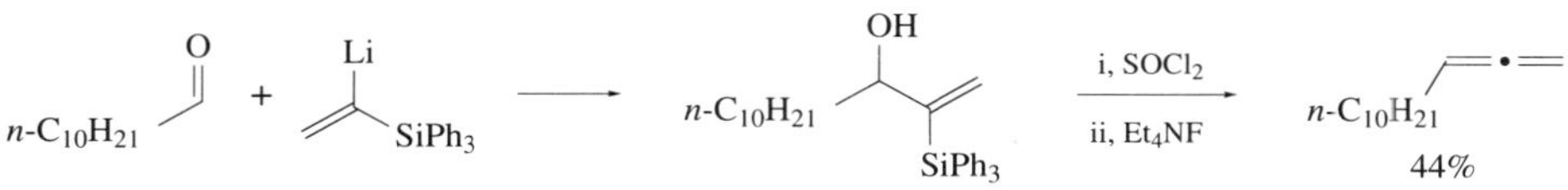

Scheme 100

(163)

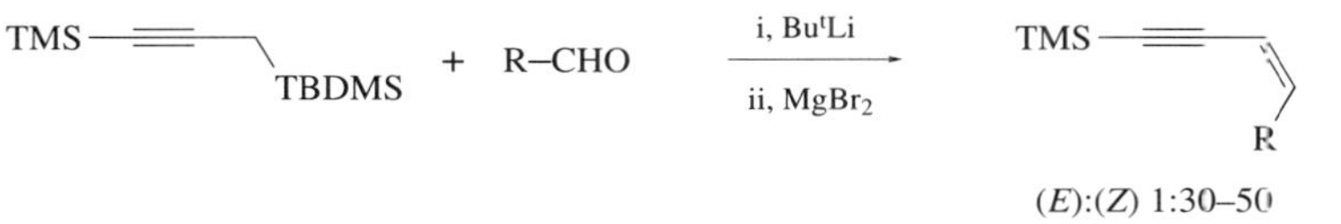

(164)

Conjugated dienes may be synthesized by reaction of an α-silylanion with an α,β-unsaturated aldehyde or ketone ⟨70TL1137⟩ or by reaction of a silylallyl anion with a carbonyl compound ⟨78TL2383, 80JA5004⟩. However, the latter reaction can lead to side products due to attack at the γ-position thereby reducing the yield of the diene. This problem has been overcome by use of boronic esters to control the condensation (Scheme 101) ⟨81TL2751⟩. Furthermore, the β-hydroxysilane is formed diastereoselectively; therefore both (Z)- and (E)-dienes may be prepared by choosing the appropriate elimination conditions. An alternative which has been investigated is the use of 1,3-bis(trimethylsilyl)propenyl anion with aldehydes or ketones to form 1-silyl-1,3-dienes stereoselectively (Scheme 102) ⟨82CC969, 83CC921, 84JOM(264)207, 87TL211⟩. Alexakis and Jachiet developed a new approach to conjugated dienes, used in pheromone synthesis, by stereoselective ring opening of epoxysilanes with (Z)-alkenyl cuprates in the presence of boron trifluoride etherate, followed by elimination under basic or acidic conditions to form the (Z)- or (E)-conjugated dienes ⟨89T381⟩. Thus stereoselective diene synthesis is possible with the Peterson reaction while the Wittig reaction is not as efficient ⟨71HCA1767⟩.

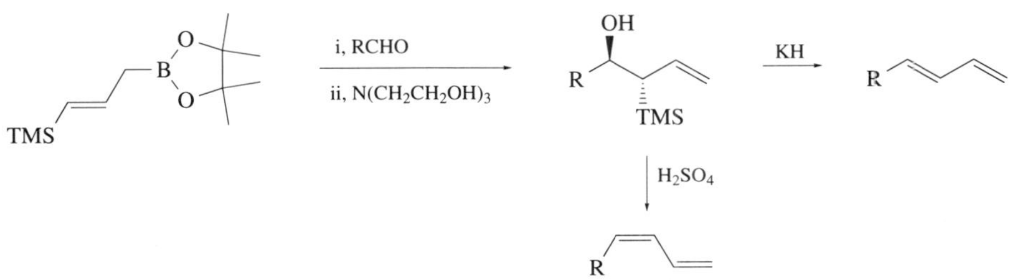

Scheme 101

The vinylogous Peterson elimination has been investigated ⟨84CC534, 90CC1575⟩; under basic conditions the elimination occurs stereospecifically *syn* to form the conjugated diene. When a *cis* double bond is relocated it is transposed to the position which originally carried the hydroxy group (Equations (165a) and (165b)).

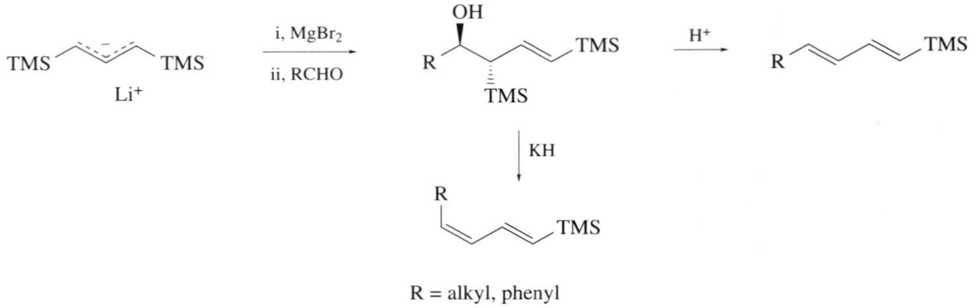

Scheme 102

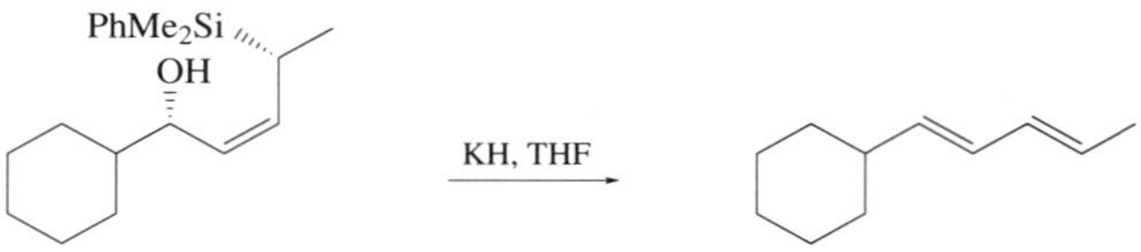

(165a)

(165b)

Allylsilane synthesis by hydroxysilane elimination has been developed by Fuchs and Anderson ⟨87SC621⟩ as well as Narayanan and Bunnelle (Scheme 103) ⟨87TL6261⟩, while Soderquist and Santiago have employed elimination of β-hydroxysilanes in a method of conversion of vinylsilanes to allylsilanes ⟨89TL5693⟩.

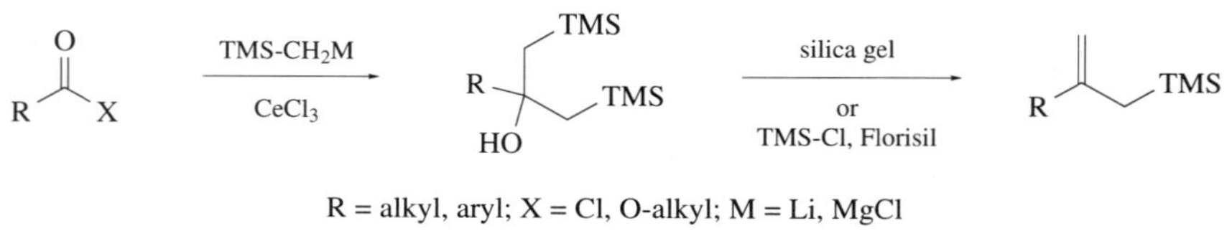

Scheme 103

β-Hydroxysilanes have been employed as alkene-protecting groups in pheromone synthesis, thereby exploiting the ease with which they may be eliminated stereospecifically to afford either the (*E*) or (*Z*) alkene ⟨93TL1593⟩.

1.14.5.1.2 Elimination of β-halosilanes

β-Halosilanes readily undergo β-elimination ⟨48JA2869⟩ to form alkenes sometimes simply on heating ⟨69JCS(B)652⟩ but frequently on treatment with fluoride ions. The elimination usually proceeds in an *anti* fashion ⟨69JCS(B)852, 76JOM(13)C11, 76JOM(121)293⟩, except on thermolysis when a four-centred intermediate appears to be involved ⟨69JCS(B)652⟩. Chan and co-workers have used this strategy to prepare reactive alkenes such as allenes, allene oxides (Equation (166)) ⟨74TL171, 74TL3511, 75TL3383, 78JOC2994, 80T2269⟩ and strained alkenes (Equation (167)) ⟨77JA936⟩. Thus *t*-butylallene oxide could be isolated in this manner; other allene oxides with less bulky substituents which were too reactive to isolate could be generated by fluoride-induced halosilane elimination and trapped *in situ*. Similarly the bridgehead alkene shown in Equation (167) was generated as an intermediate and was subsequently trapped in a cycloaddition process.

(166)

(167)

β-Silylcarbonyl compounds do not readily eliminate to form α,β-unsaturated derivatives; however, on bromination the α-bromo-β-silyl derivatives undergo facile elimination of the 1,2-halosilane either on treatment with fluoride ion or thermally ⟨56ACS633⟩. Fleming *et al.* have developed this as a method for protection of an enone as a β-silyl derivative (Scheme 104) ⟨78CC176, 78CC177, 78CC178, 80JCS(P1)1493⟩. Notably the bromination usually occurs selectively on the same side of the ketone as the silyl substituent. However, even if the other regioisomer is formed, enolization or equilibration can provide pathways for elimination to form the enone. Conjugate additions of organosilyl lithiums to enones form enolates which can be alkylated regiospecifically; as the product contains a β-silyl derivative, bromination can be used to reintroduce the double bond (Scheme 105) ⟨78CC177⟩. Fluoride ion-induced desilylchlorination has been employed in the synthesis of α-methylenecyclobutanones and α-methylenecyclopentanones from the corresponding α-chloro-β-silyl ketones (Equation (168)) ⟨84JOC1317⟩.

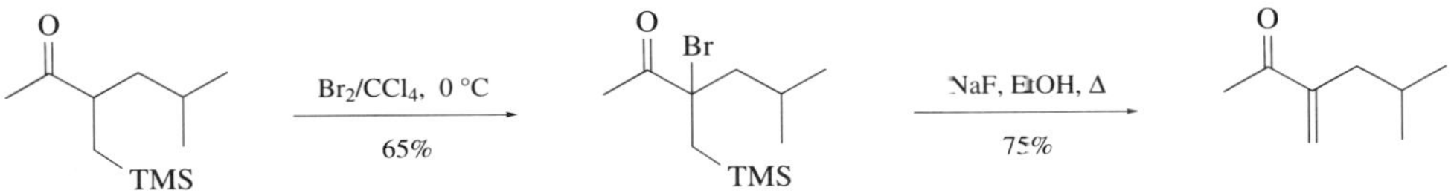

Scheme 104

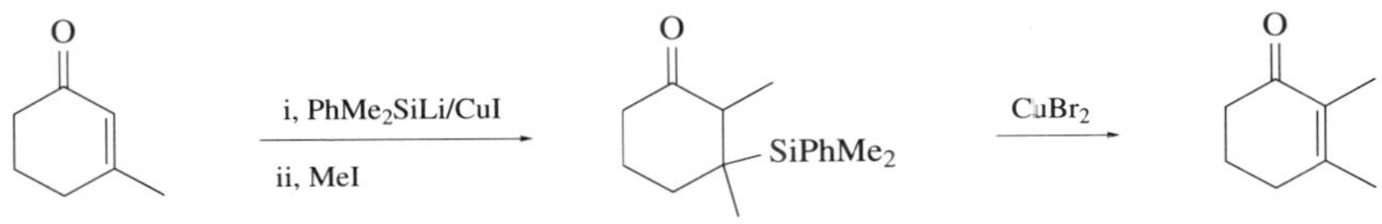

Scheme 105

(168)

Vinylsilanes can be transformed stereospecifically to vinylhalides by treatment with chlorine or bromine followed by elimination of the halosilane (Scheme 106) ⟨74TL543⟩. When iodine is employed the *cis*-vinyliodide is isolated instead. Reaction of fluorovinylsilanes with bromine involves elimination in the intermediate bromosilane to form bromofluoroalkenes ⟨90JOM(394)37⟩. Elimination of halosilanes has been employed for the synthesis of vinyl sulfides ⟨89T3163⟩ and vinyl sulfones ⟨82TL3455⟩.

1.14.5.1.3 Elimination of other β-substituted silanes

Silanes bearing alkoxy, acetoxy, sulfonyl, amino groups, etc., at the β-position undergo elimination to form alkenes, frequently on treatment with fluoride ions. The stereochemistry of the

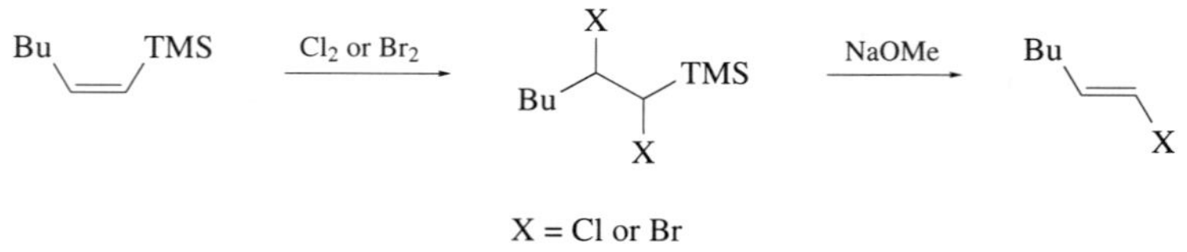

Scheme 106

elimination is usually *anti* (Equations (169) and (170)) ⟨83TL4151, 83TL4153, 84JOM(271)281⟩. Fluoride ion-induced elimination of a β-trifluoroacetylsilane was used to synthesize (*E*)-1-iodohex-1-ene ⟨74TL543⟩. Thermolysis of β-methoxysilanes to form alkenes appears to proceed via a four-centre intermediate while the acid-catalysed elimination of the same substrates proceeds in a *trans*-coplanar fashion ⟨69JA514⟩.

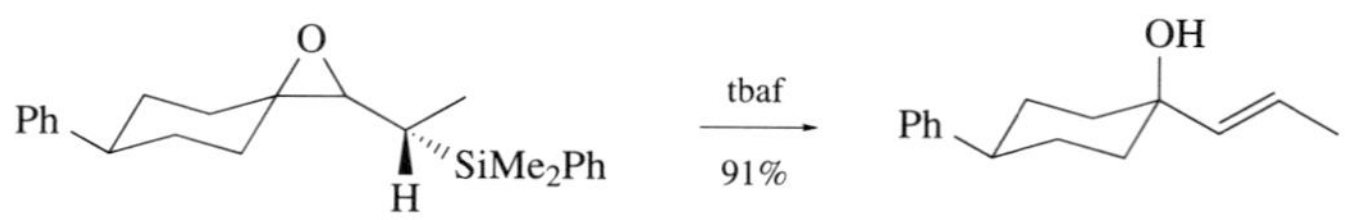

(169)

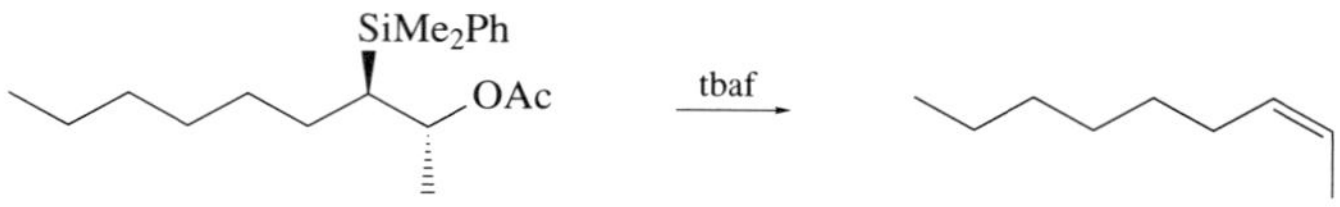

(170)

The ease of fragmentation of β-silyl-substituted compounds has been exploited in the development of new protecting groups ⟨85S817⟩. 2-Trimethylsilylethyl esters eliminate on treatment with fluoride ion to form ethene, trimethylsilylfluoride and carboxylate (Equation (171)) and are therefore a useful method of protection of carboxylic acids ⟨77HCA2711, 77HCA3039⟩, employed for example in the synthesis of a highly reactive carboxylate salt ⟨87CC106⟩. This methodology has been extended to provide a protecting group for alcohols; β-trimethylsilylethoxymethyl (SEM) ethers are removed as shown in Equation (172) ⟨80TL3343, 84JOC3671⟩. Amines can be protected in a similar fashion ⟨78CC358, 83CC630⟩.

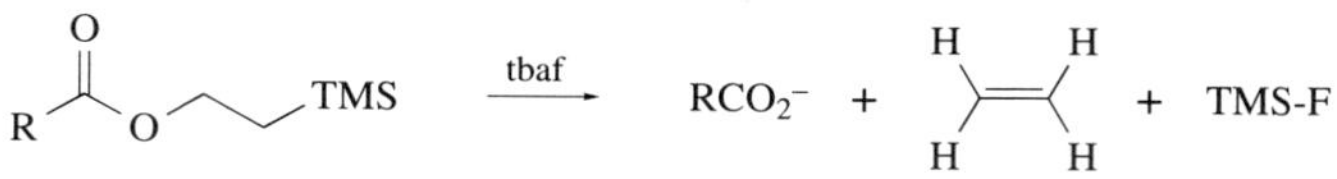

(171)

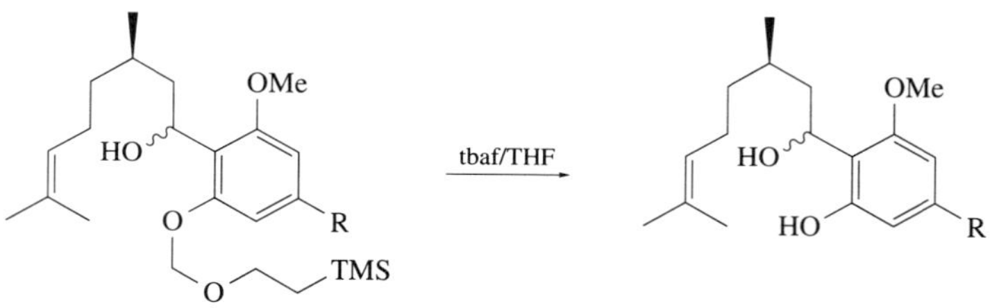

(172)

Fluoride ion-induced elimination of β-silyl sulfones (see Section 1.14.2.3.1.ii) provides a mild route to alkenes ⟨79TL2649, 88JOC2688⟩ and, in its vinylogous form, to dienes ⟨84TL1219, 88JOC2688⟩; the synthetic utility is limited by the accessibility of β-silyl sulfones. They are usually prepared by alkylation of a sulfone anion with iodomethyltrimethylsilane, thereby limiting the alkenes synthesized to monosubstituted or 1,1-disubstituted alkenes. Application of the methodology to the introduction of the C-11 methylene unit of moenocinol is illustrated in Equation (173) ⟨78JCS(P1)829⟩. An alternative route to a β-silyl sulfone, employed in the synthesis of the antibiotic bicyclomycin, involves a conjugate addition to an unsaturated β-silyl sulfone ⟨86CC620⟩; elimination to form the alkene was effected on treatment with tbaf as usual. 1-Trimethylsilyl-2-benzenesulfonylethylene can act as an alkyne equivalent in Diels–Alder cycloadditions, as elimination of the silyl-sulfone units in the adducts forms alkenes (Scheme 107) ⟨83JOC4976⟩. Furthermore alkylation of the sulfone prior to elimination on treatment with fluoride ion increases the versatility of this methodology.

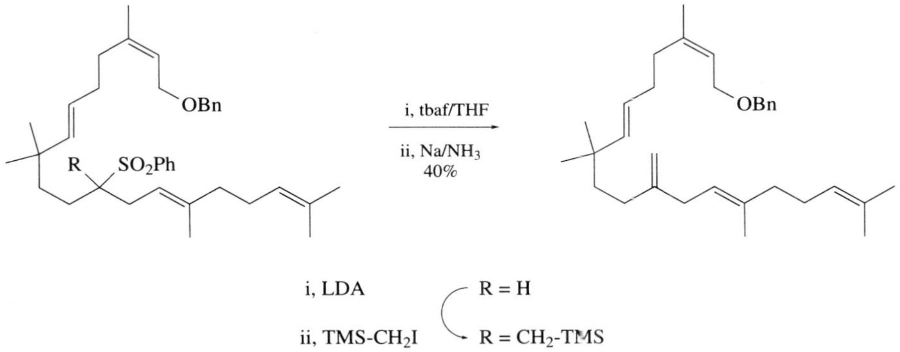

(173)

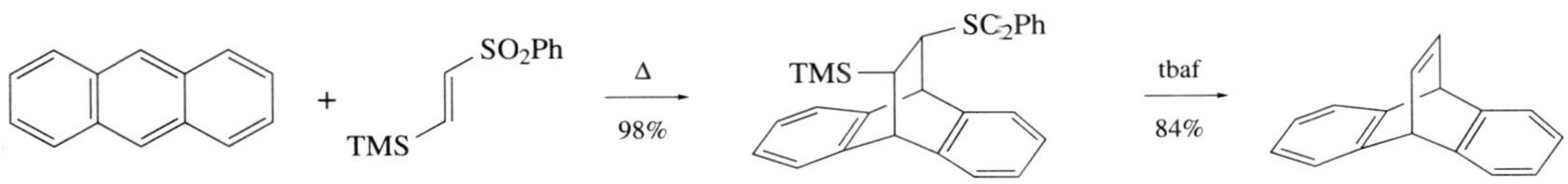

Scheme 107

β-Silyl sulfoxides in which the sulfoxide moiety does not have a β-hydrogen atom undergo *syn* elimination of silylsulfenate on heating (Equation (174)) ⟨82JCS(P1)1563⟩. Similarly sila versions of the Hoffmann and Cope eliminations are possible in β-silyl derivatives of ammonium salts and amine oxides respectively (see Sections 1.14.3.1 and 1.14.3.2.2) ⟨82JA7666⟩. Thus fluoride ion-induced fragmentation of a silyl ammonium iodide produced the (*Z*,*E*)-diene via an *anti* elimination (Equation (126)) while the corresponding amine oxide decomposed spontaneously by a *syn* elimination to form the (*Z*,*Z*)-diene (Scheme 108) (see also Equation (118)). Related 1,4-fragmentation processes effected by fluoride ion have been employed in the generation of *o*-quinodimethanes (Scheme 109) ⟨80JA6885, 82JA7609, 88JA6579⟩.

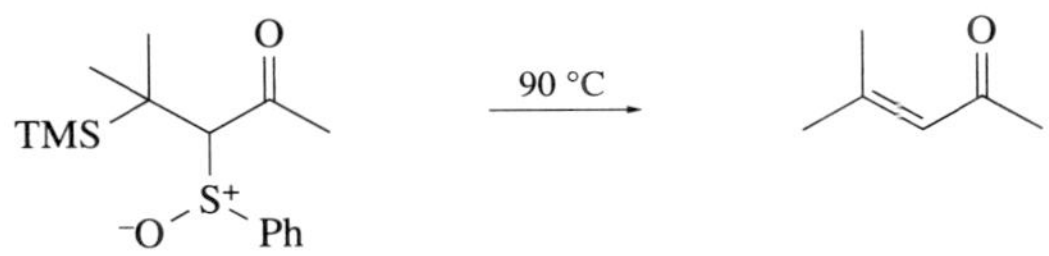

(174)

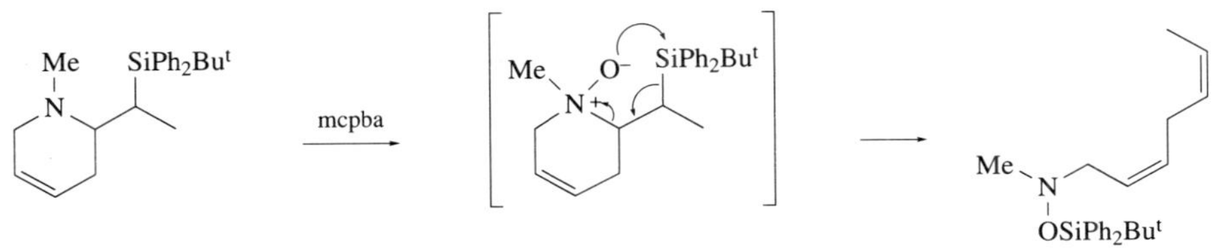

Scheme 108

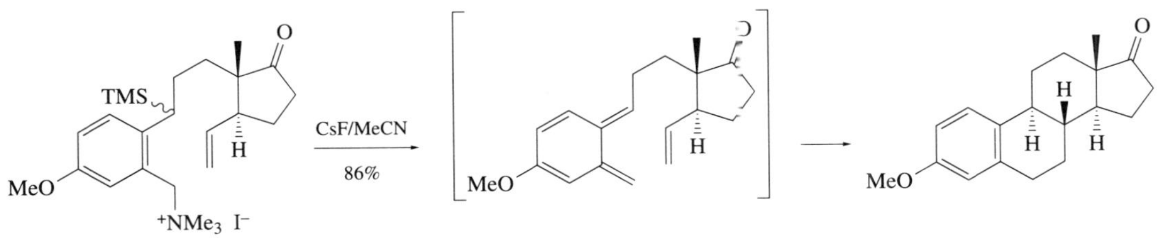

Scheme 109

Silyl-directed fragmentation reactions leading to alkenes have been extensively investigated; for example oxidative fragmentation of γ-trimethylsilyl alcohols on treatment with ceric ammonium nitrate to form keto alkenes is illustrated in Equation (175) ⟨87JOM(332)53⟩, while Equation (176) shows fragmentation of the silacyclohexane ring on treatment with trityl fluoroborate ⟨74JOM(81)41⟩.

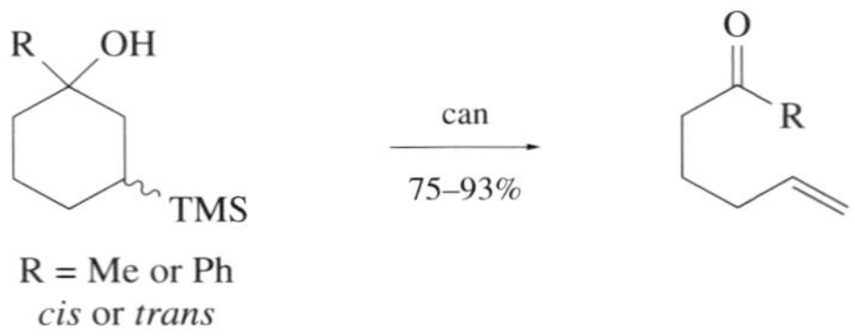

(175)

can = ceric ammonium nitrate

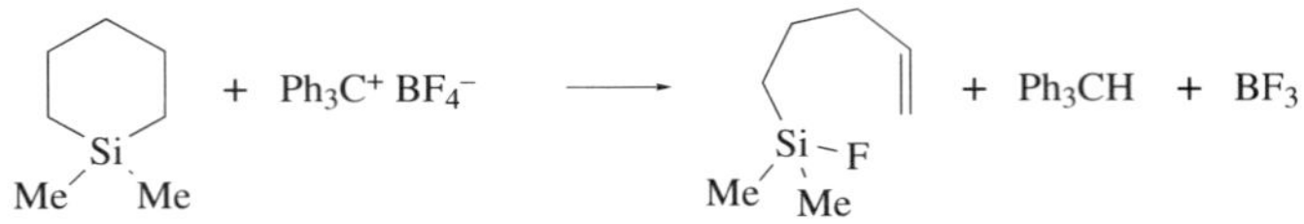

(176)

Silacyclopropanes undergo thermal cycloreversion on heating to form a silylene and an alkene as illustrated in Equation (177) ⟨75JA7162⟩.

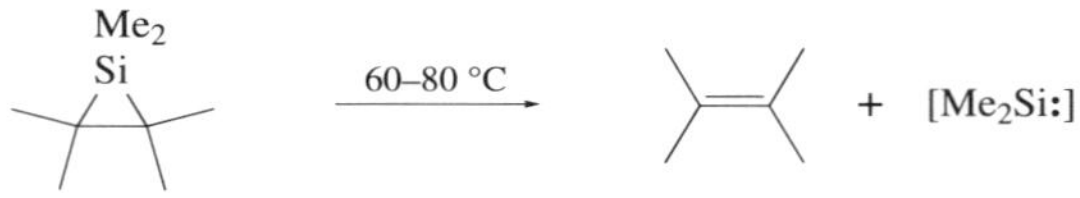

(177)

1.14.5.2 Elimination of Germanium Groups

Germanes bearing leaving groups at the β-position undergo elimination to form alkenes analogous to the behaviour of silanes; β-halogermanes decompose very readily forming ethene (Equation (178)) ⟨63MI 114-01, B-67MI 114-03⟩. The unstable heterocycles 2-oxa and 2-thia-germacyclobutanes lead to alkenes and transient germanones or germathiones via β-elimination (Scheme 110) ⟨83JOM(243)281, 90CRV283, 90JOM(400)121⟩. A similar elimination to form a butadiene has been proposed ⟨84OM1340⟩. Carbonyl alkenation, similar to the Wittig reaction, is possible with germanium-based reagents; fragmentation of four-membered intermediates is proposed for these reactions ⟨85TL4523, 90CRV283, 90JOM(400)121⟩. Treatment of aldehydes, but not ketones, with the reagent $Me_3GeCH_2TiCl_3$, leads to alkenes in good yields (Equation (179)) ⟨81TL5031⟩, presumably involving elimination from a β-oxygenated germane intermediate. Butadienes are formed during the gas-phase pyrolysis of 1,2-digermacyclohex-4-enes (Equation (180)) ⟨84JOM(267)93⟩.

(178)

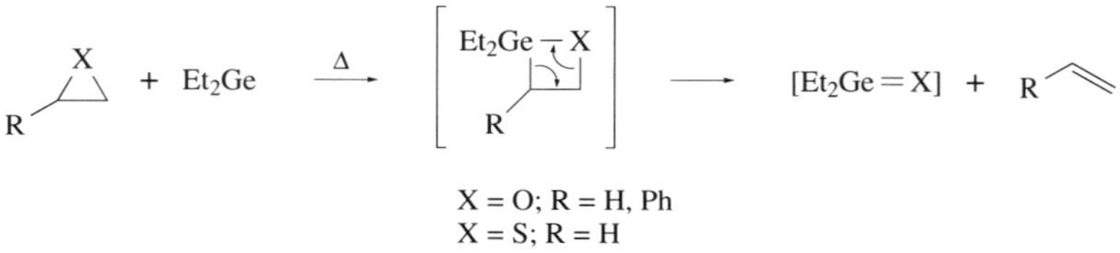

Scheme 110

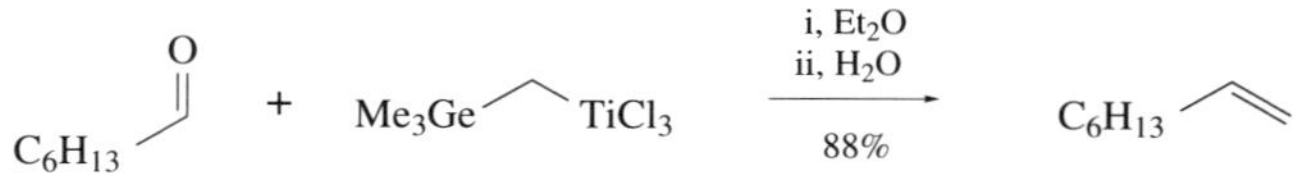

(179)

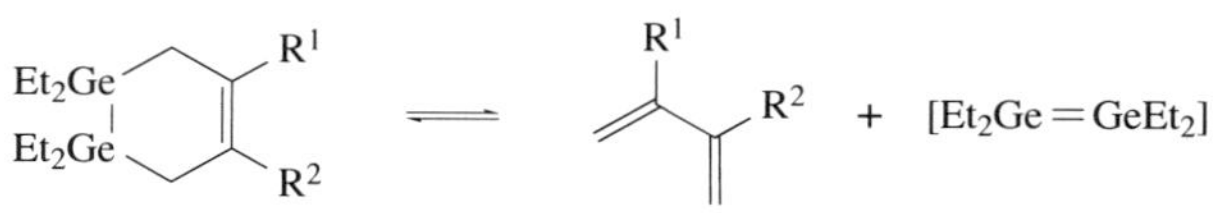

(180)

1.14.5.3 Elimination of Boron Functions

1.14.5.3.1 Elimination of alkylboranes

On heating alkylboranes, in which the alkyl group bears at least one β-hydrogen atom, elimination may occur (retrohydroboration) reversibly to give an alkene ⟨55JA5016, 68JCS(B)1044⟩; this can be driven to completion by using another (usually less volatile) alkene as a trap and by removing the alkene by distillation as it is formed (Equation (181)) ⟨67JA567, 79BCJ2942⟩. In the latter case, after 8 h at 160°C, 90% of the hexyl group was displaced by 1-decene, with only 2% of the norbornyl group undergoing displacement; the alkene isolated was almost entirely 1-hexene containing only minor amounts of 2-hexene and 3-hexene. Thus selective displacement of alkenes from alkyldi-*exo*-norbornylborane is possible under these conditions; with other trialkylboranes bearing different alkyl groups mixtures of alkenes may be formed due to less-selective elimination. Borane elimination forms the more substituted alkene wherever possible (Equation (182)) ⟨67JA567⟩.

(181)

R = *exo*-norbornyl

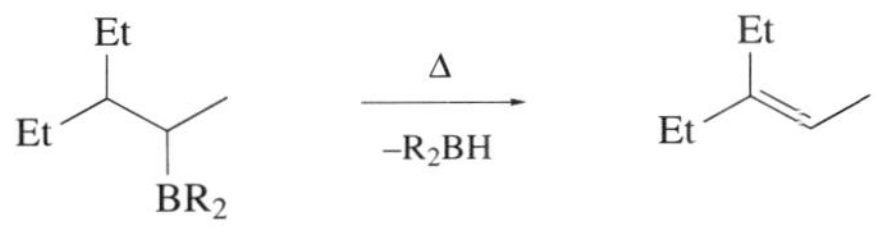

(182)

Alternatively, the B—H bond may be trapped irreversibly with alkynes, thereby liberating the alkenes and forming alkenylboranes in the process (Equation (183)) ⟨65JCS6669⟩, or with aldehydes such as formaldehyde ⟨72JA6549⟩, benzaldehyde ⟨78JOM(156)203⟩, ketones ⟨78JOM(156)203⟩ or azo compounds ⟨72JCS(P2)803, 73TL401⟩. In some cases the elimination of alkenes occurs very rapidly under mild conditions in reactions using these trapping agents as illustrated for example for the *B*-alkyl-9-BBN derivatives in Equation (184) ⟨78JOM(156)203⟩; it is possible that a cyclic transition state ⟨68DOK(179)1344⟩ involving the displacing group is involved in these reactions. The rate of elimination from the *B*-alkyl-9-BBN derivatives increases with increasing substitution at the β-carbon atom and increasing ability of the alkyl group to adopt a *syn*-planar B—C—C—H conformation. Brown and Ford have illustrated the utility of this in the transformation of cycloalkenes to methylenecycloalkenes as illustrated in Scheme 111 ⟨81JOC647⟩. The methylene derivative was formed specifically via the displacement reaction; in contrast methylenation of isopinocamphone would result in significant by-products due to epimerization or rearrangement. Alkene displacements may also be conducted with other unsaturated compounds ⟨79COC(3)791⟩ such as allyl alcohols and amines ⟨65MI 114-01⟩, vinylsilanes ⟨60ZOB3615⟩, dienes ⟨59AG520⟩ and nitriles ⟨69JCS(A)433⟩.

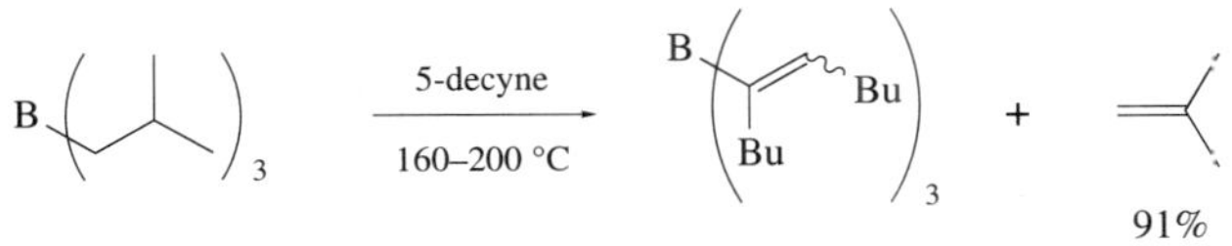

(183)

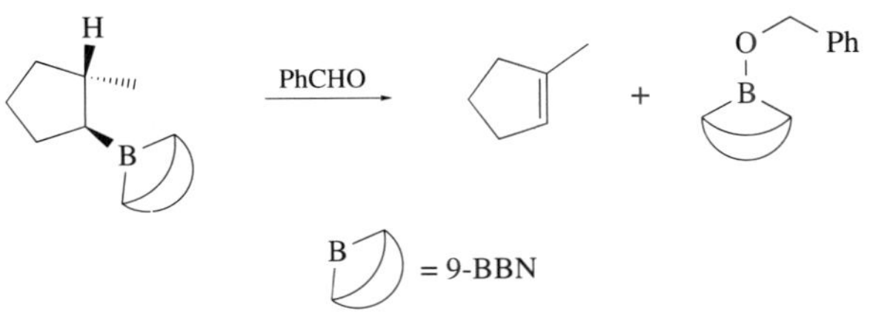

(184)

9-BBN = 9-borabicyclo[3.3.1]nonyl

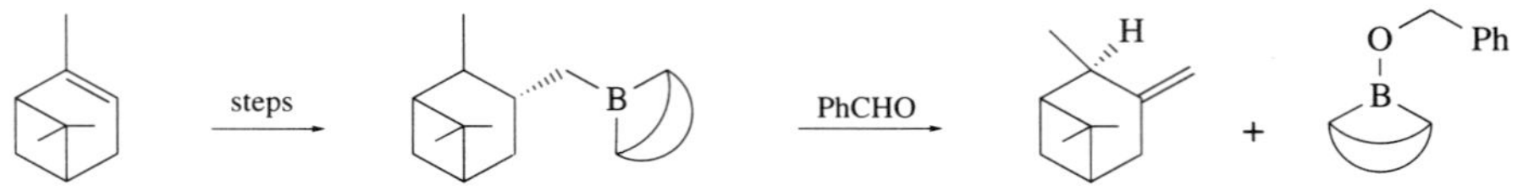

Scheme 111

The major synthetic use of borane eliminations is for contrathermodynamic alkene isomerization ⟨66JA1440, 67JA567⟩. On heating, alkylboranes undergo isomerization to the most stable isomer which is in fact the least sterically crowded isomer, via elimination and re-addition to the alkene. Therefore if an internal alkene is hydroborated then heated to effect isomerization to the terminal alkylborane, followed by displacement, this results in isomerization of an internal alkene to its less stable terminal isomer (Scheme 112), a transformation which is difficult to achieve otherwise. As di-*exo*-norbornylborane is an excellent agent for selective displacement of alkyl groups as alkenes as illustrated in Equation (181), this reagent is particularly well suited to the hydroboration–isomerization–displacement sequence to effect isomerization of alkenes from internal to terminal positions ⟨79BCJ2942⟩.

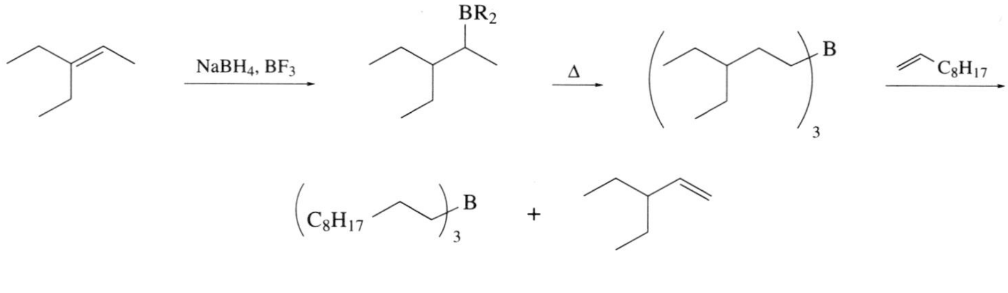

Scheme 112

Organoboron compounds in which a halo or alkylthio group is attached to boron can eliminate alkenes by using more forcing displacement conditions than those required with alkylboranes; when dialkylamino or alkoxy groups are bonded to boron elimination of alkenes is not observed ⟨68IZV548⟩.

1.14.5.3.2 Elimination of β-substituted alkylboranes

Alkylboranes bearing a heterosubstituent at the β-position can undergo elimination of R_2BX either in an uncatalysed *syn* elimination process or a catalysed (usually by acid or base) *anti* elimination (Scheme 113) ⟨79COC(3)791⟩. With β-chloroalkylboranes elimination to form the alkene and the boron halide occurs readily especially in the case of acyclic alkenes ⟨64JA1801, 66JOM(5)506, 68JA2906⟩; the β-fluoro derivatives are also quite labile ⟨68IC1014⟩ while the iodo analogues appear to be more stable ⟨70BSF2505⟩. As illustrated in Scheme 114, elimination of β-chloroalkylboranes occurs rapidly under normal hydroboration conditions, at 0 °C. This illustrates a significant feature of the hydroborations of substituted alkenes; if β-elimination occurs readily under the reaction conditions, then subsequent hydroboration of the newly formed alkene results in isolation of the product of a second hydroboration. The cyclic analogues are more stable than the acyclic β-chloroalkylboranes ⟨66JOC2777, 68JA4445⟩; *cis*-2-chlorocyclohexylboranes are quite stable while the *trans* isomer eliminates rapidly in THF at room temperature but is stable in diethyl ether, indicating that in THF the solvent acts as a nucleophile resulting in *anti*-elimination. Alkenylboranes with β-chloro ⟨70JA1427⟩ or β-alkoxy groups ⟨74JA316⟩ eliminate to form allenes (Scheme 115).

Boranes bearing other leaving groups such as acetoxy, benzyloxy and tosyloxy at the β-position behave very similarly to the chloro compounds readily eliminating to form alkenes ⟨68JA2906⟩. β-Alkoxyorganoboranes undergo uncatalysed *syn* elimination (Scheme 116) ⟨66JOC2777⟩ although they have been reported to be more stable than the chloro derivatives ⟨62IZV1373⟩. Catalysis with nucleophiles such as THF ⟨64JA1801⟩ or electrophiles such as BF_3 ⟨64JA4343, 73JOM(60)19⟩ results in *anti* elimination (Scheme 116) ⟨66JOC2777⟩. In cyclic derivatives the *trans* isomers are sufficiently stable to allow oxidation to occur ⟨68JA4439, 68JA4445⟩. Other oxygen substituents behave similarly,

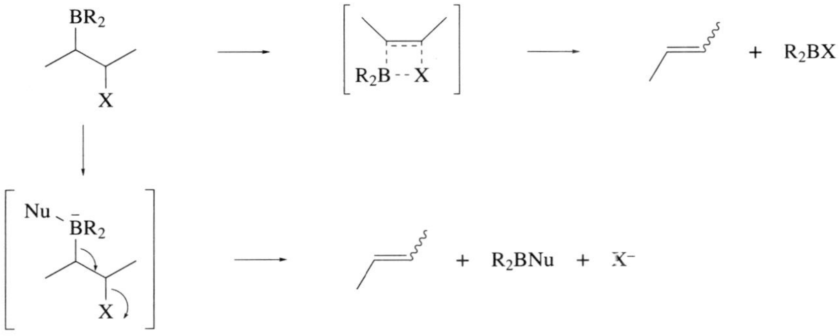

Scheme 113

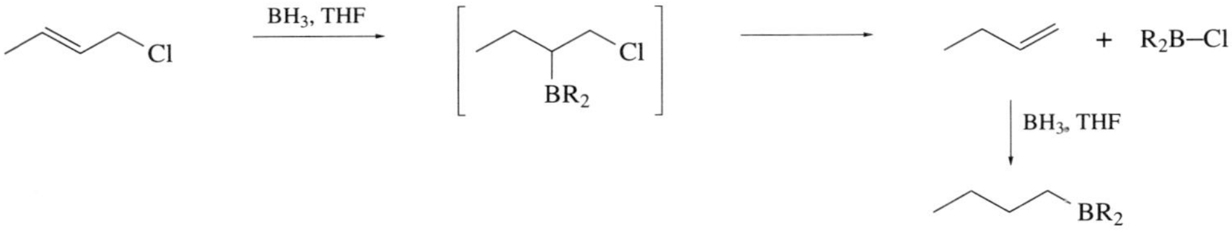

Scheme 114

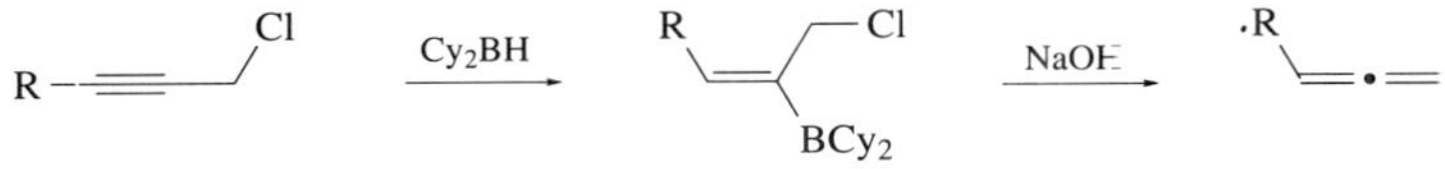

Scheme 115

e.g., phenoxy ⟨68JA2906⟩, dialkylboryloxy ⟨68JA2906, 68JA4439⟩, lithium alkoxide ⟨66TL4315⟩ and trimethylsilyloxy groups ⟨75JOM(102)123⟩. In contrast organoboranes bearing β-alkylthio or β-arylthio groups do not undergo uncatalysed elimination but on treatment with butyllithium give alkenes via *anti* elimination (Equation (185)) ⟨66JOC2777⟩.

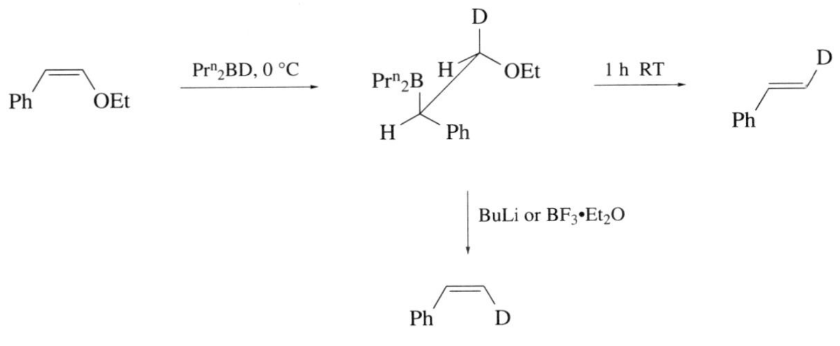

Scheme 116

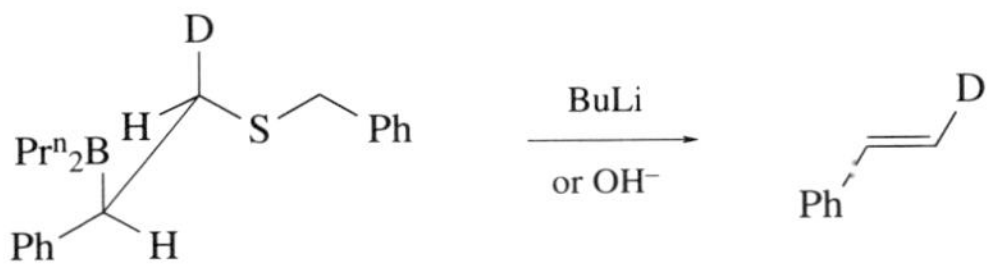

(185)

Organoboranes with β-dialkylamino groups eliminate in an *anti* manner even in the absence of a catalyst; presumably self-catalysis occurs via an acid–base B—N interaction ⟨66JOC2777⟩. Addition of a carboxylic acid facilitates the elimination and on heating this produces the alkenes in good yields (Equation (186)) ⟨69JCS(B)863⟩. As with oxygen substituents, cyclic amino derivatives are usually sufficiently stable to allow removal of the boron oxidatively without elimination occurring ⟨71BSF3978⟩. Elimination of a β-amidoborane on treatment with tetrahydroborate resulted in cleavage of the bicyclic system as shown in Equation (187) ⟨66JOC3498⟩.

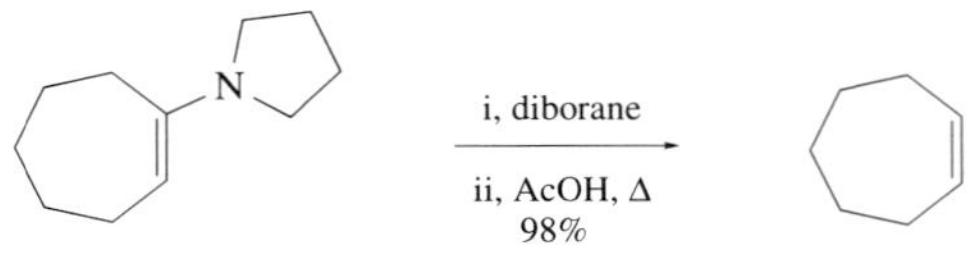

(186)

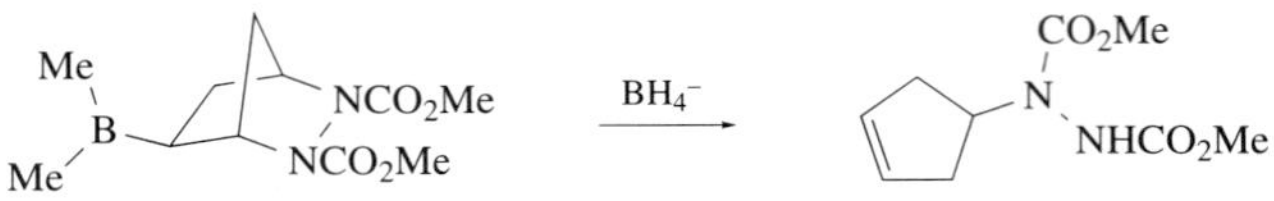

(187)

The facility with which boranes eliminate when β-acetoxy or β-amino substituents are present can be exploited in the transformation of ketones to alkenes via enol acetates or enamines as illustrated in Scheme 117 ⟨64TL2039⟩. Since the regiochemistry of enol acetate and enamine formation are complementary either the more-substituted or less-substituted alkene, respectively, can be obtained.

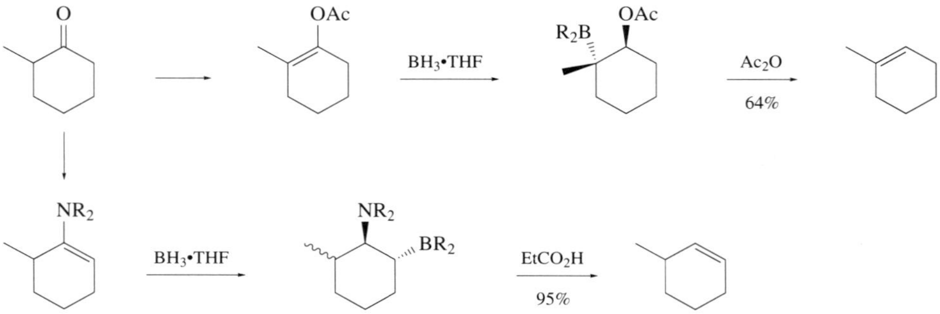

Scheme 117

Elimination of boranes with an oxygen substituent in the β-position has been employed in a boron–Wittig reaction ⟨83TL635, 87CC297⟩. The anion formed by condensation of dimesitylboryl-stabilized carbanions with aromatic aldehydes eliminates to form selectively either (*E*) or (*Z*) alkenes on treatment with TMS-Cl/HF or $(CF_3CO)_2O$, respectively (Scheme 118). The (*E*) alkene results from *anti* elimination while the (*Z*) alkene is formed by a *syn* elimination process.

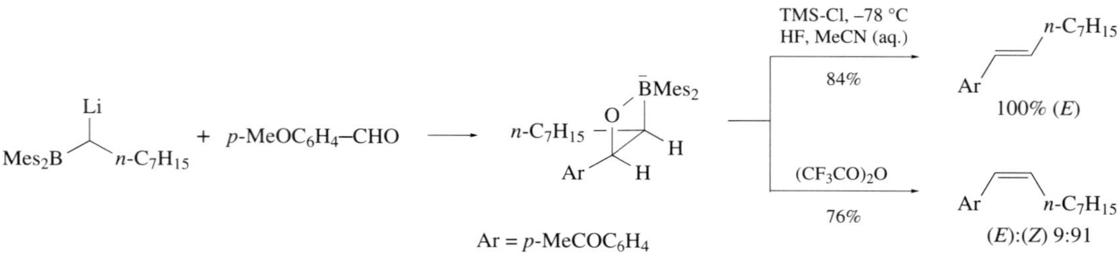

Scheme 118

Fragmentation of δ-substituted boranes with, for example, a sulfonyloxy group at the 4-position may occur as illustrated in Scheme 119 ⟨70TL3861⟩.

Alkenylboronic esters undergo deboronobromination to form vinyl bromides with excellent stereochemical control; both the addition of bromine to the alkenylborane and the elimination on treatment with base occur stereospecifically *anti* resulting in formation of the bromoalkene of opposite stereochemistry to the starting alkenylborane (Scheme 120) ⟨65JA1526, 73JA6456, 88TL21, 89T1859⟩. This methodology has been used in the stereospecific synthesis of both (*E*) and (*Z*) 1,2-

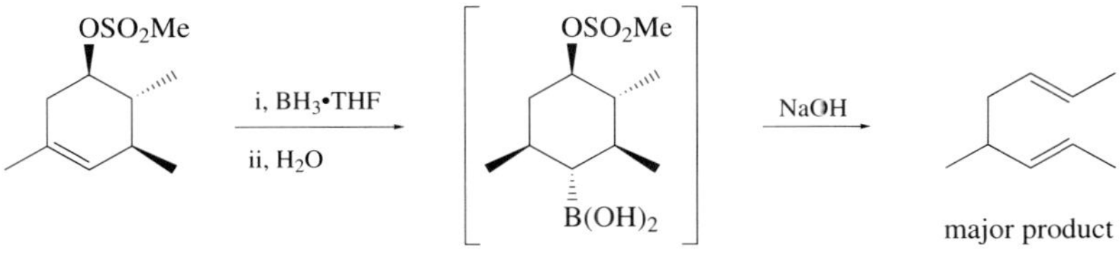

Scheme 119

disubstituted vinyl bromides ⟨88TL21⟩. The opposite stereochemical result is observed when (*E*)-alkenylboronic acids are iodinated in the presence of base ⟨73JA5786⟩ or with chloramine-T/sodium iodide ⟨88MI 114-04⟩ to form the (*E*)-iodoalkenes as shown in Equation (188). Thus alkynes can be readily transformed to *cis* and *trans* vinyl bromides and *trans* vinyl iodides via hydroboration to the *trans* borane derivatives followed by halogenation.

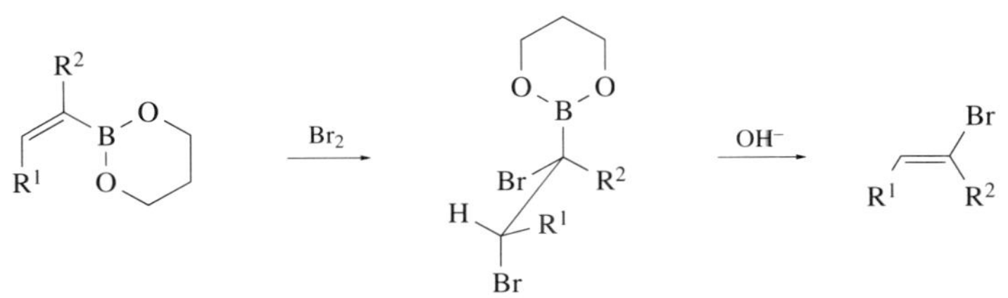

Scheme 120

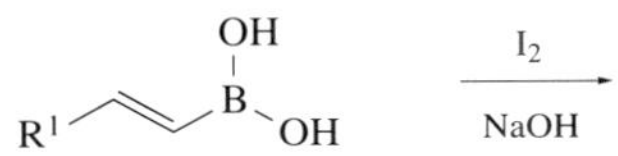

(188)

This deboronobromination may also be applied to certain alkenyldialkylboranes as illustrated in Scheme 121 ⟨67JA4531⟩; by effecting the elimination from the dibromo intermediate thermally the product of *syn* elimination is obtained whereas the base-induced process results in the product of *anti* elimination. Furthermore, phenylethenyl derivatives consistently result in the opposite stereochemical outcome to those observed with alkenyl derivatives.

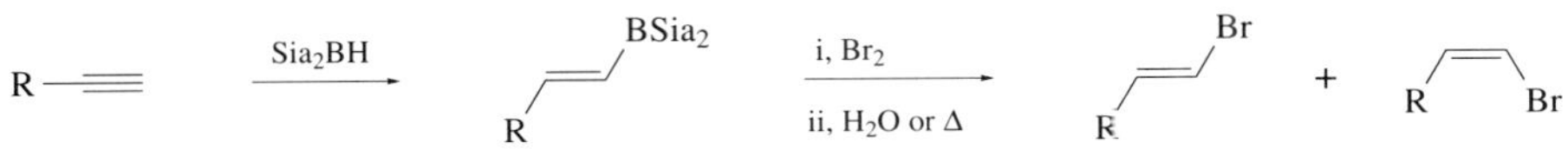

R = alkyl		
Thermal	88–100%	0–12%
Solvolysis	2–5%	95–98%
R = phenyl		
Thermal	18%	82%
Solvolysis	96%	4%

Sia_2BH = disiamylborane

Scheme 121

More frequently the reaction of alkenyldialkylboranes with bromine or iodine proceeds with migration of one of the alkyl groups on boron. This methodology, which is useful for the stereoselective transformation of alkynes to alkenes, has been extensively investigated and developed, mainly by Brown *et al.* Thus when alkynes are hydroborated to form (*E*)-alkenyldialkylboranes and this is followed by treatment with iodine and base (typically hydroxide or methoxide), migration of one of the alkyl groups from the boron occurs followed by *anti* elimination of the iodine and boron group to form the (*Z*)-alkene (Scheme 122) ⟨67JA3652⟩. The migration of the alkyl group occurs with inversion at the migration terminus and retention of configuration of the migrating group

⟨71JA6309⟩. This methodology can also be extended to the stereospecific synthesis of trisubstituted alkenes ⟨82JOC171, 82JOC754, 82JOC5407⟩ and also β-aminoalkenes ⟨82T2355⟩.

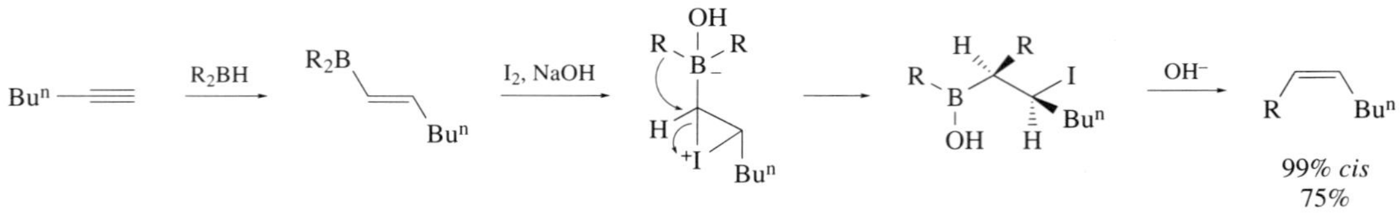

R = cyclohexyl

Scheme 122

In order to stereoselectively form (*E*)-alkenes from the (*E*)-alkenyldialkylboranes, treatment with bromocyanogen instead of iodine is employed; the strongly electron-withdrawing group on the boron enhances the tendency for *syn* elimination, thereby producing the *trans* alkene (Scheme 123) ⟨72JA6560⟩. Bromination of thexylalkylalkenylboranes has been employed to synthesize either the (*Z*) or (*E*) isomer of α,β-unsaturated esters with very high stereoselection ⟨74JOC2321⟩. Thermal elimination is employed to form the (*E*) isomer by *syn* elimination while use of sodium ethoxide to effect the elimination produces the (*Z*) isomer by *anti* elimination.

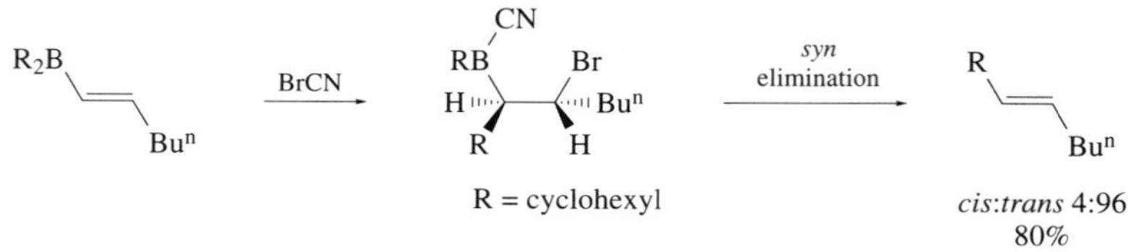

R = cyclohexyl

Scheme 123

Only one of the alkyl groups on the boron is transferred in the halogenation–elimination sequence which can be wasteful unless the alkene from which they are derived is readily available. However considerable success has been achieved in using different substituents on boron. Negishi has applied the iodination–base catalysed elimination methodology with borates to the synthesis of *trans*-enynes (Scheme 124) ⟨73CC874⟩. A leukotriene synthesis involving similar *trans*-enyne formation has also been reported ⟨86TL539⟩. Negishi's enynes are readily transformed to *cis,trans*-dienes as illustrated by application to the synthesis of the pheromone bombykol. Small amounts of migration of the thexyl group are observed in this reaction (see also ⟨72JA4013⟩). However, in other cases thexyl group migration can compete effectively ⟨68JA6243⟩. To overcome this problem, Evans *et al.* employed the iodination–base-catalysed elimination sequence with boronic ester derived ate complexes to synthesize prostaglandin model systems as illustrated in Scheme 125 ⟨76JOC3947⟩. Similarly alkenylalkylbromoboranes and alkenylalkylmethoxyboranes ⟨82JOC754, 82JOC5407⟩ can be employed thereby circumventing the migration problem as illustrated in the synthesis of the pheromone muscalure (Equation (189)) ⟨82JOC3806⟩.

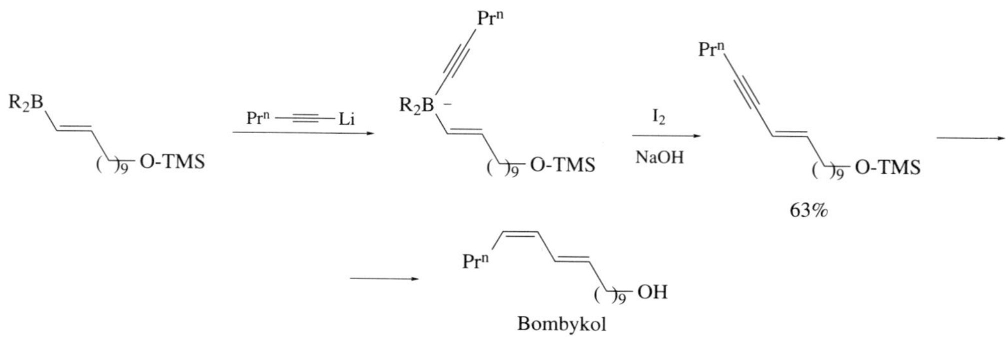

$R = Me_2CHCHMe_2$–

Scheme 124

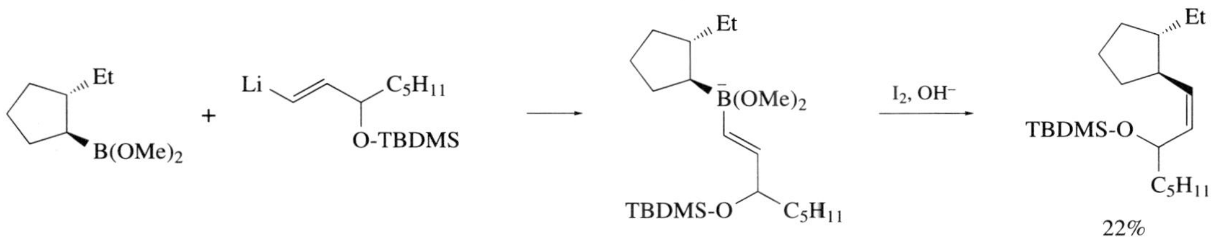

Scheme 125

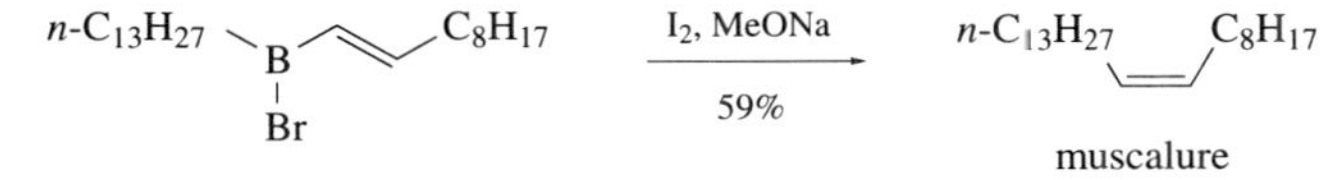
(189)

Trans-alkenylborepanes undergo elimination on treatment with iodine and base to form (*Z*)-7-alkenyl-1-boronate esters which on oxidation furnish (*Z*)-7-alkenyl-1-ols (Scheme 126) ⟨82JOC1792⟩. Yamamoto *et al.* have demonstrated that alkenyldialkylboranes are transformed stereoselectively to (*E*) alkenes on treatment with palladium acetate and triethylamine via a *cis* β-elimination ⟨77CC852, 80BCJ1670⟩. 1-Halo-1,2-dialkylalkenes may be synthesized stereoselectively from (1-halo-1-alkenyl)dialkylboranes on treatment with lead(IV) acetate or (diacetoxyiodo)benzene ⟨80BCJ1652⟩.

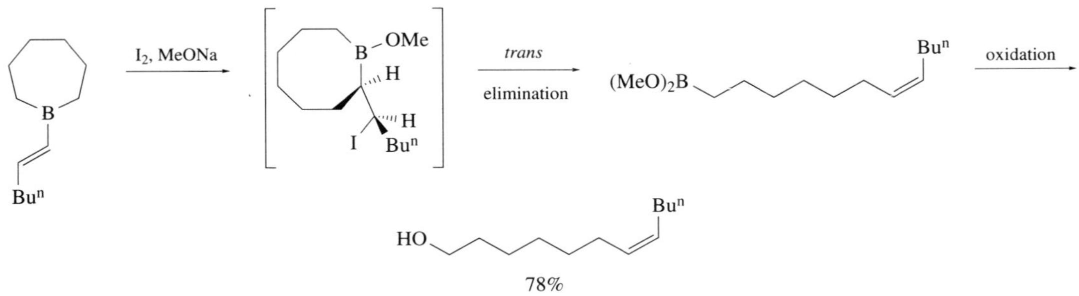

Scheme 126

1.14.6 BY ELIMINATION OF METAL FUNCTIONS

Frequently alkyl metal compounds in which there is a β-hydrogen atom are unstable and eliminate readily to form alkenes ⟨76CRV219, 76CRV243⟩. However, as highlighted in these reviews, while this decomposition pathway is important, it may be overemphasized.

1.14.6.1 Elimination of Groups 1 and 2 Metals—Li, Na, K, Mg

Organolithiums in which the alkyl group has a β-hydrogen atom decompose thermally to form alkenes and lithium hydride usually via β-elimination ⟨50LA(567)179, 65JOC1258, B-74MI 114-01⟩; occasionally the products of α-elimination and coupling are observed also. With primary alkyllithiums temperatures of over 100 °C are usually required to effect elimination but the secondary and tertiary derivatives are less stable and eliminate more easily. If aromatization occurs as a result of elimination then the decomposition occurs rapidly ⟨38JA2333⟩. As illustrated in Equation (190) *s*-butyllithium produces mainly the product of Zaitsev elimination; furthermore formation of the *cis* alkene is preferred over the *trans*-alkene ⟨65JOC1258⟩. Elimination in alkyllithiums is believed to proceed via a four-centre cyclic transition state ⟨83OM554⟩; therefore it is likely that the conformation required for elimination to form the *cis*-alkene is favoured over that for the *trans*-isomer. Elimination occurs also in organosodium and organopotassium compounds; the ease of elimination increases on going from lithium to potassium ⟨62CI(L)895, 63TL851⟩.

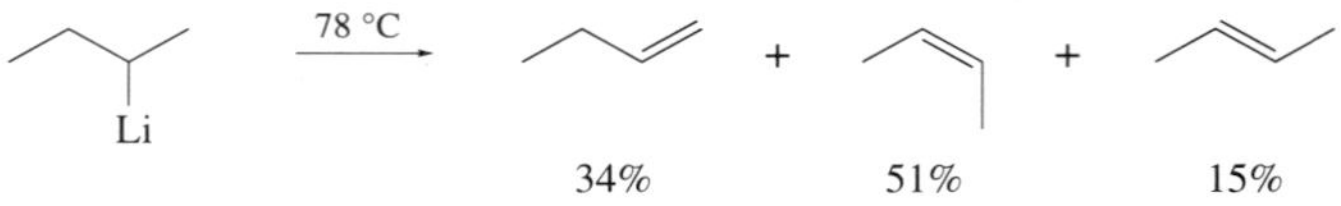

(190)

Organolithium derivatives bearing a leaving group in the *β*-position, e.g., a silyloxy group, may also undergo elimination to form alkenes ⟨88JOM(346)C1⟩. Perfluoroalkyllithiums decompose thermally via *β*-LiF elimination ⟨79JOC3431⟩. Treatment of 1,2-diiodoadamantane with *n*-butyllithium generates adamantene as a reactive intermediate via elimination of LiI (Equation (191)) ⟨76JCS(P1)234⟩. Occasionally elimination of ethylene to form an alkyllithium can occur (Equation (192)) if the carbanion corresponding to the newly formed organolithium is highly stabilized ⟨68C338⟩.

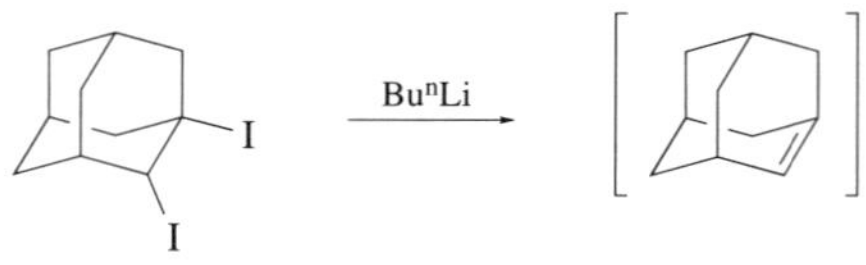

(191)

(192)

Frequently ring opening may occur in eliminations of alkyllithium reagents, for example the cleavage of cyclic ethers as shown in Equation (193) ⟨78CI(L)234⟩ and other heterocyclic compounds (Equation (194)) ⟨79JCR(S)54⟩. Ring opening is also possible when the lithium is *exo*-cyclic, for example with furans (Equation (195)) ⟨78JOC4235⟩, azoles ⟨73JA3408⟩ and the reversible opening of lithiated dihydrooxazines illustrated in Equation (196) ⟨70TL4355⟩. Similar fragmentation forming alkynes is possible with a wide variety of heterocyclic systems ⟨71JCS(C)3447, 76ACS(B30)485, 82COMC-I(7)85⟩.

(193)

(194)

(195)

(196)

On treatment with ethylene and nickel chloride organolithium and magnesium reagents undergo elimination to form alkenes (Equation (197)) ⟨80LA171⟩. Under these conditions a slight preference for formation of the less-substituted alkene is observed. Similarly on treatment with hydride acceptors such as triphenylmethyltetrafluoroborate or tricyclohexylborane, organolithium and organomagnesium compounds eliminate under mild conditions to produce selectively the alkene formed by Hofmann elimination ⟨77AG(E)44a⟩. This elimination has been employed in nucleophilic substitution of vinylic hydrogen atoms on activated alkenes by carbanions (Scheme 127) ⟨77AG(E)44b⟩.

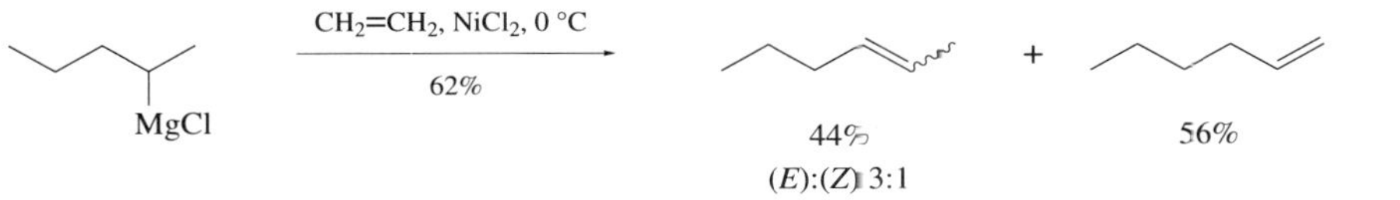

(197)

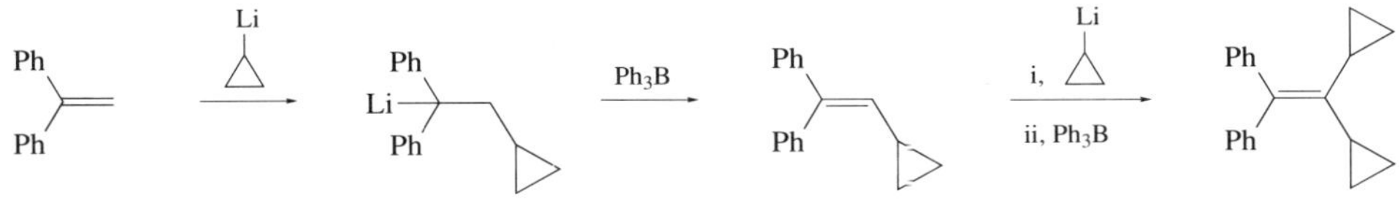

Scheme 127

Pyrolysis of organomagnesium compounds results in elimination to form alkenes, such as ethene and 1-butene, and magnesium hydride ⟨52CB593, 65JOM(4)349, 69JOM(16)7⟩. In the pyrolysis of ethylmagnesium halides, other hydrocarbons have also been identified, e.g., ethane and 1-butene ⟨69JOM(16)7⟩. Photolysis of Grignard reagents also results in alkenes via β-elimination ⟨75JOM(85)1⟩.

In reactions of Grignard reagents with carbonyl compounds, β-elimination can compete if steric hindrance by bulky alkyl groups slows the usual reaction pathway. Occasionally reduction products are observed in the product mixtures as a result of this elimination process (Equation (198)) ⟨66MI 114-01, B-80MI 114-02⟩.

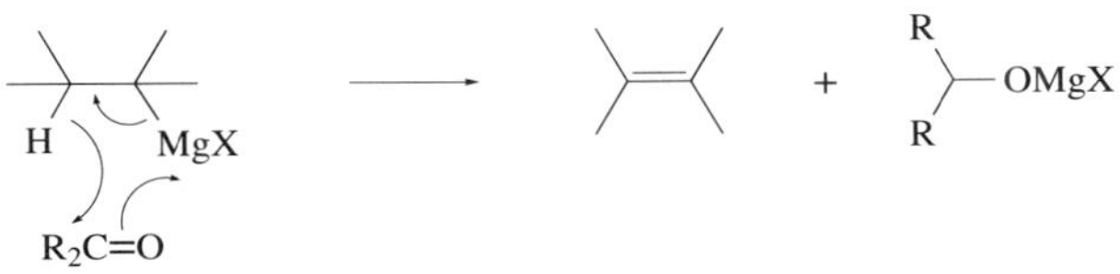

(198)

Organomagnesium compounds bearing a leaving group in the β-position undergo facile elimination, e.g., 1,2-dibromoethane forms ethylene on reaction with magnesium via the β-bromo Grignard reagent ⟨B-80MI 114-01⟩. Similarly β-alkoxy groups can be eliminated ⟨87JCS(P1)475⟩.

1.14.6.2 Elimination of Group 13 Metals—Aluminum

Organoalanes eliminate reversibly to form alkenes and an aluminohydride even more easily than organoboranes ⟨60LA(629)1, 60LA(629)53, 60LA(629)210⟩; this has been used industrially in the synthesis of long-chain alkenes mainly due to the pioneering work of Ziegler ⟨82COMC-I(7)365⟩. Elimination to form terminal alkenes is generally quite efficient at 200–300 °C; the aluminohydride formed may be trapped with ethene to form the ethylaluminum thereby displacing the alkene with ethene. Thus alkene polymerization can be conducted as illustrated in Scheme 128; firstly ethene inserts into ethylaluminum to form chains with even numbers of carbon atoms, then by using conditions which favour displacement (typically 200–300 °C) the alkene is displaced by ethene regenerating the triethylaluminum. Alternatively the displacement reaction can be catalysed by nickel ⟨60LA(629)172⟩. As an extension of this research, the development of Ziegler–Natta catalysts (involving aluminum and a transition metal such as titanium) for alkene polymerization was particularly significant ⟨55AG541, B-79MI 114-02⟩.

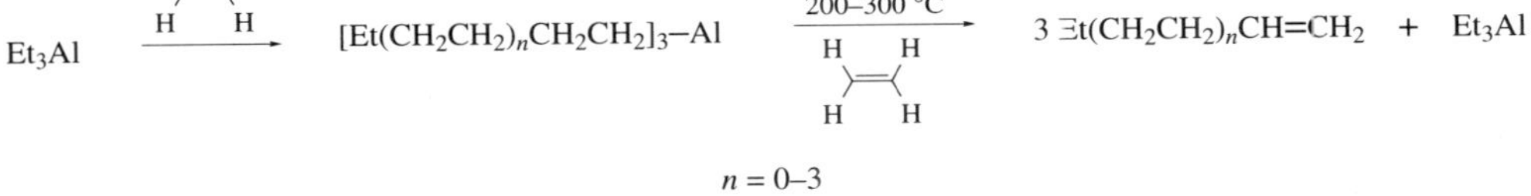

Scheme 128

Elimination of β-branched alkylaluminums, such as isobutylaluminum, is particularly facile and occurs quite rapidly at 100–150 °C (Equation (199)). When tripropylaluminum is treated with propene at 140–200 °C catalytic dimerization of the propene occurs as insertion of propene forms a β-branched alkylaluminum which rapidly eliminates at this temperature (Scheme 129). Thus monosubstituted alkenes may readily be dimerized. Equilibria can be established between organoaluminum compounds derived from various alkenes—alkenes are generally displaced by less-substituted alkenes thereby forming the more stable organoaluminum compound ⟨60LA(629)14, 60LA(629)53⟩.

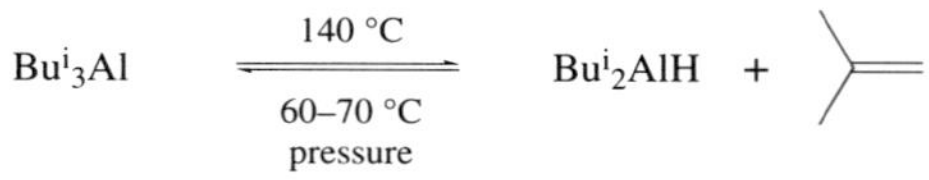

(199)

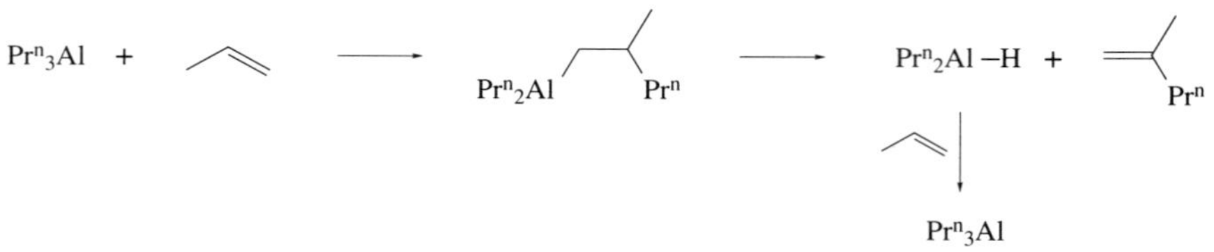

Scheme 129

Elimination to form an alkylaluminum and an alkene is possible with trineopentylaluminum and like compounds (Equation (200)) but only under extreme conditions ⟨60LA(629)207⟩.

(200)

When trialkylaluminum compounds are reacted with carbonyl compounds reduction products are often observed as well as addition products. The reduction is a result of β-elimination in the organoaluminum compound (Scheme 130) ⟨73JOC1130, 90OM3086⟩. Triisobutylaluminum is particularly effective as a reducing agent.

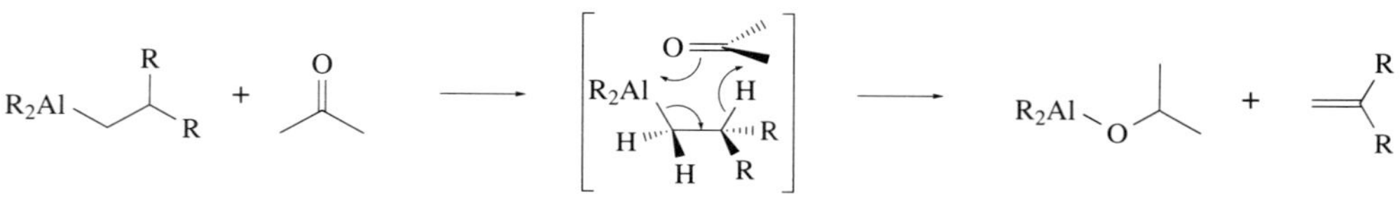

Scheme 130

1.14.6.3 Elimination of Group 14 Metals—Tin and Lead

On treatment with trityl cation, tetraalkyltin and tetraalkyllead compounds (and also dialkylmercuries) undergo elimination to form alkenes (Equation (201)) ⟨71JA6278⟩. Hydride abstraction in these reactions is facilitated by stabilization of the carbonium ion intermediate by the metal. β-Trichlorostannyl aldehydes and ketones undergo dehydrostannation on heating in DMSO or on treatment with bases such as pyridine, triethylamine, 1,4-diazabicyclo[2.2.2]octane (dabco), or TMEDA; as these β-trichlorostannyl aldehydes and ketones are readily accessible by ring opening of siloxycyclopropanes with stannic chloride, this provides a useful route from siloxycyclopropanes to α-methylene aldehydes and ketones (Scheme 131) ⟨86JOC2389, 92JOC17⟩. The elimination to form unsaturated aldehydes is much more facile than that to form methylene ketones.

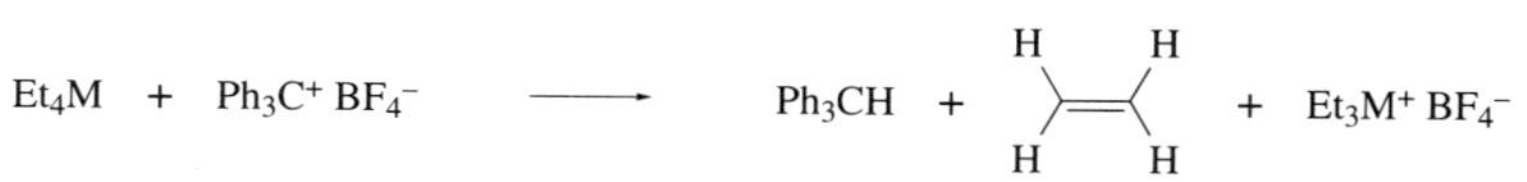

(201)

M = Sn or Pb

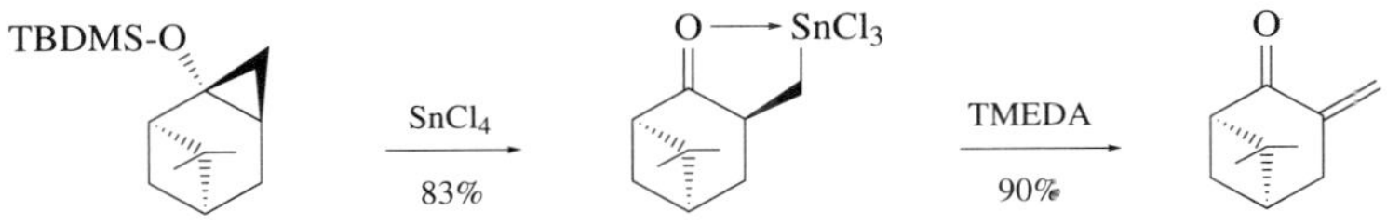

Scheme 131

Elimination of organotin and organolead units in compounds bearing a leaving group in the β-position is also possible; acid-catalysed elimination of β-hydroxyalkyl triphenylmetal compounds has been reported (Equation (202)) ⟨69JOM(18)P1, 70JOC1303⟩. Interestingly, the corresponding triphenylsilicon and triphenylgermanium compounds do not eliminate under these conditions. Thus the ease of elimination of the Group 14 elements is Pb > Sn ≫ Ge > Si. The rate of elimination increases with increasing substitution at the sites of elimination. The stereochemistry of the elimination has been established as *trans* by use of the *threo* 3-triphenylstannyl-2-butanol prepared from *cis*-2,3-butene oxide (Scheme 132). Elimination of a hydroxy and a tributylstannyl group from β-hydroxyvinylstannanes under acidic conditions has been applied to the synthesis of α-allenylsilanes (Equation (203)) ⟨87TL2751⟩. As β-stannylalcohols can be formed by reaction of trialkyl or triarylstannylmethyllithium with carbonyl compounds and subsequently eliminated under acidic conditions or thermally this can prove a useful method for methylenation of ketones (Scheme 133) ⟨77AG(E)862, 78TL4399, 84CL1897⟩. Thus methylenation of *cis*-1-decalone proceeds in excellent yield without any epimerization whereas the corresponding Wittig reaction proceeds to give a mixture of *cis* and *trans* isomers in lower yield ⟨90S259⟩. Vinyl oxiranes can be prepared in this fashion employing silica gel to effect the elimination of the hydroxy and tin groups (Scheme 134) ⟨85TL2205, 90S259⟩. Interestingly while Wittig reactions with α-chloro ketones can be complicated ⟨68JOC2975⟩, reaction of α-chloro ketones with trimethylstannylmethyllithium leads to allyl alcohols in reasonable yields ⟨84CL1897, 90S259⟩—elimination from an intermediate epoxystannane is the proposed mechanism (Scheme 135). Similarly, epoxystannanes have been proposed as intermediates in the mcpba oxidation of allylstannanes to form allyl alcohols ⟨80S1011⟩.

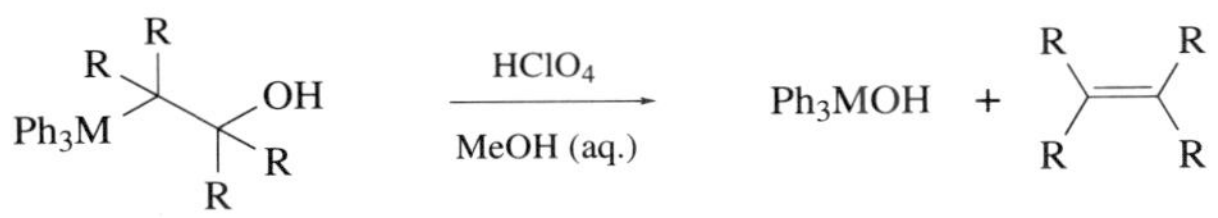

(202)

M = Sn, Pb

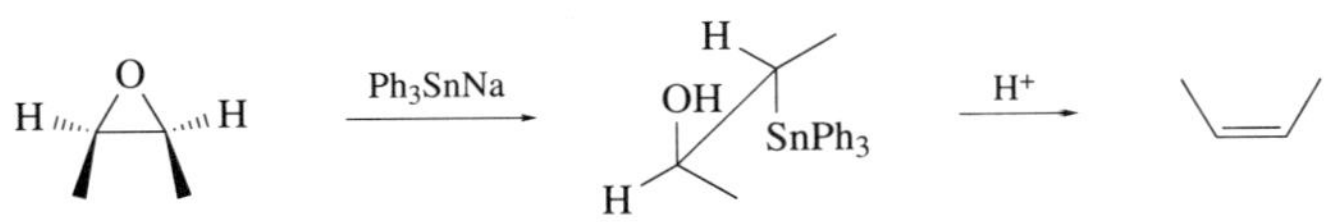

Scheme 132

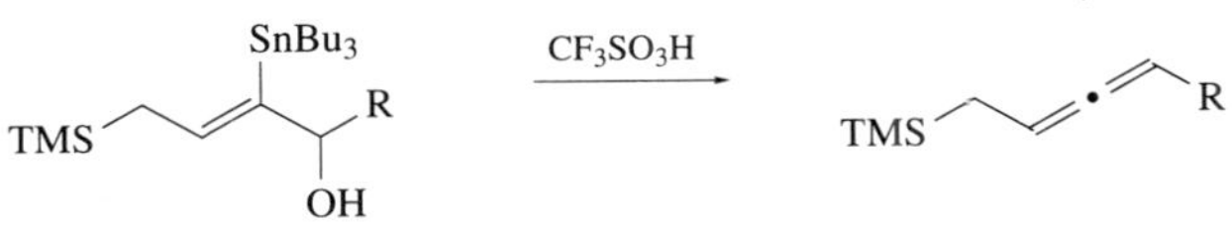

(203)

R = Ph, 92%
R = Pr, 72%

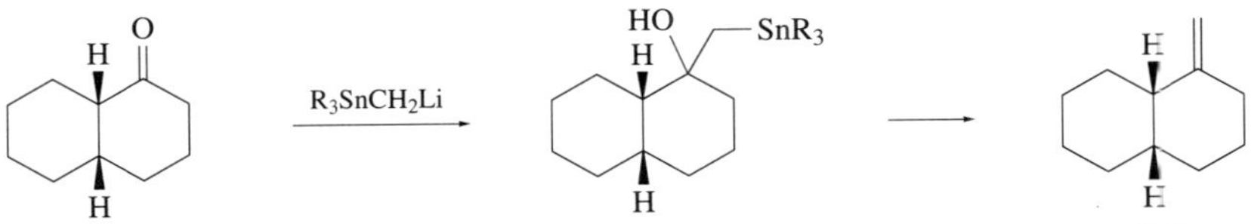

91%

Scheme 133

Elimination of stannyl halide has been postulated as a pathway in the transformation of α-haloepoxides to allyl alcohols on treatment with trialkylstannates ⟨87CJC1821⟩. Diene formation by

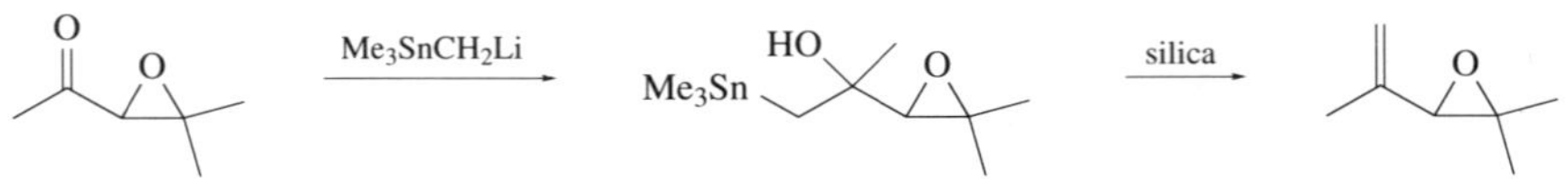

Scheme 134

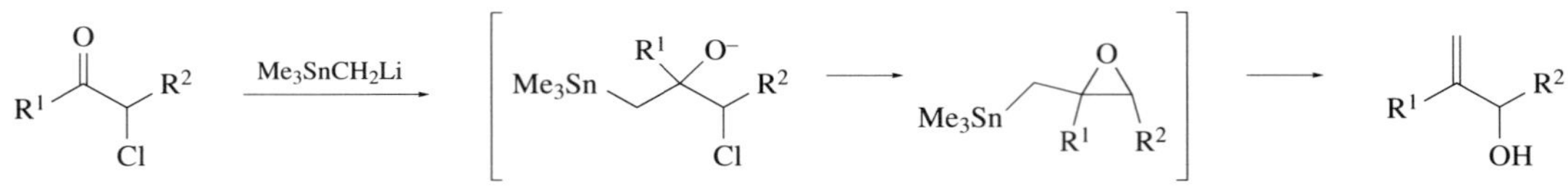

Scheme 135

elimination of β-stannyl and sulfide groups is illustrated in Scheme 3 ⟨82TL2205⟩ (Section 1.14.2.1.1). Elimination of β-stannyl sulfones is quite facile as discussed in Section 1.14.2.3.1 ⟨82TL2205, 85JOC3622⟩, occurring on silica gel or by heating to form alkenes (Scheme 136) ⟨83CC619⟩. Stereoselectivity in this example was low. Alkene formation by destannylselenylation of a β-stannyl selenide is also possible as illustrated in Equation (12) in Section 1.14.2.1.2 ⟨92JOC3145⟩.

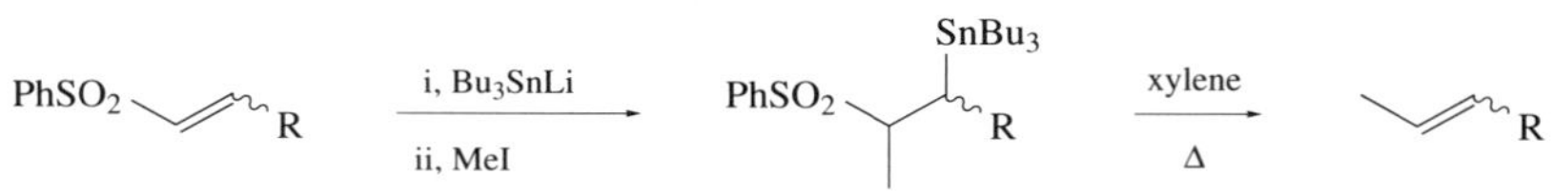

Scheme 136

The ability of the tributylstannyl group to participate in fragmentation has been employed to direct Baeyer–Villiger and Beckmann fragmentations ⟨90JA6729⟩.

1.14.6.4 Elimination of Transition Metals

Decomposition of alkyl transition metals bearing a β-hydrogen atom via β-elimination is very frequently observed ⟨76CRV219, 76CRV243, B-78MI 114-04, B-85MI 114-01, B-87MI 114-04⟩; however, as pointed out by Lappert and Schrock, there is in fact a risk that its importance may be overemphasized. The elimination is believed to proceed via a four-centre planar cyclic transition state (Scheme 137). The elimination illustrated in Equation (204) of the deuterium-labelled alkyl copper confirms that the β-hydrogen atom is transferred to the metal ⟨70JA1426⟩. Therefore, the transition-metal complex must be coordinatively unsaturated with vacant *cis* coordination sites to allow the elimination to occur. In the transition-metal series the alkene formed by the elimination may remain within the coordination sphere of the metal (Scheme 138) ⟨72JA5258⟩. Whitesides *et al.* have studied eliminations from dialkylplatinum complexes in detail (Scheme 138) ⟨73JA4451, 76JA6521, 86OM1473⟩. This elimination, which is first order, is inhibited by addition of triphenylphosphine, presumably by slowing the dissociation of phosphine from the complex. The reversibility of this elimination is illustrated by scrambling of deuterium labels (see also ⟨68JCS(A)190⟩). Once the butene elimination has occurred, *n*-butane is produced also, by reductive elimination from the resulting platinum hydride.

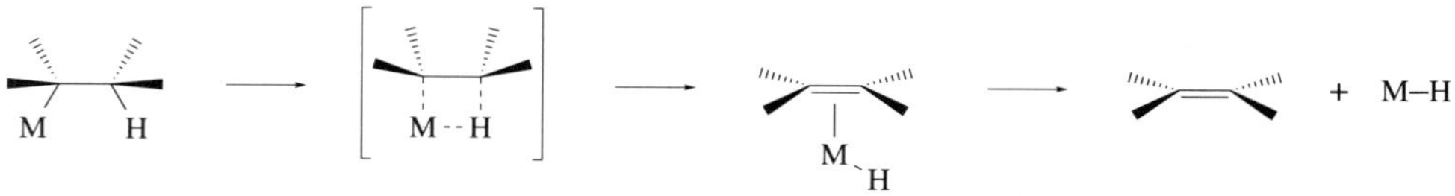

Scheme 137

(204)

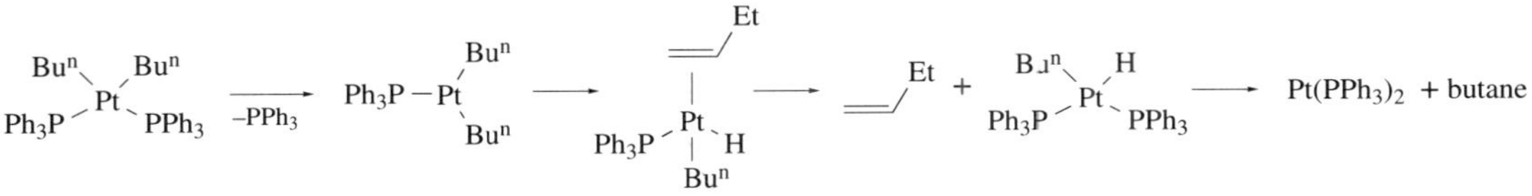

Scheme 138

Elimination to form the less-substituted alkene is normally preferred, e.g., terminal alkenes are formed preferentially when the alkyliron complexes are treated with trityl ion followed by iodide to liberate the alkene from the transition-metal complex (Scheme 139) ⟨78TL3307⟩.

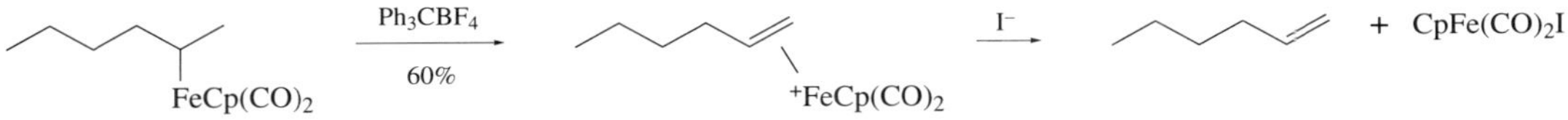

Scheme 139

Metallacyclic complexes (**13**) are much more stable thermally than their acyclic analogues as the planar four-centre transition state is not easily attained in the cyclic systems ⟨73JA4451, 76JA6521, 86OM1473⟩. Similarly use of bulky ligands can prevent the dihedral angle of the M—C—C—H bond approaching 0° as seen for example in $CrBu^t_4$ ⟨72JOM(42)C39⟩. In the alkyliridium(I) complex (**14**) the kinetic isotope effect for β-hydrogen elimination is $k_H/k_D = 2.28 \pm 0.20$ indicating that in the transition state, both C—Ir and H—Ir bond formation are important ⟨74JOM(81)C37⟩. In this complex elimination is rate determining and there are no complications of deuterium scrambling.

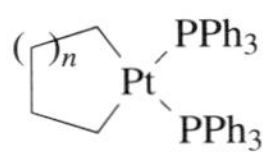

n = 1, 2
(13)

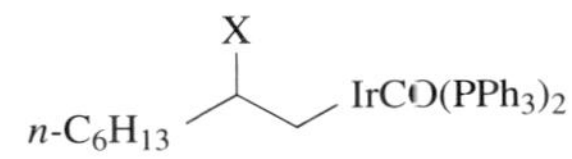

X = H/D
(14)

Isomerization of alkenes in the presence of transition metal complexes (e.g., Pd or Pt) to form an equilibrium mixture of alkenes may also proceed via a reversible alkene addition–elimination sequence ⟨64JA1776, 69CRV799⟩. The isomerization of the alkyl group in the complex shown in Equation (205) proceeds via elimination followed by alkene addition leading to the more stable alkylmetal, again illustrating the reversibility of the process ⟨74JA6140⟩. Similarly alkylzirconium complexes rearrange rapidly to place the zirconium at the least-hindered position on the alkyl chain (Equation (206)) ⟨74JA8115⟩. In general secondary and tertiary alkyls are least stable; thus on treatment of Cp_2ZrCl_2 with *t*-butyl Grignard reagent the hydride complex Cp_2ZrClH is formed, presumably via elimination from an intermediate alkyl zirconium ⟨84TL3407⟩. A further example of β-elimination in the early transition metals is illustrated in Equation (207) ⟨71JA3793⟩.

$$AuMe_2Bu^tPPh_3 \longrightarrow AuMe_2Bu^iPPh_3 \qquad (205)$$

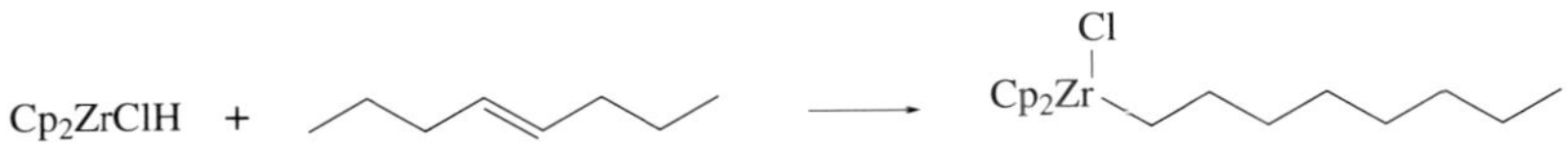

(206)

$$Cp_2NbEt(C_2H_4) \underset{}{\overset{-C_2H_4}{\rightleftharpoons}} Cp_2NbH(C_2H_4) \qquad (207)$$

Fragmentation of metallacyclobutanes to form alkenes and carbene complexes has been investigated (Equation (208)) ⟨86JA733, 86OM721⟩. Unlike most other metallacyclobutanes, titanacyclobutanes are more stable than their fragmentation products and therefore their exchange reactions with alkenes are readily studied. In this case the alkene is displaced by the highly strained cyclopropene. Carbonyl methylenation using $Cp_2TiCl_2/AlMe_3$ ⟨78JA3611⟩ or $TiCl_4/Zn/CH_2Br_2$

⟨78TL2417⟩ proceed via the reactive species $[Cp_2Ti{=}CH_2]$ formed by fragmentation of aluminum alkyl adducts or titanacyclobutanes ⟨86OM721, 91JOM(405)1⟩.

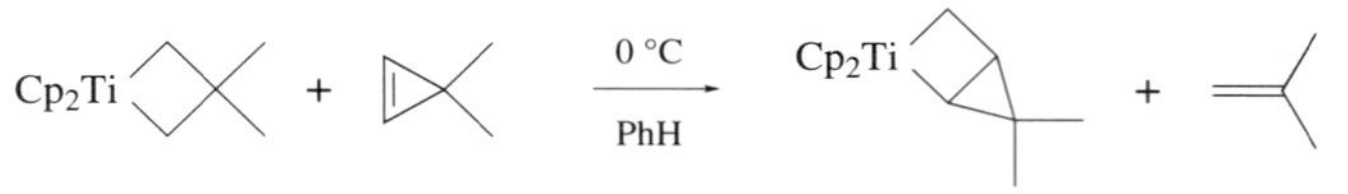

(208)

β-Elimination of alkyl transition-metal complexes frequently features as a step in catalytic cycles of reactions involving transition metal complexes, e.g., as a termination step in alkene polymerization ⟨B-80MI 114-03⟩. In the decarbonylation of acyl halides with $RhCl(PPh_3)_3$, the stereospecific formation of alkenes as illustrated in Scheme 140 and Equations (209a) and (209b) is consistent with *syn* β-hydride elimination from an intermediate alkyl rhodium species ⟨74JA1518⟩.

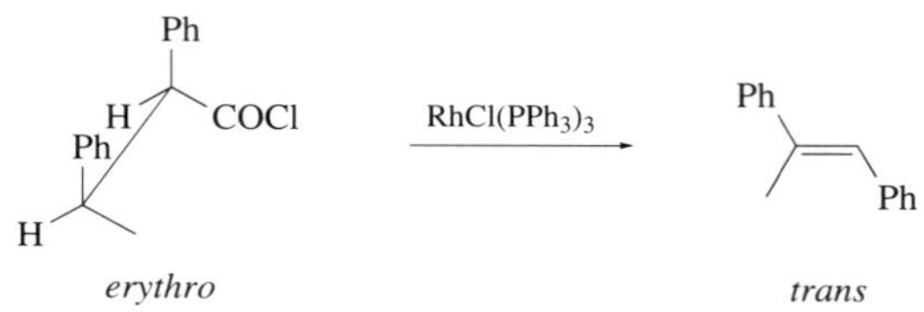

(209a)

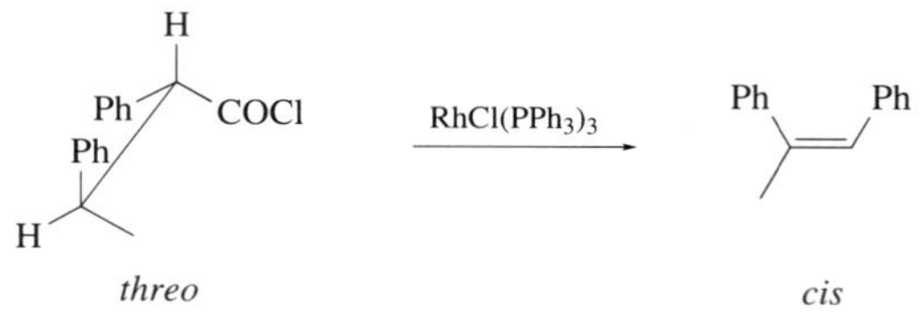

(209b)

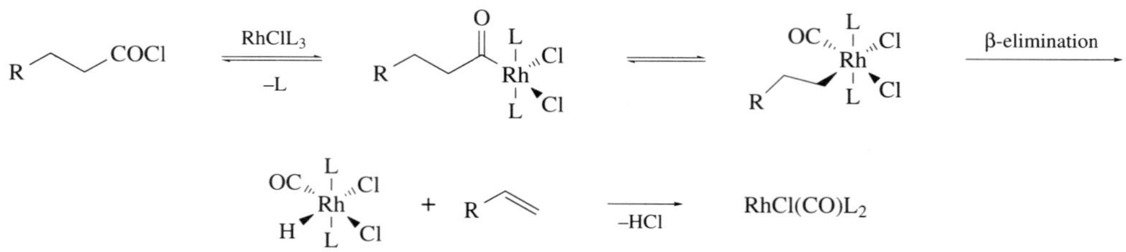

Scheme 140

Elimination of ethene or propene from titanium(III) alkyls on treatment with carbon dioxide has been reported (Equation (210)) ⟨87CC849⟩. Addition of alkyl radicals generated from alkyl cobalt 'salophen' reagents to activated alkenes followed by elimination of a cobalt hydride produces new alkenes (Scheme 141) ⟨87CC871⟩.

$$(\eta^5\text{-}C_5Me_5)_2TiCH_2CH_2R \ + \ CO_2 \longrightarrow (\eta^5\text{-}C_5Me_5)_2TiO_2CH \ + \ \text{=\!\!\diagup}^{R} \qquad (210)$$

R = H, Me

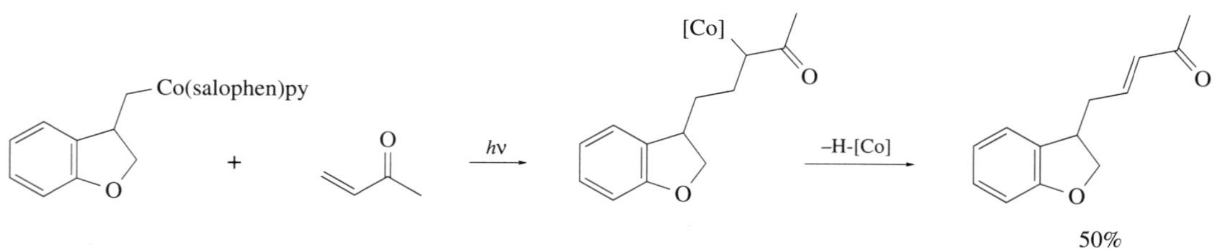

Scheme 141

Chiral allenes can be prepared by reaction of chiral propargylic ethers and acetates with organocopper reagents as shown in Scheme 142 ⟨90JA8042⟩. This process involves an *anti* elimination of the copper group and the methoxy group as shown. Extension of this process to the use of a Grignard reagent in the presence of a catalytic amount of a copper(I) salt has also been undertaken; the stereochemistry of the elimination step is *anti* if the counterion present is iodide but *syn* with chloride.

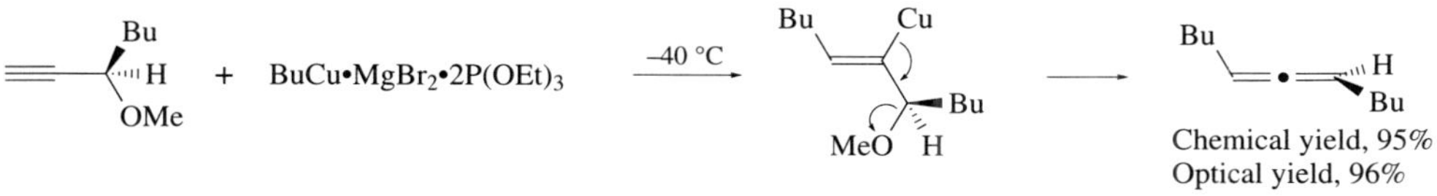

Scheme 142

Alkylmercury compounds react with chlorocarbons thermally, photochemically or by free-radical initiation to form alkenes *via* radical intermediates as shown in Equation (211) ⟨77JOM(124)327, 77JOM(124)349, 77JOM(124)371⟩. Secondary and tertiary alkylmercurials are particularly unstable thermally. Elimination of alkenes from dialkylmercury compounds on treatment with trityl cation proceeds rapidly via a hydride abstraction from the position β to the carbon—mercury bond ⟨71JA6278⟩ (see Section 1.14.6.3). Deoxymercuration under acidic conditions to form alkenes is believed to proceed via a metal bridged ionic intermediate ⟨60JA739, 65JOC1593⟩. Nitroalkenes can be prepared by base-induced elimination from β-nitro mercuric halides formed by nitromercuration of alkenes (Scheme 143) ⟨78JA6294⟩.

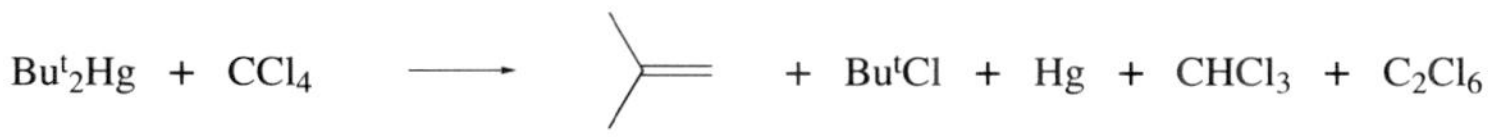

(211)

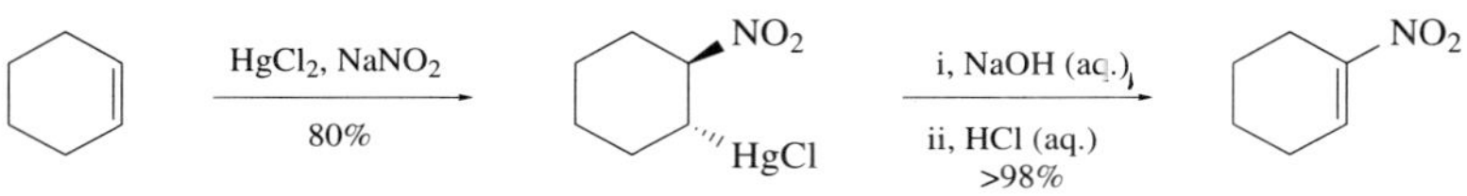

Scheme 143

1.15
One or More C=C Bond(s) Formed by Condensation: Condensation of Nonheteroatom Linked Functions, Halides, Chalcogen or Nitrogen Functions

CHRISTOPHER M. RAYNER
University of Leeds, UK

1.15.1 BY CONDENSATION FROM NONHETEROATOM LINKED FUNCTIONS

1.15.1.1 Oxidative Coupling of Hydrocarbons

The oxidative coupling of hydrocarbons to give alkenes is a reaction of relatively little use synthetically, as harsh reaction conditions are required and yields are low, and so will be mentioned only briefly. The coupling of methane to give ethylene using a manganese oxide/sodium chloride catalyst at 750 °C occurs in up to 30% yield ⟨87CC388⟩. Other catalysts based on ytterbium(III) oxide/calcium oxide are less efficient, particularly at lower temperatures ⟨90AC59⟩.

1.15.1.2 Alkene Metathesis

The metathesis of alkenes using transition metal catalysts was first reported in 1964 ⟨64IEC170⟩, and has since been the subject of a number of extensive reviews ⟨82COMC-I(8)499, 86JOM(300)249, B-86MI 115-01, 91COS(5)1115⟩. The metathesis reaction is basically a transition metal (e.g., Ti, Ta, W, Mo, Ru, Re, Os) ⟨86JOM(300)249⟩ catalysed exchange of the carbon atoms (and their substituents) which make up the double bond of an alkene (Equation (1)). The reaction has been the subject of considerable mechanistic investigation ⟨82COMC-I(8)499, 85CC874, 87JA899⟩; however, this is beyond the scope of this chapter.

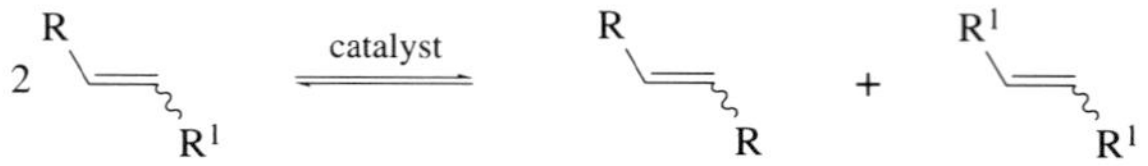

(1)

At present, one of the major limitations of the reaction is the sensitivity of the catalyst to functional groups ⟨76CL1019⟩ which, although better catalysts are being developed, has limited the synthetic applications of the standard metathesis reaction. This is particularly the case for catalysts based on early transition metals such as Ti, Ta, W and Mo. More recently, systems using the later transition metals such as Re and Ru are potentially more useful as they are able to metathesise alkenes containing basic functional groups ⟨91COS(5)1115⟩. Catalysts have been developed which will tolerate a number of functional groups including quaternary ammonium salts ⟨77CC502, 77RTC(M)91⟩, ethers ⟨74CC793, 76CL1019, 77CC198⟩, nitriles ⟨74CC793, 76CL1019, 77CC198, 81CC1081, 85CC1816⟩, esters ⟨74CC793, 76MAC1349, 77TL441, 80TL1715⟩ and sulfonates ⟨82TL2997⟩.

The reaction is an equilibrium, and often mixtures of products are obtained which can be difficult to separate. The reaction is, however, very important industrially and is used for the production of alkenes, for example, the Shell Higher Olefin Process is used to make C_{12} to C_{16} alkenes which are used in the manufacture of surfactants ⟨67HP232, 79CT494, 82COMC-I(8)499⟩.

Alkene metathesis has been used for the synthesis of a number of macrocyclic alkenes. Cycloheptene and cyclooctene undergo metathetic dimerisation using Re_2O_7/Al_2O_3. The activity of the catalyst is increased significantly by the addition of $SnMe_4$. Good yields are obtained by displacing the equilibria using a Soxhlet-type apparatus to promote metathesis of the more volatile components of the reaction mixture (Scheme 1) ⟨87AG(E)702⟩. In an approach to the synthesis of civetone (**1**), the metathesis of unsaturated ester (**2**) gave the diester (**3**) which could be isolated by chromatography as a 1.3 : 1 mixture of *cis* (*Z*)- and *trans* (*E*)-isomers in good yield. Conversion to the macrocycle by Dieckman condensation, hydrolysis and decarboxylation gave civetone as an inseparable mixture of isomers (Scheme 2) ⟨80TL2955⟩.

The use of cyclic alkenes in metathesis reactions leads to the production of polymeric systems by

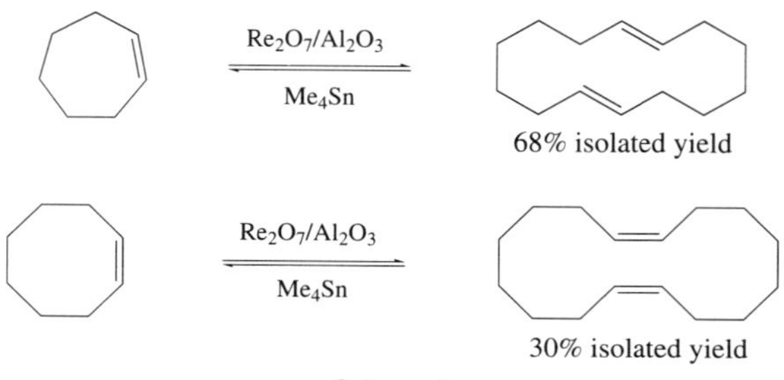

Scheme 1

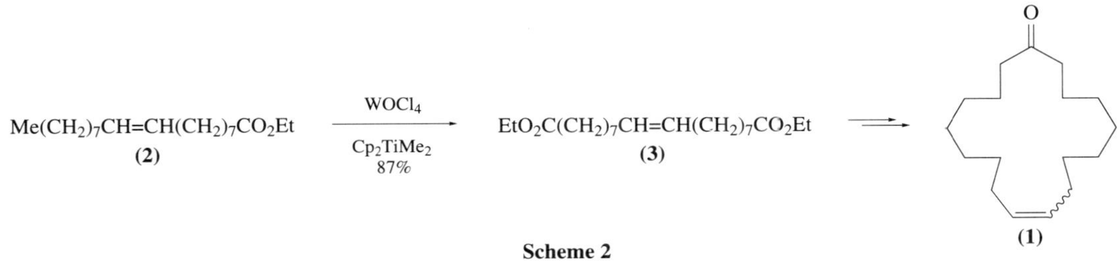

Scheme 2

a process known as ring opening metathesis polymerisation (ROMP). A full discussion of ROMP is beyond the scope of this chapter; however, extensive reviews are available ⟨82COMC-I(8)499, 86JOM(300)249, 89SCI907⟩. Metathesis of alkenes such as norbornene derivatives, using tungsten, ruthenium or osmium catalysts, gives rise to low polydispersity homopolymers (Equation (2)) ⟨87MM1169, 87MM2640⟩.

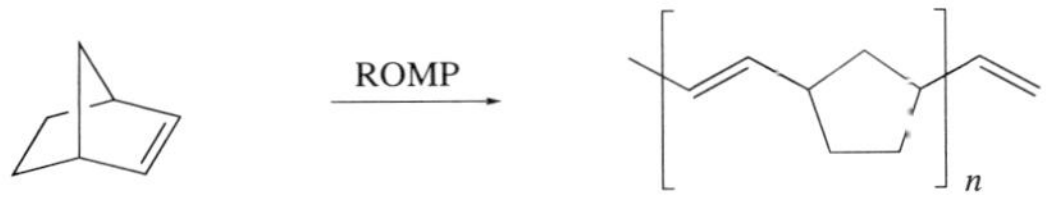

(2)

The ROMP of 7-oxabicyclo[2.2.1]hept-5-ene derivatives provides an interesting example of the choice of catalyst (Equation (3)). Highly Lewis acidic catalysts such as WCl_6 or $WOCl_4$ preferentially coordinate to, and cationically open, the bridging ether. However, less Lewis acidic catalysts such as $RuCl_3$, Ru(1,5-COD)Cl_3, and $OsCl_3$ are much more successful. The homogeneity and *cis*/*trans* ratio of the polymer can be controlled depending on the catalyst used ⟨88JA960⟩. A related reaction has also been carried out under aqueous conditions using $RuCl_3$ as a catalyst ⟨88JA7542⟩. The poly(7-oxanorbornene) products are believed to adopt a helical structure (**4**) with the tetrahydrofuranyl oxygen atoms pointing inwards, and thus have potential as ionophores and ion channels. Recently reported chiral molybdenum-based systems give all-*cis* highly tactic poly(2,3-(R)$_2$norbornadiene) (R = CF_3 or CO_2Me) with superior selectivity to that previously observed ⟨93JA4413⟩.

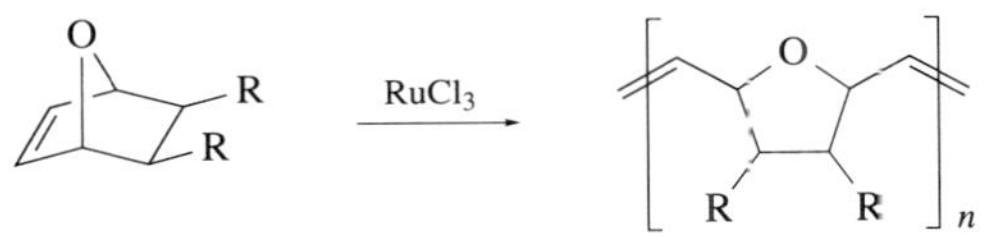

(3)

R = H, OH, OMe, CH_2O-TMS

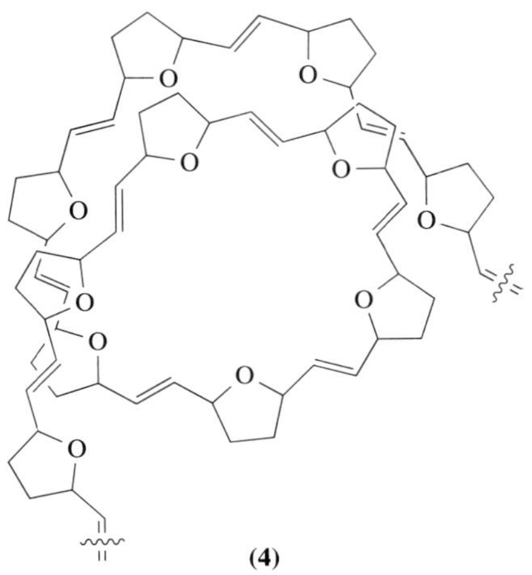

(4)

Metathesis of alkynes is also a known reaction ⟨85JA5987⟩; however, more relevant to this chapter is work recently reported by Trost *et al.* on enyne metathesis (Scheme 3) ⟨91JA1850, 93S824⟩.

Catalysts based on the palladacycle 2,3,4,5-tetrakis(methoxycarbonyl)palladacyclopentadiene (TCPC) (**5**; R = Me) ⟨88JA1636⟩ provide good yields and a high degree of selectivity for the enyne metathesis reaction, which is exploited in an approach to the botrydianes.

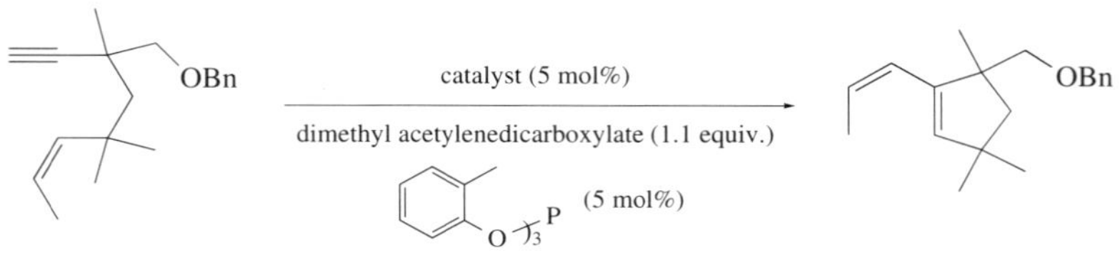

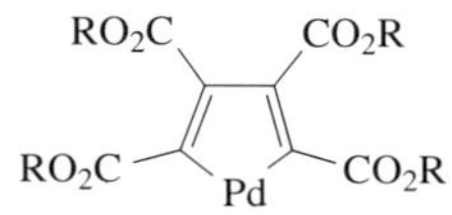

(5)

Catalyst (**5**)	*Time*	*Yield*(%)
R = Me	60h	62
R = $CH_2C_3F_7$	22h	66
R = CH_2CF_3	1h	68

Scheme 3

1.15.2 BY CONDENSATION OF HALIDES: METAL MEDIATED CONDENSATION OF *gem*-DIHALIDES AND RELATED COMPOUNDS WITH CARBONYL GROUPS

This section describes methods for the formation of alkenes by metal mediated condensation of *gem*-dihalides, or equivalent species with carbonyl compounds. The reactions are divided according to the metal involved, and approaches involving Cr, Zn, Mg, Al, Zr, Ti, Sm and Li are included. This kind of reaction has found particular application in the synthesis of terminal alkenes, although in some cases more substituted systems can be prepared. The main emphasis is on reactions where the metal complex is generally not isolated but generated and used *in situ*; reactions of more specific isolated organometallic reagents will be covered elsewhere (see Section 1.16.3).

1.15.2.1 Chromium-mediated Condensation

The condensation of a *gem*-dihalide, usually the iodide or bromide, with aldehydes and ketones mediated by a chromium(II) species provides a method for direct formation of an alkene ⟨68JCS(A)34, 87JA951, 91COS(1)807, 92S248⟩. Reduction of a dihalide by $CrCl_2$ gives the *gem*-dichromium intermediate (**6**) which reacts efficiently with aldehydes with high (*E*) selectivity (Scheme 4). Lower (*E*)/(*Z*) selectivity is observed using $CrCl_3$/Zn as the source of chromium(II) species. The reaction is most successful with 1,1-diiodoethane and aldehydes in THF. With other diiodides, dramatic increases in yields are observed in the presence of a donor such as DMF (1 equivalent), which greatly increases the reduction power of the chromium(II) species ⟨87JA951⟩ (Scheme 5). The reaction is also promoted by ultrasound ⟨84BCJ2335, 87JA951⟩. Even so, low yields are still obtained with *gem*-dibromides and chlorides, although the presence of α-trimethylsilyl or -phenylsulfenyl substituents allow use of dibromides for the preparation of vinylsilanes and vinylsulfides, respectively, again with high (*E*) selectivity. $CrCl_3$/$LiAlH_4$ can also be used in place of $CrCl_2$ as the source of chromium(II) species for these reactions ⟨87TL1443⟩. The reaction with ketones is only really successful using diiodoethane, other diiodides giving low yields. However, the low basicity of the reagent system means that easily enolisable ketones are converted efficiently to alkenes, subject to the above limitations (Equation (4)).

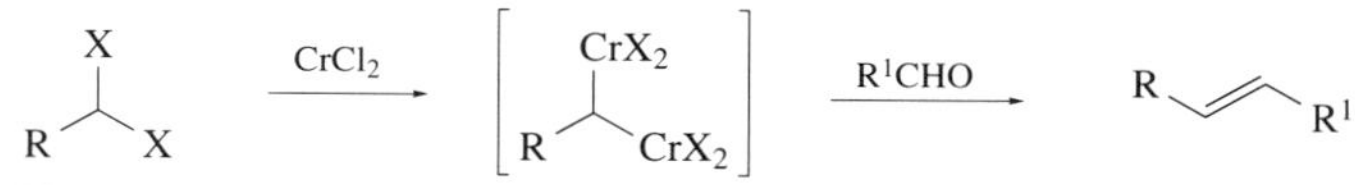

X = I, Br

(6)

Scheme 4

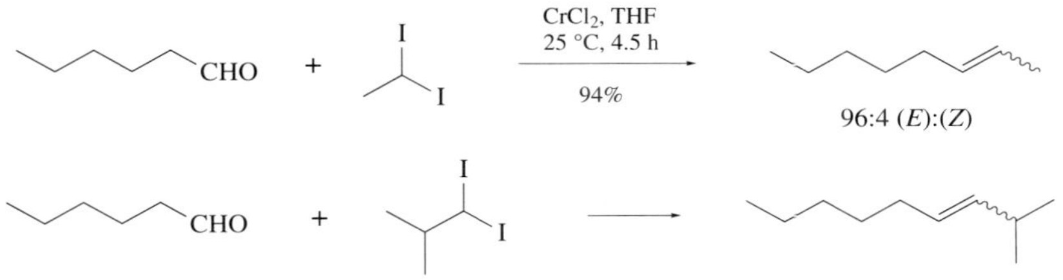

Conditions
$CrCl_2$, THF, 24 h, 12%, 72:28 (*E*):(*Z*)
$CrCl_2$, DMF (1 equiv.), THF, 1 h, 74%, 93:7 (*E*):(*Z*)

Scheme 5

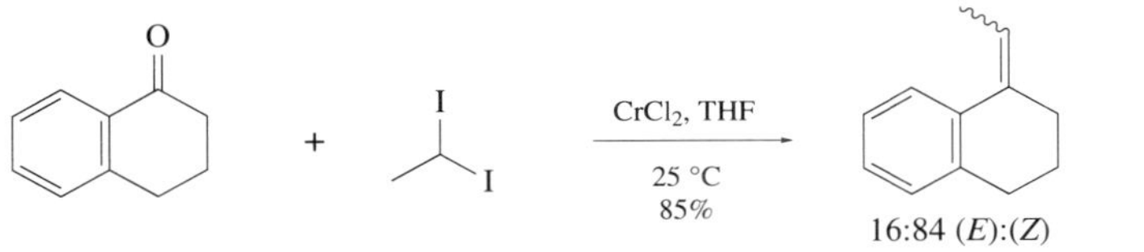

(4)

α-Acetoxybromides react with aldehydes under similar reaction conditions to give alkenes. Commercial $CrCl_2$ is reported to be unsuccessful for this reaction, instead the chromium species being generated using $CrCl_3$/Zn. A coordinating agent such as DMF is essential for an efficient reaction; however, other complexing agents such as TMEDA or tetraethylethylenediamine [1,2-bis(diethylamino)ethane] (TEEDA) are also effective, but can affect the stereoselectivity of double bond formation. Ketones and esters remain unaffected by the reaction conditions. α,β-Unsaturated and aromatic aldehydes react with little stereoselectivity and low yields. In general, use of higher temperatures can improve yields but often at the expense of (*E*)/(*Z*) selectivity. A variety of α-acetoxybromides have been used in this reaction; one example is shown in Equation (5) ⟨93SL837⟩.

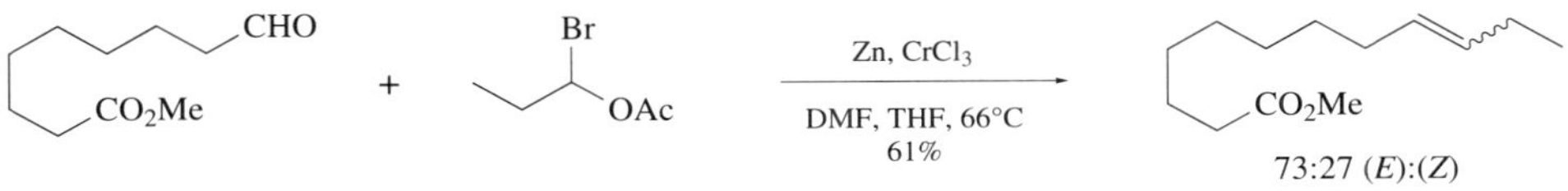

(5)

The high (*E*) selectivity and mild reaction conditions for the condensation of aldehydes and diiodides using $CrCl_2$ had led to its exploitation in natural product synthesis. In an approach to the synthesis of a segment of macbecin I, condensation of aldehyde (**7**) with 1,1-diiodoethane using $CrCl_2$ proceeds in high yield with exclusive formation of the (*E*)-alkene with no racemisation at the chiral centre adjacent to the aldehyde group (Equation (6)) ⟨89JCS(P1)190⟩. Similarly, the synthesis of unusual long chain ketones originating from algae has been accomplished using successive $CrCl_2$-mediated condensations to give the all-(*E*) pentaene (**8**) (Scheme 6) ⟨88TL2599⟩. The analogous all-(*Z*) alkene was prepared, for comparison with the natural product, using a salt-free Wittig reaction.

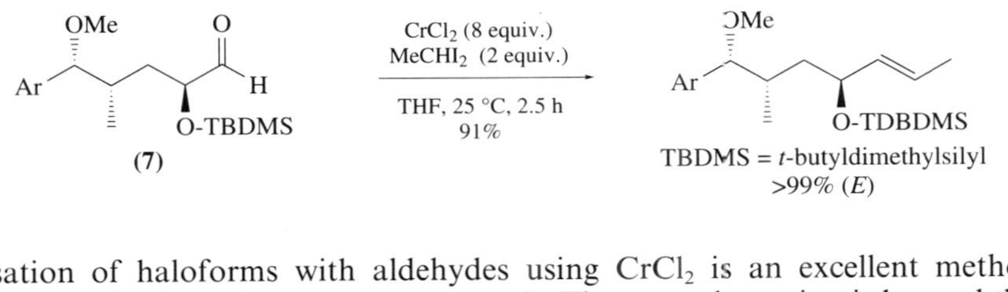

(6)

The condensation of haloforms with aldehydes using $CrCl_2$ is an excellent method for the preparation of (*E*)-vinyl halides ⟨86JA7408, 91COS(1)807⟩. The general reaction is beyond the scope of this chapter; however, it has led to the development of a one-pot procedure for the preparation of allylic alcohols by $CrCl_2$-mediated condensation of iodoform and benzaldehyde. The initially formed (*E*)-vinyl iodide, is reacted *in situ* with another aldehyde using $NiCl_2$ catalyst in DMF, to give the allylic alcohol in good overall yield (Scheme 7) ⟨83TL5281, 86JA5644, 86JA6048, 86JA7408⟩.

In a different approach to the use of chromium-based reagents for alkene synthesis, the trans-

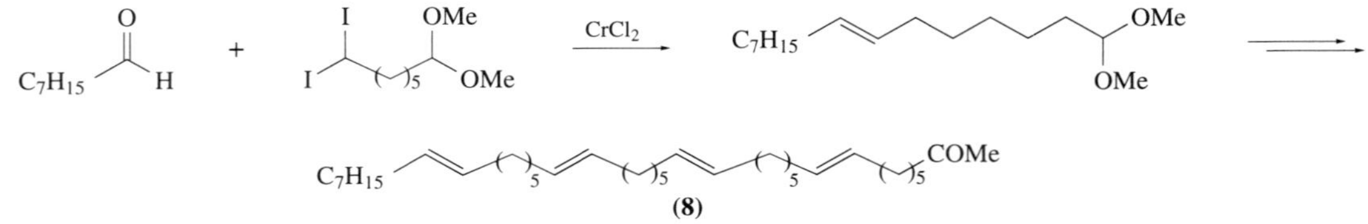

(8)

Scheme 6

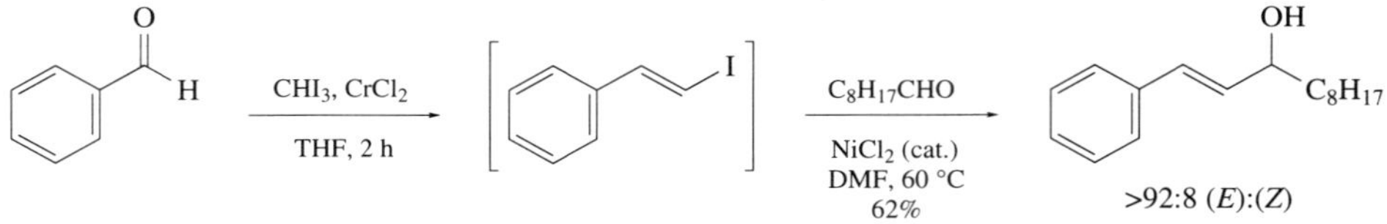

Scheme 7

metallation of the Grignard reagent derived from trimethylsilylmethyl chloride using $CrCl_3$ gives the organochromium reagent (**9**) which adds efficiently to aldehydes. The β-trimethylsilyl alcohols thus prepared can be isolated and on treatment with dilute mineral acid eliminate to give the alkene (Scheme 8) in a process related to the Peterson alkeneation (see Section 1.16.2). The reaction is reported to be unsuccessful with ketones ⟨81TL5031⟩.

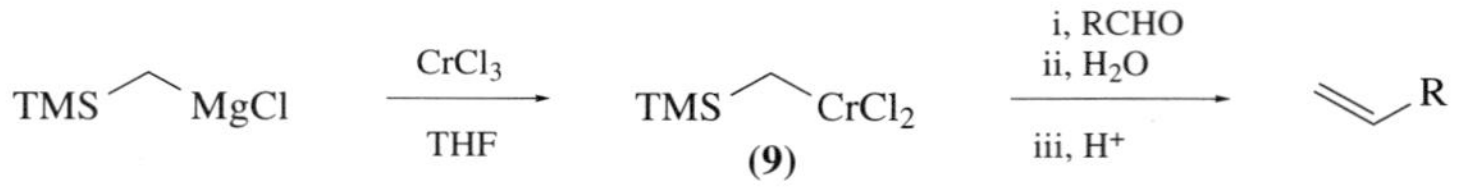

Scheme 8

1.15.2.2 Zinc-mediated Condensation

The methylenation of carbonyl compounds using organozinc reagents derived from dihalomethanes works best in the presence of Lewis acids, typically $TiCl_4$ or $AlMe_3$ ⟨78TL2417, 80BCJ1698, 91COS(1)749⟩. The reaction is successful for aldehydes, ketones and α,β-unsaturated aldehydes (Scheme 9). Because it is an electrophilic reagent, it is also successful with enolisable aldehydes and ketones ⟨85TL5579⟩. Diiodomethane works best when using $AlMe_3$ or VCl_4 as the catalyst; the stronger Lewis acid $TiCl_4$ enables dibromomethane to be used ⟨78TL2417, 80BCJ1698⟩. Detailed experimental procedures are available ⟨87OS(65)81⟩. More complex dihalides have also been used in related reactions ⟨88TL3357⟩. Excellent chemoselectivity for aldehydes in the presence of ketones can be achieved using the much milder Lewis acid, $Ti(OPr^i)_4$ (Scheme 10). Reversal of such chemoselectivity can be achieved by pretreatment of the substrate with $Ti(NEt_2)_4$ (Scheme 10) ⟨85TL5581⟩. Similarly, ketones can be methylenated in the presence of esters and free alcohols ⟨85TL5579⟩. This has been exploited in a synthesis of a gibberellin precursor (Equation (7)) ⟨82TL4293⟩.

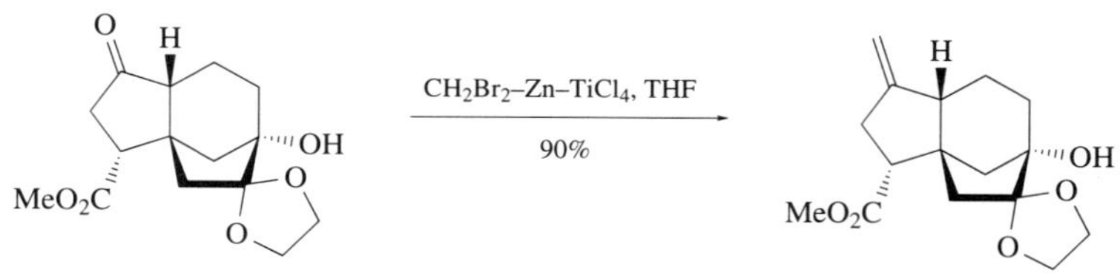

(7)

The reaction can also be carried out using dibromo- or diiodomethane in the absence of Lewis acids; however, the use of ultrasound is desirable and greatly accelerates the process (Equation (8)). Even so, yields with ketones and α,β-unsaturated aldehydes are low under these conditions ⟨84BCJ2335⟩.

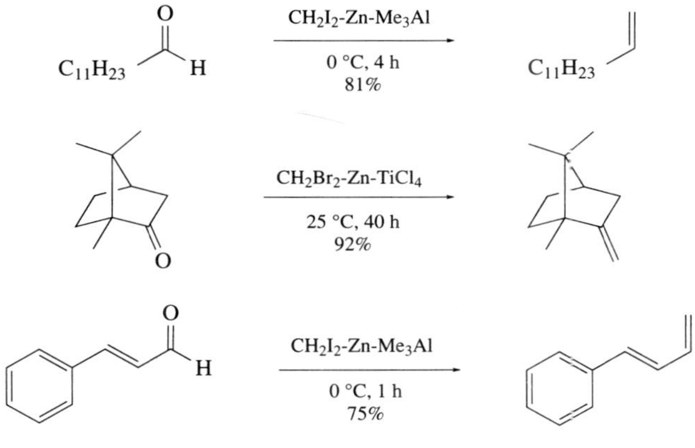

Scheme 9

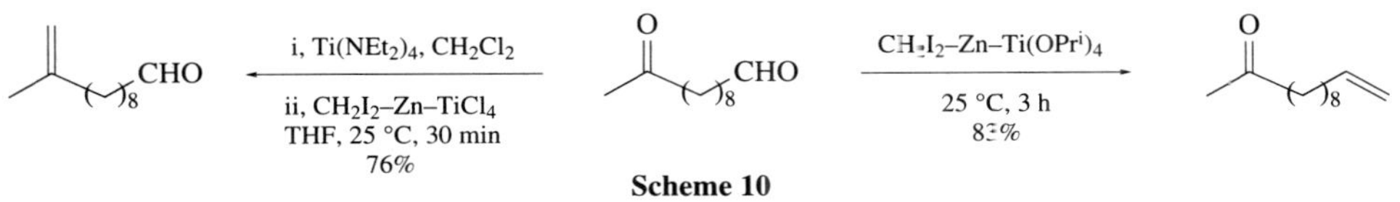

Scheme 10

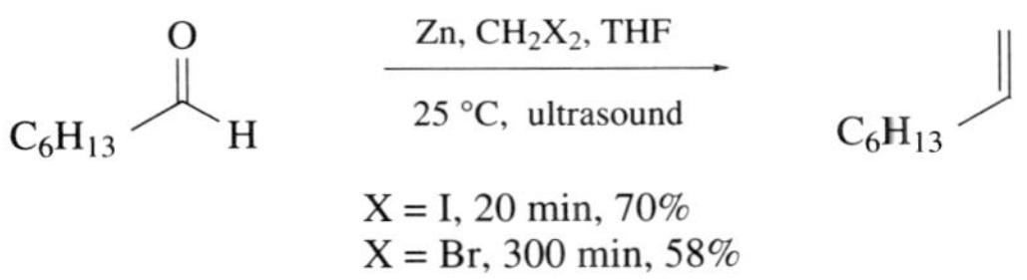

(8)

A related reaction involves the generation of a mixed bismetallic reagent from an allylic zinc bromide and a vinyl-Grignard, -lithium or -diethylaluminum reagent. Condensation with an aldehyde in the presence of $BF_3 \cdot OEt_2$ gives the 1,5-diene in good yield and with very high (*E*)-selectivity (Scheme 11) ⟨86TL1039, 86TL1043⟩.

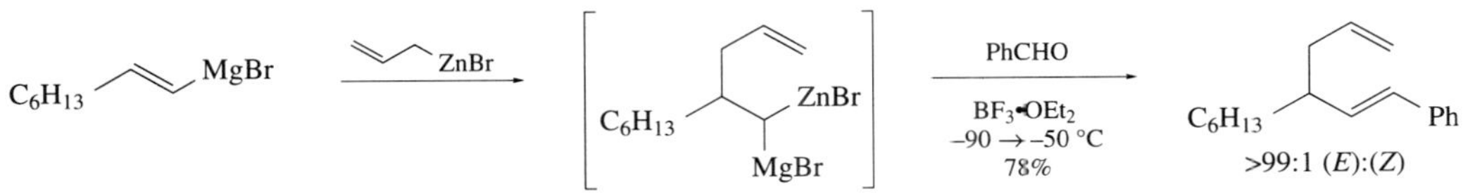

Scheme 11

1.15.2.3 Condensations Mediated by Other Metals

The condensation of *gem*-dibromides and -diiodides with carbonyl compounds can be accomplished using magnesium, either as a mercury amalgam or, less efficiently, as magnesium turnings ⟨70T1281⟩. The reaction is rapid, often being complete in a few minutes, and good yields can be obtained with both aldehydes and ketones (Scheme 12).

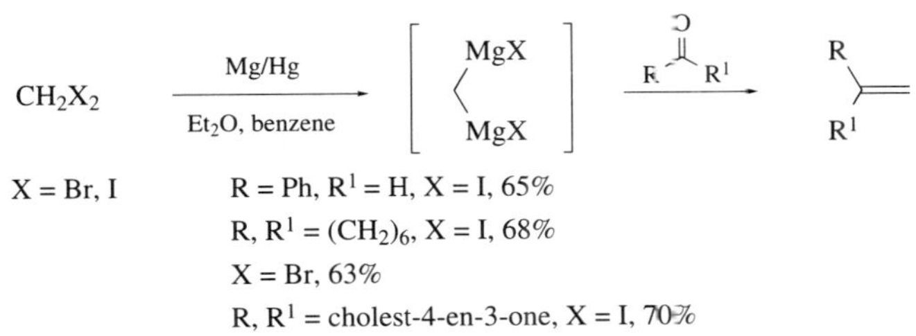

Scheme 12

A series of methylenating agents can be prepared by the action of aluminum on dichloromethane ⟨66TL6021, 67TL1581, 88JOC2829⟩. The initial complex (**10**) is relatively unreactive, but if treated with triethyl- or trimethyl-aluminium, very effective ketone methylenating agents (**11**) and (**12**) are obtained. They convert a broad range of aliphatic, alicyclic and aromatic ketones to their methylene derivatives with no competitive alkylation or reduction. Complexes of these reagents with ethers (THF or Et_2O) are particularly efficient. Unfortunately, related multicarbon alkenylating agents such as (**13**) are much less successful, inducing severe side reactions for substrates other than aromatic ketones, which themselves give only fair to good yields (Scheme 13).

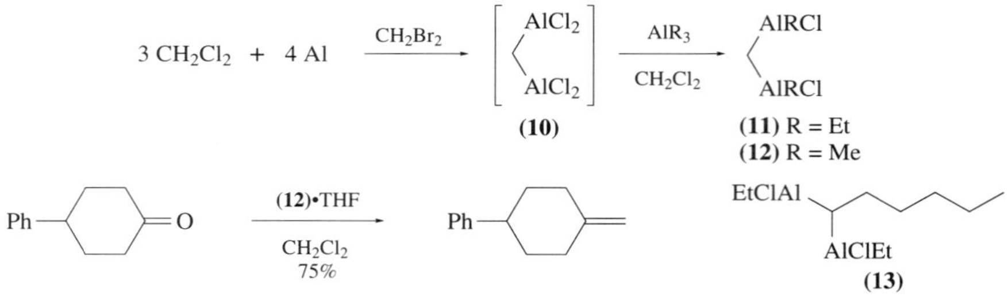

Scheme 13

Organozirconium-based methylenating agents such as (**14**) can be prepared by treatment of a dihalomethane with zirconocene dichloride and zinc. They are usually generated *in situ* and rapidly methylenate aldehydes, ketones and enones, and are reportedly particularly useful for acid-sensitive substrates (Scheme 14) ⟨89TL3927⟩. Unfortunately, homologous dihalides such as 1,1-dibromopentane are unsuccessful in this reaction probably owing to β-elimination in the organozirconium intermediate.

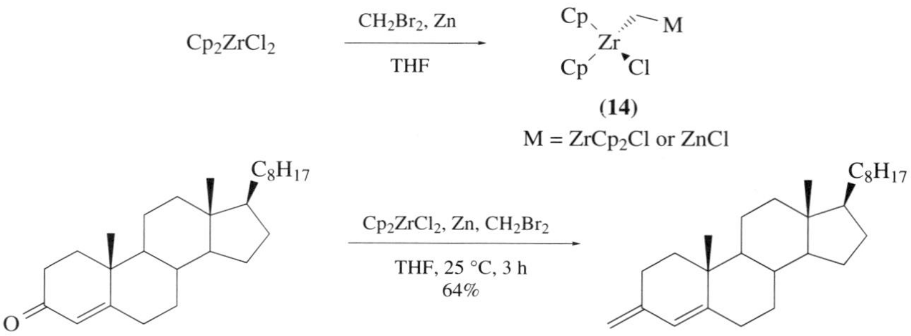

Scheme 14

Transmetallation of the Grignard reagent derived from trimethylsilylmethyl chloride with $TiCl_4$ gives a reagent (**15**) which methylenates aldehydes but not ketones (Scheme 15). The intermediate alkoxytitanium complex is not isolated but undergoes *in situ* elimination to give the corresponding alkene ⟨81TL5031⟩. This is in contrast to the related chromium-based system which needs further acid treatment before the alkene is formed (Scheme 8). Higher yields can be obtained using the trimethylgermanium reagent (**16**) under similar conditions, although the reaction is still poor with ketones.

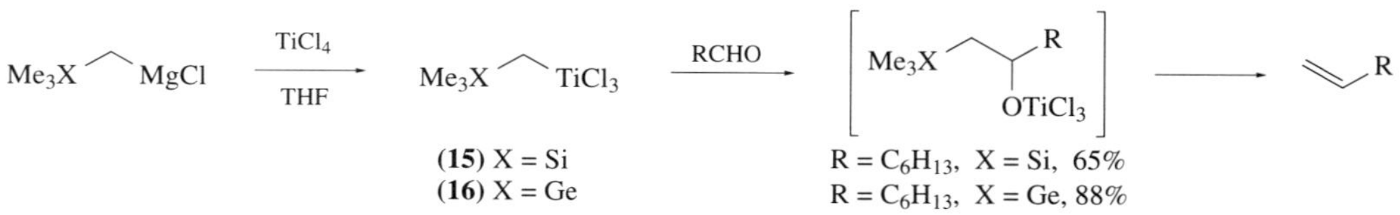

Scheme 15

The methylenation of aldehydes and ketones using samarium has been reported ⟨80JA2693, 86TL3891, 87CL2101⟩. Treatment of an aldehyde or ketone with SmI_2 and a dihalomethane initially gives an iodohydrin, which upon treatment with a further equivalent of SmI_2 in hexamethylphosphoramide (HMPA) with an additive such as *N*,*N*-dimethylaminoethanol (DMAE) or glutaric anhydride, eliminates to give the alkene in a useful one-pot procedure (Scheme 16).

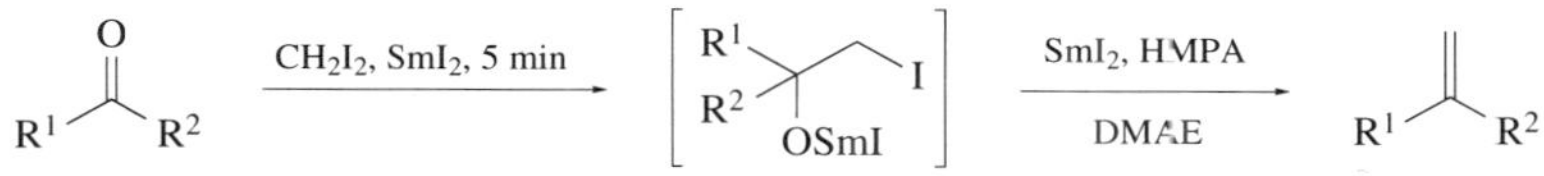

$R^1 = C_{11}H_{23}$, R^2 = H, 73%
R^1, $R^2 = (CH_2)_{11}$, 80%

Scheme 16

Finally, chloromethyllithium can be generated readily from chloroiodomethane and methyllithium or lithium powder ⟨86CC1665⟩. It is usually generated *in situ*, and reacts with aldehydes and ketones to give the methylenated products in moderate to good yields (Equation (9)).

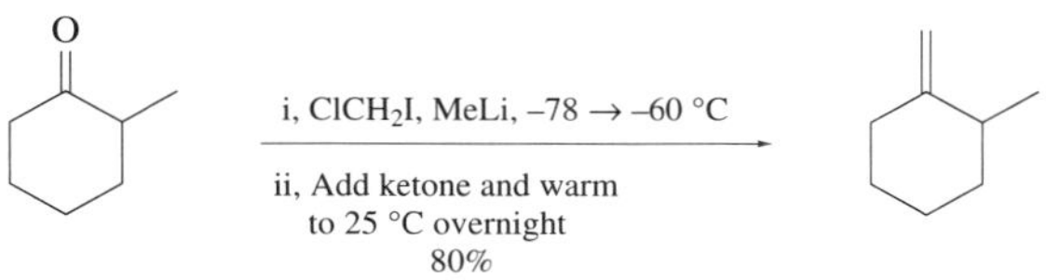

(9)

1.15.3 BY CONDENSATION OF OXYGEN FUNCTIONS

1.15.3.1 Synthesis of *β*-Lactones and Subsequent Decarboxylation

The thermal elimination of carbon dioxide from a *β*-lactone is generally a stereospecific process, with geometry of the *β*-lactone being conserved in the alkene product (Equation (10)) ⟨66JOC4043, 72JA2000, 93S441⟩. The reaction occurs at temperatures ranging from approximately 80 °C to 160 °C, although more extreme conditions are required in some cases ⟨86T1581, 91JOC1176⟩. The reaction can also be catalysed by SiO_2 ⟨91JOC1176⟩ or a carboxylic acid (e.g., trifluoroacetic acid); however, in the latter case stereoselectivity can be seriously affected (Scheme 17) ⟨81CC891⟩.

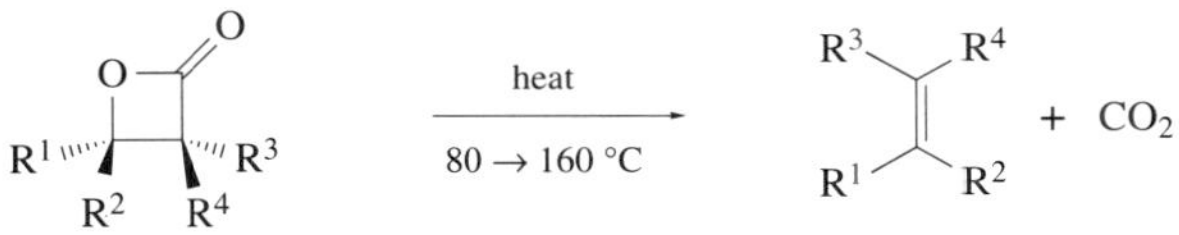

(10)

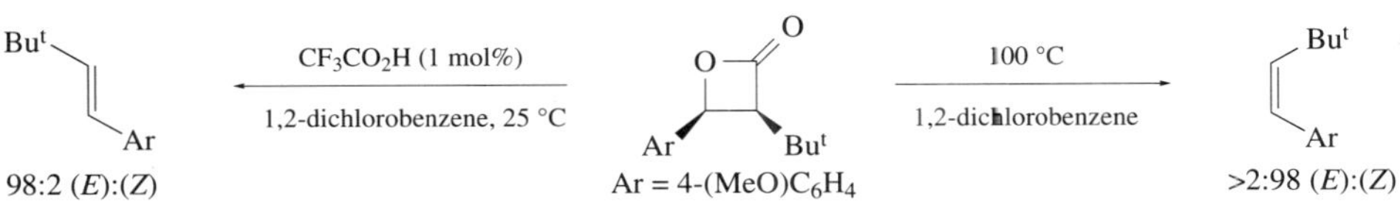

Scheme 17

The preparation of *β*-lactones for this reaction is generally carried out by condensation of a carboxylic acid enolate equivalent with an aldehyde or ketone. The dianions of carboxylic acids add to aldehydes ⟨71JOC1149, 79CC52, 81JOC3359⟩ and ketones ⟨72JA2000, 74JOC1650, 78JOC4574⟩ to give predominantly the *threo*-*β*-hydroxy acids where appropriate (Scheme 18). These products are then readily cyclised to the *β*-lactones by treatment with phenylsulfonyl chloride/pyridine, *p*-toluenesulfonyl chloride/pyridine ⟨84T1269⟩, or methanesulfonyl chloride/Na_2CO_3 ⟨76JCS(P1)884⟩. Thermolysis (180 °C) of *β*-lactones derived from *threo*-*β*-hydroxy acids gives *trans*-alkenes as the major products ⟨81JOC3359⟩. Alternatively, if the lactonisation is carried out at slightly elevated temperatures (45–55 °C), alkene products can be isolated directly ⟨72JA2000⟩. This procedure is often limited by the efficiency of the initial condensation reaction.

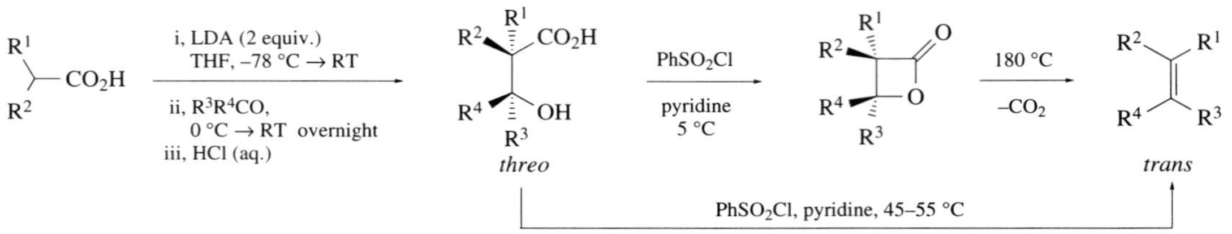

				Yields (%)			
R¹	*R²*	*R³*	*R⁴*	*Acid*	*Lactone*	*Alkene*	*Ref.*
Bu^t	H	H	Bu^t	79	82	95	81JOC3359
1-Ad[b]	H	H	1-Ad[b]	22	43	82	81JOC3359
Bu^t	H	H	1-Ad[b]	64	85	100	81JOC3359
H	Ph	Ph	Ph	81	70	100	72JA2000
H	Me	Ph	Ph	76	37[a]	100	72JA2000
Me	Me	–$(CH_2)_5$–		75	67	100	72JA2000
–$(CH_2)_5$–		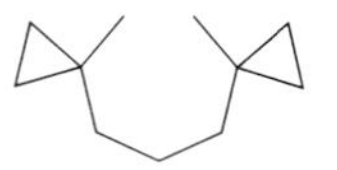		61	90	100	74JOC1650

[a] >50% alkene also isolated. [b] 1-Ad = 1-Adamantyl.

Scheme 18

The condensation of esters with aldehydes and ketones allows the preparation of β-hydroxyesters, which may also be considered as precursors to β-lactones. However, problems can be encountered as the conditions required for ester hydrolysis and/or lactone synthesis can induce a retro-Reformatsky reaction with destruction of the β-hydroxyester ⟨72JA2000⟩. Nevertheless, some β-hydroxyesters have been used in this manner. The β-hydroxyester (**17**) can be stereoselectively alkylated with alkyl halides ⟨84T1269⟩. Hydrolysis, lactonisation and thermolysis then give the *trans*-alkene selectively (Scheme 19). Alternatively, the β-hydroxyester (**18**), synthesised by nonstereoselective reduction (52 : 48 mixture) of the corresponding β-keto ester followed by chromatography, can be processed similarly to give a trisubstituted alkene (Scheme 20) ⟨68TL4569, 76JCS(P1)884⟩.

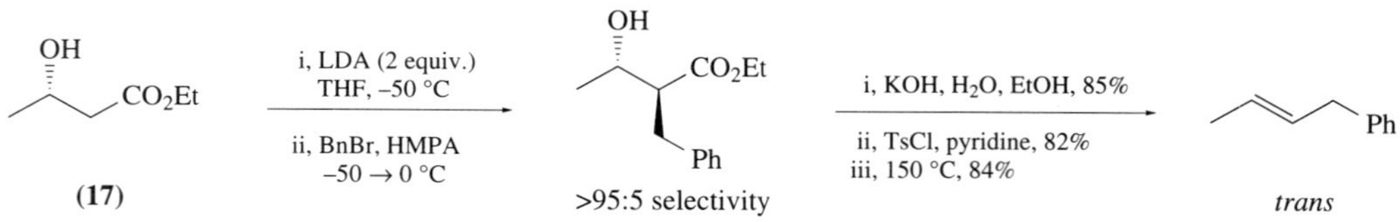

Scheme 19

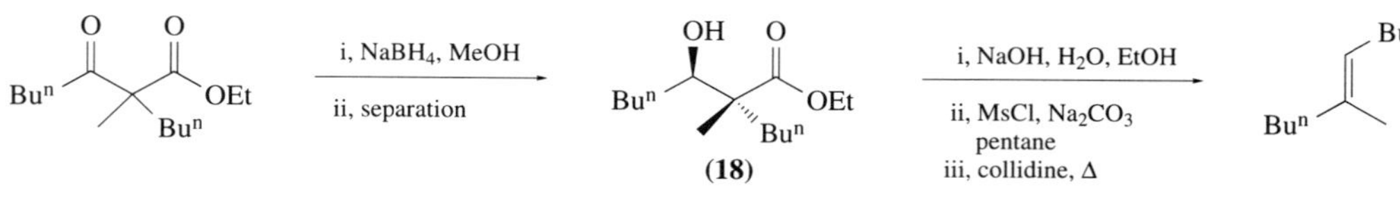

Scheme 20

The condensation reaction between a thioester enolate and a ketone occurs with high stereoselectivity; this provides a direct route to β-lactones, as the intermediate alkoxide cyclises *in situ* with loss of the thiophenolate group (Scheme 21) ⟨91JOC1176⟩. In the case of more hindered thioesters, improved yields of the β-lactone can be obtained by use of the 2,6-dimethylthiophenyl ester. Decarboxylation gives the (*E*)-alkene in high yield, the reaction occurring at 80–110 °C when catalysed by silica gel (10% w/w).

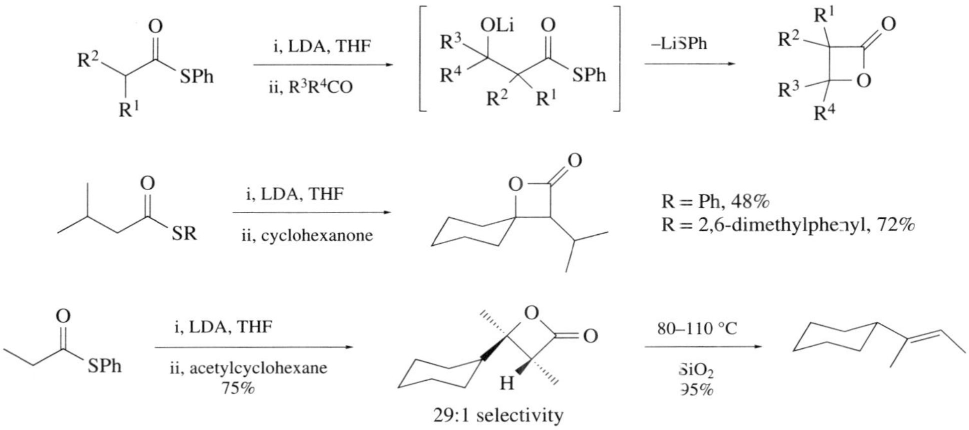

Scheme 21

The condensation of (2-alkenyl)triphenoxytitanium derivatives with unsymmetrical ketones provides a highly diastereoselective route to substituted homoallylic alcohols, which are readily converted into β-lactones, via β-hydroxy acids (Scheme 22) ⟨82HCA1972⟩. Thermolysis again provides access to the alkene with control of double bond geometry.

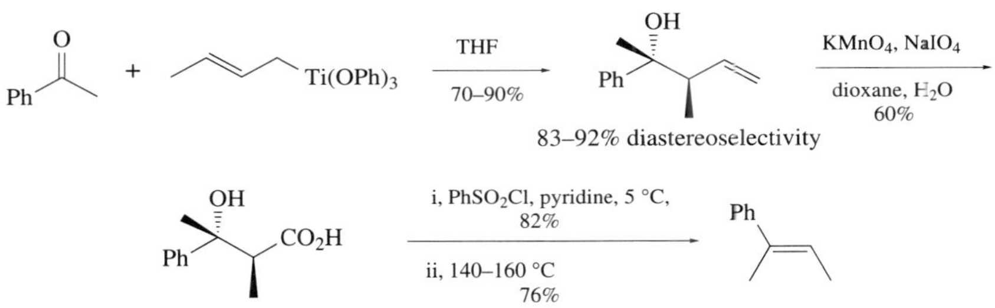

Scheme 22

Lithium phenylethynolate can be readily generated by loss of benzonitrile from the lithiated isoxazole (**19**). It condenses with carbonyl compounds to form the lithiated β-lactone (**20**) via the corresponding ketene (Scheme 23). The enolate (**20**) may be reacted with alkyl halides, aldehydes, or protonated to give highly substituted β-lactones, which extrude CO_2 on heating to 140 °C to form the expected alkenes ⟨75AG(E)765⟩.

The synthesis of β-lactones and subsequent decarboxylation provides an alternative process to the Wittig reaction. Although examples of synthetic applications are limited ⟨93S441⟩, this reaction has been used for the introduction of the isopropylidene groups in syntheses of terpinolene (**21**) and (±)-α-curcumene (**22**), by condensation of the required carboxylic acid dianion with acetone, followed by β-lactone formation and thermolysis ⟨74JOC1322⟩.

In addition to forming β-lactones which may subsequently decarboxylate, β-hydroxy acids can also undergo elimination via a Grob-type fragmentation, to give alkenes. This is particularly useful as it provides the opposite alkene geometrical isomer to that which would be produced if a β-lactone intermediate was used (Scheme 24). The alcohol is converted into a leaving group, either a phosphonium salt using Mitsunobu-type chemistry ⟨79CC52⟩, or an imidate by treatment with *N*,*N*-dimethylformamide dimethyl acetal (DMF–DMA) ⟨84T1269⟩. The zwitterionic intermediates thus formed after deprotonation of the acid then undergo stereocontrolled *anti*-elimination with loss of CO_2, generally to give (*Z*)-alkenes ⟨75HCA1450⟩. A related oxidative decarboxylation has been reported using $VOCl_3$ and other vanadium(V) complexes ⟨89JA3069⟩.

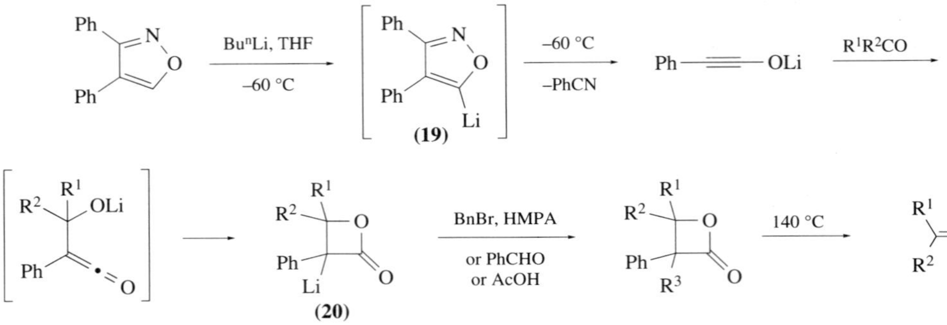

R^1	R^2	R^3	Yield (%)[a] Lactone	Alkene
Ph	H	Bn	68	95(*Z*)
Me	Me	Bn	56	93
$-(CH_2)_5-$		H	51	92
Me	Me	H	59	88
Bu[t]	H	H	65	—
Ph	H	PhCH(OH)	79[b]	—

[a] 75AG(E)765. [b] based onPhCHO.

Scheme 23

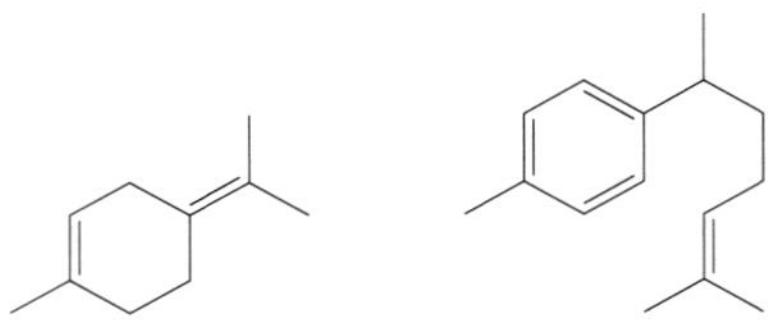

(21) **(22)**

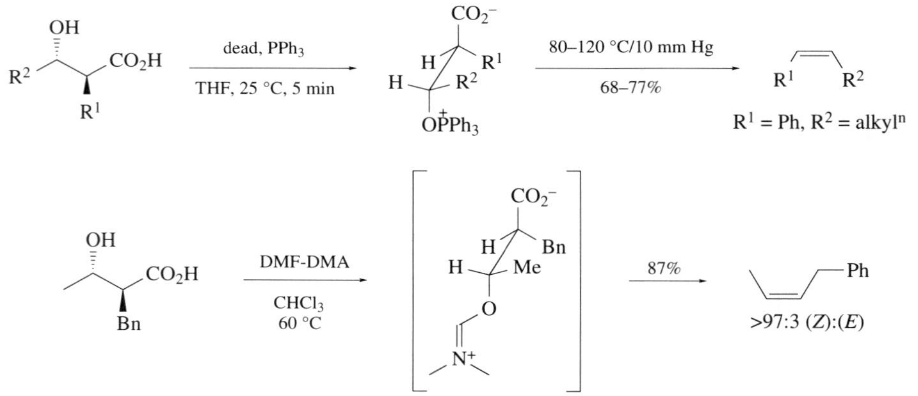

Scheme 24

1.15.3.2 Conversion of Carbonyl Compounds to 1,3,4-Thia- and 1,3,4-Selenadiazolines and Extrusion of N_2 and S or Se and Related Reactions

The two-fold extrusion of nitrogen and sulfur from 1,3,4-thiadiazolines to form alkenes, was originally developed by Barton ⟨72JCS(P1)305, 74JCS(P1)1794⟩ and also independently by Kellogg

⟨72JOC4045⟩; for a review see ⟨88T6241⟩. The reaction involves initial loss of N_2 to form a thiocarbonyl ylide which undergoes conrotatory ring closure to a thiirane which loses sulfur, often induced by a phosphine (PBu_3, PPh_3 or $P(NEt_2)_3$) (Scheme 25).

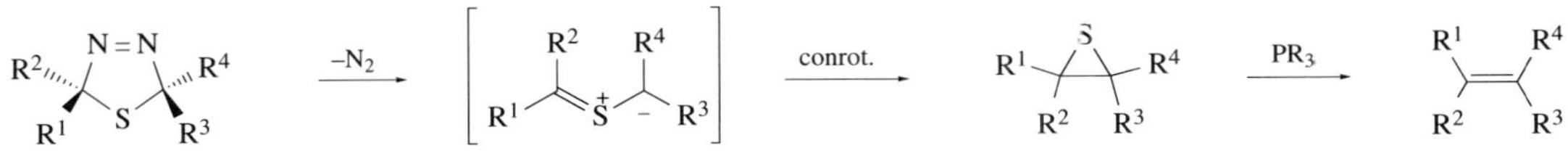

Scheme 25

The 1,3,4-thiadiazolines may be synthesised by a number of routes. Condensation of hydrazine hydrate, hydrogen sulfide, and a carbonyl compound gives a 1,3,4-thiadiazolidine (**23**) with low stereoselectivity. This can then be oxidised to the 1,3,4-thiadiazoline using diethyl azodicarboxylate (dead) ⟨72JOC4045⟩, lead tetraacetate ⟨72JCS(P1)305, 93JOC43⟩ or 2,3-dichloro-5,6-dicyano-1,4-benzoquinone ⟨72JCS(P1)305⟩. Alternatively, condensation of hydrazine and a carbonyl compound followed by chlorination and cyclisation provides the most efficient route to these systems (Scheme 26).

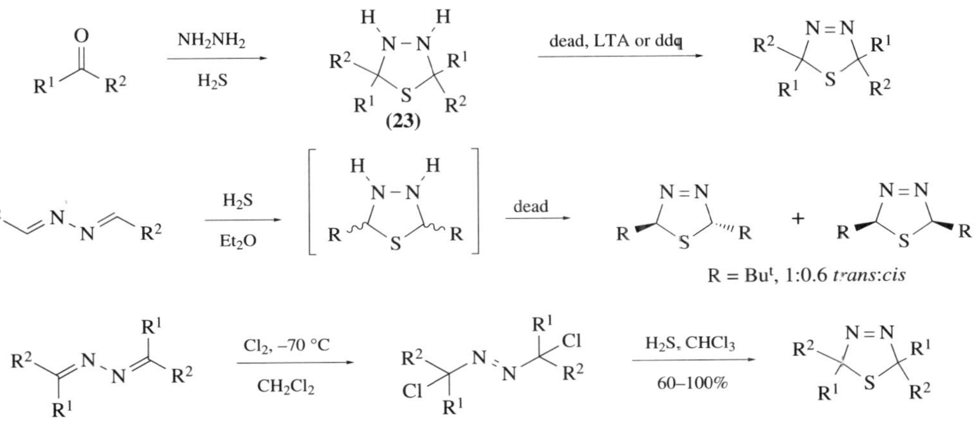

ddq = 2,3-dichloro-5,6-dicyano-1,4-benzoquinone

Scheme 26

1,3,4-Thiadiazolines can also be synthesised directly by cycloaddition of a thioketone and a diazo compound (Scheme 27) ⟨72JOC4045⟩. This has two advantages; it allows the synthesis of unsymmetrical alkenes and it alleviates the requirement of a final oxidation step. However, the dangers and unpleasantness of working with the diazo and thioketone precursors should also be taken into consideration. Both the thioketone and diazo compound are readily synthesised from the corresponding carbonyl compounds. The diazo compound is prepared via the hydrazone either via the triphenylphosphoranylhydrazone ⟨74JCS(P1)1794, 75CC539⟩ or by oxidation using barium manganate ⟨93JOC5900⟩. The intermediate 1,3,4-thiadiazolines can be isolated, or directly converted to alkenes by carrying out the cycloaddition reaction at elevated temperatures (160 °C) (Equation (11)) ⟨75CC539⟩.

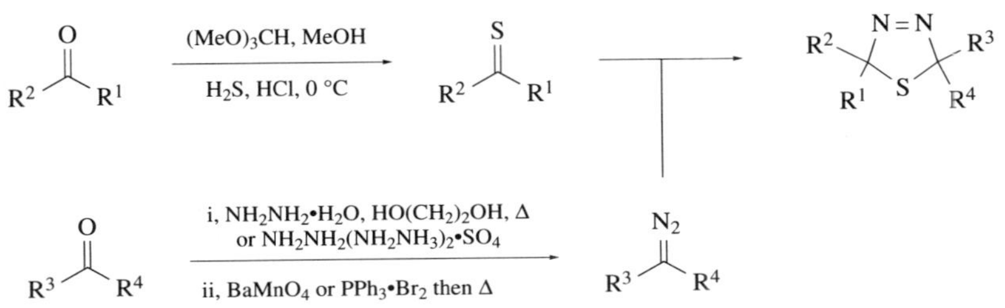

Scheme 27

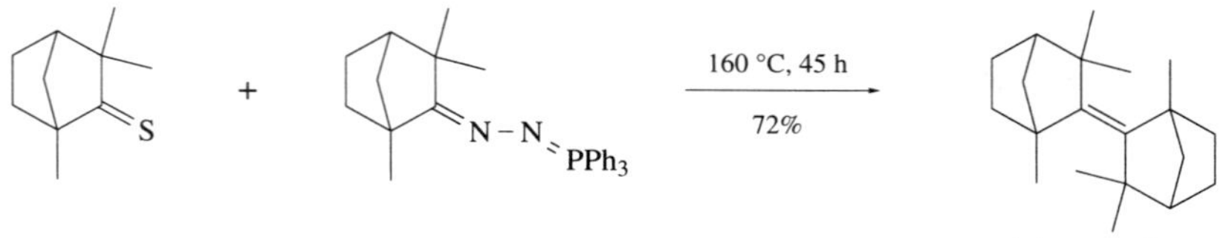

mainly *trans*-isomer (11)

1,3,4-Selenadiazolines undergo similar extrusion reactions to the corresponding sulfur systems, and have the advantage that the selenoketone precursors (also referred to as selones) show greater reactivity toward diazo compounds; thus 1,3,4-selenadiazolines may be easier to prepare by the cycloaddition route (Scheme 28) ⟨75CC539, 82JOC3563, 93JOC5900⟩. Again, it is possible to carry out the reaction at elevated temperatures without isolation of the intermediate selenadiazoline ⟨82JOC3563⟩. The selenoketones are prepared from ketones by heating the triphenylphosphoranylhydrazone derivative with selenium powder ⟨82JOC3563⟩ and a trace of base (e.g., NBu_3) if necessary (Scheme 29) ⟨76JCS(P1)2079⟩.

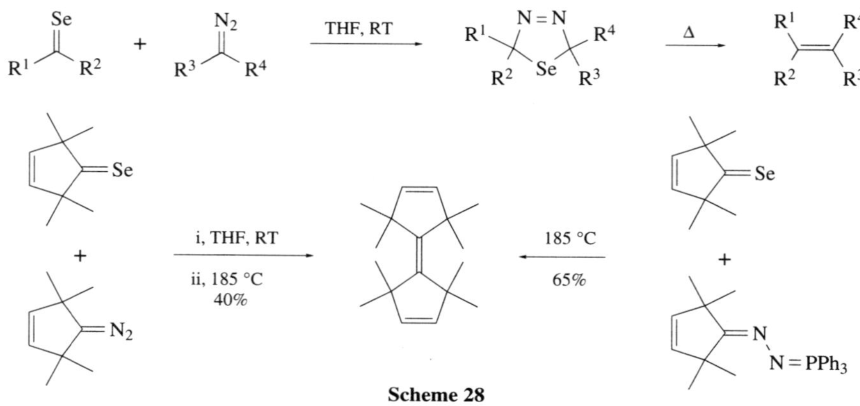

Scheme 28

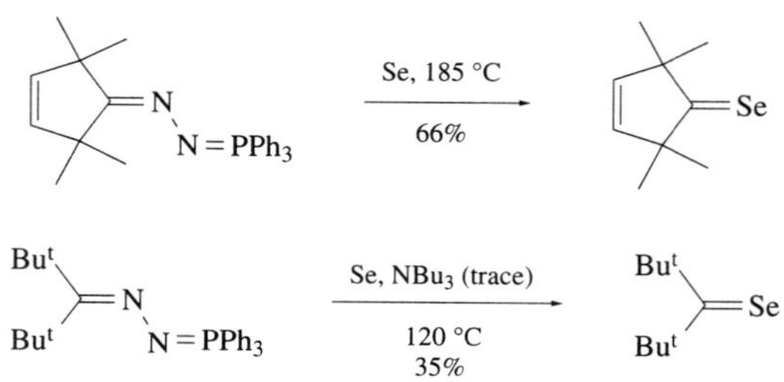

Scheme 29

This reaction has found application in the synthesis of very hindered alkenes including **(24)**, **(25)** and **(26)** ⟨74JCS(P1)1794, 75CC539, 84CB277⟩, and potential precursors of the elusive tetra-*t*-butylethylene, and tied back analogues thereof **(27)** and **(28)** ⟨81TL4937, 86T1693, 93JOC5900⟩.

More recently, this reaction has been applied to the synthesis of less sterically congested alkenes, such as those derived from aldehydes (Scheme 30). Previously, thermolysis of 1,3,4-thiadiazolines to give relatively nonhindered alkenes resulted in only low yields of thiiranes or alkene products. This is now believed to be a result of the instability of 1,3,4-thiadiazoline and 1,3,4-thiadiazolidine precursors, a problem which has been overcome by improved synthetic procedures ⟨93JOC43⟩.

A related two-fold extrusion reaction which can also be used for alkene synthesis is the thermolysis of the oxathiolan-5-ones **(31)** and **(32)**. Both these compounds lose CO_2 and sulfur on thermolysis, with the corresponding alkene formed in good yield ⟨72JCS(P1)305⟩. The substrates are readily prepared by acid-catalysed condensation of an aldehyde or ketone with mercaptodiphenylacetic acid (Scheme 31).

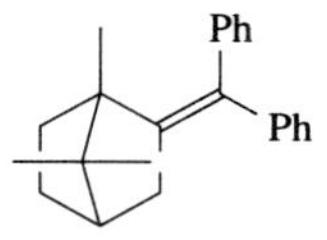

(24)

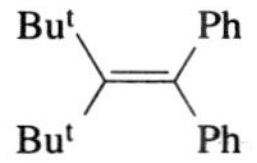

(25)

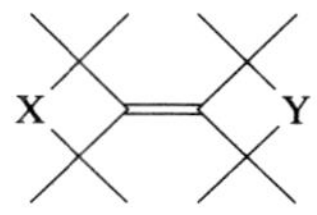

(26)

X = S, Y = Me, Me
X = CH_2CH_2, Y = CH_2CH_2

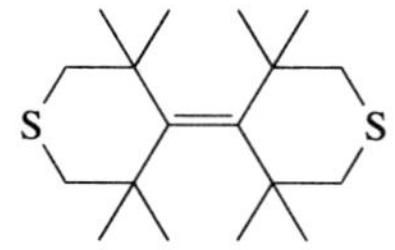

(27)

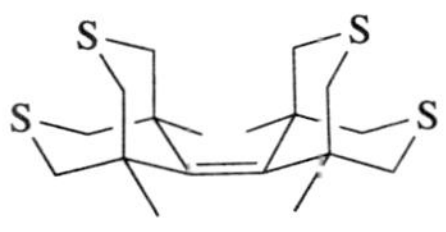

(28)

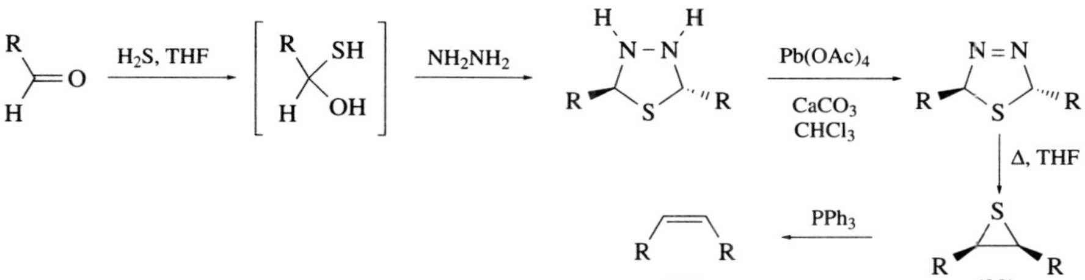

R	*(29)* *Yield* (cis:trans) (%)	*(30)* *Yield* (cis:trans) (%)
C_6H_{13}	75(4:1)	70(4:1)
C_9H_{19}	62(4:1)	56(4:1)
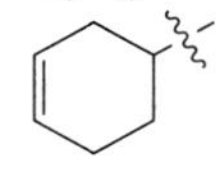	66(6:1)	63(6:1)

Scheme 30

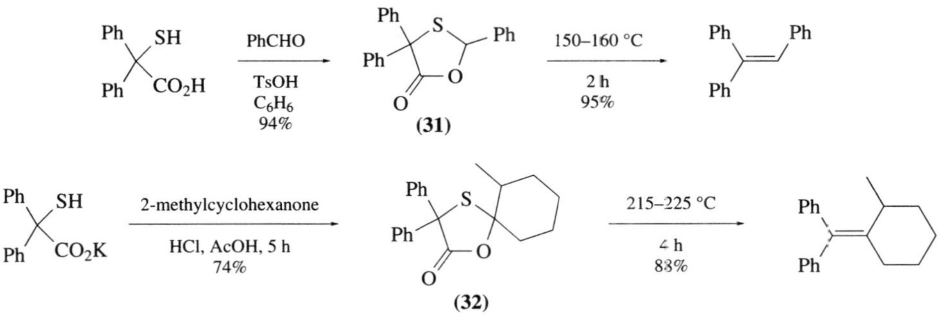

Scheme 31

1.15.3.3 McMurry Coupling of Carbonyl Compounds

The titanium induced reductive coupling of carbonyl compounds is an efficient and widely-used process for alkene preparation. It has been the subject of a number of reviews ⟨80OPP361, 83ACR405, 83MI 115-01, 88CRV733, 88T4295, 89C39, 89CRV1513, B-89MI 115-01, 89S883⟩. The reaction has general applicability and is successful for a wide variety of aldehydes and ketones, including unsaturated, aryl and alkyl, both intermolecular and intramolecular. Although the reaction is often referred to as the McMurry reaction ⟨74JA4708⟩, it was also independently reported by Mukaiyama ⟨73CL1041⟩ and Tyrlik ⟨73BSF2147⟩ around the same time, with slightly different reaction conditions.

McMurry's initial discovery of this reaction provides a nice example of serendipity in synthetic chemistry. While attempting to develop a new procedure for carbonyl reduction, McMurry investigated the use of $LiAlH_4$ modified by $TiCl_3$, and obtained the dimeric alkene (**33**) in 80% yield, rather than the desired reduction product (**34**) (Scheme 32) ⟨83ACR405⟩.

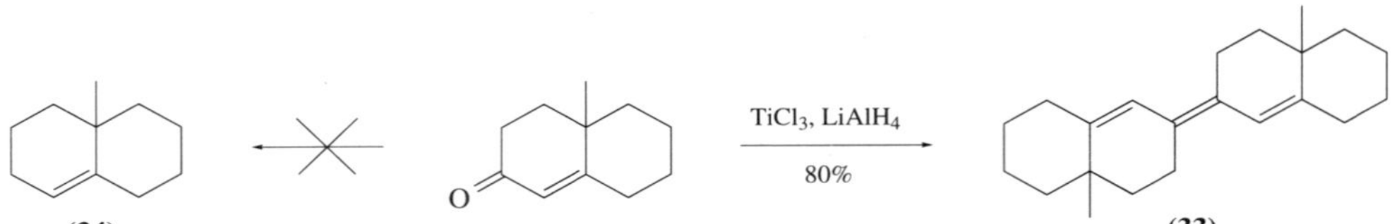

(**34**) (**33**)

Scheme 32

The reactive species in the reaction is almost certainly a Ti(0) species in a finely divided form ⟨81BSB1141, 82JOC248, 82MI 115-01⟩. There are many reported procedures for the generation of such a species, the most reactive and reproducible being formed from $TiCl_3$ and a zinc–copper couple ⟨78JOC3255⟩. In early experiments, the age, history and source of $TiCl_3$ caused problems with reproducibility; however, purification by crystallisation as the complex $TiCl_3(DME)_{1.5}$ gives reliable results, even from batches of $TiCl_3$ which were previously ineffectual (Scheme 33) ⟨89JOC3748⟩. Higher yields may be obtained if a nonaqueous filtration procedure is used rather than an aqueous work-up. The coupling is believed to occur on the surface of the active titanium particle; the highly pebbled 'Swiss cheese' type of surface generated gives a very high surface-to-mass ratio ⟨78JOC3255⟩.

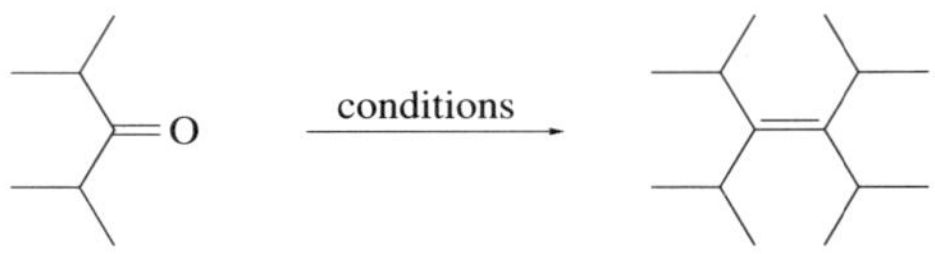

Conditions	*Yield(%)*
$TiCl_3/LiAlH_4$	12
$TiCl_3$/Zn-Cu	37
$TiCl_3(DME)_{1.5}$	87

Scheme 33

Many other procedures are also available for the production of low valent titanium species by reduction of either $TiCl_3$ or, less commonly, $TiCl_4$. These are: $TiCl_3$/K(graphite) ⟨76JOC896, 78JOC3255, 82JOC248, 85JOM(280)307, 87S1071, 88JA6914, 88JCS(P1)1729⟩; $TiCl_3$/Li ⟨78JOC3255, 82JOC248, 82TL5485, 84LA1905⟩; $TiCl_3$/Na ⟨82JOC248⟩; $TiCl_3$/Zn—Cu ⟨78JOC3255, 89JOC3748⟩; $TiCl_3/LiAlH_4$ ⟨74JA4708, 76TL3265, 78JOC3609, 78JOC3249⟩; $TiCl_3/LiAlH_4/NR_3$ ⟨76CL1127, 77S553, 83JA1660⟩; $TiCl_4$/Mg(Hg) ⟨73BSF2147, 82JOC248, 90AJC1439⟩; $TiCl_4$/Zn/pyridine ⟨77S553⟩. In the case of $TiCl_4$/Zn, the addition of an amine induces disproportionation to the active Ti(0) species ⟨87S1071⟩.

A synthetic approach to the synthesis of mevinolin (**39**) and compactin (**40**) gives a good example for comparison of some of the various procedures available for generation of the active Ti(0) (Scheme 34). Under standard conditions no alkene was formed, possibly due to the highly oxygenated nature of the substrate. Addition of a tertiary amine gave a better yield and improved the stability of the acetal unit; however, the best yield was obtained using potassium graphite as the reducing agent. Importantly, no racemisation was observed at the position α- to the aldehyde unit in the cyclisation of compound (**36**). Unfortunately this reagent system is very sensitive to atmospheric moisture and/or oxygen, which can lead to low yields with aged reagents ⟨91JOC6447⟩.

A survey of functionality compatible with reaction conditions has been published ⟨89CRV1513⟩. Acetals (best with added amine), alcohols (but not allylic alcohols), alkenes, silanes, amines, ethers, halides (but not 1,2-dihalides) and sulfides are unaffected by the reaction. Alkynes, amides, carboxylic acids, esters, ketones, nitriles and tosylates will survive in some cases. Allylic and benzylic alcohols, 1,2-diols, epoxides, enediones and quinones, halohydrins, α-haloketones, aliphatic and aromatic nitro compounds, oximes and sulfoxides are all generally incompatible with the reaction

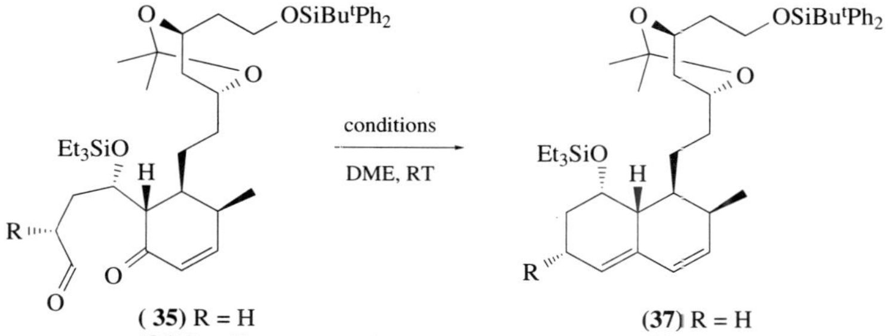

(**35**) R = H
(**36**) R = Et

(**37**) R = H
(**38**) R = Et

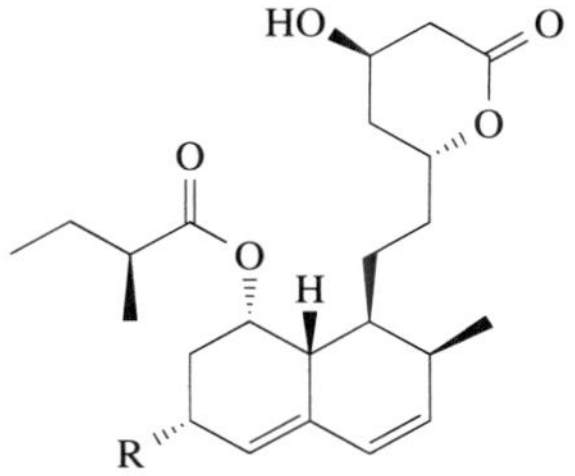

(**39**) mevinolin R = Me
(**40**) compactin R = H

Conditions	*(37) Yield (%)*
$TiCl_3$, Zn-Cu	0
$TiCl_3$, $LiAlH_4$	0
$TiCl_3$, $LiAlH_4$, NEt_3	30–35
$TiCl_3$, C_8K	71

Scheme 34

conditions. Reductive coupling of aromatic ketones can lead to further reduction of the new C═C bond forming the ethane derivative ⟨82JCS(P2)77⟩.

The reaction is usually efficient and generally successful. Symmetrical alkenes are prepared readily from aldehydes or ketones, saturated or unsaturated, using any of the conditions described above (Scheme 35) ⟨74JA4708, 77S553, 78JOC3255, 84JOC3618⟩. Hindered alkenes can also be prepared although reduced yields are observed. In extreme cases (e.g., di-*tert*-butylketone), reduction of the ketone to the secondary alcohol is observed with no evidence of coupling ⟨80JCR(S)396⟩.

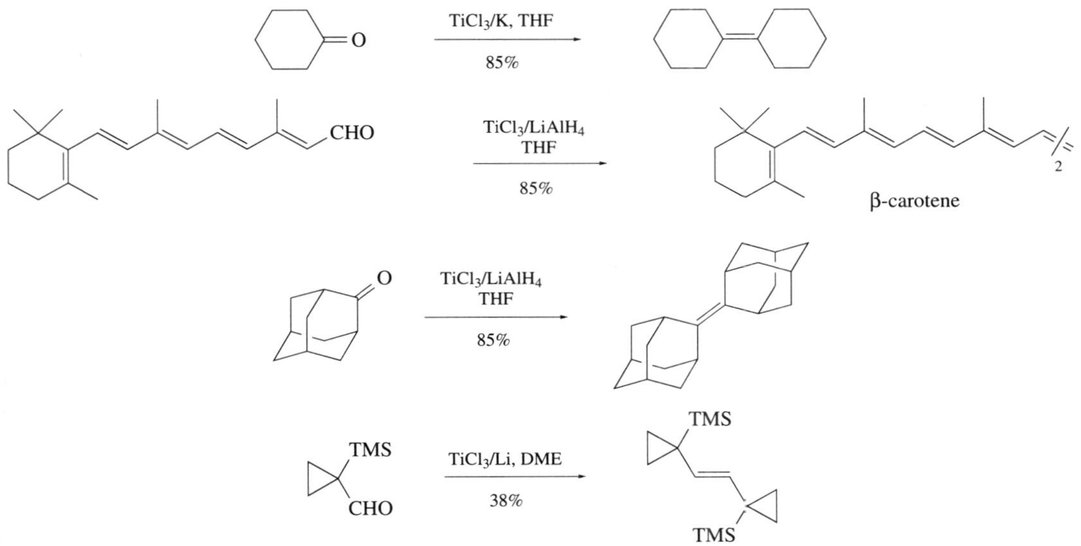

Scheme 35

(continued)

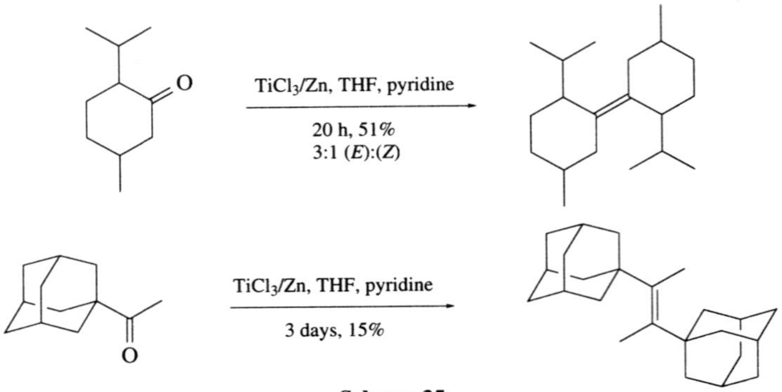

Scheme 35

Unsymmetrical alkenes can also be prepared in some cases, particularly if one carbonyl compound (often acetone) is used in excess (Scheme 36) 〈78JOC1612, 78JOC3255〉, or if one is a diaryl ketone such as benzophenone or fluorenone. Thioketones also undergo reductive coupling to give alkenes under similar conditions 〈78JOC3255〉.

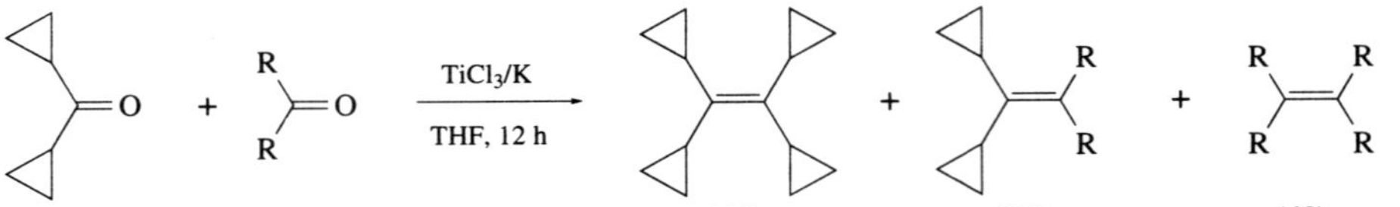

1 : 1 **(41)** **(42)** **(43)**

R = Pr^i, ratio **(41)**:**(42)**:**(43)**, 2.5:6:1.5, 29%

R = Et, ratio **(41)**:**(42)**:**(43)**, 1:6.5:2.5, 31%

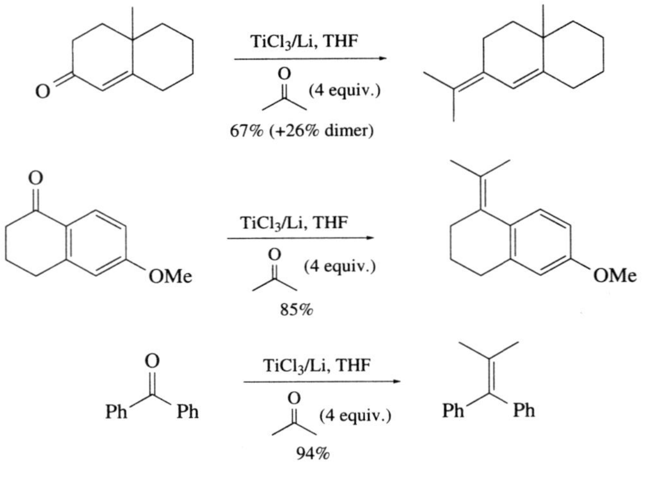

Scheme 36

Intramolecular couplings are also very efficient for formation of cyclic alkenes. As well as the more usual five and six-membered rings 〈78JOC3255〉, highly strained three and four-membered cycloalkenes have been synthesised (Scheme 37) 〈76TL3265, 78JOC3255, 78JOC3609〉. Macrocyclic alkenes can also be prepared, generally under high dilution conditions (Scheme 38) 〈78JOC3255, 78JOC3609, 83BCJ192, 84JA5018, 86BCJ2209, 86JA515, 86JA2932, 86JA3513, 88JA6914〉.

The McMurry coupling has been extensively used in the synthesis of interesting and unusual molecules 〈79JOC4729, 81SC895, 82BCJ2434, 82CB1234, 83BCJ2023, 83JA7171, 84JA723, 84JA6450, 84JOC1627, 84LA340, 85AG(E)1046, 86AG(E)720, 86JCS(P1)475, 87AG(E)928, 87CB79, 87CC381m 87TL3209, 87TL4965, 88AG(E)298, 88AG(E)411, 88BCJ1281, 88JA3578〉, and of natural products 〈76CL1127, 78CL803, 82TL1777, 82TL2723, 85CC1282, 85TL2167, 85TL2171, 85TL3393, 86BCJ1109, 87JOC2905, 87JOC4885, 87T5489, 88CC354, 88TL4369〉. Some of these have already been described (Schemes 34, 35, 37 and 38), and additional examples are shown in Scheme 39 〈77JA602, 83BCJ192, 84JA5018, 86JA515, 86JA2932〉 and Scheme 40 〈82JOC5229, 84JOC3610, 86JA3513, 86TL399, 87TL3091, 88JCS(P1)1085〉.

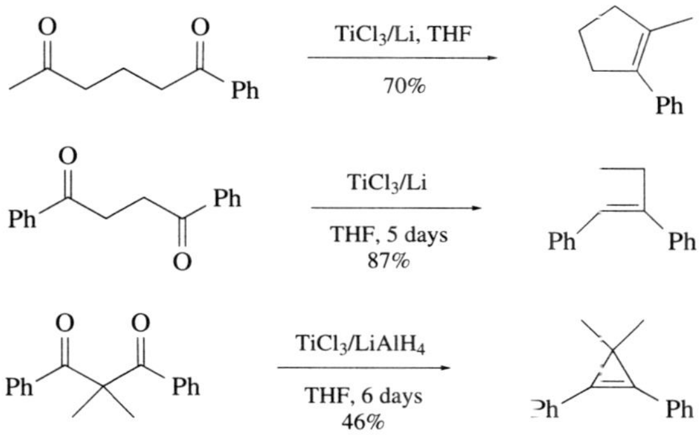

Scheme 37

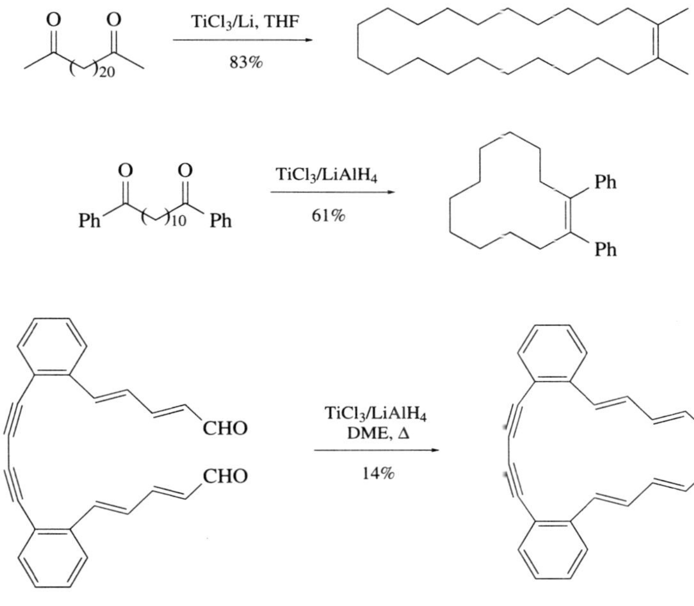

Scheme 38

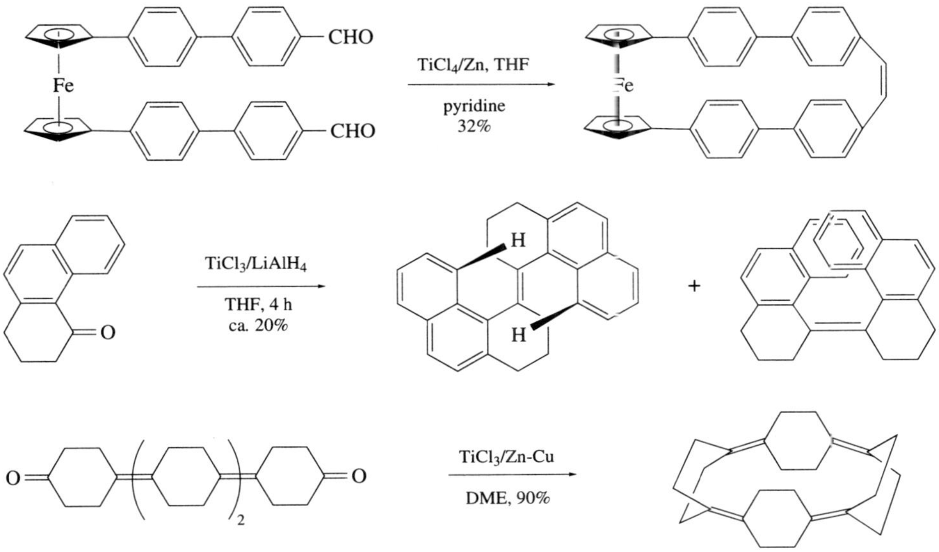

Scheme 39

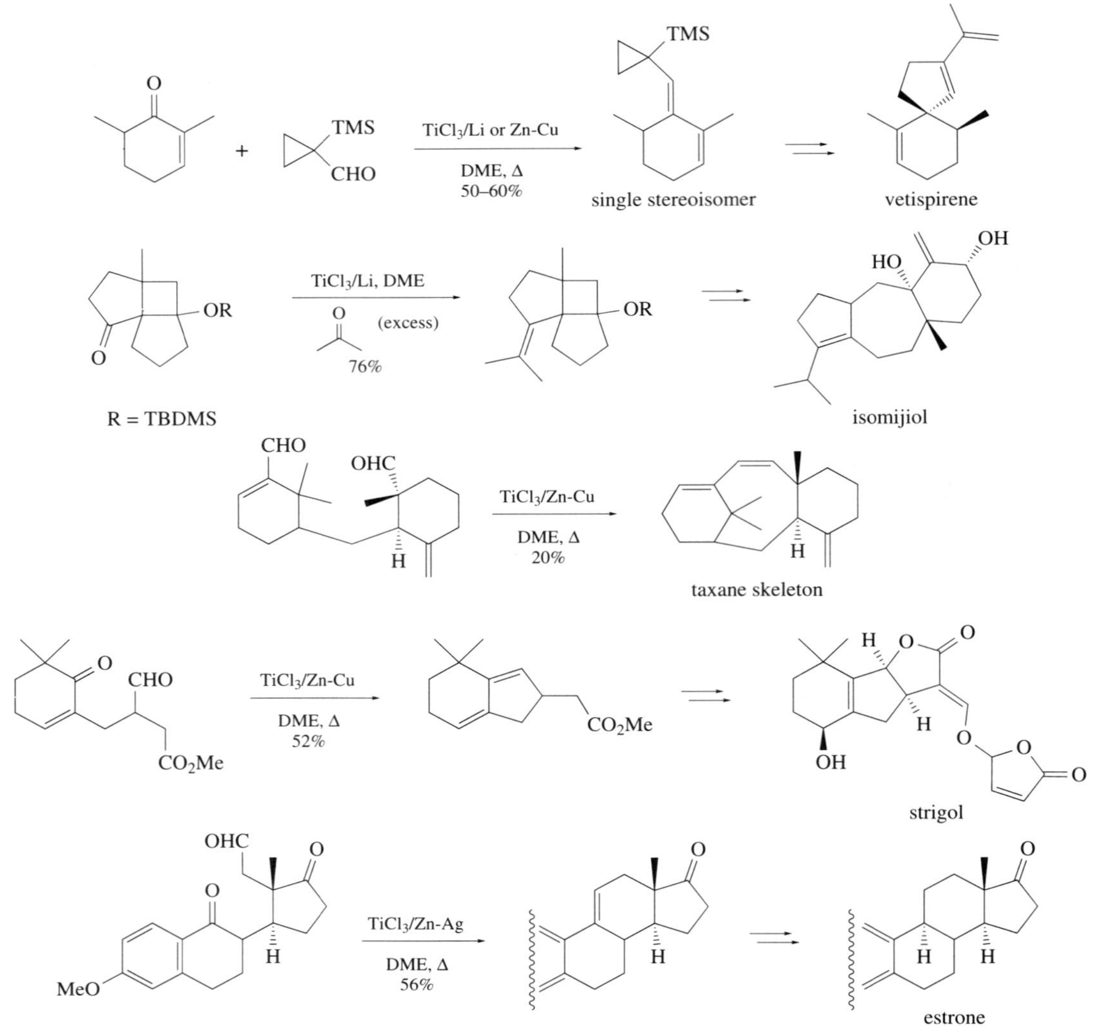

Scheme 40

McMurry-type coupling has also been reported using other metal systems ⟨88CRV733⟩, including $Cp_2Ti(CO)_2$ ⟨84JOM(268)C1⟩, $NbCl_5/NaAlH_4$ ⟨82BSB149, 82CL157⟩, $NbCl_5/K$ ⟨92CB149⟩, low valent tungsten ⟨72JA6538, 78JOC2477, 82BSB149, 84CC341, 89JA2324⟩, and molybdenum ⟨78JOC2477, 82BSB149⟩. Most of these systems are limited to the synthesis of aromatic alkenes. A reagent system based on zinc has also been described (Equation (12)) ⟨92TL3899⟩.

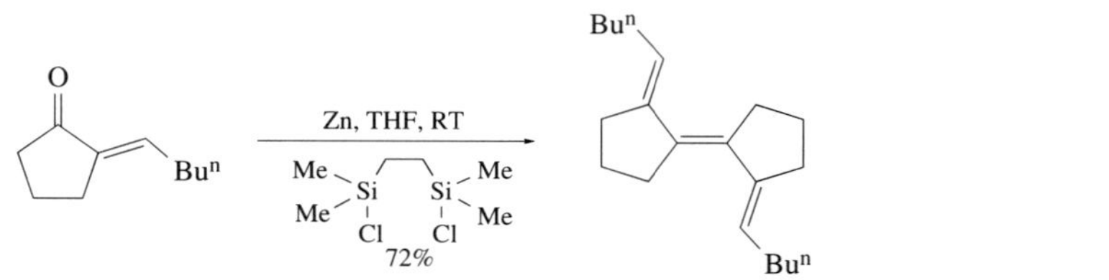

(12)

1.15.3.4 Addition of Organometallic Reagents to Carbonyl Compounds and *In Situ* Dehydration

The dehydration of alcohols is a very general method for the formation of alkenes. A full discussion of this reaction is beyond the scope of this chapter and is covered elsewhere (see Section

1.13.4). However, there are examples where an elimination reaction can be carried out *in situ* immediately after carbon—carbon bond formation, and thus represents a method for the direct synthesis of alkenes.

The preparation of crowded alkenes from hindered ketones can be difficult using common methods such as the Wittig reaction, where often no reaction or electron transfer reduction is observed. The addition of organolithium reagents to ketones generates a lithium alkoxide which is often converted to the alcohol by aqueous workup. If, however, thionyl chloride is added to the alkoxide slowly at $-78\,^{\circ}C$ (the reaction is reported to be extremely exothermic) and the resulting solution allowed to warm to room temperature, then direct formation of alkene is observed in high yield (Scheme 41) ⟨89JOC1375⟩. This reaction has been applied to the synthesis of crowded alkenes, which could otherwise not be prepared in a satisfactory manner using alternative strategies.

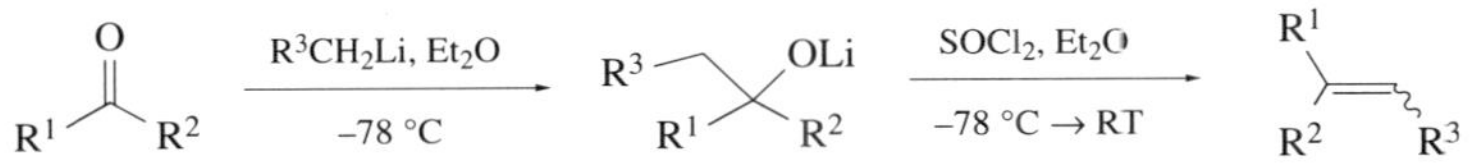

(44) **(45)** **(46)** **(47)** **(48)**

Variation of product yield on use of different alkyllithium reagents

Ketone	*MeLi*	*PrⁱLi*	*BuˢLi*
(44)	94	94	95
(45)	93	90	88
(46)	82	82	83
(47)	96	95	96
(48)	96	93	90(*E*)+(*Z*)

Scheme 41

1.15.4 BY CONDENSATION OF SULFUR, SELENIUM OR TELLURIUM FUNCTIONS

1.15.4.1 Alkylation of Sulfur(II) Stabilised Carbanions and Elimination

The addition of anions derived from thioethers to carbonyl compounds provides efficient access to β-hydroxysulfides. The anions can be generated by lithiation of the thioether ⟨67JOC1717, 66JOC4097, 78JA2140, 82JA2923⟩ or from a halide via the Grignard reagent ⟨82CL1697⟩. Conversion of the alcohol group to a benzoate ester and reductive elimination using $TiCl_4$/Zn/pyridine ⟨74CL1523⟩, $TiCl_3$/K ⟨81JOC4072⟩, or Na/naphthalene ⟨72JA4758⟩ then forms the alkene (Scheme 42). Related procedures are also successful using organoselenium chemistry (see Section 1.15.4.4). This reaction sequence is successful for hindered ketones, β,γ-unsaturated ketones which transform without double bond migration ⟨72JA4758⟩ and α,β-unsaturated ketones without overreduction (Scheme 43) ⟨81JOC4072⟩.

An alternative procedure for the conversion of β-hydroxysulfides into alkenes has also been developed ⟨72TL737⟩. Alkoxide formation and reaction with *ortho*-phenylene phosphochloridite form a phosphorus(III) ester which on thermolysis gives the desired alkene (Scheme 44).

1.15.4.2 Alkylation of Sulfur(IV) Stabilised Carbanions followed by Elimination

1.15.4.2.1 Alkylation of sulfoxides and elimination

Sulfoxides have been used extensively for alkene synthesis using the *syn*-elimination of a sulfenic acid. Similarly, alkylation of sulfoxide stabilised carbanions is also used extensively for carbon—carbon bond formation. However, the two procedures have been rarely linked together to form simple alkenes. The related reaction using selenium-based systems has been investigated more thoroughly (see Section 1.15.4.4).

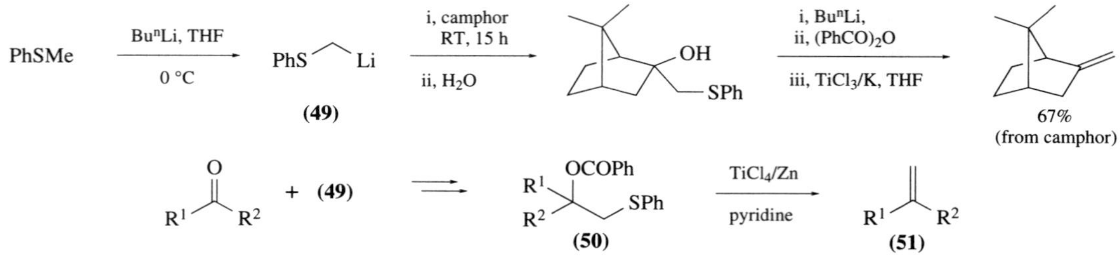

R^1	R^2	(50)–(51) Yield (%)
$PhCH_2CH_2$	Me	93
Bn	Bn	94
PhCH=CH	Me	90
Cholestan-3-one		87

Scheme 42

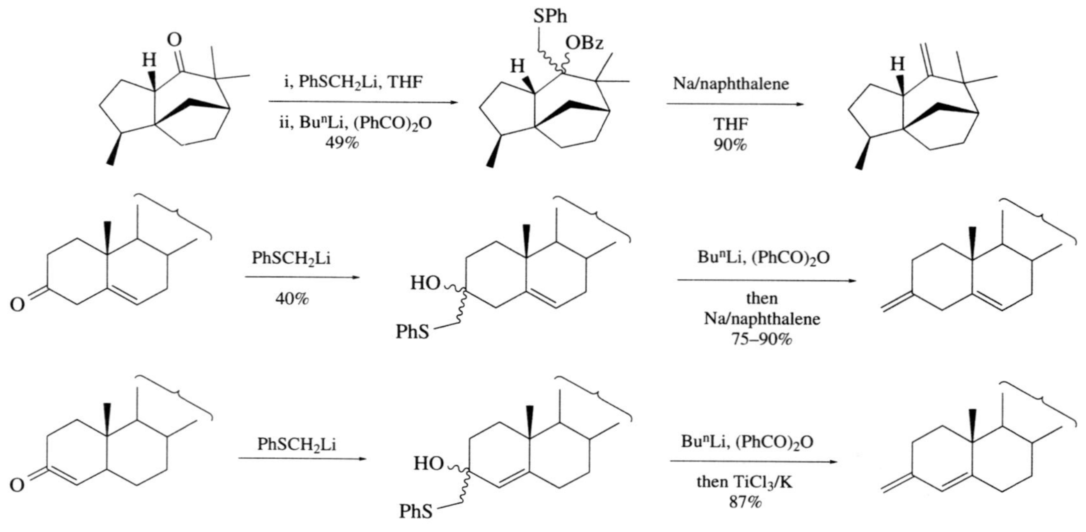

Scheme 43

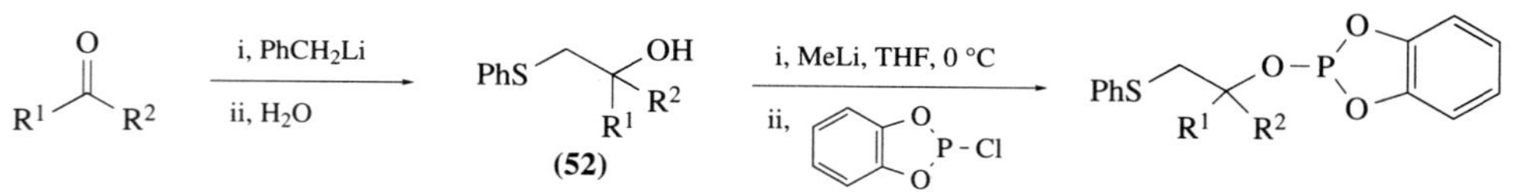

R^1	R^2	Yield from (52)
Ph	Ph	83
n-C_9H_{11}	H	80
n-$C_{11}H_{23}$	H	64

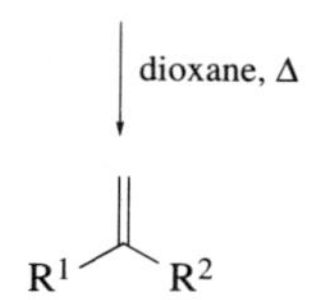

Scheme 44

In an approach to the synthesis of the elaeokanines, the sulfoxide (**53**) was alkylated and reduced to give a mixture of products which were separated. Regeneration of the sulfoxide anion of (**54**) and condensation with butanal gave the two epimeric sulfoxides which after separation and thermolysis gave (−)-elaeokanine B (Scheme 45) ⟨90JOC2128⟩.

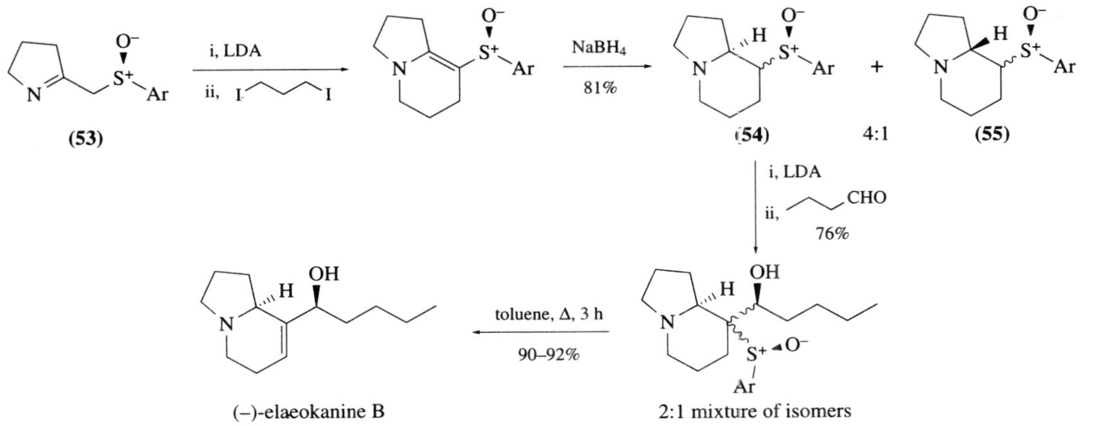

Scheme 45

1.15.4.2.2 *Alkylation of sulfinamides and elimination*

The addition of α-lithiosulfinamides to carbonyl compounds and subsequent thermal elimination of the *β*-hydroxysulfinamide intermediates provides a route to alkenes (Scheme 46) ⟨66JA5656, 67PAC19, 68JA5548, 68JA5553⟩. The sulfinamide reagent is limited in that the initial anion formed using *N*,*N*-disubstituted sulfinamides (**56**), eliminates to give a sulfine and low yields are obtained. It is for this reason that dianions of sulfinanilides are reagents of choice.

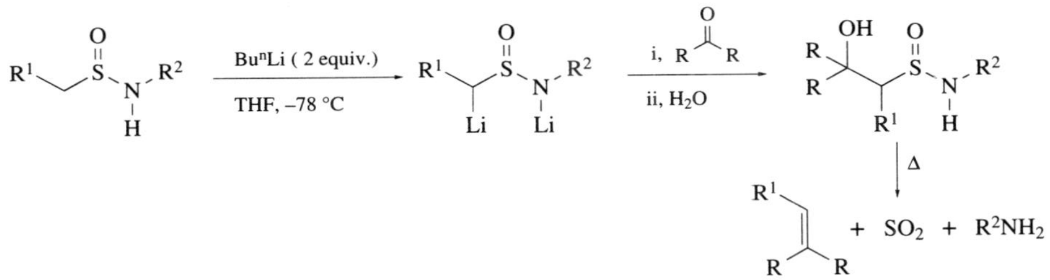

Scheme 46

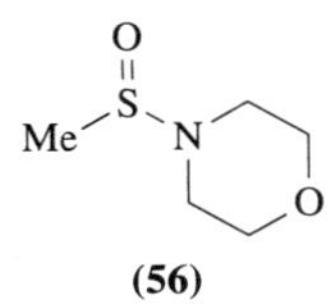

(56)

The dilithio salt of *N*-methanesulfinyl-*p*-toluidine, generated using Bu^{n}Li, adds smoothly to aldehydes and ketones to give *β*-hydroxysulfinamides. Competing enolate formation can be an important side reaction although yields, taking into account recovered ketone, are high (Equation (13)). For example, 4-*t*-butylcyclohexanone (**57**) undergoes approximately 60% addition and 40% enolate formation. Thermolysis of the *β*-hydroxysulfinamide adducts gives alkenes in high yield, although this reaction is considerably slower for adducts derived from aldehydes and moisture must be excluded to prevent deleterious side reactions occurring ⟨66JA5656, 67PAC19⟩. With *N*-ethylsulfinyl aniline, a mixture of (*Z*)- and (*E*)-alkenes is formed (Scheme 47). This is a result of the non-stereoselective *β*-hydroxysulfinamide formation. The elimination reaction itself has been shown to occur stereospecifically by a *cis* elimination pathway (Equation (14)) ⟨68JA5553⟩. Further evidence for this is provided by the *trans*-2-hydroxycyclohexylsulfinanilide (**58**). Stereospecific *cis*- elimination would form the highly unstable *trans*-cyclohexene, whereas *trans* elimination would give the *cis*-cyclohexene. Significantly, (**58**) is stable under thermolysis conditions which cause complete decomposition of other *β*-hydroxysulfinanilides indicating that only the *cis* elimination mechanism operates.

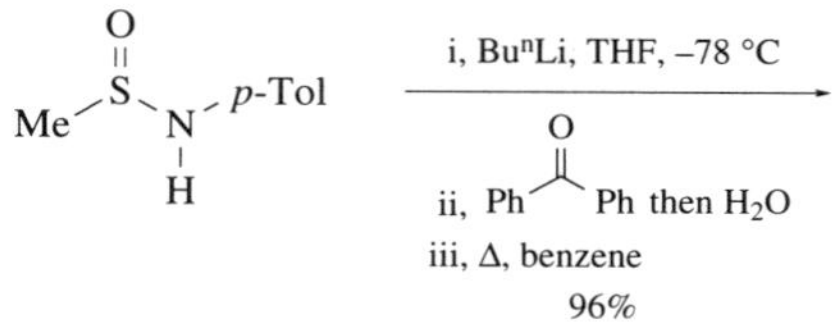

(13)

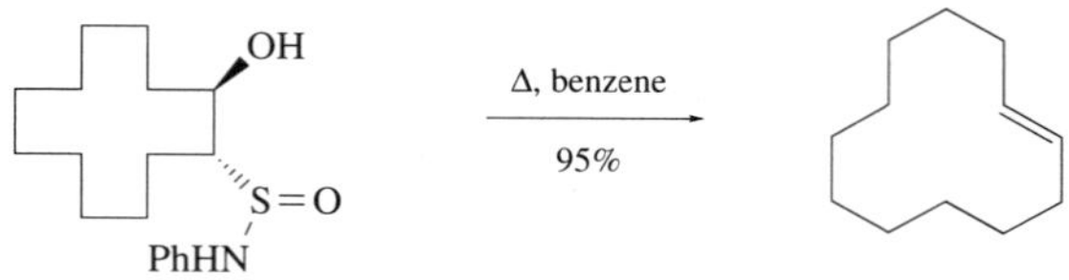

(14)

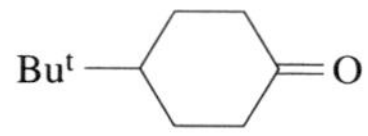

(57)

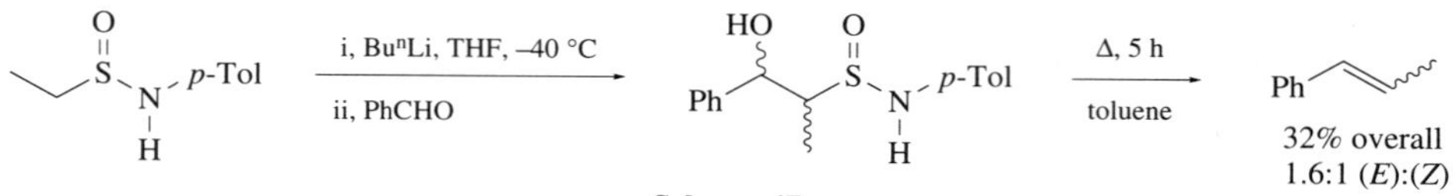

Scheme 47

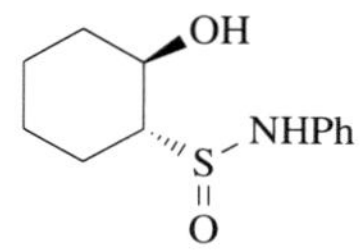

(58)

1.15.4.3 Alkylation of Sulfur(VI) Stabilised Carbanions followed by Elimination

1.15.4.3.1 The Julia reaction and related transformations

The condensation of a sulfone anion with a carbonyl compound and subsequent reductive elimination of the β-hydroxysulfone (or a derivative thereof) to form an alkene (Scheme 48), is usually referred to as the Julia reaction ⟨73TL4833⟩, although the important contributions made by Lythgoe and Kocienski ⟨78JCS(P1)829⟩ have also been recognised. It has been the subject of a number of reviews ⟨81CI(L)548, 85PAC763, 85PS(24)97, 88BCJ107, 91COS(1)792, B-93MI 115-02⟩. The reaction usually gives good *trans* selectivity, particularly if there is some branching adjacent to the newly formed double bond (Scheme 49) ⟨80JCS(P1)1045⟩.

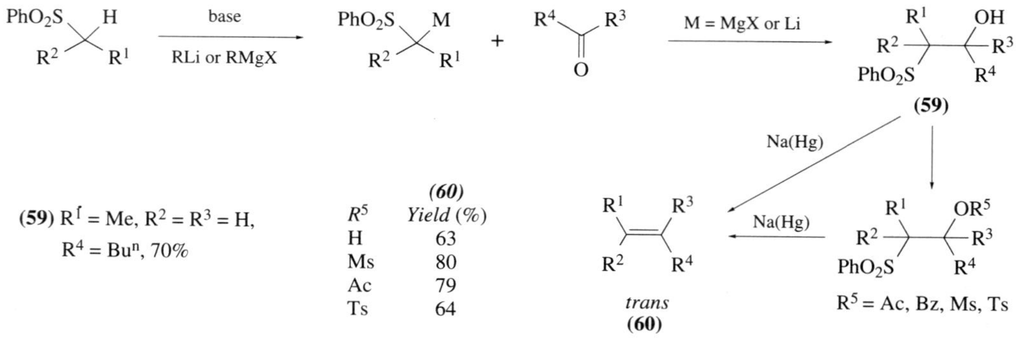

(59) R^1 = Me, R^2 = R^3 = H, R^4 = Bu^n, 70%

(60)	
R^5	*Yield* (%)
H	63
Ms	80
Ac	79
Ts	64

Scheme 48

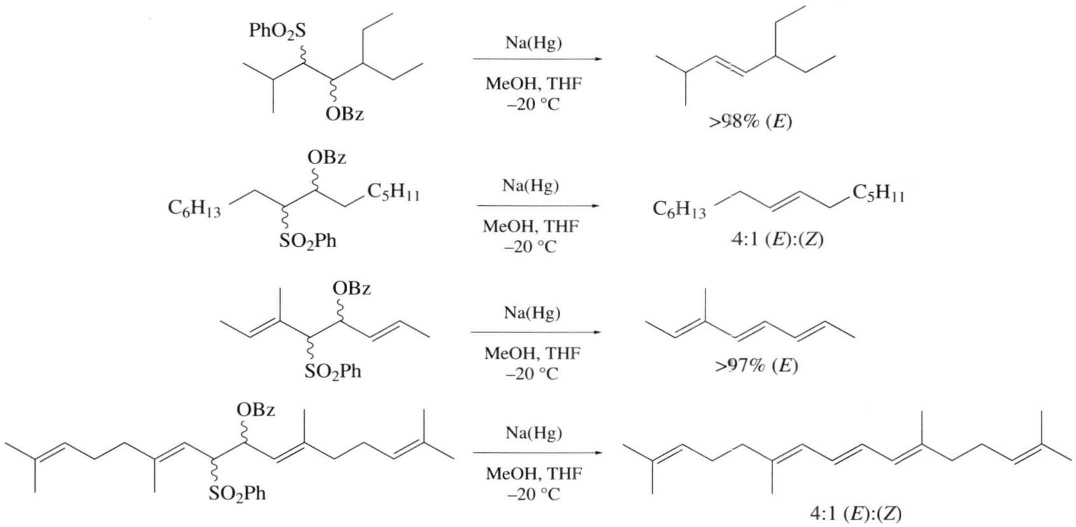

Scheme 49

The sulfone anion is usually generated using Bu^{n}Li in THF, although other reagents have also been used, including Bu^{t}Li ⟨87JCS(P1)1623, 87TL5763, 89T7161, 90TL5525⟩, MeLi ⟨86TL2095⟩, LDA ⟨86JOC3916, 88JOC4282, 88TL5169⟩, and magnesium bases, which are superior for reactions involving easily enolisable carbonyl compounds, including aldehydes ⟨73TL4833, 78JCS(P1)829⟩. An interesting case, shown in Scheme 50, required the use of phenyllithium, as bases such as LDA and Et_2NLi were not of sufficient strength to deprotonate the sulfone, whereas stronger bases (Bu^{n}Li, Bu^{t}Li) caused undesired side reactions ⟨90JA7407⟩. In some cases use of the co-solvent HMPA is required ⟨83CC630, 87T4395⟩. For sulfones with additional acidic protons, it is possible to avoid protection by the use of polyanions (Scheme 51) ⟨86TL2095, 87TL5759⟩. The condensation is successful with both aldehydes and ketones and can be promoted by Lewis acids, typically Bu^i_2AlOMe, generated *in situ* from diisobutylaluminum hydride and methanol ⟨86TL2095, 87JOC3759⟩, or $BF_3 \cdot OEt_2$ (Schemes 51 and 52) ⟨85TL5597, 88JA5198⟩.

The initial products of the condensation reaction are *β*-hydroxysulfones, which may be eliminated

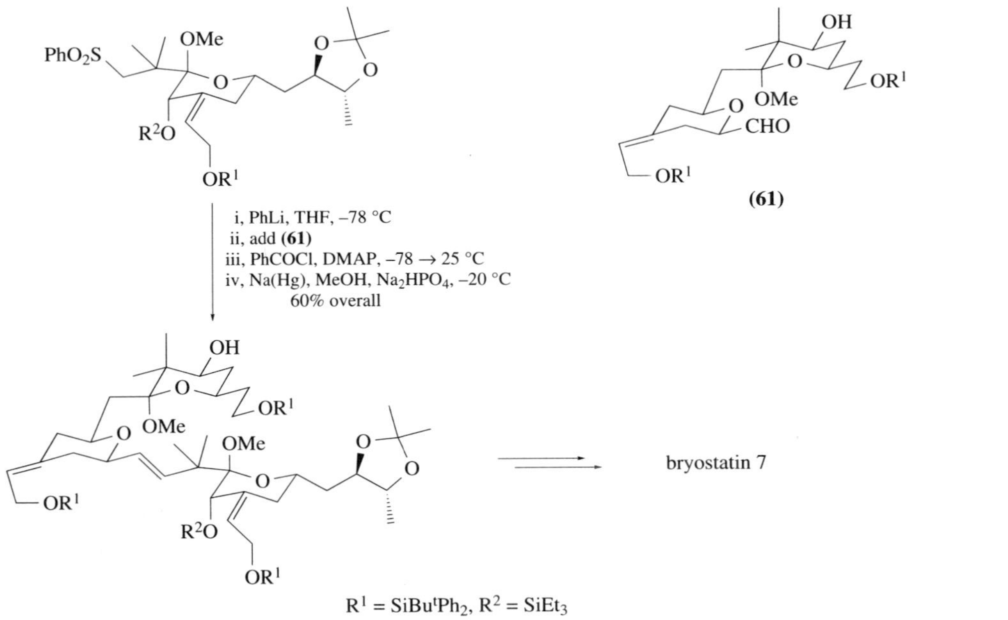

$R^1 = SiBu^tPh_2$, $R^2 = SiEt_3$

Scheme 50

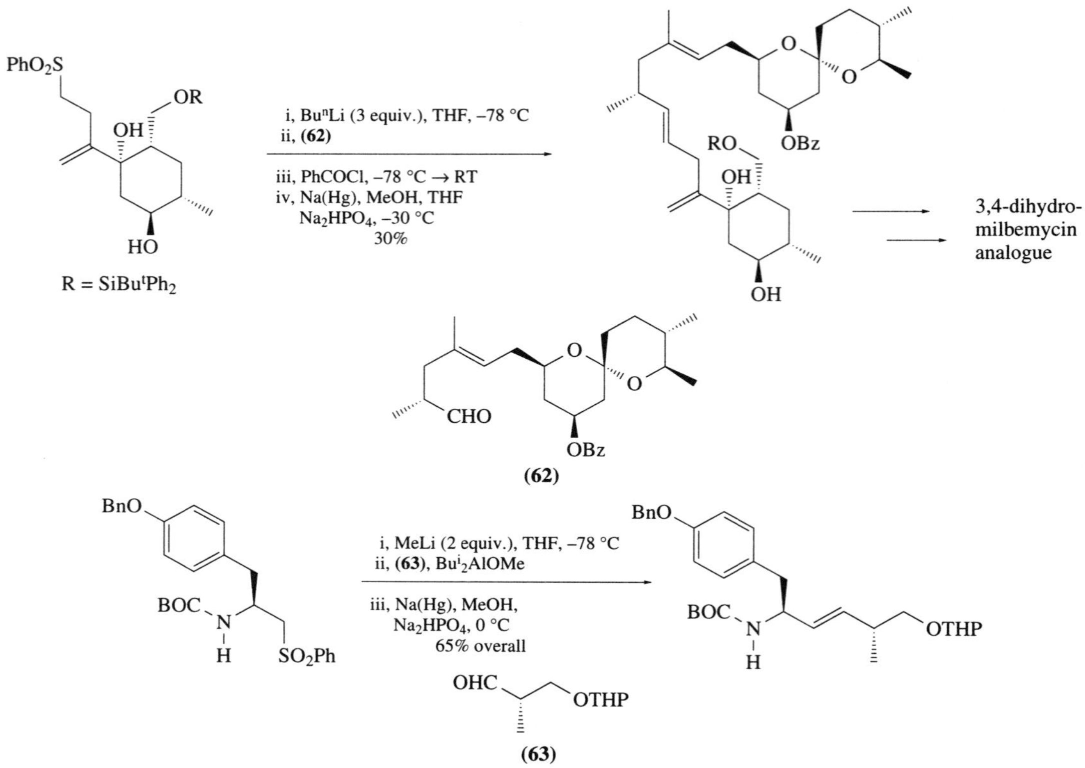

Scheme 51

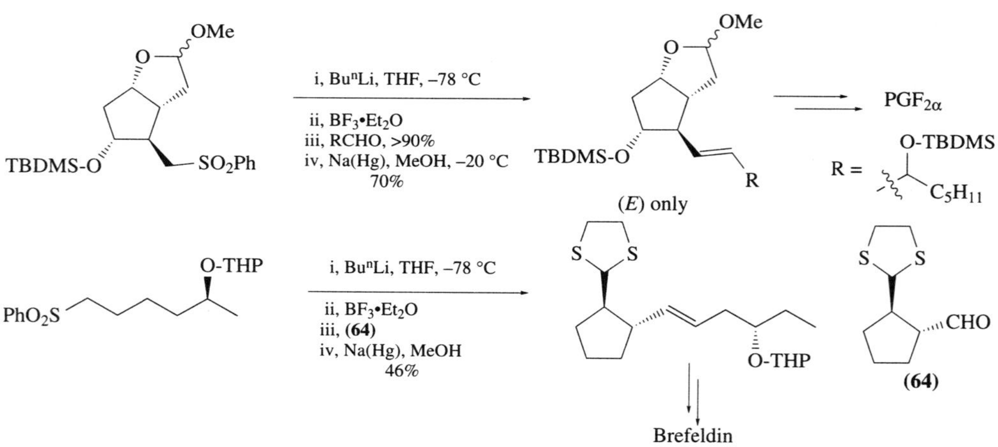

Scheme 52

directly using sodium amalgam ⟨73TL4833, 77JOC2036, 85JOC5465, 85TL5597, 88JA5198⟩; see also ⟨76TL3477, 78TL2625⟩, often in the presence of the phosphate buffers Na_2HPO_4 ⟨86TL2095, 87JOC3759, 89BCJ3132⟩, or KH_2PO_4 ⟨86JA2776⟩; electrochemically ⟨78CL69⟩; using Bu_3SnH/2,2′-azobisisobutyronitrile (AIBN) ⟨81TL2675⟩; or with SmI_2 on imidazole-derived sulfones ⟨90TL7105⟩ (Scheme 53). The double bond formed from this reduction is predominantly the *trans*-isomer, irrespective of the stereochemistry in the β-hydroxysulfone. This implies that any intermediate radical or anion produced by reduction of the sulfone is sufficiently long-lived to allow the molecule to equilibrate to an intermediate with the large groups as far apart as possible, prior to elimination ⟨78JCS(P1)829, 81CI(L)548⟩.

More commonly, the hydroxyl group of a β-hydroxysulfone will be converted into an ester prior to reduction using sodium amalgam. Again high *trans* selectivity is observed. Typical esters are

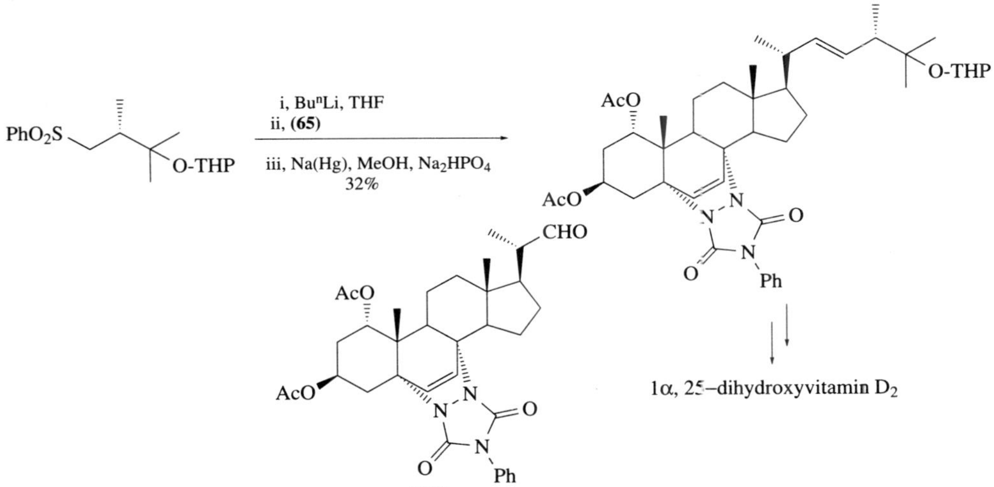

(65)

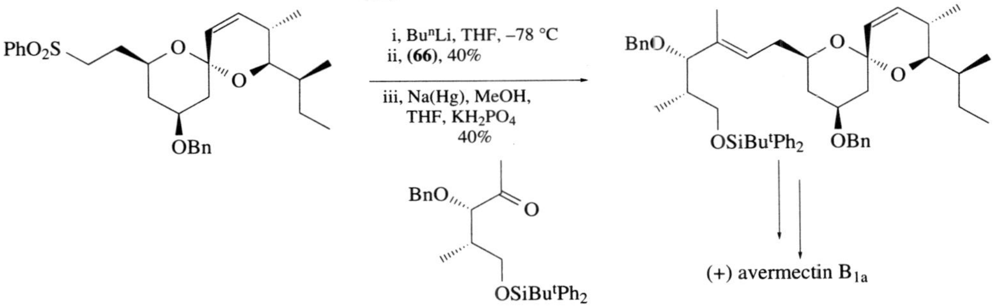

(66)

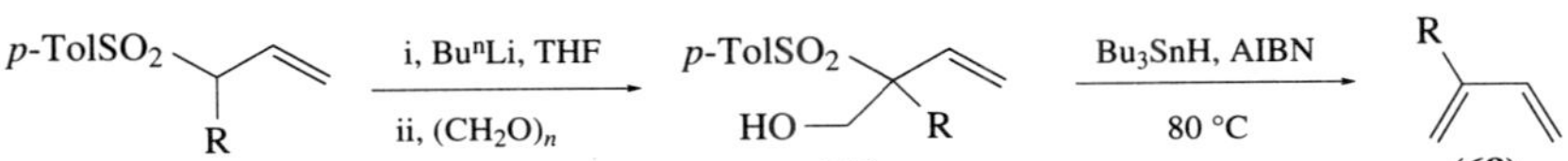

(67) **(68)**

R	*(67)* *Yield* (%)	*(68)* *Yield* (%)
C_8H_{17}	95	92
C_6H_{13}	92	63
C_4H_9	96	53
$PhCH_2$	88	100

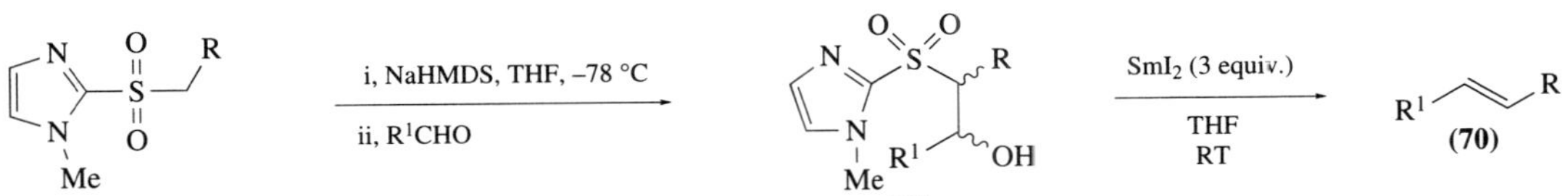

(69) **(70)**

R	*R¹*	*(69)* → *(70)* *Yield* (%)	(*E*):(*Z*)
Ph	$PhCH_2CH_2$	82	8:1
$(CH_3)_2C{=}CH$	$PhCH_2CH_2$	84	5:1
$PhCH_2CH_2$	$PhCH_2CH_2$	55	3:1
PhCH=CH	$(CH_3)_2C{=}CH$	78	(*E*) only
Ph	$(CH_3)_2C{=}CH$	85	5:1

Scheme 53

acetates, prepared using acetic anhydride ⟨78JCS(P1)834, 86JOC4840, 87JOC2838⟩ with pyridine and/or 4-dimethylaminopyridine (dmap) when necessary (Scheme 54) ⟨87T4395, 88TL5169, 89JOC12⟩, benzoates prepared using benzoyl chloride (Scheme 55) ⟨80JCS(P1)1045, 80JCS(P1)1400, 82TL4879, 83CC630, 86JA4603, 87JCS(P1)1623, 87TL5755, 87TL5759, 87TL5763, 88JOC1064, 90JA7407, 90TL5525⟩, mesylates (Scheme 56) ⟨73TL4833, 78JCS(P1)829, 81CI(L)548, 82TL1963, 85TL5597⟩ or tosylates ⟨73TL4833, 78JCS(P1)829⟩. Ester formation is often carried out *in situ* on the β-alkoxysulfone intermediate formed from the initial condensation reaction. Acetoxy and benzoyloxy groups are better than mesylate and tosylate for derivatisation of allylic hydroxyl groups leading to the formation of conjugated dienes ⟨78JCS(P1)829⟩. Methanol is the solvent of choice for this reaction being superior to EtOH and Pr^{i}OH ⟨78JCS(P1)829⟩; however, often a co-solvent such as EtOAc ⟨78JCS(P1)829, 78JCS(P1)834, 86JA4603⟩ or THF ⟨80JCS(P1)1400, 82TL4879, 83CC630, 86JOC4840, 87TL5763⟩ is added to help dissolve the substrate. The reduction reaction is usually carried out between $-20\,^\circ$C and $-40\,^\circ$C ⟨78JCS(P1)829⟩. Higher geometrical selectivity is often observed at these low temperatures; however, in some cases this can be at the expense of lower overall yield ⟨86JOC4840⟩.

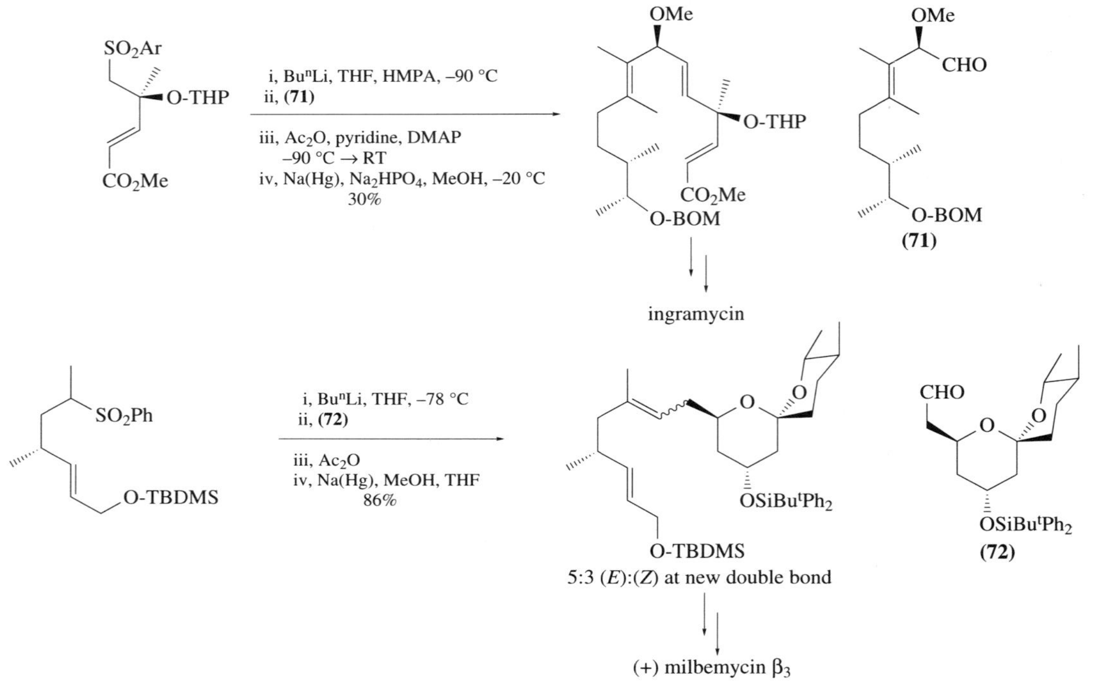

Scheme 54

Alternative procedures via xanthate esters have also been developed; these methods give good yields of alkene, even in cases where other elimination protocols are unsuccessful. The xanthates are prepared from the initially formed β-alkoxysulfone intermediate using CS_2 and methyl iodide. Reduction using Bu_3SnH (Scheme 57) ⟨77TL4223, 86JOC3916, 88JOC4282⟩, or photolysis in the presence of *N*-acetyloxy-2-thiopyridine (**76**) (Scheme 58) ⟨92T7109, 91TL2703⟩ gives alkenes with high *trans* selectivity. In the latter case, use of other radical initiators such as AIBN, Et_3B-O_2, or $(PhCO_2)_2$ is also successful.

Procedures closely related to the Julia reaction, provide additional alternative routes to alkenes. If the β-hydroxysulfone intermediates are eliminated to the vinylsulfones, usually by simple base catalysis on the derived acetates ⟨82TL2465, 87JCS(P1)2171, 87T859⟩; see also ⟨84JA3670, 86JOC3830⟩; or dehydration using $POCl_3$/pyridine ⟨73TL4833⟩, the sulfone group on these intermediates may be reductively removed with retention of double bond geometry using sodium dithionite ⟨82TL3265⟩ or sodium amalgam ⟨73TL4833⟩. Alternatively, it is also possible to use a Grignard reagent (Bu^{n}MgCl) which, in the presence of a nickel(II) or lead(II) catalyst, acts as a reducing agent (Scheme 59); use of the latter metal often leads to higher yields and better stereochemical control ⟨83TL4311, 83TL4319, 86JOC3830, 87T859⟩. In some cases, using nickel(II), or more commonly iron(III) catalysis, it is possible to replace the sulfone group with a Grignard reagent leading to the formation of trisubstituted alkenes, again with complete control of stereochemistry ⟨82TL2469, 87JCS(P1)2171⟩, see also ⟨82TL2457⟩ (Scheme 60).

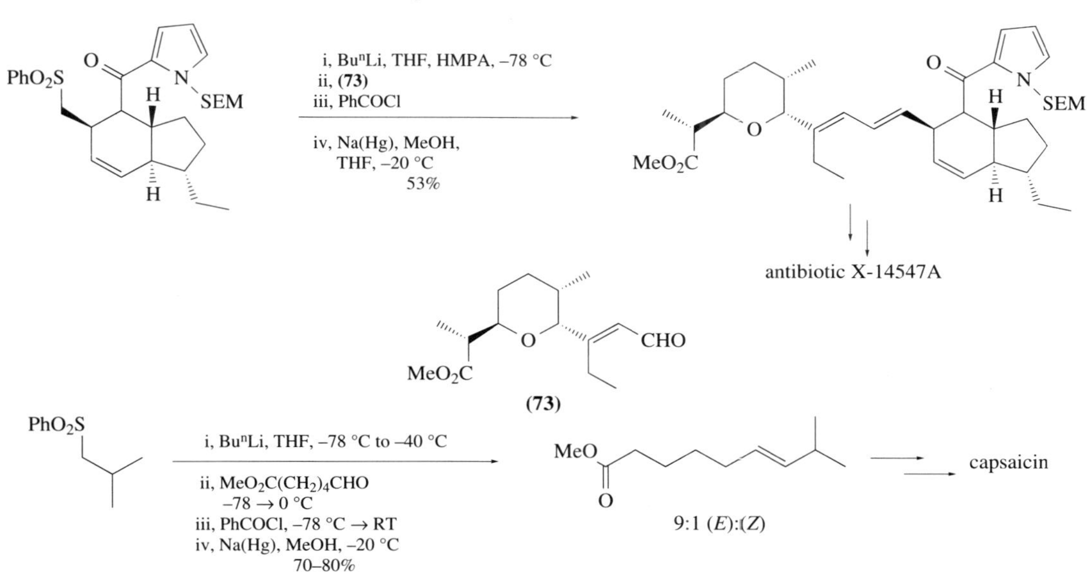

Scheme 55

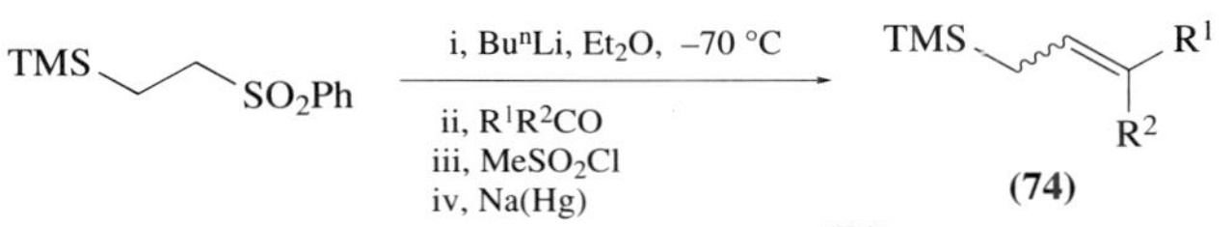

R^1	R^2	(74) Overall yield (%)
Me	Me	92
Et	Et	92
Ph	Ph	95
— $(CH_2)_5$ —		94
$(CH_3)_2C{=}CH$	H	85 (*E*)+(*Z*)
Ph	H	92 (*E*)+(*Z*)

Scheme 56

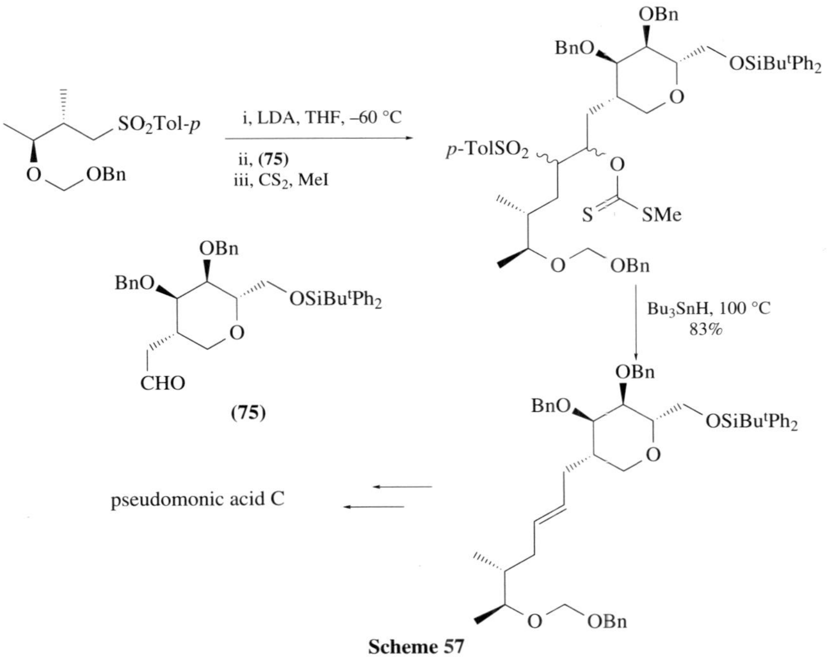

Scheme 57

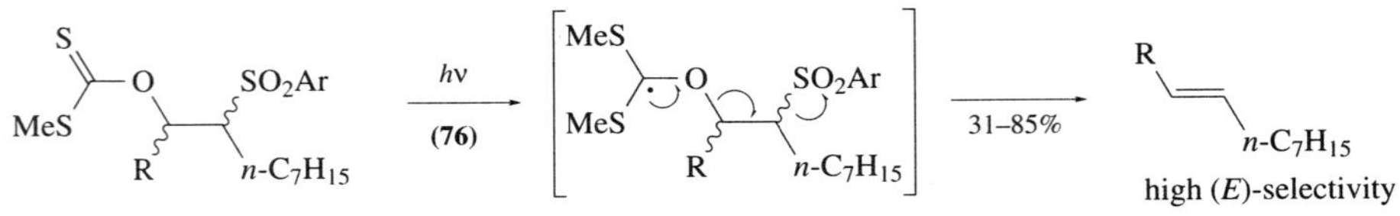

R = n-C_6H_{13} or $PhCH_2CH_2$
Ar = Ph or p-Tol

Scheme 58

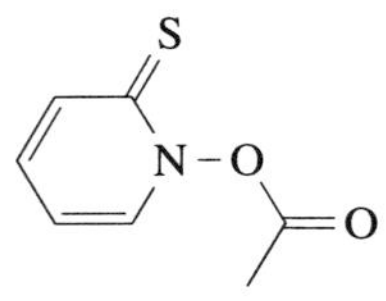

(76)

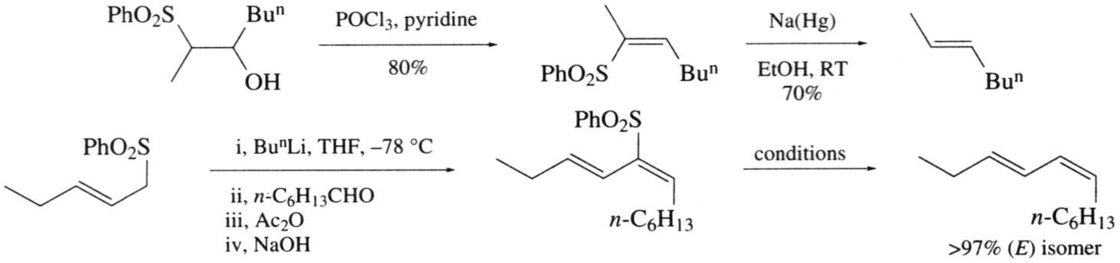

Conditions	*Yield(%)*
Bu^nMgCl, THF, RT, $Ni(acac)_2$, DABCO	46
Bu^nMgCl, THF, RT, $Pd(acac)_2$	41
$Na_2S_2O_4$, cyclohexane-H_2O, Adogen 464, $NaHCO_3$, 80 °C	65

Scheme 59

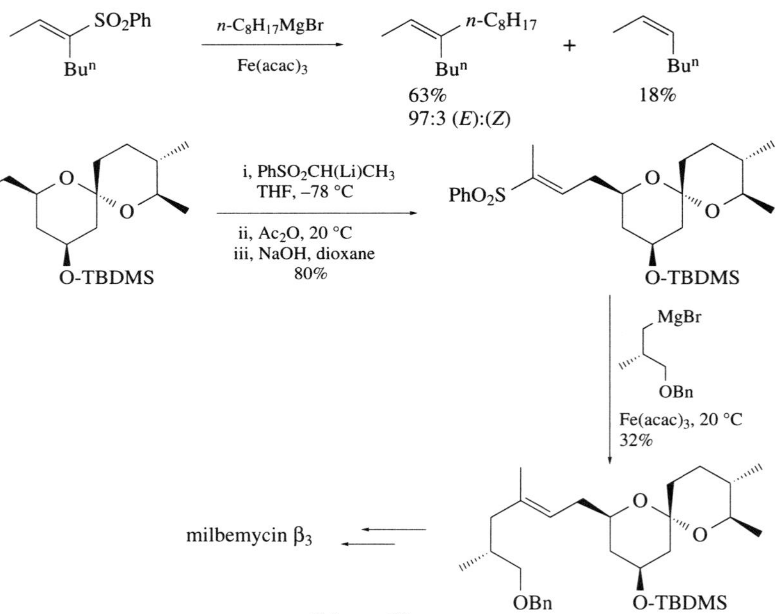

Scheme 60

The Julia reaction has found extensive use in natural product synthesis; an excellent review has been published ⟨91COS(1)792⟩. The main advantages of the reaction are good *trans* selectivity, particularly with more substituted systems, the ease with which the coupling partners can be synthesised ⟨77JOC2036, 88CJC2860, 88TL1197⟩, and the reaction's success with relatively complex molecules. Many important examples have been discussed already to illustrate a particular point (*vide supra*) and some further details are summarised in Scheme 61. These include approaches to 1α-hydroxyvitamin D_3 (Scheme 61) ⟨80JCS(P1)1400⟩, 1α,25-dihydroxyvitamin D_2 (Scheme 53)

⟨89BCJ3132⟩, (+)-milbemycin β_3 (Schemes 54, 60 and 61) ⟨85CC1326, 86JOC4840, 87JCS(P1)1623, 87JCS(P1)2171, 87TL5763, 89T7161⟩ and analogues (Scheme 51) ⟨87TL5755, 87TL5759⟩, (+)-avermectin B_{1a} (Scheme 53) ⟨86JA2776⟩, premonensin ⟨86JA2476, 86JA4603⟩, pseudomonic acid C (Scheme 57) ⟨85JOC4317, 86JOC3916, 88JOC4282⟩, methyl deoxypseudomonate B ⟨84TL2085⟩, tetronasin (Scheme 61) ⟨86JA4603, 90TL5525⟩, bryostatin 7 (Scheme 50) ⟨90JA7407⟩, brefeldin (Scheme 52) ⟨88JA5198⟩, the antibiotic X-14547A (Schemes 55 and 61) ⟨82TL4879, 83CC630⟩, $PGF_{2\alpha}$ (Scheme 52) ⟨85TL5597⟩, *trans*-D-*erythro*-sphingosine (Scheme 61) ⟨87JOC2838⟩, (±)-trinoranastreptene ⟨88TL5169⟩; FK-506 (Scheme 61) ⟨89JOC12⟩, ingramycin (Scheme 54) ⟨87T4395⟩, capsaicin (Scheme 55) ⟨88JOC1064⟩ and *trans*-alkene dipeptide isosteres (Scheme 51) ⟨86TL2095, 87JOC3759⟩.

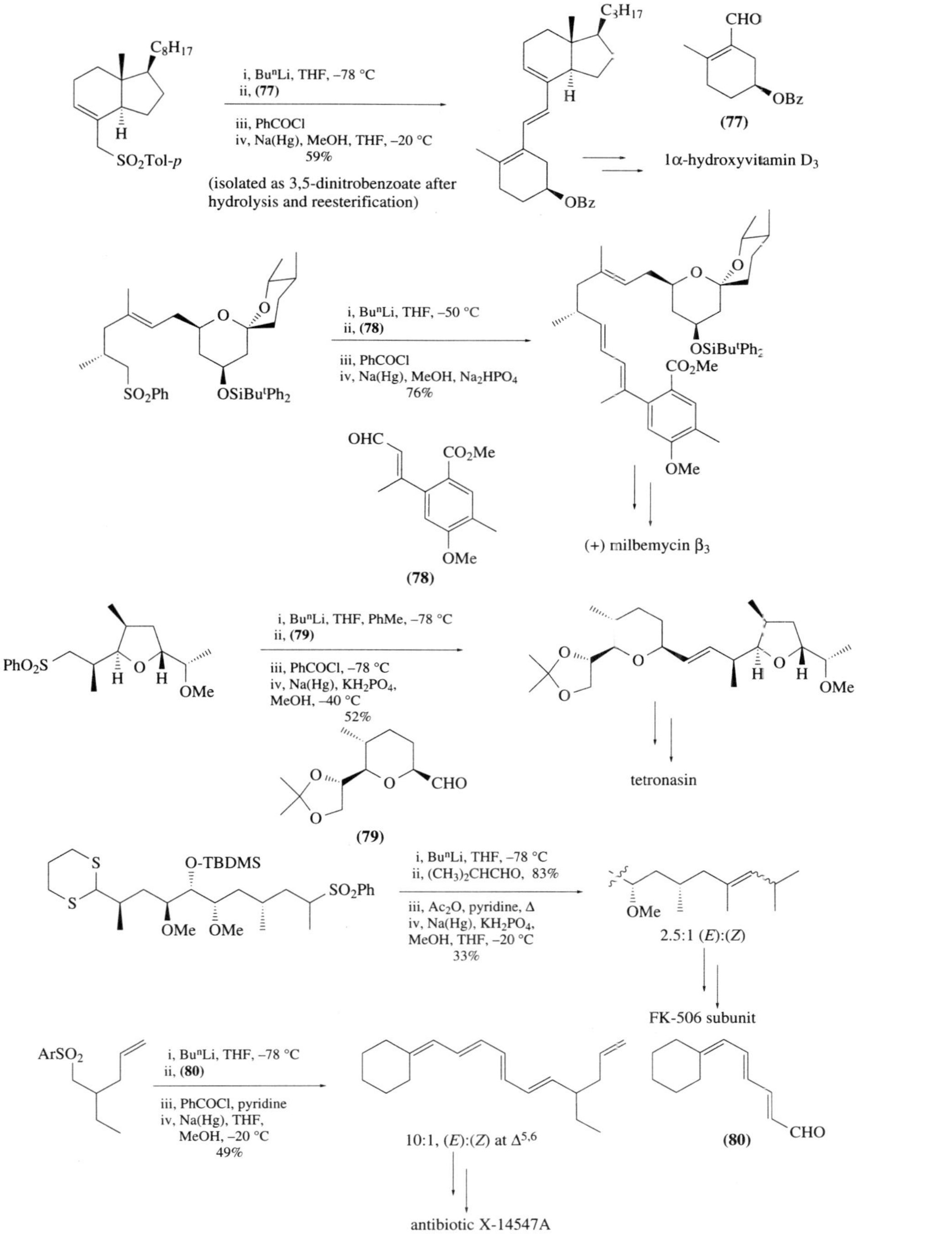

Scheme 61 *(continued)*

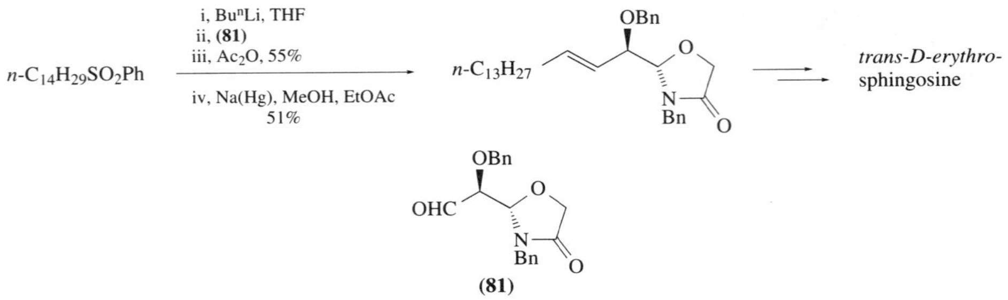

(81)

Scheme 61

1.15.4.3.2 *The Ramberg–Bäcklund reaction*

The base-induced conversion of α-halosulfones to alkenes was first reported in 1940 by Ramberg and Bäcklund ⟨40AK(13A)1⟩. They observed that, if α-bromoethylsulfone was treated with aqueous potassium hydroxide, 2-butene was formed, predominantly as the (*Z*)-isomer (Equation (15)).

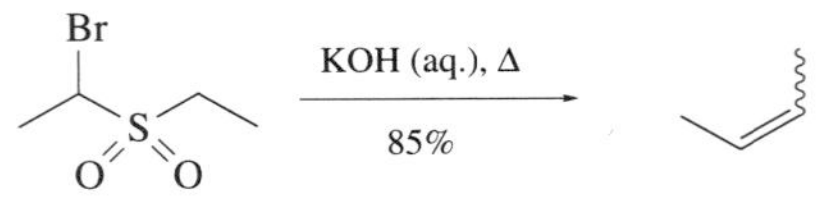

(15)

Since then the reaction has been extensively used for the synthesis of alkenes and an excellent detailed review has been published ⟨91COS(3)861⟩. Earlier reviews are also particularly useful for discussions on the mechanism ⟨B-67MI 115-01, 68ACR209, 77OR1⟩ and for methods of preparing suitable precursors for the reaction ⟨77OR1, 86T3731, B-88MI 115-01, B-93MI 115-01⟩. Other reviews are also of interest ⟨77T2019, 79COC(3)187, 82T2857, B-88MI 115-02, B-88MI 115-03, 88T6241⟩.

The reaction is believed to proceed by formation of a sulfone anion which cyclises by displacement of a leaving group (X) to give an episulfone intermediate (**82**), from which sulfur dioxide is lost to give the alkene (Scheme 62). The leaving group (X) is usually a halogen, although tosyloxy groups have also been used but give a much slower reaction; in the latter cases the process is prone to side reactions, particularly if NaOMe/MeOH is used as base rather than KOBu^t/Bu^{t}OH ⟨80JOC1719⟩. A sulfinate can also be used as the leaving group and *p*-toluenesulfinate ⟨86CL433⟩; see also ⟨80RTC39⟩ as well as trifluoromethanesulfinate (triflinate) (Schemes 63 and 64) have been used ⟨84TL4617, 85JOC2110, 86JA2358⟩. Simple alkylsulfinates require harsh reaction conditions and are of limited use ⟨52JA1225⟩.

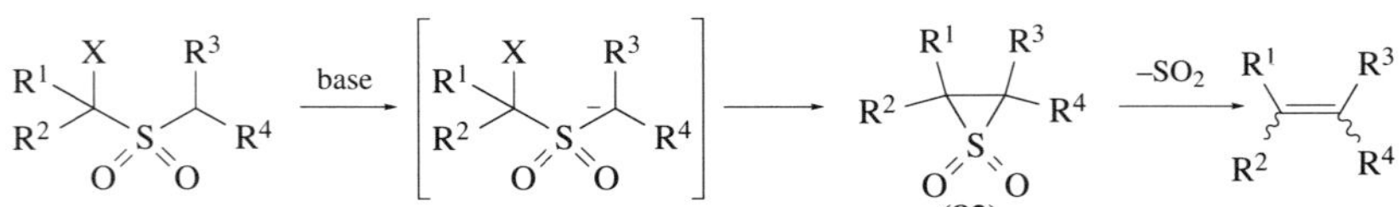

(82)

X = Cl, Br, I, OTs, SO_2CF_3, SO_2Tol-*p*

Scheme 62

(83)

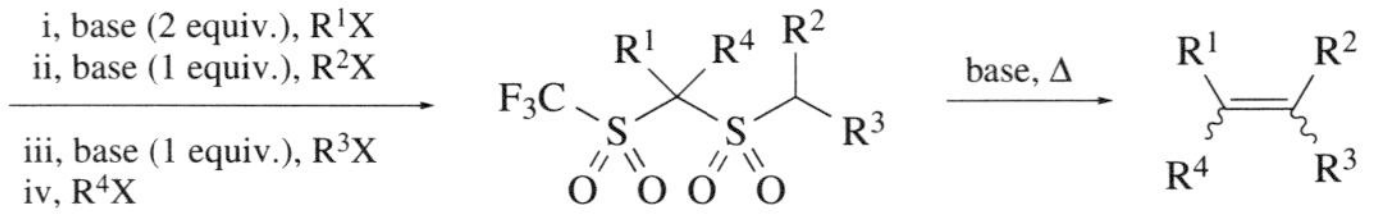

Scheme 63

The α-halosulfones can be prepared by a variety of methods (Scheme 65) ⟨77OR(25)1, 91COS(3)861⟩, most commonly from a sulfide by oxidation followed by halogenation ⟨69JA3870, 88JA7247⟩, or by halogenation ⟨86T3731⟩ and oxidation ⟨77JA2815, 77JA6946, 78JOC4527, 86JOC2397, 87JA2857⟩. Chlorination of bridgehead sulfides and sulfones has been reported (Scheme 66) ⟨69JOC3870⟩. Other miscellaneous preparative procedures are also available (Schemes 71 and 75). Reactions related to the Ramberg-Bäcklund, where the α-halosulfone moiety is replaced by an α-halosulfide ⟨73TL4395⟩, α-halo sulfoxide ⟨84CB966, 86CB1540⟩ or an α-halo *N*-*p*-toluenesulfonyl sulfoximine ⟨78JOC4140⟩, have also been reported, and in the particular case of the sulfoxide shown (Equation (16)), was superior to the sulfone system.

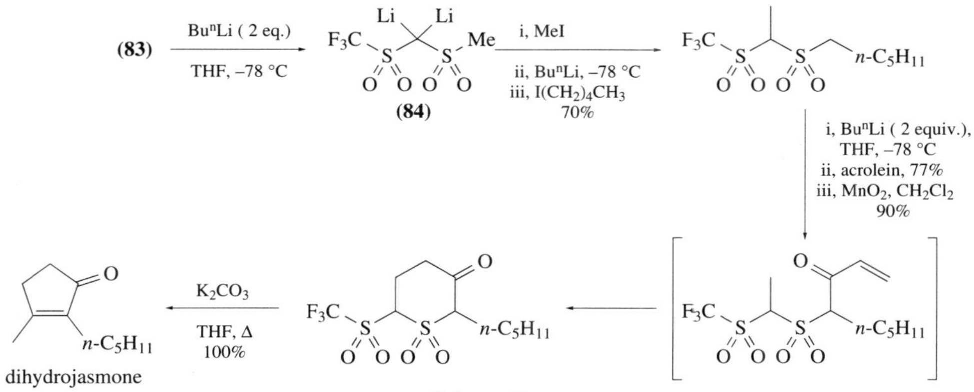

Scheme 64

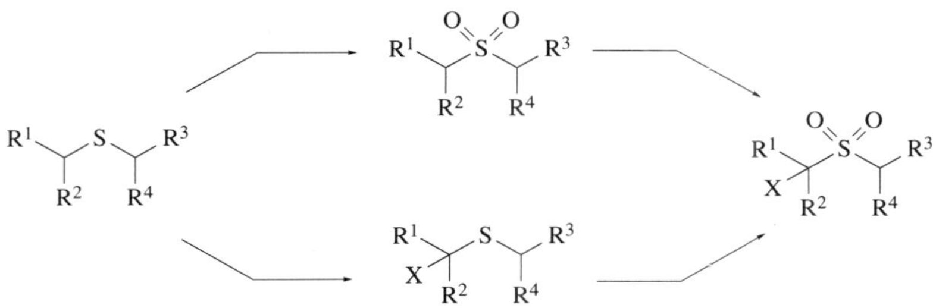

Scheme 65

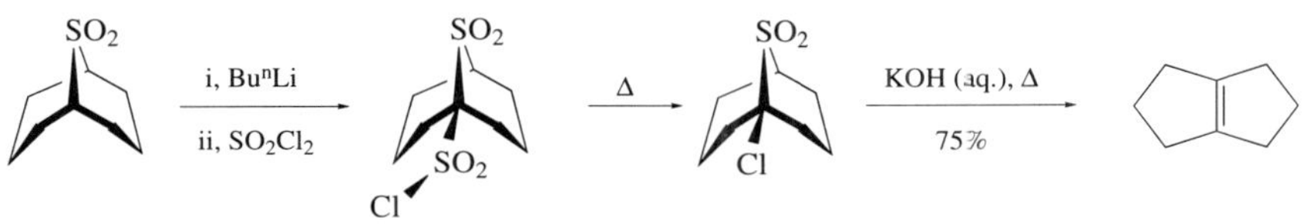

Scheme 66

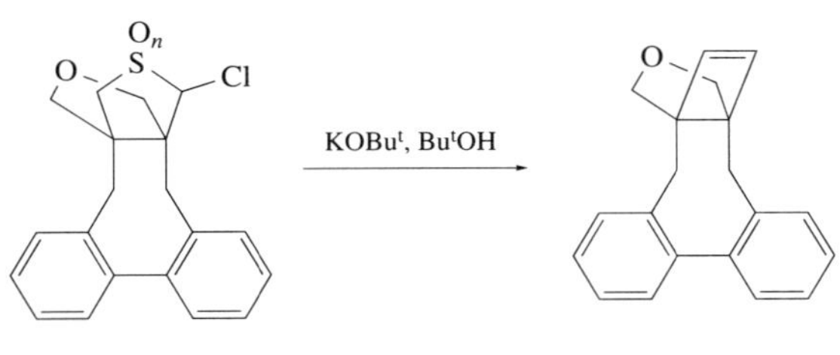

$n = 1$, 25–90%
$n = 2$, <10% (16)

The use of trifluoromethanesulfinate (triflinate) as a leaving group in the Ramberg-Bäcklund reaction has found particular application for alkene synthesis using mesyl triflone (**83**), which acts as an alkene polyanion equivalent (Scheme 67) ⟨84TL4617, 85JOC2110, 86JA2358⟩. The alkene precursors are prepared by sequential alkylation of mesyl triflone, where high regioselectivity can be achieved, but low stereoselectivity (Scheme 67). The reaction succeeds only if the precursor is fully substituted on the carbon atom between the two sulfone groups so as to prevent deprotonation at this highly activated position (note formation of dianion (**84**)), and thus is best suited to the synthesis of 1,1-di-, tri- and tetra-substituted alkenes.

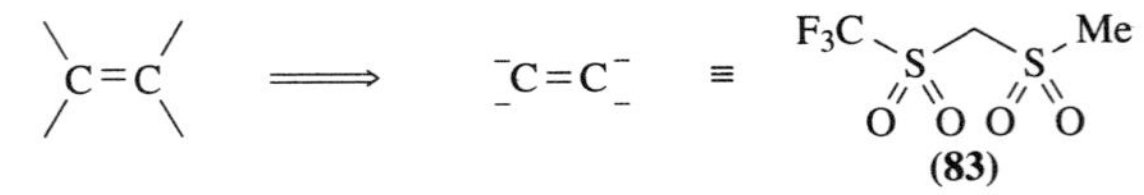

(83)

Scheme 67

One of the most useful improvements on the original Ramberg-Bäcklund reaction was developed by Meyers. Chlorination and alkene formation occur in one pot using KOH, CCl_4, Bu^tOH and H_2O. The reaction is a general procedure for the chlorination of compounds with acidic methylene groups and, for sulfones, cyclisation and extrusion of SO_2 occurs *in situ* (Scheme 68) ⟨69JA7510, 74JA3332, B-77MI 115-01, 81ACS(B)263, 82S504, 84CL833, 87JOC1703⟩; but see ⟨79CJC1958⟩; see also ⟨78JOC1985⟩. In the case of simple acyclic dialkyl sulfones, a significant amount of product resulting from dichlorocarbene addition to the alkene is formed, which can be a serious limitation on the utility of this reaction (Equation (17)) ⟨69JA7510⟩. However, these conditions, in some cases, are successful when the more conventional Ramberg-Bäcklund approach fails. In the specific example given (Equation (18)), problems were encountered with the chlorination step ⟨73CC520⟩. An alternative one-pot procedure allows formation of alkenes directly from sulfones in moderate yield by oxidation of α,α-dianions with $CuCl_2$ (Equation (19)) ⟨74TL2805⟩; see also ⟨88TL909⟩.

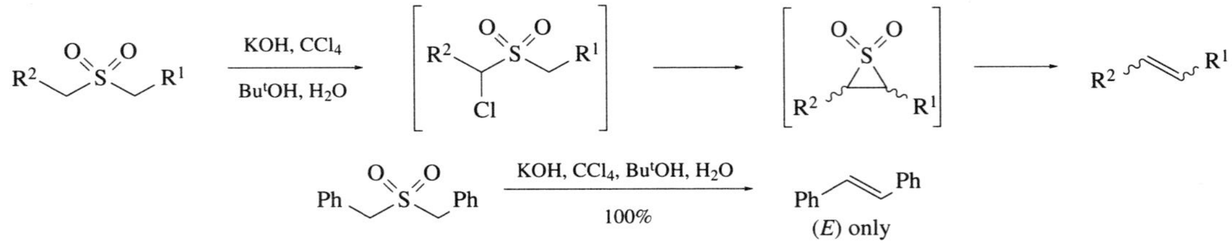

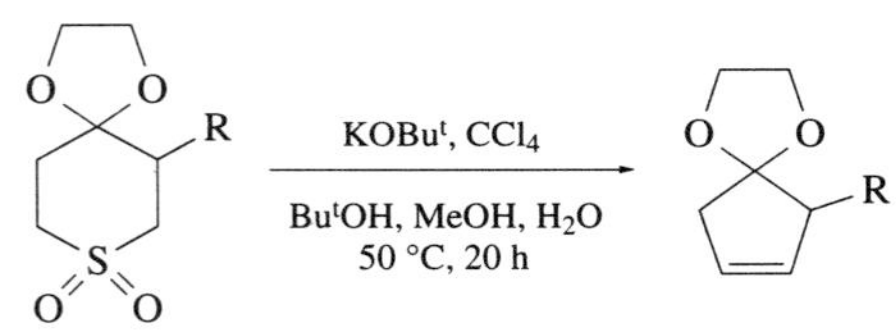

R	*Yield* (%)
Bn	84
n-C_5H_{11}	50
CH_2=$CHCH_2$	73
EtCH=$CHCH_2$	77
Pr^n	60

Scheme 68

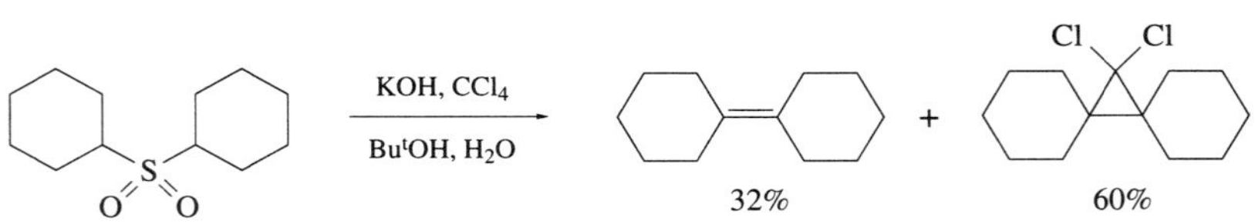 (17)

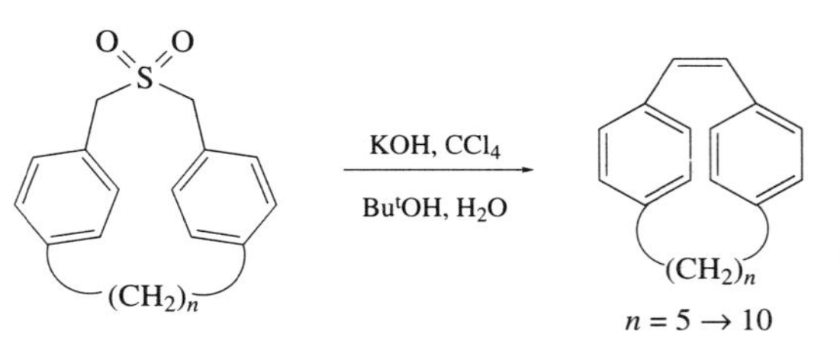 (18)

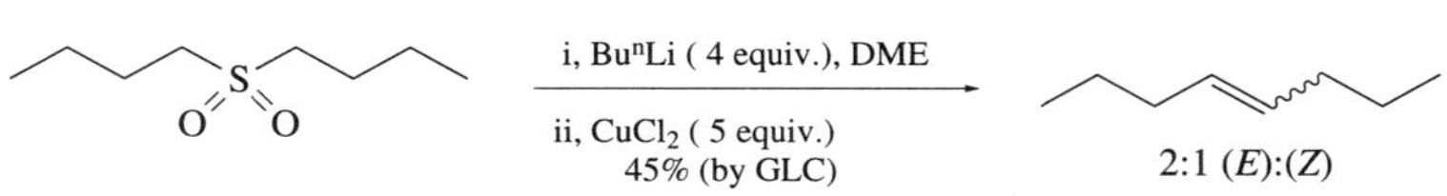 (19)

In addition to Meyers' system, a number of other bases can be used to induce a Ramberg-Bäcklund reaction on an α-halosulfone. The original literature report ⟨40AK(13A)1⟩ used aqueous

KOH; however, Bu^tOK/Bu^tOH is also now widely used. The two bases show interesting stereoselectivity, KOH favouring the (*Z*)-isomer of the alkene, whereas $KOBu^t$ favours the (*E*)-isomer (Scheme 69) ⟨81CB909⟩. Other bases have been used including $LiOBu^t$ or $CsOBu^t$, which give minor changes in stereoselectivity relative to $KOBu^t$ ⟨86JA4568⟩, $KOCEt_3$ ⟨91JA9682⟩, NaOMe/MeOH ⟨70JA2581⟩, MeLi ⟨88JA7247⟩, NaH/KH ⟨86CL433⟩ and $LiAlH_4$ (for 1,2-disubstituted cyclobutenes) (Equation (21)) ⟨74JA5801⟩. For the vinylogous Ramberg-Bäcklund reaction (Scheme 70), $KOBu^t$ can be used, the precursors for such a reaction being easily prepared by regiospecific photochemical addition of $BrCH_2SO_2Br$ to alkenes followed by elimination ⟨75IJC251, 83JA6164, 83JA6165, 84TL5469, 86JA4568, 87OS(65)90, 90JA4072⟩. Alternatively, it is possible to induce a Ramberg-Bäcklund reaction by Michael addition to an α,β-unsaturated sulfone, although stereoselectivity can be low (Scheme 71) ⟨77TL4527, 81T417⟩. The vinylogous, and Michael-induced Ramberg-Bäcklund reactions are both useful for the synthesis of conjugated dienes and polyenes (see also ⟨74JA3332⟩).

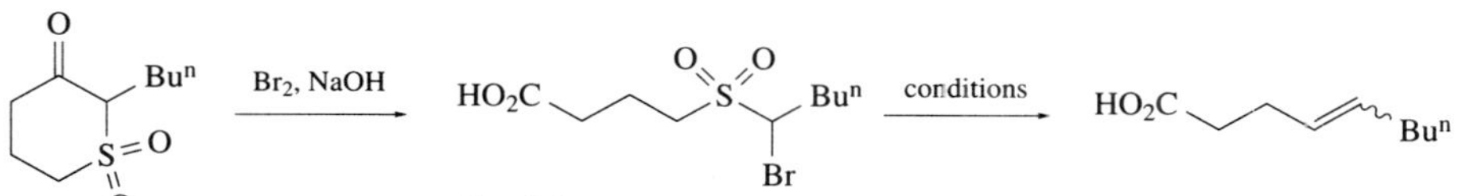

Conditions
KOH, H_2O, Δ, 65%, 82:18 (*Z*):(*E*)
$KOBu^t$, Bu^tOH, Δ, 85%, 24:76 (*Z*):(*E*)

Scheme 69

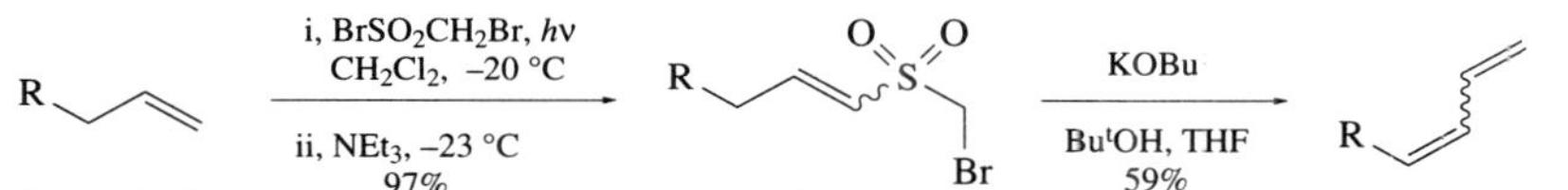

R = *n*-C_5H_{11} | **(85)** 10:1 (*E*):(*Z*) | 83:17 (*Z*):(*E*) from pure (*E*)-**(85)**

Scheme 70

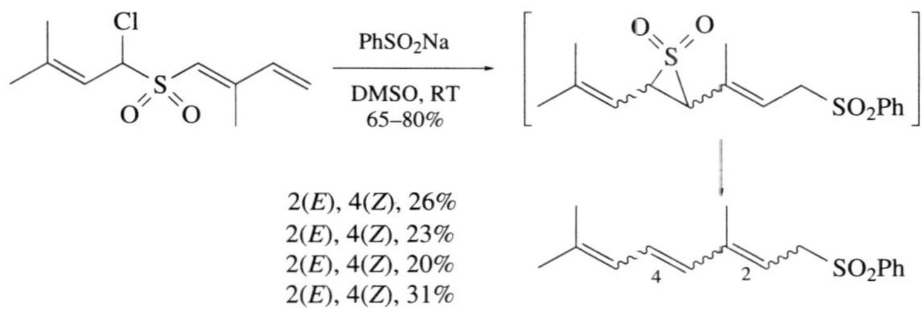

2(*E*), 4(*Z*), 26%
2(*E*), 4(*Z*), 23%
2(*E*), 4(*Z*), 20%
2(*E*), 4(*Z*), 31%

Scheme 71

The use of phase transfer catalysts allows a Ramberg-Bäcklund reaction to proceed under milder conditions, i.e. with more dilute NaOH solutions, or employing Na_2CO_3 and $NaHCO_3$ as base, which allows base sensitive functional groups such as esters to be present in the substrate ⟨82S504⟩. Typically aliquat-336 (tricaprylmethylammonium chloride) or benzyltriethylammonium chloride are used to catalyse the reaction under conditions similar to those developed by Meyers (NaOH, CCl_4, H_2O) ⟨82S504⟩, although in some cases this is less satisfactory (Scheme 72) ⟨81ACS(B)263⟩.

The main disadvantage of the Ramberg-Bäcklund reaction is its lack of stereoselectivity. Although this can be controlled to a certain extent (Scheme 69), the reaction generally produces mixtures of (*E*)- and (*Z*)- isomers whenever possible. It is known that loss of SO_2 from the episulfone occurs stereospecifically, with retention of configuration. It is thus at the cyclisation stage, which is usually the rate determining step, where the geometry of the final product is determined, and lack of stereocontrol is apparent (Scheme 73) ⟨68ACR209, 70JA2581, 76T1705, 77OR(25)1, 85CJC1, but see B-88MI 115-04⟩. For this reason, the reaction is most useful for the synthesis of alkenes where stereoisomerism is not a problem. In many cases the methodology is superior to alternative protocols available for the synthesis of strained or reactive cycloalkenes, particularly if both allylic positions are fully substituted, as this prevents elimination of HX to form a vinylsulfone which in some cases can be an important competing side reaction (Scheme 74) ⟨71JOC1015, 79CJC1958, 80LA1540, 83HCA1090⟩.

The Ramberg-Bäcklund reaction has been used extensively for ring synthesis. In such cases, it is

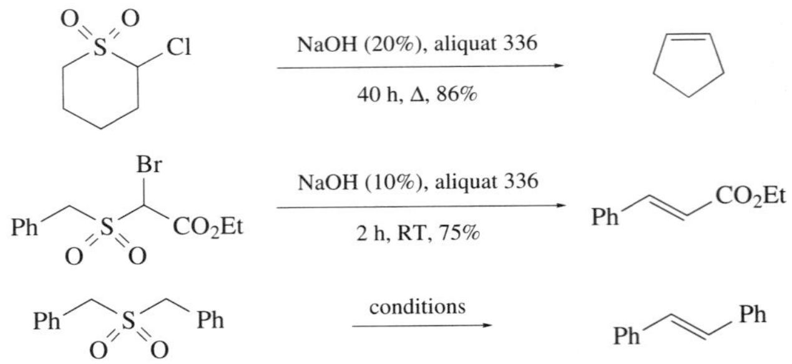

Conditions
NaOH (10%), CCl_4, Δ, no reaction
NaOH (10%), CCl_4, aliquat 336, Δ, 94%
Na_2CO_3 (sat. aq.), CCl_4, aliquat 336, Δ, 70%
KOH, CCl_4, Bu^{t}OH, H_2O, 100%

Scheme 72

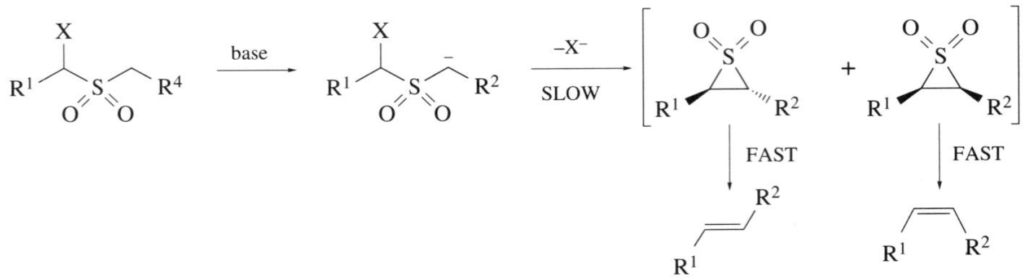

Scheme 73

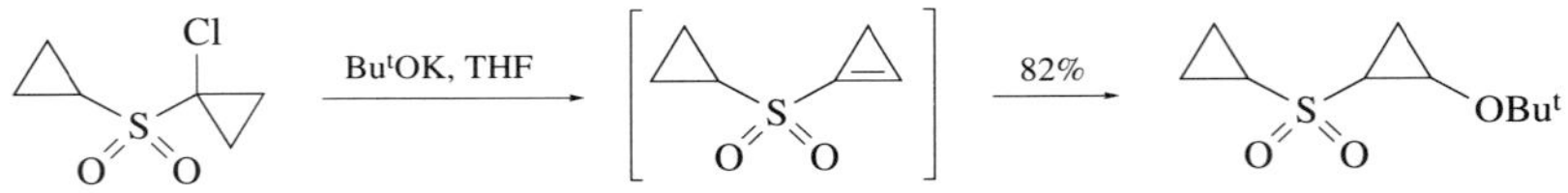

Scheme 74

actually a ring contraction rather than a cyclisation reaction, and stereoselectivity, which can be a serious problem in acyclic systems, is controlled. It has been used for the synthesis of cyclobutenes unsubstituted at positions 1 and 2 (Equation (20)) ⟨74CB1915, 74JA5801, 76CB793, 77CB2978, 77JA2815, 77JA6946, 78JOC4527, 80LA1540, 81AG(E)677, 83HCA1090, 84CB1868, 86TL1895, 87CB5, 88JA8197⟩, and also related 1,2-disubstituted systems (Equation (21)) ⟨74JA4715, 74JA5801, 76JA4936, 77OS(57)53⟩. Cyclopentenes can also be prepared (Scheme 66), including tricyclo[5.3.0.0^{2,8}]deca-3,5,9-triene (**86**), which is unstable and prone to isomerisation (Equation (22)) ⟨87JA2857; see also 75TL947, 80H(14)1739, 84HCA434, 87LA679, 88JA5479⟩. Cyclopentenones and related systems are also available (Schemes 64, 68 and 75) ⟨77TL1173, 84CL833, 85JOC2110, 86CL433, 87JOC1703, 88CC454⟩.

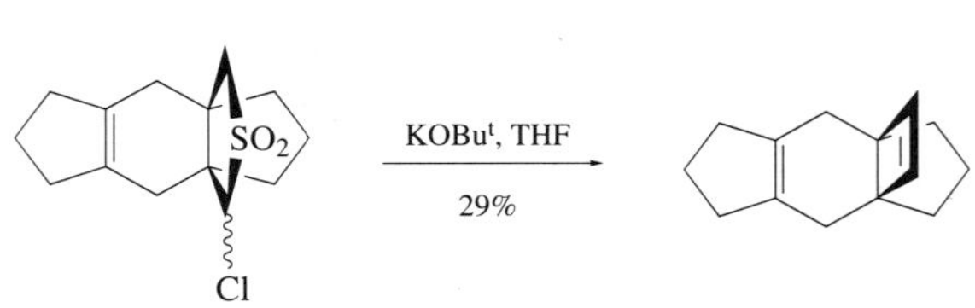

(20)

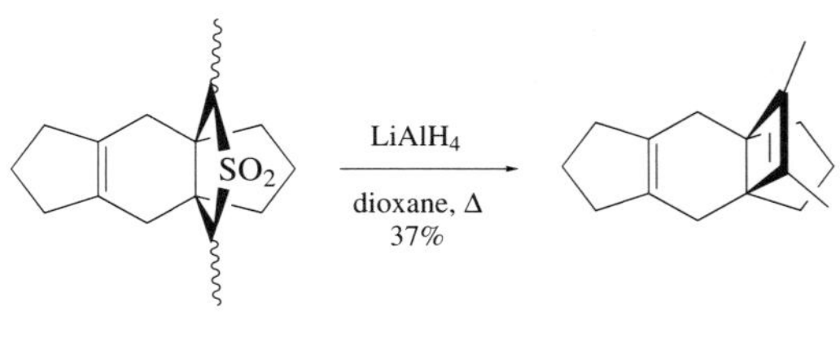

(21)

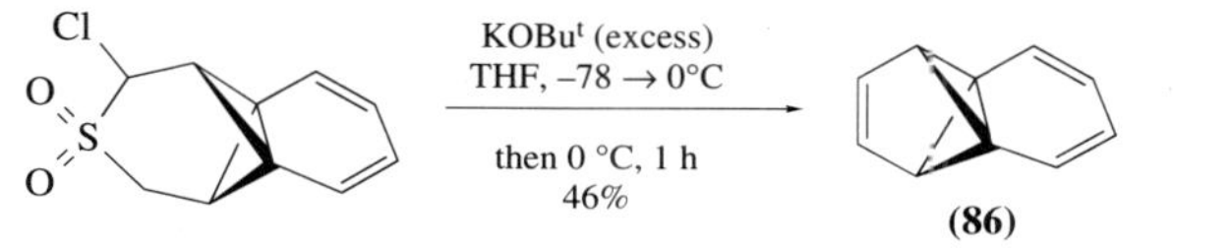

(86)

(22)

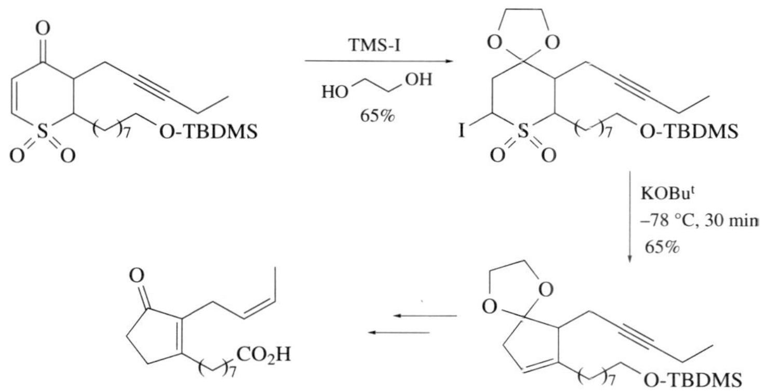

tetrahydrodicranenone B

Scheme 75

The unstable cyclohexene (**87**) (Equation (23)) ⟨83JA667, 86JOC2397, 88TL4803⟩, other cyclohexenes ⟨86JA2358⟩ and phenanthrenes ⟨64JA4085⟩ can also be prepared using this reaction. For the synthesis of larger rings (Equation (18)), the Ramberg-Bäcklund reaction suffers from low yields and has considerable drawbacks over alternative procedures ⟨73CC520, 76TL1243, 78TL4269, 81JOC1747⟩. However, the reaction has recently found application for the synthesis of analogues of the DNA cleaving enediynes related to calicheamicins (Scheme 76) ⟨88JA4866, 88JA7247⟩ and has also been used in synthesis of the nine-membered ring macrocyclic alkene (+)-eremantholide (Equation (24)) ⟨91JA9682⟩.

(87)

(23)

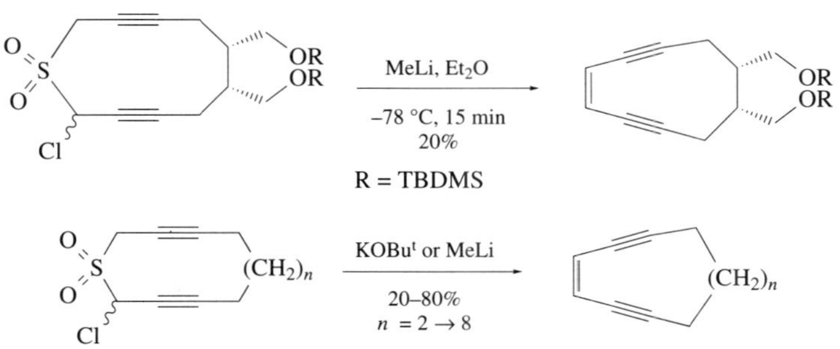

Scheme 76

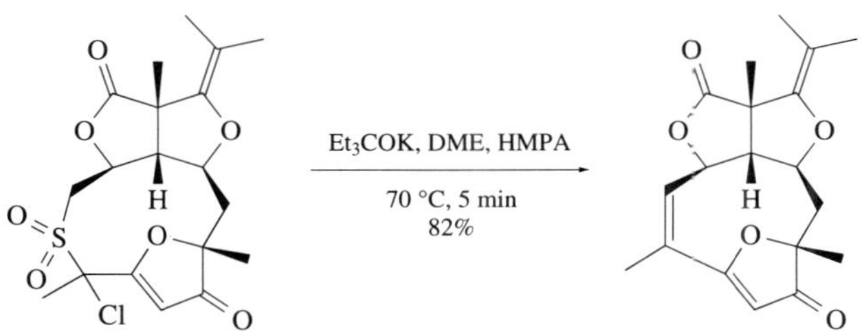

(+)-eremantholide A

(24)

1.15.4.3.3 Alkylation of sulfoximine stabilised carbanions followed by elimination – the Johnson (N-methylphenylsulfonimidoyl)methyl lithium method

The anions of alkylaryl sulfoximines (**88**) readily add to aldehydes and ketones to form *β*-hydroxysulfoximines (Scheme 77). These compounds can readily undergo reductive elimination using aluminum amalgam in aqueous THF and acetic acid to give alkenes (Scheme 77) ⟨73JA6462, 79JA3602; see also 73JA7424⟩. The reaction can be carried out without isolation of the *β*-hydroxysulfoximine intermediate ⟨79JA3602⟩. The reaction is successful for the synthesis of mono-, di-, and trisubstituted alkenes, but fails in the initial condensation reaction which would lead to tetrasubstituted systems owing to problems with enolisation. Aryl aldehydes are also poor substrates as the thiophenol biproduct from the reductive elimination reaction can add to the styrene product resulting in a significant reduction of yield. Another serious limitation is that, for the synthesis of polysubstituted alkenes using alkyl sulfoximines (**89**) and (**90**), mixtures of *cis*- and *trans*-isomers are produced in the reaction irrespective of the geometry of the *β*-hydroxysulfoximine ⟨79JA3602⟩.

(**88**) R = H
(**89**) R = Me
(**90**) R = $(CH_2)_2Me$
(**91**) (**92**)

Reaction of sulfoximines with carbonyl compounds and subsequent elimination

R¹R²CO	*Sulfoximine*	(**91**) *Yield (%)*	(**92**) *Yield (%) (E):(Z)*
$Me(CH_2)_{14}CO$ Me	(**88**)	90	90 (85[a])
$Me(CH_2)_6CHO$	(**88**)	85	60
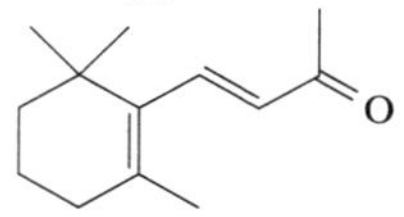	(**88**)	85	93
$Me(CH_2)_4CHO$	(**89**)	75	100 (60:40)
$Me(CH_2)_2CHO$	(**90**)	78	78 (64:36)
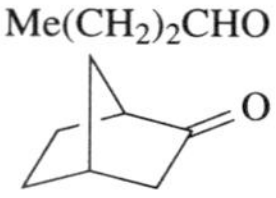	(**89**)	75	65 (78:22[b])

[a] Yield without isolation of intermediate β-hydroxysulfoximine. [b] Stereochemistry not determined.

Scheme 77

The initial sulfoximine anions can be considered as being less basic but more nucleophilic than the corresponding Wittig reagent ⟨79JOC2880⟩. As a result, the reaction using sulfoximine anions can be successful in cases where the Wittig reaction has failed, two such cases being during the total synthesis of gascardic acid (Equation (25)) and 6a-carbaprostaglandin I_2 (Equation (26)) ⟨79JA5060, 79JOC2880⟩. Sulfoximines can also show useful chemoselectivity, the product (**93**) being formed by selective attack at the least hindered ketone (Equation (27)) ⟨84JA4547⟩.

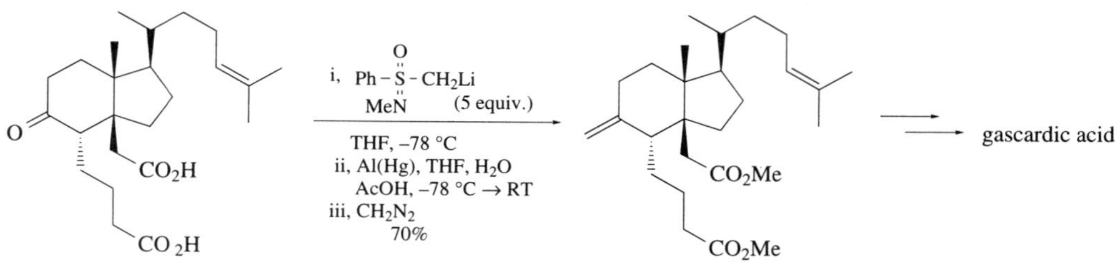

(25)

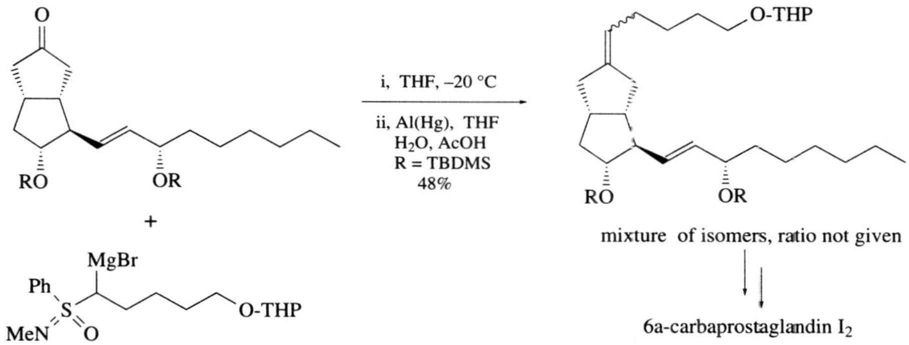

(26)

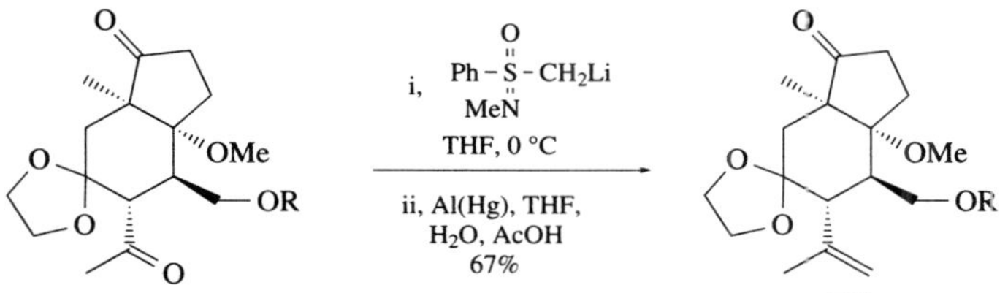

(93)

(27)

Sulfoximines are chiral, and can be readily resolved ⟨85PS(24)531⟩. If used in an enantiomerically pure form, they provide an opportunity for resolution by separation of diastereomeric β-hydroxysulfoximines prior to reductive elimination. This approach has been used for the synthesis of (–)-β-Panasinsene (**94**) and its enantiomer (Scheme 78). The resolution procedure is greatly simplified in this case, as addition of the sulfoximine anion to the ketone shows complete facial selectivity resulting in formation of only two diastereomeric β-hydroxysulfoximines, which were separated by chromatography ⟨81JA7667⟩.

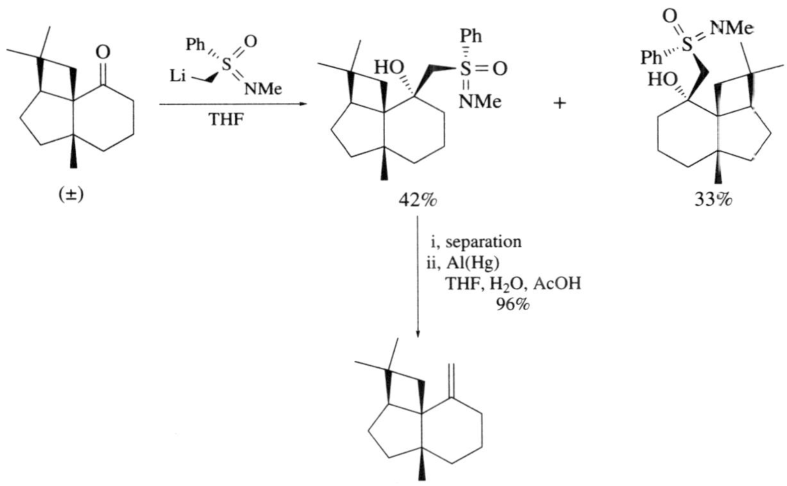

(–)-β-panasinsene **(94)**

Scheme 78

A related process involves the use of resolved sulfoximines adding to symmetrical ketones (Scheme 79 and Equation (28)). Again, addition occurs selectively to one face of the ketone to give a β-hydroxy sulfoximine, which can undergo asymmetric elimination via its trimethylsilyl ether, to give the axially chiral alkenyl sulfoximine (**95**) with high diastereoselectivity ⟨86AG(E)935⟩. The sulfoximine group can be substituted with full retention of stereochemistry using a nickel-catalysed cross-coupling reaction with an organozinc reagent, in the presence of a metal salt ($MgBr_2$, LiCl), allowing access to trisubstituted alkenes ⟨89JA1125⟩.

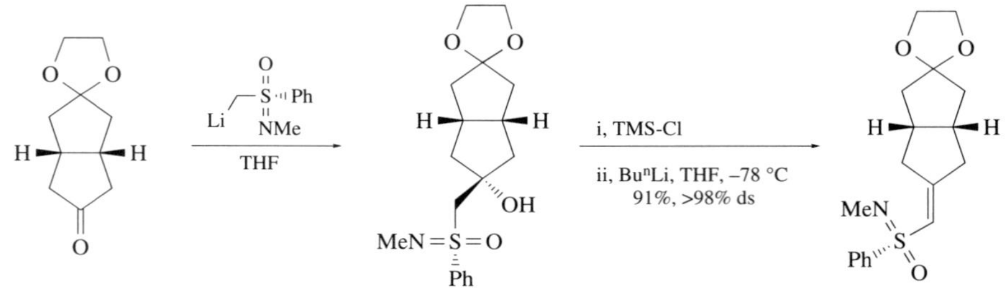

(95)

Scheme 79

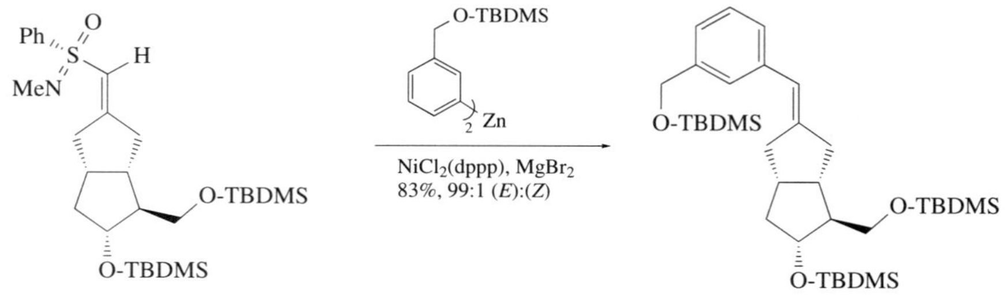

(28)

1.15.4.4 Alkylation of Selenium(II) and Selenium(IV) Stabilised Carbanions Followed By Elimination

There are a number of important ways in which alkylation of selenium stabilised anions can lead to alkene formation at the new C—C bond. First, alkylation of selenide α-anions using alkyl halides leads to the formation of substituted selenides, which, after oxidation to the corresponding selenoxides, undergo the well-known *syn*-elimination reaction to give, predominantly, the less substituted (E)-alkene (Scheme 80). A related process is also possible *via* a selenonium ylide (Scheme 81). Alternatively, selenoxide α-anions can be alkylated and eliminated directly (Scheme 82). The addition of selenide α-anions to carbonyl compounds provides an efficient method for the preparation of β-hydroxyselenides. These then undergo a reductive elimination reaction, induced by a variety of reagents (*vide infra*), with overall loss of RSeOH to give an alkene (Scheme 83). In this case, the geometry of the new double bond is controlled by the β-hydroxyselenide stereochemistry. An excellent extensive review in this area has been published ⟨91COS(1)629⟩ and other reviews are also available ⟨78T1049, 80T2531, B-86MI 115-02, B-87MI 115-01, B-87MI 115-02, 87TCC1, 89H(28)1203⟩.

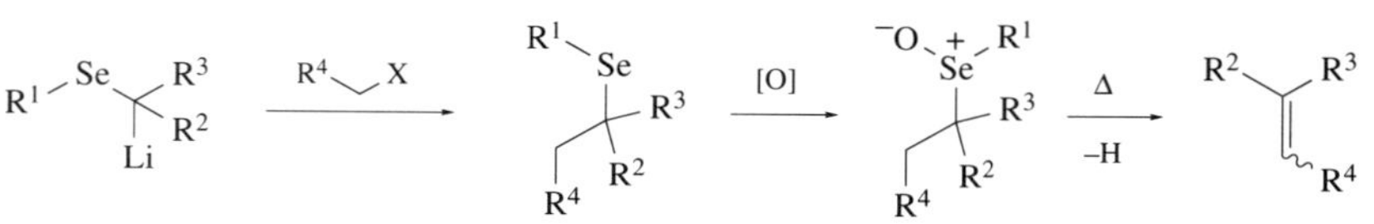

Scheme 80

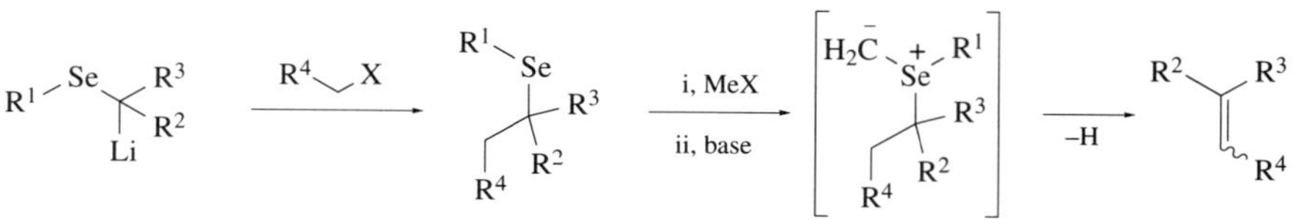

Scheme 81

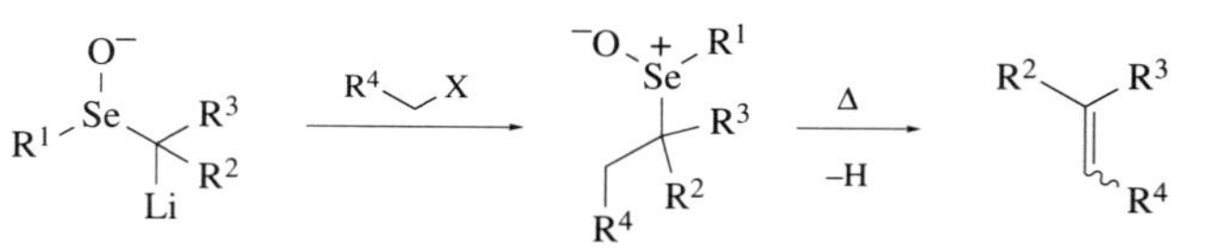

Scheme 82

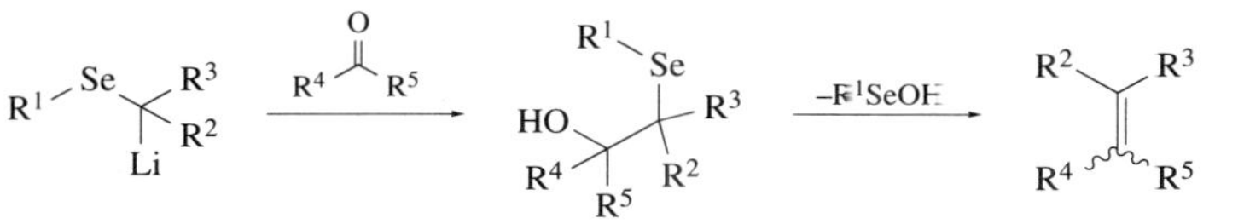

Scheme 83

As all these processes begin with the alkylation of a selenium stabilised anion, the generation of such species will be briefly mentioned here; more detail can be found in the aforementioned reviews.

The most important methods for generating selenium-stabilised anions are summarized in Scheme 84 ⟨78T1049, 79TL4233, 80T2531, 84IJ125, B-87MI 115-01, B-87MI 115-02, 89H(28)1203, 91COS(1)629⟩. These involve the nucleophile induced (usually Bu^nLi) cleavage of 1,1-selenoacetals ⟨76TL3743, 79CC1136, 80TL1997, 82CC564, 87CC1540, 89T2005⟩, deprotonation of selenides (although C—Se bond cleavage can be a serious side reaction) ⟨74CC990, 75CC790, 79JA6638, 84JOC1096, 89T2023⟩, Michael addition to vinylselenides (rarely used) ⟨78AG(E)526⟩, and oxidation of selenides to selenoxides and subsequent deprotonation (which is often carried out in one pot) ⟨75CC790, 75JA3250, 79JA6648, 83JOC2098⟩.

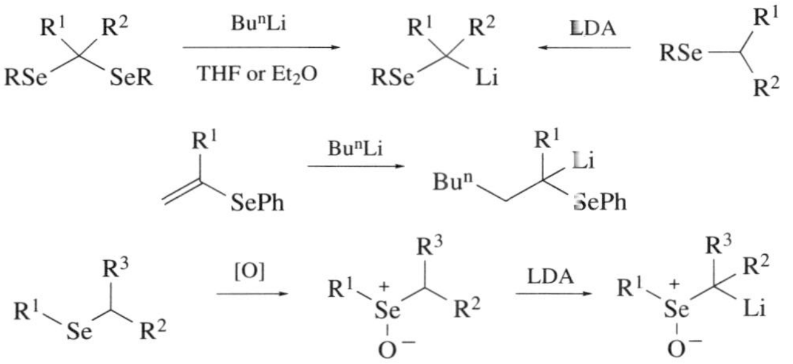

Scheme 84

In general, anions stabilised by an $SeCH_3$ group are more reactive than the corresponding SePh analogue ⟨89T2023⟩. They react well with aldehydes and ketones, including hindered ketones ⟨82CC564, 86CB1162⟩, to form *β*-hydroxyselenides (Scheme 85) ⟨75CC790, 76TL1385, 76TL3743, 79CC1136, 79JA6638, 89T2005, 89T2023, see also 77LA846, 78TL1145, 80TL1997, 81TL18[illegible]3, 82TL4385, 84TL2329, 87CC1540⟩, and will also add in a Michael-type reaction to *α*,*β*-unsaturated lactones (Scheme 86) ⟨87CC106⟩.

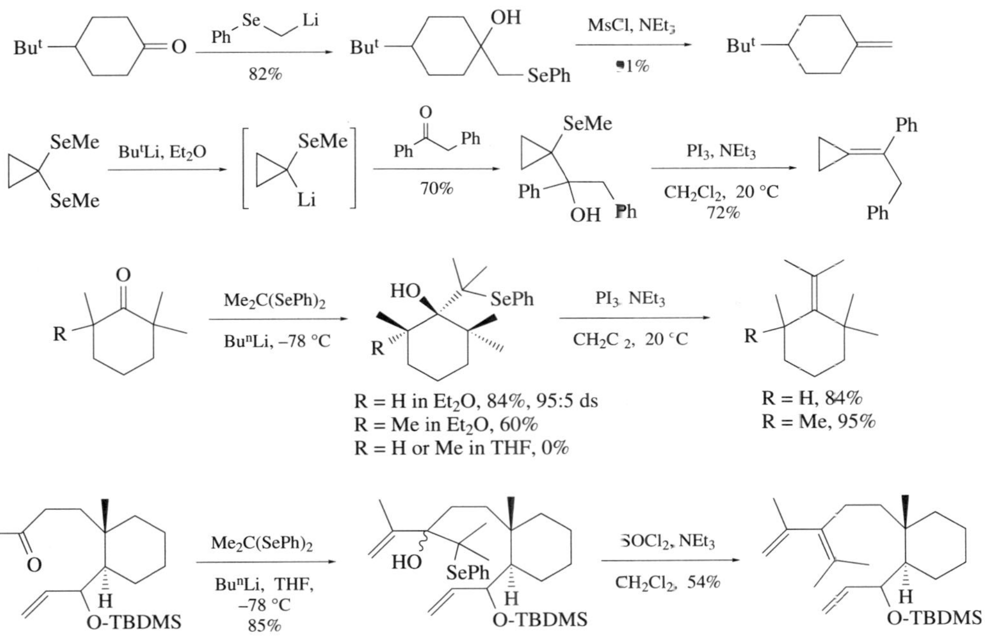

Scheme 85

For allylic anion systems, complementary regio- and enhanced stereocontrol can be obtained by transmetallation of the lithium species with triethylaluminum to form an allylic aluminum 'ate' complex, which adds to aldehydes selectively at the α-position, whereas the original organolithium reagent alkylates predominantly in the γ-position (Equation (29)) ⟨84JOC1096⟩.

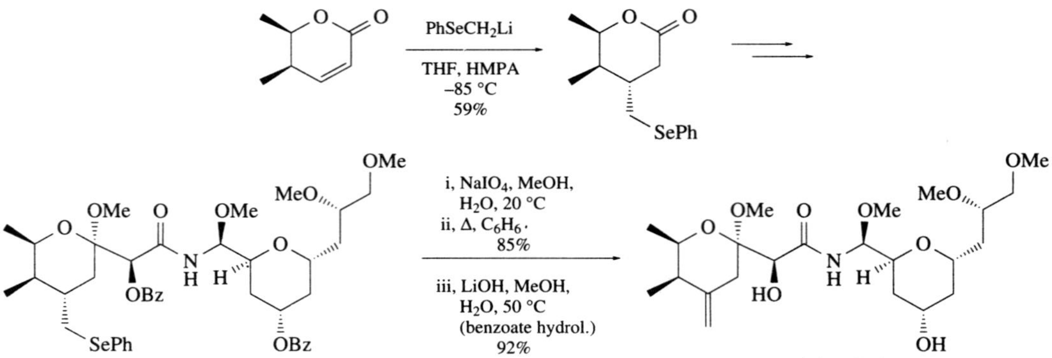

(±)-pederin

Scheme 86

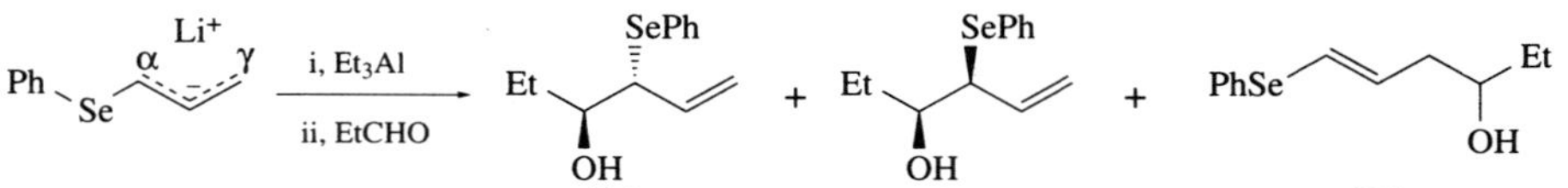

(96) **(97)** **(98)** (29)

with $AlEt_3$, 90%, ratio **(96)**:**(97)**:**(98)**, 81:19:0

without $AlEt_3$, 91%, ratio **(96)**+**(97)**:**(98)**, 12:88

Alkylation with alkyl halides is much less general (Scheme 87), with only primary alkyl bromides and iodides reacting efficiently (Scheme 88) ⟨74CC990, 79JA6638⟩ with best results being obtained using THF with HMPA as co-solvent ⟨B-87MI 115-02, 87TCC(135)1, 91COS(1)629⟩. Allylation can be achieved by using $CuBr \cdot Me_2S$ as a catalyst (Scheme 89) ⟨84IJ125, 87TCC1⟩, although this is not always necessary (Scheme 90) ⟨74CC990, 75JA3250, 79JA6648⟩.

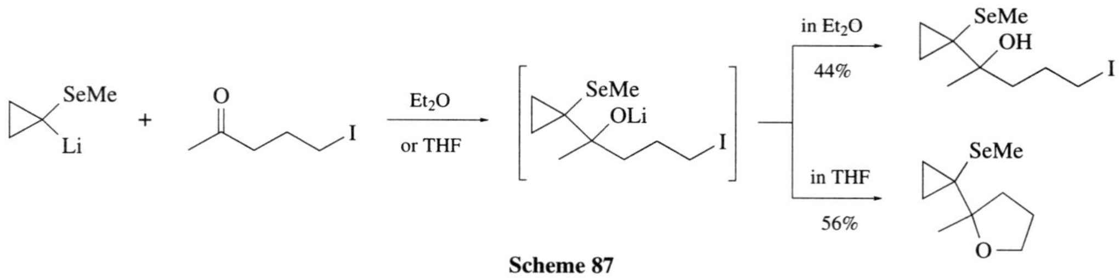

Scheme 87

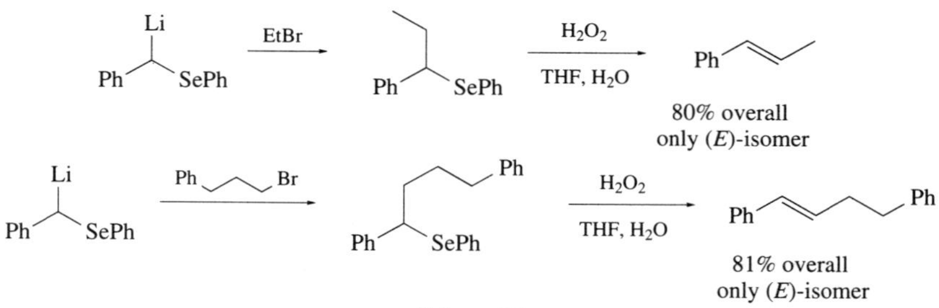

Scheme 88

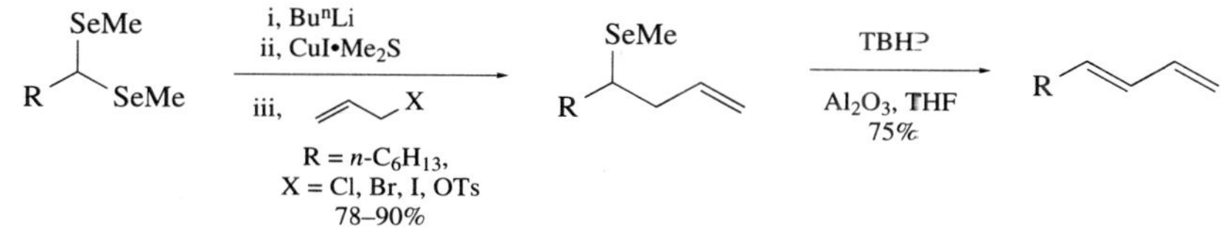

Scheme 89

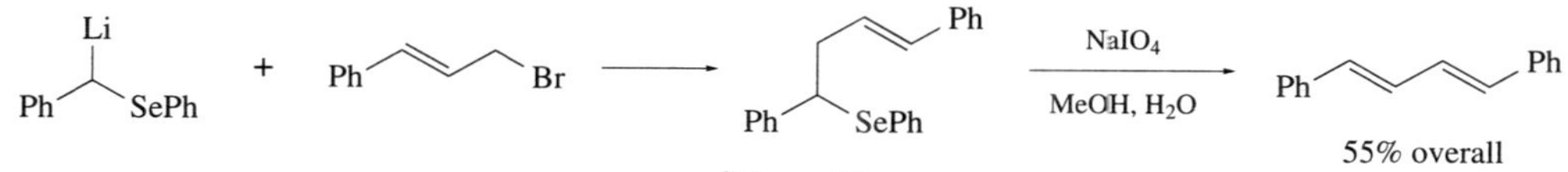

Scheme 90

Selenoxides are much easier to metallate than selenides, and are often generated by oxidation of a selenide (*m*-chloroperoxybenzoic acid) and deprotonated *in situ* using an excess of base to neutralise the *m*-chloroperoxybenzoic acid by-product. In such cases, LDA in THF at low temperatures (−78 °C) is successful ⟨75CC790, 75JA3250, 79JA6648, 83JOC2098, B-87MI 115-01, B-87MI 115-02⟩ whereas use of Bu^nLi results in C—Se bond cleavage ⟨B-87MI 115-02⟩. Alkylation can be achieved with reactive alkyl halides (Scheme 91) ⟨75JA3250, 79JA6648⟩ and carbonyl compounds (Scheme 92) (Equation (30)) ⟨75CC790, 83JOC2098⟩.

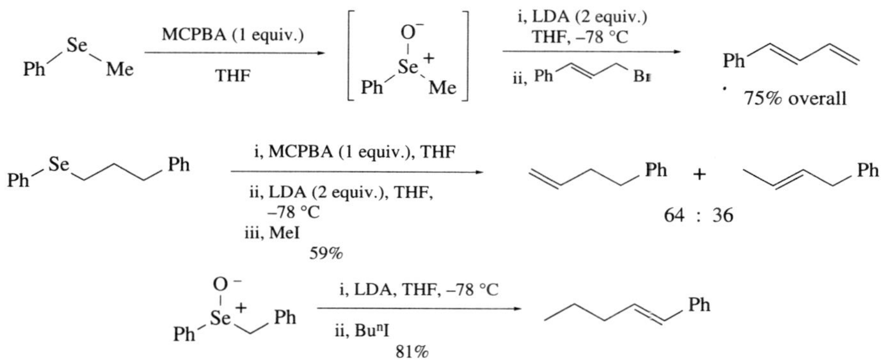

Scheme 91

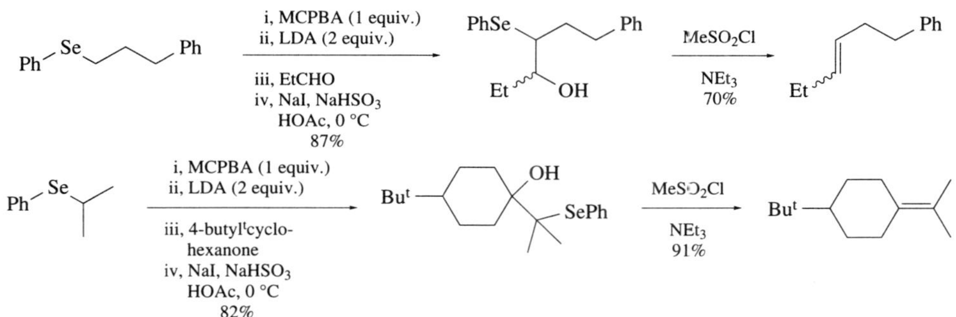

Scheme 92

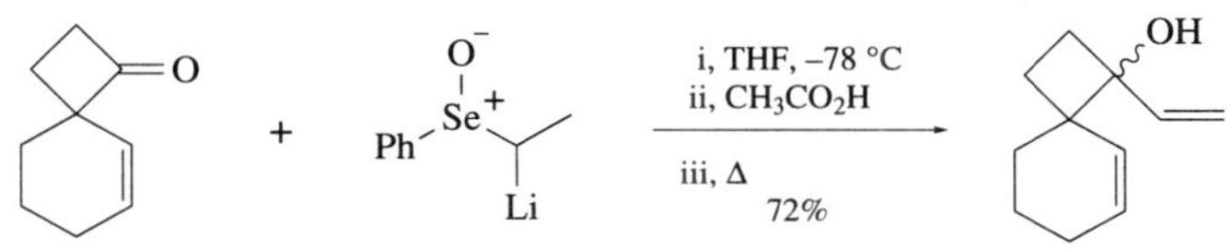 (30)

After alkylation, there are two routes which can be used for alkene formation. The first involves loss of RSeOH from a β-hydroxyselenide (Schemes 85 and 92) ⟨78T1049, 91COS(1)629⟩. This can be induced using a number of reagents including $SOCl_2/NEt_3$ ⟨76TL3227, 76TL3743, 82CC564, 86JOC3108, 87CC1540⟩, $MsCl/NEt_3$ ⟨75CC790, 79JA6638⟩, TsOH ⟨76TL1385, 76TL3227, 76TL3743, 84JOC1096⟩; but see ⟨79CC1136, 90BSF681⟩, PI_3/NEt_3 ⟨79CC1136, 80TL1997, 82CC564⟩, P_2I_4/NEt_3 ⟨81TL1833⟩, $HClO_4$ ⟨76TL1385, 76TL3743⟩, carbonyl diimidazole ⟨79CC1136, 80TL1997⟩, $POCl_3$ ⟨78TL2693, 82CC564⟩, $(CF_3CO)_2O/NEt_3$ ⟨76TL1385, 76TL3743⟩, NaH and phosphochloridite **(99)** ⟨76TL3743⟩, and TMS-Cl/NaI ⟨81JOC231⟩. β-Alkoxyselenides undergo a similar elimination using either TMS-Cl/NaI ⟨81JOC231⟩ or Na/NH_3 ⟨79CC83⟩. Note that the geometry of the alkene produced is dependent on the stereochemistry of the β-hydroxy- (or alkoxy)selenide precursor, a mechanism consistent with loss of an RSe^+ equivalent from a seleniiranium salt intermediate (Scheme 93) ⟨76TL1385, 76TL3743, 81JOC231⟩.

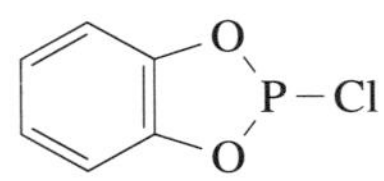

(99)

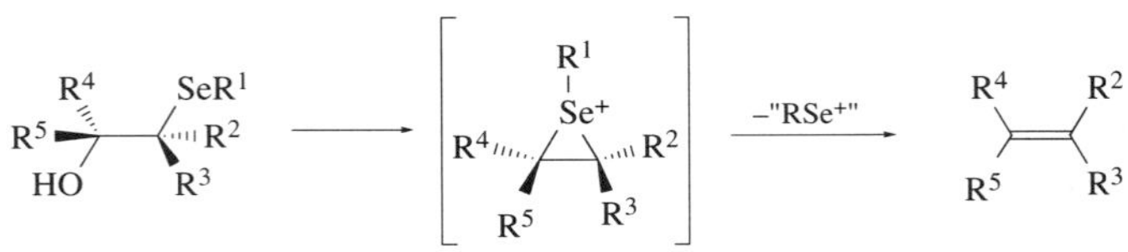

Scheme 93

β-Hydroxyselenides can also be prepared by addition of α-lithioselenoxides to carbonyl compounds and subsequent reduction back to the selenide using $NaI/NaHSO_3$ or $P(OMe)_3$ ⟨75CC790, 79JA6648⟩. Elimination (as above) then provides an alkene (Scheme 92). Alternatively, direct elimination of the β-hydroxyselenoxide intermediate generally occurs in the direction away from the alcohol group to give an allylic alcohol (Equation 30) ⟨77LA846, 78TL1145, 79JA6648, 80TL1997, 83JOC2098, 87JOC774⟩.

Substituted selenides, prepared by alkylation of α-seleno anions with alkyl and allyl halides, can be readily converted into selenoxides, and subsequently alkenes by *in situ* elimination by oxidation using many reagents (Schemes 88, 89 and 90), ⟨B-87MI 115-02, 91COS(1)629⟩ including $TBHP/Al_2O_3$ which is reported to be particularly good for primary alkylphenylselenides and dialkylselenides ⟨78TL1141, 84IJ125⟩, ozone ⟨79TL4233⟩, H_2O_2 ⟨74CC990, 85CC1037⟩, or $NaIO_4$ ⟨85CC1037, 87CC106⟩. In some cases yields can be low, and an alternative procedure involving elimination via a selenonium ylide can be used (Scheme 94). The required ylides are readily formed by alkylation of a dialkylselenide with either $MeSO_2F$ or $MeI/AgBF_4$ and treatment with $KOBu^t$ ⟨79TL4233, 84IJ125⟩.

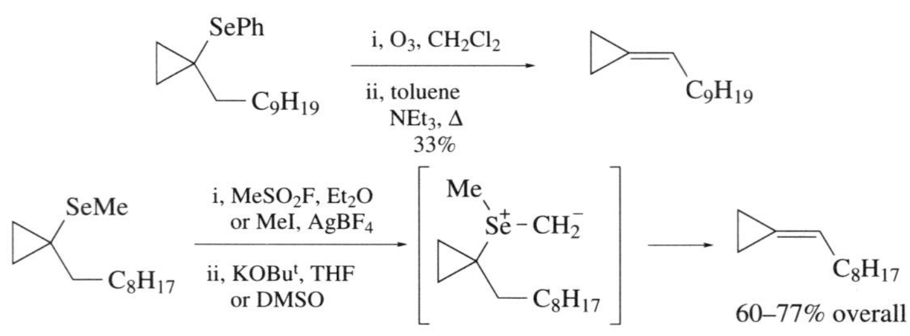

Scheme 94

1.15.4.5 Alkylation of Tellurium Stabilised Anions Followed By Elimination

Although elimination of telluroxides, analogous to that of selenoxides (see Section 1.15.4.4), is known to produce alkenes ⟨75CS9, 83JA2748, 85CC1037, 86S1, 87MI 115-01, B-87 MI115-02⟩, it is generally a much less attractive reaction for a number of reasons ⟨83TL4347⟩, including difficulty in controlling the tellurium oxidation state (overoxidation to tellurones), formation of stable hydrates, reduction of telluroxide by RTeOH by-product to give the unreactive telluride, and formation of alcohols via tellurinic esters (RTe(O)OR). In addition, relatively little is known about the reactivity of α-telluroorganometallics toward electrophiles ⟨B-87MI 115-02⟩. Similarly, unlike the analogous selenium reaction, β-hydroxytellurides do not undergo elimination of RTeOH to form alkenes in a high enough yield to be synthetically useful ⟨78TL4399⟩.

1.15.5 BY CONDENSATION OF NITROGEN FUNCTIONS

1.15.5.1 Rhodium Catalysed Dimerisation of Diazo Compounds

There are few examples of simple alkene synthesis by condensation of nitrogen functions; however, an interesting reaction involving the rhodium catalysed dimerisation of primary aryl diazoalkanes results in formation of diarylethylenes, predominantly as the (*Z*)-isomers (Scheme 95) ⟨82TL2277⟩. The (*Z*):(*E*) ratio varies according to the catalyst used, higher (*Z*) selectivity being observed using iodorhodium(III)tetraphenyl porphyrin rather than $Rh_2(OAc)_4$. The reaction is unsuccessful with secondary aryl diazoalkanes.

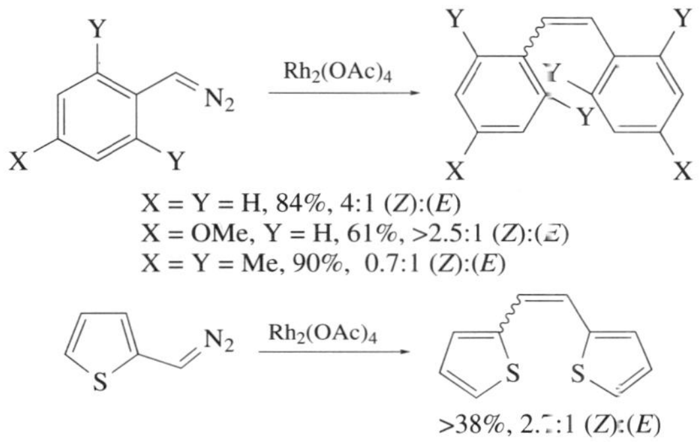

Scheme 95

1.16
One or More C=C Bond(s) Formed by Condensation: Condensation of P, As, Sb, Bi, Si or Metal Functions

IAN GOSNEY
University of Edinburgh, UK

and

DOUGLAS LLOYD
University of St. Andrews, UK

1.16.1 C=C BONDS BY CONDENSATION OF PHOSPHORUS, ARSENIC, ANTIMONY OR BISMUTH FUNCTIONS

1.16.1.1 Alkenation via the Wittig Reaction

The Wittig reaction is one of the most important processes in organic synthesis. It consists of the formation of carbon–carbon double bonds by transfer of an alkylidene group, in the form of a carbanion attached to a heteroatom that carries a high degree of positive charge, to a carbonyl compound with displacement of the carbonyl oxygen (Equation (1)).

(1)

The heteronium group X attached to the alkylidene group usually contains an atom belonging to Group 15, commonly phosphorus and arsenic, and to a lesser used extent, antimony and bismuth. Such compounds are defined as ylides which may be represented as hybrids of the dipolar (**1**) or covalent (**2**) structures. For convenience, in this chapter ylides will commonly be represented by covalent structures such as (**2**) but the dipolar contribution to such structures must be taken as understood.

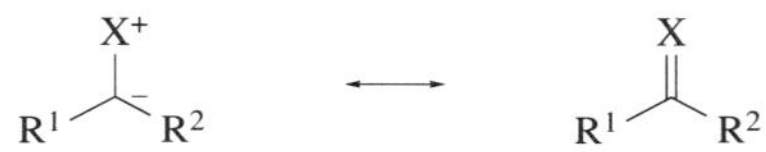
(1) **(2)**

If electron-withdrawing centres are conjugated with the ylidic carbon atom, further dipolar structures such as (**3**) and (**4**) may make major contributions to the overall structure. Delocalisation of the negative charge onto the alkylidene grouping in this way frequently leads to the ylides being isolable. Such ylides are commonly described as *stable* ylides; in this context stable is, in effect, usually synonymous with isolable. Many ylides are not, however, isolable, because of their high reactivity, in particular they are very prone to hydrolysis. In this chapter such ylides will be called *reactive* ylides. Some other ylides notably benzylides (**1**; $R^1 = Ph$, $R^2 = H$) have a reactivity intermediate between those ylides which are obviously *stable* or obviously *reactive*; these will be termed *moderately reactive* ylides. This section is organised according to the type of heteronium group X used in the Wittig reagent. Section 1.16.1.2 covers phosphorus ylides and is further subdivided depending on the nature of the alkylidene grouping and its stabilising action. The competing benefits and disadvantages of the use of arsenic, antimony and bismuth ylides for carbon–carbon double bond formation are discussed in Sections 1.16.1.3, 1.16.1.4, and 1.16.1.5, respectively.

(3) **(4)**

In general, sulfonium, selenonium and telluronium ylides either react with carbonyl compounds to form epoxides, or, in the case of the stable ylides, undergo no reaction at all, but there has been a report in 1993, of the formation of an alkene from a carbonyl compound which may involve a telluronium ylide as an intermediate (Section 1.16.2.4).

1.16.1.2 Phosphonium Ylides

While the Wittig reaction is used widely to construct one or more C=C bond(s) in a specific position, the stereochemical course of the reaction is sometimes less predictable and mixtures of (*Z*)- and (*E*)-alkenes are often produced if the phosphonium ylide and carbonyl compound are both unsymmetrically substituted. This lack of stereoselectivity was first noted by Wittig and Schöllkopf in the reaction of the allylic ylide (**5**) with benzaldehyde to give a 1 : 1 mixture of (*Z*)-(**6**) and (*E*)-phenylbutadiene (**7**) (Scheme 1) ⟨54CB1318⟩.

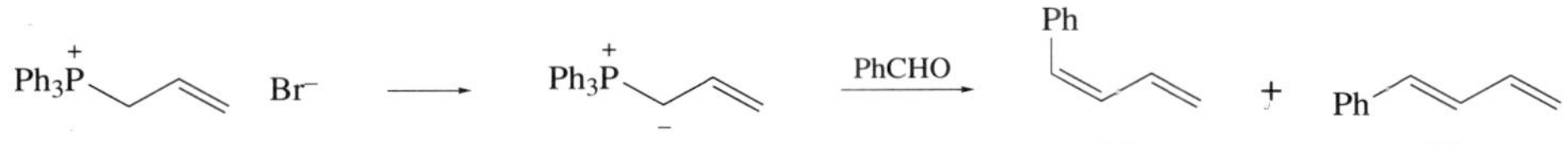

(**5**) (**6**) (**7**)

Scheme 1

Subsequent investigations revealed wide variation in the stereoselectivity of the reaction depending on the base used ⟨54CB1318, 55CB1654⟩, and the reaction temperature ⟨57LA(603)115⟩. In many cases only the (*E*)-alkene is formed, a fact that is frequently put to synthetic use. For example, Trippett ⟨56CI(L)80⟩ was able to prepare the first pure sample of all-*trans*-squalene from the condensation of two molecules of geranylacetone (**8**) with the bis ylide derived from 1,4-bis(triphenylphosphonio)butane dibromide (**9**) (Scheme 2).

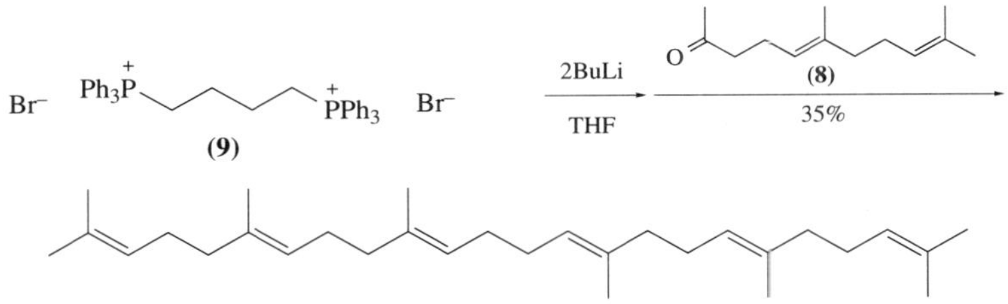

Scheme 2

Bergelson and Shemyakin were the first to explore the *cis*-selectivity in detail and to demonstrate its synthetic utility ⟨64PAC271⟩. Their pioneering work on the effect of solvents and additives on the (*Z*) and (*E*) ratio of alkenes laid the foundations for subsequent mechanistic investigations, notably by Schlosser ⟨B-70MI 116-01⟩ which revealed subtle ways to bring about a high degree of stereoselectivity in the Wittig reaction.

1.16.1.2.1 Mechanism

Any attempt to rationalise the stereochemical outcome of the Wittig reaction hinges on a full understanding of the mechanism. This aspect has been extensively studied and the reader is referred to an excellent article on the subject ⟨89CRV863⟩. Over much of its history, the Wittig reaction has been described as a stepwise ionic process initiated by nucleophilic attack of the ylide on the carbonyl carbon atom to give two possible diastereomeric betaine intermediates (**10a**) and (**10b**) (Scheme 3). Subsequent decomposition then occurs by way of *cis*- and *trans*-oxaphosphetanes (structures (**11a**) and (**11b**)) ⟨73JA5778, 75C341⟩ to form the (*Z*)- or (*E*)-alkenes by elimination of phosphine oxide in a *syn* fashion ⟨65AG(E)689, 67LA(708)1⟩. At least part of the driving force for the second stage of the reaction appears to be the formation of the very strong phosphorus–oxygen bond. Vedejs has shown that there is no longer a clear need to invoke an ionic intermediate on the way to the covalent oxaphosphetane intermediate (**11**) in the Wittig reaction of 'reactive' phosphorus ylides ⟨89JA5861⟩. For reasons that are not clearly understood there is a decided kinetic preference for oxaphosphetane formation via a one-step cycloaddition reaction that only under 'salt-free' conditions suffers negligible reversal to starting materials. In consequence, a high proportion of the less thermodynamically stable (*Z*)-alkene is formed via the incursion of an early 'puckered' transition state that gives a high ratio of *cis*- to *trans*-oxaphosphetanes. In most, if not all, other cases, oxaphosphetane formation is reversible, and consequently stable phosphorus ylides furnish predominantly (*E*)-alkenes by allowing interconversion to the thermodynamically more stable *threo*-isomer (**11b**) and collapse by a later almost planar transition state. It follows that the stereochemistry of the alkene produced will depend both on the nature of the ylide and on the reaction conditions that affect the balance between the betaine dissociation and formal cycloaddition steps ⟨89TL1361⟩. For this reason it is convenient to deal separately with the three distinct categories of ylides: stable, reactive and so-called moderately reactive.

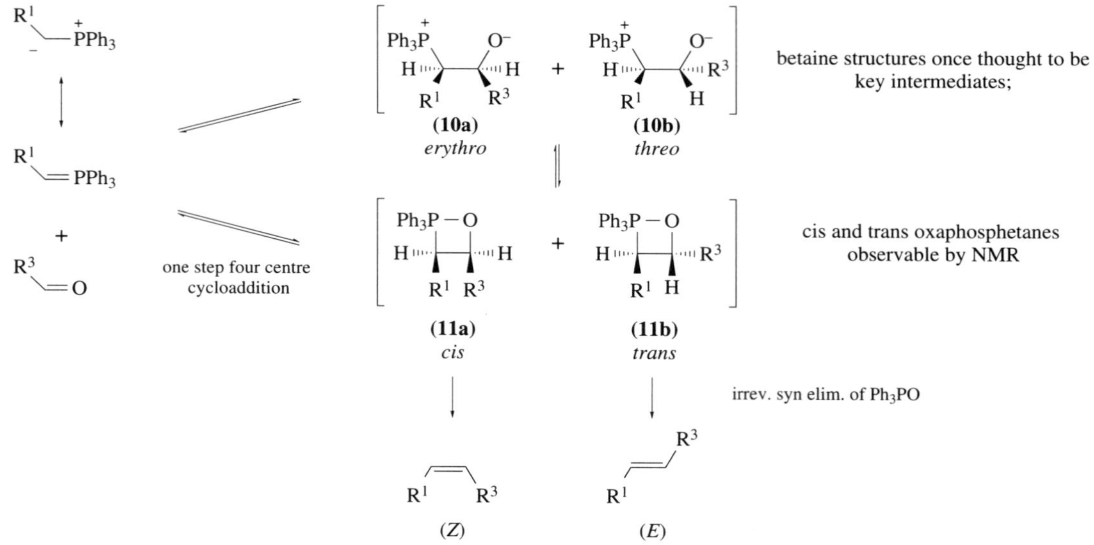

Scheme 3

1.16.1.2.2 Stable ylides

In general, reactions employing stabilised ylides favour the formation of the stereoisomer with the activating group at the α-carbon of the alkene *trans* to the larger group at the β-carbon. This stereoselectivity, as discussed above, is attributed to the preferential decomposition of the betaine intermediate in the *threo*- rather than the *erythro*-configuration since, in the transition state leading to the (*E*)-alkene, there is less steric interference to conjugative stabilisation of the incipient double bond. The condensation of the ethoxycarbonylmethylide **(12)** with benzaldehyde in Scheme 4 is illustrative ⟨63IZV1053⟩.

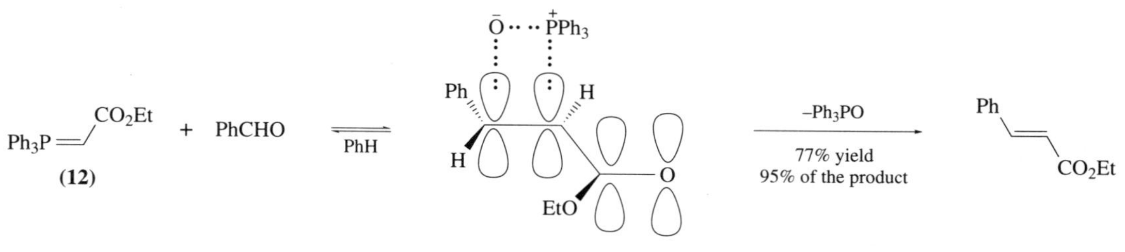

Scheme 4

Scheme 5, Equation (2) and Scheme 6 provide three other examples indicating the high degree of stereoselectivity that may be obtained with stabilised ylides ⟨72JOC2579, 75JA3512⟩. The use of epoxides as the base in a one-pot synthesis of alkenes by Buddrus is worthy of special note ⟨68AG(E)536, 72AG(E)1041, 74CB2050⟩.

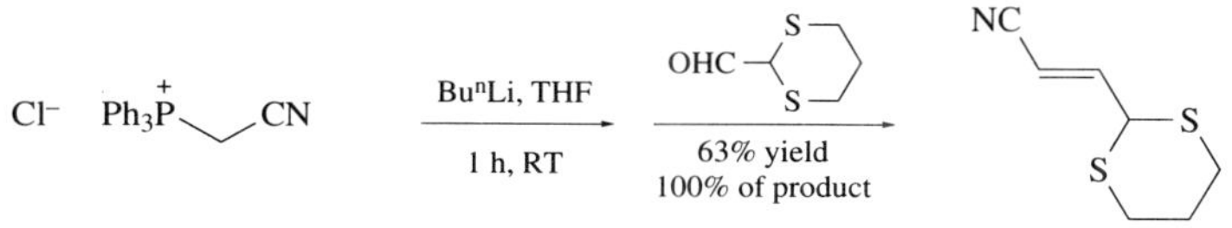

Scheme 5

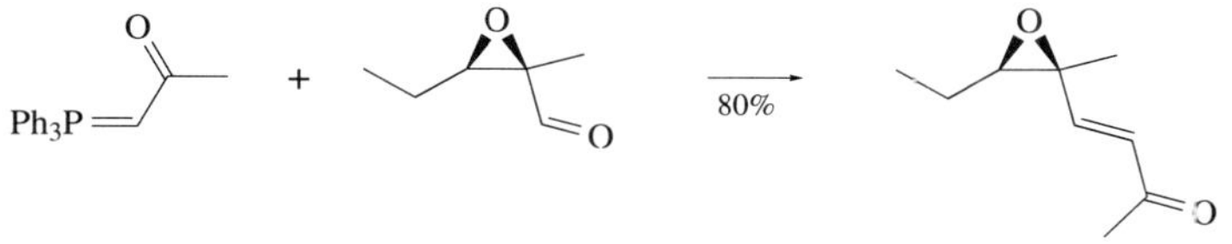

(2)

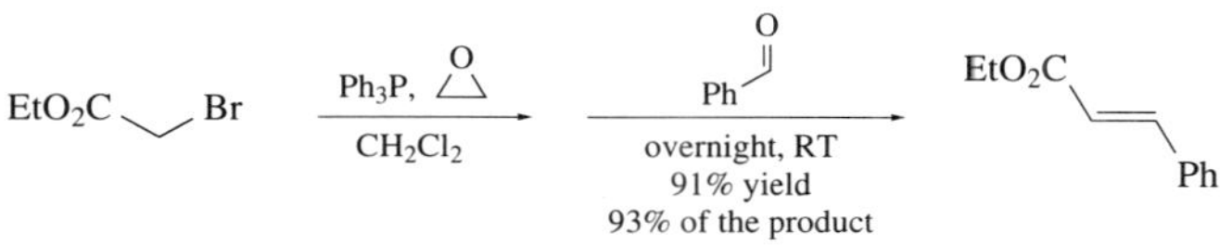

Scheme 6

As the examples in Scheme 5, Equation (2) and Scheme 6 indicate, the (*E*)-isomer is often produced with almost complete exclusion of the (*Z*)-isomer. It should be recognised that this may not be a true reflection of the stereochemistry of the reaction. In many cases, (*Z*)-alkenes having electron-withdrawing groups at the α-carbon are isomerised to the thermodynamically more stable (*E*)-isomer by the elevated temperatures required to effect the reaction ⟨B-70MI 116-02⟩. In addition, the reactants or products may catalyse the isomerisation ⟨63JA3878, 64JA5041⟩.

The degree of stereoselectivity observed may be influenced by the nature of the substituents on the ylide. A characteristic feature of stabilised ylides is that replacement of phenyl groups on phosphorus by alkyl groups such as butyl ⟨65JOC1296⟩ or cyclohexyl ⟨62CB1894, 65JOC1296⟩ shifts the isomer ratio further in favour of the (*E*)-product (Equation (3)). To account for this effect it has been suggested that alkyl groups (which are more electron-donating than phenyl) reduce the effective charge at phosphorus, thus allowing equilibration of the intermediate betaines. However, it may equally be due to increased steric interaction preventing the formation of the transition state leading to (*Z*)-alkene.

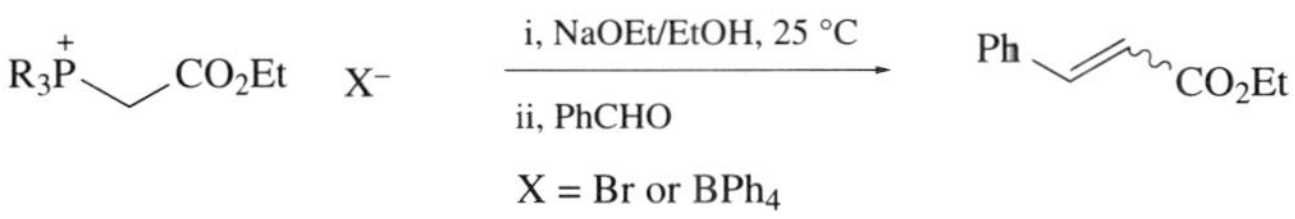

(3)

R	*Product* (*Z*)	(*E*)
Ph	15	85
Bu	5	95
n-C_6H_{13}	2	98
n-$C_{10}H_{21}$	4	96
c-C_6H_{13}	0	100

Substitution of phenyl groups on phosphorus by alkyl groups also results in an increase in the yields of alkene (Equation (4)) ⟨60T130⟩. This is because electron-releasing groups attached to the phosphorus facilitate nucleophilic attack ⟨63JA3878⟩ on the carbonyl group in the first (rate-controlling) step of the Wittig reaction by stabilising the contributing dipolar canonical form, for example, (**13b**). Thus, fluorenylidenetriphenylphosphorane (**13**; R = Ph) reacts only with the more reactive (electrophilic) aldehydes, whereas the tri-*n*-butyl ylide (**13**; R = Bu^n) gives quantitative yields of alkene with most aldehydes, and good yields with the more reactive ketones, for example, *p*-nitrobenzophenone.

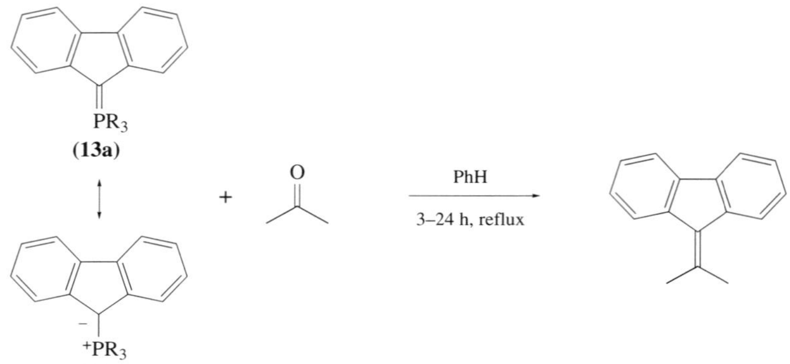

(13a) **(13b)** (4)

Carbonyl	Yield (%) R = Ph	Yield (%) R = Bu^n
p-$O_2NC_6H_4CHO$	96	99
p-ClC_6H_4CHO	93	96
PhCHO	84	96
p-$MeOC_6H_4CHO$	37	94
p-$Me_2NC_6H_4CHO$	0	94
$(p$-$O_2NC_6H_4)_2CO$	0	93
p-$O_2NC_6H_4COMe$	0	67
p-ClC_6H_4COMe	0	9

The *trans*-stereoselectivity of Wittig reactions of stabilised ylides may be adversely affected by the introduction of substituents at the position α to the phosphorus atom. If the group is sufficiently large, the energy difference between the two diastereoisomers of the intermediate betaine will be lowered. In this case appreciable amounts of (*Z*)-alkenes are formed in addition to the (*E*)-isomers. For example, reaction of phenacylidenetriphenylphosphorane (**14**; R = H) with *p*-nitrobenzaldehyde gives entirely (*E*)-alkene, whereas introduction of an α-bromine atom (**14**; R = Br) leads to the formation of 35% of the (*Z*)-isomer (Equation (5)) ⟨63JA2790⟩.

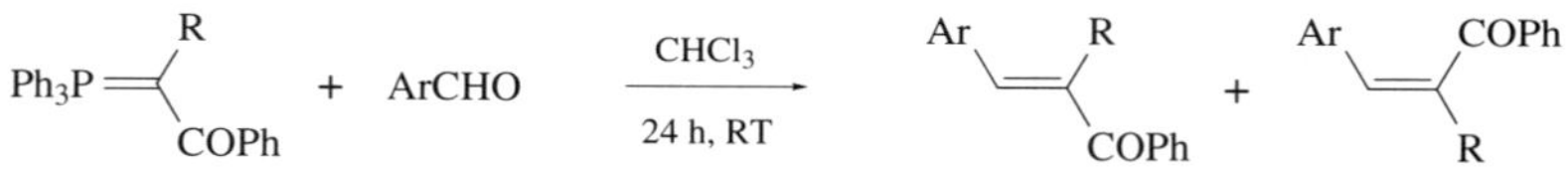

(14) (5)

R	Yield (%)	Product composition	
H	72	100	0
Br	64	35	65

Ar = p-$O_2NC_6H_4$

A similar situation prevails in the preparation of esters related to the insecticide bioresmethrin (Equation (6)) ⟨74JCS(P1)2470⟩. Condensation of the aldehyde (**16**) with the ethoxycarbonylmethylide (**15**; R = H) gives a product containing >90% of the (*E*)-alkene; alternatively, when R = Br or Cl, the condensation produces an isomer mixture in which the (*Z*)-product predominates (70–80%).

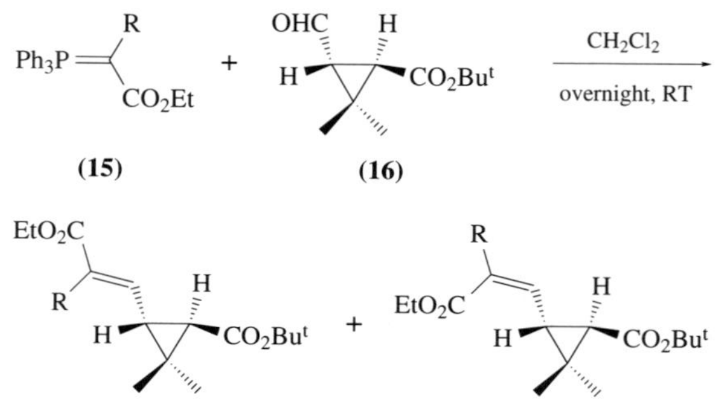

(15) **(16)** (6)

R	Product composition	
H	>90	<10
Cl	30	70
Br	20	80

The effect of α-methyl substituents is of interest (Scheme 7). As a rule, the (*E*)-alkene is formed almost exclusively as illustrated by the reaction ⟨61JOC4278⟩ of methoxycarbonylethylidenetriphenylphosphorane (**17**) with acetaldehyde to give practically pure methyl tiglate (**18**), i.e., the (*E*)-isomer, and with the alkynic aldehyde ⟨68JCS(C)1997⟩ (**19**) to give a product (**20**) which was all-*trans* as shown by NMR. Similarly, the same ylide with 2-hydroxytetrahydropyran (**21**) ⟨63T149, 64IZV2003⟩ afforded methyl 7-hydroxy-2-methyl-(*E*)-hept-2-enoate (**22**) stereospecifically. However, loss of stereospecificity has occasionally been observed ⟨67TL401, 70TL41, 73JCS(P1)1416, 76LA1475⟩.

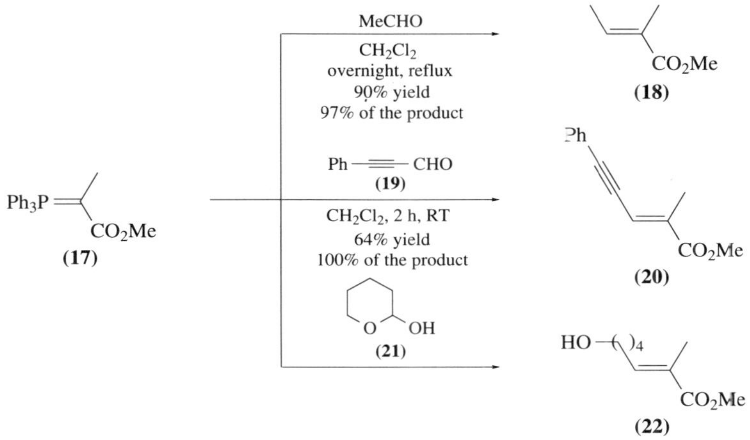

Scheme 7

A significant loss of *trans*-stereoselectivity also occurs in the reactions of stabilised ylides with aldehydes having α-alkoxy substituents although precise rules to predict the extent of loss are still difficult to formulate. In the case of the aldehydosugar (**23**) ⟨79HCA2091⟩, reaction with ethoxycarbonylmethylidenetriphenylphosphorane (**12**) in chloroform gave an isomer mixture (**24**) in which the (*Z*)-product predominates (60%) (Equation (7)). The loss in stereoselectivity is solvent dependent and (*Z*)-alkene formation can be minimised in DMF (14%) and maximised in methanol (92%). This effect is general and usually more (*Z*)-alkene is obtained when an aprotic solvent is replaced by a protic one, such as methanol, or when soluble lithium salts are added.

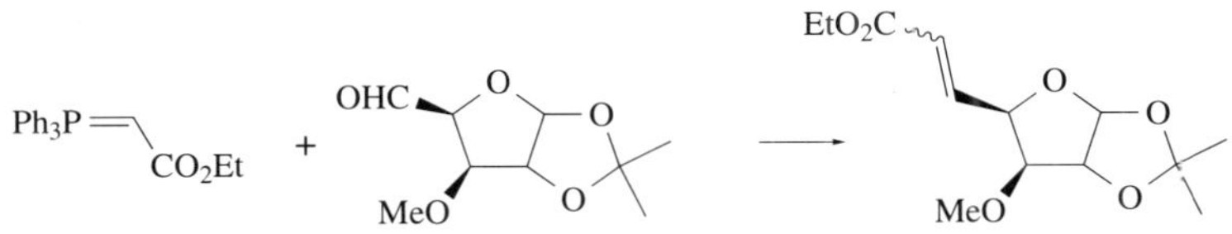

(7)

(12) **(23)** **(24)**

Solvent	*Product composition* (*Z*):(*E*)
DMF	14:86
$CHCl_3$	60:40
MeOH	92:8

Whereas in the reactions of stabilised ylides with aldehydes, the (*E*)-alkene is predominantly formed, interaction with unsymmetrical ketones almost always leads to a considerable amount of (*Z*)-isomer. However, condensation often fails unless an acid catalyst, for example, benzoic acid is present ⟨67CB1144, 69JCS(C)2728, 74CB1773⟩ (Equation (8)).

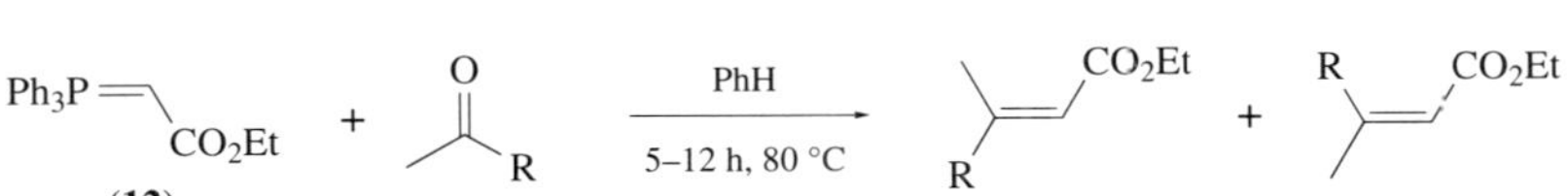

(8)

(12)

R	*Yield (%)* without $PhCO_2H$	*Yield (%)* with $PhCO_2H$	*Product composition*	
Et	5	89	66	34
Pr	12	83	60.5	39.5

An alternative, but often better procedure, to acid catalysis involves heating a mixture of the reactants without a solvent to relatively high temperatures (150–200 °C) ⟨61MI 116-01, 61TL579⟩. For example, in the synthesis of abscisic acid (Equation (9)), the condensation between ethoxycarbonylmethylidenetriphenylphosphorane (**12**) and 1-hydroxy-4-oxo-α-ionone (**25**) was practically complete after 10 min at 140–170 °C and gave approximately equal amounts of the (*E,Z*)- and (*E,E*)-esters (**26**) and (**27**) respectively ⟨68JOC3566⟩. A comparison of the stereochemical outcome of the procedure with that obtained in toluene (or ethanol) after prolonged heating revealed little difference in the isomer ratio ⟨71BSF(2)4551, 72TL2521⟩ although increasing the temperature to about 170 °C appeared to increase the amount of (*E,E*)-isomer ⟨68JOC3566⟩.

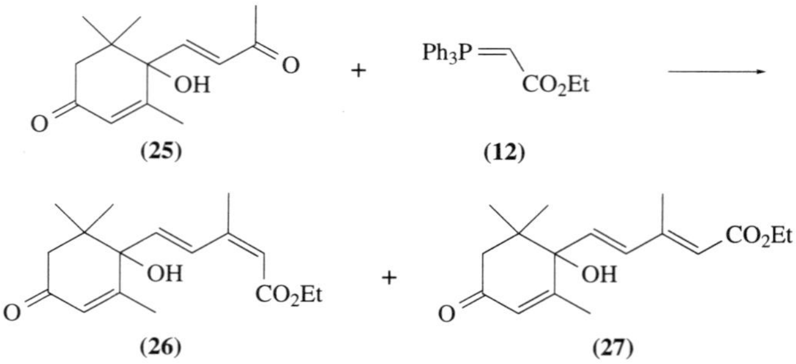

(9)

Reaction conditions	Yield (%)	Product composition (*E,Z*)	Product composition (*E,E*)
Melt (10 min, 140-170 °C)	83	50	50
Toluene (4 h, reflux)	77	50	50
EtOH		38	62

1.16.1.2.3 Reactive ylides

From Section 1.16.1.2.2 it can be assumed that, in the absence of additives or solvent interactions, the (*E*)-isomer usually forms the major product in Wittig reactions effected with stabilised ylides. This situation changes profoundly in the case of the so-called 'reactive' ylides, i.e., ylides that do not contain stabilising α-substituents. For these the thermodynamically less stable (*Z*)-isomers tend to dominate the mixture of alkenes obtained, as illustrated by the two examples ⟨63JOC372, 73HCA1176⟩ in Scheme 8. The degree of stereoselectivity varies considerably with the reaction conditions, especially with the nature of the solvent and the base used. The decisive influence of dissolved lithium salts is particularly stressed. It was originally thought that Wittig reactions of reactive ylides produced largely (*E*)-alkenes but that the (*Z*) : (*E*) ratio could be increased by the use of special polar solvents and added lithium salts.

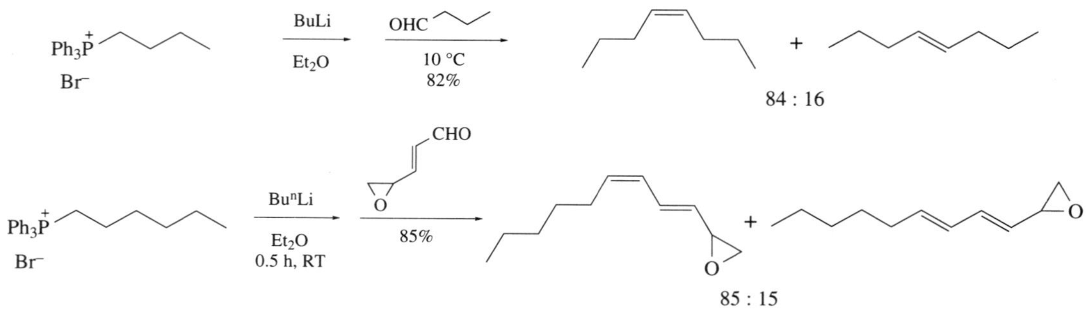

Scheme 8

While these conditions have been used to effect striking syntheses of a number of naturally occurring *cis*-unsaturated acids it has become apparent that lithium salts have little or no effect on the stereochemistry of reactions involving reactive ylides in dipolar aprotic solvents. The examples in Scheme 9 illustrate the generalisation that reactive ylides normally yield alkenes that are >90% (*Z*), irrespective of the presence or absence of lithium salts, if the reaction is carried out in very polar aprotic solvents such as dimethylformamide ⟨67T2709⟩, dimethylsulphoxide (DMSO) ⟨77JOC2783, 77TL4257⟩, or hexamethylphosphoramide (HMPA) ⟨74S798, 74TL207⟩. The addition of DMSO or HMPA as cosolvent to tetrahydrofuran also provides a suitable reaction medium for promoting the formation of (*Z*)-alkene ⟨74JOC3793, 74OPP269⟩. It should be stressed that organolithium compounds are unsuitable as bases in dimethylformamide (DMF) since they may react with this solvent to give aldehydes ⟨56JCS4691, 57CI(L)1596⟩. Sodium hydride is recommended as base if DMF is used as solvent; metal alkoxides have also enjoyed widespread use ⟨64CB3293⟩.

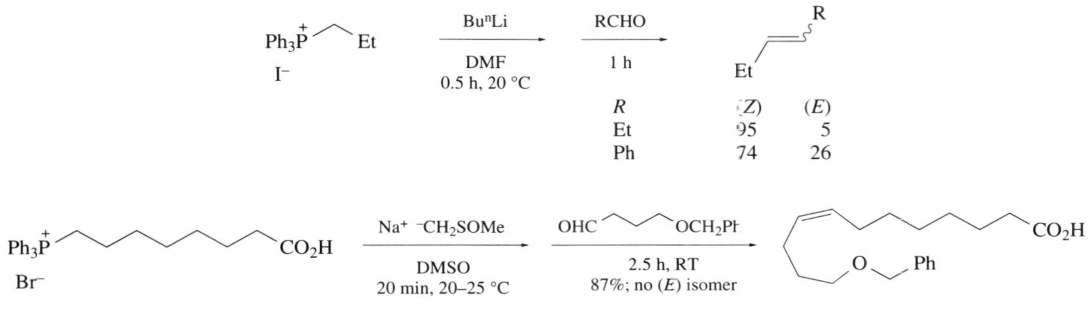

Scheme 9

A completely different situation is observed for reactions in nonpolar solvents such as benzene and ether. Reactions in solvents of this type show a salt dependence, giving increased amounts of the (*E*)-alkene on addition of relatively soluble lithium salts such as lithium iodide ⟨66AG(E)667, 67LA(708)1⟩.

In order to explain these effects in nonpolar media, it is argued that when dissolved metal cations, especially Li^+, are present in the reaction solution, the decomposition of both diastereoisomeric betaine intermediates to alkene is retarded by complex formation with the cation. As a result, reversal of the betaine formation step becomes important and more of the kinetically favoured *erythro*-betaine is converted into the thermodynamically more stable *threo*-isomer, thus reducing the proportion of (*Z*)-alkene in the product. The absence of salt effects in more polar media is regarded as being due to destabilisation of any such betaine–cation complex due to preferential solvation of the cation.

It is important, therefore, when preparing (*Z*)-alkenes by the Wittig reaction in nonpolar media, to ensure that ylide solutions that are free from dissolved salts are used. Several means of achieving this have been reported. For example, the ylide can be prepared by the use of sodamide in anhydrous ammonia and then extracted into benzene ⟨58LA(619)10, 65AG(E)583⟩. A variation of this method is to prepare the ylide with sodamide in boiling tetrahydrofuran and to remove the insoluble inorganic salts by filtration ⟨70LA(739)211⟩. Potassium *t*-butoxide ⟨75JA4327⟩ and sodium bis(trimethylsilyl)amide ⟨76CB1694⟩ have also been employed as bases in tetrahydrofuran without the tedious necessity for filtration.

In another convenient procedure ⟨70MI 116-04⟩ a mixture of the phosphonium salt and the aldehyde is treated with sodium *t*-pentoxide dissolved in benzene containing trace amounts of dimethyl sulphoxide. Typical 'salt-free' (*Z*):(*E*) distributions are also obtained when 18-crown-6-complexes of potassium carbonate or potassium *t*-butoxide are used in THF ⟨75S784⟩. Major advantages of the last two methods are that preformation of the ylide is unnecessary. Apart from the obvious advantages of a 'one-pot' process, *in situ* formation of the ylide gives high yields of alkenes and a minimum of self-condensation. Some examples of the use of the 'salt-free' Wittig reactions are shown in Schemes 10, 11 and Equation (10) ⟨70MI 116-03, 75JA4327, 75S784, 76CB1694⟩.

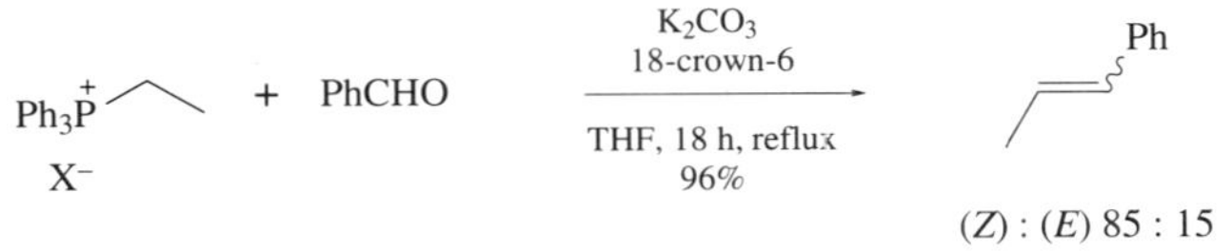

(10)

Although the choice of reaction conditions may exert a significant influence on the stereochemical composition of the product of Wittig reactions, other factors of importance are the nature of the substituents both on the carbonyl and on the ylide component. As the data in Table 1 clearly demonstrates, the reaction between homologous unbranched alkylidene ylides and primary aliphatic

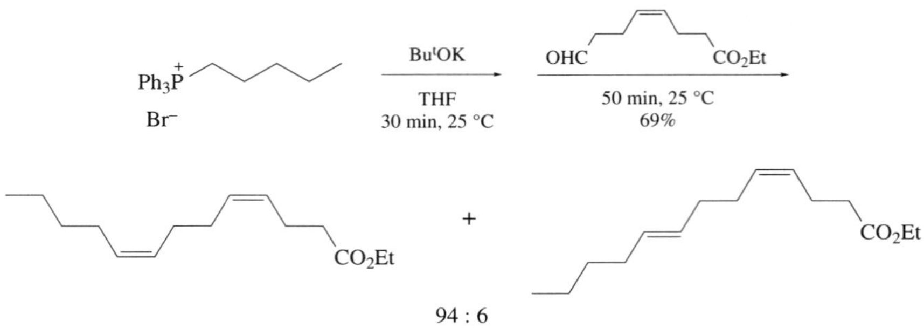

Scheme 10

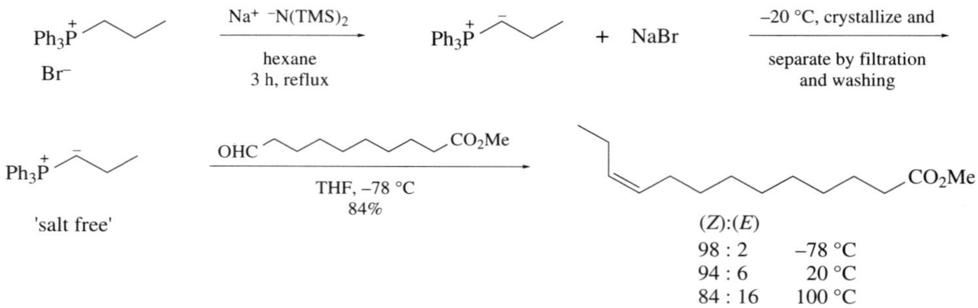

Scheme 11

aldehydes is highly stereoselective in salt-free solution, yielding products in which the proportion of the (*Z*)-isomer exceeds 90% (in most cases 95%). However, when α,β-unsaturated or aromatic aldehydes serve as the carbonyl component, some loss of stereoselectivity is observed, especially in dipolar aprotic media. This difference in behaviour may best be explained by assuming that betaine formation with aliphatic aldehyde is essentially irreversible (in the absence of lithium salts), whereas with more electrophilic carbonyl compounds such as *p*-cyanobenzaldehyde, reversibility is more pronounced and the betaine formation and hence the product ratio become thermodynamically controlled.

Table 1 Yields and (*Z*):(*E*) ratios of alkenes from 'salt-free' Wittig reactions of reactive ylides in benzene, toluene or tetrahydrofuran at 0°C ⟨66AG(E)667, 67LA(708)1, B-70MI 116-01, 71BSF(2)453⟩.

$$Ph_3P^+—^-CHR^1 + R^2CH{=}O \rightarrow R^1CH{=}CHR^2 + Ph_3P{=}O$$

R^2	(*Z*):(*E*) ratio (*yield* %) $R^1 = Me$	$R^1 = Et$	$R^1 = Pr$
Et		97:3 (70)	
Pr^n	95:5	95:5 (49)	95:5 (78)
n-C_5H_{11}	91:9 (66)	96:4	
(*E*)-PhCH=CH	87:13 (77)		
PhC≡C	73:27		
p-$MeOC_6H_4$	90:10 (78)	92:8 (95)	90:10 (80)
p-MeC_6H_4	89:11 (88)	95:5 (59)	92:8 (71)
Ph	87:13 (98)	96:4 (88)	94:6 (66)
p-ClC_6H_4	82:18 (77)	93:7 (54)	92:8 (64)
m-ClC_6H_4	74:26		
p-NCC_6H_4	74:26		
p-$F_3C\ C_6H_4$	69:31		

In the case of ketones the isomer ratio depends on the nature of the substituents at the carbonyl carbon. With alkyl aryl ketones the (*Z*)-isomer is still favoured as shown in Scheme 12

⟨76JCS(P1)1476⟩. On the other hand, *cis*-stereoselectivity is almost totally lost with unsymmetrical aliphatic ketones using a variety of reaction conditions.

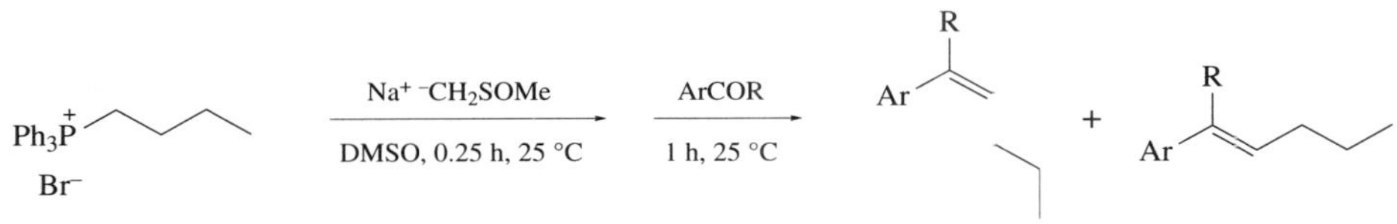

R	*Yield* (%)	*Product composition*
Me	80	9 : 1
Et	71	9 : 1

Ar = p-MeOC$_6$H$_4$

Scheme 12

An unexpectedly large tendency to form (Z)-alkenes has been observed with aliphatic aldehydes when the alkyl group is branched ⟨73JA5778, 75JA6214⟩. For example, reaction of ethylidenetriphenylphosphorane with bulky aldehydes such as pivalaldehyde and 2,2-dimethylbut-3-enal gives predominantly, if not exclusively, (Z)-alkene, even in the presence of lithium salts. Steric factors may also be responsible for the exclusive formation ⟨69CC303, 69CC785⟩ of the (Z)-alkene (**28**) from the reaction of hexylidenetriphenylphosphorane with the *endo*-aldehyde (**29**); in the case of the corresponding *exo*-aldehyde a mixture of (Z)- and (E)-alkenes is obtained in the ratio of 80:20 (Scheme 13) ⟨69JA5364⟩. A possible explanation for this surprising selectivity can be formulated ⟨69CC303, 69CC785⟩ based on initial coordination of the carbonyl oxygen atom to the phosphorus atom of the ylide before C—C bond formation takes place.

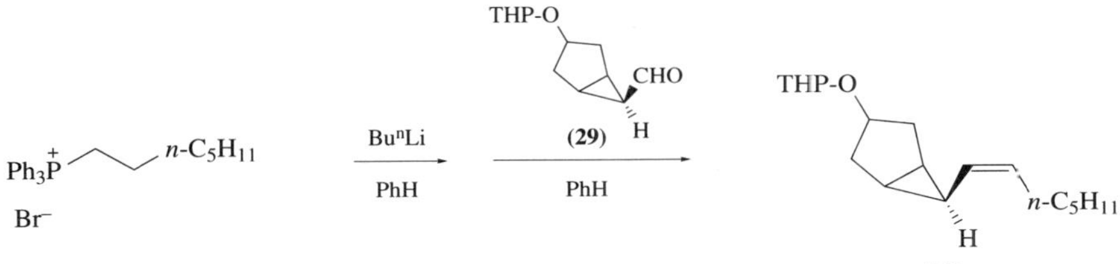

Scheme 13

This rationale may also explain the loss of *cis*-selectivity in Wittig reactions of reactive ylides having branched alkyl substituents at the α-carbon atom. The example in Scheme 14 illustrates this effect ⟨71JCS(C)3252⟩.

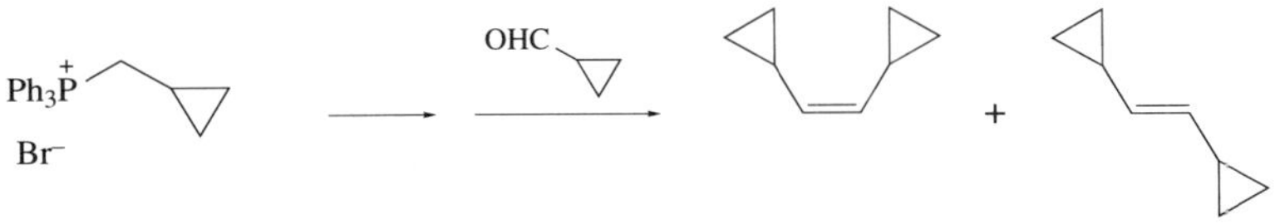

Reaction conditions	*Yield* (%)	*Product composition*
Bu^{n}Li, Et$_2$O, 4 h, 35 °C	70	1 : 1
Na$^+$ $^-$CH$_2$SOMe, DMSO, 4 h, 25 °C	71	2.4 : 1

Scheme 14

1.16.1.2.4 *Formation of (E)-alkenes via Wittig–Schlosser modification*

As mentioned previously, reactive ylides will afford large amounts of (E)-alkenes if equilibration of the *erythro* and *threo* betaines can be accelerated. A method has been described ⟨66AG(E)126, 67LA(708)1, B-76MI 116-01⟩ whereby the initially formed betaine–lithium complex, with predominantly the *erythro* configuration, is α-metallated with an equivalent of an organolithium reagent to form a new β-oxido phosphorus ylide (betaine ylide) (**31**) (Scheme 15). Whereas the initially formed *erythro*-complex (**30**) was relatively stable to inversion, the diastereoisomeric betaine ylides are rapidly interconverted, even at temperatures as low as −78 °C. Significantly, the equilibrium lies far to the side of form (**31b**), so addition of one equivalent of a proton donor occurs with remarkable stereospecificity to form the Wittig intermediate (**32**) in the *threo* configuration, which decomposes selectively to give pure (*E*)-alkene. It should be noted that protonation of *β*-oxido ylides prepared from salt-free ylides leads to mixture of *erythro*- and *threo*-betaines and hence to mixtures of (*Z*)- and (*E*)-alkenes ⟨70CB2814⟩.

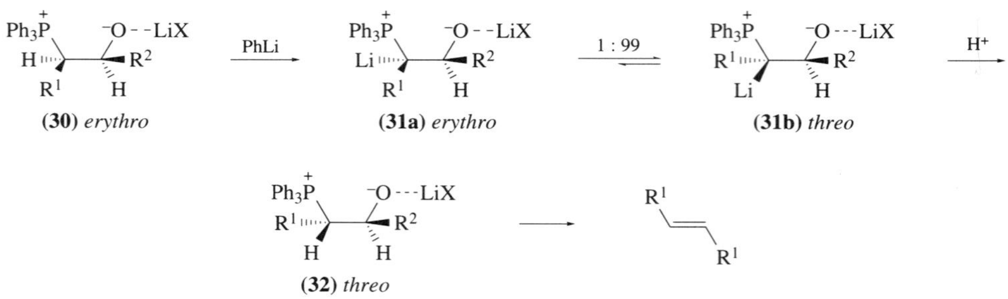

Scheme 15

Table 2 summarises, the application of this method, known as the Wittig–Schlosser reaction, to the preparation of (*E*)-alkenes using reactive ylides in combination with a variety of aliphatic, unsaturated and aromatic aldehydes. For comparison, the conventional Wittig reactions lead to a (*E*):(*Z*) distribution that varies between 75:25 (for phenylhexene) and 20:80 (for oct-2-ene).

Table 2 *trans*-Stereoselective alkenation of aldehydes $R^2CH{=}O$ with reactive ylides $[Ph_3P^+\text{-}^-CHR^1]$ via the Wittig–Schlosser reaction ⟨66AG(E)126, 67LA(708)1, B-76MI 116-01⟩.

Alkene formed	R^1	R^2	*Yield* (%)	(*E*):(*Z*) *ratio*
Oct-2-ene	Me	n-C_5H_{11}	70	99:1
Oct-2-ene	n-C_5H_{11}	Me	60	96:4
Oct-4-ene	n-C_3H_7	n-C_3H_7	72	98:2
PhCH=CHMe	Me	Ph	69	99:1
'salt-free'			82	76:24
PhCH=CHEt	Et	Ph	72	97:3
PhCH=CH$(CH_2)_3$Me	Bu^n	Ph	75	96:4
PhCH=CHCH=CHMe	Me	(*E*) PhCH=CH	63	97:3

trans-Stereoselective alkene synthesis by means of the Wittig–Schlosser reaction is not limited to aldehydes, as is shown by the reaction of the ethylidene-triphenylphosphorane with the unsymmetrical ketone acetophenone to give (*E*)- and (*Z*)-phenylbut-2-enes in the ratio 89:11 ⟨67LA(708)1⟩. When carried out in the conventional manner, the same reaction led to a ratio of 40:60 ⟨B-70MI 116-01⟩. In contrast, no detectable increase in stereoselectivity was observed ⟨70JOC565⟩ for the reactions of *n*-butylidenetriphenylphosphorane with the *aliphatic* methyl ketone, methylheptenone, under the same conditions.

Exaggerated *trans*-selectivity is also observed with oxido ylides bearing various nucleophilic groups (OH, CO_2H and $CONH_2$) in the side chain ⟨85JA217⟩, relative to a ylide that does not possess such a group, upon reaction with both benzaldehyde ((*E*):(*Z*) = 93:7) and hexanal ((*E*):(*Z*) = 52:48).

1.16.1.2.5 *Ylides of intermediate (moderate) reactivity*

As discussed previously, ylides fall into two main groups according to their reactivity. The major group are the so-called 'reactive' ylides in which the negative charge is localised on the α-carbon, whereas the second group includes the 'stable' ylides which are much less reactive owing to extensive delocalisation of negative charge through participation of resonance structures. Obviously, a whole spectrum of reactivity lies in between, and some ylides stabilised by α-substituents such as vinyl, aryl or alkynyl show an intermediate 'moderate' reactivity which is also reflected in the stereochemical outcome of their reactions with carbonyl compounds. Scheme 16, Equation (11) and

Schemes 17–19 illustrate the fact that stereoisomeric mixtures of alkenes are usually produced when this preparatively important class of ylides are employed in Wittig reactions under a variety of conditions ⟨65JOC1473, 69S75, 70T1291, 71CB2885, 72AJC545, 73S295, 74TL2587⟩. It will be noted that in most cases a slight preponderance of the (*E*)-alkene is favoured.

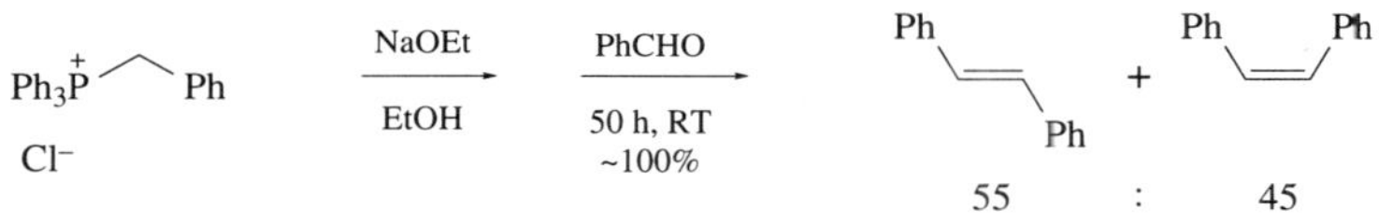

Scheme 16

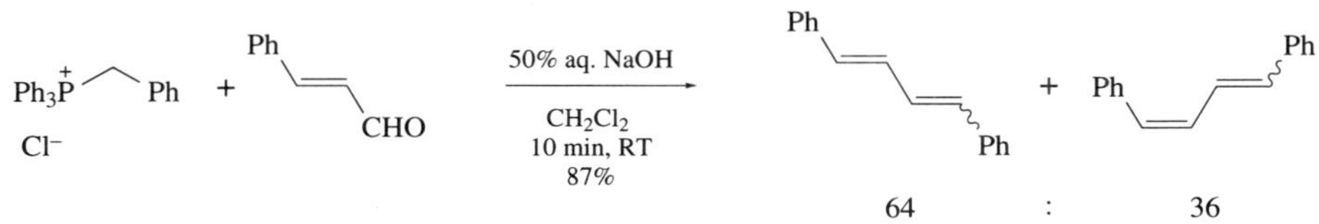

(11)

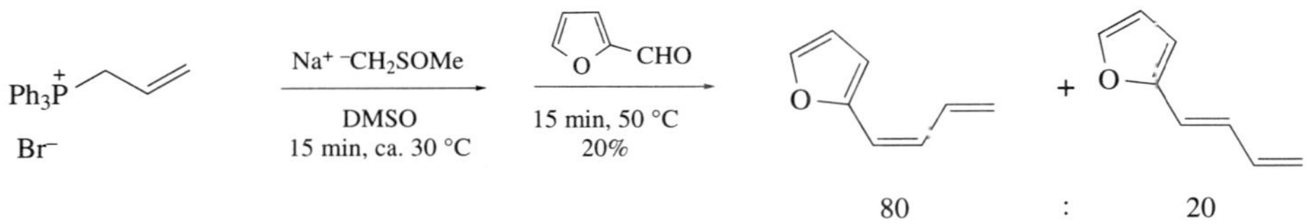

Scheme 17

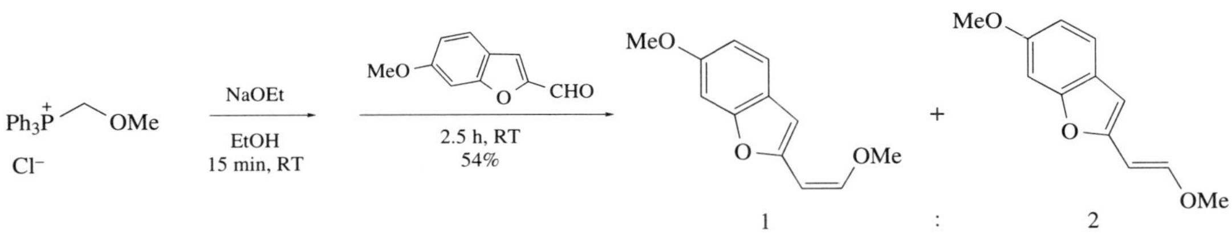

Scheme 18

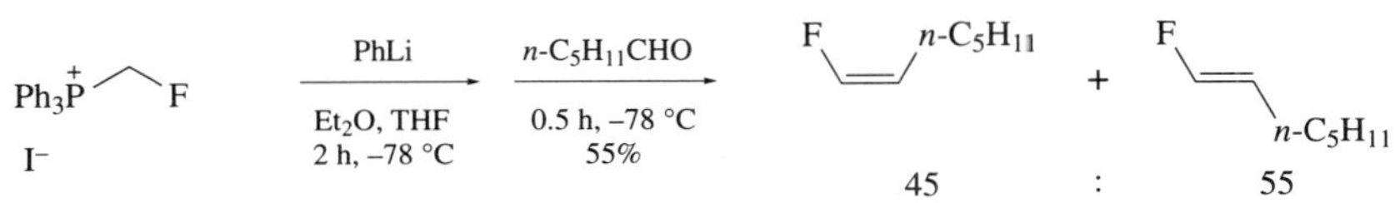

Scheme 19

Since moderately reactive ylides possess different structural features, certain of which deserve special recognition, it is convenient, when discussing stereochemical implications as well as preparative applications, to treat the important types of moderately reactive ylides; benzylic and allylic separately.

(*i*) *Benzylic ylides*

For Wittig reactions effected with triphenylphosphonium benzylides, the stereochemistry of the products depends largely on the nature of the substituents in the aryl ring. In keeping with previous stereochemical arguments, electron-releasing groups which enhance the reactivity (nucleophilicity) of the ylide reduce the *trans*-selectivity, whereas the reverse is true for electron-withdrawing groups. Important advantages may be gained in synthesis from this influence to produce a higher proportion of (*Z*)-alkene. For example, in the preparation of 4-nitro-4′-methoxystilbene, the reaction of

4-nitrobenzylidenetriphenylphosphorane (**33**) with anisaldehyde leads to the predominant formation of the (*E*)-alkene as shown in Scheme 20 ⟨62JOC4666, 66JOC334⟩. However, by interchanging the ylide and aldehyde substituents, a 1 : 1 mixture of stereoisomers is obtained (cf. ylide (**34**)).

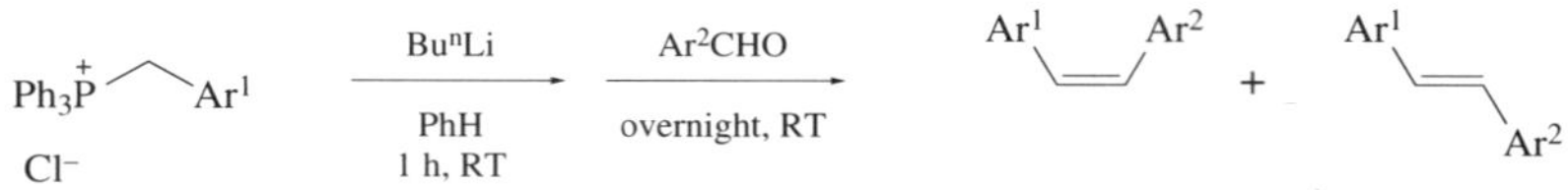

Ylide	*Ar¹*	*Ar²*	*Product composition*		
(**33**)	*p*-$O_2NC_6H_4$	*p*-$MeOC_6H_4$	26(0)	:	74(100)
(**34**)	*p*-$MeOC_6H_4$	*p*-$O_2NC_6H_4$	44(48)	:	56(52)

Scheme 20

Perhaps a more effective means of shifting the isomer ratio in favour of the (*Z*)-product is through structural alteration of the phosphonium group. As already noted, if the phosphorus atom is rendered more electrophilic by substitution of electron-withdrawing groups, the betaine intermediate has less chance to equilibrate to the more stable *threo* form. As a consequence greater selectivity for (*Z*)-alkene formation is observed ⟨66JOC334⟩. Table 3 shows that in the synthesis of 4-nitro-4′-methoxystilbene, use of 4-methoxybenzylidenetris(*p*-chlorophenyl)phosphorane (**35**) in place of (**34**) leads to an increase from 44:56 to 80:20 in the (*Z*):(*E*) ratio. In a similar manner, the (*Z*):(*E*) ratio increases in the nitrobenzylidene series as the phosphorus substituent is successively changed from *n*-butyl (**36**) to phenyl (**37**) to *p*-chlorophenyl (**38**). As indicated, the course of the reaction may be shifted overwhelmingly in favour of either the (*Z*)- or the (*E*)-isomer by the appropriate choice of reactants.

Table 3 Relative yields of (*Z*)- and (*E*)-4-nitro-4′-methoxystilbenes from the reaction of $R^1{}_3P^+CH_2C_6H_4R^2Br^-$ with $R^3C_6H_4CHO$ in benzene at room temperature using *n*-butyllithium as base ⟨66JOC334⟩.

Ylide	*R¹*	*R²*	*R³*	*(Z):(E) ratio*
34	Ph	OMe	NO_2	44:56
35	*p*-ClC_6H_4	OMe	NO_2	80:20
36	*n*-C_4H_9	NO_2	OMe	17:83
37	Ph	NO_2	OMe	26:74
38	*p*-ClC_6H_4	NO_2	OMe	52:48

Apart from the effect of the substituents, the degree of stereoselectivity observed may be influenced by changes in the reaction medium. As shown in Table 4, formation of (*Z*)-alkene is increased by the use of either protic solvents or dipolar aprotic solvents. However, under no circumstances is the effect appreciable, and (*E*)-alkene usually dominates the product composition even when salt-free solutions of the ylide are employed. This would imply that in general solvation of the betaine intermediate is not significantly affecting its formation or reversible decomposition. It has been suggested ⟨B-70MI 116-01⟩ that *threo*-betaine formation becomes more important under conditions of kinetic control if the ylide sidechain bears bulky ligands such as phenyl (or cyclohexyl). Even so, it should be noted that use of Schlosser's modification, whereby the initially formed betaine–lithium complex is α-metallated, fails to improve the *trans*-selectivity in the reaction with propionaldehyde. An interesting change in selectivity is observed ⟨74JCS(P1)2435⟩ when the reaction solvent in the Wittig condensation between 2-methoxybenzyltriphenylphosphonium bromide and the aldehyde (**39**) is changed from methanol to dimethylformamide (Equation (12)). In methanolic solution (2 h, reflux) with lithium methoxide as base the (*E*)-stilbene predominated, but in dimethylformamide (1 h, 90 °C) the (*Z*)-stilbene predominated. The reason why opposite stereoselectivities are preferred in these solvents is not apparent.

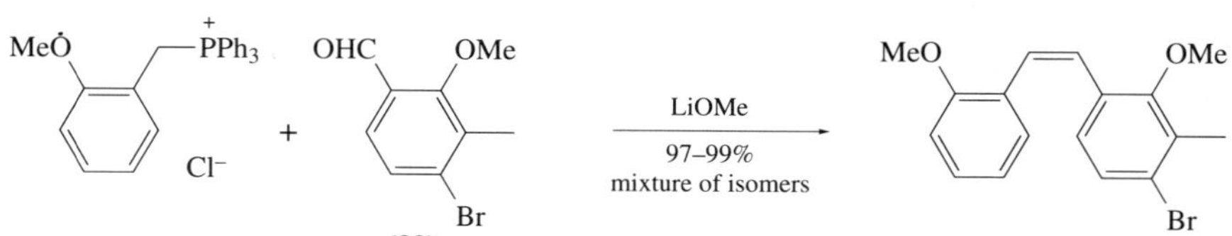

(**39**)

(12)

Table 4 Effect of the reaction medium on the stereoselectivity of the Wittig reaction between benzyltriphenylphosphonium salts and propionaldehyde.

Reaction medium (*salt present*)	*PhCH=CHEt* (*Z*):(*E*) *ratio*	*Yield* (%)	*Ref.*
$Bu^nLi—C_6H_6$ (LiBr)	27:73	57	68JCS(C)1984
(LiCl)	23:77	32–83	64JOC3327
(LiCl)	21:79		67T2709
(LiBr)	27:73		
(LiI)	41:59		
Schlosser's α-metallation product	43:57		
Bu^nLi—hexane	22:78		
'salt-free' C_6H_6	25:75		B-70MI 116-01
	18:82	61	67T2709
Et_2O—NaOEt	31:69		63T149
THF—NaOEt	33:67		
DMF—Bu^nLi	44:56	32–44	64JOC3327
	40–45:60–55	60–74	68JCS(C)1984
DMF—NaOEt	46:54	66	64JOC3327
EtOH—NaOEt	46:54	95	
K_2CO_3—18-crown-6-ether—THF	25:75	94	75S784

The isomer ratio may be shifted further in favour of the (*E*)-isomer by replacing the phenyl groups on phosphorus by alkyl groups as illustrated in Table 5 where a comparison is made of the distribution of stilbene isomers obtained in ethanol. It should be noted that substitution of alkyl groups for phenyl groups on phosphorus leads not only to almost pure (*E*)-stilbene, but also to better yields. A similar effect has been noted previously for stabilised ylides.

Table 5 Yields and (*Z*):(*E*) ratios for stilbene resulting from the interaction of different benzylphosphonium salts $R^1R^2R^3P^+CH_2PhX^-$ and benzaldehyde in ethanol–sodium ethoxide.

$R^1R^2R^3$	*PhCH=CHPh* (*Z*):(*E*) *ratio*	*Yield* (%)	*Ref.*
Ph	53:47	76	55CB1654
Ph_2Me	28:72	97	66JCS(C)1090
$PhMe_2$	13:87	91	
Bu	9:91		B-70MI 116-01
cyclohexyl	5:95	92	62CB1894

In general, structural changes in the carbonyl co-reactant exert only a minor effect on the stereochemistry of Wittig reactions of benzylidenetriphenylphosphorane. *trans*-Selectivity is preserved when using a variety of aliphatic, unsaturated and aromatic aldehydes under aprotic conditions. There are some exceptions to this observation, but one in particular deserves special attention ⟨68JCS(C)1997⟩ and involves condensation of benzylidenetriphenylphosphorane with the α,β-unsaturated aldehyde (**40**) having a carboxylate ion on the β-carbon atom (lactol in disguise) to give a mixture of pentadienoic acids (**41**) and (**42**) (Scheme 21), in which a high proportion of the newly formed double bond has the (*Z*) configuration. Molecular models suggest that this may result from preferential stabilisation of the *erythro* form of the intermediate betaine by interaction of the positively charged phosphorus atom and the carboxylate ion in the staggered conformation ⟨68JCS(C)1982, 68JCS(C)1997⟩.

As might be expected, in some instances only the (*E*)-isomer is formed, especially when it is strongly favoured on steric grounds. An example is the conversion of indole-3-aldehyde (**43**) into (*E*)-1-phenyl-2-(3-indolyl)ethylene (**44**) in 30–60% yield by condensation with benzylidenetriphenylphosphorane (Scheme 22) ⟨76IJC(B)419⟩.

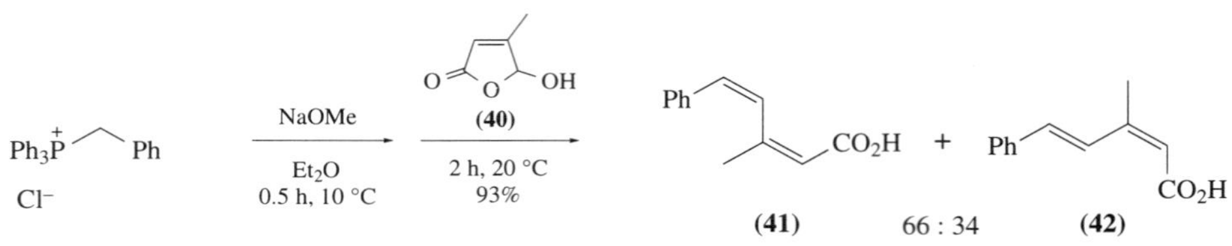

Scheme 21

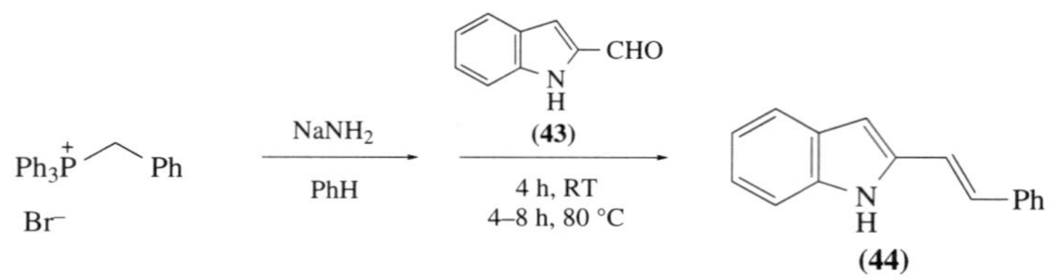

Scheme 22

(ii) Allylic ylides

In contrast to reactions of reactive ylides, the stereochemistry of Wittig reactions effected with moderately reactive ylides is little affected by the presence of dissolved lithium salts. Scheme 23 shows one example ⟨69JCS(C)1024⟩ in which the stereochemical outcome is not significantly different under 'salt-free' conditions from that obtained under normal Wittig reaction conditions ⟨77T1845⟩. Owing to the intrinsic nonstereoselectivity of allylic ylides, formation of conjugated dienes of definite configuration is best conducted by reaction of a saturated aliphatic reactive ylide with an α,β-unsaturated aldehyde, as exemplified by the synthesis of the (9*Z*, 11*E*)-diene **(46a)** ⟨75CL103⟩, a major component of the sex pheromone of the Egyptian cotton leafworm (*Spodoptera littoralis*), by the Wittig reaction of 9-acetoxynonyltriphenylphosphonium bromide **45** with (*E*)-pent-2-en-1-al in dimethyl sulphoxide using dimsylsodium as the base (Scheme 24). This gave a 74% yield of the dienes containing 85–90% of **(46a)** and 10–15% of the (9*E*, 11*E*)-isomer **(46b)**, whereas alternative Wittig condensation of (*E*)-pent-2-entyltriphenylphosphonium bromide **(47)** with 9-oxonon-1-ylacetate produced a mixture of isomers in various ratios depending on the reaction conditions.

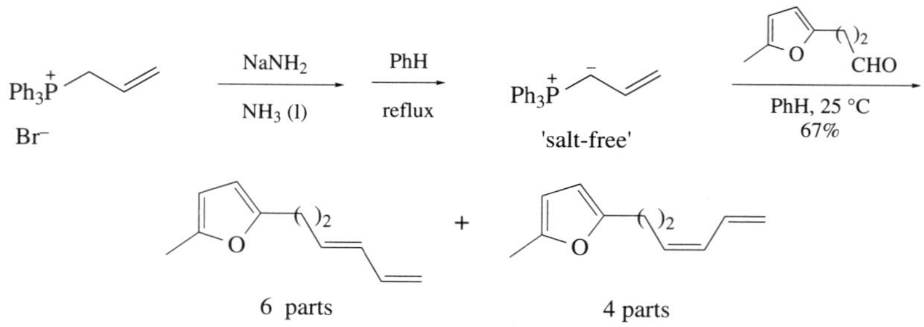

Scheme 23

A further stereochemical complication sometimes encountered during Wittig reactions with allylic ylides having a terminal substituent is loss of configuration in the allylic double bond. Thus, the reaction ⟨66JOC2907, 72HCA1828⟩ between salicylaldehyde and the ylide generated from the crotylphosphonium salt **(48)** (Equation (13)) gave a mixture of all four geometrical isomers of the diene **(49)**. The same mixture was also obtained starting from pure (*E*)-crotyl salt. However, the tendency to isomerise is not a general phenomenon. Preservation of stereochemical integrity has been demonstrated for both (*Z*)- and (*E*)-allylic phosphonium salts during ylide formation and subsequent reaction with aldehydes ⟨62LA(658)65, 75HCA1016⟩.

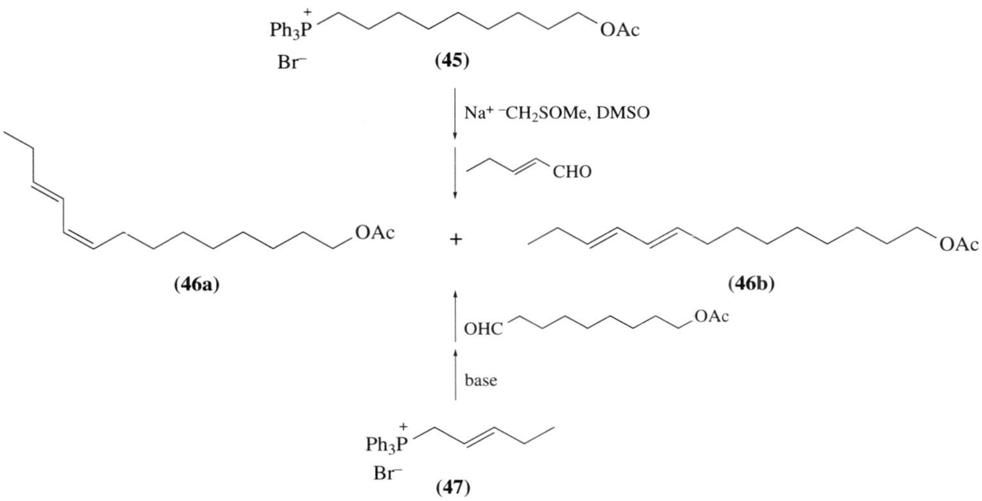

Reaction conditions	(46a) : (46b)
KOBu^t/PhH	40 : 60
GLYME/Bu^{n}Li	50 : 50
$Na^+ \ ^-CH_2SOMe$/DMSO	60 : 40

Scheme 24

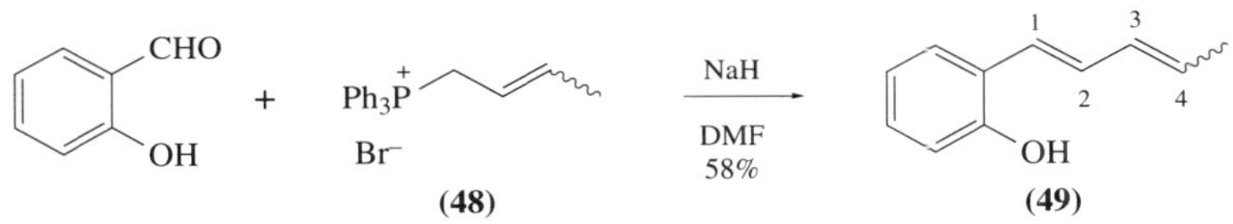

(48) (49) (13)

(*E*)-1, (*E*)-3, 34%
(*E*)-1, (*Z*)-3, 15%
(*Z*)-1, (*E*)-3, 36%
(*Z*)-1, (*Z*)-3, 15%

In some cases, at least, stereomutation may be suppressed by an appropriate choice of reaction conditions. For example, the (*Z*)-allyltriphenylphosphonium bromide (**50**) with butyllithium and cyclohexanone in tetrahydrofuran at 40 °C gave a mixture of dienes (**51a**) and (**51b**) containing 20% of the (*E*)-isomer (Scheme 25) ⟨58JCS843⟩. By preparing the ylide at −25 °C and warming to 5 °C after addition of the ketone the proportion of (*E*)-isomer was reduced to <5%.

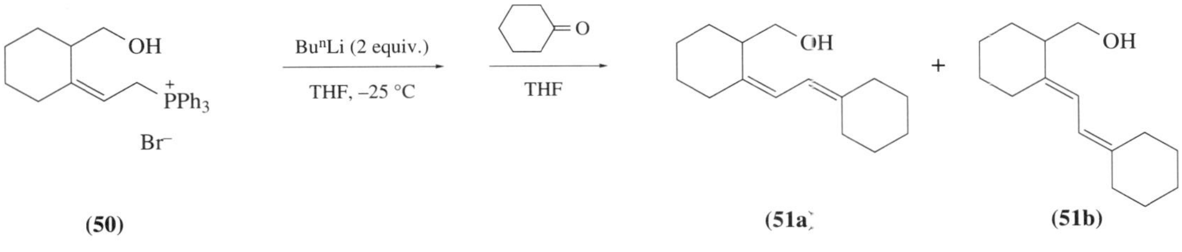

Scheme 25

Substitution of a methyl group at the *β*-position usually increases the isomerisation about the allylic double bond ⟨71JA3457⟩. Another difficulty that frequently arises in Wittig reactions of allylic ylides is concurrent condensation at the *γ*-position. For example, reaction of hexanal with the allylic ylide (**52**) furnished ⟨74JOC821⟩ not only all four geometric isomers of methyl 3-methyldeca-2,4-dienoate (**53**), the normal *α*-condensation product, but also both geometric isomers of methyl 2-isopropenyl-2-enoate (**55**), the *γ*-condensation product (Scheme 26). The latter could have been formed via the cyclic six-centre phosphorane (**54**).

However, by generating the ylide from its salt, using diisopropylamine in DMF in the presence

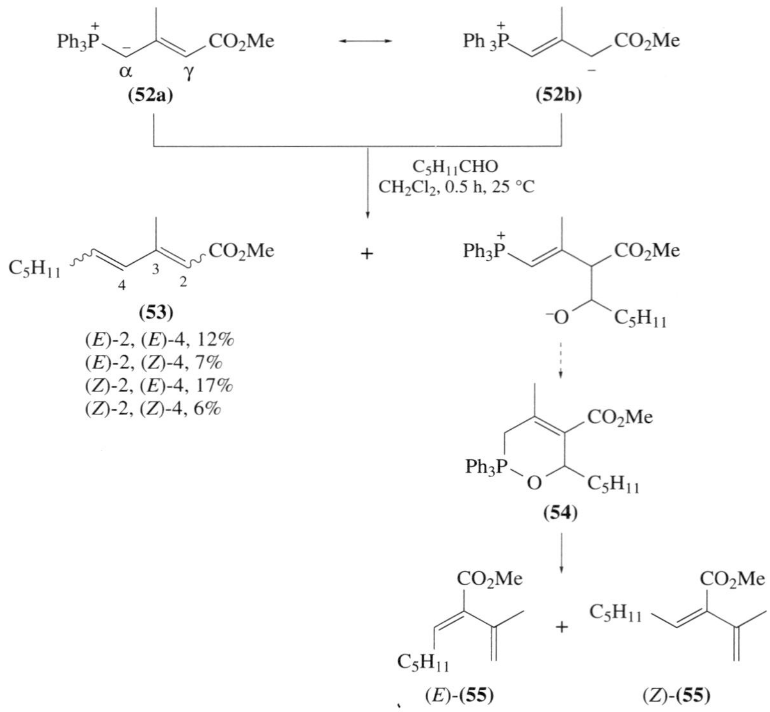

Scheme 26

of cadmium iodide, it is possible to minimise γ-condensation and obtain fair yields of the α-condensation product (**53**) (36%).

In the case of α,β-unsaturated carbonyl compounds, condensation at the γ-position is highly specific and takes place in a Michael-type fashion to provide a simple and elegant synthesis of cyclohexa-1,3-dienes ⟨71HCA1767, 73CB3779, 74JOC1318⟩. A possible mechanism for this transformation is illustrated in Scheme 27 ⟨71HCA1767⟩.

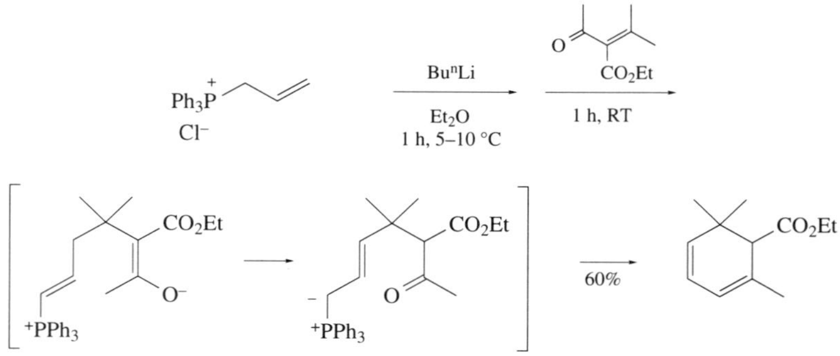

Scheme 27

As expected increased steric hindrance at the γ-carbon atom inhibits γ-condensation ⟨73JOC3625, 73TL4425⟩, although condensation of 2-methylenecyclohexanone with the ylide derived from the allylic salt (**56**) ⟨55CB1415⟩ resulted in the formation of 1-(cyclohex-1-enyl)-3-cyclohexylidenepropene (**57**) via a [1,3]-hydrogen shift as shown in Scheme 28.

Finally, mention should be made of a related 'one pot' procedure ⟨75TL1359⟩ for the synthesis of conjugated polyene esters or ketones involving the condensation of allylidenetriphenylphosphorane with α,β-unsaturated β-chloro-carbonyl compounds, for example, (**58**), in the presence of an aldehyde (Scheme 29). The reaction is believed to involve analogous conjugate addition to give intermediates such as (**59**). Subsequent loss of chloride and deprotonation by base convert (**59**) into the

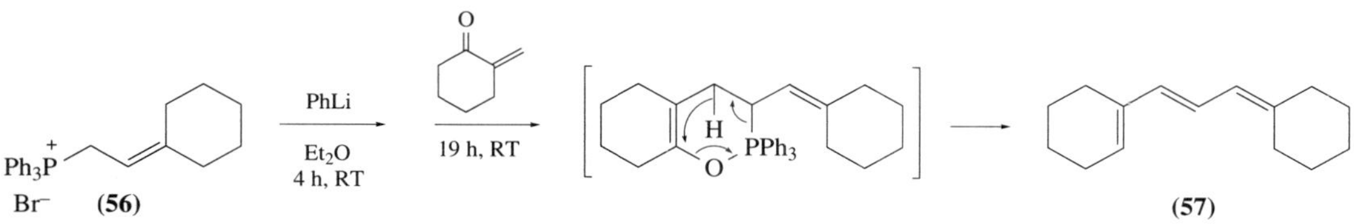

(56) (57)

Scheme 28

conjugated ylide (**60**) which is trapped by the aldehyde. It should be noted that trapping the ylide with benzaldehyde gives a high yield of the expected all-(*E*)-triene (**61a**) whereas the more bulky isobutyraldehyde affords a 40 : 60 mixture of (*Z*)- and (*E*)-isomers (**61b**) and (**61a**), respectively.

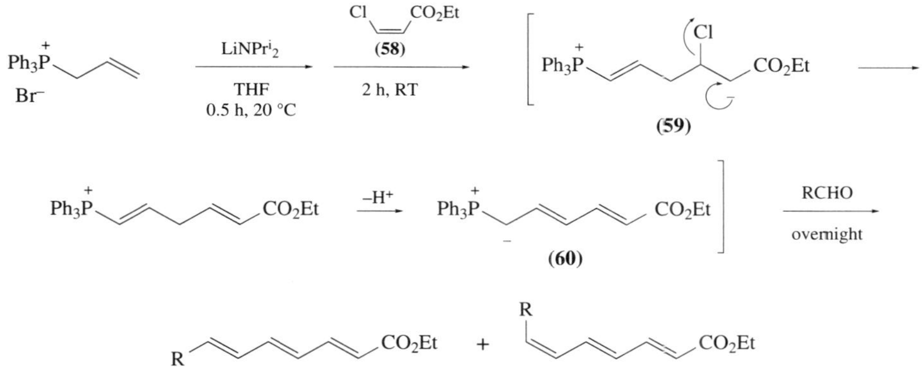

(61a) (61b)

R	*Yield* (%)	(*E*)		(*Z*)
Ph	84	100	:	0
Me_2CH	67	60	:	40

Scheme 29

1.16.1.3 Arsonium Ylides

Comparative studies, involving acylylides ⟨82OPP373, 88JCS(2)1829⟩, fluorenylides ⟨60JOC183⟩ and cyclopentadienylides ⟨72T343⟩, show that arsonium ylides are markedly more reactive than their phosphonium (and sulfonium) analogues. In many cases reactions proceed only in the case of the arsonium ylides; this is especially true the more electron-withdrawing are the substituents on the ylidic carbon atom, although some arsonium ylides with two electron-withdrawing substituents will not react even with aldehydes as reactive as 2,4-dinitrobenzaldehyde ⟨73T1697⟩. This difference has commonly been attributed (*inter alia*, ⟨64ACS655, 68CJC461, 68JOM(13)263, 71ACS1452, 72T353, 76CB473⟩) to the lower electronegativity of the arsenic atom, which leads to a lower electrostatic interaction between the arsenic and ylidic carbon atoms, and to a lower effectiveness of $p\pi$–$d\pi$ orbital overlap between these atoms because of the greater size and diffuseness of the arsenic 4*d*-orbitals compared with the 3*d*-orbitals of phosphorus or sulfur. But it has been pointed out that other factors in addition to electronegativity and $p\pi$–$d\pi$ orbital overlap must play a part in determining the relative acidity of the heteronium salts and the stability of the related ylides ⟨67LA(707)120⟩; the involvement of steric factors has been noted.

Arsonium ylides also differ in behaviour from their phosphorus counterparts in that in reactions with carbonyl compounds they may provide alkenes, or like sulfonium ylides, epoxides. The type of product formed depends on the nature of the substituents on both the ylidic carbon atom and the arsenic atom, and also to some extent on the solvent used; these aspects have been the subject of an excellent review ⟨B-94MI 116-01⟩. The first example of a reaction between an arsonium ylide and a carbonyl compound was recorded in a thesis in 1937 ⟨64MI 116-01⟩; it was reported that triphenylarsonium benzoylylide reacted with benzaldehyde to give benzylideneacetophenone. Two publications dealing with the reactions of arsonium ylides appeared in 1960. One described the formation of alkenes in high yield, starting from a fluorenylide ⟨60JOC183⟩, and the other reported that from triphenylarsonium methylide and benzophenone both 1,1-diphenylethylene and phenylacetaldehyde were obtained, with the latter predominating (1 : 3.5); it was suggested that the aldehyde arose from an initially formed epoxide during an acid workup of the reaction mixture ⟨60JA563⟩. Similarly a reaction between triphenylarsonium ethylide and *p*-tolualdehyde gave a mixture of a small amount of an alkene and, as principal product, *p*-tolylacetone, again formed by acid induced rearrangement of an initially formed epoxide ⟨64MI 116-02⟩.

Thus at an early stage in the study of arsonium ylides it was shown that either alkenes or epoxides might be formed, in contrast to the behaviour of phosphonium ylides, which gave only alkenes, and sulfonium ylides, which gave only epoxides. It was also apparent that arsonium ylides were more reactive than their phosphonium analogues, for while triphenylarsonium fluorenylide reacted with *p*-dimethylaminobenzaldehyde to give an alkene in high yield, triphenylphosphonium fluorenylide did not react with this aldehyde ⟨60JOC183⟩. This arsonium ylide reacted in high yield with a number of substituted benzaldehydes and with acetaldehyde; it did not react with acetone or acetophenone but did with the more reactive ketone *p*-nitroacetophenone ⟨82OPP373⟩.

The reaction of triphenylarsonium benzylide (**62**) with *p*-nitrobenzaldehyde (Scheme 30) provided an alkene and an epoxide in about equal amounts, together with equimolar amounts of triphenylarsine and triphenylarsine oxide ⟨65CI(L)1726⟩.

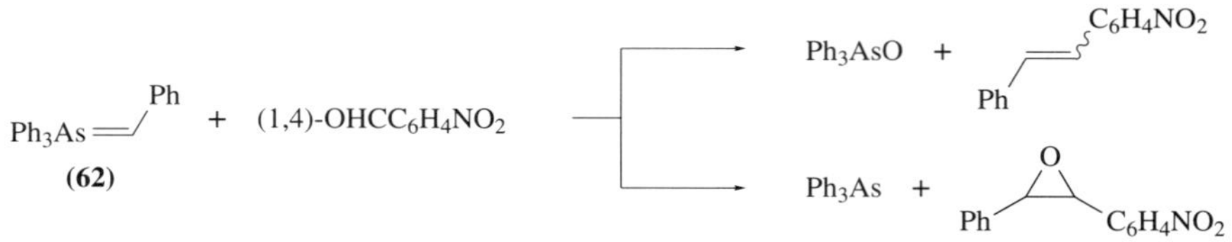

Scheme 30

By contrast, in another investigation of the reactions between arsonium ylides and aldehydes, it was found that *either* an alkene or an epoxide was formed, depending upon the identity of the ylide, but not both of them together ⟨71JCS(C)1114⟩. Both alkenes and epoxides were always *trans*. This usually appears to be the case.

1.16.1.3.1 Comparison of stable, reactive and semi-stabilised (moderately stabilised) ylides

The general pattern which has emerged is that stable arsonium ylides provide alkenes whilst reactive arsonium ylides give epoxides ⟨65BAU1434, 65CI(L)1726, 65IZV1474, 70JOC2678, 71JCS(C)1114⟩. This can be attributed to stabilisation of the transition state leading to alkene formation being provided by those same electron-withdrawing groups that stabilise the ylides ⟨65CI(L)1726, 71JCS(C)1114⟩.

Thus arsonium ylides stabilised by acyl groups ⟨70JOC2678, 75G109, 78IJC(B)623, 78S616, 79IJC(B)637, 82OPP373⟩, alkoxycarbonyl groups ⟨65MI 116-01, 73T1697, 85TL6447⟩, cyano groups ⟨81MI 116-01, 83MI 116-01⟩, and cyclopentadiene rings ⟨68CI(L)1277, 72T343, 72T353⟩ all provide alkenes, predominantly *trans*, as products from reactions with carbonyl compounds. Most ylides with two electron-withdrawing substituents do not, however, take part in Wittig reactions ⟨83T297, 88JCS(P2)1829⟩; steric factors may also sometimes inhibit reaction ⟨83T297⟩.

The arsonium salt $Ph_2(PhCH_2)AsCH_2CH_2OH\ Br^-$ requires two equivalents of butyl lithium for its conversion into an ylide $^-OCH_2CH_2(Ph_2)As{=}CHPh$; the presence of the alkoxide group does not prevent the ylide reacting with aldehydes to form a *trans*-alkene together with 2-hydroxyethyldiphenylarsine oxide ⟨90AG(E)1454⟩. In contrast, 2-hydroxyethyltriphenylarsonium bromide reacts with arylaldehydes in the presence of solid potassium hydroxide to give 2-aryl-3-hydroxymethyloxiranes ⟨88TL5295⟩.

Reactive ylides give good yields of *trans*-epoxides ⟨81JA1283⟩, although stereospecificity may vary with conditions. For example, use of an arsonium tetrafluoroborate as precursor of the ylide, and potassium bis(trimethylsilyl)amide as base gives 100% *trans*-epoxide, whereas with iodide as counterion and butyl lithium as base, there is less stereospecificity ⟨81JA1283⟩.

Semi-stabilised arsonium ylides are intermediate in behaviour between stable and reactive ylides, and may provide alkenes and/or epoxides ⟨69JCS(C)2464, 73JOM(60)247, 75JOM(85)173, 75JOM(96)237, 76JOM(108)175, 76JOM(112)279, 88BSB271, 89JOC3229⟩. Allylic arsonium ylides show such a pattern of reactivity. Ethoxycarbonylallyl ylides (**63**), wherein the ester group is conjugated with the ylidic carbon atom, afford dienes in reactions with aldehydes or ketones ⟨85S57, 87S370⟩, whereas other allylic ylides (**64**) lacking such an electron-withdrawing substituent give vinylic epoxides in high yield (Scheme 31) ⟨83SC1193, 91TL2913⟩. In the latter case it is found that the presence of hexamethylphosphoramide leads to the formation of a diene instead of the epoxide ⟨83SC1193, 91TL2913⟩. The ethynyl ylide $Ph_3As{=}CHC{\equiv}C$-TMS reacts with ketones to give conjugated enynes, giving higher yields than its phosphonium analogue ⟨88JOM(346)181⟩.

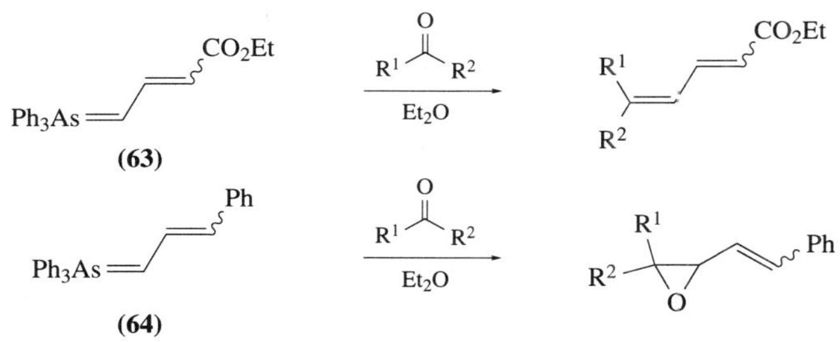

Scheme 31

In these cases, factors such as the substituent groups on arsenic, and the nature of the solvent and of the base, may become important in determining the nature of the product; this will be considered in more detail later. Small changes in the structure of the ylidic moiety may also have a marked effect; for example, whereas triphenylarsonium β-naphthylmethylide reacts to give epoxide, the presence of a bromine atom at the adjacent α-position of the naphthalene ring results instead in the formation of alkenes ⟨76JOM(108)175⟩. These results may be summarised as shown in Scheme 32.

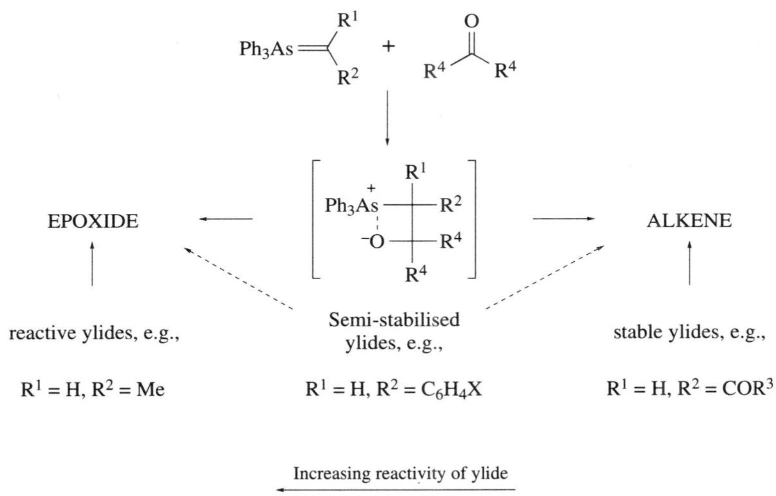

Scheme 32

(i) Effect of substituents on arsenic

The first report of such effects was in a study of the reactions of a series of tris(*p*-substituted phenyl)arsonium ylides with benzaldehyde; all these ylides gave epoxides in high yield save for the tris(*p*-dimethylamino) compound which gave instead the *trans*-alkene ⟨70MI 116-03⟩. In further experiments replacement of a triphenylarsonium group by a tris(*p*-methoxyphenyl)arsonium group was found to have little effect on the ratio of products ⟨71JCS(C)1124⟩, and inclusion of the arsenic atom in a strained ring also had no effect ⟨76JOM(110)315⟩.

A comprehensive investigation of a series of ylides (**65**) with benzaldehyde showed that as X, Y become more electron-donating, so the proportion of alkene to epoxide increases ⟨77AG(E)487⟩.

For example, the ratios of epoxide:alkene changed from ≈11:1 (X = Y = H) to ≈6.5:1 (X = Y = MeO), ≈6:1 (X = H, Y = NMe_2), ≈4:1 (X = NMe_2, Y = H), ≈1:1 (X = Y = NMe_2). Even more striking is the effect of replacing these aryl groups by alkyl groups, *viz* ratios of epoxide:alkene were Ph_3As, ≈11:1; Et_2PhAs, ≈1:5, Et_3As, ≈1:87 ⟨77AG(E)487⟩.

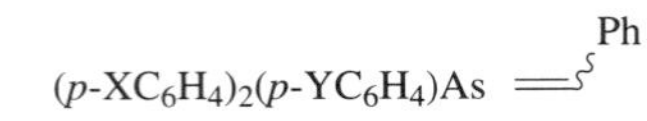

(65)

These results strongly suggest that remote control of the major pathway followed in reactions of semi-stabilised arsonium ylides with carbonyl compounds might be achieved by choice of the appropriate arsonium group. The reactivity of arsonium ylides is also affected by both electronic and steric effects on the arsenic atom.

(ii) Effects of solvent and base

Initial studies of solvent effects, on the reactions of triarylarsonium benzoylylides with *p*-nitrobenzaldehyde in *N,N*-dimethylformamide, dimethyl sulphoxide or methanol, indicated little solvent effect in these cases ⟨71ACS2541⟩, but later studies of the more finely balanced reactions of semi-stabilised ylides have provided examples of strong influences due to the effect of different base and solvent when the ylide is generated in the presence of a carbonyl compound ⟨68IC168, 83SC1193, 88BSB271, 89JOC3229, 91TL2913⟩. Thus, when benzyltriphenylarsonium bromide or *p*-chlorobenzyltriphenylarsonium bromide were treated with sodium hydride in benzene in the presence of a variety of *p*-substituted benzaldehydes the products were alkenes, but if sodium ethoxide in ethanol was used the isolated products were epoxides ⟨77TL3843, 79IJC(B)859⟩. Likewise, when triphenylarsonium benzylylide was generated by phenyllithium in the presence of either benzaldehyde or acetaldehyde, the preponderant product was the epoxide whereas use of sodium amide as base provided mostly the alkene ⟨88BSB271⟩. Similar results were obtained when an allyltriphenylarsonium salt was deprotonated using different hexamethyldisilazide salts at −65°C in tetrahydrofuran in the presence of aldehydes ⟨89JOC3229⟩. When the potassium salt was used as base the product was *trans*-alkene, but when the lithium salt was used a 2:1 mixture of *trans*- and *cis*-epoxides resulted. It was suggested that when the potassium salt is used, the reactants can equilibrate between the *cis* and *trans* form of the conjugated ylide, but that the lithium ion enhances production of a betaine and that this cyclises to give the two epoxides. Again scope for the control of the reaction, by suitable choice of solvent and/or base, at least of semi-stabilised ylides, is indicated.

(iii) Reactions with conjugated enones

Arsonium ylides may react with conjugated enones either to give dienes or by conjugate addition to provide cyclopropane derivatives ⟨80MI 116-01, 80MI 116-02⟩ as typified in Scheme 33.

Formation of a cyclopropane derivative rather than participation in a Wittig reaction appears to

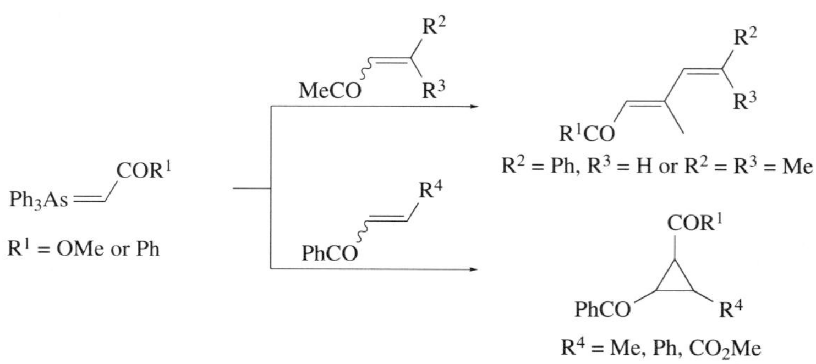

Scheme 33

be encouraged by the presence of a bulky group adjacent to the carbonyl group of the ketone ⟨80MI 116-02⟩. Enals and conjugated enals undergo Wittig reactions with arsonium ylides to give alkenes and dienes respectively ⟨87TL2155, 88TL3949⟩.

(iv) Reactions with other C=X functions

Semi-stabilised arsonium ylides react with thioketones; with cyclic thioketones triphenylarsonium benzylides gave only exocyclic alkenes and no thiirans ⟨80ZN(B)95⟩. Benzothiopyrones (and, better, benzopyrones) also react to give exocyclic alkenes, in these cases arylidenebenzopyrans ⟨80SC457⟩. Reaction of triphenylarsonium benzylide with benzylideneaniline gave 1,2,3-triphenylaziridine ⟨71JCS(C)1114⟩. This reaction is analogous to that of reactive arsonium ylides with carbonyl compounds.

(v) Mechanism

Since the behaviour of arsonium ylides in Wittig reactions appeared to be intermediate between that of sulphonium and phosphonium ylides, it was inferred that mechanisms similar to those accepted for the respective reactions of the latter ylides were involved (Scheme 34) ⟨65CI(L)1726⟩. The energetic driving force to generate an arsenic–oxygen bond is not as strong as that to form a phosphorus–oxygen bond, so that there is not the same compulsion to alkene formation in the case of arsonium ylides, allowing the alternative pathway (b) (Scheme 34) to compete. It has been suggested that in the case of stable ylides, where R^2 is an electron-withdrawing group, the presence of the latter group, which becomes conjugated with the carbon–carbon double bond in the final alkene, also stabilises the transition state leading to the formation of the alkene, thus promoting pathway (a) with respect to pathway (b) ⟨65CI(L)1726, 71JCS(C)1114⟩.

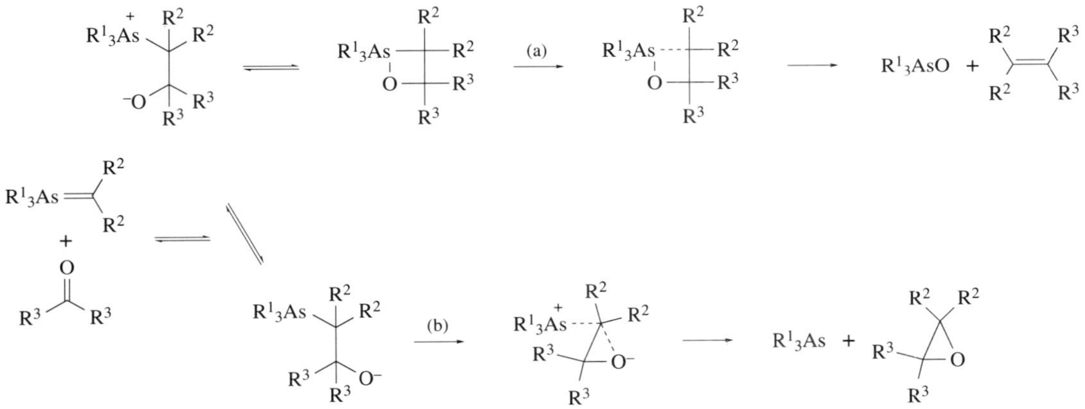

Scheme 34

The rates of the reactions of acylylides, $Ph_3As{=}CHCOX$ with ketones decreases as X is more electron-withdrawing, making the ylide less nucleophilic ⟨72BAU2056, 72IZV2113⟩. This observation, together with the relation between the reactivity and basicity of an ylide ⟨71ACS2541⟩ and the second-order character of the reaction, all suggest that the first step, which is slow and reversible, is the rate-determining step. The lack of solvent effect on the rates of reactions between benzoylylides and *p*-nitrobenzaldehyde led to the suggestion ⟨71ACS2541⟩ that in alkene formation the reaction goes directly to a four-membered ring transition state without the intermediate formation of a betaine. This was seen to be consistent with the large negative entropy of reaction and the very low activation energies observed.

Formation of an epoxide must, however, involve an intermediate betaine which reacts further by intramolecular displacement of an arsine. When pathways (a) and (b) (Scheme 34) are followed, the electrons in the arsenic–carbon bond are displaced in an opposite direction in the two mechanisms. In alkene formation displacement of electrons occurs away from the arsenic atom and in epoxide formation displacement of electrons is towards the arsenic atom. The change in pathway, depending upon the nature of the substituents at arsenic, could be associated with this, for electron-donating substituents on arsenic should assist displacement of the electrons away from arsenic and favour alkene formation as observed ⟨77AG(E)487⟩. For similar reasons electron-withdrawing substituents on the ylide carbon atom should favour alkene formation.

The observed solvent effects on the type of product formed ⟨69JCS(C)2464, 77TL3843, 79IJC(B)859⟩ could also be associated with the structure of the intermediate. Formation of alkenes or epoxides necessitates, respectively, *cisoid* and *transoid* arrangements of the arsenic and oxygen atoms, and a *transoid* structure is likely to be much more favoured in a polar protic solvent such as ethanol than in benzene. Similarly, as observed, the presence of lithium ions should favour epoxide formation compared to the effect of sodium or potassium ions since the *transoid* intermediate is likely to be stabilised by association of the oxygen with lithium but less so with sodium or potassium ⟨88BSB271, 89JOC3229⟩.

(vi) Conclusion

Reactions of arsonium ylides with carbonyl compounds take place much more readily than with phosphonium or sulphonium ylides. The nature of the products depends upon the character of the substituents on the ylide carbon atom, where electron-withdrawing substituents favour alkene formation, and of substituents on the arsenic atom, where electron-donating substituents favour alkene formation. This may be summarised as in Figure 1. However, choice of appropriate base and solvent can, in the case of less stabilised ylides, have an effect on both the product distribution and on the stereospecificity of the product. More detailed analysis is desirable, but it seems likely that, at least in the case of semi-stabilised ylides and possibly for others also, control over the product can be achieved by suitable choice of the substituents on arsenic, and of the solvent and base, thus making a valuable addition to the organic chemist's synthetic armoury.

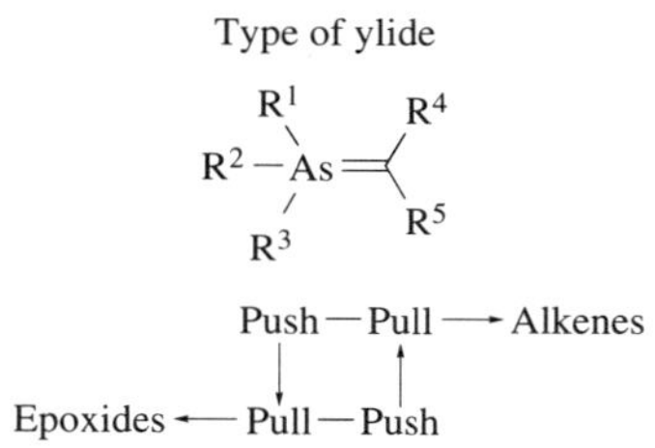

(Push, pull refer, respectively, to electron-donating or electron-withdrawing effects of substituents on the arsenic or carbon atoms)

Figure 1 Formation of alkenes or epoxides in the reaction of arsonium ylides with carbonyl compounds.

Reactions which may initially appear to involve ylides may not in fact do so. For example, a bromomalonic ester reacted with a variety of aldehydes in the presence of tributylarsine to form alkylidene– or arylidene–malonic esters, but no base was required and it was suggested that reaction proceeded not via an ylide but via an arsonium salt (**66**) formed by extraction of a bromonium ion from the malonic ester ⟨89JOM(375)45⟩. The malonate anion then makes a nucleophilic attack on the aldehyde (Scheme 35).

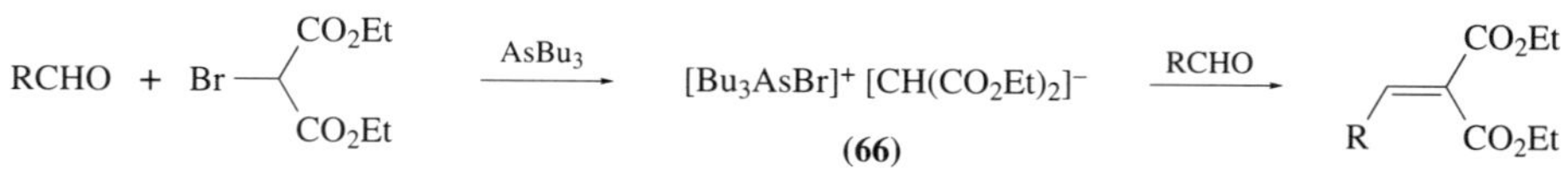

Scheme 35

1.16.1.4 Stibonium Ylides

Only limited information is available on the use of stibonium ylides in the Wittig reaction. Spectroscopic data on triphenylstibonium tetraphenylcyclopentadienylide has established that the ylide C—Sb bond is more polar and has less double-bond character than in the corresponding arsonium or phosphonium ylides. This is attributed to the less efficient overlap between the 2*p* orbitals of the ylide C-atom and the *d*-orbitals of antimony, because of the greater size and diffuseness of the *d*-orbitals in this case ⟨72T343, 72T353⟩. The greater dipolarity of the stibonium ylide compared to its arsonium, phosphonium and sulfonium analogues is also in accord with its higher reactivity towards benzaldehydes, and as a stable ylide, it gave alkenes only ⟨72T343⟩. By comparison, as might be expected for a reactive ylide, nonisolable triphenylstibonium methylylide reacted readily with benzophenone and formed an epoxide which, however, rearranged to diphenylacetaldehyde ⟨60JA563⟩. Tri-*n*-butylstibonium benzylide also reacted with aryl aldehydes, and like its arsonium analogue, gave mixtures of alkene and epoxide ⟨90CC85, 92ACR182⟩.

Tributylstibonium ylides having one electron-withdrawing substituent (e.g., (**67**)), react readily with aldehydes or ketones to give (*E*)-alkenes. Since the ylides are sensitive to air they were prepared *in situ* and not isolated. Evidence for their formation was provided by trapping the initial adduct formed from the ylide and aldehyde with trimethylsilyl chloride as shown in Scheme 36 ⟨90JCR(S)388⟩.

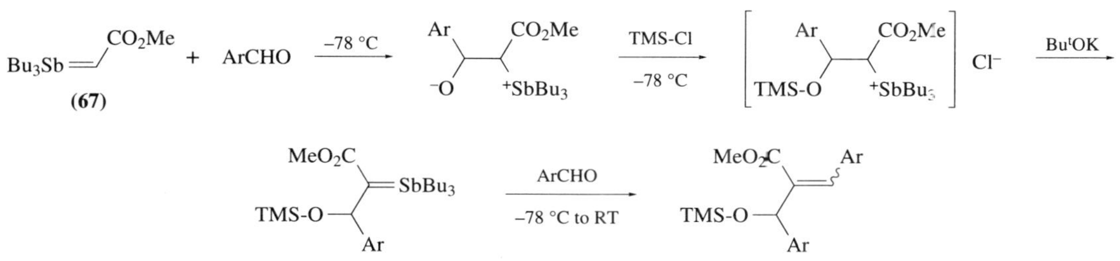

Scheme 36

Stibonium ylides bearing two electron-withdrawing substituents did not undergo Wittig reactions ⟨88JCS(P2)1829⟩.

As with comparable arsenic compounds, some α-halo esters and amides react with oxo-compounds in the presence of stibines, but without the need of base, to form arylidene or alkylidene derivatives ⟨89JOM(366)87⟩. The mechanism is thought to be the same as for the arsenic analogues, namely via initial formation of a stibonium salt (**68**) (Scheme 37).

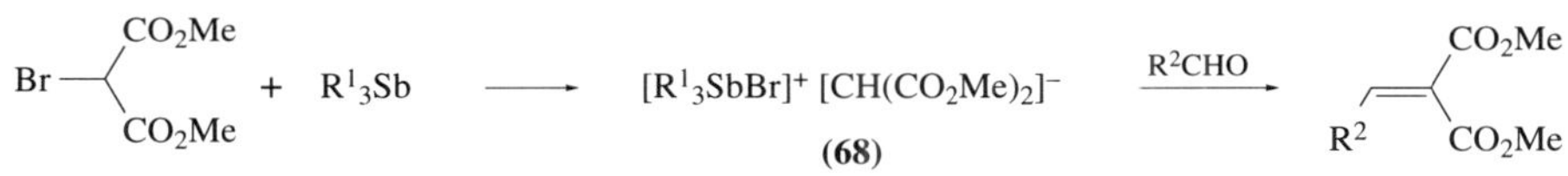

Scheme 37

It has also proved possible to make alkenes in a one-pot process from diazo-compounds N_2CR_2, where R is an electron-withdrawing group, and aldehydes or ketones by heating them in the presence of triphenylstibine; no base was required ⟨90TL5897⟩. The reaction was assumed to proceed through a stibonium ylide as intermediate, although it has proved impossible to isolate this ylide, which probably reacted further too rapidly to allow this to be accomplished.

1.16.1.5 Bismuthonium Ylides

Relatively little attention has been paid to bismuthonium ylides, and information on their effectiveness as a method of alkene formation from carbonyl compounds is limited. In the only example reported so far, the bismuthonium ylides (**69**; X = CH_2, O) reacted with aryl or alkyl aldehydes to give a variety of products, depending on the ylide, the aldehyde (RCHO) and the conditions. The products included cyclopropane derivatives (**70**), furan derivatives analogous to (**70**) but with a furan ring replacing the cyclopropane ring, and in some cases (from (**69**; X = O) and *p*-methoxybenzaldehyde or cinnamaldehyde) the straightforward product from a Wittig reaction (**71**) ⟨88CL849⟩.

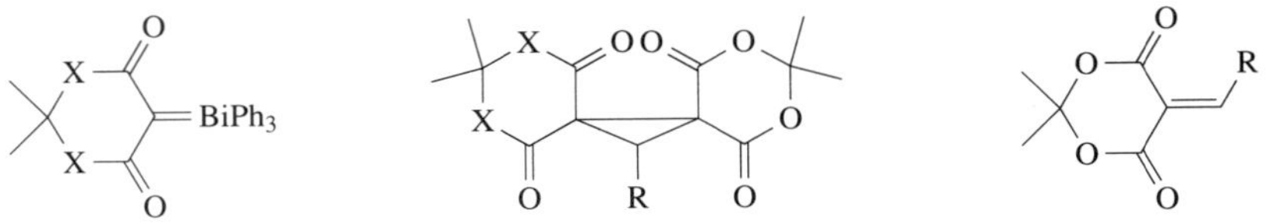

(69) **(70)** **(71)**

1.16.1.6 PO-Activated Alkene Formation

The use of PO-stabilised carbanions has come to challenge the Wittig reaction as the synthetic method of choice when a specific alkene is required. Two variants exist: the reaction of a phosphonate carbanion with a carbonyl compound to form an alkene is commonly referred to as the Wadsworth–Emmons reaction, whilst the use of a phosphine oxide stabilised anion is called the Horner reaction (Equation (14)). When compared to the Wittig reaction such PO-activated reactions offer a number of advantages, although the lesser stabilising effect of the neutral phosphoryl and phosphonate groups means that electron-withdrawing α-substituents are generally required at the carbanion centre before preparative yields of alkenes can be obtained. These advantages include: (i) greater nucleophilicity than the corresponding phosphonium ylide such that they react, often exothermically, with both aldehydes and ketones under milder conditions; (ii) the phosphinic, phosphonic and phosphoric acid derivatives obtained from PO-activated syntheses are water-soluble so separation from the alkene is easily achieved; and, (iii) considerable control of stereochemistry is possible by change of reaction conditions to yield the alkene with either the (*E*)- or the (*Z*)-geometry.

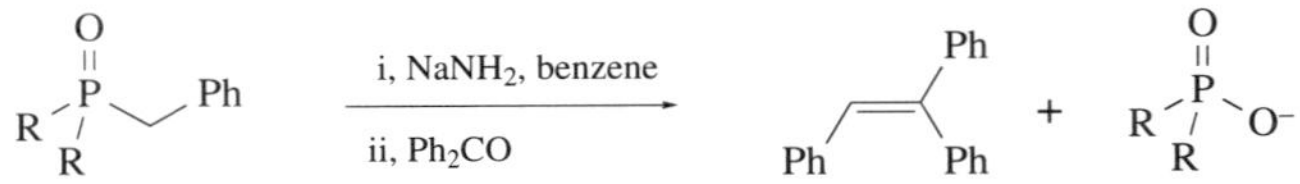

(14)

R = Ph, OEt

The scope, mechanism and stereochemistry of these types of reaction have been extensively reviewed ⟨74CRV87, B-79MI 116-01, 89CRV863, 91COS(1)761⟩, and only a summary of the mechanism will be presented here, with a more detailed account of the aspects relating to stereochemistry.

1.16.1.6.1 Mechanism

The accepted mechanism for PO-activated alkene synthesis is closely analogous to that of the Wittig reaction in that the phosphorus-stabilised carbanion reacts with the carbonyl compound in a reversible step to form a diastereomeric mixture of *erythro* (**72a**) and *threo* (**72b**) intermediate alkoxides that irreversibly decompose by *syn*-elimination of a water–phosphorus acid anion to give an alkene with specific geometry (Scheme 38). In the case of the Horner reaction, or the absence of resonance-stabilising α-substituents, the initial oxyanion intermediates may be trapped as their conjugate acids. The ratio of (*E*)- to (*Z*)-alkenes obtained in a given reaction depends largely on the relative rates of formation and decomposition of the intermediates (**72**) and to the extent of their equilibration. The development of an understanding of the various factors affecting these equilibria, including the effect of changing the base, solvent, temperature and the nature of the associated cation has led to successful attempts to control the stereochemical outcome in PO-activated alkene formation.

1.16.1.6.2 Phosphonate-stabilised carbanions (Wadsworth–Emmons Reaction)

The greater reactivity of phosphonate-stabilised carbanions over the corresponding Wittig reagent in alkene-forming reactions is ascribed to the fact that the phosphonate group has a lower net positive charge and accordingly provides less stabilisation for the adjacent carbanion by valence shell expansion, (**73a**)↔(**73b**) (Scheme 39). Thus ketones that react sluggishly, or not at all, with Wittig reagents (stabilised by an alkoxycarbonyl or acyl group) ⟨60CPB819⟩ are smoothly converted into the corresponding alkenes by their phosphonate counterparts under very mild conditions. Table 6 summarises this striking difference in reactivity for reactions of diethyl ethoxycarbonyl-

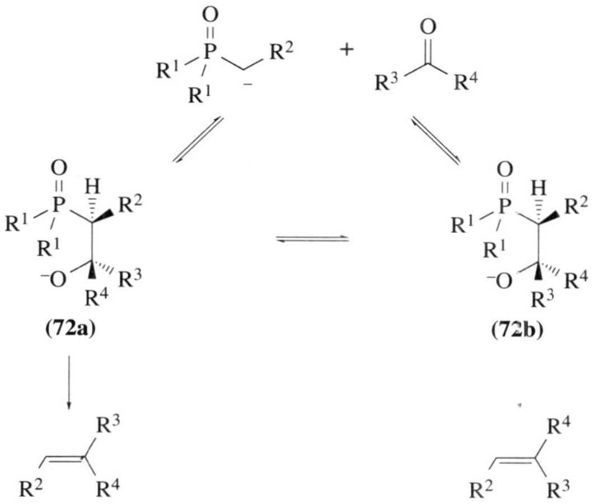

(72a) (72b)

Scheme 38

methylphosphonate and its phosphonium analogue with a variety of ketones. Ketones are generally less reactive towards phosphonate carbanions than aldehydes, and usually require much more vigorous conditions for alkene formation. This is primarily due to steric effects, and electronic effects seem less important, although trifluoromethyl ketones do show enhanced reactivity.

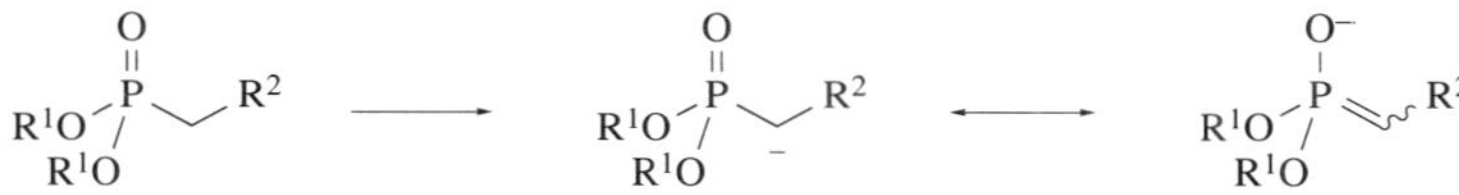

(73a) (73b)

R^1 = alkyl or phenyl; R^2 = stabilizing group, e.g. alkoxycarbonyl, acyl, nitrile, etc.

Scheme 39

A characteristic drawback to the use of phosphonate carbanions for alkene formation is that an electron-withdrawing α-substituent must be present at the carbanion centre. With hydrogen or alkyl groups as the only α-substituent ('unactivated' carbanions), reaction with carbonyl compound is rapid, but in most cases only poor yields of alkene are obtained ⟨74CRV87, B-79MI 116-01⟩, apparently because the intermediate alkoxide (**72**) requires 'activating' α-substituents (e.g., CO_2^-, CO_2Et, CN, aryl, vinyl, SO_2R, OR and NR_2), before it will decompose, often only under vigorous conditions.

*(i) Formation of (*E*)-alkenes*

Generally speaking (*E*)-alkene formation is favoured in Wadsworth–Emmons reactions when groups capable of conjugating with the incipient double bond are present. For example Royal Jelly (exclusively (*E*)) has been prepared from the reaction of diethyl ethoxycarbonylmethylphosphonate and the aldehyde (**74**) (Equation (15)) ⟨75JIC538⟩.

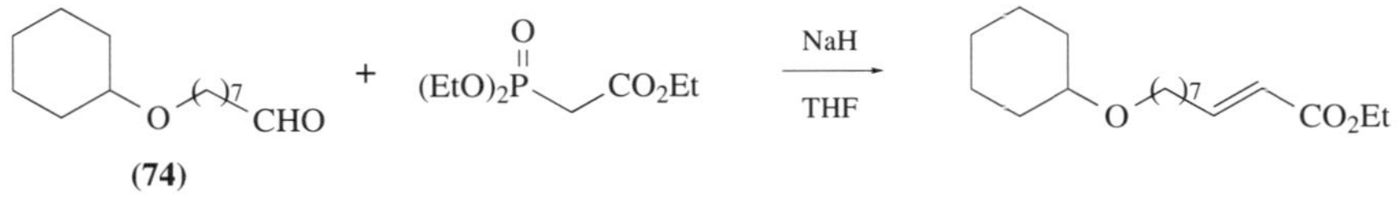

(74) (15)

Even in those cases where mixtures of (*E*)- and (*Z*)-alkenes are almost invariably obtained with the corresponding Wittig reagent ⟨64JPR143⟩, (*E*)-alkene is formed almost exclusively with phosphonate anions as in the reaction with α-naphthaldehyde to form the stilbene (**75**) (Equation (16)) ⟨69RZC95⟩; interestingly, the addition of catalytic amounts of 15-crown-5 to reactions using NaH in THF as base, is found to improve the yield of stilbenes substantially ⟨81JCS(P1)3087⟩.

Table 6 Relative yields of alkenes from the reaction of Wittig and phosphonate reagents with ketones.

Ketone	From $Ph_3P=CHCO_2Et$ Reaction conditions (Yield %)	From $(EtO_2)_2P(O)CH_2CO_2Et$ Reaction conditions (Yield %)	Alkene	Ref.
Et_2CO	xylene (reflux) (0)	NaOEt, DMF (ca. 65), 20 °C	$Et_2C=CHCO_2Et$	61CI(L)990
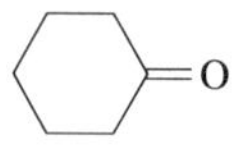	C_6H_6 (reflux) (25)	NaOEt, DMF (ca. 65), 20 °C	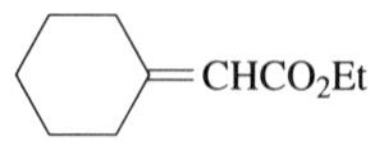	61CI(L)990
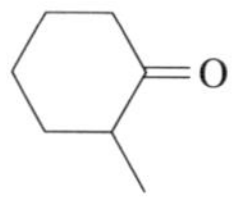	xylene (reflux) (0)	NaOEt, DMF (ca. 65), 20 °C	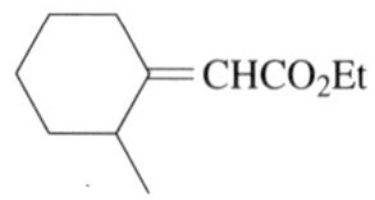	61CI(L)990
EtCOMe	xylene (reflux) (40)	$NaNH_2$, Et_2O (58)	$Et(Me)C=CHCO_2Et$	64IZV1341
PhCOMe	170 °C, 10 h (68)	$NaNH_2$, Et_2O (77)	$Ph(Me)C=CHCO_2Et$	72M1651
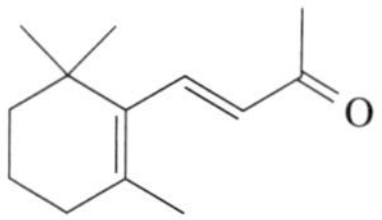	toluene (<50) 2 d (reflux)	NaOMe, MeOH (90), 40 °C, 10 h	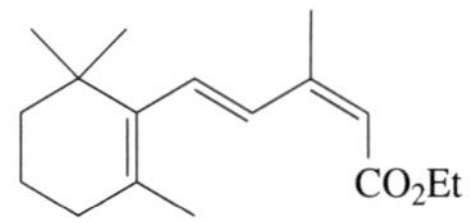 (Z) and (E)	73ACS1401
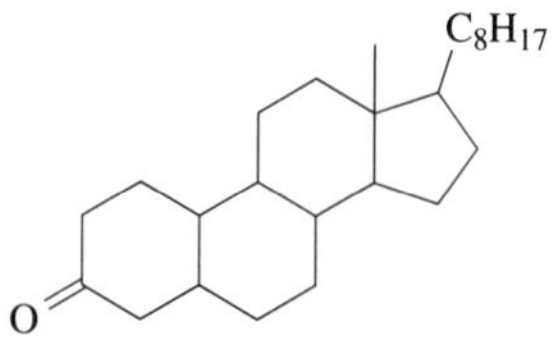	no reaction	NaOEt, DMF (93), 20 °C, 20 h	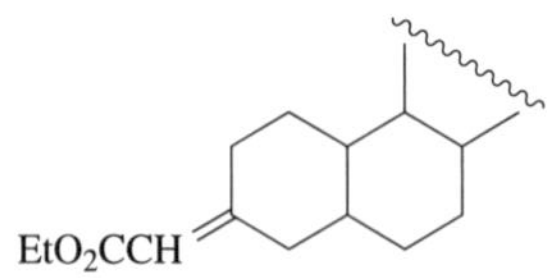	65JOC505
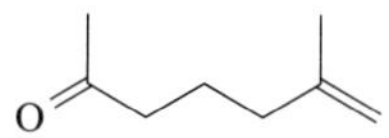	xylene, 6–14 h, reflux (15)	NaOEt, DMF (43), 30 °C, overnight	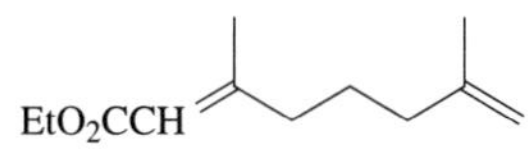	64IZV1341
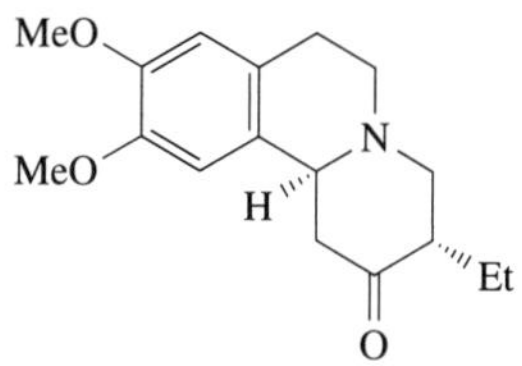	152 °C, 3 h (54)	$KOBu^t$, DMF, (99), 20 °C, 30 h	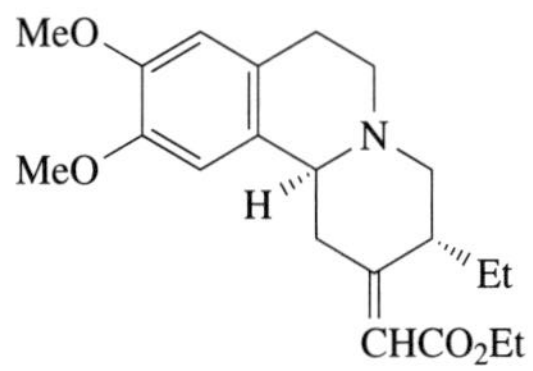	66JOC1447 63JCS1461

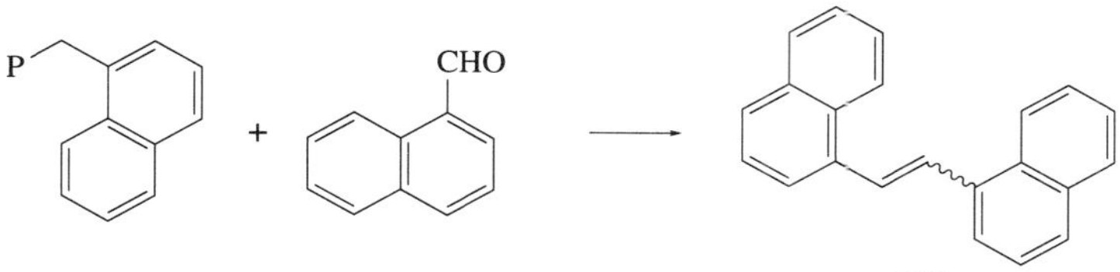

(16)

(75)

P	*Reaction conditions*	*Yield* (%)	(*Z*) : (*E*)
PH_3P^+–	NaOMe/MeOH	71	58 : 42
$(EtO)_2P(O)$	NaOMe/DMF	67	0 : 100

The use of the Wadsworth–Emmons reaction is generally the method of choice for the preparation of (*E*)-α,β-unsaturated carbonyl compounds, especially in natural product syntheses. For example, the C_8 side chain in prostaglandins, such as 10,10-dimethyl prostaglandin E_1 (**77**), can be introduced successfully by reaction of dimethyl 2-oxoheptylphosphonate (**76**) with the appropriate aldehyde followed by reduction of the keto group (Scheme 40) ⟨78JCS(P1)304⟩.

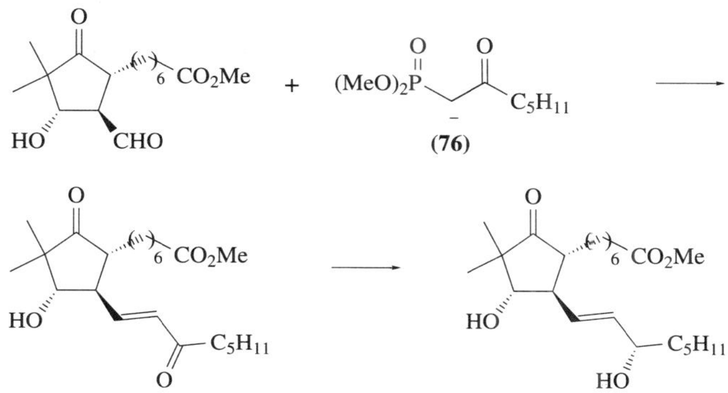

(76) **(77)**

Scheme 40

Extensive use has been made of the exclusive *trans*-stereochemistry shown by ethoxycarbonylmethyl phosphonate anions even when the conjugation is extended by unsaturation. The dehydro derivative (**80**) of the C_{18} juvenile hormone of *cecropia* has been synthesised by Corey *et al.* in 78% yield (99% stereoselectivity) from the condensation of phosphonate (**79**) with aldehyde (**78**) (Equation (17)) ⟨71TL1821⟩. A Wittig reaction between the same aldehyde and the corresponding (*E*)-triphenylphosphorane afforded non-stereoselective mixtures of (*E*)- and (*Z*)-alkene.

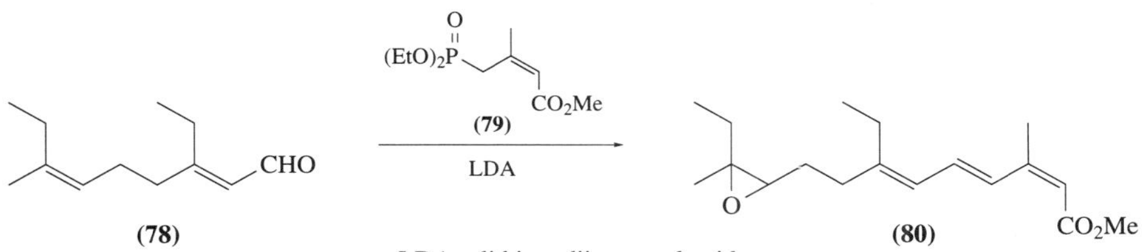

(17)

(78) **(79)** **(80)**

LDA = lithium diisopropylamide

Significant loss of *trans*-stereoselectivity will occur when steric interactions in the product become serious ⟨68BCJ1252, 68CC1699, 68CJC2225⟩. For example, in the reactions of diethyl ethoxycarbonylalkylphosphonates (**81**) with aliphatic aldehydes (**82**) (Scheme 41) ⟨68CC1699⟩, only (*E*)-ester (**83**) is formed when $R^1 = H$. With the size of R^1 increasing together with the branching of the alkyl group R^2 of the aldehyde, the (*Z*)-ester (**84**) becomes the major product of the reaction.

The reverse effect is found upon incorporation of an α-fluoro substituent even with bulky aldehydes. For example, condensation of $(EtO)_2P(O)CH(F)CO_2Me$ with isobutyraldehyde in the presence of $Bu^nLi/THF(-78\,°C)$ afforded a 90% yield of $Pr^iCH{=}C(F)CO_2Me$ with almost complete *trans*-selectivity ⟨85BSF(2)448⟩. This higher than expected *trans*-stereoselection is attributed to electronic effects ⟨84T2871⟩ and has been applied to the synthesis of fluorinated sex pheromones.

Loss of *trans*-stereoselectivity does occur in condensations with unsymmetrical ketones ⟨65IZV1504, 67JA5292, 68CC543, 69AJC1737, 69AJC2145, 71ABC1116, 73ACS1401⟩. Two examples from the synthesis of

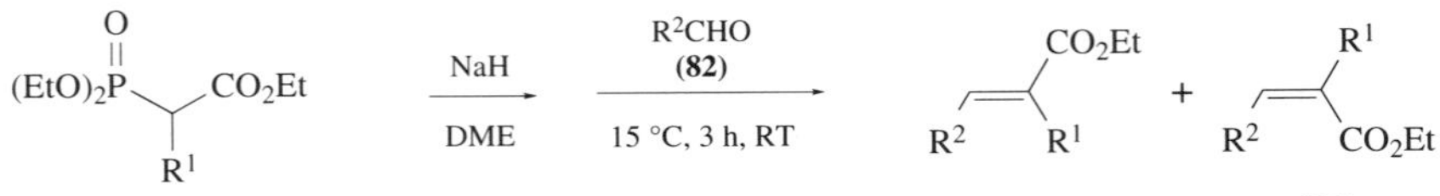

(81) (83) (84)

R¹	*R²*	*% (84) in the product*
H	Pr^i	0
H	Bu^t	0
Me	Me	0
Me	Et	16
Me	Pr^i	62–73
Me	Bu^i	67
Et	Me	18
Et	Et	41
Et	Pr^i	84

Scheme 41

natural products are illustrative of this (Equation (18) and Scheme 42). Thus, in the first step of the synthesis of racemic *β,γ*-carotene ⟨73ACS1401⟩, condensation of *β*-ionone (**85**) with diethyl methoxycarbonylmethylphosphonate in the presence of sodium methoxide gave an isomer mixture containing 35% of methyl (*Z*)-*β*-ionylidene acetate (**86b**). Likewise, during the synthesis of the juvenile hormone isolated from the giant silkworm moth *Hyalophora cecropia* ⟨69AJC1737, 69AJC2145⟩, treatment of the methyl ketone (**87**) with the sodium salt of trimethyl phosphonacetate in DME led to a mixture of the (*E,E*)- (**88a**) and (*Z,E*)-ester (**88b**) in yield in the ratio 60 : 40.

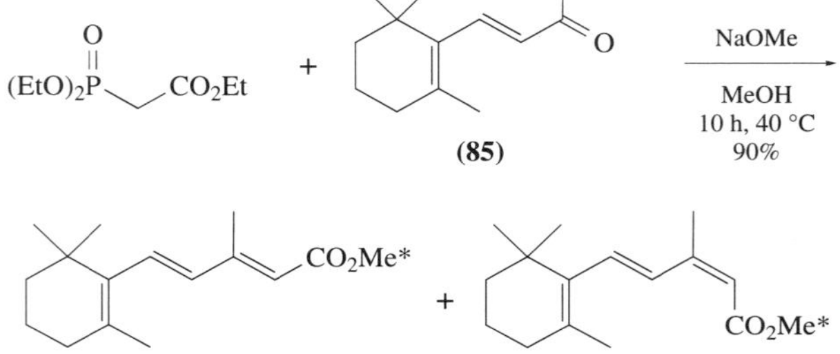

(85)

(86a) 65 : 35 (86b) (18)

* ester change

trans-Selectivity can be maximised by increasing the size of the substituents on both the phosphonate and stabilising ester groups. For example, in the reaction with 2-phenylpropanal, the α-methyl substituted ethyl phosphonate (**89**) gave a ratio of 95 : 5 of (*Z*)-(**90**) and *E*-(**90**) (Scheme 43). By using bulky isopropyl groupings instead, the stereoselectivity is dramatically reversed and the corresponding reaction with (**91**) occurs to give 5 : 95 ratio of (*Z*)-(**92**) and (*E*)-(**92**) ⟨81T3873, 82JA1109, 87JA4755⟩.

This approach has been used to ensure *trans*-alkene formation in natural product syntheses. One example of its utility is represented by the synthesis of brefeldin C in which the (*E*)-alkene (**93**) was selectively synthesised in 72% yield without epimerisation (Equation (19)) ⟨89TL4845⟩.

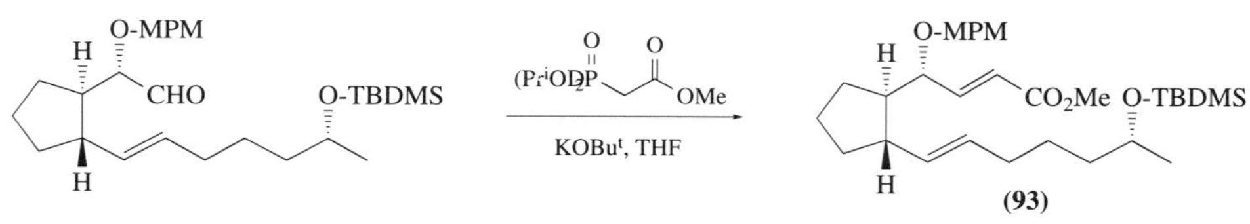

(93) (19)

The same effect is observed with α-cyanophosphonates which are usually less stereodemanding than their carbalkoxy counterparts and produce mixtures of (*Z*)- and (*E*)-alkenes in the range 1 : 4 to 2 : 1 ⟨80SC509, 80ZOB76⟩. For example, in the reaction with *β*-ionone (**85**), $(Pr^iO_2)_2P(O)CHCN^-Na^+$ is found to be more *trans*-selective ((*Z*)/(*E*) = 18 : 15) than the corresponding diethyl phosphonate

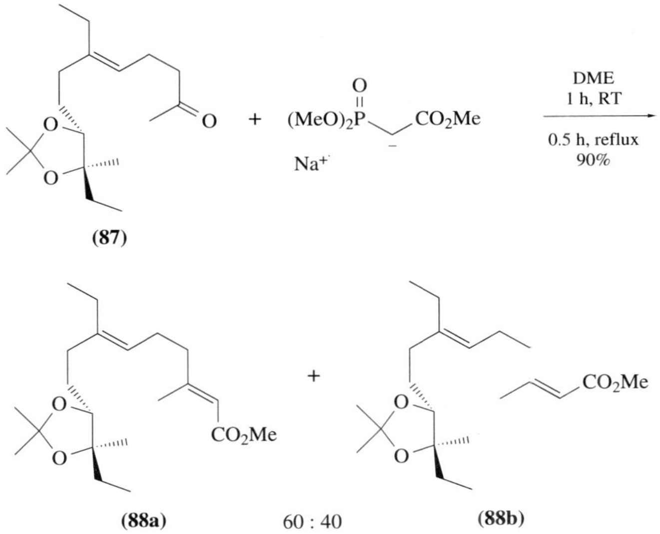

Scheme 42

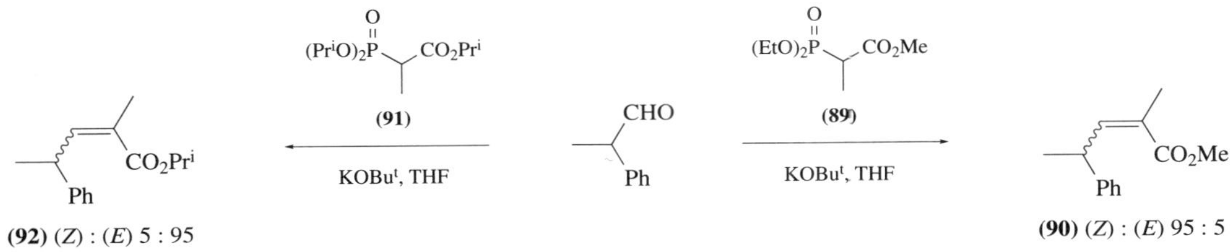

Scheme 43

((*Z*)/(*E*) = 35 : 65) ⟨80SC509⟩ (see Equation (18) and Scheme 42). Choice of solvent can also play a role in determining the final product ratios; such an effect is ascribed to the extent of intermediate alkoxide reversibility. In benzene, the reaction of diethyl cyanomethylphosphonate with the ketone **(94)** is (*E*)-selective, whereas in DMF or DMSO, a moderate level of (*Z*)-selectivity is obtained (Equation (20)).

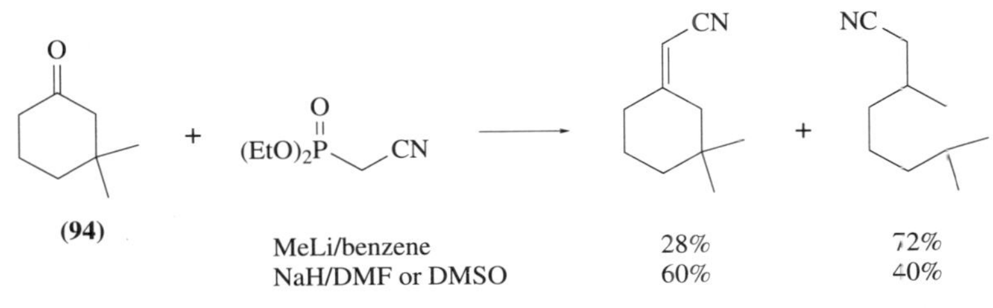

(20)

Bulky ester groups can also induce high (*E*)-selectivity even in reactions with hindered ketones. Thus, condensation of the *t*-butyl ester phosphonate **(96)** with ketone **(95)** affords the (*E*)-product **(97)** with only a trace of the (*Z*)-isomer (Scheme 44) ⟨86JOC959⟩. Notably, when the *t*-butyl group is replaced by a large alkyl chain, selectivity is lost and in the corresponding condensation with phosphonate **(98)**, a 1 : 4 mixture of (*Z*)/(*E*)-**(99)** is formed ⟨87JOC1372⟩.

For reactants that racemise easily or are base-sensitive a new protocol devised by Masamune and co-workers ⟨84TL2183⟩ exists for generating phosphonate carbanions under mild conditions. This involves use of lithium chloride and a weak base such as 1,5-diazabicyclo[5.4.0]undec-5-ene (dbu) or diisopropylamine. In one exemplary case directed at a key intermediate for the synthesis of norescurinine, Heathcock ⟨87H(75)75⟩ demonstrated that potassium *t*-butoxide gave epimerised

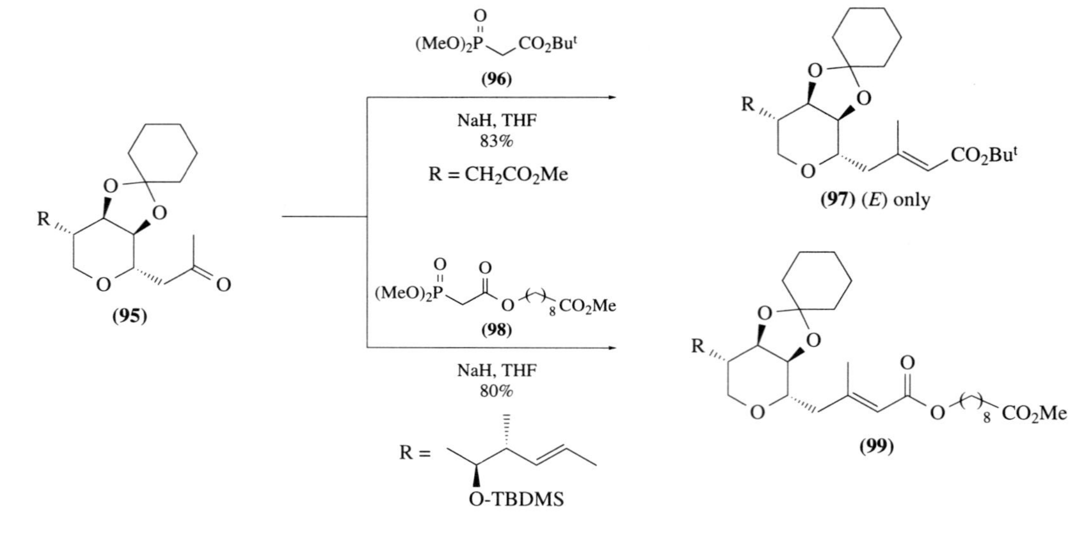

Scheme 44

material in the reaction of phosphonate anion (**100**) with the complex aldehyde (**101**), but that LiCl/dbu circumvented racemisation to produce (**102**) in 80% yield and 93% *ee* (Equation (21)).

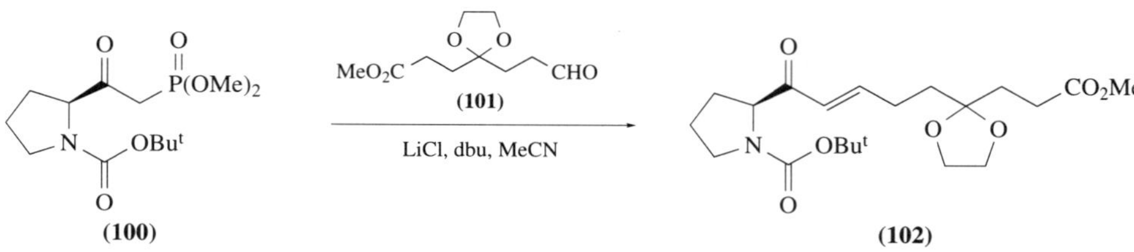

(21)

dbu = 1,5-diazabicyclo[5.4.0]undec-5-ene

Another example of the utility of this modification is the synthesis of amphotericin-B by Nicolaou ⟨88JA4685⟩.

(ii) Formation of (Z)-alkenes

One method for obtaining almost pure (*Z*)-alkene from the reaction of phosphonate carbanions with aldehydes makes use of the fact that intermediate alkoxides with magnesium as the cation are much more stable than the comparable sodium or lithium complexes (Scheme 45) ⟨72T4209, 73T4199⟩.

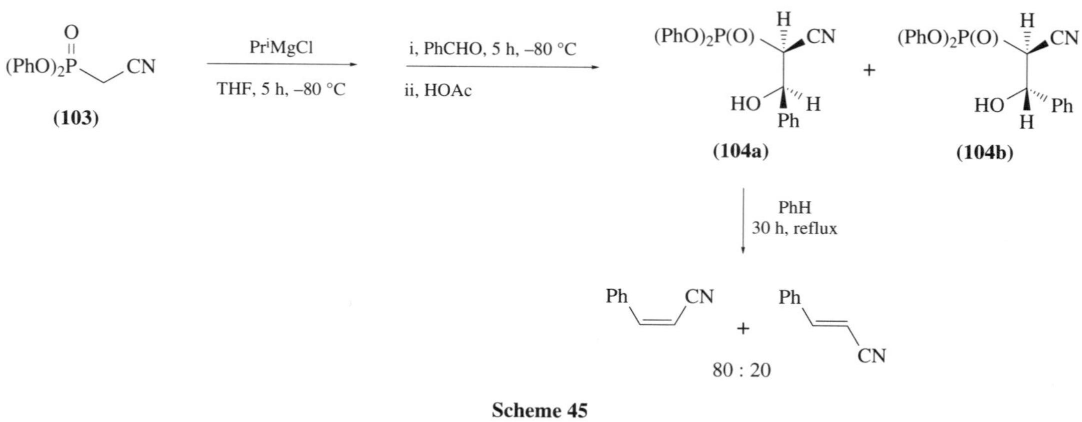

Scheme 45

When hydrolysed, these reaction mixtures afford diastereomeric β-hydroxyphosphonates from which the *erythro* component may be separated by fractional crystallisation and stereospecifically converted into the corresponding (Z)-alkene by heating. For example ⟨72T4209⟩, in the direct reaction between diphenyl cyanomethylphosphonate (**103**) and benzaldehyde under normal conditions, the ratio of (Z)- to (E)-cinnamonitriles is 15 : 85, whereas thermal decomposition of the isolated *erythro*-adduct (**104a**) gives a (Z) : (E) ratio of 80 : 20.

Incorporation of the phosphonate group into a ring as in structure (**105**) promotes (Z)-alkene selectivity in reactions with aromatic aldehydes as well as enhancing reactivity compared to its acyclic counterpart $(EtO_2)P(O)CH_2CO_2Et$ (Scheme 46) ⟨83PS(16)339⟩. This high (Z)-selectivity (65–70%) is attributed ⟨77TL1137, 78T997⟩ to a more rapid closure to the pentacoordinate intermediate (**106**) owing to release of strain in the five-membered ring coupled with an increase in the rate of elimination relative to the equilibration of intermediates. Replacement of the CO_2Et grouping in (**105**) by CN brings about a reversal in selectivity to *trans* (*vide supra*), whereas with the corresponding *t*-butyl ester, very high levels of (Z)-alkenes ($>95\%$) are obtained instead ⟨86MI 116-01⟩.

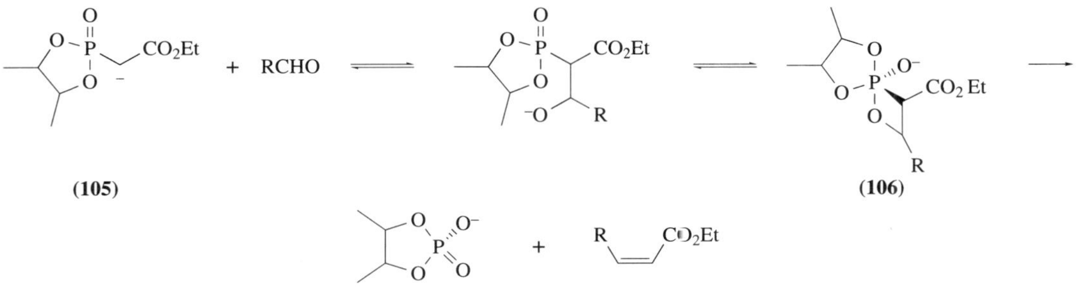

(**105**) (**106**)

Scheme 46

Dramatically higher levels of *cis*-selectivity can also be obtained by a variant and introduced by Still and Gennari ⟨83TL4405⟩ in 1983 that uses bis(trifluoroethyl)phosphonates (e.g., (**107**) and (**108**)) in reactions with aldehydes. The magnitude of selectivity is dependent on the base and solvent combination used, and in this respect, potassium hexamethyldisilazide with 18-crown-6 in THF appears to be most effective by increasing the rate of elimination relative to equilibration due to minimal complexation of the intermediate alkoxide with its counterion. For example, phosphonate (**107**) combined with benzaldehyde to give a $>50:1$ (Z)/(E)-ratio of alkenes that dropped to 12 : 1 with octanal ⟨83TL4405⟩. The improved *cis*-selectivity in such reactions is attributed to the electron-withdrawing effect of the trifluoromethyl groups that forces an increase in the rate of elimination of alkene from the initially formed β-hydroxyphosphonate adduct before equilibration can take place (cf. **72**). Still's method has been used extensively in synthesis ⟨88JA2248, 88JA3929, 88JOC4274⟩. A typical example of its utility, with an aldehyde possessing an α-ether functionality, is represented by Danishefsky's synthesis of the ester (**109**) which was contaminated with only 5% of the (E)-isomer (Equation (22)) ⟨88JA3929⟩.

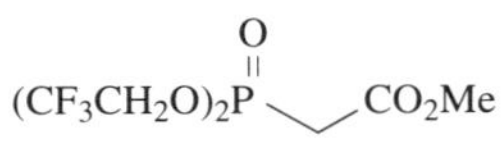

(**107**)

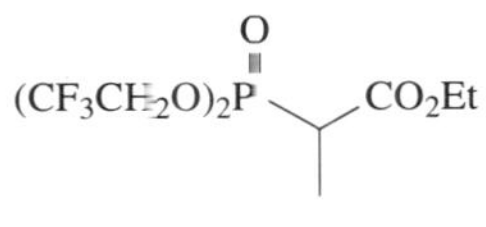

(**108**)

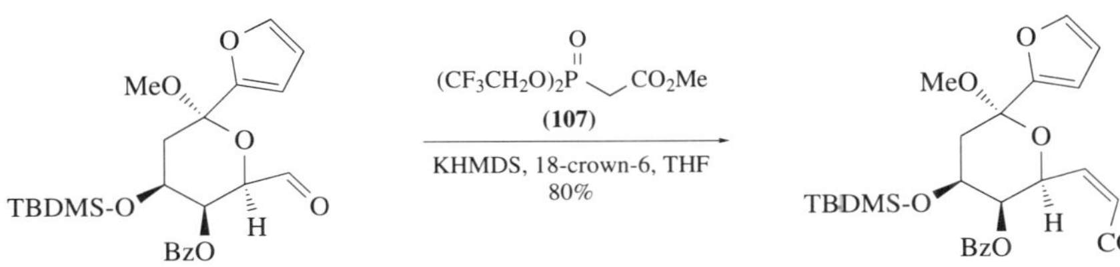

(22)

(**109**)

KHMDS = potassium hexamethyldisilazide (Z) : (E) > 95 : 5

Still's approach is also effective with a variety of carbanion stabilising groups other than esters, including cyano which normally exhibits poor stereoselectivity when used as a diethylphosphonate.

In one interesting example ⟨88TL419⟩, the conjugated cyano phosphonate (**111**) condensed with the *cis*-isomer (**110**) of β-ionone to produce the highly strained all-*cis*-retinal (**112**) (Equation (23)); the crude yield was 45% with 72% of the material having the *cis*-configuration at the newly formed double bond (C-11).

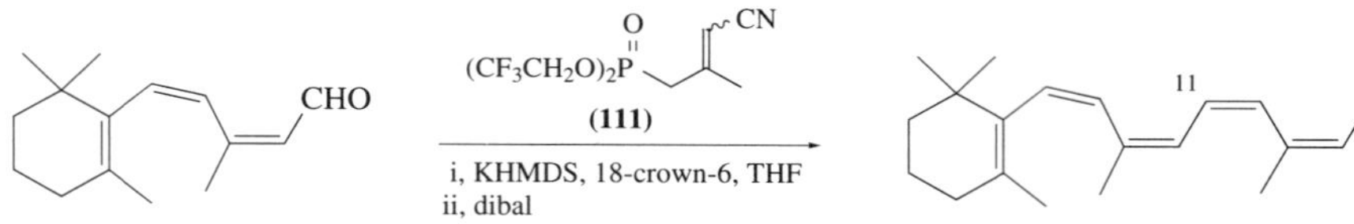

(23)

(110) **(112)**

dibal = diisobutylaluminum as a mixture of isomers

Good levels of *cis*-selectivity are also obtained even with α-methyl substituents. For example, the reaction of trifluoroethylphosphonate (**108**) with benzaldehyde afforded a 30 : 1 ratio of (*Z*)/(*E*) isomers, which is reversed to 1 : 22 in the corresponding reaction with $(EtO)_2P(O)CH_2(Me)CO_2Et$ ⟨83TL4405⟩. The effect is general, and is further exemplified by the synthesis of methyl tri-substituted alkenes from the condensation of (**108**) with aldehydes (**113**) ⟨87TL3075⟩, (**114**) ⟨88H(27)2077⟩ and (**115**) ⟨87JOC3883⟩ with virtually completely (*Z*)-selectivity. In an interesting application to natural product synthesis, Roush and Palkowitz ⟨89JOC3009⟩ have synthesised the C(1)–C(15) segment of streptovaricin by applying Still's phosphonate to form the (*Z*)-alkene (**117**) from the α-methyl substituted aldehyde (**116**) with greater than 90% selectivity (Equation (24)). Even with groups larger than methyl, *cis*-stereoselectivity is compromised to only a small extent.

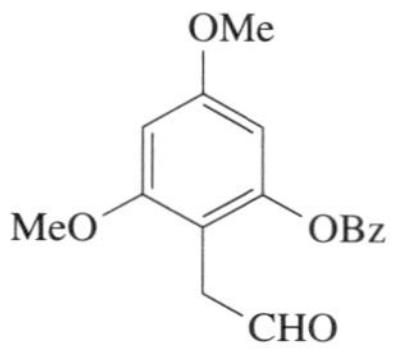

(113)

(114)

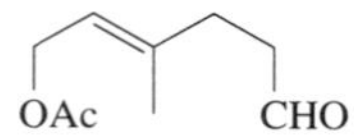

(115)

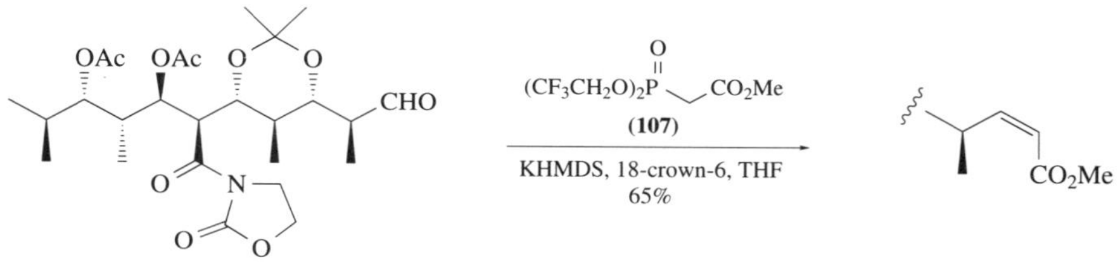

(24)

(116) **(117)**

In a further variant, Corey and Wiatkowski ⟨88JA5653⟩ have introduced the use of phosphoric acid bisamides, for example, $(Me_2N)_2P(O)CH_2R$ (R = alkyl or aryl) for alkenation reactions. For R = aryl, reactions with aldehydes are well controlled to give only *erythro*-adducts which decompose stereospecifically to (*Z*)-alkenes. Reactions with alkyl substituents are not stereoselective. For example, with R = *n*-C_7H_{15} condensation with pivalaldehyde produced a 3 : 1 (*Z*) : (*E*) mixture of alkenes compared to the corresponding reaction of the triphenylphosphonium ylide which gave a 98.5 : 1.5 mixture in favour of the (*Z*)-isomer.

1.16.1.6.3 Phosphoryl-stabilised carbanions (Horner Reaction)

As far back as 1958, Horner ⟨58CB61⟩ discovered the alkene forming reaction that bears his name by treatment of methyldiphenylphosphine oxide (**118**) with sodamide and then benzophenone to give 1,1-diphenylethene (**119**) in 70% yield together with diphenylphosphinic acid (Scheme 47). The intervention of an alkoxide intermediate corresponding to the conventional Wittig-type betaine was first proved ⟨59CB2499⟩ by isolation of the β-hydroxyphosphine oxide (**121**) from the reaction of benzyldiphenylphosphine oxide (**120**) with phenyllithium and benzaldehyde, followed by quenching with acid (Equation (25)).

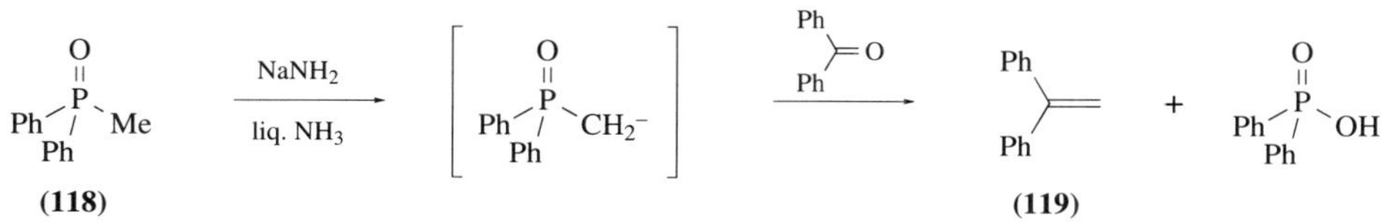

Scheme 47

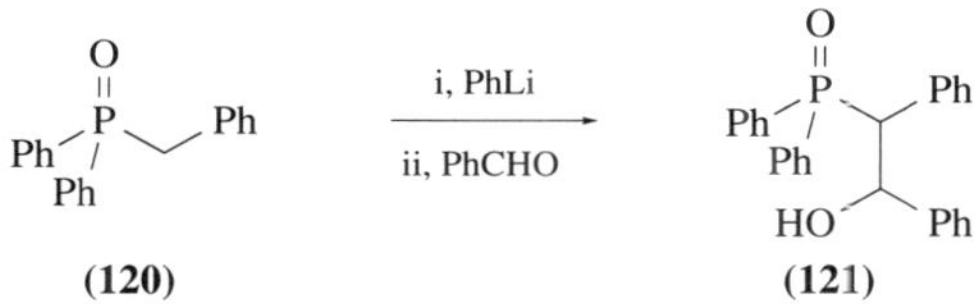

(25)

By using potassium *t*-butoxide as base, no adduct was obtained, but 1,2-diphenylethene was isolated in good yield. In a further significant discovery in 1964, Horner ⟨64TL2467⟩ showed that the *β*-hydroxyphosphine oxide (**121**) consisted of a mixture of diastereomers that could be separated by chromatography into a minor isomer (5%) with the *erythro*-configuration (**122a**) and a major isomer (90%) with the *threo*-configuration (**122b**) (Scheme 48).

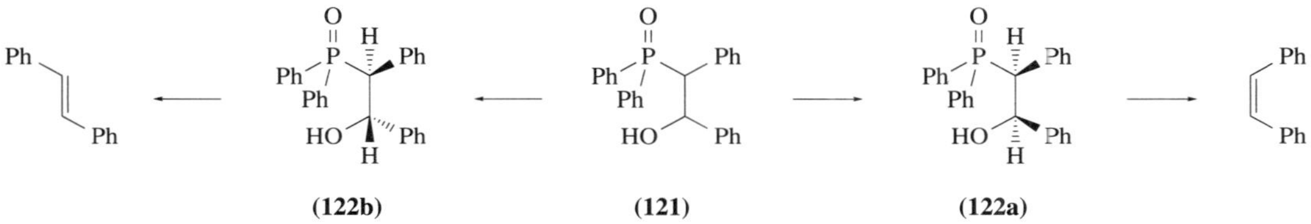

Scheme 48

Assignment of configuration was based on subsequent decomposition of the adducts by treatment with phenyllithium and lithium hydride to give (*Z*)- and (*E*)-1,2-diphenylethene, respectively, by *syn*-elimination of diphenylphosphinic acid. This finding laid the basis for the use of anions from phosphine oxides to form alkenes of specific geometry; however the intermediate *β*-hydroxyphosphine oxide can be isolated only if lithium is used as the counterion to the base for deprotonation ⟨58CB61, 59CB2499⟩. Sodium or potassium bases promote *in situ* decomposition to form alkenes directly. The fact that the Horner reaction can be stopped at the first stage and the intermediate adducts isolated and separated into pure diastereomers, that are frequently crystalline, does offer a definite advantage over the more conventional Wittig reaction in that each diastereomer (*erythro* and *threo*) offers direct access to a single geometric isomer of the corresponding alkene ⟨85JCS(P1)2307⟩. An added bonus is that the phosphine oxide reagents do not require an α-stabilising substituent (cf. phosphonate reagents) to effect alkenation. The presence of such groups, for example, carbonyl, even when protected as the ethylene acetal, does place a limitation on the Horner reaction by inhibiting the decomposition of the intermediate *β*-hydroxyphosphine oxide and thus preventing the formation of alkenes ⟨83JCS(P1)1173⟩. Furthermore, such adducts even with moderate stabilising groups next to $Ph_2P(O)$ (e.g., phenyl) have to be isolated at low temperatures otherwise the reaction proceeds stereorandomly to give a mixture of alkenes.

Other complications do exist and specifically involve loss of stereospecificity in the case of *erythro*-adducts by equilibration with the starting materials, a problem that is exacerbated with aromatic aldehydes ⟨83TL3931⟩; however specific reaction conditions have been developed to bring about greater *erythro*/*threo* ratios ⟨83TL3931, 85JCS(P1)2307⟩. It is also worth mentioning at this point of an alternative approach to the intermediate *β*-hydroxyphosphine oxides by ring-opening of epoxides. Bridges and Whitham ⟨74CC142⟩ originally used this method to carry out a stereospecific *cis* to *trans* alkene conversion by isolating pure diastereomers of *β*-hydroxyphosphone oxides (**123**) from treatment of the derived epoxide with lithium diphenylphosphide (Scheme 49). The (*E*)-alkenes were then obtained by *syn* elimination initiated by a strong base, usually NaH in DMF.

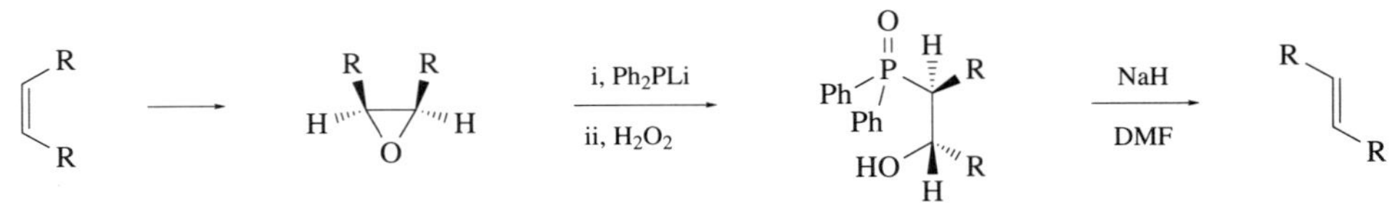

(123)

Scheme 49

In a variant to this method, epoxidation of allylphosphine oxides followed by ring-opening with nucleophiles has been employed to generate the intermediate hydroxyphosphine oxides ⟨84CL1997, 85TL2119, 85TL5709⟩. Such oxidations normally proceed in a highly stereoselective manner provided the allylphosphine oxide is trisubstituted.

(i) Stereoselective formation of (E)- and (Z)-alkenes

The reversibility of the first step in the Horner reaction coupled with the diminished rate of elimination from *erythro*-adducts can be utilised to synthesise (*E*)-alkenes without isolation of intermediates ⟨76JCS(P1)2386⟩. Steric factors are important as in the approach by Lythgoe *et al.* ⟨78JCS(P1)590⟩ to the synthesis of the diene portion of vitamin D_2 (**126**) (Equation (26)). Thus, metallation of the phosphine oxide (**124**) followed by treatment under mild conditions with the ketone (**125**) gave, after deprotection, the newly formed trisubstituted double-bond with the desired (*E*)-geometry. The same technique has been used to synthesise other conjugated dienes with specific *trans* geometry, including polyene isoprenoids. For example ⟨78TL4159⟩, treatment of the lithiated derivative of the phosphine oxide (**127**) with (*E*)-citral (**128**) gave rise to the isoprenoid (**129**) with almost complete (>98%) *trans*-selectivity (Equation (27)). By comparison, condensation of the same aldehyde with the corresponding phosphonium ylide led to a 50 : 50 mixture of (*E*)- and (*Z*)-8-polyenes. Change of base is often advantageous. For example, Nicolaou *et al.* ⟨84CC349⟩ synthesised the pentaene (**130**) with specific (*E*)-geometry by carrying out the initial addition with LDA followed by use of potassium *t*-butoxide to effect elimination (Equation (28)). In the total synthesis of milbemycin, Smith and co-workers ⟨86JA2662⟩ also found that in the condensation of the phosphine oxide (**131**) with aldehyde (**132**), the sensitivity of the alkene stereochemistry depended profoundly on the choice of base (Equation (29)). The use of NaH led to a 1 : 7 (*Z*)/(*E*) ratio, but owing to epimerisation at the aldehydic methine, the yield was only 15%. This improved to 74% with potassium hexamethyldisilazide, but the use of this base, gave only a 2 : 3 (*Z*)/(*E*) ratio. These problems were avoided by the finding that sodium hexamethyldisilazide effected the condensation almost quantitatively to give the desired (*E*)-alkene (**133**) in a 7 : 1 ratio with its (*Z*)-isomer.

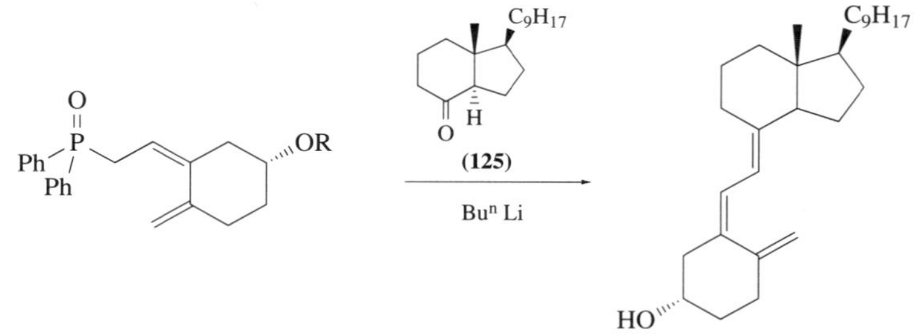

(124) (12i) (26)

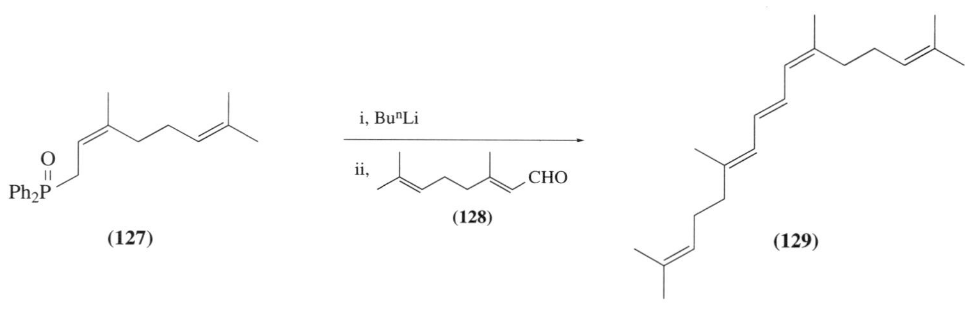

(127) (129) (27)

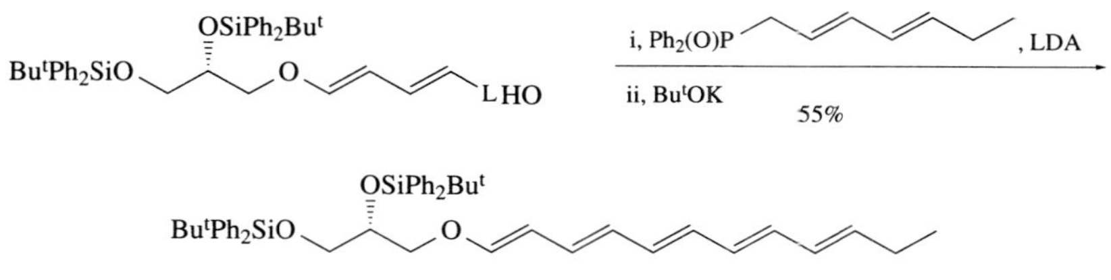

(130) (28)

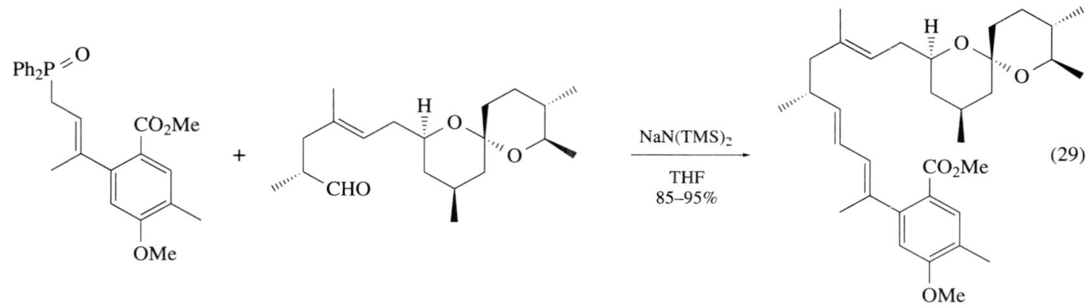

(131) **(132)** **(133)**

Solvent effects can also be important in determining the stereochemical outcome. In an ingenious synthesis of 'gossyplure', the insect pheromone that consists of a 50 : 50 mixture of (*Z*,*Z*)- and (*E*,*E*)-dienes (**136a**) and (**136b**) (Scheme 50) ⟨81JOC459⟩, reaction of the unsaturated diphenylphosphine oxide (**134**) with pentanal in a precise 1 : 1 mixture of THF and Et_2O afforded the β-hydroxyphosphine oxide (**135**) as an equal mixture of diastereomers. The isomers were not isolated but decomposed under basic conditions to give the desired mixture of dienes.

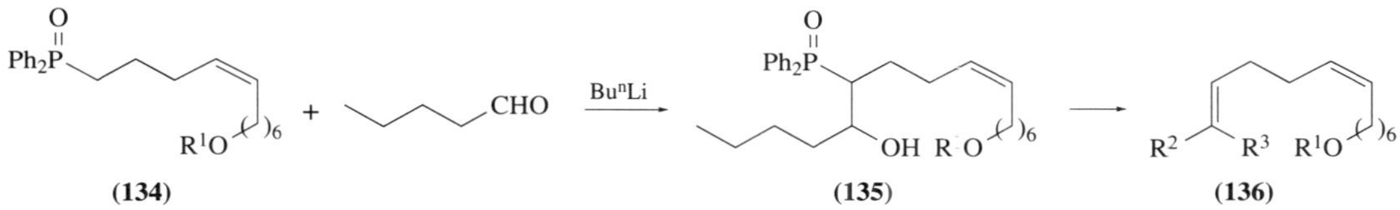

(134) **(135)** **(136)**

a; R^2 = H, R^3 = $(CH_2)_3Me$
b; R^2 = $(CH_2)_3Me$, R^3 = H

Scheme 50

In general, the Horner reaction does not proceed in one step to give essentially pure (*E*)- or (*Z*)-alkenes. For this purpose it is necessary to isolate the intermediate diasteromeric β-hydroxyphosphine oxides, usually by chromatography, and effect decomposition stereospecifically to the corresponding alkenes via *syn* elimination of diphenylphosphinic acid. This methodology has been used with allylphosphine oxides to prepare dienes and polyenes without undesired double-bond stereomutation that often accompanies such condensations ⟨86JA2662, 88TL2401⟩. For example, the lithio derivative of allylphosphine oxide (**137**) added to acetaldehyde to form the easily separated mixture of diastereomeric β-hydroxyphosphine oxides (**138a**) and (**138b**) (Scheme 51) ⟨77JCS(P1)550⟩. Independent decomposition of these diastereomers led to pure (*E*)-(**139a**)- and (*Z*)-(**139b**)-dienes. The synthetic utility of this indirect method is further demonstrated by the construction of a *cis*-double bond at C(11) of polyene (**141**) (Scheme 52) ⟨79TL5043⟩ following chromatographic separation of the diastereomeric β-hydroxy phosphine oxides and specific treatment of the *erythro*-isomer (**140**) with NaH in DMF.

When using this protocol, decomposition of *threo*-adducts affords (*E*)-alkenes stereospecifically and in high yields without exception. Unfortunately, loss of selectivity can occur with *erythro*-adducts, especially when conjugating groups are present α to the phosphorus ⟨84JCS(P1)243⟩, due to breakdown to starting materials and equilibration with the *threo*-adduct. This leads to contamination of the (*Z*)-alkene, often with predominating amounts of (*E*)-alkene. Equilibration can be minimised and greater *erythro*/*threo* ratios obtained by the use of electron-donating *ortho*-substituents in diarylphosphine oxides. This is demonstrated convincingly by the reaction of $(o\text{-}MeOC_6H_4)_2P(O)C^-HPr^iLi^+$ with benzaldehyde to form the *erythro*-adduct only ⟨86CB2150⟩.

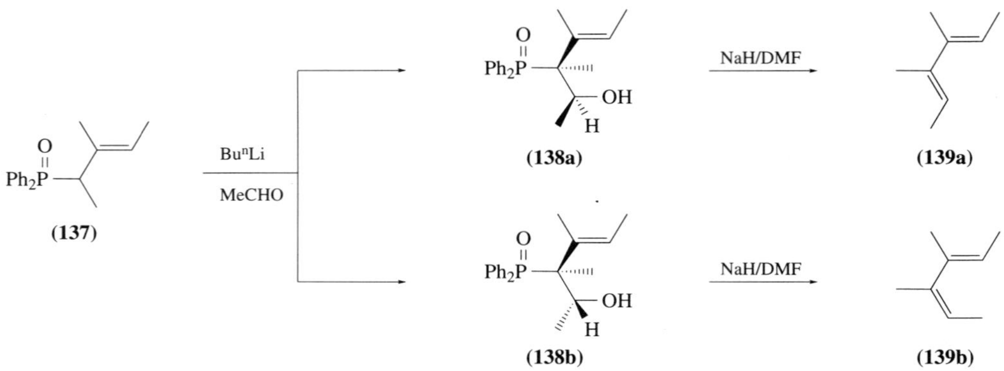

Scheme 51

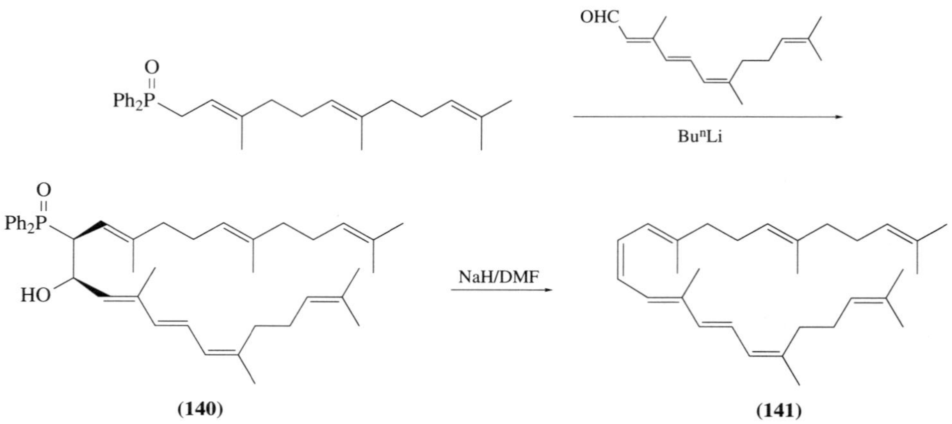

Scheme 52

erythro-Selectivity is also favoured by using very polar solvents such as DMF or DMSO; in nonpolar solvents, reaction occurs with virtually no selectivity ⟨85JCS(P1)2307⟩. An excellent measure of selectivity can also be obtained in THF, especially in conjunction with the lithium complexing agent tetramethylene diamine (TMEDA) at low temperatures ⟨85JCS(P1)2307⟩. For example, compared to the poor selectivity reported earlier (see Scheme 48), these conditions (−72°C) led to an 88 : 12 mixture of *erythro*-(**122a**) and *threo*-(**122b**). A different technique employs dibenzophosphole oxide-stabilised carbanions ⟨85JCS(P1)1953⟩. Decomposition of the *erythro*-adduct (**142a**; $R^1 = R^2 =$ Ph), (generated specifically via (*E*)-stilbene oxide) with dbu led to (*Z*)-stilbene exclusively owing to a marked increase in the rate of elimination upon incorporation of phosphorus into a five-membered ring. Even with NaH/DMSO, which led to *erythro*-(**122a**) giving mostly (*E*)-alkene, the cyclic analogue (**142a**; $R^1 = R^2 =$ Ph) decomposed to furnish an 89 : 11 mixture of (*Z*)/(*E*)-stilbenes. In a significant development, Warren ⟨86TL645, 89TL601⟩ extended this approach by introducing selective reduction of the α-ketophosphine oxide (**143**) to obtain both the *erythro*-(**142a**) and *threo*-(**142b**) adducts separately. Reduction under normal conditions with $NaBH_4$ gave rise to the *threo* isomer, but by addition of $CeCl_3$, the stereoselectivity is reversed in favour of the *erythro* isomer, due to chelation control. In an excellent example of the utility of this method ⟨86TL645⟩, both (*E*)- and (*Z*)-isosafroles were synthesised from a single ketophosphine oxide intermediate (**143**; $R^1 =$ Me, $R^2 =$ 3,4-(methylenedioxy)phenyl).

The utility of this modification is general. *trans*-Alkenes can be synthesised by acylation of diphenylphosphinoyl-stabilised anions with carboxylate esters or lactones to furnish α-ketophosphine oxides, which can be reduced with different agents, including sodium borohydride, to yield predominantly *threo*-β-hydroxyphosphine oxides ⟨83TL593, 85JCS(P1)2307⟩. This general protocol is exemplified in Scheme 53 by the synthesis of alkenol (**144**), a component of the Mediterranean fruit fly pheromone with specific (*E*)-geometry ⟨85JCS(P1)2307⟩; the (*Z*)-isomer could not be detected by GLC.

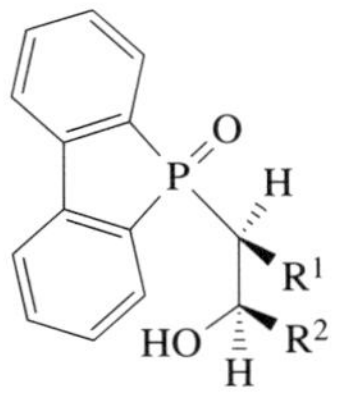

(142a)

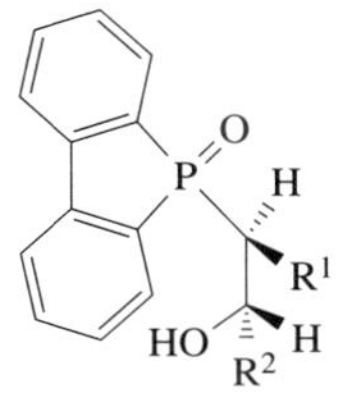

(142b)

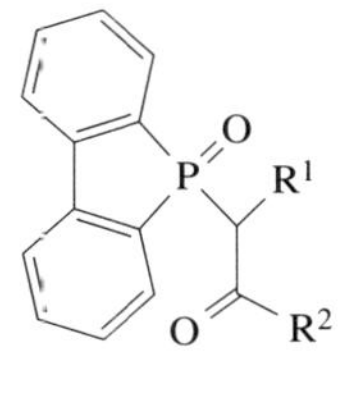

(143)

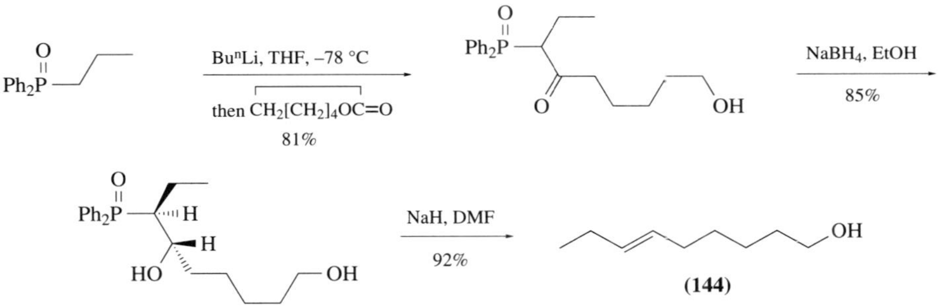

Scheme 53

By incorporating a carboxylic acid in the side chain, reduction of the ketophosphine oxide leads to lactonization. Separation and treatment with KOH to effect elimination provides a route to unsaturated acids. In a further extension of this approach, Warren ⟨88JCS(P1)2971⟩ has synthesised (*E*)-*homo*-allylic alcohols (**146**) by incorporating an intramolecular acyl transfer to the ketophosphine oxide (**145**) (Scheme 54).

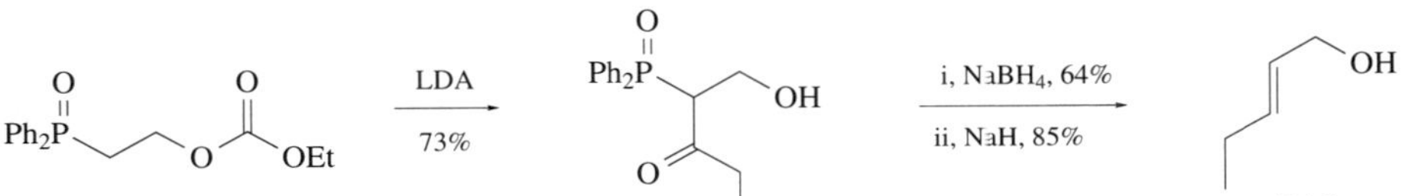

(145) **(146)**

Scheme 54

Stereospecific syntheses of trisubstituted alkenes, for example, (**149**) can be achieved by addition of butyllithium to the ketophosphine oxide (**147**) with chiral centres to induce high diastereoselectivity in the β-hydroxy intermediate (**148**) (Scheme 55) ⟨89TL4581⟩. A more common method involves reaction of branched phosphine oxides with aldehydes in preference to treatment of unbranched phosphine oxides with ketones, which leads to lower yields due to the instability of the intermediate β-hydroxyphosphine oxides ⟨83TL111, 87JCS(P1)2569⟩. In both approaches, the initial condensation occurs stereorandomly and requires separation by chromatography of the two diastereomeric phosphine oxides. Both (*Z*)-α-bisabolene (**150**), one of the main constituents of Opoponax oil ⟨83TL111⟩, and (**151**) ⟨85T3825⟩ have been prepared by these two different approaches.

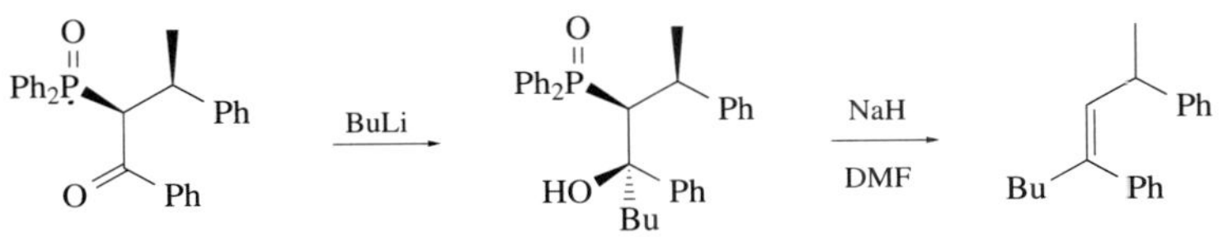

(147) **(148)** **(149)**

Scheme 55

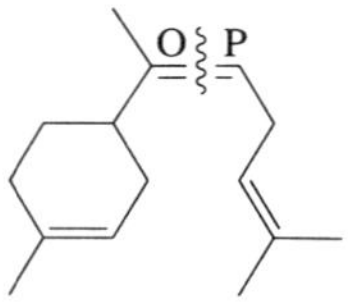

(150)

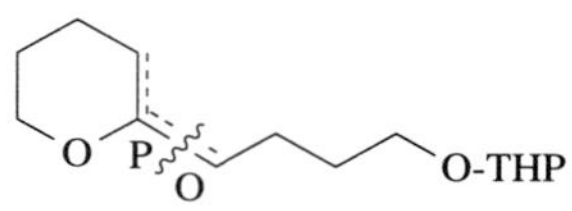

(151)

1.16.2 C=C BONDS BY CONDENSATION OF SILICON, BORON, GERMANIUM OR TELLURIUM FUNCTIONS

1.16.2.1 Silicon-Mediated Alkenation: The Peterson Reaction

A valuable alternative to the Wittig and related reactions involves α-silyl carbanions which can be generated by metallation of weakly acidic organosilanes with butyllithium. Such anions (**152**) react readily with carbonyl compounds as illustrated in Scheme 56 to form lithio derivatives of β-hydroxysilanes (**153**), which spontaneously eliminate the trimethylsilyloxy group to form alkenes in excellent yields. The major driving force for this process, commonly referred to as the Peterson reaction after its discoverer in 1968 ⟨68JOC780⟩, is provided by the formation of the strong silicon–oxygen bond ($D_{Si\text{-}O}$531 kJ mol^{-1}). As ordinarily effected, the Peterson reaction has the practical advantage in that the by-product hexamethyldisiloxane (**154**) is volatile, and consequently is easier to remove from the product alkene than triphenylphosphine oxide formed in the Wittig reaction.

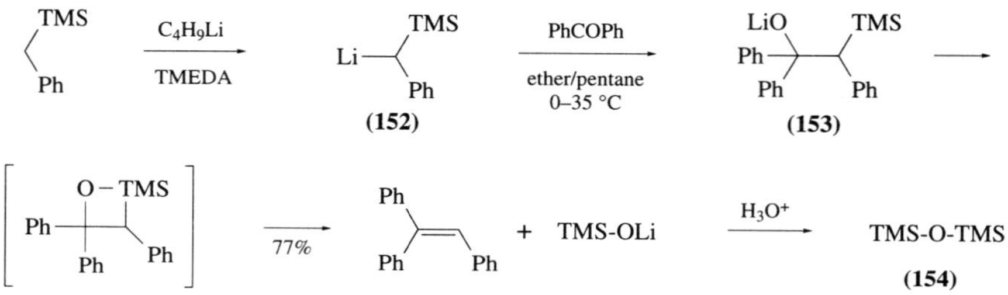

Scheme 56

The Peterson reaction has been applied extensively in organic synthesis to convert aldehydes and ketones into alkenes, but owing to difficulties in generating the necessary α-silyl carbanions for more complex applications, its greatest utility lies in the synthesis of methylene derivatives. The reaction has been reviewed by Ager ⟨84S384⟩, as well as by Colvin ⟨B-88MI 116-01⟩. A review article by Kelly ⟨91COS(1)731⟩ covers the topic in some depth.

1.16.2.1.1 Mechanism

In general, condensation of an α-silylcarbanion with a carbonyl compound proceeds in a manner analogous to the Wittig reaction. The overall transformation is depicted in Scheme 57 and will proceed directly to alkene if X is an anion-stabilising group (phenyl, ester, thio, sulphonyl, etc.). However, if X is not anion-stabilising, the elimination reaction of the β-oxido intermediate (**155**) is slow, and consequently the β-hydroxysilane (**156**) can be isolated and subsequently converted into an alkene in a separate step on treatment with either basic (NaH, KH or KOBu^t) or acidic (HOAc,H_2SO_4 or $BF_3 \cdot Et_2O$) reagents ⟨B-88MI 116-01⟩.

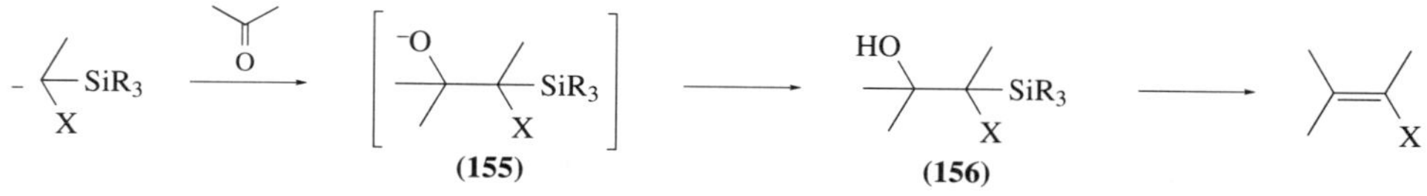

Scheme 57

1.16.2.1.2 *Preparation of silicon-stabilised carbanions*

A variety of methods exist for the generation of α-silylcarbanions for use in the Peterson reaction. Proton abstraction is only feasible provided the organosilane is functionalised with an anion-stabilising group. The bases most commonly used for deprotonation are either alkyllithiums, usually in conjunction with TMEDA, or LDA ⟨B-88MI 116-01⟩. When an anion-stabilising group is not present, access to α-silylcarbanions can be achieved by indirect methods, including the addition of an organometallic species to a vinylsilane ⟨70JA7424⟩, metal–halogen exchange ⟨70CJC561⟩, *trans*-metallation of an already α-metallated organosilane ⟨76AG(E)161⟩, and cleavage of a silicon–carbon bond in bis- or poly-silylated compounds with alkoxide or fluoride anions ⟨73TL4193⟩. The need for such procedures can be circumvented by taking advantage of alternative approaches, which offer direct access to β-hydroxysilanes, often in a stereodefined manner. These methods involve the reaction of β-ketosilanes with hydride sources ⟨74TL1133⟩ or organometallic reagents such as alkyllithiums and Grignard reagents ⟨89OS(67)125⟩, and the regiospecific ring-opening of α,β-epoxysilanes with dialkylcuprates ⟨75JOC2264⟩. Examples of these methods are given in Section 1.16.2.1.4.

1.16.2.1.3 *Methylenation reactions*

The greatest utility of the Peterson reaction is in the synthesis of methylene derivatives. For nonenolizable ketones, a particularly clean method for the generation of the required α-silylcarbanion is by reaction of a bissilylated methane with an alkoxide (Scheme 58) ⟨73TL4193⟩. The driving force for this reaction is the thermodynamically favoured formation of a silicon–oxygen bond.

Scheme 58

In other methylenation reactions, the use of α-silyl organometallic reagents, particularly lithium- and Grignard-based, has proved advantageous in comparison to the Wittig reaction for enolizible ketones, although exceptions have been reported. For example, the bridged ketone (**157**) reacted smoothly with methylenetriphenylphosphorane, whereas trimethylsilylmethylmagnesium chloride acted as a base and abstracted an α-proton with aldol products being isolated ⟨76MI 116-01⟩. Interestingly, Lewis-acid catalysis could have proved advantageous in this instance as evidenced by the reaction of β-tetralone (**158**) with trimethylsilylmethyllithium in the presence of cerium trichloride. Under these conditions, methylenation is virtually complete and occurs without isomerisation, whereas in its absence the ketone is recovered unchanged. Improved yields of methylenation are obtained even with base-sensitive aldehydes by the prior addition of cerium trichloride ⟨87JOC281⟩. For example, octanal can be methylenated to only 56% extent under normal conditions, but the cerium-modification approach afforded 86% of the desired product. Other additives such as titanium tetrachloride have been employed, albeit in conjunction with the Grignard reagent (TMS-CH_2MgCl) ⟨81TL5031⟩, but yields of methylenated product are inferior with aldehydes (octanal, 59%), and for ketones, the additive virtually prevents reaction. Typically, acetophenone gave only 3% methylenation ⟨81TL5031⟩, whilst the cerium-modification (TMS-CH_2Li/$CeCl_3$) occurred successfully in 93% yield ⟨87JOC281⟩, which dropped to 78% under normal conditions.

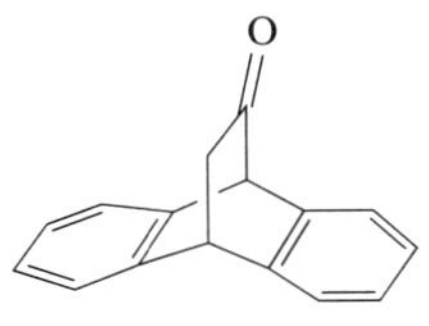

(157)

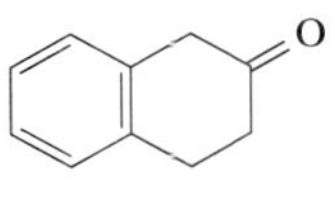

(158)

Even in the absence of additives, such α-silyl organometallic reagents possess a distinct advantage when compared to the corresponding Wittig reagent by permitting the introduction of a methylene group into a sterically hindered ketone. For example, the (±)-β-gorgonene precursor (**159**) failed to react under Wittig conditions ⟨73TL3497⟩, whereas methylenation could be achieved with trimethylsilylmethylmagnesium chloride, albeit in only 15% yield (Scheme 59). Even with less sterically demanding carbonyl compounds, the Peterson method for the synthesis of methylene derivatives offers advantages in being compatible with a variety of sensitive functionalities, including the anomeric alkoxy group in (**160**) ⟨82JOC3548⟩, the thioacetal group in (**161**) ⟨88TL4521⟩, and an aziridine moiety in (**162**) ⟨88JOC3391⟩. The latter example is of interest in that the elimination from the intermediate β-hydroxysilane was effected quantitatively by the unusual treatment with 2,3-dichloro-5,6-dicyano-1,4-benzoquinone (ddq).

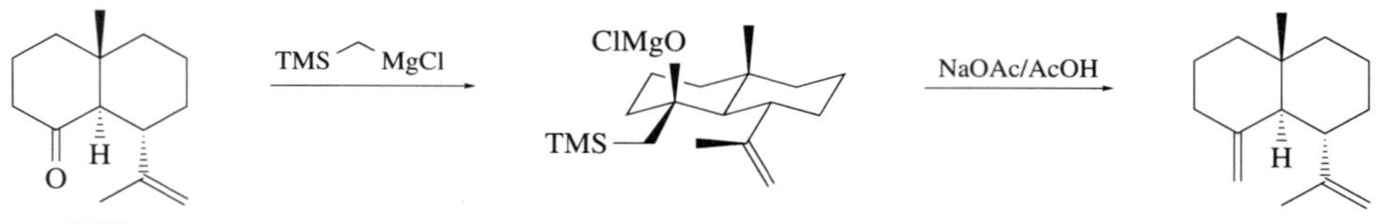

(159)

Scheme 59

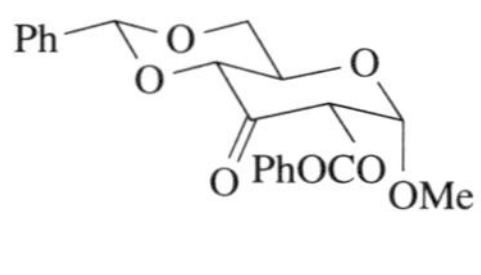

(160)

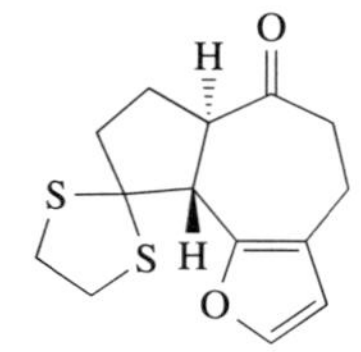

(161)

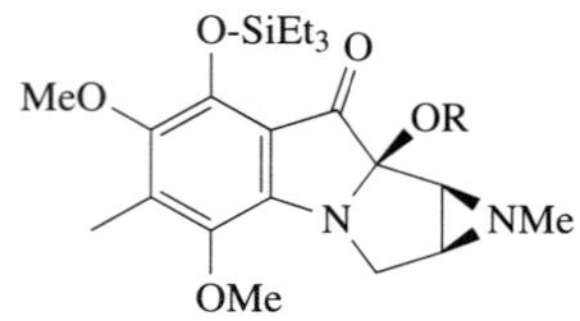

(162)

Hindered ketones can also be transformed into an ethylidene unit by the ingenious use of α-(trimethylsilyl)vinyllithium (**163**), which is readily accessible by metallation of the corresponding bromocompound ⟨80JA2463⟩. The general procedure is outlined in Scheme 60 and forms the basis of the synthesis of coronafacic acid by Jung and co-workers ⟨80JA2463⟩. The same reagent can also be utilised for the preparation of terminal allenes, although in this situation the intermediate adducts are most reluctant to eliminate, but can be isolated as the corresponding alcohol and converted into a better leaving group such as trifluoroacetate or chloride ⟨78JOC1526⟩; treatment with fluoride ion results in smooth elimination to give the desired allene as illustrated in Scheme 61 for the synthesis of 1,1-diphenylallene in 45% yield.

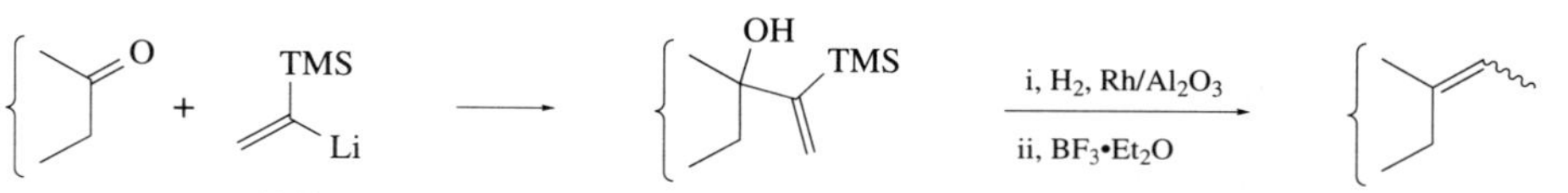

(163)

Scheme 60

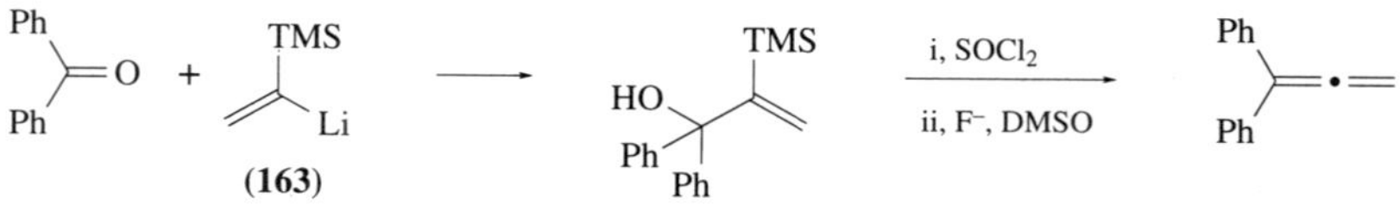

Scheme 61

1.16.2.1.4 *Stereoselective formation of alkenes*

As mentioned earlier, the presence of an anion-stabilising group facilitates the elimination step in the Peterson reaction and alkenes are formed directly. Conversely, the lack of such a group permits isolation of the intermediate β-hydroxysilane and elimination of the alkene is carried out in a separate step. For this reason, it is convenient to deal with these two categories of alkene formation separately.

(i) Formation of aliphatic alkenes

Since alkylsilanes are not readily deprotonated, access to the required α-metallated silanes is obtained by the various methods discussed in Section 1.16.2.1.2. Unlike the Wittig reaction, when such α-silylcarbanions react with carbonyl compounds where *cis*- and *trans*-alkenes can be produced, both isomers are formed in almost equal amounts. This lack of stereospecificity results from the formation, under kinetic control, of the intermediate β-hydroxysilane as a ca. 1 : 1 mixture of *threo*- and *erythro*-diastereomers, which decompose with a high degree of selectivity. The ratio is found not to be affected significantly by any change in reaction conditions, but fortunately, a preponderance of either *cis*- or *trans*-alkenes can be obtained by other routes that provide a diastereomerically highly enriched β-hydroxysilane.

In one approach, illustrated in Scheme 62, the addition of ethyllithium to trimethylvinylsilane is used to generate an α-silylcarbanion, which is quenched with carbon dioxide to form a carboxylic acid ⟨75JA1464⟩. Further steps as indicated afford the β-ketosilane (**164**), the reduction of which using dibal occurs in a manner predicted by the Felkin–Anh mode to afford predominantly the *threo*-β-hydroxysilane (**165**). In the second stage, treatment of (**165**) with potassium hydride on one hand, or with boron trifluoride etherate or sulfuric acid on the other, led, respectively, to the stereospecific formation of (*E*)- and (*Z*)-oct-4-ene (Scheme 63) ⟨75JA1464⟩. This difference in geometry of the alkene depending on the type of treatment is a reflection of the mechanistic requirement for *syn*-elimination in the former case, and the need for an *anti*-periplanar geometry of elimination in the latter.

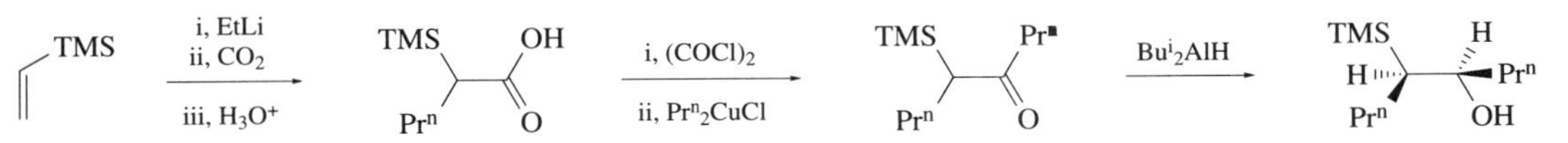

(**164**) (**165**)

Scheme 62

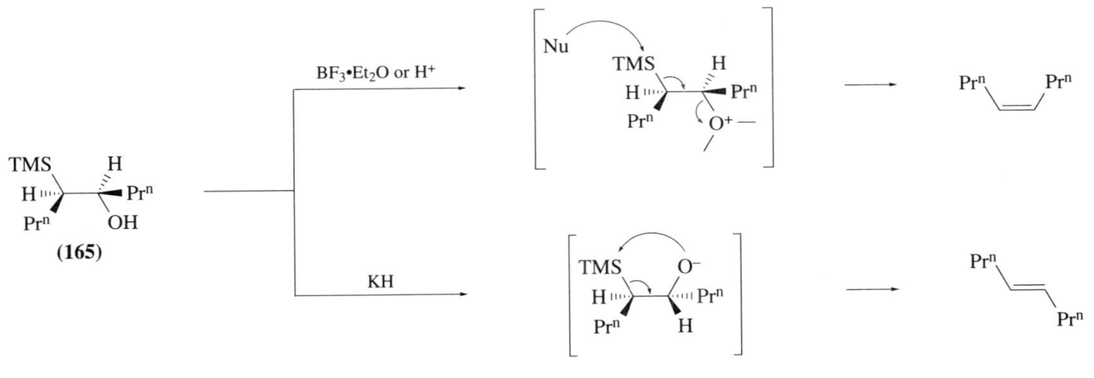

(**165**)

Scheme 63

The same method can be utilised to provide a stereoselective synthesis of trisubstituted alkenes as exemplified in Scheme 64 by the addition of methyllithium to β-ketosilane (**166**) to give predominantly one diastereomeric form of the two possible β-oxidosilanes ⟨76JOC2940, 77TL1807⟩. Workup under either basic or acidic conditions as discussed earlier allowed stereoselective generation of the trisubstituted alkene.

Unlike the Horner reaction, it is not possible to control such processes to produce predominantly the *erythro*-adduct. This is relatively unimportant provided that one of the diastereomeric β-hydroxysilanes is formed in sufficiently large excess, since the two complimentary methods of elimination allow the choice of stereoselective delivery of either geometrical isomer of the alkene.

By analogy to the Horner reaction, synthetic access to diastereomeric β-hydroxysilanes such as (**165**) and (**167**) can be obtained by regiospecific opening of α,β-epoxysilanes with dialkylcuprates (Scheme 65) ⟨75JOC2264⟩. The overall reaction is highly stereoselective and is consistent with the silyl group facilitating specific nucleophilic attack α to silicon, coupled with treatment by potassium

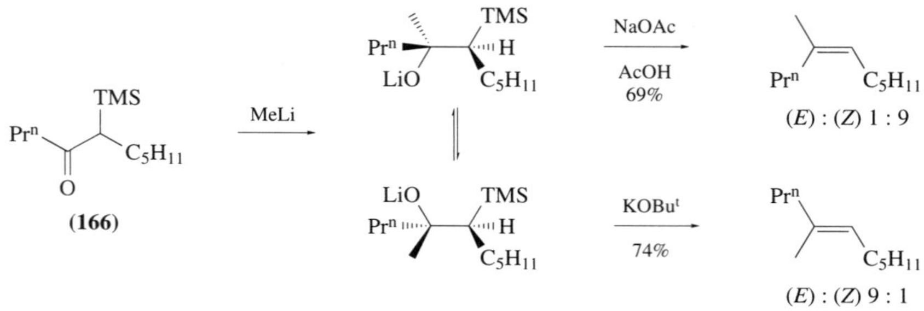

Scheme 64

hydride to effect *syn* elimination and produce alkenes in >99% geometric purity. Unfortunately, considerable limitations are placed on the synthetic usefulness of this method by the need for epoxysilanes with defined geometry.

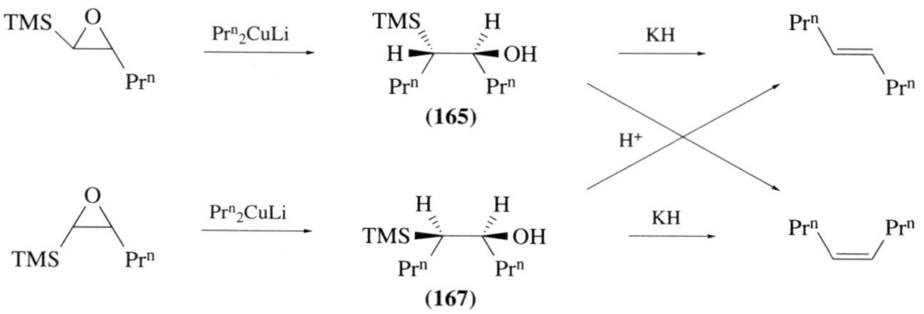

Scheme 65

(ii) Formation of functionalised alkenes

The Peterson reaction can be successfully employed with α-silylallyl anions (**168**) to synthesise conjugated alkenes directly provided the addition to the carbonyl compound is carried out in the presence of magnesium bromide to promote α-attack (Scheme 66) ⟨78TL2383⟩, otherwise the selectivity is inverted and γ-attack predominates ⟨80JA5004⟩. By incorporating a boronate function at the γ-position as in (**169**), reaction with aldehydes is also directed predominantly towards the α-position to form the *erythro*-adduct (**170**) as the major diastereomer (Scheme 67) ⟨81TL2751⟩, which undergoes base-catalysed elimination to form the mono-substituted (*Z*)-double bond isomer (**171**).

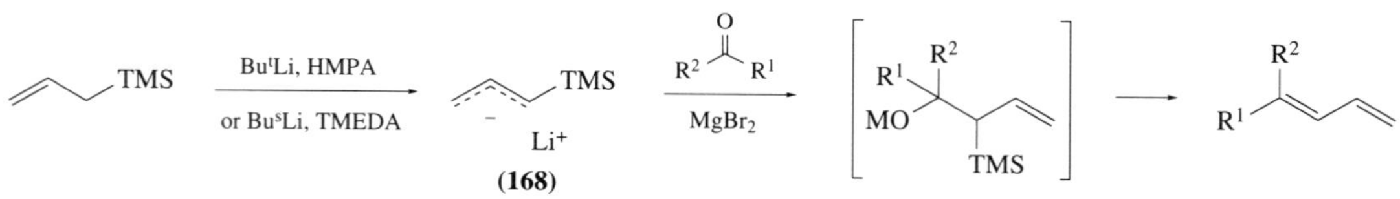

HMPA = hexamethylphosphoramide

Scheme 66

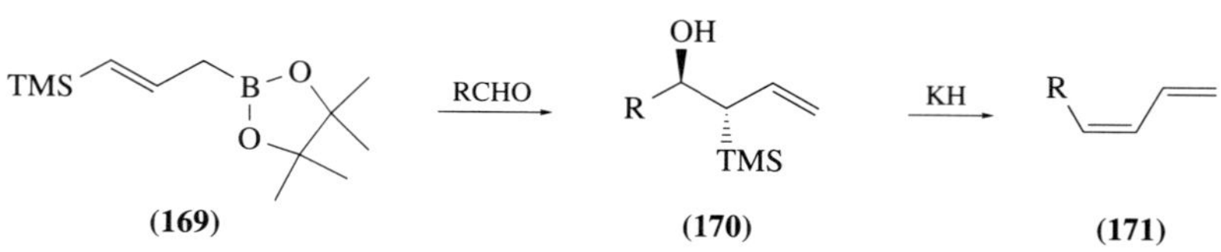

Scheme 67

The same conditions are also required for the synthesis of α,β-unsaturated esters by the addition of functionalised α-silylcarbanions such as **(172)** to aldehydes (RCHO). The elimination from the β-oxidosilane intermediate, which is formed in the presence of magnesium bromide, is directed towards the (*Z*)-isomer **(173)** by treatment with HMPT, or alternatively is carried out with $BF_3 \cdot Et_2O$ to synthesise the (*E*)-isomer **(174)** (Equation (30)) ⟨81CC877⟩. Even easily enolisable ketones, for example, cyclohexanone, which often give poor yields in the Wittig reaction, react readily with reagents such as **(172)** to give >95% yield of the desired alkene. Such reactions with unsymmetrical ketones proceed with high kinetic selectivity to form predominantly (*Z*)-alkenes, and as such are complimentary to the corresponding Wittig reaction, which produces the (*E*)-isomer as the predominant product. The (*Z*):(*E*) ratio of the derived conjugated alkenes **(175)**, **(176)** and **(177)** illustrate this selectivity ⟨81TL1575, 84TL5603⟩.

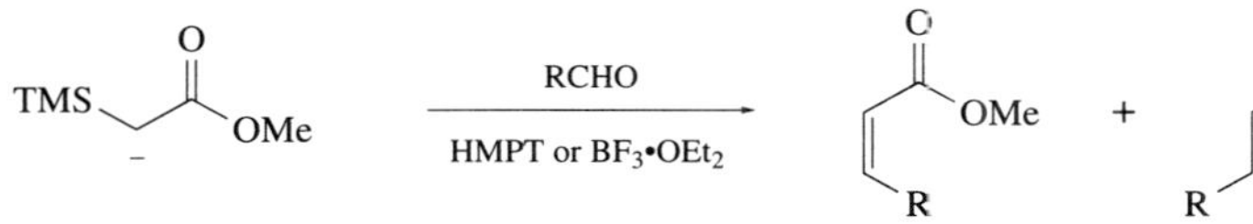

(30)

(172) **(173)** **(174)**

Conditions	*RCHO*	*(173) : (174)*
HMPT	Ph	80 : 20
	Bu^n	85 : 15
$BF_3 \cdot Et_2O$	Ph	1 : 99
	Bu^n	2 : 98

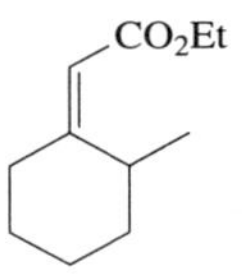

(175)
(*Z*) : (*E*) 89 : 11

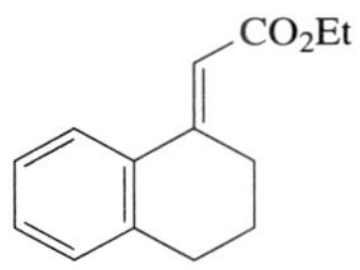

(176)
67 : 33

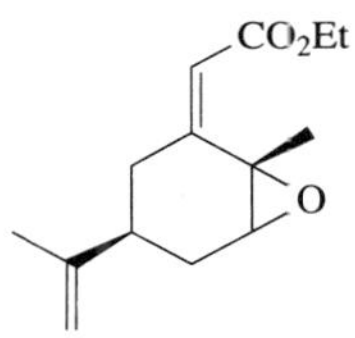

(177)
90 : 10

By increasing the size of the ester grouping, it is possible to raise the (*Z*):(*E*) ratios of alkenes as exemplified by the butenolide product **(178)** ⟨81TL1595⟩. On changing the ester group (R) from methyl to propyl and then *t*-butyl, the (*Z*):(*E*) ratio improved from 43:57 to 80:20 and 88:12 without much change in the yield (ca. 70%). Surprisingly, (*Z*)-selectivity is maintained even when the product becomes sterically crowded as in the pseudomonic acid derivative **(179)** which is produced in an (*Z*):(*E*) ratio of 88:12 ⟨85JCS(P1)541⟩.

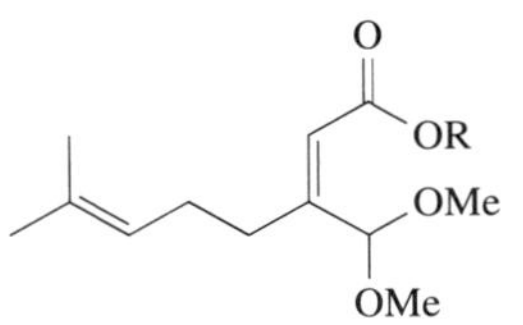

(178)

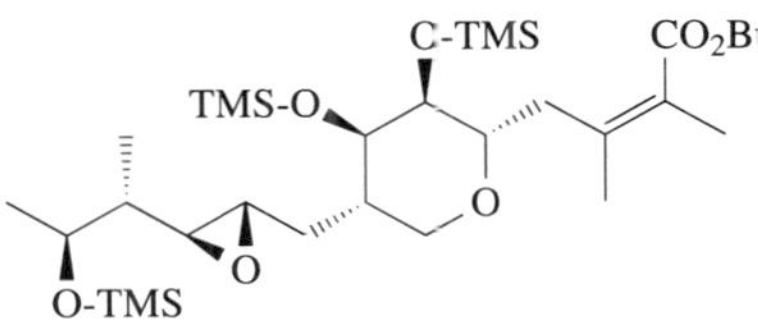

(179)

Synthesis of hetero-substituted alkenes using more complex metallated silanes such as **(180)**, **(181)**, **(182)** and **(183)** ⟨85TL4359⟩ have been outlined. The latter example is of interest in that unlike the preceding one, reaction with aldehydes led to a change in selectivity from equal amounts of (*E*)- and (*Z*)-alkenes to formation of (*E*)-alkenes almost exclusively without any racemisation of the sulfoximine. The Peterson reaction is also compatible with a phosphorus substituent as demonstrated in the original work of Gilman and Tomasi (Scheme 68) ⟨62JOC3647⟩ in which the silyl-substituted phosphorane **(184)** reacted with benzophenone to produce tetraphenylallene **(185)** by the preferential loss of TMS-O instead of elimination of triphenylphosphine oxide expected under Wittig conditions.

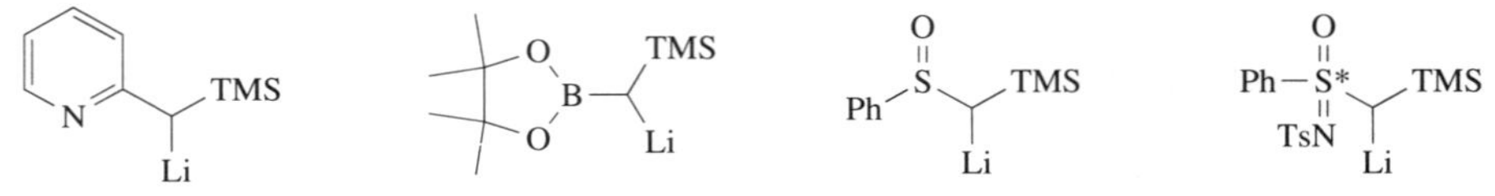

(180) **(181)** **(182)** **(183)**

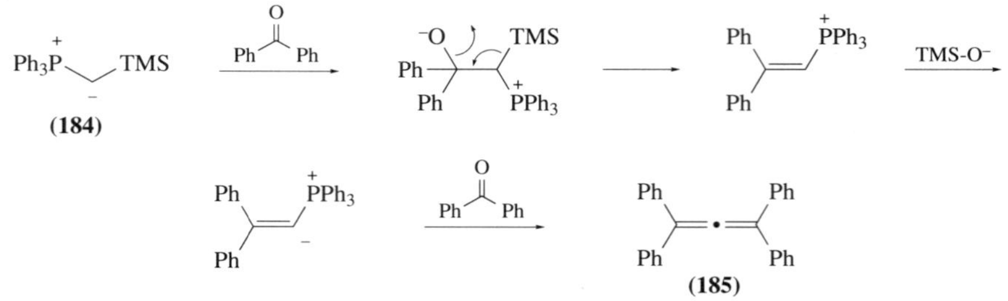

Scheme 68

1.16.2.2 Boron-mediated Alkenation

In comparison to the well-established chemistry of silicon-stabilised carbanions that of the corresponding boron reagents has received scant attention, especially in connection with the conversion of a carbonyl compound into an alkene by addition to an aldehyde or ketone. Evidence so far shows that such reactions do occur readily ⟨66TL4315⟩, but may lead to products other than alkenes depending upon the nature of the carbonyl compound. For example, the bis(mesitylboryl)stabilised carbanion (**187**), prepared conveniently by reaction of fluoride ion with the bis(mesitylboryl)(trimethylsilyl)alkane (**186**) ⟨83OM236⟩, condensed with benzophenone in a straightforward manner to afford the corresponding alkene (**188**) in 70% isolated yield (Scheme 69) ⟨83TL635⟩. On the other hand reaction with benzaldehyde at room temperature led to a mixture of products, including the expected alkene (**189**) in only 40% yield, but with 99.9% (*E*)-selectivity ⟨85TL5093⟩. The reluctance of (**187**) to be (*Z*)-selective, like the corresponding Wittig reagent, is the result of *anti*-elimination from the initially formed and favoured *erythro*-adduct (**190**; see Scheme 70), which is formed even at low temperatures (−78 °C) and decomposes only on warming to room temperature.

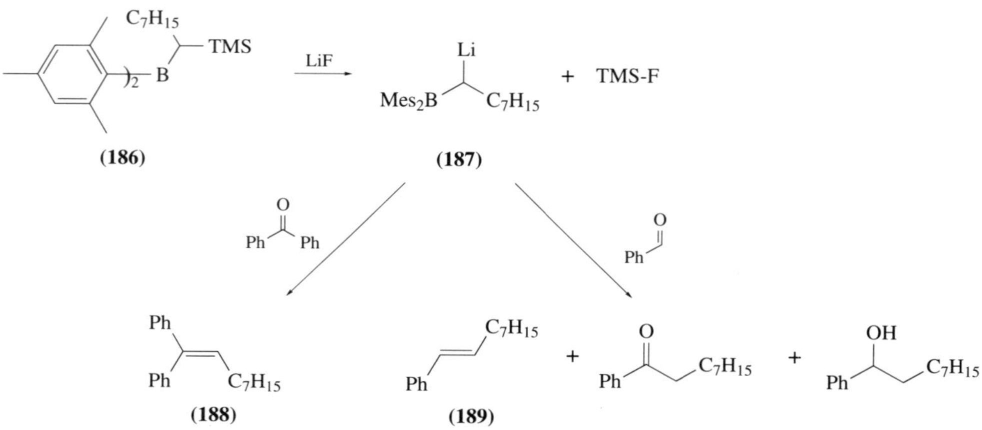

Scheme 69

By analogy to the Peterson reaction, the stability of intermediates such as (**190**) can be utilised to bring about the stereoselective formation of both geometric forms of the alkene by inducing decomposition with various reagents selected to promote either *anti*- or *syn*-elimination of the boryl function. Thus, treatment of (**190**) with trifluoroacetic anhydride (TFAA) at low temperature

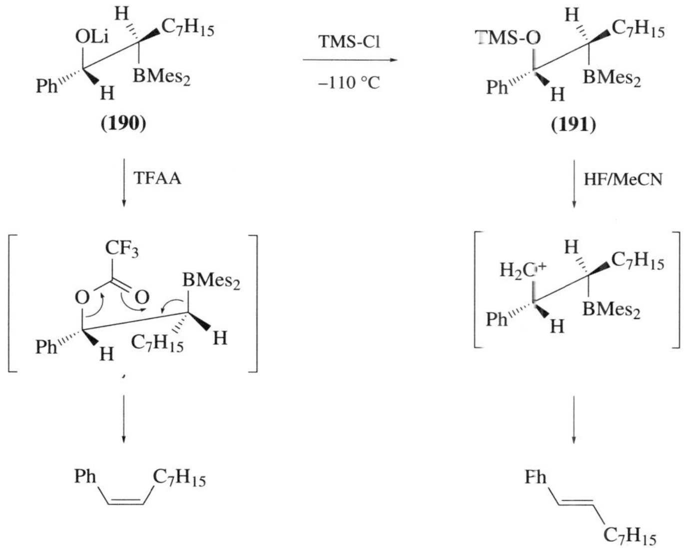

Scheme 70

($-110\,^{\circ}C$) followed by warming to room temperature resulted in decomposition by a cyclic ester-type elimination to a 9:1 mixture of (Z)/(E)-alkenes (Scheme 70) ⟨87CC297⟩. Alternatively, prior reaction of (**190**) with trimethylsilylchloride produced the silyl ether (**191**) which fragmented to a 95:5 mixture of (E)/(Z)-alkenes (73% yield) with aqueous HF in acetonitrile ⟨87CC297⟩. Interestingly, substitution of the 4-position of the aromatic aldehyde by electron-withdrawing or -donating groups caused little variation in yield or selectivity for (E)-alkenes, but led to a lowering on both accounts for (Z)-alkenes.

In the reaction of (**190**) with aliphatic aldehydes it is necessary to premix the aldehyde with a protic acid, for example, acetic acid, to ensure alkene formation ⟨89TL5643, 89TL5647⟩, which proceeds with poor selectivity unless the aldehyde is bulky.

Finally, it is worth noting that boron-stabilised carbanions like their silicon analogues can bring about methylation of ketones. For example, the condensation of (**193**) with cyclohexanone gave methylenecyclohexanone in ca. 55% yield (Scheme 71), but deprotonation of β-methyl-9-borabicyclo[3.3.1]nonane (**192**) proved difficult and only occurred to an appreciable degree with the highly hindered base, lithium piperidide ⟨72JA6854⟩.

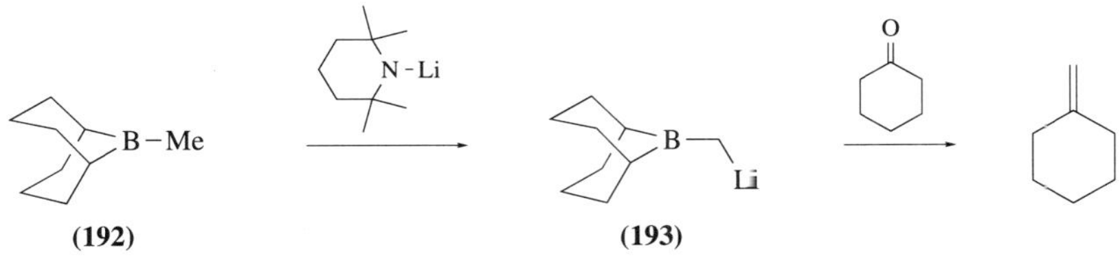

Scheme 71

1.16.2.3 Germanium-mediated Alkenation

Little information is available on pseudo-Wittig reactions of germenes, $R^1{}_2Ge{=}CR^2{}_2$, compounds with a germanium–carbon double bond, due to their transient nature and a propensity to poly-condense and form oligomers. This is demonstrated by the reaction of Ph_2Ge and N_2CHCO_2Et (Carius tube, copper powder, 60 °C) to generate the transient germene $Ph_2Ge{=}CHCO_2Et$, which is only partially trapped by benzaldehyde with the formation of a germoxane $(Ph_2GeO)_n$ and the corresponding alkene ($PhCH{=}CHCO_2Et$, 7%) ⟨80JA5413⟩; the side reaction led to oligomers.

In another report ⟨91OM1771⟩, the 'free' germene (**195**) was obtained by thermal decomposition of the stable adduct (**194**), but upon condensation with benzaldehyde, the reaction stopped at the intermediate germaoxetane (**196**) due to the large steric hindrance of the substituent (Scheme 72). In the case of acetone, reaction occurred with the enolic form, followed by a germanotropic rearrangement.

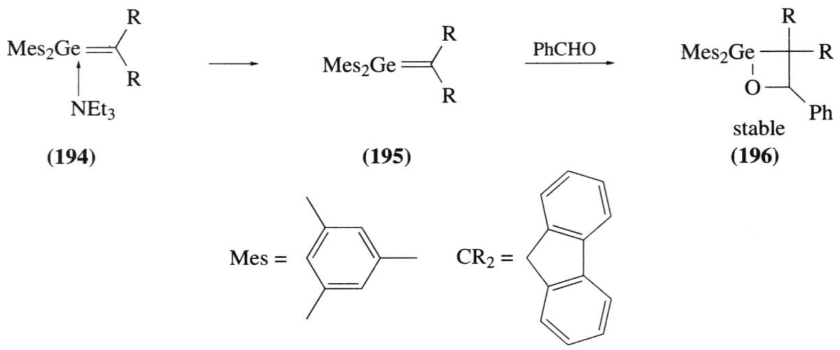

Scheme 72

1.16.2.4 Tellurium-mediated Alkenation

The formation of an alkene from a carbonyl compound which may involve a telluronium ylide as an intermediate has been reported (Equation (31)) ⟨93T6821⟩.

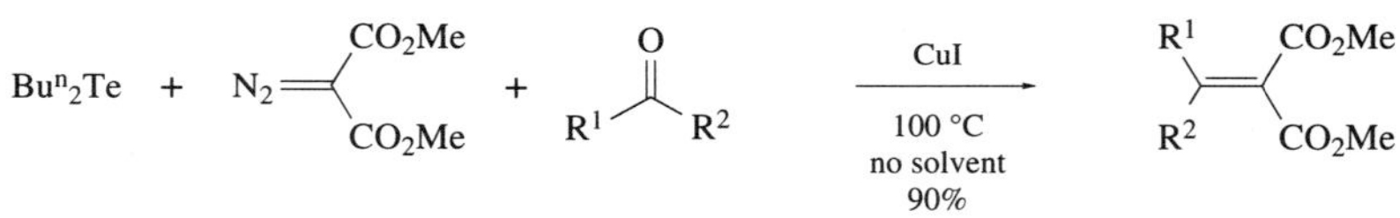

(31)

No reaction takes place in the absence of a tellurium compound, but there is no firm evidence that a tellurium ylide is indeed an intermediate; attempts to isolate an ylide were unsuccessful, but ^{1}H-NMR provided evidence for its presence. The one-pot reaction is simple and proceeds without base and in high yield.

1.16.3 METAL-INDUCED METHYLENATION

One-carbon homologation of carbonyl compounds can be achieved with certain methylene-bridged metallic species that act as CH_2 transfer agents. The most useful preparative examples of such species are the titanium methylenation complex (**197**) known as Tebbe's reagent ⟨78JA3611⟩, and the zinc divalent complex (**198**) developed by Oshima ⟨87JOC4410⟩. Both of these complexes have found widespread application in organic synthesis and offer distinct advantages over the more conventional Wittig methylenation reagent. In particular they remove the possibility of enolising sensitive carbonyl functions, whilst the more reactive Tebbe's reagent (**197**) can be used in the methylenation of esters and amides for which the conventional Wittig reaction is not suitable. The literature on these synthetically useful processes has been reviewed by Pine ⟨93OR(43)1⟩.

1.16.3.1 Tebbe's Reagent

The η^5-C_5H_5 (hereafter Cp) complex (**197**) of titanium is readily made by the reaction of red, crystalline $CpTiCl_2$ (mp 230°C) with Al_2Me_6 ⟨85JOC2386⟩. In the reaction with carbonyl compounds, the actual methylene transfer agent is $Cp_2Ti{=}CH_2$, which is formed from (**197**) by elimination of Me_2AlCl upon warming or by addition of a Lewis base such as pyridine or 4-dimethylaminopyridine (dmap). In the absence of an added Lewis base, reaction of (**197**) with an ester such as methyl benzoate (Scheme 73) ⟨80JA3270⟩ proceeds slowly in contrast to ketones and amides which react rapidly. In each case, the driving force for methylenation is the formation of titanocene oxide, $Cp_2Ti{=}O$ ⟨84AG(E)587⟩.

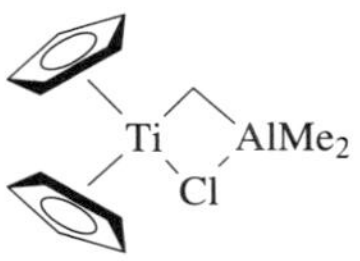

(197)

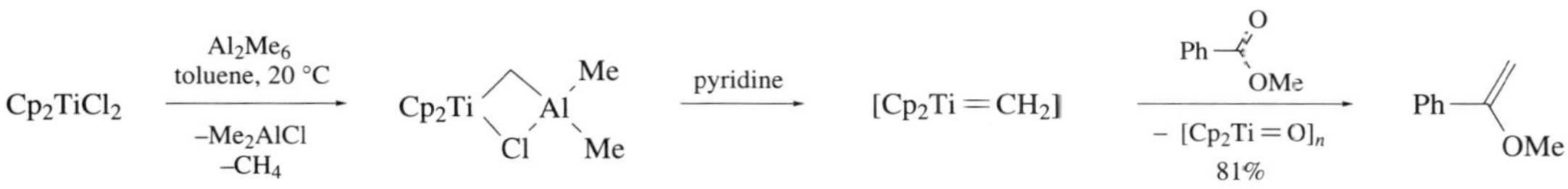

Scheme 73

Table 7 illustrates the application of this method to the preparation of methylene derivatives from a wide range of carbonyl-containing compounds, including enol ethers from esters ⟨78JA3611⟩, and enamines from amides ⟨89SC2659⟩; the reaction is not applicable to hindered ketones due to enolate formation, nor anhydrides and acid chlorides.

1.16.3.2 Oshima's Reagent

Unlike Tebbe's reagent, the use of Oshima's reagent (**198**) ⟨78TL2417⟩ as a methylene transfer-agent does not require the presence of a base, but like its counterpart, it is more effective in the methylenation of enolizable ketones than the Wittig reaction ⟨B-86MI 116-01⟩. It is readily prepared by the treatment of dibromo- or diiodo-methane and zinc with titanium tetrachloride and may be stored indefinitely at −20°C ⟨87OS(65)81⟩. The procedure is exemplified in Scheme 74 by the methylenation of the β-keto ester (**199**) in 40% yield ⟨88BCJ457⟩; the corresponding Wittig reaction failed.

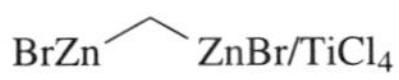

(198)

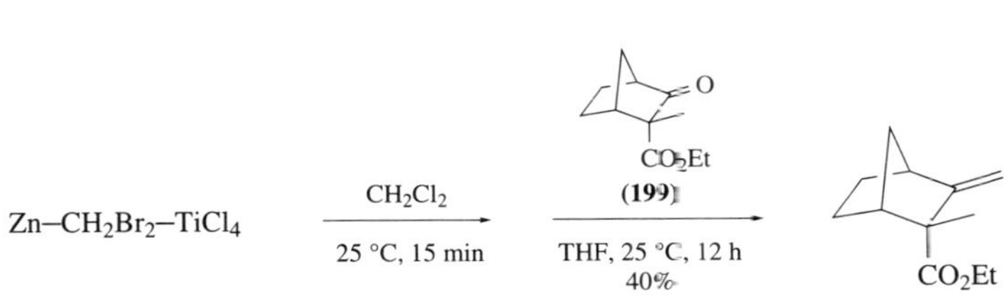

Scheme 74

Table 7 Yields from methylenation of carbonyl-containing compounds via Tebbe reagent (**197**) in toluene.

Reaction	*Conditions*	*Yield* (%)	*Ref.*
1 Aldehydes			
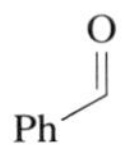	THF, 0 °C to RT, 30 min	92	78JA3611
2 Lactones			
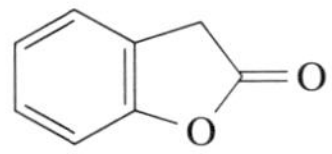	THF, 0 °C to RT, 30 min	85	80JA3270
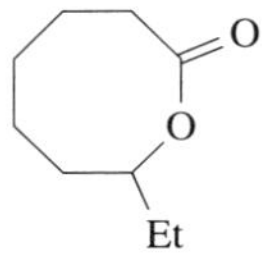	THF, tol, dmap, –40 °C, 30 min; RT, 90 min		88TL4333
3 Esters			
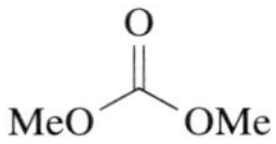	THF, 0 °C to RT, 30 min	60	83PAC1733
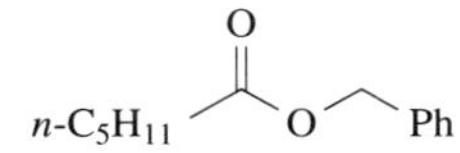	THF, 0 °C to RT, 30 min	47	88JOC1903
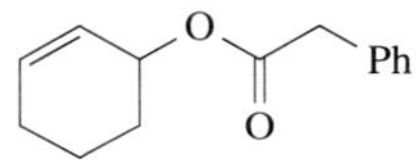	THF, 0 °C to RT, 30 min	96	80JA3270
4 Ketones			
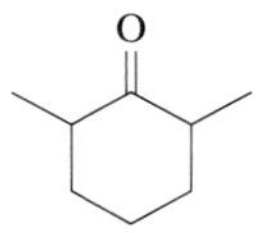	THF, 0 °C to RT, 30 min	97	91S165
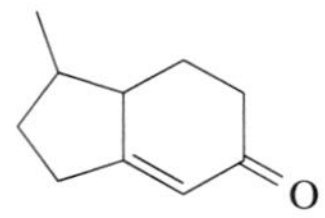	THF, 0 °C to RT, 30 min	65	82PAC1915
5 Amides			
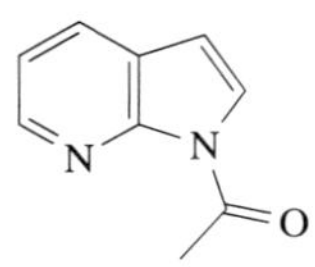	THF, 0 °C to RT, 30 min	56	89SC2659
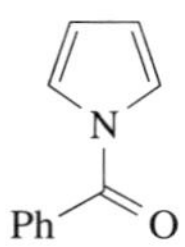	THF, 0 °C to RT, 30 min	74	89SC2659

Oshima's reagent is only suitable for the methylenation of aldehydes and ketones (cf. Tebbe's reagent), but this modified reactivity has led to a widespread application in organic synthesis due to a tolerance of a variety of other functional groups, including esters, lactones, acetates and ketals as exemplified in Table 8.

Table 8 Yields from methylenation of aldehydes and ketones via Oshima's reagent (**198**).

Reactant	*Conditions*	*Yield* (%)	*Ref.*
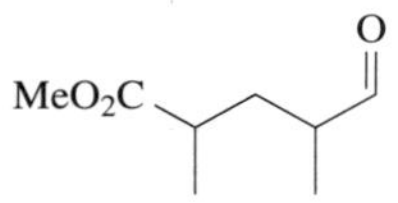	Zn, CH_2Br_2, $TiCl_4$ CH_2Cl_2, 20 °C, 1.5 h	40	88CJC1359
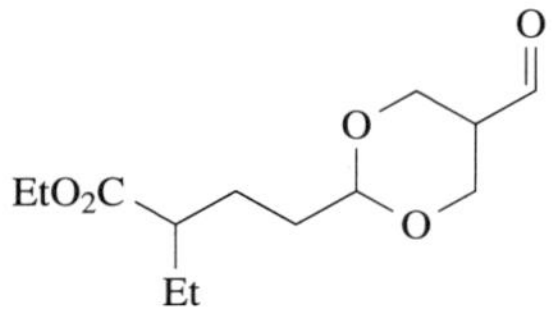	Zn, CH_2Br_2, $TiCl_4$ CH_2Cl_2, 20 °C, 1.5 h	35	84TL2035
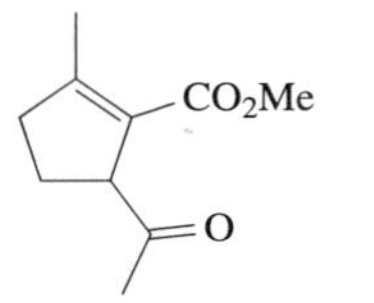	Zn, CH_2Br_2, $TiCl_4$ THF, 25 °C, 12 h	94	80JOC2005
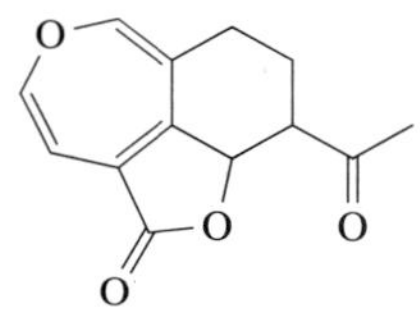	Zn, CH_2Br_2, $TiCl_4$ THF, 25 °C, 12 h	98	89TL1241
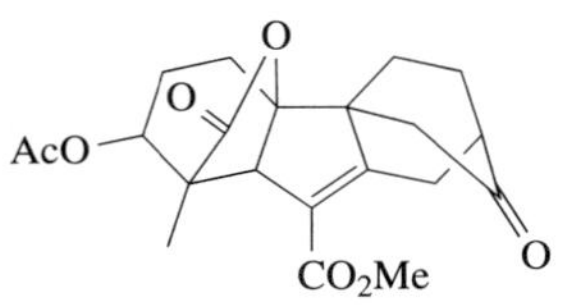	Zn, CH_2Br_2, $TiCl_4$ CH_2Cl_2, 20 °C, 1.5 h	79	88TL3339
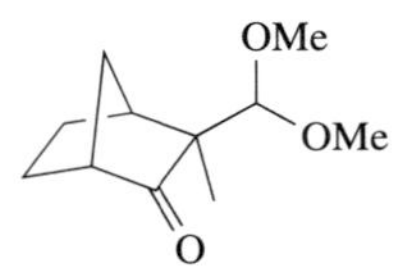	Zn, CH_2Br_2, $TiCl_4$ THF, 25 °C, 12 h	30	88BCJ467
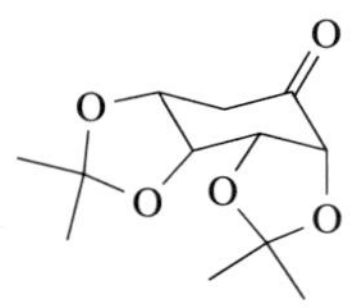	Zn, CH_2Br_2, $TiCl_4$ CH_2Cl_2, 20 °C, 1.5 h	95	87CC1008
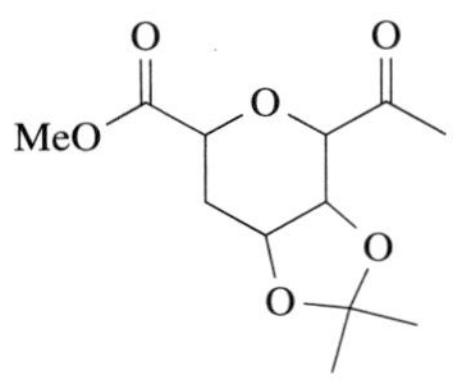	Zn, CH_2I_2, $TiCl_4$ THF, 25 °C, 15–60 min	47	90CAR(201)185

In a synthetically useful modification of (**198**), the use of $Ti(OPr^i)_4$ instead of $TiCl_4$ ⟨85TL5581⟩ allows the exclusive methylenation of an aldehyde in the presence of a ketone. Moreover, the reverse effect can be achieved by complexation of the aldehydic carbonyl with $Ti(NEt_2)_4$ prior to treatment with (**198**). Scheme 75 illustrates the benefit of this chemoselectivity.

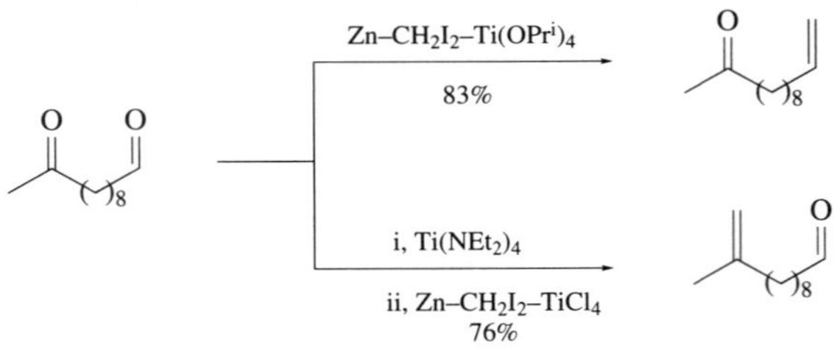

Scheme 75

Finally, it is worth mentioning a clever variation of this method that allows the general alkylidenation of esters (and lactones) by use of dibromoalkanes instead of dibromomethane. Tetramethylethylenediamine is required and the reagents are mixed in a different order to that normally used for (**198**), but yields are high and the process leads to the formation of (*Z*)-alkenyl ethers predominantly ⟨87JOC4410⟩. By contrast, Tebbe's reagent cannot be extended to the alkylidenation of esters.

1.17
One or More C=C Bond(s) by Pericyclic Processes

HAMISH McNAB
University of Edinburgh, UK

1.17.1 INTRODUCTION

This chapter is concerned with the creation of the C=C unit by retro-pericyclic cleavage reactions, with emphasis on their synthetic utility. Retro-sigmatropic-shift processes, which are considered first, are characterised by cleavage of one C—H and one C—X bond to create the alkene, whereas the retro-cycloaddition reactions which form the next section of the chapter are characterised by cleavage of two C—C (or C—X) bonds. Formation of dienes and polyenes are considered separately, and include a consideration of retro-cheletropic reactions. Concerted rearrangements are covered in Chapters 1.09 and 1.18. Within each section, the mechanism and stereochemistry of the process under consideration will be covered, together with an assessment of the scope, limitations and reaction conditions; finally, some examples of applications to the syntheses of natural and unnatural products will be given.

The majority of the relevant reactions are thermal eliminations, which may be carried out preparatively either in solution or in the gas phase. Most chemists have an intuitive 'feel' for solution temperature conditions, but the same cannot be said for gas phase reactions, where the contact time of the substrate in the hot zone is the most important parameter. This chapter will describe as 'flow pyrolyses' those reactions which are carried out at atmospheric pressure (often in a stream of nitrogen) by feeding the substrate, either neat or in solution, down a heated tube packed with glass helices. Contact times under these conditions are generally of the order of a few seconds. In contrast, 'flash vacuum pyrolysis' (FVP) involves vacuum distillation of the substrate through a heated tube, and contact times are generally a few milliseconds. Clearly, much higher temperatures are needed for FVP and, as a general rule, a reaction which proceeds at 180 °C in solution may well require in excess of 800 °C under FVP conditions, with flow pyrolyses taking place at intermediate temperatures.

1.17.2 FORMATION OF MONOENES BY RETRO-ENE AND RELATED REACTIONS

The general reactions under consideration are shown in Equations (1) and (2). Equation (1) involves hydrogen transfer via a six-membered ring transition state and is formally a $[2\sigma_s + 2\sigma_s + 2\pi_s]$ reaction. In certain cases where X and Y are carbon-containing groups, the formation of the by-product X=Y—ZH may also be synthetically useful. Equation (2) is isoelectronic with Equation (1), but involves a five-membered ring transition state $[2\sigma_s + 2\sigma_s + 2\omega_s]$. A number of general reviews are available (e.g., ⟨69AG(E)556, 93S659⟩), which include examples of the generation of C—heteroatom double bonds by this mechanism.

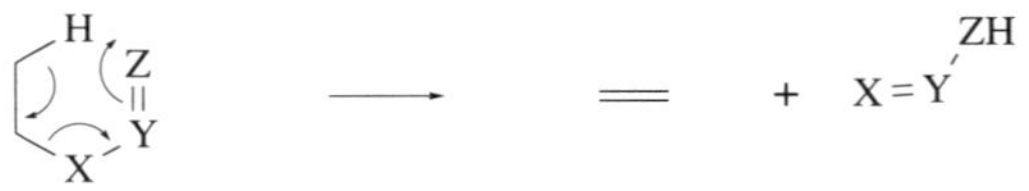

(1)

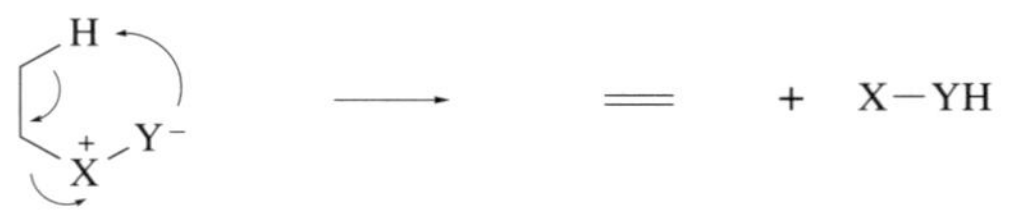

(2)

1.17.2.1 Cleavage of One C—H and One C—C Bond

Apart from the intramolecular case which results in a rearrangement, the 'all-carbon' retro-ene reaction (Equation (1); X = Y = Z = C) is of little synthetic value ⟨93S659⟩. In part, this is due to a mechanistic dichotomy in which concerted and homolytic cleavage mechanisms are in competition and which therefore leads to mixtures of products. In the gas phase pyrolysis of hex-1-ene, for example, retro-ene cleavage is most important at lower temperatures, with radical fission of the C-3—C-4 bond being predominant at high temperatures ⟨79MI 117-01⟩.

Simple ketones decompose on pyrolysis under flow conditions to give alkenes (Equation (1); Z = O), but they require temperatures ca. 150 °C higher than the corresponding acetates (see Section 1.17.2.2.1) for comparable conversion ⟨78JOC1421⟩, and therefore the reaction is seldom used synthetically.

Decarboxylation of carboxylic acids containing β,γ-unsaturation (Equation (3)) forms a special case of Equation (1), in which the 'useful' product is the HZ—Y=X fragment ⟨B-79MI 117-02, B-80MI 117-01⟩. Pyrolyses are carried out at 340 °C in solution (boiling phenanthrene) ⟨64JCS3897⟩ or at ca. 420 °C in the gas phase under flow conditions ⟨65JCS6202⟩. The reaction takes place with a wide variety of substituted acids, including cyclic examples ⟨67JCS(B)557⟩, but it does not appear to have been exploited in synthesis.

(3)

Facile decarboxylation of β-ketoacids and 1,3-diacids ⟨51QR131⟩ is familiar as a key step in the acetoacetic ester and malonic ester syntheses of methyl ketones and carboxylic acids, respectively, in which the initially formed enol tautomerises to the final product (Scheme 1). In practice, the reaction may only require heating of the acid for a few hours in refluxing aqueous acid solution ⟨43OSC(2)416⟩, or heating at 125–130 °C in the absence of solvent ⟨43OSC(2)93⟩.

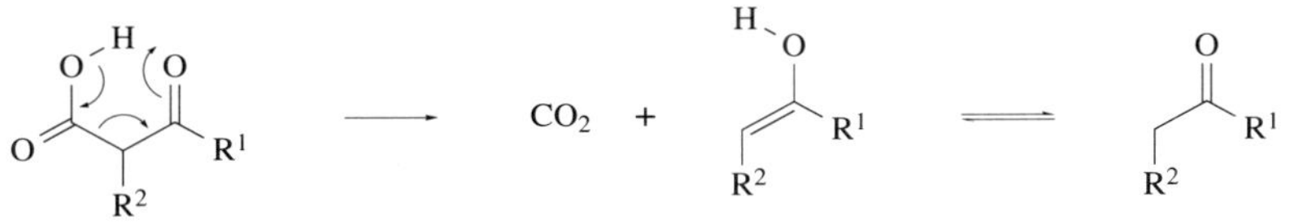

R^1 = Me or OH; R^2 = alkyl

Scheme 1

1.17.2.2 Cleavage of One C—H and One C—O Bond

These cleavage processes are an important group of preparatively useful reactions (Equation (1); X = O) which differ in the nature of the other heteroatoms (Y,Z). Most have the overall effect of accomplishing an alcohol dehydration, by first modifying the alcohol function and then performing a pyrolytic *syn*-elimination ⟨60CRV431⟩. The efficiency of both of these steps is therefore important in the synthetic applicability of the various methods.

1.17.2.2.1 Pyrolysis of acetates and related esters

Many reviews of this topic (Equation (1); X = Z = O) are available ⟨60CRV431, 71MI 117-01, B-79MI 117-02, B-80MI 117-01, 91COS1011⟩.

This reaction is traditionally carried out in the gas phase under flow conditions, with temperatures of 500–525 °C recommended for preparative purposes ⟨60CRV431⟩. FVP conditions are also effective, but require slightly higher temperatures (600–700 °C). The mechanism has been extensively studied ⟨B-79MI117-02⟩. The *syn*-nature of the elimination was suspected for many years before it was elegantly proved by pyrolyses of the specifically labelled acetates (**1**) and (**2**), which lead to deuterium incorporation into, and deuterium exclusion from, the alkene product, respectively (Scheme 2) ⟨53JA6011, 72JCS(P2)165⟩. These results have been applied to the conformational analysis of steroids (⟨51JCS1048⟩ and references therein).

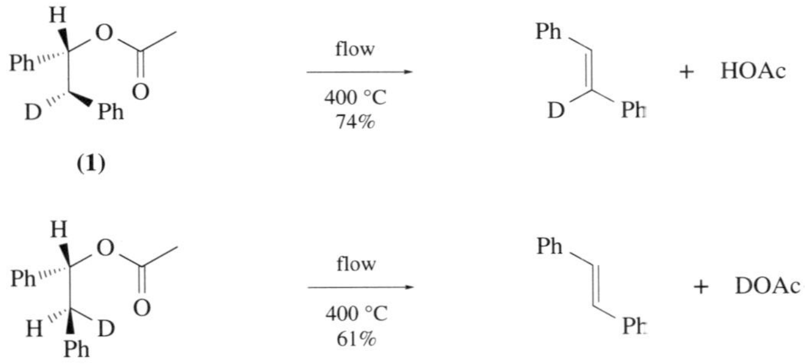

Scheme 2

The regioselectivity of the elimination is dependent on a number of factors, including statistical effects, thermodynamic stability of the products (arising from steric effects) and electronic effects, and the overall stereochemical outcome can be rather modest. Statistical effects dominate the pyrolyses of simple alkyl acetates; for example, but-2-yl acetate gives a 57 : 43 ratio of but-1-ene to but-2-enes, close to the statistical ratio of 60 : 40 ⟨63RTC1123⟩ (Equation (4)). Steric effects promote the formation of the (*E*)- rather than the (*Z*)-alkenes if either is possible (Equation (4)) and may cause a slight reduction in the amount of terminal alkene in competitive situations. Cycloalkyl acetates generally form the alkene unexceptionally; however, with large ring sizes, both (*E*)- and (*Z*)-alkenes are formed (Equation (5)) ⟨55JA1001⟩. The formation of a double bond *endo*- to a ring is generally favoured relative to the *exo*-isomer, due to the greater reduction in eclipsing interactions in the former case ⟨B-79MI 117-02⟩ (e.g., Equation (6) ⟨59JA651⟩). Conjugative stabilisation of the product appears to be unimportant; even the acetate (**3**) gives just a 1 : 1 mixture of conjugated and unconjugated dienes (Equation (7)) ⟨69MI 117-01⟩. Large ring lactones, which can accommodate the required transition state, undergo corresponding thermal rearrangement to give ω-unsaturated carboxylic acids ⟨77JOC3895⟩.

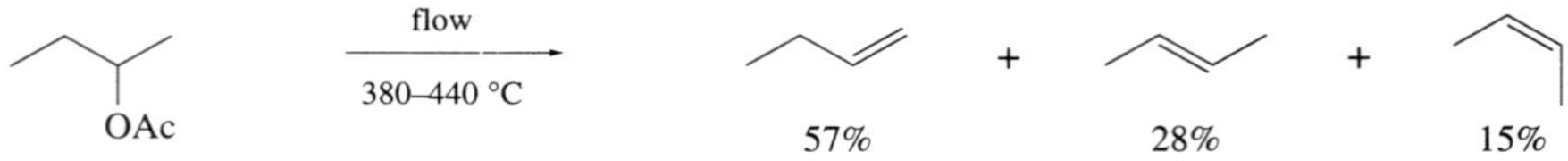

(4)

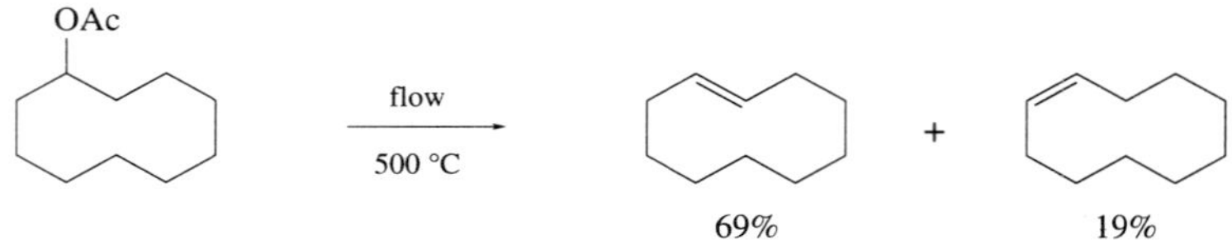

(5)

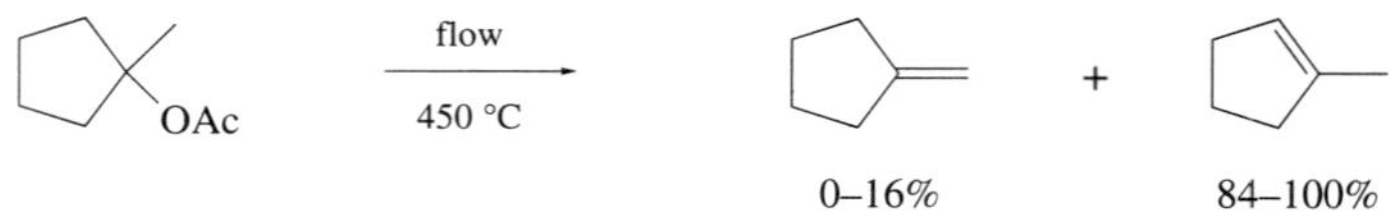

(6)

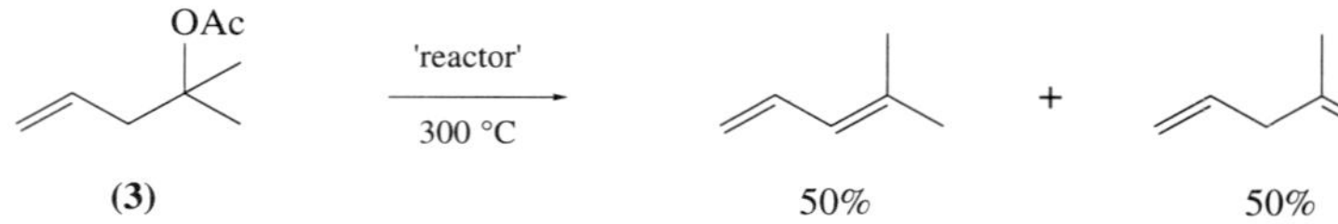

(3)

(7)

Electronic factors are now recognised as being dominant in governing the rate and direction of alkene formation. The reaction is aided by electron donation at C-1, such that the order of reactivity is tertiary > secondary > primary, even allowing for statistical correction (e.g., the relative rates of ethyl, isopropyl and *t*-butyl acetates are 1 : 28.8 : 3315 ⟨75JCS(P2)1025⟩). The reaction is aided by electron withdrawing groups at C-2, such that pyrolysis of (*E*)-methyl 2-acetoxycyclohexanecarboxylate gave the 1-alkene almost exclusively (Equation (8)) ⟨59JA2126⟩. Bulky electron donating (alkyl) substituents at this position can also cause a small rate enhancement due to 'steric acceleration' ⟨76JCS(P2)280⟩, which reaches a maximum factor of 13 for a *t*-butyl group relative to a hydrogen atom.

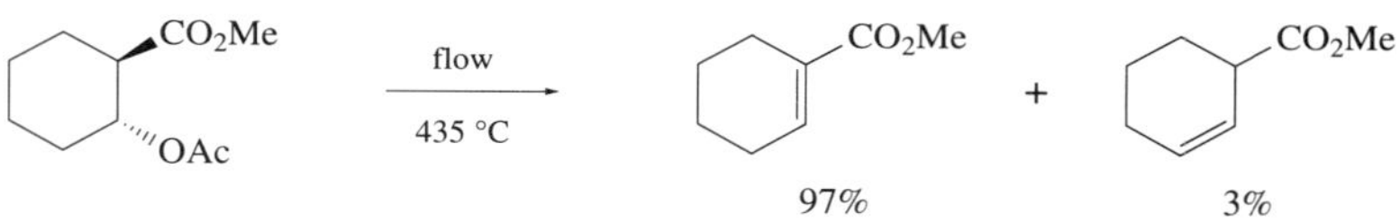

(8)

The reaction is also accelerated by electron-withdrawing groups at the carbonyl carbon atom; formates, chloroacetates, chloroformates and other functional groups such as carbonates and carbamates (see Section 1.17.2.2.3) are therefore all in principle more efficient than acetates ⟨B-79MI 117-02⟩, although the effect is often small ⟨63JCS1246⟩. However, cyclohexyl trifluoroacetate is some 19 times more reactive than the corresponding acetate ⟨69JCS(B)187, 72RTC3⟩.

From a synthetic point of view, these reactions are attractive despite the relatively high temperatures involved, because of the ready availability of starting materials, and because of the absence of any acidic or basic reagents which might promote rearrangements. At a practical level, the use of benzoates rather than acetates has the advantage that benzoic acid crystallises at the exit point of the furnace, well away from the more volatile alkene ⟨88MI 117-01⟩. The gas-phase, short contact time conditions are particularly effective for the synthesis of alkenes which readily polymerise or decompose unless kept in the cold. Examples include the generation of reactive dihydrofurans ⟨87TL1519, 93AJC1321⟩ or dihydrothiophenes ⟨86JOC113⟩ (Scheme 3); in both monocyclic cases the isolated yield from the pyrolysis step was >85%.

Acetate pyrolysis continues to be employed as a key step in natural product syntheses. It was used selectively by Klaver *et al. en route* to the alkaloid peduncularine (**4**), when anionic or cationic eliminations failed (Scheme 4) ⟨89JA2588⟩. The *exo*-cyclic methylene groups of longifolene (**5**) ⟨90JA4609, 93JOC2186⟩ and sinularene (**6**) ⟨79AJC1819⟩ were introduced by classic 'flow' pyrolysis methodology using benzene or toluene solutions of the substrates (Scheme 5).

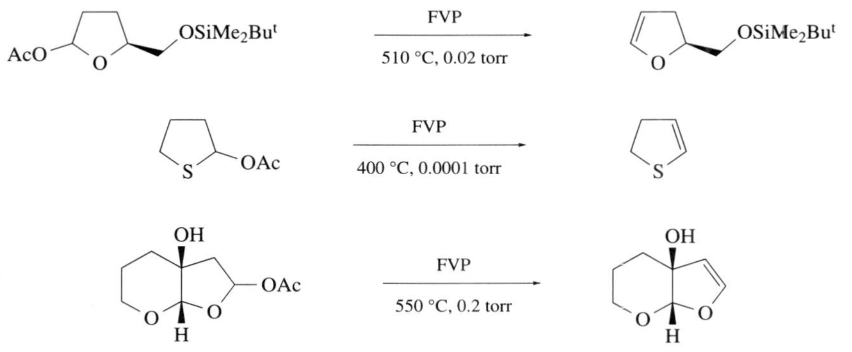

Scheme 3

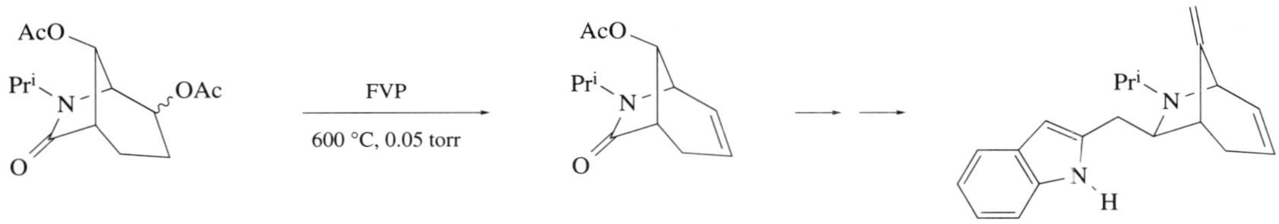

(4)

Scheme 4

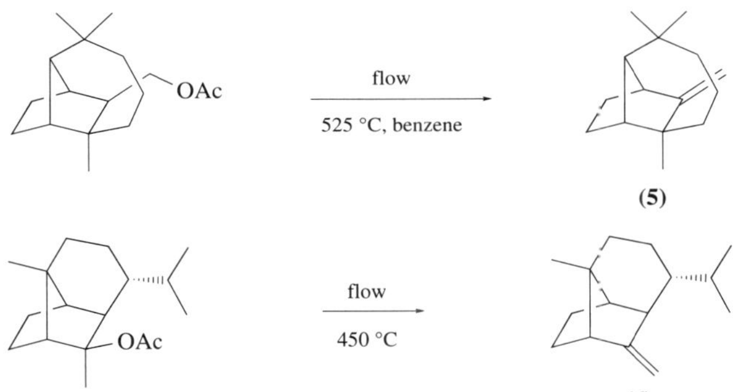

(5)

(6)

Scheme 5

1.17.2.2.2 Pyrolysis of xanthate esters

This method follows the general mechanism of Equation (1) (X = O, Y = C—SR, Z = S); invariably the *S*-methyl derivatives are used. As well as the reviews cited in the previous subsection, a specialised account of this transformation—the Chugaev reaction—is available ⟨62OR(12)57⟩.

The reaction is aided by greater nucleophilicity in the atom Z ⟨B-79MI 117-02⟩ and so, not surprisingly, xanthates are more reactive than acetates and the reaction may be carried out under relatively mild conditions. Possible thermal isomerisation of the alkene is therefore minimised. Typically, the pure xanthate is heated to ca. 150 °C for a few hours either at atmospheric pressure or under vacuum; the product may distil out or remain in the pyrolysis flask. If the xanthate distils without decomposition, the elimination can be effected by dropwise addition to a high boiling point solvent such as diphenyl ether or biphenyl ⟨62OR(12)57⟩. Two disadvantages of this method vis-à-vis ester pyrolysis are that xanthates are more difficult to prepare than acetates and that the products are frequently contaminated with sulfur-containing impurities (which are traditionally removed by treatment with metallic sodium) ⟨62OR(12)57⟩.

The *syn*-stereochemistry of the process was established by Cram ⟨49JA3883⟩ (e.g., Equation (9)), and this was applied to assign the configuration of terpene hydroxyl groups in the early days of conformational analysis ⟨49JCS2174⟩. The involvement of the C=S sulfur atom in the elimination has been proved by ^{34}S and ^{14}C isotope effects ⟨61CJC348⟩. Statistical, thermodynamic and steric factors again govern the direction of the elimination ⟨62OR(12)57⟩, and complex mixtures will result if a variety of decomposition pathways are open (cf. Section 1.17.2.2.1). Where a comparison has been made, the results have been found to be closely similar to those of the corresponding acetate, bearing in mind that the reaction conditions are substantially different ⟨60CRV431⟩. Again, the formation of (*E*)-alkenes is favoured over (*Z*)-alkenes in competitive situations. (*Z*)-Cycloalkenes are formed when the ring size is smaller than eight carbon atoms ⟨62OR(12)57⟩, but the proportion of (*E*)-isomer increases the larger the ring size ⟨55JA1001⟩. The preference for *endo*- rather than *exo*-elimination in cyclic cases is apparently less marked than for the corresponding acetate ⟨60CRV431⟩. Electron withdrawing groups attached to sulfur further accelerate the reaction ⟨53JA2118⟩.

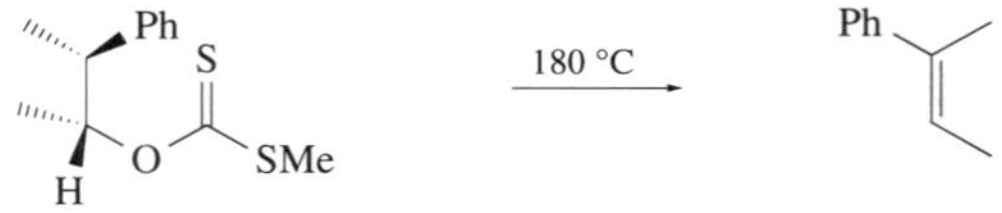

(9)

Xanthate pyrolysis is still used substantially in synthesis. At a simple level, *β*-^{13}C-labelled styrene was prepared in 20% yield by pyrolysis of the xanthate derived from labelled phenethyl alcohol ⟨87JA553⟩. A comparison of acetate, tosylate and xanthate pyrolysis as a means of preparing tens of grams of 6-chlorohex-1-ene has shown that the last method is preferred, with yields of 75–80% being obtained routinely by FVP at 525°C ⟨89JOC5811⟩. Applications in propellane ⟨87TL2795, 90JA228⟩ and triquinancene chemistry ⟨90TL3409, 92JOC5121⟩ have also been published since the late 1980s, including a spectacular triple-Chugaev process (Equation (10)) which takes place in 91% yield when carried out at 220–230°C in hexamethylphosphoramide (HMPA) ⟨90TL3409, 92JOC5121⟩.

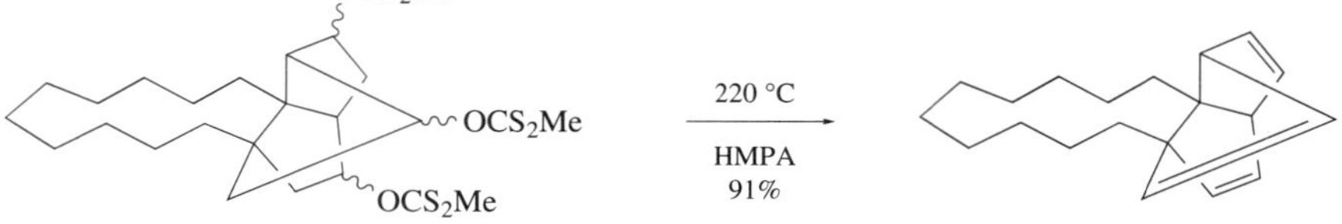

(10)

A number of applications in natural product synthesis may be cited. Pyrolysis of the xanthate (**7**) in biphenyl solution at 210°C gave an 80% yield of the alkene (**8**) *en route* to racemic lindestrene (Scheme 6) ⟨85TL6393⟩. The method is compatible with the presence of remote chiral centres, and the alkene function of the alkaloid (−)anisomycin (**9**) was introduced by xanthate pyrolysis in *o*-dichlorobenzene (Equation (11)); the 4,5-dehydropyrrolidine isomer was also obtained as a minor by-product ⟨89H(29)1861⟩. Not surprisingly, heating the xanthate (**10**) at its sublimation temperature (210°C) gave regiospecific formation of the dihydropyran (**11**) in high yield (Equation (12)); structure (**11**) has been proposed as a potential intermediate in yohimbine alkaloid synthesis ⟨86JOC3023⟩. In other cases, a comparison of xanthate and selenoxide methods has come out in favour of the latter methodology ⟨90JOC4051⟩.

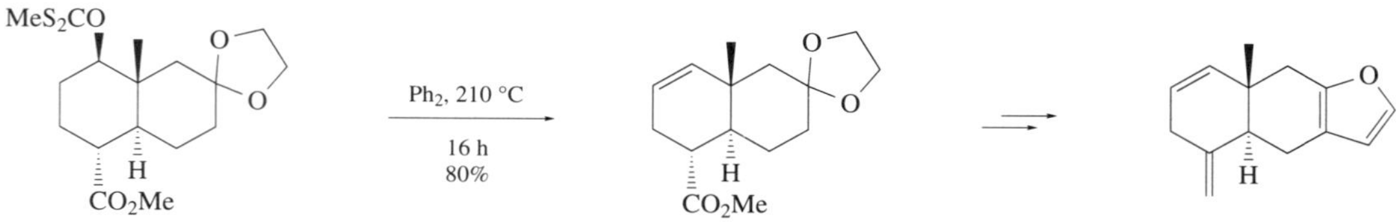

(**7**) (**8**)

Scheme 6

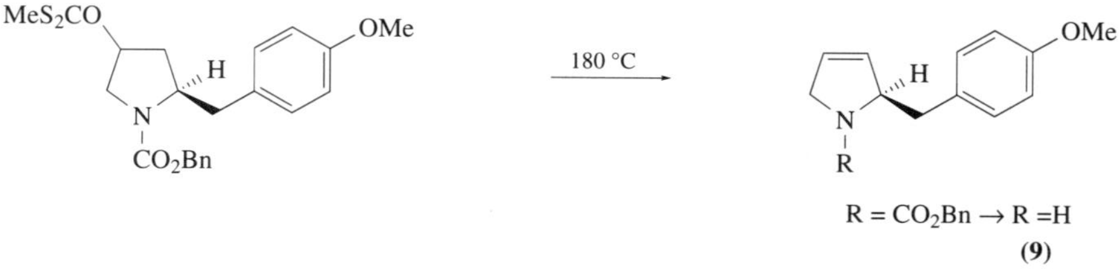

(11)

(**9**)

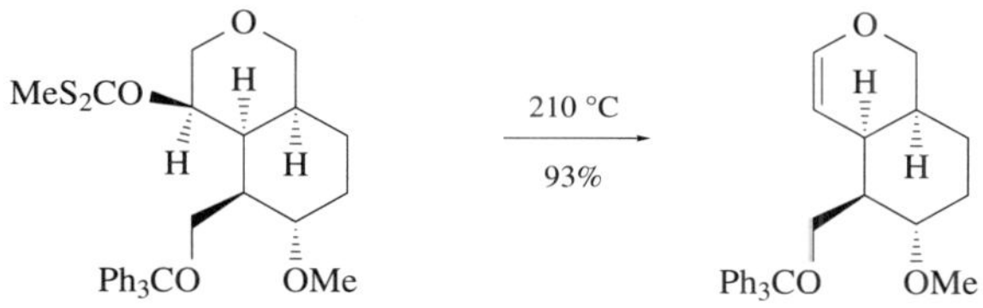

(10) **(11)** (12)

1.17.2.2.3 Other concerted pyrolytic eliminations

Wide structural variation is possible, and some examples have been summarised ⟨91JOC846⟩; Taylor has made extensive studies of the mechanistic effects of changing the heteroatoms (e.g., ⟨91JCS(P2)1703⟩ and earlier papers in the series). Important factors include the nucleophilicity of Z and the electronegativity of Y (or of atoms attached to Y), both of which aid the reaction (Equation (1)) ⟨B-79MI 117-02⟩. However, these variations have, as yet, not significantly displaced the traditional acetate or xanthate methods for preparative purposes.

The heteroatom Z is not required for a successful pyrolytic elimination, and indeed simple vinyl ethers **(12)** (Equation (1); X = O) can be pyrolysed in a flow system at atmospheric pressure ⟨77JOC3899⟩, to give alkenes at a comparable rate to the corresponding acetate, although the product distribution is slightly altered (Equation (1); X = O) ⟨82JCS(P2)1179, 88JCS(P2)737⟩.

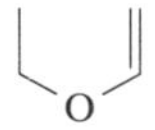

(12)

2-Alkoxypyridines **(13)** ⟨82JCS(P2)1175⟩ and related heterocycles ⟨86JCS(P2)1255⟩ undergo a similar decomposition (Equation (1); X = O, Z = N).

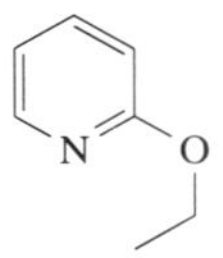

(13)

Other examples of substrates with two heteroatoms which give alkenes in a similar manner to ester pyrolyses are thionacetates **(14)** (and the isomeric thioacetates **(15)**) ⟨73JCS(P2)1293, 75JCS(P2)317⟩ and benzimidates **(16)** ⟨66TL6279⟩, whereas examples related to xanthates include the carbonates **(17)** ⟨72JCS(P2)926, 72JCS(P2)1744, 72JCS(P2)2359, 83JCS(P2)291⟩ and thiocarbonates **(18)**. Within the series of carbonates and their possible sulfur analogues, the following order of reactivity has been found ⟨88JCS(P2)177⟩: PhOCSOR > $PhOCO_2R$ > PhSCSOR > $PhSCO_2R$ > $PhOCS_2R$ > $PhSCS_2R$ > PhOCOSR > PhSCOSR. The major rate change occurs when OR is replaced by SR, with the change from carbonyl to thiocarbonyl producing a relatively small effect ⟨88JCS(P2)177⟩.

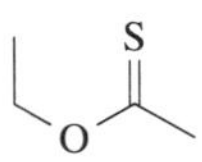

(14)

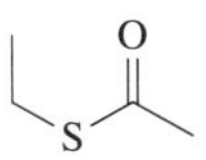

(15)

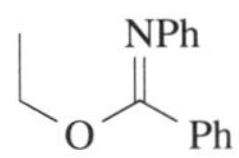

(16)

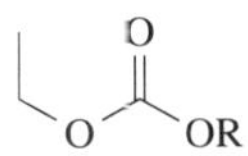

(17)

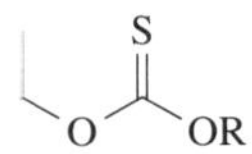

(18)

The flow pyrolysis of carbamates **(19)** ⟨73AJC1259⟩ has attracted some attention. Empirically, *N*,*N*-dimethylcarbamates require similar conditions to those of acetates, but their diphenyl analogues react at rather lower temperatures ⟨81JOC2804⟩. In one useful application, decomposition of the carbamate **(20)** at the temperature of refluxing carbon tetrachloride (10 h) gave an 80% yield of the silole **(21)** (Equation (13)) ⟨83TL3521⟩; the method was much superior to earlier methods of dehydration. A study of *N*-aryl-*O*-alkylselenocarbamate **(22)** pyrolysis has been made ⟨94T639⟩.

Conditions are mild (80 °C in chloroform solution), but the *syn*-elimination is complicated by a competing bimolecular (E2) mechanism.

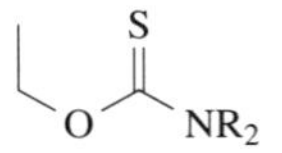

(19)

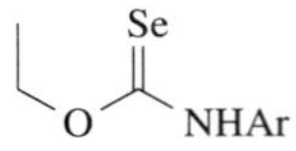

(22)

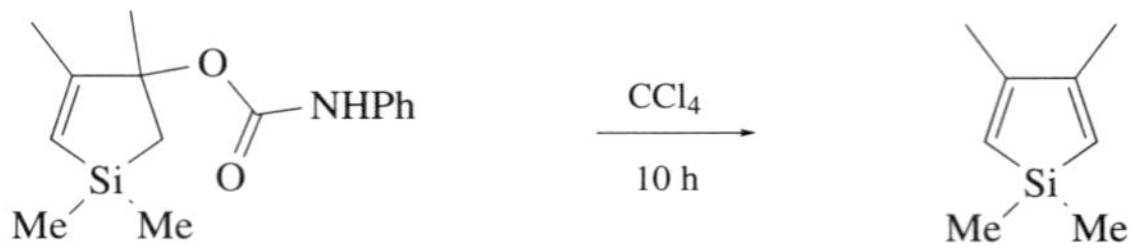

(20) **(21)** (13)

Similar eliminations have been observed from phosphate (**23**), phosphinate (**24**) or sulfonate (**25**) esters. Phosphates are much more reactive than the corresponding acetate owing to the greater electronegativity of the phosphorus atom ⟨61JOC846⟩, and the rate spread for alkyl diphenylphosphinates (primary : secondary : tertiary = 1 : 400 : 10^6) is much greater than for the corresponding acetates ⟨75JCS(P2)1025⟩.

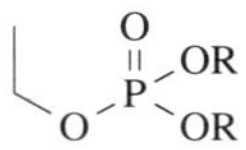

(23)

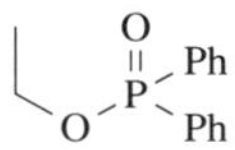

(24)

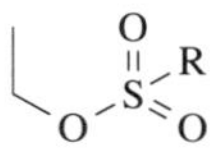

(25)

Tosylates have been found to be more reactive than either acetates or xanthates in 1,3-eliminations in the adamantane series ⟨72JCS(P1)2533⟩. Like benzoates, they have the added practical advantage for FVP that the co-product condenses at a position in the trap, well away from the alkene, allowing facile separation ⟨89JOC5811⟩. As an alternative to gas-phase methods, 8-quinolylsulfonates or 2-pyridylsulfonates have been found to decompose cleanly at ca. 150 °C to give good to excellent yields of simple alkenes ⟨89JOC389⟩.

1.17.2.3 Cleavage of One C—H and One C—S Bond, including Pyrolysis of Sulfoxides

The cleavage of C—S bonds in ester-type pyrolyses is well known (Equation (1); X = S, etc.). Examples include thioacetates (**15**) and *S*-alkyldimethylthiocarbamates (**26**), which show standard eliminations to give alkenes (80–98% on a 100 mg scale) upon FVP at 560 °C ⟨87JOC3688⟩; these compounds are less reactive than their *O*-alkyl isomers.

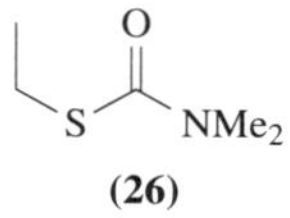

(26)

Pyrolysis of sulfoxides has long been known to give alkenes, but the reaction has only been developed as a useful synthetic method since the mid 1970s ⟨78ACR453, 78CRV363, 78S713, 91COS1011⟩. The availability of the precursors and the mild conditions required for the elimination are two major attractions of this method (and for the related application of selenoxide pyrolysis (Section 1.17.2.4)). In particular, it provides a three-step route from ketones or esters to their α,β-unsaturated analogues by sequential sulfenylation, oxidation and thermolysis (Scheme 7) ⟨78CRV363⟩. The thermolysis step usually takes place in solution in the temperature range 25–130 °C, with *S*-arylsulfoxides decomposing at the lower temperatures and *S*-alkylsulfoxides decomposing at the higher temperatures in the range ⟨78ACR453⟩. Electron-withdrawing groups on the *S*-aryl ring also accelerate the reaction

⟨78CL541⟩. The method is also useful for the preparation of the sulfenic acid co-products ⟨91COS1011⟩. Sulfinate esters (**27**) show similar elimination reactions ⟨70JCS(C)81⟩.

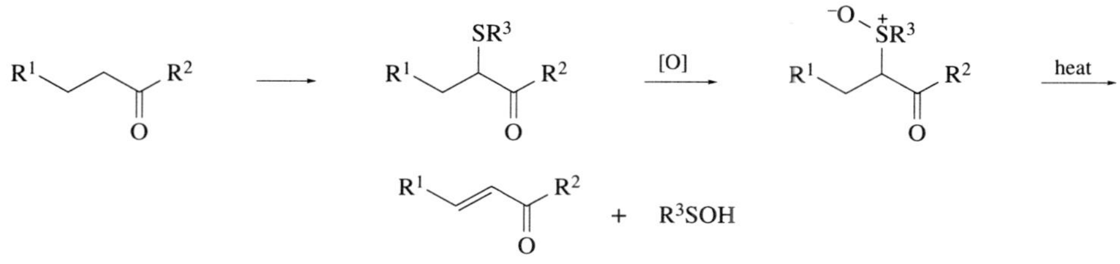

Scheme 7

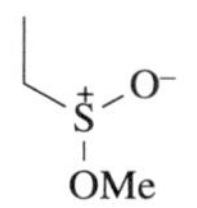

(27)

The standard *syn*-stereochemistry of the process for acyclic examples was established by Kingsbury and Cram ⟨60JA1810⟩ and the solvent independence of the reaction rate is consistent with a concerted mechanism. Isotope effect studies suggest that the hydrogen transfer occurs via a linear transition state ⟨78JA2802, 78JA3927⟩. Acyclic alkenes are produced with (*E*) stereochemistry ⟨73JA6840, 75JOC148⟩. Fragmentation takes place towards the most acidic β-hydrogen atom, and the order of reactivity quoted by Trost ⟨78ACR453⟩ is as follows: $C{=}C{-}CH_2 > C{\equiv}C{-}CH_2 > ArCH_2 \sim CH_3 > CH_2 > CH$. The elimination takes place away from β-hydroxy groups to give allylic alcohols (Scheme 8) ⟨78TL4903⟩: the equivalent reaction of selenoxides (Section 1.17.2.4) is more useful in practice. Where no other reaction is possible, pyrolysis of β-hydroxysulfoxides gives ketones via their enol tautomers ⟨75TL2841⟩. Chirality at the sulfoxide group can also have an effect on the regiochemistry, especially in sterically hindered situations where alternative β-hydrogen atoms have similar acidities ⟨67JCS(C)532, 70JCS(C)833, 76JCS(P1)459⟩. Early attempts to obtain optically active alkenes by the use of homochiral sulfoxides have met with only limited success ⟨67JOC2059⟩. As found for acetate and xanthate pyrolysis, *endo*-cyclic rather than *exo*-cyclic elimination takes place in five- and six-membered cyclic systems (Equation (14) and Scheme 9, where LDA is lithium diisopropylamide) ⟨73JA6840, 75JOC148, 77HCA2388⟩.

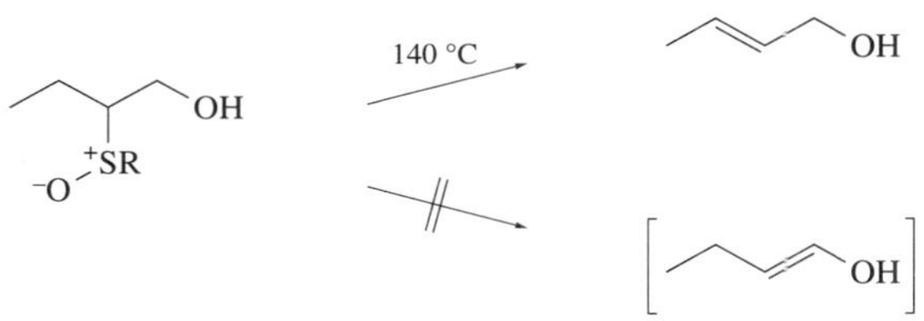

Scheme 8

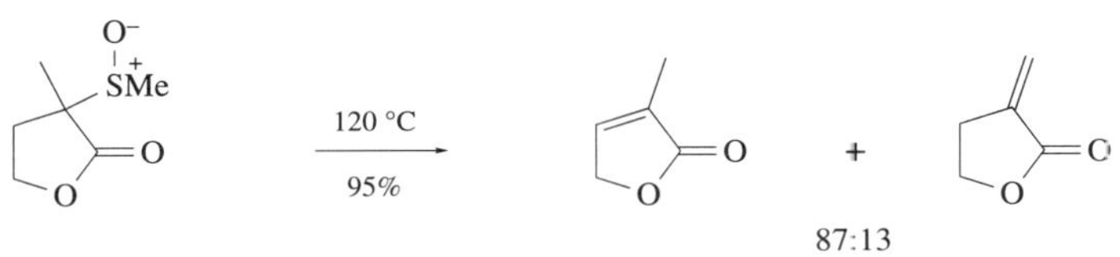

(14)

As the elimination step takes place under such mild conditions, the synthetic utility of sulfoxide eliminations is primarily dependent on the availability of the precursors. The route to α,β-unsaturated carbonyl compounds (Scheme 7) can be extended by regiospecific alkylation—either at the sulfide or the sulfoxide stage ⟨78S713⟩—to give a combined alkylation–elimination sequence (e.g., Scheme 9) ⟨77HCA2388⟩. Dimsyl anion methodology can be used as a route to isolated terminal alkenes, although the thermolysis temperatures are comparatively high (e.g., Scheme 10) ⟨64JOC2699⟩.

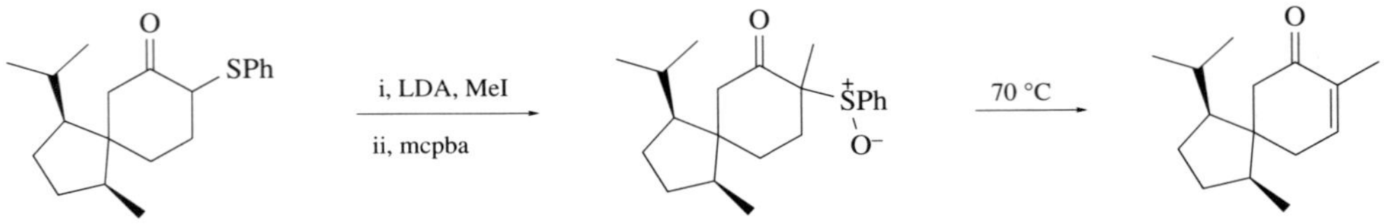

Scheme 9

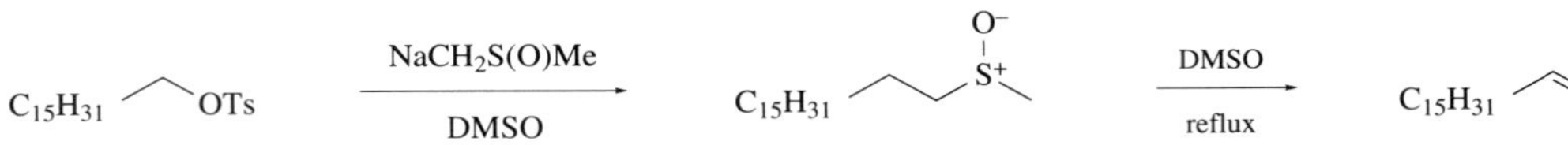

Scheme 10

In some applications from the late 1980s, vinyl fluorides have been obtained from α-fluorosulfoxides ⟨87TL3901⟩; for the corresponding β-fluoro derivatives elimination takes place away from the fluorine atom to give a high yield of an allylic fluoride (Equation (15)) ⟨92JOC714⟩. Dimethyldioxirane has been advocated for oxidising sterically hindered sulfides to the sulfoxide prior to the elimination, for which traditional reagents are ineffective ⟨93T7967⟩. In the field of natural product synthesis, sulfoxide elimination has been employed in routes to avermectin ⟨87TL4959⟩ and azadirachtin ⟨87TL221, 91T6813⟩ subunits (Scheme 11) (see also Chapters 1.13 and 1.14).

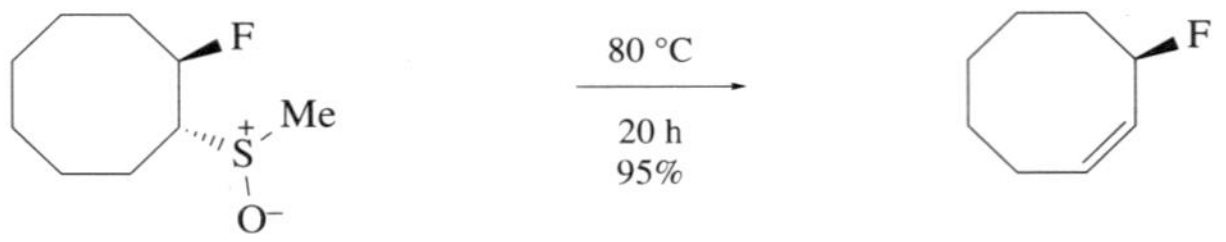

(15)

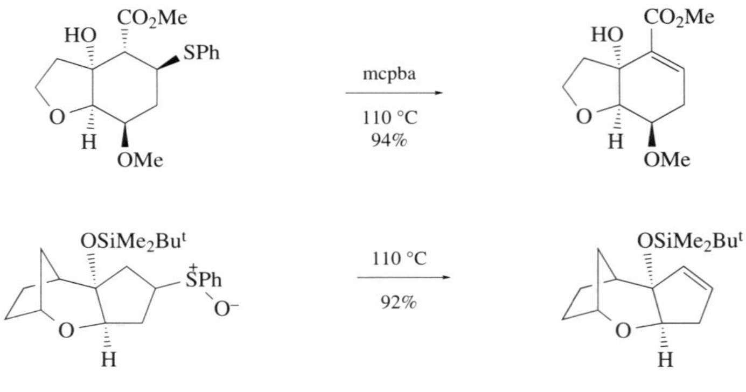

Scheme 11

1.17.2.4 Cleavage of One C—H and One C—Se Bond—Pyrolysis of Selenoxides

Selenoxide elimination is the most recently developed method to be considered in this chapter, and it occurs under the mildest conditions, with the 'pyrolysis' frequently taking place in solution between 0°C and room temperature (Equation (2); X = Se). On average, temperatures for the elimination are 50–120°C lower than those required for sulfoxide pyrolysis ⟨78T1049⟩. Alkenes which are sensitive to nucleophiles, polymerisation, enolisation and thermal degradation can be prepared efficiently by this method. Flexible routes to the selenide precursors are available from ketones, esters, epoxides or by substitution reactions, which enhance the synthetic utility of the method, and these lead to α,β-unsaturated carbonyl compounds, allylic alcohols or terminal alkenes, respectively. Reviews have been published ⟨78T1049, B-86MI 117-01, 91COS1011⟩. Detailed practical considerations are discussed by Clive ⟨78T1049⟩, and applications to natural product synthesis are summarised by Paulmier ⟨B-86MI 117-01⟩.

The reaction is carried out by oxidation of a selenide precursor (usually a phenylselenide) and the elimination then takes place directly without isolation of the selenoxide. A range of possible reagents may be employed, but hydrogen peroxide or meta-chloroperoxybenzoic acid are used most commonly ⟨78T1049⟩. The formation of terminal unconjugated alkenes from primary selenides is slow, although it can be encouraged by incorporating an electron withdrawing group into the phenyl ring of the selenide ⟨75JOC947, 80TL5037⟩. Since the precursors are easily made from alcohols ⟨76JOC1485⟩, this provides an alternative to acetate and related pyrolyses for primary alcohol dehydration. The generation of acyclic enones is particularly rapid. Often, an excess of the oxidising agent is employed to destroy the selenenic acid co-product, which could otherwise react with the alkene (particularly terminal alkenes) or undergo competing disproportionation reactions ⟨78JOC1689⟩. The selenenic acid can also be destroyed by addition of a secondary amine to convert it into a selenenamide ⟨76JOC2503⟩, a strategy which is particularly useful when cyclic enones are formed ⟨75JA5434⟩.

The mild conditions employed, compared with sulfoxides (and amine oxides), are due to the longer Se—O and Se—C bond lengths and to the greater polarisation of the Se—O bond ⟨81JA1232⟩. The *syn*-nature of the elimination was established in 1970 using a steroid example ⟨70JCS(D)86⟩, and later confirmed for acylic cases ⟨73TL1979⟩. As usual, (*E*)-alkenes are obtained predominantly ⟨74TL2279⟩, although almost equal amounts of (*E*)- and (*Z*)-α,β-unsaturated nitriles may be obtained. In most cyclic systems a (*Z*)-alkene is produced, but for large ($\geqslant$12-membered) rings either pure (*E*)- isomer or mixtures of (*E*)- and (*Z*)-isomers may be obtained ⟨78T1049⟩. As with sulfoxide- or acetate-type pyrolysis, *endo*-alkene formation is favoured over *exo*-elimination for both five- and six-membered rings, although in some cases the preference may not be very great (Scheme 12) ⟨73JA5813⟩. Where possible, the fragmentation occurs *away* from a β-electronegative substituent (particularly oxygen atoms) (e.g., Equation (16)) ⟨74JOC429⟩, and this provides an efficient and highly effective synthetic route to allylic alcohols. Decomposition takes place *towards* allylic or benzylic centres to give the conjugated product ⟨73JA5813, 73TL1979, 75JA3250⟩; conditions are particularly mild in these cases ⟨75JA5434⟩. The direction of the elimination is also affected by statistical factors, and if these are in opposition to the above effects then the overall regioselectivity may not be very good.

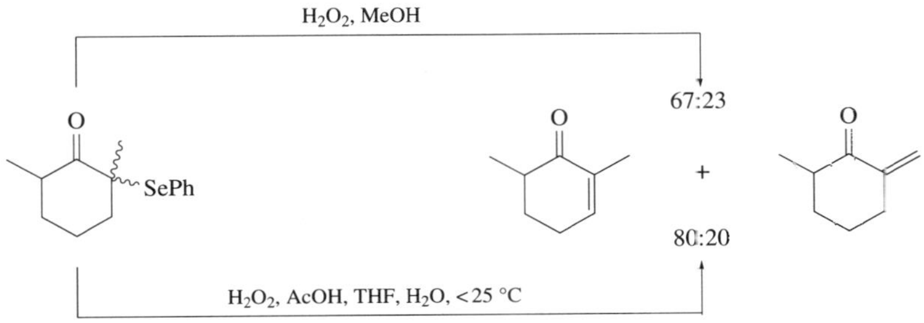

Scheme 12

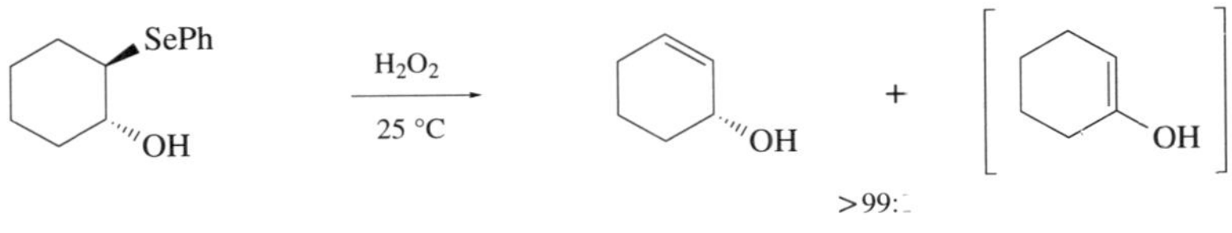

(16)

Partly because of the availability of precursors, selenoxide eliminations are particularly effective for the synthesis of α,β-unsaturated carbonyl compounds, and for allylic alcohols ⟨73JA2697⟩ and related compounds ⟨91COS1011⟩ via enolate selenenylation and epoxide ring opening, respectively. These methods are of significant importance in contemporary organic synthesis, and frequently appear in multistep syntheses of natural products. Thus the reaction has been employed in quassinoid chemistry (Equation (17)) ⟨94JOC311⟩, where it was found that removal of the phenylselenenic acid co-product was simply effected by addition of sodium bicarbonate. Cases of α,β-unsaturated carbonyl compounds being produced include formation of α-alkylidene-β-lactones (**28**) ⟨93JOC322⟩ (which can be transformed into allenes by β-lactone thermolysis (Section 1.17.3.1)) and a subunit (**29**) of the streptogramin antibiotics (Scheme 13) ⟨89JOC3988⟩. Some *exo*-methylene heterocyclic compounds have been prepared via selenoxide intermediates (e.g., Equation (18)) ⟨93JOC1349⟩. Allylic alcohol formation has been employed in key steps *en route* to symbioramide (**30**) ⟨94LA41⟩, valienamine (**31**) ⟨92TL1025⟩ and coriolin (**32**) ⟨93LA1133⟩ (Scheme 14) (see also Chapter 1.14).

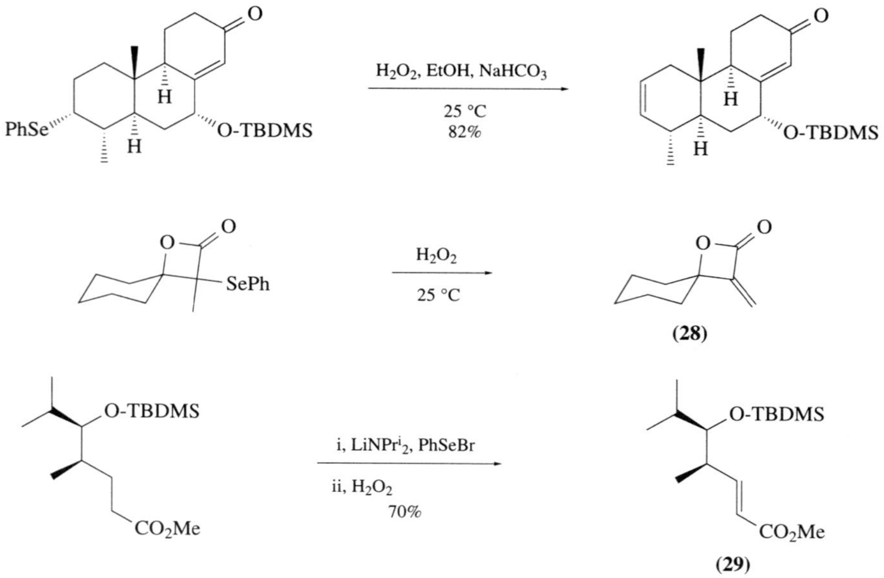

(17)

(28)

(29)

Scheme 13

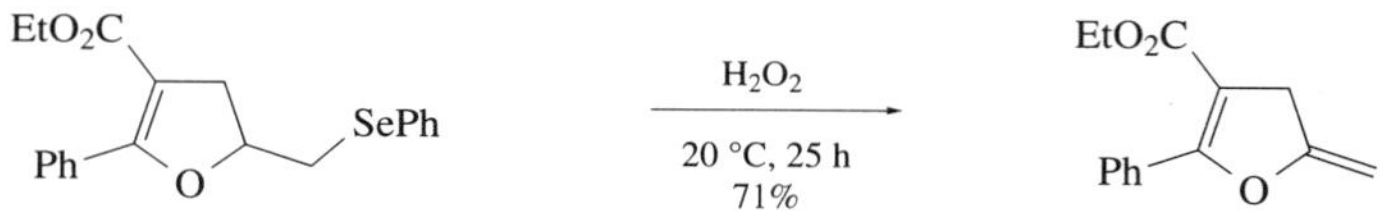

(18)

1.17.2.5 Cleavage of One C—H and One C—N Bond, including the Cope Elimination

Alkenes may be formed by pyrolysis of suitable amides (Equation (1); X = NR, Z = O). Although the temperatures required are higher (ca. 100 °C) than those for the corresponding acetate ⟨60CRV431⟩, similar isomer distributions are obtained ⟨59JA651⟩. Kinetic data for amides and the corresponding thioamides (Equation (1); X = N, Z = S) are available ⟨89JCS(P2)579, 90JCS(P2)2187⟩. In a cyclic example, facile elimination of isobutene from *N*-*t*-butylpyridazin-3-ones at 700 °C under FVP conditions has been observed ⟨82JCS(P1)1845⟩. Disulfonimides (**33**) can also be pyrolysed neat at 160–200 °C to give alkenes in up to 99% yield.

The most important reaction type in this section is the Cope elimination (Equation (2); X = NR), for which specialised reviews are available ⟨60OR317, 93S263⟩ in addition to more general surveys ⟨60CRV431, 91COS1011⟩. The method presents an alternative to the more widely used Hofmann elimination of quaternary ammonium hydroxides, and has certain advantages in terms of ease of manipulation and lack of product isomerisation. However, the Meisenheimer rearrangement to form *N*,*N*,*O*-trisubstituted hydroxylamines may compete ⟨93S263⟩.

As with the related sulfoxide and selenoxide eliminations, the reaction is normally carried out with a mixture of amine and oxidising agent without purification of the amine oxide. After destruction of the excess oxidising agent, the reaction mixture is concentrated and the crude material is pyrolysed for a few minutes at 100–140 °C (e.g., ⟨53JA3212⟩). These conditions are significantly more severe than those required for sulfoxide and particularly selenoxide eliminations. Both the alkene and the hydroxylamine by-product distil from the reaction mixture and are separated by treatment with acid. Significant rate enhancements can be observed if the elimination is carried out in DMSO solution ⟨62JA1734⟩.

The *syn*-stereochemistry of the elimination has been frequently established (e.g., Equation (19) ⟨73JOC1742⟩), but isotope effects suggest that there may be subtle differences in the mechanism in

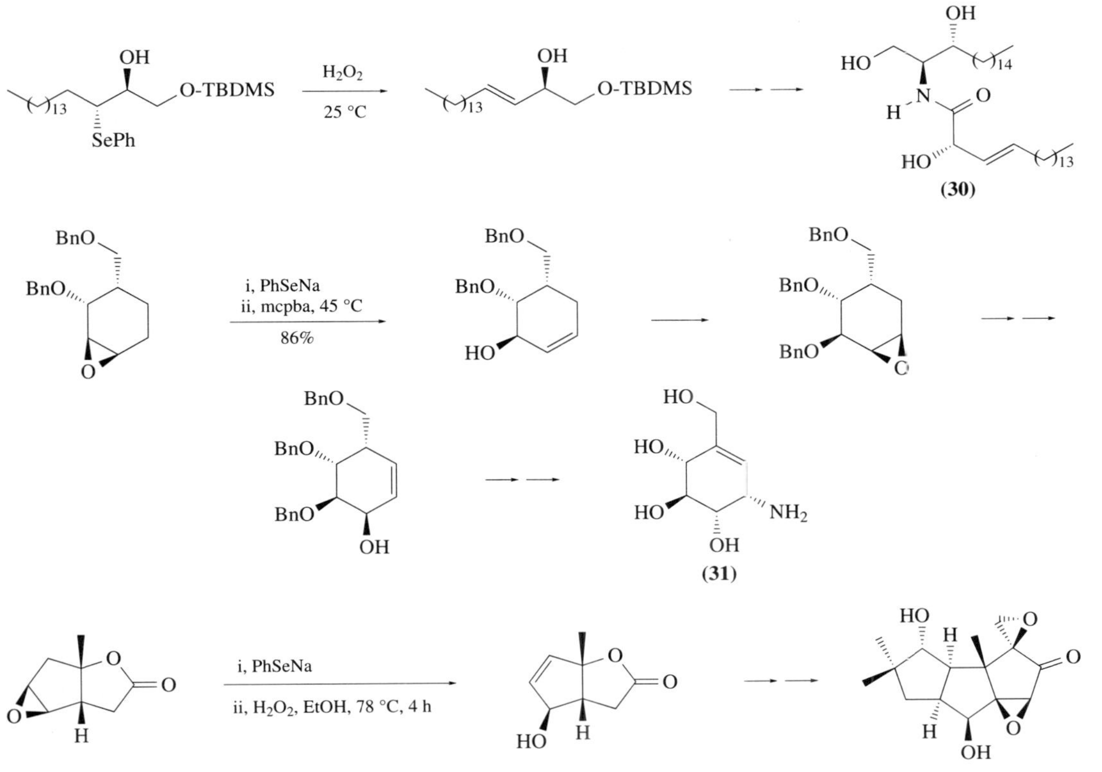

(30) (31) (32)

Scheme 14

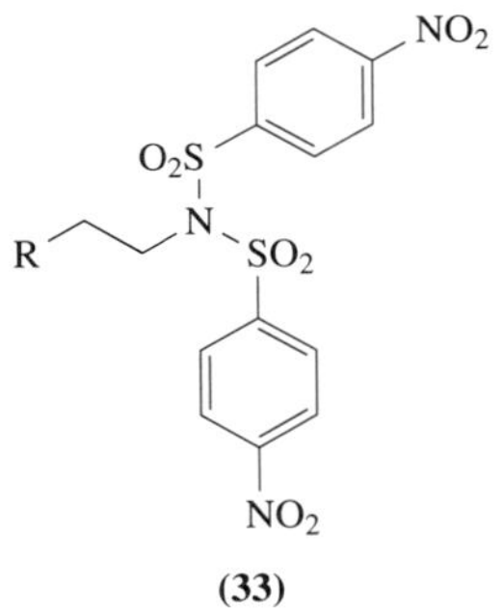

(33)

DIGLYME solution (nonlinear hydrogen transfer via a bent transition state) and DMSO solution (linear hydrogen transfer, possibly incorporating a molecule of solvent) ⟨78JA2802, 78JA3927, 81JA4650⟩. ^{14}C-Isotope effects indicate that extensive rupture of both the C—N and the C—H bonds has taken place at the transition state, with relatively little C═C character ⟨83JA3714⟩. In aliphatic systems, the direction of elimination qualitatively resembles that found for esters, with the number of hydrogen atoms on β-positions being the major factor ⟨60CRV431, 60OR317⟩; the formation of (*E*)-alkenes is again favoured if both (*E*) and (*Z*) isomers can be obtained ⟨60CRV431, 60OR317⟩. In medium-sized rings ($\geqslant$nine-membered) (*E*)-alkenes are obtained exclusively ⟨53JA3212, 55JA1628⟩. A striking difference compared with the results of ester, sulfoxide or selenoxide pyrolyses is that a double bond *exo*- to a six-membered ring is obtained in preference to the *endo*-isomer (Equation (20)) ⟨57JA4729⟩. This may be due to the more severe constraints of the five-membered ring transition state with relatively short C—N bonds, which requires a quasi-boat form in the cyclohexane ring for *endo*-elimination. Five- and seven-membered ring analogues give more of the *endo*-cyclic alkene ⟨57JA4729⟩.

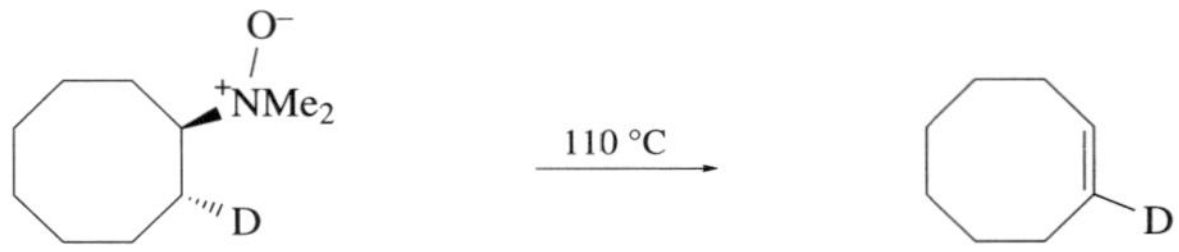

(19)

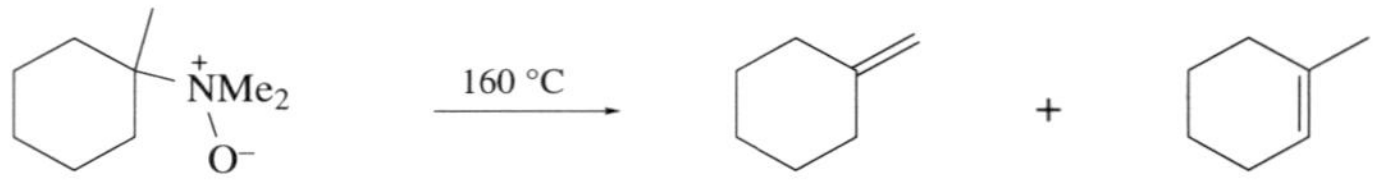

97.2:2.8

(20)

Applications of the Cope elimination to synthesis have been rare. Following earlier work in thiete dioxide chemistry ⟨64JOC3131⟩, Woolhouse *et al.* obtained the conjugated thiete (**34**) exclusively by heating a sample of the amine oxide (**35**) at 90 °C (Scheme 15) ⟨93JHC873⟩; the isomeric thiete (**36**) is thought to be formed initially, which then isomerises under the reaction conditions. In a highly stereoselective, convergent synthesis of racemic trichodiene (**37**), a Cope elimination was used in the final step to introduce the *exo*-cyclic alkene function (Equation (21)) ⟨84JOC3870⟩. Examples of intramolecular ring cleavage by Cope elimination have been published with simple heterocycles ⟨60JA4656⟩ and the work has been extended to alkaloid chemistry in the mid 1980s ⟨84JCS(P1)1701, 85H(23)3085⟩ (see also Chapters 1.13 and 1.14).

(35) **(36)** **(34)**

Scheme 15

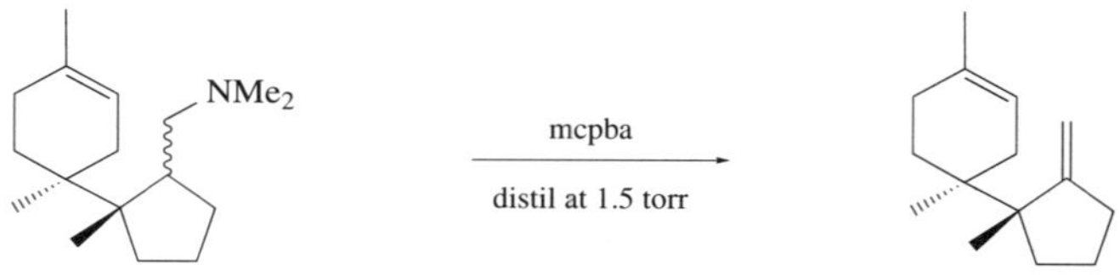

(37)

(21)

1.17.3 FORMATION OF MONOENES BY RETRO-CYCLOADDITION REACTIONS

1.17.3.1 Retro [2 + 2] Cycloadditions

Thermal cycloreversions of this type (Equation (22)) would formally be [$2\sigma_s + 2\sigma_a$] processes and hence many of these reactions occur thermally by diradical mechanisms ⟨86T2135⟩ and are not considered in detail here. The incorporation of a carbonyl group in the four-membered ring apparently increases the likelihood of a (stereospecific) concerted process taking place under thermal conditions. The topic has been reviewed ⟨82AG(E)225⟩.

(22)

Mechanistic aspects of the formation of alkenes by cyclobutane pyrolysis have been reviewed by Gajewski ⟨B-81MI 117-02⟩ and Brown ⟨B-80MI 117-01⟩, and the weight of evidence is in favour of diradical intermediates ⟨94TL2675⟩. In contrast, the thermal decomposition of simple cyclobutanones is probably concerted and proceeds with the retention of stereochemistry predicted for such a mechanism ⟨72JA7235⟩. Under photochemical conditions, even simple cyclobutanes would be expected to undergo concerted fragmentation; one example is shown in Equation (23) ⟨75TL1335⟩.

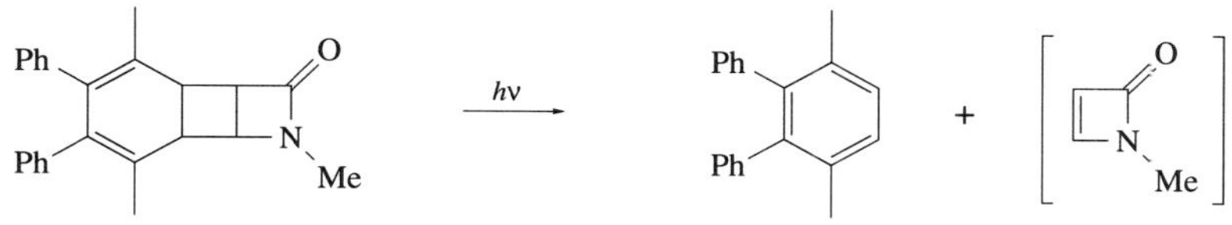

(23)

The photochemical formation of oxetanes followed by thermal cracking in the opposite sense ⟨73CC374⟩ has been employed as a synthetic method for α,ω-enones ⟨75CC206⟩; this is probably another diradical process ⟨77CRV473⟩.

The facile loss of carbon dioxide from β-lactones is well established ⟨1883CB2208⟩ and has been reviewed (e.g., ⟨84CHEC(7)363, 93S441⟩. The process is of synthetic value for the two-step conversion of 3-hydroxycarboxylic acids into alkenes, and since these precursors can be readily obtained from ketones the method is a viable alternative to the Wittig reaction for the conversion of ketones into alkenes. Experimental conditions include heating in water ⟨58JA3585⟩, or heating at 140–160 °C either neat ⟨72JA2000⟩ or in solution ⟨93JOC322⟩, or FVP at 400 °C ⟨91JOC5782⟩. The decomposition is accelerated by electron donating substituents at the 4-position, although it is affected little by substituents at the 3-position ⟨84CHEC(7)363⟩. The precise mechanism of the reaction is a matter for debate ⟨93S441⟩, but the *syn*-nature of the elimination was confirmed in 1966 (Equation (24)) ⟨66JOC4043⟩. This feature, together with the absence of isomerisation, is particularly important for synthetic applications ⟨72JA2000⟩. The method has been used as a route to enol ethers ⟨73CJC981, 79S388⟩, and applications to the synthesis of allenes are also noteworthy ⟨91JOC5782, 93JOC322⟩. In exceptional cases where a highly strained alkene would result, the cycloreversion may take place in the opposite sense to give a keto-ketene ⟨86JA7121⟩.

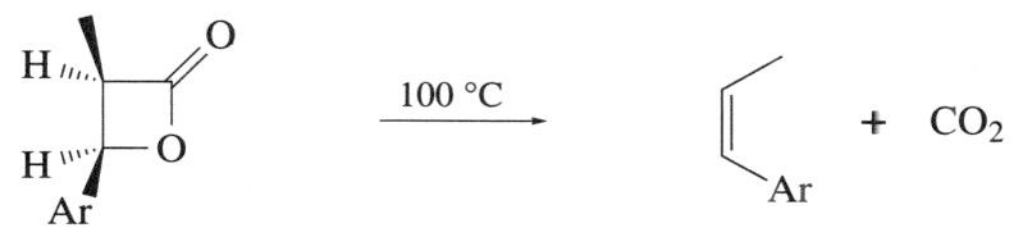

(24)

Simple β-lactams can decompose thermally to give isocyanates and alkenes ⟨59RTC551⟩, although the alternative mode of fragmentation to ketene and imine may take place with appropriate substituents ⟨63JA3524⟩ and under photochemical conditions ⟨68CB2669⟩. Upon FVP at 600 °C (12 torr) reaction takes place to give alkenes with almost complete retention of stereochemistry (Equation (25)) ⟨70JA1763⟩. This reaction does not appear to have been used synthetically.

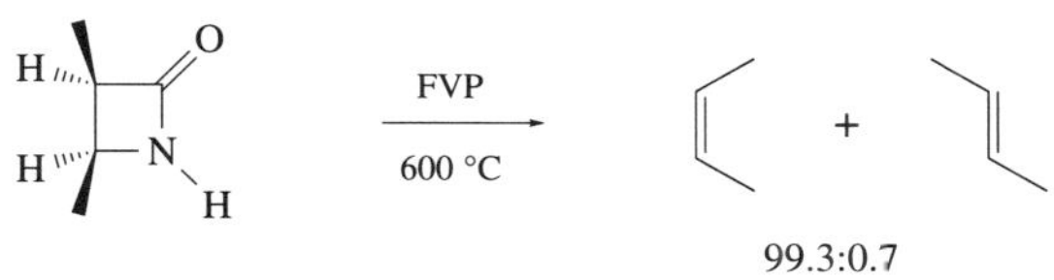

(25)

1.17.3.2 Retro [4 + 2] Cycloadditions (Retro-Diels–Alder Reactions)

This $[2\pi_s + 2\sigma_s + 2\sigma_s]$ reaction (Equation (26)) has been used extensively as a preparative route to either monoenes, dienes (Section 1.17.4.1) or other multiply bonded functional groups containing heteroatoms. For the preparation of alkenes (Equation (26); Y = Z = C), the transformation almost always involves cleavage of two C—C single bonds (i.e., U = X = C). It can be carried out in solution, usually at temperatures in excess of 150 °C ⟨91COS(5)551⟩, although in appropriate cases acceleration by cation, anion or radical substitution has been predicted ⟨78T1877⟩. Indeed, oxy-anionic substitution may cause the reaction to proceed rapidly at room temperature (Equation (27)) ⟨80TL2799⟩. Cycloreversion of *endo*-adducts may be substantially faster than that of *exo*-adducts ⟨78JOC518⟩. Increasingly, FVP methods are employed to accomplish the transformation for neutral species, using temperatures in the range 400–600 °C ⟨91COS(5)551⟩. Cyclopentadiene or anthracene adducts ((**38**) and (**39**), respectively) have been commonly used as alkene precursors. The former has the advantage that its low molecular weight causes the precursor (and the cyclopentadiene co-product) to be relatively volatile and it is commonly used for complex alkenes, although relatively high temperatures may be required. In contrast, anthracene is less volatile and condenses near the exit point of the furnace; it is often used for simple alkenes from which it is readily separated.

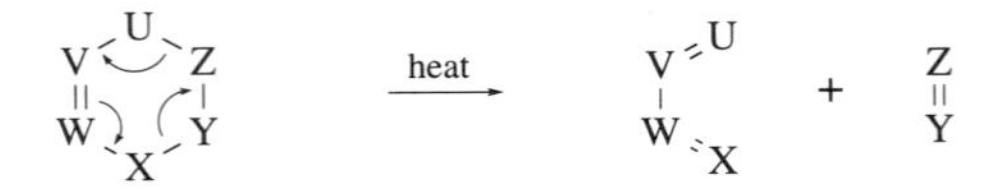
(26)

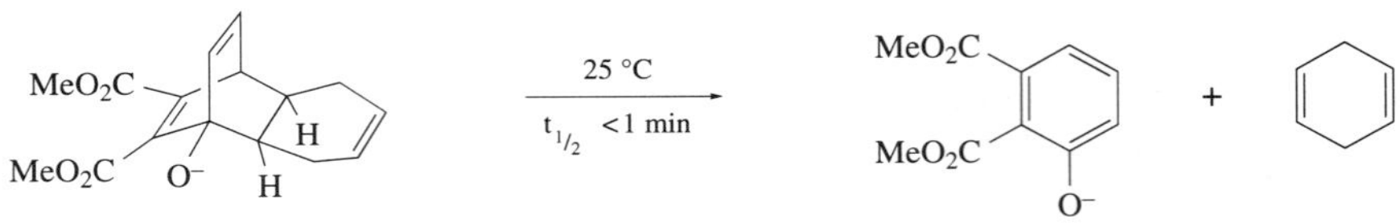
(27)

(38)

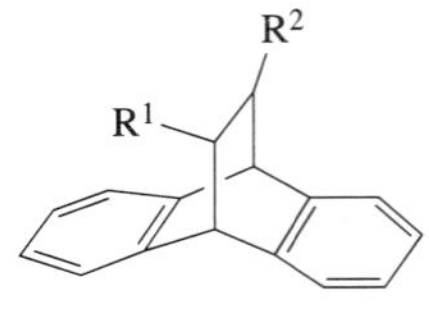
(39)

These methods have been used to prepare a wide range of both simple and functionalised alkene systems, and some typical examples are given below. In many cases reactive alkenes are released at a late stage of a synthesis and so the Diels–Alder adduct serves as a protecting group. A number of comprehensive reviews of retro-Diels–Alder processes are available, covering mechanistic features ⟨68CRV415, 80AG(E)779⟩, preparative aspects ⟨78T19, 85S121, 91COS(5)551⟩ and application to natural product synthesis ⟨87S207⟩.

The vinylimidazole (**40**) is a good example of a simple alkene best prepared by retro-Diels–Alder reactions. A yield of 85–90% is claimed for flow pyrolysis (methanol solution) at 500 °C (Equation (28)) ⟨83AG(E)560⟩. The method is also useful for releasing sensitive α,β-unsaturated carbonyl functionality. The example shown in Equation (29) is noteworthy, since the chiral centres and the alcohol protecting group are both unaffected, despite the apparently severe conditions of the retro-Diels–Alder reaction ⟨86TL3511⟩. Zwanenburg and co-workers have used FVP methodology (500 °C) to generate the alkene unit of cyclopentenones (e.g., Equation (30), which are important intermediates in the synthesis of quinane natural products; once again, optically active products can be obtained under these conditions (e.g. ⟨87TL357, 94TL2787⟩). A large number of methylidenemalonic esters have been generated *in situ* by reaction of malonic ester with formaldehyde, trapped with anthracene, and then released again by pyrolysis of the adducts (**41**) in solution at 200–250 °C in the presence of maleic anhydride (which reacts with the anthracene formed). For most esters, yields are in the 40–60% range ⟨88JOC4859⟩. Natural products containing an allylic alcohol unit—such as (**42**), a flavour component of mushrooms—have been made in high enantiomeric purity by a method which requires an FVP cycloreversion to create the alkene function (Scheme 16) ⟨87TL183⟩.

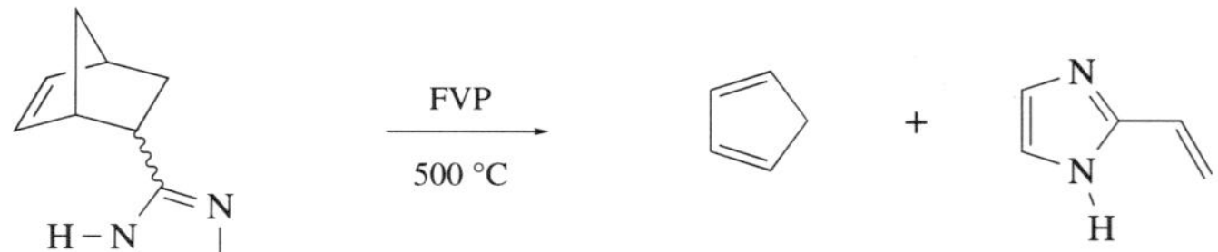
(28)

(40)

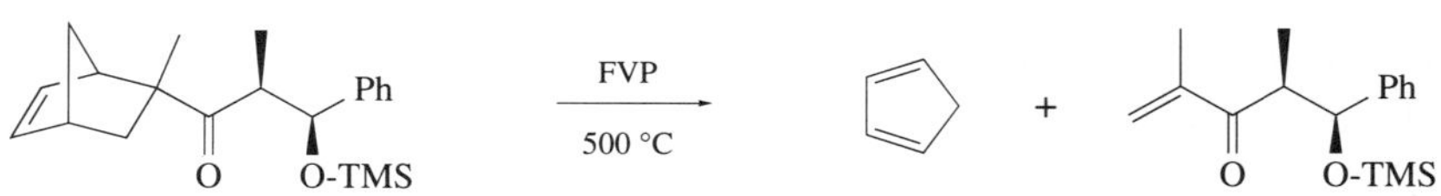
(29)

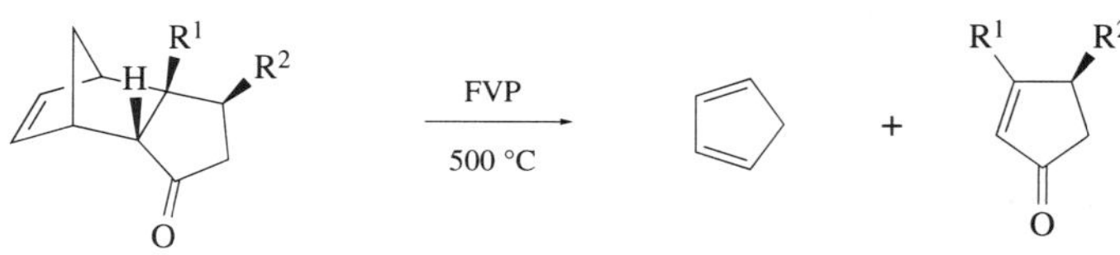
(30)

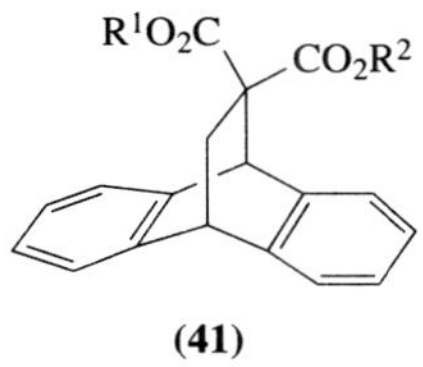

(41)

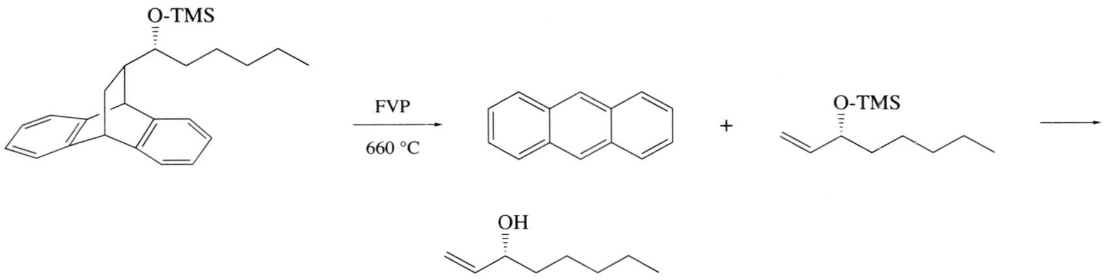

(42)

Scheme 16

The retro-Diels–Alder cleavage of anthracene cycloadducts has been used very effectively by Ripoll and co-workers ⟨85S121⟩ to generate unstable or highly reactive tautomers such as enols ⟨79NJC195⟩, simple enamines ⟨80T2497⟩ or vinyl phosphines ⟨80T2497⟩, as shown in Equation (31). In many cases, these products were trapped or were characterised by spectroscopy at low temperatures.

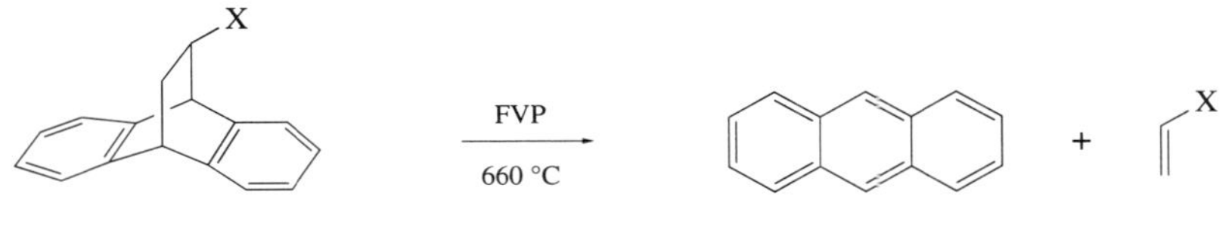

(31)

X = OH, NH_2, PH_2, etc.

In certain cases where strained alkenes may be formed, an alternative free-radical ring expansion may take place rather than the retro-Diels–Alder process. Certain cyclopropene–anthracene adducts, for example, have been shown to behave in this way ⟨94CC889⟩.

1.17.4 FORMATION OF DIENES AND POLYENES

The methods described in the previous sections have been applied—often unexceptionally—to generate dienes and polyenes, and some typical examples of each will be given below. The major advantage of these methods vis-à-vis ionic eliminations is that the reactions usually proceed with a notable lack of isomerisation, and so nonconjugated alkenes can often be obtained if desired.

Standard flow pyrolysis of an acetone solution of the diacetate (**43**) gave the diene (**44**) in up to 85% yield on a 250 g scale (Equation (32)) ⟨87JOC5034⟩. Trahanovsky and co-workers have made extensive application of ester pyrolysis under FVP conditions as a route to heteroaromatic analogues of *o*-xylylenes (e.g., ⟨81JA6691, 86JOC4208⟩). The mechanism may involve a direct δ-elimination, or a [3,3]-sigmatropic shift of the ester function prior to standard β-elimination (Scheme 17). These xylylenes are stable at $-60\,^{\circ}C$ but form dimers at room temperature.

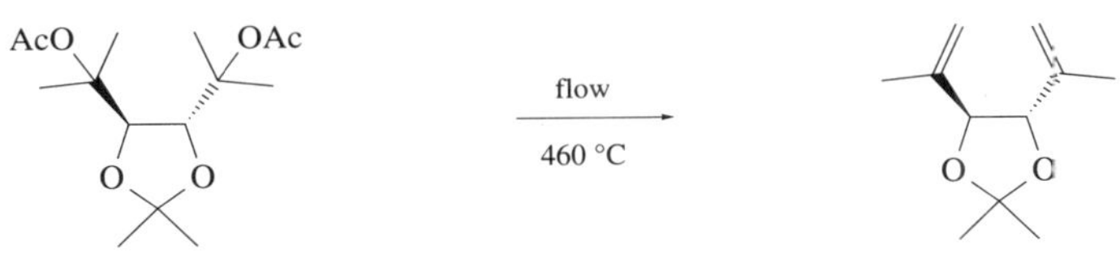

(32)

(43) **(44)**

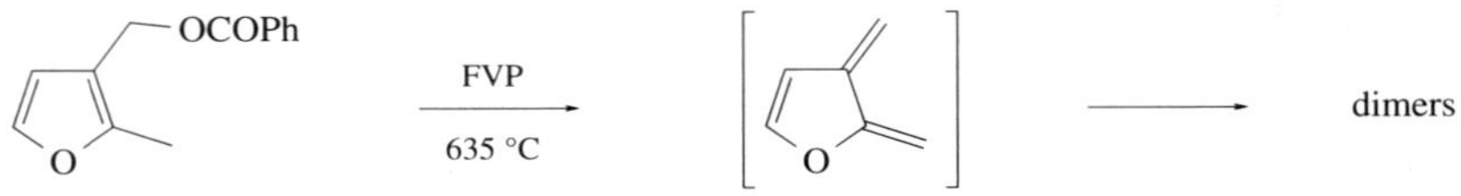

Scheme 17

Facile Chugaev eliminations can take place instead of the expected reduction when appropriate xanthate-type compounds are heated to 80–110 °C in the presence of tributyltin hydride ⟨93JA2268⟩.

Allylic alcohols have been transformed into conjugated dienoic esters by a one-pot process involving sequential *ortho*-ester exchange, [3,3]-sigmatropic shift and sulfoxide pyrolysis (Scheme 18) ⟨91JOC6981⟩. Although mixtures of isomers may be obtained, the method has been usefully applied in synthesis ⟨90JOC3967, 91JOC4339⟩. In a more esoteric application, a quadruple sulfoxide elimination was employed to generate four alkene units in the novel cyclophane (**45**) (Equation (33)) ⟨91AG(E)1173⟩. The use of sulfoxide and particularly selenoxide pyrolysis to give *conjugated* dienes is complicated by the possibility of a competitive [2,3]-sigmatropic shift to give the isomeric sulfenate or selenenate, respectively (Scheme 19), which may be trapped to give an allylic alcohol ⟨74ACR147⟩. As a result of the reversibility of the process in the case of sulfur, the diene may also be obtained if desired ⟨82JOC4801⟩, especially if the *S*-aryl ring contains electron-withdrawing groups which accelerate the elimination ⟨78CL541⟩. With selenoxides, the equilibrium lies much further towards the selenenate, and dienes have only been obtained in a few special cases and in modest yield ⟨83JA2503⟩.

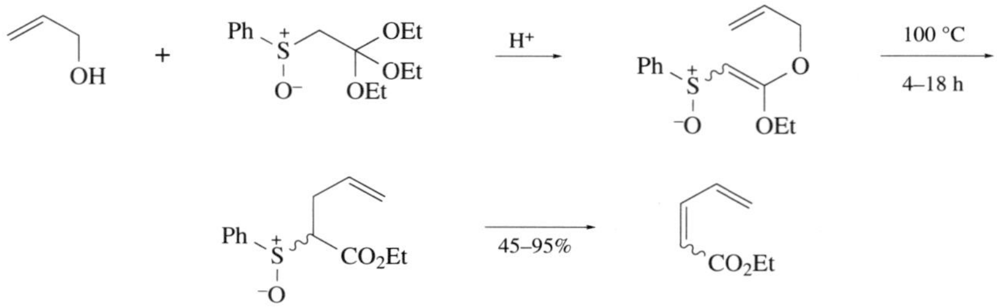

Scheme 18

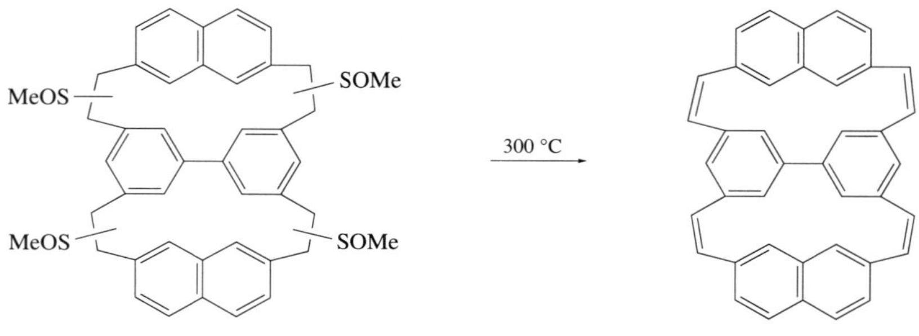

(33)

(45)

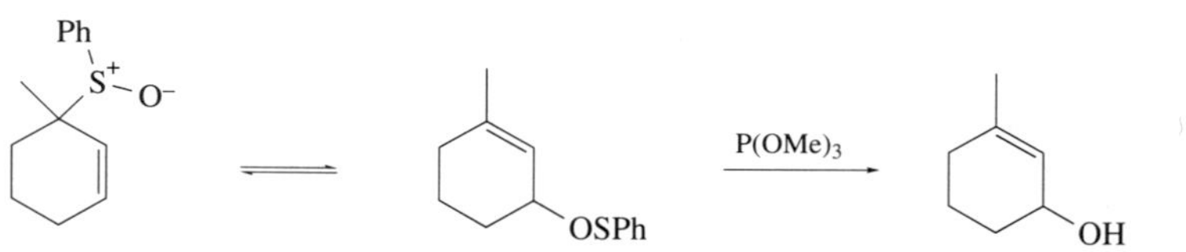

Scheme 19

The Cope elimination has been employed to prepare penta-1,4-diene and allylbenzene; both were obtained in the absence of their conjugated isomers penta-1,3-diene and *β*-methylstyrene, respectively ⟨57JA960⟩. Although the overall yield is often high, isomerisation may be a problem in amine oxide pyrolysis ⟨92TL2461⟩, and traditional Hofmann elimination can be more selective. A spontaneous 'sila-Cope' elimination, leading specifically to a (*Z*),(*Z*)-diene, has been reported to take place under particularly mild conditions ⟨82JA7666⟩.

1.17.4.1 Retro-cycloaddition Reactions

Photochemical cleavage of cyclobutanes has been used to achieve some spectacular syntheses of unnatural target molecules such as bullvalene (Equation (34)) ⟨64CB3140⟩, and—in conjunction with matrix isolation—reactive polyenes such as pentalenes ⟨73AG(E)337⟩ and cyclobutadienes ⟨74AG(E)425⟩ have been obtained.

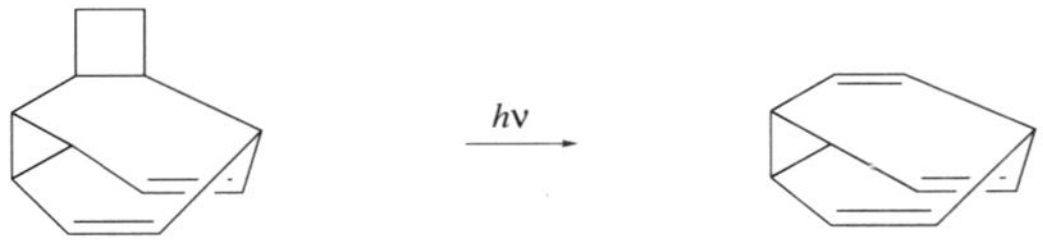

(34)

The pyrolytic cleavage of *β*-lactones has been employed as a convenient route to arene oxides and related dihydroaromatic compounds ⟨78JA352⟩, since the precursors can be easily obtained in a few steps from benzoic acid.

The retro-Diels–Alder reaction can be used to release a single alkene of a diene (or polyene) unit. The general conditions of the reaction are described in Section 1 17.3.2, and a typical example is shown in Equation (35) ⟨86TL1445⟩.

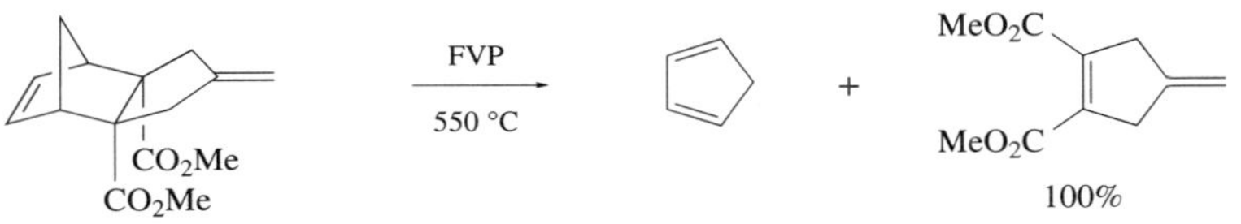

(35)

Clearly, dienes can also be generated directly by utilising the other product of the retro-Diels–Alder process (Equation (26); U = V = W = X = C). When the 'diene' released forms part of a benzenoid system, the reactions may be especially facile. Much variation in the precursor structure is possible, and useful syntheses have been developed in which the diene is formed by cleavage of either two C—C bonds, or one C—C bond together with a C—O or C—N bond, or by cleavage of two C—N bonds. By way of illustration, some typical examples of each of these classes are given below.

The retro-Diels–Alder cleavage of ethylene from 1,4-epoxytetrahydronaphthalenes (**46**) under FVP conditions has been developed into a highly efficient synthesis of the highly reactive iso-benzofuran system (Equation (36)) ⟨81MI 117-01⟩. The parent compound, which is a useful Diels–Alder diene, can be accumulated in quantitative yield at a rate of 5–10 g h^{-1} from a commercially available precursor ⟨72CC347⟩.

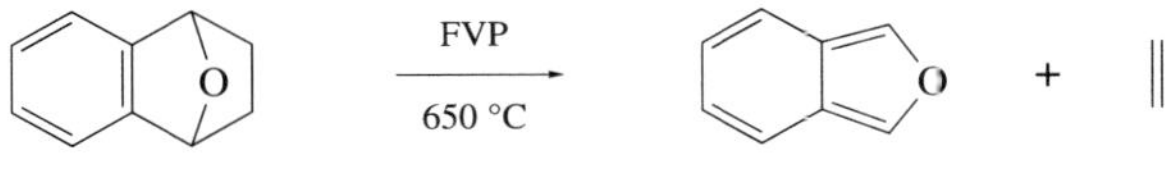

(**46**)

(36)

Cycloadducts derived from *α*-pyrones (e.g., (**47**)) acting as dienes often lose CO_2 under the conditions of their formation to generate diene products by cleavage of one C—C and one C—O bond (e.g., Scheme 20) ⟨72TL4651⟩. When the original dienophile is an alkyne, substituted benzene systems are obtained directly (e.g., ⟨87T5245⟩).

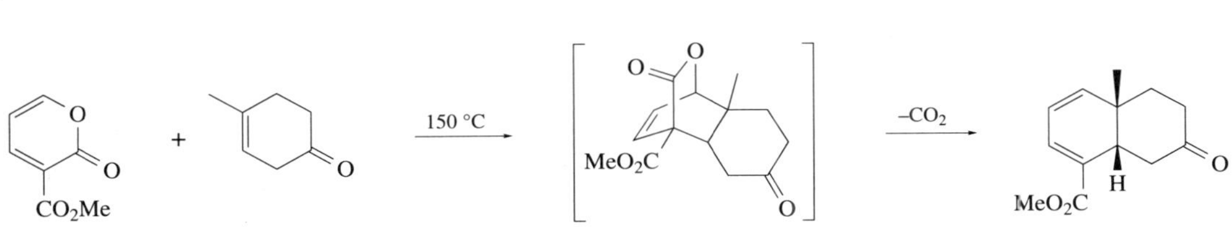

(**47**)

Scheme 20

In a similar fashion to the examples in the preceding paragraph, retro-Diels–Alder cleavage of one C—C and one C—N bond (usually as RCN), or two C—N bonds (usually as N_2) frequently takes place *in situ* following cycloaddition reactions of azines and alkynes, leaving an alternative heteroaromatic system as the final product of the reaction. Typical examples of the many possibilities are given in Scheme 21 ⟨73LA437, 74LA1190⟩. This area of heterodiene reactions has been comprehensively reviewed ⟨83T2869⟩.

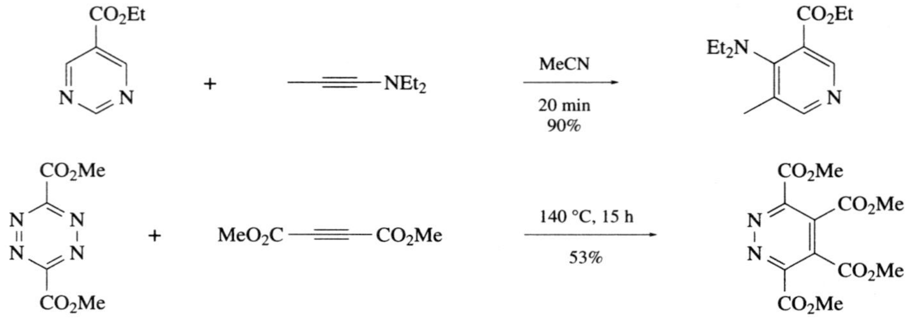

Scheme 21

1.17.4.2 Retro-cheletropic Reactions

Retro-cheletropic processes are applicable to the stereospecific formation of conjugated dienes (Equation (37)). The extruded group (X) is most often CO, N_2 or SO_2. The cleavage of SO_2 from 3-sulfolenes is of particular importance in synthesis, and this topic has been reviewed ⟨93PHC1⟩. Both the thermal extrusion of sulfur dioxide from **(48)** ⟨66JA2857, 66JA2858⟩ and of nitrogen from **(49)** ⟨66JA1335⟩ have been shown to take place stereospecifically with disrotatory motions of the terminal methylene groups (e.g., Scheme 22). Carbon monoxide extrusion is particularly important when an aromatic product is formed ⟨67TL4945⟩.

(37)

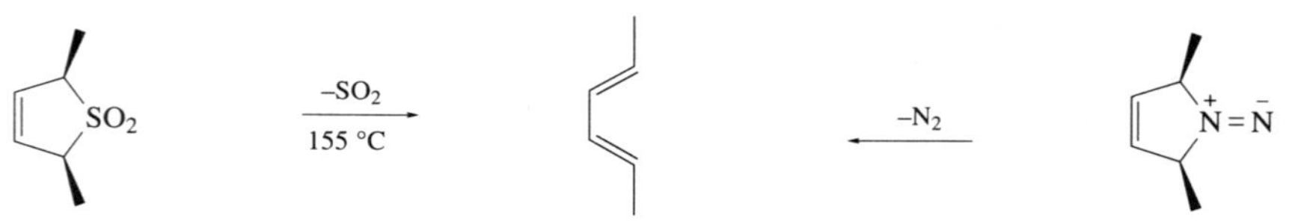

(48) **(49)**

Scheme 22

The formation of butadiene derivatives by pyrolysis of 3-sulfolenes takes place readily at 100–150 °C in solution or in a sealed tube, at 350 °C under flow conditions, or at ca. 500 °C by FVP. The first conditions may be used to generate a reactive diene *in situ* in the presence of a dienophile, whereas the last conditions are often more efficient for the isolation of the diene itself. 1-Substituted or 1,4-disubstituted butadienes can be made by alkylation of the sulfolene at its acidic α-positions prior to the pyrolysis, although special conditions must be used to avoid the ring opening of the anion to butadienyl sulfinate. Therefore, nonnucleophilic bases have been used so that the anion may be generated in the presence of the alkylating agent ⟨86JCS(P1)1039⟩. Alternatively, the protection and subsequent pyrolytic (FVP) release of the sulfolene as a formal Diels–Alder adduct has the significant advantage that alkylation(s) take place exclusively on the *exo*-face and therefore lead to a stereospecific diene synthesis (Equation (38)) ⟨82TL3277⟩. This method has been employed extensively by Bloch *et al.* as a route to insect sex pheromones (e.g., ⟨83TL1247, 86T4975⟩). 2-Trimethylsilylsulfolene can be used to achieve α,α-dialkylation and hence give a potential route to 1,1-disubstituted butadienes ⟨87CC934⟩.

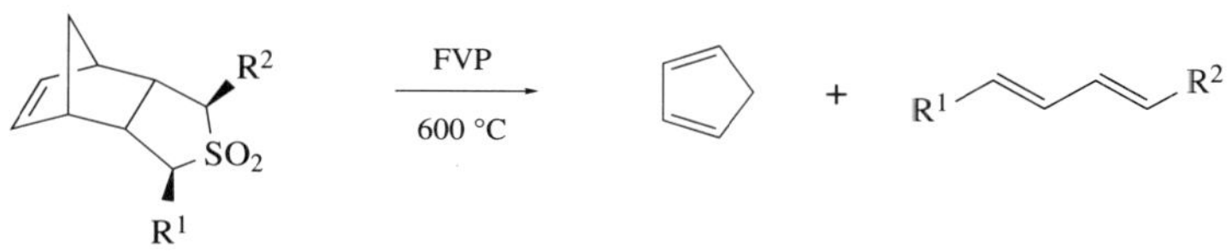

(38)

Reaction of the anion derived from 3-methyl-3-sulfolene with prenyl or geranyl bromides followed by extrusion of SO_2 has been reported as a stereo- and regioselective means of introducing an isoprene unit onto an existing skeleton ⟨84CC1323⟩. Many functionalised 2-substituted butadienes have been obtained by pyrolysis of the corresponding 3-substituted sulfolene. For example, 2-trimethylsilylbuta-1,3-diene was generated by sealed-tube (160 °C) pyrolysis of the appropriate sulfolene, and underwent *in situ* Diels–Alder reaction with dimethyl acetylenedicarboxylate (DMAD) ⟨87JOC244⟩.

In applications of sulfolene pyrolysis to alkaloid synthesis, the butadiene released is often in place to undergo an intramolecular Diels–Alder reaction and hence create a new six-membered ring. Two examples are given in Scheme 23 ⟨80JOC3372, 94TL1071⟩; in the first of these, the heteroene component was also generated by a pyrolytic method.

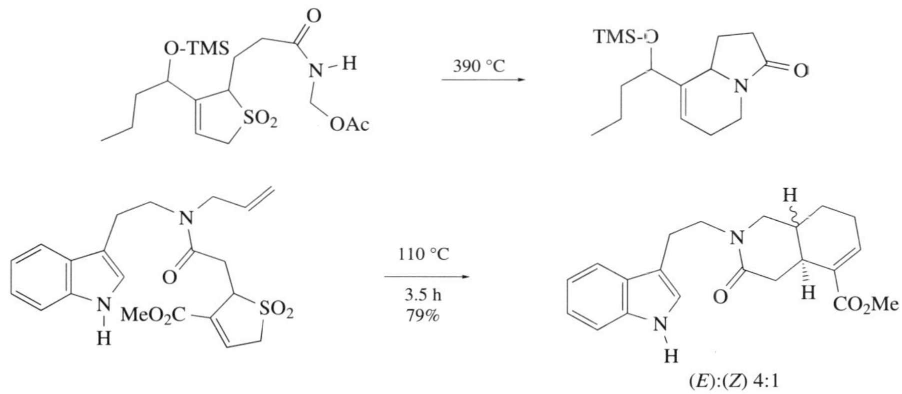

Scheme 23

Pyrolysis of 2,7-dihydrothiepin-1,1-dioxides (**50**) leads to 1,3,5-trienes by an analogous concerted process ⟨69JA5682⟩.

(50)

1.18
One or More =CH, =CC, and/or C=C Bonds Formed by Rearrangement

PATRICK J. MURPHY
University of North Wales, Bangor, UK

1.18.1 GENERAL INTRODUCTION AND METHOD OF CATEGORISATION

Rearrangement reactions lie at the heart of modern synthetic organic chemistry, and in their highest form they can effect complex but predictable manipulations of carbon frameworks and associated functionality. This chapter is complementary to other discussions in this work and the reader is specifically referred to Chapters 1.09, 1.17 and 1.20 (Sections 1.20.1.3 and 1.20.3.2.) for other information. Allylic rearrangement during substitution, aromatic rearrangements, and reductive transposition, will not be dealt with in this chapter. Two works edited by Mayo ⟨B-63MI 118-01, B-80MI 118-01⟩ and the chapter on rearrangements in the excellent annual series *Organic Reaction Mechanisms* currently edited by Knipe and Watts ⟨B-66MI 118-01⟩ provide useful background information.

The order of discussion is based upon the generalised transformation found in Equation (1) involving the formation of product (**2**) from a precursor (**1**); the organisational principles apply to the function Y.

(1)

(**1**) (**2**)

For example, the first section deals with rearrangements involving [1,*j*]-processes (*j* = odd numeral; these are not necessarily sigmatropic in their nature) and the case where Y = H hydrogen is discussed first, followed sequentially by Y = carbon-containing and finally with hetero- or chalcogen-containing substituents. Several other comments are also necessary at this point. First, hetero-substituted alkenes (enol ethers, enamines etc.) will not be covered in this section, and recourse should be made to the relevant chapters. Stereochemical and mechanistic considerations will be included as, and where, necessary.

1.18.2 REARRANGEMENTS INVOLVING 1,2-SHIFTS AND 1,*j* MIGRATION

This section forms the bulk of the chapter and deals with rearrangements involving a 1,2-shift of a double bond, together with the concomitant migration of the group X to give the new function Y (in many cases X = Y) (Equation (1)). These reactions are discussed in the following order: sigmatropic (the molecular orbital considerations for these processes are covered in general texts ⟨B-70MI 118-01, B-76MI 118-01⟩), acid-catalysed, base-catalysed, and metal-catalysed.

1.18.2.1 Where Y = H

1.18.2.1.1 [1,j] Migrations of hydrogen

(i) Sigmatropic migrations of hydrogen

Many examples of thermal or photochemical rearrangements have been reported in which a hydrogen migrates along a π-system; these have been reviewed on several occasions ⟨B-70MI 118-01, 71ACR272, 71AG(E)761, 71MI 118-01, 76CRV187, 81RCR666⟩. The most commonly encountered rearrangements of this type are [1,3] (photochemical and suprafacial), [1,5] (thermal and suprafacial) and [1,7] (thermal and antarafacial or photochemical and suprafacial) hydrogen shifts (Scheme 1). The [1,3] process is the least common, though the photochemical transformations in Scheme 2 show typical examples ⟨63HCA678, 64PAC539, 65TL3955, 76JA840⟩.

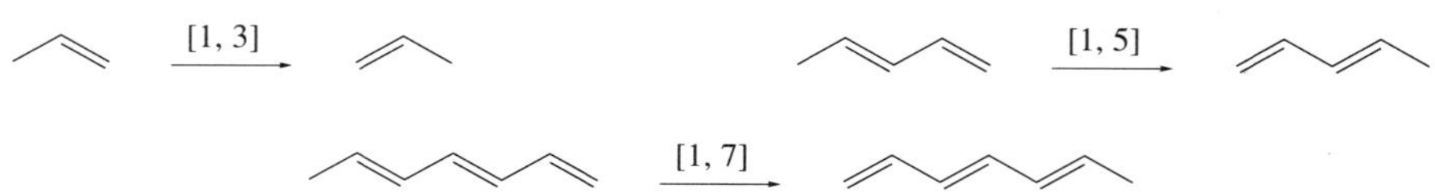

Scheme 1

In contrast, [1,5] hydrogen shifts are quite common (for excellent reviews of early work, see ⟨69CRV103, 76CRV187⟩) and are thermal in nature (Scheme 3) ⟨63RTC741, 64TL1009, 65T2329, 69CJC1555, 70CB426, 73T3781, 76CRV187, 82T567, 84MI 118-01⟩. Some photochemical antarafacial [1,5] hydrogen shifts are also known ⟨69TL993, 70JA7408⟩.

The [1,7] hydrogen shift has been studied extensively and is known to proceed antarafacially

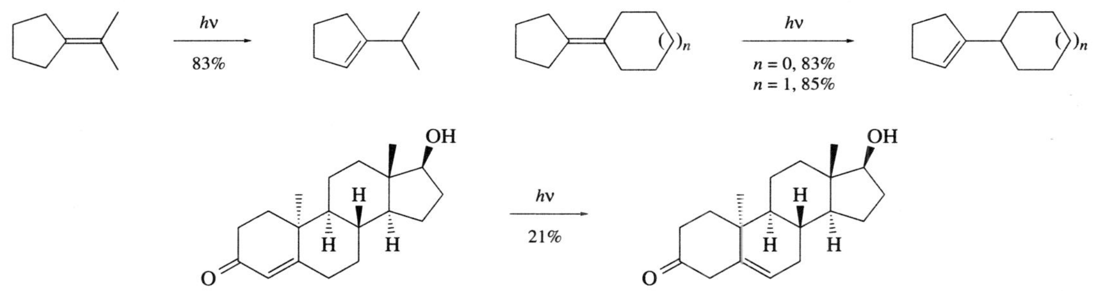

Scheme 2

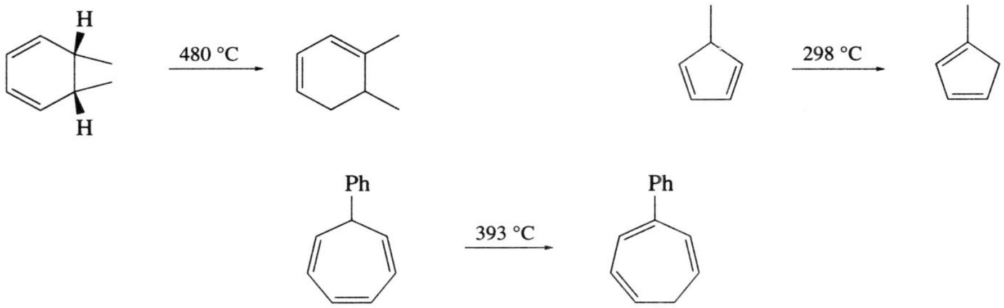

Scheme 3

under thermal conditions ⟨64RTC1173, 70HCA173, 72BSF4238, 72HCA1385, 72HCA1828, 76CRV187⟩, and suprafacially under photochemical conditions ⟨69RTC266, 70BCJ1120⟩. This rearrangement has attracted considerable interest, owing to the presence of a conjugated triene in vitamin D_3 (cholecalciferol) **(3)** which can undergo a reversible [1,7]-hydrogen shift to a minor previtamin form **(4)** (Scheme 4) ⟨64RTC1173, 83ACR81, 87TL6021⟩. Some interesting work reported in 1991 has involved the thermolysis of (7*Z*)-analogues of vitamin D ⟨91JA1355, 91JOC3572⟩.

Several similar processes exist, in particular the [1,5] hydrogen shift found in vinyl cyclopropanes (Scheme 5), which is competitive with the rearrangement of vinyl cyclopropanes to cyclopentenes (see Section 1.18.2.2.1(ii)). The [1,5] shift generally occurs at lower temperatures than cyclopentene formation and signifies the lowest energy pathway for rearrangement ⟨69CRV103, 72MI 118-01, 76RCR469⟩; the reaction is thought to proceed via a chair-like transition state with an activation energy of 30–35 kcal mol^{-1} ⟨72MI 118-01, 76RCR469, 85OR(33)247⟩; most of the reactions are stereospecific yielding only the *cis*-alkene. The process has also been shown to be reversible ⟨65CB756, 65LA(688)28, 80TL2487⟩. The reaction has many examples of theoretical and synthetic interest (Scheme 6) ⟨64JCS5578, 65LA(688)28, 67JA1404, 68CC35, 72TL125, 73RTC605⟩ and has been used in the total synthesis of sarkomycin **(5)** (Scheme 7) ⟨80TL2487, 83JOC3581⟩. 2-Vinyl cycloalkanols are also known to undergo a thermal [1,5] hydrogen shift ⟨60JA4918⟩.

(ii) Acid- and base-catalysed migrations of a hydrogen atom

Rearrangement of double bonds is possible under both proton and Lewis acid catalysis. The reaction is not generally useful as carbocations are involved as intermediates and these give rise to many side products. Also, where several double bond isomers are possible, mixtures are obtained, generally favouring the thermodynamic product ⟨70S405⟩. Examples are known of mineral acid ⟨36CB882, 63JCS4091⟩, alumina ⟨55JA5166, 65JA4107⟩, *p*-toluene sulfonic acid ⟨62BSF1325⟩, oxalic acid ⟨53JA5421⟩ and Lewis acid-catalysed isomerisations ⟨67CC397⟩. The reaction generally proceeds well when conjugation is present in the product ⟨53JA5421, 65JA3228⟩.

The base-mediated or -catalysed migration of double bonds is a well-documented procedure which has been reviewed on several occasions ⟨B-64MI 118-01, B-65MI 118-01, 68ACR231, 69S97, B-70MI 118-02, 70S405, B-77MI 118-01⟩. Excellent results can be obtained when amide bases ⟨87OS(65)224⟩ or alkali metals absorbed on to alumina are used (Scheme 8) ⟨85AG(E)320⟩; potassium *t*-butoxide has

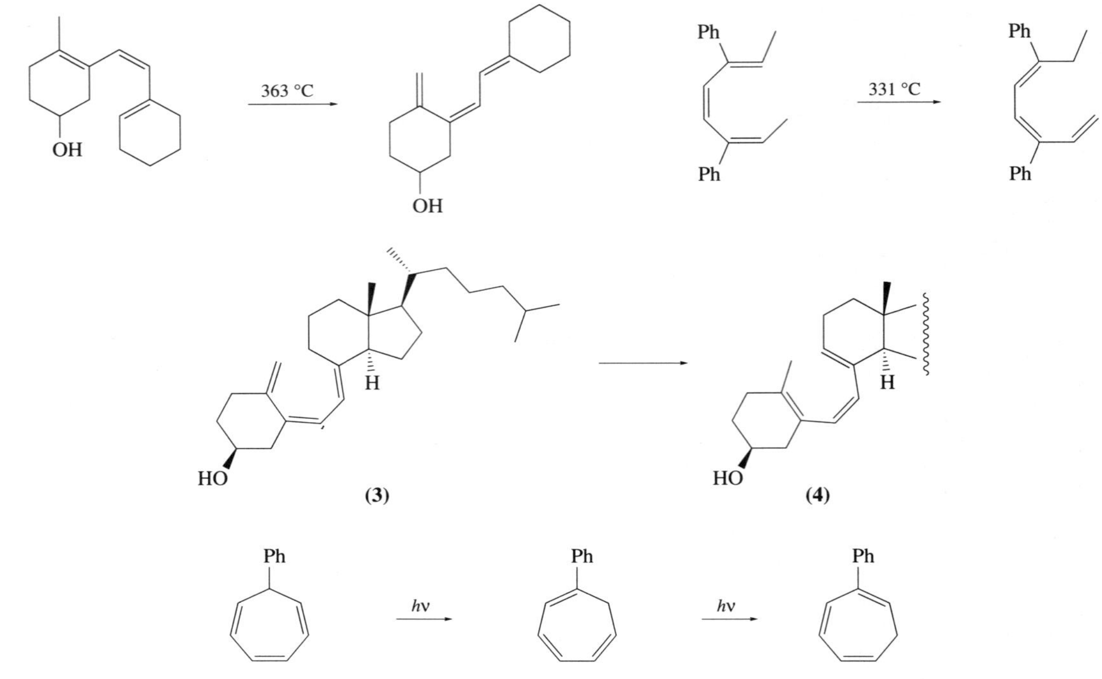

Scheme 4

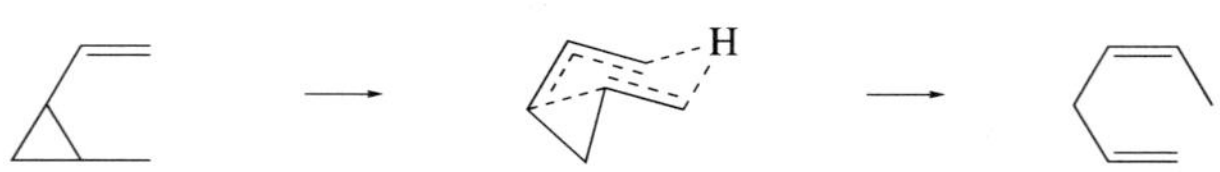

Scheme 5

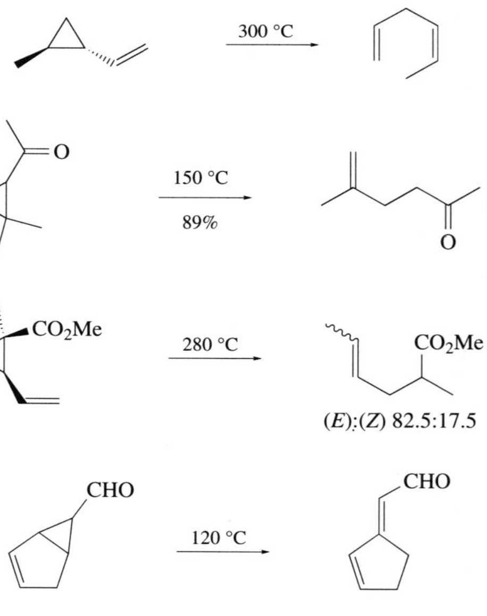

Scheme 6

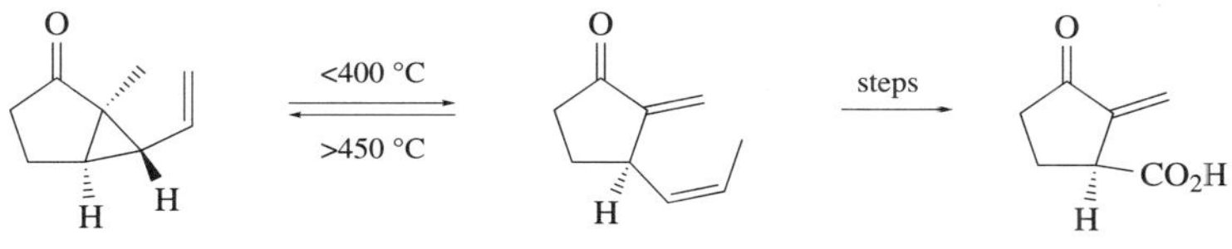

(5)

Scheme 7

also been used to good effect ⟨68JOC221, 80JOC900⟩. The process normally leads to equilibrium mixtures favouring the thermodynamic product. The mechanism of the reaction, often referred to as prototropic rearrangement, is an example of electrophilic substitution with allylic rearrangement of an intermediate carbanion. It has been demonstrated to be partially intramolecular in some instances, in which case the base 'leads' the proton from one site to another ⟨64JA5466, 69JA4459⟩.

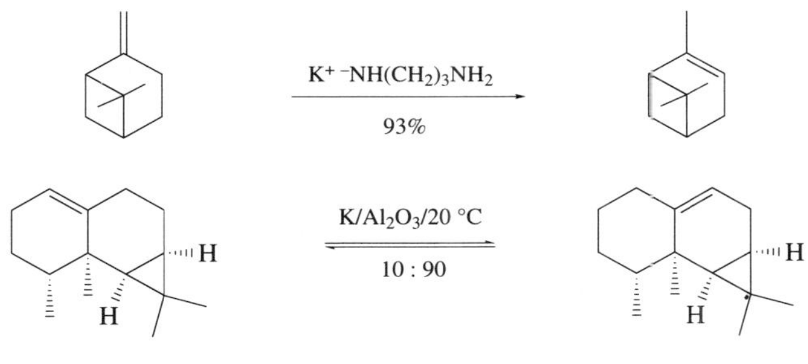

Scheme 8

Rearrangement is possible under relatively mild conditions when conjugation is possible in the product. Many examples are known using nitrogen bases and potassium *t*-butoxide and the criteria for the position of the equilibrium (shown in Scheme 9) has been studied extensively ⟨82JOC4758, 82JOC4766, 83JOC584⟩; two examples are also shown in Scheme 9 ⟨86S921, 93HCA418⟩. The isomerisation of β,γ-enones to α,β-enones under acidic, basic and enzymatic conditions has been extensively studied ⟨B-89MI 118-01, 90CR1203⟩. In the majority of cases the equilibrium lies on the side of the conjugated product but steric factors can, in some cases, reverse this trend. In small ring (five- or six-membered) enones the conjugated product also prevails ($>99:1$) but, in medium rings, a cross-over is observed eventually favouring the unconjugated product (i.e. seven-membered, 73 : 27; eight-membered, 20 : 80; nine-membered, $<0.3:>99.7$) (Scheme 10); this is probably due to unfavourable transannular interactions in the conjugated product ⟨66JCS(B)164, 76JA7319⟩. It is also possible to effect the reverse reaction, i.e. an overall rearrangement of conjugated esters via a two-step procedure, involving lithium amide base generation of an enolate which is then protonated to give the kinetic, deconjugated product; excellent stereocontrol is generally observed (Scheme 11) ⟨81HCA1023, 82JOC163, 84CL1541⟩.

A [1,3] hydrogen shift is also invoked in the Sommelet–Hauser rearrangement (see Section 1.18.2.5.2(ii)); iodine-catalysed alkene isomerisation is also known ⟨87JOC2586⟩.

(iii) Metal-catalysed migrations of hydrogen

Double-bond migrations are also possible under metal catalysis; common techniques utilise supported metals (on alumina, silica or carbon), metal salts, metal complexes and also Ziegler–Natta conditions. The most commonly used metals are palladium and rhodium ⟨68ACR186⟩, although examples are known of platinum, nickel and iridium-based catalysis ⟨66MI 118-02, 67MI 118-01, 70S405, B-74MI 118-01⟩; typical examples are shown in Scheme 12 ⟨68TL3797, 83S952, 85JA7352, 87TL5473, 94T3139⟩. The reaction proceeds either via a metal hydride addition–elimination process or a π-allyl complex ⟨73JA2248⟩. Metal carbonyls are also known to effect alkene isomerisations ⟨73JA2248, 89BSF799⟩.

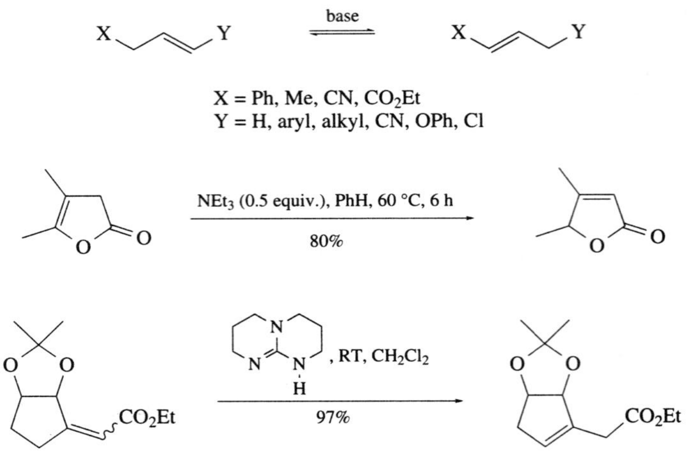

Scheme 9

Scheme 10

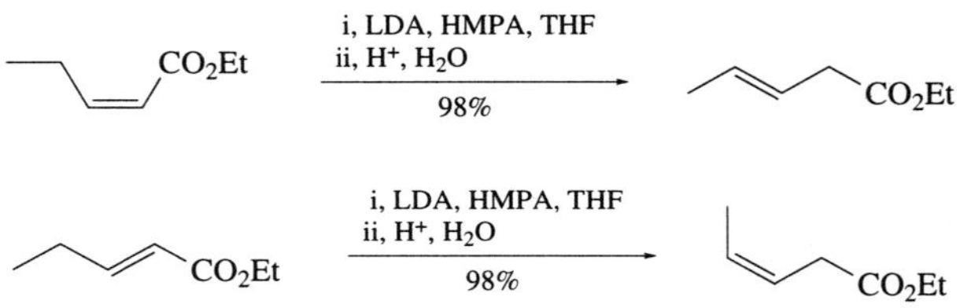

LDA = lithium diisopropylamide
HMPA = hexamethylphosphoramide

Scheme 11

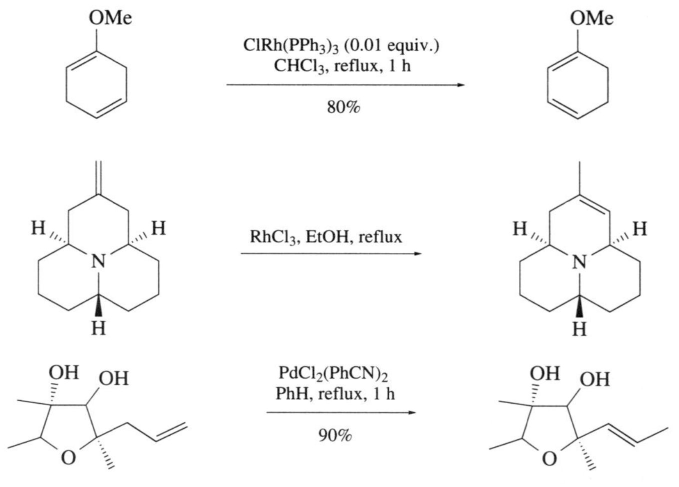

Scheme 12

(iv) Anion accelerated migrations of hydrogen atoms

Paquette demonstrated that oxyanion substituents facilitate rapid [1,5] sigmatropic shifts, with rate increases of $>10^5$; for example the rearrangement of 2,4-cycloctadienol to 3-cycloctenone proceeds rapidly at room temperature on addition of potassium hydride (Equation (2)) ⟨80JA3972⟩.

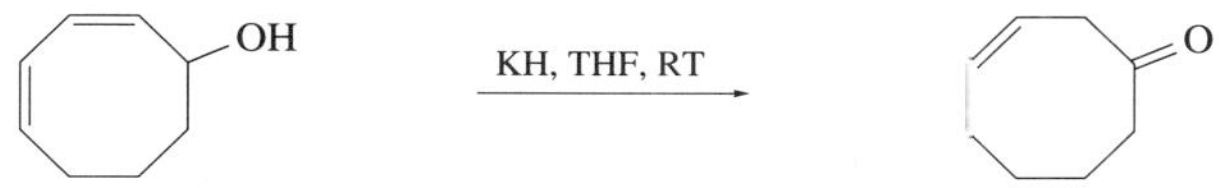

(2)

1.18.2.2 Where Y = C

1.18.2.2.1 [1,j] Rearrangements involving C migration

(i) Sigmatropic migrations of carbon centres

Sigmatropic rearrangements involving the migration of alkyl or aryl groups are well known despite evidence which suggests a lower migratory aptitude for alkyl and aryl groups when compared with the corresponding hydrogen migrations ⟨69JA1578, 71JA650⟩. The sigmatropic parameters for these reactions are known to be that the thermal suprafacial [1,3] rearrangement occurs with inversion of configuration at the migrating carbon atom, while the thermal suprafacial [1,5] shift occurs with retention of configuration; photochemical suprafacial [1,3] shifts occur with retention of configuration. Various articles have been published on this matter ⟨69TL2607, 70CC1168, 71CC107, 71CC108, 72ACR406, 76CRV187⟩ and many exceptions to these observations are known ⟨67JA5503, 68ACR152, 72ACR406, 75JA3837, 80JA2469, 80JA4841⟩.

Much work has been performed on [1,3]-rearrangements involving the migration of a carbon centre. The thermal rearrangement of substituted bicyclo[3.2.0]hept-2-enes (**6**) and bicyclo [2.1.1]hex-2-enes (**7**) are known to proceed with inversion of configuration ⟨67JA5503, 69CC1069, 69JA4322, 69TL2607, 70JA1096, 70JCS(B)1410, 72CC716⟩; studies on the rearrangement of (**8**) to (**9**) have shown the reaction to proceed with retention. The aza-anion (**8**) Li undergoes an accelerated rearrangement, the reaction being complete in one minute at 30 °C ⟨75JA3837⟩. An interesting tandem [1,3]-[3,3]-rearrangement has been reported for the 7-vinylbicyclohept-2-ene (**10**), which leads to the formation of the tetrahydroindene (**11**) (Scheme 13) ⟨86TL4107⟩. Photochemical examples of this rearrangement are also known (Scheme 14) ⟨65TL3955, 90CJC1961, 92JA5878⟩.

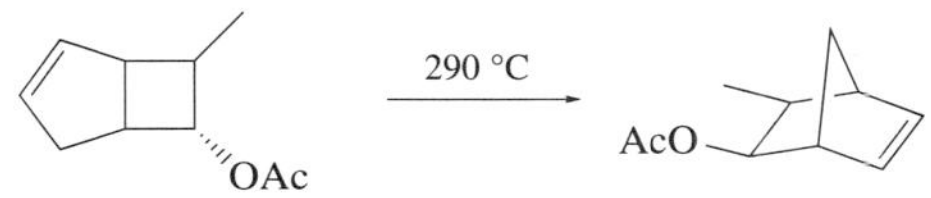

(6)

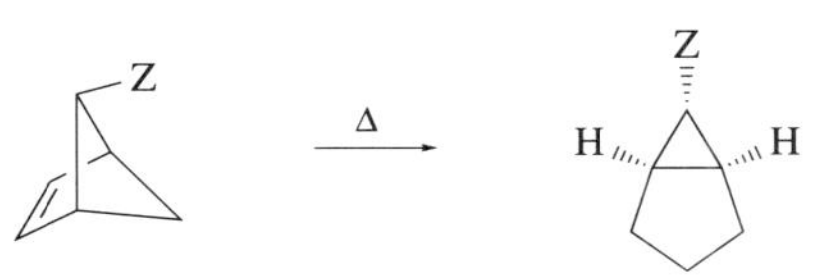

(7) Z = Me, Ac, MeO, N_3, $MeOCH_2CO_2$

(8) R = H, Me, CO_2Ph, CO_2Me, Li **(9)**

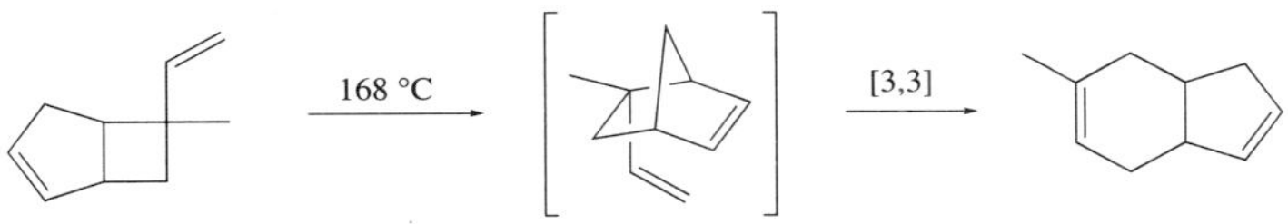

(10) **(11)**

Scheme 13

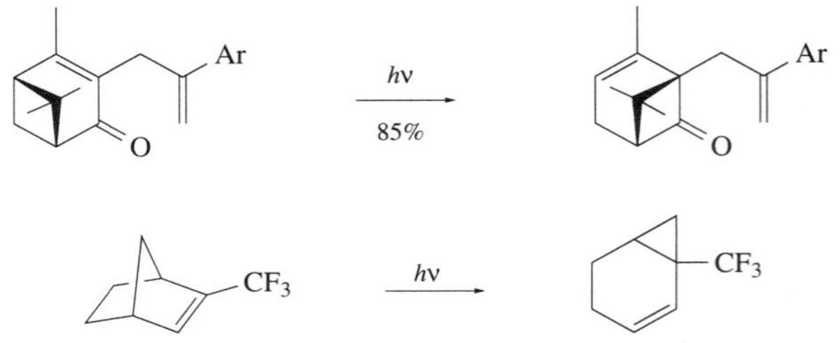

Scheme 14

Anion-accelerated examples of thermal [1,3]-rearrangements are known, including reactions of lithium, sodium, magnesium and zinc alkoxides ⟨66LA(698)80, 69JA5162, 74BSF1924, 77TL2559, 78JA2134, 82JOC4786⟩, with the best examples generally being rearrangements of potassium alkoxides (Scheme 15) ⟨79TL2779⟩.

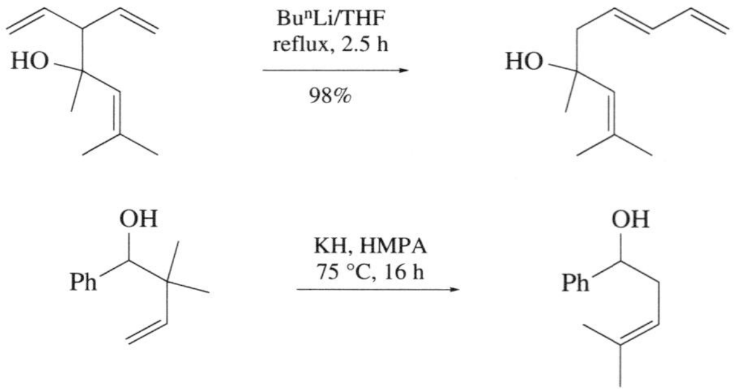

Scheme 15

A formal [1,3]-shift is also observed in the oxa-di-π-methane and aza-di-π-methane rearrangement ⟨B-83MI 118-01⟩. For example, irradiation of the oxime acetate of 2,2-dimethyl-3-phenylbut-3-enal (**12**) has been shown to afford the oxime acetate of 4-methyl-3-phenylpent-3-enal (**13**), presumably via a free radical pathway (Equation (3)) ⟨90CC934⟩. Many examples of the related photochemical 1,3-acyl migration are also known ⟨B-80MI 118-01⟩.

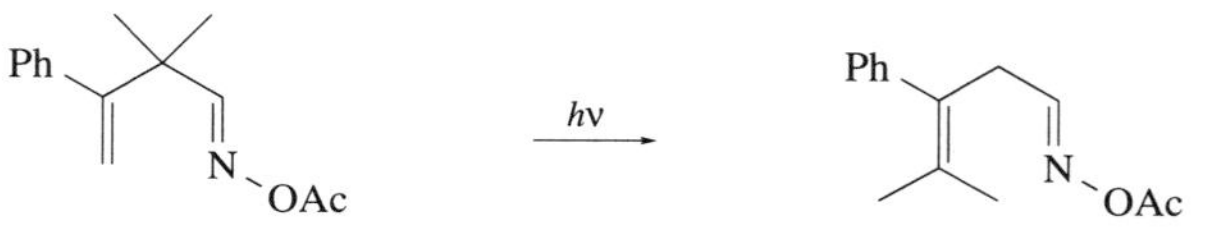

(12) **(13)** (3)

There are also many examples known of thermal [1,5]-migration of alkyl and aryl substituents particularly in cyclic systems (Scheme 16) ⟨68RTC298, 70TL2755, 72JOC2601, 73JOC487, 74JA1962, 76CRV187⟩. These reactions again require extreme conditions, which limit their synthetic application. Anion-accelerated versions of this reaction which illustrate migration of alkyl, vinyl and cyclopropyl groups have been reported (Equation (4)) ⟨84CC990⟩.

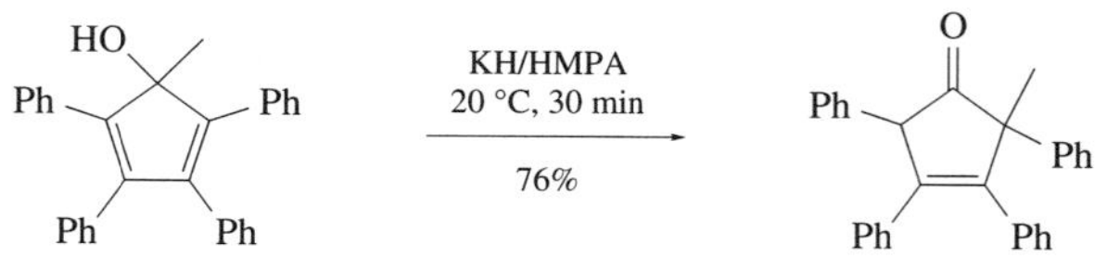

(4)

Several [1,7] migrations of carbon substituents are known, the thermal processes occurring with inversion ⟨72AG(E)832⟩ and the photochemical processes with retention of configuration ⟨90JOC1506⟩; the subject has been reviewed ⟨B-81MI 118-01⟩.

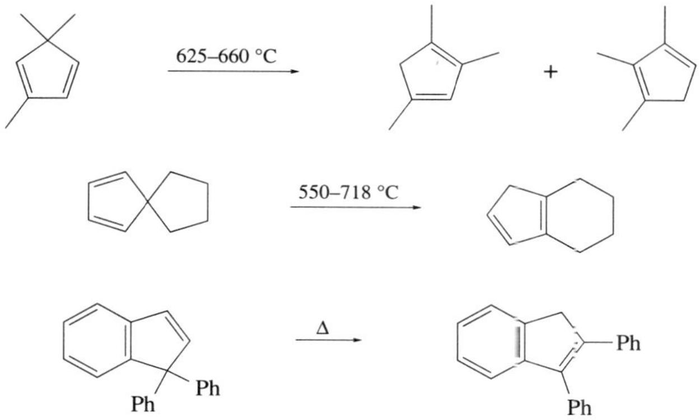

Scheme 16

(ii) Rearrangement of vinylcyclopropanes

The conversion of vinylcyclopropanes to cyclopentenes is a frequently encountered rearrangement (see also Section 1.18.2.1.1(i)), and can be considered as a formal [1,3]-migration of a carbon centre (Equation (5)) ⟨61JCS3547, 62JPC1671, 82TL236⟩. The reaction is conducted under thermal, photochemical and metal-catalysed conditions and the topic has been reviewed on several occasions ⟨85OR(33)247, B-87MI 118-01, 88CSR229, 88TCC73, 91COC(5)899, 93SL875⟩.

(5)

(a) Thermal rearrangement of vinylcyclopropanes. Two mechanisms have been invoked to explain the selectivity found in the thermal reaction, which was first reported by Neureiter ⟨59JOC2044⟩, either a concerted $[2\pi_s + 2\sigma_s]$ reaction or via diradical intermediates ⟨72MI 118-01, 76RCR469, 85OR(33)247, B-87MI 118-02, 89CRV165⟩. The majority of this experimental evidence suggests a diradical cleavage of the cyclopropane with subsequent reclosure of the allylic diradical. Despite the nature of this mechanism, the reaction is known to proceed with a high degree of stereochemical integrity; for example, the optically pure cyclopropane **(14)** rearranges to give the expected product ((−)15), as well as ((+)15), ((−)16) and ((+)16); these additional products suggest a biradical mechanism, with intermediates that cannot freely rotate (Equation (6)) ⟨76JA6705, B-87MI 118-03⟩.

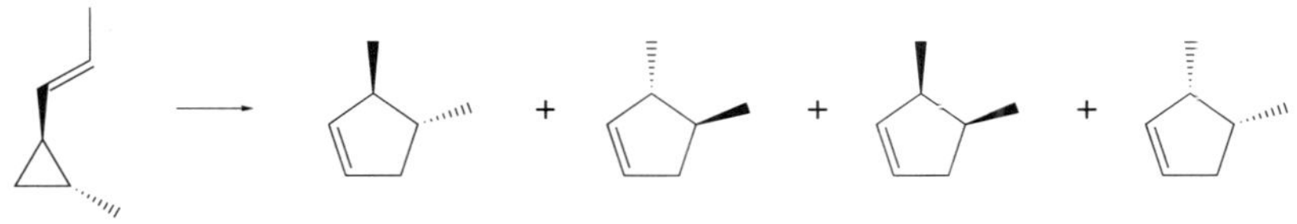

(6)

(14) ((−) **15**) 65% ((+) **15**) 8% ((−) **16**) 22% ((+) **16**) 5%

The reaction has been investigated thoroughly from a synthetic standpoint, and some general conclusions can be drawn. In general, increased carbon substitution, extended conjugation or heterosubstitution all lower the activation barrier for rearrangement to the cyclopentene, thus the transformation of vinylcyclopropanes **(17)**–**(21)** into the corresponding cyclopentenes all occur more freely than for the parent system (Scheme 17) ⟨64JCS959, 64JCS5578, 72JA2544, 72MI 118-01, 73TL2545, 76TL575, 81JOC506⟩. Other facile transformations are the interesting tandem rearrangement of 1,1-bis(cyclopropylethene) **(22)** and the rearrangement of methylene cyclopropane **(23)** (Scheme 18) ⟨66JCS(A)1342, 72JA8934, 73JA8096, 82CB2508, 82JA2498⟩.

Reports from the 1980s and 1990s have highlighted the potential for this rearrangement, and have also shown the possibility for catalysis of the reaction (Scheme 19). Davies has reported the facile thermal rearrangement of the 'push–pull' vinylcyclopropanes **(24)** and **(25)**; an even better method employed a diethylaluminum chloride-catalysed rearrangement, which proceeded at −78 °C with excellent stereocontrol and in high yield ⟨92TL453⟩. A similar rearrangement was reported by Corey in his synthesis of antheridiogen-An ⟨85JA5574, 86CL621⟩. The thermal rearrangement of spirovinylcyclopropanes **(26)** has also been reported ⟨92TL1645⟩. The transformation can also be mediated in other ways; for example by free radical ⟨88TL1543⟩ and Li/DMF ⟨91TL4591⟩ promoted rearrangements.

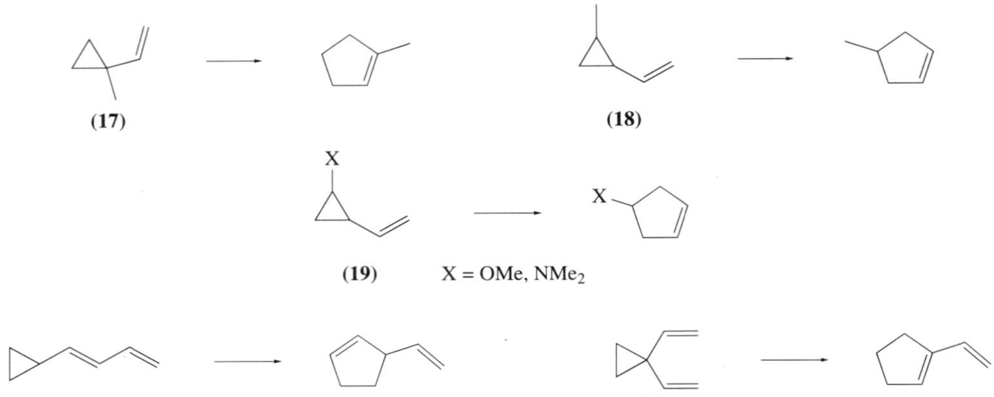

Scheme 17

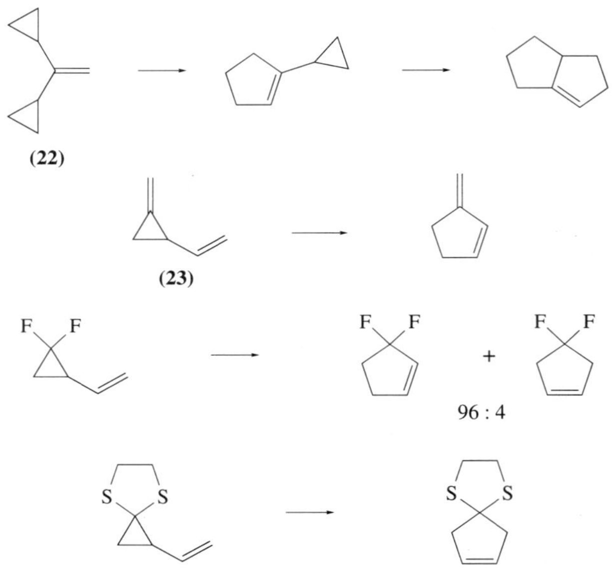

Scheme 18

(b) Photochemical rearrangement of vinylcyclopropanes. The photochemical rearrangement of vinylcyclopropanes under direct and sensitised conditions has also proved synthetically useful, and has been reviewed ⟨85OR(33)247, B-87MI 118-04, 88CSR229, 89CRV165, 93SL875⟩. Several intermediates have been proposed for the reaction including diradicals, zwitterions and carbenes ⟨65JA5264⟩. Some representative examples are given in Scheme 20. Notable examples include the formation of the novel tricyclic carbon skeletons (**27**) and (**28**) via this rearrangement, the reaction being subject to a remarkable substituent effect ⟨84TL1255⟩. A similar thermal rearrangement has also been reported in the synthesis of the structurally complex sesquiterpene longifolene ⟨85JOC915⟩. The conversion of vinylcyclopropanes (**29**) into the bicyclo[3.2.0]heptene (**30**) (a precursor for (−)-$\Delta^{9(12)}$-capnellene (**31**)) has also been reported ⟨91T8259, 91TL1107⟩. A similar transformation has also been observed for vinylcyclopropenes which rearrange to give substituted cyclopentadienes (Equation (7)) ⟨77JA2342, 77JA2344, 78JOC1493⟩.

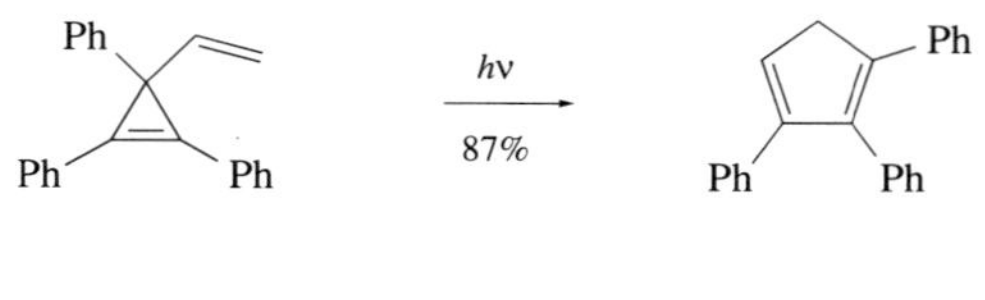

(7)

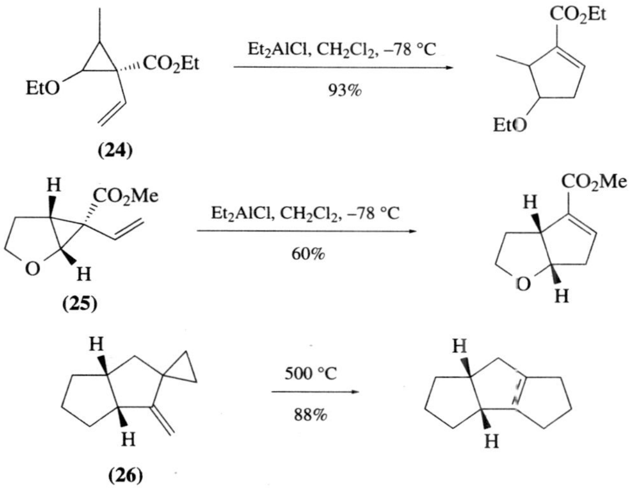

Scheme 19

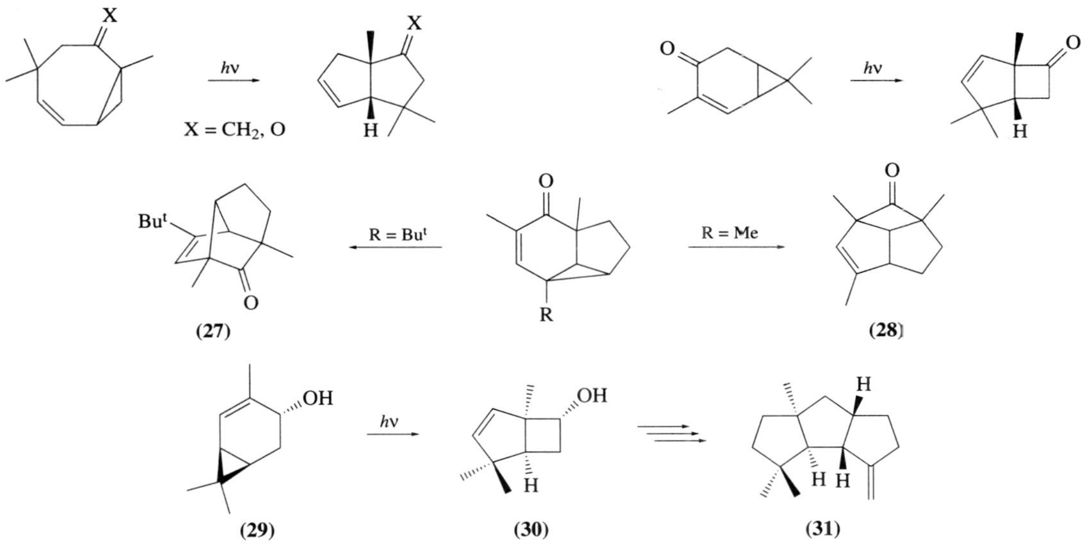

Scheme 20

(c) Metal-catalysed rearrangement of vinylcyclopropanes. The transition metal-catalysed rearrangement of vinyl cyclopropanes is of much importance, as the reaction generally occurs under mild conditions ⟨78ACR204, 88CSR229⟩. The reaction is catalysed by several transition metal complexes, the most common being those of rhodium(I) ⟨71CC1248, 76CC89, 79JCS(P2)962, 80JOC5020⟩, nickel(0) ⟨79CL927⟩, nickel(II) ⟨63T715⟩ and palladium(0) ⟨82TL2271, 84MI 118-02⟩. One prerequisite for the reaction is the presence of an additional site of coordination for the metal since, if this is not available, then β-hydride elimination is the predominant reaction pathway ⟨76CC89, 79CL927, 79JCS(P2)962⟩; representative examples of this reaction are shown in Scheme 21.

(d) Charge-accelerated rearrangement of vinylcyclopropanes. As with many rearrangement processes ⟨91COC(5)999, 93OR(43)93⟩ the introduction of a charged substituent into a vinylcyclopropane leads to an acceleration of the reaction with the effect that the transformations are generally easily performed under very mild conditions. Examples are known of oxyanion (**32**) ⟨80JOC1340, 81JA2443⟩ and carbanion (**33**) and (**34**) ⟨85JA4579⟩ acceleration and the Lewis acid-catalysed examples (Scheme 19) are essentially carbocation accelerated ⟨85JA5574, 86CL621⟩. Dinnocenzo and Conlon ⟨88JA2324⟩

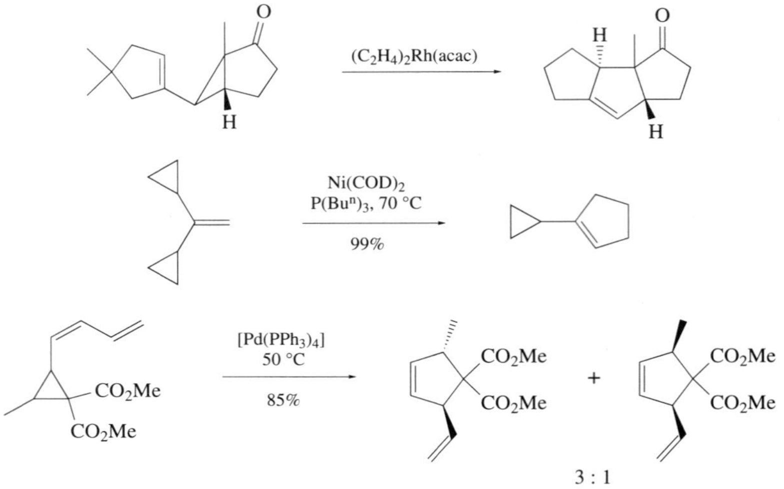

Scheme 21

have demonstrated a catalytic, one electron oxidation of vinyl cyclopropanes (**35**) which effects a rapidly accelerated rearrangement (Scheme 22).

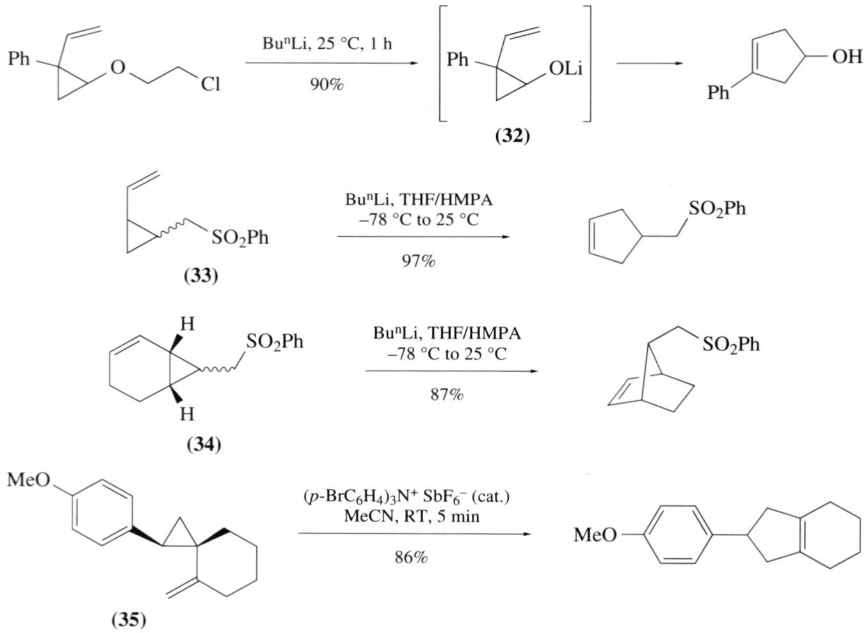

Scheme 22

(e) Natural product synthesis using the rearrangement of vinylcyclopropanes. Much work has been done on the synthesis of natural products using the vinylcyclopropane to cyclopentene rearrangement. Notable work is by Hudlicky on hirsutene ⟨80JA6351, 80JOC5020⟩, isocumene ⟨84TL2447⟩, pentalene and related structures ⟨83JOC4453, 87JOC4641⟩. The thermal rearrangement of cyclopropane (**36**), for example, leads to a 66 : 34 mixture of cyclopentene (**37**) and diene (**38**), the former being further converted to pentalene (**39**). Corey has reported the Lewis acid-catalysed conversion of cyclopropane (**40**) to the fused cyclopentene (**41**), in his synthesis of the plant hormone antheridiogen-An (Scheme 23) ⟨85JA5574⟩. Related work worthy of note is by Piers and co-workers *en route* to zizaene ⟨79CC1138⟩ and hinesol ⟨83CJC3227⟩.

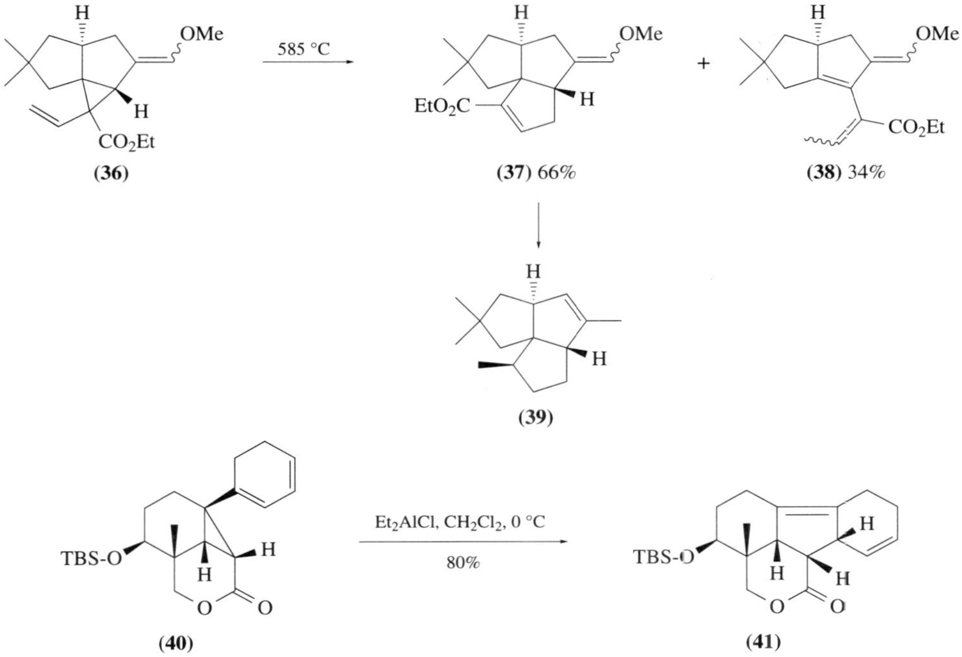

Scheme 23

(iii) Rearrangement of vinylcyclobutanes

Since its discovery ⟨60JA1007⟩, the vinylcyclobutane to cyclohexene rearrangement has been the subject of a considerable number of mechanistic studies ⟨66MI 118-03, 67JA5503, 68ACR152, 72ACR406, 72JA8917, B-81MI 118-02⟩. Despite this large quantity of work, the thermal rearrangement has found limited applications in synthesis, as high temperatures are generally required to effect the transformation, and the reaction is further limited by competitive decomposition ⟨63MI 118-02⟩. Some of the more successful examples are given in Scheme 24 ⟨77TL1779, 80JCR1551⟩.

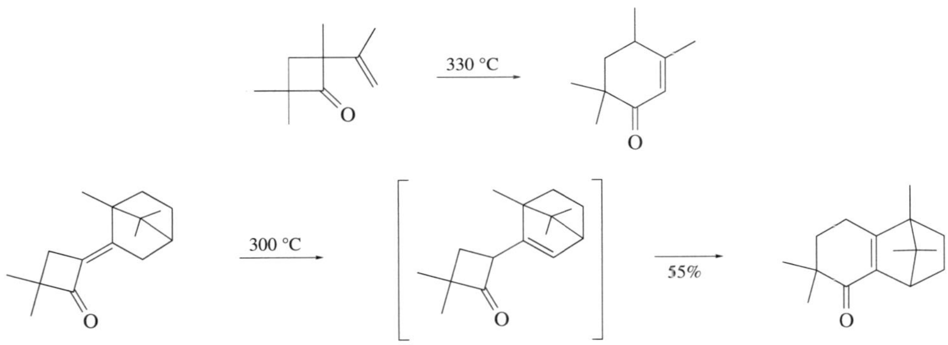

Scheme 24

The most significant contributions in vinylcyclobutane rearrangement are the anion-accelerated cases ⟨78CC479⟩. The basic reaction is the rearrangement of a 2-vinylcyclobutanol to a 3-hydroxycyclohexene, as exemplified in the rearrangement of *exo*-bicyclo[3.2.0]heptane (**42**) to a 8 : 1 mixture of *exo* and *endo*-norbornenol (**43**) when treated with potassium hydride (Equation (8)) ⟨78CC479⟩. Many examples are known in which lithium ⟨81T3943⟩ and sodium ⟨83TL2931⟩ alkoxides have also been used. The reaction has also been applied to the synthesis of several fused carbocyclic systems ⟨80JA6900, 81T3943, 83JA520, 83JA6798, 84JA2730, 88JOC1848⟩. Representative examples of these reactions are given in Scheme 25 including the formation of (**44**) which was subsequently converted into (−)β-selinene (**45**). The topic has been the subject of a short review ⟨93OR(43)93⟩.

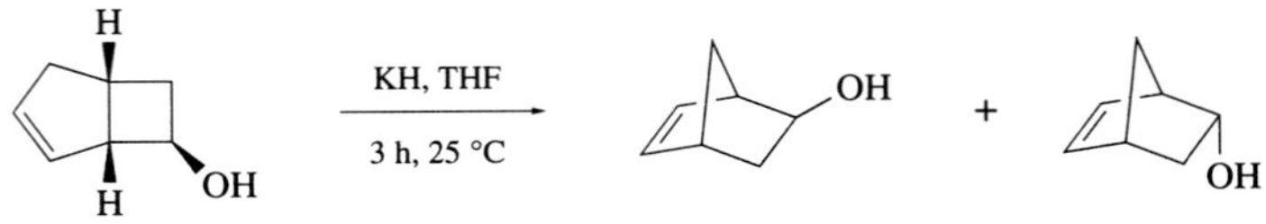

(8)

(42) *exo*-**(43)** 8 : 1 *endo*-**(43)**

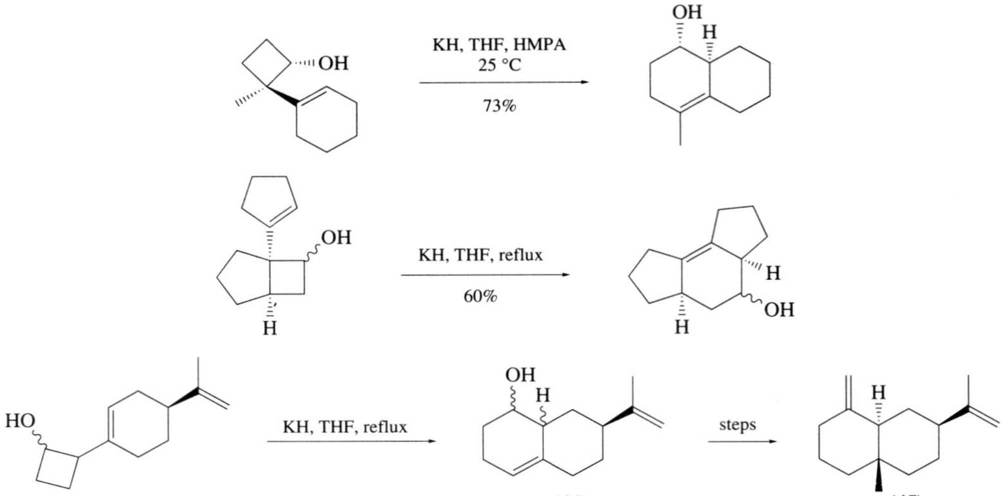

(44) **(45)**

Scheme 25

1.18.2.3 Where Y = C—C

Baldwin has reported an anionic [2,3]-rearrangement of an all-carbon system, which, depending on reaction temperature, is thought to proceed via either a sigmatropic or a dissociative mechanism, or a combination of both ⟨70CC165⟩.

1.18.2.4 Where Y = C—C—C

1.18.2.4.1 Cope rearrangement

The Cope rearrangement in its generalised form, can be considered as the interconversion of two 1,5-dienes (Equation (9)). The conditions for this conversion are numerous and include thermal, photochemical, metal-catalysed, acid-catalysed and charge-accelerated examples. The reaction has also been reviewed ⟨63AG(E)115, 75OR(22)1, 80T3, 84CRV205, B-84MI 118-03, 90AG(E)609, 90SL67, 91COC(5)785, 92OR(41)1⟩. The reaction proceeds via a six-membered cyclic transition state, with a preference for a chair conformation ⟨77JA1499, 80JA3156⟩, structural features permitting ⟨74JA3705, 85JA5739, 88CC354⟩. This, in combination with a marked preference for sp^3 carbon centres to occupy equatorial, rather than axial, positions in the transition state, leads to the formation of products with (*E*) double bond geometry. This also has the effect that transfer of chirality from a stereogenic centre in the substrate to a new centre in the product is possible with excellent levels of integrity, generally greater than 97% ⟨67CC619, 80JOC1172, 82JA7225⟩.

(9)

(i) Thermal Cope rearrangement

The thermal Cope rearrangement ⟨40JA441⟩ can be considered as the parent member of a large family of related 3,3-sigmatropic rearrangements. The thermal reaction has been used in synthesis on many occasions and selected examples serve to illustrate this point and to highlight the reaction's drawbacks. The reaction is reversible, but several factors can be applied and shift the position of the equilibrium in a predictable manner. In general terms, and in the absence of any other contributing factors, the equilibrium will tend to favour the formation of the most highly substituted alkenes (Scheme 26); this is exemplified by the equilibrium mixture obtained on heating **(46)** and **(47)** ⟨68MI 118-01⟩. Numerous further examples are known ⟨77JA1499, 80JA2392, 88CC354⟩ including a report ⟨92JA2635⟩ which demonstrates that DL-**(48)** and *meso*-**(48)** require different conditions to effect Cope rearrangement and illustrates the marked preference for a chair conformation (which DL-**(48)** is forced to adopt) over a boat conformation (which *meso*-**(48)** is forced to adopt) in the transition state of the reaction.

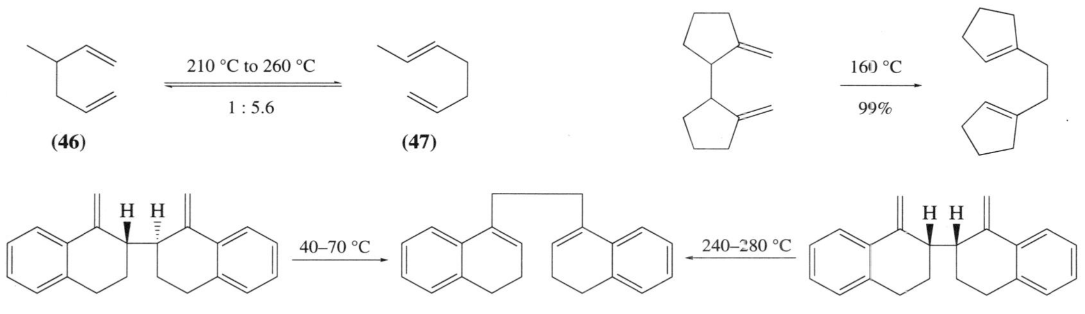

DL-**(48)** meso-**(48)**

Scheme 26

An important driving force for the rearrangement is if one or both of the double bonds enters into conjugation. Examples are known in which this conjugation is provided by ester ⟨41JA1843, 79JA1611⟩, ketone ⟨60TL29, 61T13, 62TL505, 63MI 118-03, 81TL1583⟩, nitrile ⟨80TL2125, 81TL1179, 84JA718⟩ and aryl ⟨67CC619⟩ functions (Scheme 27). In certain cases the preference for polyalkyl substitution takes precedence over the stabilisation that is possible through conjugation (Equation (10)) ⟨79JA1611, 86TL5923⟩. The rearrangement is highly favoured when one of the double bonds of the product is incorporated into an aromatic ring, and examples include the formation of indoles and pyrazoles (Equation (11)) ⟨69CC99⟩.

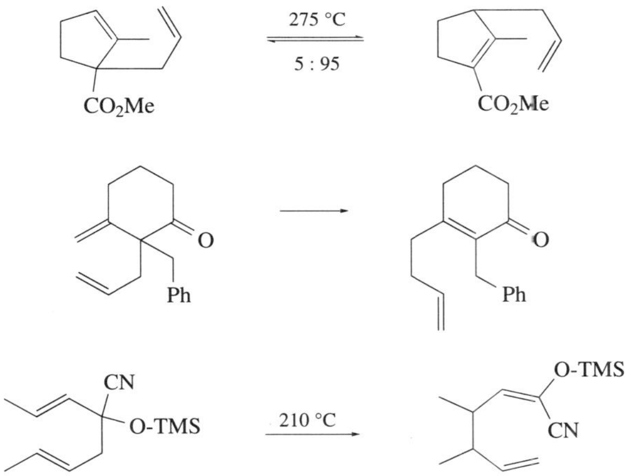

Scheme 27

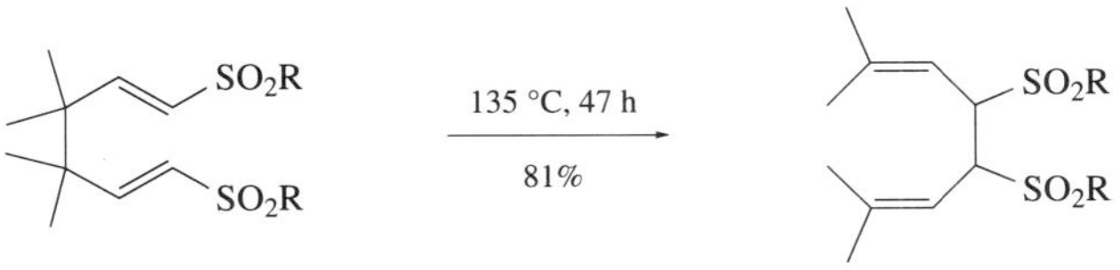

(10)

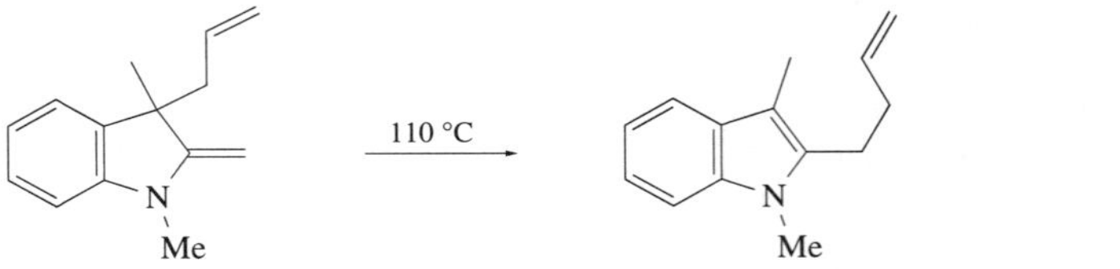

(11)

The Cope product may also be able to tautomerise to form an aromatic ring and thus move the position of equilibrium to the aromatic product. Many examples are known of this phenomenon and these include the well-known *para*-Claisen rearrangement of aryl allyl ethers (see Section 1.18.2.6.1.(i).b). For example, 5-allyl-2,4-cyclohexadieneone (**49**), rapidly rearranges at relatively low temperature to give the phenol (**50**) via its keto tautomer (Scheme 28) ⟨75ACR245⟩. With the similarities between the Cope and Claisen reactions, one might envisage a Cope analogue of the aromatic Claisen rearrangement; however, the reaction is not possible using simple homoallyl benzene derivatives, even in the presence of conjugating substituents ⟨34JA447, 66BSF281, 77MI 118-02, 79JA1793, 80JOC4793⟩. Some of the rare examples of this reaction are shown in Scheme 29 ⟨56JA2547, 56JA2551, 80CB3679, 86JOC183⟩ and include the rearrangement of the aryl vinyl cyclopropane (**51**) to give the annulated product (**52**), in which the cyclopropane assists in lowering the activation energy required for disruption of aromaticity ⟨78JA877⟩. Other factors that affect the position of equilibrium in the Cope rearrangement are ring strain (see Section 1.18.2.4.1.v) and some conformational effects ⟨59T19, 80JA2797⟩.

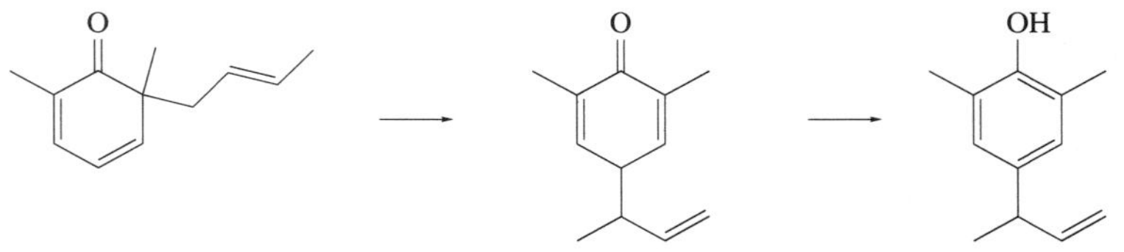

(49) **(50)**

Scheme 28

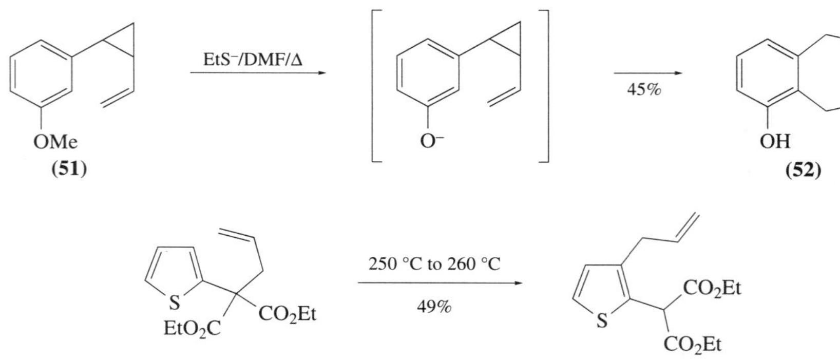

(51) **(52)**

Scheme 29

Alkynes and allenes participate in the Cope rearrangement. Cope rearrangement of 1,5-enynes provides a useful route to allenes ⟨65JCS6784, 66JA5846, 68JOC4413⟩ and the oxy-Cope variant leads to the formation of α,β-unsaturated ketones ⟨78CC651⟩. Allenes also participate readily in Cope rearrangements, leading to the formation of conjugated dienes ⟨78TL1803, 88JOC196⟩, and examples are known where the allene and isolated double bond are both incorporated into a large ring ⟨65JA4501, 65JA4506, 65TL1359⟩. Bisallenes participate in Cope rearrangements ⟨65JA4501, 65JA4506, 65TL1359⟩. Some examples are illustrated in Scheme 30.

(ii) Catalysis of the Cope rearrangement

The Cope rearrangement is often accelerated by the presence of a catalyst, particularly acids and metals. For example, the thermal rearrangement of ketone (**53**) takes several hours at 80 °C, whereas

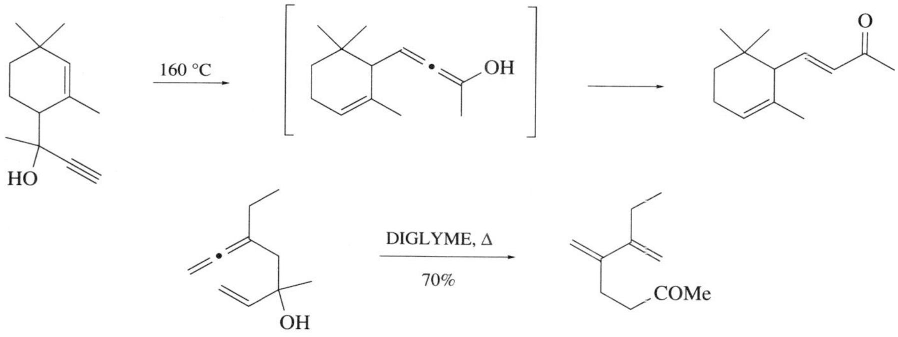

Scheme 30

under acid catalysis the reaction produces good yields in 15 minutes (Equation (12)) ⟨81TL1583⟩. This acid catalysis is largely restricted to 1,5-dienes that contain carbonyl substituents; other examples are known involving catalysis by mineral acids ⟨65JA5115, 71HCA937, 73HCA75, 75ACR245, 81TL1583⟩, Lewis acids ⟨60TL29, 61T13⟩, alumina ⟨76JOC2048⟩ and weak acids such as ammonium salts and iodine ⟨80TL1347⟩.

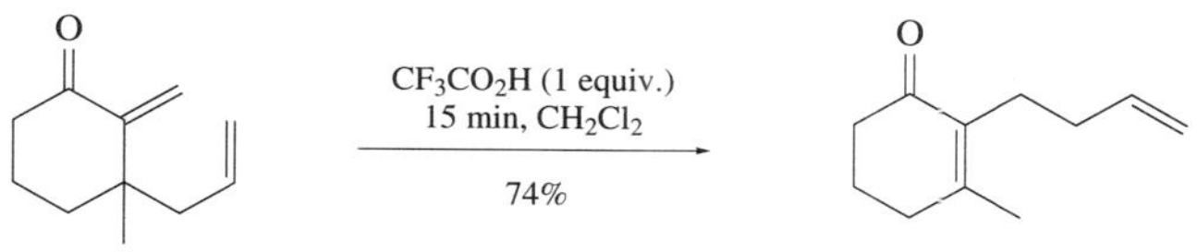

(12)

(53)

Several metal complexes are known to act as catalysts for the Cope rearrangement. Palladium dichloride was originally found to effect the rearrangement when a stoichiometric amount was used ⟨66JOM(6)412, 67AG(E)800, 73JOM(49)483⟩; however, subsequent investigation has shown that catalytic quantities of palladium(II) complexes can be used in a wide range of rearrangements (Scheme 31) ⟨82JA7225, 83TL3757, 84AG(E)579⟩. Other metals are also known to accelerate the rate of Cope rearrangement, for example $Rh_2(CO)_4Cl_2$ ⟨73JCS(P1)1913⟩ and also silver and mercury salts ⟨84T3277, 86T1333⟩. A few examples of photochemically initiated Cope rearrangements are also known, although the factors affecting the rearrangement are not thoroughly understood ⟨65TL3955, 67TL227, 69JA522, 85JOC3393⟩.

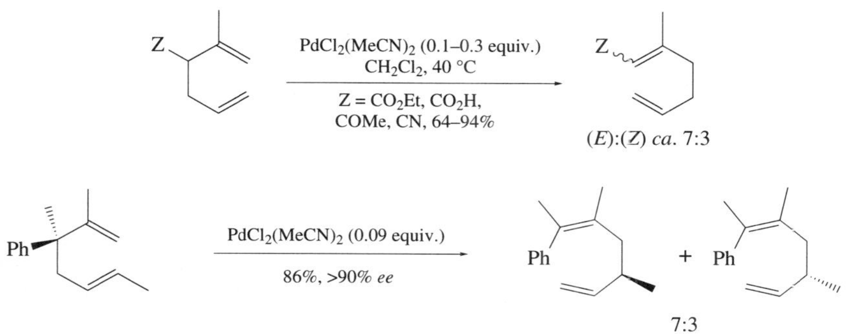

Scheme 31

(iii) Oxy-Cope rearrangement

The most important development in the synthetic applications of the Cope rearrangement was the discovery of the oxy-Cope variant ⟨64JA5019⟩. In this type of reaction, a C-3 hydroxy-1,5-diene undergoes Cope rearrangement to give an intermediate enol, which on tautomerism leads to a δ,ε-unsaturated carbonyl compound (Scheme 32). This reaction has several advantages over the normal

Cope reaction, in that the precursors to the reaction are easily prepared via a Grignard-type protocol and the product contains two easily distinguishable and useful functional groups. Another very important consideration is that, by generating an enol which will rapidly tautomerise to a carbonyl compound, the reaction removes the product from the equilibrium and this drives the rearrangement to completion (for a reversible case see ⟨91TL319⟩). This allows the formation of the thermodynamically unfavoured product in many cases and this protocol has been used on many occasions. (This factor is also evident in the anionic oxy-Cope reaction; see next section.) One major drawback of the reaction is the possibility of a competing thermal retro-ene reaction ⟨70JOC856⟩. This problem is often circumvented by the use of ether ⟨68JA4729⟩ or ester ⟨68FRP1519701⟩ derivatives. Some examples of this reaction are given in Scheme 33 ⟨67JA3462, 76TL677, 92JCS(P1)2727⟩. Palladium(II) catalysis of the oxy-Cope rearrangement is also known ⟨83TL1157⟩.

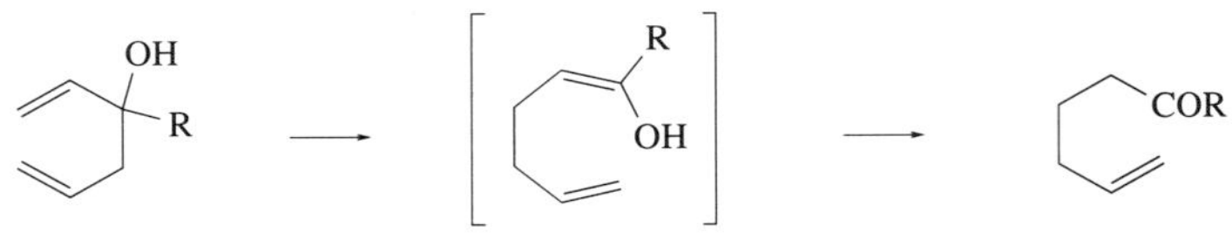

Scheme 32

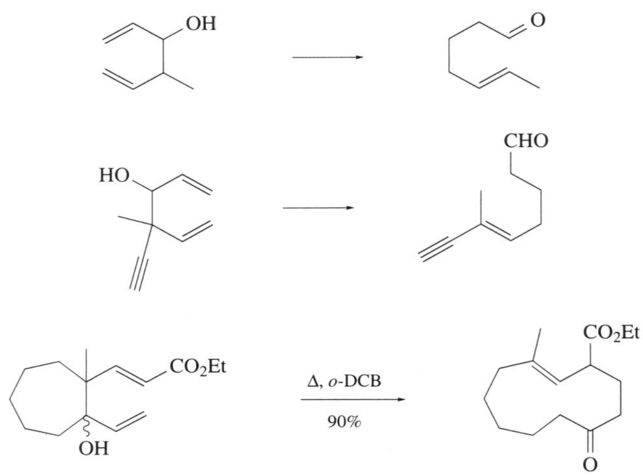

Scheme 33

(iv) Anionic oxy-Cope rearrangements

The major drawback of all the previous methods for Cope rearrangements is the requirement for extreme reaction conditions to effect the transformation. It was thus of major synthetic importance when Evans and Golob reported ⟨75JA4765⟩ that rate enhancements of 10^{10}–10^{17} were possible in oxy-Cope rearrangements if 1,5-diene alkoxides were used. This anionic oxy-Cope rearrangement (Scheme 34) of a lithium, sodium or potassium alkoxide (**54**) normally proceeds to give an intermediate enolate (**55**) as the rearranged product under considerably milder conditions than are required for the thermal variant. Additionally, the analogous anion-accelerated Cope rearrangements of 3-amino-1,5-dienes has also been demonstrated and was found to be considerably more facile than those of the corresponding 3-hydroxy-1,5-dienes ⟨93TL247⟩.

Synthetic applications illustrating the importance of this reaction include approaches to the taxane diterpenes; for example, the AB ring system model (**57**) was constructed using the oxy-Cope rearrangement of divinylcyclobutane alkoxides (**56**) ⟨91JOC321, 92HCA1755, 93JOC4952⟩. A short synthesis of the tricyclic ring system of vinigrol (**59**), has been reported using the anionic oxy-Cope rearrangement of (**58**) ⟨93JOC2349⟩, and the rearrangement of ether (**60**) has been used in the enantiospecific synthesis of ajmaline-related alkaloids (Scheme 35) ⟨93JOC661⟩; see also ⟨91CJC77, 93JOC4245, 93JOC5482, 93SC311, 93T1025⟩.

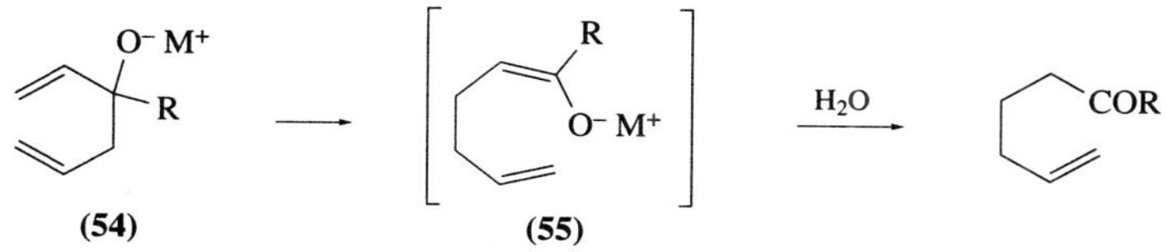

(54) **(55)**

M = Li, Na, K

Scheme 34

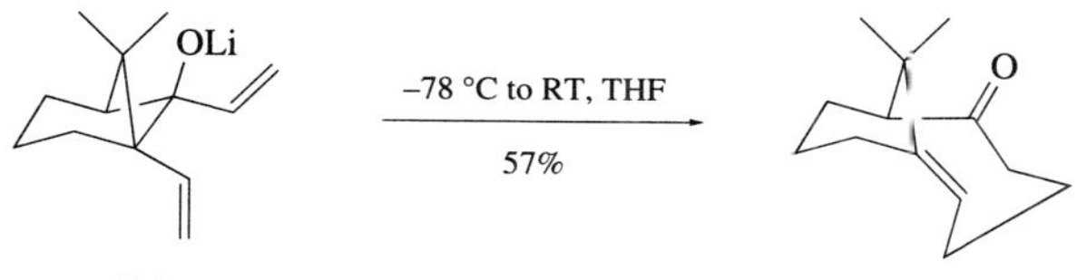

(56) **(57)**

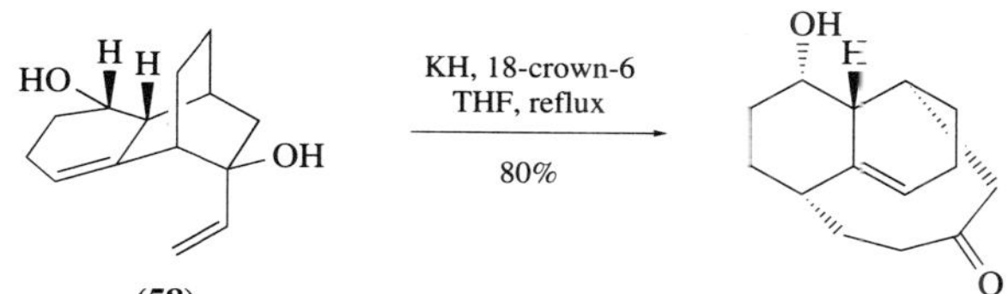

(58) **(59)**

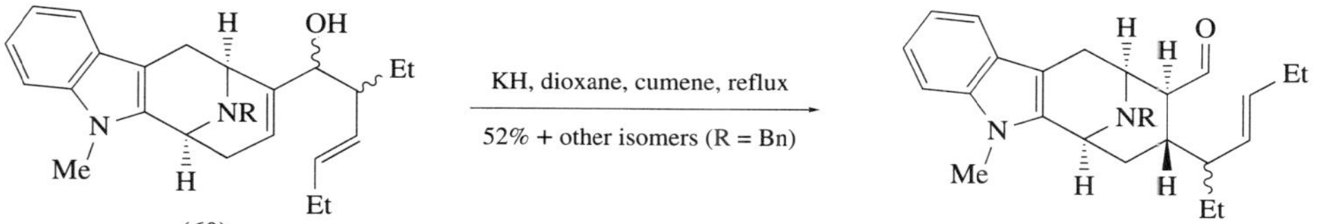

(60)

Scheme 35

(v) Cope reaction of divinyl cycloalkanes

The Cope reaction of divinyl cycloalkanes has been an area of considerable investigation. The basic conversion is shown in Equation (13) and can be represented as the equilibrium between the divinyl cycloalkane (**61**) and a cyclic diene (**62**). The primary factor that affects the position of this equilibrium is the strain associated with the ring sizes present. In the case of *cis*-1,2-divinylcyclopropanes ($n = 1$) and cyclobutanes ($n = 2$), the rearrangement lies completely on the side of the less-strained seven- and eight-membered ring dienes, and very low temperature are required for this transformation (Scheme 36) ⟨58LA(615)1, 61LA(644)172, 64TL2503, 65CC226, 70CC1519, 73CC319, 76JA5937, 82CC1164, 86JA5901⟩. However, it is known that, for bridged 1,5-cyclooctadienes, the equilibrium can lie on the side of the cyclobutane ⟨70JA703, 72JA9228, 79JA7599, 81CC785⟩.

(13)

(61) $n = 1–10$ **(62)**

The Cope rearrangements of 1,2-divinylcyclopropanes have been used extensively in organic synthesis ⟨91COC(5)785, 92OR(41)1⟩; early work concentrated on the synthesis of cycloheptane-containing natural products such as the dictyopterenes ⟨69HCA880, 69TL3459, 71JA3087, 76JOC3490, 82TL1405⟩. Many synthetic strategies have been developed to prepare the precursors: these include conjugate addition/elimination of vinylcyclopropyl organometallic reagents to β-substituted cycloalkenones ⟨76JOC3490, 76TL3245, 78JOC3630, 83CJC1226, 83CJC1239, 93T5203⟩, construction of cyclopropanes from sulfur ylides ⟨70JA2158, 70JA6641, 70JA6643, 73TL3971, 73TL3975, 74JOC3175, 81JCS(P1)3225⟩, transition metal-catalysed addition of vinylcarbenes to cyclic dienes ⟨87T4265, 87TL1853, 88TL975⟩ and

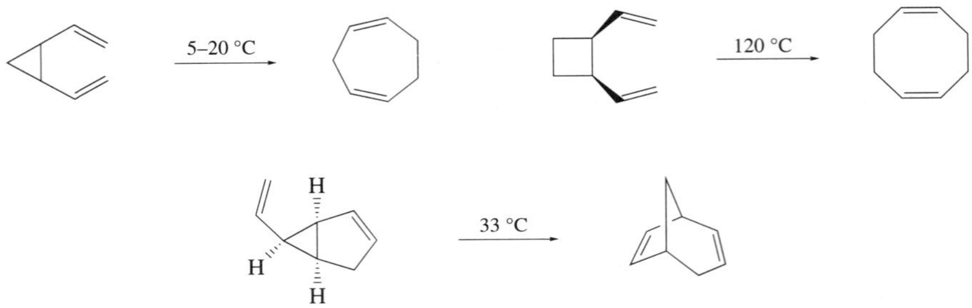

Scheme 36

generation of enol derivatives ⟨84TL3959, 85TL2735⟩. Examples of these methods are shown in Scheme 37 and include the Cope rearrangement of divinylcyclopropane (**63**), prepared via the conjugate addition/elimination protocol, which led to the cycloheptadiene (**64**), a precursor of β-himachalene ⟨83CJC1239⟩. A rhodium-catalysed intermolecular cyclopropanation–Cope rearrangement approach to the synthesis of hydroaxulenes has been described ⟨91JOC723⟩, as well as a similar sequence for the synthesis of various seven-membered carbocycles ⟨91JOC3817⟩ and tropane derivatives ⟨91JOC5696⟩, including the enantioselective route for the preparation of cyclopropane (**65**) which rearranges to tropane (**66**) ⟨92TL6935⟩. Both *cis*- and *trans*-1,2-divinylcyclopropanes give the same products on Cope rearrangement (the *trans*-isomer undergoes isomerisation via a diradical intermediate to the *cis*-isomer ⟨63AG(E)1, 72JA5910⟩ and consequently higher temperatures are necessary to effect rearrangement). A similar effect is seen with *trans*-1,2-divinylcyclobutanes, but side reactions are known to occur ⟨64JA899⟩.

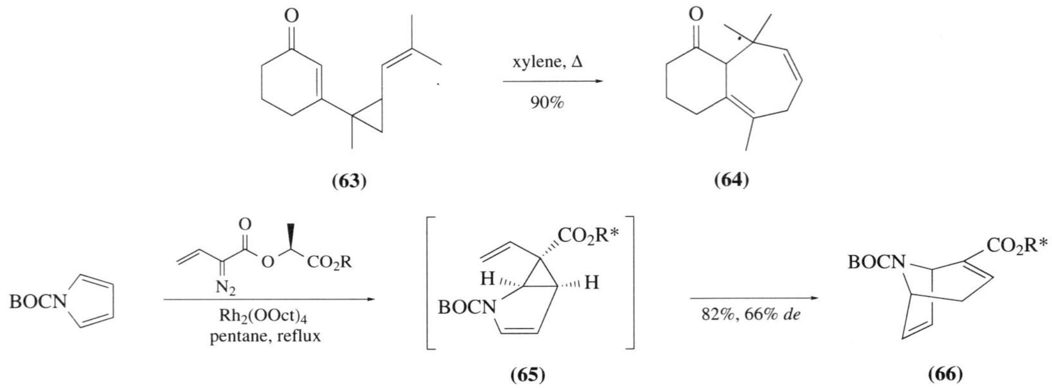

Scheme 37

Many examples of Cope rearrangement are known for divinylcyclobutanes, particularly the oxy-Cope variant, as several methods for the preparation of 2-vinylcyclobutanones are available ⟨89OS(67)210⟩. For example, addition of a vinyl group to the cyclobutanone (**67**), leads to divinylcyclobutanol (**68**), which undergoes oxy-Cope rearrangement ⟨82JOC2268, 84HCA774⟩. Many other strategies are available for the preparation of divinylcyclobutane Cope precursors including [2 + 2] cycloadditions and fluoride ion-assisted conjugate additions (Scheme 38) ⟨80TL4547, 81JOC2199, 82HCA451, 82JA7670, 85JOC201, 88TL2773, 91TL6969⟩.

The position of equilibrium is totally reversed in the rearrangement of 1,2-divinylcyclopentanes ($n = 3$) and cyclohexanes ($n = 4$) (Equation (13)), as the small amount of ring strain in five- and six-membered rings is preferred to the considerable strain found in medium-sized rings (eight- to eleven-membered). In the parent ring systems, 1,5-cyclononadiene (**69**) and cyclodecadiene (**70**) rearrange completely to the corresponding divinylcycloalkanes (Scheme 39) ⟨63AG(E)739, 63HCA483, 73JOC4117⟩; see also ⟨68JOC4258, 69LA(727)194⟩. The germacrane class of sesquiterpenes displays this characteristic rearrangement as many of them contain the 1,5-cyclodecadiene skeleton; for example, the thermal conversion of (+)-hedycaryol (**71**) to (−)-elemol (**72**) takes place readily (Equation (14)) ⟨68CC1229, 70CC892⟩. Many other examples are known ⟨59CCC2371, 59LA(625)206⟩ and information is available on the effect of substituents on the conformation adopted for rearrangement ⟨68CC1229, 70CC892,

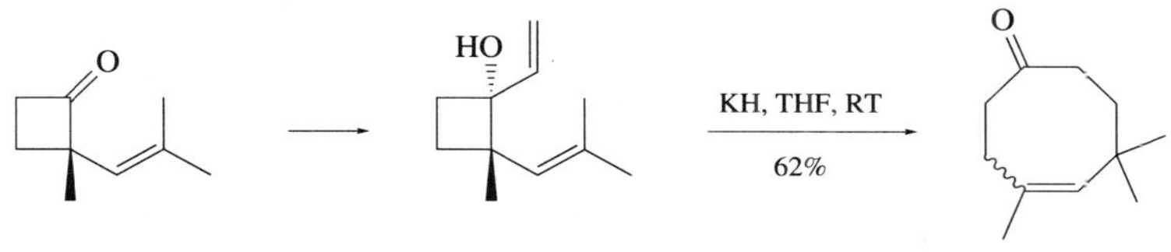

(67) **(68)** (*E*):(*Z*) 21:79

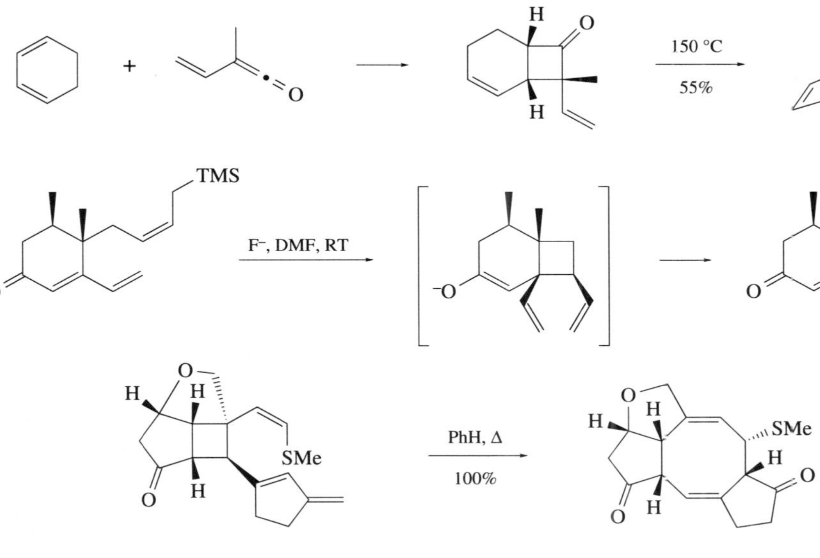

Scheme 38

70JCS(C)2697, 70JCS(C)2704, 74T1525⟩. It is also possible to observe equilibrium mixtures when highly substituted cyclodecadienes are used ⟨64JCS4578, 80JA1756⟩; and it is also possible to reverse the position of equilibrium completely ⟨87CC1196⟩.

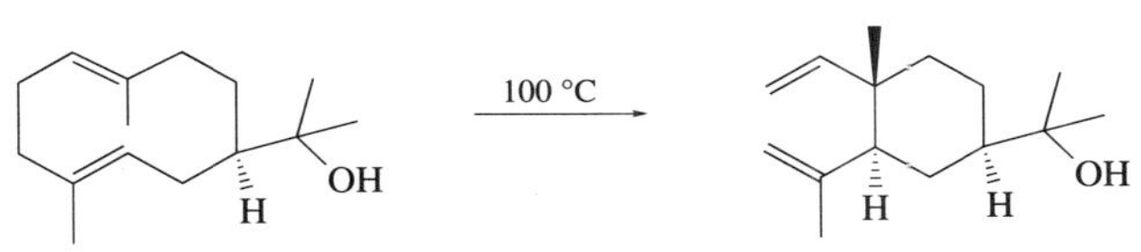

(14)

(71) **(72)**

(69) **(70)**

Scheme 39

Furthermore, it is possible to reverse the position of this equilibrium to favour the medium-sized ring if the oxy- or anionic oxy-Cope rearrangement is employed, and this methodology has been used to good effect. Several examples exist of 1,2-divinylcyclopentane to cyclononenone rearrangements under oxy-Cope conditions, many of these in the synthesis of natural products (Scheme 40) ⟨81JOC9, 81JOC2199, 86TL219⟩. In a similar manner, cyclodecenones can be prepared by oxy- or anionic oxy-Cope rearrangement of 1,2-divinylcyclohexanols ⟨70TL509⟩. The reaction has found widespread application in the synthesis of cyclodecane natural products (germacranolides ⟨85BCJ146⟩), periplanone B ⟨79JA2493, 84JA4038, 86TL1315⟩, eucannabinolide ⟨77JA4186, 83JA625⟩), as well as many other polycyclic systems (Scheme 41) ⟨81JOC2199, 85TL6137, 88JA890, 92TL923⟩.

The smaller ring product is also favoured in the thermal rearrangement of divinylcycloheptanes ($n = 5$) ⟨67TL227⟩, but again the large ring product is favoured in the oxy-Cope rearrangement ⟨80BCJ2958⟩. In the Cope rearrangement of 1,2-divinylcyclooctanes ($n = 6$) and larger rings ($n = 7$–10),

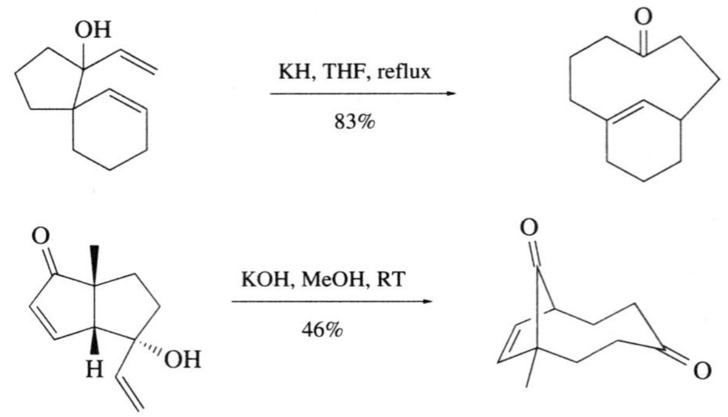

Scheme 40

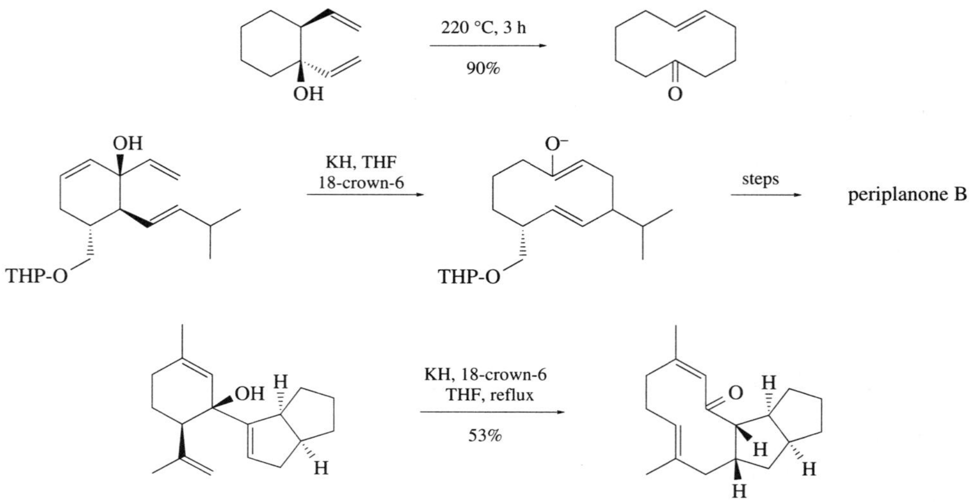

Scheme 41

the equilibrium position shifts again, leading to the preferential formation of twelve- to sixteen-membered rings. The determining factor in these equilibria is the strain present in medium-sized rings (eight- to eleven-membered). Examples are known of divinylcyclooctane rearrangements to twelve-membered rings ⟨79HCA2630, 82AG(E)686, 82JOC1632⟩ and also rearrangements of divinylcyclo-undecanes ⟨86AG(E)1026⟩ and cyclododecanes (Scheme 42) ⟨80BCJ2958, 82JOC1632⟩.

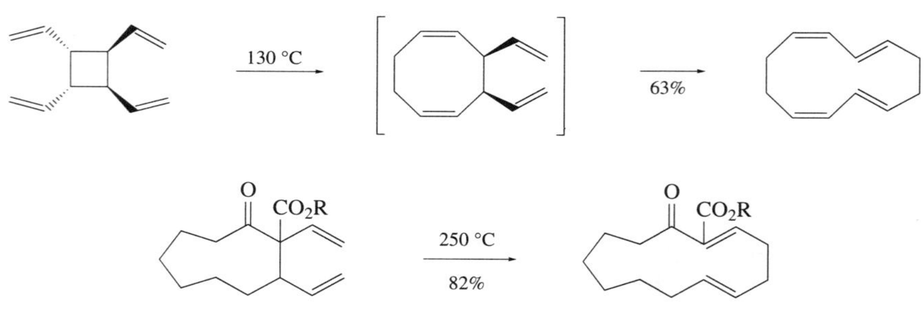

Scheme 42

(vi) Cope rearrangements of vinylbicycloalkenes

The Cope rearrangement of vinylbicycloalkenes leads to the formation of a fused bicyclic diene; the most frequently reported cases are the bicyclo[2.2.1]heptenes (**73**) and bicyclo[2.2.2]octenes (**74**), which lead to the formation of indene and decalin skeletons respectively (Scheme 43). This powerful methodology can be used to rapidly assemble complex polycyclic skeletons, and has been used in the synthesis of several natural products. Representative examples of indene ⟨76BCJ2017, 86JA7873, 88CC539⟩ and decalin formation are given in Scheme 44 ⟨64JA5019, 68JA4729, 78JA8170, 79CC106, 84TL5103, 88CC1281, 88T3139, 92TL923⟩ including the formation, via anionic oxy-Cope rearrangement, of a precursor **(75)** for the indacene subunit of ikarugamycin ⟨87TL31, 88JA879⟩. The rearrangement of 7-vinylbicyclo[3.3.1]heptenes also leads to the formation of indenes, but can be problematic in some cases ⟨74JA3468, 82JOC4815, 87JA5731⟩. Larger ring ⟨82JOC4815⟩ and tricyclic ⟨66CC775, 67TL4537, 71JCS(C)970, 77T1527⟩ variants are also known.

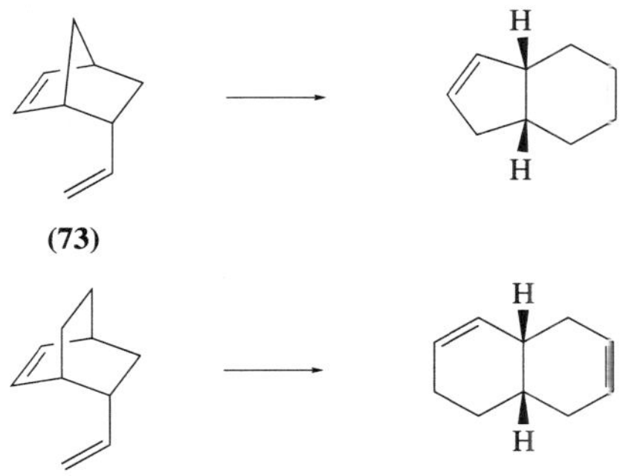

Scheme 43

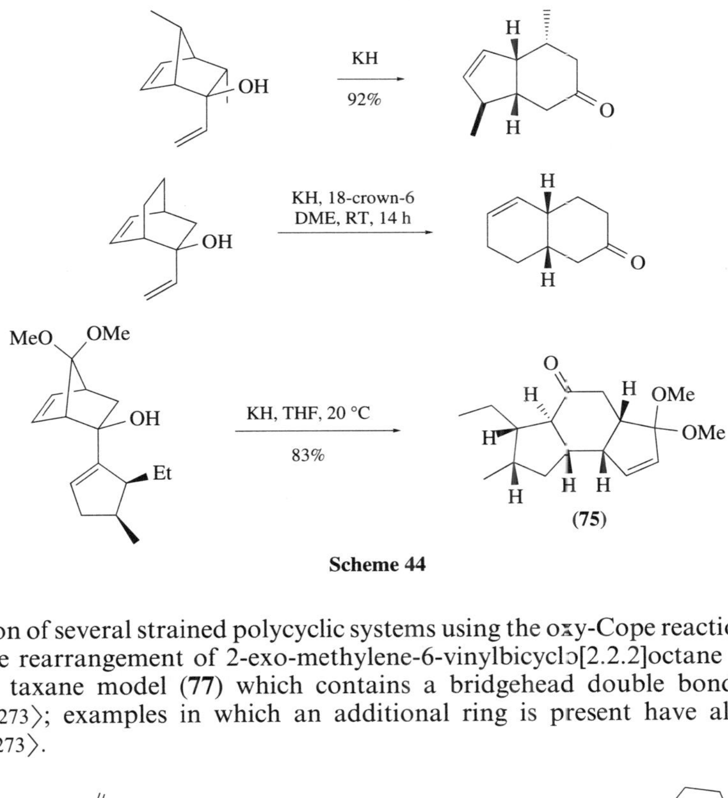

Scheme 44

The preparation of several strained polycyclic systems using the oxy-Cope reaction is also possible. For example, the rearrangement of 2-exo-methylene-6-vinylbicyclo[2.2.2]octane **(76)**, has led to a synthesis of the taxane model **(77)** which contains a bridgehead double bond (Equation (15)) ⟨82JOC3190, 88TL273⟩; examples in which an additional ring is present have also been reported ⟨88JOC4508, 88TL273⟩.

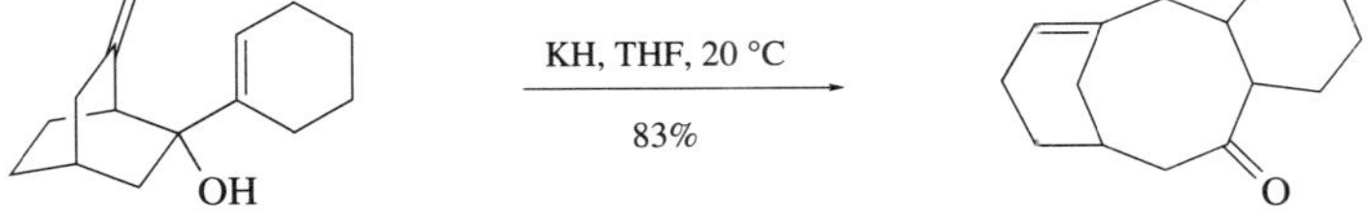

(76) **(77)** (15)

1.18.2.5 Where Y = C—Z (Z = Heteroatom)

1.18.2.5.1 Where Z = chalcogen

(i) Wittig rearrangement (Y = C—O)

The 2,3-Wittig rearrangement ⟨86CRV885, 91S594⟩ involves the rearrangement of the anion of a substituted methyl allyl ether to a 1-alkoxybut-3-ene (Equation (16)); this reaction proceeds under relatively mild conditions in generally good yield and with predictable stereochemistry ⟨79AG(E)563, 92AG(E)647⟩. Several methods are available for the generation of carbanions in these systems and much depends on the nature of the substituent E, which is employed to promote formation of the anion ⟨90JOC1421⟩. A list of literature examples of different types of group E is given in Table 1.

(16)

Table 1 Examples of group E.

Type	*Ref.*
H	78JA1927, 85TL4959, 85TL5013, 85TL5017, 85TL5021, 86JA3841, 87JA3017, 87JA6199, 87JOC2960, 87TL2099, 87TL4993, 90CB153, 90CB917, 90TL4301, 90TL4425, 93TL2597, 93TL5217
aryl/heteroaryl	66LA(698)80, 66TL6393, 70CC4, 70TL353, 71JA3556, 83JOC279, 84JOC1842, 89TL2549, 93CL1397
vinylic	51JA1437, 70CC4, 81JA6492, 82TL3931, 83JOC279, 84JOC1707, 84JOC1842, 85JA3915, 86JOC4315, 91CL43, 91JOC1185, 94T3687
alkynyl	83CL1379, 83JOC279, 84CL255, 84JOC1842, 84T2303, 84TL565, 85CL1719, 85CL1723, 85JOC1143, 85TL5799, 86JOC4316, 87JOC3860, 87JOC3883, 87TL723, 87TL3323, 88TL4587, 90JOC1421, 91CL43, 92CC350, 92T4087
oxazolines and oxamines	83TL513, 84TL6011, 86TL4185, 88JOC4631, 88JOC5189, 93TL749, 93TL753
R—C=O	74HCA2084, 85JA5572
R^1—C=N(R^2)	89JOC5415
CO_2^-	81TL69, 86JOC4090, 86TL4899
CO_2R	86TL1557, 86TL4511, 86TL4581, 87CL69, 87TL803, 88TL4763, 91TL5137
$CONR_2$	85CL1729, 86TL4577
CN	74TL2077, 77BSF925

The stereochemical implications of the reaction are governed to a large extent by the configuration of the double bond in the allyl ether; for example, rearrangement of the isomeric benzyl crotyl ethers **(78)** and **(79)**, which proceeds with excellent *syn* selectivity for the *cis*-crotyl ether **(78)**, whereas the selectivity is lower for the *trans*-isomer **(79)** (Scheme 45) ⟨70CC4, 70LA(734)106, 76CC684⟩.

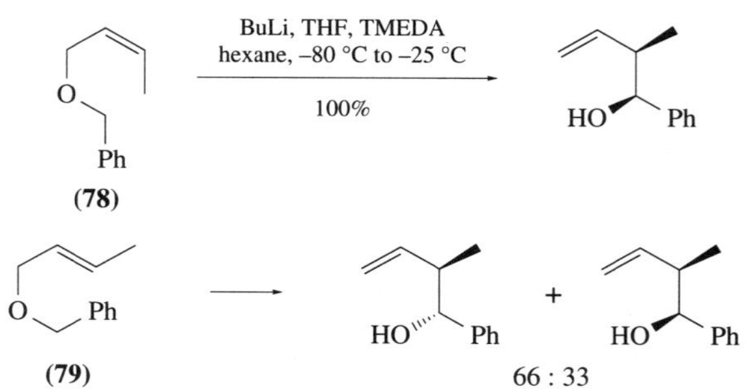

Scheme 45

A nonstabilised variant of the 2,3-Wittig rearrangement employs an unstabilised lithiomethyl allylic ether generated *in situ* through transmetallation of a (trialkylstannyl)methyl allylic ether ⟨78JA1927⟩. The reaction has been used extensively in synthesis and several examples serve to illustrate the selectivity inherent in the process (Scheme 46). In acyclic systems, the stereochemistry of the rearranged double bond depends upon the substitution pattern of the allylic ether (structures (**80**) and (**81**)) ⟨78JA1927⟩. In addition, high stereoselectivity ⟨85TL5013⟩ and absolute stereoinduction ⟨90CB153, 90CB917⟩ are also known, for example, the Still–Wittig rearrangement of the carbohydrate-derived stannane (**82**) gave the homoallylic alcohol (**83**) as a single diastereoisomer. The reaction has also found application in natural product synthesis, including the preparation of steroidal side chains ⟨85TL4959, 85TL5017, 85TL5021⟩, the sesquiterpene punctatin A ⟨86JA3841⟩, hydroxylated vitamin D_3 analogues ⟨87TL2099⟩ and dipeptide isosteres ⟨91TL5401, 93TL5217⟩.

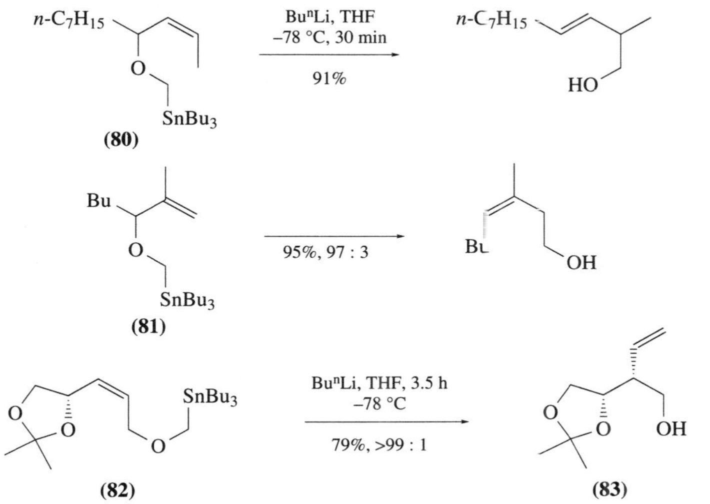

Scheme 46

Probably the most studied substrates for 2,3-Wittig rearrangement are allyl propargyl ethers; these are selectively deprotonated at the propargylic position to give anions which readily undergo 2,3-rearrangement. A high degree of diastereoselectivity is often seen in this rearrangement and, in general, (*E*)-allylic ethers afford *anti*-homoallylic alcohols (**84**), and (*Z*)-allylic ethers rearrange to *syn*-homoallylic alcohols (**85**) ⟨84TL2303⟩; however, many exceptions are known (Scheme 47) ⟨84T2303, 86TL4189⟩ (for mechanistic considerations, see ⟨83JOC279, 90JOC1421⟩). Stereocontrolled transformations have also been reported (Equation (17)) ⟨84CL255, 84JOC1842, 84TL565, 85CL1719, 85CL1723, 85JOC1143, 88TL4587⟩.

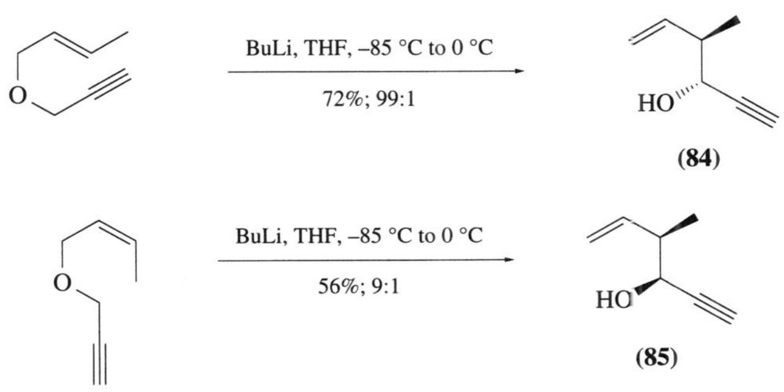

Scheme 47

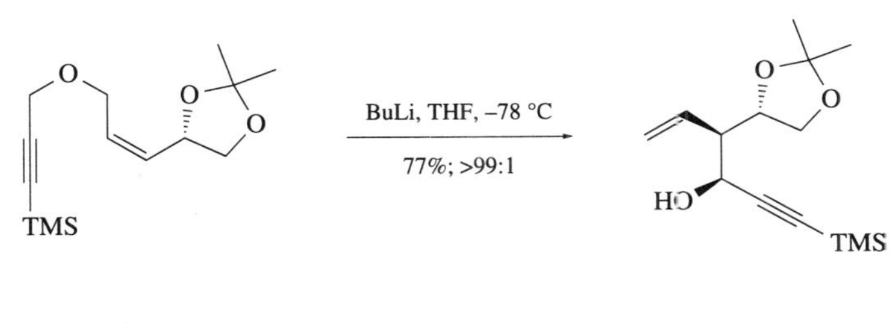

(17)

The reaction has also found extensive use in natural product synthesis, for example, Nakai has employed the propargylic 2,3-Wittig rearrangement for elaboration of side chains in steroidal systems ⟨85CL1719, 85TL5799⟩ and both he and Takahashi have reported approaches to substituted cyclopentenes as prostaglandin precursors ⟨93SL527, 93TL5923⟩. Perhaps the most synthetically important application of this reaction is found in the reports by Marshall ⟨86JOC4316, 87JOC3860⟩ of the synthesis of medium- and large-ring carbocycles by 2,3-Wittig ring contraction of allylpropargyl ethers. Thus, treatment, of the seventeen-membered allylpropargyl ether (**86**) using Bu^{n}Li in pentane–THF–TMEDA gave (**87**) as the major diastereomer in excellent yield ⟨88TL3547⟩; this reaction demonstrates the ability of remote stereocentres to effect chirality transfer in these macrocyclic systems (Equation (18)) ⟨88TL4913, 90JOC227⟩. This chemistry presents a powerful route for the synthesis of cembranoid diterpenes ⟨87JOC3883, 90JOC227⟩, with similar rearrangements to ten-membered rings having been used in a synthetic approach to the germacranolide sesquiterpene aristolactone ⟨87JOC3883, 88JOC4108⟩; homochiral lithium amide base methodology has also been applied to the latter reaction ⟨87TL3323, 88JA2925⟩. A similar strategy has been applied to the synthesis of neocarzinostatin–chromophore analogues and esperamicin–calichemicin analogues; for example, the transannular [2,3]-Wittig rearrangement of a twelve-membered cyclic ether led to the diyne (**88**) (Equation (19)) ⟨91JOC3465, 92TL4563⟩.

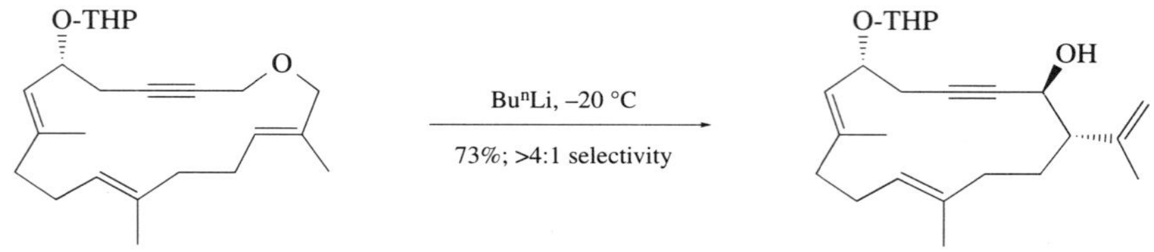

(18)

(**86**) (**87**)

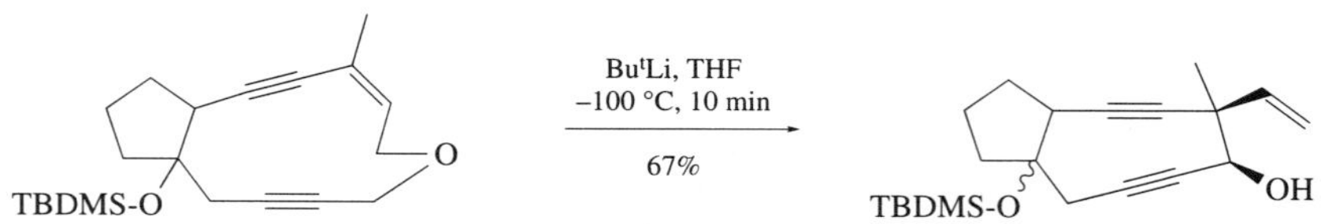

(19)

(**88**)

Similar ring contractions of diallylic macrocyclic ethers are also known; Takahashi *et al.* reported 2,3-Wittig ring contractions of thirteen-membered diallylic ethers (Equation (20)) ⟨86JOC4315, 87H(25)139⟩; a similar 2,3-Wittig ring expansion protocol has also been reported ⟨93TL8453⟩. Acyclic diallyl ethers also undergo 2,3-Wittig rearrangement and a systematic study of these reactions has been performed; the regioselectivity of the rearrangement was found to be related to the stability of the presumed allylic anion intermediates and, in general, the less-substituted allylic anion was preferred (Scheme 48) ⟨81JA6492, 83JOC279⟩; in addition, the stereoselectivity of the process is highly dependent upon the geometry of the intermediate allylic anion ⟨94T3687⟩. Tertiary allyl ethers derived from (+)-camphor and (−)-fenchone also impart a high degree of absolute stereochemical control to the [2,3]-Wittig rearrangement ⟨91JOC1185⟩. Many examples of aryl stabilised anions have also been reported ⟨70CC4, 70TL353, 71JA3556, 84JOC1842, 89TL2549⟩.

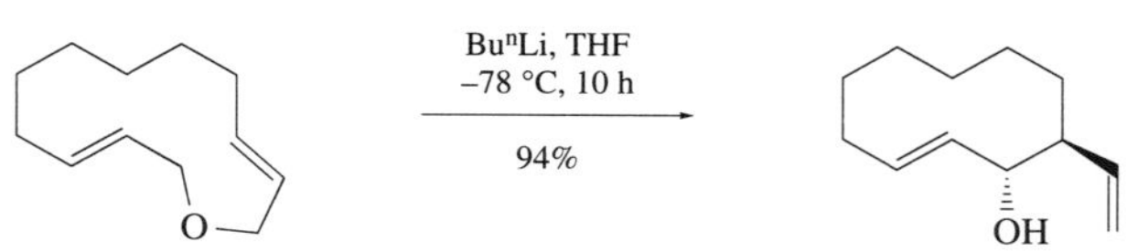

(20)

One area of considerable interest is where the anion that initiates a 2,3-Wittig rearrangement is derived from an enolate. Numerous examples have been reported and include enolates derived from α-allyloxy or α-propargyloxy ketones, hydrazones, nitriles, acids, esters, amides and also oxazoles (see Table 1). In general, such anions are easily formed under mild conditions and lead to highly functionalised products with great synthetic potential; other advantages include the possibility of incorporation of chiral auxiliary groupings and the opportunity to effect changes in the counterion. Some representative examples are shown in Scheme 49, which includes methodology for the modification of the 22-hydroxy substituent found in ecdysone and brassinolide natural products (**89**)

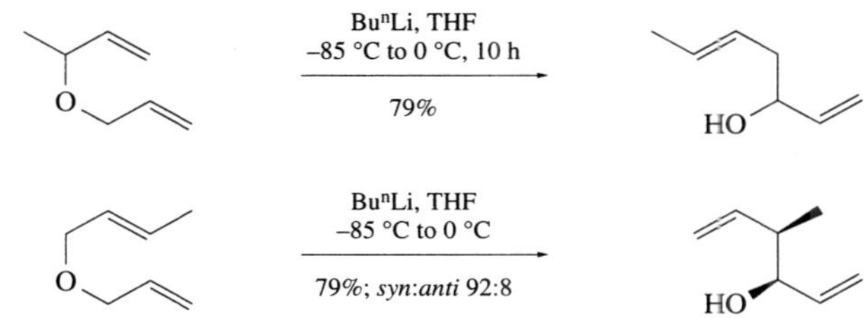

Scheme 48

⟨86JOC4090, 86TL4899⟩ and in the synthesis of an advanced intermediate for the preparation of rapamycin **(90)** ⟨93TL749, 93TL753, 93TL1103⟩. Ester enolates show a high propensity for 2,3-rearrangement, and chiral auxiliaries have been used to effect high levels of diastereoselectivity and asymmetric induction in the transformation **(91)** (Equation (21)) ⟨87CL69⟩. Examples using amides of homochiral pyrrolidines ⟨85CL1729⟩ and oxazoline derivatives ⟨84TL6011, 86TL4185, 88JOC4631⟩ are also known.

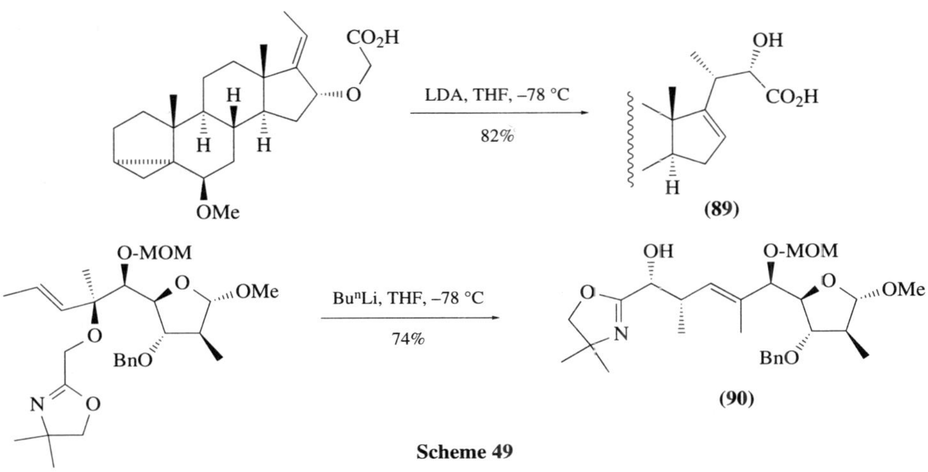

Scheme 49

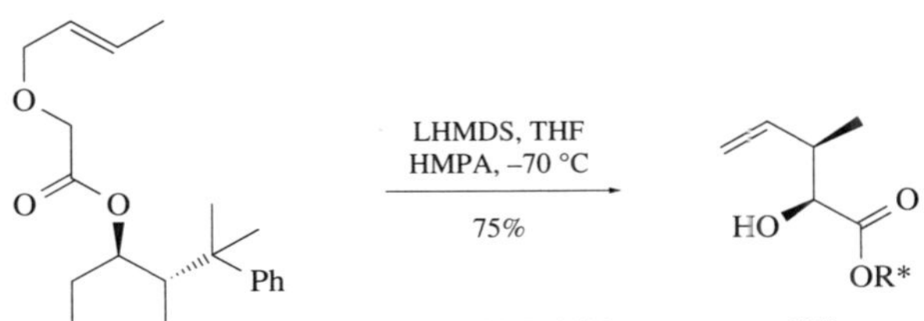

(91) (21)

Transmetallation of enolates has been used to good effect and it is well known that zirconium and titanium enolates improve the efficiency and stereoselectivity of the Wittig rearrangement of allylic glycolate esters ⟨86TL4581, 88CL4768, 88TL4763⟩; similar results are also obtained with boron or tin enolates ⟨91TL4647⟩. A high erythro-selectivity is observed when silyl ketene acetals of α-allyloxy esters are treated with $TiCl_4$ or $SnCl_4$ to induce a [2,3]-sigmatropic shift (Scheme 50) ⟨91SL629⟩. A titanium-mediated [2,3]-Wittig rearrangement of glycolate ester **(92)** gave the hexenoate **(93)** with excellent stereocontrol and diastereoselectivity (Equation (22)) ⟨92BCJ1841⟩.

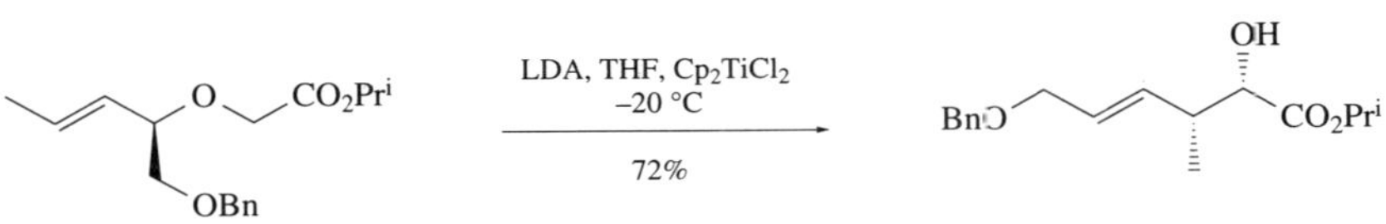

(92) *syn*:*anti* 29:1 **(93)** (22)

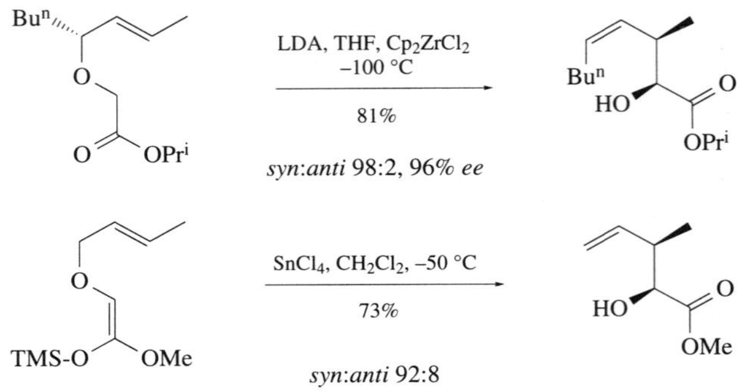

Scheme 50

(ii) Rearrangement of oxonium ylides (Y = C—O)

The 2,3-rearrangement of oxonium ylides effects a similar transformation to that found in the [2,3]-Wittig reaction. Various methods are available for the generation of the ylide ⟨91CRV263⟩; these include metal-catalysed decomposition of α-diazo ketones ⟨86JA6060, 86JA6062, 88TL5119, 92TL6193⟩ with intramolecular 2,3-rearrangement and similar intermolecular versions using α-diazomalonate- ⟨72JA3870⟩ and α-diazoacetate-based ⟨71JOC1732⟩ rearrangements of allylic ethers. Trimethylsilyloxonium ylides can also be generated *in situ* by treatment of α-allyloxyacetic esters with TMSO-Tf and triethylamine ⟨86TL4511⟩. Representative examples are given in Scheme 51.

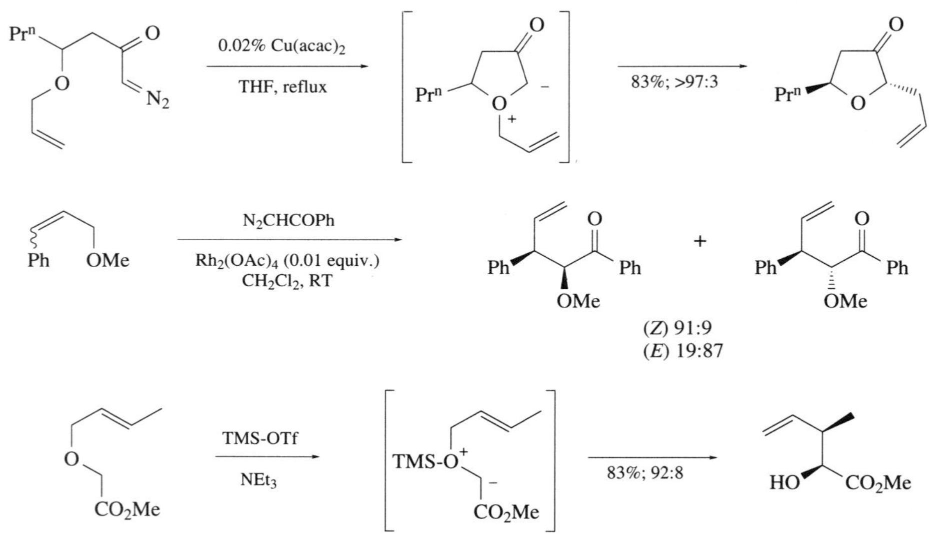

Scheme 51

(iii) Thia–Wittig rearrangement (Y = C—S)

Allylic sulfides undergo 2,3-Wittig rearrangement, but at a slower rate than their oxygen counterparts. Several anion stabilising groups have been utilised including cyano ⟨76JA7115⟩, vinyl ⟨71HCA739⟩, aryl ⟨71HCA739⟩, ester ⟨80CL1061⟩, thiolether ⟨77SC103⟩ and sulfoxide ⟨75CR(C)1231⟩; rearrangements in which no stabilising group is used ⟨92TL7323⟩ have been reported and can be considered a direct heterologue of the Still 2,3-Wittig variant (Scheme 52) ⟨78JA1927⟩.

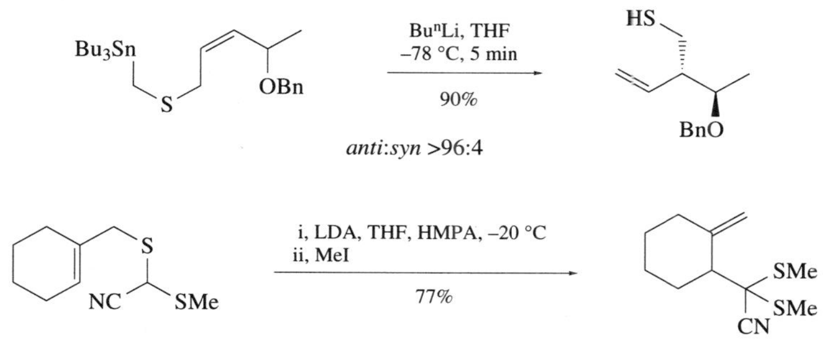

Scheme 52

(iv) Rearrangement of sulfur ylides ($Y = C—S$)

The 2,3-rearrangement of allylic sulfur ylides has been studied extensively ⟨91COC(3)913⟩. It is a facile process occurring readily under mild conditions, proceeding in a concerted fashion with inversion at the allyl centre. A high degree of stereocontrol is observed for the alkene geometry and, in the absence of any special effects, such as conformational or steric factors ⟨81JOC3315⟩, the rearrangement is (*E*)-selective ⟨69CC99, 72CC702, 73JOC2572, 79JOC4128⟩.

Several methods exist for the preparation of the intermediate ylide and a common method is the deprotonation of sulfonium salts (Scheme 53) ⟨71CC1639, 72CC757, 72TL3587, 78T317, 80CC1096⟩. This prototypic reaction has also found application in ring contraction ⟨72JA8932, 81JCS(P1)1953⟩ and expansion methodologies ⟨75JA6878, 77HCA1361, 78JOC1185, 78JOC4826, 78JOC4831, 78TL519, 79JOC4128, 84TL3555⟩ including impressive total synthetic work towards erythronolide A ⟨83JA127⟩, cytochalasin ⟨84JA4617⟩, zygosporin ⟨88JA4822⟩ and a precursor of methynolide **(94)** ⟨87JA5878⟩ performed by Vedejs and co-workers (Scheme 54). Examples are also known of asymmetric transfer using homochiral sulfonium salts ⟨73JA962, 73JOC3438, 77TL1337, 78JOC2518, 91TL335⟩ although, in the majority of cases, the reaction proceeds with low enantiomeric excess. Enantioselective deprotonation of a prochiral sulfonium salt using a chiral base has also been reported ⟨73JOC3438⟩, but asymmetric induction was low.

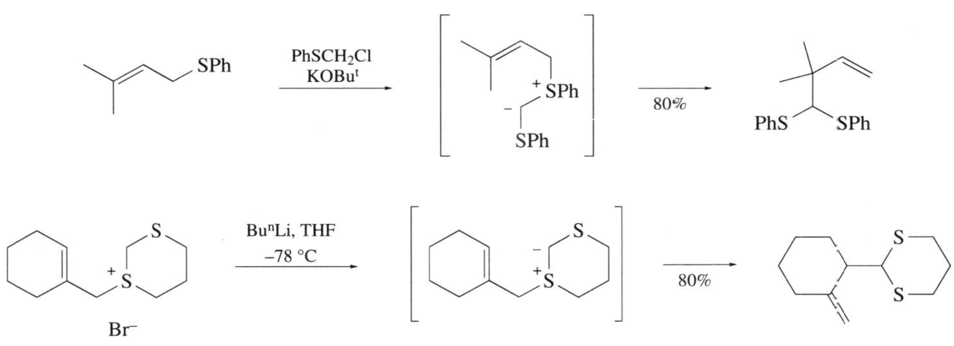

Scheme 53

Regiospecific generation of ylides using fluoride ion-catalysed desilylation of trimethylsilyl sulfonium salts has been explored ⟨79JA6452⟩; this technique has the advantage over the previous method in that 2,3-rearrangement usually takes place before any equilibration to thermodynamically more stable ylides can occur; examples include a rearrangement that was important in the total synthesis of sarkomycin (Scheme 55) ⟨85JOC2965⟩. The direct formation of sulfur ylides is also possible by the addition of a carbene or benzyne to a sulfide; this has been an area of exceptional development ⟨77ACR179, B-78MI 118-01, 86ACR348⟩. Several methods are available for the generation of carbene intermediates in these reactions, the most common being the light-promoted ⟨72JA3870, 72JOC3791⟩ or metal-catalysed generation of carbenes from diazonium compounds, the

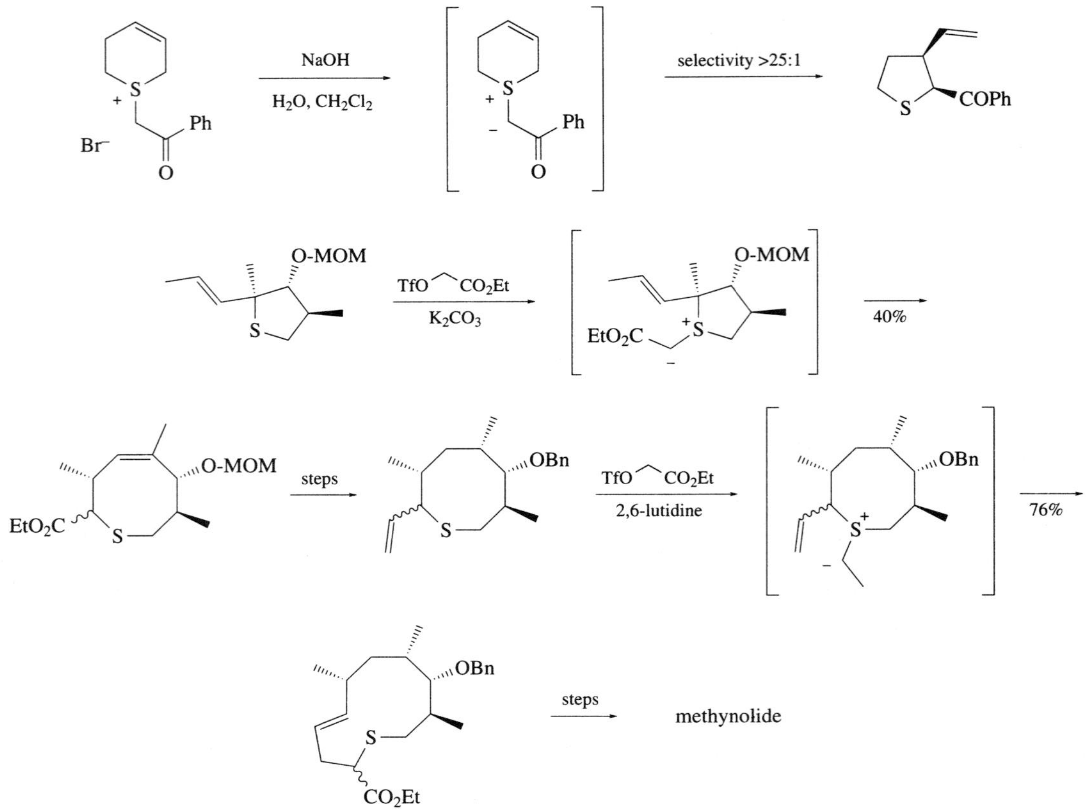

(94)

Scheme 54

latter method giving better results ⟨72JA3870⟩. Copper ⟨72TL5121, 72JA3870, 72JOC3791, 72TL2923, 80TL399, 82JCS(P1)2757⟩ or rhodium ⟨81JOC5094, 84ACR358, 87CL1569, 90CC418, 90CL613, 90T4887⟩ catalysis is generally used. Carbenes generated under phase transfer ⟨72TL5121⟩ or Simmons–Smith-type ⟨82TL3019⟩ conditions have been used, as have carbenes from the Lewis acid mediated decomposition of diazopenicillins ⟨80TL399, 82JCS(P1)2757⟩. Several examples of these methods are given in Scheme 56, including methodology by which treatment of a tosylhydrazone with hydride generates the intermediate ylide which then rearranges to the dithiolester (**95**), a synthetic precursor of the natural product bakkenolide ⟨72CC354, 77JA5453⟩. In addition, Yoshikoshi and co-workers describe a novel ring contraction strategy directed towards the preparation of highly substituted lactones (**96**) ⟨89TL1575, 90CC418, 90CL613, 90T4887, 91TL6159⟩, see also ⟨72JA3870, 72TL2923, 94CC789⟩.

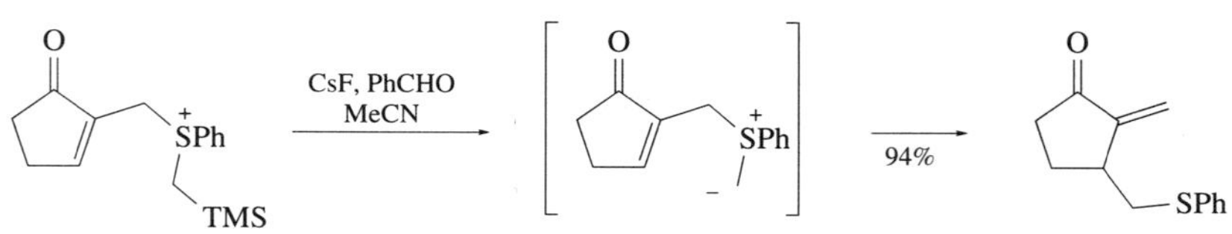

Scheme 55

Some noteworthy work has been reported using benzyne for generating sulfur ylides (including a total synthesis of squalene ⟨69CC99⟩); however, competitive rearrangements are frequently observed ⟨63LA(662)188, 68CC186⟩. Sulfoxonium ylides also undergo 2,3-sigmatropic rearrangements ⟨66TL1787, 68CPB793, 73TL2351⟩ as do selenonium ylides ⟨80TL399, 82JCS(P1)2757⟩.

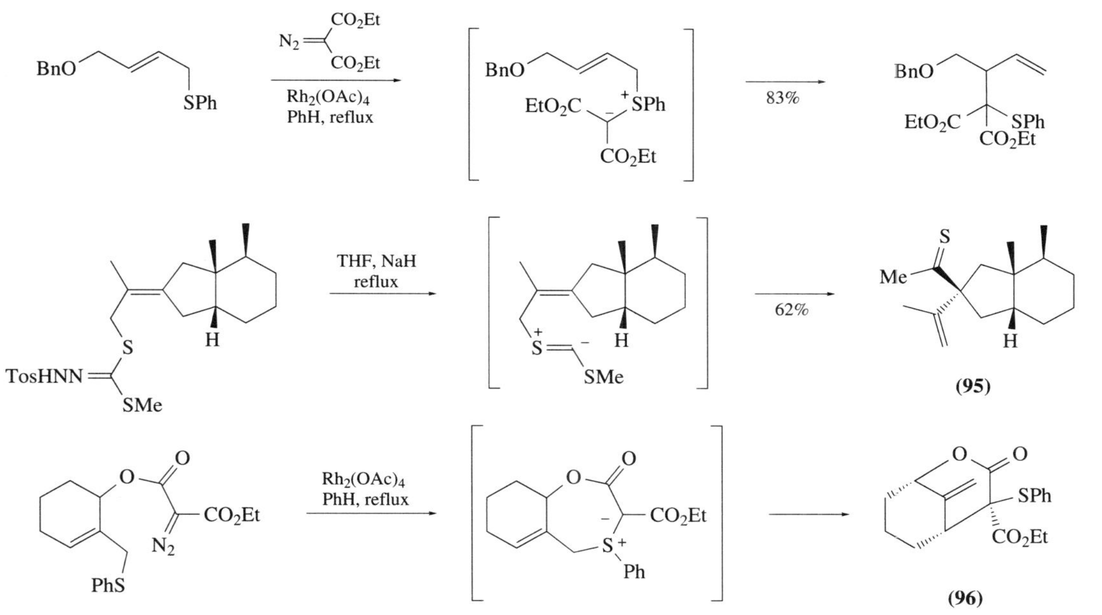

Scheme 56

1.18.2.5.2 Where Z = nitrogen

(i) Aza-Wittig rearrangement (Y = C—N)

Some examples of the 2,3-aza-Wittig rearrangement are known utilising basic conditions to generate the requisite anion (Equation (23)) 〈72JA9261, 90CL2069〉.

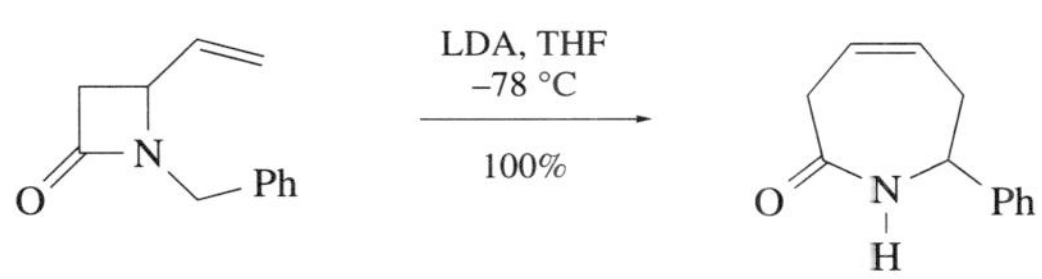

(23)

(ii) Rearrangement of nitrogen ylides (Y = C—N)

The 2,3-rearrangement of nitrogen ylides (the 2,3-Stevens rearrangement) has been used extensively in synthesis and has many parallels with the corresponding rearrangement of sulfur ylides 〈77ACR179, 86ACR348, 91COC(3)913〉. The rearrangement is a concerted, symmetry-allowed process, proceeding with inversion at the allylic centre and showing high regio-, diastereo- and enantioselectivities.

Several methods exist for the preparation of nitrogen ylides, the most commonly used being the deprotonation of quaternary ammonium salts under basic conditions 〈71JA2342, 73JOC2915, 74JA7573, 80JCS(P1)1436, 80JCS(P1)2033, 81JCS(P1)1963, 90JA1999〉. Other methods include the fluoride-catalysed desilylation of trimethylsilyl ammonium salts 〈79JA6452, 83JOC4773, 90JA1999〉, rhodium-catalysed reaction of allylic amines with diazo-compounds 〈81JOC5094, 84JOC1917〉 and silyl triflate-mediated rearrangements (Scheme 57) 〈90CL2069〉.

High levels of diastereo-control are possible in acyclic rearrangements, particularly with propargylic ammonium ylides, the stereochemistry being dependent on the double-bond geometry in the starting material 〈69CC294, 72CC62, 80JCS(P1)1436〉. In contrast, acyclic, nonpropargylic ammonium ylides give mixtures of diastereoisomers 〈81JCS(P1)1953〉 with steric factors playing a significant role (Scheme 58) 〈81JCS(P1)1963〉. Efficient transfer of configuration is also possible 〈B-84MI 118-04〉 as is asymmetric induction by remote chiral centres 〈71JA2342〉. Ring contraction 〈81JCS(P1)1953, 88TL3993〉 and ring expansion 〈78TL523〉 reactions are also known; these generally proceed with higher diastereoselectivity than is observed in acyclic cases (Scheme 59).

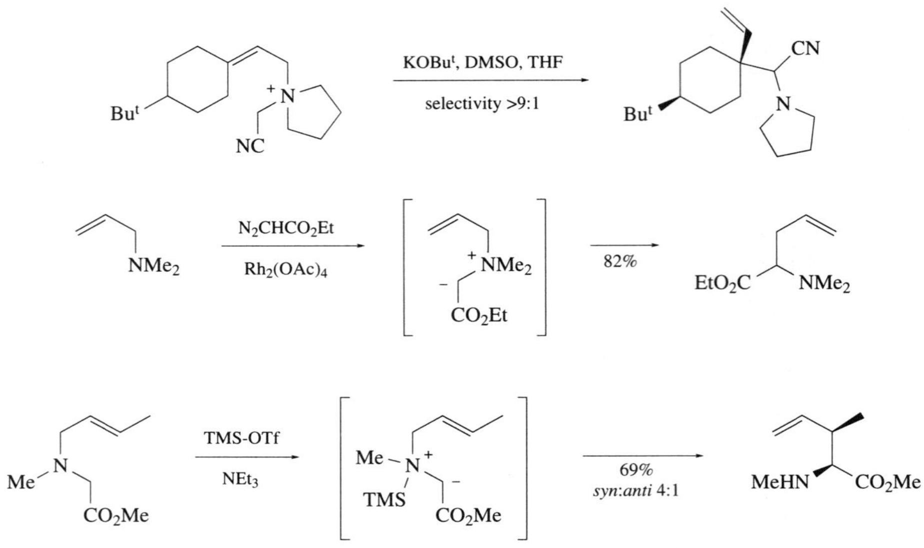

Scheme 57

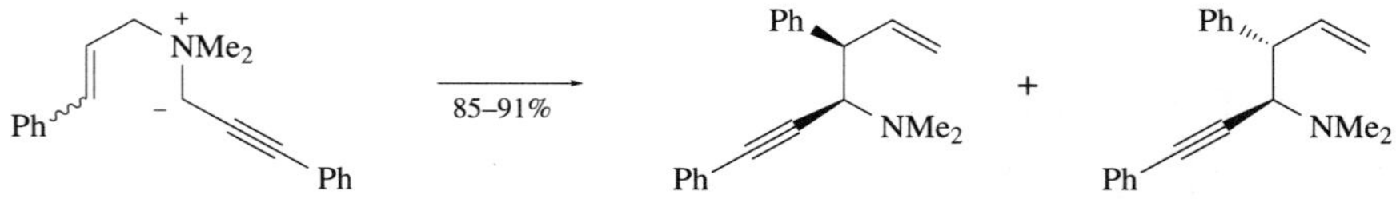

6 (*E*); ratio; 86:14
6 (*Z*); ratio; 10:90

Scheme 58

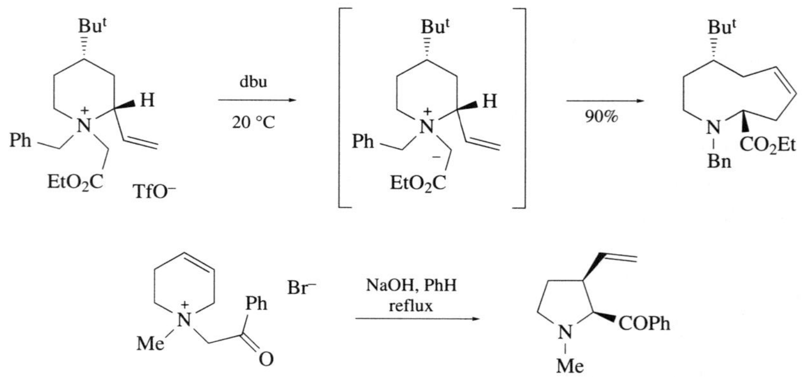

Scheme 59

The Sommelet–Hauser rearrangement ⟨37CR(205)56, 51JA4122⟩ is also a 2,3-rearrangement, but one in which a benzylic quaternary ammonium ylide is employed. The reaction proceeds via an intermediate triene (for example **(97)**) which on most occasions undergoes re-aromatisation, via a 1,3-hydrogen shift; however, on some occasions these intermediates have been intercepted and isolated (Equation (24)) ⟨88CC370, 94JCS(P1)393⟩.

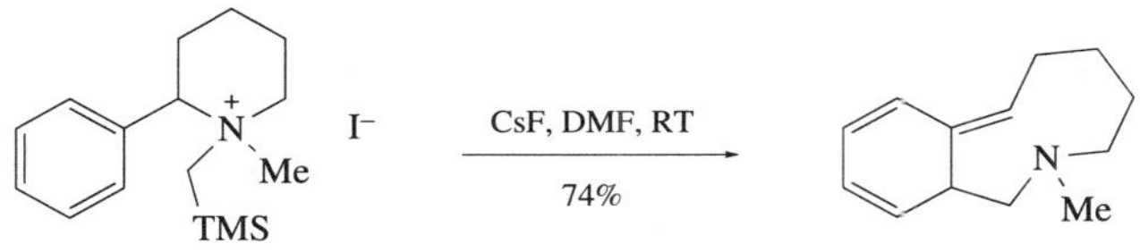

(97)

(24)

1.18.2.5.3 Miscellaneous rearrangements

(i) Y = C—Hal

Rhodium-catalysed decomposition of diazoacetates in the presence of allylic halides allows the preparation of halonium ylides; these compounds undergo sigmatropic rearrangement to homoallylic halides, with iodides giving the best results (Scheme 60) ⟨81JOC5094⟩.

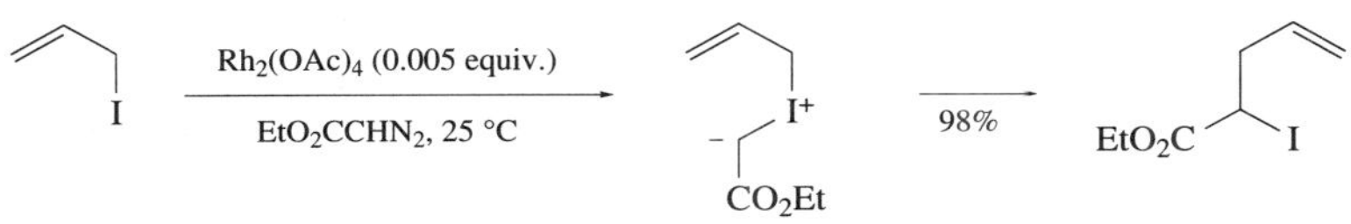

Scheme 60

1.18.2.6 Where Y = C—C—Z

1.18.2.6.1 Where Z = chalcogen

(i) Claisen rearrangement (Y = C—C—O)

The Claisen rearrangement can be represented simply as the [3,3] sigmatropic conversion of allyl vinyl ethers (**98**) into γ,δ-unsaturated carbonyl compounds (**99**) (Equation (25)). Since its discovery ⟨12CB3157⟩ the reaction has been the subject of several modifications, which have led to it becoming one of the most powerful synthetic operations available for carbon–carbon bond formation ⟨80T3, B-81MI 118-01, 84CRV205, 87AHC(42)203, 88CRV1423, 89S71⟩. The common variants of the Claisen rearrangement are shown in Equation (25) and, in addition to the parent rearrangement of allyl vinyl ethers (X = H, R, Ar), the Johnson orthoester (X = OR) and the Eschenmoser amide acetal (X = NR_2) modifications are facile processes which have increased the potential of this methodology. Possibly the most important developments in this area are the enolate Claisen rearrangement (X = OM), the Ireland rearrangement of silyl ketene acetals (X = $OSiR_3$) being the most widely used. An aromatic version of the reaction is also known, where the vinyl ether portion of the molecule is contained within an aromatic ring. In addition, anion-accelerated examples of the Claisen rearrangement are known, as are several other versions, for example where X = F ⟨81CL1289⟩, X = Si, Ge ⟨90JA7791⟩ and X = Fe ⟨87OM2249⟩. The reaction, in accord with related processes, is generally considered to be an exothermic, intramolecular and concerted process ⟨79JA2746, 88CRV1423⟩.

(98) **(99)**

X = H, R, Ar, Aliphatic
X = OR, Johnson
X = NR_2, Eschenmoser
X = OM, Ireland

(25)

(a) Claisen rearrangement of allyl vinyl ethers. The Claisen rearrangement of allyl vinyl ethers has been used extensively in synthesis; this is despite the fact that the uncatalysed thermal reaction generally requires temperatures of 100–200 °C and thus limits the process to thermally stable compounds. Thermal Claisen rearrangement of ethers normally results in high (*E*)-selective product formation ⟨59JA3379, 60JA4681⟩. The reaction has found considerable application in the total synthesis of natural products, particularly those which contain multiple contiguous quaternary carbon centres

⟨88TL1669, 93JOC7751⟩. Many ring expansion reactions have also been reported, the precursors being prepared via Tebbe alkeneations of lactones or esters ⟨85JA7352, 90TL6799, 91JA2610, 91JOC3841, 93TL1721⟩ or via selenoxide elimination ⟨90TL4041, 90TL6979⟩. High levels of stereocontrol ⟨89JOC4278, 90JA4035⟩ and chirality transfer ⟨89HCA996, 90JCS(P1)11⟩ are also observed. Specific examples of these transformations are shown in Scheme 61 and include the formation of contiguous quaternary centres in the rearrangement of allyl vinyl ether (**100**) ⟨88TL1669⟩, the two-carbon ring contraction of macrocyclic ether (**101**) (prepared using the Tebbe methodology) ⟨93TL1721⟩ and the tandem selenoxide elimination–Claisen rearrangement of substrate (**102**) ⟨90TL6979⟩ (for a related sulfoxide elimination–Claisen protocol, see ⟨90TL4041⟩). Numerous other synthetic examples are known ⟨89HCA996, 89JOC4278, 90JA3252, 90JA4035, 90T3601, 90T4487, 93TL3523⟩.

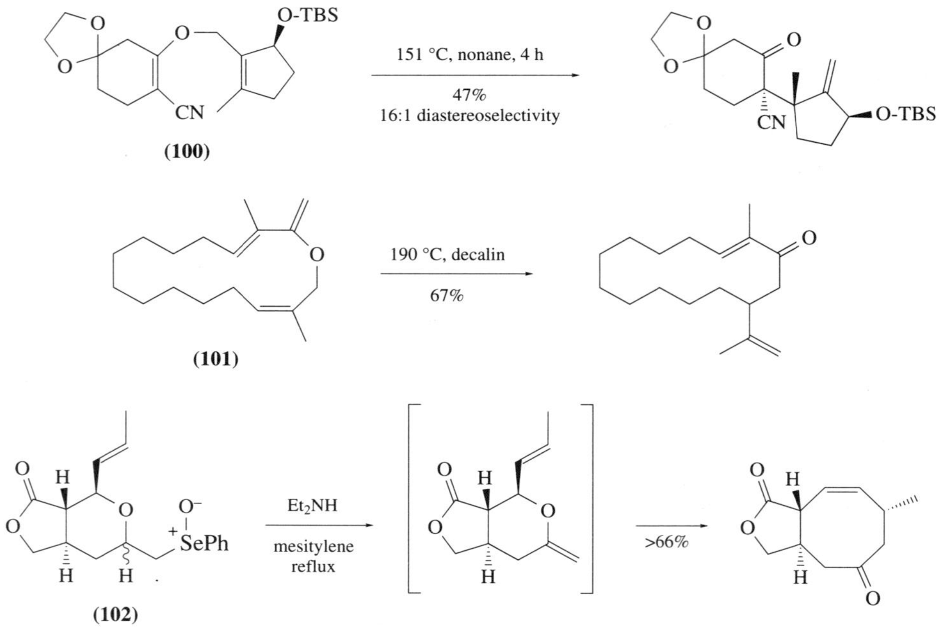

Scheme 61

The chorismate mutase catalysed [3,3]-sigmatropic rearrangement of chorismate to prephenate in the shikimic acid pathway ⟨63NAT388, 73B3492, 79AJC1921⟩ presents a biosynthetic example of this versatile process. Transition-state analogues have been prepared in order to study both enzyme inhibition ⟨85JA7792, 88JOC3195⟩ and the preparation of catalytic monoclonal antibodies ⟨88JA5593, 88SCI426⟩.

Catalysis of the Claisen reaction of allyl vinyl ethers (and of many of the Claisen variants) has received considerable attention. Catalysis by Brönsted and Lewis acids ⟨90JA316, 90JA7791, 90TL377, 90TL3241⟩, bases, metal salts ⟨89BCJ3835, 91JCR(S)172⟩, transition metal complexes ⟨84AG(E)579, 86TL6267, 87TL5879, 89SC2109⟩, and microwave methodology ⟨86TL4945⟩ have been reported and early work in this field has been reviewed ⟨84CRV205⟩. Later work has concentrated on aspects of organoaluminum catalysis of the reaction, and high levels of stereocontrol and chiral transmission have been reported ⟨88JA7922, 89TL1265, 90JA316, 91JA2762, 91JOC3841⟩. In addition, chiral organoaluminum reagents have also been shown to effect asymmetric rearrangements in enantiomeric excesses as high as 93% (Scheme 62) ⟨90JA7791⟩. The rate of reaction in the aliphatic Claisen rearrangement of allyl vinyl ethers has been shown to be highly dependent on solvent with a rate acceleration occurring in polar solvents, particularly water/alcohol mixtures ⟨93CRV2023⟩; see also ⟨70JOC3585, 87JA1160, 87JA1170, 87JA5008, 89JOC515, 89JOC5849, 90TL4147, 93T2093⟩.

Unlike the Cope rearrangement the Claisen reaction is not generally reversible. However, several examples are known of the so-called retro-Claisen rearrangement, where the allyl vinyl ether product is favoured; these cases generally occur where excessive strain is present in the ketonic system ⟨65HCA1985, 68CC587, 69JA2815, 71HCA1589, 79CC584, 82JA6129, 83CC932, 85JA4359⟩.

(b) Aromatic Claisen rearrangement. The aromatic Claisen rearrangement is the thermal [3,3]-sigmatropic rearrangement of an allyl aryl ether to give initially an *ortho*-dienone, which on

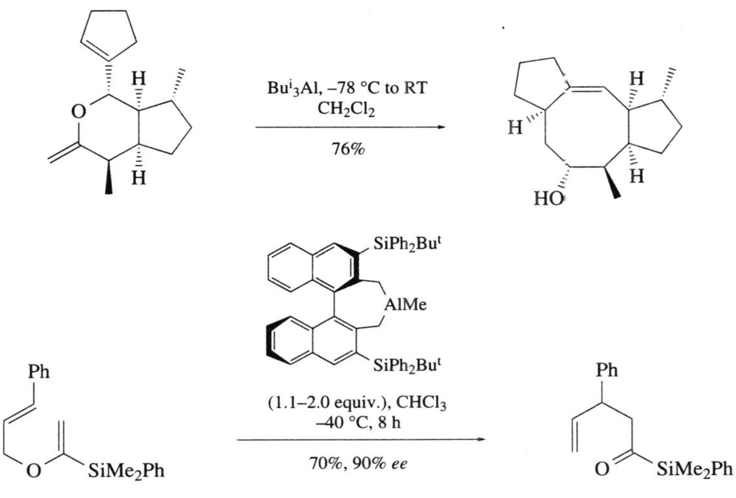

Scheme 62

enolisation yields an *ortho*-allylphenol. This is referred to as the *ortho*-Claisen rearrangement and is typified by the rearrangement of the allyl ether (**103**) (Scheme 63) ⟨93CC793⟩. When both *ortho* positions are substituted, a second 3,3-sigmatropic shift (Cope rearrangement) occurs, which, on enolisation, generates a *para*-allylphenol in the so called *para*-Claisen rearrangement (see 1.18.2.4.1.i, Scheme 28). These products are also obtained occasionally when one or both *ortho* positions are unsubstituted; however, no reaction occurs when all *ortho* and *para* positions are substituted ⟨63HCA2444, 68HCA1603⟩. These reactions have been used extensively in synthesis ⟨44OR(2)1, 84CRV205⟩ and remain the best synthetic route for the formation of allyl or phenyl substituted aromatic compounds ⟨88TL6797, 89JOC3137, 90SC1437, 91JCS(P1)2089, 91SC1455, 94CC251, 94TL1661⟩. A further complicating factor in this rearrangement is the 'abnormal' Claisen rearrangement ⟨71MI 118-02, 73JA3348⟩; this results in the formation of mixtures of isomeric allyl phenols via a homodienyl 1,5-sigmatropic H-shift ⟨75OR(22)1⟩. This process can be suppressed by the introduction of a trapping agent for the intermediate phenol, *N-O*-bis(trimethylsilyl)acetamide (bsa) and hexamethyldisilazane being particularly effective ⟨94TL2145⟩.

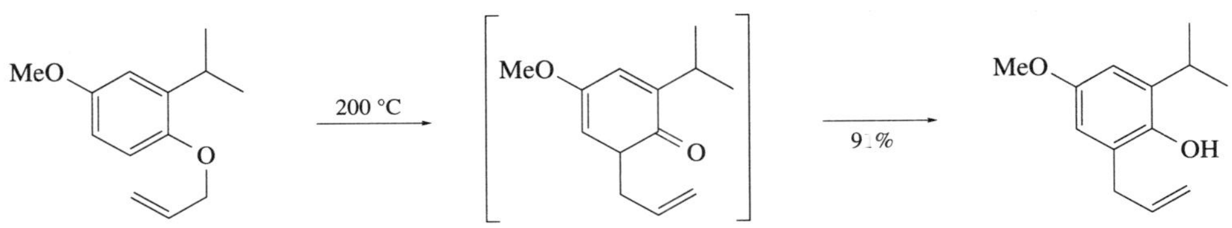

(103)

Scheme 63

(c) The Eschenmoser amide acetal and Johnson orthoester Claisen rearrangements. The discoveries of the Johnson orthoester ⟨70JA741⟩ and the Eschenmoser amide acetal ⟨64HCA2425, 79HCA1922⟩ rearrangement protocols were major developments in this area. The formation of vinyl ketene acetals and ketene *N,O*-acetals, by treatment of allylic alcohols with *ortho* or amide acetals and their subsequent *in situ* sigmatropic rearrangement facilitated the stereoselective synthesis of γ,δ-unsaturated amides (**104**) and esters (**105**) (Scheme 64). The reactions generally require high temperatures to effect conversion; these conditions are required for the alcohol exchange reaction, and not for the actual rearrangement which occurs at significantly lower temperatures ⟨85JOC5909⟩. Both reactions have been used extensively in synthesis and are known to proceed with high levels of chirality transfer and stereoselectivity ⟨70JA741, 77TL1625, 89CPB2047⟩. In addition, the Eschenmoser rearrangement produces stereochemically well-defined ketene *N,O*-acetals with a thermodynamic preference for the (*Z*)-isomer ⟨71CB3679⟩; for the formation of (*E*)-ketene *N,O*-acetals, see ⟨79JOC882⟩. This is not the case for the Johnson variant and mixtures of diastereomers are obtained where α-substituents are present in the orthoester ⟨71JCS(C)2950, 87JOC1201, 88JOC3868, 88TL655,

89JOC1223⟩. Nevertheless, the reaction has found considerable synthetic application ⟨79TL3057, 80TL4335, 81JOC1485, 83T3731, 86JOC123, 89TL3377, 89TL6271, 89TL6577, 90SC423⟩.

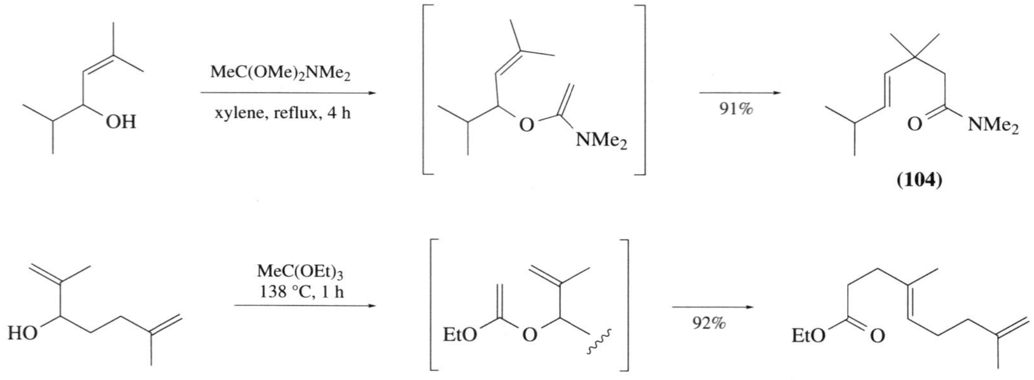

(104)

(105)

Scheme 64

The Eschenmoser rearrangement of the allylic alcohol (**106**) was a crucial step in the total synthesis of DL-stenine ⟨90JOC6236⟩ and similar transformations have been used in the synthesis of branched-chain D-myo-inositols ⟨93CC960⟩ and in the stereoselective construction of the C(21)–C(42) fragment of rapamycin ⟨93CC619⟩. Many additional examples are known, including the Eschenmoser rearrangement of alcohol (**107**) to amide (**108**), a key step in the total synthesis of thromboxane B_2 (Scheme 65) ⟨77TL1625⟩. Approaches to carbohydrate-derived carbocycles ⟨89JOC5350⟩, alkaloids ⟨91JCS(P1)561⟩ and optically active fluorine compounds ⟨91TL4267⟩ have also been reported using the Eschenmoser protocol, as have many other examples ⟨79HCA1922, 82JA7174, 82TL109, 84JOC2682, 86TL3053, 87JOC1372, 87TL2041, 89CPB2047⟩.

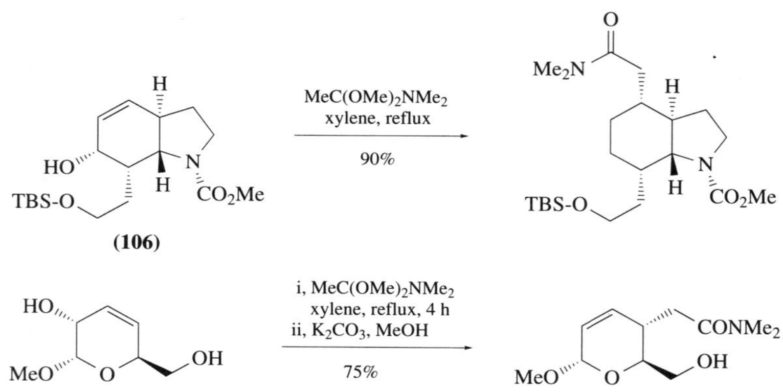

(106) **(107)** **(108)**

Scheme 65

Several modified Eschenmoser procedures are also known, including the ynamine–Claisen rearrangement ⟨66TL6425, 79JOC882, 81JOC3896, 81TL4097⟩, where the reactive *N,O*-ketene acetal intermediates are formed at ambient temperature by treatment of an ynamine with an allylic alcohol. Methods for the preparation of *N*-silylketene *N,O*-acetals from allyl *N*-phenylimidates by deprotonation and subsequent *N*-silylation have been reported ⟨90SL97⟩, as have *N,N*-dialkyl-alkoxymethyleneiminium salts which, in combination with lithium alkoxides, generate the requisite intermediates for the Claisen rearrangement (Scheme 66) ⟨85JOC5909, 87JA6716, 91TL4199⟩.

Similar progress has been made with the Johnson rearrangement, and the rearrangement of allylic alcohol (**109**), using orthoacetate (**110**), illustrates both the absolute transfer of configuration in this reaction but, due to the formation of both (*Z*)- and (*E*)-ketene acetals, the product (**111**) is formed as an epimeric mixture ⟨76JA1583⟩. The Johnson rearrangement of (**112**) has been used in the total synthesis of the alkaloid (−)-mesembranol ⟨94CC901⟩, and a similar transformation was used in an

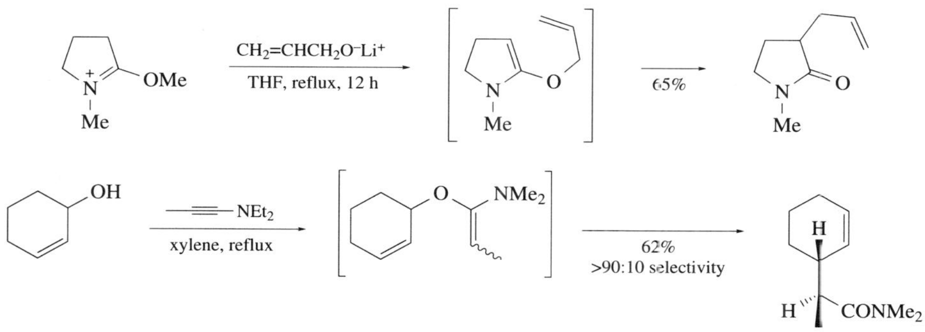

Scheme 66

approach to the thapsane sesquiterpenes ⟨93JCS(P1)667⟩. A recent synthetic approach to pyrrolizidine alkaloids uses the Johnson–Claisen rearrangement of alcohol (**113**) as a key transformation, the reaction proceeding with complete chirality transfer (Scheme 67) ⟨93S615⟩. Other examples are worthy of note ⟨87JA3025, 87JOC1201, 87JOC3541, 87TL2597, 88CC1531, 88JOC1922, 89JOC736, 90JOC2108, 91CC1579, 91TL5525, 93T4789, 93TL1851⟩.

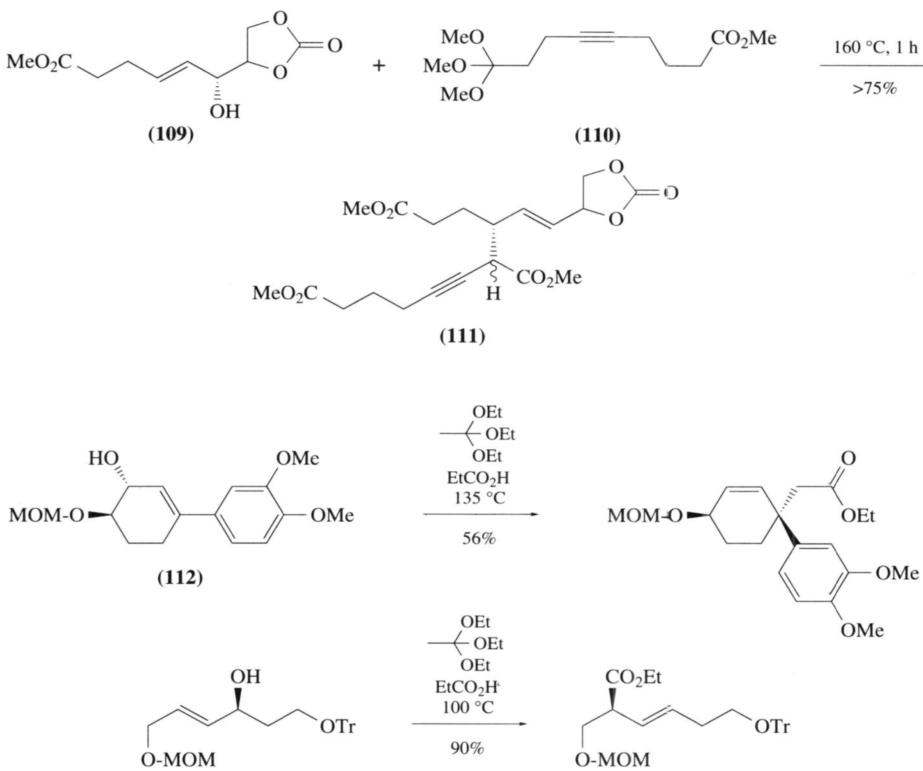

Scheme 67

The related rearrangements of vinyl ketene acetals have been reported, including a key step in the total synthesis of the marine metabolite (+)-laurencin via a tandem selenoxide elimination–Claisen rearrangement of selenide (**114**) to give lactone (**115**) ⟨93JA10400⟩; similar examples are known (Scheme 68) ⟨91T7171, 93JA5815⟩.

(d) The Ireland ester enolate Claisen rearrangement. The rearrangement of metal enolates generated from allylic esters has long been known to be a viable variant of the Claisen reaction ⟨49JA1150, 49JA2439, 60JOC1815, 72SC27⟩; however, it was not until Ireland and co-workers reported ⟨72JA5897, 75TL3975, 76JA2868⟩ the effective [3,3] sigmatropic rearrangement of silyl ketene acetals that the full

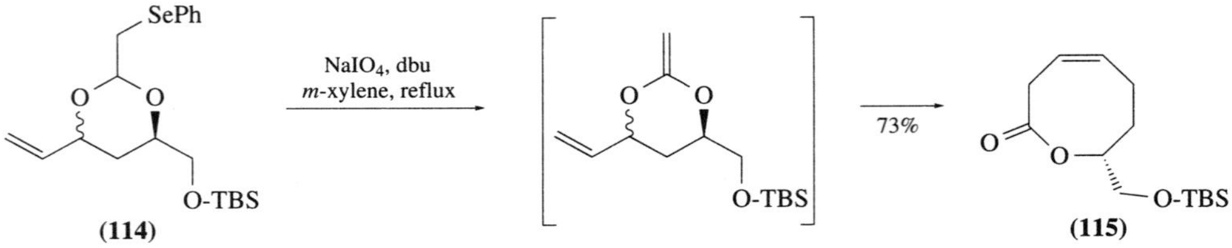

(114) **(115)**

Scheme 68

potential of the reaction was realised. Treatment of an allylic ester (**116**) with a lithium amide base, followed by *O*-silylation (normally with *t*-butyldimethylsilyl (TBDMS) chloride to limit competitive *C*-silylation ⟨72JA5897, 73SC67⟩) generates a silyl ketene acetal (**117**) which, on warming, undergoes sigmatropic rearrangement to yield a γ,δ-unsaturated silyl ester (**118**) (Scheme 69). In the 1980s and 1990s this simple procedure has been developed into a general way to effect stereocontrolled Claisen rearrangements and represents one of the most powerful methodologies available to the synthetic chemist ⟨93MI 118-01⟩.

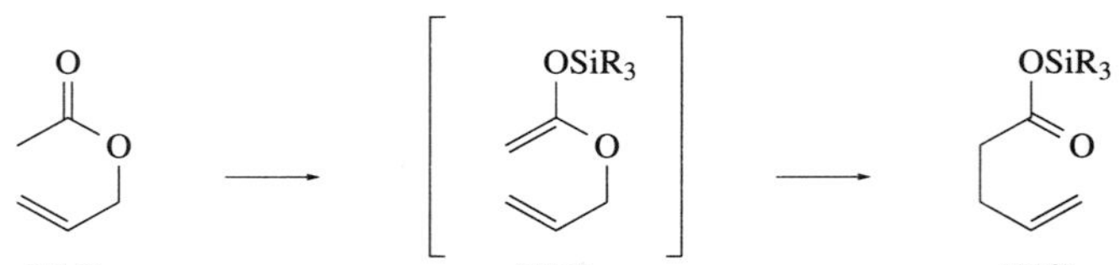

(116) **(117)** **(118)**

Scheme 69

The most important aspects of this process are those which allow control of the stereochemistry of the intermediate enolate, as the geometry of the products of the Ireland–Claisen rearrangement can be predicted by consideration of configuration of the double bonds involved in the ketene acetal rearrangement. The formation of (*E*)-silyl ketene acetal (**119**) is favoured when THF is used in the formation of the lithium enolate, while the use of 23% hexamethylphosphoramide (HMPA)/THF favours the formation of the (*Z*)-silyl ketene acetal (**120**); on rearrangement these intermediates give the *erythyro*- (**121**) and *threo*-products (**122**), respectively (Scheme 70) ⟨75TL3975⟩. Other additives such as *N,N′*-dimethylpropyleneurea (DMPU) or TMEDA can also have a profound effect ⟨91JOC650, 93MI 118-01⟩ as can other factors such as the choice of base ⟨93JOC5301, 93MI 118-01, 94T3099⟩. Several other conditions, such as treatment of the ester precursor with silyl triflates in the presence of a nitrogen base, can also be used to generate the intermediate silyl ketene acetal ⟨88JA3298, 89TL515, 93TL2251⟩. The reaction has been extensively studied from a mechanistic standpoint and the factors affecting the transition-state geometry (boat versus chair shaped six-membered rings) for both acyclic and cyclic cases have been examined ⟨91JOC3572⟩.

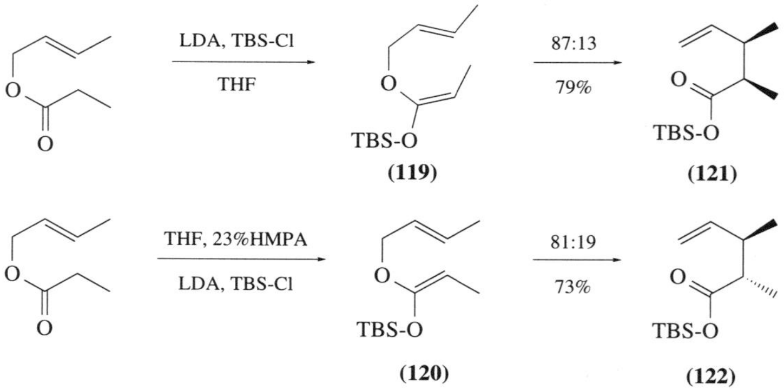

(119) **(121)**

(120) **(122)**

Scheme 70

The Ireland–Claisen rearrangement has been used extensively in synthesis. Applications include the synthesis of polyether antibiotics ⟨80JA1155, 83JA1988, 85JA3285⟩, sesquiterpenes ⟨80JA6891, 81JA2446, 84CC827, 89CC1665⟩, alkaloids ⟨93TL3985⟩, steriods ⟨90CC1001⟩, iridoids ⟨82JA4030⟩, tetronates ⟨86JOC635, 88JA854, 90TL4433⟩, marine natural products ⟨89JA2735⟩, amino acids ⟨82JOC3933, 89CC1812⟩, carbocycles ⟨80JA6889, 83HCA2608, 84JOC4320, 85TL1163, 88JOC4923, 88TL1371, 89JA8231, 89JOC9⟩, nitrogen heterocycles ⟨85JOC707, 86JOC3247, 86T2831, 87CC1220, 87TL3031, 89TL515, 91JCS(P1)705⟩, oxygen heterocycles ⟨90TL5285⟩, macrocyclic lactones ⟨83HCA2608⟩, C-glycosides ⟨79CJC1743⟩, substituted cyclohexanes ⟨85JOC5179, 89JOC9, 94TL1317⟩, chiral stannanes ⟨88JOC5189, 90TL869⟩ silanes ⟨85JOC5179, 86JCS(P1)161, 88JOC5189⟩ and numerous other examples ⟨88T3079, 90TL1977, 91JOC353, 91TL3209, 93JOC5873⟩.

Other notable applications include the total synthesis of ebelactone A, wherein the ester (**123**), is treated under the Ireland–Claisen conditions to generate an (*E*)-*O*-silylketene acetal which rearranges to give the ester (**124**), with excellent stereocontrol ⟨90TL7513⟩. The reaction has also been developed into a double Ireland–Claisen reaction for stereocontrolled two-directional chain synthesis ⟨91TL7601⟩. Complementary to this, iterative enolate Claisen rearrangements have proved to be a versatile route to synthetically useful 2,7-nonadiene-5-carboxylic acids, proceeding with high enantioselectivity ⟨93JOC4589⟩. The methodology lends itself well to the stereoselective synthesis of contiguous quaternary centres ⟨80T419, 86JOC4485, 90JA5601⟩ as illustrated by the rearrangement of ester (**125**) under Ireland conditions to give predominantly (**126**), a precursor of trichodiene ⟨93JOC6255, 94T1651⟩. The Ireland–Claisen rearrangement has also been used in syntheses involving the ring contraction of lactones ⟨88JA3298⟩. Knight and co-workers have used this strategy for the enantiospecific total synthesis of (−)kainic acid via Ireland–Claisen ring contraction of lactone (**127**) to the pyrrolidine (**128**) in 55% overall yield (Scheme 71) ⟨87CC1220, 87TL3031⟩.

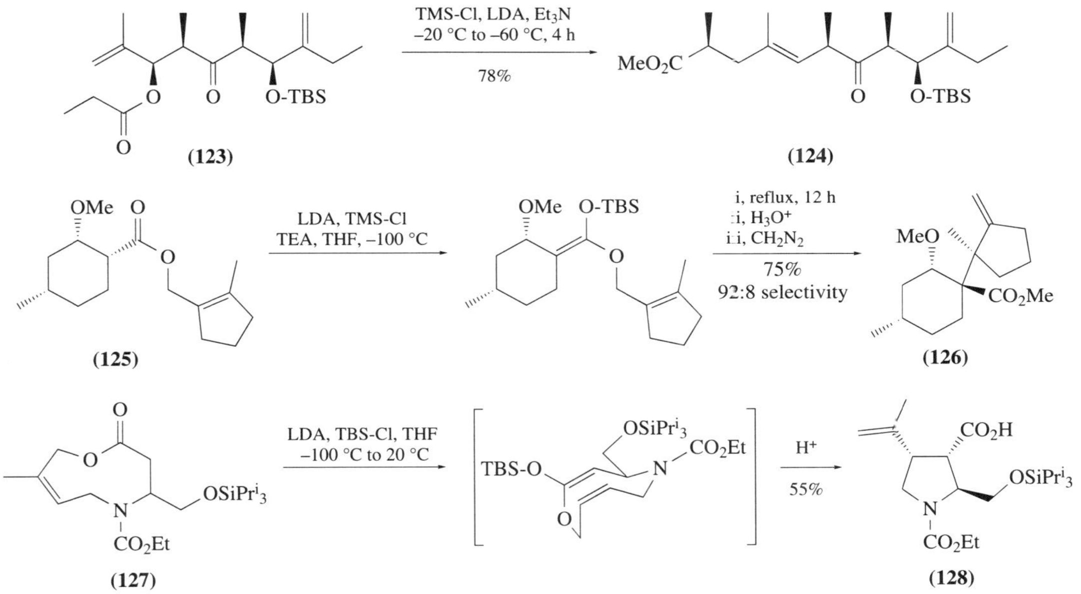

Scheme 71

Several other enolates also participate in Ireland-type [3,3]-sigmatropic rearrangements and indeed the use of zinc enolates derived from α-halo esters in the Reformatsky–Claisen rearrangement was reported shortly after Ireland's original report ⟨73CC117, 88TL3291⟩. The selective formation and rearrangement of (*E*)-enol dialkylboronates has also been reported ⟨89JA3441⟩; the process has been extended to afford an enantioselective and diastereoselective Ireland rearrangement of achiral allylic esters utilising a recyclable chiral boron reagent (**129**) (Scheme 72) ⟨91JA4026⟩. A procedure for effecting the Claisen rearrangement of allyl esters or lactones via the corresponding dialkoxyphosphinyl ketene acetals has also been reported as being more facile than the analogous Ireland rearrangement ⟨93JA8847⟩.

(e) The Carroll rearrangement. The Carroll variant of the Claisen rearrangement is the base-catalysed or thermal rearrangement of *β*-keto esters and allylic alcohols to alkenic ketones with concomitant decarboxylation (Scheme 73) ⟨40JCS704, 40JCS1266, 41JCS507, 43JA1992⟩. The intermediates

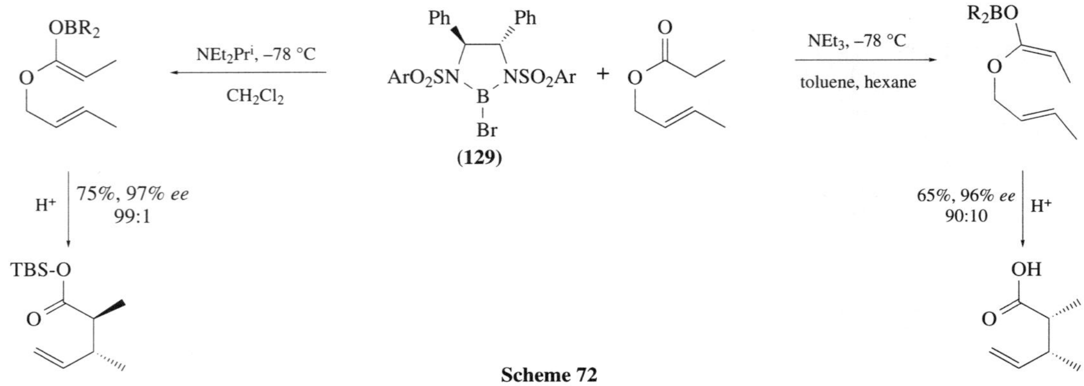

Scheme 72

in the Carroll reaction are easily prepared with well-defined double-bond configuration, and many synthetic applications of the reaction have been reported ⟨69LA(729)52, 69TL3253, 71LA(747)60, 72JCS(P1)367, 74HCA771, 80JA862, 81SC237, 82TL109, 83JOC2122⟩, including some of industrial relevance ⟨75PAC527⟩. The reaction has been improved further with the discovery of the ester enolate Carroll rearrangement ⟨84JOC722⟩, where dianions of allylic acetoacetates are rearranged at elevated temperatures to yield β-keto acids. This methodology has been used in an approach to the Prelog–Djerassi lactone framework wherein the dianion generated from β-keto ester (**130**) undergoes rearrangement and decarboxylation to give cyclopentanone (**131**) (Scheme 74) ⟨93TL1149⟩. The reaction has also been used in the synthesis of the sesquiterpene isocomene ⟨88JOC4508⟩, and a modified silyl ketene acetal variant has been used for the diastereoselective formation of contiguous quaternary centres ⟨88T7587⟩.

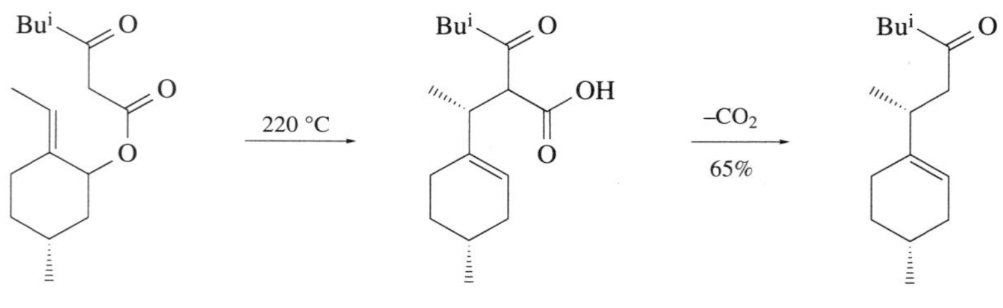

Scheme 73

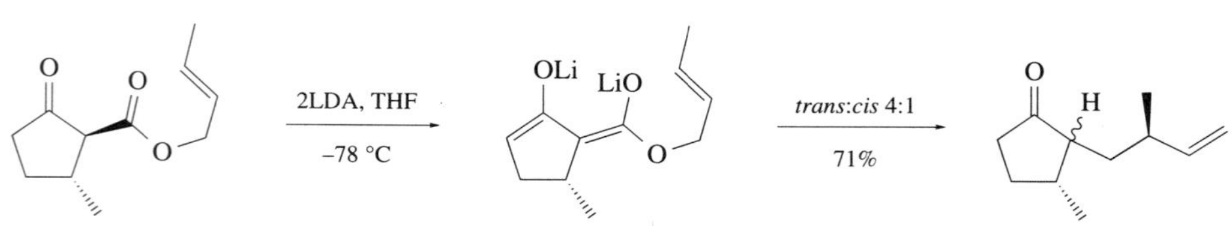

(**130**) (**131**)

Scheme 74

(*f*) *Charge-accelerated Claisen rearrangements.* Since the first report of a carbanion-accelerated Claisen rearrangement ⟨82JA4972, 83JOC3369, 84TL1543⟩ considerable work has been performed on this and other charge-accelerated rearrangements ⟨85JOC4664⟩. Several anion stabilising groups have been utilised and the most successful have proved to be arylsulfonyl groups ⟨89JA8878⟩, phosphine oxides and phosphonates ⟨81JCS(P1)2127, 91JOC1003⟩ as well as phosphonamide groups ⟨91JOC5063⟩; in contrast, ethoxycarbonyl- , cyano- , and cyanohydrin-groups do not promote the rearrangement ⟨89JA8878⟩. Examples are given in Scheme 75, including an asymmetric example utilising the phosphorus-stabilised carbanion (**132**) ⟨87JOC5742, 89TL2469⟩.

Remarkable rate accelerations are observed for the rearrangement of substituted allyl cyclohexenyl ethers (**133**). The thermal rearrangement of ketone (**133a**) is slow, having a half-life of 340 h,

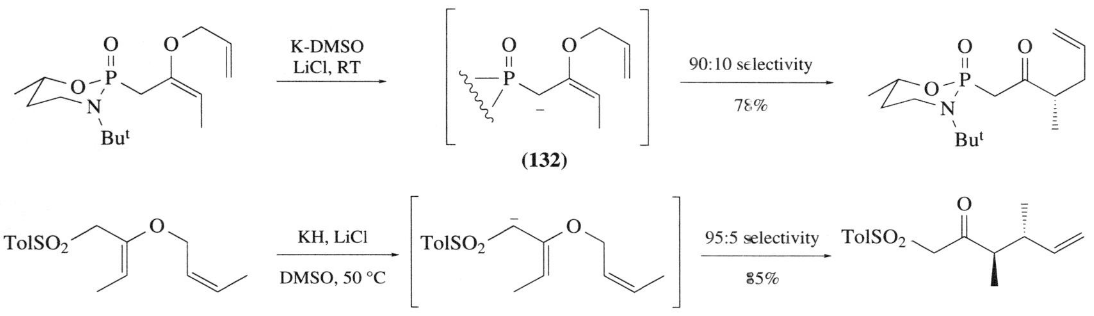

Scheme 75

whilst the rearrangement of carbomethoxyhydrazone (**133b**) and its sodium salt (**133c**) are considerably accelerated with half-lives of 22 h and 1.5 h, respectively ⟨80JOC4413, 80TL4803, 83JOC3866, 83TL3⟩. The metal enolate (**134**) ⟨85JA5572⟩ also undergoes accelerated rearrangement, and this methodology has also found synthetic application ⟨88TL4229⟩; related accelerated rearrangements of vinyl fluorides ⟨74BSF2072⟩ and enamines are also known ⟨84CC1427⟩. Though not necessarily a charge-accelerated reaction, the ketene Claisen rearrangement of allylic ethers (**135**) also proceeds with good yields (Scheme 76) ⟨78HCA3096, 83JOC860⟩.

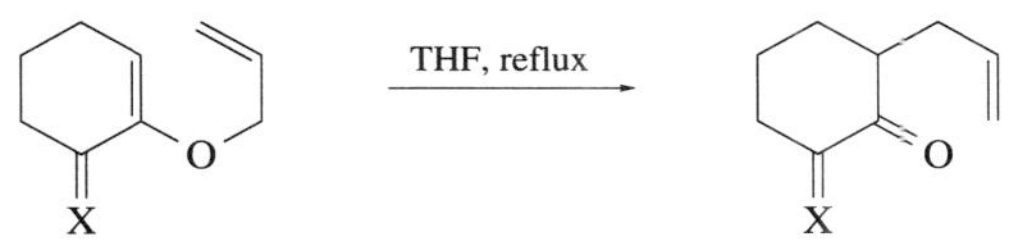

(**133**) **a**; X = O
b; X = $NNHCO_2Me$
c; X = $NN^- CO_2MeNa^+$

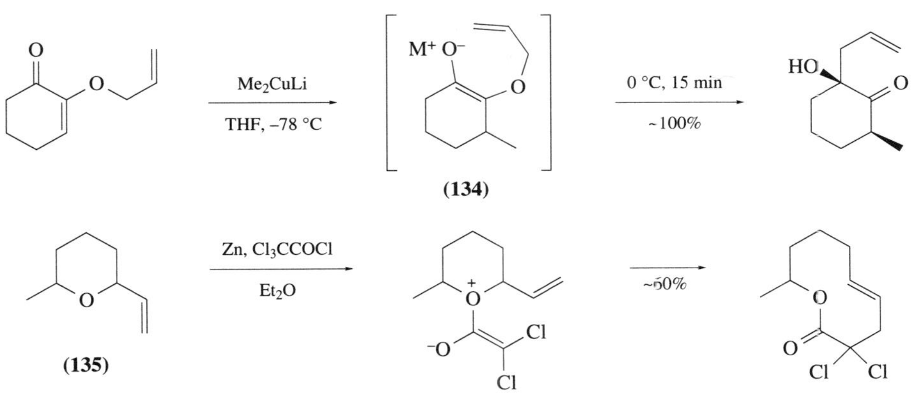

Scheme 76

(ii) Thia-Claisen rearrangement

The [3,3] sigmatropic rearrangement of allyl vinyl thioether, the thia-Claisen rearrangement (commonly referred to as the thio-Claisen rearrangement) is a facile process which has many parallels with the parent reaction ⟨79PS(7)69⟩. The acyclic and aromatic rearrangements occur under relatively mild conditions; however, the thioaldehydes that are formed are unstable and undergo further reaction to generate cyclised products. Thermal rearrangement of allyl vinyl thioether (**136**) leads to the formation of bicyclic product (**137**); if, however, the reaction is performed in the presence of an acid anhydride, the intermediate thioketone is trapped, to give (**138**), and the cyclisation step is averted (Scheme 77) ⟨87TL6461⟩. An alternative strategy is to conduct the reaction under conditions which will effect hydrolysis of the thioaldehyde to the corresponding aldehyde ⟨70JA5522, 72CC62, 73JA2693⟩; many other examples of the reaction are known ⟨74JOC1575, 77JOC172, 79PS(7)69, 81JA1757, 86H(24)3347, 87JOC370, 89CC414, 90TL523⟩. Similar reactions have been reported for allyl vinyl sulfoxide rearrangements to give γ,δ-unsaturated sulfines, which again undergo hydrolysis to furnish the corresponding aldehydes ⟨85JA6731, 91JCS(P1)3199⟩; the corresponding sulfone rearrangement to afford unstable γ,δ-unsaturated sulfenes is also known ⟨76JA3312⟩.

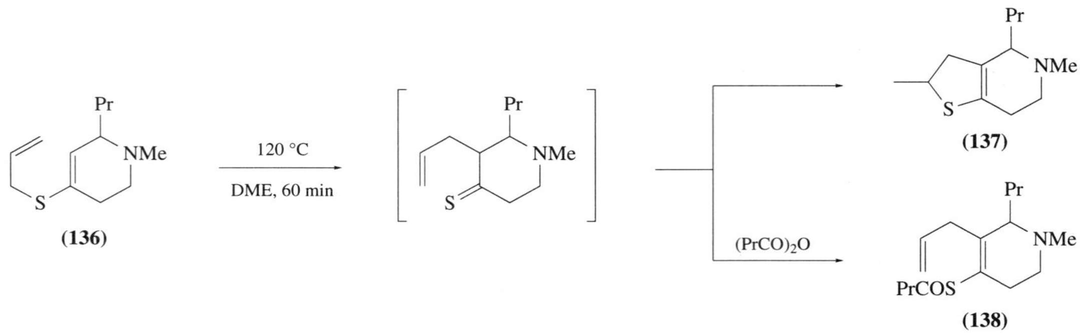

Scheme 77

Reports have highlighted chirality transfer and asymmetric induction in the reaction; for example, the diketene thia-Claisen rearrangement of allylic sulphide (**139**) proceeds with complete 1,2-induction via the transition state shown in Scheme 78 ⟨81JOC2228, 84JOC1840, 89TL3517, 91AG(E)1465⟩. Correspondingly high levels of *syn*-stereoselectivity have been observed for the rearrangement of allyl ketene dithioacetals ⟨91T6275⟩ and, in the thia-Claisen rearrangement of *S*-crotyl ketene dithioacetals such as (**140**), it is possible to control the formation of three contiguous stereogenic centres ⟨92T10315, 93T3131⟩. *S*-Allylation of thioamides, for example (**141**), followed by treatment with base effects the formation of allyl vinyl thioether (**142**) which undergoes rapid rearrangement under remarkably mild conditions (Scheme 78) ⟨80JA2392⟩; attempts at developing an asymmetric version of this reaction have also been reported ⟨87JA6716⟩.

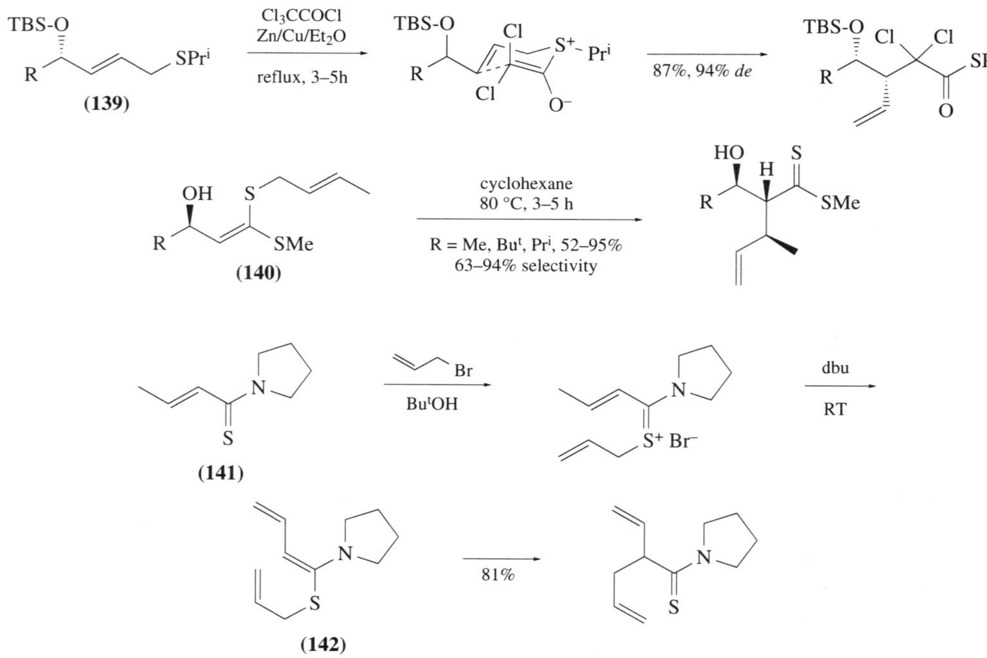

Scheme 78

1.18.2.6.2 *Where Z = N; the aza-Claisen rearrangement and related processes*

Many examples of the aza-Claisen rearrangement have been reported and have proved useful in synthesis. The aliphatic aza-Claisen reaction is known, requiring high temperatures to effect rearrangement ⟨75OR(22)1⟩; however, the use of both Brönsted and Lewis acids results in effective catalysis of the reaction. For example, the rearrangement of allylic enamines (**143**) to imines (**144**) is easily effected under mild conditions using a range of catalysts (Equation (26)) ⟨78TL4337, 83TL1021, 92JOC7188⟩. The 3-aza-Cope rearrangement of *N*-alkyl-*N*-allylanilines has also been the subject of extensive study ⟨57JOC1418, 71TL4661, 73HCA105, 77HCA978, 84TL3159⟩, and this process is again easily promoted by Lewis acid catalysis which generally lowers the temperature required for the reaction by over 100 °C ⟨93JOC5095⟩.

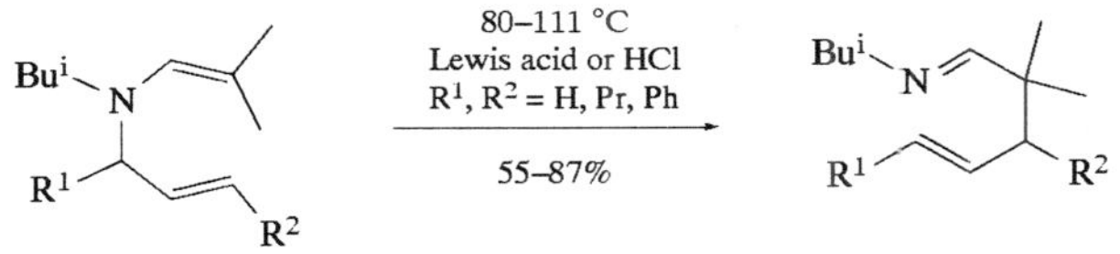

(143) **(144)** (26)

A powerful extension of the rearrangement is found in the rearrangement of homochiral *N*-alkyl-*N*,*O*-ketene acetals, for example (**145**); this intermediate undergoes rapid rearrangement under very mild conditions with excellent diastereocontrol ⟨85JA443, 85JOC5769, 86JOC1377, 87TL1031, 88S362⟩. A further variant is the zwitterionic aza-Cope rearrangement which involves the treatment of tertiary amines, for example (**146**), with propiolate esters to effect the formation of the zwitterionic enamine intermediates (**147**) which undergo facile rearrangement to give enamines (**148**) (Scheme 79) ⟨83JOC4262⟩. Many examples of this reaction are known and it has found extensive use in alkaloid synthesis ⟨84JOC2708, 84TL2303, 89JOC2893, 90JA7682, 90JCS(P1)1493⟩. The rearrangement of *N*-allyl ket-enimines (**149**), readily generated from *N*-allyl amides, provides a useful synthesis of *β*,*γ*-unsaturated nitriles ⟨91TL179⟩ and several alternative methods are also available for the generation of the intermediates required for the reaction (Scheme 80) ⟨91JOC970, 91TL4041, 91TL7233⟩. A 3-aza-Ireland–Claisen rearrangement has also been reported ⟨90TL727⟩.

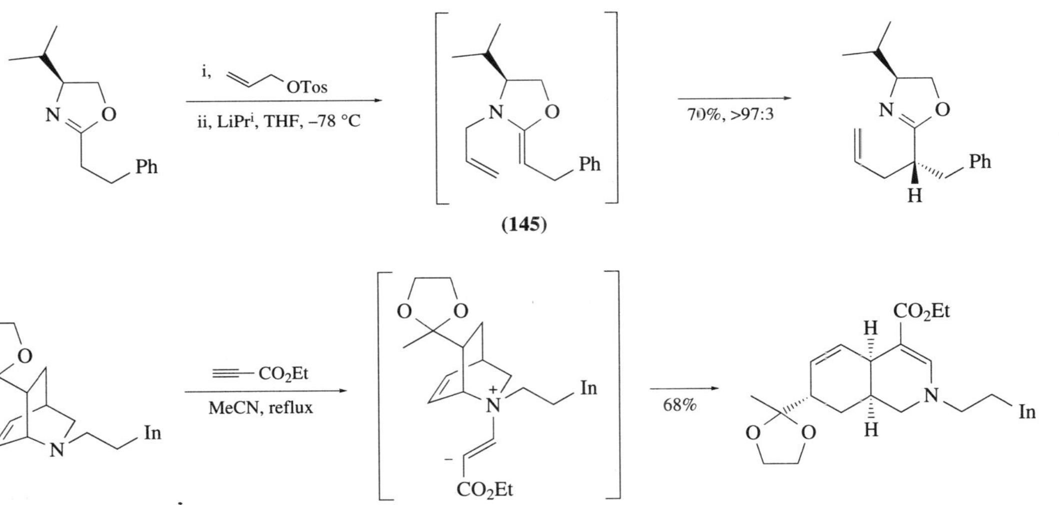

(146) **(147)** **(148)**

In = indole

Scheme 79

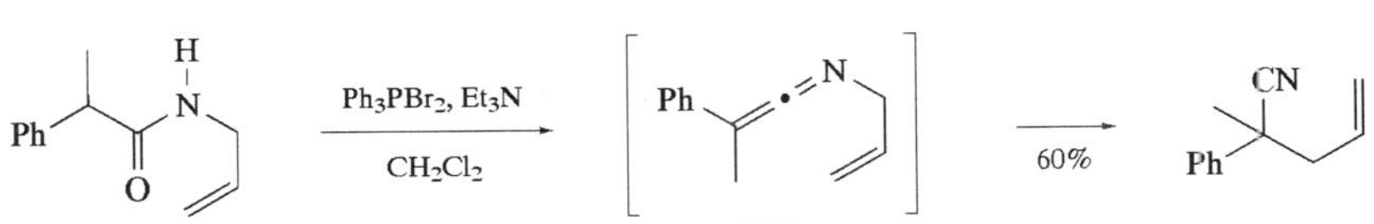

(149)

Scheme 80

1.18.2.7 Other Heteroatom Variants

Many heteroatom variants of the Claisen rearrangement (3-hetero-Cope rearrangement) have been reported. Possibly the most remarkable of these is the metallo-Cope rearrangement ⟨86TL1039, 86TL1043, 86TL4427, 86TL4431, 91TL5969⟩. For example, lithiation of *Z*-vinyl iodide (**150**) followed by treatment with crotyl magnesium bromide and transmetallation with zinc bromide generates the organozinc intermediate (**151**); this then undergoes rearrangement to the stable bimetallic species (**152**) which on hydrolysis gives the product (**153**) with excellent diastereocontrol (Scheme 81). Examples of the 3-phospha- ⟨86CL501⟩ and 3-sila-Cope ⟨73JOC3658⟩ rearrangements are also known.

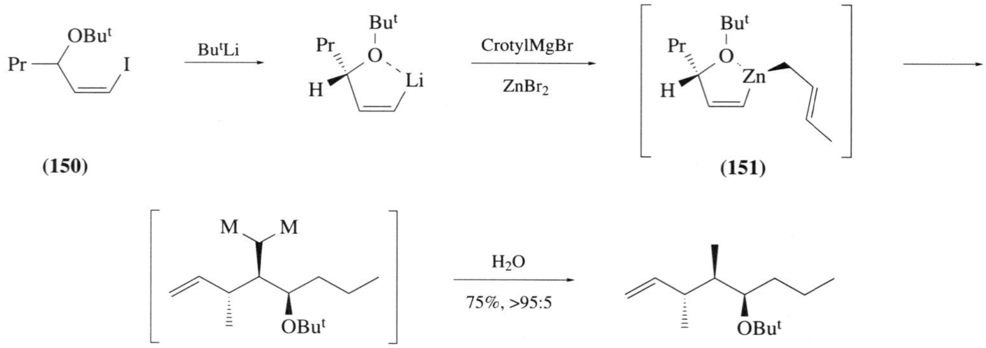

(**150**) (**151**) (**152**) (**153**)

Scheme 81

1.18.2.8 Where Y = Z—C—Z′

A variety of 1,3-diheteroanalogues of the [3,3]-sigmatropic Cope rearrangement are known and have been reviewed ⟨80T3, 84AG(E)579, 84CRV205, 88CRV1423, 89S71⟩.

1.18.2.8.1 Where Z and Z′ = chalcogen

The [3,3]-rearrangement of allylic acetates is a palladium catalysed ⟨80JA7587, 84AG(E)579, 90JOC4614, 93JCS(P1)2913⟩ or thermal process: this process normally leads to equilibrium mixtures which slightly favour primary allylic acetates where possible. Either reaction usually proceeds with complete transfer of absolute or relative stereochemistry as illustrated by the rearrangement of the acetates (**154**) or (**155**) (Scheme 82) ⟨90JOC4614, 93JCS(P1)1849⟩. The related mercury(II) catalysed rearrangements of allyl-*N*,*N*-dimethylcarbamic esters is also of considerable synthetic importance ⟨78JA4822⟩. *O*-Allylic dithiocarbonates, for example (**156**), also undergo facile rearrangement ⟨87TL5973⟩ and this particular example, which is catalysed via a *β*-cyclodextrin complex, proceeds to give the isomeric *S*-allylic dithiocarbonate (**157**) in a remarkable 46% *ee* ⟨91TL7557⟩. Examples of the reaction are also known to effect ring expansions ⟨91CPB1659⟩ and a thiolcarbonate variant has been used in pheremone synthesis ⟨93T7657⟩. The corresponding rearrangement of allylic dithiocarbamates has found considerable synthetic application and a typical example is the thermal rearrangement of (**158**) (Scheme 83) ⟨90TL4155⟩.

1.18.2.8.2 Where Z and/or Z′ = nitrogen

The rearrangement of trichloroimidates (Overman's methodology) has been used extensively in synthesis ⟨80ACR218⟩. For example, the thermal rearrangement of trichloroimidate (**159**) generated

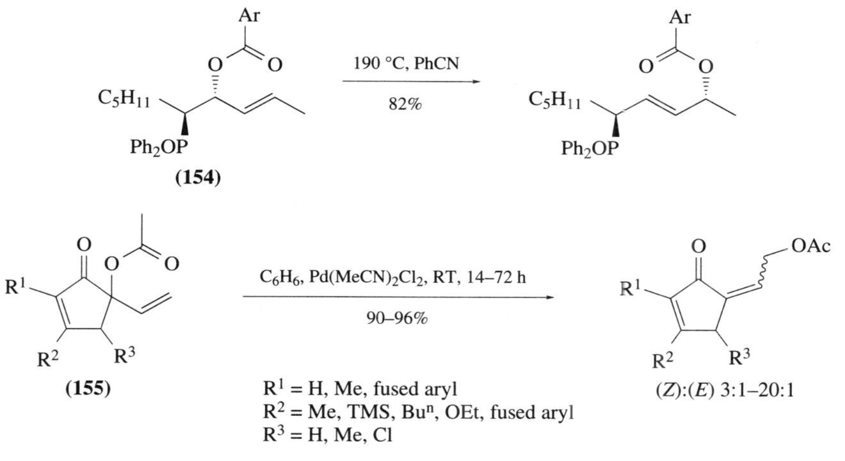

Scheme 82

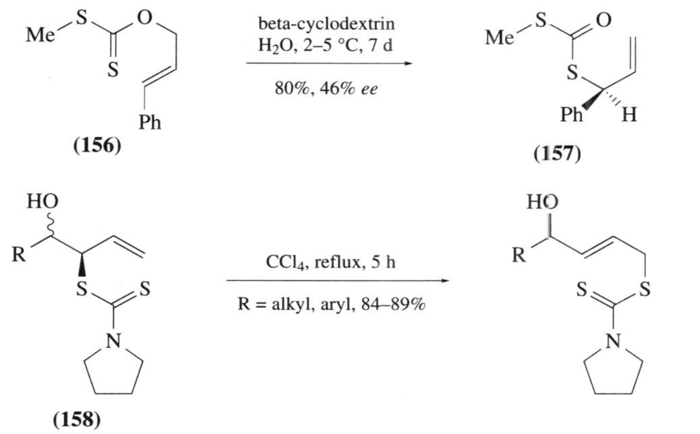

Scheme 83

the amide (**160**) with complete transfer of chirality; the product subsequently being used for the total synthesis of the alkaloid (*R*,*R*)-crinan ⟨93T5177⟩. The reaction is also facilitated by palladium catalysis ⟨84AG(E)579, 93S729⟩ and generally proceeds with complete stereocontrol, although exceptions are known ⟨92TL4313⟩. The related rearrangement of *O*-tetrahydropyranylhydroximates, for example (**161**), has also been reported (Scheme 84) ⟨94TL15⟩. *N*-Allyl-2-thiones are also formed from the [3,3]-sigmatropic rearrangement of 2-allylthioimadazoles ⟨68ZOR1114⟩. The preparation of allylic isocyanates (**163**) has been reported using the sigmatropic rearrangement of allylic cyanates (**162**) (Scheme 85) ⟨91SL238⟩; a 1,3-diphospha-Cope rearrangement has also been reported ⟨87CB131⟩.

1.18.2.9 Where Y = Z

The acid-catalysed rearrangement of allylic alcohols is a process of some importance in organic chemistry ⟨49JCS396, B-63MI 118-04, B-64MI 118-02⟩. The reaction is typified by the sulfuric acid-catalysed rearrangement of allylic alcohol (**164**) to its conjugated isomer (**165**) ⟨45JCS90⟩, as was used in the original synthesis of vitamin-A (Equation (27)) ⟨47HCA489, 47HCA1911⟩. In favourable cases the use of Brönsted or Lewis acids can effect the transformation cleanly and in high yield; however, many side reactions can also occur including elimination (to produce dienes), skeletal rearrangements, cyclisations and polymerisation. These problems have led to the development of other methodology

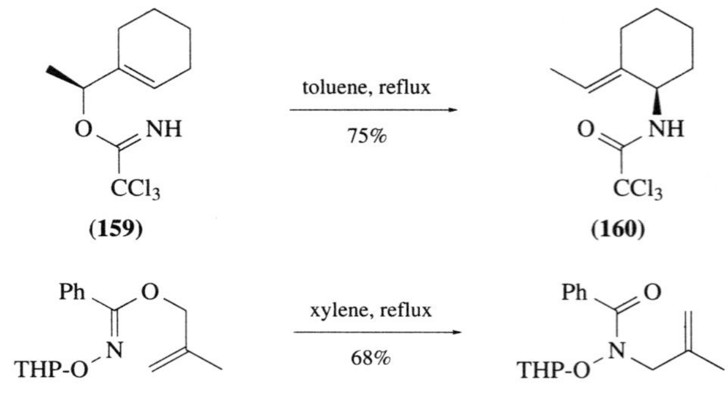

(159) (160) (161)

Scheme 84

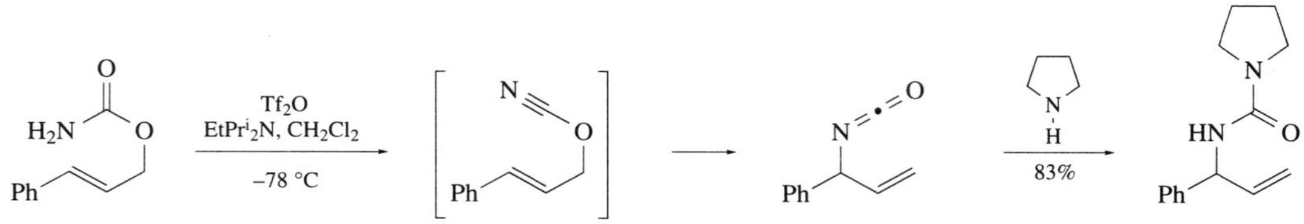

(162) (163)

Scheme 85

for effecting oxygen transposition including metal-catalysed rearrangements of allylic acetates (see Section 1.18.2.8.1) and related systems ⟨78JA4822⟩. [1,5]-sigmatropic rearrangements of oxygen (ethers) and nitrogen functions (tertiary amines) are also known ⟨76CRV187⟩.

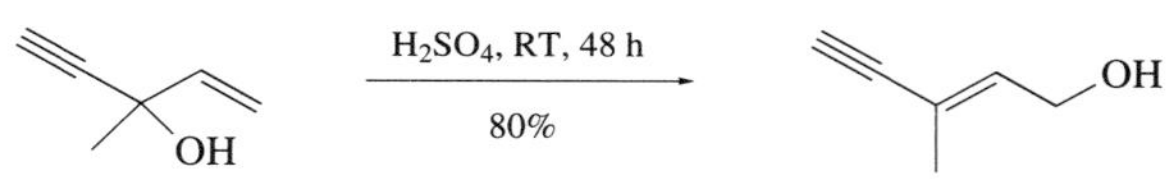

(164) (165) (27)

The [1,3]-sigmatropic migration of boron is a facile process ⟨68ACR152, B-80MI 118-02⟩ and this leads to a fleeting existence for simple allyl boranes. The rearrangement is accelerated by Lewis acids ⟨74PAC505⟩ and is inhibited by complexation of the borane with amines ⟨77JOM(132)9⟩. Brown has reported the facile rearrangement of α-substituted allyl boronate esters **(166)** to the more thermodynamically stable primary boronates **(167)** (Equation (28)) ⟨90TL7115⟩. The thermal rearrangement of allylic halides has been observed ⟨59JCS2720⟩, and some [1,5]-sigmatropic migrations are also known ⟨76CRV187⟩. The thermal [1,3]-rearrangement of allylsilanes is known to proceed with inversion at silicon, favouring the most highly substituted alkene product but leading to mixtures of stereoisomers ⟨72JA2515, 73JA8678⟩. Tin and germanium analogues of this process are also known, as are some [1,5]-rearrangements of these systems ⟨76CRV187⟩.

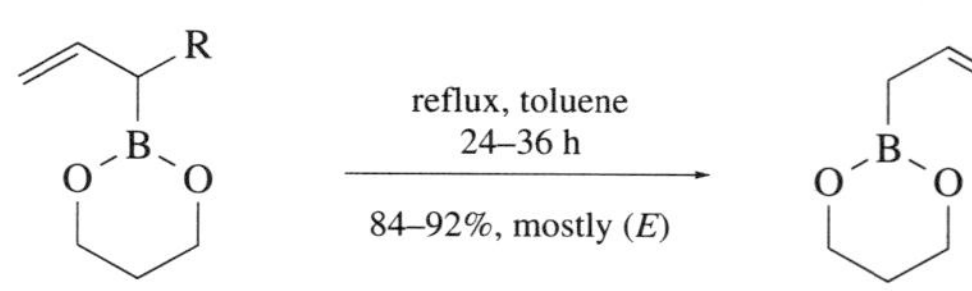

(166) (167) (28)

1.18.2.10 Where Y = Z—Z′

1.18.2.10.1 Where Z, Z′ = chalcogen

The rearrangement of allylic peroxides via intermediate peroxy radicals is catalysed by free-radical intermediates or light and has been shown to proceed by a concerted [2,3]-process ⟨90JA1266⟩;

the reaction was first observed by Schenck ⟨58LA(618)202⟩ for cholesterol hydroperoxides but has limited synthetic use.

Perhaps the most important rearrangement in this area is the reversible [2,3] sulfoxide–sulfenate rearrangement. This reaction is of considerable importance in the synthesis of allylic alcohols by *in situ* decomposition of the sulfenate to destroy the equilibrium process ⟨B-88MI 118-01⟩. A representative example of each reaction is shown in the tandem [2,3]–[2,3] rearrangement of the diester **(168)** (Scheme 86) ⟨82TL3463⟩. A similar process is also observed for the [2,3] sigmatropic selenoxide–selenate rearrangement; an integral process in the selenium dioxide oxidation of allylic carbon centres ⟨76OR(24)261⟩.

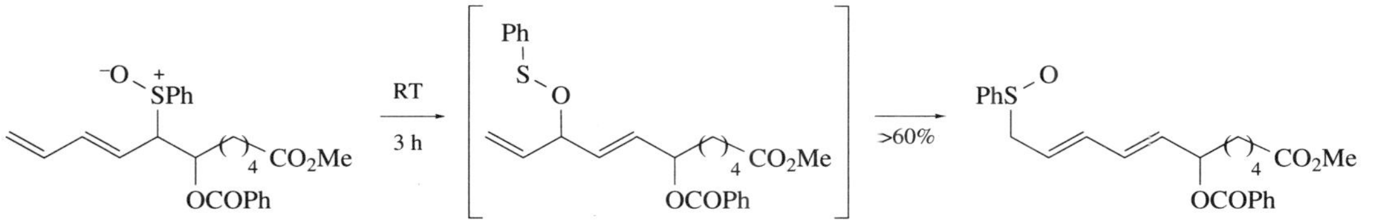

(168)

Scheme 86

1.18.2.10.2 Where Z and/or Z′ = nitrogen

The [2,3]-sigmatropic rearrangement of amine oxides to the corresponding hydroxylamine has been shown to be a particularly facile process ⟨44JA1929⟩ and proceeds with exceptionally high levels of asymmetric transfer ⟨73CC848⟩. For example, oxidation of homochiral γ-*N*,*N*-dibenzylamino-α,β-unsaturated esters **(169)** effects formation of an intermediate amine oxide **(170)**, which undergoes low temperature [2,3]-sigmatropic rearrangement with complete 1,3-transfer of chirality ⟨91TL4481⟩. The reaction has also been used in a nonasymmetric sense for the preparation of naturally occurring terpene alcohols ⟨90BCJ1328⟩. In a related process allyl ethers of oximes, for example **(171)**, are also known to undergo a reversible formal [2,3]-shift with the corresponding *N*-allyl nitrone **(172)** via radical pair intermediates, the example given being best effected using palladium catalysis ⟨91TL279⟩; a sulfur heterologue of this reaction is also known ⟨91JA3526⟩. The Stevens-like [2,3]-rearrangement of allylic aminoamidates, for example **(173)**, has also been observed (Scheme 87) ⟨83JCS(P1)1029⟩.

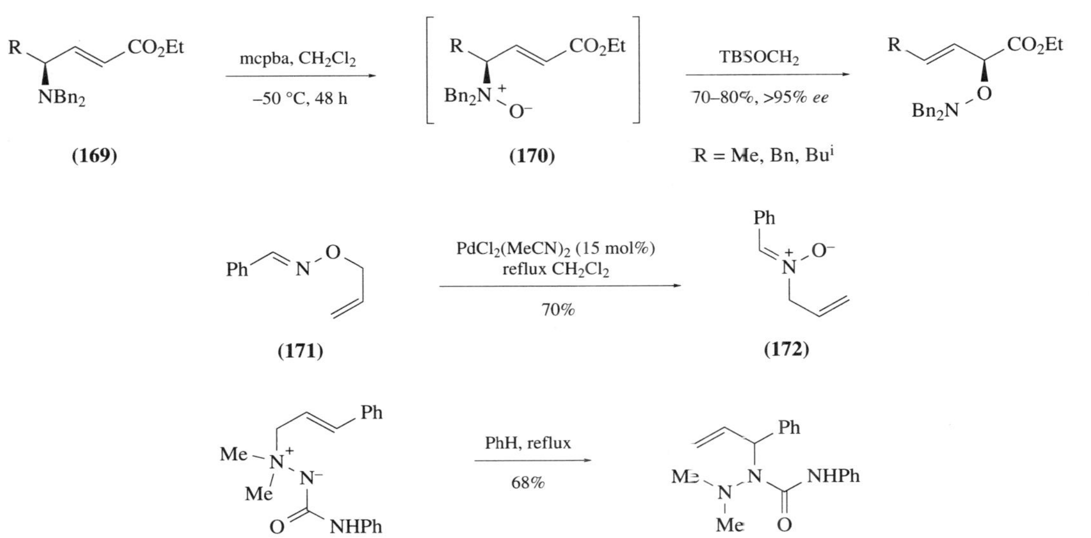

(169) **(170)** **(171)** **(172)** **(173)**

Scheme 87

1.18.2.11 Where Y = Z—Z—Z

A duality of mechanisms is possible in trihetero-[3,3]-rearrangements, for example, the rearrangement of the allylic-*p*-nitrophenylsulfonyl derivative (**174**) is thought to proceed via a concerted [3,3]-sigmatropic process involving a chair-like transition state ⟨91JOC1014⟩, whereas the related rearrangement of allylic thionophosphates involves ion-paired intermediates (Equation (29)) ⟨91JA4895⟩.

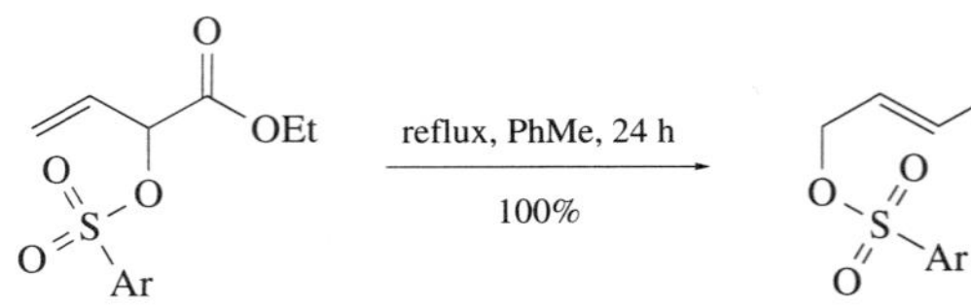

(**174**)

(29)

1.18.3 OTHER REARRANGEMENTS

1.18.3.1 Tandem and Higher Sigmatropic Rearrangements

The field of tandem sigmatropic rearrangements is one of considerable interest and represents a powerful synthetic application of rearrangement methodology ⟨91COC(5)875⟩. The *para*-Claisen rearrangement is an early example of the tandem Claisen–Cope rearrangement ⟨26LA(449)81⟩ (see also Sections 1.18.2.4.1.(i) and 1.18.2.6.1.(i)(b)), and this process has also found considerable application in aliphatic systems, as exemplified by the synthesis of *β*-sinesal (**175**) ⟨69JA3281⟩. The Cope–Claisen sequence is also known. For example, the substrate (**176**) is transformed into the aldehyde (**177**) as the major product; the latter compound was subsequently used in natural product synthesis (Scheme 88) ⟨82JA7181⟩.

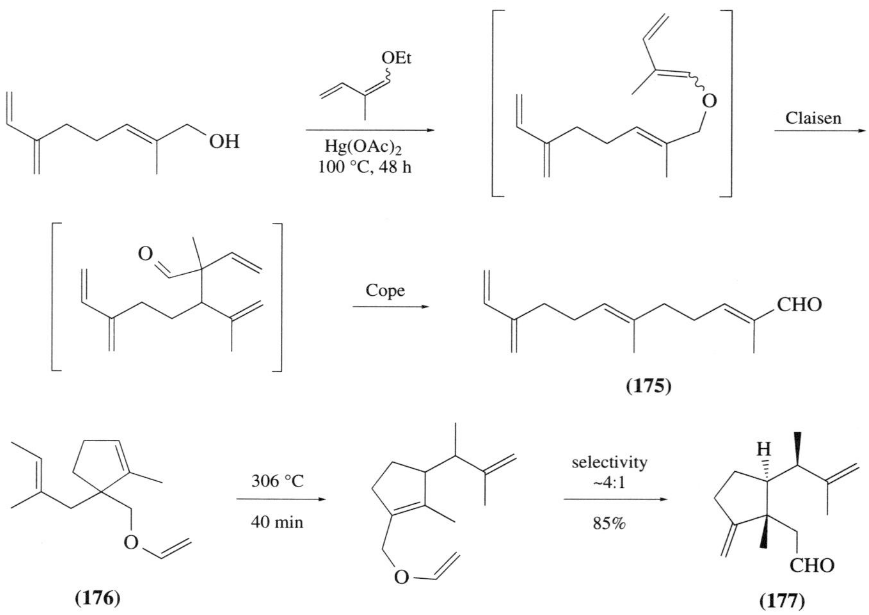

Scheme 88

Numerous higher order sigmatropic rearrangements are known and have been the subject of many theoretical studies; however, they have found limited application in synthesis ⟨B-66MI 118-01⟩. A [3,5]-sigmatropic rearrangement has been put forward to rationalise the rearrangement of vinylcyclopropane (**178**) to dihydroindene (**179**) ⟨86CC1807⟩, and a [5,5]-sigmatropic process is thought to occur in the oxy-Cope rearrangement of (**180**) (Scheme 89) ⟨81T3967, 81TL2471⟩. A [7,7] rearrangement has also been reported ⟨85TL1057⟩.

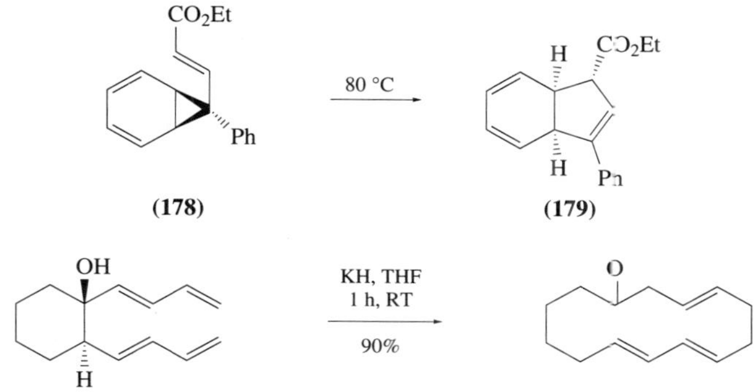

(**178**) (**179**) (**180**)

Scheme 89

1.18.3.2 Rearrangements That Involve Ring Opening

1.18.3.2.1 Ring opening of cyclopropanes

Many rearrangement processes are known in cyclopropane chemistry. These can be effected under a variety of conditions including thermal, photochemical or metal-catalysed ⟨B-87MI 118-04⟩. The photochemical irradiation of arylcyclopropanes provides an example of a typical rearrangement which in general leads to a mixture of many products, including variable amounts of substituted propenes (Scheme 90) ⟨79MI 118-01⟩. Indeed, yields of products are not generally good or predictable in such reactions, as several mechanistic pathways are known to operate ⟨83JOC2202, 88JA3601, 88JOC2706, 88JOC2711⟩. An example of a Rh(I)-catalysed rearrangement is also given in Scheme 90 ⟨72JA7757, 73TL2727⟩.

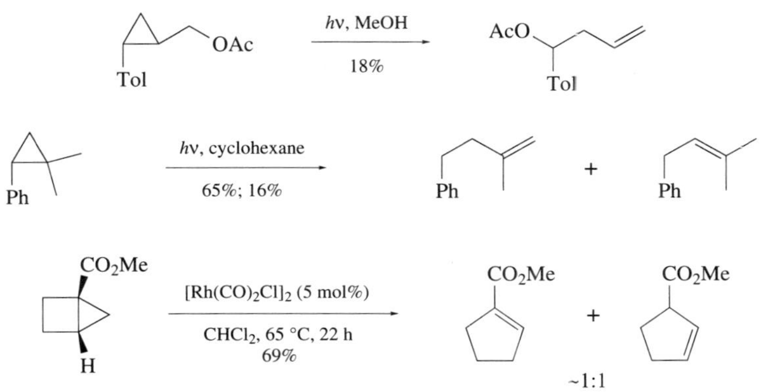

Scheme 90

1.18.3.2.2 Photochemical ring opening of cyclic ketones

The photochemical ring cleavage of cyclic ketones to yield *ω*-unsaturated aldehydes is a useful, if somewhat unpredictable, process ⟨70TL1175, B-80MI 118-03⟩. Synthetic examples of note include the ring opening of macrocyclic ketone (**181**) ⟨77TL763⟩ and the formation of medium ring lactones from epoxy ketone (**182**) ⟨73CC223⟩; a reaction using a cyclopropane in place of the epoxide has been used in the synthesis of medium ring ketones ⟨69CC1049, 72BCJ1777⟩. The reaction has also found an application in the total synthesis of insect pheromones (Scheme 91) ⟨73TL3459⟩.

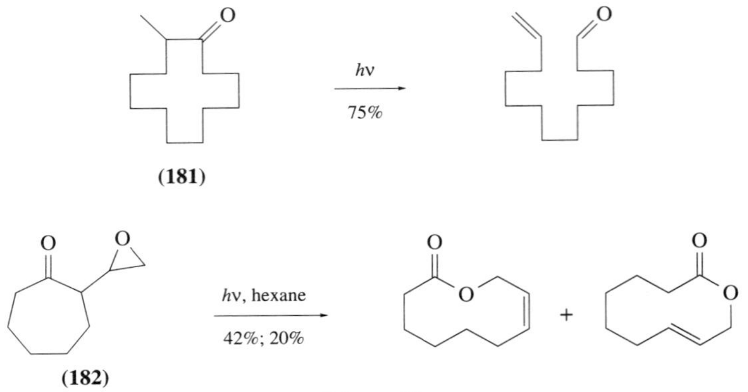

Scheme 91

1.18.3.2.3 *Rearrangements of epoxides*

Several fragmentation reactions of epoxides result in the formation of unsaturated alcohols. The rearrangement of epoxides to allylic alcohols has been well documented and occurs under a variety of conditions ⟨83T2323, 84S629⟩. Base-mediated rearrangement of epoxides is a common process and the use of the hindered bases such as LDA can result in excellent regioselectivity, for example, in the rearrangement of epoxide (**183**) (Scheme 92). Homochiral bases have also been used in the enantioselective rearrangement of meso-epoxides ⟨91TA1, 93CC884⟩. Fragmentation with ring opening is also known and perhaps the most elegant application is found in the two epoxide rearrangements reported by Holton ⟨84JA5731, 84TL4455⟩. Treatment of β-patchouline oxide (**184**) with boron trifluoride etherate effects rearrangement with a [1,2] carbon shift to give homoallylic alcohol (**185**); epoxidation of (**185**) leads to the intermediate (**186**) which rearranges *in situ* with ring fission, to give the taxane AB-ring model (**187**); extension of this chemistry led to the first total synthesis of taxol ⟨94JA1597, 94JA1599⟩.

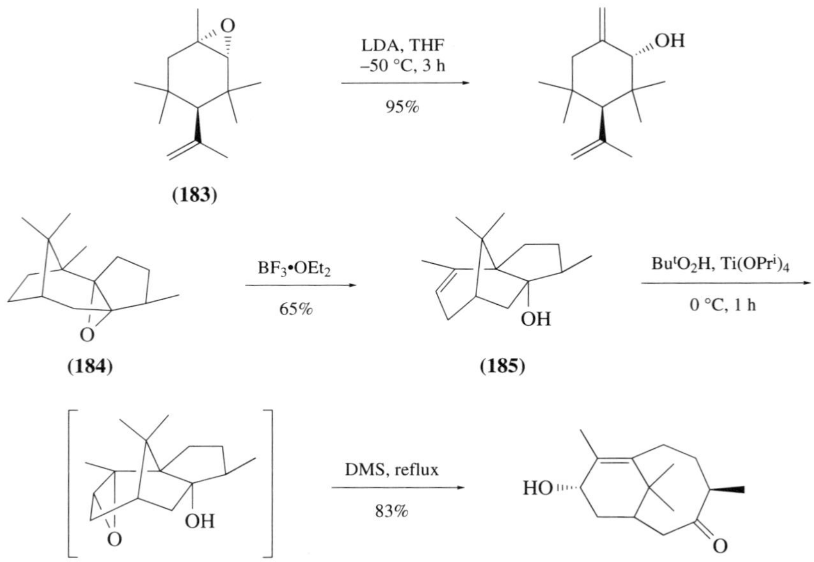

Scheme 92

1.19 Tricoordinate Anions, Cations, and Radicals

JULIAN O. WILLIAMS
University of Exeter and White Knight Chemicals Ltd, Paignton, UK

and

MICHAEL J. KELLY
University of Exeter, UK

1.19.1 TRICOORDINATE CARBANIONS

1.19.1.1 General Aspects of Carbanion Formation

1.19.1.1.1 General literature survey

Since the 1960s, a number of monographs have been published covering general aspects of carbanion chemistry ⟨B-80MI 119-01, B-83MI 119-03, B-84MI 119-01, B-87MI 119-01⟩. Cram discussed in general the relationship between carbanion structure and reactivity ⟨B-65MI 119-01⟩. Synthetic applications were specifically described by Stowell ⟨B-79MI 119-01⟩. Streitwieser and Hammons surveyed the estimation of hydrocarbon acidity by kinetic and thermodynamic methods ⟨65MI 119-06⟩. Exchange reactions in liquid ammonia have been reviewed ⟨63APO(1)155⟩. Albery ⟨67MI 119-07⟩, Eigen ⟨64AG(E)1⟩ and Grunwald ⟨65MI 119-07⟩ dealt with rapid proton-transfer reactions. Fischer and Rewicki considered the chemistry of acidic hydrocarbons ⟨68MI 119-05⟩. Carbocyclic nonbenzenoid aromatic carbanions have been reviewed by Lloyd ⟨B-66MI 119-01⟩. The subject of saturated (sp^3) carbanions adjacent to both heterocyclic and nonheterocyclic nitrogen has been reviewed ⟨88PIA187, 92T2589⟩. General reviews on carbanions have appeared frequently ⟨B-68MI 119-01, 68MI 119-06, 69FCF387, B-73MI 119-01, B-73MI 119-04, 78MI 119-01, 78RCR425, 79COC(2)411, B-78MI 119-01, 81MI 119-01, 85MI 119-01, B-92MI 119-01⟩. The subject of carbanions in general and electrophilic aliphatic substitution in particular has been reviewed at yearly intervals, the latest in the series being in 1991 ⟨B-91MI 119-01⟩.

1.19.1.1.2 *Different types of carbanion*

Carbanions may be divided broadly into two classes ⟨B-87MI 119-02⟩:

(i) Short-lived reactive intermediates in solution in basic conditions, found in base-catalysed deuterium exchange, ElcB eliminations and α-eliminations. In these cases, another substance in the solution (often the solvent) acts as proton donor; such a carbanion exists in equilibrium with its conjugate acid or is present only as a stationary-state intermediate and is probably in low concentration with a brief lifetime.

(ii) Carbanions, or the closely related organometallic compounds, which occur in solution at higher concentration with longer lifetimes. The solvent must be markedly less acidic than the conjugate acid of the carbanion and not vulnerable to nucleophilic attack. Typically anhydrous ether proves to be a suitable solvent for such carbanions.

The occurrence of carbanions in moderate to high concentration in aqueous solution is somewhat limited because a C—H bond will dissociate appreciably in aqueous solution only if substituted by two or three strongly electron-withdrawing groups. However, dipolar aprotic solvents such as acetonitrile, hexamethylphosphoramide (HMPA) or DMSO can sustain significant concentrations of less strongly stabilised carbanions. In such solvents carbanions may sometimes exist as free ions, but more often they will exist in the form of solvent-separated or contact ion pairs. In the less-acidic solvents, such as THF, 2-Me-THF, DME and the higher glymes, diethyl ether and pentane, strong ion pairing of a carbanion with its gegenion will be normal. When not stabilised by an electron-delocalising group, the anion is likely to bond covalently to the cation, forming a polar covalent compound.

In organometallic compounds a carbanion provides a σ-electron pair and is said to be a σ-donor ligand of the metal, as in for example alkyllithium and alkyl magnesium compounds. Certain carbanions also bond strongly to transition metals, particularly those in low oxidation states, through a combination of σ- and π-bonding, with the ligand carbanion donating an unshared pair to form a σ-bond (or in the case of alkenes, one or more pairs of π-electrons to give $d\pi$–$p\pi$-bonding) to the metal and receiving electron density from filled metal d-orbitals into empty π^*-orbitals to give $d\pi$–$p\pi^*$-bonding. Such complexes undergo a variety of reactions useful in synthetic organic chemistry.

1.19.1.1.3 *Simple alkyl carbanion formation*

Heterolytic cleavage of one of the bonds of a tetrahedrally coordinated carbon atom in molecule (**1**) can occur in such a way as to leave a net negative charge on the resultant tricoordinate carbon atom (**2**), (Equation (1)).

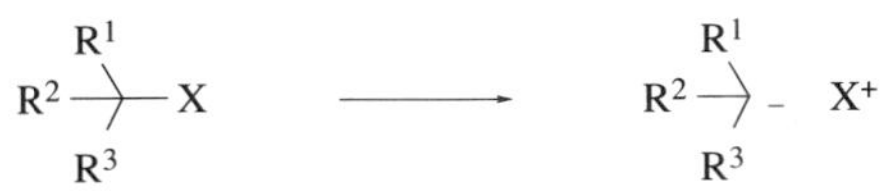

(**1**) (**2**) (1)

Although the evidence for the existence for such species as (**2**) is overwhelming, they are usually highly unstable, transient and not given to direct observation. Their existence and physical characteristics can frequently only be inferred by examination of the products or kinetics of a given reaction ⟨B-92MI 119-02⟩.

Every carbanion possesses an unshared pair of electrons and is therefore, by definition, a Lewis base. When a carbanion (e.g., **2**) (Equation (2)) is protonated, it is converted into its conjugate acid (e.g., **3**). The stability of the carbanion is proportional to the strength of the conjugate acid. The weaker the acid, the greater the strength of the base and the lower the stability of the carbanion. Determination of the strengths of the conjugate acids immediately determines the order of stability of a series of carbanions.

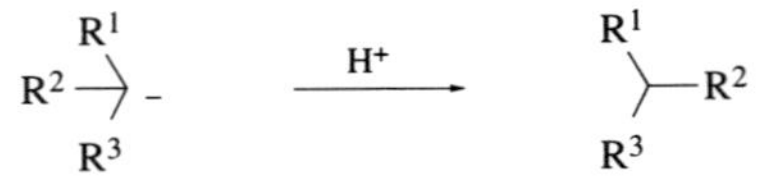

(2) **(3)** (2)

The aliphatic C—H bond is relatively strong and largely covalent ⟨B-60MI 119-01⟩. The removal of a proton from a simple alkane consequently proves difficult. Consideration may be given to the equilibrium between methane and its conjugate base in basic medium (Equation (3)). A number of factors combine to make the methyl anion an energetically unfavourable species, and this equilibrium lies overwhelmingly to the left. Not least of the factors which destabilise the methyl anion is the absence of any means to delocalise the negative charge.

$$CH_4 + B^- \rightleftharpoons CH_3^- + BH \qquad (3)$$

There is no doubt that carbanions derived from simple alkanes are very unstable in solution. There is even some evidence that simple carbanions such as ethyl and isopropyl degenerate to radicals through the loss of an electron ⟨86TL4411, 88JA607⟩. However, indirect methods have enabled a study to be made of these species. Applequist and O'Brien investigated ⟨63JA743⟩ the position of the equilibrium for the reaction in ether and ether/pentane (Equation (4)).

$$R^1Li + R^2I \rightleftharpoons R^1I + R^2Li \qquad (4)$$

The rationale underlying these experiments was that the R group which forms the more stable carbanion would be more likely to be bonded to lithium rather than iodine. Carbanion stability was found by this method to be in the order: vinyl > phenyl > cyclopropyl > ethyl > *n*-propyl > isobutyl > neopentyl > cyclobutyl > cyclopentyl. Later, using a similar method, Dessy *et al.* set up the equilibrium, shown in Equation (5), treating a number of alkyl magnesium compounds with various alkyl mercury compounds in THF solution ⟨66JA460⟩.

$$R^1{}_2Mg + R^2{}_2Hg \rightleftharpoons R^1{}_2Hg + R^2{}_2Mg \qquad (5)$$

The group of greater carbanion stability is associated with the magnesium cation. The order of carbanionic stability, determined by this method was: phenyl > vinyl > cyclopropyl > methyl > ethyl > isopropyl, which was in broad agreement with the work of Applequist and O'Brien. It may be concluded that the stability of alkyl carbanions decreases in the order: methyl > primary > secondary. Dessy *et al.* were unable to obtain data for the *t*-butyl anion, but there can be little doubt that it is even less stable. This observed order of stability may be attributed to inductive effects transmitted through the σ-framework. The $+I$ methyl groups α to the isopropyl carbanion serve to increase charge density at the central carbanionic atom, leading to a decrease in stability. The results of Applequist and O'Brien show that β-branching also reduces carbanion stability. The cyclopropyl carbanion seems anomalously stable, but this may be viewed as due to ring strain promoting a large amount of *s*-character in the C—H bond.

There is no direct evidence as to the structure of unsubstituted alkyl carbanions, since such species have not been isolated. However, it is plausible that the charge-carrying carbon atom may be regarded as sp^3-hybridised with the lone pair of electrons directed towards one vertex of the tetrahedron, as shown in structure (**4**) (Equation (6)). Such carbanions may be regarded as isoelectronic with amines. When $R^1 \neq R^2 \neq R^3$ in structure (**4**) the carbanion may be viewed as chiral. However, such species undergo lone-pair pyramidal inversion with concomitant racemisation in much the same way as tertiary amines (Equation (6)).

(4) (6)

Typically, Letsinger ⟨50JA4842⟩ found that in light petroleum/diethyl ether 2-octyllithium racemised slowly at $-70\,^\circ C$, but within minutes at $0\,^\circ C$. Contrastingly, Curtin and Koehl reported that the compound required hours to racemize at $-6°C$ in pentane ⟨60CI(L)262⟩. The catalysis of the racemisation by diethyl ether may be attributed to the coordination of the oxygen atom to the lithium atom to promote ion-pair formation by the normally covalent organolithium compound (Scheme 1).

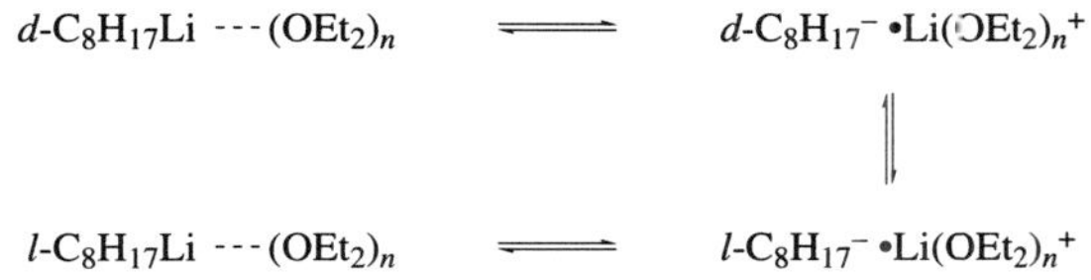

Scheme 1

Carbon atoms at bridgeheads form carbanions relatively easily and stable bridgehead carbanions are known ⟨66MI 119-07⟩. The pyramidal structure of unsubstituted alkyl carbanions is further supported by Streitwieser *et al.* ⟨69JA529⟩ and Peoples and Grutzner ⟨80JA4709⟩.

1.19.1.1.4 Carbanions stabilised by conjugation

When a multiple bond is located α to the charge-bearing carbon atom, the carbanion (**5**) (Equation (7)) is stabilised by conjugation between the unshared pair of electrons and the π-electronic system of the multiple bond.

(5a) **(5b)** (7)

This phenomenon is responsible for the stability of the allylic ⟨B-70MI 119-01⟩ and benzylic ⟨74JA5708⟩ carbanions (see Equation (8) and Scheme 2 respectively).

(8)

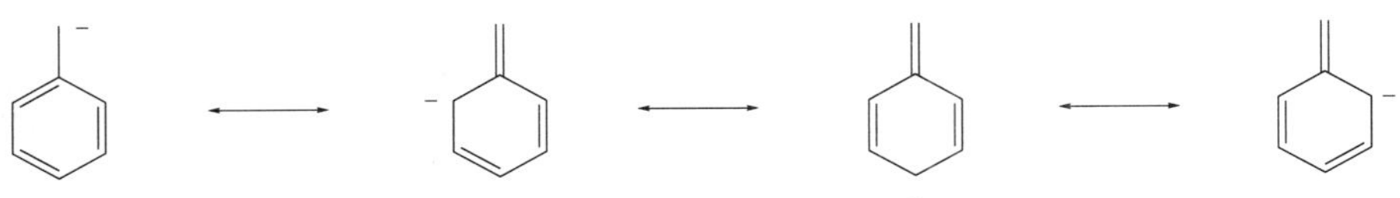

Scheme 2

Diphenylmethyl and triphenyl anions are even more stable than the benzylic anion and can be kept in solution indefinitely, provided anhydrous conditions are rigorously maintained ⟨B-80MI 119-02⟩. The diphenylmethyl and triphenyl carbanions enclosed by crown ethers have been the subject of x-ray crystallographic analyses ⟨85JA2174⟩.

Methyl and alkyl anions have proved too reactive to have any well-characterised solution chemistry, and even when the gegenion is a very electropositive metal like lithium, the carbon–metal bond is largely covalent. Most of the existing knowledge of carbanion chemistry has been derived from the study of more stable species. It is certainly of importance to quantify relative carbanion stability ⟨B-71MI 119-01⟩. A suitable fundamental approach has been to measure acid–base equilibria, in which both components are carbon acids. By selecting anions that are highly coloured, McEwen, Conant and Wheland, working in the 1930s, were able to measure their concentrations spectrometrically and even colorimetrically. They were able to draw up a hydrocarbon pK_a scale measured in ethereal solutions. Typically, a solution of the orange-coloured sodium salt of diphenylmethane, $Ph_2CH^-\cdot Na^+$, on treatment with triphenylmethane, was gradually converted into diphenylmethane by proton transfer and the solution assumed a red colour due to the predominance of the triphenylmethyl anion Ph_3C^-. Quantitatively, it was ascertained that triphenylmethane was the more acidic by at least 2 pK_a units. By taking as a standard, at one extreme, a compound whose pK_a could be measured independently, for example in aqueous solution, this spectrometric technique was used to obtain data for a range of hydrocarbon compounds with pK_a values in the range 15 to 30 ⟨32JA1212, 36JA1124, 47JA2025⟩. Simple colorimetric techniques were used to rank hydrocarbons in a series of increasing acidity in ethereal solution ⟨32JA1212⟩.

Subsequently more refined work was carried out by Streitwieser *et al.* in the polar protic solvent cyclohexylamine, which has very low acidity ⟨67JA63⟩. These results served in general to confirm the earlier studies by McEwen *et al.* ⟨36JA1124⟩.

When the negative charge of a carbanionic carbon atom is conjugated with a C=O double bond or a carbon–nitrogen multiple bond (Equation (7); Z = O or N), the carbanion is stabilised to an even greater extent than even triarylmethyl anions, since such electronegative heteroatoms are more able to accommodate the excess negative charge than carbon atoms.

For the anion **(6)** (Equation (9)) the canonical form **(6b)** is the major resonance canonical, contributing more to the hybrid than canonical **(6a)**, because of the high electron affinity of the oxygen atom. Nevertheless, for the purposes of this chapter, resonance hybrids such as that depicted in Equation (9) are regarded as tricoordinate carbanions.

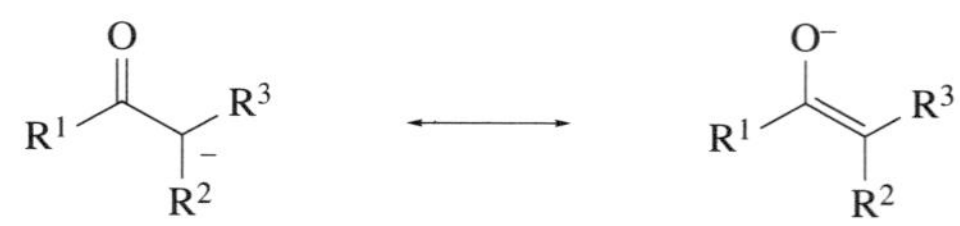

(6a) **(6b)** (9)

Using both equilibrium and kinetic data, a comparison of hydrocarbon acidities may be made (Table 1) ⟨68JA1791, B-71MI 119-02, B-78MI 119-05⟩. It should be noted, however, that even the most acidic compounds tabulated form conjugate anions which are more strongly basic than aqueous sodium hydroxide solution.

1.19.1.2 Carbanions by Proton Loss

1.19.1.2.1 Carbanion formation by proton loss stabilised by hybridization effects

The foregoing discussion in Section 1.19.1.1.3 has given an indication of the difficulties of formation of tricoordinate alkyl carbanions, which do not benefit from resonance stabilisation. There are, however, some factors which can conspire to stabilise such carbanions.

Within the context of the molecular orbital version of the quantum mechanical molecular model, the shapes and distribution of electron density for 2*s*- and 2*p*-carbon orbitals and the characteristics of various hybridised *s*/*p* mixtures may be calculated. As the percentage of *s*-character of a carbon sp^n-hybridised orbital increases, the average distance, measured over time, between electron and nucleus decreases, and the coulombic term relating to stabilisation by electron cloud with carbon nucleus electrostatic interaction increases. Accordingly, the acidity of a C—H bond is enhanced by increasing *s*-character, a phenomenon exemplified by the facile conversion of alkynes into alkynyl carbanions, and one which has wide implications in organic synthesis.

It has been found that the percentage of *s*-character in a C—H bonding orbital plays a major role in determining the NMR spin–spin coupling constant between ^{1}H and ^{13}C. By taking alkyne, ethylene and ethane as reference points, it is possible to derive an empirical relationship between percentage *s*-character and $J_{^{13}C—H}$, as shown in Equation (10).

$$J_{^{13}C\text{–}H} = 500 - 5(100 - \%s_{C\text{–}H}) \qquad (10)$$

Streitwieser *et al.* measured the rates of hydrogen-isotope exchange of a series of C3 to C8 cyclic hydrocarbons in cyclohexylamine containing caesium cyclohexylamide ⟨69JA527⟩ and showed a close correlation with $J_{^{13}C—H}$. Highly strained hydrocarbons are sufficiently acidic to be deprotonated by methyl lithium in ether ⟨65TL287⟩, and reaction rates correlate with $J_{^{13}C—H}$ values. The increase in kinetic acidity with decreasing ring size undoubtedly results from the increase in *s*-character of the carbon orbital in the C—H bond ⟨65MI 119-06⟩. Hydrogen-isotope exchange of cyclobutane with caesium cyclohexylamide in cyclohexylamine in somewhat faster than that of cyclohexane, and cyclopropane reacts faster by a factor of approximately 10^5 ⟨61JCS73⟩.

The highly strained hydrocarbon, tricyclo-[4.1.0.0^{2,7}]-heptane **(7)**, has been found to undergo

Table 1 Approximate pK_a values for various hydrocarbons.

Compound	pK_a	*Compound*	pK_a
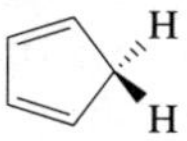	~16	Ph_2CH_2	31[b]
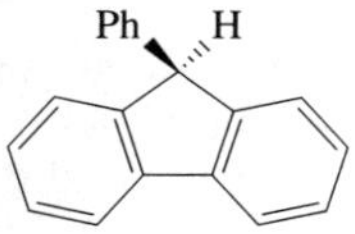	16.4[a]	PhMe	35[b]
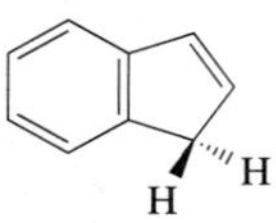	17.8[b]	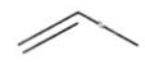	35[c]
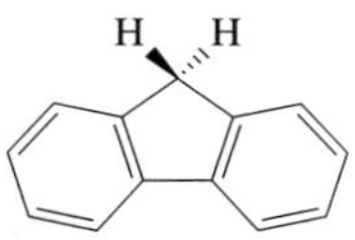	20.6[b]	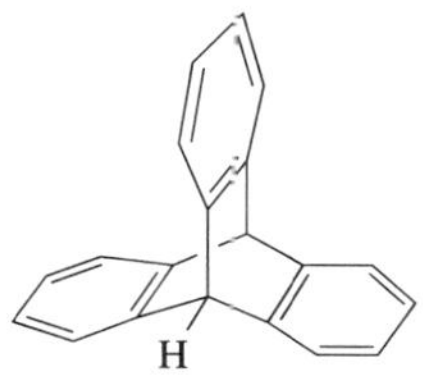	39[c]
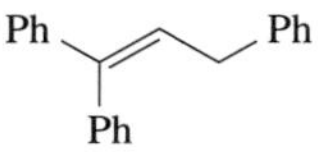	24.4[b]	$PhCHMe_2$	39[c]
Ph_3CH	29[b]	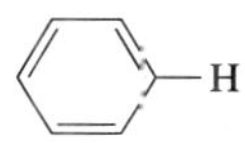	39[c]
			43[c]
		CH_4	44[c]
			48[c]

[a] 'Anchored' in aqueous solution. [b] Obtained from equilibrium data. [c] Extrapolated from isotopic exchange rates.

rapid deuterium exchange with potassium *t*-butoxide in *t*-butyl alcohol at 99.5°C ⟨60JA3652⟩. This remarkable kinetic acidity is presumably a consequence of the high degree of *s*-character in the C—H bond, estimated from the $J_{^{13}C—H}$ coupling constant to be 40%.

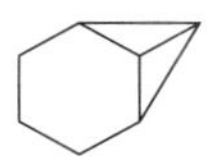

(7)

1.19.1.2.2 Carbanion formation by proton loss stabilised by conjugation

The above discussion implies that the lone-pair orbital of a localised carbanion will be most stable when it has the maximum possible *s*-character. It might be expected that an alkyl carbanion would have strong preference for sp^3-hybridisation over sp^2 in which the lone pair is forced to occupy a *p*-orbital. In fact, a pyramidal configuration with sp^3-hybridisation and C_{3v} symmetry is predicted for the CH_3^- anion theoretically by CNDO-type calculations ⟨68JA2471, 69T4117⟩. Such calculations predict an energy barrier to pyramidal inversion of approximately 15 kcal mole^{-1}. The inversion barriers of substituted, and in some cases conjugated, carbanions have been studied theoretically by Niemeyer utilising CNDO/2 and STO-3G calculations ⟨77T2267⟩, while the so-called 'gauche effect', with its stereochemical consequences for adjacent electron pairs and polar bonds, was reviewed by Wolfe, with respect to *ab initio* molecular-orbital calculations ⟨72ACR102⟩. A planar

sp^2-hybridised trigonal conformation with D_{3h} symmetry was predicted as an energy minimum for the $(N{\equiv}C)_3C^-$ carbanion ⟨76AJC1635⟩.

Consideration of the benzyl anion reveals the operation of two opposing effects. The tendency to maximise the *s*-character of the orbital carrying the negative charge is opposed by the potential energy benefits to be gained by increased delocalisation of the negative charge made possible if the carbanionic lone pair is associated with a *p*-orbital. There is no way of predicting *a priori* which effect will predominate. Recourse to the valence bond model (Scheme 2) indicates the possibility of extensive resonance stabilisation and implies at least qualitatively that $p\pi$–$p\pi$-delocalisation will be a significant stabilising factor. Physical properties of the benzyl anion are consistent with the picture of a highly delocalised species as depicted in Figure 1. Significantly, the ^{13}C-NMR chemical shift and $J_{^{13}C-H}$ (138 Hz for the $PhCH_2$ methylene group, compared with a coupling of 91 Hz for MeLi) ⟨66JA1272⟩ indicate the conformation with the atom centres coplanar is energertically favoured.

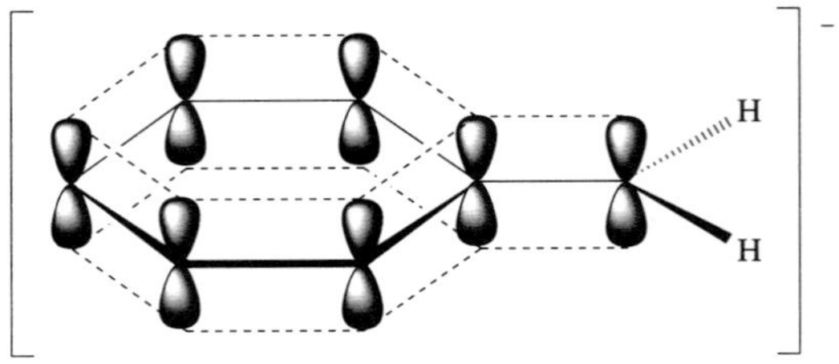

Figure 1 Stabilisation by $p\pi$–$p\pi$ delocalisation of the negative charge in the benzyl anion.

It would appear that there exists a considerable degree of stabilisation attributable to $p\pi$–$p\pi$-electron delocalisation in an sp^2 carbanion of this type. The relevant data in Table 1 indicate that the successive substitution of phenyl groups for hydrogen atoms in methane increases the acidity and decreases the pK_a value in stepwise fashion by 9, 13 and 15 pK_a units. The values imply that the triphenylmethyl anion is 21 kcal mole^{-1} ($RT\log_e 10^{15}$) more stable than the methyl anion at 25°C. It may be seen from the data that proportionately less energetic stabilisation is gained with each sequential replacement of a hydrogen atom by a phenyl group. As has been argued above, the most effective stabilisation occurs when all the atom centres are coplanar, i.e., when the *p*-orbital of the sp^2-hybridised carbanionic carbon atom lies parallel to the *p*-orbitals which constitute the π-clouds of the phenyl ring(s). This condition becomes increasingly difficult to fulfil due to unfavourable steric interactions as the number of phenyl substituents increases. In fact, the triphenylmethyl anion is thought to be propeller-shaped (**8**), a structure which reflects a compromise between energetically favourable conjugative effects and unfavourable steric interactions.

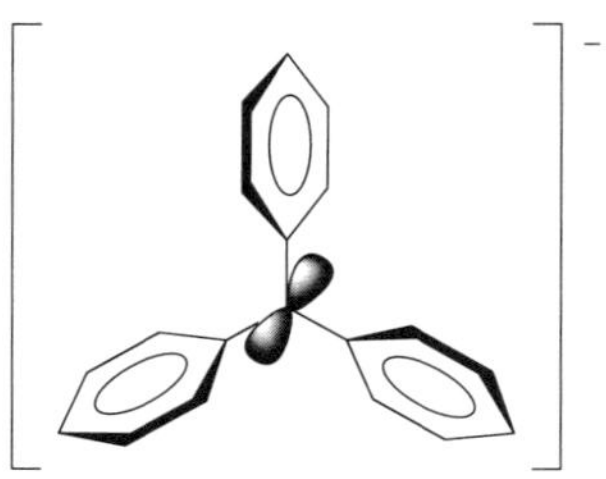

(8)

Should a carbanionic carbon atom be adjacent to a $p\pi$–$p\pi$-bond orbital whose remote conjugated atom is oxygen or nitrogen, a stable system may result. In species such as enolate anions, α-nitro-carbanions, and α-cyano-carbanions, it is to be expected that most of the negative charge will reside on the oxygen or nitrogen atom, each of which is more electronegative than carbon. The pK_a values observed for this type of system are generally much lower than those of analogous hydrocarbons. Pearson and Dillon ⟨53JA2439⟩ measured the acidities of various compounds and obtained the following series of pK_a values: $\mathit{CH_3}NO_2$ 10.2; $\mathit{CH_3}COCH_3$ 20; $\mathit{CH_3}CN$ 25; $\mathit{CH_3}CO_2Et$ 25; $\mathit{CH_3}CH{=}CH_2$ 35 (the active protons are italicised).

The significance of the role of resonance in the stabilisation of anions was quantified by Belanger *et al.* who measured the rates of exchange of α-substituted phenylacetates in alkaline D_2O (Scheme 3) ⟨69CC1067⟩. With a carbanion of this type, the degree of deformation from coplanarity of the atoms involved in conjugation will vary with the steric interactions between the *ortho* protons of the phenyl ring and the substituent R. It is clear from the experimental evidence that the rate of deuterium exchange increases with diminishing steric bulk of the group R. In the case where R = Pr^i there are considerable unfavourable steric interactions.

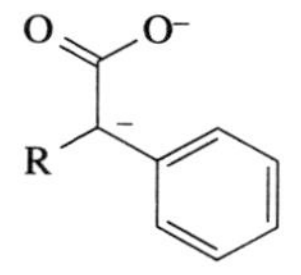

(8)

Relative rates of deuterium exchange as a function of the size of substituent R:

Pr^i = 1; Et = 18; Me = 155; H = 42 000

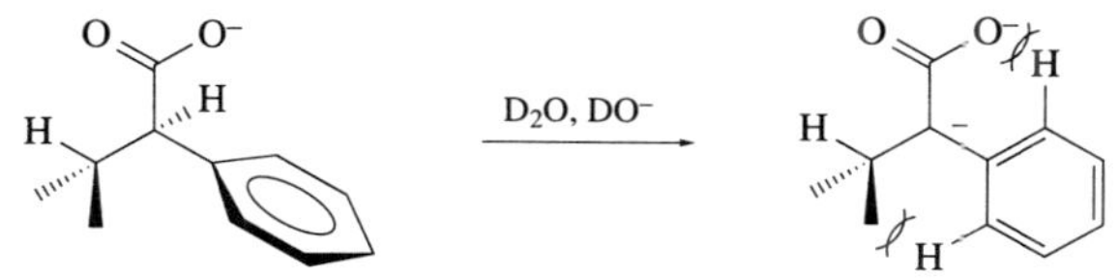

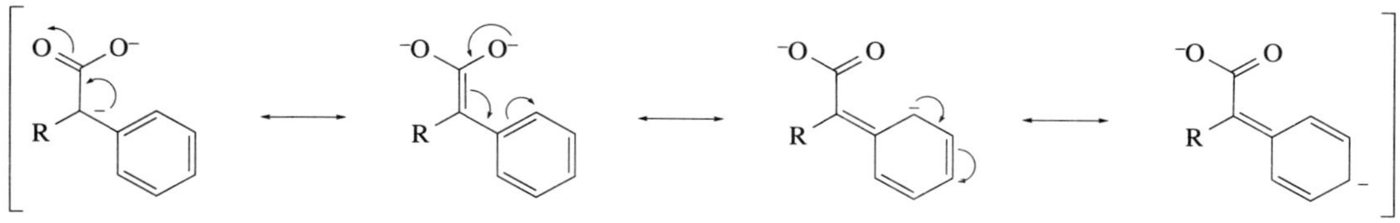

(8a) **(8b)** **(8c)** **(8d)**

Resonance stabilisation of the dianion (**10**)

Variation of steric inhibition of resonance stabilisation in the formation of planar or near-planar carbanions

Scheme 3

The base-catalysed isotope exchange of a hydrogen α to a $-M$ substituent such as a carbonyl or cyano functionality will normally result in the loss of chirality at that site since the intermediate will be planar due to sp^2-hybridisation at the carbanionic carbon centre. However, this general rule does not apply in the case of an acidic proton bonded to either a cyclopropane ⟨62JA2465⟩ or cyclopropene ⟨68JA2698⟩ ring. For these two systems, a nonplanar anion would seem to be preferred because of the extra stabilisation inherent in the small three-carbon ring (see Section 1.19.1.2.1) and also because of the additional bond angle strain needed to attain sp^2-hybridisation in a three-carbon ring. The nitrile (**9**) undergoes deuterium exchange more than 4000 times faster than it racemises in NaOMe/MeOD (Scheme 4).

Consistent with the above observation, a structurally related cyclopropyllithium was shown to be configurationally stable under conditions where optically active alkyllithium compounds rapidly racemise ⟨64JA3283⟩.

1.19.1.2.3 Carbanion formation by proton loss stabilised by induction

Almost all substituents which stabilise carbanions by conjugation also offer significant stabilisation by field and inductive effects ⟨B-71MI 119-03⟩; alkyl substituents interact with carbanions by field and inductive mechanisms alone. Thus alkyl substitution at an sp^3-hybridised carbanion causes very marked destabilisation. Methyllithium is stable for months in ethereal solution, whereas *t*-butyllithium can be kept only in hydrocarbon solvents in the absence not only of oxygen but also of nitrogen. Conversely trifluoromethyl substituents are strongly stabilizing of carbanions and $(CF_3)_3CH$ has an estimated $pK_a = 11$ ⟨B-71MI 119-04⟩. There is no doubt that in this case the nine fluorine atoms combine to stabilise the resultant carbanion by induction as shown in structure (**10**).

Additionally, carbanion (**10**) may be regarded as being stabilised by resonance in the manner

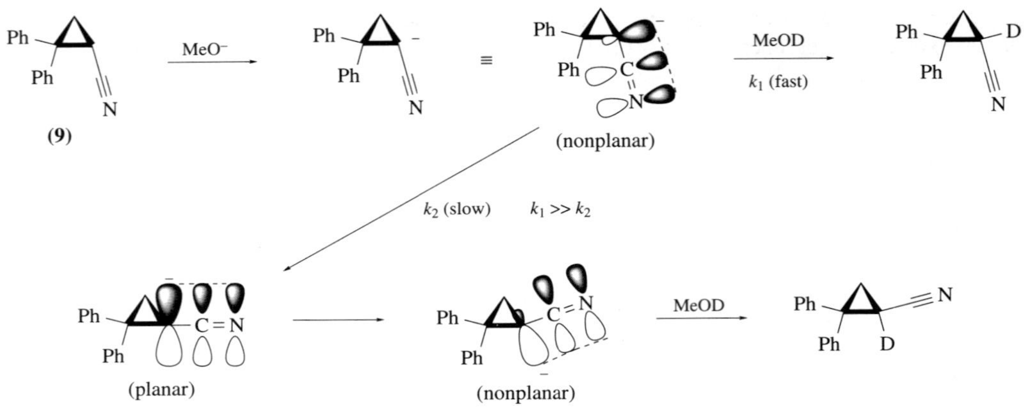

Deuterium exchange and racemisation of a cyclopropyl-nitrile in MeONa/MeOD

Scheme 4

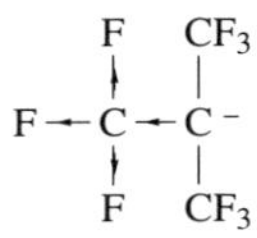

(10)

shown in Scheme 5 ⟨50JA408, B-59MI 119-01, B-60MI 119-02, 79COC(1)525, 83T1141⟩. Von R. Schleyer and Kos refer to this type of effect as 'negative (anionic) hyperconjugation' and have investigated the theoretical rationalisation using molecular orbital theory ⟨83T1141⟩.

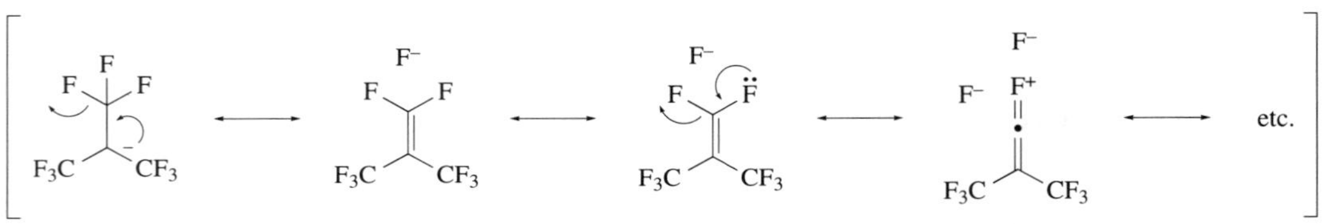

(10a) **(10b)** **(10c)** **(10d)**

Scheme 5

1.19.1.2.4 Carbanion formation by proton loss stabilised by multiple substituents

As might be expected, a multiplicity of suitably positioned substituents, of the requisite type, serve to increase the stabilisation of a carbanion over the corresponding monosubstituted case. When a dilute solution of aqueous sodium hydroxide is added to a solution of a β-diketone (e.g., acetyl acetone (**11**) (Scheme 6), or to a β-keto ester (e.g., ethyl acetoacetate (**13**) (Scheme 7)), proton abstraction from the active methylene group takes place. In both cases, ionisation by loss of a proton is facilitated by two factors: first, the presence of adjacent $-I$ substituents, caused by the high electronegativity of the oxygen atoms and second, that the $-M$ carbonyl groups are able to stabilise the resultant carbanions (**12**) and (**14**) by mesomeric delocalisation of the negative charge.

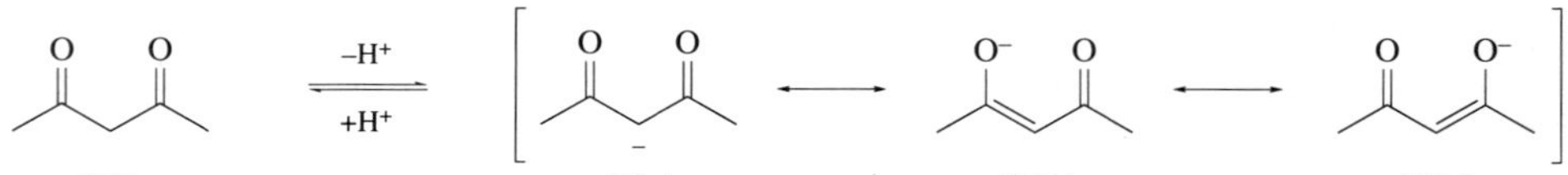

(11) **(12a)** **(12b)** **(12c)**

Scheme 6

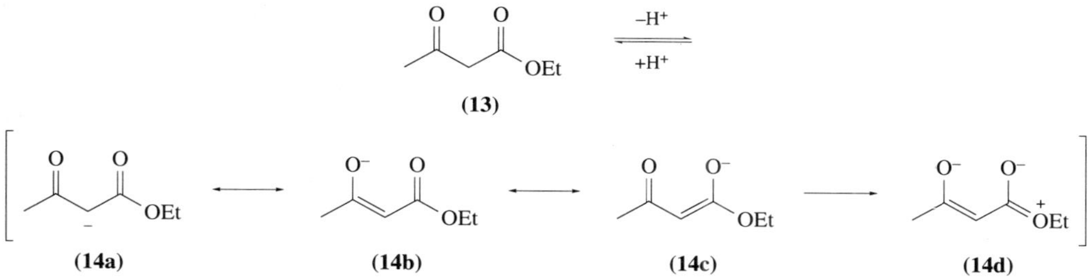

Scheme 7

Substitution of three electron-attracting fluorine atoms for methyl hydrogen atoms in acetyl acetone (**11**) increases its acidity by a factor of 4 pK_a units. Conversely, substitution of a methyl group on the centre carbon of diketone (**11**) (i.e., taking the value for 3-methyl-2,4-pentanedione) lowers its acidity by a factor of 2 pK_a units, an effect partly due to steric inhibition of resonance in the anion, which may find difficulty in attaining coplanarity of the conjugated system due to methyl–methyl interactions. Ethyl acetoacetate (**13**), in which one of the carbonyl groups is bound to an ethoxy group, is a weaker acid than acetylacetone (**11**) by 0.3 pK_a units. Since the ester carbonyl group of carbanion (**14**) acquires electron density from conjugation with the ethoxy group, as shown in canonical form (**14d**), it can absorb a correspondingly smaller amount of negative charge from the carbanionic carbon in the manner represented by canonical (**14c**). Thus the ester carbonyl group is less effective in stabilizing the carbanion than the ketone carbonyl group. Further comparisons of 'activated' dicarbonyl compounds can be made from Tables 2 and 3.

The bridged bicyclic diketone (**15**) is not appreciably more acidic than ordinary monoketones (with pK_a values approaching 19 or 20), since canonical form (**16b**) (Scheme 8) would necessitate a double bond at a bridgehead, in violation of Bredt's rule ⟨24LA(437)1, 40JA2933⟩. The preferred mode of ionisation of diketone (**15**) giving enolate carbanion (**17**) is shown in Scheme 9.

Similarly, although there are three benzene rings α to the C—H bond in trypticine (**18**), the compound is not appreciably more acidic than unsubstituted alkanes ⟨50JA1005⟩. Canonical forms such as (**19b**) (Scheme 10), which might be expected to stabilise the trypticide anion (**19**), are precluded by Bredt's rule ⟨24LA(437)1⟩.

1.19.1.2.5 Kinetics of base-catalysed formation of carbanions

The bromination of almost any enolisable ketone (e.g., (**20**) (Scheme 11) or aldehyde in the position α to the carbonyl group is accelerated by the addition of bases. Such base-promoted halogenations of carbonyl compounds are kinetically similar to reactions subject to general base catalysis ⟨B-59MI 119-02⟩. A simple example ⟨46JCS636⟩ is the bromination of acetone (**20**), promoted by the hydroxide anion giving (**21**; X = Br) via carbanion (**22**). The rate of the reaction is proportional to the concentration of ketone and base, but is independent of the concentration of bromine. Chlorination ⟨46JCS636⟩ and iodination ⟨34JA967⟩ of acetone (**20**) in the presence of hydroxide take place in a similar manner.

If the mechanism proposed for acetone also applies to the bromination of the optically active ketone (**23**) (Scheme 12), a basic solution of this ketone should undergo racemisation in the absence of bromine at the same rate as in the presence of bromine, since the presumed rate-determining step in the bromination reaction, specifically the conversion of ketone (**23**) to the carbanion (**24**), destroys the chiral centre. The rates of bromination and racemisation of ketone (**26**) have been compared in aqueous acetic acid, with acetate ion added as the base, and found to be the same ⟨36JCS623⟩. Since the concentration of anion (**24**) is small, its rate of destruction should equal its rate of formation, once the initial step is underway. When anion (**24**) is generated in D_2O it incorporates a deuterium atom α to the carbonyl group, to give (**27**). It has been found that, within the limits of experimental error, the rates of racemisation and deuteration of ketone (**23**) are equal in D_2O/dioxane mixtures catalysed by the presence of NaOD ⟨38JCS78⟩. It is worthy of note that the specific rates of deuteration of ketone (**23**) in D_2O/dioxane and of racemisation in H_2O/dioxane are found to be different. The strength of a given base is slightly different in each medium, because their solvation characteristics are not identical.

Table 2 Rate of transfer of protons to water at 25 °C ⟨53JA2439⟩.

Compound	k_1 (min^{-1})	*Compound*	k_1 (min^{-1})
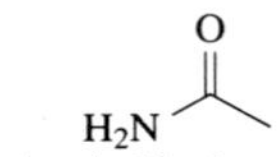 (methyl hydrogen)	~2 x 10^{-12}	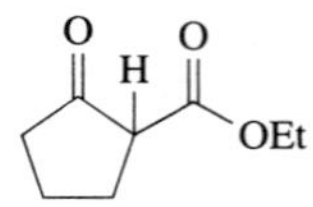	1.4 x 10^{-1}
Me—CN	~4 x 10^{-12}	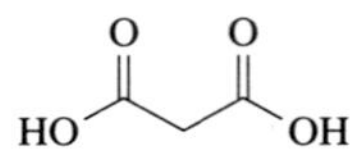 (methylene hydrogen)	~1.7 x 10^{-1}
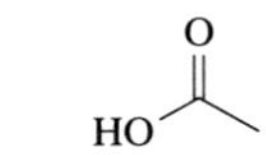 (methyl hydrogen)	2.0 x 10^{-11}	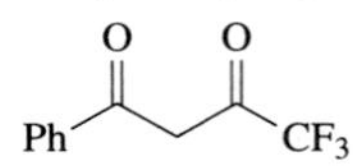	5.0 x 10^{-1}
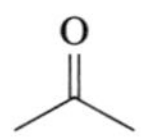	2.8 x 10^{-8}	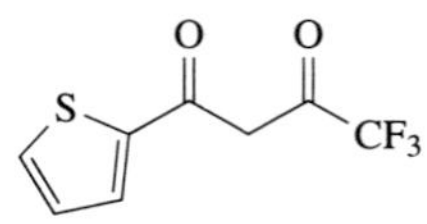	6.0 x 10^{-1}
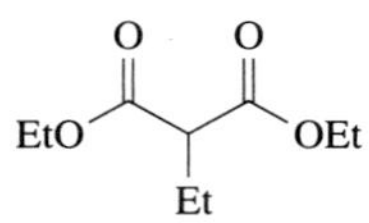	2.0 x 10^{-5}	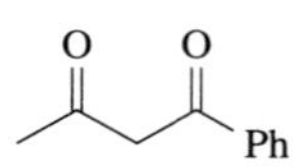	6.6 x 10^{-1}
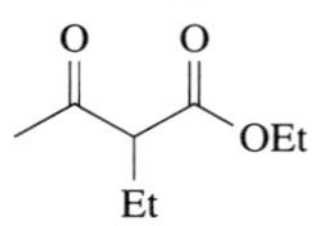	4.5 x 10^{-4}	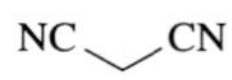	9.0 x 10^{-1}
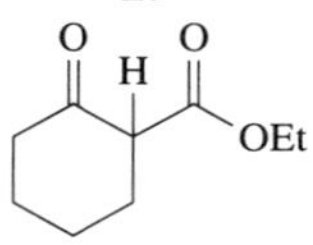	5.8 x 10^{-4}	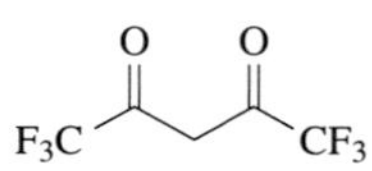	9.0 x 10^{-1}
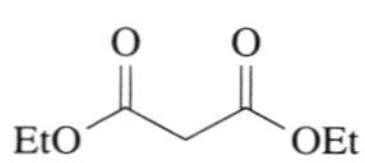	1.5 x 10^{-3}	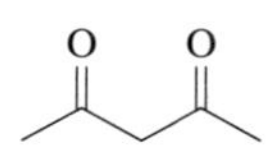	1.0
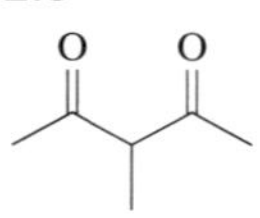	5.0 x 10^{-3}	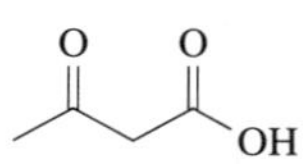 (methylene hydrogen)	7.8
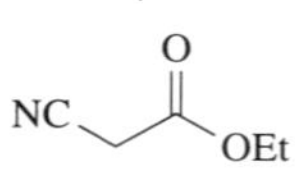	7.0 x 10^{-2}		
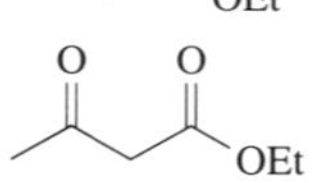	7.2 x 10^{-2}		

The mechanisms shown in Schemes 11 and 12 are classed as S_E1: initially a slow ionisation and a fast subsequent addition. First-order kinetics are observed and the IUPAC designation is $D_E + A_E$.

Carbanions generated from optically active compounds during hydrogen exchange can give products implying either inversion or retention of configuration or racemisation or an intermediate result ⟨61JA3688, 63JA3890, 64JA5445, 64JA5457, 65MI 119-08, B-65MI 119-02, 71JA2225, 71JA2231, 73MI 119-20⟩. The ratio of k_e (rate constant for isotopic exchange) to k_α (rate constant for racemisation) may be considered. When $k_e/k_\alpha \gg 1$, deuterium exchange proceeds with retention of configuration. When $k_e/k_\alpha \sim 1$ racemisation results, while $k_e/k_\alpha \sim 0.5$ leads to inversion. All three types of behaviour occur and are dependent on the structure and substitution pattern of the anion, the base and the solvent. Retention is generally found in solvents of low dielectric constant, racemisation is favoured in polar aprotic solvents and inversion in protic solvents. Finally, a fourth type of behaviour is encountered

Table 3 Ionisation constants of some carbon acids at 25 °C ⟨53JA2439⟩.

Compound	pK_a	*Compound*	pK_a
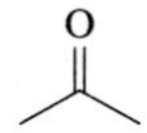	20	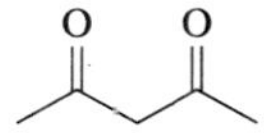	9
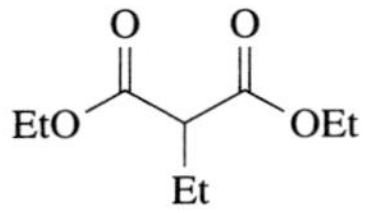	15	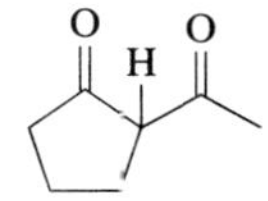	8
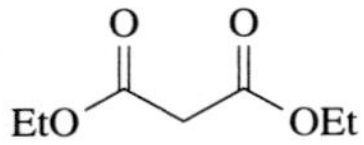	13	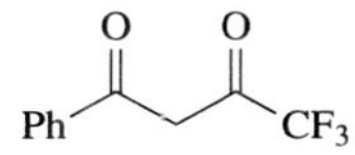	7
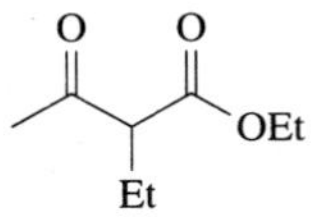	13	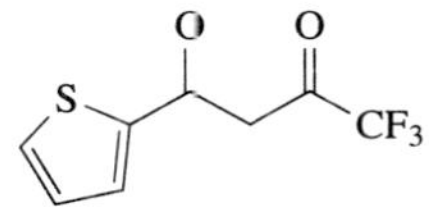	6
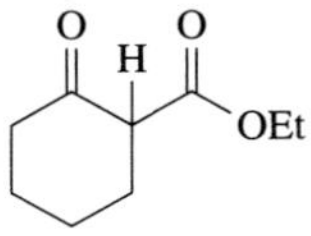	12	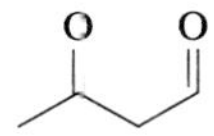	6
	11	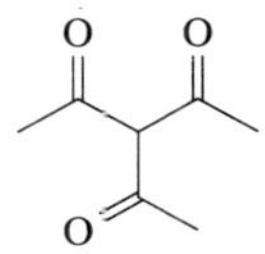	6
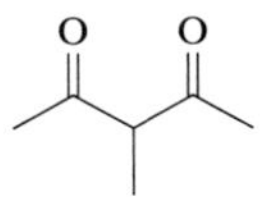	11	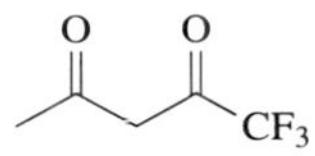	5
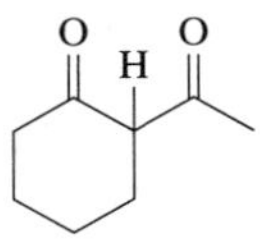	10	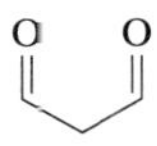	1
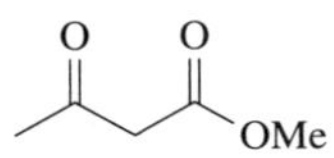	10		
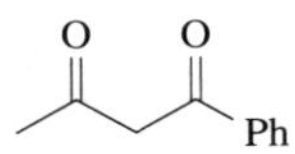	9		

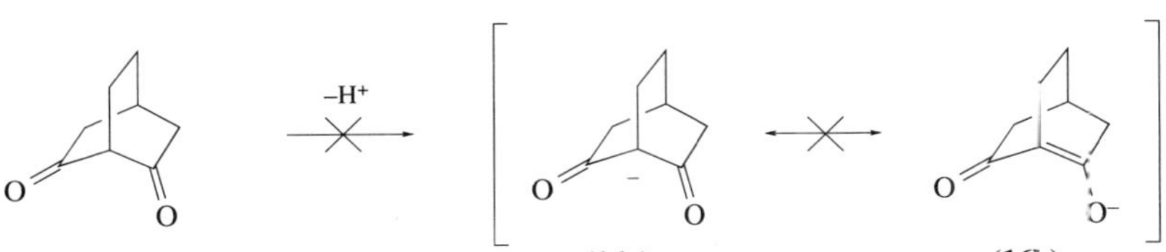

(15) **(16a)** **(16b)**

Scheme 8

in aprotic solvents with aprotic bases like tertiary amines, the ratio $k_e/k_\alpha < 0.5$, indicating that racemisation takes place faster than isotopic exchange. Cram termed this process 'isoracemisation', during which the conjugate acid of the amine remains associated with the carbanion as an ion pair ⟨B-65MI 119-02⟩. Occasionally, the ion pair dissociates long enough for the carbanion to rotate and recapture the deuteron.

For 2-carboxamido-9-deuterio-9-methylfluorene (**28**), it was found that retention of configuration ($k_e/k_\alpha = 148$) occurs in THF with ammonia or a primary amine as base ⟨63JA3890⟩. Similar

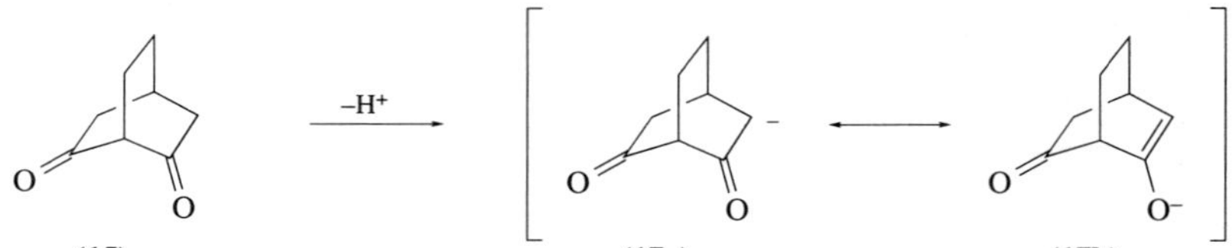

Scheme 9

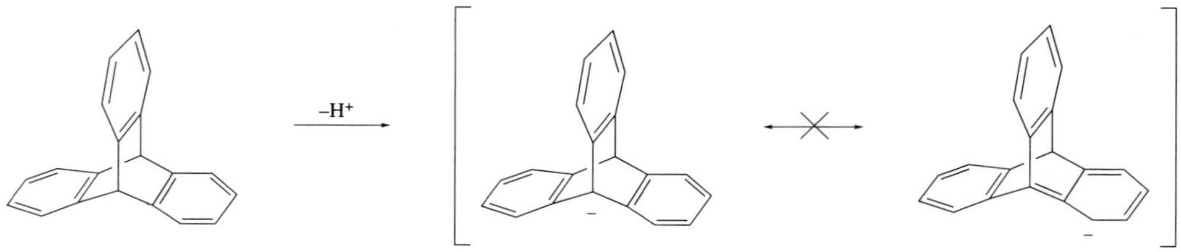

Scheme 10

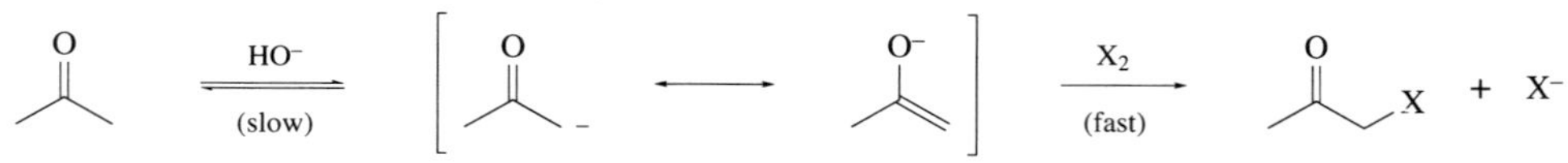

Scheme 11

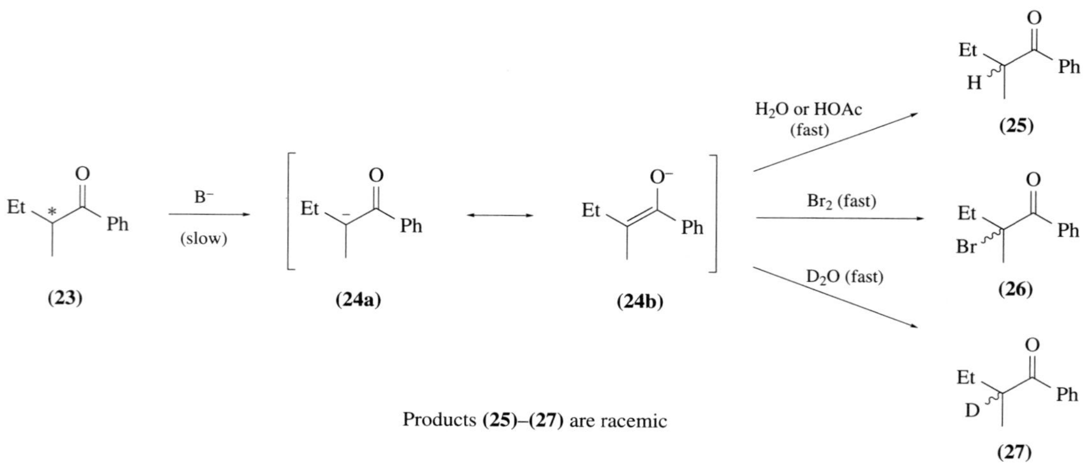

Scheme 12

results were obtained with the benzyl system in cyclohexylamine using cyclohexylamide as base ⟨65MI 119-08⟩. Cram's proposed mechanism is shown in Scheme 13. In the low-dielectric solvent, the carbanion is closely paired with the ammonium cation, so that reprotonation occurs from this same ammonium cation that was formed in the original dedeuteration and from the same face.

Another process that can lead to retention of configuration occurs with the chiral substrate **(28)** in benzene solution containing 10% phenol with potassium phenoxide as the base. The carbanion remains paired to the potassium cation, which is also coordinated to phenol molecules. Rotation of the potassium phenol complex deploys a phenol molecule into a position to protonate the anion with retention ($k_e/k_\alpha = 18$), as shown in Scheme 14.

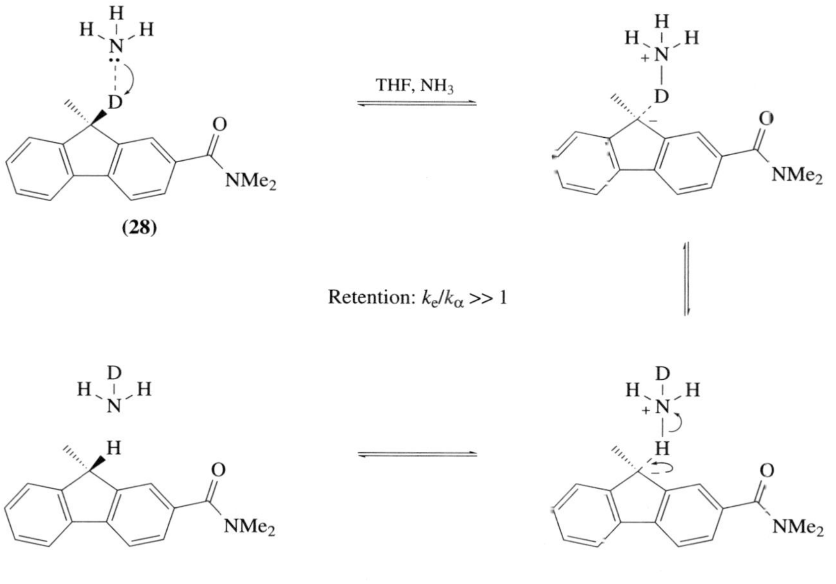

Scheme 13

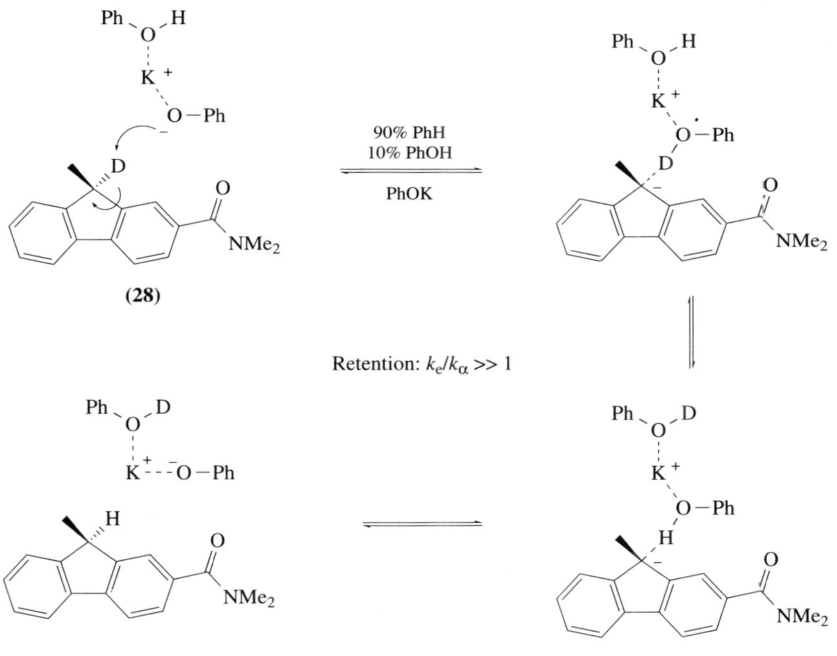

Scheme 14

The stereochemistry of the exchange reaction of benzylic systems has been studied using optically active ethylbenzene-α-*d*; each replacement of an α-H by hydrogen was found to proceed with about 82% net retention of configuration ⟨62JA258⟩. On the basis of this kinetic and stereochemical data, the mechanism of this type of exchange reaction was formulated by Streitwieser as proceeding from (**29**) (Scheme 15) through a highly polar four-centre transition state (**30**) to yield a benzyllithium intermediate, which reacts with solvent ⟨65MI 119-08⟩. The dotted lines in intermediate (**30**) denote bonds which may be largely or totally ionic. The experimental results require that the organic moieties in (**30**) embody a substantial degree of negative charge. The high primary isotope effect implies that the intermediate organolithium compound survives long enough for the deuteriated cyclohexylamine to diffuse away.

For similar benzyl systems in DMSO, the rates of deuterium exchange and racemisation are more

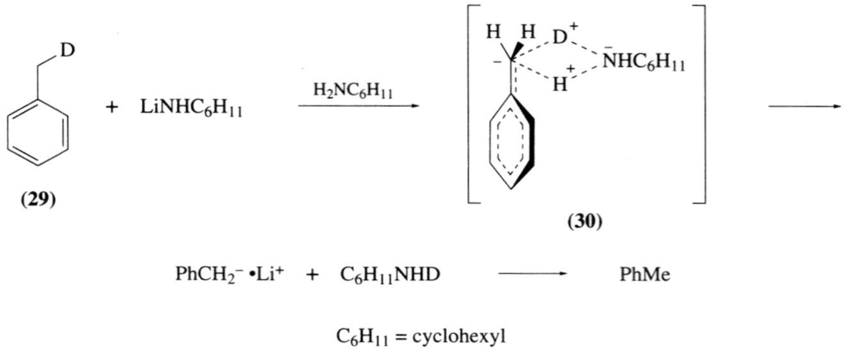

Scheme 15

nearly equal ($k_e/k_\alpha \sim 1$). The more polar solvent presumably allows dissociation of the paired ions. The newly created free carbanion then enters a symmetric solvation environment. The lone pair of the carbanion is then free to rehybridise to an sp^2 orbital, and in this symmetrical form the carbanion may be protonated from either face with equal probability ⟨65MI 119-08⟩.

In methanolic solution, as shown in Scheme 16, there are protons readily available from the surrounding solvent sheath. As the proton departs from one side of the molecule, it can be replaced concertedly from the opposite face. Predominance of this pathway dictates that a majority of the product will result from inversion at the chiral centre, as shown in Scheme 16. The exchange of nitrile (**31**) in methanol with tri-*n*-propylamine as the base gave $k_e/k_\alpha = 0.84$ ⟨64JA5457⟩.

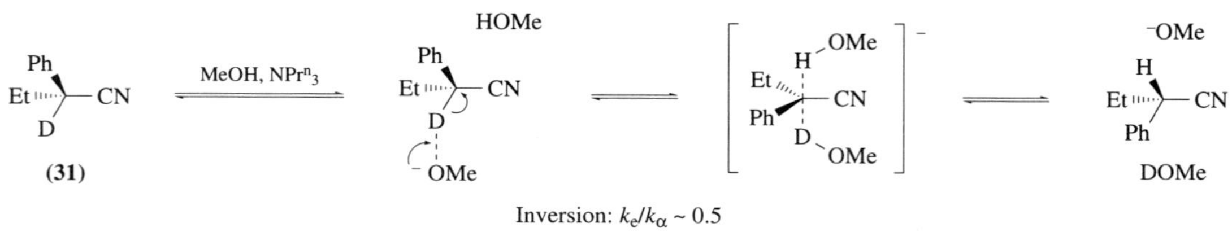

Scheme 16

When a base that possesses no exchangeable protons induces carbanion formation in a solvent of low polarity and low proton concentration, the ion pair formed will revert to base and substrate without deuterium exchange. However, the carbanion will sometimes rotate while still a partner in the ion pair. Inversion, proceeding by way of two changes in hybridisation of the orbital containing the carbanionic lone electron pair, followed by redeuteration, forms the enantiomer. This constitutes isoracemisation as shown in Scheme 17, whereby nitrile (**31**), in THF containing tri-*n*-propylamine as base and 1.5 M *t*-butyl alcohol, undergoes isoracemisation to give a mixture of (**31**) and its enantiomer (**32**) such that $k_e/k_\alpha = 0.05$ ⟨64JA5457⟩. The mechanism explains how inversion occurs without deuterium exchange leading to racemisation, which results from repeated inversion. A single act of inversion without deuterium exchange is termed 'isoinversion'.

Cram has identified a second process, which gives values of $k_e/k_\alpha < 1$ involving carbanions with a strategically placed substituent capable of undergoing tautomerism or having high electron density. Chiral 2-nitro-7-carboxamido-9-deutero-9-methylfluorene (**33**) (Scheme 18) is dedeuterated to form the plane symmetric ion pair (**34**). Redeuteration may occur, with equal probability onto either face, via either transition state (**35**), which recreates the original substrate (**33**) or via transition state (**36**), which results in the formation of enantiomer (**37**) ⟨68JA2598, 68JA2606, 68JA2612, 71JA2235, 89AG(E)73⟩. This process is termed the 'conducted tour' mechanism, since the positive species migrates in a stepwise fashion around the periphery of the substrate from one nucleophilic position to another. At 25°C in THF, chiral-substituted fluorene (**36**) with 16% *t*-butyl alcohol with 10 mole% tri-*n*-propylamine as base gives rise to $k_e/k_\alpha = 0.10$.

In conclusion, it is unlikely that free carbanions exist in solution; they are usually in ion pairs or else solvated ⟨77APO(15)153, 77T2737⟩. This assertion was supported by studying the treatment of $PhCOCHMe^- \cdot M^+$ with ethyl iodide, where M^+ was Li^+, Na^+ or K^+. The half-lives of the reaction (Li, 31×10^{-6}; Na, 0.39×10^{-6}; K, 0.0045×10^{-6}) demonstrates that the species involved were not

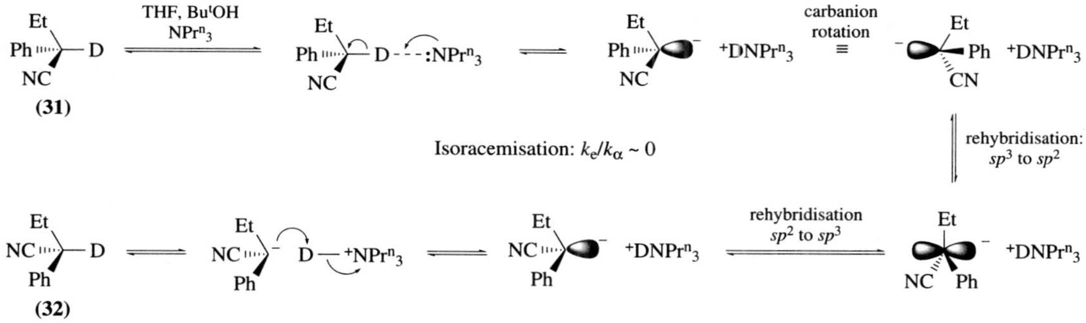

Base-catalysed isoracemisation of **(31)** involving inversion of configuration

Scheme 17

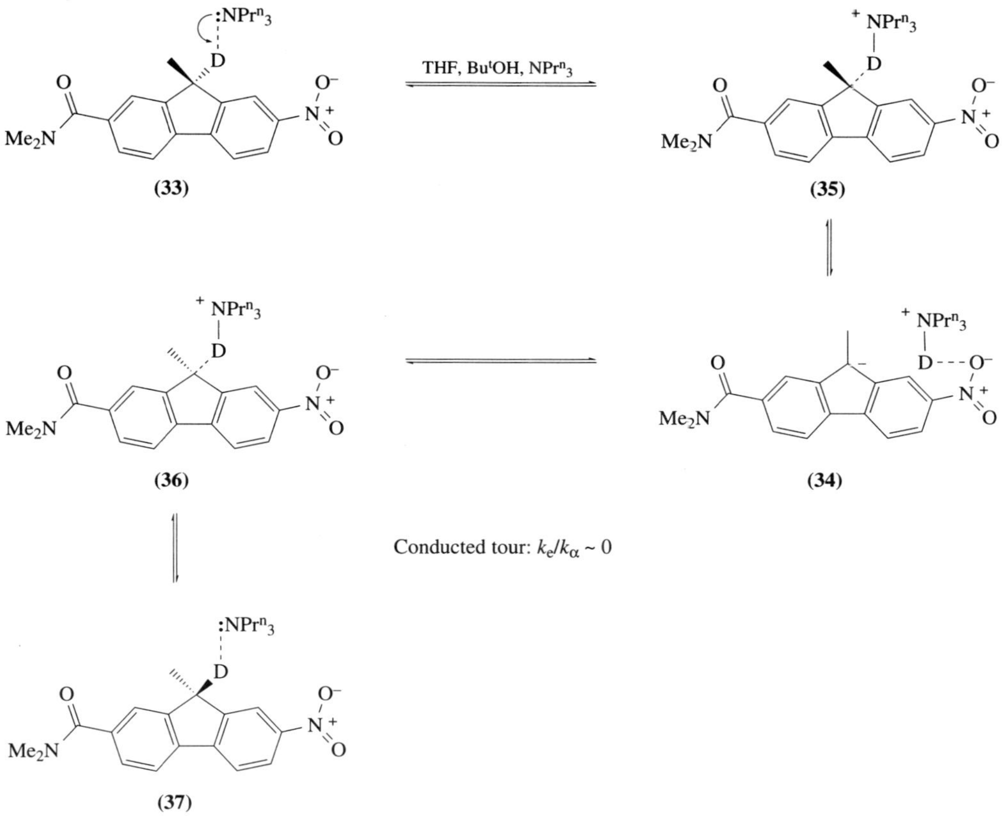

Scheme 18

identical ⟨60JA1386⟩. Similar results were reported with Li, Na and Cs triphenylmethides $Ph_3C^- \cdot M^+$ ⟨81JOU381⟩. Alternatively, where ion pairs do not play an important role, carbanions are solvated. Cram has shown evidence of solvation of carbanions in many solvents. The field of carbanionic pair formation has been extensively investigated ⟨66JA307, 66JA318, 69JA4580, 77JA8262, 77MI 119-01, 78JA3514, 79JA633, 79JOC317, 79TL327⟩.

1.19.1.2.6 Carbanion formation stabilised by conjugation with heterocyclic systems

Alkyl groups situated in the α- and γ-positions of pyridine rings show reactions which result from the relatively easy loss of a proton from the alkyl carbon atom ⟨B-67MI 119-01, 84CHEC(2)315, 88PIA187,

92T2589, B-92MI 119-03⟩. Such strategically placed alkyl groups α and γ to the ring nitrogen of a pyridine ring are significantly more reactive than alkyl groups attached to a benzene ring in base-catalysed processes involving deprotonation of the alkyl substituent ⟨79JOM129⟩. The carbanion thus produced is stabilised by charge delocalisation into the heterocyclic ring, whereby one of the canonical forms **(39b)** (Scheme 19) accommodates the negative charge on the electronegative nitrogen atom. Strong bases, such as sodamide ($NaNH_2/NH_3$, −40 °C) or organometallic compounds ($PhLi/Et_2O$, 40 °C), convert 2-alkylpyridines **(38)** essentially completely into the corresponding anions **(39)** as is the case with 4-alkylpyridines. Anions such as **(39)** react readily with electrophilic reagents (Equation (11)).

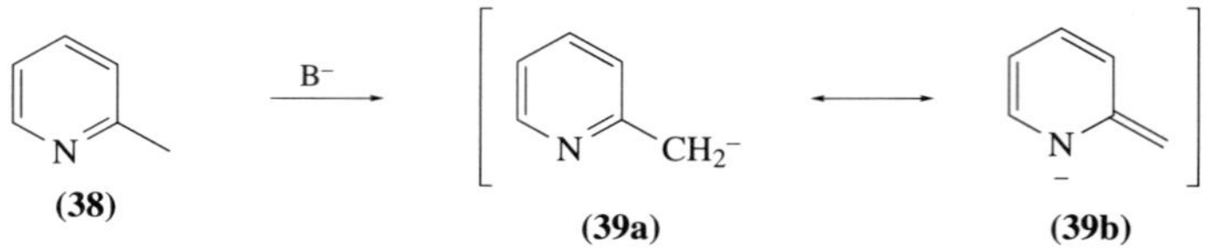

Scheme 19

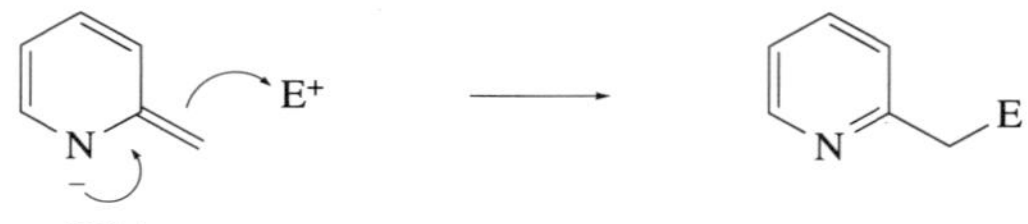

(11)

4- and 2-Alkylpyridines can undergo base-catalysed substitution at the alkyl position in the following ways (Scheme 20):

(i) Alkylation, e.g., 2-picoline **(38)** to give 2-*n*-propylpyridine, conyrine, **(40)** ⟨51JA3308, 56JA1723, 59JA2568, 60JCS4454⟩.

(ii) Reaction with carbonyl compounds to give alcohols, e.g., 2-picoline **(38)** gives the tertiary alcohol **(41)**.

(iii) Carboxylation, e.g. 2-picoline **(38)** gives 2-pyridineacetic acid **(42)**, which must be esterified before isolation.

(iv) Acylation, e.g., 2-picoline **(38)** gives ketone **(43)** ⟨51JA4301, 74JOC2006⟩.

(v) Aldol condensation followed by dehydration, e.g., 2-picoline **(38)** gives stilbazole **(44)** ⟨45JOC21, 74JA8095⟩.

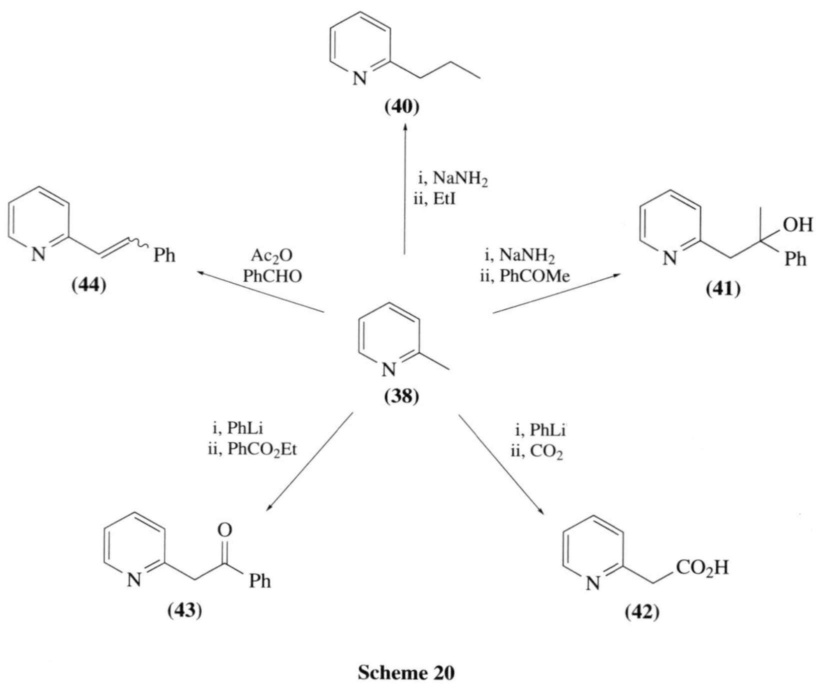

Scheme 20

Similarly, the anion derived from 4-methyl quinoline, by the action of sodamide, reacts with benzoyl chloride to give 4-phenacylquinoline.

3-Methyl pyridine is less acidic than either 2-methylpyridine (**38**) or 4-methylpyridine and will not undergo base-catalysed aldol reactions as easily as either of the latter. It can, however, be deprotonated by very strong bases such as sodamide in liquid ammonia and the anions can then be alkylated ⟨51JA3308⟩. 2-Methyl groups can be selectively functionalised in the presence of 3-methyl groups, an example of which is the selective formation of the sulfide (**46**) (Equation (12)) in high yield from 2,3-dimethylpyridine (**45**), butyllithium and diphenyl disulfide ⟨81JOC2059⟩.

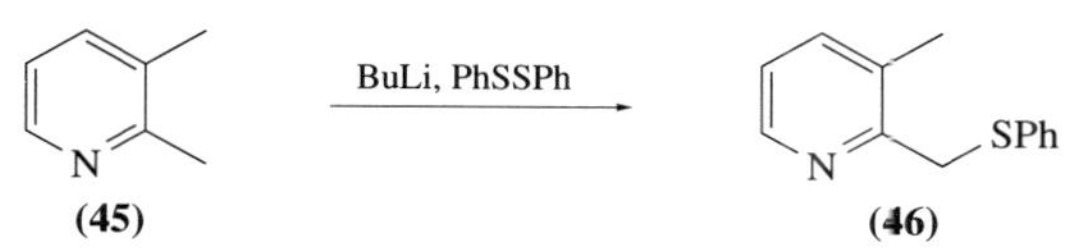

(12)

Proton removal from methyl groups on either the 2- or 4-position of the pyridine nucleus is rendered even easier when the heterocyclic nitrogen bears a formal positive charge, as in quaternary salts (**47**) (Equation (13)) and *N*-oxides (Scheme 21), both of which require only mild base catalysis to effect deprotonation.

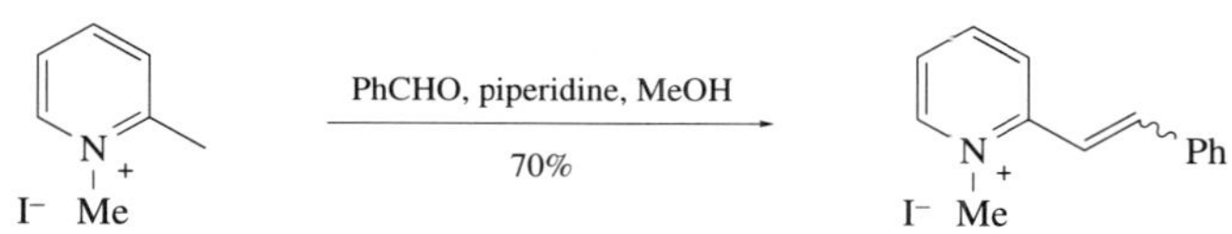

(13)

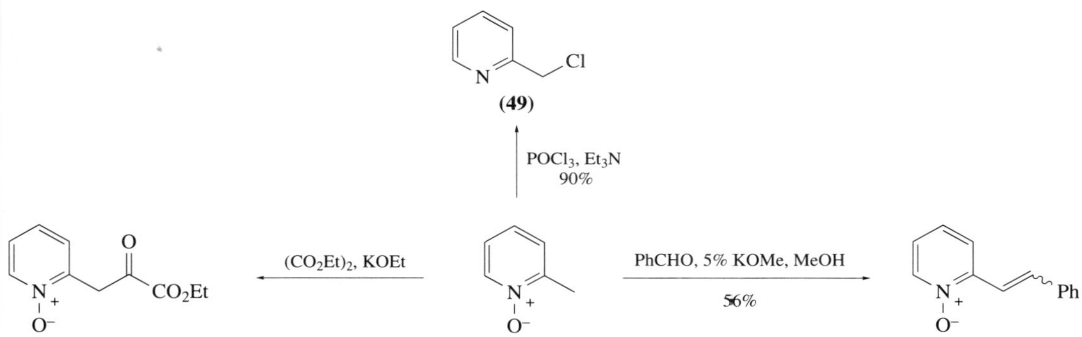

Scheme 21

Reaction of 2-methylpyridine-*N*-oxide (**48**) with *p*-toluenesulfonyl chloride or phosphorus oxychloride gives 2-pyridylmethyl chloride (**49**) (Scheme 21), and related reactions with reactive halides have also been studied ⟨62JOC3856, 81JHC939⟩. Condensation of benzaldehyde with (**48**) gives alkene (**50**). Reaction of (**48**) with diethyl oxalate under basic conditions furnishes the α-keto ester (**51**).

*1.19.1.2.7 Carbanion formation by proton loss stablised in accordance with the Hückel (4*n *+ 2) rule*

The formation of aromatic monocyclic carbanions has been reviewed ⟨79COC(2)361⟩. Cyclopentadiene (**52**) has apparently anomolously high acidic properties with $pK_a \sim 16$ since, with the loss of a methylene proton, the resultant carbanion (**53**) is stabilised by resonance and has attained a sextet of delocalised electrons thus conforming to the Hückel $(4n + 2)$ π-electron rule (Scheme 22) ⟨B-65MI 119-03⟩. The resonance stabilisation of the cyclopentadienyl anion is greater than that accorded to pyrrole, thiophene or furan, since all five canonical forms are equivalent ⟨B-92MI 119-04⟩. The resonance energy stabilisation for (**53**) has been estimated at 24–27 kcal $mole^{-1}$ ⟨81JOC632⟩. The chemistry of the cyclopentadienide anion has been reviewed by Lloyd ⟨B-66MI 119-02⟩.

Webster observed that electrophilic substitution could not ordinarily employ strongly acidic conditions because of the basicity of the cyclopentadienyl anion ⟨67JOC39⟩. However, when the negative charge was dispersed over peripheral nitrile groups, strongly acidic conditions could be

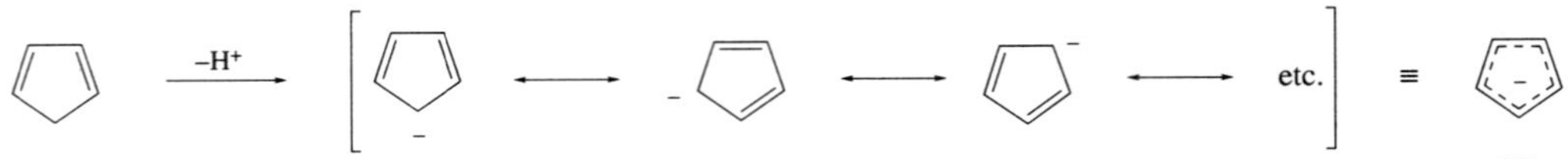

(52) **(53)**

Scheme 22

employed. Tetracarbomethoxycyclopentadiene can be isolated ⟨61PCS117⟩, but the more strongly acidic tetracyanocyclopentadiene (pK_aMeCN ≈ 0) or monosubstituted tetracyanocyclopentadienes are elusive. The tetracyanocyclopentadienyl anion (**54**) (Scheme 23) could be compared in reactivity to anisole ⟨65JA1820, 66AG(E)420, 66JA3046, 66JA4055⟩. The studies were conducted on the tetraethylammonium, potassium and silver tetracyanocyclopentadienide salts. All three salts were surprisingly soluble in organic solvents and changes in cation had little effect on the course of reaction. The most convenient cation was found to be tetraethylammonium, the salts of which were found to be in general water insoluble and crystalline.

The tetracyanocyclopentadienyl anion (**54**) (Scheme 23) reacts with concentrated nitric acid, formaldehyde, bromine, chlorine, trifluoromethylsulphenyl chloride and acetic anhydride to give nitro-, methylidene-, bromo-, chloro-, trifluoromethylthio- and acetyl-tetracyanocyclopentadienyl anions. The chlorotetracyanocyclopentadienyl anion and chlorine produce 5,5-dichloro-1,2,3,4-tetracyanocyclopentadiene, an enophile which undergoes Diels–Alder cycloaddition with ethylene at atmospheric pressure and room temperature.

Suitably positioned substituents such as the trifluoromethyl groups of 1,2,3,4,5-pentakis(trifluoromethyl)cyclopentadiene (**55**) (Scheme 24) render the compound more acidic than nitric acid ⟨80JA6633⟩.

Fulvenes such as (**56**) form cyclopentadienide salts on treatment with lithium alkyls ⟨58CI(L)1589, 58PCS289⟩, or with sodamide in liquid ammonia (Scheme 25) ⟨58PCS289, 61JCS4610⟩.

Indene (**57**) and fluorene (**58**) (Scheme 26) are also acidic, but less so than cyclopentadiene (see Table 2). The negative charge, formed initially on the five-membered ring of either indene or fluorene, is delocalised onto a fused six-membered ring only at the expense of the loss of aromatic stability of the fused six-ring electron system as depicted in canonical (**59b**) (Scheme 26). However, the tribenzopentalene derivative fluoradene (**60**) is sufficiently acidic (pK_a = 13.6), that it will dissolve in aqueous sodium hydroxide ⟨60JA934⟩.

The cyclopentadienyl anion may be stabilised by complexation with an Fe^{2+} ion to form the 'sandwich compound' (a descriptive designation originated by Dunitz and Orgel ⟨53NAT121, B-87MI 119-03⟩) dicyclopentadienyl iron, $Fe(C_6H_6)_2$ (**61**) (dubbed 'ferrocene' by Woodward *et al.* ⟨52JA3458⟩), which was first reported almost simultaneously by two groups of workers ⟨51NAT1039, 52JCS632⟩. The first preparation by Kealy and Pauson used the reaction of cyclopentadienylmagnesium bromidewith $FeCl_3$ in diethyl ether–benzene. X-ray analyses of ferrocene and its simple derivatives have revealed that this complex exists in the staggered or antiprismatic conformation shown in structure (**61**) ⟨52JA2125, 52ZN(B)377⟩. In solution the rings rotate quite freely, with the barrier to ring rotation estimated at 2–5 kcal mole^{-1} ⟨79ACR415⟩.

Ferrocene is exceptionally stable. It is not only unaffected by air, moisture and heat up to 470°C, but it can also be boiled in concentrated hydrochloric acid or 10% caustic soda without decomposition. The dicyclopentadienyl derivatives, Cp_2M, of the other metals in the first transition series from V to Ni have been synthesised ⟨59MI 119-04, 59MI 119-05⟩. Metallocenes containing two metal atoms and three cyclopentadienyl anion rings have been prepared and are known as 'triple-decker sandwich' compounds ⟨77AG(E)1⟩. Tetradecker, pentadecker and hexadecker sandwich compounds have been reported ⟨85AG(E)943⟩. For a bibliography of reviews on metallocenes see Bruce ⟨72AOC(10)273⟩.

The cyclopentadienyl anions of ferrocene, in common with other metallocenes, readily undergo substitution reactions underlining their aromatic character ⟨62AG(E)312, 62AG(E)394⟩. The first example of such a reaction was reported by Woodward *et al.* ⟨52JA3458⟩. When ferrocene in carbon disulfide was treated with acetyl chloride in the presence of aluminum chloride, it was smoothly converted to the diacetyl derivative (**62**) (Equation (14)).

Treatment of benzocycloheptratrienes (**63**) or (**64**) with potassium amide in liquid ammonia cleanly affords solutions of the benzocycloheptatrienyl anion (**65**) (Scheme 27) ⟨73JA3382⟩. Such solutions can be quenched with wet ether to produce isomers (**63**) and (**64**) in about equal proportions.

The cyclononatetraenide anion (**67**) has been prepared from bicyclo[6.1.0]nona-2,4,6-triene (**66**) in moderate yield by deprotonation with the methylsulfinyl carbanion (Equation (15)) ⟨65JA1941⟩.

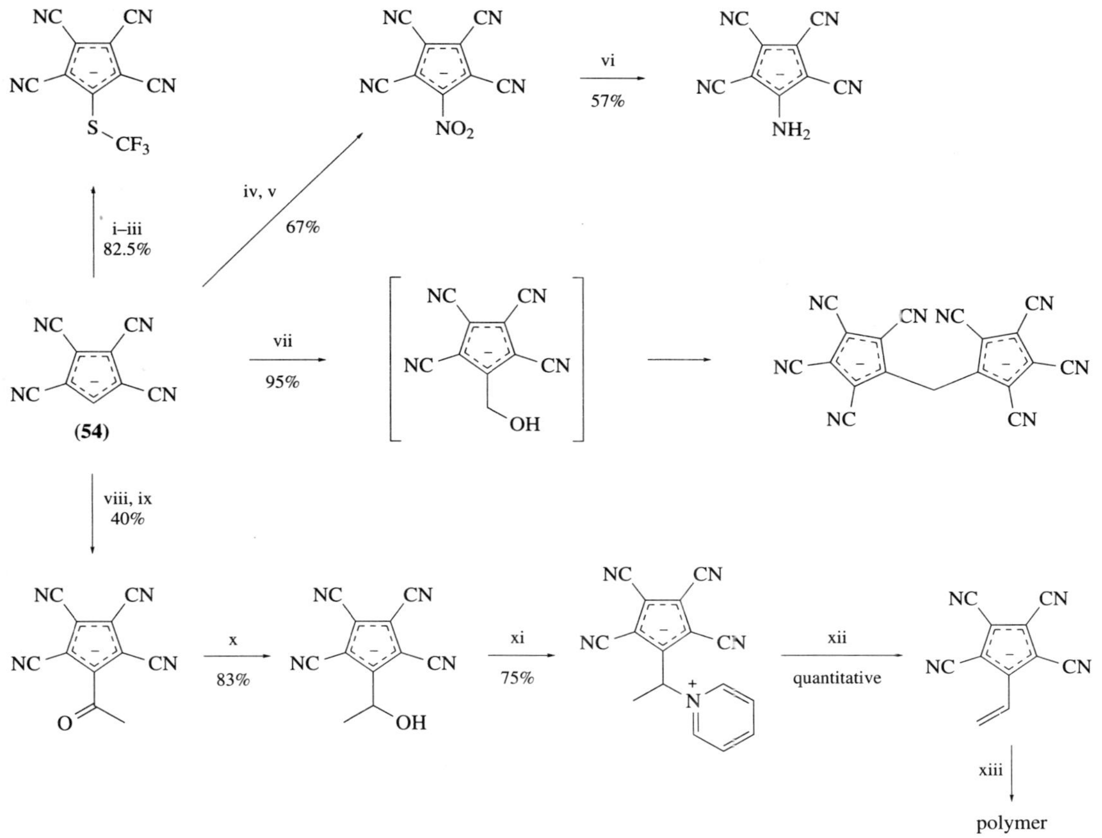

i, $[Et_4N]^+$•$[Cp(CN)_4]^-$ (starting material), $AlCl_3$, 0 °C to 50 °C; ii, CF_3SCl (g), RT to 40 °C; iii, reflux, 1.5 h, RT, overnight; iv, K^+•$[Cp(CN)_4]^-$ (starting material), HNO_3 (conc.), RT, 15 min; v, H_2O, Et_4N•Cl, 0 °C; vi, Zn, H_2O, 6M HCl, reflux, 15 min; vii, $[Et_4N]^+$•$[Cp(CN)_4]^-$ (starting material), paraformaldehyde, CF_3CO_2H, RT, 1 h; viii, K^+•$[Cp(CN)_4]^-$ (starting material), Ac_2O, CF_3CO_2H, reflux, 1 h; ix, Et_4N•Cl, H_2O; x, $NaBH_4$, EtOH, reflux, 1 h; xi, pyridine, $SOCl_2$, RT, 2.5 h; xii, pyridine, reflux, 15 h; xiii, 135–140 °C, 15 min.

Scheme 23

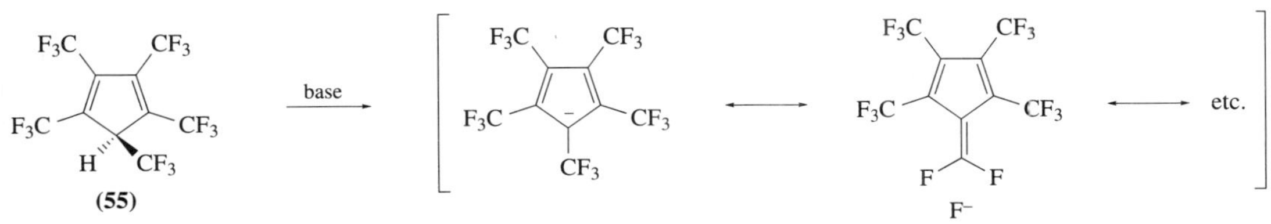

Scheme 24

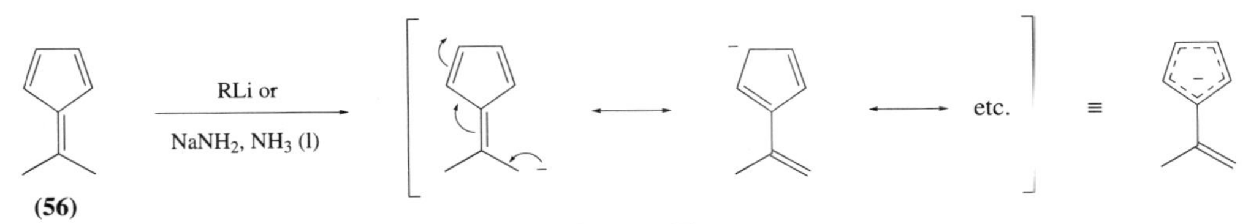

Scheme 25

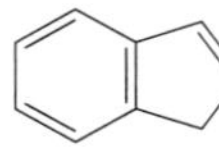

(57)

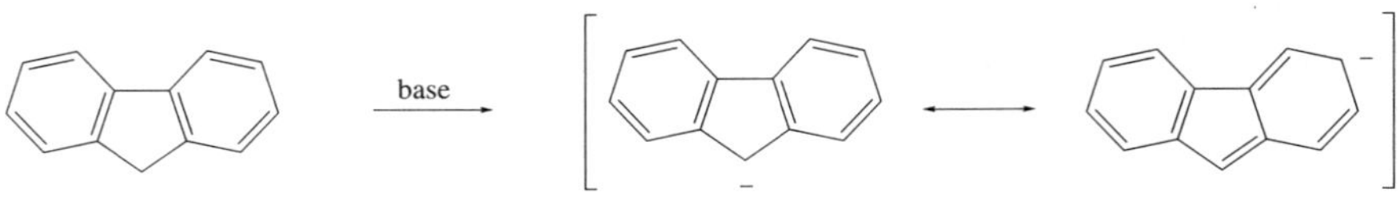

(58) **(59a)** **(59b)**

Scheme 26

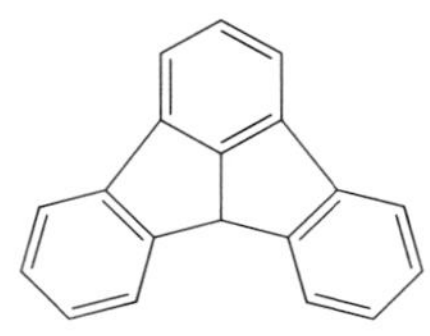

(60)

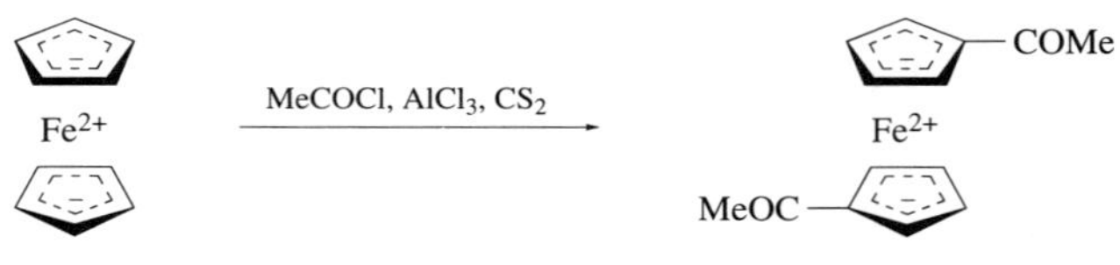

(14)

(61) **(62)**

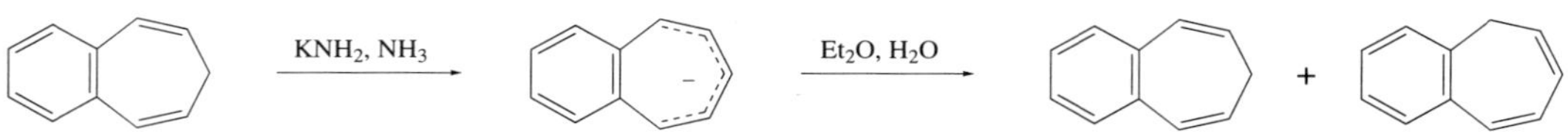

(63) **(65)** **(63)** **(64)**

Scheme 27

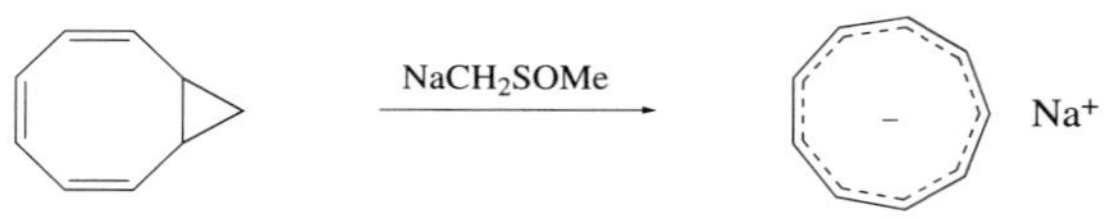

(15)

(66) **(67)**

The preparation from triene system **(68)** of the stable 1,5-methanocyclononatetraenyl 10π-electron anion **(69)** has been reported (Equation (16)) ⟨66JA3461⟩. The downfield ^{1}H-NMR signals of the ring protons ($8 \geqslant \delta \geqslant 6$) combined with the strong shielding of the methylene protons indicated the induction of a ring current and suggested that **(69)** was a 10π-electron delocalised homoaromatic system.

(16)

(68) **(69)**

1.19.1.2.8 Destabilised carbanion formation by proton loss viewed in the context of the Hückel (4n + 2) rule

If proton removal from a molecule results in formation of an anti-aromatic π-electron system, this may prove a great disincentive to carbanion formation. It has been found that **(70)** loses its proton about 6×10^3 more slowly than the saturated analogue (**71**) ⟨67JA4383⟩. For the similar pair (**72**) and (**73**) the observed ratio is approximately 10^4 ⟨68JA2698⟩. This illustrates that (**70**) and (**72**) are much more reluctant to form carbanions, which would be unstable 4π-electron antiaromatic cyclopropenyl carbanions, than (**71**) and (**73**) which form simple carbanions. Although derivatives of the cyclopropenyl anion have been inferred as transient intermediates, all attempts to prepare the ion or any of its derivatives as relatively stable intermediates have been in vain ⟨82TL795⟩. A triphenylcyclopropyl anion has been prepared in the gas phase, with a lifetime of 1–2 seconds ⟨86TL5931⟩.

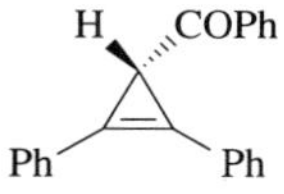

(70)

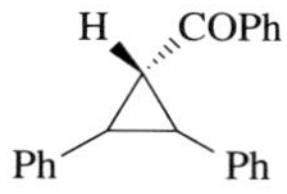

(71)

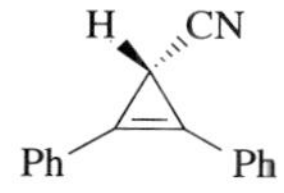

(72)

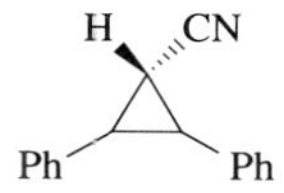

(73)

In marked contrast to cyclopentadiene (**52**) cycloheptatriene (**74**) displays no unusual acidity. The cycloheptatrienyl anion (**75**) has been prepared in solution albeit in modest yield by the action of benzyl potassium on cycloheptatriene (Scheme 28) ⟨63JA3041⟩. The anion (**75**) is less stable than the cyclopentadienyl anion (**53**), having an 8π-electron system as discussed in detail by Dauben and Rifi ⟨63JA3041⟩. The formation of antiaromatic monocyclic carbanions has been reviewed ⟨79COC(2)391⟩.

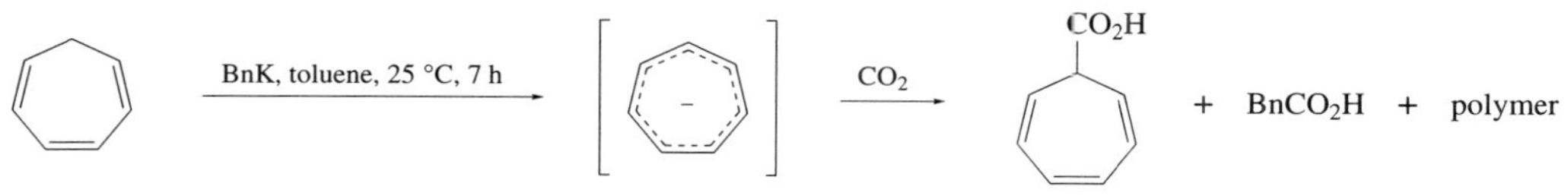

(74) **(75)**

Scheme 28

1.19.1.2.9 Formation of homoaromatic carbanions by proton loss

A bishomocyclopentadienyl anion (**77**) has been prepared, in which delocalisation of the negative charge appears to take place to a significant degree. Brown and Occolowitz originally proposed (**77**) as being involved in the deprotonation of bicyclo[3.2.1]octa-2,6-diene (**76**) (Equation (17)), which occurs $10^{4.5}$ times faster than the deprotonation of the analogous monoene (**78**) ⟨65CC376, 67CC638, 68JCS(B)411, 70JA3821⟩. Paquette has reviewed the concept of extended homoaromaticity ⟨78AG(E)106⟩. Carbanion (**79**) ⟨73JA6449⟩ and the heterocyclic carbanion (**80**) ⟨75CC201⟩ exhibit comparable ^{1}H-NMR spectral properties to (**77**), suggesting a sustained ring current.

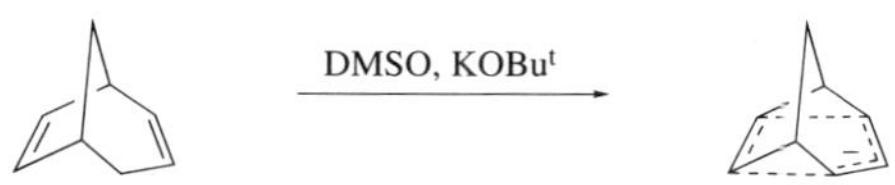

(76) **(77)** (17)

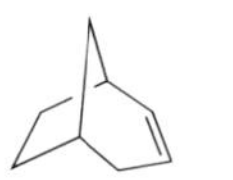

(78)

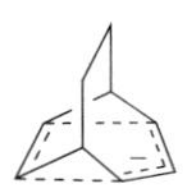

(79)

(80)

The instability of bishomocyclopentadienyl anions was underlined by the finding that the 2-bromobishomocyclopentadienyl anion (**82**) is not obtained by the treatment of the bromide (**81**) with base, the final product of the reaction being the bicyclic alkyne (**83**). Bergman and Rajadhyaksha suggested that carbanion (**82**) is formed, but that it subsequently loses Br^- and rearranges to give (**83**) (Scheme 29) ⟨70JA2163⟩.

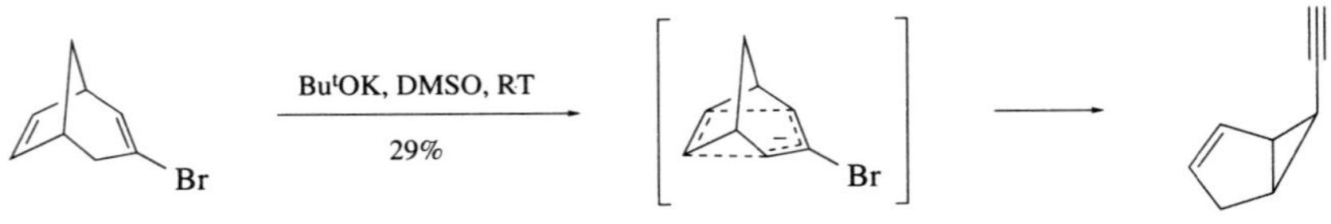

(**81**) (**82**) (**83**)

Scheme 29

It should be noted that the precise realities of carbanionic homoaromaticity is an area of controversy. However, a balanced account of the debate is given by Lee and Squires ⟨86JA5078⟩, who present evidence, derived from gas-phase experiments, supporting the bishomoaromatic charge delocalisation of carbanion (**77**).

1.19.1.2.10 Formation of the phenalene anion by proton loss

The chemistry of phenalenes has been reviewed ⟨58T339, 65QR274, 73MI 119-19⟩. Phenalene (**84**) is remarkable, in that it gives rise to a relatively stable anion (**85**), cation (**86**) and radical (**87**).

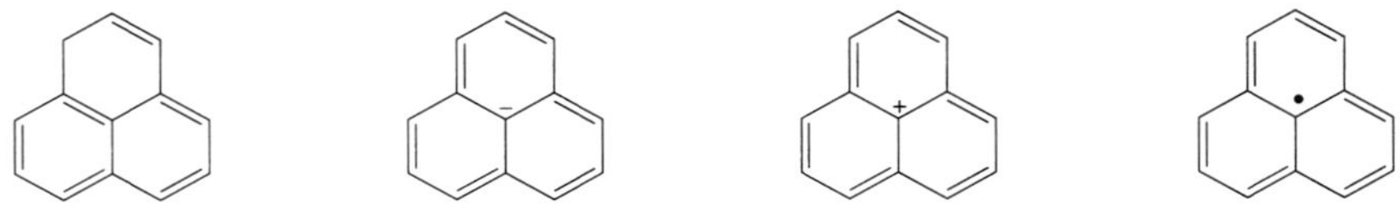

(**84**) (**85**) (**86**) (**87**)

Quantum mechanical studies have provided a satisfactory explanation for their observed stability. Hückel molecular orbital (HMO) calculations indicate that all three species possess the same π-electron delocalisation energy ⟨56CI(L)1306, 57DOK(117)1021, 60JA1972, 64CCC1932⟩. In simple terms, the phenalenyl anion (**85**) (derived from phenalene using phenyllithium ⟨50JA1245⟩ or potassium methoxide ⟨56CI(L)1504, 58T339⟩) can be represented as having a 12π-electron periphery, with the negative charge formally assigned to the central carbon atom. Phenalene is more acidic than triphenylmethane ($pK_a = 25$) and less so than cyclopentadiene ($pK_a = 16$) ⟨50JA1245⟩.

1.19.1.2.11 Carbanions derived from unsaturated acidic hydrocarbons involving stabilisation by very extensive delocalisation

7*H*-Dibenzo[*c*,*g*]fluorene (**88**) is about 10^4 times more acidic than fluorene. Tris(biphenylenevinyl)methane (**89**) had the highest acidity of any of the fluorene-substituted hydrocarbons described prior to 1966 ⟨62LA(654)64, 64AG(E)312, 65LA(689)1⟩.

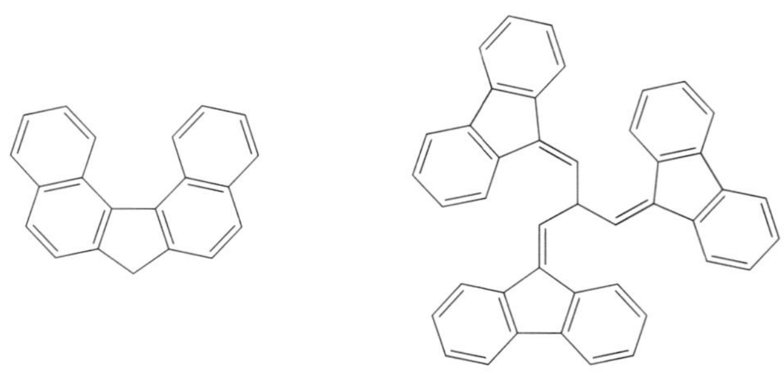

(**88**) (**89**)

In 1967, however, Kuhn and Rewicki reported the preparation of tris-(7*H*-dibenzo[*c*,*g*]-fluorenylidenemethyl)methane (**90**) and hydrocarbon (**91**) which were obtained as a tautomeric mixture by the method shown in Scheme 30 ⟨67AG(E)635⟩.

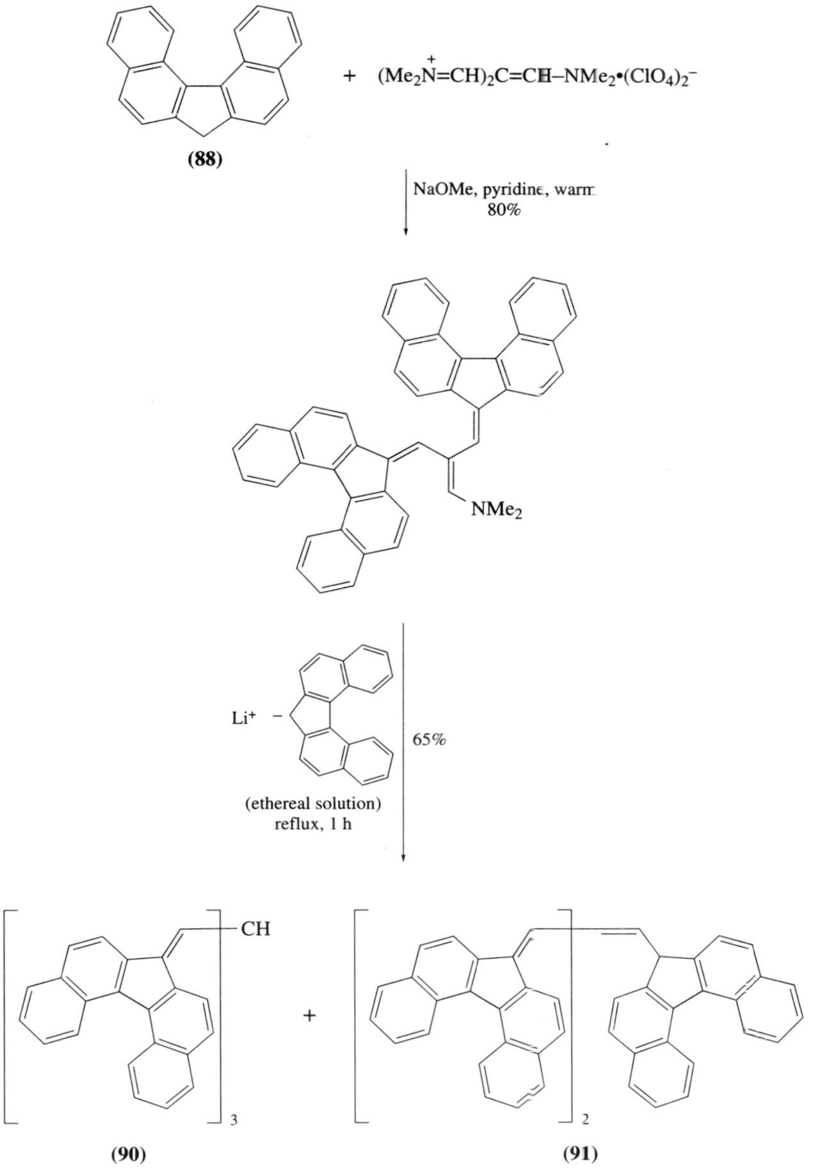

Scheme 30

When treated with DMSO containing a base (e.g., NaOH or tri-*n*-propylamine), the hydrocarbons (**90**) and (**91**) gave a deep green carbanion (**92**) now known as Kuhn's anion. As might be expected, hydrocarbons (**90**) and (**91**) are highly acidic. A 10^{-5} M solution in DMSO is almost completely dissociated even without addition of base ⟨67AG(E)635⟩.

Okamoto *et al.* have published detailed experimental techniques for the preparation of two further hydrocarbon salts both consisting of Kuhn's anion as the carbanion moiety ⟨90JOC996⟩.

1.19.1.2.12 Carbanion stabilization by a nonadjacent π-bond: the formation of homoenolate anions

Under certain circumstances (Scheme 31) a carbonyl group can activate a distant proton (e.g., on a *β*-carbon atom as in (**93**) when $n = 1$) sufficiently to allow conversion to an anion (**94**) stabilised

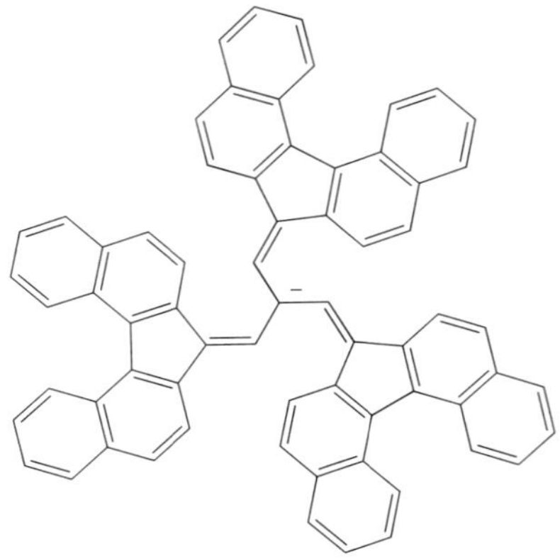

(92)

by homoconjugation. Such an anion is called a 'homoenolate carbanion' **(94)** and the cyclic alcohol derived by protonation on oxygen is the homoenol **(95)**.

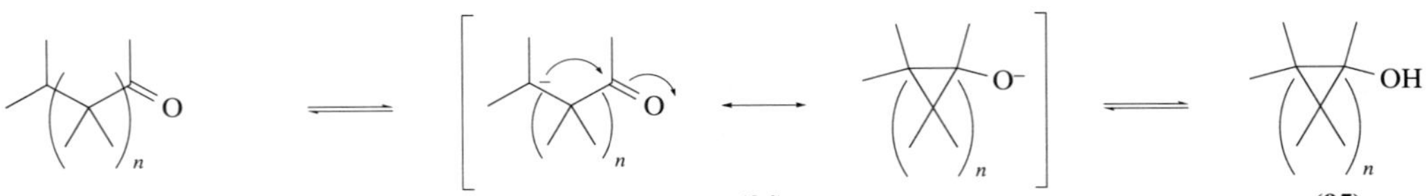

(93) **(94)** **(95)**

Scheme 31

In the simplest case (Scheme 31; $n = 1$), the homoenol formed is a cyclopropyl alcohol. The archetypal example of homoenolisation is that of (+)-camphenilone **(96)**, which, when treated with potassium *t*-butoxide, underwent racemisation (Scheme 32) ⟨66JA1905⟩. The experimental evidence pointed to the formation of the symmetrical intermediate **(97)**.

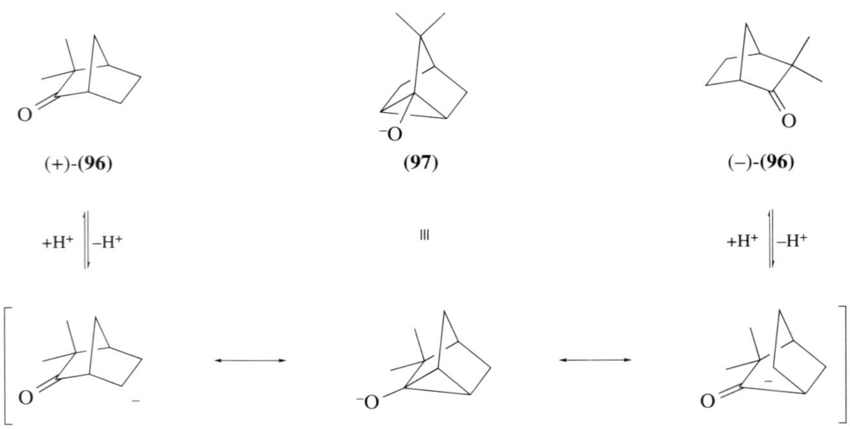

Scheme 32

γ-Homoenolate formation (Scheme 31; $n = 2$) has been observed in 'half-cage' ketones ⟨65JA914, 65JA915, 65JA916⟩. Winstein and Fukunaga both found that the hexachloro half-cage ketone **(98a)** was converted, in good yield, to the corresponding hexachlorohomoenol **(99a)** in refluxing pyridine at 100 °C ($k = 2.9 \times 10^{-5}$ s^{-1}). Winstein also studied the γ-enolisation of the parent ketone **(98b)** at a range of temperatures and found, for example, that **(98b)** rearranged to **(100)** in 0.9 M *t*-BuOK/*t*-BuOH with a rate constant of 7.58×10^{-5} at 195 °C ($k_{185\,°C} \sim 3.8 \times 10^{-5}$ s^{-1}). The subject of homoenolisation has been reviewed ⟨83T206⟩.

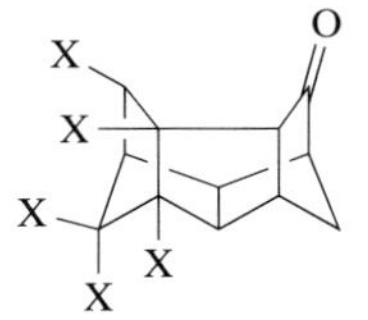

(98a) X = Cl
(98b) X = H

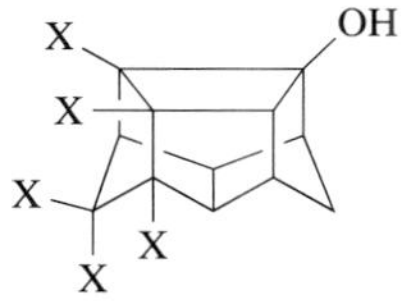

(99a) X = Cl
(99b) X = H

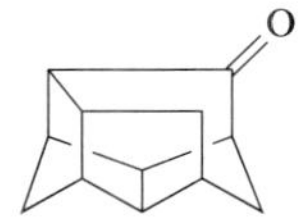

(100)

1.19.1.2.13 Kinetic versus thermodynamic factors determining the course of enolisation

The presence of certain unsaturated substituents at a saturated carbon atom, for example, carbonyl, cyano or phenyl groups renders any hydrogen atoms bonded to that carbon relatively acidic as shown in Table 4. As indicated in Table 4, the effectiveness of these activating groups follows the approximate order: $COR > CO_2R \sim CN > Ph$. The presence of two such activating groups further enhances the acidity of an active methylene compound; conversely, acidity is decreased by about 1–2 pK_a units by an alkyl substituent. The values or ranges of values given in Table 4 for weakly acidic substances should be viewed as approximate. The values listed are not corrected for deviations in pK_a values which result from the measurements being made in different solvents with a variety of bases and cations present ⟨32JA1212, 36JA1124, 53JA2439, 59JOC709, 65JA382, 65JA384, B-65MI 119-04, 65JCS(B)1752, 67JA1721, 67JA2752, 67JA2751, 67JA2960, 67JA3905, 67JCS(B)827, 68JA2821, 68JA3415, 68JOC2222, 68JOC3477, 68MI 119-06, 69JA6749, 69JA2797, 70JCS(B)173, 70JCS(B)179, 70JCS(B)194, 70JA4689, 70JA5926, 70JA5933, 70JA5939⟩.

Table 4 Approximate pK_a values of some active methylene compounds.

Compound	pK_a	*Compound*	pK_a
$CH_2(CN)CO_2Et$	9	$MeCOCH_3$	20
$CH_2(COMe)_2$	9	Indene	21
$MeCOCH_2CO_2Et$	11	Bu^tCOCH_3	21
$MeCH(COMe)_2$	11	$Bu^tCOCHMe_2$	23
$CH_2(CN)_2$	11	CH_3CO_2Et	25
$CH_2(CO_2Et)_2$	13	CH_3CN	25
CpH	15	Ph_3CH	28–23
$PhCH_2COPh$	16	Ph_2CH_2	33–35
$MeCOCH_2Cl$	17	$PhCH_3$	37
$PhCOCH_3$	19		

Much attention has been devoted to the determination of rate and equilibrium constants for enolate formation by the loss of protons α to a carbonyl group ⟨B-90MI 119-01⟩. Synthetic routes that involve the formation and subsequent reaction of carbanions (enolate anions) derived from active methylene compounds constitute very important and synthetically useful classes of organic chemical and enzymatic reactions, including eliminations, aldol condensations and alkylations ⟨75T2979, 91JA9667⟩. Although studies on the acid- and base-catalysed conversion of aldehydes and ketones to their enols were reported as early as the 1900s ⟨82APO(18)1, 89T4913⟩, the inaccessiblity of these enols in aqueous solution hampered studies of their ketonisations. Keefe and Kresge ⟨B-90MI 119-02⟩ and Capon ⟨B-90MI 119-03⟩ developed techniques for producing enols in sufficiently high concentration to enable their ketonisation reactions to be examined. Aqueous pK_a values, keto–enol equilibrium constants and rate constants for keto–enol interconversions are now available for a variety of aldehydes and ketones ⟨82APO1, 89T4913, B-90MI 119-02, B-90MI 119-04⟩.

In the case of most monoketones and esters the amount of enol present at equilibrium is small. Conversely 1,2- and 1,3-dicarbonyl compounds often contain a large proportion (>50%) of their enol tautomers at equilibrium ⟨54JOC1960, B-60MI 119-03, 61JOC3156, 63T1625, 65JOC3212, 66JCS(B)241, 67JOC1994, 67JOC2162, 70T2027, 70T2035⟩.

Scheme 33 illustrates the general rule that the enol tautomer is more acidic than the keto tautomer by 4–5 pK_a units ⟨66JCS(B)241⟩. This acidity difference is attributable to the fact that the rate of proton abstraction from a C—H bond usually proceeds slowly relative to the very fast (frequently diffusion-controlled) rate of proton removal from OH bonds or proton addition to either oxygen or a carbanionic carbon atom (Scheme 34) ⟨B-59MI 119-03, 62JA4824, 64AG(E)1, 67JA2826, 69JA1146, 69JOC4083⟩.

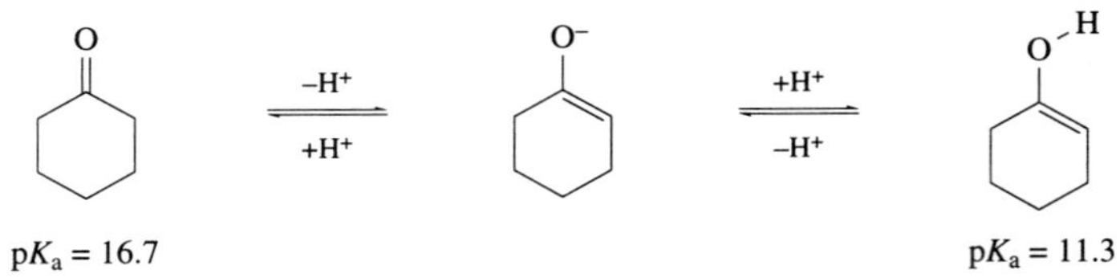

Scheme 33

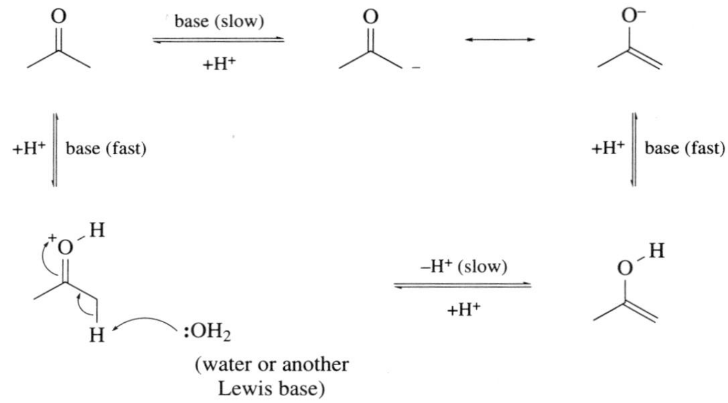

Scheme 34

The degree of enolisation and the structure and stereochemistry of the enols obtained from 1,3-dicarbonyl compounds are dependent both on the structure of the dicarbonyl compound and the solvent in which the equilibria are measured ⟨B-60MI 119-04⟩. The positions and compositions of the equilibrium mixtures involving 1,3-dicarbonyl compounds are measured from NMR experiments ⟨63JA1696, 65JA505, 66JCS(B)161, 67JA1183, 67JCS(C)1716, 67TL29, 67TL4525, 68JCS(B)132, 68JCS(C)676, 68T1651, 68T1869, 69JOC2817⟩. Furthermore, the stabilisation of dienol tautomers as their tricarbonyliron complexes has been studied ⟨68CC1225, 69TL3989⟩.

1,3-Dicarbonyl compounds in nonpolar, aprotic solvents usually favour the *cisoid* enol form, which normally exists as a monomer with an intramolecular hydrogen bond as shown in Scheme 35.

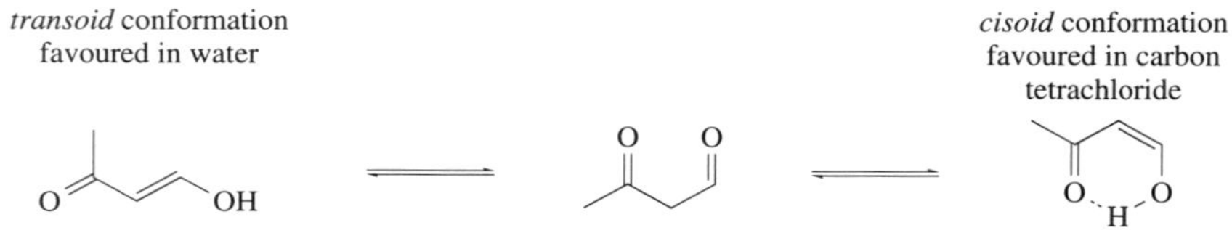

Scheme 35

Dimeric and polymeric hydrogen-bonded species are formed by cyclic 1,3-diketones where intramolecular hydrogen bonding in the enol is prevented by the geometry of the molecule ⟨67JOC2162⟩. The percentage of enol present in such cases is dependent on both the concentration and the temperature of the solution. In relatively polar solvents, and particularly in polar protic solvents, the percentage of the keto form is usually enhanced, since this tautomer is more subject to solvation through a process of intermolecular hydrogen bonding with proximate solvent molecules. It should be noted, however, that cyclic 1,3-diketones represent an exception to this generalisation.

The presence of substituents, especially sterically bulky substituents, in the 1,3-dicarbonyl system usually serves to depress the percentage of enol present at equilibrium, since formation of a planar enol tautomer incorporating an intramolecular hydrogen bond results in adjacent substituents being forced into sterically unfavourable eclipsed conformations.

The ratio at equilibrium between the pairs of isomeric enols, shown in Equation (18) and listed in Table 5 is evidently a consequence of the stabilising effects resulting from conjugation with the carbonyl group set against the requirement for minimisation of unfavourable steric interactions and torsional strains ⟨63JA1696, 69JOC2817⟩.

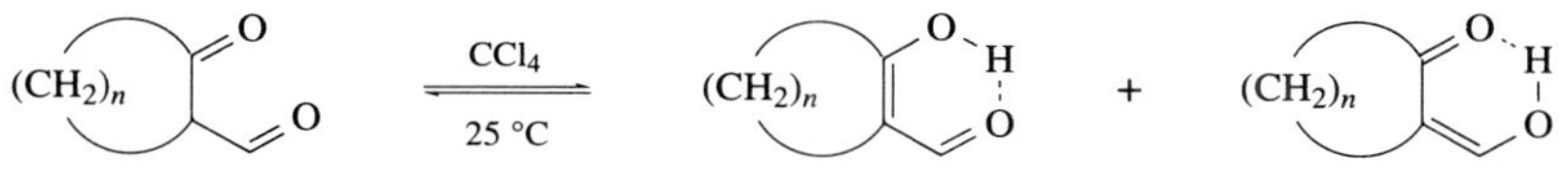

(101) **(102)** **(103)** (18)

Table 5 Percentage composition of enol mixtures derived from Equation (18).

Value of n	*Tautomer (**102**)*	*Tautomer (**103**)*
3	22	78
4	76	24
5	34	66
6	56	44
7	69	31
8	82	18

Measurements of the rates of proton removal from C—H bonds α to carbonyl groups are somewhat more plentiful in the literature than data concerning the equilibrium dissociation constants for such weak acids ⟨40JA113, 51JCS2430, 51JCS2442, 56JA6259, 59JOC709, 60HCA1915, 61JA3678, 62JA2899, 62JA2905, 62JA3164, 62JA3402, 62TL669, 63JA1538, 63JOC3362, 63MI 119-04, 64JA1997, 64JA2832, 64JA5457, 65JA3228, 65JA5492, 65JA5493, 65JA5513, 65JCS(B)1752, 65JOC1341, 65JOC2502, 66JA1332, 66JOC520, 67JA2664, 67JCS(B)827, 67JCS(B)1173, 67JOC492, 67JOC3700, 67JOC4127, 67RCP99, 68MI 119-06, 68JOC2222, 68JOC3477, 68JOC1301, 68JOC1750, 68JCS(C)338, 68T6287, 68TL2317, 68TL2921, 68JA4629, 68JOC935, 68JOC943, 68JA6749, 68JA6091, 68CRV375, 69JA2797, 69CC292, 69JA530, 69JOC3070, 69CC1498 69CC1497, 69JOC2324, 70JCS(B)194, 70JA4689, 70JA5926, 70JA5933, 70JA5939, 70JA2445, 70JA3697⟩.

In base-catalysed enolisations, alkyl substituents appear invariably to decrease the rate of deprotonation from ketones at both α-positions (Table 6) ⟨67JA2664, 67JOC3700, 67JOC4127, 68TL2317, 68TL2921, 68T6287, 68JA4629, 69CC292⟩. This phenomenon may be attributed in part to enhanced steric hindrance towards the approach of the base engaged in proton abstraction.

Table 6 Relative rates[a] of H/D exchange for a series of ketones R-CO-Me catalysed by Na_2CO_3 or NaOD in D_2O or D_2O/dioxane mixtures.

R	*Rate of exchange at Me*	*Rate of exchange at an α-C—H bond on R*
Me—	100	100
$MeCH_2$—	45–50	28
$EtCH_2$—	33–41	12–13
Me_2CHCH_2—	27	7
Me_3CCH_2—	5–21	0.3–0.7
Me_2CH—	33–45	0.1–0.9
Me_3C—	16–27	

[a] These values represent the total rate of exchange at each group and are not corrected for the number of hydrogen atoms at each α-position. From 'Modern Synthetic Reactions', 2nd edn., by Herbert O. House, Copyright (©) 1972 by W. A. Benjamin, Inc.

Furthermore, in the case of methylene groups or α-protons activated by one or more carbonyl groups, the presence of alkyl substituents at either carbon atom may interfere by steric overcrowding with the attainment of a transition state during proton removal that would otherwise permit efficient overlap of the *p*-orbitals involved (Scheme 36) ⟨B-72MI 119-01⟩.

Ketones can be totally converted to their enolate anions in a kinetically controlled process by adding the ketones slowly to an excess of a strong base in an aprotic solvent (Scheme 36).

In α,β-unsaturated ketones, nitriles and esters (e.g., (**104**) (Scheme 37)) the γ-proton assumes the acidity, which is normally a prerogative of the proton α to the carbonyl or nitrile group, especially when R^1 is other than a proton and cannot compete. Alkylation at the α-position to give (**105**),

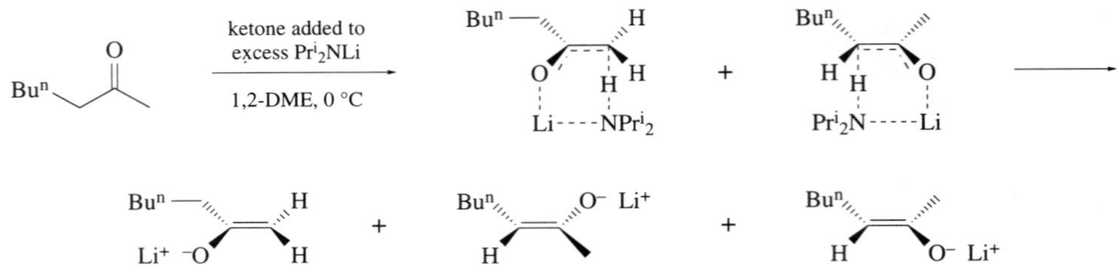

84 : 7 : 9

Scheme 36

competes with alkylation at the γ-position (leading to product (**106**), and usually predominates ⟨B-92MI 119-05⟩.

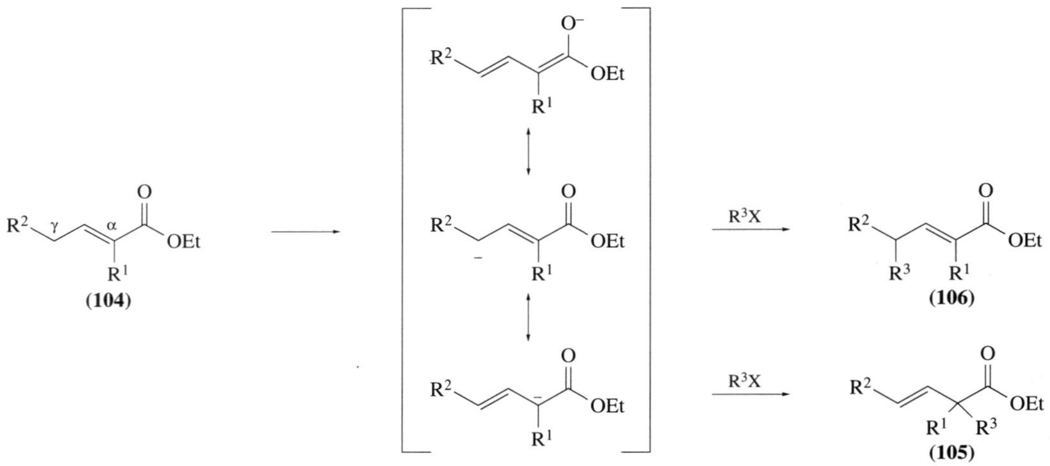

Scheme 37

For ketone enolates the question of the regioselectivity may be summarised ⟨76T2979, 84MI 119-02⟩. Under kinetic conditions (low temperature, $-78\,^{\circ}C$, excess base, small cation), the less-substituted enolate is formed. More hindered bases give higher selectivity. Under conditions designed to provide an equilibrium mixture of product enolates (relatively higher temperature, excess ketone, larger cation) the more substituted enolate is usually formed.

The required one of two possible enolates, derived from an unsymmetrical ketone, may be selectively prepared and such methods have been reviewed ⟨75PAC553, 76T2979⟩. The two isomeric carbanions (**107**) and (**108**) interconvert rapidly only in the presence of the parent ketone or any stronger acid (Scheme 38) ⟨65JOC1341⟩.

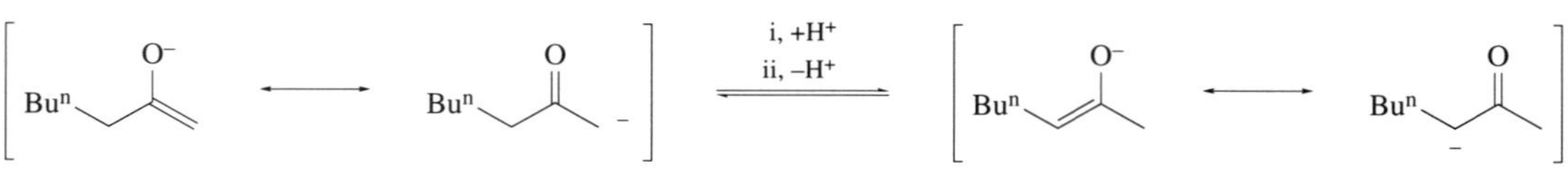

(**107**) (**108**)

Scheme 38

In the absence of such acids, it is possible to prepare either (**107**) or (**108**) and thus in some cases to carry out selective alkylation at the α-position on either flank of the ketone functionality ⟨65JOC2502, 69JOC1962, 71JOC2361⟩. The regioselective formation of enolates derived from unsymmetrical ketones and their subsequent alkylation have been the subject of extensive research. Whereas numerous effective and ingenious methods have been reported for the regioselective formation of the anion, the alkylation step has frequently encountered serious difficulties. Equilibration of enolates accompanying the alkylation process has resulted in loss of regioselectivity and in the formation of polyalkylated products ⟨76T2979, 77T2737, B-79MI 119-02⟩. An especially notorious example of this phenomenon is the benzylation of the kinetic lithium enolate mixture derived either from 2-heptanone directly or from a mixture of the corresponding silyl enol ethers (**109**) and (**110**), which gave as the major products (**111**) and (**112**) in 53% and 37% yield, respectively, along with dibenzylated isomers in 10% yield ⟨71JOC2361⟩.

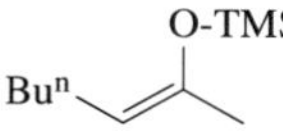

(**109**)

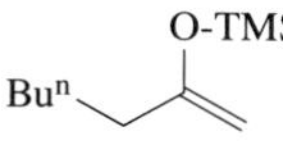

(**110**)

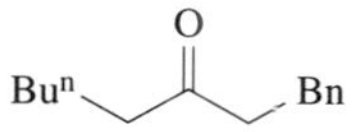

(**111**)

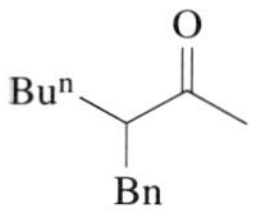

(**112**)

Liotta and Caruso argued that if the kinetic enolate mixture could be substantially activated towards alkylation, then proton transfer and the resulting equilibration and polyalkylation might no longer be an effective competing process in the presence of excess alkylating agent ⟨85TL1599⟩. They reported the use of various Li^+ complexing agents to accomplish these objectives. All the ligands reported produced more terminal than internal alkylation. The order of effectiveness was found to be HMPA (4.9 equiv.) > DMF (neat) > benzo-14-crown-4 (1.61 equiv.) > DMF (6.18 equiv.) > triglyme (1.94 equiv.). HMPA was found to give, starting from a mixture of silyl enol ethers (**109**) and (**110**) a ratio of terminal (**111**) to internal (**112**) benzylation products of 92 : 8, with an absolute yield of (**111**) of 77% (a percentage yield quite close to the absolute amount of terminal enolate (**110**) initially present). The rates of reaction in the presence of each of the ligands were enhanced; in the presence of benzo-14-crown-4, DMF or HMPA the reactions were complete in less than one minute. Since benzyl bromide may be considered to be a relatively reactive electrophile, the regiochemistry obtained using the less-reactive alkylating reagent, *n*-butyl iodide was examined. In the presence of HMPA, not only was the rate accelerated but the product mixture showed a definite preference for the terminal *n*-butylated product. In a preparative scale run using HMPA, a 40% absolute yield of mono-*n*-butylated products was obtained in which 56% was the terminal isomer. Thus, even with a poorer electrophile the presence of a suitable complexing agent gave greater regioselectivity. Liotta and Caruso discussed these observations in the light of NMR work concerning organolithium complexes formed in various ethereal solutions ⟨73JA3687, 77JA4954, 77T2737, 81HCA2617, 81JA4494⟩.

1.19.1.2.14 Factors determining the course of ketone and aldehyde enolate formation

The subject of enolate anion formation has been reviewed by Mekelburger and Wilcox ⟨91COS(2)99⟩. Sterically encumbered amide bases are employed to retard nucleophilic attack on the carbonyl group. The most frequently employed bases are lithium diisopropylamide (LDA), lithium isopropylcyclohexylamide (LICA), lithium tetramethylpiperidine (LITMP), lithium hexamethyldisilylamide (LHMDS) and lithium tetramethyldiphenyldisilylamide (LTDDS). Other useful bases include sodium hydride, potassium hydride and triphenylmethyllithium. The stereochemical course of an enolate reaction frequently depends on the geometry of the enolate anion. The structures of enolates have been studied through NMR studies and by x-ray crystallography ⟨77JA4954, 77T2737, 81JA4494, 85CB764, 85HCA1373, 85JA3345, 85JA5403, 85TL3931, 86HCA1041, 86JA462, 87JA5539⟩.

Aldehyde enolates and aldehydes are extremely reactive and therefore, to avoid undesirable side reactions, fast and quantitative conversion of aldehydes to enolates is necessary. Strong bases are required, e.g., potassium amide in liquid ammonia ⟨70RTC1153⟩ or potassium hydride in THF ⟨78TL491⟩. As a consequence of their excessive reactivity, aldehyde enolates are infrequently used in organic synthesis.

For all types of enolates either (*E*)- or (*Z*)-geometry is possible and this has been the subject of much study ⟨82TS1, B-84MI 119-03, B-84MI 119-04⟩. Such information is pertinent to predicting the outcome of aldol reactions because (*Z*)-enolates tend to give *syn* products, whereas (*E*)-enolates tend to give *anti* products with varying degrees of stereoselectivity.

Under equilibrium conditions, the (*Z*)-enolate is invariably the major product except in the case of three-to-nine membered ring systems, where the (*E*)-enolate is favoured. The investigations of Rathke and co-workers are consistent with this observation (Equation (19)), where the use of LDA gives rise to a product ratio determined largely by kinetic factors ⟨80JA3959⟩.

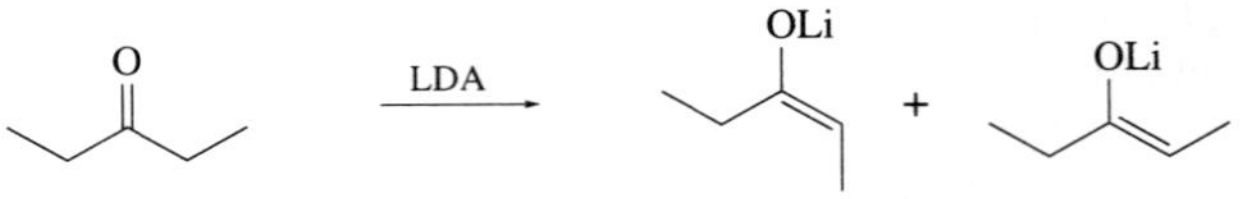

(*E*) 87% (*Z*) 13% (19)

LDA = lithium diisopropylamide

Corey and Gross' studies on enolate selectivity led to the conclusion that the use of sterically hindered bases like lithium *t*-octyl butyl amide (LOBA) gave superior selectivity to (*E*)-enolates as compared to LDA ⟨84TL495⟩. He argued that the stereochemical outcome in the presence of HMPA was not a kinetic effect, but was due to equilibration to the more thermodynamically stable (*Z*)-enolate. Corey and Gross' conclusion was founded on their experiments using TMS-Cl as an internal quenching agent during the enolisation with a lithium amide base (Equation (20)).

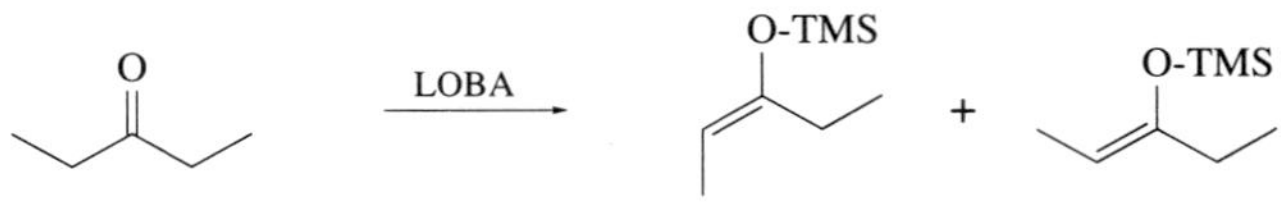

(*E*) (*Z*) (20)

Method	*Solvent*	(E) : (Z)
Internal quench	THF	98 : 2
Internal quench (8 equiv. TMS-Cl)	HMPA, THF	37 : 63
Internal quench (17 equiv. TMS-Cl)	HMPA, THF	46 : 54
Two-step procedure[a]	HMPA, THF	18 : 82

[a] Slow addition of ketone to LDA followed by silylation.

LOBA = Lithium *t*-octyl butylamide

Rathke and co-workers found that the addition of 1–4 equivalents of HMPA or TMEDA did not change the (*E*):(*Z*) ratio (Scheme 39), but the addition of 0.2 equivalents of 3-pentanone caused rapid isomerisation to an equilibrium mixture of enolates with an (*E*):(*Z*) ratio of 16:84. He suggested as explanation the mechanism illustrated in Scheme 39 ⟨80JA3959⟩. Rathke and co-workers found it was possible to control the deprotonation of 3-pentanone in THF solution so as to produce predominantly the (*E*)-isomer by addition of the ketone to 10% excess LITMP at 0°C ((*E*):(*Z*) ratio 87:13), or to produce predominantly the (*Z*)-isomer by addition of the ketone to a slight deficiency of LITMP ((*E*):(*Z*) ratio 16:84). He concluded that the formation of the (*E*)-enolate could be the result of kinetically controlled deprotonation, but the formation of the (*Z*)-enolate is thermodynamically favoured.

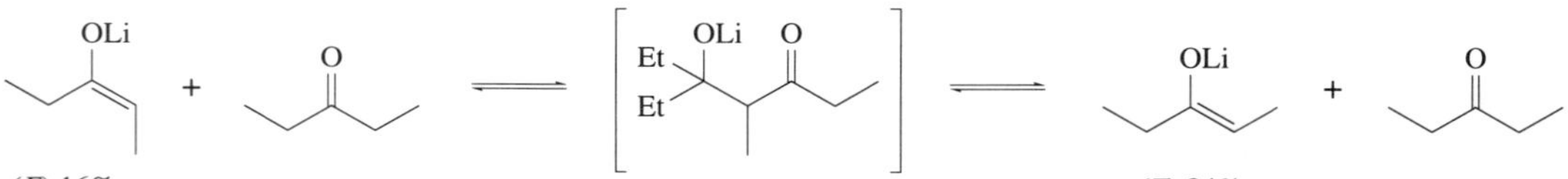

(*E*) 16% (*Z*) 84%

Scheme 39

Ketone enolate stereochemical assignments have been determined by Ireland *et al.* ⟨76JA2868⟩. The preponderance of (*E*)-enolate observed with LDA as the base in THF solution in these and related studies with esters can be rationalised by the consideration of the nonbonding interactions in the alternative pericyclic transition states (**113**) and (**114**) (Scheme 40). Unfavourable steric interactions between R^1 and R^2 disfavour transition state (**113**) and promote enolisation via the transition state (**114**) to give (*Z*)-geometry, whereas dominant R^2 versus L (R^3) nonbonded interactions should militate against transition state (**114**) in favour of (**113**).

If lithium dialkylamide bases are used in a mixture of THF and HMPA, or if silylamide bases are employed, then (*Z*)-enolates prove to be the major products. Prediction of the outcome of kinetic product control necessitates a more precise study of the reactants and bases. Reactions of ketones, esters or thioesters with lithium dialkylamides (e.g., LDA, LICA or LITMP) in THF favour the (*E*)-enolate. Larger, sterically encumbered bases give predominantly (*Z*)-selectivity (Table 7) ⟨77JA8109, 80JOC1066⟩.

However, if a bulky group (e.g. *t*-butyl or phenyl) is located α to the carbonyl group, the

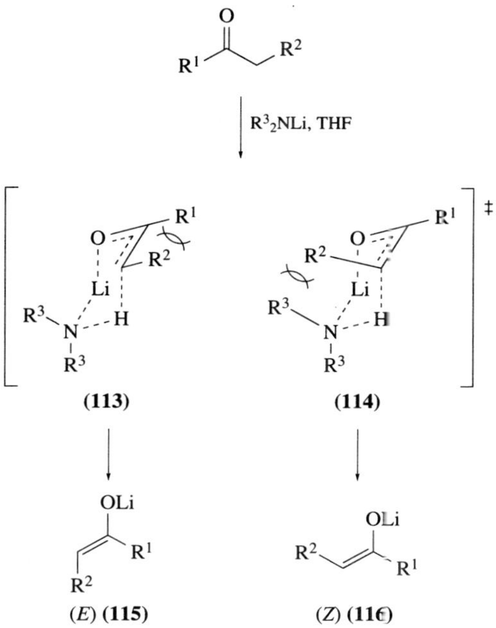

Scheme 40

Table 7 Kinetic enolisation of aldehydes and ketones where R^2 = Et (Scheme 40).

R^1	Base[a]	Enolate ratio (Z) : (E)
Et	LDA	30 : 70
Et	LCPA	35 : 65
Et	LHMDS	66 : 34
Et	LITMP	20 : 80
Pr[i]	LDA	60 : 40
Pr[i]	LCPA	59 : 41
Pr[i]	LHMDS	>98 : 2
Pr[i]	LITMP	32 : 68
Bu[t]	LDA	>98 : 2
1-Adamantyl	LDA	>98 : 2
TMS	LDA	38 : 62
Ph	LDA	>98 : 2
Ph	LCPA	>98 : 2
Ph	LHMDS	>98 : 2
Ph	LITMP	>98 : 2
Mesityl	LDA	5 : 95
Mesityl	LCPA	4 : 96
Mesityl	LHMDS	87 : 13

[a] LCPA = lithium cyclohexylisopropylamide, LHMDS = lithium hexamethyldisylamide, LDA = lithium diisopropylamide, LITMP = lithium tetramethylpiperidine.

(Z)-enolate is preferred ⟨77JA247⟩. As has been discussed, ketones enolise under kinetic conditions regioselectively towards the less-substituted carbon atom, and stereoselectively to the (Z)-enolate, if the enolates are formed by a bulky base and trapped with dialkyl boron triflates, $R_2BOSO_2CF_3$, at low temperatures ⟨79JA6120⟩. Both types of (Z)-enolates can be generated for future application in stereoselective aldol reactions as shown in Equations (21) and (22).

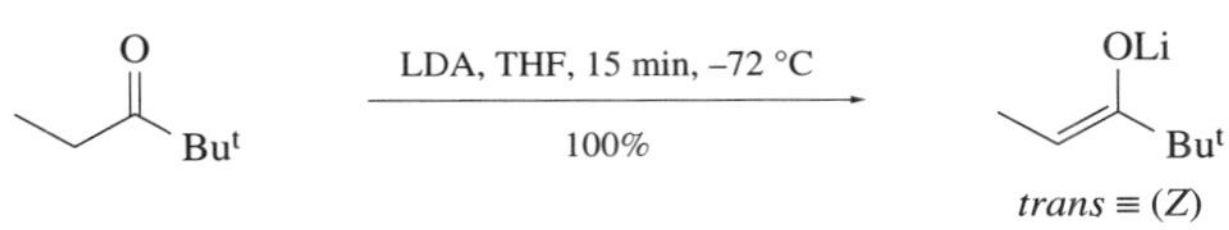

(21)

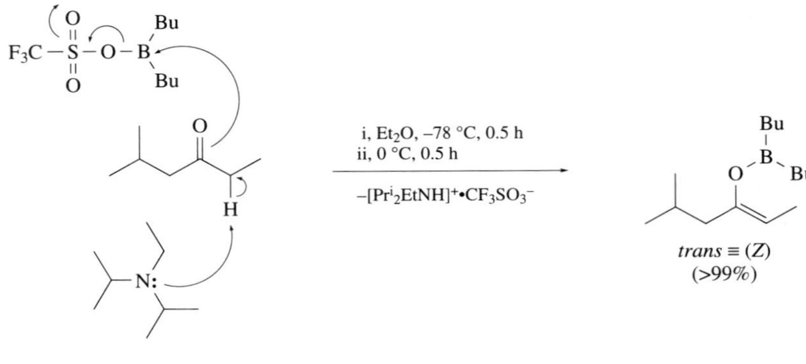

(22)

1.19.1.2.15 Factors determining the course of ester and amide enolate formation

The subject of ester enolate anion formation has been reviewed by Wipf ⟨91COS(5)827⟩. It was initially proposed that the (*E*) : (*Z*) ratio of the enolate esters was kinetically determined, regardless of whether THF or a mixture of HMPA/THF was used ⟨75TL3975, 76JA2868⟩. Two possible transition states (**113**) and (**114**) leading to lithium ester enolate formation are shown in Scheme 40, where in this case $R^1 = OR$. In the absence of the solvating ligand HMPA or DMPU the lithium atom is strongly coordinated to the ester carbonyl oxygen atom leading to strongly unfavourable interaction between R^1 and R^3 in transition state (**114**). Although transition state (**113**) is energetically disfavoured by unfavourable interaction between R^1 and R^2, this transition state leads preferentially to (*E*)-enolate formation in the absence of HMPA or DMPU (*N,N′*-dimethylolpropyleneurea). Contrastingly, in the presence of HMPA or DMPU, the lithium atom is highly solvated resulting in a weakened coordination of the Li^+ by the carbonyl oxygen atom. In this case, unfavourable steric interactions between R^1 and R^3 are reduced by such solvation, which lengthens the Li^+ to carbonyl oxygen bond, while steric repulsion between R^1 and R^2 is relieved by rotation about the methylene carbon to carbonyl carbon C—C bond. These two factors combine to lower the potential energy of transition state (**114**) relative to the unsolvated case and leads preferentially to (*Z*)-enolate formation. It should also be noted in this latter solvated situation, that a decrease in polarisation of the carbonyl oxygen also results in a significantly less reactant-like transition state, as the α-C—H bond becomes more difficult to break ⟨85JA2264, 91JOC650⟩. In the presence of a slight excess of ester in the enolisation mixture, there is a change in the (*E*) : (*Z*) ratio of the lithium enolate in favour of the (*Z*)-enolate ⟨91COS(5)827, 91JOC650⟩.

Tertiary carboxylic acid amides always give (*Z*)-enolates with high selectivity. Ireland *et al.* have proposed a cyclic six-membered transition state between the carbonyl group and the dialkylamide base to explain this phenomenon ⟨76JA2868⟩, and in the mid 1980s the validity of this model was bolstered by theoretical calculations by Moreland and Dauben ⟨85JA2264⟩. As summarised in Table 8, under ‘apparent’ ⟨82TS1⟩ kinetic control, esters and thioesters afford largely (*E*)-enolates, while dialkylamides exhibit predominant to exclusive (*Z*)-enolisation.

Evans *et al.* have postulated ⟨80TL3975, 80TL4233, 81TH⟩ that the high (*Z*)-stereoselection observed in the deprotonation of dialkylamides is a consequence of ground state allylic strain considerations ⟨66AG(E)899, 68CJC821, 68CRV375, 78JOC2705⟩ that disfavour deprotonation via transition state (**113**) (Scheme 40; $R^1 = NR_2$). Thus the formation of (*E*)-enolates (**115**) is disfavoured, whereas formation of (*Z*)-enolates (**116**) via transition state (**114**) (Scheme 40; $R^1 = NR_2$) is perforce the preferred reaction pathway. Investigations indicate that the enolisation process catalysed by LDA in THF affords enolates (**115**) and (**116**) with at least 97% (*Z*)-stereoselection ⟨80TL4233⟩. Related observations have been reported on the stereoselective enolisation of dialkylthioamides ⟨80JA7806⟩. In this latter study, the Ireland–Claisen strategy ⟨76JA2868⟩ was employed to assign enolate geometry.

Studies on the role of base structure and solvent have appeared in the literature ⟨78TL2079, 80JA3959⟩ and the Ireland enolisation model for lithium amide bases, involving a chair-configured six-centre pericyclic mechanism, has now been widely accepted. The stereoselectivity of enolisation has been well reviewed by Evans, Nelson and Taber ⟨82TS1⟩ and also, particularly with respect to the Ireland–Claisen rearrangement, by Pereira and Srebnik ⟨93MI 119-01⟩. The effect of fluorine substituents on the structure and reactivity of enolates and enolate equivalents has been reviewed ⟨B-88MI 119-01⟩. Structural investigations of lithium enolates and related species have been reviewed ⟨88AG(E)1624, 89AG(E)277⟩ and also, specifically, NMR investigations ⟨87AG(E)1212⟩.

Table 8 Kinetic enolisation of carboxylic acid esters and amides.

R^1	R^2	*Base*	*Enolate ratio* (E):(Z)	*Ref.*
OMe	Me	LDA	95:5	80JOC1066
OMe	Et	LDA	91:9	76JA2868
OMe	Ph	LDA	29:71	76JA2868
OMe	Bu^t	LDA	97:3	76JA2868
OBu^t	Me	LDA	95:5	80JOC1066
OBu^t	Et	LDA	95:5	76JA2868
SBu^t	Me	LDA	90:10	80TL3975
NEt_2	Me	LDA	<3:97	81TH
$N(CH_2)_4$	Me	LDA	<3:97	80TL4233
NPr^i_2	Me	LDA	19:81	80JOC1066

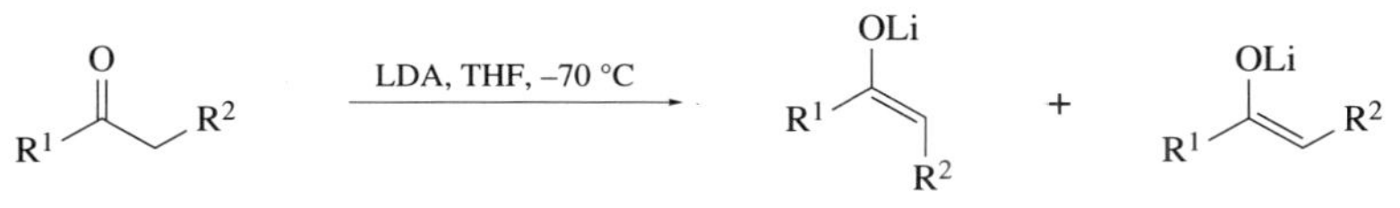

(*E*) (*Z*)

1.19.1.2.16 *Structural effects on the energetics of* s-cis *and* s-trans *dienolates*

Eldin, Whalen and Pollack have examined the acidifying effect of the positioning of either a phenyl or vinyl substituent at the α-carbon of a carbonyl compound ⟨88JA1981, 89JA6419, 90JOC1019, 91JA1344, 91JA3833⟩. Both a phenyl substituent and a vinyl substituent can increase the aqueous acidity of a simple ketone by as much as 10^7-fold. The aqueous acidities of 2-tetralone **(117)** ⟨88JA1981⟩ and 5-androstene-3,17-dione **(118)** ⟨89JA6419⟩ are comparable (Equations (23)–(26)), which implies a quantitatively similar acidifying effect for a phenyl group and a double bond on the α-carbon of carbonyl compounds. However, the double bonds in the β,γ-unsaturated ketones 5(10)-androstene-3,17-dione **(119)** and 3-cyclohexenone **(120)** are approximately 10^2–10^3 less effective in promoting the ionisation of the α-hydrogen atom than in compound **(118)**.

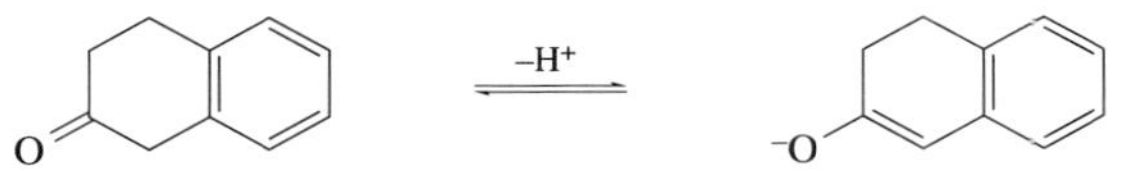

(117) **(121)** pK_a = 12.9 (23)

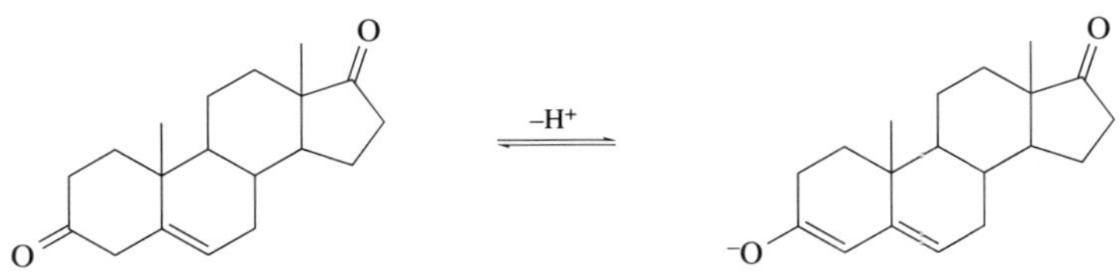

(118) **(122)** pK_a = 12.7 (24)

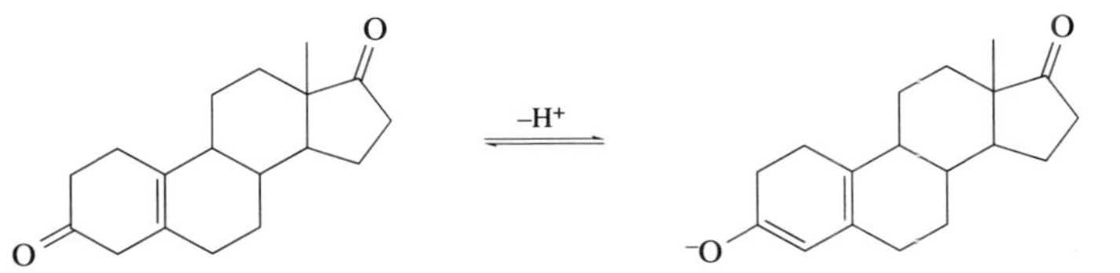

(119) **(123)** pK_a >> 14 (25)

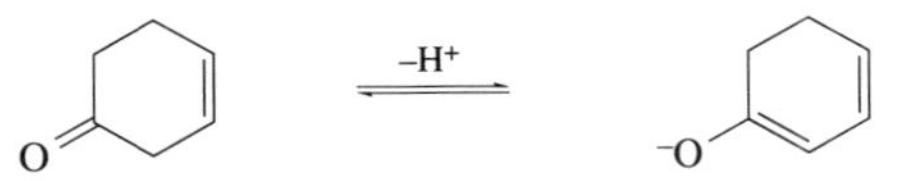

(120) **(124)** pK_a ~ 15.2 (26)

The most salient differences between enolate (**122**) and enolates (**123**) and (**124**) are firstly, that (**122**) has an s-*trans* conformation of the conjugated double bonds, whereas the anions of (**123**) and (**124**) possess an s-*cis* geometry and secondly, (**122**) has an exocyclic double bond that is not present in (**123**) and (**124**).

Pollack and co-workers have reported an ^{1}H-NMR investigation of the relative kinetic acidities of the *cis* and *trans* γ-protons of 2-cyclopentylidenecyclopentanone (**125**) (Scheme 41) ⟨93JOC7100⟩. In this system the influence of endocyclic versus exocyclic double bonds is eliminated, rendering it possible in theory to isolate the effect of orientation of the double bonds of a dienolate ion (s-*cis* versus s-*trans*) on the acidifying effect of a β,γ-double bond. Both the kinetic and quenching experiments indicated that the transition state for the formation of the s-*cis* dienolate (**126**) is higher than that for the s-*trans* dienolate (**127**) by a factor of about four-fold (ca. 0.8 kcal mol^{-1}). Pollack and co-workers estimated that thermodynamically the s-*trans* dienolate (**127**) was about 1.7 kcal mol^{-1} more stable than the s-*cis* dienolate, corresponding to a difference in pK_a of 1.2 units.

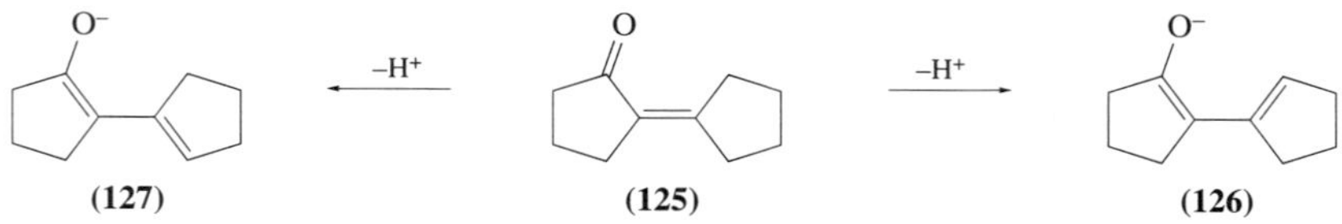

(**127**) (**125**) (**126**)

Scheme 41

Pollack and co-workers concluded that cyclic s-*cis* dienolates were considerably less stable than the corresponding s-*trans* dienolates due to two factors; first, greater strain caused by the presence of two double bonds in the same ring in the case of the s-*cis* dienolate and second, an inherent energy difference between s-*cis* and s-*trans* geometries. The authors ascribed predominance to the first effect.

1.19.1.2.17 Chiral auxiliaries conferring stereocontrol over enolate reactions

Valinol- and ephedrine-derived auxiliaries, to be found incorporated in structures (**128**) and (**129**) respectively, both possess similar characteristics as chiral controllers for a wide range of asymmetric transformations ⟨81PAC1109, 82MI 119-01⟩. (*Z*)-Enolates, derived from *N*-acyl imides of chiral oxazolidinones react with aldehydes in a highly stereoregulated fashion yielding α-substituted-β-hydroxyamides in high yields (Equations (27) ⟨81JA2127⟩ and (28) ⟨82TL807⟩).

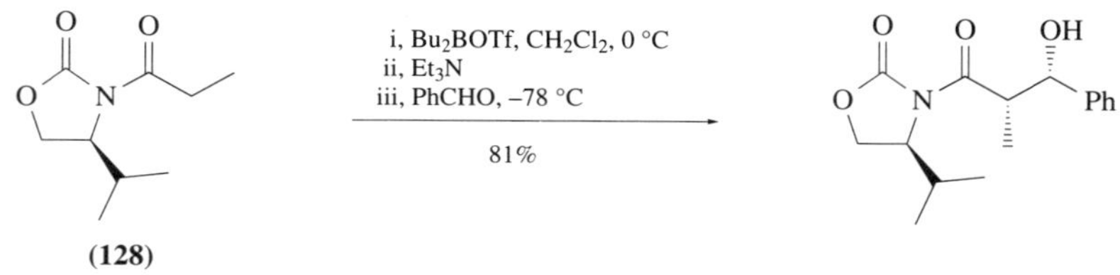

(**128**) (27)

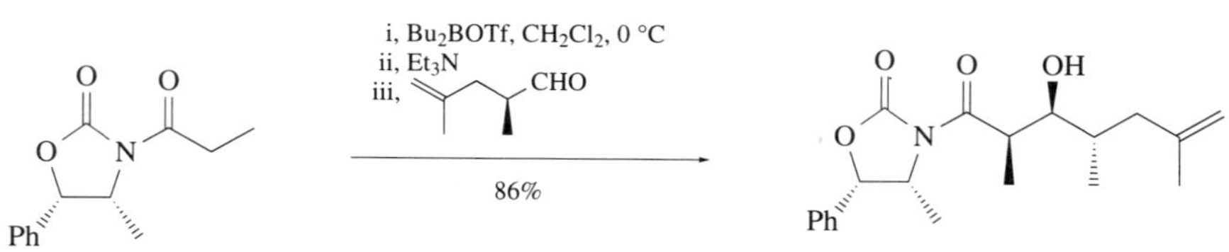

(**129**) (28)

Equations (27) and (28) are illustrative of the vast and growing literature describing the reactions of chiral enolate anions. The reader is directed to the authoritative texts of Evans ⟨B-84MI 119-02⟩ and Heathcock ⟨B-84MI 119-05⟩ for reviews of this topic.

1.19.1.2.18 Stereoelectronic effects in the formation of enolates and related functions

The subject of such stereoelectronic effects has been reviewed by Deslongchamps ⟨B-83MI 119-04⟩. Corey observed ⟨53E329⟩ that the kinetically controlled bromination of ketosteroids always gives the epimer in which the bromine atom is 'polar' (i.e. axial) ⟨54JA175⟩. In a subsequent publication, Corey and Sneen ⟨56JA6269⟩ mentioned that this was an example of a stereochemical-electronic effect.

Fraser and Champagne have observed a marked selectivity in the exchange of the conformationally fixed substrate (**130**), a bridged biaryl ketone in which the C—H^1 bond is parallel to the $p\pi$-orbitals of the carbonyl group; the $H^1:H^2$ ratio of rates of exchange was found to be 73 using sodium methoxide and 30 using sodium phenoxide ⟨76CJC3809⟩.

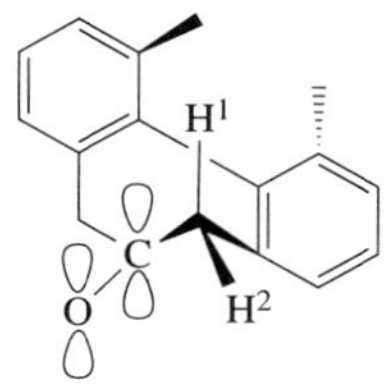

(130)

Fraser and Champagne also studied the base-catalysed H/D exchange reaction of 4-twistanone (**131**) ⟨78JA657⟩. One C—H bond alignment (H^1) gives maximum overlap with the adjacent $p\pi$-orbital of the carbonyl function, whereas the dihedral angle between the C—H^2 bond and that $p\pi$-system is 30°. Accordingly, when 4-twistanone was treated with sodium methoxide in methanol-O-d, the relative rates of exchange of the diastereotopic protons H^1 and H^2 were found to be in the ratio 290 : 1. Using *ab initio* molecular orbital calculations Wolfe and Schlegel have predicted that the enolate derived from the removal of H^1 from (**131**) should be more stable than that derived by abstraction of H^2 by 18 kcal mole^{-1} ⟨75CJC3365⟩.

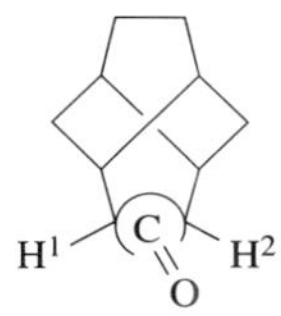

(131)

1.19.1.2.19 Dicarbanion formation by proton loss

Compounds of the type MeCO—CH_2—CO— can release two protons, if treated with 2 moles of a sufficiently strong base (e.g., LDA or BuLi), to give dicarbanions (e.g., (**132**) (Scheme 42)). Such dianions are ambident carbon nucleophiles, in addition to an attack capability involving the oxygen atoms. For such species the electrophilic attack is almost invariably on the more basic carbon atom. However, counter-examples do exist, for example the work of Trimitsis *et al.* ⟨83JOC2957⟩. Since a proton on a carbon bonded to two carbonyl groups is more acidic than that on a carbon adjacent to just one carbonyl group, it follows that the CH^- group of (**132**) is more acidic than the CH_2^- group, so it is the latter group which is attacked by an electrophile to give (**133**) ⟨58JA6360, 65JOC2934⟩. The subject has been reviewed ⟨69OR(17)155, 77S509, 91COS(2)181, 91T4223⟩.

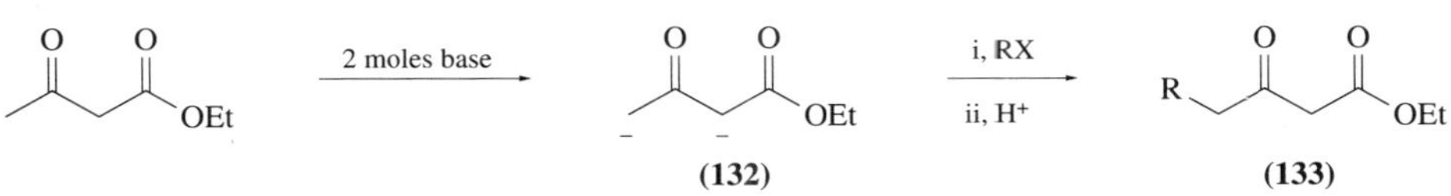

Scheme 42

The hydrocarbon analogue (**135**) of the squarate dianion (**136**) has been prepared by metallation of 1,2-dimethyl-3,4-bismethylenecyclobutene (**134**) with Bu^nLi, Bu^tOK or Bu^nLi, TMEDA (Scheme 43) ⟨89JOC773, B-89MI 119-01⟩.

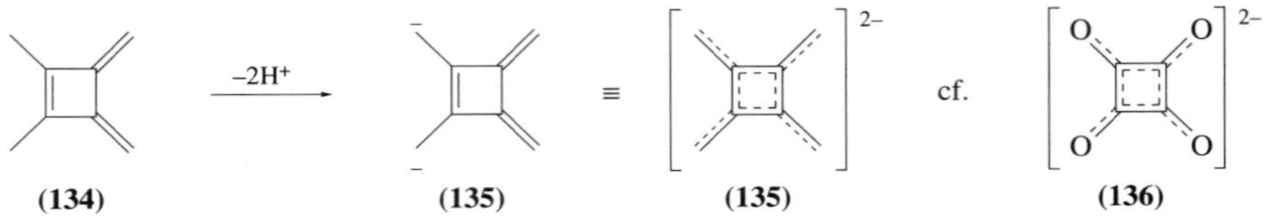

Scheme 43

Methylated biphenylene dianions have been characterised by NMR spectroscopy and their direction of ring-opening has been rationalised by molecular orbital calculations ⟨89TL159⟩.

1.19.1.2.20 Choice of base and solvent for the formation of various types of carbanion stabilised by conjugation

C—H bonds are polarised by attached unsaturated carbon substituents. Such groups activate the neighbouring CH_3, CH_2 or CH in the following order: $CR{=}NR_2^+ > COR > C{\equiv}N > CO_2R > CR{=}NR > Ph > CR{=}CR_2$.

The formation of the carbanions described in Scheme 44 depends on equilibria between the carbon acid, its conjugate base and the solvent. To ensure a substantial concentration of the carbanion in solution the pK_a of both the conjugated acid of the base and of the solvent must be higher than the pK_a value of the carbon acid. Alkali hydroxides in water ($pK_a \sim 16$), alkoxides in the corresponding alcohols ($pK_a \sim 20$), sodium amide in liquid ammonia ($pK_a \sim 35$), sodium methylsulfinylmethylide in dimethyl sulfoxide ($pK_a \sim 35$), alkali metal hydrides ⟨86AG(E)465⟩, lithium amides or lithium alkyls in ether or hydrocarbon solvents ($pK_a > 40$) are common combinations used in synthesis (Table 9).

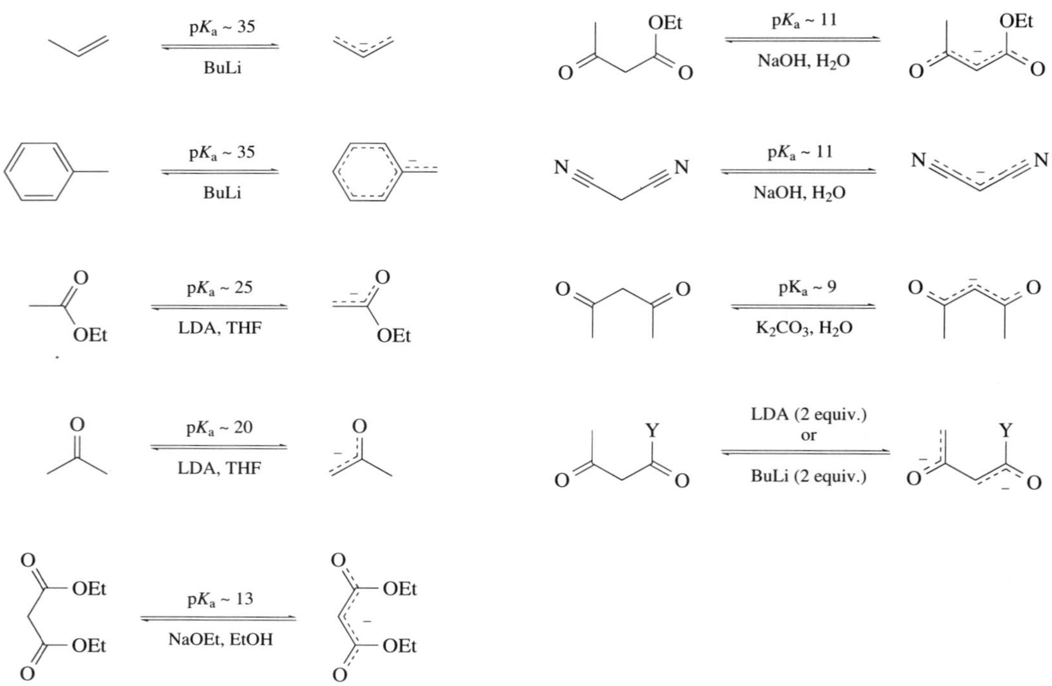

Scheme 44

Sometimes, unfortunately, the bases (e.g., methoxides, amides, lithium alkyls) react as nucleophiles, rather than carry out the desired proton abstraction, with the result that the nucleophile undergoes an addition and/or substitution reaction with the carbon compound. In order to avoid this undesirable situation, nonnucleophilic sterically hindered bases are popular, e.g., potassium *t*-butoxide, LDA, sodium bis[trimethylsilyl]amide, potassium 3-aminopropylamide (KAPA),

Table 9 Common combinations of bases and solvents used for the conversion of weakly acidic methylene compounds to their anions.

Base	*Frequently used solvents*	*References to preparation and use*
Bu^tOK	Bu^tOH, DMSO, Et_2O^a, 1,2-DME, PhH^a, THF	51JA5770, 51OR(6)41, 57JA2906, 57JOC602, 58JA5220, 58JA6296, 58JOC841, 60JA2847, 61JA3678, 62JA3164, 62JA3402, 62TL669, 63JA1538, 63OSC(4)132, 64JA1997, 64JA5457, 65JA3228, 65OS(45)89, 66JA1332, 67JCS(C)1860 [For crystal structure of Bu^tOK see 67AG(E)801]
$EtMe_2CONa$	Et_2O, PhH, PhMe	50BSF533, 50BSF537, 56BSF1040, 61BSF836, 61T45, 62TL505, 63BSF1925, 63BSF1930, 63RCP42
$NaNH_2$	liq. NH_3, Et_2O^a, PhH^a, $PhMe^a$, 1,2-DME	41JA2975, 55OSC(3)44, 55OSC(3)219, 55OSC(3)778, 60OS(40)38, 62JOC3836, 64AG(E)471, 65JOC2937, 65JOC4135, 66JOC982, 66JOC989, 67OS(47)72, 67OS(47)92, 68OS(48)53, 68OS(48)80
KNH_2	liq. NH_3, Et_2O^a	58JA6360, 59JA1154, 59JA1160, 59OS(39)73, 60JOC158, 62JOC2737, 63JOC1946, 64AG(E)471, 65JA82, 66JOC1035, 68OS(48)40
Et_2NLi	1,2-DME, Et_2O	32LA(495)84, 45JA2106, 45JA2110, 50JOC850, 51JA4158, 64CB3548, 64JOM(1)476, 65JOC1341, 65JOC2502, 68JOC935, 68JOC943, 70JA3222, 70OS(50)66
Pr^i_2NLi	1,2-DME, Et_2O, THF	67JA2500, 69JOC2324, 70JA1396, 70JA1397, 70OS(50)58, 70OS(50)66
$(TMS)_2NNa$	THF, Et_2O, PhH	64JOM(1)476, 69CC1497, 69CC1498, 70JA3222
NaH	Et_2O^a, PhH^a, $PhMe^a$, $C_6H_4Me_2{}^a$, 1,2-DME^a, DMF^a	45JA1273, 50JA3704, 52JA831, 59JOC709, 60JA1386, 60JA1684, 60JA2895, 60JA5704, 61JOC644, 61OS(41)5, 62OS(42)28, 63JOC3362, 63OSC(4)261, 65JOC1341, 65JOC2502, 67OR(15)1, 67OS(47)20, 68JOC935, 68JOC943, 68JOC2222, 68JOC3425, 68JOC3477, 70TL3903 (For the preparation of a very reactive form of sodium hydride by reaction of hydrogen with sodium naphthelenide in THF solution, see ⟨68JA4505⟩)
LiH	Et_2O^a	59JOC709, 60JA1386, 68JOC2222, 68JOC3477
$MeSOCH_2Na$	DMSO	61JOC4112, 62JA866, 64JOC221, 64PCS108, 65JA1345, 69MI 119-01
Ph_3CLi	Et_2O, 1,2-DME	59JOC229, 63JOC3362, 63JOC1725, 65JOC1341, 65JOC2502, 68JOC935, 68JOC943, 70CC1325
Ph_3CNa	Et_2O, PhH, PhMe, liq. NH_3	43OSC(2)268, 43OSC(2)606, 54JOC1490, 55JOC1531, 57JA881, 59JOC709
Ph_3CK	1,2-DME	54JOC1490, 55JOC1531, 62JOC4146, 63JOC3362, 65JOC1341, 65JOC2502, 68JOC935, 68JOC943
Sodium naphthelenide and related carbanions	1,2-DME	36JA2442, 60BSF354, 62JOC3745

[a] Solvents in which the base is either insoluble or only partially soluble. From 'Modern Synthetic Reactions', 2nd edn., by Herbert O. House, Copyright (©) 1972 by W. A. Benjamin, Inc.

lithium 1,1-bis(trimethylsilyl)-3-methylbutoxide and the weak bases 1-(dimethylamino)naphthalene and poly(4-(diisopropylaminomethyl)styrene) ⟨B-72MI 119-02, 76TL1817⟩.

1.19.1.2.21 Metallation of N*-alkylimines*

A carbonyl function may be converted into an amino group by condensation with a primary amine. The resulting *N*-alkylimine ('Schiff base') (**137**) (Scheme 45) can be metallated with LDA in ethereal solvents at 0°C. At temperatures below −70°C such lithiated Schiff bases are stable against self-addition. These carbanions are useful in selective aldol additions ⟨68AG(E)7⟩. Metallated dimethyl hydrazones have also been used extensively. They permit stereo- and regioselective carbanion formation since the dimethylamino group orients away from large substituents and tends to locate the metal ion proximate to the less substituted carbon atom ⟨78CB1337⟩.

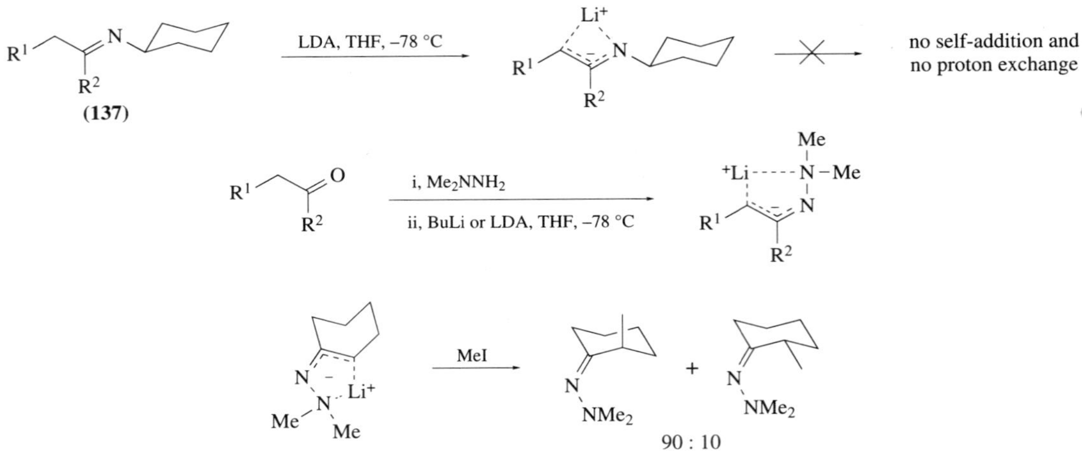

Scheme 45

1.19.1.2.22 Carbanion formation with concomitant double-bond rearrangement

The double bonds of many unsaturated compounds are shifted on treatment with strong bases ⟨B-89MI 119-02⟩ and there have been numerous reviews ⟨68ACR231, 69S97, 70RCR859, 70S405⟩. In many cases equilibrium mixtures are obtained and the thermodynamically most stable isomer predominates. Lists of which double bonds are more stable in conversions of $XCH_2CH{=}CHY$ to $XCH{=}CHCH_2Y$ are to be found in the literature ⟨82JOC4766, 83JOC584⟩. If formation of the new double bond places it in conjugation with an existing double bond or an aromatic ring, this pathway will usually be favoured, although there are counter-examples ⟨62TL669, 65JOC2513⟩. If formation of the new double bond promotes resonance stabilisation with a lone electron pair then, similarly, this pathway will be favoured ⟨64TL1191, 66JCS(C)82, 68JCS(C)1903, 72JCS(P1)1535, 73JCS(P1)1858, 87TL5887⟩.

An example of the isomerisation of alkenes by base is furnished by the synthetically useful Varrentrapp reaction ⟨60T246⟩. Despite the vigorous conditions, the reaction can be quite specific. The product (**138**) (Scheme 46) is derived in high yield and results from at least nine successive isomerisations followed by base-catalysed cleavage.

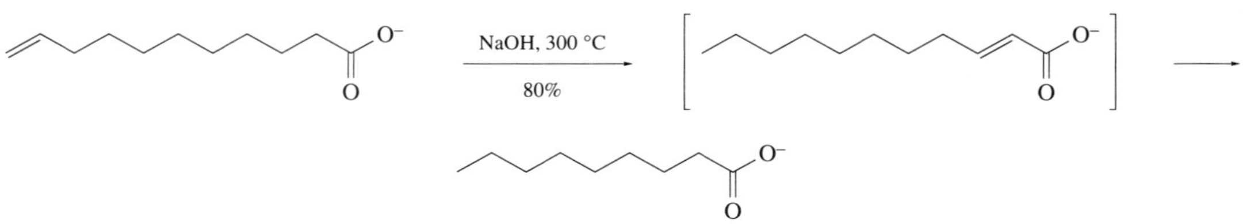

(138)

Scheme 46

Terminal alkenes react with potassium *t*-butoxide in DMSO to give internal alkenes of predominantly *cis*-configuration (e.g., (**140**) (Scheme 47)) as the favoured product via the *cis*-allylic anion (**139**) ⟨65JA3244⟩.

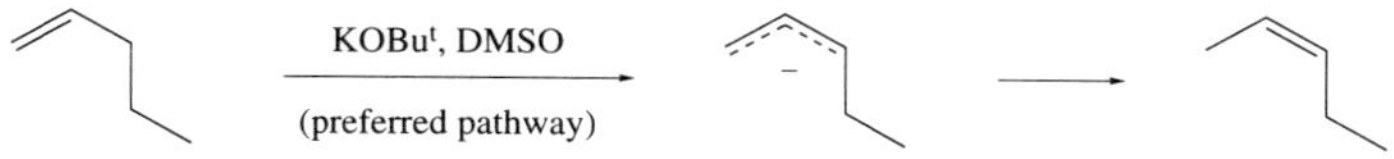

(139) **(140)**

Scheme 47

1.19.1.3 Carbanions by Scission of C—C or C—X Bonds

1.19.1.3.1 General considerations of carbanion formation by C—C bond scission

In these reactions, the substrate may be regarded as the moiety that retains the electron pair. The resultant carbanion is usually quenched by a proton. The reactions are sometimes termed 'anionic cleavages' ⟨87RCR983⟩ and usually proceed via an S_E1 carbanionic mechanism. Such reactions are facilitated kinetically by any inherent stabilisation of the nascent carbanion.

1.19.1.3.2 Formation of carbanions by decarboxylation

The elimination of carbon dioxide from a carboxylic acid is an important reaction in organic chemistry. However, decarboxylation of simple aliphatic acids requires high temperatures and is of limited synthetic value, although acetic acid in the form of the acetate salt, when heated with base, gives good yields of methane ⟨63JCE212⟩. Contrastingly, decarboxylation readily occurs when certain structural features are present. When the acid possesses a suitably situated electron-withdrawing substituent as in oxalic, nitroacetic or trichloroacetic acid, carbon dioxide is readily eliminated on heating and the resultant carbanion is protonated. The facile elimination of carbon dioxide from β-ketonic esters to yield ketones probably proceeds via an effectively concerted mechanism involving a cyclic transition state without formation of a carbanion ⟨61JA1951, 63PCS253⟩. However, monocarboxylic acids or their alkali salts on heating with sodium hydroxide, calcium hydroxide or other metallic hydroxides give an alkane and a metal carbonate (Equation (29)) ⟨1856MI 119-01⟩. This pyrolysis presumably involves transient carbanion formation.

$$(RCO_2)_2Ca + Ca(OH)_2 \xrightarrow{\text{heat}} 2RH + 2CaCO_3 \qquad (29)$$

1.19.1.3.3 Formation of carbanions during cleavage of alkoxides

Alkoxides of tertiary alcohols can be cleaved by thermolysis (Equation (30)) ⟨59JA1617, 77JOM445, 78JA2134⟩. The reaction fails when the R groups are unbranched alkyl groups. Branched alkoxides, such as those derived from diisopropylneopentylcarbinol or tri-*t*-butylcarbinol, can be cleaved ⟨78JOC815⟩. Allylic ⟨87HCA1858, 87HCA1879⟩, benzylic ⟨88JCS(P2)983⟩ and aryl groups also cleave. Gas-phase studies indicate that cleavage proceeds via carbanion formation ⟨88JA2714, 89CC161⟩. However, with some substrates ⟨89JOC5044, 89JOC5054⟩ and hindered alcohols (not the alkoxides) a radical pathway is suspected ⟨90JOC1052⟩. Mechanistic studies have examined the inversion and the subsequent degree of racemisation of carbanions thus formed ⟨59JA5740, 69JA1009, B-92MI 119-06⟩.

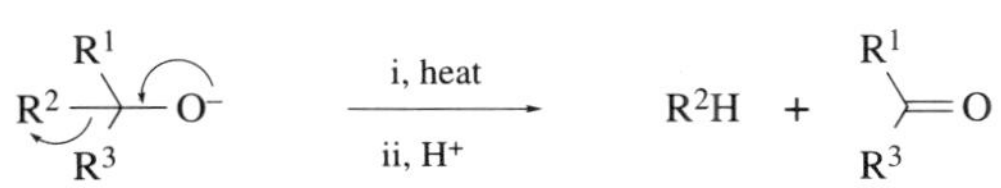

(30)

1.19.1.3.4 Carbanion formation by acyl cleavages

(i) Formation of carbanions during the retro-aldol reaction

The factors affecting reversibility of the aldol reaction have been reviewed ⟨91COS(2)133⟩. The entire aldol reaction (including the possible dehydration step giving the α,β-unsaturated carbonyl compound (**141**)) is an equilibrium (Scheme 48) and α,β-unsaturated and β-hydroxy aldehydes and ketones can be cleaved by treatment with hydroxide anion. However, there is evidence that an SET (single electron-transfer reaction) mechanism can participate when the substrate is an aromatic ketone ⟨82JA6788, 86JOC472⟩.

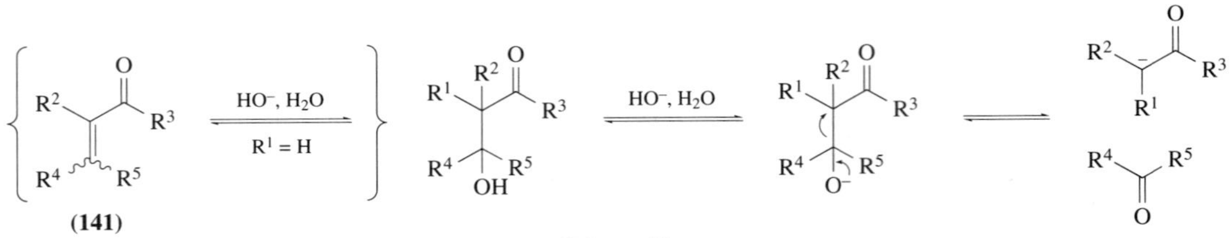

Scheme 48

Aldol reactions in which aldehydes are receptors (i.e., $R^4 = H$) are often easily reversible under mild conditions while aldol reactions in which ketones are receptors are more readily reversible than the analogous reactions with aldehyde receptors. The main structural factor that favours reversibility is steric compression of the aldol. Branching at the hydroxy carbon in the aldol favours reversibility, as does branching at the carbon situated between the hydroxy and carbonyl functions in the aldol ⟨83JA1598⟩. The degree of stereoselectivity of aldol reactions can be adversely affected by the retrograde reaction. In most cases, aldolate equilibration must be avoided, because it usually degrades the kinetically established stereoisomer ratio ⟨91COS(2)234⟩.

The cleavage of cyanohydrins is mechanistically similar to the retrograde aldol reaction ⟨55OSC(3)101⟩. The retrograde Michael reaction also gives rise to carbanions via C—C bond scission (see Section 1.19.1.4.1.1).

(ii) Base-induced cleavage of β-keto esters

When β-keto esters are treated with concentrated base, cleavage occurs via a tetrahedral intermediate with concomitant carbanion formation (Scheme 49) ⟨37JA1823⟩; see also ⟨B-92MI 119-07⟩ although there is evidence that in some cases a SET mechanism is involved ⟨83TL1667⟩.

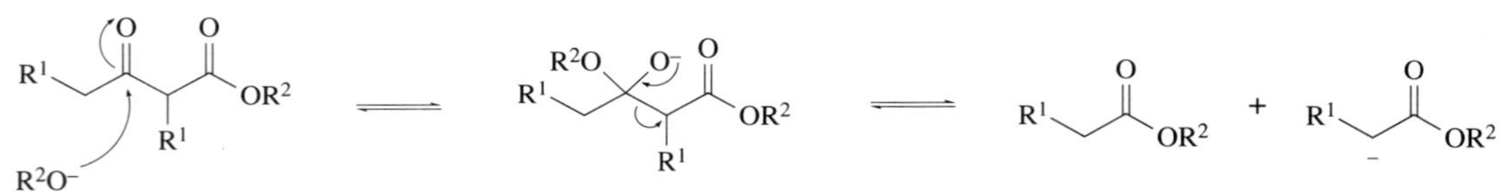

Scheme 49

The ready reversibility of the closely related Dieckmann condensation combined with the tendency to form the least hindered metal enolate usually determine the structure of the cyclised product from an unsymmetrical diester ⟨43OSC(2)116, 50JA2395, 52JA1704, 64JOC2781, 65JOC181, 65OS(45)25, 66CC880, 66T285, 67JOC2257, 67OR(15)1, 68JOC460, B-72MI 119-03⟩.

(iii) Carbanion formation resulting from cleavage of ketones

Carbanion formation resulting from the heterolytic C—C bond scission is of pivotal importance to the mechanism of the synthetically useful Favorskii rearrangement (see Section 1.09.1.2.5) ⟨70RCR732, B-92MI 119-08⟩.

(iv) Cleavage of nonenolisable ketones

Ordinary ketones are generally much more difficult to cleave than trihaloketones or β-diketones, because the carbanionic intermediates in these latter cases are more stable than simple unsubstituted carbanions ⟨B-92MI 119-09⟩. However, nonenolizable ketones can be cleaved by treatment with a 10:3 mixture of Bu^tOK/H_2O in an aprotic solvent such as ether (Scheme 50), DMSO, 1,2-DME

⟨48JCS1408, 67JA946⟩ or with solid *t*-BuOK in the absence of a solvent ⟨77JCS(P1)460⟩. In certain cases, cyclic ketones can be cleaved by treatment with base, even if they are enolisable ⟨58T88⟩.

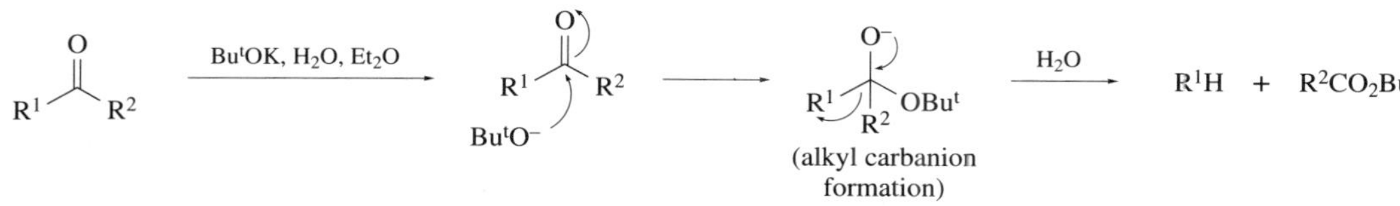

Scheme 50

Cleavage of ketones with sodium amide is called the Haller–Bauer reaction ⟨75S395, 90OPP167⟩. As in the preceding case, the reaction is usually applied only to nonenolisable ketones, and most frequently to ketones of the form $ArCOCR_3$, where the products R_3CCONH_2 are not easily attainable by other methods. However, a diversity of ketones can be cleaved by this method. It has been shown that the configuration of optically active R^1 is retained (Scheme 51) ⟨52JA4838, 62JOC4152, 88JOC4972, 88JOC4978⟩.

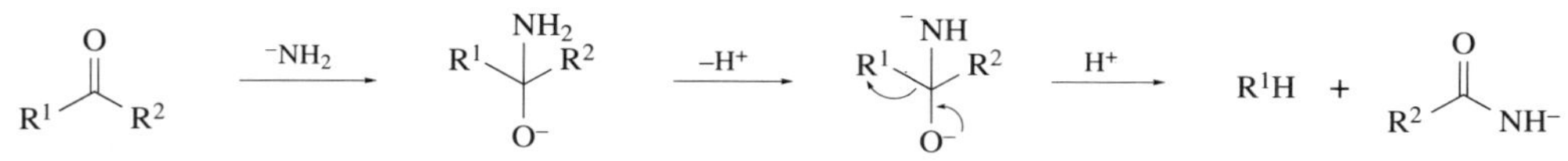

Scheme 51

1.19.1.3.5 Carbanions derived from organometallic species

The nature of the carbon—metal bond may be ionic or polar-covalent depending chiefly on the electronegativity of the metal and the structure of the organic moiety. Ionic bonding is especially favoured when the negative charge on the carbanionic carbon is stabilised by resonance or field effects. Thus the sodium salt of acetoacetic ester is more ionic than methylsodium. The enormous literature of organometallic chemistry has been reviewed ⟨B-67MI 119-02, B-85MI 119-02, B-89MI 119-03, B-92MI 119-10⟩. Organometallic chemistry involving species either π-bonded or σ-bonded to metals has been reviewed by Pearson ⟨B-85MI 119-01⟩. The structures of organodialkali compounds have been specifically reviewed ⟨B-87MI 119-04⟩. The synthesis and structural characterisation of carbanions of alkali and alkaline earth cations has been reviewed ⟨91COS(1)1⟩, along with organoaluminum reagents ⟨91COS(1)77⟩, organocopper reagents ⟨91COS(1)107⟩, organotitanium and organozirconium reagents ⟨91COS(1)139⟩, organochromium reagents ⟨91COS(1)173⟩, organozinc, organocadmium and organomercury reagents ⟨91COS(1)211⟩, organocerium reagents ⟨91COS(1)231⟩, organosamarium and organoytterbium reagents ⟨91COS(1)251⟩. In 1991 developments in synthetic organic chemistry via organometallics were reviewed by Dötz and Hoffmann ⟨B-91MI 119-02⟩.

Most organometallic bonds are polar-covalent; only the alkali metals have electronegativities low enough to form truly ionic bonds ⟨B-92MI 119-10⟩ with carbon, while even lithium alkyls show considerable covalent character. X-ray crystallographic studies of organoalkali compounds have been reviewed ⟨87AOC(27)169⟩. X-ray crystallographic examination of potassium, rubidium and caesium methyls indicates completely ionic lattices ⟨70CB265, 77CB717⟩. The simple alkyls of sodium, potassium, rubidium and caesium are insoluble in benzene or other nonpolar organic solvents, whereas alkyllithiums are soluble in such media.

Metal-associated alkyl carbanions are usually produced from the corresponding halides by metal–halogen exchange. Standard procedures may be found in, besides others, the text by Fieser and Fieser ⟨B-67MI 119-03⟩. Halides at sp^3-hybridised carbon atoms are more reactive than sp^2-hybridised carbon atoms. The scissile ionic carbon–metal bonds (e.g., C—Na, C—Li or C—Mg) may subsequently be replaced by a more covalent carbon–metal bond (e.g., C—Cu) by metal–metal interchange ⟨B-86MI 119-01⟩. The metal ion may be strongly associated with the carbanion centre and modify its behaviour. However, carbanion chemistry and the chemistry of organometallic compounds, especially of Li, Mg, Na and K are closely related ⟨66MI 119-08, B-73MI 119-03, B-74MI 119-01, B-88MI 119-02⟩. Their chemical reactions are similar. Synthetic chemistry, however, takes advantage of two important phenomena: first lithium forms more covalent structures than other alkali metals and second, carbanions with a 'soft' copper(I) gegenion tend to be highly nucleophilic. Organometallics, as either reactants/reagents, or reaction intermediates, can exhibit fascinating reactivity profiles. The assessment of the diverse reactivity patterns is of great theoretical and practical interest. In a generalisation of the Lewis acid–base concept and by dividing the acids and bases into two categories ('hard' and 'soft'), Pearson formulated the Hard and Soft Acids and Bases (HSAB) principle ⟨63JA3533, 67JA1827⟩. Tse-Lok Ho has surveyed a number of organometallic reactions, useful in organic synthesis, in terms of the HSAB principle ⟨85T1⟩.

Carbanions are much less prone to rearrangement than are carbocations ⟨B-87MI 119-05⟩. The analogue of the most common type of carbocation rearrangement, which would be migration of an alkyl group to an adjacent electron-rich centre leaving its bonding pair of electrons behind, does not occur at all readily, although a rare example is shown in Equation (31) ⟨77JA1881⟩. Pericyclic rearrangements of carbanions have been reviewed ⟨78AG(E)313⟩.

(31)

With the exception of pericyclic rearrangements, carbanion rearrangements usually occur stepwise either via intramolecular addition of a carbanion to a neighbouring unsaturated group or by nucleophilic attack by the carbanion at a neighbouring substitution site ⟨B-63MI 119-01, B-65MI 119-05, B-75MI 119-01, B-80MI 119-03⟩.

Scheme 52 shows an anionic rearrangement in which a phenyl group migrates from an adjacent centre, leaving a more stable carbanion ⟨61JA412, 61JA2537⟩. Such reactions are termed Grovenstein–Zimmerman rearrangements. The intermediate carbanion is $Ph_3C—CH_2^-$, and the phenyl group moves without its electron pair. Aryl and vinyl, but not alkyl groups, migrate in this type of rearrangement ⟨89JOC1671⟩.

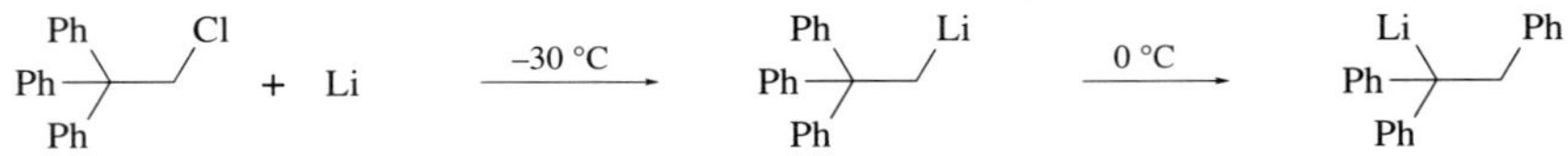

Scheme 52

Zimmerman and Zweig have demonstrated that when there is a choice between a phenyl group and a tolyl group, the phenyl group migrates preferentially (Equation (32)) ⟨61JA1196⟩. This is consistent with a mechanism in which the migrating aryl group interacts with the carbanion centre through its π-electron system to give a transition state (or perhaps an intermediate) structure (**142**) or (**143**) (Scheme 53). The phenyl migration involving (**142**) would be favoured because carbanion (**143**) is destabilised by the electron-donating methyl group. Anion (**144**) can be detected in solution at low temperature ⟨76JA7835, 82JA6681, 82JOC2928⟩.

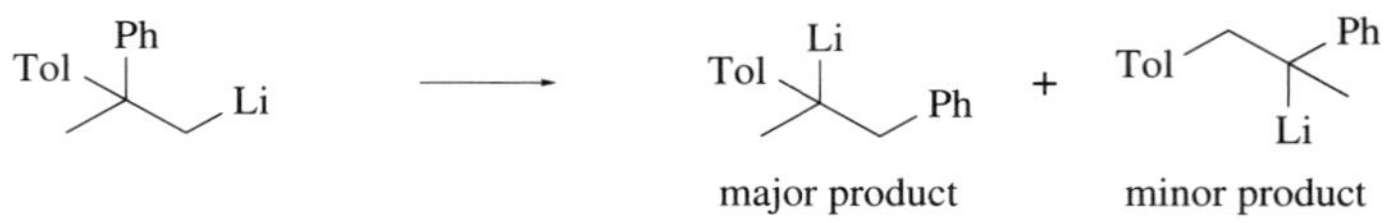

(32)

1.19.1.3.6 *Formation of carbanions by scission of C—O and C—halogen bonds*

Standard procedures for the preparation of organometallic species from halogenated carbon compounds may be found in the text by Fieser and Fieser ⟨B-67MI 119-03⟩ and are also covered in Chapter 2.11 of this work.

Cleavage of C—O and C—Cl bonds by metals dissolved in an aprotic solvent can facilitate the preparation of a variety of aromatic ⟨78CB2480⟩ and homoaromatic ⟨68JA6562⟩ carbanions and carbanionic annulenes. In 1978, Boche *et al.* found (Equations (33) and (34)) that the reaction of 9-*anti*-methoxy- or 9-*anti*-chlorobicyclo[6.1.0]nona-2,4,6-triene (**145**) or (**146**) with Li, Na or K in THF at −30 °C to −20 °C led stereoselectively to the aromatic *cis,cis,cis,trans*-[9]annulene anion (**147**) ⟨78CB2480⟩. The isomeric *syn*-9-methoxy- (**148**) or *syn*-9-chloro-*cis*-bicyclo[6.1.0]nona-2,4,6-triene (**149**) on the other hand, gave the *all-cis*-anion (**67**). Boche *et al.* concluded that 9-*syn*-lithio- and 9-*anti*-lithio-*cis*-bicyclo[6.1.0]nona-2,4,6-trienes (**148**; X = Li) and (**145**; X = Li) were not intermediates in the formation of (**67**) and (**147**). They presumed that electron transfer, giving

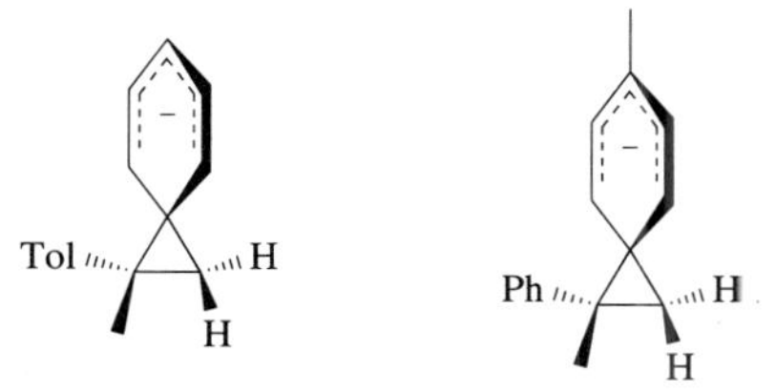

(142) **(143)**

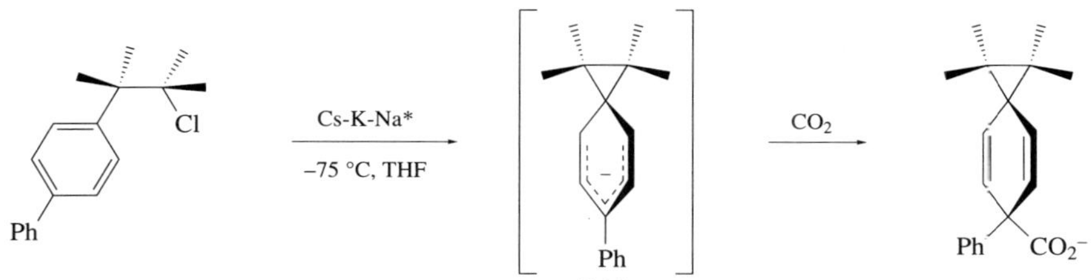

(144)

* alloy of eutectic composition

Scheme 53

bicyclo[6.1.0]nonatriene anion radicals followed by stereoselective ring-opening of these species, was responsible for the course of these reactions.

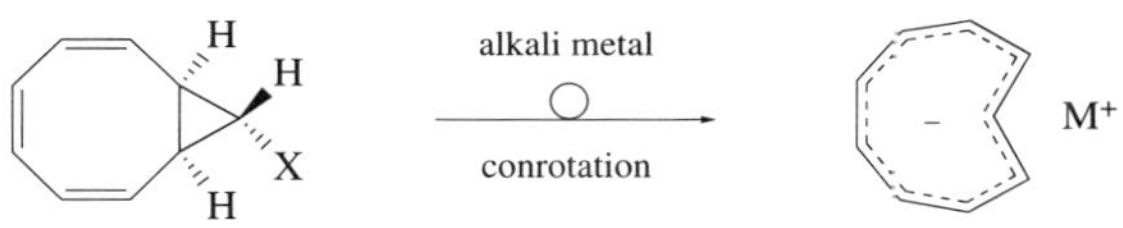

(33)

(145) X = OMe
(146) X = Cl

(147)

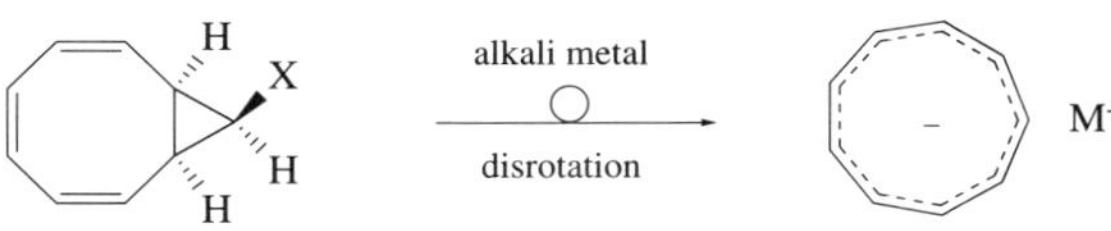

(34)

(148) X = OMe
(149) X = Cl

(67)

Substrate	*Reagent(s)*	*Product Carbanion*	*Yield(s) (%)*
(145)	Li or Na or K	**(147)**	>97%
(146)	Li	**(147)**	68%
(146)	Na	**(147)**	75%
(146)	K	**(147)**	82%
(146)	Li-naphthelenide	**(147)**	68%
(146)	Na-naphthelenide	**(147)**	67%
(148)	Li or Na or K	**(67)**	>97%
(149)	Li or Na or K	**(67)**	>97%

1.19.1.3.7 Formation of carbanions by scission of C—N bonds

The mechanism of the Wolff–Kishner reduction of aldehydes and ketones involves transient carbanion formation as a result of the cleavage of a C—N bond leading to the evolution of a molecule of nitrogen. This and related reactions have been reviewed ⟨48OR(4)378, 68AG(E)120, B-72MI 119-04, 91COS(8)327, B-92MI 119-11⟩. The Wolff–Kishner reduction of the α,β-unsaturated ketone **(150)** (Scheme 54) results in a mixture of products **(152)** and **(153)**, which demonstrates the role of the intermediate allylic carbanion **(151)**, but also highlights one of the limitations of the procedure ⟨64JA5503⟩.

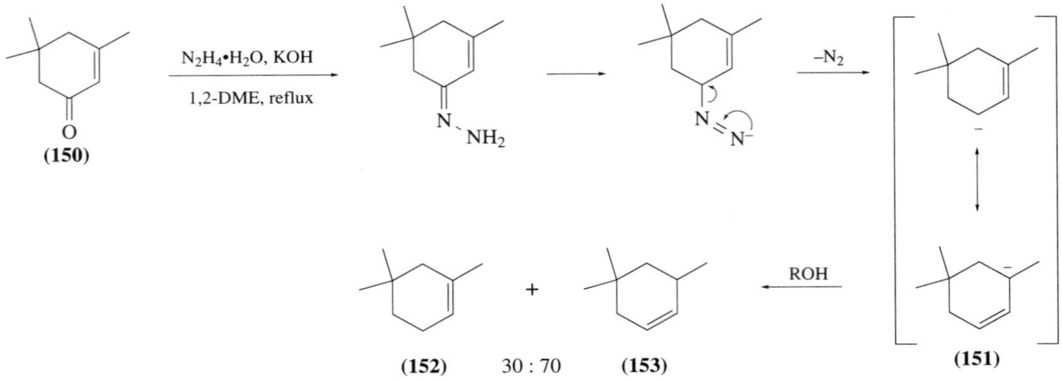

Scheme 54

1.19.1.4 Carbanions by Addition to C═C Bonds

1.19.1.4.1 Nucleophilic addition to C═C bonds resulting in carbanion formation

(i) Carbanion formation during nucleophilic addition to C═C bonds

Nucleophilic addition to C═C bonds has been reviewed ⟨B-64MI 119-01⟩. In the first step of the addition the nucleophile Nu^- attacks, with its pair of electrons, one carbon atom of the double bond, creating a carbanion (Scheme 55). The second step is combination of this carbanion with an electrophile E^+.

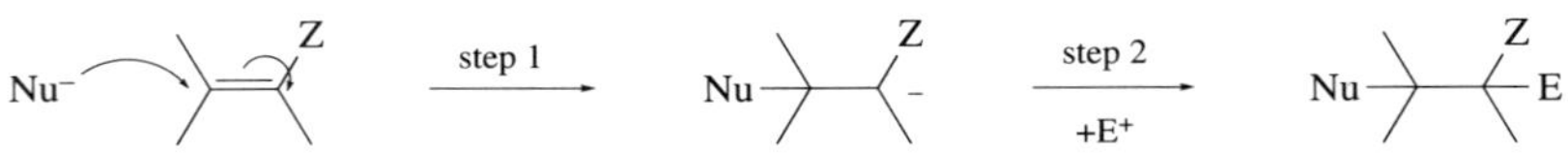

Scheme 55

(ii) Carbanion formation during the Michael-type nucleophilic addition to C═C bonds

In the special case of addition (Scheme 55) of NuE to a substrate of the form —C═C—Z, where Z is an electron-withdrawing group, e.g., CHO, COR (including quinones), CO_2R, $CONH_2$, CN, NO_2, SOR, SO_2R, etc., addition nearly always follows a mechanism involving a tetrahedral intermediate ⟨89T4017⟩, with Nu^- attacking the sp^2-hybridised carbon atom furthest from the carbonyl group (Scheme 56). For reviews of the reactions of C═C—C═O compounds see ⟨B-89MI 119-04, B-89MI 119-05⟩ and for quinones see ⟨B-74MI 119-02, B-88MI 119-03, 91RCR72⟩.

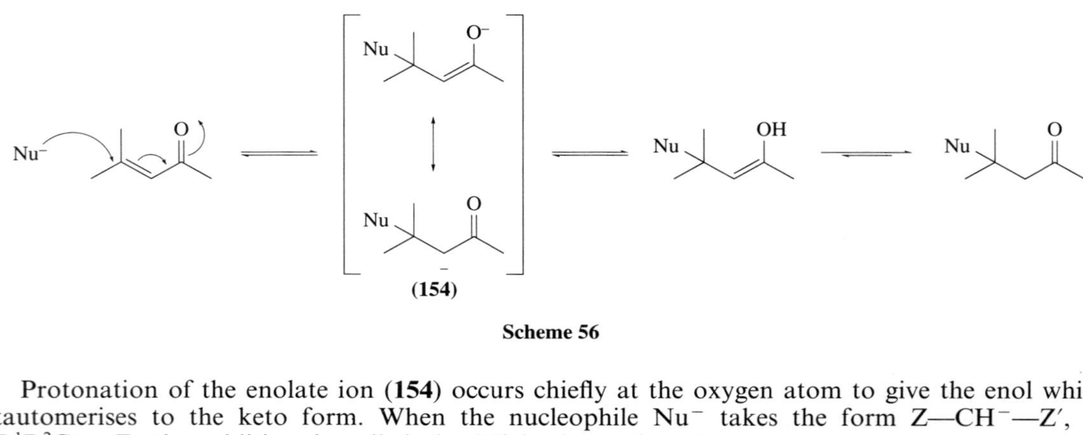

Scheme 56

Protonation of the enolate ion (**154**) occurs chiefly at the oxygen atom to give the enol which tautomerises to the keto form. When the nucleophile Nu^- takes the form Z—CH$^-$—Z′, or $R^1R^2C^-$—Z, the addition is called the Michael reaction ⟨59OR(10)179, B-72MI 119-06, 84RCR744,

91COS(4)1, B-92MI 119-12⟩, the importance of which in synthetic organic chemistry cannot be over emphasised; a simple example is shown in Scheme 57.

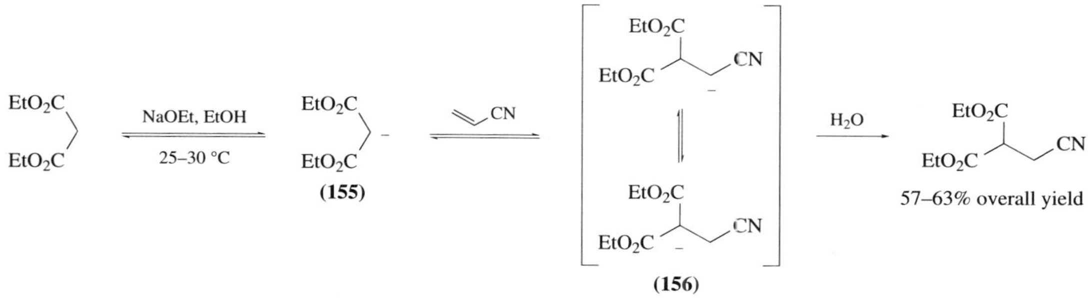

Scheme 57

The rate-limiting step in the reaction (i.e., **(155)** into **(156)**), is reversible giving rise to the possibility of retrograde Michael reactions, wherein the key step is formation of a carbanion by C—C bond scission ⟨62JCS396, 62JCS1741, 65JOC1291, 67JOC2778, 68TL4309, 70JCS(B)469⟩.

1.19.1.4.2 Carbanion formation during single electron transfer to C=C bonds

(i) Dissolving metal reduction of α,β-unsaturated ketones

Upon treatment with lithium in liquid ammonia, an α,β-unsaturated ketone is reduced to the enolate anion which may, in turn, react with an alkyl halide, frequently giving regiospecific alkylation ⟨71JA1027, 74HCA1317, 78JA292⟩. Exploitation of this synthetic strategy has been reviewed ⟨B-72MI 119-05, 76OR(23)1, B-92MI 119-13⟩ and an example is given by Scheme 58 ⟨65JA275⟩.

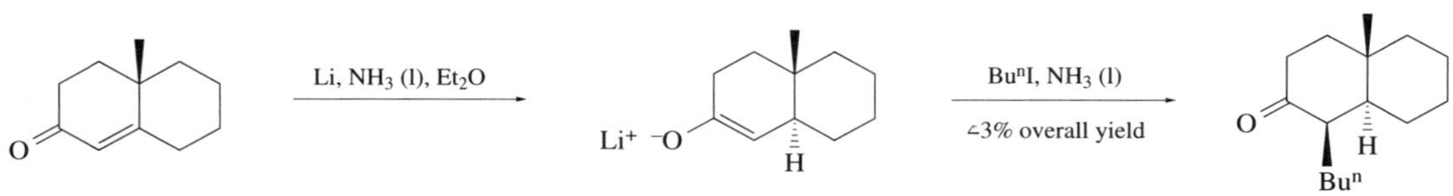

Scheme 58

(ii) Dissolving metal reduction of aromatic systems

Alkali metals are used to partially reduce aromatic systems ⟨50QR69, 58QR17, 70S161, 72S391, 89T1579⟩. Illustrative of this is the reduction of benzene to 1,4-cyclohexadiene with lithium (or sodium or potassium) and ethanol in liquid ammonia (Scheme 59) ⟨59JA3652, 59JA3658, 64JA1096, 90SC2165⟩. Such reactions are referred to as Birch reductions. One important function of the alcohol in this type of reaction is to provide a proton source that is more acidic than ammonia.

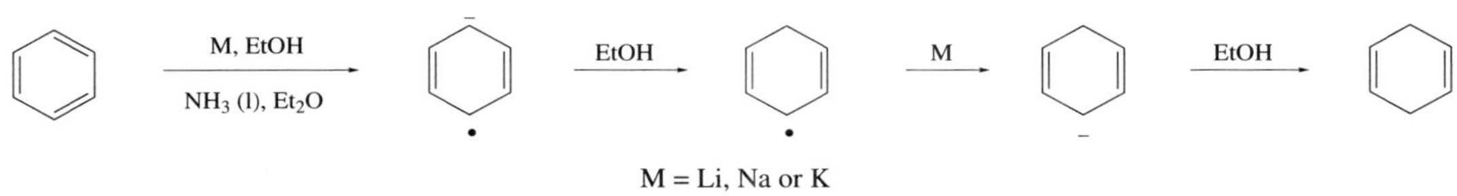

Scheme 59

A number of polycyclic aromatic systems, which have less negative reduction potentials than benzene, are reduced with an alkali metal in liquid ammonia in the absence of an alcohol, or even by less-reactive metals such as calcium or magnesium ⟨62OS(42)48, 64JOC1261, 66JCS(C)154⟩. When substituted aromatic compounds are subjected to the Birch reduction, electron-donating groups such as alkyl or alkoxyl decrease the rate of the reaction and are generally found in the nonreduced positions of the product. For example, anisole gives 1-methoxy-1,4-cyclohexadiene. Conversely, electron-withdrawing groups such as CO_2H or CO_2NH_2 increase the rate of reaction and are found in the reduced positions of the product. Such regioselectivities have been rationalised by molecular orbital theory ⟨80JA3370, 80JA4074, 80JA6430, 81JA284, 90JA1280⟩. Methods do exist, however, to reverse these regioselectivities ⟨86TL1469, 87TL2481⟩. The mechanism of Birch reductions involves solvated electrons ⟨87TCC(138)149⟩ which are transferred from the metal atom via the solvent to the aromatic ring ⟨59T148⟩. The generation of ions and diions from aromatic compounds has been reviewed ⟨74CRV243⟩. Internal alkenes are not normally reduced by metals in ammonia; however, certain nonconjugated alkenes with low-lying antibonding orbitals represent exceptions ⟨67JA3365⟩.

(iii) Dissolving metal reduction of annulenes to give aromatic dianions

The cyclooctatetraenyl dianion (**159**) can be prepared by treatment of solutions of cyclooctatetraene (**157**) in ether, THF or liquid ammonia with alkali metals. The reaction was originally reported by Reppe *et al.* ⟨48LA(560)1⟩. Their investigations showed that a dialkali metal salt was formed. The reduction of cyclooctatetraene (**157**) to form the radical anion (**158**) and then the dianion (**159**) was first achieved by Katz (Scheme 60) ⟨60JA3784, 60JA3785, 79COC(2)398⟩. Subsequent work has confirmed that the radical anion and the dianion are planar, delocalised systems ⟨74AX(B)833, 74JA1348, 81JOC3925, 84CC1654, 86CC947, 89JOM(371)279⟩. The chemistry of dianions and polyanions has been reviewed by Bates ⟨B-80MI 119-04⟩.

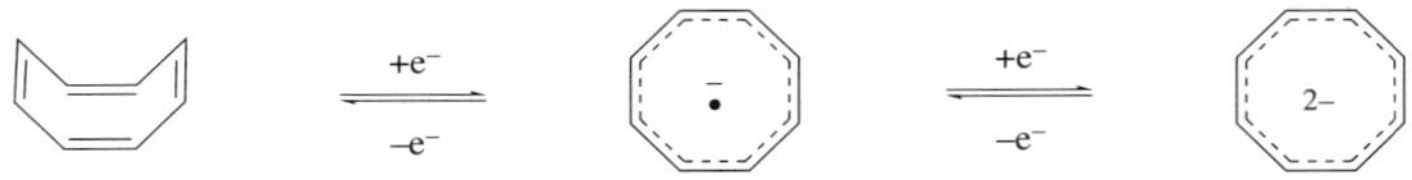

(157) **(158)** **(159)**

Scheme 60

(iv) Reversible electron transfer with C_{60}

One of the most remarkable properties associated with buckminsterfullerene C_{60}, and a property that has profoundly influenced scientific activity, is its tremendous electron affinity ⟨93CI(L)226⟩. Electrochemically, C_{60} is seen to accept six electrons ⟨92CC782, 92JA3978⟩. Electron addition to C_{60} is reversible. Multiple and reversible electron addition is possible because the resulting C_{60} anions (such as C_{60}^{-}, C_{60}^{2-} or C_{60}^{3-}) exhibit unique stability relative to other multiply charged organic ions. Such labile behaviour suggests that C_{60} should engage in reversible electron transfer with many electron donors including metals and/or metal complexes as well as organic reducing agents.

1.19.2 TRICOORDINATE CARBOCATIONS

1.19.2.1 General Aspects of Carbocation Formation

1.19.2.1.1 Nomenclature

For many years species of the type (**160**) were called 'carbonium ions', although as long ago as 1902 Gomberg had suggested that this nomenclature was inappropriate because '-onium' usually referred to a covalency higher than that of the neutral atom ⟨02CB2397⟩. Nevertheless the term 'carbonium ion' became well established ⟨86JCE930⟩ until 1971, when Olah found evidence for another type of intermediate, in which there is a positive charge at a carbon atom, but in which the formal covalency of the carbon atom is five rather than three. The simplest example is the methanonium ion CH_5^+ ⟨71CT566, 72JA808⟩. Olah proposed that the title 'carbonium ion' be reserved for such pentacoordinate positive carbon species, and that species of the type (**160**) be called 'carbenium ions'. He also proposed the term 'carbocation' to encompass both types, and these definitions have been accepted by IUPAC. While some authors still refer to **(160)** as a carbonium ion and others call it a carbenium ion, the general practice is to refer to it simply as a *carbocation* and this latter method will be employed in this chapter.

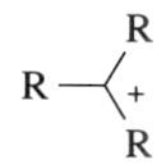

(160)

The species shown by structure **(161)** contains a two-electron, three-centre bond and is known as a 'non-classical carbonium ion', although strictly speaking it is simply a 'carbonium ion'. Harwood has clarified the situation by stating that: a *carbenium ion* only contains two-electron, two-centre bonds, whereas a *carbonium ion* contains a two-electron, three-centre bond and the cationic centre has a coordination number of five ⟨B-92MI 119-14⟩.

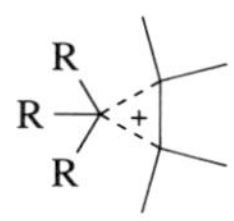

(161)

1.19.2.1.2 General literature survey

Between 1968 and 1976, a five-volume treatise edited by Olah and von R. Schleyer was published ⟨B-68MI 119-02, B-70MI 119-02, B-72MI 119-07, B-73MI 119-04, B-76MI 119-01⟩. Monographs have been written by Bethell and Gold ⟨B-67MI 119-04⟩ and Vogel ⟨B-85MI 119-03⟩ and reviews have been published ⟨66APO(4)305, 73AG(E)173, 73TS253, B-73MI 119-05, 74MI 119-04, B-74MI 119-03, 78MI 119-04, 79COC(2)411, 79TCC(80)19, 81MI 119-02, 81CS97, 87MI 119-10, 91CRV375⟩; see also the series *Advances in Carbocation Chemistry*. For a review of carbocations in superacid solutions, see Olah *et al.* ⟨B-85MI 119-04⟩. The chemistry of carbocyclic nonbenzenoid aromatic carbocations has been reviewed by Lloyd ⟨B-66MI 119-01⟩. Over the years, carbocations have been periodically reviewed by Cox, the latest being ⟨B-90MI 119-05⟩. Dicarbocations have been specifically reviewed ⟨83AG(E)390, 84T4161, 89AG(E)1321⟩. Intramolecular cationic rearrangements have been reviewed by Lowry and Richardson ⟨B-87MI 119-06⟩.

1.19.2.1.3 General features of carbocation formation and structural stability

(i) Introduction to carbocation formation by bond scission

By 1935 it had become evident through the investigations of Hughes *et al.* ⟨33JCS526, 33JCS1571, 35JCS255⟩ that nucleophilic-substitution reactions could proceed by what appeared to be two distinct paths. In the first of these, the new bond is being formed simultaneously as the old bond is breaking, the S_N2 mechanism, and this is the path by which the basic hydrolysis of methyl iodide proceeds. In the second path, the S_N1 route, the bond to the leaving group is broken *before* the new bond with the nucleophile is created, and the reaction proceeds ostensibly through the intermediacy of a free carbocation. Reactions proceeding by this mechanism may be represented in two steps, for example, the hydrolysis of *t*-butyl chloride (Scheme 61).

$$\mathrm{Bu^t{-}Cl} \xrightarrow{-\mathrm{Cl^-}} \mathrm{{}^+Bu^t} \xrightarrow{\mathrm{H_2O}} \mathrm{Bu^t{-}OH} + \mathrm{H^+}$$

Scheme 61

This path has been variously called the 'dissociation', 'heterolysis' or 'carbocation' mechanism, or 'solvolytic displacement'. In the 'dissociation mechanism', the initial step is generally the slow one. (Although some evidence has been obtained that in a few S_N reactions proceeding via preliminary dissociation, the initial step is more rapid than the subsequent attack on the carbocation ⟨54JCS2918, 55JA1122⟩. (These cases, it must be stressed, are exceptions.)

In general, nucleophilic substitutions on primary carbon atoms tend to proceed via the S_N2 mechanism, while those on tertiary carbon atoms tend to proceed via the S_N1 mechanism, and those on secondary carbon atoms often constitute borderline cases. For the S_N1 mechanism, the activated complex in the slow (bond-breaking) step assumes some of the character of an ion pair, implying its formation necessitates considerable separation of charge. Alkyl groups are more effective electron releasers than a hydrogen atom. It would seem reasonable to expect that the leaving group :L (with its pair of bonding electrons) would depart more easily from $R_3C : L$ than from $RCH_2 : L$, (where R = alkyl) and this proves to be the case. Two additional effects may account, in part, for the ease of unimolecular substitution on tertiary carbon atoms. First, since the tertiary carbon atom generally suffers from steric crowding, it may be expected that relief from such strain will be more pronounced on expulsion of the leaving group from a tertiary carbon, than from a primary carbon. Such crowded halides as $Bu^t_2Pr^iCCl$ are known to undergo unimolecular solvolysis extremely rapidly ⟨55JA2801⟩. Second a carbocation (and presumably an activated complex having much of the character of a carbocation) should be stabilised by hyperconjugation. At present, the weight of evidence is against hyperconjugation in the ground states of neutral molecules, although this is not an entirely unanimous verdict ⟨87AJC1537⟩. However, for carbocations and free radicals ⟨62T333⟩ there is evidence that hyperconjugation is important. In hyperconjugation in the ground state of neutral molecules, which Muller and Mulliken called ‘sacrificial hyperconjugation’ ⟨58JA3489⟩, the canonical forms involve not only ‘no-bond resonance’, but also charge separation not extant in the main undisturbed form. In carbocations, the canonical forms display no more charge separation than the main form, and this has been called ‘isovalent hyperconjugation’.

Such hyperconjugation would be expected to be more important if there are three alkyl groups on the α-carbon than if there is only one alkyl group. Such canonical forms as in (**162**) (Equation (35)) imply that the β C—H bonds are loosened during the formation of the activated complex. Since C—H bonds are ‘looser’ (or longer) than C—D bonds, it might be supposed that substitution of deuterium for β-hydrogen atoms would lower the extent of hyperconjugation slightly, thus retarding S_N1 substitution at the α-carbon atom. Such an effect has indeed been observed ⟨53JA2925, 54JA791, 58JA2326⟩. Further evidence of hyperconjugation being the cause of this β secondary isotope effect is provided by the observation that the equilibrium constant for the reaction depicted by Equation (36) is such that $K_{298} = 1.97 \pm 0.20$, indicating that (**164**) is more stable than (**163**) ⟨87JA7947⟩.

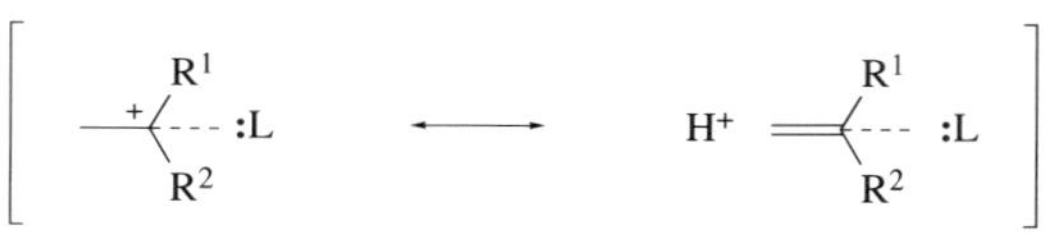

(35)

(162)

$$(CD_3)_3C^+ + Bu^tH \rightleftharpoons {}^+Bu^t + (CD_3)_3CH \qquad (36)$$

(163) **(164)**

Aryl groups linked to the α-carbon atom are even more effective in facilitating S_N1 substitution than are alkyl groups, for such groups stabilise the carbocation (and the carbocation-like activated complex) by supplying π-electron density from the ring (Scheme 62).

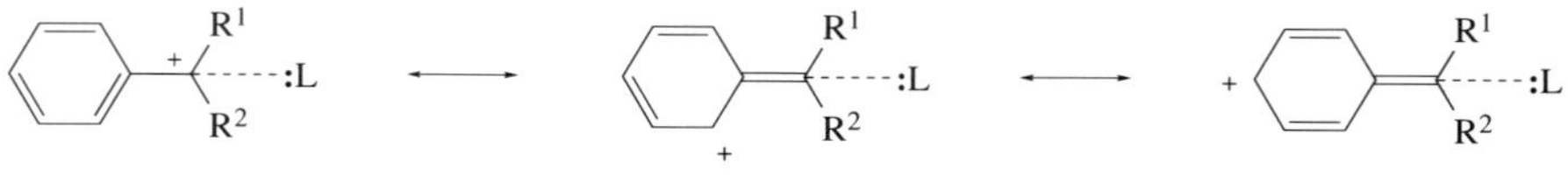

Scheme 62

The ease of both S_N1 and S_N2 reactions is affected by the polarity of the solvent. For the S_N1 reactions of alkyl halides and sulfonate esters, activation requires separation of opposite charges and, in general, such reactions are more rapid in solvents of high dielectric constant (e.g., water or formic acid) than in solvents of low dielectric constant (e.g., acetone or dioxane).

There is substantial theoretical and experimental evidence that the optimum structure for a carbocation is such that the three sp^2-orbitals lie in the same plane at angles of 120° ⟨69JA1037⟩. Any distortion of this optimum geometry will be strongly resisted by a carbocation, since both the necessary rehybridisation of the cationic carbon centre and the resultant greater steric repulsion increase the energy of the system.

Two types of distortion are possible: deformations in which the cationic carbon centre and the three attached atoms remain in the same plane, and distortions leading to nonplanar structures (Figure 2). This latter type of distortion is unavoidable during carbocation formation in some bridged systems in which, for structural reasons, planar ions cannot form. The variation in reactivity of a range of ring systems is shown in structures **(165)**–**(169)** ⟨66MI 119-01⟩.

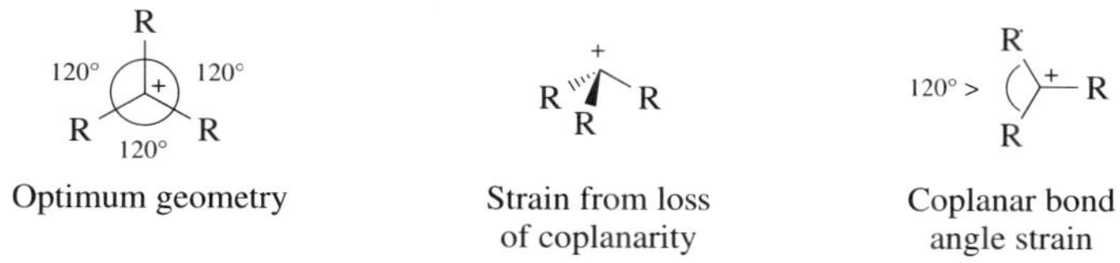

Figure 2 Sources of strain in carbocations.

1.0	0.5	10^{-3}	10^{-6}	10^{-13}
Br—Bu^t				
(165)	**(166)**	**(167)**	**(168)**	**(169)**

Both bridged bromides **(168)** and **(169)** are unreactive, with the latter yielding the corresponding alcohol only after treatment with aqueous silver nitrate at 150 °C for 2 days. With similar reagents, **(167)** produces alcohol after reaction at room temperature for 4 hours. It may be inferred that 1-bromoadamantane **(167)** can achieve a greater degree of coplanarity at the nascent cationic carbon centre, but it is still markedly less reactive than *t*-butyl bromide **(165)**. Only bromide **(166)** proves to be of comparable reactivity to **(165)** since its extra methylene group gives the molecule much greater flexibility than the more rigid adamantane skeleton. For discussions of the stereochemistry of carbocations see ⟨73CSR397, 75ACR369⟩.

Carbocations prove to be intermediates in diverse types of reaction and the more stable types can be kept in solution or in some cases isolated as solid salts. In solution the carbocation may be free, especially in polar solvents, or it may be part of an ion pair, which is more prevalent in nonpolar solvents ⟨B-72MI 119-08, B-74MI 119-05⟩.

Among simple alkyl carbocations, the order of stability is tertiary > secondary > primary ⟨71CT566, 72JA808⟩. Many examples of rearrangements of primary or secondary carbocations to tertiary carbocations are known both in solution and gaseous phases. The field of intramolecular cationic rearrangements has been reviewed by Lowry and Richardson ⟨B-87MI 119-05⟩. Simple alkyl cations are not stable in conventional strong-acid solutions, for example, H_2SO_4. However, it was discovered that many of them could be kept stable indefinitely in solution with mixtures of fluorosulfuric acid and antimony pentafluoride. Such mixtures, usually dissolved in SO_2 or SO_2ClF, are often called 'superacids'. Although the term 'superacid' was first used by Conant in 1927 ⟨27JA3047⟩, the technology was systematically developed by Olah, whose early experiments involved the reaction of alkyl fluorides with SbF_5 which might more accurately be termed a 'super Lewis acid' (Equation (37)) ⟨64JA1360, 64PCS147, 69JA4819⟩.

$$RF + SbF_5 \xrightarrow{SO_2 \text{ or } SO_2ClF} R^+ \cdot SbF_6^- \tag{37}$$

It was found later by Olah that similar carbocations could also be generated from alcohols in superacid in sulfur dioxide solution at −60 °C ⟨65JA2997, 67JA3576⟩. Indeed even alkanes can be induced to give carbocations in superacid solution by loss of H^- (Equation (38)) ⟨67JA4739⟩. In accordance with the observed stability order, H^- is abstracted from alkanes by superacid most readily from tertiary carbons and least readily from primary carbons.

$$Bu^tH \xrightarrow{FSO_3H,\ SbF_5} {}^+Bu^t + SbF_5 \cdot FSO_3^- + H_2 \tag{38}$$

As of 1995, no primary carbocation has survived long enough to be detected. Neither methyl nor ethyl fluoride gave observable Me^+ or Et^+ ions when treated with SbF_5. A number of benzylic carbocations have been obtained in solution as SbF_6^- salts ⟨67JA5687, 72JA2044⟩. Diarylmethyl and triarylmethyl carbocations are further stabilised. It has long been known that in SO_2, triphenylchloromethane ionises to give the equilibrium shown in Equation (39).

$$Ph_3C-Cl \rightleftharpoons Ph_3C^+ + Cl^- \qquad (39)$$

Both triphenylmethyl and diphenylmethyl carbocations have been isolated as solid salts ⟨63AG(E)622, 65AG(E)873⟩. Arylmethyl carbocations are further stabilised when they have electron-donating groups in the *ortho* and/or *para* positions ⟨49JCS1724, 55JA3051⟩.

Cyclopropyl groups have an even greater stabilising effect on carbocations than aryl groups. Carbocation (**170**) has been prepared by dissolving the corresponding alcohol in 96% sulfuric acid ⟨62JA2016⟩, while (**171**) and (**172**) have been similarly prepared in $FSO_3H/SO_2/SbF_5$ ⟨65JA2998, 65JA3000⟩.

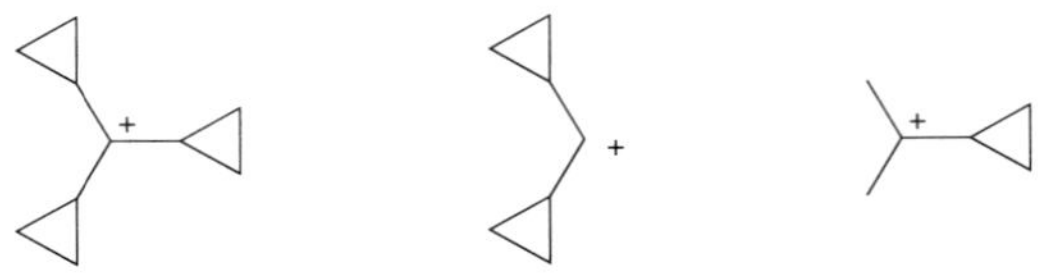

(**170**) (**171**) (**172**)

This special ability to stabilise carbocations, which increases with each additional cyclopropyl ring, results from conjugation between the bonding orbitals of the cyclopropyl rings and the empty *p*-orbital of the cationic carbon atom. NMR studies indicate that this vacant *p*-orbital lies parallel to the C3—C4 bond of the cyclopropane ring, as shown in structure (**173**), and not orthogonal to it ⟨62JA2016, 70JA1660, 70JA3234, 71JA3927, 74JA7591, 90JA8912⟩. In solvolysis of primary cyclopropylmethyl systems, the rate is enhanced because of this overlap of the σ-bonds of the ring with the empty *p*-orbital of the cationic carbon atom ⟨79JOC2860⟩. The ion that is formed initially is an unrearranged cyclopropylmethyl cation that is symmetrically stabilised as represented in Scheme 63 ⟨68JA63⟩. Evidence that (**173**) is orientated as shown is provided by the fact that substitution of one or more methyl groups in the 3- and 4-positions increases the rate of solvolysis of cyclopropylcarbinyl 3,5-dinitrobenzoates by approximately a factor of 10 for each methyl group ⟨66JA2321⟩. If only one of the σ-bonds of the cyclopropyl ring was involved in stabilisation (e.g., the 2,3 bond), then methyl substitution at the 3-position should increase the rate, and a second methyl group at the same position should enhance it further, but placing the second methyl group at the 4-position should have little effect; such is not the case. It may thus be inferred that both the C2—C3 bond and the C2—C4 bond contribute electron density towards the stabilisation of positive charge on the C1 atom.

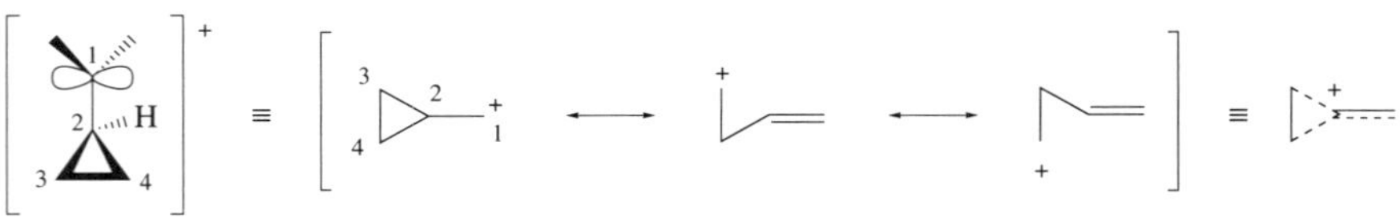

(**173**)

Scheme 63

Numerous methods have been employed for the examination of factors governing the stability of carbocations. 'Stability', however, is only a relative term, since in chemical systems, carbocations are frequently associated with anions and solvent. Intermolecular ion–solvent interactions can be eliminated if studies are undertaken in the gas phase, and scales of stabilities have been obtained by the measurement of ionisation potentials (Table 10) ⟨B-63MI 119-02⟩.

An important technique for the investigation of carbocation stability and structure is measurement of the ^{13}C-NMR chemical shift of the cationic carbon atom ⟨75APO(11)123, 79MI 119-05⟩. ^{13}C-NMR chemical shifts for a range of carbocations are given in Table 11 ⟨68JA1884, 69JA5801, 77JA5026, 77JOC2666⟩.

However, the ^{13}C-NMR chemical shifts do not always reflect the precise order of carbocation stabilities as determined by other methods; for example, the chemical shift shows that the triphenyl-

Table 10 Ionisation potentials (kcal mol^{-1}) of radicals derived from electron-impact studies.

Radical	*Ionisation potential*	*Radical*	*Ionisation potential*
Me	229.4 ± 0.7	Me_2CH	182.2 ± 1.2
$MeCH_2$	202.5 ± 1.2	$PhCH_2$	178.9 ± 1.8
$MeCH_2CH_2$	200.4 ± 1.2	$MeCH{=}CHCH_2$	177.8 ± 1.2
p-$NCC_6H_4CH_2$	197.9 ± 2.3	Me_3C	171.8 ± 1.2
$HC{\equiv}CCH_2$	190.2 ± 1.8	Ph_2CH	168.8 ± 2.3
$CH_2{=}CHCH_2$	188.2 ± 0.7	p-$MeOC_6H_4CH_2$	157.7 ± 2.3

Table 11 ^{13}C-NMR chemical shift values, in ppm from $^{13}CS_2$, for the cationic carbon in some carbocations in SO_2ClF/SbF_5, $SO_2/FSO_3H/SbF_5$, or SO_2/SbF_5 ⟨71JA464⟩.

Carbocation	*Chemical shift*	*Temp. (°C)*	*Carbocation*	*Chemical shift*	*Temp. (°C)*
Et_2MeC^+	−139.4	−20	Me_2(cyclopropyl)C^+	−86.8	−60
Me_2EtC^+	−139.2	−60	$PhMeCH^+$	−40[a]	
Me_3C^+	−135.4	−20	Ph_3C^+	−18.1	−60
Me_2CH^+	−125.0	−20	Ph_2CH^+	−5.6	−60
$PhMe_2C^+$	−61.1	−60			

methyl carbocation has a more positive cationic carbon than the diphenylmethyl carbocation, despite the fact that the former is known to be more stable. The 2-cyclopropylpropyl and 2-phenylpropyl carbocations have shifts of −86.8 ppm and −61.1 ppm, respectively, despite the fact that according to other experimental evidence a cyclopropyl group is more efficient than a phenyl group at stabilising a carbocation ⟨71JA464, 73JA2400, 74JA3548, 75JA2902, 75APO(11)123, 76CJC2839, 77JA1712, 79JOC2652, 79MI 119-07⟩.

(ii) Introduction to carbocation formation by electrophilic addition to C=C bonds

Alkenes dissolved in aqueous acid are in rapid, reversible equilibrium with a cation formed by addition of a proton ⟨52JA5372⟩. This cation has to be nonclassical in nature in view of the behaviour of the isomeric pentenes (**174**) and (**175**), (Scheme 64) ⟨53JA1253, 53JA3955⟩. Both pentenes react with dilute nitric acid to give the same tertiary alcohol (**179**). If the reaction is interrupted, when half the alkene has been converted to the alcohol, the remaining alkene has the starting structure in each case. The initial product of protonation is therefore of such a structure that loss of a proton gives only the starting alkene. The reversibly formed carbocations are the π-complexes (**176**) and (**177**), which are precursors to the hydration of the alkenes in a relatively mild acidic medium and, have the proton embedded in the π-electron cloud. They are therefore reasonable intermediates to expect from reactions in which a classical carbocation (**178**) is subsequently formed ⟨52JA5943⟩.

It has been found that if the hydration of an alkene (**179**) is conducted in D_2O but terminated in its early stages, the unreacted alkene has acquired *no* deuterium and has undergone *no* isomerisation ⟨56JA5807, 57JA3724⟩, whereas both deuteration and isomerisation should be expected if the reversible formation of a *classical* carbocation were involved.

Tertiary cycloalkyl cations, such as the 1-methyl-1-cyclopentyl cation (**180**) show high stability in superacid media and can be derived from a number of precursors, e.g., (**181**) and (**182**) ⟨68JA933⟩. Cyclohexene (**183**) also gives cation (**180**) via isomerisation of the secondary cyclohexyl cation to the more thermodynamically stable tertiary methylcyclopentyl cation. Under such extreme conditions, the cation (**180**) undergoes both carbon and hydrogen scrambling. Sorensen and co-workers have prepared tertiary cycloalkyl cations of different ring sizes, $4 \geqslant n \geqslant 20$ and these cations were found to undergo various expansion or contraction reactions ⟨78JA5134⟩.

In contrast to simple alkenes, aryl-substituted alkenes dissolve in sulfuric acid to give comparatively stable carbocations, as indicated by the spectra and the recovery of the alkene on dilution ⟨52JCS2167⟩. Because of the similarity of their spectra it is believed that the carbocations (**184**)–(**186**) are best represented by the classical structures shown ⟨52JCS2172⟩.

The addition of a proton to an unsaturated molecule can sometimes lead to the same carbocation

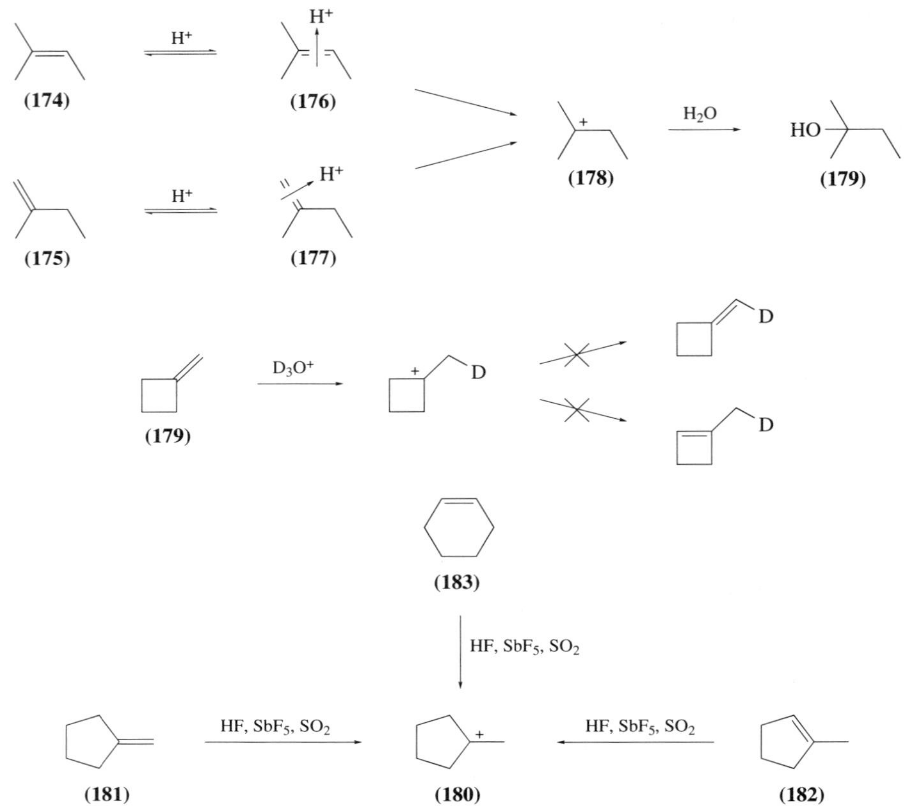

Scheme 64

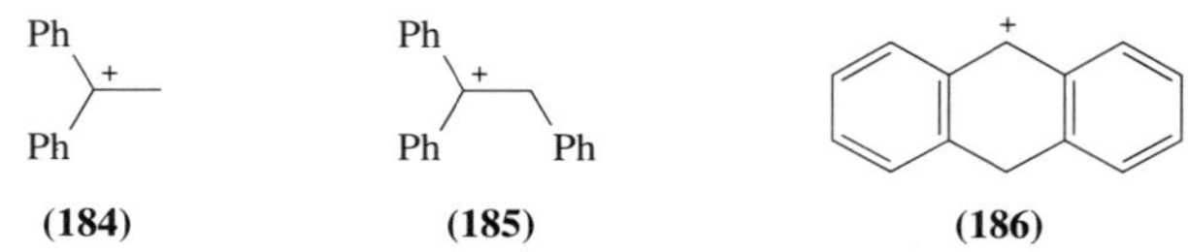

as the loss of an anionic group from a saturated species, as shown by the two routes to the 1,1-diphenylethyl cation **(187)** (Scheme 65) ⟨52JCS2172⟩.

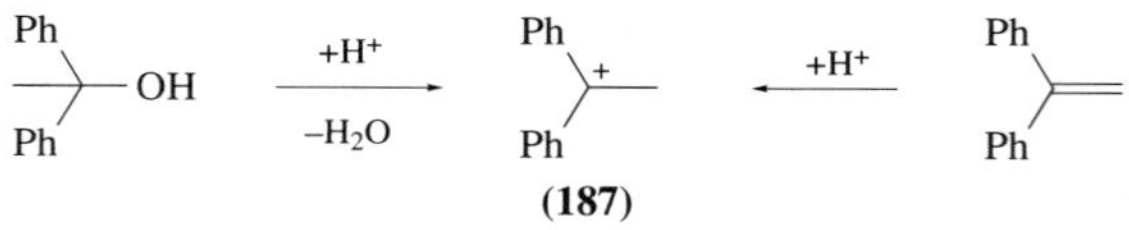

Scheme 65

The formation of transient carbocations is thought to be an intermediate step in electrophilic addition reactions, for example the hydration of certain alkenes ⟨64JA4727, 65JA2157, 65JCS6718, 66JA120⟩, though it is not the only mechanism for such reactions ⟨60JA4729, 63JA2245, 63JCS1930, 65JCS6019⟩.

Carbocation formation will clearly be favoured by the strength of the source of the protons as an acid, and this consideration should be foremost in deciding the choice of the reaction medium, in addition to the usual desiderata of power for anion stabilisation, chemical inertness and high dielectric constant ⟨B-67MI 119-05⟩. Concentrated sulfuric acid, and to a lesser extent, other strong mineral acids were therefore the first media employed in such reactions. Sulfuric acid is in many respects a very suitable solvent, but presents problems connected with its high reactivity both as a sulfonating agent and as an oxidant. For this reason MSA ⟨50JA163⟩, chlorosulfonic acid ⟨64JA934⟩, fluorosulfonic acid ⟨64CJC502⟩ and liquid hydrogen halides, especially hydrogen fluoride, have been used.

The protonating power of a Brönsted acid can be increased by stabilising the gegenion formed alongside the carbocation. If the Brönsted acid is a hydrogen halide, the proton donating power of the medium can be increased by adding a substance with a strong specific affinity for the relevant halide ion. Acid/halide combinations that have been used extensively as media for the generation of stable carbocations include $HCl/AlCl_3$, HCl/BCl_3 ⟨62JCS600⟩, HF/BF_3 ⟨50IEC2103⟩, HF/SbF_5 ⟨65JA1103⟩ and FSO_3H/SbF_5 ⟨65JA2997⟩. Such media are superior to sulfuric acid both in acidity and chemical inertness. Furthermore, they can be used at low temperatures. The solutions can be diluted with sulfur dioxide to facilitate work at even lower temperatures ⟨65JA1103⟩.

In a case where the cationic carbon centre is in conjugation with a double bond the stability of the system is enhanced due to resonance and because the positive charge is delocalized over two atoms (Equation (40)). If the substituents R are equivalent, then each of the two terminal allylic carbon atoms will bear a charge of exactly +0.5.

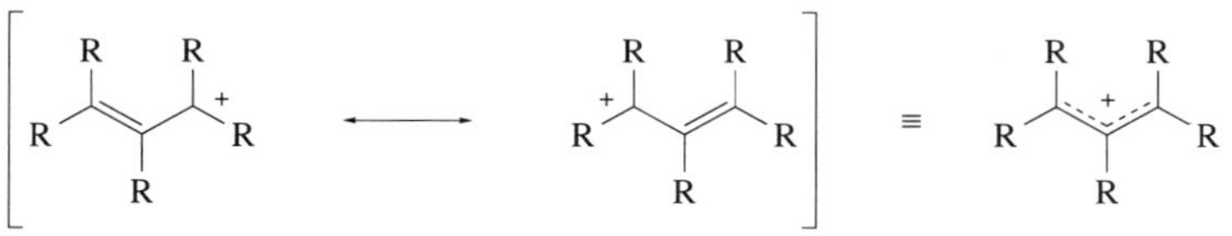

(40)

Stable allylic carbocations have been prepared by dissolving conjugated dienes in concentrated sulfuric acid (Equation (41)) ⟨62JA1498, 63JA2991⟩.

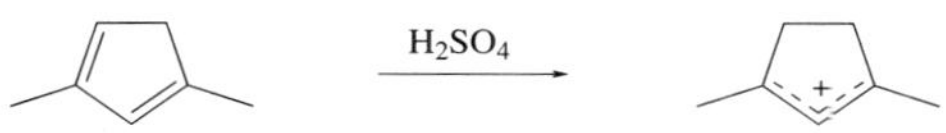

(41)

1.19.2.2 Carbocations by Bond Scission

1.19.2.2.1 Aryl-substituted carbocations by bond scission

The triphenylmethyl carbocation (**188**) represents not only the first stable carbocation to be discovered ⟨02CB1189, 02CB3031⟩, but also the first to be studied by x-ray diffraction ⟨65AX437⟩. Carbocation (**189**) has proved to be propeller-shaped with threefold (D_3) symmetry: the cationic carbon atom and the three adjacent ring carbons are coplanar, while the three benzene rings, due to steric crowding of the *ortho* hydrogens, are twisted out of coplanarity, reducing the degree of resonance stabilisation of the system and resulting in a slightly 'feathered' propeller geometry. In the perchlorate salt of (**188**), the aromatic rings of the cation are mutually twisted at an angle of 54° to each other. These twist angles are in the range 45–61° in the platinum salts ⟨72JA5511⟩.

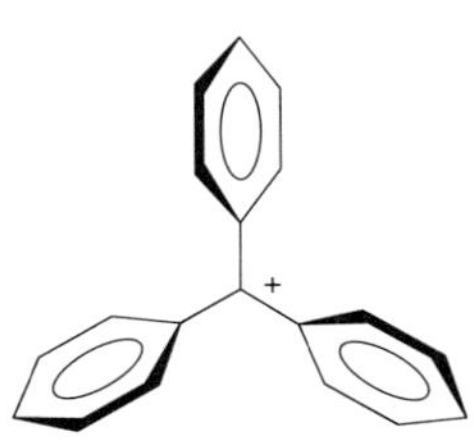

(**188**)

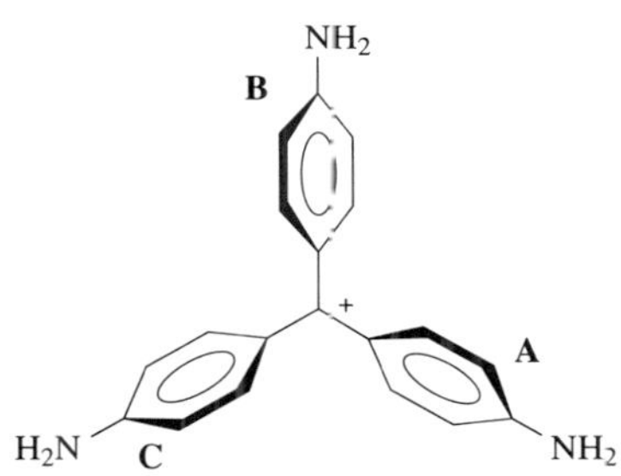

(**189**)

The tri-(*p*-aminophenyl)methyl carbocation (**189**) is propeller-shaped, but does not exhibit threefold symmetry ⟨71AX(B)1405⟩. The three exocyclic bonds adjoining the central cationic carbon are coplanar indicating strong electron delocalisation from the aryl rings. The carbocation (**189**) has the rings A, B and C twisted by 34.4°, 33.8° and 27.3°, respectively, from the plane formed by the cationic carbon and the three atoms attached to it. Despite these large angles, there is evidence of appreciable π-electron delocalisation.

para-Methoxy or amino substituents have a powerful stabilising effect on the cation, as illustrated by the various ionic triphenylmethane dyes bearing these groups, e.g., derivatives of malachite green (**190**). Extensive data on the stability of substituted triphenylmethyl carbocations have been tabulated by Freedman ⟨B-73MI 119-06⟩.

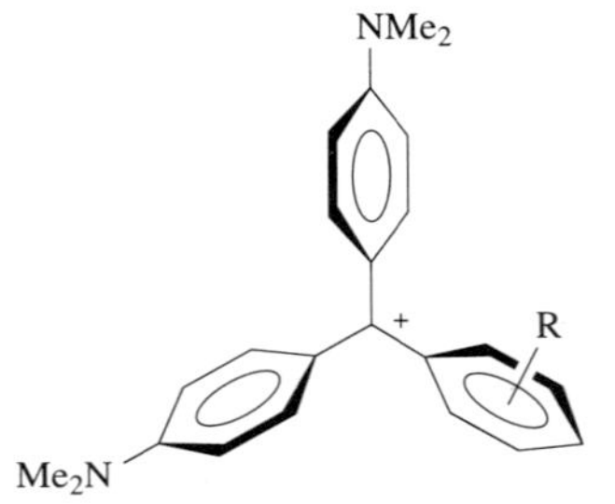

(190)

The degrees of ionisation of triphenylmethyl chloride, *p,p'*-dimethoxybenzhydryl chloride and dimesitylmethyl chloride in liquid sulfur dioxide are about the same ⟨21JPR1, 37RTC535⟩. Apparently one ring of the triphenylmethyl carbocation may be dispensed with, if the other two rings are adequately substituted. 1-Phenyl-2,2,2-trifluoroethyl cations, purportedly highly unstable, in fact have fairly long lifetimes ⟨89JA7659⟩. They are generated by the solvolysis of $YC_4H_4CH(CF_3)X$ (Y = e.g., MeS; X = OTs, OMs or Br) in 1 : 1 H_2O/F_3CCH_2OH and solvolysis rate constants are almost identical with those for the nonfluorinated analogues $YC_4H_4CH(CH_3)X$; only S_N1 mechanisms were observed, even with nucleophiles such as N_3^-, 1^- and Br^-.

The tripticyl carbocation (**191**) is a system in which the bridged structure ensures that all three rings will always deviate from coplanarity with the bonds leading to the cationic carbon atom to the maximum possible extent. Tripticyl bromide exhibits none of the tendency to ionise that is shown by the closely analogous triphenylmethyl bromide ⟨50JA1005⟩. It seems likely that the major reason for the instability of carbocation (**191**) is structural inhibition of resonance stabilisation by the aromatic rings and prevention of the cationic carbon atom adopting planar sp^2-hybridisation.

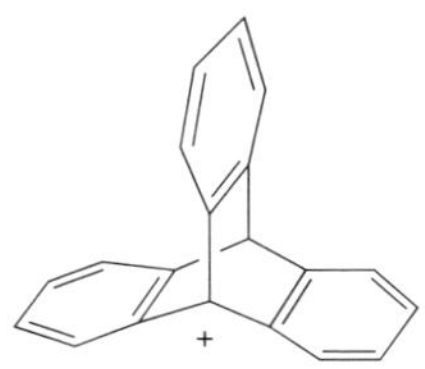

(191)

The work of Walden is remarkable for his early utilisation of a nonaqueous solvent, liquid sulfur dioxide, possessing an exceptional ionising power ⟨02CB2018⟩. The importance of the choice of solvent had also been noticed by Gomberg ⟨02CB2397⟩, who showed that in some solvents such as benzene and ether, triphenylmethyl chloride is nonionised and colourless, while in solutions of 'dissociating solvents' such as SO_2, MeCOCl and SO_2Cl_2, are conductors and exhibit a yellow colour. Both properties were correctly attributed to Ph_3C^+. Ionisation occurs also when triphenylmethyl chloride is complexed with inorganic halides, e.g., $SnCl_4$, $ZnCl_2$ or $AlCl_3$ ⟨02CB1189⟩.

Using the modern superacid technology, developed by Olah, the following stable triphenylmethyl carbocationic salts have been prepared and are commercially available: $Ph_3C \cdot SbCl_6$ ⟨B-72MI 119-09, B-90MI 119-06⟩, $Ph_3C \cdot PF_6$ ⟨B-74MI 119-06⟩, $Ph_3C \cdot SnCl_5$, $Ph_3C \cdot BF_4$ (⟨B-67MI 119-06, B-86MI 119-02⟩ and references therein).

The structures of such exotic species as the biscyclobutadienyliron tricarbonyl phenylmethyl carbocation (**192**) ⟨71JA6688⟩ have been determined by x-ray crystallography as have those of the dicarbocations (**193**) ⟨67JA3405⟩ and (**194**) ⟨71JCS(B)918⟩, and all show the propeller-type structures depicted.

The synthesis of the tri(1-azulenyl)methyl carbocation (**195**; R = H) has been reported by two methods: one by the reaction of azulene with tetraethyl *ortho*-carbonate [C(OEt)$_4$] in the absence of

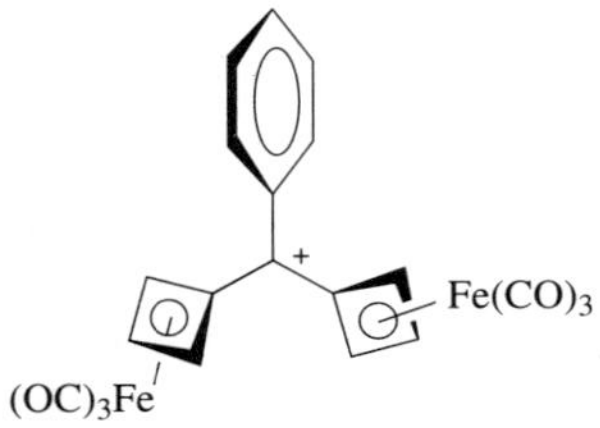

(192)

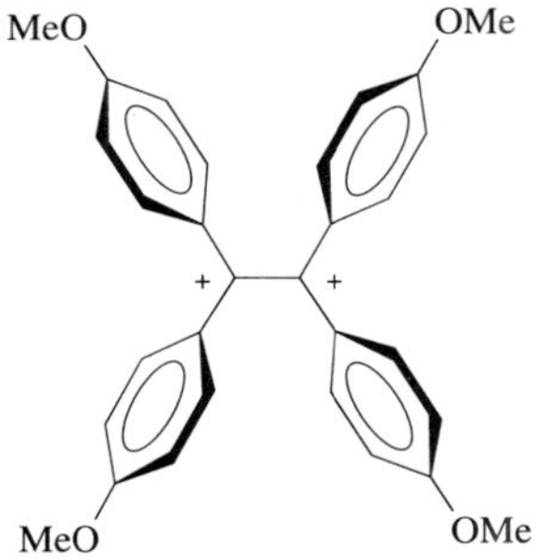

(193)

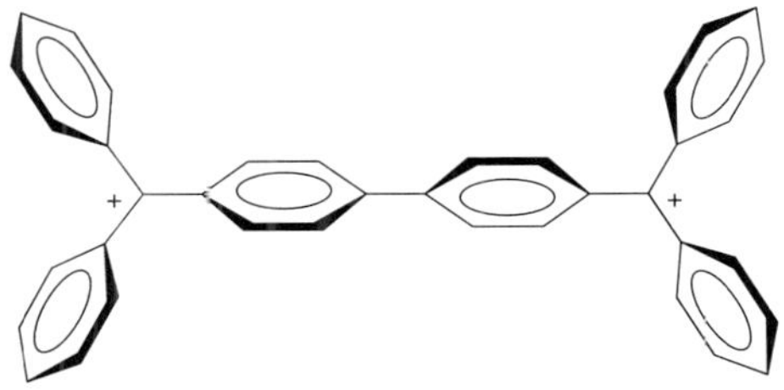

(194)

acid ⟨61LA(650)62⟩ and the other by the hydride abstraction from tri(1-azulenyl)methane by the triphenylmethyl carbocation ⟨87BCJ3721, 88NKK157⟩. Substituted tri(1-azulenyl)methyl carbocations (**195**; R = Me, OMe, CO_2Me) were obtained by Asao by treatment of the triazulenylmethanes with 2,3-dichloro-5,6-dicyano-1,4-benzoquinone (ddq) in CH_2Cl_2, followed by addition of hexafluorophosphoric acid to give the crystalline products ⟨90PAC507⟩. The chemistry of the perchlorate salt of the tri(2-thiophenyl)methyl carbocation (**196**) has been examined ⟨90TL2623⟩.

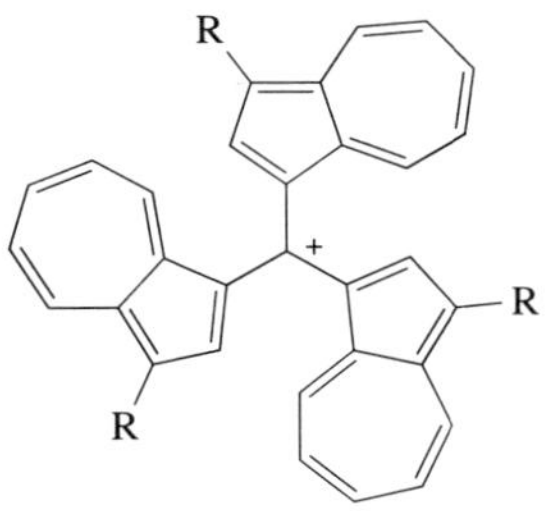

(195)

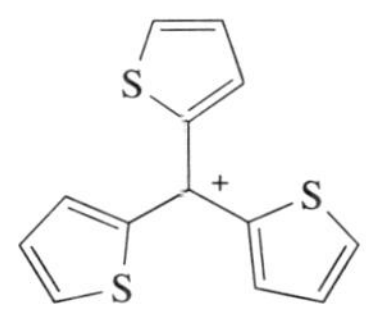

(196)

1.19.2.2.2 Alkyl carbocations by bond scission

Some carbocations are stable only as salts of complex anions, for example, $AlCl_4^-$, BF_4^-, $SnCl_6^{2-}$ or SbF_6^-. The alkylcarbocationic salts belong to this class ⟨B-68MI 119-03⟩.

Remarkably stable alkylcarbocationic hexafluoroantimonates $R^+SbF_6^-$ have been obtained by Olah and co-workers ⟨63JA1328, 64JA1360⟩. Such ionisations can be accompanied by extensive rearrangements ⟨73ACR53⟩. Through hydride and methide shifts, the thermodynamically most stable carbocations are formed. *n*-Propyl and *i*-propyl fluorides give the identical isopropyl cation (**197**) and *n*-butyl, *sec*-butyl, *i*-butyl and *t*-butyl fluorides all lead to the formation of the same tertiary carbocation complex (**198**) while each of the seven isomeric pentyl fluorides including neopentyl fluoride, gives the same tertiary carbocation complex (**199**).

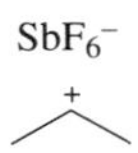

(197)

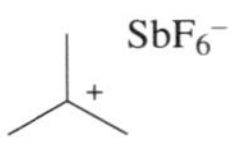

(198)

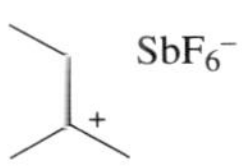

(199)

The alkyl carbocation hexafluoroantimonates are generally stable only if an excess of SbF_5 is present. The role of this excess of Lewis acid is undoubtedly to solvate the organic salt or ion pair. Such solvated tertiary alkyl cation hexafluoroantimonates are soluble without decomposition in cold SO_2. Their optical ^{1}H- and ^{13}C-NMR spectra and chemical behaviour have been studied by Olah and co-workers ⟨63JA1328, 64JA1360, 68JA1884, 69JA5801, 77JA5026, 77JOC2666, 81PNA(78)1998⟩. The crystal structure of the *t*-butyl cation has been determined by x-ray crystal analysis of its $Sb_2F_{11}^-$ salt at −80°C ⟨93JA7240⟩. The cation was found to be planar; the C—C$^+$ bonds had an average length of 1.442(5) Å and the C—C$^+$—C bond angles were 120(1)°.

Attempts to obtain alkyl carbocation complexes by dissolving alkyl chlorides in other liquid Lewis acid halides (e.g., $SnCl_4$, $TiCl_4$, $SbCl_5$) have proved unsuccessful, either because the Lewis acids are too weak to cause ionisation of the C—Cl bond or the solvating effect of the halides is not sufficient to stabilise any carbocations formed. However, the *sec*-butyl cation has been prepared by slow addition of *sec*-butyl chloride to SbF_5/SO_2ClF solution at $-110\,^{\circ}C$ ⟨68JA6882⟩. Alkyl carbocations are not stable in sulfuric acid solutions as has been observed by Deno and co-workers ⟨62JA2016⟩.

A wide range of phenyl, alkyl and cyclopropylcarbinyl cations have been prepared by Olah from alcohols by treatment with superacids FSO_3H/SBF_5 (1 : 1) or FSO_3H/SO_2ClF solution at $-78\,^{\circ}C$ ⟨77JOC2666⟩.

Carbocations are difficult or impossible to form at bridgehead carbon atoms in [2.2.1] systems ⟨B-73MI 119-07⟩ and bridgehead adamantane carbocation reactivities have been studied by von R. Schleyer *et al.* ⟨64JA4194⟩. However, von R. Schleyer *et al.* have reported the preparation of the adamantyl cation (**200**; R = H) as the hexafluoroantimonate salt ⟨64JA4195⟩ and subsequently that of the 3,3′-(1,1′-biadamantyl) dication (**201**) and the 4,9-diamantyl dication (**202**) in superacid media ⟨85JA2764⟩. In 1990, Olah prepared the tris(1-adamantyl)methyl cation (**203**), which is stable for about 30 min at $-70\,^{\circ}C$ in FSO_3H/SbF_5 (1 : 1), and is the most crowded carbocation yet synthesised ⟨90JA6422⟩. Its relatively high stability is attributed to the structural constraint of the adamantyl cage framework producing a high barrier to decomposition. The first crystal structure of an alkyl carbocation [the $Sb_2F_{11}^-$ salt of (**200**; R = Me)] was reported by Laube ⟨86AG(E)349⟩. The cationic centre was found to be almost planar.

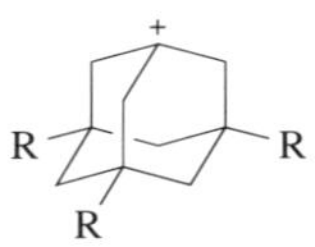

(200)

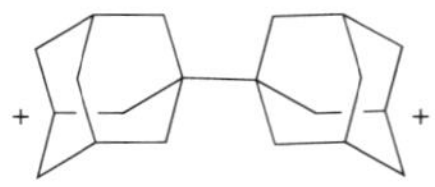

(201)

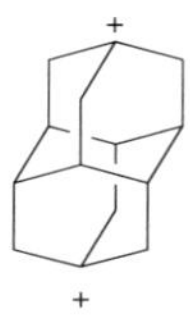

(202)

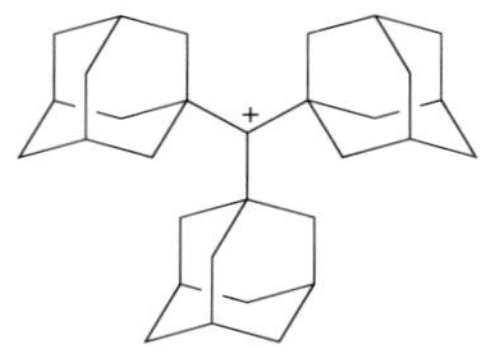

(203)

A striking example of bridgehead unreactivity is furnished by compound (**204**), which resisted hydrolysis when boiled for 21 hours in 30% KOH/EtOH or for 48 hours in $H_2O/AgNO_3$ ⟨39JA3184⟩. [2.2.2]-Bicyclic systems undergo S_N1 reactions much faster than smaller bicyclic systems, although the reactions are retarded compared to open-chain analogues. Synthetic use has been made of the former type of reaction ⟨85JOC4605⟩. The relatively larger and more flexible structure permits formation of cation (**205**), which is sufficiently stable to be kept in solution with SbF_5/SO_2ClF below $-50\,^{\circ}C$. The brominated [3.1.1]system (**206**) undergoes S_N1 reaction ⟨87CC833⟩ as does the cubyl triflate (**207**) ⟨90JA3225, 90JA3228⟩. *Ab initio* calculations show that the cubyl carbocation, though permanently constrained to be nonplanar requires less energy to form than the 1-norbornyl cation ⟨90JA3227⟩.

(204)

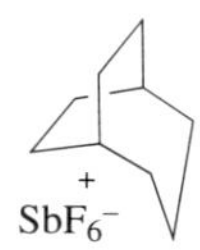

(205)

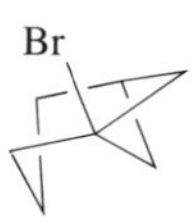

(206)

(207)

The 1-trishomobarrelyl (**208**) and 1-trishomobullvalyl (**209**) cations have been prepared in the superacid media (SbF_5/SO_2ClF) ⟨73AG(E)399⟩. In these two cations, the inherent instability of the bridgehead position is ameliorated by the stabilisation derived from conjugation with the three cyclopropyl groups. The dodecahedryl carbocation (**210**) and the 1,16-dodecahedryl dication (**211**) have been prepared in superacid media (SbF_5/SO_2ClF) by Olah and co-workers ⟨88JA8599⟩.

Timberlake and co-workers have prepared the pentacyclopropylethyl carbocation (**212**) (Equation (42)) ⟨86JOC2969⟩. ^{13}C-NMR spectra revealed all five cyclopropyl rings to be equivalent down to $-100\,^{\circ}C$. Further spectroscopic evidence showed species (**212**) to be a classical cation undergoing rapid interchange.

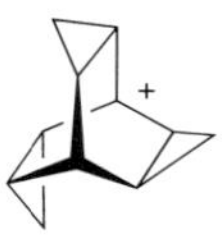

(208)

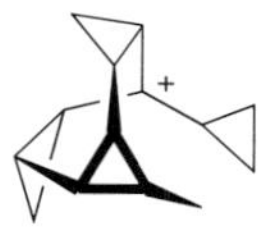

(209)

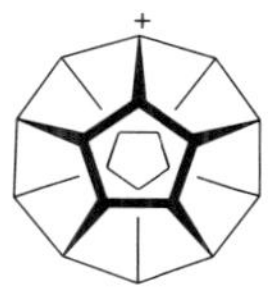

dodecahedryl cation

(210)

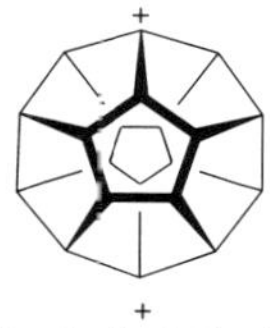

dodecahedryl dication

(211)

(42)

(212)

The C—C bonds of alkanes can be cleaved by superacids and primary, secondary and tertiary hydrogens of alkanes can be exchanged under such conditions ⟨67CC635, 68RTC319, 69JA3261, 69RTC371, 73JA4960⟩ and these topics have been reviewed ⟨71CT566, 72MI 119-13, 73AG(E)173⟩. Typically neopentane in FSO_3H/SbF_5 can cleave to give CH_4 and the *t*-butyl cation (Equation (43)). C—H cleavage is a competing reaction and, for example, neopentane can give H_2 and the *t*-pentyl cation (formed by rearrangement of the initially formed neopentyl cation) by this type of pathway.

$$Me_4C \xrightarrow{FSO_3H,\ SbF_5} Me_3C^+ + CH_4 \qquad (43)$$

Alkanes can give carbocations in superacid by loss of a hydride ion (Equation (44)) ⟨67JA4739⟩. However, Herlem has later asserted that the mechanism is somewhat more complex than mere hydride abstraction alone, invoking oxidation by both SbF_5 and SO_3 in the presence of HSO_3F and electrochemical oxidation of the hydrocarbon ⟨77PAC107⟩.

$$Bu^tH \xrightarrow{FSO_3H,\ SbF_5} Me_3C^+ + SbF_5{\bullet}FSO_3^- + H_2 \qquad (44)$$

1.19.2.2.3 Enyl carbocations by bond scission

Unlike alkyl carbocations, allylic cations are stable in concentrated sulfuric acid provided that the termini of the allylic system are completely alkylated (the contributing resonance canonical forms should be those of tertiary carbocations) ⟨B-68MI 119-03⟩. The simplest ion of this type (**213**) was obtained by Deno and co-workers ⟨62JA1498, 63JA2991, 63JA2995, 63JA2998⟩. Enyl carbocations are stable for a relatively long time in 96% sulfuric acid, but they quickly disappear in more dilute solutions, due to hydration and polymerisation. Allylic cations possessing a cyclopentane skeleton are more stable than the cyclohexane analogues.

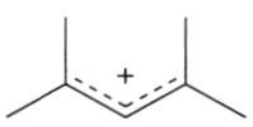

HSO_4^-

(213)

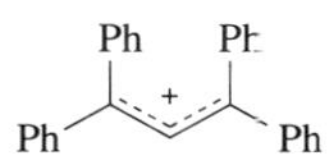

ClO_4^-

(214)

Ziegler found allylic carbocations are additionally stabilised by the introduction of phenyl groups, e.g., the 1,1,3,3-tetraphenylpropenylium cation (**214**), which is formed from tetraphenylallyl alcohol and perchloric acid ⟨23LA(434)34⟩.

A cyclobutenium carbocation (**215**) was obtained ⟨62JA3199⟩ by treating 1,2-dichloro-1,2,3,4-tetramethylcyclobutene-3 with SbF_5/SO_2 at −70 °C. The analogous tetraphenylated compound (**216**) was obtained as a crystalline $SnCl_5^-$ complex ⟨62JA4165, 64JA733⟩.

Allylic, acyclic dienyl or polyenyl cations are usually obtained by addition of a proton to the terminal carbon in the array of C═C bonds. However, by treating certain diphenylpolyvinylcarbinols with HF and BF_3, Hafner obtained crystalline fluoroborates of a series of

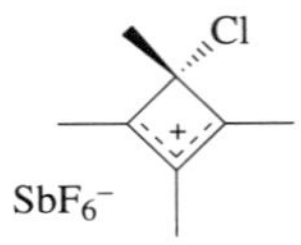

(215)

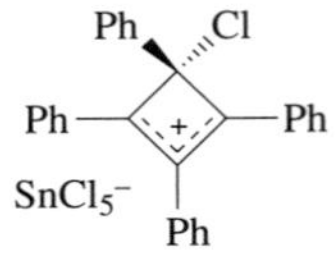

(216)

diphenylmethine carbocations (**217**), the colour of which varies from orange ($n = 0$), to blue ($n = 4$) ⟨61AG342⟩.

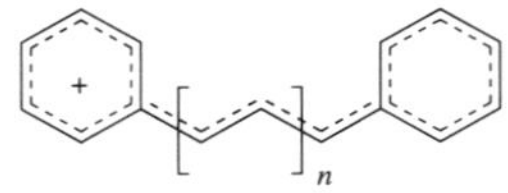

(217)

Carbocation (**219**), resulting from alcohol (**218**), has a half life of only a few seconds in 96% sulfuric acid at 25°C, quantitatively rearranging to the monoenylic cation (**220**) (Scheme 66) ⟨64JA1871⟩.

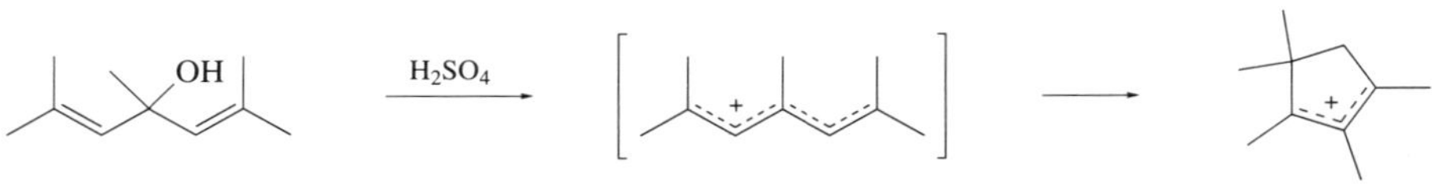

(218) **(219)** **(220)**

Scheme 66

Stable allylic cations have been prepared by Olah and co-workers in superacid media by the reaction of alkyl halides, alcohols, or alkenes (via hydride abstraction) with SbF_5 in SO_2 or SO_2ClF (Equations (45)–(47), Schemes 67 and 68, Equations (48)–(52)) ⟨64JA5682, 71JA4219, 72JA3544, 72JA6434, 75JA1539⟩.

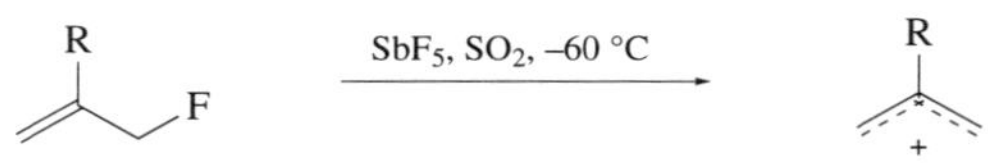

R = H, Me (45)

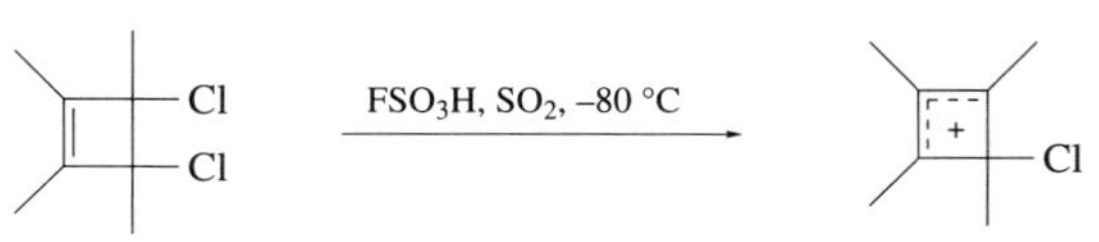

(46)

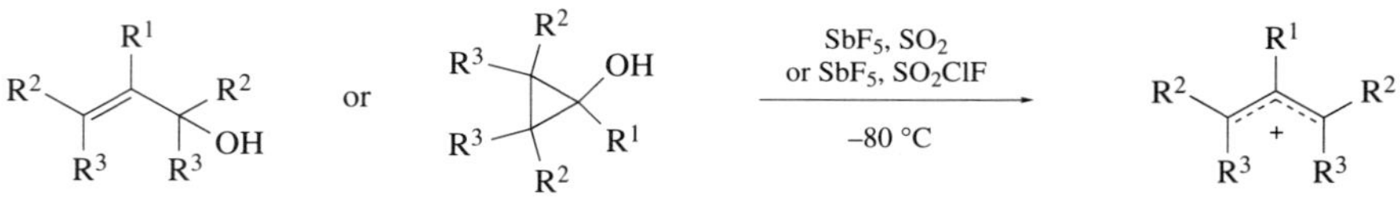

(47)

$R^1 = R^2 = R^3 = Me$
$R^1 = H$; $R^2 = R^3 = Me$
$R^1 = R^2 = Me$; $R^3 = H$
$R^1 = Ph$; $R^2 = Me$; $R^3 = H$

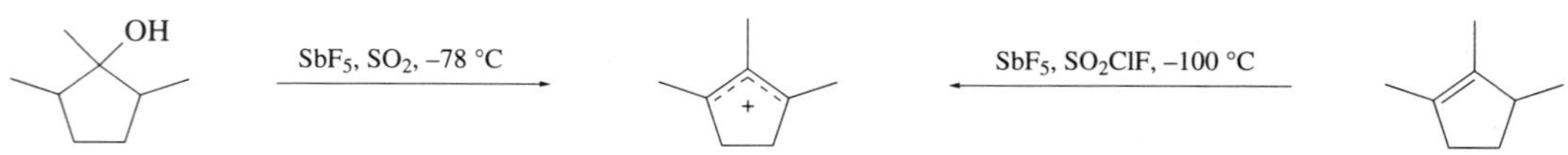

Scheme 67

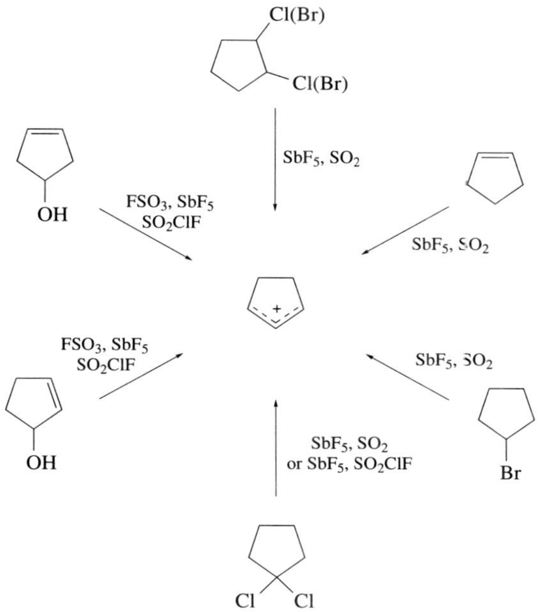

All reactions carried out at reduced temperature

Scheme 68

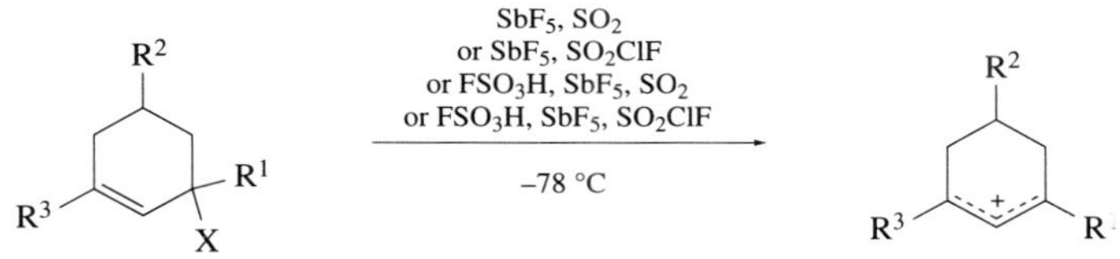

(48)

$R^1 = R^2 = R^3 = H$; X = OH, Br or Cl
$R^1 = Me$; $R^2 = R^3 = H$; X = OH
$R^1 = R^2 = Me$; $R^3 = H$; X = OH
$R^1 = R^3 = Me$; $R^2 = H$; X = OH
$R^1 = R^2 = R^3 = Me$; X = OH

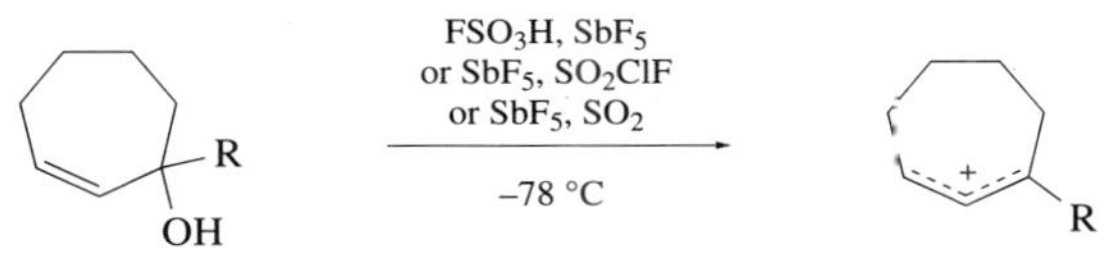

(49)

R = H or Me

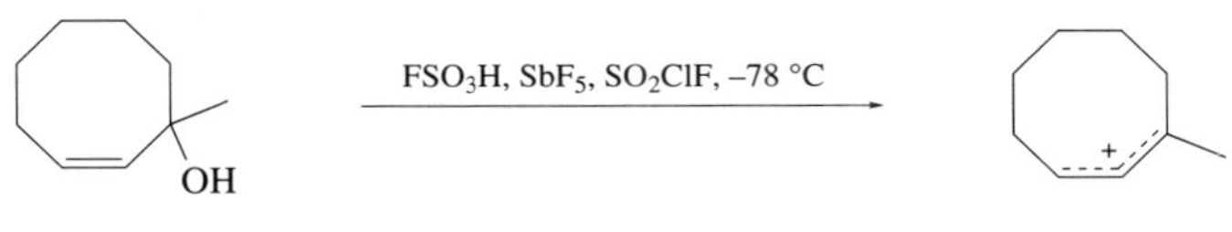

(50)

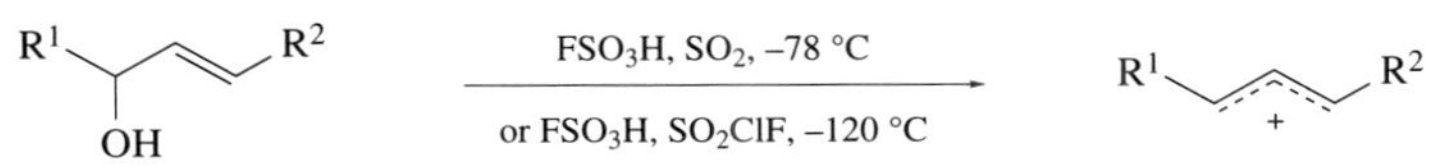

(51)

$R^1 = R^2 = Me$
$R^1 = Pr^c$; $R^2 = Me$
$R^1 = Ph$; $R^2 = Me$
$R^1 = R^2 = Pr^c$
$R^1 = Pr^c$; $R^2 = Ph$
$R^1 = R^2 = Ph$

(Pr^c = cyclopropyl)

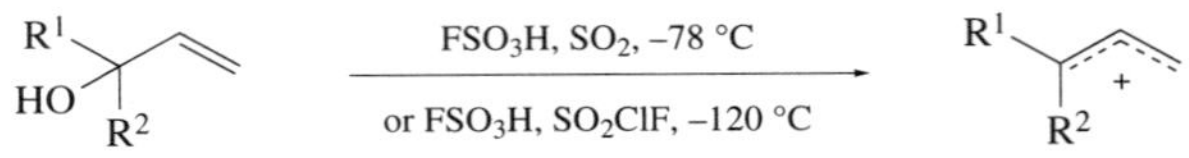

(52)

$R^1 = R^2 = Me$
$R^1 = Me; R^2 = Pr^c$
$R^1 = Me; R^2 = Ph$
$R^1 = R^2 = Pr^c$
$R^1 = Pr^c; R^2 = Ph$
$R^1 = R^2 = Ph$

1.19.2.2.4 Ynyl and cyano carbocations by bond scission

Diarylalkynyl cations (R—C≡C—C^+Ar_2) and other alkynic cations have been formed ⟨65JA1381, 65JA4017⟩. Stable propargyl cations (R—C≡C—CR_2^+) have been prepared in superacid media by Olah and co-workers ⟨65JA5632, 74JA5855⟩. The tris(ethynyl)methyl cations (**222**; R = Me, Bu^t, TMS, Ph) can be obtained by treating the parent alcohols (**221**) (Scheme 69) with FSO_3H/SO_2ClF at −78 °C. Significant participation of resonance canonicals (**222b**) with allenyl character is envisaged by Olah and co-workers, and the electronic effects due to changing R were found to be only slight ⟨90JOC6061⟩. Unsubstituted, *p*-Me- and *p*-MeO-1,1-diphenyl-2-butynyl cations (Me—C≡C—C-Ar_2^+) have also been studied by Olah *et al.* ⟨86JA1575⟩.

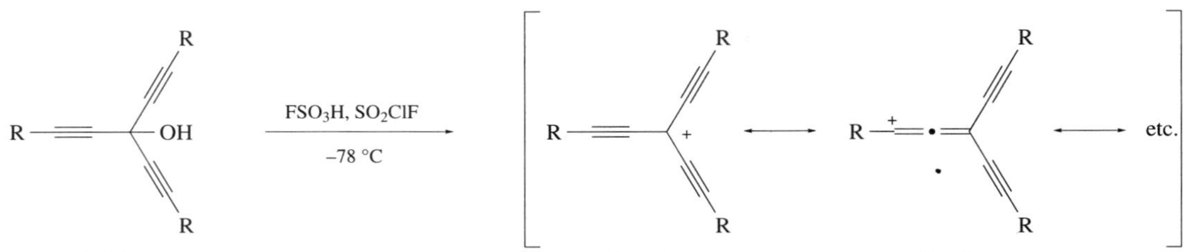

(221) **(222a)** **(222b)**

Scheme 69

Solvolytic studies ⟨80JA1214, 80JA4138, 80JA7613⟩ have shown that a cyano group bonded to a cationic carbon atom stabilises it through resonance, despite the strong −*I* effect. This model is supported by ^{15}N-NMR spectroscopy of cyanodiarylmethyl cations (N≡C—CAr_2^+) ⟨80JA6640, 82JA1628⟩.

1.19.2.2.5 Nonclassical carbocations (carbonium ions) by bond scission

The subject of nonclassical carbocations (carbonium ions) has been reviewed ⟨B-72MI 119-10, B-72MI 119-11, B-87MI 119-07, B-92MI 119-15⟩. Overlap of the vacant orbital of a carbocation with nonconjugated, distant π- or σ-bond orbitals within the same molecule may produce a stabilisation of the ion which can be recognised by kinetic and stereochemical behaviour. The extreme of stabilisation would be to ensure the existence of a carbonium ion in protic acid solutions or in the form of crystalline salts. In 1959, Leal and Pettit reported the preparation of a crystalline perchlorate (**224**) (Scheme 70) from the tertiary benzyl chloride (**223**) ⟨59JA3160⟩. The stability of the salt (**224**) was attributed to the overlap of the empty p_z-orbital of cationic carbon atom C2 with the π-molecular orbital of the double bond.

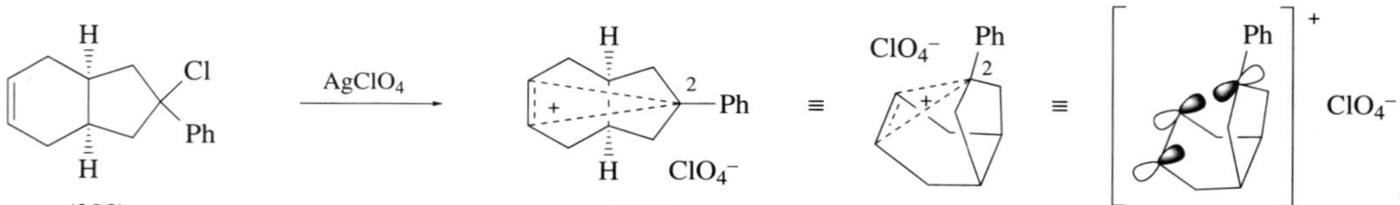

(223) **(224)**

Scheme 70

2-*p*-Anisyl-camphenilol and its epimer iso-2-*p*-anisyl-camphenilol yield in sulfuric, hydrochloric or formic acid solutions a stable nonclassical carbocation, characterised by a strong absorption in the visible and near-UV spectra and by its conductivity (in formic acid solution). On solvolysis, rearranged reaction products are formed; therefore carbocation (**225**) may be considered to be an example of the isolation of a carbonium ion acting as an intermediate in a Wagner–Meerwein rearrangement ⟨59LA(623)217⟩.

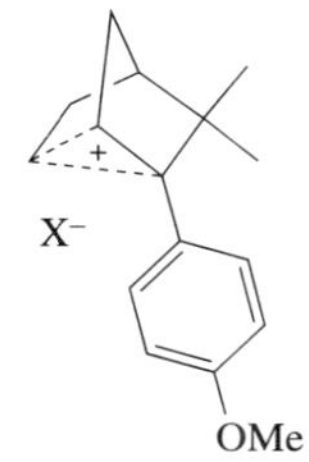

(225)

$X = HSO_4^-, Cl^-$ or HCO_2^-

A prodigious quantity of work has been published on the solvolysis of the 2-norbornyl system in an effort to determine whether the 1,6-bond participates and if the carbonium ion (**226**) (Scheme 71) is a true intermediate ⟨52JA1147, 52JA1154, 65JA376, 66PAC25, 83ACR426, 83ACR432, 83ACR448, 88JOC661⟩. The controversy was for many years dominated by the conflicting views of Winstein, who proposed the nonclassical ion, and Brown, who argued in favour of a classical carbocation ⟨85MI 119-06⟩.

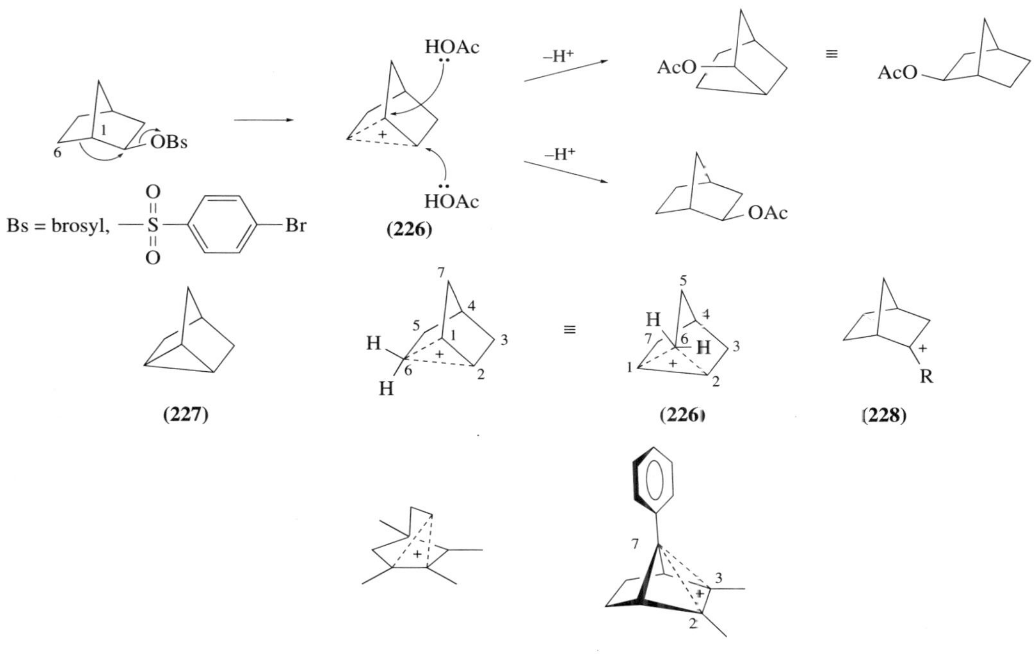

Scheme 71

A majority of scientists now acknowledge the validity of the carbonium ion (**226**) ⟨79JA522, 80JCS(P2)1447, 82TL3119, 83JA950, 83JA7123, 84CB2510, 84JA810, 84TL3403, 85ACR254, 85JA3747, 87JA7263, 88JOC661⟩, but this view is by no means unanimous. For evidence since the late 1970s against a nonclassical carbocation (**226**) see ⟨77JA377, 78JA2501, 78JCR(S)62, 78JOC3667, 79JOC133, 79JOC3536, 79TCC(80)1, 80JCR(S)152, 80JOC2113, 83CJC2403, 85JOC826, 85JOC1165, 86JOC3736⟩. The 2-norbornyl cation (**226**) has also been extensively studied at low temperatures and these studies tend to support the nonclassical model. Olah and co-workers have prepared the 2-norbornyl cation in stable solutions at temperatures below $-150°C$ in SbF_5/SO_2 and $FSO_3H/SbF_5/SO_2$, when the structure is static and hydride shifts are absent. Raman spectra, ^{13}C-NMR and x-ray analysis indicated the ion was nonclassical ⟨70JA4627, 72JA808, 73JA8698, 76ACR41, 80JA683, 80JA6867, 82JA7105, 83ACR440, 83JA3571, 88JA4112⟩. There are, however, those who still challenge the nonclassical view ⟨74JA7638, 75APO(11)177, 82JOC2089, 89MI 119-09⟩, which Olah continues to defend ⟨83JOC2146⟩. The nonclassical model is supported by the fact that ^{13}C-NMR spectra of the 2-norbornyl cation in the solid state at 77K and even 5K gave no evidence of 'freezing out' a single classical ion ⟨82BSB422, 82JA907, 82JA7380, 90JA8991⟩. Olah represents the nonclassical structure as nortricyclane (**227**) protonated at the apex. Olah concluded that all the positive charge resides on C1 and C2 with a negligible amount on C6 in carbonium ion (**226**).

Ab initio calculations show that the nonclassical model (**226**) corresponds to an energy minimum ⟨89JA1527⟩. At low temperatures the spectra of (**228**; R = Me or Et) show reduced charge delocalisation ⟨69JA3958⟩, and (**228**; R = Ph) is effectively classical ⟨74JA195⟩, as are (**228**; R = OMe) ⟨69JA6861⟩ and (**228**; R = Cl) ⟨69JOC2314⟩. ^{13}C-NMR studies show that electron-withdrawing substituents on a phenyl ring at C2 render the cation less classical, whereas electron-donating groups confer greater classical character ⟨77JA5683, 78JA3847, 82JOC3903⟩.

The first crystal structure of a norbornyl cation was reported by Laube for the $Sb_2F_{11}^-$ salt of the *anti*-1,2,4,7-tetramethyl-2-norbornyl cation (**229**), which was found to have no symmetry ⟨87AG(E)560, 91CRV375⟩. Attempting to assess the nonclassical character of 7-norbornenyl cations, Laube isolated the SbF_6^- salt of the 2,3-dimethyl-7-phenyl-2-norbornen-7-yl cation (**230**) ⟨89JA9224, 91CRV375⟩. The crystal structure of (**230**) showed the carbon skeleton surrounded by two SbF_6^- anions, with no overall symmetry. It was felt the unexpectedly strong bending of the C7 bridge toward the C2 to C3 bridge was indicative of the formation of a three-centre, two-electron bond between C2, C3 and C7.

1.19.2.2.6 Formation by bond scission of carbocations stabilised in accordance with the Hückel (4n + 2) rule

Aromatic and antiaromatic monocyclic ions have been reviewed by Garratt ⟨79COC(2)390⟩.

(i) Cyclic carbocations with (4n + 2) π-electrons, where n = 1

In 1937 Hückel ⟨37MI 119-01, 37MI 119-02⟩ 'predicted' a stable cycloheptatrienyl carbocation, unaware that such a species had already been prepared by Merling without its structure being fully understood ⟨1891CB3108⟩. Doering and Knox subsequently proved ⟨54JA3203⟩ that the crystals obtained by Merling from cycloheptatriene (tropylidene) (**74**) (by addition of bromine and elimination of hydrobromic acid) were cycloheptatrienium bromide (tropylium bromide) (**231**; X = Br) (Equation (53)). Dauben *et al.* later obtained tropylium salts by a hydride transfer from cycloheptatriene (**74**) to the triphenylcarbonium ion in acetonitrile or in liquid SO_2 solution at $-20°C$ ⟨57JA4557, 60JOC1442⟩.

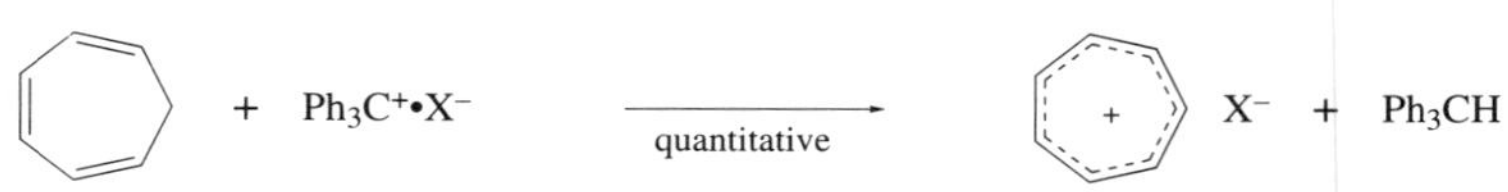

(53)

(74) **(231)**

This method is satisfactory for a wide range of substituent groups, gegenions (X^-) and solvents and generally gives nearly quantitative yields of high-purity salts, from which the by-product triphenylmethane is easily separated. Tropenylium cation chemistry has been reviewed ⟨B-66MI 119-03, 67RCR735, 73MI 119-20, 73CRV293, B-73MI 119-08⟩. The benzotropylium ion (**232**) ⟨55CI(L)415⟩, *sym*- and *asym*-dibenzotropylium ions (**233**) ⟨59JCS3139⟩ and (**234**) ⟨50LA(568)63⟩, respectively, and the tribenzotropylium ion (**235**) ⟨57JOC1243, 58HCA414⟩ have been synthesised. The relatively high stability decreases as indicated in Scheme 72, apparently because the 7-membered ring is distorted from planarity by steric crowding of phenyl protons, which increases through the series.

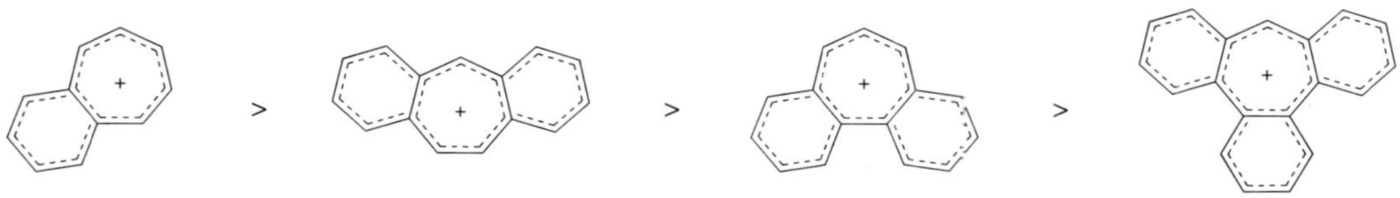

(232) **(233)** **(234)** **(235)**

Scheme 72

The heptaphenyltropylium cation (**236**) was prepared by Battiste as the bromide ⟨61CI(L)550, 61JA4101⟩. The stable 1,3,5-tricyclopropyltropylium cation (**237**) has been reported ⟨82BCJ3257⟩. One of the most stable carbocations ever produced, the 1,2 : 3,4 : 5,6-tris(bicyclo[2.2.2]octeno)tropylium ion (**238**), has been described by Komatsu *et al.* ⟨88JA633⟩. Its stability is attributed to inductive effects and enhanced hyperconjugation of the aromatic system with twelve of the C—C σ-bonds of the cyclohexyl side groups. These bonds lie almost orthogonal to the plane of the 7-membered ring and are thus almost parallel to the *p*-orbitals of the aromatic system.

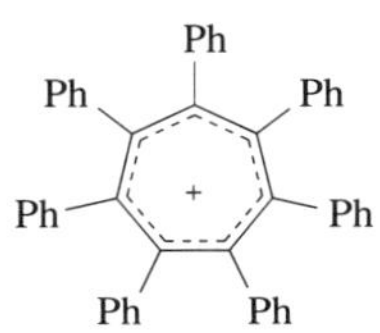

(236)

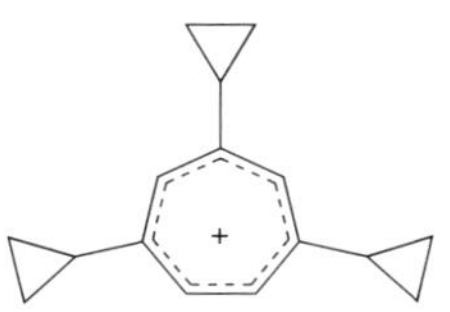

(237)

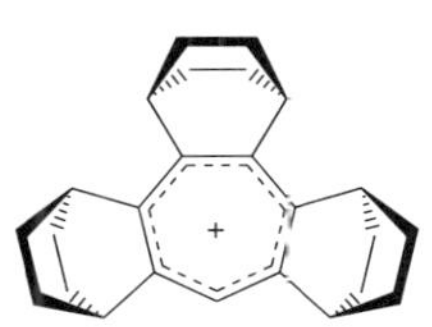

(238)

*(ii) Cyclic carbocations with (4*n *+ 2) π-electrons, where* n *= 0*

For reviews of two-electron aromatic systems see ⟨65AG(E)10, 66MI 119-09, B-66MI 119-04, 67RCR557, 73TCC47, 74CRV189, B-87MI 119-08⟩. The Hückel $(4n+2)$ rule predicts that in general cyclopropenyl cations (**239**) should possess aromatic stability, and this prediction is validated by the fact that a large number of such cations have been reported. Even the unsubstituted archetype (**240**) has been prepared (Equation (54)) ⟨67JA5048a, 67JA5048b, 70JA984⟩.

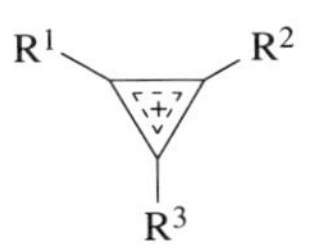

(239)

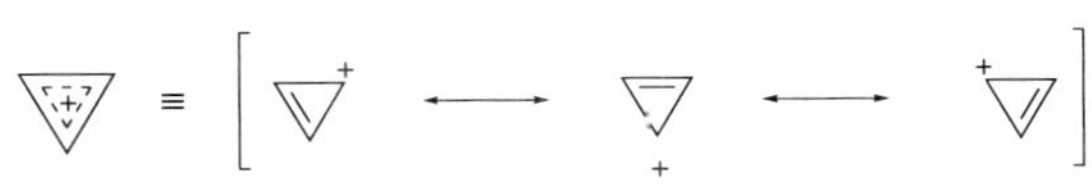

(240) (54)

The majority of preparations of cyclopropenyl salts have involved carbene addition to an acetylenic bond. The first example of such a salt (**241**) was prepared by Breslow and co-workers using a carbene generated by decomposition of a diazo compound. This synthesis and other examples are shown in Scheme 73 ⟨57JA5318, 58JA5991⟩, Scheme 74 ⟨61JA2367⟩, Scheme 75 ⟨61JA2375⟩, Equation (55) ⟨60JA2644, 60JA2651, 62JA3168⟩, Scheme 76 ⟨64JA1459⟩, Scheme 77 ⟨62JA3168⟩ and Equation (56) ⟨62JA3168⟩.

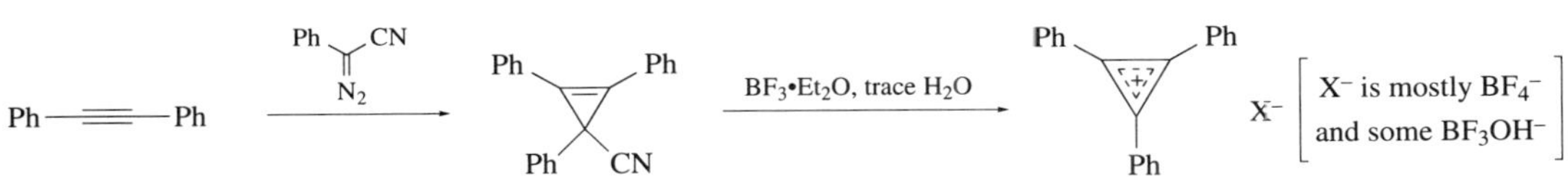

(241)

Scheme 73

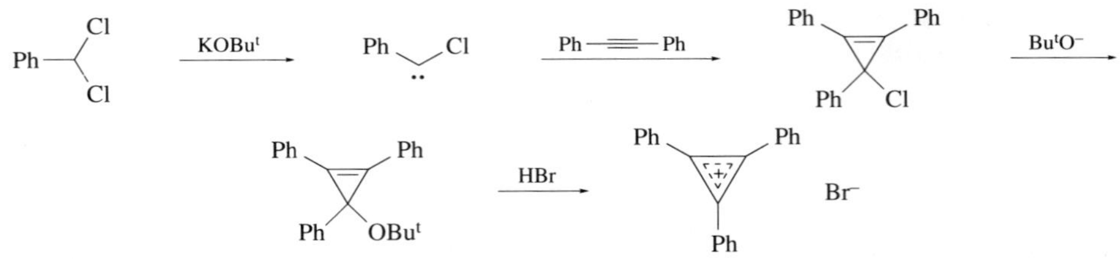

Scheme 74

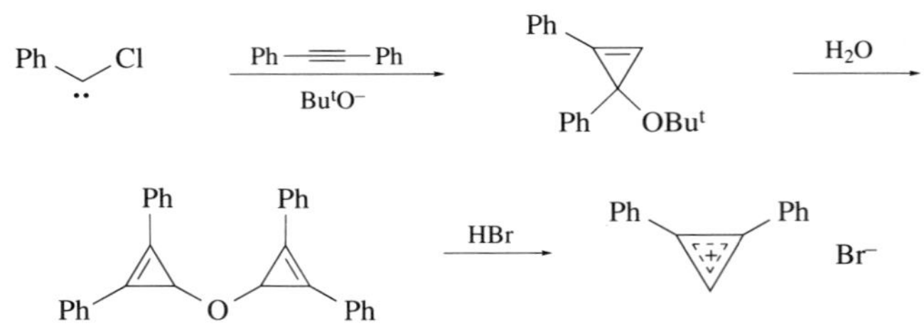

Scheme 75

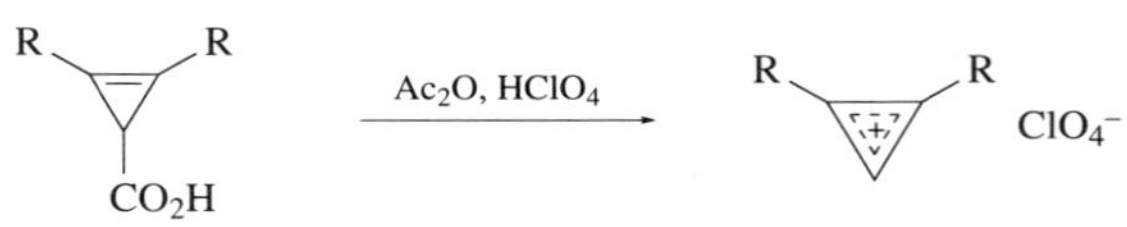

(55)

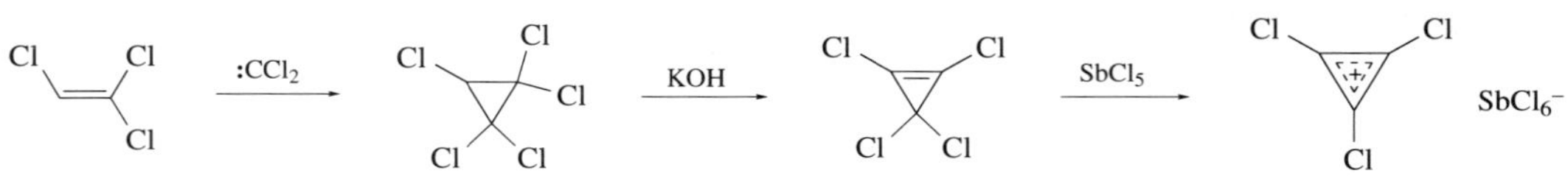

Scheme 76

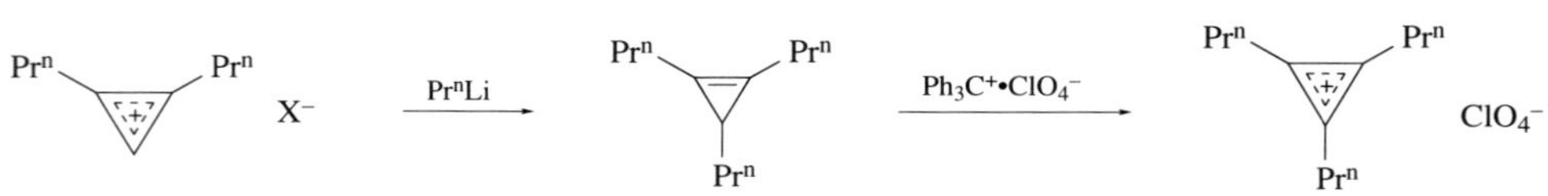

Scheme 77

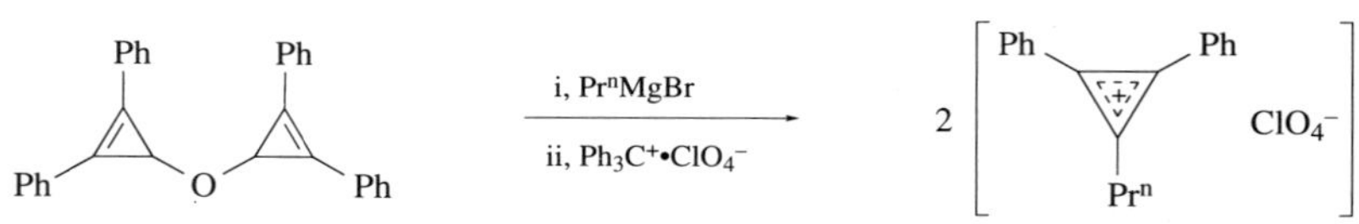

(56)

Cyclopropenium salts (**239**) are soluble and stable in acid solution, but neutral aqueous solutions become turbid owing to the equilibration with a covalent carbinol (**242**).

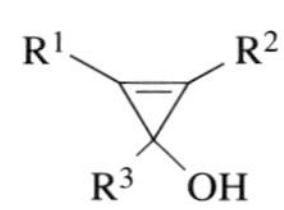

(242)

It has been suggested ⟨62JA3168⟩ that the pH value at which there is 50% ionisation of the carbinol (**242**) to the cyclopropenium ion (**239**), described as pK_{R^+}, can be taken as a criterion of the stability of the cation. Some values have been determined experimentally by potentiometric titration or spectrophotometrically (Table 12).

Table 12 pK_{R^+} values of some cyclopropenyl cations.

Compound	*Medium*	*Method*	pK_{R^+}
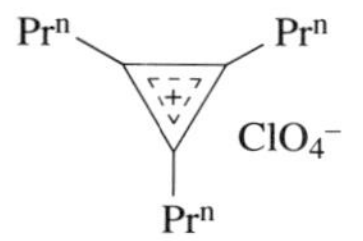	50% $MeCN/H_2O$	potentiometric titration	7.2
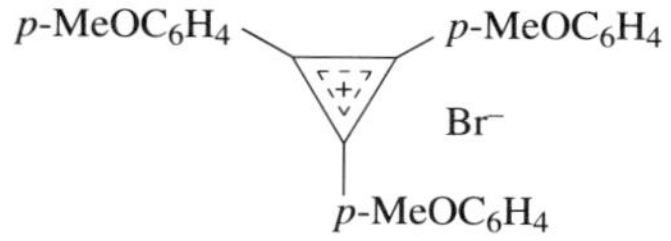	50% $MeCN/H_2O$	potentiometric titration	6.5
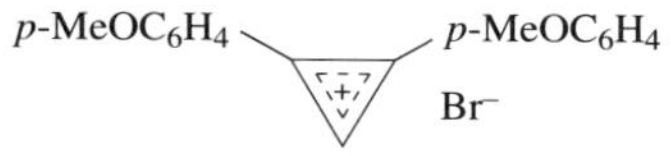	50% $MeCN/H_2O$	potentiometric titration	5.2
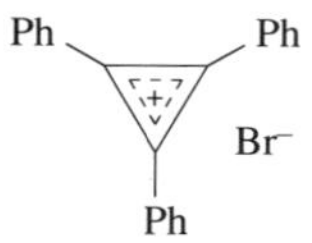	50% $MeCN/H_2O$	potentiometric titration	3.1
	50% $MeCN/H_2O$	potentiometric titration	2.7
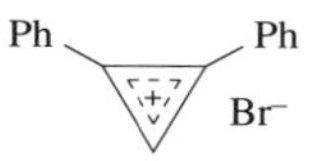	H_2O	spectrophotometric titration	0.32

A number of factors other than π-electron stabilisation affect the position of equilibrium so that the absolute values of pK_{R^+} are much less easy to interpret than their relative values. The pK_{R^+} values in Table 12 suggest that *n*-propyl groups stabilise the cation more effectively than do phenyl groups. An alternative explanation is that the covalent carbinol is more effectively stabilised by phenyl than by propyl groups. On comparing the diphenyl and diphenylpropyl compounds, the pK_{R^+} value is greater for the latter. This suggests that the *n*-propyl group does indeed have a stabilising effect on the cyclopropenium cation; it seems likely that this is an inductive rather than hyperconjugative effect.

The *syn*-triphenylcyclopropenium perchlorate (**243**) was first synthesised by Breslow and Yuan ⟨57JA5318, 58JA5991⟩. Due to the steric compression of the *ortho*-hydrogens, the phenyl rings cannot lie coplanar with the 3-ring and the molecule adopts a propeller-like arrangement with the benzene rings twisted at a shallow average angle of 13.6° ⟨63JA3302, 66JA198⟩. The tripropylcyclopropenyl ⟨62JA3168⟩ and the tricyclopropylcyclopropenyl (**244**) ⟨80TL947, 86JA134⟩ cations rank amongst the most stable carbocations known and are even stable in aqueous solution. The tri-*t*-butylcyclopropenyl cation is also very stable ⟨68JA4495⟩, as is the tris-1-(5-isopropyl-3,8-dimethylazulenyl)cyclopropenylium cation (Agranat's cation) ⟨76JOC2379, 90PAC507⟩.

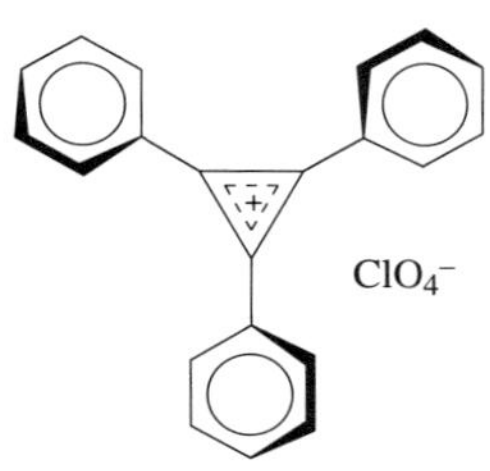

(243)

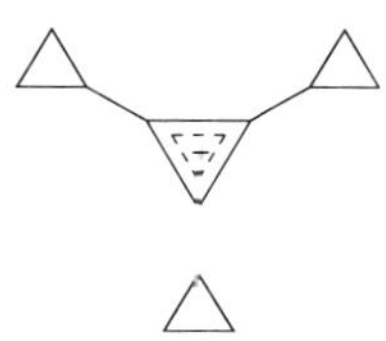

(244)

The dications (**245**; R = Me or Ph) have been prepared ⟨64JA734, 69JA3667, 70JA1430, 76JA6290⟩ and their aromaticity discussed ⟨74JOC378, 78JA4301⟩. The IGLO (individual gauge for localised molecular orbitals) method has been used by von R. Schleyer and co-workers to study the cyclobutadiene cation ⟨89JA1147⟩. These and earlier *ab initio* calculations ⟨86JA561⟩ and x-ray studies ⟨84AG(E)371⟩ of derivatives suggest that the puckered ring (**246**) is more stable than the planar configuration implied by structure (**245**).

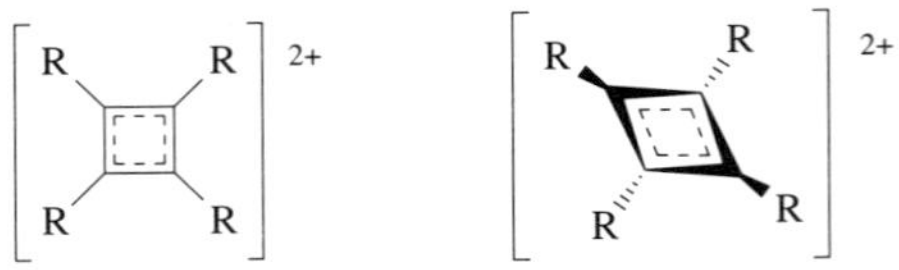

(245) **(246)**

(iii) Cyclic carbocations with (4n + 2) π-electrons, where n = 2

1,6-Methano[10]annulene (**247**) reacts with diazomethane in the presence of cuprous chloride to give bicyclo[5.4.1]dodecapentene (**248**), from which a hydride ion may be abstracted producing a bridged cation (**249**) (Scheme 78) having a delocalised system of 10π-electrons ⟨65AG(E)354, 66AG(E)604, 68HCA225, 70TL1941⟩. The observed ^{1}H-NMR ring current ⟨65AG(E)354⟩ and x-ray analysis ⟨76JA1999, 79AX(B)1846⟩ favour the delocalised 10π-electron system (**249**) rather than a benzo-homotropylium structure (**250**). The distance between C1 and C6 is long (2.3 Å) and the rather flat nature of the peripheral carbon system combine to suggest that transannular interactions in this carbocation are not important ⟨84ACR347⟩.

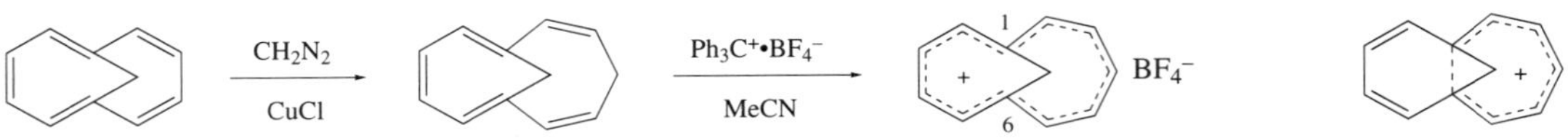

(247) **(248)** **(249)** **(250)**

Scheme 78

For discussions relating to factors affecting paramagnetic ring currents in the analogous bicyclo-[5.4.1]dodecapentaenyl anion see ⟨73JA3384⟩.

1.19.2.2.7 *Phenalene cations by bond scission*

The stability of the phenalenium cation (**86**) was mentioned in Section 1.19.1.2.10. The cation was first prepared by Pettit as its perchlorate salt ⟨56CI(L)1306, 60JA1972⟩. A general synthesis of phenalenium salts features hydride abstraction from phenalenes by high potential quinones in the presence of perchloric acid ⟨63JCS3295⟩ or by triphenylmethyl perchlorate ⟨59JCS2773⟩. Several alkyl phenalenium perchlorates have been prepared by this method (e.g., (**251**)) as well as benzo[*a*]phen-alenium perchlorate (**252**) ⟨65QR274⟩.

(251) **(252)**

Although not conforming to the (4*n* + 2) rule the perinaphthenylium cation is remarkably stable. According to molecular orbital theory ⟨B-61MI 119-02⟩ the positive charge is distributed between positions 1,3,4,6,7 and 9 (but not on the central position 13). These six positions are equivalent and vulnerable to nucleophilic attack. The solid yellow perchlorate is stable in dry nitrogen, but in the presence of water it yields perinaphthenone and perinaphthylene denoting the intermediate formation of a secondary alcohol which subsequently disproportionates ⟨56CI(L)1306, 60JA1972⟩. Certain hydrocarbons, e.g., pyrene and perilene, whose conjugate acids contain the perinaphthenylium cation, are highly reactive towards electrophiles at the positions predicted by molecular orbital theory ⟨B-61MI 119-01⟩.

1.19.2.3 Carbocations by Addition to C=C Bonds

1.19.2.3.1 Stable carbocations by protonation of alkenes

Cationic polymerisation generally occurs when treating reactive alkenes with strong acids ⟨B-63MI 119-03⟩. However, Olah and Halpern discovered conditions which allowed the formation of stable monomeric carbocations as the only or predominant product in high yields by the protonation of alkenes with superacids (Scheme 79, Equation (57), Scheme 80, Equations (58) and (59)) ⟨71JOC2354⟩. However, it was emphasised that no single general procedure could be applied to any alkene, but specific conditions had to be employed in each case.

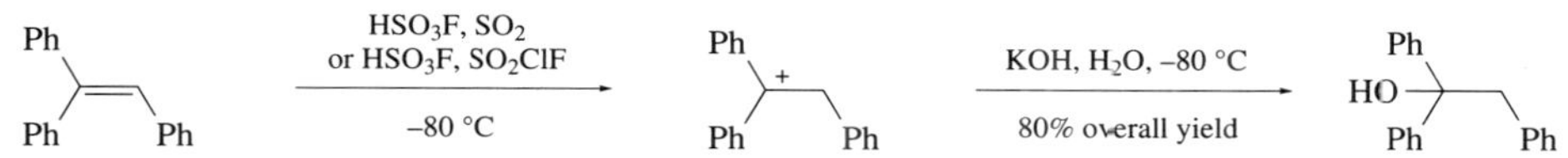

Scheme 79

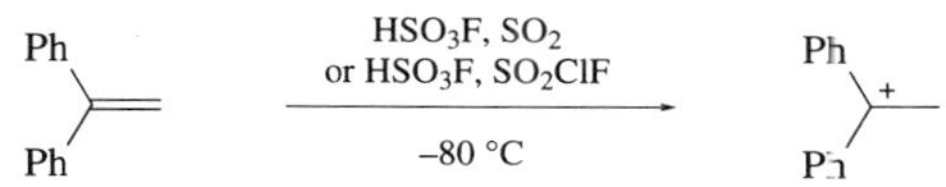

(57)

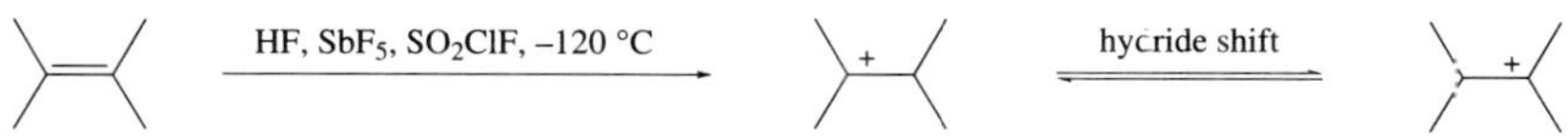

Scheme 80

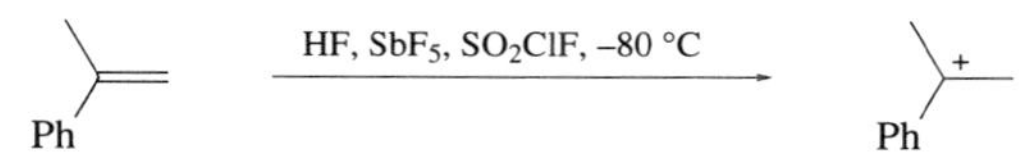

(58)

(59)

1.19.2.3.2 Stable carbocations by protonation of dienes

The formation of the allyl cation by protonation of dienes has been reviewed by Deno ⟨B-70MI 119-03⟩ and his work is referred to in Section 1.19.2.2.3, ⟨62JA1498, 63JA2991⟩. Deno emphasised the importance of the measurements of freezing-point depression in sulfuric acid leading the initial confirmation of the existence of conjugated alkenyl cations ⟨B-70MI 119-03⟩. The demonstration that 1,3,5,5-tetramethyl-cyclohexadiene (**253**) produces a twofold depression in freezing point in sulfuric acid combined with quantitative regeneration of the starting diene gave the strongest evidence for the validity of Equation (60) and the formation of carbocation (**254**).

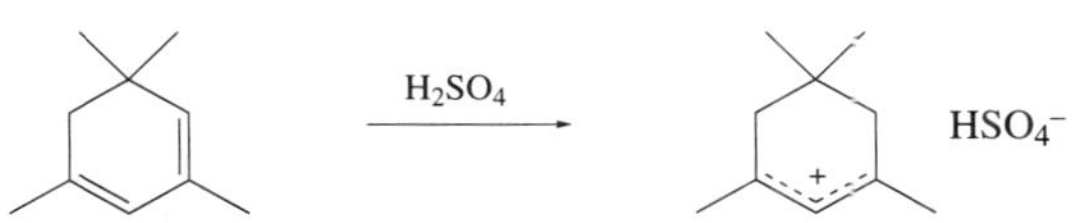

(60)

(253) **(254)**

An idea of the thermodynamic stability of various allylic cations derived from protonated dienes may be derived from Table 13. It may be noted that stability is conferred on the allylic cation by the cyclopropyl group, a phenomenon discussed by Deno *et al.* ⟨65JA4533⟩.

1.19.2.3.3 Dienylic and polyenylic cations by protonation of trienes and polyenes

This topic has been reviewed by Sorensen ⟨B-70MI 119-04⟩. The interaction of polyenes with strong acids has long been a recognised colour test, dating back to early days of the dye industry, when it was observed that the colour of many compounds could be enhanced or shifted to longer wavelengths by the addition of acid. The great majority of compounds, in which colour enhancement occurred, contained an 'auxochromic' group, but some (e.g., the purely hydrocarbon polyenes) probably derived their enhanced colours from carbocation formation ⟨38CB442⟩. Hexaene (**255**) has been protonated under strongly acidic conditions and the NMR spectrum indicates the formation of the carbocation (**256**), which gives a blue colour in dilute solution (Equation (61)) ⟨65JA5075⟩. Treatment of (**257**) with acid gave the magenta tetraenylic cation (**258**) ⟨61JA2392⟩, which is rendered particularly stable by the heavily alkylated terminal cyclopentenyl rings. Treatment of the triene (**259**) with acid also gave the dimeric cation (**258**) (Scheme 81) ⟨61JA2392⟩.

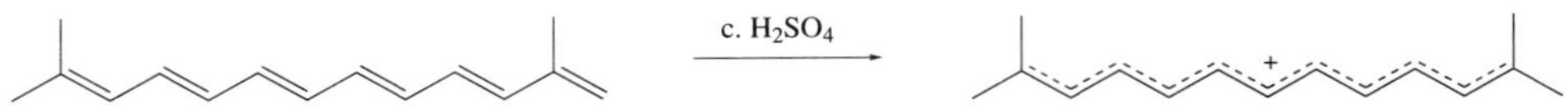

(61)

(255) **(256)**

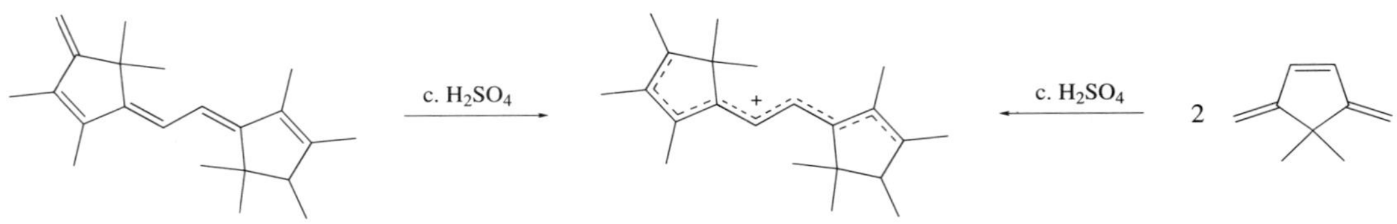

(257) **(258)** **(259)**

Scheme 81

The dienylic cations (**260**) ⟨64JA1871⟩ and (**261**) (Scheme 82 ⟨64JA1871⟩, Equation (62) ⟨65CJC2744, 65JA2153⟩) are stable in 96% H_2SO_4; compound (**261**) results from cyclisation and/or rearrangement of either (**262**) or (**263**). The first direct observation of a dienylic cation (**265**) of the type which is in equilibrium with an aliphatic triene (**264**) was made by von Doering and Saunders *et al.* who discovered the heptamethylcyclohexadienyl cation (**265**) as a product of the exhaustive methylation of benzene ⟨58T178⟩. Cation (**265**) is an example of an 'arenonium ion', the formation of which type of species is reviewed in Section 1.19.2.3.4 (Equation (63)) ⟨58T178⟩.

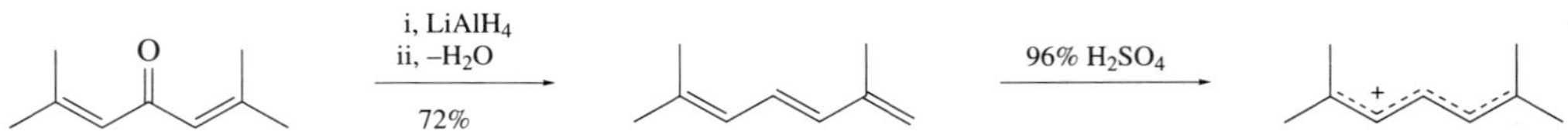

(260)

Scheme 82

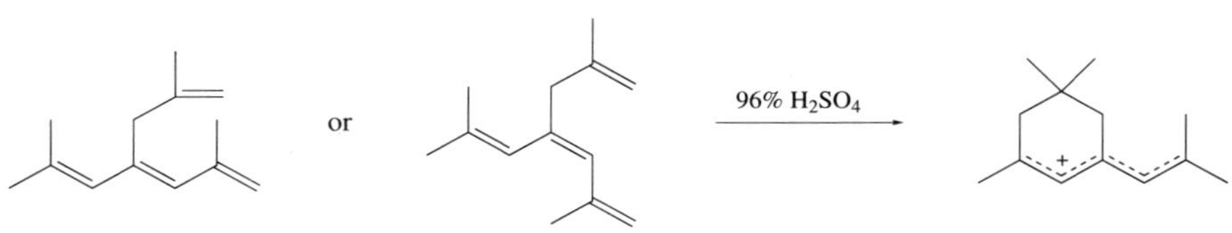

(62)

(262) **(263)** **(261)**

Table 13 p*K* values for the protonation of some dienes.

Alkenyl cation[a]	*Concentration of* H_2SO_4 *(%) required for half-protonation*	p*K*[b]	*Ref.*
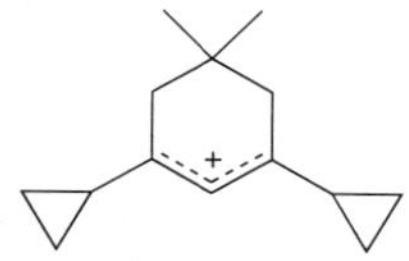	1.2	0.6	65JA4533
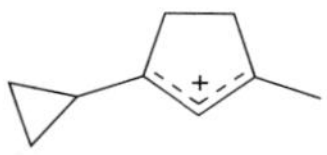	12	–0.46	65JA4533
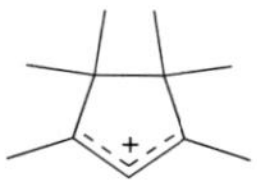	33	–1.9	63JA2998
	35	–2.1	63JA2998
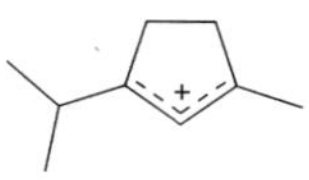	42	–2.6	63JA2998
	42	–2.6	63JA2998
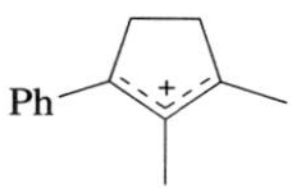	45	–2.9	65TH
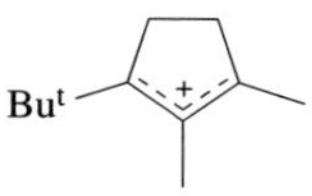	47	–3.1	65TH
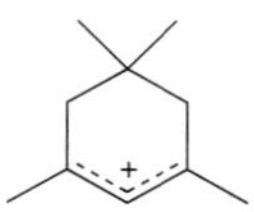	50	–3.4	63JA2998
	50	–3.4	65JA4533
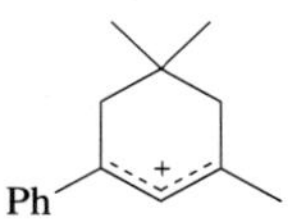	52	–3.6	65JA4533
	54.5	–3.8	65JA4533
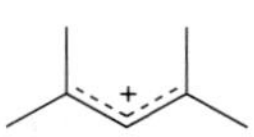	73	6.3	63JA2998
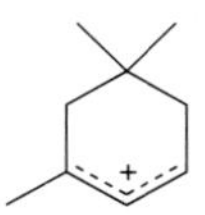	80	–7.3	63JA2998
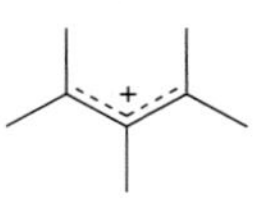	82	7.7	63JA2998

[a] The structure of the cation is given instead of that of the diene because in most cases a single cation is in equilibrium with several dienes.
[b] Values of p*K* were calculated from the equation: $H_o = pK + \log([C_B]/[C_{BH+}])$ where H_o = Hammett's acidity function.

(264) (265) (63)

1.19.2.3.4 Arenonium and phenonium ions

(i) Arenonium ions

The early history of arenonium ions was outlined by Nenitzescu ⟨B-68MI 119-03⟩, and the field has been subsequently reviewed ⟨B-70MI 119-05, 74BAU1031, 82ACR46, 84TCC1⟩. Until about 1940, the σ-complexes of aromatic compounds, also called 'Wheland intermediates', were only hypothetical entities. In 1958, Olah and Kuhn achieved the decisive breakthrough by isolating crystalline cyclohexadienyl fluoroborates (**266**)–(**269**) from the reaction between an aromatic hydrocarbon (toluene, *m*-xylene, mesitylene or isodurene), HF, and BF_3 at low temperature (see Scheme 83), which are salts of arenonium ions ⟨58JA6535⟩. The resulting coloured complexes had a high specific conductivity and definite decomposition points, varying from −65 °C for the toluene complex to −15 °C and −10 °C for the complexes of mesitylene and isodurene. Using similar techniques Olah and co-workers prepared the related PF_6^- and SbF_6^- complexes of methylbenzenes ⟨62JOC3441⟩. A classic paper by MacLean and Mackor reported the NMR spectra of a number of stable arenonium ions, including those formed by protonation of toluene, mesitylene and hexamethylbenzene ⟨61MP241⟩.

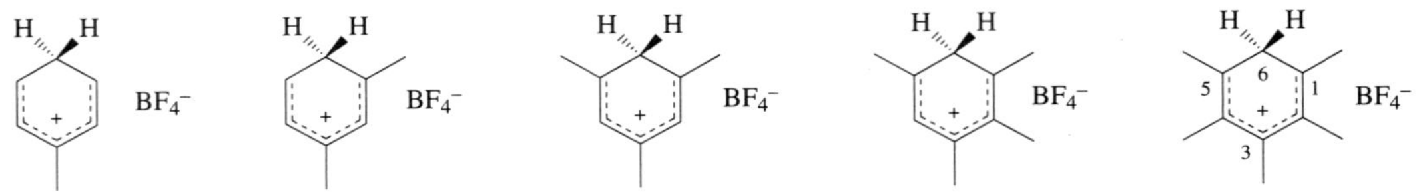

(266) (267) (268) (269) (270)

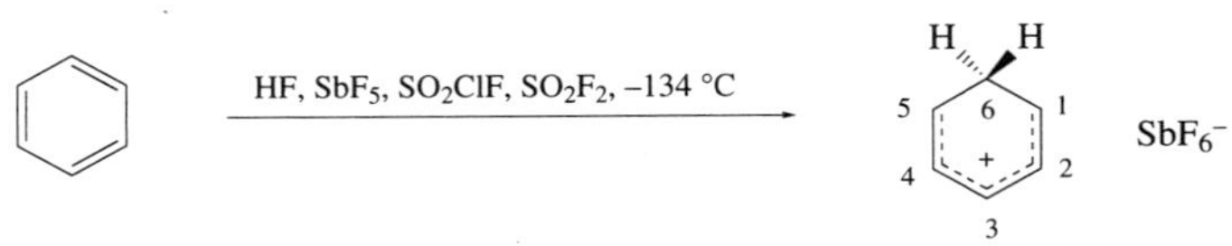

(271)

Scheme 83

The archetypal arenonium ion, the benzenonium ion (**271**), has been prepared at low temperature in superacid media (Scheme 83) ⟨72JA2034⟩. The ^{13}C-NMR spectra of the benzenonium ion (**271**) ⟨78JA6299⟩ and the pentamethylbenzenonium ion (**270**) ⟨79JA4770⟩ both indicate that the unit positive charge is distributed approximately equally between C1, C3 and C5.

Olah and Kuhn were the first to isolate typical intermediates of the Friedel–Crafts aromatic alkylation reactions, for example, the crystalline complex of the type (**271a**) obtained from the reaction of EtF/BF_3 with mesitylene (Scheme 84), along with similar results with toluene, *m*-xylene, and *asym*-tetramethylbenzene ⟨58JA6541⟩. Olah and Kuhn also isolated crystalline complexes of the type (**272**) which, according to theory, are intermediates in aromatic formylation, acetylation and nitration reactions ⟨58JA6541, 60JA2380⟩. Contemporaneously, von Doering and co-workers introduced, by exhaustive methylation of benzene with $AlCl_3$ and MeCl, a seventh methyl group to hexamethylbenzene; the resultant chloroaluminate (**273**) was obtained; this compound gave methanol and hexamethylbenzene upon hydrolysis ⟨58T178⟩.

It should be noted ⟨B-70MI 119-05⟩ that the stability of an arenonium ion is determined by two factors: (i) its resistance toward isomerisation and decomposition reactions; and (ii) its resistance towards deprotonation, which is related to the basicity of the parent molecule. Cations of the less basic isomers of the di-, tri- and tetramethylbenzenes are chemically unstable owing to their tendency to isomerise to the more stable cations of *m*-xylene, mesitylene and isodurene, respectively. These isomerisations proceed by successive intramolecular 1,2-hydrogen and 1,2-methyl shifts. In benzenonium ions, the 1,2-methyl shift is a very slow process relative to the 1,2-hydrogen shift.

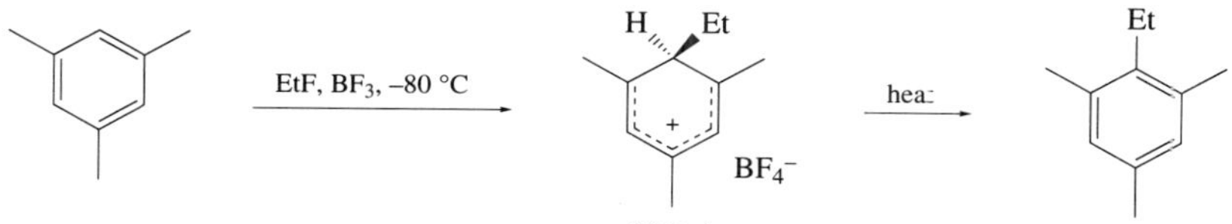

(271a)

Scheme 84

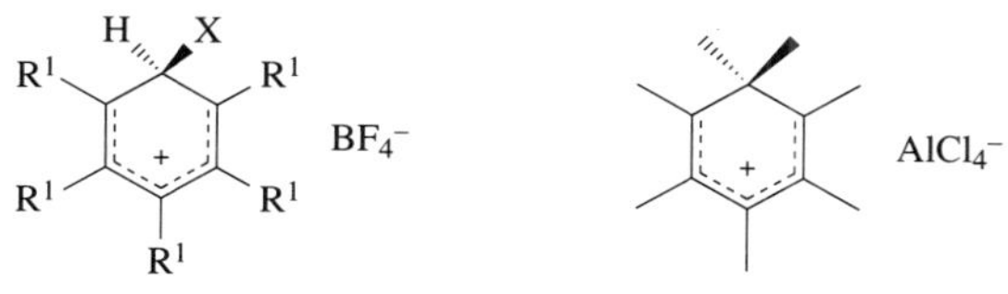

(272) **(273)**

R^1 = Me or H; X = CHO, R^2CO, NO_2

(ii) Phenonium ions

The topic of phenonium ion formation has been reviewed by Lancelot, Cram and von R. Schleyer ⟨B-72MI 119-12⟩ and by Lowry and Richardson ⟨B-87MI 119-09⟩. Several stable phenonium ions have been prepared by Olah in superacid media Scheme 85 ⟨67JA5259, 71JA6877, 76JA6284⟩, Scheme 86 ⟨69JA1458, 84JA3265⟩, whereby their NMR spectra could be studied.

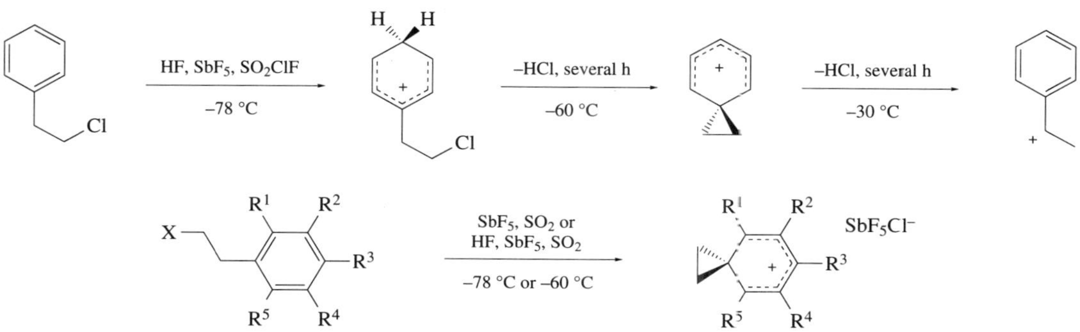

X = OH or Cl

$R^1 = R^2 = R^3 = R^4 = R^5 = H$
$R^1 = R^2 = R^4 = R^3 = H; R^5 = Me$
$R^1 = R^2 = R^4 = H; R^3 = R^5 = Me$
$R^2 = R^4 = H; R^1 = R^3 = R^5 = Me$
$R^1 = R^2 = R^3 = R^4 = R^5 = Me$

Scheme 85

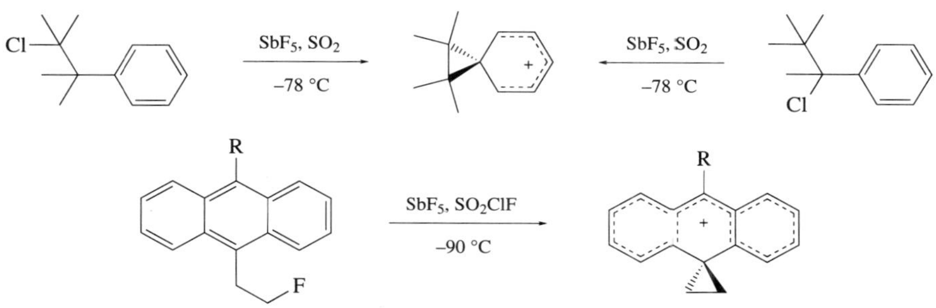

R = H or Me

Scheme 86

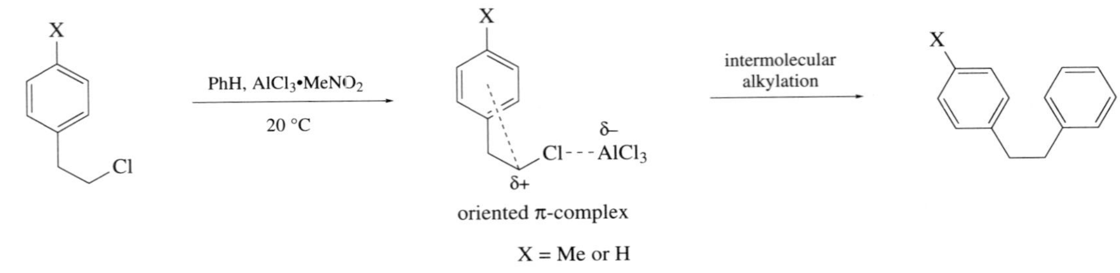

Scheme 87

Investigations by Olah *et al.* concerning the phenylethylation of benzene and toluene with β-phenylethyl chlorides, in more nearly normal Friedel–Crafts conditions, indicated π-aryl participation in polarised donor-acceptor β-phenylethylating complexes rather than the formation of phenonium ions (σ-complexes) (Scheme 87) ⟨92PNA(89)915⟩.

1.19.2.3.5 Azulenyl carbocations

The chemistry of azulene (**274**) and its derivatives has been reviewed ⟨B-66MI 119-05, 77RCR530, 85HOU(V/2c)127⟩. Azulene is an unusual hydrocarbon molecule in that it has a permanent dipole moment (1.08D). Canonicals (**274b and c**) (Scheme 88) reveal how the dipolarity occurs and may be summarised by structure (**274d**). The molecule (**274**) is readily protonated at C1, or its equivalent C3, to give carbocation (**275**) (Scheme 89), which may be portrayed by canonicals (**275a–e**). The seven-membered ring takes on the characteristics of an aromatic tropylium cation and the system may be described by structure (**275f**), which implies partial fixation of the C2=C3 double bond. It is this ability of the seven-membered ring of the azulenyl system to accommodate a positive charge, which enables the three guaiazulene substituents to stabilise the cyclopropenium ion in the tris-1-(5-isopropyl-3,8-dimethylazulenyl)cyclopropenylium cation (Agranat's cation) (Section 1.19.2.2.6.(ii)).

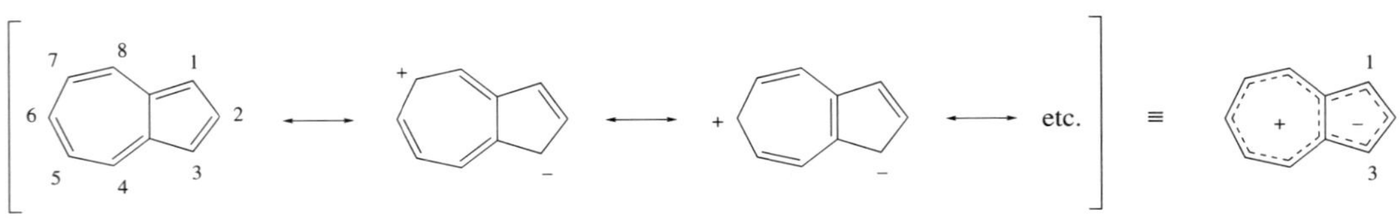

(**274a**) (**274b**) (**274c**) (**274d**)

Scheme 88

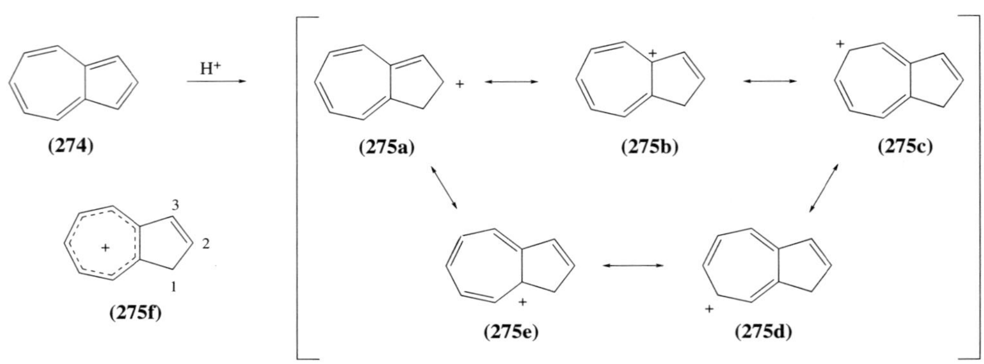

Scheme 89

An impression of the relative stability of various azulenyl cations may be derived from Table 14, which reflects the work of Schulze and Long ⟨64JA322, 64JA327, 64JA331⟩. Worthy of note is the unusual behaviour of 1-nitrotrimethylazulene, which protonates at C1 as opposed to the C3 position favoured by the other 1-substituted azulenes. It is thought that this anomolous C1 protonation relieves steric crowding between the 1-nitro group and the 8-methyl group.

Table 14 Spectrophotometric measurements of the formation of azulenyl carbocations.

Azulene	*Azulenyl cation*	*Molarity of $HClO_4$ required for half-protonation*
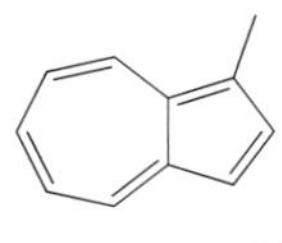	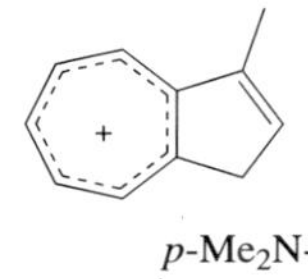	1.22
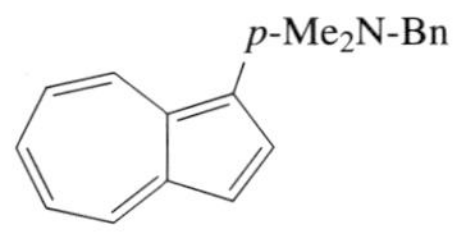	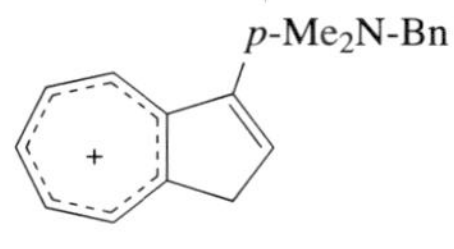	2.16
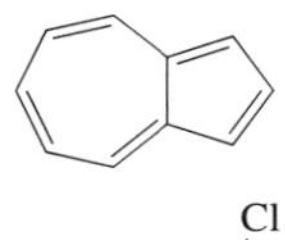	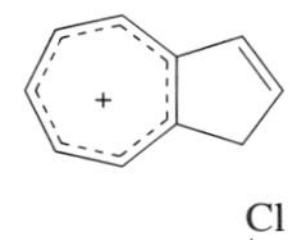	2.30
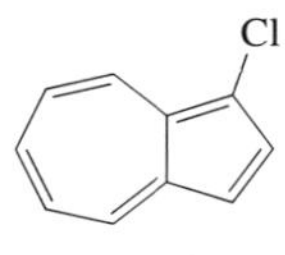	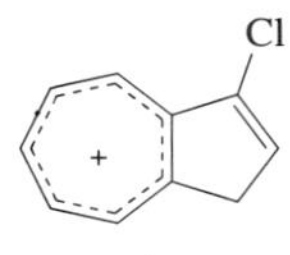	4.17
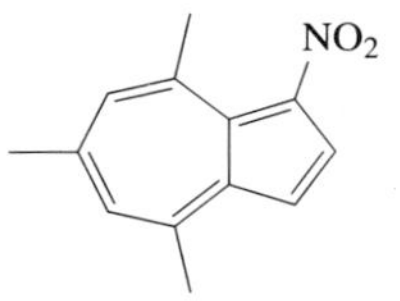	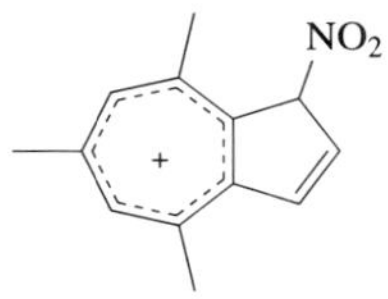	5.06
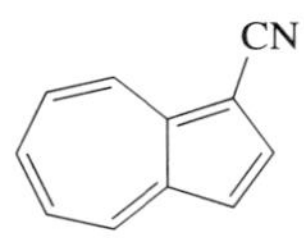	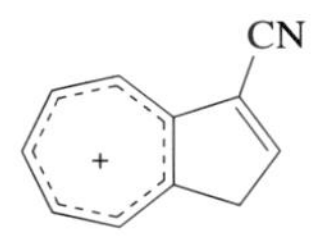	8.36

1.19.2.3.6 Homotropylium and homoaromatic carbocations

The subject of homoaromaticity has been reviewed by Childs ⟨84ACR347⟩, Paquette ⟨78AG(E)106⟩, Winstein ⟨69QR141, B-72MI 119-10⟩, Story and Clark ⟨B-72MI 119-11⟩, Olah, Prakash and Sommer ⟨B-85MI 119-04⟩ and Garratt ⟨79COC(2)460, B-86MI 119-03⟩.

Cyclooctatetraene (**157**) undergoes protonation, when dissolved in concentrated H_2SO_4. The proton adds to one of the double bonds of (**157**) to give the homotropylium cation (**276**) (Equation (64)) ⟨65JA3267, 66JA604, 66JA606, 66JA2047, 83JOC1431, 88JA1108⟩. Carbocation (**276**) may be isolated as a solid salt, $C_8H_9^+ \cdot SbCl_6^-$ ⟨62JA2842⟩. In cation (**276**) an aromatic sextet of electrons is delocalised over seven sp^2-hybridised carbons C1 to C7 somewhat analogously to the aromatic tropylium cation. The remaining carbon C8 is sp^3-hybridised and takes no part in the delocalisation of the positive charge. NMR spectra show the presence of a ring current. In order to maximise the overlap of the C1 and C7 p-orbitals, the C8 methylene group is constrained to lie almost vertically above the plane of the seven-carbon aromatic ring formed by C1 to C7.

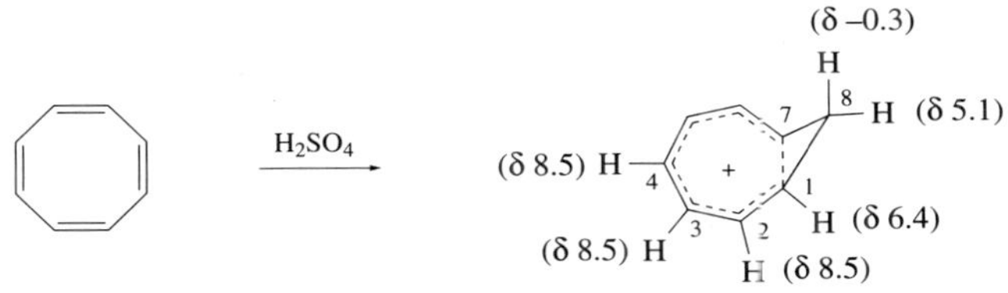

(157) **(276)** (64)

Paquette has stated ⟨78AG(E)106⟩ that (**277**) and (**278**) (Scheme 90) ⟨78AG(E)106⟩ are examples of bishomoaromatic cations, while Childs ⟨84ACR347⟩, being somewhat more stringent in his definition of homoaromaticity, is sceptical of this assertion. Jonsäll and Ahlberg have used $^{1}J_{C-C}$ values as a test for homoaromaticity in (**277**; R = H) ⟨86JA3819, 91CRV375⟩. ^{13}C-NMR spectroscopy in superacid media supports its aromatic character, although solvolytic studies suggest that (**277**; R = H) has the planar dihydroindenyl structure (**279**) ⟨72JA7063, 75JOC700⟩. Semiempirical and *ab initio* (STO-3G) calculations find both structures (**277**; R = H) and (**279**) to represent energy minima ⟨83JCS(P2)305, 85T6055⟩.

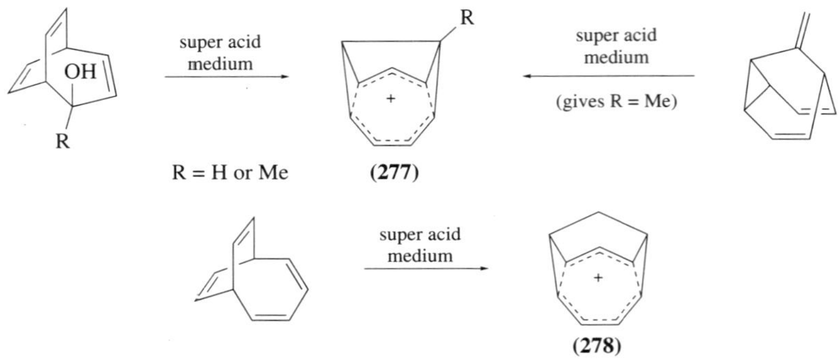

Scheme 90

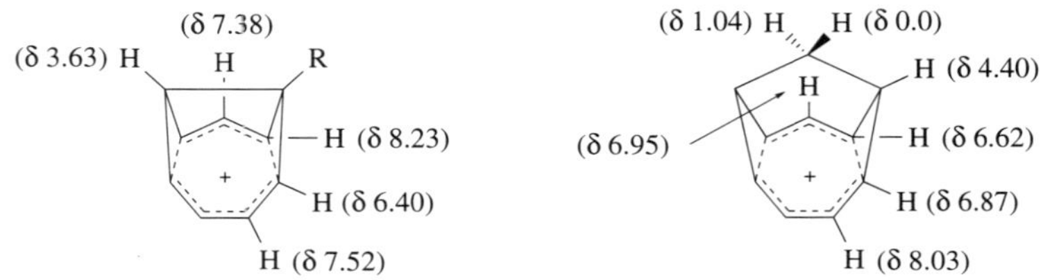

(277) **(278)**

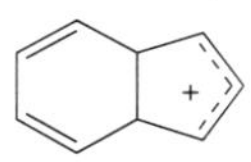

(279)

Olah and co-workers have reported the preparation of the unusual pagodyl dication (**281**) (Scheme 91) from pagodane (**280**) ⟨89JA746⟩; see also ⟨83TL5857, B-87MI 119-01⟩. The dication was described as homoaromatic.

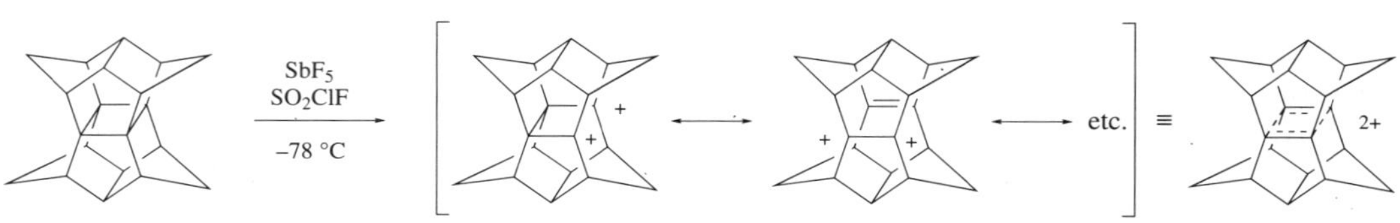

(280) **(281)**

Scheme 91

1.19.2.3.7 μ-Hydrido-bridged carbocations

Three-centre, two-electron C—H—C bonds occur only rarely in organic chemistry. However, Sorensen has shown that the 1,6-dimethyl-1-cyclodecyl cation (**282**) (Equation (65)) and related ions contain transannular three-centre C—H—C bonds, when generated from the corresponding alcohols at −120 °C in superacid media ⟨81JA588⟩. These ions are not stable, rearranging and ultimately decomposing above −70 °C. However, McMurry *et al.* have prepared the relatively stable *in*-bicyclo[4.4.4]-1-tetradecyl cation (**283**) (Scheme 92), which is comparable in acidity to acetic acid ⟨89JA8867, B-89MI 119-06, 91CRV375⟩. Sorensen and Whitworth have subsequently prepared a series of μ-hydrido-bridged tricycloalkyl cations (**284**) varying the acid media with the value of n, where $5 \geqslant n \geqslant 8$ (Equation (66)) ⟨90JA8135⟩.

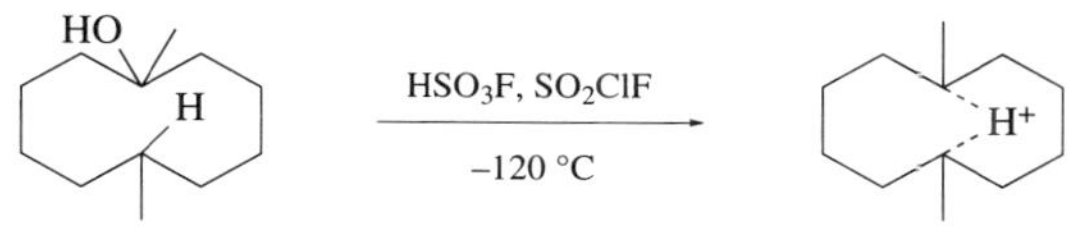

(282) (65)

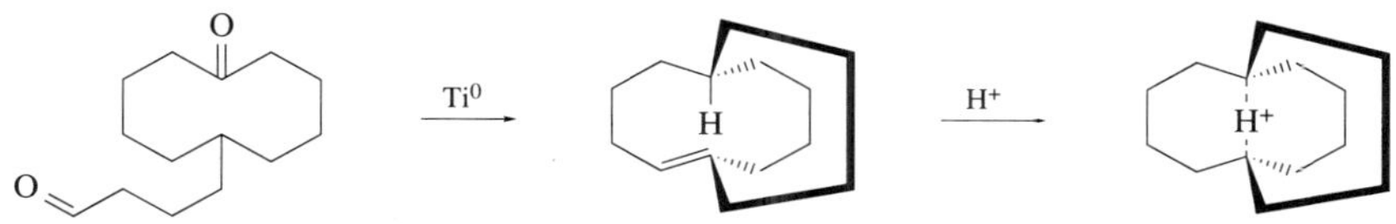

(283)

Scheme 92

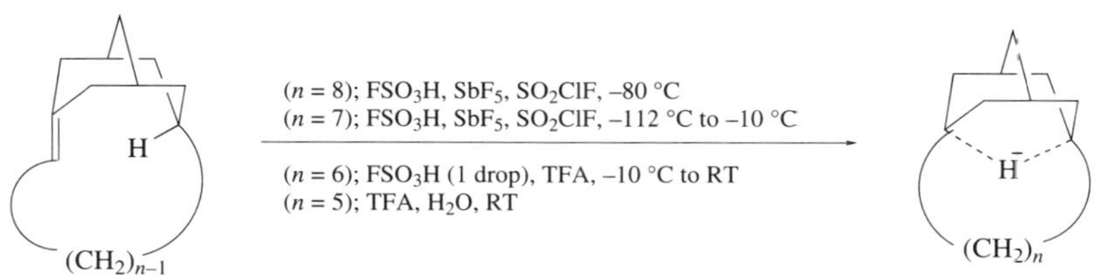

(66)

1.19.3 TRICOORDINATE RADICALS

1.19.3.1 General Features of Radical Formation

1.19.3.1.1 General literature survey

The generation of radicals has been known since 1900 when Gomberg first investigated the triphenylmethyl radical ⟨00CB3150, 00JA757⟩. The following 70 years saw a steady increase in the available information regarding these intermediates ⟨B-46MI 119-01, B-57MI 119-01, 63OR(13)91, 63OR(13)150, B-64MI 119-02, B-66MI 119-06, B-70MI 119-06, 71ACR386, B-71MI 119-05, B-71MI 119-06⟩. Research in this period culminated with an excellent two-volume work edited by Kochi ⟨B-73MI 119-09, B-73MI 119-10⟩. Since 1970 there has been a dramatic increase in the generation, understanding and use of radicals in synthesis and this is reflected in the number of literature reviews. General texts in the field include: Barton and Parekh ⟨B-92MI 119-16⟩, Motherwell and Crich ⟨B-92MI 119-17⟩, Regitz and Giese ⟨89HOU(1)1, 89HOU(2)1⟩, Curran ⟨88S417, 88S489⟩, Ramaiah ⟨87T3541⟩, Giese ⟨B-86MI 119-04⟩, Walling ⟨85T3887⟩, Tedder ⟨82AG(E)401⟩, Nonhebel *et al.* ⟨B-79MI 119-06⟩, Davies and Parrott ⟨B-78MI 119-05⟩ and Hay ⟨B-74MI 119-07⟩. Reviews on more specific topics are: bridgehead radicals ⟨89CRV975, 92CSR105⟩, the captodative—or synergistic ⟨89JA7558⟩—effect ⟨85ACR148, 88PAC1635⟩, cyclisations ⟨B-80MI 119-05, 81T3073, 84SCI883, 88PAC1645, 88BCJ149, 90T1385, 91ACR139, 91COS(4)779⟩, cyclopropylmethyl radical ⟨92MI 119-18, 93CSR347⟩, diastereofacial selection in intermolecular reactions ⟨91ACR296, 94SL1⟩, organosilane ⟨92ACR188⟩ and organotin reagents ⟨B-86MI 119-05, 87S665⟩, persistent radicals ⟨76ACR13⟩, retrosynthetic planning ⟨91SL63⟩ and ring expansions ⟨93CRV2091⟩.

1.19.3.1.2 Description of the different character of carbon-centred radicals

Much of the research in the previous sections concerning carbanion formation (see 1.19.1) and carbocation formation (see 1.19.2) was carried out from the perspective of mechanistic studies roughly during the period 1960 to 1985. However, many of the most important developments in the formation of carbon-centered radicals have been motivated by the needs of synthetic organic chemistry and this is reflected in the approach used in section 1.19.3.

Radicals are species that contain at least one unpaired electron. Although they are neutral species, radicals may be defined as nucleophilic (even though they are electron deficient), electrophilic or ambiphilic in character. These tendencies have been rationalised using the frontier molecular orbital theory ⟨B-76MI 119-02, 83AG(E)753, 91COS(4)715⟩ and are influenced by the functional group(s) attached to the carbon holding the unpaired electron. Thus, interaction of the singly occupied molecular orbital (SOMO) with the lowest unoccupied (LUMO) and/or the highest occupied molecular orbital (HOMO) of its reaction partner will be influenced by the relative energy of the SOMO.

(i) Nucleophilic radicals

Carbon-centred radicals that add more rapidly to electron-poor alkenes, such as acrylonitrile and methyl acrylate, are defined as nucleophilic. Alkyl-substituted radicals have a nucleophilic tendency and increasing the number of alkyl substituents serves to raise the relative energy of the SOMO. This reduces the energy difference between the SOMO and LUMO (Figure 3) and is reflected in the slight increase in relative reaction rates. Thus, the relative reaction rates of alkyl radicals with dimethyl fumarate, an electron-deficient alkene, follows the order tertiary > secondary > primary (Table 15) (Equation (67)) ⟨83TL703⟩. For other information see chapters 1.08 and 1.12.

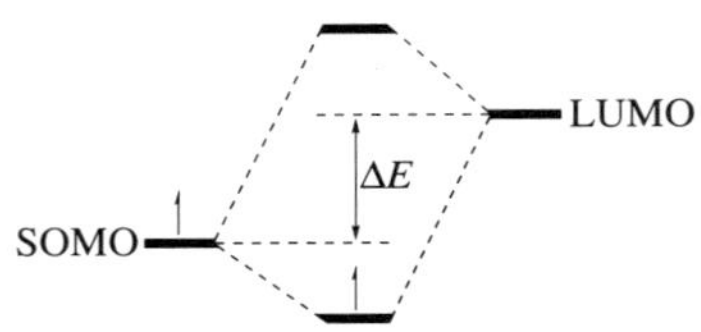

Figure 3 Interaction of nucleophilic radical SOMO with LUMO of alkene.

Table 15 Relative rates of radical addition to dimethyl fumarate (Equation (67))

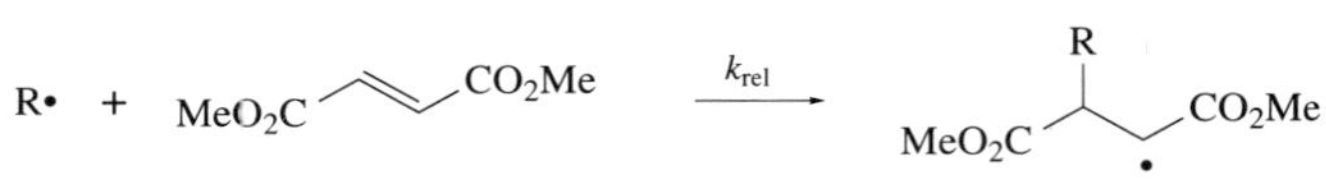

(67)

R•	k_{rel}
n-C_7H_{15}•	1
c-C_6H_{11}•	1.1
Bu^t•	1.6

(ii) Electrophilic radicals

Electrophilic radicals have a reaction profile directly opposite to that of nucleophilic radicals. Hence, these species react with electron-rich alkenes, such as enol ethers and enamines, more rapidly than with electron-deficient alkenes. Electron-withdrawing substituents enhance the electrophilic character and the frontier molecular orbital interaction is that of the SOMO with the HOMO; an energy decrease in the SOMO leads to a lower ΔE and an increased rate of reaction (Figure 4). However, unlike their nucleophilic counterparts, information relating to absolute rate constants for electrophilic radicals is scarce ⟨86JOC5007, 86CB444, 87JOC205, 89JOC308⟩.

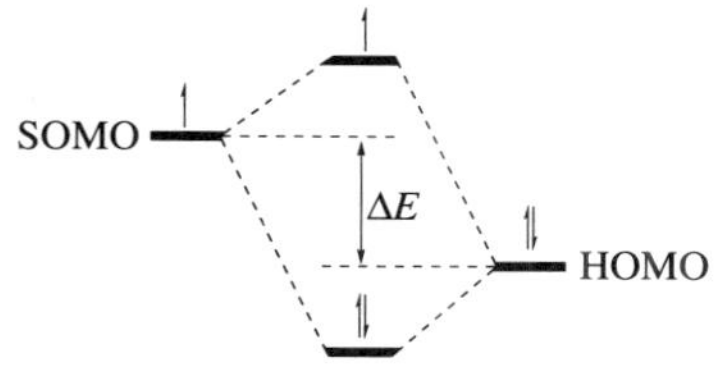

Figure 4 Interaction of electrophilic radical SOMO with HOMO of alkene.

(iii) Ambiphilic radicals

On considering the frontier molecular orbital theory, one might expect a certain group of radicals to have a SOMO of intermediate energy. In this scenario, the relative energy difference between interactions of the SOMO, with either the LUMO or HOMO of the alkene, would be similar. Thus, both electron-donating and -releasing substituents attached to the alkene would accelerate the reaction rate. Ambiphilic radicals fall into this grey category separating nucleophilic and electrophilic character. However, very little is known about these species and the borders of ambiphilic and both nucleophilic and electrophilic radicals are very poorly defined. Recent research suggests that a carbon-centred radical substituted by one weakly electron-withdrawing group has ambiphilic tendencies (Table 16) (Equation (68)) ⟨88CB2063, B-89MI 119-08⟩.

Table 16 Relative rates of ambiphilic radical addition to alkenes (Equation (68))

(68)

R	k_{rel}
CO_2Me	12
Et	1
Ph	40

1.19.3.1.3 Stability and structure of carbon-centred radicals

One of the major differences between radicals and their ionic counterparts is that radicals will undergo direct radical–radical coupling, forming a dimer (Equation (69)). For simple alkyl radicals, this reaction typically occurs at the diffusion-controlled rate, hence their lifetime is extremely short in solution. However, these transient species may be trapped in a frozen lattice and kept for relatively long periods, allowing spectral measurements to be made ⟨63JA384, 68AG(E)507, 71ARP109, 80CSR1⟩. Transient radicals are the most useful in synthesis. In analogous fashion to carbocations, the stability order of free radicals is tertiary > secondary > primary and this can be explained in terms of hyperconjugation and inductive effects (Scheme 93). The most simple and useful method of assessing radical stabilisation is that of bond dissociation energy [D(C—H)], a high value indicating lower stability ⟨82ARP493, 82JA3655, 85JA2872, 86TCC(130)1, 88JA7339, 88JA7543, 89JOU449⟩; see also ⟨90ACR375, 93ACR510, 93JOC6067⟩. However, Nicholas and Arnold have disputed the usefulness of this measure ⟨84CJC1850, 84CJC1860⟩.

(69)

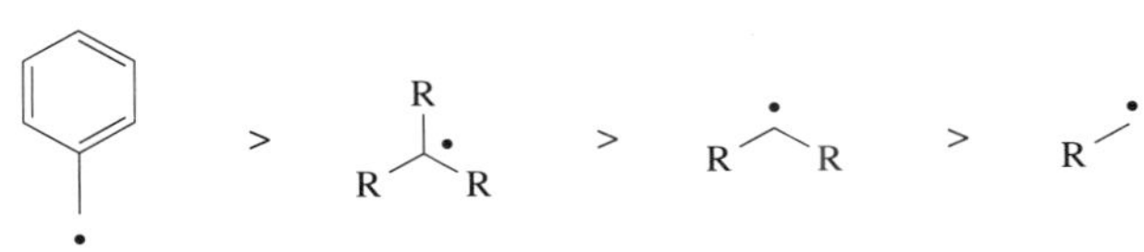

Scheme 93

On introducing conjugation at the carbon-centred radical, delocalisation of the unpaired electron can occur and this results in a more stable species ⟨B-68MI 119-04, 86JA3224⟩. Increasing the number of substituents that allow conjugation enhances the stability of the radical (Scheme 94). It has also been postulated that radicals have enhanced stability (via resonance) when substituted by both an electron-donating and electron-withdrawing group, the so-called 'captodative' or 'synergistic' effect ⟨85ACR148, 88TL6839, 88TL1437, 88PAC1635, 88JPO197, 88JA8164, 89JA7558, 90APO(25)131⟩. However, it is important to note that there is also evidence, both theoretical and experimental, against this enhancement ⟨87AG(E)770, 89JA5845, 90JOC58, 90CJC1714⟩. For example, evidence suggests that, while both mono- and difluoromethyl radicals are more stable than a methyl radical, the trifluoromethyl radical is less stable ⟨89JOC5648⟩. However, although these electronic effects increase *relative* stability, radicals generally remain transient species.

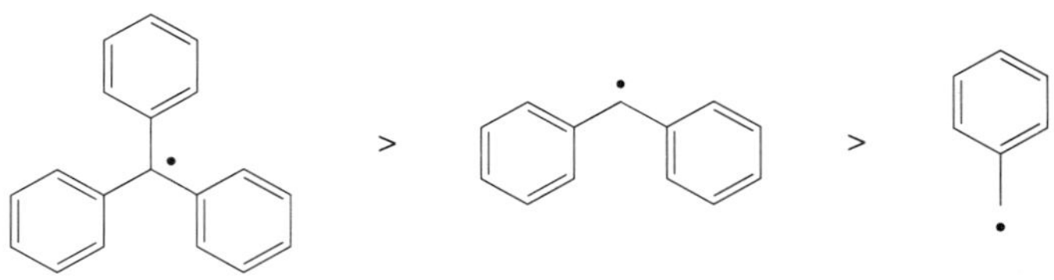

Scheme 94

With respect to radical chemistry, the opposite of a transient species is a persistent one and the important factor to consider here is not that of electronic stabilisation but of steric hindrance ⟨80TCC(88)1⟩. Hence, although the benzyl radical is more stable than the methyl radical, it remains a transient species. The triphenylmethyl radical (**285**), however, exists in solution at room temperature ⟨73RCR1011⟩. This species is one example of a *persistent* radical and occurs, at a concentration of approximately 2%, in equilibrium with its dimer (**286**) (Equation (70)) ⟨68TL249, 70T5343, 70CB1101, 74T2009⟩. The greater importance of steric hindrance to the dimerisation of persistent radicals was elegantly demonstrated by Sabacky *et al.* ⟨67JA2054⟩. This group prepared the radicals (**287**) and (**288**) and compared the ability of these compounds to dimerise with that of the triphenylmethyl radical (**285**). The carbon containing the radical in compound (**285**) is sp^2-hybridised and, to maximise delocalisation, all three phenyl rings must be simultaneously coplanar. X-ray crystallographic measurements, however, have shown that this is not the case. In reality the triphenylmethyl radical is propeller-shaped and the extent of delocalisation is similar to that in both the diphenylmethyl and phenylmethyl radicals. Compound (**287**) is a planar species and maximum interaction (resonance stabilisation) occurs between the *p*-orbital on the central carbon atom and the π-orbital systems of the three aromatic rings. Conversely, the *ortho*-methoxy groups in compound (**288**) increase the steric bulk around the central carbon atom, resulting in an increased deviation from a planar geometry and a decrease in resonance stability. Thus, compounds (**287**) and (**288**) will have similar electronic character with compound (**287**) having greater delocalisation and compound (**288**) having greater steric bulk. Thus, if resonance is the major factor influencing the stability of compound (**285**), compound (**287**) would dimerise and not compound (**288**). The opposite would occur if steric hindrance is the major influence on stability. In actuality, there is no evidence for dimerisation of compound (**288**) either in solution or the solid state. Compound (**287**), however, exists as a dimer and dissociates only slightly in solution ⟨67TL3877, 78TL4871⟩. Similar effects are observed with the di-*t*-butylmethyl radical (**289**) where steric effects alone retard the rate of dimerisation ⟨76ACR13, 78CB72, 79TL929⟩.

There are two possible shapes for simple alkyl radicals. If the unpaired electron were to be

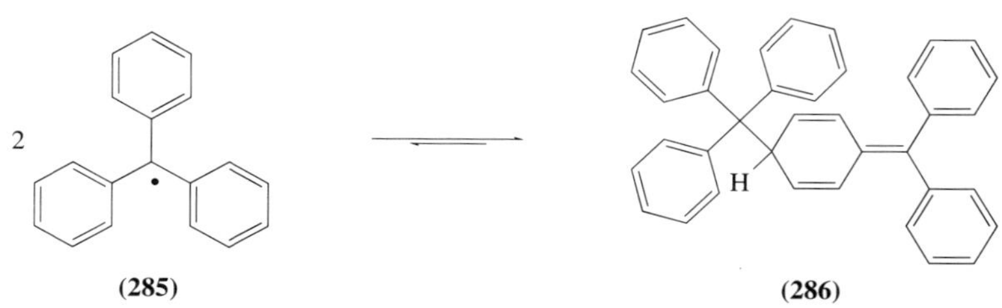

(70)

(285) **(286)**

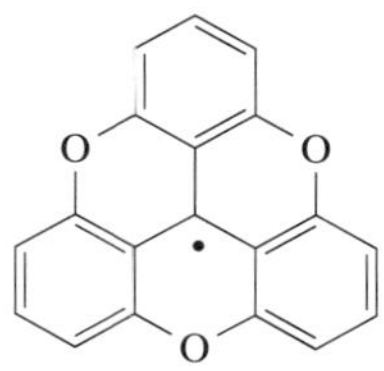

(287)

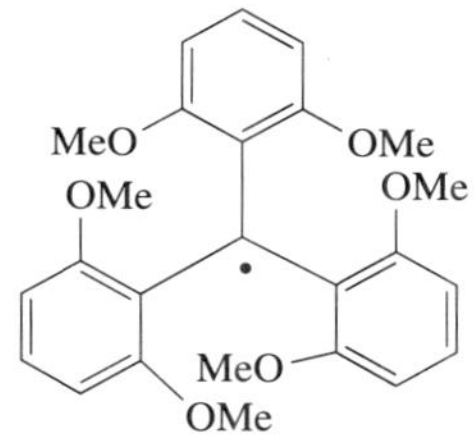

(288)

(289)

accommodated in a *p*-orbital, the carbon would have planar geometry (**290**). Alternatively, if an *sp*^{3}-orbital were occupied, the carbon would adopt a pyramidal shape (**291**) (Figure 5). Direct physical evidence was gained from both electron spin resonance (ESR) and infrared spectroscopy of the methyl radical and other simple alkyl radicals ⟨67JCP(47)3637, 67JCP(47)5146, 77JA8348, 78AG(E)594⟩. This information, and the known loss of optical activity on generation of a radical at a chiral carbon, indicate the geometry to be essentially planar. However, if the carbon-centred radical contains a highly electronegative substituent, the configuration becomes more pyramidal and this deviation increases with the number of electronegative substituents ⟨65JCP(43)2704, 67JCP(46)3193, 69JCP(51)2767, 74JA2201, 76JA230⟩. Thus, the geometry changes from essentially planar to pyramidal across the series shown in Scheme 95, though the barrier to inversion is low. In contrast to this, bridgehead radicals are pyramidal and their rigid structure prevents inversion ⟨87JOC2627, 88JCS(P2)1989, 89CRV975, 90CC1589, 92CSR105⟩. This geometry results in an increased *s*-character of the SOMO and reduced stability of the radical, when compared to their simple alkyl counterparts.

(290) **(291a)** **(291b)**

Figure 5 Pyramidal nature of tricoordinate C-centred radical.

$$Me\bullet \quad < \quad FH_2C\bullet \quad < \quad F_2HC\bullet \quad < \quad F_3C\bullet$$

Scheme 95

1.19.3.1.4 Solvent and concentration effects

In contrast to carbanion and carbocation chemistry, carbon-centred radicals are neutral species and therefore their reactions are virtually independent of solvent effects ⟨67JA2654, B-73MI 119-11, 85T3887, 91COS(4)715⟩. Thus, the rate of a radical reaction is similar over a wide range of solvent polarities. When selecting the solvent, it is more important to consider the rate of a potential radical–solvent side reaction. Typical solvents used in radical chemistry are aromatic hydrocarbons, particularly benzene, ethers and alcohols (see ⟨86JA303⟩). However, it is important to note that alkyl radicals will both abstract α-hydrogen atoms from THF and diethyl ether and add to benzene, with pseudo first-order rate constants of approximately 10^4 s^{-1}, 10^3 s^{-1} and 10^2 s^{-1}, respectively ⟨77JA7960, 85MI 119-07, 88ACR206⟩. Halocarbons should also be avoided when tin-, silicon- and phosphorus-centred radicals are involved as intermediates, since abstraction of either a halogen or a hydrogen atom may occur.

As with the choice of solvent, concentrations become very important when considering the relative rates and efficiency of radical reactions. For example, Giese *et al.* and Barclay *et al.* have both demonstrated that, on altering the concentration of tributyltin hydride (TBTH), reactions can afford different products ⟨84JA1793, 90OS(68)66, 93JOC1083⟩. Thus, on generating the anomeric radical (**292**) using a high concentration of TBTH, the product of deoxygenation (**293**) is observed. However, on slow addition of TBTH, in a solution containing 2,2′-azobisisobutyronitrile (AIBN; **302**) the initial radical undergoes rearrangement of the acetyl moiety, generating radical (**294**). This second radical is subsequently quenched with TBTH to afford the 2-deoxyglucose derivative (**295**) (Scheme 96). Phosphate groups have also been shown to rearrange in this manner ⟨93JOC1083⟩.

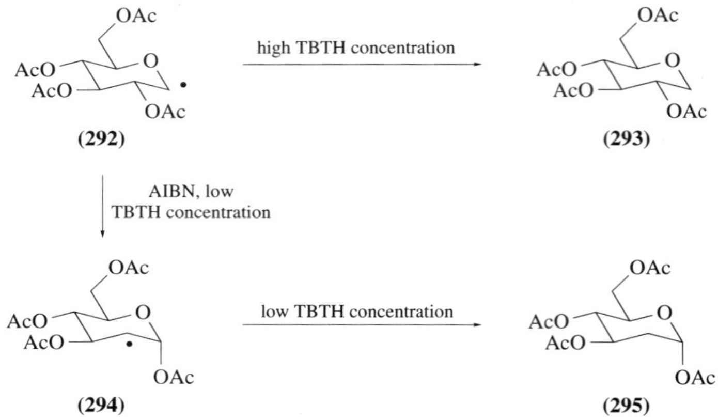

Scheme 96

1.19.3.1.5 Initiation

The homolytic cleavage of a chemical bond requires an energy input and there are four methods by which this energy may be supplied: thermolysis, photolysis, redox reactions and sonication. Thermolysis may be achieved at high temperatures in the gas phase; however, this is not a useful process for synthetic applications ⟨B-80MI 119-06⟩. Thus, to generate a radical in solution at moderate temperatures, it is usually necessary to kick-start the process by adding an initiator. The selection of an initiator is generally dictated by the operating temperature, although the character of the bond being cleaved should also be considered ⟨63OR(13)91, 85T3887⟩. Commonly utilised initiators at elevated temperatures include azo compounds ⟨65JA3413, 70JA4395, 80CRV99, 86JA7935, 87JA2750⟩, diacyl peroxides and peroxides ⟨B-79COC(1)909, 88JA2877⟩. In the late 1980s, the bulky hexa-organodistannanes (**296**, **297** and **298**), via the intermediary stannyl radical (Equation (71)) ⟨89T951, B-91MI 119-03⟩, and the bis(trialkylstannyl)benzopinacolates (**299** and **300**) ⟨88JA1631, 93TL7819⟩ have been shown to generate, under thermolytic conditions, carbon-centred radicals efficiently (**301**). The mechanism of the latter reaction is uncertain, though the initial fragmentation is thought to be that outlined in Equation (72). Although there are many thermal initiators, the most frequently used in synthetic applications is AIBN (**302**). This compound has an effective temperature range of 60–110 °C and its decomposition pathway is well understood (Equation (73)) ⟨64JA1911, 80CRV99, 85T3887⟩.

$$(R_3Sn)_2 \xrightleftharpoons{\text{heat}} 2\ R_3Sn\bullet \qquad (71)$$

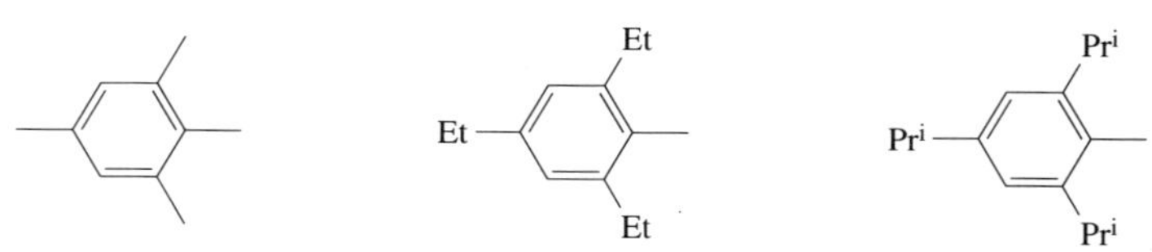

(296) **(297)** **(298)**

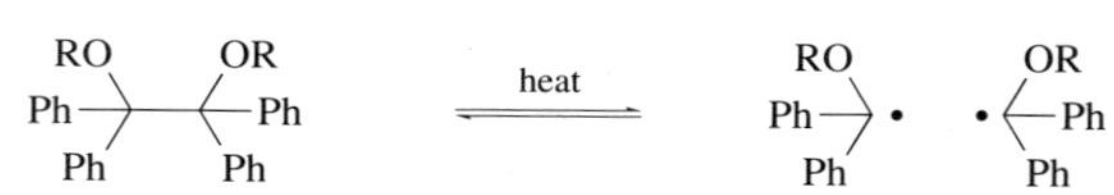

(299) R = $Bu^n{}_3Sn$
(300) R = Me_3Sn **(301)**

(72)

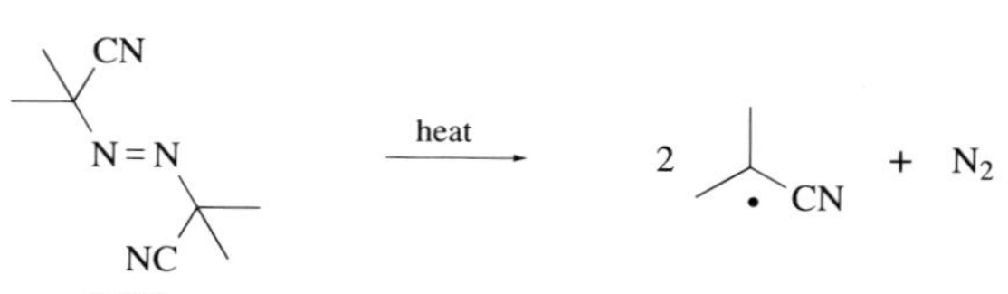

(302)

(73)

Photoinitiation is the most applicable technique for low temperature work ⟨79JOC4123, 85JOC3498⟩. However, Oshima and co-workers have reported a triethylborane-tin hydride method of generating radicals effectively even at −78 °C ⟨88TL6127, 89BCJ143⟩. The most useful photoinitiators are dibenzoyl peroxide and AIBN; however, Neumann and co-workers have reported the use of hexabutyldistannane under mild conditions, with triplet sensitisers ⟨65JA3413, 70JA4395, 80CRV99⟩. Thus, on irradiation of a solution containing hexabutyldistannane and the sensitiser, either acetone, diphenyl ether or 4-methoxy acetophenone, with a 125 W daylight lamp, selective homolytic cleavage of the tin–tin bond was observed ⟨93SL286⟩.

Redox reactions are emerging as useful processes for the generation of carbon-centred radicals. The most important techniques from a synthetic viewpoint are reactions involving transition metals, though electrolysis is also useful (e.g., Kolbe reactions; section 1.19.3.2.2(iii)). The use of transition metals has been reviewed by Iqbal *et al.* ⟨94CRV519⟩. Sono-initiation is also emerging as a potentially useful method ⟨89JOC6101, 89JA6849, 93TL1317⟩; see also ⟨87CSR239⟩.

1.19.3.1.6 Vicinal substituent

As with ionic chemistry, certain vicinal moieties may cause undesirable elimination reactions to occur (Equation (74)). Functional groups to beware of are halogens, xanthates, nitro, isonitrile, selenides, sulfides and sulfoxides (see Chapters 1.13 and 1.14).

(74)

1.19.3.2 Radicals by Bond Scission

1.19.3.2.1 Carbon–hydrogen bond scission

Simple hydrocarbons may undergo carbon–hydrogen bond scission by either thermolysis or abstraction together with a radical species generated from an initiator (e.g., Cl·, Br·, RO·, R·). These processes have been extensively studied and the order of hydrogen atom abstraction generally parallels their bond dissociation energies ⟨82ARP493⟩. Thus, the relative susceptibility for abstraction is tertiary > secondary > primary ⟨B-73MI 119-12, 82AG(E)401, 82T313⟩. However, the regioselectivity is variable and dependent on both the abstracting radical and the reaction temperature ⟨71JA2065, 73S1, 81TL1891, 89JOC3571, 93ACR219⟩. Also, on the introduction of a carbon–carbon double bond, the preferred site of abstraction is at the allylic position. This is ascribed to resonance stabilisation of the resultant radical; allylic rearrangements are frequently observed (Equation (75)) ⟨B-73MI 119-13⟩. Similarly, on introduction of an aromatic ring, the abstraction usually, but not always, occurs at the benzylic position ⟨B-73MI 119-12⟩.

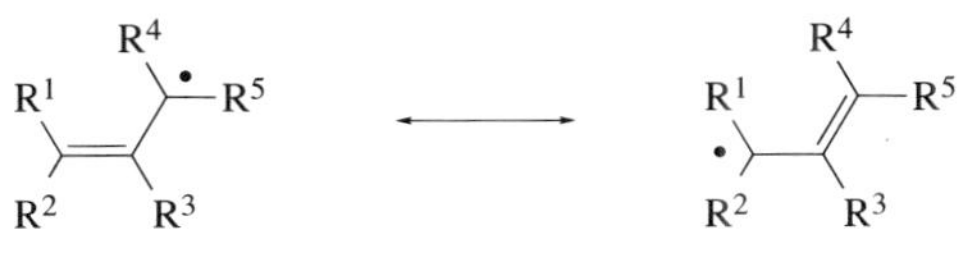

(75)

Electron-donating and withdrawing groups reduce the α-carbon–hydrogen bond strength hence radicals are generated with a high degree of regioselectivity ⟨73HCA1516, 82ARP493, 83CB1042, 83JA4701, 83TL185⟩. Thus, on treatment with di-*t*-butyl peroxide (DTBP), ketones and carboxylic acids, esters and lactones readily form the radical at the carbon atom α to the electron-withdrawing moiety (Equations (76)–(79)) ⟨60DOK(131)580, 61DOK(136)1099, 61IZV2065, 65CC7, 85JOC3220⟩. Higher regioselectivities are facilitated when two electron withdrawing groups are present (Equation (80)) ⟨62JCS4468, 71BSF461, 86JA1708⟩. Furthermore, although it is possible to generate a radical at the alkoxy carbon atom of an ester or lactone, this generally does not occur. This is in stark contrast to amides and lactams, where the radical is preferentially generated at the carbon atom α to the nitrogen atom (Equation (81)) ⟨68T4509⟩.

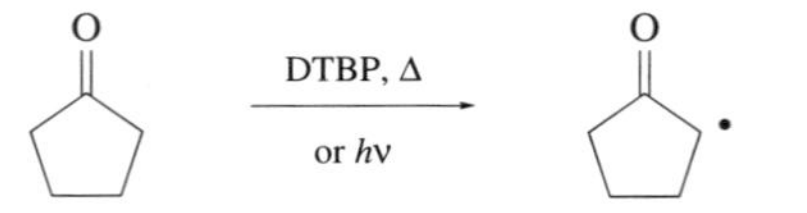
(76)

(77)

(78)

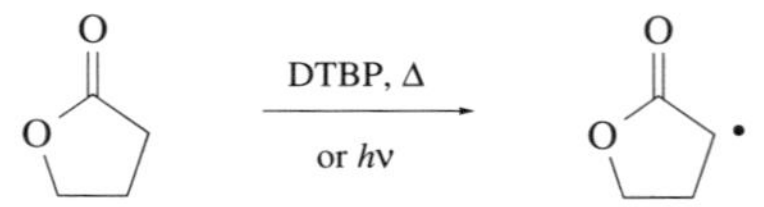
(79)

(80)

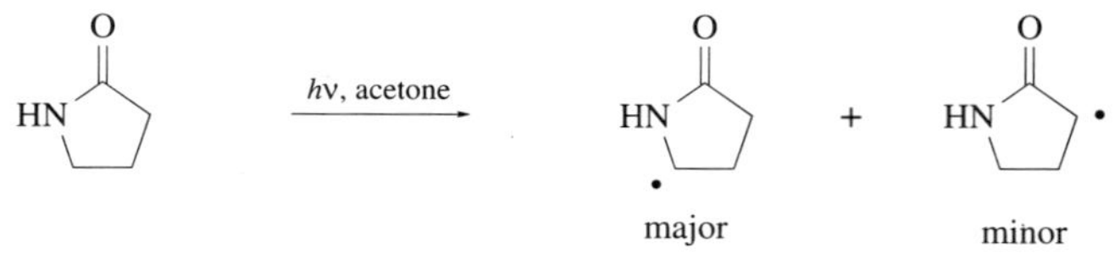
(81)

A superior procedure to the DTBP method is the use of metal salts in an oxidative process. Several metals in high oxidation states, including Ag^{2+}, Co^{3+}, Ce^{4+}, Fe^{3+}, Ir^{4+}, Mn^{3+} and V^{5+} are effective in generating a radical at the carbon atom α to an electron-withdrawing group (Equations (82) and (83)) ⟨72JOU2515, 74JA7977, 82TL5587, 88JA2210, 93TL1701, 94CRV519⟩. Of these, manganese has been studied most intensively and a simplified mechanism is depicted in Scheme 97 ⟨74JA7977, 88JOC2137, 88PAC1659, 91CRV1237⟩. A representative selection of substrates is depicted in Equations (84) and (85), Scheme 98, Equations (86–88) ⟨84SC1191, 84JOC1603, 85TL4291, 93JOC7610, 93JOC7640⟩.

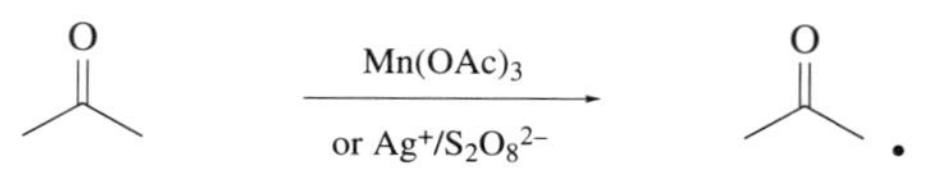
(82)

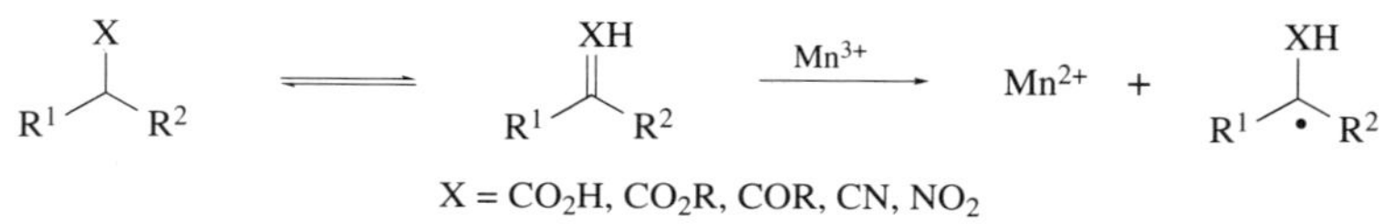

Scheme 97

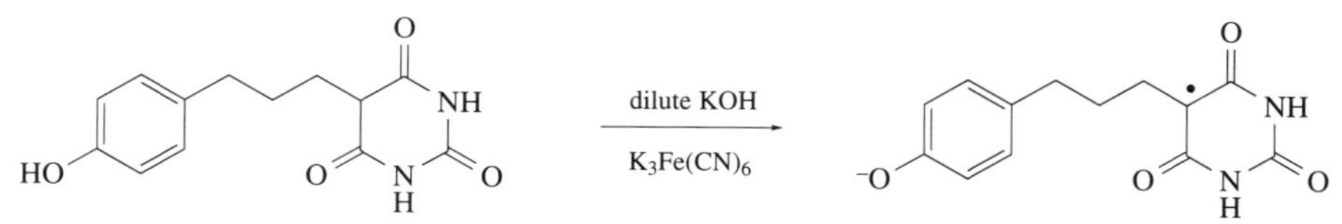
(83)

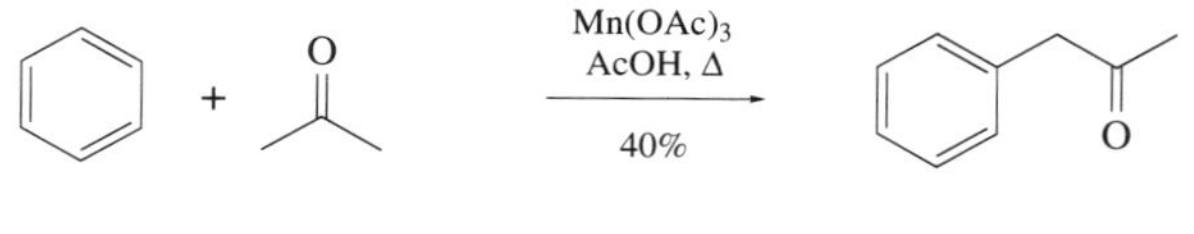
(84)

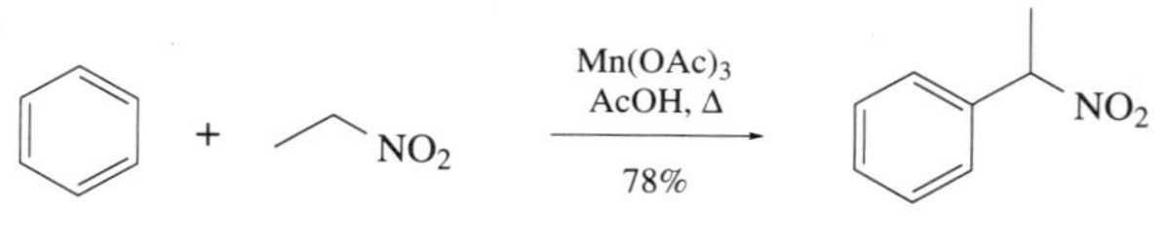
(85)

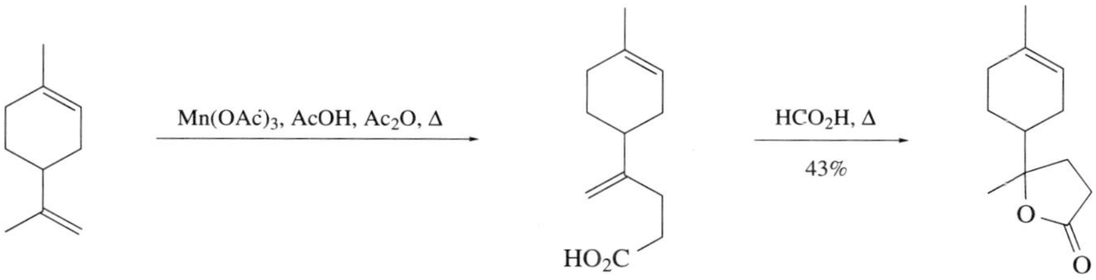

Scheme 98

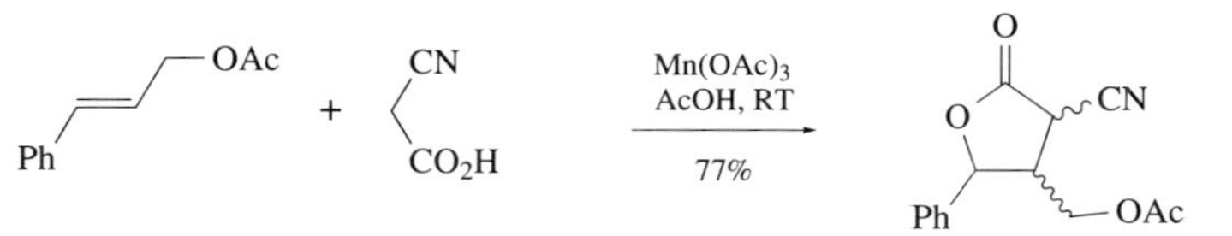

(86)

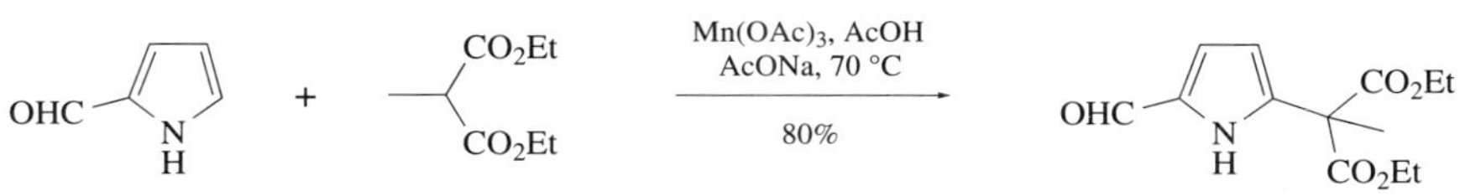

(87)

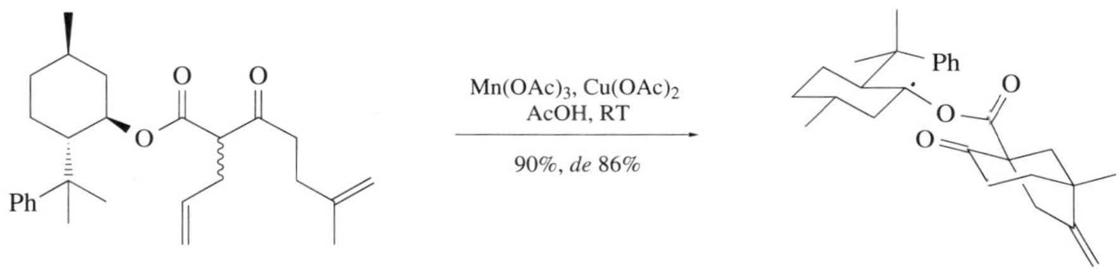

(88)

The regioselective intramolecular formation of a remote carbon-centred radical may also be achieved by initially generating, for example, a heteroatom radical, typically based on nitrogen or oxygen 〈64AG(E)525, B-73MI 119-13, B-80MI 119-05, B-92MI 119-17〉. Heteroatom-centred radicals may effect a 1,5-hydrogen atom abstraction with concomitant generation of the desired carbon radical (Equations (89) and (90)). These reactions have been the subject of many investigations and are now well understood 〈63CRV55, 68PAC1, 71S1, 86JA2381, 87JA2195, 88JCS(P1)3255, 88TL837, 88TL1441, 88TL2215, B-89MI 119-07, 92TL5913〉. Curran and co-workers developed this methodology to include the use of aryl and vinyl radicals as the initially formed reactive species (Schemes 99 and 100) 〈88JA5900, 90JA896, 91JOC4335, 93JA6051〉. It should, however, be noted that this method is not confined to 1,5-abstractions; Breslow and co-workers have elegantly demonstrated that 1,*n*-abstraction may occur in steroid systems; an example and the proposed mechanism is shown in Scheme 101 〈88MI 119-04, 90CC1074, 90CC1076, 90JA6842, 91JA8977〉.

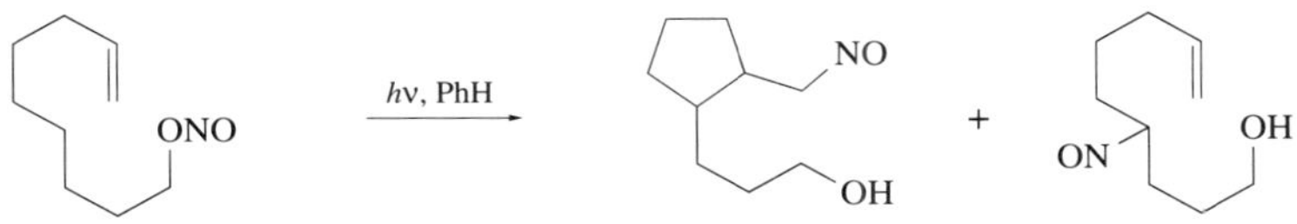

(89)

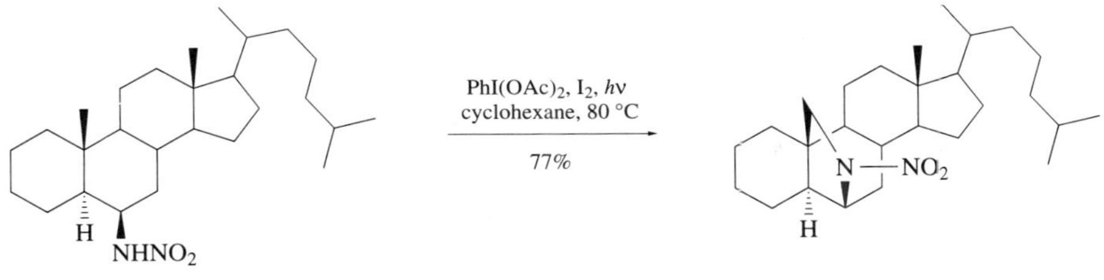

(90)

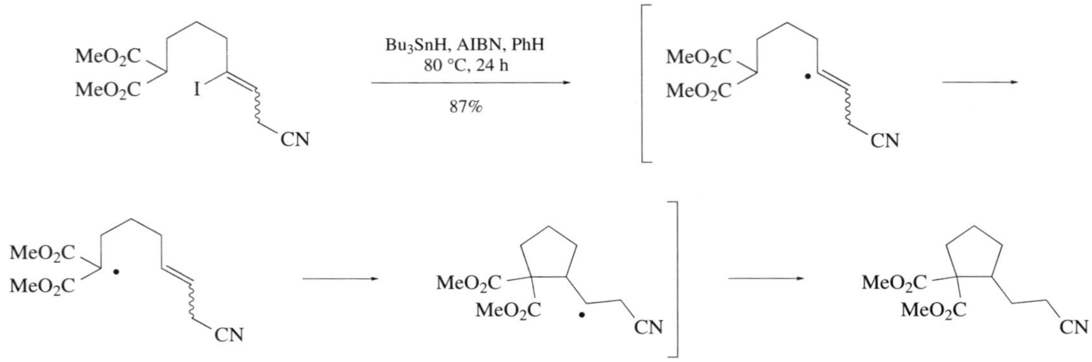

Scheme 99

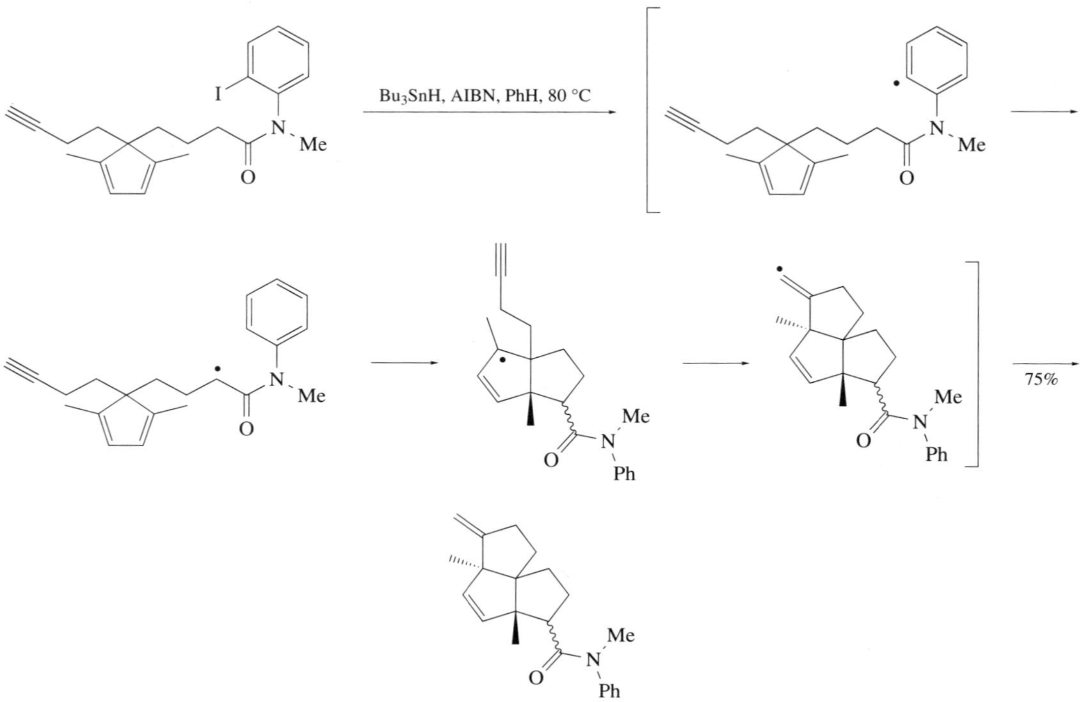

Scheme 100

1.19.3.2.2 Carbon–carbon bond scission

The homolytic fission of a carbon–carbon bond is a useful preparative method for the formation of carbon-centred radicals. Typically, this process proceeds via a heteroatom-centred radical which subsequently suffers β-carbon–carbon bond scission with simultaneous formation of the desired carbon-centred radical and expulsion of either a ketone, aldehyde or carbon dioxide. However, other methods including decarbonylation, addition to cyclopropanes as well as cyclopropyl- and cyclobutylcarbinyl fragmentation reactions have been investigated ⟨72JA301, 73JA856, 73JA6728, 76JOC523, 79ACR66, 86TL1553, 88CC1380, 88JA3300, 89CC332, 90CJC1698, 91JOC3801, 92MI 119-18, 92T8031, 92TL181, 93CRV2091, 93CSR347⟩.

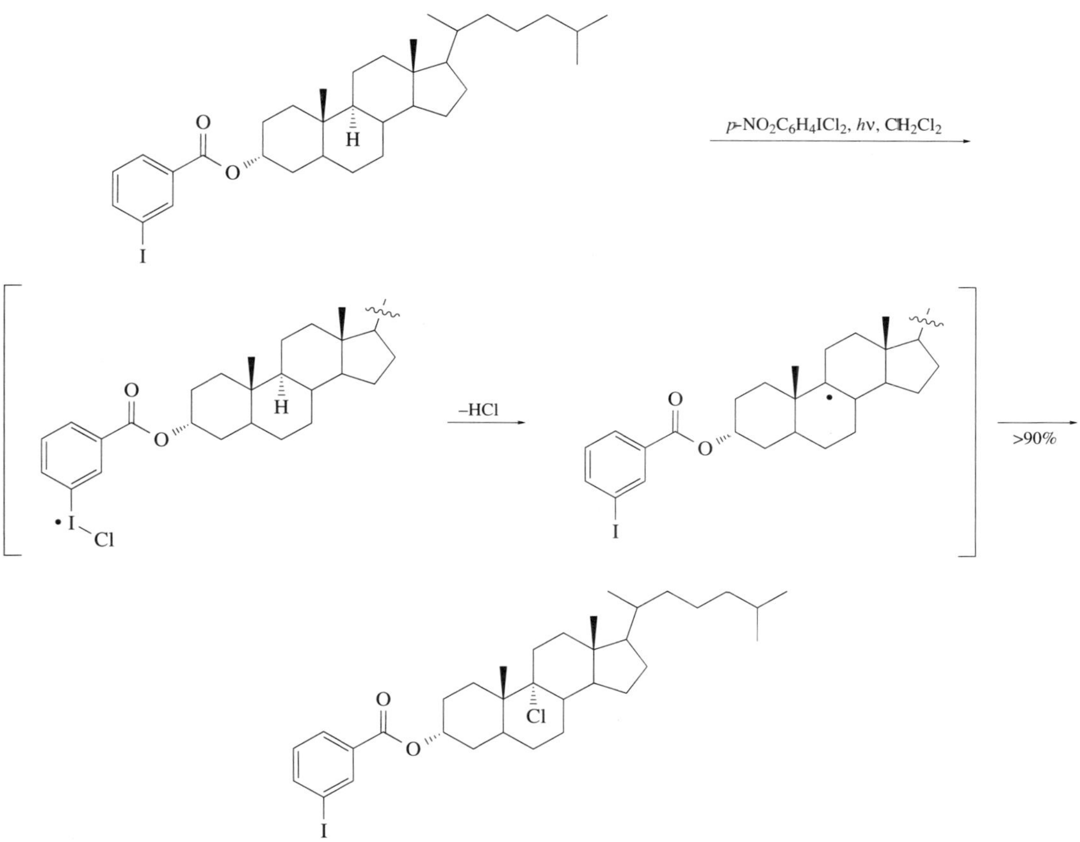

Scheme 101

(i) β-Fragmentation of oxygen-centred radicals

There are several factors including ring strain, stereoelectronic effects and relative stability of the adduct radical, that influence the direction of β-fragmentation in unsymmetrical molecules. The observed fragmentation results from a combination of these determinants and cannot be deduced by simple analysis of individual effects.

Several methods have been developed for the generation of alkoxy radicals ⟨B-73MI 119-14, B-79MI 119-03, 84TL1953, 87T3541⟩. Booker-Milburn has reported that slow addition of a solution of iron(III) chloride in DMF to the cyclopropyl silyl ether (**303**) affords the alkoxy radical (**304**) ⟨92SL809⟩. This species subsequently fragments, to afford the ring-expanded carbon-centred radical (**305**). Intramolecular addition followed by chlorine-atom abstraction generated the bicyclic compound (**306**) (Scheme 102). A similar reaction was observed by Snider *et al.* when a cyclopropyl silyl ether was treated with cupric tetrafluoroborate ⟨92JOC2399⟩.

A number of systems have been developed to generate the precursor alkoxy radical from the parent alcohol. On reacting cyclopropyl alcohol derivatives with either copper(II), iron(III), lead(IV) or manganese(IV) salts, the alkoxy radical undergoes a facile β-fragmentation with concomitant relief of ring strain ⟨68RTC1301, 73TL827, 84TL5335, 91CL1193, 93CL545⟩. Hence, on treatment of the bicyclic compound (**307**) with silyl enol ether (**308**) and manganese(III) 2-pyridinecarboxylate ($Mn(pic)_3$), diketones (**309**) and (**310**) are formed in 77 and 5% yield respectively (Scheme 103) ⟨93BCJ819⟩. Similarly cyclobutyl, cyclopentyl and cyclohexyl alcohols are suitable substrates for this reaction. Typical substrates and reagents are illustrated in Equations (91)–(94) ⟨72JA1209, 87TL3963, 93JCS(P1)3007, 93JOC7228⟩. A related method involves the facile homolytic fission of an epoxide carbon–oxygen bond at the β-position to the carbon-centred radical. The chemistry of the initial radical intermediate (**311**), generated by the tin hydride mediated dehalogenation or deoxygenation, is however dependent on the reagent concentration and substrate substitution patterns (Scheme 104) ⟨81JCS(P1)2363, 86CC1419, 87CC1238, 88TL837, 88TL955, 89T7835, 89TL3343, 91TL7493, 92T6883, 93CRV2091, 93JOC1215, 93TL867, 93TL4505, 93TL5197⟩.

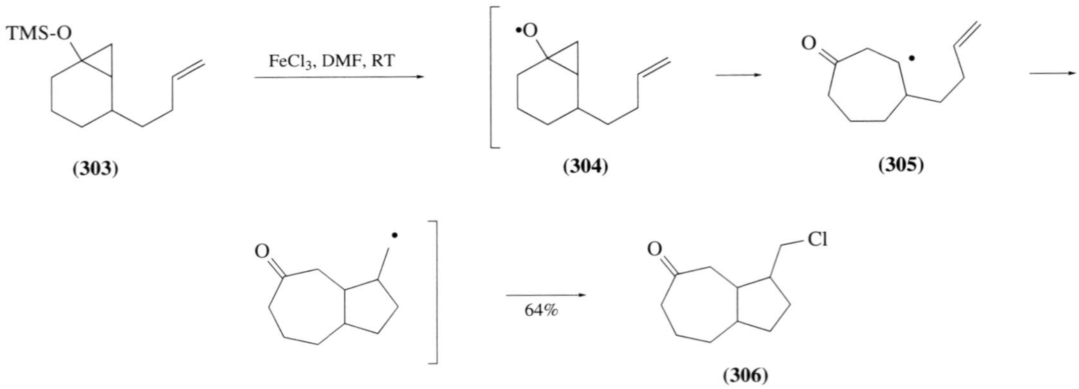

Scheme 102

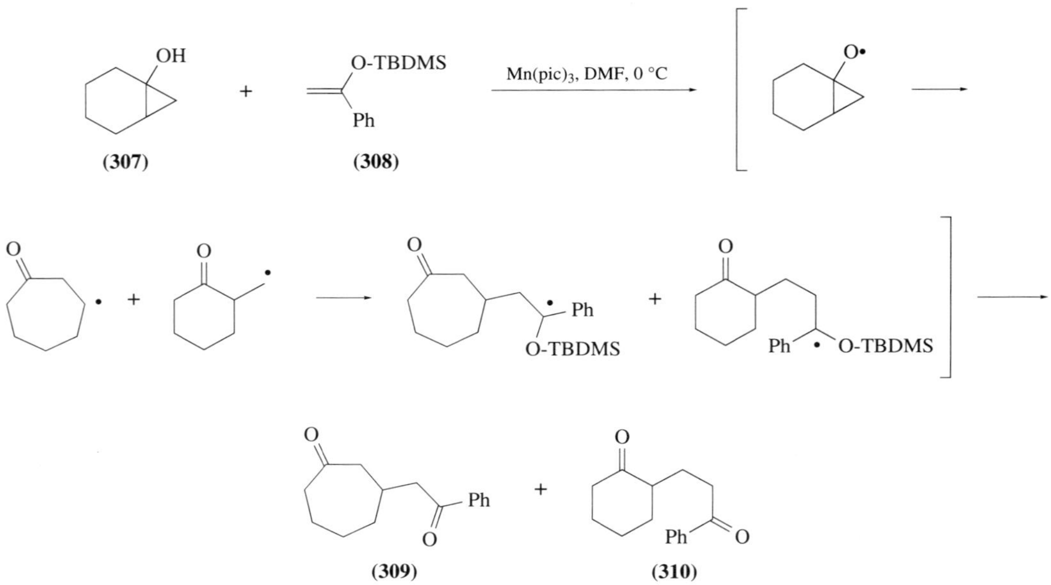

Scheme 103

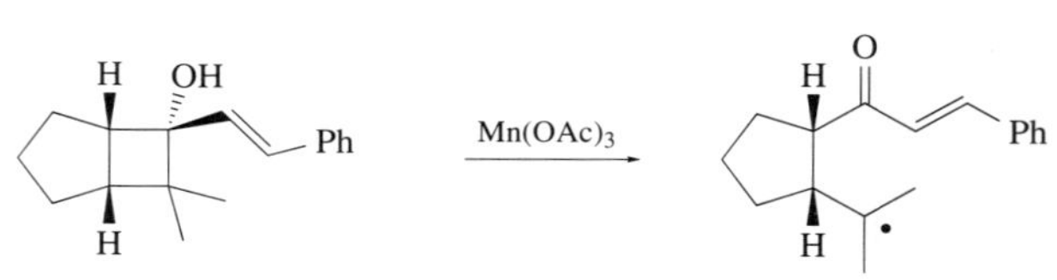

(91)

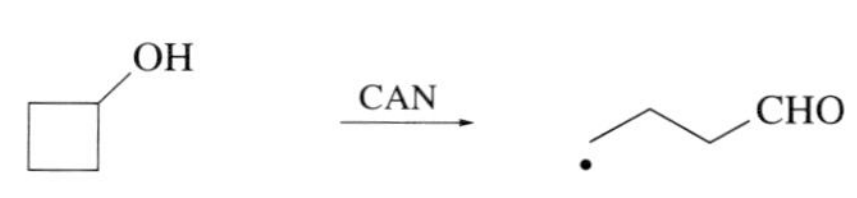

(92)

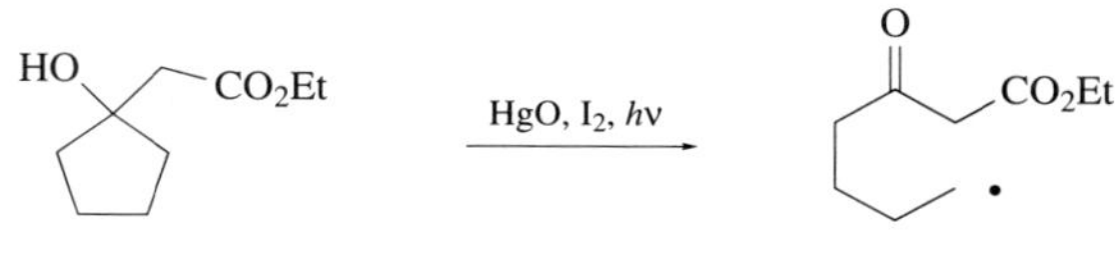

(93)

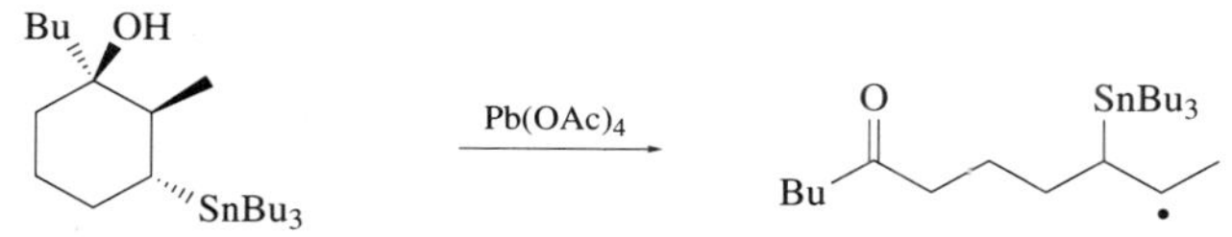

(94)

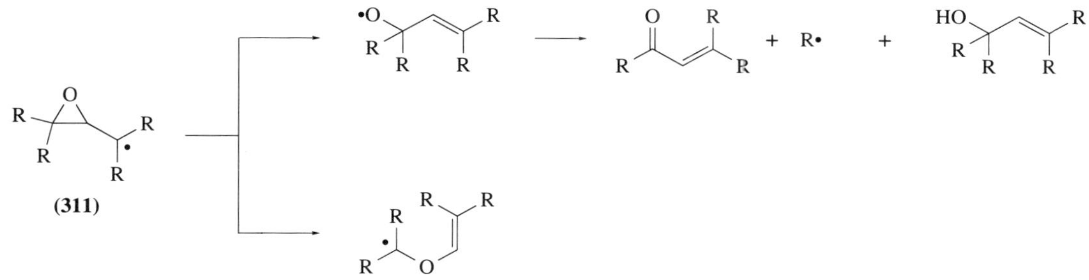

(311)

Scheme 104

A β-fragmentation reaction, analogous to that of the alkoxy version has been developed for lactols. The method generally involves the use of either a mercury(II) oxide, iodine and $h\nu$ or (diacetoxyiodo)benzene (DIB) and iodine system ⟨86TL383, 87CL133, 87TL3397, 92JCS(P1)1837, 93TL4865⟩. Thus Suginome and Yamada demonstrated that treatment of the lactol (**312**) with mercury(II) oxide and iodine followed by irradiation, produced the 10-membered iodolactone (**313**) in 79% yield (Scheme 105). This method was also demonstrated to be useful for the synthesis of 9-, 10- and 11-membered iodolactones ⟨87T3371⟩. The related method involving use of DIB and iodine was shown to be useful in generating formate esters from deoxypentose derivatives ⟨92AG(E)772⟩. Accordingly, the sugar derivative (**314**) afforded the formate ester (**315**) (Equation (95)).

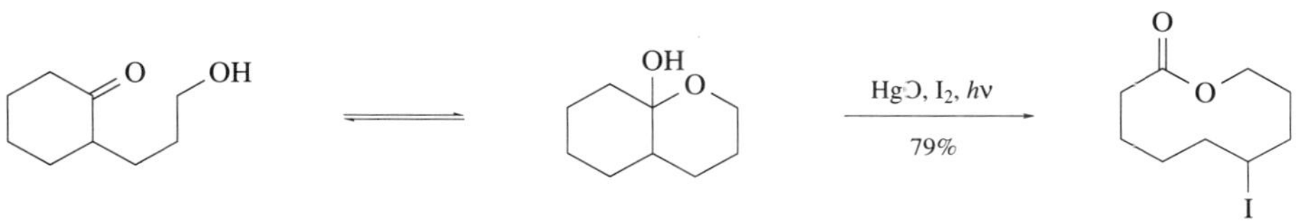

(**312**) (**313**)

Scheme 105

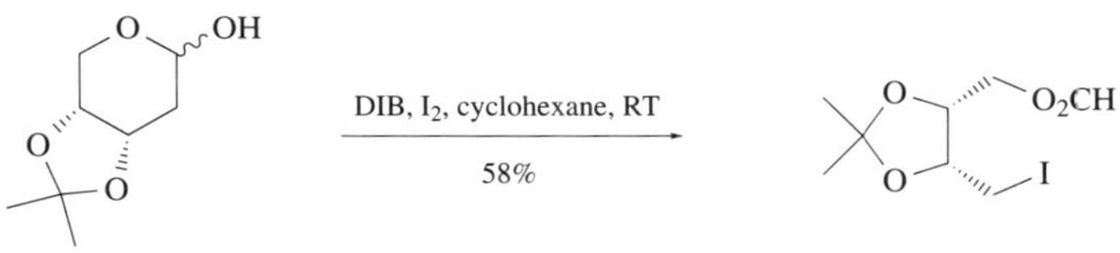

(95)

(**314**) (**315**)

Beckwith and Hay reported the use of *N*-(cyclopentyloxy)pyridine-2-thione (**316**) as a clean source of generating alkoxy radicals for a kinetic study of the reversible β-scission reaction. In this study they showed that the distribution of products (**317**) and (**318**) is dependent on the concentration of the TBTH indicating a reversible reaction (Scheme 106) ⟨88JA4415, 89JA230⟩. Hartung and Giese have also reported on the formation of carbon-centred radicals via the β-scission of alkoxy radicals, generated by photolysis of tertiary (alkylperoxy)(pyridine)cobaloximes (Scheme 107) ⟨91CB387⟩.

Hirao *et al.* have demonstrated that cyclobutanone (**319**), on treatment with $VO(OEt)Cl_2$, undergoes carbon–carbon bond scission to afford compounds consistent with the radical intermediate (**320**). The proposed mechanism is outlined in Scheme 108 ⟨91JOC2264⟩. The intramolecular addition of carbon-centred radicals to ketones, with subsequent β-scission, has been investigated in greater depth ⟨93CRV2091⟩. As can be seen in Scheme 109, this ring-expansion reaction has a wide substrate tolerance and ring expansions of one, three and four methylene units can be achieved in moderate-to-high yields. This methodology has been developed to include cyclisation onto imines ⟨93TL6787⟩; see also ⟨91JA1055⟩. Heterocyclic ring systems are also suitable substrates for the formation of carbon-centred radicals via ring expansion ⟨91T4847, 92T3385, 92T4027⟩. An interesting development

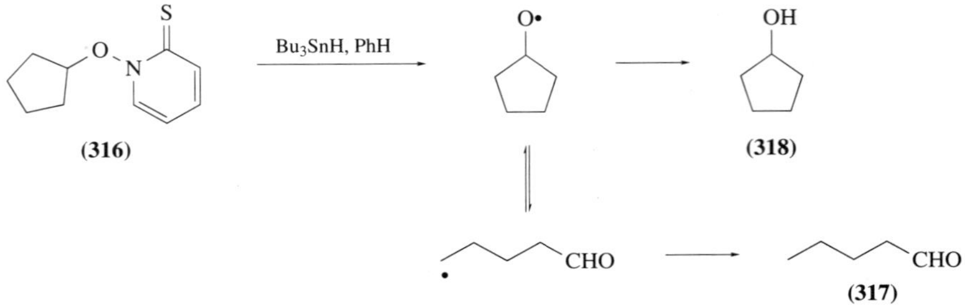

(316) **(318)** **(317)**

Scheme 106

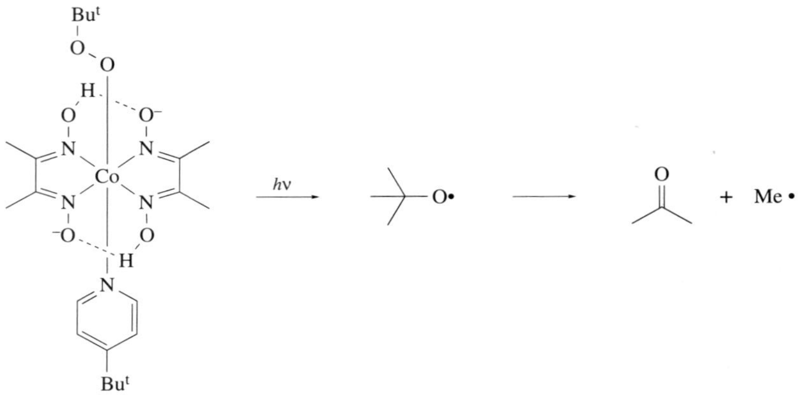

Scheme 107

is the use of the azido moiety (as in compounds (**321**)) as an aminyl radical precursor. Cyclisation followed by β-scission of the intermediate (**322**) afforded high yields of ring-expanded lactams (**323**) (Scheme 110) ⟨93JA3328⟩.

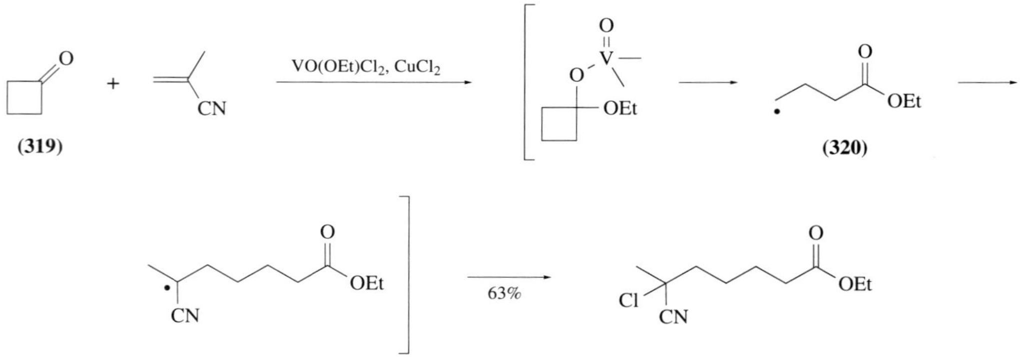

(319) **(320)**

Scheme 108

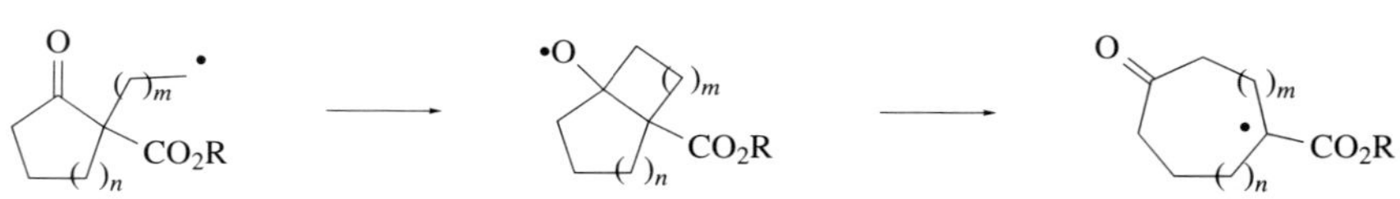

$n = 0$–11; $m = 0, 2, 3$

Scheme 109

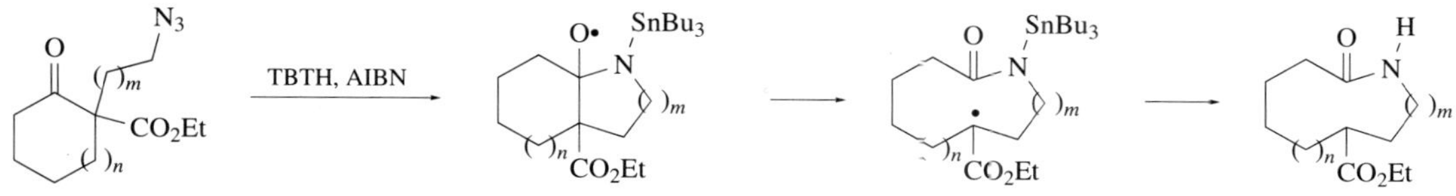

(321) (322) (323)

$n = 1,2$; $m = 1, 2$

Scheme 110

(ii) Decarbonylation

This has been demonstrated to be a preparatively useful method of generating carbon-centred radicals and three complementary procedures exist. Pfenninger *et al.* treated the phenyl selenoester **(324)** with TBTH, initiated with AIBN, in an inert aromatic solvent to afford the acyl radical **(325)**. At higher temperatures, secondary and tertiary acyl substrates undergo decarbonylation to form the alkyl radical **(326)**; primary and arylacyl substrates do not suffer decarbonylation (Scheme 111) ⟨80HCA2328, 89JOC1777⟩. Similarly, Patel and Pattenden demonstrated that, on sunlamp irradiation, arylmethyl- and allylacylcobalt salophens **(327)** undergo homolysis of the acyl–cobalt bond with subsequent decarbonylation (Scheme 112) ⟨88TL707⟩. Unfortunately, primary and secondary acyl radicals are stable to decarbonylation ⟨87TL5949, 88CSR361⟩. A noteworthy development involves the formation of acyl radicals **(329)** via dehalogenation of acid chlorides mediated by tris(trimethylsilyl)silane (TTMSS). Thus, on treatment of primary, secondary and tertiary acid chlorides **(328)** with TTMSS and AIBN in toluene at 80 °C, the nor-alkanes **(330)** are obtained in high yields (Scheme 113) ⟨79JCS(P1)1137, 90JOM(381)155, 92TL1787⟩.

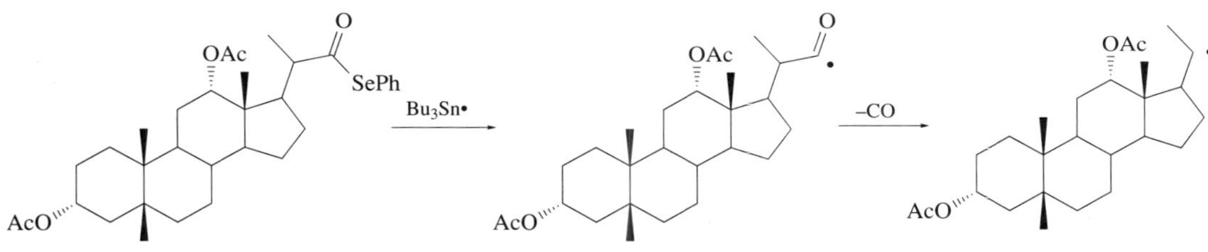

(324) (325) (326)

Scheme 111

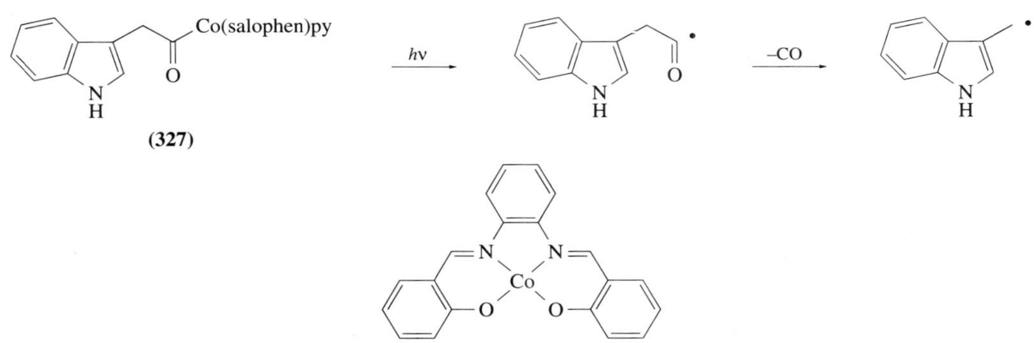

Scheme 112

(iii) Decarboxylation

The formation of alkyl radicals via the decarboxylation of carboxylic acids has been known for some time. Several methods are outlined below.

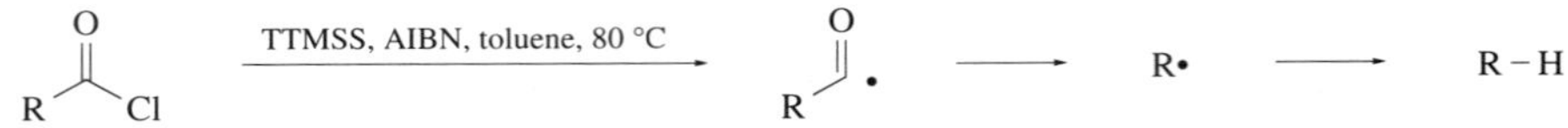

(328) **(329)** **(340)**

Scheme 113

The electrolytic Kolbe decarboxylation generates alkyl radicals which undergo recombination to form dimers (Equation (96)) ⟨63JA3419, 63MI 119-04, 78JA4903, 81AG(E)911⟩. If two different acids are employed, unsymmetrical compounds result, the yields of which are increased when an excess of one acid is used ⟨78AG(E)47, 88TL2797, 88TL2801⟩. The procedure is, however, only useful for primary and allylic carboxylic acids and overoxidation is a problem. Diacylperoxides may also be used as the source of alkyl radicals, either under thermolytic or photolytic conditions. This procedure has very limited advantages over the Kolbe electrolysis. These advantages include a greater selectivity under low-temperature photolysis in the solid state, and the fact that tertiary acids undergo decarboxylation ⟨B-73MI 119-15, 79COC(1)909, 85T4213, 88JA2877⟩.

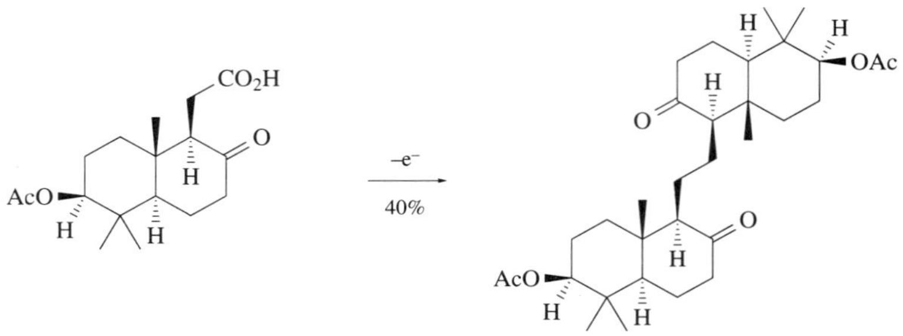

(96)

Decarboxylation may be effected in the presence of high oxidation-state metals ⟨64JOC921, B-73MI 119-16, 89H(28)489, 90JHC79, 91JOC2866⟩. Thus, in the presence of catalytic quantities of silver(I) nitrate or acetate and a peroxo compound (e.g., aroylperoxide, perborate, percarbonate and peroxydisulfate) homolytic alkylation of protonated heteroaromatic systems occurs. Primary, secondary and tertiary carboxylic acids are suitable substrates for this method and a mechanism has been proposed in Scheme 114 ⟨71T3575, 76TCC(62)1, 78TL619, 83ACR27, 85TL4803⟩. A similar method, adopting the use of hypervalent iodine compounds, has been reported by Togo *et al.* ⟨91CL1691, 91TL6559, 93JCS(P1)2417⟩. Thus, on treatment with either (bis(trifluoroacetoxy)iodo)benzene or (bis(trifluoroacetoxy)iodo)-pentafluorobenzene, tertiary (**341**) and secondary (**342**) acids decarboxylate to generate the alkyl radical which can subsequently alkylate a heteroaromatic base in good yields (**344** and **345**) (Equation (97)). Primary carboxylic acids (**343**) are poor substrates for this procedure and, with respect to the hypervalent iodine compound, a three-fold excess of both acid and base is required.

$$(PhCO_2)_2 + Ag^+ \longrightarrow PhCO_2\bullet + PhCO_2^- + Ag^{2+} \xrightarrow{RCO_2H} R\bullet + CO_2 + H^+ + Ag^+$$

Scheme 114

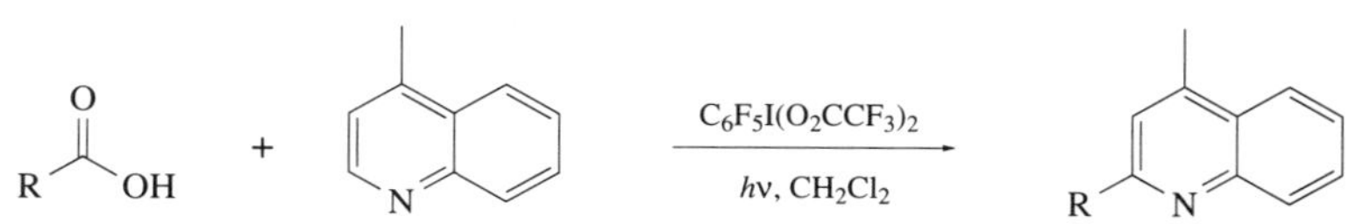

(97)

(341) R = adamantyl
(342) R = cyclohexyl
(343) R = 2-phenylethyl

(344) R = adamantyl, 85%
(345) R = cyclohexyl, 72%
(346) R = 2-phenylethyl, 28%

In 1980, Barton introduced a novel method suitable for the radical decarboxylation of primary, secondary, tertiary, α,β-unsaturated and aromatic acids ⟨80CC732⟩. The procedure involved a TBTH-mediated dehalogenation of chlorodihydrophenanthrene esters (**347**), to afford the radical species (**348**). Rapid aromatisation followed by decarboxylation afforded phenanthrene (**349**) and the desired alkyl radical (Scheme 115). Although there is a broad substrate compatibility, formation of the ester was problematic and this instigated further research. Barton *et al.* subsequently discovered that *O*-esters of thiohydroxamic acid derivatives, e.g., (**350**) were reduced, in a radical chain manner, on treatment with TBTH (Scheme 116) ⟨85T3901, 85TL5943, 91CRV477, B-92MI 119-17⟩. The commercially available sodium salt of *N*-hydroxypyridine-2-thione was chosen since both the potential for aromatisation and reduction of the thiocarbonyl moiety would assist the fragmentation process ⟨75JCS(P1)1574, 81PAC15⟩. The *O*-acyl derivatives of several thio- and selenohydroxamic acid compounds (**350**–**357**) have since been prepared and their radical fragmentation studied ⟨83TL5889, 84TL1055, 86JCS(P1)39, 92T7091, 92T7121, 93TL4505⟩. The *N*-hydroxypyridine-2-thione derivative however, remains the most popular choice and has been utilised frequently in both functional-group transformations and in synthesis ⟨86PAC675, 87MI 119-11, 87S665, 87T3541, 88S417, 88S489, 89CRV1413, 92T2529⟩. A number of simple methods are available to prepare these *O*-esters (**350**) in high yields ⟨83CC939, 84CC1298, 85T3901, 88T5479, 92T7083⟩. The radical chain process is usually initiated by either thermolysis, in refluxing benzene or toluene, or photolysis using a 250 W tungsten lamp. On photolysis at low temperature, thiohydroxamic ester derivative (**350**) produces radicals without a subsequent chain process at $-30\,^{\circ}C$, while derivative (**354**) continues a radical chain process at $-60\,^{\circ}C$ ⟨87T2733⟩. Zard and co-workers have demonstrated that a combination of both heating and irradiating leads to higher product yields in some addition reactions ⟨91TL199⟩. Sonication of thiohydroxamic esters in carbon tetrachloride has also been reported as a useful method of inducing the halogenative decarboxylation ⟨89JOC6101⟩. Ultrasound is reported to generate the trichloromethyl radical (**358**) which initiates fragmentation of the thiohydroxamic ester (**350**) (Scheme 117).

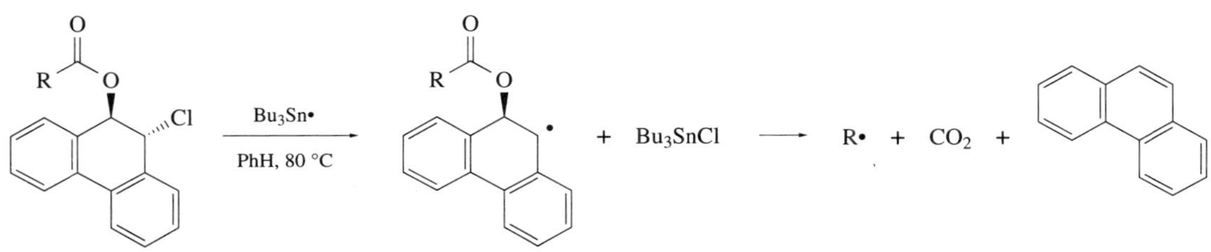

(347) **(348)** **(349)**

Scheme 115

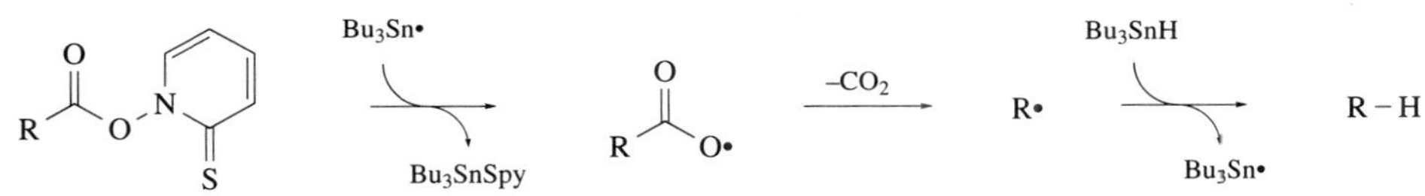

(350)

Scheme 116

1.19.3.2.3 *Carbon–halogen bond scission*

The homolysis of a carbon–halogen bond is an important method of forming carbon-centred radicals and many procedures have been described. The most important involves removal of the halogen atom by trialkyltin radicals, themselves formed via one of the initiation procedures described above (Equation (98)) ⟨68ACR299, 85AG(E)553, 86PAC675, 87S665, 88S417, 88PAC1659, 91CRV1237⟩. The halogen atom can be chlorine, bromine or iodine, fluorine being essentially inert (see ⟨68CJC633, 73JCS(P2)1570, 85JOC2149⟩). The reactivity follows the order iodine > bromine > chlorine, with rate constants in the order of $\sim 10^9$, 10^7 and $10^2\ M^{-1}\,s^{-1}$, respectively ⟨B-83MI 119-01⟩. Thus, it is possible to selectively generate a carbon radical by homolysis of a carbon–bromine bond in the presence of a carbon–chlorine bond (Equation (99)) ⟨79JOC151⟩. The nature of the alkyl group is also an influential factor and the ease of radical formation for simple alkyl substrates follows the expected order, tertiary > secondary > primary. Several tin reagents are available for this reaction; however, tributyl- and to a lesser extent, triphenyltin hydride, are most frequently used ⟨88OM2220⟩. The

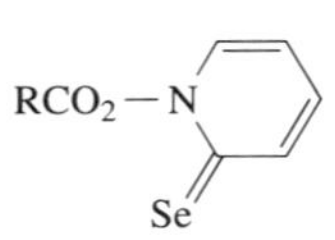

(351)

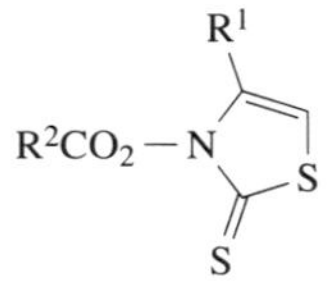

(352) R^1 = Me or Ph

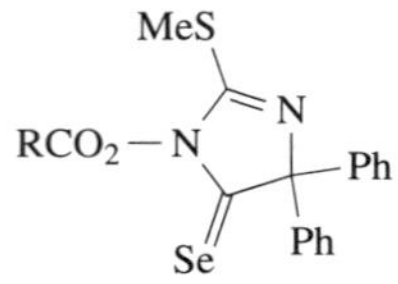

(353)

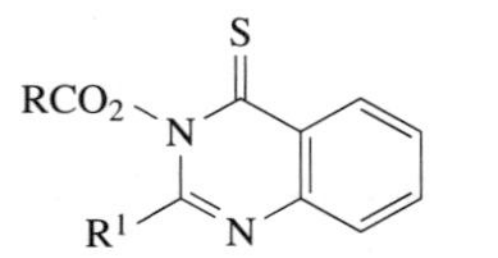

(354) R^1 = Ph, 4-MeOPh or 1-naphthyl

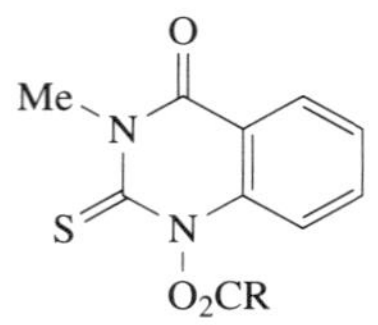

(355)

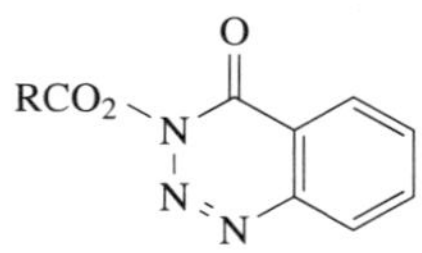

(356)

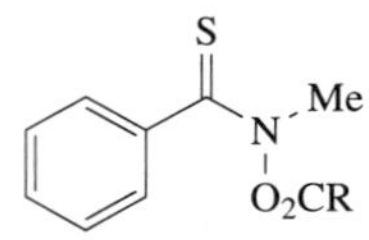

(357)

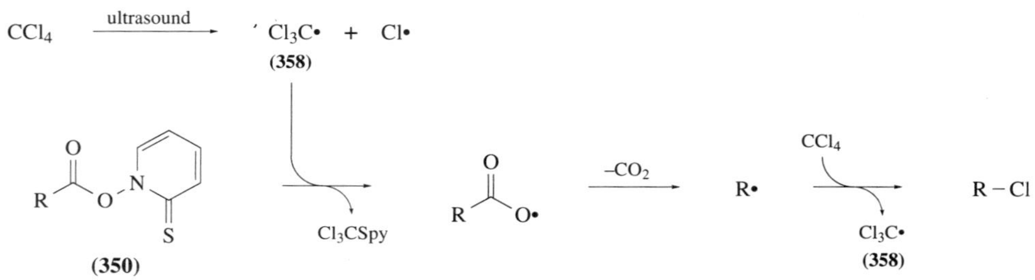

Scheme 117

reactions are generally performed in refluxing benzene or toluene solution under AIBN initiation, although other solvents, with polarities ranging from cyclohexane to methanol, have been used. Light and Breslow have developed a water-soluble tin hydride reagent which reduces alkyl halides in either water or organic solvents ⟨90TL2957, 93OS(72)199⟩. Tributylgermanium hydride is also effective in the formation of carbon-centred radicals from alkyl halides, though the germyl radical has a reduced rate of halogen atom abstraction which can result in a poor chain propagation step ⟨83JA3578, 84JA343, 85TL6289, 85JA4595, 86JA5893, 87JOC3509, 88T6295⟩. Removal of the tin residues from the reaction mixture is a major problem with the use of these reagents; however, several work-up procedures have been developed ⟨79S471, 79JOC449, 81JCS(P1)2363, 89JOC3140⟩. The use of catalytic quantities of a tin reagent, in conjunction with either lithium aluminum hydride, sodium borohydride or diisobutylaluminum hydride, has been developed to reduce this problem ⟨63JOC2165, 73JOC3055, 75JOC2554, 77TL4291, 78MI 119-03, 84CL2069, 86JA303⟩. Organotin reagents have also been immobilised on polymers, alumina and silica, thus allowing removal of the tin residues by simple filtration ⟨75JOC1966, 81AG(E)1043, 82JA5564, 91JOC5971, 92JCS(P1)3165, 93SL283⟩.

$$R_3Sn\bullet + R{-}X \xrightarrow{k} R_3Sn{-}X + R\bullet \qquad (98)$$

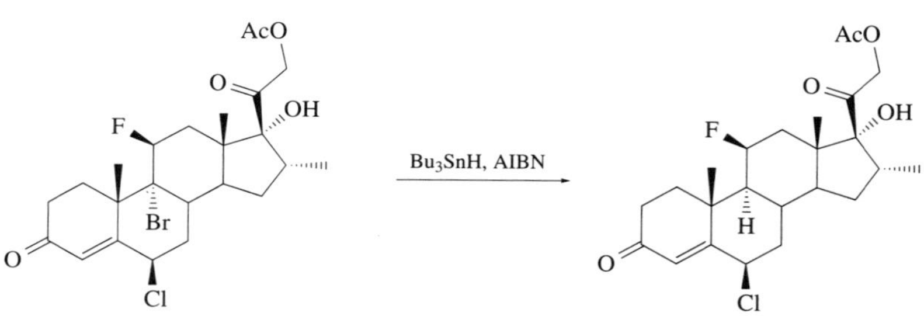

(99)

Chatgilialoglu has introduced tris(trimethylsilyl)silane (TTMSS) as an excellent alternative to tin hydrides. Although the silicon–hydrogen bond strength is greater, comparative studies have shown this reagent to be equally effective in the formation of carbon-centred radicals for use in synthesis ⟨87JA5267, 88JOC3641, 89JOC2492, 89TL2733, 89TL681, 91JOC678, 93SL143⟩. The use of both tris-(alkylthio)silane and a triethylsilane/thiol system have also been shown to be useful alternatives ⟨89CC1387, 90SL219, 92ACR188⟩. A solution developed by Barton *et al.* adopts the use of dialkyl phosphites and hypophosphorous acid. This method has been demonstrated to be useful for secondary and tertiary alkyl iodides and bromides ⟨92TL5709, 93JOC6838⟩. Giese and Thoma have shown that the dimeric metal complex dicarbonyl (η^5-cyclopentadienyl)iron (**359**) undergoes metal–metal bond cleavage on irradiation with a 250 W sunlamp to generate the iron-centred radical species (**360**). This subsequently abstracts a halide atom to afford the carbon-centred radical (Equations (100) and (101)). However, the reaction is limited to alkyl iodides and activated bromides ⟨89TL2907, 91HCA1143⟩. Decacarbonyldimanganese (**361**) also generated carbon-centred radicals via this metal–metal bond homolysis and subsequent iodine atom abstraction process ⟨91HCA1135⟩.

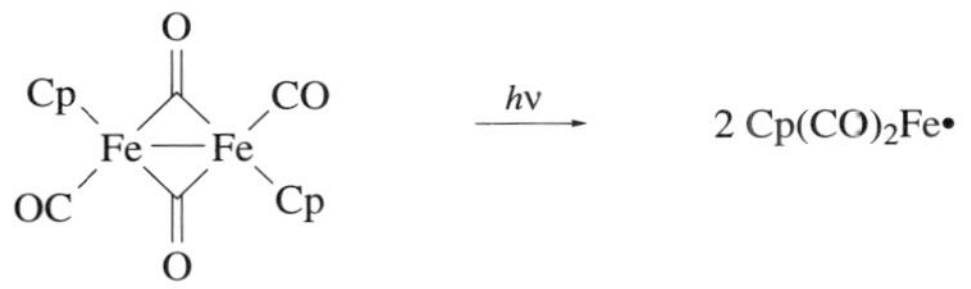

(100)

(359) **(360)**

$Cp(CO)_2Fe\bullet$ + R–X ⟶ $Cp(CO)_2FeX$ + R• (101)

(360)

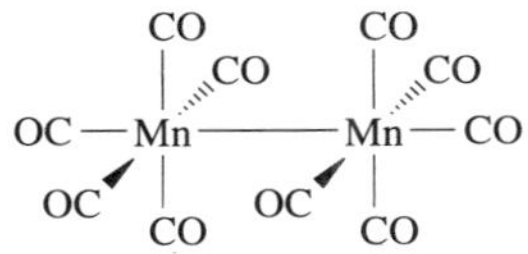

(361)

Samarium(II) iodide reacts with alkyl halides to generate radicals; however the method is extremely solvent-dependent. When the reaction is performed in THF, only primary alkyl iodides and bromides are effective substrates, but on addition of HMPA, primary, secondary and tertiary halides become efficient radical precursors ⟨92CRV29, 92SL943⟩. Some bimolecular rate constants for the reduction of primary alkyl radicals have been reported. These constants were dependent on the amount of HMPA added, varying from 5×10^5 to 7×10^6 M^{-1} s^{-1} ⟨93TL1717⟩. Low oxidation state salts of chromium ⟨92CB287, 93CB1759⟩, iron ⟨88T4671⟩ and nickel ⟨93JCS(P1)2339⟩ have also been shown to be effective in the homolytic dehalogenation of alkyl bromides and iodides. Representative examples are given in Equations (102)–(104) (see also ⟨94CRV519⟩).

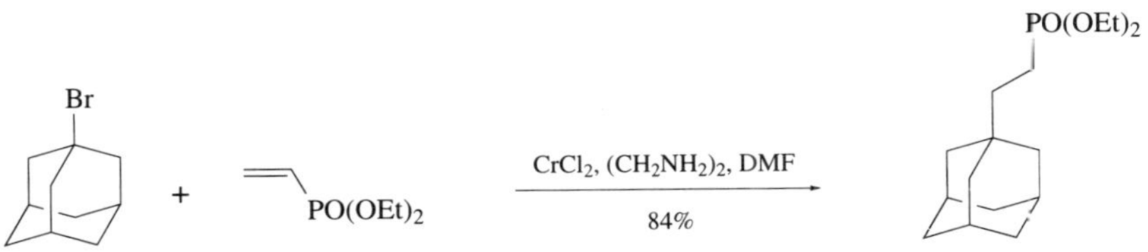

(102)

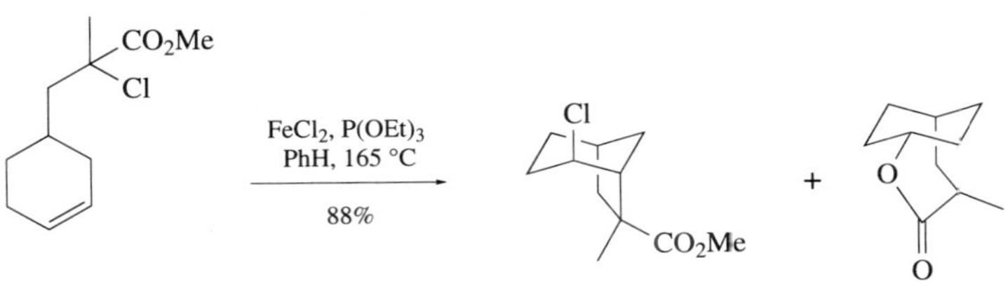

(103)

64 : 36

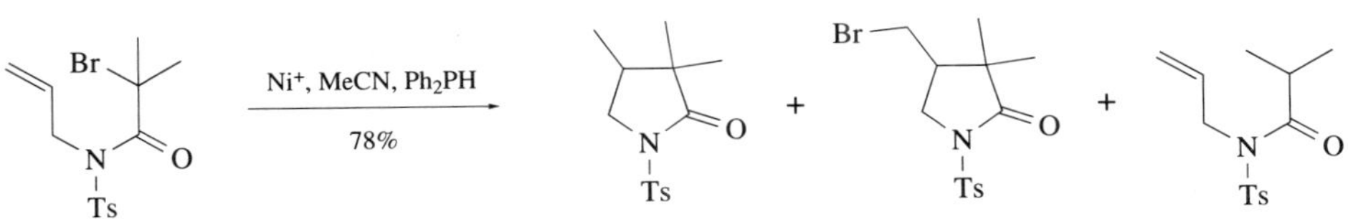

(104)

86 : 10 : 4

1.19.3.2.4 Carbon–chalcogen bond scission

(i) Carbon–oxygen bond scission

Homolytic cleavage of the carbon–oxygen bond of alcohols was initially investigated by Barton and McCombie ⟨75JCS(P1)1574, 81PAC15, B-92MI 119-17⟩. A number of secondary alcohols were converted into thiocarbonyl derivatives (**362**–**369**) which were subsequently transformed into the parent hydrocarbon, on treatment with TBTH. It has been demonstrated that the mechanism involves the reversible addition of a tributyltin radical to the thiocarbonyl moiety (e.g., **369**) to form the intermediate radical (**374**), followed by homolytic scission of the carbon–oxygen bond, generating the carbon-centred radical and a carbonyl species (**375**) (Scheme 118). The tin intermediate (**375**) subsequently fragments to the tin sulfide (**376**) and COS ⟨88JCS(P1)1517, 89CRV1413, 90TL3991, 91JCS(P1)1173⟩. Since this initial work, several alternative derivatives have been demonstrated to be useful carbon centred radical precursors (**370**–**373**) ⟨83T2609, 86PAC675, 87S665, B-92MI 119-17⟩.

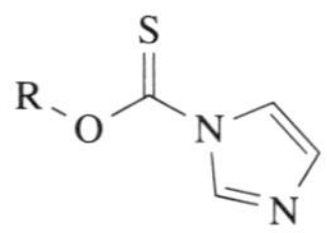

(**362**)

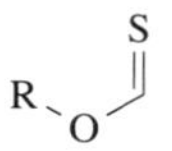

(**363**)

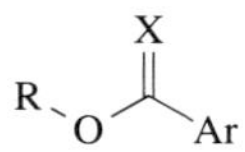

(**364**) X = S, Ar = Ph
(**365**) X = S, Ar = 4-MeOPh
(**366**) X = S, Ar = 4-$MeSO_2Ph$
(**367**) X = S, Ar = 2,4-MeO_2Ph
(**368**) X = Se, Ar = Ph

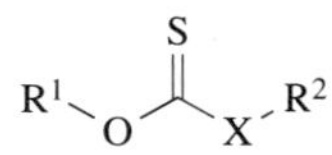

(**369**) X = S, R^2 = Me
(**370**) X = S, R^2 = Ph
(**371**) X = O, R^2 = Me, Ph or aryl

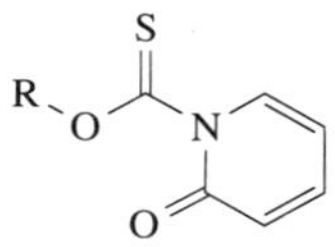

(**372**)

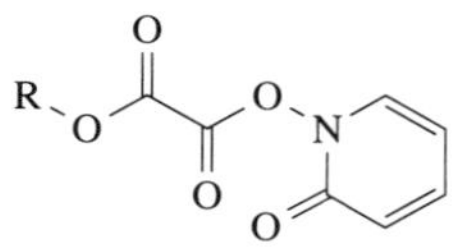

(**373**)

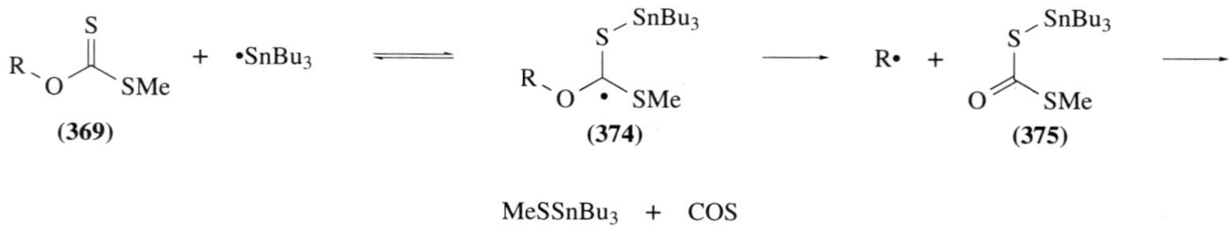

Scheme 118

Primary alcohol derivatives, as might be expected, require the most forcing conditions to fragment; hence it is necessary to heat a solution of xanthate esters (**369**) to 150 °C. Thus, on slow addition of TBTH in *p*-cymene at 150 °C, the hederagenin xanthate derivative (**377**) afforded methyl oleanolate (**378**), via the primary radical species (Scheme 119). The use of thiocarbonyl imidazolide (**362**) and thiobenzoate (**364**) derivatives, in conjunction with TBTH, however, is effective at the slightly lower temperature of 130 °C (compare ⟨81S743⟩ with ⟨90TL2565⟩). Barton *et al.* have since systematically investigated the use of a variety of silanes in alliance with thionocarbonate derivatives of type (**371**). Hence, on initiation with triethylborane/oxygen in the presence of diphenylsilane, these primary alcohol derivatives are reduced to the parent hydrocarbon in high yields at 80 °C ⟨90TL4681, 91T8969, 91SL435, 91TL7187⟩. Thus, not only is the reaction temperature lowered but the use of toxic TBTH is eliminated. AIBN and benzoyl peroxide may also be used to initiate this silane chemistry; however, large quantities are required, indicating a short radical chain process ⟨92ACR188⟩. Similarly, dialkylphosphites and hypophosphorus acid effect the reaction in refluxing dioxan solution ⟨92TL5709, 93JOC6838⟩. A cautionary note should be mentioned with regard to the generation of carbon-centred radicals from the precursory primary thiocarbonyl derivatives. If the substrate contains homoallylic unsaturation, the radical intermediate may be captured to afford a lactone. Three groups have developed synthetically useful lactonisation procedures that adopt this methodology (Equations (105) and (106)) ⟨88TL6127, 89JOC1234, 89CC1265, 91JCS(P1)1173, 92JOC6803⟩.

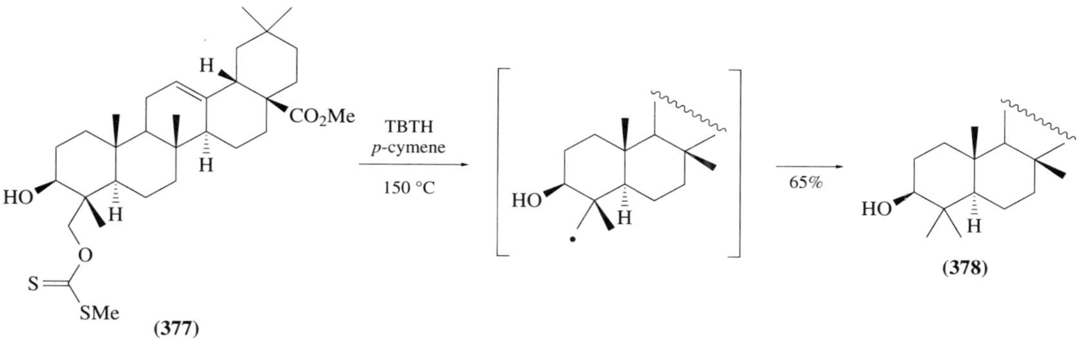

(377) (378)

Scheme 119

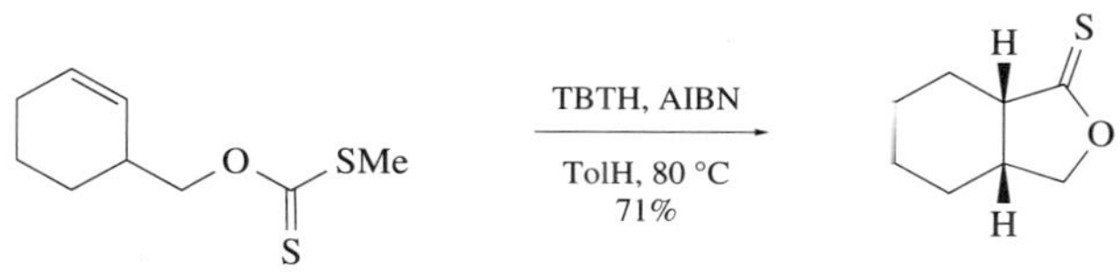

(105)

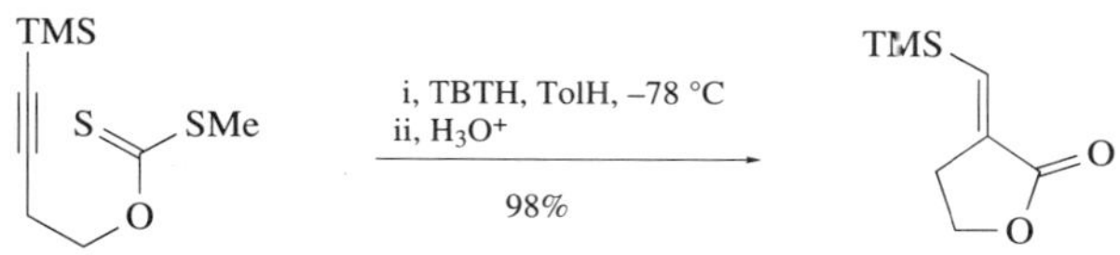

(106)

The formation of carbon-centred radicals from the corresponding secondary alkyl derivatives (**362–371**) is an exceptionally facile process that has been utilised frequently in synthesis. The reactions proceed at a lower temperature to that of their primary analogues, even at $-20\,^\circ\mathrm{C}$ with the triethylborane/oxygen initiation procedure ⟨85SC401, 86JOC2613, 88TL6125, 90JA2034, 92TL6629, 92T2529⟩. Although all of the methods available for the generation of primary radicals are applicable to the generation of secondary radicals, xanthate (**369**) and thionocarbonate derivatives (**371**) are more commonly used (Equation (107)) ⟨81JA932, 83JA4059, 89TL2619, 92T7435, B-92MI 119-17, 93JOC5028⟩. An interesting and potentially useful development is the use of *O*-acetyl thiohydroxamic ester (**379**) in place of the alkyltin or alkylsilane reagent ⟨91TL2703, 92T7109⟩. Irradiation with visible light at 40 °C afforded the methyl radical which reacts with the thiocarbonyl moiety of xanthate (**380**). Subsequent β-scission of the carbon–oxygen bond afforded the secondary cyclododecyl radical (**381**) which, on reaction with a second molecule of the thiohydroxamic ester derivative (**379**), carries the chain (Scheme 120). A polystyrene-supported organotin hydride has also been reported to effect the radical deoxygenation of secondary alcohols after conversion to either their methyl xanthate (**369**) or thionocarbonate derivatives (**371**) ⟨92SL801⟩.

(107)

The thiocarbonyl derivatives (**362, 364–372**) work well for primary and secondary carbon–oxygen bonds, but due to the facile Chugaev reaction at elevated temperatures, their use is limited with tertiary substrates. Four methods have now been reported for the formation of carbon-centred radicals from tertiary alcohols. The first solution, reported by Pfenninger *et al.*, adopted the use of a selenocarbonate derivative (**382**) ⟨80HCA2328⟩. Treatment with TBTH and AIBN in a refluxing benzene solution effected deselenylation, to afford the formyl radical (**383**) which subsequently decarboxylated to generate the carbon-centred radical (**384**). Hydrogen atom abstraction from

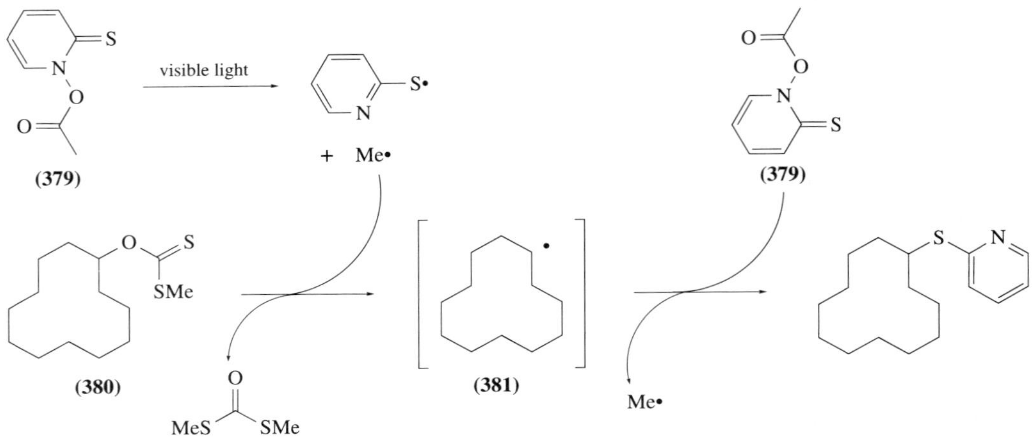

Scheme 120

TBTH gave the deoxygenated compound (**385**). Trapping of the intermediate formyl radical with TBTH also occurred, to give the formate (**386**) (Scheme 121). Barton's initial solution for the generation of carbon-centred radicals from tertiary alcohols was to use the thermally more stable thioformate ester derivatives (**363**) ⟨82TL2019⟩. However, the synthesis of these radical precursors is not easy. Thus, Barton and Crich have developed a facile method which incorporates the radical chemistry of *O*-esters of thiohydroxamic acids ⟨84CC774, 85TL757, 86JCS(P1)1603⟩. The mixed half esters of oxalic acid (**373**) are conveniently prepared from the tertiary alcohol, oxalyl chloride and the sodium salt of *N*-hydroxypyridine-2-thione. As with the radical decarboxylation procedure either heating or irradiation with a tungsten lamp is sufficient to initiate the radical chain process, affording the tertiary carbon radical, the thiyl radical and two equivalents of carbon dioxide (Equation (108)). Barton *et al.* have also demonstrated the utility of the *O*-acetyl thiohydroxamic ester (**372**) derivative in the formation of tertiary carbon-centred radicals from methyl xanthate derivatives. The fragmentation is effected by photolysis with a 500 W tungsten lamp at 0–40 °C, hence the competing Chugaev elimination is suppressed (Scheme 120) ⟨93TL2733⟩.

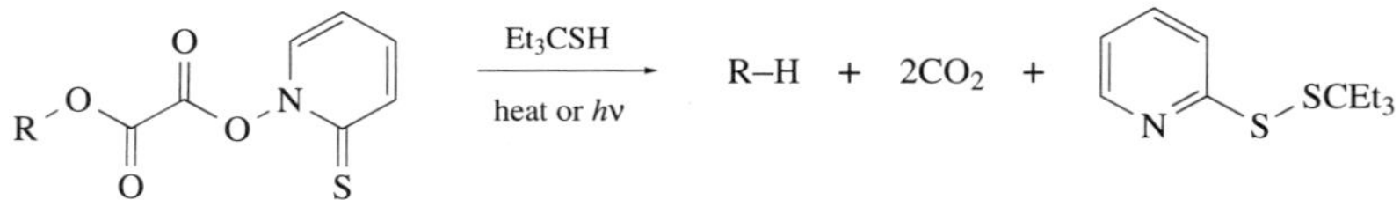

(**372**) (108)

Selective formation of carbon-centred radicals parallels the stability of the generated radical. Thus, Dolan and MacMillan have shown that deoxygenation of the secondary allylic derivative (**387**) does not occur under the conditions required to transform the tertiary allylic compound (**388**) into the deoxy compound (**389**) ⟨85CC1588⟩. A similar deoxygenation of a tertiary centre was effected in the presence of a secondary, cyclic thionocarbonate moiety (Equation (109)) ⟨88TL281⟩.

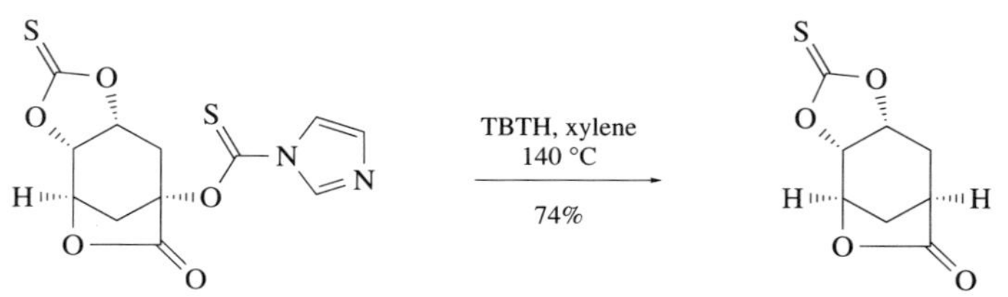

(109)

In general, when cyclic thionocarbonates are formed from a primary and either a secondary or tertiary alcohol moiety, the radical is generated at the most substituted centre ⟨77JCS(P1)1718, 85T4253, 93TL8245⟩. There are, however, exceptions to this generality and the cleavage of a primary over a secondary carbon–oxygen bond was explained in terms of the bond-angle strain energy ⟨90JOC1416⟩. With cyclic thionocarbonate systems involving secondary and either secondary or tertiary carbon–

oxygen bonds, product mixtures may arise (Equations (110)–(112)) ⟨77JCS(P1)1718, 78HCA690, 85T4253, 87TL5973, 89CRV1413⟩.

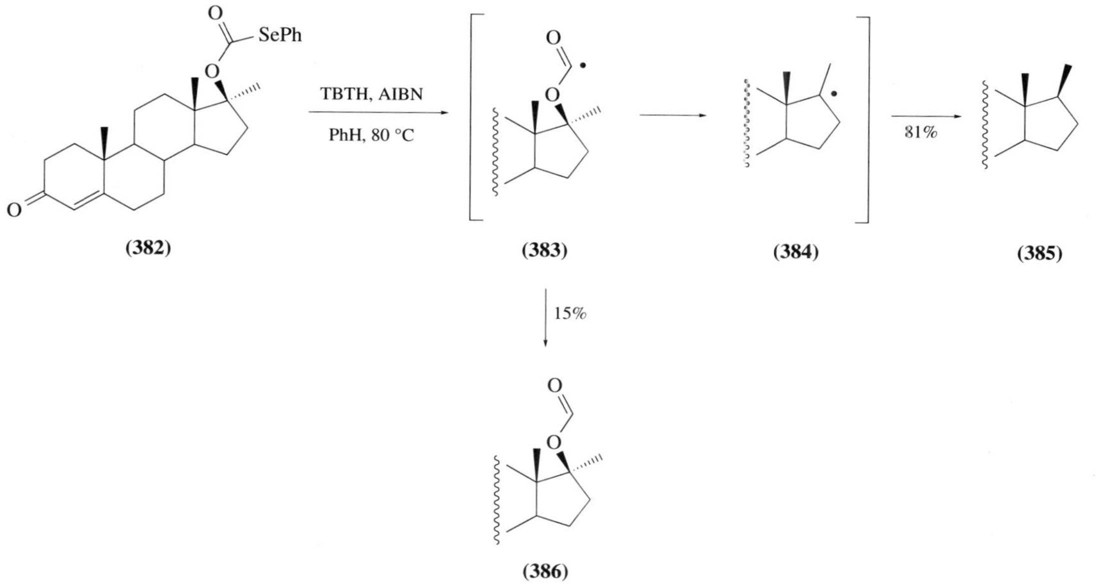

(382) **(383)** **(384)** **(385)** **(386)**

Scheme 121

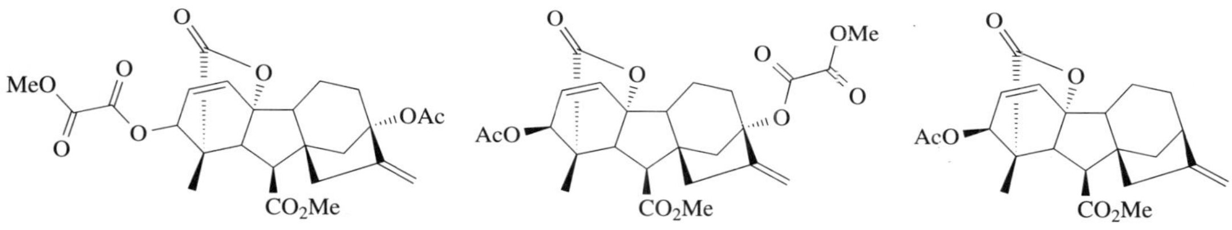

(387) **(388)** **(389)**

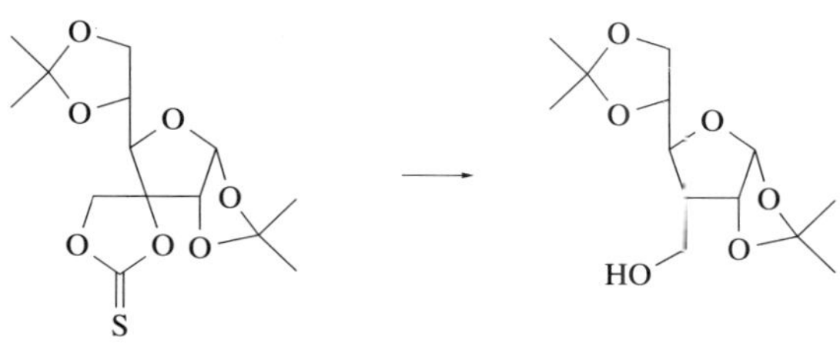

(110)

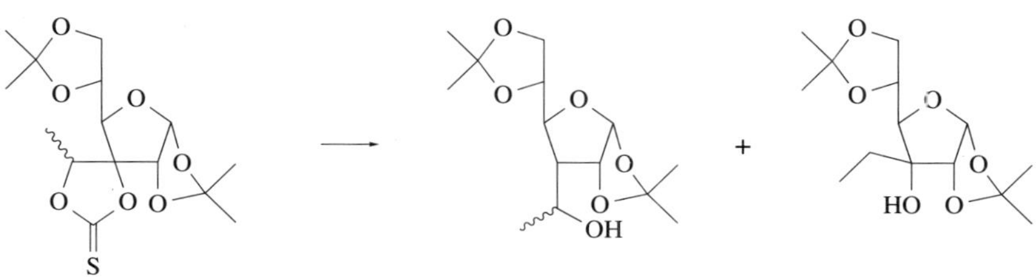

(111)

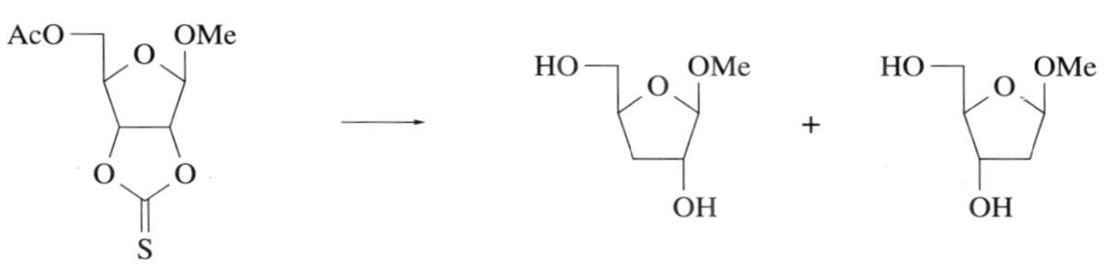

(112)

Lowinger and Weiler have developed a novel method for the generation of carbon-centred radicals from epoxides ⟨90CJC1636, 94CRV519⟩. Treatment of the epoxide (**390**), in THF solution, with two mole equivalents of bis(cyclopentadienyl)titanium(III) chloride effected homolytic scission of one carbon–oxygen bond to form a carbon radical (Scheme 122). The reaction seems to proceed, as one would predict, via the radical intermediate (**391**).

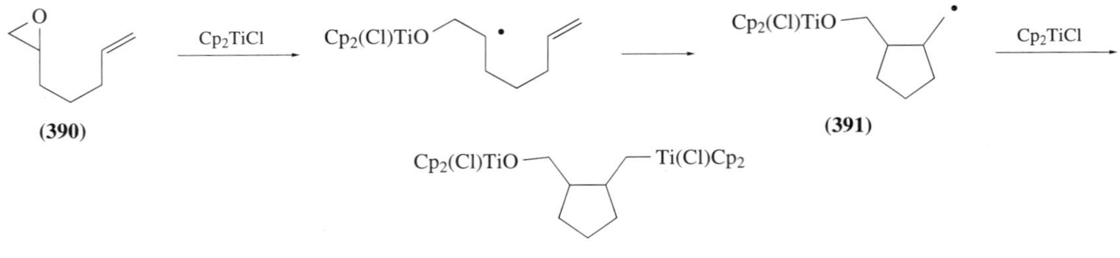

Scheme 122

(ii) Carbon–sulfur bond scission

Unlike their alcohol counterparts, thiols readily undergo desulfurisation via a radical chain mechanism. Hence, on treatment of the thiol with trialkyl phosphite, in the presence of either AIBN or light, reductive desulfurisation occurs according to the mechanism described in Scheme 123 ⟨56JA6414, 57JA5326, 59JA1243⟩. Thiols are similarly desulfurised on treatment with two equivalents of TBTH and AIBN (Equation (113)) ⟨82JA2046, 85TL135⟩. The ability of a substrate to undergo desulfurisation mirrors the relative stability of the generated radical, benzyl ≈ allyl > tertiary > secondary > primary > methyl ⟨82TL443, 84JOC5206, 86AJC77⟩. The formation of carbon-centred radicals from the corresponding sulfide is a more energetically demanding process and these compounds are rarely used as radical precursors. Phenyl sulfide derivatives, however, have been applied in a limited number of synthetically useful reactions since the phenyl–sulfur bond is not affected under the reaction conditions (Equation (114)) ⟨83JCS(P1)403, 87S665, 92CC613⟩.

$$RSH + P(OEt)_3 \xrightarrow[\Delta \text{ or } h\nu]{AIBN} RS\bullet + P(OEt)_3 \longrightarrow RS-\dot{P}(OEt)_3 \longrightarrow$$

$$R\bullet + SP(OEt)_3 \xrightarrow{RSH} RH + RS\bullet$$

Scheme 123

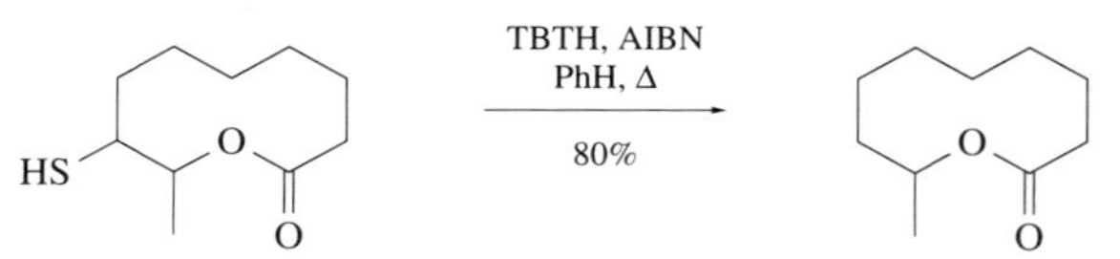

(113)

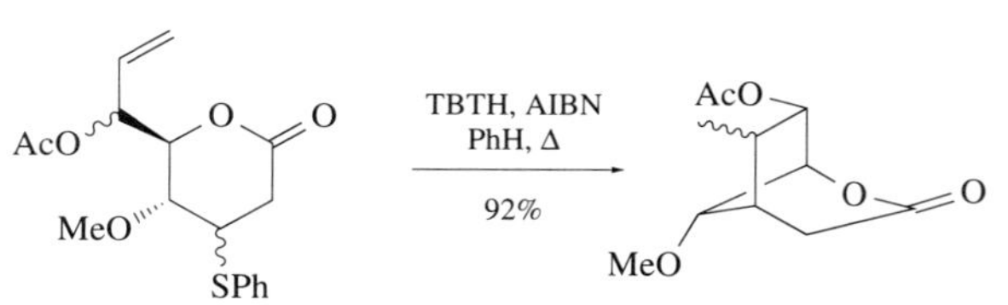

(114)

(iii) Carbon–selenium bond scission

In contrast to their sulfur analogues, the homolytic scission of carbon–selenium bonds is a relatively easy process and has frequently been adopted for the generation of carbon-centred radicals ⟨78CC41, 87S665, 88PAC1659, 91JCS(P1)1433, 93JOC6857⟩. Primary, secondary and tertiary radicals have been generated from their phenylselenyl precursors. Typically, the selenide substrate is treated with triphenyltin hydride in refluxing toluene; however, a more facile reaction occurs at lower temperatures under AIBN initiation (Equation (115)) ⟨80JA4438⟩. Tributyltin hydride and tris(trimethylsilyl)silane (TTMSS) may also be used ⟨90CC1575, 90T545, 91JCS(P1)2797, 91JOC678, 92ACR188⟩. A novel photosensitised one-electron reduction of primary and secondary phenylselenyl compounds has been reported ⟨93CC1636⟩. The selenides were irradiated with a 450W Hanovia medium pressure lamp ($\lambda > 280$ nm) in an isopropanol solution containing 1,5-dimethoxynaphthalene (DMN) as sensitiser and ascorbic acid as co-reductant. Yields of the reduced compounds were in the range 80–90% (Scheme 124). Kataoka *et al.* have also demonstrated that single electron transfer (SET) can effect the reductive cleavage of selenium– and sulfur–carbon bonds of cyclic onium salts via a carbon-centred radical (Scheme 125) ⟨90JCS(P1)3017⟩.

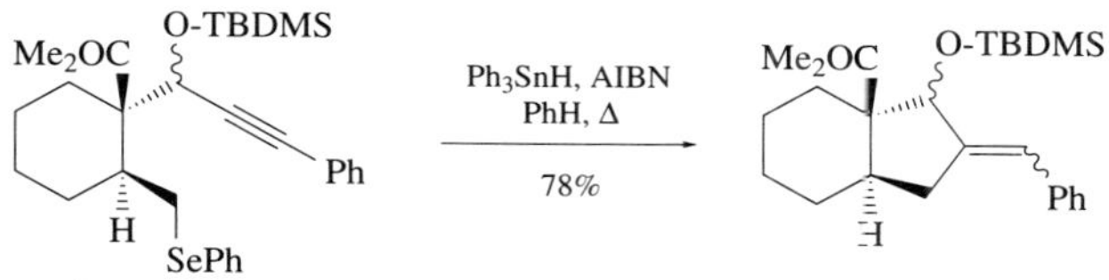

(115)

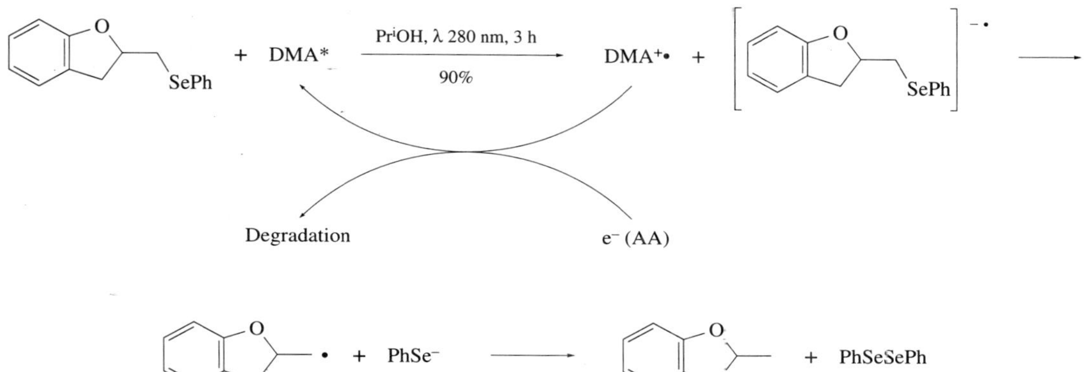

Scheme 124

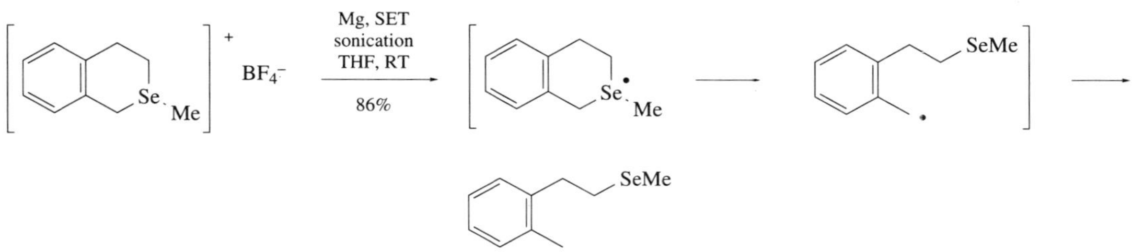

Scheme 125

(iv) Carbon–tellurium bond scission

The homolytic cleavage of a carbon–tellurium bond is an even more facile process than that with selenides. The formation of carbon-centred radicals occurs either at room temperature for secondary or at 80 °C for primary substrates, on treatment with triphenyltin hydride. Addition of any initiator to the reaction mixture is unnecessary ⟨80JA4438⟩. Although information in the literature is limited, the homolytic cleavage of a carbon–tellurium bond appears to be very selective and seems to parallel the ease of radical formation. An understanding of this selectivity has been successfully applied by Barton *et al.* in an elegant extension of the thiohydroxamic ester chemistry. A methyl radical generated from the *O*-acetyl thiohydroxamic ester derivative (**379**) generated under standard conditions, was allowed to react with anisyltelluride derivatives (**392**), liberating the more stable secondary radical (**393**). This radical subsequently undergoes typical radical addition and/or substitution reactions (Scheme 126) ⟨88TL6581, 90JA891, 92JA5904, 92T9261, 93JA8050⟩.

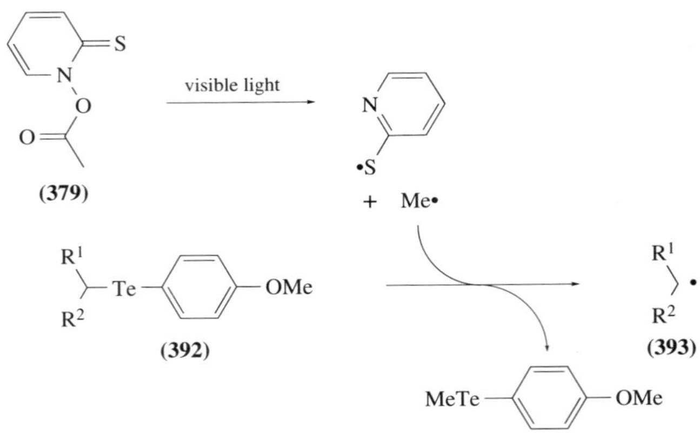

Scheme 126

1.19.3.2.5 Carbon–nitrogen bond scission

(i) Azoalkanes

Homolysis of the carbon–nitrogen bond in azoalkanes (**394**) can be effected either by thermolysis or photolysis over a wide temperature range, to afford a carbon-centred radical or biradical and nitrogen gas ⟨80CRV99⟩. They are therefore, probably the cleanest and most convenient source of radicals and are frequently used as initiators in radical reactions. The ease of homolysis, as expected, parallels the relative stability of the formed radical, benzyl > tertiary > secondary > primary. With cyclic azoalkanes, release of ring strain is also a contributory factor. There are two mechanistic possibilities for this fragmentation, either simultaneous or stepwise carbon–nitrogen bond scission. It is now believed that for symmetrical azoalkanes, the simultaneous pathway is followed while, for unsymmetrical compounds, there is a continuum between the two. The more unsymmetrical the compound, the greater the tendency to proceed via a diazenyl radical (**395**) (Scheme 127). The synthetic potential of this method for generating carbon-centred radicals has been demonstrated by Little and co-workers in the synthesis of several ring systems (Equation (116)) ⟨79JA7129, 81JA2744, 81TL4389, 85TL3417, 85JOC2400, 85JOC3940, 92SL107⟩. Taber *et al.* have proposed a radical chain mechanism for the reductive cyclisation of a ketone tosylhydrazone (Scheme 128). They propose the monoalkyl diazene (**397**) is formed on reduction of the hydrazone (**396**) with $NaBH_3CN/ZnCl_2$, which subsequently fragments to form the carbon radical intermediate (**398**) ⟨93TL6209⟩.

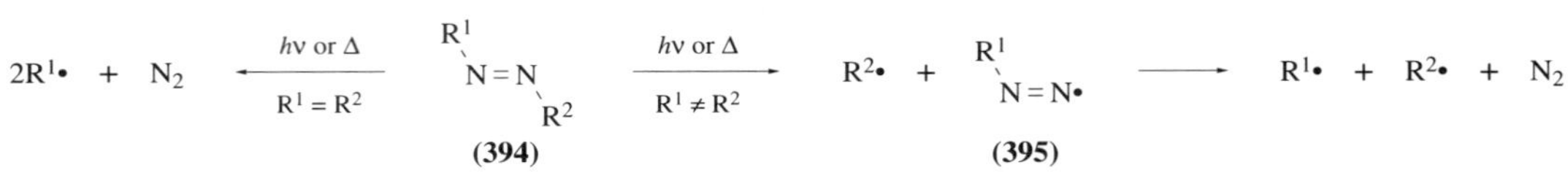

Scheme 127

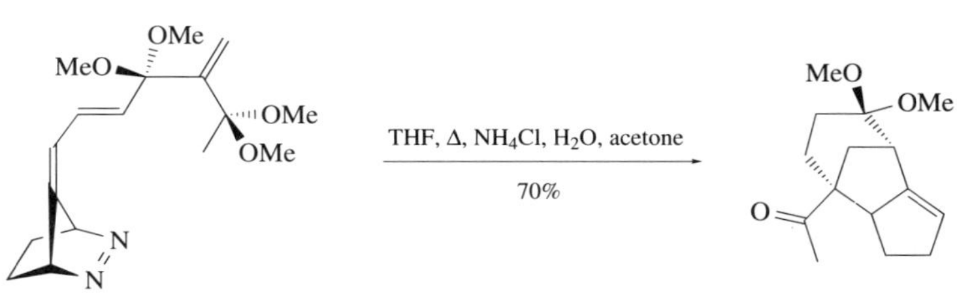

(116)

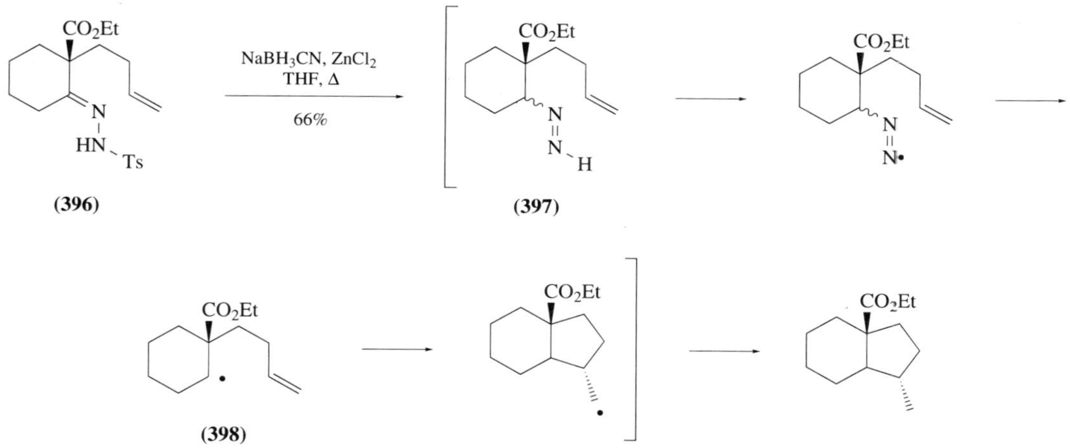

Scheme 128

(ii) Isocyanides, isothiocyanates and isoselenocyanates

Primary, secondary and tertiary isocyanides react with TBTH to generate carbon-centred radicals ⟨68JA4182, 79TL2291, 80JCS(P1)2657, 80S68⟩. This process is initiated with AIBN and is temperature-dependent; however, very little is known about the mechanism ⟨78JCS(P2)1313⟩. As noted with previous radical precursors, ease of radical formation parallels the stability of the generated radical. Consequently, while tertiary and secondary isocyanides afford radicals in either refluxing benzene or toluene solution, primary substrates require the higher temperatures afforded by a refluxing xylene solution. Barton *et al.* demonstrated that this phenomenon allowed selective formation of secondary and tertiary radicals over their primary counterparts ⟨80JCS(P1)2665⟩. In a refluxing benzene solution, the tetraisocyanide derivative (**399**) was completely deaminated, however, at 70–72 °C, the primary isocyanide moiety was unaffected (Scheme 129). It should be noted that the formation of the primary radical is at an unusually low temperature and this was attributed to the β-oxygen effect ⟨82CC447⟩. Chatgilialoglu and co-workers have investigated the ability of silanes to effect the generation of radicals from isocyanides ⟨90SL219, 90TL6013, 91JOC678, 92ACR188⟩. With these silane reagents, there is no temperature dependence and a facile reductive homolysis of primary, secondary and tertiary isocyanides occurs at about 75 °C. A tertiary isocyanide has also been reduced via a radical chain process using hypophosphorus acid ⟨93JOC6838⟩. Isothiocyanates and isoselenocyanates also form carbon-centred radicals, via the intermediary of the isocyanide ⟨80JCS(P1)2657, 80JCS(P1)2665, 81CC901⟩.

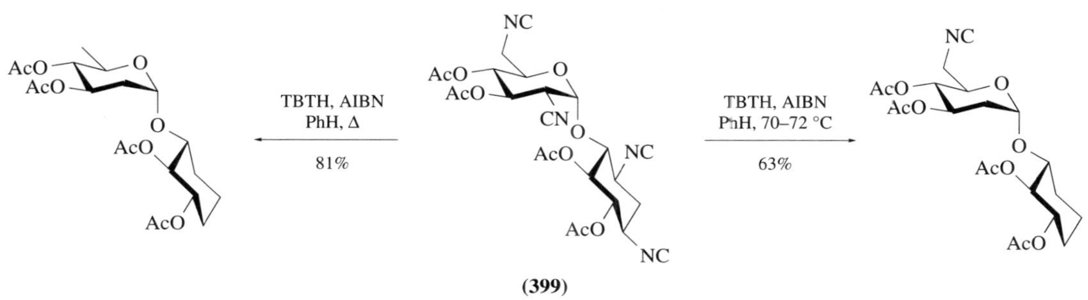

Scheme 129

(iii) Nitroalkanes

The reduction of aliphatic nitro compounds with TBTH has been reported ⟨81JA1557, 81TL1705⟩. Tertiary nitro compounds are the most effective in generating radicals, though secondary systems are also useful. However, a primary nitro moiety is inert to this TBTH-mediated reaction. Typically, tertiary and allylic substrates react with about one equivalent TBTH and AIBN at 80°C while secondary substrates require the higher temperatures afforded by toluene and about five equivalents of hydride ⟨83HCA2210, 85T4013, 86S693, 87S665, 93T5189⟩. The mechanism has been the subject of several studies and is now believed to proceed via addition of the tributyltin radical to the nitro compound (**400**) to afford the nitroxide radical (**401**). Subsequent β-scission of the carbon–nitrogen bond afforded the carbon-centred radical (Scheme 130) ⟨87CB1197, 88BCJ3629, 90JOC3321, 93T9079; compare with 81JA1557⟩.

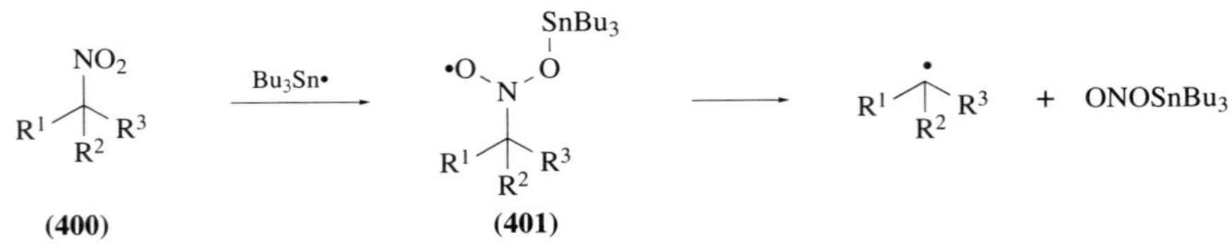

Scheme 130

1.19.3.2.6 *Carbon–boron bond scission*

Homolytic scission of a carbon–boron bond may be induced thermally, on initiation with either AIBN or diacylperoxides, photochemically and by the slow addition of air (oxygen) ⟨70JA712, 70JA714, 72AG(E)692, 79COC(3)791, B-86MI 119-04⟩. The latter method is becoming an important low-temperature method for the initiation of radical reactions associated with other functional groups ⟨88TL6125, 88TL6127, 89BCJ143, 90JA2034, 90TL3991, 90TL4681, 91T8969, 92T2529, 93JA10464, 93TL1317⟩. Again, the ease of radical formation parallels the stability of the generated radical, tertiary > secondary > primary. Thus, with *B*-alkylborinanes, it has been demonstrated that generation of either a secondary or tertiary carbon-centred radical occurs in preference to the primary radical (Scheme 131) ⟨71JA3777⟩. This phenomenon was utilised to circumvent the inefficiency of free-radical addition reactions. With simple trialkylboranes, only one of the three alkyl groups participate in the reaction, thus any reaction can only be 33% efficient ⟨68JA4165⟩. However, if two sacrificial substituents are introduced, as in *B*-alkylborinanes and alkyldiphenylboranes, the reaction can achieve a 100% efficiency with respect to the synthetically useful alkyl group ⟨71JA3777, 78JOM(156)101; see also 71JA1024, 86T5523⟩.

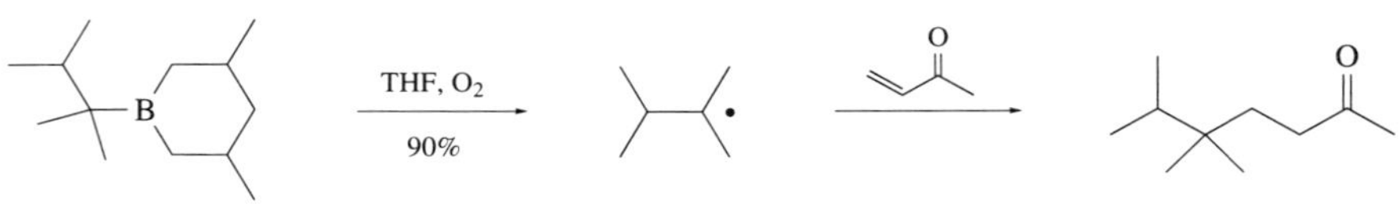

Scheme 131

1.19.3.2.7 *Carbon–cobalt bond scission*

Carbon-centred radicals are efficiently generated from a variety of alkylcobalt(III) species ⟨83ACR343, 88CSR361, 88S489, 94CRV519⟩. Pioneering work by Scheffold established the use of vitamin B_{12} catalysis in the chemical, electrochemical and photoelectrochemical reduction of alkyl halides. Homolytic scission of an intermediary cobalt–carbon bond generates the carbon-centred radical (Equations (117)–(120)) ⟨80JA3642, B-83MI 119-05, 87OM1606, 87PAC363, 90T3155⟩. This facile homolytic cleavage has now been extended to include the more readily available bis(dimethylglyoximato) (dmgH) (**402**), *N*,*N*-ethylenebis(salicylidaminato) (salen) (**403**) and *N*,*N*-phenylenebis-(salicylidaminato) (salophen) (**404**) derivatives. Typically, carbon-centred radicals are afforded on either sunlamp irradiation or thermolysis and the value of this method has been demonstrated by several groups (Equations (121) and (122)) ⟨82JOC1775, 85JOC5875, 86JA4790, 89JOC1320, 90SL166, 90JCS(P1)2691, 90JCS(P1)2703, 90JCS(P1)2709, 92T9373, 93CB1187⟩.

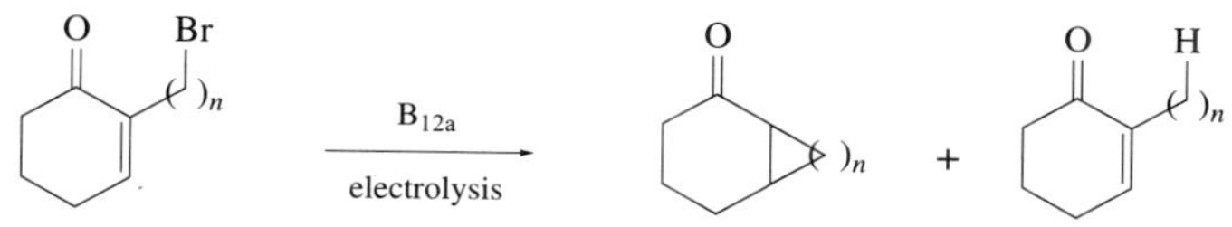

(117)

90% $n = 3$; 0:100
95% $n = 4$; 100:0
80% $n = 5$; 88:12

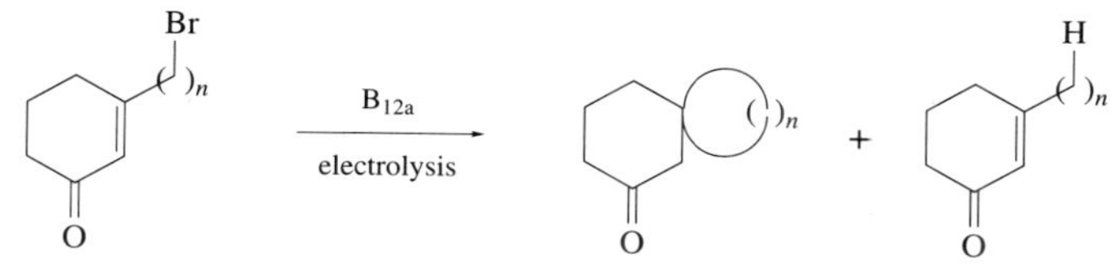

(118)

90% $n = 3$; 0:100
95% $n = 4$; 100:0
85% $n = 5$; 53:47

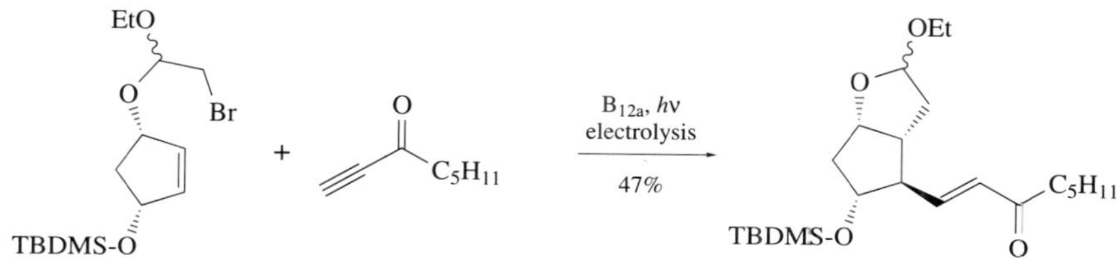

(119)

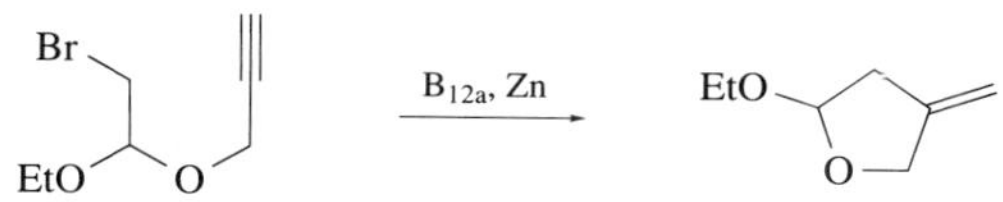

(120)

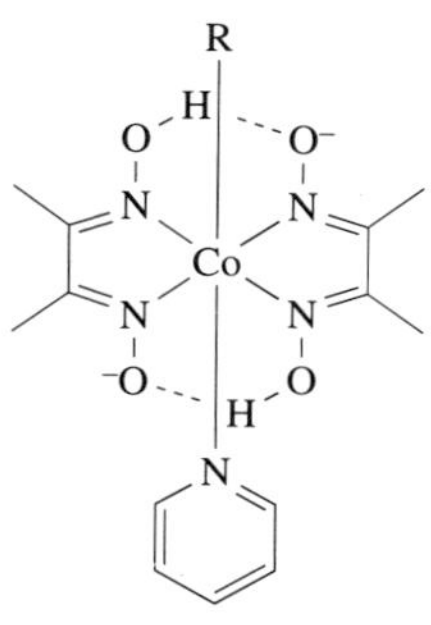

(402)

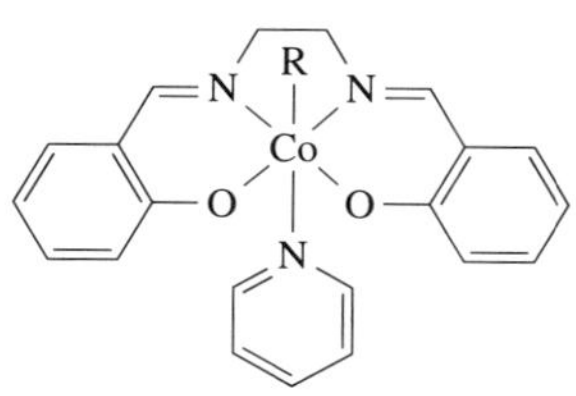

(403)

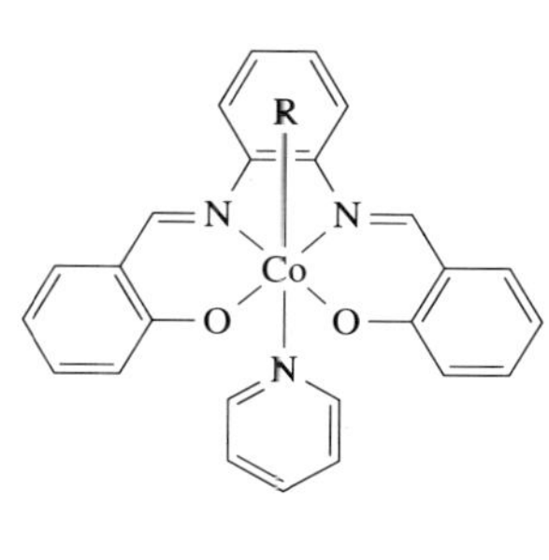

(404)

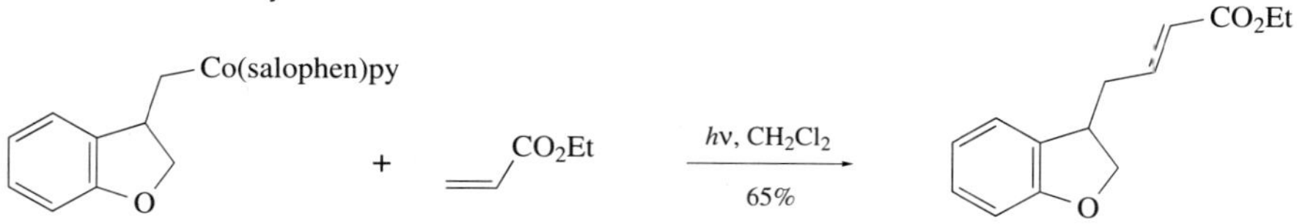

(121)

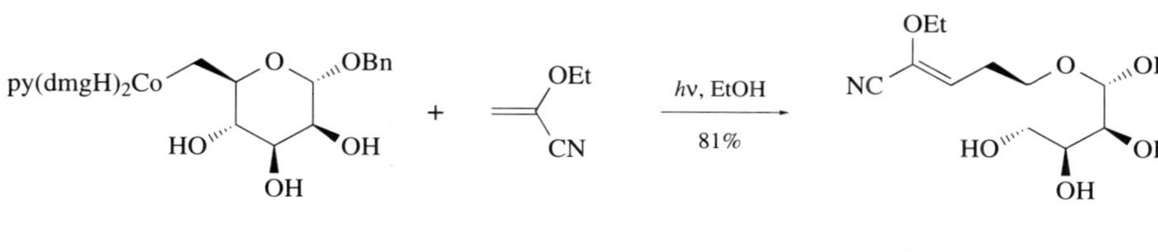

(122)

1.19.3.2.8 Carbon–mercury bond scission

The generation of free radicals via homolysis of a carbon–mercury bond, often called 'the mercury method', is a well-documented and important process; however, it has now largely been surpassed by the use of tin hydrides ⟨85AG(E)553, B-86MI 119-04, 88CRV487, 89ACR1, 91COS(4)715⟩. The importance of the former method lies in both the ease of alkylmercurial preparation, especially in relation to the solvomercuration reaction and subsequent radical formation ⟨B-86MI 119-06⟩. Alkyl radicals may be generated by both photolysis and thermolysis, on addition of AIBN; however, the most common method is reduction in the presence of hydrogen donors such as sodium borohydride or tributyltin hydride, to the unstable mercuric hydride (Scheme 132) ⟨77CB2588, 84CB3175, 84TL5239⟩. Primary, secondary and tertiary alkylmercuric halides, acetates and nitrates are compatible with this procedure. Representative examples are depicted in Equation (123), Schemes 133 and 134 ⟨82CB2526, 84TL2743, 84JCS(P1)721⟩. More specific methods that involve mercury in the generation of alkyl radicals include electrochemical, autoxidation, halodemercuration and the use of photosensitised mercury and metallic salts ⟨88CRV487⟩.

$$\mathrm{RHgX} \xrightarrow{\mathrm{NaBH_4}} \mathrm{RHgH} \longrightarrow \mathrm{RHg}\bullet \longrightarrow \mathrm{R}\bullet + \mathrm{Hg}$$

Scheme 132

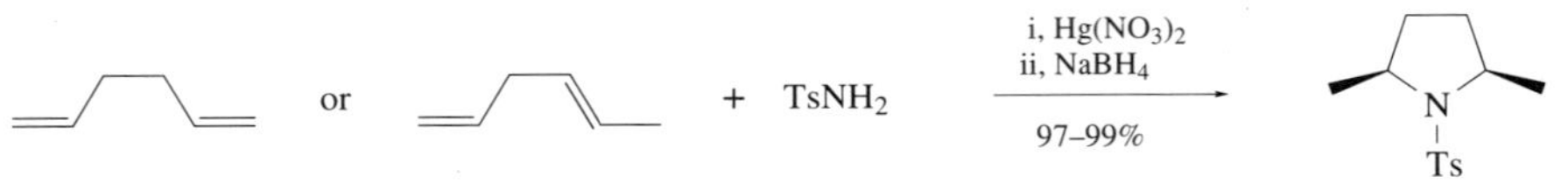

(123)

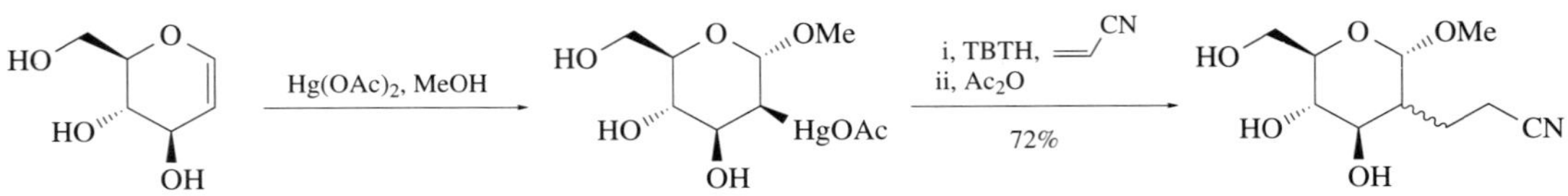

Scheme 133

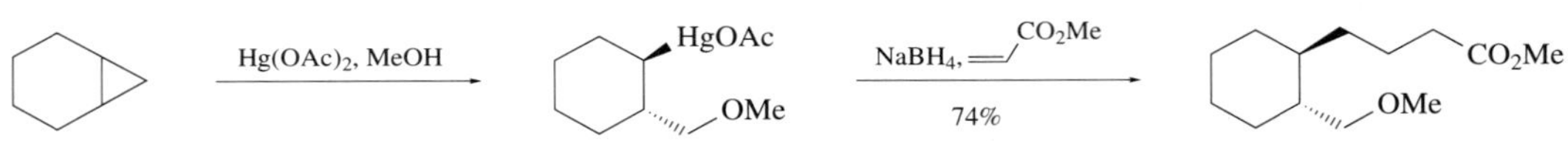

Scheme 134

1.19.3.3 Radicals by Addition to C═C

The addition of heteroatom-centred radicals to carbon–carbon double bonds, with concomitant formation of a carbon-centred radical, has been known for many years. This anti-Markovnikov addition process is well understood and there are several review articles that discuss this process ⟨63OR(13)91, 63OR(13)150, 70AG(E)273, B-73MI 119-17, 88S489, 91CRV1237, 91COS(4)715⟩. However, many of these reactions are not useful, being superseded by the hydroboration and heteroatom mercuration methods (Sections 1.19.3.2.6 and 1.19.3.2.8, respectively). Outlined below are some of the preferred methods of forming carbon-centred radicals by the addition of heteroatom radicals to carbon–carbon double bonds.

1.19.3.3.1 The addition of oxygen-centred radicals to C═C

Oxygen-centred radicals are high energy species that typically undergo β-scission and atom abstraction reactions (see Sections 1.19.3.2.1 and 1.19.3.2.2(i)). Intramolecular cyclisations, however, are possible when a C═C double bond is suitably positioned within the substrate. Thus, oxygen-centred radicals afford five- and six-membered ring systems in high yielding cyclisation reactions (Equations (124)–(126)) ⟨75JOC3614, 84JOC1345, 87TL4011, 91TL45, 91TL49⟩.

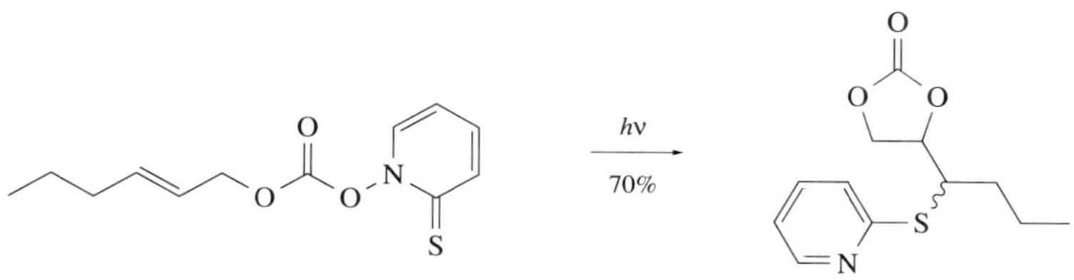

(124)

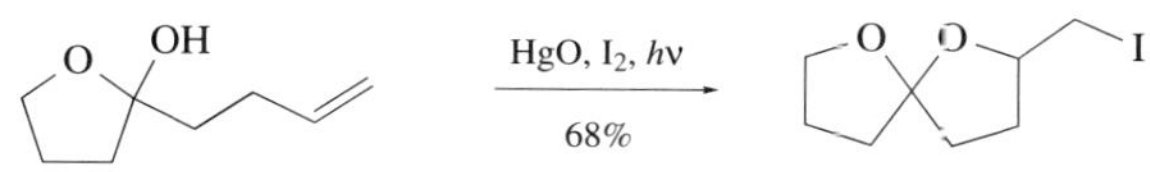

(125)

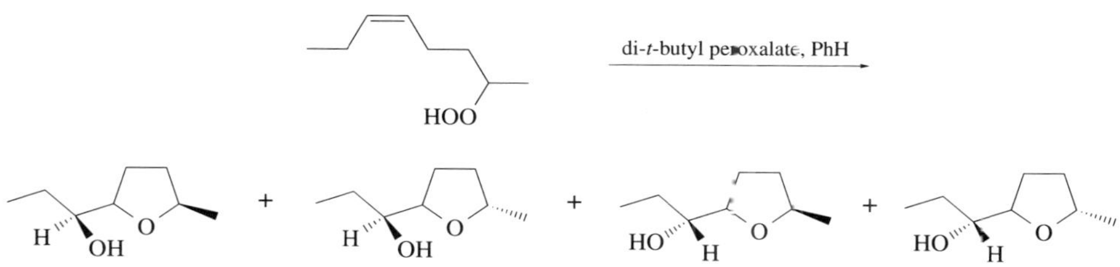

(126)

68 : 21 : 8 : 3

1.19.3.3.2 *The addition of sulfur-centred radicals to C=C*

Thiyl radicals are readily generated from thiols either on photolysis or treatment with radical initiators. These radicals are known to undergo reversible addition to alkenes at the least-hindered carbon atom. These properties have been used by the groups of Singleton and Feldman in syntheses of cyclopentane and cyclohexane systems, respectively, from cyclopropyl substituted alkenes ⟨90TL823, 90TL5551, 93SL827⟩. Feldman has also demonstrated the utility of this addition–elimination process in the preparation of the brefeldin ring system ⟨93JOC6851⟩. Sulfonyl radicals also add to carbon–carbon double bonds to afford the carbon-centred radical in preparatively useful reactions (Equations (127)–(130)) ⟨86JA4568, 90T3061, 91T5425, 91JCR(S)1, 91JOC2179, 92JOC6118⟩.

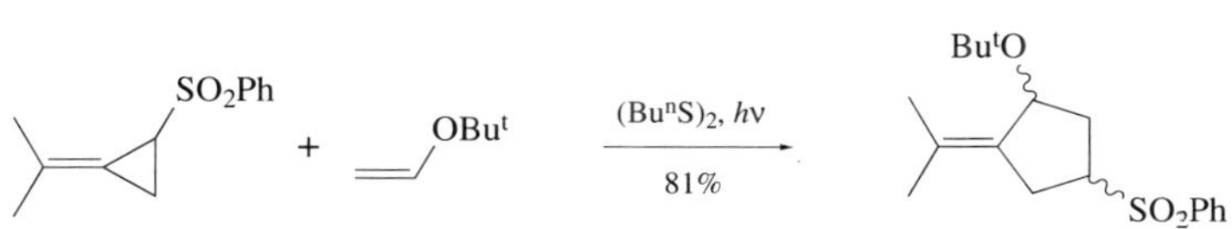

(127)

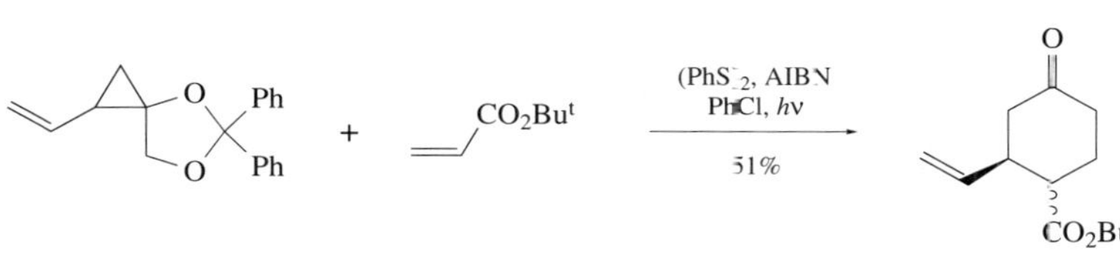

(128)

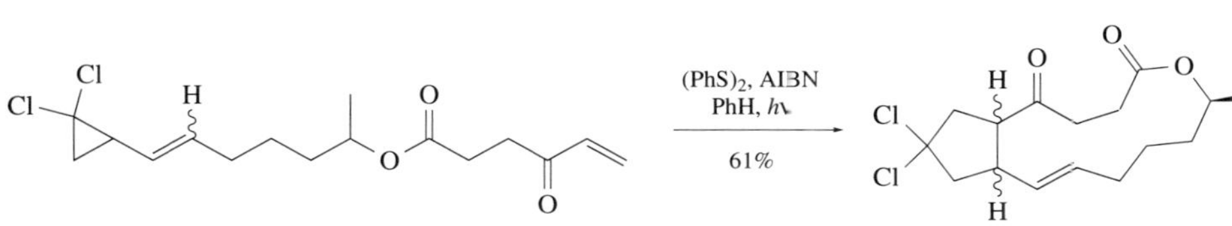

(129)

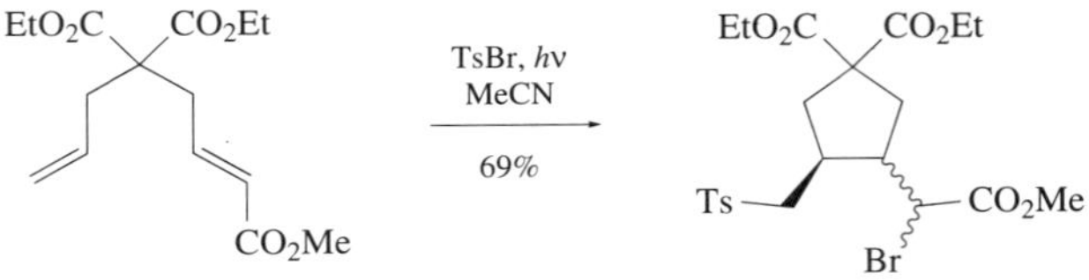

(130)

1.19.3.3.3 *The addition of nitrogen-centred radicals to C=C*

It has been demonstrated that the addition of nitrogen-centred radicals to carbon–carbon double bonds is a reversible process. The intramolecular aminyl radical cyclisation is also substrate-dependent and has been shown to require an activated carbon–carbon double bond ⟨92TL4993, 93TL5523⟩. However, dialkylamminium radicals participate in both facile and high-yielding radical addition reactions. These reactions can be either intramolecular additions (with unactivated double bonds) or intermolecular additions (with electron-rich alkenes) ⟨87JA3163, 90TL1675, 90T2317, 90T2329, 90T2345, 91JOC1309⟩. Amidyl radicals have a similar reaction profile ⟨92TL5913, 93CB1171, 93TL6877, 93JOC4933⟩. Representative examples of amminium and amidyl radical addition reactions are depicted in Equations (131)–(134).

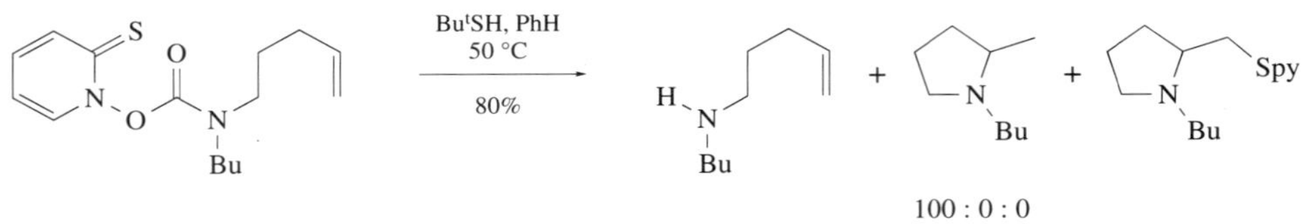

(131)

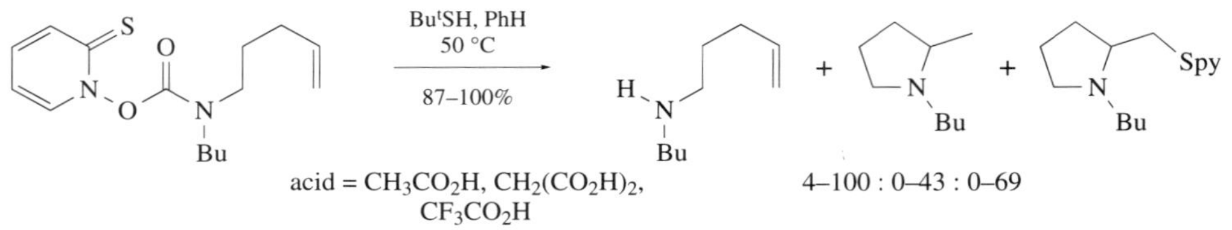

(132)

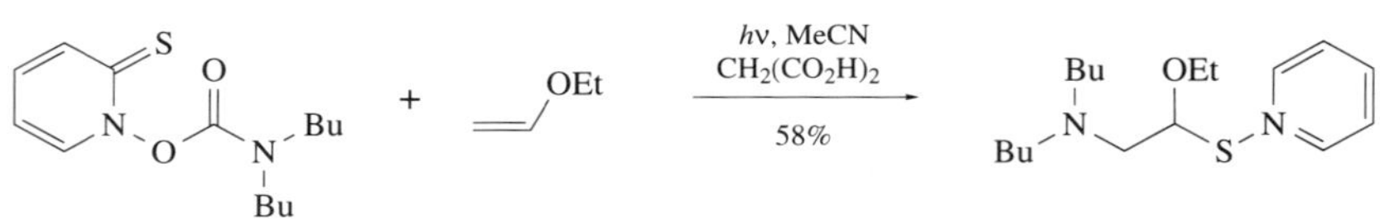

(133)

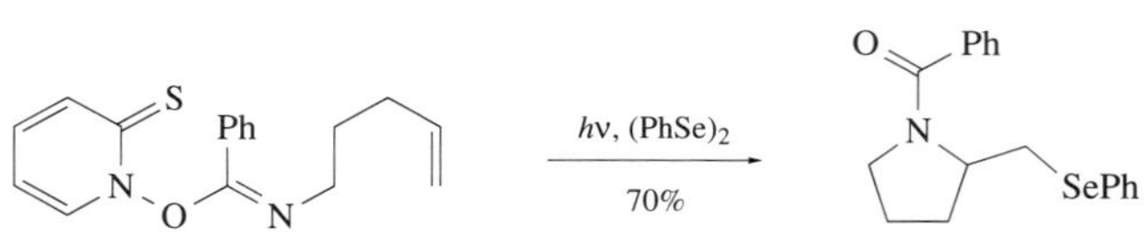

(134)

1.19.3.3.4 *The addition of group IV-centred radicals to C=C*

The addition of carbon-centred radicals to a carbon–carbon double bond is arguably the most synthetically useful of these processes and Chapter 1.08 discusses this in detail. The addition of stannyl, germyl and silyl radicals to carbon–carbon double bonds, to afford the adduct carbon-centred radical, are preparatively useful procedures ⟨63OR(13)150, B-73MI 119-17, 88JOM(340)317, 91JCR(S)216, 92JOC3994⟩. The precursory Group 4 radicals, generated by the typical initiation procedures described above, add regioselectively to the least-hindered terminus of the double bond. The addition of stannyl radicals to carbon–carbon double bonds has also been demonstrated to be reversible. With respect to silyl radicals, Chatgilialoglu *et al.* have illustrated that the addition process is extremely facile and both electron-rich and electron-deficient alkenes participate equally well in this reaction, to generate a carbon-centred radical (Equations (135)–(137)) ⟨92JOC3994⟩.

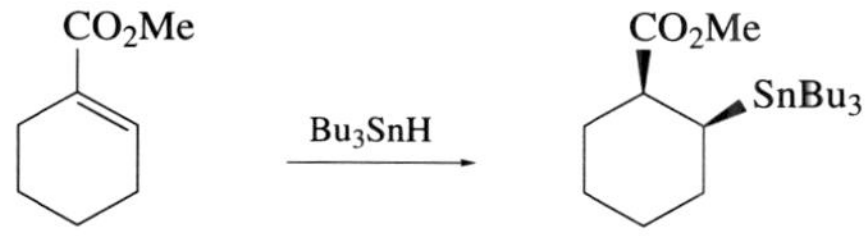
CO2Me
Bu3SnH
CO2Me
SnBu3

(135)

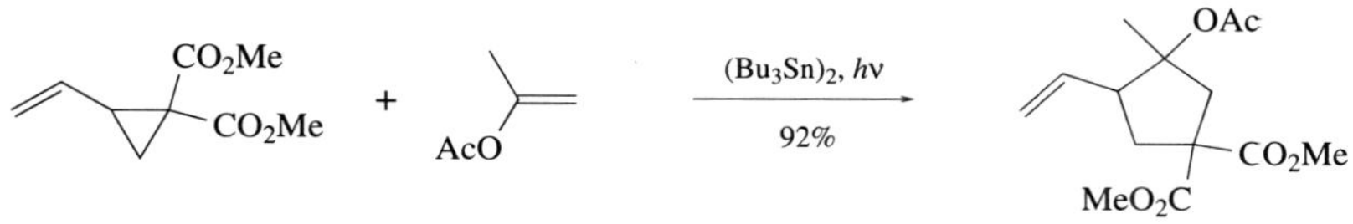
CO2Me
CO2Me
+
AcO
(Bu3Sn)2, hv
92%
OAc
CO2Me
MeO2C

(136)

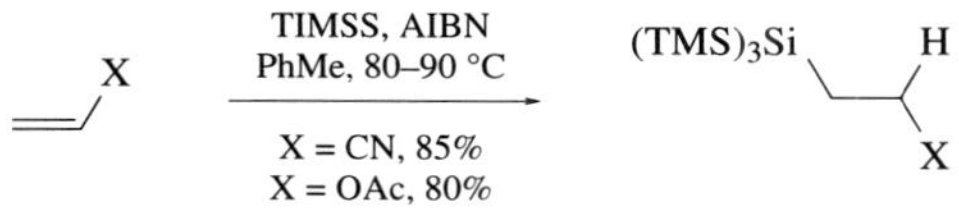
X
TIMSS, AIBN
PhMe, 80–90 °C
X = CN, 85%
X = OAc, 80%
(TMS)3Si
H
X

(137)

1.20
Allenes and Cumulenes

CHRISTIAN BRUNEAU and PIERRE H. DIXNEUF
Université de Rennes 1, France

1.20.1 INTRODUCTION

The preparation and reactivity of allenes and cumulenes have received special attention in several reviews. After the first reviews on allenes ⟨64RCR1, B-69MI 120-01⟩ and cumulenes ⟨61BSF2176⟩, general information on methods for preparation was given by Murray ⟨77HOU(5)967⟩, Hopf ⟨B-80MI 120-01⟩,

Landor ⟨B-82MI 120-01⟩, and Brandsma ⟨B-84MI 120-01⟩. More specialized reviews have appeared dealing with propargylic rearrangements ⟨B-69MI 120-02⟩, allenic ketones ⟨80T331⟩, strained cumulenes ⟨89CRV1111⟩, and reactivity of allenes ⟨B-84MI 120-02, 84T2805⟩.

1.20.2 BY C—H BOND FORMATION

1.20.2.1 Isomerization of Alkynic Compounds

1.20.2.1.1 Isomerization of alkynes and diynes

Allenes were initially prepared by isomerization of the $RCH_2—C{\equiv}C—$ moiety by a strong base catalyst such as an alkali metal amide ($NaNH_2$) at high temperature ⟨B-69MI 120-01⟩ or, when the α hydrogen is more acidic, at 20°C with a tertiary amine ⟨65CB2611⟩ or basic alumina ⟨64T2177⟩. Coordination of the triple bond to a metal like Mn can help the isomerization with Al_2O_3 ⟨79AG(E)688⟩. This method can be applied to the formation of allenyl sulfides from alkynyl thioethers ⟨70QRS1⟩. Internal alkynes containing a hydroxy group can be isomerized regioselectively into allenic alcohols with BuLi/TMEDA (Scheme 1) ⟨86TL4599⟩. For diynes in the presence of Bu^tOK the prototropic rearrangement leads to bisallenyl derivatives ⟨73TL3181⟩, while for 1,4-diynes milder conditions can be used (NaOH/EtOH) for the preparation of alkynyl allenes (Scheme 2) ⟨58JA1376⟩.

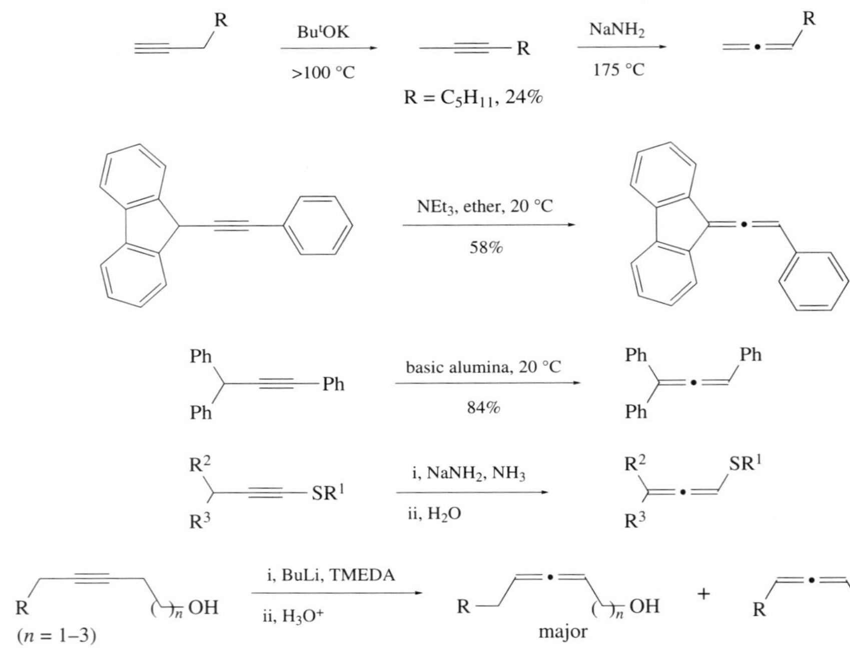

Scheme 1

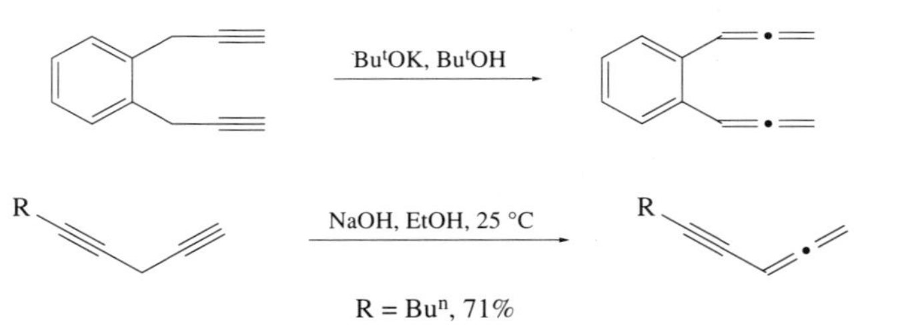

Scheme 2

1.20.2.1.2 Isomerization of enynes

1,4-Enynes can be isomerized using Bu^{t}OK to produce conjugated ene–allenes ⟨65JOC2983⟩. Cross-conjugated ene–allenes and ene–diallenes have been produced by prototropic shift from a 1,4-enediyne ⟨87TL2697⟩. 1,4-Enynes containing a hydroxy group are transformed into conjugated vinylallenyl alcohol derivatives in the presence of 5% methanolic sodium hydroxide (Scheme 3) ⟨53JA1050⟩.

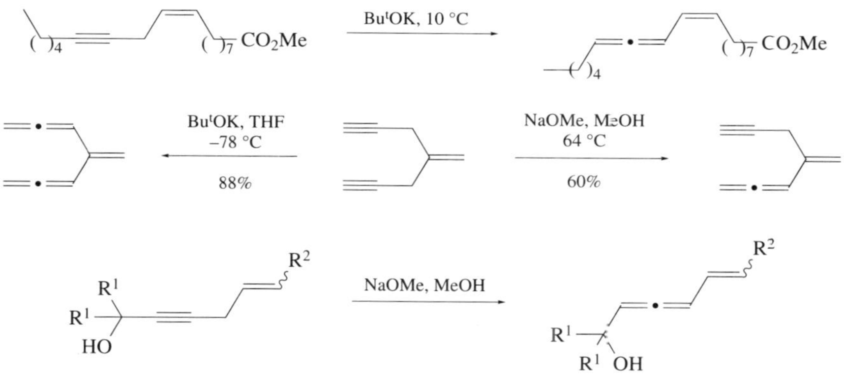

Scheme 3

1.20.2.1.3 Propargylic rearrangement

Propargyl derivatives with a heteroatom X attached to the sp^3 carbon (R^1—C≡C—CHR^3—X—R^2) easily undergo a ready base-catalyzed rearrangement leading to stable allenes R^1—CH=C=CR^3—X—R^2. Allenyl ethers, efficient building blocks in organic synthesis ⟨92AG(E)1033, 93JOC5709, 93S165, 93SL105⟩, are generally obtained in 80–100% yields by reaction of propargyl ethers with Bu^{t}OK ⟨68RTC916, 93JOC5702⟩. Catalytic amounts of Bu^{t}OK are used for the selective transformation of 2-alkynyl ethers (**1**) (containing a bromine atom) into allenyl ethers (**2**) ⟨88CC237⟩. This reaction can also be applied to the generation of alkenyl allene (**4**) (precursor for Diels–Alder cycloadditions) from the enyne (**3**) ⟨93CC270⟩. 2-Alkynyl ethers with an internal triple bond usually give complete conversion into allenyl ethers on treatment with NaNH$_2$ ⟨65RTC31⟩. Bispropargylic ethers and thioethers have been converted into bispropadienyl ethers and sulfides with various base systems (Scheme 4) ⟨76TL3181, 78TL691⟩.

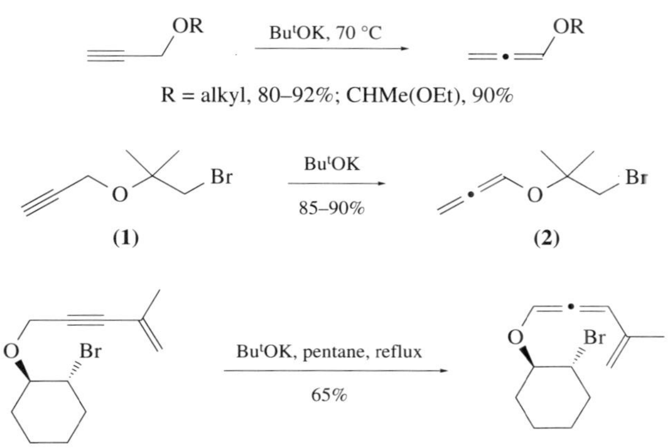

Scheme 4 *(continued)*

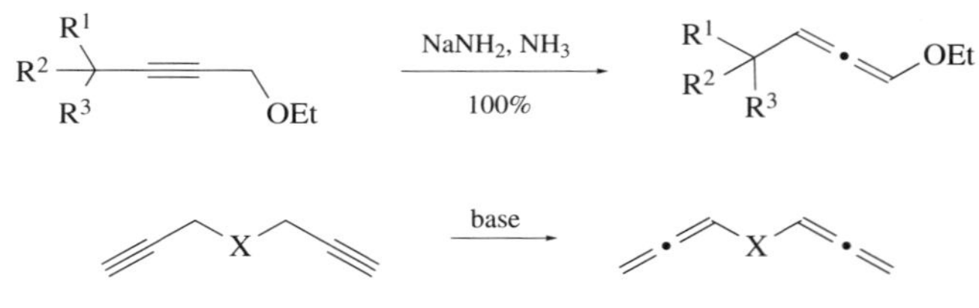

Scheme 4

The isomerization of β-alkynic alcohols is less common but occurs in the presence of $AgNO_3$ to afford α-allenic alcohols ⟨59HCA719⟩. The isomerization of propargylic derivatives into allenes containing a SR ⟨63RTC1040⟩, SeR, or TeR group is achieved with strong bases (Scheme 5). Thioether derivatives generally isomerize faster than the corresponding selenium or tellurium compounds ⟨66BSF3016, 66BSF3024⟩. Novel cyclic enyne allene sulfones (**6**) have been prepared by oxidation of cyclic enediyne sulfides (**5**) with *m*-chloroperbenzoic acid (Scheme 5) ⟨93CC1406⟩. 2-Alkynyl amines (**7**) can be converted into *N*,*N*-dialkylallenyl amines (**8**) by treatment with potassium amide or alumina; they are stable except in the case where R = H and are precursors of 1-alkynyl amines ⟨68JCS(C)228⟩. Heterocyclic allenyl amines have been prepared in only moderate yields by this method ⟨68JCS(C)606⟩. However, allenyl benzotriazole can be prepared in satisfactory yield by treatment of propargyl benzotriazole with NaOH (Scheme 5) ⟨93JOC3038⟩.

The mild isomerization of the prop-2-ynyl into the allenyl group within phosphine ligands coordinated to metal complexes can also be achieved (Scheme 6) ⟨93CB1077⟩.

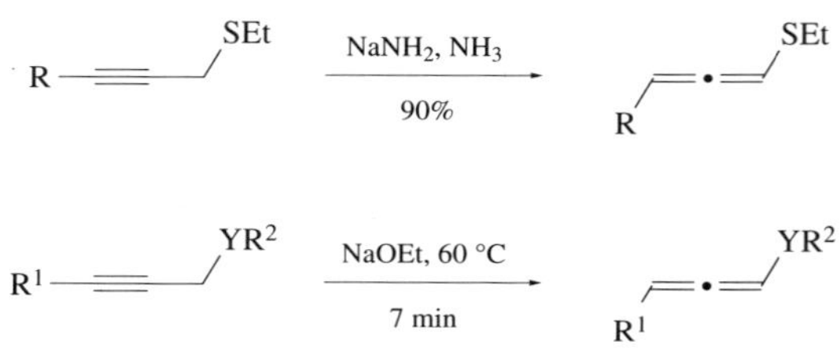

R^1 = H; R^2 = Ph; Y = S, 90%; Y = Se, 30%

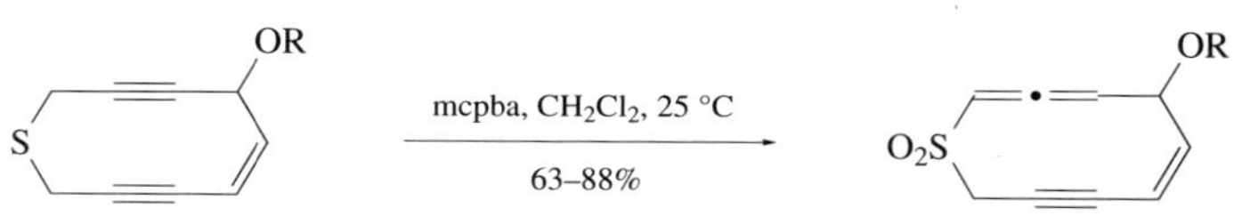

(5) **(6)**

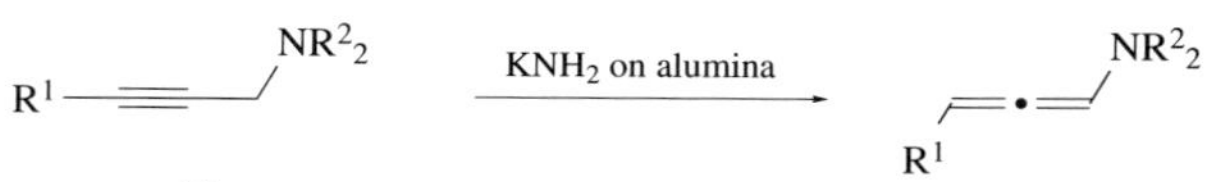

(7) **(8)**

$R^1 \neq$ H; NR^2_2 = N(alkyl)$_2$, N(cycloalkyl)
R^1 = H; NR^2_2 = imidazol-1-yl, 40%, carbazol-9-yl, 35%, pyrrol-1-yl, 20%

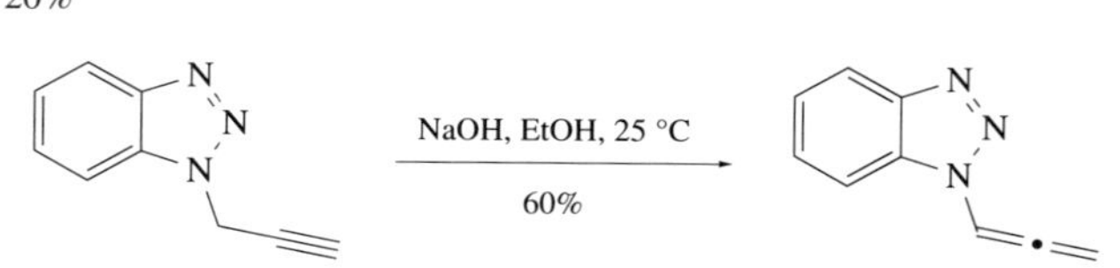

Scheme 5

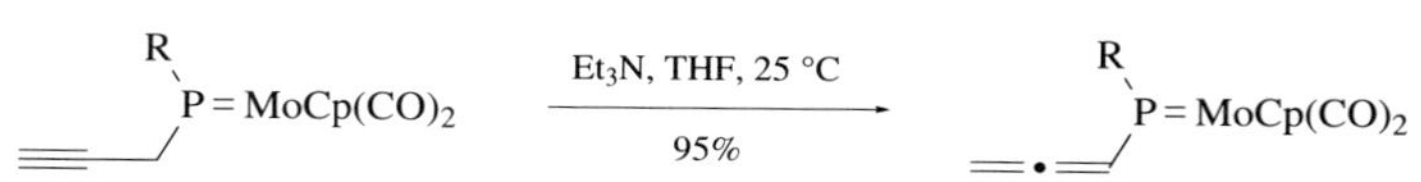

Scheme 6

1.20.2.1.4 *Rearrangement into α-allenic ketones*

The rearrangement of β-alkynic ketones into α-allenic ketones can be achieved with base as in the transformation of **(9)** to **(10)** ⟨55JCS1740⟩, or under milder conditions by using $NaHCO_3$ in water for **(11)** to **(12)** ⟨B-84MI 120-01⟩, or basic alumina for **(13)** to **(14)** and **(15)** to **(16)** ⟨86JOC2623⟩. A similar isomerization takes place with β-alkynic esters in the presence of K_2CO_3 ⟨72TL4249⟩. Homopropargylic alcohols give α-allenic ketones upon oxidation with Jones reagent via the formation of β-alkynic ketones, followed by isomerization ⟨57CR(C)1790⟩. A good method to produce allenynone **(19)** from alkynyl homopropargylic alcohol **(17)** involves first, the Swern oxidation into **(18)** followed by thermal rearrangement into **(19)**; Jones reagent converts alcohol **(17)** into the γ-pyrone **(20)** in low yield (Scheme 7) ⟨93TL449⟩. The oxidation of homopropargylic alcohols by the Dess–Martin periodinane reagent also affords allenones in good yields ⟨90JOC3450⟩.

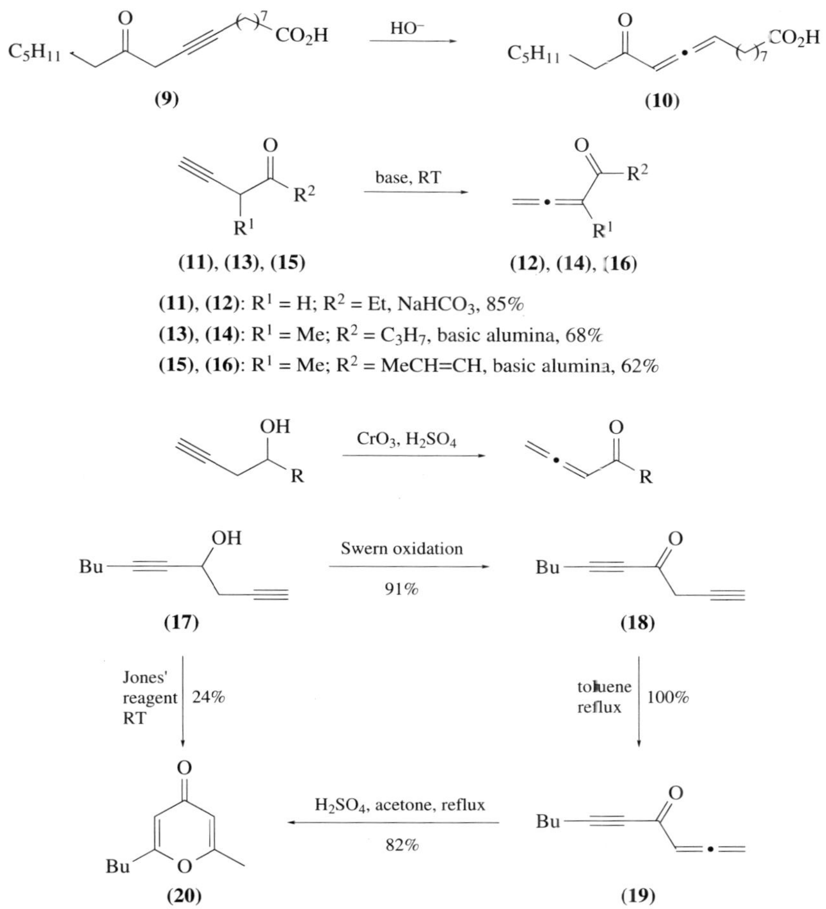

Scheme 7

1.20.2.1.5 *Photorearrangement*

Irradiation of 1,1,3-triphenylpropyne in methanol, in the presence of 1,4-dicyanonaphthalene (1,4-DCN) as photosensitizer, gives a rearrangement leading to triphenylallene as the major product ⟨85JA6615⟩. A stereoisomerization with no proton migration has been observed during the irradiation (at 230 nm) of alkenyl allenes **(21)** and **(22)** in hexane (Scheme 8) ⟨76RTC80⟩.

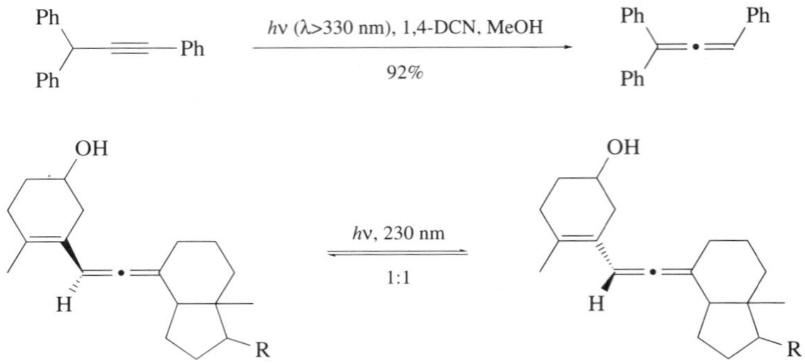

(21) **(22)**

Scheme 8

1.20.2.2 Reduction of Carbon–carbon Triple Bonds

1.20.2.2.1 Dehalogenation of alkynic halides

Hydrodehalogenation of α-alkynic halides by reducing agents such as ZnCu/ROH or $LiAlH_4$ constitutes a general means of access to allenic derivatives (see Table 1). Since the direct reduction of propargylic alcohols is not possible under these conditions, α-allenyl alcohols can be prepared from α-halo-α′-hydroxyalkynic compounds ⟨55JOC1337⟩. For example, in the synthesis of the first fluoroallenylamino acid (**25**), the fluoroallenyl group is formed by reduction of the chlorofluoropropargyl synthon (**23**) into the fluoroallenic alcohol (**24**) (Scheme 9) ⟨87JA3491⟩. The Hiyama reagents, $CrCl_2$ generated from $CrCl_3$ and $LiAlH_4$, or $CrCl_2$ in the presence of AcOH ⟨83T2185⟩, are also useful reducing systems that allow the obtention of allenes and vinyl allenes from propargylic halides ⟨78TL3801⟩ and 1-bromo-4-ene-2-ynes ⟨79NJC321⟩ (Table 1). The utilization of a chiral protonating agent like (-)menthol or (-)borneol in the reduction of 1-bromo-2-alkynes leads to the formation of enantiomerically enriched allenes ⟨81TL103⟩.

Table 1 Reduction of α-alkynic halides.

R^1	R^2	R^3	X	Reducing agent	Yield (%)	Ref.
Me	Et	H	Cl	$LiAlH_4$	63	55JOC95
H	H	CH_2OH	Cl	$LiAlH_4$	94	87TL2697
Pr	H	H	Cl	ZnCu	71	49JA1964
H	H	CH(Me)OAc	Br	Zn	45	67CR(C)1492
H	H	C_5H_{11}	Br	$CrCl_2$	75	78TL3801
H	H	CH_2=CH	Br	$CrCl_2$	40	79NJC321
H	n-C_7H_{15}	CH_2=CMe	Br	$CrCl_2$	95	79NJC321
H	Me	1-cyclohexenyl	Br	$CrCl_2$	95	79NJC321

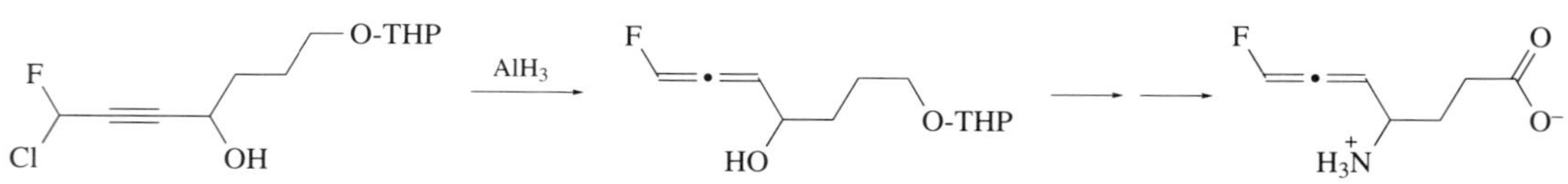

(23) **(24)** **(25)**

Scheme 9

Alkynic halides of different structures, with the halogen separated from the triple bond by an unsaturated conjugated chain, have been reduced by generation of vinylallenyl metal intermediates which release the allene on hydrolysis (Scheme 10) ⟨60JCS341, 72TL4465, 72CC866, 74BSF1119, 74CR(C)149, 76S755, 77T2813, 79TL1845⟩.

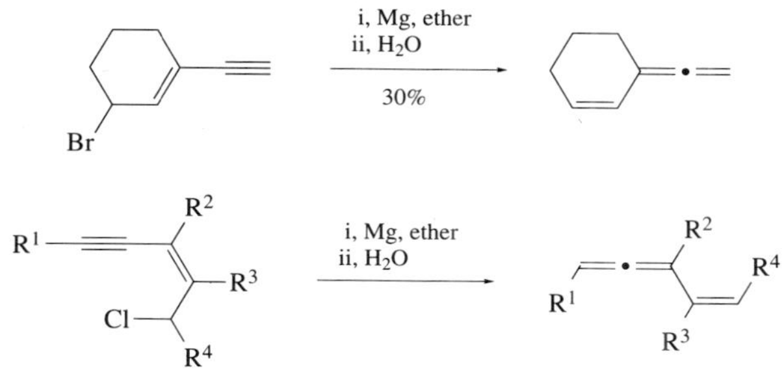

R¹	*R²*	*R³*	*R⁴*	*Yield (%)*
H	H	H	H	31
H	Me	Me	H	55
Me	H	H	H	52
H	n-C_5H_{11}	H	H	95
H	$CH_2C{\equiv}CEt$	H	H	93

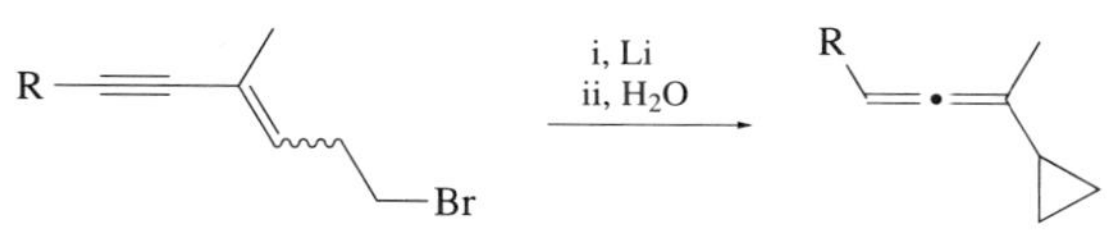

R = H, 30%
R = Me, 27%

Scheme 10

1.20.2.2.2 Reduction of polyunsaturated hydrocarbons

The formal 1,4-addition of hydrogen to conjugated enynes affords allenes as illustrated by the reduction of dicyclohexen-1-yl acetylene with hydrogen in the presence of an iron/aluminum catalyst (Scheme 11) ⟨42JA363⟩. Conjugated 2-en-4-yn-1-ols are reduced to 3,4-dien-1-ols by $LiAlH_4$ in refluxing diethyl ether ⟨75BSF1407⟩. Enynes containing an allenyl group and either an hydroxy **(26)** or an acetate **(27)** group have been reduced into diallenes **(28)** by reaction with $LiAlH_4$ **(26)** or $LiAlH_3OMe$ **(26)**–**(27)**. With this latter reagent, less competing hydroxy substitution by the hydride ion takes place. This reduction corresponds to a 1,4-addition of H^-/H^+ and the presence of a

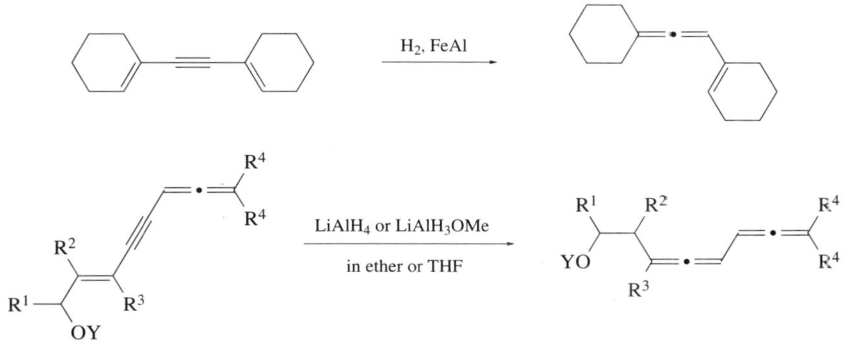

(26) Y = H
(27) Y = MeCO

(28)

Scheme 11 *(continued)*

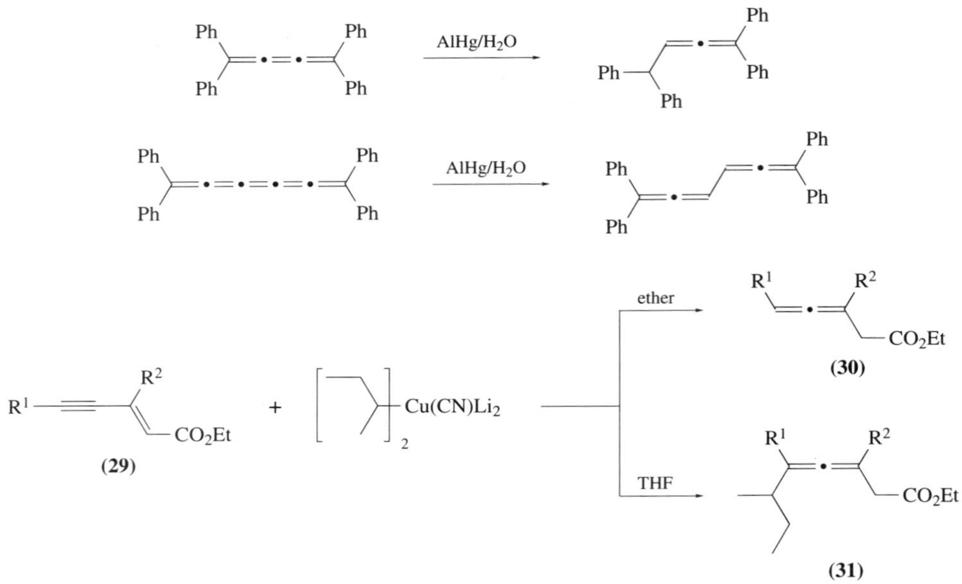

Scheme 11

hydroxy group in (**26**) is expected to facilitate the hydride transfer via an alkoxyhydridoaluminate —$OAlH_3$ intermediate ⟨74TL1593⟩. The reduction of tetraphenylbutatriene with aluminum amalgam in THF containing water gives 1,1,4,4-tetraphenyl-1,2-butadiene in 70% yield ⟨61CB3060⟩, but this strategy has been scarcely used for allene preparation (Scheme 11). A specific reduction of the enyne (**29**), with the double bond linked to an ester group, has been achieved by treatment with a cyanocuprate in ether. After protonation with pivalic acid, the allene (**30**) is obtained; the carbocupration product (**31**) is formed when the reaction is performed in THF (Scheme 11) ⟨91TL7229⟩. It has also been shown that a low temperature favours C—H rather than C—C bond formation in the reaction of propargylic acetates with organocuprates and $LiAlH_4$ ⟨76CC183⟩.

A reduction sequence, formally with R^- and H^+, is exemplified by the addition of an organolithium ⟨77HOU(5)1030⟩ or a Grignard reagent ⟨74BSF1119⟩ to enynes followed by hydrolysis (Scheme 12). A similar strategy was applied to the formation of aminomethyl allenes, by addition of amines to 1,3-enynes, or by reaction with lithium amides followed by protonation (Scheme 12) ⟨77HOU(5)1037⟩. Under similar conditions, lithium phosphide leads to dialkylphosphinomethyl allenes (Scheme 12) ⟨60ZOB1056⟩.

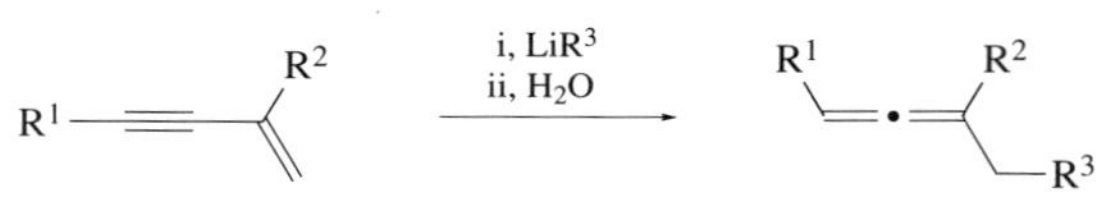

R¹	*R²*	*R³*	*Yield (%)*
Me	H	Et	72
Bu^t	H	Et	70
CH=CH$_2$	H	Et	58

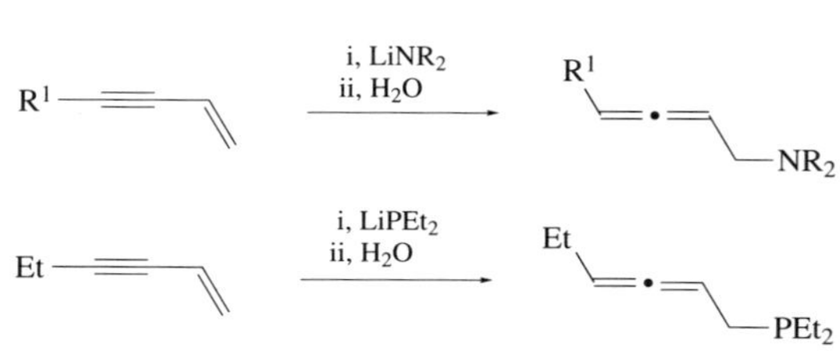

Scheme 12

1.20.2.2.3 Reduction of propargylic derivatives

α-Allenyl alcohols (**33**) are obtained in good yields from the mono-*O*-tetrahydropyranyl ethers of but-2-yn-1,4-diols (**32**) by reduction with $LiAlH_4$ under mild conditions (Scheme 13). This reaction is expected to produce first a ≡C—CH_2—O—AlH— intermediate (**34**) able to intramolecularly and stereoselectively substitute the leaving group O—THP by transfer of hydride ⟨73JCS(P1)720⟩. β-Allenic alcohols can be obtained under similar conditions by reduction of mono-*O*-tetrahydropyranyl ethers of pent-3-yn-1,5-diols with $LiAlH_4$ in THF at 65 °C ⟨74ACS(B)765⟩. α-Allenyl ethers and tertiary amines have been prepared by reduction of 1,4-dialkoxybut-2-ynes and 4-dialkylamino-1-methoxybut-2-ynes, respectively, with lithium aluminum hydride and a Lewis acid ($MgBr_2$ or $AlCl_3$) ⟨85S768⟩. An analogous transformation can be achieved from (**35**) which bears a quaternary ammonium group as the leaving group at the α′ position of the internal triple bond (Scheme 13) ⟨74S344⟩. It is, however, possible to transform propargylic alcohols with no leaving groups such as the steroid (**36**) into the corresponding allene (**37**), but the reaction requires the presence of $LiAlH_4/AlCl_3$ (3/1) in THF (Scheme 13) ⟨75CC362⟩. The regioselective reduction of aryloxymethylethynyl carbinols (**38**) into aryloxymethyl allenes (**39**) can be achieved using only $LiAlH_4$ in THF; in this case, the observed regioselectivity is thought to result from the coordination of both oxygen atoms to the aluminum centre (Scheme 13) ⟨86TL3777⟩. A versatile method to produce a variety of hydroxymethyl allenes of type (**41**) results from reduction of hydroxymethyl alkynes (**40**) with $LiAlH_4$ in THF followed by treatment with solid iodine at −78 °C. The reaction proceeds via the alanate (**42**); treatment of (**42**) with iodine in a THF solution affords allylic iodo alcohols (**43**) (Scheme 13) ⟨82TL3051⟩.

The reaction of propargylic ether or acetate (**44**) with ethylcuprate generated from an excess of EtMgBr with CuBr·Me_2S in THF gives a reaction mixture containing allenes (**45**), (**46**) resulting from reduction and alkylation, respectively, together with the alkynic isomers (Scheme 14) ⟨84JOC4120⟩.

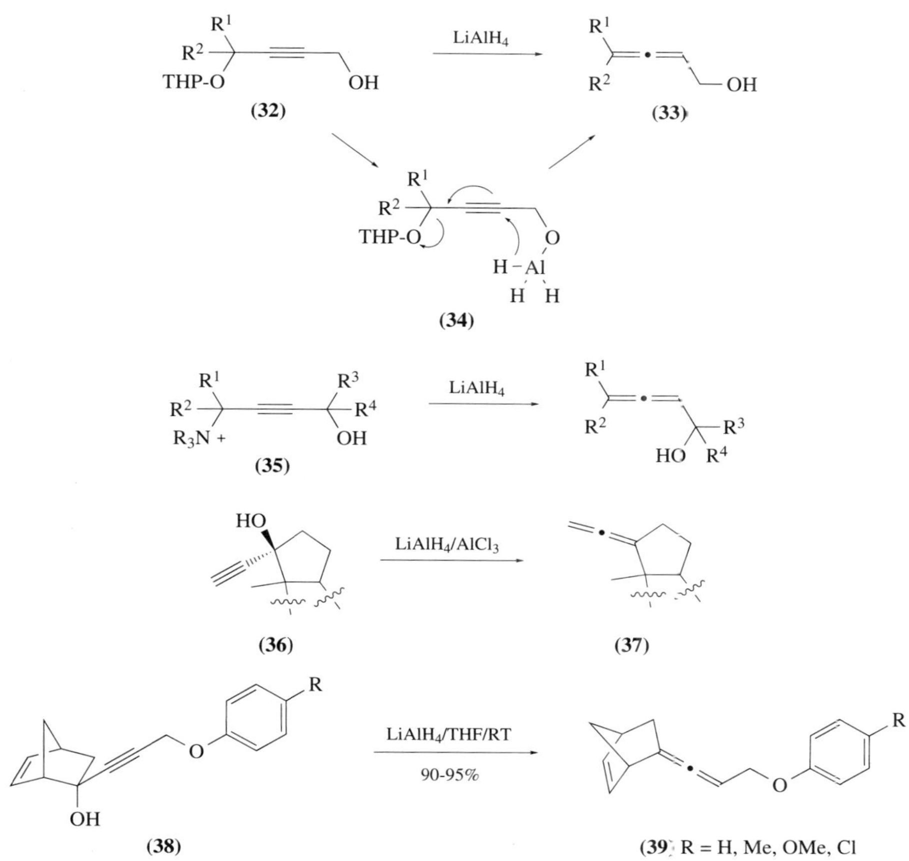

Scheme 13 *(continued)*

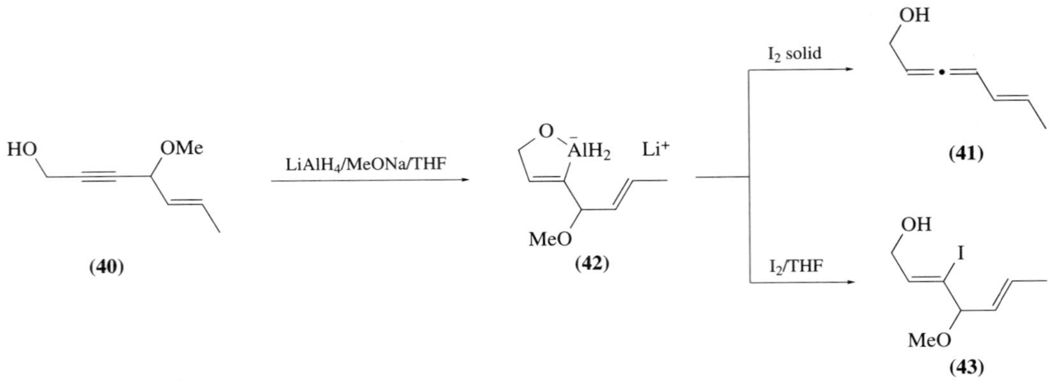

Scheme 13

Cumulenes have been obtained from 4,5-dien-2-yn-1-ols (**47**) (or the corresponding acetates), by reduction with $LiAlH_4$. The addition of hydride to the central allenyl carbon atom with elimination of the hydroxy (or acetate) group leads to the formation of cumulenes (**48**) (Scheme 14) ⟨75BSF2159, 75BSF2166⟩.

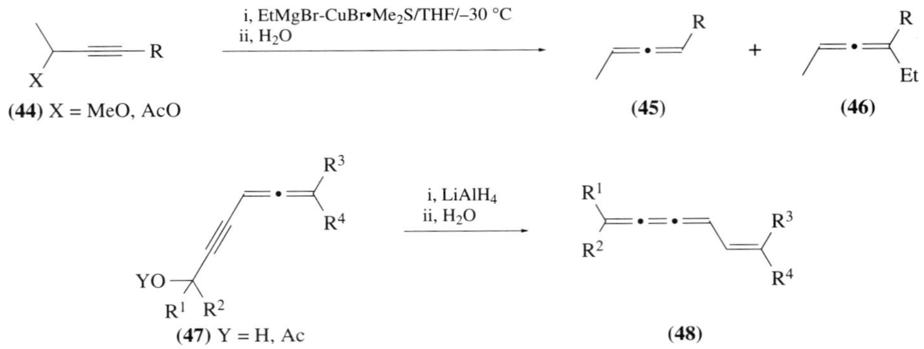

Scheme 14

1.20.2.2.4 Palladium catalyzed reduction of propargylic derivatives

Propargylic derivatives bearing a good leaving group (carboxylate, carbonate or halide) are known to be transformed by palladium(0) complexes to afford allenyl palladium intermediates. Their hydrogenolysis by an alcohol ⟨86TL5237⟩, or various sources of hydride, like metal hydride ⟨84TL845⟩ or ammonium formates ⟨87S603, 86CC922⟩ leads to the formation of allenes (Scheme 15). α-Hydroxy allenes have been obtained by a similar strategy, by reduction of alkynic epoxides with triethylammonium formate in the presence of $Pd_2(dba)_3 \cdot CHCl_3$ as the catalyst precursor, but the selectivity was not good, and homopropargylic alcohol isomers were also formed (Scheme 15) ⟨86CC922⟩. α-Hydroxy allenes are formed more selectively starting from alkynyl cyclic carbonates in the presence of $Pd(dba)_2$/1,2-bis(diphenylphosphino)ethane ⟨94SL457⟩. Palladium-catalyzed decarboxylation–hydrogenolysis of propargylic formates, which have a terminal triple bond, cleanly gives allenes at room temperature ⟨93TL2161⟩.

1.20.2.2.5 Elimination involving H migration

Rearrangement of propargylic diazene (**50**), involving the loss of nitrogen and the regioselective 1,5-migration of hydride, leads to 1,3-disubstituted allenes. The intermediates (**50**) can be formed from α-alkynic ketones, by reduction of the corresponding tosylhydrazone derivatives (**49**) with catecholborane, to afford alkyl and aryl allenes ⟨78CC726⟩, or with $NaBH_3CN$ to produce silylated allenes ⟨80JOC3925⟩. The same type of intermediate (**50**) is involved in the direct oxidation (by diethylazodicarboxylate (dead) at 0 °C) of hydrazines (**51**), obtained in two steps from α-alkynic alcohols; this process generates trienynes (**52**) with complete stereospecificity from optically active propargylic alcohols (Scheme 16) ⟨89TL5747⟩.

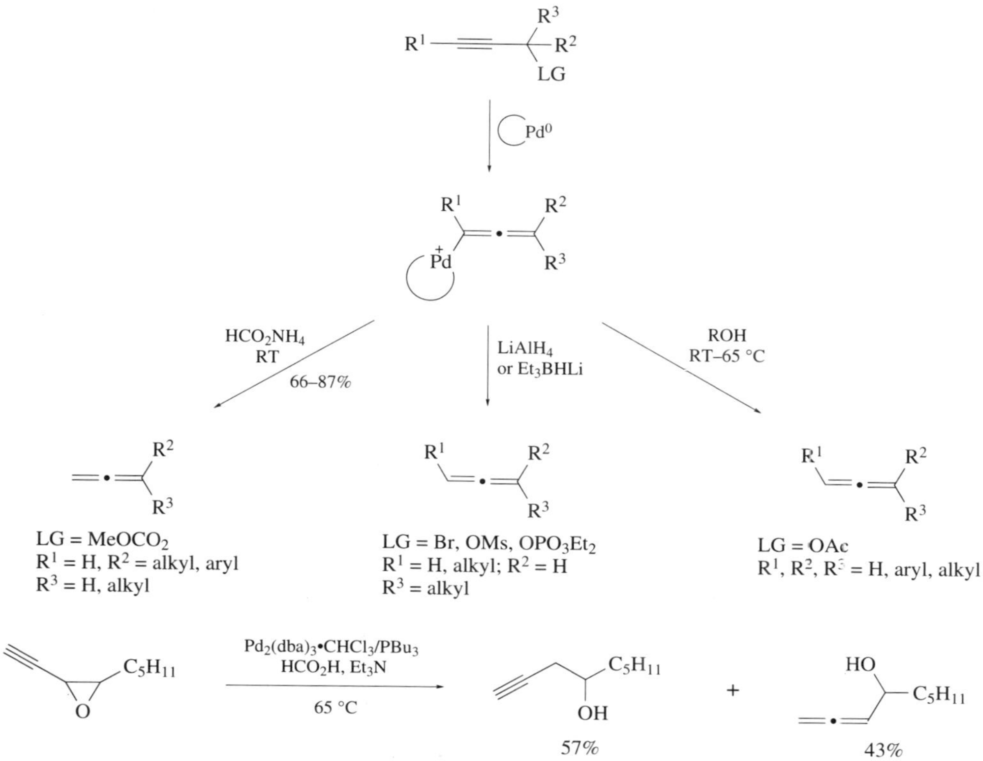

Scheme 15

Another example of elimination–rearrangement involving a sigmatropic 1,5-migration is illustrated by the formation of allenes with loss of SO_2 via sulfinic acids (**54**) generated by hydrolysis of propargylic sulfinamides (**53**) (Scheme 16) ⟨90TL213⟩.

1.20.2.2.6 Hydrolysis of alkynic silanes

Electrophilic desilylation of propargylic silanes TMS-$CH_2C{\equiv}CR$ (**55**), in the presence of carboxylic acids such as CF_3CO_2H ⟨80JOC5006⟩ or $BF_3 \cdot 2\,CH_3CO_2H$ ⟨87JOM(319)333⟩ provides a regioselective synthesis of monosubstituted allenes (**56**) under mild conditions (Scheme 17). Selective cleavage of the propargylic trimethylsilyl group from TMS-$C(Me)_2C{\equiv}C$—TMS (**57**) leads to the substituted allenyl silane (**58**) on treatment with CF_3CO_2H at 0 °C ⟨81CC1094, 81TL3401⟩ or $MeSO_3H$ at 25 °C ⟨72JOM(39)C44⟩. Alcohols (**61**), containing a terminal allenyl group, have been obtained under similar conditions from alcohols (**59**) or their trimethylsilyl derivatives (**60**) (Scheme 17). The protection of the alcohol functionality by a tetrahydropyranyl group gives rise to the formation of the two possible allenols ⟨87JOM(319)333⟩. 1-Trimethylsilylpropargyl alcohols are converted into trimethylsilyloxy allenes in 63–96% yields by reaction with BuLi, via migration of the trimethylsilyl group and protonation ⟨80TL623⟩.

The cleavage of a ≡C—Si bond leads to a terminal allene via proton transfer and protonolysis. This method has been used to access, respectively, 1,2,4,6-tetraenes (**64**) and (**65**) from a free (**63**) or coordinated (**62**) trimethylsilyldienyne by cleavage with fluoride. During this procedure, the propargylic proton transfer induces a *cis–trans* isomerization of the diene system (Scheme 18) ⟨92BSF151⟩.

1.20.2.2.7 Allenes via hydroboration

Hydroboration of 1-chloro-2-alkynes with diisoamyl or dicyclohexylborane in THF at 0 °C gives α-chloromethylvinyl boranes which undergo easy elimination on treatment with sodium hydroxide to produce terminal allenes in good yields (Scheme 19) ⟨70JA1427⟩.

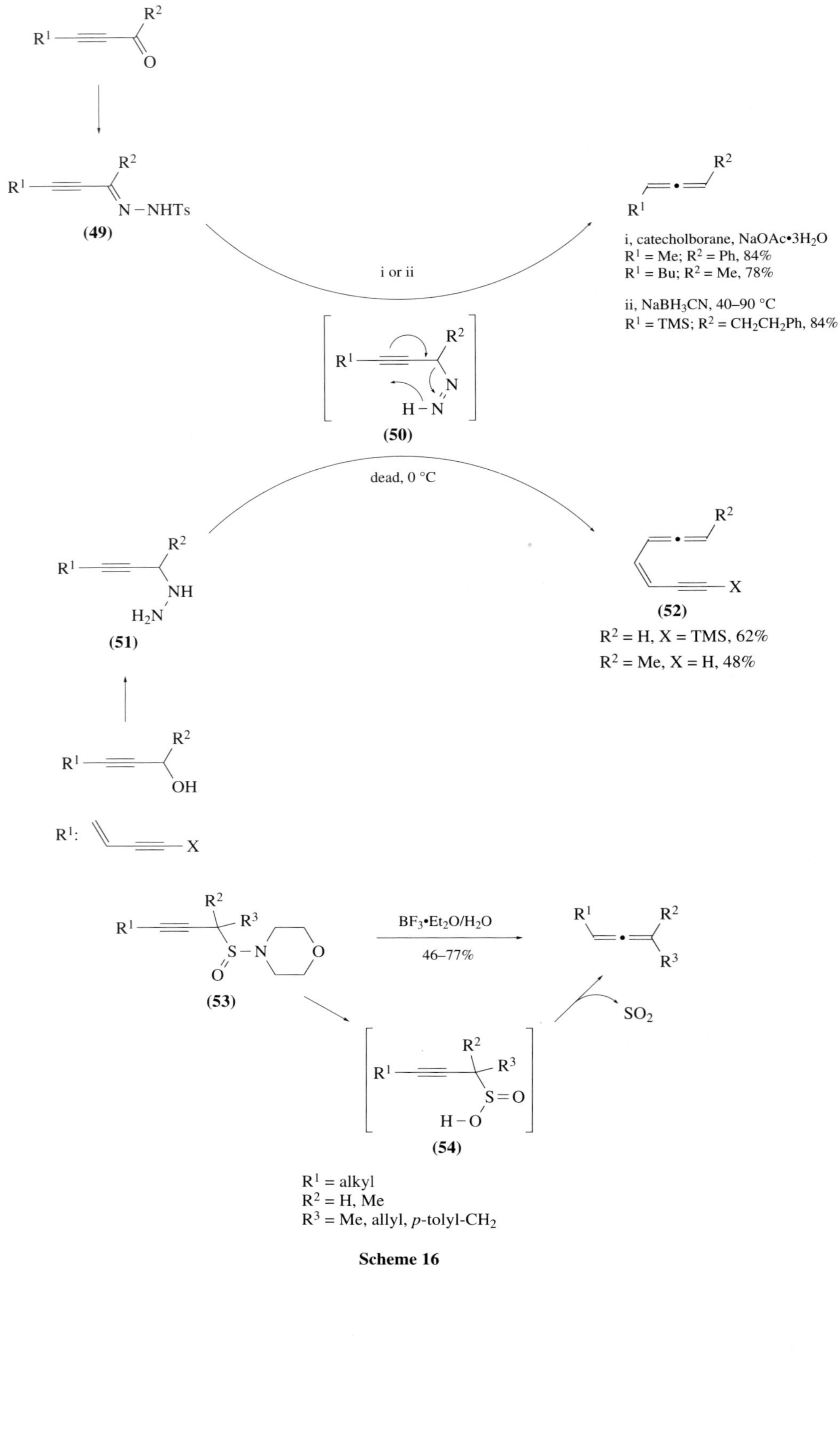
R1
R2
O
N−NHTs
(49)
i or ii
i, catecholborane, NaOAc•3H2O
R1 = Me; R2 = Ph, 84%
R1 = Bu; R2 = Me, 78%
ii, NaBH3CN, 40–90 °C
R1 = TMS; R2 = CH2CH2Ph, 84%
N
H−N
(50)
dead, 0 °C
NH
H2N
(51)
X
(52)
R2 = H, X = TMS, 62%
R2 = Me, X = H, 48%
OH
R1:
R3
S−N
O
(53)
BF3•Et2O/H2O
46–77%
SO2
S=O
H−O
(54)
R1 = alkyl
R2 = H, Me
R3 = Me, allyl, p-tolyl-CH2

Scheme 16

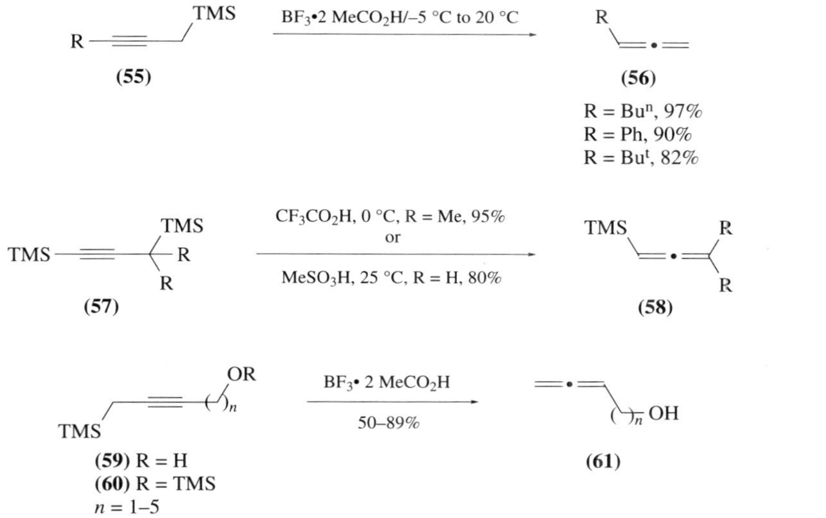

Scheme 17

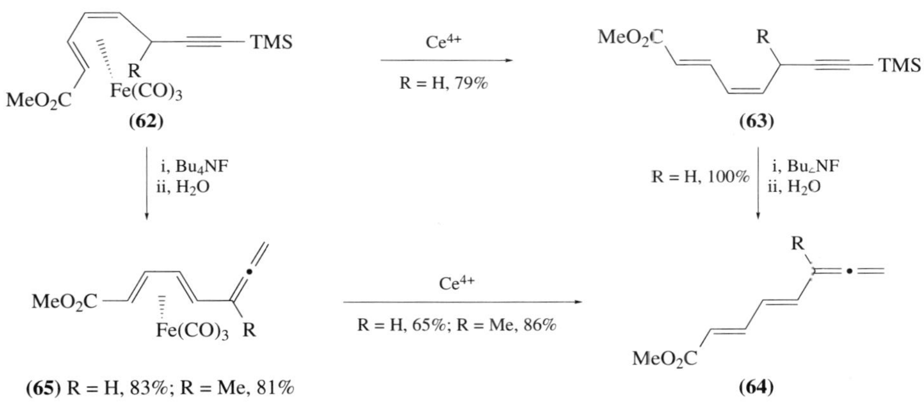

Scheme 18

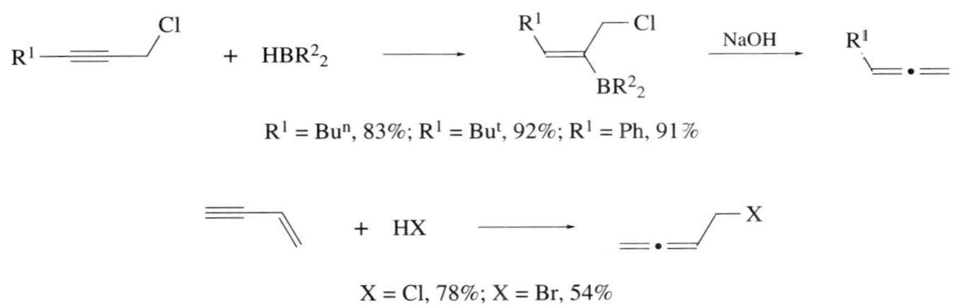

Scheme 19

1.20.2.2.8 Hydrohalogenation of enyne derivatives

1,4-Addition of HX (X = Cl, Br) to vinylalkyne or enynols leads to halogenated allenes; yields are improved in the presence of Lewis acids, e.g. $AlCl_3$ or $ZnCl_2$ (Scheme 19) ⟨32JA4066, 63IZV499, 66JCS(C)1223⟩.

1.20.3 BY C—C BOND FORMATION

1.20.3.1 Nucleophilic Substitution with Organocopper Compounds from Propynyl Derivatives

Since the first use of organocopper reagents to synthesize allenes from propargylic derivatives via selective S_N2' displacement of a leaving group ⟨68JA4733⟩, many examples of this strategy have been reported involving a variety of labile groups and copper derivatives.

Lithium cuprates allow the selective formation of allenes from propargylic acetates containing a spirocycloalkyl substituent at the propargylic carbon ⟨75TL4615⟩. Pasto *et al.* ⟨78JOC1389⟩ have shown the preferred formation of alkyl allenes from mixed methyl alkyl cuprates and alkyl allenyl cuprates, and pointed out the competition between reduction and substitution in the case of terminal chloropropargylic derivatives (Scheme 20). Chiral allenes, precursors of natural pheromones, have been obtained with lithium dialkylcuprates starting from optically active propargylic carbamates **(66)** (Scheme 21) ⟨78JOC1950, 78JOC2091⟩. α-Allenyl alcohols, precursors of dihydrofurans, have been prepared by treatment of epoxy propargylic alcohols with methylcuprate ⟨93JOC7180⟩. The reaction of the epoxy alcohol **(67)** with organocuprates leads to a mixture of *anti* and *syn* diols **(68)** and **(69)** at 35°C, whereas the presence of dimethyl sulfide at −60°C selectively directs the nucleophilic substitution to the formation of the *anti* isomer **(68)** ⟨83TL5587⟩.

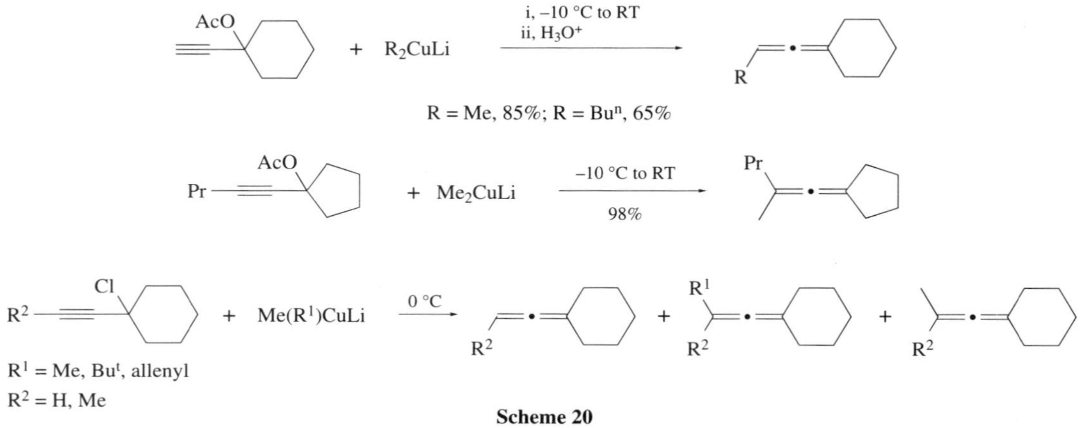

Scheme 20

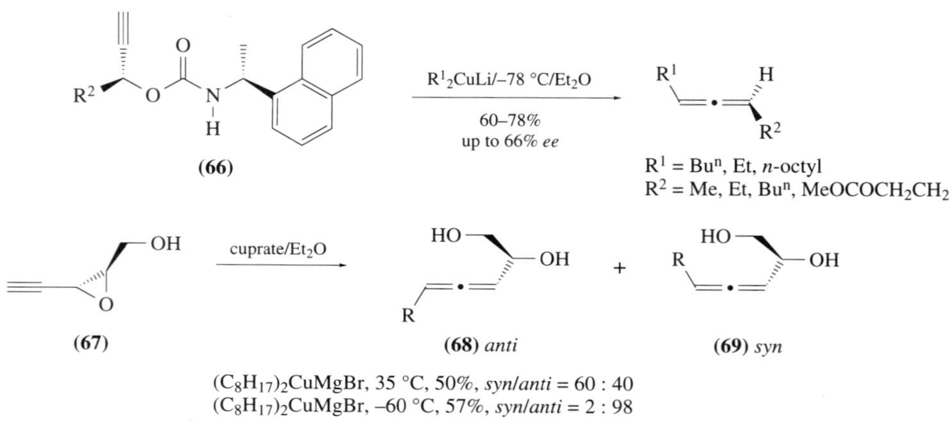

Scheme 21

However, diorganocuprates are known to racemize optically active allenes, and other organocopper derivatives have been used for the synthesis of optically active allenes from chiral propargylic derivatives. Chiral 1-phenyl-3-alkyl and 1,3-dialkyl allenes have thus been obtained from propargylic sulfinates or sulfonates and (RCuBr)MgBr·LiBr at −65°C ⟨89JOC3726⟩. Dienylallene tricarbonyl iron complexes **(71)** have been obtained from dienylpropargylic carbonate **(70)** or mesylate complexes and (RCuBr)MgBr·LiBr (R = Et, Bu^t) with very high diastereoselectivity ⟨92AG(E)224, 93T9775⟩. These organoheterocuprates are also suitable reagents for the synthesis of silylated allenes

(**72**) from trimethylsilyl propargylic sulfinates, mesylates or tosylates ⟨79S390⟩. 5-Trimethylsilyl-2-methylpenta-2,3-dien-1-ol (**74**) has been produced by opening the epoxide (**73**) with a trimethylsilylmethyl copper(I) species formed from trimethylsilylmethyl magnesium chloride and $LiCuBr_2$ ⟨85JOC5143⟩. Pure butatrienes (**76**) have been obtained from 3-bromo-alk-3-en-1-ynes (**75**), as a result of *anti*-1,3-substitution, on treatment with a copper(I) species in THF at −50 °C (Scheme 22) ⟨82JOM(240)329⟩.

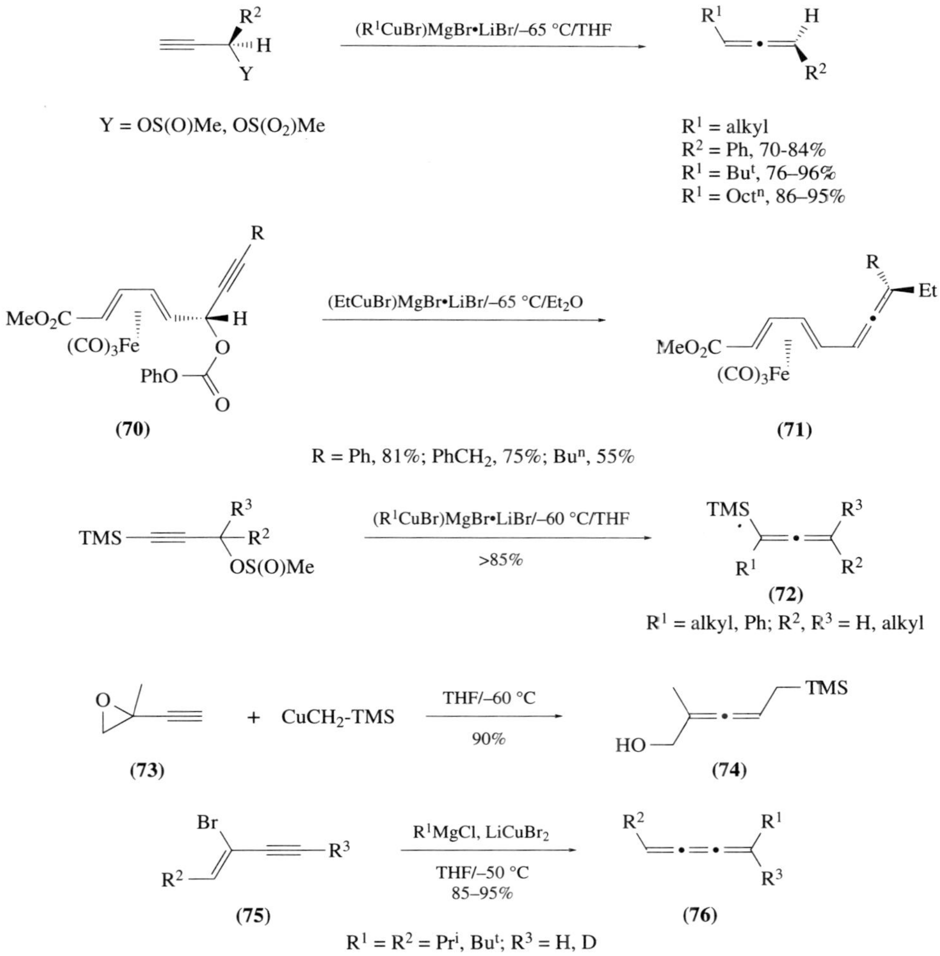

Scheme 22

Other organocopper(I) reagents have been prepared by using Grignard reagents in the presence of catalytic amounts of Cu(I) halides. This procedure usually avoids the parallel formation of alkynic derivatives resulting for simple nucleophilic substitution. Propargylic derivatives bearing an alkoxy group as the leaving group have been used to gain access to polyalkylpropa-1,2-dienes ⟨76JOM(108)159⟩, 1,2,4,6-tetraenes (**77**) ⟨79TL7⟩ and α-allenyl amines ⟨87TL2207⟩. Chiral allenes are obtained with high stereoselectivity from optically active propargylic ethers ⟨90JA8042⟩ and cyclic acetals ⟨85TL4197⟩ by using RMgBr, CuBr (5%), and $P(OEt)_3$ to inhibit racemization. Cyclic allenes (**79**) can be obtained from cyclic propargylic esters (**78**) ⟨86TL4845⟩, and the use of CuX catalyst also allows the diastereoselective synthesis of α-allenols from epoxides ⟨91T1677⟩ or cyclic carbonates (or sulfates) in the presence of a Lewis acid (Scheme 23) ⟨92TA1509⟩.

More simple organocopper(I) derivatives promote 1,3-substitution reactions with propargylic derivatives. Fluorinated allenes (**80**) and bisallenes (**81**) have been prepared from propargylic halides or tosylates and perfluoroalkyl copper(I) compounds (Scheme 24) ⟨90TL3699, 90TL3703⟩. The synthesis of α-allenic alcohols (**82**), of high optical purity, has been carried out starting from the appropriate chiral propargylic bromide or tosylate and functional Cu(I) derivatives ⟨91JOC1083⟩. Another route,

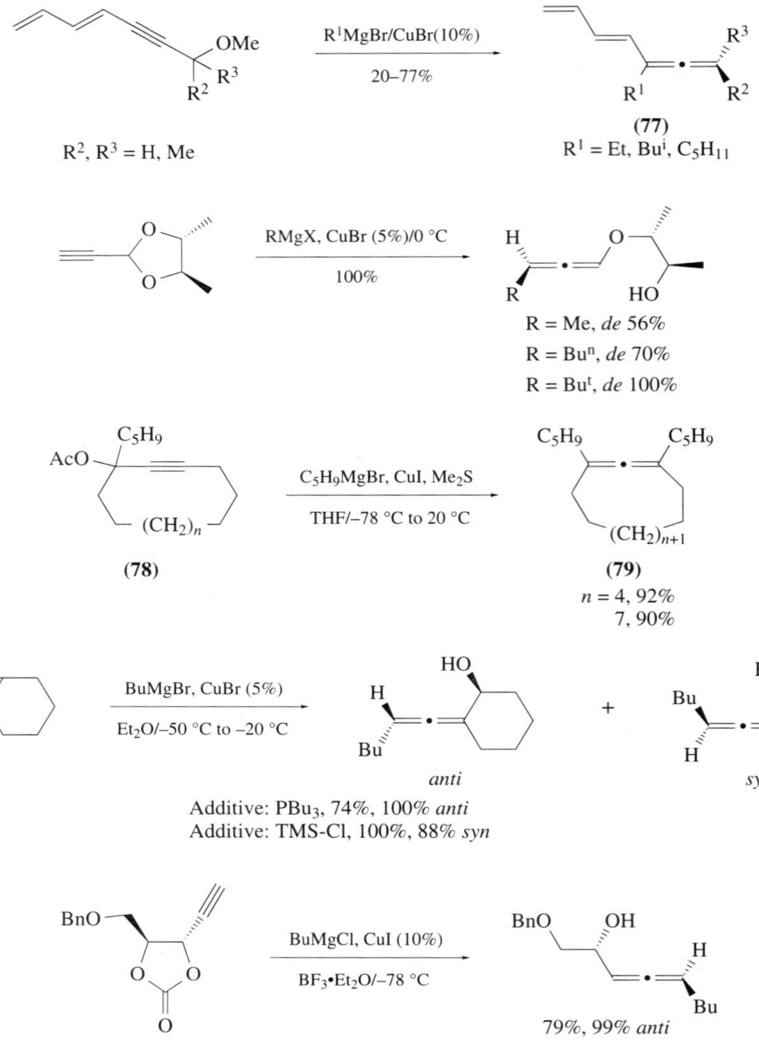

Scheme 23

based on the use of propargyl chloride and alkyl copper reagents arising from optically active amino acids, has allowed the formation of enantiomerically pure protected amino acids containing an allenyl group ⟨93SL219⟩. Similar syntheses of allenic amino acids (**83**) are based on the use of zinc–copper reagents (Scheme 24) ⟨92CC319, 93SL499⟩. Stannyl alkynes (**84**) react with alkyl copper(I) species at $-60\,^{\circ}C$ to give stannyl allenes (**85**) in high yields, but quantitative transmetallation of (**84**) takes place with MeCu, $CH_2{=}CH{-}Cu$, Ph—Cu and TMS—C≡C—Cu (Scheme 25) ⟨84TL3019⟩. The copper(I) enolate generated from the lithium enolate of acetylacetate and CuI at $-78\,^{\circ}C$ is able to substitute propargylic esters to produce β-allenyl esters ⟨78JOC555⟩.

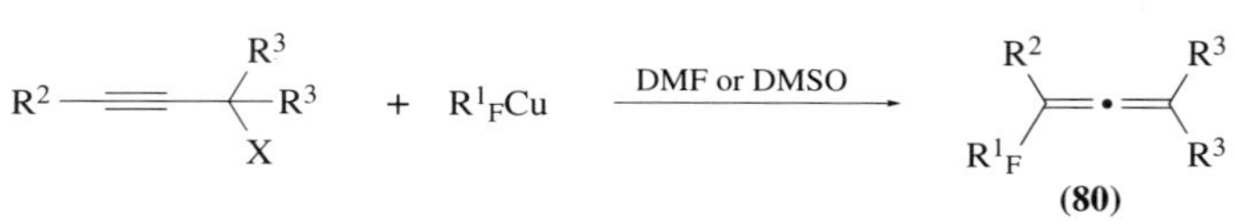

X	R^1_F	R^2	R^3	*Yield* (%)	*Ref.*
OTs	CF_3, C_3F_7, C_6F_{13}	H	H	68, 41, 49	90TL3699
Cl	CF_3, C_3F_7, C_6F_{13}, C_8F_{17}	H	Me	73, 68, 66, 60	
Cl	CF_3	EtO_2C	Me	57	90TL3699
Cl	CF_3	TMS	Me	52	90TL3699
Br	C_6F_{13}, C_8F_{17}	H	H	70–75	90TL3703

Scheme 24

(continued)

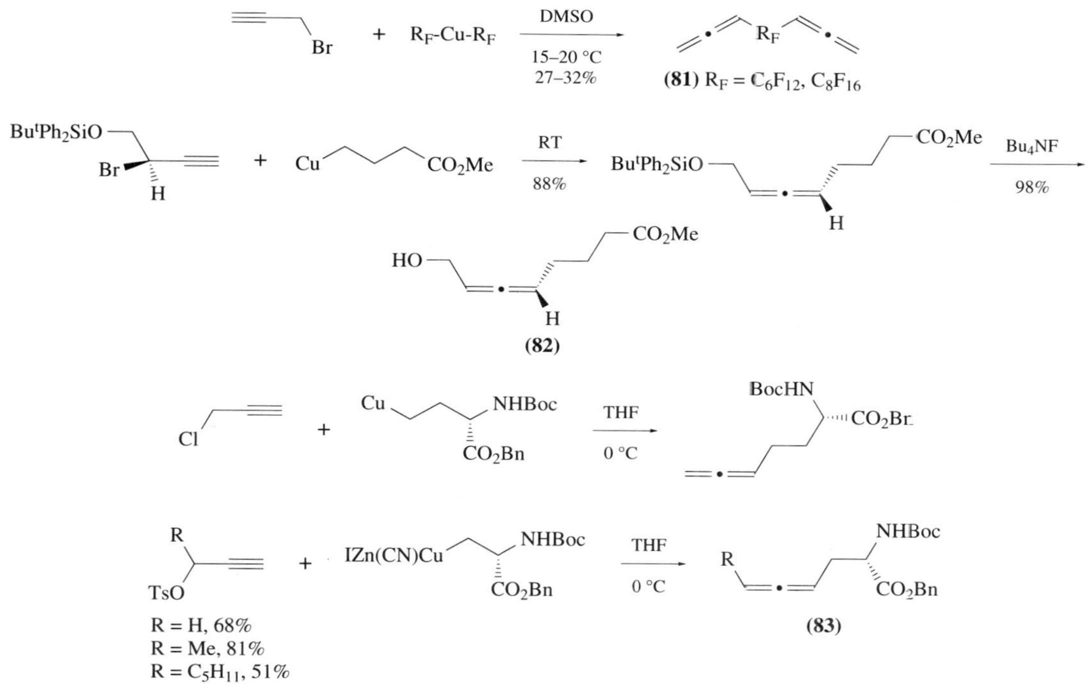

Scheme 24

The direct 1,3-substitution of the hydroxy group of propargylic alcohols has been performed successfully by an organocuprate formed *in situ* from CuI, RLi, and (methyl phenylamino) tributylphosphonium iodide (Scheme 25) ⟨80JOC4536⟩.

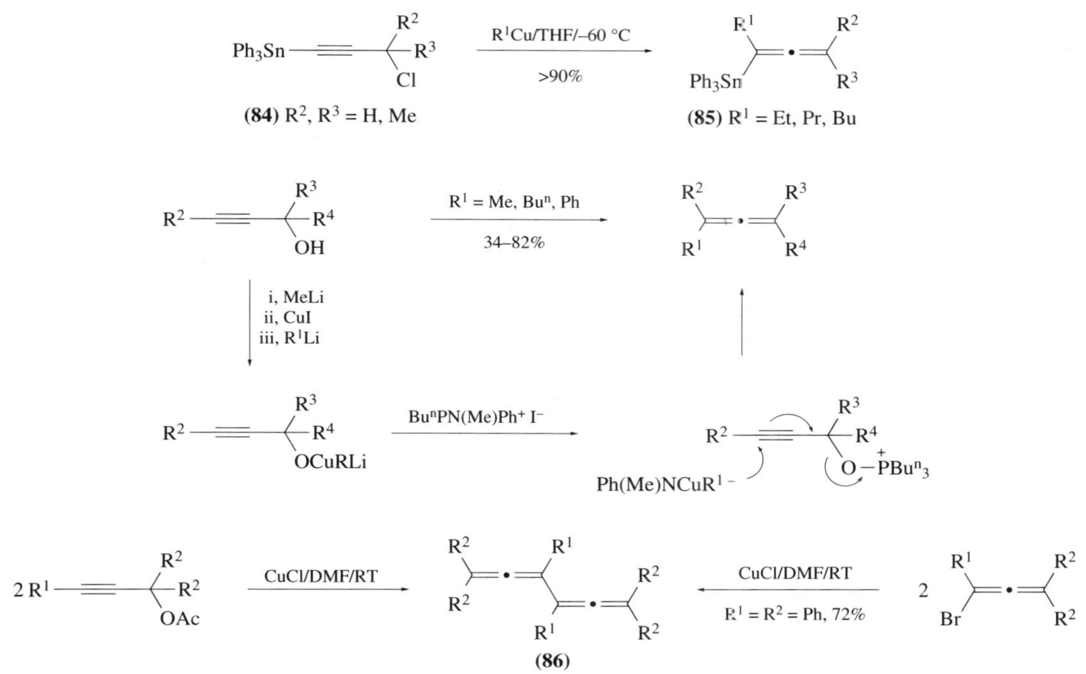

Scheme 25

Coupling either two propargylic acetates or two allenyl bromides in the presence of CuCl in DMF at room temperature, furnishes symmetrical diallenes (**86**) via allenyl radical intermediates (Scheme 25) ⟨75CC174⟩.

The 1,3-alkadien-2-yl diethyl phosphate (**87**) reacts with an organocuprate reagent formed *in situ* from RMgX and CuX to afford the allene (**88**) in moderate yield; suitable conditions have to be found to avoid nucleophilic substitution giving the 1,3-diene (**89**) (Scheme 26) ⟨83TL1297⟩.

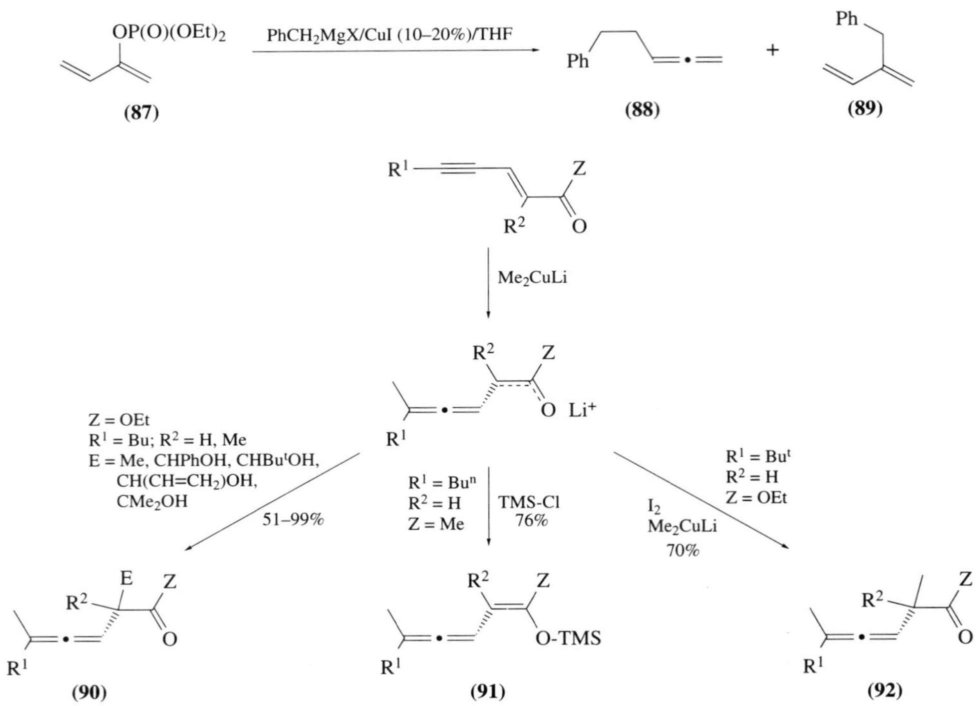

Scheme 26

The generation of allenyl enolates by 1,6-addition of lithium dimethylcuprate to conjugated functional enynes makes possible access to allenyl esters (**90**) by C—C bond formation, and to allenyl enol ethers (**91**) by C—O bond formation, on addition of electrophiles ⟨93CB251, 93LA521⟩. When the starting alkyne is substituted by a bulky group like Bu^t, these allenyl enolates can be trapped to selectively afford allene derivatives (**92**) (Scheme 26) ⟨93CB261⟩.

1.20.3.2 Nucleophilic Substitution with Organomagnesium Reagents

The reaction of propargyl derivatives with Grignard reagents is well documented, but the competition between alkyne and allene formation has brought some restriction to its utilization. However, the selectivity of the 1,3-substitution has been improved by the use of specific leaving groups or catalytic systems involving transition metals. Thus, propargylic alcohols, protected *in situ* by 1-chloro-2-methyl-*N*,*N*-tetramethylenepropenyl amine, lead to the selective formation of allenes at 0 °C with alkyl, phenyl, vinyl, and allyl Grignard reagents in very good yields (Scheme 27) ⟨84TL4007⟩.

Palladium(0) is an efficient catalyst for the cross-coupling of propargylic and allenic halides with Grignard reagents to selectively produce alkyl and aryl allenes ⟨80TL5019⟩. The nickel-catalyzed reaction of propargylic alcohols (**93**) with organomagnesium bromides selectively gives silyl allenes (**94**) (Scheme 27) ⟨85JOC1122⟩. Pasto *et al.* ⟨76JOC3496, 78JOC1382, 78JOC1385⟩ have shown that iron (especially $FeCl_3$) and cobalt derivatives were good catalysts for the production of allenes from propargylic halides and Grignard reagents.

A similar nucleophilic substitution involving triethylaluminum and tris(trimethylsilylethynyl)methyl acetate (**95**) gave compound (**96**) in low yield (20%) (Scheme 27) ⟨88CC748⟩.

It must be noted that vinylic allenes have been obtained via selective 1,5-substitution of chloride from 5-chloropent-3-en-1-ynes on treatment with MeMgI ⟨72CC866⟩.

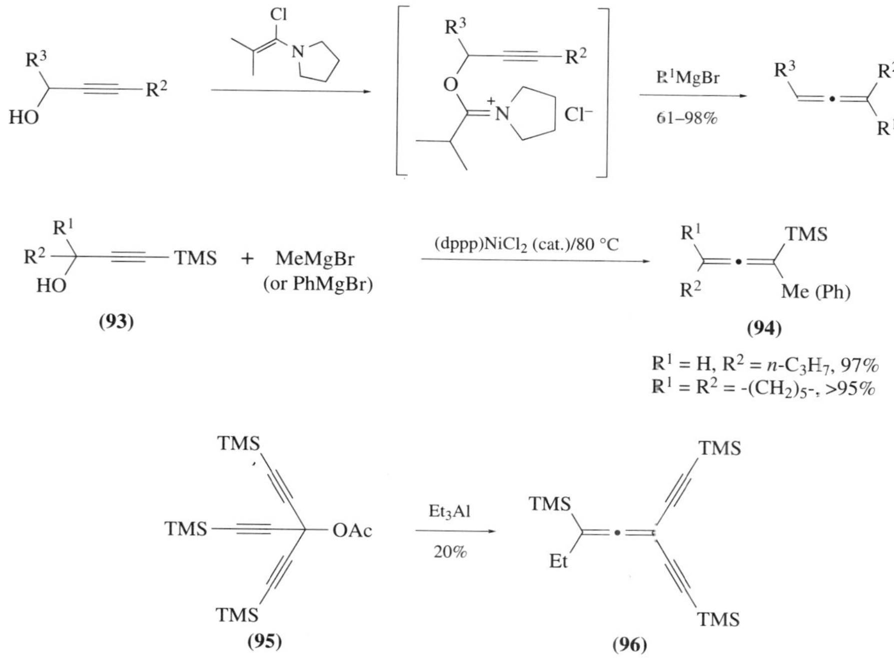

Scheme 27

1.20.3.3 Reaction of Electrophiles with Propargyl and Allenyl Organometallics

Propargylic and allenic organometallics are usually prepared by reaction of a metal or an alkyl lithium with either the corresponding halide or the hydrocarbon. Metallation of allenic hydrocarbons or halogen–lithium exchange from haloallenes with butyllithium at $-78\,^{\circ}C$ generates carbanions capable of reacting with electrophiles to produce allenyl derivatives via C—C bond formation ⟨91COS(2)81⟩. Various electrophiles have been used leading to the synthesis of various functional allenes.

The selective formation of mono-, di-, tri- and tetraalkyl allenes (84–94%) is possible with alkyl halides ⟨75CC561, 82SC739⟩. The cross-coupling between allenyl lithium compounds and aryl or vinyl halides, catalyzed by $Pd(PPh_3)_4$ makes possible the preparation of vinyl and aryl allenes ⟨82S738⟩.

The reaction of allenyl lithium with ketones and epoxides, respectively, leads to α-allenic and β-allenic alcohols ⟨81S875, 83CC1133⟩, the selective formation of the latter being improved in the presence of hexamethylphosphorotriamide. When CO_2 is used as the electrophile, allenic carboxylic acids are formed in good yields ⟨77JA7632, 81S875, 85JA6046⟩. Whereas the reaction with amides gives allenic ketones ⟨77NJC373⟩, the cross-coupling with ureas provides an efficient route to allenic amides in satisfactory yields (Scheme 28) ⟨81S875⟩. The ambident alkynic–allenic nature of 3,3-dimethylallenyl lithium has been discussed by Creary on the basis of hard and soft acidity ⟨77JA7632⟩.

Functional allenic compounds have been prepared from γ-lithio-γ-alkoxyallenes ⟨68RTC1179⟩, silyl allenes ⟨80TL3987⟩, γ-methoxy-γ-alkyl allenes ⟨78TL1137⟩, and the synthesis of the first primary helicoidal molecule (**97**) was based on the reaction of γ-methoxy-γ-lithio allene with tetrahydrofuranone moities (Scheme 29) ⟨80JA2134⟩. The use of enantiopure alkoxy allenes has allowed the preparation of enantiomerically enriched allenols by reaction with aldehydes ⟨93SL105⟩. Silyloxypropargyl lithium reagents (**98**), generated either by alkylation of silylated ketones or alkynylation of silyl ketones, react with electrophiles to produce allenol silyl ethers (**99**) (Scheme 29) ⟨80JA1423, 86JA7791⟩.

Allenyl Grignard reagents have not been used extensively because of their lack of selectivity. However, γ-allenyl ketones have been produced from allenyl magnesium bromide and esters ⟨69BSF898, 70BSF3157⟩.

The organoaluminum derivative (**100**), generated by reaction of aluminum with trimethylsilylpropargyl bromide selectively affords allenyl alcohols (**101**) and (**102**), on reaction with carbonyl compounds, whereas magnesium or zinc propargyl derivatives selectively give homopropargylic alcohols ⟨81TL1579, 91JOM(403)299⟩. Reaction of (**100**) with triethyl *ortho*-formate affords the aldehyde

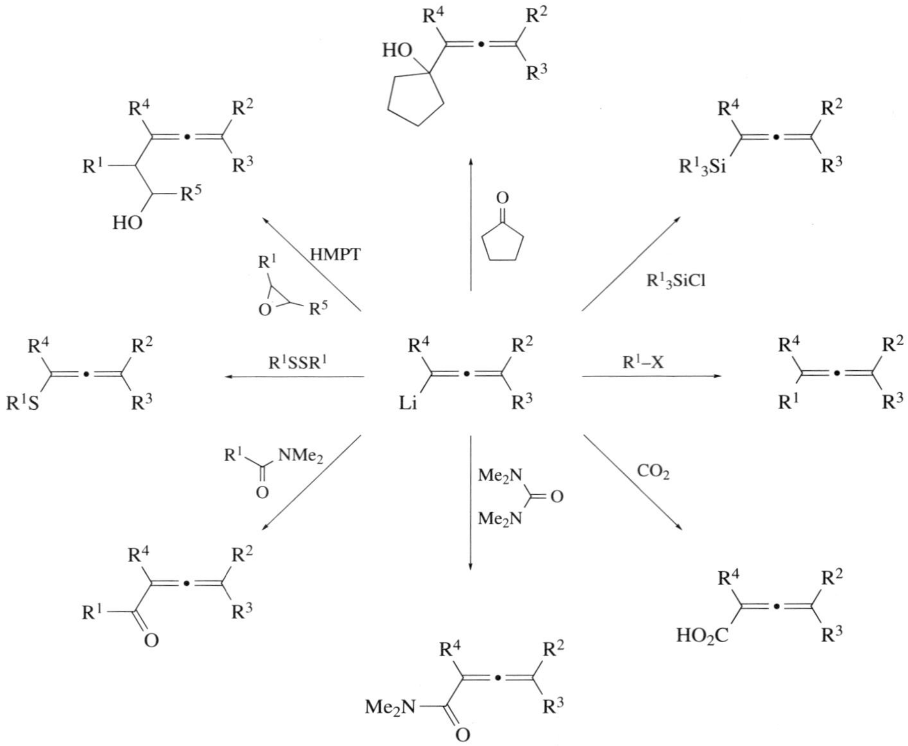

Scheme 28

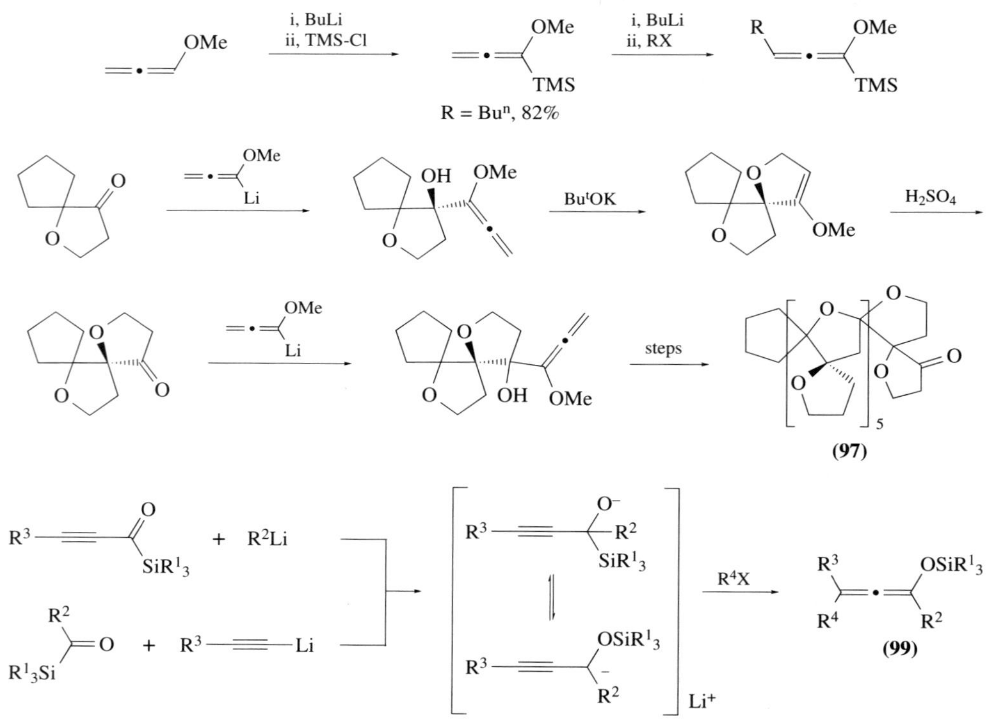

Scheme 29

(103), a useful intermediate for the access to functionalized allenic alcohols on reaction with organomagnesium, zinc, or aluminum reagents (Scheme 30) ⟨92JOM(440)277⟩. Propargylic lithium alanates generated from methylacetylenes, selectively afford allenes, α-allenic alcohols and acids by reaction with allyl halide, carbonyl compounds and carbon dioxide, respectively. Similarly, methoxyallenyl aluminum reagents react with aldehydes to produce methoxyhydroxy allenes **(104)** which cyclize into furans under acidic conditions with loss of methanol ⟨82CL1029⟩. With organoaluminum derivatives arising from substituted propargyl bromides, α-allenic tertiary amines have been produced with high selectivity (>90%) when the electrophile is an iminium salt ⟨81JOM(218)1⟩ or a *gem*-aminoether (Scheme 30) ⟨80JOM(198)1⟩.

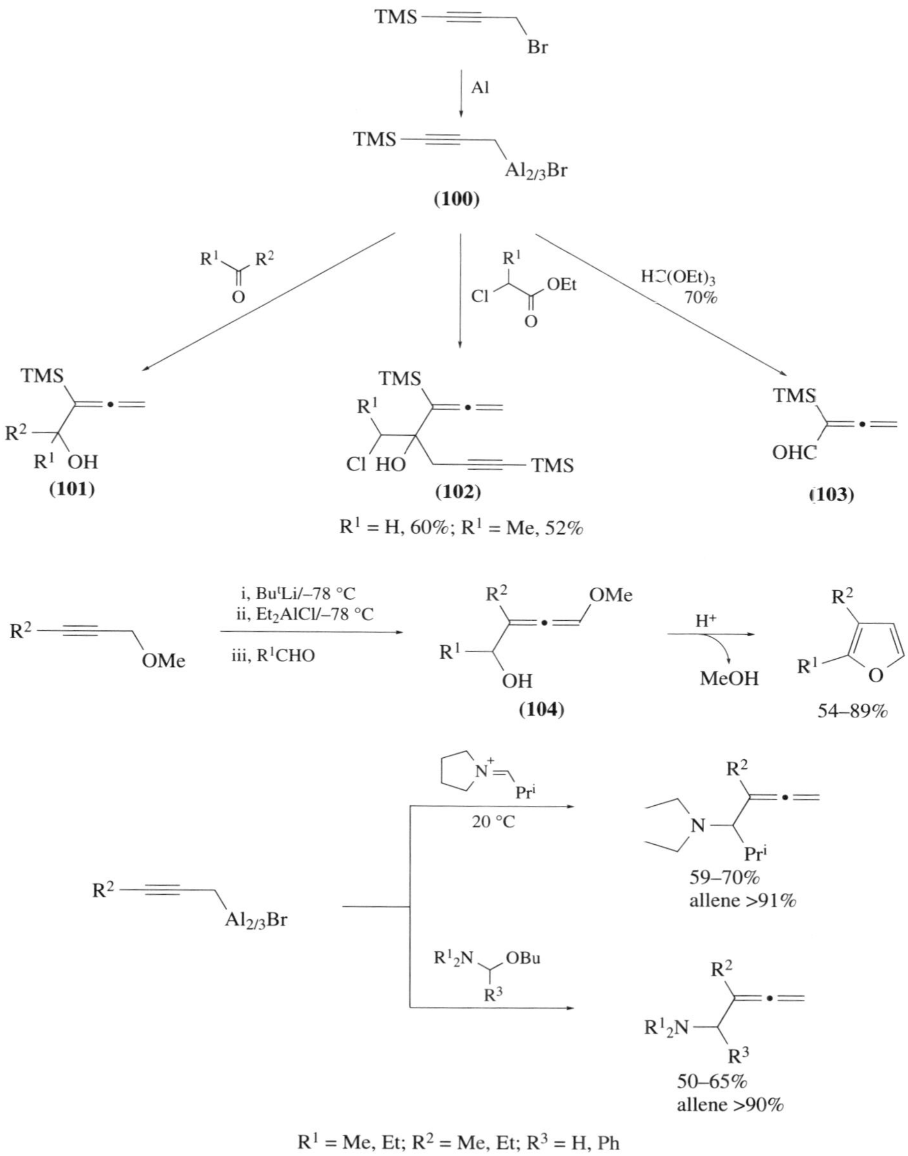

Scheme 30

Stable organotin compounds react with activated aldehydes to produce γ-allenyl alcohols in good yields ⟨73JOM(54)153⟩. The *in situ* formation of organotin derivatives by stirring stannous chloride with propargylic halides leads to allenic alcohols on reaction with aldehydes at 0°C ⟨81CL621, 93TL449⟩. The best selectivity in allenols is observed when γ-substituted propargyl iodides are used (Scheme 31).

Transmetallation of 1-trimethylsilyl (or 1-methyl) propynyl lithium with titanium isopropoxide

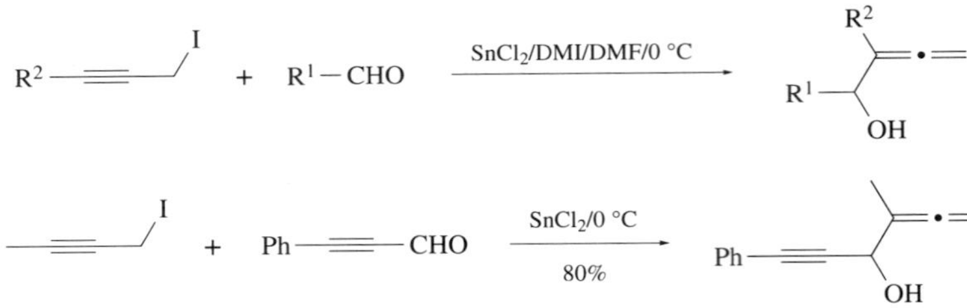

Scheme 31

at $-78\,^{\circ}C$ gives access to a titanium reagent which reacts with carbonyl compounds at $-78\,^{\circ}C$ to specifically afford allenic alcohols, whereas similar reactions from 1-trimethylsilyl-3-substituted propynyl lithium lead to alkynic alcohols (Scheme 32) ⟨82JOC2225⟩. Highly regio- and diastereoselective syntheses of 4-hydroxy-1,2-alkadienyl carbamates (**105**) have been obtained from propargylic carbamates by using the corresponding titanium derivative, whereas no selectivity was observed with the lithium reagent ⟨87T2457⟩.

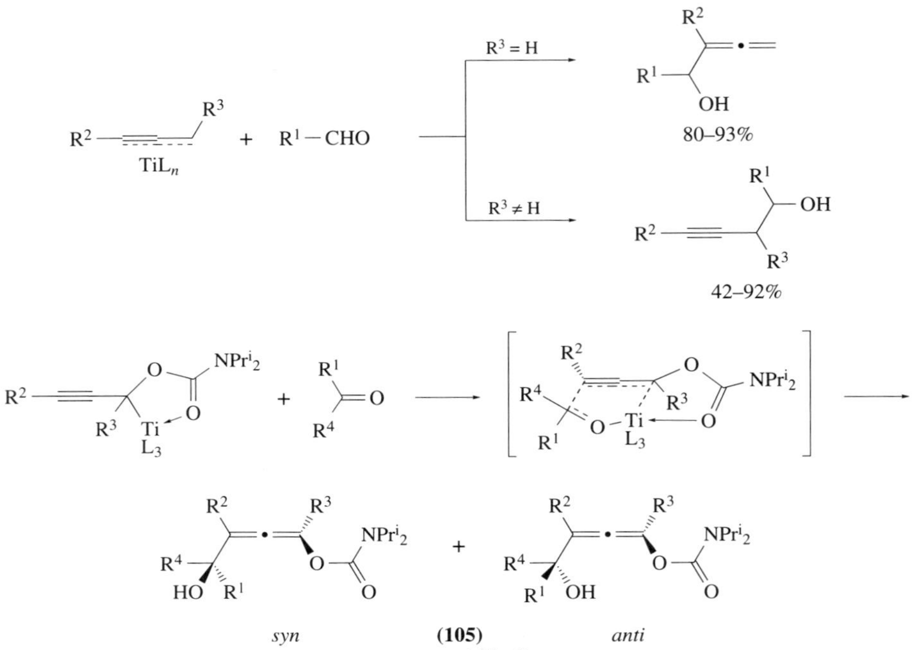

(105)
syn : *anti* 95 : 5

Scheme 32

By reaction of aldehydes or ketones in the presence of hexamethylphosphoramide (HMPA), a very selective formation of allenols has been observed when chromium(II) derivatives were used to generate the allenyl organometallic intermediate from propargylic bromides ⟨81T1359⟩. This system allows the use of propargylic halides containing other functionalities such as esters, halides and nitriles ⟨92JOC4070⟩.

The organocupration of methyl propynoate by an allenyl cuprate selectively affords *trans*-2,4,5-trienyl methyl esters in good yields (Scheme 33) ⟨76TL275⟩.

Allenic ketones have been synthesized by reaction of propargyl mercury iodide with acyl chlorides in the presence of $AlCl_3$ at $-40\,^{\circ}C$ (Scheme 33) ⟨86JOC2623⟩.

Organozinc compounds prepared from propargylic halides usually give no selectivity on reaction with carbonyl electrophiles, but after treatment with zinc, γ-substituted α-alkynic bromides react with *N*-chloromethyl-*N*-methylformamide to afford allenic methyl formamides, which are easy to convert into secondary allenyl amines with BuLi ⟨86BSF449⟩. On hydrolysis with D_2O, allenic zinc

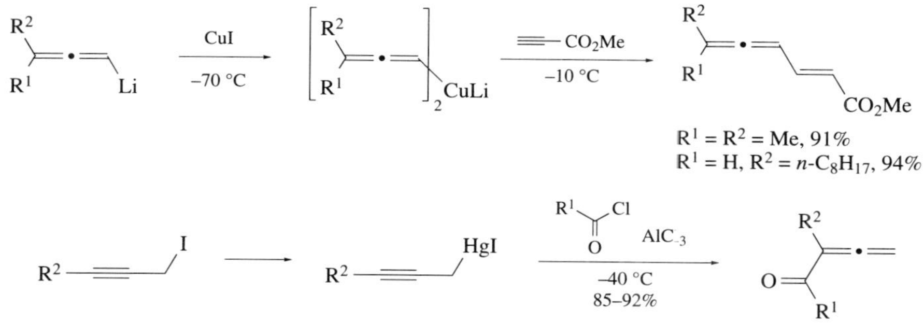

Scheme 33

reagents, generated from propargylic mesylates or chlorides and triorganozincates, give allenes in high yields via C—C and C—D bond formations ⟨93JOC6166⟩.

Nucleophilic organosamarium derivatives, formed by electron transfer from SmI_2 to allenyl palladium species, react with carbonyl compounds to afford homopropargylic and allenic alcohols, the proportions of which depend on the substitution pattern of the starting prop-2-yn-1-yl acetate ⟨87CL2275⟩.

1.20.3.4 Catalytic Formation of Allenes via Allenyl–Metal Intermediates

The reaction with palladium(0) catalysts probably proceeds via oxidative addition of a propargylic carboxylate, halide, or carbonate to produce an allenyl organopalladium(II) moiety. Subsequent reaction with electrophiles gives 1,2-diene derivatives.

Starting from propargylic halides or esters and hard nucleophiles such as R—ZnCl, in the presence of catalytic amounts of $Pd(PPh_3)_4$, functionalized allenes and bisallenes have been obtained with high regio- ⟨81TL1451⟩ and stereoselectivity ⟨83JOC1103⟩ except in the case of 1-alkynylcyclopropyl derivatives which led to alkynic cyclopropanes as major compounds ⟨92JA4051⟩. Conjugated *para*-allenyl styrenes and vinyl allenes have thus been obtained from various nucleophiles including zinc, aluminum and tin organometallics ⟨86JOC4006⟩. Allenynes have been synthesized in one step from propargylic halides or acetates and terminal alkynes in the presence of tetrakis(triphenylphosphine)palladium, copper iodide and a base ⟨93TL3853⟩. The cross-coupling is also possible from allenyl halides and organozinc ⟨84TL5571⟩, magnesium ⟨80TL5019⟩ or copper ⟨83S32⟩ species as nucleophiles, catalyzed by palladium(0) complexes (Scheme 34). Goré has recently shown the direct access to functionalized vinyl allenes (**106**) by coupling of propargylic acetates and pentynoates based on the efficient activation of both substrates by palladium complexes ⟨93TL3129⟩ (Scheme 34). Since they are easily transformed into allenyl—Pd(II) intermediates, CO_2 and alkoxide, propargylic carbonates are especially good substrates for various transformations into allenic derivatives via palladium(0) catalysis. In alcohol, under CO pressure, propargylic carbonates are easily converted into allenic esters (**107**) ⟨86TL731, 93JA5865, 93JOM(451)15⟩. Mono- and

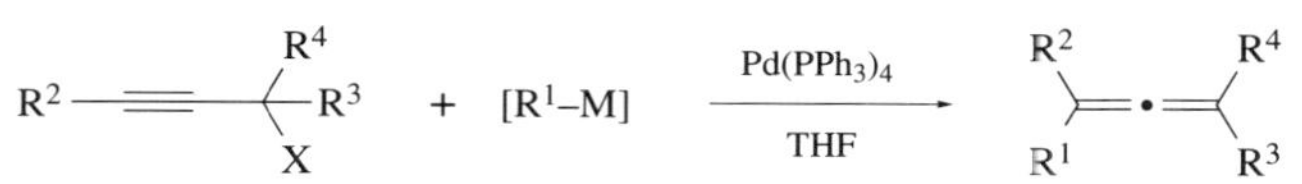

$[R^1–M] = R^1ZnCl; R^1C{\equiv}CH/CuI/base$
X = Br, Cl, OAc, OTs
R^2 = H, Me, Ph; R^3, R^4 = H, Me, Ph, $PhCH_2CH_2$, PhCH=CH, Ar
R^1 = Ph, TMS-C≡C, PhC≡C, $H_2C{=}C(Me)C{\equiv}C$, HC≡C–C≡C, $H_2C{=}CH$, $Bu^tCH{=}CH$

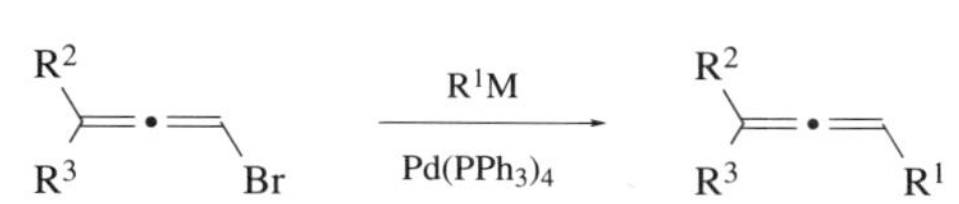

$R^1M = Ph_2Zn$; R^2 = H; R^3 = Ph, Bu^t, 20 °C, 98%
R^1M = RMgX; R^2 = H, Me; R^3 = Me, 60–98%
$R^1M = R^1C{\equiv}CH$/CuI (2%)/Et_2NH; R^2 = H, Me; R^3 = Pr^n, Me, 20 °C, 70–94%

Scheme 34 *(continued)*

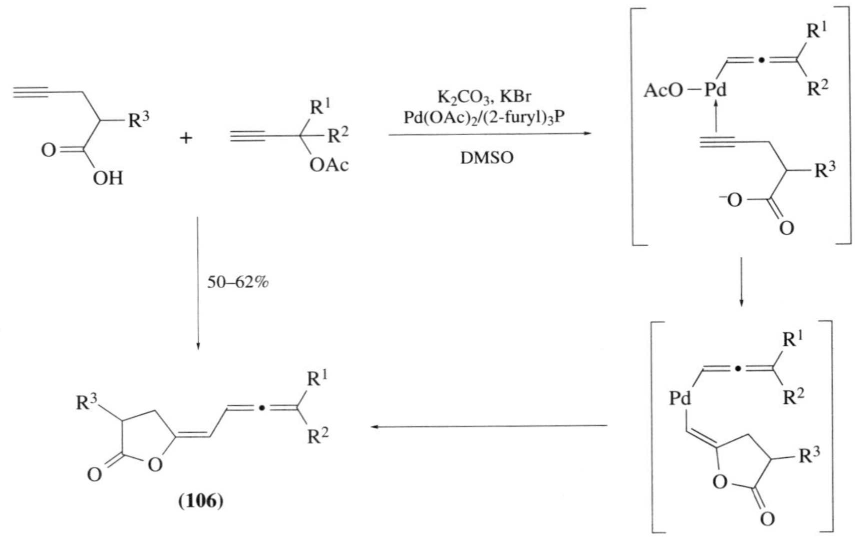

Scheme 34

dicarbonylation of propargyl halides are also possible under phase transfer conditions, with $Ni(CN)_2 \cdot 4H_2O$ as the catalyst precursor to produce allenyl mono- and dicarboxylic acids ⟨92OM493, 93OM1871⟩. The carbonylation in the presence of carbanions generated from active methylene compounds with NaH gives allenyl ketones (**108**) ⟨91SL697⟩. From 4-amino-2-alkynyl carbonates, the carbonylation leads directly to allenic β-lactams via subsequent intramolecular C—N bond formation ⟨91TL7683⟩. Based on the easy formation of alkynyl copper derivatives from terminal alkynes, CuI and a base, 1,2-dien-4-ynes, 1,2-dien-4-yn-6-ols or their tetrahydropyranyl ethers (**109**) have been obtained in good yields in the presence of palladium catalyst precursors with KBr or LiCl ⟨90TL7179, 91JOM(417)305⟩. Alkenyl allenes (**110**) have been synthesized in one step from propargylic carbonates and activated olefins via the Heck reaction, in the presence of $Pd(OAc)_2$, PPh_3 and KBr in DMF at 70 °C (Scheme 35) ⟨91TL3397⟩.

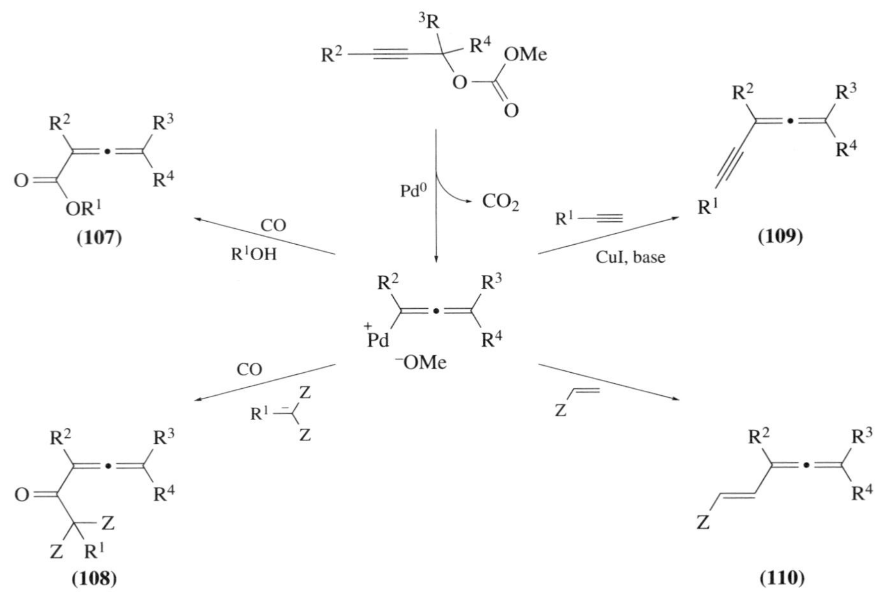

Scheme 35

1.20.3.5 Allene Synthesis via Boron Derivatives

Organoboranes react with lithium chloro- or acetatopropargylides at low temperature to produce allenyl boranes. Further treatment with acetic acid gives the allenes (**111**) (>70% yields), corresponding to the migration of an alkyl group from the boron atom to the terminal carbon of the initial alkyne (Scheme 36) 〈74JA5620, 77JOC2650〉. Propargylic organoborane intermediates react with carbonyl compounds to produce allenic alcohols in good yields 〈78JA5561, 82JOC3364, 83JOC5376〉, with CO_2 to give allenic acids, and with allylic halides to form 1,2,5-trienes 〈82JOC3364〉. 3-Alkenylpropargylic boranes (**112**) react at $-78\,^{\circ}C$ with aldehydes to afford the conjugated 1,2,4-trienols (**113**) in more than 60% yield (Scheme 36) 〈81JOC829〉. Recently, the enantioselective synthesis of chiral α-allenols via reaction of aldehydes with an enantiopure propargyl borane, generated from bromoborane (**114**), has been described 〈90JA878〉. The formation of allenols is also possible by reaction of trialkylborane with 3,4-epoxybut-1-ynes in the presence of oxygen 〈73S305〉.

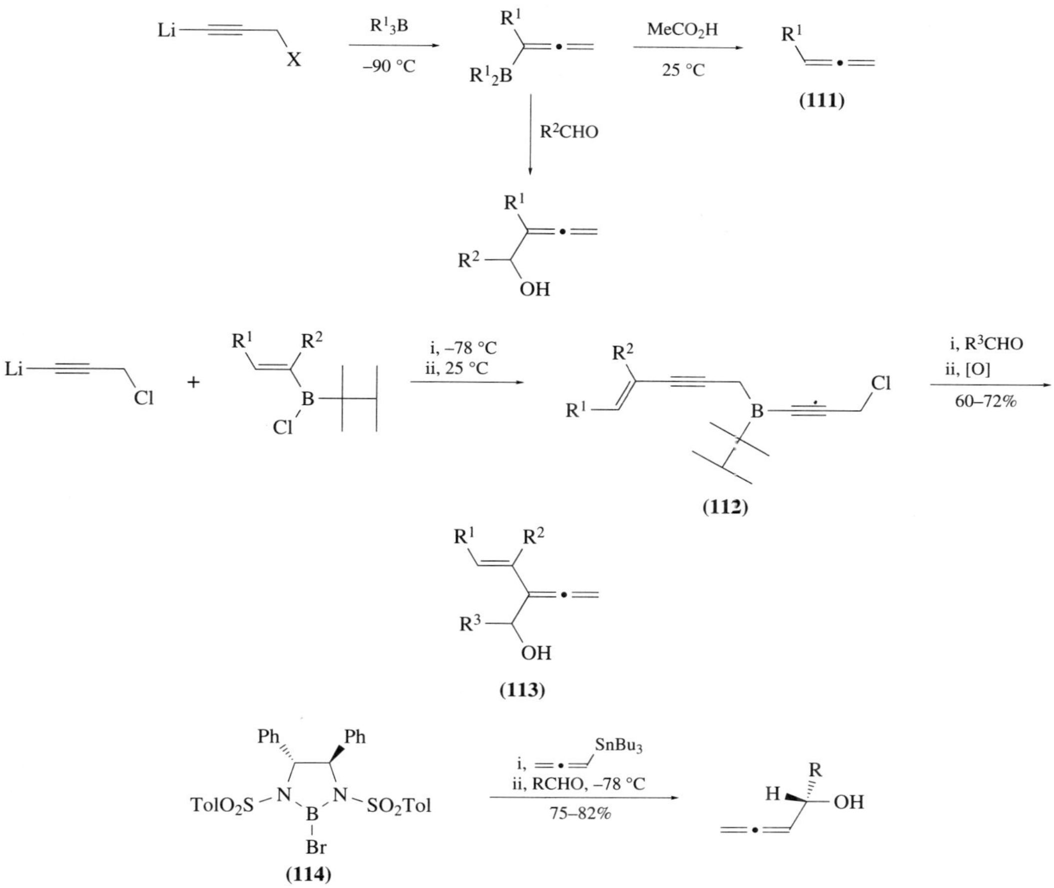

R = n-C_5H_{11}, Pr^i, c-C_6H_{11}, Ph, PhCH=CH; *ee* >91%

Scheme 36

1.20.3.6 Specific Reactivity of Propargyl Silanes

In the presence of a Lewis acid as catalyst, propargyl trialkyl silanes react regioselectively with electrophiles to afford allenic derivatives via a stabilized vinylic carbocation according to Equation (1).

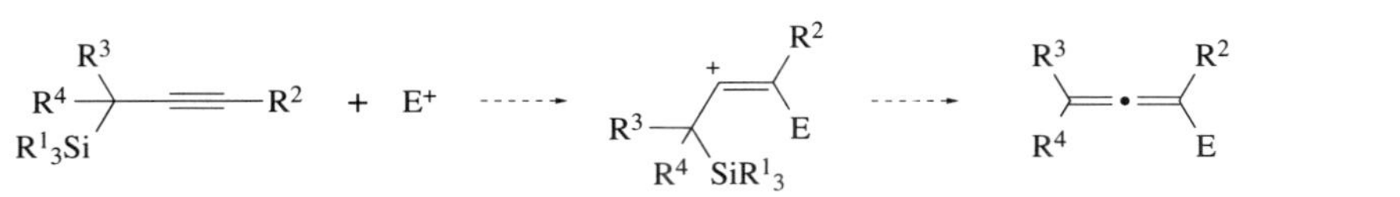

(1)

This strategy has allowed the preparation of a variety of allenes using electrophiles such as acyl chlorides ⟨80JOC5006, 81TL3401⟩, acetals ⟨81TL3609, 85OM333⟩, aldehydes ⟨75JOM(93)43, 81TL455, 81TL1327, 84TL651⟩, and ketones (Scheme 37) ⟨81TL455⟩. For propargyl trimethyl silanes containing a silyl ether or a hydroxy group, the reaction with aldehydes leads to vinylidene cyclic compounds **(115)**, **(116)**, **(117)** (Scheme 37) ⟨84TL651, 86T2017, 86T2501, 88JOM(349)43⟩.

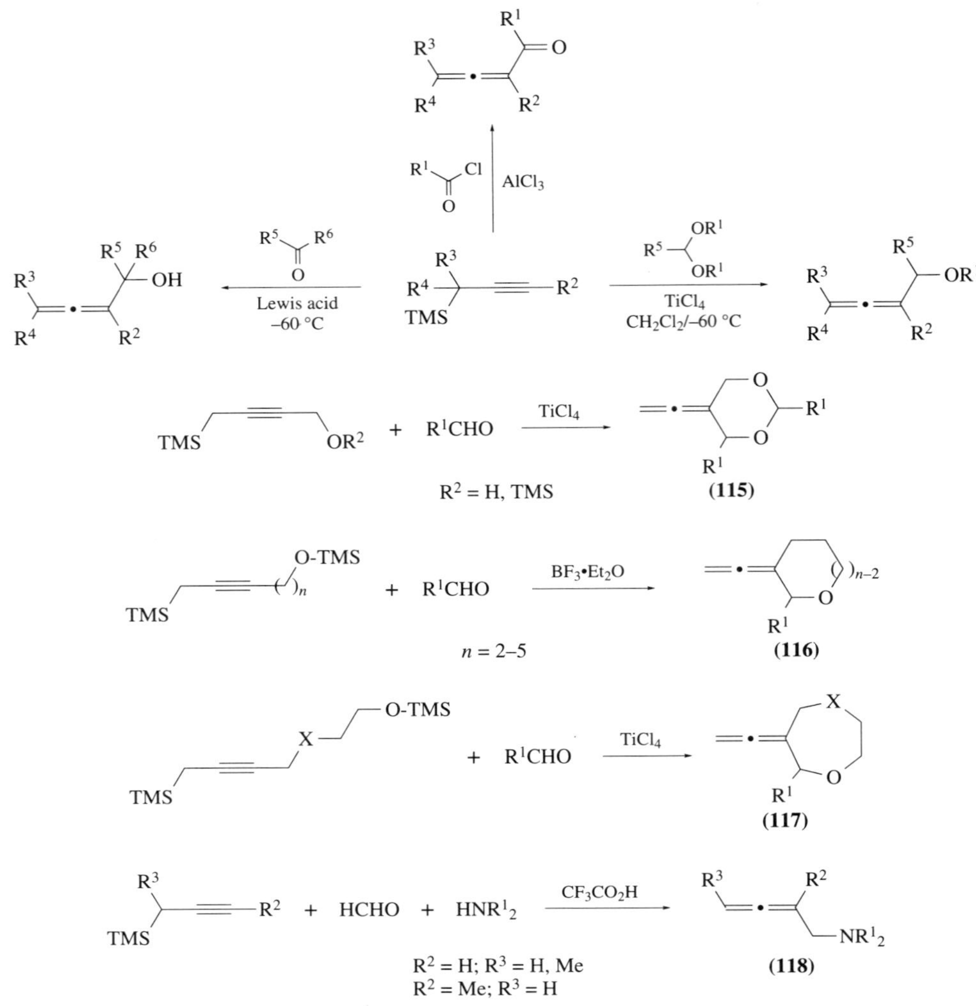

Scheme 37

Iminium salts, generated from formaldehyde and secondary amines in the presence of CF_3CO_2H, react with propargyl silanes to afford tertiary allenyl amines (**118**) (Scheme 37) ⟨90JOM(396)289⟩. Similarly, vinylidene cyclic amines have been prepared from amino propargyl trimethyl silanes via a regiospecific intramolecular reaction ⟨87TL4689⟩.

Allenyl lactams (**119**) have been produced by reaction of propargyl silanes with acyl iminium cations used as electrophiles ⟨83TL1407, 84JOC1149, 84TL3115, 86TL1411, 88TL4253, 92T3445⟩. Under similar conditions, allenyl ethers (**120**) have been obtained from oxonium derivatives, formed *in situ* with $BF_3 \cdot Et_2O$ (Scheme 38) ⟨87JOC1370, 88JOC2450⟩. The acetolysis of 8-(trimethylsilyl)-6-octyn-2-ol tosylate led to an exocyclic allene via an electrophilic intramolecular reaction ⟨80JA5120⟩.

Propargyl silanes react smoothly with the activated double bond of alkylidene malonates in the presence of $TiCl_4$ at 20 °C to produce allenes in moderate to good yields ⟨82JOM(236)177⟩. Allenyl acyl cyanides (**121**) are obtained in good yields from α-unsaturated acyl cyanides and 3-trimethylsilylprop-1-yne (Scheme 38) ⟨86JOC1199⟩. Intramolecular addition of propargyl silanes to conjugated

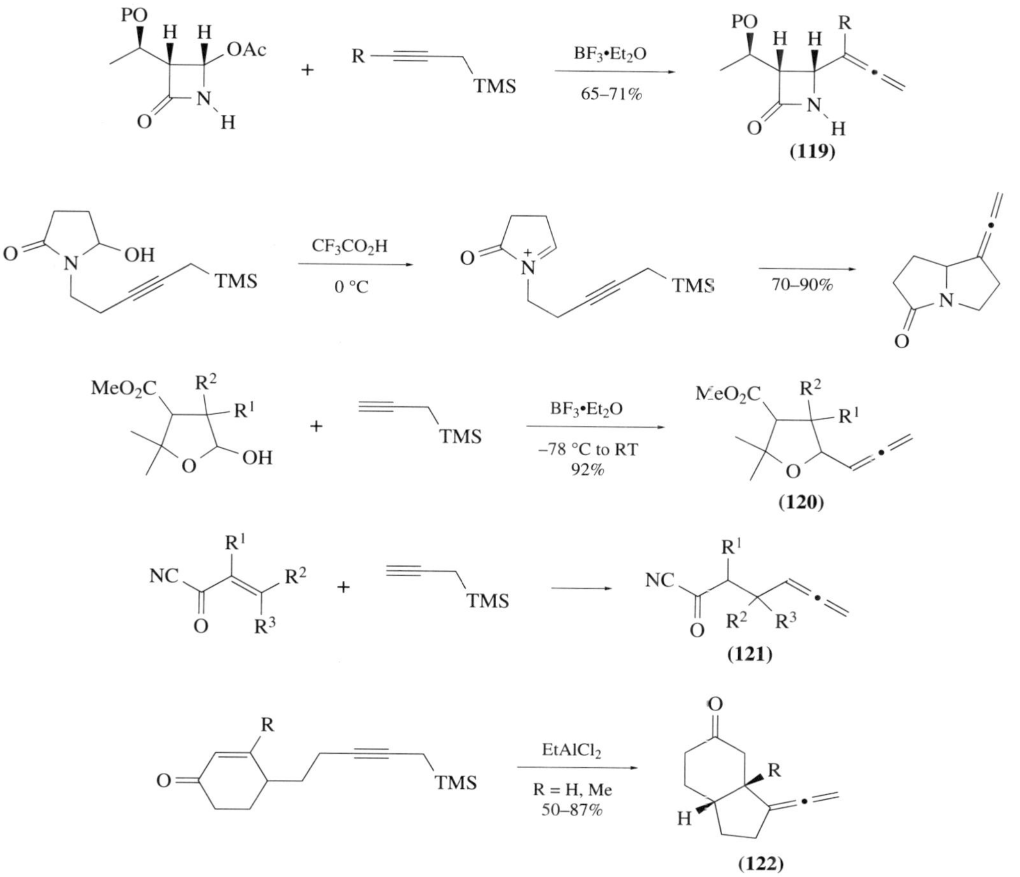

Scheme 38

enones, in the presence of $TiCl_4$ or $EtAlCl_2$ at very low temperature (−70 °C) gives rise to annelation compounds (**122**) containing the exocyclic allenic moiety (Scheme 38) ⟨85TL1831, 86CC829, 88S263⟩. Other examples of acid-catalyzed or Lewis acid-promoted cyclizations from trimethylsilyl propargyl derivatives, leading to the exocyclic terminal allenyl group, have been reported ⟨91CB247, 92TL8017, 93TL7849⟩.

1.20.3.7 C—C Bond Formation via Sigmatropic Rearrangements

Claisen–Cope rearrangement of propargyl vinyl ethers is a method of choice for the access to β-allenic aldehydes, ketones, esters and amides according to the general equation **A** → **B** (Equation (2)).

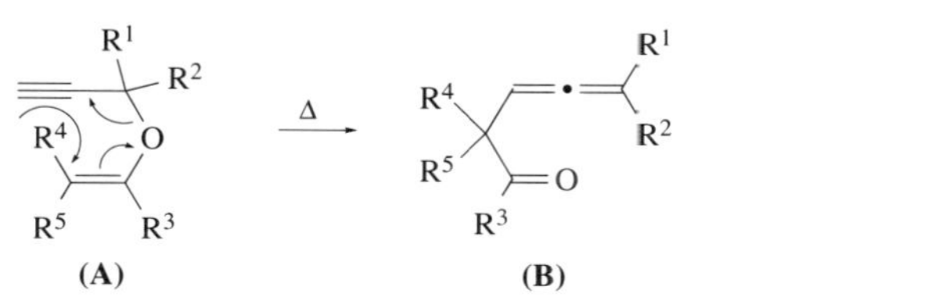

(2)

A classical route for access to intermediates of type (**A**) is the condensation, under acidic conditions, of prop-2-yn-1-ols with enol ethers ⟨67HCA1158⟩ or *ortho*-esters (Scheme 39) ⟨77JOC353, 82JOC4478, 92TL1057⟩. The good diastereoselectivity of this reaction has been used for the preparation of chiral methylmalonaldehyde derivatives ⟨88JOC4736⟩. The reaction of amide acetals with prop-2-

yn-1-ols in refluxing benzene made possible the preparation of allenic pyrrolidines (**123**) and piperidine amides in more than 54% yield (Scheme 39) ⟨82JOC389, 84JOC1204⟩. The Claisen rearrangement can also be extended to propargyl allyl ethers ⟨80JOC2080⟩ and 2-propargyloxy imines ⟨92TL4447⟩. Allenic acids (**124**) have been obtained in 47–88% yields starting from propargyl glycolates via enolates generated on treatment with TMS_2NLi (Scheme 39) ⟨85CL1457⟩. Furans undergo ring opening in the presence of BuLi, and the subsequent rearranged intermediate can be trapped by electrophiles to give allenyl ketones ⟨79JA2208⟩. Other examples of sigmatropic rearrangements involving C—C bond formation have been described, among them the aza–Cope rearrangement involving 2-(*N*-succinimidyl)-4-pentyne derivatives (**125**) ⟨84JA1877⟩, or the formation of the allenic thioether (**128**) from the allylic sulfide (**126**) and 3-chloro-3-methylbut-2-yn-1-yl lithium (**127**) ⟨74CC10⟩. Allenyl sulfones have been obtained by nucleophilic substitution of a phenylsulfonyl group from 2,3-bis(phenylsulfonyl)-1,3-butadiene by soft carbanions ⟨93JA3776⟩.

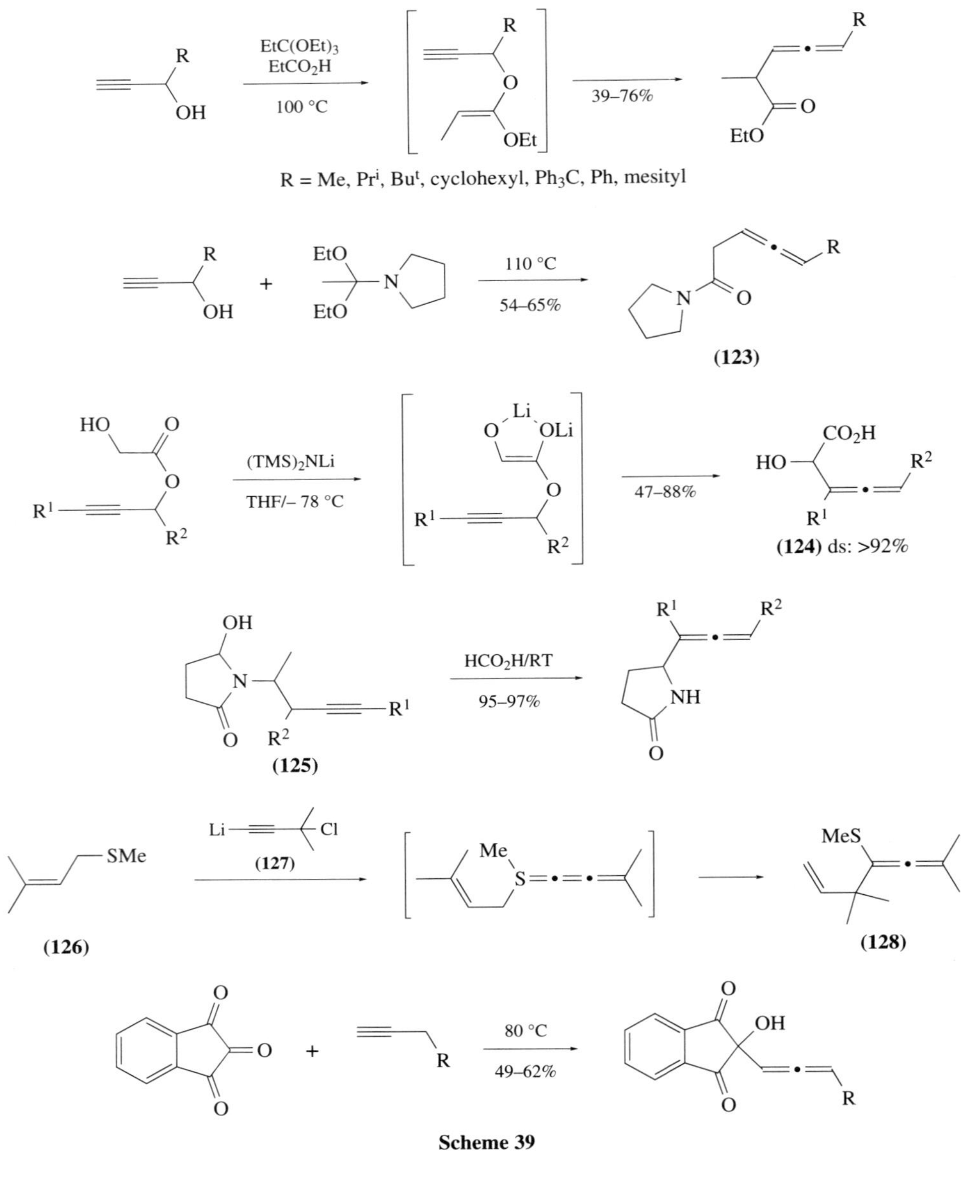

Scheme 39

The thermal ene reaction between terminal alkynes and indane-1,2,3-trione involving C—C bond formation and H migration gives rise to 2-allenyl-2-hydroxy indane-1,3-diones (Scheme 39) ⟨92JCS(P1)2355⟩.

1.20.4 BY C=C BOND FORMATION

1.20.4.1 Elimination Reactions

1,2-Elimination from an unsaturated molecule or 1,4-elimination from both sides of a triple bond, provide simple routes to allenes and cumulenes. The sterically hindered tetra-*tert*-butyl allene has been obtained by elimination of water under acid catalysis from 1,1,3,3-tetra-*tert*-butylprop-2-en-1-ol ⟨82AG(E)924⟩. However, there are only a small number of examples of direct elimination of water from alcohols. In most cases, ester type derivatives like trifluorosulfonates ⟨75JOC657⟩, trifluoroacetates ⟨78JOC1526, 88TL1355⟩, or sulfinates ⟨85AG(E)851⟩ facilitate the elimination (Scheme 40).

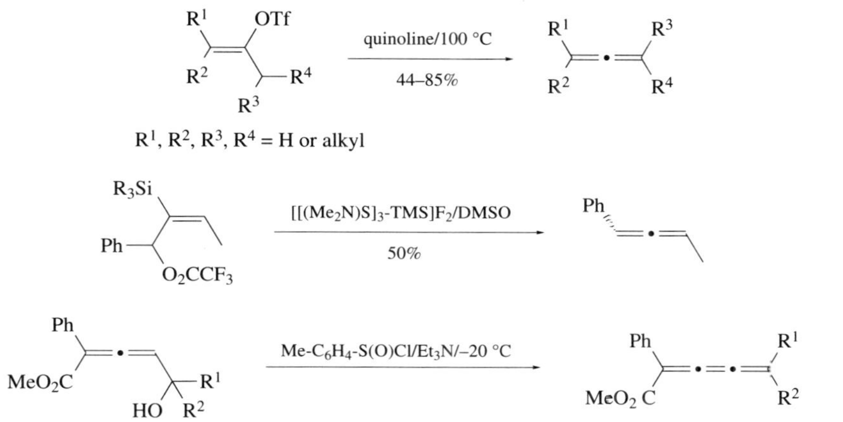

Scheme 40

1,2,3-Butatriene derivatives have been formed by 1,4-elimination of methanol, by treatment of propargylic methyl ether with two equivalents of Bu^{t}Li. The intermediate lithiobutatriene (**129**) allowed access to α-silyl α-silyloxy butatrienes (**130**) on reaction with TMS-Cl, to allenyl ketones and aldehydes (**131**) on hydrolysis ⟨81TL2827⟩, or to allenyl amines (**132**), (**133**) starting from 1-dialkylamino-4-methoxy-4-methylpent-2-ynes (Scheme 41) ⟨82JOM(233)C25⟩. Butatrienyl ethers have been obtained from 1-trimethylsilyloxy-4-alkoxybut-2-ynes by 1,4-elimination of Me_3SiOH on treatment with ethyllithium ⟨81RTC34⟩. Dehydrohalogenation by strong bases has been widely used to prepare allenes in good yields from unsaturated halopropanes with release of HX (Scheme 42) ⟨61JCS2687⟩. The reactive allenyl diazomethyl ketone (**135**) has been obtained from the bromo-derivative (**134**) on treatment with 1,5-diazabicyclo[4.3.0]non-5-ene (dbn) at −20 °C ⟨76JOC3326⟩. Elimination of HCl from the β-chlorovinyl aldehydes (**136**) gave allenyl aldehydes (**137**) ⟨70TL4315⟩, whereas the elimination of HCl from (**138**) using pyridine led to a 'push–pull' type cumulene (**139**)

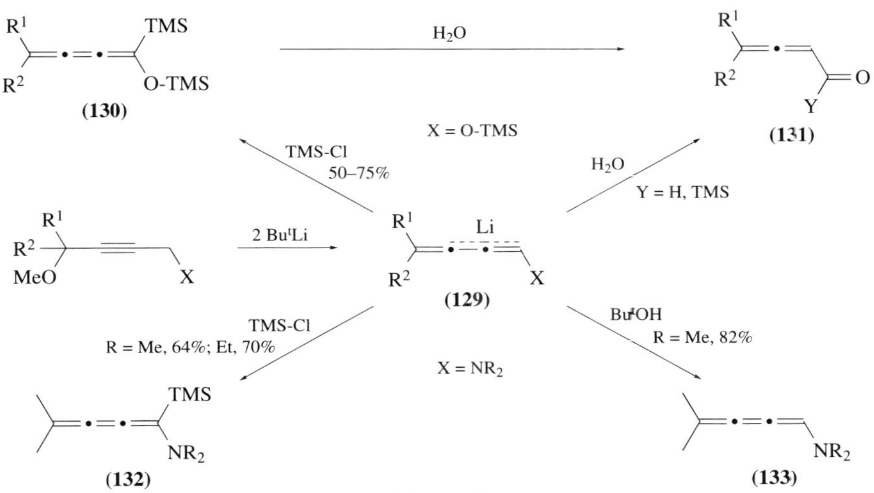

Scheme 41

containing the fulvene and heptafulvene units linked by a cumulative double bond ⟨87AG(E)335⟩. Similarly, 1,3-di-*tert*-butyl-5-vinylidene cyclopentadiene was obtained at low temperature by elimination of HCl from 6-chloro-6-methylpentafulvene using 2,2,6,6-tetramethylpiperidine lithium as a base ⟨86AG(E)466⟩. γ-Ketoallenes have been prepared in 34–40% yield by reaction of dialkyl ketones with 1,4-dibromobut-2-ene in the presence of NaH in DMSO at room temperature via a process involving elimination of HBr ⟨91CC294⟩. Tris(arylthio)butatrienes **(141)** have been obtained by elimination of HCl from 1,2-dichlorobuta-1,3-dienes **(140)**; these compounds rearrange at room temperature into enyne derivatives ⟨84LA1873⟩. Unsaturated β-chlorosilanes **(142)** can eliminate one molecule of chlorosilane in the presence of a fluoride (KF or Et_4NF) to afford allenes **(143)** ⟨74TL171, 84S384⟩. The elimination of HX (X = Cl, ClO_4) from 1,1,3,3-tetrakis(dialkylamino) allylic chlorides or perchlorates **(144)** with BuLi yields 1,1,3,3-tetrakis(dialkylamino) allenes **(145)** in good yields (Scheme 42) ⟨73AG(E)566⟩.

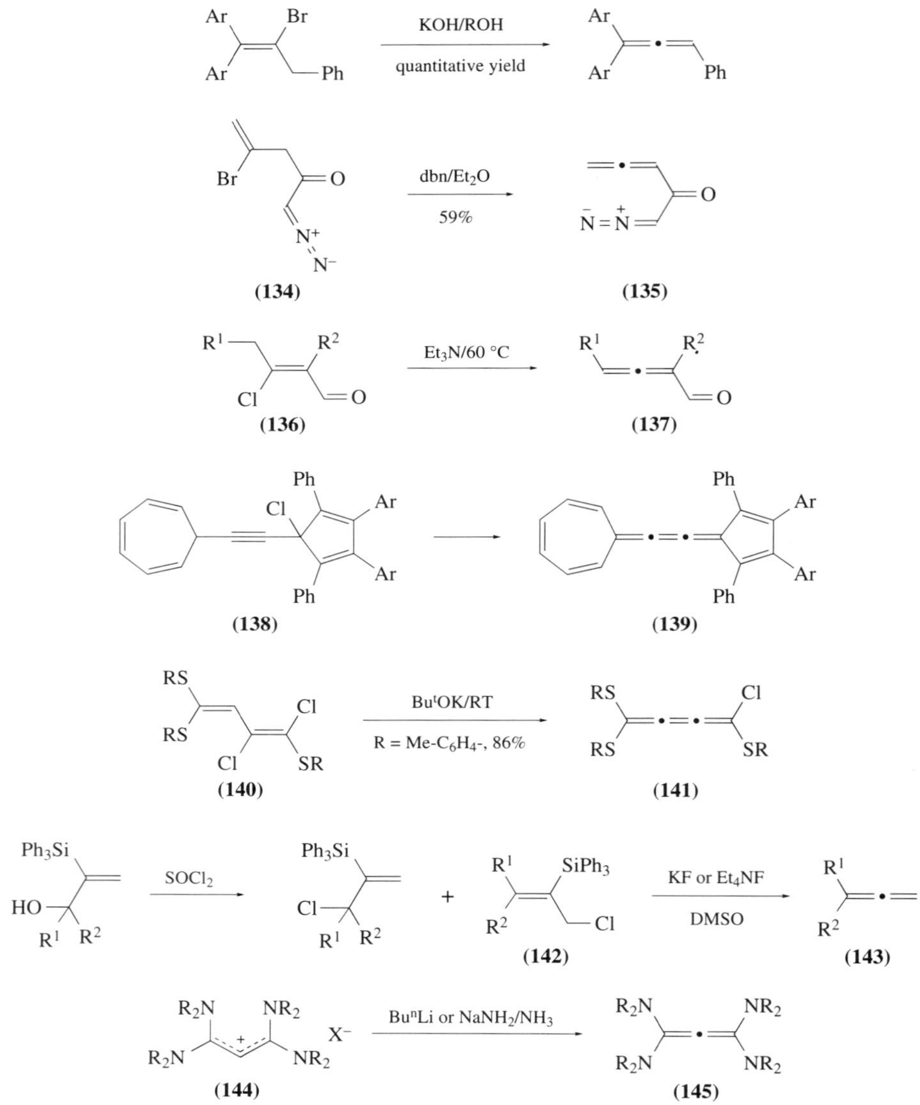

Scheme 42

The elimination of sulfur-containing units provides another route to cumulenes. Thus, desulfurization of 4-vinylidene-1,3-dithiolane-2-thione **(146)** by 1,3-dimethyl-2-phenyl-1,3,2-diazaphospholidine **(147)** gives butatriene derivatives under very mild conditions (Scheme 43) ⟨91S1151, 92AG(E)1611⟩. Treatment of episulfides **(148)** with PBu_3 at 130 °C, or with dimethyl diazomalonate in the presence of a rhodium catalyst, leads to desulfurization and gives cumulenes (Scheme 43)

⟨86T1989⟩. 1,2,3,4-Pentatetraenes (**151**) have been obtained in fair yields (> 50%) by thermolytic or photolytic desulfurization of episulfides (**149**), (**150**) (Scheme 43) ⟨87TL1803, 89TL4271⟩. Thermal elimination of sulfur dioxide from the alkylidene sulfones (**152**) at 650 °C affords vinyl allenes (**153**) in states of good purity and high yields (Scheme 43) ⟨83TL4691⟩.

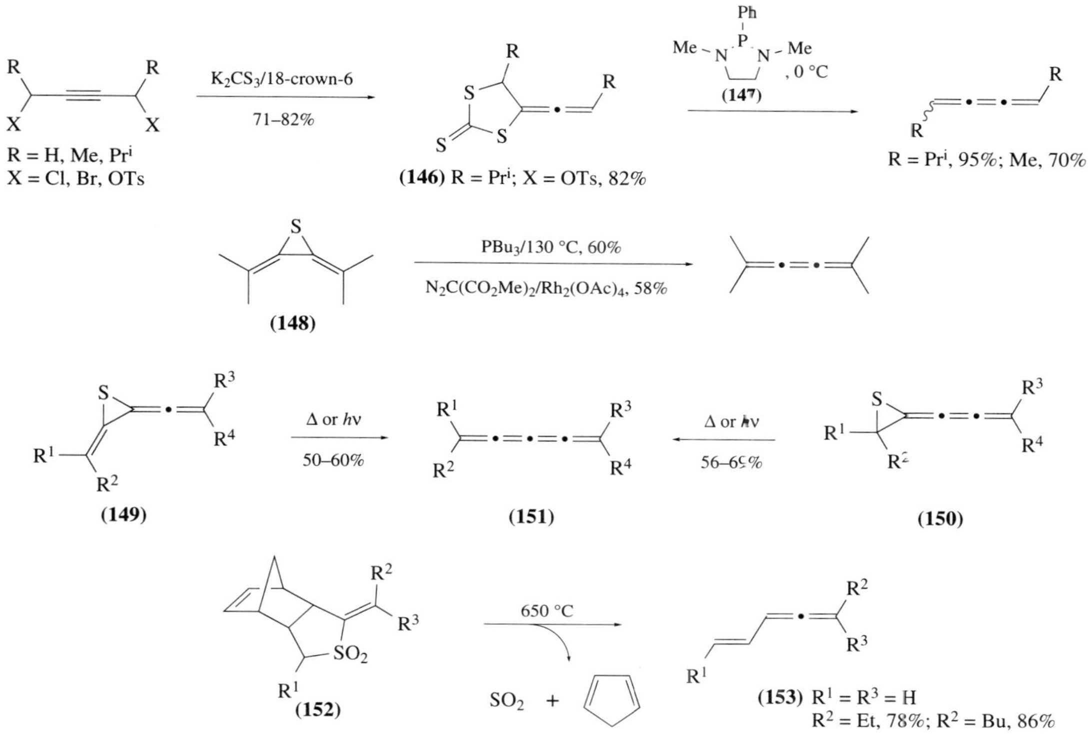

Scheme 43

Addition of a ketone or a diketone to dibromotriphenylphosphine gives access to a bromide salt; its treatment with Et_3N leads to elimination of triphenylphosphine oxide and formation of terminal allenes and allenyl ketones ⟨72TL3257⟩.

The thermal elimination of ArSeOH from vinyl selenoxides leads to allenes when the *syn*-elimination to form a carbon–carbon triple bond is not possible ⟨80JA5967⟩. Based on the same principle, the asymmetric oxidation of vinyl selenides (**154**) by David oxidants, or Bu^tO_2H in the presence of Sharpless catalysts, leads to chiral allenic sulfones (**155**) (Scheme 44) ⟨92CC46, 93JOC3697⟩.

Cyclic allenic esters (**157**) have been produced by oxidation of 3,4-polymethylene-2-pyrazolin-5-ones (**156**) by thallium nitrate followed by ring opening with MeOH (Scheme 44) ⟨80JOC3522⟩. Allenes have been formed by acylation of *N*-nitroenamines (**158**) with concomitant elimination of N_2O, but the reaction is not specific (depending on the site of deprotonation), and only low yields of allenes have been obtained (Scheme 44) ⟨79JOC4116⟩.

Treatment of allylic mesylates (**159**) activated by a chiral sulfoxide group with $Me_2CuLi \cdot LiI$ in THF at −78 °C gave allenes (**160**) corresponding to the formal elimination of MsOS(O)*p*-Tol (Scheme 45) ⟨92TL4985⟩.

Butatrienes (**162**) have been obtained from 1-iodohex-1-yne and *tert*-hexyl borane on reaction with MeONa via the intermediate divinyl boranes (**161**) (Scheme 45) ⟨91T343⟩. Oxidation of cumulenes by *m*-chloroperbenzoic acid leads to alkylidene cyclopropanones, which photochemically release carbon monoxide to produce new cumulenes containing one double bond less (Scheme 45) ⟨87JA4338, 92JA5998⟩.

The reaction of propargyl alcohol with $SOBr_2$ affords an example of intramolecular nucleophilic substitution with formation of a carbon–carbon double bond. Bromoallenes (**163**) are thus formed preferentially when the reaction is carried out in the presence of propylene oxide (Scheme 46) ⟨84TL3055⟩. Similarly, acyl allenes have been obtained by reaction of $SOCl_2$ with acyl propargylic alcohols ⟨85TL631⟩.

The rearrangement of alkynic disilanes (**164**) initiated by $FeCl_3$ or $TiCl_4$ leads to 1,1-bissilyl allenes

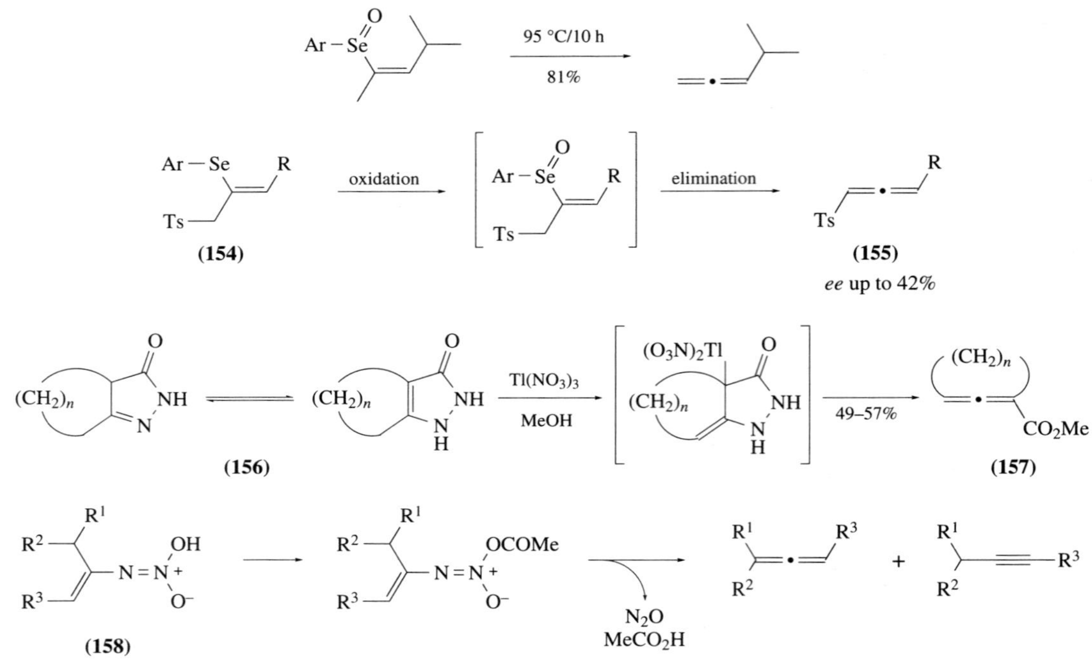

Scheme 44

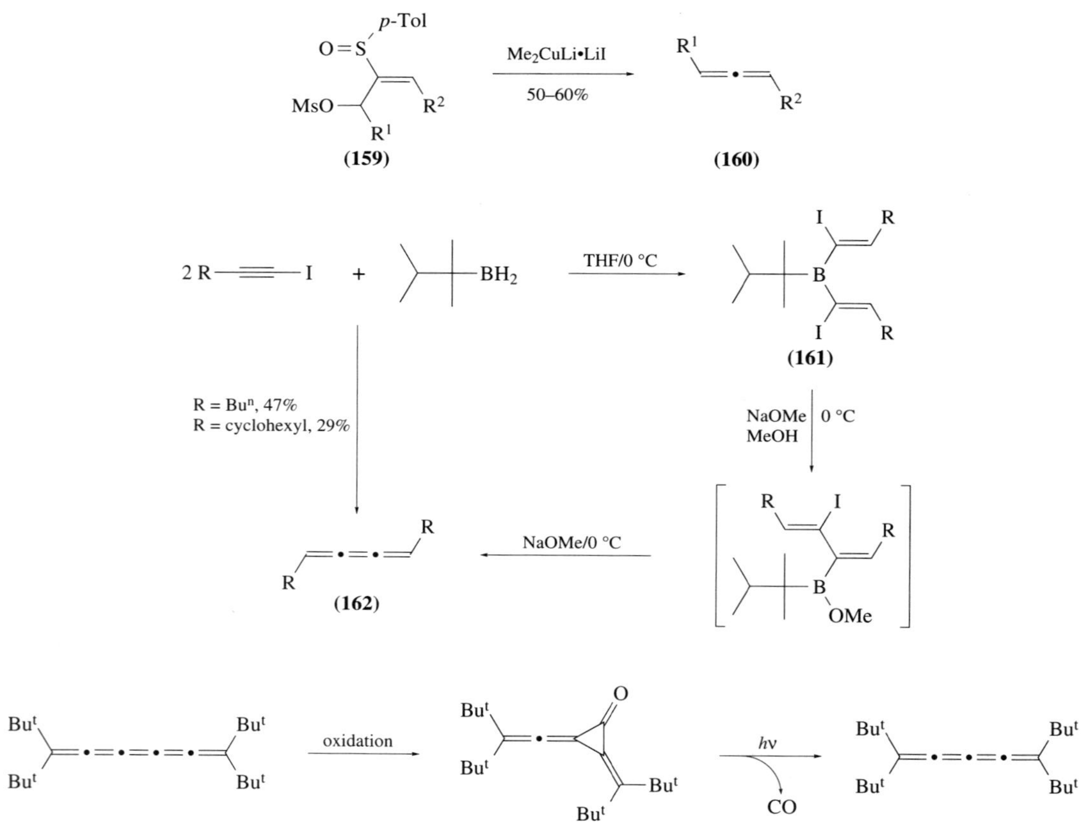

Scheme 45

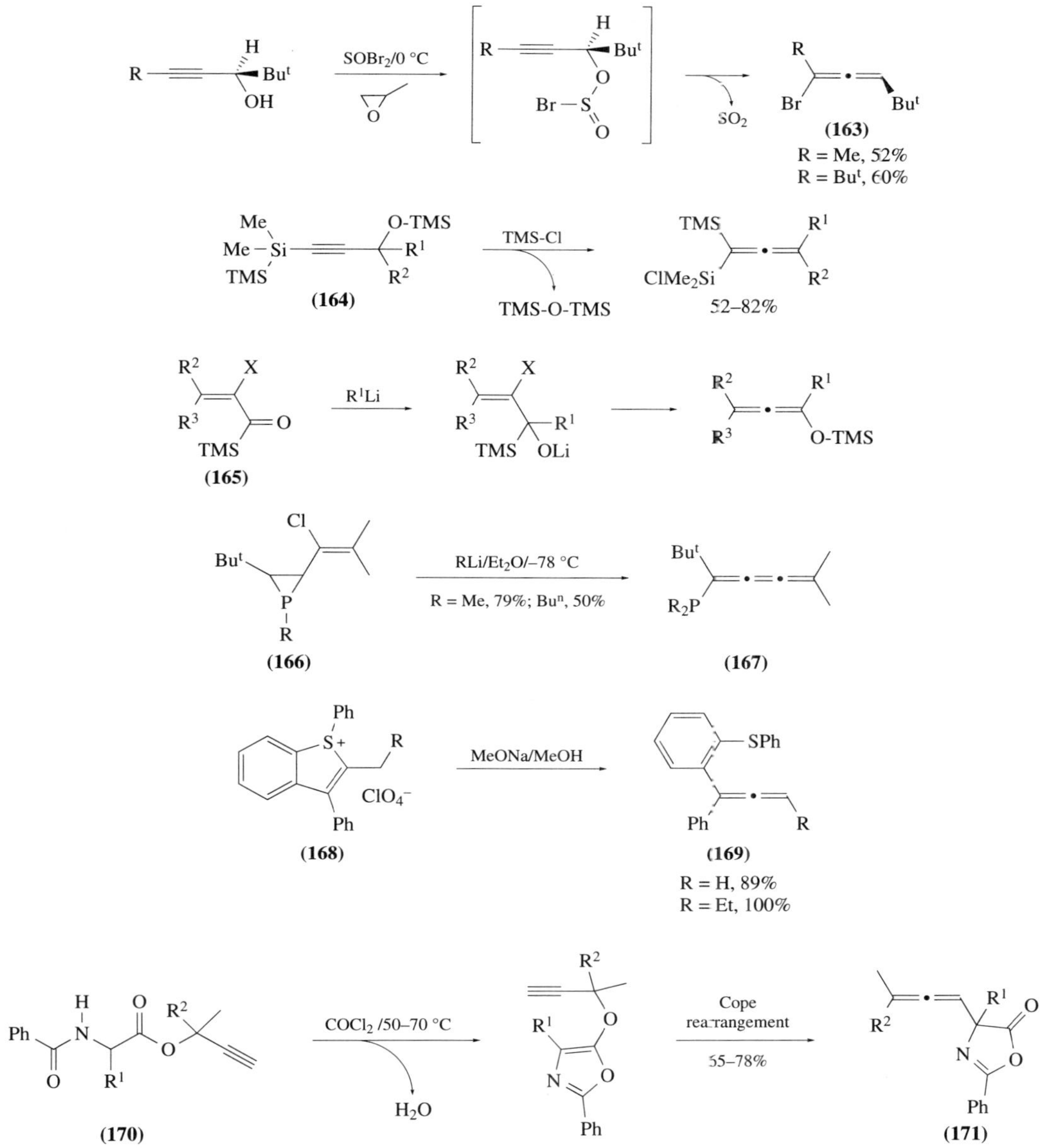

Scheme 46

on elimination of TMS-O-TMS (Scheme 46) ⟨90TL5607⟩. Treatment of 2-halo-2-alkenoyl silanes (**165**) with organolithium gives trimethylsilyloxy allenes after migration of the silyl group to the oxygen atom and elimination of LiX ⟨86JA7791⟩. A similar base-induced Brook rearrangement and elimination of an ether group is observed in the formation of 4-[(benzyloxy)(*tert*-butyl)(methyl) silyloxy]penta-1,2,3-triene on reaction of the corresponding acyl silane with lithium or bromomagnesium 3-tetrahydropyranyloxyprop-1-ynide ⟨94TL1161⟩.

Treatment of chlorinated vinylic phosphirenes (**166**) with BuLi at −78°C gives phosphinobutatrienes (**167**) in good yields via C—P bond formation, rearrangement and elimination of LiCl ⟨92SL635⟩. Deprotonation of the benzothiophenium perchlorate (**168**) by MeONa leads to ring opening and formation of the allene (**169**) (Scheme 46) ⟨92CL1357⟩.

α-Hydroxyallenes have been prepared from 3-phenylselenobuta-1,3-dienes by oxidation with H_2O_2, followed by sigmatropic rearrangement and elimination ⟨84TL1987⟩.

The hetero-Cope rearrangement involved in the reaction of propargyl esters of *N*-acyl amino acids (**170**) with phosgene leads to allenyl oxazolinones (**171**) after elimination of water through prolonged heating at 50–70°C (Scheme 46) ⟨75AG(E)58⟩.

1.20.4.2 Reduction of Unsaturated Alcohols and Halides

Reduction of 1,4-dihydroxy-2-butynes (**172**) by various reducing agents (e.g., PBr_3, $SnCl_2/HCl$) has been used to prepare tetrasubstituted butatrienes ⟨66BSF2885, 59CB1849⟩. Reaction of the diols (**172**) with EtOH in the presence of hot H_2SO_4 leads to the synthesis of butatrienes ⟨87IZV695⟩. The extension of this reaction to 1,6-dihydroxy-2,4-hexadiynes (**173**) ⟨66BSF2885⟩ has recently been used for the preparation of hexapentaenes (**174**) and bishexapentaenes as potential molecular materials for nonlinear optics ⟨93CM357⟩ or as host molecules (Scheme 47) ⟨91BCJ659⟩.

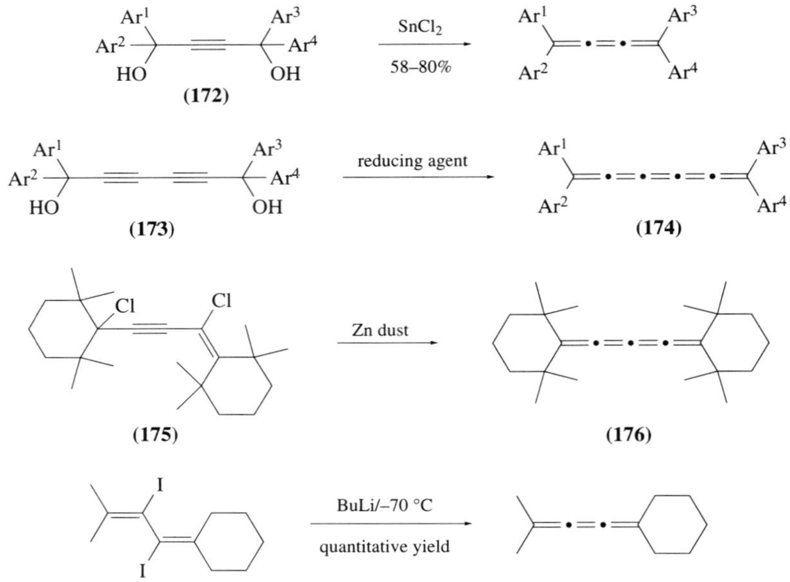

Scheme 47

Reductive dehalogenation of 1,4-dihalo-2-butynes by zinc dust or even by an excess of NaI in DMSO are simple routes to butatrienes. The reaction is possible under phase transfer conditions ⟨84AKZ116⟩, and when one halogen is replaced by an alkoxy group, a 1 : 1 mixture of an allenyl ether and an alkynyl ether is obtained ⟨67RTC129⟩. If the starting compound has a 2,5-dihalopent-1-en-3-yne structure like (**175**), the reductive cleavage of halogen–carbon bonds with zinc dust leads to pentatetraenes of type (**176**) ⟨86AG(E)340⟩ (Scheme 47).

It is also possible to generate butatrienes in good yields and high stereoselectivity by reduction of 2,3-diiodobuta-1,3-dienes using BuLi at $-70\,^{\circ}C$ ⟨84CL131⟩.

1.20.4.3 Wittig and Related Reactions

The formation of allenes from phosphorus ylides and carbonyl compounds can be considered either starting from an alkylidene phosphorane and a ketene ⟨63CB1535⟩ or from a vinylidene phosphorane and carbonyl compounds. Both these strategies have been used to gain access to a variety of functionalized allenes.

Terminal allenes have been obtained in 59–80% yields by reaction of stable phosphorus ylides with $CH_2{=}C{=}O$ at 0–5 °C (Scheme 48) ⟨70S543⟩. Electron-deficient allenes such as 1,1,3,3-(tetra-alkoxycarbonyl) allenes have also been prepared in more than 60% yield by treatment of alkylidene phosphoranes with ketenes in refluxing benzene ⟨79LA1388⟩.

A similar process, based on the use of carboxylic acids or acyl chlorides that are able to give a β-elimination reaction in place of a ketene, can be used to prepare allenes (Scheme 48). Thus, ethyl allenyl carboxylates have been obtained from [(ethoxycarbonyl)methylidene] phosphoranes (**177**) and acids, in the presence of Bu_3N and 2-chloro-1-methylpyridinium iodide ⟨85HCA2244⟩. From unsaturated acids like acrylic or propargylic acids, the formation of vinyl allenes takes place but only in low yields (6–18%). In dichloromethane at room temperature, and in the presence of Et_3N, acyl halides react with [(alkoxycarbonyl)methylidene] phosphoranes (**177**) to provide allenes ⟨80HCA438, 92CPL243⟩ and alkenyl allenes from α,β-unsaturated acyl chlorides ⟨85HCA2249⟩.

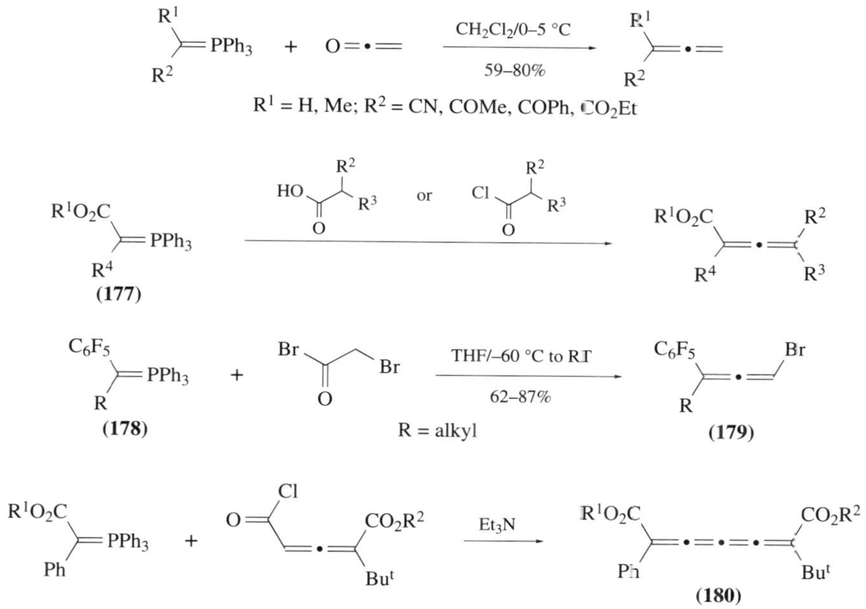

Scheme 48

1-Alkyl-3-bromo-1-pentafluorophenyl allenes (**179**) have been prepared from phosphoranes (**178**) and bromoacetyl bromide at $-20\,^{\circ}C$ in THF ⟨91JCS(P1)665⟩. Unsymmetrical pentatetraenes (**180**) have been prepared in good yields (50–75%) on reaction of a phosphorus ylide with allenyl acyl chloride in the presence of Et_3N at room temperature (Scheme 48) ⟨86CB1208⟩. Photolysis of stabilized phosphoranes ylides with alkoxycarbene chromium complexes (**181**) under a carbon monoxide atmosphere leads to allenes via formation of a ketene (Scheme 49) ⟨92JA4079⟩. The coordination of vinyl ketenes to a $Fe(CO)_3$ moiety makes possible the direct synthesis of coordinated vinyl allenes (**182**) ⟨91CC1290, 92JCS(P1)259⟩ on reaction with phosphonoacetate in a Wadsworth–Emmons type reaction ⟨61JA1733⟩ (Scheme 49). On reaction with phosphonium ylides, carbon

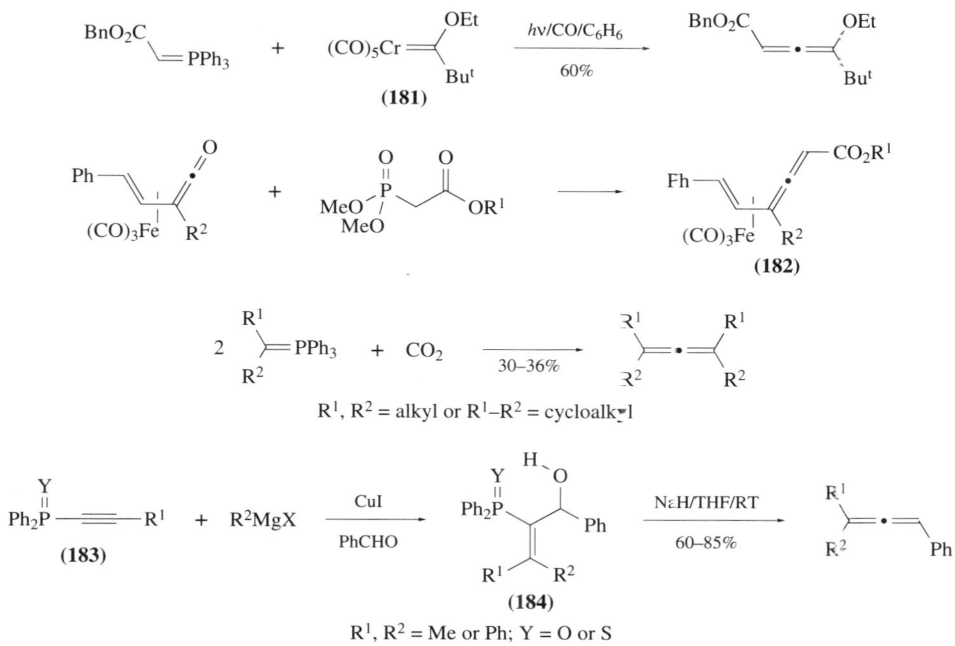

Scheme 49

dioxide allows the formal coupling of two carbenic moieties with the carbon atom of CO_2 ⟨74TL1275⟩. Alkynic phosphine oxides (or the corresponding sulfur-containing compounds) (**183**) react with organomagnesium reagents in the presence of CuI to give nucleophilic species which couple with aldehydes to afford alcohols of type (**184**). On treatment with NaH, the latter compounds give allenes via a Wittig–Horner type elimination (Scheme 49) ⟨74TL1905⟩.

The other strategy based on the use of vinylidene triphenylphosphoranes can be exemplified by the synthesis of push–pull allenes of type (**185**) from cyclic 1,2-diketones and 2,2-diethoxyvinylidene phosphorane (Scheme 50) ⟨73TL3985, 75TL4405⟩.

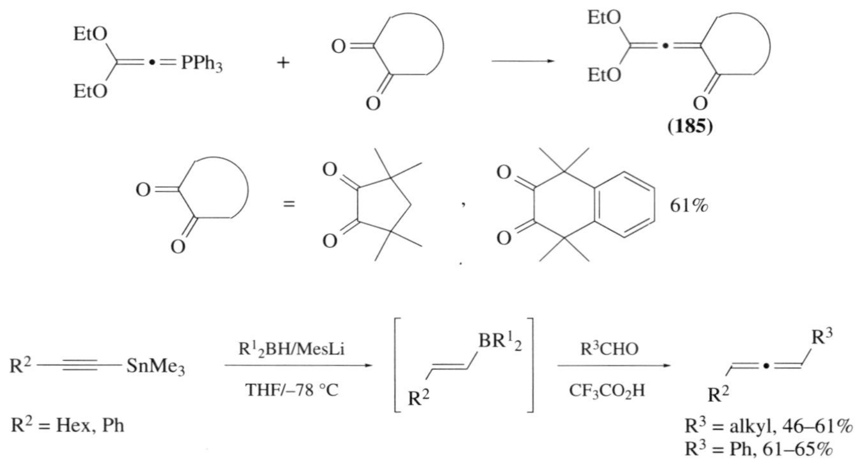

Scheme 50

A new allene synthesis involving a boron–Wittig reaction of aldehydes with boron stabilized carbanions at $-78\,^{\circ}C$ has been recently reported (Scheme 50) ⟨92JCS(P1)747⟩. Regioselective halide displacement from halogenated phosphonium salts gives rise to phosphacumulenes, which react with ketones to afford butatriene derivatives ⟨69JA6112⟩. The bromoalkenylphosphonium bromide (**186**) gave a 53% yield of the cumulene diester (**187**) ⟨87AJC1675⟩, whereas on reaction with ketones in the presence of $(Me_3Si)_2NLi$ at low temperature dibromo- or dichlorophosphonium salts (**188**) led to halogenated butatrienes (**189**) (Scheme 51) ⟨84CC152, 87JOC443, 88TL411, 90JOC2983⟩.

The symmetrical butatriene (**191**) has been obtained in 71% yield by reaction of fluorenone with a propylidene phosphorane produced from methylidene triphenyl phosphorane and the *gem*-dihaloalkene (**190**) (Scheme 51) ⟨75TL4025⟩. The Horner–Emmons type reaction starting from allenyl diphenylphosphine oxide (**192**) ⟨90TL7469⟩, or allenyl phosphonates (**193**) ⟨85IJ136, 86JA343⟩ and aldehydes or ketones, makes possible the synthesis of butatrienes derivatives and bicyclic cumulenes (**194**) via an intramolecular reaction (Scheme 52). Another strategy based on the intermediate formation of a cumulenyl diylide is exemplified by the synthesis of 1,4-diphenylbutatriene from benzaldehyde and the diphosphonium bistriflate (**195**) ⟨91T4539⟩. Higher symmetrical cumulenes have been obtained by Wittig reaction of carbon suboxide with 2 equivalents of phosphorus ylide at room temperature (Scheme 52) ⟨86AG(E)93, 87G625⟩.

1.20.4.4 Addition of Carbene Moieties to Double Bonds

Unsaturated carbenes generated *in situ* by treatment of propargylic halides or esters with Bu^tOK react with alkenes to produce alkenylidene cyclopropanes (**196**) in moderate yields (10–52%) (Scheme 53) ⟨61JA4990, 71JA4527⟩. Extended unsaturated carbenes like alkadienylidene, alkatrienylidene and alkatetraenylidene carbenes have been generated from 1-ethynylvinyl triflates ⟨79JA4772, 82ACR348⟩, (1,3-diynyl)alkyl mesylate or halide ⟨81JA4638, 88CC301, 88JOC3122⟩, and (1-butadiynyl)vinyl triflates ⟨81JA6437, 82ACR348⟩, respectively. These unsaturated carbenes give access to a variety of cyclopropyl cumulenes and dimerized cumulenic compounds ⟨87JA782⟩. Silicon–, germanium–, and tin–functionalized cumulenes have been prepared in good yields (26–88%) on interaction of alkadienylidene ⟨81JA5429⟩ and alkatetraenylidene ⟨81JA6437⟩ carbenes with Group 4 hydrides.

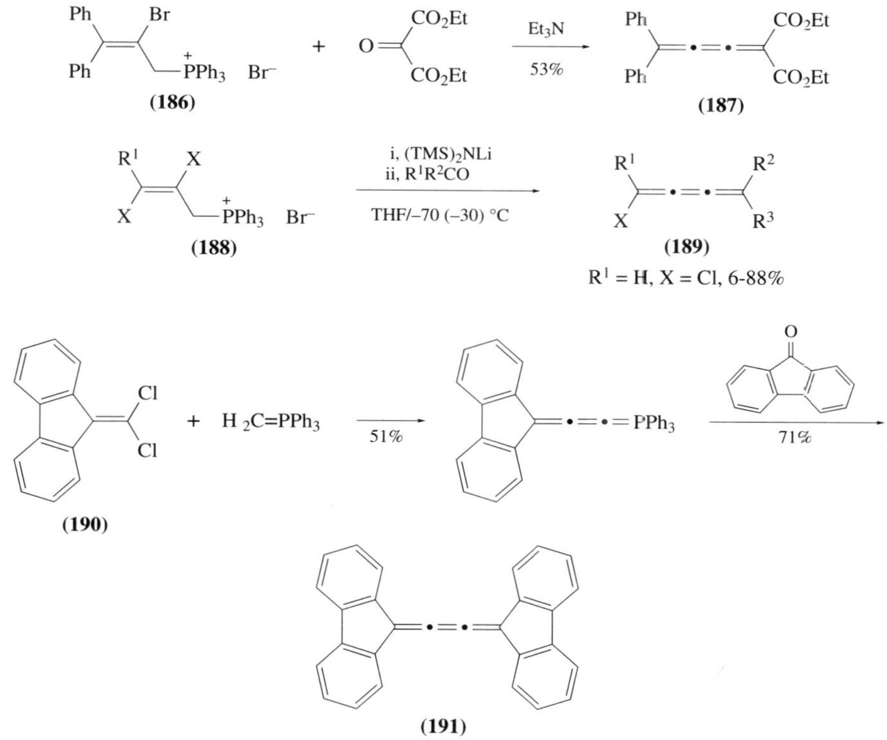

Scheme 51

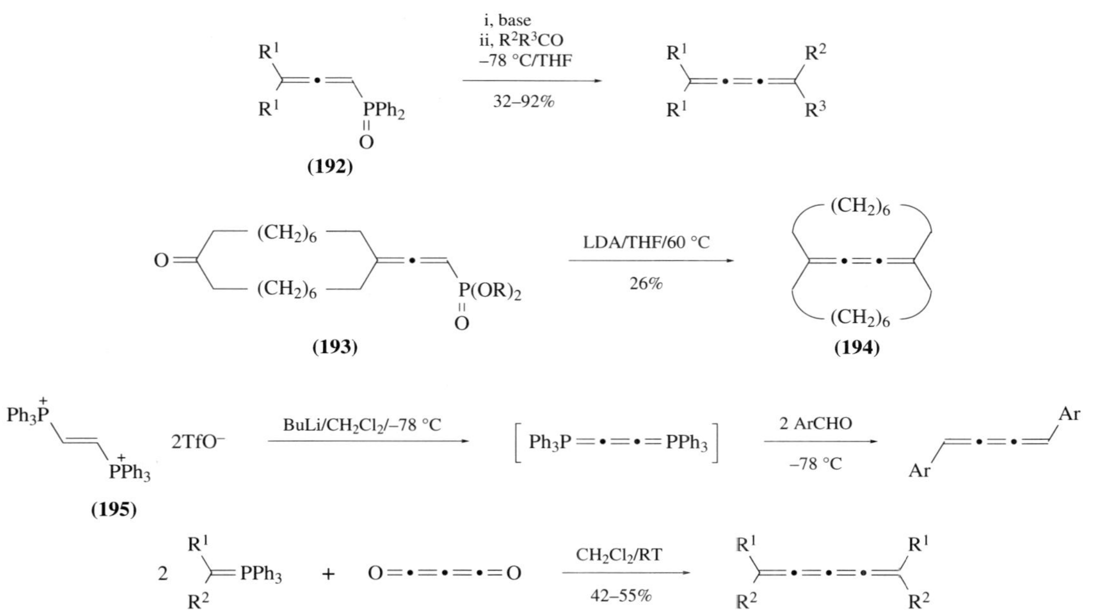

Scheme 52

cis-Cumulenes (**197**) have also been prepared in 20–50% yields via homocoupling of two carbenes generated from (α-methylthiovinyl) copper reagents and CH_3I at 0°C in THF (Scheme 53) ⟨75TL2923⟩.

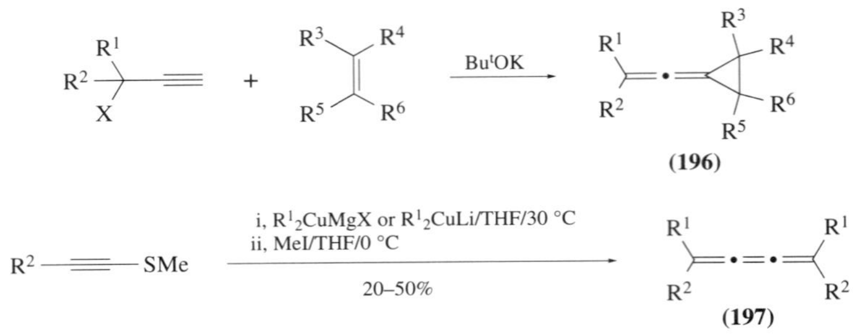

(196)

(197)

R^1 = alkyl, Ph

R^2 = H, Ph

Scheme 53

1.20.4.5 Dehalogenation

1.20.4.5.1 Dehalogenation of gem-*dihalocyclopropanes*

Treatment of *gem*-dihalocyclopropanes by alkyllithium at low temperature (−30 °C to room temperature) gives allenes in good yields ⟨63ACS1683, 84CJC1558, 84TL2887, 85CC1812⟩. Butatrienes have also been prepared in approximately 75% yields by reaction of BuLi with *gem*-dibromo methylene cyclopropanes ⟨67JCS(C)194⟩. This strategy has been used for the synthesis of cyclic allenes **(198)** ⟨73JOC864, 86TL4679⟩ and bisallenes **(199)** ⟨90TL1841⟩, bicyclic allenes **(200)** ⟨82JOC1435⟩, and small ring cyclic cumulenes such as cyclonona-1,2,3-triene (Scheme 54) ⟨84JOC2880⟩.

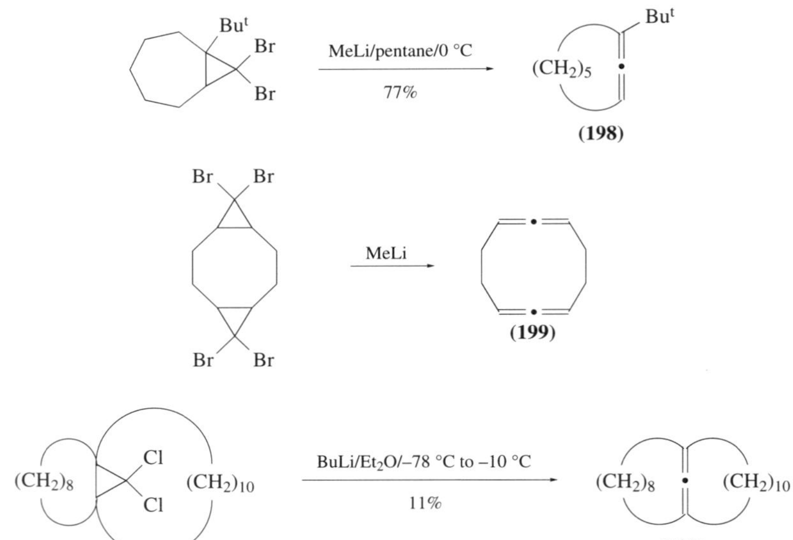

(198)

(199)

(200)

Scheme 54

1.20.4.5.2 Dehalogenation of gem-*dihaloalkenes*

Dimerization of *gem*-dihaloalkenes on treatment with alkyllithium leads to butatrienes via 1-halo-1-lithium alkene intermediates. Cumulenic thioesters **(201)** ⟨83CPB3306⟩ and tetra(2-thienyl)butatrienes **(202)** ⟨92CC778⟩ were prepared in this way (Scheme 55). From these intermediates, lithium can be exchanged with copper ⟨86JA5371⟩ to produce functionalized alkoxy **(203)** ⟨88JA5567⟩, alkynyl **(204)** ⟨93AG(E)1187⟩ and fluorinated **(205)** ⟨91CC566⟩ butatrienes (Scheme 55). 1-Halo-1-lithio alkenes intermediates have also been generated on treatment of terminal monohalogenated alkenes with Bu^nLi ⟨84JOM(264)135, 84JOM(271)181⟩. Aryl-substituted butatrienes have been conveniently prepared from 1,1-dibromoethenes using a nickel(0) catalyst at 50 °C (Scheme 55) ⟨84CL2005, 89CC1690⟩.

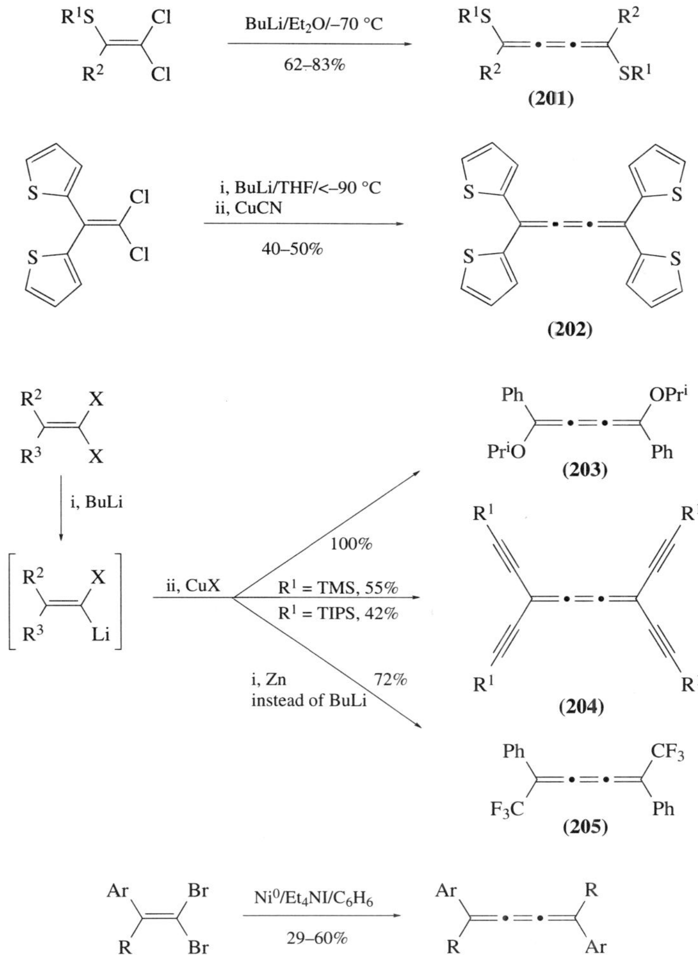

R = Me; Ar = C_6H_5, MeO-C_6H_4, NC-C_6H_4, MeO_2C-C_6H_4

Scheme 55

1.20.4.6 Reactions Involving Coordinated Organometallic Species

Coordinated cumulenes can be generated by reaction of suitable substrates with organometallic complexes. The cumulene moiety can be eliminated from the metal centre by ligand exchange from (**206**) ⟨89CB1237⟩, oxidation of (**207**) with Me_3NO ⟨93OM3971⟩ or protonation of (**208**) (Scheme 56) ⟨93AG(E)1315⟩. Cumulenic systems can also be formed from metal–vinylidene intermediates by dimerization of the vinylidene ligand on thermal treatment ⟨85IJ131, 90JOC1874, 87CC981⟩ or ligand exchange (Scheme 56) ⟨92CB2667⟩.

Titanium complexes (**209**) or mixed titanium aluminum complexes (**210**) react with aldehydes or ketones to afford allenes ⟨81JA1276, 83JA5490, 93AG(E)554, 93JOC1298⟩. Under similar conditions, 1-zinc-1-zirconium alkene reagents (**211**), (generated by reaction of zinc acetylides with $H(Cl)Zr(Cp)_2$) react with aldehydes in dichloromethane at 25 °C to smoothly produce allenes (Scheme 57) ⟨91JA9888, 94OM94⟩.

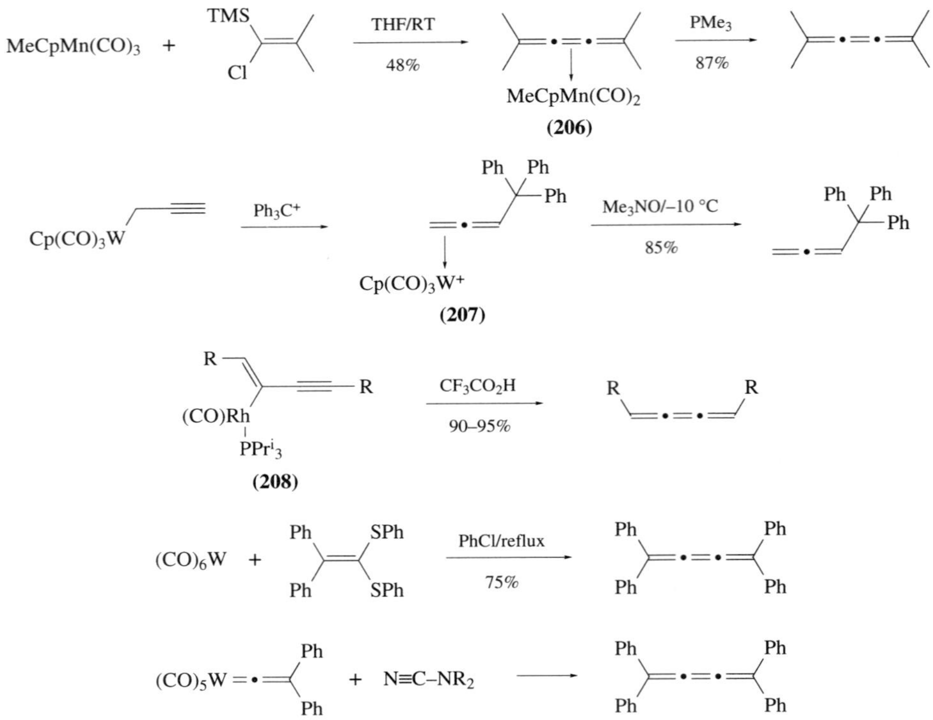

Scheme 56

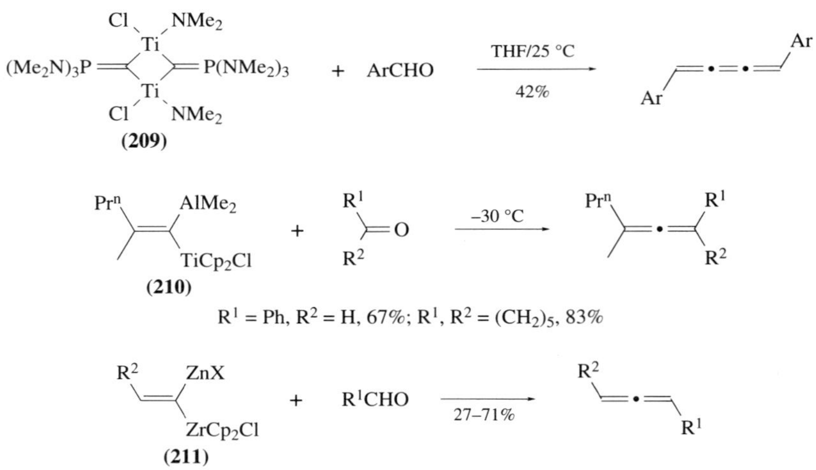

Scheme 57

1.20.4.7 Metal-catalyzed Homocoupling of Alkynes or Cumulenes

Activation of terminal alkynes by metal complexes, via a vinylidene intermediate, makes possible the catalytic synthesis of cumulenes. Ruthenium complexes such as $H_2Ru(CO)(PPh_3)_3$ ⟨76CC841⟩, or Ru(cod)(cot) ⟨93BCJ987, 91JA9604⟩ catalyze the dimerization of bulky terminal alkynes such as *tert*-butylacetylene or trimethylsilylacetylene into the corresponding 1,4-disubstituted butatrienes (Scheme 58).

Iyoda *et al.* have shown that the nickel(0)-catalyzed cyclodimerization of tetraarylhexapentaenes led to new cumulenes such as the head-to-head dimers of type **(212)** ⟨89JA3761⟩. Hexapentaenes bearing bulky alkyl substituents gave the head-to-tail dimer **(213)** in good yield using a nickel(0) catalyst in refluxing benzene (Scheme 58) ⟨90CL2149⟩.

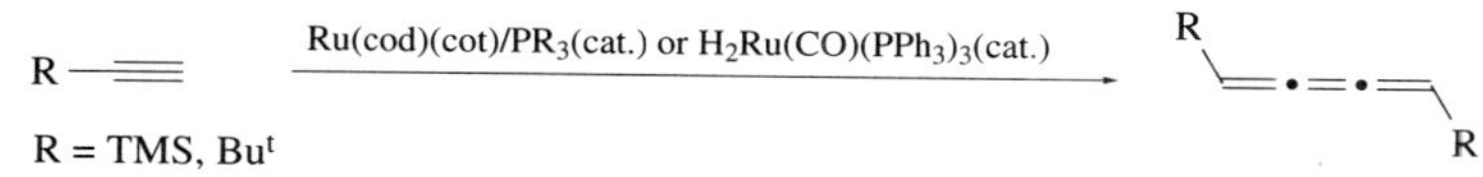

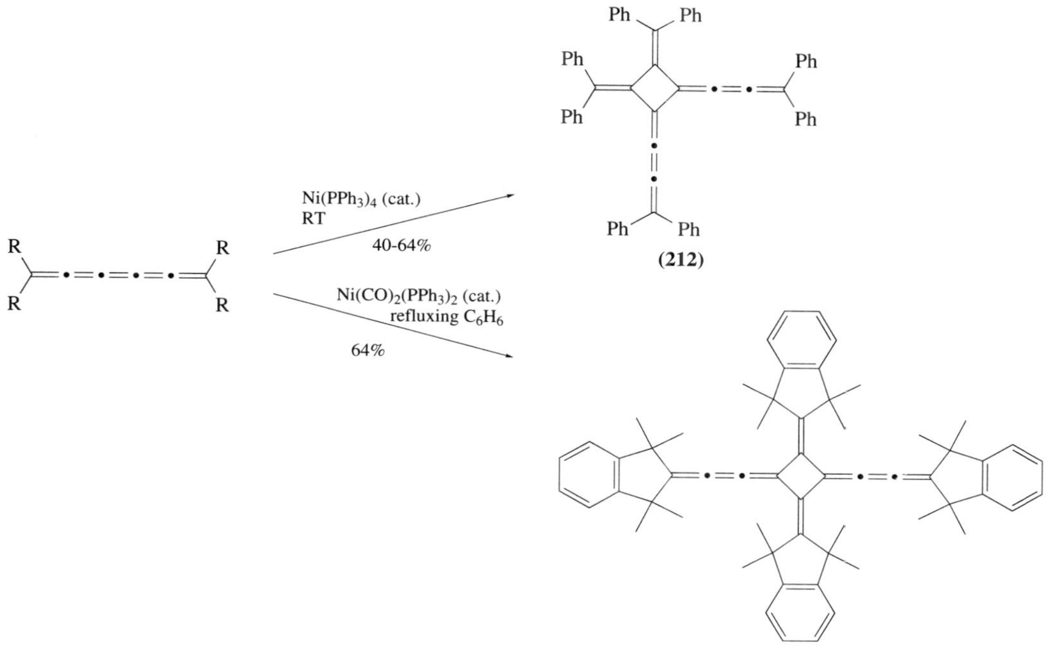

Scheme 58

1.20.4.8 Intramolecular Rearrangement with Formation of a Heteroatom–carbon Bond

Intramolecular rearrangement involving heteroatom–carbon and carbon–carbon double bond formation can take place from various functionalized propargyl derivatives. Thus, the condensation of Ph_2PCl with prop-2-yn-1-ol derivatives leads to the formation of alkynyl phosphinites (**214**), which undergo rearrangement into allenes via intramolecular addition of the phosphorus atom to the triple bond ⟨62JOC1828, 89JA1770, 89TL4995⟩.

Similarly, allenyl dihalophosphine oxides (**215**) were obtained at room temperature from propargylic alcohols and phosphorus trihalide via dihalopropargyl phosphites ⟨76JOC3191⟩; the latter compounds are useful intermediates for access to allenyl phosphonic derivatives and oxaphospholenes. Allenyl dialkyl phosphates (**216**) have been prepared in 43–62% yields on rearrangement of stable propargylic phosphates with $AgClO_4$ as catalyst (Scheme 59) ⟨77JOC1804⟩. The same type of silver-catalyzed reaction allows the efficient conversion of alkynyl esters into allenyl esters ⟨73HCA875, 77JOC1804⟩.

The reaction of propargylic alcohols with sulfenyl chloride at low temperature in the presence of a base leads to allenyl sulfoxides via unstable propargyl sulfenates ⟨71JCS(C)1530, 88JA4062, 93SL931⟩. This method has been used for the preparation of allenyl chloromethyl sulfones (**217**) as dienophiles ⟨90JA4072⟩, and in a key step for the modification of steroids ⟨79JOC1582⟩ (Scheme 60). The utilization of sulfinyl chloride allows the obtention of allenyl sulfones via a similar intramolecular rearrangement ⟨87JOC4031⟩. Treatment of propargylic diol monothionocarbonates with $(TMS)_2NLi$ has been used to produce heterocyclic allenes (**218**) ⟨94TL1255⟩.

Allenyl trichloroacetamides (**219**) are formed by thermal interconversion of propargylic trichloroacetimidate, which isomerize into trichloroacetamidodienes on prolonged heating ⟨80ACR218⟩. Propadienyl isothiocyanates (**221**) have been prepared by thermolysis of propynyl thiocyanates (**220**) via a sigmatropic rearrangement in 7–20% yield in dilute solution at 60–100 °C, or better in 95–100% yield in the gas phase at 400 °C ⟨92AG(E)90⟩.

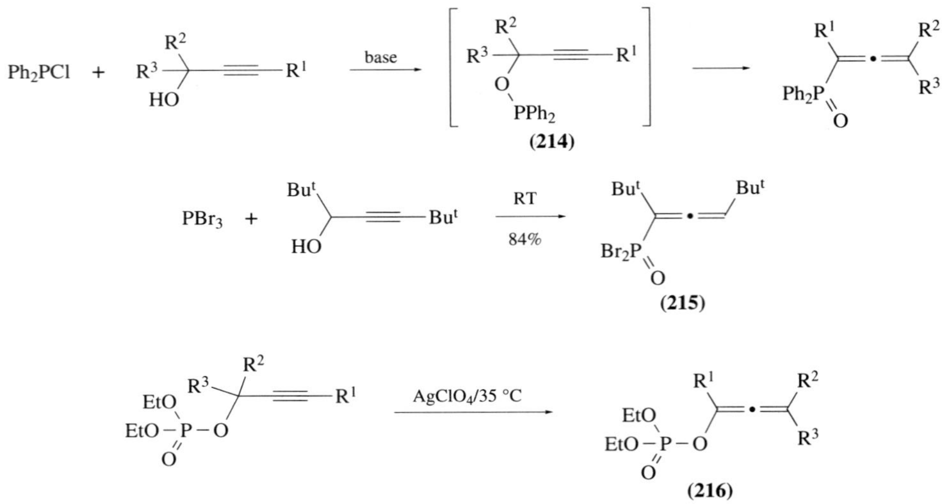

Scheme 59

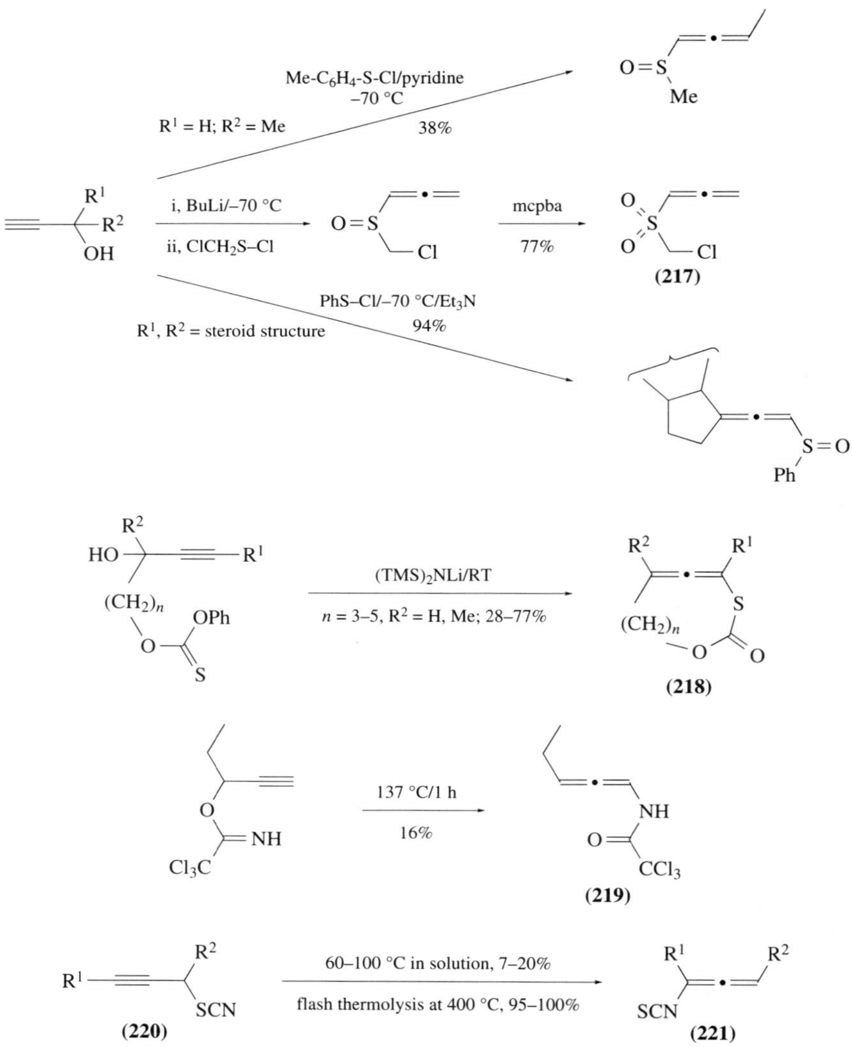

Scheme 60

1.20.4.9 Miscellaneous Reactions

Push–pull butatrienes like 1,1-bis(dimethylamino)-4,4-dicyanobutatriene (**222**) ⟨93TL1779⟩ and 1,1-bis(dimethylamino)-4,4-bis(methoxycarbonyl)butatriene (**223**) ⟨78TL4263⟩ have been synthesized by coupling 1,1-dimethylaminoethylene with 1,1-dichloroethylene compounds. Homologation of alkynic compounds to allenes has been performed with formaldehyde and diisopropylamine in the presence of CuBr as catalyst in refluxing dioxan ⟨79CC859⟩. On bromination, divinylalkyne derivatives containing naphthyl substituents lead to 1,6-dibromocumulatrienes (**224**) via 1,6-addition of bromine (Scheme 61) ⟨90PJC123⟩. Deprotonation of the bisallene (**225**) by 2,2,6,6-tetramethylpiperidine lithium, followed by oxidation with $CuCl_2$ at $-80\,°C$ gives the hexapentaene (**226**) ⟨90PAC531⟩.

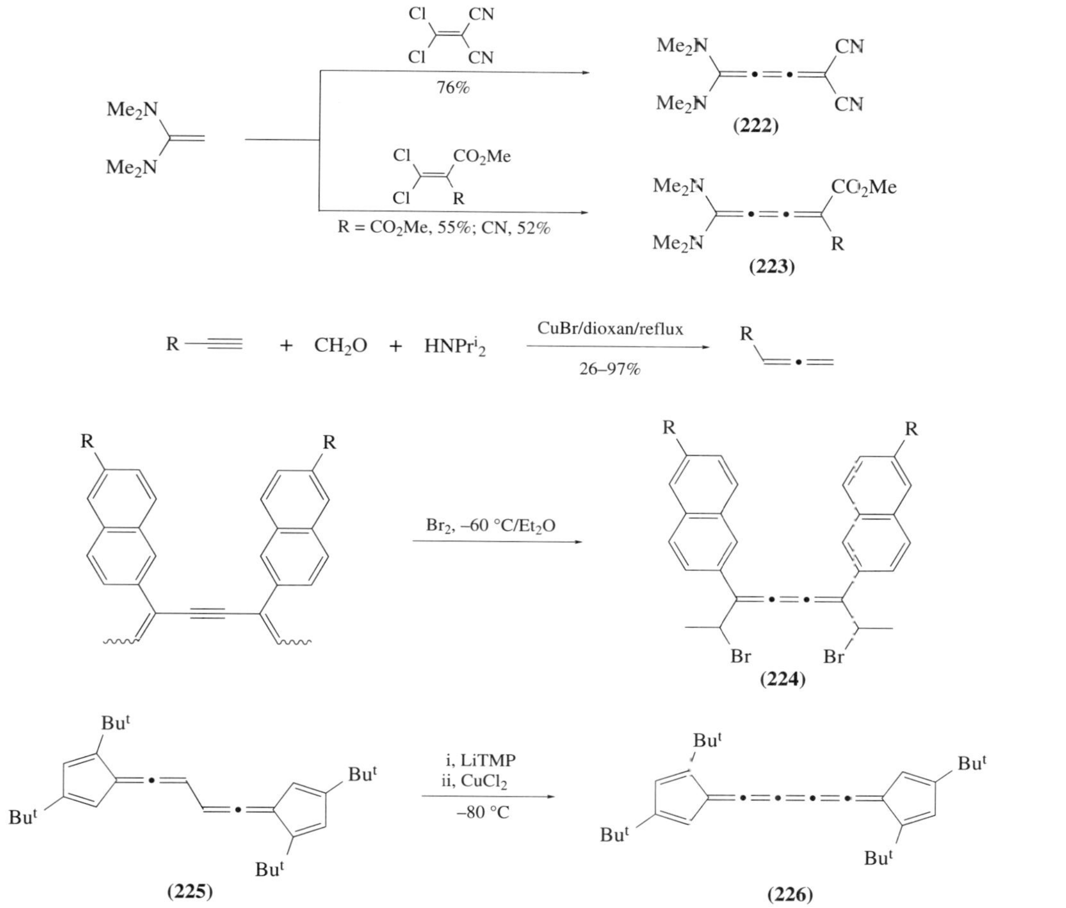

Scheme 61

1.21
Alkynes

MARK FURBER
Astra Charnwood, Loughborough, UK

INTRODUCTION

Alkynes are of enormous value both as products in their own right and, more importantly, as synthetic intermediates from which the triple bond can be readily converted into a number of other functional groups. Consequently, a number of synthetic methods have been developed for the introduction of alkyne units into organic molecules. Many of these methods have been described, for example in books by Raphael ⟨B-55MI 121-01⟩, Rutledge ⟨B-68MI 121-01⟩, Viehe ⟨B-69MI 121-01⟩ and Brandsma ⟨B-71MI 121-01, B-88MI 121-01⟩ and in reviews by Ben-Efraim ⟨B-78MI 121-03⟩, Friedrich ⟨B-83MI 121-03⟩ and Jones and Stanforth ⟨B-91MI 121-01⟩. In addition, more specialised reviews are available and these are referenced in the appropriate sections of this chapter.

For the purpose of this review the preparation of alkynes is divided into methods of C—H bond formation, C—C single bond formation and C—C triple bond formation (Figure 1). C—H bond formation is clearly only applicable to terminal alkynes and even then has little practical value aside from alkyne protection–deprotection sequences. C—C single bond formation encompasses the large class of substitution and addition reactions of terminal alkynes and their anions, and also the somewhat smaller class of substitution and addition reactions of propargyl/allenyl anions. C—C triple bond formation is dominated by elimination reactions, although fragmentation, isomerisation and electrocyclic reactions are also included within this section.

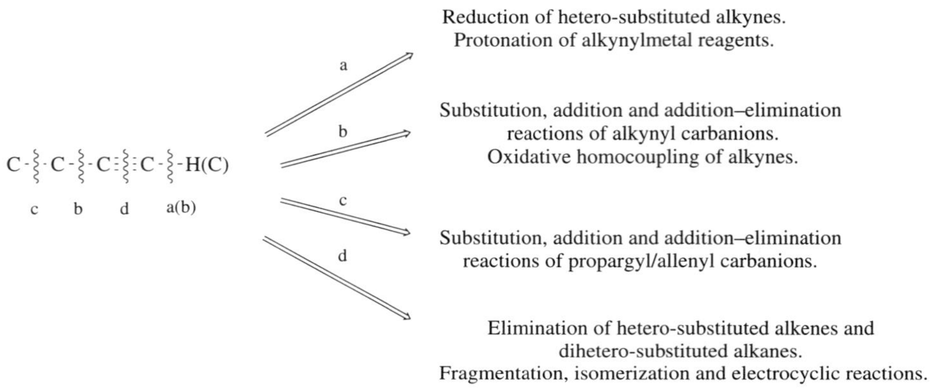

Figure 1

Figure 1 Methods of Alkyne Formation.

1.21.1 ALKYNES BY C—H BOND FORMATION

Aside from the production of ethyne and propyne by the hydrolysis of carbides, terminal alkynes can be prepared by the cleavage of alkynyl–heteroatom and alkynyl–metalloid/metal bonds (Equation (1)). In most cases the reaction is of little synthetic value since most effort has been devoted to the reverse transformation, that is, the conversion of 1-alkynes into the corresponding 1-substituted derivatives. Exceptions are the usually trivial hydrolysis of alkynylmetal compounds and the cleavage of alkynyl–silicon bonds in alkyne protection/deprotection sequences.

(1)

X = heteroatom, metalloid or metal function

1.21.1.1 Reduction of Alkynyl–Heteroatom Functions

1.21.1.1.1 Reduction of alkynyl halides

1-Alkynyl halides are most commonly converted into the corresponding free alkynes by treatment with organolithium reagents followed by hydrolysis of the intermediate alkynyllithium. The reaction is applicable to alkynyl chlorides, bromides and iodides. Similar halogen–metal exchange reactions are observed on treatment of 1-bromo- and 1-iodoalkynes with Grignard reagents or with the sodium salt of diethylmalonate ⟨63RTC183⟩. 1-Alkynyl halides have also been converted into the free alkynes by treatment with metallic sodium in ether ⟨B-69MI 121-05⟩, copper metal in acetic acid ⟨82TL2373⟩ and with a copper hydride reagent derived from potassium tri-*s*-butylborohydride and copper(I) iodide ⟨74CC762⟩.

1.21.1.1.2 Reduction of alkynyl-oxygen, -sulfur, -selenium or -tellurium functions

1-Alkynyl ethers can be reduced to 1-alkynes by diisobutylaluminum hydride, as illustrated by the conversion of 1-methoxyhexyne into hexyne in 32% yield ⟨66JOC329⟩. 1-Alkynylthioethers are most commonly cleaved by reaction with alkali metal amide and alkyllithium reagents ⟨64RTC307, B-69MI 121-06⟩. The course of the reaction depends on the structure of the thioether, but in the absence of an acidic γ-proton exclusive nucleophilic substitution at sulfur occurs to furnish the 1-alkyne on hydrolysis. 1-Alkynyl selenides display an even greater propensity to undergo substitution at the heteroatom with organolithium reagents, and generally afford 1-alkynes on hydrolysis of the intermediate alkynyllithium. In comparison, 1-alkynyl ethers undergo addition–elimination reactions with organolithium reagents (see Section 1.21.3.1.2(iii)). Alkynyl tellurides have been reduced with sodium borohydride in ethanol to afford 1-alkynes ⟨88OM84⟩.

1-Alkynyl sulfoxides afford 1-alkynes on treatment with Grignard reagents ⟨74RTC240⟩. The reaction presumably proceeds by attack of the Grignard reagent at the sulfur atom of the sulfoxide. 1-Alkynyl phenylselenoxides undergo hydrolysis with aqueous base to afford 1-alkynes and, in view of their ready removal by mcpba oxidation and hydrolysis, phenylselenoethers have been proposed as protecting groups for 1-alkynes ⟨86SC283⟩. In an isolated example phenyl (*p*-methoxyphenyltelluro)ethyne has been reported to afford phenylethyne as the sole product on oxidation with mcpba ⟨90SC2181⟩.

1.21.1.1.3 Reduction of alkynyl-nitrogen, -phosphorus, -arsenic, -antimony or -bismuth functions

1-Alkynyl amines are not readily converted into 1-alkynes. In contrast, 1-alkynyl phosphines undergo cleavage of the carbon–phosphorus bond on treatment with strong alkali ⟨63N403⟩, organolithium reagents ⟨62RTC993⟩ and lithium metal ⟨66JOM(5)205⟩. In a similar manner, the action of PhLi on the oxide of diphenylethynylphosphine leads to the formation of phenylethyne ⟨66BSF971⟩. Alkynyl derivatives of arsenic ⟨57ZAAC(290)348⟩, antimony ⟨61ZAAC(312)186⟩ and bismuth ⟨62ZAAC(317)54⟩ readily undergo cleavage of the carbon–heteroatom bond affording the 1-alkyne with water or aqueous base.

Alkynyl phosphonium and arsonium salts are readily cleaved by water ⟨62CB319, 64TL983⟩, aqueous base ⟨87IJC(B)77⟩ or basic alumina ⟨56ZAAC(287)261⟩.

1.21.1.2 Reduction of Alkynyl–Metalloid Functions

The alkynyl–boron bond of 1-alkynylboranes and 1-alkynylboronates is cleaved by a variety of proton donors, that is, proton acids, alcohols and water ⟨83COMC-I337, 88TL2631⟩. The reaction of alkynylborates with proton donors is dependent on the acidity of the reagents and the migratory aptitude of the other ligands on boron. Weak proton donors and ligands of low migratory aptitude favour protonolysis of the alkynyl–boron bond. Strong acids and ligands of high migratory aptitude favour protonation of the alkynyl group followed by 1,2-migration of an organic group from boron to the alkynyl α-carbon atom ⟨83COMC-I337, 88TL4181⟩.

The cleavage of alkynyl–silicon bonds is a valuable reaction in organic synthesis since silyl protection is commonly used for the storage, protection and selective functionalisation of terminal alkynes. Commonly employed deprotection conditions include: tetrabutylammonium fluoride in THF or DMF ⟨73TL1495⟩, potassium fluoride dihydrate in methanol ⟨67JOM(8)239⟩, silver nitrate in aqueous methanol followed by potassium cyanide ⟨67RTC1138, 68TL5041⟩ or potassium iodide ⟨87TL3923⟩, methanolic alkoxide ⟨51JA5864⟩ and carbonate ⟨92AG(E)1240⟩ solutions, aqueous solutions of sodium hydroxide with ⟨86SC983⟩ and without ⟨65JOM(4)217⟩ phase-transfer catalysis and aqueous solutions of sodium bicarbonate ⟨69JCS(C)2273⟩, potassium carbonate ⟨73T883⟩ and sodium borate ⟨72JOM(37)45⟩. The rate of protodesilylation depends on both the nature of the cleavage reagent and on the particular silicon and carbon substituents. Through appropriate choice of substituents, selective cleavage reactions can be achieved.

The cleavage of alkynyl–germanium bonds has been accomplished under aqueous base ⟨63RTC183, 65JOM(4)217⟩ and acid conditions ⟨60BSF856, 64JOM(1)420⟩.

1.21.1.3 Reduction of Alkynyl–Metal Functions

Alkynyl–metal bonds are cleaved by acid hydrolysis. For active metals such as Mg, Li, etc., water is sufficiently acidic. For alkynyl derivatives of less electropositive metals such as copper ⟨72S63⟩, tin ⟨65BSF35⟩ and lead ⟨68BSF1085⟩, stronger acid conditions are required. Alkynylmercury compounds have been cleaved with cyanide ion ⟨58JA6626⟩.

1.21.2 ALKYNES BY C—C BOND FORMATION

Alkynes are frequently prepared by substitution, addition and addition–elimination reactions. Many of these reactions can be further categorised into reactions of alkynyl carbanions and reactions of propargyl/allenyl carbanions.

1.21.2.1 Substitution Reactions of Alkynyl Carbanions

The coupling of *sp* and sp^3 centres can be achieved by the alkylation of an alkynyl carbanion or by alkynylation of an alkyl organometallic reagent (see Section 1.21.2.1.1(iii)). In practice, by far the most important process involves the use of a deprotonated alkyne as the nucleophile. Detailed procedures for the preparation of alkynyl carbanions are described in two excellent books by Brandsma ⟨B-71MI 121-01, B-88MI 121-01⟩.

1.21.2.1.1 Substitution of halogen

(i) Substitution involving alkyl halides

The alkylation of terminal alkynes has traditionally been carried out using sodamide suspended in liquid ammonia to form a sodium alkynide, which is then alkylated with the appropriate alkyl halide ⟨B-71MI 121-01⟩. Essentially the reaction is limited to primary alkyl halides with a carbon chain unbranched at the *β*-position. *β*-Branched primary alkyl halides give a mixture of substitution and elimination products while secondary and tertiary alkyl halides (and also primary alkyl halides containing an electron-withdrawing group at the *β*-position) undergo almost exclusive elimination. Even with primary alkyl halides variable yields have been observed due to the low solubilities of higher alkyl halides (greater than C_5) in liquid ammonia. To overcome this problem other solvents have been investigated both alone and as cosolvents with liquid ammonia (Scheme 1) ⟨37JOC1, 50JCS115, B-71MI 121-01, 74S441, 86TL573, 88S250⟩.

The use of lithium alkynides in liquid ammonia with THF as the cosolvent has been reported to give good yields with alkyl bromides ⟨77T1845⟩. Dipolar aprotic solvents such as DMF, hexamethylphosphoramide (HMPA) and DMSO have been used successfully. DMF has been used as solvent alone ⟨59AG245, 61T18⟩ or as a cosolvent with THF ⟨65BSF859⟩ or xylene ⟨59JOC840⟩. HMPA has been used very successfully as solvent alone ⟨73SC245, 74SC199⟩ or as a cosolvent with THF ⟨65BSF859, 65BSF1872, 72S567, 75S423⟩. Reactions occur rapidly at or below room temperature and yields are high even with alkyl chlorides. Since the early 1980s, however, HMPA has been shown to be highly carcinogenic and the HMPA substitute *N,N′*-dimethylpropyleneurea (DMPU) might prove preferable since it has been observed to promote the alkylation of lithium alkynides with equal effect ⟨88S250⟩. DMSO has been used successfully as solvent alone or as a cosolvent with liquid ammonia, but due to competing metallation of this solvent ⟨73JOC3588⟩, the use of DMSO is best restricted to alkylations using ethynyllithium itself (or the commercially available ethynyllithium–ethylenediamine complex) ⟨74S441⟩ or lithium alkynides possessing inductively stabilising electron-withdrawing groups ⟨86TL5445⟩. Dioxan has also been used as a solvent at temperatures of 100–150 °C and gives moderate yields of branched alkynes from secondary alkyl

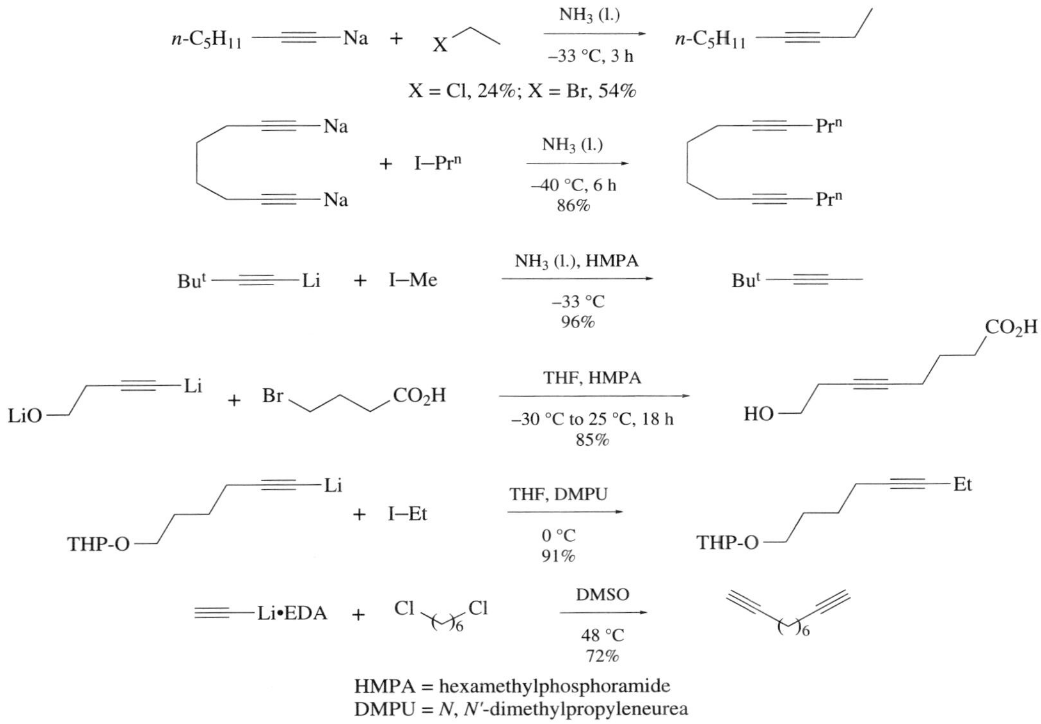

Scheme 1

bromides ⟨58LA37, 68JMC371, 68JMC373⟩. Lithium salts in refluxing THF have been recommended for the cyclisation of acyclic diterminal diynes with dihaloalkanes (Equation (2)) ⟨93S558⟩. Magnesium alkynides are inert towards even primary alkyl halides in the absence of catalysis by other metals.

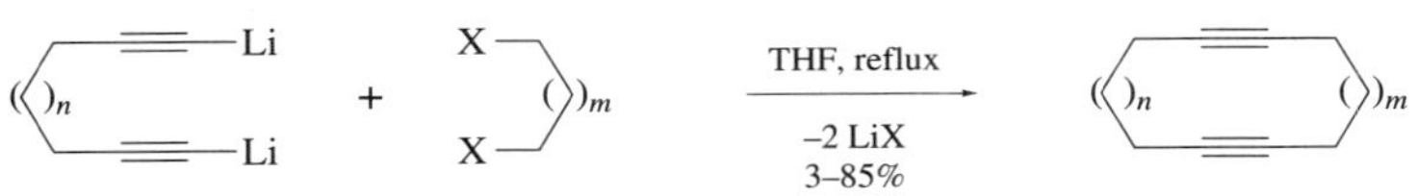 (2)

$n = 1–7$; $m = 1–7$; X = I

Unprotected ω-hydroxy-1-alkynes ⟨63JCS5889, 64BSF2477, 68AJC2989, B-71MI 121-01⟩ and ω-carboxy-1-alkynes ⟨74SC199⟩ can be selectively monoalkylated by treatment of the dilithium compounds with one equivalent of alkyl halide. Similarly, unprotected ω-haloalkanols ⟨63JCS5889⟩ and ω-haloalkanoic acids ⟨63JCS775, 63JCS5889, 65JCS894, 67TL401, 74SC199, 74TL423⟩ can be used to alkylate lithium and sodium alkynides in the presence of an extra equivalent of the metal alkynide. Carbonyl groups in either coupling partner are protected as acetal, ketal or *ortho*-ester functions. Representative examples of the alkylation of alkynyl carbanions in multifunctional substrates are given in Scheme 2 ⟨84CC350, 89CB193, 90JOC791⟩.

The alkylation of ethyne itself is complicated by disproportionation of the monoanion driven by precipitation of the dianion from many organic solvents. The monoanion is stabilised by liquid ammonia but in this solvent alkylation is poor. The commercially available complex of ethynyllithium and ethylenediamine is stable and soluble in DMSO, and alkylations occur in good yield in this solvent ⟨74S441⟩. As expected, 1-bromoalkanes and 1-iodoalkanes are more reactive and give the same or higher yields than 1-chloroalkanes, which require forcing conditions for complete reaction.

The reaction of Group 1 alkynides with allylic and propargylic halides is complicated by multiple alkylation, isomerisation and polymerisation reactions. For example, in a study of the reaction of sodium alkynides with allylic halides in liquid ammonia it was found that complex isomeric mixtures of mono-, di- and tri-allylation products were obtained ⟨75LA2130⟩. Surprisingly, addition of lithium iodide has been reported to improve matters and allows the high yield coupling of lithium hexynide with crotyl chloride ⟨80CL669⟩. This procedure is reported to be applicable to other allylic halides

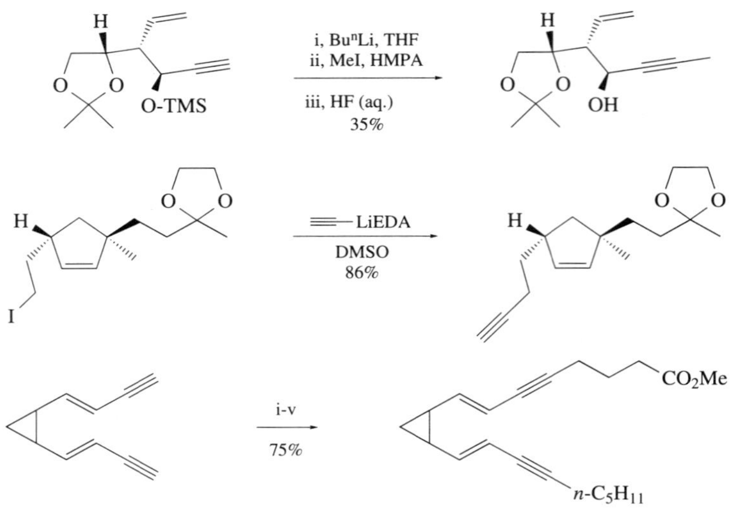

i, Bu^nLi, THF; ii, n-$C_5H_{11}I$, HMPA; iii, Bu^nLi, THF; iv, $(MeO)_3C(CH_2)_3I$, HMPA; v, H_3O^+

Scheme 2

and lithium alkynides, and has been applied in an intramolecular reaction to produce the 14-membered ring of the marine natural product crassin acetate (Equation (3)) ⟨91TL3605⟩. Group 2 alkynides show little reactivity in the absence of other metal catalysts. The use of copper(I) alkynides has, however, proven successful in these reactions ⟨B-69MI 121-03, 72S63⟩. Good results have been obtained using preformed copper(I) alkynides, or copper(I) alkynides generated *in situ* from terminal alkynes or lithium and magnesium alkynides (Scheme 3) ⟨83TL5185, 92TL135⟩. Propargylic and unsymmetrically substituted allylic halides have the potential to produce regioisomeric products and the factors governing product regiochemistry are often unclear. Propargylic halides generally afford 1,4-diynes as the main products when $R^2 =$ alkyl and allenynes when $R^2 = H$ (Scheme 4), although exceptions are known. In a similar manner, allylic halides generally undergo substitution at the least hindered end of the allyl system although the ratio of normal allylation product to isomeric allylic rearrangement product can be strongly influenced by the reaction conditions ⟨77T1845⟩. High regioselectivities have been described for the coupling of allylic and propargylic halides with 1-alkynes in the presence of copper(I) halides and 1,5-diazabicyclo[5.4.0]undec-5-ene (dbu) or 1,5-diazabicyclo[4.3.0]non-5-ene (dbn) to afford 1,4-enynes and 1,4-diynes respectively ⟨78LA658⟩. 1,4-Enynes and 1,4-diynes have also been formed in high yield from the coupling of allylic ⟨89TL2225⟩ and propargylic ⟨92TL5757⟩ halides with 1-alkynes in the presence of copper(I) iodide, sodium or potassium carbonate and tetra-n-butylammonium chloride, in DMF or acetonitrile. The mild reaction conditions (room temperature, 2–24 h) are compatible with a number of functional groups, including esters and unprotected hydroxys. Allylic bromides also undergo substitution by 1-alkynes under more classical phase-transfer conditions, using a concentrated alkali–dichloromethane biphasic system, copper(I) chloride and Aliquat-336 or triethylbenzylammonium chloride ⟨92JOC2188⟩.

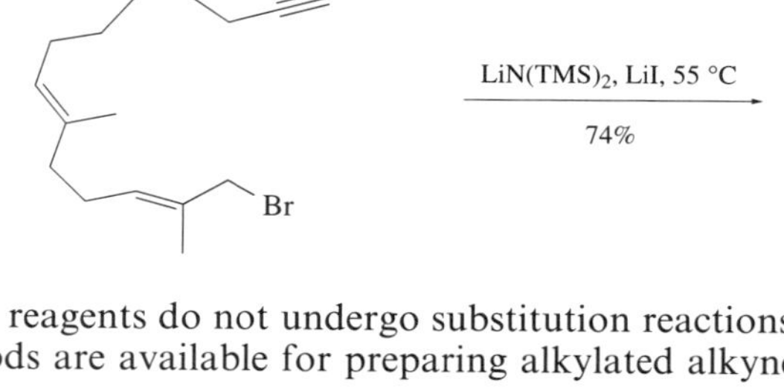

(3)

Alkynylboron reagents do not undergo substitution reactions with alkyl halides although highly successful methods are available for preparing alkylated alkynes using these reagents (see Section

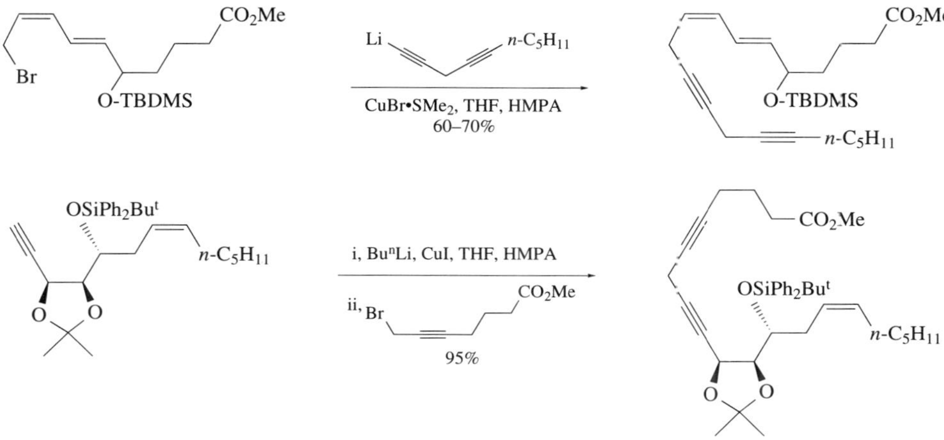

Scheme 3

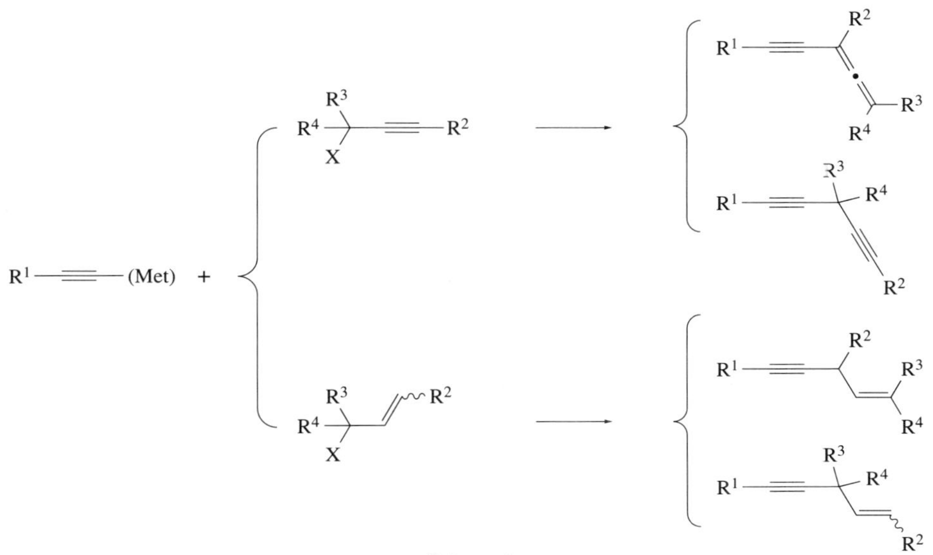

Scheme 4

1.21.3.1.5). Alkynylaluminum reagents do, however, undergo alkylation and provide one of the few satisfactory methods available for direct tertiary alkyl–alkynyl coupling. Thus, treatment of lithium alkynides with aluminum trichloride leads to trialkynylaluminum intermediates which undergo coupling with tertiary alkyl chlorides and bromides in high yield (Equation (4)) ⟨75JA7385⟩. Primary and secondary alkyl halides do not undergo substitution reactions with these reagents. The yields obtained through this procedure are much higher than those observed in the reactions of tertiary alkyl chlorides with 1-alkynes in the presence of Friedel–Crafts catalysts, where the yields of the coupling products are, in general, disappointingly low ⟨B-64MI 121-01⟩. An exception to this generalisation is the reaction of alkynylsilanes with tertiary alkyl chlorides and bromides in the presence of aluminum trichloride which can proceed in high yield ⟨82CC959⟩. This 'Friedel–Crafts like' reactivity of alkynylsilanes has been applied intramolecularly in the cyclisation of homoconjugated polyalkynes ⟨85JA6546⟩.

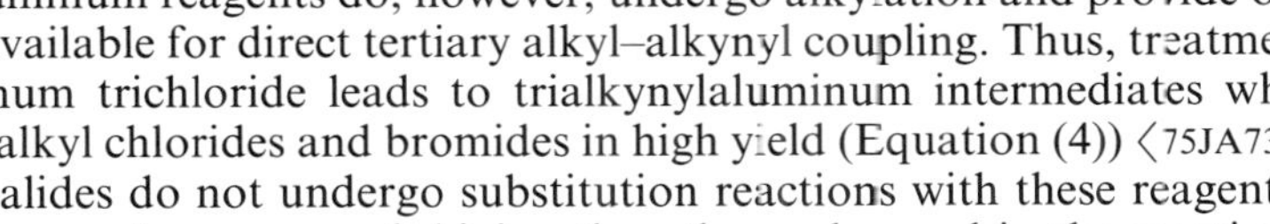

(4)

Examples of substitution by alkynyltin reagents are also known. Thus treatment of mixed halo acetals containing O, N and S heteroatoms with tri-*n*-butyltinalkynides in the presence of zinc chloride leads to the formation of α-alkynyl ethers, amines and sulfides. This methodology has been applied to the synthesis of amino acids and C-glycosides ⟨88JA2501⟩.

Transition metal mediated cross-coupling reactions have been limited to allylic and propargylic halides ⟨82ACR340⟩. The coupled product is obtained in almost quantitative yield when a copper(I) salt of phenylethyne is reacted with allyl bromide in HMPA in the presence of tetra-*n*-butylammonium iodide and $(PPh_3)_2Pd(I)Ph$ ⟨81IZV2836⟩. Alkynylzinc, copper, silver and magnesium reagents undergo palladium-catalysed cross-coupling reactions with propargyl bromides to afford high yields of conjugated allenynes ⟨81TL1451, 82RTC405⟩. Similarly, propargylic halides couple with 1-alkynes themselves in the presence of a base and $Pd(PPh_3)_4$ and copper(I) iodide catalysts to afford substituted allenynes, including 1,1-disubstituted, 1,3-disubstituted, tri- and tetrasubstituted allenynes ⟨93TL3853⟩. Alkynyltin reagents, alkynylsilanes and 1-alkynes themselves have been observed to undergo more complicated palladium-catalysed reactions with allylic halides (Equation (5)) ⟨79JOC55, 82CL1485, 90JOC5833⟩.

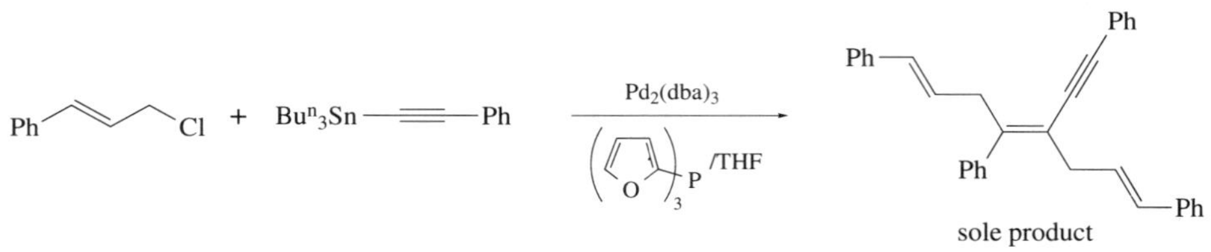

(5)

(ii) Substitution involving alkenyl, allenyl or aryl halides

The coupling between *sp* and sp^2 carbon atoms is most commonly achieved by the reaction of an alkenyl, allenyl or aryl halide with a 1-alkyne or an alkynyl organometallic reagent (Scheme 5). The reverse process, involving alkynylation of an alkenyl, allenyl or aryl organometallic, is discussed in Section 1.21.3.1.1 (iii).

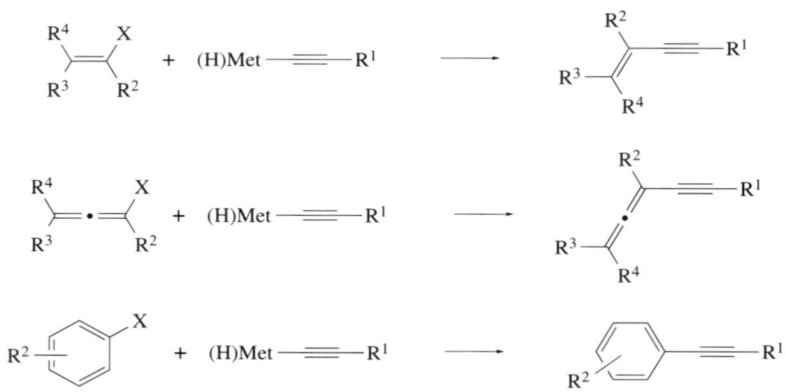

Scheme 5

Traditionally, the coupling between *sp* and sp^2 centres has been achieved through the use of alkynylcopper reagents. The Castro–Stephens reaction between an aryl iodide and a copper(I) alkynide in refluxing pyridine has been used with particular success for the preparation of diaryl alkynes (Equation (6)) ⟨63JOC3313, 64JA4358⟩. The copper(I) alkynide is usually preformed by reaction of the terminal alkyne with copper(I) chloride in an appropriate solvent ⟨72S63⟩, but it can also be generated *in situ* by reaction of the terminal alkyne with copper(I) iodide in the presence of an amine base such as dbu ⟨93JOC2694⟩, by reaction with copper powder in the presence of an inorganic base such as K_2CO_3 ⟨83JCR(M)167⟩ or by reaction of a preformed metal alkynide with a copper(I) halide ⟨77JOC2626⟩. Halogen reactivity follows the order I > Br > Cl ≫ F and, when the reaction is carried out under nitrogen to avoid oxidative coupling of the alkyne, yields are usually high. Diverse functional groups can be tolerated in both the halide and the alkyne, although nucleophilic substituents (OH, SH, NH_2, CO_2H, NO_2) in the *ortho* position of aryl halides give rise to cyclisation products ⟨66JOC4071⟩. Electron-withdrawing groups in the halide accelerate the reaction although high temperatures (ca. 100 °C) and stoichiometric quantities of copper are generally still required for the reaction. Several reviews of the reaction have been published ⟨75OR(22)253, B-74MI 121-01, B-80MI 121-01, 84T1433⟩; representative examples are provided in Scheme 6 ⟨64JA4358, 77JOC2626, 78TL3195, 83JCR(M)167⟩.

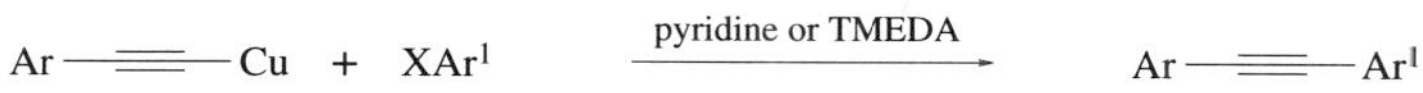

(6)

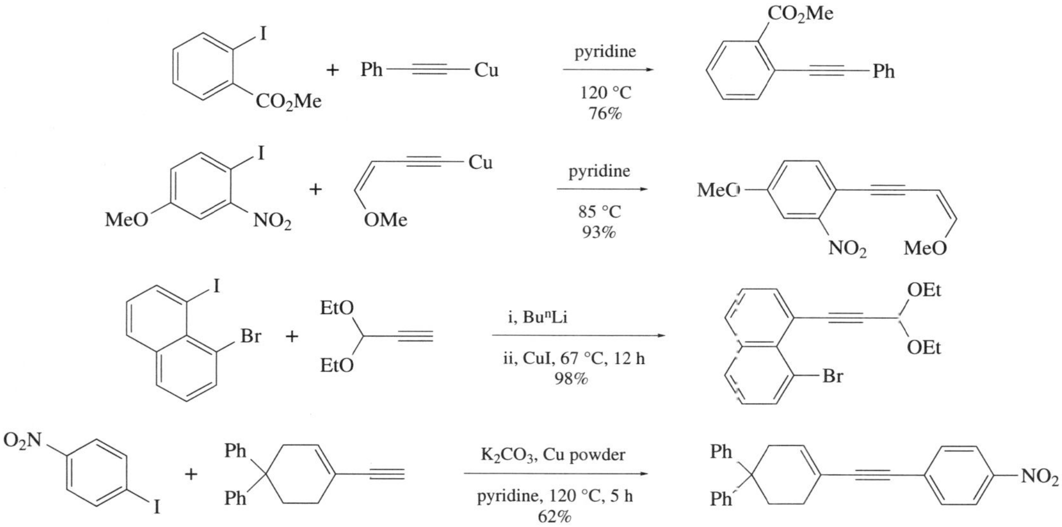

Scheme 6

Although relatively few examples have been reported, alkenyl and allenyl halides will undergo coupling reactions with copper(I) alkynides under certain conditions (Scheme 7) ⟨67CC1259, 73TL1019, 89SC2199⟩. Successful couplings have been achieved with β-bromostyrenes and with the iodine bearing carbon atoms of 1-chloro-2-iodoethylene and 1,2-diiodoethylene in refluxing pyridine or DMF ⟨67CC1259, 68JOU21, 72JCS(P1)639⟩. In *N*-methylpyrrolidone or in the absence of solvent at 220 °C, alkenyl halides have been reported not to react with copper(I) alkynides ⟨69JA6464⟩. A more generally applicable procedure is the reaction of terminal alkynes with alkenyl iodides or bromides in the presence of copper(I) iodide or bromide in HMPA ⟨89SC2199⟩. This reaction is stereospecific and is applicable to both alkyl and aryl substituted alkynes. Allenyl bromides have been reported to couple with terminal alkynes in the presence of copper(I) bromide and an amine base ⟨73TL1019, 75JCS(P1)1628⟩.

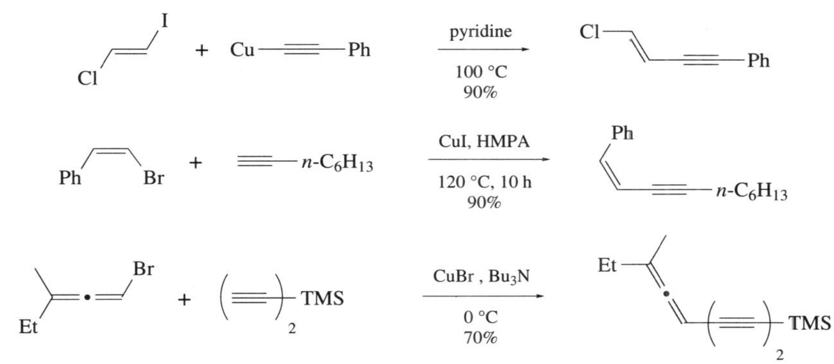

Scheme 7

Terminal alkynes can be prepared through the use of protecting groups. Ethyne, protected at one terminus, can be introduced on to the alkene, allene or aryl nucleus via coupling at the free terminus. Subsequent removal of the protecting group generates a terminal alkyne. Commonly used protecting groups are acetals ⟨77JOC1480, 77JOC2626⟩, ketones ⟨74JOU1990⟩, ketals ⟨67JCS(C)2011⟩, hydroxymethyl ⟨67JCS(C)578, 75JOM(93)253⟩, tetrahydropyran-protected hydroxymethyl ⟨67JCS(C)578, 70IZV1144⟩, dimethylcarbinol and ethyl vinyl ether-protected carbinol ⟨71IZV1534, 72IZV472⟩ and ethyl ester ⟨69JCS(C)2173⟩. The conditions required for removal of these protecting groups can be incompatible with sensitive functionality elsewhere in the product. Trimethylsilyethynyl copper is reported to be unstable under the usual Castro–Stephens conditions ⟨75JOC131⟩ while trimethylsilylbutadiyne undergoes coupling with allenyl bromides in moderate yield ⟨73TL1019⟩.

Although the use of stoichiometric quantities of copper(I) salts can be limiting, it has been reported that the coupling of alkenyl and aryl halides with terminal alkynes proceeds efficiently with catalytic copper(I) iodide in the presence of potassium carbonate and triphenylphosphine (Equation (7)) ⟨92TL5363⟩. Yields are high in the quoted examples, but the required reaction temperatures are also relatively high and it seems unlikely that this process will compete with the now widely used palladium-catalysed coupling reaction which proceeds at or below room temperature (see below).

$$R^1X \;+\; R^2{-}{\equiv} \xrightarrow[\text{DMF or DMSO, 80–120 °C, 5–26 h, 85–96\%}]{\text{CuI (cat.), 2PPh}_3\text{, K}_2\text{CO}_3} R^2{-}{\equiv}{-}R^1 \qquad (7)$$

Transition metal-catalysed cross-coupling reactions, particularly those based on palladium, have dominated the area of *sp–sp*^{2} carbon–carbon bond formation in recent years and have been used with remarkable success in natural product synthesis. Palladium mediated *sp–sp*^{2} coupling processes have emerged involving either the terminal alkyne or its metallated counterpart.

(a) Reaction of metallated alkynes with alkenyl, allenyl or aryl halides in the presence of catalytic quantities of Pd(II) or Pd(0) compounds (Equation (8)). By appropriate choice of metal, palladium catalyst and solvent, very high yields of coupling product can be obtained under mild conditions (Scheme 8) ⟨77CC683, 78JOC358, 82RTC405, 83JOM(250)551, 87JA2138⟩. With alkenyl and allenyl halides the reaction proceeds with retention of the double bond geometry, although inappropriate choice of catalyst can lead to long reaction times and to product isomerisation. Terminal alkynes can be prepared without recourse to protecting groups. With respect to the metal, the reactivity of metallated alkynes shows an interesting trend. Highly electropositive metals such as sodium, lithium and magnesium are often less effective than metals of intermediate electronegativity, possibly due to the formation of stable tetraalkynylpalladates with the former reagents ⟨87CC477⟩. In most cases alkynyltin ⟨81JOU18, 83DOK(272)1384, 83JOM(250)551, 87JA2138⟩, alkynylzinc ⟨77CC683, 78JOC358⟩ and alkynylcopper ⟨81IZV2836, 83JOM(250)551⟩ reagents appear to be optimal with respect to reactivity and toleration of functionality in both coupling partners. With organotin reagents alkyl groups have the slowest transfer rate, thus by using a trimethyl- or tributylalkynyltin reagent the alkynyl group is transferred exclusively. Although the trimethyltin reagents are more reactive and allow for easier product purification (Me_3SnI is water soluble), the tributyltin reagents are more stable and can be purified by chromatography and stored. The preparation of alkynylcopper reagents has already been described; alkynyltin and alkynylzinc reagents are generally prepared by reaction of the lithium or magnesium alkynide with a trialkyltin chloride or with zinc chloride. Alkynyltin reagents have also been prepared by the condensation of a terminal alkyne with bis(triethyltin)oxide or by reaction with an *N*-trialkylstannylamine ⟨92SL199⟩. Alkynylmagnesium reagents have been used successfully but their scope is limited ⟨84T2773⟩.

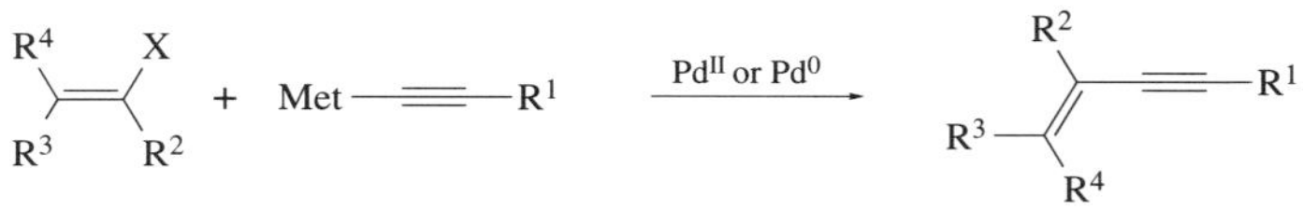

(8)

Coupling rates and yields are frequently very sensitive to the particular transition metal catalyst employed. In contrast to *sp*^{2}–*sp*^{2} and *sp*^{3}–*sp*^{2} coupling reactions, the product yields in nickel-catalysed *sp–sp*^{2} coupling processes are substantially lower than those realised under palladium catalysis, presumably because the alkyne-containing starting compounds and/or products tend to react more readily with nickel catalysts than with similar palladium catalysts to form polymers, most notably cyclic trimers and tetramers. In practice, therefore, palladium catalysts are used almost exclusively. The rate of the coupling reaction and/or the reaction temperature can also be very sensitive to the particular palladium catalyst used. Higher ratios of phosphine ligand to palladium slow the reaction and higher temperatures are generally required to achieve the same yields of products. Thus, in the

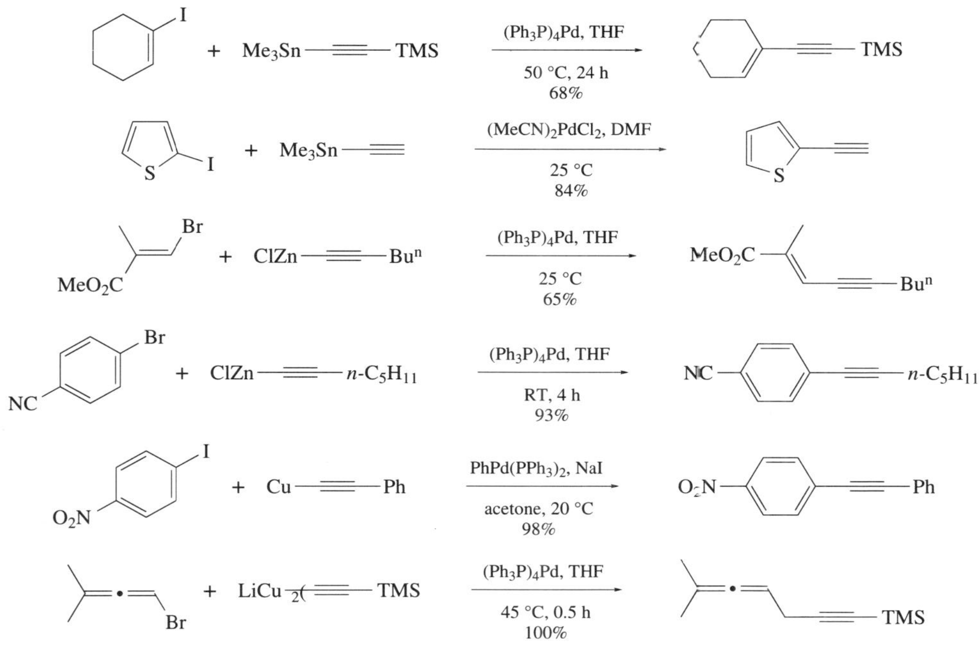

Scheme 8

reaction of alkynyltin reagents with alkenyl halides, higher reaction temperatures and/or longer reaction times are required for tetrakis(triphenylphosphine)palladium than for bis(triphenylphosphine)palladium dichloride ⟨87JA2138⟩. Bis(acetonitrile)dichloropalladium in the weakly coordinating solvent DMF is also a very active catalyst, giving high yields of coupled products instantaneously at ambient temperatures and within 3 min at $-50\,^{\circ}C$ ⟨87JA2138⟩.

Reactivity of the alkenyl, allenyl or aryl halide follows the order I > Br > Cl ≫ F and with respect to the sp^2 centre, alkenyl > allenyl > aryl. Iodides and bromides generally undergo coupling reactions under mild conditions, although unactivated aryl bromides can require elevated temperatures. Chlorides are generally unreactive unless activated by electron-withdrawing substituents. Fluorides are inert.

Marked reactivity differences between stereoisomeric alkenyl halides have been observed with the *E*-isomers showing the highest reactivity. This observation has been used very effectively in highly diastereoselective mono- and dialkynylation of a commercial mixture of *E*/*Z*-dibromoethylenes (Equation (9)) ⟨86TL4351⟩ and in the monoalkynylation of a commercial mixture of *E*/*Z*-bromotrimethylsilylethylenes ⟨88TL2239⟩.

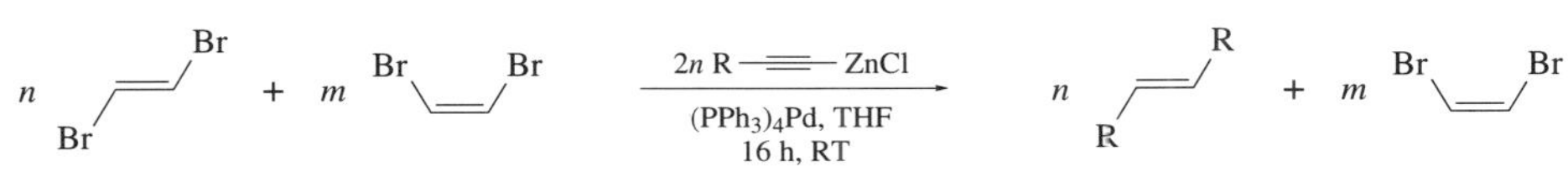 (9)

R = Bu^n, n-C_5H_{11}, Ph, TMS

A potentially very useful one-pot synthesis of stereodefined 1,5-diene-3-ynes has been described involving the sequential coupling of trimethylstannyl(trimethylsilyl)ethyne with two different alkenyl iodides (Scheme 9) ⟨89TL2403⟩. The greater reactivity of the carbon–tin bond in the presence of $Pd(PPh_3)_4$ allows selective coupling with an alkenyl iodide to give a trimethylsilyl-substituted enyne intermediate (**1**). The carbon–silicon bond is normally inert under the usual palladium-catalysed carbon–carbon bond forming conditions, but *in situ* activation with tris(diethylamino)sulfonium trimethyldifluorosilicate (tasf) enables a palladium-catalysed cross-coupling reaction with a second alkenyl iodide ⟨88JOC918⟩. Transmetallation presumably proceeds due to the enhanced nucleophilicity of a pentacoordinated silicate, or even a naked anion, generated on treatment with the fluoride ion source.

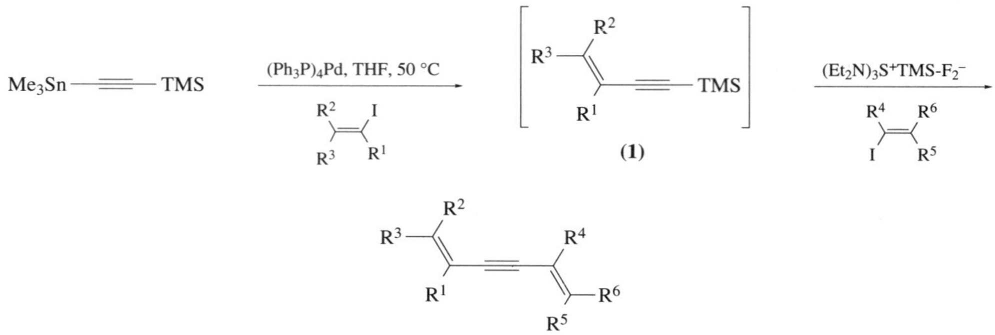

Scheme 9

Alkynylzinc reagents undergo palladium-catalysed coupling with *Z*-bromoalkenyl boranes generated *in situ* from terminal alkynes and boron tribromide (Scheme 10) ⟨88TL1811⟩. The intermediate alkenylborane can be further transformed by protonolysis or by palladium-catalysed coupling with an organic halide in the presence of a base.

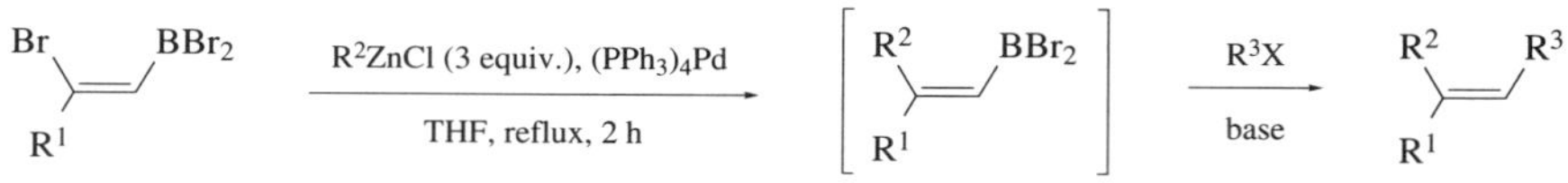

R^1 = *n*-C_6H_{13}, R^2 = $Bu^nC{\equiv}C$, R^3 = H, 61%
R^1 = *n*-C_6H_{13}, R^2 = $Bu^nC{\equiv}C$, R^3 = Ph, 54%

Scheme 10

Although the potential of this general approach to the formation of *sp*–*sp*^{2} carbon–carbon bonds is well recognised, its use for the preparation of complicated natural products and their analogues has been limited (Equation (10)) ⟨87JMC193⟩. Two reasons for this are apparent. In the first instance, the reaction requires the prior formation of a metal alkynide under conditions (strong base) which might not be compatible with functionality elsewhere in the molecule. Secondly, a milder procedure involving the palladium-catalysed coupling of terminal alkynes themselves with alkenyl, allenyl and aryl halides, and related electrophiles, is available and has been used with great success in the natural products area.

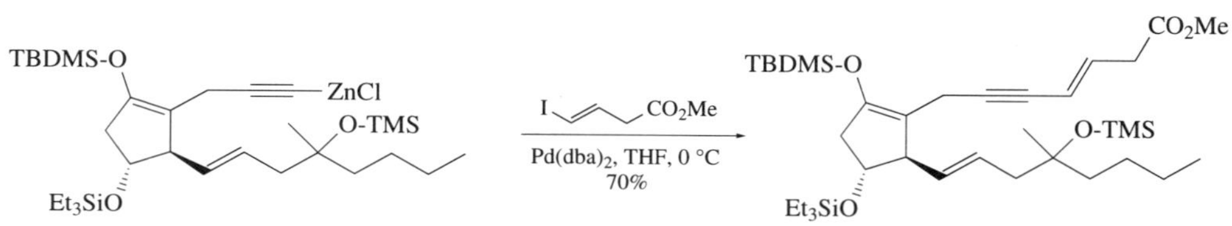

(10)

(b) Reaction of terminal alkynes with alkenyl, allenyl or aryl halides in the presence of catalytic quantities of Pd(II) or Pd(0) compounds and a base (Equation (11)). Terminal alkynes can be coupled directly to alkenyl, allenyl and aryl halides in the presence of a palladium catalyst and a base. The reaction was independently reported by three groups. Cassar demonstrated that alkenyl and aryl bromides and iodides undergo a palladium-catalysed alkynylation in the presence of sodium methoxide at 40–100°C, over 3–8 h ⟨75JOM(93)253⟩. Dieck and Heck carried out similar reactions using amine bases at 100°C, the reaction times ranging from 0.5 h to 2.25 h ⟨75JOM(93)259⟩. Sonogashira *et al.* used a $Pd(PPh_3)_2Cl_2/CuI/Et_2NH$ system at room temperature to provide a range of alkenyl and aryl substituted alkynes after 3–6 h ⟨75TL4467⟩. Both copper(I) iodide and bromide have been demonstrated to accelerate these reactions significantly, and this variant of the procedure is used in most cases. In the absence of copper(I) co-catalysis no reaction generally occurs at room temperature unless the substrate is highly activated. An additional advantage to the use of copper(I)

co-catalysis is the diminished triphenylphosphine contamination of the reaction products ⟨89JOC4453⟩.

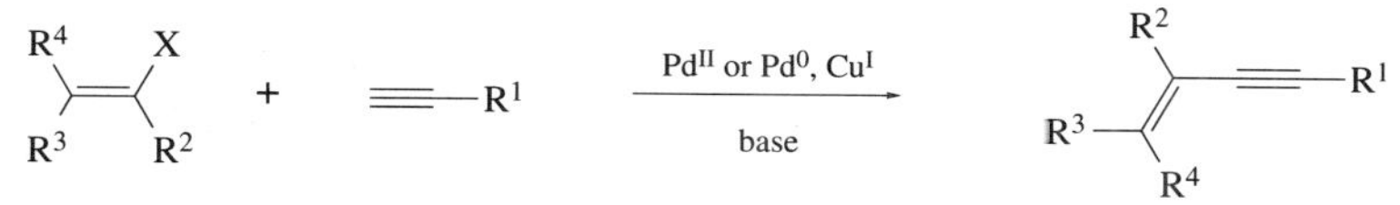

(11)

The most commonly employed palladium catalysts are $Pd(PPh_3)_4$, $Pd(PPh_3)_2Cl_2$ and $Pd(OAc)_2/PPh_3$, although other systems such as 10% Pd on carbon/PPh_3 have been used with success. Amine bases are used almost exclusively. A wide variety of amine bases including *n*-propylamine, *n*-butylamine, diethylamine, piperidine and triethylamine has been used successfully, often as the reaction solvent.

Reactivity follows the usual order I > Br > Cl ≫ F and alkenyl > allenyl > aryl. With unactivated aryl rings, iodides and bromides can be used although the iodides are substantially more reactive, requiring lower temperatures and shorter reaction times. Chlorides generally show poor reactivity unless electron-withdrawing substituents are present on or within the ring. Fluorides are inert. The reaction shows a marked sensitivity to substituent effects; the relative reaction rates for the reaction of substituted bromobenzenes with 1-heptyne catalysed by 2 mol % $Pd(PPh_3)_4$ and 3 mol % CuBr are: *p*-NO_2 (1800), *p*-CN (1190), *p*-CHO (313), *p*-H (38) and *p*-OCH_3 (1) ⟨89JOC4453⟩. With prolonged reaction times, however, even hindered and highly electron-rich aryl iodides undergo successful coupling reactions with terminal alkynes ⟨89JOC249⟩. Electron-withdrawing groups directly attached to the alkyne adversely affect the coupling reaction. For example, the reaction of iodopyrimidines with ethylpropiolate fails to give the cross-coupling product, while the reaction with an alkyl or arylethyne gives the desired ethynylpyrimidines ⟨78H(9)271⟩. Similar results in the palladium-catalysed reaction of iodobenzenes with perfluoroalkylethynes have been described ⟨90BCJ2124⟩. To overcome this problem 3,3,3-trifluoro-1-pentynylzinc chloride was used in the place of 3,3,3-trifluoro-1-pentyne itself, and was successfully coupled to iodobenzene in high yield. 3,3,3-Triethoxypropyne has been used successfully as a replacement for ethyl propiolate ⟨92S746⟩.

A wide variety of heteroarylalkynes have been prepared under palladium and copper co-catalysis (Table 1). Yields are invariably high and a wide variety of functional groups is tolerated in both the halide and the alkyne.

Aryl and heteroaryl dihalides and polyhalides can undergo selective mono-, di- and poly-substitution under appropriate conditions. As expected, 1,2-dibromo-4-nitro- and 1,2-dibromo-3-nitrobenzenes undergo alkynylation to rapidly provide the product in which the bromine *para* or *ortho* to the nitro group is replaced, whereas the corresponding dibromoaminobenzenes slowly provide the product of *meta* substitution ⟨89JOC4453⟩. 1,2-Dibromo- and diiodobenzene will undergo selective monoalkynylation under the appropriate conditions; the second alkynylation proceeds at approximately one-tenth the rate of the initial alkynylation, allowing a subsequent alkynylation using a different alkyne (Scheme 11) ⟨92TL2315⟩. Mixed bromoiodoaromatics were found to undergo alkynylation reactions more rapidly at the iodo positions, and some selective reactions have been achieved (Scheme 12) ⟨90JOC63⟩. Peralkynylation of polyhaloaromatic and polyhaloheteroaromatic systems is readily achieved. For instance, hexaethynylbenzene is prepared from either hexabromo- or hexaiodobenzene in moderate yield by cross-coupling with trimethylsilylethyne, followed by desilylation (Scheme 12) ⟨86AG(E)268, 90JOC63⟩.

Alkenyl and allenyl halides are more reactive than the corresponding aryl halides and undergo stereospecific cross-coupling reactions with terminal alkynes (Scheme 13) ⟨75JOM(93)253, 75JOM(93)259, 75TL4467, 81TL1451, 83S32⟩. Both bromides and iodides react readily at room temperature under copper(I) co-catalysis. As seen with aryl systems, electron-withdrawing groups on the alkenyl halide accelerate the reaction; thus coupling with a range of 3-bromopropenoates has been observed at room temperature in the absence of copper co-catalysis ⟨80JOC4925⟩. Although chlorides are less reactive, activated systems such as 1,1-dichloroalkenes ⟨87TL1649⟩ and (*E*)- and (*Z*)-1,2-dichloroalkenes ⟨85TL709, 85TL3811⟩ undergo facile coupling reactions (Scheme 14) ⟨81TL315, 85TL3811, 86TL5857⟩. The commercial availability of pure (*E*)- and (*Z*)-1,2-dichloroethylene and the ability to perform two successive and different stereospecific alkynylation reactions at the termini of these systems, has been utilised with great success in the synthesis of a number of polyene and eneyne containing natural products, particularly the eicosanoids ⟨89TL4951, 93SL217⟩ and the enediyne antibiotics ⟨91AG(E)1387⟩. The preparation of intermediates in the synthesis of lipoxin B_4 and in the synthesis of dynemicin and calicheamicin/esperamicin models (**2**) and (**3**) is illustrated in Scheme 15 and Equations (12) and (13) ⟨88JA1626, 91CL1271, 93SL217⟩.

Table 1 Palladium catalysed formation of heteroarylalkynes.

$$\text{HET–X} + \equiv\!\!-\text{R} \xrightarrow{\text{Pd}^{\text{II}}\text{ or Pd}^{0}\text{, CuI, amine}} \text{HET}-\!\!\equiv\!\!-\text{R}$$

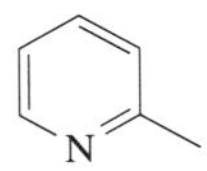	80S627 80BCJ3273 81S364 83S312	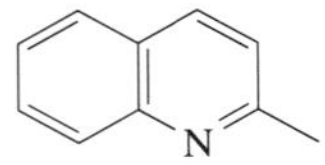	83S312 88CPB1890	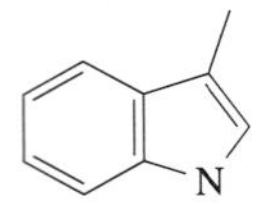	88CPB2248
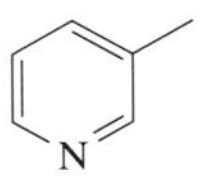	83S312 88CPB1890 92BMC1619	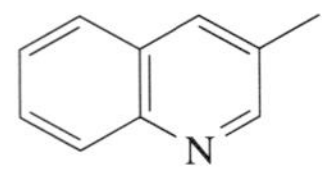	80S627 81S364 83S312 88CPB1890	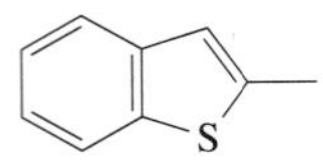	84S571
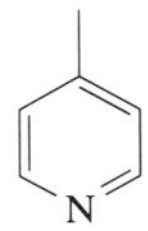	80S627 80BCJ3273 81S364 88CPB1890	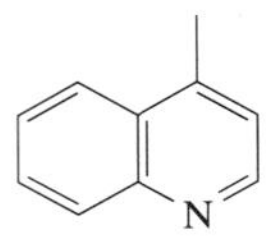	83S312 88CPB1890	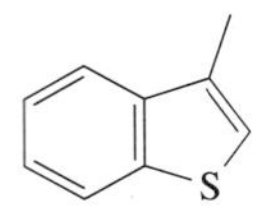	88CPB2248
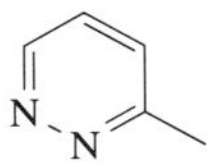	78H(9)1397 88M751 83S312	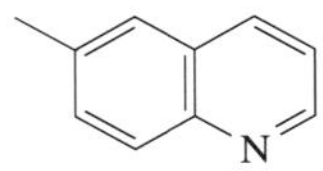	80S627 81S364	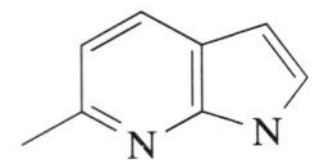	92BCJ2992
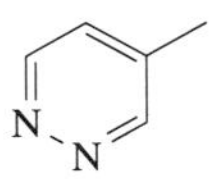	83S312	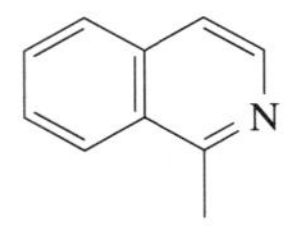	83S312 88CPB1890	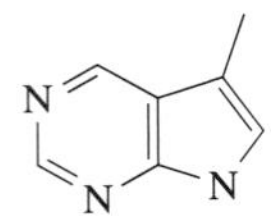	89JOC3420 90TL3731
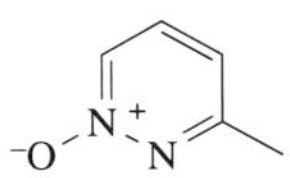	78H(9)1397	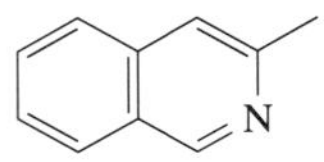	83S312 88CPB1890	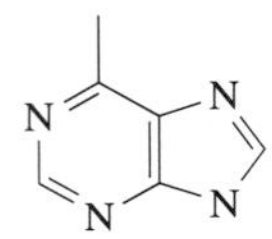	88CPB1935
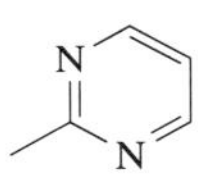	78CPB3843 78H(9)1397 83S312	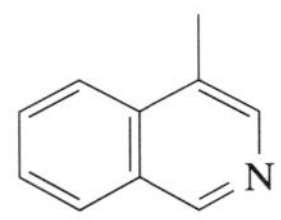	80S627 81S364 83S312 88CPB1890	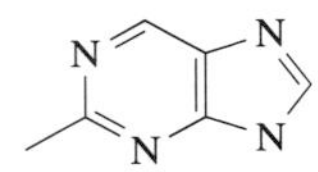	85CPB1761 92JMC241
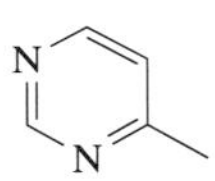	79CPB193 82CPB1865 82CPB2410 82CPB2417 83S312	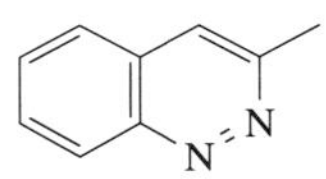	80 S627 81S364	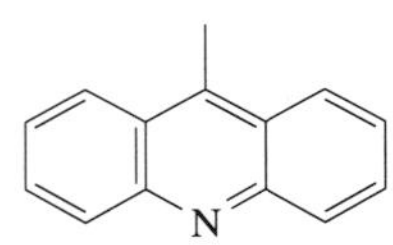	88CPB1890
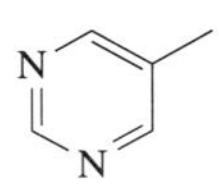	78CPB3843 78H(9)271 79CPB193 82CPB1865 82CPB2410 82CPB2417 83S312	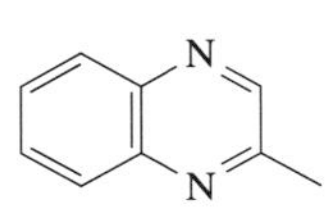	80S627 81S364	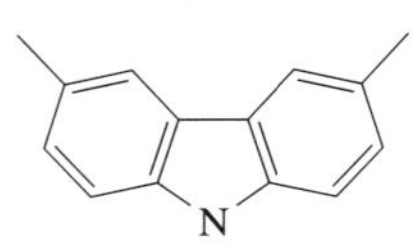	88ZOR650
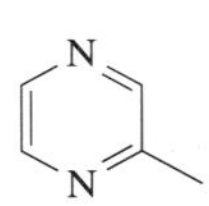	82H(19)329 86CPB1447 87JOC3997 89JOC1218	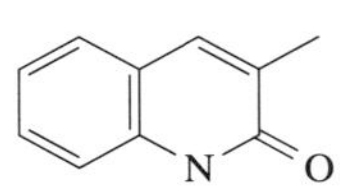	88LA69	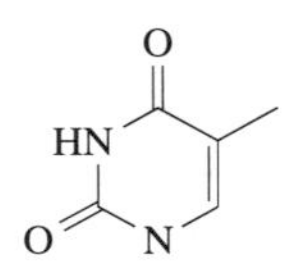	81TL421 83JOC1854 87H(26)355 89JOC3420 90TL3731
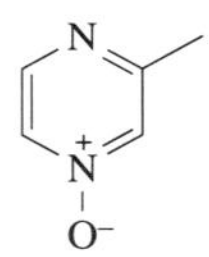	82H(19)329 86CPB1447 90JHC503	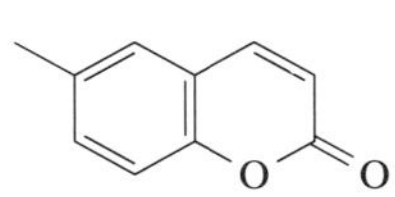	88LA543	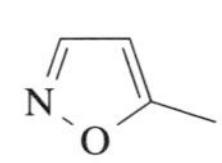	81CPB3543
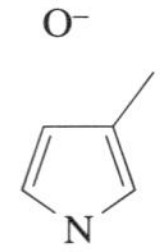	80IZV726	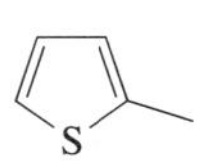	84IZV1561 85DOK(105)630 85T621	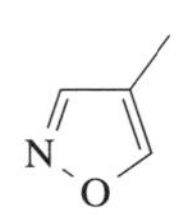	81CPB3543
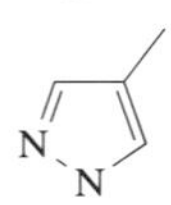	88M253	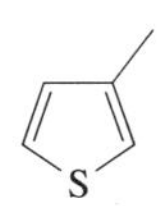	90SL427 88JOC2489	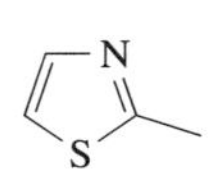	88JOC2489

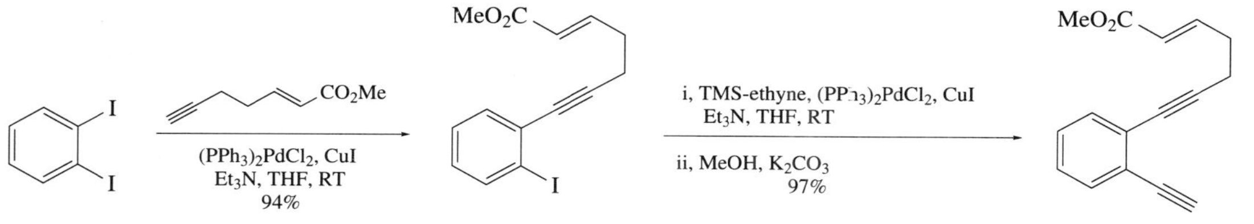

Scheme 11

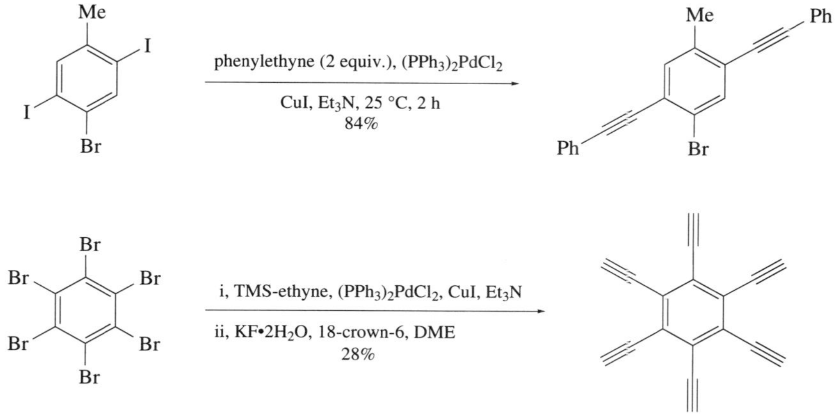

Scheme 12

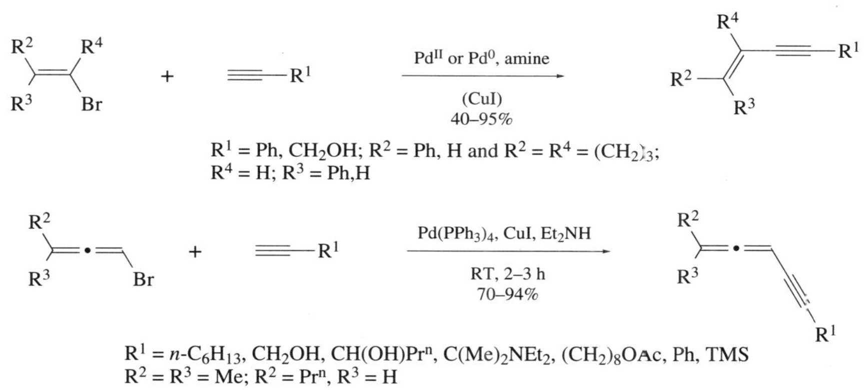

Scheme 13

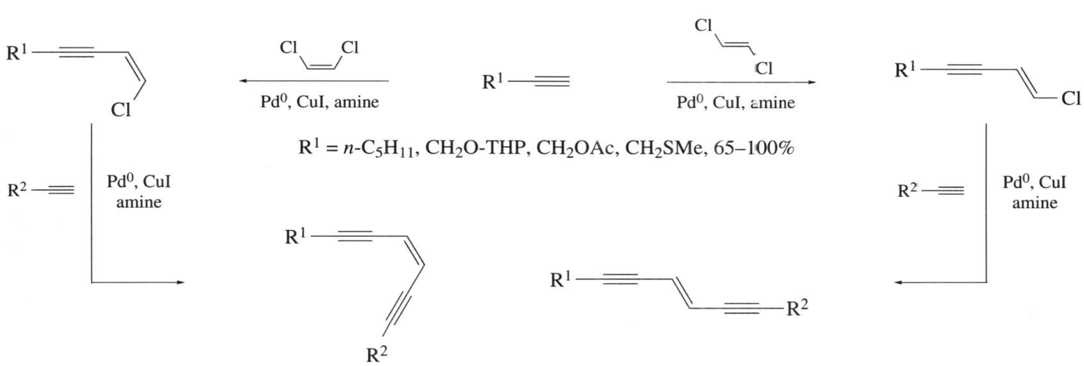

Scheme 14

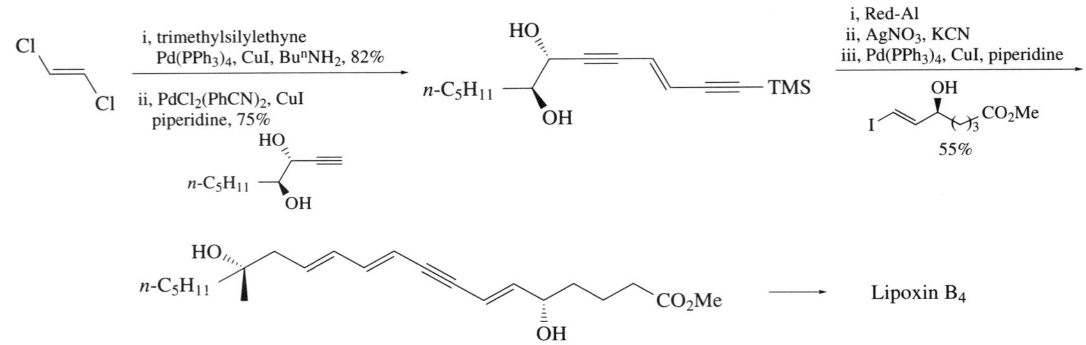

Scheme 15

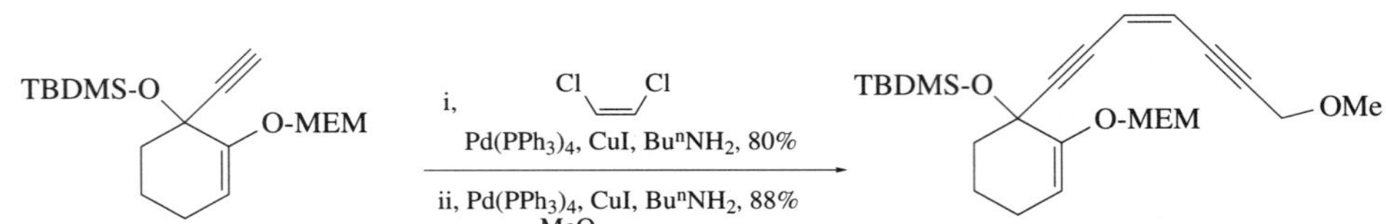

(2) (12)

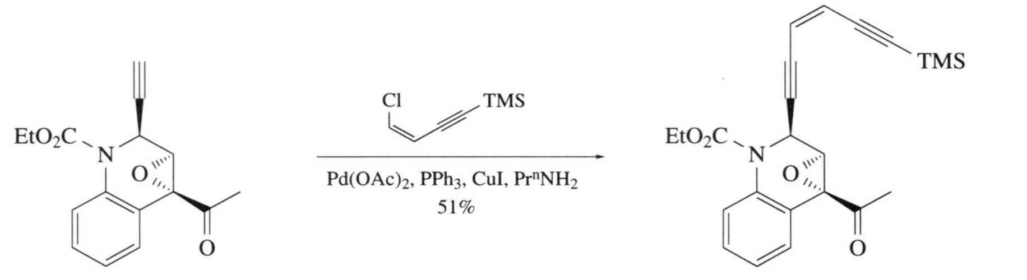

(3) (13)

Unactivated alkenyl chlorides can undergo rapid coupling reactions at room temperature using a weakly ligated palladium catalyst and the appropriate choice of amine base and solvent. Optimisation of the conditions for the coupling of heptyne with (1*E*)-chloroheptene gave a very efficient system comprising $PdCl_2(MeCN)_2$ or $PdCl_2(PhCN)_2$ (5%), CuI (10%) and piperidine as solvent ⟨91TL6109⟩. Under these conditions the coupling product was isolated in 93% yield after 1 h at room temperature. A similar yield of product was obtained after 6 h by employing 20 equivalents of piperidine in tetrahydrofuran as solvent, whereas other solvents and amine bases gave low yields of product. Alkenyl iodides under the same conditions gave lower yields of products. As an interesting comparison, when heptynyl tributylstannane was treated with (1*E*)-chloroheptene in the presence of $PdCl_2(MeCN)_2$, no reaction had occurred after 20 h at room temperature.

Terminal alkynes are most conveniently prepared by cross-coupling reactions with trimethylsilylethyne ⟨80S627⟩. Alternative procedures are available based upon coupling reactions with propargyl alcohol or 2-methylbut-3-yn-2-ol, followed by oxidative decarbonylation with MnO_2/KOH/18-crown-6 or base cleavage ⟨81S364, 84S728⟩. With ethyne itself, reaction occurs at both termini to give dialkenyl-, diallenyl- and diarylalkynes (Scheme 16) ⟨75TL4467, 88TL3651⟩.

In addition to intermolecular alkynylations, ring closures have been successfully achieved by intramolecular alkynylation (Scheme 17) ⟨88T6337, 94SL766⟩. Closure of the 11-membered ring of the calicheamicin model system (**4**) was achieved by a palladium-catalysed intramolecular cross-coupling reaction ⟨92SL766⟩. In a similar fashion, the dehydro [12] annulene (**5**) was formed by an intermolecular followed by an intramolecular palladium-catalysed alkynylation reaction employing (*Z*)-1,2-diiodoethylene; the corresponding dichloride failed to give any of the desired macrocycle ⟨88T6337⟩.

The palladium-catalysed cross-coupling of alkenyl and aryl halides with terminal alkynes has

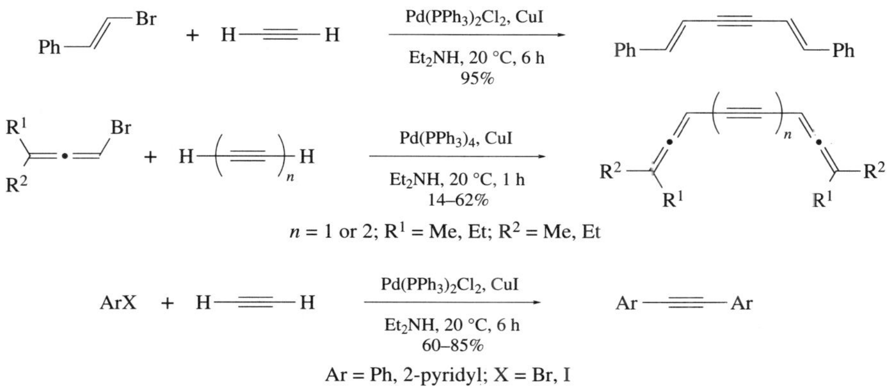

Scheme 16

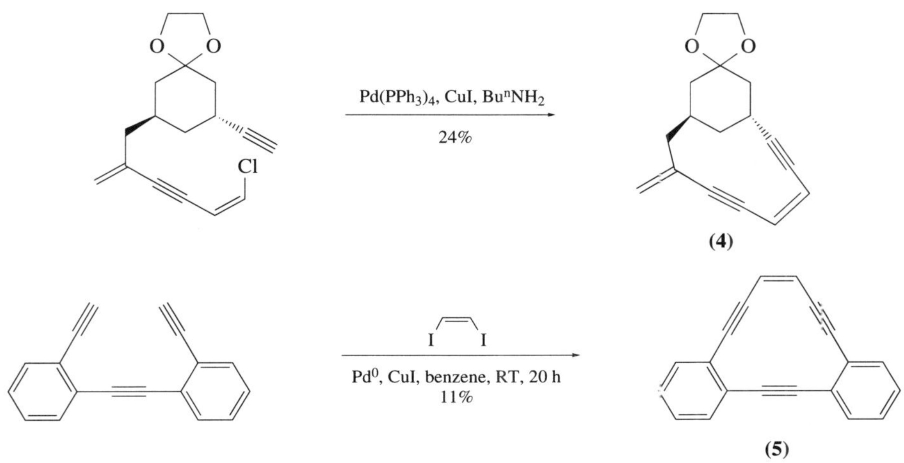

Scheme 17

been achieved both under phase-transfer catalysis and in aqueous solvent systems with water-soluble palladium catalysts. The phase-transfer conditions typically employ benzyltriethylammonium chloride (1 mol%) as the phase-transfer agent, a large excess of 2.5 N aqueous sodium hydroxide as the base, and a mixture of $Pd(PPh_3)_4$ (1 mol%) and CuI (0.5–2 mol%) as the catalyst ⟨82T631⟩. Yields are reportedly higher than those obtained under the standard anhydrous conditions, but the procedure has not been extensively used. The conditions are such that alkynyltrimethylsilanes undergo deprotection/cross coupling in a one-pot reaction ⟨84T2773⟩. In a similar manner, the monoprotected alkyne 2-methyl-3-butyn-2-ol will successfully undergo two successive arylation reactions (Scheme 18) ⟨84S571, 88T6337⟩.

Water soluble palladium catalysts based on the sulfonated phosphines $PPh_2(m\text{-}C_6H_5SO_3M)$ ($M = Na^+, K^+$) ⟨90JA4324⟩ and $P(m\text{-}C_6H_5SO_3Na)_3$ ⟨92SL715⟩ have been successfully applied to the coupling of alkenyl and aryl halides in acetonitrile/water mixtures. The reported yields are high, the reactions are carried out at room temperature, and in the case of the $Pd(OAc)_2/P(m\text{-}C_6H_5SO_3Na)_3$ system, copper(I) co-catalysis is unnecessary.

The mechanism of the palladium-catalysed cross-coupling reaction of alkenyl, allenyl and aryl halides with terminal alkynes or alkynylmetal reagents has not been determined, presumably because, in common with most other transition metal-catalysed reactions, the transient nature of the intermediates makes mechanistic studies extremely difficult. A plausible mechanistic scheme is shown in Scheme 19. A stereospecific oxidative addition of, for instance, an alkenyl iodide to Pd(0) gives an alkenylpalladium(II) iodide (**6**). The subsequent reaction between (**6**) and a terminal alkyne gives a new alkenylpalladium(II) iodide (**7**) which undergoes elimination of HPdI at elevated temperatures to produce the cross-coupled product. Alternatively, with preformed alkynylmetal reagents, or with alkynylcopper compounds generated *in situ* from terminal alkynes and copper(I)

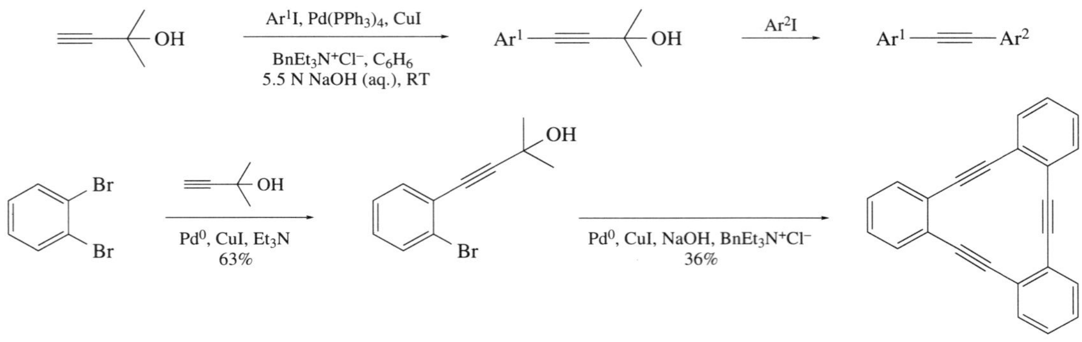

Scheme 18

halides, a preferable mechanism might involve transmetallation with the alkenylpalladium iodide (**6**), followed by reductive elimination of the cross-coupled product and regeneration of the Pd(0) catalyst.

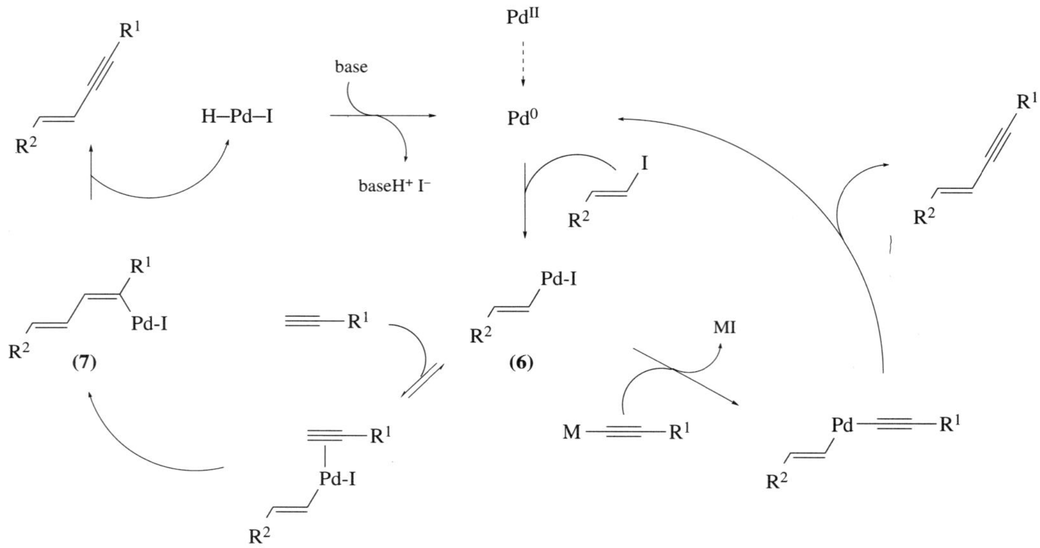

Scheme 19

It is interesting to note that in the absence of copper co-catalysis and in the presence of excess alkyne, alkenyl iodides undergo a sequence of reactions to give, not the expected enynes, but substituted fulvenes ⟨90JA9330⟩. The reaction can be understood if it is assumed that in the presence of excess alkyne the alkenylpalladium(II) iodide (**7**) is converted to the triene (**8**) (Scheme 20) ⟨90JA9330⟩. A suprafacial insertion at C-5 followed by *cis* β-hydride elimination then generates the (*E*)-fulvene. In the presence of copper(I) co-catalysis the normal cross-coupling product is obtained.

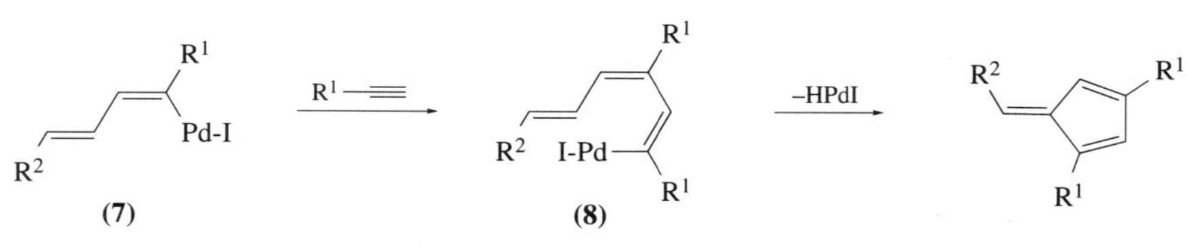

Scheme 20

In addition to these direct coupling reactions of metal alkynides and terminal alkynes, more complicated coupling reactions are known, in which insertion of an additional unsaturated function (such as carbon monoxide, an alkene, or another alkyne) into the alkenyl- or arylpalladium bond occurs prior to insertion of the terminal alkyne.

The palladium-catalysed carbonylation of alkenyl and aryl halides in the presence of a terminal alkyne gives mixed alkenyl/alkynyl and aryl/alkynyl ketones in fair to moderate yields (Equation (14)) ⟨81CC333⟩. The reaction works well with aryl iodides, but not with aryl bromides possessing strong electron-withdrawing substituents. In these cases the arylalkyne is produced exclusively. The reactions with 2-bromo-2-butene, β-bromostyrene and 2-ethoxyvinyllbromide give the phenylethynyl ketones in 47, 81 and 58% yields respectively. Isolated examples of the carbonylative coupling reaction of alkynyltin reagents with alkenyl and aryl halides have been reported. With aryl iodides the use of $(\eta^3\text{-}C_3H_5PdCl)_2$ in HMPA, DMF or acetone allows extremely mild conditions to be used; 20°C and 1 atmosphere of carbon monoxide ⟨81DOK(261)1141⟩. Cyclopentenyliodide has been coupled with pentynyltributylstannane in the presence of $(PPh_3)_2PdCl_2$ (50psi CO, 50°C, 7 h) to give the corresponding ketone in 54% yield ⟨84JA6417⟩.

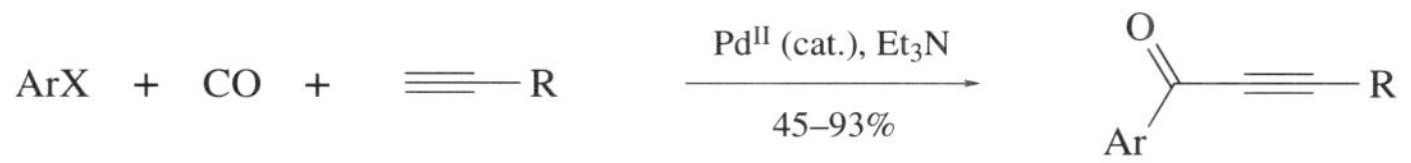

(14)

The palladium-catalysed cross-coupling of an alkenyl, allenyl or aryl halide and a terminal alkyne with the additional insertion of an alkene (Heck reaction), is a very desirable goal. In practice, the problem of rapid β-hydride elimination from the intermediate alkylpalladium species has limited this reaction to those alkenes from which β-hydride elimination is less facile, most notably norbornene and norbornadiene (Equation (15)) ⟨82TL4517, 92TL3503⟩. In contrast with the direct coupling reaction, alkenyl iodides undergo alkene isomerisation in this reaction.

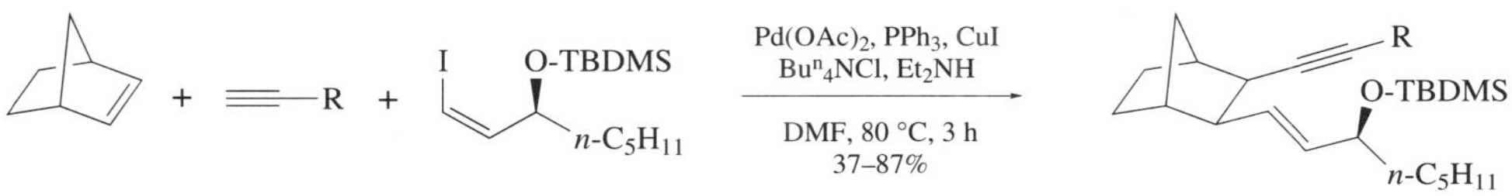

(15)

The palladium-catalysed cross-coupling of an alkenyl, allenyl or aryl halide and a terminal alkyne with the additional insertion of a second alkyne has been more successful, the crucial difference being the greater resistance of alkenylpalladium intermediates to β-hydride elimination. The palladium-catalysed reaction of alkenyliodide (**9**) with hexynylmetal reagents demonstrates a strong dependence on the metal, with the highest proportion of insertion product being obtained with hexynyltrimethylstannane (Equation (16)) ⟨90TL4393⟩. Applications of this reaction to the preparation of both carbocyclic ⟨91TL5243, 93TL2139⟩ and heterocyclic ⟨88TL5565⟩ systems have been reported. In work directed towards a synthesis of the neocarzinostatin chromophore, the dibromoalkene (**10**) was treated with an alkynyltin reagent under palladium catalysis to give the product derived from cross-coupling with insertion (Scheme 21) ⟨93TL2139⟩. In a subsequent step, the insertion product was subjected to a second cross-coupling reaction.

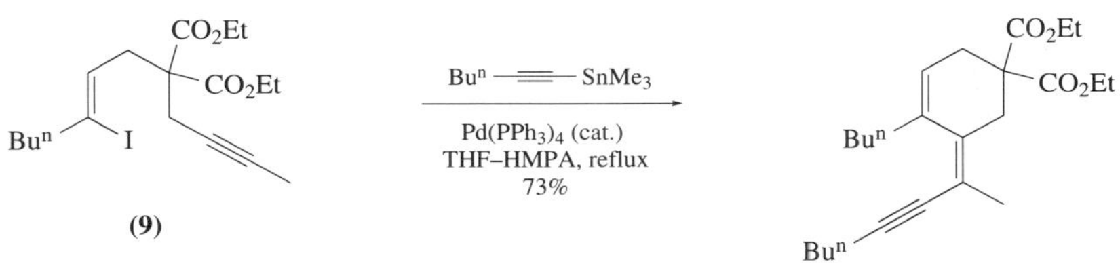

(**9**)
(16)

Finally, it should be noted that in addition to direct substitution, allenyl halides are able to undergo 1,3-substitution reactions with nucleophiles, particularly organocopper reagents, to afford alkynes (Equation (17)) ⟨81JA4618, 82TL1651, 83TL3291, 84TL3059, 86JOC2230⟩. Although the alkyne is formed in a substitution reaction, this process is quite different to those discussed so far, in that the alkyne unit is not present in the nucleophile but is introduced by a reorganisation of the allene. The regiochemistry of substitution by organocopper reagents, and hence the balance between alkynic and allenic products, is dependent on the substitution pattern of the allenyl halide and the nature of the organocopper reagent. In particular, it has been observed that 3-substituted- and 3,3-disubstituted-1-haloallenes react with lithium dialkylcuprates to afford 1-alkylallenes via direct substitution ⟨73JCS(P1)1347⟩, while cyanocuprates R(CN)CuLi react with 1-substituted- and 1,3-disubstituted-1-haloallenes to give mainly internal alkynes ⟨84TL3059⟩. With 3,3-disubstituted-1-haloallenes n-alkyl cyanocuprates afford alkynes, whereas α-branched primary, secondary and

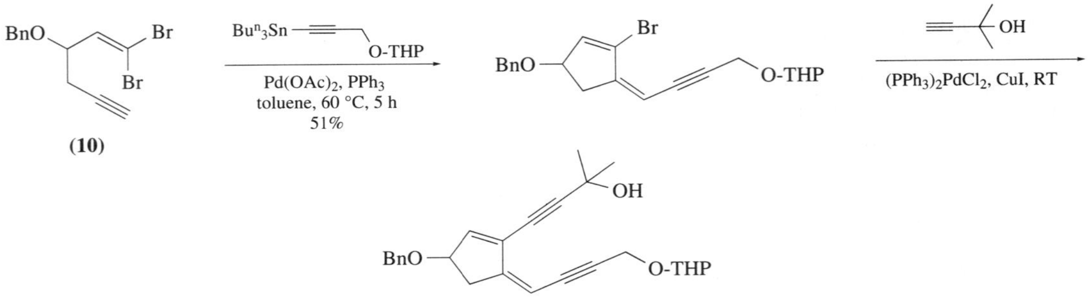

Scheme 21

tertiary alkyl cyanocuprates afford allenic products ⟨87TL6073⟩. Employing optically active 1,3-disubstituted-1-bromoallenes, S_N2' organocuprate displacements have been shown to proceed with very high *anti*-selectivity ⟨84TL3059⟩. The reaction has been applied as a key step in the synthesis of hybridalactone (Equation (18)) ⟨84JA2735⟩ and 11(*R*)-HETE ⟨81JA4618⟩, and in the synthesis of prostaglandin ⟨83TL3291⟩ and arachidonic acid ⟨82TL1651⟩ analogues.

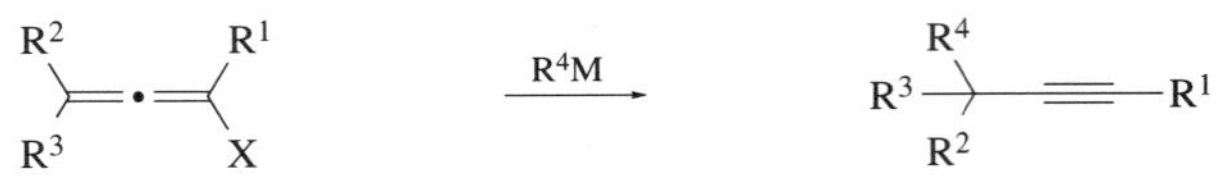

(17)

X = Br, I

R^4M = Me_3Cu_2Li, $MeCu{\bullet}Bu^n{}_3P$, RCuCNLi, ArCuCNLi, RCH=CHLi/CuCN, RCH=CHCu, RC≡CCuCNLi

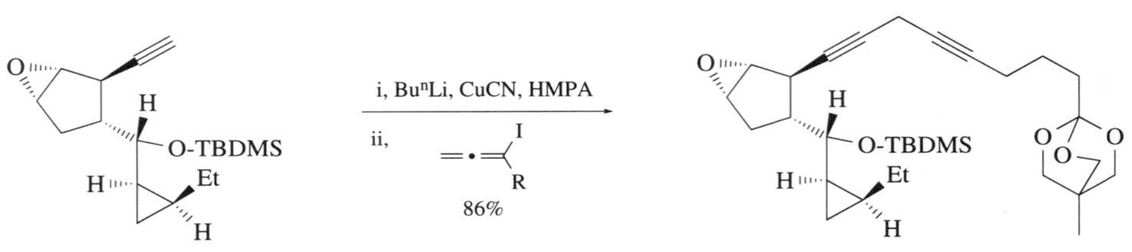

(18)

(iii) Substitution involving alkynyl halides

This section is concerned with the reaction of alkynyl halides with stabilised and unstabilised carbon nucleophiles. Although many of these reactions appear to involve electrophilic substitution of an *sp* carbon atom, simple substitution mechanisms are rarely involved. In many cases the reactions proceed through an addition–elimination mechanism. More general reviews covering the reactions of alkynyl halides with nucleophiles are available ⟨B-69MI 121-05, B-78MI 121-02⟩.

Relatively few examples of the reaction of alkynyl halides with stabilised carbon nucleophiles have been reported. Tertiary enolates undergo a substitution reaction with alkynyl chlorides bearing an anion stabilising substituent, for example, phenyl or trimethylsilyl (Equation (19)) ⟨43CB80, 84JA3551⟩; primary and secondary enolates give rise to unidentifiable product mixtures. In the absence of an anion-stabilising group no reaction occurs, supporting an addition–elimination mechanism. Alkynyl(phenyl)iodonium salts react with soft nucleophiles, such as β-dicarbonyl and aminomalonate enolates, to give the corresponding alkynylated products ⟨90CC118, 91JOC3912⟩; alkynylation of simple enolates does not occur. Alkynyl bromides react with methylene(triphenyl)phosphorane to afford propargylidene(triphenyl)phosphoranes ⟨80LA2061⟩. The ylides have been reacted with aldehydes to afford a wide range of enynes (Scheme 22).

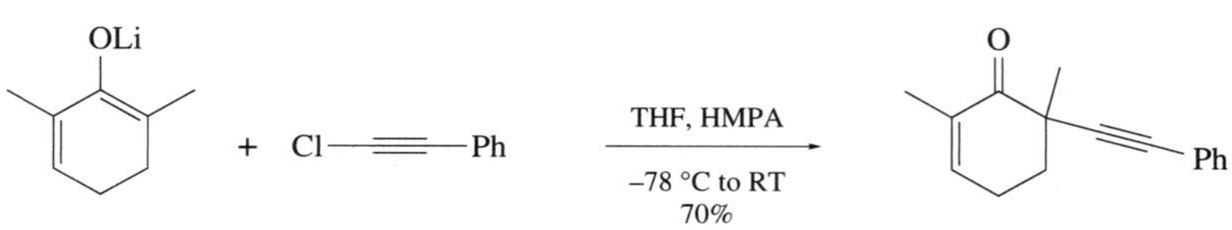

(19)

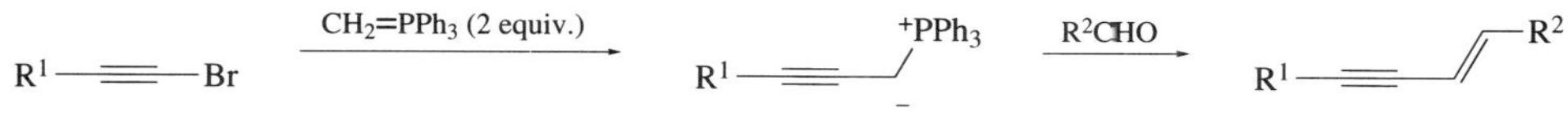

Scheme 22

The reaction of alkynyl halides with Grignard and organolithium reagents is dependent upon a number of factors including the halide, the solvent and the nature of the organometallic ligand. In the absence of catalysis by other metals, alkynyl chlorides only undergo effective coupling with Grignard reagents when activating substituents are present on the alkyne, for example, 3-alkoxychloroalkynes ⟨63BSF1876, 63BSF1888⟩. With organolithium reagents the most common reaction observed is halogen–metal interconversion with formation of an alkynyllithium. Alkynyl bromides have been observed to undergo moderately effective coupling reactions with alkylmagnesium halides in polar solvents, although competing halogen–metal interconversion does still occur. The influence of solvent polarity is demonstrated by the reaction of bromophenylethyne with pentylmagnesium bromide. In diethyl ether, phenylethyne is formed almost quantitatively. In tetrahydrofuran the yield drops to 83%, and in HMPA to 20%, with a 55% yield of substitution product being obtained in the latter solvent ⟨63BSF1876, 63BSF1888⟩. Similarly, a 50% yield of 6-dodecyne was obtained from the reaction of 1-bromoheptyne and pentylmagnesium bromide in HMPA. The reaction with Grignard reagents in polar solvents is, however, limited to alkylmagnesium halides. With either alkenyl or aryl reagents metal–halogen interchange occurs, even in HMPA. Alkynylmagnesium halides do effect substitution but the yields are inferior to those obtained with alkynyl iodides. With organolithium reagents rapid bromine–lithium exchange occurs. Alkynyl iodides generally undergo metal–halogen exchange with Grignard and organolithium reagents. An exception is the reaction with alkynylmagnesium halides which provides a convenient and high-yielding route for the preparation of symmetrical 1,3-diynes ⟨B-69MI 121-04⟩. For the preparation of unsymmetrical 1,3-diynes, the route is inferior to the Cadiot–Chodkiewicz method since it gives a mixture of both symmetrical and unsymmetrical products. Alkynyl fluorides, perhaps surprisingly, undergo very effective coupling reactions with both types of organometallic reagent. Fluoroethyne itself has been prepared from 1,1-difluoroethylene at low temperatures, and coupled *in situ* with alkyl, alkenyl and aryl Grignard and organolithium reagents (Scheme 23) ⟨82TL4325⟩. In a similar manner, aryl-substituted alkynyl fluorides have been coupled with alkyl- and aryllithium reagents (Scheme 24) ⟨82TL4329⟩.

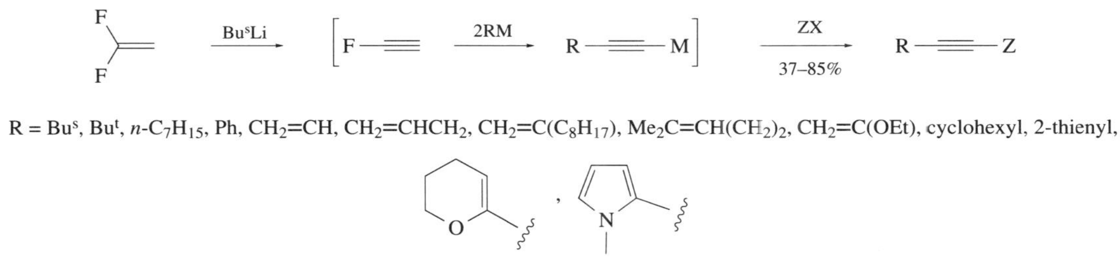

Scheme 23

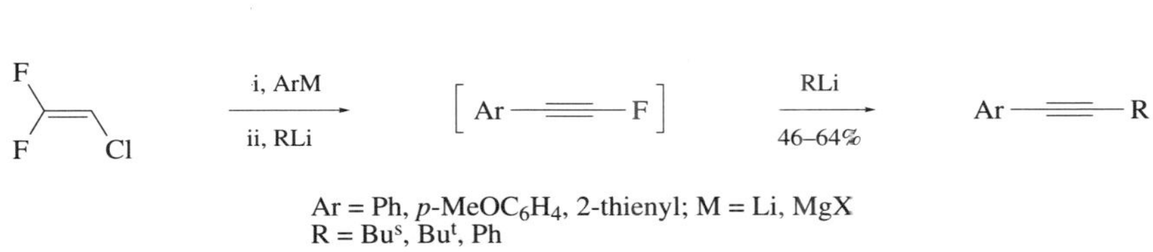

Scheme 24

Organocopper(I) compounds undergo successful substitution reactions with alkynyl halides (Scheme 25) ⟨72TL5209, 79S826, 81JOM(217)267, 89TL4799⟩. Alkylcopper reagents of the type $RCu \cdot MgX_2$ ⟨80T1215⟩ and alkylcopper/zinc reagents of the type RCu(CN)ZnX react, under mild conditions, with alkynyl bromides and iodides to afford disubstituted alkynes ⟨89TL4799⟩. Alkenyl

⟨75TL1465, 79S826, 80T1215⟩, allenyl ⟨81JOM(217)267, 88T6699⟩ and arylcopper(I) reagents ⟨72TL5209⟩ also undergo successful substitution reactions with alkynyl bromides and iodides in tetrahydrofuran or tetrahydrofuran/TMEDA mixtures; the chlorides give very low yields. Alkenylcopper(I) reagents react with retention of configuration. Interestingly, the reactions can proceed at a sufficiently fast rate that a hydroxy group in the alkynyl halide need not be protected ⟨79S826⟩. Alkenylcopper/zinc reagents have also been coupled with functionalised alkynyl halides ⟨91TL6085⟩. Alkenylcopper(I) reagents can be generated from alkenylboranes providing a method that is stereochemically complementary to the carbometallation route (Equation (20)) ⟨81JOC645⟩. The derived alkenylcopper(I) reagents couple effectively with alkynyl bromides and iodides with retention of configuration ⟨81JOC645⟩. The reaction can also be successfully achieved with bis(acetylacetonato)copper catalysis ⟨83BCJ2855, 84CL1029⟩. Terminal alkynes can be prepared using iodoethynyltrimethylsilane followed by brief treatment of the product with dilute aqueous alkali ⟨72TL5209⟩. Reactions of alkynyl halides with lithium diorganocuprate reagents or with organomagnesium halide reagents in the presence of copper(I) catalysis, proceed mainly by metal–halogen exchange and poor yields of enynes are obtained ⟨72TL5209, 79S826⟩.

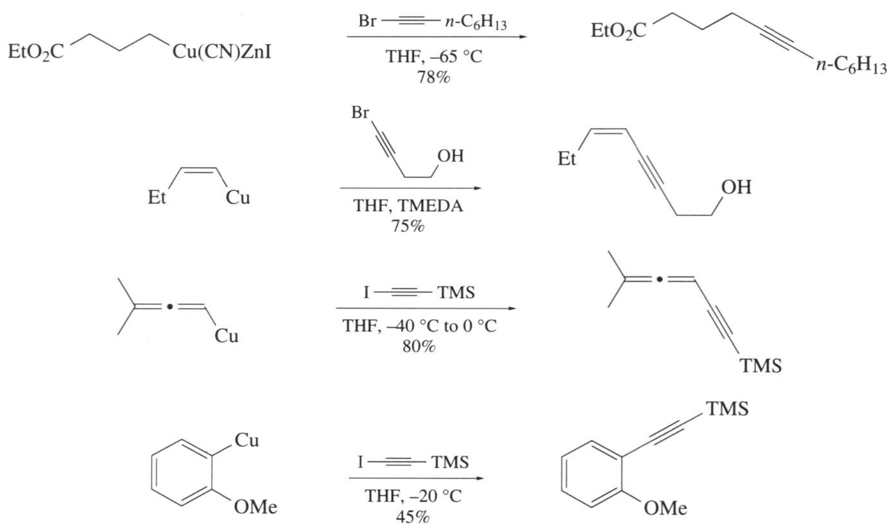

Scheme 25

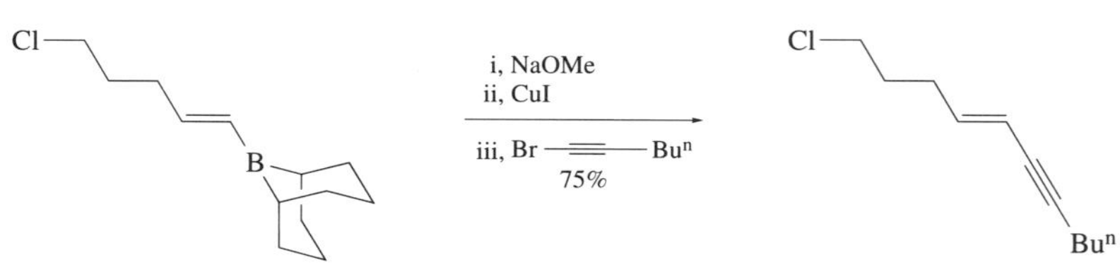

(20)

Organocopper(I) compounds also undergo successful substitution reactions with alkynyl iodonium salts. Treatment of alkynyl(phenyl)iodonium tosylates with dialkylcuprate ⟨91JCS(P1)2892⟩, alkenylcopper ⟨87JA7561⟩ and alkynylcyanocuprate ⟨91JCS(P1)2892⟩ reagents affords disubstituted alkynes, 1,3-enynes and unsymmetrical 1,3-diynes respectively (Scheme 26).

Terminal alkynes undergo coupling with alkynyl bromides in the presence of an aliphatic amine and a catalytic amount of a copper(I) salt to give unsymmetrical diynes. A reducing agent, such as hydroxylamine, is often added to maintain the level of cuprous ion. Alkynyl chlorides show low reactivity, although some chloroalkynols have been successfully employed. Alkynyl iodides have a strong tendency to undergo self-coupling to give symmetrical diynes. The reaction, commonly known as the Cadiot–Chodkiewicz coupling, is particularly valuable for the preparation of unsymmetrical diynes ⟨B-69MI 121-04⟩. Coupling almost certainly proceeds through *in situ* formation of the copper(I) alkynide and, indeed, a variant of the procedure involves the use of preformed copper(I) alkynides in pyridine ⟨68TL2919, 71JCS(C)186⟩. Some examples of the reaction are illustrated in Scheme 27 ⟨87JMC254, 92TL5355⟩. Polyalkynes are also accessible by employing the Cadiot–Chodkiewicz

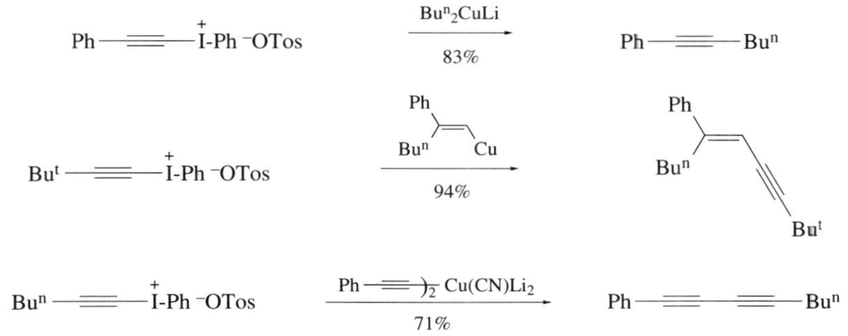

Scheme 26

reaction, although protecting group strategies are required for successful couplings (Equation (21)) ⟨90JA4054⟩. Without recourse to protecting groups polymerisation and symmetrical coupling reactions are unavoidable, employing the parent alkyne $H(C{\equiv}C)_nH$ or bromoalkyne $H(C{\equiv}C)_nBr$. Carboxylate, primary and tertiary carbinol, and trialkylsilyl protecting groups have been successfully applied to this reaction, most commonly in the bromoalkyne ⟨B-69MI 121-04⟩. A particularly convenient and high yielding procedure employs the coupling of preformed copper(I) alkynides with (bromoethynyl)trimethylsilane ⟨83S128⟩.

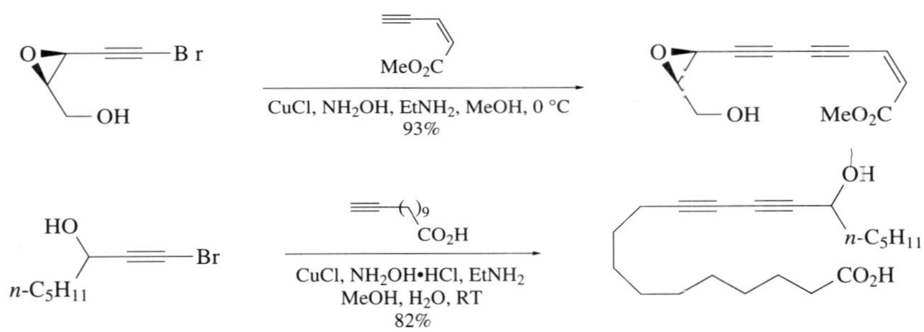

Scheme 27

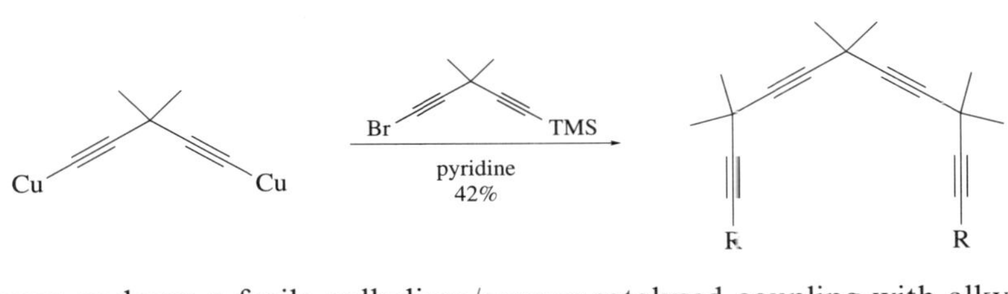

(21)

Terminal alkynes undergo a facile palladium/copper-catalysed coupling with alkynyl bromides and iodides (Equation (22)) ⟨90TL7227, 91SC977⟩. In THF, this palladium-catalysed version of the Cadiot–Chodkiewicz reaction proceeds rapidly at room temperature to give moderate to high yields of coupling products.

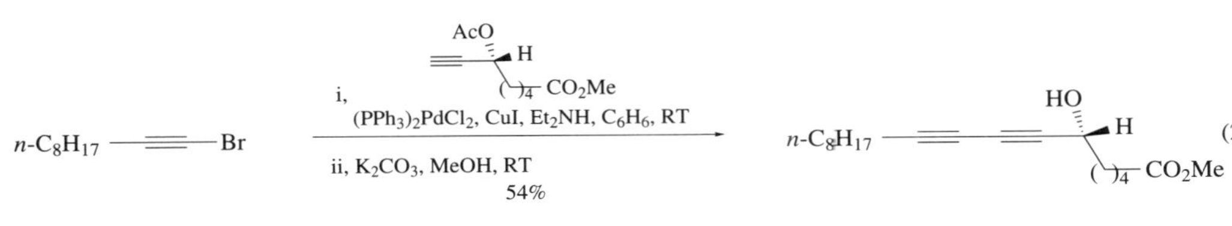

(22)

Transition-metal catalysis has also been successfully applied to the alkynylation of organometallic reagents. Alkynyl bromides react with trialkylaluminum reagents in the presence of bis(*N*-methylsalicylaldimine)nickel to give the substitution products in 42–90% yield (Equation (23)) ⟨78TL2831⟩.

Similarly, alkenyldiisobutylalanes, derived from hydroalumination of terminal alkynes, react with alkynyl halides in the presence of zinc chloride and a palladium catalyst to afford the cross-coupling product (Equation (24)) ⟨78JA2254⟩.

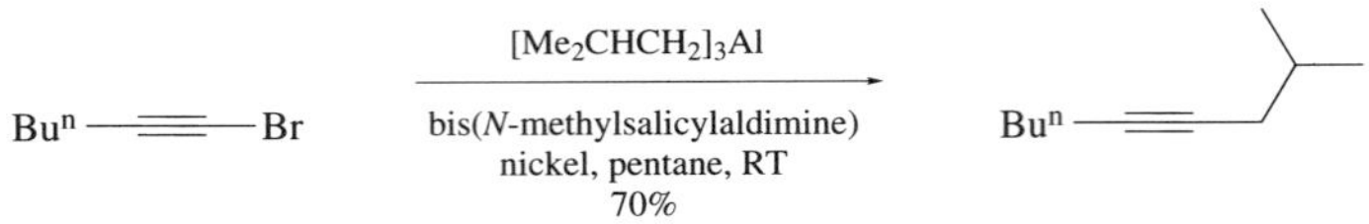

(23)

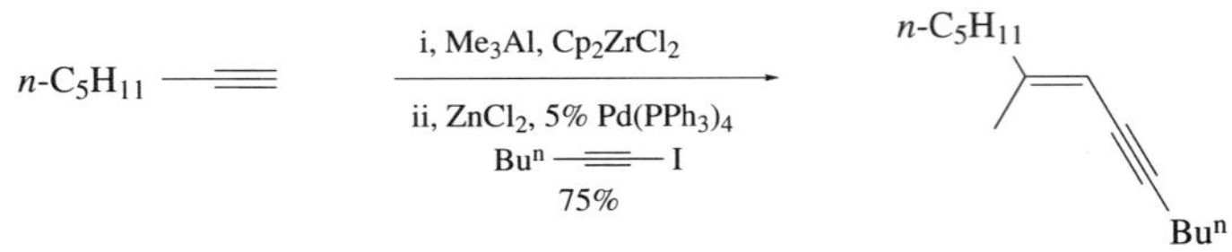

(24)

Alkenylboranes undergo palladium-catalysed cross-coupling reactions with alkynyl bromides in the presence of a base such as alkoxide, hydroxide or acetate (Equation (25)) ⟨79TL3437, 85JA972⟩. Normally transition metal transmetallation processes do not occur readily with organoboranes, due to the weak carbanionic character of organic groups in the organoborane. However, the reaction is facilitated in the presence of base through the formation of an organoborate, and 1,3-enynes can be formed in high yield.

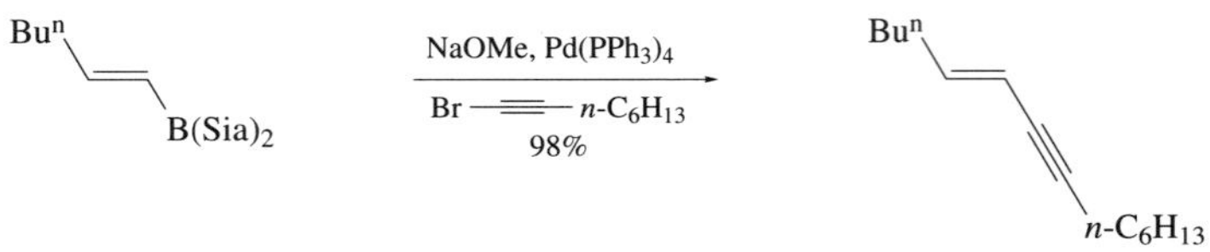

(25)

Alkenyltin compounds undergo cross-coupling reactions with alkynyl bromides and iodides in the presence of palladium(II) catalysts (Equation (26)) ⟨92TL3647⟩. Copper(I) co-catalysis has been reported to accelerate this reaction ⟨90JOC5359⟩.

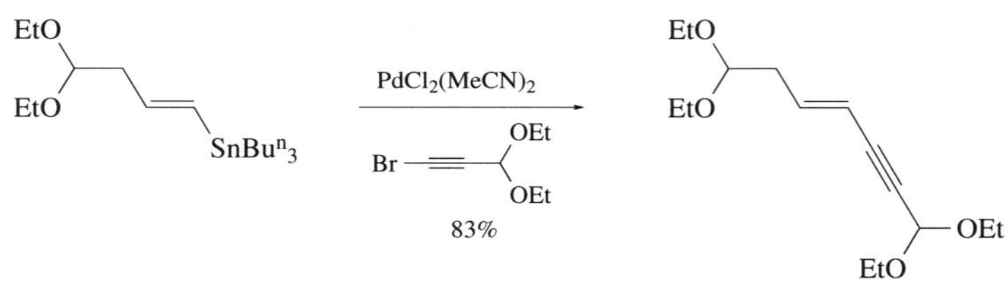

(26)

More complicated reactions have been reported in which a cyclisation process accompanies the cross-coupling reaction. Ethynylidene cyclopentanes have been formed with high stereoselectivity, in 35–71% yield, from the reaction of δ-ethynyl malonate enolates (and related enolates) with alkynyl halides (Equation (27)) ⟨91TL6541⟩. The reaction requires catalysis by both palladium(0) and copper(I) species. Ynenol lactones are formed with high stereoselectivity when the potassium salts of γ-alkynylcarboxylic acids are reacted with alkynyl bromides in the presence of a palladium/triarylphosphine catalyst (Equation (28)) ⟨92TL2811⟩. The nature of the phosphine is important; only tri-*o*-tolylphosphine and tri-(2-furyl)phosphine promote an efficient transformation.

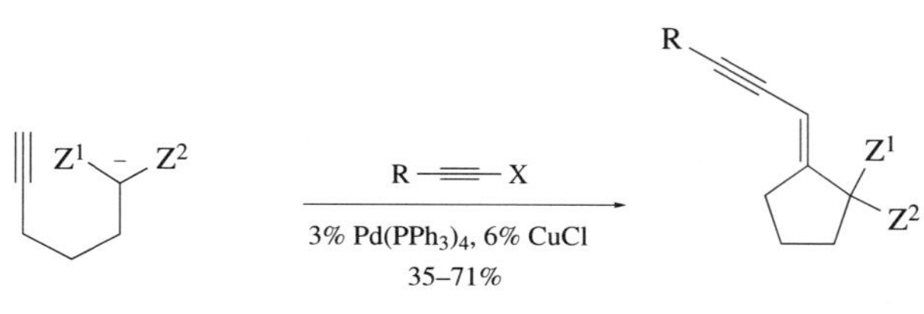

(27)

$Z^1 = CO_2Me$; $Z^2 = CO_2Me$, COMe, CN; X = Br, I; R = Bu^n, $Cl(CH_2)_3$, $MeOCH_2$, Ph

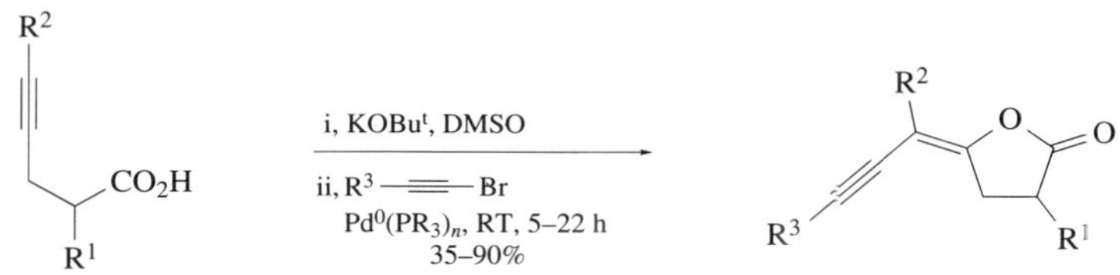

(28)

Heck-type alkenylation of alkynyl halides would represent a very simple route for the preparation of 1,3-enynes. Unfortunately, very low yields are obtained under the normal conditions. 1-Bromophenylethyne does react normally, but a large amount of diphenylbutadiyne is also produced ⟨82OR(27)345⟩. Under solid–liquid phase-transfer conditions (potassium or sodium carbonate/tetra-*n*-butylammonium chloride/DMF) alkynyl iodides do undergo a palladium-catalysed reaction with methyl-2-propenoate or 3-buten-2-one, at room temperature, to give methyl (*E*)-enynoates and (*E*)-enynones respectively in 40–70% yield (Equation (29)) ⟨87S70⟩. Application of the conditions to unactivated alkenes has not been reported.

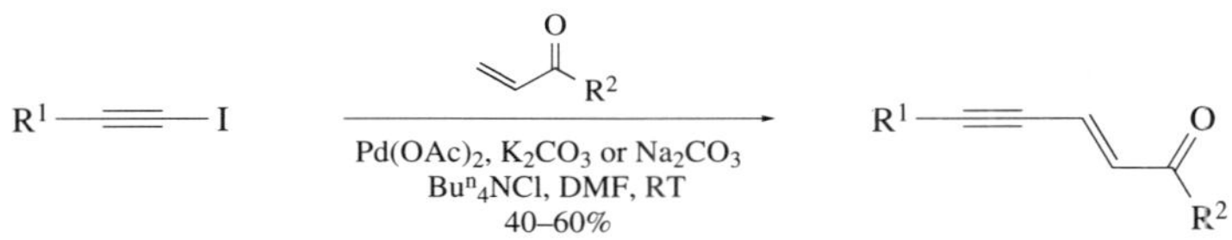

(29)

R^1 = Bu^n, *n*-C_5H_{11}, *n*-C_6H_{13}, TMS, Ph
R^2 = Me, OMe

1.21.2.1.2 *Substitution of oxygen functions*

(i) Substitution of alkyl–oxygen functions

The substitution of ester and ether functions by alkynylmetal reagents, in particular the substitution of sulfonate esters and epoxides, is an extremely important method for carbon–carbon bond formation.

Dialkyl sulfates and aryl-, alkyl- and perfluoroalkylsulfonate esters have all been successfully employed in substitution reactions with alkynylmetal reagents. Whereas the use of dialkyl sulfates is limited to lower members of the group (methyl, ethyl, propyl and isopropyl) ⟨B-69MI 121-03⟩, sulfonate esters of primary alkanols are generally applicable alkylating agents (Schemes 28 ⟨86JOC3098, 88CPB4337, 91TL53⟩ and 29 ⟨80TL4047, 86TL4913, 90TL4609⟩). The sulfonate esters are readily prepared and often give yields of substitution product comparable to, or better than, those obtained with the corresponding halide. In addition, the wide reactivity range of the various substituted sulfonate esters (detailed studies establish triflates to be some 10^4–10^5 times more reactive than the corresponding tosylates ⟨82S85⟩) can allow selective substitution reactions. For example, whereas the coupling reactions of alkynyllithium reagents with tosylates and iodides bearing *β*-oxygen functions do not generally proceed well, the corresponding triflates often allow successful substitution (Equation (30)) ⟨86TL6133, 90TL4609⟩.

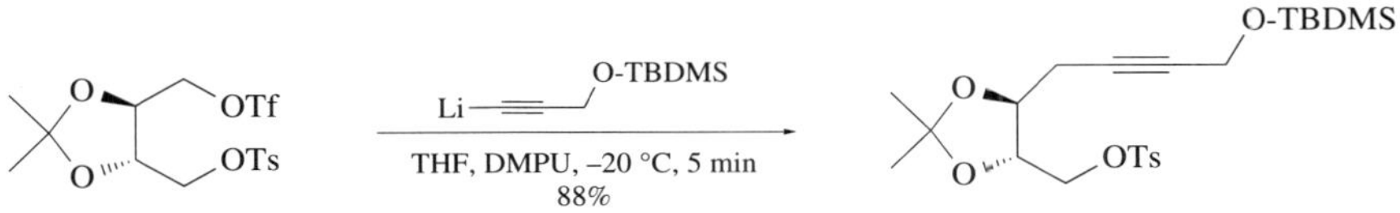

(30)

Alkynylaluminum reagents undergo substitution reactions with secondary alkyl sulfonates. Trihexynylaluminum has been reported to afford the substitution products with isopropyl and cyclopentyl mesylates in 60 and 99% yields respectively (Equation (31)) ⟨75JA7385⟩. 3-Methyl-2-butyl mesylate affords the isomerization product, 3,3-dimethyl-4-nonyne, as the only coupling product (60%). Although alkynyldialkylalanes, such as those derived from diethylaluminum chloride, were reported not to be dependable alkynylating reagents, such reagents have since been employed successfully (Equation (32)) ⟨86JMC230⟩. Interestingly, trialkynyl- and alkynyldiethylalanes have been observed to undergo direct substitution reactions with a secondary homoallylic

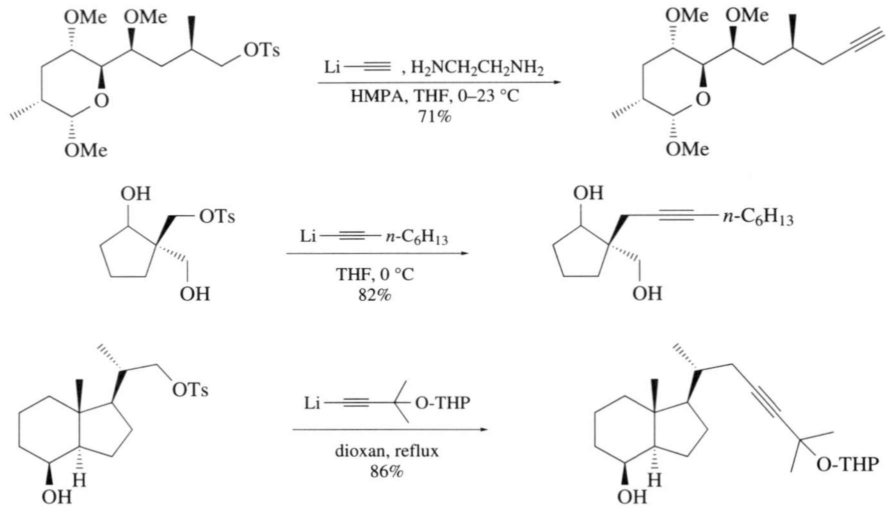

Scheme 28

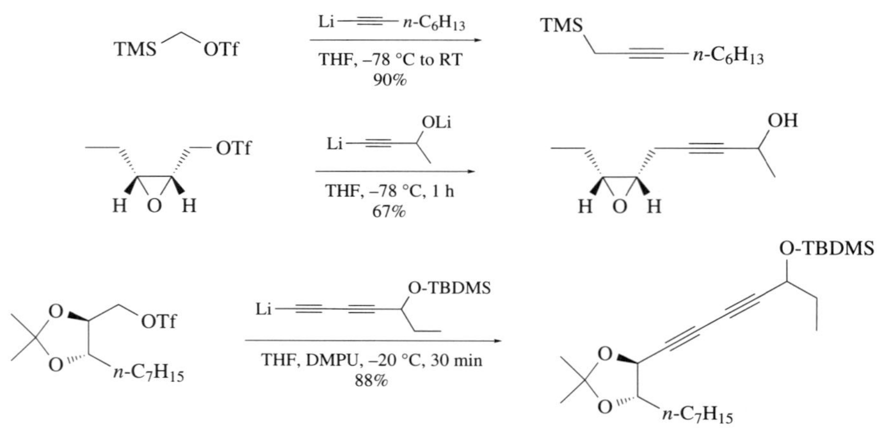

Scheme 29

tosylate whereas the corresponding alkyl- and alkenylaluminum reagents afford cyclopropanes through homoallylic rearrangement ⟨85IZV1109⟩. Reactions with tertiary alkyl sulfonates have not been reported.

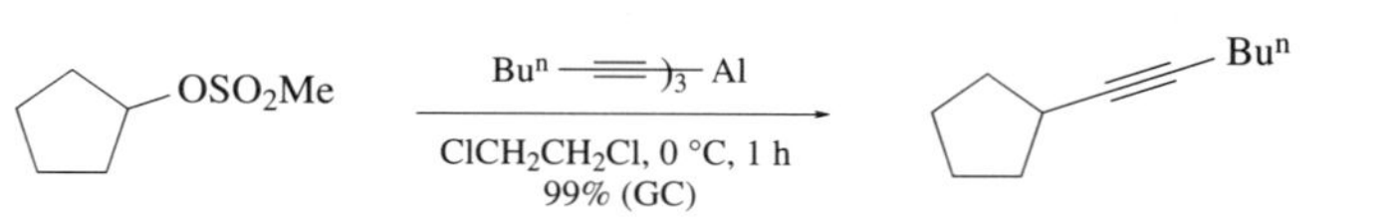

(31)

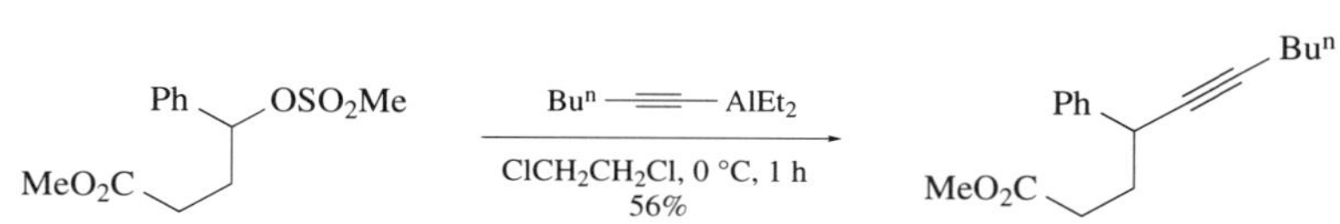

(32)

Oxiranes (epoxides) undergo substitution reactions with alkynylmetal reagents to yield *β*-hydroxyalkynes. With unsymmetrical oxiranes the ring opening by Group 1 metal alkynides usually occurs at the less substituted carbon atom ⟨78JOU1095⟩. In liquid ammonia, oxirane itself and homologues react rather slowly, although high yields of homopropargylic alcohols can be obtained when a mixture of lithium alkynide and the oxirane in ammonia is allowed to stand for 12–24 hours ⟨B-88MI 121-02⟩. With sodium alkynides, there is some risk of product *O*-alkylation when the oxirane is used in excess. Alkynylmagnesium halides also undergo substitution reactions with oxiranes, although the reaction is generally unsatisfactory since impure products are often obtained ⟨68RTC1249⟩. The Lewis acidity of the reagents can cause S_N1-like attack of the alkynide ion on both ring carbons, and can cause the formation of halohydrins. It can also cause rearrangement of the oxirane to an aldehyde or ketone which, in turn, can couple with the alkynylmetal. Since the reaction with alkali metal alkynides in liquid ammonia is reported to be limited to terminal oxiranes ⟨56CB853, 58JA4593, 60JA4918⟩, a variety of improved procedures have been developed to expand the scope of this reaction. The use of strongly coordinating solvents such as DMSO ⟨67CR(C)839, 69CC674, 78S26⟩, HMPA ⟨75TL51, 76TL1539, 81S490⟩ or DMPU ⟨92SL291⟩ enhances the nucleophilicity of the alkynide ion and can afford good yields of substitution products with nonterminal oxiranes. Examples are illustrated in Scheme 30 ⟨74TL2615, 87S139, 89TL2153⟩. The coupling of oxiranes with alkynyllithium reagents can be accelerated by the addition of boron trifluoride etherate either to the oxirane/alkynyllithium mixture ⟨84CC202, 92JOC6344, 93SL37⟩ or to the alkynyllithium reagent prior to addition of the epoxide ⟨83TL391⟩. Although these two procedures would appear to be quite different, it has been established that RLi and $BF_3 \cdot OEt_2$ are reasonably stable together at low temperatures and react independently as potent nucleophile and strong Lewis acid ⟨84JA3693⟩. Reactions with oxiranes are rapid at $-78\,°C$ and functional groups such as halogen, ester or ketal are unaffected under the reaction conditions (Scheme 31) ⟨83TL391⟩. Alkynyl groups are introduced regioselectively at the less hindered site in the case of monoalkyl-substituted oxiranes, but with low regioselectivity with styrene oxide, indicating cationic character in the transition state. Epoxy alcohols derived from (*E*)-allylic alcohols react with lithium alkynides in the presence of boron trifluoride etherate to give mixtures of rearranged and unrearranged products with the former predominating in most cases (Equation (33)) ⟨84CC202⟩.

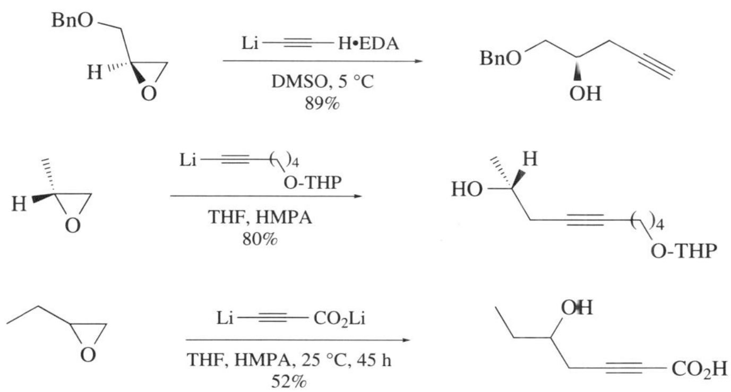

Scheme 30

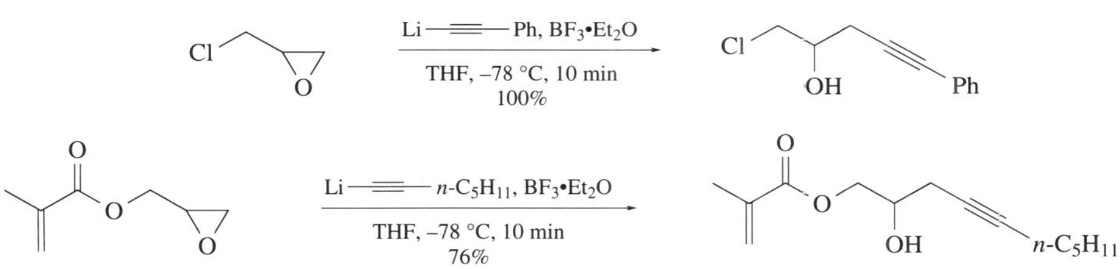

Scheme 31

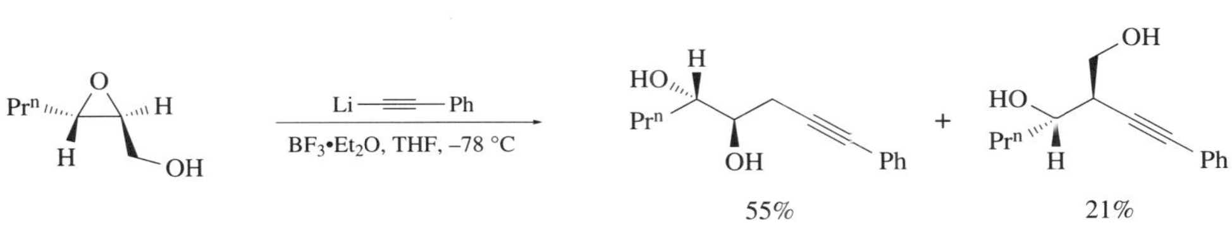

(33)

Alkynylaluminum reagents also undergo effective substitution reactions with oxiranes in nonpolar ⟨68CC634, 69TL1379, 70TL2695, 73TL3899⟩ or ethereal solvents ⟨76CC907, 78JOC353⟩. The Lewis acidic nature of the alanes and their consequent ability to coordinate oxirane oxygen atoms (initial electrophilic attack) facilitates alkynyl transfer. The choice of solvent can have a dramatic effect on the outcome of the reaction. For example, the reaction of diethylhexynylaluminum with 3,4-epoxycyclopentene in an ether solvent resulted in predominant opening of the oxirane ring by attack at the allylic carbon, while use of a hydrocarbon solvent led to rearrangement of the oxirane to cyclopent-3-enone followed by further reaction of the carbonyl group ⟨75CC287⟩. The reaction of a dimethylaluminum alkynide with the diol oxirane (**11**) has been successfully applied to the preparation of prostaglandins and the sesquiterpene vernolepin (Scheme 32) ⟨73TL3899, 76TL1769, 80JCS(P1)852⟩. Different sources of dimethylaluminum chloride were found to give variable regioselectivities in this reaction, the difference being traced to the formation of methoxymethylaluminum chloride through air oxidation. Deliberate formation of the methoxymethylaluminum alkynide gave complete regioselectivity. The opening of oxiranes with an unsubstituted ethynylalane $R_2AlC{\equiv}CH$ has been accomplished with dimethylethynylaluminum etherate ⟨76CC907, 78JOC353, 81JA7520⟩ and also with trimethylsilyl protection/deprotection ⟨76TL1769⟩. Alkynylalanate reagents will also undergo substitution reactions with oxiranes in the presence of boron trifluoride etherate ⟨89T6197, 92SL266⟩ and have been used successfully in cases where the corresponding alane has failed ⟨90TL7145⟩.

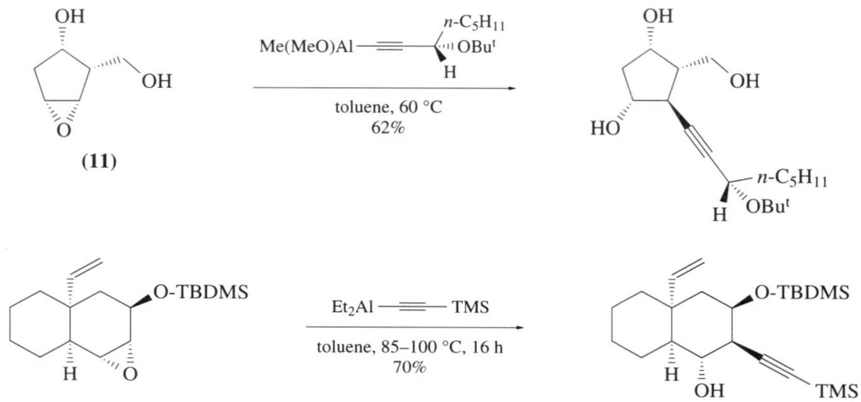

Scheme 32

Alkynylgallium reagents undergo substitution reactions with oxiranes. A reagent prepared from 1-lithio-1-octyne and equimolar trimethylgallium in THF-hexane underwent a rapid reaction with oxirane at room temperature to afford 3-decyn-1-ol in 95% yield ⟨84TL5423⟩. Furthermore, reaction with a variety of oxiranes can be accomplished in good yield employing catalytic trimethylgallium, substitution occurring at the less substituted position of the epoxide.

Triisopropoxytitanium alkynides react with aryl-, alkenyl-, alkynyl- and trimethylsilyl-substituted oxiranes to afford moderate to good yields of adduct resulting from attack at the more substituted carbon of the oxirane ring (Equation (34)) ⟨88CB1315⟩. Further substitution on the oxirane ring or at the alkenyl group increases the yield. Alkyl substituted oxiranes give very low yields of substitution product. The reaction with alkenyl- and alkynyl-substituted oxiranes is particularly interesting since these systems invariably undergo substitution with migration of the multiple bond.

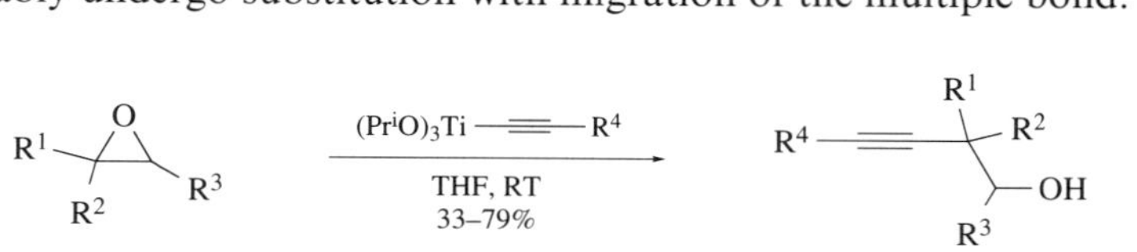

(34)

R^1 = Ph, $CH{=}CH_2$, $C(Me){=}CH_2CH{=}CHPh$, $C{\equiv}CMe$, TMS and $R^1 = R^3 = (CH_2)_4$;
R^2 = H, Me, Ph; R^3 = H and $R^3 = R^1 = (CH_2)_4$; R^4 = n-C_6H_{13}, Ph, TMS

Oxetane and substituted oxetanes undergo substitution reactions with alkynyllithium reagents in the presence of boron trifluoride etherate to afford γ-hydroxyalkynes (Equation (35)) ⟨83TL5121, 84T4261⟩. In a related reaction, β-propiolactones undergo substitution reactions with dimethylalkynylaluminum reagents to afford β-alkynylpropionic acids (Equation (36)) ⟨86TL87⟩.

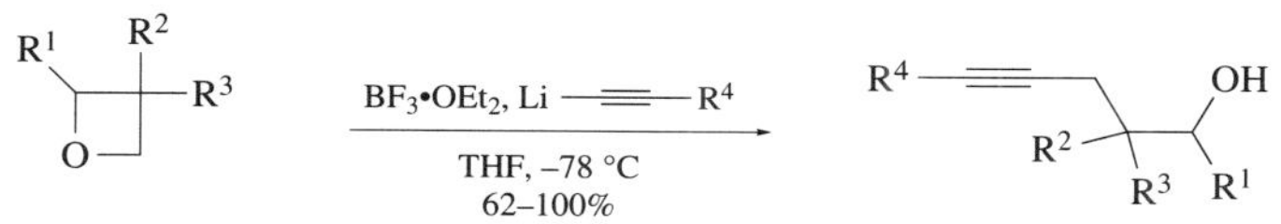

(35)

R^1 = H, *n*-C_7H_{15}, Ph, $(CH_2)_2Ph$; R^2 = H, Me, CH_2Ph; R^3 = H, CH_2Ph; R^4 = H, Et, *n*-C_5H_{11}, Ph, CH_2(O-THP)

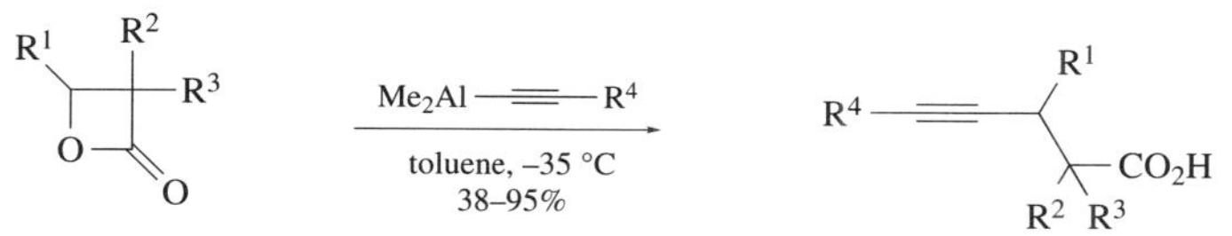

(36)

R^1 = H, Me; R^2 = H, Me; R^3 = H, Me;
R^4 = Et, Pr^n, Ph, cyclopentyl, $(CH_2)_4Cl$, CH(O-THP)CH(Pr^n)Et, $(CH_2)_2$O-TBDMS,

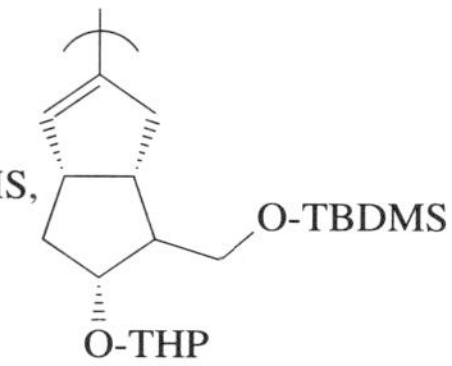

Allylic and propargylic oxygen functions undergo 1,1- and 1,3-substitution reactions with terminal alkynes and alkynylmetal reagents. As with the corresponding halides, the regiochemistry of substitution depends on a number of factors, particularly the substitution pattern of the substrate and the nature of the alkynylmetal reagent. In compound **(12)** the C—OMs carbon atom is sufficiently hindered towards direct nucleophilic attack that a regioselective copper-catalysed 1,3-substitution occurs with phenylethynylmagnesium chloride (Equation (37)) ⟨88CB1461⟩. More generally, allylic ethers of 2-hydroxybenzothiazole undergo stereo- and regioselective substitution reactions with copper(I) alkynides in THF or DMF at 40–60 °C, reaction occurring at the less substituted terminus of the allyl system (Equation (38)) ⟨79TL3873⟩. Unsubstituted propargyl toluenesulfonate undergoes a copper-catalysed 1,1-substitution reaction with alkynylmagnesium halide reagents in THF to afford 1,4-diynes in good yield ⟨79S292⟩. In contrast, propargyl acetates ⟨73TL3361, 82RTC405⟩, methanesulfinates ⟨82RTC405⟩, methanesulfonates ⟨74S573⟩ and toluenesulfonates ⟨65TL1953⟩ have been reported to undergo 1,3-substitution reactions with copper(I) alkynides to afford allenynes in good yield. The $Pd(PPh_3)_4$-catalysed reaction of propargylic acetates, diethylphosphates, methylsulfinates, methanesulfonates and epoxides affords allenynes with a wide range of alkynylmetal reagents (Scheme 33) ⟨82RTC97, 82RTC405, 86JOC4006⟩. Propargylic acetates, carbonates and toluenesulfonates also react with terminal alkynes in the presence of $Pd(PPh_3)_4$, copper(I) iodide, an amine base and lithium chloride to afford allenynes by 1,3-substitution ⟨90TL7179, 93TL3853⟩. Propargyl acetates undergo substitution reactions with trialkynylalanes when activated as their dicobalt hexacarbonyl complexes, the free 1,4-diynes being recovered from their $Co_2(CO)_6$ complexes by oxidation with ceric ammonium nitrate at −78 °C ⟨83TL2239⟩.

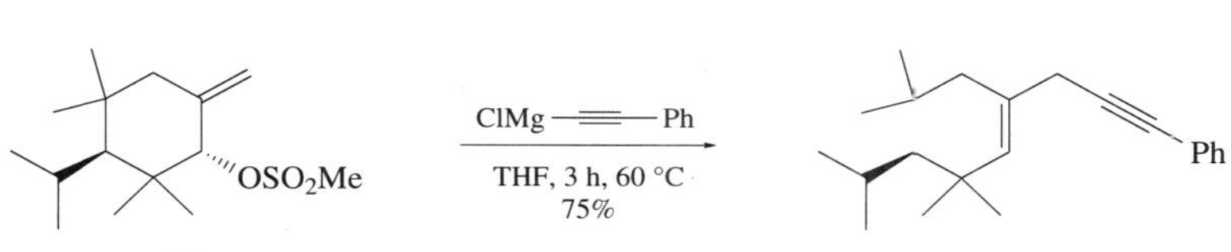

(37)

(12)

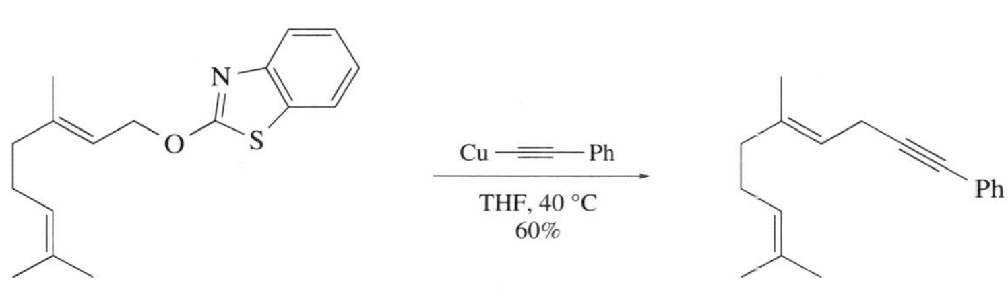

(38)

Acetals, *ortho*-esters, *N*,*N*-dialkylformamide acetals and *ortho*-carbonates have also been reacted with alkynes or alkynylmetal reagents to afford products formally derived from substitution reactions. The reaction mechanisms are more complicated in these cases, often involving Friedel–Crafts

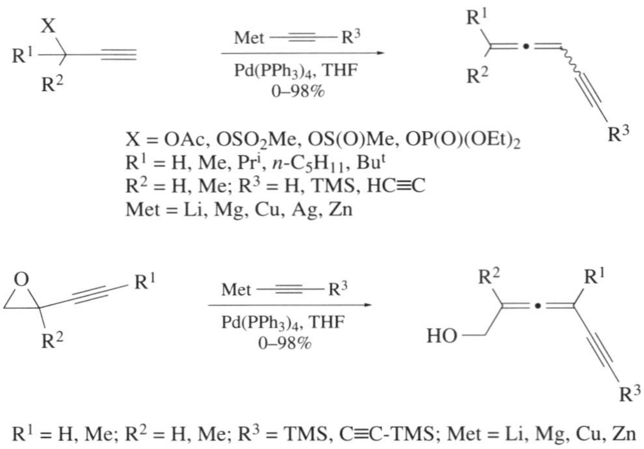

Scheme 33

type alkylation of the alkyne ⟨B-69MI 121-03⟩. Alkynyltrimethylsilanes ⟨83JA2904, 87HCA448⟩ and alkynyltributylstannanes ⟨86JA7116⟩ undergo coupling reactions with acetals in the presence of stoichiometric titanium(IV) chloride to give propargyl ethers in high yield. Very high diastereoselectivities are observed by employing acetals derived from chiral 1,3-diols (Equation (39)) ⟨83JA2904, 86JA7116⟩. Propargyl ethers are also formed from acetals and alkynyltrimethylsilanes in the presence of catalytic quantities of zinc chloride and tin(IV) chloride ⟨87CL1975⟩, and, under the same conditions, directly from aldehydes in the presence of alkoxytrimethylsilanes (Scheme 34) ⟨88BCJ4037⟩. *C*-Glycosylated alkynes have been prepared by the reaction of *D*-glucal acetates with alkynyltrimethylsilanes in the presence of tin(IV) chloride ⟨93SL413⟩, and by the reaction of an *O*-glycoside with ethynylmagnesium bromide in the presence of zinc bromide ⟨89TL6027⟩.

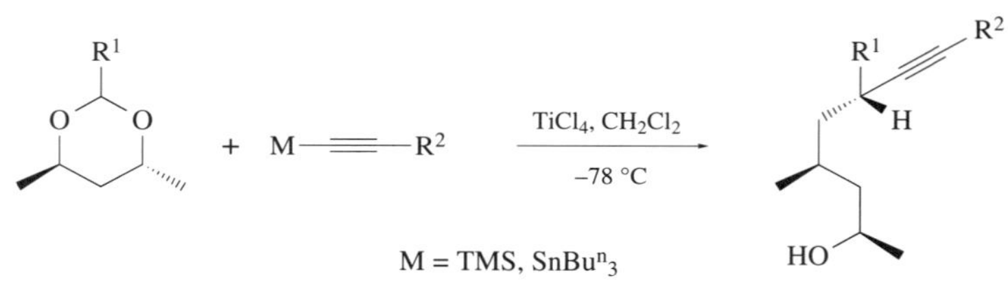

(39)

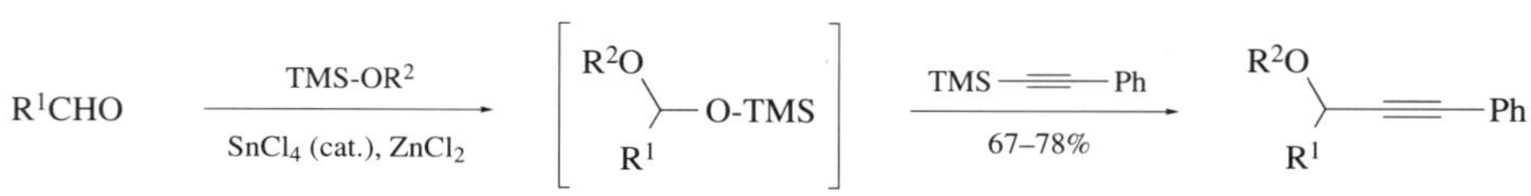

Scheme 34

Ortho-formates, *ortho*-acetates and *ortho*-carbonates react with alkynyl magnesium halides ⟨B-69MI 121-03, 84JOC2031⟩ and with 1-alkynes in the presence of a zinc halide catalyst ⟨58JA4607, 86JOC2607⟩ to afford propargyl acetals, ketals and *ortho*-esters respectively. *N*,*N*-Dialkylformamide acetals condense with 1-alkynes in the absence of catalysts simply on heating the two components together ⟨60AG836⟩.

(ii) Substitution involving alkenyl-, allenyl- and aryl-oxygen functions

The formation and trapping of an enolate, allenolate or phenolate ion, followed by substitution of the enol derivative with a terminal alkyne or alkynylmetal reagent, allows the regiospecific formation of enyne, allenyne or arenyne products (Scheme 35). In practice, enol substitution reactions of this type have only been successfully achieved under transition metal catalysis. Although a wide range of diorganocuprate reagents do undergo successful substitution reactions with alkenyl and aryl triflates ⟨80TL4313⟩, alkynyl groups are not transferred. Substitution by an alkynyl group is most successfully achieved by the palladium-catalysed cross coupling of a terminal alkyne, or alkynylmetal reagent, with an alkenyl/aryl phosphate or sulfonate. As with the corresponding halides, distinct palladium mediated sp–sp^2 coupling processes have emerged involving either the terminal alkyne or its metallated counterpart.

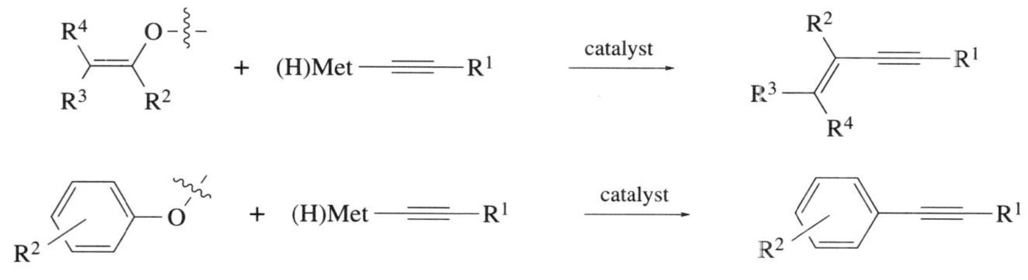

Scheme 35

(a) Reaction of metallated alkynes with alkenyl/aryl phosphates or sulfonates in the presence of catalytic quantities of Pd(II) or Pd(0) compounds. With appropriate choice of metal, palladium catalyst and solvent, very high yields of coupling product have been obtained under mild conditions with both alkenyl and aryl sulfonates, and with alkenyl phosphates. Metals of intermediate electronegativity such as Al, Sn and Zn are most successful; organolithium and magnesium reagents do not effect carbon–oxygen bond cleavage in these systems, rather they effect attack at the phosphorus or sulfur atom with regeneration of the metallic enolate ⟨80TL2531⟩. Diethylalkynylalanes undergo palladium-catalysed cross-coupling reactions with alkenyl phosphates with exclusive transfer of the alkyne (Scheme 36) ⟨84BCJ108⟩. $(PPh_3)_4Pd$ was found to be the most effective catalyst. The reaction has also been successfully applied to alkenyl phosphates derived from thiocarboxylic *S*-esters ⟨81TL1609⟩.

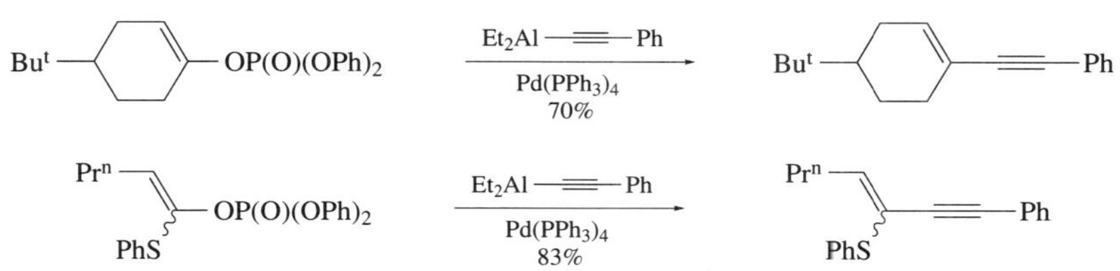

Scheme 36

The palladium-catalysed cross-coupling reaction of alkynyltrialkyltin reagents with alkenyl and aryl triflates provides a useful route to enynes and arynes (Scheme 37) ⟨84JA4630, 86JA3033, 87JA5478⟩. The reaction has been applied to both highly functionalised ⟨89JA5417⟩ and highly hindered alkenyl and aryl triflates (Scheme 38) ⟨89JOC5828, 92JOC678⟩, and to a limited number of heteroaryl systems ⟨89AJC279⟩. At least one equivalent of lithium chloride is required for the reaction to proceed to completion, indicating that the initial oxidative addition product, an alkenyl- or arylpalladium(II) triflate, is incapable of entering a catalytic cycle. In the presence of lithium chloride, an alkenyl- or arylpalladium(II) chloride is generated which reacts with the organotin reagent to afford coupled products with regeneration of the palladium(0) catalyst. Lithium bromide and iodide can also be used, but catalyst decomposition leads to longer reaction times and, in the case of lithium iodide, to incomplete reaction. Sodium and potassium salts are unsuitable for the reaction. Zinc chloride can accelerate the reaction, presumably by transmetallation to the organozinc chloride, but it cannot replace the use of lithium chloride. $(PPh_3)_4Pd$ and $(PPh_3)_2PdCl_2$ both give good results with alkenyl and aryl triflates, although different selectivities have been observed. For instance, with 4-haloaryltriflates a change in selectivity is observed on going from $(PPh_3)_4Pd$ (I > Br > OTf) to $(PPh_3)_2PdCl_2$ (I > OTf > Br). The coupling reaction works well in a number of solvents including THF and other ethers; those solvents able to solubilise lithium chloride and act as good ligands (HMPA, DMPU, DMF and DMSO) are particularly effective. Low concentrations of water do not adversely affect the reaction, nor does the presence of oxygen, although reactions are usually carried out under an inert atmosphere.

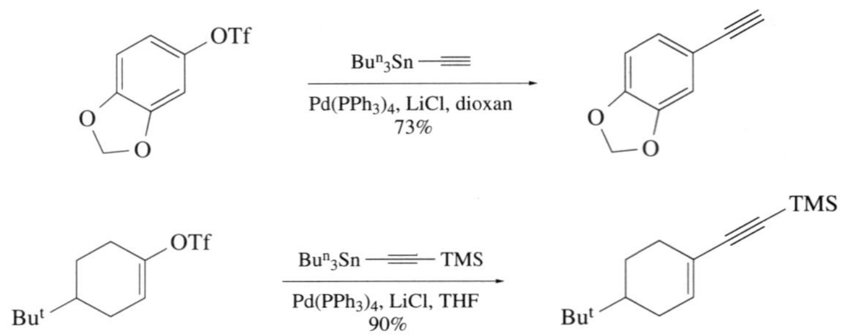

Scheme 37

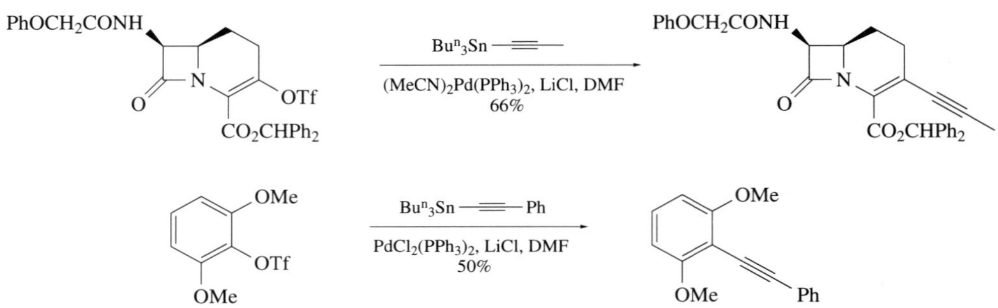

Scheme 38

Alkynylzinc chlorides undergo palladium-catalysed cross-coupling reactions with aryl fluoroalkanesulfonates in the presence of lithium chloride (Equation (40)) ⟨87TL2387⟩. Alkynylzinc chlorides also undergo nickel- or palladium-catalysed cross-coupling reactions with diketene (Equation (41)) ⟨83CPB4346⟩.

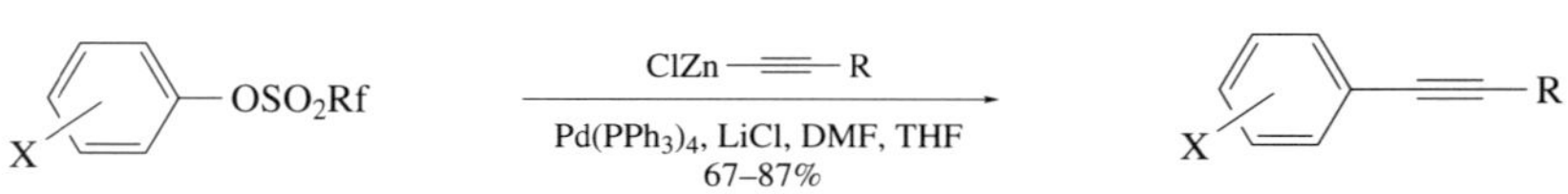

(40)

Rf = CF$_3$, CF$_2$CF$_2$OCF$_2$CHF$_2$; X = H, *p*-Cl, *o*-Cl, *m*-OMe, *p*-NO$_2$; R = Bu^n, CH$_2$OMe, Ph

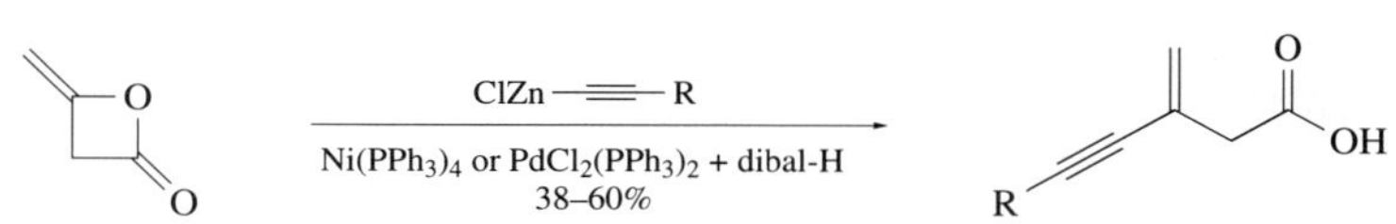

(41)

R = Pr^n, Bu^n, *n*-C$_5$H$_{11}$, Ph

dibal-H = diisobutylaluminum hydride

In the presence of fluoride ion, various types of organosilicon compounds undergo palladium-catalysed cross-coupling with alkenyl and aryl triflates under mild conditions ⟨90TL2719⟩. A single example of the coupling of an alkynyltrimethylsilane with an alkenyl triflate has been reported (Equation (42)).

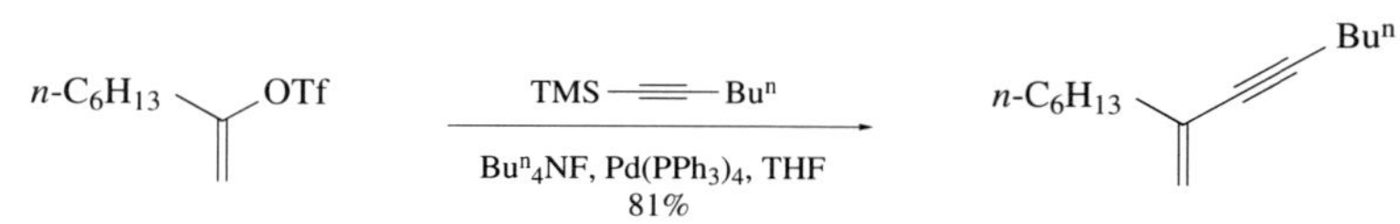

(42)

(b) Reaction of terminal alkynes with alkenyl and aryl sulfonates in the presence of catalytic amounts of Pd(II) or Pd(0) compounds and a base. Terminal alkynes can be coupled directly to

alkenyl and aryl sulfonates in the presence of a palladium catalyst and a base (Scheme 39). The palladium-catalysed cross-coupling of alkenyl triflates with terminal alkynes was independently reported by two groups. Stille *et al.* used $(PPh_3)_4Pd$ or $(PPh_3)_2PdCl_2$ in DMF at 70°C in the presence of triethylamine ⟨85JOC2302⟩. Yields are considerably improved by the addition of lithium chloride. Cacchi *et al.* carried out similar reactions using amine bases or sodium acetate in DMF at 60°C, and also using an amine base and copper(I) iodide at room temperature ⟨86S320⟩. The presence of copper co-catalysis is essential for the reaction to proceed at room temperature. Aryl fluoroalkanesulfonates, including aryl triflates, similarly undergo palladium-catalysed cross-coupling with terminal alkynes in the presence of triethylamine in DMF at 90°C ⟨86TL1171⟩ and with copper(I) iodide co-catalysis at room temperature ⟨89TL2581⟩. Heteroaryl triflates, including pyridines ⟨88JOC386, 89TL2581⟩ and isoquinolines ⟨89JMC1820⟩, undergo similar reactions. An interesting preparation of a [5.6] spiro ring system has been reported employing the palladium-catalysed reaction of 4-*t*-butyl-1-cyclohexene triflate with an excess of trimethylsilylethyne (Equation (43)) ⟨90JA9330⟩. The attempted use of the less reactive mesylate and tosylate functions, where reported, has been unsuccessful ⟨86TL1171, 87H(26)355⟩.

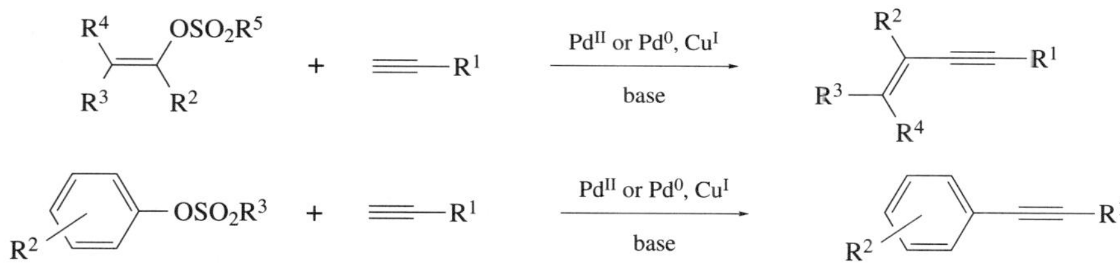

Scheme 39

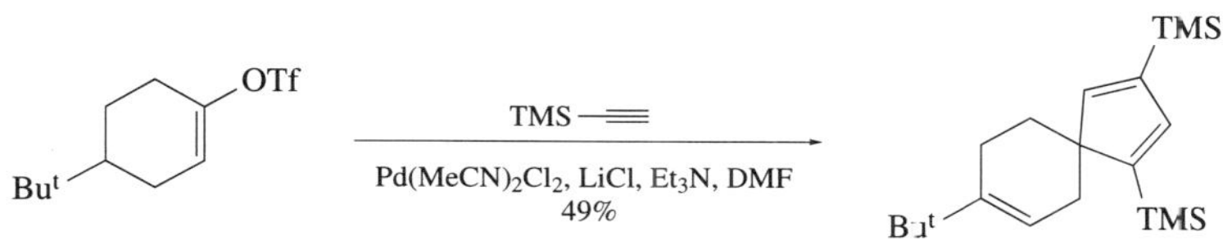

(43)

The palladium-catalysed cross-coupling of alkenyl triflates with terminal alkynes has been extensively used for the synthesis of analogues of vitamin D (calciferol) and its hydroxylated metabolites ⟨86TL1523, 88JOC1790, 88TL1203, 91JA3873, 91JMC2452⟩, for the synthesis of *Z*-enediynes related to the neocarzinostatin chromophore ⟨91TL3405, 93TL4177⟩, and in a total synthesis of ginkgolide B (Scheme 40) ⟨88JA649, 92CB271⟩.

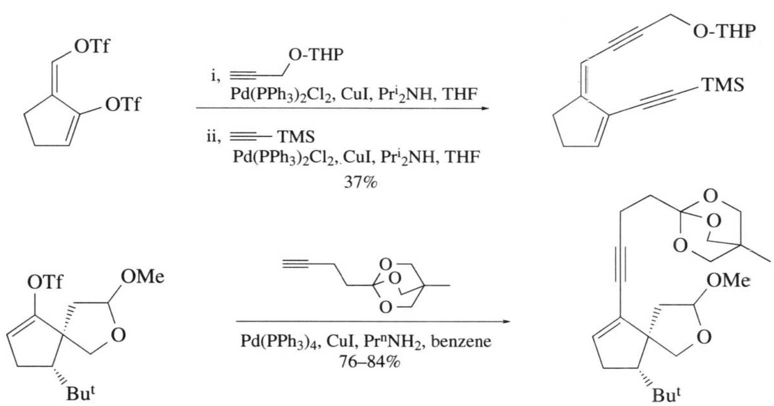

Scheme 40

Carbonylative coupling reactions of alkenyl and aryl triflates with terminal alkynes and alkynyltin reagents have been reported. Alkenyl triflates undergo carbonylative coupling with alkynyltin reagents in the presence of $(PPh_3)_4Pd$ and lithium chloride ⟨84JA7500⟩. In the presence of bidentate phosphine ligands and palladium(II) acetate, alkenyl triflates undergo carbonylative coupling with terminal alkynes to afford 1-alken-4-yn-3-ones in good yields (Equation (44)) ⟨91TL6449⟩. Monodentate phosphine ligands are not effective in the reaction, and the conditions are unsuccessful with electron deficient alkynes such as ethyl propynoate and 3,3-diethoxy-1-propyne, and with aryl triflates. Aryl triflates do, however, undergo successful carbonylative coupling reactions with alkynylstannanes in the presence of dichloro(1,1′-bis(diphenylphosphino)ferrocene)palladium(II), lithium chloride and carbon monoxide (1 atm.) ⟨88JA1557⟩.

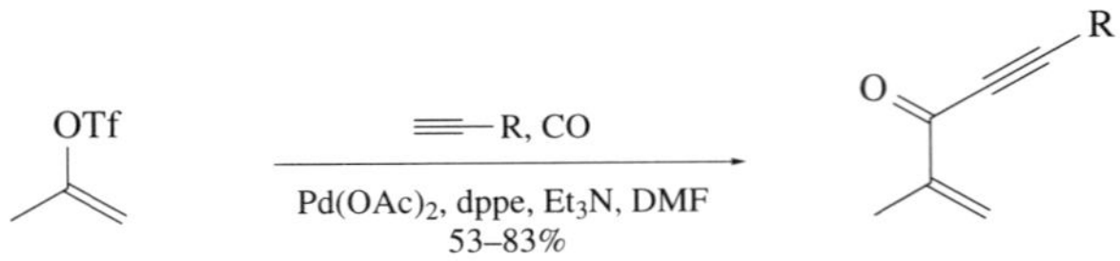

(44)

(c) Substitution involving alkynyl–oxygen functions. The use of alkynyl ethers as electrophilic alkynylation reagents has been reported although these reactions have seen little use compared to the more commonly employed nucleophilic alkynylation reagents (alkynylmetals). A range of disubstituted alkynes has been produced from the reaction of 1-ethoxy-1-hexyne with organolithium reagents (Equation (45)) ⟨68RTC69⟩. Phenylmagnesium bromide, when employed instead of phenyllithium, gave a poor result. Ethynylation is also possible, as in the preparation of phenylethyne from ethoxyethyne. Strong evidence for an addition–elimination mechanism was obtained from the reaction of ethoxyhexyne with hexynyllithium in dioxan. Prolonged heating of the solution of these compounds at 100°C and subsequent hydrolysis afforded the diyne (**14**), in addition to a small amount of the ether (**13**) (Scheme 41) ⟨68RTC69⟩. Ether (**13**) became the main product when the reaction time was shortened.

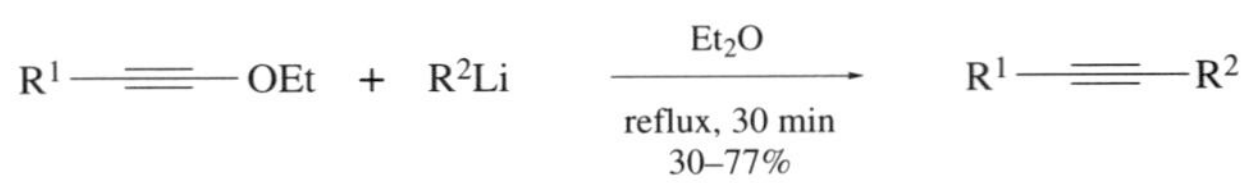

(45)

R^1 = Bu^n,H; R^2 = Me, Bu^n, 1-cyclohexenyl, Bu^{n}C≡C, Ph

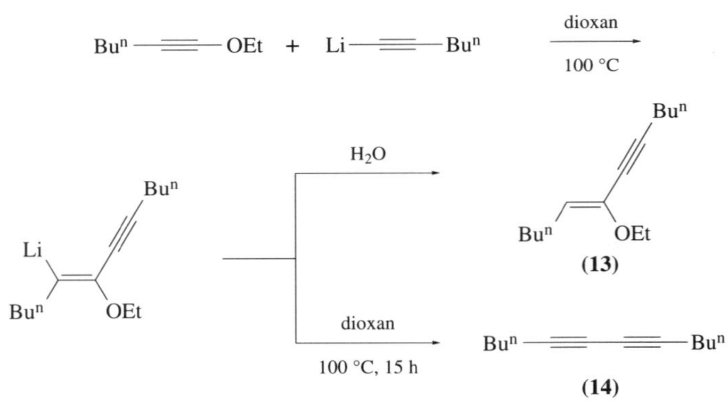

Scheme 41

Allylboranes and substituted allylboranes undergo an addition–elimination reaction with ethoxyethyne in the presence of triethylalane to afford the products of a formal substitution reaction (Scheme 42) ⟨80S904⟩. The initial addition reaction is exothermic and proceeds at a low temperature, yielding a thermally stable alkenylborane. Elimination to 1,4-pentenynes is achieved by treating the alkenylboranes with triethylaluminum at 80–110°C.

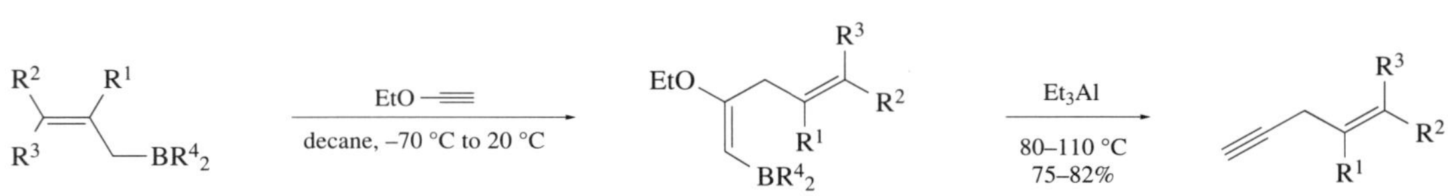

R^1 = H, Me; R^2 = H, Me; R^3 = H, Me; R^4 = Et, Pr^n, Bu^n or $R^2R^3C{=}C(R^1)CH_2$

Scheme 42

1.21.2.1.3 Substitution of sulfur, selenium or tellurium functions

The substitution of sulfur, selenium and tellurium functions by 1-alkynes or alkynylmetal reagents is generally much more difficult than substitution of the corresponding oxygen functions. In part, this is due not only to the weaker inductive effects of SR, SeR and TeR functions compared to their oxygen counterparts, but also to the increased propensity for nucleophilic substitution to occur at the chalcogen atom rather than at carbon ⟨B-67MI 121-02⟩. This is particularly true for Se and Te functions and direct replacement of selenium and tellurium groups by alkynes has not been described. Some success has been achieved employing sulfur functions, particularly sulfones.

Simple alkyl and aryl sulfides do not undergo substitution reactions with alkynylmetal reagents. Even the more activated thioacetals ($CH_2(SR)_2$) and *ortho*-thioesters are stable to alkynylmagnesium halides in contrast to their oxygen counterparts ⟨B-67MI 121-02⟩. Successful substitution reactions have been reported with episulfonium tetrafluoroborates generated *in situ* from alkenes and dimethyl(thiomethyl)sulfonium tetrafluoroborate. Alkynyldiethylalanes and lithium dialkynyldiethylaluminates were found to afford the highest yields of homopropargyl thiomethyl ethers (Scheme 43) ⟨84JA4263⟩. Alkynyldiethylalanes have also been reported to effect successful substitution of a range of allylic sulfones in the presence of aluminum trichloride (Scheme 44) ⟨86JA1098⟩. Using the more Lewis acidic alkynylethylaluminum chloride obviates the need for an exogeneous Lewis acid. The reaction is regioselective, substitution occurring at the less substituted side of the allyl system, and is compatible with ester and hydroxyl functions. The mechanism most likely involves ionisation to a carbenium ion followed by capture by the nucleophile. 2-Arylsulfonyl substituted cyclic ethers ⟨89T4293⟩ and amides ⟨90SL48⟩ react with alkynylzinc reagents to afford substitution products in high yield (Scheme 45).

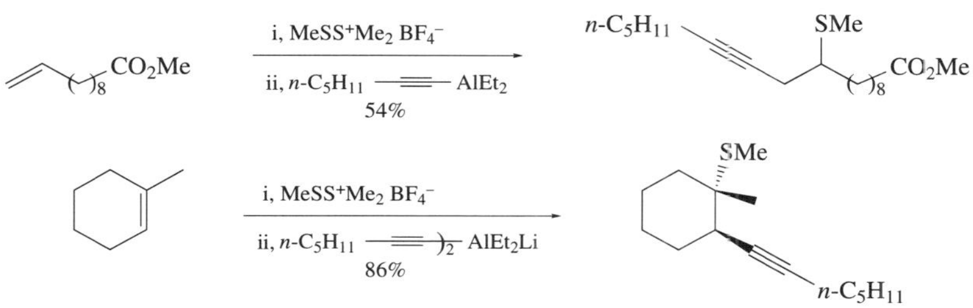

Scheme 43

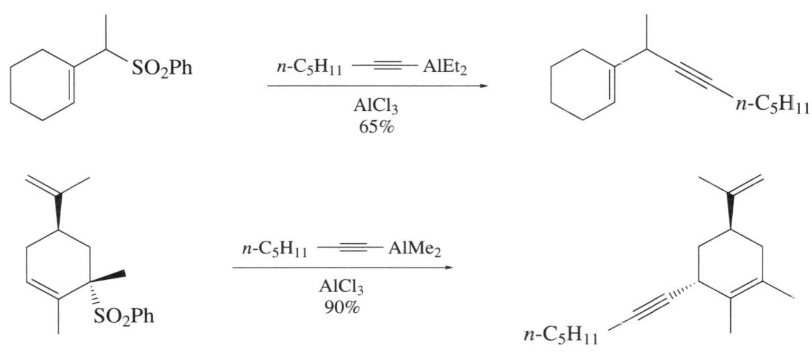

Scheme 44

Direct substitution of alkenyl and aryl sulfides or sulfones by alkynylmetal reagents has not been reported. Alkenyl sulfides are well known to undergo nickel-catalysed cross-coupling reactions with Grignard reagents, but reactions with the corresponding alkynylmagnesium halides have not been described, perhaps due to the known propensity of alkynes to undergo polymerisation under nickel catalysis. Substitution of a β-sulfonyl-α,β-unsaturated ester by 1-octynylmagnesium bromide has been reported, but the reaction proceeds by a conjugate addition–elimination mechanism (Equation (46)) ⟨89TL173⟩.

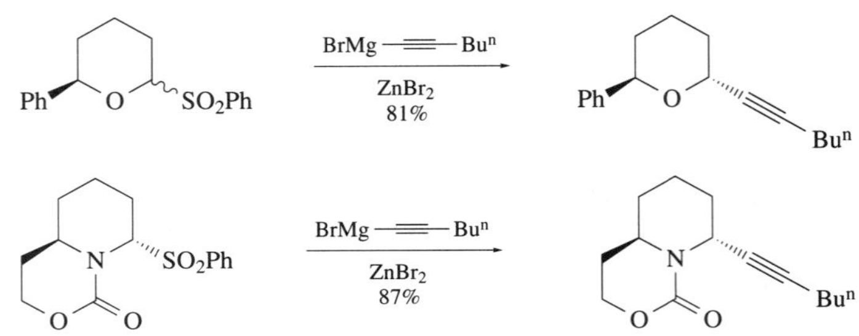

Scheme 45

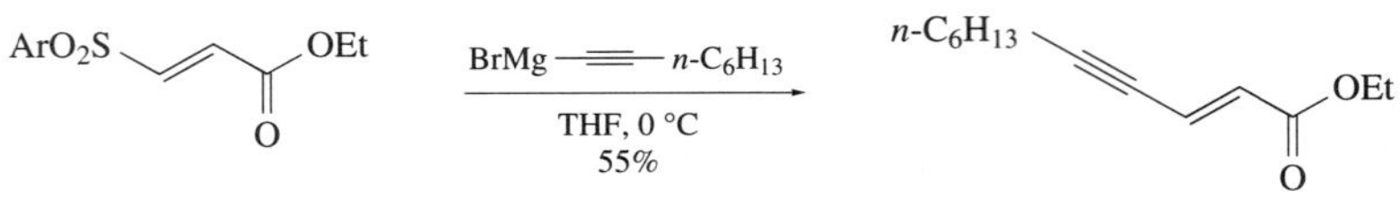

(46)

Alkynyl sulfones undergo substitution reactions with Grignard and organolithium reagents (Equation (47)) ⟨79JOC3444⟩. The reaction is limited to alkynyl sulfones not having propargylic hydrogen atoms and is believed to proceed by a mechanism involving single electron transfer (SET) ⟨84TL4851⟩.

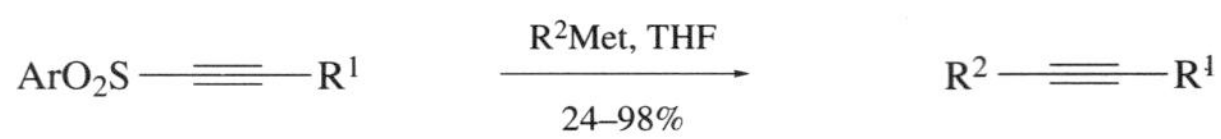

(47)

Met = MgX or Li
R^1 = H, Bu^t, Ph, TMS
R^2 = Et, Pr^i, Bu^n, Bu^t, Ph, *p*-Me-C_6H_4, PhC≡C

1.21.2.1.4 Substitution of nitrogen, phosphorus, arsenic, antimony or bismuth functions

The groups NH_2, NHR and NR_2 are extremely poor nucleofuges and direct substitution reactions with alkynide anions have not been reported. Successful substitution reactions of NR_2 functions have been achieved in aminals but these reactions presumably involve iminium ion intermediates (see Section 1.21.3.2.4).

The leaving-group ability of NH_2 and NHR can be improved by conversion to *N*-sulphonyl derivatives, but the reaction with alkynide ions has only been used scarcely. An example is the ring cleavage of bis(aziridine) (**15**) with lithium heptynide (Equation (48)) ⟨89JOC5324⟩. Even with this highly activated system no reaction is observed in the absence of HMPA. The leaving-group ability of NR_2 can also be improved by quaternisation of the nitrogen atom, although elimination reactions predominate with such substrates and substitution can only be achieved under special circumstances (Equation (49)) ⟨63JCS2990⟩.

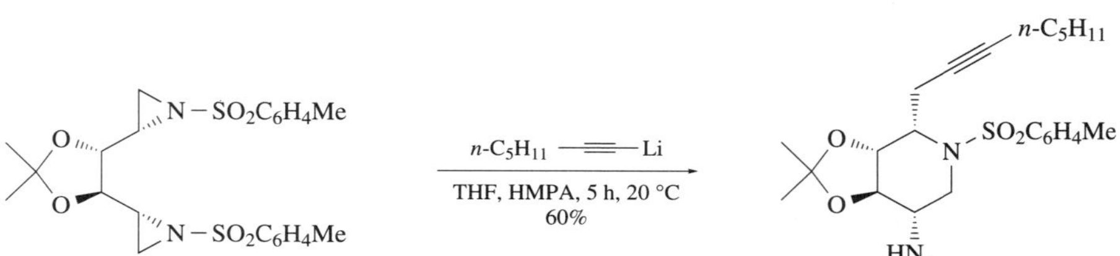

(48)

(**15**)

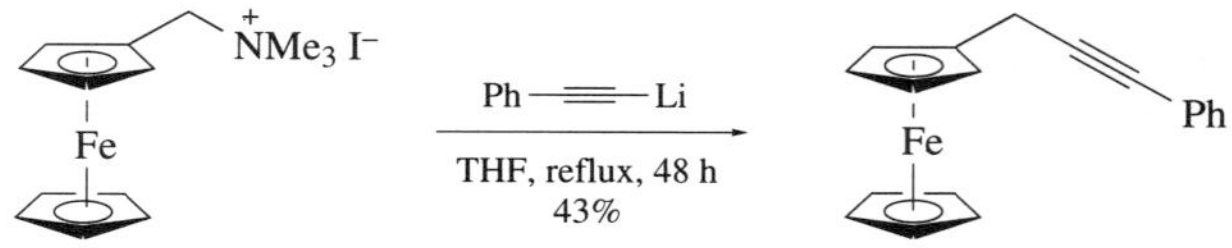

(49)

Amine diazotisation provides the very powerful leaving group (N_2) and substitution of a diazonium group is an effective method for the introduction of many functional groups into an aromatic ring. Up to 1995, alkynes have not been introduced, although palladium-catalysed substitution reactions of arenediazonium salts have been reported ⟨87JCS(P1)1511⟩, offering the potential for substitution with alkynes and alkynylmetal reagents. Substitution adjacent to a carbonyl group has been effected; thus ethyl diazoacetate reacts with trialkynylboranes to provide propargylic esters in good yield (Equation (50)) ⟨72CJC1105⟩.

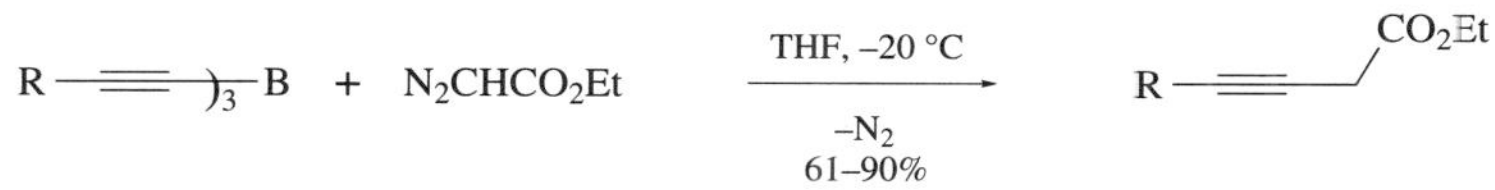

(50)

R = Bu^n, Bu^t, n-C_5H_{11}, Ph

Nitro functions do not generally undergo substitution reactions. However, α-nitroalkynes can be obtained by the reaction of alkynyllithium reagents with 2,2-dinitropropane (Equation (51)) ⟨79CC488, 82SC557⟩. The reaction with 2,2-dinitropropane has been successfully extended to alkynyllithium reagents derived from propargyl acetals and amines ⟨82SC557⟩.

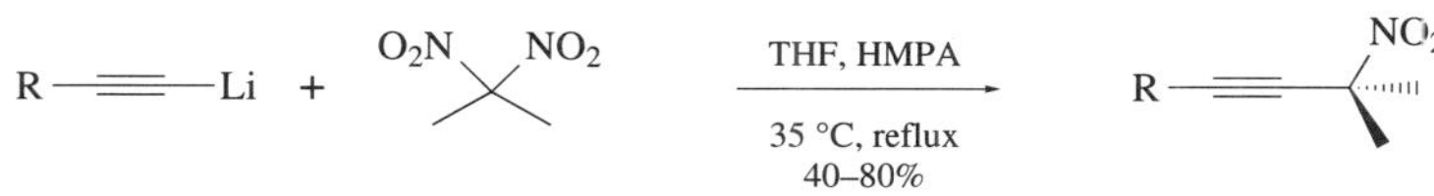

(51)

R = Pr^n, Bu^n, CH_2O-THP, CH(Me)O-THP, CH(Me)NH_2

No examples of the substitution of phosphorus, arsenic, antimony or bismuth functions by alkynes/alkynylmetal reagents have been reported.

1.21.2.1.5 Substitution of boron, silicon or germanium functions

The substitution of silicon and germanium functions is not commonly employed for the preparation of alkynes. The substitution of boron functions, however, provides a convenient and general route for the preparation of various alkyl-, alkenyl- and aryl-substituted alkynes. Lithium (1-alkynyl)organoborates (readily prepared from the reaction of an organoborane with an alkynyllithium) react with a variety of electrophiles to induce the transfer of an alkyl, alkenyl or aryl group from boron to the alkynyl C-1 carbon atom. With the proper choice of electrophile the resulting alkenyl borane can be induced to undergo β-elimination (Scheme 46) ⟨86JOC4507⟩. Although the reaction mechanism involves a complicated addition–migration–elimination sequence the overall result is the substitution of the boron function. The most commonly employed electrophile is iodine ⟨73JA3080, 86JOC4507, 91T343⟩, although methanesulphinyl chloride ⟨73TL1847, 74T2159⟩ and tributyltin chloride ⟨84JOC5175⟩ have also been used.

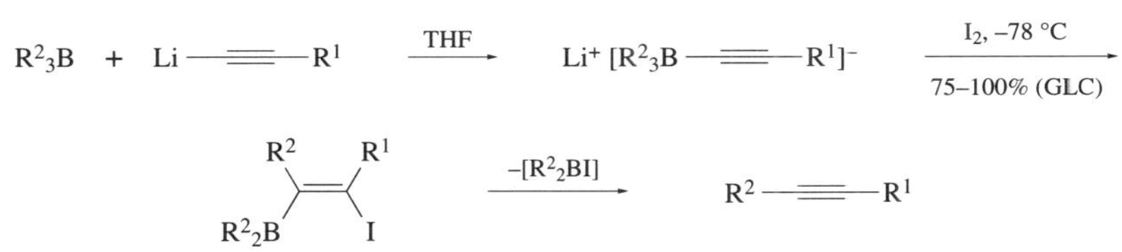

R^1 = H, Bu^n, Bu^t, Ph; R^2 = Bu^n, Bu^i, Bu^s, $MeO_2C(CH_2)_{10}$, cyclopentyl, cyclohexyl, Ph

Scheme 46

Primary alkyl groups are readily introduced as are secondary alkyl groups such as *s*-butyl, cyclopentyl and cyclohexyl, in marked contrast to their problematic introduction by conventional substitution reactions employing alkynylmetal reagents. Tertiary alkyl groups appear to undergo B to C migration with difficulty and are not generally introduced by this method. Aryl and alkenyl groups are both readily inserted, the latter being particularly convenient in view of the facile and highly stereoselective preparation of *E*-alkenyl boranes by hydroboration of alkynes (Equation (52)) ⟨91T343⟩. Conjugated 1,3-diynes can also be formed by transfer of an alkynyl group (see Section 1.21.3.1.6). Transfer of both alkyl and alkenyl groups occurs with complete retention of stereochemistry. Only one R group is transferred from organoboranes (R_3B) and a more satisfactory utilisation of the R group is achieved employing thexylmonomethoxyboron ⟨86JOC4521⟩ or bis(1,2-dimethylpropyl)boron blocking groups ⟨91T343⟩. Certain functionalities in the organoborane are compatible with the reaction provided the alkynyllithium reagent is added at low temperature. For example, methyl ester and acetate functions survive intact.

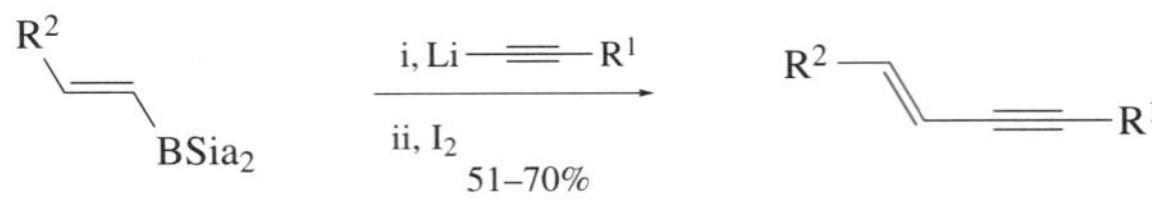

(52)

R^1 = Et, Pr^n, Bu^n, cyclohexyl
R^2 = Bu^n, TMS-O(CH_2)$_9$, AcO(CH_2)$_2$, AcO(CH_2)$_6$, cyclohexyl

Ethynyllithium itself affords low yields of 1-alkynes, although the ethylenediamine complex gives good results with most alkylboranes ⟨86JOC4507⟩. Attempted preparation of symmetrical alkynes employing dilithium ethynide is unsuccessful, but the same overall transformation can be achieved employing chloroethynyllithium, itself generated *in situ* from (*E*)-1,2-dichloroethene ⟨75TL1961⟩. Two alkyl groups are transferred from the organoborane in this reaction (Equation (53)).

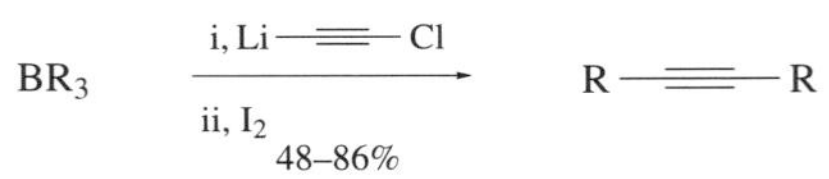

(53)

R = Bu^n, Bu^i, *n*-C_5H_{11}, *n*-C_6H_{13}

α-Chiral mono- and disubstituted alkynes have been prepared employing optically active mono-alkylthexyborinates ⟨88JA1529, 91JOC1500⟩, the latter being obtained in very high enantiomeric excess by asymmetric hydroboration of prochiral alkenes.

1.21.2.1.6 Oxidative homocoupling reactions of terminal alkynes and their derivatives

Interest in the chemistry of butadiynes and poly-ynes has led to the development of methods for the oxidative coupling of terminal alkynes and their derivatives. Reviews are available covering various aspects of oxidative coupling reactions of alkynes ⟨B-69MI 121-04, B-83MI 121-02, 91COS(5)551⟩. Most commonly, oxidative homocoupling is carried out with copper(I) (Glaser reaction) or copper(II) (Eglington reaction) salts in the presence of oxygen (Equation (54)).

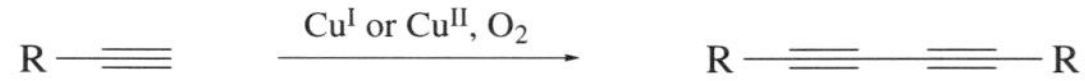

(54)

The Glaser reaction, as originally reported, is carried out in two stages. The copper(I) alkynide is formed in the first stage and then oxidised in the second stage to the diyne. Air and oxygen are most frequently employed as oxidising agents, although potassium ferricyanide, hydrogen peroxide and copper(II) salts are also used. Isolation of the copper alkynide is inconvenient and the procedure was modified to allow *in situ* formation of the copper(I) derivative. The method further evolved into that reported in 1960 by Hay ⟨60JOC1275, 62JOC3320⟩. In the Hay procedure, oxygen or air is passed through a solution of the alkyne and a catalytic quantity of a copper(I) salt in a complex forming solvent, such as pyridine or TMEDA (Scheme 47). The reaction is of wide scope and is compatible with a wide variety of functional groups in the alkyne ⟨B-69MI 121-04⟩. Yields are generally high, although the less acidic alkyl substituted alkynes require the presence of a stronger base, such as dbu, to promote the reaction ⟨B-88MI 121-01⟩.

In the Eglington reaction, the terminal alkyne is heated with a stoichiometric quantity of a copper(II) salt, most commonly copper(II) acetate, in pyridine or a similar base. Again, the reaction is of wide scope and is compatible with a variety of functional groups. The reaction is readily adapted to high dilution conditions and is particularly valuable for the preparation of large ring poly-ynes by intramolecular oxidative coupling of ω-diynes (Scheme 48) ⟨74OS(54)1, 88H(27)475⟩.

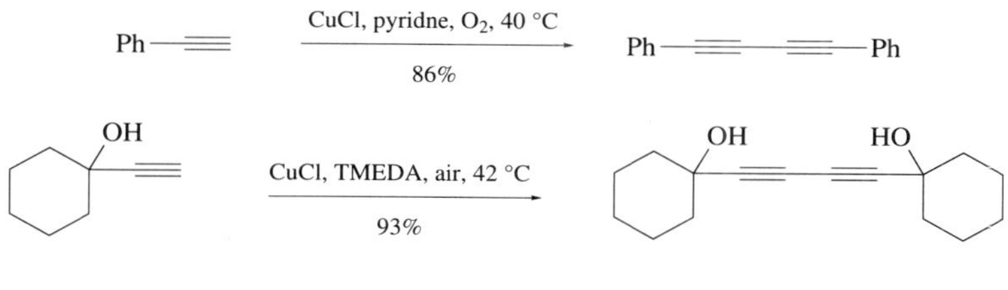

Scheme 47

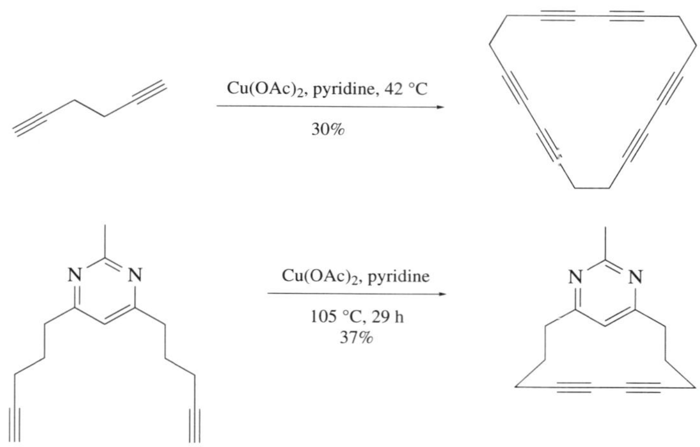

Scheme 48

Both the Glaser and Eglington oxidative coupling reactions are most suitable for the preparation of symmetrical butadiynes and poly-ynes. In general, both procedures give unsatisfactory results in unsymmetrical coupling reactions. Unsymmetrical 1,3-diynes are most conveniently prepared by the Cadiot–Chodkiewicz procedure, as described in Section 1.21.3.1.1 (iii).

Symmetrical diaryl substituted butadiynes can also be prepared in high yield by reacting benzene solutions of aryl substituted terminal alkynes with chloroacetone in the presence triethylamine and a $(PPh_3)_4Pd/CuI$ catalyst mixture ⟨85TL523⟩. Under similar conditions, aliphatic 1-alkynes afford mixtures of symmetrically disubstituted 1,4-dialkyl-1,3-butadiynes and 3-alkyl-4-(1-alkynyl)hexa-1,5-diyn-3-enes (Equation (55)). The oxidative coupling of phenylethyne has been reported employing a Pd–Cu catalyst in the presence of 4-iodo-(3*H*)-phenothiazin-3-one ⟨89JOM(364)245⟩.

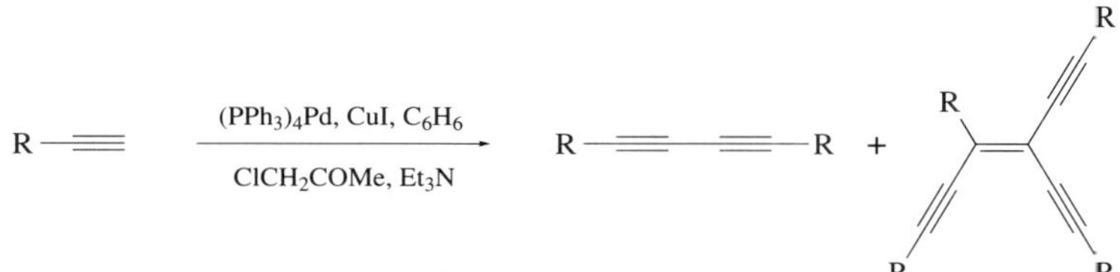

(55)

R = Ph, 2-thienyl, Pr^n, Bu^n, n-C_6H_{13}

Alkynylmetal reagents and a number of other derivatives of terminal alkynes also undergo dimerisation in the presence of a suitable oxidant. Alkynyl Grignard reagents afford conjugated diynes in the presence of iodine or other oxidants ⟨52JCS2014, 54JCS147⟩, with thionyl chloride ⟨72JOC3749⟩ or with *N*-substituted isocyanide dichlorides in the presence of a palladium catalyst ⟨88TL5379⟩. Alkynyllithium reagents undergo oxidative coupling in the presence of stoichiometric palladium and nickel complexes ⟨69TL4593, 87CC477⟩. Alkynylborates undergo oxidative coupling in the presence of iodine to afford both symmetrical and unsymmetrical 1,3-butadiynes (Equation (56)) ⟨75CC857, 76JOC1078, 76TL4385⟩. Alkynyltin ⟨87JOC4296⟩, alkynylselenium ⟨86CC1067⟩ and alkynyltellurium ⟨90SC2181⟩ compounds also undergo symmetrical coupling reactions in the presence of a suitable oxidant.

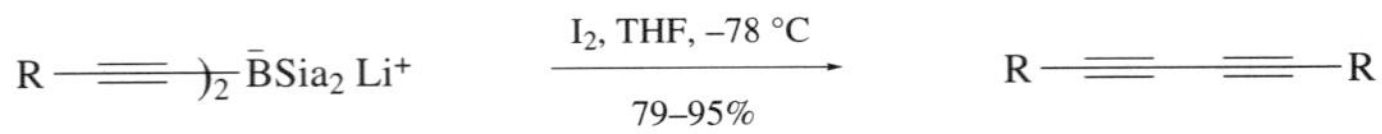

(56)

1.21.2.2 Addition Reactions of Alkynyl Carbanions

1.21.2.2.1 Addition to carbon–carbon multiple bonded functions

Terminal aliphatic alkynes react with 1,3-butadienes in the presence of catalytic quantities of a ruthenium(0)–trialkylphosphine complex to afford linear codimers in good to excellent yield (Scheme 49) ⟨85JOC565, 86BCJ3201⟩. 1,3-Butadiene itself undergoes a highly regio- and stereoselective reaction with straight chain and branched terminal alkynes to afford linear conjugated (*E*)-enynes. Phenylethyne also affords the corresponding enyne, but in low yield, the major product being a mixture of homodimers of phenylethyne. 1,3-Cyclohexadiene affords 1,5-enynes through a regioselective 1,2-addition to the cyclohexadiene. Methyl (*E*,*E*)-2,4-hexadienoate similarly undergoes highly regioselective and stereoselective reactions to give (*E*)-1,5-enynes.

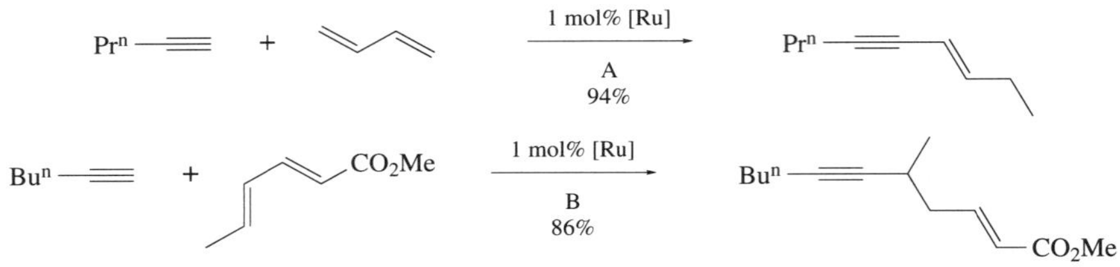

A = Ru^0(cod)(cot)-PBu^n_3; B = $RuH_2(PBu^n_3)_4$

Scheme 49

The homocoupling (Scheme 50) and cross-coupling (Scheme 51) of alkynes to furnish 1,3-enynes has been achieved in high yield employing catalytic $Pd(OAc)_2$ and the hindered phosphine ligand tris(2,6-dimethoxyphenyl)phosphine ⟨87JA3486⟩. Homocoupling to afford 1,3-enynes of opposite regiochemistry has been observed in the protonolysis of alkynylzirconocene derivatives (Scheme 52) ⟨91JA1440⟩.

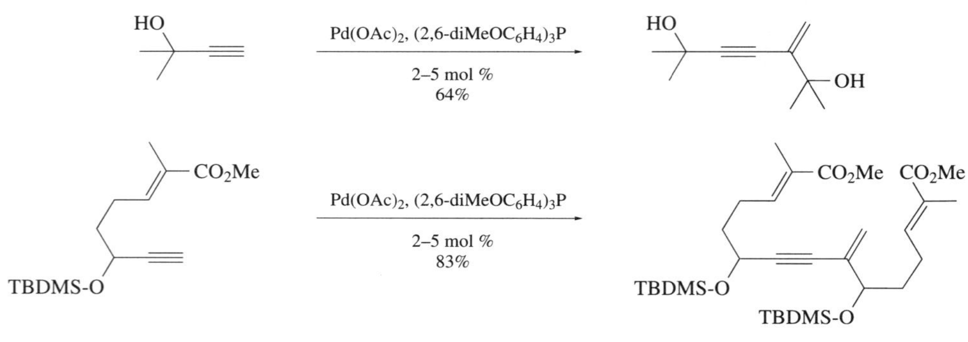

Scheme 50

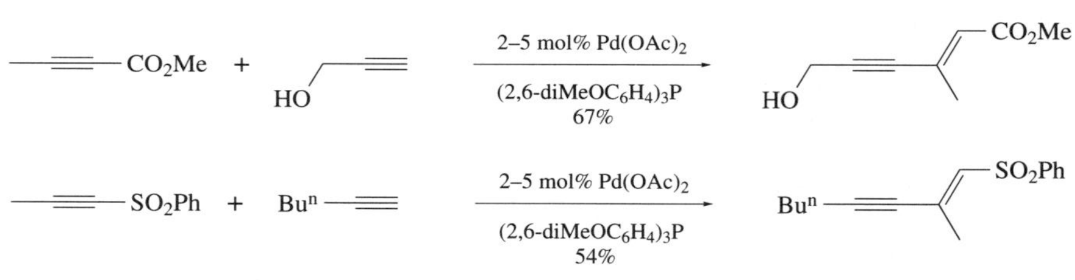

Scheme 51

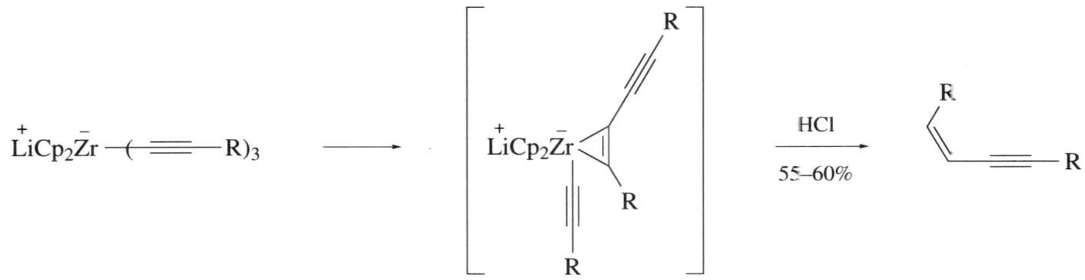

R = n-C_6H_{13}, Ph

Scheme 52

1.21.2.2.2 1,2-Addition to carbon–oxygen double bonded functions

The addition of alkynylmetal reagents to carbonyl compounds, including aldehydes, ketones, carbon dioxide and carbon monoxide affords primary, secondary or tertiary alkynols, alkynoic acids or alkynoic esters depending on the carbonyl compound employed. The literature up to 1969 has been comprehensively reviewed ⟨B-69MI 121-03⟩. Group 1 and 2 alkynylmetal reagents are most commonly employed in these reactions. Liquid ammonia and THF are the most commonly used solvents.

Aldehydes and ketones react with alkynylmetal reagents to afford propargyl alcohols (Equation (57)). Reduced yields of coupling products can be caused by enolisation of the carbonyl compound, the extent to which this occurs depending upon the acidity of the carbonyl compound, the basicity of the alkynide (kinetic basicity K > Na > Li > Mg) and the polarity of the solvent. The use of ether solvents and alkynyllithium or alkynylmagnesium halides is preferable with readily enolisable carbonyl compounds, although in general the latter offer little practical advantage over the more conveniently prepared alkynyllithium reagents. In some cases, addition of anhydrous lithium bromide has the favourable effect of solubilising poorly soluble alkynyllithium reagents ⟨81S459⟩. Addition of ethyne itself is achieved employing the readily prepared amine-free ethynyllithium ⟨75JOC2250⟩ or the commercially available complex with ethylenediamine ⟨64JOC1872⟩. High yields of addition product are obtained even with a number of readily enolised ketones. Addition of dimetalloethynides to aldehydes has been reported ⟨75JOC1131⟩. Hydroxymethylation of alkynyllithium and magnesium reagents can be achieved employing paraformaldehyde in diethyl ether or THF as solvent, depolymerisation occurring *in situ* ⟨65RTC1200⟩.

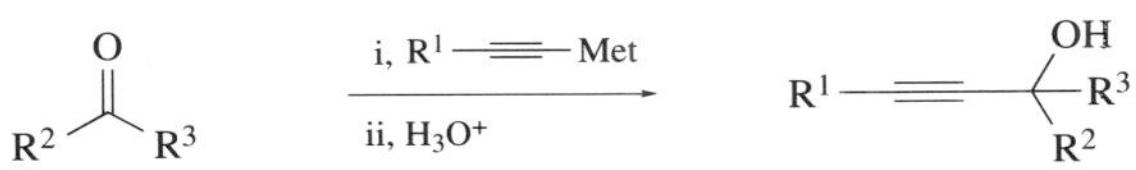

(57)

Met = Li, Na, K, MgX

Alkynylboron and aluminum reagents undergo addition reactions with aldehydes and ketones. *B*-1-Alkynyl-9-borabicyclo[3.3.1]nonanes smoothly add to aldehydes and ketones at room temperature to provide propargyl alcohols in high yield (Equation (58)) ⟨85JOC1577⟩. The reagents are nonbasic and are essentially inert at 25 °C toward alkyl halides, acid chlorides, amides, anhydrides, esters, nitriles, acetals and ketals. Furthermore, the reagents can discriminate between aldehydes and ketones and even between two ketone functions. Ethyne itself is introduced employing the trimethylsilyl protected reagent ⟨92JOC2941⟩. Alkynyldiethylalanes ⟨65CR(261)1992, 68BSF205⟩ and trialkylalanes ⟨66CR(C)289⟩ undergo addition to aldehydes and ketones although the procedure seems to offer little advantage, particularly with the latter reagents where only one of the three alkynyl groups is utilised in the reaction.

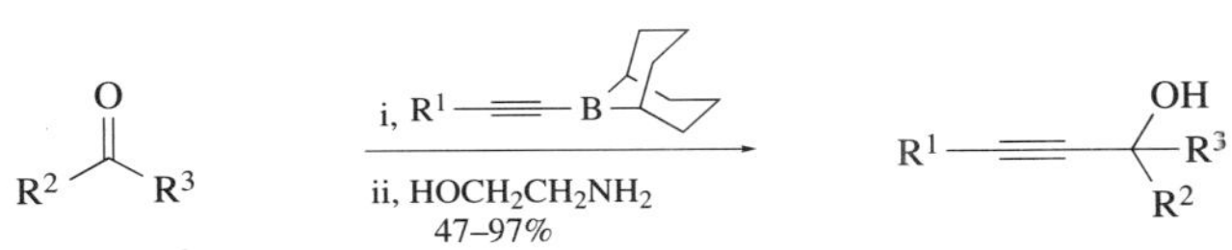

(58)

R^1 = Bu^n, Bu^t, CH_2=CH(Me), $Cl(CH_2)_3$, Ph
R^2 = H, Et, Bu^t, 3-cyclohexene, Ph; R^3 = H
R^2 = R^3 = Me, $(CH_2)_4$, $(CH_2)_5$

Alkynylcerium ⟨84TL4233⟩, titanium ⟨87CB1845⟩, chromium ⟨85TL5585⟩, vanadium ⟨86TL933⟩, manganese ⟨77TL3383⟩, ytterbium ⟨81JOM(205)4⟩ and zinc ⟨87TL1019⟩ reagents have been coupled with aldehydes and ketones. Alkynylcerium dichloride reagents have been used with particular success for readily enolised ketones such as β,γ-enones and benzyl ketones (Scheme 53) ⟨84TL4233, 91JA694, 91TL1707, 92TL4653⟩. These ketones undergo extensive enolisation even with alkynyllithium and magnesium reagents, but afford high yields of the addition product in the presence of one equivalent of cerium trichloride. Alkynyltitanium triisopropoxides react smoothly with aldehydes at $-50\,^{\circ}$C but very slowly with ketones even at $0\,^{\circ}$C (Equation (59)) ⟨87CB1845⟩. The reagents also allow highly diastereoselective reactions with some chiral aldehydes. Alkynylchromium reagents, prepared by treatment of alkynyl iodides (and less commonly alkynyl bromides) with anhydrous chromium(II) chloride, also undergo highly chemoselective reactions with aldehydes ⟨85TL5585⟩. Catalytic quantities of nickel(II) chloride are commonly employed in the addition reactions to overcome the variable results sometimes observed with different sources of $Cr(II)Cl_2$ ⟨86JA5644⟩. Alkynylchromium reagents have been applied with particular effect for cyclisations mediated by intramolecular addition to enolisable aldehydes (Scheme 54) ⟨91TL3171, 92JA9279, 92TL7511, 93TL1467⟩. Although conditions have been described for achieving such cyclisations with alkynyllithium ⟨88TL4217⟩ and alkynylcerium reagents ⟨91JA694⟩, the chromium(II)-mediated reaction of alkynyl iodides provides a more general solution.

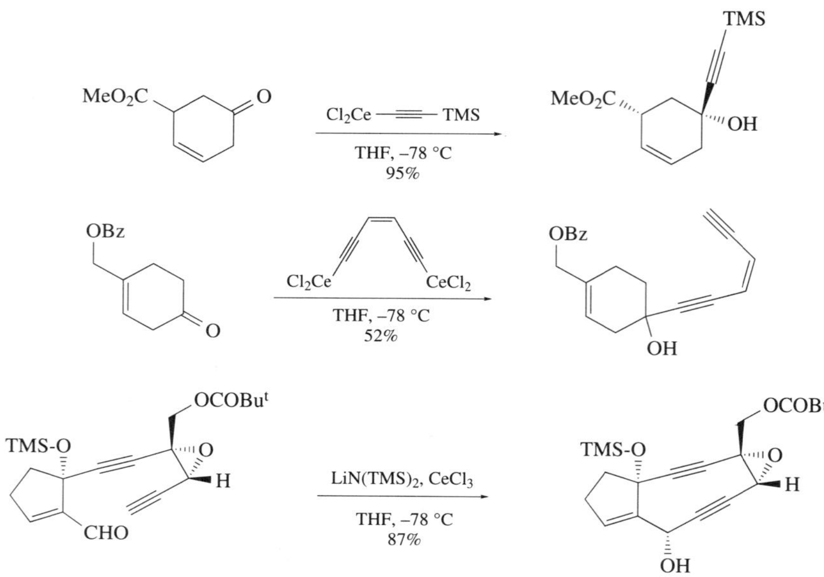

Scheme 53

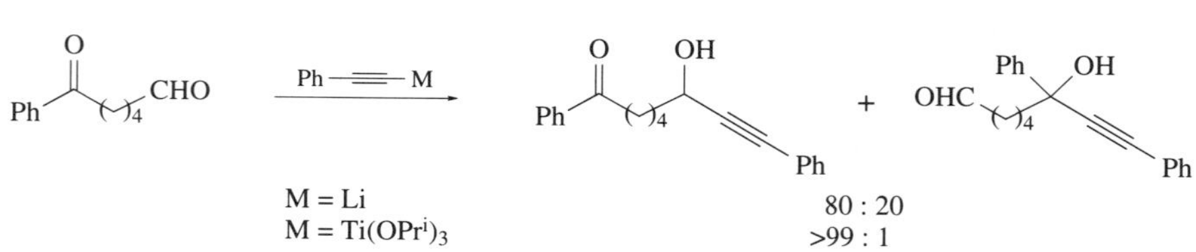

(59)

Alkynyltin ⟨67TL495, 86CC102⟩ and silicon ⟨76AG(E)498, 76TL2449, 83T975⟩ reagents undergo addition reactions with aldehydes and ketones in the presence of a Lewis acid or fluoride ion. 1-Alkynes themselves undergo addition to aldehydes and unhindered ketones in the presence of $Sn(OTf)_2$ and an amine ⟨91JOC4091⟩.

1-Alkynes will undergo addition reactions with aldehydes and ketones in the presence of basic reagents such as alkali metal hydroxides and alkoxides. The reaction is limited to less base-sensitive carbonyl compounds and is more successful with ketones than aldehydes. Phase-transfer catalysis is beneficial ⟨82LA1750⟩. The reaction has been comprehensively reviewed ⟨B-69MI 121-03⟩.

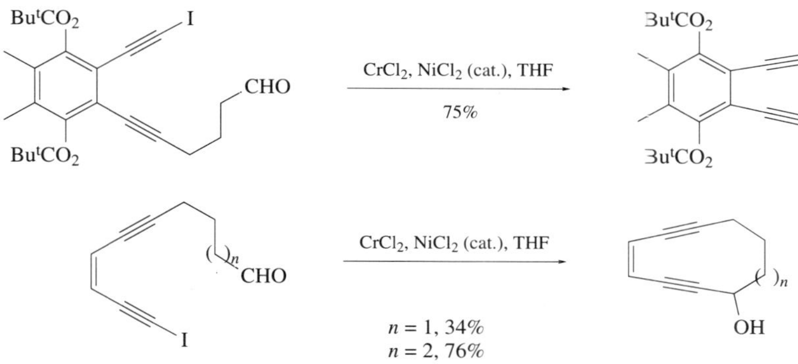

Scheme 54

In addition to chemoselectivity considerations, stereoselectivity issues including diastereoselectivity and enantioselectivity usually arise in the addition of alkynylmetal reagents to aldehydes and ketones. Diastereoselective additions to chiral aldehydes and ketones and to substituted cyclic ketones have been reported. For acyclic compounds, with a carbonyl unit as the only polar group, Cram's rule applies and stereoselectivities are generally moderate. A comparison of alkynyllithium and alkynyltitanium triisopropoxides showed little difference in diastereoselectivities for the addition to chiral aldehydes ⟨87CB1845⟩ whereas considerable improvements were observed with the $TiCl_4$-promoted addition of alkynyltin reagents ⟨86CC102⟩. Low diastereoselectivities are frequently observed in the addition of alkynyllithium and magnesium reagents to α-alkoxy aldehydes, although improved diastereoselectivities have been reported with alkynylzinc bromides and alkynyltitanium triisopropoxides in the presence of $TiCl_4$ ⟨84CL405, 87TL1019, 93JOC2468⟩. In the absence of overriding steric constraints, alkynylmetal reagents undergo moderate to complete axial additions to cyclohexanones and cyclohexenones ⟨83TL4887⟩. The presence of an axial substituent at C-3 or C-5 in either system results in a shift to complete selectivity for equatorial addition ⟨75CRV521, 89JOC4072⟩. Generalisations are less easy to make in the cyclopentanone series. 2-Alkylcyclopentanones undergo *cis*-addition with ethynylmetal reagents. 2-Methoxycyclopentanone undergoes predominantly *trans*-addition with ethynyllithium in liquid ammonia, but predominantly *cis*-addition in THF ⟨75CRV521⟩. Interestingly, a complete reversal of stereoselectivity was observed with the 3-hydroxycyclopentanone (**16**) on changing from monolithium ethynide to the bis magnesium ethynide (Scheme 55) ⟨81CC363⟩.

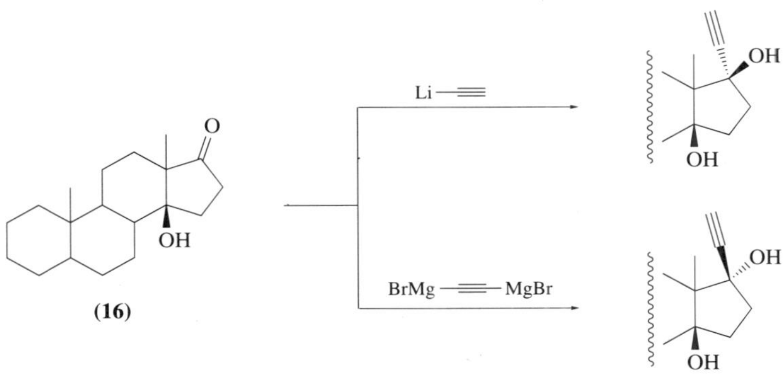

Scheme 55

Enantioselective additions to aldehydes and ketones have been achieved with trialkylsilylethynyllithium reagents which undergo addition to aromatic and aliphatic aldehydes in the presence of a chiral amino alcohol ligand (*ee* 40–92%) (Equation (60)) ⟨79CL447, 80CL255⟩. Moderate enantiomeric excesses have also been achieved in the alkynylation of alkyl and aryl aldehydes with alkynylzinc reagents in the presence of catalytic quantities of the amino alcohols (−)-*N*,*N*-dibutylnorephedrine ⟨90JCS(P1)937⟩ and (−)-*N*-methylephedrine (Equation (61)) ⟨90SL547⟩.

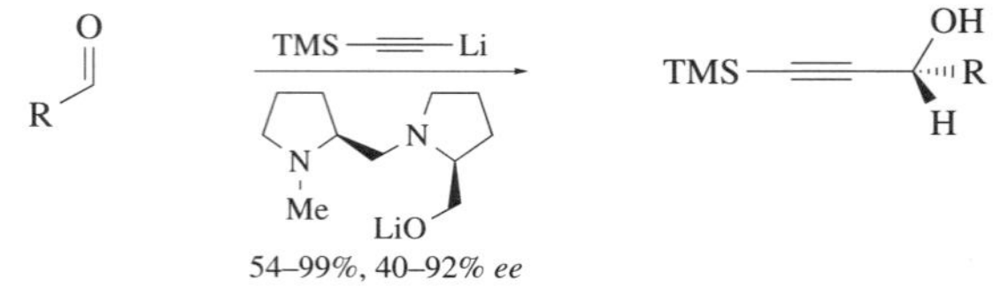

(60)

R = Et, Bu^i, n-C_5H_{11}, n-C_8H_{17}, n-$C_{11}H_{23}$, n-$C_{13}H_{27}$, Ph, $Me(CH_2)_2CH{=}CH_2$

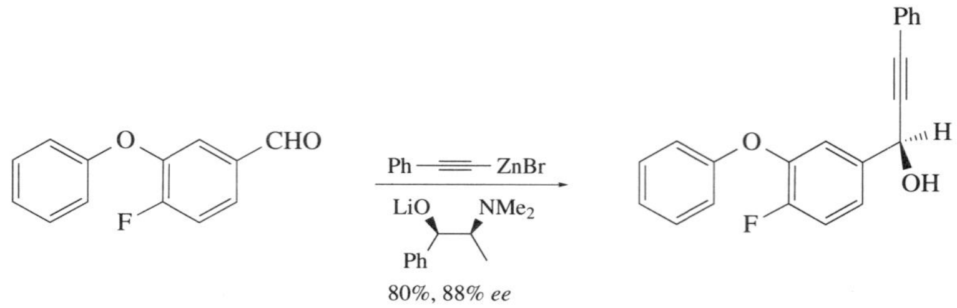

(61)

2-Alkynoic acids are prepared by the carboxylation of alkynylmetal reagents (Equation (62)) ⟨B-88MI 121-01⟩. Group 1 and 2 alkynylmetal reagents are employed almost exclusively, the reaction being performed with solid carbon dioxide ⟨54JOC1580⟩ or, preferably, with anhydrous gaseous carbon dioxide. The choice of solvent can be important; for instance, dioxan solutions of dialkynylmagnesium reagents afford higher yields of alkynoic acids than ethereal solutions of alkynylmagnesium halides ⟨56JA1221⟩. In general, the use of alkynyllithium reagents in THF is optimal providing alkynoic acids in very high, often quantitative, yields ⟨B-88MI 121-01⟩.

$$R{-}C{\equiv}C{-}Li \xrightarrow[\text{ii, } H_3O^+]{\text{i, } CO_2\text{, THF}} R{-}C{\equiv}C{-}CO_2H \qquad (62)$$

Esters of 2-alkynoic acids are conveniently obtained by the oxidative carbonylation of terminal alkynes (Equation (63)) ⟨80TL849⟩. The reaction is carried out at atmospheric pressure and room temperature in an alcoholic solvent with catalytic palladium(II) chloride, a stoichiometric quantity of copper(II) chloride and a base.

$$R^1{-}C{\equiv}CH \xrightarrow[\substack{CuCl_2\text{, NaOAc, } R^2OH \\ 58\text{–}74\%}]{CO\text{, } PdCl_2\text{ (cat.)}} R^1{-}C{\equiv}C{-}CO_2R^2 \qquad (63)$$

1.21.2.2.3 1,2-Addition to other carbon–chalcogen double bonded functions

In contrast to their carbonyl counterparts, thioaldehydes and thioketones usually undergo 1,2-addition reactions with carbanions through thiophilic attack; that is, nucleophilic attack occurs at the heteroatom end of the C=S bond ⟨B-90MI 121-01⟩. Alkynylmetal additions have been little studied but appear to conform to this mode of reactivity. Thus, thiobenzophenone reacts with ethynyl sodium to give 1,2-bis(benzhydrylthio)ethyne (Equation (64)) ⟨65CB3311⟩.

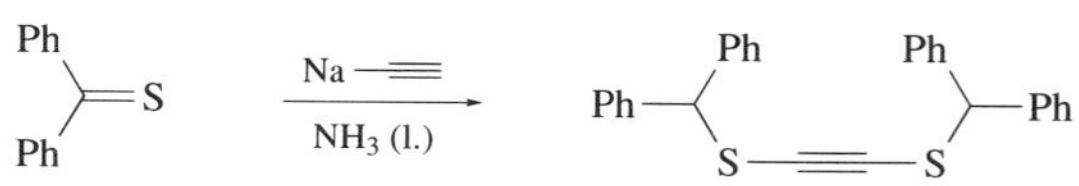

(64)

1.21.2.2.4 1,2-Addition to carbon–nitrogen multiple bonded functions

This section will deal with the reaction of 1-alkynes and alkynylmetal reagents with imines and imine derivatives, heteroaromatic nitrogen compounds (e.g., pyridines), isocyanates, isothiocyanates, carbodiimides and nitriles.

Aldimines and ketimines do not generally undergo addition reactions with alkynyllithium and magnesium reagents ⟨48JA3868⟩, although exceptions are known ⟨86AKZ306, 89S33⟩. The poor reactivity of imines is presumably due to the low electrophilicity of the imine carbon atom and competing deprotonation. Imines can be activated towards nucleophilic addition by removal of electron density from the imine carbon atom. This has been achieved by coordination of a Lewis acid to the nitrogen lone pair, by quaternisation of the nitrogen atom (iminium salts and *N*-acyliminium salts) and by incorporation of an electron-withdrawing group on the imine nitrogen (*N*-acylimines and *N*-sulphonylimines).

Aldimines undergo 1,2-addition reactions with alkynyllithium reagents, in the presence of stoichiometric boron trifluoride etherate, to afford secondary amines in reasonable yield. The reaction has been applied to acyclic aldimines (Equation (65)) ⟨84TL1083⟩ and to substituted 3-thiazolines ⟨83TL4503⟩. Ketimines do not undergo addition reactions under these conditions but are presumably intermediates in the formation of propargyl amines by trialkynylaluminum-induced Beckmann rearrangement of oxime carbonates (Equation (66)) ⟨84TL2367⟩. Other Lewis acids, for example TMS-OTf ⟨88CC439⟩ have been applied to the reaction of organometallic reagents with imines, but the use of alkynylmetal reagents has not been reported.

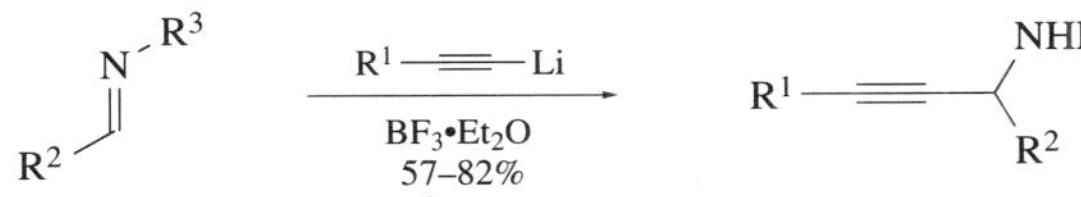

(65)

R^1 = Pr^n, Ph
R^2 = Pr^n, $CHMe_2$
R^3 = $PhCH_2$, PhCH(Me), $PhCH_2CH_2$

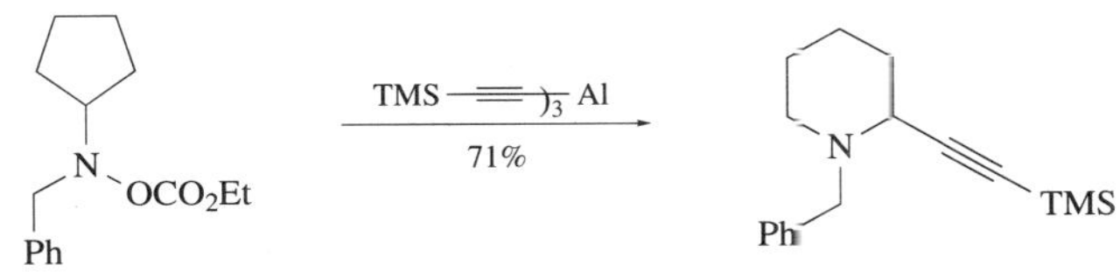

(66)

Iminium salts have long been used as activated imines. The Mannich reaction of an alkyne with formaldehyde and a secondary amine affords good yields of 1-amino-2-alkynes (Equation (67)) ⟨73S703⟩. The reaction can be viewed as a nucleophilic attack of the alkynide anion on an iminium ion generated *in situ*. Conjugated alkynes including 1,3-enynes, 1,3-diynes and arylethynes react more readily than do the less acidic alkylethynes. Copper salts markedly accelerate the reaction through the formation of an alkynylcopper(I) reagent. Unfortunately the reaction is limited to the preparation of tertiary amines in which the alkyne and *N*-atom are connected by a methylene group. This limitation is overcome by the use of 1-(1-dialkylaminoalkyl)benzotriazoles prepared from benzotriazole, an aldehyde and a secondary amine. The latter compounds react with alkynyllithium reagents to afford high yields of propargyl amines (Equation (68)) ⟨89S31⟩. An alternative to the Mannich reaction for the preparation of tertiary amines is the reaction of *N*,*N*-disubstituted aminoalkyl ethers with alkynylmagnesium halides (Scheme 56) ⟨63CPB1049⟩ or with 1-alkynes in the presence of copper(I) iodide ⟨85TL3203⟩. Primary propargyl amines are not available from the conventional Mannich reaction but have been prepared employing *N*-triphenylphosphorylidene-1-(benzotriazol-1-yl)methylamine (Scheme 57) ⟨89TL3303⟩ and *N*,*N*-bis(trimethylsilyl)methoxymethylamine ⟨84AG(E)53, 84CC794, 87S848⟩.

(67)

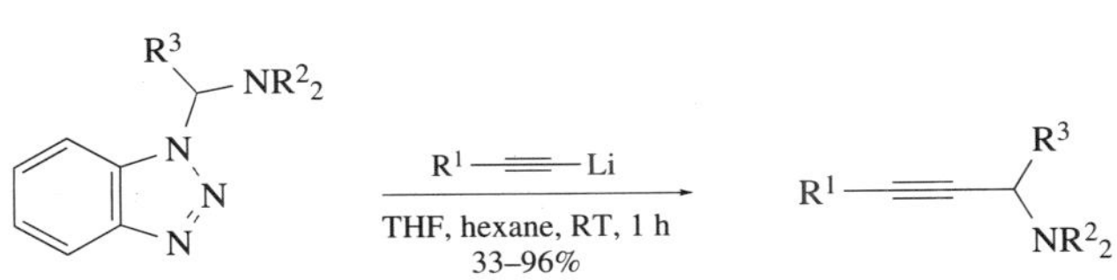

(68)

Preformed iminium salts also undergo reaction with alkynylmetal reagents although only a few examples have been reported ⟨70JA6676, 83BSF(2)21⟩. *S*-Alkylthioamidinium salts of thiolactams react with alkynyllithium reagents to afford α-alkynylazacycloalkanes after reduction with lithium aluminum hydride (Scheme 58) ⟨89JCS(P1)1211⟩.

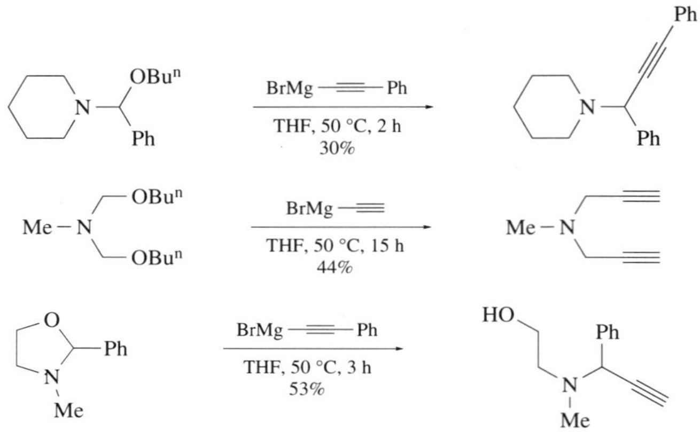

Scheme 56

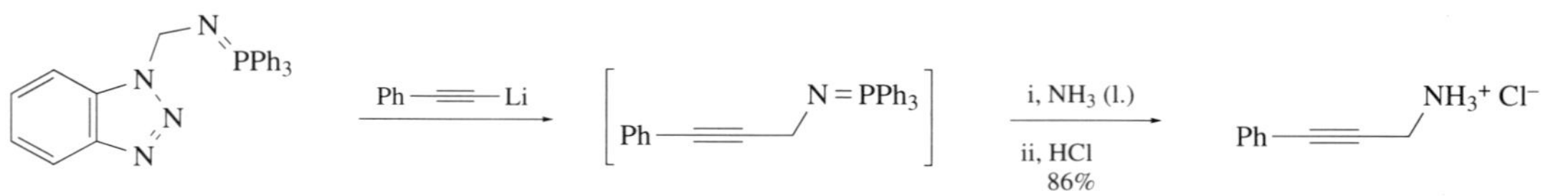

Scheme 57

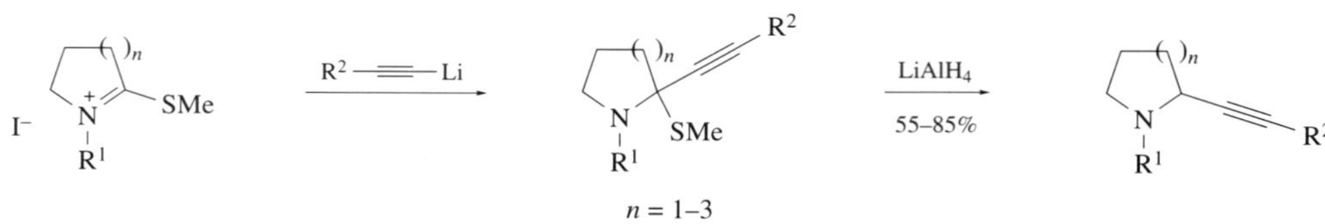

Scheme 58

N-Acylimines and *N*-acyliminium salts are also powerful electrophiles ⟨70S49, 84S85⟩. The reagents, usually generated *in situ*, undergo 1,2-addition reactions with a variety of alkynylmetal reagents ⟨80CC736, 84T811, 86S372, 86TL4241, 88T5425, 88T5451, 88TL5391, 90JOC4657, 90S115⟩. A few examples are illustrated in Scheme 59.

N-Sulphonyl aldimines undergo addition reactions with alkynyllithium reagents to generate *N*-protected primary amines ⟨90JOC393⟩. The *N*-sulphonyl imines can be generated *in situ* by treatment of aldehydes with *N*-sulphinyl sulphonamides (Scheme 60).

Oximes and hydrazones generally show low reactivity towards alkynylmetal reagents. For example, (S)-(−)-1-amino-2-methoxymethylpyrrolidine (SAMP)-hydrazones have been reported to undergo addition reactions with a wide variety of organocerium reagents (alkyl, alkenyl, aryl), but alkynylcerium reagents fail to give any addition products ⟨87JA2224⟩. Boron trifluoride-activated oxime ethers do undergo addition reactions with alkynyllithium reagents to afford *O*-alkyl hydroxylamines (Equation (69)) ⟨88TL3455⟩. Similarly, 2-isoxazolines undergo addition reactions with alkynyllithium reagents in the presence of boron trifluoride etherate provided the C-3 ring carbon atom is unsubstituted ⟨89CL1079, 91COS(1)355⟩.

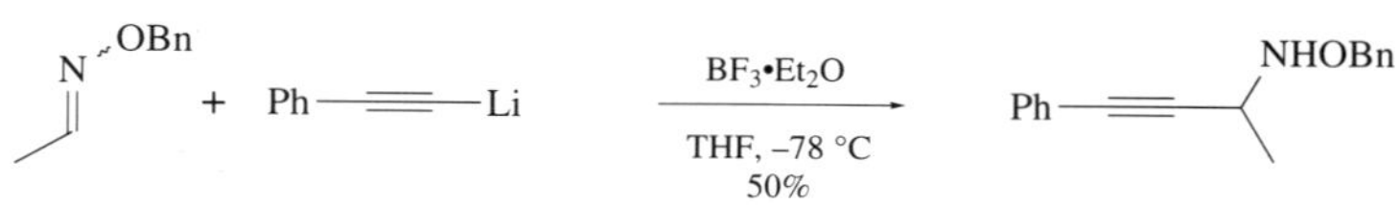

(69)

Pyridine and other heteroaromatic nitrogen compounds possessing a formal azomethine linkage undergo addition of alkynylmetal reagents when converted into iminium salts. Pyridine and substituted pyridines undergo smooth 1,2-addition reactions with alkynylmagnesium halides in the presence of methyl chloroformate (Equation (70)) ⟨83H(20)601, 83TL1801, 87JOC2094⟩. Where other

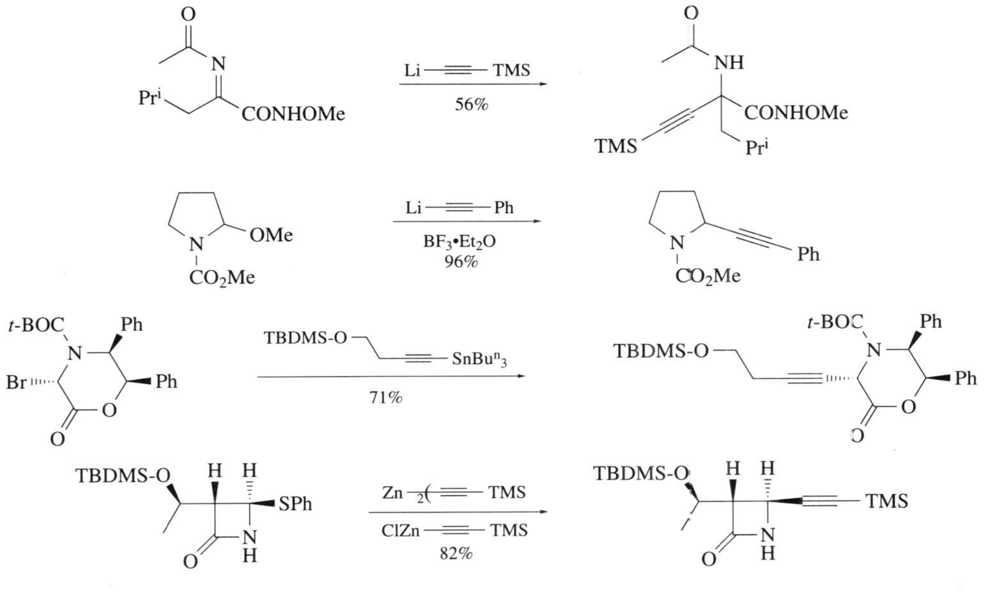

Scheme 59

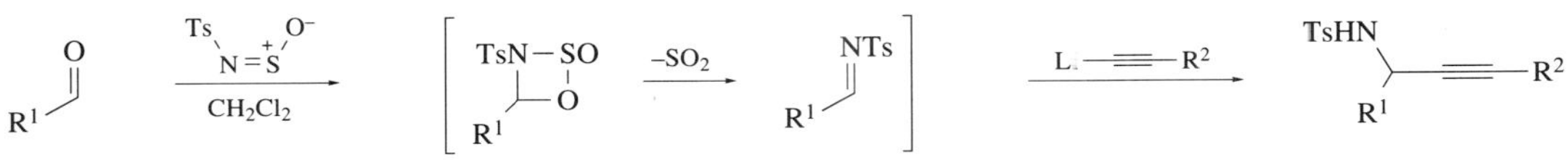

R^1 = Ph, R^2 = Bu^n, 93%
R^1 = Pr^i, R^2 = Bu^n, 76%

Scheme 60

electrophilic functions are present, such as carbonyl groups, chemoselectivity has been controlled employing alkynyltin reagents. For example, 3-acyl substituted pyridines afford reasonable yields of 2,3-disubstituted 1,2-dihydropyridines with alkynyltin reagents, whereas the corresponding magnesium reagents display poor selectivity ⟨88TL1785⟩. *N*-Benzyloxypyridinium chloride, generated *in situ* from pyridine *N*-oxide and benzoyl chloride, undergoes 1,2-addition of alkynylsilver reagents to afford the C-2 addition product ⟨89CL773⟩. The reaction is applicable to a wide range of substituted pyridine *N*-oxides and to quinoline, isoquinoline and pyrimidine *N*-oxides.

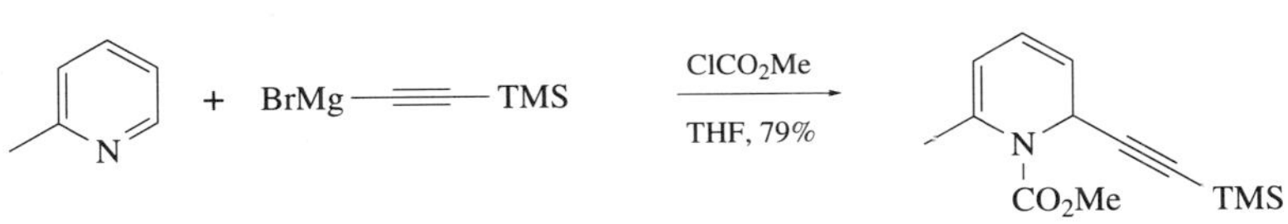

(70)

Addition of alkynylmetal reagents to heterocumulenes such as isocyanates, isothiocyanates and carbodiimides have not been studied extensively and only a few examples are available. Isocyanates have long been known to react with alkynylmetal reagents and have often been used for characterisation purposes ⟨31JA4192, 33JA1094, 59CB1950⟩; representative examples are given in Scheme 61 ⟨33JA1094, B-88MI 121-20⟩. Primary carboxylic amides are available from the reaction of alkynylmagnesium bromides with trimethylsilyl isocyanate or chloroacetyl isocyanate ⟨75TL981⟩ and from the reaction of alkynylsilanes with chlorosulfonyl isocyanate ⟨88S621⟩. A single example of the reaction of an alkynyltin reagent with phenyl isocyanate in the presence of aluminum trichloride has been reported ⟨90CB2191⟩. 9-(1-Alkynyl)-9-borabicyclo[3.3.1]nonanes react with isocyanates to afford *N*,*N*′-disubstituted-*N*-(2-alkynoyl)-ureas (Equation (71)) ⟨79S104⟩.

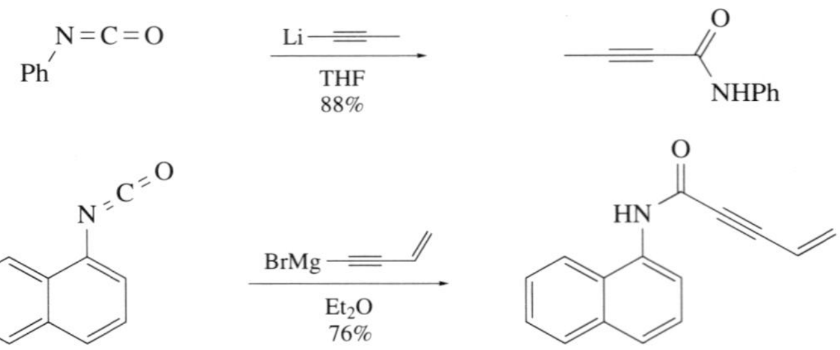

Scheme 61

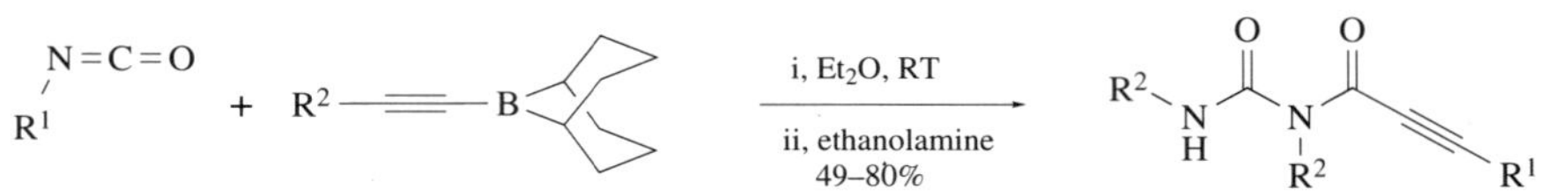

(71)

Isothiocyanates ⟨81JCS(P1)40, B-88MI 121-03⟩ and carbodiimides ⟨72BCJ1846, 84LA85⟩ react with group 1 and 2 alkynylmetal reagents to afford thioamides and amidines respectively (Scheme 62) ⟨72CPB1846, B-88MI 121-03⟩.

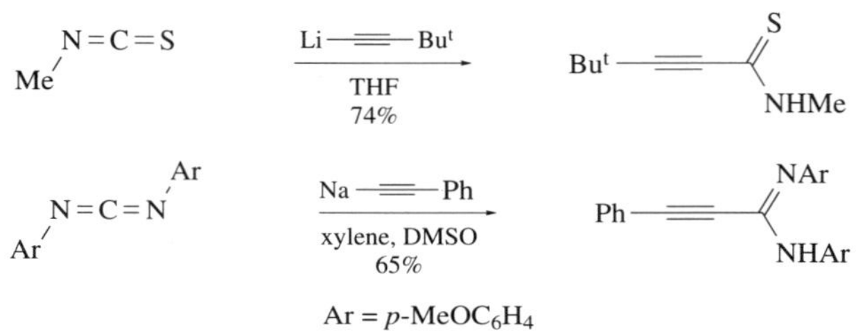

Scheme 62

Although nitriles do not undergo addition reactions with alkynyllithium reagents, and Lewis acid promoted reactions have not been reported, nitrilium ions are presumably intermediates in the Beckmann rearrangement of oxime sulfonates induced by alkynylmagnesium halides (Equation (72)) ⟨82TL3395⟩. Reduction of the intermediate imine with diisobutylaluminum hydride allows the preparation of propargyl amines in reasonable yield.

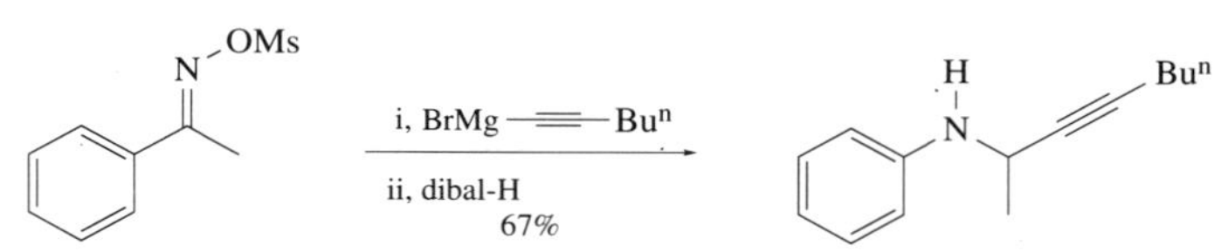

(72)

1.21.2.2.5 1,2-Addition–elimination reactions

This section deals with the reaction of alkynylmetal reagents with electrophiles of the form C(X)═Y, where both X and Y represent heteroatom functions. Substrates include carboxylic acid chlorides, anhydrides and imides, carboxylic esters, thiol esters and lactones, amides and lactams, alkyl chloroformates, alkyl chlorothionoformates and alkyl chlorodithioformates, imidoyl chlorides, *N*,*N*-dialkylcarbamoyl and *N*,*N*-dialkylthiocarbamoyl chlorides, and cyanogen chloride. Although the overall transformation for these reactions is one of substitution, the mechanism, at least of the nontransition metal-mediated reactions, is better represented as an addition–elimination reaction.

Carboxylic acid chlorides react with alkynylmetal reagents to afford conjugated alkynones (Equation (73)). Group 1 and 2 alkynylmetal reagents do not generally give satisfactory results and the less reactive alkynylcopper reagents are more successfully employed ⟨70TL2659, 75OR(22)253⟩. Although ethynylcopper itself does not undergo this reaction, trimethylsilylethynylcopper can be used successfully ⟨75JOC131⟩. Alkynylzinc halides also give very good results with saturated aliphatic acid chlorides ⟨88JOM(338)289⟩; with unsaturated acid chlorides the palladium-catalysed process is more successful. Alkynylsilver ⟨56JA1675⟩, manganese ⟨89TL3545⟩ and cadmium ⟨67ZOR219⟩ reagents have also been employed successfully. More conveniently, 1-alkynes themselves react with acid chlorides in the presence of catalytic copper(I) iodide, $(PPh_3)_2PdCl_2$ and triethylamine at room temperature to afford alkynones in high yield (Equation (74)) ⟨77S777⟩. Unfortunately the reaction is not applicable to acid chlorides such as acetyl, propanoyl and butanoyl chloride which react readily with triethylamine. Alkynylzinc ⟨83TL5181⟩ and alkynyltin ⟨86AG(E)508⟩ reagents both undergo smooth palladium-catalysed coupling reactions with acid chlorides, affording high yields of conjugated ynones under very mild conditions and in the presence of a wide variety of other functional groups. Alkynylsilanes undergo aluminum trichloride-catalysed reactions with acid chlorides ⟨63CB3280, 72JOM(37)45⟩. The intramolecular reaction has been applied to the preparation of a number of cyclic alkynones ⟨78TL2301⟩.

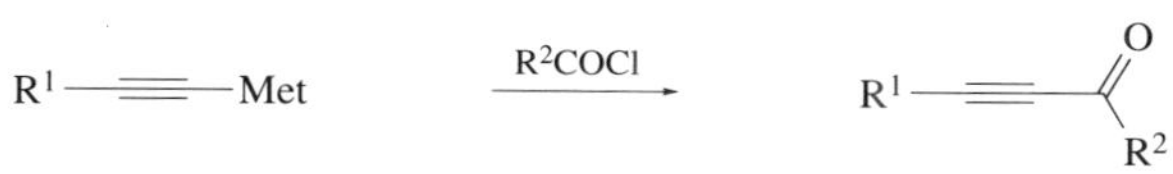

(73)

Met = Cu, Ag, MnX, Cd, ZnCl, $ZnCl_2/Pd(PPh_3)_4$, SnR_3/Pd^0

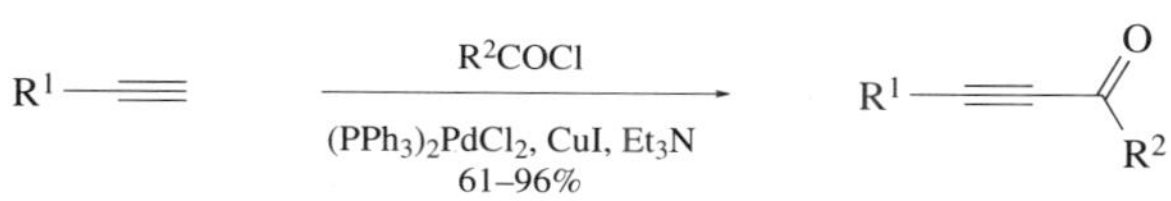

(74)

R^1 = Bu^n, Ph; R^2 = Pr^i, Bu^t, Ph, CH=CHPh, 2-furyl

Carboxylic anhydrides undergo more effective coupling reactions with group 1 and 2 alkynylmetal reagents ⟨71M214⟩. Lithium alkynyltrifluoroborates also react smoothly with carboxylic acid anhydrides to afford conjugated alkynones in high yield (Equation (75)) ⟨84TL2411⟩. Cyclic imides, such as succinimide, react with alkynylmagnesium halides to afford 4-ketoamides in yields of 40–69% ⟨71CPB391⟩.

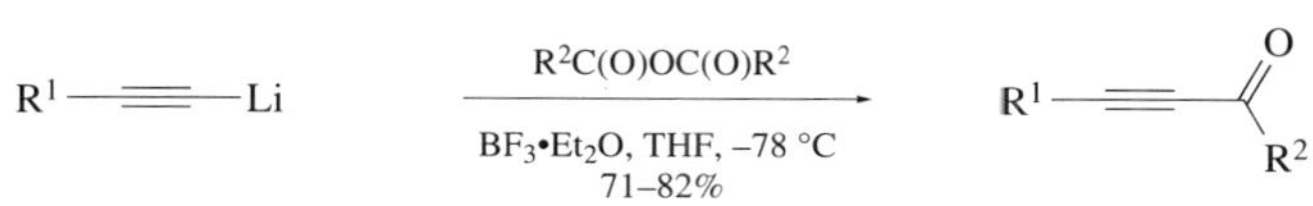

(75)

R^1 = Bu^n, Bu^t, n-C_6H_{13}, n-C_8H_{17}, Ph, cyclohexyl
R^2 = Me, Et, Ph

Carboxylic acid esters generally afford tertiary alcohols with alkynylmetal reagents; exceptions include formate, oxalate and trichloroacetate ⟨78S307, 78TL937⟩. Tertiary alcohol formation can be prevented by employing alkynyllithium reagents in the presence of boron trifluoride etherate (Equation (76)) ⟨86S421⟩. Under these conditions aliphatic esters afford conjugated alkynones in good yield, whereas aromatic and α,β-unsaturated esters do not react. Thiol esters such as *S*-ethyl carbothioates react with 1-alkynyltrimethylsilanes in the presence of silver tetrafluoroborate to give the corresponding alkynones in good yield ⟨83TL5131⟩. Lactones undergo addition–elimination reactions with alkynyllithium reagents to afford conjugated alkynones, generally in good yield, although this is dependent on the lactone ring size (Equation (77)) ⟨78TL937⟩. The products are often in equilibrium with their ring-closed hemiketal forms, the position of equilibrium depending on the ring size and substitution pattern. Successful monoaddition of the alkynylmetal reagent appears to be dependent on the stability of the intermediate ketal alkoxide which in turn is related to ring size; $6 > 5 > 7 \gg$ acyclic. 6-Membered ring lactones invariably afford conjugated alkynones in very high yield and numerous examples have been reported demonstrating the generality of this reaction (Scheme 63) ⟨81CL1415, 92JOC5818⟩. In certain cases the initial addition product can be induced to undergo a ring enlargement to an oxocene when warmed to room temperature in the presence of HMPA (Equation (78)) ⟨84TL1757⟩.

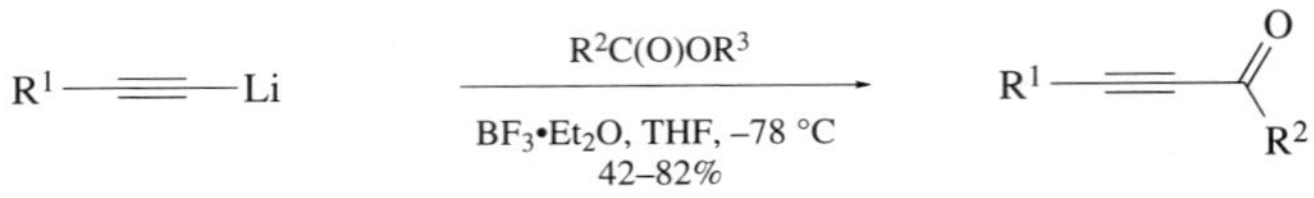

(76)

R^1 = H, n-C_5H_{11}, Ph
R^2 = Me, Pr^n, n-C_9H_{19}, $Ph(CH_2)_2$, $Ph(CH_2)_3$
R^3 = Et, Pr^i, Bu^s, Bu^t, n-C_5H_{11}, $PhCH_2$

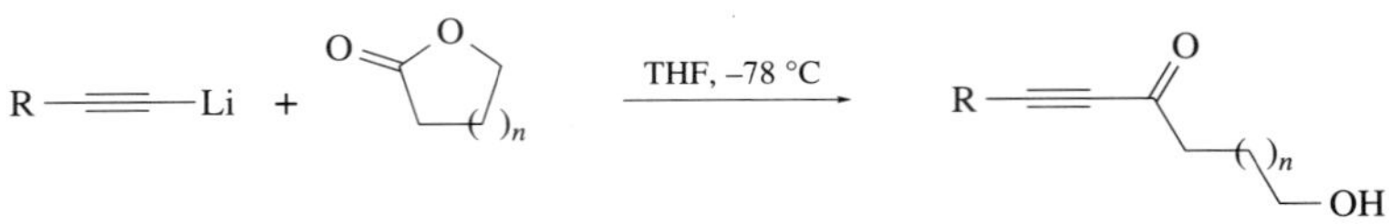

(77)

	$n = 1$	$n = 2$	$n = 3$
R = Bu^n	63%	80%	38%
R = $Ph_3COCH_2CH_2$	52%	99%	31%
R = Ph	68%	99%	31%

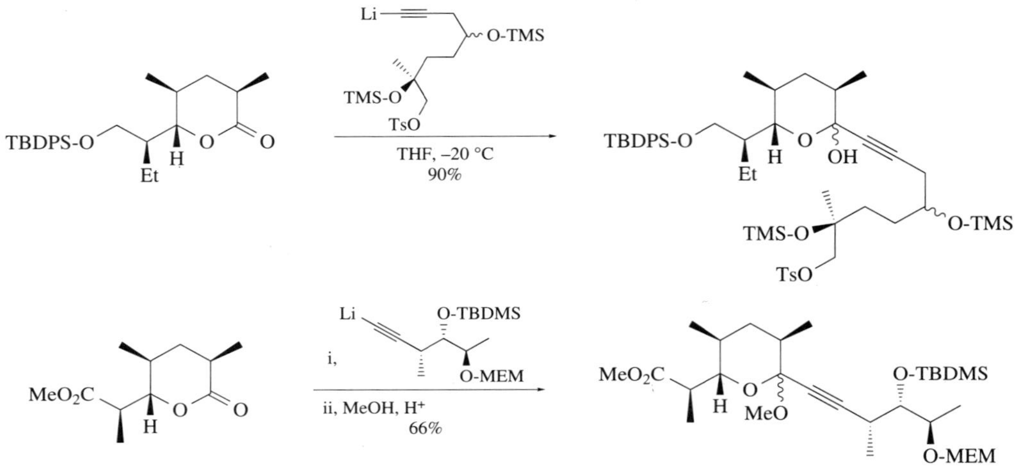

Scheme 63

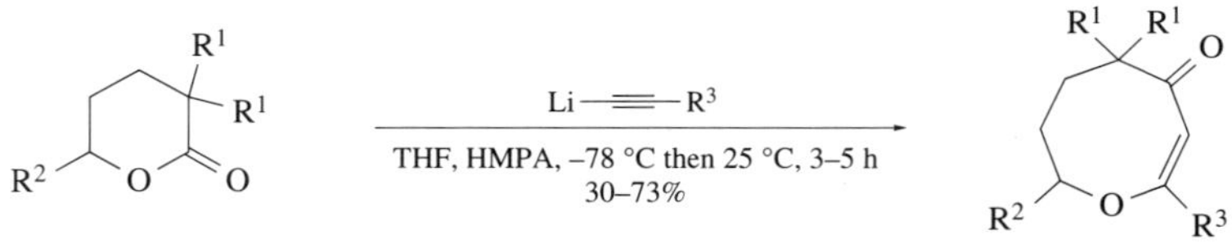

(78)

R^1 = H, Me, SMe; R^2 = H, Me
R^3 = H, CH_2OMe, $CH(OEt)_2$

Carboxylic amides acylate alkynylmetal reagents with variable success ⟨B-69MI 121-03⟩. *N,N*-Disubstituted formamides, in particular *N,N*-dimethylformamide ⟨B-88MI 121-01⟩ and formyl piperidine ⟨81AG925⟩, react with alkynyllithium reagents to afford good yields of conjugated ynals. The reaction intermediate is reasonably stable in ether solvents and further attack of the alkynyllithium reagent does not occur. Alkynylmagnesium halides give somewhat lower yields ⟨58JCS1054⟩. A number of methyl and phenyl alkynones can be prepared in high yield from an alkynyllithium reagent and *N,N*-dimethylacetamide or *N,N*-dimethylbenzamide. However, weakly basic alkynides such as phenylethynyllithium give unsatisfactory results ⟨88JOM(338)289, B-88MI 121-01⟩. The amides from other aliphatic carboxylic acids react incompletely. In the presence of boron trifluoride etherate a wide variety of *N,N*-dialkylamides react with alkynyllithium reagents to afford conjugated ynones in high yield (Equation (79)) ⟨83CL35⟩. An interesting chemoselectivity is observed under these reaction conditions with dimethylaminocarbonyl being far more reactive than methoxycarbonyl and comparable to a ketone group. Secondary amides such as acetanilide are recovered unchanged under the reaction conditions.

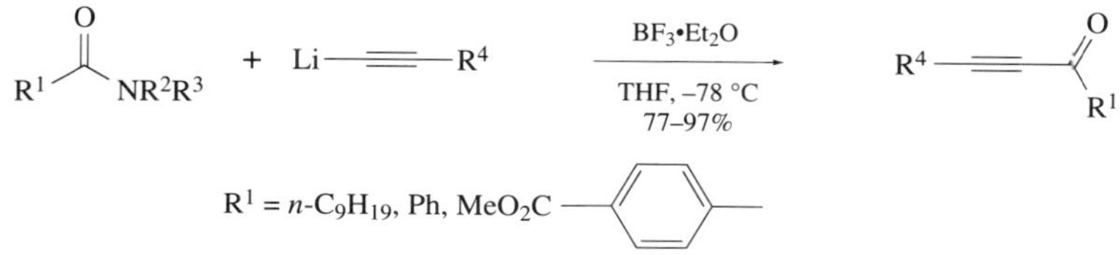

(79)

R^2 = Me, Ph; R^3 = Me; R^4 = n-C_5H_{11}, THP-OCH_2, Ph

N-Methoxy-*N*-methylamides are more effective acylating agents than the corresponding *N,N*-dialkylamides and afford high yields of conjugated ynones with alkynyllithium reagents (Equation (80)) ⟨81TL3815, 88JA2301⟩. An evaluation of a variety of carboxylic acid derivatives in the preparation of ynones showed that *N*-methoxy-*N*-methylamides are very resistant to excess nucleophilic attack as compared to acid chlorides, lithium carboxylates, 2-pyridylthioate esters and *N*-acylpyrazoles, and result in no tertiary alcohol products ⟨85JOC3972⟩. In addition, the *N,O*-dimethylhydroxylamine unit appears to enhance the reactivity of the carbonyl group such that selective displacement reactions can be achieved even in the presence of ester functions ⟨88JA2301⟩. This methodology has been usefully applied to the preparation of a number of β-lactam ynones (Equation (81)) ⟨87TL1857⟩. The symmetrical urea *N,N′*-dimethoxy-*N,N′*-dimethylurea has been used as a carbonyl dication equivalent in its reaction with phenylethynyllithium ⟨91JOC2911⟩.

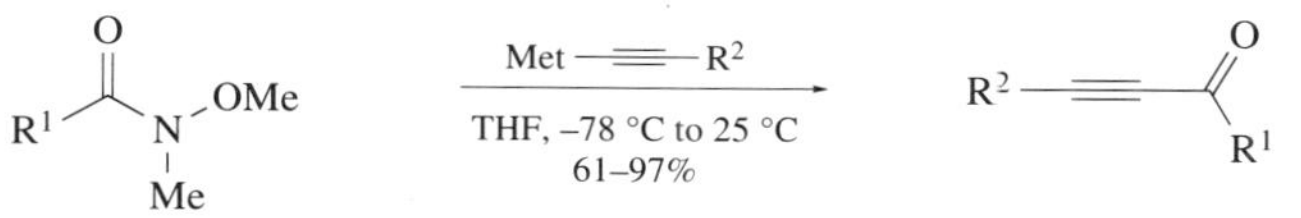

(80)

Met = Li, MgX
R^1 = Bu^t, MeCH(OAc), Ph, PhCH=CH, cyclohexyl, 4-cyclohexenyl
R^2 = Et, Pr^n, Bu^n, Ph

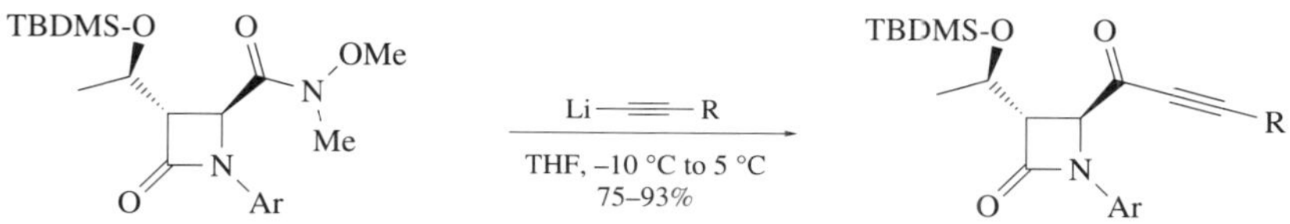

(81)

R = $MeOCH_2$, Ph, TMS
Ar = p-$MeOC_6H_4$

The reaction of lactams with alkynylmetal reagents has been little reported. Alkynyl azacycloalkanes have been prepared in good yields by the reaction of alkynyllithium reagents with lactams in the presence of boron trifluoride etherate followed by lithium aluminum hydride reduction of the adducts (Equation (82)) ⟨83TL1719⟩.

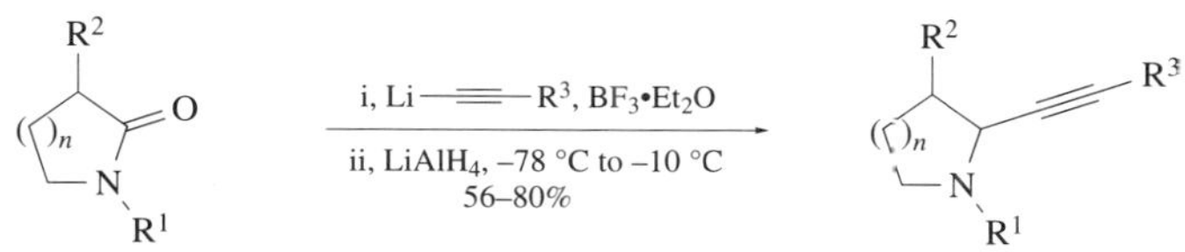

(82)

R^1 = Me, $CH_2CH=CH_2$
R^2 = H, $CH_2CH=CH_2$
R^3 = n-C_5H_{11}, Ph

Alkyl chloroformates (**17**) react with alkynyllithium reagents to form esters of 2-alkynoic acids (Equation (83)) ⟨B-69MI 121-03⟩. The reaction is normally carried out by adding the lithiated alkyne to an excess of the chloroformate in order to avoid the formation of tertiary alcohols ⟨B-88MI 121-01⟩. Esters of 2-alkynoic acids are also formed in a palladium-catalysed reaction of alkyl chloroformates with alkynylzinc chlorides ⟨83TL5181⟩. Alkyl chlorothionoformates (**18**) react with alkynylmagnesium bromides and with 1-alkynes under palladium catalysis to afford thionoesters ⟨89TL1073, 91LA903⟩. Alkyl chlorodithioformates (**19**) similarly afford dithioesters ⟨89TL1073, 91LA903⟩. Imidoyl chlorides (**20**) react with alkynylcopper reagents ⟨72S63, 91LA903⟩ and, under palladium catalysis, with 1-alkynes and alkynyltin reagents to afford alkynyl imines in high yield ⟨85TL3463, 91S235, 91TL1459⟩. In a similar manner isocyanide dichlorides afford dialkynyl imines ⟨88TL5379⟩.

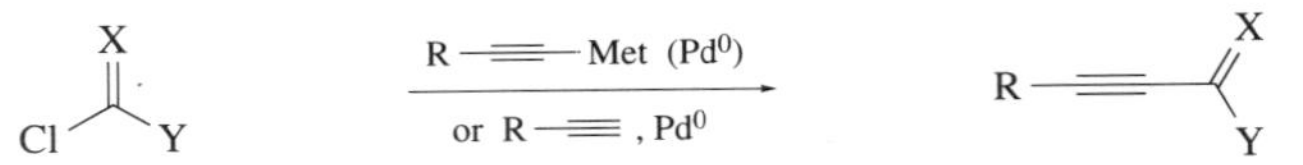

(83)

X = O	Y = OR	**(17)**
X = S	Y = OR	**(18)**
X = S	Y = SR	**(19)**
X = NR	Y = R	**(20)**
X = O	Y = NR_2	**(21)**
X = S	Y = NR_2	**(22)**

N,N-Dialkylcarbamoyl chlorides (**21**) react with alkynylmetal reagents ⟨55CR989⟩ and with 1-alkynes in the presence of a palladium catalyst ⟨77S777, 91LA903⟩ to afford amides of 2-alkynoic acids. *N,N*-Dialkylthiocarbamoyl chlorides (**22**) afford thioamides (Equation (83)) ⟨91LA903⟩.

Cyanogen chloride reacts with alkynyllithium reagents to afford conjugated alkynyl nitriles ⟨73RTC667, 80S150⟩. In the early 1990s, these compounds have been obtained from the reaction of 1-alkynes themselves with CuCN/bis(trimethylsilyl)peroxide ⟨91TL2169⟩ or with CuCN/TMS-Cl/cat. NaI (Equation (84)) ⟨93TL5911⟩.

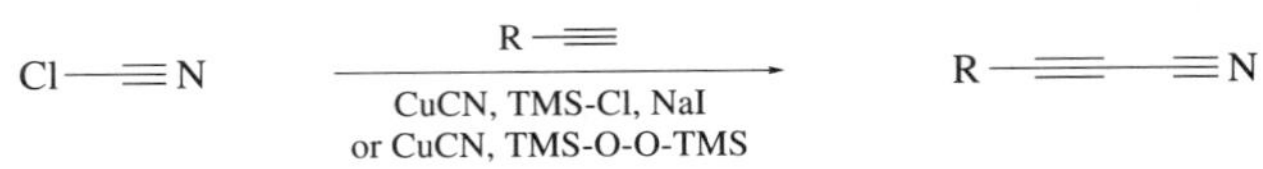

(84)

1.21.2.2.6 1,4-Addition reactions

A limited number of methods is available for the 1,4-addition of alkynylmetal reagents (or terminal alkynes themselves) to α,β-unsaturated ketones, esters, sulfones, nitro compounds, etc. Group 1 and 2 metal alkynides do not, in general, undergo successful 1,4-addition reactions. Exceptions include the purely sterically controlled addition of lithium alkynides to 1,1,1-triphenyl-3-penten-2-one ⟨81AG(E)569⟩ and the 1,4-addition of lithium and magnesium alkynides to α-silyl-α,β-unsaturated sulfones (Equation (85)) ⟨90SL701, 90TL5499⟩. The use of alkynylcopper reagents to effect the 1,4-addition of an alkynyl group is precluded due to the tenacity with which copper binds alkynyl ligands ⟨69JOC3616⟩. One possible solution to this problem employs a cuprate reagent derived from (*E*)-1,2-bis(tri-*n*-butylstannyl)ethylene as a nucleophilic ethynyl group equivalent in the conjugate addition to α,β-unsaturated ketones (Scheme 64) ⟨74JA5581⟩. Unfortunately, extension of this procedure to higher alkynes would appear to present major difficulties. More generally applicable 1,4-addition reactions of alkynylmetal reagents have to date been limited to alkynylaluminum, alkynylboron, alkynylzinc and alkynyltin reagents.

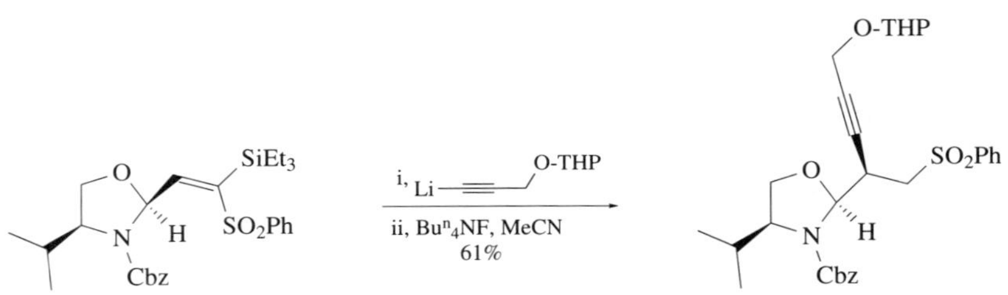

(85)

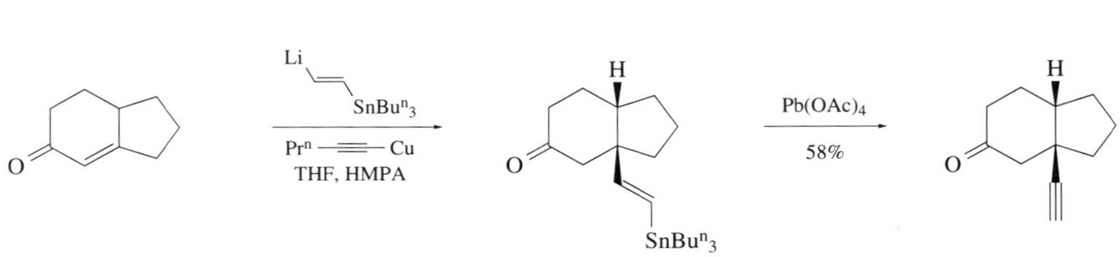

Scheme 64

Diethylalkynylalanes react with conjugated enones to give the 1,4-addition products (Scheme 65) ⟨71JA7320⟩. The reaction is believed to proceed through a cyclic six-membered transition state, and in the absence of other proximal coordinating functionalities, the reaction is limited to enones capable of adopting a *cisoid* conformation. Cyclic enones in which the enone system is rigidly constrained to a *transoid* geometry, such as 2-cyclohexenone or isophorone, react with the alane reagent to provide the tertiary carbinol derived from 1,2- rather than 1,4-addition of the alkyne unit. This constraint can be overcome by preferential coordination of the aluminum reagent to a neighbouring hydroxyl or amino group. For example, 2-(6-carbomethoxyhexyl)-4-hydroxy-2-cyclopentenone (**23**) reacts with trialkynyl- and alkynyldimethylalanes to afford the 1,4-addition product, the alkynyl group being delivered to the same face as the directing hydroxyl group (Equation (86)) ⟨72TL2627, 75TL4217⟩. Protection of the hydroxyl group of (**23**) by a tetrahydropyranyl group prevents reaction with the aluminum reagent. In a similar system, the stereochemistry of the conjugate addition has been reversed by providing a coordinating group exocyclic to the cyclopentene ring (Equation (87)) ⟨91JOC3205⟩. Conjugated enals and ynones do not afford 1,4-addition products with diethylalkynylalanes, but rather give the 1,2-addition product ⟨71JA7320⟩. Conjugated nitro alkenes undergo 1,4-addition giving γ-alkynylnitro compounds in good yield (Equation (88)) ⟨89JOC2391⟩.

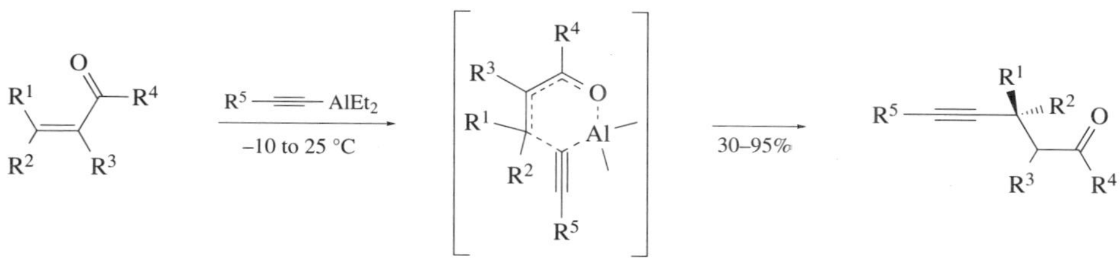

Scheme 65

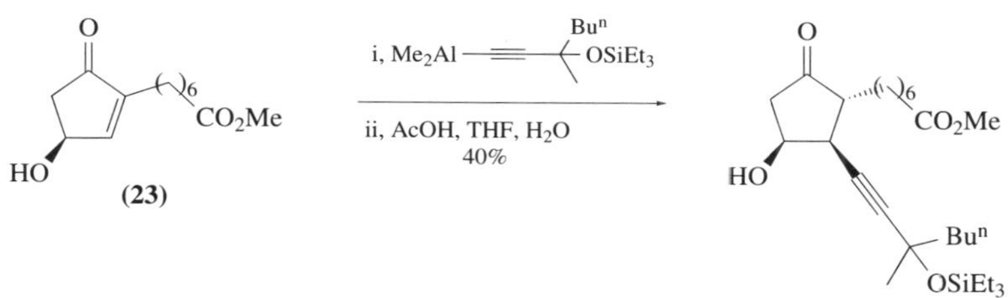

(86)

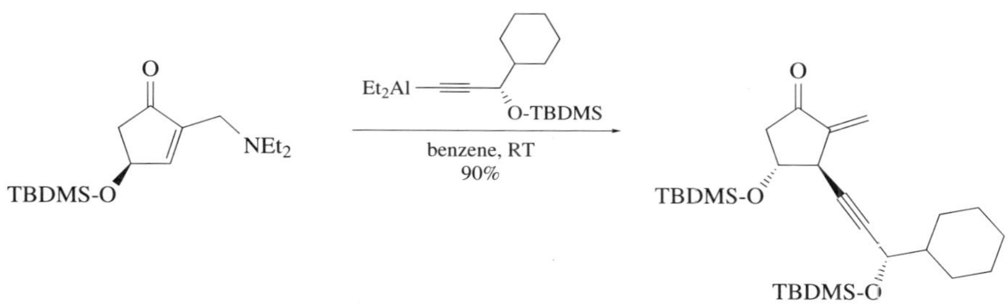

(87)

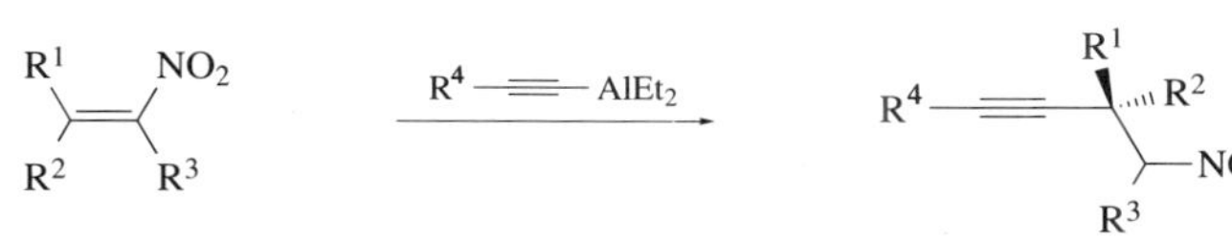

(88)

$R^1 = H$, $R^2 = R^3 = (CH_2)_4$, $R^4 = Bu^n$
$R^1 = H$, $R^2 = R^3 = (CH_2)_4$, $R^4 = Ph$
$R^1 = H$, $R^2 = Ph$, $R^3 = Me$, $R^4 = Bu^n$

A more general procedure for the 1,4-addition of an alkynyl unit to conjugated enones is the nickel-catalysed reaction with dimethylalkynylalanes (Equation (89)) ⟨78JA2244⟩. The reaction has been successfully applied to a range of cyclic *s-trans* enones using a catalyst derived from treatment of $Ni(acac)_2$ with diisobutylaluminum hydride. 1,4-Addition of an ethyne unit is most readily achieved by the use of a trimethylsilylethynyl reagent, since reaction with an unprotected aluminum ethynide is complicated by product decomposition.

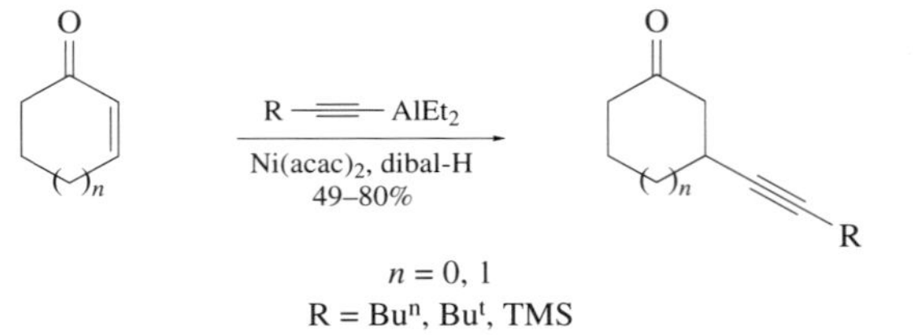

(89)

1,4-Addition of alkynylalanes followed by *in situ* alkylation is not readily achieved due to the poor reactivity of aluminum enolates. However, a protocol has been described whereby addition of alkynylalanes to alkylidenemalonates is followed by palladium-catalysed allylation, permitting vicinal β-alkynylation α-allylation in a single operation (Scheme 66) ⟨93TL2271⟩.

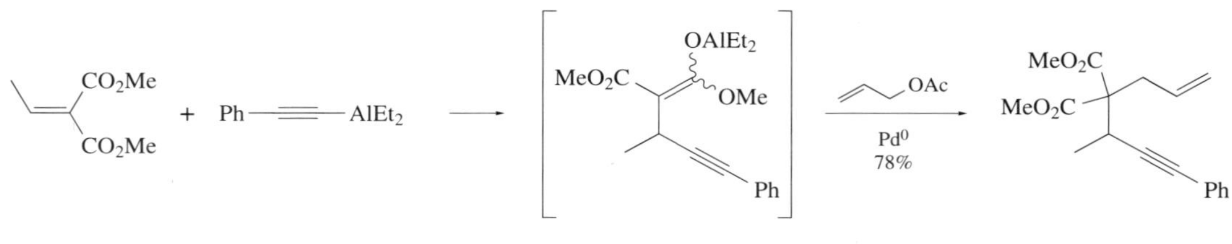

Scheme 66

Alkynylboron reagents similarly undergo 1,4-addition to conjugated enones. Thus, tri-octynylboron undergoes successful 1,4-addition to pregnenolone acetate yielding the 16-octynyl derivative ⟨72TL2627⟩. A trialkynylboron reagent has also been successfully added to methyl vinyl ketone ⟨76TL235⟩. Unfortunately, only one of the three alkynyl groups is transferred in the reaction, constituting a serious disadvantage where the alkyne moiety is valuable. Extension of this reaction to *B*-1-alkynyl-9-BBN provides a solution to this problem ⟨77JA954⟩. Reaction with conjugated enones occurs rapidly at room temperature and gives high yields of the 1,4-addition product. Analogous to the uncatalysed 1,4-addition of alkynylalanes, the reaction is believed to proceed through a cyclic transition state, consistent with the observation that the reaction is retarded by steric bulk at the β-position of the enone and fails with cyclic enones which are incapable of adopting a *cisoid* geometry. Alkynyldiisopropoxyboranes also react with conjugated enones in the presence of boron trifluoride etherate to give 1,4-addition products in good yields (Equation (90)) ⟨92CL695⟩. A disproportionation reaction is believed to occur with boron trifluoride to give the more Lewis acidic alkynylfluoroisopropoxyborane reagent; in the absence of boron trifluoride etherate no reaction occurs.

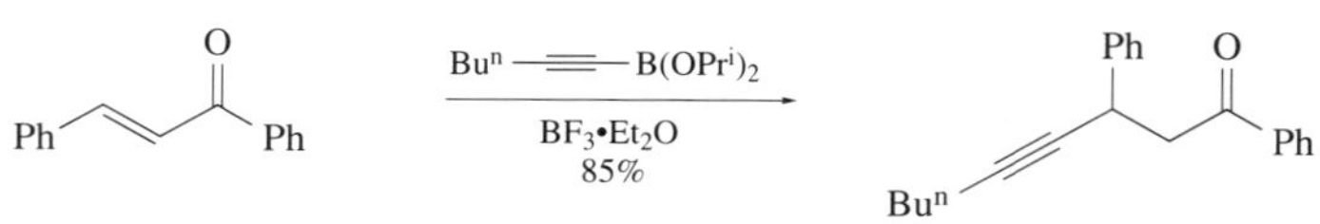

(90)

Alkynylzinc bromides undergo 1,4-addition to conjugated enones in the presence of *t*-butyldimethylsilyltriflate to give γ,δ-alkynyl silyl enol ethers (Equation (91)) ⟨90TL7627⟩. The reaction is successful with both *s-cis* and *s-trans* enones and yields are high with β-monosubstituted α,β-enones. With β-unsubstituted enones, such as methyl vinyl ketone, yields are reduced due to further reaction of the initially formed 1,4-addition product (a zinc enolate) with the enone. Furthermore, the method is unsuccessful with β,β-disubstituted enones, yielding a mixture of 1,2- and 1,4-addition products along with unreacted starting material.

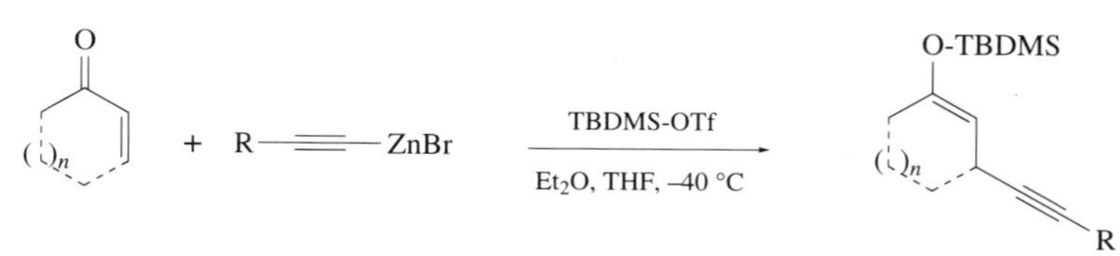

(91)

Alkynyltrichlorotin reagents, prepared from terminal alkynes and a combination of tin(IV) chloride and a trialkylamine, undergo 1,4-addition to conjugated enones (Equation (92)) ⟨92CL2479⟩.

Cyclic enones were not examined. With methyl vinyl ketone, addition of two equivalents of the alkyne has been reported under almost identical conditions (Equation (93)) ⟨93JA3362⟩.

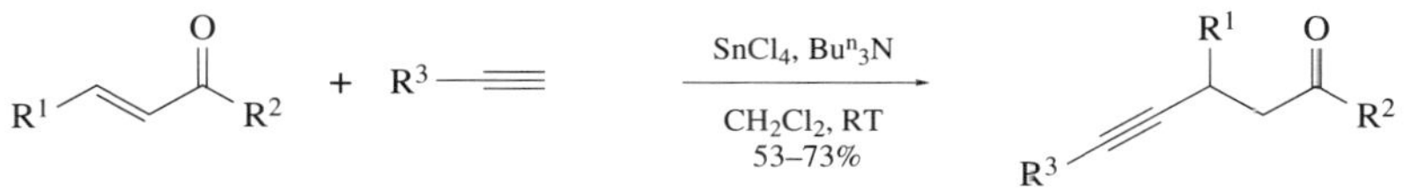

(92)

$R^1 = n\text{-}C_5H_{11}$, Ph; R^2 = Me, Ph; $R^3 = n\text{-}C_5H_{11}$, Ph

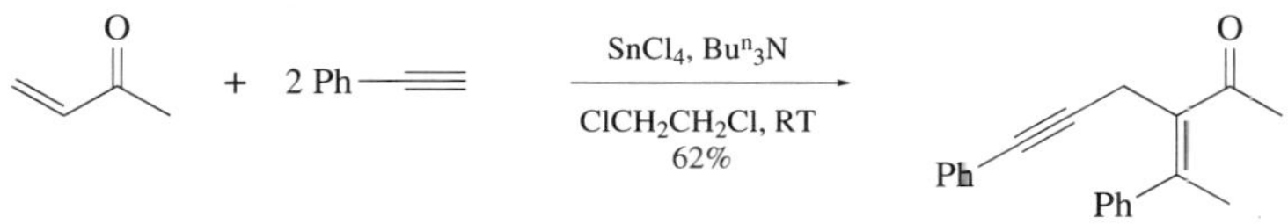

(93)

Transition metal-catalysed reactions of terminal alkynes with conjugated enones and enoates have been reported. A very efficient addition of terminal alkynes to methyl vinyl ketone has been realised with a $RhCl(PMe_3)_3$ catalyst (Equation (94)) ⟨90TL7063⟩. Very few examples were reported and it remains to be seen whether this potentially very useful reaction will prove general. Unfortunately, at temperatures significantly above 20 °C, dimerisation of the terminal alkyne occurs as a competing reaction, and this might hinder attempts to extend the reaction to β-substituted enones.

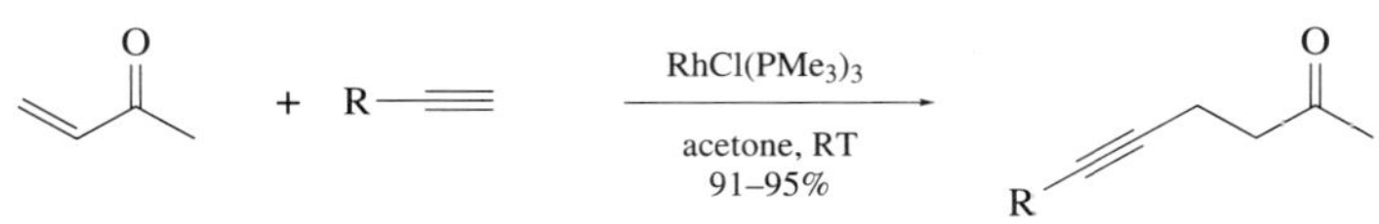

(94)

R = Pr^n, Bu^n, $n\text{-}C_6H_{13}$, Ph

A palladium-catalysed reaction of terminal alkynes with methyl 2,3-pentadienoate has been reported (Equation (95)) ⟨90JA2816⟩. An unusual dependence of product regioselectivity on the choice of catalyst is observed with the conjugated enoate (**24**) being favoured by more electron deficient catalyst systems and the nonconjugated enoate (**25**) being favoured by more electron rich catalysts. The ratio of (**24**):(**25**) was optimised at 81:19 using a catalyst comprising tetrakis(carbomethoxy)palladacyclopentadiene and tris(2,6-dimethoxyphenyl)phosphine. The ratio of (**25**):(**24**) was optimised at 91:9 using a catalyst comprising palladium acetate and triphenylphosphine. A possible explanation for the catalyst-dependent regioselectivity might relate to the preferential coordination of an active palladium intermediate to the 2,3-double bond with a more electron rich catalyst (leading to the nonconjugated ester products) and to the 3,4-double bond with a more electron deficient catalyst (leading to the conjugated ester product). Steric effects override these electronic effects such that further substitution at position 2 or 4 gives the conjugated addition product (**24**) regardless of the catalyst used.

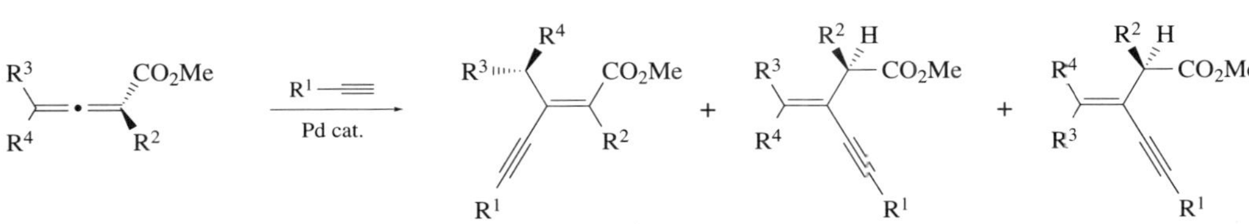

(24) **(25)**,(*E*) **(25)**,(*Z*)

(95)

1.21.2.3 Substitution Reactions of Propargyl/Allenyl Carbanions

While alkynylmetal reagents are generally used for the introduction of the carbon–carbon triple bond, a number of propargylation methods have been developed employing propargyl/allenyl metal reagents ⟨B-80MI 121-02⟩. Almost all of the reported examples of this type of substitution reaction can be categorised as substitutions of halogen or oxygen functions. In addition, the reported reactions have been limited to alkyl, allyl, propargyl and benzyl electrophiles. Nevertheless, this alternative strategy is particularly well suited for the preparation of α-branched alkynes.

1.21.2.3.1 Substitution of halogen

Propargyl and allenyl organometallic reagents are powerful nucleophiles and undergo a variety of substitution reactions. Unfortunately, the reagents frequently exist in equilibrium with each other and display ambident behaviour in their reactions with electrophiles. Substitution reactions of alkyl halides are no exception. Indeed, one of the main problems associated with the use of these reagents is controlling the balance between allenic and propargylic products (Scheme 67).

Scheme 67

Allenylmagnesium halides display poor reactivity towards unactivated alkyl halides, but couple effectively with allylic halides, propargylic halides and α-haloethers to afford mixtures of alkynes and allenes. In some cases alkynes are formed almost exclusively, as in the coupling with allyl and propargyl bromides (**26**) and (**27**) (Scheme 68) ⟨75T1846⟩. In other cases substantial quantities of allene are formed. The resulting separation problems can be overcome by silylation of the crude product, and this procedure has been successfully applied to the reaction of allenylmagnesium bromide with allylic and benzylic halides ⟨70TL2247⟩. Allenyllithium reagents generally afford substituted allenes by α-alkylation ⟨75CC561⟩, whereas allenylalanates have been reported to undergo regioselective γ-alkylation reactions with allylic bromides, affording 1,5-enynes in high yield (Equation (96)) ⟨83S883⟩.

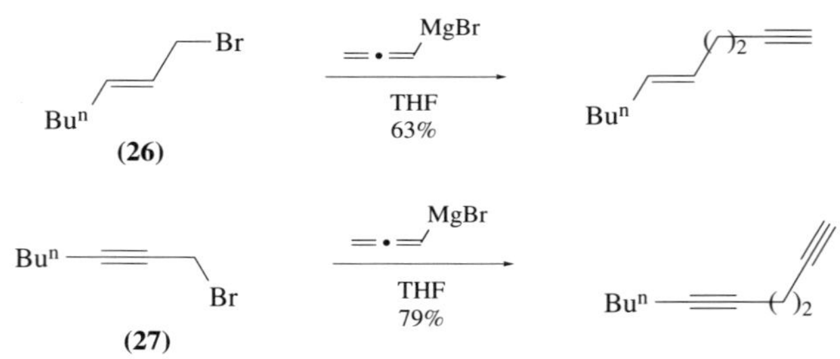

Scheme 68

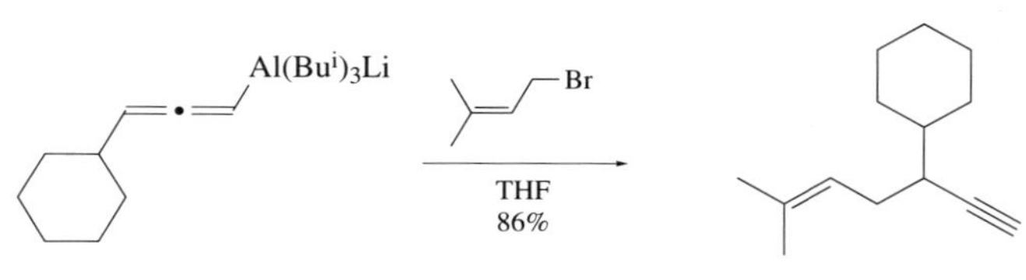

(96)

Lithium and magnesium derivatives of 1-(trimethylsilyl)propyne ⟨68TL5041⟩, 1-(triisopropylsilyl)-propyne ⟨82TL719⟩ and 1-(trimethylsilyl)butyne ⟨84S111⟩ have been used very effectively for the regioselective substitution of allyl, benzyl and unactivated alkyl halides (Scheme 69) ⟨70JA6314, 86JA1359⟩. The trialkylsilyl group functions both as a readily removable protecting group and as a steric shielding group reducing the level of substitution at C-1.

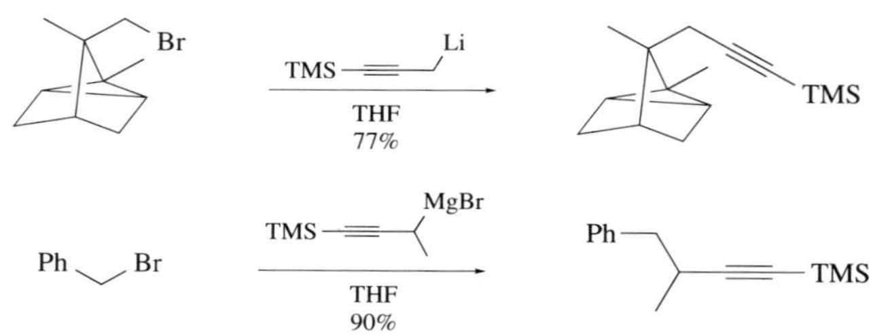

Scheme 69

An alternative and very effective solution to the regioselectivity problem is provided by the use of 1,3-dilithiopropyne. The allenyl dianion is presumably destabilised relative to the propargyl dianion allowing the reagent to undergo highly regioselective monosubstitution (Scheme 70) and 1,3-disubstitution (Scheme 71) reactions with primary alkyl bromides ⟨75CC817, 76S321⟩. The reaction can be readily extended to other terminal alkynes and provides a convenient route to α-branched alkynes, compounds which are not readily available by substitution of alkynylmetal reagents. Reactions with allylic halides have been reported to be problematic and reagents derived from dilithiation of allene ⟨85TL271⟩ and phenyl propargyl sulphide ⟨81JOC5041⟩ have been recommended as alternatives.

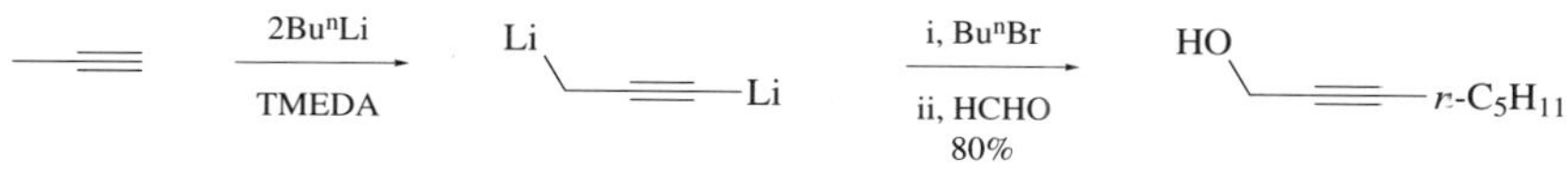

Scheme 70

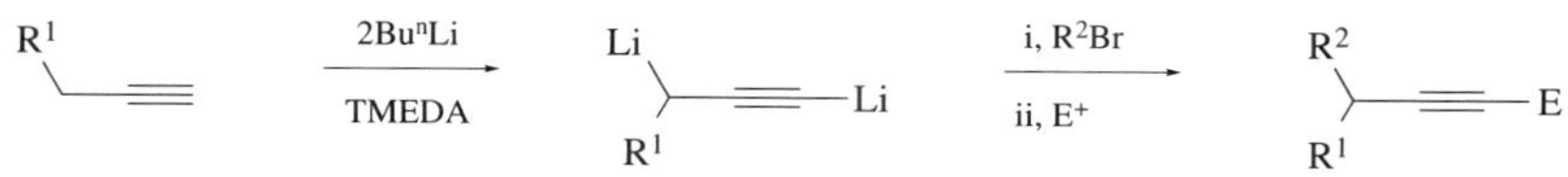

Scheme 71

1.21.2.3.2 *Substitution of oxygen functions*

Allenylmagnesium bromide reacts with 1,2-epoxybutane (Equation (97)) ⟨B-80MI 121-02⟩, whereas allenyllithium reagents afford allenes predominantly or exclusively ⟨81S875⟩. Regioselective propargylation of epoxides can be achieved employing lithiated 1-triisopropylsilylpropyne (Equation (98)) ⟨82TL719⟩. The corresponding Grignard reagent undergoes substitution reactions with allylic epoxides ⟨84JA6006⟩ and allylic phosphates ⟨87TL527⟩ in the presence of copper(I) iodide (Scheme 72). Allenyltrialkylstannanes react with allylic acetates in the presence of catalytic $Pd(PPh_3)_4$ ⟨83JOC5302⟩ and with acetals and ketals in the presence of titanium(IV) chloride ⟨87CL1345⟩ to afford 1,5-enynes and propargyl ethers respectively.

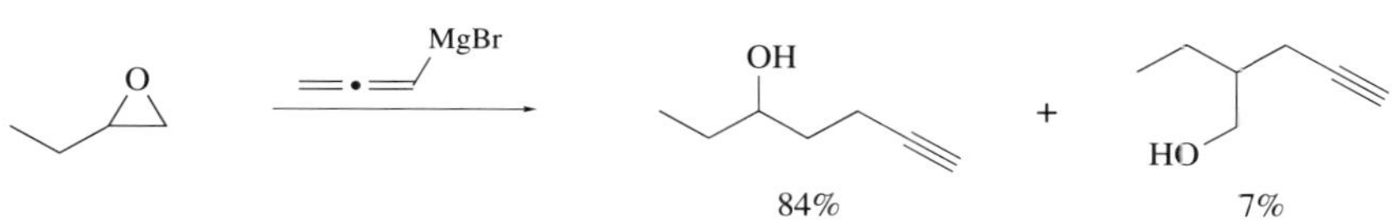 (97)

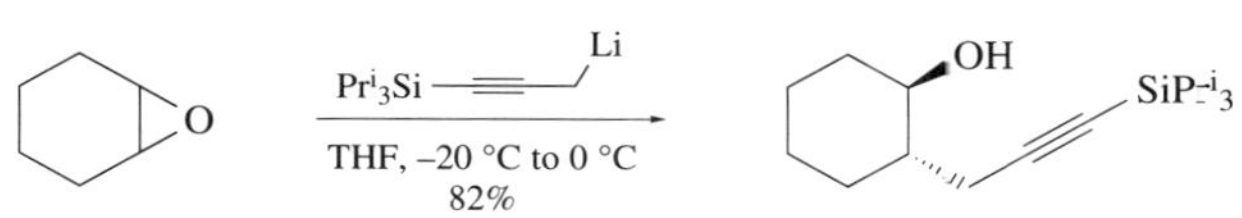 (98)

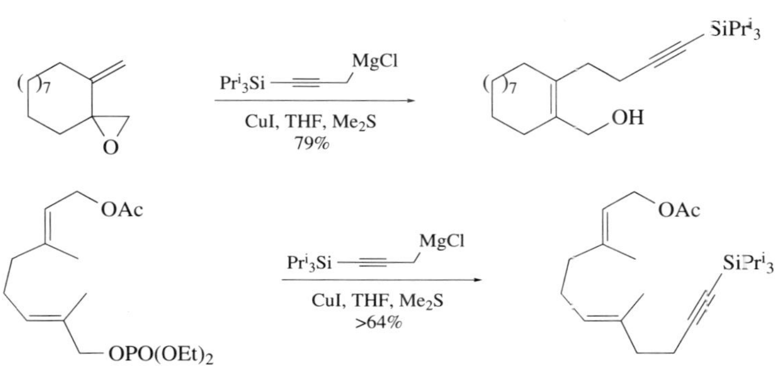

Scheme 72

1.21.2.4 Addition Reactions of Propargyl/Allenyl Carbanions

The 1,2-addition reaction of propargyl/allenyl organometallics to C=X functions provides an alternative strategy for introduction of the carbon–carbon triple bond, one which is particularly well suited for the preparation of β-functionalised alkynes ⟨B-78MI 121-01, B-80MI 121-02, 91COS(2)81⟩. The reactions can, however, be complicated by equilibration of the isomeric propargyl/allenyl organometallics with the associated formation of a mixture of propargylic and allenic adducts. In general, at least with carbonyl functions, it is believed that the reactions proceed by a cyclic S_E1' (or S_E2') process and that allenylmetals lead mainly to alkyne-containing adducts whereas the corresponding propargylic isomers lead to allenic adducts (Scheme 73). The overall regioselectivity is governed by the equilibrium concentration and relative reactivities of the organometallic nucleophiles, and is therefore highly dependent on the nature of the substituents on both the organometallic reagent and on the electrophile.

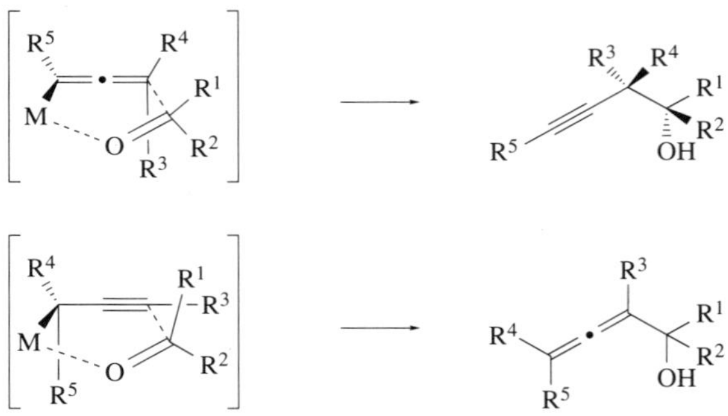

Scheme 73

In addition to the problem of regioselectivity, problems of enantioselectivity and diastereoselectivity (*syn* and *anti* isomers) can arise, depending on the substitution pattern of the organometallic reagent.

1.21.2.4.1 1,2-Addition to carbon–oxygen double bonded functions

The 1,2-addition of propargyl/allenyl metal and metalloid reagents to aldehydes and ketones provides a convenient method for the preparation of homopropargyl alcohols. Through appropriate choice of the metal and the reaction conditions, regioselective and stereoselective propargylation methods have been developed for particular substitution patterns of the organometallic reagents. It therefore seems most useful to arrange this section according to the substitution pattern of the target homopropargylic alcohol (Figure 2).

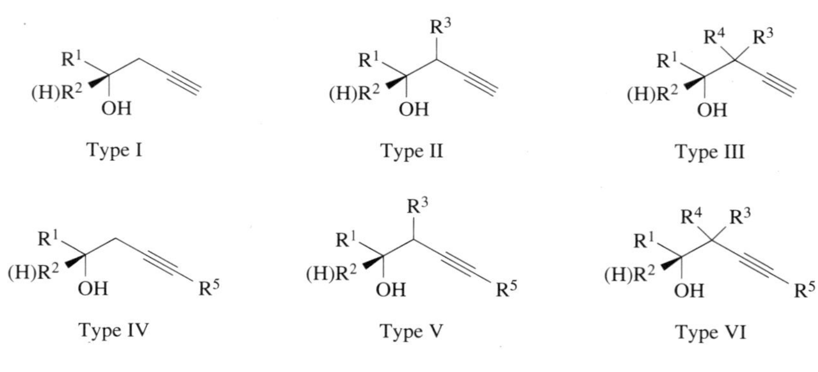

Figure 2

Figure 2 Substituted homopropargyl alcohols.

Type I alcohols can be prepared very effectively by the addition of unsubstituted allenylboron, tin and silicon reagents to aldehydes and ketones. These reagents are relatively stable towards isomerisation and generally afford predictable results in their reactions with carbonyl compounds. 9-Allenyl-9-BBN undergoes a highly regioselective addition to aldehydes and ketones, and even to relatively hindered ketones, affording high yields of type I alcohols (Scheme 74) ⟨93TL15⟩. Allenylboronic esters have also been successfully applied to the regioselective ⟨74JOM(76)297, 74JOM(76)305⟩ and enantioselective ⟨82JA7667, 86JA483, 86TL1175⟩ propargylation of aldehydes and *β*-hydroxyketones (although not to ketones in general) (Scheme 75). Enantioselective propargylation of aldehydes has also been achieved with propargyl triphenylstannane in the presence of a chiral bromoborane reagent derived from the bis-*p*-toluenesulphonamide of (*R*,*R*)-1,2-diphenyl-1,2-diaminoethane and boron tribromide (Equation (99)) ⟨90JA878⟩. The reaction proceeds through the intermediacy of an allenylborane. Diallenyltin dibromide ⟨84CL1939⟩ and allenyltrimethylsilane/titanium(IV) chloride ⟨80JOC3925⟩ also react with aldehydes to afford type I alcohols.

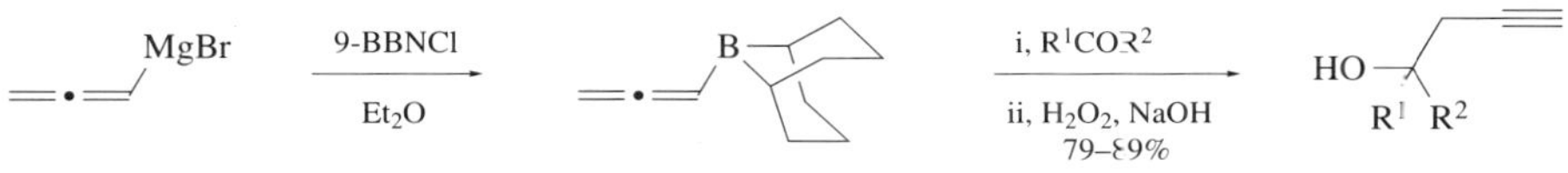

Scheme 74

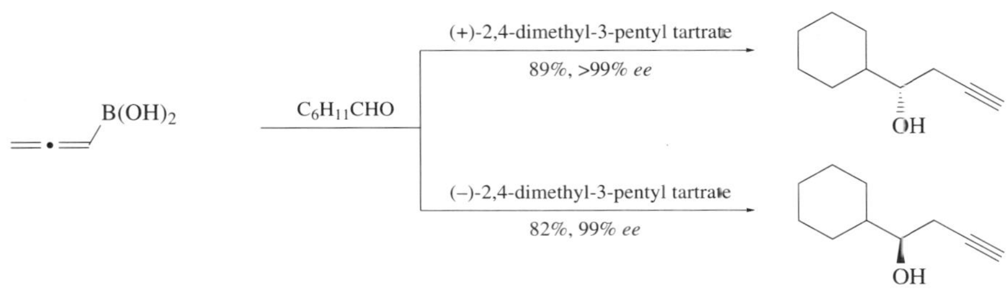

Scheme 75

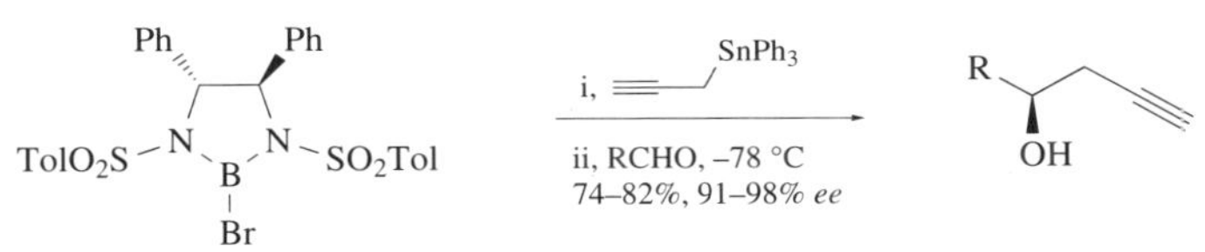

(99)

Allenyl/propargyl metal reagents derived from more electropositive metals (Li, Mg, Zn, Al) frequently undergo rapid equilibration and the outcome of their addition reactions is less predictable. However, some generalisations can be made. Allenylmagnesium halides and unsubstituted allenylaluminum reagents usually react with carbonyl derivatives to furnish predominantly or exclusively type I alcohols ⟨70BSF2175⟩, whereas, with the exception of some aromatic aldehydes and ketones ⟨89SC1705⟩, allenyllithium and zinc reagents generally afford mixtures of allenic and homopropargylic alcohols ⟨B-80MI 121-02⟩. Highly regioselective propargylation of aldehydes and ketones can be achieved employing the lithium derivative of 1-(triisopropylsilyl)propyne (Equation (100)) ⟨82TL719⟩.

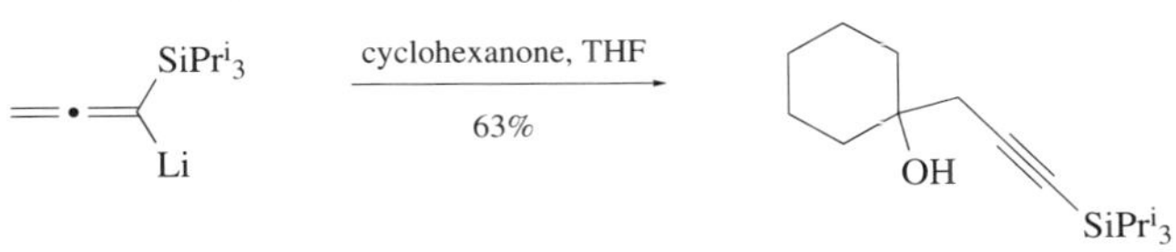

(100)

Type II homopropargyl alcohols have been prepared by the regioselective reaction of aldehydes and ketones with 3-substituted allenylmagnesium ⟨73BSF3371⟩ and aluminum ⟨83S883⟩ reagents. In the reaction of aldehydes with 1,2-butadienylmagnesium halides, the *anti*:*syn* product ratio is related to the size of the aldehyde substituent and has been rationalised as shown in Figure 3 ⟨69CR(C)342⟩. With aldehydes *anti*-diastereoselectivity can be improved to 96–98% employing the corresponding allenylzinc chlorides (Equation (101)) ⟨84JOC4565⟩. Allenyltitanium reagents also give improved *anti*-selectivity in certain cases ⟨82JOC2225⟩. Type II *syn*-homopropargyl alcohols are obtained selectively from the reaction of aldehydes and ketones with 3-substituted allenylsilanes in the presence of titanium(IV) chloride (Equation (102)) ⟨86JOC3870⟩. α-Substituted propargyl bromides are reported to undergo addition to aldehydes in the presence of chromium(II) chloride to afford homopropargyl alcohols exclusively, albeit with little diastereoselectivity ⟨92JOC4070⟩.

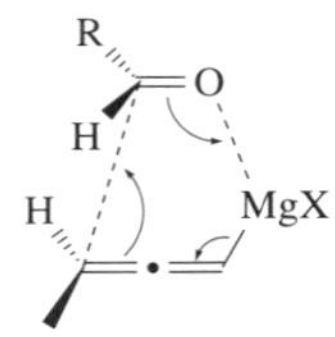

R	*syn* : *anti*
Me	76 : 24
Et	82 : 18
Pr^i	90 : 10
Bu^t	95 : 5

Figure 3

Figure 3 Steric effects in the addition of allenyl Grignard reagents to aldehydes.

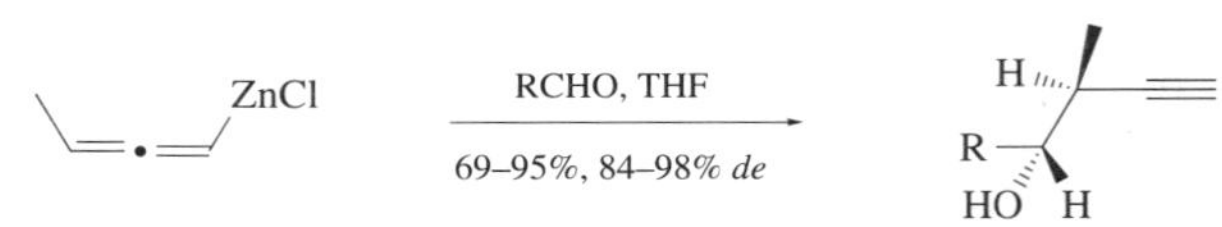

(101)

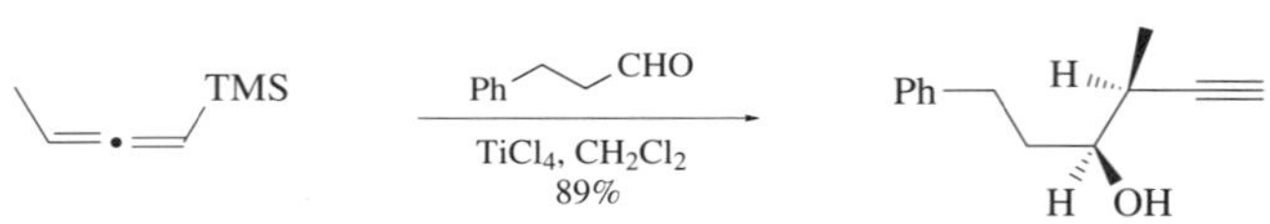

(102)

syn : *anti* 3.1 : 1

Type III homopropargylic alcohols can be formed by the reaction of 3,3-disubstituted allenyllithium and magnesium reagents with carbonyl compounds; steric factors play a large part in determining the outcome of the reaction. Thus, while the homopropargyl alcohol (**28**) was formed according to Scheme 76 ⟨62CR(255)1750⟩, an analogous Grignard derivative furnished none of the corresponding alcohol but gave (**29**) as the principal product ⟨75T3073⟩. It has since been shown that 3,3-dialkyl substituted allenyllithium ⟨75CC561, 77JA7632, 81S875⟩ and alanate ⟨83S883⟩ reagents react with aldehydes and unhindered aryl ketones to afford type III alcohols, whereas dialkyl ketones and hindered aryl ketones afford predominantly allenic alcohols. Small quantities of alkynols corresponding to (**29**) are also formed in many of these reactions, presumably by rearrangement of the allenyllithium.

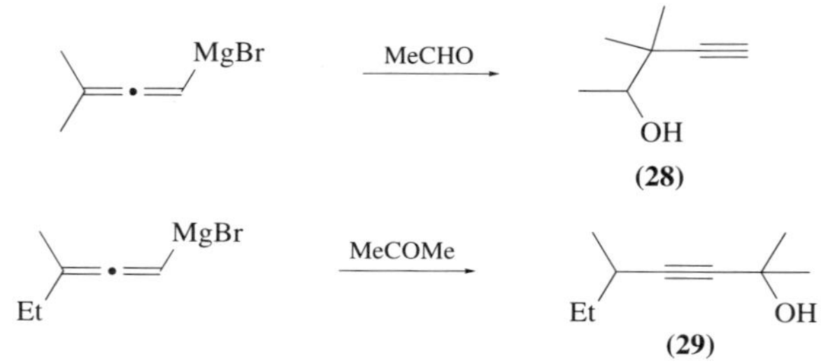

Scheme 76

Type IV homopropargylic alcohols can be prepared selectively through the reaction of aldehydes and ketones with 1-substituted allenylboranes (Scheme 77) ⟨78JA5561⟩ or with 1-substituted allenylsilanes in the presence of titanium(IV) chloride ⟨86JOC3870⟩. In contrast, 1-substituted allenyl/propargyl metal reagents generally afford allenic alcohols as the major products ⟨84BCJ2768⟩. An exception is the selective formation of homopropargyl alcohols from the reaction of hindered ketones with 1-substituted allenyltin/alkyllithium combinations ⟨90JOC441⟩. The limitation to hindered ketones can be overcome using indirect methods. Thus, acyl silanes condense with the allenyllithium reagents to afford almost exclusively the homopropargylic alcohols after desilylation with tetrabutylammonium fluoride (Equation (103)).

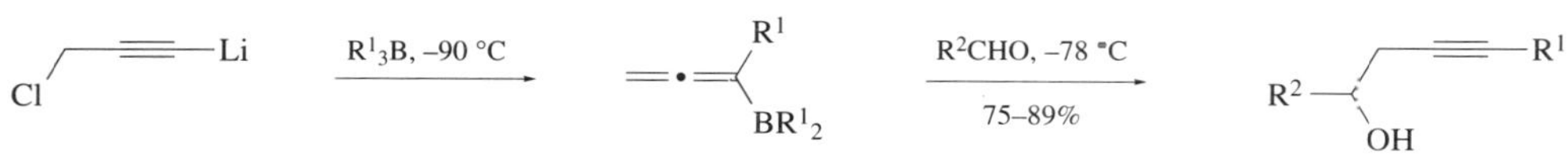

Scheme 77

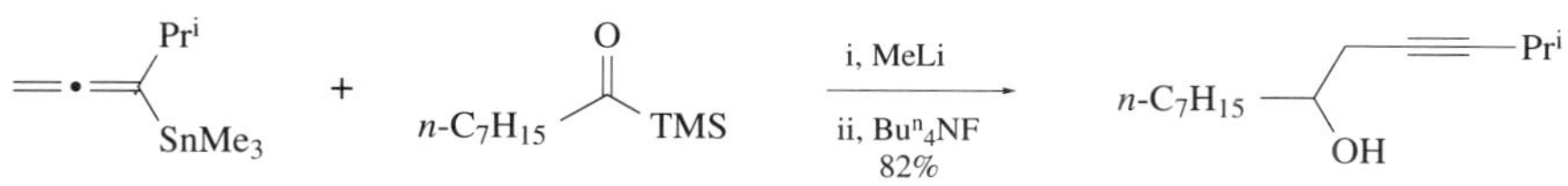

(103)

Type V homopropargylic alcohols have been prepared by the regioselective reaction of 1,3-disubstituted allenyltitanium reagents with aldehydes (Equation (104)) ⟨84BCJ2768⟩. The reaction is highly diastereoselective affording *anti*-homopropargylic alcohols. Allenylzirconium reagents, generated by treating propargylic ethers with biscyclopentadienylzirconium, have also been shown to undergo regioselective and diastereoselective reactions with aldehydes in the presence of boron trifluoride etherate to afford *anti*-homopropargylic alcohols ⟨92TL3769⟩. Complementary *syn* diastereocontrol, and also enantiocontrol, is obtained in the Lewis acid-catalysed reaction of enantiomerically enriched 1,3-disubstituted allenylsilanes ⟨93TL2383⟩ and stannanes ⟨92JOC1242⟩ with α-branched aldehydes (Scheme 78). Intramolecular cyclisations of allenylstannane aldehydes have been successfully applied to the preparation of 12–14 membered rings (Equation (105)) ⟨92JOC3387⟩.

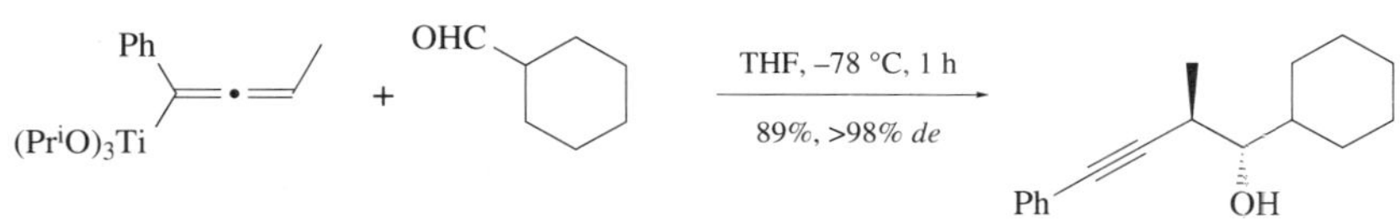

(104)

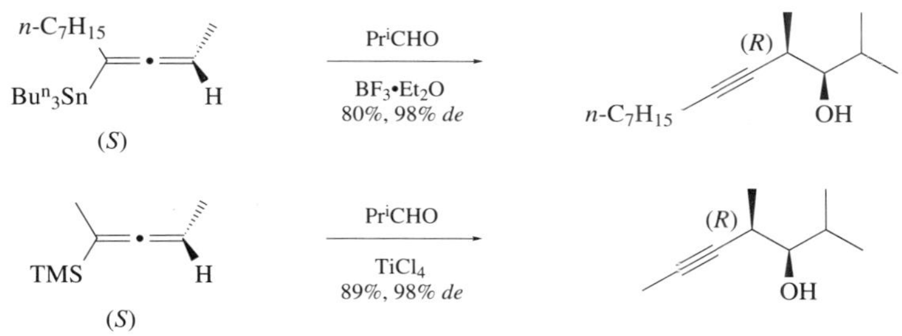

Scheme 78

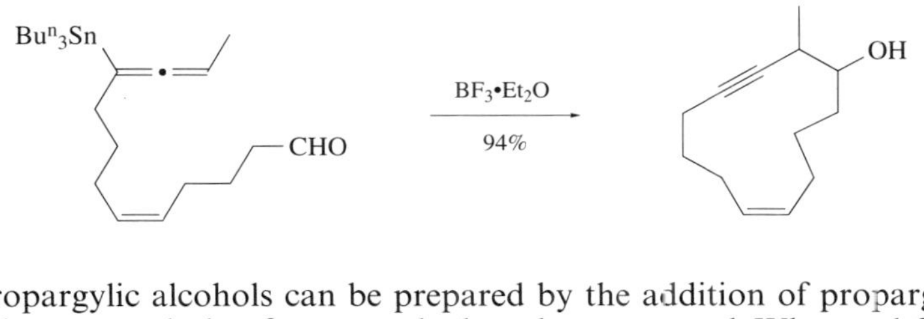

(105)

Type VI homopropargylic alcohols can be prepared by the addition of propargyl/allenyl metal reagents to carbonyl compounds, but few examples have been reported. Whereas 1,3,3-trisubstituted allenyllithium reagents seem to show a strong preference for the formation of allenic alcohols ⟨81S875⟩, the corresponding allenyl-9-BBN ⟨93JOC1651⟩ and allenylzirconium (Scheme 79) ⟨92TL3769⟩ reagents afford homopropargylic alcohols with high selectivity. 1,3,3-Trisubstituted allenylsilanes also afford type VI alcohols selectively, but the reaction is complicated by the formation of 4,5-dihydrofurans (Equation (106)) ⟨86JOC3870⟩.

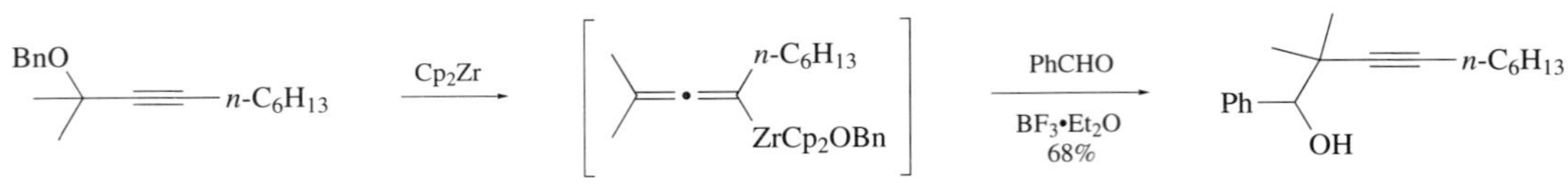

Scheme 79

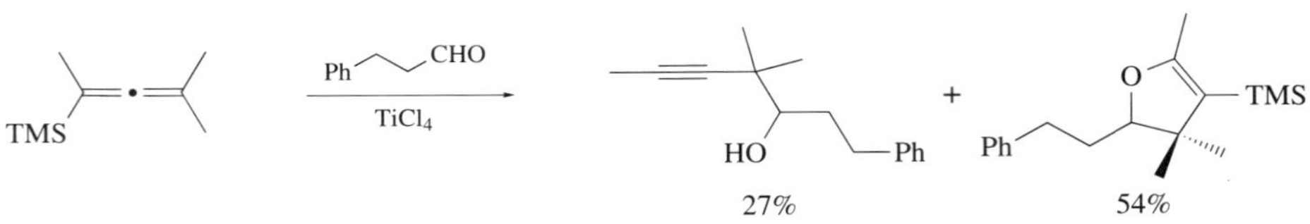

(106)

Reaction of propargyl/allenyl zinc, magnesium and lithium reagents with carbon dioxide affords mixtures of alkynoic and allenic acids, with the latter predominating ⟨B-80MI 121-02, 81S875⟩. In contrast, regioselective formation of 3-alkynoic acids has been achieved by carbonation of allenylalanates (Equation (107)) ⟨83S883⟩.

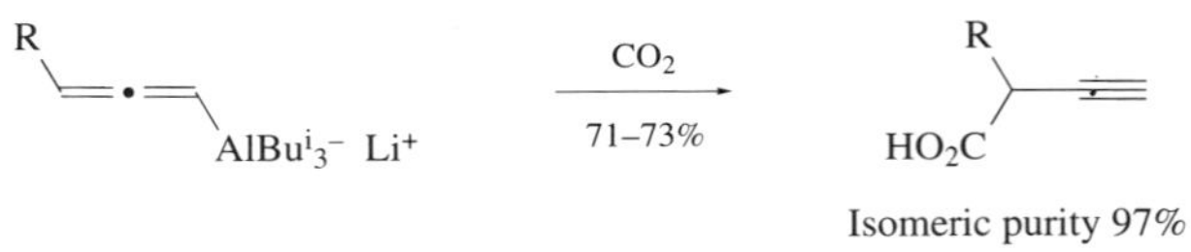

(107)

1.21.2.4.2 1,2-Addition to other carbon–chalcogen double bonded functions

Thioaldehydes are generally unstable at ambient temperatures and their reactions with propargyl/allenyl organometallics have not been reported. Di-*t*-butyl thioketone reacts with allenylmagnesium bromide to afford a mixture of homopropargylic thiol (**30**) and the heterocycle (**31**) (Equation (108)) ⟨77JOM(140)257⟩. Adamantane thione reacts similarly.

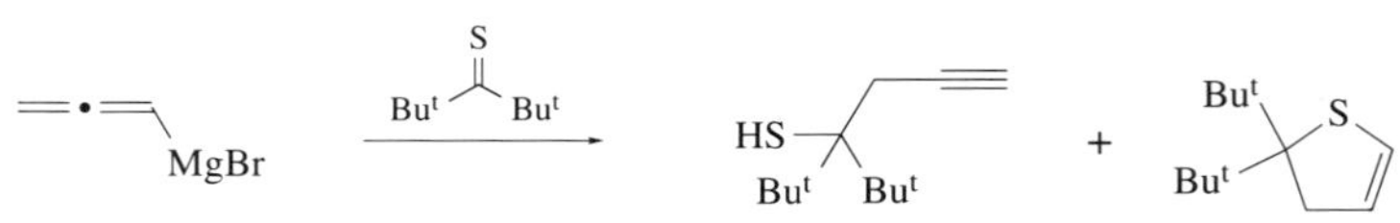

(108)

(30) **(31)**

Dithioesters react with allenylmagnesium bromide to afford dithioacetals after methylation (Equation (109)) ⟨77T2949⟩. Carbon disulfide similarly undergoes carbophilic addition with allenyllithium reagents to afford methyl 3-alkynedithioates after methylation (Equation (110)) ⟨81S551⟩.

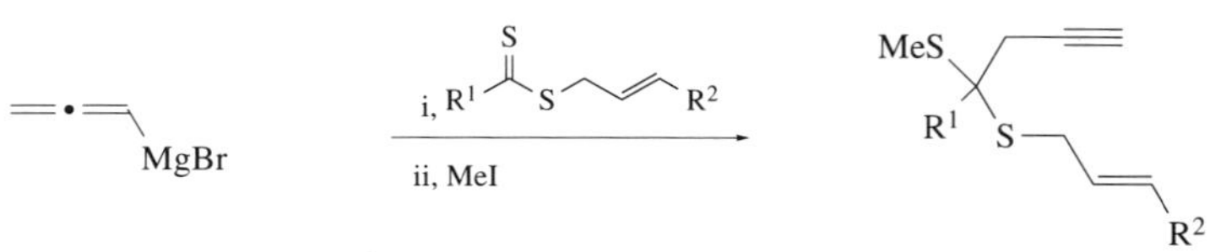

(109)

R^1 = Me, R^2 = H, 65%
R^1 = Et, R^2 = Me, 73%

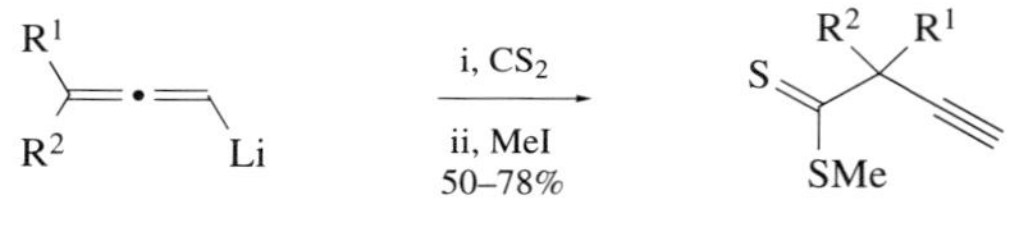

(110)

1.21.2.4.3 1,2-Addition to carbon–nitrogen multiple bonded functions

Most of the information available for the reaction of propargyl/allenyl organometallics with C–N multiple bonded functions concerns the reactions of aldimines and their iminium salts. Little or no information is available for reactions with ketimines, ketiminium salts, *N*-heterosubstituted imines (oxime ethers, phenylhydrazones, etc.) or nitriles.

The addition of unsubstituted propargyl/allenyl aluminum, zinc, magnesium and lithium reagents to aldimines has been reported (Equation (111)) ⟨64BSF952, 71BSF3071⟩. Homopropargylamines (**32**) are obtained exclusively with the aluminum reagent, whilst mixtures of alkynic (**32**) and allenic products (**33**) are formed with the zinc and lithium reagents. Propargyl/allenyl magnesium reagents do not give useful yields of (**32**).

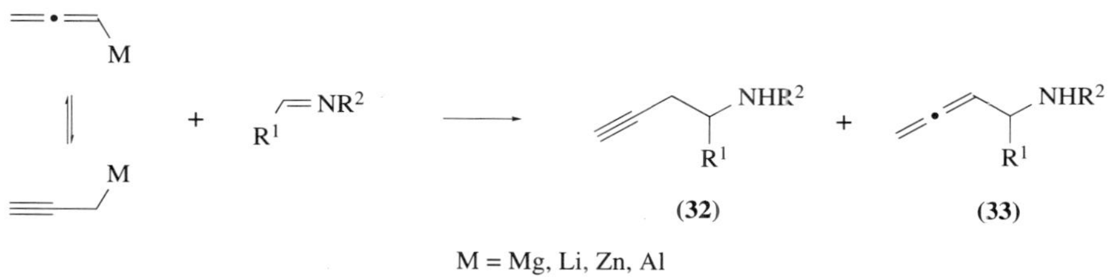

(111)

Direct additions of unsubstituted propargyl/allenyl organometallics to ketimines have not been reported. There is, however, no reason why reactions of this type should not be successful since indirect additions have been achieved in the Beckmann rearrangement of oxime sulfonates promoted by Grignard reagents ⟨82TL3395⟩. Thus, treatment of oxime sulfonates with Grignard reagents furnishes imines which can be further alkylated with allenylmagnesium bromide and lithium reagents (Scheme 80).

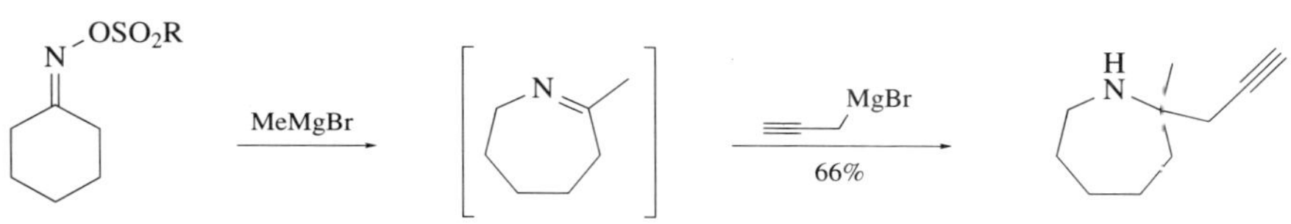

Scheme 80

Substituted propargyl/allenyl organometallics bearing a C-1 trimethylsilyl group (**34**) undergo highly regio- and diastereoselective reactions with aliphatic aldimines affording *anti*-homopropargyl amines (Equation (112)) ⟨84CC1004⟩. Yields are optimal employing the titanium reagent derived from addition of $ClTi(Pr^iO)_3$ to the lithiated allene. The stereoselectivity is readily explained by a cyclic transition state in which the α-substituent of the imine and the methyl group of the allenylmetal reagent are aligned in a less sterically congested *anti* relationship (Figure 4). α-Aryl imines are exceptions in that they react with little stereoselectivity. Related addition reactions of α-silyl-allenylboranes have been reported ⟨85JOC2193⟩.

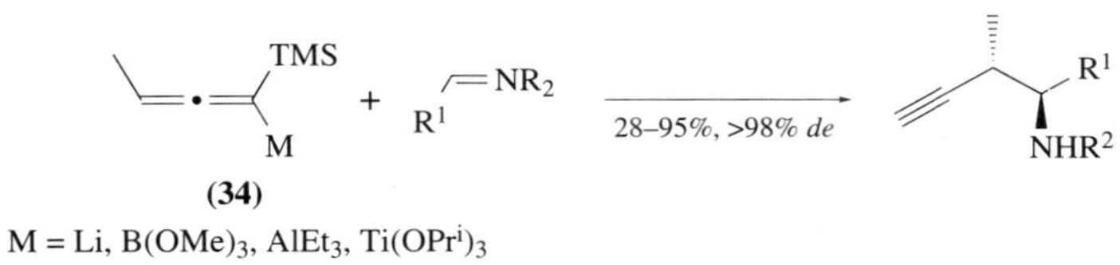

(112)

Propargyl/allenyl organometallics also undergo addition reactions with iminium salts, both preformed iminium salts ⟨81JOM(218)275⟩ and those generated *in situ* from *gem* amino ethers (Equation (113)) ⟨80JOM(198)1, 87BSF93⟩. As with the corresponding imines, aluminum reagents are more selective than the corresponding lithium, magnesium or zinc compounds. Regardless of the metal, however, the regiochemical outcome is markedly affected by the nature of R^1. Alkynic products are favoured with unsubstituted propargyl/allenyl metal reagents ($R^1 = H$) and allenic products are favoured with the corresponding substituted reagents (R^1 = alkyl).

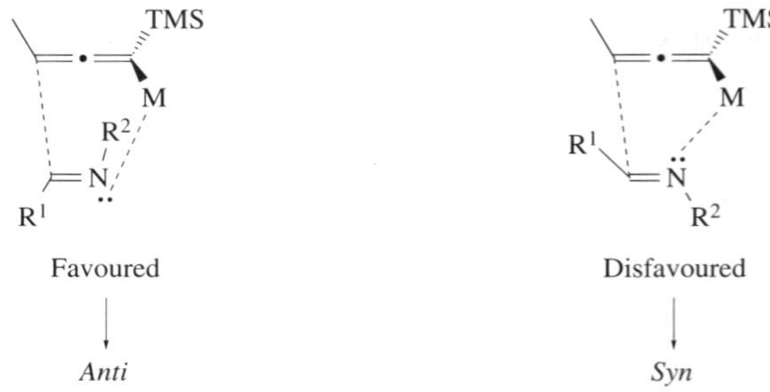

Figure 4

Figure 4 Regioselective and diastereoselective formation of anti-homopropargylamines.

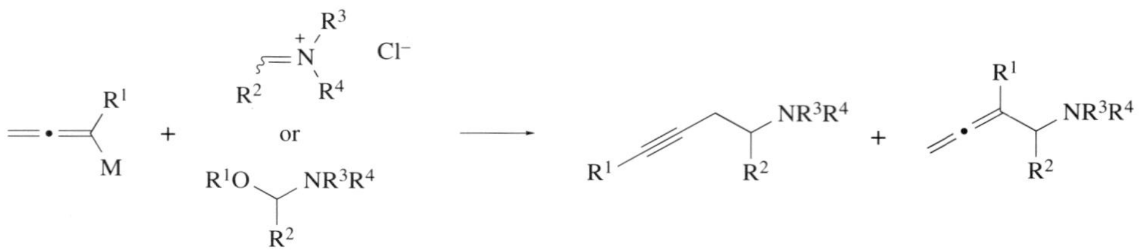

(113)

The increased electrophilicity of iminium salts can allow reactions with propargyl/allenyl organometallics derived from less electropositive metals such as tin and silicon. An example is the reaction of propargyl/allenyl tin reagents with iminium salts derived from quinoline or isoquinoline and chloroformate esters. Addition is regioselective affording 1-alkoxycarbonyl-2-(2-alkynyl)-1,2-dihydroquinolines and dihydroisoquinolines (Equation (114)) ⟨87CL1519⟩. An *N*-acyl iminium ion is also a presumed intermediate in the titanium(IV) chloride-promoted reaction of 1-methyl-1-(trimethylsilyl)allene with 5-ethoxy-2-pyrrolidinone (Equation (115)) ⟨85JA7233⟩.

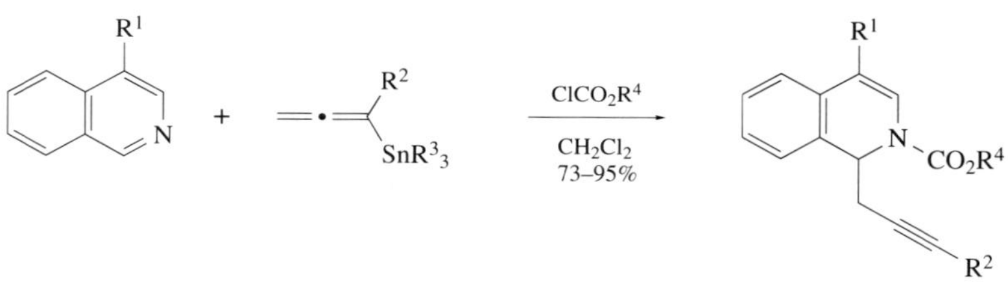

(114)

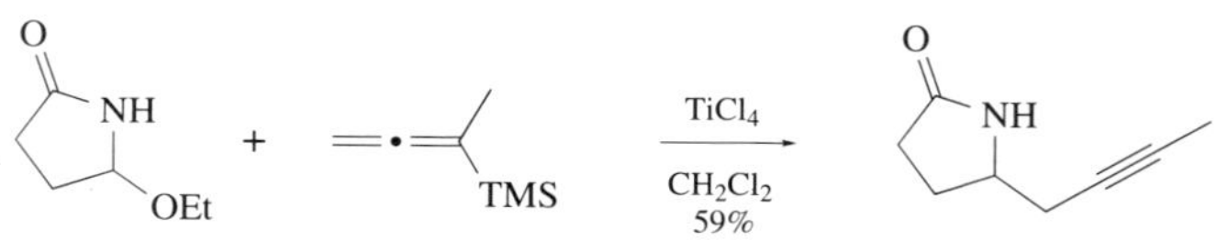

(115)

1.21.2.4.4 1,2-Addition–elimination reactions

Relatively few examples of 1,2-addition–elimination reactions of propargyl/allenyl organometallics have been reported. Carboxylic esters undergo addition–elimination reactions with propargyl/allenyl magnesium reagents to afford mixtures of the bis-addition products **(35)** and **(36)** (Equation (116)), even at low temperatures ⟨B-80MI 121-02⟩. The allenic alcohol **(36)** derives from isomerisation of the initially formed propargylic ketone. Regioselective monoaddition is observed in the reaction of allenic organomercurials with acid chlorides (Equation (117)) ⟨86JOC2623⟩; however, the propargylic ketones are relatively unstable and readily isomerise to allenones. *N*,*N*-Disubstituted carboxylic acid amides react with propargyl/allenylmagnesium reagents to afford mixtures of propargylic and allenic ketones ⟨56CR(243)1895⟩.

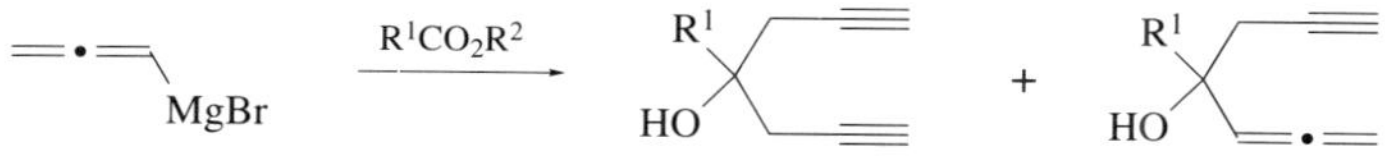

(35) (36)

(116)

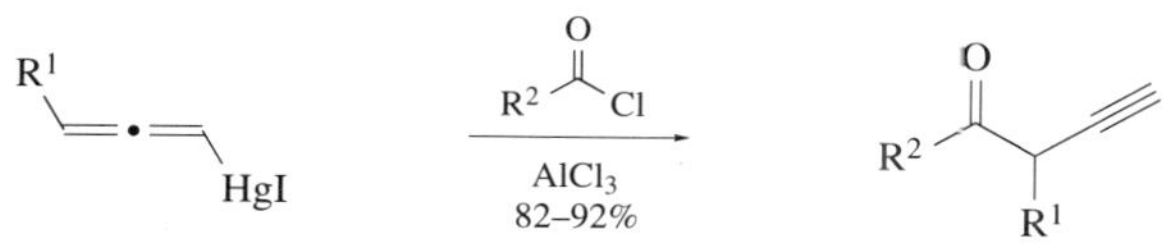

(117)

Benzoxazoles and benzothiazole react with allenylmagnesium bromide to afford products (**37**) and (**38**) by an initial addition–elimination reaction (Equation (118)) ⟨86S638⟩.

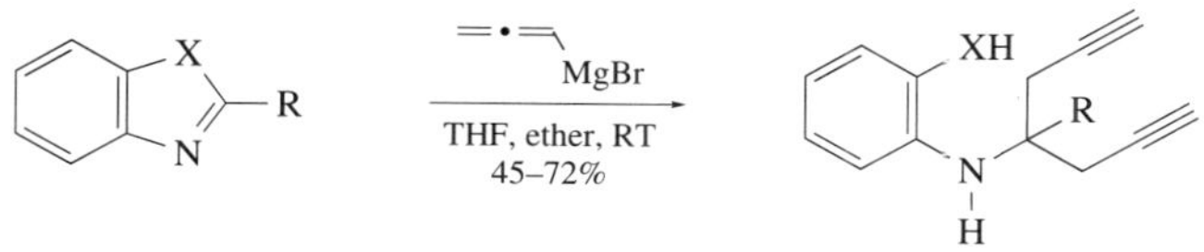

(118)

X = O, R = H, Me, Et, Pr^n, $PhCH_2$ (**37**)
X = S, R = H (**38**)

1.21.2.4.5 1,4-Addition reactions

Few examples of the 1,4-addition reactions of propargyl/allenyl organometallics have been reported, perhaps due to difficulties associated with controlling both the addition mode (1,2-addition vs. 1,4-addition) and the ambident behaviour of the reagents. Although, in general, propargyl/allenyl lithium reagents undergo 1,2-addition reactions with α,β-unsaturated carbonyl compounds ⟨77JA7632, 90JOC441⟩, lithiated 1-(triisopropylsilyl)propyne has been reported to effect regioselective conjugate propargylation of 2-cyclohexenone in a THF–HMPA solvent mixture (Equation (119)) ⟨82TL719⟩. 1,2-Addition occurs in THF alone. Organocopper reagents have met with limited success ⟨74TL4467⟩.

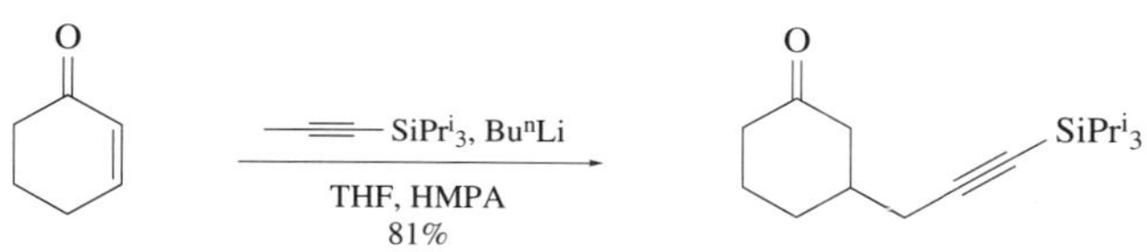

(119)

Silylallenes and stannylallenes have proven more successful. For instance, trimethylsilylallene and 3,3-dimethyl-1-trimethylsilylallene undergo successful 1,4-addition reactions with α,β-unsaturated acyl cyanides in the presence of $TiCl_4$ (Equation (120)) ⟨86JOC1199⟩. With α,β-unsaturated ketones reactions can be more complicated. For example, the reaction of trimethylsilylallene with 2-cyclohexenone in the presence of $TiCl_4$ provides a 1.6 : 1 mixture of the 1,4-adduct, 3-propargylcyclohexanone and a cyclopentene derivative (Scheme 81) ⟨83T935⟩. The minor product, presumably derived via 1,2-silyl migration in an intermediate vinyl cation, can be avoided by using triphenylstannylallene ⟨90JOC4853⟩. The weaker C–Sn bond allows a more rapid elimination of the stannyl group and the high yield formation of conjugate propargylation products. Similarly, 3-methyl-1-(triphenylstannyl)allene reacts with 2-cyclohexenone to afford the 1,4-addition product (R^1 = Me) as a 1 : 1 mixture of diastereomers.

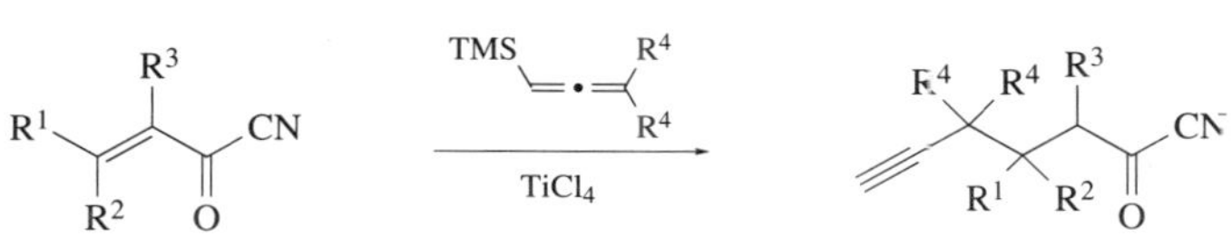

(120)

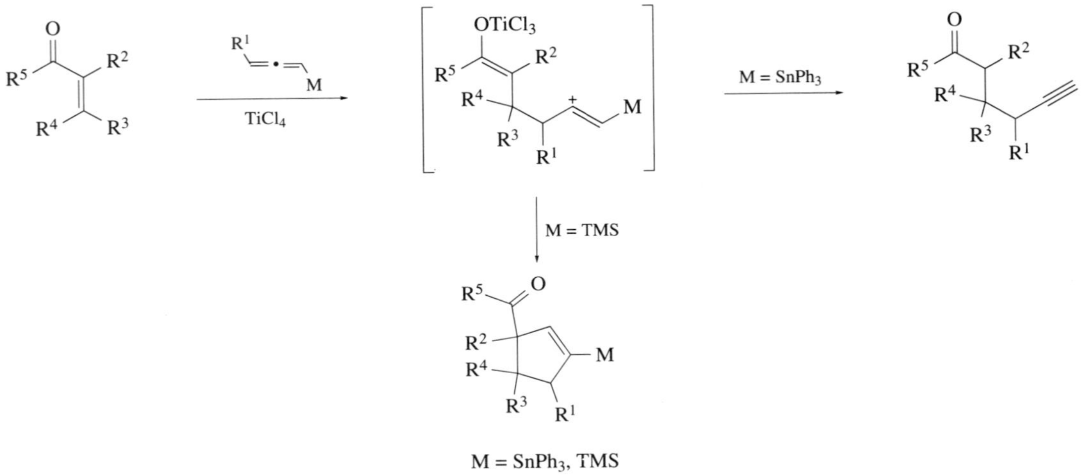

Scheme 81

The $TiCl_4$-promoted reaction of stannylallenes with α-nitro alkenes has also been reported ⟨90JOC4853⟩. Using (tri-*n*-butylstannyl)allene, two types of products, β-propargylic nitroalkanes (**39**) and α-propargylic ketones (**40**) are obtained. The selectivity depends on the presence or absence of an α-substituent on the nitro alkene (Scheme 82).

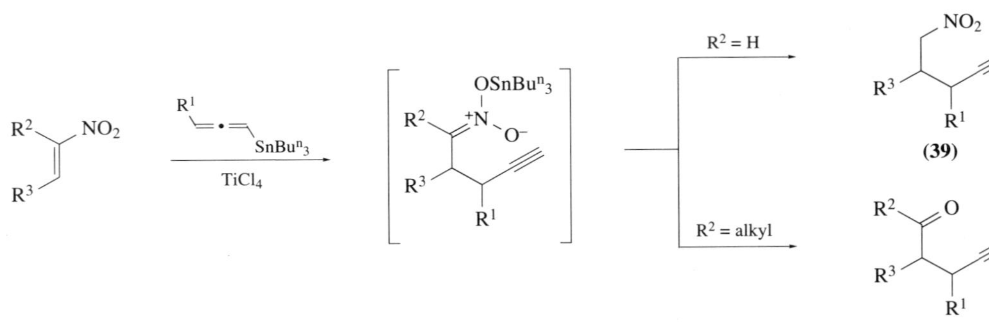

(**39**)

(**40**)

Scheme 82

1.21.3 ALKYNES BY C≡C BOND FORMATION

All alkyne carbon–carbon triple bond formation reactions can be classified as elimination reactions. However, for the purpose of this review it is convenient to make some distinctions and to classify the reactions into three categories: elimination, fragmentation and electrocyclic reactions. In addition, a fourth category of reaction, isomerisation reactions, has been added to account for those processes in which a triple bond is introduced by migration of a preexisting triple bond along a carbon chain.

1.21.3.1 Elimination Reactions

The alkyne carbon–carbon triple bond is most commonly introduced by elimination of a stable molecule or molecules from a more saturated substrate, either an alkene or an alkane. Eliminations most commonly involve the loss of groups from adjacent carbon atoms (1,2-eliminations) although alkynes can also be formed by 1,1-eliminations and, with more highly unsaturated systems, by 1,4- and 1,6-eliminations.

1.21.3.1.1 *Elimination of hydrogen*

Thermolytic elimination of hydrogen from alkenes is known to occur, but requires severe conditions and does not represent a convenient laboratory procedure for the preparation of alkynes ⟨60AG391⟩. 1-Alkenes have been observed to undergo dehydrogenation with lithium dispersion to afford lithium alkynides and lithium hydride ⟨67JOC105⟩. With 1-hexene the yield reached 65% although lower yields were obtained with other 1-alkenes. A number of diaryl substituted alkynes have been prepared in moderate yields by dehydrogenation of the corresponding alkenes with $KOBu^t$ in oxygenated DMF, but the reaction is of limited scope ⟨91CC948⟩.

In contrast to the severe conditions required for the thermal dehydrogenation of alkenes, palladium-catalysed dehydrogenation of both terminal and internal monoalkenes occurs under very mild conditions. Employing a supported palladium acetate catalyst and perchloric acid in an aerated ethanol–water solvent system, a range of straight chain 1-alkenes has been reported to undergo dehydrogenation to afford 1-alkynes in very high yields (Equation (121)) ⟨85CC1571⟩. Moreover, internal *trans* alkenes give nearly quantitative yields of disubstituted alkynes with complete regioselectivity. No isomerisation occurs and provided an aqueous ethanolic solvent system is employed, no hydration of the alkene or alkyne is observed.

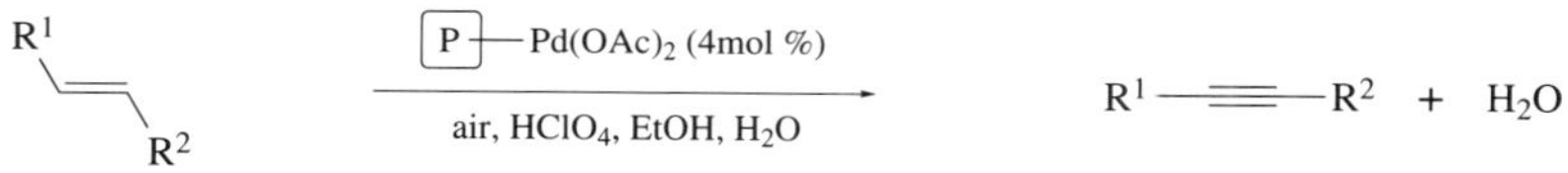

(121)

R^1 = H; R^2 = Pr^n, Bu^n, n-C_5H_{11}, n-C_6H_{13}, n-C_7H_{15}, n-C_8H_{17} 89–92%
R^1 = Me, Et, Pr^n; R^2 = Pr^n, Bu^n, n-C_5H_{11} 95–98%

1.21.3.1.2 *Elimination of carbon functions*

Although alkynes are frequently prepared by the elimination of carbon functions, these reactions are more adequately described as fragmentations or electrocyclic reactions and as such are treated in Sections 1.21.4.2 and 1.21.4.3. A unique alkene–alkyne transformation has been reported on treatment of a variety of terpenoids containing a terminal isopropylidene unit with sodium nitrite in aqueous acetic acid (Scheme 83) ⟨86TL267⟩. The reaction produces 2-alkynes in 30–98% yields and appears to involve an unprecedented elimination of a CH_4 unit from the double bond of the 2-methylpropenyl group. On closer examination, however, it has been determined that the original methyl group is lost as carbon dioxide, and a mechanism involving a complicated series of nitrations, ring closure to an isoxazoline *N*-oxide and fragmentation to the alkyne, carbon dioxide and NO has been proposed ⟨87TL4921⟩. Thus, the reaction is more adequately described as a fragmentation reaction.

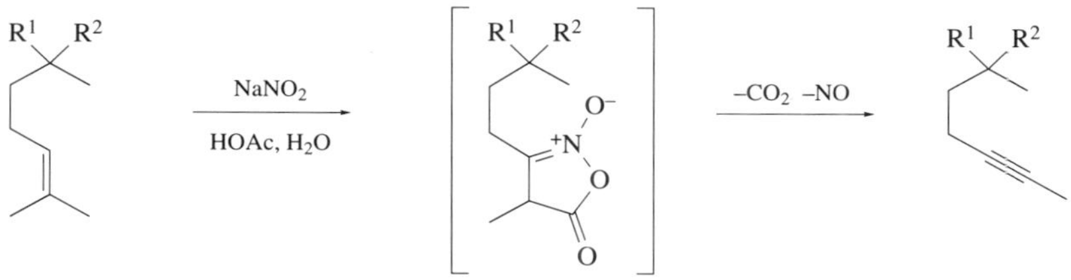

Scheme 83

1.21.3.1.3 *Elimination of halogen*

(i) *Dehydrohalogenation*

Alkynes have been prepared from organic halides by 1,1-, 1,2-, 1,4- and even 1,6- eliminations of hydrogen halide, although by far the most important type of elimination from the preparative point of view is 1,2-dehydrohalogenation.

1,2-Elimination of hydrogen halide from 1,1- and 1,2-dihaloalkanes (Equations (122) and (123)) and from haloalkenes (Equation (124)) represents the most common method for introducing the carbon–carbon triple bond. The starting materials are readily accessible; 1,2-dihaloalkanes are prepared by halogenation of alkenes and 1,1-dihaloalkanes are available from halogenation of aldehydes or ketones with, for example, phosphorus pentachloride or another acid halide. Alkenyl halides, including 1,1-dihaloalkenes, can be prepared by a variant of the Wittig reaction or by dehydrohalogenation of dihaloalkanes with bases not strong enough to accomplish full elimination to alkynes.

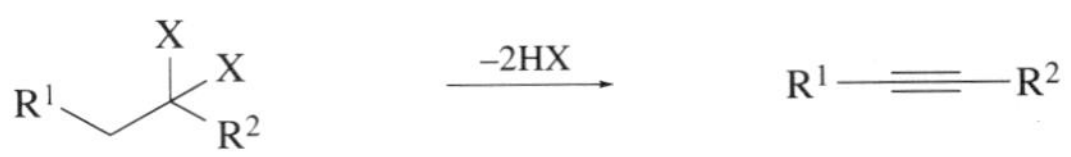

(122)

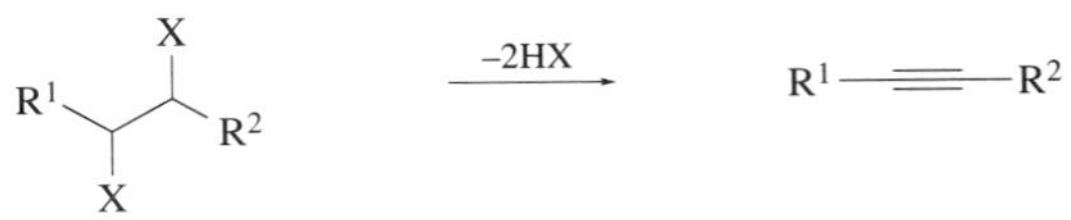

(123)

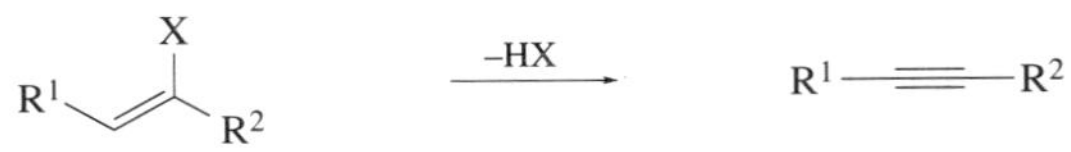

(124)

A priori, numerous complementary and opposing factors will govern the course of base-induced 1,2-dehydrohalogenations. Among these are the nature of the leaving group, the base and the solvent, and also the electronic and steric effects of the substituents in the α and β positions. For a given substrate the choice of base, solvent and temperature is often critical for successful alkyne formation, as illustrated in Scheme 84 ⟨60JCS3614, 64JCS1151⟩. This sequence demonstrates; (i) the elimination of hydrogen halide from 1,2-haloalkanes is faster than the corresponding elimination from haloalkenes; (ii) *trans* elimination from alkenyl halides is more rapid than *cis* elimination; and (iii) sodium amide in liquid ammonia is a much more effective base than potassium *t*-butoxide.

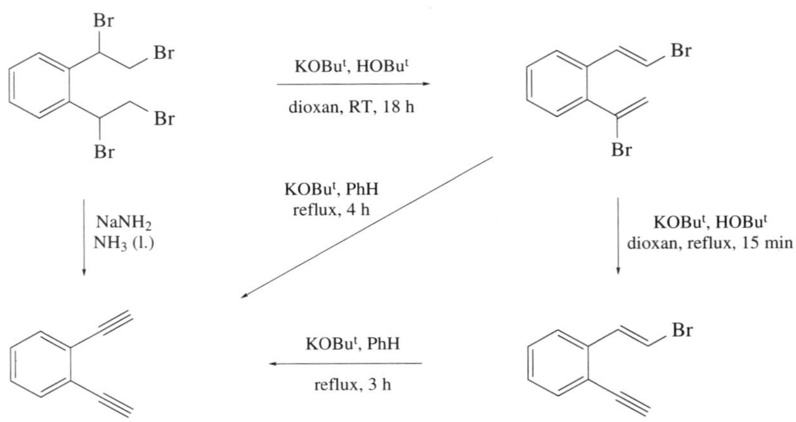

Scheme 84

Tertiary amines such as dbu and dbn which smoothly dehydrohalogenate dihaloalkanes to haloalkenes are generally inert towards the latter compounds. Similarly, alkali metal carbonates do not effect elimination in unactivated haloalkenes. Alkali metal hydroxides and alkoxides in solvents such as alcohols, dioxan and DMSO have been used with some success as illustrated in Scheme 85 ⟨76JA2667, 87CB187, 92JOC651⟩. The greater basicity of potassium *t*-butoxide and its reduced nucleophilicity due to steric encumbrance have led to its successful use in elimination reactions. In particular, the dehydrobromination of alkenyl bromides by potassium *t*-butoxide used in combination with the Wittig reaction of aldehydes with bromomethylenetriphenylphosphorane provides a convenient preparation of terminal alkynes (Equation (125)) ⟨80TL4021, 92JOC651⟩. The product alkynes can, however, be contaminated with unreacted (*E*)-alkenyl bromide. Indeed, oxygen bases generally suffer from the disadvantage that only *trans* eliminations proceed smoothly with alkenyl halides. Isomers with H and X in a *cis* relationship react very slowly, do not react at all, or evade *cis* elimination by a side reaction (Scheme 86) ⟨54JA4558, 62JOC2688⟩.

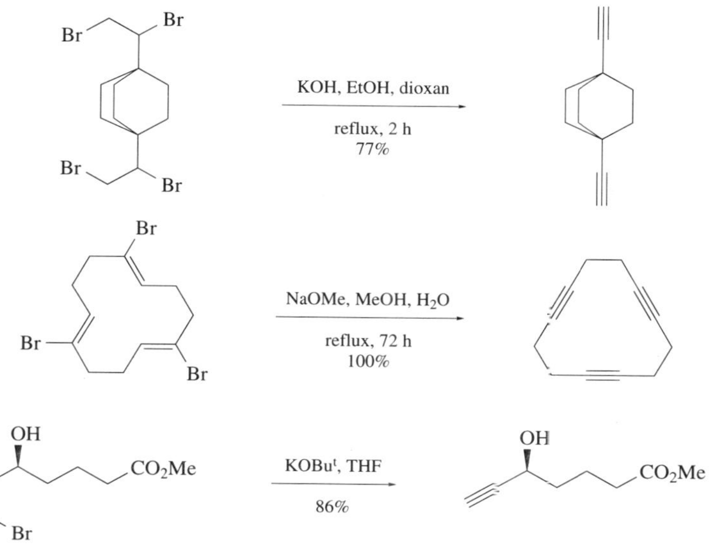

Scheme 85

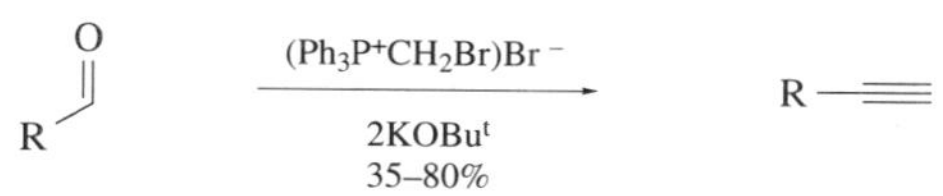

(125)

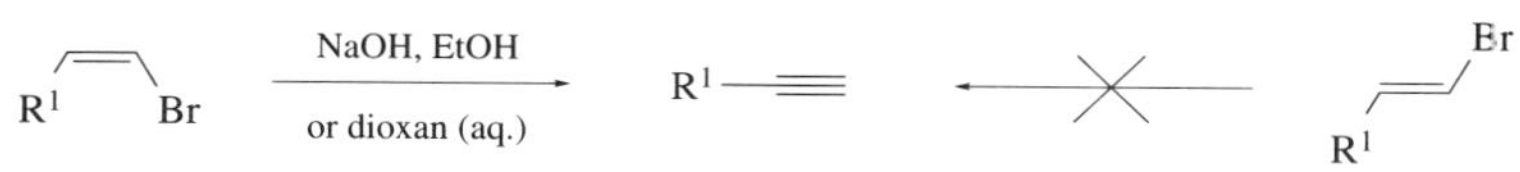

R^1 = p-$NO_2C_6H_4$
R^2 = Bu^nNHCH_2

Scheme 86

The use of oxygen bases can be further limited by isomerisation and addition reactions of the product alkyne. The preparation of straight chain terminal alkynes is particularly affected by isomerisation reactions; using sodium ethoxide in ethanol the yield rapidly drops off with increasing chain length ⟨27CB950⟩. One very effective solution to this problem is the use of potassium *t*-butoxide and catalytic amounts of 18-crown-6 in petroleum ether ⟨80LA1⟩. Very high yields of alkyl substituted alkynes are formed with no residual alkenyl halide or isomerisation/addition products being formed. 1,2-Dihalides (from alkenes) and 1,1-dihalides (from aldehydes) yield 1-alkynes and internal 1,1-dihalides (from symmetrical ketones) give internal alkynes in excellent yield (Scheme 87). 2,2-Dihalides (from methyl ketones) yield homogeneous 1-alkynes only if the 3-position is blocked. A similar high yielding procedure for the preparation of alkyl and aryl substituted alkynes from 1,2-dibromoalkanes has been reported using potassium hydroxide and catalytic amounts of lipophilic phase-transfer catalysts in petroleum ether ⟨81T1653, 86T3569⟩.

The fluoride ion can promote elimination from alkenyl halides to afford alkynes ⟨74JCS(P1)340⟩. A mild synthesis of 1,3-diynes is provided by a two step sequence involving a palladium(0)-catalysed coupling of a 1-alkyne with (Z)-1,2-dichloroethene, followed by dehydrochlorination of the intermediate (Z)-chloroenyne with tetrabutylammonium fluoride (Scheme 88) ⟨88JOC2655⟩. The corresponding (E)-chloroenynes give only traces of 1,3-diynes even under vigorous conditions, consistent with the observation that *syn* eliminations generally require a stronger base.

Alkali metal amides are stronger bases than alkoxides or fluoride ion and are frequently used in elimination reactions. Sodium amide in liquid ammonia is most commonly employed although other solvents such as DMSO and HMPT, and other amides such as lithium amide and lithium diisopropylamide are frequently used. In liquid ammonia, three equivalents of sodium amide base

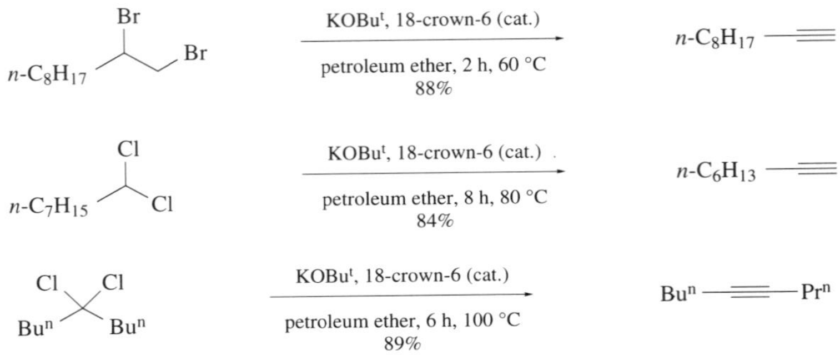

Scheme 87

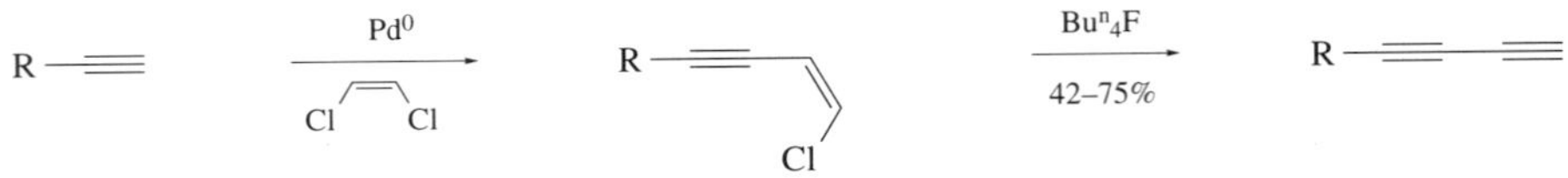

R = n-C_5H_{11}, $HOCH_2CH_2$, $MeCH_2CH(OH)$, $(EtO)_2CH$, $Pr^n{}_2NCH_2$, n-$C_5H_{11}CH{=}CH$, $MeO_2CCH_2CH_2$, $EtSCH_2$

Scheme 88

are generally employed for the preparation of 1-alkynes from dihaloalkanes. In this manner the sodium salt of the 1-alkyne is precipitated, thus avoiding prototropic isomerisations. In DMSO, the sodium salts are soluble but isomerisation to the thermodynamically more stable 2-alkynes can be prevented if prolonged reaction times are avoided ⟨70T2127⟩. A further advantage to the use of amide bases is that both (E)- and (Z)-haloalkenes are smoothly converted to the alkyne. The intermediate alkynylmetal reagent can also be used *in situ* for further reactions, for example with ketones or carbon dioxide. In spite of the frequently successful preparation of acyclic alkynes employing amide bases, the latter have not been used to a significant extent in cycloalkyne chemistry. One exception is the generation of arynes where one of the most commonly used procedures is the reaction of an aryl chloride or bromide with sodium amide or a lithium dialkylamide ⟨B-83MI 121-01⟩.

Alkali metal amides have been used very successfully for the preparation of alkynols from 1,2-haloethers. The reaction is illustrated by the preparation of 4-pentyn-1-ol from tetrahydrofurfuryl chloride and sodium amide (Equation (126)) ⟨53OS(33)68⟩. Related eliminations have been very successfully employed for the preparation of optically active 3-hydroxyalkynes (Scheme 89) ⟨89TL5455, 90SL591, 90T7033⟩. All of these eliminations appear to proceed through an initial deprotonation α to the chlorine atom followed by elimination of the ether oxygen, affording an alkenyl chloride intermediate, the alkyne-forming reaction involving a dehydrohalogenation reaction. If less than three equivalents of base are employed the alkenyl chloride intermediates can be isolated.

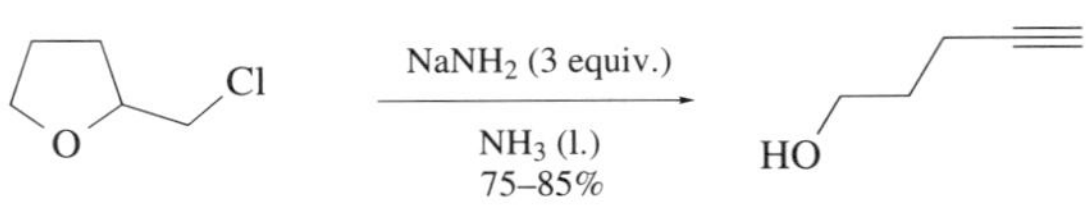

(126)

Organolithium compounds are very effective reagents for the dehydrohalogenation of dichloroalkanes and certain haloalkenes. The course of these organolithium-induced elimination reactions is governed by the relative rates of deprotonation and lithium–halogen exchange reactions in the haloalkene. With chloroalkenes, initial deprotonation geminal to the chlorine atom is followed by a rapid dehydrochlorination reaction induced by a second equivalent of organolithium (Scheme 90) ⟨73TL1495, 92TL2681⟩. With bromo- and iodoalkenes lithium–halogen exchange usually predominates and alkynes are not formed with these substrates. In a similar manner 1,1-dichloro- and 1,1-dibromoalkenes also undergo a rapid lithium–halogen exchange reaction, activated by the second halogen atom. However, with these substrates the intermediate lithiated alkenyl halides are able to undergo a further 1,2-dehydrohalogenation reaction to afford an alkynyllithium species. Very

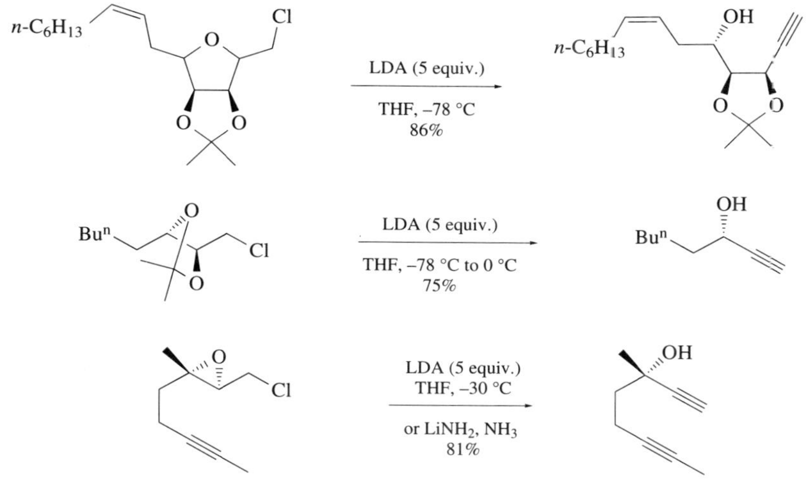

Scheme 89

successful formyl to ethynyl conversion reactions have been developed, based on the organolithium induced dehydrohalogenation of 1,1-dichloroalkanes (Scheme 91) ⟨79S502⟩, 1,1-dichloroalkenes (Scheme 92) ⟨74BSF(2)1731, 75S458⟩ and 1,1-dibromoalkenes (Scheme 93) ⟨72TL3769⟩. The combination of aldehyde alkenation employing dibromomethylenetriphenylphosphorane (prepared *in situ* from carbon tetrabromide and triphenylphosphine) and organolithium-induced dehydrohalogenation of the intermediate 1,1-dibromoalkene has been used extensively, and represents one of the most commonly employed methods for introducing the alkyne unit. The intermediate alkynylmetal reagent can be used *in situ* for further reactions, for example alkylation and carboxylation (Scheme 94) ⟨91TL4115, 92JA10654, 92JA10653, 92JOC5060⟩. Dehydrofluorination of 1,1-difluoroethene using organolithium reagents also provides a route to mono- and disubstituted alkynes ⟨82TL4325⟩.

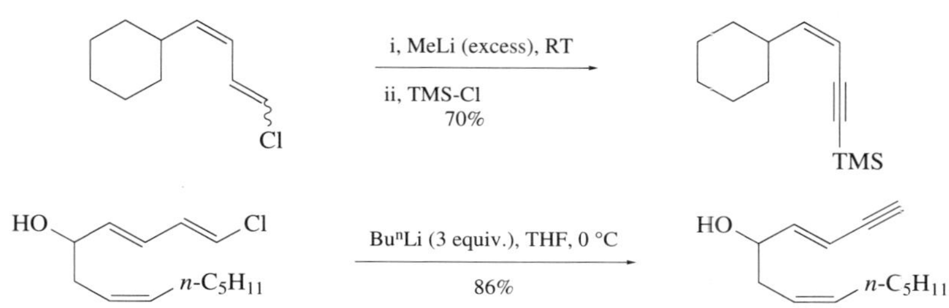

Scheme 90

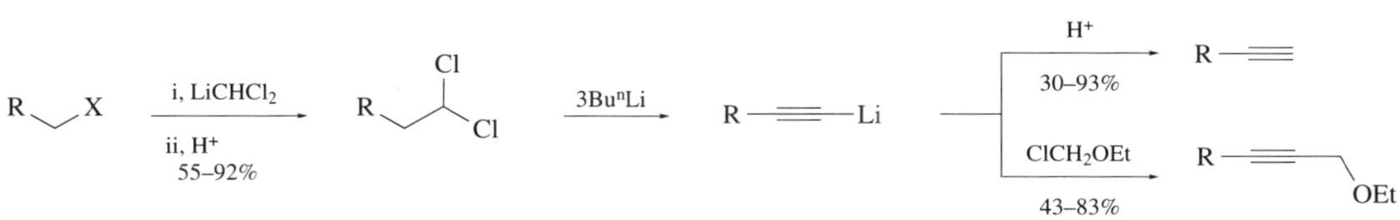

Scheme 91

The preparation of alkynes by the 1,1-elimination of hydrogen halide from alkenyl halides is commonly known as the Fritsch–Buttenberg–Wiechell rearrangement (Equation (127)). Alkynes are formed successfully from monoaryl and diaryl substituted alkenyl halides. The low tendency of alkyl groups to undergo migration usually leads to intermolecular carbenoid reactions with 2,2-dialkyl substituted alkenyl halides. Rearrangement is highly stereoselective, migration of the *trans* aryl group occurring simultaneously with elimination of the halogen in a carbenoid intermediate. Where the *trans* substituent is an alkyl group, a preferential *cis* migration of an aryl residue

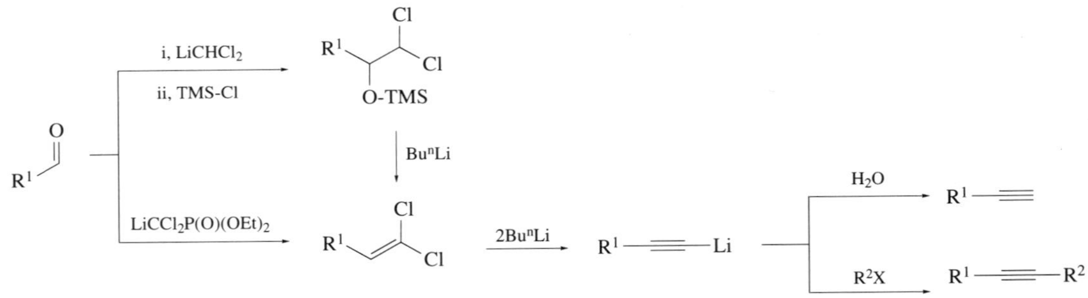

Scheme 92

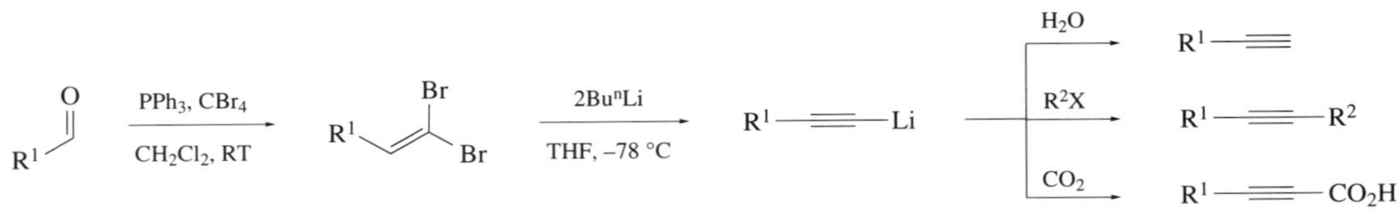

Scheme 93

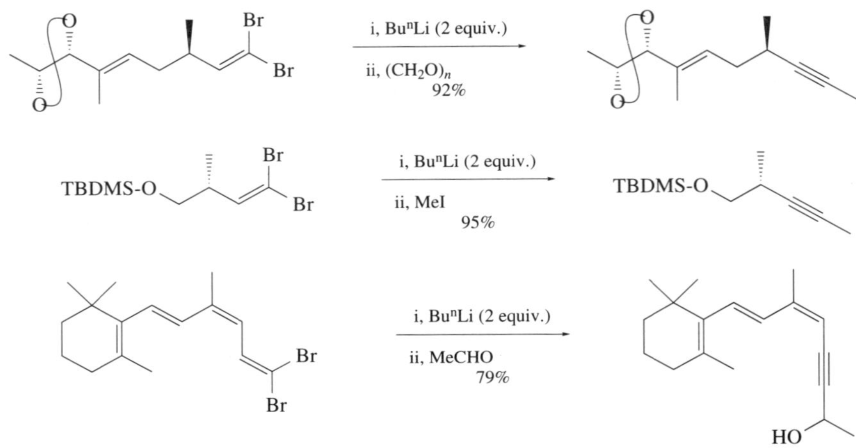

Scheme 94

occurs. The whole spectrum of bases, including hydroxide, alkoxide, amide and organolithium bases, has been employed in the Fritsch–Buttenberg–Wiechell rearrangement. The reaction has been extensively reviewed ⟨B-69MI 121-02, B-78MI 121-03⟩. In a variant of this rearrangement, alkenyliodonium tetrafluoroborates have been reported to undergo 1,1-elimination reactions with migration of the β-substituent under very mild conditions (Equation (128)) ⟨88JA6565⟩.

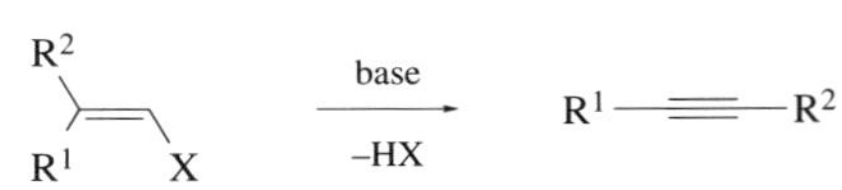 (127)

R^1, R^2 = Ar, H

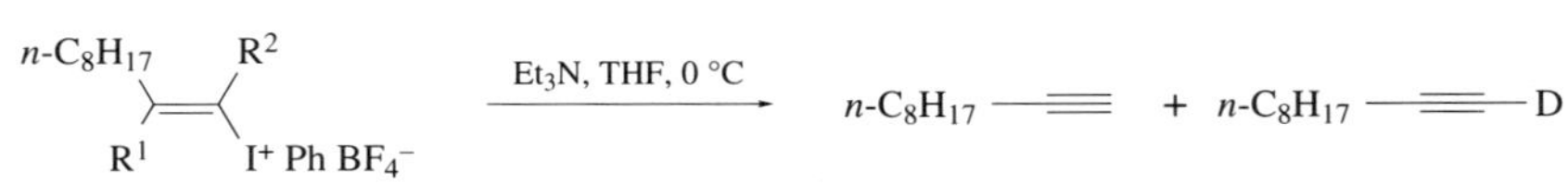 (128)

$R^1 = R^2 = H$	100 : 0	93%
R^1 = D (92%), R^2 = H	11 : 89	99%
R^1 = H, R^2 = D (100%)	90 : 10	87%

The preparation of alkynes by 1,4- and 1,6-elimination of hydrogen halides is also known where unsaturation is present between the eliminating atoms. For example, 1,3-butadiyne and 1,3-butenyne have been prepared by dehydrohalogenation of 1,4-dichloro-2-butyne and 1,4-dichloro-2-butene respectively ⟨54JA1700⟩. Similarly, 1,3,5-hexatriyne has been prepared by the 1,6-dehydrochlorination of 1,6-dichloro-2,4-hexadiyne ⟨51CB785, 52JCS2010, 53CB657, B-88MI 121-01⟩.

(ii) Dehalogenation

Dehalogenation of 1,2-dihaloalkenes and 1,1,2,2-tetrahaloalkanes has been effected with zinc or organolithium reagents, and less commonly with magnesium or sodium. Other reagents such as sodium sulfide ⟨81S879⟩ and lithium diphenylphosphide ⟨83JCS(P1)1689⟩ have also been used. Examples can be found in reviews by Ben-Efraim ⟨B-78MI 121-03⟩ and Köbrich and Buck ⟨B-69MI 121-02⟩.

Considered in isolation, the reaction is of limited preparative value since the substrates for dehalogenation are themselves generally prepared by halogenation of alkynes. Of considerably more synthetic utility is the combination of dehalogenation with C–C bond formation. An example is found in a two-step reaction sequence for the conversion of organolithium and magnesium compounds into lithium and magnesium alkynides having two more carbon atoms (Scheme 95) ⟨76JOC1487⟩. In the first step, the organometallic reagent is treated with 1,1-dichloro-2,2-difluoroethene to afford an intermediate 1,1-dichloro-2-fluoroalkene which, in a second step, is treated with *n*-butyllithium to effect dehalogenation. Although alkyllithium and alkylmagnesium halide reagents do not afford alkynes by this method, a wide variety of aryl, heteroaryl and alkynyl organometallic reagents have been successfully employed. The reaction sequence has also been employed iteratively to afford diynes, triynes and even tetraynes.

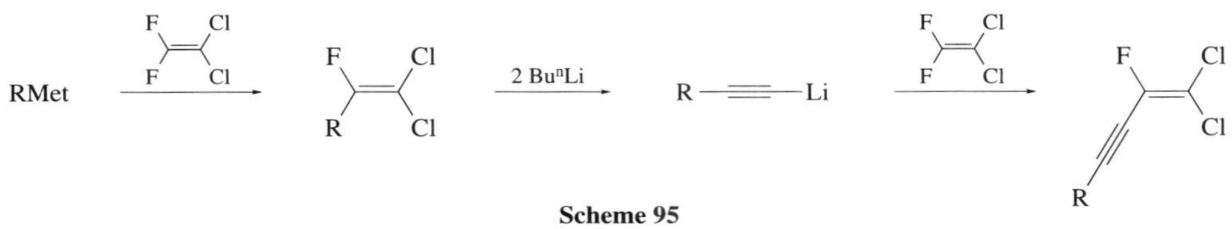

Scheme 95

Another example of C–C bond formation/dehalogenation is the preparation of cyclopropyl-substituted alkynes by thermal ring-opening of tetrachlorocyclopropene in the presence of an alkene, followed by dehalogenation of the intermediate trichloroalkene with an organolithium reagent (Scheme 96) ⟨86CB2995, 89CB377⟩.

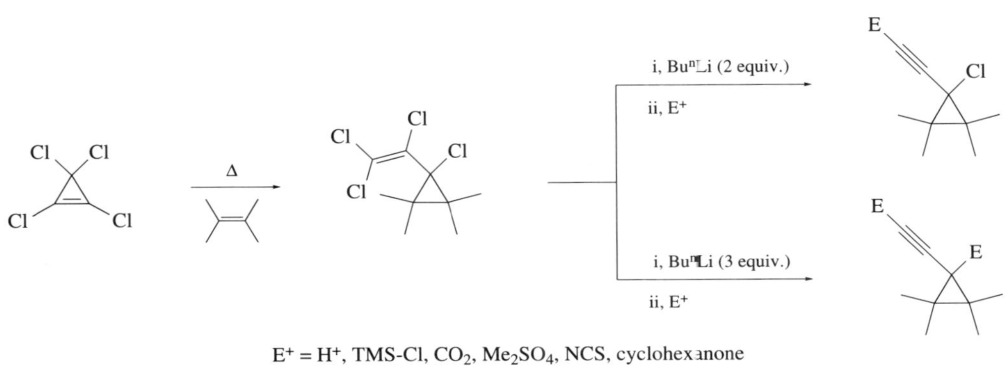

Scheme 96

1.21.3.1.4 Elimination of oxygen functions

The elimination of oxygen functions from enol derivatives represents an attractive route for the preparation of alkynes (Equation (129)), especially in view of the ready availability of enol derivatives from carbonyl compounds. Although, in 1995, the reaction appears more limited than the corresponding elimination of halogen derivatives, a number of successful methods has been developed for the 1,2-elimination of alkenyl phosphate and sulfonate esters.

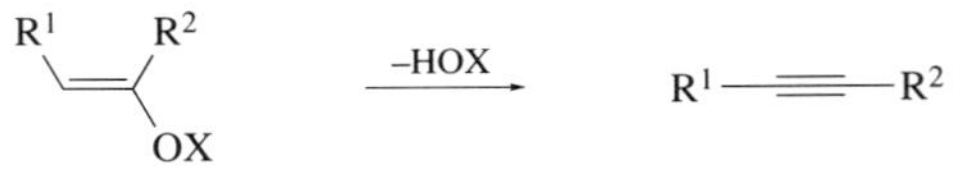

(129)

Alkenyl phosphates undergo elimination reactions with sodium amide in liquid ammonia ⟨63JCS3712⟩, and with lithium dialkylamides ⟨80JOC2526⟩, to afford alkynes. The reaction has been applied most successfully to the conversion of methyl ketones into terminal alkynes (Scheme 97) ⟨80JOC2526, 85OS(65)44, 87TL5793⟩. Using lithium diisopropylamide (LDA) as the base, good results are obtained where the group (R) of the methyl ketone ($RCOCH_3$) is either bulky or does not contain an α-methylene or methine proton. In other cases, LDA produces alkyne/allene mixtures, and the more sterically hindered lithium 2,2,6,6-tetramethylpiperidide gives superior results.

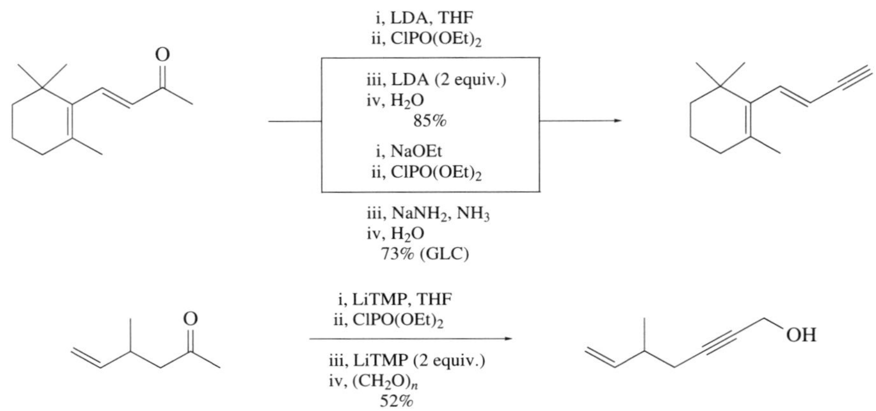

Scheme 97

Alkenyl triflates have been converted to alkynes by treatment with pyridine ⟨74JOC581⟩, NaH/DMSO ⟨84JA462⟩, potassium 2,6-di-*t*-butyl phenoxide ⟨81CC459⟩ and LDA ⟨85S962⟩ (Scheme 98). Under certain circumstances, alkyne formation has also been observed with higher-order cuprate reagents (Equation (130)) ⟨88TL1197⟩. Detailed comparisons of the effectiveness of phosphate and triflate groups have not been reported, but in a single example the triflate group afforded higher yields of alkyne (Equation (131)) ⟨93TL5257⟩.

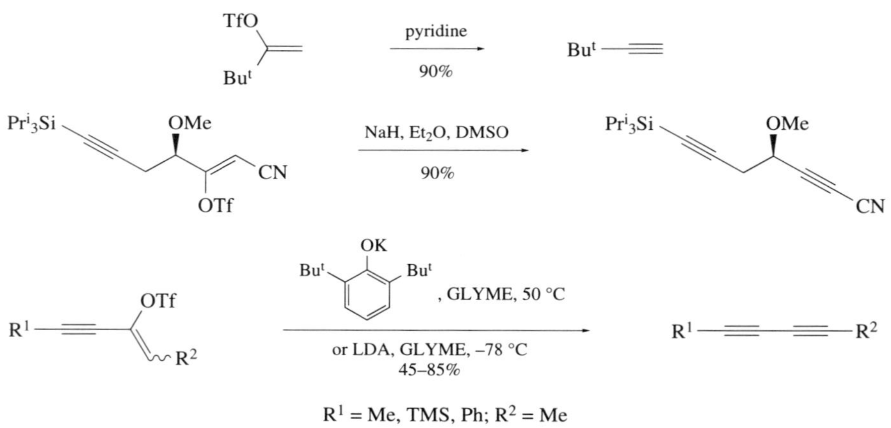

Scheme 98

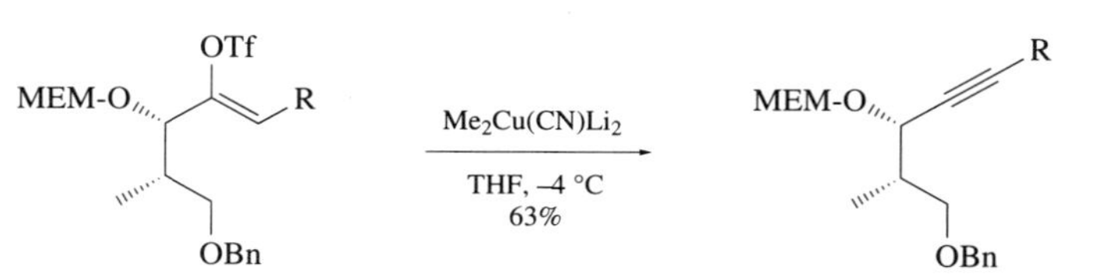

(130)

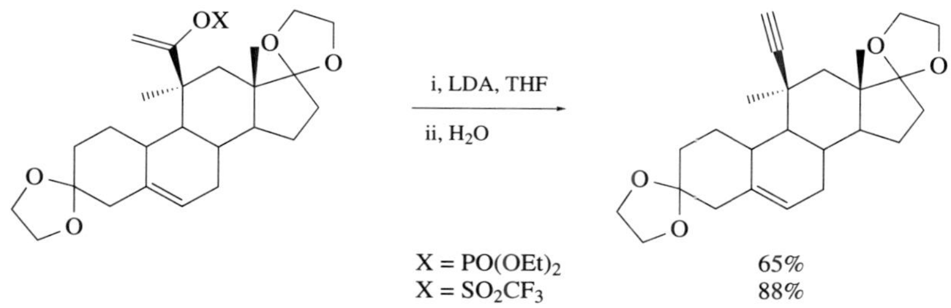

(131)

Other enol derivatives affording alkynes by elimination reactions include enol phosphonium ⟨89S217⟩ and enol benzoxazolium salts ⟨79CL481⟩. The former only undergo successful elimination reactions when the system is activated, (Equation (132)) ⟨89S217⟩. Enol benzoxazolium salts afford alkynes on treatment with a large excess of triethylamine at room temperature (Equation (133)) ⟨79CL481⟩.

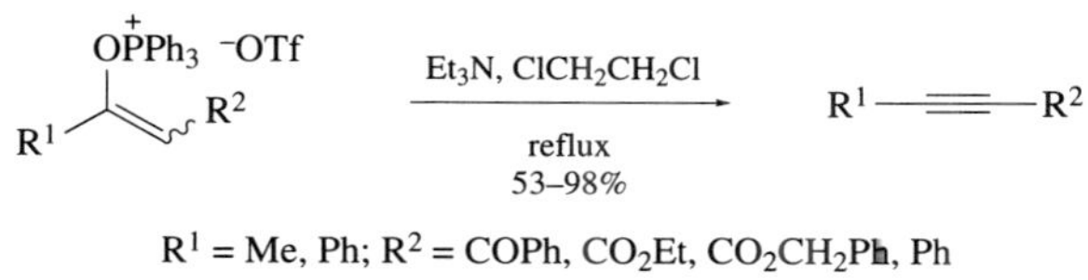

(132)

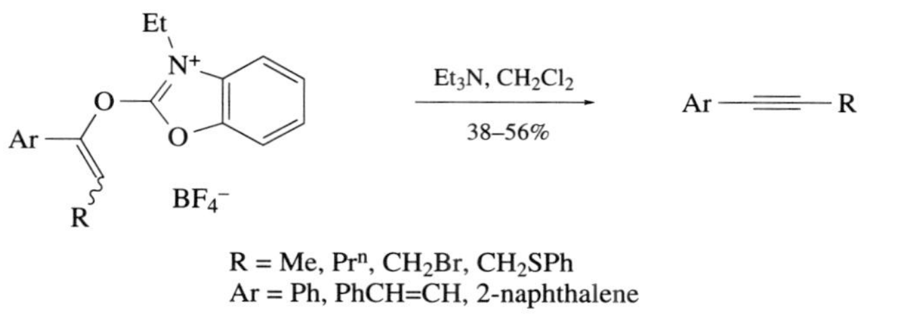

(133)

With the exception of benzyne, cycloalkynes have not generally been formed by the elimination of enol derivatives. Aryl triflates undergo elimination with LDA to afford diisopropylanilines through aryne intermediates ⟨91JOC2045⟩. *o*-Haloaryl triflates afford aryne intermediates on treatment with *n*-butyllithium ⟨91TL6735⟩.

Alkenyl triflates possessing a *β*-hydrogen or *β*-aryl group can undergo 1,1-elimination with rearrangement on treatment with a strong base such as potassium *t*-butoxide (Equation (134)) ⟨74JA4562⟩. *β,β*-Dialkyl substituted alkenyl triflates afford carbene derived products under identical conditions. Similar Fritsch–Buttenberg–Wiechell type rearrangements of alkenyl carbamates have been reported (Equation (135)) ⟨91SL611⟩.

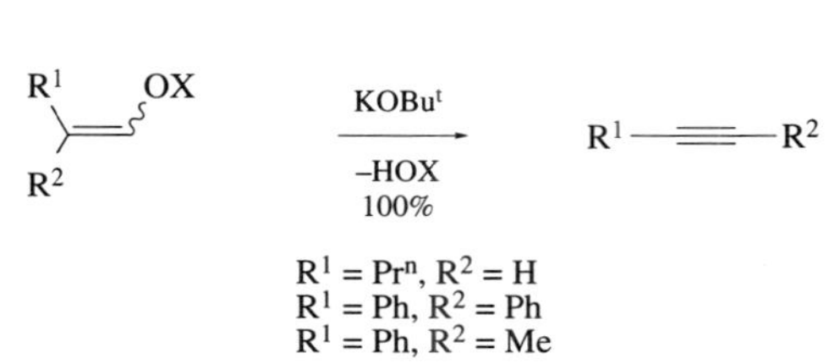

(134)

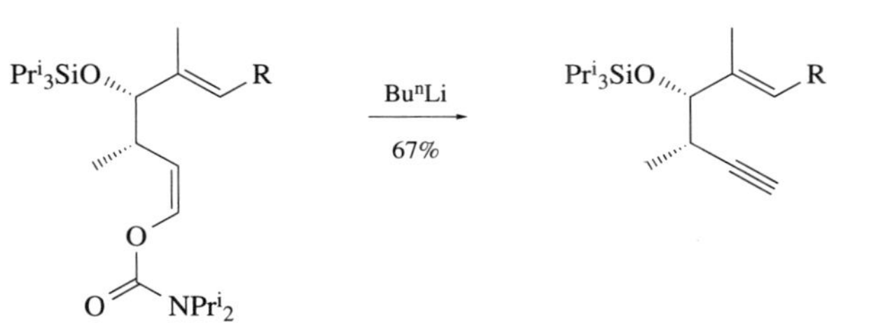

(135)

1.21.3.1.5 *Elimination of sulfur, sellenium or tellurium functions*

Although of little practical importance, alkenyl thioethers eliminate thiols on treatment with strong bases. Examples include the preparation of 1,3-diynes and 1,3,5-triynes from the corresponding alkenyl thioethers in the presence of KNH_2 in liquid ammonia (Equation (136)) ⟨71RTC265, B-88MI 121-01⟩. Alkenyl thioethers also undergo elimination and *in situ* isomerisation (Section 1.21.4.4) in the presence of potassium 3-aminopropylamide to afford 1-alkynes in high yield ⟨78JOC3083⟩. 1,4-Elimination of thiols is known, as illustrated by the preparation of 1,3-enynes from allenyl thioethers of the type $CH_3CH{=}C{=}C(R)SR'$ and potassium amide (Equation (137)) ⟨64RTC456⟩. Analogous elimination reactions of alkenyl/allenyl selenoethers and telluroethers have not been reported.

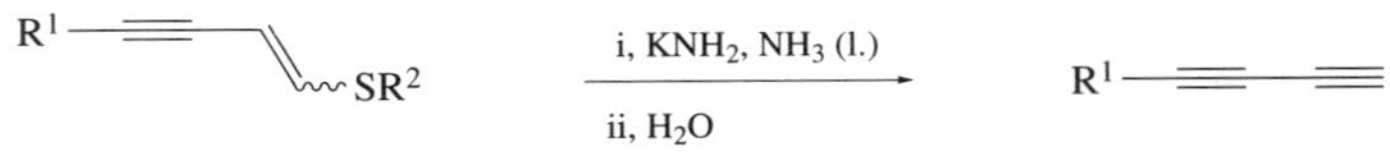

(136)

$R^1 = R^3{}_2NCH_2,\ R^4C{\equiv}C$

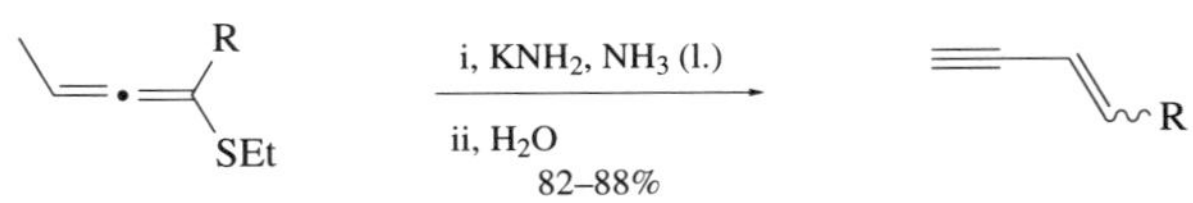

(137)

Thiophene and its 2-alkyl derivatives undergo intramolecular thiol eliminations (ring-opening) in the presence of Bu^nLi and HMPA to afford enynethiolates (Scheme 99) ⟨76RTC264⟩. The reaction proceeds by abstraction of the C-3 proton in an intermediate 2-thienyllithium. In a similar manner 3-halothiophenes undergo ring-opening in the presence of organolithium reagents. The reaction has been extensively investigated by Gronowitz and co-workers ⟨84JOC2018⟩. Representative examples are given in Scheme 100 ⟨76ACS(B)341, 83JHC729⟩. 3-Haloselenophene and tellurophene derivatives also afford alkynes, but through an alternative mechanistic pathway involving an initial attack of the organolithium reagent at the chalcogen atom ⟨79TL1509⟩.

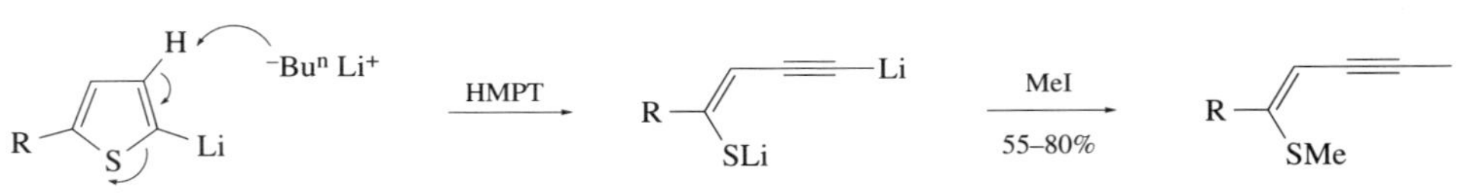

Scheme 99

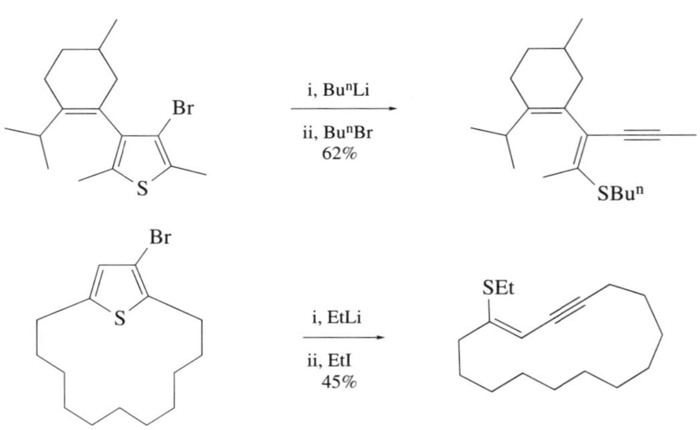

Scheme 100

Alkenyl sulfonium salts afford alkynes on treatment with aqueous alkali ⟨69ZC229⟩. Alkyl dithioacetals of 3-oxocarboxylic acids undergo elimination in the presence of excess dimethyl sulfate and aqueous alkali to afford 2-alkynoic acids (Scheme 101) ⟨65AG(E)1080⟩.

The elimination of alkenyl sulfoxides has scarcely been reported. In an isolated example the tetrahydrofuran (**41**) undergoes successive elimination and isomerisation reactions in the presence

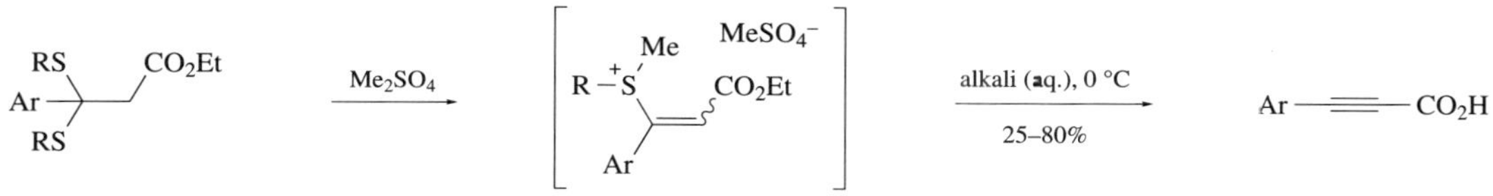

Scheme 101

of potassium 3-aminopropylamide to afford a terminal alkyne (Equation (138)) ⟨90SC2733⟩. The reaction presumably proceeds via an intermediate alkenyl sulfoxide. Analogous alkenyl selenoxides undergo elimination in the presence of bases such as 1,4-diazabicyclo[2.2.2]octane (dabco), quinuclidine and hexamethyldisilazane to afford alkynes ⟨80JA5967⟩. The utility of this elimination is enhanced when combined with the alkylation of deprotonated alkenyl selenides (Scheme 102). Elimination is believed to proceed by a pericyclic *syn* mechanism, as evidenced by the predominant formation of allenes from (*Z*)-selenoxides. An interesting elimination/[2,3]-sigmatropic rearrangement of 1,2-bis(phenylseleno)-1-alkenes is reported to occur in the presence of dabco affording propargyl alcohols (Scheme 103) ⟨91CL1805⟩.

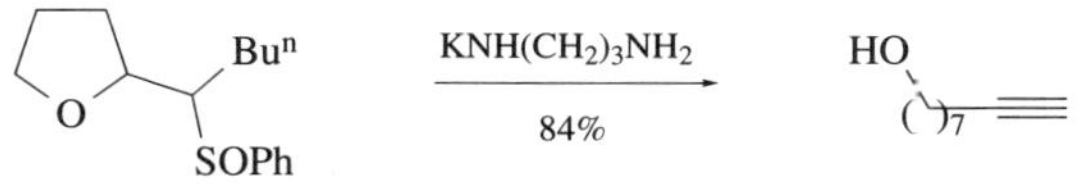

(41)

(138)

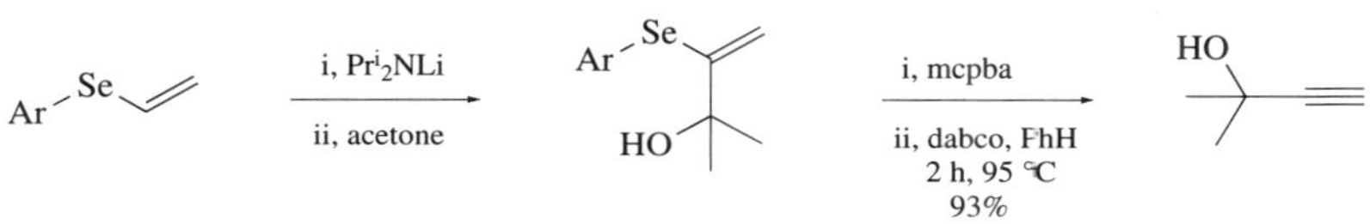

Scheme 102

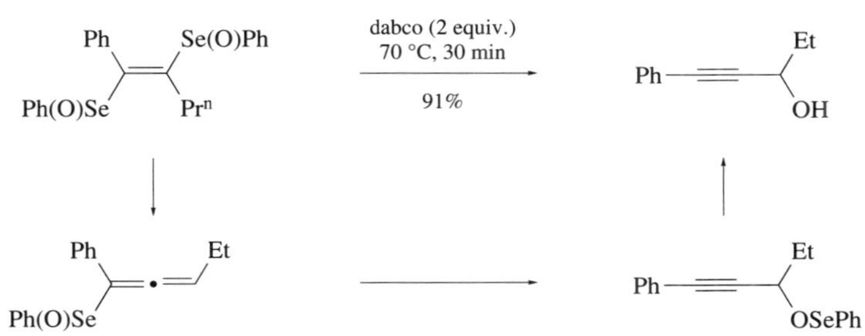

Scheme 103

β-Acetoxy and alkoxy sulfones undergo a double elimination reaction to afford enynes or polyenes depending on the substitution pattern of the substrate (Scheme 104) ⟨84JA3670⟩. The initial elimination step involves loss of the acetoxy or alkoxy group furnishing an alkenyl sulfone. Alkenyl sulfones with no allylic *γ*-hydrogen atoms undergo elimination in the presence of excess Bu^tOK to afford alkynes. Where an allylic *γ*-hydrogen atom is present 1,3-dienes are formed through isomerisation to an allylic sulfone and 1,4-elimination ⟨86JOC3830⟩. In practice, this limitation restricts the Julia coupling–double elimination sequence, (illustrated in Scheme 105) ⟨84JA3670⟩, to aldehydes R^2CHO where R^2 = aryl, heteroaryl, *β*-silyl substituted ethenyl or *β*-silyl substituted ethynyl.

Of more general applicability is the reductive elimination of enol phosphates derived from *β*-keto sulfones (Scheme 106) ⟨78JA4852, 78TL2625⟩. The reaction has been successfully applied to mono- and disubstituted alkynes bearing alkyl, functionalised alkyl, alkenyl and aryl substituents (Scheme 107) ⟨78JA4852, 78TL2625, 79JCS(P1)2429⟩. Reductive elimination can be carried out using sodium in liquid ammonia, or by using sodium amalgam in a mixture of THF and DMSO. The latter conditions are milder and are more suitable where, for example, benzyl ether or ester functions are present.

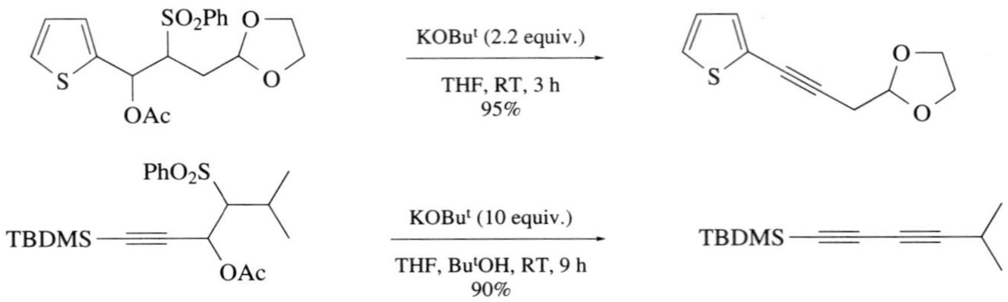

Scheme 104

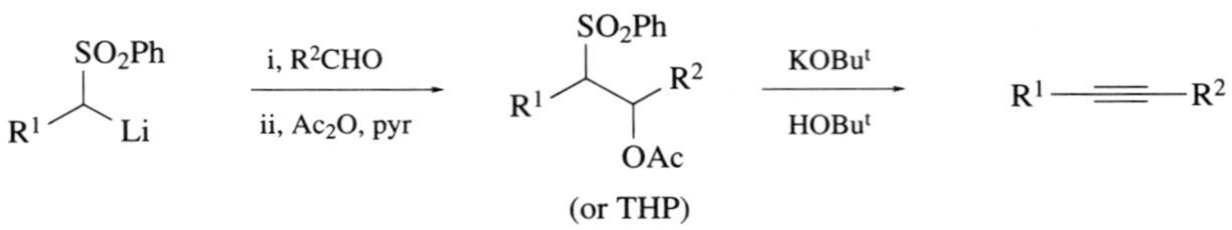

Scheme 105

Aryl substituted alkynes can be prone to overreduction, as found during the attempted synthesis of 6′-arylethynyl-6′-dethia analogues of *S*-adenosylhomocysteine ⟨90JOC3147⟩.

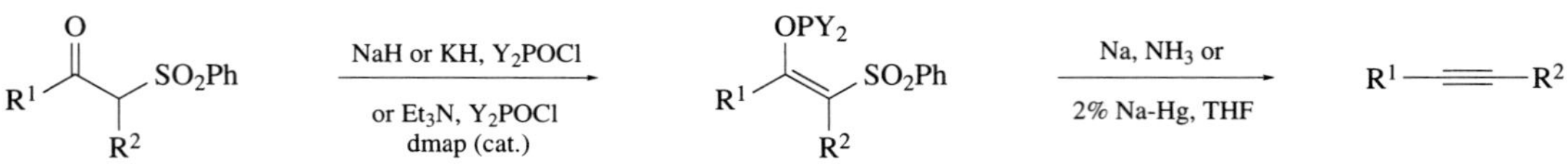

Y = OEt, OPh, NMe_2
dmap = 4-dimethylaminopyridine

Scheme 106

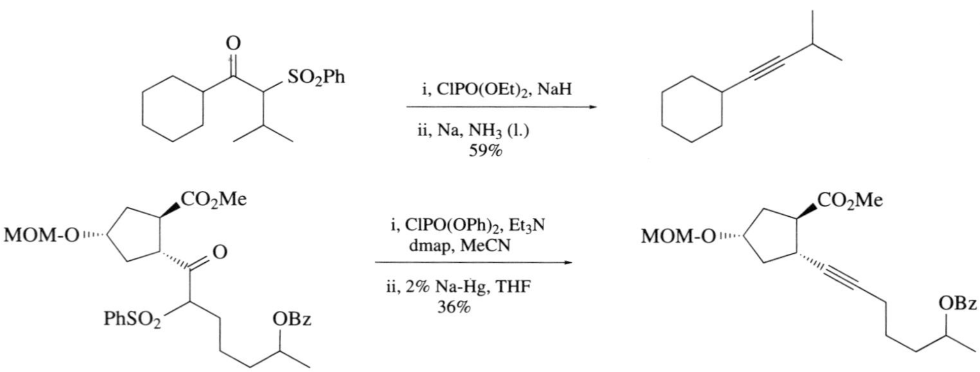

Scheme 107

α,α-Dichlorodibenzyl sulfones undergo Ramberg–Bäcklund rearrangement in the presence of dabco to afford diaryl substituted alkynes (Scheme 108) ⟨71CC22⟩. The reaction proceeds through the intermediacy of thiiren-1,1-dioxides which can be isolated if desired. α,α-Dichlorodibenzyl sulfides similarly undergo rearrangement to afford diaryl substituted alkynes in the presence of triphenylphosphine and $KOBu^t$ (Equation (139)) ⟨73CC955⟩. The mechanism is likely to be quite different from that of the Ramberg–Bäcklund rearrangement.

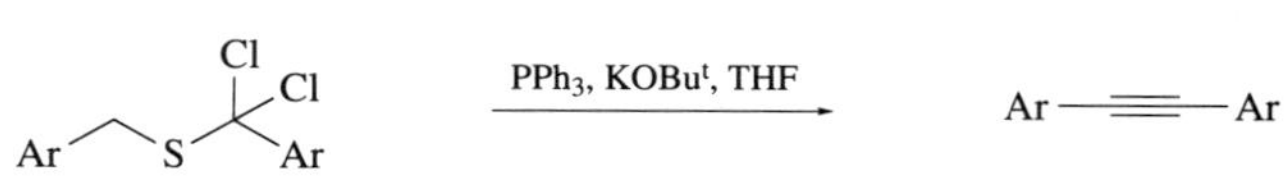 (139)

Ar = Ph, 1-naphthyl, *m*-xylyl, *p*-xylyl

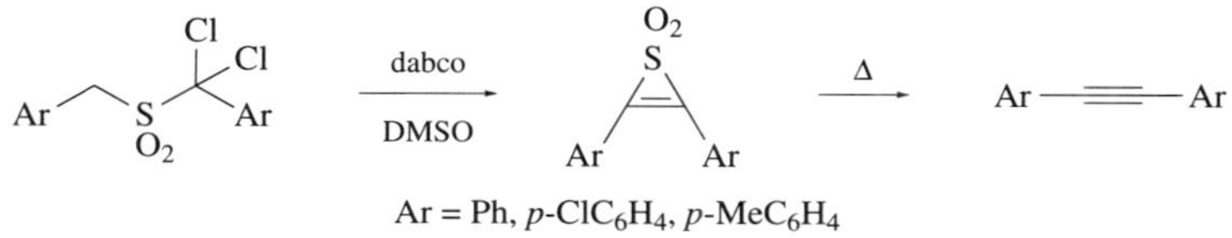

Scheme 108

1.21.3.1.6 Elimination of nitrogen functions

Alkynes have been formed by the Hofmann elimination of tertiary amines from quaternised enamines, although yields are generally low and the method is of very limited usefulness ⟨75TL1513⟩. Higher yields of alkynes have been obtained by the pyrolysis of bis quaternary ammonium salts, but again the method is of little value. Of far greater utility is the preparation of alkynes by elimination of molecular nitrogen from functional groups possessing two directly bonded nitrogen atoms. Such functional groups include hydrazones, 1,2-dihydrazones and a variety of nitrogen-containing heterocycles such as 1-amino- and 1-sulfonylamino-1,2,3-triazoles, 5-pyrazolones, 3-nitroso-2-oxazolidones, 2-nitroso-5-isoxazolones, 1,2,3-selenodiazoles and 1,2,3-thiadiazoles. Additionally, alkynes have been prepared through the elimination of molecular nitrogen from *p*-toluenesulfonylhydrazones, tetrazoles and diazo compounds possessing an adjacent leaving group, and by the base-induced condensation of carbonyl compounds with trimethylsilyldiazomethane or dimethyl (diazomethyl)phosphonate.

Hydrazones of benzyl ketones react with mercury(II) trifluoroacetate to afford alkynes in moderate yield ⟨66JOC624⟩. More generally, 1,2-dihydrazones undergo reaction with a variety of oxidants, including mercury(II) oxide ⟨63OSC377⟩, silver trifluoroacetate ⟨58JOC665⟩, lead(IV) acetate ⟨68TL4511⟩ and iodine/*N*-*t*-butyl-*N*′,*N*′,*N*″,*N*″-tetramethylguanidine ⟨88T147⟩, to afford alkynes. A very effective system employing catalytic copper(I) chloride, pyridine and oxygen has also been reported and appears to be superior to the other oxidants (Equation (140)) ⟨83T3279⟩.

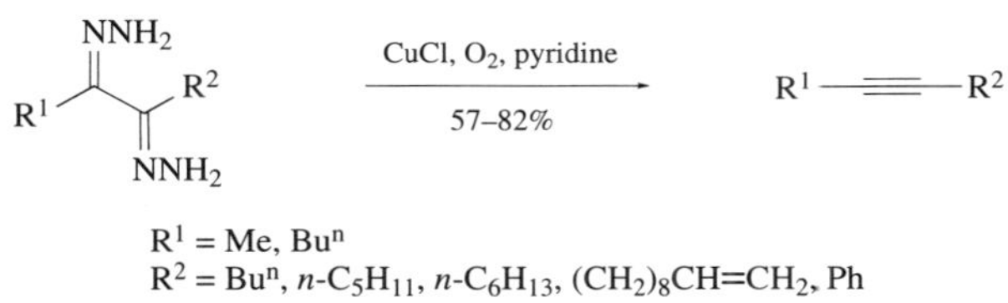

(140)

1-Amino-1,2,3-triazoles, which are sometimes observed as by-products in the mercury(II) oxide oxidation of 1,2-dihydrazones, can be oxidized with lead(IV) acetate to give alkynes. Cyclooctyne has been prepared ⟨68LA46⟩ and cycloheptyne and cyclohexyne ⟨68LA65⟩ have been generated by this method. The salts of 1-tosylamino-1,2,3-triazoles undergo photolytic decomposition with evolution of two mole equivalents of nitrogen and the formation of a metal toluenesulfinate and the corresponding alkyne (Scheme 109) ⟨64AG(E)138, 71S215⟩. The starting materials are readily available from 1,2-bistosylhydrazones and need not be isolated if the latter are photolysed in aqueous methanolic NaOH ⟨71S215⟩.

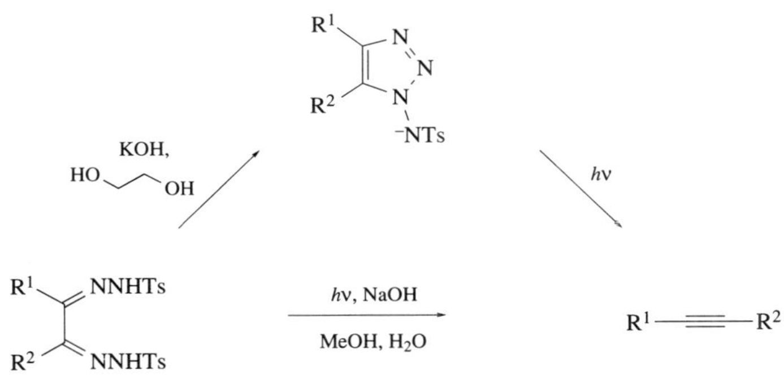

Scheme 109

4,4-Dihalo-5-pyrazolones (**42**), readily available from β-ketoesters, afford conjugated alkynoic acids on treatment with aqueous alkali (Scheme 110) ⟨66JOC2867⟩. A modification of this reaction using 3,4-disubstituted-4-monohalo-5-pyrazolones (**43**) in the presence aqueous alkali and potassium ferricyanide allows the preparation of disubstituted alkynes ⟨76JOC3650⟩. 3-Alkyl- and 3-aryl-5-pyrazolones (**44**) afford 2-alkynoic esters on treatment with thallium(III) nitrate ⟨72AG(E)48⟩, lead(IV) acetate ⟨82S1100⟩ or (diacetoxyiodo)benzene ⟨87CC711⟩.

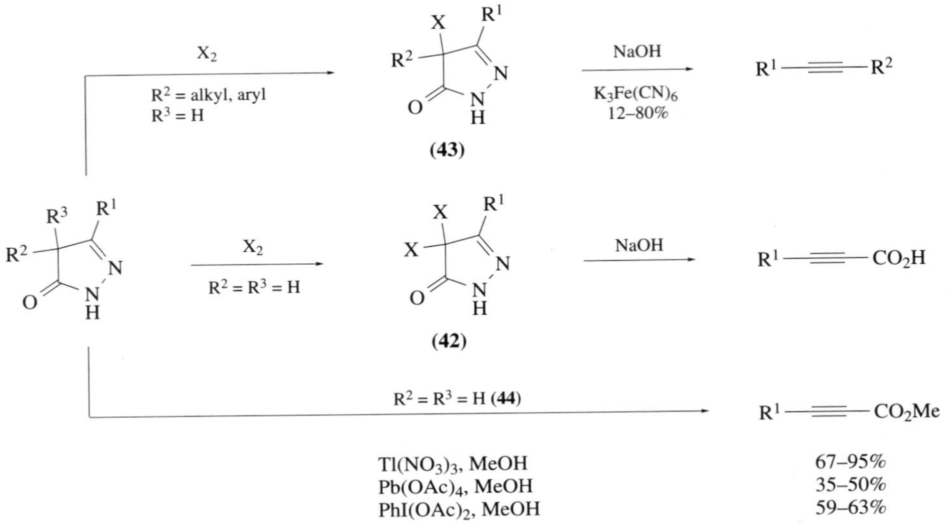

Scheme 110

5,5-Diaryl, 5,5-alkylaryl- and 5-aryl-3-nitroso-2-oxazolidones afford aryl substituted alkynes quantitatively on treatment with *n*-butylamine (Equation (141)) ⟨72JOC4466, 72JOC4468⟩. When an aryl group is not present in the 5-position little alkyne is produced. Hydroxide and alkoxide bases have also been used but afford lower yields of alkyne ⟨51JA4199, 72JOC4467⟩. 2-Nitroso-5-isoxazolones, generated *in situ* from readily available 5-isoxazolones, undergo elimination of nitrous oxide and carbon dioxide to afford alkynes (Equation (142)) ⟨91TL5321⟩. The procedure is applicable to the preparation of both alkyl and aryl substituted alkynes. 5-Isoxazolones have also been used as precursors to a variety of aryl substituted terminal alkynes through flash vacuum pyrolysis of their condensation products with aryl aldehydes (Scheme 111) ⟨78AG(E)609⟩. *N*-Nitroso derivatives of 5-aminoisoxazoles undergo decomposition with loss of nitrogen, carbon dioxide and water to afford alkynes in moderate to good yield (Equation (143)) ⟨85JOC2372⟩.

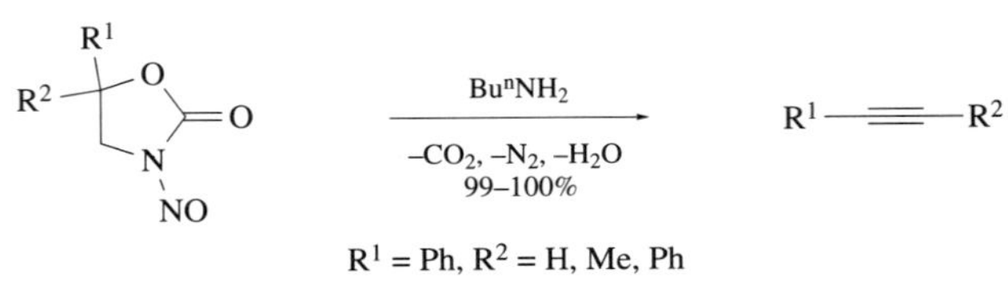

(141)

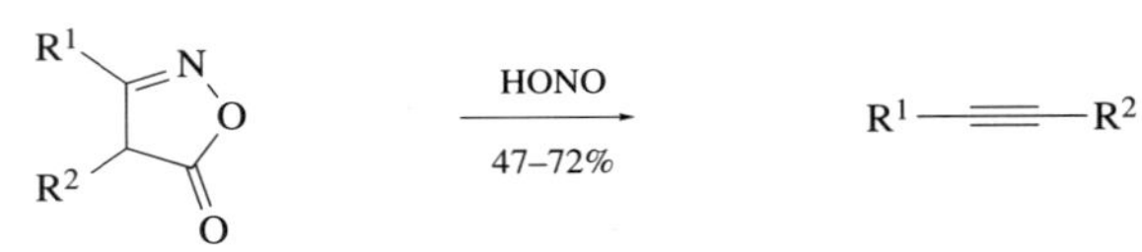

(142)

R^1 = Me, $CH_2{=}CH(CH_2)_7$, $MeO_2C(CH_2)_4$, Ph
$R^1 = R^2 = (CH_2)_{10}$
R^2 = $MeO_2C(CH_2)_2$, $MeO_2C(CH_2)_3$, $CH_2{=}CHCH_2$, $CH_2{=}CH(CH_2)_8$, $PhCH_2$, 1-naphthylmethyl

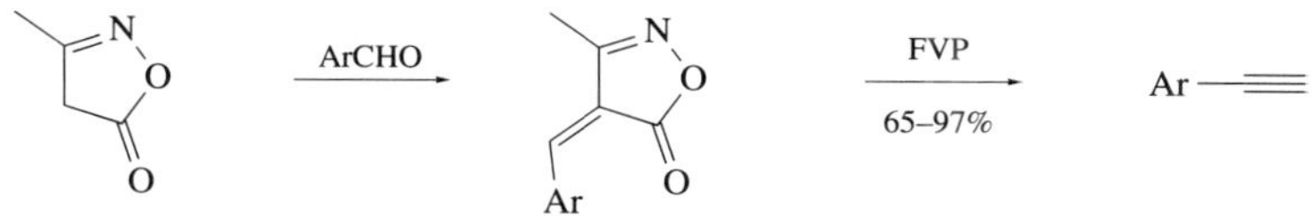

Scheme 111

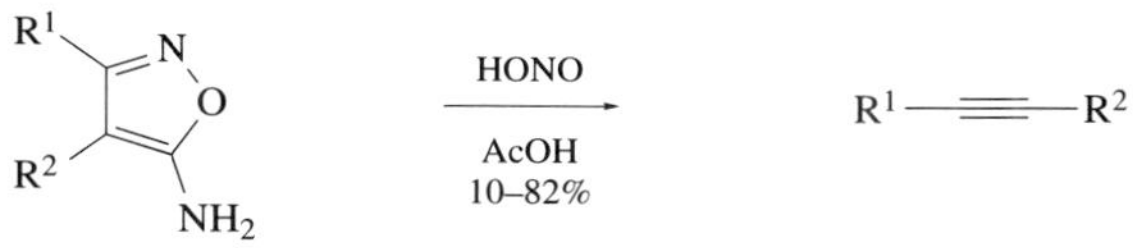

(143)

R^1 = Ph, CO_2Et, H
R^2 = CN, CO_2R, $CONH_2$, SO_2Ar, $PO(OEt)_2$, Ar

1,2,3-Selenodiazoles, readily available from oxidation of the semicarbazones of aliphatic and aromatic aldehydes and ketones with selenium dioxide, are cleaved by the action of strong bases ⟨72S235⟩ or by thermolysis ⟨71JOC2836⟩, to afford alkynes, usually in high yield (Scheme 112) ⟨70AG(E)464⟩. The reaction is generally applicable and has been used for the preparation of alkyl, aryl and heteroaryl mono- and disubstituted alkynes ⟨70AG(E)464, 75JHC801, 77JHC567⟩, non-conjugated diynes ⟨73JHC655⟩, alkynoic acids and esters ⟨73JHC655⟩ and for the preparation of strained cycloalkynes and cycloalkenynes ⟨86T1711, 88LA221⟩. Although less frequently used, 4,5-disubstituted 1,2,3-thiadiazoles also afford disubstituted alkynes on treatment with organolithium reagents ⟨71OPP163, 78S536⟩, and by photolysis (Scheme 113) ⟨78S736, 79CB3728⟩. In the absence of a 5-substituent lithium alkynylthiolates are obtained ⟨68CJC1057⟩.

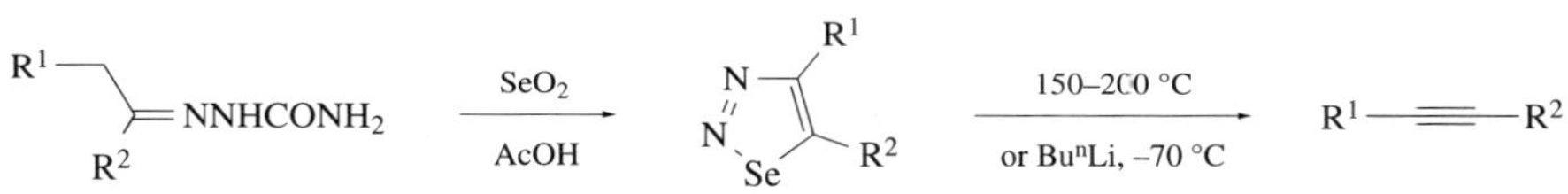

Scheme 112

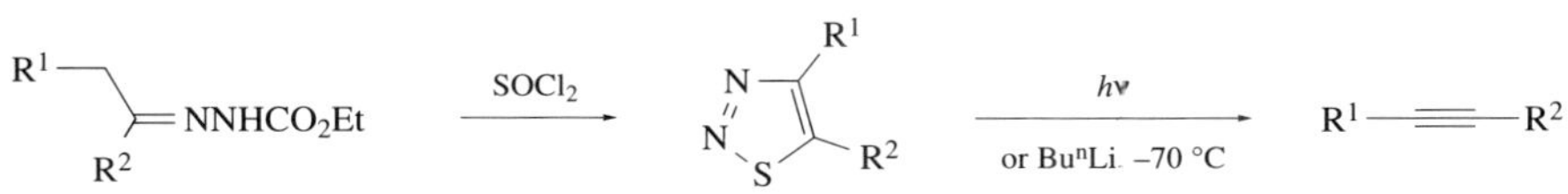

Scheme 113

Diazo compounds possessing an adjacent leaving group, such as a hydroxyl, thioether, sulfone, sulfonate, carboxylate, halogen or epoxy group, have been used for the preparation of alkynes (Scheme 114). The opening of epoxides represents a special case of the reaction in that it involves a subsequent fragmentation and as such is reviewed in Section 1.21.4.2. Most commonly, the diazo compound is generated *in situ* from an arylsulfonylhydrazone, although aminoaziridine hydrazones and *N*-acetylnitrosamides ⟨88CC830, 90JA3717⟩ have also been employed. As illustration, treatment of the tosylhydrazones of benzoin and its acetate and benzoate with alkoxides in protic or aprotic solvents affords diphenylalkyne in 13%, 94% and 98% yields respectively ⟨69TL4447⟩. Similarly, the tosylhydrazones of 20-oxo-21-fluoro- and 21-methanesulfonyloxypregnenes have been converted to pregnenynes ⟨70HCA171⟩. Tosylhydrazones of α-thiomethylketones have been converted into alkyl and aryl monosubstituted and disubstituted alkynes in high yield by treatment with methyl lithium ⟨78JOC4366, 78S305⟩. In a synthesis of tetraethynylmethane (**45**), incorporation of the final ethyne unit was achieved employing a phenylsulfone leaving group in combination with a (2,4,6-triisopropylphenyl)sulfonylhydrazone and the hindered base lithium tetramethylpiperidide Equation (144)) ⟨93JA3846⟩. Tosylhydrazones of α-furyl ketones and aldehydes undergo thermolysis as their sodium salts to afford (*E*)- and (*Z*)-2-alken-4-ynals and ynones (**46**) through cleavage of the furan ring (Equation (145)) ⟨78JA7927⟩. In an analogous reaction α-thienyl and α-oxazolyl ketones and aldehydes afford the corresponding thiocarbonyl compounds (**47**) and *N*-acyl-3-alkyn-2-one imines (**48**) respectively ⟨79TL2961⟩.

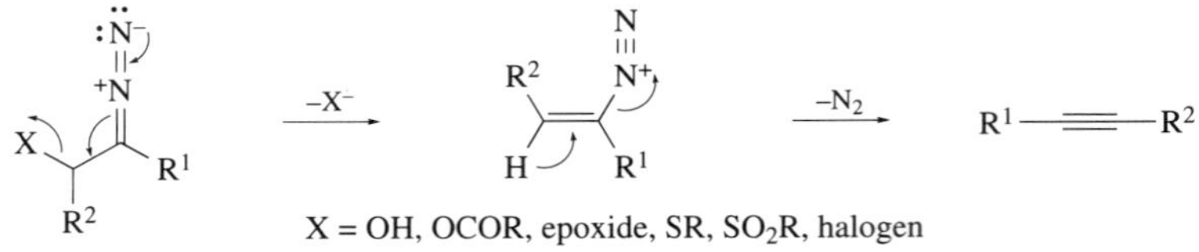

X = OH, OCOR, epoxide, SR, SO_2R, halogen

Scheme 114

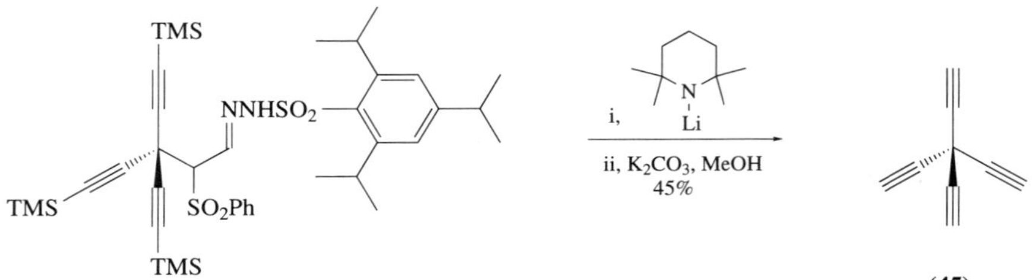

(144)

(**45**)

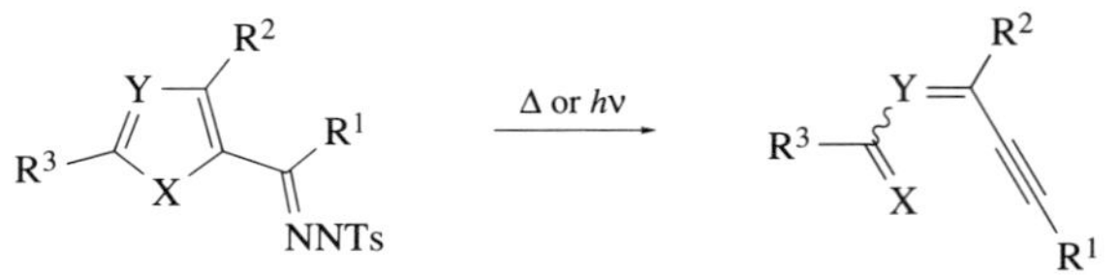

(145)

X = O, Y = C (**46**)
X = O, Y = N (**47**)
X = S, Y = C (**48**)

2-Alkynones and alkynoates are conveniently prepared by the base-catalysed condensation of diazoketones and esters with aldehydes. The intermediate α-diazo-β-hydroxycarbonyl compounds undergo elimination on treatment with boron trifluoride etherate (Scheme 115) ⟨72JA8084, 72SC331⟩.

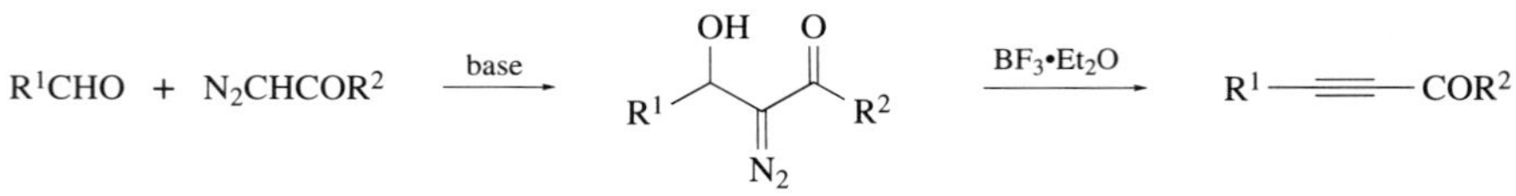

Scheme 115

Substituted tetrazoles (**49**; X = Cl, N_3, OH, NH_2) liberate nitrogen on heating at 110–200 °C to furnish alkynes in moderate to good yield (Scheme 116) ⟨66TL1663⟩.

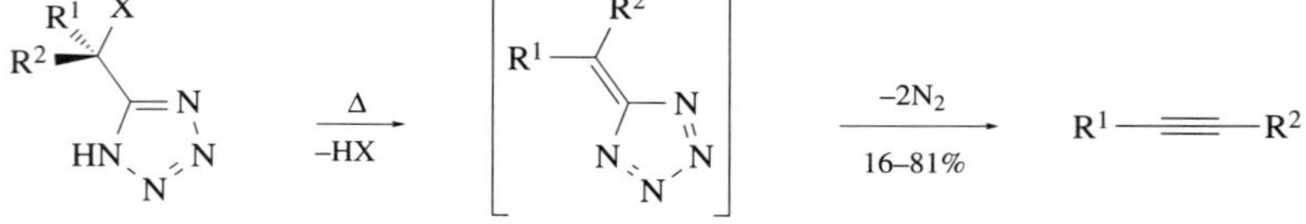

(**49**)

X = OH, NH_2, N_3, Cl
R^1, R^2 = H, aryl

Scheme 116

1-Alkynes have been prepared by the base-promoted reaction of dimethyl (diazomethyl) phosphonate and trimethylsilyldiazomethane with aldehydes and aryl ketones (Equation (146)) ⟨73CC151, 77JCS(P1)869⟩. By operating at low temperatures enolisable aldehydes, aryl ketones and enals can be transformed successfully ⟨82JOC1837⟩. Dialkyl ketones do not form alkynes but instead give cyclopentenes ⟨92CC721⟩. The reaction has been applied to the preparation of chiral propargyl amines from aminoaldehydes ⟨92TL3715⟩ and to the preparation of the immunosuppressive agent

FK506 ⟨89JOC4267, 92JOC5071⟩. The reaction mechanism is believed to involve the loss of nitrogen from diazoalkene intermediates.

(146)

R^1 = alkyl, Ar; R^2 = H, Ar

1.21.3.1.7 *Elimination of phosphorus, arsenic, antimony or bismuth functions*

Although alkynes are frequently prepared by Wittig-type reactions of phosphorus ylides with aldehydes and ketones, these invariably precede a further heteroatom elimination, and as such are treated under the appropriate heteroatom. Elimination reactions involving the loss of a phosphorus function in the alkyne forming step are more limited in number.

One commonly employed method of alkyne synthesis involves the coupling of phosphorus ylides with carboxylic acid chlorides followed by thermal extrusion of triphenylphosphine oxide from the intermediate oxoalkylidenetriphenylphosphorane (**50**) (Scheme 117). Using conventional pyrolysis techniques the elimination step is only successful when R^1 is an electron-withdrawing group (e.g., Ar ⟨59JCS3874, 62JCS2333, 64JCS543, 82TL343⟩, COR ⟨65JOC1015⟩, CO_2R ⟨61CB3005, 62JCS2333, 64JCS543, 77LA282⟩ and CN ⟨62JCS2333, 64JCS543, 81TL5283⟩. Purely aliphatic alkynes and terminal alkynes are obtained from the corresponding ylides (R^1 = alkyl or H) under flash vacuum pyrolysis (FVP) conditions (Scheme 118) ⟨85CC1140⟩. Terminal alkynes can also be prepared by acylation of ethoxycarbonylmethylenetriphenylphosphorane with acid chlorides followed by FVP at 750°C ⟨90SL211⟩. Loss of Ph_3PO is followed by loss of the ethoxycarbonyl group to afford the 1-alkyne (Scheme 118). At lower temperatures ethyl 2-alkynoates (**51**) are obtained. The reaction has been applied iteratively to the preparation of diynes and triynes ⟨90SL211, 90SL212⟩. Where ester (or thioloester) and keto (or 1,2-diketo) carbonyl groups are both present, oxygen is eliminated exclusively from the latter, giving alkynyl esters ⟨93TL5621⟩. Room temperature eliminations have been reported in the presence of organotin halides ⟨74JOM(77)C22⟩.

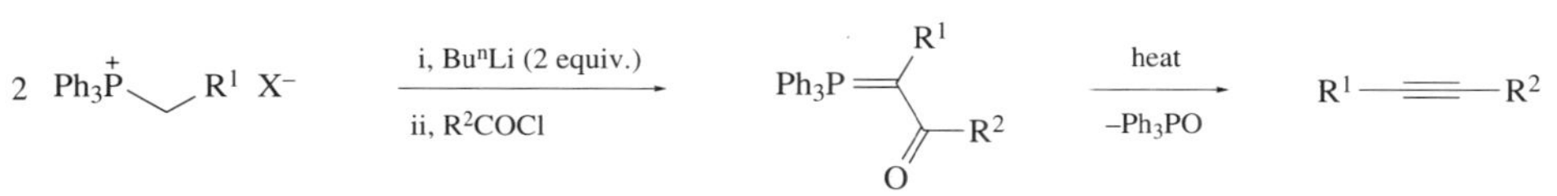

(50)

Scheme 117

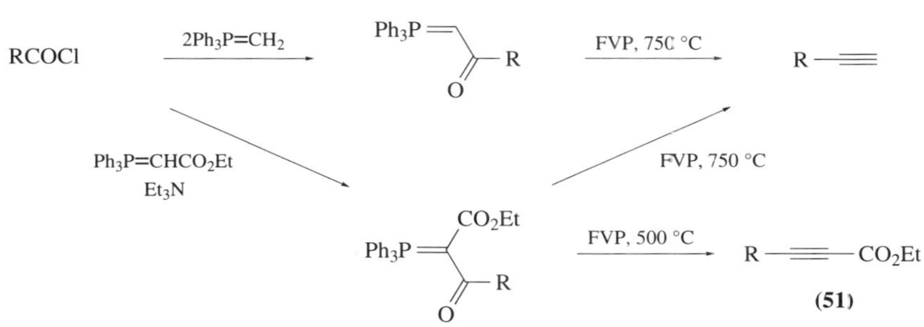

(51)

Scheme 118

Dephosphorylation of 2-fluoroalkenylphosphonates in the presence of fluoride ion can afford 1-alkynes, although the reaction has only been applied to perfluoroalkyl substituted alkynes and is of limited use ⟨84TL1377⟩.

1.21.3.1.8 Elimination of boron, silicon or germanium functions

Few elimination reactions of this type have been described for the preparation of alkynes. Those methods which have been reported invariably involve nucleophilic attack at the heteroatom with expulsion of a β-leaving group. Alkenylboranes possessing a β-halogen atom or sulfinate group undergo alkyne forming elimination reactions, but these eliminations invariably form part of a sequence of reactions in which the overall transformation can be categorised as one of substitution, at least for the purpose of this review (Section 1.21.3.1.5). Alkenylsilanes possessing a β-halogen atom also undergo elimination reactions in the presence of nucleophiles, most commonly fluoride ion. The reaction represents an important method for the preparation of arynes and other strained cycloalkynes, such as 1,3-cyclohexenyne (Scheme 119) ⟨90JA8578⟩, but has been little used for the preparation of simple alkynes. β-Silylalkenyl sulfinates undergo *syn*-stereospecific elimination reactions to form alkynes (Scheme 120) ⟨82JCS(P1)1563⟩. The reactions are analogous to the boron sulfinate elimination reaction (Section 1.21.3.1.5).

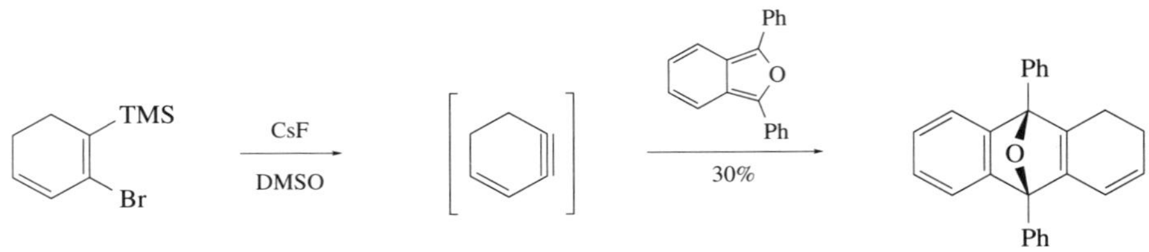

Scheme 119

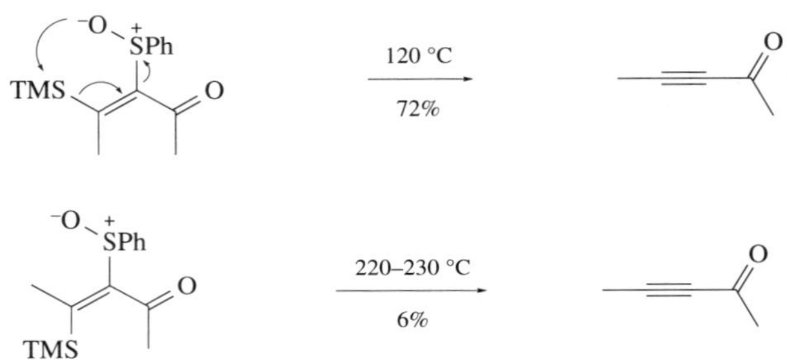

Scheme 120

Dehydrosilylation of alkenylsilanes has been achieved by employing boron trifluoride activated iodosylbenzene ⟨85CC697⟩. The reaction proceeds by halodesilylation followed by elimination of an intermediate alkenyl(phenyl)iodonium salt (Scheme 121). Both (*E*)- and (*Z*)-alkenylsilanes afford alkynes by this method, although an additional base is required in the former case.

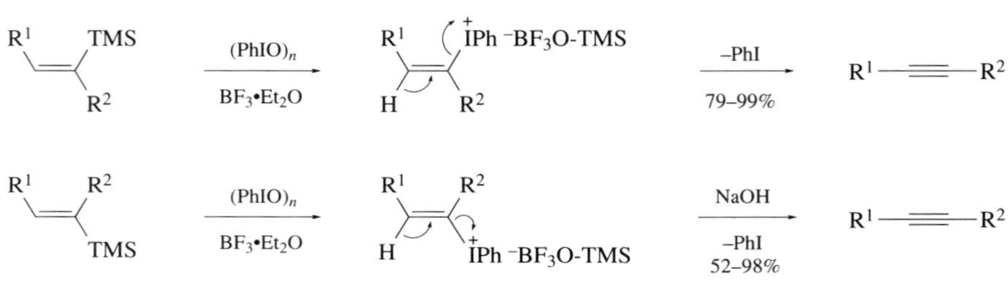

Scheme 121

1.21.3.1.9 Elimination of metal functions

While many of the elimination reactions described in Section 1.21.4 proceed through organometallic intermediates, without exception these alkyne syntheses involve the coelimination of a heteroatom function (Equation (147)). Dehydrometallation reactions (Equation (148)), which represent an important source of instability for metal alkyls, are less well described for alkenyl metals. Aside from the possible involvement of dehydropalladation processes in the catalytic coupling of unsaturated organic halides with terminal alkynes (Section 1.21.3.1.1(ii)), the only dehydrometallation reaction of practical importance for the preparation of alkynes is the lead(IV) acetate induced elimination of trialkyltin hydrides from alkenylstannanes (Scheme 122). The reaction has been used to prepare terminal alkynes and to introduce the ethynyl group into the β-position of α,β-unsaturated ketones ⟨74JA5581, 82TL4607⟩.

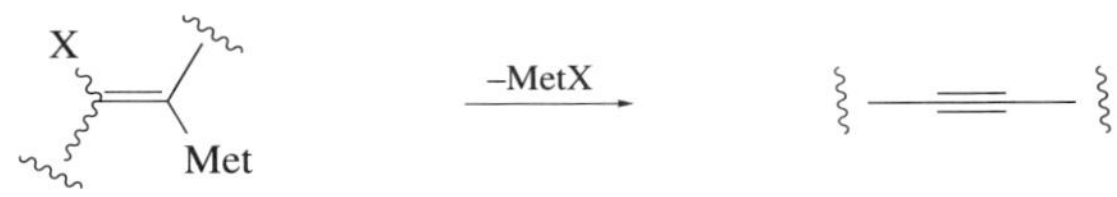

(147)

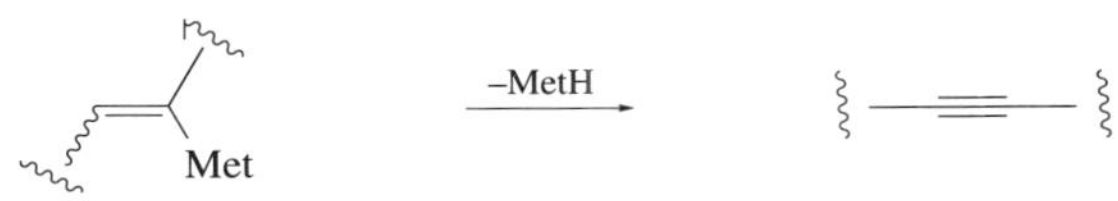

(148)

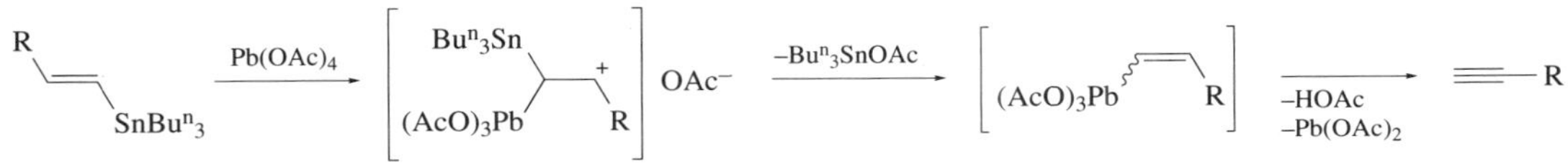

Scheme 122

1.21.3.2 Formation by Fragmentation Reactions

Fragmentation reactions, according to the definition of Grob and Schiess ⟨67AG(E)1⟩, are processes in which the reacting molecule breaks into three fragments according to Equation (149). The electrofugal group Y–B forms a stable cation or neutral molecule, the central group becomes an unsaturated fragment and the nucleofugal group X departs with an electron pair. Alkynes are most commonly formed in fragmentation reactions with substrates of the form Y-C-C=C-X, where X is the nucleofuge (e.g., halogen, OSO_2Ar, SeO_2Ar, N_2^+) and Y is a heteroatom able to stabilise the positive charge on carbon. Most commonly, the electrofugal fragment is a carbonyl compound or carbon dioxide.

(149)

Many examples of decarboxylative and decarbonylative fragmentations have been reported where 3-halo-2-alkenoic acids, esters and aldehydes are treated with aqueous base (Scheme 123) ⟨B-69MI 121-02⟩. Decarbonylative fragmentation requires the addition of a nucleophile, invariably hydroxide, to the carbonyl carbon atom prior to fragmentation. Decarboxylative fragmentation has also been successfully applied to the preparation of alkynes from enol sulfonates ⟨69AG(E)429, 71JCS(C)2013⟩. As expected, fragmentation occurs more readily when the nucleofuge and electrofuge have a *trans* orientation. Arynes have also been prepared by decarboxylative fragmentation of 2-haloarylcarboxylic acids, 2-carboxyaryl tosylates and aryldiazonium-2-carboxylic acids ⟨B-83MI 121-01, 90JOC3808, 91COS(4)483⟩ and by decarbonylative fragmentations ⟨B-67MI 121-01⟩.

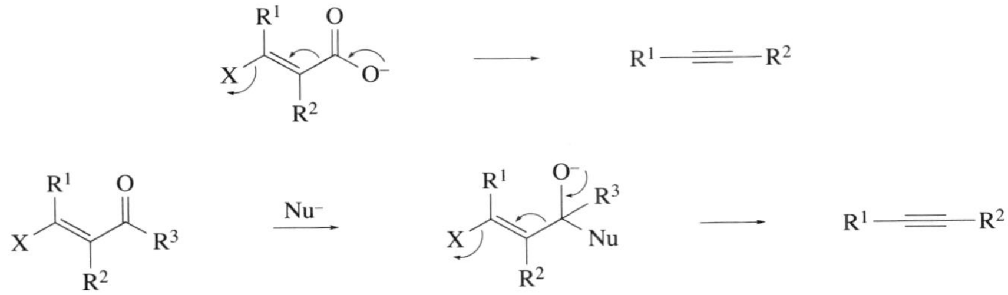

Scheme 123

Tosylhydrazones of α,β-epoxy ketones undergo fragmentation to afford an alkyne, a ketone (or aldehyde), nitrogen and toluenesulfinate ion (Eschenmoser fragmentation) (Scheme 124) ⟨67HCA2101, 67TL3739, 67TL3943, 71HCA2896⟩. The reaction is of most value with epoxyketones derived from cyclic enones, where the product alkyne and carbonyl functions are connected. In general, the reaction involves treating the cyclic enone with *p*-toluenesulfonylhydrazine in dichloromethane–acetic acid 1 : 1 to form an intermediate tosyl hydrazone which decomposes to give the alkynal or alkynone. A number of representative examples are illustrated in Scheme 125 ⟨71HCA2896, 90TL6205, 92JOC50⟩. 2,4-Dinitrobenzenesulfonylhydrazones ⟨75JOC579⟩ and mesityl- and triisopropylsulfonylhydrazones ⟨81S276⟩ have also been found to be advantageous in difficult cases, especially in instances where the product is an alkynal. Other preparative variations include the use of oximes ⟨67HCA2108⟩, 1,3,4-oxadiazolin derivatives ⟨78CJC308⟩ and 2-phenylaminoaziridine and *trans*-2,3-diphenylaminoaziridine hydrazones ⟨72HCA1276, 76OS(55)114⟩. The aminoaziridine hydrazones of α,β-epoxy ketones decompose thermally and afford alkynones in higher yield than the corresponding tosylhydrazones. In addition, only inert and volatile by-products are formed and the reaction takes place in a neutral medium.

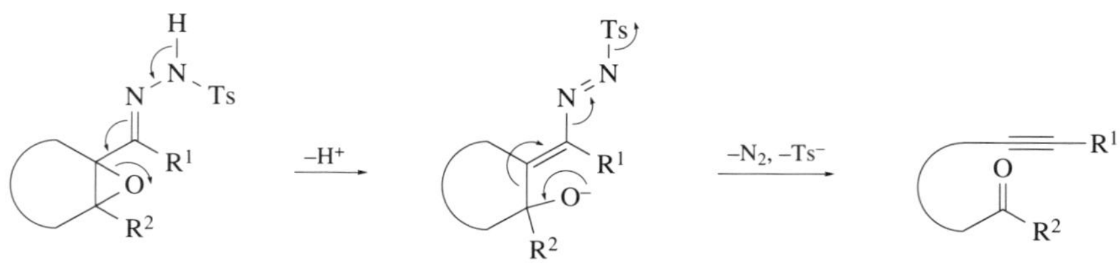

Scheme 124

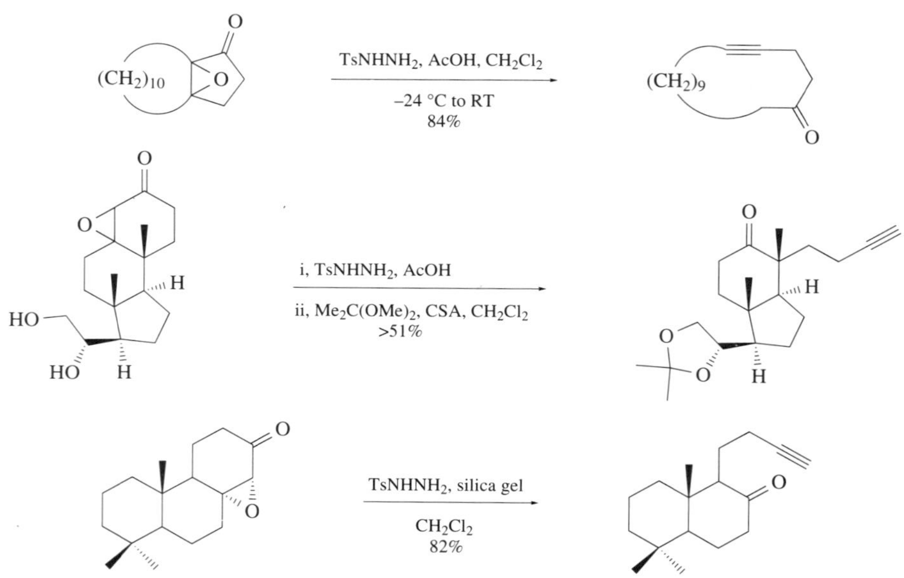

Scheme 125

The formation of epoxy ketones from sterically crowded enones is often difficult, but this step can be avoided by fragmentation of α,β-unsaturated tosylhydrazones using NBS in alcohol solvents. The reaction has been applied to the preparation of the natural products muscone and exaltone (Scheme 126) ⟨79HCA2655⟩.

A variant of the Eschenmoser fragmentation employing the cleavage of a cyclopropyl substituted tosylhydrazone has been reported to afford moderate yields of cyclic alkenynes, sometimes contaminated with the isomeric allene (Scheme 127) ⟨71HCA1789⟩. The reaction can be undesirable, as in the formation of alkyne (**52**) during a synthesis of sesquicarene ⟨70T3487⟩.

The nucleofuge in the Eschenmoser fragmentation is N_2 (and toluenesulfinate ion). Analogous fragmentations employing an allylic alkoxide with halide ⟨77JOC2380⟩, tosylate ⟨84JA2735, 86JOC3214⟩

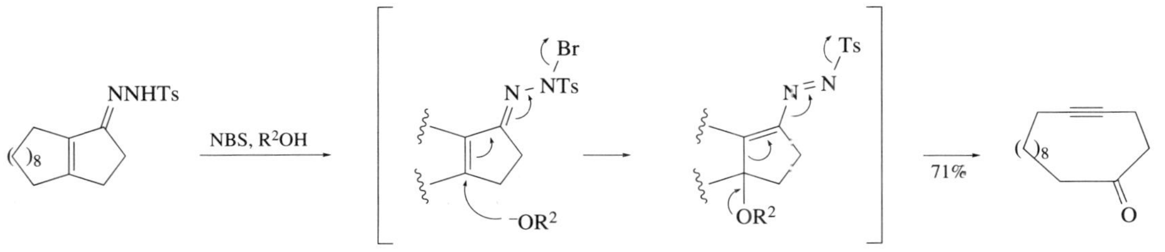

Scheme 126

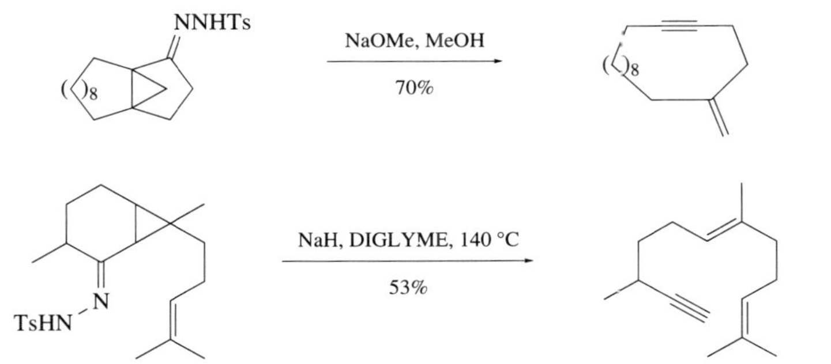

(52)

Scheme 127

or selenone ⟨84JOC1230⟩ leaving groups have been reported (Scheme 128) ⟨77JOC2380, 84JA2735, 84JOC1230⟩. The allylic alkoxide intermediate has been generated by base treatment of the alcohol or by organometallic addition to the corresponding ketone.

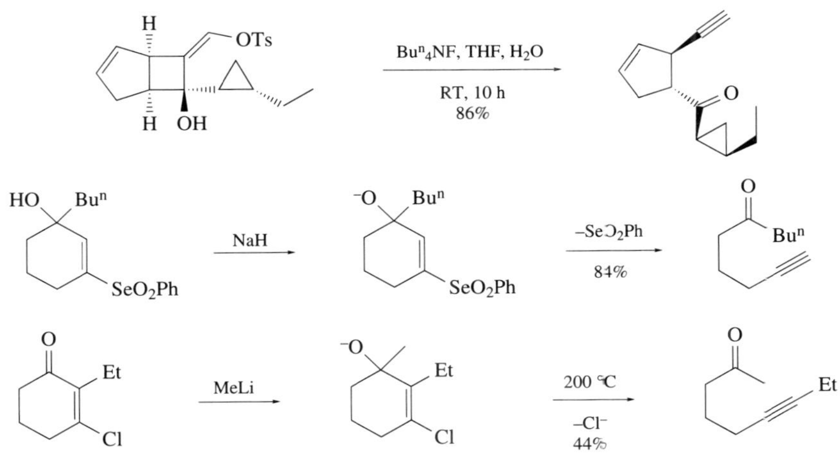

Scheme 128

1.21.3.3 Formation by Pericyclic Reactions

Alkynes have occasionally been prepared by pericyclic reactions, generally retrodiene cleavage reactions. Aside from the trivial case where an alkyne is regenerated from its cycloaddition adduct, reversal of the Diels–Alder reaction can lead to products where the diene and dienophile have exchanged substituents (Scheme 129). An example is the formation of dicyanoethyne and dimethylfurandicarboxylate from the reaction of dimethylethynedicarboxylate with 3,4-dicyanofuran at 175–180 °C ⟨62JOC3520⟩.

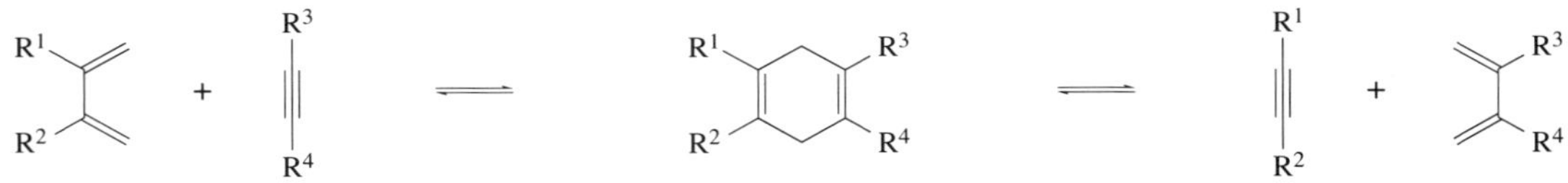

Scheme 129

The ketene–anthracene cycloaddition adduct serves as a good precursor in a synthesis of mono- and disubstituted alkynes (Scheme 130) ⟨81CL1241⟩. The key step, a retro Diels–Alder reaction, is achieved by flash vacuum pyrolysis at 500–550 °C. A retrodiene cleavage of (**53**) has also been used to generate cyclohexyne ⟨63CB342⟩.

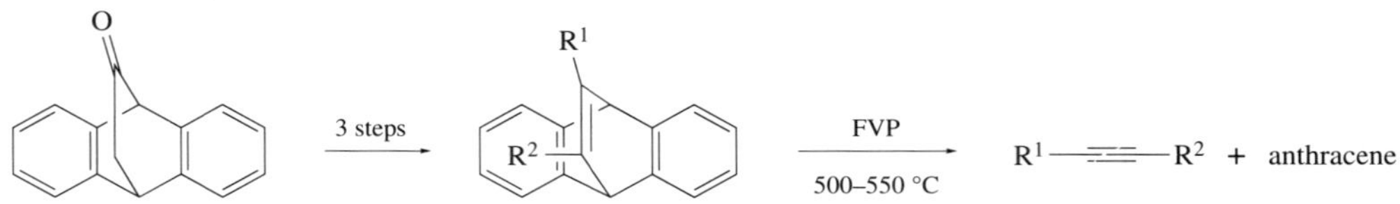

Scheme 130

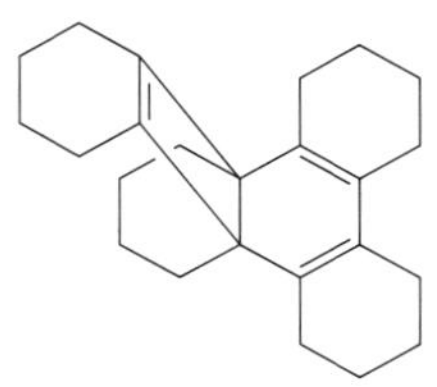

(53)

1.21.3.4 Formation by Isomerisation Reactions

Alkynes are able to undergo isomerisation reactions in the presence of strong bases as long as there is at least one hydrogen atom attached to a carbon atom adjacent to the triple bond. The equilibrium in linear alkynes has been shown to heavily favour the internal isomers and 1-alkynes can often be cleanly isomerised to the thermodynamically more stable 2-alkynes using, for example, NaH or KOBu^t in DMSO (Scheme 131) ⟨70T2127, 87CL1905, B-88MI 121-01⟩. Isomerisation to 3-alkynes can be achieved employing stronger bases although the comparable energies of 2- and 3-alkynes leads to product mixtures.

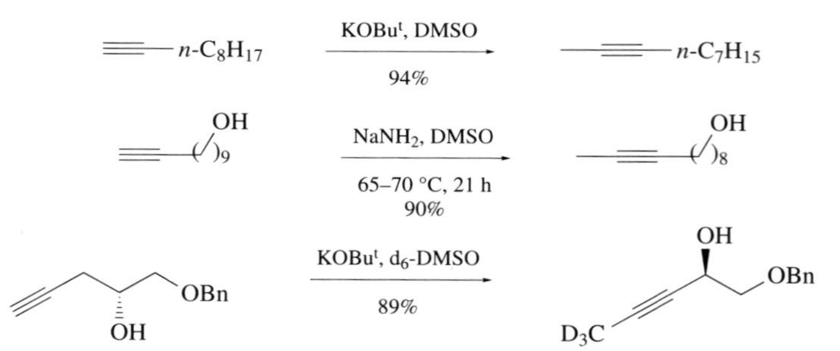

Scheme 131

Contrathermodynamic isomerisations of internal to terminal alkynes are dependent on displacement of the equilibrium by precipitation of the metal alkynide ion. The isomerisation of

2-alkynes to 1-alkynes has been achieved with strong bases such as $NaNH_2$ in aromatic hydrocarbons at high temperature, and with KNH_2/NH_3/HMPA at room temperature, but isomerisations over two or more carbon atoms are less successful. Highly effective multipositional isomerisations of alkynes are possible employing potassium 3-aminopropylamide (KAPA) in excess amine solvent (the 'Acetylene Zipper Reaction') (Equation (150)) ⟨75JA891⟩. Addition of internal alkynes to this base system at room temperature generally results in almost instantaneous precipitation of the potassium alkynide in very high yields. Alkynols ⟨76CC959⟩, alkynyl boranes ⟨77CC318⟩ and some alkynyl ethers ⟨88S707⟩ also undergo successful isomerisation reactions. The isomerisation of optically-active secondary propargyl alcohols to terminal alkynols proceeds without loss of configuration at the hydroxyl centre ⟨81TL4171⟩. A slight variation of the procedure employing potassium *t*-butoxide and the lithium salt of either 1,2-diaminoethane or 1,3-diaminopropane has been reported ⟨84CJC1333⟩.

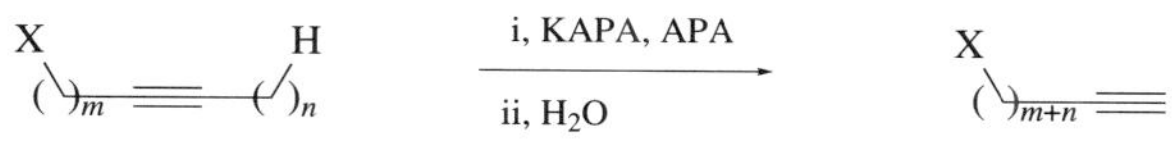

(150)

X = H	m = 2–6, n = 2–6	74–100%
X = OH	m = 1–6, n = 2–8	83–98%
X = BR_2	m = 3, n = 3	>95%

KAPA = potassium 3-aminopropylamine
APA = 3-aminopropylamine

Alkenyl sulfides undergo elimination and concomitant multipositional isomerisation to 1-alkynes in the presence of KAPA (Equation (151)) ⟨78JOC3083⟩. In a related reaction thiophenyl α-substituted tetrahydrofurans afford ω-alkyn-1-ols on treatment with KAPA in 3-aminopropylamine ⟨90SC2733⟩.

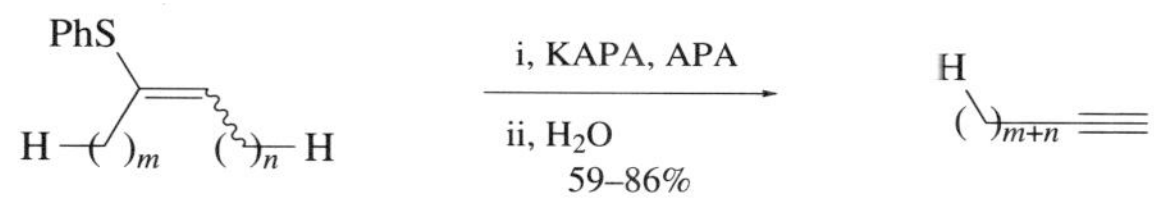

(151)

1.22
Ions, Radicals, Carbenes and Other Monocoordinated Systems

JULIA M. DICKINSON
Manchester Metropolitan University, UK

1.22.1 DICOORDINATE CARBANIONS (i.e., VINYLIC CARBANIONS)

1.22.1.1 General Aspects of Carbanion Formation

A number of reviews and monographs that discuss carbanion chemistry in general, also discuss various aspects of vinylic carbanions, such as their formation, stability, stereochemistry and reactivity ⟨B-65MI 122-01⟩. The following subsections will deal principally with methods of vinyl anion formation and reactivity.

The method of formation of vinylic (alkenyl) carbanions, and their subsequent reactivity is dependent largely on the nature of the countercation. A number of metals have been found to act as the cationic component in these species, with alkenyllithium and alkenyl Grignard reagents being the most widely studied; other metals that have been employed include copper, aluminum, boron and tin. The nature of the metal counterion, as is to be expected, markedly alters the stability and reactivity of the alkenyl carbanion, but only those organometallic reagents that have found subsequent use in synthesis will be discussed. In addition, vinylic carbanions are often implicated as intermediates in elimination reactions during the conversion of alkenes to alkynes. A number of monographs exist which deal with this aspect of vinyl carbanion chemistry; such reactions will only be discussed briefly.

No matter what the method chosen for the generation of the vinylic carbanion, or the nature of the countercation in reactions employing such anions, a number of features are general to these processes. The most important of these features from a synthetic chemist's point of view is the nonlinearity of the vinylic carbanion and its overall retention of configuration. Carbonation reactions have been used to demonstrate this latter property. For example, the reaction of (*Z*)-2-bromobut-2-ene (**1**) with lithium to give the corresponding alkenyl carbanion (**2**), followed by treatment with carbon dioxide gave (*Z*)-2-methylbut-2-enoic acid (**3**) as the major isomer in a yield of ~75%, with the geometrical isomer being formed in <5% yield (Scheme 1). The nonlinearity of vinyl anions has also been demonstrated, for example, cyclopropene is able to exchange its vinylic protons readily ⟨71RTC1110⟩. A few vinyl carbanions have been characterised by x-ray diffraction analysis ⟨85CB1504, 88JA6033⟩.

In general there are three main methods by which vinylic carbanions may be formed, and these will be discussed in more detail in the following sections.

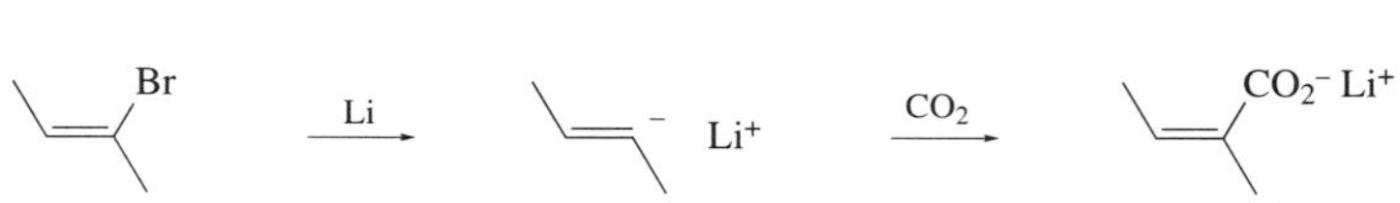

(1) (2) (3)

Scheme 1

1.22.1.2 Formation of Vinylic Carbanions by Proton Loss

The increase in *s* character of the sp^2 carbon atom of an alkene, as compared to an alkyl sp^3 carbon, leads to an increase in acidity of the attached proton. This proton can thus be removed by suitable bases to give the corresponding alkenyl carbanion. The presence of heteroatom-containing substituents on the double bond can lead to increased stabilisation of the vinylic carbanion, with some substituents stabilising a negative charge in the α position, whilst others stabilise a β carbanion. For example, methylvinyl ether reacts with *t*-butyllithium to give the α-lithio species ⟨74JA7125, 75JA3822⟩; higher homologues are also available if the methoxymethyl (MOM) analogue is employed and *n*-butyllithium in TMEDA is the lithiating species ⟨87JOC4817⟩. Cyclic analogues such as 6-lithio-2,3-dihydropyran can be obtained using the same methodology ⟨77TL4187, 84H(22)1489⟩. Sulfur and selenium containing substituents are often used to stabilise carbanionic centres but, whereas sulfur is more stabilising than selenium for sp^3 hybridized carbanionic centres, the reverse is true for sp^2 hybridized anions, i.e., the seleno moiety is more stabilising than the thio moiety ⟨80JOC5227⟩; this is attributed to the greater polarisability of the selenium atom. The stabilising effect of the heteroatom is modified further by the other groups attached ⟨83TL4825⟩; α-lithiovinylsulfones have also been prepared by direct deprotonation of the corresponding alkenes ⟨87TL4127, 87TL6709⟩. Vinyl selenide has been prepared, and alkylated in good yield with *n*-octyl bromide ⟨78AG(E)526, 82AG(E)410⟩. Vicinal- and geminal-dihalides can undergo elimination to give vinyl halides which, in the presence of a second equivalent of base, either undergo loss of a vinylic proton to give the corresponding anion (quenching with an electrophile then gives a substituted alkene) or eliminate a second halide ion to give an alkyne. An example of the former can be seen in the reaction of the carbamate (**4**) with two equivalents of lithium tetramethylpiperidine (LITMP) ⟨83JA6155⟩, followed by reaction of the intermediate α-lithiodifluorovinyl carbamate (**5**) with a series of electrophiles (Scheme 2) ⟨93TL415⟩. An example of alkyne formation via the above mentioned elimination sequence is seen in the synthesis of an analogue of 3′-azido-3′-deoxythymide (AZT) (**6**) (Scheme 3) ⟨90TL597⟩. Elimination of HX from vinyl halides via an E2 mechanism is well-documented.

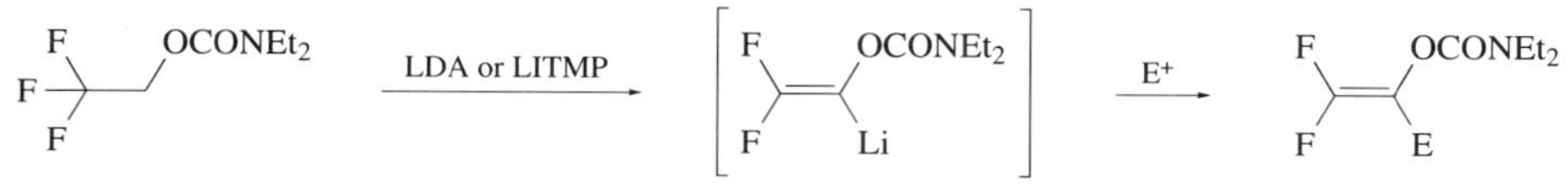

(**4**) (**5**)

E+	*E*	*Yield* (%)
TMS-Cl	TMS	80
PhSeBr	SePh	58
MeOTf	Me	71

LDA = lithium diisopropylamide
LITMP = lithium tetramethylpiperidine

Scheme 2

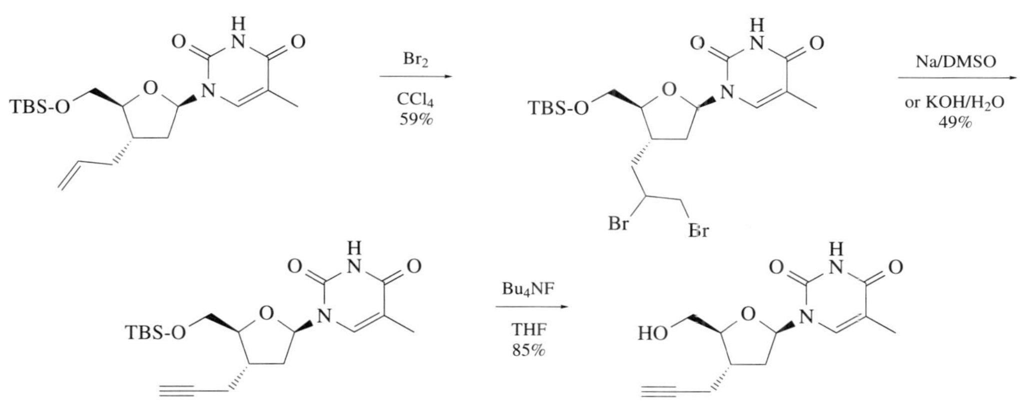

(**6**)

Scheme 3

When secondary allylic amines are treated with *t*-butyllithium, lithiation occurs at the vinylic position to give a γ-aminated organolithium compound (**7**). With tertiary amines, however, lithiation occurs at the allylic position, for example, (**8**) ⟨88CC1135, 88TL4859, 89JCR(S)200, 90CC587, 90CC1521, 91CC1001, 92TL831⟩. To overcome this problem, the ability of a sulfone group to direct metallations and stabilise the resulting sp^2 hybridised organolithium compound has been utilised. This allowed the reaction of (*E*)-*N*-(3-tosyl-2-propenyl)morpholine (**9**) with *s*-butyllithium at $-78\,^\circ$C to take place, giving the γ-functionalised organolithium compound (**10**) (Scheme 4) ⟨93TL2003⟩. Deuterolysis of (**10**) with deuterium oxide showed the decreasing stability of the anion with increasing temperature, such that at $-78\,^\circ$C, 80% of deuteriated product (**11**; E = D) was obtained; at $-20\,^\circ$C the yield had decreased to 70%, with only decomposition products being obtained at $0\,^\circ$C. Other electrophiles were also reacted with (**11**) to give, on samarium iodide reduction of the tosyl group, γ-substituted allylic amines. The regio- and stereoselective lithiation of vinyl sulfones has been reported on a number of occasions ⟨79JOC3277, 79JOC3279, 87TL989, 87TL4123, 87TL4127, 87TL6709, 89JCS(P1)1387, 89JOC1757⟩.

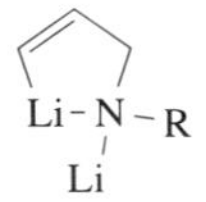

(7)

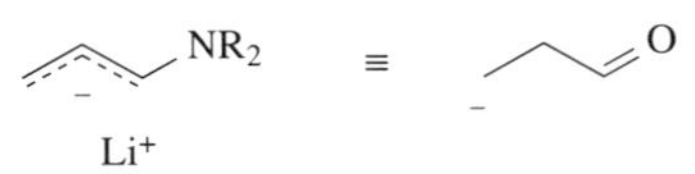

(8)

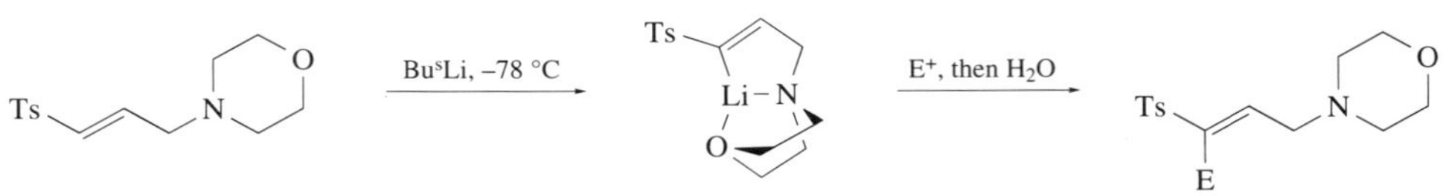

(9) **(10)** **(11)**

Scheme 4

1.22.1.3 Formation of Vinylic Carbanions by C—C and C—X Bond Scission

Vinylic halides are the common precursors to a number of alkenylmetallic reagents. However, vinylic halides are often less reactive than the corresponding alkyl halides because of the overlap between the filled *p* orbital of a halogen atom, and the *p* orbital of the double bond, leading to resonance stabilisation. Although the vinyl halides are less reactive than expected, they will undergo either metal–halogen exchange to give the corresponding alkenylmetallic reagents, or a metal may insert into the vinylic carbon–halogen bond to give an alkenylmetal halide. As in the case of carbanion formation through proton loss (Section 1.22.1.2), heteroatom-containing substituents have been found to stabilise the carbanion formed. Other alkenylmetal reagents can be prepared through metal–metal exchange, in which a preformed alkenylmetal derivative (usually the alkenyllithium or alkenyl Grignard reagent) reacts with another metal containing reagent (with or without additional organic substituents), usually a halide, to give the new alkenylmetal and the corresponding lithium or magnesium halide. These new alkenylmetallic reagents often exhibit modified reactivity as compared to the lithium or magnesium precursors, namely enhanced chemo- or regioselectivity.

1.22.1.3.1 Alkenyllithium reagents

Vinyllithium compounds may be prepared stereospecifically using dispersed lithium ⟨75S434, 76OS(55)103⟩. Certain vinyllithium reagents that are unstable towards elimination or rearrangement at ordinary temperatures can be prepared at low temperatures, with terminal vinyllithium reagents being prepared stereospecifically from the corresponding 1-iodo-1-alkenes ⟨76S245, 76S761, 77JOC260⟩. Halogen–metal exchange of vinyl halides is also achieved efficiently using *t*-butyllithium as the source of lithium ⟨76TL4839, 78CB2785⟩. The lithiated derivatives are generated using two equivalents of *t*-butyllithium at $-120\,^\circ$C in a THF–pentane solvent mixture, followed by quenching of the anion at $-78\,^\circ$C with a suitable nucleophile. Such a method has been used, for example, to generate alkenylsilanols which were then employed efficiently in a Sharpless epoxidation process which did not require an allylic alcohol for a display of stereoselectivity (Scheme 5) ⟨88CC1280⟩. Alkenyllithium reagents generated from alkenyl halides and butyllithium will undergo intramolecular cyclisations to give 3-, 4-, 5- and 6-membered rings ⟨78TL5193, 86CL459⟩ (e.g., Equation (1)). Arylsulfonyl hydrazones have also been employed as precursors to vinyl anions; treatment of hydrazone (**12**), with *s*-butyllithium gave the corresponding alkenyllithium reagent (**13**) which could be treated with various electrophiles (Scheme 6) ⟨78JOC147⟩.

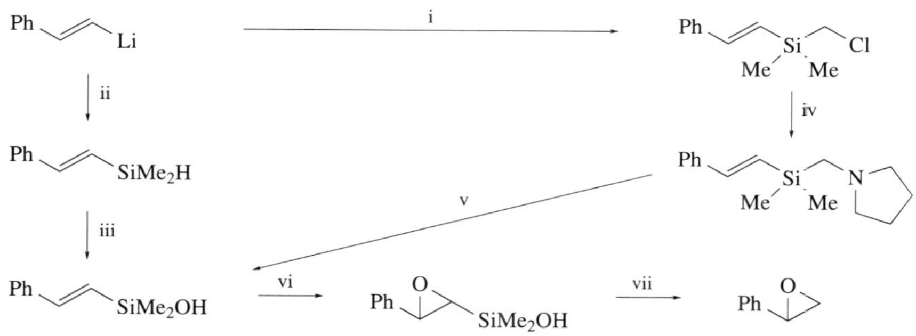

i, $Me_2Si(Cl)CH_2Cl$, Et_2O, –25 °C; ii, Me_2SiHCl, Et_2O, –20 °C; iii, $AgNO_3$, Ag_2O (1:6), THF, RT; iv, pyrrolidine, 80 °C; v, Bu^tO_2H, $VO(acac)_2$, acetone; vi, $Ti(OPr^i)_4$, (+)-DET, Bu^tO_2H, CH_2Cl_2, –20 °C; vii, Et_4NF, MeCN, RT

Scheme 5

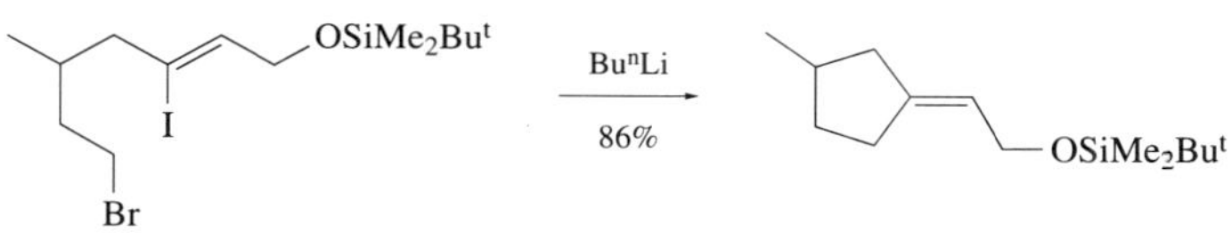

(1)

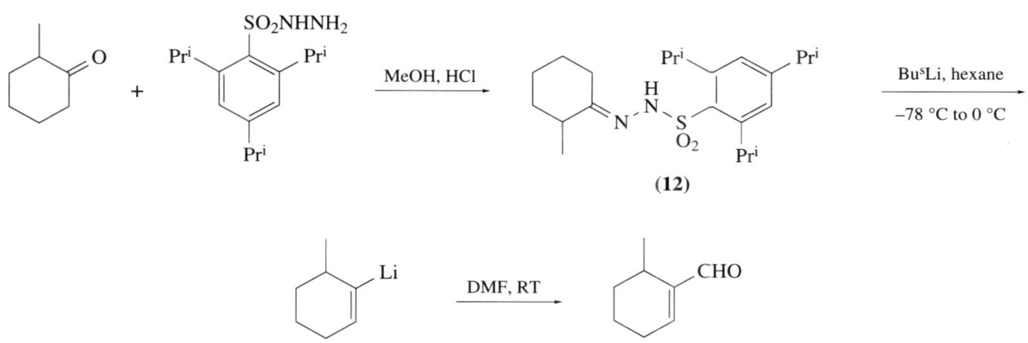

(**12**)

(**13**)

Scheme 6

Alkenyllithium reagents can also be obtained via a transmetallation process, vinyllithium itself may be prepared from the corresponding tin compounds by, for example, treatment of tetravinyltin with phenyl- or butyllithium ⟨61JA3583⟩. Substituted alkenyllithium reagents can be prepared from the products of hydrostannation of alkynes by exchange with *n*-butyllithium (Scheme 7) ⟨77JA7365, 77TL3847⟩. Exchange of tin for lithium in (**14**) generated the corresponding alkenyllithium derivative (**15**) which was alkylated in good yields by 1,8-dibromooctane (Equation (2)) ⟨75JOC2265⟩. A cyclopentenyllithium reagent has been generated via this method ⟨81TL1437⟩, whilst *trans*-1,2-bis(tri-*n*-butylstannyl)ethylene, on treatment with butyllithium, gives the corresponding alkenyllithium reagent which is readily alkylated by a range of electrophiles ⟨74JA5581⟩; such methodology has been applied to the synthesis of 5,6-dehydroarachidonic acid ⟨84TL3437⟩. The β-lithiosilane has been generated by Sn–Li exchange, and can be alkylated in good yields, without showing any tendency to undergo elimination ⟨76JOC1480⟩. Other examples of the generation of alkenyllithium reagents via tin–lithium exchange have been reported in the literature ⟨79JOC3437, 80JOC4263, 84JOC3438, 84TL3155⟩.

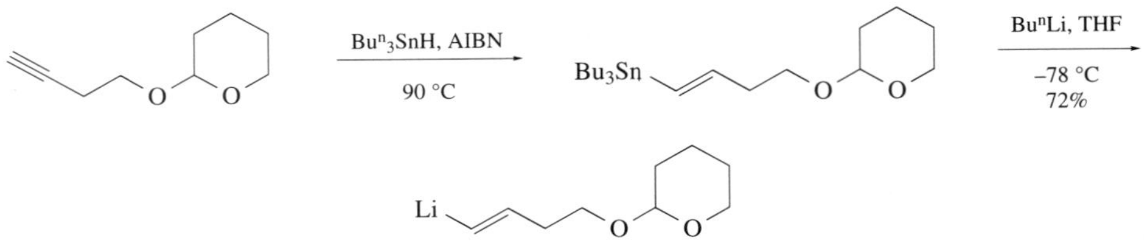

Scheme 7

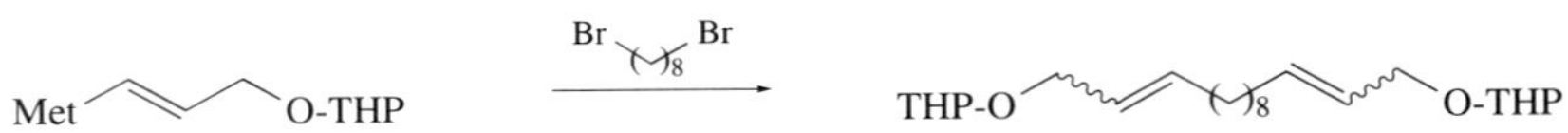

(2)

(**14**) Met = $SnBu_3$
(**15**) Met = Li

1.22.1.3.2 *Alkenylmagnesium reagents*

The usual methods of obtaining Grignard reagents, using ether as a solvent, are not applicable to the vinyl halides, because of competing elimination reactions leading to alkenes and alkynes. The use of tetrahydrofuran as solvent has greatly improved the process, due to its stronger basicity as compared to ether, helping to solvate the vinylmagnesium halides as they form. Reaction conditions are otherwise similar to the normal protocol adopted for Grignard reactions, with the best results being found with alkenyl bromides, and on utilising saturated ammonium chloride solution for workup procedures ⟨60MI 122-01⟩. As already mentioned, subsequent reactions of these vinylic carbanions proceed with retention of configuration about the double bond. For example, vinyl, 1-alkenyl and 2-alkenyl Grignard reagents couple with alkyl bromides or iodides in the presence of CuI or $AgNO_3$ to give the corresponding alkenes (Equations (3) and (4), Scheme 8 and Equation (5)) ⟨71JA1483, 76TL3225, 79TL1073, 83TL1041⟩. The 2-butadienyl Grignard reagent (**16**) couples with alkyl bromides and iodides in the presence of functional groups such as esters, in the presence of a copper catalyst (Equation (6)) ⟨83JOC1912⟩. Coupling reactions between vinyl Grignard reagents and 2-oxy secondary alkyl bromides are catalysed by copper or iron salts (Equations (7) and (8)) ⟨78TL3711, 83CB882⟩. Alkenyl, silylalkenyl and 2-butadienyl Grignard reagents have been used for the coupling of primary alkyl tosylates in the presence of a copper catalyst (Equation (9)) ⟨76TL3225, 79TL1073, 83TL1003⟩; the coupling of optically active tosylates proceeds without racemisation (Equation (10)) ⟨83TL1463⟩. Ring opening of epoxides by alkenyl Grignard reagents is greatly assisted by copper catalysis (Equations (11) and (12)) ⟨79TL1503, 82JOC3630, 82TL1267⟩.

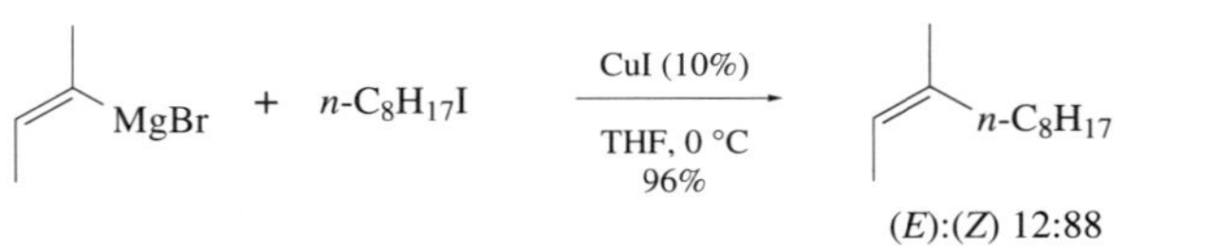

(3)

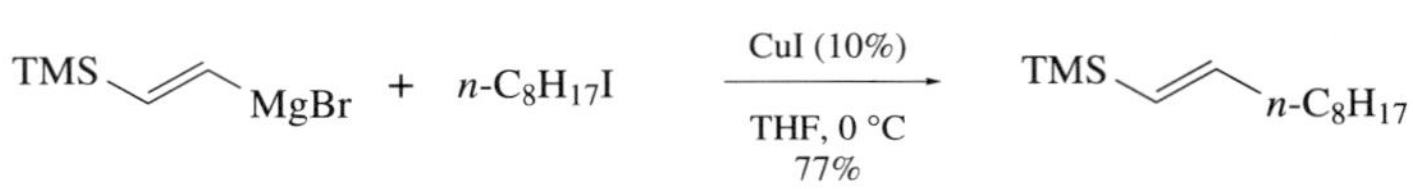

(4)

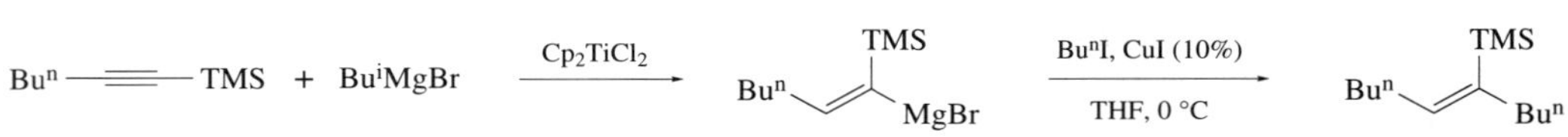

Scheme 8

(5)

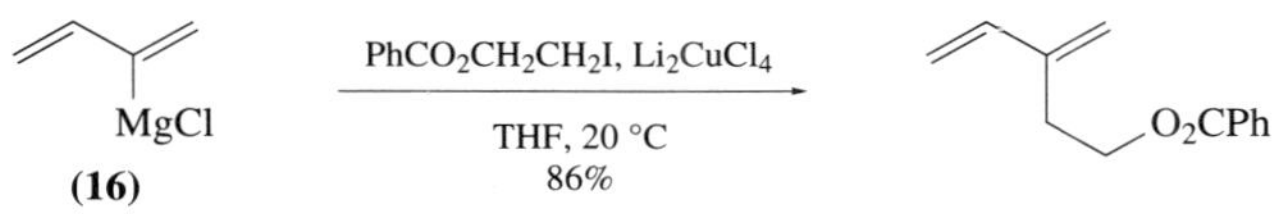

(6)

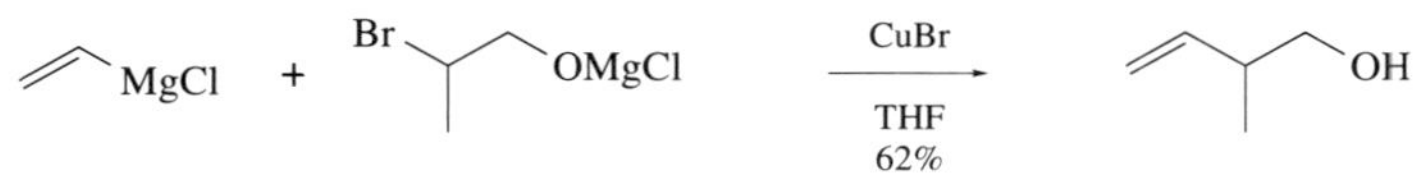

(7)

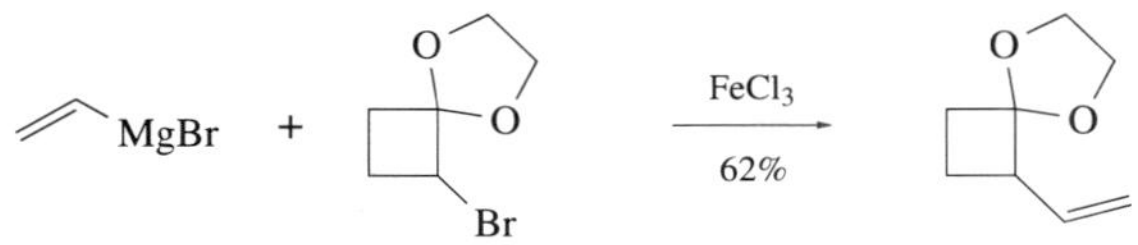

(8)

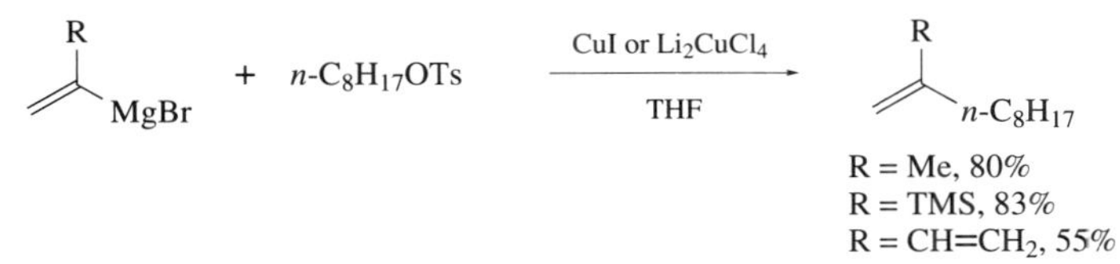

(9)

(10)

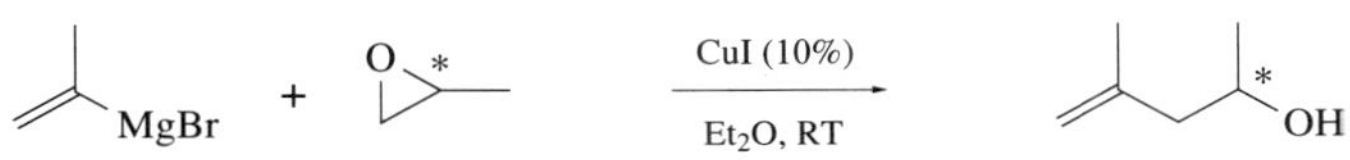

(11)

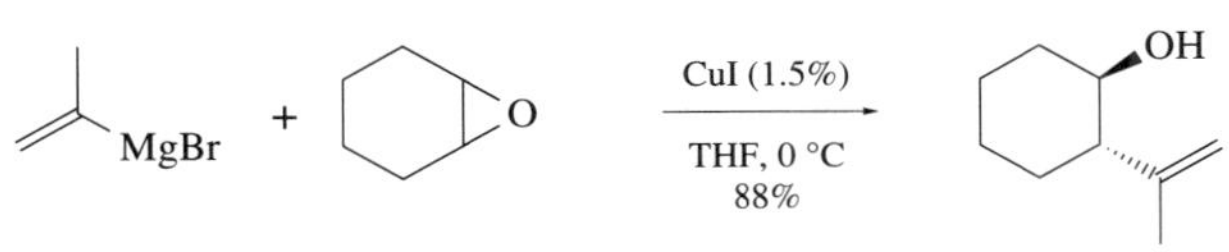

(12)

1.22.1.3.3 Alkenylcopper reagents

Alkenyllithium cuprates ⟨81S841⟩ are formed readily from reaction of the corresponding alkenyllithium reagents with copper(I) iodide, and will undergo conjugate additions to α,β-unsaturated carbonyl systems. This methodology has been employed in, for example, the synthesis of a *cis*-fused δ-lactone **(17)** (Scheme 9) ⟨75JOC2970⟩ and prostaglandin synthesis; treatment of the lithium reagent with an alkynylcopper complex gave the mixed cuprate **(18)** which was employed in a conjugate addition to cyclopentenone, for the eventual synthesis of 11-deoxy-prostaglandin E_2 **(19)** (Scheme 10) ⟨75JOC2265⟩. Ultrasonication has also been used to generate alkenyllithium cuprates from the alkenyl bromide, lithium metal and copper(I) iodide ⟨82JOC3805⟩. The corresponding magnesium cuprates can be formed from a Grignard reagent and copper(I) iodide, and will undergo similar reactions ⟨76JOC3144⟩. Lithium divinylcuprate itself has been prepared using dimethyl sulfide as solvent, giving a soluble complex with the cuprous halide. The copper complex $CuBr \cdot Me_2S$ may also be conveniently used in excess dimethyl sulfide ⟨75JOC1460, 76JOC1396⟩.

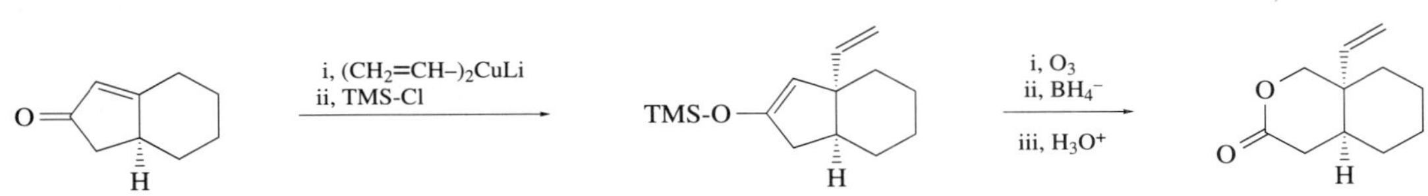

(17)

Scheme 9

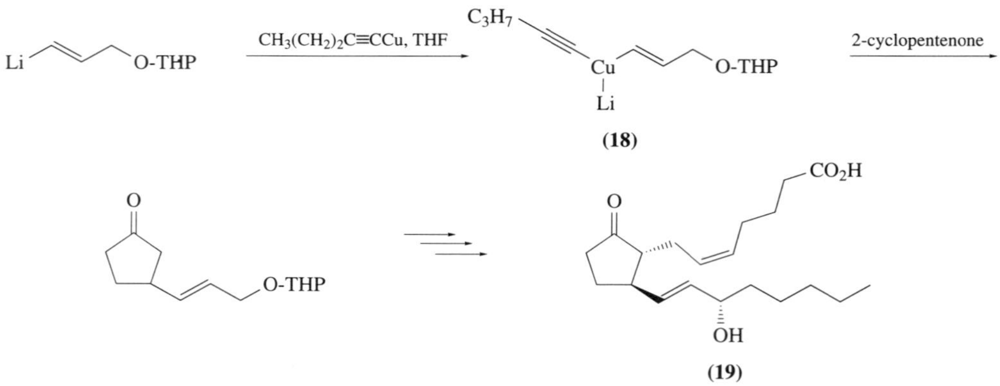

(18)

(19)

Scheme 10

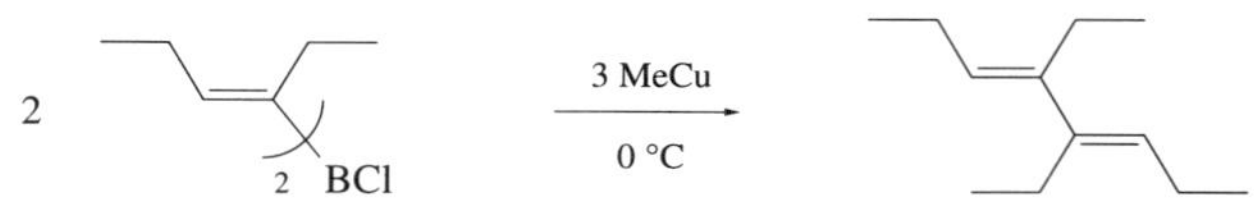

(13)

Dialkenylchloroboranes undergo coupling to dienes in excellent yields (and with a high degree of retention of configuration) in the presence of methylcopper (Equation (13)) ⟨76JA1785⟩; the reaction presumably proceeds via an alkenylcopper intermediate. Milder conditions can be employed through the use of a sodium methoxide addition compound of an alkenyldialkylborane ⟨80JOC549⟩. When such reactions are carried out at low temperatures, the vinylcopper intermediate is sufficiently stabilised for it to be trapped by allylic halides, thus allowing the stereochemically defined synthesis of 1,4-dienes (Equation (14)) ⟨80JOC550⟩; a variety of functional groups can be tolerated in these reactions. Copper reagents formed in this way also undergo 1,4-conjugate addition to α,β-unsaturated carbonyl systems to give γ,δ-unsaturated carbonyl compounds ⟨81MI 122-01⟩. Reaction between a dialkylvinylborane and $CuBr \cdot Me_2S$ also generates vinylcopper boranes ⟨80BCJ1471, 80JOC550⟩.

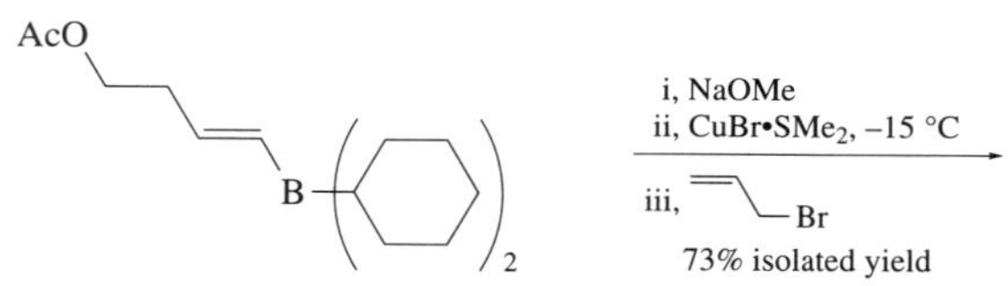

(14)

α-Silylvinylcuprates have been found to be useful synthetic intermediates. They can be prepared either by the addition of an organomagnesiocuprate reagent to a silylalkyne (Equation (15)) ⟨77TL159, 77TL1805, 77TL1823⟩ or by a boron–lithium exchange in the presence of copper(I) iodide to give the alkenylcuprate of complementary stereochemistry (Equation (16)) ⟨76JOC2941, 77T2987⟩. These silicon substituted alkenyl carbanions can be alkylated in good yields by primary alkyl halides and tosylates, and allylic halides. Intramolecular cyclizations are also possible ⟨76S801, 76TL2287, 76TL2947, 77TL4145, 78JOC147, 79S44, 81JOC4600, 81JOC4608, 83ACR55, 83JA6344, 84TL4901, 86CL459, 87TL5793⟩, with one example being illustrated (Equation (17)) ⟨85TL5671⟩. Reactions of vinylic carbanions with epoxides usually rely on either catalytic or stoichiometric amounts of copper to be present, THF or ether being the solvents of choice. Copper(I) iodide or copper(I) cyanide can be employed in the catalytic reactions, whilst organolithiocuprates are used for stoichiometric reactions ⟨75T2735, 84T641, 87JOC4412⟩. An example of such an epoxide opening is the addition of the vinylsilane (**20**) to an epoxide in the presence of copper(I) iodide, to give the homoallylic alcohol (**21**) (Equation (18)) ⟨78CL773⟩. A number of vinyllithium reagents react with epoxides in good yields, if converted to the lithium divinylcuprate first (Equation (19)) ⟨83TL3905, 84CC151⟩.

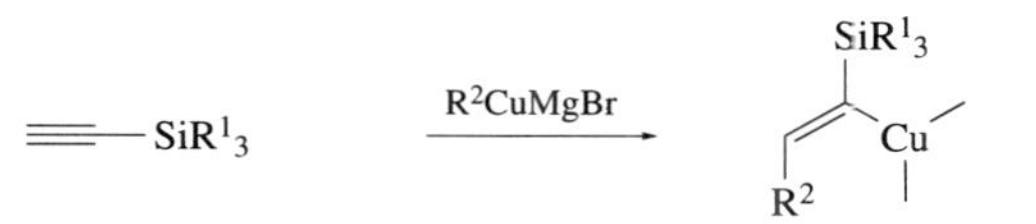

(15)

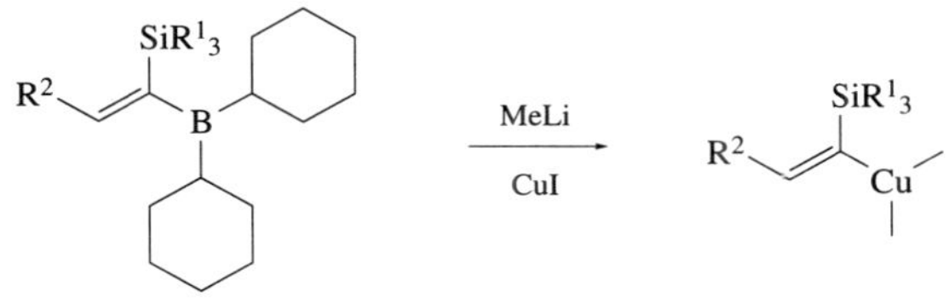

(16)

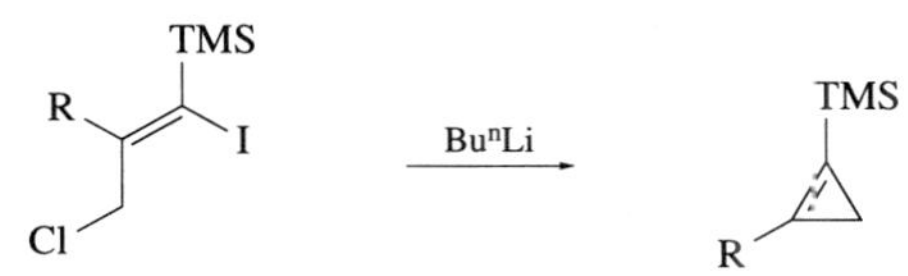

(17)

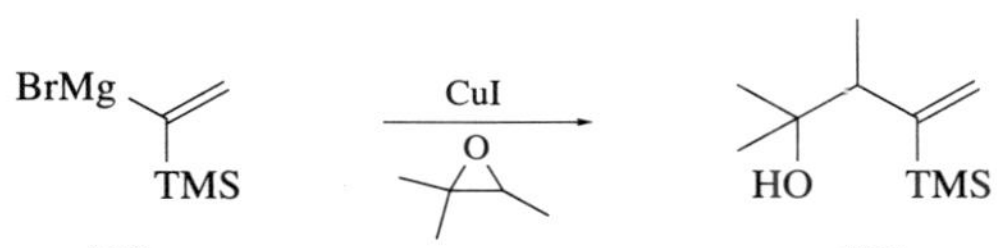

(18)

(20) **(21)**

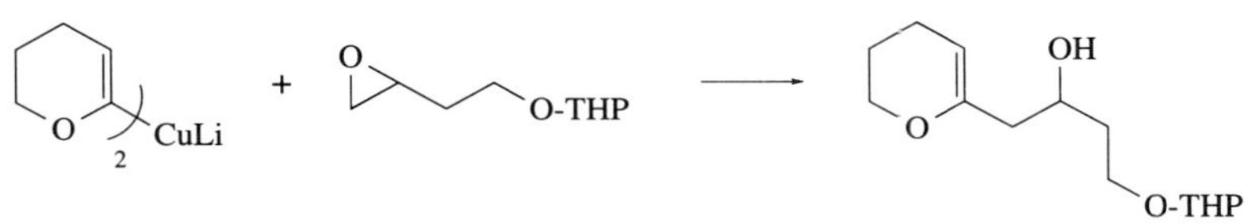

(19)

1.22.1.3.4 Alkenylmercury reagents

Alkenylboranes can be mercurated with, for example, mercuric acetate to give the corresponding alkenylmercurials, with retention of configuration about the double bond (Equation (20)) ⟨72JA4371, 72JOM(36)1⟩. The resultant organomercurials can be utilised in a number of synthetic processes such as carbon–carbon bond forming reactions ⟨78AG(E)27⟩.

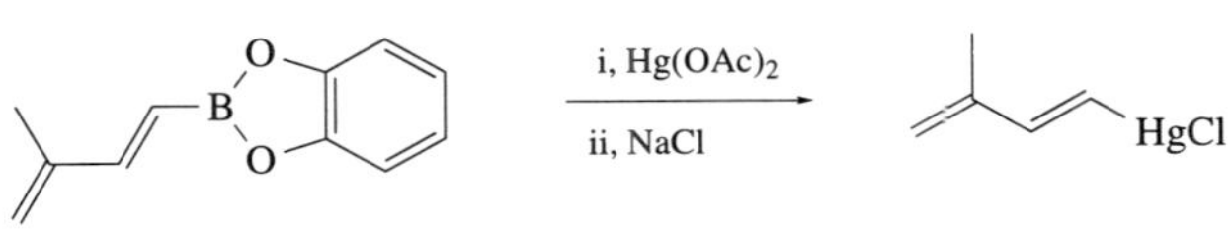

(20)

Vinylic mercurials react in a similar manner to the arylmercurials ⟨82T1713, B-85MI 122-01⟩. For example, in the presence of $PdCl_2$, coupling occurs to give symmetrical 1,3-dienes, presumably via a divinylpalladium species which reductively eliminates the diene (Equation (21)) ⟨76JOC2241⟩. Lithium chloride is a necessary component for such reactions, which are stoichiometric in palladium(II), and are usually carried out in hexamethylphosphoramide (HMPA) solution at 0°C. As has been seen in other cases the stereochemistry of the original alkene is retained in the final coupling product. However, the reaction is of limited use because alkenylboranes, from which alkenylmercurials are obtained, also undergo such reactions, the reaction is not catalytic, and there is the inherent toxicity of the mercurials and HMPA to be considered. These coupling reactions may be carried out catalytically with some rhodium catalysts, however, with mercury(II) acting as the reoxidant ⟨77JOC1680⟩. Another useful procedure is the bis(1,5-cyclooctadiene)-nickel(0) coupling of alkenyl halides ⟨72JA9234⟩.

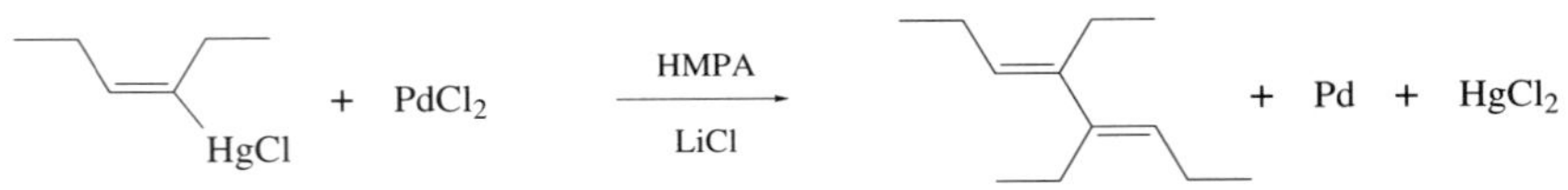

(21)

HMPA = hexamethylphosphoramide

In the absence of added nucleophiles, and in the presence of a rhodium catalyst, alkenylmercurials dimerise onto carbon monoxide to give divinyl ketones, which are otherwise very difficult to prepare (Equation (22)) ⟨80JOC3840⟩. Enol esters can be obtained in reasonable yields on reaction of alkenylmercurials with mercuric carboxylates (Equation (23)) ⟨71JOC1886⟩.

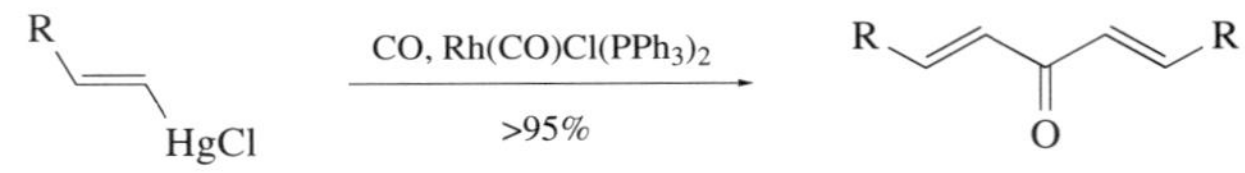

(22)

R = Bu^t, cyclohex-2-enyl, Ph

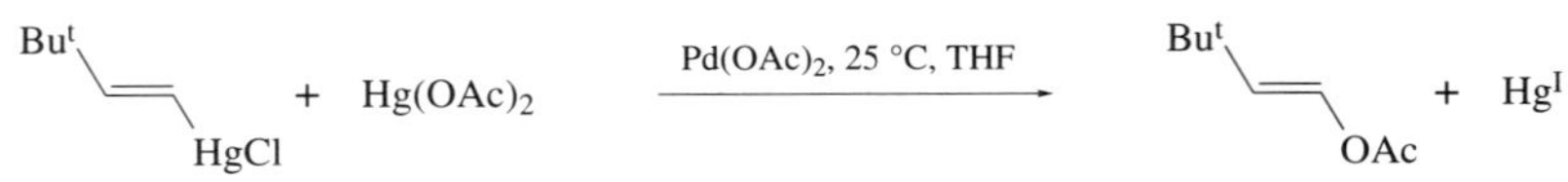

(23)

1.22.1.3.5 *Alkenylpalladium reagents*

Vinyl halides will undergo Heck-type coupling reactions, exhibiting the same high stereoselectivity and generality as found in the reactions of aryl halides ⟨86AG(E)508⟩. A number of palladium catalysts can be utilised, although typically Pd(OAc)$_2$ is used in conjunction with tri-*o*-tolylphosphine; triethylamine is added to scavenge the HX produced. Other triarylphosphines can also be used in these reactions, but the aforementioned phosphine is found to give faster reaction rates ⟨78JOC2941⟩. The reaction is thought to proceed via an alkenylpalladium halide species which undergoes *syn* addition to an unsaturated system, followed by *syn* β-elimination of palladium hydride, leading to overall retention of configuration of both the alkenes employed (Equations (24)–(26)) ⟨81JOC1061, 81JOC1067⟩. The regiochemistry of addition of the organopalladium species to the alkene is mainly sterically controlled, with addition taking place at the less substituted carbon atom. However, in some cases electronic factors may also become important, and occasionally oppose the steric effect, leading to mixtures of regioisomers. If an electron-withdrawing group is present on the alkene, addition is usually β to this moiety, with the product obtained being the (*E*) isomer. Coupling with allylic alcohols, in the presence of piperidine or morpholine, gives 4-enones or 4-enals ⟨82JOC1267⟩. Reaction with acrolein acetals under similar conditions gives 2,4-dienals, mainly as the (*E*,*E*)-isomers (Scheme 11), whilst reaction with 3-buten-2-one ethylene ketal gives the corresponding dienones ⟨81JOC1061⟩.

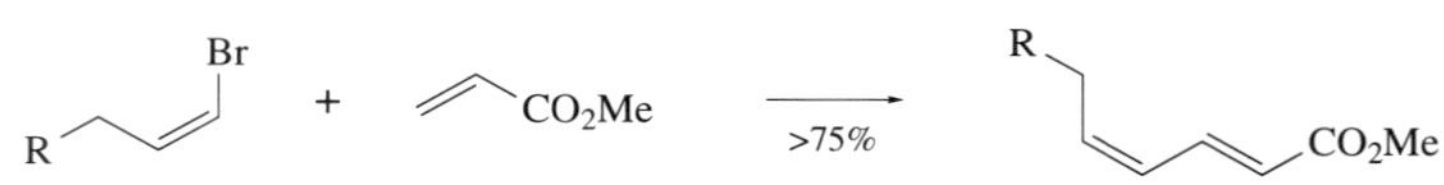

(24)

(25)

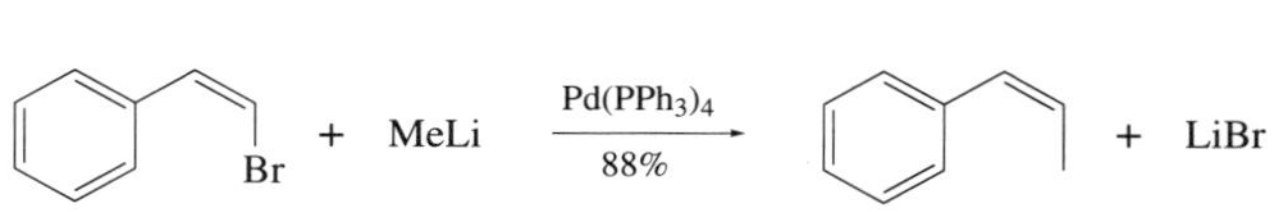

(26)

Vinylic halides will also couple with various vinylic organometallic reagents in the presence of a palladium catalyst. Again, the reaction is thought to proceed by way of oxidative addition of the vinylic halide to a palladium(0) catalyst to give the alkenylpalladium halide, followed by vinylation by the organometallic (Scheme 12). Vinylic Grignard reagents, alanes, boranes, copper and zirconium reagents have all been employed in such reactions, and some will have been dealt with in the appropriate sections.

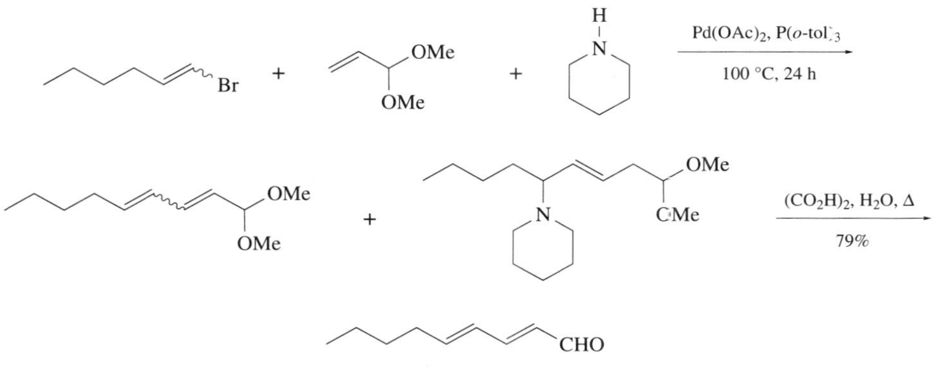

Scheme 11

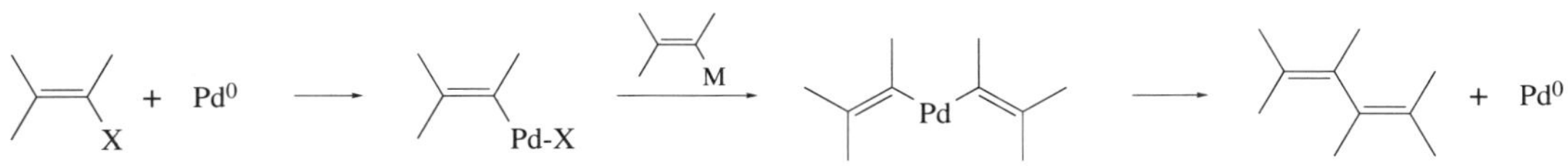

Scheme 12

Vinylic halides will couple directly with alkyllithium and alkyl Grignard reagents at reflux temperature in benzene solution in the presence of $Pd(PPh_3)_4$ catalyst to give alkenes in good yield, and with retention of configuration about the double bond ⟨79JOC2408⟩. Again the reaction is thought to proceed via an alkenylpalladium halide species. The reaction has been applied to the synthesis of several pheromones ⟨83S130⟩. Chiral coupling products can be formed from secondary alkyl Grignard reagents and vinylic halides when a chiral ferrocenylphosphine palladium chloride (or nickel chloride) catalyst is employed; enantiomeric excesses of up to 68% have been reported ⟨82JA180⟩. Vinylic iodides couple with vinylic Grignard reagents with high stereoselectivity in the presence of $Pd(PPh_3)_4$ catalyst, allowing the preparation of (*E*,*E*)- and (*E*,*Z*)-dienes ⟨78TL191, 83S130⟩.

Vinyl halides will undergo palladium-catalysed coupling with alkenylboranes (prepared from alkynes, Section 1.22.1.4.1) to give dienes. This approach has been employed in the synthesis of (*S*)-(−)-solanone (**22**) (Equation (27)) ⟨81CC951⟩. The alkene geometry of the intermediate alkenylborane is retained in the coupling reaction.

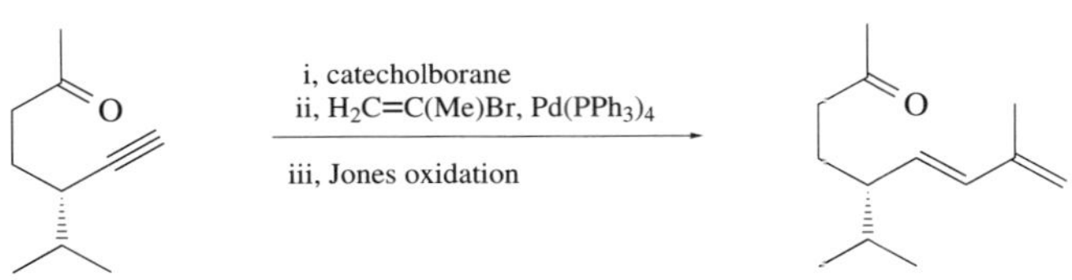

(27)

(22)

1.22.1.3.6 Alkenylzinc reagents

Grignard reagents have been used extensively to prepare alkenyl and dialkenylzinc reagents (Equation (28)) ⟨88TL5645⟩. Alkenylzinc compounds, obtained from the corresponding alkenylaluminum or alkenylzirconium compounds and zinc chloride, couple with aryl iodides in the presence of a $Pd(PPh_3)_4$ catalyst to give the corresponding aryl alkenes (Equations (29) and (30)) ⟨78JA2254⟩. Lithium dialkenylcuprates will couple with vinyl halides in the presence of zinc bromide and a $Pd(PPh_3)_4$ catalyst, with high yield and stereoselectivity, presumably via the alkenylzinc reagent (Equation (31)). Alkenylzinc reagents have also been prepared from the corresponding alkenylcuprates *in situ*, and coupled to aryl iodides stereospecifically using $Pd(PPh_3)_4$ as catalyst (Scheme 13) ⟨81TL3851⟩.

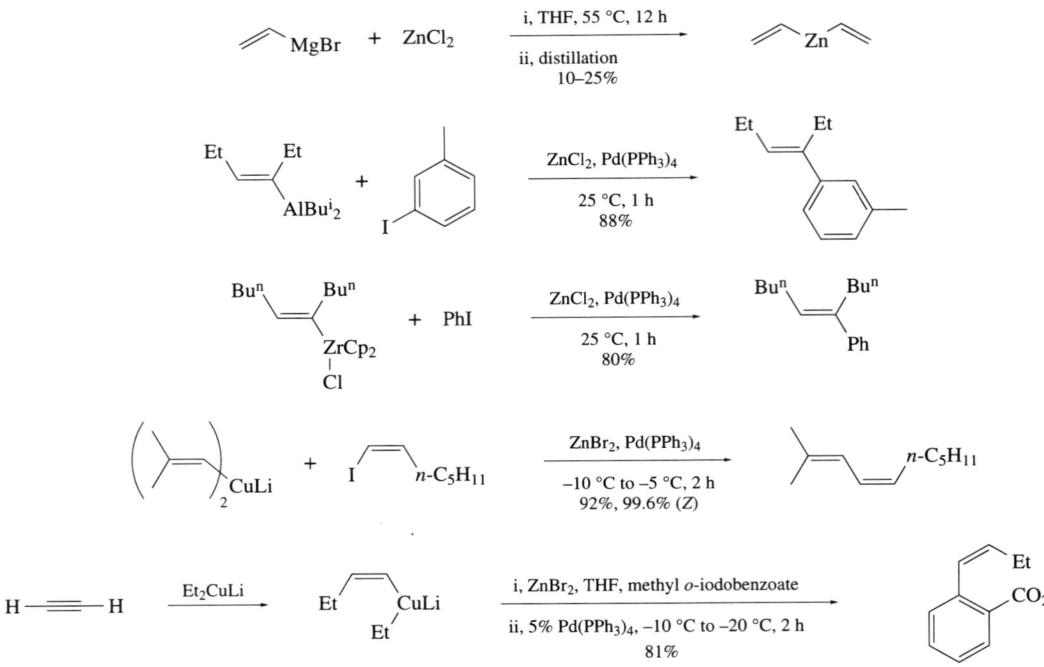

(28)

(29)

(30)

(31)

Scheme 13

1.22.1.3.7 Alkenylcerium reagents

Alkenylcerium reagents are prepared by treating the corresponding organolithium compounds with cerium chloride or cerium iodide (Equation (32)). The chloride is used in preference, as it is more readily available. The organocerium reagent obtained is represented as $RCeX_2$, although the exact structure is unknown ⟨82CC1042, 84JOC3904, 84TL4233⟩.

$$RLi + CeX_3 \xrightarrow[X = Cl,\ I]{THF} RCeX_2 + LiX \qquad (32)$$

Alkenylcerium reagents add to ketones to generate the corresponding alcohols with high diastereoselectivity (Equation (33)) ⟨84JOC3904, 84TL4233⟩, and have been employed in the stereoselective synthesis of polycyclic molecules (Scheme 14) ⟨88JA879, 89JOC2278, 90JA277, 90JA5562⟩. α-Trimethylsilylvinylcerium reagents add to readily enolisable ketones in high yields; this property has allowed the synthesis of (−)-eldanolide (**23**) (Scheme 15) ⟨85TL861⟩. β-Silylvinylcerium reagents have also been employed in synthesis (Equation (34)) ⟨89TL647⟩; in the example given, the (*E*)-cerium reagent adds to the ketone (**24**) as shown to give the alcohol (**25**) as a single adduct, whereas the (*Z*)-cerium reagent does not react, presumably due to steric effects.

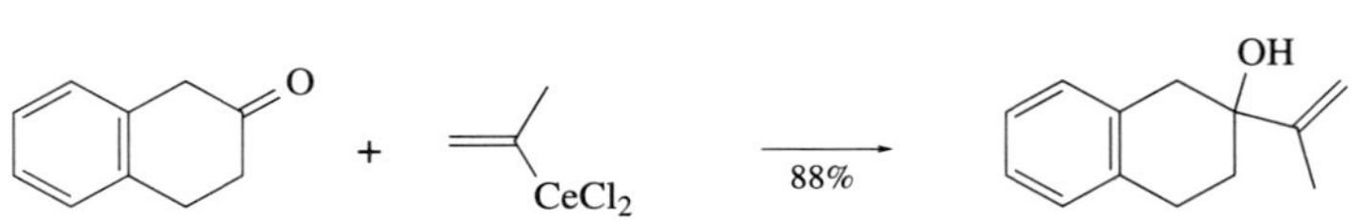

(33)

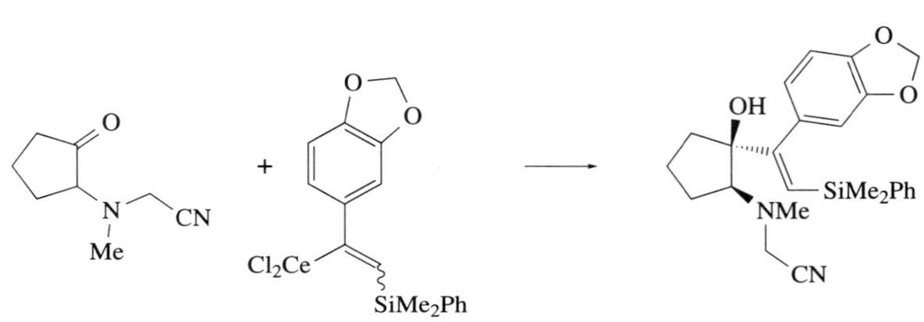

(34)

(**24**) (*E*):(*Z*) 50:50 (**25**)

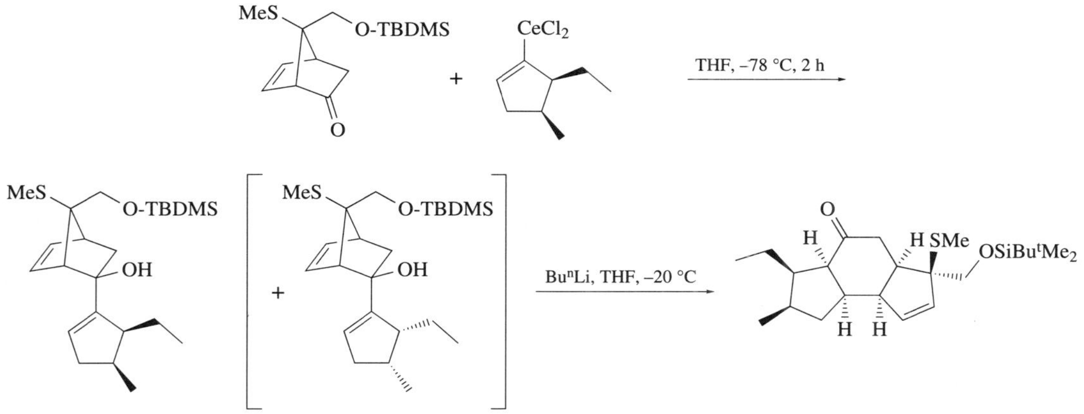

Scheme 14

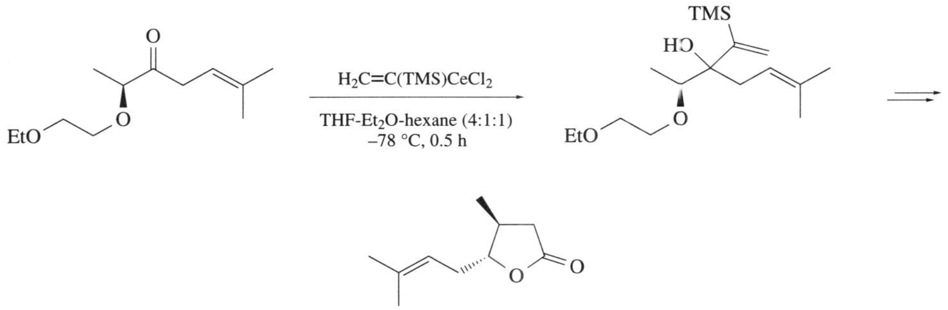

(23)

Scheme 15

1.22.1.3.8 Alkenylchromium reagents

Anhydrous chromium(II) chloride transforms vinyl halides to the corresponding alkenylchromium compound which undergoes addition reactions to aldehydes (Equation (35)) ⟨83TL5281⟩. Aldehydes react more readily than ketones; the stereochemistry of the original alkenyl halide is retained. Alkenylchromium reagents can also be obtained from enol triflates in the presence of a nickel catalyst (Equation (36)) ⟨86JA5644, 86JA6048⟩; a catalytic cycle for such reactions has been proposed ⟨72JA9234, 79JA7547⟩. When vinyl iodides are employed, palladium(II) salts also catalyse the reaction. The use of vinyl triflates as precursors to vinyl anions is unusual, and so this method is particularly useful ⟨84TL2271⟩. Alkenylchromium reagents have been employed in the synthesis of palytoxin ⟨87CS573⟩, and glycosides and C-methyl glycoside analogues (Equation (37)) ⟨87JOC4819, 87JOC4823, 87JOC4825, 88JOC3383, 88JOC4151, 88JOC5580⟩, with intramolecular addition reactions being studied for the synthesis of brefeldin analogues ⟨88JA5198⟩ and ophibolins (Equation (38)) ⟨88TL4909⟩.

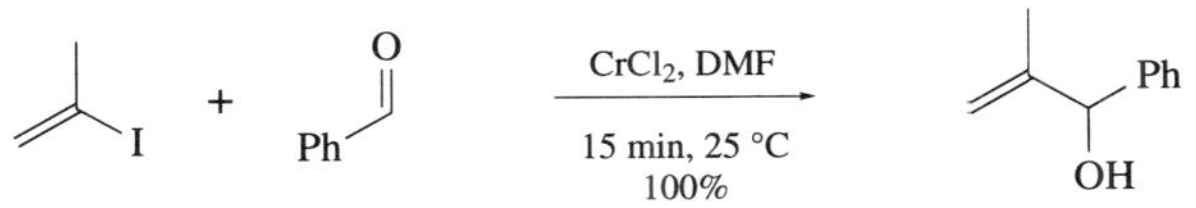

(35)

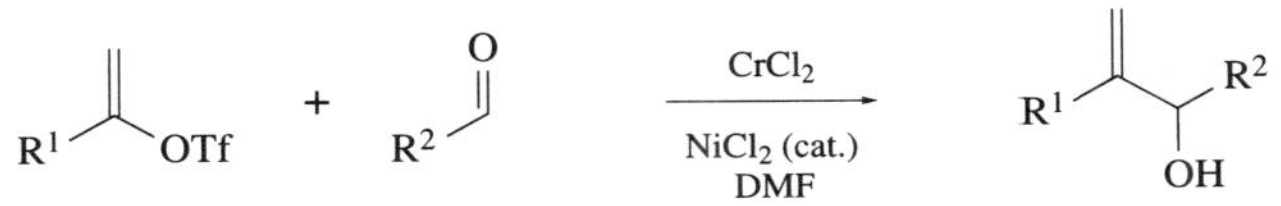

(36)

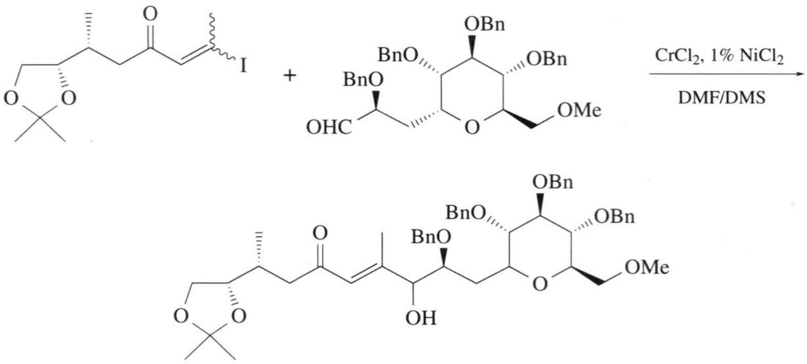

(37)

(38)

1.22.1.3.9 *Alkenylboron reagents*

Alkenyl Grignard reagents react with trialkylboranes to give an intermediate borate; decomposition of this complex results in overall alkylation of the vinylic carbanion centre by B→C alkyl migration (Scheme 16) ⟨77T1945⟩.

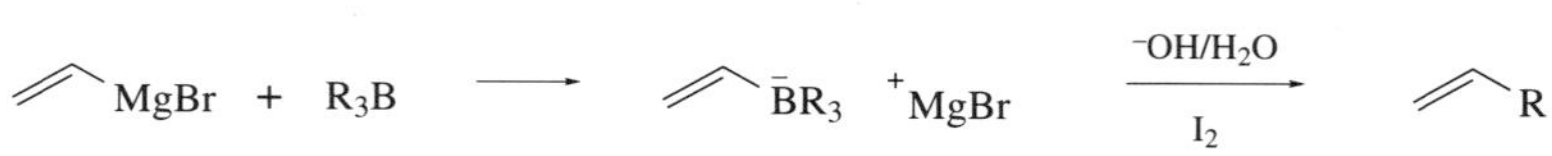

Scheme 16

1.22.1.4 Formation of Vinylic Carbanions by Addition to Acetylenes and Allenes

1.22.1.4.1 *Alkenylboron reagents*

The hydroboration of alkynes with diborane to give vinylboranes is of limited use, with internal alkynes giving only moderate yields of trialkenylboranes, whilst terminal alkynes give little or none of the required borane ⟨61JA3834⟩. However, the use of more sterically demanding hydroborating agents, for example, 9-borabicyclo[3.3.1]nonane (9-BBN-H), or heterosubstituted boranes, for example, catecholborane, has allowed the preparation of a vast array of vinylboranes from the corresponding alkynes ⟨81MI 122-01⟩.

When diisoamylborane is used as the hydroborating agent, reaction with either internal or terminal alkynes proceeds rapidly at 0 °C, allowing hydroboration to occur selectively, even in the presence of alkenes (Equation (39)) ⟨63JA2063⟩. Reaction of thexylborane with either 1-chloro- or 1-bromoalkynes gives a thexyl-1-halo-1-alkenylborane (**26**) which can then either react with an alkyne to yield a mixed thexyldialkenylborane (**27**), or with an alkene to give a thexylalkylalkenylborane (**28**) (Scheme 17) ⟨73CC606⟩. *trans*-α-Halovinylboranes have also been prepared from the reaction of 1-bromo- or 1-iodoalkynes with dicyclohexylborane; the alkenylborane undergoes a number of transformations (Scheme 18) ⟨67JA5086⟩.

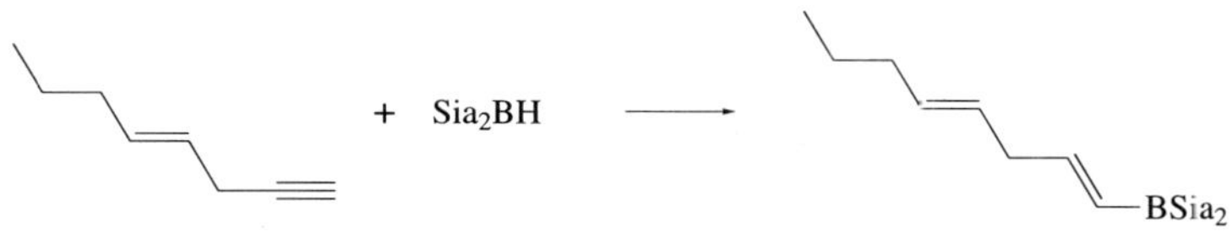

(39)

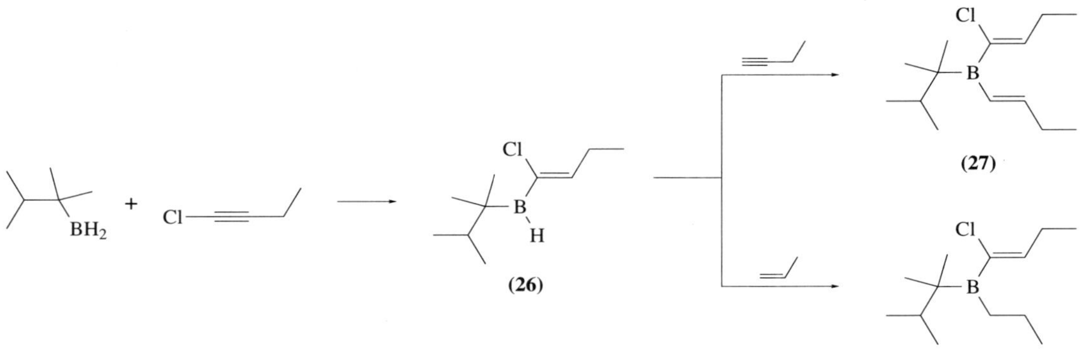

(26)

(27)

(28)

Scheme 17

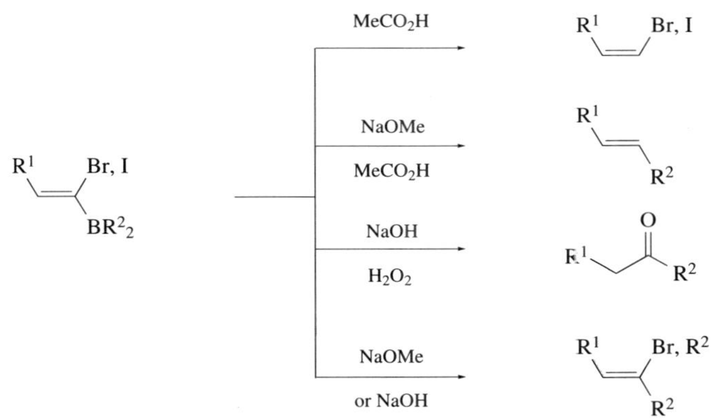

Scheme 18

9-BBN reacts with internal alkynes to give the corresponding alkenyl derivatives in good yields ⟨79JA96⟩. However, terminal alkynes can undergo further reaction to give 1,1-diboraalkanes. This problem can be overcome by using an excess of the alkyne (usually 100% excess), the unreacted alkyne being recovered at the end of the reaction. The resulting alkenylboranes are stable, and can be recovered by vacuum distillation. Unlike the previous examples, 9-BBN shows increased reactivity towards terminal alkenes which react more readily than alkynes; terminal alkynes react more readily than internal alkynes (Equation (40)) ⟨79JOC2328⟩. Catecholborane reacts with alkynes in refluxing THF to give alkenylcatecholboranes in excellent yields (Equation (41)) ⟨72JA4370, 75JA5249⟩. These alkenylboranes are air-stable and can be purified by distillation or recrystallisation. Yields of hydroboration can be increased by the use of ultrasonication in place of the prolonged heating that is often required ⟨85TL2187⟩.

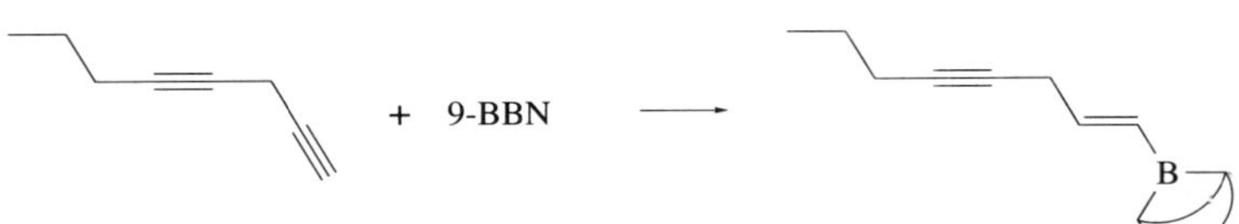

(40)

9-BBN = 9-borabicyclo[3.3.1]nonyl

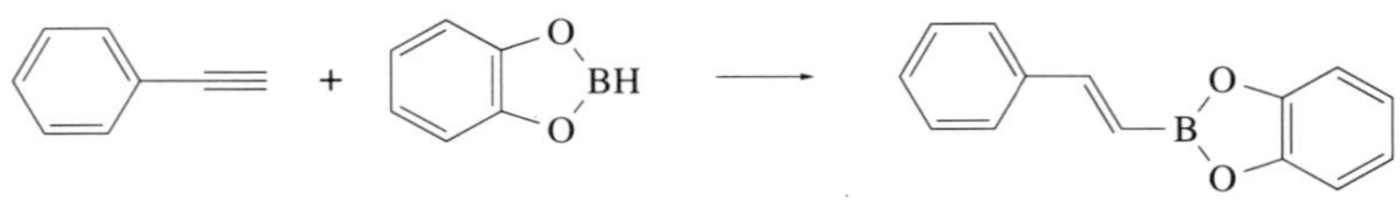

(41)

Dialkenylchloroboranes can be obtained from the reaction of monochloroborane-etherate with two equivalents of an internal alkyne (Equation (42)). Reaction with 1-alkynes is only possible if an excess (~40%) of the alkyne is used ⟨76JA1785⟩. The rate of the hydroboration reaction can be increased by addition of one equivalent of boron trichloride to the reaction mixture ⟨56JA2173, 76JA1798⟩. Hydrolysis or alcoholysis of the dialkenylchloroborane provides dialkylborinic acids or esters.

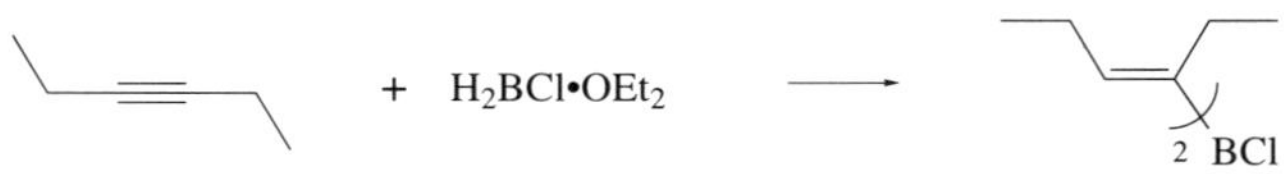

(42)

The dibromoborane-methyl sulfide complex reacts cleanly with 1-alkynes to give the corresponding alkenyldibromoboranes in excellent yields, without the problems of competing dihydroboration (Equation (43)) ⟨80JOC389⟩. Reaction of dibromoborane-dimethyl sulfide is unusually rapid with internal alkynes, allowing their selective hydroboration in the presence of terminal alkynes or alkenes. This selectivity in hydroboration complements that exhibited by 9-BBN, and so a judicious choice of hydroborating reagent should allow the preparation of either 1-substituted or *n*-substituted alkenylboranes (Equation (44)).

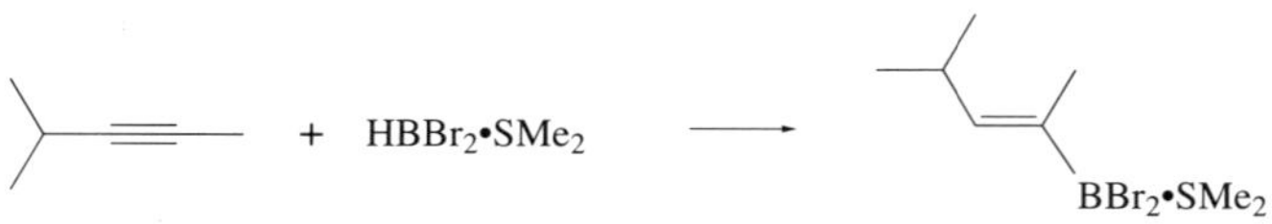

(43)

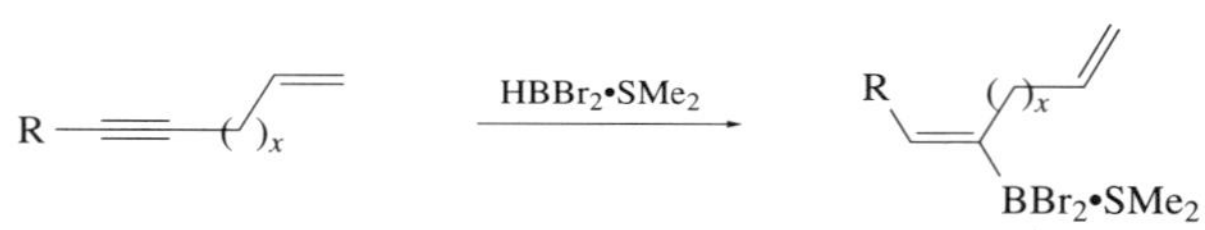

(44)

Due to the hydroboration of alkynes occurring to give exclusively the *trans*-alkenylborane many subsequent reactions can be carried out in a stereodefined manner, allowing the precise prediction of the stereochemistry of the product. Examples of the reactions of alkenylboranes will be discussed later in this section. The regioselectivity of hydroboration is, however, influenced by stereo and electronic effects imposed by the alkyne, and on the nature of the hydroborating agent employed. However, appropriate choice of reagent should allow regiospecific hydroborations to take place as required. When 1-alkynes are used, nearly all of the available reagents react such that the boron adds to the terminal carbon atom ⟨81MI 122-01⟩. With internal alkynes, nearly all the reagents have been shown to be influenced to some extent by steric effects, with boron preferring to add to the less sterically hindered position; 9-BBN and $HBBr_2 \cdot SMe_2$ appear to be the more selective than hindered reagents such as diisoamylborane and dicyclohexylborane. Electronic effects are also significant; when 1-cyclohexyl-1-propyne is hydroborated, all the reagents show a tendency to add boron to the least hindered position of the alkyne, i.e., β to the ring. This tendency is reversed in 1-phenyl-1-propyne, when 9-BBN and $HBBr_2 \cdot SMe_2$ are the reagents used, with addition to the position α to the ring being preferred. However, the more hindered reagents still show a preference to add to the β position, steric influences remaining the dominant factor ⟨81MI 122-01⟩.

The alkenylboranes have found a number of applications as intermediates in organic synthesis. Their conversion to other organometallic reagents has already been discussed (Section 1.22.1.3), but the boron atom can also be displaced by other electrophiles. Addition of acetic acid to an alkenylborane results in protonolysis of the carbon–boron bond to give an alkene ⟨61JA3834⟩. Retention of configuration is observed, and the reaction thus provides a method for the noncatalytic hydrogenation of an alkyne to a *cis*-alkene, with a number of other functional groups being tolerated. This has been exploited in the synthesis of the sex pheromone of *Lobesia botrana* (Equation (45)) ⟨77TL411⟩. The use of deuterioacetic acid provides a method for the stereospecific preparation of deuteriated alkenes ⟨71JA3395⟩.

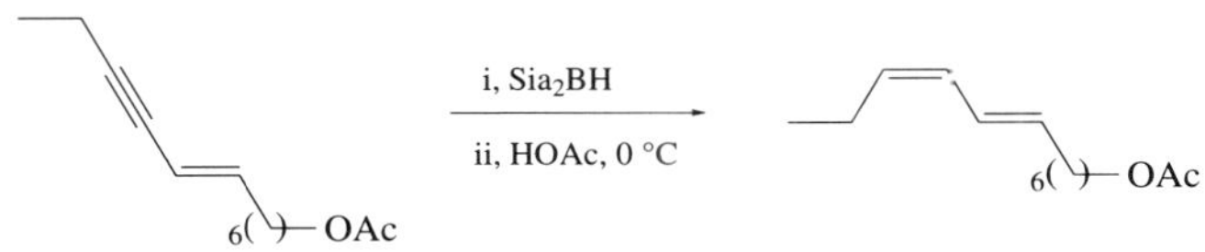

(45)

Stereochemically pure *cis*-alkenyl halides can be prepared on reaction of the alkenylborane with halogen followed by the addition of base (Equation (46)) ⟨73JA6456⟩. For the formation of *trans*-alkenyl halides base is added to the reaction mixture prior to addition of the halogen (Equation (47)) ⟨73JA5786, 80JOC389⟩.

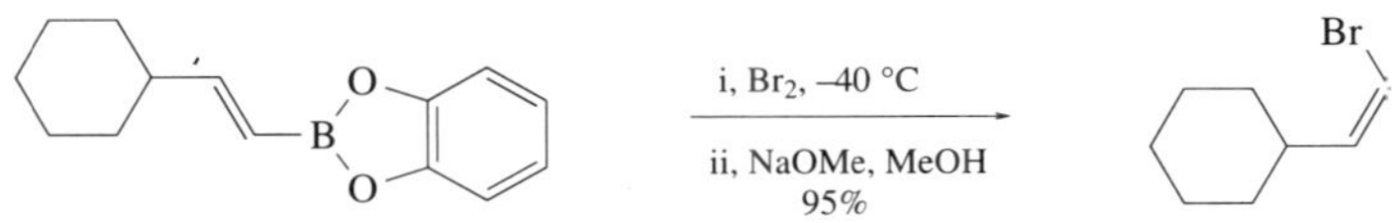

(46)

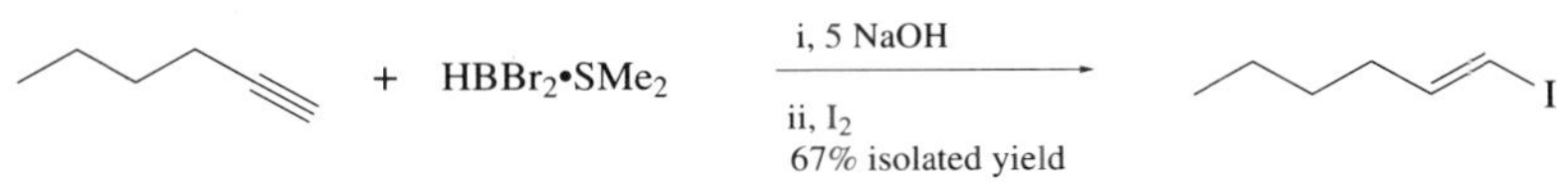

(47)

Methods for the preparation of *cis*-alkenes from alkenylboranes have been reported ⟨67JA3652⟩ and applied to the synthesis of prostaglandin analogues ⟨76JOC3947⟩. The utilisation of alkenylboranes in the preparation of *trans*-alkenes is also the subject of a number of publications ⟨67JA5086, 72JA4013, 72JA6560, 72S555⟩. Alkenylboranes have also been used to prepare conjugated *trans,trans*-dienes ⟨73CC606⟩, *cis,trans*-dienes ⟨68JA6243, 73JOC1617, 78JOM(156)159⟩, and *cis,cis*-dienes ⟨65OS(45)39, 70JA4068, 76JOC1078, 76TL4385⟩.

Alkenylboranes will add to aldehydes to give allylic alcohols of known stereochemistry about the double bond (Equation (48)) ⟨77JOC579⟩. Functional groups such as esters, halogens and nitriles are tolerated by hydroboration, and so this 1,2-addition to carbonyl groups provides a useful alternative to the Grignard reaction, in that reactive substituents can be present in both the substrate and the organometallic component. 1,4-Addition of alkenylboranes to α,β-unsaturated ketones occurs to give γ,δ-unsaturated ketones as products, with the configuration about the vinylic double bond being retained (Equation (49)). Transoid enones such as 2-cyclohexenone do not react, implying that the reaction proceeds through a cyclic transition state; on the other hand cisoid enones react readily ⟨76JA7832⟩.

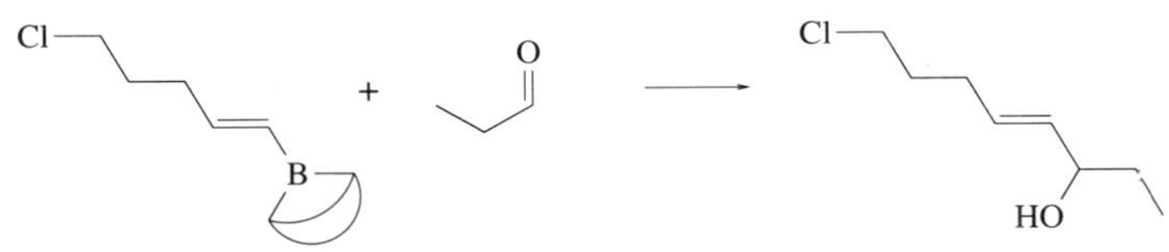

(48)

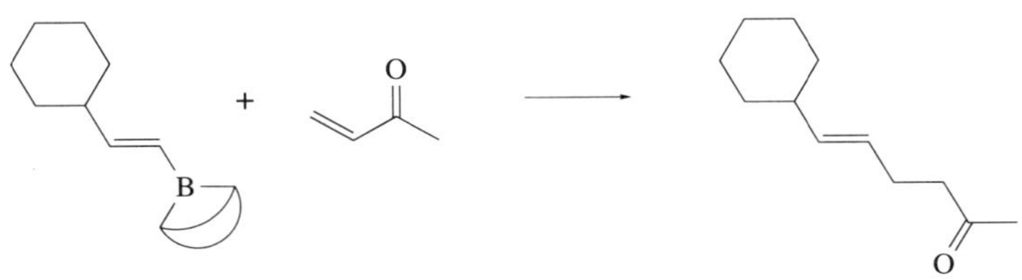

(49)

Alkenylboranes are readily oxidised by alkaline hydrogen peroxide to give the corresponding carbonyl compounds (Equation (50)) ⟨79JA96⟩. When aldehydes are the expected products (i.e., from 1-alkenylboranes), a pH 7 buffer prevents base-catalysed condensation reactions occurring ⟨80JOC389⟩. 1-Alkynes can be converted into carboxylic acids by formation of the 1-trimethylsilyl derivative, hydroboration of the resultant 1-trimethylsilylalkyne, and then oxidation (Scheme 19) ⟨77JA7097⟩.

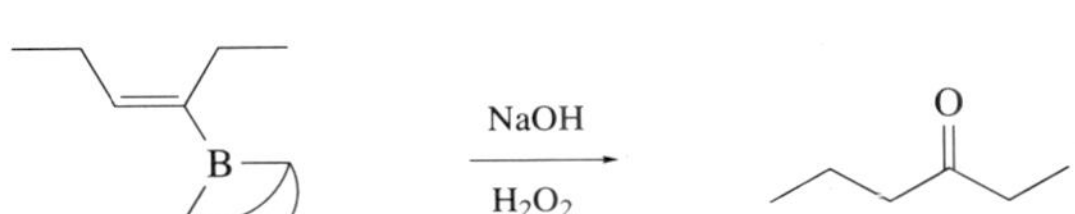

(50)

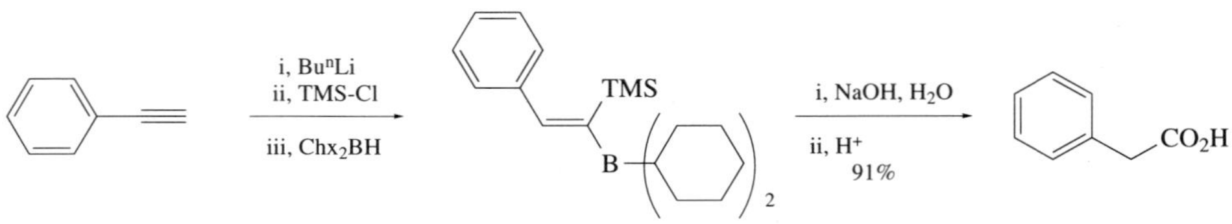

Scheme 19

Haloboration of alkynes has also been reported, and employed in a carboxyolefination reaction (Scheme 20) ⟨89TL5153⟩.

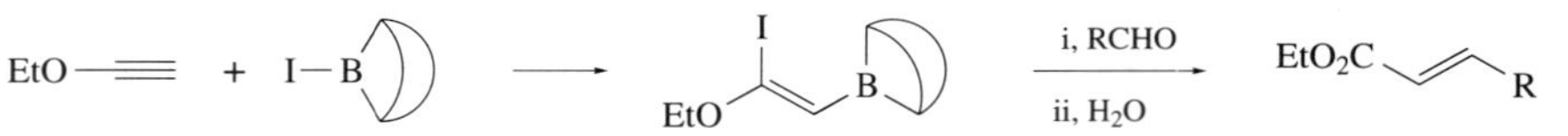

Scheme 20

Alkenylboranes will undergo a number of reactions mediated by palladium catalysis. For example, (*E*)-disubstituted alkenes can be obtained from alkenyldialkylboranes in the presence of $Pd(OAc)_2$ and triethylamine (Scheme 21) ⟨80BCJ1670⟩, whilst $Pd(PPh_3)_4$ catalysed coupling with aryl halides yields arylated alkenes (Equation (51)) ⟨79CC866⟩. Coupling with allylic chlorides with $Pd(OAc)_2$ as the catalyst leads to 1,4-dienes (Equation (52)) ⟨80BCJ1670⟩; allylic bromides require $Pd(PPh_3)_4$ as the catalyst and aqueous sodium hydroxide ⟨80TL2865⟩. Alkenylboranes will couple with benzyl bromides in the presence of $Pd(PPh_3)_4$ and sodium hydroxide, with retention of configuration (Equation (53)) ⟨80TL2865⟩.

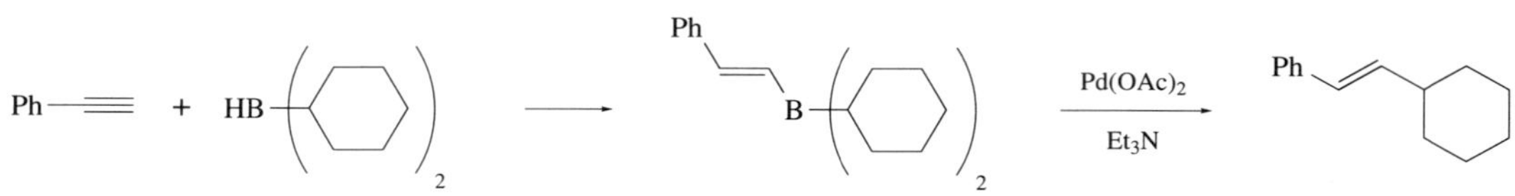

Scheme 21

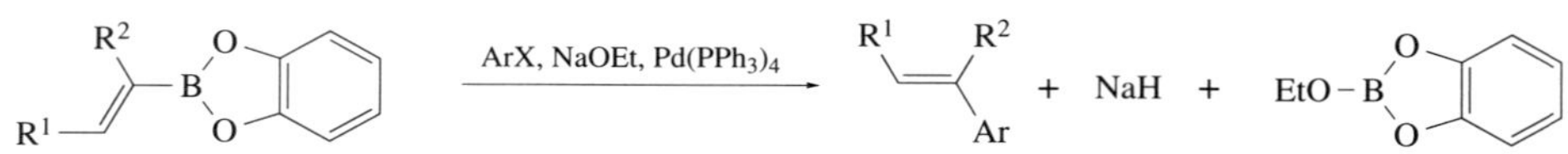 (51)

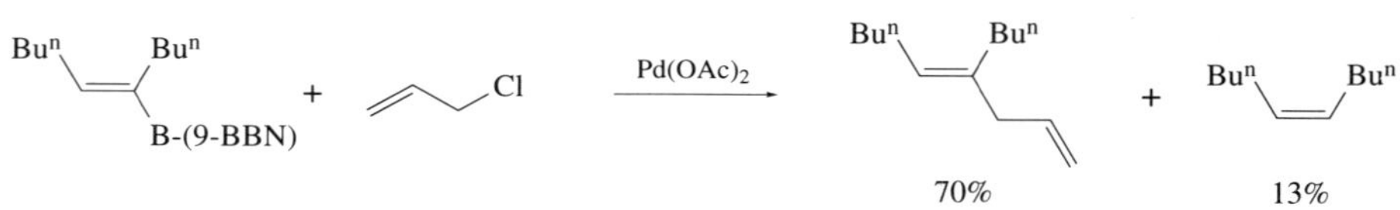 (52)

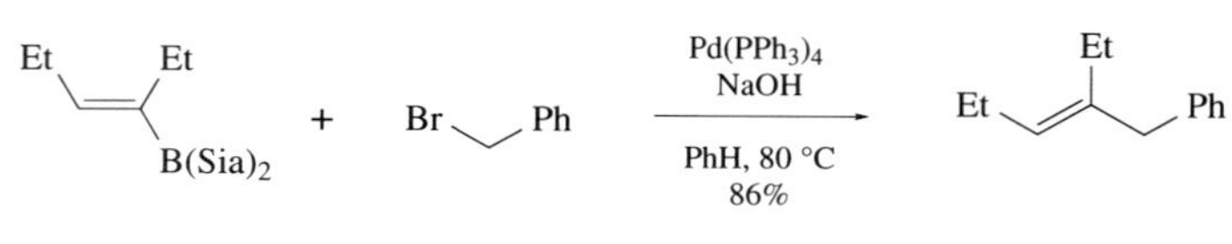 (53)

1.22.1.4.2 *Alkenylzirconium reagents*

Alkenylzirconium reagents offer some advantages over their magnesium, aluminum and mercury counterparts in terms of preparation, stability and functional group compatibility ⟨78JA2254, 78TL1027⟩, and are comparable to alanes with regard to reaction rates. Alkenylzirconium reagents are capable of discriminating between aldehydes and ketones, and may also differentiate between similar functional groups that differ only in their steric and/or electronic environment. Epoxides, ethers, carboxylic acid esters and thioesters, nitriles and alkyl halides are all tolerated. Because of the lower basicity of alkenylzirconium reagents, they can be used for addition to readily enolisable carbonyl compounds. Alkyltitanium reagents show similar reactivity to their zirconium counterparts; however, alkenyltitanium reagents cannot be exploited for nucleophilic additions because of their propensity to undergo oxidative coupling reactions.

Alkenylzirconium compounds are commonly prepared from an alkyne via hydrozirconation, the zirconium reagent most commonly used being Cp_2ZrClH (Equation (54)) ⟨71JA1379, 75JA679⟩. Vinylzirconium(IV) complexes are stable, and can be used in conjugate additions to α,β-unsaturated ketones, usually in the presence of a nickel(0) catalyst (Scheme 22) ⟨77JA8045, 81JA4466⟩ and this methodology has been used in the preparation of prostaglandin precursors ⟨78TL4381⟩. A mechanistic investigation of the addition reaction of alkenylzirconium reagents to carbonyl compounds has been reported ⟨81JA4466⟩. They have also been employed in the formation of α-(phenylselenenyl)ketones (Scheme 23) ⟨80TL1497⟩.

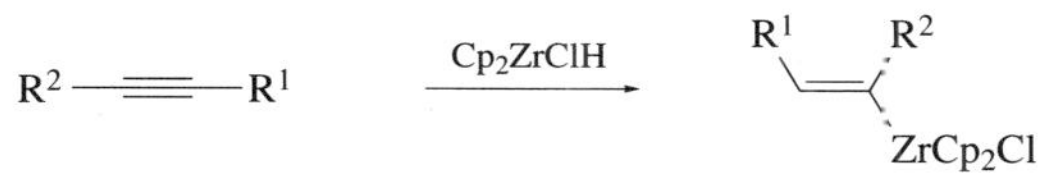

(54)

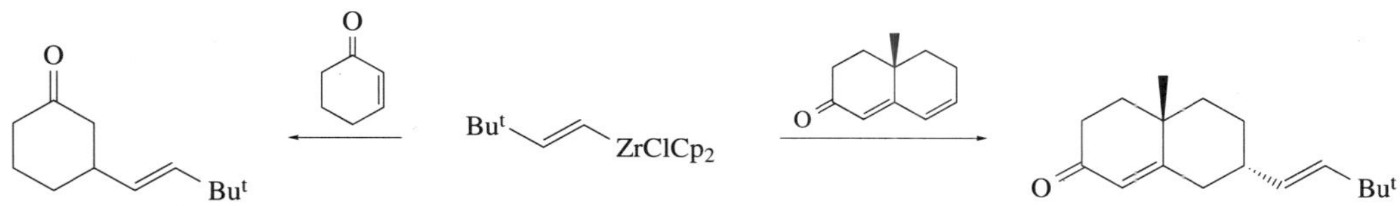

Scheme 22

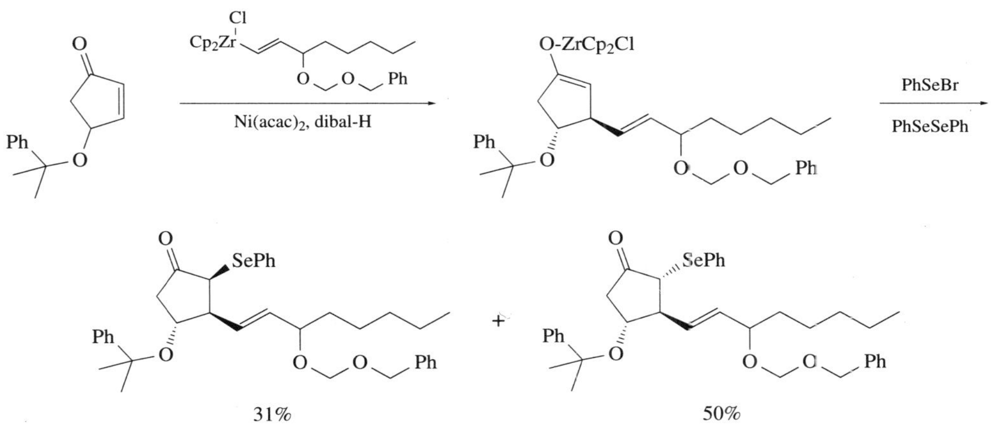

Scheme 23

1.22.1.4.3 Alkenylmagnesium reagents

Alkylmagnesium bromides add across an alkyne under nickel(0) catalysis to give vinylmagnesium bromides which then react as previously discussed. This approach has been used in a synthesis of geraniol (**29**) (Scheme 24) ⟨78JA4624⟩. α-Silylvinylmagnesium bromides have also been prepared by the addition of an alkyl Grignard reagent to a silylalkyne under nickel catalysis, and subsequently undergo alkylation by methyl iodide or allyl bromide (Equation (55)) ⟨78JA4624⟩. α- and β-silylvinyl Grignard reagents can be alkylated under palladium catalysis ⟨79JOC155, 79TL1073, 83TL1041⟩.

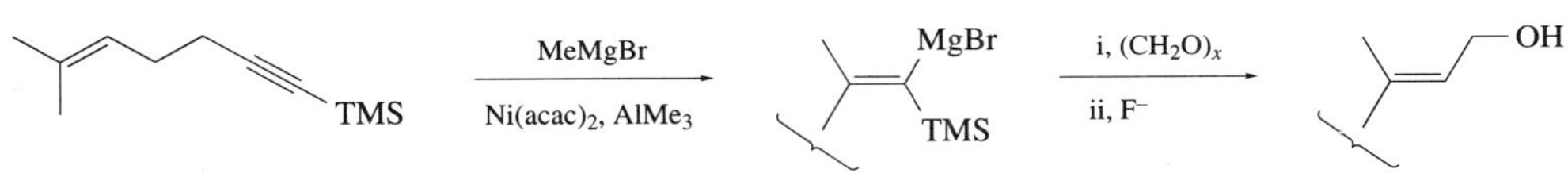

(29)

Scheme 24

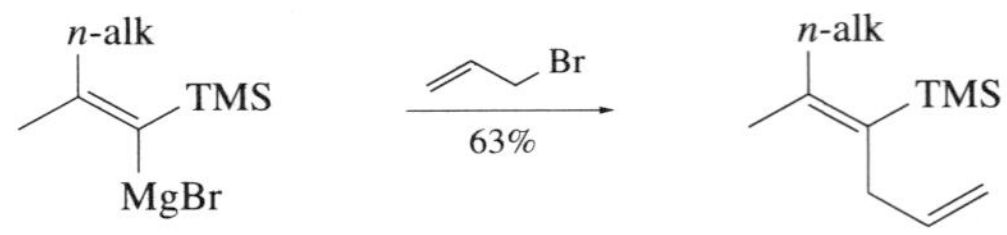

(55)

1.22.1.4.4 *Alkenylaluminum reagents*

The preparation and reactions of alkenylaluminum reagents have been reviewed ⟨81S841, 84OR375, 85AG(E)668⟩. Alkenylalanes prepared by the addition of diisobutylaluminum hydride (dibal-H) to a terminal alkyne ⟨76CC17⟩ can be alkylated in respectable yields, after conversion to an 'ate' complex using, for example, *n*-butyllithium (Equation (56)) ⟨84OR(32)375⟩. Addition of trimethylaluminum to alkynes will take place under zirconium catalysis to give the corresponding alkenylaluminum compounds ⟨80S1034⟩. α-Silylalkenylalanes and β-stannylalkenylalanes have also been prepared ⟨76JOC2215, 86TL6161⟩, and require conversion to an 'ate' complex before alkylation can occur. The corresponding alkenylaluminum complex, on the other hand, alkylates a number of primary and secondary halides without prior conversion to the 'ate' complex ⟨87JOC1624⟩, although problems arise with substrates that can undergo both S_N2 and S_N2' displacements, for example, crotyl bromide. Vinylic alanes couple with vinylic halides in the presence of $PdCl_2(PPh_3)_2$ catalyst, that had been prereduced with dibal-H, at 25 °C in THF. Nickel also catalysed the reaction, but was not as good at stereocontrol as the palladium catalyst, for which the reaction was greater than 99% stereoselective (Equation (57)) ⟨76JA6729⟩. Alkenylalanes, prepared as described above, will react with epoxides if converted to their 'ate' complexes, by treatment with *n*-butyllithium (Scheme 25) ⟨76CC17, 80S1034⟩.

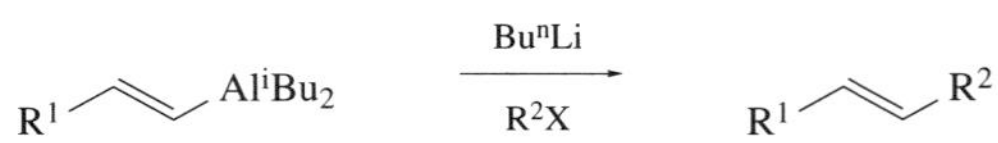

(56)

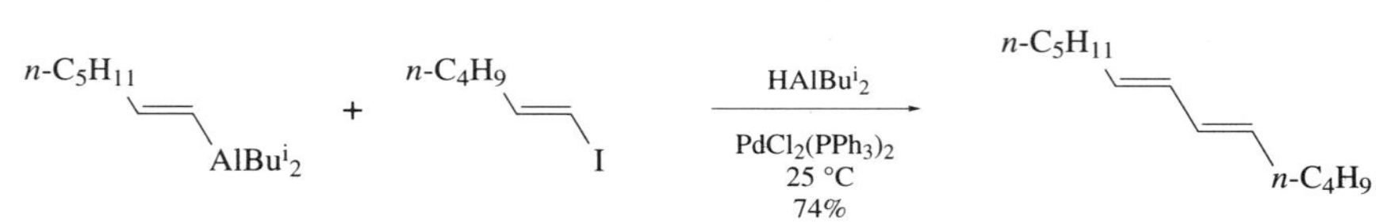

(57)

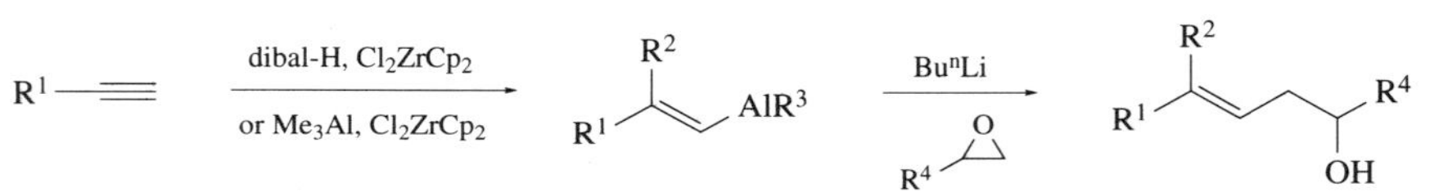

Scheme 25

1.22.1.4.5 *Alkenylsilicon reagents*

Palladium-catalysed hydrosilylation of alkynes has been used for the preparation of trichlorosilylalkenes, which were then converted to the fluoro derivatives prior to coupling with an allylic chloride, again under palladium catalysis. The product was a 1,4-diene (Scheme 26) ⟨78TL2161⟩. Trimethylsilylalkenes will also undergo palladium-catalysed coupling with allylic halides to give 1,4-dienes (Equation (58)) ⟨71TL4701, 79JOM(179)C7⟩. Although the palladium(II) used as catalyst for this reaction should not be reduced in a formal sense, side reactions can cause reduction to the metal, and so cupric chloride is added as a reoxidant.

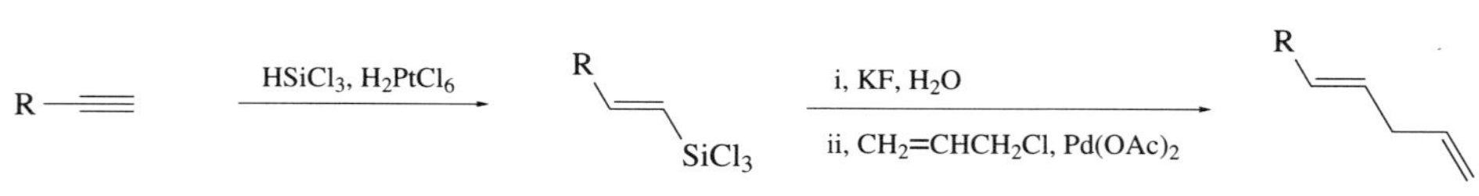

Scheme 26

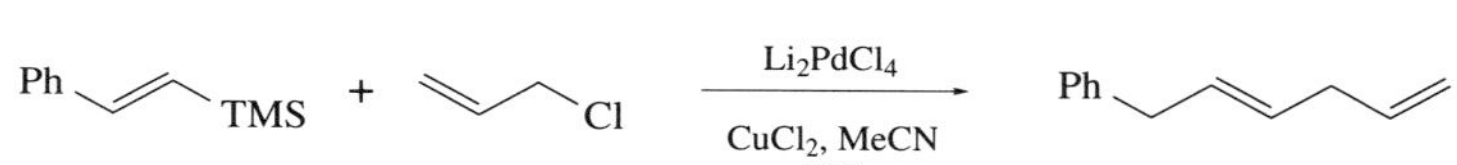

(58)

1.22.1.4.6 *Alkenylpalladium reagents*

The formation of alkenylpalladium reagents from the addition of a palladium salt across an alkyne has been reported, and used in the formation of 1,4-dienes through coupling to an allylic halide (Scheme 27) ⟨74TL1067, 79JOC55⟩. $PdX_2(PhCN)_2$ can be used as the catalyst and, if an excess of the allylic halide is also used, the products are mainly the (*Z*)-isomers.

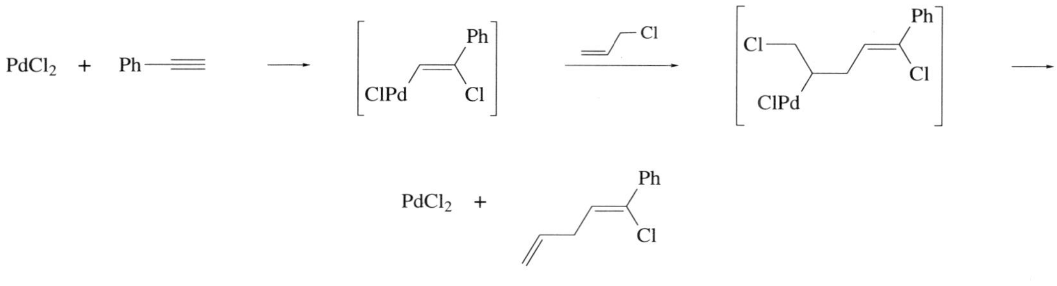

Scheme 27

1.22.1.4.7 *Alkenylcopper reagents*

The preparation of alkenylcopper reagents has been reported ⟨78JOC3418, 85CC1636⟩. Dialkyllithium cuprates will add across an alkyne to give the corresponding alkenylcuprates (Scheme 28) ⟨81TL3851⟩. Grignard reagents also add across alkynes in the presence of stoichiometric amounts of copper(I) to give the corresponding alkenylmagnesiocuprates, an approach that has been employed in the synthesis of insect pheromones (Scheme 29) ⟨79JOC3888, 82TL3587⟩. (*Z*)-Divinylcuprates derived from such alkylcuprate additions to alkynes can be alkylated by amino sulfides to give (*Z*)-allylamines (Equation (59)) ⟨80TL3763⟩. Alkenylmagnesiocuprates obtained via the addition of Grignard reagents to alkynes in the presence of copper(I), react well with ethylene oxide and monosubstituted epoxides to give homoallylic alcohols, although yields can be improved by prior conversion to the 'ate' complexes using a lithium acetylide (Equation (60)) ⟨77JA253, 78TL1363, 78TL2465, 78TL2973, 79JOC1345, 79JOC3888⟩.

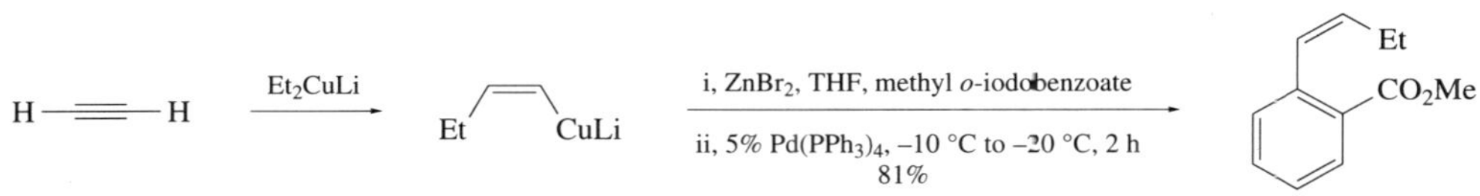

Scheme 28

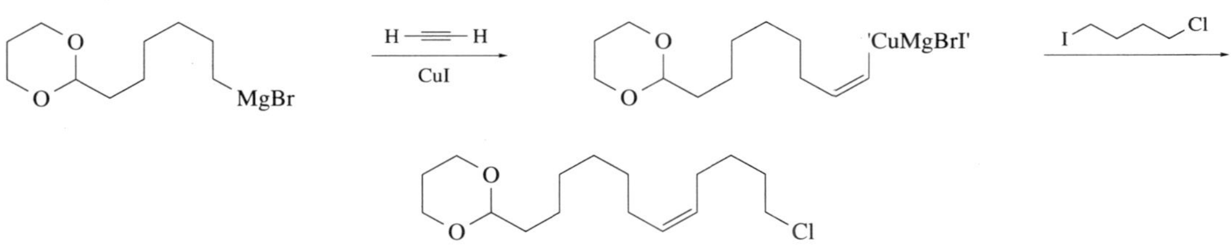

Scheme 29

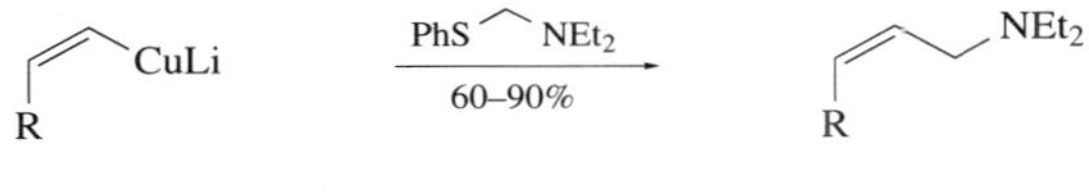

(59)

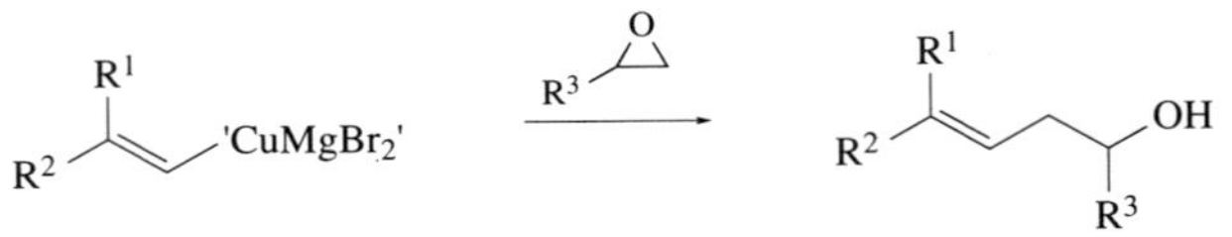
(60)

Primary alkylcopper reagents add across terminal alkynes both regio- and stereospecifically in ether in the presence of magnesium bromide or iodide. The alkenylcopper reagents thus obtained can then be protonated, alkylated, halogenated, carbonated or oxidised to the dimeric diene (Scheme 30) ⟨74JOM(77)269, 74JOM(77)281, 77TL3545⟩.

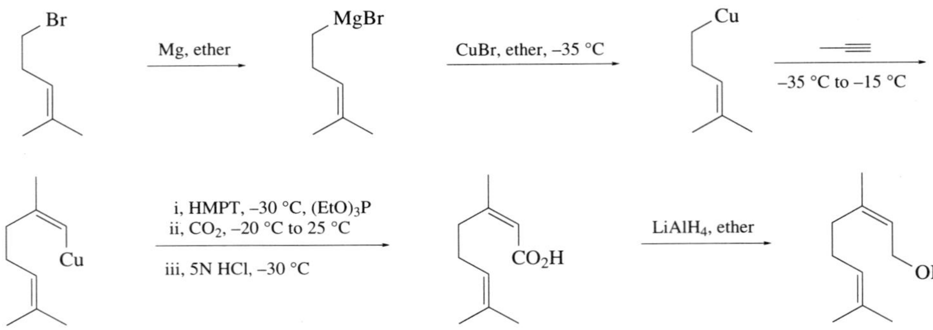

Scheme 30

1.22.1.4.8 *Alkenylmercury reagents*

Mercuric chloride adds across alkynes to give the corresponding alkenylmercury chloride. This methodology has been employed in the synthesis of β-chlorolactones, in which a propargylic alcohol was the alkyne component, and the intermediate alkenylmercury derivative was then carbonylated in the presence of a palladium salt (Scheme 31) ⟨78JOC131⟩. The reaction was limited to relatively low molecular weight propargylic alcohols, but could be made catalytic in palladium by adding cupric chloride without affecting the yields.

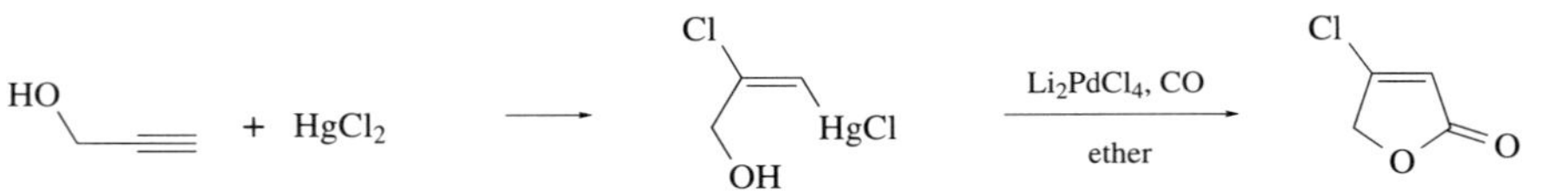

Scheme 31

1.22.1.4.9 *Alkenyllithium reagents*

Tertiary lithium enolates add to haloalkynes in special cases, i.e., when the other alkyne substituent is halo, aryl, or similar, and can stabilise a negative charge. The intermediate vinylic carbanion thus obtained eliminates halide ion to regenerate an alkyne (Equations (61) and (62)) ⟨43CB80, 84JA3551⟩.

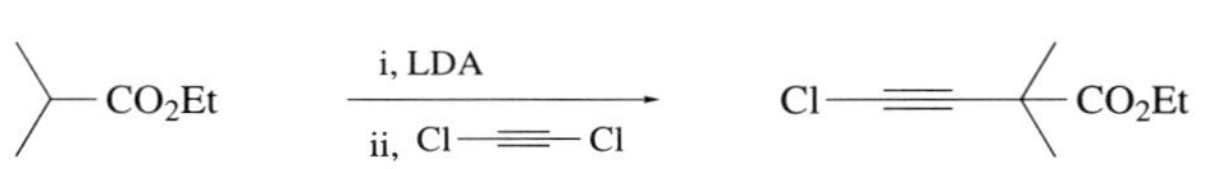
(61)

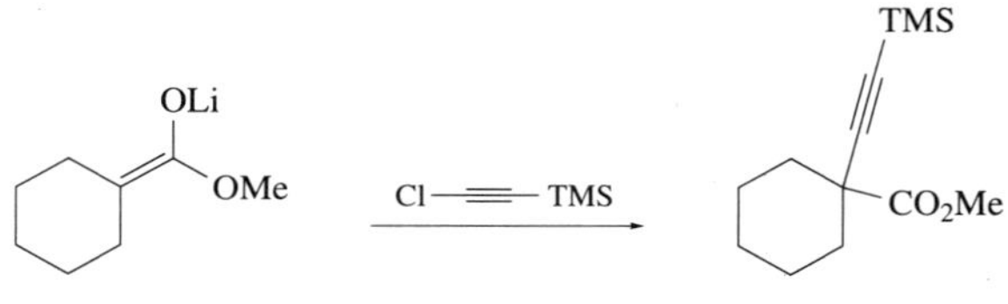
(62)

1.22.2 DICOORDINATE CARBOCATIONS (i.e., VINYLIC CATIONS)

1.22.2.1 General Aspects of Carbocation Formation

Reviews and monographs dealing with general aspects of carbocations also discuss vinyl cations, including theoretical aspects, preparations and reactions, as do general reviews on alkene chemistry ⟨B-64MI 122-01, B-67MI 122-01, 69APO(7)1, 71ACR73⟩. A monograph devoted entirely to the discussion of vinyl cations has also appeared ⟨B-79MI 122-01⟩.

1.22.2.1.1 Structures of vinyl cations

Three types of structures have been considered for vinyl cations: (i) a 'bent' geometry involving sp^2 hybridisation at C(1), which retains the structural features of the solvolysis precursors; (ii) a linear geometry involving *sp* hybridisation at C(1), which retains the features of electrophilic addition precursors; and, (iii) a bridged species with three substituents attached to C(1). Two dynamic equilibria also need to be considered; bent cations may rapidly interconvert (Equation (63)), thus showing behaviour similar to that of the linear cation, and rapid rearrangement of an R group across the double bond of a linear cation must be considered (Scheme 32).

(63)

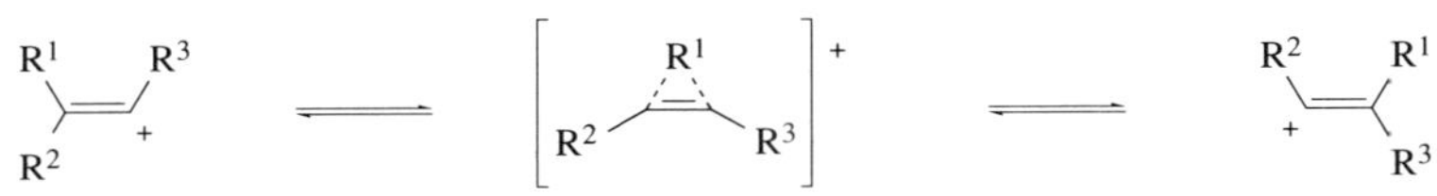

Scheme 32

Many MO calculations have been performed on the parent vinyl cation. In all cases, the bent structure was excluded as a possibility, and so the main argument has been as to whether the linear or bridged structure is preferred ⟨64JCP(40)2480, 68JA1239, 69JA5350, 70JA2141, 71JA808, 71JCP(55)3835, 71TCA65, 72CPL385, 73CPL309, 76JA875, 76JCP(64)4159, 77JA7432, 78JA5297⟩. The general conclusion is that the parent vinyl cation is probably bridged. α-Substituted vinyl cations are nearly linear at C(1), although some deviation from linearity is to be expected when the two β substituents are different. The relative stabilities of a β-thiovinyl cation and its bridged isomer have attracted special interest in view of the evidence for β-sulfur participation in solvolysis reactions (Section 1.22.2.2). The bridged structure was calculated to be more stable than the linear structure ⟨72CC8, 74JCS(P2)1808, 77JCS(P2)542⟩. Addition of methyl groups either to the sulfur atom, or to the vinylic carbon atoms increases the stability of the thiirenium ion still further ⟨78G543⟩. Similar calculations carried out on the oxygen analogues have shown that, on the contrary, it is an open ion that is more stable than the bridged ion ⟨77JCS(P2)542⟩. Bridged selenirenium ions and probably iodonium ions are more stable than the open chain ions, whereas the reverse is true for the β-fluoro- and β-chloro-substituted vinyl cations. The β-bromovinyl cation is bridged or linear depending on other substituents present. In all cases, a bent vinyl cation may be formed under certain geometrical constraints, such as in cyclic structures.

1.22.2.1.2 Relative stabilities of vinylic and trigonal cations

Relative gas phase stabilities have been examined for the heats of formation (ΔH^0_f) of several vinyl cations using mass spectrometric and ion cyclotron resonance spectroscopic techniques, and the appropriate thermochemical cycles; the parent vinyl cation was determined to be 25 kcal mol^{-1} more stable than the methyl cation and 14.5 kcal mol^{-1} less stable than the ethyl cation, whilst the 2-propenyl ion is 15 kcal mol^{-1} less stable than the secondary isopropyl cation ⟨69CRV279, 70CJC955, 71CJC357, 72CJC3973, 76JA6700⟩. Along with the other ΔH^0 values obtained, it was found that the differences in stability between analogous vinylic and trigonal cations is much lower when the ions are generated by addition of an electrophile to a multiple bond, than in the solvolysis reaction, owing to the higher ground state energies of alkynes. The relative stabilities of vinylic and trigonal cations in solution have also been determined through comparison of solvolysis rates ⟨70JA5457⟩, or of addition rates to multiple bonds of the appropriate precursors ⟨66JA4990, 73JA160, 74JA3574, 74TL2203, 75JA1977, 75TL2771, 75TL4555, 76JOC2331, 77JA290, 77JOC2021, 77JOC2689, 78JOC1521, 79JOC3022, B-79MI 122-01⟩.

1.22.2.1.3 Routes competing with vinyl cation formation

Vinyl cations are now accepted as intermediates in many addition and solvolysis reactions and the transformations discussed in the following sections are all postulated to proceed via these reactive intermediates. However, the relative unreactivity of many vinylic systems in solvolytic reactions increases the probability of alternative reaction mechanisms operating. Although the products obtained are frequently identical to those from the expected S_N1 and E1 routes, it is essential to exclude these other mechanisms, by kinetic and isotope labelling methods, before it is safe to assume vinyl cation formation. Simply changing the geometry or substituents in the vinylic system, and/or the medium and/or the nucleophile being employed may be sufficient to divert the course of the reaction from the S_N1/E1 pathways ⟨69APO(7)1, 71ACR73⟩. Some of the principle pathways will now be discussed.

An important vinylic substitution route proceeds via nucleophilic addition–elimination (Scheme 33) ⟨69APO(7)1, 71ACR73, 77T1211, 79JA5095⟩. This process is strongly dependent on the nature of the nucleophile, proceeds mainly with retention of configuration, and is the most probable route when the vinylic system contains *β*-electron-withdrawing substituents ⟨68G367, 71JCS(B)871, 71JCS(B)1461, 72JOC1174, 73JCS(P2)2078, 80JOC4309⟩.

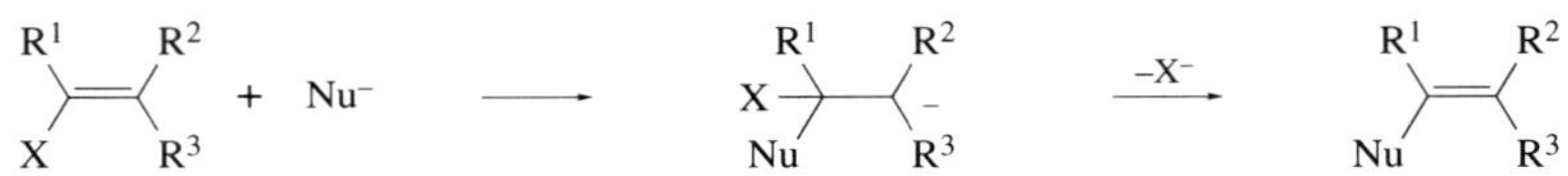

Scheme 33

In carboxylic acids or acidic media the electrophilic addition–elimination (Ad_E-E) route may compete with the S_N1 route (Scheme 34). Because most solvolysis reactions are carried out in acidic media, HX is a reaction product which may donate a proton for the addition step. The structural features that enhance the S_N1 route also enhance the Ad_E-E route, so this particular mechanism is of importance when considering vinyl cation formation. A number of examples of its co-occurrence have been reported ⟨68JA6515, 69JA6194, 70JA207, 70JA4985, 71HCA2060, 71JA1513, 73JCS(P2)301, 75JA821, 78CB1253⟩. A radical addition–elimination route has been suggested for the silver nitrate assisted solvolysis of *β*-bromostyrenes in aqueous acetonitrile (Scheme 35) ⟨69JOC1495⟩.

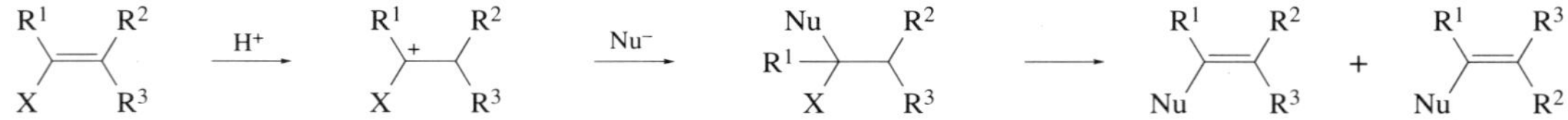

Scheme 34

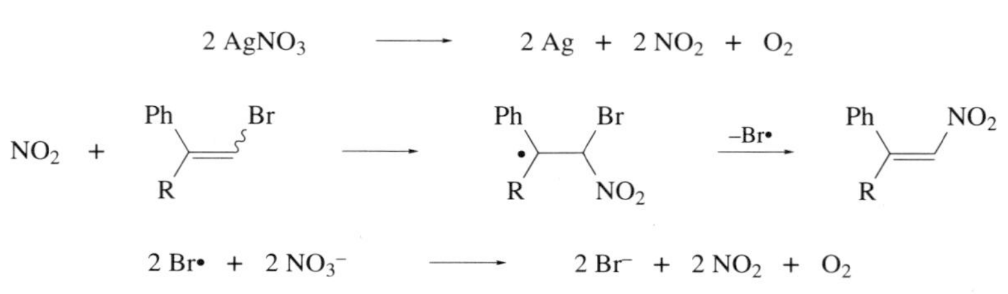

Scheme 35

When a vinyl sulfonate or carboxylate is employed in a solvolysis reaction, attack on the leaving group by the solvent or its conjugate base may occur which, when followed by S–O or C–O bond cleavage gives ketones of identical structure to those obtained via the vinyl cation route (Schemes

36 and 37). This solvolysis mechanism may be operating in those cases where the vinylic systems is unactivated ⟨69CC1044, 70JCS(C)2625, 72AG(E)714, 74JA1100, 76TL3507, 77JOC174, 78LA1894⟩.

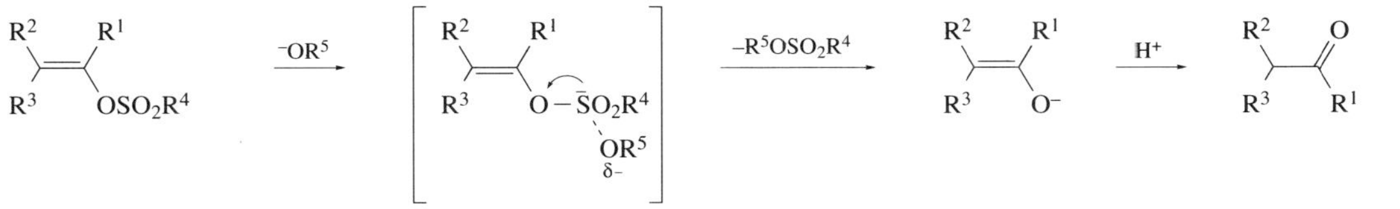

Scheme 36

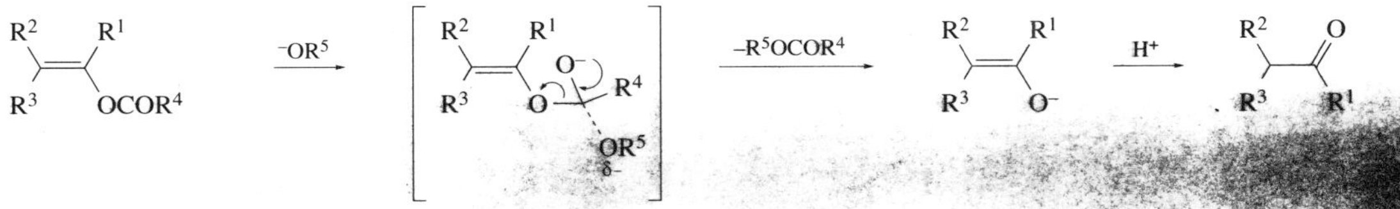

Scheme 37

Vinylic solvolysis products may be obtained in some cases by a process involving an initial rate-determining solvolysis at a nonvinylic position ⟨79JA1343, 79JA3967⟩. Substitution products formed by the S_N1 route from vinylic systems carrying β-hydrogen atoms are often accompanied by acetylenic products, the latter being formed by an E1 process on deprotonation of the intermediate cation. In certain cases, however, when the proper geometry, substituents and base are [illegible] reactions may also compete ⟨64HCA194, 69JA4600, 71HCA2060, 72JA3629, 74JCS(P2)843⟩.

1.22.2.2 Formation of Vinylic Cations by Bond Scission

There are a number of methods involving bond scission by which vinyl cations can be generated, including deamination, fragmentation, photolysis, decarboxylation and electrooxidation; these will be discussed towards the end of this section. However, by far the most widely studied method for vinyl cation formation is solvolysis, in which a suitable vinyl derivative is reacted with a solvent such as methanol, ethanol or trifluoroethanol under acid or base catalysis.

1.22.2.2.1 Vinyl cation formation via solvolysis

Vinylic systems are fairly resistant to solvolysis, their low reactivity being attributed to a combination of the high energy of the vinyl cation and the transition states leading to it, and to the sp^2 hybridisation at C(1) which makes the C–X bond stronger than is found in aliphatic sp^3-hybridised analogues ⟨68JA7282⟩. However, consideration of a number of factors, such as the nature of the leaving group, and the presence of other substituents on the vinylic system, can affect the rate of the reaction and the final product distribution. In general, simple alkylvinyl halides do not react readily, and so extremely good leaving groups such as triflate or nonafluorobutanesulfonate are required in place of the halide. These derivatives can be prepared from the corresponding carbonyl compound, via the enol or enolate. The reaction is also aided by the presence of strongly activating substituents, especially π donors (aryl, vinyl, allyl) on C(1), or by substituents at C(2) that are capable of positive charge dispersal. The presence of such activating substituents is essential for solvolysis of the less reactive vinylic derivatives, such as the halides or even tosylates. Kinetic studies on the rates of solvolysis of vinylic derivatives have been carried out giving much information on the relative effects of activating substituents and/or leaving groups. Many of these results have been discussed in detail in an excellent review, and so only those general points which need to be considered when generating vinyl cations by the solvolysis method will be considered here.

The solvolysis of α-bromostyrenes in aqueous ethanol buffered by triethylamine (Scheme 38) represents the first example of vinyl cation generation by this method ⟨64HCA194⟩. The intermediacy of the vinyl cation was indicated by a number of factors, such as the first-order kinetics, the strong effect of the aromatic substituent Y, the faster reaction rate in 50% ethanol as compared with 80%

ethanol, the independence of base concentration, and the products obtained, i.e., acetophenones. The stabilisation of the vinyl cation by the aryl group is nicely illustrated in this reaction, the corresponding unsubstituted or alkylvinyl bromides being unreactive under these conditions. The presence of other substituents at the C(2) or β position can further enhance the reactivity of the vinylic derivative by inductive, steric and participation effects. Charge dispersal onto the β carbon atom of a β substituted vinyl cation has been shown to be quite considerable by *ab initio* calculations ⟨77JOC3004⟩, and this is also true for the α-aryl-β-substituted vinyl cation, although to a smaller extent. The steric and inductive effects of β substituents are not easily separated. Anchimeric assistance by the β-substituents (e.g., sulfur derivatives) has been found in the solvolysis of α-arylvinyl derivatives; this effect has been studied extensively ⟨68CC1363, 68CC1676, 68TL4039, 71JCS(B)374, 72JCS(P2)2026⟩. The intermediacy of the thiirenium ion is implicated in, for example, the solvolysis of the vinyl ethers (**30**) and (**31**). Reaction with methanol of (**30**) : (**31**) in the ratios of 3.1 : 1 or 1 : 2.1 gave an identical mixture of products (**32**) (82%) and (**33**) (18%), indicating that the reaction had proceeded via a thiirenium ion (**34**) (Scheme 39) ⟨69CC1520, 71JCS(B)374, 71JCS(B)381⟩. Solvolysis of *trans*-1,2-diphenyl-2-(phenylthio)vinyl tosylate (**35**) in acetic acid would be expected to proceed via the formation of a vinyl cation intermediate (**36**), to give the corresponding vinyl acetate. However, studies in which the tosylate was ^{13}C labelled at the 1 position showed that ^{13}C was distributed evenly between the two alkenic carbons of the corresponding acetates (**37**) and (**38**), leading to the postulated mechanism shown, in which a thiirenium ion (**39**) is involved (Scheme 40) ⟨82CJC735⟩. Further examples have been reported in which the intermediacy of thiirenium ions is implicated ⟨71JCS(B)381, 71JCS(C)3018⟩. In contrast, there is no evidence for nitrogen or oxygen anchimeric assistance on the basis of kinetic data; such reactions are thought to proceed via the open chain vinyl cations ⟨72JCS(P1)216, 72JCS(P1)1136, 76BCJ2560, 76CC612⟩. The effect of the leaving group on the rate of solvolysis of α-arylvinyl derivatives is similar in most cases to its effect on the solvolysis of saturated systems ⟨69JA5246, 70TL3233, 71JCS(B)1569, 71TL4039, 72JOC1174, 73JCS(P2)301, 74JA586, 74JA4518, 75JA821, 75JA836, 76BCJ2560⟩.

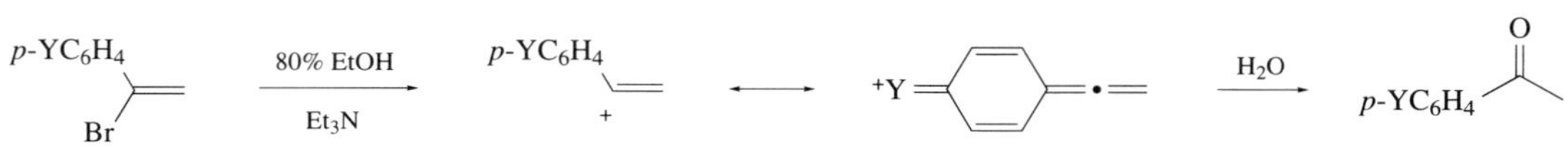

Scheme 38

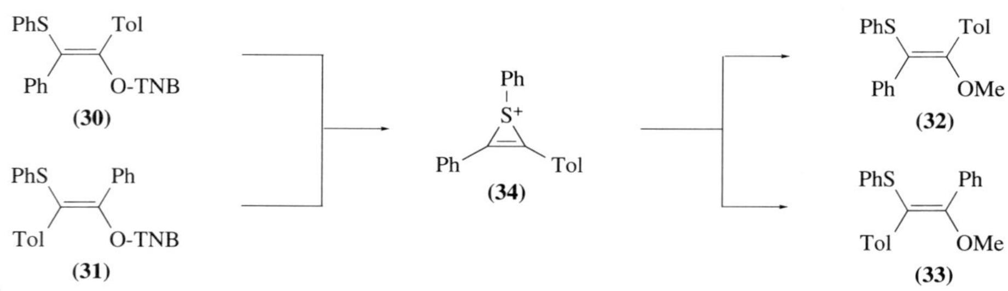

Scheme 39

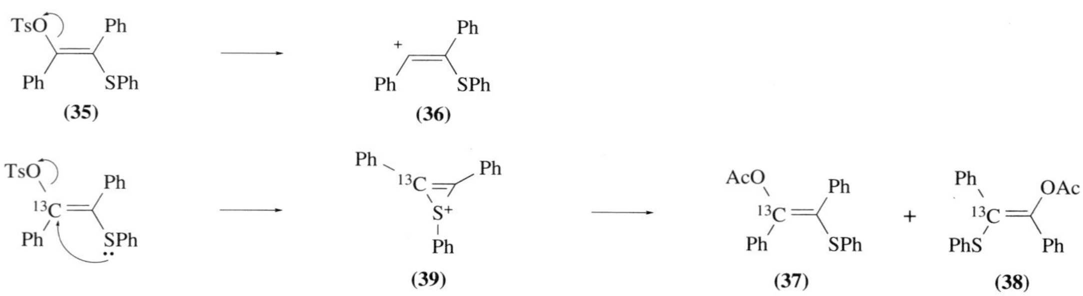

Scheme 40

When a double or triple bond is attached to the vinyl system at either C(1) or C(2), two opposing effects on solvolysis are observed. On the one hand, unsaturated groups have a negative inductive effect thereby slowing the rate of reaction, whilst on the other hand they exhibit a positive resonance (mesomeric) effect through allylic type delocalisation and so enhance reaction rates; the greater the efficiency of overlap between the π system and the vacant p orbital, then the more dominant is this latter effect. Solvolysis of these systems is therefore strongly geometry dependent.

The first example that could be considered under this category is where the vinylic system is substituted at the C(1) position by another vinyl group (Scheme 41). Here, stabilisation of the intermediate hybrid cation depends on the deconjugation of the two double bonds and on their *ortho*-gonality in the cation. Solvolysis of acyclic unsubstituted and methyl substituted 2-bromo-1,3-dienes in 80% ethanol gave three products ⟨69TL1439, 70HCA2119⟩ (Equation (64)). Each can be derived formally from the hybrid cation in Scheme 41. Successive methyl substitution enhances the reactivity of the system appreciably, through a combination of steric and electronic effects. When considering small ring systems such as 2-substituted 1,3-cyclohexadienes where the double bonds are incorporated into the ring itself, overlap of the π and the p orbitals is geometrically restricted. Under these conditions the negative inductive effect of the double bond predominates, and the vinyl cation is not formed ⟨70HCA2130, 78LA1894⟩. As ring size increases, however, approach to co-planarity of the vinyl cation is possible, allowing its formation from 1,3-cycloheptadienyl derivatives ⟨73CB3216⟩, 1,3-cyclooctadienyl derivatives ⟨73CB3216⟩, and cyclooctatetraenyl derivatives ⟨78JA7066⟩. *Ab initio* calculations provide further evidence for the importance of geometry in vinyl cation formation ⟨77JOC3004⟩. When the two double bonds are orthogonal, the α-vinyl substituent stabilises the vinyl cation (**40**) by 14.3 kcal mol^{-1} as compared with the 2-propenyl cation (**41**) (Equation (65)). The conformation with the π(C=C) and $p(C^+)$ orbital orthogonal is, by contrast, 22.2 kcal mol^{-1} less stable than the conformation when the two orbitals are in the same plane, thus reflecting the negative inductive effect of the double bond.

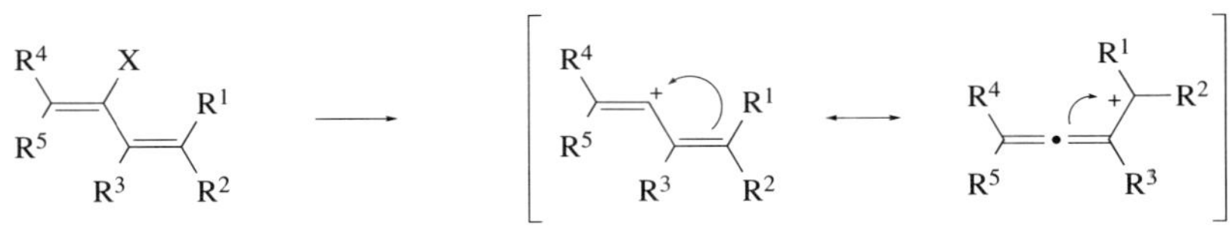

Scheme 41

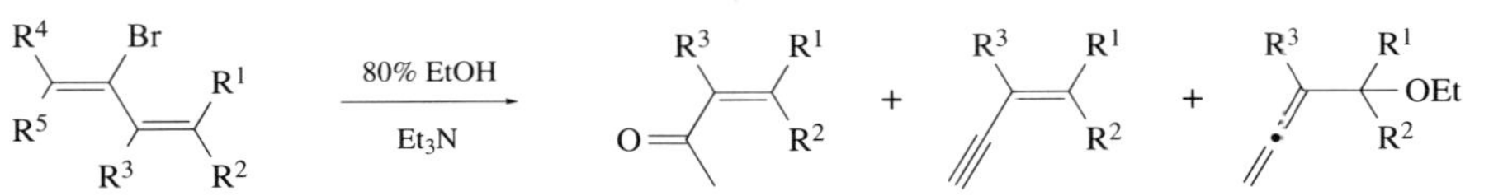

(64)

(40) (41) (65)

The α-ethynylvinyl system provides the second example of π-bond stabilisation of the vinyl cation (Scheme 42). *Ab initio* calculations have shown that the stability of the α-ethynyl substituted vinyl cation (**42**) in the gas phase is nearly the same as that of the 2-propenyl cation (**43**), reflecting the compensation of the electron-donating π character and the electron-withdrawing inductive effect of the triple bond (Equation (66)) ⟨77JOC3004⟩. These results have been verified experimentally ⟨80CC103, 80TL503⟩.

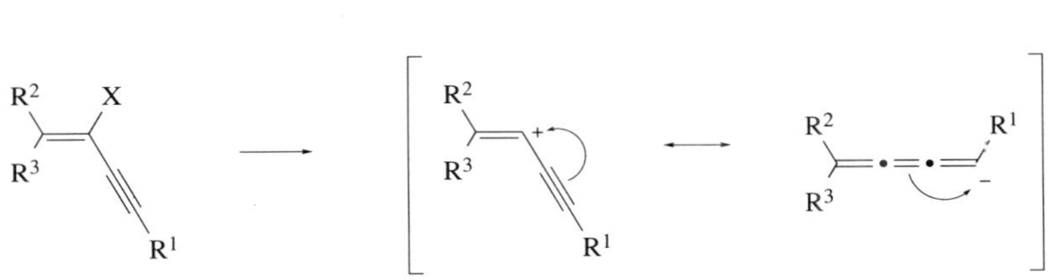

Scheme 42

(42) (43) (66)

The third example of π-bond stabilisation of the vinyl cation is found when the vinyl group is part of an allenyl system (Scheme 43). In this system, the two double bonds are *ortho*-gonal to one another and overlap of the C(2)—C(3) bond with the vacant orbital is at a maximum in the hybrid ion. (The generation of vinylic cations by addition of electrophiles to allenyl groups will be considered in Section 1.22.2.3.2). The allenic system has a high ground state energy which, coupled with the perfect disposition of the double bond for overlap with the vacant p orbital, makes the allenyl system highly reactive. The solvolysis of α-haloallenes has been studied ⟨70JA1082, 71JA6989, 72JA5061, 73JOC3054, 73JA7916, 76JOC681, 77JA3769⟩ showing that reaction rates are reasonable, with even a primary allenyl cation being generated solvolytically. Allylic stabilisation by the additional double bond in these systems is similar regardless of whether it is attached to C(1) or C(2). Solvolysis of haloallenes gives both propargylic and allenic products, with the latter tautomerising in aqueous organic media to give the α,β-unsaturated ketones (Scheme 44) ⟨65JOC1808, 70JA1082, 71JA6989, 72JA5061, 78JOC763⟩.

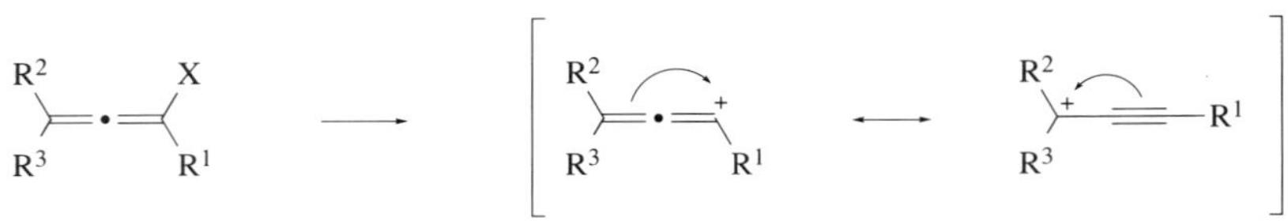

Scheme 43

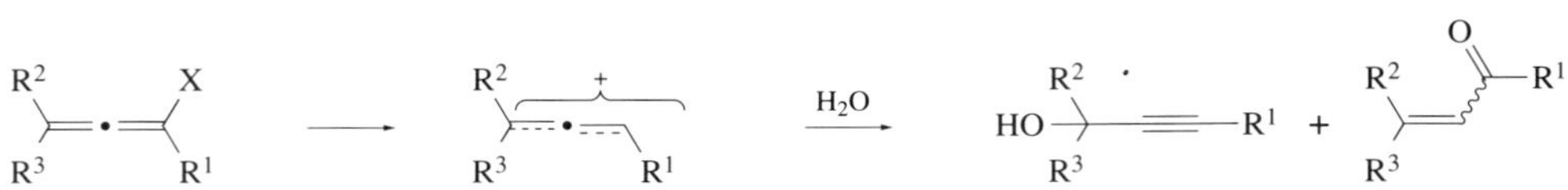

Scheme 44

Relative solvolytic reactivities of vinylic and saturated halides and triflates for all three systems (vinyl-substituted, ethynyl-substituted and allenyl) have been determined ⟨48JA846, 51JA2700, 57JA4146, 69JA5246, 70HCA2119, 70JA2538, 70JA2540, 70JA3789, 71JA6989, 71TL2171, 71TL2519, 72JA992, 72JOC1174, 74JA1100, 76JA7658, 76JA7667, 76JOC681, 80CC103, 80TL503⟩.

Closely related to the above systems are those vinyl derivatives that are substituted by a cyclopropyl group, either as a substituent at the α position, or directly attached to the β position. Silver salt promoted acetolysis of cyclopropylvinyl iodide is several orders of magnitude more rapid than that of the corresponding acyclic analogue ⟨69JA2117⟩. α-Cyclopropylvinyl chloride reacts similarly ⟨69JA2115⟩. Cyclopropyl vinyl cations can rearrange to give cyclobutyl cations; α-cyclopropylvinyl halides, on solvolysis, give cyclobutenyl and methylenecyclobutyl derivatives amongst the products ⟨71JA1925⟩. Thus acetolysis of the isomeric vinyl iodides (**44**) and (**45**) gave identical mixtures of products (Scheme 45) ⟨71JA1941⟩. The (E) isomer (**45**) is more reactive by a factor of 10.3 than the (Z) isomer (**44**). This is a general trend ⟨73CC589⟩. This difference in reactivity is attributed to the relief of steric strain between the α-cyclopropyl and the β-alkyl group. The distribution of the cyclopropyl, cyclobutyl and homoallyl products depends on the substituents both on C(2) and on the cyclopropyl ring ⟨73CC589, 74T243, 75JA1531⟩.

When cyclopropylidenemethyl bromide (**46**) is solvolysed in buffered 50% methanol, cyclobutanone (**47**) was obtained as the sole product (Scheme 46) ⟨71TL2171⟩. The formation of the vinyl cation (**48**) is indicated by the independence of the reaction rate on base concentration and on solvent effects ⟨74JA6686⟩. A methyl substituent on the cyclopropyl ring changes both the products formed and the reaction rate. Thus, when the isomeric bromides (**49**) and (**50**) are solvolysed in 50% ethanol the same product mixture distribution is observed, in which the open chain alcohol (**51**) predominates ⟨74CB1676⟩ (Scheme 47). The stabilising effect of the methyl group is reflected in the 10-fold rate increase of (**49**) and the 27-fold rate increase of (**50**) over (**46**) in 60% ethanol ⟨74CB1676⟩. When the methyl substituent is in the α position, as in vinyl bromide (**52**) solvolysis in various solvents (aqueous ethanol, TFE, DMF) gives (**53**) and (**54**) as the main products, with cyclopropyl methyl ketone (**55**) a minor component (Scheme 48) ⟨74JA6686⟩. In contrast, aryl-

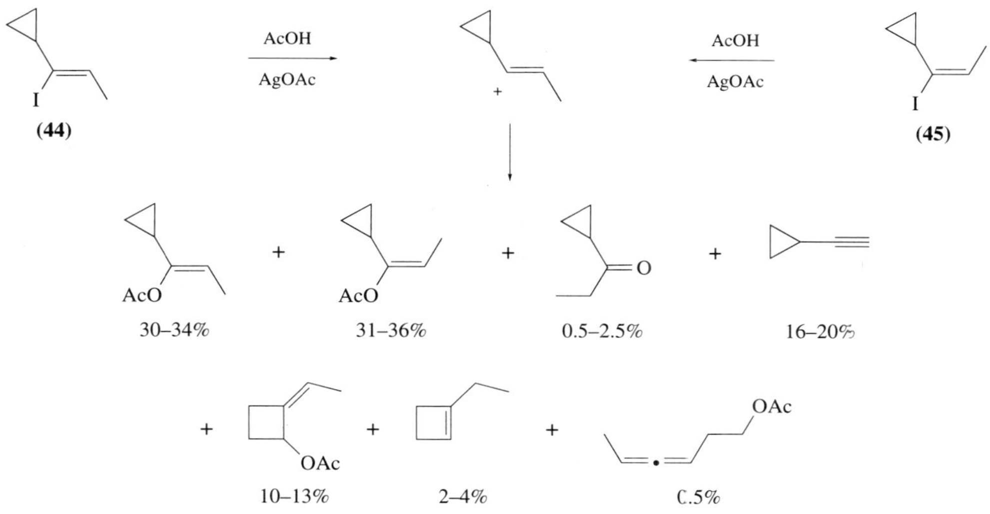

Scheme 45

substituted vinyl bromides (**56**; Ar = Ph, Tol, An) give predominantly arylcyclopropyl ketones (**57**) and 4-arylbut-1-en-3-ynes (**58**) with only a few percent of any cyclobutyl derivatives (Scheme 49) ⟨74JA6686, 75JOC1994⟩. This product distribution is due to the extensive charge delocalisation on the aryl group in vinyl cation (**59**) which is absent in vinyl cation (**60**). The same behaviour is observed in α-cyclopropyl substituted vinyl bromides ⟨79CB2453⟩.

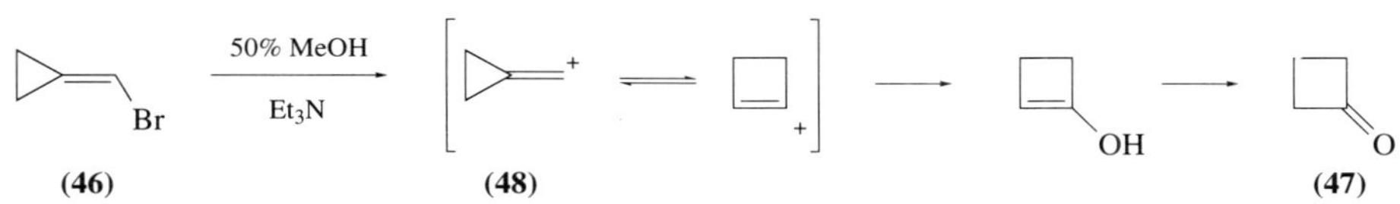

Scheme 46

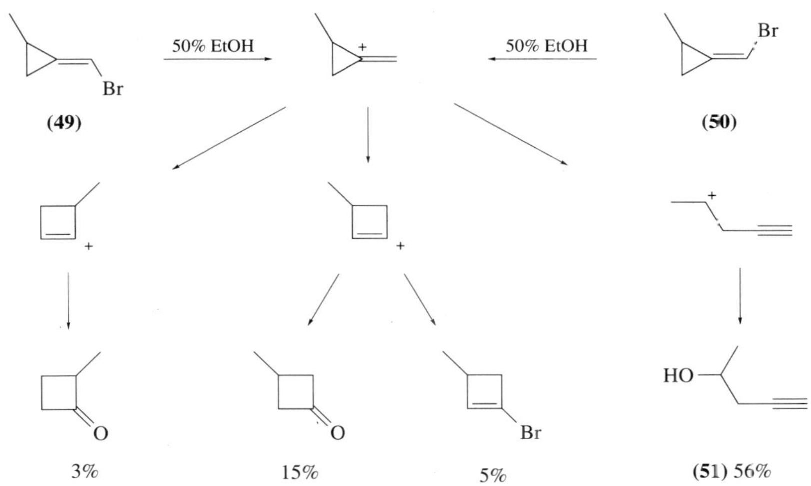

Scheme 47

In all the previous examples, the vinyl cation has been stabilised by a suitable α substituent, thus increasing the rate of reaction. It is when investigating α-alkylvinyl systems that the relatively low solvolytic reactivity of vinylic derivatives is found. Both acyclic and cyclic vinyl halides are unreactive, whilst the brosylates and tosylates react slowly ⟨69JA6194⟩. It is only when highly nucleofugal leaving groups are employed that reaction rates are convenient for kinetic and product distribution

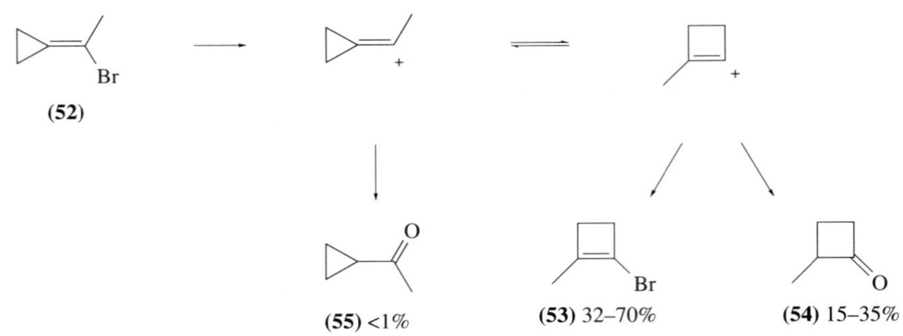

Scheme 48

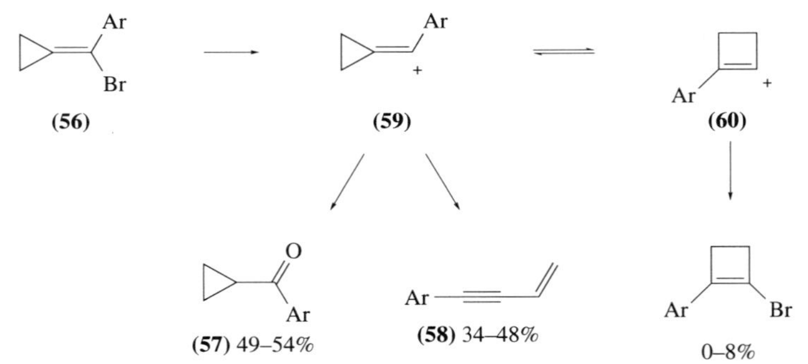

Scheme 49

studies. The effect of various alkyl groups (R = Me, Et, Bu^n, Bu^t, Pr^i) on C(1) and C(2) was investigated for substituted vinyl triflates in aqueous ethanol, 97% TFE, and AcOH with the main products in almost all cases being the alkynes, along with smaller quantities of ketones and allenes (e.g., Equation (67)). An ionic transition state is implicated in these reactions by the effect of the substituents, by the increase in solvolysis rate on addition of $LiClO_4$ or LiBr, and by capture experiments ⟨74JA1100⟩. Although E2 and S_N2 routes were excluded, there is a solvent effect indicating that the solvent plays some role in the solvolysis ⟨48JA846, 51JA2700, 57JA4146, 72JA992, 76JA7658, 76JA7667⟩. Larger substituent effects are exhibited in systems where two alkyl groups are *cis* to each other. When a better stabilising group is present on C(2) of an α-alkylvinyl system, rearrangements may occur across the double bond such that the more stable vinyl cation is obtained. For example, when a β-cyclopropyl substituent is present as in the (*E*)- and (*Z*)-isomers of (**61**), although the major products are derived from the unrearranged cation (**62**), 11–17% of cyclopropyl ethyl ketone (**63**) is also formed, presumably from hydride migration across the double bond of (**62**) to give the more stable cation (**64**) (Scheme 50) ⟨74TL1637⟩. A similar extent of hydride migration is found in solvolysis of the β-phenyl substituted vinyl derivative (Equation (68)) ⟨77CB199⟩. In the β,β-diphenyl substituted vinyl derivative (**65**), phenyl migration occurs to such an extent as to give exclusively the rearranged ketone (**66**) (Scheme 51) ⟨70JA3802⟩.

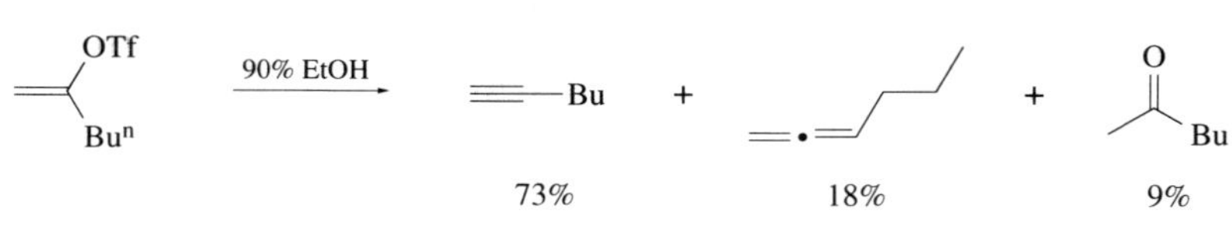
(67)

(68)

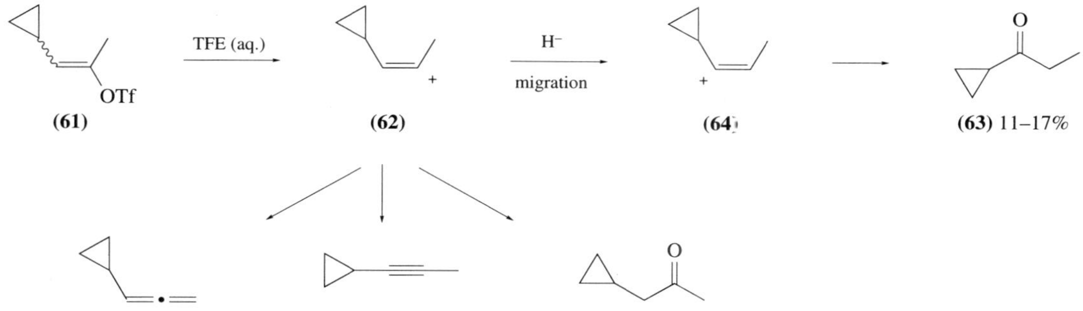

Scheme 50

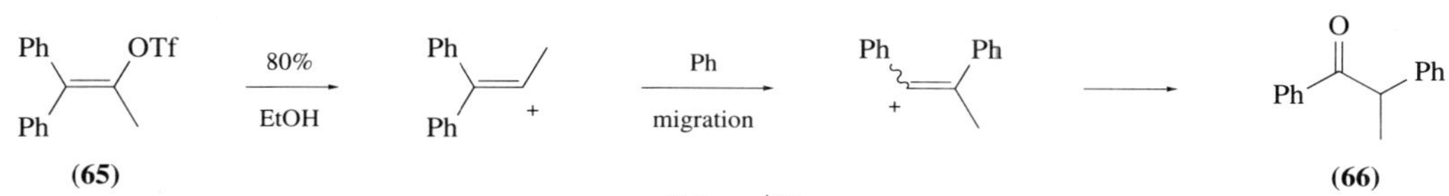

Scheme 51

In cyclic vinyl triflates, the α-alkyl group is part of the ring system (Equation (69)). When the ring is small ($n = 2$–4), the cation derived from solvolysis would be strained, leading to decreased reaction rates. This has been found to be the case ⟨72AG(E)714, 77JOC174⟩. Increasing the ring size leads to an increase in solvolytic reactivity up to $n = 9$, after which the relative reaction rates are roughly the same. This is the case whether 50% ethanol or 70% TFE is used for the solvolysis ⟨71JA1513, 72CB3789, 76T37⟩. For the smaller ring sizes, the lower reaction rates are attributed to the fact that the vinyl cation cannot achieve the preferred linear geometry at C(1). The larger rings, however, are more flexible allowing them to achieve linearity, thus giving rates similar to or greater than acyclic systems, strong evidence for the sp hybridisation of vinyl cations. The ring strain found in smaller ring systems such as cyclohexyl cations leads to alkyl rearrangement to the unsaturated centre from both the saturated position ⟨71JA1513⟩ (Scheme 52) and vinylic positions (Scheme 53) ⟨B-79MI 122-01⟩. Ring strain can also be relieved in cyclic vinyl cations by rearrangement to a cation within a larger ring. Thus solvolysis of **(67)** in 60% MeOH gives both **(68)** and **(69)** the latter presumably being derived from rearrangement of cyclohexenyl cation **(70)** to cycloheptenyl ion **(71)** (Scheme 54) ⟨79AG(E)870, B-79MI 122-01, 80CB2015⟩.

(69)

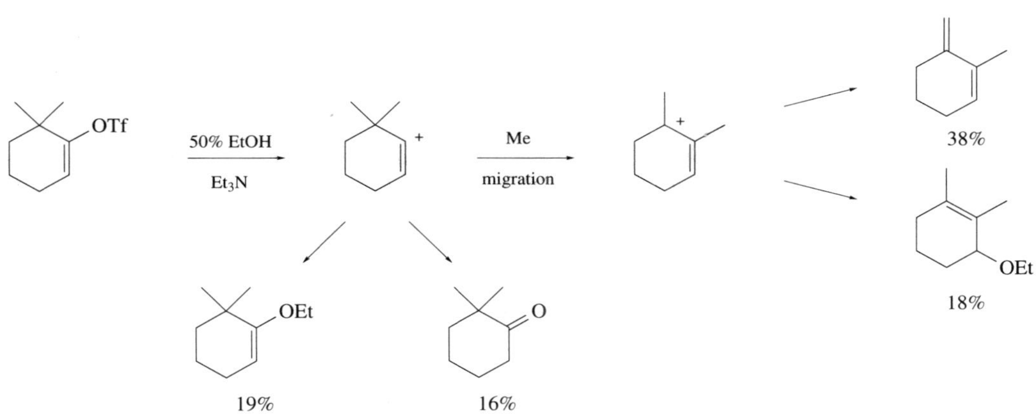

Scheme 52

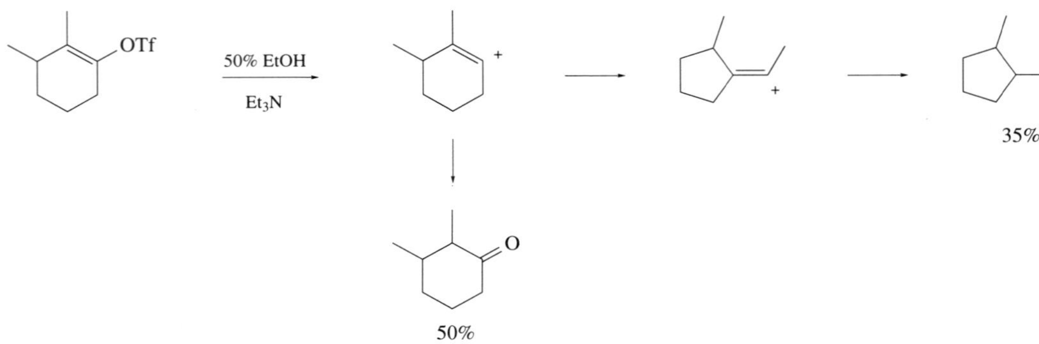

Scheme 53

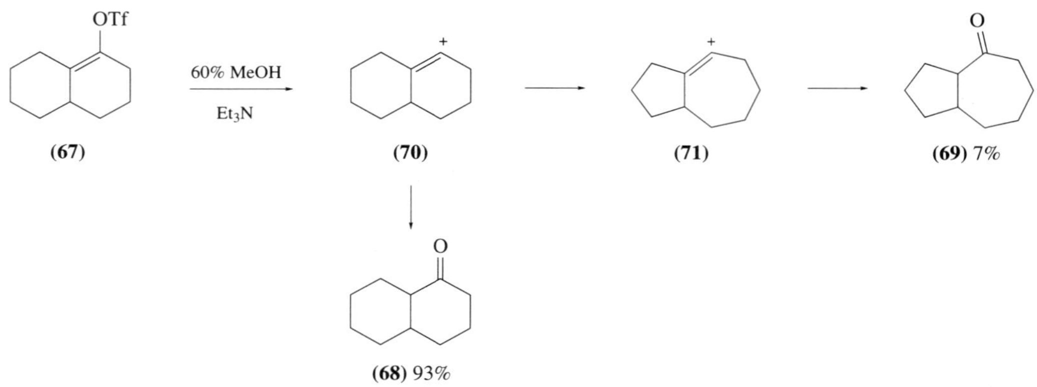

Scheme 54

1.22.2.2.2 Vinyl cation formation via deamination of diazonium salts

There are three main precursors that allow vinyl cation formation via deamination of a diazonium ion; amines, 3-nitroso-2-oxazolidones, and triazenes. One of the first attempts to generate a vinyl cation involved treatment of 2,2-diphenylvinyl amine (**72**) with nitrosyl chloride in methylene chloride (Scheme 55) ⟨65JA863⟩. The products obtained can be rationalised as being formed from the diazonium ion (**73**), expelling nitrogen to form the vinyl cation (**74**), which then undergoes phenyl migration (see 1.22.2.2.1) to give the more stable vinyl cation (**75**). It is also possible, however, that the reaction proceeds by α-elimination to the vinylidene carbene rather than via a vinyl cation ⟨78CRV383⟩.

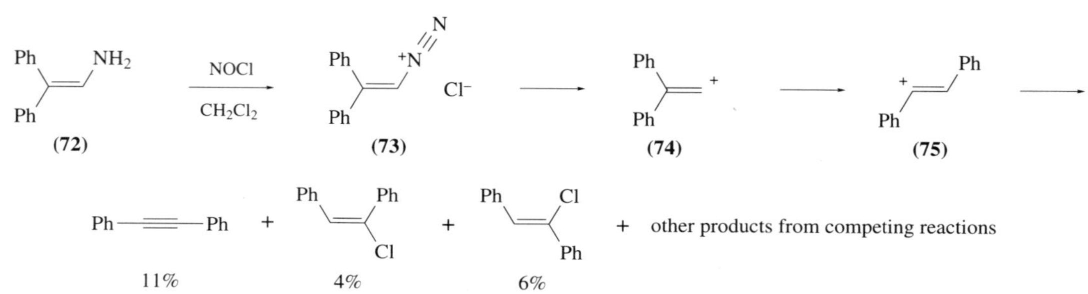

Scheme 55

3-Nitroso-2-oxazolidones (**76**) decompose in aqueous or alcoholic media under basic conditions to give aldehydes, ketones, alkynes and vinyl ethers ⟨51JA4199, 56JA4654, 69JA5677, 70JA7564⟩. The reaction can be envisaged as proceeding via the diazonium ion (**77**) which then expels nitrogen to yield vinyl cation (**78**) (Scheme 56). In aprotic media, however, it is again thought likely that the reaction proceeds via unsaturated carbenes ⟨69JOC1220⟩.

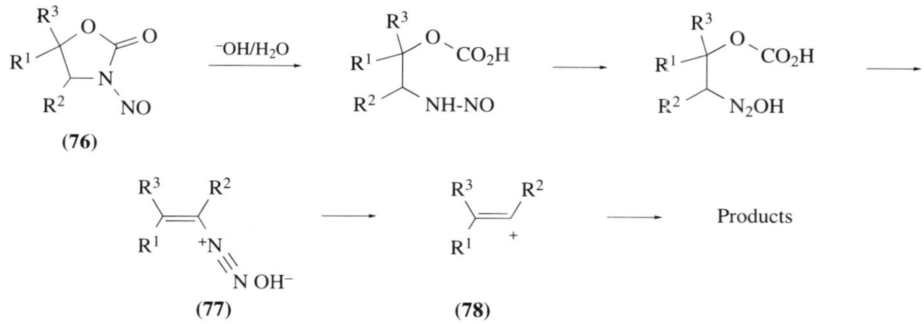

Scheme 56

Aliphatic triazenes react with acid to give products which are derived from the corresponding diazonium salts, and so the same method has been used with vinylic systems. Thus, reaction of (**79**) with acetic acid gives products which can be rationalised to have been derived from vinyl cations (**80**) and (**81**) (Scheme 57) ⟨67JA1960⟩.

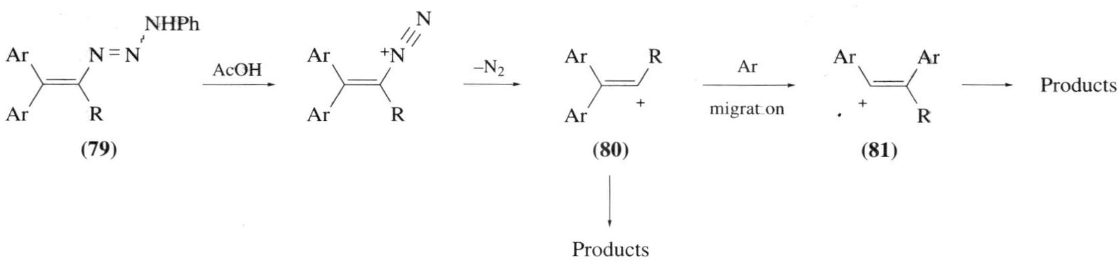

Scheme 57

1.22.2.2.3 *Vinyl cation formation via fragmentation*

Thermal decarboxylation of (*Z*)-β-bromo-α,β-unsaturated acids (**82**) gives alkynes (**83**) and ketones (**84**) as the products. The zwitterion (**85**) has been suggested to be the probable intermediate (Scheme 58) ⟨64HCA1590⟩.

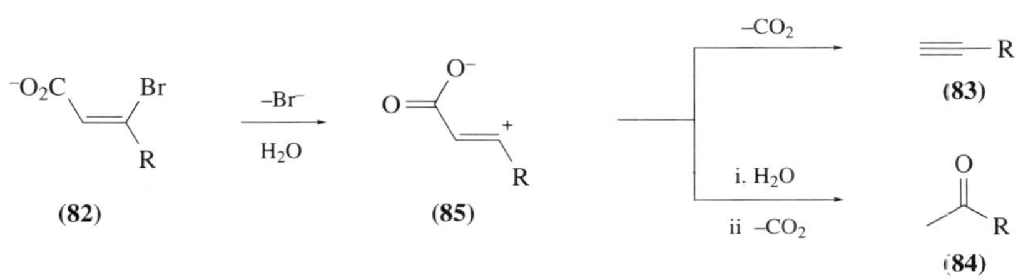

Scheme 58

Reaction of oxime tosylates (**86**) with 80% ethanol gave the ketone (**87**) and the enol ethers (**88**) and (**89**) ((*E*):(*Z*) 1:1) (Scheme 59) ⟨76TL4191, 76TL4195⟩. The latter were postulated to be derived from vinyl cation (**90**) via a fragmentation pathway. Electron-donating substituents on the aryl ring increased the fragmentation rate, and hence the yields of (**88**) and (**89**), such that it is the predominant pathway when Ar = *p*-$Me_2NC_6H_4$.

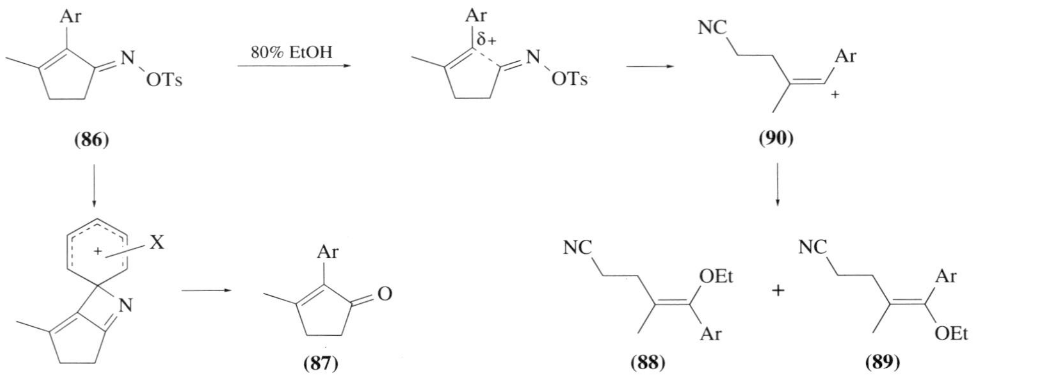

Scheme 59

1.22.2.2.4 *Vinyl cation formation via photolysis*

Vinyl cations have been suggested as intermediates in a number of photolytic reactions. For example, photolysis of pyrazoline (**91**) is postulated to proceed via a vinyl cation to give allene (**92**) and diene (**93**) as products (Scheme 60) ⟨67JCS(B)991⟩.

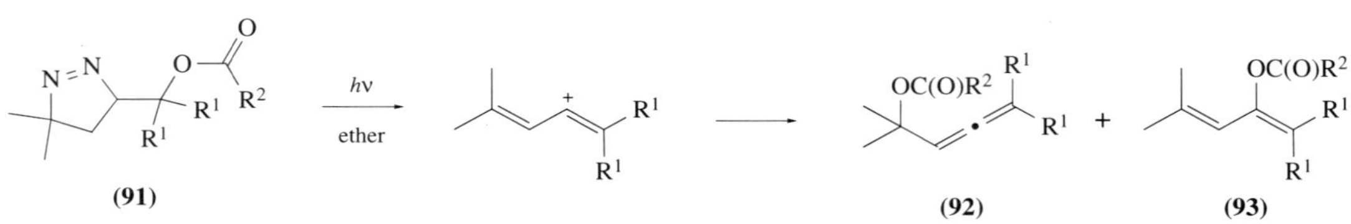

Scheme 60

Photolysis of vinyl iodide (**94**) in methylene chloride was suggested to form initially a radical pair which then underwent electron transfer to give a vinyl cation iodide ion pair (**95**). Alkyl shift followed by abstraction of chloride from the solvent gives cycloheptenyl chloride (**96**), one of the products from this photolysis reaction (Scheme 61) ⟨76JA4319⟩. A similar mechanism has been invoked to explain the formation of alkynes (**98**) and benzofurans (**99**) on irradiation of 1,1-diarylethylenes (**97**) (Equation (70)) ⟨76CC180⟩, and for alkyne formation from (*E*)-1,2-dianisylvinyl bromide (**100**) (Equation (71)) ⟨79TL1619⟩. Evidence that these reactions proceed via a vinyl cation has been obtained from capture experiments ⟨79TL1619⟩, and from flash photolysis ⟨80T3229⟩.

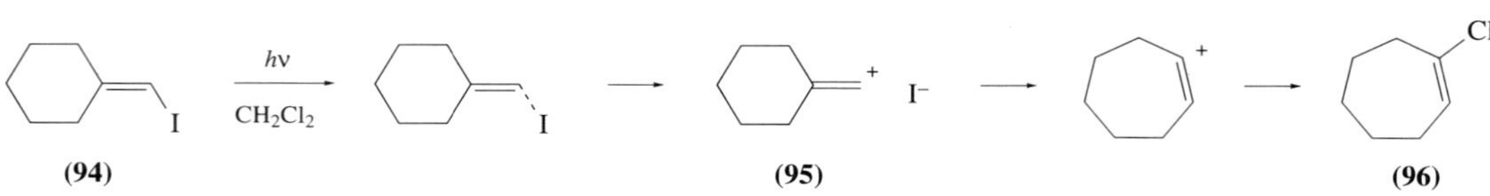

Scheme 61

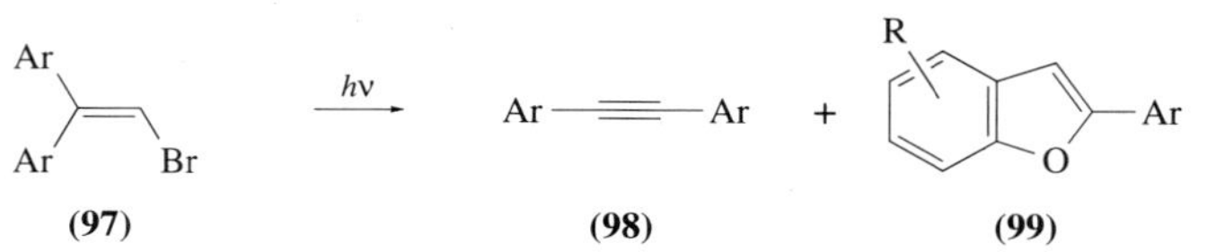

(70)

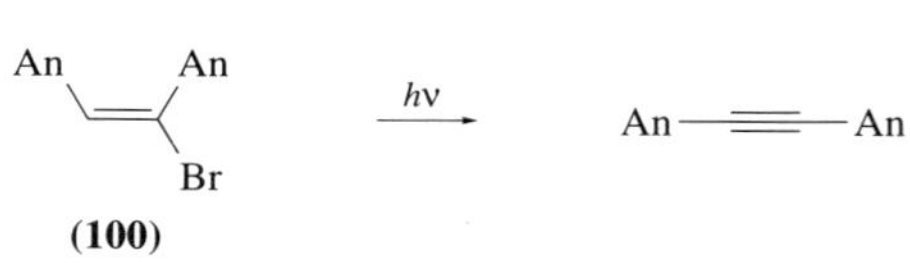

(71)

1.22.2.2.5 *Vinyl cation formation via oxidative decarboxylation*

Oxidative decarboxylation of β-anisyl-α,β-unsaturated acids (**101**) by lead tetraacetate is thought to proceed in a similar manner to that discussed above for photolysis reactions (Scheme 62) ⟨73CI(L)1067⟩.

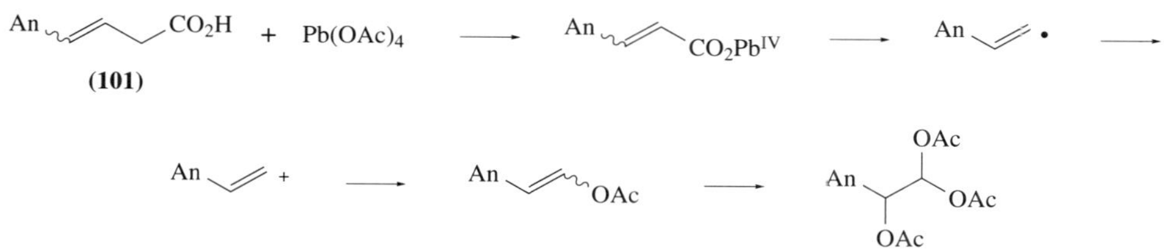

Scheme 62

1.22.2.2.6 *Vinyl cation formation via electrooxidation*

Electrooxidation of (**102**) in aqueous *t*-butanol was suggested to involve formation of the cation radical (**103**), followed by disproportionation to vinyl cation (**104**) and cation (**105**), yielding the ketone (**106**), benzoin (**107**) and benzil (**108**) as products (Scheme 63) ⟨71TL1385⟩.

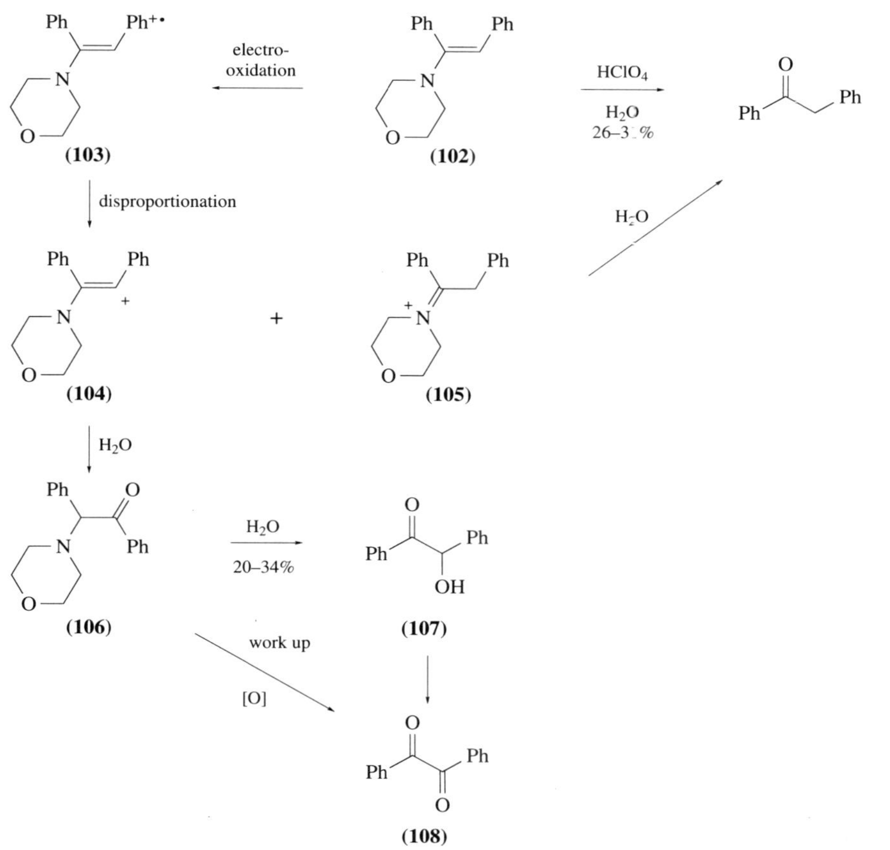

Scheme 63

1.22.2.2.7 *Vinyl cation formation via mass spectrometry*

The mass spectra of many vinyl halides and tosylates show peaks with m/e values corresponding to those of the derived vinyl cations ⟨68JA2432, 81JA2770⟩.

1.22.2.3 Formation of Vinylic Cations by Addition to Alkynes and Allenes

1.22.2.3.1 *By addition to alkynes*

The addition of an electrophile to an alkyne results in the formation of a new σ bond and a vinyl cation (Scheme 64). Attack can occur at either carbon atom of the alkyne to give, in theory, two possible products. In practice, however, it is found that one product predominates often to the exclusion of the other; the product ratios can usually be rationalised by considering the stability of the intermediate carbocations. In most cases, R^1 and R^2 are of sufficiently different stabilising ability that only one product is obtained. For terminal alkynes, for example, it is the internal carbon atom which bears the positive charge (Schemes 65 and 66 and Equation (72)) ⟨44JA686, 60RTC1002, 61RTC797, 63RTC375, 63RTC394, 63RTC410, 65JCS384, 68JA1020, 68JA1023, 76JA3295⟩. When $R^1 \neq R^2$, the product obtained can be rationalised by assuming the more stable carbocation is formed; thus, alkynyl ethers and thioethers give the corresponding esters and thioesters, and not the isomeric ketoethers ⟨61RTC1285, 63RTC405, 65JCS384, 68JA1020, 68JA1023⟩ (Equation (73)). Ynamines yield amides as the only product (Equation (74)) ⟨74JA6692⟩. *t*-Butyl and benzyl cations, generated under Friedel–Crafts conditions, add to alkylphenylalkynes to give the products derived from α-phenylvinyl cations (Scheme 67) ⟨76JA3295⟩. When R is electron withdrawing, an α-arylvinyl cation is formed preferentially ⟨68JA372⟩. The direction of addition of electrophiles to disubstituted alkynes is consistent with the relative rates of addition to the individual monosubstituted alkynes, although caution should be exercised when applying rate data for the prediction of the regiospecificity of a reaction ⟨65JA2295, 74JA1110, 74JOC1438, 75JA1845, 79JOC3076⟩. The importance of substituents and their ability to control carbocation formation can be seen in the synthesis of steroids (Scheme 68) ⟨73JA4416, 78TL2549⟩. When R = Me, five-membered ring formation is observed via a linear vinylic carbocation; this is attributed to the higher stability of the linear over the bent vinyl carbocations and to higher nonbonded interactions in the transition state. When R = TMS, however, six-membered ring formation is observed, attributable possibly to the tendency of a silicon atom to stabilise a positive charge at a β-position ⟨78TL2549⟩.

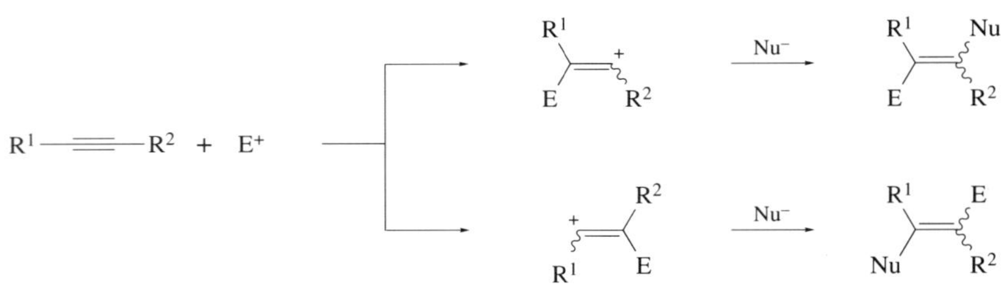

Scheme 64

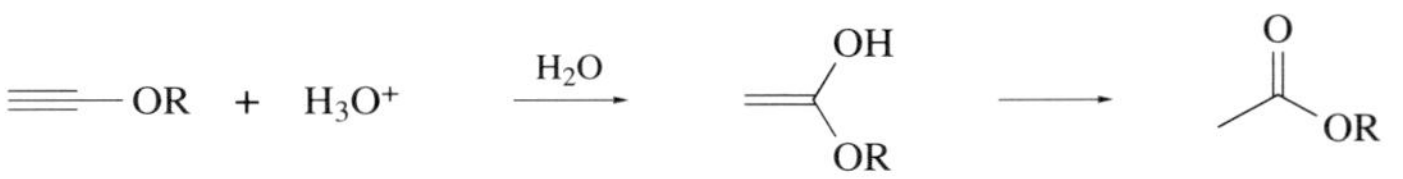

Scheme 65

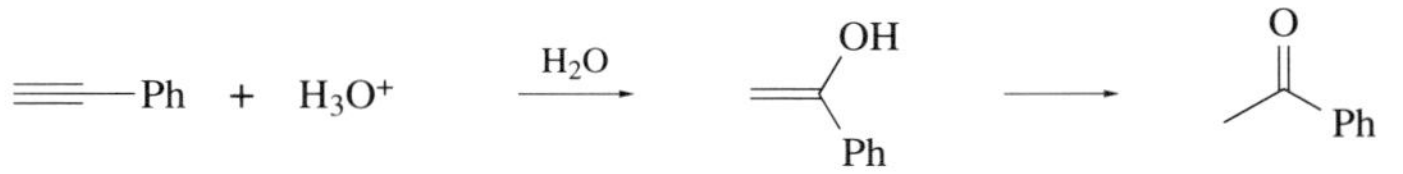

Scheme 66

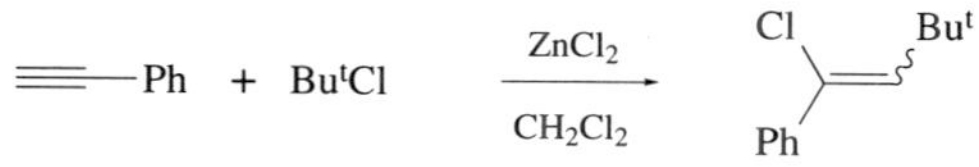

(72)

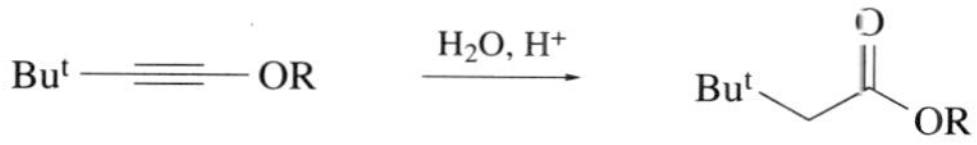

(73)

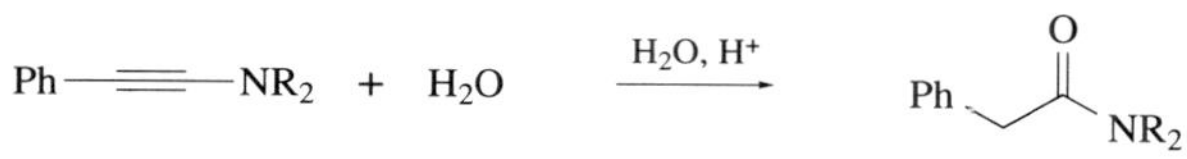

(74)

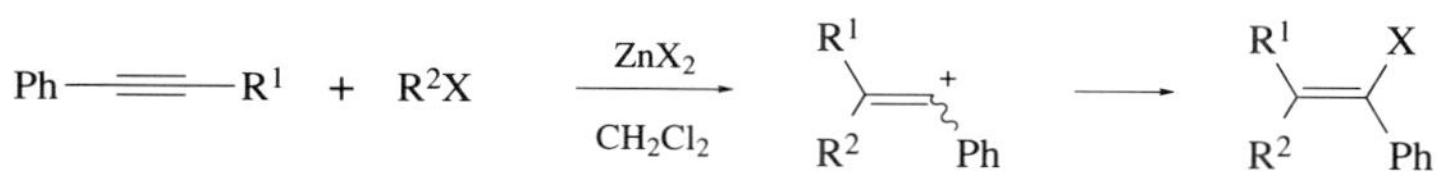

Scheme 67

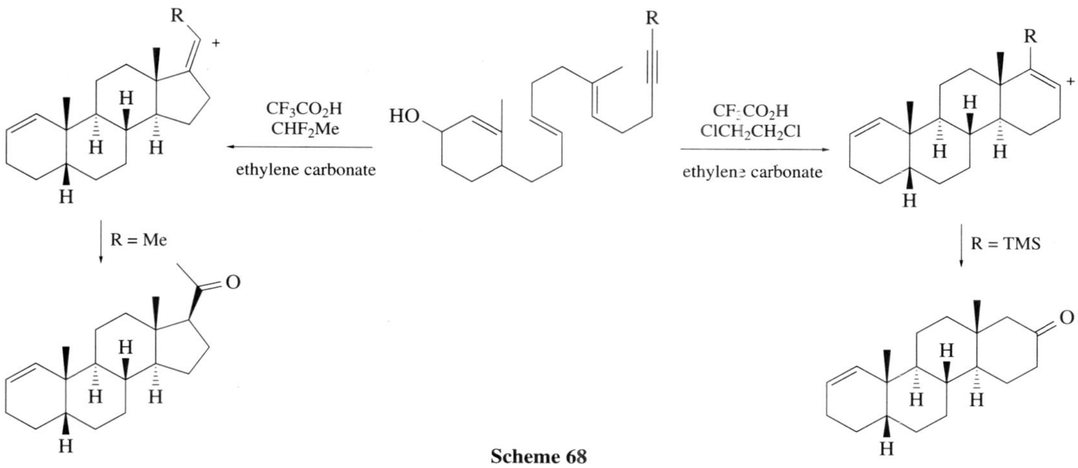

Scheme 68

The addition of halogens and of sulfenyl and selenenyl halides to alkynes is thought to involve the formation of a bridged species. Evidence for this has been obtained from the reaction of sulfenyl halides with an unsymmetrical alkyne to give two regioisomers, presumably via a thiirenium ion (Scheme 69) 〈65TL4399, 68JA3466〉. Such ring openings are subject to electronic and steric effects as well as to solvent effects 〈65TL4399, 65TL4405, 76CJC3045, 76CS76, 76JOC2331〉.

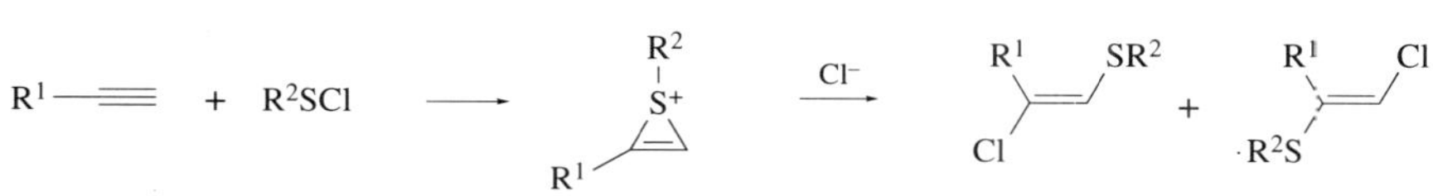

Scheme 69

The attacking electrophile does not necessarily have to be an external species, but could be a carbocation located in the alkyl chain containing the triple bond. The intermediate vinylic carbocations formed in these reactions often undergo rearrangements 〈61LA(746)211, 65TL875, 66TL3847, 67TL3871, 70JA2542, 70LA(733)5〉. For example, homopropargyl derivatives such as deuteriated 3-pentyn-1-yl *m*-nitrobenzenesulfonate, when solvolyzed in buffered trifluoroacetic acid, give cyclobutanone derivatives deuterated equally at C(3) and C(4); this result suggests the intermediacy of the cyclopropylidenevinyl cation which undergoes bond migration, rather than direct cyclization to yield a four-membered ring (Scheme 70) 〈68TL4617〉. A number of reports of this homopropargyl rearrangement can be found in the literature 〈69JA3238, 72TL2983, 73TL4767, 74JA6686, 74TL2457, 76JA862, 77JA1669, 77JOC174〉.

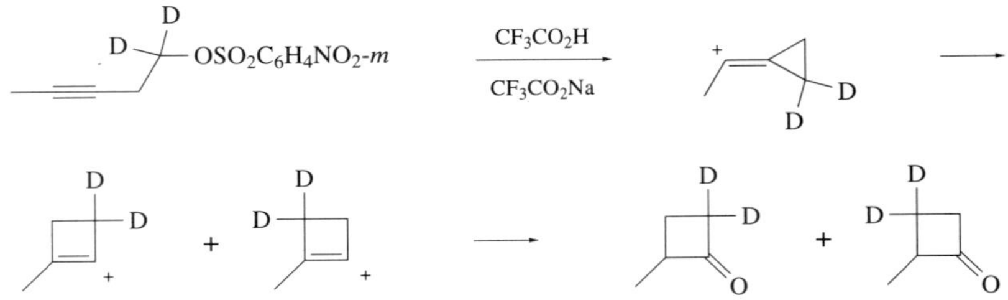

Scheme 70

Cyclisation to give larger rings has been observed in systems where the triple bond is at an even more remote position. Which end of the triple bond is attacked is dependent on the substitution of the alkyne, the intermediate vinyl cation then being either *endo*- or *exo*-cyclic (Scheme 71). For example, when 5-hexyn-1-yl triflate (**109**; R = H, X = OTf) is solvolyzed in buffered TFE, the open chain ether (**110**) is obtained in 58% yield, along with the cyclohexenyl derivatives (**113**; X = OTf, 12%) and (**113**; X = OCH_2CF_3, 24%). The latter compounds are presumably derived from the intermediate *endo*-cyclic vinyl cation (**112**). In this instance, there is no trace of products derived from the *exo*-cyclic carbocation (**111**). On the other hand, when 5-heptyn-1-yl triflate (**109**; R = Me, X = OTf) is solvolysed, both six-membered and five-membered ring products are obtained with the cyclopentene derivatives predominating. When the substituent is a phenyl group as in 6-phenyl-5-hexyn-1-yl acetate (**109**; R = Ph, X = OAc) the only cyclic products obtained are cyclopentene derivatives presumably derived from the *exo*-cyclic vinyl cation (**111**; R = Ph). Similar behaviour is shown by analogous systems 〈66JA3152, 69JA4521〉. Seven-membered rings have been formed from heptyn-6-yl derivatives, although to a much lesser extent. Transannulation reactions involving cyclodec-5-ynols have also been reported in the literature, the cyclisations proceeding by way of a vinyl cation intermediate (Equations (75)–(77)) 〈71CJC3135, 71TL927, 72CB421, 77TL4377〉.

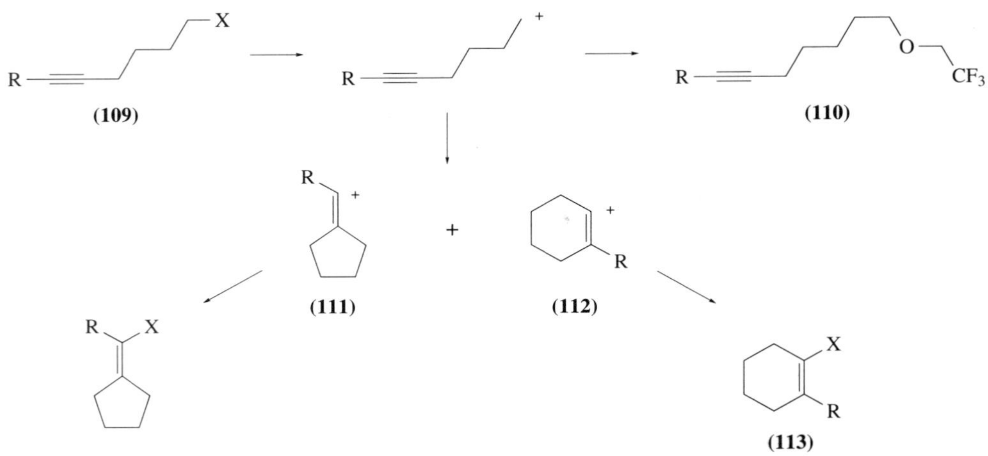

Scheme 71

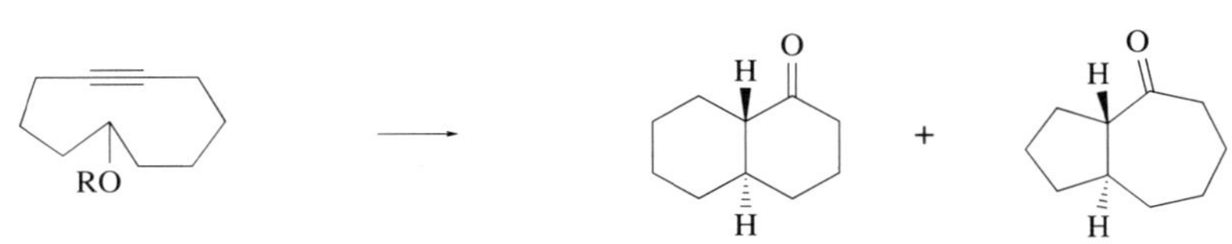

(75)

R = H	HCl, MeOH		75%	
R = Ts	80% EtOH, 2,6-lutidine	93%		7%
R = PNB	50% EtOH, 2,6-lutidine	92%		8%

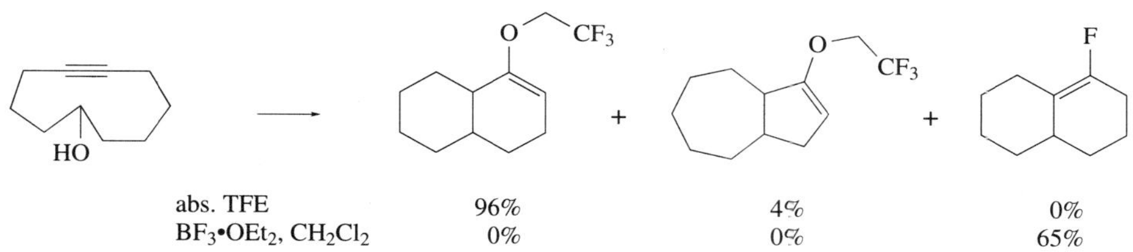

(76)

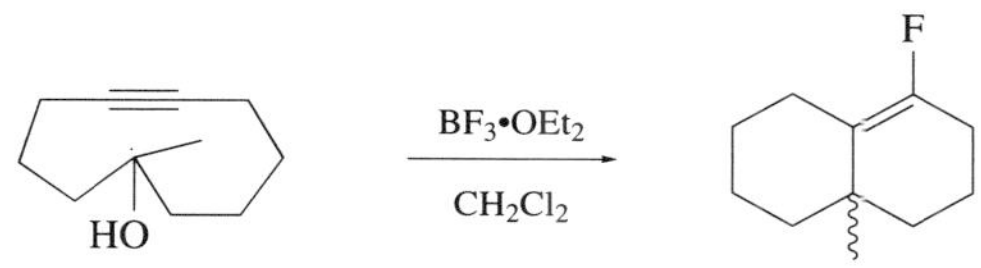

(77)

1.22.2.3.2 *By addition to allenes*

Addition of an electrophile to an unsymmetrical allene can occur at any one of the three unsaturated carbon atoms. Attack at C(1) or C(3) results in vinyl cation formation, whilst attack at C(2) generates an allylic cation. The position of attack is dependent mainly on the nature of the substituents R^1, R^2, R^3, R^4, and their ability to stabilise the intermediate transition state. When both the terminal carbon atoms are disubstituted, that is when R^1, R^2, R^3, $R^4 \neq H$, then electrophilic attack occurs preferentially at the central carbon atom of the allene system generating a tertiary allylic carbocation (Equation (78)) ⟨69CC122⟩. Aryl substituted allenes also give products arising from attack at the central carbon atom yielding, eventually, the more stable cinnamyl derivatives (Scheme 72) ⟨49JA1964⟩. Initial protonation of the central carbon atom has been reported for a number of other substrates, for example, R^1, $R^2 = Me$ ⟨60JA6397⟩, R^2, $R^4 = Me$ ⟨69CC122⟩, $R^1 =$ alkyl, $R^4 = OEt$ ⟨65RTC31⟩, and $R^1 = Me$, $R^3 = Et$ ⟨68BSF2120⟩. It should be pointed out that, although allyl cation formation is eventually achieved in some cases, the initially formed allyl cation is not stabilised by resonance owing to the orthogonality of the two π orbitals and the perpendicular direction of attack of E^+ on the double bond. The vacant p orbital and the double bond in the initially formed ion (and in the transition state leading to it) are orthogonal; the stabilised allyl cation is achieved upon rotation by 90° about the σ bond.

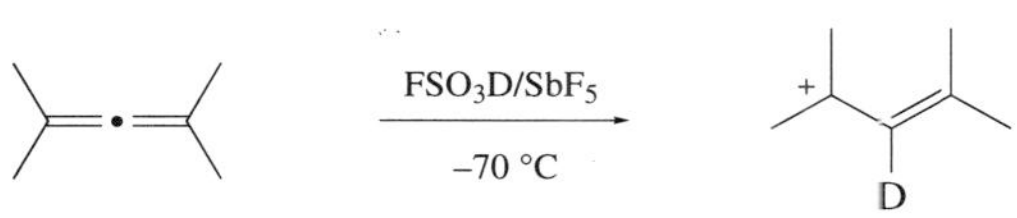

(78)

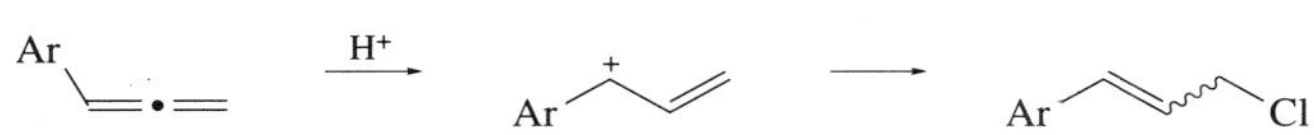

Scheme 72

Vinyl cation formation is observed in the reaction of hydrogen chloride with 1,2-butadiene (Scheme 73) ⟨60JA6397⟩, whilst in other systems the reaction products suggest competing vinyl/allyl cation formation ⟨63JGU3508⟩. Addition of halogens to allenic systems is reported to proceed via a bridged halonium ion intermediate ⟨68JA6741, 71JA5758, 74JOC2255⟩ (e.g., Scheme 74). Bridged intermediates have also been suggested for the addition of arylsulfenyl halides to allenes ⟨73JA4090⟩.

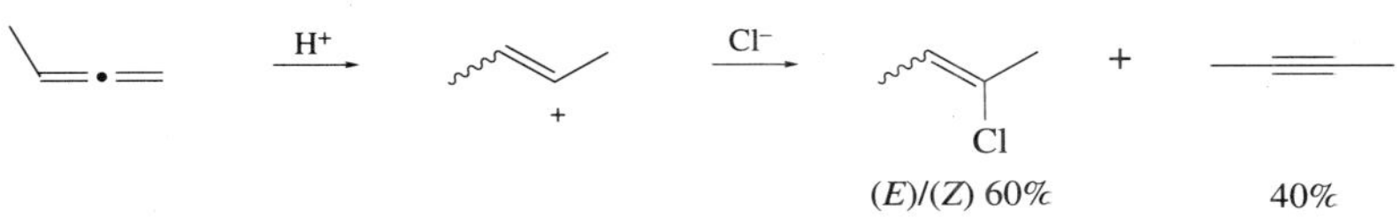

Scheme 73

As in the formation of vinyl cations by intramolecular addition of a carbocation to an alkyne, similar processes have been observed for allenes ⟨64CR(259)2251, 64TL2191, 69JA4824⟩. The product distribution is strongly dependent on the substituents present in the system, and on the ionising power

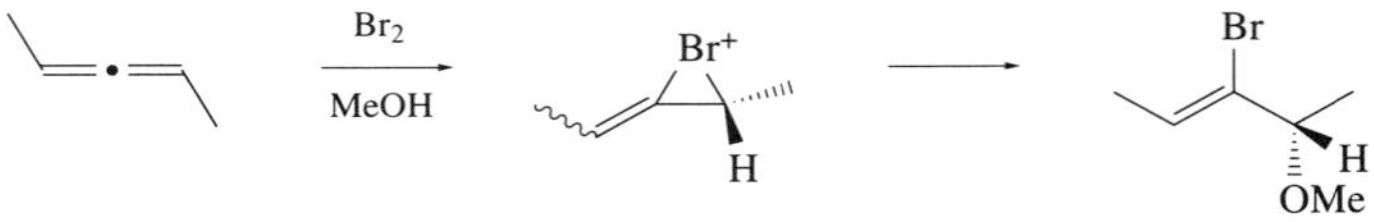

Scheme 74

of the solvent. The enol esters or ethers which are the initial products frequently tautomerise to give the corresponding ketone. Further examples of participation of allenes in solvolysis reactions, including kinetic studies, have been reported ⟨68CC718, 69JA3292, 69JA3299, 69TL3699, 74T227, 74T235⟩. When the leaving group is suitably positioned on a cyclic system, bicyclic products may be obtained (Equation (79)) ⟨69JA4824, 69TL3699⟩. Bicyclic products are also obtained when the allenic group is part of the cyclic system, presumably via a vinylic cation (Scheme 75) ⟨73BSF1800⟩.

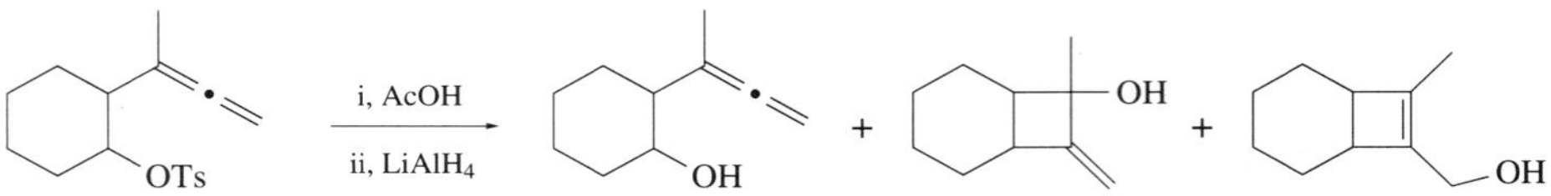

(79)

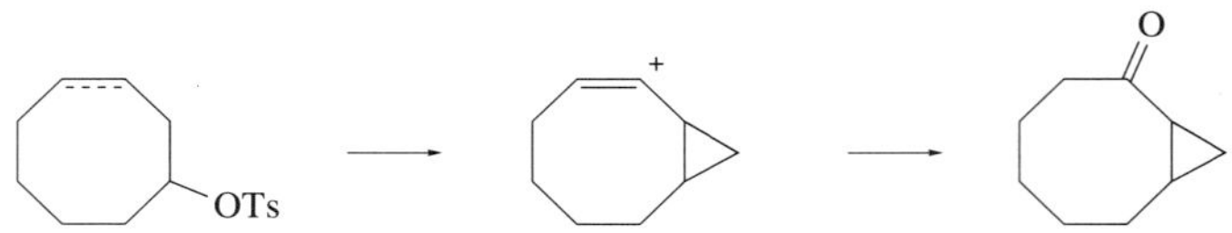

Scheme 75

When additional methylene groups are placed between the allenyl group and the leaving group, cyclisation occurs, but usually via the allyl cation (attack at sp allenyl carbon), rather than via a vinyl cation (attack at terminal sp^2 allenyl carbon atoms) ⟨73CC679⟩. There are, however, exceptions to this generalisation; the intermediacy of a vinyl cation has been reported in the cyclisation of cyclohexene (**114**) to give the *cis*-octalone (**115**) (Scheme 76) ⟨78JA993⟩.

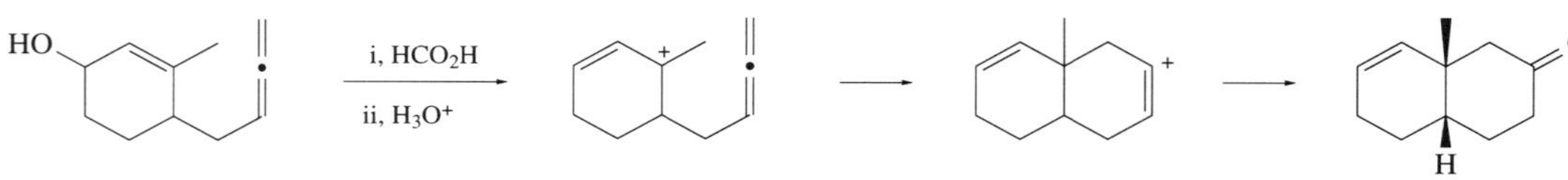

(114) **(115)**

Scheme 76

1.22.3 DICOORDINATE RADICALS (i.e., VINYLIC RADICALS)

1.22.3.1 General Aspects of Radical Formation

A number of general reviews and monographs on free radical chemistry, and the generation of carbon-centred radicals have been published, with many of these making reference to vinyl radicals ⟨85AG(E)553, B-86MI 122-01, 87T3541, 88S417⟩. The aim of this discussion, therefore, will be to highlight those methods of vinyl radical formation which have found subsequent use in organic synthesis.

The intermediacy of vinyl radicals has been demonstrated by ESR studies, with the large ^{13}C coupling constants observed demonstrating that the unpaired electron is in an orbital possessing considerable *s*-character. They also suggest that vinyl radicals adopt a bent structure (i.e., **116**) rather than a linear structure (**117**). Inversion of configuration of these radicals is rapid even at temperatures as low as $-180\,^{\circ}C$ (Equation (80)). Some vinyl radicals have been shown to adopt a linear structure, but these are sterically congested.

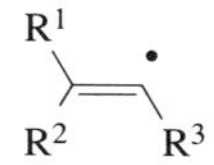

(116)

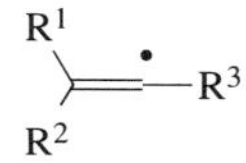

(117)

(80)

Vinyl radicals and their reaction with O_2 play a crucial role in the combustion of hydrocarbons ⟨84JA4356, 86JPC402, 92JPC2151⟩. A number of computational predictions of possible mechanisms, based on quantum mechanical calculations, have been made ⟨93JA9806⟩.

In synthesis, vinyl radicals can be formed by C—X bond scission, most commonly where X is a halogen, or by addition of another radical to a multiple bond as found in an alkyne or an allene. Vinyl radicals are higher in energy than the corresponding alkyl radicals; the formation of vinyl radicals from the corresponding vinyl halides requires higher reaction temperatures, with the bromides or iodides being the necessary halogen precursors. Once formed, the vinyl radicals can undergo a number of reactions, depending on the nature of the other reactants present. Intermolecular additions to unsaturated systems have been reported (Equation (81)), as have intramolecular cyclizations onto a suitably positioned multiple bond (Equation (82)); intramolecular hydrogen atom abstraction is also a possibility (Equation (83)). Examples of these processes will be given in the following sections. In the case of intramolecular cyclisations, the 5-*exo*-trig mode usually predominates to give methylenecyclopentanes (**118**) as the major product. However, methylenecyclohexanes (**119**) can also form, as a result of what at first glance appears to be an overall 6-*endo*-trig process; the latter become the major products under appropriate conditions (Equation (84)) ⟨82JA2321⟩. Substituting the internal position of the alkene acting as the radical acceptor with a radical stabilising group is one method of obtaining high yields of the methylenecyclohexanes (Equation (85)) ⟨86TL1355, 89CC480⟩. The formation of the methylenecyclohexanes has been rationalised as proceeding via kinetic cyclisation in the expected *exo*-mode followed by rapid rearrangement (Scheme 77) ⟨86TL4525, 86TL4529, 87JA2829, 88JA2565⟩. Cyclohexane formation can be eliminated, however, if a high concentration of stannane is used; because vinyl cyclisations of this type are fast, the reduced (open chain) product is not obtained in significant amounts even under these conditions ⟨86TL4525, 86TL4529⟩. The cyclopropane intermediate in these reactions can be trapped when a suitable substrate is employed (Equation (86)) ⟨81JOC2327, 93TL2091⟩. When the vinyl radical generated is part of a diene system, cyclisation onto a pendent alkene generates a homoallylic radical which can then give either the cyclopropane ring or the cyclohexane ring, the final product being determined by the nature of substituents attached to the diene (Scheme 78) ⟨93JOC5574⟩. Whether it is the five- or six-membered ring that is obtained, vinyl radical cyclisations of this type offer the advantage over alkyl radical cyclisations in that there is the retention of some functionality in the product in the form of an *exo*-cyclic double bond; this has been exploited on a number of occasions ⟨82JA2321, 83TL1871, 84CL1437, 85TL957, 86TL1355, 87JCS(P1)1237, 87TL2895, 89JCS(CC)1090⟩.

(81)

(82)

(83)

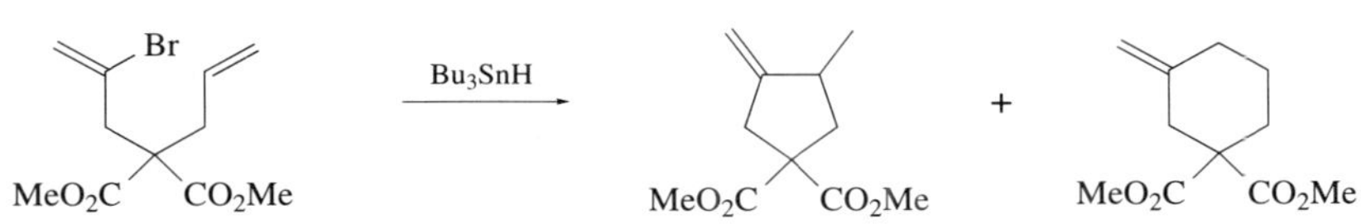

(**118**) 65% (**119**) 22%

(84)

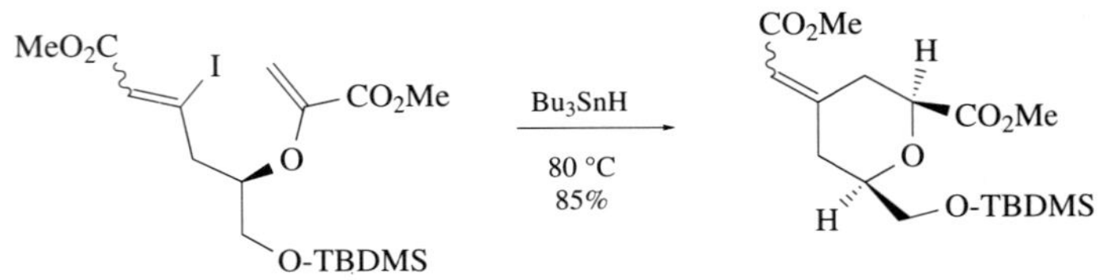

(85)

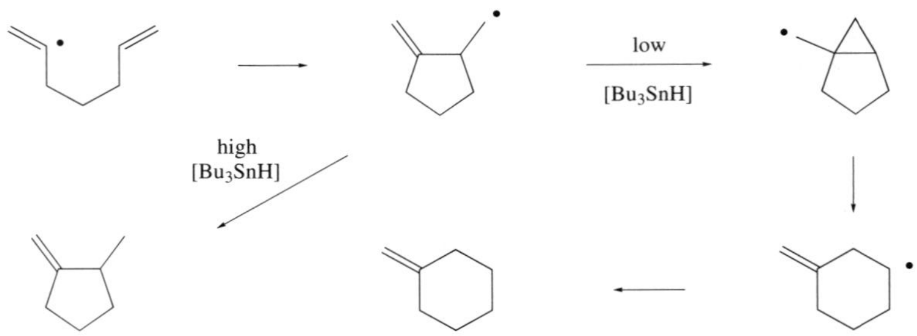

Scheme 77

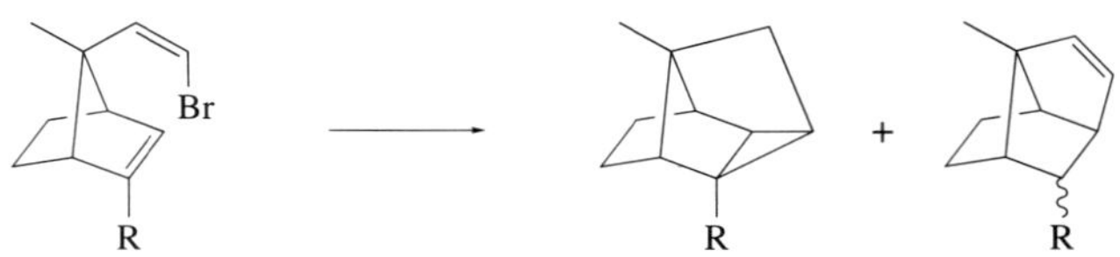

(86)

R = H	100%	
R = O-TBDMS	100%	
R = Ph	39%	45%

(Yields based on NMR spectroscopy)

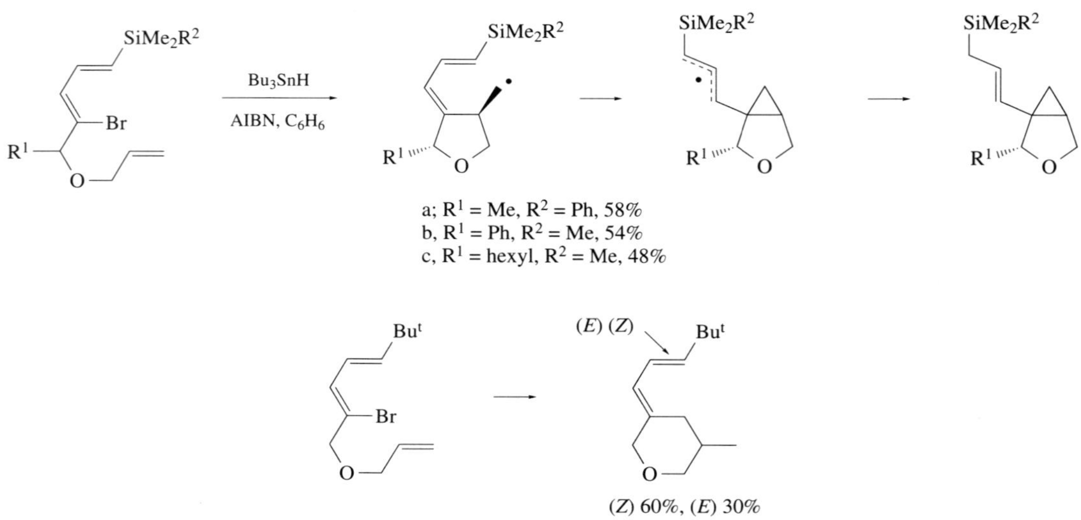

Scheme 78

Vinyl radicals are much more reactive towards cyclisations than the corresponding alkyl radicals; in 5-*exo* cyclisations the rate difference is ⩾1000 ⟨82JA2321, 86TL4525, 86TL4529⟩.

1.22.3.2 Formation of Vinylic Radicals by Bond Scission

A number of reagents can be utilised in the homolytic cleavage of vinyl halides to give vinylic radicals, with tri-*n*-butyltin hydride being the most widely used. The reaction conditions employed reflect those already well-established in alkyl radical chemistry, for example, use of 2,2′-azobisisobutyronitrile (AIBN), thermolysis, or photolysis as initiators; the choice of solvent and operating temperature varies considerably, but in all cases it is the vinyl bromide or iodide, and not the chloride, that acts as the precursor to the vinyl radical; the stereochemistry of the vinyl halide is unimportant as inversion about the radical centre occurs rapidly at the operating temperatures normally employed. Other radical precursors include, for example, the vinylselenides (Equation (87)) ⟨87CC786⟩, whilst trialkylsilanes are just one example of a number of other reagents to have been utilised. The following examples have been chosen to illustrate methods of generation of vinyl radicals by bond scission, and their subsequent utility in synthesis.

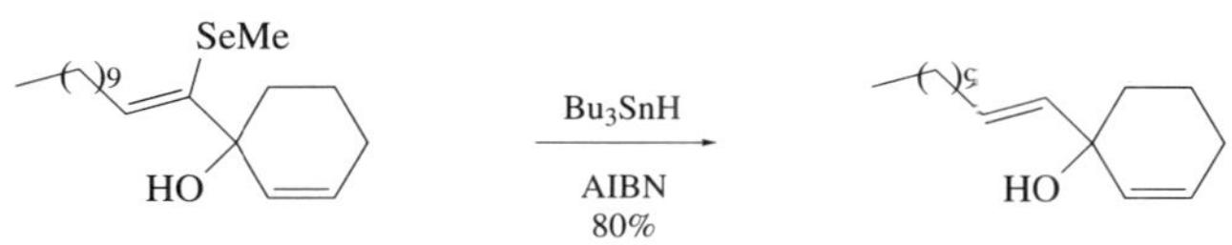

(87)

AIBN = 2,2'-azobisisobutyronitrile

There are a number of reports in which vinyl halides can be used to generate vinyl radicals that subsequently undergo 6-*exo* cyclisation to give a six-membered ring (cf. the 6-*endo* cyclisations mentioned previously); examples include the formation of (**120**) and (**121**) (Equations (88) and (89)) ⟨82JA2321, 85TL5927⟩.

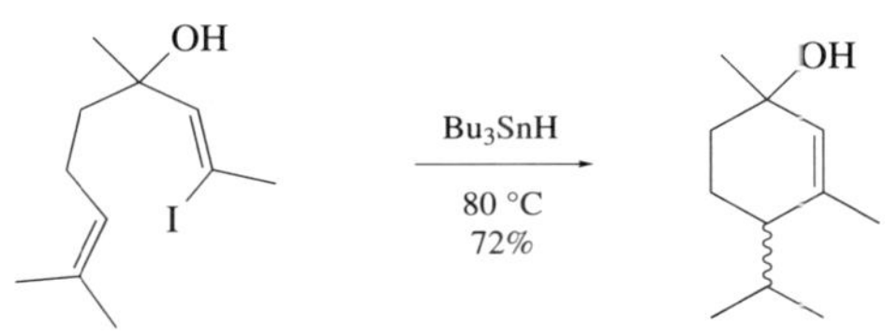

(88)

(**120**)

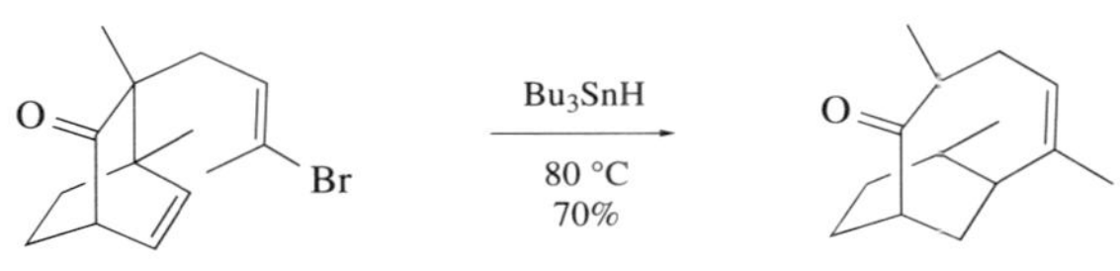

(89)

(**121**)

The tendency of vinyl radicals to abstract a hydrogen atom rather than undergo cyclisation is illustrated in Scheme 79. In this case, the vinyl radical (**123**) was generated from the corresponding vinyl iodide (**122**) and tri-*n*-butyltin hydride under photolytic conditions ⟨88JCS(CC)81⟩. Hydrogen atom abstraction from the δ position leads to the formation of an allylic radical (**124**) which cyclises back onto the original vinylic double bond to generate the tricyclic system (**125**).

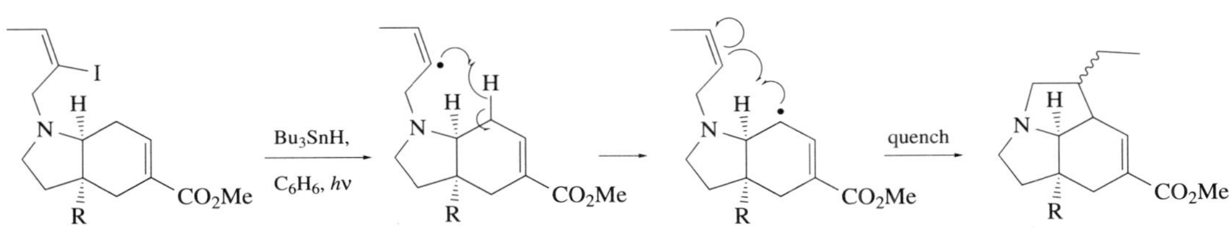

(**122**) (**123**) (**124**) (**125**)

Scheme 79

Vinyl radicals have been generated in a number of tandem radical cyclisation sequences; the synthesis of silphiperfil-6-ene (**127**) (from the bromoketone (**126**)) (Scheme 80) ⟨82JA2321, 87T5663⟩, angular triquinanes ⟨86JA1106⟩ and the southern part of the milbemycin system (Equation (90)) ⟨88CC283⟩ are just a few examples that illustrate this type of process.

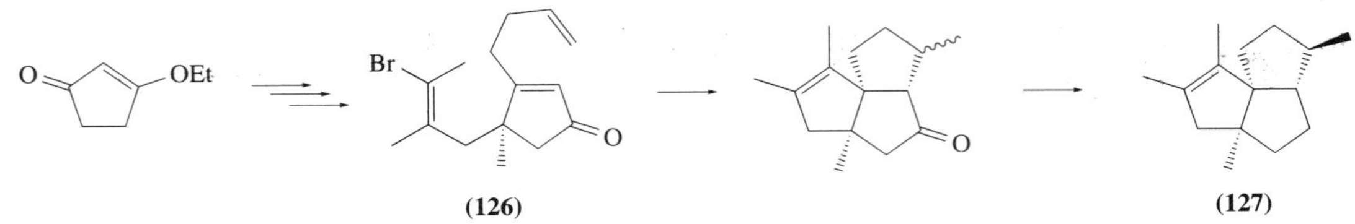

Scheme 80

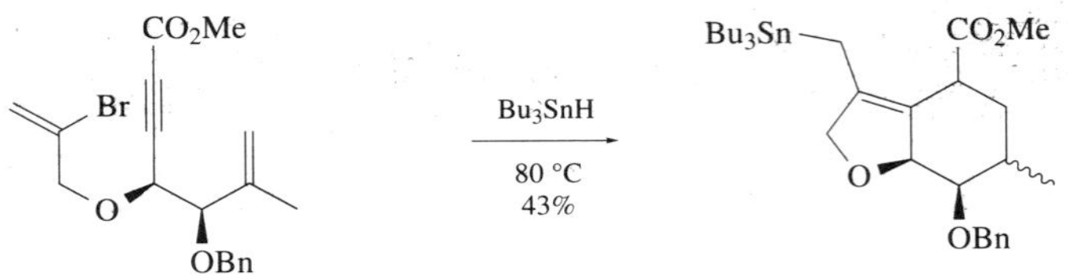

(90)

1.22.3.3 Formation of Vinylic Radicals by Addition to Alkynes and Allenes

The formation of vinyl radicals from alkynes has been employed on a number of occasions in synthesis; quite simply, a preformed radical adds to a triple bond forming a σ bond, with the vinyl radical being generated at the β position. The radical that initially attacks can be carbon centred (alkyl, aryl or vinyl), or heteroatom centred (e.g., tin, sulfur or selenium). Once formed, the vinyl radical generated in this way can undergo the same type of reactions as seen in the previous section, i.e., intermolecular additions, intramolecular cyclisations or hydrogen atom abstraction. The following examples serve to illustrate the utility of alkynes as precursors to vinyl radicals in synthesis.

The principles applied to the addition of radicals to alkenes can also be extended to alkynes, with nucleophilic radicals adding preferentially to electron poor alkynes, whilst electrophilic radicals add to electron rich alkynes. Addition to mono- and disubstituted alkynes can lead to mixtures of both *cis*-alkenes (from *syn*-addition to the alkyne), and *trans*-alkenes (*anti*-addition), although it has been reported that in intermolecular processes, *syn*-addition is preferred unless the incoming radical is sterically bulky, for example, Bu^t ⟨89TL827⟩. A number of different methods for the intermolecular addition of radicals to alkynes have been reported ⟨82TL2575, 85TL3311, 85TL6001, 88TL391, 89TL827, 89JOC1238⟩ to give alkenes as products. Allenes have also been formed from appropriately substituted alkynes, via an intermediate vinyl radical (Equations (91) and (92)) ⟨84CL1437, 86TL2833⟩, and from addition of radicals to cyclopropylacetylene ⟨86JCS(P1)39⟩.

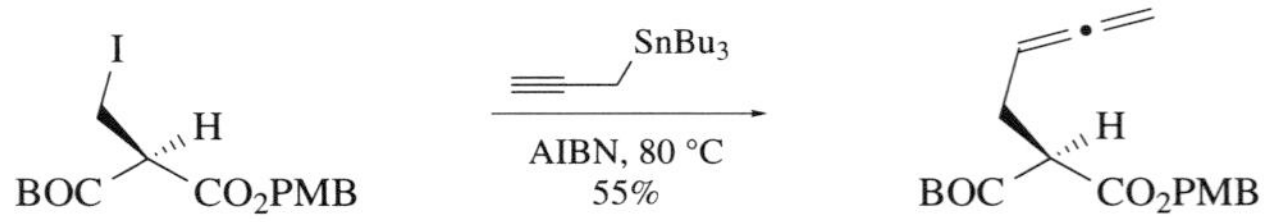

(91)

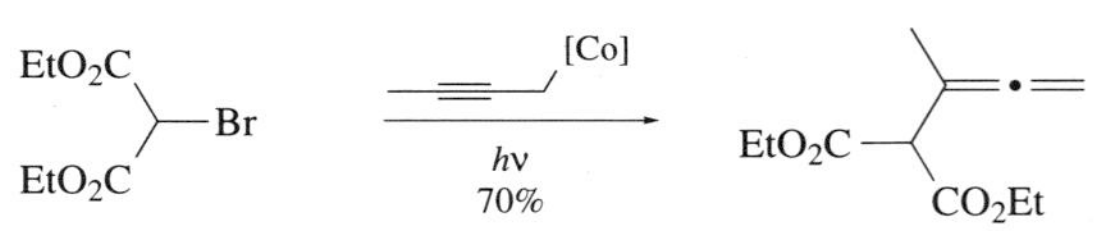

(92)

The formation of perfluoroalkyl analogues is of considerable importance for a number of synthetic targets, with the perfluoroalkylation of alkynes and alkenes being a major process for the introduction of a perfluoroalkyl group. Methods that have been utilised to achieve this include photolysis ⟨53JCS1199, 54JCS1634⟩, pyrolysis ⟨51JCS588⟩, electrolysis ⟨82CC433⟩, free radical initiation ⟨82JOC2251⟩ and transition-metal catalysis ⟨84TL303, 85CL1689, 85JA5186, 86CL1895⟩. Triethylborane-induced radical addition has been employed in synthesis on a number of occasions ⟨87JA2547, 88BCJ2693, 88TL6125,

88TL6127⟩, and was found to exhibit extremely good catalytic activity in the intermolecular addition of perfluoroalkyl iodides to both terminal and internal alkynes. The reaction proceeded with high regio- and stereoselectivity, with the *trans* product being formed exclusively, whilst the mild reaction conditions employed allowed the use of the more volatile fluorides (Equation (93)) ⟨89TL3159⟩.

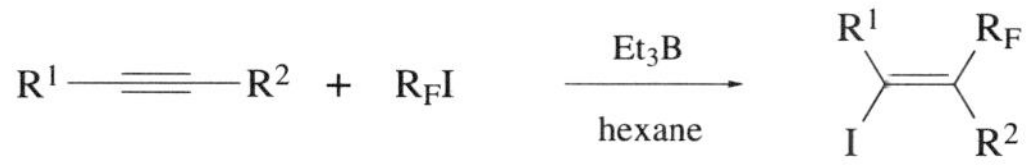

(93)

5-Hexynyl cyclizations (K_{exo} (25°C) = 2.8×10^4 s^{-1}; K_{endo} (25°C) < 6×10^2 s^{-1}) are slower than the corresponding 5-hexenyl cyclisations but, nevertheless, have found extensive use in synthesis. As previously mentioned, one advantage of these types of reactions is that the double bond formed in the process can be manipulated further, thus permitting the introduction of other functionalities ⟨69TL1653, 84CL1437, 85CC1205, 85T3959, 85T4013, 86CC878, 86TL641, 87TL2009, 87TL2637, 87TL2887, 88TL57, 88TL1085, 89TL5017⟩. The addition of carbon-centred radicals to a pendent alkyne has also been employed in a number of 6-*exo* cyclisations, the resultant vinyl radical abstracting a hydrogen atom from the tin hydride to continue the chain process (Equations (94) and (95)) ⟨79JOC546, 84JOC1313, 87CJC104, 87JA2829⟩. Tri-*n*-butyltin hydride is most commonly used in these cyclisation processes with, in general, a solution of the tin hydride and a catalytic amount of AIBN (~5 mol%) in the chosen solvent being added dropwise to a solution of the radical precursor at reflux temperature under nitrogen. Syringe pump addition to maintain low concentrations of tin hydride has been employed in those cases where the cyclisation is slow.

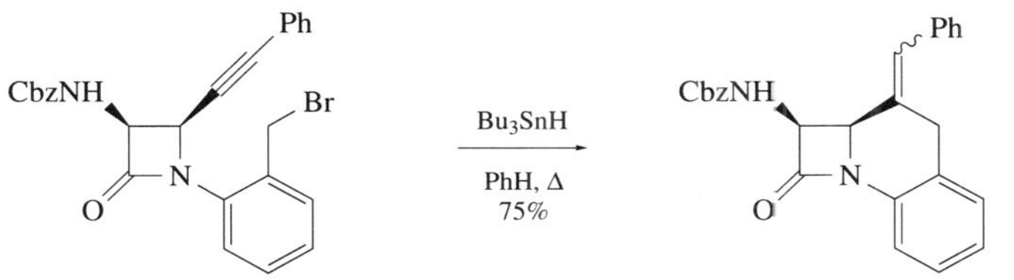

(94)

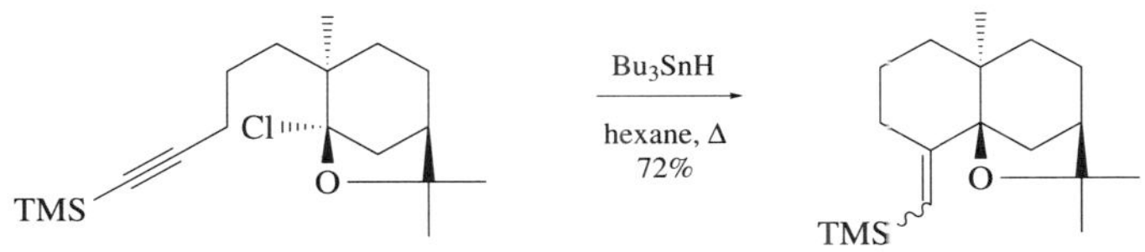

(95)

The addition of a carbon-centred radical onto an alkyne directly attached to a sulfur atom has been applied to the synthesis of cephalosporins (Scheme 81) ⟨89H(28)47⟩. Although fragmentation and reduction products were obtained as by-products in this process, there was no evidence for penem/penam formation, suggesting that *endo*-addition was favoured over *exo*-addition; this was attributed to the relative ring strains (4/5 versus 4/6 fused rings) found in such systems.

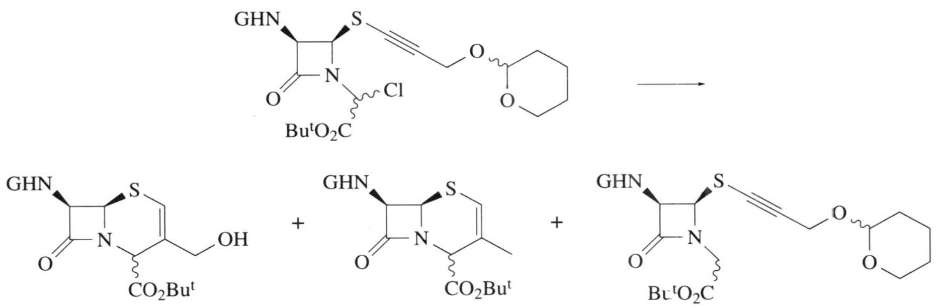

Scheme 81

Vinyl radicals generated from alkynes have found an important role in tandem cyclisation processes. In the synthesis of hirsutene (**128**), for example, the vinyl radical (**129**) is formed in the last cyclisation step, and acquires a hydrogen atom from the tri-*n*-butyltin hydride to generate the tin radical needed for halogen atom abstraction (Scheme 82) ⟨85JA1448, 85T3943⟩. The same methodology has been applied to the synthesis of $\Delta^{9(12)}$-capnellene ⟨85TL4991⟩. In both cases, the second cyclisation step involves the addition of an alkyl radical onto an alkyne to generate a vinyl radical. Alkyl radicals are more stable than vinyl radicals and so, at first glance, this appears to be an unfavourable process; however, a σ bond is formed in this cyclisation at the expense of a π bond, and so the transformation is exothermic. Vinyl radical cyclisation as the ultimate step in a tandem process has also been utilised in the synthesis of linear triquinanes (Equation (96)) ⟨83T3943⟩. In the construction of the C/D ring of steroidal butenolides through a tandem cyclisation process a vinyl radical is generated in the first cyclisation step, and subsequently adds to an adjacent alkene (Equation (97)) ⟨83JA3720⟩.

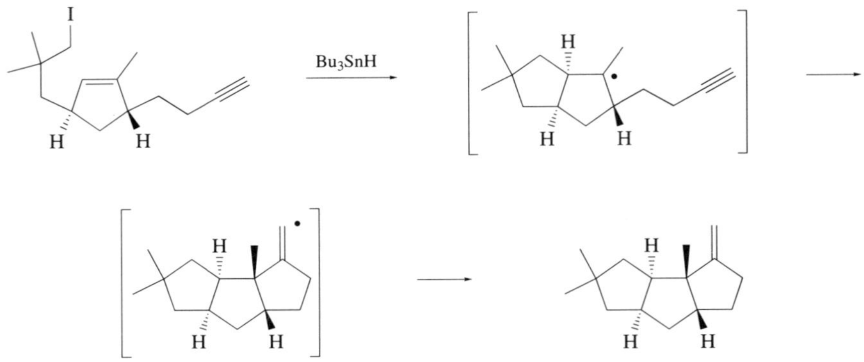

(129) **(128)**

Scheme 82

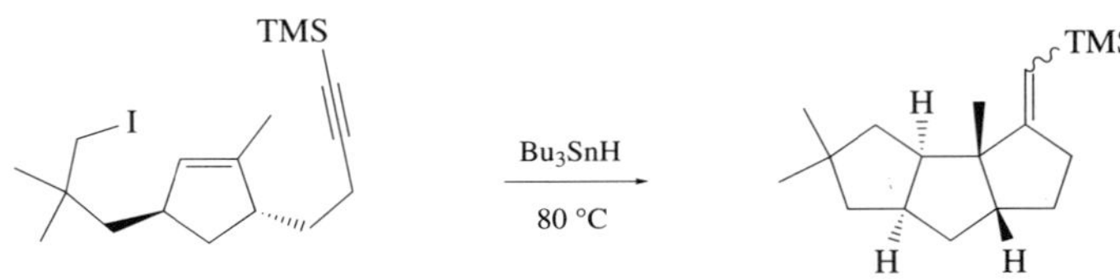

(96)

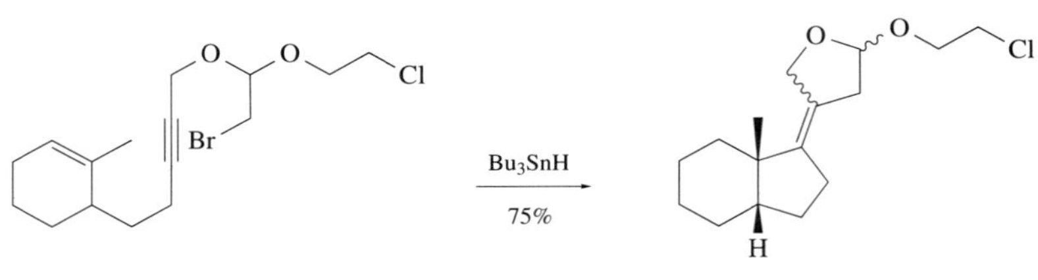

(97)

When tri-*n*-butyltin hydride is employed as the reagent in radical-forming reactions, and when no halogen atom is present in the substrate, the tin radicals generated as a result of either thermolysis or photolysis can add to a multiple bond; in the case of alkynes, vinyl radicals result. This tin radical addition has been utilised in the synthesis of natural products such as seychellene **(130)** (Scheme 83) ⟨89TL225⟩, amongst others ⟨87JA2547, 87JA2829, 89TL827, 89TL1997⟩. The stannyl group can be removed after cyclisation either by protodestannylation with silica gel and dichloromethane or by transmetallation with butyllithium and subsequent quenching.

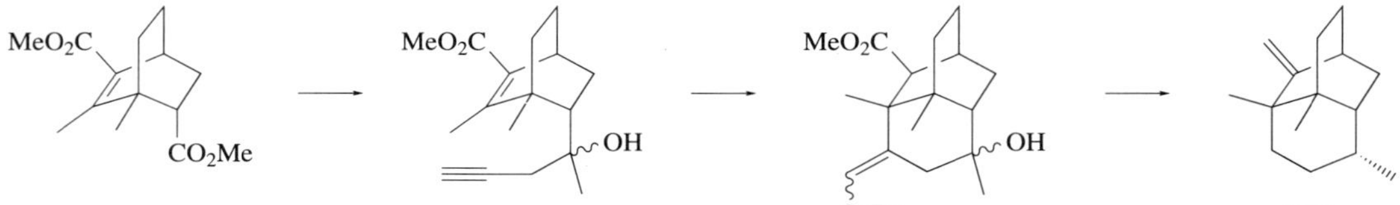

(130)

Scheme 83

1.22.4 MONOCOORDINATED ALKYNIC IONS AND RADICALS

1.22.4.1 Alkynic Carbanions

As in the case of vinylic carbanions, there are a number of reviews and monographs dealing with general aspects of carbanion chemistry, which include discussions on acetylide anion formation; monographs concerned specifically with alkyne chemistry also discuss acetylide anions ⟨B-69MI 122-01, B-83MI 122-01, B-78MI 122-02, B-88MI 122-01⟩.

The increasing *s* character of the hybrid orbital involved in the σ bond to H in the *sp* orbital of an alkyne, as compared to sp^2 (alkenes) and sp^3 (alkanes), results in a corresponding increase in acidity of the alkynic proton. Thus, alkynes RC≡CH, and ethyne HC≡CH ($pK_a = 25$), can be converted by strong bases such as $^-NH_2$ in liquid ammonia, into the corresponding anions, which are somewhat more nucleophilic than cyanide ions. The low acidity of 1-alkynes means that strong bases must be used to form the corresponding anions and that water is not a suitable solvent; a very low concentration of alkynide ions is present in aqueous solutions. However, some transition metal alkynides can be prepared by precipitation from aqueous solution because of their low solubility in water. In general, the solvent of choice for the preparation of alkynide ions should be less acidic than the alkyne, and allow the solvation of both the alkyne and the alkynide ion; liquid ammonia, tetrahydrofuran, ether and hydrocarbons have all been employed Ammonia is most often used, because the alkynide ion is readily formed by metal amides. Acetylide anions do not require the presence of electron-withdrawing groups for stabilisation, the carbanion being stabilised sufficiently as a result of the high *s* character of the carbon atom. Addition of acetylide anions to electrophilic centres constitutes an important method in synthetic chemistry, with the added advantage that the C≡C linkage present in the product can be modified further, for example, reduction by hydrogen in the presence of Lindlar catalyst to give the alkene. However, acetylide ions are also strong bases and so, in certain cases, can induce elimination reactions as well as nucleophile displacement. For example, reaction with tertiary halides is often found to result in alkenes being formed, rather than in alkylation of the acetylide. As in the case of vinylic carbanions, acetylide anion reactivity can be tailored to meet specific needs by the judicious choice of the counter cation. A number of metallated alkynes have been reported in the literature, with those that have been found to be of some synthetic utility being discussed in this article. The most widely reported alkynides are those of Group 1 and 2 metals, particularly lithium and magnesium, and these have been employed principally in alkylation reactions. Other metals utilised include copper(I), silver, gold(I), zinc, mercury and aluminum. In general, these metallated acetylides are prepared via transmetallation of the corresponding lithium or magnesium acetylide with the halide of the metal required. Palladium-catalysed reactions have also been employed, particularly in the reactions of alkynides of metals other than those of Groups 1 and 2. Alkynic carbanion structures incorporating the *t*-butylacetylide or phenylacetylide anion have been characterized by x-ray diffraction analysis ⟨83CB3212⟩.

In addition to the metallated acetylides mentioned above, acetylides have also been prepared through electrochemical reduction of the corresponding alkyne; when the reaction is carried out in the presence of an alkyl halide, the disubstituted alkynes can be obtained in high yields (Scheme 84) ⟨76CC606⟩.

Scheme 84

1.22.4.1.1 Lithium acetylides

Monometallation of ethyne itself can be achieved under special conditions. Addition of pieces of lithium or sodium metal to boiling ammonia whilst introducing ethyne gives ammoniacal solutions of LiC≡CH and NaC≡CH, respectively. The dianion can be prepared as a fine suspension by introducing ethyne into a solution of BuⁿLi in ether or THF at 0 °C. Ammonia or other amines stabilise the monoanion. Thus, when ethyne is introduced into a BuLi-TMEDA solution at low temperatures, a clear solution of the monolithium salt is obtained. Other alkynyllithium complexes are most readily prepared by treating the corresponding 1-alkyne with butyllithium, in a solvent such as THF or DIGLYME. The lithium reagents thus obtained have been employed in alkylation reactions, to give a range of natural products such as pheromones (Scheme 85) ⟨72AG(E)60⟩, macrolides ⟨77CL959, 78CC754, 78JA4618, 78JA4620, 79TL3021, 80HCA2426, 83AG(E)780⟩ (e.g., Scheme 86), sesquiterpenes ⟨69JA4318, 70TL2245, 90JOC791⟩ (e.g., Equation (98)), and leukotrienes and prostaglandins ⟨79JCS(P1)2954, 80JA5425, 80JCS(P1)852, 83JA3656, 83TL5185, 89S898⟩ (e.g., Scheme 87).

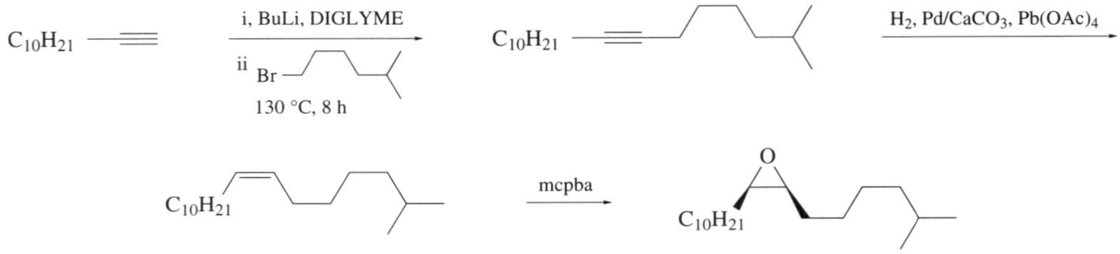

Scheme 85

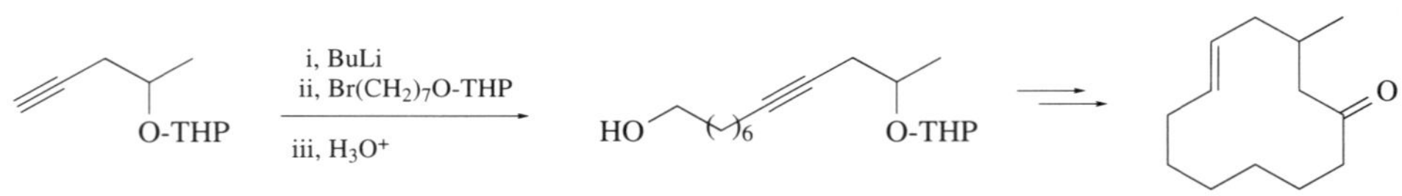

Scheme 86

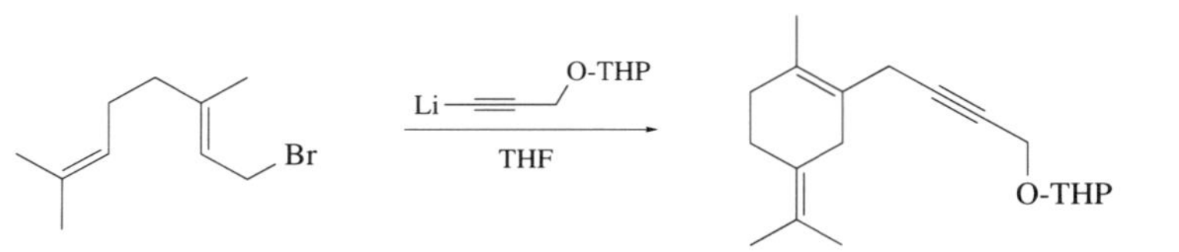

(98)

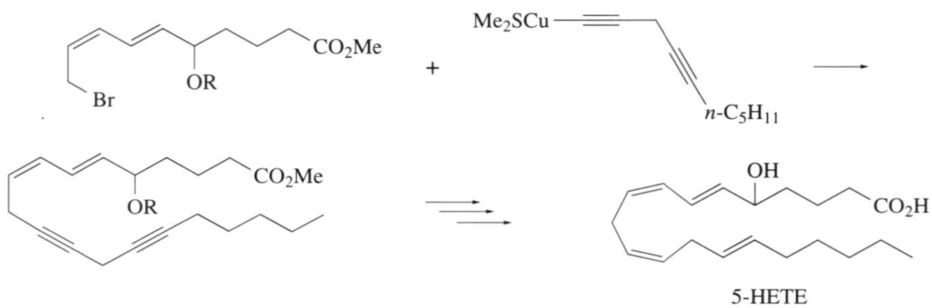

Scheme 87

The tetrahydropyranyl ether of propargyl alcohol (**131**) is readily metallated at the alkynic carbon by *n*-butyllithium, or ethylmagnesium bromide to give the corresponding anion (**132**), which reacts with a variety of electrophiles (Scheme 88) ⟨82OS(60)81, 84JOC1898, 84T2537, 85T2369, 85TL2237⟩. Lithium amide in liquid ammonia has also been used as a base for the generation of an anion from a substituted propargyl alcohol; subsequent addition to ethylene oxide allowed the eventual synthesis of the C-12 to C-18 segment of rhizoxin, an antitumour macrolide antibiotic ⟨93TL707⟩.

1.22.4.1.2 Magnesium acetylides

The most common method for the generation of magnesium acetylides is to treat the corresponding alkyne with ethylmagnesium bromide, resulting in deprotonation and metallation, with ethane being produced as a side-product ⟨82OS81⟩. The preparation of the dianion of ethyne can be achieved simply by introducing ethyne into a solution of ethylmagnesium bromide in THF at

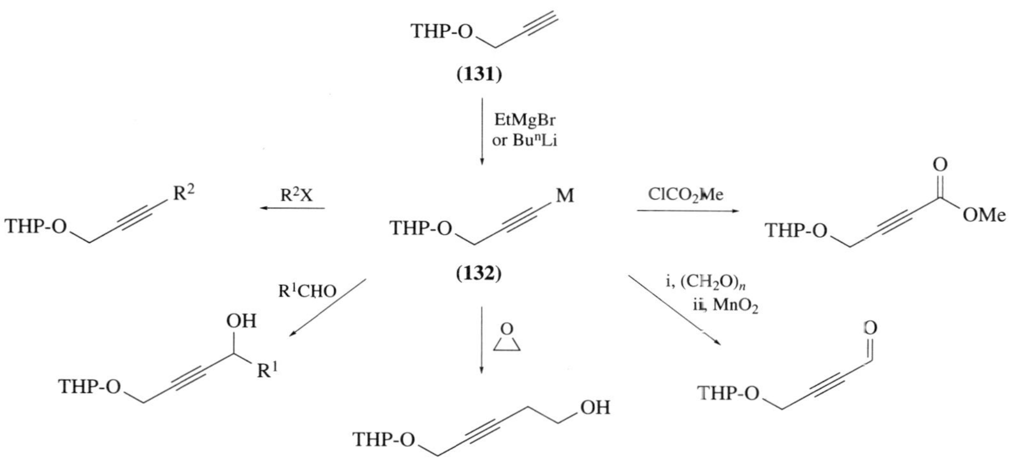

Scheme 88

>40 °C. Preparation of the monoanion is more difficult, although BrMgC≡CH can be prepared by adding a solution of EtMgBr to a solution of ethyne in THF kept below 30 °C.

Alkynyl Grignard reagents react stereospecifically with alkenyl iodides in the presence of a palladium catalyst, $Pd(PPh_3)_4$ (Equation (99)) ⟨78TL191⟩.

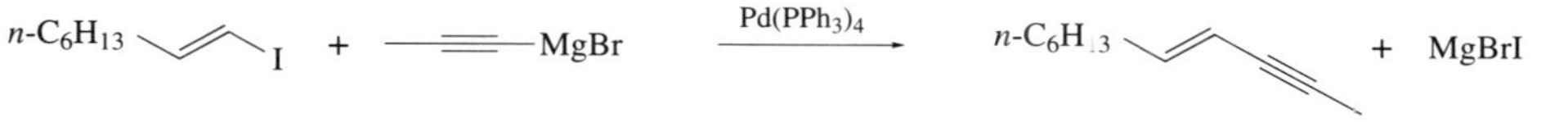

(99)

1.22.4.1.3 *Tin acetylides*

Trialkyltin acetylides have been used in the synthesis of chiral β,γ-unsaturated amino acids, a class of compounds some of which have been shown to exhibit neuroexcitatory, antimicrobial and enzyme inhibitory properties (e.g., ⟨75ACR281, 76ACR313⟩). An oxazinone (Scheme 89) was brominated and then condensed with the required trialkyltin acetylide in the presence of zinc chloride in warm carbon tetrachloride. The alkyne thus obtained is reduced directly to provide exclusively the (*E*)-vinylglycine derivatives in good to excellent *ee* ⟨88JA2501, 88T5425⟩. Trialkyltin acetylides will also couple with acid chloride in the presence of a palladium catalyst, $PdCl_2(PPh_3)_2$, with 1,2-dichloroethane as solvent, to give alkynyl ketones as products (Equation (100)) ⟨82JOC2549⟩.

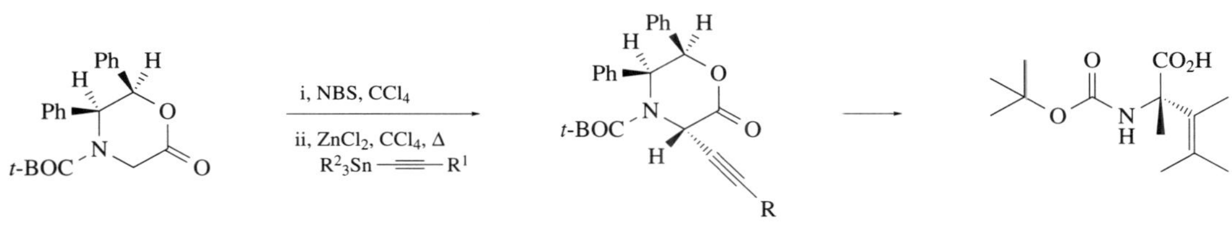

Scheme 89

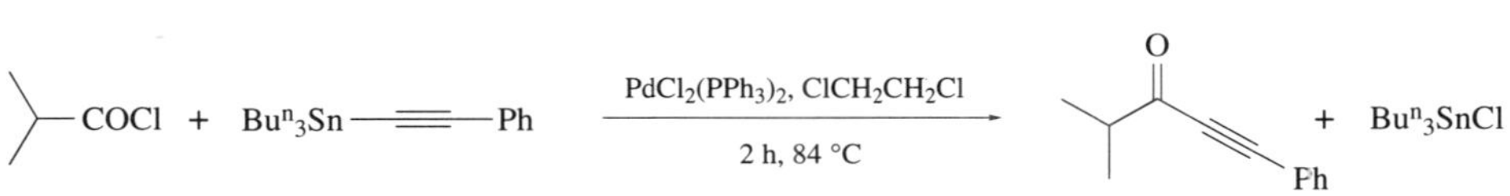

(100)

1.22.4.1.4 *Copper acetylides*

One of the drawbacks often encountered in the use of organocuprates in synthesis is that one of the organic ligands is invariably 'wasted', (i.e., only one of the ligands is transferred during the reaction). The comparative strength of the sp carbon–copper bond has led to the use of copper acetylides in the preparation of mixed organocuprates, in which it is the other organic group, and not the acetylide, which participates in reactions. This has allowed the preparation of a number of organocuprates in which a synthetically more valuable and less readily available organic group, usually as its lithium salt, is coupled with copper acetylide to give the corresponding alkylethynyllithium cuprate, which then reacts with electrophiles in a similar manner to that described elsewhere for dialkyllithium cuprates. The alkyl component can also be vinylic or aryl, and the corresponding Grignard reagents can be used in place of the organolithium reagents. Thus enamidines (**133**) can be metallated at the 2-position to furnish the alkenyllithio derivatives (**134**), which are then transformed into the mixed organocuprate (**135**) prior to alkylation by an allylic halide (Scheme 90) ⟨85JOC1019⟩. This methodology has been applied to other systems. A mixed organocuprate was employed in the synthesis of a *trans*-fused α-methylene-γ-lactone (**136**), via a conjugate addition (Scheme 91) ⟨75JOC1932⟩, and in an epoxide ring opening to give the precursor to another α-methylene-γ-lactone (**137**) (Scheme 92) ⟨76JOC3213⟩.

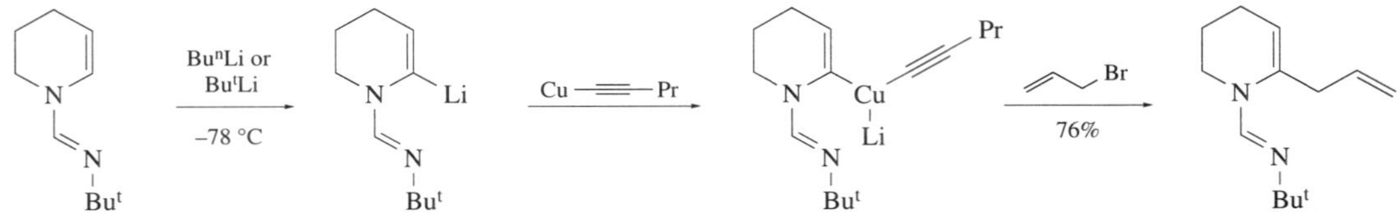

(**133**) (**134**) (**135**)

Scheme 90

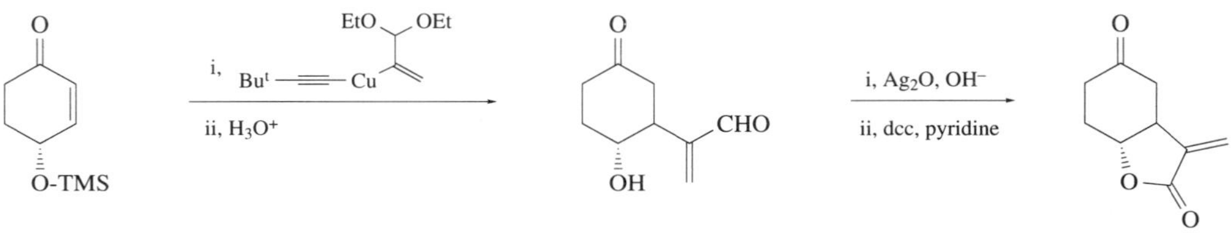

(**136**)

Scheme 91

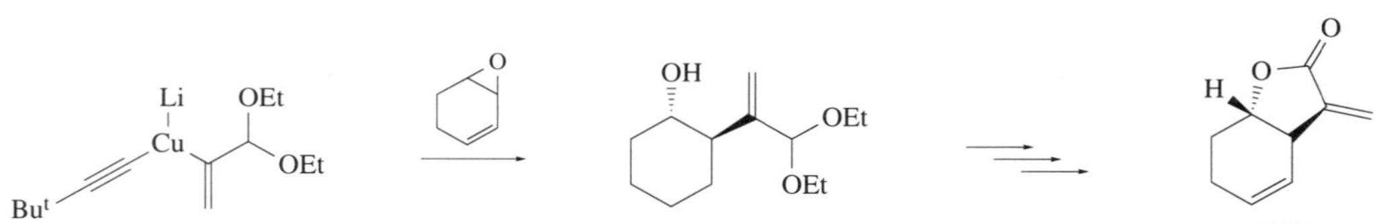

(**137**)

Scheme 92

Copper acetylides prepared from copper(II) sulfate/hydroxylamine hydrochloride have been utilised in coupling reactions to aryl halides. Both euparin (**138**) (Scheme 93) ⟨77JCS(P1)90⟩ and the antibiotic frustulosin ⟨79CC124⟩ have been synthesized via this method.

Copper acetylides may also be prepared from the corresponding terminal alkyne and copper

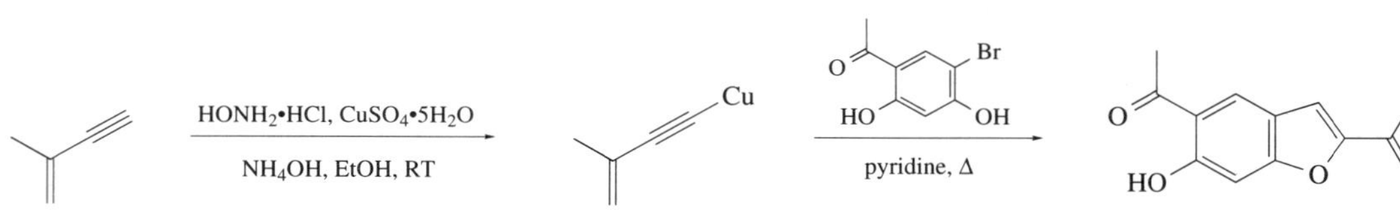

(**138**)

Scheme 93

t-butoxide; the anions prepared in this way will react with carbon dioxide to give copper alkynoates which can be converted into their methyl esters (e.g., **139**) on treatment with methyl iodide (Scheme 94) ⟨74CC380⟩.

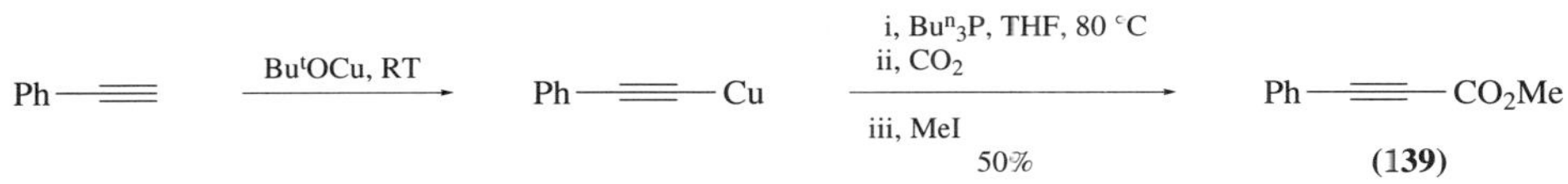

(139)

Scheme 94

1.22.4.1.5 *Silver acetylides*

Silver acetylides can be prepared from the corresponding terminal alkynes and silver nitrate ⟨56JA1675⟩. These silver salts are of low nucleophilic character, as evidenced by their inability to react with either acetone or *t*-butyl bromide; however, they will react with acyl halides in either carbon tetrachloride, chloroform, or benzene to give α,β-alkynic ketones in reasonable yields (Equation (101)) ⟨56JA1675⟩. In the presence of pyridine, the expected acylation does not occur; instead, a dihydropyridine addition product (**140**) is obtained (Scheme 95), which can undergo a number of reactions ⟨61JA449⟩. Silver acetylides generated from silver nitrate and phenyl ethyne in aqueous ethanol react with purified arenediazonium salts (as their chlorides or bromides) to give phenylazoethynylaromatic compounds (Equation (102)) ⟨75JOC124⟩.

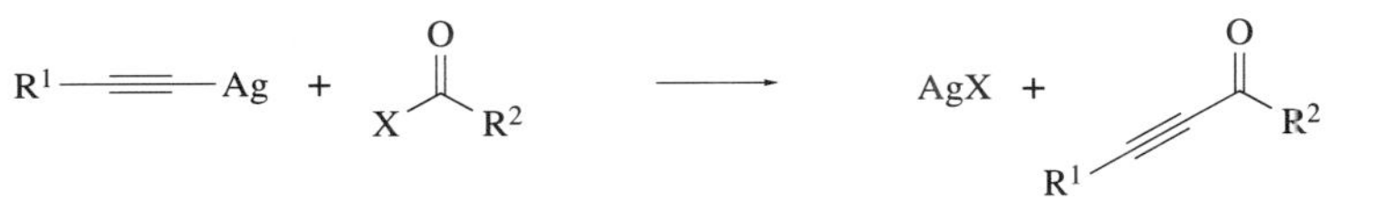

(101)

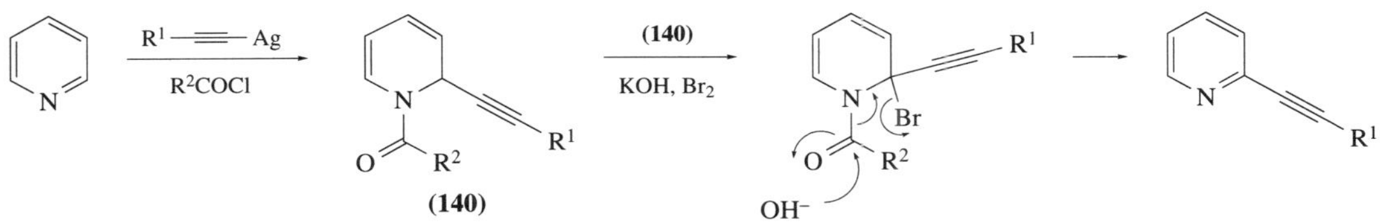

(140)

Scheme 95

(102)

Silver acetylides can also be prepared from a terminal alkyne and silver *t*-butoxide. Subsequent reaction with carbon dioxide gives the corresponding silver alkynoate which can be methylated with methyl iodide to yield the ester ⟨74CC380⟩.

1.22.4.1.6 *Palladium acetylides*

Organic halides will couple with ethyne in the presence of a palladium catalyst and base to give disubstituted alkynes (Equation (103)) ⟨75JOM(93)253, 75JOM(93)259, 75TL4467⟩. As the reaction does not proceed in the absence of base, attack of acetylide ion on an intermediate organopalladium halide is implicated. Vinyl bromides react to give enynes, whilst bromoallenes will react with terminal alkynes to give allenynes (Equation (104)) ⟨83S32⟩. Monosubstituted alkynes can be obtained by reaction of trimethylsilylethyne with an organic halide followed by removal of the silyl group (Scheme 96) ⟨80S627, 81JOC2280⟩. 1-Alkynes can also be coupled to heterocyclic halides in the presence of a palladium catalyst, $Pd(PPh_3)_2Cl_2$, CuI and triethylamine; coupling to bromopyridines was used in the design of ruthenium(II) complexes which spectrochemically recognised Group 1 and 2 metal cations ⟨93JCS(D)2629⟩.

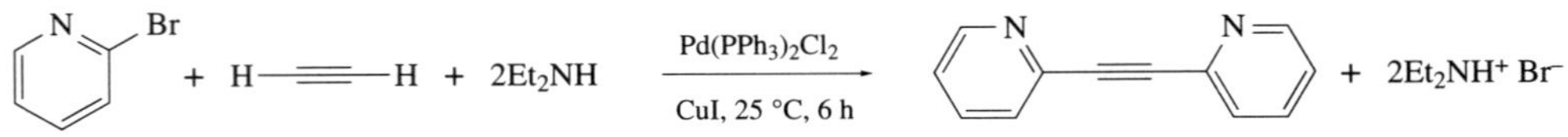

(103)

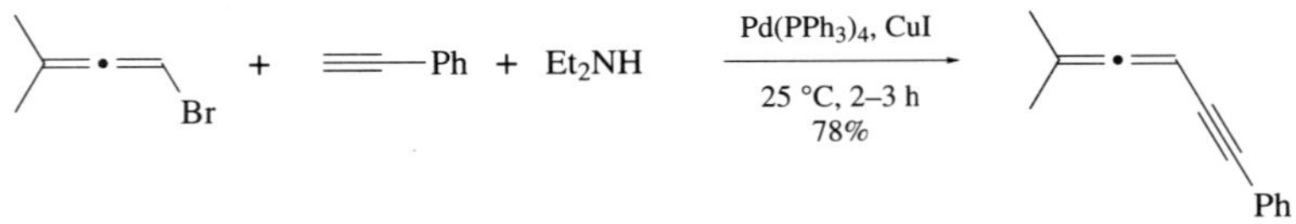

(104)

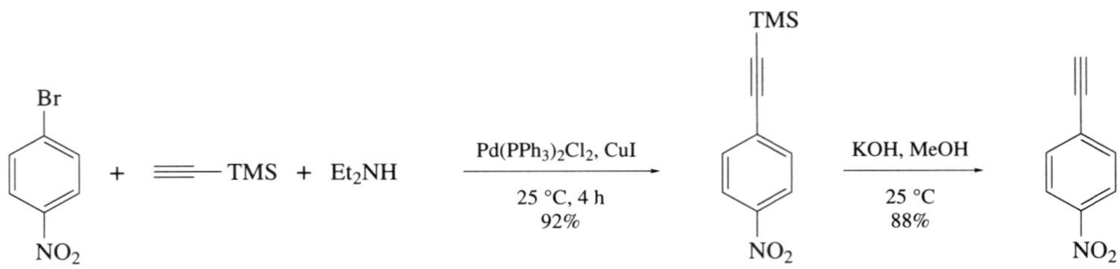

Scheme 96

Alkynylpalladium complexes are implicated when aryl or vinyl halides are reacted with terminal alkynes and norbornene ⟨82TL4517⟩. It is proposed that the aryl- or vinylpalladium halide formed initially adds across the alkene in norbornene to give an alkylpalladium adduct which then undergoes substitution of the bromide ligand by acetylide. Reductive elimination of the complex then gives the alkyne derivative (Scheme 97). This reaction sequence has also been applied to norbornadiene.

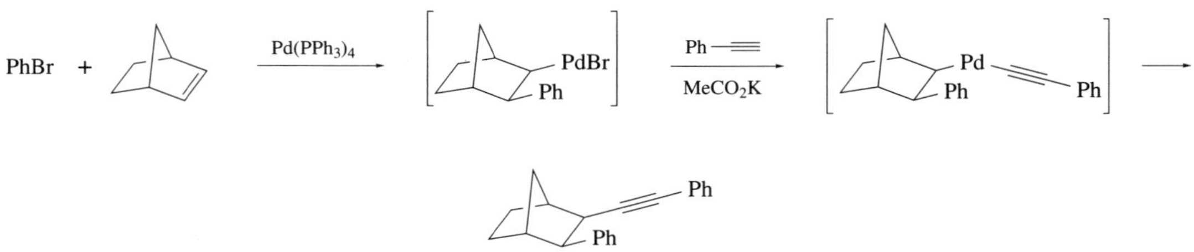

Scheme 97

1.22.4.1.7 Silicon acetylides

Alkynylsilanes react with electrophiles such that the electrophile attaches itself to the carbon atom bearing silicon, the vinylic carbocation thus forming β to the silicon (Scheme 98). Subsequent loss of the silyl group regenerates an alkyne. A number of examples of this type of reaction are to be found in the literature ⟨63CB3280, 72JOM(37)45, 73JOC2254, 76TL2449, 78TL1581, 78TL2301⟩, including the formation of a D-homosteroid from a trienyne (Scheme 68). The directing effect of the silicon group in such reactions is demonstrated very nicely in this example, for if the silyl group is replaced by methyl then cyclisation occurs to give the five-membered D-ring normally encountered in steroids ⟨78TL2549⟩.

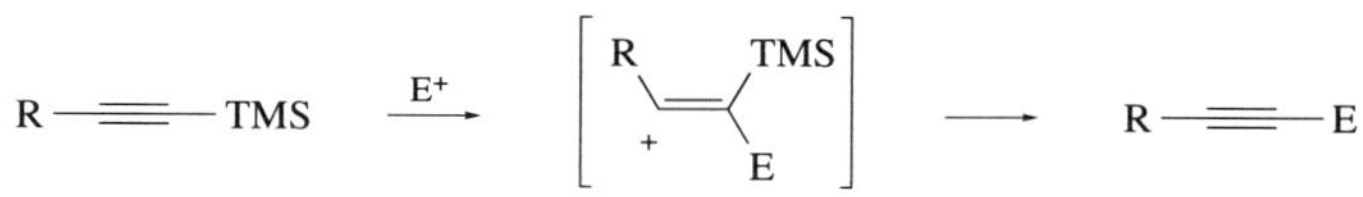

Scheme 98

1.22.4.1.8 Aluminum acetylides

Reviews discussing the preparation and reactions of aluminum acetylides have been published ⟨81S841, 84OR(32)375⟩. Aluminum acetylides can be prepared by the reaction of lithium acetylides with a dialkylaluminum halide. Dimethylaluminum chloride has been used in the preparation of acetylides for conjugate addition to α,β-unsaturated ketones (Scheme 99) ⟨78JA2244, 80JOC3053⟩, whilst methoxymethylaluminum chloride was used in a route to α-(phenylselenenyl)ketones (**141**) (Equation (105)) ⟨80TL1497⟩. Although organocuprates are more commonly employed in conjugate additions to enones, the strength of the C≡C-Cu bond precludes the use of copper acetylides in conjugate alkynylation reactions. Aluminum acetylides thus offer the same advantages of chemoselectivity over lithium reagents as would be expected from an organocuprate, and are a useful source of γ-alkynones from the corresponding α,β-enones ⟨71JA7320⟩. Alkynylalanes undergo conjugate addition to enones (e.g., **142**) under nickel(0) catalysis (Equation (106)) ⟨80JOC3053⟩.

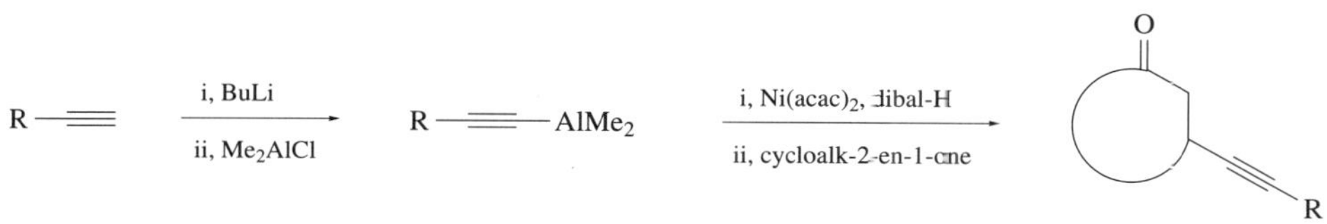

Scheme 99

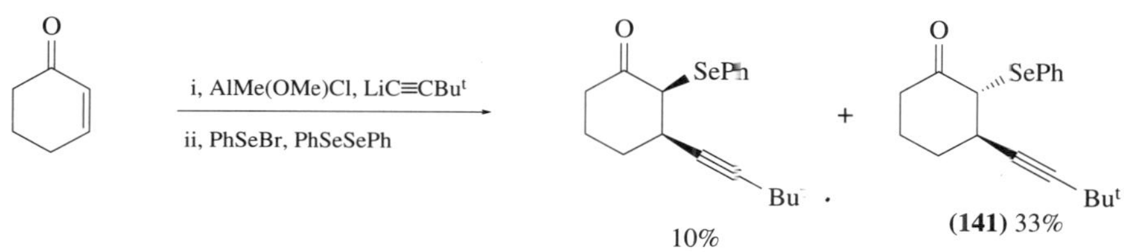

(105)

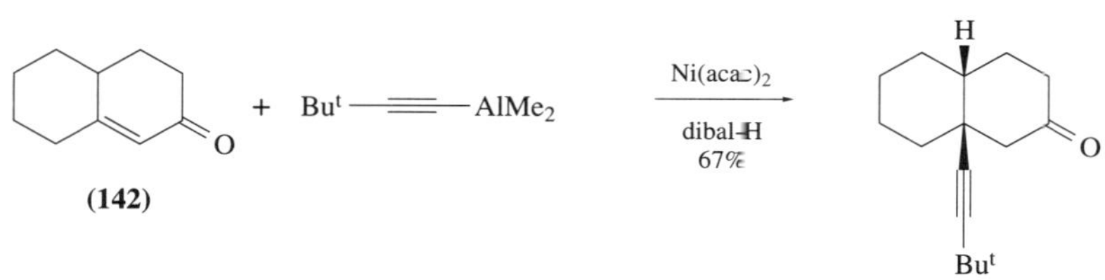

(106)

The selectivity of aluminum acetylides has been demonstrated in the ring opening of cyclohexane oxide; when the lithium salt of ethoxyethyne was employed, the required product was not obtained. However, conversion of this lithium acetylide to the corresponding diethylaluminum acetylide *in situ*, followed by addition of epoxide gave the alkynol in excellent yield (Scheme 100) ⟨76JOC1669⟩. The ring opening of an epoxide by a dialkylaluminum acetylide has been exploited in the synthesis of a *cis*-fused hydroxy-γ-lactone (**143**) (Scheme 101) ⟨75TL51, 76TL1769⟩.

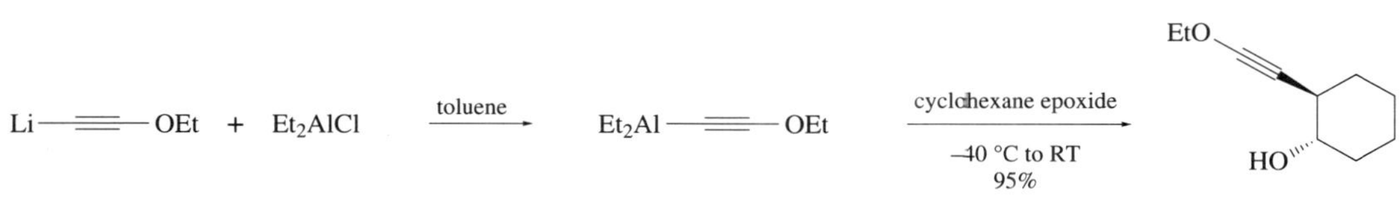

Scheme 100

1.22.4.1.9 Zinc acetylides

Alkynylzinc reagents are obtained from the reaction of zinc chloride with the corresponding lithium acetylide. Zinc acetylides react with vinyl iodides or bromides in the presence of a palladium catalyst, $Pd(PPh_3)_4$, to give alkenynes in which the stereochemistry about the original double bond

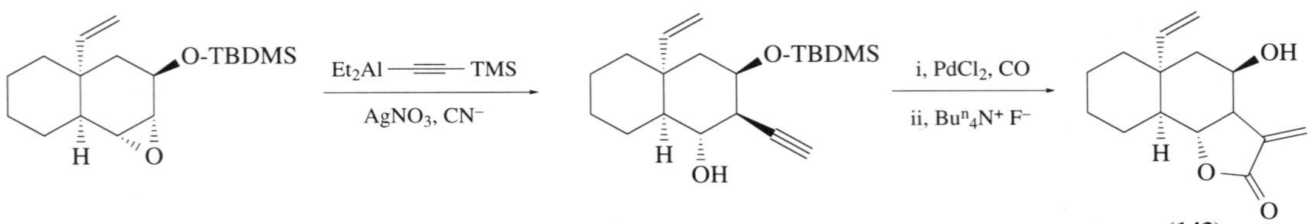

(143)

Scheme 101

is retained (Equation (107)) ⟨77CC683⟩. Because zinc acetylides are less reactive than their Grignard counterparts, they can also be used in palladium-catalysed coupling reactions with vinyl halides that contain an ester group (Equation (108)).

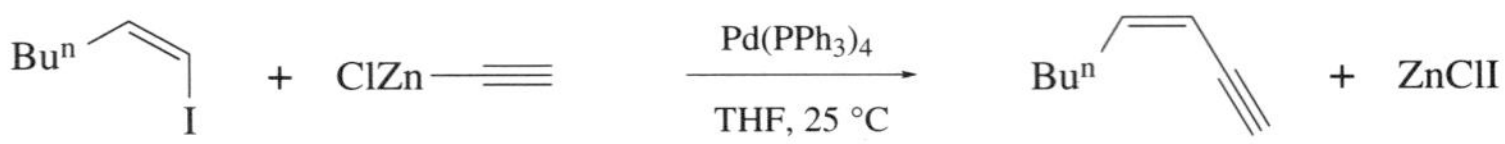

(107)

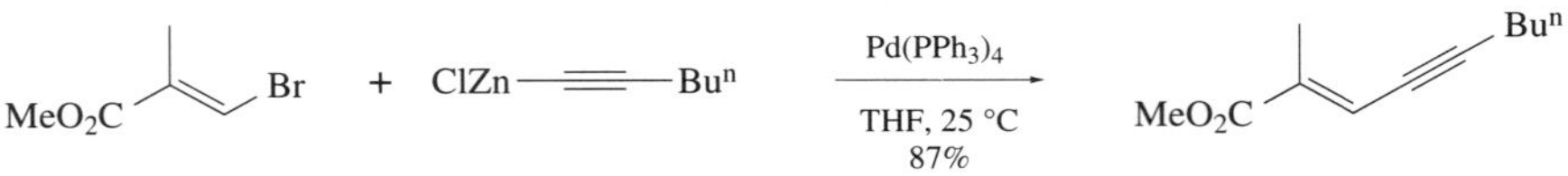

(108)

1.22.4.1.10 *Cerium acetylides*

Cerium acetylides are prepared by the reaction of the corresponding lithium reagents with cerium chloride or cerium iodide ⟨82CC1042, 84JOC3904, 84TL4233⟩. As with the alkenylcerium reagents, cerium acetylides add to ketones and aldehydes to yield the corresponding alcohols in good yields (much improved over those obtained using the lithium acetylide directly) (Equations (109) and (110)) ⟨84JOC3904, 84TL4233, 88TL2923, 88TL5017, 88TL6951, 89TL2627, 90CL1411⟩. The trimethylsilylethynylcerium reagent has been found to be particularly useful for the addition of ethynyl groups to carbonyl compounds ⟨84CL1543⟩, and has been employed successfully in a number of syntheses (Scheme 102) ⟨86CPB1531, 86TL195, 87CPB1405, 87TL4583⟩; readily enolisable ketones are also alkynylated by this reagent to give homoallylic alcohols in good yields (Equation (111)) ⟨85TL1549, 88CPB3897, 91TL1707⟩.

(109)

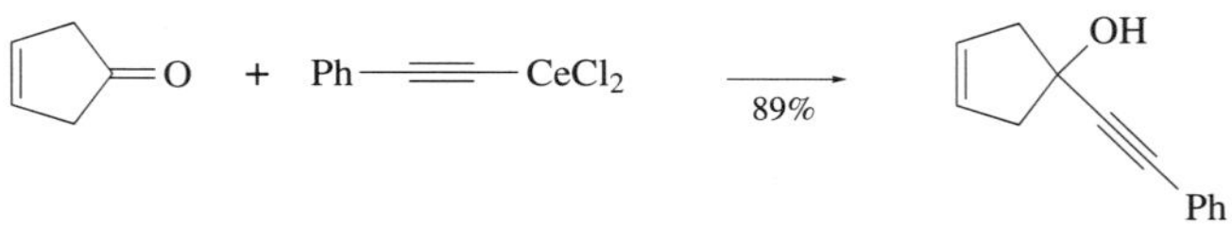

(110)

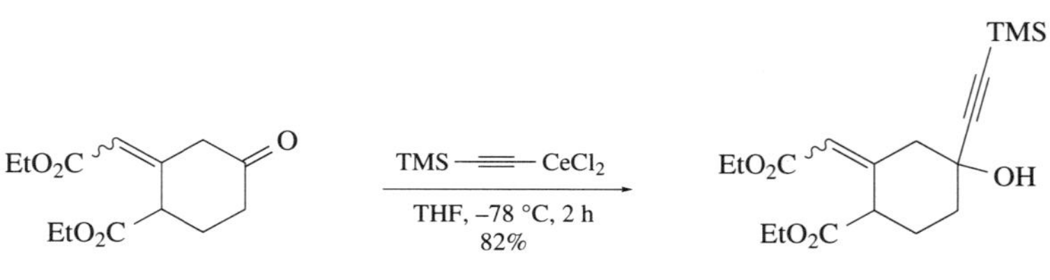

(111)

1.22.4.1.11 *Chromium acetylides*

Alkynylchromium reagents can be prepared by the treatment of the corresponding haloalkyne with chromium(II) chloride in dimethylformamide ⟨85TL5585⟩. The carbanions thus obtained react readily with ketones and aldehydes to give the corresponding alcohols (Equations (112)–(114)).

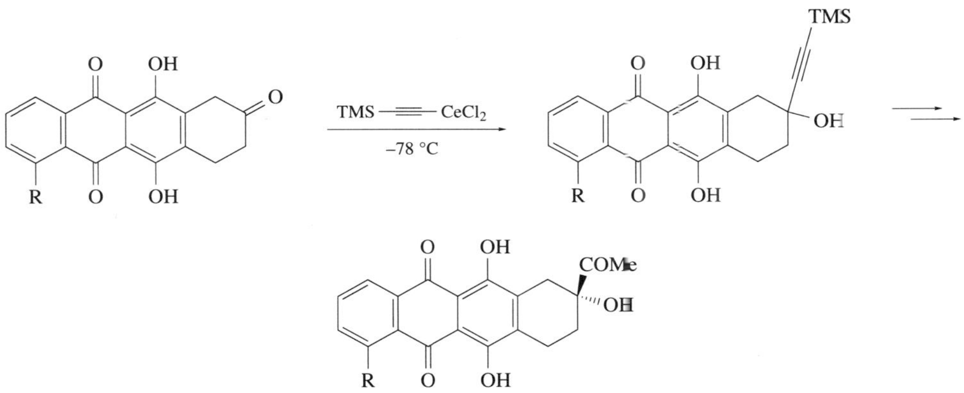

Scheme 102

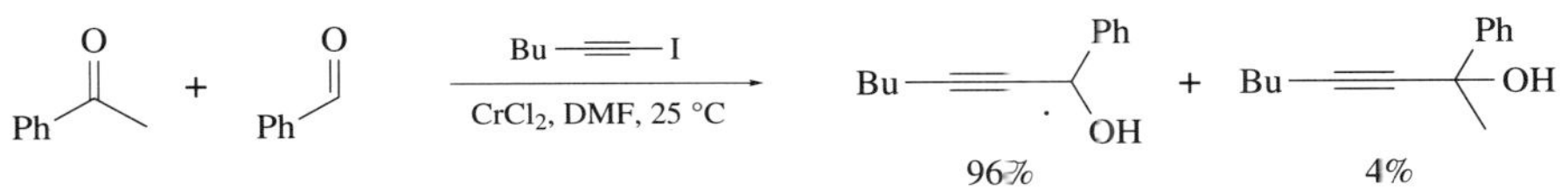

(112)

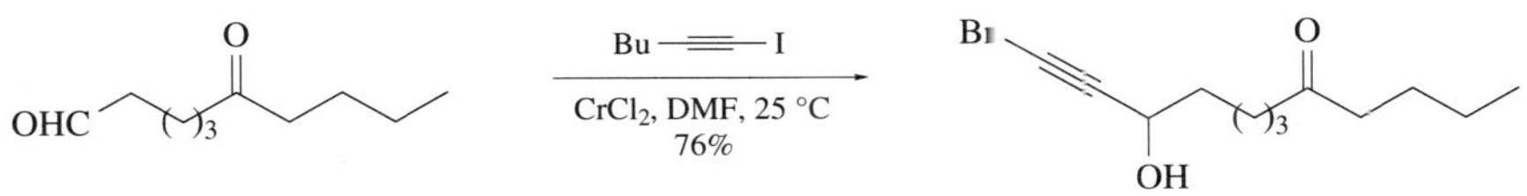

(113)

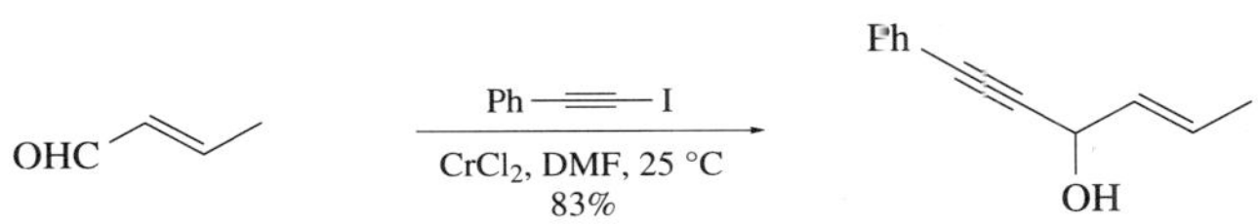

(114)

1.22.4.1.12 *Nickel acetylides*

Bromoalkynes treated with trialkylaluminum reagents in the presence of bis(*N*-methylsalicylaldimine)nickel are converted to the corresponding alkyl-substituted alkynes (Equation (115)) ⟨78TL2831⟩. It is suggested that the reaction proceeds via insertion of nickel into the alkyne halide bond to generate an alkynyl nickel complex which undergoes exchange of the halide for an alkyl group; the final step is elimination of the nickel, yielding the alkylated product.

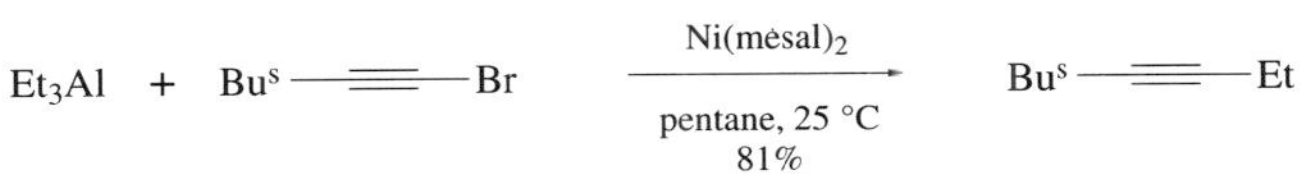

(115)

1.22.4.2 Alkynic Carbocations

As has already been discussed, vinyl cations are accepted organic intermediates generated by, for example, solvolytic processes; some have been characterised by ^{1}H and ^{13}C NMR spectroscopy ⟨85TL1493⟩. On the other hand, alkynyl cations are rare; some have been observed by mass spectrometry ⟨75JOC124, 84JA521⟩, but attempts to generate these species by solvolysis have been unsuccessful, due mainly to the alkynyl cation $HC{\equiv}C^{+}$ being ca. 130 kcal mol^{-1} less stable than the H_3C^{+} cation. Even the use of an excellent leaving group, N_2, as in the decomposition of the alkynyldiazonium salts has not yielded the required carbocation. *Ab initio* calculations have provided a rationale for this situation ⟨87JA4237⟩.

The tritium decay method ⟨93CRV2933⟩ has, however, allowed products to be isolated which suggest the intermediacy of an alkynyl carbocation. Thus, when 1,4-bis(tritioethynyl)benzene was allowed to decay in the presence of a nucleophile NuA, in both liquid and gaseous phase, for six months, significant yields of tritiated derivatives were obtained (Scheme 103) ⟨75MI 122-01, 77JCP(66)4671, 78MI 122-03⟩.

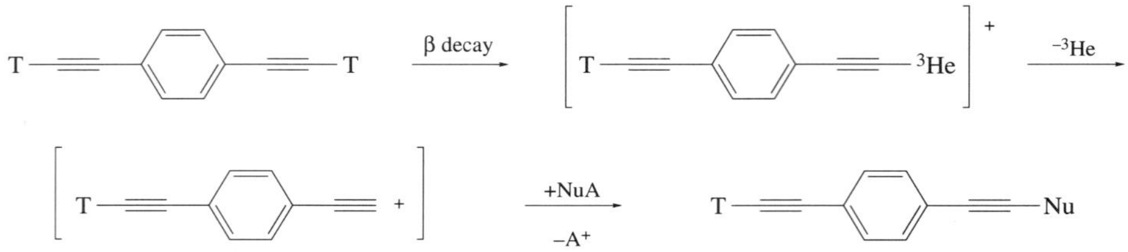

Nu-A	*Yield* (%)
Ph–H	96
Br–$(CH_2)_4$Br	36
Br–Bu^t	31

Scheme 103

1.22.4.3 Alkynic Radicals

Although the formation of alkyne radicals is nowhere near as well documented as vinyl radical formation, there are nevertheless some methods of alkyne radical formation worthy of mention. These principally occur under copper catalysis and have been employed in a number of coupling reactions. The two classical examples of these are the Glaser reaction, and the Cadiot–Chodkiewicz coupling.

In the Glaser reaction an alkyne such as 1-phenylpropyne (**144**) reacts with basic $CuCl_2$ to give a diyne (**145**) upon oxidation in air (Equation (116)) ⟨1869CB422, 1870LA137⟩. Other copper derivatives and bases can be employed in place of the copper(II) chloride and hydroxide used in the original procedures. For example, (**146**) was converted to a 1 : 2 mixture of (**147**) and (**148**) in 38% yield using copper acetate and pyridine (Equation (117)) ⟨89JA3640⟩.

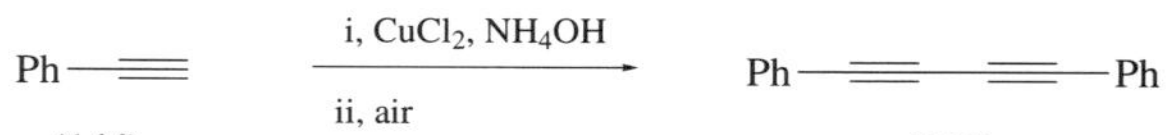

(144) **(145)** (116)

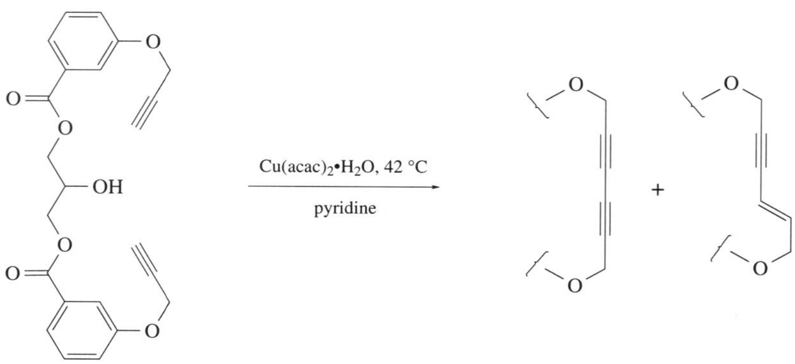

(146) **(147)** **(148)** (117)

In the Cadiot–Chodkiewicz coupling, a bromoalkyne (**150**) reacts with a monoalkyne (**149**) in the presence of cupric chloride and an amine to yield a diyne (Equation (118)) ⟨55CR1055, 57AC(P)819⟩. Alkynyl copper derivatives will also couple with aryl halides or other aromatic derivatives ⟨63JOC2163, 63JOC3313, 66CB1642, 67CB1193, 67CB1200, 73JOC3893, B-80MI 122-01⟩.

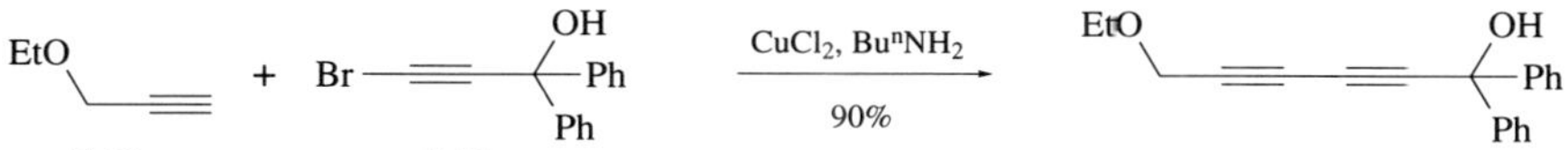

(118)

(149) **(150)**

1.22.5 VINYLIDENE CARBENES

The simplest unsaturated carbene is vinylidene, $H_2C{=}C{:}$, which is tautomeric with ethyne. Reviews concerning this and related species have appeared in the literature ⟨B-73MI 122-01, B-75MI 122-01, 78ACR107, 78CRV383, 78MI 122-03, 82ACR348, 91CRV197⟩. Theoretical studies of the ethyne/vinylidene system have been made on numerous occasions ⟨73TCA183, 75JCP(63)3238, 76JCP(64)4544, 77JA2919, 78JA1378, 78JA1928, 79ACR288, 80JA3334, 81CPL412, 82CPL493, 83CPL124⟩, and the results have been summarised in other reviews ⟨81CPL408, 81JA1904⟩.

Vinylidenes have been generated via a number of different routes, the three most important being: (i) alkoxide or alkyllithium-induced decomposition of vinyl halides (Equation (119)) ⟨57JOC1248, 59JA4719, 63JA3522, 71JA1544, 72CB2290⟩; (ii) alkoxide-induced decomposition of 5,5-disubstituted *N*-nitrosooxazolidones (Equation (120)) ⟨68JA4189, 69JA6461, 69JOC1220, 73JOC547, 74JOC761⟩; (iii) from vinyl triflates and potassium *t*-butoxide, this method being the superior choice in terms of yield, simplicity, and flexibility (Equation (121)) ⟨74JA4562, 75CC17, 75JA1459, 75JA3854, 75JA6478⟩. The required triflates can be obtained from the corresponding aldehydes via their trimethylsilyl enol ethers. The geometry of the starting triflate does not affect the product distribution, suggesting an identical and symmetrical intermediate is being formed. Small amounts of the *t*-butyl ethers always accompany the major products. Deuterium incorporation ⟨74JA4562⟩ and trapping experiments ⟨75CC17⟩ also provide evidence for the intermediacy of a methylene carbene in the above reactions. The ground state has been assigned a singlet multiplicity ⟨75JA1459, 77JA2919, 78JA6930⟩, and the carbenes generated by the above methods appear electrophilic in character ⟨72JOC1553, 74JA4562⟩, comparable to the dihalocarbenes.

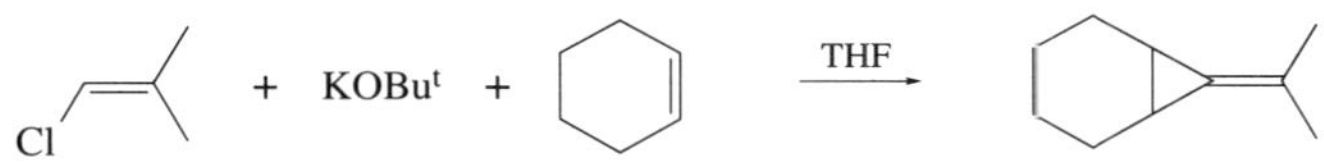

(119)

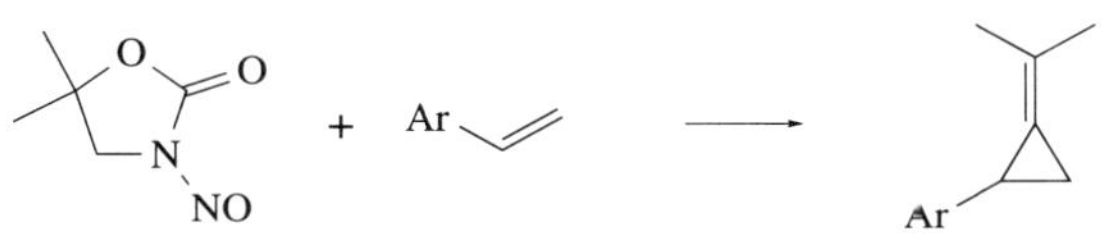

(120)

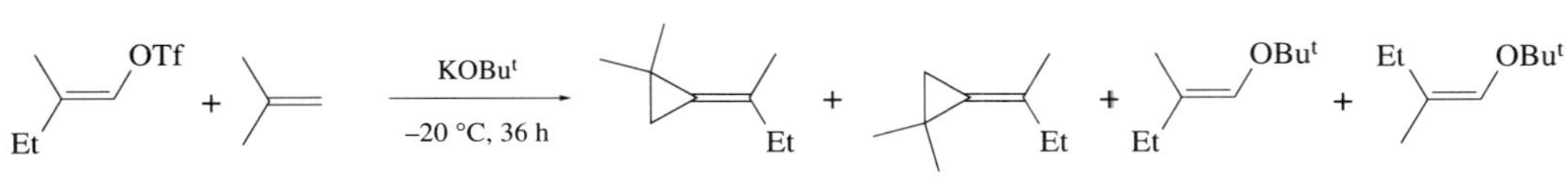

(121)

42% 32% 19% 7%

Photolysis of arylsubstituted methylenecyclopropanes has been found to generate methylene carbenes by fragmentation (Equation (122)) ⟨70JA7493, 70JA7606⟩. Methylenecyclopropane will undergo mercury-sensitised photofragmentation to ethene and ethyne ⟨68JPC321⟩, whilst organomercurial (**151**) undergoes photolysis to generate the dichloromethylene carbene (**152**) (Scheme 104) ⟨68TL503⟩.

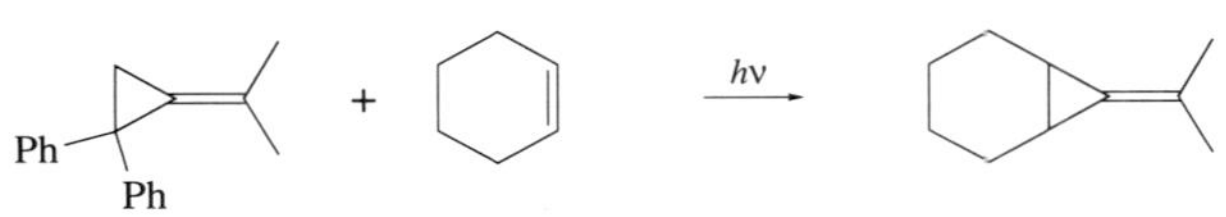

(122)

Methylene carbenes, once formed, undergo a number of reactions, of which the most important are alkyne formation, insertion reactions, addition to alkenes and coupling reactions.

Alkyne formation occurs readily when aryl substituents are present, and hydrogen migration is also very rapid ⟨72JA1578, 72JA7862⟩, although with aliphatic and halogen substituents the reaction

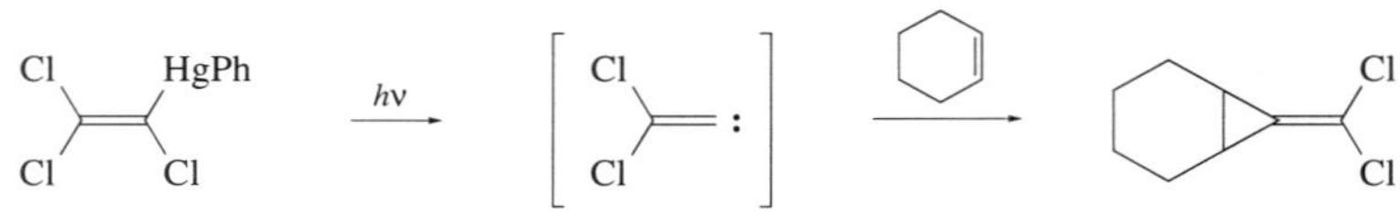

(151) **(152)**

Scheme 104

is slowed to such an extent that only minor amounts of alkynic products are obtained. The cyclopropyl group is intermediate in its migratory aptitude (Scheme 105) ⟨72JOC3220⟩.

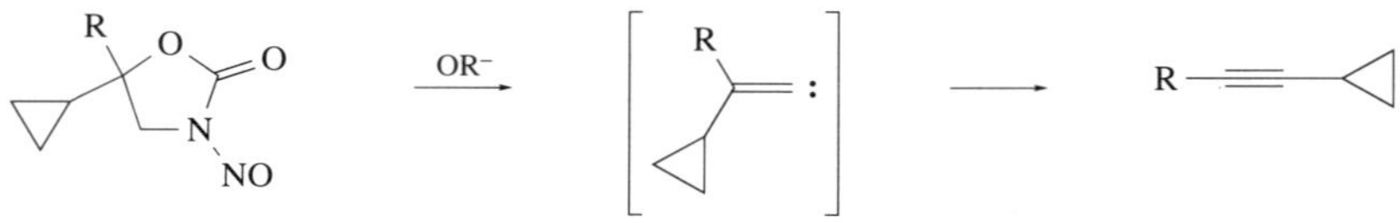

Scheme 105

In the Fritsch–Buttenberg–Wiechell rearrangement ⟨65AG(E)49⟩, a 1,1-diaryl-2-chloroethylene (**153**) rearranges to a diarylethyne on treatment with sodium ethoxide in ethanol at 180–200 °C; the mechanism was originally proposed to proceed via a vinylidene carbene (**154**) (Scheme 106) ⟨1894LA(279)319, 1894LA(279)324, 1894LA(279)337⟩. Alkali metal amides in liquid ammonia ⟨36JA2310⟩, and metal alkyls in nonpolar solvents ⟨59JA4714, 67CB2011⟩ also induce this rearrangement. Work in the early 1960s has demonstrated that a vinyl carbanion intermediate is more probable ⟨55JA3293, 60JPS1271, 61G889, 63G757, 63JA2273⟩; in fact, in diethyl ether, halovinyllithium compounds generated by this method can be trapped as their carboxylation derivatiyes on treatment with CO_2 ⟨63ZN(B)1125⟩. Metal–halogen exchange may compete with metal–hydrogen exchange, but only when the halogen is bromine or iodine ⟨59JA4714⟩. Other factors which govern metal–hydrogen vs. metal–halogen exchange are the choice of metal alkyl, salt effects and the substituents present in the aryl groups ⟨59JA4714, 59JA4719⟩.

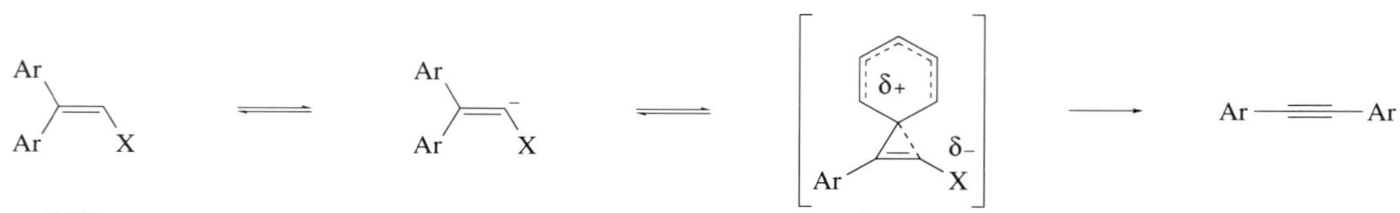

(153) **(154)**

Scheme 106

Insertion reactions of methylene carbenes occur on reaction with trialkylsilanes; insertion into the Si—H bond gives vinyl silanes as products (Scheme 107) ⟨70JA4309, 70JA4312⟩. Insertion into CH bonds is also possible in some cases ⟨65JA1142, 68TL503, 70JA7493⟩, although intramolecular insertion reactions appear to occur more readily into C—H bonds α to heteroatoms ⟨70JOC1086⟩.

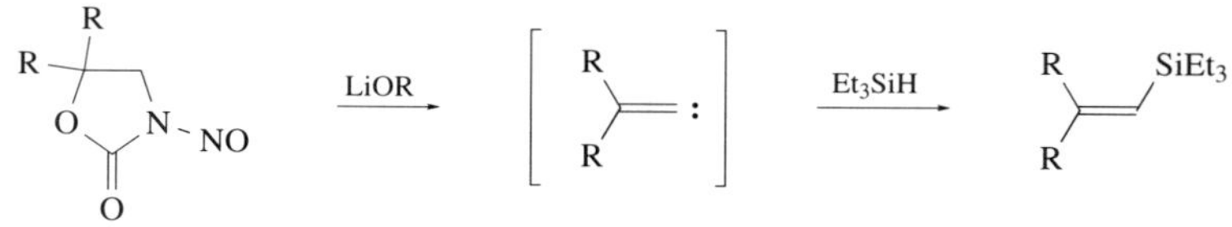

Scheme 107

A number of examples of additions of methylene carbenes to alkenes to give methylenecyclopropanes have already been mentioned, as has the apparent electrophilic character of these intermediates. Higher yields of addition products are obtained with vinyl ethers than with hydrocarbon alkenes ⟨64JA526, 69JOC1220⟩. In addition to electronic effects, steric effects also influence the addition reaction, with cyclohexene reacting 50 times faster than tetramethylethene. Coupling reactions are found to occur in those methylene carbenes where rearrangements to alkynes are sterically unfavourable (Scheme 108) ⟨57JOC1248, 59JA4719, 64AG(E)651, 65AG95, 67T565⟩. The enolate anions derived from β-dicarbonyl compounds react with either ethynyl(phenyl)iodonium tetrafluoroborate or with ethynyl-lead tetraacetate to give the corresponding α-ethynyl-1,3-dicarbonyl compounds ⟨89JCS(P1)333, 90CC118⟩. When the tetrafluoroborate reagent is employed, the reaction is thought to proceed by way of a 1,2-hydrogen migration of an alkylidenecarbene intermediate (Scheme 109).

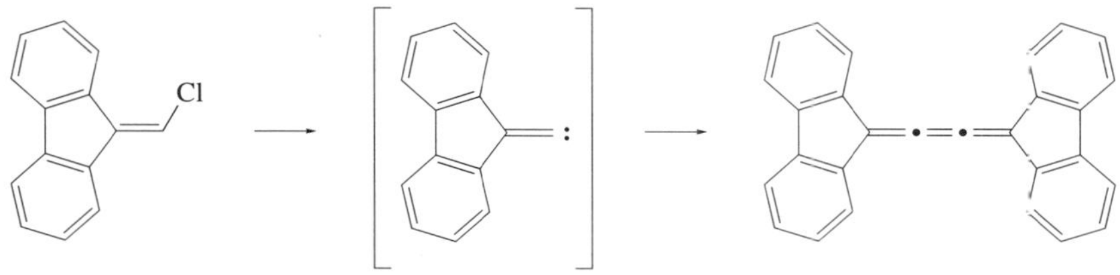

Scheme 108

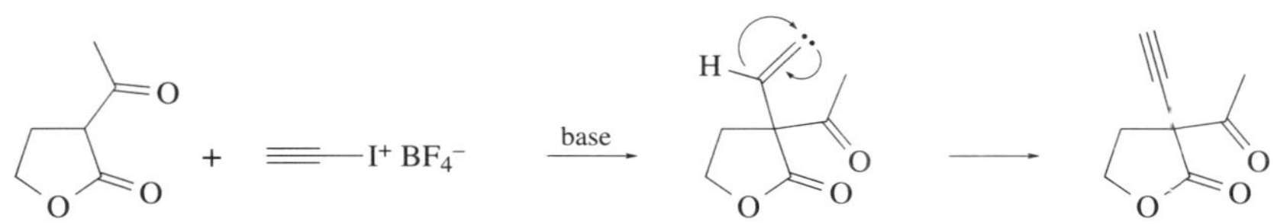

Scheme 109

The stabilisation of a reactive intermediate by complexation to a metal centre has been exploited for many systems over the years, and this is also the case for vinylidene carbenes ⟨91CRV197⟩. The most commonly exploited metal centres include Mo, Fe, Pd, Rh, Ru, Ir and Os. Several methods have been employed for the preparation of these complexes: (i) from 1-alkynes via a formal 1,2-hydrogen shift; (ii) by addition of electrophiles to metal alkynyl complexes; (iii) by deprotonation of carbyne complexes; and (iv) by formal dehydration of acyl complexes. Alkenes, preformed vinyl complexes, and disubstituted alkynes have been utilised as precursors; transfer of vinylidene ligands between metal centres and modification of existing vinylidene ligands are reactions that have also been reported ⟨91CRV197⟩. The metal complexes once formed, can undergo a variety of reactions, including electrophilic or nucleophilic attack, oxidations, cycloaddition reactions or disproportionation.

Finally, mention should be made of extended unsaturated carbenes or cumylenylidenes which have received attention as reactive intermediates. They are generated in a similar way, by base-catalysed α-elimination of triflates with $KOBu^t$ as was mentioned for the parent methylene carbenes ⟨79JA4992, 80JA6813, 81JA5429⟩, and they undergo similar reactions to that of the parent compound, such as insertion reactions and addition to alkenes.

References

EXPLANATION OF THE REFERENCE SYSTEM

Throughout this work, references are designated by a number–lettering coding of which the first two numbers denote tens and units of the year of publication, the next one to three letters denote the journal, and the final numbers denote the page. This code appears in the text each time a reference is quoted; the advantages of this system are outlined in the Introduction. The system has been used previously in "Comprehensive Heterocyclic Chemistry," eds A. R. Katritzky and C. W. Rees, Pergamon, Oxford, 1984 and is based on that used in the following two monographs: (a) A. R. Katritzky and J. M. Lagowski, "Chemistry of the Heterocyclic N-Oxides," Academic Press, New York, 1971; (b) J. Elguero, C. Marzin, A. R. Katritzky and P. Linda, "The Tautomerism of Heterocycles," in "Advances in Heterocyclic Chemistry," Supplement 1, Academic Press, New York, 1976.

The following additional notes apply:

1. A list of journal codes in alphabetical order, together with the journals to which they refer, is given immediately following these notes. Journal names are abbreviated throughout using the CASSI (Chemical Abstracts Service Source Index) system.
2. Each volume contains all the references cited *in that volume*; no separate lists are given for individual chapters.
3. The list of references is arranged in order of (a) year, (b) journal in alphabetical order of journal code, (c) part letter or number if relevant, (d) volume number if relevant, (e) page number.
4. In the reference list the code is followed by (a) the complete literature citation in the conventional manner and (b) the number(s) of the page(s) on which the reference appears, whether in the text or in tables, schemes, etc.
5. For nontwentieth-century references the year is given in full in the code.
6. For journals which are published in separate parts, the part letter or number is given (when necessary) in parentheses immediately after the journal code letters.
7. Journal volume numbers are *not* included in the code numbers unless more than one volume was published in the year in question, in which case the volume number is included in parentheses immediately after the journal code letters.
8. Patents are assigned appropriate three-letter codes.
9. Frequently cited books are assigned codes.
10. Less common journals and books are given the code "MI" for miscellaneous with the whole code for books prefixed by the letter "B-".
11. Where journals have changed names, the same code is used throughout, e.g. CB refers to both *Chem. Ber.* and to *Ber. Dtsch. Chem. Ges.*

Journal Codes

AAC	Antimicrob. Agents Chemother.
ABC	Agric. Biol. Chem.
AC	Appl. Catal.
AC(P)	Ann. Chim. (Paris)
AC(R)	Ann. Chim. (Rome)
ACH	Acta Chim. Acad. Sci. Hung.

ACR	Acc. Chem. Res.
ACS	Acta Chem. Scand.
ACS(A)	Acta Chem. Scand., Ser. A
ACS(B)	Acta Chem. Scand., Ser. B
AF	Arzneim.-Forsch.
AFC	Adv. Fluorine Chem.
AG	Angew. Chem.
AG(E)	Angew. Chem., Int. Ed. Engl.
AHC	Adv. Heterocycl. Chem.
AHCS	Adv. Heterocycl. Chem. Supplement
AI	Anal. Instrum.
AJC	Aust. J. Chem.
AK	Ark. Kemi
AKZ	Arm. Khim. Zh.
AM	Adv. Mater. (Weinheim, Ger.)
AMLS	Adv. Mol. Spectrosc.
AMS	Adv. Mass. Spectrom.
ANC	Anal. Chem.
ANL	Acad. Naz. Lncei
ANY	Ann. N. Y. Acad. Sci.
AOC	Adv. Organomet. Chem.
AP	Arch. Pharm. (Weinheim, Ger.)
APO	Adv. Phys. Org. Chem.
AQ	An. Quim.
AR	Annu. Rep. Prog. Chem.
AR(A)	Annu. Rep. Prog. Chem., Sect. A
AR(B)	Annu. Rep. Prog. Chem., Sect. B
ARP	Annu. Rev. Phys. Chem.
ASI	Acta Chim. Sin. Engl. Ed.
ASIN	Acta Chim. Sin.
AX	Acta Crystallogr.
AX(A)	Acta Crystallogr., Part A
AX(B)	Acta Crystallogr., Part B
B	Biochemistry
BAP	Bull. Acad. Pol. Sci., Ser. Sci. Chim.
BAU	Bull. Acad. Sci. USSR, Div. Chim. Sci.
BBA	Biochim. Biophys. Acta
BBR	Biochim. Biophys. Res. Commun.
BCJ	Bull. Chem. Soc. Jpn.
BEP	Belg. Pat.
BJ	Biochem. J.
BJP	Br. J. Pharmacol.
BMC	Bioorg. Med. Chem. Lett.
BP	Biochem. Biopharmacol.
BPJ	Br. Polym. J.
BRP	Br. Pat.
BSB	Bull. Soc. Chim. Belg.
BSF	Bull. Soc. Chim. Fr.
BSF(2)	Bull. Soc. Chim. Fr., Part 2
C	Chimia
CA	Chem. Abstr.
CAN	Cancer
CAR	Carbohydr. Res.
CAT	Chim. Acta Turc.
CB	Chem. Ber.

CBR	Chem. Br.
CC	J. Chem. Soc., Chem. Commun.
CCA	Croat. Chem. Acta
CCC	Collect. Czech. Chem. Commun.
CCR	Coord. Chem. Rev.
CE	Chem. Express
CEN	Chem. Eng. News
CHE	Chem. Heterocycl. Compd. (Engl. Transl.)
CHEC	Comp. Heterocycl. Chem.
CI(L)	Chem. Ind. (London)
CI(M)	Chem. Ind. (Milan)
CJC	Can. J. Chem.
CJS	Can. J. Spectrosc.
CL	Chem. Lett.
CLY	Chem. Listy
CM	Chem. Mater.
CMC	Comp. Med. Chem.
COC	Comp. Org. Chem.
COMC-I	Comp. Organomet. Chem., 1st edn.
COS	Comp. Org. Synth.
CP	Can. Pat.
CPB	Chem. Pharm. Bull.
CPH	Chem. Phys.
CPL	Chem. Phys. Lett.
CR	C. R. Hebd. Seances Acad. Sci.
CR(A)	C. R. Hebd. Seances Acad. Sci., Ser. A
CR(B)	C. R. Hebd. Seances Acad. Sci., Ser. B
CR(C)	C. R. Hebd. Seances Acad. Sci., Ser. C
CRAC	Crit. Rev. Anal. Chem.
CRV	Chem. Rev.
CS	Chem. Scr.
CSC	Cryst. Struct. Commun.
CSR	Chem. Soc. Rev.
CT	Chem. Tech.
CZ	Chem.-Ztg.
CZP	Czech. Pat.
DIS	Diss. Abstr.
DIS(B)	Diss. Abstr. Int. B.
DOK	Dokl. Akad. Nauk SSSR
DP	Dyes Pigm.
E	Experientia
EC	Educ. Chem.
EF	Energy Fuels
EGP	Ger. (East) Pat.
EJM	Eur. J. Med. Chem.
EUP	Eur. Pat.
FCF	Forschr. Chem. Forsch.
FCR	Fluorine Chem. Rev.
FES	Farmaco Ed. Sci.
FOR	Forschr. Chem. Org. Naturst.
FRP	Fr. Pat.
G	Gazz. Chim. Ital.
GAK	Gummi Asbest Kunstst.
GEP	Ger. Pat.
GEP(O)	Ger. Pat. Offen.

GSM	Gen. Synth. Methods
H	Heterocycles
HAC	Heteroatom Chem.
HC	Chem. Heterocycl. Compd.
HCA	Helv. Chim. Acta
HOU	Methoden Org. Chem. (Houben-Weyl)
HP	Hydrocarbon Process
IC	Inorg. Chem.
ICA	Inorg. Chim. Acta
IEC	Ind. Eng. Chem. Res.
IJ	Isr. J. Chem.
IJC	Indian J. Chem.
IJC(A)	Indian J. Chem., Sect. A
IJC(B)	Indian J. Chem., Sect. B
IJM	Int. J. Mass Spectrom. Ion Phys.
IJQ	Int. J. Quantum Chem.
IJS	Int. J. Sulfur Chem.
IJS(A)	Int. J. Sulfur Chem., Part A
IJS(B)	Int. J. Sulfur Chem., Part B
IS	Inorg. Synth
IZV	Izv. Akad. Nauk SSSR Ser. Khim.
JA	J. Am. Chem. Soc.
JAN	J. Antibiot.
JAP	Jpn. Pat.
JAP(K)	Jpn. Kokai
JBC	J. Biol. Chem.
JC	J. Chromatogr.
JCC	J. Coord. Chem.
JCE	J. Chem. Ed.
JCED	J. Chem. Eng. Data
JCI	J. Chem. Inf. Comput. Sci.
JCP	J. Chem. Phys.
JCPB	J. Chim. Phys. Physico-Chim. Biol.
JCR(M)	J. Chem. Res. (M)
JCR(S)	J. Chem. Res. (S)
JCS	J. Chem. Soc.
JCS(A)	J. Chem. Soc. (A)
JCS(B)	J. Chem. Soc. (B)
JCS(C)	J. Chem. Soc. (C)
JCS(D)	J. Chem. Soc., Dalton Trans.
JCS(F1)	J. Chem. Soc., Faraday Trans. 1
JCS(F2)	J. Chem. Soc., Faraday Trans. 2
JCS(P1)	J. Chem. Soc., Perkin Trans. 1
JCS(P2)	J. Chem. Soc., Perkin Trans. 2
JCS(S2)	J. Chem. Soc. (Suppl. 2)
JEC	J. Electroanal. Chem. Interfacial Electrochem.
JEM	J. Energy Mater.
JES	J. Electron. Spectrosc.
JFA	J. Sci. Food. Agri.
JFC	J. Fluorine Chem.
JGU	J. Gen. Chem. USSR (Engl. Transl.)
JHC	J. Heterocycl. Chem.
JIC	J. Indian Chem. Soc.
JINC	J. Inorg. Nucl. Chem.
JLC	J. Liq. Chromatogr.

JMAS	J. Mat. Sci.
JMC	J. Med. Chem.
JMOC	J. Mol. Catal.
JMR	J. Magn. Reson.
JMS	J. Mol. Sci.
JOC	J. Org. Chem.
JOM	J. Organomet. Chem.
JOU	J. Org. Chem. USSR (Engl. Transl.)
JPC	J. Phys. Chem.
JPJ	J. Pharm. Soc. Jpn.
JPO	J. Phys. Org. Chem.
JPP	J. Pharm. Pharmacol.
JPR	J. Prakt. Chem.
JPS	J. Pharm. Sci.
JPS(A)	J. Polym. Sci., Polym. Chem., Part A
JPU	J. Phys. Chem. USSR (Engl. Transl.)
JSC	J. Serbochem. Soc.
JSP	J. Mol. Spectrosc.
JST	J. Mol. Struct.
K	Kristallografiya
KFZ	Khim. Farm. Zh.
KGS	Khim. Geterotsikl. Soedin.
KO	Kirk-Othmer Encyc.
KPS	Khim. Prir. Soedin.
L	Langmuir
LA	Liebigs Ann. Chem.
LC	Liq. Cryst.
LS	Life Sci.
M	Monatsh. Chem.
MAC	Macromol. Chem.
MC	Mendeleev Chem. J. (Engl. Transl.)
MCLC	Mol. Cryst. Liq. Cryst.
MI	Miscellaneous [book/journal]
MIP	Miscellaneous Pat.
MM	Macromolecules
MP	Mol. Phys.
MRC	Magn. Reson. Chem.
N	Naturwissenschaften
NAT	Nat.
NEP	Neth. Pat.
NJC	Nouv. J. Chim.
NKK	Nippon Kagaku Kaishi (J. Chem. Soc. Jpn.)
NKZ	Nippon Kagaku Zasshi
NZJ	N. Z. J. Sci. Technol.
OCS	Organomet. Synth.
OM	Organometallics
OMR	Org. Magn. Reson.
OMS	Org. Mass Spectrom.
OPP	Org. Prep. Proced. Int.
OR	Org. React.
OS	Org. Synth.
OSC	Org. Synth., Coll. Vol.
P	Phytochemistry
PA	Polym. Age
PAC	Pure Appl. Chem.

PAS	Pol. Acad. Sci.
PB	Polym. Bull.
PC	Personal Communication
PCS	Proc. Chem. Soc.
PHA	Pharmazi
PHC	Prog. Heterocycl. Chem.
PIA	Proc. Indian Acad. Sci.
PIA(A)	Proc. Indian Acad. Sci., Sect. A
PJC	Pol. J. Chem.
PJS	Pak. J. Sci. Ind. Res.
PMH	Phys. Methods Heterocycl. Chem.
PNA	Proc. Natl. Acad. Sci. USA
POL	Polyhedron
PP	Polym. Prepr.
PRS	Proceed. Roy. Soc.
PS	Phosphorus Sulfur
QR	Q. Rev., Chem. Soc.
QRS	Quart. Rep. Sulfur. Chem.
QSAR	Quant. Struct. Act. Relat. Pharmacol. Chem. Biol.
RC	Rubber Chem. Technol.
RCM	Rapid Commun. Mass Spectrom.
RCP	Rec. Chem. Prog.
RCR	Russ. Chem. Rev. (Engl. Transl.)
RHA	Rev. Heteroatom Chem.
RJ	Rubber J.
RP	Rev. Polarogr.
RRC	Rev. Roum. Chim.
RS	Ric. Sci.
RTC	Recl. Trav. Chim. Pays-Bas
RZC	Rocz. Chem.
S	Synthesis
SA	Spectrochim. Acta
SA(A)	Spectrochim. Acta, Part A
SAP	S. Afr. Pat.
SC	Synth. Commun.
SCI	Science
SL	Synlett
SM	Synth. Met.
SR	Sulfur Reports
SRI	Synth. React. Inorg. Metal-Org. Chem.
SS	Sch. Sci. Rev.
SST	Org. Compd. Sulphur, Selenium, Tellurium [R. Soc. Chem. series]
SUL	Sulfur Letters
SZP	Swiss Pat.
T	Tetrahedron
T(S)	Tetrahedron, Suppl.
TA	Tetrahedron Asymmetry
TAL	Talanta
TCA	Theor. Chim. Acta
TCC	Top. Curr. Chem.
TCM	Tetrahedron, Comp. Method
TFS	Trans. Faraday Soc.
TH	Thesis
TL	Tetrahedron Lett.
TS	Top. Stereochem.

UK	Usp. Khim.
UKZ	Ukr. Khim. Zh. (Russ. Ed.)
UP	Unpublished Results
URP	USSR Pat.
USP	US Pat.
WCH	Wiadom. Chem.
YGK	Yuki Gosei Kagaku Kyokaishi
YZ	Yakugaku Zasshi (J. Pharm. Soc. Jpn.)
ZAAC	Z. Anorg. Allg. Chem.
ZAK	Zh. Anal. Khim.
ZC	Z. Chem.
ZN	Z. Naturforsch.
ZN(A)	Z. Naturforsch., Teil A
ZN(B)	Z. Naturforsch., Teil B
ZOB	Zh. Obshch. Khim.
ZOR	Zh. Org. Khim.
ZPC	Hoppe–Seyler's Z. Physiol. Chem.
ZPK	Zh. Prikl. Khim.

VOLUME 1 REFERENCES

1851LA(78)253	A. W. Hofmann; *Justus Liebigs Ann. Chem.*, 1851, **78**, 253.	627
1856MI 119-01	R. Piria; *Ann. Chim. Phys*, 1856, **48**, 113.	883
1860LA(114)54	R. Fittig; *Justus Liebigs Ann. Chem.*, 1860, **114**, 54.	384
1869CB422	C. Glaser; *Ber. Dtsch. Chem. Ges.*, 1869, **2**, 422.	1142
1870LA(154)137	C. Glaser; *Justus Liebigs Ann. Chem.*, 1870, **154**, 137.	104
1874CB352	M. P. von Wilde; *Ber. Dtsch. Chem. Ges.*, 1874, **7**, 352.	71
1883CB2208	A. Einhorn; *Ber. Dtsch. Chem. Ges.*, 1883, **16**, 2208.	785
1883LA(218)145	T. Komnenos; *Justus Liebigs Ann. Chem.*, 1883, **218**, 145.	306
1884AC(P)449	C. Friedel and J. M. Crafts; *Ann. Chim. Phys.* 1884, **1**, 449.	462
1887JPR349	A. Michael; *J. Prakt. Chem.*, 1887, **35**, 349.	306
1887MI 107-01	A. Michael; *Am. Chem. J.*, 1887, **9**, 115.	306
1891CB3108	G. Merling; *Ber. Dtsch. Chem. Ges.*, 1891, **24**, 3108.	906
1894LA(279)319	P. Fritsch; *Justus Liebigs Ann. Chem.*, 1894, **279**, 319.	1144
1894LA(279)324	W. P. Buttenberg; *Justus Liebigs Ann. Chem.*, 1894, **279**, 324.	1144
1894LA(279)337	H. Wiechell; *Justus Liebigs Ann. Chem.*, 1894, **279**, 337.	1144
1894MI 109-01	A. E. Favorskii; *Zh. Russ. Fiz.-Khim. O-va.*, 1894, **26**, 556.	391
1895MI 103-01	F. B. Ahrens; *Z. Elektrochem.*, 1895, **2**, 577.	103
1897MI 103-01	P. Sabatier and J. B. Senderens; *Compt. Rend.*, 1897, **124**, 1358.	71
1898CB1553	W. Wernick and R. Wolffenstein; *Ber. Dtsch. Chem. Ges.*, 1898, **31**, 1553.	623
1899CB2307	G. Wagner and W. Brickner; *Ber. Dtsch. Chem. Ges.*, 1899, **32**, 2307.	380
00CB3150	M. Gomberg; *Ber. Dtsch. Chem. Ges.*, 1900, **33**, 3150.	919
00JA757	M. Gomberg; *J. Am. Chem. Soc.*, 1900, **22**, 757.	919
02CB2018	P. Walden; *Ber. Dtsch. Chem. Ges.*, 1902, **35**, 2018.	898
02CB2397	M. Gomberg; *Ber. Dtsch. Chem. Ges.*, 1902, **35**, 2397.	890, 898
02CB3031	A. Baeyer and V. Villiger; *Ber. Dtsch. Chem. Ges.*, 1902, **35**, 3031.	897

04BSF101	P. Sabatier and J. B. Senderens; *Bull. Soc. Chim. Fr.*, 1904, **31**, 101.	87
05MI 107-01	E. P. Kohler and G. L. Heritage; *Am. Chem. J.*, 1905, **33**, 21.	310
07CB4959	N. J. Demjanow; *Ber. Dtsch. Chem. Ges.*, 1907, **40**, 4959.	383
10CB1528	H. Finkelstein; *Ber. Dtsch. Chem. Ges.*, 1910, **43**, 1528.	107
11CB2134	H. Thoms and W. Siebling; *Ber. Dtsch. Chem. Ges.*, 1911, **44**, 2134.	432
12CB3157	L. Claisen; *Ber. Dtsch. Chem. Ges.*, 1912, **45**, 3157.	414, 825
12LA(394)23	L. Wolff; *Justus Liebigs Ann. Chem.*, 1912, **394**, 23.	393
13CB1837	E. Clemmensen; *Ber. Dtsch. Chem. Ges.*, 1913, **46**, 1837.	18
14BSF228	P. Sabatier and L. Espil; *Bull. Soc. Chim. Fr.*, 1914, **15**, 228.	87
16CB25	H. Rope and M. Iselin; *Ber. Dtsch. Chem. Ges.*, 1916, **49**, 25.	469
16CB2222	W. Borsche; L. Stackmann and J. Makaroff-Semijanski; *Ber. Dtsch. Chem. Ges.*, 1916, **49**, 2222.	474
20MI 104-01	E. E. Turner; *J. Proc. Roy. Soc. N.S.W.*, 1920, **54**, 37.	117
20MI 110-01	F. Swarts; *Bull. Acad. Res. Belg.*, 1920, 399.	426
21JPR1	F. Straus and A. Dützmann; *J. Prakt. Chem.*, 1921, **103**, 1.	898
22CB1960	B. Nybergh; *Ber. Dtsch. Chem. Ges.*, 1922, **55**, 1960.	384
22CB2500	H. Meerwein and K. Van Emster; *Ber. Dtsch. Chem. Ges.*, 1922, **55**, 2500.	380
23LA(434)34	K. Ziegler; *Justus Liebigs Ann. Chem.*, 1923, **434**, 34.	901
24CB1634	P. Schorigin; *Ber. Dtsch. Chem. Ges.*, 1924, **57**, 1634.	396, 397
24LA(437)1	J. Bredt; *Justus Liebigs Ann. Chem.*, 1924, **437**, 1.	853
25CB2028	P. Schorigin; *Ber. Dtsch. Chem. Ges.*, 1925, **58**, 2028.	396
26LA(449)81	L. Claisen and E. Tietze; *Justus Liebigs Ann. Chem.*, 1926, **449**, 81.	840
26MI 104-01	B. V. Tronov and E. A. Krüger; *J. Russ. Phys.-Chem. Soc.*, 1926, **58**, 1270; *Chem. Abstr.*, 1927, **21**, 3887.	107
27CB950	J. Loevenich, J. Losen and A. Dierichs; *Ber. Dtsch. Chem. Ges.*, 1927, **60**, 950.	1065
28CB1750	K. Freudenberg, H. Toepffer and C. C. Andersen; *Ber. Dtsch. Chem. Ges.*, 1928, **61**, 1750.	21
28JCS3046	G. Newberry and M. A. Phillips; *J. Chem. Soc.*, 1928, 3046.	446
28JCS3193	T. S. Stevens, E. M. Creighton, A. B. Gordon and M. MacNicol; *J. Chem. Soc.*, 1928, 3193.	398
29JA594	C. R. Noller; *J. Am. Chem. Soc.*, 1929, **51**, 594.	127
29JA1491	F. C. Whitmore and E. N. Thurman; *J. Am. Chem. Soc.*, 1929, **51**, 1491.	128
29JCS2342	G. W. Fenton and C. K. Ingold; *J. Chem. Soc.*, 1929, 2342.	49, 640
30JCS705	G. W. Fenton and C. K. Ingold; *J. Chem. Soc.*, 1930, 705.	612
30JCS2107	T. S. Stevens; *J. Chem. Soc.*, 1930, 2107.	398
30JCS2171	D. A. Ainley and F. G. Challenger; *J. Chem. Soc.*, 1930, 2171.	447
31CB2543	E. Benary; *Ber. Dtsch. Chem. Ges.*, 1931, **64**, 2543 (*Chem. Abstr.*, 1932, **26**, 420).	486
31JA4192	H. Gilman, A. P. Hewlett and G. F. Wright; *J. Am. Chem. Soc.*, 1931, **53**, 4192.	1043
32JA1212	J. B. Conant and G. W. Wheland; *J. Am. Chem. Soc.*, 1932, **54**, 1212.	848, 869
32JA4066	W. H. Carothers, G. J. Berchet and A. M. Collins; *J. Am. Chem. Soc.*, 1932, **54**, 4066.	965
32JCS55	T. Thomson and T. S. Stevens; *J. Chem. Soc.*, 1932, 55.	398
32JCS69	T. Thomson and T. S. Stevens; *J. Chem. Soc.*, 1932, 69.	399
32JCS1926	J. L. Dunn and T. S. Stevens; *J. Chem. Soc.*, 1932, 1926.	398
32JCS1932	T. Thomson and T. S. Stevens; *J. Chem. Soc.*, 1932, 1932.	407, 628
32LA(495)84	K. Ziegler and H. Ohlinger; *Justus Liebigs Ann. Chem.*, 1932, **495**, 84.	881
33JA1094	W. H. Carothers and G. J. Berchet; *J. Am. Chem. Soc.*, 1933, **55**, 1094.	1043
33JCS526	E. D. Hughes, C. K. Ingold and C. S. Patel; *J. Chem. Soc.*, 1933, 526.	891
33JCS1571	E. D. Hughes and C. K. Ingold; *J. Chem. Soc.*, 1933, 1571.	891
34CB317	A. N. Nesmejanow and K. A. Kozeschkow; *Ber. Dtsch. Chem. Ges.*, 1934, **67**, 317.	446
34JA447	C. T. Hurd and H. T. Bollman; *J. Am. Chem. Soc.*, 1934, **56**, 447.	808
34JA967	P. D. Bartlett; *J. Am. Chem. Soc.*, 1934, **56**, 967.	853
34MI 103-01	I. Horiuti and M. Polanyi; *Trans. Faraday Soc.*, 1934, **30**, 1164.	71

35CB200	F. Arndt and B. Eistert; *Ber. Dtsch. Chem. Ges.*, 1935, **68**, 200.	394
35JCS255	T. L. Gleave, E. D. Hughes and C. K. Ingold; *J. Chem. Soc.*, 1935, 255.	891
36CB882	A. Butenandt, J. Schmidt-Thome; *Ber. Dtsch. Chem. Ges.*, 1936, **69**, 882.	795
36CB2106	C. Mannich and H. Davidsen; *Ber. Dtsch. Chem. Ges.*, 1936, **69**, 2106.	82
36JA816	C. B. Linn and C. R. Noller; *J. Am. Chem. Soc.*, 1936, **58**, 816.	117
36JA1124	W. K. Mcewen; *J. Am. Chem. Soc.*, 1936, **58**, 1124.	848, 869
36JA2310	G. H. Coleman, W. H. Holst and R. D. Maxwell; *J. Am. Chem. Soc.*, 1936, **58**, 2310.	1144
36JCS584	J. S. McFadyen and T. S. Stevens; *J. Chem. Soc.*, 1936, 584.	39
36JCS623	S. K. Hsü and C. L. Wilson; *J. Chem. Soc.*, 1936, 623.	853
36JPR251	*J. Prakt. Chem.*, 1936, **147**, 251.	58
36RTC518	H. Gilman and J. F. Nelson; *Recl. Trav. Chim. Pays-Bas*, 1936, **55**, 518.	127
37CR(205)56	M. Sommelet; *C. R. Hebd. Seances Acad. Sci.*, 1937, **205**, 56.	824
37JA596	C. B. Wooster and K. L. Godfrey; *J. Am. Chem. Soc.*, 1937, **59**, 596.	95
37JA1823	C. R. Hauser and W. B. Renfrew, Jr.; *J. Am. Chem. Soc.*, 1937, **59**, 1823.	884
37JCS1136	R. P. Linstead, A. B. Wang, J. H. Williams and K. D. Errington; *J. Chem. Soc.*, 1937, 1136.	78
37JCS1177	E. D. Hughes and U. G. Shapiro; *J. Chem. Soc.*, 1937, 1177.	107
37JCS1183	K. A. Cooper and E. D. Hughes; *J. Chem. Soc.*, 1937, 1183.	107
37JOC1	T. H. Vaughn, G. F. Hennion, R. R. Vogt and J. A. Nieuwland; *J. Org. Chem.*, 1937, **2**, 1.	1000
37MI 119-01	E. Hückel; *Z. Elektrochem.*, 1937, **43**, 752.	906
37MI 119-02	E. Hückel; *Z. Elektrochem.*, 1937, **43**, 827.	906
37RTC535	W. T. Nauta and P. J. Wuis; *Recl. Trav. Chim. Pays-Bas*, 1937, **56**, 535.	898
38CB442	R. Kuhn and C. Grundmann; *Ber. Dtsch. Chem. Ges.*, 1938, **71B**, 442.	912
38JA115	J. R. Johnson, H. R. Snyder and M. G. Van Campen; *J. Am. Chem. Soc.*, 1938, **60**, 115.	58
38JA121	J. R. Johnson and M. G. van Campen; *J. Am. Chem. Soc.*, 1938, **60**, 121.	447
38JA1905	C. D. Hurd and M. A. Pollack; *J. Am. Chem. Soc.*, 1938, **60**, 1905.	414
38JA2333	H. Gilman and C. W. Bradley; *J. Am. Chem. Soc.*, 1938, **60**, 2333.	663
38JCS78	S. K. Hsü, C. K. Ingold and C. L. Wilson; *J. Chem. Soc.*, 1938, 78.	853
39JA2418	R. Q. Brewster and J. A. Poje; *J. Am. Chem. Soc.*, 1939, **61**, 2418.	442
39JA3184	P. D. Bartlett and L. H. Knox; *J. Am. Chem. Soc.*, 1939, **61**, 3184.	900
40AK(13A)1	L. Ramberg and B. Bäcklund; *Ark. Kemi*, 1940, **13A**, No. 27, 1 (*Chem. Abstr.*, 1940, **34**, 4725).	704, 706
40CA(34)6407	S. Orlic; *Chem. Abstr.*, 1940, **34**, 6407.	446
40JA113	W. G. Brown and K. Eberly; *J. Am. Chem. Soc.*, 1940, **62**, 113.	871
40JA441	A. C. Cope and E. M. Hardy; *J. Am. Chem. Soc.*, 1940, **62**, 441.	411, 806
40JA2590	J. G. Aston and R. B. Greenburg; *J. Am. Chem. Soc.*, 1940, **62**, 2590.	391
40JA2933	P. D. Bartlett and G. F. Woods; *J. Am. Chem. Soc.*, 1940, **62**, 2933.	853
40JCS207	H. H. Hodgson and E. Marsden; *J. Chem. Soc.*, 1940, 207.	442
40JCS704	M. F. Carroll; *J. Chem. Soc.*, 1940, 704.	416, 831
40JCS1266	M. F. Carroll; *J. Chem. Soc.*, 1940, 1266.	416, 831
40MI 114-01	L. Ramberg and B. Bäcklund; *Arkiv Kemi, Mineral. Geol.*, 1940, **13A**, No. 27 (*Chem. Abstr.*, 1940, **34**, 4725).	617
41JA216	K. N. Campbell and L. T. Eby; *J. Am. Chem. Soc.*, 1941, **63**, 216.	456, 508
41JA758	F. C. Whitmore and G. W. Pedlow, Jr.; *J. Am. Chem. Soc.*, 1941, **63**, 758.	301
41JA1563	H. Adkins and H. L. Coonradt; *J. Am. Chem. Soc.*, 1941, **63**, 1563.	90, 92
41JA1674	J. F. Lane and E. S. Wallis; *J. Am. Chem. Soc.*, 1941, **63**, 1674.	394
41JA1843	A. C. Cope, K. E. Hoyle and D. Heyl; *J. Am. Chem. Soc.*, 1941, **63**, 1843.	411, 807
41JA1852	A. C. Cope, C. M. Hofmann and E. M. Hardy; *J. Am. Chem. Soc.*, 1941, **63**, 1852.	411
41JA2308	M. S. Kharasch and P. O. Tawney; *J. Am. Chem. Soc.*, 1941, **63**, 2308.	311
41JA2316	M. S. Kharasch and E. K. Fields; *J. Am. Chem. Soc.*, 1941, **63**, 2316.	117
41JA2975	G. Wash, B. Shive and H. L. Lochte; *J. Am. Chem. Soc.*, 1941, **63**, 2975.	104
41JCS507	M. F. Carroll; *J. Chem. Soc.*, 1941, 507.	416, 831
42CRV77	K. N. Campbell and B. K. Campbell; *Chem. Rev.*, 1942, **31**, 77.	502
42IZV392	I. N. Nazarov and A. I. Kuznetsova; *Izv. Akad. Nauk SSSR, Ser. Khim.*, 1942, 392 (*Chem. Abstr.*, 1945, **39**, 1620).	421
42JA363	A. F. Thompson and E. N. Shaw; *J. Am. Chem. Soc.*, 1942, **64**, 363.	959
42JA2365	S. Wawzonek and H. Laitinen; *J. Am. Chem. Soc.*, 1942, **64**, 2365.	103
42JCS748	H. H. Hodgson and H. S. Turner; *J. Chem. Soc.*, 1942, 748.	442
42LA(550)260	G. Wittig and L. Löhmann; *Justus Liebigs Ann. Chem.*, 1942, **550**, 260.	396, 397
42OR(1)38	W. E. Bachmann and W. S. Struve; *Org. React.*, 1942, **1**, 38.	393
42OR(1)155	E. L. Martin; *Org. React.*, 1942, **1**, 155.	18
43CB80	E. Ott and G. Dittus; *Ber. Dtsch. Chem. Ges.*, 1943, **76**, 80.	1016, 1108
43CB1252	G. Lock and K. Stach; *Ber. Dtsch. Chem. Ges.*, 1943, **76**, 1252.	42
43HCA800	P. Karrer and E. Schick; *Helv. Chim. Acta*, 1943, **26**, 800.	46
43IZV389	I. N. Nazarov and Ya. M. Yanbikov; *Izv. Akad. Nauk SSSR, Ser. Khim.*, 1943, 389 (*Chem. Abstr.*, 1945, **39**, 503).	421

43JA1516	M. L. Wolfrom and R. L. Brown; *J. Am. Chem. Soc.*, 1943, **65**, 1516.	45
43JA1992	W. Kimel and A. C. Cope; *J. Am. Chem. Soc.*, 1943, **65**, 1992.	416, 831
43JA2020	A. L. Henne and K. W. Greenlee; *J. Am. Chem. Soc.*, 1943, **65**, 2020.	508
43JCS86	H. H. Hodgson and H. S. Turner; *J. Chem. Soc.*, 1943, 86.	442
43OSC(2)93	C. S. Marvel and V. du Vigneaud; *Org. Synth., Coll. Vol.*, 1943, **2**, 93.	772
43OSC(2)116	P. S. Pinkney; *Org. Synth., Coll. Vol.*, 1943, **2**, 116.	884
43OSC(2)268	C. R. Hauser and W. B. Renfrew, Jr.; *Org. Synth., Coll. Vol.*, 1943, **2**, 268.	881
43OSC(2)416	E. B. Vliet, C. S. Marvel and C. M. Hsueh; *Org. Synth., Coll. Vol.*, 1943, **2**, 416.	772
43OSC(2)491	R. M. Herbst and D. Shemin; *Org. Synth. Coll. Vol.*, 1943, **2**, 491.	80
43OSC(2)566	D. Starr and R. M. Hixon; *Org. Synth., Coll. Vol.*, 1943, **2**, 566.	91
43OSC(2)606	H. Adkins and W. Zartman; *Org. Synth., Coll. Vol.*, 1943, **2**, 606.	881
44BSF1944	Buu-Hoi and P. Cagniant; *Bull. Soc. Chim. Fr.*, 1944, **11**, 349.	104
44HCA1297	P. Karrer and H. Rentschler; *Helv. Chim. Acta*, 1944, **27**, 1297.	502
44IZV65	I. N. Nazarov and I. I. Zaretskaya; *Izv. Akad. Nauk SSSR Ser. Khim.*, 1944, 65 (*Chem. Abstr.*, 1945, **39**, 1620).	421, 422
44JA686	T. L. Jacobs and S. Searles; *J. Am. Chem. Soc.*, 1944, **66**, 686.	1122
44JA1389	G. Calingaert, H. Soroos, V. Hnizda and H. Shapiro; *J. Am. Chem. Soc.*, 1944, **66**, 1389.	114
44JA1515	H. Gilman and A. H. Haubein; *J. Am. Chem. Soc.*, 1944, **66**, 1515.	110
44JA1684	H. Levy and A. C. Cope; *J. Am. Chem. Soc.*, 1944, **66**, 1684.	411
44JA1929	R. F. Kleinschmidt and A. C. Cope; *J. Am. Chem. Soc.*, 1944, **66**, 1929.	839
44JCS322	J. W. Cook, W. Graham, A. Cohen, R. W. Lapsley and C. A. Lawrence; *J. Chem. Soc.*, 1944, 322.	637
44JCS430	A. J. Birch; *J. Chem. Soc.*, 1944, 430.	95, 98
44JOC310	E. D. Amsutz; *J. Org. Chem.*, 1944, **9**, 310.	591
44OR(2)1	D. S. Tarbell; *Org. React.*, 1944, **2**, 1.	827
44OR(2)262	N. Kornblum; *Org. React.*, 1944, **2**, 262.	36
45JA1273	B. Riegel and W. M. Lilienfeld; *J. Am. Chem. Soc.*, 1945, **67**, 1273.	881
45JA2106	H. Gilman, N. N. Crounse, S. P. Massie, Jr., R. A. Benkeser and S. M. Spatz; *J. Am. Chem. Soc.*, 1945, **67**, 2106.	881
45JA2110	C. H. Horning and F. W. Bergstrom; *J. Am. Chem. Soc.*, 1945, **67**, 2110.	881
45JCS90	J. Cymerman, I. M. Heilbron and E. R. H. Jones; *J. Chem. Soc.*, 1945, 90.	837
45JOC21	M.-C. Chiang and W. H. Hartung; *J. Org. Chem.*, 1945, **10**, 21.	860
46IZV633	I. N. Nazarov and L. N. Pinkina; *Izv. Akad. Nauk SSSR, Ser. Khim.*, 1946, 633 (*Chem. Abstr.*, 1948, **42**, 7731).	422
46JA261	R. Baltzly and A. P. Phillips; *J. Am. Chem. Soc.*, 1946, **68**, 261.	2, 6, 8, 12, 427
46JA502	D. S. Tarbell, H. R. Frank and P. E. Fanta; *J. Am. Chem. Soc.*, 1946, **68**, 502.	637
46JA1813	R. E. Lutz, P. S. Bailey, M. T. Clark, J. F. Codington, A. J. Deinet, J. A. Freek, G. H. Harnest, N. H. Leake, T. A. Martin, R. J. Rowlett, Jr., J. M. Salsbury, N. H. Shearer, Jr., J. D. Smith and J. W. Wilson, III; *J. Am. Chem. Soc.*, 1946, **68**, 1813.	45
46JA2487	Huang-Minlon; *J. Am. Chem. Soc.*, 1946, **68**, 2487.	11, 18, 19, 41, 45
46JCS52	K. Bowden and E. R. H. Jones; *J. Chem. Soc.*, 1946, 52.	519
46JCS396	E. A. Braude, E. R. H. Jones and E. S. Stern; *J. Chem. Soc.*, 1946, 396.	104
46JCS636	R. P. Bell and H. C. Longuet-Higgins; *J. Chem. Soc.*, 1946, 636.	853
B-46MI 119-01	W. A. Waters; "The Chemistry of Free Radicals," Clarendon Press, Oxford, 1946.	919
47CB202	H. Kindler and W. Hückel; *Chem. Ber.*, 1947, **80**, 202.	57
47HCA1911	O. Isler, W. Haber, A. Ronco and M. Kofler; *Helv. Chim. Acta*, 1947, **30**, 1911.	502, 837
47IZV205	I. N. Nazarov and S. S. Bukhmutskaya; *Izv. Akad. Nauk SSSR Ser. Khim.*, 1947, 205 (*Chem. Abstr.*, 1948, **42**, 7733).	422
47JA1551	F. C. Whitmore, L. H. Sommer, J. Gold and R. E. Van Strien; *J. Am. Chem. Soc.*, 1947, **69**, 1551.	648
47JA2025	R. S. Stearns and G. W. Wheland; *J. Am. Chem. Soc.*, 1947, **69**, 2025.	848
47JA2916	S. Winstein and F. H. Seubold, Jr.; *J. Am. Chem. Soc.*, 1947, **69**, 2916.	396
47JCS93	A. Campbell, A. H. J. Houston and J. Kenyon; *J. Chem. Soc.*, 1947, 93.	398
47JCS1190	E. B. Knott; *J. Chem. Soc.*, 1947, 1190.	256
47JCS1270	A. J. Birch; *J. Chem. Soc.*, 1947, 1270.	99
47LA(557)205	G. Wittig and W. Happe; *Justus Liebigs Ann. Chem.*, 1947, **557**, 205.	396
48BSF811	J. Colonge and P. Rochas; *Bull. Soc. Chim. Fr.*, 1948, 811 (*Chem. Abstr.*, 1949, **43**, 2182).	475
48BSF818	J. Colonge and P. Rochas; *Bull. Soc. Chim. Fr.*, 1948, 818.	212
48BSF822	J. Colonge and P. Rochas; *Bull. Soc. Chim. Fr.*, 1948, 822.	212
48BSF825	J. Colonge and P. Rochas; *Bull. Soc. Chim. Fr.*, 1948, 825.	212
48BSF827	J. Colonge and P. Rochas; *Bull. Soc. Chim. Fr.*, 1948, 827.	212
48JA232	H. R. Snyder and C. Weaver; *J. Am. Chem. Soc.*, 1948, **70**, 232.	447
48JA846	E. Grunwald and S. Winstein; *J. Am. Chem. Soc.*, 1948, **70**, 846.	1114, 1116
48JA1490	R. H. Baker, K. H. Cornell and M. J. Cron; *J. Am. Chem. Soc.*, 1948, **70**, 1490.	15
48JA2869	L. H. Sommer, D. L. Bailey and F. C. Whitmore; *J. Am. Chem. Soc.*, 1948, **70**, 2869.	652
48JA3770	H. R. Snyder and F. J. Pilgrim; *J. Am. Chem. Soc.*, 1948, **70**, 3770.	257
48JA3868	K. N. Campbell, C. H. Helbing, M. P. Florkowski and B. K. Campbell; *J. Am. Chem. Soc.*, 1948, **70**, 3868.	1040

48JA4045	J. J. Ritter and P. P. Minieri; *J. Am. Chem. Soc.*, 1948, **70**, 4045.	637
48JCS1408	G. A. Swan; *J. Chem. Soc.*, 1948, 1408.	885
48JOC763	A. L. Wilds and A. L. Meader, Jr.; *J. Org. Chem.*, 1948, **13**, 763.	394
48LA(560)1	W. Reppe, O. Schlichting, K. Klager and T. Toepel; *Justus Liebigs Ann. Chem.*, 1948, **560**, 1.	530, 890
48LA(560)116	G. Wittig, R. Mangold and G. Felletschin; *Justus Liebigs Ann. Chem.*, 1948, **560**, 116.	399
48OR(4)378	D. Todd; *Org. React.*, 1948, **4**, 378.	11, 18, 19, 41, 887
49BSF501	L. Velluz, G. Amiard and A. Petit; *Bull. Soc. Chim. Fr.*, 1949, 501.	403
49CR(228)1501	I. Marszak, A. Marszak-Fleury and M. Koulkes; *C. R. Hebd. Seances Acad. Sci.* 1949, **228**, 1501.	506
49HCA489	O. Isler, A. Ronco, W. Geux, N. C. Hindley, W. Huber, K. Digler and M. Kofler; *Helv. Chim. Acta*, 1949, **32**, 489.	104
49HCA1414	F. Fichter and C. Simon; *Helv. Chim. Acta*, 1949, **32**, 1414.	103
49JA663	H. R. Snyder and E. L. Eliel; *J. Am. Chem. Soc.*, 1949, **71**, 663.	257
49JA1150	R. T. Arnold and S. Searles, Jr.; *J. Am. Chem. Soc.*, 1949, **71**, 1150.	417, 829
49JA1675	L. W. Trevoy and W. G. Brown; *J. Am. Chem. Soc.*, 1949, **71**, 1675.	3, 4, 7, 8, 10
49JA1767	H. B. Hass and M. L. Bender; *J. Am. Chem. Soc.*, 1949, **71**, 1767.	6
49JA1964	G. F. Hennion and J. J. Sheehan; *J. Am. Chem. Soc.*, 1949, **71**, 1964.	958, 1125
49JA2137	N. Kornblum and D. C. Iffland; *J. Am. Chem. Soc.*, 1949, **71**, 2137.	442
49JA2439	R. T. Arnold, W. E. Parham and R. M. Dodson; *J. Am. Chem. Soc.*, 1949, **71**, 2439.	417, 829
49JA2870	M. L. Wolfrom, R. U. Lemieux and S. M. Olin; *J. Am. Chem. Soc.*, 1949, **71**, 2870.	24
49JA2953	S. Winstein and D. Trifan; *J. Am. Chem. Soc.*, 1949, **71**, 2953.	381
49JA3301	Huang-Minlon; *J. Am. Chem. Soc.*, 1949, **71**, 3301.	11, 18, 19, 41
49JA3883	D. J. Cram; *J. Am. Chem. Soc.*, 1949, **71**, 3883.	775
49JA3929	A. C. Cope, T. T. Foster and P. H. Towle; *J. Am. Chem. Soc.*, 1949, **71**, 3929.	624
49JA4154	A. Weizmann; *J. Am. Chem. Soc.*, 1949, **71**, 4154.	89
49JCS282	C. C. J. Culvenor, W. Davies and N. S. Heath; *J. Chem. Soc*, 1949, 282.	592
49JCS737	I. Heilbron, E. R. H. Jones, R. W. Richardson and F. Sondheimer; *J. Chem. Soc.*, 1949, 737.	506
49JCS1074	J. W. Cook, G. T. Dickson, D. Ellis and J. D. Loudon; *J. Chem. Soc.*, 1949, 1074.	637
49JCS1724	R. J. Goldacre and J. N. Phillips; *J. Chem. Soc.*, 1949, 1724.	894
49JCS1827	I. Heilbron, E. R. H. Jones, J. B. Toogood and B. C. L. Weedon; *J. Chem. Soc.*, 1949, 1827.	506
49JCS2028	I. Heilbron, E. R. H. Jones, J. B. Toogood and B. C. L. Weedon; *J. Chem. Soc.*, 1949, 2028.	506
49JCS2174	D. H. R. Barton; *J. Chem. Soc.*, 1949, 2174.	775
49LA(562)177	G. Wittig and M. Rieber; *Justus Liebigs Ann. Chem.*, 1949, **562**, 177.	157
50AP(283)36	H. W. Bersch; *Arch. Pharm.*, 1950, **283**, 36.	628
50BSF35	R. Jacquier; *Bull. Soc. Chim. Fr.*, 1950, 35.	391
50BSF121	R. Paul, G. Roy, M. Fluchaire and G. Collardeau; *Bull. Soc Chim. Fr.*, 1950, 121.	414
50BSF533	J. M. Conia; *Bull. Soc. Chim. Fr.*, 1950, 533.	881
50BSF537	J. M. Conia; *Bull. Soc. Chim. Fr.*, 1950, 537.	881
50BSF1305	I. Marszak and A. Marszak-Fleury; *Bull. Soc. Chim. Fr.*, 1950, 1305.	506
50IEC2103	D. A. McCaulay and A. P. Lien; *Ind. Eng. Chem. Res.*, 1950, **42**, 2103.	897
50JA163	R. A. Craig, A. B. Garrett and M. S. Newman; *J. Am. Chem. Soc.*, 1950, **72**, 163.	896
50JA362	P. L. Julian and W. J. Karpel; *J. Am. Chem. Soc.*, 1950, **72**, 362.	10
50JA408	J. D. Roberts, R. L. Webb and E. A. McElhill; *J. Am. Chem. Soc.*, 1950, **72**, 408.	852
50JA494	R. B. Woodward and T. Singh; *J. Am. Chem. Soc.*, 1950, **72**, 494.	538
50JA849	R. L. Letsinger and J. G. Traynham; *J. Am. Chem. Soc.*, 1950, **72**, 849.	113, 114
50JA1005	P. D. Bartlett and E. S. Lewis; *J. Am. Chem. Soc.*, 1950, **72**, 1005.	853, 898
50JA1245	V. Boekelheide and C. E. Larrabee; *J. Am. Chem. Soc.*, 1950, **72**, 1245.	866
50JA2134	V. Boekelheide and C. Ainsworth; *J. Am. Chem. Soc.*, 1950, **72**, 2134.	257
50JA2395	W. S. Johnson, A. R. Jones and W. P. Schneider; *J. Am. Chem. Soc.*, 1950, **72**, 2395.	884
50JA2884	R. B. Wagner and J. A. Moore; *J. Am. Chem. Soc.*, 1950, **72**, 2884.	393
50JA3539	R. G. Jones, E. C. Kornfeld and K. C. McLaughlin; *J. Amer. Chem. Soc.*, 1950, **72**, 3539.	81
50JA3704	M. D. Soffer, R. A. Stewart, J. C. Cavagnol, H. E. Gellerson and E. A. Bowler; *J. Am. Chem. Soc.*, 1950, **72**, 3704.	881
50JA4296	M. S. Newman and H. M. Walborsky; *J. Am. Chem. Soc.*, 1950, **72**, 4296.	21, 24
50JA4842	R. L. Letsinger; *J. Am. Chem. Soc.*, 1950, **72**, 4842.	846
50JA5127	S. E. Ulrich, F. H. Gentes, J. F. Lane and E. S. Wallis; *J. Am. Chem. Soc.*, 1950, **72**, 5127.	113
50JA5566	O. H. Johnson and D. M. Harris; *J. Am. Chem. Soc.*, 1950, **72**, 5566.	449
50JCS115	R. A. Raphael and F. Sondheimer; *J. Chem. Soc.*, 1950, 115.	1000
50JCS3000	P. Bladon, W. G. Overend, L. N. Owen and L. F. Wiggins; *J. Chem. Soc.*, 1950, 3000.	21
50JGU450	B. N. Dolgov and N. A. Larin; *J. Gen. Chem. USSR*, 1950, **20**, 450 (*Chem. Abstr.*, 1951, **45**, 566).	463
50JOC850	J. Cason, G. Sumrell and R. S. Mitchell; *J. Org. Chem.*, 1950, **15**, 850.	881
50LA(567)179	K. Ziegler and H.-G. Gellert; *Justus Liebigs Ann. Chem.*, 1950, **567**, 179.	663
50LA(568)63	H. Fernholz; *Justus Liebigs Ann. Chem.*, 1950, **568**, 63.	906
50QR69	A. J. Birch; *Q. Rev., Chem. Soc.*, 1950, **4**, 69.	889
51BSFC13	M. Julia; *Bull. Soc. Chim. Fr.*, 1951, C13 (*Chem. Abstr.*, 1952, **46**, 414).	469
51CB785	F. Bohlmann; *Chem. Ber.*, 1951, **84**, 785.	1069

51CRV273	J. F. Bunnet and R. E. Zahler; *Chem. Rev.*, 1951, **49**, 273.	474, 483
51JA1066	C. S. Marvel and C. H. Young; *J. Am. Chem. Soc.*, 1951, **73**, 1066.	503
51JA1437	C. R. Hauser and S. W. Kantor; *J. Am. Chem. Soc.*, 1951, **73**, 1437.	396, 397, 404, 816
51JA1528	J. Romo, M. Romero, C. Djerassi and G. Rosenkranz; *J. Am. Chem. Soc.*, 1951, **73**, 1528.	23
51JA2700	S. Winstein, E. Grunwald and H. W. Jones; *J. Am. Chem. Soc.*, 1951, **73**, 2700.	1114, 1116
51JA2718	H. Rapoport, T. P. King and J. B. Lavigne; *J. Am. Chem. Soc.*, 1951, **73**, 2718.	34
51JA2880	H. Gilman and H. L. Yale; *J. Am. Chem. Soc.*, 1951, **73**, 2880.	446
51JA3308	H. C. Brown and W. A. Murphey; *J. Am. Chem. Soc.*, 1951, **73**, 3308.	860, 861
51JA3900	J. G. Aston and J. D. Newkirk; *J. Am. Chem. Soc.*, 1951, **73**, 3900.	392
51JA4033	H. D. Hartough; *J. Am. Chem. Soc.*, 1951, **73**, 4033.	42
51JA4122	S. W. Kantor and C. R. Hauser; *J. Am. Chem. Soc.*, 1951, **73**, 4122.	824
51JA4158	A. C. Cope and B. D. Tiffany; *J. Am. Chem. Soc.*, 1951, **73**, 4158.	881
51JA4199	M. S. Newman and A. Kutner; *J. Am. Chem. Soc.*, 1951, **73**, 4199.	1076, 1118
51JA4301	N. N. Goldberg, L. B. Barkley and R. Levine; *J. Am. Chem. Soc.*, 1951, **73**, 4301.	860
51JA4707	R. B. Loftfield; *J. Am. Chem. Soc.*, 1951, **73**, 4707.	392
51JA5073	E. B. Hershberg, E. P. Oliveto, C. Gerold and L. Johnson; *J. Am. Chem. Soc.*, 1951, **73**, 5073.	503
51JA5187	F. G. Bordwell and G. D. Cooper; *J. Am. Chem. Soc.*, 1951, **73**, 5187.	618
51JA5444	G. E. Benedict and R. R. Russell; *J. Am. Chem. Soc.*, 1951, **73**, 5444.	513
51JA5770	H. Rinderknecht; *J. Am. Chem. Soc.*, 1951, **73**, 5770.	881
51JA5864	C. R. Hausner and C. R. Hance; *J. Am. Chem. Soc.*, 1951, **73**, 5864.	999
51JCS556	W. Baker and W. D. Ollis; *J. Chem. Soc.*, 1951, 556.	21
51JCS588	R. N. Haszeldine; *J. Chem. Soc.*, 1951, 588.	1130
51JCS1048	D. H. R. Barton and W. J. Rosenfelder; *J. Chem. Soc.*, 1951, 1048.	773
51JCS1945	A. J. Birch, A. R. Murray and H. Smith; *J. Chem. Soc.*, 1951, 1945.	98
51JCS2430	H. M. E. Cardwell and A. E. H. Kilner; *J. Chem. Soc.*, 1951, 2430.	871
51JCS2442	H. M. E. Cardwell; *J. Chem. Soc.*, 1951, 2442.	871
51JCS2556	S. F. Birch and D. T. McAllan; *J. Chem. Soc.*, 1951, 2556.	100
51NAT1039	T. J. Kealy and P. L. Pauson; *Nat.*, 1951, **168**, 1039.	862
51OR(6)41	W. S. Johnson and G. H. Daub; *Org. React.*, 1951, **6**, 41.	881
51QR131	B. R. Brown; *Q. Rev., Chem. Soc.*, 1951, **5**, 131.	772
52CB593	E. Wiberg and R. Bauer; *Chem. Ber.*, 1952, **85**, 593.	665
52JA295	W. Oroshnik, G. Karmas and A. D. Mebane; *J. Am. Chem. Soc.*, 1952, **74**, 295.	503
52JA831	G. S. Fonken and W. S. Johnson; *J. Am. Chem. Soc.*, 1952, **74**, 831.	881
52JA1147	S. Winstein and D. Trifan; *J. Am. Chem. Soc.*, 1952, **74**, 1147.	905
52JA1154	S. Winstein and D. Trifan; *J. Am. Chem. Soc.*, 1952, **74**, 1154.	381, 905
52JA1225	M. W. Cronyn; *J. Am. Chem. Soc.*, 1952, **74**, 1225.	704
52JA1704	N. J. Leonard and R. C. Sentz; *J. Am. Chem. Soc.*, 1952, **74**, 1704.	884
52JA2125	G. Wilkinson, M. Rosenblum, M. C. Whiting and R. B. Woodward; *J. Am. Chem. Soc.*, 1952, **74**, 2125.	862
52JA2137	D. J. Cram; *J. Am. Chem. Soc.*, 1952, **74**, 2137.	624
52JA3074	N. Kornblum, A. E. Kelley and G. D. Cooper; *J. Am. Chem. Soc.*, 1952, **74**, 3074.	36
52JA3458	R. B. Woodward, M. Rosenblum and M. C. Whiting; *J. Am. Chem. Soc.*, 1952, **74**, 3458.	862
52JA4126	G. B. Diamond and M. D. Soffer; *J. Am. Chem. Soc.*, 1952, **74**, 4126.	103
52JA4582	L. F. Cason and H. G. Brooks; *J. Am. Chem. Soc.*, 1952, **74**, 4582.	310
52JA4867	A. C. Cope, A. C. Haren, Jr., F. L. Ramp and E. R. Trumbull; *J. Amer. Chem. Soc.*, 1952, **74**, 4867.	78, 421
52JA5179	J. H. Brewster and M. W. Kline; *J. Am. Chem. Soc.*, 1952, **74**, 5179.	398
52JA5372	R. W. Taft, Jr.; *J. Am. Chem. Soc.*, 1952, **74**, 5372.	895
52JA5381	D. Y. Curtin and M. J. Hurwitz; *J. Am. Chem. Soc.*, 1952, **74**, 5381.	128, 396
52JA5943	J. D. Roberts and J. A. Yancey; *J. Am. Chem. Soc.*, 1952, **74**, 5943.	895
52JCS153	N. L. McNiven and J. Read; *J. Chem. Soc.*, 1952, 153.	628
52JCS632	S. A. Miller, J. A. Tebboth and J. F. Tremaine; *J. Chem. Soc.*, 1952, 632.	862
52JCS1430	E. A. Braude and J. A. Coles; *J. Chem. Soc.*, 1952, 1430.	422
52JCS2010	J. B. Armitage, E. R. H. Jones, C. L. Cook and M. C. Whiting; *J. Chem. Soc.*, 1952, 2010.	1069
52JCS2014	J. B. Armitage, E. R. H. Jones and M. C. Whiting; *J. Chem. Soc.*, 1952, 2014.	1035
52JCS2127	R. H. Glauert and F. G. Mann; *J. Chem. Soc.*, 1952, 2127.	41
52JCS2167	V. Gold, B. W. V. Hawes and F. L. Tye; *J. Chem. Soc.*, 1952, 2167.	895
52JCS2172	V. Gold and F. L. Tye; *J. Chem. Soc.*, 1952, 2172.	895, 896
52JCS4735	W. R. Bamford and T. S. Stevens; *J. Chem. Soc.*, 1952, 4735.	39, 630, 631
52JOC1630	H. Gilman, R. G. Jones and L. A. Woods; *J. Org. Chem.*, 1952, **17**, 1630.	117
52LA(577)26	G. Wittig and K. Clauss; *Justus Liebigs Ann. Chem.*, 1952, **577**, 26.	446
52LA(578)136	G. Wittig and K. Clauss; *Justus Liebigs Ann. Chem.*, 1952, **578**, 136.	446
52QR101	L. Crombie; *Q. Rev., Chem. Soc.*, 1952, **6**, 101.	502
52ZN(B)377	E. O. Fischer and W. Z. Pfab; *Z. Naturforsch., Teil B*, 1952, **7**, 377.	862
53CB167	K. Kindler, H. Oelschläger and P. Henrick; *Chem. Ber.*, 1953, **86**, 167.	89
53CB657	F. Bohlmann; *Chem. Ber.*, 1953, **86**, 657.	1069
53CR(236)1044	R. Romanet; *C. R. Hebd. Seances Acad. Sci.*, 1953, **236**, 1044.	503
53E329	E. J. Corey; *Experientia*, 1953, **9**, 329.	878
53HCA1809	E. Urech, E. Tagmann, E. Sury and K. Hoffmann; *Helv. Chim. Acta*, 1953, **36**, 1809.	537
53JA246	C. G. Swain and C. B. Scott; *J. Am. Chem. Soc.*, 1953, **75**, 246.	107

53JA1050 W. Oroshnik, A. D. Mebane and G. Karmas; *J. Am. Chem. Soc.*, 1953, **75**, 1050. 955
53JA1128 E. M. Schultz and J. B. Bicking; *J. Am. Chem. Soc.*, 1953, **75**, 1128. 34
53JA1253 J. B. Levy, R. W. Taft, Jr. and L. P. Hammett; *J. Am. Chem. Soc.*, 1953, **75**, 1253. 895
53JA2118 G. L. O'Connor and H. R. Nace; *J. Am. Chem. Soc.*, 1953, **75**, 2118. 776
53JA2153 A. T. Blomquist and L. H. Liu; *J. Am. Chem. Soc.*, 1953, **75**, 2153. 635
53JA2439 R. G. Pearson and R. L. Dillon; *J. Am. Chem. Soc.*, 1953, **75**, 2439. 155, 850, 855, 869
53JA2925 V. J. Shiner, Jr.; *J. Am. Chem. Soc.*, 1953, **75**, 2925. 892
53JA3212 A. C. Cope, R. A. Pike and C. F. Spencer; *J. Am. Chem. Soc*, 1953, **75**, 3212. 624, 782, 783
53JA3589 E. L. Eliel and N. J. Murphy; *J. Am. Chem. Soc.*, 1953, **75**, 3589. 256
53JA3955 J. B. Levy, R. W. Taft, Jr., D. Aaron and L. P. Hammett; *J. Am. Chem. Soc.*, 1953, **75**, 3955. 895
53JA3975 H. D. Zook and R. N. Goldey; *J. Am. Chem. Soc.*, 1953, **75**, 3975. 112
53JA5421 L. F. Fieser; *J. Am. Chem. Soc.*, 1953, **75**, 5421. 795
53JA6011 D. Y. Kellom and D. B. Kellom; *J. Am. Chem. Soc.*, 1953, **75**, 6011. 773
53JCS124 G. Baddeley, H. T. Taylor and W. Pickles; *J. Chem. Soc.*, 1953, 124. 422
53JCS922 R. N. Haszeldine; *J. Chem. Soc.*, 1953, 922. 431
53JCS1199 R. N. Haszeldine and B. R. Steele; *J. Chem. Soc.*, 1953, 1199. 1130
53JCS2247 A. M. Islam and R. A. Raphael; *J. Chem. Soc.*, 1953, 2247. 422
53JCS3156 B. B. Elsner and P. F. M. Paul; *J. Chem. Soc.*, 1953, 3156. 503
53JCS3294 P. Mildner and B. C. L. Weedon; *J. Chem. Soc.*, 1953, 3294. 503
53LA(580)44 G. Wittig and G. Geissler; *Justus Liebigs Ann. Chem.*, 1953, **580**, 44. 638
53NAT121 J. P. Dunitz and L. E. Orgel; *Nat.*, 1953, **171**, 121. 862
53OR(7)263 W. H. Hartung and R. Simonoff; *Org. React.*, 1953, **7**, 263. 11, 12, 18, 19, 21
53OS(33)68 E. R. H. Jones, G. Eglington and M. C. Whiting; *Org. Synth.*, 1953, **33**, 68. 1066

54CB1318 G. Wittig and U. Schöllkopf; *Chem. Ber.*, 1954, **87**, 1318. 639, 721
54JA35 R. B. Loftfield and L. Schaad; *J. Am. Chem. Soc.*, 1954, **76**, 35. 393
54JA175 E. J. Corey; *J. Am. Chem. Soc.*, 1954, **76**, 175. 879
54JA494 D. Y. Curtin and W. R. Proops; *J. Am. Chem. Soc.*, 1954, **76**, 494. 397
54JA791 E. S. Lewis and C. E. Boozer; *J. Am. Chem. Soc.*, 1954, **76**, 791. 892
54JA1613 L. H. Sommer, D. L. Bailey, G. M. Goldberg, C. E. Buck, T. S. Bye, F. J. Evans and F. C. Whitmore; *J. Am. Chem. Soc.*, 1954, **76**, 1613. 450
54JA1700 W. J. Croxall and J. O. Van Hook; *J. Am. Chem. Soc.*, 1954, **76**, 1700. 1069
54JA1706 G. E. Wiseman and E. S. Gould; *J. Am. Chem. Soc.*, 1954, **76**, 1706. 437
54JA2753 D. J. Cram and H. Steinberg; *J. Am. Chem. Soc.*, 1954, **76**, 2753. 384
54JA3203 W. von E. Doering and L. H. Knox; *J. Am. Chem. Soc.*, 1954, **76**, 3203. 906
54JA3972 G. B. Bachman, H. B. Hass and G. O. Platau; *J. Am. Chem. Soc.*, 1954, **76**, 3972. 156
54JA4137 F. Bergmann and A. Kalmus; *J. Am. Chem. Soc.*, 1954, **76**, 4137. 108
54JA4446 D. Papa, F. J. Villani and H. F. Ginsberg; *J. Am. Chem. Soc.*, 1954, **76**, 4446. 503
54JA4558 S. J. Cristol, A. Begoon, W. P. Norris and P. S. Ramey; *J. Am. Chem. Soc.*, 1954, **76**, 4558. 1065
54JA5740 D. J. Cram and J. E. McCarty; *J. Am. Chem. Soc.*, 1954, **76**, 5740. 624
54JA6040 E. J. Corey and W. J. Wechter; *J. Am. Chem. Soc.*, 1954, **76**, 6040. 108
54JCS113 A. R. Pinder and H. Smith; *J. Chem. Soc.*, 1954, 113. 96
54JCS147 J. B. Armitage, E. Entwistle, E. R. H. Jones and M. C. Whiting; *J. Chem. Soc.*, 1954, 147. 1035
54JCS1634 R. N. Haszeldine and K. Leedham; *J. Chem. Soc.*, 1954, 1634. 1130
54JCS1854 E. B. Bates, E. R. H. Jones and M. C. Whiting; *J. Chem. Soc.*, 1954, 1854. 513
54JCS2918 E. Gelles, E. D. Hughes and C. K. Ingold; *J. Chem. Soc.*, 1954, 2918. 892
54JCS3045 D. H. R. Barton and C. H. Robinson; *J. Chem. Soc.*, 1954, 3045. 13
54JCS3197 G. Eglinton, E. R. H. Jones, G. H. Mansfield and M. C. Whiting; *J. Chem. Soc.*, 1954, 3197. 80
54JCS3217 B. L. Shaw and M. C. Whiting; *J. Chem. Soc.*, 1954, 3217. 513
54JCS3617 R. D. Haworth, H. K. Pindred and P. R. Jefferies; *J. Chem. Soc.*, 1954, 3617. 45
54JOC1490 D. F. Thompson, P. L. Bayless and C. R. Hauser; *J. Org. Chem.*, 1954, **19**, 1490. 881
54JOC1580 J. H. Wotiz and E. S. Hudak; *J. Org. Chem.*, 1954, **19**, 1580. 1040
54JOC1960 A. Gero; *J. Org. Chem.*, 1954, **19**, 1960. 869
54MI 102-01 Z. Welvart; *Compt. Rend.*, 1954, **238**, 2536. 55
54MI 102-02 Z. Welvart; *Compt. Rend.*, 1954, **239**, 1299. 55
B-54MI 104-01 M. S. Kharasch and O. Reinmuth; in "Grignard Reactions of Nonmetallic Substances," Prentice-Hall, New York, 1954, Chapter 16, pp. 1046–1132. 114

55AG541 K. Ziegler, E. Holzkamp, H. Breil and H. Martin; *Angew. Chem.*, 1955, **67**, 541. 665
55AG710 J. Goubeau, R. Epple, D. Ulmschneider and H. Lehman; *Angew. Chem.*, 1955, **67**, 710. 58
55BSF1341 L. Velluz, G. Amiard and B. Goffinet; *Bull. Soc. Chim. Fr.*, 1955, 1341. 419
55CB934 L. Horner and E. Spietschka; *Chem. Ber.*, 1955, **88**, 934. 395
55CB1415 H. H. Inhoffen, K. Brückner, G. F. Domagk and H.-M. Erdmann; *Chem. Ber.*, 1955, **88**, 1415. 736
55CB1654 G. Wittig and W. Haag; *Chem. Ber.*, 1955, **88**, 1654. 640, 721, 733
55CI(L)415 H. H. Rennhard, E. Heilbronner and A. Eschenmoser; *Chem. Ind.* (*London*), 1955, 415. 906
55CR(240)989 R. Epsztein; *C. R. Hebd. Seances Acad. Sci.*, 1955, **240**, 989 (*Chem. Abstr.*, **50**, 5520). 1048
55CR(241)1055 W. Chodkiewicz and P. Cadiot; *C. R. Hebd. Seances Acad Sci.*, 1955, **241**, 1055. 1142
55JA73 W. J. Bailey and J. Rosenberg; *J. Am. Chem. Soc.*, 1955, **77**, 73. 521
55JA1001 A. T. Blomquist and A. Goldstein; *J. Am. Chem. Soc.*, 1955, **77**, 1001. 773, 776
55JA1122 C. G. Swain and M. M. Kreevoy; *J. Am. Chem. Soc.*, 1955, **77**, 1122. 892
55JA1628 A. C. Cope, D. C. McClean and N. A. Nelson; *J. Am. Chem. Soc.*, 1955, **77**, 1628. 625, 783

55JA2801	P. D. Bartlett and M. S. Swain; *J. Am. Chem. Soc.*, 1955, **77**, 2801.	892
55JA3051	N. C. Deno and A. Schriesheim; *J. Am. Chem. Soc.*, 1955, **77**, 3051.	894
55JA3230	D. M. Sauve, O. H. Thomas, R. A. Benkeser and R. E. Robinson; *J. Am. Chem. Soc.*, 1955, 77, 3230.	51, 52, 101
55JA3293	A. A. Bothner-By; *J. Am. Chem. Soc.*, 1955, **77**, 3293.	1144
55JA3378	R. A. Benkeser, G. Schroll and D. M. Sauve; *J. Am. Chem. Soc.*, 1955, **77**, 3378.	457, 508
55JA4262	L. D. Freeman, G. O. Doak and E. L. Petit; *J. Am. Chem. Soc.*, 1955, **77**, 4262.	88, 89
55JA4367	R. S. Rosenfeld and T. F. Gallagher; *J. Am. Chem. Soc.*, 1955, **77**, 4367.	16
55JA5016	L. Rosenblum; *J. Am. Chem. Soc.*, 1955, **77**, 5016.	657
55JA5083	H. O. House; *J. Am. Chem. Soc.*, 1955, **77**, 5083.	390
55JA5166	R. H. Beutel, D. F. Hinkley and P. J. Pollak; *J. Am. Chem. Soc.*, 1955, **77**, 5166.	795
55JA5517	C. J. Collins; *J. Am. Chem. Soc.*, 1955, **77**, 5517.	385
55JA5533	F. M. Beringer and H. S. Schultz; *J. Am. Chem. Soc.*, 1955, **77**, 5533.	108
55JA5960	H. Conroy; *J. Am. Chem. Soc.*, 1955, **77**, 5960.	8
55JA6026	H. M. Walborsky and F. M. Hornyak; *J. Am. Chem. Soc.*, 1955, **77**, 6026.	155
55JCS522	G. W. Kenner and N. R. Williams; *J. Chem. Soc.*, 1955, 522.	432
55JCS769	M. A. T. Rogers; *J. Chem. Soc.*, 1955, 769.	626
55JCS1740	L. Crombie and A. G. Jacklin; *J. Chem. Soc.*, 1955, 1740.	957
55JCS2044	M. Lamchen and A. M. Stephen; *J. Chem. Soc.* 1955, 2044.	41
55JOC9	C. S. Rondestvedt, Jr., R. M. Scribner and C. E. Wulfman; *J. Org. Chem.*, 1955, **20**, 9.	447
55JOC95	W. J. Bailey and C. R. Pfeifer; *J. Org. Chem.*, 1955, **20**, 95.	958
55JOC1119	C. R. Hauser, R. M. Manyik, W. R. Brasen and P. L. Bayless; *J. Org. Chem.*, 1955, **20**, 1119.	399
55JOC1337	W. J. Bailey and C. R. Pfeifer; *J. Org. Chem.*, 1955, **20**, 1337.	958
55JOC1531	C. R. Hauser, D. S. Hoffenberg, W. H. Puterbaugh and F. C. Frostick; *J. Org. Chem.*, 1955, **20**, 1531.	881
55LA177	A. Dornow and M. Gellrich; *Justus Liebigs Ann. Chem.*, 1955, 177.	57
B-55MI 121-01	R. A. Raphael; "Acetylenic Compounds in Organic Synthesis," Butterworths, London, 1955.	998
55OSC(3)44	C. A. Vanderwerf and L. F. Lemmerman; *Org. Synth., Coll. Vol.*, 1955, **3**, 44.	881
55OSC(3)101	G. Braun; *Org. Synth., Coll. Vol.*, 1955, **3**, 101.	884
55OSC(3)219	E. M. Hancock and A. C. Cope; *Org. Synth., Coll. Vol*, 1955, **3**, 219.	881
55OSC(3)778	F. W. Bergstrom; *Org. Synth., Coll. Vol.*, 1955, **3**, 778.	881
56ACS633	L. Eberson; *Acta Chem. Scand.*, 1956, **10**, 633.	653
56BSF1040	J. M. Conia; *Bull. Soc. Chim. Fr.*, 1956, 1040.	881
56BSF1653	R. Jacquier, M. Mousseron and S. Boyer; *Bull. Soc. Chim. Fr.*, 1956, 1653.	422
56CB853	H. H. Inhoffen, K. Weissermel, G. Quinkert and D. Bartling; *Chem. Ber.*, 1956, **89**, 853.	1023
56CB2105	W. Hückel and E. Vevera; *Chem. Ber.*, 1956, **89**, 2105.	98
56CI(L)80	S. Trippett; *Chem. Ind. (London)*, 1956, 80.	721
56CI(L)1306	R. Pettit; *Chem. Ind. (London)*, 1956, 1306.	866, 910
56CI(L)1504	D. H. Reid; *Chem. Ind. (London)*, 1956, 1504.	866
56CR(243)1895	F. Bardone-Gaudemar; *C. R. Hebd. Seances Acad. Sci.*, 1956, **243**, 1895.	1060
56JA82	C. R. Hauser and W. R. Brassen; *J. Am. Chem. Soc.*, 1956, **78**, 82.	560
56JA92	L. Field; *J. Am. Chem. Soc.*, 1956, **78**, 92.	146
56JA790	D. J. Cram, F. D. Greene and C. H. Depuy; *J. Am. Chem. Soc.*, 1956, **78**, 790.	627, 628
56JA1221	J. H. Wotiz, C. A. Hollingsworth and R. E. Dessy; *J. Am. Chem. Soc.*, 1956, **78**, 1221.	1040
56JA1675	R. B. Davis and D. H. Scheiber; *J. Am. Chem. Soc.*, 1956, **78**, 1675.	1045, 1137
56JA1723	C. Osuch and R. Levine; *J. Am. Chem. Soc.*, 1956, **78**, 1723.	860
56JA2173	H. C. Brown and R. R. Holmes; *J. Am. Chem. Soc.*, 1956, **78**, 2173.	1102
56JA2518	D. J. Cram and N. L. Allinger; *J. Am. Chem. Soc.*, 1956, **78**, 2518.	503, 505
56JA2547	A. C. Cope, L. D. Field, D. W. MacDowell and M. E. Wright; *J. Am. Chem. Soc.*, 1956, **78**, 2547.	808
56JA2551	A. C. Cope, J. E. Meili and D. W. MacDowell; *J. Am. Chem. Soc.*, 1956, **78**, 2551.	808
56JA3484	F. L. M. Pattison, W. J. Cott, W. C. Howell and R. W. White; *J. Am. Chem. Soc.*, 1956, **78**, 3484.	107
56JA4604	G. Stork and W. N. White; *J. Am. Chem. Soc.*, 1956, **78**, 4604.	99
56JA4654	M. S. Newman and A. E. Weinberg; *J. Am. Chem. Soc.*, 1956, **78**, 4654.	1118
56JA5447	W. von E. Doering and P. LaFlamme; *J. Am. Chem. Soc.*, 1956, **78**, 5447.	299
56JA5807	E. L. Purlee and R. W. Taft, Jr.; *J. Am. Chem. Soc.*, 1956, **78**, 5807.	895
56JA6269	E. J. Corey and R. A. Sneen; *J. Am. Chem. Soc.*, 1956, **78**, 6269.	871, 879
56JA6414	F. W. Hoffmann, R. J. Ess, T. C. Simmons and R. S. Hanzel; *J. Am. Chem. Soc.*, 1956, **78**, 6414.	942
56JCS2419	G. Brownlie, F. S. Spring, R. Stevenson and W. S. Strachan; *J. Chem. Soc.*, 1956, 2419.	383
56JCS4691	E. A. Evans; *J. Chem. Soc.*, 1956, 4691.	727
56JOC918	P. A. Cantor, R. Lancaster and C. A. Vander Werf; *J. Org. Chem.*, 1956, **21**, 918.	42
56JPC1454	J. R. Lacher, A. Kianpour and J. D. Park; *J. Phys. Chem.*, 1956, **60**, 1454.	6
B-56MI 102-01	N. G. Gaylord; "Reduction with Complex Metal Halides," Interscience, New York, 1956.	28, 32
56MI 103-01	S. Akabori, S. Sakurai, Y. Izumi and Y. Fujii; *Nature*, 1956, **178**, 323.	76
56ZAAC(287)261	H. Hartmann, C. Beermann and H. Czempik; *Z. Anorg. Allg. Chem.*, 1956, **287**, 261.	999
56ZOB1175	I. N. Nazarov, S. N. Ananchenko and I. V. Torgov; *Zh. Obshch. Khim.*, 1956, **26**, 1175.	79
57AC(P)819	W. Chodkiewicz; *Ann. Chim. (Paris)*, 1957, **2**, 819.	1142

57CB1774 A. Dornow and K. J. Fust; *Chem. Ber.*, 1957, **90**, 1774. 57

57CB2733 D. Ulmschneider and J. Goubeau; *Chem. Ber.*, 1957, **90**, 2733. 58

57CI(L)1596 E. A. Evans; *Chem. Ind. (London)*, 1957, 1596. 727

57CR(C)1790 M. Bertrand; *C. R. Hebd. Seances Acad. Sci. Ser. C*, 1957, **244**, 1790. 957

57CRV895 R. L. Burwell; *Chem. Rev.*, 1957, **57**, 895. 89

57DOK(117)1021 M. E. Dyatkina and E. M. Shustorovich; *Dokl. Akad. Nauk SSSR*, 1957, **117**, 1021. 866

57HCA1250 M. Montavon, H. Lindlar, R. Marbet, R. Rüegg, G. Ryser, G. Saucy, P. Zeller and O. Isler; *Helv. Chim. Acta*, 1957, **40**, 1250. 156

57JA881 H. D. Zook and W. L. Rellahan; *J. Am. Chem. Soc.*, 1957, **79**, 881. 881

57JA960 A. C. Cope and C. L. Bumgardner; *J. Am. Chem. Soc.*, 1957, **79**, 960. 624, 625, 788

57JA1546 P. J. Elving and J. T. Leone; *J. Am. Chem. Soc.*, 1957, **79**, 1546. 6

57JA2217 H. R. Snyder and D. S. Matteson; *J. Am. Chem. Soc.*, 1957, **79**, 2217. 258

57JA2311 F. L. M. Pattison and J. J. Norman; *J. Am. Chem. Soc.*, 1957, **79**, 2311. 107

57JA2828 W. H. Watanabe and L. E. Conlon; *J. Am. Chem. Soc.*, 1957, **79**, 2828. 414

57JA2906 F. Sondheimer and Y. Mazur; *J. Am. Chem. Soc.*, 1957, **79**, 2906. 881

57JA3567 W. J. Bailey and S. A. Buckler; *J. Am. Chem. Soc.*, 1957, **79**, 3567. 51

57JA3724 P. Riesz, R. W. Taft, Jr. and R. H. Boyd; *J. Am. Chem. Soc.*, 1957, **79**, 3724. 895

57JA4146 S. Winstein, A. H. Fainberg and E. Grunwald; *J. Am. Chem. Soc.*, 1957, **79**, 4146. 1114, 1116

57JA4557 H. J. Dauben, Jr., F. A. Gadecki, K. M. Harmon and D. L. Pearson; *J. Am. Chem. Soc.*, 1957, **79**, 4557. 906

57JA4720 A. C. Cope, N. A. LeBel, H.-H. Lee and W. R. Moore; *J. Am. Chem. Soc.*, 1957, **79**, 4720. 624, 628

57JA4729 A. C. Cope, C. L. Bumgardner and E. E. Schweizer; *J. Am. Chem. Soc.*, 1957, **79**, 4729. 625, 783

57JA4985 E. Grovenstein; *J. Am. Chem. Soc.*, 1957, **79**, 4985. 378

57JA5279 N. J. Leonard and F. P. Hauck, Jr.; *J. Am. Chem. Soc.*, 1957, **79**, 5279. 55, 91

57JA5318 R. Breslow; *J. Am. Chem. Soc.*, 1957, **79**, 5318. 907, 909

57JA5326 C. Walling and R. Rabinowitz; *J. Am. Chem. Soc.*, 1957, **79**, 5326. 942

57JA5455 H. E. Zimmerman and F. J. Smentowski; *J. Am. Chem. Soc.*, 1957, **79**, 5455. 378

57JA5540 R. S. Rosenfeld; *J. Am. Chem. Soc.*, 1957, **79**, 5540. 16

57JA6210 N. J. Leonard and R. R. Sauers; *J. Amer. Chem. Soc.*, 1957, **79**, 6210. 82

57JCS498 F. B. Deans and C. Eaborn; *J. Chem. Soc.*, 1957, 498. 447

57JCS1622 L. Crombie and A. G. Jacklin; *J. Chem. Soc.*, 1957, 1622. 506

57JGU693 I. N. Nazarov and I. I. Zaretskaya; *J. Gen. Chem. USSR (Engl. Transl.)*, 1957, **27**, 693. 421

57JIC169 S. Dev; *J. Indian Chem. Soc.*, 1957, **34**, 169. 422

57JOC602 H. J. Ringold and G. Rosenkrantz; *J. Org. Chem.*, 1957, **22**, 602. 881

57JOC1118 C. B. Scott; *J. Org. Chem.*, 1957, **22**, 1118. 646

57JOC1243 M. Stiles and A. J. Libbey; *J. Org. Chem.*, 1957, **22**, 1243. 906

57JOC1248 C. R. Hauser and D. Lednicer; *J. Org. Chem.*, 1957, **22**, 1248. 1143, 1145

57JOC1326 J. Cason and R. J. Fessenden; *J. Org. Chem.*, 1957, **22**, 1326 127, 128

57JOC1418 C. D. Hurd and W. W. Jenkins; *J. Org. Chem.*, 1957, **22**, 1418. 835

57LA(603)115 A. Mondon; *Justus Liebigs Ann. Chem.*, 1957, **603**, 115. 721

57LA(612)102 G. Wittig and R. Polster; *Justus Liebigs Ann. Chem.*, 1957, **612**, 102. 628

B-57MI 119-01 C. Walling; "Free Radicals in Solution," Wiley, New York, 1957. 919

57RTC689 A. Verloop, A. L. Koevoet and E. Havinga; *Recl. Trav. Chim. Pays-Bas*, 1957, **76**, 689. 403

57ZAAC(290)348 H. Hartmann and G. Nowak; *Z. Anorg. Allg. Chem.*, 1957, **290**, 348. 445, 999

58CB61 L. Horner, H. Hoffman and H. G. Wippel; *Chem. Ber.*, 1958, **91**, 61. 639, 752, 753

58CB750 G. Drefahl and H. Crahmer; *Chem. Ber.*, 1958, **91**, 750. 57

58CB2383 L. Birkofer and L. Erlenbach; *Chem. Ber.*, 1958, **91**, 2383. 43

58CI(L)1589 R. C. Koestler and W. K. Little; *Chem. Ind. (London)*, 1958, 1589. 862

58DOK31 I. P. Tsukervanik and C. S. Kadyrov; *Dokl. Akad. Nauk SSSR*, 1958, 31 (*Chem. Abstr.*, 1959, **53**, 16 083). 475

58HCA414 D. Meuche, H. Strauss and E. Heilbronner; *Helv. Chim. Acta*, 1958, **41**, 414. 906

58IEC1677 T. F. Doumani; *Ind. Eng. Chem.*, 1958, **50**, 1677. 56

58JA355 A. C. Cope and E. M. Acton; *J. Am. Chem. Soc.*, 1958, **80**, 355. 625, 628

58JA751 H. N. Cripps, J. K. Williams and W. H. Sharkey; *J. Am. Chem. Soc.*, 1958, **80**, 751. 498

58JA1376 W. J. Gensler and J. Casella Jr.; *J. Am. Chem. Soc.*, 1958, **80**, 1376. 954

58JA1654 J. H. Boothe, G. E. Bonvicino, C. W. Waller, J. P. Petisi, R. W. Wilkinson and R. B. Broschard; *J. Am. Chem. Soc.*, 1958, **80**, 1654. 34

58JA2326 A. Streitwieser, Jr., R. H. Jagow, R. C. Fahey and S. Suzuki; *J. Am. Chem. Soc.*, 1958, **80**, 2326. 892

58JA2896 R. F. Nystrom and C. R. A. Berger; *J. Am. Chem. Soc.*, 1958, **80**, 2896. 20

58JA3489 N. Muller and R. S. Mulliken; *J. Am. Chem. Soc.*, 1958, **80**, 3489. 892

58JA3585 P. D. Bartlett and H. T. Liang; *J. Am. Chem. Soc.*, 1958, **80**, 3585. 785

58JA3625 W. E. Truce, W. J. Ray, O. L. Norman and D. B. Eickemeyer; *J. Am. Chem. Soc.*, 1958, **80**, 3625. 379

58JA4386 R. Mechoulam and F. Sondheimer; *J. Am. Chem. Soc.*, 1958, **80**, 4386. 268

58JA4405 G. E. Sunko and G. Wolf; *J. Am. Chem. Soc.*, 1958, **80**, 4405. 255

58JA4588 H. E. Baumgarten, F. A. Bower, R. A. Setterquist and R. E. Allen; *J. Am. Chem. Soc.*, 1958, **80**, 4588. 637

58JA4593 W. J. Gensler and C. B. Abrahams; *J. Am. Chem. Soc.*, 1958, **80**, 4593. 1023

58JA4607 B. W. Howk and J. C. Sauer; *J. Am. Chem. Soc.*, 1958, **80**, 4607. 1026

58JA5220 Y. Mazur and F. Sondheimer; *J. Am. Chem. Soc.*, 1958, **80**, 5220. 881

58JA5270	M. Freifelder and R. G. Stone; *J. Am. Chem. Soc.*, 1958, **80**, 5270.	88
58JA5867	B. S. Friedman, F. L. Morritz, C. J. Morrissey and R. Koncos; *J. Am. Chem. Soc.* 1958, **80**, 5867.	462
58JA5991	R. Breslow and C. Yuan; *J. Am. Chem. Soc.*, 1958, **80**, 5991.	907, 909
58JA6296	Y. Mazur and F. Sondheimer; *J. Am. Chem. Soc.*, 1958, **80**, 6269.	881
58JA6360	C. R. Hauser and T. M. Harris; *J. Am. Chem. Soc.*, 1958, **80**, 6360.	879, 881
58JA6412	H. K. Hall, Jr.; *J. Am. Chem. Soc.*, 1958, **80**, 6412.	91
58JA6535	G. A. Olah and S. J. Kuhn; *J. Am. Chem. Soc.*, 1958, **80**, 6535.	914
58JA6541	G. A. Olah and S. J. Kuhn; *J. Am. Chem. Soc.*, 1958, **80**, 6541.	914
58JA6626	N. A. Nelson and J. C. Wollensak; *J. Am. Chem. Soc.*, 1958, **80**, 6626.	1000
58JCS403	C. A. Bunton, T. Hadwick, D. R. Llewellyn and Y. Pocker; *J. Chem. Soc.*, 1958, 403.	384
58JCS843	I. T. Harrison and B. Lythgoe; *J. Chem. Soc.*, 1958, 843.	735
58JCS1054	E. R. H. Jones, L. Skattenbol and M. C. Whiting; *J. Chem. Soc.*, 1958, 1054.	1046
58JCS1451	E. W. Abel, W. Gerrard and M. F. Lappert; *J. Chem. Soc.*, 1958, 1451.	447
58JOC133	G. N. Walker; *J. Org. Chem.*, 1958, **23**, 133.	19
58JOC665	M. S. Newman and D. E. Reid; *J. Org. Chem.*, 1958, **23**, 665.	1075
58JOC841	M. Yanagita, M. Hirakura and F. Seki; *J. Org. Chem.*, 1958, **23**, 841.	881
58JOC996	W. J. Bailey and C. N. Bird; *J. Org. Chem.*, 1958, **23**, 996.	637
58JOC1281	J. H. Burckhalter and J. H. Short; *J. Org. Chem.*, 1958, **23**, 1281.	256
58JOC1938	L. J. Sargent and J. H. Ager; *J. Org. Chem.*, 1958, **23**, 1938.	2, 8
58LA(614)37	H. H. Schlubach and K. Repenning; *Justus Liebigs Ann. Chem.*, 1958, **614**, 37.	104
58LA(615)1	E. Vogel; *Justus Liebigs Ann. Chem.*, 1958, **615**, 1.	811
58LA(618)202	G. O. Schenk, O.-A. Neumuller and W. Eisfeld; *Justus Liebigs Ann. Chem.*, 1958, **618**, 202.	839
58LA(619)10	G. Wittig, H. Eggers and P. Duffner; *Justus Liebigs Ann. Chem.*, 1958, **619**, 10.	727
58PCS289	G. R. Knox and P. L. Pauson; *Proc. Chem. Soc.*, 1958, 289.	862
58QR17	A. J. Birch and H. Smith; *Q. Rev., Chem. Soc.*, 1958, **12**, 17.	889
58T88	S. Swaminathan and M. S. Newman; *Tetrahedron*, 1958, **2**, 88.	885
58T178	W. von E. Doering, M. Saunders, H. G. Boyton, H. W. Earhart, E. F. Wadley and W. R. Edwards; *Tetrahedron*, 1958, **4**, 178.	912, 914
58T339	D. H. Reid; *Tetrahedron*, 1958, **3**, 339.	866
59AC(P)541	G. Martin; *Ann. Chim. Paris*, 1959, **4**, 541 (*Chem. Abstr.*, 1960, **54**, 1278).	469
59AG245	E. F. Jenny and K. D. Meier; *Angew. Chem.*, 1959, **71**, 245.	1000
59AG520	R. Köster; *Angew. Chem.*, 1959, **71**, 520.	657
59AG726	D. Staschewski; *Angew. Chem.*, 1959, **71**, 726.	18
59BCJ861	K. Hata and K. Watanabe; *Bull. Chem. Soc. Jpn.*, 1959, **32**, 861.	48
59CB528	F. Weygand and H. J. Bestmann; *Chem. Ber.*, 1959, **92**, 528.	395
59CB916	E. Buchta and F. Güllich; *Chem. Ber.*, 1959, **92**, 916.	42
59CB1700	L. Horner, L. Schläfer and H. Kämmerer; *Chem. Ber.*, 1959, **92**, 1700.	2, 6, 8, 10
59CB1849	R. Kuhn and H. Fischer; *Chem. Ber.*, 1959, **92**, 1849.	986
59CB1950	H. G. Viehe; *Chem. Ber.*, 1959, **92**, 1950.	1043
59CB2499	L. Horner, H. Hoffman, H. G. Wippel and G. Klahre; *Chem. Ber.*, 1959, **92**, 2499.	639, 752, 753
59CCC2371	I. Ognjanov, V. Herout, M. Horak and F. Somm; *Collect. Czech. Chem. Commun.*, 1959, **24**, 2371.	812
59HCA401	E. F. Jenny and J. Druey; *Helv. Chim. Acta*, 1959, **42**, 401.	513
59HCA719	H. Gutmann, O. Isler, G. Ryser, P. Zeller and B. Pellmont; *Helv. Chim. Acta*, 1959, **42**, 719.	956
59JA481	W. E. Truce and W. J. Ray; *J. Am. Chem. Soc.*, 1959, **81**, 481.	379
59JA578	N. P. Neureiter and F. G. Bordwell; *J. Am. Chem. Soc.*, 1959, **81**, 578.	592, 646
59JA651	W. J. Bailey and W. F. Hale; *J. Am. Chem. Soc.*, 1959, **81**, 651.	637, 773, 782
59JA1154	C. R. Hauser and T. M. Harris; *J. Am. Chem. Soc.*, 1959, **81**, 1154.	881
59JA1160	C. R. Hauser and T. M. Harris; *J. Am. Chem. Soc.*, 1959, **81**, 1160.	881
59JA1243	C. Walling and R. Rabinowitz; *J. Am. Chem. Soc.*, 1959, **81**, 1243.	942
59JA1497	M. Stiles and R. P. Mayer; *J. Am. Chem. Soc.*, 1959, **81**, 1497.	385
59JA1512	H. C. Brown and G. Zweifel; *J. Am. Chem. Soc.*, 1959, **81**, 1512.	449, 509
59JA1617	H. D. Zook, J. March and D. F. Smith; *J. Am. Chem. Soc.*, 1959, **81**, 1617.	883
59JA2126	W. J. Bailey and R. A. Baylouny; *J. Am. Chem. Soc.*, 1959, **81**, 2126.	774
59JA2568	H. Pines and D. Wunderlich; *J. Am. Chem. Soc.*, 1959, **81**, 2568.	860
59JA2572	L. Field, J. R. Holsten and R. D. Clark; *J. Am. Chem. Soc.*, 1959, **81**, 2572.	144
59JA2723	H. N. Cripps, J. K. Williams and W. H. Sharkey; *J. Am. Chem. Soc.*, 1959, **81**, 2723.	498
59JA2799	A. C. Cope, E. Ciganek and N. A. LeBel; *J. Am. Chem. Soc.*, 1959, **81**, 2799.	625
59JA3160	G. Leal and R. Pettit; *J. Am. Chem. Soc.*, 1959, **81**, 3160.	904
59JA3379	K. C. Brannock; *J. Am. Chem. Soc.*, 1959, **81**, 3379.	825
59JA3383	R. C. Woodworth and P. S. Skell; *J. Am. Chem. Soc.*, 1959, **81**, 3383.	299
59JA3634	H. E. Zimmerman and W.-H. Chang; *J. Am. Chem. Soc.*, 1959, **81**, 3634.	8
59JA3644	H. E. Zimmerman and A. Mais; *J. Am. Chem. Soc.*, 1959, **81**, 3644.	8
59JA3652	J. F. Eastham and D. R. Larkin; *J. Am. Chem. Soc.*, 1959, **81**, 3652.	889
59JA3658	A. P. Krapcho and A. A. Bothner-By; *J. Am. Chem. Soc.*, 1959, **81**, 3658.	95, 96, 889
59JA4108	H. C. Brown and K. J. Murray; *J. Amer. Chem. Soc.*, 1959, **81**, 4108.	75, 85
59JA4266	M. P. Cava and A. A. Deana; *J. Am. Chem. Soc.*, 1959, **81**, 4266.	418
59JA4390	R. H. Mazur, W. N. White, D. A. Semenov, C. C. Lee, M. S. Silver and J. D. Roberts; *J. Am. Chem. Soc.*, 1959, **81**, 4390.	383
59JA4714	D. Y. Curtin and E. W. Flynn; *J. Am. Chem. Soc.*, 1959, **81**, 4714.	1144
59JA4719	D. Y. Curtin and W. H. Richardson; *J. Am. Chem. Soc.*, 1959, **81**, 4719.	1144, 1145

59JA4733	H. O. House and E. J. Grubbs; *J. Am. Chem. Soc.*, 1959, **81**, 4733.	384, 388
59JA4850	E. Grovenstein and R. W. Stevenson; *J. Am. Chem. Soc.*, 1959, **81**, 4850.	34
59JA5740	D. J. Cram, A. Langemann, J. Allinger and K. R. Kopecky; *J. Am. Chem. Soc.*, 1959, **81**, 5740.	883
59JA6330	D. B. Denney and M. J. Boskin; *J. Am. Chem. Soc.*, 1959, **81**, 6330.	268
59JCS2720	G. Valkanus and E. S. Waight; *J. Chem. Soc.*, 1959, 2720.	838
59JCS2773	W. Bonthrone and D. H. Reid; *J. Chem. Soc.*, 1959, 2773.	910
59JCS2902	F. J. McQuillin and W. O. Ord; *J. Chem. Soc.*, 1959, 2902.	506
59JCS3139	M. J. S. Dewar and C. R. Ganellin; *J. Chem. Soc.*, 1959, 3139.	906
59JCS3874	S. Trippett and D. M. Walker; *J. Chem. Soc.*, 1959, 3874.	1079
59JCS4017	N. Jones and H. T. Taylor; *J. Chem. Soc.*, 1959, 4017.	422
59JOC229	P. Tomboulian; *J. Org. Chem.*, 1959, **24**, 229.	881
59JOC709	W. L. Rellahan, W. L. Gumby and H. D. Zook; *J. Org. Chem.*, 1959, **24**, 709.	869, 871, 881
59JOC840	T. F. Rutledge; *J. Org. Chem.*, 1959, **24**, 840.	1000
59JOC843	A. S. Hussey and R. R. Herr; *J. Org. Chem.*, 1959, **24**, 843.	108, 464
59JOC2044	N. P. Neureiter; *J. Org. Chem.*, 1959, **24**, 2044.	801
59LA(623)217	P. D. Bartlett, E. R. Webster, C. E. Dills and H. G. Richey, Jr.; *Justus Liebigs Ann. Chem.*, 1959, **623**, 217.	905
59LA(625)206	G. Ohloff, H. Farnow, W. Phillipp and G. Schade; *Justus Liebigs Ann. Chem.*, 1959, **625**, 206.	812
B-59MI 119-01	E. S. Gould; "Mechanism and Structure in Organic Chemistry," Holt, Rinehart and Winston, New York, 1959, p. 218.	852
B-59MI 119-02	E. S. Gould; "Mechanism and Structure in Organic Chemistry," Holt, Rinehart and Winston, New York, 1959, p. 372.	853
B-59MI 119-03	R. P. Bell; "The Proton in Chemistry," Cornell University Press, Ithaca, New York, 1959.	870
59MI 119-04	G. Wilkinson and F. A. Cotton; *Prog. Inorg. Chem.*, 1959, **1**, 1.	862
59MI 119-05	E. O. Fischer and H. P. Fritz; *Adv. Inorg. Chem. Radiochem.*, 1959, **1**, 55.	862
59OR(10)179	E. D. Bergmann, D. Ginsberg and R. Pappo; *Org. React.*, 1959, **10**, 179–560.	888
59OS(39)37	D. W. Goheen and W. R. Vaughan; *Org. Synth.*, 1959, **39**, 37.	393
59OS(39)68	B. C. McKusick and L. R. Melby; *Org. Synth.*, 1959, **39**, 68.	560
59OS(39)73	C. R. Hauser and W. R. Dunnavant; *Org. Synth.*, 1959, **39**, 73.	881
59RTC551	A. M. van Leusen and J. F. Arens; *Recl. Trav. Chim. Pays-Bas*, 1959, **78**, 551.	785
59T19	R. B. Woodward and T. J. Katz; *Tetrahedron*, 1959, **5**, 19.	808
59T148	A. J. Birch and D. Nasipuri; *Tetrahedron*, 1959, **6**, 148.	99, 890
59T305	J. W. Powell and M. C. Whiting; *Tetrahedron*, 1959, **7**, 305.	631
60AG391	W. Franke, W. Ziegenbein and H. Meister; *Angew. Chem.*, 1960, **72**, 391.	1063
60AG535	F. Weygand and H. J. Bestmann; *Angew. Chem.*, 1960, **72**, 535.	393
60AG836	H. Eilingsfeld, M. Seefelder and H. Weidinger; *Angew. Chem.*, 1960, **72**, 836.	1026
60BSF856	P. Mazerolles; *Bull. Soc. Chim. Fr.*, 1960, 856.	1000
60CB2819	V. Franzen and C. Mertz; *Chem. Ber.*, 1960, **93**, 2819.	623
60CI(L)262	D. Y. Curtin and W. J. Koehl; *Chem. Ind.* (*London*), 1960, 262.	847
60CI(L)405	P. Bruck, D. Thompson and S. Winstein; *Chem. Ind.* (*London*), 1960, 405.	6
60CPB819	S. Sugasawa and H. Matsuo; *Chem. Pharm. Bull.* (*Japan*), 1960, **8**, 819 (*Chem. Abstr.*, 1961, **55**, 20 901).	744
60CRV431	C. H. DePuy and R. W. King; *Chem. Rev.*, 1960, **60**, 431.	624, 637, 773, 776, 782, 783
60DOK(131)27	C. S. Kadyrov and I. P. Tsukervanik; *Dokl. Akad. Nauk SSSR*, 1960, **131**, 27 (*Chem. Abstr.*, 1963, **58**, 2420).	475
60DOK(131)580	A. D. Petrov, G. I. Nikishin and Y. N. Ogibin; *Dokl. Akad. Nauk SSSR*, 1960, **131**, 580.	925
60HCA1915	A. K. Mills and A. E. Wilder Smith; *Helv. Chim. Acta*, 1960, **43**, 1915.	871
60JA563	M. C. Henry and G. Wittig; *J. Am. Chem. Soc.*, 1960, **82**, 563.	271, 647, 738, 743
60JA634	C. H. DePuy and D. H. Froemsdorf; *J. Am. Chem. Soc.*, 1960, **82**, 634.	632, 633
60JA739	M. M. Kreevoy and F. R. Kowitt; *J. Am. Chem. Soc.*, 1960, **82**, 739.	671
60JA934	H. Rapoport and G. Smolinsky; *J. Am. Chem. Soc.*, 1960, **82**, 934.	862
60JA1007	C. G. Overberger and A. E. Borchert; *J. Am. Chem. Soc.*, 1960, **82**, 1007.	805
60JA1386	H. D. Zook and W. L. Gumby; *J. Am. Chem. Soc.*, 1960, **82**, 1386.	859, 881
60JA1684	J. H. Fried, G. E. Arth and L. H. Sarett; *J. Am. Chem. Soc.*, 1960, **82**, 1684.	881
60JA1810	C. A. Kingsbury and D. J. Cram; *J. Am. Chem. Soc.*, 1960, **82**, 1810.	595, 779
60JA1972	R. Pettit; *J. Am. Chem. Soc.*, 1960, **82**, 1972.	866, 910
60JA2380	G. A. Olah and S. J. Kuhn; *J. Am. Chem. Soc.*, 1960, **82**, 2380.	914
60JA2644	R. Breslow and H. Höver; *J. Am. Chem. Soc.*, 1960, **82**, 2644.	907
60JA2651	D. G. Farnum and M. Burr; *J. Am. Chem. Soc.*, 1960, **82**, 2651.	907
60JA2847	N. W. Atwater; *J. Am. Chem. Soc.*, 1960, **82**, 2847.	881
60JA2895	H. E. Zaugg, B. W. Horrom and S. Borgwardt; *J. Am. Chem. Soc.*, 1960, **82**, 2895.	881
60JA3652	L. A. Errede and J. P. Cassidy; *J. Am. Chem. Soc.*, 1960, **82**, 3652.	849
60JA3784	T. J. Katz; *J. Am. Chem. Soc.*, 1960, **82**, 3784.	890
60JA3785	T. J. Katz; *J. Am. Chem. Soc.*, 1960, **82**, 3785.	890
60JA4307	G. Stork and I. J. Borowitz; *J. Am. Chem. Soc.*, 1960, **82**, 4307.	392
60JA4656	A. C. Cope and N. A. LeBel; *J. Am. Chem. Soc.*, 1960, **82**, 4656.	625, 784
60JA4663	A. C. Cope, E. Ciganek, C. F. Howell and E. E. Schweizer; *J. Am. Chem. Soc.*, 1960, **82**, 4663.	625
60JA4681	A. W. Burgstahler; *J. Am. Chem. Soc.*, 1960, **82**, 4681.	415, 825
60JA4692	S. J. Cristol and F. R. Stermitz; *J. Am. Chem. Soc.*, 1960, **82**, 4692.	628

60JA4729	R. H. Boyd, R. W. Taft, Jr., A. P. Wolf and D. R. Christman; *J. Am. Chem. Soc.*, 1960, **82**, 4729.	896
60JA4918	R. T. Arnold and G. Smolinsky; *J. Am. Chem. Soc.*, 1960, **82**, 4918.	795, 1023
60JA5704	J. H. Fried, A. N. Nutile and G. E. Arth; *J. Am. Chem. Soc.*, 1960, **82**, 5704.	881
60JA6397	T. L. Jacobs and R. N. Johnson; *J. Am. Chem. Soc.*, 1960, **82**, 6397.	1125
60JCS179	C. Eaborn and D. E. Webster; *J. Chem. Soc.*, 1960, 179.	447
60JCS298	P. W. Kent, A. Morris and N. F. Taylor; *J. Chem. Soc.*, 1960, 298.	2, 6
60JCS341	E. R. H. Jones, H. H. Lee and M. C. Whiting; *J. Chem. Soc.*, 1960, 341.	959
60JCS1045	M. J. Allen, J. A. Siragusa and W. Pierson; *J. Chem. Soc.*, 1960, 1045.	548
60JCS1566	C. Eaborn and K. C. Pande; *J. Chem. Soc.*, 1960, 1566.	448, 449
60JCS3521	J. Cast, T. S. Stevens and J. Holmes; *J. Chem. Soc.*, 1960, 3521.	397
60JCS3614	O. M. Behr, G. Eglinton, A. R. Galbraith and R. A. Raphael; *J. Chem. Soc.*, 1960, 3614.	1064
60JCS4454	C. T. Kyte, G. H. Jeffery and A. I. Vogel; *J. Chem. Soc.*, 1960, 4454.	860
60JGU765	I. N. Nazarov, I. I. Zaretskaya and T. I. Sorkina; *J. Gen. Chem. USSR* (*Engl. Transl.*), 1960, **30**, 765.	421
60JOC158	R. B. Meyer and C. R. Hauser; *J. Org. Chem.*, 1960, **25**, 158.	881
60JOC183	A. W. Johnson; *J. Org. Chem.*, 1960, **25**, 183.	646, 737, 738
60JOC463	A. Fischer, R. D. Topsom and J. Vaughan; *J. Org. Chem.*, 1960, **25**, 463.	258
60JOC1275	A. S. Hay; *J. Org. Chem.*, 1960, **25**, 1275.	1034
60JOC1442	H. J. Dauben, Jr., L. R. Honnen and K. M. Harmon; *J. Org. Chem.*, 1960, **25**, 1442.	906
60JOC1815	K. C. Brannock, H. S. Pridgen and B. Thompson; *J. Org. Chem.*, 1960, **25**, 1815.	417, 829
60JOC2203	L. A. Rothman and E. I. Becker; *J. Org. Chem.*, 1960, **25**, 2203.	428
60JOC2217	S. M. Kupchan and A. Afonso; *J. Org. Chem.*, 1960, **25**, 2217.	431
60JPS1271	J. G. Pritchard and A. A. Bothner-By; *J. Phys. Chem.*, 1960, **64**, 1271.	1144
60LA(629)1	K. Ziegler, H.-G. Gellert, H. Lehmkuhl, W. Pfohl and K. Zosel; *Justus Liebigs Ann. Chem.*, 1960, **629**, 1.	665
60LA(629)14	K. Ziegler, H. Martin and F. Krupp; *Justus Liebigs Ann. Chem.*, 1960, **629**, 14.	666
60LA(629)53	K. Ziegler, W.-R. Kroll, W. Larbig and O.-W. Steudel; *Justus Liebigs Ann. Chem.*, 1960, **629**, 53.	665, 666
60LA(629)172	K. Ziegler, H.-G. Gellert, E. Holzkamp, G. Wilke, E. W. Duck and W.-R. Kroll; *Justus Liebigs Ann. Chem.*, 1960, **629**, 172.	665
60LA(629)207	W. Pfohl; *Justus Liebigs Ann. Chem.*, 1960, **629**, 207.	666
60LA(629)210	K. Ziegler, K. Nagel and W. Pfohl; *Justus Liebigs Ann. Chem.*, 1960, **629**, 210.	665
60LA(629)222	G. Wilke and H. Müller; *Justus Liebigs Ann. Chem.*, 1960, **629**, 222.	511, 534, 542
60MI 105-01	Ch. Sh. Kadyrov and I. P. Tsukervanik; *Dokl. Akad. Nauk Uzb. SSR*, 1960, 27; (*Chem. Abstr.*, 1963, **58**, 2420).	212
B-60MI 109-01	G. W. Wheland; "Advanced Organic Chemistry," Wiley, New York, 1960, p. 550.	379, 380
B-60MI 119-01	L. Pauling; "The Nature of the Chemical Bond," 3rd edn., Cornell University Press, Ithaca, New York, 1960, p. 90.	846
B-60MI 119-02	L. Pauling; "The Nature of the Chemical Bond," 3rd edn., Cornell University Press, Ithaca, New York, 1960, p. 314.	852
B-60MI 119-03	L. Pauling; "The Nature of the Chemical Bond," 3rd edn., Cornell University Press, Ithaca, New York, 1960, p. 488.	869
B-60MI 119-04	G. W. Wheland; "Advanced Organic Chemistry," Wiley, New York, 1960, pp. 663–730.	870
60MI 122-01	H. Normant; *Adv. Org. Chem.*, 1960, **2**, 1.	1092
60OR(11)157	P. A. S. Smith and D. R. Baer; *Org. React.*, 1960, **11**, 157.	389
60OR(11)261	A. S. Kende; *Org. React.*, 1960, **11**, 261.	391
60OR(11)317	A. C. Cope and E. R. Trumbull; *Org. React.*, 1960, **11**, 317.	624, 627, 628, 637, 640, 782, 783
60OS(40)38	C. R. Hauser and W. R. Dunnavant; *Org. Synth.*, 1960, **40**, 38.	881
60QR357	C. J. Collins; *Quart. Rev., Chem. Soc.*, 1960, **14**, 357.	384
60RTC1002	W. Drenth and H. Hogeveen; *Rec. Trav. Chim. Pays-Bas*, 1960, **79**, 1002.	1122
60T130	A. W. Johnson and R. B. LaCount; *Tetrahedron*, 1960, **9**, 130.	104
60T246	R. A. Dytham and B. C. L. Weedon; *Tetrahedron*, 1960, **9**, 246.	882
60TL29	R. C. Cookson, J. Hudec and R. O. Williams; *Tetrahedron Lett.*, 1960, **1**, 29.	807, 809
60ZOB1056	A. A. Petrov and V. A. Kormer; *Zh. Obshch. Khim.*, 1960, **30**, 1056. (*Chem. Abstr.*, 1961, **55**, 358f).	960
60ZOB3615	B. M. Mikhailov and A. N. Blokhina; *Zh. Obshch. Khim.*, 1960, **30**, 3615 (*Chem. Abstr.*, 1961, **55**, 20 920).	657

61AG342	K. Hafner and H. Pelster; *Angew. Chem.*, 1961, **73**, 342.	902
61BSF836	J.-M. Conia and P. Gosselin; *Bull. Soc. Chim. Fr.*, 1961, 836.	881
61BSF2176	P. Cadiot, W. Chodkiewicz and J. Rauss-Godineau; *Bull. Soc. Chim. Fr.*, 1961, 2176.	953
61CA(55)14340	T. Hashimoto and M. Seki; *Chem. Abstr.*, 1961, **55**, 14 340.	447
61CB1157	I. Ugi and F. Bodesheim; *Chem. Ber.*, 1961, **94**, 1157.	38
61CB1447	E. Amberger; *Chem. Ber.*, 1961, **94**, 1447.	446
61CB3005	G. Märkl; *Chem. Ber.*, 1961, **94**, 3005.	1079
61CB3060	R. Kuhn and H. Fischer; *Chem. Ber.*, 1961, **94**, 3060.	960
61CI(L)550	M. A. Battiste; *Chem. Ind.* (*London*), 1961, 550.	907
61CI(L)990	S. Trippett and D. M. Walker; *Chem. Ind.* (*London*), 1961, 990.	747
61CJC348	R. F. W. Bader and A. N. Bourns; *Can. J. Chem.*, 1961, **39**, 348.	776
61DOK(136)1099	G. I. Nikishin, G. V. Somov and A. D. Petrov; *Dokl. Akad. Nauk SSSR*, 1961, **136**, 1099.	925
61G889	P. Beltrame and S. Carra; *Gazz. Chim. Ital.*, 1961, **91**, 889.	1144

61IZV1099 G. I. Nikishin, G. V. Somov and A. D. Petrov; *Izv. Akad. Nauk SSSR, Ser. Khim.*, 1961, **136**, 1099. 331

61IZV2065 G. I. Nikishin, G. V. Somov and A. D. Petrov; *Izv. Akad. Nauk SSSR, Ser. Khim.*, 1961, 2065. 925

61JA198 A. W. Burghstahler and I. C. Nordin; *J. Am. Chem. Soc.*, 1961, **83**, 198. 104

61JA412 E. Grovenstein, Jr. and L. P. Williams, Jr.; *J. Am. Chem. Soc.*, 1961, **83**, 412. 378, 886

61JA417 A. W. Johnson and R. B. LaCount; *J. Am. Chem. Soc.*, 1961, **83**, 417. 646

61JA449 T. Agawa; *J. Am. Chem. Soc.*, 1961, **83**, 449. 1137

61JA927 G. Büchi, R. E. Erickson and N. Wakabayashi; *J. Am. Chem. Soc.*, 1961, **83**, 927. 561

61JA1196 H. E. Zimmerman and A. Zweig; *J. Am. Chem. Soc.*, 1961, **83**, 1196. 378, 886

61JA1251 J. B. Hendrickson; *J. Am. Chem. Soc.*, 1961, **83**, 1251. 442

61JA1705 E. Grovenstein Jr., D. V. Rao and J. W. Taylor; *J. Am. Chem. Soc.*, 1961, **83**, 1705. 562

61JA1733 W. S. Wadsworth, Jr. and W. D. Emmons; *J. Am. Chem. Soc.*, 1961, **83**, 1733. 639, 987

61JA1951 C. G. Swain, R. F. W. Bader, R. M. Esteve, Jr. and R. N. Griffin; *J. Am. Chem. Soc.*, 1961, **83**, 1951. 883

61JA2159 H. G. Kuivala and K. V. Nahabedian; *J. Am. Chem. Soc.*, 1961, **83**, 2159, 2164, 2167. 447

61JA2367 R. Breslow and H. W. Chang; *J. Am. Chem. Soc.*, 1961, **83**, 2367. 907

61JA2375 R. Breslow, J. Lockhart and H. W. Chang; *J. Am. Chem. Soc.*, 1961, **83**, 2375. 907

61JA2392 L. de Vries; *J. Am. Chem. Soc.*, 1961, **83**, 2392. 912

61JA2537 E. Grovenstein, Jr. and L. P. Williams, Jr.; *J. Am. Chem. Soc.*, 1961, **83**, 2537. 886

61JA2944 J. R. Leto and M. F. Leto; *J. Am. Chem. Soc.*, 1961, **83**, 2944. 530

61JA2957 E. J. Corey, D. J. Pasto and W. L. Mock; *J. Am. Chem. Soc.*, 1961, **83**, 2957. 74

61JA3160 D. V. Young and H. R. Snyder; *J. Am. Chem. Soc.*, 1961, **83**, 3160. 88

61JA3583 D. Seyferth and M. A. Weiner; *J. Am. Chem. Soc.*, 1961, **83**, 3583. 1091

61JA3678 D. J. Cram, B. Rickborn, C. A. Kingsbury and P. Haberfield; *J. Am. Chem. Soc.*, 1961, **83**, 3678. 871, 881

61JA3688 D. J. Cram, C. A. Kingsbury and B. Rickborn; *J. Am. Chem Soc.*, 1961, **83**, 3688. 854

61JA3834 H. C. Brown and G. Zweifel; *J. Am. Chem. Soc.*, 1961, **83**, 3834. 509, 1100, 1102

61JA3861 A. C. Cope, N. A. LeBel, P. T. Moore and W. R. Moore; *J. Am. Chem. Soc.*, 1951, **83**, 3861. 627

61JA4101 M. A. Battiste; *J. Am. Chem. Soc.*, 1961, **83**, 4101. 907

61JA4990 H. D. Hartzler; *J. Am. Chem. Soc.*, 1961, **83**, 4990. 988

61JCS73 A. L. J. Beckwith and M. J. Thompson; *J. Chem. Soc.*, 1961, 73. 848

61JCS2687 W. Tadros, A. B. Sakla and A. A. Helmy; *J. Chem. Soc.*, 1961, 2687. 981

61JCS3452 E. D. Bergmann, S. Cohen, E. Hoffman and Z. Rand-Meir; *J. Chem. Soc.*, 1961, 3452. 108

61JCS3547 M. C. Flowers and R. M. Frey; *J. Chem. Soc.*, 1961, 3547. 801

61JCS4610 G. R. Knox and P. L. Pauson; *J. Chem. Soc.*, 1961, 4610. 862

61JOC631 H. C. Brown and K. J. Murray; *J. Org. Chem.*, 1961, **26**, 631. 58

61JOC644 H. E. Zaugg, D. A. Dunnigan, R. J. Michaels, L. R. Swett, T. S. Wang, A. H. Sommers and R. W. De Net; *J. Org. Chem.*, 1961, **26**, 644. 881

61JOC818 S. L. Shapiro, H. Soloway and L. Freedman; *J. Org. Chem.*, 1961, **26**, 818. 91

61JOC846 C. E. Higgins and W. H. Baldwin; *J. Org. Chem.*, 1961, **26**, 846. 778

61JOC1073 G. F. Hennion and A. C. Perrino; *J. Org. Chem.*, 1961, **26**, 1073. 506

61JOC1835 M. Freifelder; *J. Org. Chem.*, 1961, **26**, 1835. 88

61JOC2934 D. Seyferth, G. Raab and K. A. Brandle; *J. Org. Chem.*, 1961, **26**, 2934. 455

61JOC3156 A. Gero; *J. Org. Chem.*, 1961, **26**, 3156. 869

61JOC3772 N. R. Easton, R. D. Dillard, W. J. Doran, M. Livezey and D. E. Morrison; *J. Org. Chem.*, 1961, **26**, 3772. 506

61JOC4112 J. J. Bloomfield; *J. Org. Chem.*, 1961, **26**, 4112. 881

61JOC4278 H. O. House and G. H. Rasmusson; *J. Org. Chem.*, 1961, **26**, 4278. 725

61LA(641)1 H. Meerwein, W. Florian, N. Schön and G. Stopp; *Justus Liebigs Ann. Chem.*, 1961, **641**, 1. 415

61LA(642)1 U. Schöllkopf and W. Fabian; *Justus Liebigs Ann. Chem.*, 1961, **642**, 1. 396

61LA(644)172 E. Vogel, K.-H. Ott and K. Gajek; *Justus Liebigs Ann. Chem.*, 1961, **644**, 172. 811

61LA(650)62 K. Hafner, H. Pelster and J. Schneider; *Justus Liebigs Ann. Chem.*, 1961, **650**, 62. 899

61LA(746)211 K. Hummel and M. Hanack; *Justus Liebigs Ann. Chem.*, 1951, **746**, 211. 1123

B-61MI 110-01 W. Gerrard; "The Organic Chemistry of Boron," Academic Press, London, 1961. 447

61MI 116-01 H. T. Openshaw and N. Whittaker; *Proc. Chem. Soc.*, 1961, 454. 726

B-61MI 119-01 C. A. Coulson; "Valence," Oxford University Press, 2nd edn., 1961, p. 215. 911

B-61MI 119-02 A. Streitwieser, Jr.; "Molecular Orbital Theory for Organic Chemists," Wiley, New York, 1961, p. 358. 910

61MP241 C. MacLean and E. L. Mackor; *Mol. Phys.*, 1961, **4**, 241. 914

61OS(41)5 R. E. Ireland and M. Chaykovsky; *Org. Synth.*, 1961, **41**, 5. 881

61PCS117 R. C. Cookson, J. Hudec and B. Whitear; *Proc. Chem. Soc.*, 1961, 117. 862

61RTC797 E. J. Stamhuis and W. Drenth; *Rec. Trav. Chim. Pays-Bas*, 1961, **66**, 797. 1122

61RTC1285 G. L. Hekkert and W. Drenth; *Rec. Trav. Chim. Pays-Bas*, 1961, **80**, 1285. 1122

61T13 P. Yates and P. Eaton; *Tetrahedron*, 1961, **12**, 13. 807, 809

61T16 N. A. Dobson, G. Eglington, M. Krishnamurti, R. A. Raphael and R. G. Willis; *Tetrahedron*, 1961, **16**, 16. 504

61T18 G. I. Fray, R. H. Jaeger, E. D. Morgan, R. Robinson and A. D. B. Sloan; *Tetrahedron*, 1961, **15**, 18. 1000

61T45 J.-M. Conia and F. Rouessac; *Tetrahedron*, 1961, **16**, 45. 881

61T146 E. Havinga and J. L. M. A. Schlatmann; *Tetrahedron*, 1961, **16**, 146. 419

61TL347 E. J. Corey, W. L. Mock and D. J. Pasto; *Tetrahedron Lett.*, 1961, **2**, 347. 458

61TL353 S. Hünig, R. Müller and W. Thier; *Tetrahedron Lett.*, 1961, **2**, 353. 74

61TL579 G. Fodor and I. Tomoskozi; *Tetrahedron Lett.*, 1961, **2**, 579. 726

61ZAAC(312)186 H. Hartmann and G. Kuhl; *Z. Anorg. Allg. Chem.*, 1961, **312**, 186. 999

62ACS(B)523 B. Smith and U. Persmark; *Acta Chem. Scand., Ser. B*, 1962, **16**, 523. 560

62AG(E)116 H. J. Bestmann and F. Seng; *Angew Chem., Int. Ed. Engl.*, 1962, **1**, 116. 268

62AG(E)155 E. F. Jenny and J. Druey; *Angew. Chem., Int. Ed. Engl.*, 1962, **1**, 155. 398

62AG(E)312 K. Plesske; *Angew. Chem., Int. Ed. Engl.*, 1962, **1**, 312. 862

62AG(E)394 K. Plesske; *Angew. Chem., Int. Ed. Engl.*, 1962, **1**, 394. 862

62AG(E)415 G. Wittig; *Angew. Chem., Int. Ed. Engl.*, 1962, **1**, 415. 519

62BSF1325 H. Christol and F. Plenat; *Bull. Chim. Soc. Fr.*, 1962, 1325. 795

62BSF1882 C. Brocquet and R. Quelet; *Bull. Soc. Chim. Fr.*, 1962, 1882. 181

62CB319 H. G. Viehe and E. Franchimont; *Chem. Ber.*, 1962, **95**, 319. 999

62CB1894 H. J. Bestmann and O. Kratzer; *Chem. Ber.*, 1962, **95**, 1894. 723, 733

62CHE430 D. J. Brown; *Chem. Heterocycl. Compd.* (*Engl. Transl.*), 1962, **16**, 430. 90

62CI(L)484 F. Aylward and M. Sawistowska; *Chem. Ind.* (*London*), 1962, 484. 74

62CI(L)895 R. A. Finnegan; *Chem. Ind.* (*London*), 1962, 895. 663

62CI(L)1236 D. A. Lightner and C. Djerassi; *Chem. Ind.* (*London*), 1962, 1236. 592

62CR(254)1817 M. Pfau, R. Dulou and M. Vilkas; *C. R. Hebd. Seances Acad. Sci.*, 1962, **254**, 1817. 546

62CR(255)959 M. Julia and C. James; *C. R. Hebd. Seances Acad. Sci.*, 1962, **255**, 959. 548

62CR(255)1750 Y. Pasternak; *C. R. Hebd. Seances Acad. Sci.*, 1962, **255**, 1750. 1056

62CRV347 H. Hauptmann and W. F. Walter; *Chem. Rev.*, 1962, **62**, 347. 21, 23, 24, 435, 438

62IZV1373 B. M. Mikhailov and A. N. Blokhina; *Izv. Akad. Nauk SSSR, Ser. Khim.*, 1962, 1373 (*Chem. Abstr.*, 1963, **58**, 5707). 658

62JA258 A. Streitwieser, Jr., D. E. Van Sickle and L. Reif; *J. Am. Chem. Soc.*, 1962, **84**, 258. 857

62JA676 N. E. McEwen and A. P. Wolf; *J. Am. Chem. Soç.*, 1962, **84**, 676. 268

62JA677 N. E. McEwen, A. Bladé-Font and C. A. Vanderwerf; *J. Am. Chem. Soc.*, 1962, **84**, 677. 268

62JA679 L. Vaska and J. W. Diluzio; *J. Am. Chem. Soc.*, 1962, **84**, 679. 73

62JA866 E. J. Corey and M. Chaykovsky; *J. Am. Chem. Soc.*, 1962, **84**, 866. 881

62JA1015 K. R. Kopecky, G. S. Hammond and P. A. Leermakers; *J. Am. Chem. Soc.*, 1962, **84**, 1015. 299

62JA1498 N. C. Deno, H. G. Richey, Jr., J. D. Hodge and M. J. Wisotsky; *J. Am. Chem. Soc.*, 1962, **84**, 1498. 897, 901, 911

62JA1734 D. J. Cram, M. R. V. Sahyun and G. R. Knox; *J. Am. Chem. Soc.*, 1962, **84**, 1734. 19, 43, 624, 782

62JA2016 N. C. Deno, H. G. Richey, Jr., J. S. Liu, J. D. Hodge, J. J. Houser and M. J. Wisotsky; *J. Am. Chem. Soc.*, 1962, **84**, 2016. 894, 900

62JA2465 H. M. Walborsky, A. A. Youssef and J. M. Motes; *J. Am. Chem. Soc.*, 1962, **84**, 2465. 851

62JA2842 J. L. von Rosenberg, Jr., J. E. Mahler and R. Pettit; *J. Am. Chem. Soc.*, 1962, **84**, 2842. 917

62JA2899 R. E. Dessy, Y. Okuzumi and A. Chen; *J. Am. Chem. Soc.*, 1962, **84**, 2899. 871

62JA2905 H. Shechter, M. J. Collis, R. Dessy, Y. Okuzumi and A. Chen; *J. Am. Chem. Soc.*, 1962, **84**, 2905. 871

62JA3164 A. Schriesheim, R. J. Muller and C. A. Rowe, Jr.; *J. Am. Chem. Soc.*, 1962, **84**, 3164. 871, 881

62JA3168 R. Breslow, H. Höver and H. W. Chang; *J. Am. Chem. Soc.*, 1962, **84**, 3168. 907, 908, 909

62JA3199 T. J. Katz, J. R. Hall and W. C. Neikam; *J. Am. Chem. Soc.*, 1962, **84**, 3199. 901

62JA3402 H. J. Ringold and S. K. Malhotra; *J. Am. Chem. Soc.*, 1962, **84**, 3402. 871

62JA3744 E. C. Taylor and C. W. Jefford; *J. Am. Chem. Soc.*, 1962, **84**, 3744. 47

62JA3944 D. B. Denney, J. J. Vill and M. J. Boskin; *J. Am. Chem. Soc.*, 1962, **84**, 3944. 268

62JA4141 R. Srinivasan; *J. Am. Chem. Soc.*, 1962, **84**, 4141. 418, 419

62JA4165 H. H. Freedman and A. M. Frantz, Jr.; *J. Am. Chem. Soc.*, 1962, **84**, 4165. 901

62JA4660 P. S. Skell and G. P. Bean; *J. Am. Chem. Soc.*, 1962, **84**, 4660. 114

62JA4824 V. J. Shiner, Jr. and B. Martin; *J. Am. Chem. Soc.*, 1962, **84**, 4824. 870

62JA4838 F. J. Impastato and H. M. Walborsky; *J. Am. Chem. Soc.*, 1962, **84**, 4838. 885

62JCS396 S. Patai and Z. Rappoport; *J. Chem. Soc.*, 1962, 396. 889

62JCS600 M. E. Peach and T. C. Waddington; *J. Chem. Soc.*, 1962, 600. 897

62JCS1741 S. Patai, S. Weinstein and Z. Rappoport; *J. Chem. Soc.*, 1962, 1741. 889

62JCS1801 D. J. Alsop, J. Burdon and J. C. Tatlow; *J. Chem. Soc.*, 1962, 1801. 426

62JCS2333 S. Trippett and S. T. D. Gough; *J. Chem. Soc.*, 1962, 2333. 1079

62JCS3253 G. M. Brooke, J. Burdon and J. C. Tatlow; *J. Chem. Soc.*, 1962, 3253. 426

62JCS4468 J. C. Allen, J. I. G. Cadogan, B. W. Harris and D. H. Hey; *J. Chem. Soc.*, 1962, 4468. 925

62JOC293 S. J. Cristol and D. I. Davies; *J. Org. Chem.*, 1962, **27**, 293. 628

62JOC1406 M. S. Raasch; *J. Org. Chem.*, 1962, **27**, 1406. 91

62JOC1828 A. P. Boisselle and N. A. Meinhardt; *J. Org. Chem.*, 1962, **27**, 1828. 993

62JOC1910 W. G. Dauben and R. L. Cargill; *J. Org. Chem.*, 1962, **27**, 1910. 419

62JOC1933 P. T. Lansbury and V. A. Pattison; *J. Org. Chem.*, 1962, **27**, 1933. 397

62JOC2288 J. H. Stocker; *J. Org. Chem.*, 1962, **27**, 2288. 88

62JOC2291 O. L. Chapman and S. L. Smith; *J. Org. Chem.*, 1962, **27**, 2291. 404

62JOC2398 R. J. Tedeschi; *J. Org. Chem.*, 1962, **27**, 2398. 503, 506

62JOC2737 W. G. Kofron, W. R. Dunnavant and C. R. Hauser; *J. Org. Chem.*, 1962, **27**, 2737. 881

62JOC3320 A. S. Hay; *J. Org. Chem.*, 1962, **27**, 3320. 1034

62JOC3441 G. A. Olah, W. S. Tolgyesi and R. E. A. Dear; *J. Org. Chem.*, 1962, **27**, 3441. 914

62JOC3520 C. D. Weis; *J. Org. Chem.*, 1962, **27**, 3520. 521, 1084

62JOC3647 H. Gilman and R. A. Tomasi; *J. Org. Chem.*, 1962, **27**, 3647. 649, 763

62JOC3688 A. T. Bottini, B. J. King and J. M. Lucas; *J. Org. Chem.*, 1962, **27**, 3688. 1065

62JOC3819 G. B. Payne; *J. Org. Chem.*, 1962, **27**, 3819. 223

62JOC3836 J. F. Bunnett and J. A. Skorcz; *J. Org. Chem.*, 1962, **27**, 3836. 881

62JOC3856	J. F. Vozza; *J. Org. Chem.*, 1962, **27**, 3856.	861
62JOC4046	M. Freifelder; *J. Org. Chem.*, 1962, **27**, 4046.	91
62JOC4146	H. O. House and V. Kramar; *J. Org. Chem.*, 1962, **27**, 4146.	881
62JOC4152	J. F. Bunnett and B. F. Hrutfiord; *J. Org. Chem.*, 1962, **27**, 4152.	885
62JOC4323	R. J. Tedeschi and G. Clark, Jr.; *J. Org. Chem.*, 1962, **27**, 4323.	503, 506
62JOC4666	R. Ketcham, D. Jambotkar and L. Martinelli; *J. Org. Chem.*, 1962, **27**, 4666.	732
62JPC1671	C. A. Wellington; *J. Phys. Chem.*, 1962, **66**, 1671.	801
62LA(651)141	K. Alder and H. von Brachel; *Justus Liebigs Ann. Chem.*, 1962, **651**, 141.	524
62LA(654)64	R. Kuhn, H. Fischer, F. A. Neugebauer and H. Fischer; *Justus Liebigs Ann. Chem.*, 1962, **654**, 64.	866
62LA(658)65	E. Truscheit and K. Eiter; *Justus Liebigs Ann. Chem.*, 1962, **658**, 65.	734
62OR(12)57	H. R. Nace; *Org. React.*, 1962, **12**, 57.	775, 776
62OR(12)356	G. R. Pettit and E. E. van Tamelen; *Org. React.*, 1962, **12**, 356.	21, 23, 435
62OS(42)28	S.-O. Lawesson, S. Gronwall and R. Sandberg; *Org. Synth.*, 1962, **42**, 28.	881
62OS(42)48	K. C. Bass; *Org. Synth.*, 1962, **42**, 48.	889
62PCS334	K. J. Crowley; *Proc. Chem. Soc.*, 1962, 334.	418
62RTC993	W. Voskuil and J. F. Arens; *Recl. Trav. Chim. Pays-Bas*, 1962, **81**, 993.	999
62T67	W. von E. Doering and W. R. Roth; *Tetrahedron*, 1962, **18**, 67.	412
62T333	M. C. R. Symons; *Tetrahedron*, 1962, **18**, 333.	892
62TL505	J.-M. Conia and A. Sandré-Le Craz; *Tetrahedron Lett.*, 1962, **3**, 505.	807, 881
62TL549	G. J. Fonken; *Tetrahedron Lett.*, 1962, **3**, 549.	420
62TL669	H. J. Ringold and S. K. Malhotra; *Tetrahedron Lett.*, 1962, **3**, 669.	871, 881, 882
62TL1261	L. Caglioti and M. Magi; *Tetrahedron Lett.*, 1962, **3**, 1261.	630
62ZAAC(317)54	H. Hartmann, G. Habenicht and W. Reiss; *Z. Anorg. Allg. Chem.*, 1962, **317**, 54.	999
62ZN(B)787	H. J. Bestmann and H. Häberlein; *Z. Naturforsch., Teil B*, 1962, **17**, 787.	157
63ACS1683	L. Skattebol; *Acta Chem. Scand.*, 1963, **17**, 1683.	990
63AFC63	R. E. Parker; *Adv. Fluorine Chem.*, 1963, **3**, 63.	108
63AG(E)1	E. Vogel; *Angew. Chem., Int. Ed. Engl.*, 1963, **2**, 1.	419, 812
63AG(E)115	W. von E. Doering and W. R. Roth; *Angew. Chem., Int. Ed. Engl.*, 1963, **2**, 115.	411, 806
63AG(E)153	G. Märkl; *Angew. Chem., Int. Ed. Engl.*, 1963, **2**, 153.	157
63AG(E)622	H. Volz; *Angew. Chem., Int. Ed. Engl.*, 1963, **2**, 622.	894
63AG(E)688	W. R. Roth; *Angew. Chem., Int. Ed. Engl.*, 1963, **2**, 688.	404
63AG(E)739	E. Vogel, W. Grimme and E. Dinné; *Angew. Chem., Int. Ed. Engl.*, 1963, **2**, 739.	812
63AP(296)324	H. Oelschager and P. Schmersahl; *Arch. Pharm.* (*Weinheim, Ger.*), 1963, **296**, 324.	426
63APO(1)155	A. I. Shatenshtein; *Adv. Phys. Org. Chem.*, 1963, **1**, 155.	844
63BSF798	J. Schmitt, J. J. Panouse, P.-J. Cornu, P. Comoy, A. Hallut and H. Pluchet; *Bull. Soc. Chim. Fr.*, 1963, 798.	82
63BSF1428	G. Charvière, B. Tchoubar and Z. Welvart; *Bull. Soc. Chim. Fr.*, 1963, 1428.	55
63BSF1876	J. F. Normant; *Bull. Soc. Chim. Fr.*, 1963, 1876.	1017
63BSF1888	J. F. Normant; *Bull. Soc. Chim. Fr.*, 1963, 1888.	1017
63BSF1925	J.-M. Conia and F. Rouessac; *Bull. Soc. Chim. Fr.*, 1963, 1925.	881
63BSF1930	J.-M. Conia and F. Rouessac; *Bull. Soc. Chim. Fr.*, 1963, 1930.	881
63CA(59)2732	K. M. Shadmanov; *Chem. Abstr.*, 1963, **59**, 2732.	463
63CB342	G. Wittig and U. Mayer; *Chem. Ber.*, 1963, **96**, 342.	1084
63CB1535	G. Wittig and A. Haag; *Chem. Ber.*, 1963, **96**, 1535.	986
63CB3280	L. Birkofer, A. Ritter and H. Uhlenbrauck; *Chem. Ber.*, 1963, **96**, 3280.	1045, 1138
63CI(L)1203	R. M. Dodson and G. Klose; *Chem. Ind.* (*London*), 1963, 1203.	234
63CPB1049	I. Iwai and Y. Yura; *Chem. Pharm. Bull.*, 1963, **11**, 1049.	1041
63CRV55	M. E. Wolff; *Chem. Rev.*, 1963, **63**, 55.	927
63G757	P. Beltrame and G. Favini; *Gazz. Chim. Ital.*, 1963, **93**, 757	1144
63HCA483	C. A. Grob, H. Link and P. Schiess; *Helv. Chim. Acta*, 1963, **46**, 483.	812
63HCA678	H. Wehrli, R. Wenger, K. Schaffner and O. Jeger; *Helv. Chim. Acta*, 1963, **46**, 678.	794
63HCA1792	M. Winter; *Helv. Chim. Acta*, 1963, **46**, 1792.	178
63HCA2444	J. Borgulya, H.-J. Hansen, R. Barner and H. Schmid; *Helv. Chim. Acta*, 1963, **46**, 2444.	827
63IC1111	M. M. Fein, J. Bobinski, N. Mayes, N. Schwartz and M. S. Cohen; *Inorg. Chem.*, 1963, **2**, 1111.	168
63IZV499	S. A. Vartanyan and A. I. Tosunyan; *Izv. Akad. Nauk SSSR, Ser. Khim.*, 1963, **16**, 499 (*Chem. Abstr.*, 1964, **60**, 13 125h).	965
63IZV1053	L. D. Bergelson, V. A. Vaver, L. I. Barsukov and M. M. Shemyakin; *Izv. Akad. Nauk SSSR, Otd. Khim. Nauk*, 1963, 1053 (*Chem. Abstr.*, 1963, **59**, 8783).	722
63JA207	G. Stork, A. Brizzolara, H. Landesman, J. Szmuszkovicz and R. Terrell; *J. Am. Chem. Soc.*, 1963, **85**, 207.	307
63JA384	P. J. Sullivan and W. S. Koski; *J. Am. Chem. Soc.*, 1963, **85**, 384.	921
63JA743	D. E. Applequist and D. F. O'Brien; *J. Am. Chem. Soc.*, 1963, **85**, 743.	846
63JA1108	D. J. Cram and J. S. Bradshaw; *J. Am. Chem. Soc.*, 1963, **85**, 1108.	35, 39
63JA1328	G. A. Olah, W. S. Tolgyesi, S. J. Kuhn, M. E. Moffatt, I. J. Bastien and E. B. Baker; *J. Am. Chem. Soc.*, 1963, **85**, 1328.	899
63JA1538	S. K. Malhotra and H. J. Ringold; *J. Am. Chem. Soc.*, 1963, **85**, 1538.	871, 881
63JA1548	W. A. Mosher and N. D. Heindel; *J. Am. Chem. Soc.*, 1963, **85**, 1548.	384
63JA1609	J. Bornstein, M. R. Borden, F. Nunes and H. I. Tarlin; *J. Am. Chem. Soc.*, 1963, **85**, 1609.	2, 3, 6
63JA1696	E. W. Garbisch, Jr.; *J. Am. Chem. Soc.*, 1963, **85**, 1696.	870, 871
63JA1949	A. C. Cope and A. S. Mehta; *J. Am. Chem. Soc.*, 1963, **85**, 1949.	627

63JA2063 H. C. Brown and A. W. Moerikofer; *J. Am. Chem. Soc.*, 1963, **85**, 2063. 1100
63JA2171 J. F. Eastham and G. W. Gibson; *J. Am. Chem. Soc.*, 1963, **85**, 2171. 111
63JA2236 P. T. Lansbury and J. O. Peterson; *J. Am. Chem. Soc.*, 1963, **85**, 2236. 102
63JA2245 M. J. S. Dewar and R. C. Fahey; *J. Am. Chem. Soc.*, 1963, **85**, 2245. 896
63JA2273 W. M. Jones and R. Damico; *J. Am. Chem. Soc.*, 1963, **85**, 2273. 1144
63JA2739 T. J. Wallace, J. E. Hofmann and A. Schriesheim; *J. Am. Chem. Soc.*, 1963, **85**, 2739. 612
63JA2743 H. W. Heine; *J. Am. Chem. Soc.*, 1963, **85**, 2743. 266
63JA2790 A. J. Speziale and K. W. Ratts; *J. Am. Chem. Soc.*, 1963, **85**, 2790. 724
63JA2881 E. F. Meyer and R. L. Burwell, Jr.; *J. Am. Chem. Soc.*, 1963, **85**, 2881. 459
63JA2991 N. C. Deno, H. G. Richey, Jr., N. Friedman, J. D. Hodge, J. J. Houser and C. U. Pittman, Jr.; *J. Am. Chem. Soc.*, 1963, **85**, 2991. 897, 901, 911
63JA2995 N. C. Deno, N. Friedman, J. D. Hodge and J. J. Houser; *J. Am. Chem. Soc.*, 1963, **85**, 2995. 901
63JA2998 N. C. Deno, J. Bollinger, N. Friedman, K. Hafer, J. D. Hodge and J. J. Houser; *J. Am. Chem. Soc.*, 1963, **85**, 2998. 901, 913
63JA3041 H. J. Dauben, Jr. and M. R. Rifi; *J. Am. Chem. Soc.*, 1963, **85**, 3041. 865
63JA3302 M. Sundaralingam and L. H. Jensen; *J. Am. Chem. Soc.*, 1963, **85**, 3302. 909
63JA3419 G. Stork, A. Meisels and J. E. Davies; *J. Am. Chem. Soc.*, 1963, **85**, 3419. 934
63JA3522 M. Tanabe and R. A. Walsh; *J. Am. Chem. Soc.*, 1963, **85**, 3522. 104
63JA3524 K. R. Henery-Logan and J. V. Rodricks; *J. Am. Chem. Soc.*, 1963, **85**, 3524. 785
63JA3533 R. G. Pearson; *J. Am. Chem. Soc.*, 1963, **85**, 3533. 886
63JA3878 A. J. Speziale and D. E. Bissing; *J. Am. Chem. Soc.*, 1963, **85**, 3878. 723
63JA3890 D. J. Cram and L. Gosser; *J. Am. Chem. Soc.*, 1963, **85**, 3890. 854, 855
63JA4033 E. J. Corey and A. G. Hortmann; *J. Am. Chem. Soc.*, 1963, **85**, 4033. 421
63JA4045 R. Srinivasan; *J. Am. Chem. Soc.*, 1963, **85**, 4045. 418
63JCE212 J. March; *J. Chem. Ed.*, 1963, **40**, 212. 883
63JCS577 D. Bryce-Smith and A. C. Skinner; *J. Chem. Soc.*, 1963, 577. 115
63JCS775 D. E. Ames and A. N. Covell; *J. Chem. Soc.*, 1963, 775. 1001
63JCS1246 E. U. Emovon; *J. Chem. Soc.*, 1963, 1246. 774
63JCS1461 H. T. Openshaw and N. Whittaker; *J. Chem. Soc.*, 1963, 1461. 747
63JCS1930 V. Gold and R. S. Satchell; *J. Chem. Soc.*, 1963, 1930. 896
63JCS2990 P. L. Pauson and W. E. Watts; *J. Chem. Soc.*, 1963, 2990. 1032
63JCS3295 D. H. Reid and R. G. Sutherland; *J. Chem. Soc.*, 1963, 3295. 910
63JCS3397 B. J. Millard and T. S. Stevens; *J. Chem. Soc.*, 1963, 3397. 398
63JCS3712 J. C. Craig and M. Moyle; *J. Chem. Soc.*, 1963, 3712. 1070
63JCS4091 A. J. Hubert and J. Dale; *J. Chem. Soc.*, 1963, 4091. 795
63JCS4522 F. G. Baddar, S. Farid and N. A. Starkowsky; *J. Chem. Soc.*, 1963, 4522. 181
63JCS5889 D. E. Ames, A. N. Covell and T. G. Goodburn; *J. Chem. Soc.*, 1963, 5889. 1001
63JGU3508 A. V. J. Fedorova; *J. Gen. Chem. USSR* (*Engl. Transl.*), 1963, **33**, 3508. 1125
63JOC6 R. E. Ireland and P. W. Shiess; *J. Org. Chem.*, 1963, **28**, 6. 89
63JOC372 C. F. Hauser, T. W. Brooks, M. L. Miles, M. A. Raymond and G. B. Butler; *J. Org. Chem.*, 1963, **28**, 372. 726
63JOC421 J. A. Marshall and W. S. Johnson; *J. Org. Chem.*, 1963, **28**, 421. 82
63JOC568 K. G. Rutherford and W. A. Redmond; *J. Org. Chem.*, 1963, **28**, 568. 442
63JOC878 H. Gurien; *J. Org. Chem.*, 1963, **28**, 878. 636
63JOC1254 W. J. Gensler and J. J. Bruno; *J. Org. Chem.*, 1963, **28**, 1254. 453, 512
63JOC1395 J. H. Jordaan and W. S. Serfontein; *J. Org. Chem.*, 1963, **28**, 1395. 21
63JOC1725 H. Gilman and B. J. Gaj; *J. Org. Chem.*, 1963, **28**, 1725. 881
63JOC1946 K. G. Hampton, T. M. Harris and C. R. Hauser; *J. Org. Chem.*, 1963, **28**, 1946. 881
63JOC2163 W. R. Hatchard; *J. Org. Chem.*, 1963, **28**, 2163. 1142
63JOC2165 H. G. Kuivila and L. W. Menapace; *J. Org. Chem.*, 1963, **28**, 2165. 330, 341, 936
63JOC2347 H. S. Broadbent and D. W. Seegmiller; *J. Org. Chem.*, 1963, **28**, 2347. 21
63JOC3313 R. D. Stephens and C. E. Castro; *J. Org. Chem.*, 1963, **28**, 3313. 1004, 1142
63JOC3362 H. O. House and V. Kramar; *J. Org. Chem.*, 1963, **28**, 3362. 871, 881
63LA(662)188 H. Hellman and D. Eberle; *Justus Liebigs Ann. Chem.*, 1963, **662**, 188. 822
63LA(663)22 U. Schöllkopf and H. Schäfer; *Justus Liebigs Ann. Chem.*, 1963, **663**, 22. 396
63MI 103-01 F. Hartog and L. Zweitering; *J. Catalysis*, 1963, **2**, 79. 88
63MI 103-02 I. M. Coggiola; *Nature*, 1963, **200**, 954. 100
B-63MI 109-01 P. de Mayo (ed.); "Molecular Rearrangements," Wiley-Interscience, New York, 1963, vol. 1. 377, 380, 382, 395, 414
63MI 114-01 V. F. Mironov and N. G. Dzhurinskaya; *Izv. Akad. Nauk SSSR, Otdel. Khim. Nauk*, 1963, 75 (*Chem. Abstr.*, 1963, **58**, 10 225). 656
B-63MI 118-01 P. De Mayo (ed.); "Molecular Rearrangements," Interscience, New York, 1963. 794
63MI 118-02 R. J. Ellis and H. M. Frey; *Trans. Faraday Soc.*, 1963, **59**, 2076. 805
63MI 118-03 J. M. Conia; *Rec. Chem. Prog.*, 1963, **24**, 61. 807
B-63MI 118-04 P. B. O. de la Mare; in "Molecular Rearrangements," ed. P. De Mayo, Interscience, New York, 1963, vol. 1, chapter 2. 837
B-63MI 119-01 H. E. Zimmerman; in "Molecular Rearrangements," ed. P. deMayo, Wiley, New York, 1963, vol. 1, pp. 345–372. 886
B-63MI 119-02 F. P. Lossing; in "Mass Spectrometry," ed. C. A. McDowell, McGraw-Hill, New York, 1963, p. 442. 894
B-63MI 119-03 S. Bywater; in "Chemistry of Cationic Polymerization," ed. P. H. Plesch, MacMillan, New York, 1963, p. 311. 911

63MI 119-04 B. C. L. Weedon; *Adv. Org. Chem.*, 1963, **1**, 1. 871, 934
63N403 H. Hartmann and A. Meixner; *Naturwissenschaften*, 1963, **50**, 403. 999
63NAT388 F. Gibson and L. M. Jacman; *Nature*, 1963, **198**, 388. 826
63OR(13)91 C. S. Walling and E. S. Huyser; *Org. React.*, 1963, **13**, 91. 919, 924, 948
63OR(13)150 F. W. Stacey and J. F. Harris, Jr.; *Org. React.*, 1963, **13**, 150. 919, 948, 950
63OSC(4)132 W. S. Johnson and W. P. Schneider; *Org. Synth., Coll. Vol.*, 1963, **4**, 132. 881
63OSC(4)136 G. A. Page and D. S. Tarbell; *Org Synth. Coll. Vol.*, 1963, **4**, 136. 80
63OSC(4)261 A. L. McCloskey, G. S. Fonken, R. W. Kluiber and W. S. Johnson; *Org. Synth., Coll. Vol.*, 1963, **4**, 261. 881
63OSC(4)304 A. C. Cope and E. C. Herrick; *Org. Synth. Coll. Vol.*, 1963, **4**, 304. 78
63OSC(4)377 A. C. Cope, D. S. Smith and R. J. Cotter; *Org. Synth., Coll. Vol.*, 1963, **4**, 377. 104
63OSC(4)585 W. R. Brasen and C. R. Hauser; *Org. Synth., Coll. Vol.*, 1963, **4**, 585. 399
63OSC(4)612 A. C. Cope and E. Ciganek; *Org. Synth., Coll. Vol.*, 1963, **4**, 612. 624
63OSC(4)980 W. S. Emerson and T. M. Patrick, Jr.; *Org. Synth., Coll. Vol.*, 1963, **4**, 980. 629
63PCS19 J. W. Lewis and P. P. Lynch; *Proc. Chem. Soc.*, 1963, 19. 443
63PCS219 D. Bryce-Smith, B. J. Wakefield and E. T. Blues; *Proc. Chem. Soc.*, 1963, 219. 3, 6
63PCS253 F. H. Westheimer; *Proc. Chem. Soc.*, 1963, 253. 883
63QR406 S. Trippett; *Q. Rev., Chem. Soc.*, 1963, **17**, 406. 638
63RCP42 J.-M. Conia; *Rec. Chem. Progr.*, 1963, **24**, 42. 881
63RCR59 B. I. Belov and V. V. Kozlov; *Russ. Chem. Rev.* (*Engl. Transl.*), 1963, **32**, 59. 36
63RTC183 J. F. Arens; *Recl. Trav. Chim. Pays-Bas*, 1963, **82**, 183. 999, 1000
63RTC375 H. Hogeveen and W. Drenth; *Rec. Trav. Chim. Pays-Bas*, 1963, **82**, 375. 1122
63RTC394 E. J. Stamhuis and W. Drenth; *Rec. Trav. Chim. Pays-Bas*, 1963, **82**, 394. 1122
63RTC405 G. L. Hekkert and W. Drenth; *Rec. Trav. Chim. Pays-Bas*, 1963, **82**, 405. 1122
63RTC410 H. Hogeveen and W. Drenth; *Rec. Trav. Chim. Pays-Bas*, 1963, **82**, 410. 1122
63RTC741 A. P. ter Borg and H. Kloosterziel; *Recl. Trav. Chim. Pays-Bas*, 1963, **82**, 741. 794
63RTC1040 L. Brandsma, H. E. Wijers and J. H. Arens; *Recl. Trav. Chim. Pays-Bas*, 1963, **82**, 1040. 956
63RTC1123 J. C. Scheer, E. C. Kooyman and F. L. J. Sixma; *Recl. Trav. Chim. Pays-Bas*, 1963, **82**, 1123. 773
63T149 L. D. Bergelson and M. M. Shemyakin; *Tetrahedron*, 1963, **19**, 149. 725, 733
63T715 W. von E. Doering and W. R. Roth; *Tetrahedron*, 1963, **19**, 715. 803
63T1127 L. Caglioti and M. Magi; *Tetrahedron*, 1963, **19**, 1127. 18, 39
63T1625 S. J. Rhoads, J. C. Gilbert, A. W. Decora, T. R. Garland, R. J. Spangler and M. J. Urbigkit; *Tetrahedron*, 1963, **19**, 1625. 869
63TL851 R. A. Finnegan; *Tetrahedron Lett.*, 1963, **4**, 851. 663
63ZN(B)84 G. Märkl; *Z. Naturforsch., Teil B*, 1963, **18**, 84. 157
63ZN(B)1125 G. Köbrich and H. Trapp; *Z. Naturforsch., Teil B*, 1963, **18**, 1125. 1144

64ACS655 G. Asknes and J. Songstad; *Acta Chem. Scand.*, 1964, **18**, 655. 737
64AG(E)1 M. Eigen; *Angew. Chem., Int. Ed. Engl.*, 1964, **3**, 1. 844, 870
64AG(E)138 F. G. Willey; *Angew. Chem., Int. Ed., Engl.*, 1964, **3**, 138. 1075
64AG(E)250 L. D. Bergelson and M. M. Shemyakin; *Angew. Chem., Int. Ed. Engl.*, 1964, **3**, 250. 639
64AG(E)287 M. Schlosser; *Angew. Chem., Int. Ed. Engl.*, 1964, **3**, 287. 113
64AG(E)312 C. Jutz and H. Amschler; *Angew. Chem., Int. Ed. Engl.*, 1964, **3**, 312. 866
64AG(E)362 M. Schlosser; *Angew. Chem., Int. Ed. Engl.*, 1964, **3**, 362. 113
64AG(E)471 R. Juza; *Angew. Chem., Int. Ed. Engl.*, 1964, **3**, 471. 881
64AG(E)525 K. Heusler and J. Kalvoda; *Angew. Chem., Int. Ed. Engl.*, 1964, **3**, 525. 927
64AG(E)651 W. Kirmse and G. Wächterhäuser; *Angew. Chem., Int. Ed. Engl.*, 1964, **3**, 651. 1145
64BSF952 J. Huet; *Bull. Soc. Chim. Fr.*, 1964, 952. 1059
64BSF2477 M. D. D'Engenieres, M. Miocque and J. A. Gautier; *Bull. Soc. Chim. Fr.*, 1964, 2477. 1001
64BSF2485 C. Prévost, P. Miginiac and L. Miginiac-Groizeleau; *Bull. Soc. Chim. Fr.*, 1964, 2485. 127
64CB769 G. Wittig and D. Hellwinkel; *Chem. Ber.*, 1964, **97**, 769. 445
64CB2451 G. N. Schrauzer and P. Glockner; *Chem. Ber.*, 1964, **97**, 2451. 517
64CB3140 G. Schröder, *Chem. Ber.*, 1964, **97**, 3140. 789
64CB3293 N. Petragnani and G. Schill; *Chem. Ber.*, 1964, **97**, 3293. 727
64CB3548 G. Wittig and H.-D. Frommeld; *Chem. Ber.*, 1964, **97**, 3548. 881
64CCC1932 R. Zahradník, J. Michl and J. Koutecký; *Collect. Czech. Chem. Commun.*, 1964, **29**, 1932. 866
64CI(L)153 L. Caglioti and P. Grasselli; *Chem. Ind.* (*London*), 1964, 153. 18, 19
64CI(L)1689 A. N. de Belder and H. Weigel; *Chem. Ind.* (*London*), 1964. 1689. 18
64CJC502 T. Birchall and R. J. Gillespie; *Can. J. Chem.*, 1964, **42**, 502. 897
64CJC572 G. R. Pettit and R. L. Smith; *Can. J. Chem.*, 1964, **42**, 572. 3, 4, 6
64CR(259)2251 M. Bertrand and M. Santelli; *C. R. Hebd. Seances Acad. Sci.*, 1964, **259**, 2251. 1125
64G537 L. Caglioti, P. Grasselli and A. Selva; *Gazz. Chim. Ital.*, 1964, **94**, 537. 630
64HCA194 C. A. Grob and G. Cseh; *Helv. Chim. Acta*, 1964, **47**, 194. 1111
64HCA1590 C. A. Grob, J. Csapilla and G. Cseh; *Helv. Chim. Acta*, 1964, **47**, 1590. 1119
64HCA2425 A. E. Wick, D. Felix, K. Steen and A. Eschenmoser; *Helv. Chim. Acta*, 1964, **47**, 2425. 414, 415, 827
64IEC170 R. L. Banks and G. C. Bailey; *Ind. Eng. Chem. Res.*, 1964, **3**, 170. 674
64IZV1341 L. A. Yanovskaya and V. F. Kucherov; *Izv. Akad. Nauk SSSR, Ser. Khim.*, 1964, 1341 (*Chem. Abstr.*, 1964, **61**, 11 887). 747
64IZV2003 L. D. Bergelson, E. V. Dyatlovitskaya and M. M. Shemyakin; *Izv. Akad. Nauk SSSR, Ser. Khim.*, 1964, 2003 (*Chem. Abstr.*, 1965, **62**, 8998). 725
64JA118 F. Johnson, N. A. Starkovsky, A. C. Paton and A. A. Carlson; *J. Am. Chem. Soc.*, 1964, **86**, 118. 89

64JA322 J. Schulze and F. A. Long; *J. Am. Chem. Soc.*, 1964, **86**, 322. 916

64JA327 F. A. Long and J. Schulze; *J. Am. Chem. Soc.*, 1964, **86**, 327. 916

64JA331 J. Schulze and F. A. Long; *J. Am. Chem. Soc.*, 1964, **86**, 331. 916

64JA478 E. J. Corey, M. Ohno, R. B. Mitra and P. A. Vatakencherry; *J. Am. Chem. Soc.*, 1964, **86**, 478. 389

64JA526 H. D. Hartzler; *J. Am. Chem. Soc.*, 1964, **86**, 526. 1144

64JA733 R. F. Bryan; *J. Am. Chem. Soc.*, 1964, **86**, 733. 901

64JA734 H. H. Freedman and A. E. Young; *J. Am. Chem. Soc.*, 1964, **86**, 734. 909

64JA899 G. S. Hammond and C. D. DeBoer; *J. Am. Chem. Soc.*, 1964, **86**, 899. 812

64JA934 D. G. Farnum; *J. Am. Chem. Soc.*, 1964, **86**, 934. 896

64JA1096 A. P. Krapcho and M. E. Nadel; *J. Am. Chem. Soc.*, 1964, **86**, 1096. 889

64JA1112 V. D. Parker and C. R. Noller; *J. Am. Chem. Soc.*, 1964, **86**, 1110. 117

64JA1360 G. A. Olah, E. B. Baker, J. C. Evans, W. S. Tolgyesi, J. S. McIntyre and I. J. Bastien; *J. Am. Chem. Soc.*, 1964, **86**, 1360. 893, 899

64JA1459 S. W. Tobey and R. West; *J. Am. Chem. Soc.*, 1964, **86**, 1459. 907

64JA1561 J. E. Hofmann, T. J. Wallace and A. Schriesheim; *J. Am. Chem. Soc.*, 1964, **86**, 1561. 600, 612

64JA1639 E. J. Corey and M. Chaykovsky; *J. Am. Chem. Soc.*, 1964, **86**, 1639. 51

64JA1652 E. J. Corey and S. Nozoe; *J. Am. Chem. Soc.*, 1964, **86**, 1652. 382

64JA1776 J. F. Harrod and A. J. Chalk; *J. Am. Chem. Soc.*, 1964, **86**, 1776. 669

64JA1801 H. C. Brown and O. J. Cope; *J. Am. Chem. Soc.*, 1964, **86**, 1801. 658

64JA1871 N. C. Deno and C. U. Pittman, Jr.; *J. Am. Chem. Soc.*, 1964, **86**, 1871. 902, 912

64JA1911 H. P. Waits and G. S. Hammond; *J. Am. Chem. Soc.*, 1964, **86**, 1911. 924

64JA1997 S. K. Malhotra and H. J. Ringold; *J. Am. Chem. Soc.*, 1964, **86**, 1997. 871, 881

64JA2722 P. von R. Schleyer, W. E. Watts and C. A. Cupas; *J. Am. Chem. Soc.*, 1964, **86**, 2722. 380

64JA2832 R. Beugelmans, R. H. Shapiro, L. J. Durham, D. H. Williams, H. Budzikiewicz and C. Djerassi; *J. Am. Chem. Soc.*, 1964, **86**, 2832. 871

64JA3283 H. M. Walborsky, F. J. Impastato and A. E. Young; *J. Am. Chem. Soc.*, 1964, **86**, 3283. 851

64JA4085 L. A. Paquette; *J. Am. Chem. Soc.*, 1964, **86**, 4085. 709

64JA4194 R. C. Fort, Jr. and P. von R. Schleyer; *J. Am. Chem. Soc.*, 1964, **86**, 4194. 900

64JA4195 P. von R. Schleyer, R. C. Fort, Jr., W. E. Watts, M. B. Comisarow and G. A. Olah; *J. Am. Chem. Soc.*, 1964, **86**, 4195. 900

64JA4343 D. J. Pasto and C. C. Cumbo; *J. Am. Chem. Soc.*, 1964, **86**, 4343. 658

64JA4358 C. E. Castro and R. D. Stephens; *J. Am. Chem. Soc.*, 1964, **86**, 4358. 457, 515, 1004, 1005

64JA4727 W. M. Schubert, B. Lamm and J. R. Keeffe; *J. Am. Chem. Soc.*, 1964, **86**, 4727. 896

64JA5017 J. A. Berson and M. Jones, Jr.; *J. Am. Chem. Soc.*, 1964, **86**, 5017. 412

64JA5019 J. A. Berson and M. Jones, Jr.; *J. Am. Chem. Soc.*, 1964, **86**, 5019. 412, 809, 815

64JA5041 A. T. Blomquist and V. J. Hruby; *J. Am. Chem. Soc.*, 1964, **86**, 5041. 723

64JA5361 D. J. Burton and R. L. Johnson; *J. Am. Chem. Soc.*, 1964, **86**, 5361. 429

64JA5445 D. J. Cram and L. Gosser; *J. Am. Chem. Soc.*, 1964, **86**, 5445. 854

64JA5457 D. J. Cram and L. Gosser; *J. Am. Chem. Soc.*, 1964, **86**, 5457. 854, 858, 871, 881

64JA5466 D. J. Cram and R. T. Uyeda; *J. Am. Chem. Soc.*, 1964, **86**, 5466. 797

64JA5503 R. A. Sneen and N. P. Matheny; *J. Am. Chem. Soc.*, 1964, **86**, 5503. 887

64JA5570 E. J. Corey, J. D. Bass, R. LeMahieu and R. B. Mitra; *J. Am. Chem. Soc.*, 1964, **86**, 5570. 498

64JA5682 G. A. Olah and M. B. Comisarow; *J. Am. Chem. Soc.*, 1964, **86**, 5682. 902

64JCP(40)213 E. L. Cochran, F. J. Adrian and V. A. Bowers; *J. Chem. Phys.*, 1964, **40**, 213. 322

64JCP(40)2480 R. Hoffmann; *J. Chem. Phys.*, 1964, **40**, 2480. 1109

64JCS543 S. Trippett and S. T. D. Gough; *J. Chem. Soc.*, 1964, 543. 1079

64JCS959 R. J. Ellis and R. M. Frey; *J. Chem. Soc.*, 1964, 959. 801

64JCS1151 O. M. Behr, G. Eglinton, I. A. Lardy and R. A. Raphael; *J. Chem. Soc.*, 1964, 1151. 1064

64JCS3080 K. E. Lewis and H. Steiner; *J. Chem. Soc.*, 1964, 3080. 419

64JCS3897 D. B. Bigley; *J. Chem. Soc.*, 1964, 3897. 772

64JCS4578 K. Takeda, H. Minato and M. Ishikawa; *J. Chem. Soc.*, 1964, 4578. 813

64JCS4804 R. W. Bott, C. Eaborn and P. M. Greasley; *J. Chem. Soc.*, 1964, 4804. 448

64JCS5578 R. J. Ellis and R. M. Frey; *J. Chem. Soc.*, 1964, 5578. 795, 801

64JOC221 V. J. Traynelis and W. L. Hergenrother; *J. Org. Chem.*, 1964, **29**, 221. 881

64JOC423 E. Le Goff and R. B. LaCount; *J. Org. Chem.*, 1964, **29**, 423. 537

64JOC921 J. W. Wilt and J. A. Lundquist; *J. Org. Chem.*, 1964, **29**, 921. 934

64JOC940 K. C. Brannock, R. D. Burpitt and J. G. Thweatt; *J. Org. Chem.*, 1964, **29**, 940. 355

64JOC1261 A. R. Utke and R. T. Sanderson; *J. Org. Chem.*, 1964, **29**, 1261. 889

64JOC1316 R. D. Clark and G. K. Helmkamp; *J. Org. Chem.*, 1964, **29**, 1316. 636

64JOC1872 O. F. Beumel and R. F. Harris; *J. Org. Chem.*, 1964, **29**, 1872. 1037

64JOC2317 G. A. Olah and S. J. Kuhn; *J. Org. Chem.*, 1964, **29**, 2317. 462

64JOC2325 P. E. Peterson and C. Casey; *J. Org. Chem.*, 1964, **29**, 2325. 12

64JOC2699 C. Walling and L. Bollyky; *J. Org. Chem.*, 1964, **29**, 2699. 595, 779

64JOC2781 K. Sisido, K. Utimoto and T. Isida; *J. Org. Chem.*, 1964, **29**, 2781. 884

64JOC2788 J. W. Wilt and W. J. Wagner; *J. Org. Chem.*, 1964, **29**, 2788. 632

64JOC2870 Z. B. Papanastassiou and R. J. Bruni; *J. Org. Chem.*, 1964, **29**, 2870. 3, 6

64JOC3042 D. G. Holland, G. J. Moore and C. Tamborski; *J. Org. Chem.*, 1964, **29**, 3042. 426

64JOC3131 D. C. Dittmer and F. A. Davis; *J. Org. Chem.*, 1964, **29**, 3131. 784

64JOC3258 G. K. Helmkamp and D. J. Pettitt; *J. Org. Chem.*, 1964, **29**, 3258. 592

64JOC3327 H. O. House, V. K. Jones and G. A. Frank; *J. Org. Chem.*, 1964, **29**, 3327. 733

64JOM(1)420 R. W. Bott, C. Eaborn and D. R. M. Walton; *J. Organomet. Chem.*, 1964, **1**, 420. 1000

64JOM(1)476 C. R. Krüger and E. G. Rochow; *J. Organomet. Chem.*, 1964, **1**, 476. 881

64JPR143	G. Drefahl, D. Lorenz and G. Schnitt; *J. Prakt. Chem.*, 1964, **23**, 143.	745
64MI 103-01	G. C. Bond and P. B. Wells; *Adv. Catalysis*, 1964, **15**, 92.	80
B-64MI 107-01	L. Schmerling; in "Friedel–Crafts and Related Reactions," ed. G. A. Olah, Wiley, New York, 1964, vol. 2, p. 1075.	294
64MI 111-01	J. E. Hofman and A. Schriesheim; in "Friedel–Crafts and Related Reactions," ed. G. A. Olah, Interscience, New York, 1964, vol II, pt 1, ch 19.	475
64MI 112-01	G. C. Bond and P. B. Wells; *Adv. Catal.*, 1964, **15**, 91.	505
64MI 116-01	W. Heffe; Dissertation, University of Berlin, 1937 (quoted by G. Wittig; *Pure Appl. Chem.*, 1964, **9**, 249).	737
64MI 116-02	A. Maccioni and M. Secci; *Rend. Seminario Fac. Sci. Univ. Cagliari*, 1964, **34**, 328.	738
B-64MI 118-01	K. Mackenzie; in "The Chemistry of Alkenes," ed. S. Patai, Interscience, New York, 1964, vol. 1, p. 415.	795
B-64MI 118-02	R. H. Dewolfe and W. G. Young; in "The Chemistry of Alkenes," ed. S. Patai, Interscience, New York, 1964, vol. 1, chapter 10.	837
B-64MI 119-01	S. Patai and Z. Rappoport; in "The Chemistry of Alkenes," ed. S. Patai, Wiley, New York, 1964, pp. 469–584.	888
B-64MI 119-02	G. Sosnovsky; "Free Radical Reactions in Preparative Organic Chemistry," MacMillan, New York, 1964.	919
B-64MI 121-01	L. Schmerling; in "Friedel–Crafts and Related Reactions," ed. C. A. Olah, Interscience, New York, 1964, vol. 2, ch. 25.	1003
B-64MI 122-01	S. Patai and Z. Rappoport; in "The Chemistry of Alkenes," Vol. 1, ed., S. Patai, Wiley-Interscience, New York, 1964.	1109
64PAC271	L. D. Bergelson and M. M. Shemyakin; *Pure Appl. Chem.*, 1964, **9**, 271.	721
64PAC539	W. G. Dauben and W. T. Wipke; *Pure Appl. Chem.*, 1964, **9**, 539.	794
64PCS108	A. Ledwith and N. McFarlane; *Proc. Chem. Soc.*, 1964, 108.	881
64PCS147	D. M. Brouwer and E. L. Mackor; *Proc. Chem. Soc.*, 1964, 147.	893
64QR406	S. Trippett; *Q. Rev., Chem. Soc.*, 1964, **17**, 406.	104
64RCR1	A. A. Petrov and A. V. Fedorova; *Russ. Chem. Rev.*, 1964, **33**, 1.	953
64RTC307	L. Brandsma; *Recl. Trav. Chim. Pays-Bas*, 1964, **83**, 307.	999
64RTC456	P. P. Montijn and L. Brandsma; *Recl. Trav. Chim. Pays-Bas*, 1964, **83**, 456.	1072
64RTC1173	J. L. M. A. Schlatmann, J. Pot and E. Havinga; *Recl. Trav. Chim. Pays-Bas*, 1964, **83**, 1173.	403, 795
64T1259	W. G. Dauben and P. Lang; *Tetrahedron*, 1964, **20**, 1259.	259
64T1695	Y. Inoue, T. Sugata and H. M. Walborsky; *Tetrahedron*, 1964, **20**, 1695.	268
64T2177	T. L. Jacobs, D. Dankner and S. Singer; *Tetrahedron*, 1964, **20**, 2177.	954
64TL983	H. Hoffmann and H. Förster; *Tetrahedron Lett.*, 1964, **5**, 983.	999
64TL1009	W. R. Roth; *Tetrahedron Lett.*, 1964, **5**, 1009.	794
64TL1191	J. Cunningham, R. Gigg and C. D. Warren; *Tetrahedron Lett*, 1964, **5**, 1191.	882
64TL2039	J. W. Lewis and A. A. Pearce; *Tetrahedron Lett.*, 1964, **5**, 2039.	443, 660
64TL2191	M. Hanack and J. Häffner; *Tetrahedron Lett.*, 1964, **5**, 2191.	1125
64TL2467	L. Horner and W. Klink; *Tetrahedron Lett.*, 1964, **5**, 2467.	753
64TL2503	C. Cupas, W. E. Watts and P. von R. Schleyer; *Tetrahedron Lett.*, 1964, **5**, 2503.	811
64TL3695	P. E. Eaton; *Tetrahedron Lett.*, 1964, **5**, 3695.	498

65AG95	G. Köbrich and W. Drischel; *Angew. Chem.*, 1965, **77**, 95.	1145
65AG967	K. Bott; *Angew. Chem.*, 1965, **77**, 967.	186
65AG(E)10	A. W. Krebs; *Angew. Chem., Int. Ed. Engl.*, 1965, **4**, 10.	907
65AG(E)49	G. Köbrich; *Angew. Chem. Int. Ed. Engl.*, 1965, **4**, 49.	1144
65AG(E)354	W. Grimme, H. Hoffmann and E. Vogel; *Angew. Chem., Int. Ed. Engl.*, 1965, **4**, 354.	910
65AG(E)583	H. J. Bestmann; *Angew. Chem., Int. Ed. Engl.*, 1965, **4**, 583.	727
65AG(E)689	M. Schlosser and K. F. Christmann; *Angew. Chem. Int. Ed. Engl.*, 1965, **4**, 689.	721
65AG(E)873	H. Volz and H. W. Schnell; *Angew. Chem., Int. Ed. Engl.*, 1965, **4**, 873.	894
65AG(E)1075	E. J. Corey and D. Seebach; *Angew. Chem., Int. Ed. Engl.*, 1965, **4**, 1075.	210
65AG(E)1080	J. Gosselck, L. Béress, H. Schenk and G. Schmidt; *Angew. Chem., Int. Ed., Engl.*, 1965, **4**, 1080.	1072
65AX437	A. H. Gomes de Mesquita, C. H. MacGillavry and K. Eriks; *Acta Crystallogr.*, 1965, **18**, 437.	897
65BAU1434	N. A. Nesmeyanov, V. V. Pravdina and O. A. Reutov; *Bull. Acad. Sci. USSR, Div. Chim. Sci.*, 1965, 1434.	738
65BSF35	P. Cadiot and M. Le Quan; *Bull. Soc. Chim. Fr.*, 1965, 35.	1000
65BSF859	J. F. Normant; *Bull. Soc. Chim. Fr.*, 1965, 859.	1000
65BSF1872	H. Normant and T. Cuvigny; *Bull. Soc. Chim. Fr.*, 1965, 1872.	1000
65CB756	W. Grimme; *Chem. Ber.*, **98**, 756, 1965.	795
65CB824	W. Ziegenbein and W. M. Schneider; *Chem. Ber.*, 1965, **98**, 824.	453
65CB2611	R. Kuhn and D. Rewicki; *Chem. Ber.*, 1965, **98**, 2611.	954
65CB3236	M. Fischer, Z. Pelah, D. H. Williams and C. Djerassi; *Chem. Ber.*, 1965, **98**, 3236.	630
65CB3311	A. Schönberg, E. Singer, E. Frese and K. Praefche; *Chem. Ber.*, 1965, **98**, 3311.	1040
65CB4022	W. Kirmse and K.-H. Pook; *Chem. Ber.*, 1965, **98**, 4022.	631
65CC7	D. Elad and R. D. Youssefyeh; *J. Chem. Soc., Chem. Commun.*, 1965, 7.	925
65CC29	I. D. Entwhistle and R. A. W. Johnstone; *J. Chem. Soc., Chem. Commun.*, 1965, 29.	142, 595
65CC71	E. E. Van Tamelen, M. Davis and M. F. Deem; *J. Chem. Soc., Chem. Commun.*, 1965, 71.	82
65CC131	J. F. Young, J. A. Osborn, F. H. Jardine and G. Wilkinson; *J. Chem. Soc. Chem. Commun.*, 1965, 131.	73

65CC226	J. M. Brown; *J. Chem. Soc., Chem. Commun.*, 1965, 226.	811
65CC376	J. M. Brown and J. L. Occolowitz; *J. Chem. Soc., Chem. Commun.*, 1965, 376.	865
65CI(L)560	J. A. Osborn, F. H. Jardine, G. F. Young and G. Wilkinson; *Chem. Ind. (London)*, 1965, 560.	73
65CI(L)1726	A. W. Johnson and J. O. Martin; *Chem. Ind. (London)*, 1965, 1726.	738, 741
65CI(L)1768	T. J. Wallace, J. E. Hofmann and A. Schriesheim; *Chem. Ind. (London)*, 1965, 1768.	590
65CJC1689	F. L. M. Pattison, D. A. V. Peters and F. H. Dean; *Can. J. Chem.*, 1965, **43**, 1689.	2, 6
65CJC2744	T. S. Sorensen; *Can. J. Chem.*, 1965, **43**, 2744.	912
65CR(261)1992	H. Demarne; *C. R. Hebd. Seances Acad. Sci.*, 1965, **261**, 1992.	1037
65HCA1985	M. Rey and A. S. Dreiding; *Helv. Chim. Acta*, 1965, **48**, 1985.	826
65IZV1474	N. A. Nesmeyanov, V. V. Pravdina and O. A. Reutov; *Izv. Akad. Nauk SSSR, Ser. Khim.*, 1965, 1474.	738
65IZV1504	L. A. Yanovskaya and V. F. Kucherov; *Izv. Akad. Nauk SSSR, Ser. Khim.*, 1965, 1504 (*Chem. Abstr.*, 1965, **63**, 16 387).	747
65JA82	S. Boatman, T. M. Harris and C. R. Hauser; *J. Am. Chem. Soc.*, 1965, **87**, 82.	881
65JA275	G. Stork, P. Rosen, N. Goldman, R. V. Coombs and J. Tsuji; *J. Am. Chem. Soc.*, 1965, **87**, 275.	130, 889
65JA376	S. Winstein, E. Clippinger, R. Howe and E. Vogelfanger; *J. Am. Chem. Soc.*, 1965, **87**, 376.	905
65JA382	E. C. Steiner and J. M. Gilbert; *J. Am. Chem. Soc.*, 1965, **87**, 382.	869
65JA384	A. Streitwieser, Jr., J. I. Brauman, J. H. Hammons and A. H. Pudjaatmaka; *J. Am. Chem. Soc.*, 1965, **87**, 384.	869
65JA395	R. B. Woodward and R. Hoffmann; *J. Am. Chem. Soc.*, 1965, **87**, 395.	417, 418, 419, 422
65JA505	E. W. Garbisch, Jr.; *J. Am. Chem. Soc.*, 1965, **87**, 505.	870
65JA659	J. A. Smith, H. Shechter, J. Bayless and L. Friedman; *J. Am. Chem. Soc.*, 1965, **87**, 659.	631
65JA863	D. Y. Curtin, J. A. Kampmeier and B. R. O'Connor; *J. Am. Chem. Soc.*, 1965, **87**, 863.	1118
65JA914	P. Carter, R. Howe and S. Winstein; *J. Am. Chem. Soc.*, 1965, **87**, 914.	868
65JA915	R. Howe and S. Winstein; *J. Am. Chem. Soc.*, 1965, **87**, 915.	868
65JA916	T. Fukunaga; *J. Am. Chem. Soc.*, 1965, **87**, 916.	868
65JA935	L. Friedman and H. Shecter; *J. Am. Chem. Soc.*, 1965, **87**, 935.	39
65JA1103	G. A. Olah; *J. Am. Chem. Soc.*, 1965, **87**, 1103.	897
65JA1142	K. L. Erickson and J. Wolinsky; *J. Am. Chem. Soc.*, 1965, **87**, 1142.	1144
65JA1150	A. Viola and L. Levasseur; *J. Am. Chem. Soc.*, 1965, **87**, 1150.	412
65JA1345	E. J. Corey and M. Chaykovsky; *J. Am. Chem. Soc.*, 1965, **87**, 1345.	142, 881
65JA1353	E. J. Corey and M. Chaykovsky; *J. Am. Chem. Soc.*, 1965, **87**, 1353.	304
65JA1381	H. G. Richey, Jr., J. C. Philips and L. E. Rennick; *J. Am. Chem. Soc.*, 1965, **87**, 1381.	904
65JA1526	D. S. Matteson and J. D. Liedtke; *J. Am. Chem. Soc.*, 1965, **87**, 1526.	660
65JA1820	O. W. Webster; *J. Am. Chem. Soc.*, 1965, **87**, 1820.	862
65JA1941	E. A. LaLancette and R. E. Benson; *J. Am. Chem. Soc.*, 1965, **87**, 1941.	863
65JA1984	E. C. Taylor and A. McKillop; *J. Am. Chem. Soc.*, 1965, **87**, 1984.	81
65JA2153	N. C. Deno, C. U. Pittman, Jr. and J. O. Turner; *J. Am. Chem. Soc.*, 1965, **87**, 2153.	912
65JA2157	N. C. Deno, F. A. Kish and H. J. Peterson; *J. Am. Chem. Soc.*, 1965, **87**, 2157.	896
65JA2295	D. S. Noyce, M. A. Matesich, M. D. Schiavelli and P. E. Peterson; *J. Am. Chem. Soc.*, 1965, **87**, 2295.	1122
65JA2511	R. B. Woodward and R. Hoffmann; *J. Am. Chem. Soc.*, 1965, **87**, 2511.	400
65JA2997	G. A. Olah, M. B. Comisarow, C. A. Cupas and C. U. Pittman, Jr.; *J. Am. Chem. Soc.*, 1965, **87**, 2997.	893, 897
65JA2998	C. U. Pittman, Jr. and G. A. Olah; *J. Am. Chem. Soc.*, 1965, **87**, 2998.	894
65JA3000	N. C. Deno, J. S. Liu, J. O. Turner, D. N. Lincoln and R. E. Fruit, Jr.; *J. Am. Chem. Soc.*, 1965, **87**, 3000.	894
65JA3228	S. K. Malhotra and H. J. Ringold; *J. Am. Chem. Soc.*, 1965, **87**, 3228.	795, 871, 881
65JA3244	S. Bank, A. Schriesheim and C. A. Rowe, Jr.; *J. Am. Chem. Soc.*, 1965, **87**, 3244.	882
65JA3267	S. Winstein, H. D. Kaesz, C. G. Kreiter and E. C. Friedrich; *J. Am. Chem. Soc.*, 1965, **87**, 3267.	917
65JA3277	E. E. van Tamelen and M. A. Schwartz; *J. Amer. Chem. Soc.*, 1965, **87**, 3277.	174
65JA3413	C. Walling and M. J. Gibian; *J. Am. Chem. Soc.*, 1965, **87**, 3413.	924, 925
65JA3996	K. M. Shumate, P. N. Neuman and G. J. Fonken; *J. Am. Chem. Soc.*, 1965, **87**, 3996.	418
65JA4017	H. G. Richey, Jr., L. E. Rennick, A. S. Kushner, J. M. Richey and J. C. Philips; *J. Am. Chem. Soc.*, 1965, **87**, 4017.	904
65JA4107	J. Herling, J. Shabtai and E. Gil-Av; *J. Am. Chem. Soc.*, 1965, **87**, 4107.	795
65JA4501	K. G. Untch and D. J. Martin; *J. Am. Chem. Soc.*, 1965, **87**, 4501.	808
65JA4506	L. Skattebol and S. Solomon; *J. Am. Chem. Soc.*, 1965, **87**, 4506.	808
65JA4533	N. C. Deno, H. G. Richey, Jr., J. S. Liu, D. N. Lincoln and J. O. Turner; *J. Am. Chem. Soc.*, 1965, **87**, 4533.	912, 913
65JA4585	M. F. Hawthorne and M. Reintjes; *J. Am. Chem. Soc.*, 1965, **87**, 4585.	273
65JA5075	T. S. Sorensen; *J. Am. Chem. Soc.*, 1965, **87**, 5075.	912
65JA5115	B. Miller; *J. Am. Chem. Soc.*, 1965, **87**, 5115.	809
65JA5264	M. J. Jorgenson and C. H. Heathcock; *J. Am. Chem. Soc.*, 1965, **87**, 5264.	802
65JA5492	F. Johnson and S. K. Malhotra; *J. Am. Chem. Soc.*, 1965, **87**, 5492.	871
65JA5493	S. K. Malhotra and F. Johnson; *J. Am. Chem. Soc.*, 1965, **87**, 5493.	871
65JA5513	S. K. Malhotra and F. Johnson; *J. Am. Chem. Soc.*, 1965, **87**, 5513.	871
65JA5632	C. U. Pittman, Jr. and G. A. Olah; *J. Am. Chem. Soc.*, 1965, **87**, 5632.	904
65JCP(43)2704	R. W. Fessenden and R. H. Schuler; *J. Chem. Phys.*, 1965, **43**, 2704.	923
65JCS384	R. W. Bott, C. Eaborn and D. R. M. Walton; *J. Chem. Soc.*, 1965, 384.	1122

65JCS894 D. E. Ames, A. N. Covell and T. G. Goodburn; *J. Chem. Soc.*, 1965, 894. 1001

65JCS2513 M. Chérest, H. Felkin, J. Sicher, F. Sipos and M. Tichy; *J. Chem. Soc.*, 1965, 2513. 389

65JCS4120 F. G. Mann and M. J. Pragrell; *J. Chem. Soc.*, 1965, 4120. 445

65JCS4355 P. Bamfield, A. Crabtree and A. W. Johnson; *J. Chem. Soc.*, 1965, 4355. 537

65JCS5045 R. D. Chambers, F. G. Drakesmith and W. K. R. Musgrave; *J. Chem. Soc.*, 1965, 5045. 428

65JCS6019 A. Gandini and P. H. Plesch; *J. Chem. Soc.*, 1965, 6019. 896

65JCS6202 D. B. Bigley and J. C. Thurman; *J. Chem. Soc.*, 1965, 6202. 772

65JCS6669 A. J. Hubert; *J. Chem. Soc.*, 1965, 6669. 657

65JCS6718 V. Gold and M. A. Kessick; *J. Chem. Soc.*, 1965, 6718. 896

65JCS6784 D. K. Black and S. R. Landor; *J. Chem. Soc.*, 1965, 6784. 808

65JCS(B)1752 J. A. Feather and V. Gold; *J. Chem. Soc. (B)*, 1965, 1752. 869, 871

65JOC181 W. L. Meyer, A. P. Lobo and E. T. Marquis; *J. Org. Chem.*, 1965, **30**, 181. 884

65JOC505 A. K. Bose and R. T. Dahill, Jr.; *J. Org. Chem.*, 1965, **30**, 505. 747

65JOC1015 P. A. Chopard, R. J. G. Searle and F. H. Devitt; *J. Org. Chem.*, 1965, **30**, 1015. 1079

65JOC1154 J. N. Burnett and A. L. Underwood; *J. Org. Chem.*, 1965, **30**, 1154. 103

65JOC1258 W. H. Glaze, J. Lin and E. G. Felton; *J. Org. Chem.*, 1965, **30**, 1258. 663

65JOC1291 J. H. Burckhalter and B. A. Brown; *J. Org. Chem.*, 1965, **30**, 1291. 889

65JOC1296 D. E. Bissing; *J. Org. Chem.*, 1965, **30**, 1296. 723

65JOC1341 H. O. House and B. M. Trost; *J. Org. Chem.*, 1965, **30**, 1341. 871, 872, 881

65JOC1473 O. H. Wheeler and H. N. Batlle de Pabon; *J. Org. Chem.*, 1965, **30**, 1473. 731

65JOC1593 K. Ichikawa, K. Nishimura and S. Takayama; *J. Org. Chem.*, 1965, **30**, 1593. 671

65JOC1752 J. Martel, E. Toromanoff and C. Huynh; *J. Org. Chem.*, 1965, **30**, 1752. 503

65JOC1808 T. L. Jacobs and D. M. Fenton; *J. Org. Chem.*, 1965, **30**, 1808. 1114

65JOC1939 D. J. Peterson and H. R. Hays; *J. Org. Chem.*, 1965, **30**, 1939. 157

65JOC2502 H. O. House and B. M. Trost; *J. Org. Chem.*, 1965, **30**, 2502. 130, 871, 872, 881

65JOC2513 H. O. House, B. M. Trost, R. W. Magin, R. G. Carlson, R. W. Franck and G. H. Rasmusson; *J. Org. Chem.*, 1965, **30**, 2513. 882

65JOC2934 K. G. Hampton and C. R. Hauser; *J. Org. Chem.*, 1965, **30**, 2934. 879

65JOC2937 W. G. Kenyon, E. M. Kaiser and C. R. Hauser; *J. Org. Chem.*, 1965, **30**, 2937. 881

65JOC2983 K. L. Mikolajczak, M. O. Bagby, R. B. Bates and I. A. Wolff; *J. Org. Chem.*, 1965, **30**, 2983. 955

65JOC3212 S. J. Rhoads and C. Pryde; *J. Org. Chem.*, 1965, **30**, 3212. 869

65JOC3888 A. R. Lepley and R. H. Becker; *J. Org. Chem.*, 1965, **30**, 3888. 399

65JOC3897 W. Z. Heldt; *J. Org. Chem.*, 1965, **30**, 3897. 484

65JOC4135 W. G. Kenyon, E. M. Kaiser and C. R. Hauser; *J. Org. Chem.*, 1965, **30**, 4135. 881

65JOC4230 J. J. Beereboom; *J. Org. Chem.*, 1965, **30**, 4230. 381

65JOM(4)217 C. Eaborn and D. R. M. Walton; *J. Organomet. Chem.*, 1965, **4**, 217. 999, 1000

65JOM(4)349 L. I. Zakharkin, O. Y. Okhlobystin and B. N. Strunin; *J. Organomet. Chem.*, 1965, **4**, 349. 665

65JPC1278 A. Amano and M. Uchiyama; *J. Phys. Chem.*, 1965, **69**, 1278. 411

65LA(681)196 K. Schreiber and H. Rönsch; *Justus Liebigs Ann. Chem.*, 1965, **681**, 196. 2, 10

65LA(688)28 W. R. Roth and J. König; *Justus Liebigs Ann. Chem.*, 1965, **688**, 28. 795

65LA(689)1 R. Kuhn, Herbert Fischer, D. Rewicki and Hans Fischer; *Justus Liebigs Ann. Chem.*, 1965, **689**, 1. 866

65MI 102-01 L. I. Zamyshlyaeva, T. A. Slovokhotova and A. A. Balandin; *Vestn. Mosk. Univ., Ser. II, Khim.*, 1965, **20(4)**, 39 (*Chem. Abstr.*, 1965, **63**, 17 944b). 56

B-65MI 103-01 R. L. Augustine; "Catalytic Hydrogenation," Marcel Dekker, New York, 1965. 71, 87

65MI 103-02 C. L. Perrin; *Prog. Phys. Org. Chem.*, 1965, **3**, 165. 103

B-65MI 109-01 R. Kh. Friedlina; in "Advances in Free-Radical Chemistry," ed. G. H. Williams, Academic Press, 1965, vol. 1, p. 211. 395

B-65MI 112-01 R. L. Augustine; "Catalytic Hydrogenation," Arnold, London, 1965, pp. 69–71. 502

65MI 114-01 B. M. Mikhailov, V. A. Dorokhov and M. V. Mostovoi; *Probl. Organ. Sinteza, Akad. Nauk SSSR, Otdel. Obshchei, Tekhn. Khim.*, 1965, 228 (*Chem. Abstr.*, 1965, **64**, 14 204). 657

65MI 116-01 Y. T. Huang, W. Y. Ting and H. Sheng; *Acta. Chim. Sinica Engl. Ed.*, 1965, **31**, 38. 738

B-65MI 118-01 D. J. Cram; "Fundamentals of Carbanion Chemistry," Academic Press, New York, 1965, chapter V. 795

B-65MI 119-01 D. J. Cram; "Fundamentals of Carbanion Chemistry," Academic Press, New York, 1965. 844

B-65MI 119-02 D. J. Cram; "Fundamentals of Carbanion Chemistry," Academic Press, New York, 1965, pp. 85–105. 854, 855

B-65MI 119-03 J. N. Murrell, S. F. A. Kettle and J. M. Tedder; "Valence Theory," 2nd edn., Wiley, London, 1965, p. 283. 860

B-65MI 119-04 D. J. Cram; "Fundamentals of Carbanion Chemistry," Academic Press, New York, 1965, pp. 1–84. 869

B-65MI 119-05 D. J. Cram; "Fundamentals of Carbanion Chemistry," Academic Press, New York, 1965, chapter 6. 886

65MI 119-06 A. Streitwieser, Jr. and J. H. Hammons; *Prog. Phys. Org. Chem.*, 1965, **3**, 41. 844, 848

65MI 119-07 E. Grunwald; *Prog. Phys. Org. Chem.*, 1965, **3**, 317. 844

65MI 119-08 A. Streitwieser, Jr.; *Prog. Phys. Org. Chem.*, 1965, **3**, 41. 854, 856, 857, 858

B-65MI 122-01 D. J. Cram; "Fundamentals of Carbanion Chemistry," Academic Press, New York, 1965. 1088

65OR(14)270 A. Maercker; *Org. React.*, 1965, **14**, 270. 638

65OS(45)25 A. T. Nielsen and W. R. Carpenter; *Org. Synth.*, 1965, **45**, 25. 884

65OS(45)39 I. D. Campbell and G. Eglington; *Org. Synth.*, 1965, **45**, 39. 1103

65OS(45)89 M. R. V. Sahyun and D. J. Cram; *Org. Synth.*, 1965, **45**, 89. 881

65QR274 D. H. Reid; *Q. Rev., Chem. Soc.*, 1965, **19**, 274. 866, 910

65RTC31 J. H. van Boom, P. P. Montijn, L. Brandsma and J. F. Arens; *Rec. Trav. Chim. Pays-Bas*, 1965, **84**, 31. 955, 1125
65RTC241 A. P. Ter Borg and H. Kloosterziel; *Recl. Trav. Chim. Pays-Bas*, 1965, **84**, 241. 404
65RTC1200 L. Brandsma, A. Schaap and J. F. Arens; *Recl. Trav. Chim. Pays-Bas*, 1965, **84**, 1200. 1037
65T1001 K. J. Crowley; *Tetrahedron*, 1965, **21**, 1001. 418
65T2329 S. McLean and P. Haynes; *Tetrahedron*, 1965, **21**, 2329. 402, 794
65TH N. Friedman; Ph.D. Thesis, Pennsylvania State University, 1965. 913
65TL287 G. L. Closs and R. B. Larrabee; *Tetrahedron Lett.*, 1965, **6**, 287. 848
65TL377 D. S. Glass, J. W. H. Watthey and S. Winstein; *Tetrahedron Lett.*, 1965, **6**, 377. 421
65TL385 E. N. Marvell, G. Caple and B. Schatz; *Tetrahedron Lett.*, 1965, **6**, 385. 419, 420
65TL391 E. Vogel, W. Grimme and E. Dinné; *Tetrahedron Lett.*, 1965, **6**, 391. 419, 421
65TL875 M. Hanack, J. Häffner and I. Herterich; *Tetrahedron Lett.*, 1965, **6**, 875. 1123
65TL1359 J. F. Harris, Jr.; *Tetrahedron Lett.*, 1965, **6**, 1359. 808
65TL1953 P. Cadiot, W. Chodkiewicz and A. Sevin; *Tetrahedron Lett.*, 1965, **6**, 1953. 1025
65TL2899 S. Searles, Jr. and R. E. Wann; *Tetrahedron Lett.*, 1965, **6**, 2899. 623
65TL3955 R. C. Cookson, V. N. Gogte, J. Hudec and N. A. Mirza; *Tetrahedron Lett.*, 1965, **6**, 3955. 794, 799, 809
65TL4399 V. Calò, G. Melloni, G. Modena and G. Scorrano; *Tetrahedron Lett.*, 1965, **6**, 4399. 1123
65TL4405 L. DiNunno, G. Melloni, G. Modena and G. Scorrano; *Tetrahedron Lett.*, 1965, **6**, 4405. 1123

66AG1057 H. Prinzbach, M. Arguëlles and E. Druckrey; *Angew. Chem.*, 1966, **78**, 1057. 522
66AG(E)126 M. Schlosser and K. F. Christmann; *Angew. Chem., Int. Ed. Engl.*, 1966, **5**, 126. 729, 730
66AG(E)420 K. Friedrich; *Angew. Chem., Int. Ed. Engl.*, 1966, **5**, 420. 862
66AG(E)604 W. Grimme, M. Kaufhold, U. Dettmeier and E. Vogel; *Angew. Chem., Int. Ed. Engl.*, 1966, **5**, 604. 910
66AG(E)667 M. Schlosser, G. Müller and K. F. Christmann; *Angew. Chem., Int. Ed. Engl.*, 1966, **5**, 667. 727, 728
66AG(E)899 H. Paulsen and K. Todt; *Angew. Chem., Int. Ed. Engl.*, 1966, **5**, 899. 876
66AG(E)1039 H. Prinzbach, M. Arguëlles and E. Druckrey; *Angew. Chem., Int. Ed. Engl.*, 1966, **5**, 1039. 521
66AJC521 T. J. Broxton, Y. C. Mac, A. J. Parker and M. Ruane; *Aust. J. Chem.*, 1966, **19**, 521. 178
66APO(4)305 G. A. Olah and C. U. Pittman, Jr.; *Adv. Phys. Org. Chem.*, 1966, **4**, 305–347. 891
66BCJ1178 K. Sonogashira and N. Hagihara; *Bull. Chem. Soc. Jpn.*, 1966, **39**, 1178. 456
66BSF281 J. M. Conia and P. Le Perchec; *Bull. Chem. Soc. Fr.*, 1966, 281. 808
66BSF971 W. Smadja, M. Simalty-Siemiatycki and C. Prevost; *Bull. Soc. Chim. Fr.*, 1966, 971. 999
66BSF1888 J. Normant; *Bull. Soc. Chim. Fr.*, 1963, 1888. 464
66BSF2885 J. Rauss-Godineau, W. Chodkiewicz and P. Cadiot; *Bull. Soc. Chim. Fr.*, 1966, 2885. 986
66BSF3016 G. Pourcelot and P. Cadiot; *Bull. Soc. Chim. Fr.*, 1966, 3016. 956
66BSF3024 G. Pourcelot and P. Cadiot; *Bull. Soc. Chim. Fr.*, 1966, 3024. 956
66C229 W. R. Roth; *Chimia*, 1966, **20**, 229. 402
66CB227 K. Kindler and K. Lührs; *Chem. Ber.*, 1966, **99**, 227. 47
66CB1393 R. Mayer, S. Scheithauer and D. Kunz; *Chem. Ber.*, 1966, **99**, 1393. 24
66CB1642 F. Bohlmann, K.-M. Klein and C. Arndt; *Chem. Ber.*, 1966, **99**, 1642. 1142
66CC757 R. Baker and M. J. Spillett; *J. Chem. Soc., Chem. Commun.*, 1966, 757. 240
66CC775 W. M. Horspool, J. M. Tedder and Z. U. Din; *J. Chem. Soc., Chem. Commun.*, 1966, 775. 815
66CC880 C. G. Chin, H. W. Cuts and S. Masamune; *J. Chem. Soc., Chem. Commun.*, 1966, 880. 884
66CR(C)289 H. Demarne and M. G. Champetier; *C. R. Hebd. Seances Acad. Sci., Ser. C.*, 1966, 289. 1037
66CRV161 A. E. Pohland and W. R. Benson; *Chem. Rev.*, 1966, **66**, 161. 469
66HCA1145 W. Büchner and R. Dufaux; *Helv. Chim. Acta.*, 1966, **49**, 1145. 38
66JA78 P. T. Lansbury, V. A. Pattison, J. D. Sidler and J. B. Bieber; *J. Am. Chem. Soc.*, 1966, **88**, 78. 396, 397
66JA120 W. M. Schubert and B. Lamm; *J. Am. Chem. Soc.*, 1966, **88**, 120. 896
66JA198 M. Sundaralingam and L. H. Jensen; *J. Am. Chem. Soc.*, 1966, **88**, 198. 909
66JA307 T. E. Hogen-Esch and J. Smid; *J. Am. Chem. Soc.*, 1966, **88**, 307. 859
66JA318 T. E. Hogen-Esch and J. Smid; *J. Am. Chem. Soc.*, 1966, **88**, 318. 859
66JA460 R. E. Dessy, W. Kitching, T. Psarras, R. Salinger, A. Chen and T. Chivers; *J. Am. Chem. Soc.*, 1966, **88**, 460. 846
66JA604 C. E. Keller and R. Pettit; *J. Am. Chem. Soc.*, 1966, **88**, 604. 917
66JA606 C. E. Keller and R. Pettit; *J. Am. Chem. Soc.*, 1966, **88**, 606. 917
66JA1073 K. M. Shumate and G. J. Fonken; *J. Am. Chem. Soc.*, 1966, **88**, 1073. 419
66JA1272 R. Waack, M. A. Doran, E. B. Baker and G. A. Olah; *J. Am. Chem. Soc.*, 1966, **88**, 1272. 850
66JA1332 G. Subrahmanyam, S. K. Malhotra and H. J. Ringold; *J. Am. Chem. Soc.*, 1966, **88**, 1332. 871, 881
66JA1335 D. M. Lemal and S. D. McGregor; *J. Am. Chem. Soc.*, 1966, **88**, 1335. 790
66JA1440 H. C. Brown and M. V. Bhatt; *J. Am. Chem. Soc.*, 1966, **88**, 1440. 658
66JA1473 H. M. Bell and H. C. Brown; *J. Am. Chem. Soc.*, 1966, **88**, 1473. 4, 7
66JA1905 A. Nickon and J. L. Lambert; *J. Am. Chem. Soc.*, 1966, **88**, 1905. 868
66JA2047 S. Winstein, C. G. Kreiter and J. I. Brauman; *J. Am. Chem. Soc.*, 1966, **88**, 2047. 917
66JA2257 R. R. Sauers, R. A. Parent and S. B. Damle; *J. Amer. Chem. Soc.*, 1966, **88**, 2257. 214
66JA2321 P. von R. Schleyer and G. W. Van Dine; *J. Am. Chem. Soc.*, 1966, **88**, 2321. 894
66JA2857 W. L. Mock; *J. Am. Chem. Soc.*, 1966, **88**, 2857. 620, 790
66JA2858 S. D. McGregor and D. M. Lemal; *J. Am. Chem. Soc.*, 1966, **88**, 2858. 790
66JA3027 S. A. Kandil and R. E. Dessy; *J. Am. Chem. Soc.*, 1966, **88**, 3027. 539
66JA3046 O. W. Webster; *J. Am. Chem. Soc.*, 1966, **88**, 3046. 862
66JA3152 P. E. Peterson and R. J. Kamat; *J. Am. Chem. Soc.*, 1966, **88**, 3152. 1124

66JA3461 P. Radlick and W. Rosen; *J. Am. Chem. Soc.*, 1966, **88**, 3461. 864
66JA4055 O. W. Webster; *J. Am. Chem. Soc.*, 1966, **88**, 4055. 862
66JA4271 W. J. Musliner and J. W. Gates, Jr.; *J. Am. Chem. Soc.*, 1966, **88**, 4271. 432
66JA4964 C. E. Castro, R. D. Stephens and S. Mojé; *J. Am. Chem. Soc.*, 1966, **88**, 4964. 516
66JA4990 P. E. Peterson and J. E. Duddey; *J. Am. Chem. Soc.*, 1966, **88**, 4990. 1110
66JA5652 E. J. Corey and G. T. Kwiatkowski; *J. Am. Chem. Soc.*, 1966, **88**, 5652. 643, 644
66JA5653 E. J. Corey and G. T. Kwiatkowski; *J. Am. Chem. Soc.*, 1966, **88**, 5653. 643, 644
66JA5654 E. J. Corey and G. T. Kwiatkowski; *J. Am. Chem. Soc.*, 1966, **88**, 5654. 639, 644, 645, 646
66JA5656 E. J. Corey and T. Durst; *J. Am. Chem. Soc.*, 1966, **88**, 5656. 621, 695
66JA5663 G. A. Russell and W. C. Danen; *J. Am. Chem. Soc.*, 1966, **88**, 5663. 156
66JA5846 W. D. Huntsman, J. A. DeBoer and M. H. Woosley; *J. Am. Chem. Soc.*, 1966, **88**, 5846. 808
66JCE398 R. J. Baumgarten; *J. Chem. Ed.*, 1966, **43**, 398. 32, 35
66JCS(A)453 J. Kollonitsch; *J. Chem. Soc. (A)*, 1966, 453. 127
66JCS(A)1342 G. R. Branton and H. M. Frey; *J. Chem. Soc. A.*, 1966, 1342. 801
66JCS(B)161 G. Allen and R. A. Dwek; *J. Chem. Soc. (B)*, 1966, 161. 870
66JCS(B)164 N. Heap and G. H. Whitham; *J. Chem. Soc. B.*, 1966, 164. 797
66JCS(B)241 R. P. Bell and P. W. Smith; *J. Chem. Soc. (B)*, 1966, 241. 869
66JCS(C)41 W. Cocker, P. V. R. Shannon and P. A. Staniland; *J. Chem. Soc. (C)*, 1966, 41. 72
66JCS(C)82 J. Gigg and R. Gigg; *J. Chem. Soc. (C)*, 1966, 82. 882
66JCS(C)154 D. Bryce-Smith and A. C. Skinner; *J. Chem. Soc. (C)*, 1966, 154. 889
66JCS(C)425 I. Fleming and J. Harley-Mason; *J. Chem. Soc. (C)*, 1966, 425. 45
66JCS(C)1090 M. E. Jones and S. Trippett; *J. Chem. Soc. (C)*, 1966, 1090. 733
66JCS(C)1223 S. R. Landor, A. N. Patel, P. F. Whiter and P. M. Greaves; *J. Chem. Soc. (C)*, 1966, 1223. 965
66JCS(C)1894 A. J. Birch and K. A. M. Walker; *J. Chem. Soc. (C)*, 1966, 1894. 82
66JOC321 P. Pino and G. P. Lorenzi; *J. Org. Chem.*, 1966, **31**, 321. 434
66JOC329 P. Pino and G. P. Lorenzi; *J. Org. Chem.*, 1966, **31**, 329. 999
66JOC334 A. W. Johnson and V. L. Kyllingstad; *J. Org. Chem.*, 1966, **31**, 334. 732
66JOC520 J. Fishman; *J. Org. Chem.*, 1966, **31**, 520. 871
66JOC624 R. J. Theis and R. E. Dessy; *J. Org. Chem.*, 1966, **31**, 624. 1075
66JOC646 H. O. House, W. L. Roelofs and B. M. Trost; *J. Org. Chem.*, 1966, **31**, 646. 307
66JOC870 E. E. Schweizer and K. K. Light; *J. Org. Chem.*, 1966, **31**, 870. 90
66JOC908 D. B. Miller; *J. Org. Chem.*, 1966, **31**, 908. 128
66JOC982 P. G. Gassman and B. L. Fox; *J. Org. Chem.*, 1966, **31**, 982. 881
66JOC989 H. L. Needles and R. E. Whitfield; *J. Org. Chem.*, 1966, **31**, 989. 881
66JOC1035 K. G. Hampton, T. M. Harris and C. R. Hauser; *J. Org. Chem.*, 1966, **31**, 1035. 881
66JOC1447 C. Szántay, L. Töke and P. Kolonits; *J. Org. Chem.*, 1966, **31**, 1447. 747
66JOC2047 A. R. Lepley, W. A. Khan, A. B. Giumanini and A. G. Giumanini; *J. Org. Chem.*, 1966, **31**, 2047. 148
66JOC2061 A. R. Lepley and W. A. Khan; *J. Org. Chem.*, 1966, **31**, 2061. 148
66JOC2373 D. J. Peterson and J. H. Collins; *J. Org. Chem.*, 1966, **31**, 2373. 157
66JOC2376 L. H. Klemm, D. Hsu Lee, K. W. Gopinath and C. E. Klopfenstein; *J. Org. Chem.*, 1966, **31**, 2376. 523
66JOC2777 D. J. Pasto and R. Snyder; *J. Org. Chem.*, 1966, **31**, 2777. 658, 659
66JOC2867 L. A. Carpino, P. H. Terry and S. D. Thatte; *J. Org. Chem.*, 1966, **31**, 2867. 1076
66JOC2907 E. E. Schweizer, E. T. Shaffer, C. T. Hughes and C. J. Berninger; *J. Org. Chem.*, 1966, **31**, 2907. 734
66JOC3128 H. O. House, W. L. Respess and G. M. Whitesides; *J. Org. Chem.*, 1966, **31**, 3128. 311
66JOC3498 E. L. Allred and R. L. Smith; *J. Org. Chem.*, 1966, **31**, 3498. 660
66JOC3595 D. L. Ransley; *J. Org. Chem.*, 1966, **31**, 3595. 463
66JOC4043 D. S. Noyce and E. H. Banitt; *J. Org. Chem.*, 1966, **31**, 4043. 681, 785
66JOC4071 C. E. Castro, E. J. Gaughan and D. C. Owsley; *J. Org. Chem.*, 1966, **31**, 4071. 1004
66JOC4097 E. J. Corey and D. Seebach; *J. Org. Chem.*, 1966, **31**, 4097. 140, 693
66JOM(5)205 A. M. Aguiar, H. J. Aguiar, T. G. Archibald and D. Daigle; *J. Organomet. Chem.*, 1966, **5**, 205. 999
66JOM(5)506 F. Joy, M. F. Lappert and B. Prokai; *J. Organomet. Chem.*, 1966, **5**, 506. 658
66JOM(6)412 J. C. Trebellas, J. R. Olechowski and H. B. Jonassen; *J. Organomet. Chem.*, 1966, **6**, 412. 413, 809
66LA(691)15 F. Klages and F. E. Malecki; *Justus Liebigs Ann. Chem.*, 1966, **691**, 15. 28
66LA(691)41 W. Kirmse, B.-G. von Bülow and H. Schepp; *Justus Liebigs Ann. Chem.*, 1966, **691**, 41. 630
66LA(698)80 U. Schollkopf and K. Fellenberger; *Justus Liebigs Ann. Chem.*, 1966, **698**, 80. 800, 816
66MI 103-01 S. Siegel; *Adv. Catal.*, 1966, **16**, 123. 72
B-66MI 109-01 C. J. Collins and J. F. Eastham; in "The Chemistry of the Carbonyl Group," ed. S. Patai, Wiley-Interscience, New York, 1966, ch. 15. 384
66MI 114-01 B. J. Wakefield; *Organometallic Chem. Rev.*, 1966, **1**, 131. 665
B-66MI 118-01 Molecular Rearrangements; in "Organic Reaction Mechanisms," Wiley-Interscience, 1966–1993, 27 vols. 794, 840
66MI 118-02 M. Orchin; *Adv. Catal.*, 1966, **16**, 1. 797
66MI 118-03 H. M. Frey; *Adv. Phys. Org. Chem.*, 1966, **4**, 147. 805
B-66MI 119-01 D. Lloyd; "Carbocyclic Non-Benzenoid Aromatic Compounds," Elsevier, Amsterdam, 1966. 845, 891
B-66MI 119-02 D. Lloyd; "Carbocyclic Non-Benzenoid Aromatic Compounds," Elsevier, Amsterdam, 1966, chapter 4, pp. 55–97. 861
B-66MI 119-03 D. Lloyd; "Carbocyclic Non-Benzenoid Aromatic Compounds," Elsevier, Amsterdam, 1966, chapter 5, pp. 98–116. 906

B-66MI 119-04	D. Lloyd; "Carbocyclic Non-Benzenoid Aromatic Compounds," Elsevier, Amsterdam, 1966, chapter 2, pp. 16–34.	907
B-66MI 119-05	D. Lloyd; "Carbocyclic Non-Benzenoid Aromatic Compounds," Elsevier, Amsterdam, 1966, 190–202.	916
B-66MI 119-06	W. A. Pryor; "Free Radicals," McGraw Hill, New York, 1966.	919
66MI 119-07	R. C. Fort, Jr. and P. von R. Schleyer; *Adv. Alicyclic Chem.*, 1966, **1**, 283.	847, 853
66MI 119-08	B. J. Wakefield; *Organomet. Chem. Rev.*, 1966, 131–156.	885
66MI 119-09	G. L. Closs; *Adv. Alicyclic Chem.*, 1966, **1**, 53–127.	907
66OR(12)356	G. R. Pettit and E. E. van Tamelen; *Org. React.*, 1966, **12**, 356.	438
66OS(46)89	H. Lindlar and R. Dubuis; *Org. Synth.*, 1966, **46**, 89.	502, 503
66PAC25	G. E. Gream; *Pure Appl. Chem.*, 1966, **16**, 25.	905
66T285	J. L. Baas, A. Davies-Fidder and H. O. Huisman; *Tetrahedron*, 1966, **22**, 285.	884
66T487	L. Caglioto; *Tetrahedron*, 1966, **22**, 487.	39
66TL1605	J. Blum; *Tetrahedron Lett.*, 1966, **7**, 1605.	73
66TL1663	H. Behringer and M. Matner; *Tetrahedron Lett.*, 1966, **7**, 1663.	1078
66TL1787	J. Ide and Y. Kishida; *Tetrahedron Lett.*, 1966, **7**, 1787.	822
66TL3847	M. Hanack and I. Herterich; *Tetrahedron Lett.*, 1966, **7**, 3847.	1123
66TL4315	G. Cainelli, G. Dal Bello and G. Zubiani; *Tetrahedron Lett.*, 1966, **7**, 4315.	659, 764
66TL6021	G. Zweifel and R. B. Steele; *Tetrahedron Lett.*, 1966, **7**, 6021.	680
66TL6115	E. Brown, P. Leriverend and J.-M. Conia; *Tetrahedron Lett.*, 1966, **7**, 6115.	412
66TL6279	N. P. Marullo, C. D. Smith and J. F. Terapane; *Tetrahedron Lett.*, 1966, **7**, 6279.	777
66TL6393	Y. Makisumi and S. Notzumoto; *Tetrahedron Lett.*, 1966, **7**, 6393.	816
66TL6425	J. Ficini and C. Barbara; *Tetrahedron Lett.*, 1966, **7**, 6425.	416, 828
66ZAAC(343)196	K. Issleib and B. Hamann; *Z. Anorg. Allg. Chem.*, 1966, **343**, 196.	446
67AG(E)1	C. A. Grob and P. W. Schiess; *Angew. Chem., Int. Ed., Engl.*, 1967, **6**, 1.	1081
67AG(E)124	G. P. Chiusoli and L. Cassar; *Angew. Chem., Int. Ed. Engl.*, 1967, **6**, 124.	526
67AG(E)402	G. Maier; *Angew. Chem., Int. Ed. Engl.*, 1967, **6**, 402.	420
67AG(E)635	R. Kuhn and D. Rewicki; *Angew. Chem., Int. Ed. Engl.*, 1967, **6**, 635.	867
67AG(E)800	P. Heimbach and W. Brenner; *Angew. Chem., Int. Ed. Engl.*, 1967, **6**, 800.	809
67AG(E)801	E. Weiss, H. Alsdorf and H. Kühr; *Angew. Chem., Int. Ed. Engl.*, 1967, **6**, 801.	104
67ANY(145)91	R. J. Tedeschi, H. C. McMahon and M. S. Pawlak; *Ann. N. Y. Acad. Sci.*, 1967, **145**, 91.	506
67BSF675	J. F. Biellmann and J. P. Pète; *Bull. Soc. Chim. Fr.*, 1967, 675.	630
67BSF2641	M. Julia, M. Maumy and L. Mion; *Bull. Soc. Chim. Fr.*, 1967, 2641.	343
67CB1144	C. Rüchardt, P. Panse and S. Eichler; *Chem. Ber.*, 1967, **100**, 1144.	725
67CB1193	F. Bohlmann, C. Zdero and W. Gordon; *Chem. Ber.*, 1967, **100**, 1193.	1142
67CB1200	F. Bohlmann, P.-H. Bonnet and H. Hofmeister; *Chem. Ber.*, 1967, **100**, 1200.	1142
67CB2011	G. Köbrich and F. Ansari; *Chem. Ber.*, 1967, **100**, 2011.	1144
67CC397	K. H. Bell; *J. Chem. Soc., Chem. Commun.*, 1967, 397.	795
67CC619	R. K. Hill and N. W. Gilman; *J. Chem. Soc., Chem. Commun.*, 1967, 619.	806, 807
67CC635	H. Hogeveen and A. F. Bickel; *J. Chem. Soc., Chem. Commun.*, 1967, 635.	901
67CC638	J. M. Brown; *J. Chem. Soc., Chem. Commun.*, 1967, 638.	865
67CC1042	D. Lloyd and M. I. C. Singer; *Chem. Commun.*, 1967, 1042.	648
67CC1259	J. Burdon, P. L. Coe, C. R. Marsh and J. C. Tatlow; *J. Chem. Soc., Chem. Commun.*, 1967, 1259.	1005
67CI(L)787	D. Lloyd and M. I. C. Singer; *Chem. Ind.* (*London*), 1967, 787.	54
67CR(C)839	E. Casadevall, J. C. Jallageas, L. Mion, M. Mion and P. Moreau; *C. R. Hebd. Seances Acad. Sci., Ser. C.*, 1967, **264**, 839.	1023
67CR(C)1492	M. Gaudemar and P. Pierrot; *C. R. Hebd. Seances Acad. Sci., Ser. C*, 1967, **264**, 1492.	958
67CR(C)1204	M. Pereyre, G. Colin and J. Valade; *C. R. Hebd. Séances Acad. Sci., Ser. C.*, 1967, **264**, 1204.	132
67HCA1158	G. Saucy and R. Marbet; *Helv. Chim. Acta*, 1967, **50**, 1158.	979
67HCA2101	A. Eschenmoser, J. Schreiber, D. Felix, M. Winter, F. Gautschi, K. H. Schulte-Elte, E. Sundt, G. Ohloff, J. Kalvoda, H. Kaufmann, P. Wieland and G. Anner; *Helv. Chim. Acta*, 1967, **50**, 2101.	1082
67HCA2108	P. Wieland, H. Kaufmann and A. Eschenmoser; *Helv. Chim. Acta*, 1967, **50**, 2108.	1082
67HP232	R. L. Banks; *Hydrocarbon Process.*, 1967, **46**, 232.	674
67JA63	A. Streitwieser, Jr., E. Ciuffarin and J. H. Hammons; *J. Am. Chem. Soc.*, 1967, **89**, 63.	848
67JA112	R. S. H. Liu; *J. Am. Chem. Soc.*, 1967, **89**, 112.	419
67JA291	G. Zweifel and H. Arzoumanian; *J. Am. Chem. Soc.*, 1967, **89**, 291.	509
67JA471	R. H. Shapiro, J. H. Duncan and J. C. Clopton; *J. Am. Chem. Soc.*, 1967, **89**, 471.	631
67JA567	H. C. Brown, M. V. Bhatt, T. Munekata and G. Zweifel; *J. Am. Chem. Soc.*, 1967, **89**, 567.	657, 658
67JA946	P. G. Gassman, J. T. Lumb and F. V. Zalar; *J. Am. Chem. Soc.*, 1967, **89**, 946.	885
67JA1044	J. Fried, N. A. Abraham and T. S. Santhanakrishnan; *J. Am. Chem. Soc.*, 1967, **89**, 1044.	99
67JA1183	M. Gorodetsky, Z. Luz and Y. Mazur; *J. Am. Chem. Soc.*, 1967, **89**, 1183.	870
67JA1404	R. M. Roberts, R. G. Landolt, R. N. Greene and E. W. Heyer; *J. Am. Chem. Soc.*, **89**, 1404, 1967.	795
67JA1442	R. H. Shapiro, J. H. Duncan and J. C. Clopton; *J. Am. Chem. Soc.*, 1967, **89**, 1442.	631
67JA1721	C. D. Ritchie and R. E. Uschold; *J. Am. Chem. Soc.*, 1967, **89**, 1721.	869
67JA1753	R. Huisgen, W. Scheer and H. Huber; *J. Am. Chem. Soc.*, 1967, **89**, 1753.	524
67JA1761	J. P. Freeman and W. H. Graham; *J. Am. Chem. Soc.*, 1967, **89**, 1761.	636
67JA1827	R. G. Pearson and J. Sangstad; *J. Am. Chem. Soc.*, 1967, **89**, 1827.	886
67JA1880	L. B. Jones and V. K. Jones; *J. Am. Chem. Soc.*, 1967, **89**, 1880.	404

67JA1960	W. M. Jones and F. W. Miller; *J. Am. Chem. Soc.*, 1967, **89**, 1960.	1119
67JA2054	M. J. Sabacky, C. S. Johnson, Jr., R. G. Smith, H. S. Gutowsky and J. C. Martin; *J. Am. Chem. Soc.*, 1967, **89**, 2054.	922
67JA2500	P. L. Creger; *J. Am. Chem. Soc.*, 1967, **89**, 2500.	881
67JA2654	F. R. Mayo; *J. Am. Chem. Soc.*, 1967, **89**, 2654.	923
67JA2664	J. Hine, K. G. Hampton and B. C. Menon; *J. Am. Chem. Soc.*, 1967, **89**, 2664.	871
67JA2751	E. C. Steiner and J. D. Starkey; *J. Am. Chem. Soc.*, 1967, **89**, 2751.	869
67JA2752	C. D. Ritchie and R. E. Uschold; *J. Am. Chem. Soc.*, 1967, **89**, 2752.	869
67JA2754	G. Zweifel and R. B. Steele; *J. Am. Chem. Soc.*, 1967, **89**, 2754.	534
67JA2755	E. J. Corey and M. F. Semmelhack; *J. Am. Chem. Soc.*, 1967, **89**, 2755.	127
67JA2826	J. Stuehr; *J. Am. Chem. Soc.*, 1967, **89**, 2826.	870
67JA2960	C. D. Ritchie and R. E. Uschold; *J. Am. Chem. Soc.*, 1967, **89**, 2960.	869
67JA3365	B. R. Ortiz de Montellano, B. A. Loving, T. C. Shields and P. D. Gardner; *J. Am. Chem. Soc.*, 1967, **89**, 3365.	890
67JA3405	N. C. Baenziger, R. E. Buckles and T. D. Simpson; *J. Am. Chem. Soc.*, 1967, **89**, 3405.	898
67JA3462	A. Viola, E. J. Iorio, K. K. Chen, G. M. Glover, U. Nayak and P. J. Kocienski; *J. Am. Chem. Soc.*, 1967, **89**, 3462.	412, 810
67JA3576	G. A. Olah, J. Sommer and E. Namanworth; *J. Am. Chem. Soc.*, 1967, **89**, 3576.	893
67JA3600	H. W. Whitlock and G. L. Smith; *J. Am. Chem. Soc.*, 1967, **89**, 3600.	266
67JA3652	G. Zweifel, H. Arzoumanian and C. C. Whitney; *J. Am. Chem. Soc.*, 1967, **89**, 3652.	278, 487, 661, 1103
67JA3905	F. G. Bordwell, R. H. Imes and E. C. Steiner; *J. Am. Chem. Soc.*, 1967, **89**, 3905.	869
67JA3911	E. J. Corey and G. H. Posner; *J. Am. Chem. Soc.*, 1967, **89**, 3911.	117, 118
67JA4383	R. Breslow, J. Brown and J. J. Gajewski; *J. Am. Chem. Soc.*, 1967, **89**, 4383.	865
67JA4531	H. C. Brown, D. H. Bowman, S. Misumi and M. K. Unni; *J. Am. Chem. Soc.*, 1967, **89**, 4531.	661
67JA4739	G. A. Olah and J. Lukas; *J. Am. Chem. Soc.*, 1967, **89**, 4739.	893, 901
67JA5048a	R. Breslow, J. T. Groves and G. Ryan; *J. Am. Chem. Soc.*, 1967, **89**, 5048.	907
67JA5048b	D. G. Farnum, G. Mehta and R. G. Silberman; *J. Am. Chem. Soc.*, 1967, **89**, 5048.	907
67JA5085	G. Zweifel and R. B. Steele; *J. Am. Chem. Soc.*, 1967, **89**, 5085.	514, 534
67JA5086	G. Zweifel and H. Arzoumanian; *J. Am. Chem. Soc.*, 1967, **89**, 5086.	487, 1100, 1103
67JA5259	G. A. Olah, M. B. Comisarow, E. Namanworth and B. Ramsey; *J. Am. Chem. Soc.*, 1967, **89**, 5259.	915
67JA5292	K. H. Dahm, B. M. Trost and H. Röller; *J. Am. Chem. Soc.*, 1967, **89**, 5292.	747
67JA5421	J. I. Brauman and A. J. Pandell; *J. Am. Chem. Soc.*, 1967, **89**, 5421.	476
67JA5503	J. A. Berson and G. L. Nelson; *J. Am. Chem. Soc.*, 1967, **89**, 5503.	400, 799, 805
67JA5687	J. M. Bollinger, M. B. Comisarow, C. A. Cupas and G. A. Olah; *J. Am. Chem. Soc.*, 1967, **89**, 5687.	894
67JA5708	A. Suzuki, A. Arase, H. Matsumoto, M. Itoh, H. C. Brown, M. M. Rogic and M. W. Rathke; *J. Am. Chem. Soc.*, 1967, **89**, 5708.	273
67JA5709	A. Suzuki and H. C. Brown; *J. Am. Chem. Soc.*, 1967, **89**, 5709.	273
67JA5710	H. C. Brown, M. M. Rogic, M. W. Rathke and G. W. Kabalka; *J. Am. Chem. Soc.*, 1967, **89**, 5710.	273
67JA5734	R. H. Shapiro and M. J. Heath; *J. Am. Chem. Soc.*, 1967, **89**, 5734.	630, 631, 632, 633
67JA5736	G. Kaufman, F. Cook, H. Shechter, J. Bayless and L. Friedman; *J. Am. Chem. Soc.*, 1967, **89**, 5736.	630
67JA5850	R. M. Gerkin and B. Rickborn; *J. Amer. Chem. Soc.*, 1967 **89**, 5850.	212
67JA6701	J. L. Coke and M. P. Cooke, Jr.; *J. Am. Chem. Soc.*, 1967, **89**, 6701.	627
67JCP(46)3193	M. T. Rogers and L. D. Kispert; *J. Chem. Phys.*, 1967, **46**, 3193.	923
67JCP(47)3637	L. Andrews and G. C. Pimentel; *J. Chem. Phys.*, 1967, **47**, 3637.	923
67JCP(47)5146	D. E. Milligan and M. E. Jacox; *J. Chem. Phys.*, 1967, **47**, 5146.	923
67JCS(A)1574	F. H. Jardine, J. A. Osborn and G. Wilkinson; *J. Chem. Soc.* (*A*), 1967, 1574.	73, 79, 80
67JCS(B)509	M. A. Baldwin, D. V. Banthorpe, A. G. Loudon and F. D. Waller; *J. Chem. Soc.* (*B*), 1967, 509.	627
67JCS(B)557	D. B. Bigley and R. W. May; *J. Chem. Soc.* (*B*), 1967, 557.	772
67JCS(B)827	R. P. Bell, G. R. Hillier, J. W. Mansfield and D. G. Street; *J. Chem. Soc.* (*B*), 1967, 827.	869, 871
67JCS(B)991	A. C. Day and M. C. Whiting; *J. Chem. Soc.* (*B*), 1967, 991.	1120
67JCS(B)1173	J. R. Jones and R. Stewart; *J. Chem. Soc.* (*B*), 1967, 1173.	871
67JCS(C)194	W. J. Ball, S. R. Landor and N. Punja; *J. Chem. Soc.* (*C*), 1967, 194.	990
67JCS(C)270	F. H. Jardine and G. Wilkinson; *J. Chem. Soc.* (*C*), 1967, 270.	85
67JCS(C)532	D. N. Jones and M. J. Green; *J. Chem. Soc.* (*C*), 1967, 532.	779
67JCS(C)578	R. E. Atkinson, R. F. Curtis and J. A. Taylor; *J. Chem. Soc. C*, 1967, 578.	1006
67JCS(C)1099	A. R. Hands and A. J. H. Mercer; *J. Chem. Soc.* (*C*), 1967, 1099.	647
67JCS(C)1120	D. C. Curry, B. C. Uff and N. D. Ward; *J. Chem. Soc.* (*C*), 1967, 1120.	74, 82, 84
67JCS(C)1625	J. A. Barltrop and B. Hesp; *J. Chem. Soc.* (*C*), 1967, 1625.	517
67JCS(C)1716	D. C. Nonhebel; *J. Chem. Soc.* (*C*), 1967, 1716.	870
67JCS(C)1720	A. B. Turner and H. J. Ringold; *J. Chem. Soc.* (*C*), 1967, 1720.	555
67JCS(C)1860	J. I. G. Cadogan, J. K. A. Hall and J. T. Sharp; *J. Chem. Soc.* (*C*), 1967, 1860.	881
67JCS(C)1964	R. K. Bartlett and T. S. Stevens; *J. Chem. Soc.* (*C*), 1967, 1964.	630
67JCS(C)1986	R. C. Cookson, J. B. Henstock, J. Hudec and B. R. D. Whitear; *J. Chem. Soc.* (*C*), 1967, 1986.	537
67JCS(C)2011	R. E. Atkinson, R. F. Curtis and G. T. Phillips; *J. Chem. Soc. C*, 1967, 2011.	1006
67JOC39	O. W. Webster; *J. Org. Chem.*, 1967, **32**, 39.	861

67JOC102	D. W. Emerson, A. P. Craig and I. W. Potts, Jr.; *J. Org. Chem.*, 1967, **32**, 102.	595
67JOC105	D. L. Skinner, D. J. Peterson and T. J. Logan; *J. Org. Chem.*, 1967, **32**, 105.	1063
67JOC492	A. A. Bothner-By and C. Sun; *J. Org. Chem.*, 1967, **32**, 492.	871
67JOC855	D. Brett, I. M. Downie and J. B. Lee; *J. Org. Chem.*, 1967, **32**, 855.	185, 200
67JOC869	T. Inukai and T. Kojima; *J. Org. Chem.*, 1967, **32**, 869.	373
67JOC893	J. W. Wilt, G. Gutman, W. J. Ranus, Jr. and A. R. Zigman; *J. Org. Chem.*, 1967, **32**, 893.	81
67JOC1155	K. P. Klein, D. N. Van Eenam and C. R. Hauser; *J. Org. Chem.*, 1967, **32**, 1155.	399
67JOC1631	J. L. Kice and J. D. Campbell; *J. Org. Chem.*, 1967, **32**, 1631.	595
67JOC1717	D. J. Peterson; *J. Org. Chem.*, 1967, **32**, 1717.	140, 693
67JOC1994	N. L. Allinger, L. W. Chow and R. A. Ford; *J. Org. Chem.*, 1967, **32**, 1994.	869
67JOC2013	M. Brown and L. W. Piszkiewicz; *J. Org. Chem.*, 1967, **32**, 2013.	85
67JOC2059	S. I. Goldberg and M. S. Sahli; *J. Org. Chem.*, 1967, **32**, 2059.	779
67JOC2162	A. Yogev and Y. Mazur; *J. Org. Chem.*, 1967, **32**, 2162.	869, 870
67JOC2257	R. L. Augustine, Z. S. Zelawski and D. H. Malarek; *J. Org. Chem.*, 1967, **32**, 2257.	884
67JOC2627	A. M. Agviar and T. G. Archibald; *J. Org. Chem.*, 1967, **32**, 2627.	445
67JOC2701	J. F. Bunnett and D. A. R. Happer; *J. Org. Chem.*, 1967, **32**, 2701.	445
67JOC2778	P. Beak and B. M. Munroe; *J. Org. Chem.*, 1967, **32**, 2778.	889
67JOC2794	S. D. Darling and K. D. Wills; *J. Org. Chem.*, 1967, **32**, 2794.	100
67JOC3700	C. Rappe and W. H. Sachs; *J. Org. Chem.*, 1967, **32**, 3700.	871
67JOC4126	C. C. Price and J. R. Sowa; *J. Org. Chem.*, 1967, **32**, 4126.	62
67JOC4127	C. Rappe and W. H. Sachs; *J. Org. Chem.*, 1967, **32**, 4127.	871
67JOC4143	R. Vaidyanathaswamy and D. Devaprabhakara; *J. Org. Chem.*, 1967, **32**, 4143.	459
67JOM(8)239	C. S. Kraihanzel and J. E. Poist; *J. Organomet. Chem.*, 1967, **8**, 239.	999
67LA(707)120	K. Isslieb and R. Lindner; *Justus Liebigs Ann. Chem.*, 1967, **707**, 120.	737
67LA(708)1	M. Schlosser and K. F. Christmann; *Justus Liebigs Ann. Chem.*, 1967, **708**, 1.	721, 727, 728, 729, 730
B-67MI 103-01	P. N. Rylander; "Catalytic Hydrogenation over Platinum Metals," Academic Press, New York, 1967.	71, 87, 88, 89
B-67MI 104-01	S. T. Ioffe and A. N. Nesmeyanov; in "The Organic Compounds of Magnesium, Beryllium, Calcium, Strontium, and Barium, Series: Methods of Elemento-Organic Chemistry," eds A. N. Nesmeyanov and K. A. Kocheshkov, North-Holland, Amsterdam, 1967, vol. 2, Chapter 17, pp. 474–506.	114
B-67MI 104-02	L. G. Makarova and A. N. Nesmeyanov; in "The Organic Compounds of Mercury, Series: Methods of Elemento-Organic Chemistry," eds A. N. Nesmeyanov and K. A. Kocheshkov, North-Holland, Amsterdam, 1967, vol. 4, pp. 375–381.	128
B-67MI 112-01	L. F. Fieser and M. Fieser; "Reagents for Organic Synthesis," Wiley, New York, 1967, vol. 1, p. 556.	503
B-67MI 112-02	R. W. Hoffmann; "Dehydrobenzene and Cycloalkynes," VCH, Weinheim, 1967.	519
B-67MI 114-01	F. G. Bordwell; in "Organosulphur Chemistry," ed. M. J. Janssen, Interscience, New York, 1967, p. 271.	617
67MI 114-02	J. L. Huguet; *Adv. Chem. Ser.*, 1967, **76**, 345 (*Chem. Abstr.*, 1968, **69**, 105 823y).	601
B-67MI 114-03	V. F. Mironov and T. K. Gar; "Organic Compounds of Germanium," Nauka, Moscow, 1967, p. 363 (*Chem. Abstr.*, 1968, **68**, 114 732z).	656
B-67MI 115-01	F. G. Bordwell; "Organosulfur Chemistry," ed. M. J. Janssen, Wiley, New York, 1967, p. 271.	704
67MI 118-01	C. W. Bird; "Transition Metal Intermediates in Organic Synthesis," Academic Press, New York, 1967.	797
B-67MI 119-01	A. R. Katritzky and J. M. Lagowski; "The Principles of Heterocyclic Chemistry," Chapman and Hall, London, 1967, p. 55.	859
B-67MI 119-02	G. E. Coates, M. L. H. Green and K. Wade; "Organometallic Compounds," 3rd edn., Methuen, London, 1967, vol. 1.	885
B-67MI 119-03	M. Fieser and L. F. Fieser; "Reagents for Organic Synthesis," Wiley, New York, 1967, vol. 1.	885, 886
B-67MI 119-04	D. Bethel and V. Gold; "Carbonium Ions," Academic Press, New York, 1967.	891
B-67MI 119-05	D. Bethel and V. Gold; "Carbonium Ions," Academic Press, New York, 1967, p. 53.	896
B-67MI 119-06	M. Fieser and L. F. Fieser; "Reagents for Organic Synthesis," Wiley, New York, 1967, vol. 1, p. 1256.	898
67MI 119-07	W. J. Albery; *Prog. Reaction Kinetics*, 1967, **4**, 353.	844
B-67MI 121-01	R. W. Hoffmann; in "Dehydrobenzene and Cycloalkynes," Academic Press, New York, 1967, p. 33.	1081
B-67MI 121-02	L. Brandsma and J. F. Arens; in "The Chemistry of the Ether Linkage," ed. S. Patai, Wiley, 1967, p. 553.	1031
B-67MI 122-01	D. Bethell and V. Gold; "Carbonium Ions. An Introduction," Academic Press, London, 1967.	1109
67OR(15)1	J. P. Schaefer and J. J. Bloomfield; *Org. React.*, 1967, **15**, 1.	881, 884
67OS(47)20	A. P. Krapcho, J. Diamanti, C. Cayen and R. Bingham; *Org. Synth.*, 1967, **47**, 20.	881
67OS(47)72	E. M. Kaiser, W. G. Kenyon and C. R. Hauser; *Org. Synth.*, 1967, **47**, 72.	881
67OS(47)92	K. G. Hampton, T. M. Harris and C. R. Hauser; *Org. Synth.*, 1967, **47**, 92.	881
67PAC19	E. J. Corey; *Pure Appl. Chem.*, 1967, **14**, 19.	695
67RCP99	H. O. House; *Rec. Chem. Prog.*, 1967, **28**, 99.	104
67RCR260	L. L. Rodina and I. K. Korobitsyna; *Russ. Chem. Rev.* (*Engl. Transl.*), 1967, **36**, 260.	393
67RCR557	I. A. D'yakonov and R. R. Kostikov; *Russ. Chem. Rev.* (*Engl. Transl.*), 1967, **36**, 557.	907
67RCR735	G. D. Kolomnikova and Z. N. Parnes; *Russ. Chem. Rev.* (*Engl. Transl.*), 1967, **36**, 735.	906
67RTC129	P. P. Montijn, L. Brandsma and J. F. Arens; *Recl. Trav. Chim. Pays-Bas*, 1967, **86**, 129.	986

67RTC1138	J. F. Arens and H. M. Schmidt; *Rec. Trav. Chim. Pays-Bas*, 1967, **86**, 1138.	999
67T215	M. L. Mihailovic, R. I. Mamuzić, L. Žigić-Mamuzić, J. Bošnjak and Ž. Ceković; *Tetrahedron*, 1967, **23**, 215.	91
67T565	G. Köbrich, H. Heinemann and W. Zündorf; *Tetrahedron*, 1967, **23**, 565.	1145
67T793	J. R. Hanson; *Tetrahedron*, 1967, **23**, 793.	382
67T2235	D. N. Kursanov, Z. N. Parnes, G. I. Bassova, N. M. Loim and V. I. Zdanovich; *Tetrahedron*, 1967, **23**, 2235.	76
67T2709	L. D. Bergelson, L. I. Barsukov and M. M. Shemyakin; *Tetrahedron*, 1967, **23**, 2709.	727, 733
67TL29	E. W. Garbisch, Jr. and J. G. Russell; *Tetrahedron Lett.*, 1967, **8**, 29.	870
67TL227	N. H. Subba Rao, N. P. Damodaran and S. Dev; *Tetrahedron Lett.*, 1967, **8**, 227.	809, 813
67TL401	R. G. Lewis, D. H. Gustafson and W. F. Erman; *Tetrahedron Lett.*, 1967, **8**, 401.	725, 1001
67TL495	K. König and W. P. Neumann; *Tetrahedron Lett.*, 1967, **8**, 495.	1038
67TL857	A. J. Birch and G. S. R. Subba Rao; *Tetrahedron Lett.*, 1967, **8**, 857.	99
67TL1581	G. Cainelli, F. Bertini, P. Grasselli and G. Zubiani; *Tetrahedron Lett.*, 1967, **8**, 1581.	680
67TL2329	L. P. Reiff and H. S. Aaron; *Tetrahedron Lett.*, 1967, **8**, 2329.	19
67TL3457	A. J. Birch and K. A. M. Walker; *Tetrahedron Lett.*, 1967, **8**, 3457.	85
67TL3739	M. Tanabe, D. F. Crowe, R. L. Dehn and G. Detre; *Tetrahedron Lett.*, 1967, **8**, 3739.	1082
67TL3871	M. Hanack, I. Herterich and V. Vött; *Tetrahedron Lett.*, 1967, **8**, 3871.	104
67TL3877	E. Müller, A. Moosmayer, A. Rieker and K. Scheffler; *Tetrahedron Lett.*, 1967, **8**, 3877.	922
67TL3943	M. Tanabe, D. F. Crowe and R. L. Dehn; *Tetrahedron Lett.*, 1967, **8**, 3943.	1082
67TL4525	S. T. Yoffe, P. V. Petrovskii, E. I. Fedin, K. V. Vatsuro P. S. Burenko and M. I Kabachnick; *Tetrahedron Lett.*, 1967, **8**, 4525.	870
67TL4537	M. F. Ansell, A. F. Gosden and V. J. Leslie; *Tetrahedron Lett.*, 1967, **8**, 4537.	815
67TL4901	P. Radlick and W. Fenical; *Tetrahedron Lett.*, 1967, **8**, 4901.	418
67TL4945	S. Yankelevich and B. Fuchs; *Tetrahedron Lett.*, 1967 **8**, 4945.	790
67ZOR219	O. G. Yashina, T. V. Zarva and L. I. Vereschagin; *Zh. Org. Khim.*, 1967, **3**, 219 (*Chem. Abstr.*, **66**, 46 070p).	1045
67ZOR598	N. A. Nesmeyanov, V. V. Pravdina and O. A. Reutov *Zh. Org. Khim.*, 1967, **3**, 598 (*Chem. Abstr.*, 1967, **67**, 11 553s).	647
68ACR152	J. A. Berson; *Acc. Chem. Res.*, 1968, **1**, 152.	400, 799, 805, 838
68ACR186	R. Cramer; *Acc. Chem. Res.*, 1968, **1**, 186.	797
68ACR209	L. A. Paquette; *Acc. Chem. Res.*, 1968, **1**, 209.	144, 617, 704, 707
68ACR231	C. D. Broaddus; *Acc. Chem. Res.*, 1968, **1**, 231.	795, 882
68ACR299	H. G. Kuivila; *Acc. Chem. Res.*, 1968, **1**, 299.	935
68ACS1883	J. Bergman; *Acta Chem. Scand.*, 1968, **22**, 1883.	484
68AG(E)7	G. Wittig and H. Reiff; *Angew. Chem., Int. Ed. Engl.*, 1968, **7**, 7.	881
68AG(E)67	H. Prinzbach, R. Fuchs and R. Kitzing; *Angew. Chem., Int. Ed. Engl.*, 1968, **7**, 67.	521
68AG(E)120	H. H. Szmant; *Angew. Chem., Int. Ed. Engl.*, 1968, **7**, 120.	887
68AG(E)456	J. Gosselck and G. Schmidt; *Angew. Chem., Int. Ed. Engl.*, 1968, **7**, 456.	241
68AG(E)507	B. Mile; *Angew. Chem., Int. Ed. Engl.*, 1968, **7**, 507.	921
68AG(E)536	J. Buddrus; *Angew. Chem., Int. Ed. Engl.*, 1968, **7**, 536	722
68AG(E)650	M. Schlosser; *Angew. Chem., Int. Ed. Engl.*, 1968, **7**, 650.	639
68AG(E)942	L. Horner, H. Siegel and H. Büthe; *Angew. Chem., Int Ed. Engl.*, 1968, **7**, 942.	76
68AJC1375	P. M. Pojer, E. Ritchie and W. C. Taylor; *Aust. J. Chem.*, 1968, **21**, 1375.	36
68AJC2989	M. Rasmussen, D. D. Ridley, E. Ritchie and W. C. Taylor; *Aust. J. Chem.*, 1968, **21**, 2989.	1001
68BCJ1252	K. Sasaki; *Bull. Chem. Soc. Japan*, 1968, **41**, 1252, and references cited therein	747
68BCJ2921	A. Misono, T. Osa, T. Yamagichi; *Bull. Chem. Soc. Jpn.*, 1968, **41**, 2921.	103
68BSF205	H. Demarne and P. Cadiot; *Bull. Soc. Chim. Fr.*, 1968, 205.	1037
68BSF1085	J. C. Masson, M. Le Quan and P. Cadiot; *Bull. Soc. Chim. Fr.*, 1968, 1085.	1000
68BSF2120	J. P. Blanchini and A. Guillemonat; *Bull. Soc. Chim. Fr.*, 1968, 2120.	1125
68BSF3401	P. Courtot and O. Le Goff-Hays; *Bull. Soc. Chim. Fr.*, 1968, 3401.	420
68C338	H. P. Fischer, E. Kaplan and P. Neuenschwander; *Chimia*, 1968, **22**, 338.	664
68CB2669	M. Fischer; *Chem. Ber.*, 1968, **101**, 2669.	785
68CC35	F. Bickelhaupt, W. L. de Graff and G. W. Klumpp; *J. Chem. Soc., Chem. Commun.*, 1968, 35.	795
68CC186	G. M. Blackburn, W. D. Ollis, J. D. Plackett, C. Smith and I. O. Sutherland; *J. Chem. Soc., Chem. Commun.*, 1968, 186.	822
68CC543	G. Jones and R. F. Maisey; *J. Chem. Soc., Chem. Commun.*, 1968, 543.	747
68CC552	R. P. Gandhi and V. K. Chadha; *J. Chem. Soc., Chem. Commun.*, 1968, 552.	521
68CC587	M. T. Hughes and R. O. Williams; *J. Chem. Soc., Chem. Commun.*, 1968, 587.	826
68CC634	J. Fried, S. Heim, S. J. Etheridge, P. Sunder-Plassmann, T. S. Santhanakrishan, J. Himizu and C. H. Lin; *J. Chem. Soc., Chem. Commun.*, 1968 634.	1023
68CC718	M. Bertrand and M. Santelli; *J. Chem. Soc., Chem. Commun.*, 1968, 718.	1126
68CC934	J. H. Stocker and R. M. Jenevein; *J. Chem. Soc., Chem. Commun.*, 1968, 934.	6
68CC982	B. D. Challand and P. de Mayo; *J. Chem. Soc., Chem. Commun.*, 1968, 982.	356
68CC1225	C. H. DePuy, R. N. Greene and T. E. Schroer; *J. Chem. Soc., Chem. Commun.*, 1968, 1225.	870
68CC1229	R. V. H. Jones and M. D. Sutherland; *J. Chem. Soc., Chem. Commun.*, 1968, 1229.	812
68CC1363	G. Modena, U. Tonellato and F. Naso; *J. Chem. Soc., Chem. Commun.*, 1968, 1363.	1112
68CC1445	W. S. Knowles and M. J. Sabacky; *J. Chem. Soc., Chem. Commun.*, 1968, 1445	76
68CC1549	H. C. Brown and N. M. Yoon; *J. Chem. Soc., Chem. Commun.*, 1968, 1549.	14
68CC1676	G. Modena and U. Tonellato; *J. Chem. Soc., Chem. Commun.*, 1968, 1676.	1112
68CC1699	T. H. Kinstle and B. Y. Mandanas; *J. Chem. Soc., Chem. Commun.*, 1968, 1699.	639, 644, 747

68CI(L)1277	D. Lloyd and M. I. C. Singer; *Chem. Ind.* (*London*), 1968, 1277.	738
68CJC84	S. N. Sharma, R. K. Srivastava and D. Devaprabhakara; *Can. J. Chem.*, 1968, **46**, 84.	459
68CJC461	A. W. Johnson and R. T. Amel; *Can. J. Chem.*, 1968, **46**, 461.	737
68CJC633	M. Akhtar and H. C. Clark; *Can. J. Chem.*, 1968, **46**, 633.	935
68CJC1057	R. Raap and R. G. Micetich; *Can. J. Chem.*, 1968, **46**, 1057.	1077
68CJC2225	D. E. McGreer and N. W. K. Chiu; *Can. J. Chem.*, 1968, **46**, 2225.	747
68CJC2821	Y. L. Chow, C. J. Colón and J. N. S. Tam; *Can. J. Chem.*, 1968, **46**, 2821.	104
68CPB793	J. Ide and Y. Kishida; *Chem. Pharm. Bull.*, 1968, **16**, 793.	822
68CR(C)1162	J.-F. Fauvarque, B. Meklati and C. Dearing; *C. R. Hebd. Séances Acad. Sci., Ser. C.*, 1968, **267**, 1162.	155
68CR(C)1614	C. Bouchoule and P. Miginiac; *C. R. Hebd. Séances Acad. Sci., Ser. C.*, 1968, **266**, 1614.	127
68CR(C)1834	M. Nicolas and R. Pallaud; *C. R. Hebd. Seances Acad. Sci., Ser. C*, 1962, **267**, 1834.	104
68CRV375	F. Johnson; *Chem. Rev.*, 1968, **68**, 375.	871, 876
68CRV415	H. Kwart and K. King; *Chem. Rev.*, 1968, **68**, 415.	786
68DOK(179)1344	B. M. Mikhailov, M. E. Kuimova and E. A. Shagova; *Dokl. Akad. Nauk SSSR*, 1968, **179**, 1344 (*Chem. Abstr.*, 1968, **69**, 76 149u).	657
68FCR315	F. J. Mettille and D. J. Burton; *Fluorine Chem. Rev.*, 1968, **1**, 315.	426
68FRP1519701	A. Bader and F. Weiss, *Fr. Pat.*, 1 519 701 (1968) (*Chem. Abstr.*, 1969, **70**, 96 177).	810
68G367	P. Beltrame and P. L. Beltrame; *Gazz. Chim. Ital.*, 1968, **98**, 367.	1110
68HCA225	W. Grimme, E. Heilbronner, E. Hohlneicher, E. Vogel and J. P. Weber; *Helv. Chim. Acta*, 1968, **51**, 225.	910
68HCA371	P. Kniel; *Helv. Chim. Acta*, 1968, **51**, 371.	441
68HCA1457	P. Vittorelli, T. Winkler, H.-J. Hansen and H. Schmid; *Helv. Chim. Acta*, 1968, **51**, 1457.	415
68HCA1603	F. Scheinmann, R. Barner and H. Schmid; *Helv. Chim. Acta*, 1968, **51**, 1603.	827
68IC168	H. Schmidbaur and W. Tronich; *Inorg. Chem.*, 1968, **7**, 168.	740
68IC1014	T. D. Coyle, J. Cooper and J. J. Ritter; *Inorg. Chem.*, 1968, **7**, 1014.	658
68IJ691	H. C. Brown and E. F. Knights; *Isr. J. Chem.*, 1968, **6**, 691.	82
68IS101	K. Vrieze, J. P. Collman, C. T. Sears and M. Kubota; *Inorg. Synth.*, 1968, **11**, 101.	73
68IZV548	B. M. Mikhailov, M. E. Kuimova and E. A. Shagova; *Izv. Akad. Nauk SSSR Ser. Khim.*, 1968, 548 (*Chem. Abstr.*, 1968, **69**, 67 447).	658
68JA63	K. B. Wiberg and A. J. Ashe, III; *J. Am. Chem. Soc.*, 1968, **90**, 63.	894
68JA113	C. M. Cimarusti and J. Wolinsky; *J. Am. Chem. Soc.*, 1968, **90**, 113.	563
68JA372	D. S. Noyce and K. E. DeBruin; *J. Am. Chem. Soc.*, 1968, **90**, 372.	1122
68JA429	F. G. Bordwell, J. M. Williams, Jr., E. B. Hoyt, Jr. and B. B. Jarvis; *J. Am. Chem. Soc.*, 1968, **90**, 429.	620
68JA435	F. G. Bordwell and J. M. Williams, Jr.; *J. Am. Chem. Soc.*, 1968, **90**, 435.	620
68JA531	J. H. Bayless, L. Friedman, F. B. Cook and H. Shechter; *J. Am. Chem. Soc.*, 1968, **90**, 531.	631
68JA800	J. Meinwald and F. Uno; *J. Am. Chem. Soc.*, 1968, **90**, 800.	631
68JA818	H. C. Brown, M. M. Rogic, M. W. Rathke and G. W. Kabalka; *J. Am. Chem. Soc.*, 1968, **90**, 818.	125, 280
68JA933	G. A. Olah and J. Lukas; *J. Am. Chem. Soc.*, 1968, **90**, 933.	895
68JA1020	D. S. Noyce and M. D. Schiavelli; *J. Am. Chem. Soc.*, 1968, **90**, 1020.	1122
68JA1023	D. S. Noyce and M. D. Schiavelli; *J. Am. Chem. Soc.*, 1968, **90**, 1023.	1122
68JA1239	T. Yonezawa, H. Nakatsuji and H. Kato; *J. Am. Chem. Soc.*, 1968, **90**, 1239.	1109
68JA1791	D. J. Cram and W. D. Kollmeyer; *J. Am. Chem. Soc.*, 1968, **90**, 1791.	848
68JA1884	G. A. Olah and A. M. White; *J. Am. Chem. Soc.*, 1968, **90**, 1884.	894, 899
68JA1911	H. C. Brown, M. M. Rogic, M. W. Rathke and G. W. Kabalka; *J. Am. Chem. Soc.*, 1968, **90**, 1911.	280
68JA2432	S. J. Huang and M. V. Lessard; *J. Am. Chem. Soc.*, 1968, **90**, 2432.	1122
68JA2471	M. S. Gordon and H. Fischer; *J. Am. Chem. Soc.*, 1968, **90**, 2471.	849
68JA2598	D. J. Cram, W. T. Ford and L. Gosser; *J. Am. Chem. Soc.*, 1968, **90**, 2598.	858
68JA2606	W. T. Ford and D. J. Cram; *J. Am. Chem. Soc.*, 1968, **90**, 2606.	858
68JA2612	W. T. Ford and D. J. Cram; *J. Am. Chem. Soc.*, 1968, **90**, 2612.	858
68JA2698	R. Breslow and M. Douek; *J. Am. Chem. Soc.*, 1968, **90**, 2698.	851, 865
68JA2821	C. D. Ritchie and R. E. Uschold; *J. Am. Chem. Soc.*, 1968, **90**, 2821.	869
68JA2906	H. C. Brown and R. M. Gallivan, Jr.; *J. Am. Chem. Soc.*, 1968, **90**, 2906.	658, 659
68JA2927	N. M. Yoon and H. C. Brown; *J. Am. Chem. Soc.*, 1968, **90**, 2927.	14
68JA3415	C. D. Ritchie and R. E. Uschold; *J. Am. Chem. Soc.*, 1968, **90**, 3415.	869
68JA3466	G. H. Schmid and M. Heinola; *J. Am. Chem. Soc.*, 1968, **90**, 3466.	1123
68JA3590	E. W. Garbisch, Jr. and M. G. Griffith; *J. Am. Chem. Soc.*, 1968, **90**, 3590.	95
68JA4165	H. C. Brown, G. W. Kabalka, M. W. Rathke and M. M. Rogić; *J. Am. Chem. Soc.*, 1968, **90**, 4165.	274, 946
68JA4166	H. C. Brown, G. W. Kabalka, M. W. Rathke and M. M. Rogic; *J. Am. Chem. Soc.*, 1960, **90**, 4166.	274
68JA4182	T. Saegusa, S. Kobayashi, Y. Ito and N. Yasuda; *J. Am. Chem. Soc.*, 1968, **90**, 4182.	38, 945
68JA4187	E. F. Kiefer and M. Y. Okamura; *J. Am. Chem. Soc.*, 1968, **90**, 4187.	498
68JA4189	M. S. Newman and A. O. M. Okorodudu; *J. Am. Chem. Soc.*, 1968, **90**, 4189.	1143
68JA4439	H. C. Brown and E. F. Knights; *J. Am. Chem. Soc.*, 1968, **90**, 4439.	658, 659
68JA4445	D. J. Pasto and J. Hickman; *J. Am. Chem. Soc.*, 1968, **90**, 4445.	658
68JA4462	G. Stork and P. F. Hudrlik; *J. Am. Chem. Soc.*, 1968, **90**, 4462.	130
68JA4464	G. Stork and P. F. Hudrlik; *J. Am. Chem. Soc.*, 1968, **90**, 4464.	130
68JA4495	J. Ciabattoni and E. C. Nathan, III; *J. Am. Chem. Soc.*, 1968, **90**, 4495.	909
68JA4498	R. Srinivasan; *J. Am. Chem. Soc.*, 1968, **90**, 4498.	419

68JA4505	S. Bank and T. A. Lois; *J. Am. Chem. Soc.*, 1968, **90**, 4505.	881
68JA4629	J. Warkentin and C. Barnett; *J. Am. Chem. Soc.*, 1968, **90**, 4629.	871
68JA4729	J. A. Berson and E. J. Walsh, Jr.; *J. Am. Chem. Soc.*, 1968, **90**, 4729.	810, 815
68JA4733	P. Rona and P. Crabbé; *J. Am. Chem. Soc.*, 1968, **90**, 4733.	966
68JA4762	W. G. Dauben, M. E. Lorber, N. D. Vietmeyer, R. H. Shapiro, J. H. Duncan and K. Tomer; *J. Am. Chem. Soc.*, 1968, **90**, 4762.	633
68JA5518	R. F. Heck; *J. Am. Chem. Soc.*, 1968, **90**, 5518.	493
68JA5531	R. F. Heck; *J. Am. Chem. Soc.*, 1968, **90**, 5531.	128
68JA5548	E. J. Corey and T. Durst; *J. Am. Chem. Soc.*, 1968, **90**, 5548.	621, 695
68JA5553	E. J. Corey and T. Durst; *J. Am. Chem. Soc.*, 1968, **90**, 5553.	621, 644, 695
68JA5561	J. L. Coke and M. C. Mourning; *J. Am. Chem. Soc.*, 1968, **90**, 5561.	628
68JA5615	E. J. Corey and G. H. Posner; *J. Am. Chem. Soc.*, 1968, **90**, 5615.	117, 118, 466
68JA6091	H. E. Zimmerman and P. S. Mariano; *J. Am. Chem. Soc.*, 1968, **90**, 6091.	871
68JA6171	J. D. White and D. N. Gupta; *J. Am. Chem. Soc.*, 1968, **90**, 6171.	357
68JA6243	G. Zweifel, N. L. Polston and C. C. Whitney; *J. Am. Chem. Soc.*, 1968, **90**, 6243.	278, 488, 531, 662, 1103
68JA6515	P. E. Peterson and J. M. Indelicato; *J. Am. Chem. Soc.*, 1968, **90**, 6515.	1110
68JA6562	J. B. Grutzner and S. Winstein; *J. Am. Chem. Soc.*, 1968, **90**, 6562.	886
68JA6741	W. L. Waters, W. S. Linn and M. Caserio; *J. Am. Chem. Soc.*, 1968, **90**, 6741.	1125
68JA6749	F. G. Bordwell and R. G. Scamehorn; *J. Am. Chem. Soc.*, 1968, **90**, 6749.	871
68JA6816	E. J. Corey and G. T. Kwiatkowski; *J. Am. Chem. Soc.*, 1968, **90**, 6816.	643, 644
68JA6882	M. Saunders, E. L. Hagen and J. Rosenfeld; *J. Am. Chem. Soc.*, 1968, **90**, 6882.	900
68JA7139	G. Zweifel, J. T. Snow and C. C. Whitney; *J. Am. Chem. Soc.*, 1968, **90**, 7139.	534
68JA7271	P. G. Gassman, D. H. Aue and D. S. Patton; *J. Am. Chem. Soc.*, 1968, **90**, 7271.	6
68JA7282	L. L. Miller and D. A. Kaufman; *J. Am. Chem. Soc.*, 1968, **90**, 7282.	1111
68JCS(A)34	D. Dodd and M. D. Johnson; *J. Chem. Soc. (A)*, 1968, 34.	676
68JCS(A)190	J. Chatt, R. S. Coffey, A. Gough and D. T. Thompson; *J. Chem. Soc. (A)*, 1968, 190.	668
68JCS(A)2665	C. O'Connor and G. Wilkinson; *J. Chem. Soc. (A)*, 1968, 2665.	73, 74
68JCS(B)132	W. O. George and V. G. Mansell; *J. Chem. Soc. (B)*, 1968, 132.	870
68JCS(B)411	J. M. Brown and J. L. Occolowitz; *J. Chem. Soc. (B)*, 1968, 411.	865
68JCS(B)1044	E. Abuin, J. Grotewold, E. A. Lissi and M. C. Vara; *J. Chem. Soc. (B)*, 1968, 1044.	657
68JCS(C)228	A. J. Hubert and H. G. Viehe; *J. Chem. Soc. (C)*, 1968, 228.	956
68JCS(C)338	G. H. Whitman and J. A. F. Wickramasinghe; *J. Chem. Soc. (C)*, 1968, 338.	871
68JCS(C)606	A. J. Hubert and H. Reimlinger; *J. Chem. Soc. (C)*, 1968, 606.	956
68JCS(C)676	D. C. Nonhebel; *J. Chem. Soc. (C)*, 1968, 676.	870
68JCS(C)1474	A. G. Davies and B. P. Roberts; *J. Chem. Soc. (C)*, 1968, 1474.	75
68JCS(C)1903	R. Gigg and C. D. Warren; *J. Chem. Soc. (C)*, 1968, 1903.	882
68JCS(C)1982	C. F. Garbers, D. F. Schneider and J. P. van der Merwe; *J. Chem. Soc. (C)*, 1968, 1982.	733
68JCS(C)1984	G. Pattenden and B. C. L. Weedon; *J. Chem. Soc. (C)*, 1968, 1984.	733
68JCS(C)1997	G. Pattenden and B. C. L. Weedon; *J. Chem. Soc. (C)*, 1968, 1997.	725, 733
68JHC337	N. Milstein; *J. Heterocycl. Chem.*, 1968, **5**, 337.	212
68JMC371	D. Warthen; *J. Med. Chem.*, 1968, **11**, 371.	1001
68JMC373	D. Warthen and M. Jacobson; *J. Med. Chem.*, 1968, **11**, 373.	1001
68JOC70	K. W. Ratts and A. N. Yao; *J. Org. Chem.*, 1968, **33**, 70.	399
68JOC221	S. Bank, C. A. Rowe, Jr., A. Schriesheim and L. A. Naslund *J. Org. Chem.*, 1968, **33**, 221.	797
68JOC460	G. Büchi and E. C. Roberts; *J. Org. Chem.*, 1968, **33**, 460.	884
68JOC747	E. Wenkert, K. G. Dave, F. Haglid, R. G. Lewis, T. Oishi, R. V. Stevens and M. Terashima; *J. Org. Chem.*, 1968, **33**, 747.	91
68JOC780	D. J. Peterson; *J. Org. Chem.*, 1968, **33**, 780.	648, 649, 758
68JOC935	H. O. House, B. A. Tefertiller and H. D. Olmstead; *J. Org. Chem.*, 1968, **33**, 935.	871, 881
68JOC943	H. O. House and T. M. Bare; *J. Org. Chem.*, 1968, **33**, 943.	871, 881
68JOC1301	J. Warkentin and R. A. Cox; *J. Org. Chem.*, 1968, **33**, 1301.	871
68JOC1421	R. Hiatt, T. Mill, K. C. Irwin and J. K. Castleman; *J. Org. Chem.*, 1968, **33**, 1421.	37
68JOC1675	P. R. Jones and J. R. Young; *J. Org. Chem.*, 1968, **33**, 1675.	128
68JOC1750	G. Kruger; *J. Org. Chem.*, 1968, **33**, 1750.	871
68JOC2203	R. A. Benkeser, T. E. Johnston and W.-H. Tong; *J. Org. Chem.*, 1968, **33**, 2203.	115
68JOC2222	H. D. Zook, T. J. Russo, E. F. Ferrand and D. S. Stotz; *J. Org. Chem.*, 1968, **33**, 2222.	130, 869, 871, 881
68JOC2975	D. J. Pasto, K. Garves and J. P. Sevenair; *J. Org. Chem.*, 1968, **33**, 2975.	667
68JOC3425	J. S. McConaghy, Jr. and J. J. Bloomfield; *J. Org. Chem.*, 1968, **33**, 3425.	881
68JOC3477	H. D. Zook, W. L. Kelly and I. Y. Posey; *J. Org. Chem.*, 1968, **33**, 3477.	869, 871, 881
68JOC3566	D. L. Roberts, R. A. Heckmann, B. P. Hege and S. A. Bellin; *J. Org. Chem.*, 1968, **33**, 3566.	726
68JOC3679	K. J. Crowley; *J. Org. Chem.*, 1968, **33**, 3679.	420
68JOC3948	A. Carbonaro, A. Greco and G. Dall'Asta; *J. Org. Chem.*, 1968, **33**, 3948.	519
68JOC4258	P. S. Wharton and R. A. Kretchmer; *J. Org. Chem.*, 1968, **33**, 4258.	812
68JOC4413	D. T. Manning, H. A. Coleman and R. A. Langdale-Smith; *J. Org. Chem.*, 1968, **33**, 4413.	808
68JOM(11)P7	J. J. Eisch and M. W. Foxton; *J. Organomet. Chem.*, 1968, **11**, P7.	511
68JOM(12)P33	J. J. Eisch and M. W. Foxton; *J. Organomet. Chem.*, 1968, **12**, P33.	511
68JOM(13)263	N. A. Nesmeyanov, V. V. Mikulshina and O. A. Reutov; *J. Organomet. Chem.*, 1968, **13**, 263.	737
68JOU21	L. Y. Ukhin, A. M. Sladkov and U. N. Gorshkov; *J. Org. Chem. USSR (Engl. Transl.)*, 1968, **4**, 21 (*Chem. Abstr.*, **68**, 78 330f).	1005

68JPC321 R. K. Brinton; *J. Phys. Chem.*, 1968, **72**, 321. 1143

68LA(711)46 G. A. Wittig and H. Dorsch; *Justus Liebigs Ann. Chem.*, 1968, **711**, 46. 104

68LA(711)65 G. Wittig and J. Meske-Schüller; *Justus Liebigs Ann. Chem.*, 1968, **711**, 65. 104

B-68MI 102-01 E. H. White and D. J. Woodcock; in "The Chemistry of the Amino Group," ed. S. Patai, Wiley, London, 1968, vol. 1, p. 409. 28, 32, 35

68MI 103-01 J. P. Candlin and A. R. Oldham; *Discuss. Far. Soc.*, 1968, **46**, 60. 82, 84

B-68MI 106-01 S. Patai (ed.); "The Chemistry of the Amino Group," Wiley-Interscience, 1968. 258

68MI 110-01 J. P. Candlin and A. R. Oldman; *Faraday Discuss. Chem. Soc.*, 1968, **46**, 60. 456

68MI 112-01 J. P. Candlin and A. R. Oldham; *Discuss. Faraday Soc.*, 1968, **46**, 60. 507

B-68MI 114-01 E. H. White and D. J. Woodcock; in "The Chemistry of the Amino Group," ed. S. Patai, Interscience, New York, 1968, pp. 423–425. 624

B-68MI 114-02 E. H. White and D. J. Woodcock; in "The Chemistry of the Amino Group," ed. S. Patai, Interscience, New York, 1968, pp. 409–416. 627

B-68MI 114-03 E. H. White and D. J. Woodcock; in "The Chemistry of the Amino Group," ed. S. Patai, Interscience, New York, 1968, pp. 440–445. 637

68MI 114-04 L. A. Paquette; *Mech. Mol. Migr.*, 1968, **1**, 121. 617

68MI 118-01 H. M. Frey and R. K. Solly; *Trans. Faraday Soc.*, 1968, **64**, 1858. 807

B-68MI 119-01 O. A. Reutov and I. P. Beletskaya; "Reaction Mechanisms of Organometallic Compounds," North-Holland Publishing Co., Amsterdam, 1968, p. 1. 845

B-68MI 119-02 G. A. Olah and P. von R. Schleyer (eds.); "Carbonium Ions," Wiley, New York, 1968, vol. 1. 891

B-68MI 119-03 C. D. Nenitzescu; in "Carbonium Ions," eds. G. A. Olah and P. von R. Schleyer, Wiley, New York, 1968, vol. 1, chapter 1. 899, 901; 914

B-68MI 119-04 A. R. Forrester and J. M. Hay and R. H. Thomson; "Organic Chemistry of Stable Free Radicals," Academic Press, New York, 1968. 922

68MI 119-05 H. Fischer and D. Rewicki; *Prog. Org. Chem.*, 1968, **7**, 116. 844

68MI 119-06 D. J. Cram; *Surv. Prog. Chem.*, 1968, **4**, 45. 845, 869, 871

B-68MI 121-01 T. F. Rutledge; "Acetylenic Compounds," Reinhold, New York, 1968. 998

68OS(48)40 S. Boatman, T. M. Harris and C. R. Hauser; *Org. Synth.*, 1968, **48**, 40. 881

68OS(48)53 J. A. Skorcz and F. E. Kaminski; *Org. Synth.*, 1968, **48**, 53. 881

68OS(48)68 P. G. Gassman and J. L. Marshall; *Org. Synth.*, 1968, **48**, 68. 430

68OS(48)80 W. S. Murphy, P. J. Hamrick and C. R. Hauser; *Org. Synth.*, 1968, **48**, 80. 881

68PAC1 D. H. R. Barton; *Pure Appl. Chem.*, 1968, **16**, 1. 927

68PAC79 O. A. Reutov; *Pure Appl. Chem.*, 1968, **17**, 79. 65, 66

68QR338 G. B. Gill; *Q. Rev., Chem. Soc.*, 1968, **22**, 338. 400, 402, 403

68QR549 J. A. Kerr and A. C. Lloyd; *Q. Rev. Chem. Soc.*, 1968, **22**, 549. 321

68RCR789 V. S. Fel'dblyum and N. V. Obeshchalova, *Russ. Chem. Rev.* (*Engl. Trans.*), 1968, **37**, 789. 294

68RTC69 L. Brandsma, J. F. Arens, J. G. A. Kooyman, H. P. G. Hendricks and P. P. Montijn; *Recl. Trav. Chim. Pays-Bas*, 1968, **87**, 69. 1030

68RTC298 J. W. de Haan and H. Kloosterziel; *Recl. Trav. Chim. Pays-Bas*, 1968, **87**, 298. 403, 800

68RTC319 H. Hogeveen and C. J. Gaasbeek; *Recl. Trav. Chim. Pays-Bas*, 1968, **87**, 319. 901

68RTC916 S. Hoff, L. Brandsma and J. F. Arens; *Recl. Trav. Chim. Pays-Bas*, 1968, **87**, 916. 955

68RTC1179 S. Hoff, L. Brandsma and J. F. Arens; *Recl. Trav. Chim. Pays-Bas*, 1968, **87**, 1179. 971

68RTC1249 A. Schaap and J. F. Arens; *Recl. Trav. Chim. Pays-Bas*, 1968, **87**, 1249. 1023

68RTC1301 S. E. Schaafsma, E. J. F. Molenaar, H. Steinberg and J. Th. de Boer; *Recl. Trav. Chim. Pays-Bas*, 1968, **87**, 1301. 929

68T545 S. B. Kulkarni and S. Dev; *Tetrahedron*, 1968, **24**, 545. 422

68T1651 K. M. Baker and J. P. Bartley; *Tetrahedron*, 1968, **24**, 1651. 870

68T1869 D. C. Nonhebel; *Tetrahedron*, 1968, **24**, 1869. 870

68T2215 J. D. Bacha and J. K. Kochi; *Tetrahedron*, 1968, **24**, 2215. 561

68T4285 T. Metler, A. Uchida and S. I. Miller; *Tetrahedron*, 1968, **24**, 4285. 538

68T4489 J. M. Coulter, J. W. Lewis and P. P. Lynch; *Tetrahedron*, 1968, **24**, 4489. 443

68T4509 J. Sinnreich and D. Elad; *Tetrahedron*, 1968, **24**, 4509. 925

68T6287 C. Rappe and W. H. Sachs; *Tetrahedron*, 1968, **24**, 6287. 871

68TL249 H. Lankamp, W. Th. Nauta and C. MacLean; *Tetrahedron Lett.*, 1968, **9**, 249. 922

68TL345 R. H. Shapiro; *Tetrahedron Lett.*, 1968, **9**, 345. 633

68TL503 T. Sakakibara, Y. Odaira and S. Tsutsumi; *Tetrahedron Lett.*, 1968, **9**, 503. 1143, 1144

68TL1457 H. G. Richey, Jr. and A. M. Rothman; *Tetrahedron Lett.*, 1968, **9**, 1457. 539

68TL2101 J.-M. Conia and M.-L. Leriverend; *Tetrahedron Lett.*, 1968, **9**, 2101. 422

68TL2317 W. H. Sachs and C. Rappe; *Tetrahedron Lett.*, 1968, **9**, 2317. 871

68TL2809 H. Schäfer, U. Schöllkopf and D. Walter; *Tetrahedron Lett.*, 1968, **9**, 2809. 396, 397

68TL2919 R. F. Curtis and J. A. Taylor; *Tetrahedron Lett.*, 1968, **9**, 2919. 1018

68TL2921 W. H. Sachs and C. Rappe; *Tetrahedron Lett.*, 1968, **9**, 2921. 871

68TL3635 H. Nozaki, M. Kurita and R. Noyori; *Tetrahedron Lett.*, 1968, **9**, 3635. 422

68TL3797 A. J. Birch and G. S. R. Subba Rao; *Tetrahedron Lett.*, 1968, **9**, 3797. 797

68TL4039 G. Capozzi, G. Melloni, G. Modena and M. Piscitelli; *Tetrahedron Lett.*, 1968, **9**, 4039. 1112

68TL4309 H. Junek and H. Sterk; *Tetrahedron Lett.*, 1968, **9**, 4309. 889

68TL4511 A. Krebs; *Tetrahedron Lett.*, 1968, **9**, 4511. 1075

68TL4569 M. U. S. Sultanbawa; *Tetrahedron Lett.*, 1968, **9**, 4569. 682

68TL4617 M. Hanack, V. Vött and H. Ehrhardt; *Tetrahedron Lett.*, 1968, **9**, 4617. 1123

68TL5041 E. J. Corey and H. A. Kirst; *Tetrahedron Lett.*, 1968, **9**, 5041. 999, 1052

68TL5215 R. Huisgen, G. Boche, A. Dahmen and W. Hechtl; *Tetrahedron Lett.*, 1968, **9**, 5215. 421

68TL5381 Y. Hayashi and R. Oda; *Tetrahedron Lett.*, 1968, **9**, 5381. 399

68ZOR1114 K. M. Krivozheiko and A. V. El'tsov; *Zh. Org. Khim.*, 1968, **4**, 1114. 837

69AG496 H. Fernholz and H. J. Schmidt; *Angew. Chem.*, 1969, **81**, 496. 181
69AG(E)429 J. C. Craig, M. D. Bergenthal, I. Fleming and J. Harley-Mason; *Angew. Chem., Int. Ed. Engl.*, 1969, **8**, 429. 1081
69AG(E)556 H. M. R. Hoffmann; *Angew. Chem., Int. Ed. Engl.*, 1969, **8**, 556. 524, 772
69AG(E)604 R. Huisgen, W. Scheer, H. Mader and E. Brunn; *Angew. Chem., Int. Ed. Engl.*, 1969, **8**, 604. 524
69AG(E)781 R. B. Woodward and R. Hoffmann; *Angew. Chem., Int. Ed. Engl.*, 1969, **8**, 781. 349, 400, 417, 418, 419, 422
69AJC1321 A. J. Birch, R. W. Rickards and K. J. S. Stapleford; *Aust. J. Chem.*, 1969, **22**, 1321. 100
69AJC1737 G. W. K. Cavill and P. J. Williams; *Aust. J. Chem.*, 1969, **22**, 1737. 747, 748
69AJC2145 G. W. K. Cavill, D. G. Liang and P. J. Williams; *Aust. J. Chem.*, 1969, **22**, 2145. 747, 748
69APO(7)1 Z. Rappoport; *Adv. Phys. Org. Chem.*, 1969, **7**, 1. 1109, 1110
69BSF842 L. Hilaire and F. G. Gault; *Bull. Soc. Chim. Fr.*, 1969, 842. 430
69BSF898 R. Couffignal and M. Gaudemar; *Bull. Soc. Chim. Fr.*, 1969, 898. 971
69BSF2427 M. Julia and M. Maumy; *Bull. Soc. Chim. Fr.*, 1969, 2427. 343
69BSF3665 P. Courtot and R. Rumin; *Bull. Soc. Chim. Fr.*, 1969, 3665. 420
69CB2966 T. Severin, D. Scheel and P. Adhikary; *Chem. Ber.*, 1969, **102**, 2966. 486
69CB3877 M. Regitz and J. Rüter; *Chem. Ber.*, 1969, **102**, 3877. 395
69CC99 G. M. Blackburn, W. D. Ollis, C. Smith and I. O. Sutherland; *J. Chem. Soc., Chem. Commun.*, 1969, 99. 807, 821, 822
69CC122 C. U. Pittman, Jr.; *J. Chem. Soc., Chem. Commun.*, 1969, 122. 1125
69CC292 J. E. Dubois and J. Toullec; *J. Chem. Soc., Chem. Commun.*, 1969, 292. 871
69CC293 R. W. C. Cose, A. M. Davies, W. D. Ollis, C. Smith and I. O. Sutherland; *J. Chem. Soc., Chem. Commun.*, 1969, 293. 407, 408
69CC294 R. W. Jemison and W. D. Ollis; *J. Chem. Soc., Chem. Commun.*, 1969, 294. 823
69CC303 U. Axen, F. H. Lincoln and J. L. Thompson; *J. Chem. Soc., Chem. Commun.*, 1969, 303. 729
69CC674 R. G. Carlson and D. E. Henton; *J. Chem. Soc., Chem. Commun.*, 1969, 674. 1023
69CC785 W. P. Schneider; *J. Chem. Soc., Chem. Commun.*, 1969, 785. 729
69CC1044 N. Frydman, R. Bixon, M. Sprecher and Y. Mazur; *J. Chem Soc., Chem. Commun.*, 1969, 1044. 1111
69CC1049 R. G. Carlson and E. L. Biersmith; *J. Chem. Soc., Chem. Commun.*, 1969, 1049. 842
69CC1067 P. Belanger, J. G. Atkinson and R. S. Stuart; *J. Chem. Soc., Chem. Commun.*, 1969, 1067. 850
69CC1069 H. M. Frey, R. G. Hopkins, H. E. O'Neal and F. T. Bond; *J. Chem. Soc., Chem. Commun.*, 1969, 1069. 799
69CC1225 J. E. Baldwin and U. V. Roy; *J. Chem. Soc., Chem. Commun.*, 1969, 1225. 498
69CC1365 R. E. Harmon, J. L. Parsons and S. K. Gupta; *J. Chem. Soc., Chem. Commun.*, 1969, 1365. 80
69CC1395 J. E. Herz and E. González; *J. Chem. Soc., Chem. Commun.*, 1969, 1395. 630
69CC1497 D. H. R. Barton, R. H. Hesse, G. Tarzia and M. M. Pechet; *J. Chem. Soc., Chem. Commun.*, 1969, 1497. 871, 881
69CC1498 M. Tanabe and D. F. Crowe; *J. Chem. Soc., Chem. Commun.*, 1969, 1498. 871, 881
69CC1503 P. Abley and F. J. McQuillin; *J. Chem. Soc., Chem. Commun.*, 1969, 1503. 507
69CC1520 G. Capozzi, G. Melloni, G. Modena and U. Tonellato; *J. Chem. Soc., Chem. Commun.*, 1969, 1520. 1112
69CI(L)1595 B. Heath-Brown; *Chem. Ind. (London)*, 1969, 1595. 100
69CJC1555 S. McLean, C. J. Webster and R. J. D. Rutherford; *Can. J. Chem.*, 1969, **47**, 1555. 794
69CJC2391 R. C. Bansal, A. W. McCulloch and A. G. McInnes; *Can. J. Chem.*, 1969, **47**, 2391. 522
69CR(C)342 W. Chodkiewicz, M. Karila and M. L. Capmau; *C. R. Hebd. Seances Acad. Sci., Ser. C*, 1969, **269**, 342. 1055
69CR(C)567 M. Pedoussaut and J. Coudane; *C. R. Hebd. Seances Acad. Sci., Ser. C.*, 1969, **269**, 567. 388
69CRV103 H. M. Frey and R. Walsh; *Chem. Rev.*, 1969, **69**, 103. 402, 794, 795
69CRV279 S. W. Benson, F. R. Cruickshank, D. M. Golden, G. R. Haugen, H. E. O'Neal, A. S. Rodgers, R. Shaw and R. Walsh; *Chem. Rev.*, 1969, **69**, 279. 1109
69CRV799 F. R. Hartley; *Chem. Rev.*, 1969, **69**, 799. 669
69FCF387 H. F. Ebel; *Forschr. Chem. Forsch.*, 1969, **12**, 387. 104
69HCA880 G. Ohloff and W. Pickenhagen; *Helv. Chim. Acta*, 1969, **52**, 880. 811
69HCA1030 D. Felix, K. Gschwend-Steen, A. E. Wick and A. Eschenmoser; *Helv. Chim. Acta*, 1969, **52**, 1030. 415
69JA514 W. K. Musker and G. L. Larson; *J. Am. Chem. Soc.*, 1969, **91**, 514. 654
69JA522 H. R. Ward and E. Karafiath; *J. Am. Chem. Soc.*, 1969, **91**, 522. 809
69JA527 A. Streitwieser, Jr., R. A. Caldwell and W. R. Young; *J. Am. Chem. Soc.*, 1969, **91**, 527. 848
69JA529 A. Streitwieser, Jr., R. A. Caldwell and W. R. Young; *J. Am. Chem. Soc.*, 1969, **91**, 529. 847
69JA530 H. W. Amburn, K. C. Kauffman and H. Shechter; *J. Am. Chem. Soc.*, 1969, **91**, 530. 871
69JA676 F. E. Henoch, K. G. Hampton and C. R. Hauser; *J. Am. Chem. Soc.*, 1969, **91**, 676. 154
69JA1009 T. D. Hoffman and D. J. Cram; *J. Am. Chem. Soc.*, 1969, **91**, 1009. 883
69JA1037 J. E. Williams, Jr., R. Sustmann, L. C. Allen and P. von R. Schleyer; *J. Am. Chem. Soc.*, 1969, **91**, 1037. 892
69JA1146 G. E. Lienhard and T. C. Wang; *J. Am. Chem. Soc.*, 1969, **91**, 1146. 870
69JA1237 A. R. Lepley; *J. Am. Chem. Soc.*, 1969, **91**, 1237. 399
69JA1458 G. A. Olah, M. B. Comisarow and C. J. Kim; *J. Am. Chem. Soc.*, 1969, **91**, 1458. 915
69JA1578 L. L. Miller, R. Greisinger and R. F. Boyer; *J. Am. Chem. Soc.*, 1969, **91**, 1578. 799
69JA1853 J. B. Siddall, M. Biskup and J. H. Fried; *J. Am. Chem. Soc.*, 1969, **91**, 1853. 540
69JA2115 S. A. Sherrod and R. G. Bergmann; *J. Am. Chem. Soc.*, 1969, **91**, 2115. 1114

69JA2117 M. Hanack and T. Bässler; *J. Am. Chem. Soc.*, 1969, **91**, 2117. 1114

69JA2147 H. C. Brown, M. M. Rogić, H. Nambu and M. W. Rathke; *J. Am. Chem. Soc.*, 1969, **91**, 2147. 125

69JA2150 H. C. Brown, M. M. Rogić, M. W. Rathke and G. W. Kabalka; *J. Am. Chem. Soc.*, 1969, **91**, 2150. 126

69JA2797 P. W. K. Flanagan, H. W. Amburn, H. W. Stone, J. G. Traynham and H. Shechter; *J. Am. Chem. Soc.*, 1969, **91**, 2797. 869, 871

69JA2815 S. J. Rhoads and R. D. Cockroft; *J. Am. Chem. Soc.*, 1969, **91**, 2815. 826

69JA3238 J. W. Wilson; *J. Am. Chem. Soc.*, 1969, **91**, 3238. 1124

69JA3261 G. A. Olah, G. Klopman and R. H. Schlosberg; *J. Am. Chem. Soc.*, 1969, **91**, 3261. 901

69JA3281 A. F. Thomas; *J. Am. Chem. Soc.*, 1969, **91**, 3281. 840

69JA3292 R. S. Bly and S. U. Koock; *J. Am. Chem. Soc.*, 1969, **91**, 3292. 1126

69JA3299 R. S. Bly and S. U. Koock; *J. Am. Chem. Soc.*, 1969, **91**, 3299. 1126

69JA3667 G. A. Olah, J. M. Bollinger and A. M. White; *J. Am. Chem. Soc.*, 1969, **91**, 3667. 909

69JA3870 L. A. Paquette and R. W. Houser; *J. Am. Chem. Soc.*, 1969, **91**, 3870. 618, 704

69JA3958 G. A. Olah, J. R. DeMember, C. Y. Lui and A. M. White; *J. Am. Chem. Soc.*, 1969, **91**, 3958. 906

69JA4318 E. J. Corey, K. Achiwa and J. A. Katzenellenbogen; *J. Am. Chem. Soc.*, 1969, **91**, 4318. 1134

69JA4322 S. Masamune, S. Takada, N. Nakatsuka, R. Vukov and E. N. Cain; *J. Am. Chem. Soc.*, 1969, **91**, 4322. 400, 799

69JA4459 J. Almy and D. J. Cram; *J. Am. Chem. Soc.*, 1969, **91**, 4459. 797

69JA4521 P. E. Peterson and R. J. Kamat; *J. Am. Chem. Soc.*, 1969, **91**, 4521. 1124

69JA4580 T. E. Hogen Esch and J. Smid; *J. Am. Chem. Soc.*, 1969, **91**, 4580. 859

69JA4600 P. J. Stang and R. Summerville; *J. Am. Chem. Soc.*, 1969, **91**, 4600. 1111

69JA4819 G. M. Kramer; *J. Am. Chem. Soc.*, 1969, **91**, 4819. 893

69JA4824 T. L. Jacobs and R. S. Macomber; *J. Am. Chem. Soc.*, 1969, **91**, 4824. 1125, 1126

69JA4871 G. M. Whitesides, W. F. Fischer, Jr., J. San Filippo, Jr., R. W. Bashe and H. O. House; *J. Am. Chem. Soc.*, 1969, **91**, 4871. 118, 119, 469, 473

69JA5162 R. A. Benkeser and W. E. Broxterman; *J. Am. Chem. Soc.*, 1969, **91**, 5162. 401, 800

69JA5246 Z. Rappoport and A. Gal; *J. Am. Chem. Soc.*, 1969, **91**, 5246. 1112, 1114

69JA5350 R. Sustmann, J. E. Williams, M. J. S. Dewar, L. C. Allen and P. v. R. Schleyer; *J. Am. Chem. Soc.*, 1969, **91**, 5350. 1109

69JA5364 G. Just, C. Simonovitch, F. H. Lincoln, W. P. Schneider, U. Axen, G. B. Spero and J. E. Pike; *J. Am. Chem. Soc.*, 1969, **91**, 5364. 729

69JA5677 M. S. Newman and C. D. Beard; *J. Am. Chem. Soc.*, 1969, **91**, 5677. 1118

69JA5682 W. L. Mock; *J. Am. Chem. Soc.*, 1969, **91**, 5682. 620, 791

69JA5801 G. A. Olah and A. M. White; *J. Am. Chem. Soc.*, 1969, **91**, 5801. 894, 899

69JA5901 C. A. Brown; *J. Am. Chem. Soc.*, 1969, **91**, 5901. 506

69JA6112 K. W. Ratts and R. D. Partos; *J. Am. Chem. Soc.*, 1969, **91**, 6112. 988

69JA6194 P. E. Peterson and J. M. Indelicato; *J. Am. Chem. Soc.*, 1969, **91**, 6194. 1110, 1115

69JA6461 M. S. Newman and T. B. Patrick; *J. Am. Chem. Soc.*, 1969, **91**, 6461. 1143

69JA6464 C. E. Castro, R. Havlin, V. K. Honwad, A. Malte and S. Mojé; *J. Am. Chem. Soc.*, 1969, **91**, 6464. 1005

69JA6707 R. F. Heck; *J. Am. Chem. Soc.*, 1969, **91**, 6707. 493

69JA6749 C. D. Ritchie; *J. Am. Chem. Soc.*, 1969, **91**, 6749. 869

69JA6852 H. C. Brown, H. Nambu and M. M. Rogic; *J. Am. Chem. Soc.*, 1969, **91**, 6852. 464

69JA6854 H. C. Brown, H. Nambu and M. M. Rogić; *J. Am. Chem. Soc.*, 1969, **91**, 6854. 125

69JA6861 A. Nickon and Y.-i. Lin; *J. Am. Chem. Soc.*, 1969, **91**, 6861. 906

69JA7510 C. Y. Meyers, A. M. Malte and W. S. Matthews; *J. Am. Chem. Soc.*, 1969, **91**, 7510. 618, 706

69JCP(51)2767 L. Pauling; *J. Chem. Phys.*, 1969, **51**, 2767. 923

69JCS(A)433 V. A. Dorokhov and M. F. Lappert; *J. Chem. Soc. (A)*, 1969, 433. 657

69JCS(B)187 T. O. Bamkole and E. U. Emovon; *J. Chem. Soc. (B)*, 1969, 187. 774

69JCS(B)652 D. Graham, R. N. Haszeldine and P. J. Robinson; *J. Chem. Soc. (B)*, 1969, 652. 652

69JCS(B)852 A. W. P. Jarvie, A. Holt and J. Thompson; *J. Chem. Soc. (B)*, 1969, 852. 652

69JCS(B)863 J. W. Lewis and A. A. Pearce; *J. Chem. Soc. (B)*, 1969, 863. 659

69JCS(B)1071 D. Dodd and M. D. Johnson; *J. Chem. Soc. B*, 1969, 1071. 66

69JCS(C)1024 L. Crombie, P. Hemesley and G. Pattenden; *J. Chem. Soc. (C)*, 1969, 1024. 734

69JCS(C)2173 R. E. Atkinson, R. F. Curtis, D. M. Jones and J. A. Taylor; *J. Chem. Soc. (C)*, 1969, 2173. 1006

69JCS(C)2273 D. W. Burt and P. Simpson; *J. Chem. Soc. (C)*, 1969, 2273. 999

69JCS(C)2464 D. Lloyd and N. W. Preston; *J. Chem. Soc. (C)*, 1969, 2464. 738, 742

69JCS(C)2728 A. K. Bose, M. S. Manhas and R. M. Ramer; *J. Chem. Soc. (C)*, 1969, 2728. 725

69JOC4 F. A. Carey and H. S. Tremper; *J. Org. Chem.*, 1969, **34**, 4. 13

69JOC520 S. P. Pappas, B. C. Pappas and N. A. Portnoy; *J. Org. Chem.*, 1969, **34**, 520. 517

69JOC638 J. D. Citron, J. E. Lyons and L. H. Sommer; *J. Org. Chem.*, 1969, **34**, 638. 428

69JOC641 W. H. Glaze, C. M. Selman, A. L. Ball, Jr. and L. E. Bray; *J. Org. Chem.*, 1969, **34**, 641. 110

69JOC1220 M. S. Newman and A. O. M. Okorodudu; *J. Org. Chem.*, 1969, **34**, 1220. 1119, 1143, 1144

69JOC1233 E. J. Corey and E. Block; *J. Org. Chem.*, 1969, **34**, 1233. 232

69JOC1495 D. Kaufman and L. L. Miller; *J. Org. Chem.*, 1969, **34**, 1495. 1110

69JOC1936 J. E. Mulvaney and D. J. Newton; *J. Org. Chem.*, 1969, **34**, 1936. 539

69JOC1962 H. W. Whitlock, Jr. and L. E. Overman; *J. Org. Chem.*, 1969, **34**, 1962. 872

69JOC2314 A. J. Fry and W. B. Farnham; *J. Org. Chem.*, 1969, **34**, 2314. 906

69JOC2324 H. O. House, L. J. Czuba, M. Gall and H. D. Olmstead; *J. Org. Chem.*, 1969, **34**, 2324. 871, 881

69JOC2817 D. H. Sardella, D. H. Heinert and B. L. Shapiro; *J. Org. Chem.*, 1969, **34**, 2817. 870, 871

69JOC3053	E. J. Corey and D. E. Cane; *J. Org. Chem.*, 1969, **34**, 3053.	643, 644
69JOC3070	B. J. L. Huff, F. N. Tuller and D. Caine; *J. Org. Chem.*, 1969, **34**, 3070.	871
69JOC3615	H. O. House and W. F. Fischer, Jr.; *J. Org. Chem.*, 1969, **34**, 3615.	311
69JOC3616	H. O. House and W. F. Fischer, Jr.; *J. Org. Chem.*, 1969, **34**, 3616.	1048
69JOC3667	E. J. Corey and K. Achiwa; *J. Org. Chem.*, 1969, **34**, 3667.	17
69JOC3918	H. C. Brown and S. Krishnamurthy; *J. Org. Chem.*, 1969, **34**, 3918.	428
69JOC3923	H. M. Bell, C. W. Vanderslice and A. Spehar; *J. Org. Chem.*, 1969, **34**, 3923.	4, 7, 9, 10, 17, 429
69JOC3974	Y. Ogata and K. Aoki; *J. Org. Chem.*, 1969, **34**, 3974.	108
69JOC3978	Y. Ogata and K. Aoki; *J. Org. Chem.*, 1969, **34**, 3978.	108
69JOC4083	J. Hine, B. C. Menon, J. Mulders and R. L. Flachskam Jr.; *J. Org. Chem.*, 1969, **34**, 4083.	870
69JOC4134	V. Boekelheide and J. E. Nottke; *J. Org. Chem.*, 1969, **34**, 4134.	537
69JOM(16)7	M. Lefrançois and Y. Gault; *J. Organomet. Chem.*, 1969, **16**, 7.	665
69JOM(16)503	D. Seyferth, S. S. Woshurne and T. F. Jula; *J. Organometallic Chem.*, 1969, **16**, 503.	291
69JOM(18)263	W. L. Respess and C. Tamborski; *J. Organomet. Chem.*, 1969, **18**, 263.	426
69JOM(18)P1	D. D. Davis and C. E. Gray; *J. Organomet. Chem.*, 1969, **18**, P1.	667
69LA(727)194	W. Brenner, P. Heimbach and G. Wilke; *Justus Liebigs Ann. Chem.*, 1969, **727**, 194.	812
69LA(729)52	W. Hoffmann, H. Pasedach and H. Pommer; *Justus Liebigs Ann. Chem.*, 1969, **729**, 52.	832
69M724	F. Asinger, H. Offermanns and A. Saus; *Monatsh. Chem.*, 1969, **100**, 724.	46
69MI 102-01	D. S. Matteson; *Organometal. Rev., Sect. A*, 1969, 263.	65
69MI 103-01	G. K. Koch; *J. Labelled Comp. Radiopharm.*, 1969, **5**, 99.	74
B-69MI 104-01	A. T. Nielsen; in "The Chemistry of the Nitro and Nitroso Groups," *Part* 1, ed. H. Feuer; Interscience, New York, 1969, Chapter 7, pp. 349–486.	156
B-69MI 112-01	T. F. Rutledge; "Acetylenes and Allenes," Reinhold, New York, 1969, p. 232.	537
B-69MI 112-02	T. F. Rutledge; "Acetylenes and Allenes," Reinhold, New York, 1969, p. 173.	549
69MI 117-01	K. K. Lum and G. G. Smith; *Int. J. Chem. Kinet.*, 1969, **1**, 401.	773
69MI 119-01	T. Durst; *Adv. Org. Chem.*, 1969, **6**, 285.	881
B-69MI 120-01	T. F. Rutledge; "Acetylenes and Allenes," Rheinhold Book Corp., New York, 1969.	953, 954
B-69MI 120-02	J. H. Wotiz; in "Chemistry of Acetylenes," eds. H. G. Viehe, M. Dekker, New York, 1969, p. 365.	954
B-69MI 121-01	H. G. Viehe (ed.); "Chemistry of Acetylenes," Marcel Dekker, New York, 1969	998
B-69MI 121-02	G. Köbrich and P. Buck; in "Chemistry of Acetylenes," ed. H. G. Viehe, Marcel Decker, New York, 1969, ch. 2, p. 99.	1068, 1069, 1081
B-69MI 121-03	W. Ziegenbein; in "Chemistry of Acetylenes," ed. H. G. Viehe, Marcel Decker, New York, 1969, ch. 3, p. 169.	1002, 1021, 1026, 1037, 1038, 1046, 1047
B-69MI 121-04	P. Cadiot and W. Chodkiewicz; in "Chemistry of Acetylenes," ed. H. G. Viehe, Marcel Dekker, New York, 1969, ch. 9, p. 597.	1017, 1018, 1019, 1034
B-69MI 121-05	S. Y. Delavarenne and H. G. Viehe; in "Chemistry of Acetylenes," ed. H. G. Viehe, Marcel Dekker, New York, 1969, ch. 10, p. 651.	999, 1016
B-69MI 121-06	L. Brandsma, H. J. T. Bos and J. F. Arens; in "Chemistry of Acetylenes," ed. H. G. Viehe, Marcel Dekker, New York, 1969, ch. 11, p. 751.	999
B-69MI 122-01	W. Ziegenbein; in "Chemistry of Acetylenes," ed. H. G. Viehe, Dekker, New York, 1969, chap. 3.	1133
69NAT1212	J. W. Cornforth, J. W. Redmond, H. Eggerer, W. Buckel and C. Gutschow; *Nat.*, 1969, **221**, 1212.	458
69OR(17)155	T. M. Harris and C. M. Harris; *Org. React.*, 1969, **17**, 155.	879
69PAC1	W. Kitching; *Pure Appl. Chem.*, 1969, **19**, 1.	65
69QR141	S. Winstein; *Q. Rev., Chem. Soc.*, 1969, **23**, 141.	917
69RTC266	A. P. Ter Borg and H. Kloosterziel; *Recl. Trav. Chim. Pays-Bas*, 1969, **88**, 266.	402, 795
69RTC371	H. Hogeveen and A. F. Bickel; *Recl. Trav. Chim. Pays-Bas*, 1969, **88**, 371.	901
69RZC95	W. E. Witold and J. Zimmicki; *Rocz. Chem.*, 1969, **43**, 95 (*Chem. Abstr.*, 1969, **71**, 3169p).	745
69S75	M. Schlosser and M. Zimmermann; *Synthesis*, 1969, 75	731
69S97	A. J. Hubert and H. Reimlinger; *Synthesis*, 1969, 97.	795, 882
69T1479	B. N. Blackett, J. M. Coxon, M. P. Hartshorn, B. L. J. Jackson and C. N. Muir; *Tetrahedron*, 1969, **25**, 1479.	387
69T2647	J. Nishimura, N. Kawabata and J. Furukawa; *Tetrahedron*, 1969, **25**, 2647.	300
69T3107	J. M. Coxon, M. P. Hartshorn, D. N. Kirk and M. A. Wilson; *Tetrahedron*, 1969, **25**, 3107.	631
69T4117	T. P. Lewis; *Tetrahedron*, 1969, **25**, 4117.	849
69T4445	J. M. Coxon, M. P. Hartshorn, A. J. Lewis, K. E. Richards and W. H. Swallow; *Tetrahedron*, 1969, **25**, 4445.	387
69T5771	T. L. Kwa and C. Boelhouwer; *Tetrahedron*, 1969, **25**, 5771.	464
69TL505	M. Pereyre and Y. Odic; *Tetrahedron Lett.*, 1969, **10**, 505	132
69TL707	H. Felkin, G. Swierczewski and A. Tambuté; *Tetrahedron Lett.*, 1969, **10**, 707.	316
69TL821	H. Felkin and A. Tambuté; *Tetrahedron Lett.*, 1969, **10**, 821.	397
69TL993	E. F. Kiefer and J. Y. Fukunaga; *Tetrahedron Lett.*, 1969, **10**, 993.	794
69TL1117	P. A. Tardella; *Tetrahedron Lett.*, 1969, **10**, 1117.	132
69TL1379	J. Fried, C. Lin and S. H. Ford; *Tetrahedron Lett.*, 1969, **10**, 1379.	1023
69TL1439	C. A. Grob and R. Spaar; *Tetrahedron Lett.*, 1969, **10**, 1439.	1113
69TL1653	J. K. Crandall and D. J. Keyton; *Tetrahedron Lett.*, 1969, **10**, 1653.	1131
69TL1739	M. Černý and J. Málek; *Tetrahedron Lett.*, 1969, **10**, 1739.	18, 21
69TL2145	R. E. Ireland and G. Pfister; *Tetrahedron Lett.*, 1969, **10**, 2145.	434
69TL2313	H. G. Richey, Jr. and W. C. Kossa, Jr.; *Tetrahedron Lett.*, 1969, **10**, 2313.	495
69TL2607	W. R. Roth and A. Friedrich; *Tetrahedron Lett.*, 1969, **10**, 2607.	400, 799
69TL2619	U. Schöllkopf, G. Ostermann and J. Schossig; *Tetrahedron Lett.*, 1969, **10**, 2619.	399

69TL3101	S. Yamada, M. Kuramoto and Y. Kikugawa; *Tetrahedron Lett.*, 1969, **10**, 3101.	102
69TL3105	S.-I. Yamada and H. Akimoto; *Tetrahedron Lett.*, 1969, **10**, 3105.	55
69TL3253	N. Wakabayashi, R. M. Waters and J. P. Church; *Tetrahedron Lett.*, 1969, **10**, 3253.	832
69TL3415	U. Schöllkopf, U. Ludwig, G. Ostermann and M. Patsch; *Tetrahedron Lett.*, 1969, **10**, 3415.	398, 399
69TL3459	K. C. Das and B. Weinstein; *Tetrahedron Lett.*, 1969, **10**, 3459.	811
69TL3699	M. Santelli and M. Bertrand; *Tetrahedron Lett.*, 1969, **10**, 3699.	1126
69TL3989	C. H. DePuy and C. R. Jablonski; *Tetrahedron Lett.*, 1969, **10**, 3989.	870
69TL4001	R. M. Carlson and S. Y. Lee; *Tetrahedron Lett.*, 1969, **10**, 4001.	636
69TL4447	T. Iwadare, I. Adachi, M. Hayashi, A. Matsunaga and T. Kitai; *Tetrahedron Lett.*, 1969, **10**, 4447.	1077
69TL4593	I. Rhee, M. Ryang and S. Tsutsumi; *Tetrahedron Lett.*, 1969, **10**, 4593.	1035
69ZC229	W. Schroth, F. Billig and G. Reinhold; *Z. Chem.*, 1969, **9**, 229.	1072
70ACR281	F. G. Bordwell; *Acc. Chem. Res.*, 1970, **3**, 281.	617
70AG(E)273	K. Griesbaum; *Angew. Chem., Int. Ed. Engl.*, 1970, **9**, 273.	948
70AG(E)374	H. W. Heine, J. D. Myers and E. T. Peltzer, III; *Angew. Chem., Int. Ed. Engl.*, 1970, **9**, 374.	636
70AG(E)464	I. Lalezari, A. Shafiee and M. Yalpani; *Angew. Chem., Int. Ed., Engl.*, 1970, **9**, 464.	1077
70AG(E)763	U. Schöllkopf; *Angew. Chem., Int. Ed. Engl.*, 1970, **9**, 763.	196, 233, 396
70AJC1641	A. J. Birch and G. S. R. Subba Rao; *Aust. J. Chem.*, 1970, 1641.	434
70BCJ1120	T. Tezuka, M. Kimura, A. Sato and T. Mukai; *Bull. Chem. Soc. Jpn.*, 1970, **43**, 1120.	795
70BCJ2506	Y. Hayasi, T. Okada and M. Kawanisi; *Bull. Chem. Soc. Jpn.*, 1970, **43**, 2506.	11, 109
70BSF991	E. Elkik and H. Assadi-Far; *Bull. Soc. Chim. Fr.*, 1970, 991.	108
70BSF2175	J. Moreau and M. Gaudemar; *Bull. Soc. Chim. Fr.*, 1970, 2175.	1055
70BSF2505	H. Mongeot; *Bull. Soc. Chim. Fr.*, 1970, 2505.	658
70BSF2981	J.-M. Conia and M.-L. Leriverend; *Bull. Soc. Chim. Fr.*, 1970, 2981.	422
70BSF3157	R. Couffignal and M. Gaudemar; *Bull. Soc. Chim. Fr.*, 1970, 3157.	971
70CB265	E. Weiss, G. Sauermann; *Chem. Ber.*, 1970, **103**, 265.	885
70CB426	W. R. Roth, J. König and K. Stein; *Chem. Ber.*, 1970, **103**, 426.	402, 794
70CB1101	H. A. Staab, H. Brettschneider and H. Brunner; *Chem. Ber.*, 1970, **103**, 1101.	922
70CB2814	M. Schlosser, K. F. Christmann and A. Piskala; *Chem. Ber.*, 1970, **103**, 2814.	730
70CC4	V. Rautenstrauch; *J. Chem. Soc., Chem. Commun.*, 1970, 4.	396, 405, 816, 818
70CC86	D. N. Jones, D. Mundy and R. D. Whitehouse; *J. Chem. Soc., Chem. Commun.*, 1970, 86.	601
70CC165	J. E. Baldwin and F. J. Urban; *J. Chem. Soc., Chem. Commun.*, 1970, 165.	806
70CC338	J. Jacobus; *J. Chem. Soc., Chem. Commun.*, 1970, 338.	4
70CC495	R. L. Augustine and J. F. Van Peppen; *J. Chem. Soc., Chem. Commun.*, 1970, 495.	73
70CC497	R. L. Augustine and J. F. Van Peppen; *J. Chem. Soc., Chem. Commun.*, 1970, 497.	73
70CC571	R. L. Augustine and J. F. Van Peppen; *J. Chem. Soc., Chem. Commun.*, 1970, 571.	73
70CC576	J. E. Baldwin, W. F. Erickson, R. E. Hackler and R. M. Scott; *J. Chem. Soc., Chem. Commun.*, 1970, 576.	399
70CC738	J. P. Coleman, H. G. Gilde, J. H. P. Utley and B. C. L. Weedon; *J. Chem. Soc., Chem. Commun.*, 1970, 738.	6
70CC892	R. V. H. Jones and M. D. Sutherland; *J. Chem. Soc., Chem. Commun.*, 1970, 892.	812
70CC1168	B. A. M. Boersma, J. W. de Haan, H. Kloosterziel and L. J. M. van de Ven; *J. Chem. Soc., Chem. Commun.*, 1970, 1168.	403, 799
70CC1325	G. Brieger and D. W. Anderson; *J. Chem. Soc., Chem. Commun.*, 1970, 1325.	881
70CC1519	C. B. Reese and M. S. Baird; *J. Chem. Soc., Chem. Commun.*, 1970, 1519.	811
70CC1683	D. H. R. Barton, D. G. T. Greig, G. Lucente, P. G. Sammes, M. V. Taylor, C. M. Cooper, G. Hewitt and W. G. E. Underwood; *J. Chem. Soc., Chem. Commun.*, 1970, 1683.	595
70CJC561	A. G. Brook, J. M. Duff and D. G. Anderson; *Can. J. Chem.*, 1970, **48**, 561.	759
70CJC955	F. P. Lossing and G. P. Semeluk; *Can. J. Chem.*, 1970, **48**, 955.	1109
70CJC2110	D. V. Gardner and D. E. McGreer; *Can. J. Chem.*, 1970, **48**, 2110.	637
70CR(C)100	Y. Odic and M. Pereyre; *C. R. Hebd. Séances Acad. Sci., Ser. C.*, 1970, **270**, 100.	132
70HCA171	P. Wieland; *Helv. Chim. Acta*, 1970, **53**, 171.	1077
70HCA173	H. Heimgartner, H.-J. Hansen and H. Schmid; *Helv. Chim. Acta*, 1970, **53**, 173.	403, 795
70HCA485	P. Schiess and H. L. Chia; *Helv. Chim. Acta*, 1970, **53**, 485.	506
70HCA1713	P. Schiess, R. Seeger and C. Suter; *Helv. Chim. Acta*, 1970, **53**, 1713.	502
70HCA2119	C. A. Grob and R. Spaar; *Helv. Chim. Acta*, 1970, **53**, 2119.	1113, 1114
70HCA2130	C. A. Grob and H. R. Pfaendler; *Helv. Chim. Acta*, 1970, **53**, 2130.	1113
70IZV1144	M. S. Shvartsberg, I. L. Kotlyarevskii, A. N. Kozheynikova and V. N. Andrievskii; *Izv. Akad. Nauk SSSR, Ser. Khim.*, 1970, 1144 (*Chem. Abstr.*, **73**, 130 851f).	1006
70JA207	W. M. Schubert and G. W. Barfknecht; *J. Am. Chem. Soc.*, 1970, **92**, 207.	1110
70JA703	E. A. Chandross and C. J. Dempster; *J. Am. Chem. Soc.*, 1970, **92**, 703.	811
70JA712	H. C. Brown and G. W. Kabalka; *J. Am. Chem. Soc.*, 1970, **92**, 712.	946
70JA714	H. C. Brown and G. W. Kabalka; *J. Am. Chem. Soc.*, 1970, **92**, 714.	946
70JA741	W. S. Johnson, L. Werthemann, W. R. Bartlett, T. J. Brocksom, T. Li, D. J. Faulkner and M. R. Petersen; *J. Am. Chem. Soc.*, 1970, **92**, 741.	414, 416, 827
70JA984	R. Breslow and J. T. Groves; *J. Am. Chem. Soc.*, 1970, **92**, 984.	907
70JA1082	M. D. Schiavelli, S. C. Hixon and H. W. Moran; *J. Am. Chem. Soc.*, 1970, **92**, 1082.	1114
70JA1096	J. A. Berson and G. L. Nelson; *J. Am. Chem. Soc.*, 1970, **92**, 1096.	799
70JA1396	P. L. Creger; *J. Am. Chem. Soc.*, 1970, **92**, 1396.	881
70JA1397	P. L. Creger; *J. Am. Chem. Soc.*, 1970, **92**, 1397.	881
70JA1426	G. M. Whitesides, E. R. Stedronsky, C. P. Casey and J. San Filippo, Jr.; *J. Am. Chem. Soc.*, 1970, **92**, 1426.	668

70JA1427	G. Zweifel, A. Horng and J. T. Snow; *J. Am. Chem. Soc.*, 1970, **92**, 1427.	658, 963
70JA1430	G. A. Olah and G. D. Mateescu; *J. Am. Chem. Soc.*, 1970, **92**, 1430.	909
70JA1660	B. R. Ree and J. C. Martin; *J. Am. Chem. Soc.*, 1970, **92**, 1660.	894
70JA1763	L. A. Paquette, M. J. Wyvratt and G. R. Allen, Jr.; *J. Am. Chem. Soc.*, 1970, **92**, 1763.	785
70JA1769	G. F. Hennion and M. J. Shoemaker; *J. Am. Chem. Soc.*, 1970, **92**, 1769.	399
70JA1784	L. A. Carpino and R. K. Kirkley; *J. Am. Chem. Soc.*, 1970, **92**, 1784.	636
70JA2139	E. H. Axelrod, G. M. Milne and E. E. Van Tamelen; *J. Am. Chem. Soc.*, 1970, **92**, 2139.	49, 51, 157
70JA2141	J. E. Williams, Jr., V. Buss, L. C. Allen, P. v. R. Schleyer, W. A. Lathan, W. J. Hehre and J. A. Pople; *J. Am. Chem. Soc.*, 1970, **92**, 2141.	1109
70JA2158	E. Vedejs, R. A. Shepherd and R. P. Steiner; *J. Am. Chem. Soc.*, 1970, **92**, 2158.	811
70JA2163	R. G. Bergman and V. J. Rajadhyaksha; *J. Am. Chem. Soc.*, 1970, **92**, 2163.	866
70JA2445	H. M. Walborsky and J. M. Motes; *J. Am. Chem. Soc.*, 1970, **92**, 2445.	871
70JA2538	J. L. Fry, C. J. Lancelot, L. K. M. Lam, J. M. Harris, R. C. Bingham, D. J. Raber, R. E. Hall and P. v. R. Schleyer; *J. Am. Chem. Soc.*, 1970, **92**, 2538.	1114
70JA2540	J. L. Fry, J. M. Harris, R. C. Bingham and P. v. R Schleyer; *J. Am. Chem. Soc.*, 1970, **92**, 2540.	1114
70JA2542	P. v. R. Schleyer, J. L. Fry, L. K. M. Lam and C. J. Lancelot; *J. Am. Chem. Soc.*, 1970, **92**, 2542.	104
70JA2581	F. G. Bordwell, E. Doomes and P. W. R. Corfield; *J. Am. Chem. Soc.*, 1970, **92**, 2581.	707
70JA3222	M. W. Rathke; *J. Am. Chem. Soc.*, 1970, **92**, 3222.	130, 881
70JA3232	R. A. Benkeser, K. M. Foley, J. M. Gaul and G. S. Li; *J. Am. Chem. Soc.*, 1970, **92**, 3232.	21
70JA3234	D. S. Kabakoff and E. Namanworth; *J. Am. Chem. Soc.*, 1970, **92**, 3234.	894
70JA3503	A. Suzuki, S. Nozawa, M. Itoh, H. C. Brown, G. W. Kabalka and G. W. Holland; *J. Am. Chem. Soc.*, 1970, **92**, 3503.	274, 546
70JA3697	J. M. Motes and H. M. Walborsky; *J. Am. Chem. Soc.*, 1970, **92**, 3697.	871
70JA3789	S. H. Liggero, J. J. Harper, P. v. R. Schleyer, A. P. Krapcho and D. E. Horn; *J. Am. Chem. Soc.*, 1970, **92**, 3789.	1114
70JA3802	M. A. Imhoff, R. H. Summerville, P. v. R. Schleyer, A. G. Martinez, M. Hanack, T. E. Dueber and P. J. Stang; *J. Am. Chem. Soc.*, 1970, **92**, 3802.	1116
70JA3821	J. M. Brown and E. N. Cain; *J. Am. Chem. Soc.*, 1970, **92**, 3821.	865
70JA4068	G. Zweifel and N. L. Polston; *J. Am. Chem. Soc.*, 1970, **92**, 4068.	509, 1103
70JA4309	M. S. Newman and C. D. Beard; *J. Am. Chem. Soc.*, 1970, **92**, 4309.	1144
70JA4312	M. S. Newman and T. B. Patrick; *J. Am. Chem. Soc.*, 1970, **92**, 4312.	1144
70JA4395	R. A. Sheldon and J. K. Kochi; *J. Am. Chem. Soc.*, 1970, **92**, 4395.	924, 925
70JA4627	G. A. Olah, A. M. White, J. R. DeMember, A. Commeyras and C. Y. Lui; *J. Am. Chem. Soc.*, 1970, **92**, 4627.	906
70JA4689	M. Fukuyama, P. W. K. Flanagan, F. T. Williams, Jr., L. Frainier, S. A. Miller and H. Shechter; *J. Am. Chem. Soc.*, 1970, **92**, 4689.	869, 871
70JA4898	W. T. Borden; *J. Am. Chem. Soc.*, 1970, **92**, 4898.	513
70JA4978	R. J. Anderson; *J. Am. Chem. Soc.*, 1970, **92**, 4978.	226, 479
70JA4979	R. W. Herr and C. R. Johnson; *J. Am. Chem. Soc.*, 1970, **92**, 4979.	226
70JA4985	Z. Rappoport, T. Bässler and M. Hanack; *J. Am. Chem. Soc.*, 1970, **92**, 4985.	1110
70JA5457	W. M. Jones and D. D. Maness; *J. Am. Chem. Soc.*, 1970, **92**. 5457.	1110
70JA5522	E. J. Corey and J. I. Shulman; *J. Am. Chem. Soc.*, 1970, **92**, 5522.	834
70JA5790	H. Nambu and H. C. Brown; *J. Am. Chem. Soc.*, 1970, **92**, 5790.	126
70JA5926	F. G. Bordwell, W. J. Boyle, Jr. and K. C. Yee; *J. Am. Chem. Soc.*, 1970, **92**, 5926.	869, 871
70JA5933	F. G. Bordwell and K. C. Yee; *J. Am. Chem. Soc.*, 1970, **92**, 5933.	869, 871
70JA5939	F. G. Bordwell and K. C. Yee; *J. Am. Chem. Soc.*, 1970, **92**, 5939.	869, 871
70JA6314	E. J. Corey, H. A. Kirst and J. A. Katzenellenbogen; *J. Am. Chem. Soc.*, 1970, **92**, 6314.	1052
70JA6611	G. M. Whitesides and J. San Fillipo, Jr.; *J. Am. Chem. Soc.*, 1970, **92**, 6611.	67
70JA6641	K. Hojo, R. T. Seidner and S. Masamune; *J. Am. Chem. Soc.*, 1970, **92**, 6641.	811
70JA6643	T. J. Katz, J. J. Cheung and N. Acton; *J. Am. Chem. Soc.*, 1970, **92**, 6643.	811
70JA6676	A. I. Meyers and E. W. Collington; *J. Am. Chem. Soc.*, 1970, **92**, 6676.	1041
70JA6678	G. Zweifel and R. L. Miller; *J. Am. Chem. Soc.*, 1970, **92**, 6678.	491, 534
70JA7408	W. G. Dauben, C. D. Poulter and C. Suter; *J. Am. Chem. Soc.*, 1970, **92**, 7408.	794
70JA7424	G. R. Buell, R. Corriu, C. Guerin and L. Spialter; *J. Am. Chem. Soc.*, 1970, **92**, 7424.	759
70JA7493	J. C. Gilbert and J. R. Butler; *J. Am. Chem. Soc.*, 1970, **92**, 7493.	1143, 1144
70JA7564	M. S. Newman and C. D. Beard; *J. Am. Chem. Soc.*, 1970, **92**, 7564.	1118
70JA7606	A. S. Kende, Z. Goldschmidt and R. F. Smith; *J. Am. Chem. Soc.*, 1970, **92**, 7606.	1143
70JCS2365	W. Nagata, M. Yoshioka and T. Okumura; *J. Chem. Soc.*, 1970, 2365.	208
70JCS(B)173	K. Bowden and A. F. Cockerill; *J. Chem. Soc. (B)*, 1970, 173.	869
70JCS(B)179	K. Bowden, A. F. Cockerill and J. R. Gilbert; *J. Chem. Soc. (B)*, 1970, 179.	869
70JCS(B)194	R. P. Bell and B. G. Cox; *J. Chem. Soc. (B)*, 1970, 194.	869, 871
70JCS(B)469	H. Shenhav, Z. Rappoport and S. Patai; *J. Chem. Soc. (B)*, 1970, 469.	889
70JCS(B)1410	H. M. Frey and R. G. Hopkins; *J. Chem. Soc. (B)*, 1970, 1410.	799
70JCS(C)81	D. N. Jones and W. Higgins; *J. Chem. Soc. (C)*, 1970, 81.	779
70JCS(C)833	D. N. Jones, E. Helmy and A. C. F. Edmonds; *J. Chem. Soc. (C)*, 1970, 833.	779
70JCS(C)2625	G. Capozzi, G. Melloni and G. Modena; *J. Chem. Soc. (C)*, 1970, 2625.	1111
70JCS(C)2697	K. Takeda, K. Tori, I. Horibe, M. Ohtsuru and H. Minato; *J. Chem. Soc. (C)*, 1970, 2697.	813
70JCS(C)2704	K. Takeda, I. Horibe and H. Minato; *J. Chem. Soc. (C)*, 1970, 2704.	813
70JOC22	L. H. Sommer and W. D. Korte; *J. Org. Chem.*, 1970, **35**, 22.	112
70JOC347	G. Nagendrappa, R. K. Srivastava and D. Devaprabhakara; *J. Org. Chem.*, 1970, **35**, 347.	459
70JOC532	J. P. Kennedy; *J. Org. Chem.*, 1970, **35**, 532.	128

70JOC565	K. Sato, S. Inoue and S. Ota; *J. Org. Chem.*, 1970, **35**, 565.	730
70JOC856	A. Viola and E. J. Iorio; *J. Org. Chem.*, 1970, **35**, 856.	810
70JOC1086	R. A. Walsh and A. T. Bottini; *J. Org. Chem.*, 1970, **35**, 1086.	1144
70JOC1237	M. P. Servé and H. M. Rosenberg; *J. Org. Chem.*, 1970, **35**, 1237.	516
70JOC1303	D. D. Davis and C. E. Gray; *J. Org. Chem.*, 1970, **35**, 1303.	667
70JOC1828	W. E. Truce and W. W. Brand; *J. Org. Chem.*, 1970, **35**, 1828.	379
70JOC2678	A. W. Johnson and H. Schubert; *J. Org. Chem.*, 1970, **35**, 2678.	738
70JOC3273	K. Garves; *J. Org. Chem.*, 1970, **35**, 3273.	494
70JOC3510	E. J. Glamkowski, G. Gal, R. Purick, A. J. Davidson and M. Sletzinger; *J. Org. Chem.*, 1970, **35**, 3510.	80
70JOC3585	W. N. White and E. F. Wolfarth; *J. Org. Chem.*, 1970, **35**, 3585.	826
70JOM(24)263	D. S. Matteson and J. R. Thomas; *J. Organomet. Chem.*, 1970, **24**, 263.	163
70JPR150	R. Mueller, M. Dressler and C. Dathe; *J. Prakt. Chem.*, 1970, **312**, 150.	490
70KGS495	E. E. Mikhlina, V. V. Vorob'eva and L. N. Yakhontov; *Khim. Geterotsikl. Soedin.*, 1970, 495.	45
70LA(733)5	M. Hanack, S. Bocher, I. Herterich, K. Hummel and V. Vött; *Justus Liebigs Ann. Chem.*, 1970, **733**, 5.	1123
70LA(734)106	U. Schöllkopf, K. Fellenberger and M. Rizk; *Justus Liebigs Ann. Chem.*, 1970, **734**, 106.	816
70LA(737)158	U. Schöllkopf, J. Schossig and G. Ostermann; *Justus Liebigs Ann. Chem.*, 1970, **737**, 158.	399
70LA(739)211	R. Köster, D. Simić and M. A. Grassberger; *Justus Liebigs Ann. Chem.*, 1970, **739**, 211.	727
B-70MI 102-01	M. Lemal; in "Nitrenes," ed. W. Lwowski, Interscience, New York, 1970, vol. 8, p. 347.	61
70MI 102-02	Yu. I. Baukov and I. F. Lutsenko; *Organometal. Chem. Rev.*, 1970, **A6**, 355.	62
B-70MI 109-01	R. B. Woodward and R. Hoffmann; "Conservation of Orbital Symmetry," Academic Press, New York, 1970.	417, 422
B-70MI 114-01	G. O. Doak and L. D. Freedman; "Organometallic Compounds of Arsenic, Antimony and Bismuth," Wiley, New York, 1970, pp. 223–228.	646
B-70MI 114-02	G. O. Doak and L. D. Freedman; "Organometallic Compounds of Arsenic, Antimony and Bismuth," Wiley, New York, 1970, p. 343.	647
B-70MI 116-01	M. Schlosser; in "Topics in Stereochemistry," ed. E. L. Eliel and N. L. Allinger, Interscience, New York, 1970, vol. 5, p. 1.	721, 728, 730, 732, 733
B-70MI 116-02	K. Mackenzie; in "The Chemistry of Alkenes," ed. S. Patai, ch. 7, Interscience, New York, 1964, vol. 2, ch. 3, 1970.	723
70MI 116-03	M. Fétizon and J. Schalbar; *Fr. Ses Parfums*, 1969, **12**, 330 (*Chem. Abstr.*, 1970, **72**, 42 886).	727, 739
70MI 116-04	S. G. Dwyer; Ph.D. Thesis, University of North Dakota, 1970.	727
B-70MI 118-01	R. B. Woodward and R. Hoffmann; "The Conservation of Orbital Symmetry," Verlag Chemie/Academic Press, New York, 1970.	794
B-70MI 118-02	K. Mackenzie; in "The Chemistry of Alkenes," ed. J. Zabicky, Interscience, New York, 1970, vol. 2, p. 132.	797
B-70MI 119-01	H. G. Richey, Jr.; in "The Chemistry of Alkenes," ed. J. Zabicky, Wiley, New York, 1970, vol. 2, p. 67.	847
B-70MI 119-02	G. A. Olah and P. von R. Schleyer (eds.); "Carbonium Ions," Wiley, New York, 1970, vol. 2.	891
B-70MI 119-03	N. C. Deno; in "Carbonium Ions," eds. G. A. Olah and P. von R. Schleyer, Wiley, New York, 1970, vol. 2, chapter 18, pp. 783–835.	911
B-70MI 119-04	T. S. Sorensen; in "Carbonium Ions," eds. G. A. Olah and P. von R. Schleyer, Wiley, New York, 1970, vol. 2, chapter 18, pp. 783–835.	912
B-70MI 119-05	D. M. Brouwer, E. L. Mackor and C. MacLean; in "Carbonium Ions," eds. G. A. Olah and P. von R. Schleyer, Wiley, New York, 1970, vol. 2, chapter 20.	914
B-70MI 119-06	E. S. Huyser; "Free Radical Chain Reactions," Wiley-Interscience, New York, 1970.	919
70OR(18)99	W. E. Truce, E. M. Kreider and W. W. Brand; *Org. React.*, 1970, **18**, 99.	379
70OR(18)403	S. H. Pine; *Org. React.*, 1970, **18**, 403.	398, 407
70OS(50)58	P. L. Creger; *Org. Synth.*, 1970, **50**, 58.	881
70OS(50)66	G. Wittig and A. Hesse; *Org. Synth.*, 1970, **50**, 66.	881
70QRS1	J. F. Arens, L. Brandsma, P. J. W. Schuijk and H. W. Wijers; *Quart. Rep. Sulfur Chem.*, 1970, **5**, 1.	954
70RCR732	A. A. Akhrem, T. K. Ustynyuk and Y. A. Titov; *Russ. Chem. Rev.* (*Engl. Transl.*), 1970, **39**, 732.	391, 884
70RCR859	L. A. Yanovskaya and Kh. Shakhidayatov; *Russ. Chem. Rev.* (*Engl. Transl.*), 1970, **39**, 859.	882
70RTC1153	G. J. Heiszwolf and H. Kloosterziel; *Recl. Trav. Chim. Pays-Bas*, 1970, **89**, 1153.	873
70S49	H. E. Zaugg; *Synthesis*, 1970, 49.	1042
70S161	R. G. Harvey; *Synthesis*, 1970, 161.	95, 889
70S405	A. J. Hubert and H. Reimlinger; *Synthesis*, 1970, 405.	795, 882
70S543	Z. Hamlet and W. D. Barker; *Synthesis*, 1970, 543.	986
70T1281	F. Bertini, P. Grasselli, G. Zubiani and G. Cainelli; *Tetrahedron*, 1970, **26**, 1281.	679
70T1291	J. W. Van Reijendam, G. J. Heeres and M. J. Janssen; *Tetrahedron*, 1970, **26**, 1291.	731
70T2027	W. Hänsel and R. Haller; *Tetrahedron*, 1970, **26**, 2027.	869
70T2035	R. Haller and W. Hänsel; *Tetrahedron*, 1970, **26**, 2035.	869
70T2127	J. Klein and E. Gurfinkel; *Tetrahedron*, 1970, **26**, 2127.	1066, 1084
70T2683	A. J. M. Reuvers, H. van Bekkum and B. M. Wepster; *Tetrahedron*, 1970, **26**, 2683.	115
70T3487	R. M. Coates and R. M. Friedinger; *Tetrahedron*, 1970, **26**, 3487.	1082
70T4609	J. Nakayama, M. Yoshida and O. Simamura; *Tetrahedron*, 1970, **26**, 4609.	442
70T5343	H. Volz, W. Lotsch and H.-W. Schnell; *Tetrahedron*, 1970, **26**, 5343.	922
70TL41	H. C. Kretschmar and W. F. Erman; *Tetrahedron Lett.*, 1970, **11**, 41.	725

70TL311 E. J. Corey and R. Noyori; *Tetrahedron Lett.*, 1970, **11**, 311. 218
70TL315 E. J. Corey and G. H. Posner; *Tetrahedron Lett.*, 1970, **11**, 315. 125
70TL353 J. E. Baldwin, J. DeBernardis and J. E. Patrick; *Tetrahedron Lett.*, 1970, **11**, 353. 816, 818
70TL509 E. N. Marvell and W. Whalley; *Tetrahedron Lett.*, 1970, **11**, 509. 412, 813
70TL761 A. Krebs and H. Kimling; *Tetrahedron Lett.*, 1970, **11**, 761. 635, 635
70TL1137 T. H. Chan, E. Chang and E. Vinokur; *Tetrahedron Lett.*, 1970, **11**, 1137. 651
70TL1175 T. Mori, K. Matsui and H. Nozaki; *Tetrahedron Lett.*, 1970, **11**, 1175. 841
70TL1941 E. Vogel, R. Feldmann and H. Düwel; *Tetrahedron Lett.*, 1970, **11**, 1941. 910
70TL2245 E. J. Corey and K. Achiwa; *Tetrahedron Lett.*, 1970, **11**, 2245. 1134
70TL2247 R. E. Ireland, M. I. Dawson and C. A. Lipinski; *Tetrahedron Lett.*, 1970, **11**, 2247. 1052
70TL2659 J. F. Normant and M. Bourgain; *Tetrahedron Lett.*, 1970, **11**, 2659. 1045
70TL2695 J. Fried, M. M. Mehra, W. L. Kao and C. H. Lin; *Tetrahedron Lett.*, 1970, **11**, 2695. 1023
70TL2755 L. M. Dané, J. W. de Haan and H. Kloosterziel; *Tetrahedron Lett.*, 1970, **11**, 2755. 800
70TL3233 Z. Rappoport and A. Gal; *Tetrahedron Lett.*, 1970, **11**, 3233. 1112
70TL3861 J. A. Marshall and J. H. Babler; *Tetrahedron Lett.*, 1970, **11**, 3861. 660
70TL3903 J. M. Lalancette and A. Lachance; *Tetrahedron Lett.*, 1970, **11**, 3903. 881
70TL4243 G. Nagendrappa and D. Devaprabhakara; *Tetrahedron Lett.*, 1970, **11**, 4243. 459
70TL4315 H. Schelhorn, H. Frischleder and S. Hauptmann; *Tetrahedron Lett.*, 1970, **11**, 4315. 981
70TL4355 A. I. Meyers and E. M. Smith; *Tetrahedron Lett.*, 1970, **11**, 4355. 664
70TL5139 H. P. Figeys and M. Gelbecke; *Tetrahedron Lett.*, 1970, **11**, 5139. 503

71ABC1116 K. Mori, T. Mitsui, J. Fukami and T. Ohtaki; *Agric. Biol. Chem.*, 1971, **35**, 1116. 747
71ACR57 K. Fukui; *Acc. Chem. Res.*, 1971, **4**, 57. 417
71ACR73 G. Modena; *Acc. Chem. Res.*, 1971, **4**, 73. 1109, 1110
71ACR193 E. M. Kosower; *Acc. Chem. Res.*, 1971, **4**, 193. 35
71ACR272 H. E. Zimmerman; *Acc. Chem. Res.*, 1971, **4**, 272. 794
71ACR386 M. Julia; *Acc. Chem. Res.*, 1971, **4**, 386. 337, 548, 919
71ACS1452 A. J. Dale, P. Frøyen and D. G. Morris; *Acta Chem. Scand.*, 1971, **25**, 1452. 737
71ACS2541 P. Frøyen; *Acta Chem. Scand.*, 1971, **25**, 2541. 740, 741
71AG(E)653 W. Steglich, P. Gruber, G. Höfle and W. König; *Angew. Chem., Int. Ed. Engl.*, 1971, **10**, 653. 537
71AG(E)761 M. S. Dewar; *Angew. Chem., Int. Ed. Engl.*, 1971, **10**, 761. 794
71AJC513 A. J. Birch and K. A. M. Walker; *Aust. J. Chem.*, 1971, **24**, 513. 82
71AJC2325 P. J. Collin and W. H. F. Sasse; *Aust. J. Chem.*, 1971, **24**, 2325. 517
71ARP109 L. Andrews; *Annu. Rev. Phys. Chem.*, 1971, **22**, 109. 921
71AX(B)1405 L. L. Koh and K. Eriks; *Acta Crystallogr., Part B*, 1971, **27**, 1405. 897
71BCJ166 S. Nishimura, M. Katagiri, T. Watanabe and M. Uramoto; *Bull. Chem. Soc. Jpn.*, 1971, **44**, 166. 72, 81
71BSF461 M. Cazaux and R. Lalande; *Bull. Soc. Chim. Fr.*, 1971, **2**, 461. 925
71BSF(2)453 M. Schlosser; *Bull. Soc. Chim. France*, 1971, 453. 728
71BSF(2)4551 J.-C. Bonnafous and M. Mosseron-Canet; *Bull. Soc. Chim. Fr., Part* 2, 1971, 4551. 726
71BSF3071 J. Moreau and M. Gaudemar; *Bull. Soc. Chim. Fr.*, 1971, 3071. 1059
71BSF3978 J.-J. Barieux and J. Gore; *Bull. Soc. Chim. Fr.*, 1971, 3978. 660
71CB2885 M. Schlosser and M. Zimmermann; *Chem. Ber.*, 1971, **104**, 2385. 731
71CB3679 W. Sucrow and W. Richter; *Chem. Ber.*, 1971, **104**, 3679. 415, 827
71CC22 J. C. Philips, J. V. Swisher, D. Haidukewych and O. Morales; *J. Chem. Soc., Chem. Commun.*, 1971, 22. 620, 1074
71CC107 R. C. Cookson, J. Hudec and M. Sharma; *J. Chem. Soc., Chem. Commun.*, 1971, 107. 799
71CC108 R. C. Cookson, J. Hudec and M. Sharma; *J. Chem. Soc., Chem. Commun.*, 1971, 108. 799
71CC465 N. W. Gilman and L. H. Sternbach; *J. Chem. Soc., Chem. Commun.*, 1971, 465. 3, 4, 6
71CC840 P. Abley, I. Jardine and F. J. McQuillen; *J. Chem. Soc., Chem. Commun.*, 1971, 840. 73, 74
71CC1144 H. A. Chia, B. E. Kirk and D. R. Taylor; *J. Chem. Soc., Chem. Commun.*, 1971, 1144. 518
71CC1248 R. Grigg, R. Hayes and A. Sweeney; *J. Chem. Soc., Chem. Commun.*, 1971, 1248. 803
71CC1639 B. M. Trost, P. Conway and J. Stanton; *J. Chem. Soc., Chem. Commun.*, 1971, 1639. 821
71CJC357 F. P. Lossing; *Can. J. Chem.*, 1971, **49**, 357. 1109
71CJC3135 R. J. Balf, B. Rao and L. Weiler; *Can. J. Chem.*, 1971, **49**, 3135. 1124
71CPB391 M. Sekiya and Y. Terao; *Chem. Pharm. Bull.*, 1971, **19**, 391. 1045
71CT566 G. A. Olah; *Chem. Tech.*, 1971, **1**, 566. 890, 893, 901
71HCA739 V. Rautenstrauch; *Helv. Chim. Acta*, 1971, **54**, 739. 820
71HCA937 M. Schmid, H.-J. Hansen and H. Schmid; *Helv. Chim. Acta*, 1971, **54**, 937. 809
71HCA1589 B. C. Baumann, M. Rey, J. Market, H. Prinzbach and A. S. Dreiding; *Helv. Chim. Acta*, 1971, **54**, 1589. 826
71HCA1767 G. Büchi and H. Wüest; *Helv. Chim. Acta*, 1971, **54**, 1767. 651, 736
71HCA1789 G. Ohloff and W. Pickenhagen; *Helv. Chim. Acta*, 1971, **54**, 1789. 1082
71HCA1939 F. Näf and P. Degen; *Helv. Chim. Acta*, 1971, **54**, 1939. 465
71HCA2060 C. A. Grob and H. R. Pfaendler; *Helv. Chim. Acta*, 1971, **54**, 2060. 1110, 1111
71HCA2896 A. Eschenmoser, J. Schreiber, D. Felix and G. Ohloff; *Helv. Chim. Acta*, 1971, **54**, 2896. 1082
71IZV1534 M. S. Shvartsberg, I. L. Kotlyarevskii and L. N. Bizhan; *Izv. Akad. Nauk. SSSR, Ser. Khim.*, 1971, 1534 (*Chem. Abstr.*, **75**, 98 497p). 1006
71JA464 G. A. Olah, R. D. Porter and D. P. Kelly; *J. Am. Chem. Soc*, 1971, **93**, 464. 895
71JA476 L. A. Carpino, L. V. McAdams, III, R. H. Rynbrandt and J. W. Spiewak; *J. Am. Chem. Soc.*, 1971, **93**, 476. 620
71JA650 L. L. Miller and R. F. Boyer; *J. Am. Chem. Soc.*, 1971, **93**, 650. 403, 799

71JA676 B. M. Trost, W. L. Schinski, F. Chen and I. B. Mantz; *J. Am. Chem. Soc.*, 1971, **93**, 676. 232, 234
71JA782 D. C. Owsley and J. J. Bloomfield; *J. Am. Chem. Soc.*, 1971, **93**, 782. 516
71JA808 W. A. Lathan, W. J. Hehre and J. A. Pople; *J. Am. Chem. Soc.*, 1971, **93**, 808. 1109
71JA1024 H. C. Brown, M. M. Midland and G. W. Kabalka; *J. Am. Chem. Soc.*, 1971, **93**, 1024. 946
71JA1027 R. M. Coates and R. L. Sowerby; *J. Am. Chem. Soc.*, 1971, **93**, 1027. 889
71JA1308 G. M. Clark, K. G. Hancock and G. Zweifel; *J. Am. Chem. Soc.*, 1971, **93**, 1308. 509
71JA1379 G. M. Whitesides, C. P. Casey and J. K. Krieger; *J. Am. Chem. Soc.*, 1971, **93**, 1379. 491, 1105
71JA1483 M. Tamura and J. K. Kochi; *J. Am. Chem. Soc.*, 1971, **93**, 1483. 117, 1092
71JA1506 H. C. Brown and M. M. Midland; *J. Am. Chem. Soc.*, 1971, **93**, 1506. 274
71JA1508 H. C. Brown, M. M. Midland, A. Suzuki, S. Nozawa, M. Harada and M. Itoh; *J. Am. Chem. Soc.*, 1971, **93**, 1508. 274
71JA1513 W. D. Pfeifer, C. A. Bahn, P. v. R. Schleyer, S. Bocher, C. E. Harding, K. Hummel, M. Hanack and P. J. Stang; *J. Am. Chem. Soc.*, 1971, **93**, 1513. 1110, 1117
71JA1532 R. S. H. Liu and Y. Butt; *J. Am. Chem. Soc.*, 1971, **93**, 1532. 420
71JA1544 J. A. Berson, R. J. Bushby, J. M. McBride and M. Tremelling; *J. Am. Chem. Soc.*, 1971, **93**, 1544. 104
71JA1693 B. Rickborn and R. M. Gerkin; *J. Am. Chem. Soc.*, 1971, **93**, 1693. 386
71JA1925 S. A. Sherrod and R. G. Bergmann; *J. Am. Chem. Soc.*, 1971, **93**, 1925. 1114
71JA1941 D. R. Kelsey and R. G. Bergman; *J. Am. Chem. Soc.*, 1971, **93**, 1941. 1114
71JA2065 N. C. Deno, R. Fishbein and J. C. Wyckoff; *J. Am. Chem. Soc.*, 1971, **93**, 2065. 925
71JA2225 J. N. Roitman and D. J. Cram; *J. Am. Chem. Soc.*, 1971, **93**, 2225. 854
71JA2231 J. N. Roitman and D. J. Cram; *J. Am. Chem. Soc.*, 1971, **93**, 2231. 854
71JA2235 S. M. Wong, H. P. Fisher and D. J. Cram; *J. Am. Chem. Soc.*, 1971, **93**, 2235. 858
71JA2342 G. V. Kaiser, C. W. Ashbrook and J. E. Baldwin; *J. Am. Chem. Soc.*, 1971, **93**, 2342. 410, 823
71JA2738 B. Frydman, S. Reil, A. Valasinas, R. B. Frydman and H. Rapoport; *J. Am. Chem. Soc.*, 1971, **93**, 2738. 259
71JA2792 A. Suzuki, N. Miyaura, H. C. Brown, M. Itoh, G. W. Holland and E.-I. Negishi; *J. Am. Chem. Soc.*, 1971, **93**, 2792. 277
71JA3047 D. M. Wieland and C. R. Johnson; *J. Am. Chem. Soc.*, 1971, **93**, 3047. 226
71JA3077 T. Durst, R. Viau and M. R. McClory; *J. Am. Chem. Soc.*, 1971, **93**, 3077. 143
71JA3087 J. A. Pettus, Jr. and R. E. Moore; *J. Am. Chem. Soc.*, 1971, **93**, 3087. 811
71JA3395 G. Zweifel, G. M. Clark and N. L. Polston; *J. Am. Chem. Soc.*, 1971, **93**, 3395. 449, 509, 1102
71JA3457 R. K. Howe; *J. Am. Chem. Soc.*, 1971, **93**, 3457. 735
71JA3556 J. E. Baldwin and J. E. Patrick; *J. Am. Chem. Soc.*, 1971, **93**, 3556. 816, 818
71JA3777 H. C. Brown and E. Negishi; *J. Am. Chem. Soc.*, 1971, **93**, 3777. 333, 946
71JA3793 F. N. Tebbe and G. W. Parshall; *J. Am. Chem. Soc.*, 1971, **93**, 3793. 669
71JA3828 W. J. Gensler, J. J. Langone and M. B. Floyd; *J. Am. Chem. Soc.*, 1971, **93**, 3828. 549
71JA3927 V. Buss, R. Gleiter and P. von R. Schleyer; *J. Am. Chem. Soc.*, 1971, **93**, 3927. 894
71JA4070 E. Vedejs and P. L. Fuchs; *J. Am. Chem. Soc.*, 1971, **93**, 4070. 640, 641
71JA4219 G. A. Olah, P. R. Clifford, Y. Halpern and R. G. Johanson; *J. Am. Chem. Soc.*, 1971, **93**, 4219. 902
71JA4316 N. Kornblum, S. D. Boyd, H. W. Pinnick and R. G. Smith; *J. Am. Chem. Soc.*, 1971, **93**, 4316. 637, 638
71JA4318 F. E. Ziegler and P. A. Wender; *J. Am. Chem. Soc.*, 1971, **93**, 4318. 267
71JA4527 H. D. Hartzler; *J. Am. Chem. Soc.*, 1971, **93**, 4527. 988
71JA5758 L. R. Byrd and M. C. Caserio; *J. Am. Chem. Soc.*, 1971, **93**, 5758. 1125
71JA6205 W. Egan, R. Tang, G. Zon and K. Mislow; *J. Am. Chem. Soc.*, 1971, **93**, 6205. 53
71JA6278 J. M. Jerkunica and T. G. Traylor; *J. Am. Chem. Soc.*, 1971, **93**, 6278. 666, 671
71JA6309 G. Zweifel, R. P. Fisher, J. T. Snow and C. C. Whitney; *J. Am. Chem. Soc.*, 1971, **93**, 6309. 278, 487, 662
71JA6688 R. E. Davis, H. D. Simpson, N. Grice and R. Pettit; *J. Am. Chem. Soc.*, 1971, **93**, 6688. 898
71JA6989 M. D. Schiavelli, S. C. Hixon, H. W. Moran and J. C. Boswell; *J. Am. Chem. Soc.*, 1971, **93**, 6989. 1114
71JA7118 R. D. Bach and D. Andrzejewski; *J. Am. Chem. Soc.*, 1971, **93**, 7118. 624, 629
71JA7320 J. Hooz and R. B. Layton; *J. Am. Chem. Soc.*, 1971, **93**, 7320. 314, 1048, 1049, 1139
71JCP(55)3835 A. C. Hopkinson, K. Yates and I. G. Csizmadia; *J. Chem. Phys.*, 1971, **55**, 3835. 1109
71JCS(A)3224 R. W. Mitchell, J. D. Ruddick and G. Wilkinson; *J. Chem. Soc. (A)*, 1971, 3224. 507
71JCS(B)374 G. Modena and U. Tonellato; *J. Chem. Soc. (B)*, 1971, 374. 1112
71JCS(B)381 G. Modena and U. Tonellato; *J. Chem. Soc. (B)*, 1971, 381. 1112
71JCS(B)871 Z. Rappoport and R. Ta-Shma; *J. Chem. Soc. (B)*, 1971, 871. 1110
71JCS(B)918 J. S. McKechnie and I. C. Paul; *J. Chem. Soc. (B)*, 1971, 918. 898
71JCS(B)1461 Z. Rappoport and R. Ta-Shma; *J. Chem. Soc. (B)*, 1971, 1461. 1110
71JCS(B)1569 G. Modena and U. Tonellato; *J. Chem. Soc. (B)*, 1971, 1569. 1112
71JCS(C)186 R. F. Curtis and J. A. Taylor; *J. Chem. Soc. (C)*, 1971, 186. 1018
71JCS(C)970 D. M. Bratby and G. I. Fray; *J. Chem. Soc. (C)*, 1971, 970. 815
71JCS(C)1114 S. Trippett and M. A. Walker; *J. Chem. Soc. (C)*, 1971, 1114. 738, 739, 741
71JCS(C)1327 D. A. Archer; *J. Chem. Soc. (C)*, 1971, 1327. 627
71JCS(C)1530 G. Smith and C. J. M. Stirling; *J. Chem. Soc. (C)*, 1971, 1530. 993
71JCS(C)2013 I. Fleming and C. R. Owen; *J. Chem. Soc. (C)*, 1971, 2013. 1081
71JCS(C)2950 I. J. Bolton, R. G. Harrison and B. Lythgoe; *J. Chem. Soc. (C)*, 1971, 2950. 827
71JCS(C)3018 G. Capozzi, G. Melloni and G. Modena; *J. Chem. Soc. (C)*, 1971, 3018. 1112
71JCS(C)3173 B. D. Cuddy, D. Grant and M. A. McKervey; *J. Chem. Soc. (C)*, 1971, 3173. 631
71JCS(C)3252 T. Teraji, I. Moritani, E. Tsuda and S. Nishida; *J. Chem. Soc. (C)*, 1971, 3252. 729

71JCS(C)3447	R. P. Dickinson and B. Iddon; *J. Chem. Soc. (C)*, 1971 3447.	664
71JFC(1)121	A. L. Anderson, R. T. Bogan and D. J. Burton; *J. Fluorine Chem.*, 1971, **1**, 121.	431
71JINC1157	D. B. Bigley and D. W. Payling; *J. Inorg. Nucl. Chem.*, 1971, **33**, 1157.	75
71JOC279	W. A. Remers, G. J. Gibs, C. Pidacks and M. J. Weiss; *J. Org. Chem.*, 1971, **36**, 279.	100
71JOC737	J. W. Burnham and E. J. Eisenbraun; *J. Org. Chem.*, 1971, **36**, 737.	43
71JOC758	F. A. Carey and H. S. Tremper; *J. Org. Chem.*, 1971, **36**, 758.	13
71JOC1015	L. A. Paquette and R. W. Houser; *J. Org. Chem.*, 1971 **36**, 1015.	707
71JOC1112	H. D. Zook and J. A. Miller; *J. Org. Chem.*, 1971, **36**, 1112.	130
71JOC1149	G. W. Moersch and A. R. Burkett; *J. Org. Chem.*, 1971, **36**, 1149.	681
71JOC1568	R. O. Hutchins, R. J. Bertsch and D. Hoke; *J. Org. Chem.*, 1971, **36**, 1568.	4, 7
71JOC1732	W. Ando, T. Yagihara, S. Kondo, K. Nakayma, H. Yamato, S. Nakaido and T. Migita; *J. Org. Chem.*, 1971, **36**, 1732.	820
71JOC1886	P. M. Henry; *J. Org. Chem.*, 1971, **36**, 1886.	1096
71JOC1898	A. K. Colter and R. E. Miller, Jr.; *J. Org. Chem.*, 1971 **36**, 1898.	612
71JOC2354	G. A. Olah and Y. Halpern; *J. Org. Chem.*, 1971, **36**, 2354.	911
71JOC2361	H. O. House, M. Gall and H. D. Olmstead; *J. Org. Chem.*, 1971, **36**, 2361.	872, 873
71JOC2379	E. E. Schweizer and W. S. Creasy; *J. Org. Chem.*, 1971 **36**, 2379.	268
71JOC2561	R. Walter and J. Roy; *J. Org. Chem.*, 1971, **36**, 2561.	601
71JOC2588	S. S. Hall, S. D. Lipsky, F. J. McEnroe and A. P. Bartels; *J. Org. Chem.*, 1971. **36**, 2588.	97
71JOC2836	I. Lalezari, A. Shafiee and M. Yalpani; *J. Org. Chem.*, 1971, **36**, 2836.	1077
71JOM(33)25	G. W. Kabalka; *J. Organomet. Chem.*, 1971, **33**, C25.	275
71JOM(33)321	E. Rosenberg and J. J. Zuckerman; *J. Organomet. Chem.*, 1971, **33**, 321.	450
71LA(747)60	W. Hoffmann, H. Pasedach, H. Pommer and W. Reif; *Justus Liebigs Ann. Chem.*, 1971, **747**, 60.	525, 832
71LA(748)123	E. Velarde, L. H. Knox, A. D. Cross and P. Crabbé. *Liebigs Ann. Chem.*, 1971, **748**, 123.	2, 4, 6
71M214	R. Finding, G. Zimmerman and U. Schmidt; *Monatsh. Chem.*, 1971, **102**, 214.	1045
71MI 102-01	G. L. Rabinovich, G. N. Maslyanskii and L. M. Treiger; *Kinet. Katal.*, 1971, **12(6)**, 1567 (*Chem. Abstr.*, 1972, **76**, 45 839r)	56
B-71MI 103-01	M. Freifelder; "Practical Catalytic Hydrogenation: Techniques and Applications," Wiley Interscience, New York, 1971.	89, 90
B-71MI 109-01	J. J. Gajewski; in "Mechanisms of Molecular Migrations," ed. B. S. Thyagarajan, Wiley-Interscience, 1971, vol. 4, p. 1.	402
B-71MI 109-02	A. R. Lepley and A. G. Giumanini; in "Mechanisms of Molecular Migrations," ed. B. S. Thyagarajan, Wiley-Interscience, 1971, vol. 2, p. 297.	407
B-71MI 112-01	M. Freifelder; "Practical Catalytic Hydrogenation," Wiley, New York, 1971, pp. 84–127.	548
71MI 117-01	G. G. Smith and F. W. Kelly; *Prog. Phys. Org. Chem.*, 1971, **8**, 75.	773
71MI 118-01	J. J. Gajewski; *Mech. Mol. Migr.*, 1971, **4**, 1.	794
71MI 118-02	H.-J. Hansen; *Mech. Mol. Migr.*, 1971, **3**, 177.	827
B-71MI 119-01	R. W. Alder, R. Baker and J. M. Brown; "Mechanism in Organic Chemistry," Wiley, London, 1971, p. 120.	847
B-71MI 119-02	R. W. Alder, R. Baker and J. M. Brown; "Mechanism in Organic Chemistry," Wiley, London, 1971, p. 124.	848
B-71MI 119-03	R. W. Alder, R. Baker and J. M. Brown; "Mechanism in Organic Chemistry," Wiley, London, 1971, p. 133.	851
B-71MI 119-04	R. W. Alder, R. Baker and J. M. Brown; "Mechanism in Organic Chemistry," Wiley, London, 1971, p. 134.	851
B-71MI 119-05	K. U. Ingold and B. P. Roberts; "Free-Radical Substitution Reactions, Bimolecular Homolytic Substitutions (S_H2 Reactions) at Saturated Multivalent Atoms," Wiley-Interscience, New York, 1971.	919
B-71MI 119-06	W. A. Pryor; "Introduction to Free Radical Chemistry," Prentice-Hall, Englewood Cliffs, 1971.	919
B-71MI 121-01	L. Brandsma; "Preparative Acetylene Chemistry," Elsevier, Amsterdam, 1971.	998, 1000, 1001
71OPP163	R. G. Micetich; *Org. Prep. Proced. Int.*, 1971, **3**, 163.	1077
71OS(51)103	A. I. Meyers, W. N. Beverung and R. Gault; *Org. Synth.*, 1971, **51**, 103.	89
71RTC265	L. Brandsma; *Recl. Trav. Chim. Pays-Bas*, 1971, **90**, 265.	1072
71RTC961	A. Steenhoek, B. H. von Wijngaarden and H. J. J. Pabon; *Recl. Trav. Chim. Pays-Bas*, 1971, **90**, 961.	505
71RTC1110	A. J. Schipperijn; *Rec. Trav. Chim., Pays-Bas*, 1971, **90**, 1110.	1088
71S1	S. R. Neale; *Synthesis*, 1971, 1.	927
71S139	J. B. Koster, G. J. Timmermans and H. Van Bekkum; *Synthesis*, 1971, 139.	522
71S203	G. Sosnovsky, E. H. Zaret and B. Boehnel; *Synthesis*, 1971, 203.	181
71S215	H. Meier and I. Menzel; *Synthesis*, 1971, 215.	1075
71S303	M. Tamura and J. F. Kochi; *Synthesis*, 1971, 303.	117, 119, 465
71T3575	F. Minisci, R. Bernardi, F. Bertini, R. Galli and M. Perchinummo; *Tetrahedron*, 1971, **27**, 3575.	934
71T4921	K. Tsujihara, N. Furukawa and S. Oae; *Tetrahedron*, 1971, **27**, 4921.	621
71T5523	E. H. Braye, I. Caplier and R. Saussez; *Tetrahedron*, 1971, **27**, 5523.	53
71T6101	K. Tsujihara, K. Harada, N. Furukawa and S. Oae; *Tetrahedron*, 1971, **27**, 6101.	621
71TCA65	H. Kollmar and H. O. Smith; *Theor. Chim. Acta*, 1971, **20**, 65.	1109
71TL137	J. M. Townsend and T. A. Spencer; *Tetrahedron Lett.* 1971, **12**, 137.	5, 8, 9, 10
71TL927	B. Rao and L. Weiler; *Tetrahedron Lett.*, 1971, **12**, 927.	1124
71TL1385	S. J. Huang and E. T. Hsu; *Tetrahedron Lett.*, 1971, **12**, 1385.	1121

71TL1821 E. J. Corey, J. A. Katzenellenbogen, S. A. Roman and N. W. Gilman; *Tetrahedron Lett.*, 1971, **12**, 1821. 747

71TL2075 R. Vaidyanathaswamy, G. C. Joshi and D. Devaprabhakara; *Tetrahedron Lett.*, 1971, **12**, 2075. 459

71TL2171 T. Bässler and M. Hanack; *Tetrahedron Lett.*, 1971, **12**, 2171. 1114

71TL2519 C. V. Lee, R. J. Hargrove, T. E. Dueber and P. J. Stang; *Tetrahedron Lett.*, 1971, **12**, 2519. 1114

71TL2583 J. F. Normant and M. Bourgain; *Tetrahedron Lett.*, 1971, **12**, 2583. 540

71TL2603 B. M. Trost and F. Chen; *Tetrahedron Lett.*, 1971, **12**, 2603. 564

71TL3777 F. W. Von Rein and H. G. Richey, Jr.; *Tetrahedron Lett.*, 1971, **12**, 3777. 539

71TL3785 H. G. Richey, Jr. and S. S. Szucs; *Tetrahedron Lett.*, 1971, **12**, 3785. 495

71TL3791 D. G. Martin, L. J. Hanka and L. M. Reineke; *Tetrahedron Lett.*, 1971, **12**, 3791. 241

71TL4039 Z. Rappoport and J. Kaspi; *Tetrahedron Lett.*, 1971, **12**, 4039. 1112

71TL4661 N. Takamatsu, S. Inoue and Y. Kishi; *Tetrahedron Lett.*, 1971, **12**, 4661. 835

71TL4701 W. P. Weber, R. A. Felix, A. K. Willard and K. E. Koenig; *Tetrahedron Lett.*, 1971, **12**, 4701. 490, 1106

72ACR102 S. Wolfe; *Acc. Chem. Res.*, 1972, **5**, 102. 849

72ACR406 J. A. Berson; *Acc. Chem. Res.*, 1972, **5**, 406. 799, 805

72AG(E)48 E. C. Taylor, R. L. Robey and A. McKillop; *Angew. Chem., Int. Ed., Engl.*, 1972, **11**, 48. 1076

72AG(E)60 K. Eiter; *Angew. Chem., Int. Ed. Engl.*, 1972, **11**, 60. 1134

72AG(E)473 G. Köbrich, *Angew. Chem., Int. Ed. Engl.*, 1972, **11**, 473. 164

72AG(E)692 H. C. Brown and M. M. Midland; *Angew. Chem., Int. Ed. Engl.*, 1972, **11**, 692. 333, 946

72AG(E)714 L. R. Subramanian and M. Hanack; *Angew. Chem., Int. Ed. Engl.*, 1972, **11**, 714. 1111, 1117

72AG(E)832 F. G. Klärner; *Angew. Chem., Int. Ed. Engl.*, 1972, **11**, 832. 800

72AG(E)1041 J. Buddrus; *Angew. Chem., Int. Ed. Engl.*, 1972, **11**, 1041. 722

72AJC545 J. D. Brewer and J. A. Elix; *Aust. J. Chem.*, 1972, **25**, 545, and references cited therein. 731

72AOC(10)273 M. I. Bruce; *Adv. Organomet. Chem.*, 1972, **10**, 273. 862

72BAU2056 N. A. Nesmeyanov, E. V. Binshtok, O. A. Rebrova and O. A. Reutov; *Bull. Acad. Sci. USSR, Div. Chem. Sci.*, 1972, 2056. 741

72BCJ264 M. Toda, M. Hayashi, Y. Hirata and S. Yamamura; *Bull. Chem. Soc. Jpn.*, 1972, **45**, 264. 19

72BCJ1120 M. Tamura and J. K. Kochi; *Bull. Chem. Soc. Jpn.*, 1972, **45**, 1120. 491

72BCJ1777 A. Sonota, I. Moritani, J. Miki, T. Tsuji and S. Nishidu; *Bull. Chem. Soc. Jpn.*, 1972, **45**, 1777. 842

72BCJ1846 H. Fujita, R. Endo, A. Aoyama and T. Ichii; *Bull. Chem. Soc. Jpn.*, 1972, **45**, 1846. 1044

72BCJ2582 K. Onuma and H. Hashimoto; *Bull. Chem. Soc. Jpn.*, 1972, **45**, 2582. 465

72BCJ2723 T. Funabiki, M. Matsumoto and K. Tarama; *Bull. Chem. Soc. Jpn.*, 1972, **45**, 2723. 73, 74, 79, 80

72BSF4238 P. Courtot and R. Rumin; *Bull. Soc. Chim. Fr.*, 1972, 4238. 795

72CB421 M. Hanack, C. E. Harding and J. L. Derocque; *Chem. Ber.*, 1972, **105**, 421. 1124

72CB2290 H. A. Staab, E. Wehinger and W. Thorwart; *Chem. Ber.*, 1972, **105**, 2290. 1143

72CB3789 E. Lamparter and M. Hanack; *Chem. Ber.*, 1972, **105**, 3789. 1117

72CC8 A. S. Denes, I. G. Csizmadia and G. Modena; *J. Chem. Soc., Chem. Commun.*, 1972, 8. 1109

72CC62 K. Kondo and I. Ojima; *J. Chem. Soc., Chem. Commun.*, 1972, 62. 823, 834

72CC347 U. E. Wiersum and W. J. Mijs; *J. Chem. Soc., Chem. Commun.*, 1972, 347. 789

72CC354 J. E. Baldwin and J. A. Walker; *J. Chem. Soc., Chem. Commun.*, 1972, 354. 822

72CC541 J. R. du Manoir and J. F. King; *J. Chem. Soc., Chem. Commun.*, 1972, 541. 232

72CC595 A. Meisters and T. Mole; *J. Chem. Soc., Chem. Commun.*, 1972, 595. 186

72CC702 P. A. Grieco; *J. Chem. Soc., Chem. Commun.*, 1972, 702. 821

72CC716 F. Scheidt and W. Kirmse; *J. Chem. Soc., Chem. Commun.*, 1972, 716. 799

72CC757 E. Hunt and B. Lythgoe; *J. Chem. Soc., Chem. Commun.*, 1972, 757. 821

72CC860 K. Kondo and I. Ojima; *J. Chem. Soc., Chem. Commun.*, 1972, 860. 400

72CC866 J. Goré and J. P. Dulcère; *J. Chem. Soc., Chem. Commun.*, 1972, 866. 959, 970

72CC1034 B. J. Arnold and P. G. Sammes; *J. Chem. Soc., Chem. Commun.*, 1972, 1034. 420

72CJC1105 J. Hooz and R. B. Layton; *Can. J. Chem.*, 1972, **50**, 1105. 1033

72CJC3973 F. P. Lossing; *Can. J. Chem.*, 1972, **50**, 3973. 1109

72CPL385 P. C. Hariharan, W. A. Lathan and J. A. Pople; *Chem. Phys. Lett.*, 1972, **14**, 385. 1109

72HCA1276 D. Felix, R. K. Müller, U. Horn, R. Joos, J. Schreiber and A. Eschenmoser; *Helv. Chim. Acta*, 1972, **55**, 1276. 1082

72HCA1385 H. Heimgartner, H.-J. Hansen and H. Schmid; *Helv. Chim. Acta*, 1972, **55**, 1385. 795

72HCA1828 R. Hug, H.-J. Hansen and H. Schmid; *Helv. Chim. Acta.*, 1972, **55**, 1828. 734, 795

72HCA2233 V. Rautenstrauch; *Helv. Chim. Acta*, 1972, **55**, 2233. 104

72HCA3005 W. Sieber, H. Heimgartner, H.-J. Hansen and H. Schmid; *Helv. Chim. Acta*, 1972, **55**, 3005. 420

72HOU(5/1b)180 M. Schlosser; *Methoden Org. Chem.* (*Houben-Weyl*), 1972, **5/1b**, 180. 568

72IZV472 M. S. Shvartsberg, I. L. Kotlyarevskii, L. N. Bizhan and E. E. Zaev; *Izv. Akad. Nauk, SSSR Ser. Khim.*, 1972, 472 (*Chem. Abstr.*, **77**, 34 422u). 1006

72IZV2113 N. A. Nesmeyanov, E. V. Binshtok, O. A. Rebrova and O. A. Reutov; *Izv. Akad. Nauk SSSR, Ser. Khim*, 1972, 2113. 741

72JA301 J. H. Incremona and C. J. Upton; *J. Am. Chem. Soc.*, 1972, **94**, 301. 928

72JA808 G. A. Olah; *J. Am. Chem. Soc.*, 1972, **94**, 808. 890, 893, 906

72JA992 T. W. Bentley, F. L. Schadt and P. v. R. Schleyer; *J. Am. Chem. Soc.*, 1972, **94**, 992. 1114, 1116

72JA1578 P. S. Skell and P. W. Owen; *J. Am. Chem. Soc.*, 1972, **94**, 1578. 1143

72JA2000 W. Adam, J. Baeza and J. C. Liu; *J. Am. Chem. Soc.*, 1972, **94**, 2000. 681, 682, 785

72JA2034 G. A. Olah, R. H. Schlosberg, R. D. Porter, Y. K. Mo, D. P. Kelly and G. D. Mateescu; *J. Am. Chem. Soc.*, 1972, **94**, 2034. 914

72JA2044	G. A. Olah, R. D. Porter, C. L. Jeuell and A. M. White; *J. Am. Chem. Soc.*, 1972, **94**, 2044.	894
72JA2128	F. Sondheimer and P. J. Beeby; *J. Am. Chem. Soc.*, 1972, **94**, 2128.	104
72JA2515	H. Kwart and J. Slutsky; *J. Am. Chem. Soc.*, 1972, **94**, 2515.	838
72JA2544	W. R. Dolbier, Jr. and J. H. Alonso; *J. Am. Chem. Soc.*, 1972, **94**, 2544.	801
72JA3544	G. A. Olah, G. Liang and Y. Mo; *J. Am. Chem. Soc.*, 1972, **94**, 3544.	902
72JA3629	R. H. Summerville and P. v. R. Schleyer; *J. Am. Chem. Soc.* 1972, **94**, 3629.	1111
72JA3870	W. Ando, S. Kondo, K. Nakayama, K. Ichibori, H. Kohoda, H. Yamato, I. Imai, S. Nakaido and T. Migita; *J. Am. Chem. Soc.*, 1972, **94**, 3870.	408, 409, 410, 820, 821, 822
72JA4013	E. J. Corey and T. Ravindranathan; *J. Am. Chem. Soc.*, 1972, **94**, 4013.	450, 662, 1103
72JA4370	H. C. Brown and S. K. Gupta; *J. Am. Chem. Soc.*, 1972, **94**, 4370.	510, 1101
72JA4371	R. C. Larock, S. K. Gupta and H. C. Brown; *J. Am. Chem. Soc.*, 1972, **94**, 4371.	1095
72JA4758	R. L. Sowerby and R. M. Coates; *J. Am. Chem. Soc.*, 1972, **94**, 4758.	591, 693
72JA5003	R. J. Arhat and J. C. Martin; *J. Am. Chem. Soc.*, 1972, **94**, 5003.	576
72JA5061	M. D. Schiavelli, R. P. Gilbert, W. A. Boynton and C. J. Boswell; *J. Am. Chem. Soc.*, 1972, **94**, 5061.	1114
72JA5098	R. E. Ireland, D. C. Muchmore and U. Hengartner; *J. Am. Chem. Soc.*, 1972, **94**, 5098.	17, 434
72JA5258	G. M. Whitesides, J. F. Gaasch and E. R. Stedronsky; *J. Am. Chem. Soc.*, 1972, **94**, 5258.	668
72JA5495	H. O. House and M. J. Umen; *J. Am. Chem. Soc.*, 1976, **94**, 5495.	275
72JA5511	P. M. Cook, L. F. Dahl and D. W. Dickerhoof; *J. Am. Chem. Soc.*, 1972, **94**, 5511.	897
72JA5897	R. E. Ireland and R. H. Mueller; *J. Am. Chem. Soc.*, 1972, **94**, 5897.	414, 416, 829, 830
72JA5910	C. Ullenius, P. W. Ford and J. E. Baldwin; *J. Am. Chem. Soc.*, 1972, **94**, 5910.	812
72JA6229	T. J. Barton, J. L. Witiak and C. L. McIntosh; *J. Am. Chem. Soc.*, 1972, **94**, 6229.	522
72JA6434	G. A. Olah and G. Liang; *J. Am. Chem. Soc.*, 1972, **94**, 6434.	902
72JA6538	K. B. Sharpless, M. A. Umbreit, M. T. Nieh and T. C. Flood; *J. Am. Chem. Soc.*, 1972, **94**, 6538.	692
72JA6549	N. Miyaura, M. Itoh, A. Suzuki, H. C. Brown, M. M. Midland and P. Jacob, III; *J. Am. Chem. Soc.*, 1972, **94**, 6549.	657
72JA6560	G. Zweifel, R. P. Fisher, J. T. Snow and C. C. Whitney; *J. Am. Chem. Soc.*, 1972, **94**, 6560.	278, 531, 662, 1103
72JA6854	M. W. Rathke and R. Kow; *J. Am. Chem. Soc.*, 1972, **94**, 6854.	162, 765
72JA6867	L. G. Cannell; *J. Am. Chem. Soc.*, 1972, **94**, 6867.	357
72JA7063	P. Ahlberg, D. L. Harris, M. Roberts, P. Warner, P. Seidl, M. Sakai, D. Cook, A. Diaz, J. P. Dirlam, H. Hamberger and S. Winstein; *J. Am. Chem. Soc.*, 1972, **94**, 7063.	918
72JA7081	A. Nickon and N. H. Werstiuk; *J. Am. Chem. Soc.*, 1972, **94**, 7081.	631
72JA7154	K. B. Sharpless and R. F. Lauer; *J. Am. Chem. Soc.*, 1972, **94**, 7154.	608
72JA7235	J. Metcalfe, H. A. J. Carless and E. K. C. Lee; *J. Am. Chem. Soc.*, 1972, **94**, 7235.	784
72JA7448	G. A. Olah, S. Kobayashi, and M. Tashiro; *J. Am. Chem. Soc.*, 1972, **94**, 7448.	462
72JA7859	G. A. Olah, P. W. Westerman, Y. K. Mo and G. Klopman; *J. Am. Chem. Soc.*, 1972, **94**, 7859.	58, 841
72JA7862	P. S. Skell, F. A. Fagone and K. J. Klabunde; *J. Am. Chem. Soc.*, 1972, **94**, 7862.	1143
72JA8084	E. Wenkert and C. A. McPherson; *J. Am. Chem. Soc.*, 1972, **94**, 8084.	1078
72JA8225	E. E. van Tamelen and R. J. Anderson; *J. Am. Chem. Soc.*, 1972, **94**, 8225.	216
72JA8608	R. D. Bach, K. W. Bair and D. Andrzejewski; *J. Am. Chem. Soc.*, 1972, **94**, 8608.	624, 629
72JA8917	J. A. Berson and L. Salem; *J. Am. Chem. Soc.*, 1972, **94**, 8917.	805
72JA8932	E. J. Corey and S. W. Walinsky; *J. Am. Chem. Soc.*, 1972, **94**, 8932.	821
72JA8934	E. J. Corey and S. W. Walinsky; *J. Am. Chem. Soc.*, 1972, **94**, 8934.	801
72JA9228	N. C. Yang and J. Libman; *J. Am. Chem. Soc.*, 1972, **94**, 9228.	811
72JA9234	M. F. Semmelhack, P. M. Helquist and J. D. Gorzynski; *J. Am. Chem. Soc.*, 1972, **94**, 9234.	1095, 1099
72JA9256	A. F. Kluge, K. G. Untch and J. H. Fried; *J. Am. Chem. Soc.*, 1972, **94**, 9256.	458
72JA9261	T. Durst, R. Van Den Elzen and M. J. LeBelle; *J. Am. Chem. Soc.*, 1972, **94**, 9261.	823
72JA9268	K. Tamao, Y. Kiso, K. Sumitani and M. Kumada; *J. Am. Chem. Soc.*, 1972, **94**, 9268.	471
72JCS(P1)216	G. Capozzi and G. Modena; *J. Chem. Soc., Perkin Trans. 1*, 1972, 216.	1112
72JCS(P1)305	D. H. R. Barton and B. J. Willis; *J. Chem. Soc., Perkin Trans. 1*, 1972, 305.	684, 685, 686
72JCS(P1)367	B. A. McAndrew and G. Riezebos; *J. Chem. Soc., Perkin Trans. 1*, 1972, 367.	832
72JCS(P1)639	J. Burdon, P. L. Coe, C. R. Marsh and J. C. Tatlow; *J. Chem. Soc., Perkin Trans. 1*, 1972, 639.	1005
72JCS(P1)1136	G. Capozzi, G. Modena and L. Ronzini; *J. Chem. Soc., Perkin Trans. 1*, 1972, 1136.	1112
72JCS(P1)1274	J. MacMillan and E. R. H. Walker; *J. Chem. Soc., Perkin Trans. 1*, 1972, 1274.	382
72JCS(P1)1535	P. A. Gent, R. Gigg and R. Conant; *J. Chem. Soc., Perkin Trans. 1*, 1972, 1535.	882
72JCS(P1)2271	C. W. Shoppee and B. J. A. Cooke; *J. Chem. Soc., Perkin Trans. 1*, 1972, 2271.	421
72JCS(P1)2533	J. Boyd and K. H. Overton; *J. Chem. Soc., Perkin Trans. 1*, 1972, 2533.	778
72JCS(P2)165	R. Taylor; *J. Chem. Soc., Perkin Trans. 2*, 1972, 165.	773
72JCS(P2)172	C. Eaborn, Z. S. Salih and D. R. M. Walton; *J. Chem. Soc., Perkin Trans. 2*, 1972, 172.	447
72JCS(P2)803	A. G. Davies, B. P. Roberts and J. C. Scaiano; *J. Chem. Soc., Perkin Trans. 2*, 1972, 803.	657
72JCS(P2)926	D. B. Bigley and C. M. Wren; *J. Chem. Soc., Perkin Trans. 2*, 1972, 926.	777
72JCS(P2)1744	D. B. Bigley and C. M. Wren; *J. Chem. Soc., Perkin Trans. 2*, 1972, 1744.	777
72JCS(P2)2026	A. Burighel, G. Modena and U. Tonellato; *J. Chem. Soc., Perkin Trans* 2, 1972, 2026.	1112
72JCS(P2)2359	D. B. Bigley and C. M. Wren; *J. Chem. Soc., Perkin Trans. 2*, 1972, 2359.	777
72JOC950	S. H. Dai and W. H. Dolbier, Jr.; *J. Org. Chem.*, 1972, **37**, 950.	497
72JOC1174	Z. Rappoport and A. Gal; *J. Org. Chem.*, 1972, **37**, 1174.	1110, 1112, 1114
72JOC1182	W. K. Baird, Jr. and J. H. Surridge; *J. Org. Chem.*, 1972, **37**, 1182.	72
72JOC1321	F. W. Fowler; *J. Org. Chem.*, 1972, **37**, 1321.	102

72JOC1553	T. B. Patrick, E. C. Haynie and W. J. Probst; *J. Org. Chem.*, 1972, **37**, 1553.	1143
72JOC1721	W. Ando, T. Yagihara, S. Tozune, I. Imai, J. Suzuki, T. Toyama, S. Nakaido and T. Migita; *J. Org. Chem.*, 1972, **37**, 1721.	233, 400
72JOC2363	P. A. Grieco; *J. Org. Chem.*, 1972, **37**, 2363.	2, 6
72JOC2579	A. I. Meyers and R. C. Strickland; *J. Org. Chem.*, 1972, **37**, 2579.	722
72JOC2601	A. K. Youssef and M. A. Ogliaruso; *J. Org. Chem.*, 1972, **37**, 2601.	800
72JOC2957	R. N. McDonald and H. E. Petty; *J. Org. Chem.*, 1972, **37**, 2957.	212
72JOC2989	E. N. Marvell, T. Gosink, P. Churchley and T. H. Li; *J. Org. Chem.*, 1972, **37**, 2989.	506
72JOC3220	M. S. Newman and S. J. Gromelski; *J. Org. Chem.*, 1972, **37**, 3220.	1144
72JOC3749	A. Uchida, T. Nakazawa, I. Kondo, N. Iwata and S. Matsuda; *J. Org. Chem.*, 1972, **37**, 3749.	1035
72JOC3791	W. Ando, M. Yamada, E. Matsuzaki and T. Migita; *J. Org. Chem.*, 1972, **37**, 3791.	409, 821, 822
72JOC3881	L. R. Byrd and M. C. Caserio; *J. Org. Chem.*, 1972, **37**, 3881.	497
72JOC4045	J. Buter, S. Wassenaar and R. M. Kellogg; *J. Org. Chem.*, 1972, **37**, 4045.	685
72JOC4090	A. J. Sisti and A. C. Vitale; *J. Org. Chem.*, 1972, **37**, 4090.	115
72JOC4236	J. K. Crandall and A. C. Clark; *J. Org. Chem.*, 1972, **37**, 4236.	316
72JOC4341	F. R. Jenson, J. J. Miller, S. J. Cristol and R. S. Beckley; *J. Org. Chem.*, 1972, **37**, 4341.	69
72JOC4466	H. P. Hogan and J. Seehafer; *J. Org. Chem.*, 1972, **37**, 4466.	1076
72JOC4467	T. B. Patrick, J. M. Disher and W. J. Probst; *J. Org. Chem.*, 1972, **37**, 4467.	1076
72JOC4468	M. S. Newman and L. F. Lee; *J. Org. Chem.*, 1972, **37**, 4468.	1076
72JOM(34)1	H. Lehmkuhl and D. Reinehr; *J. Organomet. Chem.*, 1972, **34**, 1.	495
72JOM(36)1	R. C. Larock and H. C. Brown; *J. Organomet. Chem.*, 1972, **36**, 1.	1095
72JOM(37)45	D. R. M. Walton and F. Waugh; *J. Organomet. Chem.*, 1972, **37**, 45.	1000, 1045, 1138
72JOM(39)C44	P. Bourgeois and G. Merault; *J. Organomet. Chem.*, 1972, **39**, C44.	963
72JOM(42)C21	T. J. Barton and E. Kline; *J. Organomet. Chem.*, 1972, **42**, C21.	522
72JOM(42)C39	W. Kruse; *J. Organomet. Chem.*, 1972, **42**, C39.	669
72JOM(42)163	W. Strohmeier and R. Fleischmann; *J. Organomet. Chem.*, 1972, **42**, 163.	73
72JOM(42)205	M. Tamura and J. K. Kochi; *J. Organomet. Chem.*, 1972, **42**, 205.	119
72JOM(44)69	B. Mauzé, C. Nivert and L. Miginiac; *J. Organomet. Chem.*, 1972, **44**, 69.	544
72JOU2515	M. G. Vinogradov, S. P. Verenchikov and G. I. Nikishin; *J. Org. Chem. USSR (Engl. Transl.)*, 1972, **8**, 2515.	926
72JOU2530	K. G. Golodova and S. I. Yakimovich; *J. Org. Chem. USSR (Engl. Transl.)*, 1972, 2530 (*Chem. Abstr.*, 1973, **79**, 53107c).	225
72M1651	M. Kirilov and G. Petrov; *Monatsh. Chem.*, 1972, 1651.	747
B-72MI 102-01	M. Luckner; "Secondary Metabolism in Plants and Animals," Academic Press, New York, 1972, p. 343.	35
B-72MI 102-02	L. G. Makarova; in "Organometallic Reactions," ed. E. I. Becker and M. Tsutsui, Wiley, New York, 1972, vol. 1, p. 325.	65
B-72MI 102-03	G. Bruno; "The Use of Aluminium Alkyls in Organic Synthesis," Ethyl Corp., Baton Rouge, 1970 and 1972.	69
B-72MI 103-01	A. J. Fry; "Synthetic Organic Electrochemistry," Harper and Row, New York, 1972.	87
B-72MI 110-01	L. Maier; in "Organophosphorus Compounds," eds. G. M. Kosolapoff and L. Maier, Wiley, New York, 1972, vol. 1, p. 23.	444
72MI 118-01	M. R. Willcott III, R. L. Cargill and A. B. Sears; *Prog. Phys. Org. Chem.*, 1972, **9**, 25.	795, 801
B-72MI 119-01	H. O. House; "Modern Synthetic Reactions," 2nd edn., W. A. Benjamin, Menlo Park, CA, 1972, p. 500.	871
B-72MI 119-02	H. O. House; "Modern Synthetic Reactions," 2nd edn., W. A. Benjamin, Menlo Park, CA, 1972, p. 547.	881
B-72MI 119-03	H. O. House; "Modern Synthetic Reactions," 2nd edn., W. A. Benjamin, Menlo Park, CA, 1972, p. 742.	884
B-72MI 119-04	H. O. House; "Modern Synthetic Reactions," 2nd edn., W. A. Benjamin, Menlo Park, CA, 1972, chap. 4.	887
B-72MI 119-05	H. O. House; "Modern Synthetic Reactions," 2nd edn., W. A. Benjamin, Menlo Park, CA, 1972, pp. 595–621.	889
B-72MI 119-06	H. O. House; "Modern Synthetic Reactions," 2nd edn., W. A. Benjamin, Menlo Park, CA, 1972, pp. 564–566.	888
B-72MI 119-07	G. A. Olah and P. von R. Schleyer (eds.); "Carbonium Ions," Wiley, New York, 1972, vol. 3.	891
B-72MI 119-08	M. Szwarc; "Ions and Ion Pairs in Organic Reactions," Wiley, New York, 1972, vol. 1.	893
B-72MI 119-09	M. Fieser and L. F. Fieser; "Reagents for Organic Synthesis," Wiley, New York, 1972, vol. 3, p. 401.	898
B-72MI 119-10	S. Winstein; in "Carbonium Ions," eds. G. A. Olah and P. von R. Schleyer, Wiley, New York, 1972, vol. 3, chap. 22.	904, 917
B-72MI 119-11	P. R. Story and B. C. Clark, Jr.; in "Carbonium Ions," eds. G. A. Olah and P. von R. Schleyer, Wiley, New York, 1972, vol. 3, chap. 23.	904, 917
B-72MI 119-12	C. J. Lancelot, D. J. Cram and P. von R. Schleyer; in "Carbonium Ions," eds. G. A. Olah and P. von R. Schleyer, Wiley, New York, 1972, vol. 3, chap. 27.	915
72MI 119-13	D. M. Brouwer and H. Hogeveen; *Prog. Phys. Org. Chem.*, 1972, **9**, 179–240.	901
72OR(19)1	G. H. Posner; *Org. React.*, 1972, **19**, 1.	311, 539, 540
72OR(19)115	M. F. Semmelhack; *Org. React.*, 1972, **19**, 115.	127
72OR(19)279	R. A. Sheldon and J. R. Kochi; *Org. React.*, 1972, **19**, 279.	560
72OS(52)128	D. C. Owsley and C. E. Castro; *Org. Synth.*, 1972, **52**, 128.	472
72RTC3	A. Tinkelberg, E. C. Kooyman and R. Louw; *Recl. Trav. Chim. Pays-Bas*, 1972, **91**, 3.	774

72RTC1057 F. van Rantwijk, Th. G. Spec and H. van Bekkum; *Recl. Trav. Chim. Pays-Bas*, 1972, **91**, 1057. 80

72RTC1069 Th. J. Nieuwstad and H. van Bekkum; *Recl. Trav. Chim. Pays-Bas*, 1972, **91**, 1069. 96

72S63 J. F. Normant; *Synthesis*, 1972, 63. 1000, 1002, 1004, 1047

72S142 G. Sosnovsky, E. H. Zaret and W. Mertz; *Synthesis*, 1972, 142. 181

72S235 H. Meier; *Synthesis*, 1972, 235. 1077

72S391 E. M. Kaiser; *Synthesis*, 1972, 391. 101, 889

72S483 G. L. Grady and S. K. Chokshi; *Synthesis*, 1972, 483. 9

72S555 E. Negishi, J.-J. Katz and H. C. Brown; *Synthesis*, 1972, 555. 1103

72S567 M. Schwarz and R. M. Waters; *Synthesis*, 1972, 567. 1000

72S591 H. Oediger, F. Möller and K. Eiter; *Synthesis*, 1972, 591. 573

72SC27 R. T. Arnold and C. Hoffmann; *Synth. Commun.*, 1972, **2**, 27. 417, 829

72SC331 E. Wenkert and C. A. McPherson; *Synth. Commun.*, 1972, **2**, 331. 1078

72T343 B. H. Freeman, D. Lloyd and M. I. C. Singer; *Tetrahedron*, 1972, **28**, 343. 647, 648, 737, 738, 743

72T353 D. Lloyd and M. I. C. Singer; *Tetrahedron*, 1972, **28**, 353. 737, 738, 743

72T4209 B. Deschamps, G. Lefebvre and J. Seyden-Penne; *Tetrahedron*, 1972, **28**, 4209. 750, 751

72TL125 J. C. Gilbert, K. R. Smith, G. W. Klumpp and M. Schakel; *Tetrahedron Lett.*, 1972, **13**, 175. 795

72TL649 I. Kuwajima and M. Uchida; *Tetrahedron Lett.*, 1972, **13**, 649. 600

72TL737 I. Kuwajima, S. Sato and Y. Kurata; *Tetrahedron Lett.*, 1972, **13**, 737. 590, 693

72TL1189 M. Bertrand, J.-L. Gras and J. Goré; *Tetrahedron Lett.*, 1972, **13**, 1189. 498

72TL1279 M. Wada and O. Mitsunobu; *Tetrahedron Lett.*, 1972, **13**, 1279. 179

72TL1391 M. M. Bhagwat and D. Devaprabhakara; *Tetrahedron Lett.*, 1972, **13**, 1391. 459

72TL2013 P. Naegeli and R. Kaiser; *Tetrahedron Lett.*, 1972, **13**, 2013. 525

72TL2521 T. Oritani and K. Yamashita; *Tetrahedron Lett.*, 1972, **13**, 2521. 726

72TL2627 R. Pappo and P. W. Collins; *Tetrahedron Lett.*, 1972, **13**, 2627. 1049, 1050

72TL2923 M. Yoshimoto, S. Ishihara, N. Soma and E. Nakayama; *Tetrahedron Lett.*, 1972, **13**, 2923. 822

72TL2983 R. Garry and R. Vessière; *Tetrahedron Lett.*, 1972, **13**, 2983. 1124

72TL3257 G. Buono; *Tetrahedron Lett.*, 1972, **13**, 3257. 983

72TL3587 S. Julia, C. Huynh and D. Michlot; *Tetrahedron Lett.*, 1972, **13**, 3587. 821

72TL3769 E. J. Corey and P. L. Fuchs; *Tetrahedron Lett.*, 1972, **13**, 3769. 1067

72TL3827 Y. Hata and M. Watanabe; *Tetrahedron Lett.*, 1972, **13**, 3827. 636

72TL4083 K. F. Bernady and M. J. Weiss; *Tetrahedron Lett.*, 1972, **13**, 4083. 534

72TL4249 R. W. Rathke and D. Sullivan; *Tetrahedron Lett.*, 1972, **13**, 4249. 957

72TL4465 J. P. Dulcère, M.-L. Roumestant and J. Goré; *Tetrahedron Lett.*, 1972, **13**, 4465. 959

72TL4605 G. I. Tsuchihashi, S. Iriuchijima and M. Ishibashi; *Tetrahedron Lett.*, 1972, **13**, 4605. 210

72TL4651 D. S. Watt and E. J. Corey; *Tetrahedron Lett.*, 1972, **13**, 4651. 789

72TL4659 Y. Hata and M. Watanabe; *Tetrahedron Lett.*, 1972, **13**, 4659. 636

72TL5121 G. Andrews and D. A. Evans; *Tetrahedron Lett.*, 1972, **13**, 5121. 822

72TL5209 R. Oliver and D. R. M. Walton; *Tetrahedron Lett.*, 1972, **13**, 5209. 1017, 1018

72TL5269 D. N. Reinhoudt, H. C. Volger and C. G. Kouwenhoven; *Tetrahedron Lett.*, 1972, **13**, 5269. 518

72ZOR1484 S. V. Zavgorodnii, G. K. Bereznitskii and D. A. Pisanenko; *Zh. Org. Khim.*, 1972, **8**, 1484 (*Chem. Abstr.*, 1972, **77**, 113 919). 492

73ACR53 M. Saunders, P. Vogel, E. L. Hagen and J. Rosenfeld; *Acc. Chem. Res.*, 1973, **6**, 53. 899

73ACS1401 A. G. Andrewes and S. Liaan-Jensen; *Acta Chem. Scand.*, 1973, **27**, 1401. 747, 748

73ACS3944 P. O. I. Virtanen, S. Peltonen and J. Hyyppa; *Acta Chem. Scand.*, 1973, **27**, 3944. 229

73AG(E)173 G. A. Olah; *Angew. Chem., Int. Ed. Engl.*, 1973, **12**, 173. 891, 901

73AG(E)323 U. Schöllkopf and R. Jentsch; *Angew. Chem., Int. Ed. Engl.*, 1973, **12**, 323. 209

73AG(E)337 K. Hafner, R. Dönges, E. Goedecke and R. Kaiser; *Angew. Chem., Int. Ed. Engl.*, 1973, **12**, 337. 789

73AG(E)399 A. de Meijere and O. Schallner; *Angew. Chem., Int. Ed. Engl.*, 1973, **12**, 399. 900

73AG(E)566 H. G. Viehe, Z. Janousek, R. Gompper and D. Lach; *Angew. Chem., Int. Ed. Engl.*, 1973, **12**, 566. 982

73AJC1259 N. J. Daly, G. M. Heweston and F. Ziolkowski; *Aust. J. Chem.*, 1973, **26**, 1259. 777

73B3492 P. R. Andrews, G. D. Smith and I. G. Young; *Biochemistry*, 1973, **12**, 3492. 826

73BSF746 M. Julia and D. Arnould; *Bull. Soc. Chim. Fr.*, 1973, 746. 612

73BSF1800 M. Bertrand and C. Santelli-Rouvier; *Bull. Soc. Chim. Fr.*, 1973, 1800. 1126

73BSF2147 S. Tyrlik and I. Wolochowicz; *Bull. Soc. Chim. Fr.*, 1973, 2147. 687, 688

73BSF3371 M. Karila, M. L. Capmau and W. Chodkiewicz; *Bull. Soc. Chim. Fr.*, 1973, 3371. 1055

73C277 J. Bruhin, W. Kneubueler and W. Jenny; *Chimia*, 1973, **27**, 277. 104

73CB674 H. J. Kuhn and K. Gollnick; *Chem. Ber.*, 1973, **106**, 674. 521

73CB1376 W. Kirmse, W. Gruber and J. Knist; *Chem. Ber.*, 1973, **106**, 1376. 383

73CB3216 E. Lamparter and M. Hanack; *Chem. Ber.*, 1973, **106**, 3216. 1113

73CB3779 F. Bohlmann and C. Zdero; *Chem. Ber.*, 1973, **106**, 3779. 736

73CC117 J. E. Baldwin and J. A. Walker; *J. Chem. Soc., Chem. Commun.*, 1973, 117. 831

73CC151 E. W. Colvin and B. J. Hamill; *J. Chem. Soc., Chem. Commun.*, 1973, 151. 1078

73CC223 R. G. Carlson, J. H.-A. Huber and D. E. Henton; *J. Chem. Soc., Chem. Commun.*, 1973, 223. 841

73CC319 J. M. Brown, B. T. Golding and J. J. Stofko Jr.; *J. Chem. Soc., Chem. Commun.*, 1973, 319. 411, 811

73CC351 V. Pascali and A. Umani-Ronchi; *J. Chem. Soc., Chem. Commun.*, 1973, 351. 439

73CC374 G. Jones, II, S. B. Schwartz and M. T. Marton; *J. Chem. Soc., Chem. Commun.*, 1973, 374. 785

73CC520 S. E. Potter and I. O. Sutherland; *J. Chem. Soc., Chem. Commun.*, 1973, 520. 706, 709

73CC544	A. Pelter, C. R. Harrison and D. Kirkpatrick; *J. Chem. Soc., Chem. Commun.*, 1973, 544.	126
73CC589	D. R. Kelsey and R. G. Bergman; *J. Chem. Soc., Chem. Commun.*, 1973, 589.	1114
73CC606	E. Negishi and T. Yoshida; *J. Chem. Soc., Chem. Commun.*, 1973, 606.	532, 1100, 1103
73CC679	M. H. Sekera, B.-A. Weissman and R. G. Bergman; *J. Chem. Soc., Chem. Commun.*, 1973, 679.	1126
73CC695	D. L. J. Clive; *J. Chem. Soc., Chem. Commun.*, 1973, 695.	601
73CC848	Y. Yamamoto, J. Oda and Y. Inouye; *J. Chem. Soc., Chem. Commun.*, 1973, 848.	839
73CC874	E. Negishi, G. Lew and T. Yoshida; *J. Chem. Soc., Chem. Commun.*, 1973, 874.	532, 662
73CC907	G. H. Posner and D. J. Brunelle; *J. Chem. Soc., Chem. Commun.*, 1973, 907.	482, 485
73CC955	R. H. Mitchell; *J. Chem. Soc., Chem. Commun.*, 1973, 955.	1074
73CI(L)1067	B. Danieli, F. Ronchetti and G. Russo; *Chem. Ind.* (*London*), 1973, 1067.	1121
73CJC981	G. Caron and J. Lessard; *Can. J. Chem.*, 1973, **51**, 981.	785
73CL291	T. Mukaiyama, M. Hayashi and K. Narasaka; *Chem. Lett.*, 1973, 291.	427, 428, 430
73CL667	M. Nishino, H. Kondo, and A. Miyake; *Chem. Lett.*, 1973, 667.	464
73CL1041	T. Mukaiyama, T. Sato and J. Hanna; *Chem. Lett.*, 1973, 1041.	687
73CL1097	S. Kobayashi, H. Takei and T. Mukaiyama; *Chem. Lett.*, 1973, 1097.	485
73CPL309	B. Zurawski, R. Ahlrichs and W. Kutzelnigg; *Chem. Phys. Lett.*, 1973, **21**, 309.	1109
73CRV293	F. Pietra; *Chem. Rev.*, 1973, **73**, 293.	906
73CSR397	J. W. Henderson; *Chem. Soc. Rev.*, 1973, **2**, 397.	893
73HCA75	U. Widmer, J. Zsindely, H.-J. Hansen and H. Schmid; *Helv. Chim. Acta*, 1973, **56**, 75.	809
73HCA105	M. Schmid, H.-J. Hansen and H. Schmid; *Helv. Chim. Acta*, 1973, **56**, 105.	835
73HCA875	H. Schlossarczyk, W. Sieber, M. Hesse, H.-J. Hansen and H. Schmid; *Helv. Chim. Acta*, 1973, **56**, 875.	993
73HCA1176	G. Ohloff and M. Pawlak; *Helv. Chim. Acta*, 1973, **56**, 1176.	726
73HCA1516	K. W. Egger and A. T. Cocks; *Helv. Chim. Acta*, 1973, **56**, 1516.	925
73IJS205	J. R. Shelton and K. E. Davis; *Int. J. Sulfur Chem.*, 1973, **3**, 205.	595
73JA160	K. Yates, G. H. Schmid, T. W. Regulski, D. G. Garratt, H.-W. Leung and R. McDonald; *J. Am. Chem. Soc.*, 1973, **95**, 160.	1110
73JA822	E. Vedejs and P. L. Fuchs; *J. Am. Chem. Soc.*, 1973, **95**, 822.	640, 641
73JA856	G. G. Maynes and D. E. Applequist; *J. Am. Chem. Soc.*, 1973, **95**, 856.	928
73JA962	B. M. Trost and R. F. Hammen; *J. Am. Chem. Soc.*, 1973, **95**, 962.	821
73JA967	T. Mukaiyama, K. Inomata and M. Muraki; *J. Am. Chem. Soc.*, 1973, **95**, 967.	333
73JA1547	P. S. Skell, K. J. Klambunde, J. H. Plonka, J. S. Roberts and D. L. Williams-Smith; *J. Am. Chem. Soc.*, 1973, **95**, 1547.	232
73JA2248	C. P. Casey and C. R. Cyr; *J. Am. Chem. Soc.*, 1973, **95**, 2248.	797
73JA2400	H. C. Brown and E. N. Peters; *J. Am. Chem. Soc.*, 1973, **95**, 2400.	895
73JA2693	K. Oshima, H. Takahashi, H. Yamamoto and H. Nozaki; *J. Am. Chem. Soc.*, 1973, **95**, 2693.	141, 834
73JA2694	K. Oshima, K. Shimoji, H. Takahashi, H. Yamamoto and H. Nozaki; *J. Am. Chem. Soc.*, 1973, **95**, 2694.	141
73JA2697	K. B. Sharpless and R. F. Lauer; *J. Am. Chem. Soc.*, 1973, **95**, 2697.	601, 608, 781
73JA2715	R. Kow and M. W. Rathke; *J. Am. Chem. Soc.*, 1973, **95**, 2715.	162, 163
73JA3050	M. W. Rathke and D. F. Sullivan; *J. Am. Chem. Soc.*, 1973, **95**, 3050.	130
73JA3068	B. M. Trost, D. Keeley and M. J. Bogdanowicz; *J. Am. Chem. Soc.*, 1973, **95**, 3068.	390
73JA3078	T. J. Barton, E. A. Kline and P. M. Garvey; *J. Am. Chem. Soc.*, 1973, **95**, 3078.	64, 522
73JA3080	A. Suzuki, N. Miyaura, S. Abiko, M. Itoh, H. C. Brown, J. A. Sinclair and M. M. Midland; *J. Am. Chem. Soc.*, 1973, **95**, 3080.	279, 1033
73JA3348	J. M. Watson, J. L. Irvine and R. M. Roberts; *J. Am. Chem. Soc.*, 1973, **95**, 3348.	827
73JA3382	S. W. Staley and A. W. Orvedal; *J. Am. Chem. Soc.*, 1973, **95**, 3382.	862
73JA3384	S. W. Staley and A. W. Orvedal; *J. Am. Chem. Soc.*, 1973, **95**, 3384.	910
73JA3408	A. I. Meyers and G. N. Knaus; *J. Am. Chem. Soc.*, 1973, **95**, 3408.	664
73JA3416	K. E. Koenig and W. P. Weber; *J. Am. Chem. Soc*, 1973, **95**, 3416.	450, 451
73JA3420	F. Jung, N. K. Sharma and T. Durst; *J. Am. Chem. Soc.*, 1973, **95**, 3420.	600
73JA3687	L. M. Jackman and R. C. Haddon; *J. Am. Chem. Soc.*, 1973, **95**, 3687.	873
73JA3932	W. G. Dauben, R. G. Williams and R. D. McKelvey; *J. Am. Chem. Soc.*, 1973, **95**, 3932.	421
73JA4090	K. Izawa, T. Okuyama and T. Fueno; *J. Am. Chem. Soc.*, 1973, **95**, 4090.	1125
73JA4346	R. L. Cargill, A. B. Sears, J. Boehm and M. R. Willcott; *J. Am. Chem. Soc.*, 1973, **95**, 4346.	74
73JA4416	B. E. McCarry, R. L. Markezich and W. S. Johnson; *J. Am. Chem. Soc.*, 1973, **95**, 4416.	1122
73JA4451	J. X. McDermott, J. F. White and G. M. Whitesides; *J. Am. Chem. Soc.*, 1973, **95**, 4451.	668, 669
73JA4960	G. A. Olah, Y. Halpern, J. Shen and Y. K. Mo; *J. Am. Chem. Soc.*, 1973, **95**, 4960.	901
73JA5778	E. Vedejs and K. A. J. Snoble; *J. Am. Chem. Soc.*, 1973, **95**, 5778.	721, 729
73JA5786	H. C. Brown, T. Hamaoka and N. Ravindran; *J. Am. Chem. Soc.*, 1973, **95**, 5786.	661, 1103
73JA5813	H. J. Reich, I. L. Reich and J. M. Renga; *J. Am. Chem. Soc.*, 1973, **95**, 5813.	601, 602, 604, 605, 781
73JA6137	K. B. Sharpless, R. F. Lauer and A. Y. Teranishi; *J. Am. Chem. Soc.*, 1973, **95**, 6137.	601, 604, 605, 606, 607
73JA6152	G. Stork and B. Ganem; *J. Am. Chem. Soc.*, 1973, **95**, 6152.	307
73JA6286	W. J. Baron, M. E. Hendrick and M. Jones, Jr.; *J. Am. Chem. Soc.*, 1973, **95**, 6286.	549
73JA6379	H. W. Thompson and R. E. Naipawer; *J. Amer. Chem. Soc.*, 1973, **95**, 6379.	72
73JA6449	M. V. Moncur and J. B. Grutzner; *J. Am. Chem. Soc.*, 1973, **95**, 6449.	865
73JA6456	H. C. Brown, T. Hamaoka and N. Ravindran; *J. Am. Chem. Soc.*, 1973, **95**, 6456.	660, 1103
73JA6462	C. R. Johnson, J. R. Shanklin and R. A. Kirchhoff; *J. Am. Chem. Soc.*, 1973, **95**, 6462.	622, 710
73JA6728	K. J. Shea and P. S. Skell; *J. Am. Chem. Soc.*, 1973, **95**, 6728.	928
73JA6840	B. M. Trost and T. N. Salzmann; *J. Am. Chem. Soc.*, 1973, **95**, 6840.	598, 779

73JA7418	C. R. Johnson and C. W. Schroeck; *J. Am. Chem. Soc.*, 1973, **95**, 7418.	304
73JA7424	C. R. Johnson, C. W. Schroeck and J. R. Shanklin; *J. Am. Chem. Soc.*, 1973, **95**, 7424.	710
73JA7777	C. R. Johnson and G. A. Dutra; *J. Am. Chem. Soc.*, 1973, **95**, 7777.	476
73JA7788	G. H. Posner, C. E. Whitten and J. J. Sterling; *J. Am. Chem. Soc.*, 1973, **95**, 7788.	120
73JA7916	M. D. Schiavelli and D. E. Ellis; *J. Am. Chem. Soc.*, 1973, **95**, 7916.	1114
73JA8096	W. E. Billups, K. H. Leavell, E. S. Lewis and S. Vanderpool; *J. Am. Chem. Soc.*, 1973, **95**, 8096.	801
73JA8678	H. Kwart and J. Slutsky; *J. Am. Chem. Soc.*, 1973, **95**, 8678.	838
73JA8698	G. A. Olah, G. Liang, G. D. Mateescu and J. L. Riemenschneider; *J. Am. Chem. Soc.*, 1973, **95**, 8698.	906
73JCS(P1)137	W. A. Asomaning, C. Eaborn and D. R. M. Walton; *J. Chem. Soc., Perkin Trans. 1*, 1973, 137.	452
73JCS(P1)541	J. I. G. Cadogan and G. A. Molina; *J. Chem. Soc., Perkin Trans. 1*, 1973, 541.	36, 442
73JCS(P1)654	R. B. Boar, D. W. Hawkins, J. F. McGhie and D. H. R. Barton; *J. Chem. Soc., Perkin Trans. 1*, 1973, 654.	438
73JCS(P1)720	J. S. Cowie, P. D. Landor and S. R. Landor; *J. Chem. Soc., Perkin Trans 1*, 1973, 720.	961
73JCS(P1)975	I. U. Khand, G. R. Knox, P. L. Pauson and W. E. Watts; *J. Chem. Soc., Perkin Trans. 1*, 1973, 975.	528
73JCS(P1)977	I. U. Khand, G. R. Knox, P. L. Pauson, W. E. Watts and M. I. Foreman; *J. Chem. Soc., Perkin Trans. 1*, 1973, 977.	528
73JCS(P1)1347	M. Kalli, S. R. Landor and P. D. Landor; *J. Chem. Soc., Perkin Trans. 1*, 1973, 1347.	1015
73JCS(P1)1416	Y. Badar, W. J. S. Lockley, T. P. Toube, B. C. L. Weedon and L. R. Guy Valadon; *J. Chem. Soc., Perkin Trans. 1*, 1973, 1416.	725
73JCS(P1)1858	P. A. Gent, R. Gigg and R. Conant; *J. Chem. Soc., Perkin Trans. 1*, 1973, 1858.	882
73JCS(P1)1913	R. K. Bramley, J. Caldwell and R. Grigg; *J. Chem. Soc., Perkin Trans. 1*, 1973, 1913.	809
73JCS(P1)2509	C. J. Love and F. J. McQuillen; *J. Chem. Soc., Perkin Trans. 1*, 1973, 2509.	74
73JCS(P1)2818	R. M. Acheson, N. D. Wright and P. A. Tasker; *J. Chem. Soc., Perkin Trans. 1*, 1973, 2818.	518
73JCS(P2)301	Z. Rappoport and A. Gal; *J. Chem. Soc., Perkin Trans. 2*, 1973, 301.	1110, 1112
73JCS(P2)1293	D. B. Bigley and R. E. Gabbott; *J. Chem. Soc., Perkin Trans. 2*, 1973, 1293.	777
73JCS(P2)1570	D. A. Coates and J. M. Tedder; *J. Chem. Soc., Perkin Trans. 2*, 1973, 1570.	935
73JCS(P2)1971	J. Klein and R. Levene; *J. Chem. Soc., Perkin Trans. 2*, 1973, 1971.	540
73JCS(P2)2078	D. Capocasale, L. D. Nunno, S. Florio and F. Naso; *J. Chem. Soc., Perkin Trans. 2*, 1973, 2078.	1110
73JHC655	I. Lalezari, A. Shafiee and H. Golgolab; *J. Heterocycl. Chem.*, 1973, **10**, 655.	1077
73JOC26	E. M. Burgess, H. R. Penton Jr. and E. A. Taylor; *J. Org. Chem.*, 1973, **38**, 26.	575
73JOC487	A. K. Youssef and M. A. Ogliaruso; *J. Org. Chem.*, 1973, **38**, 487.	800
73JOC547	M. S. Newman and Z. U. Din; *J. Org. Chem.*, 1973, **38**, 547.	1143
73JOC864	P. J. Garratt, K. C. Nicolaou and F. Sondheimer; *J. Org. Chem.*, 1973, **38**, 864.	990
73JOC1130	T. Araki, K. Hayakawa, T. Aoyagi, Y. Nakano and H. Tanida; *J. Org. Chem.*, 1973, **38**, 1130.	666
73JOC1348	S. B. Kadin; *J. Org. Chem.*, 1973, **38**, 1348.	28, 29
73JOC1483	D. R. Coulson; *J. Org. Chem.*, 1973, **38**, 1483.	495
73JOC1617	H. C. Brown and N. Ravindran; *J. Org. Chem.*, 1973, **38**, 1617.	488, 510, 531, 1103
73JOC1742	R. D. Bach, D. Andrzejewski and L. R. Dusold; *J. Org. Chem.*, 1973, **38**, 1742.	624, 782
73JOC1775	G. Stork and R. L. Danheiser; *J. Org. Chem.*, 1973, **38**, 1775.	482
73JOC2226	C. A. Brown and V. K. Ahuja; *J. Org. Chem.*, 1973, **38**, 2226.	503
73JOC2254	H. Newman; *J. Org. Chem.*, 1973, **38**, 2254.	1138
73JOC2572	P. A. Grieco, D. Boxler and K. Hiroi; *J. Org. Chem.*, 1973, **38**, 2572.	409, 821
73JOC2574	J. Hooz, J. N. Bridson, J. G. Calzada, H. C. Brown, M. M. Midland and A. B. Levy; *J. Org. Chem.*, 1973, **38**, 2574.	486
73JOC2664	C. Trindle, J.-T. Hwang and F. A. Carey; *J. Org. Chem.*, 1973, **38**, 2664.	649
73JOC2731	K. Matsumoto, M. Suzuli, T. Iwasaki and M. Miyoshi; *J. Org. Chem.*, 1973, **38**, 2731.	34
73JOC2747	G. H. Posner and D. J. Brunelle; *J. Org. Chem.*, 1973, **38**, 2747.	484
73JOC2915	L. N. Mander and J. V. Turner; *J. Org. Chem.*, 1973, **38**, 2915.	823
73JOC3020	R. A. Rossi and J. F. Bunnett; *J. Org. Chem.*, 1973, **38**, 3020.	474
73JOC3054	M. D. Schiavelli, P. L. Timpanaro and R. Brewer; *J. Org. Chem.*, 1973, **38**, 3054.	1114
73JOC3055	A. Sanders and W. P. Giering; *J. Org. Chem.*, 1973, **38**, 3055.	936
73JOC3189	P. R. Jones and S. J. Costanzo; *J. Org. Chem.*, 1973, **38**, 3189.	127, 128, 466
73JOC3438	B. M. Trost and W. G. Biddlecom; *J. Org. Chem.*, 1973, **38**, 3438.	821
73JOC3588	W. N. Smith and E. D. Kuehn; *J. Org. Chem.*, 1973, **38**, 3588.	1000
73JOC3625	E. Vedejs, J. P. Bershas and P. L. Fuchs; *J. Org. Chem.*, 1973, **38**, 3625.	736
73JOC3652	A. W. Burgstahler and G. N. Widiger; *J. Org. Chem.*, 1973, **38**, 3652.	503
73JOC3658	J. Slutsky and H. Kwart; *J. Org. Chem.*, 1973, **38**, 3658.	836
73JOC3660	R. A. Benkeser and D. F. Ehler; *J. Org. Chem.*, 1973, **38**, 3660.	21
73JOC3893	H. O. House and M. J. Umen; *J. Org. Chem.*, 1973, **38**, 3893.	1142
73JOC4117	P. S. Wharton and D. W. Johnson; *J. Org. Chem.*, 1973, **38**, 4117.	812
73JOC4263	C. R. Johnson, R. W. Herr and D. M. Wieland; *J. Org. Chem.*, 1973, **38**, 4263.	217, 226
73JOC4268	N. Kawabata, A. Matsumura and S. Yamashita; *J. Org. Chem.*, 1973, **38**, 4268.	115
73JOC4346	B. C. Hartman, T. Livinghouse and B. Rickborn; *J. Org. Chem.*, 1973, **38**, 4346.	217
73JOC4367	J. E. McMurry and J. Melton; *J. Org. Chem.*, 1973, **38**, 4367.	156, 309
73JOM(47)417	G. Marr and J. Ronayne; *J. Organomet. Chem.*, 1973, **47**, 417.	155
73JOM(49)477	P. Heimbach and M. Molin; *J. Organomet. Chem.*, 1973, **49**, 477.	413
73JOM(49)483	P. Heimbach and M. Molin; *J. Organomet. Chem.*, 1973, **49**, 483.	809
73JOM(50)9	L. Lochmann and D. Lím; *J. Organomet. Chem.*, 1973, **50**, 9.	130

73JOM(50)C12 Y. Kiso, K. Tamao and M. Kumada; *J. Organomet. Chem.*, 1973, **50**, C12 466
73JOM(54)153 M. Le Quan and G. Guillerm; *J. Organomet. Chem.*, 1973, **54**, 153. 973
73JOM(55)273 Y. Odic and M. Pereyre; *J. Organomet. Chem.*, 1973, **55**, 273. 132
73JOM(59)53 E. M. Kaiser, R. D. Beard and C. R. Hauser; *J. Organomet. Chem.*, 1973, **59**, 53. 144
73JOM(60)19 D. J. Pasto and P. E. Timony; *J. Organomet. Chem.*, 1973, **60**, 19. 658
73JOM(60)247 P. S. Kendurkar and R. S. Tewari; *J. Organomet. Chem.*, 1973, **60**, 247. 738
73JOM(61)167 J. D. Kennedy, H. G. Kuivila, F. L. Pelczar, R. Y. Tien and J. L. Considine; *J. Organomet. Chem.*, 1973, **61**, 167. 83
73LA437 H. Neunhoeffer and G. Werner; *Justus Liebigs Ann. Chem.*, 1973, 437. 790
B-73MI 102-01 R. Luckenbach; "Dynamic Stereochemistry of Pentacoordinated Phosphorus and Related Elements," Georg Theime, Stuttgart, 1973. 51
B-73MI 108-01 L. Kaplan; in "Free Radicals," ed. J. Kochi, Wiley, New York, 1973, vol. 2, p. 361. 321
B-73MI 108-02 K. U. Ingold; in "Free Radicals," ed. J. Kochi, Wiley, New York, 1973, vol. 1, p. 37. 329
B-73MI 109-01 T. S. Stevens and W. E. Watts; "Selected Molecular Rearrangements," Van Nostrand Reinhold, London, 1973. 377
B-73MI 114-01 K. W. Bentley; in "Elucidation of Organic Structures by Physical and Chemical Methods," 2nd edn., eds. K. W. Bentley and G. W. Kirby, Wiley, Chichester, 1973, pp. 255–289. 627
B-73MI 119-01 N. S. Isaacs; in "Organic Reactive Intermediates," ed. S. P. McManus, Academic Press, New York, 1973, p. 234. 844
B-73MI 119-02 E. M. Kaiser and D. W. Slocum; in "Organic Reactive Intermediates," ed. S. P. McManus, Academic Press, New York, 1973, p. 337. 844
B-73MI 119-03 M. Schlosser; "Struktur und Reaktivität polarer Organometalle," Springer, Berlin, 1973. 885
B-73MI 119-04 G. A. Olah and P. von R. Schleyer (eds.); "Carbonium Ions," Wiley, New York, 1973, vol. 4. 891
B-73MI 119-05 S. P. McManus and C. U. Pittman, Jr.; in "Organic Reactive Intermediates," ed. S. P. McManus, Academic Press, New York, 1973, pp. 193–355. 891
B-73MI 119-06 H. H. Freedman; in "Carbonium Ions," eds. G. A. Olah and P. von R. Schleyer, Wiley, New York, 1973, vol. 4, pp. 1532–1534. 898
B-73MI 119-07 R. C. Fort, Jr.; in "Carbonium Ions," eds. G. A. Olah and P. von R. Schleyer, Wiley, New York, 1973, vol. 4, p. 1783–1835. 900
B-73MI 119-08 K. M. Harmon; in "Carbonium Ions," eds. G. A. Olah and P. von R. Schleyer, Wiley, New York, 1973, vol. 4, chap. 29. 906
B-73MI 119-09 J. K. Kochi (ed.); "Free Radicals," Wiley-Interscience, New York, 1973, vol. 1, p. 1. 919
B-73MI 119-10 J. K. Kochi (ed.); "Free Radicals," Wiley-Interscience, New York, 1973, vol. 2, p. 1. 919
B-73MI 119-11 J. C. Martin; in "Free Radicals," ed. J. K. Kochi, Wiley-Interscience, New York, 1973, vol. 2, p. 493. 923
B-73MI 119-12 G. A. Russell; in "Free Radicals," ed. J. K. Kochi, Wiley-Interscience, New York, 1973, vol. 1, p. 275. 925
B-73MI 119-13 J. W. Wilt; in "Free Radicals," ed. J. K. Kochi, Wiley-Interscience, New York, 1973, vol. 1, p. 333. 925, 927
B-73MI 119-14 J. K. Kochi; in "Free Radicals," ed. J. K. Kochi, Wiley-Interscience, New York, 1973, vol. 2, p. 665. 929
B-73MI 119-15 T. Koenig; in "Free Radicals," ed. J. K. Kochi, Wiley-Interscience, New York, 1973, vol. 1, p. 113. 934
B-73MI 119-16 J. K. Kochi; in "Free Radicals," ed. J. K. Kochi, Wiley-Interscience, New York, 1973, vol. 1, p. 591. 934
B-73MI 119-17 P. I. Abell; in "Free Radicals," ed. J. K. Kochi, Wiley-Interscience, New York, 1973, vol. 2, p. 63. 948, 950
73MI 119-18 D. J. Cram and J. M. Cram; *Intra-Sci. Chem. Rep.*, 1973, **7**, 1. 844
73MI 119-19 I. Murata; *Top. Nonbenzenoid Aromat. Chem.*, 1973, **1**, 159. 866
73MI 119-20 D. J. Bertelli; *Top. Nonbenzenoid Aromat. Chem.*, 1973, **1**, 29. 854, 906
B-73MI 122-01 R. A. Moss; in "Carbenes," eds. M. Jones, Jr. and R. A. Moss, Wiley-Interscience, New York, 1973, Vol. I. 1143
73OR(20)1 H. E. Simmons, T. L. Cairns, S. A. Vladuchick and C. M. Hoiness; *Org. React.*, 1973, **20**, 1. 299
73OS(53)13 D. H. Wadsworth; *Org. Synth.*, 1973, **53**, 13. 635
73OSC(4)880 H. Lindlar and R. Dubuis; *Org. Synth., Coll. Vol.*, 1973, **4**, 880. 318
73OSC(5)149 S. Gronowitz and T. Raznikiewicz; *Org. Synth., Coll. Vol.*, 1973, **5**, 149. 428
73OSC(5)315 A. C. Cope and R. D. Bach; *Org. Synth., Coll. Vol.*, 1973, **5**, 315. 628
73OSC(5)346 B. A. Fox and T. L. Threlfall; *Org. Synth., Coll. Vol.*, 1973, **5**, 346. 428
73OSC(5)398 K. C. Bass; *Org. Synth., Coll. Vol.*, 1973, **5**, 398. 101
73OSC(5)400 M. E. Kuehne and B. F. Lambert; *Org. Synth., Coll. Vol.*, 1973, **5**, 400. 97
73OSC(5)467 L. A. Paquette and J. H. Barrett; *Org. Synth., Coll. Vol.*, 1973, **5**, 467. 96
73OSC(5)608 J. C. H. Hwa and H. Sims; *Org. Synth., Coll. Vol.*, 1973, **5**, 608. 628
73OSC(5)880 H. Lindlar and R. Dubuis; *Org. Synth., Coll. Vol.*, 1973, **5**, 880. 502, 503, 504
73PAC197 W. G. Dauben, M. S. Kellogg, J. I. Seeman, N. D. Vietmeyer and P. H. Wendschuh; *Pure Appl. Chem.*, 1973, **33**, 197. 420, 421
73RCR1011 V. D. Sholle and E. G. Rozantsev; *Russ. Chem. Rev.*, 1973, **42**, 1011. 922
73RTC605 M. Schakel and G. W. Klumpp; *Recl. Trav. Chim. Pays-Bas*, 1973, **92**, 605. 795
73RTC667 R. A. van der Welle and L. Brandsma; *Recl. Trav. Chim. Pays-Bas*, 1973, **92**, 667. 1048
73S1 F. Minisci; *Synthesis*, 1973, 1. 925
73S37 G. Zweifel, R. P. Fisher and A. Horng; *Synthesis*, 1973, 37. 163
73S295 G. Märkl and A. Merz; *Synthesis*, 1973, 295. 731

73S305 A. Suzuki, N. Miyaura, M. Itoh, H. C. Brown, P. Jacob, III *Synthesis*, 1973, 305. 277, 977

73S457 E. N. Marvell and T. Li; *Synthesis*, 1973, 457. 456, 502, 504, 505

73S703 M. Tramontini; *Synthesis*, 1973, 703. 1041

73S744 M. Tramontini; *Synthesis*, 1973, 744. 318

73SC67 M. W. Rathke, and D. F. Sullivan; *Synth. Commun.*, 1973, **3**, 67. 830

73SC245 C. H. Heathcock and D. N. Brattesani; *Synth. Commun.*, 1973, **3**, 245. 1000

73SC281 G. H. Posner and J.-S. Ting; *Synth. Commun.*, 1973, **3**, 281. 119

73T883 H. Chwastek, R. Epsztein and N. LeGoff; *Tetrahedron Lett.*, 1973, **29**, 883. 999

73T1069 N. Kawabata, A. Matsumura and S. Yamashita; *Tetrahedron*, 1973, **29**, 1069. 115, 452

73T1393 P. J. Vroegop, J. Lugtenburg and E. Havinga; *Tetrahedron*, 1973, **29**, 1393. 420

73T1697 I. Gosney and D. Lloyd; *Tetrahedron*, 1973, **29**, 1697. 737, 738

73T3497 S. Mannen and H. A. Itano; *Tetrahedron*, 1973, **29**, 3497. 318

73T3781 E. N. Marvell, G. Caple, B. Schatz and W. Pippin; *Tetrahedron*, 1973, **29**, 3781. 419, 420, 794

73T3797 E. N. Marvell, G. Caple, C. Delphey, J. Platt, N. Polston and J. Tashiro; *Tetrahedron*, 1973, **29**, 3797. 420

73T4199 T. Bottin-Strzalko; *Tetrahedron*, 1973, **29**, 4199. 750

73TCA183 O. P. Strausz, R. J. Norstrom, A. C. Hopkinson, M. Schoenborn and I. G. Csizmadia; *Theor. Chim. Acta*, 1973, **29**, 183. 1143

73TCC47 Z. Yoshida; *Top. Curr. Chem.*, 1973, **40**, 47. 907

73TL9 M. D. Gheórghiu, C. Draghici, L. Stănescu and M. Avram; *Tetrahedron Lett.*, 1973, **14**, 9. 518

73TL401 A. Haag and H. Baudisch; *Tetrahedron Lett.*, 1973, **14**, 401. 657

73TL447 S. J. Nelson, G. Detre and M. Tanabe; *Tetrahedron Lett.*, 1973, **14**, 447. 430

73TL787 K. Utimoto, T. Tanaka, T. Furubayashi and H. Nozaki; *Tetrahedron Lett.*, 1973, **14**, 787. 274

73TL827 S. E. Schaafsma, R. Jorritsmas, H. Steinberg and J. Th. de Boer; *Tetrahedron Lett.*, 1973, **14**, 827. 929

73TL1019 P. D. Landor, S. R. Landor and J. P. Leighton; *Tetrahedron Lett.*, 1973, **14**, 1019. 1005, 1006

73TL1375 R. B. Herbert; *Tetrahedron Lett.*, 1973, **14**, 1375. 452

73TL1495 E. J. Corey and R. A. Ruden; *Tetrahedron Lett.*, 1973, **14**, 1495. 999, 1066

73TL1513 J. B. Hendrickson and J. R. Sufrin; *Tetrahedron Lett.*, 1973, **14**, 1513. 318

73TL1611 E. J. Corey and R. H. K. Chen; *Tetrahedron Lett.*, 1973, **14** 1611. 540

73TL1843 C. Y. Meyers, L. L. Ho, G. J. McCollum and J. Branca; *Tetrahedron Lett.*, 1973, **14**, 1843. 618

73TL1847 M. Naruse, K. Utimoto and H. Nozaki; *Tetrahedron Lett.*, 1973, **14**, 1847. 1033

73TL1979 K. B. Sharpless, M. W. Young and R. F. Lauer; *Tetrahedron Lett.*, 1973, **14**, 1979. 601, 602, 603, 607, 608, 781

73TL2351 Y. Tamura, T. Miyamoto, T. Nishimura and Y. Kita; *Tetrahedron Lett.*, 1973, **14**, 2351. 822

73TL2517 D. J. Pasto and J. K. Borchardt; *Tetrahedron Lett.*, 1973, **14**, 2517. 459

73TL2545 J. M. Simpson and H. G. Richey; *Tetrahedron Lett.*, 1973, **14**, 2545. 801

73TL2595 R. J. Cregge, J. L. Herrmann, J. E. Richman, R. F. Romanet and R. H. Schlessinger; *Tetrahedron Lett.*, 1973, **14**, 2595. 309

73TL2727 K. B. Wiberg and K. C. Bishop III; *Tetrahedron Lett.*, 1973, **14**, 2727. 841

73TL3181 C. M. Bowes, D. F. Montecalvo and F. Sondheimer; *Tetrahedron Lett.*, 1973, **14**, 3181. 954

73TL3361 R. Baudouy and J. Gore; *Tetrahedron Lett.*, 1973, **14**, 3361 1025

73TL3459 J. Kossanyi, B. Forth and J.-P. Monzu; *Tetrahedron Lett.*, 1973, **14**, 3459. 842

73TL3497 R. K. Boeckman, Jr. and S. M. Silver; *Tetrahedron Lett.*, 1973, **14**, 3497. 443, 649, 759

73TL3899 J. Fried and J. C. Sih; *Tetrahedron Lett.*, 1973, **14**, 3899. 1023, 1024

73TL3971 J. P. Marino and T. Kaneko; *Tetrahedron Lett.*, 1973, **14**, 3971. 811

73TL3975 J. P. Marino and T. Kaneko; *Tetrahedron Lett.*, 1973, **14**, 3975. 811

73TL3985 R. W. Saalfrank; *Tetrahedron Lett.*, 1973, **14**, 3985. 988

73TL4193 H. Sakurai, K. Nishiwaki and M. Kira; *Tetrahedron Lett.*, 1973, **14**, 4193. 759

73TL4395 R. H. Mitchell; *Tetrahedron Lett.*, 1973, **14**, 4395. 704

73TL4425 W. G. Dauben, D. J. Hart, J. Ipaktschi and A. P. Kozikowski; *Tetrahedron Lett.*, 1973, **14**, 4425. 736

73TL4477 E. J. Corey and R. L. Danheiser; *Tetrahedron Lett.*, 1973, **14**, 4477. 523

73TL4491 A. Pelter, C. R. Harrison and D. Kirkpatrick; *Tetrahedron Lett.*, 1973, **14**, 4491. 126

73TL4767 H. Yamataka, S.-G. Kim, T. Ando and Y. Yukawa; *Tetrahedron Lett.*, 1973, **14**, 4767. 1124

73TL4833 M. Julia and J. M. Paris; *Tetrahedron Lett.*, 1973, **14**, 4833. 144, 614, 615, 696, 697, 698, 700

73TS253 V. Buss, P. von R. Schleyer and L. C. Allen; *Top. Stereochem.*, 1973, **7**, 253–293. 891

74ACR147 D. A. Evans and G. C. Andrews; *Acc. Chem. Res.*, 1974, **7**, 147. 140, 143, 609, 788

74ACS(B)765 L.-I. Olsson, A. Claesson and C. Bogentoft; *Acta Chem. Scand., Ser. B*, 1974, **28**, 765. 961

74AG50 G. Fouquet and M. Schlosser; *Angew. Chem.*, 1974, **86**, 50 177

74AG(E)82 G. Fouquet and M. Schlosser; *Angew. Chem. Int. Ed. Engl.*, 1974, **13**, 82. 465

74AG(E)202 D. Seebach and K.-H. Geiss; *Angew. Chem., Int. Ed. Engl.*, 1974, **13**, 202. 140

74AG(E)407 K. Kondo, A. Negishi and D. Tunemoto; *Angew. Chem., Int. Ed. Engl.*, 1974, **13**, 407. 159

74AG(E)425 G. Maier; *Angew. Chem., Int. Ed. Engl.*, 1974, **13**, 425. 789

74AG(E)468 W. Bannwarth, R. Eidenschink and T. Kauffmann; *Angew. Chem., Int. Ed. Engl.*, 1974, **13**, 468. 363

74AG(E)479 K. Geiss, B. Seuring, R. Pieter and D. Seebach; *Angew. Chem., Int. Ed. Engl.*, 1974, **13**, 479. 140, 141

74AG(E)627 T. Kauffmann; *Angew. Chem., Int. Ed. Engl.*, 1974, **13**, 627. 371, 372

74AOC83 J. Furukawa and N. Kawabata; *Adv. Organomet. Chem.*, 1974, **12**, 83. 122

74AOC215 A. E. Jukes; *Adv. Organomet. Chem.*, 1974, **12**, 215. 116

74AX(B)833 H. Noordik, T. E. M. van den Hark, J. J. Mooji and A. A. K. Klaassen; *Acta Crystallogr., Part B*, 1974, **30**, 833. 890

74BAU1031 V. A. Koptyug; *Bull. Acad. Sci. USSR, Div. Chem. Sci.*, 1974, **23**, 1031. 914
74BCJ503 N. Miyamoto, D. Fukuoka, K. Utimoto and H. Nozaki; *Bull. Chem. Soc. Jpn.*, 1974, **47**, 503. 485
74BCJ2529 H. Taguchi, K. Shimoji, H. Yamamoto and H. Nozaki; *Bull. Chem. Soc. Jpn.*, 1974, **47**, 2529. 651
74BCJ2663 S. Uemura, H. Tara, M. Okano and K. Ichikawa; *Bull. Chem. Soc. Jpn.*, 1974, **47**, 2663. 454
74BSF1119 J. P. Dulcère, J. Goré and M.-L. Roumestant; *Bull. Soc. Chim. Fr.*, 1974, 1119. 959, 960
74BSF1656 J. F. Normant, G. Cahiez, M. Bourgain, C. Chuit and J. Villieras; *Bull. Soc. Chim. Fr.*, 1974, 1656. 540
74BSF1924 F. Gerard and P. Miginiac; *Bull. Soc. Chim. Fr.*, 1974, 1924. 800
74BSF2072 J. F. Normant, O. Reboul, R. Sauvetre, H. Deshayes, D. Masure and J. Villieras; *Bull. Soc. Chim. Fr.*, 1974, 2072. 833
74BSF(2)1731 J. Villieras, C. Bacquet and J. F. Normant; *Bull. Soc. Chim. Fr.*, Part 2, 1974, 1731. 1067
74CA(81)120222t W. Himmele, H. Siegel, W. Aquila and H. Fröhlich; *Chem. Abstr.*, 1974, **81**, 120 222t. 296
74CB1654 H.-L. Elbe and G. Köbrich; *Chem. Ber.*, 1974, **107**, 1654. 491
74CB1676 G. Hammen, T. Bäßler and M. Hanack; *Chem. Ber.*, 1974, **107**, 1676. 1114
74CB1773 F. Bohlmann and H.-J. Bax; *Chem. Ber.*, 1974, **107**, 1773. 725
74CB1915 K. Weinges and K. Klessing; *Chem. Ber.*, 1974, **107**, 1915. 708
74CB2050 J. Buddrus; *Chem. Ber.*, 1974, **107**, 2050. 722
74CC10 D. Michelot, G. Linstrumelle and S. Julia; *J. Chem. Soc., Chem. Commun.*, 1974, 10. 980
74CC100 D. L. J. Clive; *J. Chem. Soc., Chem. Commun.*, 1974, 100. 592, 608
74CC142 A. J. Bridges and G. H. Whitham; *J. Chem. Soc., Chem. Commun.*, 1974, 142. 640, 641, 753
74CC380 T. Tsuda, K. Ueda and T. Saegusa; *J. Chem. Soc., Chem. Commun.*, 1974, 380. 1137
74CC472 A. L. J. Beckwith and G. Moad; *J. Chem. Soc., Chem. Commun.*, 1974, 472. 339
74CC762 E. Negishi and T. Yoshida; *J. Chem. Soc., Chem. Commun.*, 1974, 762. 999
74CC793 E. Verkuijlen and C. Boelhouwer; *J. Chem. Soc., Chem. Commun.*, 1974, 793. 674
74CC819 R. D. Bach, K. W. Bair and D. Andrzejewski; *J. Chem. Soc., Chem. Commun.*, 1974, 819. 629
74CC820 R. D. Bach, D. Andrzejewski and K. W. Bair; *J. Chem. Soc., Chem. Commun.*, 1974, 820. 629
74CC990 R. H. Mitchell; *J. Chem. Soc., Chem. Commun.*, 1974, 990. 713, 714, 716
74CJC761 T. Durst, M. J. LeBelle, R. van den Elzen and K.-C. Tin; *Can. J. Chem.*, 1974, **52**, 761. 142
74CL165 T. Imamoto and Y. Yukawa; *Chem. Lett.*, 1974, 165. 636
74CL385 K. Iwai, M. Kawai, H. Kosugi and H. Uda; *Chem. Lett.*, 1974, 385. 209
74CL1523 T. Mukaiyama, Y. Watanabe and M. Shiono; *Chem. Lett.*, 1974, 1523. 590, 693
74CR(C)149 J. P. Dulcère, G. Gil and M. Bertrand; *C. R. Hebd. Seances Acad. Sci., Ser C*, 1974, **278**, 149. 959
74CRV87 J. Boutagy and R. Thomas; *Chem. Rev.*, 1974, **74**, 87. 638, 639, 744, 745
74CRV189 K. T. Potts and J. S. Baum; *Chem. Rev.*, 1974, **74**, 189. 907
74CRV243 N. L. Holy; *Chem. Rev.*, 1974, **74**, 243. 890
74CRV567 G. Brieger and T. J. Nestrick; *Chem. Rev.*, 1974, **74**, 567. 85, 457
74HCA771 T. A. Narwid, K. E. Cooney and M. R. Uskokovic; *Helv. Chim. Acta*, 1974, **57**, 771. 832
74HCA1317 F. Näf and R. Decorzant; *Helv. Chim. Acta*, 1974, **57**, 1317. 889
74HCA2084 A. F. Thomas and R. Dubini; *Helv. Chim. Acta*, 1974, **57**, 2084. 816
74JA112 R. A. Rossi and J. F. Bunnett; *J. Am. Chem. Soc.*, 1974, **96**, 112. 483
74JA195 G. A. Olah and G. Liang; *J. Am. Chem. Soc.*, 1974, **96**, 195. 906
74JA316 G. Zweifel, A. Horng and J. E. Plamondon; *J. Am. Chem. Soc.*, 1974, **96**, 316. 658
74JA586 Z. Rappoport and J. Kaspi; *J. Am. Chem. Soc.*, 1974, **96**, 586. 1112
74JA870 C. L. Hill and G. M. Whitesides; *J. Am. Chem. Soc.*, 1974, **96**, 870. 67
74JA1100 R. H. Summerville, C. A. Senkler, P. v. R. Schleyer, T. E. Dueber and P. J. Stang; *J. Am. Chem. Soc.*, 1974, **96**, 1100. 1111, 1114, 1116
74JA1110 R. H. Summerville and P. v. R. Schleyer; *J. Am. Chem. Soc.*, 1974, **96**, 1110. 1122
74JA1133 H. A. Dieck and R. F. Heck; *J. Am. Chem. Soc.*, 1974, **96**, 1133. 493
74JA1230 C. P. Casey and R. L. Anderson; *J. Am. Chem. Soc.*, 1974, **96**, 1230. 165
74JA1348 S. Z. Goldberg, K. N. Raymond, C. Harmon and D. H. Templeton; *J. Am. Chem. Soc.*, 1974, **96**, 1348. 890
74JA1518 J. K. Stille, F. Huang and M. T. Regan; *J. Am. Chem. Soc.*, 1974, **96**, 1518. 670
74JA1620 K. Shimoji, H. Taguchi, K. Oshima, H. Yamamoto and H. Nozaki; *J. Am. Chem. Soc.*, 1974, **96**, 1620. 651
74JA1962 J. J. McCullogch and M. R. McClory; *J. Am. Chem. Soc.*, 1974, **96**, 1962. 800
74JA2201 K. S. Chen, D. Y. H. Tang, L. K. Montgomery and J. K. Kochi; *J. Am. Chem. Soc.*, 1974, **96**, 2201. 321, 923
74JA3332 G. Buchi and R. M. Freidinger; *J. Am. Chem. Soc.*, 1974, **96**, 3332. 706, 707
74JA3468 J. A. Berson, T. Miyashi and G. Jones II; *J. Am. Chem. Soc.*, 1974, **96**, 3468. 815
74JA3548 G. A. Olah, P. W. Westerman and J. Nishimura; *J. Am. Chem. Soc.*, 1974, **96**, 3548. 895
74JA3574 G. A. Olah and T. R. Hockswender, Jr.; *J. Am. Chem. Soc.*, 1974, **96**, 3574. 1110
74JA3705 J. J. Gajewski, L. K. Hoffman and C. N. Shih; *J. Am. Chem. Soc.*, 1974, **96**, 3705. 806
74JA3711 H. M. Walborsky and M. P. Periasamy; *J. Am. Chem. Soc.*, 1974, **96**, 3711. 155
74JA4027 D. K. Bohme, G. I. Mackay and J. D. Payzant; *J. Am. Chem. Soc.*, 1974, **96**, 4027. 107
74JA4063 E. L. Muetterties and F. J. Hirsekorn; *J. Am. Chem. Soc.*, 1974, **96**, 4063. 93
74JA4518 Z. Rappoport and J. Kaspi; *J. Am. Chem. Soc.*, 1974, **96**, 4518. 1112
74JA4562 P. J. Stang, M. G. Mangum, D. P. Fox and P. Haak; *J. Am. Chem. Soc.*, 1974, **96**, 4562. 1071, 1143
74JA4708 J. E. McMurry and M. P. Fleming; *J. Am. Chem. Soc.*, 1974, **96**, 4708. 687, 688, 689
74JA4715 J. M. Photis and L. A. Paquette; *J. Am. Chem. Soc.*, 1974, **96**, 4715. 708
74JA5032 T. Gillbro and F. Williams; *J. Am. Chem. Soc.*, 1974, **96**, 5032. 445

74JA5254	S. Tanaka, H. Yamamoto, H. Nozaki and K. B. Sharpless; *J. Amer. Chem. Soc.*, 1974, **96**, 5254.	206
74JA5268	G. Stork, L. D. Cama and D. R. Coulson; *J. Am. Chem. Soc.*, 1974, **96**, 5268.	214
74JA5270	G. Stork and J. F. Cohen; *J. Am. Chem. Soc.*, 1974, **96**, 5270.	214
74JA5272	G. Stork and L. Maldonado; *J. Am. Chem. Soc.*, 1974, **96**, 5272.	309
74JA5290	E. E. van Tamelen and J. A. Gladysz; *J. Am. Chem. Soc.*, 1974, **96**, 5290.	13, 18, 21
74JA5560	D. A. Evans, G. C. Andrews and B. Buckwalter; *J. Am. Chem. Soc.*, 1974, **96**, 5560.	136
74JA5561	W. C. Still and T. L. Macdonald; *J. Am. Chem. Soc.*, 1974, **96**, 5561.	136
74JA5581	E. J. Corey and R. H. Wollenberg; *J. Am. Chem. Soc.*, 1974, **96**, 5581.	1048, 1081, 1091
74JA5620	T. Leung and G. Zweifel; *J. Am. Chem. Soc.*, 1974, **96**, 5620.	449, 977
74JA5708	E. Bockrath and L. M. Dorfman; *J. Am. Chem. Soc.*, 1974, **96**, 5708.	847
74JA5801	L. A. Paquette, R. E. Wingard Jr. and J. M. Photis; *J. Am. Chem. Soc.*, 1974, **96**, 5801.	707, 708
74JA5855	G. A. Olah, R. J. Spear, P. W. Westerman and J.-M. Denis; *J. Am. Chem. Soc.*, 1974, **96**, 5855.	904
74JA6140	A. Tamaki, S. A. Magennis and J. K. Kochi; *J. Am. Chem. Soc.*, 1974, **96**, 6140.	669
74JA6179	R. K. Boeckman, Jr.; *J. Am. Chem. Soc.*, 1974, **96**, 6179.	307
74JA6181	G. Stork and J. Singh; *J. Am. Chem. Soc.*, 1974, **96**, 6181.	307
74JA6686	M. Hanack, T. Bässler, W. Eymann, W. E. Heyd and R. Kopp; *J. Am. Chem. Soc.*, 1974, **96**, 6686.	1114, 1115, 1124
74JA6692	W. F. Verhelst and W. Drenth; *J. Am. Chem. Soc.*, 1974, **96**, 6692.	1122
74JA7101	L. Friedman and A. Shani; *J. Am. Chem. Soc.*, 1974, **96**, 7101.	119, 464
74JA7125	J. E. Baldwin, G. A. Höfle and O. W. Lever, Jr.; *J. Am. Chem. Soc.*, 1974, **96**, 7125.	1089
74JA7165	B. M. Trost, W. P. Conway, P. E. Strege and T. J. Dietsche; *J. Am. Chem. Soc.*, 1974, **96**, 7165.	599
74JA7573	G. Buchi and H. Wuest; *J. Am. Chem. Soc.*, 1974, **96**, 7573.	823
74JA7591	C. D. Poulter and C. J. Spillner; *J. Am. Chem. Soc.*, 1974, **96**, 7591.	894
74JA7638	F. K. Fong; *J. Am. Chem. Soc.*, 1974, **96**, 7638.	906
74JA7977	E. I. Heiba, R. M. Dessau and P. G. Rodewald; *J. Am. Chem. Soc.*, 1974, **96**, 7977.	336, 926
74JA8095	E. C. Taylor and S. F. Martin; *J. Am. Chem. Soc.*, 1974, **96**, 8095.	860
74JA8115	D. W. Hart and J. Schwartz; *J. Am. Chem. Soc.*, 1974, **96**, 8115.	669
74JCS(P1)340	F. Naso and L. Ronzini; *J. Chem. Soc., Perkin Trans. 1*, 1974, 340.	1065
74JCS(P1)433	M. Jones, P. Temple, E. J. Thomas and G. H. Whitham; *J. Chem. Soc., Perkin Trans. 1*, 1974, 433.	591
74JCS(P1)1209	H.-A. Chia, B. E. Kirk and D. R. Taylor; *J. Chem. Soc., Perkin Trans. 1*, 1974, 1209.	498, 518
74JCS(P1)1794	D. H. R. Barton, F. S. Guziec Jr. and I. Shahak; *J. Chem. Soc., Perkin Trans. 1*, 1974, 1794.	684, 685, 686
74JCS(P1)1844	B. E. Kirk and D. R. Taylor; *J. Chem. Soc., Perkin Trans. 1*, 1974, 1844.	518
74JCS(P1)1938	C. R. Eck, G. L. Hodgson, D. F. MacSweeney, R. W. Mills and T. Money; *J. Chem. Soc., Perkin Trans. 1*, 1974, 1938.	381
74JCS(P1)2435	T. M. Cresp, R. G. F. Giles, M. V. Sargent, C. Brown and D. O'N. Smith; *J. Chem. Soc., Perkin Trans. 1*, 1974, 2435.	732
74JCS(P1)2470	M. Elliot, N. F. Janes and D. A. Pulman; *J. Chem. Soc., Perkin Trans. 1*, 1974, 2470.	724
74JCS(P2)596	C. Eaborn, I. D. Jenkins and D. R. M. Walton; *J. Chem. Soc., Perkin Trans. 2*, 1974, 596.	64
74JCS(P2)843	P. J. Stang, R. J. Hargrove and T. E. Dueber; *J. Chem. Soc., Perkin Trans. 2*, 1974, 843.	1111
74JCS(P2)1808	I. G. Csizmadia, A. J. Duke, V. Lucchini and G. Modena; *J. Chem. Soc., Perkin Trans. 2*, 1974, 1808.	1109
74JOC120	P. A. Grieco and M. Miyashita; *J. Org. Chem.*, 1974, **39**, 120.	601, 602, 607
74JOC378	C. U. Pittman, Jr., A. Kress and L. D. Kispert; *J. Org. Chem.*, 1974, **39**, 378.	909
74JOC400	W. H. Mandeville and G. M. Whitesides; *J. Org. Chem.*, 1974, **39**, 400.	119, 312
74JOC428	H. J. Reich; *J. Org. Chem.*, 1974, **39**, 428.	592, 608
74JOC429	K. B. Sharpless and R. F. Lauer; *J. Org. Chem.*, 1974, **39**, 429.	603, 608, 781
74JOC581	P. J. Stang and R. J. Hargrove; *J. Org. Chem.*, 1974, **39**, 581.	1070
74JOC732	P. A. Grieco and C. S. Pogonowski; *J. Org. Chem.*, 1974, **39**, 732.	210
74JOC747	H. O. House and E. F. Kinloch; *J. Org. Chem.*, 1974, **39**, 747.	457, 508
74JOC761	M. S. Newman and M. C. Vander Zwan; *J. Org. Chem.*, 1974, **39**, 761.	1143
74JOC821	E. J. Corey and B. W. Erickson; *J. Org. Chem.*, 1974, **39**, 821.	735
74JOC968	B. Grant and C. Djerassi; *J. Org. Chem.*, 1974, **39**, 968.	513
74JOC1168	W. D. Korte, K. Cripe and R. Cooke; *J. Org. Chem.*, 1974, **39**, 1168.	112
74JOC1318	A. Padwa and L. Brodsky; *J. Org. Chem.*, 1974, **39**, 1318.	736
74JOC1322	A. P. Krapcho and E. G. E. Jahngen Jr.; *J. Org. Chem.*, 1974, **39**, 1322.	683
74JOC1438	D. Kaufman and R. Kupper; *J. Org. Chem.*, 1974, **39**, 1438.	1122
74JOC1575	H. Kwart and J. L. Schwartz; *J. Org. Chem.*, 1974, **39**, 1575.	834
74JOC1650	A. P. Krapcho and E. G. E. Jahngen Jr.; *J. Org. Chem.*, 1974, **39**, 1650.	681, 682
74JOC2006	J. F. Wolfe, D. E. Portlock and D. J. Feuerbach; *J. Org. Chem.*, 1974, **39**, 2006.	860
74JOC2133	H. J. Reich, J. M. Renga and I. L. Reich; *J. Org. Chem.*, 1974, **39**, 2133.	606, 610
74JOC2255	T. Okuyama, K. Ohashi, K. Izawa and T. Fueno; *J. Org. Chem.*, 1974, **39**, 2255.	1125
74JOC2302	R. H. Shapiro and E. C. Hornaman; *J. Org. Chem.*, 1974, **39**, 2302.	632
74JOC2321	E. Negishi, G. Lew and T. Yoshida; *J. Org. Chem.*, 1974, **39**, 2321.	532, 662
74JOC2486	S. Torii, I. Okamoto and H. Tanaka; *J. Org. Chem.*, 1974, **39**, 2486.	561
74JOC3174	W. E. Truce and M. J. Lusch; *J. Org. Chem.*, 1974, **39**, 3174.	470
74JOC3175	J. P. Marino and T. Kaneko; *J. Org. Chem.*, 1974, **39**, 3175.	811
74JOC3195	R. Annunziata, R. Fornasier and F. Montanari; *J. Org. Chem.*, 1974, **39**, 3195.	85
74JOC3264	T. H. Chan and E. Chang; *J. Org. Chem.*, 1974, **39**, 3264.	648

74JOC3456 E. I. Heiba and R. M. Dessau; *J. Org. Chem.*, 1974, **39**, 3456. 336

74JOC3525 P. J. DeChristopher, J. P. Ademek, G. D. Lyon, S. A. Klein and R. J. Baumgarten; *J. Org. Chem.*, 1974, **39**, 3525. 28

74JOC3793 P. E. Sonnet; *J. Org. Chem.*, 1974, **39**, 3793. 727

74JOM(67)467 R. Levene, J. Y. Becker and J. Klein; *J. Organomet. Chem.*, 1974, **67**, 467. 540

74JOM(70)89 G. F. Pregaglia, G. F. Ferrari, A. Andreetta, G. Capparella, F. Genoni and R. Ugo; *J. Organomet. Chem.*, 1974, **70**, 89. 79

74JOM(72)163 I. S. Akhrem, N. M. Christovalova, E. I. Mysov and M. E. Vol'Pin; *J. Organomet. Chem.*, 1974, **72**, 163. 493

74JOM(73)359 H. Alper; *J. Organomet. Chem.*, 1974, **73**, 359. 437

74JOM(76)297 E. Favre and M. Gaudemar; *J. Organomet. Chem.*, 1974, **76**, 297. 1055

74JOM(76)305 E. Favre and M. Gaudemar; *J. Organomet. Chem.*, 1974, **76**, 305. 1055

74JOM(77)269 J. F. Normant, G. Cahiez, C. Chuit and J. Villieras; *J. Organomet. Chem.*, 1974, **77**, 269. 491, 1108

74JOM(77)281 J. F. Normant, G. Cahiez, C. Chuit and J. Villieras; *J. Organomet. Chem.*, 1974, **77**, 281. 1108

74JOM(77)C22 J. Buckle and P. G. Harrison; *J. Organomet. Chem.*, 1974, **77**, C22. 1079

74JOM(81)41 S. S. Washburne and J. B. Simolike; *J. Organomet. Chem.*, 1974, **81**, 41. 656

74JOM(81)C37 J. Evans, J. Schwarts and P. W. Urquhart; *J. Organomet. Chem.*, 1974, **81**, C37. 669

74JOU1990 L. P. Kirillova, S. R. Buzilova, E. S. Serebryakova and L. I. Vereshchagin; *J. Org. Chem. USSR (Engl. Transl.)*, 1974, **10**, 1990. 1006

74LA369 K. Kuhlein and H. Jensen; *Justus Liebigs Ann. Chem.*, 1974, 369. 178

74LA1190 H. Neunhoeffer and G. Werner; *Justus Liebigs Ann. Chem.*, 1974, 1190. 790

74MI 102-01 L. D. Freedman, B. R. Ezzell, R. N. Jenkins and R. M. Harris; *Phosphorus*, 1974, **4**, 199. 53

B-74MI 104-01 B. J. Wakefield; in "The Chemistry of Organolithium Compounds," Pergamon, Oxford, 1974, Chapter 10, pp. 144–159. 110

B-74MI 112-01 J. M. Bruce; "The Chemistry of Quinonoid Compounds," ed. S. Patai, Wiley, New York, 1974, part 1, ch. 9, p. 465. 517

B-74MI 114-01 B. J. Wakefield; "The Chemistry of Organolithium Compounds," Pergamon, Oxford, 1974, p. 204. 663

B-74MI 118-01 R. F. Heck; "Organotransition Metal Chemistry," Academic Press, New York, 1974, vol 76. 797

B-74MI 119-01 B. J. Wakefield; "The Chemistry of Organolithium Compounds," Pergamon, Oxford, 1974. 885

B-74MI 119-02 K. T. Finley; in "The Chemistry of Quinonoid Compounds," ed. S. Patai, Wiley, New York, 1974, part 2, pp. 877–1144. 888

B-74MI 119-03 G. A. Olah; "Carbocations and Electrophilic Reactions," Wiley, New York, 1974. 891

B-74MI 119-04 N. S. Isaacs; "Reactive Intermediates in Organic Chemistry," Wiley, New York, 1974, pp. 92–199. 891

B-74MI 119-05 M. Szwarc; "Ions and Ion Pairs in Organic Reactions," Wiley, New York, 1974, vol. 2. 893

B-74MI 119-06 M. Fieser and L. F. Fieser; "Reagents for Organic Synthesis," Wiley, New York, 1974, vol. 4, p. 548. 898

B-74MI 119-07 J. M. Hay; "Reactive Free Radicals," Academic Press, New York, 1974. 919

B-74MI 121-01 A. E. Jukes; "Adv. Organomet. Chem.," 1974, **12**, 215. 318

74OPP269 P. E. Sonnet; *Org. Prep. Proced. Int.*, 1974, **6**, 269. 727

74OS(54)1 F. Sondheimer and K. Stockel; *Org. Synth.*, 1974, **54**, 1. 1034

74PAC505 B. M. Mikhailov; *Pure Appl. Chem.*, 1974, **39**, 505. 838

74PAC553 M. Julia; *Pure Appl. Chem.*, 1974, 553. 343

74RTC240 P. Vermeer, J. Meijer and C. Eylander; *Recl. Trav. Chim. Pays-Bas*, 1974, **93**, 240. 999

74S309 H. Pines; *Synthesis*, 1974, 309. 294

74S328 G. O. Doak and L. D. Freedman; *Synthesis*, 1974, 328. 445

74S339 G. Zweifel and R. P. Fisher; *Synthesis*, 1974, 339. 449

74S344 E. Galantay, I. Bacso and R. V. Coombs; *Synthesis*, 1974, 344. 961

74S441 W. N. Smith and O. F. Beumel, Jr.; *Synthesis*, 1974, 441. 1000, 1001

74S539 L. J. Kricka and A. Ledwith; *Synthesis*, 1974, 539. 357

74S573 R. Baudouy and J. Gore; *Synthesis*, 1974, 573. 515, 1025

74S633 D. N. Kursanov, Z. N. Parnes and N. M. Loim; *Synthesis*, 1974, 201. 13, 281

74S658 R. A. Lynd and G. Zweifel; *Synthesis*, 1974, 658. 534

74S798 H. J. Bestmann and W. Stransky; *Synthesis*, 1974, 798. 727

74SC199 H. Gilman and B. Holland; *Synth. Commun.*, 1974, **4**, 199. 1000, 1001

74T227 M. Santelli and M. Bertrand; *Tetrahedron*, 1974, **30**, 227. 1126

74T235 M. Santelli and M. Bertrand; *Tetrahedron*, 1974, **30**, 235. 1126

74T243 M. Santelli and M. Bertrand; *Tetrahedron*, 1974, **30**, 243. 1114

74T1525 K. Takeda; *Tetrahedron*, 1974, **30**, 1525. 813

74T1959 H.-J. Hansen and H. Schmid; *Tetrahedron*, 1974, **30**, 1959. 415

74T2009 J. M. McBride; *Tetrahedron*, 1974, **30**, 2009. 922

74T2159 M. Naruse, K. Utimoto and H. Nozaki; *Tetrahedron*, 1974, **30**, 2159. 1033

74T3037 M. Naruse, K. Utimoto and H. Nozaki; *Tetrahedron*, 1974, **30**, 3037. 207

74TL171 T. H. Chan and W. Mychajlowskij; *Tetrahedron Lett.*, 1974, **15**, 171. 648, 652, 982

74TL207 H. J. Bestmann and O. Vostrowsky; *Tetrahedron Lett.*, 1974, **15**, 207. 727

74TL423 C. J. Sih, R. Sood and M. Nagasawa; *Tetrahedron Lett.*, 1974, **15**, 423. 1001

74TL543 R. B. Miller and T. Reichenbach; *Tetrahedron Lett.*, 1974, **15**, 543. 653, 654

74TL925 C. P. Casey and D. F. Marten; *Tetrahedron Lett.*, 1974, **15**, 925. 482

74TL1067 K. Kaneda, F. Kawamoto, Y. Fujiwara, T. Imanaka and S. Teranishi; *Tetrahedron Lett.*, 1974, **15**, 1067. 1107

74TL1097 P. A. Grieco and J. J. Reap; *Tetrahedron Lett.*, 1974, **15**, 1097. 598, 601

74TL1105	C. Y. Meyers, W. S. Matthews, G. J. McCollum and J. C. Branca; *Tetrahedron Lett.*, 1974, **15**, 1105.	618
74TL1133	P. F. Hudrlik and D. Peterson; *Tetrahedron Lett.*, 1974, **15**, 1133.	650, 759
74TL1275	H. J. Bestmann, T. Denzel and H. Salbaum; *Tetrahedron Lett.*, 1974, **15**, 1275.	988
74TL1377	S. Brenner and M. Bovete; *Tetrahedron Lett.*, 1974, **15**, 1377.	155
74TL1451	J. J. Riehl, A. Smolikiewicz and L. Thil; *Tetrahedron Lett.*, 1974, **15**, 1451.	108, 115
74TL1593	R. Baudouy and J. Goré; *Tetrahedron Lett.*, 1974, **15**, 1593.	960
74TL1637	K.-P. Jäckel and M. Hanack; *Tetrahedron Lett.*, 1974, **15**, 1637.	1116
74TL1905	M. Baran Marszak, M. Simalty and A. Seuleiman; *Tetrahedron Lett.*, 1974, **15**, 1905.	988
74TL2077	B. Cazes and S. Julia; *Tetrahedron Lett.*, 1974, **15**, 2077.	816
74TL2161	L.-I. Olsson and A. Claesson; *Tetrahedron Lett.*, 1974, **15**, 2161.	539
74TL2203	J. M. Kornprobst and J. E. Dubois; *Tetrahedron Lett.* 1974, **15**, 2203.	
74TL2279	D. N. Brattesani and C. H. Heathcock; *Tetrahedron Lett.*, 1974, **15**, 2279.	608, 781
74TL2395	A. J. Birch and P. G. Lehman; *Tetrahedron Lett.*, 1974, **15**, 2395.	99
74TL2457	H. Stutz and M. Hanack; *Tetrahedron Lett.*, 1974, **15**, 2457.	1124
74TL2587	W. Tagaki, I. Inoue, Y. Yano and T. Okonogi; *Tetrahedron Lett.*, 1974, **15**, 2587.	731
74TL2615	R. M. Carlson and A. R. Oyler; *Tetrahedron Lett.*, 1974, **15**, 2615.	1023
74TL2805	J. S. Grossert, J. Buter, E. W. H. Asveld and R. M. Kellogg; *Tetrahedron Lett.*, 1974, **15**, 2805.	706
74TL3207	E. Vedejs and P. D. Weeks; *Tetrahedron Lett.*, 1974, **15**, 3207.	491
74TL3391	H. C. Bell, J. R. Kalman, J. T. Pinhey and S. Sternhell; *Tetrahedron Lett.*, 1974, **15**, 3391.	490
74TL3479	T. H. Chan and W. Mychajlowskij; *Tetrahedron Lett.*, 1974, **15**, 3479.	63, 451
74TL3511	T. H. Chan, M. P. Li, W. Mychajlowskij and D. N. Harpp; *Tetrahedron Lett.*, 1974, **15**, 3511.	
74TL3809	G. Linstrumelle; *Tetrahedron Lett.*, 1974, **15**, 3809.	111
74TL3955	R. M. Coates, H. D. Pigott and J. Ollinger; *Tetrahedron Lett.*, 1974, **15**, 3955.	236
74TL3987	E. Vilsmaier, R. Tropitzsch and O. Vostrowsky; *Tetrahedron Lett.*, 1974, **15**, 3987.	620
74TL4005	I. Ojima and M. Kumagai; *Tetrahedron Lett.*, 1974, **15**, 4005.	651
74TL4467	B. Ganem; *Tetrahedron Lett.*, 1974, **15**, 4467.	1061
75ACR245	B. Miller; *Acc. Chem. Res.*, 1975, **8**, 245.	808, 809
75ACR281	R. R. Rando; *Acc. Chem. Res.*, 1975, **8**, 281.	1135
75ACR369	W. J. Hehre; *Acc. Chem. Res.*, 1975, **8**, 369.	893
75AG(E)15	D. Seebach and D. Enders; *Angew. Chem., Int. Ed. Engl.*, 1975, **14**, 15.	148
75AG(E)32	H. Meier and K.-P. Zeller; *Angew. Chem., Int. Ed. Engl.*, 1975, **14**, 32.	393
75AG(E)58	B. Kübel, G. Höfle and W. Steglich; *Angew. Chem., Int. Ed. Engl.*, 1975, **14**, 58.	985
75AG(E)559	F. Vögtle and J. Grütze; *Angew. Chem. Int. Ed. Engl.*, 1975, **14**, 559.	232
75AG(E)709	J. H. Lukas, A. P. Kouwenhoven and F. Baardman; *Angew. Chem., Int. Ed. Engl.*, 1975, **14**, 709.	498
75AG(E)734	N. Kornblum; *Angew. Chem., Int. Ed. Engl.*, 1975, **14**, 734.	250
75AG(E)765	U. Schöllkopf and I. Hoppe; *Angew. Chem., Int. Ed. Engl.*, 1975, **14**, 765.	683
75APO(11)123	D. G. Farnum; *Adv. Phys. Org. Chem.*, 1975, **11**, 123.	894, 895
75APO(11)177	G. M. Kramer; *Adv. Phys. Org. Chem.*, 1975, **11**, 177.	906
75BCJ960	T. Nakajima, Y. Nakamoto and S. Suga; *Bull. Chem. Soc. Jpn*, 1975, **48**, 960.	212, 475
75BCJ1865	T. Kinoshita, K. Miyano and T. Miwa; *Bull. Chem. Soc. Jpn.*, 1975, **48**, 1865.	100
75BCJ1892	M. Ohashi, T. Takahashi, S. Inoue and K. Sato; *Bull. Chem. Soc. Jpn.*, 1975, **48**, 1892.	34
75BCJ2939	S. A. M. Metwally and K. Bredereck; *Bull. Chem. Soc. Jpn.*, 1975, **48**, 2939.	201
75BSF1407	G. Markarian, B. Ragonnet, M. Santelli and M. Bertrand; *Bull. Soc. Chim. Fr.*, 1975, 1407.	959
75BSF2159	R. Baudouy and J. Goré; *Bull. Soc. Chim. Fr.*, 1975, 2159.	962
75BSF2166	R. Baudouy and J. Goré; *Bull. Soc. Chim. Fr.*, 1975, 2166.	962
75C341	M. Schlosser, A. Piskala, C. Tarchini and H. B. Tuong; *Chimia (Switz.)*, 1975, **29**, 341.	721
75CB314	D. Seebach and A. K. Beck; *Chem. Ber.*, 1975, **108**, 314.	441
75CB803	R. Luckenbach; *Chem. Ber.*, 1975, **108**, 803.	53
75CC17	P. J. Stang, J. Davis and D. P. Fox; *J. Chem. Soc., Chem. Commun.*, 1975, 17.	1143
75CC174	F. Toda and Y. Takehira; *J. Chem. Soc., Chem. Commun.*, 1975, 174.	969
75CC201	A. G. Anastassiou and H. Kasmai; *J. Chem. Soc., Chem. Commun.*, 1975, 201.	865
75CC206	G. Jones II, M. A. Acquadro and M. A. Carmody; *J. Chem. Soc., Chem. Commun.*, 1975, 206.	785
75CC287	G. A. Crosby and R. A. Stephenson; *J. Chem. Soc., Chem. Commun.*, 1975, 287.	1024
75CC362	M. Morisaki, N. Awata, Y. Fujimoto and N. Ikekawa; *J. Chem. Soc., Chem. Commun.*, 1975, 362.	961
75CC480	A. J. Birch and E. A. Karakhanov; *J. Chem. Soc., Chem. Commun.*, 1975, 480.	99
75CC537	P. A. Grieco, C.-L. J. Wang and S. D. Burke; *J. Chem. Soc., Chem. Commun.*, 1975, 537.	209, 651
75CC539	T. G. Back, D. H. R. Barton, M. R. Britten-Kelly and F. S. Guziec Jr.; *J. Chem. Soc., Chem. Commun.*, 1975, 539.	685, 686
75CC561	G. Linstrumelle and D. Michelot; *J. Chem. Soc., Chem. Commun.*, 1975, 561.	111, 464, 971, 1052, 1056
75CC790	H. J. Reich and F. Chow; *J. Chem. Soc., Chem. Commun.*, 1975, 790.	592, 713, 715, 716
75CC817	F. Scheinmann and S. Bhanu; *J. Chem. Soc., Chem. Commun.*, 1975, 817.	1053
75CC857	A. Pelter, K. Smith and M. Tabata; *J. Chem. Soc., Chem. Commun.*, 1975, 857.	1035
75CC940	D. R. Marshall, P. J. Thomas and C. J. M. Stirling; *J. Chem. Soc., Chem. Commun.*, 1975, 940.	612
75CJC3365	S. Wolfe, H. B. Schlegel, I. G. Csizmadia and F. Bernadi; *Can. J. Chem.*, 1975, **53**, 3365.	879

75CL103 G. Goto, T. Shima, H. Masuya, Y. Masuoka and K. Hiraga; *Chem. Lett.*, 1975, 103. 734
75CL523 A. Suzuki, N. Miyaura, M. Itoh and M. Tokuda; *Chem. Lett.*, 1975, 523. 276
75CL871 Y. Watanabe, M. Shiono and T. Mukaiyama; *Chem. Lett.*, 1975, 871. 590
75CPB1917 Y. Sawa, T. Kato, T. Masuda, M. Hori and H. Fujimura; *Chem. Pharm. Bull.*, 1975, **23**, 1917. 181
75CPB2620 J. F. Bunnett and J. E. Sundberg; *Chem. Pharm. Bull.*, 1975, **23**, 2620. 474
75CR(C)791 M.-L. Leriverend and P. Leriverend; *C. R. Hebd. Seances Acad. Sci., Ser. C.*, 1975, **280**, 791. 387
75CR(C)999 M.-T. Bertrand, G. Courtois and L. Miginac; *C. R. Hebd. Seances Acad. Sci., Ser. C*, 1975, **280**, 999. 543
75CR(C)1231 S. Julia; *C. R. Hebd. Seances Acad. Sci., Ser. C.*, 1975, **280**, 1231. 820
75CRV521 E. C. Ashby and J. T. Laemmle; *Chem. Rev.*, 1975, **75**, 521. 1039
75CS9 K. B. Sharpless, K. M. Gordon, R. F. Lauer, D. W. Patrick, S. P. Singer and M. W. Young; *Chem. Scr.*, 1975, **8A**, 9. 601, 610, 611, 621, 646
75CS9 K. B. Sharpless, K. M. Gordon, R. F. Lauer, D. W. Patrick, S. P. Singer and M. W. Young; *Chem. Scr.*, 1975, **8A**, 9. 716
75G109 P. Bravo, C. Ticozzi and A. Cezza; *Gazz. Chim. Ital.*, 1975, **105**, 109. 738
75HCA153 A. Rüttimann, A. Wick and A. Eschenmoser; *Helv. Chim. Acta*, 1975, **58**, 153. 566
75HCA1016 F. Näf, R. Decorzant, W. Thommen, B. Willhalm and G. Ohloff; *Helv. Chim. Acta*, 1975, **58**, 1016. 403, 734
75HCA1293 P. Vittorelli, H.-J. Hansen and H. Schmid; *Helv. Chim. Acta*, 1975, **58**, 1293. 415
75HCA1450 A. Rüttimann, A. Wick and A. Eschenmoser; *Helv. Chim. Acta*, 1975, **58**, 1450. 683
75IJC251 R. B. Mitra, M. V. Natekar and S. D. Virkari; *Indian J. Chem.*; 1975, **13**, 251. 707
75IZV888 V. S. Mikhailov, V. P. Gul'tyai, S. G. Mairanovskii, S. Z. Taits, I. V. Proskurovskaya and Y. G. Dubovik; *Izv. Akad. Nauk SSSR, Ser. Khim.*, 1975, 888. 103
75JA237 F. J. Hirsekorn, M. C. Rakowski and E. L. Mutterties; *J. Am. Chem. Soc.*, 1975, **97**, 237. 93
75JA380 S. Danishefsky, P. Cain and A. Nagel; *J. Am. Chem. Soc.*, 1975, **97**, 380. 99
75JA679 D. W. Hart, T. F. Blackburn and J. Schwartz; *J. Am. Chem. Soc.*, 1975, **97**, 679. 1105
75JA821 Z. Rappoport and Y. Apeloig; *J. Am. Chem. Soc.*, 1975, **97**, 821. 1110, 1112
75JA836 Z. Rappoport and Y. Apeloig; *J. Am. Chem. Soc.*, 1975, **97**, 836. 1112
75JA891 C. A. Brown and A. Yamashita; *J. Am. Chem. Soc.*, 1975, **97**, 891. 1085
75JA928 R. Noyori, Y. Ohnishi and M. Katô; *J. Am. Chem. Soc.*, 1975, **97**, 928. 422
75JA1459 P. J. Stang and M. G. Mangum; *J. Am. Chem. Soc.*, 1975, **97**, 1459. 1143
75JA1464 P. F. Hudrlik and D. Peterson; *J. Am. Chem. Soc.*, 1975, **97**, 1464. 649, 650, 761
75JA1531 T. von Lehman and R. S. Macomber; *J. Am. Chem. Soc.*, 1975, **97**, 1531. 1114
75JA1539 G. A. Olah and R. J. Spear; *J. Am. Chem. Soc.*, 1975, **97**, 1539. 902
75JA1597 P. A. Grieco, Y. Masaki and D. Boxler; *J. Am. Chem. Soc.*, 1975, **97**, 1597. 601, 604
75JA1845 G. A. Olah and R. J. Spear; *J. Am. Chem. Soc.*, 1975, **97**, 1845. 1122
75JA1977 M.-F. Ruasse and J. E. Dubois; *J. Am. Chem. Soc.*, 1975, **97**, 1977. 318
75JA2553 Y. Hata, M. Watanabe, S. Inoue and S. Oae; *J. Am. Chem. Soc.*, 1975, **97**, 2553. 592
75JA2902 J. F. Wolf, P. G. Harch, R. W. Taft and W. J. Hehre; *J. Am. Chem. Soc.*, 1975, **97**, 2902. 895
75JA3250 H. J. Reich and S. K. Shah; *J. Am. Chem. Soc.*, 1975, **97**, 3250. 601, 602, 610, 713, 714, 715, 781
75JA3257 I. Kuwajima and E. Nakamura; *J. Am. Chem. Soc.*, 1975, **97**, 3257. 281
75JA3512 S. E. Masamune, C. U. Kim, K. E. Wilson, G. O. Spessard, P. E. Georghiou and G. S. Bates; *J. Am. Chem. Soc.*, 1975, **97**, 3512. 722
75JA3822 C. G. Chavdarian and C. H. Heathcock; *J. Am. Chem. Soc.*, 1975, **97**, 3822. 1089
75JA3837 G. R. Krow and J. Reilly; *J. Am. Chem. Soc.*, 1975, **97**, 3837. 799
75JA3854 P. J. Stang and M. G. Mangum; *J. Am. Chem. Soc.*, 1975, **97**, 3854. 550, 1143
75JA4269 D. Griller, J. W. Cooper and K. U. Ingold; *J. Am. Chem. Soc.*, 1975, **97**, 4269. 322
75JA4327 R. J. Anderson and C. A. Henrick; *J. Am. Chem. Soc.*, 1975, **97**, 4327. 727
75JA4405 K. Tamao, S. Kodama, T. Nakatsuka, Y. Kiso and M. Kumada; *J. Am. Chem. Soc.*, 1975, **97**, 4405. 471
75JA4409 A. Demoulin, H. Gorissen, A.-M. Hesbain-Frisque and L. Ghosez; *J. Am. Chem. Soc.*, 1975, **97**, 4409. 522
75JA4413 T. Sasaki, S. Eguchi and T. Ogawa; *J. Am. Chem. Soc.*, 1975, **97**, 4413. 518
75JA4765 D. A. Evans and A. M. Golob; *J. Am. Chem. Soc.*, 1975, **97**, 4765. 413, 810
75JA5249 H. C. Brown and S. K. Gupta; *J. Am. Chem. Soc.*, 1975, **97**, 5249. 1101
75JA5434 H. J. Reich, J. M. Renga and I. L. Reich; *J. Am. Chem. Soc.*, 1975, **97**, 5434. 601, 602, 604, 605, 606, 607, 608, 609, 610, 781
75JA5603 K. Tamao, N. Miyake, Y. Kiso and M. Kumada; *J. Am. Chem. Soc.*, 1975, **97**, 5603. 451
75JA5606 Y. Yamamoto, H. Yatagai and I. Moritani; *J. Am. Chem. Soc.*, 1975, **97**, 5606. 531
75JA6214 R. G. Salomon and N. E. Sanadi; *J. Am. Chem. Soc.*, 1975, **97**, 6214. 729
75JA6478 P. J. Stang and M. G. Mangum; *J. Am. Chem. Soc.*, 1975, **97**, 6478. 1143
75JA6873 R. A. Bartsch, J. R. Allaway, D. D. Ingram and J.-G. Lee; *J. Am. Chem. Soc.*, 1975, **97**, 6873. 28
75JA6878 E. Vedejs and J. P. Hagan; *J. Am. Chem. Soc.*, 1975, **97**, 6878. 410, 821
75JA7006 W. S. Matthews, J. E. Bares, J. E. Bartmess, F. G. Bordwell, F. J. Cornforth, G. E. Drucker, Z. Margolin, R. J. McCallum, G. J. McCollum and N. R. Vanier; *J. Am. Chem. Soc.*, 1975, **97**, 7006. 142, 155
75JA7162 D. Seyferth and D. C. Annarelli; *J. Am. Chem. Soc.*, 1975, **97**, 7162. 656
75JA7193 J. B. Trimitsis and A. Tuncay; *J. Am. Chem. Soc.*, 1975, **97**, 7193. 318
75JA7372 C. E. Sacks and P. L. Fuchs; *J. Am. Chem. Soc.*, 1975, **97**, 7372. 464

75JA7385 E. Negishi and S. Baba; *J. Am. Chem. Soc.*, 1975, **97**, 7385. 1003, 1021

75JCP(63)3238 W. Lindinger, D. L. Albritton, F. C. Fehsenfeld and E. E. Ferguson; *J. Chem. Phys.*, 1975, **63**, 3238. 1143

75JCS(P1)470 C. Eaborn, R. A. Jackson and R. Pearce; *J. Chem. Soc., Perkin Trans. 1*, 1975, 470. 99

75JCS(P1)1574 D. H. R. Barton and S. W. McCombie; *J. Chem. Soc., Perkin Trans. 1*, 1975, 1574. 11, 16, 935, 938

75JCS(P1)1628 P. D. Landor, S. R. Landor and J. P. Leighton; *J. Chem. Soc., Perkin Trans. 1*, 1975, 1628. 1005

75JCS(P2)317 D. B. Bigley and R. E. Gabbott; *J. Chem. Soc., Perkin Trans. 2*, 1975, 317. 777

75JCS(P2)1025 R. Taylor, *J. Chem. Soc., Perkin Trans. 2*, 1975, 1025. 774, 778

75JHC801 H. Golgolab and I. Lalezari; *J. Heterocycl. Chem.*, 1975, **12**, 801. 1077

75JIC538 O. P. Vig, A. K. Vig, J. S. Mann and K. C. Gupta; *J. Indian Chem. Soc.*, 1975, **52**, 538. 745

75JOC124 S. J. Huang, V. Paneccasio, F. DiBattista, D. Picker and G. Wilson; *J. Org. Chem.*, 1975, **40**, 124. 1137, 1141

75JOC131 M. W. Logue and G. L. Moore; *J. Org. Chem.*, 1975, **40**, 131. 1006, 1045

75JOC148 B. M. Trost and T. N. Salzmann; *J. Org. Chem.*, 1975, **40**, 148. 779

75JOC542 P. A. Grieco, C. S. Pogonowski and S. Burke; *J. Org. Chem.*, 1975, **40**, 542. 602, 607

75JOC579 E. J. Corey and H. S. Sachdev; *J. Org. Chem.*, 1975, **40**, 579. 1082

75JOC657 P. J. Stang and R. J. Hargrove; *J. Org. Chem.*, 1975, **40**, 657. 981

75JOC700 L. A. Paquette, M. Oku, W. B. Farnham, G. A. Olah and G. Liang; *J. Org. Chem.*, 1975, **40**, 700. 918

75JOC812 T. Cohen, G. Herman, J. R. Falck and A. J. Mura, Jr.; *J. Org. Chem.*, 1975, **40**, 812. 590

75JOC923 R. O. Hutchins, M. Kacher and L. Rua; *J. Org. Chem.*, 1975, **40**, 923. 635

75JOC947 K. B. Sharpless and M. W. Young; *J. Org. Chem.*, 1975, **40**, 947. 601, 604, 781

75JOC1083 H. A. Dieck and R. F. Heck; *J. Org. Chem.*, 1975, **40**, 1083. 493

75JOC1131 W. B. Sudweeks and H. S. Broadbent; *J. Org. Chem.*, 1975, **40**, 1131. 1037

75JOC1191 J. Z. Ginos; *J. Org. Chem.*, 1975, **40**, 1191. 90

75JOC1460 H. O. House, C.-Y. Chu, J. M. Wilkins and M. J. Umen; *J. Org. Chem.*, 1975, **40**, 1460. 1093

75JOC1588 S. S. Klioze and F. P. Darmory; *J. Org. Chem.*, 1975, **40**, 1588. 506

75JOC1834 G. W. Kabalka and J. D. Baker, Jr.; *J. Org. Chem.*, 1975, **40**, 1834. 19, 41

75JOC1932 M. Kato, M. Kageyama, R. Tanaka, K. Kuwahara and A. Yoshikoshi; *J. Org. Chem.*, 1975, **40**, 1932. 1136

75JOC1966 N. M. Weinshenker, G. A. Crosby and J. Y. Wong; *J. Org. Chem.*, 1975, **40**, 1966. 936

75JOC1994 J. Salaun and M. Hanack; *J. Org. Chem.*, 1975, **40**, 1994. 1115

75JOC2009 M. Aratani, L. V. Dunkerton, T. Fukuyama, Y. Kishi, H. Kakoi, S. Sugiura and S. Inoue; *J. Org. Chem.*, 1975, **40**, 2009. 263

75JOC2014 B. M. Trost and A. J. Bridges; *J. Org. Chem.*, 1975, **40**, 2014. 142, 599

75JOC2165 D. N. Brattesani and C. H. Heathcock; *J. Org. Chem.*, 1975, **40**, 2165. 607, 608

75JOC2250 M. M. Midland; *J. Org. Chem.*, 1975, **40**, 2250. 1037

75JOC2263 P. F. Hudrlik, D. Peterson and R. J. Rona; *J. Org. Chem.*, 1975, **40**, 2263. 648, 650

75JOC2264 P. F. Hudrlik, D. Peterson and R. J. Rona; *J. Org. Chem.*, 1975, **40**, 2264. 759, 761

75JOC2265 E. J. Corey and R. H. Wollenberg; *J. Org. Chem.*, 1975, **40**, 2265. 535, 1091, 1093

75JOC2554 E. J. Corey and J. W. Suggs; *J. Org. Chem.*, 1975, **40**, 2554. 5, 9, 10, 936

75JOC2570 H. J. Reich; *J. Org. Chem.*, 1975, **40**, 2570. 609

75JOC2694 H. Alper; *J. Org. Chem.*, 1975, **40**, 2694. 24

75JOC2845 M. M. Midland and H. C. Brown; *J. Org. Chem.*, 1975, **40**, 2845. 449

75JOC2963 P. Hudrlik and C.-N. Wan; *J. Org. Chem.*, 1975, **40**, 2963. 229

75JOC2970 C. G. Chavdarian and C. H. Heathcock; *J. Org. Chem.*, 1975, **40**, 2970. 1093

75JOC3073 R. K. Sehgal, R. U. Koenigsberger and T. J. Howard; *J. Org. Chem.*, 1975, **40**, 3073. 72

75JOC3151 G. H. Small, A. E. Minnella and S. S. Hall; *J. Org. Chem.*, 1975, **40**, 3151. 13

75JOC3313 H. J. Reich and J. M. Renga; *J. Org. Chem.*, 1975, **40**, 3313. 603

75JOC3599 S. Siegel, M. Foreman, R. P. Fisher and S. E. Johnson; *J. Org. Chem.*, 1975, **40**, 3599. 74, 82

75JOC3606 S. Danishefsky and P. Cain; *J. Org. Chem.*, 1975, **40**, 3606. 99

75JOC3614 N. A. Porter and M. O. Funk; *J. Org. Chem.*, 1975, **40**, 3614. 949

75JOC3619 M. F. Semmelhack and R. D. Stauffer; *J. Org. Chem.*, 1975, **40**, 3619. 515

75JOC3649 T. Cohen and I. Cristea; *J. Org. Chem.*, 1975, **40**, 3649. 468

75JOC3686 W. L. Meyer, G. B. Clemans and R. A. Manning; *J. Org. Chem.*, 1975, **40**, 3686. 46

75JOM(85)1 B. O. Wagner and G. S. Hammond; *J. Organomet. Chem.*, 1975, **85**, 1. 665

75JOM(85)173 P. S. Kendurkar and R. S. Tewari; *J. Organomet. Chem.*, 1975, **85**, 173. 738

75JOM(88)175 K. J. Irgolic, P. J. Busse and R. A. Grigsby; *J. Organomet. Chem.*, 1975, **88**, 175. 438

75JOM(93)253 H. A. Dieck and F. R. Heck; *J. Organomet. Chem.*, 1975, **93**, 253. 1006, 1008, 1009, 1137

75JOM(93)259 F. R. Heck and H. A. Dieck; *J. Organomet. Chem.*, 1975, **93**, 259. 1008, 1009, 1137

75JOM(93)43 G. Deleris, J. Dunogues and R. Calas; *J. Organomet. Chem.*, 1975, **93**, 43. 978

75JOM(96)C47 J.-G. Duboudin and B. Jousseaume; *J. Organomet. Chem.*, 1975, **96**, C47. 539

75JOM(96)225 J. Grignon, C. Servens and M. Pereyre; *J. Organomet. Chem.*, 1975, **96**, 225. 128

75JOM(96)237 N. Kumari, P. S. Kendurkar and R. S. Tewari; *J. Organomet. Chem.*, 1975, **96**, 237. 738

75JOM(96)471 A. Alexakis, J. F. Normant and J. Villieras; *J. Organomet. Chem.*, 1975, **96**, 471. 540

75JOM(99)C8 T. Spencer and F. G. Thorpe; *J. Organomet. Chem.*, 1975, **99**, C8. 493

75JOM(100)43 C. Eaborn; *J. Organomet. Chem.*, 1975, **100**, 43. 447

75JOM(102)123 G. L. Larson and A. Hernández; *J. Organomet. Chem.*, 1975, **102**, 123. 659

75LA1113 H. Neunhoeffer and B. Lehmann; *Justus Liebigs Ann. Chem.*, 1975, 1113. 522

75LA2130 R. Köster, A. Bussman and G. Schroth; *Justus Liebigs Ann. Chem.*, 1975, 2130. 1001

B-75MI 105-01 B. M. Trost and L. S. Melvin, Jr.; "Sulfur Ylides," Academic Press, New York, 1975, ch. 7. 233

B-75MI 109-01 B. M. Trost and L. S. Melvin, Jr.; "Sulfur Ylides," Academic Press, New York, 1975. 398

B-75MI 110-01	H. C. Brown, G. W. Kramer, A. B. Levy and M. M. Midland; "Organic Syntheses via Boranes," Wiley, New York, 1975.	449
B-75MI 119-01	E. Buncel; "Carbanions: Mechanistic and Isotopic Aspects," Elsevier, Amsterdam, 1975.	886
B-75MI 122-01	R. A. Moss and M. Jones (eds.); "Carbenes," Wiley-Interscience, New York, 1975, Vol. II.	1143
75MI 122-01	S. Ikuta, K. Okuno, K. Yoshihara and T. Shiokawa; *Radiochem. Radioanal. Lett.*, 1975, **23**, 213.	1142
75OPP189	M. Tashiro, H. Watanabe and K. Oe; *Org. Prep. Proced. Int.*, 1975, **7**, 189.	57
75OR(22)1	S. J. Rhoads and N. R. Raulins; *Org. React.*, 1975, **22**, 1.	411, 414, 806, 827, 835
75OR(22)253	G. H. Posner; *Org. React.*, 1975, **22**, 253.	116, 118, 206, 540, 1005, 1045
75OR(22)401	E. Vedejs; *Org. React.*, 1975, **22**, 401.	18
75PAC553	G. Stork; *Pure Appl. Chem.*, 1975, **43**, 553.	130, 872
75S1	J. M. Conia and P. Le Perchec; *Synthesis*, 1975, 1.	526
75S194	S. N. Moorthy, R. Vaidyanathaswamy and D. Devaprabhakara; *Synthesis*, 1975, 194.	459
75S256	M. Larcheveque, G. Valette, T. Cuvigny and H. Normant; *Synthesis*, 1975, 256.	212
75S376	G. Zweifel and R. P. Fischer; *Synthesis*, 1975, 376.	487
75S395	E. M. Kaiser and C. D. Warner; *Synthesis*, 1975, 395.	885
75S423	W. Beckmann, G. Doerjer, E. Logemann, C. Merkel and S. C. Zürcher; *Synthesis*, 1975, 423.	1000
75S434	J. Millon, R. Lorne and G. Linstrumelle; *Synthesis*, 1975, 434.	464, 1090
75S448	Y. Sasson and G. L. Rempel; *Synthesis*, 1975, 448.	79, 80
75S458	J. Villieras, P. Perriot and J. F. Normant; *Synthesis*, 1975, 458.	1067
75S632	S. Warwel, G. Schmitt and B. Ahlfaenger; *Synthesis*, 1975, 632.	534
75S784	R. M. Boden; *Synthesis*, 1975, 784.	727, 733
75S807	M. Brink; *Synthesis*, 1975, 807.	232
75SC15	R. A. Ruden and B. L. Gaffney; *Synth. Commun.*, 1975, **5**, 15.	650
75SC359	P. F. Hudrlik, D. Peterson and D. Chou; *Synth. Commun.*, 1975, **5**, 359.	651
75T1035	J. Casanova and B. Waegell; *Tetrahedron*, 1975, **31**, 1035.	633
75T1846	K. Mori, M. Tominaga and M. Matsui; *Tetrahedron*, 1975, **31**, 1846.	503, 1052
75T2735	H. Felkin and G. Swierczewski; *Tetrahedron*, 1975, **31**, 2735.	1094
75T2855	L. A. Paquette; *Tetrahedron*, 1975, **31**, 2855.	517
75T2979	J. d'Angelo; *Tetrahedron*, 1975, **32**, 2979.	869
75T3073	P. M. Greaves, S. R. Landor and M. M. Lwanga; *Tetrahedron*, 1975, **31**, 3073.	1056
75TL51	J. R. Norton, K. E. Shenton and J. Schwartz; *Tetrahedron Lett.*, 1975, **16**, 51.	204, 1023, 1139
75TL219	R. H. Mitchell, T. Otsubo and V. Boekelheide; *Tetrahedron Lett.*, 1975, **16**, 219.	233
75TL363	J. C. Blazejewski, D. Cantacuzène and C. Wakselman; *Tetrahedron Lett.*, 1975, **16**, 363.	522
75TL585	J. Y. Lallemand and M. Onanga; *Tetrahedron Lett.*, 1975, **16**, 585.	214
75TL947	R. G. Carlson and K. D. May; *Tetrahedron Lett.*, 1975, **16**, 947.	708
75TL981	K. A. Parker and E. G. Gibbons; *Tetrahedron Lett.*, 1975, **16**, 981.	1043
75TL1007	K. Kondo and D. Tunemoto; *Tetrahedron Lett.*, 1975, **16**, 1007.	144
75TL1335	G. Kretschmer and R. N. Warrener; *Tetrahedron Lett.*, 1975, **16**, 1335.	784
75TL1359	E. Vedejs and J. P. Bershas; *Tetrahedron Lett.*, 1975, **16**, 1359.	470, 736
75TL1465	J. F. Normant, A. Commercon and J. Villieras; *Tetrahedron Lett.*, 1975, **16**, 1465.	1018
75TL1513	A. V. Fletcher and K. Mackenzie; *Tetrahedron Lett.*, 1975, **16**, 1513.	1075
75TL1545	S. Hara, H. Taguchi, H. Yamamoto and H. Nozaki; *Tetrahedron Lett.*, 1975, **16**, 1545.	565
75TL1633	A. Pelter, C. Subrahmanyam, R. J. Laub, K. J. Gould and C. R. Harrison; *Tetrahedron Lett.*, 1975, **16**, 1633.	126
75TL1811	R. H. Shapiro, M. F. Lipton, K. J. Kolonko, R. L. Buswell and L. A. Capuano; *Tetrahedron Lett.*, 1975, **16**, 1811.	630, 632
75TL1815	J. E. Stemke and F. T. Bond; *Tetrahedron Lett.*, 1975, **16**, 1815.	633
75TL1893	E. Cuthbertson and D. D. MacNicol; *Tetrahedron Lett.*, 1975, **16**, 1893.	243
75TL1961	K. Yamada, N. Miyaura, M. Itoh and A. Suzuki; *Tetrahedron Lett.*, 1975, **16**, 1961.	1034
75TL2771	F. Marcuzzi and G. Melloni; *Tetrahedron Lett.*, 1975, **16**, 2771.	318
75TL2825	K. Utimoto, M. Kitai and H. Nozaki; *Tetrahedron Lett.*, 1975, **16**, 2825.	451
75TL2841	J. Nokami, N. Kunieda and M. Kinoshita; *Tetrahedron Lett.*, 1975, **16**, 2841.	779
75TL2923	H. Westmijze, J. Meijer and P. Vermeer; *Tetrahedron Lett.*, 1975, **16**, 2923.	989
75TL3151	J. Ficini, J. Besseyre and C. Barbara; *Tetrahedron Lett.*, 1975, **16**, 3151.	518
75TL3383	T. H. Chan and D. Massuda; *Tetrahedron Lett.*, 1975, **16**, 3383.	318
75TL3555	K. J. Crowley and S. G. Traynor; *Tetrahedron Lett.*, 1975, **16**, 3555.	420
75TL3617	A. J. Pearson; *Tetrahedron Lett.*, 1975, **16**, 3617.	292
75TL3741	J. M. G. Bonfrèr, H. J. C. Jacobs and E. Havinga; *Tetrahedron Lett.*, 1975, **16**, 3741.	420
75TL3975	R. E. Ireland and A. K. Willard; *Tetrahedron Lett.*, 1975, **16**, 3975.	416, 417, 829, 830, 876
75TL4005	G. L. Larson, E. Hernández, C. Alonso and I. Nieves; *Tetrahedron Lett.*, 1975, **16**, 4005.	434
75TL4025	H. J. Bestmann and G. Schmid; *Tetrahedron Lett.*, 1975, **16**, 4025.	988
75TL4027	T. Nakai, H. Shiono and M. Okawara; *Tetrahedron Lett.*, 1975, **16**, 4027.	210
75TL4197	B. M. Trost and K. K. Leung; *Tetrahedron Lett.*, 1975, **16**, 4197.	598
75TL4217	P. W. Collins, E. Z. Dajani, M. S. Bruhn, C. H. Brown, J. R. Palmer and R. Pappo; *Tetrahedron Lett.*, 1975, **16**, 4217.	1049
75TL4405	R. W. Saalfrank; *Tetrahedron Lett.*, 1975, **16**, 4405.	988
75TL4467	K. Sonogashira, Y. Tohda and N. Hagihara; *Tetrahedron Lett.*, 1975, **16**, 4467.	472, 1008, 1009, 1012, 1137
75TL4555	M.-F. Ruasse and J.-E. Dubois; *Tetrahedron Lett.*, 1975, **16**, 4555.	318
75TL4615	J.-L. Luche, E. Barreiro, J. M. Dollat and P. Crabbé; *Tetrahedron Lett.*, 1975, **16**, 4615.	966
75ZAAC(418)158	K. Issleib and A. Balzuweit; *Z. Anorg. Allg. Chem.*, 1975, **418**, 158.	446

76ACR13 D. Griller and K. U. Ingold; *Acc. Chem. Res.*, 1976, **9**, 13. 919, 922

76ACR41 G. A. Olah; *Acc. Chem. Res.*, 1976, **9**, 41. 906

76ACR59 H. O. House; *Acc. Chem. Res.*, 1976, **9**, 59. 313

76ACR257 T. S. Sorensen; *Acc. Chem. Res.*, 1976, **9**, 257. 381

76ACR313 R. H. Abeles and A. L. Maycock; *Acc. Chem. Res.*, 1976, **9**, 313. 1135

76ACS(B)341 S. Gronowitz and T. Frejd; *Acta Chem. Scand., Ser. B*, 1976, **30**, 341. 1072

76ACS(B)485 S. Gronowitz and T. Frejd; *Acta Chem. Scand. Ser. B*, 1976, **30**, 485. 318

76AG(E)9 W. S. Johnson; *Angew. Chem., Int. Ed. Engl.*, 1976, **15**, 9. 295

76AG(E)161 W. Dumont and A. Krief; *Angew. Chem., Int. Ed. Engl.*, 1976, **15**, 161. 759

76AG(E)251 W. Kirmse; *Angew. Chem., Int. Ed. Engl.*, 1976, **15**, 251. 258

76AG(E)417 G. N. Schrauzer; *Angew. Chem., Int. Ed. Engl.*, 1976, **16**, 417. 335, 345

76AG(E)498 I. Kuwajima and E. Nakamura; *Angew. Chem., Int. Ed. Engl.*, 1976, **15**, 498. 1038

76AG(E)505 D. Seebach and F. Lehr; *Angew. Chem., Int. Ed. Engl.*, 1976, **15**, 505. 157

76AJC1635 L. Radom; *Aust. J. Chem.*, 1976, **29**, 1635. 850

76BCJ779 K. Narasaka, K. Soai, Y. Aikawa and T. Mukaiyama; *Bull. Chem. Soc. Jpn.*, 1976, **49**, 779. 308

76BCJ1437 S. Uemura, A. Onoe, H. Okazaki, M. Okano and K. Ichikawa; *Bull. Chem. Soc. Jpn.*, 1976, **49**, 1437. 84

76BCJ1958 K. Tamao, K. Sumitani, Y. Kiso, M. Zembayashi, A. Fujioka, S.-I. Kodama, I. Nakajima, A. Minato and M. Kumada; *Bull. Chem. Soc. Jpn.*, 1976, **49**, 1958. 466, 470

76BCJ2017 S. Iwase, M. Nakata, S. Hamanaka and M. Ogawa; *Bull. Chem. Soc. Jpn.*, 1976, **49**, 2017. 815

76BCJ2560 T. Sonoda, S. Kobayashi and H. Taniguchi; *Bull. Chem. Soc., Japan*, 1976, **49**, 2560. 1112

76BSF513 M. Julia and D. Uguen; *Bull. Soc. Chim. Fr.*, 1976, 513. 24

76BSF525 M. Julia and B. Badet; *Bull. Soc. Chim. Fr.*, 1976, 525. 24

76CB473 K-H. A. O. Starzewski, W. Richter and H. Schmidbaur; *Chem. Ber.*, 1976, **109**, 473. 737

76CB793 K. Weinges and K. Klessing; *Chem. Ber.*, 1976, **109**, 793. 708

76CB1694 H. J. Bestmann, W. Stransky and O. Vostrowsky; *Chem. Ber.*, 1976, **109**, 1694. 727

76CB2628 U. Schmidt, J. Gombos, E. Haslinger and H. Zak; *Chem. Ber.*, 1976, **109**, 2628. 91

76CB3375 H. J. Bestmann, O. Vostrowsky and W. Stransky; *Chem. Ber.*, 1976, **109**, 3375. 229

76CB3598 H. Paulsen, H. Salzburg and H. Redlich; *Chem. Ber.*, 1976, **109**, 3598. 4, 7

76CC17 E. Negishi, S. Baba and A. O. King; *J. Chem. Soc., Chem. Commun.*, 1976, 17. 479, 534, 1106

76CC89 M. F. Solomon and R. G. Solomon; *J. Chem. Soc., Chem. Commun.*, 1976, 89. 803

76CC180 T. Suzuki, T. Sonoda, S. Kobayashi and H. Taniguchi; *J. Chem. Soc., Chem. Commun.*, 1976, 180. 1120

76CC183 P. Crabbé, E. Barreiro, J.-M. Dollat and J.-L. Luche; *J. Chem. Soc., Chem. Commun.*, 1976, 183. 960

76CC582 P. A. Grieco and M. Nishizawa; *J. Chem. Soc., Chem. Commun.*, 1976, 582. 607

76CC596 E. Negishi and S. Baba; *J. Chem. Soc., Chem. Commun.*, 1976, 596. 471, 533

76CC606 M. Tokuda, T. Taguchi, O. Nishio and M. Itoh; *J. Chem. Soc., Chem. Commun.*, 1976, 606. 1133

76CC612 T. Sonoda, M. Kawakami, T. Ikeda, S. Kobayashi and H. Taniguchi; *J. Chem. Soc., Chem. Commun.*, 1976, 612. 1112

76CC684 L. Crombie, G. Darnbrough and G. Pattenden; *J. Chem. Soc., Chem. Commun.*, 1976, 684. 816

76CC697 C. M. Robbins and G. H. Whitham; *J. Chem. Soc., Chem. Commun.*, 1976, 697. 649

76CC734 J. E. Baldwin; *J. Chem. Soc., Chem. Commun.*, 1976, 734. 213, 304, 336

76CC841 H. Yamazaki; *J. Chem. Soc., Chem. Commun.*, 1976, 841. 992

76CC846 R. W. Thies and E. P. Seitz; *J. Chem. Soc., Chem. Commun.*, 1976, 846. 401

76CC907 T. F. Murray, V. Varma and J. R. Norton; *J. Chem. Soc., Chem. Commun.*, 1976, 907. 1023, 1024

76CC959 C. A. Brown and A. Yamashita; *J. Chem. Soc., Chem. Commun.*, 1976, 959. 1085

76CJC1643 K. Tanaka, G. I. Mackay, J. D. Payzant and D. K. Bohme; *Can. J. Chem.*, 1976, **54**, 1643. 107

76CJC2839 S. Fliszár; *Can. J. Chem.*, 1976, **54**, 2839. 895

76CJC3045 G. H. Schmid, A. Modro, D. G. Garratt and K. Yates; *Can. J. Chem.*, 1976, **54**, 3045. 1123

76CJC3809 R. R. Fraser and P. J. Champagne; *Can. J. Chem.*, 1976, **54**, 3809. 879

76CL163 K. Saigo, M. Osaki and T. Mukaiyama; *Chem. Lett.*, 1976, 163. 308

76CL1019 R. Nakamura, S. Matsumoto and E. Echigoya; *Chem. Lett.*, 1976, 1019. 674

76CL1127 A. Ishida and T. Mukaiyama; *Chem. Lett.*, 1976, 1127. 688, 690

76CL1237 K. Tamao, M. Zembayashi and M. Kumada; *Chem. Lett.*, 1976, 1237, 1239. 467

76CRV187 C. W. Spangler; *Chem. Rev.*, 1976, **76**, 187. 400, 402, 403, 794, 795, 799, 800, 838

76CRV219 P. J. Davidson, M. F. Lappert and R. Pearce; *Chem. Rev.*, 1976, **76**, 219. 663, 668

76CRV243 R. R. Schrock and G. W. Parshall; *Chem. Rev.*, 1976, **76**, 243. 663, 668

76CS76 G. H. Schmid and D. G. Garratt; *Chem. Scr.*, 1976, **10**, 76. 1123

76HCA453 J. Hartmann and M. Schlosser; *Helv. Chim. Acta*, 1976, **59**, 453. 113

76HCA1832 H. G. W. Leuenberger, W. Boguth, E. Widmer and R. Zell; *Helv. Chim. Acta*, 1976, **59**, 1832. 77

76HCA2100 H. Loibner and E. Zbiral; *Helv. Chim. Acta*; 1976, **59**, 2100. 179

76HCA2213 R. Henning, F. Lehr and D. Seebach; *Helv. Chim. Acta*, 1976, **59**, 2213. 156

76IC2379 M. C. Rakowski, F. J. Hirsekorn, L. S. Stuhl and E. L. Muetterties; *Inorg. Chem.*, 1976, **15**, 2379. 93

76IJC(B)419 R. S. Tewari and K. C. Gupta; *Indian J. Chem. Sect. B*, 1976, **14**, 419. 733

76JA230 P. J. Krusic and R. C. Bingham; *J. Am. Chem. Soc.*, 1976, **98**, 230. 923

76JA840 P. J. Kropp, H. G. Fravel and T. R. Fields; *J. Am. Chem. Soc.*, 1976, **98**, 840. 794

76JA862 V. J. Shiner, Jr. and R. C. Sei; *J. Am. Chem. Soc.*, 1976, **98**, 862. 1124

76JA875 J. Weber and A. D. McLean; *J. Am. Chem. Soc.*, 1976, **98**, 875. 1109

76JA951 T. E. Boothe, J. L. Greene, Jr. and P. B. Shevlin; *J. Am. Chem. Soc.*, 1976, **98**, 951. 321, 591

76JA1265 P. B. Dervan and M. A. Shippey; *J. Am. Chem. Soc.*, 1976, **98**, 1265. 650

76JA1583	G. Stork and S. Raucher; *J. Am. Chem. Soc.*, 1976, **98**, 1583.	828
76JA1612	P. A. Grieco, M. Nishizawa, S. D. Burke and N. Marinovic; *J. Am. Chem. Soc.*, 1976, **98**, 1612.	604
76JA1785	H. C. Brown and N. Ravindran; *J. Am. Chem. Soc.*, 1976, **98**, 1785.	1094, 1102
76JA1798	H. C. Brown and N. Ravindran; *J. Am. Chem. Soc.*, 1976, **98**, 1798.	1102
76JA1806	D. E. James, L. F. Hines and J. K. Stille; *J. Am. Chem. Soc.*, 1976, **98**, 1806.	298
76JA1810	D. E. James and J. K. Stille; *J. Am. Chem. Soc.*, 1976, **98**, 1810.	298
76JA1997	J. B. Trimitsis and A. Tuncay; *J. Am. Chem. Soc.*, 1976, **98**, 1997.	318
76JA1999	R. Destro, T. Pilati and M. Simonetta; *J. Am. Chem. Soc.*, 1976, **98**, 1999.	910
76JA2275	E. J. Taylor and C. Djerassi; *J. Am. Chem. Soc.*, 1976, **98**, 2275.	635
76JA2580	H. E. Zieger, I. Angres and D. Mathisen; *J. Am. Chem. Soc.*, 1976, **98**, 2580.	108
76JA2667	K. P. C. Vollhardt and A. J. Barkovich; *J. Am. Chem. Soc.*, 1976, **98**, 2667.	1064
76JA2868	R. E. Ireland, R. H. Mueller and A. K. Willard; *J. Am. Chem. Soc.*, 1976, **100**, 2868.	131, 416, 829, 874, 876
76JA3022	S. Takano, S. Hatakeyama and K. Ogasawara; *J. Am. Chem. Soc.*, 1976, **98**, 3022.	596, 597
76JA3295	F. Marcuzzi and G. Melloni; *J. Am. Chem. Soc.*, 1976, **98**, 3295.	1122
76JA3312	J. F. King and D. R. K. Harding; *J. Am. Chem. Soc.*, 1976, **98**, 3312.	834
76JA3732	R. B. Banks and H. M. Walborsky; *J. Am. Chem. Soc.*, 1976, **98**, 3732.	491
76JA4319	S. A. McNeely and P. J. Kropp; *J. Am. Chem. Soc.*, 1976, **98**, 4319.	1120
76JA4805	E. P. Kyba; *J. Am. Chem. Soc.*, 1976, **98**, 4805.	157
76JA4887	B. M. Trost, T. N. Salzmann and K. Hiroi; *J. Am. Chem. Soc.*, 1976, **98**, 4887.	595, 597, 598, 601, 621
76JA4936	L. A. Paquette and J. M. Photis; *J. Am. Chem. Soc.*, 1976, **98**, 4936.	708
76JA5017	B. M. Trost and A. J. Bridges; *J. Am. Chem. Soc.*, 1976, **98**, 5017.	598
76JA5937	J. A. Berson, P. B. Dervan, R. Malherbe and J. A. Jenkins; *J. Am. Chem. Soc.*, 1976, **98**, 5937.	811
76JA5973	R. P. Quirk and R. E. Lea; *J. Am. Chem. Soc.*, 1976, **98**, 5973.	67
76JA6284	G. A. Olah, R. J. Spear and D. A. Forsyth; *J. Am. Chem. Soc.*, 1976, **98**, 6284.	915
76JA6290	G. A. Olah and J. S. Staral; *J. Am. Chem. Soc.*, 1976, **98**, 6290.	909
76JA6521	J. X. McDermott, J. F. White and G. M. Whitesides; *J. Am. Chem. Soc.*, 1976, **98**, 6521.	668, 669
76JA6700	D. H. Aue, W. R. Davidson and M. T. Bowers; *J. Am. Chem. Soc.*, 1976, **98**, 6700.	1109
76JA6705	G. D. Andrews and J. E. Baldwin; *J. Am. Chem. Soc.*, 1976, **98**, 6705.	801
76JA6729	S. Baba and E. Negishi; *J. Am. Chem. Soc.*, 1976, **98**, 6729.	468, 533, 1106
76JA6747	G. Stork and G. Kraus; *J. Am. Chem. Soc.*, 1976, **98**, 6747.	525
76JA7115	B. B. Snider, N. J. Hrib and L. Fuzesi; *J. Am. Chem. Soc.*, 1976, **98**, 7115.	820
76JA7319	D. L. Whalen, J. F. Weimaster, A. M. Ross and R. Radhe; *J. Am. Chem. Soc.*, 1976, **98**, 7319.	797
76JA7658	T. W. Bentley and P. v. R. Schleyer; *J. Am. Chem. Soc.*, 1976, **98**, 7658.	1114, 1116
76JA7667	F. L. Schadt, T. W. Bentley and P. v. R. Schleyer; *J. Am. Chem. Soc.*, 1976, **98**, 7667.	1114, 1116
76JA7835	J. A. Bertrand, E. Grovenstein, Jr., P. Lu and D. VanDerveer; *J. Am. Chem. Soc.*, 1976, **98**, 7835.	886
76JCP(64)4159	J. Weber, M. Yoshimine and A. D. McLean; *J. Chem. Phys.*, 1976, **64**, 4159.	1109
76JCP(64)4544	I. Dotan, W. Lindinger and D. L. Albritton; *J. Chem. Phys.*, 1976, **64**, 4544.	1143
76JCS(P1)45	R. M. Acheson and G. Paglietti; *J. Chem. Soc., Perkin Trans. 1*, 1976, 45.	102
76JCS(P1)234	W. Burns, D. Grant, M. A. McKervey and G. Step; *J. Chem. Soc., Perkin Trans. 1*, 1976, 234.	664
76JCS(P1)359	P. Cachia, N. Darby, C. R. Eck and T. Money; *J. Chem. Soc., Perkin Trans. 1*, 1976, 359.	381
76JCS(P1)410	G. A. MacAlpine, R. A. Raphael, A. Shaw, A. W. Taylor and H.-J. Wild; *J. Chem. Soc., Perkin Trans. 1*, 1976, 410.	422
76JCS(P1)459	D. N. Jones, A. C. F. Edmonds and S. D. Knox; *J. Chem. Soc., Perkin Trans. 1*, 1976, 459.	595, 779
76JCS(P1)884	S. Mageswaran and M. U. S. Sultanbawa; *J. Chem. Soc., Perkin Trans. 1*, 1976, 884.	681, 682
76JCS(P1)1384	J. M. Midgley, J. S. Millership and W. B. Whalley; *J. Chem. Soc., Perkin Trans. 1*, 1976, 1384.	197
76JCS(P1)1476	B. G. James and G. Pattenden; *J. Chem. Soc., Perkin Trans 1*, 1976, 1476.	729
76JCS(P1)2079	T. G. Back, D. H. R. Barton, M. R. Britten-Kelly and F. S. Guziec Jr.; *J. Chem. Soc., Perkin Trans. 1*, 1976, 2079.	686
76JCS(P1)2386	B. Lythgoe, T. A. Moran, M. E. N. Nambudiry and S. Ruston; *J. Chem. Soc., Perkin Trans. 1*, 1976, 2386.	754
76JCS(P2)280	S. de Burgh Norfolk and R. Taylor; *J. Chem. Soc., Perkin Trans. 2*, 1976, 280.	774
76JCS(P2)925	C. Eaborn and G. Seconi; *J. Chem. Soc., Perkin Trans. 2*, 1976, 925.	449
76JOC265	J. B. Melpolder and R. F. Heck; *J. Org. Chem.*, 1976, **41**, 265.	318
76JOC439	W. G. Kofron and M.-K. Yeh; *J. Org. Chem.*, 1976, **41**, 439.	154
76JOC502	R. S. Smith and J. K. Kochi; *J. Org. Chem.*, 1976, **41**, 502.	467
76JOC523	C. J. Upton and J. H. Incremona; *J. Org. Chem.*, 1976, **41**, 523.	928
76JOC574	G. W. Kabalka, D. T. C. Yang and J. D. Baker, Jr.; *J. Org. Chem.*, 1976, **41**, 574.	41, 635
76JOC636	R. D. Clark and C. H. Heathcock; *J. Org. Chem.*, 1976, **41**, 636.	469
76JOC681	M. D. Schiavelli, T. C. Germroth and J. W. Stubbs; *J. Org. Chem.*, 1976, **41**, 681.	1114
76JOC896	J. E. McMurry and M. P. Fleming; *J. Org. Chem.*, 1976, **41**, 896.	688
76JOC1052	E. L. Eliel, R. O. Hutchins, R. Mebane and R. L. Willer; *J. Org. Chem.*, 1976, **41**, 1052.	241
76JOC1078	J. A. Sinclair and H. C. Brown; *J. Org. Chem.*, 1976, **41**, 1078.	1035, 1103
76JOC1184	W. E. Parham, L. D. Jones and Y. A. Sayad; *J. Org. Chem.*, 1976, **41**, 1184.	452
76JOC1396	R. D. Clark and C. H. Heathcock; *J. Org. Chem.*, 1976, **41**, 1396.	601, 1093
76JOC1480	R. F. Cunico and J. F. Clayton; *J. Org. Chem.*, 1976, **41**, 1480.	1092

76JOC1485	P. A. Grieco, S. Gilman and M. Nishizawa; *J. Org. Chem.*, 1976, **41**, 1485.	781
76JOC1487	K. Okuhara; *J. Org. Chem.*, 1976, **41**, 1487.	1069
76JOC1669	S. Danishefsky, T. Kitahara, M. Tsai and J. Dynak; *J. Org. Chem.*, 1976, **41**, 1669.	1139
76JOC1785	H. C. Brown and N. Ravindran; *J. Am. Chem. Soc.*, 1976, **98**, 1785.	488
76JOC2048	R. P. Lutz, H. A. J. Berg and P. J. Wang; *J. Org. Chem.*, 1976, **41**, 2048.	809
76JOC2214	J. J. Eisch and G. A. Damasevitz; *J. Org. Chem.*, 1976, **41**, 2214.	128, 161
76JOC2215	K. Uchida, K. Utimoto and H. Nozaki; *J. Org. Chem.*, 1976, **41**, 2215.	161, 450, 1106
76JOC2241	R. C. Larock; *J. Org. Chem.*, 1976, **41**, 2241.	491, 1095
76JOC2331	G. H. Schmid, A. Modro, F. Lenz, D. G. Garratt and K. Yates; *J. Org. Chem.*, 1976, **41**, 2331.	1123
76JOC2379	I. Agranat and E. Aharon-Shalom; *J. Org. Chem.*, 1976, **41**, 2379.	909
76JOC2503	H. J. Reich and J. E. Trend; *J. Org. Chem.*, 1976, **41**, 2503.	601, 781
76JOC2615	J. J. Eisch and J. E. Galle; *J. Org. Chem.*, 1976, **41**, 2615.	650
76JOC2661	R. A. Kretchmer and R. Glowinski; *J. Org. Chem.*, 1976, **41**, 2661.	492
76JOC2940	K. Utimoto, M. Obayashi and H. Nozaki; *J. Org. Chem.*, 1976, **41**, 2940.	650, 761
76JOC2941	K. Uchida, K. Utimoto and H. Nozaki; *J. Org. Chem.*, 1976, **41**, 2941.	161, 1094
76JOC3061	B. B. Snider; *J. Org. Chem.*, 1976, **41**, 3061.	519, 524
76JOC3144	P. M. Wege, R. D. Clark and C. H. Heathcock; *J. Org. Chem.*, 1976, **41**, 3144.	1093
76JOC3191	R. S. Macomber and E. R. Kennedy; *J. Org. Chem.*, 1976, **41**, 3191.	993
76JOC3213	J. P. Marino and J. S. Farina; *J. Org. Chem.*, 1976, **41**, 3213.	1136
76JOC3218	T. Cohen, A. J. Mura, Jr., D. W. Shull, E. R. Fogel, R. J. Ruffner and J. R. Falck; *J. Org. Chem.*, 1976, **41**, 3218.	590
76JOC3326	N. R. Rosenqûist and O. L. Chapman; *J. Org. Chem.*, 1976, **41**, 3326.	981
76JOC3484	E. Negishi and K.-W. Chiu; *J. Org. Chem.*, 1976, **41**, 3484.	450, 510
76JOC3490	P. A. Wender and M. P. Filosa; *J. Org. Chem.*, 1976, **41**, 3490.	811
76JOC3496	D. J. Pasto, G. F. Hennion, R. H. Shults, A. Waterhouse and S.-K. Chou; *J. Org. Chem.*, 1976, **41**, 3496.	970
76JOC3497	K.-K. Chan, N. Cohen, J. P. De Noble, A. C. Specian, Jr. and G. Saucy; *J. Org. Chem.*, 1976, **41**, 3497.	514
76JOC3648	B. Corbel and T. Durst; *J. Org. Chem.*, 1976, **41**, 3648.	214
76JOC3650	P. J. Kocienski, J. M. Ansell and B. E. Norcross; *J. Org. Chem.*, 1976, **41**, 3650.	1076
76JOC3682	I. Granoth, Y. Segall, H. Leader and R. Alkabets; *J. Org. Chem.*, 1976, **41**, 3682.	80
76JOC3947	D. A. Evans, C. C. Thomas, R. C. Thomas and J. A. Walker; *J. Org. Chem.*, 1976, **4**, 3947.	278, 487, 662, 1103
76JOC4044	A. G. Schultz and Y. K. Lee; *J. Org. Chem.*, 1976, **41**, 4044.	303
76JOC4089	J. K. Crandall and R. Collonges; *J. Org. Chem.*, 1976, **41**, 4089.	516
76JOM(107)139	B. Bøe; *J. Organomet. Chem.*, 1976, **107**, 139.	70
76JOM(108)159	J.-L. Moreau and M. Gaudemar; *J. Organomet. Chem.*, 1976, **108**, 159.	967
76JOM(108)175	P. S. Kendurkar and R. S. Tewari; *J. Organomet. Chem.*, 1976, **108**, 175.	268, 738, 739
76JOM(110)139	D. Mansuy, J. Pusset and J. C. Chottard; *J. Organomet. Chem.*, 1976, **110**, 139.	63
76JOM(110)315	D. W. Allen and G. Jackson; *J. Organomet. Chem.*, 1976, **110**, 315.	739
76JOM(112)279	R. S. Tewari and U. C. Gupta; *J. Organomet. Chem.*, 1976, **112**, 279.	738
76JOM(113)C11	A. G. Brook, J. M. Duff, P. Hitchcock and R. Mason; *J. Organomet. Chem.*, 1976, **113**, C11.	
76JOM(113)99	G. Cahiez, D. Bernard and J. F. Normant; *J. Organomet. Chem.*, 1976, **113**, 99.	491
76JOM(114)1	D. S. Matteson and P. K. Jesthi; *J. Organomet. Chem.*, 1976, **114**, 1.	163
76JOM(121)293	A. G. Brook, J. M. Duff and W. F. Reynolds; *J. Organomet. Chem.*, 1976, **121**, 293.	652
76LA1475	H. Plieninger, E. Meyer, F. Sharif-Nassirian and E. Weidmann; *Justus Liebigs Ann. Chem.*, 1976, 1475.	725
76MAC1349	W. Ast, G. Rheinwald and R. Kerber; *Macromol. Chem.*, 1976, **177**, 1349.	674
B-76MI 101-01	M. Hudlicky; "Chemistry of Organic Fluorine Compounds," Ellis Horwood, England, 1976, pp. 170–180.	6
76MI 102-01	V. D. Bianco and S. Doronzo; *Inorg. Synth.*, 1976, **16**, 155.	52
B-76MI 107-01	I. Fleming; "Frontier Orbitals and Organic Chemical Reactions," Wiley, Chichester, 1976.	301
B-76MI 108-01	I. Fleming; "Frontier Orbitals and Organic Chemical Reactions," Wiley, Chichester, 1976.	323, 350, 351
B-76MI 108-02	G. W. Griffin and A. Padwa; in "Photochemistry of Heterocyclic Compounds," ed. O. Buchart, Wiley, New York, 1976, chap. 2, p. 41.	365
B-76MI 109-01	I. Zugravescu and M. Petrovanu; "N-Ylid Chemistry," McGraw-Hill, New York, 1976.	398
B-76MI 110-01	M. Hudlicky; "Chemistry of Organic Fluorine Compounds," 2nd edn., Horwood, Chichester, 1976, p. 174.	426, 441
B-76MI 116-01	M. Schlosser; in "Methodicum Chimicum," ed. F. Korte, K. Niedenzu and H. Zimmer, Thieme, New York, 1976, vol. 7, p. 551.	729, 730
76MI 116-01	P. F. Hudrlik; *J. Organometal. Chem. Library*, 1976, **1**, 127.	759
B-76MI 118-01	I. Fleming; "Frontier Orbitals and Organic Chemical Reactions," Wiley-Interscience, New York, 1976.	794
B-76MI 119-01	G. A. Olah and P. von R. Schleyer (eds.); "Carbonium Ions," Wiley, New York, 1976, vol. 5.	891
B-76MI 119-02	I. Fleming; "Frontier Orbitals and Organic Chemical Reactions," Wiley-Interscience, New York, 1976.	920
76OR(23)1	D. Caine; *Org. React.*, 1976, **23**, 1–258.	889
76OR(23)405	R. H. Shapiro; *Org. React.*, 1976, **23**, 405.	630, 631, 632
76OR(24)1	A. J. Birch and D. H. Williamson; *Org. React.*, 1976, **24**, 1.	73, 79, 456

76OR(24)225	C. S. Rondestvedt, Jr.; *Org. React.*, 1976, **24**, 225.	486
76OR(24)261	N. Rabjohn; *Organic Reactions*, 1976, **24**, 261.	839
76OS(55)3	C. Kaiser and J. Weinstock; *Org. Synth.*, 1976, **55**, 3.	627
76OS(55)48	L. F. Elsom, A. McKillop and E. C. Taylor; *Org. Synth.*, 1976, **55**, 48.	491
76OS(55)103	G. Linstrumelle, J. K. Krieger and G. M. Whitesides; *Org. Synth.*, 1976, **55**, 103.	465, 1090
76OS(55)114	R. K. Müller, R. Juus, D. Felix, J. Schreiber, C. Wintner and A. Eschenmoser; *Org. Synth.*, 1976, **55**, 52.	1082
76RCR469	E. M. Mil'vitskaya, A. V. Tarakanova and A. F. Plate; *Russ. Chem. Rev.*, 1976, **45**, 469.	795, 801
76RTC80	J. A. van Koeveringe and J. Lugtenburg; *Recl. Trav. Chim. Pays-Bas*, 1976, **95**, 80.	957
76RTC264	L. Brandsma and R. Grafing; *Recl. Trav. Chim. Pays-Bas*, 1976, **95**, 264.	1072
76S199	M. T. Reetz and M. Plachky; *Synthesis*, 1976, 199.	650
76S200	V. Calo, L. Lopez, A. Mincuzzi and G. Pesce; *Synthesis*, 1976, 200.	592
76S245	G. Cahiez, D. Bernard and J. F. Normant; *Synthesis*, 1976, 245.	1090
76S321	F. Scheinmann and A. J. G. Sagar; *Synthesis*, 1976, 321.	1053
76S625	G. Zweifel and R. A. Lynd; *Synthesis*, 1976, 625	465
76S737	M. Grigna-Dubois, J. Dunogues and R. Calas; *Synthesis*, 1976, 737.	284
76S755	M.-L. Roumestant, M. Malacria, J. Goré, J. Grimaldi and M. Bertrand; *Synthesis*, 1976, 755.	959
76S761	D. Masure, R. Sauvetre, J. F. Normant and J. Villieras; *Synthesis*, 1976, 761.	1090
76S777	R. E. Gawley; *Synthesis*, 1976, 777.	306
76S801	T. H. Chan, A. Baldassarre and D. Massuda; *Synthesis*, 1976, 801.	1094
76S816	G. Zweifel and R. A. Lynd; *Synthesis*, 1976, 816.	533
76SC357	K. Iwai, H. Kosugi, A. Miyazaki and H. Ude; *Synth. Commun.*, 1976, **6**, 357.	599
76T3	M. E. Jung; *Tetrahedron*, 1976, **32**, 3.	306
76T37	R. J. Hargrove and P. J. Stang; *Tetrahedron*, 1976, **32**, 37.	1117
76T829	C. Rivalle, J. Andre-Louisfert and E. Bisagni; *Tetrahedron*, 1976, **32**, 829.	42
76T981	C. F. Lane and G. W. Kabalka; *Tetrahedron*, 1976, **32**, 981.	510
76T1705	K. Tatsumi, Y. Yoshioka, K. Yamaguchi and T. Fueno; *Tetrahedron*, 1976, **32**, 1705.	707
76T2157	N. J. Cusack, C. B. Reese, A. C. Risius and B. Roozpeikar; *Tetrahedron*, 1976, **32**, 2157.	74
76T2165	R. M. Kellogg; *Tetrahedron*, 1976, **32**, 2165.	367
76T2979	J. d'Angelo; *Tetrahedron*, 1976, **32**, 2979.	129, 872, 873
76TCC1	F. Minisci; *Top. Curr. Chem.*, 1976, **62**, 1.	318
76TL3	E. J. Corey and D. Enders; *Tetrahedron Lett.*, 1976, **17**, 3.	153
76TL11	E. J. Corey and D. Enders; *Tetrahedron Lett.*, 1976, **17**, 11.	212
76TL15	P. W. Chum and S. E. Wilson; *Tetrahedron Lett.*, 1976, **17**, 15.	514
76TL57	S.-I. Yamada, K. Tomioka and K. Koga; *Tetrahedron Lett.*, 1976, **17**, 57.	55
76TL61	S.-I. Yamada, K. Tomioka and K. Koga; *Tetrahedron Lett.*, 1976, **17**, 61.	55
76TL235	M. Bruhn, C. H. Brown, P. W. Collins, E. Z. Dajani, J. R. Palmer and R. Pappo; *Tetrahedron Lett.*, 1976, **17**, 235.	1050
76TL255	A. Suzuki, N. Miyaura and M. Itoh; *Tetrahedron Lett.*, 1976, **17**, 255.	276
76TL275	D. Michelot and G. Linstrumelle; *Tetrahedron Lett.*, 1976, **17**, 275.	974
76TL575	H. G. Richey, Jr. and D. W. Shull; *Tetrahedron Lett.*, 1976, **17**, 575.	801
76TL677	M. L. Roumestant, P. Place and J. Gore; *Tetrahedron Lett.*, 1976, **17**, 677.	810
76TL881	I. Fleming, F. L. Gianni and T. Mah; *Tetrahedron Lett.*, 1976, **17**, 881.	519
76TL947	H. Klein, H. Eijsinga, H. Westmijze, J. Meijer and P. Vermeer; *Tetrahedron Lett.*, 1976, **17**, 947.	496
76TL1243	E. Doomes and R. M. Beard; *Tetrahedron Lett.*, 1976, **17**, 1243.	709
76TL1385	J. Rémion, W. Dumont and A. Krief; *Tetrahedron Lett.*, 1976, **17**, 1385.	592, 593, 713, 716
76TL1439	M. E. Jung, P. A. Blair and J. A. Lowe; *Tetrahedron Lett.*, 1976, **17**, 1439.	154
76TL1539	A. B. Holmes, R. A. Raphael and N. K. Wellard; *Tetrahedron Lett.*, 1976, **17**, 1539.	1023
76TL1699	E. Nakamura, M. Shimizu and I. Kuwajima; *Tetrahedron Lett.*, 1976, **17**, 1699.	283
76TL1769	C. G. Chavdarian, S. L. Woo, R. D. Clark and C. H. Heathcock; *Tetrahedron Lett.*, 1976, **17**, 1769.	1024, 1139
76TL1817	I. Kuwajima, T. Sato, M. Arai and N. Minami; *Tetrahedron Lett.*, 1976, **17**, 1817.	881
76TL1927	S. Baba, D. E. Van Horn and E. Negishi; *Tetrahedron Lett.*, 1976, **17**, 1927.	534
76TL2287	P. C. Traas, H. Boelens and H. J. Takken; *Tetrahedron Lett.*, 1976, **17**, 2287.	633, 1094
76TL2449	G. Deleris, J. Dunogués and R. Calas; *Tetrahedron Lett.*, 1976, **17**, 2449.	1038, 1138
76TL2455	T. Kurihara, Y. Nakajima and O. Mitsunobu; *Tetrahedron Lett.*, 1976, **17**, 2455.	179
76TL2643	M. Sevrin, D. Van Ende and A. Krief; *Tetrahedron Lett.*, 1976, **17**, 2643.	21, 25, 437
76TL2947	J. E. Stemke, A. R. Chamberlin and F. T. Bond; *Tetrahedron Lett.*, 1976, **17**, 2947.	633, 1094
76TL3083	F. Bernardou and L. Miginiac; *Tetrahedron Lett.*, 1976, **17**, 3083.	543
76TL3181	S. Braverman, Y. Duar and D. Segev; *Tetrahedron Lett.*, 1976, **17**, 3181.	955
76TL3225	F. Derguini-Boumechal and G. Linstrumelle; *Tetrahedron Lett.*, 1976, **17**, 3225.	1092
76TL3227	A. M. Léonard-Coppens and A. Krief; *Tetrahedron Lett.*, 1976, **17**, 3227.	592, 593, 716
76TL3245	J. P. Marino and L. J. Browne; *Tetrahedron Lett.*, 1976, **17**, 3245.	811
76TL3265	A. L. Baumstark, E. J. H. Bechara and M. J. Semigran; *Tetrahedron Lett.*, 1976, **17**, 3265.	688, 690
76TL3477	B. M. Trost, H. C. Arndt, P. E. Strege and T. R. Verhoeven; *Tetrahedron Lett.*, 1976, **17**, 3477.	698
76TL3507	W. Eymann and M. Hanack; *Tetrahedron Lett.*, 1976, **17**, 3507.	1111
76TL3585	M. Baumann, W. Hoffmann and N. Müller; *Tetrahedron Lett.*, 1976, **17**, 3585.	422
76TL3743	J. Rémion and A. Krief; *Tetrahedron Lett.*, 1976, **17**, 3743.	592, 593, 713, 716
76TL4191	C. A. Grob and P. Wenk; *Tetrahedron Lett.*, 1976, **17**, 4191.	1119
76TL4195	C. A. Grob and P. Wenk; *Tetrahedron Lett.*, 1976, **17**, 4195.	1119

76TL4385	A. Pelter, R. Hughes, K. Smith and M. Tabata; *Tetrahedron Lett.*, 1976, **17**, 4385.	1035, 1103
76TL4405	L. Blaszczak, J. Winkler and S. O'Kuhn; *Tetrahedron Lett.*, 1976, **17**, 4405.	479
76TL4705	E. J. Corey and R. H. Wollenberg; *Tetrahedron Lett.*, 1976, **17**, 4705.	535
76TL4839	H. Neumann and D. Seebach; *Tetrahedron Lett.*, 1976, **17**, 4839.	111, 1090
77ACR179	W. Ando; *Acc. Chem. Res.*, 1977, **10**, 179.	400, 821, 823
77ACR434	J. K. Stille and K. S. Y. Lau; *Acc. Chem. Res.*, 1977, **10**, 434.	471
77ACS(B)667	B. Gustafsson, M. Nilsson and C. Ullenius; *Acta Chem. Scand. Ser. B*, 1977, **31**, 667.	311
77AG(E)1	H. Werner; *Angew. Chem., Int. Ed. Engl.*, 1977, **16**, 1.	862
77AG(E)44a	M. T. Reetz and W. Stephan; *Angew. Chem., Int. Ed. Engl.*, 1977, **16**, 44.	664
77AG(E)44b	M. T. Reetz and D. Schinzer; *Angew. Chem., Int. Ed. Engl.*, 1977, **16**, 44.	664
77AG(E)264	D. Seebach, B. Seuring, H.-O. Kalinowski, W. Lubosch and B. Renger; *Angew. Chem., Int. Ed. Engl.*, 1977, **16**, 264.	210
77AG(E)339	U. Schöllkopf; *Angew. Chem., Int. Ed. Engl.*, 1977, **16**, 339	155
77AG(E)399	R. L. Hilard and K. P. C. Vollhardt; *Angew. Chem., Int. Ed. Engl.*, 1977, **16**, 399.	64
77AG(E)487	I. Gosney, T. J. Lillie and D. Lloyd; *Angew. Chem., Int. Ed. Engl.*, 1977, **16**, 487.	739, 740, 742
77AG(E)862	T. Kauffmann, R. Kriegesmann and A. Woltermann; *Angew. Chem., Int. Ed. Engl.*, 1977, **16**, 862.	667
77AOC(16)131	E. A. Hill; *Adv. Organomet. Chem.*, 1977, **16**, 131.	316
77APO(15)153	T. E. Hogen-Esch; *Adv. Phys. Org. Chem.*, 1977, **15**, 153.	858
77BCJ2199	N. Miyaura, M. Itoh and A. Suzuki; *Bull. Chem. Soc. Jpn.*, 1977, **50**, 2199.	125
77BCJ3288	K. Kitatani, T. Hiyama and H. Nozaki; *Bull. Chem. Soc. Jpn.*, 1977, **50**, 3288.	164
77BSF765	J. Villieras, P. Perriot and J. F. Normant; *Bull. Soc. Chim. Fr.*, 1977, 765.	164
77BSF925	B. Cazes and S. Julia; *Bull. Soc. Chim. Fr.*, 1977, 925.	816
77CB154	E. V. Dehmlow, S. S. Dehmlow and F. Marschner; *Chem. Ber.*, 1977, **110**, 154.	550
77CB199	K. P. Jäckel and M. Hanack; *Chem. Ber.*, 1977, **110**, 199.	1116
77CB717	E. Weiss and H. Köster; *Chem. Ber.*, 1977, **110**, 717.	885
77CB1823	W. Boland and L. Jaenicke; *Chem. Ber.*, 1977, **110**, 1823.	5, 9
77CB1833	K.-H. Geiß, D. Seebach and B. Seuring; *Chem. Ber.*, 1977, **110**, 1833.	140
77CB2127	H. Paulsen and W. Koebernick; *Chem. Ber.*, 1977, **110**, 2127.	587
77CB2588	B. Giese and J. Meister; *Chem. Ber.*, 1977, **110**, 2588.	948
77CB2978	K. Weinges, H. Baake, H. Distler, K. Klessing, R. Kolb and G. Schilling; *Chem. Ber.*, 1977, **110**, 2978.	708
77CC198	E. Verkuijlen, F. Kapteijn, J. C. Mol and C. Boelhouwer; *J. Chem. Soc., Chem. Commun.*, 1977, 198.	674
77CC233	J. E. Baldwin and L. I. Kruse; *J. Chem. Soc., Chem. Commun.*, 1977, 233.	130, 304
77CC318	C. A. Brown and E. Negishi; *J. Chem. Soc., Chem. Commun.*, 1977, 318.	1085
77CC321	N. R. A. Beeley and J. K. Sutherland; *J. Chem. Soc., Chem. Commun.*, 1977, 321.	108
77CC427	M. J. Russell, C. White and P. M. Maitlis; *J. Chem. Soc., Chem. Commun.*, 1977, 427.	93
77CC502	J.-P. Laval, A. Lattes, R. Mutin and J. M. Basset; *J. Chem. Soc., Chem. Commun.*, 1977, 502.	674
77CC683	A. O. King, N. Okukado and E.-I. Negishi; *J. Chem. Soc., Chem. Commun.*, 1977, 683.	1006, 1140
77CC852	H. Yatagai, Y. Yamamoto, K. Maruyama, A. Sonoda and S.-I. Murahashi; *J. Chem. Soc., Chem. Commun.*, 1977, 852.	531, 663
77CJC3392	P. J. Elving; *Can. J. Chem.*, 1977, **55**, 3392.	103
77CL103	K. Itoh, T. Yogo and Y. Ishii; *Chem. Lett.*, 1977, 103.	481
77CL301	M. Kosugi, K. Sasazawa, Y. Shimizu and T. Migita; *Chem. Lett.*, 1977, 301.	471
77CL959	K. Narasaka, M. Yamaguchi and T. Mukaiyama; *Chem. Lett.*, 1977, 959.	1134
77CRV69	R. D. Howells and J. D. Mc Cown; *Chem. Rev.*, 1977, **77**, 69.	178
77CRV473	S. Braslavsky and J. Heicklen; *Chem. Rev.*, 1977, **77**, 473.	785
77CRV691	A. Efraty; *Chem. Rev.*, 1977, **77**, 691.	517
77HCA978	S. Jolidon and H.-J. Hansen; *Helv. Chim. Acta*, 1977, **60**, 978.	835
77HCA1361	R. Schmid and H. Schmid; *Helv. Chim. Acta*, 1977, **60**, 1361.	234, 821
77HCA2388	W. Oppolzer, K. K. Mahalanabis and K. Bättig; *Helv. Chim. Acta*, 1977, **60**, 2388.	599, 779
77HCA2711	P. Sieber; *Helv. Chim. Acta*, 1977, **60**, 2711.	654
77HCA3039	H. Gerlach; *Helv. Chim. Acta*, 1977, **60**, 3039.	654
77HOU(5)967	M. Murray; *Methoden Org. Chem. (Houben-Weyl)*, 1977, **5**, 967.	953
77HOU(5)1030	M. Murray; *Methoden Org. Chem. (Houben-Weyl)*, 1977, **5**, 1030.	960
77HOU(5)1037	M. Murray; *Methoden Org. Chem. (Houben-Weyl)*, 1977, **5**, 1037.	960
77JA247	W. A. Kleschick, C. T. Buse and C. T. Heathcock; *J. Am. Chem. Soc.*, 1977, **99**, 247.	875
77JA253	A. Marfat, P. R. McGuirk, R. Kramer and P. Helquist; *J. Am. Chem. Soc.*, 1977, **99**, 253.	1107
77JA290	S. DeYoung, S. Ehrlich and E. Berliner; *J. Am. Chem. Soc.*, 1977, **99**, 290.	318
77JA377	M. J. S. Dewar, R. C. Haddon, A. Komornicki and H. Rzepa; *J. Am. Chem. Soc.*, 1977, **99**, 377.	905
77JA602	B. Feringa and H. Wynberg; *J. Am. Chem. Soc.*, 1977, **99** 602.	690
77JA642	M. P. Cooke, Jr. and R. Goswami; *J. Am. Chem. Soc.*, 1977, **99**, 642.	303
77JA936	T. H. Chan and D. Massuda; *J. Am. Chem. Soc* , 1977, **99**, 936.	652
77JA954	G. A. Molander, J. A. Sinclair and H. C. Brown; *J. Am. Chem. Soc.*, 1977, **99**, 954.	1050
77JA1058	A. Stefani, C. Consiglio, C. Botteghi and P. Pino; *J. Am. Chem. Soc.*, 1977, **99**, 1058.	297
77JA1499	J. H. Davis, K. J. Shea and R. G. Bergman; *J. Am. Chem. Soc.*, 1977, **99**, 1499.	806, 807
77JA1669	C. J. Collins, B. M. Benjamin, M. Hanack and H. Stutz; *J. Am. Chem. Soc.*, 1977, **99**, 1669.	1124
77JA1673	A. Hosomi and H. Sakurai; *J. Am. Chem. Soc.*, 1977, **99**, 1673.	314
77JA1712	H. C. Brown and E. N. Peters; *J. Am. Chem. Soc.*, 1977, **99**, 1712.	895

77JA1881	E. Grovenstein, Jr. and A. B. Cottingham; *J. Am. Chem. Soc.*, 1977, **99**, 1881.	886
77JA2342	H. E. Zimmerman and S. M. Aasen; *J. Am. Chem. Soc.*, 1977, **99**, 2342.	802
77JA2344	A. Padwa, T. J. Blacklock, D. Getman and N. Hatanaka; *J. Am. Chem. Soc.*, 1977, **99**, 2344.	802
77JA2631	M. P. Periasamy and H. M. Walborsky; *J. Am. Chem. Soc.*, 1977, **99**, 2631.	155
77JA2815	L. A. Paquette, T. G. Wallis, K. Hirotsu and J. Clardy; *J. Am. Chem. Soc.*, 1977, **99**, 2815.	704, 708
77JA2919	J. H. Davis, W. A. Goddard, III and L. B. Harding; *J. Am. Chem. Soc.*, 1977, **99**, 2919.	1143
77JA3168	E. Negishi and D. E. Van Horn; *J. Am. Chem. Soc.*, 1977, **99**, 3168.	468
77JA3196	D. S. Matteson and R. J. Moody; *J. Am. Chem. Soc.*, 1977, **99**, 3196.	163
77JA3414	W. G. Dauben, G. T. Rivers and W. T. Zimmerman; *J. Am. Chem. Soc.*, 1977, **99**, 3414.	632
77JA3504	B. Sket and M. Zupan; *J. Am. Chem. Soc.*, 1977, **99**, 3504.	517
77JA3769	D. Scheffel, P. J. Abbott, G. J. Fitzpatrick and M. D. Schiavelli; *J. Am. Chem. Soc.*, 1977, **99**, 3769.	1114
77JA4186	W. C. Still; *J. Am. Chem. Soc.*, 1977, **99**, 4186.	813
77JA4417	M. J. S. Dewar and L. E. Wade; *J. Am. Chem. Soc.*, 1977, **99**, 4417.	355
77JA4833	A. A. Millard and M. W. Rathke; *J. Am. Chem. Soc.*, 1977, **99**, 4833.	467
77JA4954	L. M. Jackman and N. M. Szeverenyi; *J. Am. Chem. Soc.*, 1977, **99**, 4954.	873, 873
77JA5026	G. A. Olah and D. J. Donovan; *J. Am. Chem. Soc.*, 1977, **99**, 5026.	894, 899
77JA5210	P. A. Grieco and Y. Yokoyama; *J. Am. Chem. Soc.*, 1977, **99**, 5210.	608
77JA5453	D. A. Evans, C. L. Sims and G. C. Andrews; *J. Am. Chem. Soc.*, 1977, **99**, 5453.	822
77JA5483	R. L. Funk and K. P. C. Vollhardt; *J. Am. Chem. Soc.*, 1977, **99**, 5483.	448
77JA5652	Y. Yamamoto, H. Yatagai, K. Maruyama, A. Sonoda and S.-I. Murahashi; *J. Am. Chem. Soc.*, 1977, **99**, 5652.	488
77JA5683	G. A. Olah, G. K. S. Prakash and G. Liang; *J. Am. Chem. Soc.*, 1977, **99**, 5863.	906
77JA5946	B. D. Vineyard, W. S. Knowles, M. J. Sabacky, G. L. Bachman and D. J. Weinkauff; *J. Am. Chem. Soc.*, 1977, **99**, 5946.	76
77JA6946	L. A. Paquette, T. G. Wallis, T. Kempe, G. G. Christoph, J. P. Springer and J. Clardy; *J. Am. Chem. Soc.*, 1977, **99**, 6946.	704, 708
77JA7097	H. C. Brown and N. Ravindran; *J. Am. Chem. Soc.*, 1977, **99**, 7097.	1103
77JA7365	R. H. Wollenberg, K. F. Albizati and R. Peries; *J. Am. Chem. Soc.*, 1977, **99**, 7365.	1091
77JA7432	M. J. S. Dewar and H. S. Rzepa; *J. Am. Chem. Soc.*, 1977, **99**, 7432.	1109
77JA7632	X. Creary; *J. Am. Chem. Soc.*, 1977, **99**, 7632.	111, 971, 1056, 1061
77JA7960	A. Citterio, F. Minisci, O. Porta and G. Sesana; *J. Am. Chem. Soc.*, 1977, **99**, 7960.	923
77JA8045	M. J. Loots and J. Schwartz; *J. Am. Chem. Soc.*, 1977, **99**, 8045.	536, 1105
77JA8116	B. M. Trost, J. Ippen and W. C. Vladuchick; *J. Am. Chem. Soc.*, 1977, **99**, 8116.	598
77JA8262	A. G. Abatjoglou, E. L. Eliel and L. F. Kuyper; *J. Am. Chem. Soc.*, 1977, **99**, 8262.	859
77JA8348	L. Bonazzola, N. Leray and J. Roncin; *J. Am. Chem. Soc.*, 1977, **99**, 8348.	923
77JA8356	L. Duhamel and J. M. Poirier; *J. Am. Chem. Soc.*, 1977, **99**, 8356.	453
77JCP(66)4671	S. Ikuta, S. Iwata and M. Imamura; *J. Chem. Phys.*, 1977, **66**, 4671.	1142
77JCR(M)153	P. Bladon, I. U. Khand and P. L. Pauson; *J. Chem. Res. (M)*, 1977, 153.	529
77JCR(M)168	I. U. Khand and P. L. Pauson; *J. Chem. Rev.*, 1977, 168.	529
77JCR(M)2101	M. Bourgain-Commerçon, J. F. Normant and J. Villieras; *J. Chem. Res. M*, 1977, 2101.	477
77JCS(P1)90	F. G. Schreiber and R. Stevenson; *J. Chem. Soc., Perkin Trans. 1*, 1977, 90.	1136
77JCS(P1)123	D. Savoia, C. Tromboni and A. Umani-Ronchi; *J. Chem. Soc., Perkin Trans. 1*, 1977, 123.	439
77JCS(P1)443	I. D. Entwistle, A. E. Jackson, R. A. W. Johnstone and R. P. Telford; *J. Chem. Soc., Perkin Trans. 1*, 1977, 443.	74, 441
77JCS(P1)460	J. March and W. Plankl; *J. Chem. Soc., Perkin Trans. 1*, 1977, 460.	885
77JCS(P1)550	A. H. Davidson, I. Fleming, J. I. Grayson, A. Pearce, R. L. Snowden and S. Warren; *J. Chem. Soc., Perkin Trans. 1*, 1977, 550.	643, 755
77JCS(P1)869	E. W. Colvin and B. J. Hamill; *J. Chem. Soc., Perkin Trans. 1*, 1977, 869.	1078
77JCS(P1)1452	A. H. Davidson, C. Earnshaw, J. I. Grayson and S. Warren; *J. Chem. Soc., Perkin Trans. 1*, 1977, 1452.	643
77JCS(P1)1574	D. N. Jones, D. R. Hill, D. A. Lewton and C. Sheppard; *J. Chem. Soc., Perkin Trans. 1*, 1977, 1574.	595
77JCS(P1)1718	D. H. R. Barton and R. Subramanian; *J. Chem. Soc., Perkin Trans. 1*, 1977, 1718.	940, 941
77JCS(P2)542	I. G. Csizmadia, F. Bernardi, V. Lucchini and G. Modena; *J. Chem. Soc., Perkin Trans. 2*, 1977, 542.	1109
77JHC567	A. Shafiee, I. Lalezari, M. Mirrashed and D. Nercesian; *J. Heterocycl. Chem.*, 1977, **14**, 567.	1077
77JHC681	P. A. Zoretic and P. Soja; *J. Heterocycl. Chem.*, 1977, **14**, 681.	607
77JHC947	L. Garanti, G. Padova and G. Zecchi; *J. Heterocycl. Chem.*, 1977, **14**, 947.	524
77JOC82	R. O. Hutchins, D. Kandasamy, C. A. Maryanoff, D. Masilamani and B. E. Maryanoff; *J. Org. Chem.*, 1977, **42**, 82.	4, 6, 7, 9, 10, 11, 17
77JOC135	P. N. Confalone, G. Pizzolato and M. R. Uskoković; *J. Org. Chem.*, 1977, **42**, 135.	91
77JOC172	H. Kwart, N. A. Johnson, T. Eggerichs and T. J. George; *J. Org. Chem.*, 1977, **42**, 172.	834
77JOC174	L. R. Subramanian and M. Hanack; *J. Org. Chem.*, 1977, **42**, 174.	1111, 1117, 1124
77JOC260	W. E. Parham and D. W. Boykin; *J. Org. Chem.*, 1977, **42**, 260.	1090
77JOC353	P. J. Kocienski, G. Cernigliaro and G. Feldstein; *J. Org. Chem.*, 1977, **42**, 353.	979
77JOC394	S. Danishefsky, M.-Y. Tsai and T. Kitahara; *J. Org. Chem.*, 1977, **42**, 394.	218, 221
77JOC579	P. Jacob, III and H. C. Brown; *J. Org. Chem.*, 1977, **42**, 579.	531, 1103
77JOC1349	J. Fayos, J. Clardy, L. J. Dolby and T. Farnham; *J. Org. Chem.*, 1977, **42**, 1349.	613
77JOC1680	R. C. Larock and J. C. Bernhardt; *J. Org. Chem.*, 1977, **42**, 1680.	492, 1095
77JOC1804	D. G. Oelberg and M. D. Schiavelli; *J. Org. Chem.*, 1977, **42**, 1804.	993
77JOC2021	G. H. Schmid, A. Modro and K. Yates; *J. Org. Chem.*, 1977, **42**, 2021.	318

77JOC2036	B. M. Trost and Y. Matsumura; *J. Org. Chem.*, 1977, **42**, 2036.	615, 698, 702
77JOC2380	J. L. Coke, H. J. Williams and S. Natarajan; *J. Org. Chem.*, 1977, **42**, 2380.	1082, 1083
77JOC2431	M. P. Doyle, B. Siegfried, R. C. Elliott and J. F. Dellaria, Jr.; *J. Org. Chem.*, 1977, **42**, 2431.	36, 486
77JOC2626	J. S. Kiely, L. L. Nelson and P. Boudjouk; *J. Org. Chem.*, 1977, **42**, 2626.	1004, 1005, 1006
77JOC2650	M. M. Midland; *J. Org. Chem.*, 1977, **42**, 2650.	977
77JOC2666	G. A. Olah and G. K. S. Prakash; *J. Org. Chem.*, 1977, **42**, 2666.	894, 899, 900
77JOC2689	M.-F. Ruasse and J.-E. Dubois; *J. Org. Chem.*, 1977, **42**, 2689.	318
77JOC2783	M. I. Dawson and M. Vasser; *J. Org. Chem.*, 1977, **42**, 2783.	727
77JOC2944	N. Kornblum and L. Cheng; *J. Org. Chem.*, 1977, **42**, 2944.	638
77JOC3004	Y. Apeloig, P. v. R. Schleyer and J. A. Pople; *J. Org. Chem.*, 1977, **42**, 3004.	1112, 1113
77JOC3494	M. P. Doyle, J. F. Dellaria, Jr., B. Siegfried and S. W. Bishop; *J. Org. Chem.*, 1977, **42**, 3494.	442
77JOC3895	W. J. Bailey and C. N. Bird; *J. Org. Chem.*, 1977, **42**, 3895.	773
77JOC3899	W. J. Bailey and J. Di Pietro; *J. Org. Chem.*, 1977, **42**, 3899.	777
77JOC3961	M. E. Jung, Y.-G. Pan, M. W. Rathke, D. F. Sullivan and R. P. Woodbury; *J. Org. Chem.*, 1977, **42**, 3961.	556
77JOC3981	I. Knox, S.-C. Chang and A. H. Andrist; *J. Org. Chem.*, 1977, **42**, 3981.	511
77JOM(124)327	W. A. Nugent and J. K. Kochi; *J. Organomet. Chem.*, 1977, **124**, 327.	671
77JOM(124)349	W. A. Nugent and J. K. Kochi; *J. Organomet. Chem.*, 1977, **124**, 349.	671
77JOM(124)371	W. A. Nugent and J. K. Kochi; *J. Organomet. Chem.*, 1977, **124**, 371.	671
77JOM(127)371	C. Chuit, H. Felkin, C. Frajerman, G. Roussi and G. Swierczewski; *J. Organomet. Chem.*, 1977, **127**, 371.	477
77JOM(128)1	A. Commerçon, J. F. Normant and J. Villieras; *J. Organomet. Chem.*, 1977, **128**, 1.	466
77JOM(132)445	F. Barbot and P. Miginiac; *J. Organomet. Chem.*, 1977, **132**, 445.	318
77JOM(132)9	G. W. Kramer and H. C. Brown; *J. Organomet. Chem.*, 1977, **132**, 9.	838
77JOM(135)1	T. Iwayanagi, M. Matsuo and Y. Saito; *J. Organomet. Chem.*, 1977, **135**, 1.	66
77JOM(140)257	D. Paquer and M. Vazeux; *J. Organomet. Chem.*, 1977, **140**, 257.	1058
77JOM(141)113	R. H. Crabtree, A. Gautier, G. Giordano and T. Khan; *J. Organomet. Chem.*, 1977, **141**, 113.	456, 507
77JOM(142)1	Y. Gendreau, J. F. Normant and J. Villieras; *J. Organomet. Chem.*, 1977, **142**, 1.	238, 483
77JOU1033	D. E. Prikule and O. Neilands; *J. Org. Chem. USSR* (*Engl. Transl.*), 1977, **13**, 1033.	11
77LA40	U. Schöllkopf, K.-W. Henneke, K. Madawinata and R. Harms; *Justus Liebigs Ann. Chem.*, 1977, 40.	155
77LA282	H. J. Bestmann and C. Geismann; *Justus Liebigs Ann. Chem.*, 1977, 282.	1079
77LA846	D. Seebach, N. Meyer and A. K. Beck; *Justus Liebigs Ann. Chem.*, 1977, 846.	716
B-77MI 115-01	C. Y. Meyers, W. S. Matthews, L. L. Ho, V. M. Kolb and T. E. Parady; "Catalysis in Organic Syntheses," ed. G. V. Smith, Academic Press, New York, 1977, p. 197.	706
B-77MI 118-01	H. Pines and W. M. Stalick; "Base Catalysed Reactions of Hydrocarbons and Related Compounds," Academic Press, New York, 1977.	795
77MI 118-02	E. Wenkert and H. Sliwa; *Bioorg. Chem.*, 1977, **6**, 443.	808
77MI 119-01	A. A. Solov'yanov, A. D. Karpyuk, I. P. Beletskaya and O. A. Reutov; *Dokl. Chem.* (*Engl. Transl.*), 1977, **237**, 668.	859
77NJC373	J.-C. Clinet and G. Linstrumelle; *Nouv. J. Chim.*, 1977, **1**, 373.	971
77OR(25)1	L. A. Paquette; *Org. React.*, 1977, **25**, 1.	144, 617, 704, 707
77OR(25)73	W. S. Wadsworth, Jr.; *Org. React.*, 1977, **25**, 73.	638, 639, 640
77OS(57)22	J. C. Martin, R. J. Arhart, J. A. Franz, E. F. Perozzi and L. J. Kaplan; *Org. Synth.*, 1977, **57**, 22.	576
77OS(57)53	J. M. Photis and L. A. Paquette; *Org. Synth.*, 1977, **57**, 53.	613
77PAC107	M. Herlem; *Pure Appl. Chem.*, 1977, **49**, 107.	901
77RCR530	V. B. Mochalin and Yu. N. Porshnev; *Russ. Chem. Rev.*, 1977, **46**, 530.	916
77RTC(M)91	R. Nouguier, R. Mutin, J. P. Laval, G. Chapelet, J. Basset and A. Lattes; *Recl. Trav. Chim. Pays-Bas*, 1977, **96**, M91.	674
77S52	G. Zweifel, R. A. Lynd and R. E. Murray; *Synthesis*, 1977, 52.	514
77S124	G. W. Kabalker, D. T. C. Yang, J. H. Chandler and J. D. Baker, Jr.; *Synthesis*, 1977, 124.	41
77S172	G. W. Gribble, R. M. Leese and B. E. Evans; *Synthesis*, 1977, 172.	13
77S339	K. Dimroth and W. Tüncher; *Synthesis*, 1977, 339.	443
77S509	E. M. Kaiser, J. D. Petty and P. L. A. Knutson; *Synthesis*, 1977, 509.	879
77S553	D. Lenoir; *Synthesis*, 1977, 553.	688, 689
77S561	R. Rossi and A. Carpita; *Synthesis*, 1977, 561.	514
77S777	N. Hagihara, K. Sonogashira and Y. Tohda; *Synthesis*, 1977, 777.	1045, 1048
77S859	G. W. Gribble and J. H. Hoffman; *Synthesis*, 1977, 859.	102
77SC103	C. Huynh and S. Julia; *Synth. Commun.*, 1977, **7**, 103.	820
77T457	J. Shenkenburger, D. Heber and E. Herber-Brunschweiger; *Tetrahedron*, 1977, **33**, 457.	455
77T1211	S. I. Miller; *Tetrahedron*, 1977, **33**, 1211.	1110
77T1527	D. M. Bratby, J. C. Chadwick, G. I. Fray and R. G. Saxton; *Tetrahedron*, 1977, **33**, 1527.	815
77T1845	C. A. Henrick; *Tetrahedron*, 1977, **33**, 1845.	456, 502, 504, 505, 506, 734, 1000, 1002
77T1945	K. Utimoto, K. Uchida, M. Yamaya and H. Nozaki; *Tetrahedron*, 1977, **33**, 1945.	207, 1100
77T2019	P. D. Magnus; *Tetrahedron*, 1977, **33**, 2019.	140, 146, 436, 704
77T2737	L. M. Jackman and B. C. Lange; *Tetrahedron*, 1977, **33**, 2737.	129, 130, 858, 873
77T2813	M. Malacria and M.-L. Roumestant; *Tetrahedron*, 1977, **33**, 2813.	959
77T2949	S. Masson, M. Saquet and A. Thuillier; *Tetrahedron*, 1977, **33**, 2949.	1058
77T2987	K. Uchida, K. Utimoto and H. Nozaki; *Tetrahedron*, 1977, **33**, 2987.	1094
77TL109	D. Tunemoto, N. Araki and K. Kondo; *Tetrahedron Lett.*, 1977, **18**, 109.	299, 354

77TL135	E. Vedejs and W. T. Stolle; *Tetrahedron Lett.*, 1977, **18**, 135.	630
77TL159	R. T. Taylor, C. R. Degenhardt, W. P. Melega and L. A. Paquette; *Tetrahedron Lett.*, 1977, **18**, 159.	1094
77TL167	R. Mornet and L. Gouin; *Tetrahedron Lett.*, 1977, **18**, 167.	539
77TL411	E. Negishi and A. Abramovitch; *Tetrahedron Lett.*, 1977, **18**, 411.	1102
77TL441	R. Baker and M. J. Crimmin; *Tetrahedron Lett.*, 1977, **18**, 441.	674
77TL763	D. S. Weiss and P. M. Kochanek; *Tetrahedron Lett.*, 1977, **18**, 763.	841
77TL1069	J.-J. Brunet, L. Mordenti, B. Loubinoux and P. Caubère; *Tetrahedron Lett.*, 1977, **18**, 1069.	458, 515
77TL1137	B. Deschamps, J. P. Lampin, F. Mathey and J. Seyden-Penne; *Tetrahedron Lett.*, 1977, **18**, 1137.	751
77TL1161	D. Seebach, R. Henning, F. Lehr and J. Gonnermann; *Tetrahedron Lett.*, 1977, **18**, 1161.	156, 157
77TL1173	J. Kattenberg, E. R. de Waard and H. O. Huisman; *Tetrahedron Lett.*, 1977, **18**, 1173.	708
77TL1303	M. Yoshifuji, M. J. Loots and J. Schwartz; *Tetrahedron Lett.*, 1977, **18**, 1303.	491, 536
77TL1337	T. Numata and S. Oae; *Tetrahedron Lett.*, 1977, **18**, 1337.	821
77TL1625	E. J. Corey, M. Shibasaki and J. Knolle; *Tetrahedron Lett.*, 1977, **18**, 1625.	827, 828
77TL1779	M. Bertrand, G. Gil, A. Junino and R. Maurin; *Tetrahedron Lett.*, 1977, **18**, 1779.	805
77TL1805	M. Obayashi, K. Utimoto and H. Nozaki; *Tetrahedron Lett.*, 1977, **18**, 1805.	160, 1094
77TL1807	M. Obayashi, J. Utimoto and H. Nozaki; *Tetrahedron Lett.*, 1977, **18**, 1807.	761
77TL1823	H. Westmijze, J, Meijer and P. Vermeer; *Tetrahedron Lett.*, 1977, **18**, 1823.	1094
77TL1961	T. M. Dolak and T. A. Bryson; *Tetrahedron Lett.*, 1977, **18**, 1961.	141
77TL2015	E. Brown and J.-P. Robin; *Tetrahedron Lett.*, 1978, **18**, 2015.	473
77TL2023	H. Westmijze, H. Kleijn and P. Vermeer; *Tetrahedron Lett.*, 1977, **18**, 2023.	541
77TL2441	T. Mitsuhashi, S. Otsuka and M. Oki; *Tetrahedron Lett.*, 1977, **18**, 2441.	265
77TL2559	S. R. Wilson, D. T. Mao, K. M. Jernberg and S. T. Ezmirly; *Tetrahedron Lett.*, 1977, **18**, 2559.	401, 414, 800
77TL2831	B. B. Snider, L. A. Brown, R. S. E. Conn and T. A. Killinger; *Tetrahedron Lett.*, 1977, **18**, 2831.	525
77TL3187	A. Debal, T. Cuviguy and M. Larchevêque; *Tetrahedron Lett.*, 1977, **18**, 3187.	560
77TL3215	G. H. Posner and C. M. Lentz; *Tetrahedron Lett.*, 1977, **18**, 3215.	304
77TL3263	J. F. Normant, J. Villieras, and F. Scott; *Tetrahedron Lett.*, 1977, **18**, 3263.	465
77TL3317	T. H. Chan, P. W. K. Lau and W. Mychajlowskij; *Tetrahedron Lett.*, 1977, **18**, 3317.	535
77TL3369	M. Miyaura, N. Sasaki, M. Itoh and A. Suzuki; *Tetrahedron Lett.*, 1977, **18**, 3369.	469
77TL3383	G. Cahiez and J. F. Normant; *Tetrahedron Lett.*, 1977, **18**, 3383.	1038
77TL3407	R.-D. Acker; *Tetrahedron Lett.*, 1977, **18**, 3407.	204, 206, 478
77TL3485	H. Felkin and C. Frajerman; *Tetrahedron Lett.*, 1977, **18**, 3485.	397
77TL3545	A. B. Levy, P. Talley and J. A. Dunford; *Tetrahedron Lett.*, 1977, **18**, 3545.	1108
77TL3641	K. Utimoto, K. Uchida, M. Yamaya and H. Nozaki; *Tetrahedron Lett.*, 1977, **18**, 3641.	511
77TL3791	M. E. Jung and R. B. Blum; *Tetrahedron Lett.*, 1977, **18**, 3791.	132, 135
77TL3843	R. S. Tewari and S. C. Chaturvedi; *Tetrahedron Lett.*, 1977, **18**, 3843.	740, 742
77TL3847	E. J. Corey and D. R. Williams; *Tetrahedron Lett.*, 1977, **18**, 3847.	535, 1091
77TL4089	M. Zembayashi, K. Tamao, J. Yoshida and M. Kumada; *Tetrahedron Lett.*, 1977, **18**, 4089.	472
77TL4145	C. A. Hergrueter, P. D. Brewer, J. Tagat and P. M. Helquist; *Tetrahedron Lett.*, 1977, **18**, 4145.	1094
77TL4187	R. K. Boeckman, Jr. and K. J. Bruza; *Tetrahedron Lett.*, 1977, **18**, 4187.	1089
77TL4223	B. Lythgoe and I. Waterhouse; *Tetrahedron Lett.*, 1977, **18**, 4223.	590, 591, 615, 700
77TL4257	C.-L. Yeh and M. Dawson; *Tetrahedron Lett.*, 1977, **18**, 4257.	727
77TL4291	J. T. Groves and S. Kittisopikul; *Tetrahedron Lett.*, 1977, **18**, 4291.	936
77TL4305	G. Büchi and H. Wuest; *Tetrahedron Lett.*, 1977, **18**, 4305.	450
77TL4377	M. J. Chandy and M. Hanack; *Tetrahedron Lett.*, 1977, **18**, 4377.	1124
77TL4481	E. C. Ashby and J. J. Lin; *Tetrahedron Lett.*, 1977, **18**, 4481.	514
77TL4528	T. B. R. A. Chen, J. J. Burger and E. R. de Waard; *Tetrahedron Lett.*, 1977, **18**, 4527.	318
78ACR107	P. J. Stang; *Acc. Chem. Res.*, 1978, **11**, 107.	1143
78ACR204	S. Sarel; *Acc. Chem. Res.*, 1978, **11**, 204.	803
78ACR446	J. A. Berson; *Acc. Chem. Res.*, 1978, **11**, 446.	358, 359
78ACR453	B. M. Trost; *Acc. Chem. Res.*, 1978, **11**, 453.	596, 597, 778, 779
78AG(E)27	R. C. Larock; *Angew. Chem., Int. Ed. Engl.*, 1978, **17**, 27.	128, 1095
78AG(E)47	H. Klünenberg and H. J. Schäfer; *Angew. Chem., Int. Ed. Engl.*, 1978, **17**, 47.	934
78AG(E)106	L. A. Paquette; *Angew. Chem., Int. Ed. Engl.*, 1978, **17**, 106.	865, 917, 918
78AG(E)150	G. Kaupp; *Angew. Chem., Int. Ed. Engl.*, 1978, **17**, 150.	418, 420
78AG(E)274	J. Gosselck and I. Stahl; *Angew. Chem., Int. Ed. Engl.*, 1978, **17**, 274.	241
78AG(E)313	E. Grovenstein, Jr.; *Angew. Chem., Int. Ed. Engl.*, 1978, **17**, 313.	886
78AG(E)476	W. Oppolzer and V. Snieckus; *Angew. Chem., Int. Ed. Engl.*, 1978, **17**, 476.	524
78AG(E)526	M. Sevrin, J. N. Denis and A. Krief; *Angew. Chem., Int. Ed. Engl.*, 1978, **17**, 526.	713, 1089
78AG(E)594	B. Giese and H.-D. Beckhaus; *Angew. Chem., Int. Ed. Engl.*, 1978, **17**, 594.	923
78AG(E)609	C. Wentrup and H. W. Winter; *Angew. Chem., Int. Ed., Engl.*, 1978, **17**, 609.	1076
78AG(E)690	M. Franck-Neumann, D. Martina and F. Brion; *Angew. Chem., Int. Ed. Engl.*, 1978, **17**, 690.	70
78BCJ174	M. Inoue, T. Sugita and K. Ichikawa; *Bull. Chem. Soc. Jpn.*, 1978, **51**, 174.	212, 475
78BCJ339	Y. Takahashi, K. Yuasa, M. Tokuda, M. Itoh and A. Suzuki; *Bull Chem. Soc. Jpn.*, 1978, 339.	276
78BCJ2745	T. Sato and R. Noyori; *Bull. Chem. Soc. Jpn.*, 1978, **51**, 2745.	2, 8

78CB72	H. D. Beckhaus, G. Hellmann and C. Rüchardt; *Chem. Ber.*, 1978, **111**, 72.	922
78CB1253	L. Eckes and M. Hanack; *Chem. Ber.*, 1978, **111**, 1253.	1110
78CB1337	E. J. Corey and D. Enders; *Chem. Ber.*, 1978, **111**, 1337.	882
78CB2480	G. Boche, H. Weber, D. Martens and A. Bieberbach; *Chem. Ber.*, 1978, **111**, 2480.	886
78CB2785	H. Neumann and D. Seebach; *Chem. Ber.*, 1978, **111**, 2785.	1090
78CC30	E. C. Ashby and T. Smith; *J. Chem. Soc., Chem. Commun.*, 1978, 30.	515
78CC41	D. L. J. Clive, G. Chittattu and C. K. Wong; *J. Chem. Soc., Chem. Commun.*, 1978, 41.	942
78CC176	I. Fleming and J. Goldhill; *J. Chem. Soc., Chem. Commun.*, 1978, 176.	653
78CC177	D. J. Ager and I. Fleming; *J. Chem. Soc., Chem. Commun.*, 1978, 177.	653
78CC178	I. Fleming and A. Percival; *J. Chem. Soc., Chem. Commun.*, 1978, 178.	653
78CC358	L. A. Carpino, J.-H. Tsao, H. Ringsdorf, E. Fell and G. Hettrich; *J. Chem. Soc., Chem. Commun.*, 1978, 358.	654
78CC414	S. Kano, T. Yokomatsu, T. Ono, S. Hibino and S. Shibuya; *J. Chem. Soc., Chem. Commun.*, 1978, 414.	41
78CC479	S. R. Wilson and D. T. Mao; *J. Chem Soc., Chem. Commun.*, 1978, 479.	805
78CC518	P. H. Bentley and E. Hunt; *J. Chem. Soc., Chem. Commun.*, 1978, 518.	604
78CC596	R. Gräfing, H. D. Verkruijsse and L. Brandsma; *J. Chem. Soc., Chem. Commun.*, 1978, 596.	141
78CC621	A. Cleasson and L.-I. Olsson; *J. Chem. Soc., Chem. Commun.*, 1978, 621.	513
78CC651	T. Onishi, Y. Fujita and T. Nishida; *J. Chem. Soc., Chem. Commun*, 1978, 651.	808
78CC702	H. Yatagai, Y. Yamamoto and K. Maruyama; *J. Chem. Soc., Chem. Commun.*, 1978, 702.	450, 510
78CC726	G. W. Kabalka, R. J. Newton, Jr. and J. H. Chandler; *J. Chem. Soc., Chem. Commun.*, 1978, 726.	962
78CC754	T. Livinghouse and R. V. Stevens; *J. Chem. Soc., Chem. Commun.*, 1978, 754.	1134
78CC770	D. L. J. Clive, G. Chittattu, N. J. Curtis and S. M. Menchen; *J. Chem. Soc., Chem. Commun.*, 1978, 770.	609
78CC876	R. S. Brinkmeyer and T. L. Macdonald; *J. Chem. Soc., Chem. Commun.*, 1978, 876.	197
78CC972	Y. Fujita, T. Onishi and T. Nishida; *J. Chem. Soc., Chem. Commun.*, 1978, 972.	413
78CC1085	P. Barsantai, V. Calo, L. Lopez, G. Marchese, F. Naso and G. Pesce; *J. Chem Soc., Chem. Commun.*, 1978, 1085.	187
78CC1089	K. Yamada, N. Itoh and T. Iwakuma; *J. Chem. Soc., Chem. Commun.*, 1978, 1089.	33
78CI(L)234	A. C. Ranade and S. Jayalakshmi; *Chem. Ind.* (*London*), 1978, 234.	664
78CJC308	G. A. MacAlpine and J. Warkentin; *Can. J. Chem.*, 1978, **56**, 308.	1082
78CJC419	G. Dionne and C. R. Engel; *Can. J. Chem.*, 1978, **56**, 419.	470
78CJC512	T. Durst, J. C. Huang, N. K. Sharma and D. J. H. Smith; *Can. J. Chem.*, 1978, **56**, 512.	232
78CJC1933	F. Y. Khalil and G. Aksnes; *Can. J. Chem.*, 1978, **56**, 1933.	51
78CL69	T. Shono, Y. Matsumura and S. Kashimura; *Chem. Lett.*, 1978, 69.	615, 698
78CL413	T. Mukaiyama and M. Imaoka; *Chem. Lett.*, 1978, 413.	590, 591
78CL541	M. Isobe, H. Iio, M. Kitamura and T. Goto; *Chem. Lett.*, 1978, 541.	779, 788
78CL649	K. Yamamoto, S. Suzuki and J. Tsuji; *Chem. Lett.*, 1978, 649.	521
78CL773	I. Matsuda; *Chem. Lett.*, 1978, 773.	206, 1094
78CL803	G. Gapski, A. Kini and R. S. H. Liu; *Chem. Lett.*, 1978, 803.	690
78CL1007	H. Ishihara and Y. Yoshio; *Chem. Lett.*, 1978, 1007.	246
78CPB3843	H. Yamanaka, K. Edo and T. Sakamoto; *Chem. Pharm. Bull.*, 1978, **26**, 3843.	1010
78CR(C)265	G. Courtois, A. Masson and L. Miginiac; *C. R. Hebd. Seances Acad. Sci., Ser. C*, 1978, **286**, 265.	544
78CRV317	P. P. Fu and R. G. Harvey; *Chem. Rev.*, 1978, **78**, 317.	554
78CRV363	B. M. Trost; *Chem. Rev.*, 1978, **78**, 363.	436, 596, 598, 599, 778
78CRV383	P. J. Stang; *Chem. Rev.*, 1978, **78**, 383.	1118, 1143
78CSR15	E. W. Colvin; *Chem. Soc. Rev.*, 1978, **7**, 15.	648
78G543	I. G. Csizmadia, V. Lucchini and G. Modena; *Gazz. Chim. Ital.*, 1978, **108**, 543.	1109
78H(9)271	H. Yamanaka, K. Edo and T. Sakamoto; *Heterocycles*, 1978, **9**, 271.	1009, 1010
78H(9)1397	Y. Abe, A. Ohsawa, H. Arai and H. Igeta; *Heterocycles*, 1978, **9**, 1397.	1010
78HCA449	F. Kienzle; *Helv. Chim. Acta*, 1978, **61**, 449.	495
78HCA512	R. Schlecker, D. Seebach and W. Lubosch; *Helv. Chim. Acta*, 1978, **61**, 512.	137
78HCA690	J. P. Kutney, T. Honda, A. V. Joshua, N. G. Lewis and B. R. Worth; *Helv. Chim. Acta*, 1978, **61**, 690.	941
78HCA998	P. Dubs and R. Stüssi; *Helv. Chim. Acta*, 1978, **61**, 998.	595
78HCA3096	R. Malherbe and D. Bellus; *Helv. Chim. Acta*, 1978, **61**, 3096.	833
78IJC(B)623	R. S. Tewari and K. C. Gupta; *Indian J. Chem., Sect. B.*, 1978, **16**, 623.	738
78JA292	P. A. Wender and M. A. Eissenstat; *J. Am. Chem. Soc.*, 1978, **100**, 292.	889
78JA341	G. A. Doldouras and J. Kollonitsch; *J. Am. Chem. Soc.*, 1978, **100**, 341.	35
78JA346	P. C. Conrad and P. L. Fuchs; *J. Am. Chem. Soc.*, 1978, **100**, 346.	144
78JA352	G. W. Holbert and B. Ganem; *J. Am. Chem. Soc.*, 1978, **100**, 352.	789
78JA657	R. R. Fraser and P. J. Champagne; *J. Am. Chem. Soc.*, 1978, **100**, 657.	879
78JA877	E. N. Marvell and C. Lin; *J. Am. Chem. Soc.*, 1978, **100**, 877.	808
78JA993	K. E. Harding, J. L. Copper and P. M. Puckett; *J. Am. Chem. Soc.*, 1978, **100**, 993.	1126
78JA1378	C. E. Dykstra and H. F. Schaefer, III; *J. Am. Chem. Soc.*, 1978, **100**, 1378.	1143
78JA1481	W. C. Still; *J. Am. Chem. Soc.*, 1978, **100**, 1481.	138
78JA1548	D. J. Hart, P. A. Cain and D. A. Evans; *J. Am. Chem. Soc*, 1978, **100**, 1548.	651
78JA1558	S. Lavielle, S. Bory, B. Moreau, M. J. Luche and A. Marquet; *J. Am. Chem. Soc.*, 1978, **100**, 1558.	143
78JA1605	J. M. Domagala and R. D. Bach; *J. Am. Chem. Soc.*, 1978. **100**, 1605.	387
78JA1765	H. Takaya, S. Makino, Y. Hayakawa and R. Noyori; *J. Am. Chem. Soc.*, 1978, **100**, 1765.	2

78JA1786 Y. Hayakawa, Y. Baba, S. Makino and R. Noyori; *J. Am. Chem. Soc.*, 1978, **100**, 1786. 2, 3, 8

78JA1791 Y. Hayakawa, K. Yokoyama and R. Noyori; *J. Am. Chem. Soc.*, 1978, **100**, 1791. 363, 364

78JA1927 W. C. Still and A. Mitra; *J. Am. Chem. Soc.*, 1978, **100**, 1927. 405, 406, 816, 817, 820

78JA1928 J. H. J. Dawson and N. M. M. Nibbering; *J. Am. Chem. Soc.*, 1978, **100**, 1928. 1143

78JA2134 R. A. Benkeser, M. P. Siklosi and E. C. Mozdzen; *J. Am. Chem. Soc.*, 1978, **100**, 2134. 401, 800, 883

78JA2140 A. G. Schultz, W. Y. Fu, R. D. Lucci, B. G. Kurr, K. M. Lo and M. Boxer; *J. Am. Chem. Soc.*, 1978, **100**, 2140. 693

78JA2244 R. T. Hansen, D. B. Carr and J. Schwartz; *J. Am. Chem. Soc.*, 1978, **100**, 2244. 1049, 1139

78JA2252 D. E. Van Horn and E. Negishi; *J. Am. Chem. Soc.*, 1978, **100**, 2252. 468, 543

78JA2254 E. Negishi, N. Okukado, A. O. King, D. E. Van Horn and B. I. Spiegel; *J. Am. Chem. Soc.*, 1978, **100**, 2254. 468, 471, 533, 1020, 1097, 1104

78JA2405 L. S. Stuhl, M. Rakowski DuBois, F. J. Hirsekorn, J. R. Bleeke, A. E. Stevens and E. L. Muetterties; *J. Am. Chem. Soc.*, 1978, **100**, 2405. 93

78JA2501 J. B. Lambert and H. W. Mark; *J. Am. Chem. Soc.*, 1978, **100**, 2501. 905

78JA2802 W.-B. Chiao and W. H. Saunders, Jr.; *J. Am. Chem. Soc.*, 1978, **100**, 2802. 779, 783

78JA3514 V. M. DePalma and E. M. Arnett; *J. Am. Chem. Soc.*, 1978, **100**, 3514. 859

78JA3611 F. N. Tebbe, G. W. Parshall and G. S. Reddy; *J. Am. Chem. Soc.*, 1978, **100**, 3611. 669, 766, 767, 769

78JA3847 D. G. Farnum, R. E. Botto, W. T. Chambers and B. Lam; *J. Am. Chem. Soc.*, 1978, **100**, 3847. 906

78JA3874 T. E. Boothe, J. L. Greene, Jr., P. B. Shevlin, M. R. Wilcott, III, R. R. Inners and A. Cornelis; *J. Am. Chem. Soc.*, 1978, **100**, 3874. 600

78JA3927 H. Kwart, T. J. George, R. Louw and W. Ultee; *J. Am. Chem. Soc.*, 1978, **100**, 3927. 624, 779, 783

78JA4301 K. Krogh-Jespersen, P. von R. Schleyer, J. A. Pople and D. Cremer; *J. Am. Chem. Soc.*, 1978, **100**, 4301. 909

78JA4618 E. J. Corey, E. J. Trybulski, L. S. Melvin, Jr., K. C. Nicolaou, J. A. Secrist, R. Lett, P. W. Sheldrake, J. R. Falck, D. J. Brunelle, M. F. Haslanger, S. Kim and S.-e. Yoo; *J. Am. Chem. Soc.*, 1978, **100**, 4618. 1134

78JA4620 E. J. Corey, S. Kim, S.-e. Yoo, K. C. Nicolaou, L. S. Melvin, Jr., D. J. Brunelle, J. R. Falck, E. J. Trybulski, R. Lett and P. W. Sheldrake; *J. Am. Chem. Soc.*, 1978, **100**, 4620. 1134

78JA4624 B. B. Snider, M. Karras and R. S. E. Conn; *J. Am. Chem. Soc.*, 1978, **100**, 4624. 1105

78JA4822 L. E. Overman, C. B. Campbell and F. M. Knoll; *J. Am. Chem. Soc.*, 1978, **100**, 4822. 836, 838

78JA4852 P. A. Bartlett, F. R. Green, III and E. H. Rose; *J. Am. Chem. Soc.*, 1978, **100**, 4852. 1073

78JA4888 H. J. Reich and S. L. Peake; *J. Am. Chem. Soc.*, 1978, **100**, 4888. 108

78JA4903 B. Kraeutler, C. D. Jaeger and A. J. Bard; *J. Am. Chem. Soc.*, 1978, **100**, 4903. 934

78JA5134 R. P. Kirchen, T. S. Sorensen and K. E. Wagstaff; *J. Am. Chem. Soc.*, 1978, **100**, 5134. 895

78JA5297 H. Lischka and H.-J. Köhler; *J. Am. Chem. Soc.*, 1978, **100**, 5297. 1109

78JA5561 G. Zweifel, S. J. Backlund and T. Leung; *J. Am. Chem. Soc.*, 1978, **100**, 5561. 977, 1056

78JA6294 E. J. Corey and H. Estreicher; *J. Am. Chem. Soc.*, 1978, **100**, 6294. 671

78JA6299 G. A. Olah, J. S. Staral, G. Asencio, G. Liang, D. A. Forsyth and G. D. Mateescu; *J. Am. Chem. Soc.*, 1978, **100**, 6299. 914

78JA6483 B. Ganem, G. W. Holbert, L. B. Weiss and K. Ishizumi; *J. Am. Chem. Soc.*, 1978, **100**, 6483. 573

78JA6930 J. W. Kenny, III, J. Simons, G. D. Purvis and R. J. Bartlett; *J. Am. Chem. Soc.*, 1978, **100**, 6930. 1143

78JA7066 M. Hanack and L. Sproesser; *J. Am. Chem. Soc.*, 1978, **100**, 7066. 1113

78JA7312 G. L. Keldsen and W. E. McEwen; *J. Am. Chem. Soc.*, 1957, **100**, 3567. 51

78JA7425 M. Y. Darensbourg and E. L. Muetterties; *J. Am. Chem. Soc.*, 1978, **100**, 7425. 93

78JA7927 R. V. Hoffman, G. G. Orphanides and H. Shechter; *J. Am. Chem. Soc.*, 1978, **100**, 7927. 1077

78JA8170 D. A. Evans, A. M. Golob, N. S. Mandel and G. S. Mandel; *J. Am. Chem. Soc.*, 1978, **100**, 8170. 815

78JCR(M)4401 I. U. Khand, P. L. Pauson and M. J. A. Habib; *J. Chem. Res. (M)*, 1978, 4401. 529

78JCR(M)4454 I. U. Khand, C. A. L. Mahaffy and P. L. Pauson; *J. Chem. Res. (M)*, 1978, 4454. 529

78JCR(S)62 H. Christol, J. Coste, F. Pietrasanta, F. Plénat and G. Renard; *J. Chem. Res. (S)*, 1978, 62. 905

78JCR(S)416 S. Arseniyadis, J. Goré, A. Laurent and M.-L. Roumestant; *J. Chem. Res. (S)*, 1978, 416. 495

78JCS(D)617 D. S. Gill, C. White and P. M. Maitlis; *J. Chem. Soc., Dalton Trans.*, 1978, 617. 93

78JCS(D)657 B. G. Gowenlock, W. E. Lindswell and B. Singh; *J. Chem. Soc., Dalton Trans.*, 1978, 657. 452

78JCS(P1)304 O. G. Plantema, H. de Koning and H. O. Huisman; *J. Chem. Soc., Perkin Trans. 1*, 1978, 304. 747

78JCS(P1)590 B. Lythgoe, T. A. Moran, M. E. N. Nambudiry, J. Tideswell and P. W. Wright; *J. Chem. Soc., Perkin Trans. 1*, 1978, 590. 754

78JCS(P1)829 P. J. Kocienski, B. Lythgoe and S. Ruston; *J. Chem. Soc., Perkin Trans. 1*, 1978, 829. 615, 654, 696, 697, 698, 700

78JCS(P1)834 P. J. Kocienski, B. Lythgoe and D. A. Roberts; *J. Chem. Soc., Perkin Trans. 1*, 1978, 834. 700

78JCS(P1)1133 I. Degani and R. Fochi; *J. Chem. Soc., Perkin Trans. 1*, 1978, 1133. 48

78JCS(P1)1479 S. G. Davies and G. H. Whitham; *J. Chem. Soc., Perkin Trans. 1*, 1978, 1479. 97

78JCS(P2)1313 P. M. Blum and B. P. Roberts; *J. Chem. Soc., Perkin Trans. 2*, 1978, 1313. 945

78JOC131 R. C. Larock, B. Riefling and C. A. Fellows; *J. Org. Chem.*, 1978, **43**, 131. 1108

78JOC147 A. R. Chamberlin, J. E. Stemke and F. T. Bond; *J. Org. Chem.*, 1978, **43**, 147. 630, 632, 633, 1091, 1094

78JOC353 T. F. Murray, V. Varma and J. R. Norton; *J. Org. Chem.*, 1978, **43**, 353. 1023, 1024

78JOC358 E. Negishi, A. O. King, F. J. Villani, Jr. and A. Silveira, Jr.; *J. Org. Chem.*, 1978, **43**, 358. 1006

78JOC518 M. W. Lee and W. C. Herndon; *J. Org. Chem.*, 1978, **43**, 518. 785

78JOC555 R. A. Amos and J. A. Katzenellenbogen; *J. Org. Chem.*, 1978, **43**, 555. 968

78JOC700 H. O. House, W. V. Phillips, T. S. B. Sayer and C.-C. Yau; *J. Org. Chem.*, 1978, **43**, 700. 130

78JOC757 E. C. Ashby, J. J. Lin and A. B. Goel; *J. Org. Chem.*, 1978, **43**, 757. 458, 515

78JOC763 D. Mirejovsky, W. Drenth and F. B. van Duijneveldt; *J. Org. Chem.*, 1978, **43**, 763. 1114

78JOC788 G. R. Kieczykowski, M. R. Roberts and R. H. Schlessinger; *J. Org. Chem.*, 1978, **43**, 788. 209

78JOC815 E. M. Arnett, L. E. Small, R. T. McIver, Jr. and J. S. Miller; *J. Org. Chem.*, 1978, **43**, 815. 883

78JOC1007 L. Dickson, C. A. Matuszak and A. M. Qasi; *J. Org. Chem.*, 1978, **43**, 1007. 97

78JOC1011 Y. Ito, T. Hirao and T. Saegusa; *J. Org. Chem.*, 1978, **43**, 1011. 558

78JOC1050 R. W. Thies and E. P. Seitz; *J. Org. Chem.*, 1978, **43**, 1050. 401

78JOC1185 E. Vedejs, J. P. Hagen, B. L. Roach and K. L. Spear; *J. Org. Chem.*, 1978, **43**, 1185. 141, 234, 821

78JOC1208 P. B. Hopkins and P. L. Fuchs; *J. Org. Chem.*, 1978, **43**, 1208. 572

78JOC1263 E. C. Ashby and J. J. Lin; *J. Org. Chem.*, 1978, **43**, 1263. 4, 7, 8, 10, 17

78JOC1279 N. J. Lalima, Jr. and A. B. Levy; *J. Org. Chem.*, 1978, **43**, 1279. 278

78JOC1382 D. J. Pasto, R. H. Shults, J. A. McGrath and A. Waterhouse; *J. Org. Chem.*, 1978, **43**, 1382. 970

78JOC1385 D. J. Pasto, S.-K. Chou, A. Waterhouse, R. H. Shults and G. F. Hennion; *J. Org. Chem.*, 1978, **43**, 1385. 970

78JOC1389 D. J. Pasto, S. K. Chou, E. Fritzen, R. H. Shults, A. Waterhouse and G. F. Hennion; *J. Org. Chem.*, 1978, **43**, 1389. 966

78JOC1404 K. J. Kolonko and R. H. Shapiro; *J. Org. Chem.*, 1978, **43**, 1404. 630, 633

78JOC1409 M. F. Lipton and R. H. Shapiro; *J. Org. Chem.*, 1978, **43**, 1409. 633

78JOC1421 W. J. Bailey and F. Cesare; *J. Org. Chem.*, 1978, **43**, 1421. 772

78JOC1468 R. C. Larock and B. Riefling; *J. Org. Chem.*, 1978, **43**, 1468. 492

78JOC1493 H. E. Zimmerman and S. M. Aasen; *J. Org. Chem.*, 1978, **43**, 1493. 802

78JOC1521 G. Modena, F. Rivetti and U. Tonellato; *J. Org. Chem.*, 1978, **43**, 1521. 318

78JOC1526 T. H. Chan, W. Mychajlowskij, B. S. Ong and D. N. Harpp; *J. Org. Chem.*, 1978, **43**, 1526. 651, 760, 981

78JOC1612 S. Nishida and F. Kataoka; *J. Org. Chem.*, 1982, **47**, 5229. 690

78JOC1689 T. Hori and K. B. Sharpless; *J. Org. Chem.*, 1978, **43**, 1689. 601, 608, 781

78JOC1697 H. J. Reich, S. Wollowitz, J. E. Trend, F. Chow and D. F. Wendelborn; *J. Org. Chem.*, 1987, **43**, 1697. 601, 602, 604

78JOC1947 R. P. Woodbury and M. N. Rathke; *J. Org. Chem.*, 1978, **43**, 1947. 283

78JOC1950 W. H. Pirkle and C. W. Boeder; *J. Org. Chem.*, 1978, **43**, 1950. 966

78JOC1985 C. Y. Meyers and V. M. Kolb; *J. Org. Chem.*, 1978, **43**, 1985. 706

78JOC2091 W. H. Pirkle and C. W. Boeder; *J. Org. Chem.*, 1978, **43**, 2091. 966

78JOC2153 H. O. House, T. S. B. Sayer and C.-C. Yau; *J. Org. Chem.*, 1978, **43**, 2153. 131

78JOC2161 B. B. Snider and T. A. Killinger; *J. Org. Chem.*, 1978, **43**, 2161. 525

78JOC2259 R. O. Hutchins, D. Kandasamy, F. Dux III, C. A. Maryanoff, D. Rotstein, B. Goldsmith, W. Burgoyne, F. Cistone, J. Dalessandro and J. Puglis; *J. Org. Chem.*, 1978, **43**, 2259. 4, 7, 9, 17, 28

78JOC2299 R. O. Hutchins and N. R. Natale; *J. Org. Chem.*, 1978, **43**, 2299. 635

78JOC2396 G. E. Niznik and H. M. Walborsky; *J. Org. Chem.*, 1978, **43**, 2396. 38

78JOC2477 Y. Fujiwara, R. Ishikawa, F. Akiyama and S. Teranishi; *J. Org. Chem.*, 1978, **43**, 2477. 692

78JOC2518 M. Mikoiaczyk, A. Zatorski, S. Grzejszczak, B. Costisella and W. Midura; *J. Org. Chem.*, 1978, **43**, 2518. 821

78JOC2567 E. C. Ashby and J. J. Lin; *J. Org. Chem.*, 1978, **43**, 2567. 75, 457, 514

78JOC2676 G. F. Koser, S.-M. (Yu) Linden and Y.-J. Shih; *J. Org. Chem.*, 1978, **43**, 2676. 11

78JOC2705 J. Quick, C. Mondello, M. Humora and T. Brennan; *J. Org. Chem.*, 1978, **43**, 2705. 876

78JOC2739 G. Zweifel and W. Lewis; *J. Org. Chem.*, 1978, **43**, 2739. 511

78JOC2941 C. B. Ziegler, Jr. and R. F. Heck; *J. Org. Chem.*, 1978, **43**, 2941. 494, 1096

78JOC2994 T. H. Chan and B. S. Ong; *J. Org. Chem.*, 1978, **43**, 2994. 318

78JOC3083 C. A. Brown; *J. Org. Chem.*, 1978, **43**, 3083. 590, 1072, 1085

78JOC3095 F. G. Bordwell, J. E. Bartmess and J. A. Hautala; *J. Org. Chem.*, 1978, **43**, 3095. 155

78JOC3101 F. G. Bordwell and J. E. Bartmess; *J. Org. Chem.*, 1978, **43**, 3101. 142, 165

78JOC3249 J. E. McMurry, M. G. Silvestri, M. P. Fleming, T. Hoz and M. W. Grayston; *J. Org. Chem.*, 1978, **43**, 3249. 173, 587, 688

78JOC3255 J. E. McMurry, M. P. Fleming, K. L. Kees and L. R. Krepski; *J. Org. Chem.*, 1978, **43**, 3255. 688, 689, 690

78JOC3418 E. J. Corey, D. Floyd and B. H. Lipshutz; *J. Org. Chem.*, 1978, **43**, 3418. 120, 1107

78JOC3450 S.-M. L. Chen, R. E. Schaub and C. V. Grudzinskas; *J. Org. Chem.*, 1978, **43**, 3450. 535

78JOC3609 A. L. Baumstark, C. J. McCloskey and K. E. Witt; *J. Org. Chem.*, 1978, **43**, 3609. 688, 690

78JOC3630 E. Piers, I. Nagakura and H. E. Morton; *J. Org. Chem.*, 1978, **43**, 3630. 811

78JOC3667 H. C. Brown, M. Ravindranathan, C. G. Rao, F. J. Chloupek and M.-H Rei; *J. Org. Chem.*, 1978, **43**, 3667. 905

78JOC3800 C. K. Bradsher and D. C. Reames; *J. Org. Chem.*, 1978, **43** 3800. 213, 478

78JOC3898 B. A. Patel and R. F. Heck; *J. Org. Chem.*, 1978, **43**, 3898. 494

78JOC3985 N. A. Cortese and R. F. Heck; *J. Org. Chem.*, 1978, **43**, 3985. 79, 85

78JOC4140 C. R. Johnson and H. G. Corkins; *J. Org. Chem.*, 1978, **43**, 4140. 705

78JOC4207 M. A. Calcagno and E. E. Schweizer; *J. Org. Chem.*, 1978, **43**, 4207. 267

78JOC4235 M. J. Taschner and G. A. Kraus; *J. Org. Chem.*, 1978, **43**, 4235. 664

78JOC4248 R. F. Abdulla and K. H. Fuhr; *J. Org. Chem.*, 1978, **43**, 4248. 486

78JOC4283 P. W. Rabideau and E. G. Burkholder; *J. Org. Chem.*, 1978, **43**, 4283. 95

78JOC4366 S. Kano, T. Yokomatsu and S. Shibuya; *J. Org. Chem.*, 1978, **43**, 4366. 1077

78JOC4527 E. Chamot and L. A. Paquette; *J. Org. Chem.*, 1978, **43**, 4527. 704, 708

78JOC4555 J. D. White and G. L. Larson; *J. Org. Chem.*, 1978, **43**, 4555. 457, 508

78JOC4574	W. Adam and H. H. Fick; *J. Org. Chem.*, 1978, **43**, 4574.	681
78JOC4684	A. B. Levy; *J. Org. Chem.*, 1978, **43**, 4684.	278
78JOC4750	E. C. Ashby and G. F. Willard; *J. Org. Chem.*, 1978, **43**, 4750.	636
78JOC4797	S. C. Welch and M. E. Walters; *J. Org. Chem.*, 1978, **43**, 4797.	432
78JOC4826	V. Ceré, C. Paolucci, S. Pollicino, E. Sandri and A. Fava; *J. Org. Chem.*, 1978, **43**, 4831.	821
78JOC4831	E. Vedejs, M. J. Arco, D. W. Powell, J. M. Renga and S. P. Singer; *J. Org. Chem.*, 1978, **43**, 4831.	410, 821
78JOC4884	E. Vedejs and S. P. Singer; *J. Org. Chem.*, 1978, **43**, 4884.	619
78JOC5018	B. A. Patel, J. E. Dickerson and R. F. Heck; *J. Org. Chem.*, 1978, **43**, 5018.	494
78JOM(156)45	R. C. Larock, J. C. Bernhardt and R. J. Driggs; *J. Organomet. Chem.*, 1978, **156**, 45.	128
78JOM(156)101	P. Jacob; *J. Organometal. Chem.*, 1978, **156**, 101.	946
78JOM(156)159	G. Zweifel and S. J. Backlund; *J. Organomet. Chem.*, 1978, **156**, 159.	488, 1103
78JOM(156)203	M. M. Midland, A. Tramontano and S. A. Zderic; *J. Organomet. Chem.*, 1978, **156**, 203.	657
78JOU1095	V. R. Denisov, S. N. Lysenko, Z. A. Shevchenko and I. A. Favorskaya; *J. Org. Chem. USSR (Engl. Transl.)*, 1978, 1095.	1022
78LA658	K. Eiter, F. Lieb, H. Disselnkötter and H. Oediger; *Justus Liebigs Ann. Chem.*, 1978, 658.	1002
78LA1894	M. Hanack, H. Bentz, R. Märkl and L. R. Subramanian; *Justus Liebigs Ann. Chem.*, 1978, 1894.	1111, 1113
78M609	P. K. Claus, W. Reider and F. W. Vierhapper; *Monatsh. Chem.*, 1978, **109**, 609.	242
B-78MI 103-01	M. Freifelder; "Catalytic Hydrogenation in Organic Synthesis, Procedures and Commentary," Wiley-Interscience, New York, 1978.	71, 87
B-78MI 106-01	S. Patai (ed.); "The Chemistry of the Diazo and Diazonium Groups," Wiley-Interscience, New York, 1978, chapter 14.	258
B-78MI 110-01	F. Hegarty; in "Chemistry of the Diazo and Diazonium Groups, Part 2," ed. S. Patai, Wiley-Interscience, New York, 1978, p. 511.	442
B-78MI 112-01	M. Freifelder; "Catalytic Hydrogenation in Organic Synthesis: Procedures and Commentary," Wiley-Interscience, New York, 1978, pp. 10–14.	502
B-78MI 112-02	P. F. Hudrlik and A. M. Hudrlik; in "The Chemistry of the Carbon–Carbon Triple Bond," ed. S. Patai, Wiley, New York, 1978, part 1, p. 199.	509, 511, 531
B-78MI 112-03	J. Bastide and O. Henri-Rousseau; in "The Chemistry of the Carbon–Carbon Triple Bond," ed. S. Patai, Wiley, New York, 1978, part 1, p. 447.	516, 549
B-78MI 112-04	J. D. Coyle; in "The Chemistry of the Carbon–Carbon Triple Bond," ed. S. Patai, Wiley, New York, 1978, part 2, p. 523.	516
B-78MI 112-05	J. I. Dickstein and S. I. Miller; in "The Chemistry of the Carbon–Carbon Triple Bond," ed. S. Patai, Wiley, New York, 1978, part 2, p. 813.	536
B-78MI 114-01	E. Block; "Reactions of Organosulfur Compounds," Academic Press, New York, 1978, p. 77.	617
78MI 114-01	S. Gambino, P. Martigny, G. Mousset and J. Simonet; *J. Electroanal. Chem.*, 1978, **90**, 105 (*Chem. Abstr.*, 1978, **89**, 107 250).	615
B-78MI 114-02	E. Block; "Reactions of Organosulfur Compounds," Academic Press, New York, 1978, pp. 93–94.	623
B-78MI 114-03	E. Block; "Reactions of Organosulfur Compounds," Academic Press, New York, 1978, p. 112.	623
B-78MI 114-04	J. K. Kochi; "Organometallic Mechanisms and Catalysis," Academic Press, New York, 1978, pp. 247–258.	668
B-78MI 118-01	W. Ando; in "The Chemistry of Diazonium and Diazo groups," ed. S. Patai, Wiley-Interscience, New York, 1978, chapter 9.	821
78MI 119-01	W. J. le Noble; *React. Intermed.* (*Wiley*), 1978, **1**, 27.	891
78MI 119-02	D. Bethell; *React. Intermed.* (*Wiley*), 1978, **1**, 117–161.	891
78MI 119-03	A. Holy; *Nucl. Acid Chem.*, 1978, **2**, 1007.	936
B-78MI 119-04	O. A. Reutov, I. P. Beletskaya and K. P. Butin; "CH-Acids," Pergamon, Oxford, 1978.	844
B-78MI 119-05	D. I. Davies and M. J. Parrott; "Free Radicals in Organic Synthesis," Springer-Verlag, Berlin, 1978.	848, 919
B-78MI 121-01	J. Klein; in "Chemistry of the Carbon–Carbon Triple Bond," Part 1, ed. S. Patai, Wiley, New York, 1978, p. 343.	1054
B-78MI 121-02	J. I. Dickstein and S. I. Miller; in "The Chemistry of the Carbon–Carbon Triple Bond," Part 2, ed. S. Patai, Wiley, New York, 1978, p. 813.	1016
B-78MI 121-03	D. A. Ben-Efraim; in "The Chemistry of the Carbon–Carbon Triple Bond," ed. S. Patai, Wiley, New York, 1978, p. 786.	998, 1068, 1069
B-78MI 122-01	M. Jones, Jr. and R. A. Moss (eds.); "Reactive Intermediates," Vol. 1, Wiley-Interscience, New York, 1978.	1102
B-78MI 122-02	G. H. Schmid; in "The Chemistry of the Carbon–Carbon Triple Bond," Part 1, ed., S. Patai, Wiley, New York, 1978.	318
78MI 122-03	K. Okuno, K. Yoshihara and T. Shiokawa; *Radiochem. Acta*, 1978, **25**, 21.	1142, 1143
78OS(58)127	M. Kumada, K. Tamao and K. Sumitani; *Org. Synth.*, 1978, **58**, 127.	466, 473
78PAC709	J. F. Normant; *Pure Appl. Chem.*, 1978, **50**, 709.	116, 453, 539
78RCR425	A. A. Solov'yanov and I. P. Beletskaya; *Russ. Chem. Rev.* (*Engl. Transl.*), 1978, **47**, 425.	318
78S26	M. Hanack, E. Kunzmann and W. Schumacher; *Synthesis*, 1978, 26.	204, 1023
78S128	R. Sauvêtre, D. Masure, C. Chuit and J. F. Normant; *Synthesis*, 1978, 128.	429
78S221	S. Murata and I. Matsuda; *Synthesis*, 1978, 221.	209
78S305	S. Kano, T. Yokomatsu, T. Ono, S. Hibino and S. Shibuya; *Synthesis*, 1978, 305.	1077
78S307	H. Hauptmann and M. Mader; *Synthesis*, 1978, 307.	1045

78S528	G. Cahiez, A. Alexakis and J. F. Normant; *Synthesis*, 1978, 528.	479
78S536	H. Bühl, H. Gugel, H. Kolshorn and H. Meier; *Synthesis*, 1978, 536.	1077
78S579	J. D. Findlay, D. J. H. Smith and T. Durst; *Synthesis*, 1978, 579.	232
78S616	R. S. Tewari and S. C. Chaturvedi; *Synthesis*, 1978, 616.	738
78S713	L. Field; *Synthesis*, 1978, 713.	778, 779
78S746	S. Kano, T. Ebata, K. Funaki and S. Shibuya; *Synthesis*, 1978, 746.	651
78S755	D. Habich and F. Effenberger; *Synthesis*, 1978, 755.	448
78S763	G. W. Gribble, W. J. Kelly and S. E. Emery; *Synthesis*, 1978, 763.	19
78S5397	G. A. Olah and G. K. S. Prakash; *Synthesis*, 1978, 397.	80
78SC285	T. Otsubo and S. Misumi; *Synth. Commun.*, 1978, **8**, 285.	232
78T19	J. L. Ripoll, A. Rouessac and F. Rouessac; *Tetrahedron*, 1978, **34**, 19.	786
78T317	C. Fleischmann and E. Zbiral; *Tetrahedron*, 1978, **34**, 317.	821
78T997	E. Breuer and D. M. Bannet; *Tetrahedron*, 1978, **34**, 997.	751
78T1049	D. L. J. Clive; *Tetrahedron*, 1978, **34**, 1049.	592, 601, 712, 713, 716, 780, 781
78T1261	A. A. Allen and F. Kurzer; *Tetrahedron*, 1978, **34**, 1261.	78
78T1877	B. K. Carpenter; *Tetrahedron*, 1978, **34**, 1877.	785
78T2057	J. R. Beck; *Tetrahedron*, 1978, **34**, 2057.	486
78T2783	K. J. Crowley and S. G. Traynor; *Tetrahedron*, 1978, **34**, 2783.	420
78T2827	O. A. Reutov; *Tetrahedron*, 1978, **34**, 2827.	65, 454
78TL187	M. Sevrin and A. Krief; *Tetrahedron Lett.*, 1978, **19**, 187.	623
78TL191	H. P. Dang and G. Linstrumelle; *Tetrahedron Lett.*, 1978, **19**, 191.	468, 1097, 1135
78TL491	P. Groenewegen, H. Kallenberg and A. van der Gen; *Tetrahedron Lett.*, 1978, **19**, 491.	873
78TL519	E. Vedejs, M. J. Mullins, J. M. Renga and S. P. Singer; *Tetrahedron Lett.*, 1978, **19**, 519.	234, 821
78TL523	E. Vedejs, M. J. Arco and J. M. Renga; *Tetrahedron Lett.*, 1978, **19**, 523.	410, 823
78TL619	G. Heinisch, A. Jentzsch and I. Kirchner; *Tetrahedron Lett.*, 1978, **19**, 619.	934
78TL691	Y. S. P. Cheng, E. Dominguez, P. J. Garratt and S. B. Neoh; *Tetrahedron Lett.*, 1978, **19**, 691.	955
78TL763	N. Ono, R. Tamura, J. Hayami and A. Kaji; *Tetrahedron Lett.*, 1978, **19**, 763.	617
78TL937	J. C. Chabala and J. E. Vincent; *Tetrahedron Lett.*, 1978, **19**, 937.	1045
78TL1027	N. Okukado, D. E. Van Horn, W. L. Klima and E. Negishi; *Tetrahedron Lett.*, 1978, **19**, 1027.	468, 536, 1104
78TL1137	J.-C. Clinet and G. Linstrumelle; *Tetrahedron Lett.*, 1978, **19**, 1137.	971
78TL1141	D. Labar, L. Hevesi, W. Dumont and A. Krief; *Tetrahedron Lett.*, 1978, **19**, 1141.	601, 716
78TL1145	D. Labar, W. Dumont, L. Hevesi and A. Krief; *Tetrahedron Lett.*, 1978, **19**, 1145.	716
78TL1239	N. Petiniot, A. J. Anciaux, A. F. Noels, A. J. Hubert and Ph. Teyssie; *Tetrahedron Lett.*, 1978, **19**, 1239.	549
78TL1363	A. Marfat, P. R. McGuirk and P. M. Helquist; *Tetrahedron Lett.*, 1978, **19**, 1363.	1107
78TL1581	P. Casara and B. W. Metcalf; *Tetrahedron Lett.*, 1978, **19**, 1581.	1138
78TL1803	A. Doutheau, G. Balme, M. Malacria and J. Goré; *Tetrahedron Lett.*, 1978, **19**, 1803.	808
78TL2027	G. Cahiez, A. Alexakis and J. F. Normant; *Tetrahedron Lett.*, 1978, **19**, 2027.	479
78TL2057	M. Van Audenhove, D. Termont, D. de Keukeleire and M. Vandewalle; *Tetrahedron Lett.*, 1978, **19**, 2057.	356
78TL2079	E. Nakamura, K. Hashimoto and I. Kuwajima; *Tetrahedron Lett.*, 1978, **19**, 2079.	876
78TL2161	J.-I. Yoshida, K. Tamao, M. Takahashi and M. Kumada; *Tetrahedron Lett.*, 1978, **19**, 2161.	1106
78TL2301	K. Utimoto, M. Tanaka, M. Kitai and H. Nozaki; *Tetrahedron Lett.*, 1978, **19**, 2301.	1045, 1138
78TL2357	N. Okukado and E. Negishi; *Tetrahedron Lett.*, 1978, **19**, 2357.	534, 542, 543
78TL2383	P. W. K. Lau and T. H. Chan; *Tetrahedron Lett.*, 1978, **19**, 2383.	648, 651, 762
78TL2399	R. D. Acker; *Tetrahedron Lett.*, 1978, **19**, 2399.	204
78TL2417	K. Takai, Y. Hotta, K. Oshima and H. Nozaki; *Tetrahedron Lett.*, 1978, **19**, 2417.	670, 678, 767
78TL2445	G. Stork, D. F. Taber and M. Marx; *Tetrahedron Lett.*, 1978, **19**, 2445.	307
78TL2465	A. Marfat, P. R. McGuirk, P. M. Helquist; *Tetrahedron Lett.*, 1978, **19**, 2465.	479, 1107
78TL2549	W. S. Johnson, T. M. Yarnell, R. F. Myers and D. R. Morton; *Tetrahedron Lett.*, 1978, **19**, 2549.	1122, 1138
78TL2625	B. Lythgoe and I. Waterhouse; *Tetrahedron Lett.*, 1978, **19**, 2625.	698, 1073
78TL2693	J. Lucchetti and A. Krief; *Tetrahedron Lett.*, 1978, **19**, 2693.	592, 594, 716
78TL2807	T. Shono, Y. Matsumura, S. Kashimura and H. Kyutoku; *Tetrahedron Lett.*, 1978, **19**, 2807.	590
78TL2831	G. Giacomelli and L. Lardicci; *Tetrahedron Lett.*, 1978, **19**, 2831.	1019, 1141
78TL2973	P. R. McGuirk, A. Marfat and P. M. Helquist; *Tetrahedron Lett.*, 1978, **19**, 2973.	1107
78TL3043	A. Hosomi, A. Shirahata and H. Sakurai; *Tetrahedron Lett.*, 1978, **19**, 3043.	314
78TL3195	G. A. Kraus and K. Frazier; *Tetrahedron Lett.*, 1978, **19**, 3195.	1005
78TL3307	D. E. Laycock and M. C. Baird; *Tetrahedron Lett.*, 1978, **19**, 3307.	669
78TL3455	I. Ryu, S. Murai, Y. Hatayama and N. Sonada; *Tetrahedron Lett.*, 1978, **19**, 3455.	556
78TL3483	L. Crombie, P. J. Maddocks and G. Pattenden; *Tetrahedron Lett.*, 1978, **19**, 3483.	460
78TL3711	J. F. Normant, T. Mulamba, F. Scott, A. Alexakis and G. Cahiez; *Tetrahedron Lett.*, 1978, **19**, 3711.	1092
78TL3801	P. Place, F. Delbecq and J. Goré; *Tetrahedron Lett.*, 1978, **19**, 3801.	958
78TL4069	G. Linstrumelle, R. Lorne and H. P. Dang; *Tetrahedron Lett.*, 1978, **19**, 4069.	206, 478
78TL4159	J. M. Clough and G. Pattenden; *Tetrahedron Lett.*, 1978, **19**, 4159.	754
78TL4263	R. Gompper and U. Wolf; *Tetrahedron Lett.*, 1978, **19**, 4263.	995
78TL4269	M.-K. Au, T.-W. Siu, T. C. W. Mak and T.-L. Chan; *Tetrahedron Lett.*, 1978, **19**, 4269.	709
78TL4337	R. K. Hill and H. N. Khatri; *Tetrahedron Lett.*, 1978, **19**, 4337.	835
78TL4349	H. D. Hodes and K. M. Nicholas; *Tetrahedron Lett.*, 1978, **19**, 4349.	199
78TL4381	M. J. Loots and J. Schwartz; *Tetrahedron Lett.*, 1978, **19**, 4381.	1105

78TL4399 T. Kauffmann, H. Ahlers, R. Joußen, R. Kriegesmann, A. Vahrenhorst and A. Woltermann; *Tetrahedron Lett.*, 1978, **19**, 4399. 647, 667, 717

78TL4727 P. Müller and N. T. M. Phuong; *Tetrahedron Lett.*, 1978, **19**, 4727. 257

78TL4871 F. A. Neugebauer, D. Hellwinkel and G. Aulmich; *Tetrahedron Lett.*, 1978, **19**, 4871. 922

78TL4903 J. Nokami, K. Ueta and R. Okawara; *Tetrahedron Lett.*, 1978, **19**, 4903. 779

78TL5193 G. R. Chambers and M. Jones, Jr.; *Tetrahedron Lett.*, 1978, **19**, 5193. 1091

79ACR22 H. J. Reich; *Acc. Chem. Res.*, 1979, **12**, 22. 592, 593, 601

79ACR66 S. Danishefsky; *Acc. Chem. Res.*, 1979, **12**, 66. 928

79ACR288 H. F. Schaefer, III; *Acc. Chem. Res.*, 1979, **12**, 288. 1143

79ACR324 E. L. Muetterties and J. R. Bleeke; *Acc. Chem. Res.*, 1979, **12**, 324. 87, 93

79ACR415 A. Haaland; *Acc. Chem. Res.*, 1979, **12**, 415. 862

79AG(E)563 R. W. Hoffmann; *Angew. Chem., Int. Ed. Engl.*, 1979, **18**, 563. 404, 407, 816

79AG(E)587 J. L. Dye; *Angew. Chem., Int. Ed. Engl.*, 1979, **18**, 587. 101

79AG(E)688 M. Franck-Neumann and F. Brion; *Angew. Chem., Int. Ed. Engl.*, 1979, **18**, 688. 954

79AG(E)782 H. Schmidbaur, E. Weiss and B. Zimmer-Gasser; *Angew. Chem., Int. Ed. Engl.*, 1979, **18**, 782. 158

79AG(E)870 M. Hanack and U. Michel; *Angew. Chem. Int. Ed. Engl.*, 1979, **18**, 870. 1117

79AG(E)939 W. F. Maier, P. Grubmuller, I. Thie, P. M. Stein, M. A. McKervey and P. von Rogué Schleyer; *Angew. Chem., Int. Ed. Engl.*, 1979, **18**, 939. 48

79AJC1819 P. A. Collins and D. Wege; *Aust. J. Chem.*, 1979, **32**, 1819. 774

79AJC1921 P. R. Andrews and R. C. Haddon; *Aust. J. Chem.*, 1979, **32**, 1921. 826

79AX(B)1846 R. Destro and M. Simonetta; *Acta Crystallogr., Part B*, 1979, **35**, 1846. 910

79BCJ1107 K. Fukunaga and M. Kimura; *Bull. Chem. Soc. Jpn.*, 1979, **52**, 1107. 638

79BCJ1241 H. Takagaki, N. Yasuda, M. Asaoka and H. Takei; *Bull. Chem. Soc. Jpn.*, 1979, **52**, 1241. 133

79BCJ1865 T. Sotoyama, S. Hara and A. Suzuki; *Bull Chem. Soc. Jpn.*, 1979, **52**, 1865. 278

79BCJ2377 T. Nakajima, S. Masuda, S. Nakashima, T. Kondo, Y. Nakamoto and S. Suga; *Bull. Chem. Soc. Jpn.*, 1979, **52**, 2377. 182

79BCJ2389 I. Matsuda, S. Murata and Y. Izumi; *Bull. Chem. Soc. Jpn.*, 1979, **52**, 2389. 209

79BCJ2942 H. Taniguchi; *Bull. Chem. Soc. Jpn.*, 1979, **52**, 2942. 657, 658

79CB28 H. J. Bestmann, R. Engler, H. Hartung and K. Roth; *Chem. Ber.*, 1979, **112**, 28. 446

79CB234 D. Seebach, R. Henning and J. Gonnermann; *Chem. Ber.*, 1979, **112**, 234. 156

79CB355 G. Saleh, T. Minami, Y. Ohshiro and T. Agawa; *Chem. Ber.*, 1979, **112**, 355. 608

79CB2453 R. Kopp and M. Hanack; *Chem. Ber.*, 1979, **112**, 2453. 1115

79CB3728 H. Meier, H. Bühl and U. Timm; *Chem. Ber.*, 1979, **112**, 3728. 1077

79CC52 J. Mulzer, A. Pointner, A. Chucholowski and G. Bruntrup; *J. Chem. Soc., Chem. Commun.*, 1979, 52. 681, 683

79CC83 K. C. Nicolaou, W. J. Sipio, R. L. Magolda and D. A. Claremon; *J. Chem. Soc., Chem. Commun.*, 1979, 83. 591, 594, 716

79CC100 C. A. Brown and A. Yamaichi; *J. Chem. Soc., Chem. Commun.*, 1979, 100. 308

79CC106 A. P. Kozikowski and R. Schmiesing; *J. Chem. Soc., Chem. Commun.*, 1979, 106. 815

79CC124 R. C. Ronald and J. M. Lansinger; *J. Chem. Soc., Chem. Commun.*, 1979, 124. 1136

79CC300 A. R. Katritzky, K. Horvath and B. Plau; *J. Chem. Soc., Chem. Commun.*, 1979, 300. 30

79CC345 D. I. John, E. J. Thomas and N. D. Tyrrell; *J. Chem. Soc., Chem. Commun.*, 1979, 345. 38

79CC488 M. Jawdosiuk, M. Makosza, B. Mudryk and G. A. Russell; *J. Chem. Soc., Chem. Commun.*, 1979, 488. 1033

79CC584 F. Bourelle-Wargnier, M. Vincent and J. Chuche; *J. Chem. Soc., Chem. Commun.*, 1979, 584. 826

79CC637 E. Wenkert, T. W. Ferreira and E. L. Michelotti; *J. Chem. Soc., Chem. Commun.*, 1979, 637. 484

79CC859 P. Crabbé, H. Fillion, D. André and J.-L Luche; *J. Chem. Soc., Chem. Commun.*, 1979, 859. 995

79CC866 N. Miyaura and A. Suzuki; *J. Chem. Soc., Chem. Commun.*, 1979, 866. 472, 532, 1104

79CC959 R. Pellicciari, R. Fringuelli, P. Ceccherelli and E. Sisani; *J. Chem. Soc., Chem. Commun.*, 1979, 959. 44

79CC1136 S. Halazy and A. Krief; *J. Chem. Soc., Chem. Commun.*, 1979, 1136. 713, 716

79CC1138 E. Piers and J. Banville; *J. Chem. Soc., Chem. Commun.*, 1979, 1138. 804

79CJC1431 F.-W. Sum and L. Weiler; *Can. J. Chem.*, 1979, **57**, 1431. 482

79CJC1743 R. E. Ireland, C. S. Wilcox, S. Thaisrivongs and N. R. Vanier; *Can. J. Chem.*, 1979, **57**, 1743. 831

79CJC1958 J.-M. Gaillot, Y. Gelas-Mialhe and R. Vessiere; *Can. J. Chem.*, 1979, **57**, 1958. 706, 707

79CJC1977 S. K. Tsui and J. D. Wood; *Can. J. Chem.*, 1979, **57**, 1977. 91, 92

79CL447 T. Mukaiyama, K. Suzuki, K. Soai and T. Sato; *Chem. Lett.*, 1979, 447. 1039

79CL481 T. Tsuji, Y. Watanabe and T. Mukaiyama; *Chem. Lett.*, 1979, 481. 1071

79CL927 M. Murakami and S. Nishida; *Chem. Lett.*, 1979, 927. 803

79CL1003 N. Yoneda, T. Fukuhara, Y. Takahashi and A. Suzuki; *Chem. Lett.*, 1979, 1003. 462

79COC(1)493 R. D. Chambers and S. R. James; in *Comprehensive Organic Chemistry*, volume ed. J. F. Stoddart, series eds D. Barton and W. D. Ollis; Pergamon Press, Oxford, 1979, vol. 1, Chapter 3, pp. 493–575. 107

79COC(1)525 R. D. Chambers and S. R. James; *Comp. Org. Chem.*, 1979, **1**, 525. 852

79COC(1)909 D. Swern; *Comp. Org. Chem.*, 1979, **1**, 909. 330, 934

79COC(1)1043 A. J. Waring; *Comp. Org. Chem.*, 1979, **1**, 1043. 306

79COC(2)361 P. J. Garratt; *Comp. Org. Chem.*, 1979, **2**, 361. 861

79COC(2)390 P. J. Garratt; *Comp. Org. Chem.*, 1979, **2**, 390. 906

79COC(2)391	P. J. Garratt; *Comp. Org. Chem.*, 1979, **2**, 391.	865
79COC(2)398	P. J. Garratt; *Comp. Org. Chem.*, 1979, **2**, 398.	890
79COC(2)411	D. Bethell; *Comp. Org. Chem.*, 1979, **2**, 411.	891
79COC(2)460	P. J. Garratt; *Comp. Org. Chem.*, 1979, **2**, 460.	917
79COC(3)187	T. Durst; *Comp. Org. Chem.*, 1979, **3**, 187.	704
79COC(3)541	I. Fleming; *Comp. Org. Chem.*, 1979, **3**, 541.	648, 649
79COC(3)791	A. Pelter and K. Smith; *Comp. Org. Chem.*, 1979, **1**, 909.	657, 658, 946
79CPB193	H. Yamanaka, K. Edo and T. Sakamoto; *Chem. Pharm. Bull.*, 1979, **27**, 193.	1010
79CT494	R. L. Banks; *Chem. Tech.*, 1979, **9**, 494.	674
79HCA1922	P. R. Jenkins, R. Gut, H. Wetter and A. Eschenmoser; *Helv. Chim. Acta*, 1979, **62**, 1922.	827, 828
79HCA2091	J. M. J. Tronchet and B. Gentile; *Helv. Chim. Acta*, 1979, **62**, 2091.	725
79HCA2630	J. Bruhn, H. Heimgartner and H. Schmid; *Helv. Chim. Acta*, 1979, **62**, 2630.	814
79HCA2655	G. Büchi, C. Fehr and G. Ohloff; *Helv. Chim. Acta*, 1979, **62**, 2655.	1082
79IJC(B)637	R. S. Tewari and K. C. Gupta; *Indian J. Chem., Sect. B*, 1979, **17**, 637.	738
79IJC(B)859	R. S. Tewari and S. C. Chaturvedi; *Indian J. Chem., Sect. B*, 1979, **18**, 859.	740, 742
79JA96	H. C. Brown, C. G. Scouten and R. Liotta; *J. Am. Chem. Soc.*, 1979, **101**, 96.	510, 1101, 1103
79JA522	E. M. Arnett, C. Petro and P. von R. Schleyer; *J. Am. Chem. Soc.*, 1979, **101**, 522.	905
79JA633	D. H. O'Brien, C. R. Russel and A. J. Hart; *J. Am. Chem. Soc.*, 1979, **101**, 633.	859
79JA1284	B. M. Trost, C. D. Shuey, F. DiNinno, Jr. and S. S. McElvain; *J. Am. Chem. Soc.*, 1979, **101**, 1284.	538
79JA1343	Z. Rappoport and J. Greenblatt; *J. Am. Chem. Soc.*, 1979, **101**, 1343.	1111
79JA1599	T. Hiyama, M. Shinoda and H. Nozaki; *J. Am. Chem. Soc.*, 1979, **101**, 1599.	422
79JA1611	F. E. Ziegler and J. J. Piwinski; *J. Am. Chem. Soc.*, 1979, **101**, 1611.	807
79JA1793	J. B. Lambert, D. M. Fabricius and J. J. Napoli; *J. Am. Chem. Soc.*, 1979, **101**, 1793.	808
79JA1906	B. B. Snider and D. M. Roush; *J. Am. Chem. Soc.*, 1979, **101**, 1906.	525
79JA2207	H. E. Zieger and D. Mathisen; *J. Am. Chem. Soc.*, 1979, **101**, 2207.	112
79JA2208	K. Atsumi and I. Kuwajima; *J. Am. Chem. Soc.*, 1979, **101**, 2208.	980
79JA2246	E. Wenkert, E. L. Michelotti and C. S. Swindell; *J. Am. Chem. Soc.*, 1979, **101**, 2246.	482
79JA2493	W. C. Still; *J. Am. Chem. Soc.*, 1979, **101**, 2493.	813
79JA2746	K. D. McMichael and G. L. Korver; *J. Am. Chem. Soc.*, 1979, **101**, 2746.	825
79JA3283	J.-F. Biellmann, H. d'Orchymont and J.-L. Schmitt; *J. Am. Chem. Soc.*, 1979, **101**, 3283.	590
79JA3602	C. R. Johnson and R. A. Kirchhoff; *J. Am. Chem. Soc.*, 1979, **101**, 3602.	622, 710
79JA3704	K. C. Nicolaou, D. A. Claremon, W. E. Barnette and S. P. Seitz; *J. Am. Chem. Soc.*, 1979, **101**, 3704.	604
79JA3967	Z. Rappoport and J. Greenblatt; *J. Am. Chem. Soc.*, 1979, **101**, 3967.	1111
79JA4003	A. E. Greene and J.-P. Depres; *J. Am. Chem. Soc.*, 1979, **101**, 4003.	2, 6
79JA4236	C. P. Casey and M. C. Cesa; *J. Am. Chem. Soc.*, 1979, **101**, 4236.	313
79JA4420	W. E. Fristad, T. R. Bailey, L. A. Paquette, R. Gleiter and M. C. Bohm; *J. Am. Chem. Soc.*, 1979, **101**, 4420.	451
79JA4426	M. P. Schneider and A. Rau; *J. Am. Chem. Soc.*, 1979, **101**, 4426.	
79JA4770	J. R. Lyerla, C. S. Yannoni, D. Bruck and C. A. Fyfe; *J. Am. Chem. Soc.*, 1979, **101**, 4770.	914
79JA4772	P. J. Stang and T. E. Fisk; *J. Am. Chem. Soc.*, 1979, **101**, 4772.	988
79JA4992	D. Milstein and J. K. Stille; *J. Am. Chem. Soc.*, 1979, **101**, 4992.	471, 473, 1145
79JA5060	R. K. Boeckman Jr., D. M. Blum and S. D. Arthur; *J. Am. Chem. Soc.*, 1979, **101**, 5060.	710
79JA5095	Y. Apeloig and Z. Rappoport; *J. Am. Chem. Soc.*, 1979, **101**, 5095.	1110
79JA5283	B. B. Snider, D. J. Rodini, R. S. E. Conn and S. Sealfon; *J. Am. Chem. Soc.*, 1979, **101**, 5283.	524, 525
79JA6120	D. A. Evans, E. Vogel and J. V. Nelson; *J. Am. Chem. Soc.*, 1979, **101**, 6120.	875
79JA6429	B. M. Trost and D. M. T. Chan; *J. Am. Chem. Soc.*, 1979, **101**, 6429.	287
79JA6452	E. Vedejs and G. R. Martinez; *J. Am. Chem. Soc.*, 1979, **101**, 6452.	408, 524, 821, 823
79JA6638	H. J. Reich, F. Chow and S. K. Shah; *J. Am. Chem. Soc.*, 1979, **101**, 6638.	592, 713, 714, 716
79JA6648	H. J. Reich, S. K. Shah and F. Chow; *J. Am. Chem. Soc.*, 1979, **101**, 6648.	593, 713, 714, 715, 716
79JA7001	S. Danishefsky, C.-F. Yan, R. K. Singh, R. B. Gammill, P. M. McCurry, Jr., N. Fritsch and J. Clardy; *J. Am. Chem. Soc.*, 1979, **101**, 7001.	521
79JA7107	G. Stork, R. K. Boeckmann, Jr., D. F. Taber, W. C. Still and J. Singh; *J. Am. Chem. Soc.*, 1979, **101**, 7107.	548
79JA7129	R. D. Little and G. W. Muller; *J. Am. Chem. Soc.*, 1979, **101**, 7129.	944
79JA7547	T. T. Tsou and J. K. Kochi; *J. Am. Chem. Soc.*, 1979, **101**, 7547.	1099
79JA7599	L. A. Paquette, C. W. Doecke and G. Klein; *J. Am. Chem. Soc.*, 1979, **101**, 7599.	811
79JCR(S)54	H. Grund and V. Jäger; *J. Chem. Res. (S)*, 1979, 54.	664
79JCS(P1)26	I. Matsuda, S. Murata and Y. Ishii; *J. Chem. Soc., Perkin Trans. 1*, 1979, 26.	209
79JCS(P1)442	A. R. Katritzky, J. Lewis and P.-L. Nie; *J. Chem. Soc., Perkin Trans. 1*, 1979, 442.	30
79JCS(P1)1137	N. C. Billingham, R. A. Jackson and F. Malek; *J. Chem. Soc., Perkin Trans. 1*, 1979, 1137.	933
79JCS(P1)1290	P. J. Kocienski, B. Lythgoe and S. Ruston; *J. Chem. Soc., Perkin Trans. 1*, 1979, 1290.	615, 616
79JCS(P1)2409	K. Hanaya, T. Muramatsu, H. Kudo and Y. L. Chow; *J. Chem. Soc., Perkin Trans. 1*, 1979, 2409.	442
79JCS(P1)2429	B. Lythgoe and I. Waterhouse; *J. Chem. Soc., Perkin Trans. 1*, 1979, 2429.	1073
79JCS(P1)2954	R. J. Cave, C. C. Howard, G. Klinkert, R. F. Newton, D. P. Reynolds, A. H. Wadsworth and S. M. Roberts; *J. Chem. Soc., Perkin Trans. 1*, 1979, 2954.	1134
79JCS(P2)810	C. W. Spangler, B. Keys and D. C. Bookbinder; *J. Chem. Soc., Perkin Trans. 2*, 1979, 810.	420
79JCS(P2)962	N. W. Alcock, J. M. Brown, J. A. Conneely and D. H. Williamson; *J. Chem. Soc., Perkin Trans. 2*, 1979, 962.	803
79JCS(P2)1535	A. L. J. Beckwith and T. Lawrence; *J. Chem. Soc., Perkin Trans. 2*, 1979, 1535.	547

79JOC55	K. Kaneda, T. Uchiyama, Y. Fujiwara, T. Imanaka and S. Teranishi; *J. Org. Chem.*, 1979, **44**, 55.	1004, 1107
79JOC133	H. C. Brown and C. G. Rao; *J. Org. Chem.*, 1979, **44**, 133.	905
79JOC151	H. Parnes and J. Pease; *J. Org. Chem.*, 1979, **44**, 151.	5, 9, 935
79JOC155	P. F. Hudrlik, R. H. Schwartz and J. C. Hogan; *J. Org. Chem.*, 1979, **44**, 155.	451, 1105
79JOC228	L. M. Sayre and F. R. Jensen; *J. Org. Chem.*, 1979, **44**, 228.	68
79JOC239	N. L. Holy; *J. Org. Chem.*, 1979, **44**, 239.	93
79JOC248	R. D. Clark and K. G. Untch; *J. Org. Chem.*, 1979, **44**, 248.	517
79JOC253	R. D. Clark and K. G. Untch; *J. Org. Chem.*, 1979, **44**, 253.	517
79JOC317	E. Buncel and B. Menon; *J. Org. Chem.*, 1979, **44**, 317.	859
79JOC449	J. E. Leibner and J. Jacobus; *J. Org. Chem.*, 1979, **44**, 449.	936
79JOC546	G. Büchi and H. Wüest; *J. Org. Chem.*, 1979, **44**, 546.	1131
79JOC641	T.-Y. Luh, C. H. Lai, K. L. Lei and S. W. Tam; *J. Org. Chem.*, 1979, **44**, 641.	2, 8
79JOC882	P. A. Bartlett and W. F. Hahne; *J. Org. Chem.*, 1979, **44**, 882.	827, 828
79JOC1014	S.-K. Chung; *J. Org. Chem.*, 1979, **44**, 1014.	75, 458
79JOC1345	A. Marfat, P. R. McGuirk and P. M. Helquist; *J. Org. Chem.*, 1979, **44**, 1345.	1107
79JOC1582	V. Van Rheenen and K. P. Shephard; *J. Org. Chem.*, 1979, **44**, 1582.	993
79JOC1744	Y. Yamamoto, H. Yatagai and K. Maruyama; *J. Org. Chem.*, 1979, **44**, 1744.	540
79JOC2185	R. A. Benkeser, E. C. Mozdzen and C. L. Muth; *J. Org. Chem.*, 1979, **44**, 2185.	21
79JOC2328	C. A. Brown and R. A. Coleman; *J. Org. Chem.*, 1979, **44**, 2328.	509, 1101
79JOC2348	C. L. Bumgardner, H.-B. Hsu, F. Afghahi, W. L. Roberts and S. T. Purrington; *J. Org. Chem.*, 1979, **44**, 2348.	399
79JOC2408	S.-I. Murahashi, M. Yamamura, K. Yanagisawa, N. Mita and K. Kondo; *J. Org. Chem.*, 1979, **44**, 2408.	466, 1097
79JOC2652	W. Kitching, W. Adcock and G. Aldous; *J. Org. Chem.*, 1979, **44**, 2652.	895
79JOC2802	H. Fienemann and H. M. R. Hoffmann; *J. Org. Chem.*, 1979, **44**, 2802.	524
79JOC2860	D. D. Roberts and R. C. Snyder, Jr.; *J. Org. Chem.*, 1979, **44**, 2860.	894
79JOC2880	D. R. Morton Jr. and F. C. Brokaw; *J. Org. Chem.*, 1979, **44**, 2880.	710
79JOC3022	F. Marcuzzi, G. Melloni and G. Modena; *J. Org. Chem.*, 1979, **44**, 3022.	318
79JOC3076	D. Kaufman, R. Kupper and T. Neal; *J. Org. Chem.*, 1979, **44**, 3076.	1122
79JOC3277	J. J. Eisch and J. E. Galle; *J. Org. Chem.*, 1979, **44**, 3277.	144, 1090
79JOC3279	J. J. Eisch and J. E. Galle; *J. Org. Chem.*, 1979, **44**, 3279.	145, 310, 1090
79JOC3428	F. E. Ziegler and C. C. Tam; *J. Org. Chem.*, 1979, **44**, 3428.	187
79JOC3431	J. L. Adcock and E. B. Renk; *J. Org. Chem.*, 1979, **44**, 3431.	664
79JOC3437	E. Piers and H. E. Morton; *J. Org. Chem.*, 1979, **44**, 3437.	1092
79JOC3442	B. El Amin, G. M. Anantharamaiah, G. P. Royer and G. E. Means; *J. Org. Chem.*, 1979, **44**, 3442.	74
79JOC3444	R. L. Smorada and W. E. Truce; *J. Org. Chem.*, 1979, **44**, 3444.	485, 1032
79JOC3445	R. D. Rieke and L. D. Rhyne; *J. Org. Chem.*, 1979, **44**, 3445.	472
79JOC3471	J. K. Crandall and W. R. Heitmann; *J. Org. Chem.*, 1979, **44**, 3471.	516
79JOC3536	H. C. Brown and C. G. Rao; *J. Org. Chem.*, 1979, **44**, 3536.	905
79JOC3737	R. A. Benkeser and J. Kang; *J. Org. Chem.*, 1979, **44**, 3737.	101
79JOC3888	A. Marfat, P. R. McGuirk and P. Helquist; *J. Org. Chem.*, 1979, **44**, 3888.	541, 1107
79JOC4078	J. E. Plevyack, J. E. Dickerson and R. F. Heck; *J. Org. Chem.*, 1979, **44**, 4078.	494
79JOC4116	G. Büchi and H. Wüest; *J. Org. Chem.*, 1979, **44**, 4116.	983
79JOC4123	L. Thijs, S. N. Gupta and D. C. Neckers; *J. Org. Chem.*, 1979, **44**, 4123.	925
79JOC4128	V. Ceré, C. Paolucci, S. Pollicino, E. Sandri and A. Fava; *J. Org. Chem.*, 1979, **44**, 4128.	821
79JOC4508	T. Shono, Y. Matsumura, K. Tsubata and Y. Sugihara; *J. Org. Chem.*, 1979, **44**, 4508.	432
79JOC4729	J. W. Timberlake and Y. M. Jun; *J. Org. Chem.*, 1979, **44**, 4729.	690
79JOC4813	M. Konieczny and R. G. Harvey; *J. Org. Chem.*, 1979, **44**, 4813.	432
79JOM(166)1	M. Bellassoued, Y. Frangin and M. Gaudemar; *J. Organomet. Chem.*, 1979, **166**, 1.	544
79JOM(177)129	G. Seconi, C. Eaborn and A. Fischer; *J. Organomet. Chem.*, 1979, **177**, 129.	318
79JOM(177)273	J. F. Fauvarque and A. Jutand; *J. Organomet. Chem.*, 1979, **177**, 273.	471
79JOM(178)43	C. U. Pittman, Jr., R. C. Ryan, J. McGee and J. P. O'Conner; *J. Organomet. Chem.*, 1979, **178**, C43.	456
79JOM(179)C7	V. V. R. Rao, C. V. Kumar and D. Devaprabhakara; *J. Organomet. Chem.*, 1979, **179**, C7.	1106
79LA451	U. Schöllkopf, R. Jentsch and K. Madawinata; *Justus Liebigs Ann. Chem.*, 1979, 451.	242
79LA1388	R. Gompper and U. Wolf; *Liebigs Ann. Chem.*, 1979, 1388.	986
79LA1925	S. Hünig, M. Öller and G. Wehner; *Justus Liebigs Ann. Chem.*, 1979, 1925.	627
B-79MI 102-01	J. H. Bestman and R. Zimmermann; in "Carbon–Carbon Bond Formation," ed. R. L. Augustin, Marcel Dekker, New York, 1979, vol. 1.	52
79MI 103-01	P. N. Rylander; *Aldrichimica Acta*, 1979, **12**, 53.	71, 87
79MI 110-01	M. Terasawa, K. Kameda, Y. Yamamoto, T. Imanaka and S. Terashima; *J. Catal.*, 1979, **57**, 315.	455
B-79MI 112-01	P. N. Rylander; "Catalytic Hydrogenation in Organic Synthesis," Academic Press, New York, 1979.	502
B-79MI 114-01	J. I. G. Cadogan; "Organophosphorus Reagents in Organic Synthesis," Academic Press, New York, 1979.	638
B-79MI 114-02	J. Boor, Jr.; "Ziegler–Natta Catalysts and Polymerizations," Academic Press, New York, 1979.	665
B-79MI 116-01	B. J. Walker; in "Organophosphorus Reagents in Organic Synthesis," ed. J. I. G. Cadogan, Academic Press, New York, 1979, pp 155–205.	744, 745
79MI117-01	K. D. King; *Int. J. Chem. Kinet.*, 1979, **11**, 1071.	772

B-79MI 117-02 R. Taylor; in "The Chemistry of Acid Derivatives," ed. S. Patai, Wiley, Chichester, 1979, Supplement 2, Part 2, ch. 15, p. 860. 772, 773, 774, 775, 777

79MI 118-01 S. S. Hixson; *Org. Photochem.*, 1979, **4**, 191. 841

B-79MI 119-01 J. C. Stowell; "Carbanions in Organic Synthesis," Wiley, New York, 1979. 844

B-79MI 119-02 D. Caine; in "Carbon–Carbon Bond Formation," ed. R. L. Augustine, Marcel Dekker, New York, 1979, vol. 1, pp. 85–352. 873

B-79MI 119-03 P. Brun and B. Waegell; in "Reactive Intermediates," ed. R. A. Abromovitch, Plenum Press, New York, 1979, vol. 3, p. 367. 929

B-79MI 119-04 D. C. Nonhebel, J. M. Tedder and J. C. Walton; "Radicals,' Cambridge University Press, Cambridge, 1979. 919

79MI 119-05 R. N. Young; *Prog. Nucl. Magn. Reson. Spectrosc.*, 1979, **12**, 261–268. 894

B-79MI 122-01 P. J. Stang, Z. Rappoport, M. Hanack and L. R. Subramanian; "Vinyl Cations," Academic Press, New York, 1979. 1109, ,1110, 1117

79MIP15154 E. Mago-Karacsony and L. Toldy; *Hung. Pat.*, 15 154 (1979) (*Chem. Abs.* 1979, **90**, 55 151). 85

79NJC195 J. L. Ripoll; *Nouv. J. Chim.*, 1979, **3**, 195. 787

79NJC269 G. A. Olah and D. Meidar; *Nouv. J. Chim.*, 1979, **3**, 269. 462

79NJC321 F. Delbecq, R. Baudouy and J. Goré; *Nouv. J. Chim.*, 1979, **3**, 321. 958

79OPP201 R. O. Hutchins and N. R. Natale; *Org. Prep. Proced. Int.*, 1979, **11**, 201. 635

79PS(7)69 L. Morin, J. Lebaud, D. Paguer, R. Chaussin and Z. Barillier; *Phosphorus Sulfur*, 1979, **7**, 69. 833, 834

79S44 A. R. Chamberlin and F. T. Bond; *Synthesis*, 1979, 44. 1094

79S59 G. A. Olah, Y. D. Vankar and A. P. Fung; *Synthesis*, 1979, 59. 5, 8, 9

79S104 G. A. Molander and H. C. Brown; *Synthesis*, 1979, 104. 1043

79S117 S. Wolfe and J. R. Campbell; *Synthesis*, 1979, 117. 562

79S127 H. Ahlbrecht, W. Raab and C. Vonderheid; *Synthesis*, 1979, 127. 560

79S292 H. D. Verkruijsse and M. Hasselaar; *Synthesis*, 1979, 292. 1025

79S372 E. von Dehmlow and M. Lissel; *Synthesis*, 1979, 372. 572

79S388 W. Adam and L. A. Arias Encarnacion; *Synthesis*, 1979, 388. 785

79S390 H. Westmijze and P. Vermeer; *Synthesis*, 1979, 390. 967

79S430 H. Westmijze, H. Kleijn and P. Vermeer; *Synthesis*, 1979, 430. 514

79S446 I. Fleming and I. Paterson; *Synthesis*, 1979, 446. 213

79S471 J. M. Berge and S. M. Roberts; *Synthesis*, 1979, 471. 936

79S502 J. Villieras, P. Perriot and J. F. Normant; *Synthesis*, 1979, 502. 1067

79S508 G. Balme, A. Doutheau, J. Goré and M. Malacria; *Synthesis*, 1979, 508. 225

79S736 I. Fleming and I. Paterson; *Synthesis*, 1979, 736. 556

79S761 T. H. Chan and I. Fleming; *Synthesis*, 1979, 761. 189, 450, 535

79S826 A. Alexakis, G. Cahiez and J. F. Normant; *Synthesis*, 1979, 826. 465, 1017, 1018

79S891 M. C. Benhamou, G. Etemad-Moghadam, V. Speziale and A. Lattes; *Synthesis*, 1979, 891. 68, 454

79SC275 G. W. Kabalka and J. H. Chandler; *Synth. Commun.*, 1979, **9**, 275. 41

79SC809 T. Yogo, J. Koshino and A. Suzuki; *Synth. Commun.*, 1979, **9**, 809. 198

79T77 E. P. Serebryakov, S. D. Kulomzina-Pletneva and A. Kh. Margaryan; *Tetrahedron*, 1977, **35**, 77. 516

79T885 T. C. Jain, C. M. Banks and J. E. McCloskey; *Tetrahedron*, 1979, **35**, 885. 2

79T929 M. Saunders and S. P. Budiansky; *Tetrahedron*, 1979, **35**, 929.

79T2039 R. Rossi, P. A. Salvadori, A. Carpita and A. Niccoli; *Tetrahedron*, 1979, **35**, 2039. 506

79TCC1 H. C. Brown; *Top. Curr. Chem.*, 1979, **80**, 1.

79TCC19 G. A. Olah; *Top. Curr. Chem.*, 1979, **80**, 19.

79TL7 G. Balme, M. Malacria and J. Goré; *Tetrahedron Lett.*, 1979, **20**, 7. 967

79TL43 H. Okamura, M. Miura and H. Takei; *Tetrahedron Lett.*, 1979, **20**, 43. 484

79TL327 A. Streitwieser, Jr. and C. C. C. Shen; *Tetrahedron Lett.*, 1979, **20**, 327. 859

79TL349 H. J. Cristau, A. Long and H. Christol; *Tetrahedron Lett.*, 1979, **20**, 349. 50

79TL845 E.-i. Negishi, M. J. Idacavage, F. DiPasquale and A. Silveira, Jr.; *Tetrahedron Lett.*, 1979, **20**, 845. 133

79TL849 J. L. Fry and T. J. Mraz; *Tetrahedron Lett.*, 1979, **20**, 849. 13, 18

79TL929 K. Schlüter and A. Berndt; *Tetrahedron Lett.*, 1979, **20**, 929. 922

79TL937 R. Baudouy, F. Delbecq, and J. Gore; *Tetrahedron Lett.*, 1979, **20**, 937. 464

79TL993 I. Paterson and I. Fleming; *Tetrahedron Lett.*, 1979, **20**, 993. 599

79TL995 I. Paterson and I. Fleming; *Tetrahedron Lett.*, 1979, **20**, 995. 599

79TL1073 C. Huynh and G. Linstrumelle; *Tetrahedron Lett.*, 1979, **20**, 1073. 1092, 1105

79TL1137 K. Tamao, H. Matsumoto, T. Kakui and M. Kumada; *Tetrahedron Lett.*, 1979, **20**, 1137. 490

79TL1141 J. Yoshida, K. Tamao, T. Kakui and M. Kumada; *Tetrahedron Lett.*, 1979, **20**, 1141. 490, 515

79TL1181 V. Nair and S. G. Richardson; *Tetrahedron Lett.*, 1979, **20**, 1181. 36

79TL1503 C. Huynh, F. Derguini-Boumechal and G. Linstrumelle; *Tetrahedron Lett.*, 1979, **20**, 1503. 229, 478, 1092

79TL1509 J. Bergman and L. Engman; *Tetrahedron Lett.*, 1979, **20**, 1509. 1072

79TL1519 I. Paterson; *Tetrahedron Lett.*, 1979, **20**, 1519. 133

79TL1619 T. Kitamura, S. Kobayashi and H. Taniguchi; *Tetrahedron Lett.*, 1979, **20**, 1619. 1120

79TL1641 W. F. Berkowitz, I. Sasson, P. S. Sampathkumar, J. Hrabie, S. Choudry and D. Pierce; *Tetrahedron Lett.*, 1979, **20**, **20**, 1641. 84

79TL1845 M. Bertrand, J.-P. Dulcère, G. Gil and M.-L. Roumestant; *Tetrahedron Lett.*, 1979, **20**, 1845. 959

79TL1963 A. W. McCulloch and A. G. McInnes; *Tetrahedron Lett.*, 1979, **20**, 1963. 525

79TL1995 F. Cooke, J. Schwindeman and P. Magnus; *Tetrahedron Lett.*, 1979, **20**, 1995. 422

79TL1999	W. E. Fristad, D. S. Dime, T. R. Bailey and L. A. Paquette; *Tetrahedron Lett.*, 1979, **20**, 1999.	422
79TL2157	T. Shono, Y. Matsumara, K. Tsubata and Y. Sugihara; *Tetrahedron Lett.*, 1979, **20**, 2157.	17
79TL2179	I. Paterson and I. Fleming; *Tetrahedron Lett.*, 1979, **20**, 2179.	599
79TL2291	D. H. R. Barton, G. Bringmann, G. Lamotte, R. S. H. Motherwell and W. B. Motherwell; *Tetrahedron Lett.*, 1979, **20**, 2291.	38, 945
79TL2313	E. Marinelli and A. B. Levy; *Tetrahedron Lett.*, 1979, **20**, 2313.	278
79TL2317	E. J. Corey and G. Schmidt; *Tetrahedron Lett.*, 1979, **20**, 2317.	438
79TL2625	G. Mehta and A. V. Reddy; *Tetrahedron Lett.*, 1979, **20**, 2625.	524
79TL2649	P. J. Kocienski; *Tetrahedron Lett.*, 1979, **20**, 2649.	616, 654
79TL2697	J. N. Denis, J. Vicens and A. Krief; *Tetrahedron Lett.*, 1979, **20**, 2697.	603
79TL2779	E. N. Marvell and S. W. Almond; *Tetrahedron Lett.*, 1979, **20**, 2779.	401, 800
79TL2961	S. Hayashi, M. Nair, D. J. Hauser and H. Schechter; *Tetrahedron Lett.*, 1979, **20**, 2961.	1078
79TL3021	T. Kitahara, K. Mori and M. Matsui; *Tetrahedron Lett.*, 1979, **20**, 3021.	1134
79TL3057	S. Raucher, K.-J. Hwang and J. E. Macdonald; *Tetrahedron Lett.*, 1979, **20**, 3057.	828
79TL3145	W. Granitzer and A. Stütz; *Tetrahedron Lett.*, 1979, **20**, 3145.	512
79TL3327	H. Westmijze, H. Kleijn and P. Vermeer; *Tetrahedron Lett.*, 1979, **20**, 3327.	544
79TL3375	G. W. Gokel, H. M. Gerdes, D. E. Miles, J. M. Hufnal and G. A. Zerby; *Tetrahedron Lett.*, 1979, **20**, 3375.	166
79TL3425	H. Okamura and H. Takei; *Tetrahedron Lett.*, 1979, **20**, 3425.	238, 483
79TL3437	A. Suzuki, N. Miyaura and K. Yamada; *Tetrahedron Lett.*, 1979, **20**, 3437.	532, 1020
79TL3815	R. D. Bach and J. W. Knight; *Tetrahedron Lett.*, 1979, **20**, 3815.	629
79TL3873	V. Calò, L. Lopez, G. Marchese and G. Pesce; *Tetrahedron Lett.*, 1979, **20**, 3873.	1025
79TL4111	J. N. Denis, W. Dumont and A. Krief; *Tetrahedron Lett.*, 1979, **20**, 4111.	590
79TL4233	S. Halazy and A. Krief; *Tetrahedron Lett.*, 1979, **20**, 4233.	623, 713, 716
79TL4343	M. R. Johnson, T. Nakata and Y. Kishi; *Tetrahedron Lett.*, 1979, **20**, 4343.	478
79TL4553	S. Hacini, R. Pardo and M. Santelli; *Tetrahedron Lett.*, 1979, **20**, 4553.	422
79TL4631	J. A. Aimetti, E. S. Hamanaka, D. A. Johnson and M. S. Kellogg; *Tetrahedron Lett.*, 1979, **20**, 4631.	4, 9
79TL4985	W. G. Blenderman, M. M. Joullié and G. Preti; *Tetrahedron Lett.*, 1979, **20**, 4985.	100
79TL5043	J. M. Clough and G. Pattenden; *Tetrahedron Lett.*, 1979, **20**, 5043.	755
80ACR218	L. E. Overman; *Acc. Chem. Res.*, 1980, **13**, 218.	836, 993
80ACR385	B. M. Trost; *Acc. Chem. Res.*, 1980, **13**, 385.	192
80ACR426	B. B. Snider; *Acc. Chem. Res.*, 1980, **13**, 426.	524
80AG(E)178	H. Siegel and W. Himmele; *Angew. Chem., Int. Ed. Engl.*, 1980, **19**, 178.	296
80AG(E)463	R. Gompper and A. Schmidt, *Angew., Chem. Int. Ed. Engl.*, 1980, **19**, 463.	563
80AG(E)779	J. Sauer and R. Sustmann; *Angew. Chem., Int. Ed. Engl.*, 1980, **19**, 779.	786
80BCJ169	S. Hirano, S. Takagi, T. Hiyama and H. Nozaki; *Bull. Chem. Soc. Jpn.*, 1980, **53**, 169.	422
80BCJ553	S. Uemura, H. Miyoshi, M. Wakasugi, M. Okano, O. Itoh, T. Izumi and K. Ichikawa; *Bull. Chem. Soc. Jpn.*, 1980, **53**, 553.	493
80BCJ1471	N. Miyaura, T. Yano and A. Suzuki; *Bull. Chem. Soc. Jpn.*, 1980, **53**, 1471.	532, 1094
80BCJ1652	Y. Masuda, A. Arase and A. Suzuki; *Bull. Chem. Soc. Jpn.*, 1980, **53**, 1652.	663
80BCJ1670	H. Yatagai; *Bull. Chem. Soc., Japan*, 1980, **53**, 1670.	510, 531, 532, 663, 1104
80BCJ1698	K. Takai, Y. Hotta, K. Oshima and H. Nozaki; *Bull. Chem. Soc. Jpn.*, 1980, **53**, 1698.	678
80BCJ2958	T. Kato, H. Kondo, M. Nishino, M. Tanaka, G. Hata and A. Miyake; *Bull. Chem. Soc. Jpn*, 1980, **53**, 2958.	813, 814
80BCJ3273	A. Ohsawa, Y. Abe and H. Igeta; *Bull. Chem. Soc. Jpn.*, 1980, **53**, 3273.	1010
80C265	I. Fleming; *Chimia*, 1980, **34**, 265.	130
80CB2015	M. Hanack and W. Spang; *Chem. Ber.*, 1980, **113**, 2015.	1117
80CB3679	G. Maas and C. Hummel; *Chem. Ber.*, 1980, **113**, 3679.	808
80CC86	S. Masuda, M. Segi, T. Nakajima and S. Suga; *J. Chem. Soc., Chem. Commun.*, 1980, 86.	463
80CC103	S. Kobayashi, T. Nishi, I. Koyama and H. Taniguchi; *J. Chem. Soc., Chem. Commun.*, 1980, 103.	1113, 1114
80CC169	H. Alper and C. Blais; *J. Chem. Soc., Chem. Commun.*, 1980, 169.	22, 437
80CC732	D. H. R. Barton, H. A. Dowlatshahi, W. B. Motherwell and D. Villemin; *J. Chem. Soc., Chem. Commun.*, 1980, 732.	934
80CC736	T. Kobayashi, N. Ishida and T. Hiraoka; *J. Chem. Soc., Chem. Commun.*, 1980, 736.	1042
80CC748	Z. N. Parnes, G. I. Bolestova, I. S. Akhrem, M. E. Volpin and D. N. Kursanov; *J. Chem. Soc., Chem. Commun.*, 1980, 748.	281
80CC1096	P. J. Kocienski; *J. Chem. Soc., Chem. Commun.*, 1980, 1096.	821
80CC1195	A. M. Birch and G. Pattenden; *J. Chem. Soc., Chem. Commun.*, 1980, 1195.	357
80CC1202	M. T. Reetz, B. Wenderoth, R. Peter, R. Steinbach and J. Westermann; *J. Chem. Soc., Chem. Commun.*, 1980, 1202.	127
80CJC716	G. S. Bates and S. Ramaswamy; *Can. J. Chem.*, 1980, **58**, 716.	178, 265
80CL255	T. Mukaiyama and K. Suzuki; *Chem. Lett.*, 1980, 255.	1039
80CL331	M. Isobe, M. Kitamura and T. Goto; *Chem. Lett.*, 1980, 331.	310
80CL669	Y. Yamamoto, H. Yatagai and K. Maruyama; *Chem. Lett.*, 1980, 669.	1001
80CL1061	Y. Masaki, K. Sakuma and K. Kaji; *Chem. Lett.*, 1980, 1061.	820
80CPB282	S. Inayama, T. Tawamata and N. Shimizu; *Chem. Pharm. Bull.*, 1980, **28**, 282.	561
80CRV99	P. S. Engel; *Chem. Rev.*, 1980, **80**, 99.	330, 924, 925, 944
80CSR1	I. R. Dunkin; *Chem. Soc. Rev.*, 1980, **9**, 1.	921

80H(14)1441 T. Wakamatsa, H. Inaki, A. Ogawa, M. Watanabe and Y. Ban; *Heterocycles*, 1980, **14**, 1441. 102

80H(14)1739 J. J. Burger, T. B. R. A. Chen, E. R. de Waard and H. O. Huisman; *Heterocycles*, 1980, **14**, 1739. 708

80HCA438 R. W. Lang and H.-J. Hansen; *Helv. Chim. Acta*, 1980, **63**, 438. 986

80HCA1198 W. Oppolzer and R. D. Wylie; *Helv. Chim. Acta*, 1980, **63**, 1198. 357

80HCA2328 J. Pfenninger, C. Heuberger and W. Graf; *Helv. Chim. Acta*, 1980, **63**, 2328. 933, 939

80HCA2426 A. B. Shenvi and H. Gerlach; *Helv. Chim. Acta*, 1980, **63**, 2426. 1134

80IZV726 T. A. Sundukova, S. F. Vasilevski, M. S. Shvartsberg and I. L. Kotlyarevski; *Izv. Akad. Nauk SSSR, Ser. Khim.*, 1980, 726 (*Chem. Abstr.*, **93**, 71446f). 1010

80JA337 P. A. Bartlett and J. L. Adams; *J. Am. Chem. Soc.*, 1980, **102**, 337. 68

80JA683 P. von R. Schleyer, D. Lenoir, P. Mison, G. Liang, G. K. S. Prakash and G. A. Olah; *J. Am. Chem. Soc.*, 1980, **102**, 683. 906

80JA774 D. A. Evans and J. V. Nelson; *J. Am. Chem. Soc.*, 1980, **102**, 774. 414

80JA790 F. E. Ziegler, I. Chliwner, K. W. Fowler, S. J. Kanfer, S. J. Kuo and N. D. Sinha; *J. Am. Chem. Soc.*, 1980, **102**, 790. 473

80JA862 M. Tanabe and K. Hayashi; *J. Am. Chem. Soc.*, 1980, **102**, 862. 832

80JA865 L. E. Overman and F. M. Knoll; *J. Am. Chem. Soc.*, 1980, **102**, 865. 413

80JA994 G. de Lauzon, B. Deschamps, J. Fischer, F. Mathey and A. Mitschler; *J. Am. Chem. Soc.*, 1980, **102**, 994. 53

80JA1064 D. S. Grierson, M. Harris and H.-P. Husson; *J. Am. Chem. Soc.*, 1980, **102**, 1064. 238

80JA1155 R. E. Ireland, S. Thaisrivongs and C. S. Wilcox; *J. Am. Chem. Soc.*, 1980, **102**, 1155. 831

80JA1165 A. P. Kozikowski and M. N. Greco; *J. Amer. Chem. Soc.*, 1980, **102**, 1165. 83

80JA1201 W. C. Still and C. Sreekumar; *J. Am. Chem. Soc.*, 1980, **102**, 1201. 139

80JA1214 P. G. Gassman and J. J. Talley; *J. Am. Chem. Soc.*, 1980, **102**, 1214. 904

80JA1333 J. Schwartz, M. J. Loots and H. Kosugi; *J. Am. Chem. Soc.*, 1980, **102**, 1333. 318

80JA1423 H. J. Reich, R. E. Olson and M. C. Clark; *J. Am. Chem. Soc.*, 1980, **102**, 1423. 971

80JA1611 X. Creary; *J. Am. Chem. Soc.*, 1980, **102**, 1611. 500

80JA1756 A. Gopalan and P. Magnus; *J. Am. Chem. Soc.*, 1980, **102**, 1756. 813

80JA2125 K. Ogura, S. Furukawa and G. Tsuchihashi; *J. Am. Chem. Soc.*, 1980, **102**, 2125. 408

80JA2134 D. Gange, P. Magnus, L. Bass, E. V. Arnold and J. Clardy; *J. Am. Chem. Soc.*, 1980, **102**, 2134. 971

80JA2318 Y. Yamamoto, S. Yamamoto, H. Yatagai and K. Maruyama; *J. Am. Chem. Soc.*, 1980, **102**, 2318. 119

80JA2392 Y. Tamaru, T. Harada and Z. Yoshida; *J. Am. Chem. Soc.*, 1980, **102**, 2392. 807, 834

80JA2463 M. E. Jung and J. P. Hudspeth; *J. Am. Chem. Soc.*, 1980, **102**, 2463. 760

80JA2469 F. G. Klärner and B. Brassel; *J. Am. Chem. Soc.*, 1980, **102**, 2469. 799

80JA2693 P. Girard, J. L. Namy and H. B. Kagan; *J. Am. Chem. Soc.*, 1980, **102**, 2693. 3, 6, 8, 10, 17, 587, 680

80JA2797 A. Padwa and T. J. Blacklock; *J. Am. Chem. Soc.*, 1980, **102**, 2797. 808

80JA3040 J. M. Brown and P. A. Chaloner; *J. Amer. Chem. Soc.*, 1980, **102**, 3040. 77

80JA3156 K. J. Shea and R. B. Phillips; *J. Am. Chem. Soc.*, 1980, **102**, 3156. 806

80JA3270 S. H. Pine, R. Zahler, D. A. Evans and R. H. Grubbs; *J. Am. Chem. Soc.*, 1980, **102**, 3270. 767, 769

80JA3298 E. Negishi, L. F. Valente and M. Kobayashi; *J. Am. Chem. Soc.*, 1980, **102**, 3298. 467

80JA3334 D. J. DeFrees, R. T. McIver, Jr. and W. J. Hehre; *J. Am. Chem. Soc.*, 1980, **102**, 3334. 1143

80JA3370 A. J. Birch, A. L. Hinde and L. Radom; *J. Am. Chem. Soc.*, 1980, **102**, 3370. 96, 890

80JA3642 R. Scheffold, M. Dike, S. Dike, T. Herold and L. Walder; *J. Am. Chem. Soc.*, 1980, **102**, 3642. 345, 946

80JA3959 Z. A. Fataftah, I. E. Kopka and M. W. Rathke; *J. Am. Chem. Soc.*, 1980, **102**, 3959. 873, 874, 876

80JA3972 L. A. Paquette, G. D. Crouse and A. K. Sharma; *J. Am. Chem. Soc.*, 1980, **102**, 3972. 402, 799

80JA4074 A. J. Birch, A. L. Hinde and L. Radom; *J. Am. Chem. Soc.*, 1980, **102**, 4074. 96, 890

80JA4138 P. G. Gassman and J. J. Talley; *J. Am. Chem. Soc.*, 1980, **102**, 4138. 904

80JA4438 D. L. J. Clive, G. Chittattu, V. Farina, W. A. Kiel, S. M. Menchen, C. G. Russell, A. Singh, C. K. Wong and N. J. Curtis; *J. Am. Chem. Soc.*, 1980, **102**, 4438. 21, 25, 437, 438, 943

80JA4709 P. R. Peoples and J. B. Grutzner; *J. Am. Chem. Soc.*, 1980, **102**, 4709. 847

80JA4841 W. T. Borden, J. G. Lee and S. D. Young; *J. Am. Chem. Soc.*, 1980, **102**, 4841. 799

80JA4973 L. S. Hegedus, R. E. Williams, M. A. McGuire and T. Hayashi; *J. Am. Chem. Soc.*, 1980, **102**, 4973. 315

80JA4980 L. S. Hegedus and W. H. Darlington; *J. Am. Chem. Soc.*, 1980, **102**, 4980. 315

80JA5004 E. Ehlinger and P. Magnus; *J. Am. Chem. Soc.*, 1980, **102**, 5004. 651, 762

80JA5120 A. D. Despo, S. K. Chiu, T. Flood and P. E. Peterson; *J. Am. Chem. Soc.*, 1980, **102**, 5120. 978

80JA5253 R. L. Funk and K. P. C. Vollhardt; *J. Am. Chem. Soc.*, 1980, **102**, 5253. 64

80JA5413 P. Riviere, A. Castel, and J. Satgé; *J. Am. Chem. Soc.*, 1980, **102**, 5413. 766

80JA5425 M. Rosenberger and C. Neukom; *J. Am. Chem. Soc.*, 1980, **102**, 5425. 1134

80JA5967 H. J. Reich and W. W. Willis, Jr.; *J. Am. Chem. Soc.*, 1980, **102**, 5967. 605, 983, 1073

80JA5979 B. M. Trost, N. R. Schmuff and M. J. Miller; *J. Am. Chem. Soc.*, 1980, **102**, 5979. 238, 613

80JA6163 S. L. Schreiber; *J. Am. Chem. Soc.*, 1980, **102**, 6163. 212

80JA6351 T. Hudlicky, T. M. Kutchan, S. R. Wilson and D. T. Mayo; *J. Am. Chem. Soc.*, 1980, **102**, 6351. 804

80JA6430 A. J. Birch, A. L. Hinde and L. Radom; *J. Am. Chem. Soc.*, 1980, **102**, 6430. 96, 890

80JA6633 E. D. Laganis and D. M. Lemal; *J. Am. Chem. Soc.*, 1980, **102**, 6633. 862

80JA6640 G. A. Olah, G. K. S. Prakash and M. Arvanaghi; *J. Am. Chem. Soc.*, 1980, **102**, 6640. 904

80JA6813 P. J. Stang and T. E. Fisk; *J. Am. Chem. Soc.*, 1980, **102**, 6813. 1145

80JA6867 M. Saunders and M. R. Kates; *J. Am. Chem. Soc.*, 1980, **102**, 6867. 906

80JA6885 S. Djuric, T. Sarkar and P. Magnus; *J. Am. Chem. Soc.*, 1980, **102**, 6885. 655

80JA6889 S. Danishefsky, R. L. Funk and J. F. Kerwin, Jr.; *J. Am. Chem. Soc.*, 1980, **102**, 6889. 831

80JA6891 S. Danishefsky, and D. Tsuzuki; *J. Am. Chem. Soc.*, 1980, **102**, 6891. 831

80JA6894 P. F. Hudrlik, A. M. Hudrlik, G. Nagendrappa, T. Yimenu, E. T. Zellers and E. Chin; *J. Am. Chem. Soc.*, 1980, **102**, 6894. 650

80JA6900 T. Cohen and J. R. Matz; *J. Am. Chem. Soc.*, 1980, **102**, 6900. 805

80JA7381 J. S. Temple and J. Schwartz; *J. Am. Chem. Soc.*, 1980, **102**, 7381. 536

80JA7587 P. A. Grieco, T. Takigawa, S. L. Bongers and H. Tanaka; *J. Am. Chem. Soc.*, 1980, **102**, 7587. 836

80JA7613 P. G. Gassman, K. Saito and J. J. Talley; *J. Am. Chem. Soc.*, 1980, **102**, 7613. 904

80JA7760 T. L. Macdonald, N. Narasimhan and L. T. Burka; *J. Am. Chem. Soc.*, 1980, **102**, 7760. 108

80JCR1551 M. E. Bertrand, G. Gil, A. Junino and R. Maurin; *J. Chem. Res.*, 1980, 1551. 805

80JCR(S)98 M. Bertrand, G. Gil, A. Junino and R. Maurin; *J. Chem. Res.* (*S*), 1980, 98. 498

80JCR(S)152 K. Liu, C. Yen and H. Hwang; *J. Chem. Res.* (*S*), 1980, 152. 905

80JCR(S)396 D. Lenoir and H. Burghard; *J. Chem. Res.* (*S*), 1980, 396. 689

80JCS(P1)714 D. J. Field and D. W. Jones; *J. Chem. Soc., Perkin Trans. 1*, 1980, 714. 403

80JCS(P1)852 C. C. Howard, R. F. Newton, D. P. Reynolds, A. H. Wadsworth, D. R. Kelly and S. M. Roberts; *J. Chem. Soc., Perkin Trans. 1*, 1980, 852. 1024, 1134

80JCS(P1)1045 P. J. Kocienski, B. Lythgoe and I. Waterhouse; *J. Chem. Soc., Perkin Trans. 1*, 1980, 1045. 616, 696, 700

80JCS(P1)1098 S. Uemura, H. Miyoshi and M. Okano; *J. Chem. Soc., Perkin Trans. 1*, 1980, 1098. 66, 454

80JCS(P1)1400 P. J. Kocienski and B. Lythgoe; *J. Chem. Soc., Perkin Trans. 1*, 1980, 1400. 700, 702

80JCS(P1)1420 J. Barluenga, P. J. Campos, M. A. Roy and G. Asensio; *J. Chem. Soc., Perkin Trans. 1*, 1980, 1420. 68

80JCS(P1)1436 R. W. Jemison, T. Laird, W. D. Ollis, C. Smith and I. O. Sutherland; *J. Chem. Soc., Perkin Trans. 1*, 1980, 1436. 407, 823

80JCS(P1)1450 R. W. Jemison, T. Laird, W. D. Ollis and I. O. Sutherland; *J. Chem. Soc., Perkin Trans. 1*, 1980, 1450. 408

80JCS(P1)1458 R. W. Jemison, T. Laird, W. D. Ollis and I. O. Sutherland; *J. Chem. Soc., Perkin Trans. 1*, 1980, 1458. 401

80JCS(P1)1493 I. Fleming and J. Goldhill; *J. Chem. Soc., Perkin Trans. 1*, 1980, 1493. 653

80JCS(P1)2554 A. R. Katritsky, K. Horvath and B. Plau; *J. Chem. Soc., Perkin Trans. 1.*, 1980, 2554. 30

80JCS(P1)2657 D. H. R. Barton, G. Bringmann, G. Lamotte, W. B. Motherwell, R. S. Hay-Motherwell and A. E. A. Porter; *J. Chem. Soc., Perkin Trans. 1*, 1980, 2657. 945

80JCS(P1)2665 D. H. R. Barton, G. Bringmann and W. B. Motherwell; *J. Chem. Soc., Perkin Trans. 1*, 1980, 2665. 945

80JCS(P2)1447 C. Albano and S. Wold; *J. Chem. Soc., Perkin Trans. 2*, 1980, 1447. 905

80JOC1 H. C. Brown, S. C. Kim and S. Krishnamurthy; *J. Org. Chem.*, 1980, **45**, 1. 14

80JOC274 M. R. Detty; *J. Org. Chem.*, 1980, **45**, 274. 611

80JOC389 H. C. Brown and J. B. Campbell, Jr.; *J. Org. Chem.*, 1980, **45**, 389. 488, 510, 1102, 1103

80JOC549 J. B. Campbell, Jr. and H. C. Brown; *J. Org. Chem.*, 1980, **45**, 549. 488, 1094

80JOC550 H. C. Brown and J. B. Campbell, Jr.; *J. Org. Chem.*, 1980, **45**, 550. 532, 1094

80JOC805 M. J. Gula and T. A. Spencer; *J. Org. Chem.*, 1980, **45**, 805. 72

80JOC849 S. Krishnamurthy and H. C. Brown; *J. Org. Chem.*, 1980, **45**, 849. 3

80JOC900 S. G. Traynor, B. J. Kane, J. B. Coleman and C. G. Cárdenas; *J. Org. Chem.*, 1980, **45**, 900. 797

80JOC922 K. L. Dhawan, B. D. Gowland and T. Durst; *J. Org. Chem.*, 1980, **45**, 922. 213, 478

80JOC1046 F. Cooke, R. Moerck, J. Schwindeman and P. Magnus; *J. Org. Chem.*, 1980, **45**, 1046. 422

80JOC1066 C. H. Heathcock, C. T. Buse, W. A. Kleschick, M. C. Pirrung, J. E. Sohn and J. Lampe; *J. Org. Chem.*, 1980, **45**, 1066. 874

80JOC1172 M. Koreeda, Y. Tanaka and A. Schwartz; *J. Org. Chem.*, 1980, **45**, 1172. 806

80JOC1340 R. L. Danheiser, C. Martinez-Davila and J. M. Morin; *J. Org. Chem.*, 1980, **45**, 1340. 803

80JOC1486 E. Gipstein, C. G. Willson and H. S. Sachdev; *J. Org. Chem.*, 1980, **45**, 1486. 608

80JOC1604 T. Okuyama, K. Toyoshima and T. Fueno; *J. Org. Chem.*, 1980, **45**, 1604. 459

80JOC1640 H. Yatagai; *J. Org. Chem.*, 1980, **45**, 1640. 532

80JOC1719 C. Y. Meyers, D. H. Hua and N. J. Peacock; *J. Org. Chem.*, 1980, **45**, 1719. 618, 704

80JOC2005 T. Imagawa, T. Sonobe, H. Ishiwari, T. Akiyama and M. Kawanisi; *J. Org. Chem.*, 1980, **45**, 2005. 769

80JOC2036 J.-P. Deprés and A. E. Greene; *J. Org. Chem.*, 1980, **45**, 2036. 571

80JOC2080 G. Saucy, N. Cohen, B. L. Banner and D. P. Trullinger; *J. Org. Chem.*, 1980, **45**, 2080. 980

80JOC2113 H. C. Brown and C. G. Rao; *J. Org. Chem.*, 1980, **45**, 2113. 905

80JOC2247 P. A. Grieco, T. Takigawa and W. J. Schillinger; *J. Org. Chem.*, 1980, **45**, 2247. 604

80JOC2526 E. Negishi, A. O. King, W. L. Klima, W. Patterson and A. Silveira, Jr.; *J. Org. Chem.*, 1980, **45**, 2526. 1070

80JOC2550 S. Krishnamurthy; *J. Org. Chem.*, 1980, **45**, 2550. 4, 8, 10, 17

80JOC2773 B. B. Snider, D. M. Roush, D. J. Rodini, D. Gonzalez and D. Spindell; *J. Org. Chem.*, 1980, **45**, 2773. 525

80JOC3017 L. A. Paquette, W. E. Fristad, D. S. Dime and T. R. Bailey; *J. Org. Chem.*, 1980, **45**, 3017. 422

80JOC3053 J. Schwartz, D. B. Carr, R. T. Hansen and F. M. Dayrit; *J. Org. Chem.*, 1980, **45**, 3053. 1139

80JOC3372 H. F. Schmitthenner and S. M. Weinreb; *J. Org. Chem.*, 1980, **45**, 3372. 791

80JOC3513 S. K. Chung; *J. Org. Chem.*, 1980, **45**, 3513. 431

80JOC3522 A. Silveira Jr., M. Angelastro, R. Israel, F. Totino and P. Williamsen; *J. Org. Chem.*, 1980, **45**, 3522. 983

80JOC3531 G. A. Olah, M. Arvanaghi and Y. D. Vankar; *J. Org. Chem.*, 1980, **45**, 3531. 5, 8, 9

80JOC3537	P. A. Grieco, E. Williams, H. Tanaka and S. Gilman; *J. Org. Chem.*, 1980, **45**, 3537.	604
80JOC3549	T. R. Hoye and M. J. Kurth; *J. Org. Chem.*, 1980, **45**, 3549.	590, 591
80JOC3840	R. C. Larock and S. S. Hershberger; *J. Org. Chem.*, 1980, **45**, 3840.	1096
80JOC3925	R. L. Danheiser and D. J. Carini; *J. Org. Chem.*, 1980, **45**, 3925.	962, 1055
80JOC4097	Y. Naruta; *J. Org. Chem.*, 1980, **45**, 4097.	23
80JOC4263	E. Piers and H. E. Morton; *J. Org. Chem.*, 1980, **45**, 4263.	1092
80JOC4309	Z. Rappoport and N. Pross; *J. Org. Chem.*, 1980, **45**, 4309.	1110
80JOC4413	W. G. Dauben, A. A. Ponaras and A. Chollet; *J. Org. Chem.*, 1980, **45**, 4413.	833
80JOC4534	J. J. Eisch and J. E. Galle; *J. Org. Chem.*, 1980, **45**, 4534.	145
80JOC4536	Y. Tanigawa and S.-I. Murahashi; *J. Org. Chem.*, 1980, **45**, 4536.	969
80JOC4702	K. F. Bernady, J. F. Poletto, J. Nocera, P. Mirando, R. E. Schant and M. J. Weiss; *J. Org. Chem.*, 1980, **45**, 4702.	556
80JOC4740	T. L. Macdonald, D. R. Reagan and R. S. Brinkmeyer; *J. Org. Chem.*, 1980, **45**, 4740.	197
80JOC4793	M. Newcomb and R. S. Vieta; *J. Org. Chem.*, 1980, **45**, 4793.	808
80JOC4926	J. R. Weir, B. A. Patel and R. F. Heck; *J. Org. Chem.*, 1980, **45**, 4926.	32, 457, 507, 1009
80JOC5006	T. Flood and P. E. Peterson; *J. Org. Chem.*, 1980, **45**, 5006.	963, 978
80JOC5020	T. Hudlicky, F. U. Koszyk, T. M. Kutchan and J. D. Sheth; *J. Org. Chem.*, 1980, **45**, 5020.	803, 804
80JOC5223	M. Kobayashi and E. Negishi; *J. Org. Chem.*, 1980, **45**, 5223.	543
80JOC5227	H. J. Reich and W. W. Willis, Jr.; *J. Org. Chem.*, 1980, **45**, 5227.	1089
80JOC5378	H. Park and L. A. Paquette; *J. Org. Chem.*, 1980, **45**, 5378.	563
80JOM(186)155	M. F. Lipton, C. M. Sorenson, A. C. Sadler and R. H. Shapiro; *J. Organomet. Chem.*, 1980, **186**, 155.	633
80JOM(186)297	J. W. Wilson; *J. Organomet. Chem.*, 1980, **186**, 297.	163
80JOM(193)37	T. Nishiguchi, K. Tanaka and K. Fukuzumi; *J. Organomet. Chem.*, 1980, **193**, 37.	444
80JOM(198)1	G. Courtois, M. Harama and L. Miginiac; *J. Organomet. Chem.*, 1980, **198**, 1.	973, 1059
80JOM(201)233	A. Alvanipour, C. Eaborn and D. R. M. Walton; *J. Organomet. Chem.*, 1980, **201**, 233.	455
80JPR503	B. N. Ghose; *J. Prakt. Chem.*, 1980, **322(3)**, 503.	61
80LA1	E. V. Dehmlow and M. Lissel; *Justus Liebigs Ann. Chem.*, 1980, 1.	1065
80LA171	M. T. Reetz and W. Stephan; *Justus Liebigs Ann. Chem.*, 1980, 171.	664
80LA1540	R. Neidlein and H. Dörr; *Justus Liebigs Ann. Chem.*, 1980, 1540.	707, 708
80LA2061	H. J. Bestmann and H. Frey; *Justus Liebigs Ann. Chem.*, 1980, 2061.	1016
B-80MI 102-01	E-I. Negishi; "Organometallics in Organic Synthesis," Wiley, New York, 1980.	66, 69, 70
B-80MI 105-01	J. Tsuji; "Organic Synthesis With Palladium Compounds," Springer-Verlag, Heidelberg, 1980.	192
B-80MI 105-02	H. Bahrmann and B. Cornils; in "New Syntheses with Carbon Monoxide," ed. J. Falbe, Springer-Verlag, Berlin, 1980.	202
B-80MI 107-01	J. Falbe (ed.); "New Syntheses with Carbon Monoxide," Springer-Verlag, Berlin, 1980.	296, 297
B-80MI 109-01	E. N. Marvell; "Thermal Electrocyclic Reactions," Academic Press, New York, 1980.	418, 419, 420
B-80MI 109-02	W. G. Dauben, E. L. McInnis and D. M. Michno; in "Rearrangements in Ground and Excited States," ed. P. de Mayo, Academic Press, New York, 1980, vol. 3, p. 91.	420
B-80MI 109-03	K. Schaffner and M. Demuth; in "Rearrangements in Ground and Excited States," ed. P. de Mayo, Academic Press, New York, 1980, vol. 3, p. 281.	422
B-80MI 110-01	S. Patai (ed.); "The Chemistry of Ketenes, Allenes and Related Compounds," Wiley, New York, 1980.	458
B-80MI 114-01	E. Negishi; "Organometallics in Organic Synthesis," Wiley, New York, 1980, vol. 1, p. 101.	665
B-80MI 114-02	E. Negishi; "Organometallics in Organic Synthesis," Wiley, New York, 1980, vol. 1, p. 108.	665
B-80MI 114-03	G. W. Parshall; "Homogeneous Catalysis," Wiley, New York, 1980.	670
80MI 116-01	Huang Yao-tseng, Shen Yan-Chang, Ma Jing-ji and Xin Yuan-kong; *Acta Chim. Sinica Engl. Ed.*, 1980, **38**, 185.	740
80MI 116-02	Huang Yao-tseng, Shen Yan-Chang, Xin Yuan-Kong and Ma Jing-ji; *Acta Chim. Sinica Engl. Ed.*, 1980, **23**, 1396.	740, 741
B-80MI 117-01	R. F. C. Brown; "Pyrolytic Methods in Organic Chemistry," Academic Press, New York, 1980.	772, 784
B-80MI 118-01	P. De Mayo (ed); "Rearrangements in Ground and Excited States," Academic Press, New York, 1980, 3 vols.	794, 800
B-80MI 118-02	A. Pelter; in "Rearrangements in Ground and Excited States," ed. P. De Mayo, Academic Press, New York, 1980, vol. 2, p. 95.	838
B-80MI 118-03	P. J. Wagner; in "Rearrangements in Ground and Excited States," ed. P. De Mayo, Academic Press, New York, 1980, vol. 3, p. 381.	841
B-80MI 119-01	E. Buncel and T. Durst (eds.); "Comprehensive Carbanion Chemistry, Part A," Elsevier, New York, 1980.	844
B-80MI 119-02	E. Buncel and B. Menon; in "Comprehensive Carbanion Chemistry, Part A," ed. E. Buncel and T. Durst, Elsevier, New York, 1980, p. 97.	847
B-80MI 119-03	D. H. Hunter and J. B. Stothers; in "Rearrangements in Ground and Excited States," ed. P. de Mayo, Academic Press, New York, 1980, vol. 1, p. 392.	886
B-80MI 119-04	R. B. Bates; in "Comprehensive Carbanion Chemistry, Part A," ed. E. Buncel and T. Durst, Elsevier, New York, 1980, chap. 1, pp. 1–53.	890
B-80MI 119-05	A. L. J. Beckwith and K. U. Ingold; in "Rearrangements in Ground and Excited States," ed. P. de Mayo, Academic Press, New York, 1980, vol. 1, p. 162.	919, 927
B-80MI 119-06	R. F. C. Brown; "Pyrolytic Methods in Organic Chemistry," Academic Press, New York, 1980, p. 44.	924
B-80MI 120-01	H. Hopf; in "The Chemistry of Ketenes, Allenes and Related Compounds," ed. S. Patai, Wiley, New York, 1980, p. 779.	953

B-80MI 121-01	G. H. Posner; "An Introduction to Synthesis Using Organocopper Reagents," Wiley, New York, 1980.	1005
B-80MI 121-02	J. Moreau; in "Chemistry of Ketenes, Allenes and Related Compounds," ed. S. Patai, Wiley Interscience, New York, 1980, p. 363.	1051, 1053, 1054, 1055, 1058, 1060
B-80MI 122-01	G. H. Posner; "An Introduction to Synthesis Using Organocopper Reagents," Wiley, New York, 1980, p. 41.	1142
80OPP361	Y.-H. Lai; *Org. Prep. Proceed. Int.*, 1980, **12**, 361.	687
80OS(59)42	G. W. Kabalka, R. Hutchins, N. R. Natale, D. T. C. Yang and V. Broach; *Org. Synth.*, 1980, **59**, 42.	635
80OS(59)202	T. Cohen, R. J. Ruffner, D. W. Shull, E. R. Fogel and J. R. Falck; *Org. Synth.*, 1980, **59**, 202.	590
80PAC669	M. Kumada; *Pure Appl. Chem.*, 1980, **52**, 669.	279
80PS(8)243	Y. Segall, E. Shirin and I. Granoth; *Phosphorus Sulphur*, 1980, **8**, 243.	445
80RTC39	J. A. Loontjes, M. van der Leij and B. Zwanenburg; *Recl. Trav. Chim. Pays-Bas*, 1980, **99**, 39.	704
80S68	D. H. R. Barton, G. Bringmann and W. B. Motherwell; *Synthesis*, 1980, 68.	38, 945
80S150	R. E. Murray and G. Zweifel; *Synthesis*, 1980, 150.	1048
80S389	K. G. Bilyard, P. J. Garratt and R. Zahler; *Synthesis*, 1980, 389.	184
80S425	A. R. Pinder; *Synthesis*, 1980, 425.	2, 6, 10, 425, 429
80S627	S. Takahashi, N. Hagihara, K. Sonogashira and Y. Kuroyama; *Synthesis*, 1980, 627.	1010, 1012, 1137
80S904	Y. N. Bubnov, A. V. Tsyban and B. M. Mikhailov; *Synthesis*, 1980, 904.	1030
80S945	A. B. Levy, R. Angelastro and E. R. Marinelli; *Synthesis*, 1980, 945.	487
80S1007	A. Mizuno, Y. Hamada and T. Shioiri; *Synthesis*, 1980, 1007.	185, 200
80S1011	Y. Ueno, H. Sano and M. Okawara; *Synthesis*, 1980, 1011.	667
80S1034	M. Kobayashi, L. F. Valente and E.-I. Negishi; *Synthesis*, 1980, 1034.	1106
80SC49	M. Larcheveque and A. Debal; *Synth. Commun.*, 1980, **10**, 49.	209
80SC457	R. S. Tewari, S. K. Suri and K. C. Gupta; *Synth. Commun.*, 1980, **10**, 457.	741
80SC509	R. W. Dugger and C. H. Heathcock; *Synth. Commun.*, 1980, **10**, 509.	748, 749
80SC595	O. Tetsuo, O. Fumio, Y. Hachiro, H. Higuchi and S. Misumi; *Synth. Commun.*, 1980, **10**, 595.	243
80SC627	P. W. Rabideau, E. G. Burkholder and M. J. Yates; *Synth. Commun.*, 1980, **10**, 627.	97
80T3	P. A. Bartlett; *Tetrahedron*, 1980, **36**, 3.	414, 806, 825, 836
80T331	M. Huché; *Tetrahedron*, 1980, **36**, 331.	954
80T343	T. Bally and S. Masamune; *Tetrahedron*, 1980, **36**, 343.	517
80T419	S. F. Martin; *Tetrahedron*, 1980, **36**, 419.	831
80T557	P. E. Sonnet; *Tetrahedron*, 1980, **36**, 557.	592, 636, 640, 644, 650
80T679	A. R. Katritzky; *Tetrahedron*, 1980, **36**, 679.	30, 31
80T719	G. Marini-Bettòlo, S. P. Sahoo, G. Poulton, T. Y. R. Tsai and K. Wiesner; *Tetrahedron*, 1980, **36**, 719.	498
80T1215	J. F. Normant, A. Commercon and J. Villieras; *Tetrahedron*, 1980, **36**, 1215.	1017, 1018
80T1901	R. M. Magid; *Tetrahedron*, 1980, **36**, 1901.	186
80T1961	A. Alexakis, G. Cahiez and J. F. Normant; *Tetrahedron*, 1980, **36**, 1961.	478, 479
80T2269	T. H. Chan and B. S. Ong; *Tetrahedron*, 1980, **36**, 2269.	652
80T2305	C. Chuit, J. P. Foulon and J. F. Normant; *Tetrahedron*, 1980, **36**, 2305.	312
80T2497	J. L. Ripoll, H. Lebrun and A. Thuillier; *Tetrahedron*, 1980, **36**, 2497.	787
80T2531	A. Krief; *Tetrahedron*, 1980, **36**, 2531.	301, 712, 713
80T3229	W. Schnabel, I. Naito, T. Kitamura, S. Kobayashi and H. Taniguchi; *Tetrahedron*, 1980, **36**, 3229.	1120
80TCC1	C. Rüchardt; *Top. Curr. Chem.*, 1980, **88**, 1.	
80TL51	M. Montury and J. Goré; *Tetrahedron Lett.*, 1980, **21**, 51.	459
80TL67	G. Dressaire and Y. Langlois; *Tetrahedron Lett.*, 1980, **21**, 67.	253
80TL87	H. Okamura, M. Miura, K. Kosugi and H. Takei; *Tetrahedron Lett.*, 1980, **21**, 87.	483, 484
80TL287	J. L. Stein, L. Stella and J. M. Surzur; *Tetrahedron Lett.*, 1980, **21**, 287.	7
80TL395	P. J. Giddings, D. I. John and E. J. Thomas; *Tetrahedron. Lett.*, 1980, **21**, 395.	235, 246
80TL399	P. J. Giddings, D. I. John and E. J. Thomas; *Tetrahedron Lett.*, 1980, **21**, 395, 399.	822
80TL503	J. R. Hassdenteufel and M. Hanack; *Tetrahedron Lett.*, 1980, **21**, 503.	1113, 1114
80TL623	I. Kuwajima and M. Kato; *Tetrahedron Lett.*, 1980, **21**, 623.	963
80TL845	A. Minato, K. Tamao, T. Hayashi, K. Suzuki and M. Kumada; *Tetrahedron Lett.*, 1980, **21**, 845.	473
80TL849	J. Tsuji, M. Takahashi and T. Takahashi; *Tetrahedron Lett.*, 1980, **21**, 849.	1040
80TL935	J. F. Normant, A. Alexakis and G. Cahiez; *Tetrahedron Lett.*, 1980, **21**, 935.	230, 476
80TL947	K. Komatsu, I. Tomioka and K. Okamoto; *Tetrahedron Lett.*, 1980, **21**, 947.	909
80TL965	J. T. Pinhey and B. A. Rowe; *Tetrahedron Lett.*, 1980, **21**, 965.	490
80TL1069	B. L. Sondengam, G. Charles and T. M. Akam; *Tetrahedron Lett.*, 1980, **21**, 1069.	457, 508
80TL1347	Y. Fujita, T. Onishi, K. Hino and T. Nishida; *Tetrahedron Lett.*, 1980, **21**, 1347.	809
80TL1497	J. Schwartz and Y. Hayasi; *Tetrahedron Lett.*, 1980, **21**, 1497.	1105, 1139
80TL1595	J. E. O'Boyle and K. M. Nicholas; *Tetrahedron Lett.*, 1980, **21**, 1595.	199
80TL1715	D. Villemin; *Tetrahedron Lett.*, 1980, **21**, 1715.	674
80TL1739	J. P. Vigneron and J. M. Blanchard; *Tetrahedron Lett.*, 1980, **21**, 1739.	503, 505
80TL1877	M. Sevrin, J. N. Denis and A. Krief; *Tetrahedron Lett.*, 1980, **21**, 1877.	594
80TL1997	S. Halazy and A. Krief; *Tetrahedron Lett.*, 1980, **21**, 1997.	623, 713, 716
80TL2125	F. E. Zeigler, R. V. Nelson and T.-F. Wang; *Tetrahedron Lett.*, 1980, **21**, 2125.	807

80TL2181	T. Fujisawa, T. Sato, T. Kawara, M. Kawashima, H. Shimizu and Y. Ito; *Tetrahedron Lett.*, 1980, **21**, 2181.	230
80TL2477	J. E. McMurry and W. Choy; *Tetrahedron Lett.*, 1980, **21**, 2477.	357
80TL2487	T. Hudlicky and F. J. Koszyk; *Tetrahedron Lett.*, 1980, **21**, 2487.	795
80TL2531	K. Oshima, K. Takai and H. Nozaki; *Tetrahedron Lett.*, 1980, **21**, 2531.	1027
80TL2591	B. M. Trost and E. Keinan; *Tetrahedron Lett.*, 1980, **21**, 2591.	194
80TL2637	R. H. Mitchell and Y.-H. Lai; *Tetrahedron Lett.*, 1980, **21**, 2637.	11, 18, 19
80TL2799	O. Papies and W. Grimme; *Tetrahedron Lett.*, 1980, **21**, 2799.	785
80TL2865	N. Miyaura, T. Yano and A. Suzuki; *Tetrahedron Lett.*, 1980, **21**, 2865.	532, 1104
80TL2955	J. Tsuji and S. Hashiguchi; *Tetrahedron Lett.*, 1980, **21**, 2955	674
80TL3195	A. L. Gemal and J. L. Luche; *Tetrahedron Lett.*, 1980, **21**, 3195.	5, 9, 10
80TL3323	J. A. Finkelstein and C. D. Perchonock; *Tetrahedron Lett.*, 1980, **21**, 3323.	521
80TL3343	B. H. Lipshutz and J. J. Pegram; *Tetrahedron Lett.*, 1980, **21**, 3343.	654
80TL3763	C. Germon, A. Alexakis and J. F. Normant; *Tetrahedron Lett.*, 1980, **21**, 3763.	533, 1107
80TL3915	T. Hayashi, Y. Katsuro and M. Kumada; *Tetrahedron Lett.*, 1980, **21**, 3915.	480
80TL3975	D. A. Evans and L. R. McGee; *Tetrahedron Lett.*, 1980, **21**, 3975.	876
80TL3987	J.-C. Clinet and G. Linstrumelle; *Tetrahedron Lett.*, 1980, **21**, 3987.	971
80TL4021	M. Matsumoto and K. Kuroda; *Tetrahedron Lett.*, 1980, **21**, 4021.	1064
80TL4031	G. W. J. Fleet, P. J. C. Harding and M. J. Whitcombe; *Tetrahedron Lett.*, 1980, **21**, 4031.	41
80TL4047	P. E. Peterson and S. K. Chiu; *Tetrahedron Lett.*, 1980, **21**, 4047.	1021
80TL4233	D. A. Evans and J. M. Takacs; *Tetrahedron Lett.*, 1980, **21**, 4233.	876
80TL4313	J. E. McMurry and W. J. Scott; *Tetrahedron Lett.*, 1980, **21**, 4313.	480, 1027
80TL4335	S. Raucher, R. F. Lawrence and J. E. Macdonald; *Tetrahedron Lett.*, 1980, **21**, 4335.	828
80TL4365	Y. Kojima and N. Kato; *Tetrahedron Lett.*, 1980, **21**, 4365.	206
80TL4547	M. Kahn; *Tetrahedron Lett.*, 1980, **21**, 4547.	812
80TL4787	D. J. Pasto; *Tetrahedron Lett.*, 1980, **21**, 4787.	499
80TL4803	A. A. Ponaras; *Tetrahedron Lett.*, 1980, **21**, 4803.	833
80TL4811	M. E. Garst and A. T. Johnson; *Tetrahedron Lett.*, 1980, **21**, 4811.	241
80TL5019	T. Jeffery-Luong and G. Linstrumelle; *Tetrahedron Lett.*, 1980, **21**, 5019.	970, 975
80TL5037	A. Toshimitsu, H. Owada, S. Uemura and M. Okano; *Tetrahedron Lett.*, 1980, **21**, 5037.	602, 781
80ZN(B)95	R. S. Tewari, S. K. Suri and K. C. Gupta; *Z. Naturforsch., Teil B*, 1980, **35**, 95.	741
80ZOB76	N. I. Nesterov, N. N. Belyaev, M. D. Stadnichuk, K. S. Mingaleva and Y. F. Sigolaev; *Zh. Obshch. Khim.*, 1980, **50**, 76.	748
81ACS(B)263	S. E. Lauritzen, C. Rømming and L. Skattebøl; *Acta Chem. Scand., Ser. B*, 1981, **35**, 263.	706, 707
81AG925	G. A. Olah and M. Arvanaghi; *Angew. Chem.*, 1981, **93**, 925.	1046
81AG(E)569	R. Locher and D. Seebach; *Angew. Chem., Int. Ed. Engl.*, 1981, **20**, 569.	1048
81AG(E)585	J.-C. Pommelet, C. Nyns, F. Lahousse, R. Merényi and H. G. Viehe; *Angew. Chem., Int. Ed. Engl.*, 1981, **20**, 585.	592
81AG(E)677	H.-D. Martin, B. Mayer, M. Pütter and H. Höchstetter; *Angew. Chem. Int. Ed. Engl.*, 1981, **20**, 677.	708
81AG(E)911	H. J. Schäfer; *Angew. Chem., Int. Ed. Engl.*, 1981, **20**, 911.	934
81AG(E)965	B. Giese and K. Kretzschmar; *Angew. Chem., Int. Ed. Engl.*, 1981, **20**, 965.	332
81AG(E)1043	H. Schumann and B. Pachaly; *Angew. Chem., Int. Ed. Engl.*, 1981, **20**, 1043.	936
81BCJ2747	T. Hiyama, M. Shinoda, H. Saimoto and H. Nozaki; *Bull. Chem. Soc. Jpn.*, 1981, **54**, 2747.	422
81BSB1141	R. Dams, M. Malinowski and H. J. Geise; *Bull. Soc. Chim. Belg.*, 1981, **90**, 1141.	688
81CB292	U. Jensen and H.-J. Schäfer; *Chem. Ber.*, 1981, **114**, 292.	514
81CB909	D. Scholz; *Chem. Ber.*, 1981, **114**, 909.	707
81CC65	R. M. Adlington and A. G. M. Barrett; *J. Chem. Soc., Chem. Commun.*, 1981, 65.	634
81CC96	A. R. Katritzky and A. M. El-Mowafy; *J. Chem. Soc., Chem. Commun.*, 1981, 96.	629
81CC100	A. D. Buss and S. Warren; *J. Chem. Soc., Chem. Commun.*, 1981, 100.	641
81CC237	M. T. Reetz, J. Westermann and R. Steinbach; *J. Chem. Soc., Chem. Commun.*, 1981, 237.	186
81CC261	T. Sakakibara, I. Takai, E. Ohara and R. Sudoh; *J. Chem. Soc., Chem. Commun.*, 1981, 261.	608
81CC277	H. Lee and M. P. Cava; *J. Chem. Soc., Chem. Commun.*, 1981, 277.	611, 620
81CC313	T. Hayashi, M. Konishi, K. Yokota and M. Kumada; *J. Chem. Soc., Chem. Commun.*, 1981, 313.	477
81CC333	M. Tanaka and T. Kobayashi; *J. Chem. Soc., Chem. Commun.*, 1981, 333.	1015
81CC363	J. Beloeil, M. Bertranne, M. Fétizon and T. Prangé; *J. Chem. Soc., Chem. Commun.*, 1981, 363.	1039
81CC459	P. J. Stang and M. Ladika; *J. Chem. Soc., Chem. Commun.*, 1981, 459.	1070
81CC786	Y. Tobe, F. Hirata, K. Nishida, H. Fujita, K. Kimura and Y. Odaira; *J. Chem. Soc., Chem. Commun.*, 1981, 786.	811
81CC877	M. Larchevêque and A. Debal; *J. Chem. Soc., Chem. Commun.*, 1981, 877.	763
81CC891	J. Mulzer and M. Zippel; *J. Chem. Soc., Chem. Commun.*, 1981, 891.	681
81CC901	D. I. John, N. D. Tyrrell and E. J. Thomas; *J. Chem. Soc., Chem. Commun.*, 1981, 901.	945
81CC951	A. Kohda and T. Sato; *J. Chem. Soc., Chem. Commun.*, 1981, 951.	1097
81CC1081	R. H. A. Bosma, A. P. Kouwenhoven and J. C. Mol; *J. Chem. Soc., Chem. Commun.*, 1981, 1081.	674
81CC1094	B. Bennetau, J.-P. Pillot, J. Dunogues and R. Calas; *J. Chem. Soc., Chem. Commun.*, 1981, 1094.	963
81CC1121	J. E. Baldwin and J. C. Bottaro; *J. Chem. Soc., Chem. Commun.*, 1981, 1121.	634

81CI(L)548 P. J. Kocienski; *Chem. Ind. (London)*, 1981, 548. 614, 616, 696, 698, 700
81CJC1415 T. Durst, J. M Decesare, B. Corbel and J. F. Blount; *Can. J. Chem.*, 1981, **59**, 1415. 214
81CL315 K. Tanaka, M. Terauchi and A. Kaji; *Chem. Lett.*, 1981, 315. 235
81CL447 T. Otsubo, F. Ogura, H. Yamaguchi, H. Higuchi, Y. Sakata and S. Misumi; *Chem. Lett.*, 1981, 447. 594, 611, 621
81CL621 T. Mukaiyama and T. Harada; *Chem. Lett.*, 1981, 621. 973
81CL627 H. Higuchi, Y. Sakata, S. Misumi, T. Otsubo, F. Ogura and H. Yamaguchi; *Chem. Lett.*, 1981, 627. 243
81CL879 A. Miyaura and A. Suzuki; *Chem. Lett.*, 1981, 879. 531
81CL1139 N. Ono, H. Miyake, R. Tamura, I. Hamamoto and A. Kaji; *Chem. Lett.*, 1981, 1139. 617, 638
81CL1241 B. Tarnchompoo, Y. Thebtaranonth, S. Utamapanya and P. Kasemsri; *Chem. Lett.*, 1981, 1241. 1084
81CL1289 T. Nakai, K. Tanaka, K. Ogasawara and N. Ishikawa; *Chem. Lett.*, 1981, 1289. 825
81CL1307 T. Fujisawa, T. Sato, M. Kawashima and M. Nakagawa; *Chem. Lett.*, 1981, 1307. 476
81CL1415 J. Inanaga, Y. Kawanami and M. Yamaguchi; *Chem. Lett.*, 1981, 1415. 1045
81CPB3489 K. Nozawa, M. Yamada and Y. Tsuda; *Chem. Pharm. Bull.*, 1981, **27**, 3489. 45
81CPB3543 H. Yamanaka, M. Shiraiwa, E. Yamamoto and T. Sakamoto; *Chem. Pharm. Bull.*, 1981, **29**, 3543. 1010
81CPL408 R. Krishnan, M. J. Frisch, J. A. Pople and P. v. R. Schleyer; *Chem. Phys. Lett.*, 1981, **79**, 408. 1143
81CPL412 Y. Osamura and H. F. Schaeffer, III; *Chem. Phys. Lett.*, 1981, **79**, 412. 1143
81CS97 G. A. Olah; *Chem. Scr.*, 1981, **18**, 97. 891
81DOK(261)1141 N. A. Bumagin, I. G. Bumagina, A. N. Kashin and I. P. Beletskeya; *Dokl. Akad. Nauk SSSR (Engl. Transl.)*, 1981, **261**, 1141. 1015
81HCA1023 E.-P. Krebs; *Helv. Chim. Acta*, 1981, **64**, 1023. 797
81HCA2617 R. Amstutz, W. B. Schweizer, D. Seebach and J. D. Dunitz; *Helv. Chim. Acta*, 1981, **64**, 2617. 873
81ICA1 E. L. Muetterties; *Inorg. Chem. Acta*, 1981, **50**, 1. 456
81IZV2836 N. A. Bumagin, I. O. Kalinovskii and I. P. Beletskaya; *Izv. Akad. Nauk SSSR, Ser. Khim. (Engl. Transl.)*, 1981, **12**, 2836. 1004, 1006
81JA82 M. C. Pirrung; *J. Am. Chem. Soc.*, 1981, **103**, 82. 382
81JA284 A. J. Birch, A. L. Hinde and L. Radom; *J. Am. Chem. Soc.*, 1981, **103**, 284. 96, 890
81JA588 R. P. Kirchen, K. Ranganayakulu, A. Rauk, B. P. Singh and T. S. Sorensen; *J. Am. Chem. Soc.*, 1981, **103**, 588. 918
81JA932 M. J. Robins and J. S. Wilson; *J. Am. Chem. Soc.*, 1981, **103**, 932. 939
81JA1232 L. D. Kwart, A. G. Horgan and H. Kwart; *J. Am. Chem. Soc.*, 1981, **103**, 1232. 781
81JA1276 T. Yoshida and E.-I. Negishi; *J. Am. Chem. Soc.*, 1981, **103**, 1276. 991
81JA1283 W. C. Still and V. J. Novack; *J. Am. Chem. Soc.*, 1981, 103, 1283. 269, 738
81JA1501 A. B. Smith, III, P. A. Levenberg, P. J. Jerris, R. M. Scarborough, Jr. and P. M. Wovkulich; *J. Am. Chem. Soc.*, 1981, **103**, 1501. 606
81JA1557 D. D. Tanner, E. V. Blackburn and G. E. Diaz; *J. Am. Chem. Soc.*, 1981, **103**, 1557. 945, 946
81JA1757 H. Kwart, W. H. Miles, A. G. Horgan and L. D. Kwart; *J. Am. Chem. Soc.*, 1981, **103**, 1757. 834
81JA1904 Y. Osamura, H. F. Schaefer, III, S. K. Gray and W. H. Miller; *J. Am. Chem. Soc.*, 1981, **103**, 1904. 1143
81JA2127 D. A. Evans, J. Bartroli and T. L. Shih; *J. Am. Chem. Soc.*, 1981, **103**, 2127. 878
81JA2443 R. L. Danheiser, C. Martinez-Davila, R. J. Auchus and J. T. Kadonaga; *J. Am. Chem. Soc.*, 1981, **103**, 2443. 803
81JA2446 R. E. Ireland, J. D. Godfrey and S. Thaisrivongs; *J. Am. Chem. Soc.*, 1981, **103**, 2446. 831
81JA2744 R. D. Little and G. W. Muller; *J. Am. Chem. Soc.*, 1981, **103**, 2744. 359, 944
81JA2770 Y. Apeloig, W. Franke, Z. Rappoport, H. Schwarz and D. Stahl; *J. Am. Chem. Soc.*, 1981, **103**, 2770. 1122
81JA2886 G. H. Posner, J. P. Mallamo and K. Miura; *J. Am. Chem. Soc.*, 1981, **103**, 2886. 23, 236
81JA3002 J. M. Huggins and R. G. Bergman; *J. Am. Chem. Soc.*, 1981, **103**, 3002. 453, 545
81JA3081 A. I. Meyers, D. R. Williams, G. W. Erickson, S. White and M. Druelinger; *J. Am. Chem. Soc.*, 1981, **103**, 3081. 152
81JA3088 A. I. Meyers, D. R. Williams, S. White and G. W. Erickson; *J. Am. Chem. Soc.*, 1981, **103**, 3088. 152
81JA3486 K. C. Nicolaou, W. E. Barnette and R. L. Magolda; *J. Am. Chem. Soc.*, 1981, **103**, 3486. 603
81JA4466 F. M. Dayrit and J. Schwartz; *J. Am. Chem. Soc.*, 1981, **103**, 4466. 1105
81JA4494 L. M. Jackman and B. C. Lange; *J. Am. Chem. Soc.*, 1981, **103**, 4494. 873
81JA4618 E. J. Corey and J. Kang; *J. Am. Chem. Soc.*, 1981, **103**, 4618. 1015, 1016
81JA4638 W. J. Le Noble, S. Basak and S. Srivastava; *J. Am. Chem. Soc.*, 1981, **103**, 4638. 988
81JA4650 H. Kwart and M. Brechbiel; *J. Am. Chem. Soc.*, 1981, **103**, 4650. 624, 783
81JA4985 T. Yoshida and E. Negishi; *J. Am. Chem. Soc.*, 1981, **103**, 4985. 453
81JA5096 A. Sekiya and J. K. Stille; *J. Am. Chem. Soc.*, 1981, **103**, 5096. 94
81JA5253 M. D. Ward and J. Schwartz; *J. Am. Chem. Soc.*, 1981, **103**, 5253. 94
81JA5314 D. A. Lightner, T. D. Bouman, J. K. Gawronski, K. Gawronska, J. L. Chappuis, B. V. Crist and A. E. Hansen; *J. Am. Chem. Soc.*, 1981, **103**, 5314. 634
81JA5429 P. J. Stang and M. R. White; *J. Am. Chem. Soc.*, 1981, **103**, 5429. 988, 1145
81JA5568 Y. Yamakado, M. Ishiguro, N. Ikeda and H. Yamamoto; *J. Am. Chem. Soc.*, 1981, **103**, 5568. 651
81JA6437 P. J. Stang and M. Ladika; *J. Am. Chem. Soc.*, 1981, **103**, 6437. 988

81JA6460 M. F. Semmelhack, P. Helquist, L. D. Jones, L. Keller, L. Mendelson, L. S. Ryono, J. G. Smith and R. D. Stauffer; *J. Am. Chem. Soc.*, 1981, **103**, 6460. 468, 472
81JA6492 T. Nakai, K. Mikami, S. Taya and Y. Fujita; *J. Am. Chem. Soc.*, 1981, **103**, 6492. 404, 405, 816, 818
81JA6691 W. S. Trahanovsky, T. J. Cassady and T. L. Woods; *J. Am. Chem. Soc.*, 1981, **103**, 6691. 787
81JA7520 T. F. Murray, E. G. Samsel, V. Varma and J. R. Norton; *J. Am. Chem. Soc.*, 1981, **103**, 7520. 546, 1024
81JA7661 M. T. Zoeckler and B. K. Carpenter; *J. Am. Chem. Soc.*, 1981, **103**, 7661. 401
81JA7667 C. R. Johnson and N. A. Meanwell; *J. Am. Chem. Soc.*, 1981, **103**, 7667. 711
81JA7672 B. H. Lipshutz, R. S. Wilhelm and D. M. Floyd; *J. Am. Chem. Soc.*, 1981, **103**, 7672. 118, 465
81JA7677 W. D. Wulff, P.-C. Tang and J. S. McCallum; *J. Am. Chem. Soc.*, 1981, **103**, 7677. 551
81JC288 A. W. Hanson, A. W. McCulloch and A. G. McInnes; *Can. J. Chem.*, 1981, **59**, 288. 318
81JCR(S)278 R. Baudouy and J. Gore; *J. Chem. Res. (S)*, 1981, 278. 197
81JCS(P1)40 R. C. Cambie, D. Chambers, P. S. Rutledge and P. D. Woodgate; *J. Chem. Soc., Perkin Trans. 1*, 1981, 40. 1044
81JCS(P1)270 G. D. Annis, S. V. Ley, C. R. Self and R. Sivaramakrishnan; *J. Chem. Soc., Perkin Trans. 1*, 1981, 270. 226
81JCS(P1)991 S. Uemura, M. Miyoshi, M. Okano and K. Ichikawa; *J. Chem. Soc., Perkin Trans. 1*, 1981, 991. 454
81JCS(P1)1501 A. G. M. Barrett, C. R. A. Godfrey, D. M. Hollinshead, P. A. Prokopiou, D. H. R. Barton, R. B. Boar, L. Joukhadar, J. F. McGhie and S. C. Misra; *J. Chem. Soc., Perkin Trans. 1*, 1981, 1501. 16
81JCS(P1)1516 W. P. Jackson and S. V. Ley; *J. Chem. Soc., Perkin Trans. 1*, 1981, 1516. 526
81JCS(P1)1846 T. Durst, M. Lancaster and D. J. H. Smith; *J. Chem. Soc., Perkin Trans. 1*, 1981, 1846. 232
81JCS(P1)1953 S. Mageswaran, W. D. Ollis and I. O. Sutherland; *J. Chem. Soc., Perkin Trans. 1*, 1981, 1953. 400, 408, 410, 821, 823
81JCS(P1)1963 W. D. Ollis, I. O. Sutherland and Y. Thebtaranonth; *J. Chem. Soc., Perkin Trans. 1*, 1981, 1963. 823
81JCS(P1)2127 D. Cooper and S. Trippett; *J. Chem. Soc., Perkin Trans. 1*, 1981, 2127. 832
81JCS(P1)2363 D. H. R. Barton, R. S. Hay-Motherwell and W. B. Motherwell; *J. Chem. Soc., Perkin Trans. 1*, 1981, 2363. 929, 936
81JCS(P1)2848 R. M. Adlington and A. G. M. Barrett; *J. Chem. Soc., Perkin Trans. 1*, 1981, 2848. 633
81JCS(P1)3087 R. Baker and R. J. Sims; *J. Chem. Soc., Perkin Trans. 1*, 1981, 3087. 745
81JCS(P1)3225 R. H. Bradbury, T. L. Gilchrist and C. W. Rees; *J. Chem. Soc., Perkin. Trans. 1.*, 1981, 3225. 811
81JHC939 M. L. Ash and R. G. Pews; *J. Heterocycl. Chem.*, 1981, **18**, 939. 861
81JMOC397 M. D. Ward and J. Schwartz; *J. Mol. Catal.*, 1981, **11**, 397. 94
81JOC1 C. P. Mak and G. Büchi; *J. Org. Chem.*, 1981, **46**, 1. 554
81JOC9 C. M. Tice and C. H. Heathcock; *J. Org. Chem.*, 1981, **46**, 9. 813
81JOC197 C. J. Kowalski and K. W. Fields; *J. Org. Chem.*, 1981, **46**, 197. 482
81JOC231 D. L. J. Clive and V. N. Kalè; *J. Org. Chem.*, 1981, **46**, 231. 591, 593, 594, 716
81JOC459 J. M. Muchowski and M. C. Venuti; *J. Org. Chem.*, 1981, **46**, 459. 755
81JOC506 B. M. Trost and P. H. Scudder; *J. Org. Chem.*, 1981, **46**, 506. 801
81JOC531 H. C. Brown and G. J. Lynch; *J. Org. Chem.*, 1981, **46**, 531. 67
81JOC563 W. Kitching, A. R. Atkins, G. Wickham and V. Alberts; *J. Org. Chem*, 1981, **46**, 563. 67, 69
81JOC632 F. G. Bordwell, G. E. Drucker and H. E. Fried; *J. Org. Chem.*, 1981, **46**, 632. 861
81JOC645 H. C. Brown and G. A. Molander; *J. Org. Chem.*, 1981, **46** 645. 1018
81JOC647 H. C. Brown and T. M. Ford; *J. Org. Chem.*, 1981, **46**, 647. 657
81JOC687 W. G. Dauben, D. M. Michno and E. G. Olsen; *J. Org. Chem.*, 1981, **46**, 687. 421
81JOC807 M. D. Schiavelli, J. J. Plunkett and D. W. Thompson; *J. Org. Chem.*, 1981, **45**, 807. 543
81JOC829 G. Zweifel and N. R. Pearson; *J. Org. Chem.*, 1981, **46**, 829. 977
81JOC1061 B. A. Patel, J.-I. I. Kim, D. D. Bender, L.-C. Kao and R. F. Heck; *J. Org. Chem.*, 1981, **46**, 1061. 1096
81JOC1067 J.-I. I. Kim, B. A. Patel and R. F. Heck; *J. Org. Chem.*, 1981, **46**, 1067. 1096
81JOC1270 R. Vanderesse, J.-J. Brunet and P. Caubère; *J. Org. Chem.*. 1981, **46**, 1270. 430, 431
81JOC1315 F. T. Bond and R. A. DiPietro; *J. Org. Chem.*, 1981, **46**, 1315. 633
81JOC1410 B. H. Lipshutz and R. W. Hungate; *J. Org. Chem.*, 1981, **46**, 1410. 212
81JOC1485 G. W. Daub, D. H. Teramua, K. E. Bryant and M. T. Burch; *J. Org. Chem.*, 1981, **46**, 1485. 828
81JOC1593 P. W. Rabideau, N. K. Peters and D. L. Huser; *J. Org. Chem.*, 1981, **46**, 1593. 96
81JOC1693 A. Pross, D. J. DeFrees, B. A. Levi, S. K. Pollack, L. Radom and W. J. Hehre; *J. Org. Chem.*, 1981, **46**, 1693. 129, 162
81JOC1747 M. P. Cooke Jr.; *J. Org. Chem.*, 1981, **46**, 1747. 233, 709
81JOC1846 M. Miyano; *J. Org. Chem*., 1981, **46**, 1846. 603
81JOC1874 R. T. Luibrand, B. M. Broline, K. A. Charles and R. W. Dimes; *J. Org. Chem.*, 1981, **46**, 1874. 563
81JOC2059 E. Ghera, Y. Ben-David and H. Rapoport; *J. Org. Chem.*, 1981, **36**, 2059. 861
81JOC2065 P. A. Jacobi, D. G. Walker and I. M. A. Odeh; *J. Org. Chem.*, 1981, **46**, 2065. 523
81JOC2199 S. G. Levine and R. L. McDaniel, Jr.; *J. Org. Chem.*, 1981, **46**, 2199. 812, 813
81JOC2228 G. Rosini, G. G. Spineti, E. Foresti and G. Pradella; *J. Org. Chem.*, 1981, **46**, 2228. 834
81JOC2280 W. B. Austin, N. Bilow, W. J. Kelleghan and K. S. Y. Lau; *J. Org. Chem.*, 1981, **46**, 2280. 1137
81JOC2327 P. C. Wong and D. Griller; *J. Org. Chem.*, 1982, **46**, 2327. 1127
81JOC2363 P. Beak and L. G. Carter; *J. Org. Chem.*, 1981, **46**, 2363. 137
81JOC2596 R. Seshadri, W. J. Pegg and M. Israel; *J. Org. Chem.*, 1981, **46**, 2596. 10

81JOC2622 S. K. Pradham, S. R. Khadam, J. N. Kolhe, T. V. Radhakrishnan, S. V. Sohani and V. B. Thaker; *J. Org. Chem.*, 1981, **45**, 2622. 548

81JOC2709 S. K. Taylor, G. L. Lilley, K. J. Lilley and P. A. McCoy; *J. Org. Chem.*, 1981, **46**, 2709. 475

81JOC2804 R. F. Atkinson, T. W. Balko, T. R. Westman, G. C. Sypniewski, M. A. Carmody, C. T. Pauler, C. L. Schade, D. E. Coulter, H. T. Pham, F. Barea and A. Hassner; *J. Org. Chem.*, 1981, **46**, 2804. 777

81JOC2831 F. Sauriol-Lord and T. B. Grindley; *J. Org. Chem.*, 1981, **46**, 2831. 209

81JOC2985 R. Davis and K. G. Untch; *J. Org. Chem.*, 1981, **46**, 2985. 179

81JOC3293 R. M. Wilson, R. A. Farr and D. J. Burlett; *J. Org. Chem.*, 1981, **46**, 3293. 253

81JOC3315 V. Cere, C. Paolucci, S. Pollicino, E. Sandri and A. Fava; *J. Org. Chem.*, 1981, **46**, 3315. 821

81JOC3359 W. Adam, G. Martinez, J. Thompson and F. Yang; *J. Org. Chem.*, 1981, **46**, 3359. 566, 681, 682

81JOC3376 J. D. White, T. Matsui and J. A. Thomas; *J. Org. Chem.*, 1981, **46**, 3376. 382

81JOC3696 J. P. Marino and R. J. Linderman; *J. Org. Chem.*, 1981, **45**, 3696. 540

81JOC3721 I. L. Reich and H. J. Reich; *J. Org. Chem.*, 1981, **46**, 3721. 603

81JOC3896 P. A. Bartlett and C. F. Pizzo; *J. Org. Chem.*, 1981, **46**, 3896. 828

81JOC3909 F. Rolla; *J. Org. Chem.*, 1981, **46**, 3909. 4, 6, 7, 9, 10

81JOC3925 W. J. Evans, D. J. Wink, A. L. Wayda and D. A. Little; *J. Org. Chem.*, 1981, **46**, 3925. 890

81JOC4072 S. C. Welch and J.-P. Loh; *J. Org. Chem.*, 1981, **46**, 4072. 693

81JOC4093 C. L. Rand, D. E. Van Horn, M. W. Moore and E. Negishi; *J. Org. Chem.*, 1981, **45**, 4093. 542, 543

81JOC4600 C. K. Bradsher and K. J. Edgar; *J. Org. Chem.*, 1981, **46**, 4600. 1094

81JOC4608 C. K. Bradsher and D. A. Hunt; *J. Org. Chem.*, 1981, **46**, 4608. 1094

81JOC4727 A. Toshimitsu, T. Aoai, H. Owada, S. Uemura and M. Okano; *J. Org. Chem.*, 1981, **46**, 4727. 603

81JOC4843 J. R. Rasmussen, C. J. Slinger, R. J. Kordish and D. D. Newman-Evans; *J. Org. Chem.*, 1981, **46**, 4843. 16

81JOC4885 K. Kikukawa, K. Maemura, Y. Kiseki, F. Wada, T. Matsuda and C. S. Giam; *J. Org. Chem.*, 1981, **46**, 4885. 486

81JOC5027 M. R. Britten-Kelly, B. J. Willis and D. H. R. Barton; *J. Org. Chem.*, 1981, **46**, 5027. 240

81JOC5041 E. Negishi, C. L. Rand and K. P. Jadhav; *J. Org. Chem.*, 1981, **46**, 5041. 142, 1053

81JOC5094 M. P. Doyle, W. H. Tamblyn and V. Bagheri; *J. Org. Chem.*, 1981, **46**, 5094. 400, 408, 409, 822, 823, 825

81JOC5214 R. O. Hutchins, I. M. Taffer and W. Burgoyne; *J. Org. Chem.*, 1981, **46**, 5214. 14

81JOC5242 W. C. Still and H. J. Ohmizu; *J. Org. Chem.*, 1981, **46**, 5242. 217

81JOC5246 M. Shimizu, R. Ando and I. Kuwajima; *J. Org. Chem.*, 1981, **46**, 5246. 620

81JOC5379 J. P. Marino and H. Abe; *J. Org. Chem.*, 1981, **46**, 5379. 226

81JOC5436 N. E. Schore and M. C. Croudace; *J. Org. Chem.*, 1981, **45**, 5436. 530

81JOM(204)345 B. Meunier; *J. Organomet. Chem.*, 1981, **204**, 345. 429

81JOM(205)4 G. B. Deacon and T. D. Tuong; *J. Organomet. Chem.*, 1981, **205**, 4. 1038

81JOM(206)257 H. Kleijn, H. Westmijze, J. Meijer and P. Vermeer; *J. Organomet. Chem.*, 1981, **206**, 257. 544

81JOM(209)109 J. F. Fauvarque and A. Jutand; *J. Organomet. Chem.*, 1981, **209**, 109. 467

81JOM(213)C53 N. Miyaura and A. Suzuki; *J. Organomet. Chem.*, 1981, **213**, C53. 472, 489, 531

81JOM(214)145 L. E. Gusel'nikov, V. M. Sokolova, E. A. Volina, V. G. Zalkin, N. S. Nametkin, M. G. Voronko V, S. V. Kirpichenko and V. V. Keiko; *J. Organomet. Chem.*, 1981, **214**, 145. 63

81JOM(217)43 M. Ishikawa, H. Sugisawa, O. Harata and M. Kumada; *J. Organomet. Chem.*, 1981, **217**, 43. 450

81JOM(217)267 K. Ruitenberg, J. Meijer, R. J. Bullee and P. Vermeer; *J. Organomet. Chem.*, 1981, **217**, 267. 1017, 1018

81JOM(218)1 G. Courtois, M. Harama and P. Miginiac; *J. Organomet. Chem.*, 1981, **218**, 1. 973

81JOM(218)275 G. Courtois, M. Harama and L. Miginiac; *J. Organomet. Chem.*, 1981, **218**, 275. 1059

81JOU18 I. P. Beletskaya, N. A. Bumagin, A. N. Kashin and I. G. Bumagins; *J. Org. Chem. USSR* (*Engl. Transl.*), 1981, **47**, 18. 1006

81JOU381 A. A. Solov'yanov, A. D. Karpyuk, I. P. Beletskaya and O. A. Reutov; *J. Org. Chem. USSR*, 1981, **17**, 381. 859

81LA1215 H. Redlich and W. Roy; *Liebigs Ann. Chem.*, 1981, 1215. 5, 7

81MI 103-01 J. P. Coleman and J. H. Wagenknecht; *J. Electrochem. Soc.*, 1981, **128**, 322. 103

B-81MI 105-01 E. W. Colvin; "Silicon in Organic Synthesis," Butterworths, London, 1981. 216

B-81MI 107-01 E. W. Colvin; "Silicon in Organic Synthesis," Butterworth, London, 1981, p. 104. 295

B-81MI 109-01 J. J. Gajewski; "Hydrocarbon Thermal Isomerizations," Academic Press, New York, 1981. 419

B-81MI 110-01 E. W. Colvin; "Silicon in Organic Synthesis," Butterworths, London, 1981. 447, 450

81MI 110-01 F. H. Jardine; *Prog. Inorg. Chem.*, 1981, **28**, 63. 456

B-81MI 114-01 A. C. Knipe; in "The Chemistry of the Sulfonium Group," pt. 1, ed. C. J. M. Stirling and S. Patai, Wiley, New York, 1981, pp. 334–347. 623

B-81MI 114-02 E. W. Colvin; "Silicon in Organic Synthesis," Butterworths, London, 1981, pp. 141–164. 648

81MI 116-01 Y. Z. Huang, N. T. Hsing, L. L. Shi, F. L. Ling and Y. Y. Xu; *Acta Chim. Sinica Engl. Ed.*, 1981, **39**, 348. 738

81MI 117-01 U. E. Wiersum; *Aldrichimica Acta*, 1981, **14**, 53. 789

B-81MI 117-02 J. J. Gajewski; "Organic Chemistry: Hydrocarbon Thermal Isomerisations," Academic Press, New York, 1981, vol. 44, p. 61. 784

B-81MI 118-01 J. J. Gajewski; "Hydrocarbon Thermal Isomerisations," Academic Press, New York, 1981. 800, 825

B-81MI 118-02 J. J. Gajewski; in "Hydrocarbon Thermal Isomerisations," Academic Press, New York, 1981, p. 177. 805

81MI 119-01 S. W. Staley and C. K. Dustman; *React. Intermed.* (*Wiley*), 1981, **2**, 15. 891

81MI 119-02 D. Bethell and D. Whittaker; *React. Intermed.* (*Wiley*), 1981, **2**, 211. 891

81MI 122-01 H. C. Brown and J. B. Campbell, Jr.; *Aldrichimica Acta*, 1981, **14**, 3. 1094, 1100, 1102
81OS(60)29 M. A. Umbreit and K. B. Sharpless; *Org. Synth.*, 1981, **60**, 29. 587
81OS(60)53 A. Padwa, M. J. Pulver and T. J. Blacklock; *Org. Synth.*, 1981, **60**, 53. 571
81OS(60)108 J. L. Fry and S. B. Silverman; *Org. Synth.*, 1981, **60**, 108. 20
81PAC15 D. H. R. Barton and W. B. Motherwell; *Pure Appl. Chem.*, 1981, **53**, 15. 935, 938
81PAC1109 D. A. Evans, J. M. Takacs, L. R. McGee, M. D. Ennis, D. J. Mathre and J. Bartroli; *Pure Appl. Chem.*, 1981, **53**, 1109. 878
81PAC2333 E. Negishi; *Pure Appl. Chem.*, 1981, **53**, 2333. 453, 542
81PNA(78)1998 G. A. Olah, A. L. Berrier and G. K. S. Prakash; *Proc. Natl. Acad. Sci. USA*, 1981, **78**, 1998. 899
81RCR666 V. A. Mironov, A. D. Fedorovich and A. A. Akhrem; *Russ. Chem. Rev.*, 1981, **50**, 666. 794
81RTC34 R. G. Visser, H. J. T. Bos and L. Brandsma; *Recl. Trav. Chim. Pays-Bas.*, 1981, **100**, 34. 981
81RTC337 H. Kleijn, M. Tigchelaar, J. Meijer and P. Vermeer; *Recl. Trav. Chim. Pays-Bas*, 1981, **100**, 337. 544
81S1 O. Mitsunobu; *Synthesis*, 1981, 1. 179
81S32 T. Morita, Y. Okamoto and H. Sakurai; *Synthesis*, 1981, 32. 13, 15, 17
81S55 I. H. Sánchez and M. A. Aguilar; *Synthesis*, 1981, 55. 24
81S125 A. Burmester and H. B. Stegmann; *Synthesis*, 1981, 125. 201
81S185 G. Solladié; *Synthesis*, 1981, 185. 140
81S276 C. B. Reese and H. P. Sanders; *Synthesis*, 1981, 276. 1082
81S364 D. E. Ames, D. Bull and C. Takundwa; *Synthesis*, 1981, 364. 1010, 1012
81S459 P. E. van Rijn, S. Mommers, R. G. Visser, H. D. Verkruijsse and L. Brandsma; *Synthesis*, 1981, 459. 1037
81S490 W. Schumacher and M. Hanack; *Synthesis*, 1981, 490. 204, 1023
81S551 J. Meijer, K. Ruitenberg, H. Westmijze and P. Vermeer; *Synthesis*, 1981, 551. 1058
81S743 D. H. R. Barton, W. B. Motherwell and A. Stange; *Synthesis*, 1981, 743. 16, 938
81S753 M. Petrzilka and J. I. Grayson; *Synthesis*, 1981, 753. 520, 523
81S841 J. F. Normant and A. Alexakis; *Synthesis*, 1981, 841. 538, 539, 540, 541, 542, 1093, 1106, 1139
81S875 J.-C. Clinet and G. Linstrumelle; *Synthesis*, 1981, 875. 204, 971, 1053, 1056, 1057, 1058
81S879 K. Fukunaga and H. Yamaguchi; *Synthesis*, 1981, 879. 1069
81S999 H. P. On, W. Lewis and G. Zweifel; *Synthesis*, 1981, 999. 451
81S1001 T. Hayashi, T. Fujiwa, Y. Okamoto, Y. Katsuro and M. Kumada; *Synthesis*, 1981, 1001. 479
81SC237 T.-L. Ho; *Synth. Commun.*, 1981, **11**, 237. 832
81SC281 D. E. Seitz, R. A. Milius and H. El-Wakil; *Synth. Commun.*, 1981, 281. 452
81SC513 N. Miyaura, T. Yanagi and A. Suzuki; *Synth. Commun.*, 1981, **11**, 513. 473, 489
81SC723 S. Julia and E. Guittet; *Synth. Commun.*, 1981, **11**, 723. 225
81SC895 W. H. Richardson; *Synth. Commun.*, 1981, **11**, 895. 690
81T31 K. Kikukawa, K. Nagira, F. Wada and T. Matsuda; *Tetrahedron*, 1981, **37**, 31. 486
81T173 B. J. Brisdon, P. Nair and S. F. Dyke; *Tetrahedron*, 1981, **37**, 173. 493
81T263 J. E. Baldwin, N. V. Reed and E. J. Thomas; *Tetrahedron*, 1981, **37**, Supplement 1, 263. 601
81T417 J. J. Burger, T. B. R. A. Chen, E. R. de Waard and H. O. Huisman; *Tetrahedron*, 1981, **37**, 417. 619, 707
81T1359 P. Place, C. Vernière and J. Goré; *Tetrahedron*, 1981, **37**, 1359. 974
81T1653 E. V. Dehmlow and M. Lissel; *Tetrahedron*, 1981, **37**, 1653. 1065
81T2617 R. Rossi, A. Carpita and M. G. Quirici; *Tetrahedron*, 1981, **37**, 2617. 509, 532
81T2797 M. Hayashi, S. Terashima and K. Koga; *Tetrahedron*, 1981, **37**, 2797. 33, 34
81T3073 A. L. J. Beckwith; *Tetrahedron*, 1981, **37**, 3073. 919
81T3263 W. G. Dauben and D. M. Michno; *Tetrahedron*, 1981, **37**, 3263. 421
81T3615 R. Brambilla, R. Friary, A. Ganguly, A. McPhail, K. D. Onan, M. S. Puar, B. R. Sunday and J. J. Wright; *Tetrahedron*, 1981, **37**, 3615 91
81T3873 H. Nagaoka and Y. Kishi; *Tetrahedron*, 1981, **37**, 3873. 748
81T3935 R. M. Adlington and A. G. M. Barrett; *Tetrahedron*, 1981, **37**, 3935. 634
81T3943 R. L. Danheiser, C. Martinez-Davila and H. Sard; *Tetrahedron*, 1981, **37**, 3943. 805
81T3957 M. F. Semmelhack, G. R. Clark, J. L. Garcia, J. J. Harrison, Y. Thebtaranonth, W. Wulff and A. Yamashita; *Tetrahedron*, 1981, **37**, 3957. 474
81T3967 P. A. Wender, S. M. Sieburth, J. J. Petraitis and S. K. Singh; *Tetrahedron*, 1981, **37**, 3967. 841
81T4097 K. C. Nicolaou; *Tetrahedron*, 1981, **37**, 4097. 25
81TH J. M. Takacs; Ph.D. Thesis, California Institute of Technology, Pasadena, 1981. 876
81TL69 T. Nakai, K. Mikami, S. Taya, Y. Kimura and T. Mimura; *Tetrahedron Lett.*, 1981, **22**, 69. 404, 406, 816
81TL103 C. Vernière, B. Cazes and J. Goré; *Tetrahedron Lett.*, 1981, **22**, 103. 958
81TL127 N. Miyaura, H. Suginome and A. Suzuki; *Tetrahedron Lett.*, 1981, **22**, 127. 279, 532
81TL199 V. A. Curtis, F. J. Knutson and R. J. Baumgarten; *Tetrahedron Lett.*, 1981, **22**, 199. 636
81TL315 V. Ratovelomanana and G. Linstrumelle; *Tetrahedron Lett.*, 1981, **22**, 315. 467, 1009
81TL421 M. J. Robins and P. J. Barr; *Tetrahedron Lett.*, 1981, **22**, 421. 1010
81TL455 J. Pornet; *Tetrahedron Lett.*, 1981, **22**, 455. 978
81TL959 N. Jabri, A. Alexakis and J. F. Normant; *Tetrahedron Lett.*, 1981, **22**, 959. 468
81TL1179 F. E. Zeigler and T.-F. Wang; *Tetrahedron Lett.*, 1981, **22**, 1179. 807
81TL1327 J. Pornet and B. Randrianoelina; *Tetrahedron Lett.*, 1981, **22**, 1327. 978
81TL1437 M. Gill, H. P. Bainton and R. W. Rickards; *Tetrahedron Lett.*, 1981, **22**, 1437. 1091
81TL1451 K. Ruitenberg, H. Kleijn, C. J. Elsevier, J. Meijer and P. Vermeer; *Tetrahedron Lett.*, 1981, **22**, 1451. 975, 1004, 1009
81TL1575 G. L. Larson, J. A. Prieto and A. Hernández; *Tetrahedron Lett.*, 1981, **22**, 1575. 763
81TL1579 R. G. Daniels and L. A. Paquette; *Tetrahedron Lett.*, 1981, **22**, 1579. 971

81TL1583 W. G. Dauben and A. Chollet; *Tetrahedron Lett.*, 1981, **22**, 1583. 807, 809

81TL1595 M. Larchevêque, C. Legueut, A. Debal and J. Y. Lallemand; *Tetrahedron Lett.*, 1981, **22**, 1595. 763

81TL1609 K. Oshima, K. Takai, M. Sato and H. Nozaki; *Tetrahedron Lett.*, 1981, **22**, 1609. 1027

81TL1623 J. Lucchetti and A. Krief; *Tetrahedron Lett.*, 1981, **22**, 1623. 25

81TL1675 A. Krebs, E. Franken and S. Müller; *Tetrahedron Lett.*, 1981, **22**, 1675. 522

81TL1705 N. Ono, H. Miyake, R. Tamura and A. Kaji; *Tetrahedron Lett.*, 1981, **22**, 1705. 37, 945

81TL1751 N. J. Barnes, A. H. Davidson, L. R. Hughes, G. Procter and V. Rajcoomar; *Tetrahedron Lett.*, 1981, **22**, 1751. 215

81TL1809 M. Suzuki, T. Kawagashi and R. Noyori; *Tetrahedron Lett.*, 1981, **22**, 1809. 607

81TL1833 S. Halazy and A. Krief; *Tetrahedron Lett.*, 1981, **22**, 1833. 716

81TL1891 J. Metzger, J. Hartmanns and P. Köll; *Tetrahedron Lett.*, 1981, **22**, 1891. 925

81TL1919 J. E. Bäckvall and L. Engman; *Tetrahedron Lett.*, 1981, **22**, 1919. 595

81TL2243 D. A. Holt; *Tetrahedron Lett.*, 1981, **22**, 2243. 464

81TL2471 P. A. Wender and S. M. Sieburth; *Tetrahedron Lett.*, 1981, **22**, 2471. 841

81TL2583 T. Ohsawa, T. Takagaki, A. Haneda and T. Oishi; *Tetrahedron Lett.*, 1981, **22**, 2583. 3, 6

81TL2629 Y. Hayashi, M. Riediker, J. S. Temple and J. Schwartz; *Tetrahedron Lett.*, 1981, **22**, 2629. 536

81TL2675 Y. Ueno, H. Sano, S. Aoki and M. Okawara; *Tetrahedron Lett.*, 1981, **22**, 2675. 615, 698

81TL2751 D. J. S. Tsai and D. S. Matteson; *Tetrahedron Lett.*, 1981, **22**, 2751. 651, 762

81TL2827 R. G. Visser, L. Brandsma and H. J. T. Bos; *Tetrahedron Lett.*, 1981, **22**, 2827. 981

81TL3097 M. Mikołajczyk, S. Grzejszczak and K. Korbacz; *Tetrahedron Lett.*, 1981, **22**, 3097. 608

81TL3401 J.-P. Pillot, B. Bennetau, J. Dunogues and R. Calas; *Tetrahedron Lett.*, 1981, **22**, 3401. 963, 978

81TL3463 B. M. Trost and P. L. Ornstein; *Tetrahedron Lett.*, 1981, **22**, 3463. 438

81TL3551 R. Caputo, L. Mangoni, O. Neri and G. Palumbo; *Tetrahedron Lett.*, 1981, **22**, 3551. 587

81TL3609 J. Pornet and N. Kolani; *Tetrahedron Lett.*, 1981, **22**, 3609. 978

81TL3737 E. Negishi, S. Chatterjee and H. Matsushita; *Tetrahedron Lett.*, 1981, **22**, 3737. 476

81TL3815 S. Nahm and S. M. Weinreb; *Tetrahedron Lett.*, 1981, **22**, 3815. 1047

81TL3851 N. Jabri, A. Alexakis and J. F. Normant; *Tetrahedron Lett.*, 1981, **22**, 3851. 471, 1097, 1107

81TL3985 K. Takai, I. Mori, K. Oshima and H. Nozaki; *Tetrahedron Lett.*, 1981, **22**, 3985. 415

81TL4009 J. N. Denis, S. Desauvage, L. Hevesi and A. Krief; *Tetrahedron Lett.*, 1981, **22**, 4009. 590

81TL4069 E. Wistuba and C. Rüchardt; *Tetrahedron Lett.*, 1981, **22**, 4069. 388

81TL4097 T. Nakai, H. Setoi and Y. Kageyama; *Tetrahedron Lett.*, 1981, **22**, 4097. 828

81TL4137 A. de Groot, B. J. M. Jansen, J. T. A. Reuvers and E. M. Tedjo; *Tetrahedron Lett.*, 1981, **22**, 4137. 600

81TL4163 N. Ikota, N. Takamura, S. D. Young and B. Ganem; *Tetrahedron Lett.*, 1981, **22**, 4163. 637

81TL4171 M. M. Midland, R. L. Halterman, C. A. Brown and A. Yamaichi; *Tetrahedron Lett.*, 1981, **22**, 4171. 1085

81TL4313 Y. Murata, T. Ohtsuka, H. Shirahama and T. Matsumoto; *Tetrahedron Lett.*, 1981, **22**, 4313. 382

81TL4341 S. Halazy and A. Krief; *Tetrahedron Lett.*, 1981, **22**, 4341. 245

81TL4389 R. D. Little and G. L. Carroll; *Tetrahedron Lett.*, 1981, **22**, 4389. 944

81TL4449 T. Hayashi, Y. Katsuro, Y. Okamoto and M. Kumada; *Tetrahedron Lett.*, 1981, **22**, 4449. 482

81TL4491 M. Ochiai, M. Arimoto and E. Fujita; *Tetrahedron Lett.*, 1981, **22**, 4491. 490

81TL4495 H. Kurosawa, H. Okada and T. Hattori; *Tetrahedron Lett.*, 1981, **22**, 4495. 69

81TL4655 M. Riediker and J. Schwartz; *Tetrahedron Lett.*, 1981, **22**, 4655. 536

81TL4937 A. Krebs, W. Rüger and W.-U. Nickel; *Tetrahedron Lett.*, 1981, **22**, 4937. 686

81TL5031 T. Kauffmann, R. König, C. Pahde and A. Tannert; *Tetrahedron Lett.*, 1981, **22**, 5031. 649, 656, 678, 680, 759

81TL5111 T. G. Back and S. Collins; *Tetrahedron Lett.*, 1981, **22**, 5111. 605

81TL5231 R. C. Larock, K. Takagi, S. S. Hershberger and M. A. Mitchell; *Tetrahedron Lett.*, 1981, **22**, 5231. 493

81TL5283 Y. Z. Huang, Y. Shen, W. Ding and J. Zheng; *Tetrahedron Lett.*, 1981, **22**, 5283. 1079

81ZOB420 N. M. Salimgareeva, O. Zh. Zhebarov, N. G. Bogatova and V. P. Yur'ev; *Zh. Obshch. Khim.*, 1981, **51**, 420. 61

82ACR46 D. Farcasiu; *Acc. Chem. Res.*, 1982, **15**, 46–51. 914

82ACR178 A. Suzuki; *Acc. Chem. Res.*, 1982, **15**, 178. 198

82ACR340 E. Negishi; *Acc. Chem. Res.*, 1982, **15**, 340. 468, 542, 1004

82ACR348 P. J. Stang; *Acc. Chem. Res.*, 1982, **15**, 348. 988, 1143

82AG(E)71 A. Pfaltz and A. Mattenberger; *Angew. Chem., Int. Ed. Engl.*, 1982, **21**, 71. 220

82AG(E)96 M. T. Reetz; *Angew. Chem., Int. Ed. Engl.*, 1982, **21**, 96. 133

82AG(E)225 E. Schaumann and R. Ketcham; *Angew. Chem., Int. Ed. Engl.*, 1982, **21**, 225. 784

82AG(E)401 J. M. Tedder; *Angew. Chem., Int. Ed. Engl.*, 1982, **21**, 401. 919, 925

82AG(E)410 T. Kauffmann; *Angew. Chem., Int. Ed. Engl.*, 1982, **21**, 410. 1089

82AG(E)686 R. Gleiter and K. Gubermator; *Angew. Chem., Int. Ed. Engl.*, 1982, **21**, 686. 814

82AG(E)768 B. Giese and S. Lachhein; *Angew. Chem., Int. Ed. Engl.*, 1982, **21**, 768. 546

82AG(E)820 M. Demuth and K. Schaffner; *Angew. Chem., Int. Ed. Engl.*, 1982, **21**, 820. 96

82AG(E)924 R. Bolze, H. Eierdanz, K. Schlüter, W. Massa, W. Grahn and A. Berndt; *Angew. Chem., Int. Ed. Engl.*, 1982, **21**, 924. 981

82AG(S)161 A. Pfaltz and A. Mattenberger; *Angew. Chem. Suppl.*, 1982, 161. 220

82AJC1895 E. Dimitriadis and R. A. Massy-Westropp; *Aust. J. Chem.*, 1982, **35**, 1895. 17

82AOC(20)115 Y. Huang and Y. Shen; *Adv. Organomet. Chem.*, 1982, **20**, 115. 646

82APO(18)1 J. Toullec; *Adv. Phys. Org. Chem.*, 1982, **18**, 1. 869

82ARP493 D. F. McMillen and D. M. Golden; *Annu. Rev. Phys. Chem.*, 1982, **33**, 493. 921, 925

82BCJ167 M. Segi, M. Takebe, S. Masuda, T. Nakajima and S. Suga; *Bull. Chem. Soc. Jpn.*, 1982, **55**, 167. 475

82BCJ182 H. Higuchi, T. Otsubo, F. Ogura, H. Yamaguchi, Y. Sakata and S. Misumi; *Bull. Chem. Soc. Jpn.*, 1982, **55**, 182. 243

82BCJ643 T. Imamoto, T. Kusumoto and M. Yokoyama; *Bull. Chem. Soc. Jpn.*, 1982, **55**, 643. 202

82BCJ2434 A. Kasahara, T. Izumi, I. Shimizu, M. Satou and T. Katou; *Bull. Chem. Soc. Jpn.*, 1982, **55**, 2434. 690

82BCJ2965 K. Tanaka, H. Uneme, S. Matsui and A. Kaji; *Bull. Chem. Soc. Jpn.*, 1982, **55**, 2965. 240

82BCJ3257 K. Komatsu, K. Takeuchi, M. Arima, Y. Waki, S. Shirai and K. Okamoto; *Bull. Chem. Soc. Jpn.*, 1982, **55**, 3257. 907

82BCJ3555 T. Fujisawa, T. Sato, Y. Gotoh, M. Kawashima and T. Kawara; *Bull. Chem. Soc. Jpn.*, 1982, **55**, 3555. 481

82BSB149 R. Dams, M. Malinowski and H. J. Giese; *Bull. Soc. Chim. Belg*, 1982, **91**, 149. 692

82BSB422 C. S. Yannoni, V. Macho and P. C. Myhre; *Bull. Soc. Chim. Belg.*, 1982, **91**, 422. 906

82BSF269 M. Dagonneau; *Bull. Soc. Chim. Fr.*, 1982, 269. 495

82CB804 G. Maier and W. A. Jung; *Chem. Ber.*, 1982, **115**, 804. 521

82CB1234 J. Janssen and W. Lüttke; *Chem. Ber.*, 1982, **115**, 1234. 690

82CB2508 W. R. Roth, W. Kirmse, W. Hoffmann and H.-W. Lennartz; *Chem. Ber.*, 1982, **115**, 2508. 801

82CB2526 B. Giese and W. Zwick; *Chem. Ber.*, 1982, **115**, 2526. 948

82CC56 Y. Sato and Y. Niinomi; *J. Chem. Soc., Chem. Commun.*, 1982, 56. 451

82CC164 L. Engman and M. P. Cava; *J. Chem. Soc., Chem. Commun.*, 1982, 164. 620

82CC281 M. Ochiai, S. Tada, K. Sumi and E. Fujita; *J. Chem. Soc., Chem. Commun.*, 1982, 281. 600

82CC433 P. Calas, P. Moreau and A. Commeyras; *J. Chem. Soc., Chem. Commun.*, 1982, 433. 1130

82CC447 D. H. R. Barton, W. Hartwig and W. B. Motherwell; *J. Chem. Soc., Chem. Commun.*, 1982, 447. 945

82CC564 D. Labar and A. Krief; *J. Chem. Soc., Chem. Commun.*, 1982, 564. 592, 593, 713, 716

82CC589 D. N. Jones, T. P. Kogan, R. F. Newton and S. Smith; *J. Chem. Soc., Chem.* Commun., 1982, 589. 234

82CC821 N. Ono, I. Hamamoto and A. Kaji; *J. Chem. Soc., Chem. Commun.*, 1982, 821. 250

82CC840 E. Wenkert and T. W. Ferreira; *J. Chem. Soc., Chem. Commun.*, 1982, 840. 436, 438, 484

82CC871 S. Tomoda, Y. Takeuchi and Y. Nomura; *J. Chem. Soc., Chem. Commun.*, 1982, 871. 603

82CC959 G. Capozzi, G. Romeo and F. Marcuzzi; *J. Chem. Soc., Chem. Commun.*, 1982, 959. 1003

82CC969 T.-H. Chan and J.-S. Li; *J. Chem. Soc., Chem. Commun.*, 1982, 969. 651

82CC1042 T. Imamoto, T. Kusumoto and M. Yokoyama; *J. Chem. Soc., Chem. Commun.*, 1982, 1042. 1098, 1140

82CC1164 R. A. Aitken, J. I. G. Cadogan, I. Gosney, B. J. Hamill and L. M. McLaughlin; *J. Chem. Soc., Chem. Commun.*, 1982, 1164. 811

82CC1231 W. R. Jackson and C. G. Lovel; *J. Chem. Soc., Chem. Commun.*, 1982, 1231. 545

82CC1292 K. C. Nicolaou and J. Uenishi; *J. Chem. Soc., Chem. Commun.*, 1982, 1292. 14

82CC1331 M. F. Schlecht; *J. Chem. Soc., Chem. Commun.*, 1982, 1331. 227

82CJC425 N. Berenjian, P. de Mayo, M.-E. Sturgeon, L. K. Sydnes and A. C. Weedon; *Can. J. Chem.*, 1982, **60**, 425. 356

82CJC735 C. C. Lee, C. Y. Fiakpui and J. W. Quail; *Can. J. Chem.*, 1982, **60**, 735. 1112

82CJC1256 E. Piers, K. F. Cheng and I. Nagakura; *Can. J. Chem.*, 1982, **60**, 1256. 469

82CJC1363 A. R. Sanger; *Can. J. Chem.*, 1982, **60**, 1363. 456

82CL157 M. Sato and K. Oshima; *Chem. Lett.*, 1982, 157. 514, 587, 692

82CL165 T. Yoshida and S. Saito; *Chem. Lett.*, 1982, 165. 613

82CL351 K. Tanaka, M. Terauchi and A. Kaji; *Chem. Lett.*, 1982, 351. 235

82CL603 T. Okano, K. Tsukiyama, H. Konishi and J. Kiji; *Chem. Lett.*, 1982, 603. 94

82CL1029 M. Ishiguro, N. Ikeda and H. Yamamoto; *Chem. Lett.*, 1982, 1029. 973

82CL1485 R. Yamaguchi, H. Kawasaki, T. Yoshitome and M. Kawanisi; *Chem. Lett.*, 1982, 1485. 1004

82CL1697 K. Ogura, M. Fujita, K. Takahashi and H. Iida; *Chem. Lett.*, 1982, 1697. 693

82CL1871 K. Hiroi and S. Sato; *Chem. Lett.*, 1982, 1871. 235

82COMC-I(1)121 N. A. Bell; *Comp. Organomet. Chem., 1st edn.*, 1982, **1**, 121. 452, 456

82COMC-I(4)377 A. J. Deeming; *Comp. Organomet. Chem., 1st edn.*, 1982, **4**, 377. 196

82COMC-I(7)365 J. R. Zeitz, Jr., G. C. Robinson and K. L. Lindsay; *Comp. Organomet. Chem., 1st edn.*, 1982, **7**, 365. 665

82COMC-I(7)85 B. J. Wakefield; *Comp. Organomet. Chem., 1st edn.*, 1982, **7**, 85. 664

82COMC-I(8)101 I. Tkatchenko; *Comp. Organomet. Chem., 1st edn.*, 1982, **8**, 101. 228, 296, 297

82COMC-I(8)499 R. H. Grubbs; *Comp. Organomet. Chem. 1st edn.*, 1992, **8**, 499. 674, 675

82COMC-I(8)713 P. W. Jolly; *Comp. Organomet. Chem., 1st edn.*, 1982, **8**, 713. 195

82COMC-I(8)773 P. W. Jolly; *Comp. Organomet. Chem., 1st edn.*, 1982, **8**, 773. 195

82COMC-I(8)799 B. M. Trost and T. R. Verhoeven; *Comp. Organomet. Chem., 1st edn.*, 1982, **8**, 799. 192, 212

82COMC-I(8)939 A. J. Pearson; *Comp. Organomet. Chem., 1st edn.*, 1982, **8**, 939. 196

82CPB1865 H. Yamanaka, K. Tanji and T. Sakamoto; *Chem. Pharm. Bull.*, 1982, **30**, 1865. 1010

82CPB2410 T. Sakamoto, H. Yamanaka and Y. Kondo; *Chem. Pharm. Bull.*, 1982, **30**, 2410. 1010

82CPB2417 T. Sakamoto, H. Yamanaka and Y. Kondo; *Chem. Pharm. Bull.*, 1982, **30**, 2417. 1010

82CPL493 J. Chandrasekhar, R. A. Khan and P. v. R. Schleyer; *Chem. Phys. Lett.*, 1982, **85**, 493. 1143

82H(18)187 C. A. Brown, O. Chapa and A. Yamaichi; *Heterocycles*, 1982, **18**, 187. 308

82H(19)329 Y. Akita and A. Ohta; *Heterocycles*, 1982, **19**, 329. 318

82HCA371 D. Bravetti, R. M. Bettolo and A. Lupi; *Helv. Chim. Acta*, 1982, **65**, 371. 382

82HCA451 R. Huston, M. Rey and A. S. Dreiding; *Helv. Chim. Acta*, 1982, **65**, 451. 812

82HCA1972 D. Seebach and L. Widler; *Helv. Chim. Acta*, 1982, **65**, 1972. 683
82JA180 T. Hayashi, M. Konishi, M. Fukushima, T. Mise, M. Kagotani, M. Tajika and M. Kumada; *J. Am. Chem. Soc.*, 1982, **104**, 180. 1097
82JA907 C. S. Yannoni, V. Macho and P. C. Myhre; *J. Am. Chem. Soc.*, 1982, **104**, 907. 906
82JA1025 I. Kuwajima, E. Nakamura and M. Shimizu; *J. Am. Chem. Soc.*, 1982, **104**, 1025. 133
82JA1109 N. Minami, S. S. Ko and Y. Kishi; *J. Am. Chem. Soc.*, 1982, **104**, 1109. 748
82JA1428 F. Sainte, B. Serckx-Poncin, A.-M. Hesbain-Frisque and L. Ghosez; *J. Am. Chem. Soc.*, 1982, **104**, 1428. 375
82JA1628 G. A. Olah, M. Arvanaghi and G. K. S. Prakash; *J. Am. Chem. Soc.*, 1982, **104**, 1628. 904
82JA1737 D. A. Evans, M. D. Ennis and D. J. Mathre; *J. Am. Chem. Soc.*, 1982, **104**, 1737. 134
82JA1742 L. M. Tolbert and M. B. Ali; *J. Am. Chem. Soc.*, 1982, **104**, 1742. 357
82JA2046 E. Vedejs and D. W. Powell; *J. Am. Chem. Soc.*, 1982, **104**, 2046. 22, 942
82JA2305 B. H. Lipshutz, J. Kozlowski and R. S. Wilhelm; *J. Am. Chem. Soc.*, 1982, **104**, 2305. 204, 207
82JA2321 G. Stork and N. H. Baine; *J. Am. Chem. Soc.*, 1982, **104**, 2321. 320, 322, 1127, 1129, 1130
82JA2498 W. R. Dolbier, Jr. and S. F. Sellers; *J. Am. Chem. Soc.*, 1982, **104**, 2498. 801
82JA2560 R. H. Mitchell, R. V. Williams and T. W. Dingle; *J. Am. Chem. Soc.*, 1982, **104**, 2560. 233
82JA2642 S. E. Denmark and T. K. Jones; *J. Am. Chem. Soc.*, 1982, **104**, 2642. 422
82JA2923 A. P. Kozikowski and E. M. Huie; *J. Am. Chem. Soc.*, 1982, **104**, 2923. 693
82JA3165 J. P. Marino and J. C. Jaén; *J. Am. Chem. Soc.*, 1982, **104**, 3165. 479
82JA3655 A. L. Castelhano and D. Griller; *J. Am. Chem. Soc.*, 1982, **104**, 3655. 921
82JA3676 D. J. Pasto, P. F. Heid and S. E. Warren; *J. Am. Chem. Soc.*, 1982, **104**, 3676. 498
82JA3727 R. Tamura and L. S. Hegedus; *J. Am. Chem. Soc.*, 1982, **104**, 3727. 250
82JA3733 B. M. Trost and D. M. T. Chan; *J. Am. Chem. Soc.*, 1982, **104**, 3733. 289
82JA4030 M. M. Abelman, R. L. Funk and J. D. Munger; *J. Am. Chem. Soc.*, 1982, **104**, 4030. 831
82JA4696 B. H. Lipshutz and R. S. Wilhelm; *J. Am. Chem. Soc.*, 1982, **104**, 4696. 119
82JA4972 S. E. Denmark and M. A. Harmata; *J. Am. Chem. Soc.*, 1982, **104**, 4972. 832
82JA5564 Y. Ueno, K. Chino, M. Watanabe, O. Moriya and M. Okawara; *J. Am. Chem. Soc.*, 1982, **104**, 5564. 936
82JA5568 E. R. Koft and A. B. Smith, III; *J. Am. Chem. Soc.*, 1982, **104**, 5568. 516
82JA5829 G. E. Keck and J. B. Yates; *J. Am. Chem. Soc.*, 1982, **104**, 5829. 173, 334
82JA6129 E. J. Corey and J. E. Munroe; *J. Am. Chem. Soc.*, 1982, **104**, 6129. 826
82JA6609 D. M. Grove, G. van Koten, J. N. Louwen, J. G. Noltes, A. L. Spek and H. J. C. Ubbels; *J. Am. Chem. Soc.*, 1982, **104**, 6609. 127
82JA6668 B. M. Trost and P. Renaut; *J. Am. Chem. Soc.*, 1982, **104**, 6668. 287
82JA6681 E. Grovenstein, Jr. and P.-C. Lu; *J. Am. Chem. Soc.*, 1982, **83**, 6681. 886
82JA6788 E. C. Ashby, J. N. Argyropoulos, G. R. Meyer and A. B. Goel; *J. Am. Chem. Soc.*, 1982, **104**, 6788. 884
82JA6809 P. F. Hudrlik, A. M. Hudrlik and A. K. Kulkarni; *J. Am. Chem. Soc.*, 1982, **104**, 6809. 62
82JA6831 I. Kuwajima and H. Urabe; *J. Am. Chem. Soc.*, 1982, **104**, 6831. 471
82JA6874 B. Matuszewski, A. W. Burgstahler and R. S. Givens; *J. Am. Chem. Soc.*, 1982, **104**, 6874. 421
82JA6879 B. M. Trost and B. P. Coppola; *J. Am. Chem. Soc.*, 1982, **104**, 6879. 289
82JA7041 C. R. Johnson and R. C. Elliott; *J. Am. Chem. Soc.*, 1982, **104**, 7041. 643
82JA7051 H. J. Reich and S. Wollowitz; *J. Am. Chem. Soc.*, 1982, **104**, 7051. 609
82JA7105 G. A. Olah, G. K. S. Prakash, M. Arvanaghi and F. A. L. Anet; *J. Am. Chem. Soc.*, 1982, **104**, 7105. 906
82JA7174 F. E. Ziegler, J. M. Fany and C. C. Tam; *J. Am. Chem. Soc.*, 1982, **104**, 7174. 828
82JA7181 F. E. Ziegler and J. J. Piwinski; *J. Am. Chem. Soc.*, 1982, **104**, 7181. 840
82JA7225 L. E. Overman and E. J. Jacobsen; *J. Am. Chem. Soc.*, 1982, **104**, 7225. 806, 809
82JA7549 J.-R. Hamon, J.-Y. Saillard, A. Le Beuze, M. J. McGlinchey and D. Astruc; *J. Am. Chem. Soc.*, 1982, **104**, 7549. 165
82JA7609 Y. Ito, M. Nakatsuka and T. Saegusa; *J. Am. Chem. Soc.*, 1982, **104**, 7609. 655
82JA7666 N. V. Bac and Y. Langlois; *J. Am. Chem. Soc.*, 1982, **104**, 7666. 626, 630, 655, 789
82JA7667 R. Haruta, M. Ishiguro, N. Ikeda and H. Yamamoto; *J. Am. Chem. Soc.*, 1982, **104**, 7667. 1055
82JA7670 R. L. Danheiser, S. K. Gee and H. Sard; *J. Am. Chem. Soc.*, 1982, **104**, 7670. 812
82JCR(S)300 J.-L. Gras; *J. Chem. Res. (S)*, 1982, 300. 499
82JCS(P1)545 S. W. Russell and H. J. J. Pabon; *J. Chem. Soc., Perkin Trans. 1*, 1982, 545. 506
82JCS(P1)875 G. S. R. Subba Rao and N. S. Sundar; *J. Chem. Soc., Perkin Trans. 1*, 1982, 875. 96
82JCS(P1)1381 N. S. Sundar and G. S. R. Subba Rao; *J. Chem. Soc., Perkin Trans. 1*, 1982, 1381. 98
82JCS(P1)1563 I. Fleming, J. Goldhill and D. A. Perry; *J. Chem. Soc., Perkin Trans. 1*, 1982, 1563. 655, 1080
82JCS(P1)1797 A. J. Bloodworth and J. L. Courtneidge; *J. Chem. Soc., Perkin Trans. 1*, 1982, 1797. 68
82JCS(P1)1845 H. McNab and I. Stobie; *J. Chem. Soc., Perkin Trans. 1*, 1982, 1845. 782
82JCS(P1)2299 J. Cornforth, A. F. Sierakowski and T. W. Wallace; *J. Chem. Soc., Perkin Trans. 1*, 1982, 2299. 472
82JCS(P1)2347 A. R. Katritzky and J. M. Lloyd; *J. Chem. Soc., Perkin Trans. 1*, 1982, 2347. 629
82JCS(P1)2757 P. J. Giddings, D. I. John, E. J. Thomas and D. J. Williams; *J. Chem. Soc., Perkin Trans. 1*, 1982, 2757. 410, 822
82JCS(P2)77 F. A. Bottino, P. Finocchiaro, E. Libertini, A. Reale and A. Recca; *J. Chem. Soc., Perkin Trans. 2*, 1982, 77. 689
82JCS(P2)1175 N. Al-Awadi, J. Ballam, P. R. Hemblade and R. Taylor; *J. Chem. Soc., Perkin Trans. 2*, 1982, 1175. 777
82JCS(P2)1179 I. McEwen and R. Taylor; *J. Chem. Soc., Perkin Trans. 2*, 1982, 1179. 777
82JFC(20)157 D. J. Burton and F. J. Mettile; *J. Fluorine Chem.*, 1982, **20**, 157. 431
82JFC(20)313 N. O. Brace; *J. Fluorine Chem.*, 1982, **20**, 313. 10

82JOC119 Y. Yamamoto, S. Yamamoto, H. Yatagai, Y. Ishihara and K. Maruyama; *J. Org. Chem.*, 1982, **47**, 119. 312
82JOC161 W. S. Johnson, T. A. Lyle and G. W. Daub; *J. Org. Chem.*, 1982, **47**, 161. 295
82JOC163 A. S. Kende and B. H. Toder; *J. Org. Chem.*, 1982, **47**, 163. 797
82JOC171 H. C. Brown, D. Basavaiah and S. U. Kulkarni; *J. Org. Chem.*, 1982, **47**, 171. 531, 662
82JOC248 R. Dams, M. Malinowski, I. Westdorp and H. Y. Geise; *J. Org. Chem.*, 1982, **47**, 248. 688
82JOC276 S. Krishnamurthy and H. C. Brown; *J. Org. Chem.*, 1982, **47**, 276. 4, 7, 8, 10
82JOC389 K. A. Parker, J. J. Petraitis, R. W. Kosley Jr. and S. L. Buchwald; *J. Org. Chem.*, 1982, **47**, 389. 980
82JOC409 R. J. Ardecky, D. Dominguez and M. P. Cava; *J. Org. Chem.*, 1982, **47**, 409. 569
82JOC754 H. C. Brown and D. Basavaiah; *J. Org. Chem.*, 1982. **47**, 754. 662
82JOC798 R. W. Thies and J. R. Pierce; *J. Org. Chem.*, 1982, **47**, 798. 401
82JOC1047 R. K. Norris and D. Randles; *J. Org. Chem.*, 1982, **47**, 1047 37
82JOC1193 C. R. Johnson and C. J. Stark, Jr.; *J. Org. Chem.*, 1982, **47**, 1193. 24
82JOC1221 A. A. Jakubowski, F. S. Guziec, Jr., M. Sugiura, C. C. Tam, M. Tishler and S. Omura; *J. Org. Chem.*, 1982, **47**, 1221. 503
82JOC1267 L.-C. Kao, G. Stakem, B. A. Patel and R. F. Heck; *J. Org. Chem.*, 1982, **47**, 1267. 1096
82JOC1378 P. Ma, V. S. Martin, S. Masamune, K. B. Sharpless and S. M. Viti; *J. Org. Chem.*, 1982, **47**, 1378. 14
82JOC1435 M. Nakazaki, K. Yamamoto, M. Maeda, O. Sato and T. Tsutsui; *J. Org. Chem.*, 1982, **47**, 1435. 990
82JOC1615 E. C. Friedrich and G. Biresaw; *J. Org. Chem.*, 1982, **47**, 1615. 352
82JOC1632 D. L. J. Clive, C. G. Russell and S. C. Suri; *J. Org. Chem.*, 1982, **47**, 1632. 814
82JOC1641 D. L. J. Clive, P. C. Anderson, N. Moss and A. Singh; *J. Org. Chem.*, 1982, **47**, 1641. 243
82JOC1775 M. Okabe, M. Abe and M. Tada; *J. Org. Chem.*, 1982, **47**, 1775. 946
82JOC1792 D. Basavaiah and H. C. Brown; *J. Org. Chem.*, 1982, **47**, 1792. 531, 663
82JOC1837 J. C. Gilbert and U. Weerasooriya; *J. Org. Chem.*, 1982, **47**, 1837. 1078
82JOC1904 R. I. Davidson and P. J. Kropp; *J. Org. Chem.*, 1982, **47**, 1904. 108
82JOC2089 H. C. Brown, M. Periasamy, D. P. Kelly and J. J. Giansiracusa; *J. Org. Chem.*, 1982, **47**, 2089. 906
82JOC2117 N. Miyaura, K. Maeda, H. Suginome and A. Suzuki; *J. Org. Chem.*, 1982, **47**, 2117. 532
82JOC2125 J. A. Cella; *J. Org. Chem.*, 1982, **47**, 2125. 200
82JOC2225 H. Amamoto, M. Ishiguro and N. Ikeda; *J. Org. Chem.*, 1982, **47**, 2225. 974, 1056
82JOC2231 S. J. Danishefsky, S. Chakalamannil and B.-J. Uang; *J. Org. Chem.*, 1982, **47**, 2231. 343
82JOC2251 K. Baum, C. D. Bedford and R. J. Hunadi; *J. Org. Chem.*, 1982, **47**, 2251. 1130
82JOC2268 R. C. Gadwood and R. M. Lett; *J. Org. Chem.*, 1982, **47**, 2268. 812
82JOC2549 M. W. Logue and K. Teng; *J. Org. Chem.*, 1982, **47**, 2549. 1135
82JOC2564 Y. Gaoni; *J. Org. Chem.*, 1982, **47**, 2564. 184, 215
82JOC2590 N. Pourahmady, E. H. Vickery and E. J. Eisenbraun; *J. Org. Chem.*, 1982, **47**, 2590. 7, 19
82JOC2622 I. Colon; *J. Org. Chem.*, 1982, **47**, 2622. 428
82JOC2928 E. Grovenstein, Jr. and P.-C. Lu; *J. Org. Chem.*, 1982, **47**, 2928. 886
82JOC3190 S. F. Martin, J. B. White and R. Wagner; *J. Org. Chem.*, 1982, **47**, 3190. 815
82JOC3211 L. A. Last, E. R. Fretz and R. M. Coates; *J. Org. Chem.*, 1982, **47**, 3211. 214
82JOC3297 D. Becker, Z. Harel, M. Nagler and A. Gillon; *J. Org. Chem.*, 1982, **47**, 3297. 498
82JOC3348 S. D. Burke, W. F. Forbare and D. M. Armistead; *J. Org. Chem.*, 1982, **47**, 3348. 243
82JOC3364 N. R. Pearson, G. Hahn and G. Zweifel; *J. Org. Chem.*, 1982, **47**, 3364. 977
82JOC3506 A. R. Katritzky and A. M. El-Mowafy; *J. Org. Chem.*, 1982, **47**, 3506. 629
82JOC3521 D. A. Bright, D. E. Mathisen and H. E. Zieger; *J. Org. Chem.*, 1982, **47**, 3521. 110
82JOC3548 F. A. Carey and W. C. Frank; *J. Org. Chem.*, 1982, **47**, 3548. 760
82JOC3563 E. R. Cullen, F. S. Guziec, Jr. and C. J. Murphy; *J. Org. Chem.*, 1982, **47**, 3563. 686
82JOC3630 L. D. Martin and J. K. Stille; *J. Org. Chem.*, 1982, **47**, 3630. 1092
82JOC3707 R. D. Bach, R. A. Woodard, T. J. Anderson and M. D. Glick; *J. Org. Chem.*, 1982, **47**, 3707. 66, 69, 454
82JOC3805 J. L. Luche, C. Pétrier, A. L. Gemal and N. Zikra; *J. Org. Chem.*, 1982, **47**, 3805. 1093
82JOC3806 H. C. Brown and D. Basavaiah; *J. Org. Chem.*, 1982, **47**, 3806. 531, 662
82JOC3903 G. A. Olah, A. L. Berrier and G. K. S. Prakash; *J. Org. Chem.*, 1982, **47**, 3903. 906
82JOC3933 P. A. Bartlett and J. F. Barstow; *J. Org. Chem.*, 1982, **47**, 3933. 831
82JOC4005 T. Cohen and Z. Korsarych; *J. Org. Chem.*, 1982, **47**, 4005. 235
82JOC4254 F. Johnson, K. G. Paul and D. Favara; *J. Org. Chem.*, 1982, **47**, 4254. 503, 505
82JOC4304 J. W. Lown, S. M. Sondhi, S. B. Mandal and J. Murphy; *J. Org. Chem.*, 1982, **47**, 4304. 11, 19
82JOC4319 L. N. Pridgeon; *J. Org. Chem.*, 1982, **47**, 4319. 471, 473
82JOC4380 R. O. Hutchins and K. Learn; *J. Org. Chem.*, 1982, **47**, 4380. 15, 17, 23, 24, 25
82JOC4478 S. Tsuboi, T. Masuda and A. Takeda; *J. Org. Chem.*, 1982, **47**, 4478. 979
82JOC4577 F. Marcuzzi, G. Modena and G. Melloni; *J. Org. Chem.*, 1982, **47**, 4577. 200
82JOC4713 T. Ohnuma, N. Hata, H. Fujiwara and Y. Ban; *J. Org. Chem*, 1982, **47**, 4713. 439, 450
82JOC4758 J. Hine and M. J. Skoglund; *J. Org. Chem.*, 1982, **47**, 4758. 797
82JOC4766 J. Hine and M. J. Skoglund; *J. Org. Chem.*, 1982, **47**, 4766. 797, 882
82JOC4779 R. W. Kaesler and E. LeGoff; *J. Org. Chem.*, 1982, **47**. 4779. 521
82JOC4786 P. A. Christensen, B. J. Willis, F. W. Wehrli and S. Wehrli; *J. Org. Chem.*, 1982, **47**, 4786. 800
82JOC4801 J. H. Babler and R. A. Haack; *J. Org. Chem.*, 1982, **47**, 4801. 600, 788
82JOC4815 J. H. Rigby, J.-M. Sage and J. Raggon; *J. Org. Chem.*, 1982, **47**, 4815. 815
82JOC4963 M. P. Cooke, Jr.; *J. Org. Chem.*, 1982, **47**, 4963. 51
82JOC5041 K. E. Coblens, V. B. Muralidharan and B. Ganem; *J. Org. Chem.*, 1982, **47**, 5041. 5, 9

82JOC5229	F. E. Ziegler and H. Lim; *J. Org. Chem.*, 1982, **47**, 5229.	690
82JOC5407	H. C. Brown and D. Basavaiah; *J. Org. Chem.*, 1982, **47**, 5407.	531, 662
82JOM(225)C1	S. U. Kulkarni, D. Basavaiah and H. C. Brown; *J. Organomet. Chem.*, 1982, **225**, C1.	531
82JOM(225)63	H. C. Brown, D. Basavaiah and S. U. Kulkarni; *J. Organomet. Chem.*, 1982, **225**, 63.	509
82JOM(226)C55	D. Masure, Ph. Coutrot and J. F. Normant; *J. Organomet. Chem.*, 1982, **226**, C55.	511
82JOM(226)115	L. S. Vasilyev, V. V. Veselovskii, M. I. Struchkova and B. M. Mikhailov; *J. Organomet. Chem.*, 1982, **226**, 115.	60
82JOM(233)C25	P. E. van Rijn and L. Brandsma; *J. Organomet. Chem.*, 1982, **233**, C25.	981
82JOM(233)13	N. Miyaura, Y. Tanabe, H. Suginome and A. Suzuki; *J. Organomet. Chem.* 1982, **233**, 13.	228
82JOM(233)267	H. U. Blaser and A. Spencer; *J. Organomet. Chem.*, 1982, **233**, 267.	494
82JOM(236)177	J. Pornet, N. Kolani, D. Mesnard, L. Miginiac and K. Jaworski; *J. Organomet. Chem.*, 1982, **236**, 177.	978
82JOM(240)329	H. Kleijn, M. Tigchelaar, R. J. Bullee, C. J. Elsevier, J. Meijer and P. Vermeer; *J. Organomet. Chem.*, 1982, **240**, 329.	967
82LA1750	E. V. Dehmlow and A. R. Shamout; *Liebigs Ann. Chem.*, 1982, 1750.	1038
B-82MI 102-01	R. J. Baumgarten and V. A. Curtis; in "The Chemistry of Amino, Nitroso and Nitro Compounds and Their Derivatives," ed. S. Patai, Wiley, London, 1982, Supp. F, p. 931.	28, 35
B-82MI 110-01	S. R. Landor (ed.); "The Chemistry of the Allenes," Academic Press, London, 1982.	458
82MI 111-01	J. P. Kennedy and E. Maréchal; in "Carbocationic Polymerisation," Wiley, Chichester, 1982, p. 82.	462
82MI 115-01	R. Dams, M. Malinowski and H. J. Geise; *Transition Met. Chem.*, 1982, **7**, 37.	688
82MI 119-01	D. A. Evans; *Aldrichimica Acta*, 1982, **15(2)**, 23.	878
B-82MI 120-01	S. R. Landor; "The Chemistry of the Allenes," Academic Press, London, vol 1, 1982.	954
82OM20	D. S. Matteson and R. J. Moody; *Organometallics*, 1982, **1**, 20.	163
82OM397	M. Rosenblum and D. Scheck; *Organometallics*, 1982, **1**, 397.	519
82OM542	J. Yoshida, K. Tamao, H. Yamamoto, T. Kakui, T. Uchida and M. Kumada; *Organometallics*, 1982, **1**, 542.	490, 535
82OM586	J. C. Cochran; *Organometallics*, 1982, **1**, 586.	455
82OM1366	D. M. Grove, G. van Koten and H. J. C. Ubbels; *Organometallics*, 1982, **1**, 1366.	127
82OM1449	R. G. Daniels and L. A. Paquette; *Organometallics*, 1982, **1**, 1449.	633
82OPP373	Yaozeng Huang, Yuanyao Xu and Shong Li; *Org. Prep. Proced. Int.*, 1982, **14**, 373.	737, 738
82OR(27)1	J.-F. Biellmann and J.-B. Ducep; *Org. React.*, 1982, **27**, 1.	107
82OR(27)345	R. F. Heck; *Org. React.*, 1982, **27**, 345.	1021
82OS(60)81	R. K. Duke and R. W. Rickards; *Org. Synth.*, 1982, **60**, 81.	1134
82PAC1	H. Sakurai; *Pure Appl. Chem.*, 1982, **54**, 1.	314
82PAC197	J. Tsuji; *Pure Appl. Chem.*, 1982, **54**, 197.	192
82PAC1915	M. Martin and J. Clardy; *Pure Appl. Chem.*, 1982, **54**, 1915.	769
82RTC97	H. Kleijn, J. Meijer, G. C. Overbeek and P. Vermeer; *Recl. Trav. Chim. Pays-Bas*, 1982, **101**, 97.	1025
82RTC405	K. Ruitenberg, H. Kleijn, H. Westmijze, J. Meijer and P. Vermeer; *Recl. Trav. Chim. Pays-Bas*, 1982, **101**, 405.	1004, 1006, 1025
82S85	P. J. Stang, M. Hanack and L. R. Subramanian; *Synthesis*, 1982, 85.	177, 224, 433, 1021
82S504	G. D. Hartman and R. D. Hartman; *Synthesis*, 1982, 504.	618, 706, 707
82S738	T. Jeffery-Luong and G. Linstrumelle; *Synthesis*, 1982, 738.	971
82S876	P. N. Pandley and M. L. Purkayastha; *Synthesis*, 1982, 876.	428
82S940	K. Burdeska; *Synthesis*, 1982, 940.	19
82S1036	G. R. Brown and A. J. Foubister; *Synthesis*, 1982, 1036.	47
82S1100	B. Myrboh, H. Ila and H. Junjappa; *Synthesis*, 1982, 1100.	1076
82SC557	A. Doutheau, A. Saba and J. Goré; *Synth. Commun.*, 1982, **12**, 557.	1033
82SC739	D. Michelot, J.-C. Clinet and G. Linstrumelle; *Synth. Commun.*, 1982, **12**, 739.	971
82SC795	Y. F. Zhou and N. Z. Huang; *Synth. Commun.*, 1982, **12**, 795.	482
82SC983	A. K. Sinhababu and R. T. Borchardt; *Synth. Commun.*, 1982, **12**, 983.	11, 15
82T313	J. M. Tedder; *Tetrahedron*, 1982, **38**, 313.	925
82T567	R. F. Childs; *Tetrahedron*, **38**, 567, 1982.	794
82T631	R. Rossi, A. Carpita, M. G. Quiric and M. L. Gaudenzi; *Tetrahedron*, 1982, **38**, 631.	472, 1013
82T693	J. D. Fourneron, L. M. Harwood and M. Julia; *Tetrahedron*, 1982, **38**, 693.	240
82T1713	R. C. Larock; *Tetrahedron*, 1982, **38**, 1713.	128, 1095
82T1975	P. W. Hickmott; *Tetrahedron*, 1982, **38**, 1975.	129
82T2355	J. L. Torregrosa, M. Baboulene, V. Speziale and A. Lattes; *Tetrahedron*, 1982, **38**, 2355.	531, 662
82T2857	E. Vedejs and G. A. Krafft; *Tetrahedron*, 1982, **38**, 2857.	617, 704
82T2939	J. E. Baldwin and K. J. Lusch; *Tetrahedron*, 1982, **38**, 2939.	304
82TL27	E. Negishi, F.-T. Luo and C. L. Rand; *Tetrahedron Lett.*, 1982, **23**, 27.	542
82TL63	P. Vincent, J.-P. Beaucourt and L. Pichat; *Tetrahedron Lett.*, 1982, **23**, 63.	536
82TL109	G. W. J. Fleet and C. R. C. Spensley; *Tetrahedron Lett.*, 1982, **23**, 109.	828, 832
82TL263	L. A. Paquette, G. J. Wells, K. A. Horn and T. H. Yan; *Tetrahedron Lett.*, 1983, **23**, 263.	248
82TL323	R. D. Miller and D. R. McKean; *Tetrahedron Lett.*, 1982, **23**, 323.	579
82TL343	Y. Kobayashi, T. Yamashita, K. Takahashi, H. Kuroda and I. Kumadaki; *Tetrahedron Lett.*, 1982, **23**, 343.	1079
82TL443	Y. Ueno, T. Miyano and M. Okawara; *Tetrahedron Lett.*, 1982, **23**, 443.	942
82TL631	M. Marsi and J. A. Gladysz; *Tetrahedron Lett.*, 1982, **23**, 631.	579
82TL689	M. Node, K. Nishide, K. Ohta and E. Fujita; *Tetrahedron Lett.*, 1982, **23**, 689.	426, 432
82TL719	E. J. Corey and C. Rücker; *Tetrahedron Lett.*, 1982, **23**, 719.	651, 1052, 1053, 1055, 1061
82TL795	R. Breslow, D. A. Cortés, B. Juan and R. D. Mitchell; *Tetrahedron Lett.*, 1982, **23**, 795.	865

82TL807 D. A. Evans and J. Bartroli; *Tetrahedron Lett.*, 1982, **23**, 807. 878

82TL1035 G. L. Larson and D. Hernández; *Tetrahedron Lett.*, 1982, **23**, 1035. 650

82TL1177 S. Uemura, S. Fukuzawa, A. Toshimitsu and M. Okano; *Tetrahedron Lett.*, 1982, **23**, 1177. 611

82TL1181 S. Uemura and S. Fukuzawa; *Tetrahedron Lett.*, 1982, **23**, 1181. 437, 545

82TL1267 H. Nishiyama, H. Yokoyama, S. Narimatsu and K. Itoh; *Tetrahedron Lett.*, 1982, **23**, 1267. 206, 289, 1092

82TL1343 M. Majewski and V. Snieckus; *Tetrahedron Lett.*, 1982, **23**, 1343. 215, 576

82TL1405 P. M. Cairns, L. Crombie and G. Pattenden; *Tetrahedron Lett.*, 1982, **23**, 1405. 811

82TL1651 E. J. Corey and J. Kang; *Tetrahedron Lett.*, 1982, **23**, 1651. 1015, 1016

82TL1777 J. E. McMurry, J. R. Matz, K. L. Kees and P. A. Bock; *Tetrahedron Lett.*, 1982, **23**, 1777. 690

82TL1963 C.-N. Hsiao and H. Shechter; *Tetrahedron Lett.*, 1982, **23**, 1963. 615, 700

82TL1979 E. J. Corey and P. B. Hopkins; *Tetrahedron Lett.*, 1982, 1979. 585

82TL2019 D. H. R. Barton, W. Hartwig, R. S. H. Motherwell, W. B. Motherwell and A. Stange; *Tetrahedron Lett.*, 1982, **23**, 2019. 940

82TL2087 A. P. Kozikowski and Y. Kitigawa; *Tetrahedron Lett.*, 1982, **23**, 2087. 534

82TL2105 A. Toshimitsu, H. Owada, S. Uemura and M. Okano; *Tetrahedron Lett.*, 1982, **23**, 2105. 604

82TL2205 M. Ochiai, S. Tada, K. Sumi and E. Fujita; *Tetrahedron Lett.*, 1982, **23**, 2205. 592, 616, 668

82TL2261 M. P. Doyle, R. L. Dorow, W. H. Tamblyn and W. E. Buhro; *Tetrahedron Lett.*, 1982, **23**, 2261. 353

82TL2277 B. K. R. Shankar and H. Shechter; *Tetrahedron Lett.*, 1982, **23**, 2277. 717

82TL2373 A. S. Kende and P. Fludzinski; *Tetrahedron Lett.*, 1982, **23**, 2373. 999

82TL2379 C. L. Bumgardner, J. R. Lever and S. T. Purrington; *Tetrahedron Lett.*, 1982. **23**, 2379. 210

82TL2457 M. Julia and J.-N. Verpeaux; *Tetrahedron Lett.*, 1982, **23**, 2457. 700

82TL2465 M. Julia, M. Launay, J. P. Stacino and J.-N. Verpeaux; *Tetrahedron Lett.*, 1982, **23**, 2465. 700

82TL2469 J.-L. Fabre, M. Julia and J.-N. Verpeaux; *Tetrahedron Lett.*, 1982, **23**, 2469. 485, 700

82TL2505 M. D. Bachi and C. Hoornaert; *Tetrahedron Lett.*, 1982, **23**, 2505. 547

82TL2575 Y. Ueno, K. Chino and M. Okawara; *Tetrahedron Lett.*, 1982, **23**, 2575. 1130

82TL2627 P. T. Lansbury and T. E. Nickson; *Tetrahedron Lett.*, 1982, **23**, 2627. 311

82TL2719 J. M. Finan and Y. Kishi; *Tetrahedron Lett.*, 1982; **23**, 2719. 14

82TL2723 J. E. McMurry and J. R. Matz; *Tetrahedron Lett.*, 1982, **23**, 2723. 690

82TL2797 H. Westmijze, K. Ruitenberg, J. Meijer and P. Vermeer; *Tetrahedron Lett.*, 1982, **23**, 2797. 536

82TL2871 Y. Morizawa, K. Oshima and H. Nozaki; *Tetrahedron Lett.*, 1982, **23**, 2871. 248

82TL2997 D. G. Daly and M. A. McKervey; *Tetrahedron Lett.*, 1982, **23**, 2997. 674

82TL3019 Z. Korsarych and T. Cohen; *Tetrahedron Lett.*, 1982, **23**, 3019. 235, 408, 409, 822

82TL3051 G. E. Keck and R. R. Webb, II; *Tetrahedron Lett.*, 1982, **23**, 3051. 961

82TL3119 C. F. Wilcox, Jr. and W. J. Tuszynski; *Tetrahedron Lett.*, 1982, 3119. 905

82TL3171 H. Paulsen and M. Stubbe; *Tetrahedron Lett.*, 1982, **23**, 3171. 310

82TL3265 J. Bremner, M. Julia, M. Launay and J. P. Stacino; *Tetrahedron Lett.*, 1982, **23**, 3265. 439, 700

82TL3277 R. Bloch and J. Abecassis; *Tetrahedron Lett.*, 1982, **23**, 3277. 790

82TL3395 K. Hattori, K. Maruoka and H. Yamamoto; *Tetrahedron Lett.*, 1982, **23**, 3395. 1044, 1059

82TL3411 J. N. Denis and A. Krief; *Tetrahedron Lett.*, 1982, **23**, 3411. 440

82TL3455 C.-N. Hsiao and H. Shechter; *Tetrahedron Lett.*, 1982, **23**, 3455. 653

82TL3463 E. J. Corey and D. J. Hoover; *Tetrahedron Lett.*, 1982, **23**, 3463. 839

82TL3583 T. Fujisawa, T. Sato, M. Kawashima, K. Naruse and K. Tamai; *Tetrahedron Lett.*, 1982, 3583. 476

82TL3587 T. Sato, K. Naruse and T. Fujisawa; *Tetrahedron Lett.*, 1982, **23**, 3587. 230, 1107

82TL3597 T. Suzuki, H. Saimoto, H. Tomioka, K. Oshima and H. Nozaki; *Tetrahedron Lett.*, 1982, **23**, 3597. 14, 221

82TL3733 P. A. Brown and P. R. Jenkins; *Tetrahedron Lett.*, 1982, **23**, 3733. 633

82TL3931 N. Sayo, Y. Kimura and T. Nakai; *Tetrahedron Lett.*, 1982, **23**, 3931. 816

82TL4293 L. Lombardo; *Tetrahedron Lett.*, 1982, **23**, 4293. 678

82TL4325 R. Sauvêtre and J. F. Normant; *Tetrahedron Lett.*, 1982, **23**, 4325. 1017, 1067

82TL4329 R. Sauvêtre, J. F. Normant and S. Martin; *Tetrahedron Lett.*, 1982, **23**, 4329. 1017

82TL4385 S. Halazy, F. Zutterman and A. Krief; *Tetrahedron Lett.*, 1982, **23**, 4385. 248

82TL4517 M. Catellani and G. P. Chiusoli; *Tetrahedron Lett.*, 1982, **23**, 4517. 1015, 1138

82TL4607 M. Shibasaki, Y. Torisawa and S. Ikegami; *Tetrahedron Lett.*, 1982, **23**, 4607. 1081

82TL4733 T. Hayama, S. Tomoda, Y. Takeuchi and Y. Nomura; *Tetrahedron Lett.*, 1982, **23**, 4733. 603

82TL4879 W. R. Roush and S. M. Peseckis; *Tetrahedron Lett.*, 1982, **23**, 4879. 700, 703

82TL5005 C. R. Johnson, R. C. Elliott and N. A. Meanwell; *Tetrahedron Lett.*, 1982, **23**, 5005. 643

82TL5043 F. Näf, R. Decorzant and S. D. Escher; *Tetrahedron Lett.*, 1982, **23**, 5043. 618

82TL5151 A. Alexakis and J. F. Normant; *Tetrahedron Lett.*, 1982, **23**, 5151. 541

82TL5229 A. Pelter and R. Al-Bayati; *Tetrahedron Lett.*, 1982, **23**, 5229. 204

82TL5485 J. E. Pauw and A. C. Weedon; *Tetrahedron Lett.*, 1982, **23**, 5485. 688

82TL5549 Y. Tanigawa, K. Nishimura, A. Kawasaki and S.-I. Murahashi; *Tetrahedron Lett.*, 1982, **23**, 5549. 193

82TL5587 A. Citterio, A. Gentile and F. Minisci; *Tetrahedron Lett.*, 1982, **23**, 5587. 926

82TS1 D. A. Evans, J. V. Nelson and T. R. Taber; *Top. Stereochem.*, 1982, **13**, 1–115. 873, 876

83ACR27 F. Minisci, A. Citterio and C. Giordano; *Acc. Chem. Res.*, 1983, **16**, 27. 934

83ACR55 R. M. Adlington and A. G. M. Barrett; *Acc. Chem. Res.*, 1983, **16**, 55. 630, 632, 1094

83ACR81 W. H. Okamura; *Acc. Chem. Res.*, 1983, **16**, 81. 795

83ACR343 M. D. Johnson; *Acc. Chem. Res.*, 1983, **16**, 343. 321, 334, 946

83ACR405 J. E. McMurry; *Acc. Chem. Res.*, 1983, **16**, 405. 687, 688
83ACR426 C. A. Grob; *Acc. Chem. Res.*, 1983, **16**, 426. 905
83ACR432 H. C. Brown; *Acc. Chem. Res.*, 1983, **16**, 432. 905
83ACR440 G. A. Olah, G. K. S. Prakash and M. Saunders; *Acc. Chem. Res.*, 1983, **16**, 440. 906
83ACR448 C. Walling; *Acc. Chem. Res.*, 1983, **16**, 448. 905
83ACS(B)141 S. Brandäge, O. Dahlman and J. Ölund; *Acta Chem. Scand., Ser. B*, 1983, **37**, 141. 4, 6
83ACS(B)173 L. Jalander and M. Broms; *Acta Chem. Scand., Ser. B.*, 1983, **37**, 173. 469
83AG(E)390 G. K. S. Prakash, T. N. Rawdah and G. A. Olah; *Angew. Chem., Int. Ed. Engl.*, 1983, **22**, 390. 891
83AG(E)560 A. S. Rothenberg, D. L. Dauplaise and H. P. Panzer; *Angew. Chem., Int. Ed. Engl.*, 1983, **22**, 560. 786
83AG(E)562 S. Kim, C. Y. Hong and S. Yang; *Angew. Chem., Int. Ed. Engl.*, 1983, **22**, 562. 4, 7, 9
83AG(E)599 P. Caubère; *Angew. Chem., Int. Ed. Engl.*, 1983, **22**, 599. 428
83AG(E)753 B. Giese; *Angew. Chem., Int. Ed. Engl.*, 1983, **22**, 753. 320, 323, 324, 920
83AG(E)778 H. tom Dieck and R. Diercks; *Angew. Chem., Int. Ed. Engl.*, 1983, **22**, 778. 519
83AG(E)780 H. J. Bestmann and R. Schobert; *Angew. Chem., Int. Ed. Engl.*, 1983, **22**, 780. 1134
83AJC545 A. L. J. Beckwith, C. J. Easton, T. Lawrence and A. K. Serebis; *Aust. J. Chem.*, 1983, 545. 337
83AJC2243 D. P. G. Hamon and K. R. Richards; *Aust. J. Chem.*, 1983, **36**, 2243. 4, 9
83BCJ192 I. Shimizu, Y. Kamei, T. Tezuka, T. Izumi and A. Kasahara; *Bull. Chem. Soc. Jpn.*, 1983, **56**, 192. 690
83BCJ353 N. Suzuki, Y. Ayaguchi, T. Tsukanaka and Y. Izawa; *Bull. Chem. Soc. Jpn.*, 1983, **56**, 353. 505, 507
83BCJ1089 S. Masuda, T. Nakajima and S. Suga; *Bull. Chem. Soc. Jpn.*, 1983, **56**, 1089. 463
83BCJ2023 I. Shimizu, H. Umezawa, T. Kanno, T. Izumi and A. Kasahara; *Bull. Chem. Soc. Jpn.*, 1983, **56**, 2023. 690
83BCJ2539 Y. Murata, K. Inomata, H. Kinoshita and H. Kotake; *Bull. Chem. Soc. Jpn.*, 1983, **56**, 2539. 503
83BCJ2855 A. Arase, M. Hoshi and Y. Masuda; *Bull. Chem. Soc. Jpn.*, 1983, **56**, 2855. 1018
83BSF332 N. Jabri, A. Alexakis and J. F. Normant; *Bull. Soc. Chim. Fr.*, 1983, 332. 468
83BSF(2)21 G. Courtois and P. Miginiac; *Bull. Soc. Chim. Fr., Part 2*, 1983, 21. 1041
83CB473 T. Kauffmann, R. Jouben, N. Klas and A. Vahrenhorst; *Chem. Ber.*, 1983, **116**, 473. 271
83CB882 U. H. Brinker and L. König; *Chem. Ber.*, 1983, **116**, 882. 117, 1092
83CB1042 W. Barbe, H. D. Beckhaus and C. Rüchardt; *Chem. Ber.*, 1983, **116**, 1042. 925
83CB1336 M. Yalpani, R. Köster and G. Wilke; *Chem. Ber.*, 1983, **116**, 1336. 60
83CB2378 H. J. Bestmann, K. Kumar and L. Kisielowski; *Chem. Ber.*, 1983, **116**, 2378. 248
83CB3212 B. Schubert and E. Weiss; *Chem. Ber.*, 1983, **116**, 3212. 1133
83CC365 H. Alper and J. Heveling; *J. Chem. Soc., Chem. Commun.*, 1983, 365. 80
83CC502 B. C. Berris, G. H. Hovakeemian and K. P. C. Vollhardt; *J. Chem. Soc., Chem. Commun.*, 1983, 502. 448
83CC619 M. Ochiai, T. Ukita and E. Fujita; *J. Chem. Soc., Chem. Commun.*, 1983, 619. 440, 616, 668
83CC630 M. P. Edwards, S. V. Ley, S. G. Lister and B. D. Palmer; *J. Chem. Soc., Chem. Commun.*, 1983, 630. 654, 697, 700, 703
83CC875 N. Ono, H. Miyake and A. Kaji; *J. Chem. Soc., Chem. Commun.*, 1983, 875. 37
83CC907 A. L. J. Beckwith and S. H. Goh; *J. Chem. Soc., Chem. Commun.*, 1983, 907. 426
83CC921 F. Sato, H. Uchiyama, K. Iida, Y. Kobayashi and M. Sato; *J. Chem. Soc., Chem. Commun.*, 1983, 921. 651
83CC932 A. D. Baxter, S. M. Roberts, F. Scheinmann, B. J. Wakefield and R. F. Newton; *J. Chem. Soc., Chem. Commun.*, 1983, 932. 826
83CC939 D. H. R. Barton, D. Crich and W. B. Motherwell; *J. Chem. Soc., Chem. Commun.*, 1983, 939. 935
83CC1040 R. M. Adlington, J. E. Baldwin, J. C. Bottaro and M. W. D. Perry; *J. Chem. Soc., Chem. Commun.*, 1983, 1040. 154
83CC1133 C. Huynh and G. Linstrumelle; *J. Chem. Soc., Chem. Commun.*, 1983, 1133. 971
83CC1270 H. Alper, J. B. Woell, B. Despeyroux and D. J. H. Smith; *J. Chem. Soc., Chem. Commun.*, 1983, 1270. 297
83CJC214 R. J. Armstrong and L. Weiler; *Can. J. Chem.*, 1983, **61**, 214. 216
83CJC757 D. N. Harpp and D. F. Mullins; *Can. J. Chem.*, 1983, **61**, 757. 232
83CJC1226 E. Piers, H. E. Morton, I. Nagakura and R. W. Thies; *Can. J. Chem.*, 1983, **61**, 1226. 811
83CJC1239 E. Piers and E. H. Ruediger; *Can. J. Chem.*, 1983, **61**, 1239. 811, 812
83CJC2403 N. H. Werstiuk, D. Dhanoa and G. Timmins; *Can. J. Chem.*, 1983, **61**, 2403. 905
83CJC2415 B. J. Demopoulos, H. J. Anderson, C. E. Loader and K. Faber; *Can. J. Chem.*, 1983, **61**, 2415. 19
83CJC3227 E. Piers, C. K. Law and I. Nagakura; *Can. J. Chem.*, 1983, **61**, 3227. 804
83CL35 M. Yamaguchi, T. Waseda and I. Hirao; *Chem. Lett.*, 1983, 35. 1046
83CL1379 K. Mikami, K. Azuma and T. Nakai; *Chem. Lett.*, 1983, 1379. 816
83CL1457 M. Ochiai, T. Ukita and E. Fujita; *Chem. Lett.*, 1983, 1457. 440
83COMC-I337 E. Negishi; *Comp. Organomet. Chem., 1st edn.*, 1983, **7**, 337. 999
83CPB3306 E. Nagashima, K. Suzuki and M. Sekiya; *Chem. Pharm. Bull.*, 1983, **31**, 3306. 990
83CPB4346 M. Sato, Y. Abe, H. Goto, R. Sugawara, E. Takahashi and T. Kato; *Chem. Pharm. Bull.*, 1983, **31**, 4346. 481, 1028
83CPL124 S. M. Burnett, A. E. Stevens, C. S. Feigerle and W. C. Lineberger; *Chem. Phys. Lett.*, 1983, **100**, 124. 1143
83DOK(272)1384 I. P. Beletskaya, N. A. Bumagin and I. G. Bumagins; *Dokl. Akad. Nauk SSSR*, 1983, **272**, 1384. 1006
83H(20)601 M. Natsume and M. Ogawa; *Heterocycles*, 1983, **20**, 601. 1042

83HCA1090 K. B. Becker and M. P. Labhart; *Helv. Chim. Acta*, 1983, **66**, 1090. 618, 707, 708

83HCA2210 F. Baumberger and A. Vasella; *Helv. Chem. Acta*, 1983, 2210. 321, 946

83HCA2377 T. K. Jones and S. E. Denmark; *Helv. Chim. Acta*, 1983, **66**, 2377. 422

83HCA2397 T. K. Jones and S. E. Denmark; *Helv. Chim. Acta*, 1983, **66**, 2397. 422

83HCA2608 R. K. Brunner and H. J. Borshberg; *Helv. Chim. Acta*, 1983, **66**, 2608. 831

83JA117 A. I. Meyers and L. M. Fuentes; *J. Am. Chem. Soc.*, 1983, **105**, 117. 149

83JA127 E. Vedejs, J. M. Dolphin and H. Mastalerz; *J. Am. Chem. Soc.*, 1983, **105**, 127. 821

83JA520 T. Cohen, M. Bhupathy and J. R. Matz; *J. Am. Chem. Soc.*, 1983, **105**, 520. 805

83JA625 W. C. Still, S. Murata, G. Revial and K. Yoshihara; *J. Am. Chem. Soc.*, 1983, **105**, 625. 813

83JA650 D. J. Burton and D. G. Cox; *J. Am. Chem. Soc.*, 1983, **105**, 650. 446

83JA667 P. G. Gassman and S. M. Bonser; *J. Am. Chem. Soc.*, 1983, **105**, 667. 709

83JA950 W. Kirmse and R. Siegfried; *J. Am. Chem. Soc.*, 1983, **105**, 950. 905

83JA1072 G. Stork and D. E. Kahne; *J. Amer. Chem. Soc.*, 1983, **105**, 1072. 84

83JA1383 J. A. Miller and G. Zweifel; *J. Am. Chem. Soc.*, 1983, **105**, 1383. 512

83JA1398 G. A. Russell and H. Tashtoush; *J. Am. Chem. Soc.*, 1983, **105**, 1398. 68, 454

83JA1598 R. Noyori, I. Nishida and J. Sakata; *J. Am. Chem. Soc.*, 1983, **105**, 1598. 884

83JA1660 J. E. McMurry and D. D. Miller; *J. Am. Chem. Soc.*, 1983, **105**, 1660. 688

83JA1988 R. E. Ireland, R. C. Anderson, R. Badoud, B. J. Fitzsimmons, G. J. McGarvey, S. Thaisrivongs and C. S. Wilcox; *J. Am. Chem. Soc.*, 1983, **105**, 1988. 218, 831

83JA2080 M. Al-Aseer, P. Beak, D. Hay, D. J. Kempf, S. Mills and S. G. Smith; *J. Am. Chem. Soc.*, 1983, **105**, 2080. 148

83JA2082 A. I. Meyers, W. F. Rieker and L. M. Fuentes; *J. Am. Chem. Soc.*, 1983, **105**, 2082. 149

83JA2435 A. E. Greene, M.-J. Luche and J.-D. Deprés; *J. Am. Chem. Soc.*, 1983, **105**, 2435. 6

83JA2503 H. J. Reich, K. E. Yelm and S. Wollowitz; *J. Am. Chem. Soc.*, 1983, **105**, 2503. 788

83JA2748 S. Uemura and S. Fukuzawa; *J. Am. Chem. Soc.*, 1983, **105**, 2748. 611, 716

83JA2904 W. S. Johnson, R. Elliot and J. D. Elliot; *J. Am. Chem. Soc.*, 1983, **105**, 2904. 1026

83JA2923 A. E. Asato, A. Kini, M. Denny and R. S. H. Liu; *J. Am. Chem. Soc.*, 1983, **105**, 2923. 505

83JA3571 M. Saunders and M. R. Kates; *J. Am. Chem. Soc.*, 1983, **105**, 3571. 906

83JA3578 J. Lusztyk, B. Maillard, D. A. Lindsay and K. U. Ingold; *J. Am. Chem. Soc.*, 1983, **105**, 3578. 936

83JA3656 M. Rosenberger, C. Newkom and E. R. Aig; *J. Am. Chem. Soc.*, 1983, **105**, 3656. 1134

83JA3714 D. R. Wright, L. B. Sims and A. Fry; *J. Am. Chem. Soc.*, 1983, **105**, 3714. 783

83JA3720 G. Stork and R. Mook, Jr.; *J. Am. Chem. Soc.*, 1983, **105**, 3720. 548, 1132

83JA3741 G. Stork, R. Mook, S. A. Biller and S. D. Rychnovsky; *J. Am. Chem. Soc.*, 1983, **105**, 3741. 342

83JA4017 N. Ona, R. Tamura and A Kaji; *J. Am. Chem. Soc.*, 1983, **105**, 4017. 37, 441

83JA4059 M. J. Robins, J. S. Wilson and F. Hansske; *J. Am. Chem. Soc.*, 1983, **105**, 4059. 939

83JA4701 T. J. Burkey, A. L. Castelhano, D. Griller and F. P. Lossing; *J. Am. Chem. Soc.*, 1983, **105**, 4701. 925

83JA5075 B. M. Trost and A. C. Lavoie; *J. Am. Chem. Soc.*, 1983, **105**, 5075. 439

83JA5373 L. E. Overman, D. Lesuisse and M. Hashimoto; *J. Am. Chem. Soc.*, 1983, **105**, 5373. 505

83JA5490 S. L. Buchwald and R. H. Grubbs; *J. Am. Chem. Soc.*, 1983, **105**, 5490. 991

83JA5946 R. A. Volkmann, J. T. Davis and C. N. Meltz; *J. Am. Chem. Soc.*, 1983, **105**, 5946. 204

83JA6155 T. D. Krizan and J. C. Martin; *J. Am. Chem. Soc.*, 1983, **105**, 6155. 1089

83JA6164 E. Block and M. Aslam; *J. Am. Chem. Soc.*, 1983, **105**, 6164. 619, 707

83JA6165 E. Block, M. Aslam, V. Eswarakrishnan and A. Wall; *J. Am. Chem. Soc.*, 1983, **105**, 6165. 707

83JA6273 T. Yoshida, W. J. Youngs, T. Sakaeda, T. Ueda, S. Otsuka and J. A. Ibers; *J. Am. Chem. Soc.*, 1983, **105**, 6273. 456, 507

83JA6344 E. Negishi, L. D. Boardman, J. M. Tour, H. Sawada and C. L. Rand; *J. Am. Chem. Soc.*, 1983, **105**, 6344. 1094

83JA6798 M. Bhupathy and T. Cohen; *J. Am. Chem. Soc.*, 1983, **105**, 6798. 805

83JA6882 D. B. Collum, F. Mohamadi and J. S. Hallock; *J. Am. Chem. Soc.*, 1983, **105**, 6882. 68

83JA7123 X. Creary and C. C. Geiger; *J. Am. Chem. Soc.*, 1983, **105**, 7123. 905

83JA7171 K. Yamamoto, T. Harada and M. Nakazaki; *J. Am. Chem. Soc.*, 1983, **105**, 7171. 690

83JA7200 R. Scheffold and R. Orlinski; *J. Am. Chem. Soc.*, 1983, **105**, 7200. 335, 546

83JA7407 S. Thaisrivongs and D. Seebach; *J. Am. Chem. Soc.*, 1983, **105**, 7407. 67

83JCR(M)167 K. Eichninger, H. Berbalk, E. Machat and J. Wimmer; *J. Chem. Res. (M)*, 1983, 167. 1004, 1005

83JCS(P1)87 O. Ribeiro. S. T. Hadfield, A. F. Clayton, C. W. Vose and M. M. Coombs; *J. Chem. Soc., Perkin Trans. 1*, 1983, 87. 15

83JCS(P1)387 R. Brettle and I. A. Jafri; *J. Chem. Soc., Perkin Trans. 1*, 1983, 387. 503

83JCS(P1)403 H. Natsugari, Y. Matsushita, N. Tamura, K. Yoshioka and M. Ochiai; *J. Chem. Soc., Perkin Trans. 1*, 1983, 403. 942

83JCS(P1)1009 W. D. Ollis, M. Rey and I. O. Sutherland; *J. Chem. Soc., Perkin Trans. 1*, 1983, 1009. 398, 399

83JCS(P1)1029 K. Chantrapromma, W. D. Ollis and I. O. Sutherland; *J. Chem. Soc., Perkin Trans. 1*, 1983, 1029. 839

83JCS(P1)1131 D. J. Ager; *J. Chem. Soc., Perkin Trans. 1*, 1983, 1131. 248

83JCS(P1)1169 R. S. Torr and S. Warren; *J. Chem. Soc., Perkin Trans. 1*, 1983, 1169. 643

83JCS(P1)1173 R. S. Torr and S. Warren; *J. Chem. Soc., Perkin Trans. 1*, 1983, 1173. 753

83JCS(P1)1689 D. G. Gillespie and B. J. Walker; *J. Chem. Soc., Perkin Trans. 1*, 1983, 1689. 1069

83JCS(P1)2387 S. J. Field and D. W. Young; *J. Chem. Soc., Perkin Trans. 1*, 1983, 2387. 394

83JCS(P1)2953 V. Calò, L. Lopez and W. F. Carlucci; *J. Chem. Soc., Perkin Trans. 1*, 1983, 2953. 238

83JCS(P2)291 R. Taylor; *J. Chem. Soc., Perkin Trans. 2*, 1983, 291. 777

83JCS(P2)305 M. B. Huang, O. Goscinski, G. Jonsäll and P. Ahlberg; *J. Chem. Soc., Perkin Trans. 2*, 1983, 305. 918

83JCS(P2)1923 J. G. Dawber, J. C. Tebby and A. A. C. Waite; *J. Chem. Soc., Perkin Trans. 2*, 1983, 1923. 51

83JHC729 A. Svensson, J. O. Karlsson and A. Hallberg; *J. Heterocycl. Chem.*, 1983, **20**, 729. 1072

83JOC273 R. O. Angus, Jr. and R. P. Johnson; *J. Org. Chem.*, 1983, **48**, 273. 522

83JOC279 K. Mikami, Y. Kimura, N. Kishi and T. Nakai; *J. Org. Chem.*, 1983, **48**, 279. 404, 405, 816, 817, 818

83JOC409 R. S. Matthews, E. D. Mihelich, L. S. McGowan and K. Daniels; *J. Org. Chem.*, 1983, **48**, 409. 221

83JOC494 M. R. Johnson, M. J. Fazio, D. L. Ward and L. R. Sousa; *J. Org. Chem.*, 1983, **48**, 494. 232

83JOC584 J. Hine and S.-M. Linden; *J. Org. Chem.*, 1983, **48**, 584. 797, 882

83JOC644 H. C. Brown and J. Chandrasekharan; *J. Org. Chem.*, 1983, **48**, 644. 510

83JOC860 R. Malherbe, G. Rist and D. Belluš; *J. Org. Chem.*, 1983, **48**, 860. 7, 833

83JOC1103 C. J. Elsevier, P. M. Stehouwer, H. Westmijze and P. Vermeer; *J. Org. Chem.*, 1983, **48**, 1103. 975

83JOC1262 L. A. Paquette, G. J. Hefferon, R. Samodral and Y. Hanzawa; *J. Org. Chem.*, 1983, **48**, 1262. 517

83JOC1293 Y. Tamaru, Z. Yoshida, Y. Yamada, K. Mukai and H. Yoshioka; *J. Org. Chem.*, 1983, **48**, 1293. 238

83JOC1333 K. Kikukawa, K. Kono, F. Wada and T. Matsuda; *J. Org. Chem.*, 1983, **48**, 1333. 486

83JOC1362 M. E. Garst and B. J. McBride; *J. Org. Chem.*, 1983, **48**, 1362. 241

83JOC1373 M. E. Kurz, L. Reif and T. Tantrarat; *J. Org. Chem.*, 1983, **48**, 1373. 336, 346

83JOC1431 R. F. Childs, D. L. Mulholland, A. Varadarajan and S. Yeroushalmi; *J. Org. Chem.*, 1983, **48**, 1431. 917

83JOC1455 C. E. Coburn, D. K. Anderson and J. S. Swenton; *J. Org. Chem.*, 1983, **48**, 1455. 209

83JOC1638 D. C. Tabor, F. H. White, L. W. Collier, IV and S. A. Evans; *J. Org. Chem.*, 1983, **48**, 1638. 74

83JOC1767 R. A. Volkmann, J. T. Davis and C. N. Meltz; *J. Org. Chem.*, 1983, **48**, 1767. 119

83JOC1854 M. J. Robins and P. J. Barr; *J. Org. Chem.*, 1983, **48**, 1854. 1010

83JOC1912 S. Nunomoto, Y. Kawakami and Y. Yamashita; *J. Org. Chem.*, 1983, **48**, 1912. 119, 471, 1092

83JOC1919 J. J. Brunet, C. Sidot and P. Caubere; *J. Org. Chem.*, 1983, **48**, 1919. 254

83JOC1924 R. A. Kjonaas and D. C. Shubert; *J. Org. Chem.*, 1983, **48**, 1924. 491

83JOC1991 C. H. Kuo, A. A. Patchet and N. L. Wendler; *J. Org. Chem.*, 1983, **49**, 1991. 79

83JOC2098 R. C. Gadwood; *J. Org. Chem.*, 1983, **48**, 2098. 713, 715, 716

83JOC2120 K. Tamao, N. Ishida and M. Kumada; *J. Org. Chem.*, 1983, **48**, 2120. 160

83JOC2122 M. Koreeda and L. Brown; *J. Org. Chem.*, 1983, **48**, 2122. 832

83JOC2146 G. A. Olah, G. K. S. Prakash, D. G. Farnum and T. P. Clausen; *J. Org. Chem.*, 1983, **48**, 2146. 906

83JOC2202 J. J. M. Lamberts and W. H. Laarhoven; *J. Org. Chem.*, 1983, **48**, 2202. 841

83JOC2266 S. C. Jain, D. E. Dussourd, W. E. Conner, T. Eisner, A. Guerrero and J. Meinwald; *J. Org. Chem.*, 1983, **48**, 2266. 503, 504

83JOC2270 W. Huang, S. P. Pulaski and J. Meinwald; *J. Org. Chem.*, 1983, **48**, 2270. 504

83JOC2274 S. C. Jain, W. L. Roelofs and J. Meinwald; *J. Org. Chem.*, 1983, **48**, 2274. 503

83JOC2499 C. A. Broka, S. Chan and B. Peterson; *J. Org. Chem.*, 1983, **48**, 2499. 216

83JOC2569 T. Minami, H. Sako, T. Ikehira, T. Hanamoto and I. Hiroa; *J. Org. Chem.*, 1983, **48**, 2569. 444

83JOC2796 R. A. Benkeser, F. G. Belmonte and J. Kang; *J. Org. Chem.*, 1983, **48**, 2796. 101

83JOC2957 G. B. Trimitsis, J. M. Hinkley, R. TenBrink, A. L. Faburada, R. Anderson, M. Poli, B. Christian, G. Gustafson, J. Erdman and D. Rop; *J. Org. Chem.*, 1983, **48**, 2957. 879

83JOC3085 S. Krishnamurthy and H. C. Brown; *J. Org. Chem.*, 1983, **48**, 3085. 4, 7, 9, 10, 17

83JOC3091 H. C. Brown, S. Narasimhan and V. Somayaji; *J. Org. Chem.*, 1983, **48**, 3091. 14

83JOC3162 J. R. Williams, J. F. Callahan and C. Lin; *J. Org. Chem.*, 1983, **48**, 3162. 78

83JOC3334 B. H. Lipshutz, D. Parker, J. A. Kozlowski and R. D. Miller; *J. Org. Chem.*, 1983, **48**, 3334. 177

83JOC3369 S. E. Denmark and M. A. Harmata; *J. Org. Chem.*, 1983, **48**, 3369. 832

83JOC3370 T. C. McMorris, P. H. Le, M. W. Preus, S. R. Schow and G. R. Weihe; *J. Org. Chem.*, 1983, **48**, 3370. 85

83JOC3382 A. Hassner and J. L. Dillon, Jr.; *J. Org. Chem.*, 1983, **48**, 3382. 518

83JOC3581 S. V. Govindan, T. Hudlicky and F. J. Koszyk; *J. Org. Chem.*, 1983, **48**, 3581. 795

83JOC3667 G. A. Olah, A. Husain, B. P. Singh and A. K. Mehrotra; *J. Org. Chem.*, 1983, **48**, 3667. 5, 9

83JOC3761 J. E. Wrobel and B. Ganem; *J. Org. Chem.*, 1983, **48**, 3761. 451

83JOC3866 A. A. Ponaras; *J. Org. Chem.*, 1983, **48**, 3866. 42, 833

83JOC3918 J. E. Bäckvall, J. Bergman and L. Engman; *J. Org. Chem.*, 1983, **48**, 3918. 595

83JOC4131 M. A. Tius and A. H. Fauq; *J. Org. Chem.*, 1983, **48**, 4131. 221, 478

83JOC4262 F.-A. Kunng, J.-M. Gu, S. Chao, Y. Chen and P. S. Mariano; *J. Org. Chem.*, 1983, **48**, 4262. 835

83JOC4272 E. N. Marvell, C. Hilton and M. Cleary; *J. Org. Chem.*, 1983, **48**, 4272. 420

83JOC4370 B. B. Snider and B. W. Burbaum; *J. Org. Chem.*, 1983, **48**, 4370. 499

83JOC4419 W. A. Lindley, D. W. H. MacDowell and J. L. Peterson; *J. Org. Chem.*, 1983, **48**, 4419. 436

83JOC4453 R. P. Short, J. M. Revol, B. C. Ranu and T. Hudlicky; *J. Org. Chem.*, 83, **48**, 4453. 804

83JOC4572 S. P. Tanis and P. M. Herrinton; *J. Org. Chem.*, 1983, **48**, 4572. 216

83JOC4621 J. P. Marino and R. J. Linderman; *J. Org. Chem.*, 1983, **48**, 4621. 479

83JOC4773 E. Vedejs and F. G. West; *J. Org. Chem.*, 1983, **48**, 4773. 823

83JOC4776 T. G. Back, S. Collins, U. Gokhale and K.-W. Law; *J. Org. Chem.*, 1983, **48**, 4776. 605

83JOC4976 R. V. C. Carr, R. V. Williams and L. A. Paquette; *J. Org. Chem.*, 1983, **48**, 4976. 654

83JOC4986 W. A. Kinney, G. D. Crouse and L. A. Paquette; *J. Org. Chem.*, 1983, **48**, 4986. 603

83JOC5080 H. C. Brown and J. Chandrasekharan; *J. Org. Chem.*, 1983, **48**, 5080. 510

83JOC5133 Y. Tanaka, H. Ohta, S. Kaji and K. Okuma; *J. Org. Chem.*, 1983, **48**, 5133. 210

83JOC5221	S. D. Burke, W. F. Fobare and G. J. Pacofsky; *J. Org. Chem.*, 1983, **48**, 5221.	504
83JOC5246	A. Toshimitsu, S. Uemura, M. Okano and N. Watanabe; *J Org. Chem.*, 1983, **48**, 5246.	603
83JOC5251	J. C. Gilbert, D. H. Giamalva and U. Weerasooriya; *J. Org. Chem.*, 1983, **48**, 5251.	495
83JOC5302	E. Keinan and M. Peretz; *J. Org. Chem.*, 1983, **48**, 5302.	194, 1053
83JOC5376	K. K. Wang, S. S. Nikam and C. D. Ho; *J. Org. Chem.*, 1983, **48**, 5376.	977
83JOM(243)281	J. Barrau, M. Bouchaut, H. Lavayssiere, G. Dousse and J. Satge; *J. Organomet. Chem.*, 1983, **243**, 281.	656
83JOM(250)203	S. Uemura, S. Fukuzawa and A. Toshimitsu; *J. Organomet. Chem.*, 1983, **250**, 203.	611
83JOM(250)551	I. P. Beletskaya; *J. Organomet. Chem.*, 1983, **250**, 551.	1006
83JOM(253)131	J. V. Comasseto; *J. Organomet. Chem.*, 1983, **253**, 131.	440
83JOM(255)135	H. C. Brown and N. C. Hebert; *J. Organomet. Chem.*, 1983, **255**, 135.	58
83JOM(256)C1	P. Bras, A. Van der Gen and J. Wolters; *J. Organomet. Chem.*, 1983, **256**, C1.	248
83LA226	M. Enteroth, T. Kunczik and L. Jaenicke; *Liebigs Ann. Chem.*, 1983, 226.	212
83LA705	K. Annen, H. Hofmeister, H. Laurent and R. Wiechert; *Liebigs Ann. Chem.*, 1983, 705.	5, 9
B-83MI 101-01	K. D. Asmus, M. Bonifacic, K. U. Ingold and B. P. Roberts; in "Radical Reaction Rates in Liquids," ed. H. Fischer, Springer, Heidelberg, 1983, Vol. II, 13b, p. 5 and 13c, p. 323.	5
B-83MI 104-01	G. F. Koser; in "Supplement D: The Chemistry of Halides, Pseudo-Halides and Azides," Part 1, eds S. Patai and Z. Rappoport; Wiley, Chichester, 1983, chap. 18, pp. 721–811.	108
B-83MI 104-02	R. Verhé and N. De Kimpe; in "Supplement D: The Chemistry of Halides, Pseudo-Halides and Azides," Part 1, eds S. Patai and Z. Rappoport; Wiley, Chichester, 1983, chap. 19, pp. 813–931.	115
B-83MI 110-01	M. M. Baizer and H. Lund; "Organic Electrochemistry," ed. M. Dekker, Wiley, New York, 1983, ch. 7.	427
B-83MI 112-01	H. Nozaki; "Current Trends in Organic Synthesis," Pergamon, Oxford, 1983, pp. 291–302.	539
B-83MI 114-01	H. J. Reich; "Proceedings of the Fourth International Conference on the Organic Chemistry of Selenium and Tellurium," ed. F. J. Berry and W. R. McWinnie, 1983, p. 258.	620
B-83MI 114-02	W. P. Weber; "Silicon Reagents for Organic Synthesis," Springer-Verlag, New York, 1983, pp. 58–78.	648
83MI 115-01	P. Welzel; *Nachr. Chem. Tech. Lab.*, 1983, **31**, 814 (*Chem. Abstr.*, 1983, **99**, 193959v).	687
83MI 116-01	Huang Yao-zeng, Shi Li-lan, Li Bin-quan and Ling Fang-le; *Acta Chim. Sinica Engl. Ed.*, 1983, **41**, 269.	738
B-83MI 118-01	W. Adams, N. Caballeira, O. De Lucchi and K. Hill; in "Stereochemistry and reactivity of systems containing *p*-electrons," ed. W. H. Watson, Verlag Chemie, 1983, p. 281.	800
B-83MI 119-01	K. D. Asmus, M. Bonifacic, K. U. Ingold and B. P. Roberts; in "Radical Reaction Rates in Liquids," ed. H. Fischer, Springer Verlag, Heidelberg, 1983, vol. II, **13b**, p. 5.	935
B-83MI 119-02	K. D. Asmus, M. Bonifacic, K. U. Ingold and B. P. Roberts; in "Radical Reaction Rates in Liquids," ed. H. Fischer, Springer Verlag, Heidelberg, 1983, vol. II, **13c**, p. 323.	937
B-83MI 119-03	R. B. Bates and C. A. Ogle; "Carbanion Chemistry," Springer, New York, 1983.	844
B-83MI 119-04	P. Deslongchamps; "Stereoelectronic Effects in Organic Chemistry," Pergamon, Oxford, 1983, 274.	878
B-83MI 119-05	R. Scheffold, G. Rytz and L. Walder; in "Modern Synthetic Methods," ed. R. Scheffold, Wiley, New York, 1983, vol. 3, p. 355.	946
B-83MI 121-01	T. L. Gilchrist; in "The Chemistry of Functional Groups, Supplement C: The Chemistry of Triple-Bonded Functional Groups" ed. S. Patai and Z. Rappoport, Wiley, New York, 1983, p. 383.	1066, 1081
B-83MI 121-02	L. I. Simándi; in "The Chemistry of Functional Groups, Supplement C: The Chemistry of Triple-Bonded Functional Groups," ed. S. Patai and Z. Rappoport, Wiley, New York, 1983, p. 510.	1034
B-83MI 121-03	K. Friedrich; in "The Chemistry of Functional Groups, Supplement C: The Chemistry of Triple-Bonded Functional Groups," ed. S. Patai and Z. Rappoport, Wiley, New York, 1983, p. 1380.	998
B-83MI 122-01	S. Patai and Z. Rappoport (eds.); "The Chemistry of Triple-bonded Functional Groups," Supplement C, Part 2, Wiley, Chichester, 1983.	1133
83OM236	D. J. S. Tsai and D. S. Matteson; *Organometallics*, 1983, **2**, 236.	764
83OM554	M.-Y. Li and J. San Filippo, Jr.; *Organometallics*, 1983, **2**, 554.	663
83OM563	E. Negishi, D. E. Van Horn, T. Yoshida and C. L. Rand; *Organometallics*, 1983, **2**, 563.	544
83OM598	T. Klingstedt and T. Frejd; *Organometallics*, 1983, **2**, 598.	465
83OM779	H. C. Brown, S. C. Kim and S. Krishnamurthy; *Organometallics*, 1983, **2**, 779.	485
83OM1055	K. R. Januszkiewicz and H. Alper; *Organometallics*, 1983, **2**, 1055.	94
83OR163	R. Noyori and Y. Hayakawa; *Org. React.*, 1983, **29**, 163.	2
83PAC1733	K. A. Brown-Wensley, S. L. Buchwald, L. Cannizzo, L. Clawson, S. Ho, D. Meinhardt, J. R. Stille, D. Straus and R. H. Grubbs; *Pure Appl. Chem.*, 1983, **55**, 1733.	769
83PS(15)359	H.-J. Christeau, L. Labaudinière and H. Christol; *Phosphorous and Sulfur*, 1983, **15**, 359.	444
83PS(16)339	R. O. Larsen and G. Aksnes; *Phosphorus Sulfur*, 1983, **16**, 339.	751
83RCR61	V. A. Mironov, A. D. Fedorovich and A. A. Akhrem; *Russ. Chem. Rev. (Engl. Transl.)*, 1983, **52**, 61.	248
83S1	P. Brownbridge; *Synthesis*, 1983, 1.	308
83S32	G. Linstrumelle and T. Jeffery-Luong; *Synthesis*, 1983, 32.	469, 472, 975, 1009, 1137
83S128	G. Zweifel and J. A. Miller; *Synthesis*, 1983, 128.	1019
83S130	D. Michelot; *Synthesis*, 1983, 130.	1097
83S137	G. Rosini, R. Ballini and V. Zanotti; *Synthesis*, 1983, 137.	37
83S284	M. Visnich, L. Strenkowski and M. Battiste; *Synthesis*, 1983, 284.	226
83S304	D. G. Norman and C. B. Reese; *Synthesis*, 1983, 304.	5, 9

83S310 C. G. Shankar, B. V. Mallaiah and G. Srimannarayana; *Synthesis*, 1983, 310. 555
83S312 H. Yamanaka, T. Sakamoto, M. Shiraiwa and Y. Kondo; *Synthesis*, 1983, 312. 1010
83S429 C. Santelli-Rouvier and M. Santelli; *Synthesis*, 1983, 429. 421
83S701 E. Wenkert, J. B. Fernandes, E. L. Michelott and C. S. Swindell; *Synthesis*, 1983, 701. 195
83S801 M. A. Boaventura, J. Drouin and J. M. Conia; *Synthesis*, 1983, 801. 544
83S804 S. Suzuki, M. Shiono and Y. Fujita; *Synthesis*, 1983, 804. 187
83S883 G. Hahn and G. Zweifel; *Synthesis*, 1983, 883. 1052, 1055, 1056, 1058
83S952 B. W. Disanayaka and A. C. Weedon; *Synthesis*, 1983, 952. 797
83S1006 F. Ogura, T. Otsubo, N. Ohira and H. Yamaguchi; *Synthesis*, 1983, 1006. 611, 621
83SC237 R. M. Carlson and L. L. White; *Synth. Commun.*, 1983, **13**, 237. 205
83SC677 A. K. Sinhababu and R. T. Borchardt; *Synth. Commun.*, 1983, **13(8)**, 677. 34
83SC1103 R. A. Benkeser, F. G. Belmonte and M.-H. Yang; *Synth. Commun.*, 1983, **13**, 1103. 101
83SC1193 J. B. Ousset, C. Mioskowski and G. Solladie; *Synth. Commun.*, 1983, **13**, 1193. 739, 740
83T206 N. H. Werstiuk; *Tetrahedron*, 1983, **39**, 206. 868
83T297 G. S. Harris, D. Lloyd, W. A. MacDonald and I. Gosney; *Tetrahedron*, 1983, **39**, 297. 738
83T485 R. G. Salomon; *Tetrahedron*, 1983, **39**, 485. 357
83T935 R. L. Danheiser, D. J. Carini, D. M. Fink and A. Basak; *Tetrahedron*, 1983, **39**, 935. 450, 527, 1061
83T975 I. Kuwajima and E. Nakamura and K. Hashimoto; *Tetrahedron*, 1983, **39**, 975. 1038
83T1141 P. von R. Schleyer and A. J. Kos; *Tetrahedron*, 1983, **39**, 1141. 852
83T2185 B. Le Doussal, A. Le Coq, A. Gorgues and A. Meyer; *Tetrahedron*, 1983, **39**, 2185. 958
83T2289 M. Tiecco, L. Testaferri, M. Tingoli and E. Wenkert; *Tetrahedron*, 1983, **39**, 2289. 484
83T2323 A. S. Rao, S. K. Pakinkar and J. G. Kirtane; *Tetrahedron*, 1983, **39**, 2323. 842
83T2609 W. Hartwig; *Tetrahedron*, 1983, **39**, 2609. 16, 938
83T2863 J. Barluenga, J. López-Prado, P. J. Campos and G. Asensio; *Tetrahedron*, 1983, **39**, 2863. 68
83T2869 D. L. Boger; *Tetrahedron*, 1983, **39**, 2869. 790
83T3207 J. D. Albright; *Tetrahedron*, 1983, **39**, 3207. 165
83T3279 J. Tsuji, H. Kezuka, Y. Toshida, H. Takayanagi and K. Yamamoto; *Tetrahedron*, 1983, **39**, 3279. 635, 1075
83T3283 M. Julia, A. Righini-Tapie and J. N. Verpeaux; *Tetrahedron*, 1983, **39**, 3283. 144, 237, 483
83T3731 S. Raucher and R. F. Lawrence; *Tetrahedron*, 1983, **39**, 3731. 828
83T3943 M. T. Caproiu, N. Negoita and A. T. Balaban; *Tetrahedron*, 1983, **39**, 3943. 1132
83T4225 H. J. Meeuwissen, T. A. Van der Knaap and F. Bickelhaupt; *Tetrahedron*, 1983, **39**, 4225. 445
83TL3 A. A. Ponaras; *Tetrahedron Lett.*, 1983, **24**, 3. 833
83TL111 A. D. Buss and S. Warren; *Tetrahedron Lett.*, 1983, **24**, 111. 641, 642, 757
83TL185 H. Birkhofer, H. D. Beckhaus and C. Rüchardt; *Tetrahedron Lett.*, 1983, **24**, 185. 925
83TL391 M. Yamaguchi and I. Hirao, *Tetrahedron Lett.*, 1983, **24**, 391. 204, 1023
83TL409 L. S. Mills and P. C. North; *Tetrahedron Lett.*, 1983, **24**, 409. 505
83TL513 K. Mikama, K. Fujimoto and T. Nakai; *Tetrahedron Lett.*, 1983, **24**, 513. 405, 816
83TL555 D. D. Miller, K. E. Moorthy and A. Hamada; *Tetrahedron Lett.*, 1983, **24**, 555. 556
83TL593 A. D. Buss, R. Mason and S. Warren; *Tetrahedron Lett.*, 1983, **24**, 593. 756
83TL623 A. Pelter, B. Singaram, L. Williams and J. W. Wilson; *Tetrahedron Lett.*, 1983, **24**, 623. 163
83TL627 A. Pelter, L. Williams and J. W. Wilson; *Tetrahedron Lett.*, 1983, **24**, 627. 163
83TL631 A. Pelter, B. Singaram and J. W. Wilson; *Tetrahedron Lett.*, 1983, **24**, 631. 163
83TL635 A. Pelter, B. Singaram and J. W. Wilson; *Tetrahedron Lett.*, 1983, **24**, 635. 660, 764
83TL703 B. Giese, J. Dupius, T. Hasskerl and J. Meixner; *Tetrahedron Lett.*, 1983, **24**, 703. 323, 920
83TL1003 K. J. Shea and P. Q. Pham; *Tetrahedron Lett.*, 1983, **24**, 1003. 1092
83TL1021 Y. Chen, P. L. Huesmann and P. S. Mariano; *Tetrahedron Lett.*, 1983, **24**, 1021. 835
83TL1041 F. Sato, H. Watanabe, Y. Tanaka, T. Yamaji and M. Sato; *Tetrahedron Lett.*, 1983, **24**, 1041. 161, 450, 1092, 1105
83TL1129 B. M. Trost and R. Remuson; *Tetrahedron Lett.*, 1983, **24**, 1129. 522
83TL1147 R. Hittich and K. Griesbaum; *Tetrahedron Lett.*, 1983, **24**, 1147. 9
83TL1157 N. Bluthe, M. Malacria and J. Gore; *Tetrahedron Lett.*, 1983, **24**, 1157. 413, 810
83TL1247 R. Bloch and J. Abecassis; *Tetrahedron Lett.*, 1983, **24**, 1247. 790
83TL1297 A. Claesson, A. Quader and C. Sahlberg; *Tetrahedron Lett.*, 1983, **24**, 1297. 970
83TL1341 E.-i. Negishi and S. Chatterjee; *Tetrahedron Lett.*, 1983, **24**, 1341. 133
83TL1345 M. E. Krafft and R. A. Holton; *Tetrahedron Lett.*, 1983, **24**, 1345. 130
83TL1357 R. R. Webb, Jr. and S. Danishefsky; *Tetrahedron Lett.*, 1983, **24**, 1357. 243
83TL1377 W. R. Roush, M. A. Adam and S. M. Peseckis; *Tetrahedron Lett.*, 1983, **24**, 1377. 221
83TL1407 H. Hiemstra and W. N. Speckamp; *Tetrahedron Lett.*, 1983, **24**, 1407. 978
83TL1411 P. R. Singh and R. K. Khanna; *Tetrahedron Lett.*, 1983, **24**, 1411. 68
83TL1433 A. Pelter, S. Singaram and H. C. Brown; *Tetrahedron Lett.*, 1983, **24**, 1433. 509, 510
83TL1463 J. Leder, H. Fujioka and Y. Kishi; *Tetrahedron Lett.*, 1983, **24**, 1463. 1092
83TL1475 J. L. Belletire and D. R. Walley; *Tetrahedron Lett.*, 1983, **24**, 1475. 568
83TL1657 J. R. Schauder, J. N. Denis and A. Krief; *Tetrahedron Lett.*, 1983, **24**, 1657. 592
83TL1667 E. C. Ashby and W.-S. Park; *Tetrahedron Lett.*, 1983, 1667. 884
83TL1675 M. E. Garst, B. J. McBride and J. G. Douglass, III; *Tetrahedron Lett.*, 1983, **24**, 1675. 524
83TL1719 M. Yamaguchi and I. Hirao; *Tetrahedron Lett.*, 1983, **24**, 1719. 1047
83TL1801 R. Yamaguchi, Y. Nakazono and M. Kawanisi; *Tetrahedron Lett.*, 1983, **24**, 1801. 1042
83TL1871 N. N. Marinovic and H. Ramanathan; *Tetrahedron Lett.*, 1983, **24**, 1871. 338, 342, 1127
83TL1897 P. A. Grieco, P. Garner and Z.-M. He; *Tetrahedron Lett.*, 1983, **24**, 1897. 373
83TL1909 G. Majetich, A. M. Casares, D. Chapman and M. Behnke; *Tetrahedron Lett.*, 1983, **24**, 1909. 314
83TL2231 W. R. Roush and B. M. Lesur; *Tetrahedron Lett.*, 1983, **24**, 2231. 305

83TL2239 K. M. Nicholas and S. Padmanabhan; *Tetrahedron Lett.*, 1983, **24**, 2239. 199, 1025
83TL2311 M. Taddei, A. Papini, M. Fiorenza and A. Ricci; *Tetrahedron Lett.*, 1983, **24**, 2311. 242
83TL2473 M. Saindane and W. Ott; *Tetrahedron Lett.*, 1983, **24**, 2473. 523
83TL2599 A. Osuka, Y. Mori, H. Shimizu and H. Suzuki; *Tetrahedron Lett.*, 1983, **24**, 2599. 623
83TL2723 J. E. McMurry and S. Mohanraj; *Tetrahedron Lett.*, 1983, 2723. 482
83TL2821 E. J. Corey and S. G. Pyne; *Tetrahedron Lett.*, 1983, **24**, 2821. 548
83TL2837 M. T. Reetz and M. Sauerwald; *Tetrahedron Lett.*, 1983, **24**, 2837. 202
83TL2931 M. E. Jung and G. L. Hatfield; *Tetrahedron Lett.*, 1983, **24**, 2931. 805
83TL3165 E. J. Corey and T. M. Eckrich; *Tetrahedron Lett.*, 1983, **24**, 3165. 136
83TL3213 G. Helmchen, A. Selim, D. Dorsch and I. Taufer; *Tetrahedron Lett.*, 1983, **24**, 3213. 134
83TL3291 E. J. Corey, S. G. Pyne and A. I. Schafer; *Tetrahedron Lett.*, 1983, **24**, 3291. 1015, 1016
83TL3477 N. Ono, H. Miyake, M. Fujii and A. Kaji; *Tetrahedron Lett.*, 1983, **24**, 3477. 37
83TL3521 A. Laporterie, H. Iloughmane and J. Dubac; *Tetrahedron Lett.*, 1983, **24**, 3521. 777
83TL3757 L. E. Overman and A. F. Renaldo; *Tetrahedron Lett.*, 1983, **24**, 3757. 809
83TL3905 P. Kocienski and C. Yeates; *Tetrahedron Lett.*, 1983, **24**, 3905. 1094
83TL3931 A. D. Buss and S. Warren; *Tetrahedron Lett.*, 1983, **24**, 3931. 753
83TL3997 S. Arseniyadis and J. Gore; *Tetrahedron Lett.*, 1983, **24**, 3997. 68, 454
83TL4139 J. Blum, I. Amer, A. Zoran and Y. Sasson; *Tetrahedron Lett.*, 1983, **24**, 4139. 94
83TL4151 I. Fleming and N. K. Terrett; *Tetrahedron Lett.*, 1983, **24**, 4151. 654
83TL4153 I. Fleming and N. K. Terrett; *Tetrahedron Lett.*, 1983, **24**, 4153. 654
83TL4181 L. Xingya and R. Huisgen; *Tetrahedron Lett.*, 1983, **24**, 4181. 367
83TL4311 J.-L. Fabre and M. Julia; *Tetrahedron Lett.*, 1983, **24**, 4311. 439, 700
83TL4319 T. Cuvigny, J.-L. Fabre, C. Herve du Penhoat and M. Julia; *Tetrahedron Lett.*, 1983, **24**, 4319. 700
83TL4347 S. Uemura and S. Fukuzawa; *Tetrahedron Lett.*, 1983, **24**, 4347. 611, 620, 716
83TL4405 W. C. Still and C. Gennari; *Tetrahedron Lett.*, 1983, **24**, 4405. 751, 752
83TL4419 J. B. Ousset, C. Mioskowski and G. Solladie; *Tetrahedron Lett.*, 1983, **24**, 4419. 269, 647
83TL4503 C. N. Meltz and R. A. Volkmann; *Tetrahedron Lett.*, 1983, **24**, 4503. 1041
83TL4691 R. Bloch, D. Hassan and X. Mandard; *Tetrahedron Lett.*, 1983, **24**, 4691. 983
83TL4825 C. B. B. Ekogha, O. Ruel and S. A. Julia; *Tetrahedron Lett.*, 1983, **24**, 4825. 1089
83TL4833 J.-M. Lancelin, P. H. A. Zollo and P. Sinaÿ; *Tetrahedron Lett.*, 1983, **24**, 4833. 513
83TL4883 E. J. Corey, S. G. Pyne and W. Su; *Tetrahedron Lett.*, 1983, **24**, 4883. 504, 513
83TL4887 G. Stork and J. M. Stryker; *Tetrahedron Lett.*, 1983, **24**, 4887. 1039
83TL4899 R. Zamboni, S. Milette and J. Rokach; *Tetrahedron Lett.*, 1983, **24**, 4899. 513
83TL4923 R. K. Boeckman, Jr. and C. J. Flann; *Tetrahedron Lett.*, 1983, **24**, 4923. 67
83TL4993 T. Mandai, K. Hara, T. Nakajima, M. Kawada and J. Otera; *Tetrahedron Lett.*, 1983, **24**, 4993. 613
83TL4997 K. Suzuki, E. Katayama and G. Tsuchihashi; *Tetrahedron Lett.*, 1983, **24**, 4997. 389
83TL5089 F. Barbot, A. Kadib-Elban and P. Miginiac; *Tetrahedron Lett.*, 1983, **24**, 5089. 311
83TL5121 M. Yamaguchi, Y. Nobayashi and I. Hirao, *Tetrahedron Lett.*, 1983, **24**, 5121. 204, 1024
83TL5131 Y. Kawanami, T. Katsuki and M. Yamaguchi; *Tetrahedron Lett.*, 1983, **24**, 5131. 1045
83TL5137 C. Sahlberg, A. Quader and A. Claesson; *Tetrahedron Lett.*, 1983, 5137. 480
83TL5181 E. Negishi, V. Bagheri, S. Chatterjee, F. Luo, J. A. Miller and A. T. Stoll; *Tetrahedron Lett.*, 1983, **24**, 5181. 1045, 1047
83TL5185 J. Rokach, J. Adams and R. Perry; *Tetrahedron Lett.*, 1983, **24**, 5185. 1002, 1134
83TL5281 K. Takai, K. Kimura, T. Kuroda, T. Hiyama and H. Nozaki; *Tetrahedron Lett.*, 1983, **24**, 5281. 677, 1099
83TL5293 A. D. Buss, R. Mason and S. Warren; *Tetrahedron Lett.*, 1983, **24**, 5293. 641, 642
83TL5329 H. Alper, F. Sibtain and J. Heveling; *Tetrahedron Lett.*, 1983, **24**, 5329. 437
83TL5343 R. Henning and H. Urbach; *Tetrahedron Lett.*, 1983, **24**, 5343. 325, 326
83TL5587 A. C. Oehlschlager and E. Czyzewska; *Tetrahedron Lett.*, 1983, **24**, 5587. 966
83TL5635 J. Tsuji, L. Minami and I. Shimizu; *Tetrahedron Lett.*, 1983, **24**, 5635. 558
83TL5745 T. Fujisawa, S. Iida, H. Yukizaki and T. Sato; *Tetrahedron Lett.*, 1983, 5745. 476
83TL5857 W.-D. Fessner, H. Prinzbach and G. Rihs; *Tetrahedron Lett.*, 1983, **24**, 5857. 918
83TL5889 D. H. R. Barton and G. Kretzschmar; *Tetrahedron Lett.*, 1983, **24**, 5889. 935
83ZN(B)1697 W. Roth and P. Von R. Schleyer; *Z. Naturforsch., Teil B*, 1983, **38**, 1697. 427
83ZOB806 M. G. Voronkov, S. N. Adamovich, L. V. Sherstyannikova and V. B. Putchnarevich; *Zh. Obshch. Khim.*, 1983, **53**, 806. 61

84ACR347 R. F. Childs; *Acc. Chem. Res.*, 1984, **17**, 347. 910, 917, 918
84ACR358 E. Vedejs; *Acc. Chem. Res.*, 1984, **17**, 358. 234, 410, 822
84AG(E)53 H. J. Bestmann and G. Wölfel; *Angew. Chem., Int. Ed. Engl.*, 1984, **23**, 53. 1041
84AG(E)69 B. Giese, J. A. González-Gómez and T. Witzel; *Angew. Chem. Int. Ed. Engl.*, 1984, **23**, 69. 172, 325, 330
84AG(E)224 A. Maercker and R. Dujardin; *Angew. Chem., Int. Ed. Engl.*, 1984, **23**, 224. 68, 69, 454
84AG(E)371 M. Hilderbrand, H. Pritzkow, U. Zenneck and W. Siebert; *Angew. Chem., Int. Ed. Engl.*, 1984, **23**, 371. 910
84AG(E)413 C. Giordano, G. Castaldi and F. Uggeri; *Angew. Chem., Int. Ed. Engl.*, 1984, **23**, 413. 391
84AG(E)420 A. R. Katritzky and C. M. Marson; *Angew. Chem., Int. Ed. Engl.*, 1984, **23**, 420. 30
84AG(E)435 U. Nagel; *Angew. Chem., Int. Ed. Engl.*, 1984, **23**, 435. 76
84AG(E)539 K. P. C. Vollhardt; *Angew. Chem., Int. Ed. Engl.*, 1984, **23**, 539. 447, 448, 530
84AG(E)579 L. E. Overman; *Angew. Chem., Int. Ed. Engl.*, 1984, **23**, 579. 413, 809, 826, 836, 837
84AG(E)587 K. H. Dötz; *Angew. Chem., Int. Ed. Engl.*, 1984, **23**, 587. 767

84AG(E)847 R. Noyori and M. Suzuki; *Angew. Chem., Int. Ed. Engl.*, 1984, **23**, 847. 504

84AHC(37)167 J. E. Toomey; *Adv. Heterocycl. Chem.*, 1984, **37**, 167. 87

84AKZ116 G. A. Chukhadzhyan, R. G. Karapetyan and K. N. Babayan; *Arm. Khim. Zh.*, 1984, **37**, 116 (*Chem. Abstr.*, 1984, **101**, 6565p). 986

84BCJ108 K. Oshima, K. Takai, M. Sato and H. Nozaki; *Bull. Chem. Soc. Jpn.*, 1984, **57**, 108. 480, 1027

84BCJ2335 J. Yamashita, Y. Inoue, T. Kondo and H. Hashimoto; *Bull. Chem. Soc. Jpn.*, 1984, **57**, 2335. 676, 678

84BCJ2768 K. Furuta, M. Ishiguro, R. Haruta, N. Ikeda and H. Yamamoto; *Bull. Chem. Soc. Jpn.*, 1984, **57**, 2768. 1056, 1057

84BSB511 R. Huisgen, C. Fulka, I. Kalwinsch, X. Li, G. Mloston, J. R. Moran and A. Pröbstl; *Bull. Soc. Chim. Belg.*, 1984, **93**, 511. 367

84CAR(130)103 I. Cerny, M. Budesinsky, T. Trnka and M. Cerny; *Carbohydr. Res.*, 1984, **130**, 103. 4, 10

84CAR(130)125 K. Bock, I. Lundt and C. Pedersen; *Carbohydr. Res.*, 1984, **130**, 125. 2, 4, 8, 9

84CB277 A. Krebs, W. Ruger, B. Ziegenhagen, M. Hebold, I. Hardtke, R. Muller, M. Schutz, M. Wietzke and M. Wilke; *Chem. Ber.*, 1984, **117**, 277. 686

84CB322 M. T. Reetz, P. Walz, F. Hübner, S. H. Hüttenhain, H. Heimbach and K. Schwellnus; *Chem. Ber.*, 1984, **117**, 322. 189, 200

84CB966 K. Weinges, W. Kasel, J. Klein, G. Hubertus, H. Irngartinger, U. Huber-Patz and H. Rodewald; *Chem. Ber.*, 1984, **117**, 966. 705

84CB1868 K. Weinges, W. Kasel, U. Huber-Patz, H. Rodewald and H. Irngartinger; *Chem. Ber.*, 1984, **117**, 1868. 708

84CB2132 B. Giese, K. Heuck, H. Lenhardt and U. Lüning; *Chem. Ber.*, 1984, **117**, 2132. 325, 326

84CB2510 W. Kirmse and S. Brandt; *Chem. Ber.*, 1984, **117**, 2510. 905

84CB3160 B. Giese and G. Kretzschmar; *Chem. Ber.*, 1984, **117**, 3160. 332

84CB3169 K. Raab and W. Beck; *Chem. Ber.*, 1984, **117**, 3169. 517

84CB3175 B. Giese and G. Kretzschmar; *Chem. Ber.*, 1984, **117**, 3175. 948

84CC151 P. Kocienski and C. Yeates; *J. Chem. Soc., Chem. Commun.*, 1984, 151. 1094

84CC152 R. D. Arnold, J. E. Baldwin and C. B. Ziegler, Jr.; *J. Chem. Soc., Chem. Commun.*, 1984, 152. 988

84CC202 M. Yamaguchi and I. Hirao; *J. Chem. Soc., Chem. Commun.*, 1984, 202. 223, 1023

84CC274 N. Ono, I. Hamamoto and A. Kaji; *J. Chem. Soc., Chem. Commun.*, 1984, 274. 486

84CC341 M. Petit, A. Mortreux and F. Petit; *J. Chem. Soc., Chem. Commun.*, 1984, 341. 692

84CC349 K. C. Nicolaou, R. Zipkin and D. Tanner; *J. Chem. Soc., Chem. Commun.*, 1984, 349. 754

84CC350 K. C. Nicolaou and S. Webber; *J. Chem. Soc., Chem. Commun.*, 1984, 350. 1001

84CC534 A. G. Angoh and D. L. J. Clive; *J. Chem. Soc., Chem. Commun.*, 1984, 534. 505, 651

84CC585 T. S. Tan, A. N. Mather, G. Procter and A. H. Davidson; *J. Chem. Soc., Chem. Commun.*, 1984, 585. 216

84CC617 E. Wenkert, M. H. Leftin and E. L. Michelotti; *J. Chem. Soc., Chem. Commun.*, 1984, 617. 484

84CC735 M. H. P. J. Aerssens and L. Brandsma; *J. Chem. Soc., Chem. Commun.*, 1984, 735. 457, 508

84CC774 D. H. R. Barton and D. Crich; *J. Chem. Soc., Chem. Commun.*, 1984, 774. 940

84CC794 T. Morimoto, T. Takahashi and M. Sekiya; *J. Chem. Soc., Chem. Commun.*, 1984, 794. 1041

84CC827 M. J. Begley, A. G. Cameron and D. W. Knight; *J. Chem. Soc., Chem. Commun.*, 1984, 827. 831

84CC862 W. Sas; *J. Chem. Soc., Chem. Commun.*, 1984, 862. 68

84CC990 P. J. Battye and D. W. Jones; *J. Chem. Soc., Chem. Commun.*, 1984, 990. 403, 800

84CC1004 Y. Yamamoto, W. Ito and K. Maruyama; *J. Chem. Soc., Chem. Commun.*, 1984, 1004. 1059

84CC1273 D. Wang and T. H. Chang; *J. Chem. Soc., Chem. Commun.*, 1984, 1273. 216

84CC1287 T. Jeffery; *J. Chem. Soc., Chem. Commun.*, 1984, 1287. 494

84CC1298 D. H. R. Barton, Y. Hervé, P. Potier and J. Thierry; *J. Chem. Soc., Chem. Commun.*, 1984, 1298. 935

84CC1323 T. Chou, H.-H. Tso and L.-J. Chang; *J. Chem. Soc., Chem. Commun.*, 1984, 1323. 791

84CC1345 P. A. Grieco, H. L. Sham, J. Inanaga, H. Kim and P. A. Tuthill; *J. Chem. Soc., Chem. Commun.*, 1984, 1345. 606

84CC1417 T. G. Back; *J. Chem. Soc., Chem. Commun.*, 1984, 1417. 25, 440

84CC1427 J. Barluenga, F. Aznar, R. Liz and M. Bayod; *J. Chem. Soc., Chem. Commun.*, 1984, 1427. 833

84CC1654 M. J. Goldstein and T. T. Wenzel; *J. Chem. Soc., Chem. Commun.*, 1984, 1654. 890

84CHEC(2)315 B. C. Uff; *Comp. Heterocycl. Chem.*, 1984, **2**, 315. 859

84CHEC(7)131 D. C. Dittmer; *Comp. Heterocycl. Chem.*, 1984, **7**, 131. 242

84CHEC(7)363 S. Searles; *Comp. Heterocycl. Chem.*, 1984, **7**, 363. 785

84CJC1333 S. R. Abrams; *Can. J. Chem.*, 1984, **62**, 1333. 1085

84CJC1558 G. Bérubé and P. Deslongchamps; *Can. J. Chem.*, 1984, **62**, 1558. 990

84CJC1850 A. M. de P. Nicholas and D. R. Arnold; *Can. J. Chem.*, 1984, **62**, 1850. 921

84CJC1860 A. M. de P. Nicholas and D. R. Arnold; *Can. J. Chem.*, 1984, **62**, 1860. 921

84CL131 M. Iyoda, K. Nishioka, M. Nose, S. Tanaka and M. Oda; *Chem. Lett.*, 1984, 131. 986

84CL255 N. Sayo, F. Shirai and T. Nakai; *Chem. Lett.*, 1984, 255. 816, 817

84CL405 T. Mukaiyama, F. Tabusa, T. Yamada and K. Suzuki; *Chem. Lett.*, 1984, 405. 1039

84CL833 H. Matsuyama, Y. Miyazawa, Y. Takei and M. Kobayashi; *Chem. Lett.*, 1984, 833. 706, 708

84CL1029 A. Arase, M. Hoshi, Y. Nunokawa and Y. Masuda; *Chem. Lett.*, 1984, 1029. 1018

84CL1225 M. Kosugi, T. Sumiya, T. Ogata, H. Sano and T. Migita; *Chem. Lett.*, 1984, 1225. 471

84CL1437 O. Moriya, M. Okawara and Y. Ueno; *Chem. Lett.*, 1984, 1437. 1127, 1130, 1131

84CL1505 H. Sugimura and H. Yakei; *Chem. Lett.*, 1984, 1505. 191, 238

84CL1541 S. Tsuboi, A. Kuroda, T. Masuda and A. Takeda; *Chem. Lett.*, 1984, 1541. 797

84CL1543 M. Suzuki, Y. Kimura and S. Terashima; *Chem. Lett.*, 1984, 1543. 1140

84CL1897 E. Murayama, T. Kikuchi, K. Sasaki, N. Sootome and T. Sato; *Chem. Lett.*, 1984, 1897. 667

84CL1939 J. Nokami, T. Tamaoka, T. Koguchi and R. Okawara; *Chem. Lett.*, 1984, 1939. 1055

84CL1997	T. Nagase, T. Kawashima and N. Inamoto; *Chem. Lett.*, 1984, 1997.	754
84CL2005	M. Iyoda, M. Sakaitani, T. Miyazaki and M. Oda; *Chem. Lett.*, 1984, 2005.	991
84CL2069	T. Oriyama and T. Mukaiyama; *Chem. Lett.*, 1984, 2069.	936
84CRV205	R. P. Lutz; *Chem. Rev.*, 1984, **84**, 205.	411, 413, 414, 415, 806, 825, 826, 827, 836
84CRV471	P. Beak, W. J. Zajdel and D. B. Reitz; *Chem. Rev.*, 1984, **84**, 471.	148
84H(22)1489	R. Amouroux; *Heterocycles*, 1984, **22**, 1489.	1089
84HCA261	D. Seebach and P. Knochel; *Helv. Chim. Acta*, 1984, **67**, 261.	37, 191
84HCA434	H.-R. Sliwka and H.-J. Hansen; *Helv. Chim. Acta*, 1984, **67**, 434.	708
84HCA774	T. A. Lyle, H. B. Mereyala, A. Pascual and B. Frei; *Helv. Chim. Acta*, 1984, **67**, 774.	812
84IJ113	A. Alexakis and J. F. Normant; *Isr. J. Chem.*, 1984, **24**, 113.	496
84IJ125	M. Clarembeau, J. L. Bertrand and A. Krief; *Isr. J. Chem.*, 1984, **24**, 125.	713, 714, 716
84IZV1561	I. P. Beletskaya, N. A. Bumagin and A. B. Ponomarev; *Izv. Akad. Nauk SSSR, Ser. Khim.*, 1984, 1561 (*Chem. Abstr.*, **102**, 5601f).	1010
84IZV2580	N. A. Bumagin, I. O. Kalinovskii and I. P. Beletskaya; *Izv. Akad. Nauk SSSR, Ser. Khim.*, 1984, **33**, 2580.	492
84JA158	T. Hayashi, M. Konishi, Y. Kobori, M. Kumada, T. Higuchi and K. Hirotsu; *J. Am. Chem. Soc.*, 1984, **106**, 158.	279, 466
84JA343	K. U. Ingold, J. Lusztyk and J. C. Scaiano; *J. Am. Chem. Soc.*, 1984, **106**, 343.	330, 341, 936
84JA462	E. J. Corey and A. Tramonanto; *J. Am. Chem. Soc.*, 1984, **106**, 462.	532, 1070
84JA521	P. C. Burgers, J. L. Holmes, A. A. Mommers and J. E. Szulejko; *J. Am. Chem. Soc.*, 1984, **106**, 521.	1141
84JA718	F. E. Zeigler and T.-F. Wang; *J. Am. Chem. Soc.*, 1984, **106**, 718.	807
84JA721	W. R. Roush and A. E. Walts; *J. Am. Chem. Soc.*, 1984, **106**, 721.	434
84JA723	J. A. Marshall and K. E. Flynn; *J. Am. Chem. Soc.*, 1984, **106**, 723.	690
84JA810	S. Chang and W. J. le Noble; *J. Am. Chem. Soc.*, 1984, **106**, 810.	905
84JA1079	Y. Tamura, M. Mizutani, Y. Furukawa, S. Kawamura, Z. Yoshida, K. Yanagi and M. Minobe; *J. Am. Chem. Soc.*, 1984, **106**, 1079.	187
84JA1518	L. A. Paquette, H. Jendralla and G. DeLucca; *J. Am. Chem. Soc.*, 1984, **106**, 1518.	575
84JA1793	L. R. C. Barclay, J. Lusztyk and K. U. Ingold; *J. Am. Chem. Soc.*, 1984, **106**, 1793.	923
84JA1877	A. L. Castelhano and A. Krantz; *J. Am. Chem. Soc.*, 1984, **106**, 1877.	980
84JA2115	E. R. Koft and A. B. Smith, III; *J. Am. Chem. Soc.*, 1984, **106**, 2115.	516
84JA2443	P. P. Policastro, K. G. Au, C. T. Walsh and G. A. Berchtold; *J. Am. Chem. Soc.*, 1984, **106**, 2443.	520
84JA2730	R. A. Pabon, Jr., D. J. Bellville and N. L. Bauld; *J. Am. Chem. Soc.*, 1984, **106**, 2730.	805
84JA2735	E. J. Corey and B. De; *J. Am. Chem. Soc.*, 1984, **106**, 2735.	1016, 1082, 1083
84JA3265	G. A. Olah and B. P. Singh; *J. Am. Chem. Soc.*, 1984, **106**, 3265.	915
84JA3270	A. I. Meyers, P. D. Edwards, W. F. Rieker and T. R. Bailey; *J. Am. Chem. Soc.*, 1984, **106**, 3270.	149
84JA3353	D. G. Batt, N. Takamura and B. Ganem; *J. Am. Chem. Soc.*, 1984, **106**, 3353.	434, 635
84JA3376	J. S. Sawyer, T. L. Macdonald and G. J. McGarvey; *J. Am. Chem. Soc.*, 1984, **106**, 3376.	138, 139
84JA3551	A. S. Kende, P. Fludzinski, J. H. Hill, W. Swenson and J. Clardy; *J. Am. Chem. Soc.*, 1984, **106**, 3551.	1016, 1108
84JA3668	R. E. Ireland and M. D. Varney; *J. Am. Chem. Soc.*, 1984, **106**, 3668.	450
84JA3670	T. Mandai, T. Yanagi, K. Araki, Y. Morisaki, M. Kawada and J. Otera; *J. Am. Chem. Soc.*, 1984, **106**, 3670.	613, 700, 1073
84JA3693	B. Ganem, M. J. Eis and J. E. Wrobel; *J. Am. Chem. Soc.*, 1984, **106**, 3693.	204, 478, 1023
84JA3703	D. J. Bencivengo, M. L. Brownawell, M. Y. Li and J. San Filippo, Jr.; *J. Am. Chem. Soc.*, 1984, **106**, 3703.	65, 66
84JA3866	D. A. Evans and M. M. Morrissey; *J. Am. Chem. Soc.*, 1984, **106**, 3866.	84
84JA4038	S. L. Schreiber and C. Santini; *J. Am. Chem. Soc.*, 1984, **106**, 4038.	813
84JA4218	R. C. Larock and L. W. Harrison; *J. Am. Chem. Soc.*, 1984, **106**, 4218.	66
84JA4263	B. M. Trost and S. J. Martin; *J. Am. Chem. Soc.*, 1984, **106**, 4263.	1031
84JA4356	I. R. Slagle, J.-Y. Park, M. C. Heaven and D. Gutman; *J. Am. Chem. Soc.*, 1984, **106**, 4356.	1127
84JA4547	H. Niwa, K. Wakamatsu, T. Hida, K. Niiyama, H. Kigoshi, M. Yamada, H. Nagase, M. Suzuki and K. Yamada; *J. Am. Chem. Soc.*, 1984, **106**, 4547.	710
84JA4617	E. Vedejs and J. G. Reig; *J. Am. Chem. Soc.*, 1984, **106**, 4617.	821
84JA4630	J. K. Stille, W. J. Scott and G. T. Crisp; *J. Am. Chem. Soc.*, 1984, **106**, 4630.	1027
84JA5018	J. E. McMurry, G. J. Haley, J. R. Matz, J. C. Clardy, G. Van Duyne, R. Gleiter, W. Schafer and D. H. White; *J. Am. Chem. Soc.*, 1984, **106**, 5018.	690
84JA5274	R. C. Larock, S. Varaprath, H. H. Lau and C. A. Fellows; *J. Am. Chem. Soc.*, 1984, **106**, 5274.	493
84JA5731	R. A. Holton; *J. Am. Chem. Soc.*, 1984, **106**, 5731.	842
84JA5754	S. Hanessian, D. Delorme, S. Beaudoin and Y. Leblanc; *J. Am. Chem. Soc.*, 1984, **106**, 5754.	645
84JA6006	J. A. Marshall, J. C. Peterson and L. Lebioda; *J. Am. Chem. Soc.*, 1984, **106**, 6006.	633, 1053
84JA6093	T. Murai, S. Kato, S. Murai, T. Toki, S. Suzuki and N. Sonoda; *J. Am. Chem. Soc.*, 1984, **106**, 6093.	212
84JA6105	L. D. Boardman, V. Bagheri, H. Sawada and E.-i. Negishi; *J. Am. Chem. Soc.*, 1984, **106**, 6105.	544
84JA6417	J. K. Stille, W. F. Goure, M. E. Wright, P. D. Davis and S. S. Labadie; *J. Am. Chem. Soc.*, 1984, **106**, 6417.	1015
84JA6442	L. A. Paquette, C. Blankenship and G. J. Wells; *J. Am. Chem. Soc.*, 1984, **106**, 6442.	284
84JA6450	J. E. McMurry and C. N. Hodge; *J. Am. Chem. Soc.*, 1984, **106**, 6450.	690

84JA6467 P. v. R. Schleyer, T. Clark, A. J. Kos, G. W. Spitznagel, C. Rohde, D. Arad, K. N. Houk and N. G. Rondan; *J. Am. Chem. Soc.*, 1984, **106**, 6467. 135, 159

84JA6690 L. A. Paquette, R. A. Roberts and G. J. Drtina; *J. Am. Chem. Soc.*, 1984, **106**, 6690. 19

84JA6702 L. A. Paquette, A. Varadarajan and E. Bay; *J. Am Chem. Soc.*, 1984, **106**, 6702. 18, 564

84JA6837 B. M. Trost and M.-H. Hung; *J. Am. Chem. Soc.*, 1984, **106**, 6837. 196

84JA7260 B. M. Trost and M. R. Ghadiri; *J. Am. Chem. Soc.*, 1984, **106**, 7260. 239, 484

84JA7264 M. Marsi and M. Rosenblum; *J. Am. Chem. Soc.*, 1984, **106**, 7264. 481

84JA7500 J. K. Stille, G. T. Crisp and W. J. Scott; *J. Am. Chem. Soc.*, 1984, **106**, 7500. 1029

84JA8209 D. J. Hart and Y.-M. Tsai; *J. Am. Chem. Soc.*, 1984, **106**, 8209. 341

84JA8313 F. Paqet and P. Sinaÿ; *J. Am. Chem. Soc.*, 1984, **106**, 8313. 68

84JCS(D)1477 S. Acharya, G. Neogi, R. K. Panda and D. Ramaswamy; *J. Chem. Soc., Dalton Trans.*, 1984, 1477. 74

84JCS(D)1747 S. L. Grundy, A. V. Smith, H. Adams and P. M. Maitlis; *J. Chem. Soc., Dalton Trans.*, 1984, 1747. 94

84JCS(P1)243 A. D. Buss, W. B. Cruse, O. Kennard and S. Warren; *J. Chem. Soc., Perkin Trans. 1*, 1984, 243. 641, 755

84JCS(P1)721 J. Barluenga, C. Jiménez, C. Nájera and M. Yus; *J. Chem. Soc., Perkin Trans. 1*, 1984, 721. 948

84JCS(P1)1593 G. Lesma, G. Palmisano and S. Tollari; *J. Chem. Soc., Perkin Trans 1*, 1984, 1593. 601

84JCS(P1)1671 A. R. Katritzky, S. Bravo-Borja, A. M. El-Mowafy and M. L. Lopez-Rodriguez; *J. Chem. Soc., Perkin Trans. 1*, 1984, 1671. 30

84JCS(P1)1689 R. J. Ferrier and S. R. Haines; *J. Chem. Soc., Perkin Trans. 1*, 1984, 1689. 68

84JCS(P1)1701 H. G. Theuns, R. H. A. M. Janssen, H. W. A. Biessels, F. Menichini and C. A. Salemink; *J. Chem. Soc., Perkin Trans. 1*, 1984, 1701. 784

84JMC1539 A. Stütz and G. Petranyi; *J. Med. Chem.*, 1984, **27**, 1539. 512

84JMC1575 H. Schmidhammer, L. Aeppli, L. Atwell, F. Fritsch, A. E. Jacobson, M. Nebuchla and G. Sperk; *J. Med. Chem.*, 1984, **27**, 1575. 432

84JOC596 O. de Lucchi, V. Lucchini, L. Pasquato and G. Modena; *J. Org. Chem.*, 1984, **49**, 596. 617

84JOC670 A. P. Marchand, S. C. Suri, A. D. Earlywine, D. R. Powell and D. van der Helm; *J. Org. Chem.*, 1984, **49**, 670. 637

84JOC722 S. R. Wilson and M. F. Price; *J. Org. Chem.*, 1984, **49**, 722. 832

84JOC737 M. P. Sibi, M. A. J. Miah and V. Snieckus; *J. Org. Chem.*, 1984, **49**, 737. 464

84JOC747 J. A. Marshall, M. J. Coghlan and M. Watanabe; *J. Org. Chem.*, 1984, **49**, 747. 505

84JOC931 A. E. Greene, J.-P. Lansard, J.-L. Luche and C. Petrier; *J. Org. Chem.*, 1984, **49**, 931. 314

84JOC1096 Y. Yamamoto, H. Yatagai, Y. Saito and K. Maruyama; *J. Org. Chem.*, 1984, **49**, 1096. 713, 714, 716

84JOC1147 B. H. Lipshutz and J. A. Kozlowski; *J. Org. Chem.*, 1984, **49**, 1147. 207

84JOC1149 H. Hiemstra, W. J. Klaver and W. N. Speckamp; *J. Org. Chem.*, 1984, **49**, 1149. 978

84JOC1204 S. Tsuboi, Y. Nooda and A. Takeda; *J. Org. Chem.*, 1984, **49**, 1204. 980

84JOC1230 I. Kuwajima, M. Shimizu and R. Ando; *J. Org. Chem.*, 1984, **49**, 1230. 485, 1083

84JOC1246 S. Cavicchioli, D. Savoia, C. Trombini and A. Umani-Ronchi; *J. Org. Chem.*, 1984, **49**, 1246. 24

84JOC1313 D. L. J. Clive, P. L. Beaulieu and L. Set; *J. Org. Chem.*, 1984, **49**, 1313. 174, 547, 1131

84JOC1317 L. A. Paquette, R. S. Valpey and G. D. Annis; *J. Org. Chem.*, 1984, **49**, 1317. 653

84JOC1345 N. A. Porter and P. J. Zuraw; *J. Org. Chem.*, 1984, **49**, 1345. 949

84JOC1353 A. Padwa, W. F. Rieker and R. J. Rosenthal; *J. Org. Chem.*, 1984, **49**, 1353. 355

84JOC1460 P. A. Magriotis and F. Johnson; *J. Org. Chem.*, 1984, **49**, 1460. 238

84JOC1522 K. M. Pietrusiewicz, M. Zablocka and J. Monkiewicz; *J. Org. Chem.*, 1984, **49**, 1522. 159

84JOC1544 P. W. Rabideau, D. M. Wetzel and D. M. Young; *J. Org. Chem.*, 1984, **49**, 1544. 97

84JOC1574 E. D. Sternberg and K. P. C. Vollhardt; *J. Org. Chem.*, 1984, **49**, 1574. 541

84JOC1603 M. E. Kurz, V. Baru and P.-N. Nguyen; *J. Org. Chem.*, 1984, **49**, 1603. 926

84JOC1627 J. Tirado-Rives, M. A. Oliver, F. R. Fronczek and R. D. Gandour, *J. Org. Chem.*, 1984, **49**, 1627. 690

84JOC1662 R. A. Benkeser and F. G. Belmonte; *J. Org. Chem.*, 1984, **49**, 1662. 457, 508

84JOC1700 M. E. Thompson; *J. Org. Chem.*, 1984, **49**, 1700. 147

84JOC1707 J. A. Marshall and T. M. Jenson; *J. Org. Chem.*, 1984, **49**, 1707. 816

84JOC1840 E. Vedejs and R. A. Buchanan; *J. Org. Chem.*, 1984, **49**, 1840. 834

84JOC1842 D. J.-S. Tsai and M. M. Midland; *J. Org. Chem.*, 1984, **49**, 1842. 405, 816, 817, 818

84JOC1898 R. K. Duke and R. W. Rickards; *J. Org. Chem.*, 1984, **49**, 1898. 520, 1134

84JOC1917 M. P. Doyle, J. H. Griffin, M. S. Chinn and D. van Leusen; *J. Org. Chem.*, 1984, **49**, 1917. 400, 823

84JOC1985 K. Hayakawa, Y. Kamikawaji, A. Wakita and K. Kanematsu; *J. Org. Chem.*, 1984, **49**, 1985. 235

84JOC2018 J. O. Karlsson, A. Svensson and S. Gronowitz; *J. Org. Chem.*, 1984, **49**, 2018. 1072

84JOC2031 L. Poncini; *J. Org. Chem.*, 1984, **49**, 2031. 1026

84JOC2079 M. Koreeda and J. I. Luengo; *J. Org. Chem.*, 1984, **49**, 2079. 566

84JOC2125 M. Apparu and J. K. Crandall; *J. Org. Chem.*, 1984, **49**, 2125. 497

84JOC2510 J.-M. Tornare and P. Vogel; *J. Org. Chem.*, 1984, **49**, 2510. 519

84JOC2682 M. Majewski and V. Snieckus; *J. Org. Chem.*, 1984, **49**, 2682. 828

84JOC2708 S. Chao, F.-A. Kunng, J.-M. Gu, H. L. Ammon and P. S. Mariano; *J. Org. Chem.*, 1984, **49**, 2708. 835

84JOC2838 K. E. Harding and T. H. Marman; *J. Org. Chem.*, 1984, **49**, 2838. 68

84JOC2875 M. J. Guttieri and W. F. Maier; *J. Org. Chem.*, 1984, **49**, 2875. 32, 48

84JOC2880 R. O. Angus, Jr. and R. P. Johnson; *J. Org. Chem.*, 1984, **49**, 2880. 990

84JOC2937 J. W. Wilt, V. A. Curtis, L. N. Congson and R. Palmer; *J. Org. Chem.*, 1984, **49**, 2937. 382

84JOC3314	A. Padwa, G. Haffmanns and M. Tomas; *J. Org. Chem.*, 1984, **49**, 3314.	524
84JOC3411	R. C. Larock and K. Narayanan; *J. Org. Chem.*, 1984, **49**, 3411.	454
84JOC3438	H. J. Reich, K. E. Yelm and I. L. Reich; *J. Org. Chem.*, 1984, **49**, 3438.	1092
84JOC3610	L. A. Paquette, T.-H. Yan and G. J. Wells; *J. Org. Chem.*, 1984, **49**, 3610.	690
84JOC3618	L. A. Paquette, G. J. Wells and G. Wickham; *J. Org. Chem.*, 1984, **49**, 3618.	
84JOC3641	M. Node, T. Kawabata, K. Ohta, M. Fujimoto, E. Fujita and K. Fuji; *J. Org. Chem.*, 1984, **49**, 3641.	428
84JOC3671	J. P. Marino and S. L. Dax; *J. Org. Chem.*, 1984, **49**, 3671.	654
84JOC3796	A. Toshimitsu, H. Owada, K. Terao, S. Uemura and M. Okano; *J. Org. Chem.*, 1984, **49**, 3796.	602
84JOC3870	K. E. Harding and K. S. Clement; *J. Org. Chem.*, 1984, **49**, 3870.	422, 784
84JOC3904	T. Imamoto, T. Kusumoto, Y. Tawarayama, Y. Sugiura, T. Mita, Y. Hatanaka and M. Yokoyama; *J. Org. Chem.*, 1984, **49**, 3904.	1098, 1140
84JOC3928	B. H. Lipshutz, R. S. Wilhelm, J. A. Kozlowski and D. Parker; *J. Org. Chem.*, 1984, **49**, 3928.	217, 478
84JOC4094	A. B. Smith and J. P. Konopelski; *J. Org. Chem.*, 1984, **49**, 4094.	382
84JOC4120	C. Sahlberg and A. Claesson; *J. Org. Chem.*, 1984, **49**, 4120.	961
84JOC4203	H.-P. Kaiser and J. M. Muchowski; *J. Org. Chem.*, 1984, **49**, 4203.	92
84JOC4244	J. K. Crandall and W. J. Michaely; *J. Org. Chem.*, 1984, **49**, 4244.	547
84JOC4320	S. D. Burke, D. M. Armistead and F. J. Schoenen; *J. Org. Chem.*, 1984, **49**, 4320.	831
84JOC4565	G. Zweifel and G. Hahn; *J. Org. Chem.*, 1984, **49**, 4565.	1056
84JOC4894	E. Wenkert, E. L. Michelotti, C. S. Swindell and M. Tingoli; *J. Org. Chem.*, 1984, **49**, 4894.	480
84JOC5175	K. K. Wang and K. Chu; *J. Org. Chem.*, 1984, **49**, 5175.	455, 1033
84JOC5206	C. G. Gutierrez and L. R. Summerhays; *J. Org. Chem.*, 1984, **49**, 5206.	942
84JOC5277	J. A. Marshall, J. E. Audia and J. Grote; *J. Org. Chem.*, 1984, **49**, 5277.	374
84JOC5280	G. W. Ebert and R. D. Rieke; *J. Org. Chem.*, 1984, **49**, 5280.	121
84JOM(260)17	M. E. Gurskii, D. G. Pershin and B. M. Mikhailov; *J. Organomet. Chem.*, 1984, **260**, 17.	58
84JOM(264)135	R. Muenstedt, U. Wannagat and D. Wrobel; *J. Organomet. Chem.*, 1984, **264**, 135.	991
84JOM(264)207	R. Corriu, N. Escudie and C. Guerin; *J. Organomet. Chem.*, 1984, **264**, 207.	651
84JOM(265)323	A. Spencer; *J. Organomet. Chem.*, 1984, **265**, 323.	494
84JOM(266)203	H. Hoberg, D. Schaefer, G. Burkhart, C. Krüger and M. J. Romão; *J. Organomet. Chem.*, 1984, **226**, 203.	545
84JOM(267)93	A. Marchand, P. Gerval, F. Duboudin, M.-H. Gaufryau, M. Joanny and P. Mazerolles; *J. Organomet. Chem.*, 1984, **267**, 93.	656
84JOM(268)C1	T. L. Chen, T. H. Chan and A. Shaver; *J. Organomet. Chem.*, 1984, **268**, C1.	692
84JOM(270)C15	H. Hoberg and B. Apotecher; *J. Organomet. Chem.*, 1984, **270**, C15.	545
84JOM(270)9	M. E. Gurskii, S. V. Baranin and B. M. Mikhailov; *J. Organomet. Chem.*, 1984, **270**, 9.	58
84JOM(271)181	R. Muenstedt, D. Wrobel and U. Wannagat; *J. Organomet. Chem.*, 1984, **271**, 181.	991
84JOM(271)281	W. Bernhard and I. Fleming; *J. Organomet. Chem.*, 1984, **271**, 281.	654
84LA85	G. Himbert and W. Schwickerath; *Liebigs Ann. Chem.*, 1984, 85.	1044
84LA340	B. Hagenbruch and S. Hünig; *Liebigs Ann. Chem.*, 1984, 340.	248
84LA1165	W. Helbig; *Liebigs Ann. Chem.*, 1984, 1165.	222
84LA1873	C. Ibis; *Liebigs Ann. Chem.*, 1984, 1873.	982
84LA1905	S. Hünig and B. Ort; *Liebigs Ann. Chem.*, 1984, 1905.	688
B-84MI 101-01	M. Hudlicky; *Reductions in Organic Chemistry*, Ellis Horwood, Chichester, 1984.	21
B-84MI 104-01	D. A. Evans; in "Asymmetric Synthesis, Stereodifferentiating Addition Reactions, Part B," ed. J. D. Morrison; Academic Press, Orlando, 1984, vol. 3, chap. 1, pp. 1–110.	129
B-84MI 104-02	D. Enders; in "Asymmetric Synthesis, Stereodifferentiating Addition Reactions, Part B," ed. J. D. Morrison; Academic Press, Orlando, 1984, vol. 3, chap. 4, pp. 275–341.	153
B-84MI 105-01	R. M. Roberts and A. A. Khalaf; "Friedel-Crafts Alkylation Chemistry. A Century of Discovery," Dekker, New York, 1984.	180, 212
B-84MI 107-01	P. A. Bartlett; in "Asymmetric Synthesis," ed. J. D. Morrison, Academic Press, New York, 1984, vol. 3, p. 341.	295
B-84MI 108-01	A. Padwa (ed.); "1,3-Dipolar Cycloaddition Chemistry," Wiley-Interscience, New York, 1984.	365
B-84MI 109-01	R. K. Hill; in "Asymmetric Synthesis," ed. J. D. Morrison, Academic Press, New York, 1984, vol. 3, p. 503.	377
B-84MI 110-01	J. Czombos; "Stereochemistry of Heterogenous Metal Catalysis," Wiley, New York, 1985, ch. 4.	455
B-84MI 110-02	H. F. Schuster and G. M. Coppola; "Allenes in Organic Synthesis," Wiley, New York, 1984.	459
B-84MI 110-03	M. Hudlický; "Reductions in Organic Synthesis," Wiley, New York, 1984.	425, 429
B-84MI 112-01	A. Padwa (ed.); "1,3-Dipolar Cycloaddition Chemistry," Wiley, New York, 1984.	523
B-84MI 114-01	K. C. Nicolaou and N. A. Petasis; "Selenium in Natural Product Synthesis," CIS, Philadelphia, 1984, chap. 4.	601, 604
84MI 118-01	F. Klarner; *Topics in Stereochemistry*, 1984, **15**, 1.	794
84MI 118-02	Y. Morizawa, K. Oshima and H. Nozaki; *Isr. J. Chem.*, 1984, **24**, 149.	803
B-84MI 118-03	R. K. Hill; in "Asymmetric Synthesis," ed. J. Morrison, Academic Press, New York, 1984, vol. 3, chap. 8.	806
B-84MI 118-04	R. K. Hill; "Asymmetric Synthesis," ed. J. Morrison, Academic Press, New York, 1984, vol. 3, p. 503.	823
B-84MI 119-01	E. Buncel and T. Durst (eds.); "Comprehensive Carbanion Chemistry, Part B," Elsevier, New York, 1984.	844

B-84MI 119-02	D. A. Evans; in "Asymmetric Synthesis," ed. J. D. Morrison, Academic Press, New York, 1984, vol. 3, chap. 1, p. 1.	878
B-84MI 119-03	C. H. Heathcock; in "Asymmetric Synthesis," ed. J. D. Morrison, Academic Press, New York, 1984, vol. 3, chap. 2, p. 111.	873
B-84MI 119-04	C. H. Heathcock; in "Comprehensive Carbanion Chemistry, Part B," ed. E. Buncel and T. Durst, Elsevier, Amsterdam, 1984, chap. 4, p. 177.	873
B-84MI 119-05	C. H. Heathcock; in "Asymmetric Synthesis," ed. J. D. Morrison, Academic Press, New York, 1984, vol. 3, chap. 2.	878
B-84MI 120-01	L. Brandsma and H. D. Verkruijsse; in "Synthesis of Acetylenes, Allenes and Cumulenes a Laboratory Manual," Elsevier, Amsterdam, 1981.	954, 957
B-84MI 120-02	H. F. Schuster and G. M. Coppola; in "Allenes in Organic Synthesis," Wiley, New York, 1984.	954
84OM44	M. P. Doyle, R. L. Dorow, W. E. Buhro, J. H. Griffin, W. H. Tamblyn and M. L. Trudell; *Organometallics*, 1984, **3**, 44.	299, 353
84OM630	G. S. Reddy and W. Tam; *Organometallics*, 1984, **3**, 630.	471
84OM1340	G. Manuel, G. Bertrand, W. P. Weber and S. A. Kazoura; *Organometallics*, 1984, **3**, 1340.	656
84OM(3)923	A. G. Abatjoglou, E. Billig and D. R. Bryant; *Organometallics*, 1984, **3**, 923.	52
84OR(30)457	E. Block; *Org. React.*, 1984, **30**, 457.	585
84OR(31)1	S. Arseniyadis, K. S. Kyler and D. S. Watt; *Org. React.*, 1984, **31**, 1.	155
84OR(32)1	E. Ciganek; *Org. React.*, 1984, **32**, 1.	453, 523
84OR(32)375	G. Zweifel and J. A. Miller; *Org. React.*, 1984, **32**, 375.	207, 465, 511, 512, 533, 534, 542, 543, 1106, 1139
84OS(62)1	A. Alexakis, G. Cahiez and J. F. Normant; *Org. Synth.*, 1984, **62**, 1.	541
84OS(62)14	M. Alderdice, F. W. Sum and L. Weiler; *Org. Synth.*, 1984, **62**, 14.	482
84PAC91	A. Alexakis, C. Chuit, M. Commerçon-Bourgain, J. P. Foulon, N. Jabri, P. Mangeney and J. F. Normant; *Pure Appl. Chem.*, 1984, **56**, 91.	312
84RCR744	L. A. Yanovskaya, G. V. Kryshtal and V. V. Kul'ganek; *Russ. Chem. Rev.* (*Engl. Transl.*), 1984, **53**, 744–756.	888
84RTC220	C. G. M. Janssen, A. H. M. Hendriks and E. M. Godefroi; *Recl. Trav. Chim. Pays-Bas*, 1984, **103**, 220.	24
84S85	H. E. Zaugg; *Synthesis*, 1984, 85.	1042
84S111	G. Zweifel and S. Rajagopalan; *Synthesis*, 1984, 111.	1052
84S113	S. Rajagopalan and G. Zweifel; *Synthesis*, 1984, 113.	512
84S384	D. J. Ager; *Synthesis*, 1984, 384.	160, 648, 758, 982
84S481	L. R. Subramanian, A. G. Martinez, A. H. Fernandez and R. M. Alvarez; *Synthesis*, 1984, 481.	433
84S529	M. Ramaiah; *Synthesis*, 1984, 529.	421
84S571	R. Rossi, A. Carpita and A. Lezzi; *Synthesis*, 1984, 571.	1010, 1013
84S629	J. G. Smith; *Synthesis*, 1984, 629.	204, 580, 842
84S728	N. A. Bumagin, A. B. Ponomaryov and I. P. Beletskaya; *Synthesis*, 1984, 728.	1012
84S730	R. E. Doolittle; *Synthesis*, 1984, 730.	512
84S885	G. A. Kraus and K. Landgrebe; *Synthesis*, 1984, 885.	574
84S949	S. Takano, S. Nishizawa, M. Akiyama and K. Ogasawara; *Synthesis*, 1984, 949.	5, 7
84S1009	J. Tsuji, M. Nisar, I. Shimizu and I. Minami; *Synthesis*, 1984, 1009.	561
84SC1191	C. Gardrat; *Synth. Commun.*, 1984, **14**, 1191.	926
84SCI883	D. J. Hart; *Science*, 1984, **223**, 883.	320, 919
84T641	E. Erdik; *Tetrahedron*, 1984, **40**, 641.	116, 539, 1094
84T811	T. Shono; *Tetrahedron*, 1984, **40**, 811.	1042
84T1245	A. Miyashita, H. Takaya, T. Souchi and R. Noyori; *Tetrahedron*, 1984, **40**, 1245.	76
84T1269	G. Fráater, U. Müller and W. Günther; *Tetrahedron*, 1984, **40**, 1269.	565, 681, 682, 683
84T1345	D. Enders, H. Eichenauer, U. Baus, H. Schubert and K. A. M. Kremer; *Tetrahedron*, 1984, **40**, 1345.	153
84T1361	A. I. Meyers, L. M. Fuentes and Y. Kubota; *Tetrahedron*, 1984, **40**, 1361.	149
84T1433	J. Lindley; *Tetrahedron*, 1984, **40**, 1433.	1005
84T2303	K. Mikami, K. Azuma and T. Nakai; *Tetrahedron*, 1984, **40**, 2303.	405, 816, 817
84T2317	P. A. Bartlett, D. P. Richardson and J. Myerson; *Tetrahedron*, 1984, **40**, 2317.	67, 68
84T2415	L. S. Hegedus; *Tetrahedron*, 1984, **40**, 2415.	493
84T2537	F. Bohlmann, P. Wegner, J. Jakupovic and R. M. King; *Tetrahedron*, 1984, **40**, 2537.	1134
84T2585	O. De Lucchi and G. Modena; *Tetrahedron*, 1984, **40**, 2585.	519, 562
84T2699	J. Tsuji and H. Nagashima; *Tetrahedron*, 1984, **40**, 2699.	493
84T2741	M. Gardette, N. Jabri, A. Alexakis and J. F. Normant; *Tetrahedron*, 1984, **40**, 2741.	468
84T2773	R. Rossi, A. Carpita and A. Lezzi; *Tetrahedron*, 1984, **40**, 2773.	1006, 1013
84T2805	D. J. Pasto; *Tetrahedron*, 1984, **40**, 2805.	954
84T2871	F. Camps, J. Coll, G. Fabrias and A. Guerrero; *Tetrahedron*, 1984, **40**, 2871.	747
84T3277	N. Bluthe, M. Malacria and J. Gore; *Tetrahedron*, 1984, **40**, 3277.	809
84T4161	R. M. Pagni; *Tetrahedron*, 1984, **40**, 4161–4215.	891
84T4261	M. Yamaguchi, Y. Nobayashi and I. Hirao; *Tetrahedron*, 1984, **40**, 4261.	1024
84T5153	G. Etemad-Moghadam and J. Seyden-Penne; *Tetrahedron*, 1984, **40**, 5153.	640
84TCC1	V. A. Koptyug; *Top. Curr. Chem.*, 1984, **122**, 1–245.	914
84TL131	F. E. Ziegler and K. Mikami; *Tetrahedron Lett.*, 1984, **25**, 131.	161
84TL223	H. Nishiyama, K. Sakuta and K. Itoh; *Tetrahedron Lett.*, 1984, **25**, 223.	132
84TL263	R. E. Gawley and T. Nagy; *Tetrahedron Lett.*, 1984, **25**, 263.	154
84TL303	T. Fuchikami and I. Ojima; *Tetrahedron Lett.*, 1984, **25**, 303.	1130

84TL357	A. D. Buss, N. Greeves, D. Levin, P. Wallace and S. Warren; *Tetrahedron Lett.*, 1984, **25**, 357.	643
84TL495	E. J. Corey and A. W. Gross; *Tetrahedron Lett.*, 1984, 495.	874
84TL565	N. Sayo, K. Azuma, K. Mikami and T. Nakai; *Tetrahedron Lett.*, 1984, **25**, 565.	405, 816, 817
84TL651	J. Pornet, B. Randrianoelina and L. Miginiac; *Tetrahedron Lett.*, 1984, **25**, 651.	978
84TL845	Y. Colas, B. Cazes and J. Goré; *Tetrahedron Lett.*, 1984, **25**, 845.	962
84TL1055	D. H. R. Barton, D. Crich and G. Kretzschmar; *Tetrahedron Lett.*, 1984, **25**, 1005.	935
84TL1083	M. Wada, Y. Sakurai and K. Akiba; *Tetrahedron Lett.*, 1984, **25**, 1083.	1041
84TL1159	M. Yamaguchi, K. Shibato and I. Hirao; *Tetrahedron Lett.*, 1984, **25**, 1159.	229
84TL1219	C.-N. Hsiao and H. Shechter; *Tetrahedron Lett.*, 1984, **25**, 1219.	616, 654
84TL1255	A. G. Schultz, J. P. Dittami and K. K. Eng; *Tetrahedron Lett.*, 1984, **25**, 1255.	802
84TL1279	H. Felkin, T. Fillebeen-Khan, Y. Gault, R. Holmes-Smith and J. Zakrewski; *Tetrahedron Lett.*, 1984, **25**, 1279.	554
84TL1353	H. Ahlbrecht and H. Dollinger; *Tetrahedron Lett.*, 1984, **25**, 1353.	148
84TL1377	T. Ishihara, T. Maekawa and T. Ando; *Tetrahedron Lett.*, 1984, **25**, 1377.	1079
84TL1543	S. E. Denmark and M. A. Harmata; *Tetrahedron Lett.*, 1984, **25**, 1543.	832
84TL1555	S. Pilard and M. Vaultier; *Tetrahedron Lett.*, 1984, **25**, 1555.	514
84TL1655	R. L. Funk, C. J. Mossman and W. E. Zeller; *Tetrahedron Lett.*, 1984, **25**, 1655.	510
84TL1757	S. L. Schreiber and S. E. Kelly; *Tetrahedron Lett.*, 1984, **25**, 1757.	1045
84TL1905	K. Tamao, R. Kanatani and M. Kumada; *Tetrahedron Lett.*, 1984, **25**, 1905.	160
84TL1909	K. Tamao, T. Iwahara, R. Kanatani and M. Kumada; *Tetrahedron Lett.*, 1984, **25**, 1909.	160
84TL1913	K. Tamao, R. Kanatani and M. Kumada; *Tetrahedron Lett.*, 1984, **25**, 1913.	160
84TL1953	J. I. Concepción, C. G. Francisco, R. Hernández, J. A. Salazar and E. Suárez; *Tetrahedron Lett.*, 1984, **25**, 1953.	929
84TL1987	P. Lerouge and C. Paulmier; *Tetrahedron Lett.*, 1984, **25**, 1987.	985
84TL1999	S. Fujikura, M. Inoue, K. Utimoto and H. Nozaki; *Tetrahedron Lett.*, 1984, **25**, 1999.	160, 451
84TL2035	I. T. Kay and D. Bartholomew; *Tetrahedron Lett.*, 1984, **25**, 2035.	769
84TL2085	A. P. Kozikowski and K. L. Sorgi; *Tetrahedron Lett.*, 1984, **25**, 2085.	703
84TL2089	R. A. Benkeser, J. A. Laugal and A. Rappa; *Tetrahedron Lett.*, 1984, **25**, 2089.	101
84TL2135	G. Deleris, J. Dunogues and A. Gadras; *Tetrahedron Lett.*, 1985, **25**, 2135.	238
84TL2179	J. Adams, S. Milette, J. Rokach and R. Zamboni; *Tetrahedron Lett.*, 1984, **25**, 2179.	514
84TL2183	M. A. Blanchette, W. Choy, J. T. Davis, A. P. Essenfeld, S. Masamune, W. R. Roush and T. Sakai; *Tetrahedron Lett.*, 1984, **25**, 2183.	749
84TL2271	S. Cacchi, E. Morera and G. Ortar; *Tetrahedron Lett.*, 1984, **25**, 2271.	1099
84TL2303	J. C. Gilbert and K. P. A. Senaratne; *Tetrahedron Lett.*, 1984, **25**, 2303.	817, 835
84TL2329	L. Fitjer, H.-J. Scheuermann and D. Wehle; *Tetrahedron Lett.* 1984, **25**, 2329.	248
84TL2367	J. Fujiwara, H. Sano, K. Maruoka and H. Yamamoto; *Tetrahedron Lett.*, 1984, **25**, 2367.	1041
84TL2411	H. C. Brown, U. S. Racherla and S. M. Singh; *Tetrahedron Lett.*, 1984, **25**, 2411	1045
84TL2447	B. C. Ranu, M. Kavka, L. A. Higgs and T. Hudlicky; *Tetrahedron Lett.*, 1984, **25**, 2447.	804
84TL2487	H. Nishiyama, K. Sakuta and K. Itoh; *Tetrahedron Lett.*, 1984, **25**, 2487.	132
84TL2713	J. L. Laboureur and A. Krief; *Tetrahedron Lett.*, 1984, **25**, 2713.	390
84TL2743	B. Giese and K. Gröninger; *Tetrahedron Lett.*, 1984, **25**, 2743.	326, 948
84TL2887	N. O. Nilsen, L. Skattebol, M. S. Baird, S. R. Buxton and P. D. Slowey; *Tetrahedron Lett.*, 1984, **25**, 2887.	990
84TL3019	K. Ruitenberg and P. Vermeer; *Tetrahedron Lett.*, 1984, **25**, 3019.	968
84TL3055	E. J. Corey and N. W. Boaz; *Tetrahedron Lett.*, 1984, **25**, 3055.	983
84TL3059	E. J. Corey and N. W. Boaz; *Tetrahedron Lett.*, 1984, **25**, 3059.	1015, 1016
84TL3075	A. Ghribi, A. Alexakis and J. F. Normant; *Tetrahedron Lett.*, 1984, 3075.	479
84TL3115	H. Hiemstra, H. P. Fortgens and W. N. Speckamp; *Tetrahedron Lett.*, 1984, **25**, 3115.	978
84TL3137	S. Cacchi, M. Felici and B. Pietroni; *Tetrahedron Lett.*, 1984, **25**, 3137.	546
84TL3155	E. Piers and H. L. A. Tse; *Tetrahedron Lett.*, 1984, **25**, 3155.	1092
84TL3159	S. J. Danishefsky and G. B. Philips; *Tetrahedron Lett.*, 1984, **25**, 3159.	835
84TL3221	H. Oda, Y. Morizawa, K. Oshima and H. Nozaki; *Tetrahedron Lett.*, 1984, **25**, 3221.	452
84TL3403	C. F. Wilcox, Jr. and B. Brungardt; *Tetrahedron Lett.*, 1984, 3403.	905
84TL3407	E.-I. Negishi, J. A. Miller and T. Yoshida; *Tetrahedron Lett.*, 1984, **25**, 3407.	669
84TL3437	W. Brennenstuhl and M. Hanack; *Tetrahedron Lett.*, 1984, **25**, 3437.	1091
84TL3441	M. Meier and C. Rüchardt; *Tetrahedron Lett.*, 1984, **25**, 3441.	266
84TL3555	A. Nickon, A. Rodriquez, V. Shirhatti and R. Ganguly; *Tetrahedron Lett.*, 1984, **25**, 3555.	821
84TL3959	E. Piers, G. L. Jung and N. Moss; *Tetrahedron Lett.*, 1984, **25**, 3959.	812
84TL4007	T. Fujisawa, S. Iida and T. Sato; *Tetrahedron Lett.*, 1984, **25**, 4007.	970
84TL4011	W. Ando, T. Furuhata, Y. Hanyu and T. Takata; *Tetrahedron Lett.*, 1984, **25**, 4011.	242
84TL4233	T. Imamoto, Y. Sugiura and N. Takiyama; *Tetrahedron Lett.*, 1984, **25**, 4233.	1038, 1098, 1140
84TL4317	M. Ladlow and G. Pattenden; *Tetrahedron Lett.*, 1984, **25**, 4317.	345
84TL4349	C. Rücker; *Tetrahedron Lett.*, 1984, **25**, 4349.	236
84TL4357	R. R. Schmidt and R. Hirsenkorn; *Tetrahedron Lett.*, 1984, **25**, 4357.	520
84TL4417	Y. Masuyama, M. Takahashi and Y. Kurusu; *Tetrahedron Lett.*, 1984, **25**, 4417.	37
84TL4455	R. A. Holton and R. M. Kennedy; *Tetrahedron Lett.*, 1984, **25**, 4455.	842
84TL4549	M. Yamaguchi and I. Hirao; *Tetrahedron Lett.*, 1984, **25**, 4549.	229
84TL4617	J. B. Hendrickson, G. J. Boudreaux and P. S. Palumbo; *Tetrahedron Lett.*, 1984, **25**, 4617.	704, 705
84TL4821	S. Cacchi, E. Morera and G. Ortar; *Tetrahedron Lett.*, 1984, **25**, 4821.	435
84TL4851	J. J. Eisch, M. Behrooz and J. E. Galle; *Tetrahedron Lett.*, 1984, **25**, 4851.	485, 1032
84TL4867	D. L. Comins, A. H. Abdullah and N. B. Mantlo; *Tetrahedron Lett.*, 1984, **25**, 4867.	455
84TL4881	J. I. Luengo and M. Koreeda; *Tetrahedron Lett.*, 1984, **25**, 4881.	566

84TL4901 A. R. Chamberlin and S. H. Bloom; *Tetrahedron Lett.*, 1984, **25**, 4901. 1094
84TL5043 C. Schmit, S. Sahraoui-Taleb, E. Differding, C. G. Dehasse-De Lombaert and L. Ghosez; *Tetrahedron Lett.*, 1984, **25**, 5043. 518
84TL5103 I. Fleming and N. K. Terrett; *Tetrahedron Lett.*, 1984, **25**, 5103. 815
84TL5157 J. Tsuji, I. Minami and I. Shimizu; *Tetrahedron Lett.*, 1984, **25**, 5157. 196
84TL5239 G. A. Russell and D. Guo; *Tetrahedron Lett.*, 1984, **25**, 5239. 67, 948
84TL5319 N. Ono, A. Kamimura and A. Kaji; *Tetrahedron Lett.*, 1984, **25**, 5319. 592, 638
84TL5335 K. Nakatani and S. Isoe; *Tetrahedron Lett.*, 1984, **25**, 5335. 929
84TL5423 K. Utimoto, C. Lambert, Y. Fukuda, H. Shiragami and H. Nozaki; *Tetrahedron Lett.*, 1984, **25**, 5423. 205, 1024
84TL5469 E. Block, V. Eswarakrishnan and K. Gebreyes; *Tetrahedron Lett.*, 1984, **25**, 5469. 707
84TL5571 C. J. Elsevier, H. H. Mooiweer, H. Kleijn and P. Vermeer; *Tetrahedron Lett.*, 1984, **25**, 5571. 975
84TL5603 L. Strekowski, M. Visnick and M. A. Battiste; *Tetrahedron Lett.*, 1984, **25**, 5603. 763
84TL5769 G. Solladie and R. G. Zimmermann; *Tetrahedron Lett.*, 1984, **25**, 5769. 466
84TL5781 M. Kodama, K. Okumura, Y. Kobayashi, T. Tsunoda and S. Itô; *Tetrahedron Lett.*, 1984, **25**, 5781. 215
84TL5863 J. A. Miller and E. Negishi; *Tetrahedron Lett.*, 1984, **25**, 5863. 543
84TL5921 T. Takahashi, Y. Jinbo, K. Kitamura and J. Tsuji; *Tetrahedron Lett.*, 1984, **25**, 5921. 193
84TL5959 B. H. Lipshutz, D. A. Parker, J. A. Kozlowski and S. L. Nguyen; *Tetrahedron Lett.*, 1984, **25**, 5959. 207, 312
84TL6001 V. Ratovelomanana and G. Linstrumelle; *Tetrahedron Lett.*, 1984, 6001. 468
84TL6011 K. Mikami, K. Fujimoto, T. Kasuga and T. Nakai; *Tetrahedron Lett.*, 1984, **25**, 6011. 816, 819

85ACR148 H. G. Viehe, Z. Janousek and R. Merényi; *Acc. Chem. Res.*, 1985, **18**, 148. 919, 922
85ACR181 I. Kuwajima and E. Nakamura; *Acc. Chem. Res.*, 1985, **18**, 181. 133
85ACR254 M. Lajunen; *Acc. Chem. Res.*, 1985, **18**, 254–258. 905
85ACS(B)187 A. Nilsson and R. Carlson; *Acta. Chem. Scand., Ser. B*, 1985, **39**, 187. 82
85AG(E)316 P. Binger, Q.-H. Lü and P. Wedemann; *Angew. Chem., Int. Ed. Engl.*, 1985, **24**, 316. 527
85AG(E)320 R. Rienäcker and J. Graefe; *Angew. Chem., Int. Ed. Engl.*, 1985, **24**, 320. 795
85AG(E)553 B. Giese; *Angew. Chem., Int. Ed. Engl.*, 1985, **24**, 553. 320, 330, 331, 332, 341, 343, 935, 948, 1126
85AG(E)668 K. Maruoka and H. Yamamoto; *Angew. Chem., Int. Ed. Engl.*, 1985, **24**, 668. 1106
85AG(E)851 F. W. Nader and C.-D. Wacker; *Angew. Chem., Int. Ed. Engl.*, 1985, **24**, 851. 981
85AG(E)943 W. Siebert; *Angew. Chem., Int. Ed. Engl.*, 1985, **24**, 943. 862
85AG(E)973 M. Demuth and W. Hinsken; *Angew., Chem. Int. Ed. Engl.*, 1985, **24**, 973. 563
85AG(E)1046 E. Vogel, B. Neumann, W. Klug, H. Schmickler and J. Lex; *Angew. Chem., Int. Edn. Engl.*, 1985, **24**, 1046. 248
85AOC(24)353 W. N. Setzer and P. von R. Schleyer; *Adv. Organomet. Chem.*, 1985, **24**, 353. 110
85BCJ146 C. Kuroda, T. Nakamura, H. Hirota, K. Enomoto and T. Takahashi; *Bull. Chem. Soc. Jpn.*, 1985, **58**, 146. 813
85BCJ1067 H. Suzuki, K. Takaoka and A. Osuka; *Bull. Chem. Soc. Jpn.*, 1985, **58**, 1067. 37
85BCJ2421 H. Suzuki, H. Tani and S. Takeuchi; *Bull. Chem. Soc. Jpn.*, 1985, **58**, 2421. 24
85BCJ2812 M. Yamaura, T. Suzuki, H. Hashimoto, J. Yoshimura and C. Shin; *Bull. Chem. Soc. Jpn.*, 1985, **58**, 2812. 648
85BSF75 J. Perthuis and P. Poisson; *Bull. Soc. Chim. Fr., Part S*, 1985, 75. 69
85BSF762 J. L. Fabre, M. Julia and J. N. Verpeaux; *Bull. Soc. Chim. Fr.*, 1985, 762. 485
85BSF772 J. L. Fabre, M. Julia and J. N. Verpeaux; *Bull. Soc. Chim. Fr.*, 1985, 772. 485
85BSF(2)448 G. Etemad-Moghadam and J. Seyden-Penne; *Bull. Soc. Chim. Fr. Part* 2, 1985, 448. 747
85CB764 W. Bauer, T. Laube and D. Seebach; *Chem. Ber.*, 1985, **118**, 764. 873
85CB1039 T. Kauffman, F. Steinseifer and N. Klas; *Chem. Ber.*, 1985, **118**, 1039. 54
85CB1230 H. A. Staab, W. R. K. Reibel and C. Krieger; *Chem. Ber.*, 1985, **118**, 1230. 233
85CB1345 B. Giese, H. Harnisch and U. Lüning; *Chem. Ber.*, 1985, **118**, 1345. 325
85CB1504 W. Neugebauer, G. A. P. Geiger, A. J. Kos, J. J. Stezowski and P. von R. Schleyer; *Chem. Ber.*, 1985, **118**, 1504. 1088
85CC83 M. Larchevêque and J. Lalande; *J. Chem. Soc., Chem. Commun.*, 1985, 83. 165
85CC209 S. G. Davies and J. C. Walker; *J. Chem. Soc., Chem. Commun.*, 1985, 209. 134
85CC697 M. Ochiai, K. Sumi, Y. Nagao, E. Fujita, M. Arimoto and H. Yamaguchi; *J. Chem. Soc., Chem. Commun.*, 1985, 697. 1080
85CC874 J. Kress, J. A. Osborn, R. M. E. Greene, K. J. Ivin and J. J. Rooney; *J. Chem. Soc., Chem. Commun.*, 1985, 874. 674
85CC897 T. A. K. Smith and G. H. Whitham; *J. Chem. Soc., Chem. Commun.*, 1985, 897. 231
85CC971 R. Grigg, P. Stevenson and T. Worakun; *J. Chem. Soc., Chem. Commun.*, 1985, 971. 468
85CC980 A. G. Angoh and D. L. J. Clive; *J. Chem. Soc., Chem. Commun.*, 1985, 980. 547
85CC1037 S. Uemura, Y. Hirai, K. Ohe and N. Sugita; *J. Chem. Soc., Chem. Commun.*, 1985, 1037. 611, 716
85CC1140 R. A. Aitken and J. I. Atherton; *J. Chem. Soc., Chem. Commun.*, 1985, 1140. 1079
85CC1168 M. J. O'Donnell and J.-B. Falmagne; *J. Chem. Soc., Chem. Commun.*, 1985, 1168. 280
85CC1205 L. Set, D. R. Cheshire and D. L. J. Clive; *J. Chem. Soc., Chem. Commun.*, 1985, 1205. 1131
85CC1282 B. W. Disanayaka and A. C. Weedon; *J. Chem. Soc., Chem. Commun.*, 1985, 1282. 690
85CC1326 R. Baker, M. J. O'Mahony and C. J. Swain; *J. Chem. Soc., Chem. Commun.*, 1985, 1326. 703
85CC1452 T. Ibuka, T. Aoyagi and F. Yoneda; *J. Chem. Soc., Chem. Commun.*, 1985, 1452. 188
85CC1571 G. Cum, R. Gallo, S. Ipsale and A. Spadaro; *J. Chem. Soc., Chem. Commun.*, 1985, 1571. 1063
85CC1588 S. C. Dolan and J. MacMillan; *J. Chem. Soc., Chem. Commun.*, 1985, 1588. 940

85CC1610 M.-C. Chan, K.-M. Cheng, M. K. Li and T.-Y. Luh; *J. Chem. Soc., Chem. Commun.*, 1985, 1610. 436
85CC1636 F. Sato, Y. Kobayashi, O. Takahashi, T. Chiba, Y. Takeda and M. Kusakabe; *J. Chem. Soc., Chem. Commun.*, 1985, 1636. 1107
85CC1812 U. H. Brinker, H. Wüster and G. Maas; *J. Chem. Soc., Chem. Commun.*, 1985, 1812. 990
85CC1816 F. Quignard, M. Leconte and J.-M. Basset; *J. Chem. Soc., Chem. Commun.*, 1985, 1816. 674
85CJC1 J. F. King, J. H. Hillhouse and K. C. Khemani; *Can. J. Chem.*, 1985, **63**, 1. 707
85CL53 H. Oda, K. Oshima and H. Nozaki; *Chem. Lett.*, 1985, 53. 452
85CL1137 T. Sakai, K. Miyata and A. Takeda; *Chem. Lett.*, 1985, 1137. 422
85CL1149 T. Takeda, K. Ando, A. Mamada and T. Fujiwara; *Chem. Lett.*, 1985, 1149. 236
85CL1457 T. Fujisawa, E. Maehata, H. Kohama and T. Sato; *Chem. Lett.*, 1985, 1457. 980
85CL1477 Y. Tsuji, T. Mukai, T. Kondo and Y. Watanabe; *Chem. Lett.*, 1985, 1477. 196
85CL1491 T. Imamoto, T. Takeyama and T. Kusumoto; *Chem. Lett.*, 1985, 1491. 426, 427
85CL1689 K. Maruoka, H. Sano, Y. Fukutani and H. Yamamoto; *Chem. Lett.*, 1985, 1689. 1130
85CL1719 K. Mikami, K. Kawamoto and T. Nakai; *Chem. Lett.*, 1985, 1719. 816, 817, 818
85CL1723 N. Sayo, E. Nakai and T. Nakai; *Chem. Lett.*, 1985, 1723. 405, 816, 817
85CL1729 K. Mikami, O. Takahashi, T. Kasuga and T. Nakai; *Chem. Lett.*, 1985, 1729. 405, 816, 819
85CL1829 S. Kiyooka and A. Muyanchi; *Chem. Lett.*, 1985, 1829. 290
85CL1851 M. Kijima, Y. Nambu and T. Endo; *Chem. Lett.*, 1985, 1851. 514
85CPB1761 A. Matsuda, M. Shinozaki, T. Miyasaka, H. Machida and T. Abiru; *Chem. Pharm. Bull.*, 1985, **33**, 1761. 1010
85CRV129 R. A. W. Johnstone, A. H. Wilby and I. D. Entwistle; *Chem. Rev.*, 1985, **85**, 129. 74, 455, 457
85CRV171 P. E. Garrou; *Chem. Rev.*, 1985, **85**, 171. 444
85DOK(105)630 I. P. Beletskaya, N. A. Bumagin and A. B. Ponomarev; *Dokl. Akad. Nauk SSSR*, 1985, 630 (*Chem. Abstr.*, **105**, 78 575x). 1010
85H(23)3085 S.-T. Lu and Y.-C. Wu; *Heterocycles*, 1985, **23**, 3085. 784
85HCA1373 T. Laube, J. D. Dunitz and D. Seebach; *Helv. Chim. Acta*, 1985, **68**, 1373. 873
85HCA2244 E. Kohl-Mines and H.-J. Hansen; *Helv. Chim. Acta*, 1985, **68**, 2244. 986
85HCA2249 R. W. Lang, E. Kohl-Mines and H.-J. Hansen; *Helv. Chim. Acta*, 1985, **68**, 2249. 986
85HOU(V/2c)127 K.-P. Zeller; *Methoden Org. Chem.* (*Houben-Weyl*), 1985, **V/2c**, p. 127. 916
85IC3680 R. T. Boeré, C. D. Montgomery, N. C. Payne and C. J. Willis; *Inorg. Chem.*, 1985, **24**, 3680. 157
85IJ131 J. R. Fritch and K. P. C. Vollhardt; *Isr. J. Chem.*, 1985, **26**, 131. 991
85IJ136 R. S. Macomber and T. C. Hemling; *Isr. J. Chem.*, 1985, **26**, 136. 988
85IJC(B)59 I. Sharma and S. Ray; *Indian J. Chem., Sect. B*, 1985, **24**, 59. 46
85IZV1109 G. A. Tolstikov, A. Y. Spivak, S. I. Lomakina and A. V. Kuchin; *Izv. Akad. Nauk SSSR, Ser. Khim.*, 1985, **5**, 1109 (*Chem. Abstr.*, **104**, 33 779e). 1022
85JA217 B. E. Maryanoff, A. B. Reitz and B. A. Duhl-Emswiler; *J. Am. Chem. Soc.*, 1985, **107**, 217. 730
85JA443 M. J. Kurth, O. H. W. Decker, H. Hope and M. D. Yanuck; *J. Am. Chem. Soc.*, 1985, **107**, 443. 835
85JA682 A. I. Meyers and R. J. Himmelsbach; *J. Am. Chem. Soc.*, 1985, **107**, 682. 483
85JA972 A. Suzuki, N. Miyaura, H. Suginome and K. Yamada; *J. Am. Chem. Soc.*, 1985, **107**, 972. 532, 1020
85JA1293 B. M. Trost and T. N. Nanninga; *J. Am. Chem. Soc.*, 1985, **107**, 1293. 632
85JA1379 R. L. Halterman, N. H. Nguyen and K. P. C. Vollhardt; *J. Am. Chem. Soc.*, 1985, **107**, 1379. 448
85JA1421 S. J. Danishefsky, P. J. Harrison, R. R. Webb, II and B. J. O'Neill, *J. Am. Chem. Soc.*, 1985, **107**, 1421. 575
85JA1424 G. H. Posner and S.-B. Lu; *J. Am. Chem. Soc.*, 1985, **107**, 1424. 444
85JA1448 D. P. Curran and D. M. Rakiewicz; *J. Am. Chem. Soc.*, 1985, **107**, 1448. 347, 1132
85JA1488 D. P. Curran and D. M. Rakiewicz; *J. Am. Chem. Soc.*, 1985, **107**, 1448. 248
85JA1691 R. E. Dolle and K. C. Nicolaou; *J. Am. Chem. Soc.*, 1985, **107**, 1691. 581
85JA1726 J. Drouin, M.-A. Boaventura and J.-M. Conia; *J. Am. Chem. Soc.*, 1985, **107**, 1726. 66, 454, 526
85JA1781 B. M. Trost and M. Lautens; *J. Am. Chem. Soc.*, 1985, **107**, 1781. 525
85JA2078 R. A. Wanat and D. B. Collum; *J. Am. Chem. Soc.*, 1985, **107**, 2078. 153
85JA2174 M. M. Olmstead and P. P. Power; *J. Am. Chem. Soc.*, 1985, **107**, 2174. 847
85JA2264 D. W. Moreland and W. G. Dauben; *J. Am. Chem. Soc.*, 1985, **107**, 2264. 876
85JA2495 K. J. Stone and R. D. Little; *J. Am. Chem. Soc.*, 1985, **107**, 2495. 359
85JA2568 E. Negishi, S. J. Holmes, J. M. Tour and J. A. Miller; *J. Am. Chem. Soc.*, 1985, **107**, 2568. 536
85JA2764 G. A. Olah, G. K. Surya Prakash, J. G. Shih, V. V. Krishnamurthy, G. D. Mateescu, G. Liang, G. Sipos, V. Buss, T. M. Gund and P. von R. Schleyer; *J. Am. Chem. Soc.*, 1985, **107**, 2764. 900
85JA2811 D. G. Cox, N. Gurusamy and D. J. Burton; *J. Am. Chem. Soc.*, 1985, **107**, 2811. 446
85JA2872 W. Tsang; *J. Am. Chem. Soc.*, 1985, **107**, 2872. 921
85JA3285 R. E. Ireland, D. W. Norbeck, G. S. Mandel and N. S. Mandel; *J. Am. Chem. Soc.*, 1985, **107**, 3285. 831
85JA3343 A. Misumi, K. Iwanaga, K. Furuta and H. Yamamoto; *J. Am. Chem. Soc.*, 1985, **107**, 3343. 184
85JA3345 P. G. Williard and G. B. Carpenter; *J. Am. Chem. Soc.*, 1985, **107**, 3345. 873
85JA3348 M. Suzuki, A. Yanagisawa and R. Noyori; *J. Am. Chem. Soc.*, 1985, **107**, 3348. 132
85JA3626 W. R. Dolbier, Jr. and G. E. Wicks; *J. Am. Chem. Soc.*, 1985, **107**, 3626. 510
85JA3747 R. B. Sharma, D. K. Sen Sharma, K. Hiraoka and P. Kebarle; *J. Am. Chem. Soc.*, 1985, **107**, 3747. 905
85JA3891 S. Danishefsky, E. M. Berman, M. Ciufolini, S. J. Etheredge and B. E. Segmuller; *J. Am. Chem. Soc.*, 1985, **107**, 3891. 603
85JA3915 D. J.-S. Tsai and M. M. Midland; *J. Am. Chem. Soc.*, 1985, **107**, 3915. 816

85JA4332	J. Dupuis, B. Giese and J. Hartung; *J. Am. Chem. Soc.*, 1985, **107**, 4332.	325
85JA4359	R. K. Boekman, Jr., C. J. Flann and K. M. Poss; *J. Am. Chem. Soc.*, 1985, **107**, 4359.	826
85JA4579	R. L. Danheiser, J. J. Bronson and K. Okano; *J. Am. Chem. Soc.*, 1985, **107**, 4579.	803
85JA4594	L. H. Johnston, J. Lusztyk, D. D. M. Wayner, A. N. Abeywickregma, A. L. J. Beckwith, J. C. Scaiano and K. U. Ingold; *J. Am. Chem. Soc.*, 1985, **107**, 4594.	331
85JA4595	L. J. Johnston, J. Lusztyk, D. D. M. Wayner, A. N. Abeywickreyma, A. L. J. Beckwith, J. C. Scaiano and K. U. Ingold; *J. Am. Chem. Soc.*, 1985, **107**, 4594.	936
85JA4679	R. K. Dieter, L. A. Silks, III, J. R. Fishpaugh and M. E. Kastner; *J. Am. Chem. Soc.*, 1985, **107**, 4679.	485
85JA4964	P. S. Engel, D. E. Keys and A. Kitamura; *J. Am. Chem. Soc.*, 1985, **107**, 4964.	576
85JA5186	T. Kitazume and N. Ishikawa; *J. Am. Chem. Soc.*, 1985, **107**, 5186.	1130
85JA5301	T. Imamoto, T. Kusumoto, N. Suzuki and K. Sato; *J. Am. Chem. Soc.*, 1985, **107**, 5301.	158
85JA5403	D. Seebach, R. Amstutz, T. Laube, W. B. Schweizer and J. D. Dunitz; *J. Am. Chem. Soc.*, 1985, **107**, 5403.	873
85JA5572	M. Koreeda and J. I. Luegno; *J. Am. Chem. Soc.*, 1985, **107**, 5572.	816, 833
85JA5574	E. J. Corey and A. G. Myers; *J. Am. Chem. Soc.*, 1985, **107**, 5574.	801, 803, 804
85JA5732	V. B. Rao, C. F. George, S. Wolff and W. C. Agosta; *J. Am. Chem. Soc.*, 1985, **107**, 5732.	19
85JA5739	W. Von Eggers Doering and C. A. Troise; *J. Am. Chem. Soc.*, 1985, **107**, 5739.	806
85JA5987	L. G. McCullough, R. R. Schrock, J. C. Dewan and J. C. Murdzek; *J. Am. Chem. Soc.*, 1985, **107**, 5987.	675
85JA6046	J. M. Schwab and D. C. T. Lin; *J. Am. Chem. Soc.*, 1985, **107**, 6046.	971
85JA6121	J. D. Wuest and B. Zacharie; *J. Am. Chem. Soc.*, 1985, **107**, 6121.	24
85JA6137	D. K. Hutchinson and P. L. Fuchs; *J. Am. Chem. Soc.*, 1985, **107**, 6137.	486
85JA6546	L. T. Scott, G. J. DeCicco, J. L. Hyun and G. Reinhardt; *J. Am. Chem. Soc.*, 1985, **107**, 6546.	1003
85JA6615	M. W. Klett and R. P. Johnson; *J. Am. Chem. Soc.*, 1985, **107**, 6615.	957
85JA6729	E. Block and M. Aslam; *J. Am. Chem. Soc.*, 1985, **107**, 6729.	141
85JA6731	E. Block and S. Ahmad; *J. Am. Chem. Soc.*, 1985, **107**, 6731.	834
85JA7233	R. L. Danheiser, C. A. Kwasigroch and Y. Tsai; *J. Am. Chem. Soc.*, 1985, **107**, 7233.	1060
85JA7352	W. A. Kinney, M. J. Coghlan and L. A. Paquette; *J. Am. Chem. Soc.*, 1985, **107**, 7352.	632, 797, 826
85JA7792	P. A. Bartlett and C. R. Johnson; *J. Am. Chem. Soc.*, 1985, **107**, 7792.	826
85JA7974	A. I. Meyers, D. A. Dickman and T. R. Bailey; *J. Am. Chem. Soc.*, 1985, **107**, 7974.	150
85JA7978	J. B. Lambert and D. E. Marko; *J. Am. Chem. Soc.*, 1985, **107**, 7978.	14
85JCR(S)18	T.-S. Chou, J.-J. Yuan and C.-H. Tsao; *J. Chem. Res. (S)*, 1985, 18.	444
85JCS(P1)373	A. Toshimitsu, H. Owada, K. Terao, S. Uemura and M. Okano; *J. Chem. Soc., Perkin Trans. 1*, 1985, 373.	602
85JCS(P1)471	S. Uemura and S. Fukuzawa; *J. Chem. Soc., Perkin Trans. 1*, 1985, 471.	620
85JCS(P1)493	R. Pellicciari, B. Natalini, S. Cecchetti and R. Fringuelli; *J. Chem. Soc., Perkin Trans. 1*, 1985, 493.	44
85JCS(P1)541	M. T. Crimmin, P. J. O'Hanlon and N. H. Rogers; *J. Chem. Soc., Perkin Trans. 1*, 1985, 541.	763
85JCS(P1)1115	V. Fiandanese, G. Marchese, F. Naso and L. Ronzini; *J. Chem. Soc., Perkin Trans. 1*, 1985, 1115.	467
85JCS(P1)1171	Y. Masaki, K. Sakuma and K. Kaji; *J. Chem. Soc., Perkin Trans. 1*, 1985, 1171.	237
85JCS(P1)1289	G. Singh, M. L. Purkayastha, H. Ila and H. Junjappa; *J. Chem. Soc., Perkin Trans. 1*, 1985, 1289.	485
85JCS(P1)1589	R. C. Cookson and N. J. Liverton; *J. Chem. Soc., Perkin Trans. 1*, 1985, 1589.	207
85JCS(P1)1953	T. G. Roberts and G. H. Whitham; *J. Chem. Soc., Perkin Trans. 1*, 1985, 1953.	756
85JCS(P1)2307	A. D. Buss and S. Warren; *J. Chem. Soc., Perkin Trans. 1*, 1985, 2307.	641, 753, 756
85JCS(P1)2399	J. Buck, N. G. Clemo and G. Pattenden; *J. Chem. Soc., Perkin Trans. 1*, 1985, 2399.	450
85JFC(29)273	W. Dmowski; *J. Fluorine Chem.*, 1985, **29**, 273.	429
85JMOC213	M. O. Albers, E. Singleton and M. M. Viney; *J. Mol. Catal.*, 1985, **30**, 213.	456
85JOC147	S. C. Shim, S. Antebi and H. Alper; *J. Org. Chem.*, 1985, **50**, 147.	239
85JOC201	L. A. Paquette, J. A. Colapret and D. R. Andrews; *J. Org. Chem.*, 1985, **50**, 201.	812
85JOC546	C. S. Wilcox and L. M. Thomasco; *J. Org. Chem.*, 1985, **50**, 546.	174
85JOC565	T. Mitsudo, Y. Nakagawa, K. Watanabe, Y. Hori, H. Misawa, H. Watanabe and Y. Watanabe; *J. Org. Chem.*, 1985, **50**, 565.	1036
85JOC657	K. Fuji, M. Ueda, K. Sumi, K. Kajiwara, E. Fujita, T. Iwashita and I. Miura; *J. Org. Chem.*, 1985, **50**, 657.	141
85JOC673	M. S. Mubarak and D. G. Peters; *J. Org. Chem.*, 1985, **50**, 673.	68
85JOC707	R. L. Funk and J. D. Munger, Jr.; *J. Org. Chem.*, 1985, **50**, 707.	831
85JOC826	H. C. Brown, F. J. Chloupek and K. Takeuchi; *J. Org. Chem.*, 1985, **50**, 826.	905
85JOC915	A. G. Schultz and S. Puigh; *J. Org. Chem.*, 1985, **50**, 915.	802
85JOC1019	A. I. Meyers, P. D. Edwards, T. R. Bailey and G. E. Jagdmann, Jr.; *J. Org. Chem.*, 1985, **50**, 1019.	1136
85JOC1122	E. Wenkert, M. H. Leftin and E. L. Michelotti; *J. Org. Chem.*, 1985, **50**, 1122.	970
85JOC1125	E. Wenkert; J. M. Hanna, Jr., M. H. Leftin, E. L. Michelotti, K. T. Potts and D. Usifer; *J. Org. Chem.*, 1985, 1125.	436
85JOC1143	M. M. Midland and J. Gabriel; *J. Org. Chem.*, 1985, **50**, 1143.	816, 817
85JOC1147	M. Sodeoka and M. Shibasaki; *J. Org. Chem.*, 1985, **50**, 1147.	456, 507
85JOC1165	H. C. Brown, S. Ikegami and D. L. Vander Jagt; *J. Org. Chem.*, 1985, **50**, 1165.	905
85JOC1523	J. Tsuji, I. Shimizu, I. Minami, Y. Ohashi, T. Sugiura and K. Takahashi; *J. Org. Chem.*, 1985, **50**, 1523.	193

85JOC1577 H. C. Brown, G. A. Molander, S. M. Singh and U. S. Racherla; *J. Org. Chem.*, 1985, **50**, 1577. 1037
85JOC1749 A. C. Brown and L. A. Carpino; *J. Org. Chem.*, 1985, **50**, 1749. 617
85JOC1927 S. Kim, C. H. Oh, J. S. Ko, K. H. Ahn and Y. J. Kim; *J. Org. Chem.*, 1985, **50**, 1927. 39
85JOC2110 J. B. Hendrickson and P. S. Palumbo; *J. Org. Chem.*, 1985, **50**, 2110. 619, 704, 705, 708
85JOC2124 J. C. Ewing, G. S. Ferguson, D. W. Moore, F. W. Schultz and D. W. Thompson; *J. Org. Chem.*, 1985, **50**, 2124. 543
85JOC2149 D. D. Tanner, G. E. Diaz and A. Potter; *J. Org. Chem.*, 1985, **50**, 2149. 935
85JOC2193 S. S. Nikam and K. K. Wang; *J. Org. Chem.*, 1985, **50**, 2193. 1059
85JOC2302 J. K. Stille, W. J. Scott, M. R. Peña, K. Swärd and S. J. Stoessel; *J. Org. Chem.*, 1985, **50**, 2302. 494, 1029
85JOC2309 P. DeShong, D. A. Kell and D. R. Sidler; *J. Org. Chem.* 1985, **50**, 2309. 524
85JOC2372 E. M. Beccalli, A. Manfredi and A. Marchesini; *J. Org. Chem.*, 1985, **50**, 2372. 1076
85JOC2386 L. F. Cannizzo and R. H. Grubbs; *J. Org. Chem.*, 1985, **50**, 2386. 767
85JOC2400 R. D. Little, H. Bode, K. J. Stone, O. Wallquist and R. Dannecker; *J. Org. Chem.*, 1985, **50**, 2400. 527, 944
85JOC2589 G. H. Posner and E. Asirvatham; *J. Org. Chem.*, 1985, **50**, 2589. 23
85JOC2607 L. Anzalone and J. A. Hirsch; *J. Org. Chem.*, 1985, **50**, 2607. 19, 24
85JOC2668 S. C. Welch, C.-Y. Chou, J. M. Gruber and J.-M. Assercq; *J. Org. Chem.*, 1985, **50**, 2668. 434
85JOC2782 S. A. Carr and W. P. Weber; *J. Org. Chem.*, 1985, **50**, 2782. 229
85JOC2798 M. G. Silvestri, P. J. Bednarski and E. Kho; *J. Org. Chem.*, 1985, **50**, 2798. 635
85JOC2965 T. Cohen, Z. Kosarych, K. Suzuki and L.-C. Yu; *J. Org. Chem.*, 1985, **50**, 2965. 408, 821
85JOC3042 C. J. Elsevier and P. Vermeer; *J. Org. Chem.*, 1985, **50**, 3042. 466
85JOC3220 J. Protasiewicz and G. D. Mendenhall; *J. Org. Chem.*, 1985, **50**, 3220. 925
85JOC3246 M. Sodeoka and M. Shibasaki; *J. Org. Chem.*, 1985, **50**, 3246. 507
85JOC3283 S. Yasui, K. Nakamura, M. Fujii and A. Ohno; *J. Org. Chem.*, 1985, **50**, 3283. 431
85JOC3393 K. B. Wiberg, P. J. Okarma and W. P. Dailey; *J. Org. Chem.*, 1985, **50**, 3393. 809
85JOC3408 S. K. Boyer, J. Bach, J. McKenna and E. Jagdmann, Jr.; *J. Org. Chem.*, 1985, **50**, 3408. 2, 6
85JOC3498 P. Gottschalk and D. C. Neckers; *J. Org. Chem.*, 1985, **50**, 3498. 925
85JOC3622 B. A. Pearlman, S. R. Putt and J. A. Fleming; *J. Org. Chem.*, 1985, **50**, 3622. 594, 616
85JOC3674 J. J. Eisch, S. K. Dua and M. Behrooz; *J. Org. Chem.*, 1985, **50**, 3674. 146
85JOC3676 J. Butera, J. Rini and P. Helquist; *J. Org. Chem.*, 1985, **50**, 3676. 140
85JOC3692 N. Ono, A. Kamimura, H. Mikaye, I. Hamamoto and A. Kaji; *J. Org. Chem.*, 1985, **50**, 3692. 37
85JOC3828 M. Tiecco, M. Tingoli and E. Wenkert; *J. Org. Chem.*, 1985, **50**, 3828. 484
85JOC3846 M. J. Kurth and M. J. O'Brien; *J. Org. Chem.*, 1985, **50**, 3846. 236
85JOC3923 R. S. Matthews and D. J. Eickhoff; *J. Org. Chem.*, 1985, **50**, 3923. 218
85JOC3940 L. Van Hijfte and R. D. Little; *J. Org. Chem.*, 1985, **50**, 3940. 944
85JOC3945 O. Piccolo, F. Spreafico, G. Visentin and E. Valoti; *J. Org. Chem.*, 1985, **50**, 3945. 182
85JOC3957 A. E. Greene, M.-J. Luche and A. A. Serra; *J. Org. Chem.*, 1985, **50**, 3957. 2, 6
85JOC3972 T. L. Cupps, R. H. Boutin and H. Rapoport; *J. Org. Chem.*, 1985, **50**, 3972. 1047
85JOC4317 G. E. Keck, D. F. Kachensky and E. J. Enholm; *J. Org. Chem.*, 1985, **50**, 4317. 173, 703
85JOC4329 T. Chou, C.-H. Tsao and S. C. Hung; *J. Org. Chem.*, 1985, **50**, 4329. 53
85JOC4605 G. A. Kraus and Y.-S. Hon; *J. Org. Chem.*, 1985, **50**, 4605. 900
85JOC4664 G. Büchi and D. E. Vogel; *J. Org. Chem.*, 1985, **50**, 4664 832
85JOC5045 N. Kamigata, J. Ozaki and M. Kobayashi; *J. Org. Chem.*, 1985, **50**, 5045. 485
85JOC5143 H. Kleijn and P. Vermeer; *J. Org. Chem.*, 1985, **50**, 5143. 967
85JOC5179 P. A. Bartlett, K. H. Holm and A. Morimoto; *J. Org. Chem.*, 1985, **50**, 5179. 831
85JOC5200 L. A. Paquette, J. R. Peterson and R. J. Ross; *J. Org. Chem.*, 1985, **50**, 5200. 246
85JOC5215 B. E. La Belle, M. J. Knudsen, M. M. Olmstead, H. Hope, M. D. Yanuck and N. E. Schore; *J. Org. Chem.*, 1985, **50**, 5215. 530
85JOC5413 T.-Y. Luh and C. S. Wong; *J. Org. Chem.*, 1985, **50**, 5413. 22, 23, 437
85JOC5451 D. M. Ketcha and G. W. Gribble; *J. Org. Chem.*, 1985, **50**, 5451. 19
85JOC5460 P. E. Bauer, D. A. Nelson, D. S. Watt, J. H. Reibenspies, O. P. Anderson, W. K. Seifert and J. M. Moldowan; *J. Org. Chem.*, 1985, **50**, 5460. 630
85JOC5465 W. R. Roush and S. Russo-Rodriguez; *J. Org. Chem.*, 1985, **50**, 5465. 698
85JOC5620 A. Padwa, H. Nimmesgern and G. S. K. Wong; *J. Org. Chem.*, 1985, **50**, 5620. 342
85JOC5678 C. H. Behrens, S. Y. Ko, K. B. Sharpless and F. J. Walker; *J. Org. Chem.*, 1985, **50**, 5678. 14
85JOC5769 M. J. Kurth and O. H. W. Decker; *J. Org. Chem.*, 1985, **50**, 5769. 835
85JOC5875 S. Torii, T. Inokuchi and T. Yukawa; *J. Org. Chem.*, 1985, **50**, 5875. 946
85JOC5909 J. T. Welch and S. Eswarakrishnan; *J. Org. Chem.*, 1985, **50**, 5909. 415, 827, 828
85JOM(280)307 G. P. Boldrini, D. Savoia, E. Tagliavini, C. Trombini and A. Umani-Ronchi; *J. Organomet. Chem.*, 1985, **280**, 307. 688
85JOM(285)C13 S. J. Ladoulis and K. M. Nicholas; *J. Organomet. Chem.*, 1985, **285**, C13. 196
85JOM(285)43 U. M. Dzhemilev and O. S. Vostrikova; *J. Organomet. Chem.*, 1985, **285**, 43. 316
85JOM(285)437 B. H. Lipshutz, J. A. Kozlowski, D. A. Parker, S. L. Nguyen and K. E. McCarthy; *J. Organomet. Chem.*, 1985, **285**, 437. 207, 312
85JOM(296)269 I. Minami, I. Shimizu and J. Tsuji; *J. Organomet. Chem.*, 1985, **296**, 269. 196
B-85MI 101-01 P. N. Rylander; "Hydrogenation Methods," Academic Press, New York, 1985. 2
B-85MI 102-01 J. C. March; "Advanced Organic Chemistry," Wiley, New York, 1985. 32, 34, 36
B-85MI 102-02 R. C. Larock; "Organomercury Compounds in Organic Synthesis," Springer-Verlag, Heidelberg, 1985. 67
B-85MI 103-01 P. N. Rylander; "Hydrogenation Methods," Academic Press, New York, 1985. 71, 72, 87, 89, 92

B-85MI 103-02	A. J. Pearson; "Metallo-Organic Chemistry," Wiley, Chichester, 1985.	98
B-85MI 105-01	R. F. Heck; "Palladium Reagents in Organic Synthesis," Academic Press, London, 1985.	192
B-85MI 108-01	C. J. Evans and S. Karpel; "Organic Compounds in Modern Technology," Elsevier, New York, 1985, ch. 10.	331, 342
B-85MI 110-01	P. N. Rylander; "Hydrogenation Methods," Academic Press, London, 1985.	454, 455
B-85MI 110-02	R. C. Larock; "Organomercury Compounds in Organic Synthesis," Springer-Verlag, Berlin, 1985.	454
85MI 111-01	G. A. Olah, G. K. S. Prakash and J. Sommer; in "Superacids," Wiley, Chichester, 1985, p. 24.	462
85MI 111-02	R. F. Heck; "Palladium Reagents in Organic Syntheses," Academic Press, London, 1985.	493
B-85MI 112-01	P. N. Rylander; "Hydrogenation Methods," Wiley, New York, 1985, pp. 53–65.	502
B-85MI 112-02	M. Bartók; "Stereochemistry of Heterogeneous Metal Catalysis," Wiley, Chichester, 1985.	502, 505, 506
B-85MI 112-03	M. J. Winter; in "The Chemistry of the Metal–Carbon Bond," eds. F. R. Hartley and S. Patai, Wiley, New York, 1985, vol. 3, p. 259.	517, 530
85MI 112-04	E. Negishi and T. Takahashi; *Aldrichimica Acta*, 1985, **18**, 31.	536, 542
B-85MI 114-01	R. J. Cross; in "The Chemistry of the Metal–Carbon Bond," ed. F. R. Hartley and S. Patai, Wiley, Chichester, 1985, vol. 2, ch 8.	668
B-85MI 119-01	J. P. Oliver; "The Chemistry of the Metal–Carbon Bond," ed. F. R. Hartley and S. Patai, Wiley, New York, 1985, Vol. 2., pp. 789–826.	885
B-85MI 119-02	A. J. Pearson; "Metallo-organic Chemistry," Wiley, New York, 1985.	885
B-85MI 119-03	P. Vogel; "Carbocation Chemistry," Elsevier, New York, 1985.	891
B-85MI 119-04	G. A. Olah, G. K. S. Prakash and J. Sommer; "Superacids," Wiley, New York, 1985, chap. 3.	891, 917
85MI 119-05	S. W. Staley; *React. Intermed.* (*Wiley*), 1985, **3**, 19–43.	844
85MI 119-06	E. M. Arnett, T. C. Hofelich and G. W. Schriver; *React. Intermed.* (*Wiley*), 1985, **3**, 189–226.	905
85MI 119-07	K. Münger and H. Fischer; *Int. J. Chem. Kinet.*, 1985, **17**, 809.	923
B-85MI 122-01	R. C. Larock; "Organomercury Compounds in Organic Synthesis," Springer-Verlag, Heidelberg, 1985.	1095
85OM333	J. Pornet, L. Miginiac, K. Jaworski and B. Randrianoelina; *Organometallics*, 1985, **4**, 333.	978
85OM1690	D. S. Matteson and J. W. Wilson; *Organometallics*, 1985, **4**, 1690.	163
85OM584	A. Sekiguchi, H. Tanikawa and W. Ando; *Organometallics*, 1985, **4**, 584.	63
85OM1911	H. Nishiyama, M. Matsuto, T. Matsukara, R. Miura and K. Itoh; *Organometallics*, 1985, **4**, N10, 1911.	63
85OR(33)247	T. Hudlicky, T. M. Kutchan and S. M. Naqvi; *Org. React.*, 1985, **33**, 247.	402, 795, 801, 802
85OS(64)44	E. Negishi, A. O. King and J. M. Tour; *Org. Synth.*, 1985, **64**, 44.	1070
85PAC763	M. Julia; *Pure Appl. Chem.*, 1985, **57**, 763.	696
85PAC1749	A. Suzuki; *Pure Appl. Chem.*, 1985, **57**, 1749.	532
85PAC1839	T. Aratani; *Pure Appl. Chem.*, 1985, **57**, 1839.	353
85PS(24)97	P. J. Kocienski; *Phosphorous Sulfur*, 1985, **24**, 97.	144, 696
85PS(24)477	P. Kocienski; *Phosphorus Sulfur*, 1985, **24**, 477.	614
85PS(24)531	C. R. Johnson, M. R. Barbachyn, N. A. Meanwell and C. J. Stark Jr.; *Phosphorous Sulphur*, 1985, **24**, 531.	711
85S57	Yaozeng Huang, Yanchang Shen, Jianhua Zheng and Shixiang Zhang; *Synthesis*, 1985, 57.	739
85S121	M.-C. Lasne and J.-L. Ripoll; *Synthesis*, 1985, 121.	786, 787
85S269	G. Rosini, R. Ballini and M. Petrini; *Synthesis*, 1985, 269.	37
85S364	R. J. K. Taylor; *Synthesis*, 1985, 364.	130, 304
85S493	S. Randriamahefa, P. Deschamps and R. Gallo; *Synthesis*, 1985, 493.	186
85S505	G. Castaldi, C. Giordano and F. Uggeri; *Synthesis*, 1985, 505.	391
85S676	I. Yamamoto, T. Sakai, S. Yamamoto, K. Ohta and K. Matsuzaki; *Synthesis*, 1985, 676.	430
85S717	R. Anderson; *Synthesis*, 1985, 717.	648
85S756	C. F. Nutaitis and G. W. Gribble; *Synthesis*, 1985, 756.	475
85S768	F. Barbot, B. Dauphin and P. Miginiac; *Synthesis*, 1985, 768.	961
85S817	M. Lalonde and T. H. Chan; *Synthesis*, 1985, 817.	654
85S962	P. J. Stang and V. Dixit; *Synthesis*, 1985, 962.	1070
85S1069	P. Ongoka, B. Mauzé and L. Miginiac; *Synthesis*, 1985, 1069.	242
85S1159	F. Donotrio and A. Scettri; *Synthesis*, 1985, 1159.	578
85SC401	E. J. Prisbe and J. C. Martin; *Synth. Commun.*, 1985, **15**, 401.	939
85SC663	S. Manna, J. R. Falck and C. Mioskowski; *Synth. Commun.*, 1985, **15**, 663.	179
85SC1123	S. Suzuki, T. Onishi, Y. Fujita and J. Otera; *Synth. Commun.*, 1985, **15**, 1123.	156
85T621	A. Carpita, A. Lezzi, R. Rossi, F. Marchetti and S. Merlino; *Tetrahedron*, 1985, **41**, 621.	1010
85T837	M. Reuman and A. I. Meyers; *Tetrahedron*, 1985, **41**, 837.	473, 539
85T2369	K. Mori and Y. Funaki; *Tetrahedron*, 1985, **41**, 2369.	1134
85T2387	N. Suzuki, Y. Kaneko, T. Tsukanaka, T. Nomoto, Y. Ayaguchi and Y. Izawa; *Tetrahedron*, 1985, **41**, 2387.	515
85T2751	K. Mori, T. Uematsu, K. Yanagi and M. Minobe; *Tetrahedron*, 1985, **41**, 2751.	165
85T2759	M. Bertrand, G. Gil, A. Junino and R. Maurin; *Tetrahedron*, 1985, **41**, 2759.	498
85T3331	C. Lambert, B. Caillaux and H. G. Viehe; *Tetrahedron*, 1985, **41**, 3331.	571
85T3547	P. Armstrong, R. Grigg, M. W. Jordan and J. F. Malone; *Tetrahedron*, 1985, **41**, 3547.	369, 524
85T3663	K. Mori, H. Watanabe, K. Yanagi and M. Minobe; *Tetrahedron*, 1985, **41**, 3663.	165
85T3825	S. V. Ley, B. Lygo, H. M. Organ and A. Wonnacott; *Tetrahedron*, 1985, **41**, 3825.	757

85T3887	C. Walling; *Tetrahedron*, 1985, **41**, 3887.	919, 923, 924
85T3901	D. H. R. Barton, D. Crich and W. B. Motherwell; *Tetrahedron*, 1985, **41**, 3901.	935
85T3925	A. L. J. Beckwith and C. H. Schiesser; *Tetrahedron*, 1985, **41**, 3925.	336, 337, 338, 547
85T3943	D. P. Curran and D. M. Rakiewicz; *Tetrahedron*, 1985, **41**, 3943.	1132
85T3959	J.-K. Choi and D. J. Hart; *Tetrahedron*, 1985, **41**, 3959.	547, 1131
85T3979	J. W. Wilt; *Tetrahedron*, 1985, **41**, 3979.	337
85T4001	G. Pattenden and G. M. Robertson; *Tetrahedron*, 1985, **41**, 4001.	497
85T4013	N. Ono, H. Mikaye, A. Kamimura, I. Hamamoto, R. Tamura and A. Kaji; *Tetrahedron*, 1985, **41**, 4013.	37, 330, 443, 548, 946, 1131
85T4079	G. E. Keck, E. J. Enholm, J. B. Yates and M. R. Wiley; *Tetrahedron*, 1985, **41**, 4079.	173, 321, 334
85T4213	M. Feldhues and H. J. Schäfer; *Tetrahedron*, 1985, **41**, 4213.	934
85T4253	H. Redlich, W. Sudau and H. Paulsen; *Tetrahedron*, 1985, **41**, 4253.	940, 941
85T4347	D. H. R. Barton, D. Crich, Y. Hervé, P. Potier and J. Thierry; *Tetrahedron*, 1985, **41**, 4347.	602
85T5121	A. Arcadi, S. Cacchi and F. Marinelli; *Tetrahedron*, 1985, **41**, 5121.	546
85T5855	P. L. Pauson; *Tetrahedron*, 1985, **41**, 5855.	528
85T5995	A.-M. Montaña, A. Moyano, M. A. Pericas and F. Serratosa; *Tetrahedron*, 1985, **41**, 5995.	529
85T6055	M. B. Huang and G. Jonsäll; *Tetrahedron*, 1985, **41**, 6055.	918
85TL97	A. Mete, J. B. Hobbs, D. I. C. Scopes and R. F. Newton; *Tetrahedron Lett.*, 1985, **26**, 97.	218
85TL135	G. A. Krafft and P. T. Meinke; *Tetrahedron Lett.*, 1985, **26**, 135.	942
85TL271	J. Hooz, J. G. Calzada and D. McMaster; *Tetrahedron Lett.*, 1985, **26**, 271.	1053
85TL505	D. Levin and S. Warren; *Tetrahedron Lett.*, 1985, **26**, 505.	643
85TL523	R. Rossi, A. Carpita and C. Bigelli; *Tetrahedron Lett.*, 1985, **26**, 523.	1035
85TL631	F. Toda, M. Yamamoto, K. Tanaka and T. C. W. Mak; *Tetrahedron Lett.*, 1985, **26**, 631.	983
85TL709	K. P. C. Vollhardt and L. S. Winn; *Tetrahedron Lett.*, 1985, **26**, 709.	1009
85TL757	D. H. R. Barton and D. Crich; *Tetrahedron Lett.*, 1985, **26**, 757.	172, 940
85TL767	M. Uemura, K. Isobe and Y. Hayashi; *Tetrahedron Lett.*, 1985, **26**, 767.	202
85TL861	K. Suzuki, T. Ohkuma and G.-i. Tsuchihashi; *Tetrahedron Lett.*, 1985, **26**, 861.	1098
85TL895	S. Uemura, K. Ohe and S. Fukuzawa; *Tetrahedron Lett.*, 1985, **26**, 895.	611
85TL921	S. Uemura, S. Fukuzawa and K. Ohe; *Tetrahedron Lett.*, 1985, **26**, 921.	611
85TL957	A. Padwa, H. Nimmesgern and G. S. K. Wong; *Tetrahedron Lett.*, 1985, **26**, 957.	1127
85TL985	R. D. Bach and R. C. Klix; *Tetrahedron Lett.*, 1985, **26**, 985.	388
85TL989	W. G. Dauben and G. Shapiro; *Tetrahedron Lett.*, 1985, 989.	498
85TL1057	R. Huisgen and J. R. Moran; *Tetrahedron Lett.*, 1985, **26**, 1057.	841
85TL1093	M. Clarembeau and A. Krief; *Tetrahedron Lett.*, 1985, **26**, 1093.	245
85TL1153	M. J. Eis and B. Ganem; *Tetrahedron Lett.*, 1985, **26**, 1153.	263
85TL1163	S. D. Burke, D. M. Armistead and J. M. Fevig; *Tetrahedron Lett.*, 1985, **26**, 1163.	831
85TL1333	S. Takano, S. Hatakeyama, K. Sakurai and K. Saijo; *Tetrahedron Lett.*, 1985, **26**, 1333.	248
85TL1381	M. K. Anwer and A. F. Spatola; *Tetrahedron Lett.*, 1985, **26**, 1381.	427
85TL1429	W. G. Dauben, G. Shapiro and L. Luders; *Tetrahedron Lett.*, 1985, 1429.	498
85TL1493	E.-W. Koch, H.-U. Siehl and M. Hanack; *Tetrahedron Lett.*, 1985, **26**, 1493.	1141
85TL1519	M. Mori, Y. Kubo and Y. Ban; *Tetrahedron Lett.*, 1985, 1519.	494
85TL1549	Y. Tamura, M. Sasho, H. Ohe, S. Akai and Y. Kita; *Tetrahedron Lett.*, 1985, **26**, 1549.	1140
85TL1599	C. L. Liotta and T. C. Caruso; *Tetrahedron Lett.*, 1985, **26**, 1599.	873
85TL1831	D. Schinzer, S. Solyom and M. Becker; *Tetrahedron Lett.*, 1985, **26**, 1831.	979
85TL2119	A. B. McElroy and S. Warren; *Tetrahedron Lett.*, 1985, **26**, 2119.	754
85TL2167	J. E. McMurry and G. K. Bosch; *Tetrahedron Lett.*, 1985, **26**, 2167.	690
85TL2171	J. E. McMurry and P. Kocovsky; *Tetrahedron Lett.*, 1985, **26**, 2171.	690
85TL2187	H. C. Brown and U. S. Racherla; *Tetrahedron Lett.*, 1985, **26**, 2187.	1101
85TL2205	T. Sato, T. Kikuchi, N. Sootome and E. Murayama; *Tetrahedron Lett.*, 1985, **26**, 2205.	667
85TL2237	H. H. Wasserman and B. C. Pearce; *Tetrahedron Lett.*, 1985, **26**, 2237.	1134
85TL2609	S. Antebi and H. Alper; *Tetrahedron Lett.*, 1985, **26**, 2609.	239
85TL2735	E. Piers and N. Moss; *Tetrahedron Lett.*, 1985, **26**, 2735.	812
85TL2913	P. Bravo, G. Resnati and F. Viani; *Tetrahedron Lett.*, 1985, **26**, 2913.	143
85TL3011	Y. Terao, M. Tanaka, N. Imai and K. Achiwa; *Tetrahedron Lett.*, 1985, **26**, 3011.	367
85TL3075	L. S. Liebeskind, R. W. Fengl and M. E. Welker; *Tetrahedron Lett.*, 1985, **26**, 3075.	134
85TL3121	F. Barrière, J.-C. Barrière, D. H. R. Barton, J. Cleophax, A. Gateau-Olesker, S. D. Géro and F. Tadj; *Tetrahedron Lett.*, 1985, **26**, 3121.	223
85TL3203	R. Epsztein and N. Le Goff; *Tetrahedron Lett.*, 1985, **26**, 3203.	1041
85TL3259	T. Hiyama and N. Wakasa; *Tetrahedron Lett.*, 1985, **26**, 3259.	195, 477
85TL3285	M. Furber, R. J. K. Taylor and S. C. Burford; *Tetrahedron Lett.*, 1985, **26**, 3285.	541
85TL3311	G. E. Keck and E. J. Enholm; *Tetrahedron Lett.*, 1985, **26**, 3111.	344, 1130
85TL3393	C. B. Jackson and G. Pattenden; *Tetrahedron Lett.*, 1985, **26**, 3393.	690
85TL3417	K. D. Moeller and R. D. Little; *Tetrahedron Lett.*, 1985, **26**, 3417.	944
85TL3463	T. Kobayashi, T. Sakakura and M. Tanaka; *Tetrahedron Lett.*, 1985, **26**, 3463.	1047
85TL3643	M. P. Cooke, Jr. and I. N. Houpis; *Tetrahedron Lett.*, 1985, **26**, 3643.	214
85TL3811	D. Guillerm and G. Linstrumelle; *Tetrahedron Lett.*, 1985, **26**, 3811.	1009
85TL3931	P. G. Williard and J. M. Salvino; *Tetrahedron Lett.*, 1985, **26**, 3931.	873
85TL4111	P. J. Hodges and G. Procter; *Tetrahedron Lett.*, 1985, **26**, 4111.	218
85TL4197	A. Alexakis, P. Mangeney and J.-F. Normant; *Tetrahedron Lett.*, 1985, **26**, 4197.	967
85TL4291	E. J. Corey and A. W. Gross; *Tetrahedron Lett.*, 1985, **26**, 4291.	926
85TL4359	I. Erdelmeier and H.-J. Gais; *Tetrahedron Lett.*, 1985, **26**, 4359.	763
85TL4495	T.-S. Chou and M.-L. You; *Tetrahedron Lett.*, 1985, **26**, 4495.	24

85TL4523 W. Ando, T. Tsumuraya and A. Sekiguchi; *Tetrahedron Lett.*, 1985, **26**, 4523. 656

85TL4683 J. M. Chong and K. B. Sharpless; *Tetrahedron Lett.*, 1985, **26**, 4683. 217, 478

85TL4707 G. L. Lange, C. P. Decicco, S. L. Tan and G. Chamberlain; *Tetrahedron Lett.*, 1985, **26**, 4707. 357

85TL4747 K. Tanaka, H. Yoda and A. Kaji; *Tetrahedron Lett.*, 1985, **26**, 4747. 210

85TL4803 F. Minisci, E. Vismara and U. Romano; *Tetrahedron Lett.*, 1985, **26**, 4803. 934

85TL4887 B. M. Trost and M. Lautens; *Tetrahedron Lett.*, 1985, **26**, 4887. 525

85TL4959 L. Castedo, J. R. Granja and A. Mouriño; *Tetrahedron Lett.*, 1985, **26**, 4959. 816, 817

85TL4991 D. P. Curran and M.-H. Chen; *Tetrahedron Lett.*, 1985, **26**, 4991. 548, 1132

85TL5013 M. M. Midland and Y. C. Kwon; *Tetrahedron Lett.*, 1985, **26**, 5013. 406, 816, 817

85TL5017 M. M. Midland and Y. C. Kwon; *Tetrahedron Lett.*, 1985, **26**, 5017. 816, 817

85TL5021 M. M. Midland and Y. C. Kwon; *Tetrahedron Lett.*, 1985, **26**, 5021. 816, 817

85TL5093 A. Pelter, D. Buss and A. Pitchford; *Tetrahedron Lett.*, 1985, **26**, 5093. 764

85TL5393 J. E. Nystrom, T. D. McCanna, P. Helquist and R. S. Iyer; *Tetrahedron Lett.*, 1985, **26**, 5393. 523

85TL5579 J. Hibino, T. Okazoe, K. Takai and H. Nozaki; *Tetrahedron Lett.*, 1985, **26**, 5579. 678

85TL5581 T. Okazoe, J. Hibino, K. Takai and H. Nozaki; *Tetrahedron Lett.*, 1985, **26**, 5581. 678, 769

85TL5585 K. Takai, T. Kuroda, S. Nakatsukasa, K. Oshima and H. Nozaki; *Tetrahedron Lett.*, 1985, **26**, 5585. 1038, 1140

85TL5597 B. Achmatowicz, E. Baranowska, A. R. Daniewski, J. Pankowski and J. Wicha; *Tetrahedron Lett.*, 1985, **26**, 5597. 697, 698, 700, 703

85TL5671 A. T. Stoll and E.-i. Negishi; *Tetrahedron Lett.*, 1985, **26**, 5671. 1094

85TL5675 B. B. Snider and Y. S. Kulkarni; *Tetrahedron Lett.*, 1985, **26**, 5675. 176

85TL5709 A. B. McElroy and S. Warren; *Tetrahedron Lett.*, 1985, **26**, 5709. 754

85TL5799 K. Mikami, K. Kawamoto and T. Nakai; *Tetrahedron Lett.*, 1985, **26**, 5799. 816, 818

85TL5923 G. Büchi and R. M. Freidinger; *Tetrahedron Lett.*, 1985, **26**, 5923. 248

85TL5927 G. Stork and N. H. Baine; *Tetrahedron Lett.*, 1985, **26**, 5927. 1129

85TL5943 D. H. R. Barton, D. Crich and P. Potier; *Tetrahedron Lett.*, 1985, **26**, 5943. 332, 935

85TL6001 K. Shankaran, C. P. Sloan and V. Snieckus; *Tetrahedron Lett.*, 1985, **26**, 6001. 547, 1130

85TL6115 A. Krief and M. J. De Vos; *Tetrahedron Lett.*, 1985, **26**, 6115. 240

85TL6137 V. T. Ravikumar, K. Rajagopalan and S. Swaminathan; *Tetrahedron Lett.*, 1985, **26**, 6137. 813

85TL6163 G. L. Lange and M. Lee; *Tetrahedron Lett.*, 1985, **26**, 6163. 357

85TL6289 P. Pike, S. Hershberger and J. Hershberger; *Tetrahedron Lett.*, 1985, **26**, 6289. 936

85TL6301 S. W. McCombie, B. B. Shankar and A. K. Ganguly; *Tetrahedron Lett.*, 1985, **26**, 6301. 214

85TL6393 M. C. Honan; *Tetrahedron Lett.*, 1985, **26**, 6393. 776

85TL6397 F. Camps, J. Coll, J. M. Moretó and J. Torras; *Tetrahedron Lett.*, 1985, **26**, 6397. 527

85TL6431 W. R. Leonard and T. Livinghouse; *Tetrahedron Lett.*, 1985, **26**, 6431. 244

85TL6447 Y. Huang, L. Shi and J. Yang; *Tetrahedron Lett.*, 1985, **26**, 6447. 647, 738

86ACR348 M. P. Doyle; *Acc. Chem. Res.*, 1986, **19**, 348. 352, 400, 821, 823

86AG(E)1 B. M. Trost; *Angew. Chem., Int. Ed. Engl.*, 1986, **25**, 1. 193, 287, 358, 360, 361, 362

86AG(E)93 F. W. Nader and A. Brecht; *Angew. Chem., Int. Ed. Engl.*, 1986, **25**, 93. 988

86AG(E)167 A. Maercker, T. Graule and U. Girreser; *Angew. Chem., Int. Ed. Engl.*, 1986, **25**, 167. 508

86AG(E)268 K. P. C. Vollhardt, R. Diercks, J. C. Armstrong and R. Boese; *Angew. Chem., Int. Ed. Engl.*, 1986, 268. 1009

86AG(E)340 H. Irngartinger and W. Götzmann; *Angew. Chem., Int. Ed. Engl.*, 1986, **25**, 340. 986

86AG(E)349 T. Laube; *Angew. Chem., Int. Ed. Engl.*, 1986, **25**, 349. 900

86AG(E)465 P. A. A. Klusener, L. Brandsma, H. D. Verkruijsse, P. von R. Schleyer, T. Friedl and R. Pi; *Angew. Chem., Int. Ed. Engl.*, 1986, **25**, 465. 880

86AG(E)466 B. Stowasser and K. Hafner; *Angew. Chem., Int. Ed. Engl.*, 1986, **25**, 466. 982

86AG(E)508 J. K. Stille; *Angew. Chem. Int. Ed. Engl.*, 1986, **25**, 508. 468, 1045, 1096

86AG(E)653 E. Bartmann; *Angew. Chem. Int. Ed. Engl.*, 1986, **25**, 653. 14

86AG(E)720 E. Vogel, W. Puttmann, W. Duchatsch, T. Schieb, H. Schmickler and J. Lex; *Angew. Chem., Int. Edn. Engl.*, 1986, **25**, 720. 248

86AG(E)754 U. Schöllkopf, B. Hupfeld and R. Gull; *Angew. Chem., Int. Ed. Engl.*, 1986, **25**, 754. 183, 209

86AG(E)760 I. Mukerji, A. Wayda, G. Debbagh and S. H. Bertz; *Angew. Chem., Int. Ed. Engl.*, 1986, **25**, 760. 205

86AG(E)935 I. Erdelmeier, H.-J. Gais and H. J. Lindner, *Angew. Chem., Int. Ed. Engl.*, 1986, **25**, 935. 146, 711

86AG(E)947 Y. Yamamoto; *Angew. Chem. Int. Ed. Engl.*, 1986, **25**, 947. 312, 478, 540

86AG(E)1026 N. Eisen and F. Vögtle; *Angew. Chem., Int. Ed. Engl.*, 1986, **25**, 1026. 814

86AHC(39)1 J. G. Keay; *Adv. Heterocycl. Chem.*, 1986, **39**, 1. 87

86AJC77 A. L. J. Beckwith and P. E. Pigou; *Aust. J. Chem.*, 1986, **39**, 77. 942

86AJC2095 D. J. Collins and H. A. Jacobs; *Aust. J. Chem.*, 1986, **39**, 2095. 200

86AKZ306 D. V. Grigoryan, A. M. Galoyan, F. S. Kinoyan and A. T. Babayan; *Arm. Khim. Zh.*, 1986, **39**, 306 (*Chem. Abstr.*, **107**, 6717p). 1040

86BCJ1109 N. Kato, K. Nakanishi and H. Takeshita; *Bull. Chem. Soc. Jpn.*, 1986, **59**, 1109. 690

86BCJ1363 K. Fujimoto, Y. Iwano and K. Hirai; *Bull. Chem. Soc. Jpn.*, 1986, **59**, 1363. 208

86BCJ1656 T. Amano, T. Ota, K. Yoshikawa, T. Sano, Y. Ohuchi, F. Sato, M. Shiono and Y. Fujita; *Bull. Chem. Soc. Jpn.*, 1986, **59**, 1656. 469

86BCJ2044 S. Miyano, S.-I. Okada, T. Suzuki, S. Handa and H. Hashimoto; *Bull. Chem. Soc. Jpn.*, 1986, **59**, 2044. 473

86BCJ2209 J. Ojima, K. Yamamoto, T. Kato, K. Wada, Y. Yoneyama and E. Ejiri; *Bull. Chem. Soc. Jpn.*, 1986, **59**, 2209. 690

86BCJ2321 Y. Taguchi and Y. Suhara; *Bull. Chem. Soc. Jpn.*, 1986, **59**, 2321. 242

86BCJ2801 Y. Iitaka, A. Itai, N. Tomioka, Y. Kodama, K. Ichikawa, K. Nishihata, M. Nishio, M. Izumi and K. Doi; *Bull. Chem. Soc. Jpn.*, 1986, **59**, 2801. 143

86BCJ2807 M. Kobayashi and M. Kobayashi; *Bull. Chem. Soc. Jpn.*, 1986, **59**, 2807. 64, 449

86BCJ3097 T. Kumamoto, K. Hosoya, S. Kanzaki, K. Masuko, M. Watanabe and K. Shirai; *Bull. Chem. Soc. Jpn.*, 1986, **59**, 3097. 178

86BCJ3201 T. Mitsudo, Y. Hori and Y. Watanabe; *Bull. Chem. Soc. Jpn.*, 1986, **59**, 3201. 1036

86BCJ3617 T. Yamauchi, K. Hattori, S. Mizutaki, K. Tamaki and S. Uemura; *Bull. Chem Soc. Jpn.*, 1986, **59**, 3617. 180

86BSF449 G. Courtois, D. Mesnard, J. R. Mahoungou and L. Miginiac; *Bull. Soc. Chim. Fr.*, 1986, 449. 974

86CB244 H. Wingert and M. Regitz; *Chem. Ber.*, 1986, **119**, 244. 522

86CB444 B. Giese, H. Horler and M. Leising; *Chem. Ber.*, 1986, **119**, 444. 325, 331, 920

86CB575 U. Brändli, M. Eyer and D. Seebach; *Chem. Ber.*, 1986, **119**, 575. 156

86CB1162 L. Fitjer, D. Wehle and H.-J. Scheuermann; *Chem. Ber.*, 1986, **119**, 1162. 713

86CB1208 F. W. Nader, A. Brecht and S. Kreisz; *Chem. Ber.*, 1986, **119**, 1208. 987

86CB1540 K. Weinges, J. Klein, W. Sipos, P. Gunther, U. Huber-Patz, H. Rodewald, J. Deuter and H. Irngartinger; *Chem. Ber.*, 1986, **119**, 1540. 705

86CB2150 T. Kauffmann and P. Schwartze; *Chem. Ber.*, 1986, **119**, 2150. 755

86CB2995 T. Liese and A. de Meijere; *Chem. Ber.*, 1986, **119**, 2995. 1069

86CC102 Y. Yamamoto, S. Nishii and K. Maruyama; *J. Chem. Soc., Chem. Commun.*, 1986, 102. 1038, 1039

86CC176 J. E. Baldwin, R. M. Adlington and I. M. Newington; *J. Chem. Soc., Chem. Commun.*, 1986, 176. 154

86CC451 J. R. Bull and R. I. Thomson; *J. Chem. Soc., Chem. Commun.*, 1986, 451. 25

86CC457 A. Krief, M. Clarembeau and P. Barbeaux; *J. Chem. Soc., Chem. Commun.*, 1986, 457. 245

86CC495 S. G. Davies and J. C. Walker; *J. Chem. Soc., Chem. Commun.*, 1986, 495. 134

86CC620 I. M. Dawson, J. A. Gregory, R. B. Herbert and P. G. Sammes; *J. Chem. Soc., Chem. Commun.*, 1986, 620. 654

86CC829 D. Schinzer, J. Steffen and S. Solyom; *J. Chem. Soc., Chem. Commun.*, 1986, 829. 979

86CC855 W. Kitching, J. A. Lewis, M. T. Fletcher, J. J. De Voss, R. A. I. Drew and C. J. Moore; *J. Chem. Soc., Chem. Commun.*, 1986, 855. 68

86CC878 S. M. Bennett and D. L. J. Clive; *J. Chem. Soc., Chem. Commun.*, 1986, 878. 1131

86CC922 J. Tsuji, T. Sugiura, M. Yuhara and I. Minami; *J. Chem. Soc., Chem. Commun.*, 1986, 922. 962

86CC947 W. Heinz, P. Langensee and K. Müllen; *J. Chem. Soc., Chem. Commun.*, 1986, 947. 890

86CC1067 J. V. Comasseto, V. Catani, J. T. B. Farreira and A. L. Braga; *J. Chem. Soc., Chem. Commun.*, 1986, 1067. 1035

86CC1201 M. D. Jones and R. D. W. Kemmitt; *J. Chem. Soc., Chem. Commun.*, 1986, 1201. 194, 289

86CC1285 N. Ono, T. Yanai, A. Kamimura and A. Kaji; *J. Chem. Soc., Chem. Commun.*, 1986, 1285. 250, 251, 486

86CC1317 R. Choukroun, M. Basso-Bert and D. Gervais; *J. Chem. Soc., Chem. Commun.*, 1986, 1317. 456

86CC1419 M. Cook, O. Hares, A. Johns, J. A. Murphy and C. W. Patterson; *J. Chem. Soc., Chem. Commun.*, 1986, 1419. 929

86CC1452 Q.-Y. Chen, Y.-B. Ho and Z.-Y. Yang; *J. Chem. Soc., Chem. Commun.*, 1986, 1452. 433

86CC1665 J. Barluenga, J. L. Fernandez-Simon, J. M. Concellon and M. Yus; *J. Chem. Soc., Chem. Commun.*, 1986, 1665. 681

86CC1688 S. G. Pyne and S. L. Chapman; *J. Chem. Soc., Chem. Commun*, 1986, 1688. 21, 23

86CC1807 P. J. Battye and D. W. Jones; *J. Chem. Soc., Chem. Commun.*, 1986, 1807. 841

86CJC584 R. J. Armstrong and L. Weiler; *Can. J. Chem.*, 1986, **64**, 584. 216

86CJC793 J. L. Charlton, M. M. Alauddin and G. H. Penner; *Can. J. Chem.*, 1986, **64**, 793. 418

86CL357 N. Mine, Y. Fujiwara and H. Taniguchi; *Chem. Lett.*, 1986, 357. 462

86CL407 J. Yamashita, Y. Inoue, T. Kondo and H. Hashimoto; *Chem. Lett.*, 1986, 407. 483

86CL433 H. Matsuyama, Y. Miyazawa and M. Kobayashi, *Chem. Lett.*, 1986, 433. 704, 707, 708

86CL459 S. Hyuga, S. Takinami, S. Hara and A. Suzuki; *Chem. Lett.*, 1986, 459. 1091, 1094

86CL501 T. Kawashima, Y. Miki, T. Tomita and N. Inamoto; *Chem. Lett.*, 1986, 501. 836

86CL621 Y. Sakito and G. Suzukamo; *Chem. Lett.*, 1986, 621. 801, 803

86CL1009 T. Mukaiyama, H. Nagaoka, M. Ohshima and M. Murakami; *Chem. Lett.*, 1986, 1009. 189

86CL1329 M. Satoh, N. Miyaura and A. Suzuki; *Chem. Lett.*, 1986, 1329. 489, 531

86CL1473 M. Kusakabe and K. Sato; *Chem. Lett.*, 1986, 1473. 451

86CL1895 T. Ishihara, M. Kuroboshi and Y. Okada; *Chem. Lett.*, 1986, 1895. 1130

86CPB1447 Y. Akita, A. Inoue and A. Ohta; *Chem. Pharm. Bull.*, 1986, **34**, 1447. 1010

86CPB1531 M. Suzuki, Y. Kimura and S. Terashima; *Chem. Pharm. Bull.*, 1986, **34**, 1531. 1140

86CPB3599 M. Hori, T. Kataoka, H. Shimizu, E. Imai, T. Iwamura and K. Maeda; *Chem. Pharm. Bull.*, 1986, **34**, 3599. 108, 115

86CRV353 S. Muthukumara Pillai, M. Ravindranathan and S. Sivaram; *Chem. Rev.*, 1986, **86**, 353. 294

86CRV763 B. Ganem and J. O. Osby; *Chem. Rev.*, 1986, **86**, 763. 72, 75

86CRV875 R. D. Little; *Chem. Rev.*, 1986, **86**, 875. 359, 360

86CRV885 T. Nakai and K. Mikami; *Chem. Rev.*, 1986, **86**, 885. 196, 404, 816

86CRV919 M. P. Doyle; *Chem. Rev.*, 1986, **86**, 919. 234, 299, 352

86H(24)3149 M. Matsumoto and N. Watanabe; *Heterocycles*, 1986, **24**, 3149. 178

86H(24)3337 N. Furukawa, M. Tsuruoka and H. Fujihara; *Heterocycles*, 1986, **24**, 3337. 483

86H(24)3347 H. Takahata, Y. Banba, M. Mozumi and T. Yamazaki; *Heterocycles*, 1986, **24**, 3347. 834

86HCA1041 P. Strazewski and C. Tamm; *Helv. Chim. Acta.*, 1986, **69**, 1041. 873

86JA134 R. A. Moss, S. Shen, K. Krogh-Jespersen, J. A. Potenza, H. J. Schugar and R. C. Munjal; *J. Am. Chem. Soc.*, 1986, **108**, 134. 909

86JA303 G. Stork and P. M. Sher; *J. Am. Chem. Soc.*, 1986, **108**, 303. 330, 341, 923, 936

86JA306 A. I. Meyers and S. A. Fleming; *J. Am. Chem. Soc.*, 1986, **108**, 306. 357

86JA343 R. S. Macomber and T. C. Hemling; *J. Am. Chem. Soc.*, 1986, **108**, 343. 988

86JA462 P. G. Williard and G. B. Carpenter; *J. Am. Chem. Soc.*, 1986, **108**, 462. 873

86JA483 N. Ikeda, I. Arai and H. Yamamoto; *J. Am. Chem. Soc.*, 1986, **108**, 483. 1055

86JA515 J. E. McMurry, G. J. Haley, J. R. Matz, J. C. Clardy and J. Mitchell; *J. Am. Chem. Soc.*, 1986, **108**, 515. 690

86JA561 P. H. M. Budzelaar, E. Kraka, D. Cremer and P. von R. Schleyer; *J. Am. Chem. Soc.*, 1986, **108**, 561. 910

86JA733 L. R. Gilliom and R. H. Grubbs; *J. Am. Chem. Soc.*, 1986, **108**, 733. 669

86JA1039 R. V. Stevens, N. Beaulieu, W. Hong Chan, A. R. Daniewski, T. Takeda, A. Waldner, P. G. Williard and U. Zutter; *J. Am. Chem. Soc.*, 1986, **108**, 1039. 624, 626

86JA1098 B. M. Trost and M. R. Ghadiri; *J. Am. Chem. Soc.*, 1986, **108**, 1098. 237, 484, 1031

86JA1106 D. P. Curran and S.-C. Kuo; *J. Am. Chem. Soc.*, 1986, **108**, 1106. 1130

86JA1265 S. F. Nelsen and D. L. Kapp; *J. Am. Chem. Soc.*, 1986, **108**, 1265. 3, 6

86JA1359 B. Eaton, J. A. King and K. P. C. Vollhardt; *J. Am. Chem. Soc.*, 1986, **108**, 1359. 451

86JA1575 V. V. Krishnamurthy, G. K. S. Prakash, P. S. Iyer and G. A. Olah; *J. Am. Chem. Soc.*, 1986, **108**, 1575. 904

86JA1708 J. D. Winkler and V. Sridar; *J. Am. Chem. Soc.*, 1986, **108**, 1708. 925

86JA1940 D. L. Reger, E. Mintz and L. Lebioda; *J. Am. Chem. Soc.*, 1986, **108**, 1940. 545

86JA2094 D. B. Collum, W. C. Still and F. Mohamadi; *J. Am. Chem. Soc.*, 1986, **108**, 2094. 68

86JA2358 J. B. Hendrickson, G. J. Boudreaux and P. S. Palumbo; *J. Am. Chem. Soc.*, 1986, **108**, 2358. 618, 704, 705, 709

86JA2381 M. M. Green, B. A. Boyle, M. Vairamani, T. Mukhopadhyay, W. H. Saunders, Jr., P. Bowen and N. L. Allinger; *J. Am. Chem. Soc.*, 1986, **108**, 2381. 927

86JA2476 D. A. Evans and M. DiMare; *J. Am. Chem. Soc.*, 1986, **108**, 2476. 703

86JA2659 K. N. Houk and M. N. Paddon-Row; *J. Am. Chem. Soc.*, 1986, **108**, 2659. 130

86JA2662 S. R. Schow, J. D. Bloom, A. S. Thompson, K. N. Winzenberg and A. B. Smith, III; *J. Am. Chem. Soc.*, 1986, **108**, 2662. 754, 755

86JA2769 W. H. Pearson, M. A. Walters and K. D. Oswell; *J. Am. Chem. Soc.*, 1986, **108**, 2769. 372

86JA2776 S. Hanessian, A. Ugolini, D. Dube, P. J. Hodges and C. Andre; *J. Am. Chem. Soc.*, 1986, **108**, 2776. 615, 698, 703

86JA2787 N. A. Porter, D. R. Magnin and B. T. Wright; *J. Am. Chem. Soc.*, 1986, **108**, 2787. 340, 341

86JA2932 J. E. McMurry, G. J. Haley, J. R. Matz, J. C. Clardy, G. Van Duyne, R. Gleiter, W. Schäfer and D. H. White; *J. Am. Chem. Soc.*, 1986, **108**, 2932. 690

86JA3033 J. K. Stille and W. J. Scott; *J. Am. Chem. Soc.*, 1986, **108**, 3033. 435, 480, 1027

86JA3102 C. S. Wilcox and J. J. Gaudino; *J. Am. Chem. Soc.*, 1986, **108**, 3102. 174, 341

86JA3128 S. L. Schreiber, T. Sammakia and W. E. Crowe; *J. Am. Chem. Soc.*, 1986, **108**, 3128. 199

86JA3224 D. A. Robaugh and S. E. Stein; *J. Am. Chem. Soc.*, 1986, **108**, 3224. 922

86JA3385 A. R. Matlin, C. F. George, S. Wolff and W. C. Agosta; *J. Am. Chem. Soc.*, 1986, **108**, 3385. 596

86JA3466 E. J. Madaj, Jr., D. M. Snyder and W. E. Truce; *J. Am. Chem. Soc.*, 1986, **108**, 3466. 379

86JA3513 A. S. Kende, S. Johnson, P. Sanfilippo, J. C. Hodges and L. N. Jungheim; *J. Am. Chem. Soc.*, 1986, **108**, 3513. 690

86JA3516 L. E. Overman, T. A. Blumenkopf, A. Castenada and A. S. Thompson; *J. Am. Chem. Soc.*, 1986, **108**, 3516. 450

86JA3739 L. A. Paquette, L. Waykole, H. Jendralla and C. E. Cottrell; *J. Am. Chem. Soc.*, 1986, **108**, 3739. 576

86JA3819 G. Jonsäll and P. Ahlberg; *J. Am. Chem. Soc.*, 1986, **108**, 3819. 918

86JA3835 D. H. Hua; *J. Am. Chem. Soc.*, 1986, **108**, 3835. 17, 530

86JA3841 L. A. Paquette and T. Sugimura; *J. Am. Chem. Soc.*, 1986, **108**, 3841. 816, 817

86JA4132 M. Newcomb and S. U. Park; *J. Am. Chem. Soc.*, 1986, **108**, 4132. 331, 342

86JA4568 E. Block, M. Aslam, V. Eswarakrishnan, K. Gebreyes, J. Hutchinson, R. Iyer, J.-A. Laffitte and A. Wall; *J. Am. Chem. Soc.*, 1986, **108**, 4568. 619, 707, 949

86JA4603 D. V. Patel, F. VanMiddlesworth, J. Donaubauer, P. Gannett and C. J. Sih; *J. Am. Chem. Soc.*, 1986, **108**, 4603. 700, 703

86JA4683 G. A. Molander and D. C. Shubert; *J. Am. Chem. Soc.*, 1986, **108**, 4683. 360

86JA4790 E. G. Samsel and J. K. Kochi; *J. Am. Chem. Soc.*, 1986, **108**, 4790. 946

86JA5078 R. E. Lee and R. R. Squires; *J. Am. Chem. Soc.*, 1986, **108**, 5078. 866

86JA5371 M. Iyoda, H. Otani and M. Oda; *J. Am. Chem. Soc.*, 1986, **108**, 5371. 990

86JA5644 H. Jin, J. Uenishi, W. J. Christ and Y. Kishi; *J. Am. Chem. Soc.*, 1986, **108**, 5644. 677, 1038, 1099

86JA5893 A. L. J. Beckwith and D. H. Roberts; *J. Am. Chem. Soc.*, 1986, **108**, 5893. 936

86JA5901 K. J. Shea, A. C. Greeley, S. Nguyen, P. D. Beauchamp, D. H. Aue and J. S. Witzeman; *J. Am. Chem. Soc.*, 1986, **108**, 5901. 811

86JA6048 K. Takai, M. Tagashira, T. Kuroda, K. Oshima, K. Utimoto and H. Nozaki; *J. Am. Chem. Soc.*, 1986, **108**, 6048. 677, 1099

86JA6051 B. M. Trost, M. Mignani and T. N. Manninga; *J. Am. Chem. Soc.*, 1986, **108**, 6051. 288

86JA6060 M. C. Pirrung and J. A. Werner; *J. Am. Chem. Soc.*, 1986, **108**, 6060. 196, 410, 820

86JA6062 E. J. Roskamp and C. R. Johnson; *J. Am. Chem. Soc.*, 1986, **108**, 6062. 400, 410, 820

86JA6276 H. E. Zimmerman and R. D. Solomon; *J. Am. Chem. Soc.*, 1986, **108**, 6276. 635

86JA6826 G. Stork and M. J. Sofia; *J. Am. Chem. Soc.*, 1986, **108**, 6826 62

86JA7116 Y. Yamamoto, S. Nishii and J. Yamada; *J. Am. Chem. Soc.*, 1986, **108**, 7116. 1026

86JA7117 R. Noyori, M. Ohta, Y. Hsiao, M. Kitamura, T. Ohta and H. Takaya; *J. Am. Chem. Soc.*, 1986, **108**, 7117. 77

86JA7121 G. E. Renzoni, T.-K. Yin and W. T. Borden; *J. Am. Chem. Soc.*, 1986, **108**, 7121. 785

86JA7408 K. Takai, K. Nitta and K. Utimoto; *J. Am. Chem. Soc.*, 1986, **108**, 7408. 677

86JA7686 D. F. Taber and R. E. Ruckle, Jr.; *J. Am. Chem. Soc.*, 1986, **108**, 7686. 635

86JA7763 J. J. Eisch, L. E. Hallenbeck and K. I. Han; *J. Am. Chem. Soc.*, 1986, **108**, 7763. 436

86JA7791 H. J. Reich, E. K. Eisenhart, R. E. Olson and M. J. Kelly; *J. Am. Chem. Soc.*, 1986, **108**, 7791. 971, 985

86JA7873 L. A. Paquette and K. S. Learn; *J. Am. Chem. Soc.*, 1986, **108**, 7873. 815

86JA7935 J. S. Adams, K. A. Burton, B. K. Andrews, R. B. Weisman and P. S. Engel; *J. Am. Chem. Soc.*, 1986, **108**, 7935. 924

86JCE930 J. G. Traynham; *J. Chem. Ed.*, 1986, **63**, 930. 890

86JCS(D)997 J. Ellerman, L. Brehm, E. Lindner, W. Hiller, R. Fawzi, F. L. Dickert and M. Waidhas; *J. Chem. Soc., Dalton Trans.*, 1986, 997. 445

86JCS(P1)39 D. H. R. Barton, D. Crich and G. Kretzschmar; *J. Chem. Soc., Perkin Trans. 1*, 1986, 39. 332, 935, 1130

86JCS(P1)161 A. G. Cameron and D. W. Knight; *J. Chem. Soc., Perkin Trans. 1*, 1986, 161. 831

86JCS(P1)475 P. L. Coe and C. E. Scriven; *J. Chem. Soc., Perkin Trans. 1*, 1986, 475. 248

86JCS(P1)873 M. D. Threadgill and A. P. Gledhill; *J. Chem. Soc., Perkin Trans. 1*, 1986, 873. 442

86JCS(P1)1039 T. Chou, L.-J. Chang and H.-H. Tso; *J. Chem. Soc., Perkin Trans. 1*, 1986, 1039. 790

86JCS(P1)1439 N. Ono, I. Hamamoto and A. Kaji; *J. Chem. Soc., Perkin Trans. 1*, 1986, 1439. 250

86JCS(P1)1603 D. H. R. Barton and D. Crich; *J. Chem. Soc., Perkin Trans. 1*, 1986, 1603. 940

86JCS(P1)1613 D. H. R. Barton and D. Crich; *J. Chem. Soc., Perkin Trans. 1*, 1986, 1613. 231

86JCS(P1)1959 K. Ikenaga, K. Kikukawa and T. Matsuda; *J. Chem. Soc., Perkin Trans. 1*, 1986, 1959. 490

86JCS(P1)1983 S. Uemura, S. Fukuzawa, T. Yamauchi, K. Hattori, S. Mizutaki and K. Tamaki; *J. Chem. Soc., Perkin Trans. 1*; 1986, 1983. 246

86JCS(P2)1255 N. Al-Awadi and R. Taylor; *J. Chem. Soc., Perkin Trans. 2*, 1986, 1255. 777

86JMC112 A. Stütz, A. Georgopoulos, W. Granitzer, G. Petranyi and D. Berney; *J. Med. Chem.*, 1986, **29**, 112. 512

86JMC230 J. A. Katzenellenbogen and M. J. Sofia; *J. Med. Chem.*, 1986, **29**, 230. 1021

86JOC113 N. N. Sauer, R. J. Angelici, Y. C. J. Huang and W. S. Trahanovsky; *J. Org. Chem.*, 1986, **51**, 113. 774

86JOC123 S. Raucher and P. Klein; *J. Org. Chem.*, 1986, **51**, 123. 828

86JOC183 D. W. H. MacDowell and J. M. Purpura; *J. Org. Chem.*, 1986, **51**, 183. 808

86JOC253 M. G. Constantino, P. M. Donate and N. Petragnani; *J. Org. Chem.*, 1986, **51**, 253. 516

86JOC472 E. C. Ashby and J. N. Argyropoulos; *J. Org. Chem.*, 1986, **51**, 472. 884

86JOC537 T. Tsuda, T. Hayashi, H. Satomi, T. Kawamoto and T. Saegusa; *J. Org. Chem.*, 1986, **51**, 537. 86

86JOC558 H. Fritz, P. Sutter and C. D. Weis; *J. Org. Chem.*, 1986, **51**, 558. 651

86JOC635 R. E. Ireland and M. D. Varney; *J. Org. Chem.*, 1986, **51**, 635 831

86JOC734 I. Pri-Bar and O. Buchman; *J. Org. Chem.*, 1986, **51**, 734. 8

86JOC858 J. A. Marshall and D. G. Cleary; *J. Org. Chem.*, 1986, **51**, 858. 24

86JOC959 M. Inove and T. C. Bruice; *J. Org. Chem.*, 1986, **51**, 959. 749

86JOC1124 T. L. Macdonald, K. J. Natalie, Jr., G. Prasad and J. S. Sawyer; *J. Org. Chem.*, 1986, **51**, 1124. 413

86JOC1135 G. A. Molander and G. Hahn; *J. Org. Chem.*, 1986, **51**, 1135. 2, 7, 8, 13, 15, 16, 17, 23, 24

86JOC1159 M. Westling, R. Smith and T. Livinghouse; *J. Org. Chem.*, 1986, **51**, 1159. 524

86JOC1199 M. Santelli, D. E. Abed and A. Jellal; *J. Org. Chem.*, 1986, **51**, 1199. 978, 1061

86JOC1377 M. J. Kurth and O. H. W. Decker; *J. Org. Chem.*, 1986, **51**, 1377. 835

86JOC2004 G. J. M. Vos, P. H. Benders, D. N. Reinhoudt, R. J. M. Egberink, S. Harkema and G. J. van Hummel; *J. Org. Chem.*, 1986, **51**, 2004. 518

86JOC2024 F. H. Gouzoules and R. A. Whitney; *J. Org. Chem.*, 1986, **51**, 2024. 67

86JOC2230 P. A. Bartlett and P. C. Ting; *J. Org. Chem.*, 1986, **51**, 2230. 1015

86JOC2389 I. Ryu, S. Murai and N. Sonoda; *J. Org. Chem.*, 1986, **51**, 2389. 666

86JOC2397 P. G. Gassman and K. Mlinaric-Majerski; *J. Org. Chem.*, 1986, **51**, 2397. 704, 709

86JOC2596 G. A. Molander and G. Hahn; *J. Org. Chem.*, 1986, **51**, 2596. 14

86JOC2607 B. Byrne, L. M. L. Lawter and K. J. Wengenroth; *J. Org. Chem.*, 1986, **51**, 2607. 1026

86JOC2613 S. Kim and K. Y. Yi; *J. Org. Chem.*, 1986, **51**, 2613. 939

86JOC2623 R. C. Larock, M. Chow and S. J. Smith; *J. Org. Chem.*, 1986, **51**, 2623. 957, 974, 1060

86JOC2627 I. Colon and D. R. Kelsey; *J. Org. Chem.*, 1986, **51**, 2627. 472

86JOC2832 N. Ono, I. Hamamoto and A. Kaji; *J. Org. Chem.*, 1986, **51**, 2832. 37

86JOC2859 M. Uemura, T. Kobayashi, K. Isobe, T. Minami and Y. Hayashi; *J. Org. Chem.*, 1986, **51**, 2859. 202, 203

86JOC2874 K. N. Houk, M. N. Paddon-Row, D. C. Spellmeyer, N. G. Rordan and S. Nagase; *J. Org. Chem.*, 1986, **51**, 2874. 323

86JOC2884 C. C. Tseng, S. D. Paisley and H. L. Goering; *J. Org. Chem.*, 1986, **51**, 2884. 187

86JOC2969 M. Bruch, Y. M. Jun, A. E. Luedtke, M. Schneider and J. W. Timberlake; *J. Org. Chem.*, 1986, **51**, 2969. 900

86JOC3023 R. P. Polniaszek and R. V. Stevens; *J. Org. Chem.*, 1986, **51**, 3023. 776

86JOC3038 C. K. Lau, C. Dufrésne, P. C. Bélanger, S. Piétré and J. Scheigetz; *J. Org. Chem.*, 1986, **51**, 3038. 13, 18, 20

86JOC3098 E. G. Baggiolini, J. A. Iacobelli, B. M. Hennessy, A. D. Batcho, J. F. Sereno and M. R. Uskokovic; *J. Org. Chem.*, 1986, **51**, 3098. 576, 1021

86JOC3108 A. I. Meyers, T. Sohda and M. F. Loewe; *J. Org. Chem.*, 1986, **51**, 3108. 149, 594, 716

86JOC3214 R. Friary and V. Seidl; *J. Org. Chem.*, 1986, **51**, 3214. 1082

86JOC3247 R. L. Funk and M. M. Abelman; *J. Org. Chem.*, 1986, **51**, 3247. 831

86JOC3398 H. C. Brown, N. G. Bhat and J. B. Campbell, Jr.; *J. Org. Chem.*, 1986, **51**, 3398. 488

86JOC3502 H. Lee and R. G. Harvey; *J. Org. Chem.*, 1986, **51**, 3502. 4, 7

86JOC3643 B. B. Snider and E. Ron; *J. Org. Chem.*, 1986, **51**, 3643. 498

86JOC3726 D. B. Gerth and B. Giese; *J. Org. Chem.*, 1986, **51**, 3726. 330, 341

86JOC3736 A. Nickon, T. D. Swartz, D. M. Sainsbury and B. R. Toth; *J. Org. Chem.*, 1986, **51**, 3736. 905

86JOC3830 J. Otera, H. Misawa and K. Sugimoto; *J. Org. Chem.*, 1986, **51**, 3830. 613, 700, 1073

86JOC3834 J. Otera, H. Misawa, T. Onishi, S. Suzuki and Y. Fujita; *J. Org. Chem.*, 1986, **51**, 3834. 613

86JOC3870 R. L. Danheiser, D. J. Carini and C. A. Kwasigroch; *J. Org. Chem.*, 1986, **51**, 3870. 1056, 1058

86JOC3896 T. Moriyama, T. Mandai, M. Kawada, J. Otera and B. M. Trost; *J. Org. Chem.*, 1986, **51**, 3896. 613

86JOC3916 D. R. Williams, J. L. Moore and M. Yamada; *J. Org. Chem.*, 1986, **51**, 3916. 697, 700, 703

86JOC4006 E. Keinan and E. Bosch; *J. Org. Chem.*, 1986, **51**, 4006. 975, 1025

86JOC4080 E. Negishi, Y. Zhang, F. E. Cederbaum and M. B. Webb; *J. Org. Chem.*, 1986, **51**, 4080. 536

86JOC4090 M. Koreeda and D. J. Ricca; *J. Org. Chem.*, 1986, **51**, 4090. 816, 819

86JOC4208 C.-H. Chou and W. S. Trahanovsky; *J. Org. Chem.*, 1986, **51**, 4208. 787

86JOC4315 T. Takahashi, H. Nemoto, Y. Kanda, J. Tsuji and Y. Fujise; *J. Org. Chem.*, 1986, **51**, 4315. 816, 818

86JOC4316 J. A. Marshall, T. M. Jenson and B. S. DeHoff; *J. Org. Chem.*, 1986, **51**, 4316. 816, 818

86JOC4375 R. Tamura, Y. Kai, M. Kakihana, K. Hayashi, M. Tsuji, T. Nakamura and D. J. Oda; *J. Org. Chem.*, 1986, **51**, 4375. 250

86JOC4485 J. C. Gilbert and T. A. Kelly; *J. Org. Chem.*, 1986, **51**, 4485. 831

86JOC4507 A. Suzuki, N. Miyaura, S. Abiko, M. Itoh, J. A. Sinclair, M. M. Midland and H. C. Brown; *J. Org. Chem.*, 1986, **51**, 4507. 1033, 1034

86JOC4512 H. C. Brown and G. A. Molander; *J. Org. Chem.*, 1986, **51**, 4512. 449, 510

86JOC4521 J. A. Sikorski, N. G. Bhat, T. E. Cole, K. K. Wang and H. C. Brown; *J. Org. Chem.*, 1986, **51**, 4521. 1034

86JOC4627 P. Beak and K. D. Wilson; *J. Org. Chem.*, 1986, **51**, 4627. 363

86JOC4687 R. K. Dieter and L. A. Silks, III; *J. Org. Chem.*, 1986, **51**, 4687. 485

86JOC4840 A. G. M. Barrett, R. A. E. Carr, S. V. Attwood, G. Richardson and N. D. A. Walshe; *J. Org. Chem.*, 1986, **51**, 4840. 700, 703

86JOC5007 V. Ghodoussi, G. J. Gleicher and M. Kravetz; *J. Org. Chem.*, 1986, **51**, 5007. 325, 920

86JOC5286 K. Karabelas and A. Hallberg; *J. Org. Chem.*, 1986, **51**, 5286. 490

86JOC5400 H. Chikashita, H. Ide and K. Itoh; *J. Org. Chem.*, 1986, **51**, 5400. 10

86JOM(299)119 M. Uemura, T. Minami and Y. Hayashi; *J. Organomet. Chem.*, 1986, **299**, 119. 202

86JOM(300)249 R. R. Schrock; *J. Organomet. Chem.*, 1986, **300**, 249. 674, 675

86JOM(310)135 C. Rücker; *J. Organomet. Chem.*, 1986, **310**, 135. 178

86JOM(312)53 T.-S. Chou, C. H. Tsao and S. C. Hung; *J. Organomet. Chem.*, 1986, **312**, 53. 445

86JOM(312)343 V. Fiandanese, G. Miccoli, F. Naso and G. Ronzini; *J. Organomet. Chem.*, 1986, **312**, 343. 466

86JOM(317)373 G. Salerno, C. Gallo, G. P. Chiusoli and M. Costa; *J. Organomet. Chem.*, 1986, **317**, 373. 496

86JOM(317)383 T. Cuvigny and M. Julia; *J. Organomet. Chem.*, 1986, **317**, 383. 195, 238

86JPC402 I. R. Slagle, E. Ratajczak and D. Gulman; *J. Phys. Chem.*, 1986, 402. 1127

86LA1578 H. Spreitzer, C. Schiffer and G. Buchbauer; *Liebigs Ann. Chem.*, 1986, 1578. 564

86MI 102-01 W. Ding, W. Cai and X. Peng; *Huaxue Xuebao*, 1986, **44**, 261 (*Chem. Abstr.*, **105(21)**, 191 291q). 54

B-86MI 105-01 B. Giese; "Radicals in Organic Synthesis; Formation of Carbon–Carbon Bonds," Pergamon, Oxford, 1986. 172, 230, 243

B-86MI 105-02 C. Paulmier; "Selenium Reagents and Intermediates in Organic Synthesis," Pergamon, Oxford, 1986. 243

B-86MI 106-01 M. Regitz and G. Maas; in "Diazo Compounds: Properties and Synthesis," Academic Press, 1986, chap. 5. 258

B-86MI 108-01 B. Giese; "Radicals in Organic Synthesis: Formation of Carbon–Carbon Bonds," Pergamon, Oxford, 1986. 320, 323, 329, 330, 331, 332, 341, 342

86MI 111-01 A. Suzuki and R. S. Dhillon; "Selective Hydroboration and Synthetic Utility of Organoboranes thus Obtained" in "Topics in Current Chemistry," vol. 130, Springer-Verlag, Berlin, 1986, p. 23. 488

86MI 111-02 R. C. Larock; "Solvomercuration/Demercuration Reactions in Organic Synthesis," Springer-Verlag, Berlin, 1986. 491

B-86MI 112-01 B. Giese; "Radicals in Organic Synthesis: Formation of Carbon–Carbon Bonds," Pergamon, Oxford, 1986. 546

B-86MI 114-01 C. Paulmier; "Selenium Reagents and Intermediates in Organic Synthesis," Pergamon, Oxford, 1986, ch. 5. 601

B-86MI 114-02 N. Sonoda and A. Ogawa; in "The Chemistry of Organic Selenium and Tellurium Compounds," ed. S. Patai, Wiley, Chichester, 1986, vol. 1, p. 619. 601

B-86MI 114-03 F. Fringuelli and A. Taticchi; in "The Chemistry of Organic Selenium and Tellurium Compounds," ed. S. Patai, Wiley, Chichester, 1986, vol. 1, p. 559. 601

B-86MI 114-04 M. Regitz and G. Maas; "Diazo Compounds—Properties and Synthesis," Academic Press, London, 1986, p. 114. 637

B-86MI 115-01 P. M. Chaloner; "Handbook of Coordination Catalysis in Organic Chemistry," Butterworths, London, 1986, chap. 9. 674

B-86MI 115-02 C. Paulmier; "Selenium Reagents and Intermediates in Organic Synthesis," Pergamon, Oxford, 1986. 712

86MI 116-01 J. Tessier; *L'actualité Chim.*, 1986, 5. 751

B-86MI 116-01 M. T. Reetz; "Organotitanium Reagents in Organic Synthesis," Springer-Verlag, Berlin, 1986, p. 223. 767

B-86MI 117-01 C. Paulmier; "Selenium Reagents and Intermediates in Organic Synthesis," Pergamon, Oxford, 1986. 780

B-86MI 119-01 J. Fuhrhop and G. Penzlin; "Organic Synthesis," VCH, Weinheim, 1986, chap. 1, p. 5. 885

B-86MI 119-02 M. Fieser and L. F. Fieser; "Reagents for Organic Synthesis," Wiley, New York, 1986, vol. 12, p. 550. 898

B-86MI 119-03 P. J. Garratt; "Aromaticity," Wiley, New York, 1986, pp. 5–45. 917

B-86MI 119-04 B. Giese; "Radicals in Organic Synthesis, Formation of Carbon–Carbon Bonds," Pergamon, Oxford, 1986. 919, 946, 948

B-86MI 119-05 M. Pereyre, J. P. Quintard and A. Rahm; "Tin in Organic Synthesis," Butterworths, London, 1987. 919

B-86MI 119-06 R. C. Larock; "Solvomercuration/Demercuration Reactions in Organic Synthesis," Springer, Berlin, 1986. 948

B-86MI 122-01 B. Giese; "Radicals in Organic Synthesis: Formation of Carbon–Carbon Bonds," Pergamon, Oxford, 1986. 1126

86NPR35 J. M. Hook and L. N. Mander; *Nat. Prod. Rep.*, 1986, **3**, 35. 87

86OM721 L. R. Gilliom and R. H. Grubbs; *Organometallics*, 1986, **5**, 721. 669, 670

86OM1473 T. M. Miller and G. M. Whitesides; *Organometallics*, 1986, **5**, 1473. 668, 669

86OS(64)1 R. S. Iyer and P. Helquist; *Org. Synth.*, 1986, **64**, 1. 541

86OS(64)182 T. K. Jones and S. E. Denmark; *Org. Synth.*, 1986, **64**, 182. 514

86PAC675 D. H. R. Barton and S. Z. Zard; *Pure Appl. Chem.*, 1986, **58**, 675. 935, 938

86S1 N. Petragnani and J. V. Comasseto; *Synthesis*, 1986, 1. 595, 611, 716

86S184 A. B. Smith, III, P. A. Levenberg and J. Z. Suits; *Synthesis*, 1986, 184. 516

86S320 S. Cacchi, E. Morera and G. Ortar; *Synthesis*, 1986, 320. 1029

86S372 T. Schrader, R. Kober and W. Steglich; *Synthesis*, 1986, 372. 1042

86S386 M. Uemura, T. Kobayashi and Y. Hayashi; *Synthesis*, 1986, 386. 202

86S421 M. Yamaguchi, K. Shibato, S. Fujiwara and I. Hirao; *Synthesis*, 1986, 421. 1045

86S570 A. Ono, E. Fujimoto and M. Ueno; *Synthesis*, 1986, 570. 5, 8

86S638 F. Babudri and S. Florio; *Synthesis*, 1986, 638. 1061

86S689 T. Hiyama and M. Inoue; *Synthesis*, 1986, 689. 200

86S693 N. Ono and A. Kaji; *Synthesis*, 1986, 693. 946

86S776 H. Sano, Y. Mikazaki, M. Okawaza and Y. Veno; *Synthesis*, 1986, 776. 290

86S921 E. Guntrum, W. Kuhn, W. Spönlein and V. Jäger; *Synthesis*, 1986, 921. 797

86S1032 S. Cardani, L. Prati and O. Tinti; *Synthesis*, 1986, 1032. 208

86SC283 J. V. Comasseto, C. C. Silveira, J. T. B. Ferreira and V. Catani; *Synth. Commun.*, 1986, **16**, 283. 999

86SC779 M. R. Euerby and R. D. Waigh; *Synth. Commun.*, 1986, **16**, 779. 436

86SC983 W. Eberbach and J. Roser; *Synth. Commun.*, 1986, **16**, 983. 999

86T295 K. Mori and H. Watanabe; *Tetrahedron*, 1986, **42**, 295. 165

86T1333 N. Bluthe, J. Gore and M. Malacria; *Tetrahedron*, 1986, **42**, 1333. 809

86T1399 J. P. Foulon, M. Bourgain-Commerçon and J. F. Normant; *Tetrahedron*, 1986, **42**, 1399. 491

86T1581 G. E. Renzoni, T. Yin, F. Miyaki and W. T. Borden; *Tetrahedron*, 1986, **42**, 1581. 681

86T1693 A. Krebs, B. Kaletta, W. Nickel, W. Rüger and L. Tikwe; *Tetrahedron*, 1986, **42**, 1693. 686

86T1711 H. Meier, N. Hanold, T. Molz, H. J. Bissinger, H. Kolshorn and J. Zountsas; *Tetrahedron*, 1986, **42**, 1711. 1077

86T1989 W. Ando, Y. Hanyu, Y. Kumamoto and T. Takata; *Tetrahedron*, 1986, **42**, 1989. 983

86T2017 J. Pornet, D. Damour and L. Miginiac; *Tetrahedron*, 1986, **42**, 2017. 978

86T2043 G. Consiglio, O. Piccolo, L. Roncetti and F. Morandini; *Tetrahedron*, 1986, **42**, 2043. 195, 238

86T2111 D. A. Widdowson and Y.-Z. Zhang; *Tetrahedron*, 1986, **42**, 2111. 473

86T2135 J. I. G. Cadogan, C. L. Hickson and H. McNab; *Tetrahedron*, 1986, **42**, 2135. 784

86T2429 A. Abad, C. Aqullo, M. Arno and E. Seoane; *Tetrahedron*, 1986, **42**, 2429. 49

86T2475 M. Julia, H. Lauron, J.-P. Stacino, J.-N. Verpeaux, Y. Jeannin and Y. Dromzee; *Tetrahedron*, 1986, **42**, 2475. 439

86T2501 J. Pornet, D. Damour, B. Randrianoelina and L. Miginiac; *Tetrahedron*, 1986, **42**, 2501. 978

86T2803 T. Cohen and B.-S. Guo; *Tetrahedron*, 1986, **42**, 2803. 236

86T2821 S. E. Denmark, K. L. Habermas, G. A. Hite and T. K. Jones; *Tetrahedron*, 1986, **42**, 2821. 422

86T2831 R. L. Funk, M. M. Ableman and J. D. Munger, Jr.; *Tetrahedron*, 1986, **42**, 2831. 831

86T2873 B. H. Lipshutz, D. A. Parker, S. L. Nguyen, K. E. McCarthy, J. C. Barton, S. E. Whitney and H. Kotsuki; *Tetrahedron*, 1986, **42**, 2873. 120

86T2971 I. Minami, K. Takahashi, I. Shimizu, T. Kimura and J. Tsuji; *Tetrahedron*, 1986, **42**, 2971. 558

86T2979 R. S. Garigipati, R. Cordova, M. Parvez and S. M. Weinreb; *Tetrahedron*, 1986, **42**, 2979. 408

86T3029 R. K. Dieter; *Tetrahedron*, 1986, **42**, 3029. 140, 485

86T3569 E. V. Dehmlow, R. Thieser, Y. Sasson and R. Neumann; *Tetrahedron*, 1986, **42**, 3569. 1065

86T3731 B. M. Dilworth and M. A. McKervey; *Tetrahedron*, 1986, **42**, 3731. 704

86T3887 C. Glidewell, D. Lloyd and S. Metcalfe; *Tetrahedron*, 1986, **42**, 3887. 54

86T4361 J. Tsuji; *Tetrahedron*, 1986, **42**, 4361. 248

86T4611 J. Mann; *Tetrahedron*, 1986, **42**, 4611. 363, 364

86T4807 C. Herve du Penhoat and M. Julia; *Tetrahedron*, 1986, **42**, 4807. 612

86T4975 R. Bloch and D. Hassan-Gonzales; *Tetrahedron*, 1986, **42**, 4975. 790

86T5497	H. C. Brown and K. J. Murray; *Tetrahedron*, 1986, **42**, 5497.	58, 59
86T5523	H. C. Brown, M. M. Midland and G. W. Kabalka; *Tetrahedron*, 1986, **42**, 5523.	946
86T5607	A. Alexakis, D. Jachiet and J. F. Normant; *Tetrahedron*, 1986, **42**, 5067.	479
86TCC1	C. Rüchardt and H.-D. Beckhaus; *Top. Curr. Chem.*, 1986, **130**, 1.	321
86TCC3	B. M. Trost; *Top. Curr. Chem.*, 1986, **133**, 3.	248
86TL87	M. Shinoda, K. Iseki, T. Oguri, Y. Hayasi, S. Yamada and M. Shibasaki; *Tetrahedron Lett.*, 1986, **27**, 87.	1024
86TL195	Y. Tamura, M. Sasho, S. Akai, H. Kishimoto, J. Sekihachi and Y. Kita; *Tetrahedron Lett.*, 1986, **27**, 195.	1140
86TL219	T. Ohnuma, N. Hata, N. Miyachi, T. Wakamatsu and Y. Ban; *Tetrahedron Lett.*, 1986, **27**, 219.	813
86TL267	S. L. Abidi; *Tetrahedron Lett.*, 1986, **27**, 267.	1063
86TL299	P. Mosset, S. Manna, J. Viala and J. R. Falck; *Tetrahedron Lett.*, 1986, **27**, 299.	580
86TL383	R. Freire, J. J. Marrero, M. S. Rodríguez and E. Suárez; *Tetrahedron Lett.*, 1986, **27**, 383.	931
86TL399	G. Pattenden and G. M. Robertson; *Tetrahedron Lett.*, 1986, **27**, 399.	690
86TL539	R. N. Young, E. Champion, J. Y. Gauthier, T. R. Jones, S. Leger and R. Zamboni; *Tetrahedron Lett.*, 1986, **27**, 539.	532, 662
86TL573	J. Cossy and J. P. Pete; *Tetrahedron Lett.*, 1986, **27**, 573.	1000
86TL641	M. D. Bachi and E. Bosch; *Tetrahedron Lett.*, 1986, **27**, 641.	548, 1131
86TL645	J. Elliott and S. Warren; *Tetrahedron Lett.*, 1986, **27**, 645.	756
86TL731	J. Tsuji, T. Sugiura and I. Minami; *Tetrahedron Lett.*, 1986, **27**, 731.	975
86TL933	T. Hirao, D. Misu and T. Agawa; *Tetrahedron Lett.*, 1986, **27**, 933.	1038
86TL943	S. H. Ahn, D. Kim, M. W. Chun and W.-K. Chung; *Tetrahedron Lett.*, 1986, **27**, 943.	183
86TL967	M. Uemura, T. Minami, K. Isobe, T. Kobayashi and Y. Hayashi; *Tetrahedron Lett.*, 1986, **27**, 967.	202
86TL1039	P. Knochel and J. F. Normant; *Tetrahedron Lett.*, 1986, **27**, 1039.	679, 836
86TL1043	P. Knochel and J. F. Normant; *Tetrahedron Lett.*, 1986, **27**, 1043.	679
86TL1047	A. Alexakis, J. Berlan and Y. Besace; *Tetrahedron Lett.*, 1986, **27**, 1047.	313
86TL1171	Q. Chen and Z. Yang; *Tetrahedron Lett.*, 1986, **27**, 1171.	483, 494, 1029
86TL1175	N. Ikeda, K. Omori and H. Yamamoto; *Tetrahedron Lett.*, 1986, **27**, 1175.	1055
86TL1195	T. Tabuchi, J. Inanaga and M. Yamaguchi; *Tetrahedron Lett.*, 1986, **27**, 1195.	195
86TL1209	N. S. Isaaca and O. H. Abed; *Tetrahedron Lett.*, 1986, **27**, 1209.	51
86TL1241	W. A. Smit, A. S. Gybin, A. S. Shashkov, Y. T. Strychkov, L. G. Kyz'mina, G. S. Mikaelian, R. Caple and E. D. Swanson; *Tetrahedron Lett.*, 1986, **27**, 1241.	530
86TL1315	H. Hauptmann and G. Mühlbauer; *Tetrahedron Lett.*, 1986, **27**, 1315.	813
86TL1343	T. Kitihara, M. Mori, K. Koseki and K. Mori; *Tetrahedron Lett.*, 1986, **27**, 1343.	591
86TL1355	H. Urabe and I. Kuwajima; *Tetrahedron Lett.*, 1986, **27**, 1355.	1127
86TL1411	H. Hiemstra, W. J. Klaver and W. N. Speckamp; *Tetrahedron Lett.*, 1986, **27**, 1411.	978
86TL1445	B. M. Trost, J. M. Balkovec and S. R. Angle; *Tetrahedron Lett.*, 1986, **27**, 1445.	287, 789
86TL1449	W. L. Meyer, M. J. Brannon, A. Merritt and D. Seebach; *Tetrahedron Lett.*, 1986, **27**, 1449.	578
86TL1469	G. A. Epling and E. Florio; *Tetrahedron Lett.*, 1986, 1469.	890
86TL1523	L. Castedo, A. Mouriño and L. A. Sarandeses; *Tetrahedron Lett.*, 1986, **27**, 1523.	1029
86TL1553	K. B. Wiberg, S. T. Waddell and K. Laidig; *Tetrahedron Lett.*, 1986, **27**, 1553.	928
86TL1557	S. Raucher and L. M. Gustavson; *Tetrahedron Lett.*, 1986, **27**, 1557.	816
86TL1723	M. Clarembeau and A. Krief; *Tetrahedron Lett.*, 1986, **27**, 1723.	245
86TL1895	L. A. Paquette and M. P. Trova; *Tetrahedron Lett.*, 1986, **27**, 1895.	708
86TL2015	I. Uchida, Y. Itoh, T, Namiki, M. Nishikawa and M. Hashimoto; *Tetrahedron Lett.*, 1986, **27**, 2015.	217
86TL2095	A. Spaltenstein, P. A. Carpino, F. Miyake and P. B. Hopkins; *Tetrahedron Lett.*, 1986, **27**, 2095.	697, 698, 703
86TL2187	T. G. Back, J. R. Proudfoot and C. Djerassi; *Tetrahedron Lett.*, 1986, **27**, 2187.	439, 545
86TL2243	J. K. Crandall and M. Mualla; *Tetrahedron Lett.*, 1986, **27**, 2243.	248
86TL2265	D. Levin and S. Warren; *Tetrahedron Lett.*, 1986, **27**, 2265.	643
86TL2303	V. F. Patel, G. Pattenden and J. J. Russell; *Tetrahedron Lett.*, 1986, **27**, 2303.	346
86TL2409	I. K. Youn, G. H. Yon and C. S. Pak; *Tetrahedron Lett.*, 1986, **27**, 2409.	86
86TL2479	M. Uemura, T. Kobayashi, T. Minami and Y. Hayashi; *Tetrahedron Lett.*, 1986, **27**, 2479.	202
86TL2529	R. Rossi and A. Carpita; *Tetrahedron Lett.*; 1986, 2529.	467
86TL2829	E. Negishi, F. E. Cederbaum and T. Takahashi; *Tetrahedron Lett.*, 1986, **27**, 2829.	536
86TL2833	K. A. Parker, D. M. Spero and K. C. Inman; *Tetrahedron Lett.*, 1986, **27**, 2833.	339, 1130
86TL2881	A. Hosomi, K. Otaka and H. Sakurai; *Tetrahedron Lett.*, 1986, 2881.	468
86TL2903	Y. Huang, Y. Shen and C. Chen; *Tetrahedron Lett.*, 1986, **27**, 2903.	648
86TL2949	S. Hanessian, P. J. Hodges, S. P. Sahoo and P. J. Roy; *Tetrahedron Lett.*, 1986, **27**, 2949.	608
86TL3025	H. Miyake and K. Yamamura; *Tetrahedron Lett.*, 1986, **27**, 3025.	252
86TL3053	G. W. J. Fleet, K. James and R. J. Lunn; *Tetrahedron Lett.*, 1986, **27**, 3053.	828
86TL3337	A. E. Asato and R. S. H. Liu; *Tetrahedron Lett.*, 1986, **27**, 3337.	613
86TL3403	S. Ikegami, T. Hayama, T. Katsuki and M. Yamaguchi; *Tetrahedron Lett.*, 1986, **27**, 3403.	180
86TL3511	R. Bloch and L. Gilbert; *Tetrahedron Lett.*, 1986, **27**, 3511.	786
86TL3733	J. Vidal and F. Huet; *Tetrahedron Lett.*, 1986, **27**, 3733.	613
86TL3745	N. Miyaura, M. Satoh and A. Suzuki; *Tetrahedron Lett.*, 1986, 3745.	489
86TL3777	T. Rajamannar and K. K. Balasubramanian; *Tetrahedron Lett.*, 1986, **27**, 3777.	961
86TL3891	T. Tabuchi, J. Inanaga and M. Yamaguchi; *Tetrahedron Lett.*, 1986, **27**, 3891.	680
86TL4029	E. Nakamura, S. Matsuzawa, Y. Horiguchi and I. Kuwajima; *Tetrahedron Lett.*, 1986, **27**, 4029.	313

86TL4039 M. Aono, C. Hyodo, Y. Terao and K. Achiwa; *Tetrahedron Lett.*, 1986, **27**, 4039. 367
86TL4107 M. A. Forman and P. A. Leber; *Tetrahedron Lett.* 1986, **27**, 4107. 799
86TL4185 K. Mikami, T. Kasuga, K. Fujimoto and T. Nakai; *Tetrahedron Lett.*, 1986, **27**, 4185. 816, 819
86TL4189 K. Mikami, T. Maeda and T. Nakai; *Tetrahedron Lett.*, 1986, **27**, 4189. 817
86TL4241 B. H. Lipshutz, B. Huff and W. Vaccaro; *Tetrahedron Lett.*, 1986, **27**, 4241. 1042
86TL4343 L. Dai, B. Lou, Y. Zhang and G. Guo; *Tetrahedron Lett.*, 1986, **27**, 4343. 14
86TL4351 R. Rossi and A. Carpita; *Tetrahedron Lett.*, 1986, **27**, 4351. 1007
86TL4411 P. von R. Schleyer, G. W. Spitznagel and J. Chandrasekhar; *Tetrahedron Lett.*, 1986, **27**, 4411. 846
86TL4427 P. Knochel and J. F. Normant; *Tetrahedron Lett.*, 1986, **27**, 4427. 248
86TL4431 P. Knochel and J. F. Normant; *Tetrahedron Lett.*, 1986, **27**, 4431. 248
86TL4511 K. Mikami, O. Takahashi, T. Tabei and T. Nakai; *Tetrahedron Lett.*, 1986, **27**, 4511. 404, 406, 816, 820
86TL4525 A. L. J. Beckwith and D. M. O'Shea; *Tetrahedron Lett.*, 1986, **27**, 4525. 1127, 1129
86TL4529 G. Stork and R. Mook, Jr.; *Tetrahedron Lett.*, 1986, **27**, 4529. 338, 1127, 1129
86TL4573 J. P. Genêt, D. Ferroud, S. Juge and J. Ruiz-Montes; *Tetrahedron Lett.*, 1986, **27**, 4573. 194
86TL4577 M. Uchikawa, T. Hanamoto, T. Katsuki and M. Yamaguchi; *Tetrahedron Lett.*, 1986, **27**, 4577. 816
86TL4581 M. Uchikawa, T. Katsuki and M. Yamaguchi; *Tetrahedron Lett.*, 1986, **27**, 4581. 404, 406, 816, 819
86TL4583 Y. Wang, J. Li, Y. Wu, Y. Huang, L. Shi and J. Yang; *Tetrahedron Lett.*, 1986, **27**, 4583. 647
86TL4599 M. Enomoto, T. Katsuki and M. Yamaguchi; *Tetrahedron Lett.*, 1986, **27**, 4599. 954
86TL4679 J. D. Price and R. P. Johnson; *Tetrahedron Lett.*, 1986, **27**, 4679. 990
86TL4813 P. A. Grieco and R. P. Nargund; *Tetrahedron Lett.*, 1986, **27**, 4813. 17
86TL4845 J. A. Marshall and S. D. Rothenberger; *Tetrahedron Lett.*, 1986, **27**, 4845. 967
86TL4853 D. Fletcher, F. J. Ablenas, A. C. Hopkinson and E. Lee-Ruff; *Tetrahedron Lett.*, 1986, **27**, 4853. 106
86TL4899 K. Mikami, K. Kawamoto and T. Nakai; *Tetrahedron Lett.*, 1986, **27**, 4899. 816, 819
86TL4913 H. H. Wasserman and T. Oku; *Tetrahedron Lett.*, 1986, **27**, 4913. 1021
86TL4945 R. J. Giguere, T. L. Bray, S. M. Duncan and G. Majetich; *Tetrahedron Lett.*, 1986, **27**, 4945. 525, 826
86TL5237 T. Tabuchi, J. Inanaga and M. Yamaguchi; *Tetrahedron Lett.*, 1986, **27**, 5237. 962
86TL5445 J. M. Chong and S. Wong; *Tetrahedron Lett.*, 1986, **27**, 5445. 1000
86TL5455 J.-F. Lavallée, G. Berthiaume, P. Deslongchamps and F. Grein; *Tetrahedron Lett.*, 1986, **27**, 5455. 538
86TL5541 S. Cacchi, P. G. Ciattini, E. Morera and G. Ortar; *Tetrahedron Lett.*, 1986, **27**, 5541. 433
86TL5791 R. E. Babine; *Tetrahedron Lett.*, 1986, **27**, 5791. 206
86TL5821 D. P. Curran and D. Kim; *Tetrahedron Lett.*, 1986, **27**, 5821. 344
86TL5833 D. Schomburg, M. Thielmann and E. Winterfeldt; *Tetrahedron Lett.*, 1986, **27**, 5833. 519
86TL5857 D. Guillerm and G. Linstrumelle; *Tetrahedron Lett.*, 1986, **27**, 5857. 1009
86TL5893 Z. Čeković and R. Saičić; *Tetrahedron Lett.*, 1986, **27**, 5893. 348
86TL5931 J. E. Bartmess, J. Kester, W. T. Borden and H. G. Köser; *Tetrahedron Lett.*, 1986, 5931. 865
86TL6133 A. B. Holmes and R. W. Carling; *Tetrahedron Lett.* 1986, **27**, 6133. 1021
86TL6161 S. Sharma and A. C. Oehlschlager; *Tetrahedron Lett.*, 1986, **27**, 6161. 1106
86TL6267 J. L. van der Baan and F. Bickelhaupt; *Tetrahedron Lett.*, 1986, **27**, 6267. 415, 826
86TL6369 N. Miyaura, T. Ishiyama, M. Ishikawa and A. Suzuki; *Tetrahedron Lett.*, 1986, **27**, 6369. 279
86UKZ1090 M. Y. Belik, Y. B. Kozlikovskii and V. A. Koshchii; *Ukr. Khim. Zh.*, (*Russ. Ed.*) 1986, **52**, 1090. 492

87ACR65 E. Negishi; *Acc. Chem. Res.*, 1987, **20**, 65. 207
87ACR72 G. H. Posner; *Acc. Chem. Res.*, 1987, **20**, 72. 305
87ACR207 K. M. Nicholas; *Acc. Chem. Res.*, 1987, **20**, 207. 198
87ACR371 N. L. Bauld, D. J. Bellville, B. Harirchian, K. T. Lorenz, R. A. Pabon, Jr., D. W. Reynolds, D. D. Wirth, H.-S. Chiou and B. K. Marsh; *Acc. Chem. Res.* 1987, **20**, 371. 248
87ACR389 S. J. Tauster; *Acc. Chem. Res.*, 1987, **20**, 389. 88
87AG(E)142 F. Effenberger and T. Weber; *Angew. Chem., Int. Ed. Engl.*, 1987, **26**, 142. 182, 475
87AG(E)204 K. Müllen; *Angew. Chem., Int. Ed. Engl.*, 1987, **26**, 204. 101
87AG(E)335 T. Toda, N. Shimazaki and T. Mukai; *Angew. Chem., Int. Ed. Engl.*, 1987, **26**, 335. 982
87AG(E)357 R. Aumann and J. Uphoff; *Angew. Chem., Int. Ed. Engl.*, 1987, **26**, 357. 500
87AG(E)560 T. Laube; *Angew. Chem., Int. Ed. Engl.*, 1987, **26**, 560. 906
87AG(E)683 U. Schöllkopf, S. Grüttner, R. Anderskewitz, E. Egert and M. Dyrbusch; *Angew. Chem., Int. Ed. Engl.*, 1987, **26**, 683. 463
87AG(E)702 S. Warwel, H. Kätker and C. Rauenbusch; *Angew. Chem., Int. Ed. Engl.*, 1987, **26**, 702. 674
87AG(E)770 H.-D. Beckhaus and C. Rüchardt; *Angew. Chem., Int. Ed. Engl.*, 1987, **26**, 770. 922
87AG(E)928 E. Vogel, M. Balci, K. Pramod, P. Koch, J. Lex and O. Ermer; *Angew. Chem. Int. Ed. Engl.*, 1987, **26**, 928. 248
87AG(E)1035 R. Boese, H.-J. Knölker and K. P. C. Vollhardt; *Angew. Chem., Int. Ed. Engl.*, 1987, **26**, 1035. 448
87AG(E)1167 R. Schwesinger and H. Schlemper; *Angew., Chem., Int. Ed. Engl.*, 1987, **26**, 1167. 574
87AG(E)1187 T. Hiiro, N. Kambe, A. Ogawa, N. Miyoshi, S. Murai and N. Sonoda; *Angew. Chem., Int. Ed. Engl.*, 1987, **26**, 1187. 245, 437
87AG(E)1212 H. Günther, D. Moskau, P. Bast, D. Schmalz; *Angew. Chem., Int. Ed. Engl.*, 1987, **26**, 1212–1220. 876
87AHC(42)203 C. J. Moody, *Adv. Heterocycl. Chem.*, 1987, **42**, 203. 825

87AJC1537 B. T. Cooney and D. A. R. Happer; *Aust. J. Chem.*, 1987, **40**, 1537. 892

87AJC1675 N. R. Browne, R. F. C. Brown, F. W. Eastwood and G. D. Fallon; *Aust. J. Chem.*, 1987, **40**, 1675. 988

87AOC(27)169 C. Schade and P. v. R. Schleyer; *Adv. Organomet. Chem.*, 1987, **27**, 169. 110, 885

87BAU327 L. I. Zakharkin and E. V. Anikina; *Bull. Acad. Sci. USSR, Div. Chim. Sci.*, 1987, **36**, 327. 462

87BCJ1525 Y. Masuyama, H. Hirai, Y. Kurusu and K. Segawa; *Bull. Chem. Soc. Jpn.*, 1987, **60**, 1525. 196

87BCJ2917 S. Tsuboi, H. Wada, K. Muranaka and A. Takeda; *Bull. Soc. Chem. Jpn.*, 1987, **60**, 2917. 568

87BCJ3307 T. Umemoto and Y. Gotoh; *Bull. Chem. Soc. Jpn.*, 1987, **60**, 3307. 108

87BCJ3721 S. Takekuma, Y. Matsubara, H. Yamamoto and T. Nozoe; *Bull. Chem. Soc. Jpn.*, 1987, **60**, 3721. 899

87BCJ3823 T. Umemoto and Y. Gotoh; *Bull. Chem. Soc. Jpn.*, 1987, **60**, 3823. 109

87BCJ4027 H. Higuchi, K. Tani, T. Otsubo, Y. Sakata and S. Misumi; *Bull.. Chem. Soc. Jpn.*, 1987, **60**, 4027. 243

87BCJ4468 Y. Hoshino and N. Takeno; *Bull. Soc. Chem. Jpn.*, 1987, **60**, 4468 554

87BSF93 G. Courtois, D. Mesnard, B. Dugue and L. Miginiac; *Bull. Soc. Chim. Fr.*, 1987, 93. 1059

87BSF462 M. Vincens, R. Fadal and M. Vidal; *Bull. Soc. Chim. Fr.*, 1987, 462. 84

87BSF487 M. Julia, M. Nel and D. Uguen; *Bull. Soc. Chem. Fr.*, 1987, 487. 483

87BSF1006 M.-A. Boaventura, J. Drouin, F. Theobald and N. Rodier; *Bull. Soc. Chim. Fr.*, 1987, 1006. 526

87BSF1015 M.-A. Boaventura and J. Drouin; *Bull. Soc. Chim. Fr.*, 1987, 1015. 526

87CB5 K. Weinges, W. Sipos, J. Klein, J. Deuter and H. Irngartinger; *Chem. Ber.*, 1987, **120**, 5. 708

87CB79 H. Schwager and G. Wilke; *Chem. Ber.*, 1987, **120**, 79. 248

87CB131 R. Appel, J. Kochta and F. Knoch; *Chem. Ber.*, 1987, **120**, 131. 837

87CB187 E. Honegger, E. Heilbronner, N. Hess and H. Martin; *Chem. Ber.*, 1987, **120**, 187. 1064

87CB531 W. Adam, O. de Lucchi, L. Pasquato and B. Will; *Chem. Ber.*, 1987, **120**, 531. 520

87CB1049 H. Quast and G. Meichsner; *Chem. Ber.*, 1987, **120**, 1049. 63

87CB1197 H.-G. Korth, R. Sustmann, J. Dupuis and B. Giese; *Chem. Ber.*, 1987, **120**, 1197. 946

87CB1403 H. Schmidbaur, T. Pollok, G. Reber and G. Müller; *Chem. Ber.*, 1987, **120**, 1403. 51, 446

87CB1695 A. Maercker and R. Stötzel; *Chem. Ber.*, 1987, **120**, 1695. 229

87CB1845 D. Seebach and N. Krause; *Chem. Ber.*, 1987, **120**, 1845. 1038, 1039

87CB1933 P. Binger, E. Sternberg and U. Wittig; *Chem. Ber.*, 1987, **120**, 1933. 363

87CC106 T. Willson, P. Kocienski, A. Faller and S. Campbell; *J. Chem. Soc., Chem. Commun.*, 1987, 106. 654, 713, 716

87CC166 J. E. Baldwin and C.-S. Li; *J. Chem. Soc., Chem. Commun.*, 1987, 166. 345

87CC207 G. M. P. Giblin and N. S. Simpkins; *J. Chem. Soc., Chem. Comm.*, 1987, 207. 482

87CC297 A. Pelter, D. Buss and E. Colclough; *J. Chem. Soc., Chem. Commun.*, 1987, 297. 660, 765

87CC358 H. Iio, M. Monden, K. Okada and T. Tokoroyama; *J. Chem. Soc., Chem. Commun.*, 1987, 358. 180

87CC381 I. Agranat, M. R. Suissa, S. Cohen, R. Isaksson, J. Sandstrom, J. Dale and D. Grace; *J. Chem. Soc., Chem. Commun.*, 1987, 381. 248

87CC388 K. Otsuka and T. Komatsu; *J. Chem. Soc., Chem. Commun.*, 1987, 388. 674

87CC449 K. Fuji, M. Node, Y. Usami and Y. Kiryu; *J. Chem. Soc., Chem. Commun.*, 1987, 449. 241

87CC477 E. Negishi, K. Akiyoshi and T. Takahashi; *J. Chem. Soc., Chem. Commun.*, 1987, 477. 1006, 1035

87CC490 M. Numazawa, K. Hoshi and K. Kimura; *J. Chem. Soc., Chem. Commun.*, 1987, 490. 578

87CC711 R. M. Moriarty, R. K. Vaid and P. Farid; *J. Chem. Soc., Chem. Commun.*, 1987, 711. 1076

87CC786 A. J. Bloodworth, D. Crich and T. Melvin; *J. Chem. Soc., Chem. Commun.*, 1987, 786. 1129

87CC833 E. W. Della, P. E. Pigou and J. Tsanaktsidis; *J. Chem. Soc., Chem. Commun.*, 1987, 833. 900

87CC849 G. A. Luinstra and J. H. Teuben; *J. Chem. Soc., Chem. Commun.*, 1987, 849. 670

87CC871 V. F. Patel and G. Pattenden; *J. Chem. Soc., Chem. Commun.*, 1987, 871. 670

87CC880 S. V. Mortlock, N. A. Stacey and E. J. Thomas; *J. Chem. Soc., Chem. Commun.*, 1987, 880. 602

87CC934 H.-H. Tso, T. Chou and W.-C. Lee; *J. Chem. Soc., Chem. Commun.*, 1987, 934. 790

87CC981 L. L. Yeung, Y. C. Yip and T.-Y. Luh; *J. Chem. Soc., Chem. Commun.*, 1987, 981. 991

87CC1008 R. Blattner and R. J. Ferrier; *J. Chem. Soc., Chem. Commun.*, 1987, 1008. 769

87CC1196 H. Magari, H. Hirota and T. Takahashi; *J. Chem. Soc., Chem. Commun.*, 1987, 1196. 813

87CC1214 A. Krief and P. Barbeaux; *J. Chem. Soc., Chem. Commun.*, 1987, 1214. 246

87CC1220 J. Cooper, D. W. Knight and P. T. Gallagher; *J. Chem. Soc., Chem. Commun.*, 1987, 1220. 831

87CC1238 A. Johns, J. A. Murphy, C. W. Patterson and N. F. Wooster; *J. Chem. Soc., Chem. Commun.*, 1987, 1238. 929

87CC1342 E. Piers and J. S. M. Wai; *J. Chem. Soc., Chem. Commun.*, 1987, 1342. 615

87CC1540 R. V. Bonnert and P. R. Jenkins; *J. Chem. Soc., Chem. Commun.*, 1987, 1540. 593, 713

87CC1572 Y. Yamamoto, S. Nishii and T. Ibuka; *J. Chem. Soc., Chem. Commun.*, 1987, 1572. 305

87CC1755 W. J. Scott; *J. Chem. Soc., Chem. Commun.*, 1987, 1755. 471

87CJC104 G. Just and G. Cacripante; *Can. J. Chem.*, 1987, **65**, 104. 1131

87CJC1821 L. D. Corey, S. M. Singh and A. C. Oehlschlager; *Can. J. Chem.*, 1987, **65**, 1821. 667

87CL5 S. Kanemoto, S. Matsubara, K. Oshima, K. Utimoto and H. Nozaki; *Chem. Lett.*, 1987, 5. 491

87CL19 D. J. Cram, J. A. Bryant and K. M. Doxsee; *Chem. Lett.*, 1987, 19. 473

87CL25 T. Ishiyama, N. Miyaura and A. Suzuki; *Chem. Lett.*, 1987, 25. 279

87CL69 O. Takahashi, K. Mikami and T. Nakai; *Chem. Lett.*, 1987, 69. 816, 819

87CL133 M. Ochiai, S. Iwaki, T. Ukita and Y. Nagao; *Chem. Lett.*, 1987, 133. 931

87CL495 S. Kiyooka, F. Shiota and S. Shibuya; *Chem. Lett.*, 1987, 495. 290

87CL809 K. Tanaka, K. Minami and A. Kaji; *Chem. Lett.*, 1987, 809. 240

87CL963 Y. Naruta and K. Maruyama; *Chem. Lett.*, 1987, 963. 226

87CL1163 G. Yamamoto and M. Oki; *Chem. Lett.*, 1987, 1163. 258, 259

87CL1167 T. Mukaiyama, T. Kato and M. Murakami; *Chem. Lett*, 1987, 1167. 191

87CL1237 M. Kosugi, K. Ohashi, K. Akuzawa, T. Kawazoe, H. Sano and T. Migita; *Chem. Lett.*, 1987, 1237. 194
87CL1345 T. Takeda, H. Oshima, M. Inoue, A. Togo and T. Fujiwara. *Chem. Lett.*, 1987, 1345. 1053
87CL1519 R. Yamaguchi, M. Moriyasu, I. Takase, M. Kawanisi and S. Kozima; *Chem. Lett.*, 1987, 1519. 1060
87CL1569 S. Takano, S.-I. Tomita, M. Takahashi and K. Ogasawara; *Chem. Lett.*, 1987, 1569. 234, 822
87CL1691 Y. Hashimoto, T. Mukaiyama, H. Sugumi and T. Okauchi; *Chem. Lett.*, 1987, 1691. 239
87CL1695 T. Okauchi, Y. Hashimoto, T. Mukaiyama and H. Sugumi; *Chem. Lett.*, 1987, 1695. 239
87CL1775 S. Kiyooka and T. Yamashita; *Chem. Lett.*, 1987, 1775. 486
87CL1905 S. Takano, Y. Sekiguchi, Y. Shimazaki, M. Takahashi and K. Ogasawara; *Chem. Lett.*, 1987, 1905. 1084
87CL1975 T. Mukaiyama, M. Hayashi and A. Inubushi; *Chem. Lett.*, 1987, 1975. 1026
87CL2101 M. Matsukawa, T. Tabuchi, J. Inanaga and M. Yamaguchi; *Chem. Lett.*, 1987, 2101. 680
87CL2203 K. Fugami, K. Oshima and K. Utimoto; *Chem. Lett.*, 1987, 2203. 480
87CL2275 T. Tabuchi, J. Inanaga and M. Yamaguchi; *Chem. Lett.*, 1987, 2275. 975
87CL2347 K. Mikami, K. Takahashi and T. Nakai; *Chem. Lett.*, 1987, 2347. 525
87CPB1405 Y. Tamura, M. Sasho, S. Akai, H. Kishimoto, J. Sekihachi and Y. Kita; *Chem. Pharm. Bull.*, 1987, **35**, 1405. 1140
87CS515 L. Engman; *Chem. Scr.*, 1987, **27**, 515. 610
87CS573 Y. Kishi; *Chem. Scr.*, 1987, **27**, 573. 1099
87CSR45 D. Lloyd, I. Gosney and R. A. Ormiston; *Chem. Soc. Rev.*, 1987, **16**, 45. 646, 647, 648
87CSR239 J. P. Lorimer and T. J. Mason; *Chem. Soc. Rev.*, 1987, **16**, 239. 925
87G625 L. Bonsignore, G. Loy, D. Secci and S. Cabiddu; *Gazz. Chim. Ital.*, 1987, **117**, 625. 988
87H(25)75 C. H. Heathcock and T. W. VonGeldern; *Heterocycles*, 1987, **25**, 75. 248
87H(25)139 T. Takahashi, H. Nemoto, Y. Kanda and J. Tsuji; *Heterocycles*, 1987, **25**, 139. 818
87H(26)355 Y. Isobe, N. Hirota, Y. Kitade and Y. Maki; *Heterocycles*, 1987, **26**, 355. 1029
87H(26)1291 R. D. Clark, J. Berger, C. H. Lee and J. M. Muchowski; *Heterocycles*, 1987, **26**, 1291. 177
87HCA448 D. Seebach, R. Imwinkelried and G. Stucky; *Helv. Chim. Acta.*, 1987, **70**, 448. 1026
87HCA1326 A. Griesbeck and D. Seebach; *Helv. Chim. Acta*, 1987, **70**, 1326. 230
87HCA1477 W. Oppolzer and J.-M. Gaudin; *Helv. Chim. Acta*, 1987, **70**, 1477. 195
87HCA1858 R. L. Snowden, S. M. Linder, B. L. Muller and K. H. Schulte-Elte; *Helv. Chim. Acta*, 1987, **70**, 1858. 883
87HCA1879 R. L. Snowden, S. M. Linder, B. L. Muller and K. H. Schulte-Elte; *Helv. Chim. Acta*, 1987, **70**, 1879. 883
87HCA2201 W. Oppolzer, G. Poli, A. J. Kingma, C. Starkemann and G. Bernardinelli; *Helv. Chim. Acta*, 1987, **70**, 2201. 305
87IJC(B)77 M. N. I. Khandker and A. Ahmad; *Ind. J. Chem., Sect. B*, 1987, **26B(1)**, 77. 999
87IZV695 A. S. Medvedeva, G. G. Chichkareva, M. M. Demina and N. S. Vyazankin. *Izv. Akad. Nauk SSSR, Ser. Khim.*, 1987, 3, 695 (*Chem. Abstr.* 1988, **108**, 37 283u). 986
87JA553 O. L. Chapman, U.-P. E. Tsou and J. W. Johnson; *J. Am. Chem. Soc.*, 1987, **109**, 553. 776
87JA615 B. M. Trost and P. R. Seoane; *J. Am. Chem. Soc.*, 1987, **109**, 615. 288
87JA782 M. Kaftory, I. Agmon, M. Ladika and P. J. Stang; *J. Am. Chem. Soc.*, 1987, **109**, 782. 988
87JA813 J. K. Stille and B. L. Groh; *J. Am. Chem. Soc.*, 1987, **109**, 813. 468
87JA899 J. Kress, J. A. Osborn, R. M. E. Greene, K. J. Ivin and J. J. Rooney; *J. Am. Chem. Soc.*, 1987, **109**, 899. 674
87JA951 T. Okazoe, K. Takai and K. Utimoto; *J. Am. Chem. Soc.*, 1987, **109**, 951. 676
87JA1160 R. M. Coates, B. D. Rogers, S. J. Hobbs, D. R. Peck and D. P. Curran; *J. Am. Chem. Soc.*, 1987, **109**, 1160. 826
87JA1170 J. J. Gajewski, J. Jurayj, D. R. Kimbrough, M. E. Gande, B. Ganem and B. K. Carpenter; *J. Am. Chem. Soc.*, 1987, **109**, 1170. 826
87JA1257 A. Minato, K. Suzuki and K. Tamao; *J. Am. Chem. Soc.*, 1987, **109**, 1257. 467
87JA1596 H. Takaya, T. Ohta, N. Sayo, H. Kumobayashi, S. Akutagawa, S. Inoue, I. Kasahara and R. Noyori; *J. Am. Chem. Soc.*, 1987, **109**, 1596. 77
87JA1746 C. R. Landis and J. Halpern; *J. Am. Chem. Soc.*, 1987, **109**, 1746. 77
87JA2138 J. K. Stille and J. H. Simpson; *J. Am. Chem. Soc.*, 1987, **109**, 2138. 468, 1006, 1007
87JA2176 B. M. Trost, G. B. Tometzki and M.-H. Hung; *J. Am. Chem. Soc.*, 1987, **109**, 2176. 196
87JA2195 A. E. Dorigo and K. N. Houk; *J. Am. Chem. Soc.*, 1987, **109**, 2195. 927
87JA2224 S. E. Denmark, T. Weber and D. W. Piotrowski; *J. Am. Chem. Soc.*, 1987, **109**, 2224. 1042
87JA2393 E. Negishi, T. Takahashi, S. Baba, D. E. Van Horn and N. Okukado; *J. Am. Chem. Soc.*, 1987, **109**, 2393. 468, 533, 536
87JA2442 W. F. Bailey, T. T. Nurmi, J. J. Patricia and W. Wang; *J. Am. Chem. Soc.*, 1987, **109**, 2442. 316
87JA2547 K. Nozaki, K. Oshima and K. Utimoto; *J. Am. Chem. Soc.*, 1987, **109**, 2547. 1130, 1132
87JA2717 D. L. Boger and R. S. Coleman; *J. Am. Chem. Soc.*, 1987, **109**, 2717. 523
87JA2750 M. Schmittel and C. Rüchardt; *J. Am. Chem. Soc.*, 1987, **109**, 2750. 924
87JA2829 G. Stork and R. Mook, Jr.; *J. Am. Chem. Soc.*, 1987, **109**, 2829. 548, 1127, 1131, 1132
87JA2857 J. Dressel and L. A. Paquette; *J. Am. Chem. Soc.*, 1987, **109**, 2857. 704, 708
87JA3017 T. Sugimura and L. A. Paquette; *J. Am. Chem. Soc.*, 1987, **109**, 3017. 816
87JA3025 L. A. Paquette and W. H. Ham; *J. Am. Chem. Soc.*, 1987, **109**, 3025. 25, 829
87JA3161 B. M. Trost and F. Rise; *J. Am. Chem. Soc.*, 1987, **109**, 3161. 546
87JA3163 M. Newcomb and T. M. Deeb; *J. Am. Chem. Soc.*, 1987, **109**, 3163. 950
87JA3174 L. Waykole and L. A. Paquette; *J. Am. Chem. Soc.*, 1987, **109**. 3174. 634
87JA3483 B. M. Trost and D. T. MacPherson; *J. Am. Chem. Soc.*, 1987, **109**, 3483. 194, 288, 546

87JA3486	B. M. Trost, C. Chan and G. Ruhter; *J. Am. Chem. Soc.*, 1987, **109**, 3486.	546, 1036
87JA3491	A. L. Castelhano and A. Krantz; *J. Am. Chem. Soc.*, 1987, **109**, 3491.	958
87JA3730	L. A. Paquette, T.-Z. Wang and C. E. Cottrell; *J. Am. Chem. Soc.*, 1987, **109**, 3730.	613
87JA4129	H. Takaya, T. Ohta, N. Sayo, H. Kumobayashi, S. Akutagawa, S. Inoue, I. Kasahara and R. Noyori; *J. Am. Chem. Soc.*, 1987, **109**, 4129.	77
87JA4237	R. Glaser; *J. Am. Chem. Soc.*, 1987, **109**, 4237.	1141
87JA4338	J. K. Crandall, G. E. Salazar and R. J. Watkins; *J. Am. Chem. Soc.*, 1987, **109**, 4338.	983
87JA4424	J. D. White and T. C. Somers; *J. Am. Chem. Soc.*, 1987, **109**, 4424.	13
87JA4626	W. D. Fessner, G. Sedelmeier, P. R. Spurr, G. Riks and H. Prinzbach; *J. Am. Chem. Soc.*, 1987, **109**, 4626.	564
87JA4649	L. D. Arnold, J. C. G. Drover and J. C. Vederas; *J. Am. Chem. Soc.*, 1987, **109**, 4649.	230
87JA4710	B.-S. Guo, W. Doubleday and T. Cohen; *J. Am. Chem. Soc.*, 1987, **109**, 4710.	236
87JA4755	D. K. Hutchinson and P. L. Fuchs; *J. Am. Chem. Soc.*, 1987, **109**, 4755.	748
87JA4756	J. Uenishi, J.-M. Beau, R. W. Armstrong and Y. Kishi; *J. Am. Chem. Soc.*, 1987, **109**, 4756.	489
87JA5008	S. D. Copley and J. R. Knowles; *J. Am. Chem. Soc.*, 1987, **109**, 5008.	826
87JA5167	A. P. Kozikowski, T. R. Nieduzak, T. Konoike and J. P. Springer; *J. Am. Chem. Soc.*, 1987, **109**, 5167.	521
87JA5267	J. M. Kanabus-Kaminska, J. A. Hawari, D. Griller and C. Chatgilialoglu; *J. Am. Chem. Soc.*, 1987, **109**, 5267.	331, 936
87JA5478	A. M. Echavarren and J. K. Stille; *J. Am. Chem. Soc.*, 1987, **109**, 5478.	482, 1027
87JA5539	P. G. Williard and M. J. Hintze; *J. Am. Chem. Soc.*, 1987, **109**, 5539.	873
87JA5731	L. A. Paquette, F. Pierre and C. E. Cottrell; *J. Am. Chem. Soc.*, 1987, **109**, 5731.	815
87JA5749	S. L. Schreiber, M. T. Kilmas and T. Sammakia; *J. Am. Chem. Soc.*, 1987, **109**, 5749.	199, 200
87JA5878	E. Vedejs, R. A. Buchanan, P. Conrad, G. P. Meier, M. J. Mullins and Y. Watanabe; *J. Am. Chem. Soc.*, 1987, **109**, 5878.	821
87JA5996	S. Moran, H. B. Ellis, Jr., D. J. DeFrees, A. D. McLean and G. B. Ellison; *J. Am. Chem. Soc.*, 1987, **109**, 5996.	154
87JA6004	S. Moran, H. B. Ellis, Jr., D. J. DeFrees, A. D. McLean, S. E. Paulson and G. B. Ellison; *J. Am. Chem. Soc.*, 1987, **109**, 6004.	155
87JA6199	M. T. Crimmins and L. D. Gould; *J. Am. Chem. Soc.*, 1987, **109**, 6199.	816
87JA6396	J.-E. Bäckvall and S. K. Juntunen; *J. Am. Chem. Soc.*, 1987, **109**, 6396.	237, 238
87JA6558	D. P. Curran and M.-H. Chen; *J. Am. Chem. Soc.*, 1987, **109**, 6558.	348
87JA6716	J. T. Welch and S. Eswarakrishnan; *J. Am. Chem. Soc.*, 1987, **109**, 6716.	828, 834
87JA7230	G. Mehta and S. Padma; *J. Am. Chem. Soc.*, 1987, **109**, 7230.	3, 8
87JA7263	K. L. Servis, R. L. Domenick, D. A. Forsyth and Y. Pan; *J. Am. Chem. Soc.*, 1987, **109**, 7263.	905
87JA7561	P. J. Stang and T. Kitamura; *J. Am. Chem. Soc.*, 1987, **107**, 7561.	541, 1018
87JA7923	T. Cohen, W. D. Abraham and M. Myers; *J. Am. Chem. Soc.*, 1987, **109**, 7923.	302, 303
87JA7947	M. Meot-Ner; *J. Am. Chem. Soc.*, 1987, **109**, 7947.	892
87JA8025	M. J. Burk and R. H. Crabtree; *J. Am. Chem. Soc.*, 1987, **109**, 8025.	554
87JA8051	J. E. Baldwin and V. P. Reddy; *J. Am. Chem. Soc.*, 1987, **109**, 8051.	403, 420
87JA8056	E. Nakamura, S. Aoki, K. Sekiya, H. Oshino and I. Kuwajima; *J. Am. Chem. Soc.*, 1987, **109**, 8056.	122
87JCPB765	E. Sánchez Marcos, J. Maraver, J. Anguiano and J. Bertrán; *J. Chim. Phys. Physico-Chim. Biol.*, 1987, **84**, 765.	130
87JCR(S)353	D. Crich and L. B. L. Lim; *J. Chem. Res. (S)*, 1987, 353.	595, 599
87JCS(P1)475	P. M. Fischer and M. E. H. Howden; *J. Chem. Soc., Perkin Trans. 1*, 1987, 475.	665
87JCS(P1)805	A. R. Katritzky, S. Rachwal and B. Rachwal; *J. Chem. Soc., Perkin Trans. 1*, 1987, 805.	260
87JCS(P1)1011	R. McCague; *J. Chem. Soc. Perkin Trans. 1*, 1987, 1011.	576
87JCS(P1)1237	J. Knight and P. Parsons; *J. Chem. Soc., Perkin Trans. 1*, 1987, 1237.	1127
87JCS(P1)1511	K. Kikukawa, T. Idemoto, A. Katayama, K. Kono, F. Wada and T. Matsuda; *J. Chem. Soc., Perkin Trans. 1*, 1987, 1511.	1033
87JCS(P1)1623	R. Baker, M. J. O'Mahony and C. J. Swain; *J. Chem. Soc., Perkin Trans. 1*, 1987, 1623.	697, 700, 703
87JCS(P1)1951	N. Suzuki, T. Azuma, K. Watanabe, T. Nomoto, Y. Izawa and H. Tomioka; *J. Chem. Soc., Perkin Trans. 1*, 1987, 1951.	442
87JCS(P1)2171	P. J. Kocienski, S. D. A. Street, C. Yeates and S. F. Campbell; *J. Chem. Soc., Perkin Trans. 1*, 1987, 2171.	700, 703
87JCS(P1)2259	I. Fleming and M. Rowley; *J. Chem. Soc., Perkin Trans. 1*, 1987, 2259	566
87JCS(P1)2371	J. G. Buchanan, A. R. Edgar and B. D. Hewitt; *J. Chem. Soc., Perkin Trans. 1*, 1987, 2371.	43
87JCS(P1)2569	A. D. Buss, N. Greeves, R. Mason and S. Warren; *J. Chem. Soc., Perkin Trans. 1*, 1987, 2569.	757
87JCS(P2)1221	J. S. L. Ibaceta-Lizana, A. H. Jackson, N. Prasitpan and P. V. R. Shannon; *J. Chem. Soc., Perkin Trans. 2*, 1987, 1221.	184
87JFC(36)63	H. H. Coenen and S. M. Moerlein; *J. Fluorine Chem.*, 1987, **36**, 63.	449
87JMC193	P. W. Collins, S. W. Kramer, A. F. Gasiecki, R. M. Weier, P. H. Jones, G. W. Gullikson, R. G. Bianchi and R. F. Bauer; *J. Med. Chem.*, 1987, **30**, 193.	1008
87JMC254	F. Haviv, J. D. Ratajczyk, R. W. DeNet, Y. C. Martin, R. D. Dyer and G. W. Carter; *J. Med. Chem.*, 1987, **30**, 254.	1018
87JMC1505	J. A. Kepler, A. Philip, Y. W. Lee, H. A. Musallam and F. I. Carroll; *J. Med. Chem.*, 1987, **30**, 1505.	84
87JOC34	J. H. Rigby and J. A. Z. Wilson; *J. Org. Chem.*, 1987, **52**, 34.	218, 221
87JOC205	K. Riemenschneider, H. M. Bartels, R. Dornow, E. Drechsel-Grau, W. Eichel, H. Luthe, Y. M. Matter, W. Michaelis and P. Boldt; *J. Org. Chem.*, 1987, **52**, 205.	325, 920

87JOC218	P. Beak and K. D. Wilson; *J. Org. Chem.*, 1987, **52**, 218.	240
87JOC244	T. Chou, H.-H. Tso, Y.-T. Tao and L. C. Lin; *J. Org. Chem.*, 1987, **52**, 244.	791
87JOC281	C. R. Johnson and B. D. Tait; *J. Org. Chem.*, 1987, **52**, 281.	648, 649, 759
87JOC300	T. Ishihara, T. Maekawa, Y. Yamasaki and T. Ando; *J. Org. Chem.*, 1987, **52**, 300.	434
87JOC307	B. B. Snider and Y. S. Kulkarni; *J. Org. Chem.*, 1987, **52**, 307.	7
87JOC370	R. Plate, A. W. G. Theunisse and H. C. J. Ottenheijm; *J. Org. Chem.*, 1987, **52**, 370.	834
87JOC422	D. R. McKean, G. Parrinello, A. F. Renaldo and J. K. Stille; *J. Org. Chem.*, 1987, **52**, 422.	471
87JOC443	C. B. Ziegler, Jr., S. M. Harris and J. E. Baldwin; *J. Org. Chem.*, 1987, **52**, 443.	988
87JOC473	D. E. Bergbreiter and J. R. Blanton; *J. Org. Chem.*, 1987, **52**, 473.	5, 9
87JOC664	S. M. Moerlein; *J. Org. Chem.*, 1987, **52**, 664.	449
87JOC774	R. C. Gadwood, I. M. Mallick and A. J. DeWinter; *J. Org. Chem.*, 1987, **52**, 774.	716
87JOC897	T. L. Underiner and H. L. Goering; *J. Org. Chem.*, 1987, **52**, 897.	187
87JOC959	D. C. Spellmeyer and K. N. Houk; *J. Org. Chem.*, 1987, **52**, 959.	336
87JOC974	Y. Xu and B. Zhou; *J. Org. Chem.*, 1987, **52**, 974.	196
87JOC1013	K. Imi, N. Yanagihara and K. Utimoto; *J. Org. Chem.*, 1987, **52**, 1013.	208
87JOC1161	A. K. Jhingan and W. F. Maier; *J. Org. Chem.*, 1987, **52**, 1161.	530
87JOC1201	K. Tadano, Y. Idogaki, H. Yamada and T. Suami; *J. Org. Chem.*, 1987, **52**, 1201.	827, 829
87JOC1218	O. L. Chapman and C. J. Abelt; *J. Org. Chem.*, 1987, **52**, 1218.	597
87JOC1370	S. A. Babirad, Y. Wang and Y. Kishi; *J. Org. Chem.*, 1987, **52**, 1370.	978
87JOC1372	J. C. Barrish, H. L. Lee, E. G. Baggiolini and M. R. Uskokovic; *J. Org. Chem.*, 1987, **52**, 1372.	749, 828
87JOC1414	K. Nakamura, M. Higaki, S. Adachi, S. Oka and A. Ohno; *J. Org. Chem.*, 1987, **52**, 1414.	143
87JOC1624	T. Tsuda, T. Yoshida, T. Kawamoto and T. Saegusa; *J. Org. Chem.*, 1987, **52**, 1624.	511, 534, 1106
87JOC1703	H. Matsuyama, Y. Miyazawa, Y. Takei and M. Kobayashi; *J. Org. Chem.*, 1987, **52**, 1703.	706, 708
87JOC1885	C. R. Johnson and D. S. Dhanoa; *J. Org. Chem.*, 1987, **52**, 1885.	120, 465
87JOC1984	O. D. Dailey, Jr.; *J. Org. Chem.*, 1987, **52**, 1984.	20
87JOC2094	R. Yamaguchi, E. Hata, T. Matsuki and M. Kawanisi; *J. Org. Chem.*, 1987, **52**, 2094.	1042
87JOC2170	C. R. Johnson and T. Imamoto; *J. Org. Chem.*, 1987, **52**, 2170.	158
87JOC2224	T.-S. Chou and M.-L. Lou; *J. Org. Chem.*, 1987, **52**, 2224.	439
87JOC2317	R. A. Holton, D. J. Crouse, A. D. Williams and R. M. Kennedy; *J. Org. Chem.*, 1987, **52**, 2317.	23
87JOC2523	O. Tsuge, S. Kanemasa, T. Yamada and K. Matsuda; *J. Org. Chem.*, 1987, **52**, 2523.	372
87JOC2576	E. Keinan and D. Perez; *J. Org. Chem.*, 1987, **52**, 2576.	86
87JOC2586	H. W. Kircher and F. U. Rosenstein; *J. Org. Chem.*, 1987, **52**, 2586.	797
87JOC2627	J. S. Lomas; *J. Org. Chem.*, 1987, **52**, 2627.	923
87JOC2644	R. Gleiter, H. Zimmermann, W. Sander and M. Hauck; *J. Org. Chem.*, 1987, **52**, 2644.	634
87JOC2804	J. Blum, I. Amer, K. P. C. Vollhardt, H. Schwarz and G. Höhne; *J. Org. Chem.*, 1987, **52**, 2804.	95
87JOC2838	M. A. Findeis and G. M. Whitesides; *J. Org. Chem.*, 1987, **52**, 2838.	700, 703
87JOC2905	B. W. Disanayaka and A. C. Weedon; *J. Org. Chem.*, 1987, **52**, 2905.	690
87JOC2947	A. Hosomi, K. Hoashi, Y. Tominaga, K. Otaka and H. Sakurai; *J. Org. Chem.*, 1987, **52**, 2947.	253, 486
87JOC2958	G. E. Keck and D. A. Burnett; *J. Org. Chem.*, 1987, **52**, 2958.	348
87JOC2960	L. A. Paquette, J. Wright, G. J. Drtina and R. A. Roberts; *J. Org. Chem.*, 1987, **52**, 2960.	816
87JOC3126	J. G. Ulan, E. Kuo, W. F. Maier, R. S. Rai and G. Thomas; *J. Org. Chem.*, 1987, **52**, 3126.	502
87JOC3174	T. Ohta, H. Takaya, M. Kitamura, K. Nagai and R. Noyori; *J. Org. Chem.*, 1987, **52**, 3174.	77
87JOC3205	R. Sangaiah and A. Gold; *J. Org. Chem.*, 1987, **52**, 3205.	11, 19
87JOC3346	T. Kawabata, P. A. Grieco, H.-L. Sham, H. Kim, J. Y. Jaw and S. Tu; *J. Org. Chem.*, 1987, **52**, 3346.	25, 630
87JOC3509	J. Lusztyk, B. Maillard, S. Deycard, D. A. Lindsay and K. U. Ingold; *J. Org. Chem.*, 1987, **52**, 3509.	331, 336, 342, 936
87JOC3541	A. P. Kozikowski and C.-S. Li; *J. Org. Chem.*, 1987, **52**, 3541.	633, 829
87JOC3558	Y. Huang, L. Shi, J. Yang and Z. Cai; *J. Org. Chem.*, 1987, **52**, 3558.	647
87JOC3688	D. E. Johnson, J. C. DeLane and E. C. Wisted; *J. Org. Chem.*, 1987, **52**, 3688.	778
87JOC3697	M. W. Rathke and D. Vogiazoglou; *J. Org. Chem.*, 1987, **52**, 3697.	464
87JOC3706	S. F. Martin, M. S. Dappen, B. Dupre and C. J. Murphy; *J. Org. Chem.*, 1987, **52**, 3706.	16
87JOC3759	A. Spaltenstein, P. A. Carpino, F. Miyake and P. B. Hopkins; *J. Org. Chem.*, 1987, **52**, 3759.	697, 698, 703
87JOC3860	J. A. Marshall, T. M. Jenson and B. S. DeHoff; *J. Org. Chem.*, 1987, **52**, 3860.	816, 818
87JOC3883	J. A. Marshall, J. Lebreton, B. S. DeHoff and T. M. Jenson; *J Org. Chem.*, 1987, **52**, 3883.	752, 816, 818
87JOC3997	E. C. Taylor and P. S. Ray; *J. Org. Chem.*, 1987, **52**, 3997.	1010
87JOC4014	C. T. Goralski, B. Singaram and H. C. Brown; *J. Org. Chem.*, 1987, **52**, 4014.	60
87JOC4031	S. E. Denmark, M. A. Harmata and K. S. White; *J. Org. Chem.*, 1987, **52**, 4031.	993
87JOC4130	M. M. Abelman, T. Oh and L. E. Overman; *J. Org. Chem.*, 1987, **52**, 4130.	494
87JOC4133	N. Ono, A. Kamimura, H. Sasatani and A. Kaji; *J. Org. Chem.*, 1987, **52**, 4133.	252
87JOC4258	T. G. Back, S. Collins, M. V. Krishna and K.-W. Law; *J. Org. Chem.*, 1987, **52**, 4258.	486, 545
87JOC4265	T. G. Back and M. V. Krisha; *J. Org. Chem.*, 1987, **57**, 4265.	248
87JOC4296	S. Ghosal, G. P. Luke and K. S. Kyler; *J. Org. Chem.*, 1987, **52**, 4296.	491, 1035
87JOC4299	R. Pi, T. Friedl, P. von R. Schleyer, P. Klusener and L. Brandsma; *J. Org. Chem.*, 1987, **52**, 4299.	9
87JOC4303	M. D. Wittman and J. Kallmerten; *J. Org. Chem.*, 1987, **52**, 4303.	643

87JOC4410 T. Okazoe, K. Takai, K. Oshima and K. Utimoto; *J. Org. Chem.*, 1987, **52**, 4410. 766, 770

87JOC4412 K. Tamao, E. Nakajo and Y. Ito; *J. Org. Chem.*, 1987, **52**, 4412. 1094

87JOC4418 H. Ochai, Y. Tamaru, K. Tsubaki and Z.-i. Yoshida; *J. Org. Chem.*, 1987, **52**, 4418. 122, 123, 187

87JOC4641 T. Hudlicky, M. G. Natchus and G. Sinai-Zingde; *J. Org. Chem.*, 1987, **52**, 4641. 804

87JOC4745 D. L. Hughes, U.-H. Dolling, K. M. Ryan, E. F. Schoenewaldt and E. J. J. Grabowski; *J. Org. Chem.*, 1987, **52**, 4745. 135

87JOC4784 A. P. Marchand, G. V. M. Sharma, G. S. Annapurna and P. R. Pednekar; *J. Org. Chem.*, 1987, **52**, 4784. 6

87JOC4817 P. G. McDougal and J. G. Rico; *J. Org. Chem.*, 1987, **52**, 4817. 1089

87JOC4819 T.-C. Wu, P. G. Geokjian and Y. Kishi; *J. Org. Chem.*, 1987, **52**, 4819. 1099

87JOC4823 P. G. Geokjian, T.-C. Wu, H.-Y. Kang and Y. Kishi; *J. Org. Chem.*, 1987, **52**, 4823. 1099

87JOC4825 S. A. Babirad, Y. Wang, P. G. Geokjian and Y. Kishi; *J. Org. Chem.*, 1987, **52**, 4825. 1099

87JOC4829 S. M. Cairns and W. E. McEwen; *J. Org. Chem.*, 1987, **52**, 4829. 50

87JOC4859 K. Ohe, H. Takahashi, S. Uemura and N. Sugita; *J. Org. Chem.*, 1987, **52**, 4859. 545

87JOC4875 T. N. Wheeler and J. Ray; *J. Org. Chem.*, 1987, **52**, 4875. 549

87JOC4878 M. Nishizawa, H. Yamada and Y. Hayashi; *J. Org. Chem.*, 1987, **52**, 4878. 16

87JOC4885 J. E. McMurry and G. K. Bosch; *J. Org. Chem.*, 1987, **52**, 4885. 690

87JOC4898 J. P. Marino, R. F. de la Pradilla and E. Laborde; *J. Org. Chem.*, 1987, **52**, 4898. 14, 226, 227

87JOC4943 D. L. J. Clive, T. L. B. Boivin and A. G. Angoh; *J. Org. Chem.*, 1987, **52**, 4943. 244

87JOC5034 D. S. Matteson, E. C. Beedle and A. A. Kandil; *J. Org. Chem.*, 1987, **52**, 5034. 787

87JOC5044 S. Nagarajan and B. Ganem; *J. Org. Chem.*, 1987, **52**, 5044. 514

87JOC5056 R. M. Wehmeyer and R. D. Rieke; *J. Org. Chem.*, 1987, **52**, 5056. 121

87JOC5057 T.-C. Wu, R. M. Wehmeyer and R. D. Rieke; *J. Org. Chem.*, 1987, **52**, 5057. 478

87JOC5063 Y. Takeuchi, K. Nagata and T. Koizumi; *J. Org. Chem.*, 1987, **52**, 5063. 248

87JOC5064 R. A. Outten and G. D. Davies; *J. Org. Chem.*, 1987, **52**, 5064. 494

87JOC5111 N. Ono, A. Kamimura and A. Kaji; *J. Org. Chem.*, 1987, **52**, 5111. 617, 638

87JOC5198 J. J. Gajewski, C. W. Benner and C. M. Hawkins; *J. Org. Chem.*, 1987, **52**, 5198. 412

87JOC5484 G. Zweifel, G. R. Hahn and T. M. Shoup; *J. Org. Chem.*, 1987, **52**, 5484. 178

87JOC5695 T. Imamoto, T. Mita and M. Yokoyama; *J. Org. Chem.*, 1987, **52**, 5695. 515

87JOC5742 S. E. Denmark and J. E. Marlin; *J. Org. Chem.*, 1987, **52**, 5742. 832

87JOM(319)333 J. Pornet, D. Damour and L. Miginiac; *J. Organomet. Chem.*, 1987, **319**, 333. 963

87JOM(332)53 J. R. Hwu and B. A. Gilbert; *J. Organomet. Chem.*, 1987, **332**, 53. 656

87JOM(334)9 K. S. Chan, G. A. Peterson, T. A. Brandvold, K. L. Faron, C. A. Challenger, C. Hyldahl and W. P. Wulff; *J. Organomet. Chem.*, 1987, **334**, 9. 551

87JOM(334)181 E. Negishi, T. Takahashi and K. Akiyoshi; *J. Organomet. Chem.*, 1987, **334**, 181. 491

87JOM(334)243 H. Kurosawa; *J. Organomet. Chem.*, 1987, **334**, 243. 248

87JOM(336)143 S. Top and G. Jaouen; *J. Organomet. Chem.*, 1987, **336**, 143. 202

87LA679 K. Weinges and S. Haremsa; *Liebigs Ann. Chem.*, 1987, 679. 708

B-87MI 104-01 T. H. Lowry and K. S. Richardson; in "Mechanism and Theory in Organic Chemistry, 3rd Edition," Harper and Row, New York, 1987, chap. 4, pp. 327–423. 107

B-87MI 104-02 J. L. Wardell; in "The Chemistry of the Metal–Carbon Bond, Volume 4: The Use of Organometallic Compounds in Organic Synthesis," ed. F. R. Hartley; Wiley, Chichester, 1987, pp. 99–110. 110, 111, 166

B-87MI 104-03 C. L. Raston and G. Salem; in "The Chemistry of the Metal–Carbon Bond, Volume 4: The Use of Organometallic Compounds in Organic Synthesis," ed. F. R. Hartley; Wiley, Chichester, 1987, pp. 223–225. 114

B-87MI 104-04 D. S. Matteson; in "The Chemistry of the Metal–Carbon Bond, Volume 4: The Use of Organometallic Compounds in Organic Synthesis," ed. F. R. Hartley; Wiley, Chichester, 1987, chap. 2, pp. 307–409. 126

B-87MI 104-05 M. Pereyre, J.-P. Quintard and A. Rahm; in "Tin in Organic Synthesis," Butterworths, London, 1987, chap. 10, pp. 185–258. 128

87MI 108-01 D. Crich; *Aldrichimica Acta*, 1987, 35. 332, 344

B-87MI 110-01 D. S. Matteson; in "The Chemistry of the Metal–Carbon Bond," ed. F. R. Hartley, Wiley, Chichester, 1987, vol. 4, ch. 3, p. 307. 447

B-87MI 110-02 J. L. Wardell; "The Chemistry of the Metal–Carbon Bond," ed. F. R. Hartley, Wiley, Chichester, 1987, vol. 4, ch. 1, p. 1. 452, 452, 453

B-87MI 110-03 C. L. Raston and G. Salem; in "The Chemistry of the Metal–Carbon Bond," ed. F. R. Hartley, Wiley, Chichester, 1987, vol. 4, ch. 2, p. 159. 452, 453

B-87MI 110-04 S. Uemura; in "The Chemistry of the Metal–Carbon Bond," ed. F. R. Hartley, Wiley, Chichester, 1987, vol. 4, ch. 6, p. 453. 452

B-87MI 110-05 M. Pereyre, J.-P. Quintard and A. Rahm; "Tin in Organic Synthesis," Butterworths, London, 1987, p. 130. 452

B-87MI 112-01 B. Halton and M. G. Banwell; in "The Chemistry of the Cyclopropyl Group," ed. S. Patai, 1987, part 2, p. 1223. 549

87MI 114-01 S. P. Choudhari; *Curr. Sci.*, 1987, **56**, 947 (*Chem. Abstr.*, 1989, **110**, 74 511). 612

B-87MI 114-01 T. G. Back; in "Organoselenium Chemistry," ed. D. Liotta, Wiley, New York, 1987, ch. 1. 601

B-87MI 114-02 T. G. Back; in "The Chemistry of Organic Selenium and Tellurium Compounds," ed. S. Patai, Wiley, Chichester, 1987, vol. 2, p. 91. 601, 603, 607

B-87MI 114-03 W. R. McWhinnie; in "The Chemistry of Organic Selenium and Tellurium Compounds," ed. S. Patai, Wiley, Chichester, 1987, vol. 2, p. 495. 601

B-87MI 114-04 J. P. Collman, L. S. Hegedus, J. R. Norton and R. G. Finke; "Principles and Applications of Organotransition Metal Chemistry," University Science Books, Mill Valley, CA, 1987, p. 386. 668

B-87MI 115-01 T. G. Back; in "The Chemistry of Organic Selenium and Tellurium Compounds," ed. S. Patai, Z. Rappoport and C. J. M. Stirling, Wiley, New York, 1987, vol. 2, p. 91. 712, 713, 715

B-87MI 115-02 A. Krief; in "The Chemistry of Organic Selenium and Tellurium Compounds," ed. S. Patai, Z. Rappoport and C. J. M. Stirling, Wiley, New York, 1987, vol. 2, p. 675. 712, 713, 714, 715, 716, 717

B-87MI 118-01 H.-U. Reissing; in "The Chemistry of the Cyclopropyl Group," ed. Z. Rappaport, Wiley, New York, 1987, part 1, p. 375. 801

B-87MI 118-02 Z. Rappoport (ed.); "The Chemistry of the Cyclopropyl Group," Wiley, New York, 1987, parts 1 and 2. 801

B-87MI 118-03 B. K. Carpenter; in "The Chemistry of the Cyclopropyl Group," ed. Z. Rappaport, Wiley, New York, 1987, part 2, p. 1027. 801

B-87MI 118-04 J. R. Y. Salaun; in "The Chemistry of the Cyclopropyl Group," ed. Z. Rappaport, Wiley, New York, 1987, part 2, p. 809. 802, 841

B-87MI 119-01 E. Buncel and T. Durst (eds.); "Comprehensive Carbanion Chemistry, Part C," Elsevier, New York, 1987. 844, 918

B-87MI 119-02 T. H. Lowry and K. S. Richardson; "Mechanism and Theory in Organic Chemistry," 3rd edn., Harper, New York, 1987, chap. 6, pp. 517–545. 845

B-87MI 119-03 A. Nickon and E. F. Silversmith; "Organic Chemistry: The Name Game," Pergamon, Oxford, 1987, p. 106. 862

B-87MI 119-04 E. Grovenstein, Jr.; in "Comprehensive Carbanion Chemistry, Part C," ed. E. Buncel and T. Durst, Elsevier, New York, 1987, pp. 175–221. 885

B-87MI 119-05 T. H. Lowry and K. S. Richardson; "Mechanism and Theory in Organic Chemistry," 3rd edn., Harper, New York, 1987, p. 517. 886, 893

B-87MI 119-06 T. H. Lowry and K. S. Richardson; "Mechanism and Theory in Organic Chemistry," 3rd edn., Harper, New York, 1987, pp. 425–515. 891

B-87MI 119-07 T. H. Lowry and K. S. Richardson; "Mechanism and Theory in Organic Chemistry," 3rd edn., Harper, New York, 1987, p. 448. 904

B-87MI 119-08 W. E. Billups and A. W. Moorehead; in "The Chemistry of the Cyclopropyl Group, Part 2," Wiley, New York, 1987, pp. 1533–1574. 907

B-87MI 119-09 T. H. Lowry and K. S. Richardson; "Mechanism and Theory in Organic Chemistry," 3rd edn., Harper, New York, 1987, Chap. 5, p. 434. 915

B-87MI 119-10 A. Nickon and E. F. Silversmith; "Organic Chemistry: The Name Game," Pergamon, Oxford, 1987, p. 235. 248

87MI 119-10 E. M. Arnett, T. C. Hofelich and G. W. Schriver; *React. Intermed.* (*Wiley*), 1987, **3**, 189–226. 891

87MI 119-11 D. Crich; *Aldrichimica Acta*, 1987, **20**, 35. 935

87MM1169 R. R. Schrock, J. Feldman, L. F. Cannizzo and R. H. Grubbs; *Macromolecules*, 1987, **20**, 1169. 675

87MM2640 J. S. Murdzek and R. R. Schrock; *Macromolecules*, 1987, **20**, 2640. 675

87OM1606 L. Walder and R. Orlinski; *Organometallics*, 1987, **6**, 1606. 946

87OM1625 S. Calet, H. Alper, J.-F. Petrignani and H. Arzoumanian; *Organometallics*, 1987, **6**, 1625. 242

87OM2249 A. G. M. Barrett and N. E. Carpenter; *Organometallics*, 1987, **6**, 2249. 825

87OPP442 M. Tashiro, H. Nakamura and K. Nakamura; *Org. Prep. Proced. Int.*, 1987, **19**, 442. 426, 428

87OS(65)81 L. Lombardo; *Org. Synth.*, 1987, **65**, 81. 678, 767

87OS(65)90 E. Block and M. Aslam; *Org. Synth.*, 1987, **65**, 90. 619, 707

87OS(65)224 C. A. Brown and P. K. Jadhav; *Org. Synth.*, 1987, **65**, 224. 795

87PAC363 R. Scheffold, S. Abrecht, R. Orlinski, H.-R. Ruf, P. Stamouli, O. Tinembart, L. Walder and C. Weymuth; *Pure. Appl. Chem.*, 1987, **59**, 363. 546, 946

87RCR983 G. A. Artamkina and I. P. Beletskaya; *Russ. Chem. Rev.* (*Engl. Transl.*), 1987, **56**, 983. 883

87RTC613 A. M. Brouwer, L. Bezemer, J. Cornelisse and H. J. C. Jacobs; *Recl. Trav. Chim. Pays-Bas*, 1987, **106**, 613. 420

87S40 T. Frejd and T. Klingstedt; *Synthesis*, 1987, 40. 465

87S70 T. Jeffery; *Synthesis*, 1987, 70. 1021

87S139 S. Takano, Y. Sekiguchi, N. Sato and K. Ogasawara; *Synthesis*, 1987, 139. 1023

87S207 A. Ichihara; *Synthesis*, 1987, 207. 786

87S325 B. H. Lipshutz; *Synthesis*, 1987, 325. 116, 120, 206, 465

87S370 Lilan Shi, Wengjuan Xiao, Xueqing Wen and Yaozeng Huang; *Synthesis*, 1987, 370. 739

87S373 M. Satoh, N. Miyaura and A. Suzuki; *Synthesis*, 1987, 373. 279

87S603 J. Tsuji, T. Sugiura and I. Minami; *Synthesis*, 1987, 603. 962

87S665 W. P. Neumann; *Synthesis*, 1987, 665. 4, 5, 9, 16, 172, 230, 330, 341, 428, 919, 935, 938, 942, 946

87S719 F. Clerici, A. Di Mare, M. L. Gelmi and D. Pocar; *Synthesis*, 1987, 719. 45

87S736 A. Ono, N. Suzuki and J. Kamimura; *Synthesis*, 1987, 736. 20

87S816 M. I. Al-Hassan; *Synthesis*, 1987, 816. 471

87S821 N. Ono, T. X. Jun and A. Kaji; *Synthesis*, 1987, 821. 37

87S846 M. Yokoyama, S. Watanabe and H. Hatanaka; *Synthesis*, 1987, 846. 47

87S848 H. J. Bestmann, G. Wölfel and K. Mederer; *Synthesis*, 1987, 848. 1041

87S992 I. Minami, M. Nisar, M. Yuhara, I. Shimizu and J. Tsuji; *Synthesis*, 1987, 992. 558, 561

87S1071 A. Furstner and H. Weidmann; *Synthesis*, 1987, 1071. 688

87S1142 M. Makosza and A. Tyrala; *Synthesis*, 1987, 1142. 613, 622

87SC621 M. B. Anderson and P. L. Fuchs; *Synth. Commun.*, 1987, **17**, 621. 652

87SC703 R. M. Moriarty, R. K. Vaid and M. P. Duncan; *Synth. Commun.*, 1987, **17**, 703. 74

87SC863 J. M. Muchowski and M. E. Scheller; *Synth. Commun.*, 1987, **17**, 863. 256

87SC1839 C. Xu, G. Liu and Z. Zhang; *Synth. Commun.*, 1987, **17** 1839. 599

87T853 M. Frank-Neuman, M. Miesch and H. Kempf; *Tetrahedron*, 1987, **43**, 853. 576
87T859 T. Cuvigny, C. Herve du Penhoat and M. Julia; *Tetrahedron*, 1987, **43**, 859. 439, 700
87T1969 W. Oppolzer; *Tetrahedron*, 1987, **43**, 1969. (Erratum: 1987, **43**, 4057) 134, 305
87T2457 D. Hoppe, C. Gonschorrek, D. Schmidt and E. Egert; *Tetrahedron*, 1987, **43**, 2457. 974
87T2609 Y. G. Gololobov, A. N. Nesmeyanov, V. P. Lysenko and I. E. Boldeskul; *Tetrahedron*, 1987, **43**, 2609. 304
87T2701 J. Soulie, M. Lampilas and J. Y. Lallemand; *Tetrahedron*, 1987, **43**, 2701. 223, 224
87T2733 D. H. R. Barton, D. Bridon, I. Fernandez-Picot and S. Z. Zard; *Tetrahedron*, 1987, **43**, 2733. 935
87T3371 H. Suginome and S. Yamada; *Tetrahedron*, 1987, **43**, 3371. 931
87T3541 M. Ramaiah; *Tetrahedron*, 1987, **43**, 3541. 320, 336, 919, 929, 935, 1126
87T3817 E. O. Orlemans, A. H. Schreuder, P. G. M. Conti, W. Verboom and D. N. Reinhoudt; *Tetrahedron*, 1987, **43**, 3817. 47
87T4265 H. M. L. Davies, D. M. Clark, D. B. Alligood and G. R. Eiband; *Tetrahedron*, 1987, **43**, 4265. 811
87T4395 D. Tanner and P. Somfai; *Tetrahedron*, 1987, **43**, 4395. 697, 700, 703
87T4481 P. Gramatica, P. Manitto, D. Monti and G. Speranza; *Tetrahedron*, 1987, **43**, 4481. 77
87T4817 B. M. Trost and M. Lautens; *Tetrahedron*, 1987, **43**, 4817. 196
87T5095 A. I. Meyers, D. A. Dickman and M. Boes; *Tetrahedron*, 1987, **43**, 5095. 149
87T5245 A. Ichihara, S. Sakamura and K. Murakami; *Tetrahedron*, 1987, **43**, 5245. 789
87T5489 J. E. McMurry, J. R. Matz and K. L. Kees; *Tetrahedron*, 1987, **43**, 5489. 690
87T5663 A. I. Meyers and B. A. Lefker; *Tetrahedron*, 1987, **43**, 5663. 1130
87TCC1 A. Krief; *Top. Curr. Chem.*, 1987, **135**, 1. 712, 714
87TCC75 G. Maas; *Top. Curr. Chem.*, 1987, **137**, 75. 352, 637
87TCC77 P. Binger and H. M. Büch; *Top. Curr. Chem.*, 1987, **135**, 77. 527
87TCC149 N. M. Alpatova, L. I. Krishtalik and Y. V. Pleskov; *Top. Curr. Chem.*, 1987, **138**, 149–219. 248
87TL31 L. A. Paquette, J. L. Romine and H.-S. Lin; *Tetrahedron Lett.*, 1987, **28**, 31. 434, 815
87TL175 E. J. Corey and A. K. Ghosh; *Tetrahedron Lett.*, 1987, **28**, 175. 336, 346
87TL183 G. Helmchen, K. Ihrig and H. Schindler; *Tetrahedron Lett.*, 1987, **28**, 183. 786
87TL211 J. Yoshida, T. Murata and S. Isoe; *Tetrahedron Lett.*, 1987, **28**, 211. 651
87TL221 S. V. Ley, D. Santafianos, W. M. Blaney and M. S. J. Simmonds; *Tetrahedron Lett.*, 1987, **28**, 221. 596, 780
87TL295 S. L. Buchwald and S. J. LaMaire; *Tetrahedron Lett.*, 1987, **28**, 295. 536
87TL337 E. Nakamura, K. Sekiya and I. Kuwajima; *Tetrahedron Lett.*, 1987, **28**, 337. 122
87TL357 B. Zwanenburg, A. J. H. Klunder, W. B. Huizinga and P. J. M. Sessink; *Tetrahedron Lett.*, 1987, **28**, 357. 786
87TL527 J. A. Marshall, B. S. DeHoff and S. L. Crooks; *Tetrahedron Lett.*, 1987, **28**, 527. 1053
87TL723 J. A. Marshall, J. Lebreton, B. S. DeHoff and T. M. Jenson; *Tetrahedron Lett.*, 1987, **28**, 723. 816
87TL801 X. Huang, L. Xie and H. Wu; *Tetrahedron Lett.*, 1987, **28**, 801. 623
87TL803 S. Kuroda, T. Katsuki and M. Yamaguchi; *Tetrahedron Lett.*, 1987, **28**, 803. 816
87TL917 E. Negishi, D. R. Swanson, F. E. Cederbaum and T. Takahashi; *Tetrahedron Lett.*, 1987, **28**, 917. 536
87TL945 B. H. Lipshutz, M. Koerner and D. A. Parker; *Tetrahedron Lett.*, 1987, **28**, 945. 312
87TL989 N. S. Simpkins; *Tetrahedron Lett.*, 1987, **28**, 989. 145, 1090
87TL1019 K. T. Mead; *Tetrahedron Lett.*, 1987, **28**, 1019. 1038, 1039
87TL1031 M. J. Kurth and C. J. Soares; *Tetrahedron Lett.*, 1987, **28**, 1031. 835
87TL1043 S. L. Schreiber and M. T. Goulet; *Tetrahedron Lett.*, 1987, **28**, 1043. 397
87TL1413 D. H. R. Barton, J. Guilhem, Y. Hervé, P. Potier and J. Thierry; *Tetrahedron Lett.*, 1987, **28**, 1413. 344
87TL1443 K. Takai, Y. Kataoka, T. Okazoe and K. Utimoto; *Tetrahedron Lett.*, 1987, **28**, 1443. 676
87TL1463 L. Engman; *Tetrahedron Lett.*, 1987, **28**, 1463. 610
87TL1519 D. Pflieger, B. Muckensturm, P. C. Robert, M.-T. Simonis and J.-C. Kienlen; *Tetrahedron Lett.*, 1987, **28**, 1519. 774
87TL1549 A. Krief, J. L. Laboureur and W. Dumont; *Tetrahedron Lett.*, 1987, **28**, 1549. 246
87TL1611 B. M. Trost and D. J. Jebaratnam; *Tetrahedron Lett.*, 1987, **28**, 1611. 525
87TL1649 V. Ratovelomanana, A. Hammoud and G. Linstrumelle; *Tetrahedron Lett.*, 1987, **28**, 1649. 468, 533, 1009
87TL1737 T. G. Back, M. V. Krishna and K. R. Muralidharan; *Tetrahedron Lett.*, 1987, **28**, 1737. 605
87TL1781 M. Larchevêque and S. Henrot; *Tetrahedron Lett.*, 1987, 1781. 479
87TL1803 W. Ando, H. Hayakawa and N. Tokitoh; *Tetrahedron Lett.*, 1987, **28**, 1803. 983
87TL1853 H. M. L. Davies, H. D. Smith and O. Korkor; *Tetrahedron Lett.*, 1987, **28**, 1853. 811
87TL1857 J. S. Prasad and L. S. Liebeskind; *Tetrahedron Lett.*, 1987, **28**, 1857. 1047
87TL1881 J. Font, A. Virgili and M. Figueredo; *Tetrahedron Lett.*, 1987, **43**, 1881. 209
87TL1985 R. Gleiter, G. Müller, U. Huber-Patz, H. Rodewald and H. Irngartinger; *Tetrahedron Lett.*, 1987, **28**, 1985. 634
87TL1997 J. Ardisson, J. P. Férézou, M. Julia, L. Lenglet and A. Pancrazi; *Tetrahedron Lett.*, 1987, **28**, 1997.
87TL2009 J. P. Dulcere, J. Rodriguez, M. Santelli and J. P. Zahra; *Tetrahedron Lett.*, 1987, **28**, 2009. 1131
87TL2041 A. T. Russell and G. Procter; *Tetrahedron Lett.*, 1987, **28**, 2041. 828
87TL2061 T. K. Sarkar and S. K. Ghosh; *Tetrahedron Lett.*, 1987, **28**, 2061. 593
87TL2099 L. Castedo, J. L. Mascareñas and A. Mouriño; *Tetrahedron Lett.*, 1987, **28**, 2099. 816, 817
87TL2155 L. Shi, W. Xia, J. Yang, X. Wen and Y. Huang; *Tetrahedron Lett.*, 1987, **28**, 2155. 647, 741

87TL2159	Y. Huang, L. Shi, J. Yang and J. Zhang; *Tetrahedron Lett.*, 1987, **28**, 2159.	647
87TL2207	J. R. McCarthy, C. L. Barney, D. P. Matthews and T. M. Bargar; *Tetrahedron Lett.*, 1987, **28**, 2207.	967
87TL2221	P. A. Wender and M. L. Snapper; *Tetrahedron Lett.*, 1987, **28**, 2221.	542
87TL2387	Q.-Y. Chen and Y.-B. He; *Tetrahedron Lett.*, 1987, 2387.	483, 1028
87TL2481	P. W. Rabideau and G. L. Karrick; *Tetrahedron Lett.*, 1987, **28**, 2481.	890
87TL2597	H. Okawara, Y. Nii, A. Miwa and M. Sakakibara; *Tetrahedron Lett.*, 1987, **28**, 2597.	829
87TL2637	M. D. Bachi, A. De Mesmaeker and N. Stevenart-De Mesmaeker; *Tetrahedron Lett.*, 1987, **28**, 2637.	1131
87TL2649	G. Pandey, G. Kumaraswamy and A. Krishna; *Tetrahedron Lett.*, 1987, **28**, 2649.	286
87TL2697	F. Lehrich and H. Hopf; *Tetrahedron Lett.*, 1987, **28**, 2697.	955, 958
87TL2751	C. Nativi, A. Ricci and M. Taddei; *Tetrahedron Lett.*, 1987, **28**, 2751.	667
87TL2753	L. Pettersson, T. Frejd and G. Magnusson; *Tetrahedron Lett.*, 1987, **28**, 2753.	216
87TL2795	L. A. Paquette and M. P. Trova; *Tetrahedron Lett.*, 1987, **28**, 2795.	776
87TL2833	G. L. Lange, C. P. Decicco and M. Lee; *Tetrahedron Lett.*, 1987, **28**, 2833.	357
87TL2887	M. D. Bachi, A. De Mesmaeker and N. Stevenart-De Mesmaeker; *Tetrahedron Lett.*, 1987, **28**, 2887.	1131
87TL2895	D. Crich and S. M. Fortt; *Tetrahedron Lett.*, 1987, **28**, 2895.	1127
87TL2937	P. A. Jacobi and R. F. Frechette; *Tetrahedron Lett.*, 1987, **28**, 2937.	20
87TL3031	J. Cooper, D. W. Knight and P. T. Gallagher; *Tetrahedron Lett.*, 1987, **28**, 3031.	831
87TL3075	K. S. Reddy, O.-H. Ko, D. Ho, P. E. Persons and J. M. Cassady; *Tetrahedron Lett.*, 1987, **28**, 3075.	752
87TL3091	U. Berlage, J. Schmidt, U. Peters and P. Welzel; *Tetrahedron Lett.*, 1987, **28**, 3091.	690
87TL3209	J. E. McMurry and R. Swenson; *Tetrahedron Lett.*, 1987, **28**, 3209.	
87TL3323	J. A. Marshall and J. Lebreton; *Tetrahedron Lett.*, 1987, **28**, 3323.	816, 818
87TL3397	C. G. Francisco, R. Freire, M. S. Rodríguez and E. Suárez; *Tetrahedron Lett.*, 1987, **28**, 3397.	931
87TL3457	J. F. Lavallée and P. Deslongchamps; *Tetrahedron Lett.*, 1987, **28**, 3457.	538
87TL3497	Y. Tamura, T. Bando, M. Hojo and Z. Yoshida; *Tetrahedron Lett.*, 1987, **28**, 3497.	192
87TL3791	D. Seebach and B. Herradon; *Tetrahedron Lett.*, 1987, **28**, 3791.	311
87TL3883	P. Kulanthaivel and S. W. Pelletier; *Tetrahedron Lett.*, 1987, **28**, 3883.	4, 7
87TL3901	S. T. Purrington and J. H. Pittman; *Tetrahedron Lett.*, 1987, **28**, 3901.	780
87TL3923	J. Drouin and M. A. Boaventura; *Tetrahedron Lett.*, 1987, **28**, 3923.	526, 999
87TL3963	H. Suginome and S. Yamada; *Tetrahedron Lett.*, 1987, **28**, 3963.	929
87TL4011	G. A. Kraus and J. Thurston; *Tetrahedron Lett.*, 1987, **28**, 4011.	949
87TL4123	S. W. McCombie, B. B. Shanker and A. K. Ganguly; *Tetrahedron Lett.*, 1987, **28**, 4123.	1090
87TL4127	S. W. McCombie, B. B. Shankar, A. K. Ganguly, A. Padwa, W. H. Bullock and A. D. Dyszlewski; *Tetrahedron Lett.*, 1987, **28**, 4127.	1089, 1090
87TL4437	K. Otsubo, J. Inanaga and M. Yamaguchi; *Tetrahedron Lett.*, 1987, **28**, 4437.	14
87TL4455	P. C. B. Page, S. Rosenthal and R. V. Williams; *Tetrahedron Lett.*, 1987, **28**, 4455.	284
87TL4547	J. Cossy, D. Belotti and J. P. Pete; *Tetrahedron Lett.*, 1987, **28**, 4547.	195, 548
87TL4583	Y. Tamura, S. Akai, H. Kishimoto, M. Kirihara, M. Sasho and Y. Kita; *Tetrahedron Lett.*, 1987, **28**, 4583.	1140
87TL4653	A. B. Pierini, M. T. Baumgartner and R. A. Rossi; *Tetrahedron Lett.*, 1987, 4653.	474
87TL4689	D. Damour, J. Pornet and L. Miginiac; *Tetrahedron Lett.*, 1987, **28**, 4689.	978
87TL4745	F. Camps, J. Coll, J. M. Moretó and J. Torras; *Tetrahedron Lett.*, 1987, **28**, 4745.	527
87TL4759	E. Van der Eycken, H. De Wilde, L. Deprez and M. Vandewalle; *Tetrahedron Lett.*, 1987, **28**, 4759.	19
87TL4849	Y. Ito, M. Sawamura, M. Matsuoko, Y. Matsumoto and T. Hayashi; *Tetrahedron Lett.*, 1987, **28**, 4849.	193
87TL4875	L. A. Crombie, A. J. W. Hobbs and M. A. Horsham; *Tetrahedron Lett.*, 1987, **28**, 4875.	533, 536
87TL4879	L. A. Crombie, M. A. Horsham and R. J. Blade; *Tetrahedron Lett.*, 1987, **28**, 4879.	536
87TL4917	E. J. Corey, R. Nagata and S. W. Wright; *Tetrahedron Lett.*, 1987, **28**, 4917.	601
87TL4921	E. J. Corey, W. L. Seibel and J. C. Kappos; *Tetrahedron Lett.*, 1987, **28**, 4921.	1063
87TL4959	D. M. Armistead and S. J. Danishefsky; *Tetrahedron Lett.*, 1987, **28**, 4959.	780
87TL4965	L. A. Paquette, J. Dressel and P. D. Pansegrau; *Tetrahedron Lett.*, 1987, **28**, 4965.	248
87TL4993	C. A. Broka, L. Hu, W. J. Lee and T. Shen; *Tetrahedron Lett.*, 1987, **28**, 4993.	816
87TL5093	M. J. Sharp, W. Cheng and V. Snieckus; *Tetrahedron Lett.*, 1987, 5093.	473
87TL5097	W. Cheng and V. Snieckus; *Tetrahedron Lett.*, 1987, 5097.	473
87TL5129	M. D. Lewis and R. Menes; *Tetrahedron Lett.*, 1987, **28**, 5129.	534
87TL5279	A. I. Meyers, J. D. Brown and D. Laucher; *Tetrahedron Lett.*, 1987, **28**, 5279.	310
87TL5287	T. Hudlicky, G. Sinai-Zingde and M. G. Natchus; *Tetrahedron Lett.*, 1987, **28**, 5287.	86
87TL5339	J. M. Mellor and A. M. Wagland; *Tetrahedron Lett.*, 1987, **28**, 5339.	7
87TL5395	S. W. McCombie and T. L. Nagabhushan; *Tetrahedron Lett.*, 1987, **28**, 5395.	508
87TL5415	J. W. Dieter, Z. Li and K. M. Nicholas; *Tetrahedron Lett.*, 1987, **28**, 5415.	196
87TL5473	J. K. Cha and R. J. Cooke; *Tetrahedron Lett.*, 1987, **28**, 5473.	797
87TL5509	J. C. Clinet and G. Balavoine; *Tetrahedron Lett.*, 1987, **28**, 5509.	597
87TL5513	A. E. Vougioukas and H. B. Kagan; *Tetrahedron Lett.*, 1987, **28**, 5513.	208
87TL5521	H. Amri and J. Villieras; *Tetrahedron Lett.*, 1987, **28**, 5521.	191
87TL5755	N. J. Anthony, T. Clarke, A. B. Jones and S. V. Ley; *Tetrahedron Lett.*, 1987, **28**, 5755.	700, 703
87TL5759	C. Greck, P. Grice, A. B. Jones and S. V. Ley; *Tetrahedron Lett.*, 1987, **28**, 5759.	697, 700, 703
87TL5763	N. J. Anthony, P. Grice and S. V. Ley; *Tetrahedron Lett.*, 1987, **28**, 5763.	697, 700, 703
87TL5793	E. Negishi, Y. Zhang and V. Bagheri; *Tetrahedron Lett.*, 1987, **28**, 5793.	1070, 1094

87TL5879 K. Mikami, K. Takahashi and T. Nakai; *Tetrahedron Lett.*, 1987, **28**, 5879. 415, 826

87TL5887 T. Wakamiya, K. Yamanoi, K. Kanou and T. Shiba; *Tetrahedron Lett.*, 1987, **28**, 5887. 882

87TL5949 D. J. Coveney, V. F. Patel and G. Pattenden; *Tetrahedron Lett.*, 1987, **28**, 5949. 933

87TL5973 F. E. Ziegler and Z.-I. Zheng; *Tetrahedron Lett.*, 1987, **28**, 5973. 836, 941

87TL6021 R. A. Gibbs and W. H. Okamura; *Tetrahedron Lett.*, 1987, **28**, 6021. 795

87TL6073 A. M. Caporusso, C. Polizzi and L. Lardicci; *Tetrahedron Lett.*, 1987, **28**, 6073. 1016

87TL6159 R. Gleiter and U. Steuerle; *Tetrahedron Lett.*, 1987, **28**, 6159. 634

87TL6261 B. A. Narayanan and W. H. Bunnelle; *Tetrahedron Lett.*, 1987, **28**, 6261. 652

87TL6461 F. Tubéry, D. S. Grierson and H.-P. Husson; *Tetrahedron Lett.*, 1987, **28**, 6461. 833

87TL6565 P. G. Theobald and W. H. Okamura; *Tetrahedron Lett.*, 1987, **28**, 6565. 439

87TL6709 C. Najera and M. Yus; *Tetrahedron Lett.*, 1987, **28**, 6709. 1089, 1090

87ZOR1660 A. A. Gevorkyan, A. S. Arakelyan and S. M. Kosyan; *Zh. Org. Khim.*, 1987, **23**, 1660 (*Chem. Abstr.*, 1988, **109**, 6058). 629

87ZOR2545 N. S. Zefirov, K. A. Lukin and A. Y. Timofeeva; *Zh. Org. Chem.*, 1987, **23**, 2545. 500

88ACR47 W. J. Scott and J. E. McMurry; *Acc. Chem. Res.*, 1988, **21**, 47. 480

88ACR206 M. Newcomb and D. P. Curran; *Acc. Chem. Res.*, 1988, **21**, 206. 923

88ACR229 D. L. Reger; *Acc. Chem. Res.*, 1988, **21**, 229. 545

88ACS(B)259 W. Stensen, J. S. Svendsen, O. Hofer and L. K. Sydnes; *Acta Chem. Scand., Ser. B*, 1988, **42**, 259. 498

88AG(E)298 M. Brudermüller and H. Musso; *Angew. Chem., Int. Ed. Engl.*, 1988, **27**, 298. 510

88AG(E)411 E. Vogel, M. Sicken, P. Rohrig, H. Schmickler, J. Lex and O. Ermer; *Angew. Chem., Int. Ed. Engl.*, 1988, **27**, 411. 248

88AG(E)1624 D. Seebach; *Angew. Chem., Int. Ed. Engl.*, 1988, **27**, 1624. 876

88BCJ107 B. M. Trost; *Bull. Chem. Soc. Japan*, 1988, **61**, 107. 696

88BCJ149 G. Stork; *Bull. Chem. Soc. Jpn.*, 1988, **61**, 149. 919

88BCJ467 Y. Arai, M. Yamamoto and T. Koizumi; *Bull. Chem. Soc. Jpn.*, 1988, **61**, 467. 767, 769

88BCJ571 Y. Ukaji and K. Narasaka; *Bull. Chem. Soc. Jpn.*, 1988, **61**, 571. 178

88BCJ1231 M. Kameyama, H. Shimezawa, T. Satoh and N. Kamigata; *Bull. Chem. Soc. Jpn.*, 1988, **61**, 1231. 485

88BCJ1281 K. Yamamoto, J. Ojima, N. Morita and T. Asao; *Bull. Chem. Soc. Jpn.*, 1988, **61**, 1281. 248

88BCJ1625 K. Kodaira and K. Okuhara; *Bull. Chem. Soc. Jpn.*, 1988, **61**, 1625. 466

88BCJ2693 Y. Ichinose, K. Oshima and K. Utimoto; *Bull. Chem. Soc., Japan*, 1988, **61**, 2693. 1130

88BCJ2859 H. Naora, T. Ohnuki and A. Nakamura; *Bull. Chem. Soc. Jpn.*, 1988, **61**, 2859. 504

88BCJ3211 R. Tanikaga, N. Konya, K. Hamamura and A. Kaji; *Bull. Chem. Soc. Jpn.*, 1988, **61**, 3211. 635

88BCJ3629 A. Kamimura and N. Ono; *Bull. Chem. Soc. Jpn.*, 1988, **61**, 3629. 946

88BCJ4037 T. Mukaiyama, M. Hayashi and A. Inubushi; *Bull. Chem. Soc. Jpn.*, 1988, **61**, 4037. 1026

88BSB271 R. Broos and M. J. Anteunis; *Bull. Soc. Chim. Belg.*, 1988, **97**, 271. 647, 738, 740, 742

88BSF846 H. B. Kagan; *Bull. Soc. Chim. Fr.*, 1988, 846. 76

88C106 K. Fritzsche, H.-D. Beckhaus and C. Rüchardt; *Chimia*, 1988, **42**, 106. 638

88CB421 F. Effenberger and T. Weber; *Chem. Ber.*, 1988, **121**, 421. 182

88CB1039 W. Sucrow, R. Heider and N. Joraschek; *Chem. Ber.*, 1988, **121**, 1039. 39

88CB1151 D. Hermeling and H. J. Schäfer; *Chem. Ber.*, 1988, **121**, 1151. 561

88CB1315 D. Seebach and N. Krause; *Chem. Ber.*, 1988, **121**, 1315. 1024

88CB1461 W. Hasel and H. M. R. Hoffmann; *Chem. Ber.*, 1988, **121**, 1461. 187, 1025

88CB2063 B. Giese, J. He and W. Mehl; *Chem. Ber.*, 1988, **121**, 2063. 325, 921

88CB2239 M. Brudermüller, H. Musso and A. Wagner; *Chem. Ber.*, 1988, **121**, 2239. 504, 623

88CC44 Y. Masuyama, R. Hayashi, K. Otake and Y. Kurusu; *J. Chem. Soc., Chem. Commun.*, 1988, 44. 195

88CC81 D. C. Lathbury, P. J. Parsons and I. Pinto; *J. Chem. Soc., Chem. Commun.*, 1988, 81. 248

88CC203 C. Combellas, H. Gautier, J. Simon, A. Thiebault, F. Tournilhac, M. Barzoukas, D. Josse, I. Ledoux, C. Amatore and J. N. Verpeaux; *J. Chem. Soc., Chem. Commun.*, 1988, 203. 474

88CC237 J. P. Dulcère, M. N. Mihoubi and J. Rodriguez; *J. Chem. Soc., Chem. Commun.*, 1988, 237. 955

88CC283 P. J. Parsons, P. A. Willis and S. C. Eyley; *J. Chem. Soc., Chem. Commun.*, 1988, 283. 1130

88CC285 D. H. R. Barton, E. da Silva and S. Z. Zard; *J. Chem. Soc., Chem. Commun.*, 1988, 285. 348

88CC301 P. J. Stang and A. E. Learned; *J. Chem. Soc., Chem. Commun.*, 1988, 301. 988

88CC354 N. Kato, H. Kataoka, S. Ohbuchi, S. Tanaka and H. Takeshita; *J. Chem. Soc., Chem. Commun.*, 1988, 354. 690, 806, 807

88CC370 N. Shirai, F. Sumiya, Y. Sato and M. Hori; *J. Chem. Soc., Chem. Commun.*, 1988, 370. 824

88CC439 D. B. Repke, Jahangir, R. D. Clark and D. B. MacLean; *J. Chem. Soc., Chem. Commun.*, 1988, 439. 1041

88CC449 S. Takano, Y. Sekiguchi and K. Ogasawara; *J. Chem. Soc., Chem. Commun.*, 1988, 449. 188

88CC454 G. Casy and R. J. K. Taylor; *J. Chem. Soc., Chem. Commun.*, 1988, 454. 708

88CC539 L. Crombie and K. M. Mistry; *J. Chem. Soc., Chem. Commun.*, 1988, 539. 815

88CC748 A. H. Alberts and H. Wynberg; *J. Chem. Soc., Chem. Commun.*, 1988, 748. 970

88CC760 D. Kim, Y. M. Jang, I. O. Kim and S. W. Park; *J. Chem. Soc., Chem. Commun.*, 1988, 760. 183

88CC830 S. Neunlist and M. Rohmer; *J. Chem. Soc., Chem. Commun.*, 1988, 830. 1077

88CC975 E. Wenkert, A.-L. Han and C.-J. Jenny; *J. Chem. Soc., Chem. Commun.*, 1988, 975. 487

88CC1044 H. Takayama and T. Suzuki; *J. Chem. Soc., Chem. Commun.*, 1988, 1044. 620

88CC1110 H. Bhandal and G. Pattenden; *J. Chem. Soc., Chem. Commun.*, 1988, 1110. 335

88CC1135 J. Barluenga, F. J. Fañanás, F. Foubelo and M. Yus; *J. Chem. Soc., Chem. Commun.*, 1988, 1135. 1090

88CC1278 A. Kamimura and N. Ono; *J. Chem. Soc, Chem. Commun.*, 1988, 1278. 592, 638

88CC1280	T. H. Chan, L. M. Chen and D. Wang; *J. Chem. Soc., Chem. Commun.*, 1988, 1280.	1091
88CC1281	J. H. Hutchinson, D. L. Kuo, T. Money and B. Yokoyama; *J. Chem. Soc., Chem. Commun.*, 1988, 1281.	815
88CC1380	J. D. Harling and W. B. Motherwell; *J. Chem. Soc., Chem. Commun.*, 1988, 1380.	928
88CC1531	T. Suzuki, E. Sato, Y. Matsuda, H. Tada, S. Koizumi, K. Unno and T. Kametani; *J. Chem. Soc., Chem. Commun.*, 1988, 1531.	829
88CJC1359	K. P. Shelly and L. Weiler; *Can. J. Chem.*, 1988, **66**, 1359.	769
88CJC2860	J. S. Grossert, H. R. W. Dharmaratne, T. S. Cameron and B. R. Vincent; *Can. J. Chem.*, 1988, **66**, 2860.	702
88CL645	M. Hashimoto, M. Yanagiya and H. Shirahama; *Chem. Lett.*, 1988, 645.	210
88CL809	S. Hyuga, N. Yamashina, S. Hara and A. Suzuki; *Chem. Lett.*, 1988, 809.	489
88CL849	T. Ogawa, T. Murafuji and H. Suzuki; *Chem. Lett.*, 1988, 849.	743
88CL873	K. Ikenaga, S. Matsumoto, K. Kikukawa and T. Matsuda; *Chem. Lett.*, 1988, 873.	490
88CL2009	K. Inomata, T. Hirata, H. Suhara, H. Kinoshita, H. Kotake and H. Senda; *Chem. Lett.*, 1988, 2009.	619
88CPB1890	H. Yamanaka, T. Sakamoto, M. Annaka, Y. Kondo and T. Araki; *Chem. Pharm. Bull.*, 1988, **36**, 1890.	1010
88CPB1935	K. Tanji and T. Higashino; *Chem. Pharm. Bull.*, 1988, **36**, 1935.	1010
88CPB2248	T. Sakamoto, H. Yamanaka, Y. Kondo and T. Nagano; *Chem. Pharm. Bull.*, 1988, **36**, 2248.	1010
88CPB3897	Y. Tamura, S. Akai, H. Kishimoto, M. Sasho, M. Kirihara and Y. Kita; *Chem. Pharm. Bull.*, 1988, **36**, 3897.	1140
88CPB4337	K. Sakai, H. Suemune, T. Harabe and Z. Xie; *Chem. Pharm. Bull.*, 1988, **36**, 4337.	1021
88CRV487	J. Barluenga and M. Yus; *Chem. Rev.*, 1988, **88**, 487.	332, 343, 948
88CRV733	B. E. Kahn and R. D. Rieke; *Chem. Rev.*, 1988, **88**, 733.	687, 692
88CRV1081	N. E. Schore; *Chem. Rev.*, 1988, **88**, 1081.	516, 517, 530, 551
88CRV1423	F. E. Ziegler; *Chem. Rev.*, 1988, **88**, 1423.	414, 825, 836
88CSR229	Z. Goldschmidt and B. Crammer; *Chem. Soc. Rev.*, 1988, **17**, 229.	801, 802, 803
88CSR361	G. Pattenden; *Chem. Soc. Rev.*, 1988, **17**, 361.	320, 335, 345, 933, 946
88G577	L. Pandolfo and G. Paiaro; *Gazz. Chim. Ital.*, 1988, **118**, 577.	65
88G857	A. M. Caporusso, C. Consoloni and L. Lardicci; *Gazz. Chim. Ital.*, 1988, **118**, 857.	198
88H(27)261	A. Ohta, K. Okimura, Y. Tonomura, M. Ohta, N. Yasumuro, R. Fujita and M. Shimazaki; *Heterocycles*, 1988, **27**, 261.	595, 597
88H(27)475	T. Sakamoto, S. Nishimura, Y. Kondo and H. Yamanaka; *Heterocycles*, 1988, **27**, 475.	1034
88H(27)751	S. T. Lu and I. L. Tsai; *Heterocycles*, 1988, **27**, 751.	627
88H(27)981	Y. Terao, M. Aono and K. Achiwa; *Heterocycles*, 1988, **27**, 981.	368
88H(27)1575	E. Zeller and D. S. Grierson; *Heterocycles*, 1988, **27**, 1575.	
88H(27)2077	T. Fujii, M. Ohba and M. Sakari; *Heterocycles*, 1988, **27**, 2077.	752
88H(27)2367	E. Tojo, D. Dominguez and L. Castedo; *Heterocycles*, 1988, **27**, 2367.	626
88HCA47	C. Tamm and L. Rosslein; *Helv. Chim. Acta*, 1988, **71**, 47.	178
88HCA168	S. E. Denmark, K. L. Habermas and G. A. Hite; *Helv. Chim. Acta*, 1988, **71**, 168.	422
88IC4429	C. Bianchini, A. Meli, F. Laschi, J. A. Ramirez, P. Zanello and A. Vacca; *Inorg. Chem.*, 1988, **27**, 4429.	507
88JA291	W. S. Mahoney, D. M. Brestensky and J. M. Stryker; *J. Am. Chem. Soc.*, 1988, **110**, 291.	86
88JA607	S. T. Graul and R. R. Squires; *J. Am. Chem. Soc.*, 1988, **110**, 607.	846
88JA633	K. Komatsu, H. Akamatsu, Y. Jinbu and K. Okamoto; *J. Am. Chem. Soc.*, 1988, **110**, 633.	907
88JA649	E. J. Corey, M. Kang, M. C. Desai, A. K. Ghosh and I. N. Houpis; *J. Am. Chem. Soc.*, 1988, **110**, 649.	1029
88JA842	J. S. Sawyer, A. Kucerovy, T. L. Macdonald and G. J. McGarvey; *J. Am. Chem. Soc.*, 1988, **110**, 842.	139
88JA854	R. E. Ireland and M. G. Smith; *J. Am. Chem. Soc.*, 1988, **110**, 854.	831
88JA879	L. A. Paquette, K. S. Learn, J. L. Romine and H.-S. Lin; *J. Am. Chem. Soc.*, 1988, **110**, 879.	815, 1098
88JA890	L. A. Paquette, D. T. DeRussy and C. E. Cottrell; *J. Am. Chem. Soc.*, 1988, **110**, 890.	813
88JA960	B. M. Novak and R. H. Grubbs; *J. Am. Chem. Soc.*, 1988, **110**, 960.	675
88JA1108	R. C. Haddon; *J. Am. Chem. Soc.*, 1988, **110**, 1108.	917
88JA1288	K. Mizuno, M. Ikeda, S. Toda and Y. Otsuji; *J. Am. Chem. Soc.*, 1988, **110**, 1288.	348
88JA1529	H. C. Brown, R. K. Bakshi and B. Singaram; *J. Am. Chem. Soc.*, 1988, **110**, 1529.	1034
88JA1557	J. K. Still and A. M. Echavarren; *J. Am. Chem. Soc.*, 1988, **110**, 1557.	1030
88JA1602	B. M. Trost, S. M. Mignani and T. M. Manninga; *J. Am. Chem. Soc.*, 1988, **110**, 1602.	287
88JA1626	P. Magnus and P. A. Carter; *J. Am. Chem. Soc.*, 1988, **110**, 1626.	1009
88JA1631	D. J. Hart and F. L. Seely; *J. Am. Chem. Soc.*, 1988, **110**, 1631.	924
88JA1633	P. A. Bartlett, K. L. McLaren and P. C. Tring; *J. Am. Chem. Soc.*, 1988, **110**, 1633.	175
88JA1636	B. M. Trost and G. J. Tanoury; *J. Am. Chem. Soc.*, 1988, **110**, 1636.	676
88JA1981	A. M. Ross, D. L. Whalen, S. Eldin and R. M. Pollack; *J. Am. Chem. Soc.*, 1988, **110**, 1981.	877
88JA2066	L. J. De Koning and N. M. M. Nibbering; *J. Am. Chem. Soc.*, 1988, **110**, 2066.	590
88JA2210	A. S. Kende, K. Koch and C. A. Smith; *J. Am. Chem. Soc.*, 1988, **110**, 2210.	926
88JA2218	P. A. Wender and A. W. White; *J. Am. Chem. Soc.*, 1988, **110**, 2218.	120
88JA2248	L. E. Overman and A. S. Thompson; *J. Am. Chem. Soc.*, 1988, **110**, 2248.	751
88JA2301	B. M. Trost and T. Schmidt; *J. Am. Chem. Soc.*, 1988, **110**, 2301.	1047
88JA2314	R. L. Vance, N. G. Rondan, K. N. Houk, F. Jensen, W. T. Borden, A. Komornicki and E. Wimmer; *J. Am. Chem. Soc.*, 1988, **110**, 2314.	415
88JA2324	J. P. Dinnocenzo and D. A. Conlon; *J. Am. Chem. Soc.*, 1988, **110**, 2324.	803

88JA2501 D. Zhai, W. Zhai and R. M. Williams; *J. Am. Chem. Soc.*, 1988, **110**, 2501. 1004, 1135

88JA2565 A. L. J. Beckwith, D. M. O'Shea and S. W. Westwood; *J. Am. Chem. Soc.*, 1988, **110**, 2565. 245, 1127

88JA2641 J. R. Behling, K. A. Babiak, J. S. Ng, A. L. Campbell, R. Moretti, M. Koerner and B. H. Lipshutz; *J. Am. Chem. Soc.*, 1988, **110**, 2641. 120

88JA2655 J. J. Bozell and C. E. Vogt; *J. Am. Chem. Soc.*, 1988, **110**, 2655. 494

88JA2714 W. Tumas, R. F. Foster and J. I. Brauman; *J. Am. Chem. Soc.*, 1988, **110**, 2714. 883

88JA2877 J. Chateauneuf, J. Lusztyk and K. U. Ingold; *J. Am. Chem. Soc.*, 1988, **110**, 2877. 924, 934

88JA2925 J. A. Marshall and J. Lebreton; *J. Am. Chem. Soc.*, 1988, **110**, 2925. 818

88JA3296 S. Aoki, T. Fujimura, E. Nakamura and I. Kuwajima; *J. Am. Chem. Soc.*, 1988, **110**, 3296. 483

88JA3298 R. L. Funk, T. A. Olmstead and M. Parvez; *J. Am. Chem. Soc.*, 1988, **110**, 3298. 830, 831

88JA3300 K. S. Feldman, A. L. Romanelli, R. E. Ruckle, Jr. and R. F. Miller; *J. Am. Chem. Soc.*, 1988, **110**, 3300. 928

88JA3318 A. Bhattacharya, L. M. Di Michele, U.-H. Dolling, A. W. Douglas and E. J. J. Grabowski; *J. Am. Chem. Soc.*, 1988, **110**, 3318. 558

88JA3578 K. Yamamoto, T. Harada, Y. Okamoto, H. Chikamatsu, M. Nakazaki, Y. Kai, T. Nakao, M. Tanaka, S. Harada and N. Kasai; *J. Am. Chem. Soc.*, 1988, **110**, 3578. 690

88JA3601 S. S. Hixson, L. A. Franke, J. A. Gere and Y. Xing; *J. Am. Chem. Soc.*, 1988, **110**, 3601. 841

88JA3687 B. M. Trost and M. L. Miller; *J. Am. Chem. Soc.*, 1988, **110**, 3687. 288

88JA3929 S. J. Danishefsky, M. P. DeNinno and S. Chen; *J. Am. Chem. Soc.*, 1988, **110**, 3929. 751

88JA4062 R. A. Gibbs and W. H. Okamura; *J. Am. Chem. Soc.*, 1988, **110**, 4062. 993

88JA4112 S. A. Johnson and D. T. Clark; *J. Am. Chem. Soc.*, 1988, **110**, 4112. 906

88JA4346 J. W. L. Martin, F. S. Stephens, K. D. V. Weerasuria and S. B. Wild; *J. Am. Chem. Soc.*, 1988, **110**, 4346. 53

88JA4415 A. L. J. Beckwith and B. P. Hay; *J. Am. Chem. Soc.*, 1988, **110**, 4415. 931

88JA4473 Y. Masuyama, J. P. Takahara and Y. Kurusu; *J. Am. Chem. Soc.*, 1988, **110**, 4473. 195

88JA4685 K. C. Nicolaou, R. A. Daines, T. K. Chakraborty and Y. Ogawa; *J. Am. Chem. Soc.*, 1988, **110**, 4685. 750

88JA4718 M. Suzuki, A. Yanagisawa and R. Noyori; *J. Am. Chem. Soc.*, 1988, **110**, 4718. 132

88JA4778 A. I. Meyers, D. B. Miller and F. H. White; *J. Am. Chem. Soc.*, 1988, **110**, 4778. 149

88JA4788 A. R. Chamberlin, S. H. Bloom, L. A. Cervini and C. H. Fotsch; *J. Am. Chem. Soc.*, 1988, **110**, 4788. 316

88JA4815 T. Toru, Y. Yamada, T. Ueno, E. Maekawa and Y. Ueno; *J. Am. Chem. Soc.*, 1988, **110**, 4815. 243

88JA4822 E. Vedejs, J. D. Rodgers and S. J. Wittenberger; *J. Am. Chem. Soc.*, 1988, **110**, 4822. 821

88JA4866 K. C. Nicolaou, G. Zuccarello, Y. Ogawa, E. J. Schweiger and T. Kumazawa; *J. Am. Chem. Soc.*, 1988, **110**, 4866. 617, 709

88JA5064 T. L. Fevig, R. L. Elliot and D. P. Curran; *J. Am. Chem. Soc.*, 1988, **110**, 5064. 548

88JA5075 J. C. Fishbein and W. P. Jencks; *J. Am. Chem. Soc.*, 1988, **110**, 5075. 590

88JA5198 S. L. Schreiber and H. V. Meyers; *J. Am. Chem. Soc.*, 1988, **110**, 5198. 697, 698, 703, 1099

88JA5216 B. M. Trost and C. A. Merlic; *J. Am. Chem. Soc.*, 1988, **110**, 5216. 237

88JA5231 B. M. Trost and J. M. Tour; *J. Am. Chem. Soc.*, 1988, **110**, 5231. 496

88JA5383 E. Negishi, L. D. Boardman, H. Sawada, V. Bagheri, A. T. Stoll, J. M. Tour and C. L. Rand; *J. Am. Chem. Soc.*, 1988, **110**, 5383. 464

88JA5479 J. Dressel, K. L. Chasey and L. A. Paquette; *J. Am. Chem. Soc.*, 1988, **110**, 5479. 708

88JA5567 J. Barluenga, M. A. Rodríguez and P. J. Campos; *J. Am. Chem. Soc.*, 1988, **110**, 5567. 990

88JA5593 D. Hilvert and K. D. Nared; *J. Am. Chem. Soc.*, 1988, **110**, 5593. 826

88JA5653 E. J. Corey and G. K. Wiatkowski; *J. Am. Chem. Soc.*, 1966, **88**, 5653. 752

88JA5900 D. P. Curran, D. Kim, H. T. Liu and W. Shen; *J. Am. Chem. Soc.*, 1988, **110**, 5900. 339, 927

88JA5932 S. D. Larsen; *J. Am. Chem. Soc.*, 1988, **110**, 5932. 234

88JA6033 W. Bauer, M. Feigel, G. Müller and P. von R. Schleyer; *J. Am. Chem. Soc.*, 1988, **110**, 6033. 1088

88JA6565 M. Ochiai, Y. Takaoka and Y. Nagao; *J. Am. Chem. Soc.*, 1988, **110**, 6565. 431, 1068

88JA6579 W. S. Trahanovsky, C.-H. Chou, D. R. Fischer and B. C. Gerstein; *J. Am. Chem. Soc.*, 1988, **110**, 6579. 655

88JA6911 G. Stork and M. E. Reynolds; *J. Am. Chem. Soc.*, 1988, **110**, 6911. 338

88JA6914 D. L. J. Clive, K. S. K. Murthy, A. G. H. Wee, J. S. Prasad, G. V. J. da Silva, M. Majewski, P. C. Anderson, R. D. Haugen and L. D. Heerze; *J. Am. Chem. Soc.*, 1988, **110**, 6914. 688, 690

88JA7178 D. Pansegrau, W. F. Rieker and A. I. Meyers; *J. Am. Chem. Soc.*, 1988, **110**, 7178. 539

88JA7185 K. Tanaka, I. Funaki, A. Kaji, K. Minami, M. Sawada and T. Tanaka; *J. Am. Chem. Soc.*, 1988, **110**, 7185. 240

88JA7247 K. C. Nicolaou, Y. Ogawa, G. Zuccarello and H. Kataoka; *J. Am. Chem. Soc.*, 1988, **110**, 7247. 704, 707, 709

88JA7339 J. L. Holmes, F. P. Lossing and A. Maccoll; *J. Am. Chem. Soc.*, 1988, **110**, 7339. 921

88JA7343 J. L. Holmes and F. P. Lossing; *J. Am. Chem. Soc.*, 1988, **110**, 7343. 921

88JA7538 Y. Gao and K. B. Sharpless; *J. Am. Chem. Soc.*, 1988, **110**, 7538. 17, 185

88JA7542 B. M. Novak and R. H. Grubbs; *J. Am. Chem. Soc.*, 1988, **110**, 7542. 675

88JA7906 W. D. Harman and H. Taube; *J. Am. Chem. Soc.*, 1988, **110**, 7906. 94

88JA7922 K. Maruoka, K. Nonoshita, H. Banno and H. Yamamoto; *J. Am. Chem. Soc.*, 1988, **100**, 7922. 415, 826

88JA8145 D. R. Hay, Z. Song, S. G. Smith and P. Beak; *J. Am. Chem. Soc.*, 1988, **110**, 8145. 148

88JA8153 T. Hayashi, K. Hayashizaki, T. Kiyoi and Y. Ito; *J. Am. Chem. Soc.*, 1988, **110**, 8153. 473

88JA8164 D. J. Pasto; *J. Am. Chem. Soc.*, 1988, **110**, 8164. 922

88JA8197	L. A. Paquette and M. P. Trova; *J. Am. Chem. Soc.*, 1988, **110**, 8197.	708
88JA8599	G. A. Olah, G. K. Surya Prakash, W.-D. Fessner, T. Kobayashi and L. A. Paquette; *J. Am. Chem. Soc.*, 1988, **110**, 8599.	900
88JCS(P1)53	U. C. Dyer and J. A. Robinson; *J. Chem. Soc., Perkin Trans. 1*, 1988, 53.	614
88JCS(P1)575	T. Toru, T. Seko, E. Maekawa and Y. Ueno; *J. Chem. Soc., Perkin Trans. 1*, 1988, 575.	603
88JCS(P1)813	M. D. Brown, C. J. Meunier and G. H. Whitham; *J. Chem. Soc., Perkin Trans. 1*, 1988, 813.	591
88JCS(P1)817	M. D. Brown and G. H. Whitham; *J. Chem. Soc., Perkin Trans. 1*, 1988, 817.	591
88JCS(P1)839	C. A. Townsend, S. B. Christensen and S. G. Davis; *J. Chem. Soc., Perkin Trans. 1*, 1988, 839.	539
88JCS(P1)1085	M. J. Begley, G. Pattenden and G. M. Robertson; *J. Chem. Soc., Perkin Trans. 1*, 1988, 1085.	690
88JCS(P1)1517	M. D. Bachi and E. Bosch; *J. Chem. Soc., Perkin Trans. 1*, 1988, 1517.	938
88JCS(P1)1729	A. Fürstner, R. Csuk, C. Rohrer and H. Weidmann; *J. Chem. Soc., Perkin Trans. 1*, 1988, 1729.	688
88JCS(P1)1799	D. Levin and S. Warren; *J. Chem. Soc., Perkin Trans. 1*, 1988, 1799.	641, 643
88JCS(P1)2847	M. G. Moloney and J. T. Pinhey; *J. Chem. Soc., Perkin Trans. 1*, 1988, 2847.	490
88JCS(P1)2971	P. Wallace and S. Warren; *J. Chem. Soc., Perkin Trans. 1*, 1988, 2971.	643, 757
88JCS(P1)3061	C. Fuganti, P. Grasselli, S. Servi and H. E. Högberg; *J. Chem. Soc., Perkin Trans. 1*, 1988, 3061.	78
88JCS(P1)3255	P. de Armas, C. G. Francisco, R. Hernández, J. A. Salazar and E. Suárez; *J. Chem. Soc., Perkin Trans. 1*, 1988, 3255.	927
88JCS(P2)177	N. Al-Awadi and R. Taylor; *J. Chem. Soc., Perkin Trans. 2*, 1988, 177.	777
88JCS(P2)737	R. Taylor; *J. Chem. Soc., Perkin Trans. 2*, 1988, 737.	777
88JCS(P2)983	S. M. Partington and C. I. F. Watt; *J. Chem. Soc., Perkin Trans. 2*, 1988, 983.	883
88JCS(P2)1829	G. Ferguson, C. Glidewell, I. Gosney, D. Lloyd, S. Metcalfe, and H. Lumbroso; *J. Chem. Soc., Perkin Trans. 2*, 1988, 1829.	648, 738, 743
88JCS(P2)1989	J. C. Walton; *J. Chem. Soc., Perkin Trans. 2*, 1988, 1989.	923
88JOC50	G. Majetich, J. Defauw and C. Ringold; *J. Org. Chem.*, 1988, **53**, 50.	314
88JOC139	C. Jaime, R. M. Ortuño and J. Font; *J. Org. Chem.*, 1988, **53**, 139.	217, 225
88JOC196	J. A. Duncan, L. Y. Aki, M. J. Absalon, K. S. Kwong and R. T. Hendricks; *J. Org. Chem.*, 1988, **53**, 196.	808
88JOC239	H. C. Brown, D. Basavaiah, S. U. Kulkarni, N. G. Bhat and J. V. N. V. Prasad; *J. Org. Chem.*, 1988, **53**, 239.	487, 488
88JOC251	N. Ono, A. Kamimura and A. Kaji; *J. Org. Chem.*, 1988, **53**, 251.	600, 617, 638
88JOC386	J. W. Tilley and S. J. Zawoiski; *J. Org. Chem.*, 1988, **53**, 386.	1029
88JOC432	M. E. Krafft and W. J. Crooks, III; *J. Org. Chem.*, 1988, **53**, 432.	12
88JOC477	L. A. Paquette, M. E. Okazaki and J.-C. Caille; *J. Org. Chem.*, 1988, **53**, 477.	49
88JOC661	D. Lenoir, Y. Apeloig, D. Arad and P. von R. Schleyer; *J. Org. Chem.*, 1988, **53**, 661.	905
88JOC918	H. Hatanaka and T. Hiyama; *J. Org. Chem.*, 1988, **53**, 918.	490, 1007
88JOC1064	P. M. Gannett, D. L. Nagel, P. J. Reilly, T. Lawson, J. Sharpe and B. Toth; *J. Org. Chem.*, 1988, **53**, 1064.	700, 703
88JOC1432	M. Vaman Rao and M. Nagarajan; *J. Org. Chem.*, 1988, **53**, 1432.	330
88JOC1475	N. Kornblum, P. Ackermann, J. W. Manthey, M. T. Musser, H. W. Pinnick, S. Singaram and P. A. Wade; *J. Org. Chem.*, 1988, **53**, 1475.	200, 238
88JOC1733	C. Lillocci; *J. Org. Chem.*, 1988, **53**, 1733.	629
88JOC1790	W. H. Okamura, S. A. Barrack and R. A. Gibbs; *J. Org. Chem.*, 1988, **53**, 1790	1029
88JOC1823	A. F. Cunningham and E. P. Kündig; *J. Org. Chem.*, 1988, **53**, 1823.	25
88JOC1848	S. Y. Lee, Y. S. Kulkarni, B. W. Burbaum, M. I. Johnston and B. B. Snider; *J. Org. Chem.*, 1988, **53**, 1848.	805
88JOC1903	P. E. Peterson and M. Stepanian; *J. Org. Chem.*, 1988, **53**, 1903.	769
88JOC1922	C. H. Heathcock, B. L. Finkelstein, E. T. Jarvi, P. A. Radel and C. R. Hadley; *J. Org. Chem.*, 1988, **53**, 1922.	829
88JOC2137	B. B. Snider, J. J. Patricia and S. A. Kates; *J. Org. Chem.*, 1988, **53**, 2137.	336, 346, 926
88JOC2450	C. Brückner, H. Holzinger and H.-U. Reissig; *J. Org. Chem.*, 1988, **53**, 2450.	978
88JOC2489	G. M. Whitesides and T. X. Neenan; *J. Org. Chem.*, 1988, **53**, 2489.	1010
88JOC2655	A. S. Kende and C. A. Smith; *J. Org. Chem.*, 1988, **53**, 2655.	1065
88JOC2688	C.-N. Hsiao and H. Shechter; *J. Org. Chem.*, 1988, **53**, 2688.	616, 654
88JOC2706	S. S. Hixson and L. A. Franke; *J. Org. Chem.*, 1988, **53**, 2706.	841
88JOC2711	S. S. Hixson and C. R. Gallucci; *J. Org. Chem.*, 1988, **53**, 2711.	841
88JOC2829	A. M. Piotrowski, D. B. Malpass, M. P. Boleslawski and J. J. Eisch; *J. Org. Chem.*, 1988, **53**, 2829.	680
88JOC3122	A. E. Learned, A. M. Arif and P. J. Stang; *J. Org. Chem.*, 1988, **53**, 3122.	988
88JOC3195	P. A. Bartlett, Y. Nakagawa, C. R. Johnson, S. H. Reich and A. Luis; *J. Org. Chem.*, 1988, **53**, 3195.	826
88JOC3306	H. Alper and F. Sibtain; *J. Org. Chem.*, 1988, **53**, 3306.	202
88JOC3383	U. C. Dyer and Y. Kishi; *J. Org. Chem.*, 1988, **53**, 3383.	1099
88JOC3391	G. B. Feigelson and S. J. Danishefsky; *J. Org. Chem.*, 1988, **53**, 3391.	648, 760
88JOC3641	C. Chatgilialoglu, D. Griller and M. Lesage; *J. Org. Chem.*, 1988, **53**, 3641.	5, 7, 9, 10, 331, 936
88JOC3815	T. G. Back, V. I. Birss, M. Edwards and M. V. Krishna; *J. Org. Chem.*, 1988, **53**, 3815.	440
88JOC3868	K. Tadano, S. Kanazawa and S. Ogawa; *J. Org. Chem.*, 1988, **53**, 3868.	827
88JOC4108	J. A. Marshall and J. Lebreton; *J. Org. Chem.*, 1988, **53**, 4108.	818
88JOC4151	Y. Wang, P. G. Goekjian, D. M. Ryckman and Y. Kishi; *J. Org. Chem.*, 1988, **53**, 4151.	1099

88JOC4274	J. A. Marshall, J. D. Trometer, B. E. Blough and T. D. Crute; *J. Org. Chem.*, 1988, **53**, 4274.	226, 227, 751
88JOC4282	J. C. Barrish, H. L. Lee, T. Mitt, G. Pizzolato, E. G. Baggiolini and M. R. Usković; *J. Org. Chem.*, 1988, **53**, 4282.	697, 700, 703
88JOC4419	B. Zhou and Y. Xu; *J. Org. Chem.*, 1988, **53**, 4419.	196
88JOC4466	M.-C. Chan, K.-M. Cheng, K. M. Ho, C. T. Ng, T. M. Yam, B. S. L. Wang and T.-Y. Luh; *J. Org. Chem.*, 1988, **53**, 4466.	436
88JOC4472	Y.-H. Lai and S.-M. Lee; *J. Org. Chem.*, 1988, **53**, 4472.	233
88JOC4482	G. W. Ebert and R. D. Rieke; *J. Org. Chem.*, 1988, **53**, 4482.	121
88JOC4508	B. B. Snider and R. B. Beal; *J. Org. Chem.*, 1988, **53**, 4508.	815, 832
88JOC4522	T. V. RajanBabu; *J. Org. Chem.*, 1988, **53**, 4522.	176
88JOC4631	M. D. Wittman and J. Kallmerten; *J. Org. Chem.*, 1988, **53**, 4631.	816, 819
88JOC4736	M. A. Henderson and C. H. Heathock; *J. Org. Chem.*, 1988, **53**, 4736.	979
88JOC4807	D. J. Pasto and W. Kong; *J. Org. Chem.*, 1988, **53**, 4807.	498
88JOC4859	J.-L. De Keyser, C. J. C. De Cock, J. H. Poupaert and P. Dumont; *J. Org. Chem.*, 1988, **53**, 4859.	786
88JOC4862	X. Huang, L. Xie and H. Wu; *J. Org. Chem.*, 1988, **53**, 4862.	623
88JOC4892	P. DeShong and D. R. Sidler; *J. Org. Chem.*, 1988, **53**, 4892.	208
88JOC4923	M. H. Block and D. E. Cane; *J. Org. Chem.*, 1988, **53**, 4923.	831
88JOC4972	L. A. Paquette and J. P. Gilday; *J. Org. Chem.*, 1988, **53**, 4972.	885
88JOC4978	L. A. Paquette and C. S. Ra; *J. Org. Chem.*, 1988, **53**, 4978.	885
88JOC5189	L. T. Rossano, D. J. Plata, J. Kallmerten; *J. Org. Chem.*, 1988, **53**, 5189.	816, 831
88JOC5576	P. G. Gassman and S. J. Burns; *J. Org. Chem.*, 1988, **53**, 5576.	579
88JOC5580	W. H. Miller, D. M. Ryckman, P. G. Goekjian, Y. Wang and Y. Kishi; *J. Org. Chem.*, 1988, **53**, 5580.	1099
88JOC5584	P. C.-M. Chan and J. M. Chong; *J. Org. Chem.*, 1988, **53**, 5584.	138
88JOC5854	A. R. Katritzky, M. Drewniak and P. Lue; *J. Org. Chem.*, 1988, **53**, 5854.	260
88JOC6009	H. C. Brown and N. G. Bhat; *J. Org. Chem.*, 1988, **53**, 6009.	487, 488
88JOM(25)348	N. Kanemoto, Y. Sato and S. Inoue; *J. Organomet. Chem.*, 1988, **25**, 348.	290
88JOM(25)351	L. D. Freedmann and G. O. Doak; *J. Organomet. Chem.*, 1988, **25**, 351.	271
88JOM(338)289	H. D. Verkruijsse, Y. A. Heus-Kloss and L. Brandsma; *J. Organomet. Chem.*, 1988, **338**, 289.	1045, 1046
88JOM(339)199	E. Alvarez, T. Cuvigny and M. Julia; *J. Organomet. Chem.*, 1988, **339**, 199.	195
88JOM(340)317	A. D. Ayala, N. Giagante, J. C. Podestá and W. P. Neumann; *J. Organomet. Chem.*, 1988, **340**, 317.	950
88JOM(346)181	Yanchang Shen and Qimu Liao; *J. Organomet. Chem.*, 1988, **346**, 181.	739
88JOM(346)C1	R. Tarhouni, B. Kirschleger and J. Villieras; *J. Organomet. Chem.*, 1988, **346**, C1.	664
88JOM(348)199	K. Laali, H. Y. Chen and R. J. Gerzina; *J. Organomet. Chem.*, 1988, **348**, 199.	247
88JOM(349)43	D. Damour, J. Pornet and L. Miginiac; *J. Organomet. Chem.*, 1988, **349**, 43.	978
88JOM(354)233	D. C. Billington, I. M. Helps, P. L. Pauson, W. Thompson and D. Willison; *J. Organomet. Chem.*, 1988, **354**, 233.	529
88JOM(355)449	P. Bladon, P. L. Pauson, H. Brunner and R. Eder; *J. Organomet. Chem.*, 1988, **355**, 449.	530
88JOU530	A. A. Mironova, I. V. Soloshonok, I. I. Maletina, V. V. Orda and L. M. Yagupol'skii; *J. Org. Chem. USSR* (*Engl. Transl.*), 1988, **24**, 530.	109
88JPO197	H. Sakurai, S. Kyushin, Y. Nakadaira and M. Kira; *J. Phys. Org. Chem.*, 1988, **1**, 197.	922
88JPR309	S. Sólyom, K. Szilagyi and I. Toldy; *J. Prakt. Chem.*, 1988, **330**, 309.	578
88LA69	J. Reisch and A. Bathe; *Liebigs Ann. Chem.*, 1988, 69.	1010
88LA221	H. J. Bissinger, H. Detert and H. Meier; *Liebigs Ann. Chem.*, 1988, 221.	1077
88LA377	L. Shi, J. Yang, M. Li and Y.-Z. Huang; *Liebigs Ann. Chem.*, 1988, 377.	647
88LA543	J. Reisch and A. Bathe; *Liebigs Ann. Chem.*, 1988, 543.	576, 1010
88M253	G. Heinisch, W. Holzer and C. Obala; *Monatsh. Chem.*, 1988, **119**, 253.	1010
88M751	W. Dostal, G. Heinish and G. Lotsh; *Monatsh. Chem.*, 1988, **119**, 751.	1010
B-88MI 103-01	A. Pelter, K. Smith and H. C. Brown; "Borane Reagents," Academic Press, London, 1988.	75
88MI 103-02	*Tetrahedron*, 1988, **44**, issue 10, v-xviii.	95
B-88MI 104-01	S. Oae and Y. Uchida; in "The Chemistry of Sulphones and Sulphoxides," eds S. Patai, Z. Rappoport and C. Stirling; Wiley, New York, 1988, pp. 583–664.	140
B-88MI 104-02	J. S. Grossert; in "The Chemistry of Sulphones and Sulphoxides," eds S. Patai, Z. Rappoport and C. Stirling; Wiley, New York, 1988, pp. 925–968.	144
B-88MI 108-01	E. Vedejs; in "Advances in Cycloadditions," ed. D. P. Curran, JAI, London, 1988, vol. 1, p. 33.	321
B-88MI 110-01	A. Pelter, K. Smith and H. C. Brown; "Borane Reagents," Academic Press, London, 1988.	449
B-88MI 112-01	A. Pelter, K. Smith and H. C. Brown; "Borane Reagents," Academic Press, London, 1988.	509
B-88MI 112-02	A. Pelter, K. Smith and H. C. Brown; "Borane Reagents," Academic Press, London, 1988, p. 187.	509
88MI 114-01	O. G. Vyglazov and E. N. Manukov; *Vestsi Akad. Nauk BSSR, Set. Khim. Navuk*, 1988, 111 (*Chem. Abstr.*, 1989, **111**, 134 490).	626
B-88MI 114-02	E. W. Colvin; "Silicon Reagents in Organic Synthesis," Academic Press, New York, 1988, pp. 63–75.	648
B-88MI 114-03	P. F. Hudrlik and E. L. O. Agwaramgbo; in "Silicon Chemistry," eds E. R. Corey, J. Y. Corey and P. P. Gaspar, Ellis Horwood, Chichester, 1988, pp. 95–104.	648
88MI 114-04	G. W. Kabalka, S. J. Lambert and V. K. Jinaraj; *Appl. Radiat. Isot.*, 1988, **39**, 1113 (*Chem. Abstr.*, 1989, **110**, 57 914).	661

B-88MI 115-01	K. Schank; in "The Chemistry of Sulphones and Sulfoxides," eds S. Patai, Z Rappoport and C. J. M. Stirling, Wiley, New York, 1988, p. 165.	704
B-88MI 115-02	S. Oae and Y. Uchida; in "The Chemistry of Sulphones and Sulfoxides," ed. S. Patai, Z. Rappoport and C. J. M. Stirling, Wiley, New York, 1988, p. 583.	704
B-88MI 115-03	S. Braverman; in "The Chemistry of Sulphones and Sulfoxides," eds S. Patai, Z. Rappoport and C. J. M. Stirling, Wiley, New York, 1988, p. 665.	704
B-88MI 115-04	U. Zoller; "The Chemistry of Sulphones and Sulfoxides," eds S. Patai, Z. Rappoport and C. J. M. Stirling, Wiley, New York, 1988, p. 379.	707
B-88MI 116-01	E. W. Colvin; "Silicon Reagents in Organic Synthesis," Academic Press, New York, 1988.	758, 759
88MI 117-01	L. W. Jenneskens and U. E. Wiersum; *Aldrichimica Acta*, 1988, **21**, 58.	774
B-88MI 118-01	S. Braverman; in "The Chemistry of Sulphones and Sulphoxides," eds. S. Patai, Z. Rappoport and C. Stirling, Wiley, New York, 1988, p. 717.	839
B-88MI 119-01	J. T. Welch and S. Eswarakrishnan; in "Fluorine-Containing Molecules," eds. J. F. Liebman, A. Greenberg and W. R. Dolbier, Jr., VCH, Weinheim, 1988.	876
B-88MI 119-02	B. J. Wakefield; "Organolithium Methods," Academic Press, New York, 1988.	885
B-88MI 119-03	K. T. Finley; in "The Chemistry of Quinonoid Compounds," eds. S. Patai and Z. Rappoport, Wiley, New York, 1988, part 1, pp. 537–717.	888
88MI 119-04	R. Breslow; *Chemtracts Org. Chem.*, 1988, **1**, 333.	927
B-88MI 121-01	L. Brandsma; "Preparative Acetylenic Chemistry," 2nd edn., Elsevier, Amsterdam, 1988.	998, 1000, 1034, 1040, 1046, 1047, 1069, 1072, 1084
B-88MI 121-02	L. Brandsma; "Preparative Acetylenic Chemistry," 2nd edn., Elsevier, Amsterdam, 1988, p. 40.	1022
B-88MI 121-03	L. Brandsma; "Preparative Acetylenic Chemistry," 2nd edn., Elsevier, Amsterdam, 1988, p. 109.	1044
B-88MI 122-01	L. Brandsma; "Preparative Acetylenic Chemistry," 2nd edn., Elsevier, Amsterdam, 1988.	1133
88NKK157	S. Takekuma, Y. Matsubara, H. Yamamoto and T. Nozoe; *Nippon Kagaku Kaishi* (*J. Chem. Soc. Jpn.*), 1988, 157 (*Chem. Abstr.*, 1988, **109**, 73 113).	899
88OM84	M. J. Dabdoub and J. V. Comasseto; *Organometallics*, 1988, **7**, 84.	999
88OM739	S. Inoue, Y. Sato and T. Suzuki; *Organometallics*, 1988, **7**, 739.	290
88OM1869	K. Mochida, M. Wakasa, Y. Nakadaira, Y. Sakaguchi and H. Hayashi; *Organometallics*, 1988, **7**, 1869.	449
88OM2220	P. W. Pike, V. Gilliatt, M. Ridenour and J. W. Hershberger; *Organometallics*, 1988, **7**, 2220.	935
88OS(66)60	E. Negishi, T. Takahashi and S. Baba; *Org. Synth.*, 1988, **66**, 60.	468
88OS(66)67	E. Negishi, T. Takahashi and A. O. King; *Org. Synth.*, 1988, **66**, 67.	473
88OS(66)75	R. Mook, Jr. and P. M. Sher; *Org. Synth.*, 1988, **66**, 75.	548
88OS(66)95	T. C. T. Chang, M. Rosenblum and N. Simms; *Org. Synth.*, 1988, **66**, 95.	481
88OSC(6)210	W. Weyler, W. G. Duncan, M. B. Liewen and H. W. Moore; *Org. Synth. Coll. Vol.*, 1988, **6**, 210.	571
88OSC(6)327	Y. Ito, S. Fujii, M. Nakatsuka, F. Kawamoto and T. Saegusa; *Org. Syn. Coll. Vol.*, 1988, **6**, 327.	571
88OSC(6)427	D. Bellus, H. Sauter and C. D. Weis; *Org. Syn. Coll. Vol.*, 1988, **6**, 427.	571
88OSC(6)852	R. A. Benkeser and E. M. Kaiser; *Org. Synth., Coll. Vol.*, 1988, **6**, 852.	101
88OSC(6)936	A. J. Birch and K. B. Chamberlain; *Org. Synth., Coll. Vol.*, 1988, **6**, 936.	98
88PAC7	T. Hayashi; *Pure Appl. Chem.*, 1988, **60**, 7.	194
88PAC79	F. Naso; *Pure Appl. Chem.*, 1988, **60**, 79.	484
88PAC107	E. Piers; *Pure Appl. Chem.*, 1988, **60**, 107.	360
88PAC1635	H. G. Viehe, R. Merényi and Z. Janousek; *Pure Appl. Chem.*, 1988, **60**, 1635.	919, 922
88PAC1645	D. L. J. Clive; *Pure Appl. Chem.*, 1988, **60**, 1645.	919
88PAC1659	J.-M. Surzur and M. P. Bertrand; *Pure Appl. Chem.*, 1988, **60**, 1659.	926, 935, 942
88PIA187	A. R. Katritzky and S. Sengupta; *Proc. Indian Acad. Sci.*, 1988, **100**, 187.	845, 859
88PS(40)69	I. Petrova, M. Kirilov, S. Momchilova and K. Kossev; *Phosphorus Sulfur*, 1988, **40**, 69.	644
88S1	E. Negishi and T. Takahashi; *Synthesis*, 1988, 1.	544
88S144	J. Barluenga, F. Aznar and M. Bayod; *Synthesis*, 1988, 144.	68
88S250	M. Bengtsson and T. Liljefors; *Synthesis*, 1988, 250.	1000
88S263	D. Schinzer; *Synthesis*, 1988, 263.	979
88S319	C. Glidewell, D. Lloyd and S. Metcalfe; *Synthesis*, 1988, 319.	648
88S321	Y. Shen and Q. Liao; *Synthesis*, 1988, 321.	270
88S362	M. J. Kurth and E. G. Brown; *Synthesis*, 1988, 362.	835
88S417	D. P. Curran; *Synthesis*, 1988, 417.	172, 230, 320, 336, 338, 341, 546, 919, 935, 1126
88S489	D. P. Curran; *Synthesis*, 1988, 489.	172, 230, 320, 334, 336, 344, 546, 919, 935, 946, 948
88S621	P. C. B. Page, S. Rosenthal and R. V. Williams; *Synthesis*, 1988, 621.	1043
88S707	C. Almansa, A. Moyano, M. A. Pericàs and F. Serratosa; *Synthesis*, 1988, 707.	1085
88S854	S. Wershofen and H. D. Scharf; *Synthesis*, 1988, 854.	184
88S896	Q.-Y. Chen and Y.-B. He; *Synthesis*, 1988, 896.	433
88S943	D. Habich and W. Barth; *Synthesis*, 1988, 943.	208
88S983	B. Cazes, D. Djahanbini, J. Goré, J.-P. Genêt and J.-M. Gaudin; *Synthesis*, 1988, 983.	193
88SC89	K. E. Kim, S. B. Park and N. M. Yoon; *Synth. Commun.*, 1988, **18**, 89.	86
88SC151	S. H.-u-H. Kazmi, Z. Ahmed, A. Q. Khan and A. Malik; *Synth. Commun.*, 1988, **18**, 151.	218
88SC343	L. Jalander; *Synth. Commun.*, 1988, **18**, 343.	482
88SC445	F. Camps, V. Gasol and A. Guerrero; *Synth. Commun.*, 1988, **18**, 445.	179
88SC1159	H. Takahata, T. Yamazaki and E. C. Wang; *Synth. Commun.*, 1988, **18**, 1159.	209
88SC1207	H. S.-I. Chao; *Synth. Commun.*, 1988, **18**, 1207.	33
88SC1795	S. Abdel-Baky and A. Moussa; *Synth. Commun.*, 1988, **18**, 1795.	578

88SC1905 M. A. Tius and M. A. Kerr; *Synth. Commun.*, 1988, **18**, 1905. 580
88SCI426 P. G. Schultz; *Science*, 1988, **240**, 426. 826
88T111 E. Alvarez, T. Cuvigny, C. Hervé du Penhoat and M. Julia; *Tetrahedron*, 1988, **44**, 111. 485
88T127 M.-C. Carre, B. Jamart-Gregoire, P. Geoffroy, P. Caubere, S. Ianelli and M. Nardelli; *Tetrahedron*, 1988, **44**, 127. 539
88T147 D. H. R. Barton, G. Bashiardes and J. L. Fourrey; *Tetrahedron*, 1988, **44**, 147. 1075
88T1299 P. Gramatica, P. Manitto, D. Monti and G. Speranza; *Tetrahedron*, 1988, **44**, 1299. 77
88T1567 C. W. Bird and K. Naidoo; *Tetrahedron*, 1988, **44**, 1567. 181
88T2855 S. Matsubara, H. Matsuda T. Hamatani and M. Schlosser; *Tetrahedron*, 1988, **44**, 2855. 570
88T2913 B. Badet, M. Julia, J. M. Mallet and C. Schmitz; *Tetrahedron*, 1988, **44**, 2913. 623
88T3079 B. H. Kim, P. B. Jacobs, R. L. Elliott and D. P. Curran; *Tetrahedron*, 1988, **44**, 3079. 831
88T3139 L. A. Paquette, D. T. DeRussy and R. D. Rogers; *Tetrahedron*, 1988, **44**, 3139. 815
88T3781 G. L. Larson, J. A. Prieto and E. Ortiz; *Tetrahedron*, 1988, **44**, 3781. 649
88T3899 H. M. R. Hoffmann, U. Eggert, U. Gibbels, K. Giesel, O. Koch, R. Lies and J. Rabe; *Tetrahedron*, 1988, **44**, 3899. 364
88T4095 M. Ochiai, K. Sumi, Y. Takaoka, M. Kunishima, Y. Nagao, M. Shiro and E. Fujita; *Tetrahedron*, 1988, **44**, 4095. 431
88T4135 M. Fujita, M. Obayashi and T. Hiyama; *Tetrahedron*, 1988, **44**, 4135. 429
88T4295 J. M. Pons and M. Santelli; *Tetrahedron*, 1988, **44**, 4295. 687
88T4339 T. Kitahara, H. Kurata and K. Mori; *Tetrahedron*, 1988, **44**, 4339. 649
88T4671 G. M. Lee, M. Parvez and S. M. Weinreb; *Tetrahedron*, 1988, **44**, 4671. 937
88T5425 R. M. Williams and W. Zhai; *Tetrahedron*, 1988, **44**, 5425. 1042, 1135
88T5451 A. L. Castelhano, S. Horne, G. J. Taylor, R. Billedeau and A. Krantz; *Tetrahedron*, 1988, **44**, 5451. 1042
88T5479 D. H. R. Barton, Y. Hervé, P. Potier and J. Thierry; *Tetrahedron*, 1988, **44**, 5479. 935
88T5495 L. Duhamel, P. Duhamel, S. Fouquay, J. J. Eddine, O. Peschard, J.-C. Plaquevent, A. Ravard, R. Solliard, J.-Y. Valnot and H. Vincens; *Tetrahedron*, 1988, **44**, 5495. 180
88T6241 F. S. Guziec, Jr. and L. J. SanFilippo; *Tetrahedron*, 1988, **44**, 6241. 231, 243, 620, 685, 704
88T6295 P. Pike, S. Hershberger and J. Hershberger; *Tetrahedron*, 1988, **44**, 6295. 331, 936
88T6337 G. Linstrumelle and C. Huynh; *Tetrahedron*, 1988, **44**, 6337. 1012, 1013
88T6699 J. Boersma, W. de Graaf, A. Smits, G. van Koten and W. P. M. Hoekstra; *Tetrahedron*, 1988, **44**, 6699. 1018
88T6855 E. P. Kündig and A. F. Cunningham; *Tetrahedron*, 1988, **44**, 6855. 25
88T7037 J. M. Dener and D. J. Hart; *Tetrahedron*, 1988, **44**, 7037. 497
88T7587 J. C. Gilbert and T. A. Kelly; *Tetrahedron*, 1988, **44**, 7587. 832
88TCC73 H. U. Reissig; *Top. Curr. Chem.*, 1988, **144**, 73. 801
88TCC137 M. S. Baird; *Top. Curr. Chem.*, 1988, **144**, 137. 549
88TL21 H. C. Brown and N. G. Bhat; *Tetrahedron Lett.*, 1988, **29**, 21. 660, 661
88TL57 E. J. Corey and M. M. Mehrotra; *Tetrahedron Lett.*, 1988, **29**, 57. 1131
88TL97 Y. Hatanaka and T. Hiyama; *Tetrahedron Lett.*, 1988, 97. 490
88TL107 Y. K. Rao and M. Nagarajan; *Tetrahedron Lett.*, 1988, **29**, 107. 176
88TL167 B. P. Branchaud, M. S. Meier and Y. Choi; *Tetrahedron Lett.*, 1988, 167. 335
88TL273 L. A. Paquette and M.-A. Poupart; *Tetrahedron Lett.*, 1988, **29**, 273. 815
88TL281 S. Mills, R. Desmond, R. A. Reamer, R. P. Volante and I. Shinkai; *Tetrahedron Lett.*, 1988, **29**, 281. 940
88TL343 J. Tsuji, M. Yuhara, M. Minato, H. Yamada, F. Sato and Y. Kobayashi; *Tetrahedron Lett.*, 1988, **29**, 343. 228
88TL391 J. K. MacLeod and L. C. Monahan; *Tetrahedron Lett.*, 1988, **29**, 391. 1130
88TL411 C. B. Ziegler, Jr.; *Tetrahedron Lett.*, 1988, **29**, 411. 988
88TL419 A. Trehan and R. S. H. Liu; *Tetrahedron Lett.*, 1988, **29**, 419. 752
88TL655 K. Tadano, H. Yamada, Y. Idogaki, S. Ogawa and T. Suami; *Tetrahedron Lett.*, 1988, **29**, 655. 827
88TL659 M. Shimagaki, M. Shiokawa, K. Sugai, T. Teranaka, T. Nakata and T. Oishi; *Tetrahedron Lett.*, 1988, **29**, 659. 590, 591
88TL707 V. F. Patel and G. Pattenden; *Tetrahedron Lett.*, 1988, **29**, 707. 933
88TL815 C. Bonini and R. Di Fabio; *Tetrahedron Lett.*, 1988, **29**, 815. 215
88TL837 A. Johns and J. A. Murphy; *Tetrahedron Lett.*, 1988, **29**, 837. 927, 929
88TL893 B. H. Lipshutz, E. L. Ellsworth, J. R. Behling and A. L. Campbell; *Tetrahedron Lett.*, 1988, **29**, 893. 207
88TL905 R. Larock and B. E. Baker; *Tetrahedron Lett.*, 1988, 905. 494
88TL909 P. A. Wender, M. Harmata, D. Jeffrey, C. Mukai and J. Suffert; *Tetrahedron Lett.*, 1988, **29**, 909. 706
88TL955 J. A. Murphy, C. W. Patterson and N. F. Wooster; *Tetrahedron Lett.*, 1988, **29**, 955. 929
88TL975 H. M. L. Davies, C. E. M. Oldenburg, M. J. McAfee, J. G. Nordahl, J. P. Henretta and K. R. Romines; *Tetrahedron Lett.*, 1988, **29**, 975. 811
88TL999 M. E. Krafft; *Tetrahedron Lett.*, 1988, **29**, 999. 528, 529
88TL1041 K. Nozaki, K. Oshima and K. Utimoto; *Tetrahedron Lett.*, 1988, **29**, 1041. 333
88TL1085 G. Haaima and R. T. Weavers; *Tetrahedron Lett.*, 1988, **29**, 1085. 1131
88TL1197 M. Hirama, T. Nakamine and S. Itô; *Tetrahedron Lett.*, 1988, **29**, 1197. 702, 1070
88TL1203 L. Castedo, J. L. Mascareñas, A. Mouriño and L. A. Sarandeses; *Tetrahedron Lett.*, 1988, **29**, 1203. 1029
88TL1289 R. Beugelmans and M. Bois-Choussy; *Tetrahedron Lett.*, 1988, 1289. 474
88TL1315 D. S. Middleton, N. S. Simpkins and N. K. Terrett; *Tetrahedron Lett.*, 1988, **29**, 1315. 244

88TL1355 E. Torres, G. L. Larson and G. J. McGarvey; *Tetrahedron Lett.*, 1988, **29**, 1355. 981

88TL1371 S. J. Danishefsky and J. E. Audia; *Tetrahedron Lett.*, 1988, **29**, 1371. 831

88TL1437 C. J. Rhodes and E. Roduner; *Tetrahedron Lett.*, 1988, **29**, 1437. 922

88TL1441 Z. Citterio and R. Santi; *Tetrahedron Lett.*, 1988, **29**, 1441. 927

88TL1529 H.-J. Gais and J. Vollhardt; *Tetrahedron Lett.*, 1988, **29**, 1529. 146

88TL1543 K. Miura, K. Fugami, K. Oshima and K. Utimoto; *Tetrahedron Lett.*, 1988, **29**, 1543. 801

88TL1669 F. E. Ziegler, A. Nangia and G. Schulte; *Tetrahedron Lett.*, 1988, **29**, 1669. 826

88TL1763 H. Alper and S. Calet; *Tetrahedron Lett.*, 1988, **29**, 1763. 225

88TL1785 R. Yamaguchi, E. Hata and K. Utimoto; *Tetrahedron Lett.*, 1988, **29**, 1785. 1043

88TL1811 A. Suzuki, Y. Satoh, H. Serizawa, N. Miyaura and S. Hara; *Tetrahedron Lett.*, 1988, **29**, 1811. 542, 1008

88TL2059 J. C. Carretero and L. Ghosez; *Tetrahedron Lett.*, 1988, **29**, 2059. 210, 225, 613

88TL2195 S.-E. Yoo, S.-H. Lee and N.-J. Kim; *Tetrahedron Lett.*, 1988, **29**, 2195. 307

88TL2215 K. Furuta, T. Nagata and H. Yamamoto; *Tetrahedron Lett.*, 1988, **29**, 2215. 927

88TL2239 R. Rossi, A. Carpita and B. P. Andreini; *Tetrahedron Lett.*, 1988, **29**, 2239. 1007

88TL2257 R. Hunter and C. D. Simon; *Tetrahedron Lett.*, 1988, **29**, 2257. 254

88TL2395 M. C. P. Yeh and P. Knochel; *Tetrahedron Lett.*, 1988, **29**, 2395. 123

88TL2401 V. E. Amoo, S. De Bernardo and M. Weigele; *Tetrahedron Lett.*, 1988, **29**, 2401. 755

88TL2571 D. H. R. Barton, L. Bohe and X. Lusinchi; *Tetrahedron Lett.*, 1988, **29**, 2571. 46

88TL2599 J. A. Rechka and J. R. Maxwell; *Tetrahedron Lett.*, 1988, **29**, 2599. 677

88TL2613 G. L. Lange and C. P. Decicco; *Tetrahedron Lett.*, 1988, **29**, 2613. 357

88TL2631 H. C. Brown, N. G. Bhat and M. Srebnik; *Tetrahedron Lett.*, 1988, **29**, 2631. 999

88TL2773 G. Majetich and K. Hull; *Tetrahedron Lett.*, 1988, **29**, 2773. 812

88TL2797 L. Becking and H. J. Schäfer; *Tetrahedron Lett.*, 1988, **29**, 2797. 934

88TL2801 L. Becking and H. J. Schäfer; *Tetrahedron Lett.*, 1988, **29**, 2801. 934

88TL2805 K. Fritzsche, H.-D. Beckhaus and C. Rüchardt; *Tetrahedron Lett.*, 1988, **29**, 2805 637, 638

88TL2923 C. M. J. Fox, R. N. Hiner, U. Warrier and J. D. White; *Tetrahedron Lett.*, 1988, **29**, 2923. 1140

88TL2963 S. Becker, Y. Fort, R. Vanderesse and P. Caubère; *Tetrahedron Lett.*, 1988, **29**, 2963. 437

88TL3255 J.-B. Baudin and S. A. Julia; *Tetrahedron Lett.*, 1988, **29**, 3255. 621

88TL3291 H. Greuter, R. W. Lang and A. J. Romann; *Tetrahedron Lett.*, 1988, **29**, 3291. 831

88TL3339 M. Furber and L. N. Mander; *Tetrahedron Lett.*, 1988, **29**, 3339. 769

88TL3357 M. Mortimore and P. Kocieński; *Tetrahedron Lett.*, 1988, **29**, 3357. 678

88TL3407 P. G. Gassman and S. P. Chavan; *Tetrahedron Lett.*, 1988, **29**, 3407. 521

88TL3455 K. E. Rodriques, A. Basha, J. B. Summers and D. W. Brooks; *Tetrahedron Lett.*, 1988, **29**, 3455. 1042

88TL3533 D. H. R. Barton, N. Ozbalik and M. Ramesh; *Tetrahedron Lett.*, 1988, **29**, 3533. 247

88TL3547 J. A. Marshall, E. D. Robinson and J. Lebreton; *Tetrahedron Lett.*, 1988, **29**, 3547. 818

88TL3579 J. van der Louw, J. L. van der Baan, H. Stichter G. J. J. Out, F. Bickelhaupt and G. W. Klumpp; *Tetrahedron Lett.*, 1988, **29**, 3579. 194

88TL3651 G. Märkl, P. Attenberger and J. Kellner; *Tetrahedron Lett.*, 1988, **29**, 3651. 1012

88TL3741 S. Ram and L. D. Spicer; *Tetrahedron Lett.*, 1988, **29**, 3741. 18, 19

88TL3949 L. Shi, J. Yang, X. Wen and Y. Huang; *Tetrahedron Lett.*, 1988, **29**, 3949. 647, 741

88TL3993 S. J. Neeşon and P. J. Stevenson; *Tetrahedron Lett.*, 1988, **29**, 3993. 823

88TL4105 K. Takeda, K. Tsuboyama, K. Torii, M. Murata and H. Ogura; *Tetrahedron Lett.*, 1988, **29**, 4105. 238

88TL4181 A. Pelter and R. A. Drake; *Tetrahedron Lett.*, 1988, **29**, 4181. 999

88TL4217 A. S. Kende and C. A. Smith; *Tetrahedron Lett.*, 1988, **29**, 4217. 1038

88TL4229 J. J. Kirchner, D. V. Pratt and P. B. Hopkins; *Tetrahedron Lett.*, 1988, **29**, 4229. 833

88TL4253 J. S. Prasad and L. S. Liebeskind; *Tetrahedron Lett.*, 1988, **29**, 4253. 978

88TL4313 S. A. Rao and M. Periasamy; *Tetrahedron Lett.*, 1988, **29**, 4313. 541

88TL4333 J. S. Clark and A. B. Holmes; *Tetrahedron Lett.*, 1988, **29**, 4333. 769

88TL4369 Y.-J. Wu and D. J. Burnell; *Tetrahedron Lett.*, 1988, **29**, 4369. 690

88TL4521 S. P. Tanis, G. M. Johnson and M. C. McMills; *Tetrahedron Lett.*, 1988, **29**, 4521. 760

88TL4577 C. Larpent and H. Patin; *Tetrahedron Lett.*, 1988, **29**, 4577. 446

88TL4587 E. Nakai and T. Nakai; *Tetrahedron Lett.*, 1988, **29**, 4587. 405, 816, 817

88TL4705 W. Oppolzer, J.-M. Gaudin and T. N. Birkinshaw; *Tetrahedron Lett.*, 1988, **29**, 4705. 195

88TL4709 W. Oppolzer, J.-M. Gaudin, M. Bedoya-Zurita, J. Huesc-Rodriguez, T. M. Raynham and C. Robyr; *Tetrahedron Lett.*, 1988, **29**, 4709. 195

88TL4763 S. Kuroda, S. Sakaguchi, S. Ikegami, T. Hanamoto, T. Katsuki and M. Yamaguchi; *Tetrahedron Lett.*, 1988, **29**, 4763. 404, 406, 816, 819

88TL4803 P. G. Gassman and K. Mlinarić-Majerski; *Tetrahedron Lett.*, 1988, **29**, 4803. 709

88TL4859 J. Barluenga, F. J. Fañanás, F. Foubelo and M. Yus; *Tetrahedron Lett.*, 1988, **29**, 4859. 1090

88TL4909 M. Rowley and Y. Kishi; *Tetrahedron Lett.*, 1988, **29**, 4909. 1099

88TL4913 J. A. Marshall, E. D. Robinson and R. D. Adams; *Tetrahedron Lett.*, 1988, **29**, 4913. 818

88TL5017 L. M. Harwood, S. A. Leeming, N. S. Isaacs, G. Jones, J. Pickard, R. M. Thomas and D. Watkin; *Tetrahedron Lett.*, 1988, **29**, 5017. 1140

88TL5069 R. C. Larock and S. K. Stolz-Dunn; *Tetrahedron Lett.*, 1988, **29**, 5069. 230

88TL5119 M. P. Doyle, V. Bagheri and N. K. Harn; *Tetrahedron Lett.*, 1988, **29**, 5119. 820

88TL5169 S. H. Kang, W. J. Kim and Y. B. Chae; *Tetrahedron Lett.*, 1988, **29**, 5169. 697, 700, 703

88TL5209 J. E. Merritt, M. Sasson, S. A. Kates and B. B. Snider; *Tetrahedron Lett.*, 1988, **29**, 5209. 346

88TL5295 Lilan Shi, Weibo Wang and Yao-zeng Huang; *Tetrahedron Lett.*, 1988, **29**, 5295. 269, 738

88TL5379 Y. Ito, M. Inouye and M. Murakami; *Tetrahedron Lett.*, 1988, **29**, 5379. 1035, 1047

88TL5391 S. Mori, H. Iwakura and S. Takechi; *Tetrahedron Lett.*, 1988, **29**, 5391. 1042

88TL5553 R. A. W. Johnstone and W. N. McLean; *Tetrahedron Lett.*, 1988, 5553. 482

88TL5565 B. Burns, R. Grigg, P. Ratananakul, V. Sridharan, P. Stevenson, S. Sukirthalingam and T. Worakun; *Tetrahedron Lett.*, 1988, **29**, 5565. 1015

88TL5645 W. Oppolzer and R. N. Radinov; *Tetrahedron Lett.*, 1988, **29**, 5645. 1097

88TL5661 A. E. Greene, F. Coelho, J.-P. Deprés and T. J. Brocksom; *Tetrahedron Lett.*, 1988, **29**, 5661. 187

88TL5733 A. G. M. Barrett and C. D. Spilling; *Tetrahedron Lett.*, 1988, **29**, 5733. 156

88TL6005 C. J. Moody and R. J. Taylor; *Tetrahedron Lett.*, 1988, **29**, 6005. 233, 235

88TL6125 K. Nozaki, K. Oshima and K. Utimoto; *Tetrahedron Lett.*, 1988, **29**, 6125. 939, 946, 1130

88TL6127 K. Nozaki, K. Oshima and K. Utimoto; *Tetrahedron Lett.*, 1988, **29**, 6127. 925, 938, 946, 1131

88TL6357 D. S. Holmes, U. C. Dyer, S. Russell, J. A. Sherringham and J. A. Robinson; *Tetrahedron Lett.*, 1988, **29**, 6357. 614

88TL6433 W. Oppolzer, M. Bedoya-Zurita and C. Y. Switzer; *Tetrahedron Lett.*, 1988, **29**, 6433. 195

88TL6487 A. Srikrishna and G. V. R. Sharma; *Tetrahedron Lett.*, 1988, **29**, 6487. 547

88TL6581 D. H. R. Barton, N. Ozbalik and J. C. Sarma; *Tetrahedron Lett.*, 1988, **29**, 6581. 944

88TL6797 R. S. Subramanian and K. K. Balasubramanian; *Tetrahedron Lett.*, 1988, **29**, 6797. 827

88TL6839 X. Creary, A. F. Sky and M. E. Mehrsheikh-Mohammadi; *Tetrahedron Lett.*, 1988, **29**, 6839. 922

88TL6951 K. Takeda, S. Yano and E. Yoshii; *Tetrahedron Lett.*, 1988, **29**, 6951. 1140

88ZAAC(560)141 D. Naumann, J. Kischkewitz and B. Wilkes; *Z. Anorg. Allg. Chem.*, 1988, **560**, 141. 54

88ZOR650 L. F. Kovaleva, I. O. Reutova, T. A. Sarycheva, O. A. Sheludyakova, A. N. Novikov and V. D. Filimonov; *Zh. Org. Khim.*, 1988, **24**, 650 (*Chem. Abstr.*, **110**, 23 669d). 1010

89ACR1 G. A. Russell; *Acc. Chem. Res.*, 1989, **22**, 1. 948

89ACR152 T. Cohen and M. Bhupathy; *Acc. Chem. Res.*, 1989, **22**, 152. 236

89AG215 B. M. Trost and S. Schneider; *Angew. Chem.*, 1989, **101**, 215. 287

89AG(E)73 S. Buchholz, K. Harms, W. Massa and G. Boche; *Angew. Chem., Int. Ed. Engl.*, 1989, **28**, 73. 858

89AG(E)277 G. Boche; *Angew. Chem., Int. Ed. Engl.*, 1989, **28**, 277. 140, 146, 876

89AG(E)969 B. Giese; *Angew. Chem., Int. Ed. Engl.*, 1989, **28**, 969. 320, 325, 331

89AG(E)1025 H.-J. Gais, G. Hellmann, H. Günther, F. Lopez, H. J. Lindner and S. Braun; *Angew. Chem., Int. Ed. Engl.*, 1989, **28**, 1025. 146

89AG(E)1321 K. Lammertsma, P. von R. Schleyer and H. Schwarz; *Angew. Chem. Int. Ed. Engl.*, 1989, **28**, 1321. 552

89AHC(45)231 O. Tsuge and S. Kanemasa; *Adv. Heterocycl. Chem.*, 1989, **45**, 231. 552

89AJC279 G. T. Crisp and S. Papadopoulos; *Aust. J. Chem.*, 1989, **42**, 279. 1027

89BCJ143 K. Miura, Y. Ichinose, K. Nozaki, K. Fugami, K. Oshima and K. Utimoto; *Bull. Chem. Soc. Jpn.*, 1989, **62**, 143. 330, 925, 946

89BCJ164 T. Otsubo, Y. Aso, F. Ogura, S. Misumi, A. Kawamoto and J. Tanaka; *Bull. Chem. Soc. Jpn.*, 1989, **62**, 164. 232

89BCJ1682 Y. Nishiyama, Y. Makino, S. Hamanaka, A. Ogawa and N. Sonoda; *Bull. Chem. Soc. Jpn.*, 1989, **62**, 1682. 86

89BCJ2348 S. Fukuzawa, M. Fukushima, T. Fujinami and S. Sakai; *Bull. Chem. Soc. Jpn.*, 1989, **62**, 2348. 364

89BCJ2913 Y. Masuyama, K. Yamada, S.-H. Shimizu and Y. Kurusu; *Bull. Chem. Soc. Jpn.*, 1989, **62**, 2913. 238, 246

89BCJ3132 M. Tsuji, S. Yokoyama and Y. Tachibana; *Bull. Chem. Soc. Japan*, 1989, **62**, 3132. 698, 703

89BCJ3835 T. Hayashi, Y. Okada and T. Inaba; *Bull. Chem. Soc. Jpn*, 1989, **62**, 3835. 826

89BSF409 H. Benhaoua, R. Danion-Bougot and R. Carrie; *Bull. Soc. Chim. Fr.*, 1989, 409. 637

89BSF799 J. Rodriguez, P. Brun and B. Waegell; *Bull. Soc. Chim. Fr.*, 1989, 799. 797

89C39 C. Betschart and D. Seebach; *Chimia*, 1989, **43**, 39. 687

89CA(111)23052 S. Petrova, E. Siimer and O. Eisen; *Chem. Abstr.*, 1989, **111**, 23 052. 552

89CB123 K. Banert; *Chem. Ber.*, 1989, **122**, 123. 519

89CB193 R. Brückner; *Chem. Ber.*, 1989, **122**, 193. 1001

89CB259 H. Schmidbaur, C. M. Frazão, G. Reber and G. Müller; *Chem. Ber.*, 1989, **122**, 259. 446

89CB347 K. J. Przybilla and F. Vögtle; *Chem. Ber.*, 1989, **122**, 347. 232

89CB377 H. Hopf, R. Walsh, A. de Meijere, G. Wachholz and S. Teichmann; *Chem. Ber.*, 1989, **122**, 377. 1069

89CB389 W. Kirmse and G. Hellwig; *Chem. Ber.*, 1989, **122**, 389. 637

89CB703 R. Brückner; *Chem. Ber.*, 1989, **122**, 703. 236

89CB1113 S. Billen and F. Vögtle; *Chem. Ber.*, 1989, **122**, 1113. 232

89CB1237 U. Schubert and J. Grönen; *Chem. Ber.*, 1989, **122**, 1237. 991

89CC161 S. I. Brahim, C. I. F. Watt, J. M. Wilson and C. Moore; *J. Chem. Soc. Chem. Commun.*, 1989, 161. 883

89CC332 D. L. J. Clive and S. Daigneault; *J. Chem. Soc., Chem. Commun.*, 1989, 332. 594, 928

89CC414 P. Beslin and S. Perrio; *J. Chem. Soc., Chem. Commun.*, 1989, 414. 834

89CC480 S. P. Munt and E. J. Thomas; *J. Chem. Soc., Chem. Commun.*, 1989, 480. 1127

89CC614 F. Moulines and D. Astruc; *J. Chem. Soc., Chem. Commun.*, 1989, 614. 164

89CC1090 A. P. Neary and P. J. Parsons; *J. Chem. Soc., Chem. Commun.*, 1989, 1090. 552

89CC1265 M. Yamamoto, T. Uruma, S. Iwasa, S. Kohmoto and K. Yamada; *J. Chem. Soc., Chem. Commun.*, 1989, 1265. 938

89CC1387 R. P. Allen, B. P. Roberts and C. R. Willis; *J. Chem. Soc., Chem. Commun.*, 1989, 1387. 331, 936

89CC1464 J. Le Roux and M. Le Corre; *J. Chem. Soc., Chem. Commun*, 1989, 1464. 640

89CC1665 E. Piers and F. F. Fleming; *J. Chem. Soc., Chem. Commun.*, 1989, 1665. 831

89CC1690 M. Iyoda, A. Mizusuna, H. Kurata and M. Oda; *J. Chem. Soc., Chem. Commun.*, 1989, 1690. 991

89CC1749 T. Ogawa, T. Murafuji and H. Suzuki; *J. Chem. Soc., Chem. Commun.*, 1989, 1749. 648

89CC1812 C. P. Dell, K. M. Khan and D. W. Knight; *J. Chem. Soc., Chem. Commun.*, 1989, 1812. 831

89CC1852 J. E. Baldwin, R. M. Adlington, I. A. O'Neil, G. Schofield, A. C. Spivey and J. B. Sweeney; *J. Chem. Soc., Chem. Commun.*, 1989, 1852. 264

89CL607 K. Sasaki, Y. Aso, T. Otsubo and F. Ogura; *Chem. Lett.*, 1989, 607. 611

89CL773 N. Nishiwaki, S. Minakata, M. Komatsu and Y. Ohshiro; *Chem. Lett.*, 1989, 773. 1043

89CL1079 H. Uno, T. Terakawa and H. Suzuki; *Chem. Lett.*, 1989, 1079. 1042

89CL1711 Y. Hatanaka, S. Fukushima and T. Hiyama; *Chem. Lett.*, 1989, 1711. 490

89CPB36 M. Watanabe, M. Date, M. Tsukazaki and S. Furukawa; *Chem. Pharm. Bull.*, 1989, **37**, 36. 448

89CPB1120 K. Tomioka, M. Shindo and K. Koga; *Chem. Pharm. Bull.*, 1989, **37**, 1120. 135

89CPB2047 H. Fujioka, H. Annoura, K. Murano, Y. Kita and Y. Tamura; *Chem. Pharm. Bull.*, 1989, **37**, 2047. 827, 828

89CPB2882 H. Miyaoka, H. Nagaoka, T. Okamura and Y. Yamada; *Chem. Pharm. Bull.*, 1989, **37**, 2882. 582

89CRV165 H. N. C. Wong, M.-Y. Hon, C.-W. Tse, Y.-C. Yip, J. Tanko and T. Hudlicky; *Chem. Rev.*, 1989, **89**, 165. 801, 802

89CRV863 B. E. Maryanoff and A. B. Reitz; *Chem. Rev.*, 1989, **89**, 863. 638, 639, 647, 721, 744

89CRV975 K. B. Wiberg; *Chem. Rev.*, 1989, **89**, 975. 919, 923

89CRV1111 R. P. Johnson; *Chem. Rev.*, 1989, **89**, 1111. 954

89CRV1413 D. Crich and L. Quintero; *Chem. Rev.*, 1989, **89**, 1413. 935, 938, 941

89CRV1503 J. A. Marshall; *Chem. Rev.*, 1989, **89**, 1503. 226

89CRV1513 J. E. McMurry; *Chem. Rev.*, 1989, **89**, 1513. 687, 688

89H(28)47 V. M. Girijavallabhan and A. K. Ganguly; *Heterocycles*, 1989, **28**, 47. 1131

89H(28)171 F. Mestre, C. Tailhan and S. Z. Zard; *Heterocycles*, 1989, **28**, 171. 231

89H(28)295 A. I. Meyers and J. Guiles; *Heterocycles*, 1989, **28**, 295. 149

89H(28)321 B. M. Trost and K. D. Moeller; *Heterocycles*, 1989, **28**, 321. 289

89H(28)489 F. Minisci, E. Vismara and F. Fontana; *Heterocycles*, 1989, **28**, 489. 934

89H(28)1203 A. Krief, W. Dumont, D. Van Ende, S. Halazy, D. Labar, J.-L. Laboureur and T. Q. Le; *Heterocycles*, 1989, **28**, 1203. 712, 713

89H(29)249 S. Takano, M. Yanase and K. Ogasawara; *Heterocycles*, 1989, **29**, 249. 776

89H(29)1861 S. Takano, Y. Iwabuchi and K. Ogasawara; *Heterocycles*, 1989, **29**, 1861. 776

89HCA996 G. Buchi and H. Wuest; *Helv. Chim. Acta*, 1989, **72**, 996. 826

89HCA1301 M. C. Pirrung, S. E. Dunlap and U. P. Trinks; *Helv. Chim. Acta*, 1989, **72**, 1301. 224

89HCA1537 C. Fehr, J. Galindo, R. Haubrichs and R. Perret; *Helv. Chim. Acta*, 1989, **72**, 1537. 185

89HOU(1)1 M. Regitz and B. Giese; *Methoden Org. Chem. (Houben-Weyl)*, 1989, **1**, 1. 919

89HOU(2)1 M. Regitz and B. Giese; *Methoden Org. Chem. (Houben-Weyl)*, 1989, **2**, 1. 919

89HOU(E19a) B. Giese and M. Regitz; *Methoden Org. Chem. (Houben-Weyl)*, 1989, **E19a**. 320

89IZV495 A. L. Veretenov, A. S. Gybin and V. A. Smit; *Izv. Akad. Nauk. SSSR, Ser. Khim.*, 1990, 495 (*Chem. Abst.*, 1991, **114**, 6893n). 529

89IZV2623 A. N. Kasatkin, T. Y. Romanova, O. Y. Tsypyshev, G. A. Tolstikov and S. I. Lomakina; *Izv. Akad. Nauk SSSR Ser. Khim.*, 1989, **38**, 2623. 469

89JA230 A. L. J. Beckwith and B. P. Hay; *J. Am. Chem. Soc.*, 1989, **111**, 230. 931

89JA314 N. Miyaura, T. Ishiyama, H. Sasaki, M. Ishikawa, M. Satoh and A. Suzuki; *J. Am. Chem. Soc.*, 1989, **111**, 314. 489

89JA384 B. Singaram, C. T. Goralski, M. V. Rangaishenvi and H. C. Brown; *J. Am. Chem. Soc.*, 1989, **111**, 384. 443

89JA746 G. K. S. Prakash, W.-D. Fessner, G. A. Olah, G. Lutz and H. Prinzbach; *J. Am. Chem. Soc.*, 1989, **111**, 746. 918

89JA1125 I. Erdelmeier and H. Gais; *J. Am. Chem. Soc.*, 1989, **111**, 1125. 711

89JA1147 M. Bremer, P. von R. Schleyer and U. Fleischer; *J. Am. Chem Soc.*, 1989, **111**, 1147. 910

89JA1514 G. C. Hirst, P. N. Howard and L. E. Overman; *J. Am. Chem. Soc.*, 1989, **111**, 1514. 391

89JA1527 W. Koch, B. Liu and D. J. DeFrees; *J. Am. Chem. Soc.*, 1989, **111**, 1527. 906

89JA1759 T. V. RajanBabu, T. Fukunaga and G. S. Reddy; *J. Am. Chem. Soc.*, 1989, **111**, 1759. 338

89JA1770 J. A. Palenzuela, H. Y. Elnagar and W. H. Okamura; *J. Am. Chem. Soc.*, 1989, **111**, 1770. 993

89JA2258 J. W. Bunting and R. G. Moors; *J. Am. Chem. Soc.*, 1989, **111**, 2258. 629

89JA2324 M. H. Chisholm and J. A. Klang; *J. Am. Chem. Soc.*, 1989, **111**, 2324. 692

89JA2325 K. Meerholtz and J. Heinze; *J. Am. Chem. Soc.*, 1989, **111**, 2325. 103

89JA2363 P. Kocienski, S. Wadman and K. Cooper; *J. Am. Chem. Soc.*, 1989, **111**, 2363. 481

89JA2588 W. J. Klaver, H. Hiemstra and W. N. Speckamp; *J. Am. Chem. Soc.*, 1989, **111**, 2588. 774

89JA2735 M. Rowley, M. Tsukamoto and Y. Kishi; *J. Am. Chem. Soc.*, 1989, **111**, 2735. 831

89JA2981 C. A. Broka and T. Shen; *J. Am. Chem. Soc.*, 1989, **111**, 2981. 405

89JA3069 I. K. Meier and J. Schwartz; *J. Am. Chem. Soc.*, 1989, **111**, 3069. 683

89JA3091 E. Nakamura, K. Sekiya, M. Arai and S. Aoki; *J. Am. Chem. Soc.*, 1989, **111**, 3091. 122

89JA3441 H. C. Brown, R. K. Dhar, R. K. Bakshi, P. K. Pandiarajan and B. Singaram; *J. Am. Chem. Soc.*, 1989, **111**, 3441. 831

89JA3640 A. B. Brown and H. W. Whitlck, Jr.; *J. Am. Chem. Soc.*, 1989, **111**, 3640. 1142

89JA3717 R. A. Gibbs, K. Bartels, R. W. K. Lee and W. H. Okamura; *J. Am. Chem. Soc.*, 1989, **111**, 3717. 498

89JA3761 M. Iyoda, Y. Kuwatani and M. Oda; *J. Am. Chem. Soc.*, 1989, **111**, 3761. 992

89JA4399 D. S. Matteson, P. B. Tripathy, A. Sarkar and K. M. Sadhu; *J. Am. Chem. Soc.*, 1989, **111**, 4399. 138

89JA4864 T. Ibuka, M. Tanaka, S. Nishii and Y. Yamamoto; *J. Am. Chem. Soc.*, 1989, **111**, 4864. 188

89JA4981 I. Starý and P. Kočovský; *J. Am. Chem. Soc.*, 1989, **111**, 4981. 194

89JA5417 J. K. Stille and M. R. Peña; *J. Am. Chem. Soc.*, 1989, **111**, 5417. 1027

89JA5493 E. J. Corey, R. Imwinkelried, S. Pikul and Y. B. Xiang; *J. Am. Chem. Soc.*, 1989, **111**, 5493. 374

89JA5845 W. P. Neumann, A. Penenory, U. Stewen and M. Lehnig; *J. Am. Chem. Soc.*, 1989, **111**, 5845. 922

89JA5861 E. Vedejs and T. J. Fleck; *J. Am. Chem. Soc.*, 1989, **111**, 5861. 721

89JA5902 B. M. Trost, S. A. King and T. Schmidt; *J. Am. Chem. Soc.*, 1989, **111**, 5902. 289

89JA5948 K. A. Parker and S. M. Ruder; *J. Am. Chem. Soc.*, 1989, **111**, 5948. 522

89JA6257 Y. Horiguchi, E. Nakamura and I. Kuwajima; *J. Amer. Chem. Soc.*, 1989, **111**, 6257. 86

89JA6265 D. P. Curran, M.-H. Chen and D. Kim; *J. Am. Chem. Soc.*, 1989, **111**, 6265. 334, 347

89JA6280 M. Oshima, H. Yamazaki, I. Shimizu, M. Nisar and J. Tsuji; *J. Am. Chem. Soc.*, 1989, **111**, 6280. 14

89JA6419 R. M. Pollack, B. F. Zeng, J. P. G. Mack and S. Eldin; *J. Am. Chem. Soc.*, 1989, **111**, 6419. 877

89JA6432 P. A. Wender and T. E. Jenkins; *J. Am. Chem. Soc.*, 1989, **111**, 6432. 523

89JA6443 R. M. Moriarty, O. Prakash, R. K. Vaid and L. Zhao; *J. Am. Chem. Soc.*, 1989, **111**, 6443. 109

89JA6482 B. M. Trost, B. Yang and M. L. Miller; *J. Am. Chem. Soc.*, 1989, **111**, 6482. 194, 288, 289

89JA6849 E. Nakamura, D. Machii and T. Inubushi; *J. Am. Chem. Soc.*, 1989, **111**, 6849. 925

89JA7285 S. Yamago and E. Nakamura; *J. Am. Chem. Soc.*, 1989, **111**, 7285. 360

89JA7487 B. M. Trost, P. R. Secane, M. Acemoglu and S. Mignani; *J. Am. Chem. Soc.*, 1989, **111**, 7487. 193, 288

89JA7558 F. G. Bordwell and T.-Y. Lynch; *J. Am. Chem. Soc.*, 1989, **111**, 7558. 919, 922

89JA7659 U. S. Racherla, T. Daniel, P. R. Rajamohanan and N. R. Ayyangar; *J. Am. Chem. Soc.*, 1989, **111**, 7659. 898

89JA7938 T. Murai, E. Yasui, S. Kato, Y. Hatayama, S. Suzuki, Y. Yamasaki, N. Sonoda, H. Kurosawa, Y. Kawasaki and S. Murai; *J. Am. Chem. Soc.*, 1989, **111**, 7938. 186, 225, 229

89JA8231 E. Turos, J. E. Audia and S. J. Danishefsky; *J. Am. Chem. Soc.*, 1989, **111**, 8231. 831

89JA8309 N. A. Porter, B. Lachen, V. H.-T. Chang and D. R. Magnin; *J. Am. Chem. Soc.*, 1989, **111**, 8309. 340

89JA8311 N. A. Porter, D. M. Scott, B. Lacher, B. Giese, H.-G. Zeitz and H. J. Lindner; *J. Am. Chem. Soc.*, 1989, **111**, 8311. 326

89JA8737 T. H. Chan and P. Pellon; *J. Am. Chem. Soc.*, 1989, **111**, 8737. 162

89JA8867 J. E. McMurry, T. Leckta and C. N. Hodge; *J. Am. Chem. Soc.*, 1989, **111**, 8867. 919

89JA8878 S. E. Denmark, M. A. Harmata and K. S. White; *J. Am. Chem. Soc.*, 1989, **111**, 8878. 832

89JA9224 T. Laube; *J. Am. Chem. Soc.*, 1989, **111**, 9224. 906

89JCR(S)30 M. A. Khan and B. Al-Saleh; *J. Chem. Res. (S)*, 1989, 30. 595

89JCR(S)200 J. Barluenga, F. Foubelo, F. J. Fañanás and M. Yus; *J. Chem. Res. (S)*, 1989, 200. 1090

89JCR(S)275 J.-M. Fang, J.-M. Duh and C.-T. Chen; *J. Chem. Res. (S)*, 1989, 275. 620

89JCS(P1)190 R. Baker and J. L. Castro; *J. Chem. Soc., Perkin Trans. 1*, 1989, 190. 677

89JCS(P1)225 A. R. Katritzky, K. Yannakopoulou, P. Lue, D. Rasala and L. Urogdi; *J. Chem. Soc., Perkin Trans. 1*, 1989, **2**, 225. 260

89JCS(P1)333 M. G. Moloney, J. T. Pinhey and E. G. Roche; *J. Chem. Soc., Perkin Trans. 1*, 1989, 333. 1145

89JCS(P1)1211 H. Takahata, K. Takahashi, E. Wang and T. Yamazaki; *J. Chem. Soc., Perkin Trans. 1*, 1989, 1211. 1041

89JCS(P1)1387 C. Nájera and M. Yus; *J. Chem. Soc., Perkin Trans. 1*, 1989, 1387. 1090

89JCS(P1)1775 N. X. Hu, Y. Aso, T. Otsubo and F. Ogura; *J. Chem. Soc., Perkin Trans. 1*, 1989, 1775. 611

89JCS(P1)2297 A. R. Katritzky and M. S. C. Rao; *J. Chem. Soc., Perkin Trans. 1*, 1989, 2297. 260

89JCS(P1)2513 K. Hirota, Y. Isobe and Y. Maki; *J. Chem. Soc., Perkin Trans 1.*, 1989, 2513. 482

89JCS(P2)579 N. A. Al-Awadi, R. F. Al-Bashir and O. M. E. El-Dusouqui; *J. Chem. Soc., Perkin Trans. 2*, 1989, 579. 782

89JFC(44)161 G. Shi and Y. Xu; *J. Fluorine Chem.*, 1989, **44**, 161. 634

89JFC(44)1637 D. J. Burton, T. D. Spawn, P. L. Heinze, A. R. Bailey and S. Shiniya; *J. Fluorine Chem.*, 1989, **44**, 1637. 429

89JMC1820 R. W. Guthrie, G. L. Kaplan, F. A. Mennona, J. W. Tilley, R. W. Kierstead, J. G. Mullin, R. A. LeMahieu, S. Zawoiski, M. O'Donnell, H. Crowley, B. Yaremko and A. F. Welton; *J. Med. Chem.*, 1989, **32**, 1820. 1029

89JOC9 S. L. Schreiber and D. B. Smith; *J. Org. Chem.*, 1989, **54**, 9. 831

89JOC12 A. Villalobos and S. J. Danishefsky; *J. Org. Chem.*, 1989, **54**, 12. 700, 703

89JOC20 M.-D. Wang, S. Calet and H. Alper; *J. Org. Chem.*, 1989, **54**, 20. 242

89JOC175 R. E. Gawley, G. C. Hart and L. J. Bartolotti; *J. Org. Chem.*, 1989, **54**, 175. 152

89JOC249 M. E. Garst and B. J. McBride; *J. Org. Chem.*, 1989, **54**, 249. 1009

89JOC268 Y. Hatanaka and T. Hiyama; *J. Org. Chem.*, 1989, **54**, 268. 490

89JOC308 G. J. Gleicher, B. Mahiou and A. J. Aretakis; *J. Org. Chem.*, 1989, **54**, 308. 920

89JOC317 R. F. Horvath and T. H. Chan; *J. Org. Chem.*, 1989, **54**, 317. 161, 162

89JOC389 E. J. Corey, G. H. Posner, R. F. Atkinson, A. K. Wingard, D. J. Halloran, D. M. Radzik and J. J. Nash; *J. Org. Chem.*, 1989, **54**, 389. 778

89JOC458 P. Beak and B. Lee; *J. Org. Chem.*, 1989, **54**, 458. 148

89JOC469 M. Uemura, T. Minami, K. Hirotsu and Y. Hayashi; *J. Org. Chem.*, 1989, **54**, 469. 195

89JOC515 E. Brandes, P. A. Grieco and J. J. Gajewski; *J. Org. Chem.*, 1989, **54**, 515. 826

89JOC733 K. Okabe, T. Ohwada, T. Ohta and K. Shudo; *J. Org. Chem.*, 1989, **54**, 733. 492

89JOC736 J. D. White and J. C. Amedio; *J. Org. Chem.*, 1989, **54**, 736. 829

89JOC773 W. T. Thorstad, N. S. Mills, D. Q. Buckelew and L. S. Govea; *J. Org. Chem.*, 1989, **54**, 773. 879

89JOC868 S.-S. P. Chou, H.-L. Kuo, C.-J. Wang, C.-Y. Tsai and C.-M. Sun; *J. Org. Chem.*, 1989, **54**, 868. 489

89JOC1218 M. J. Chapdelaine, P. J. Warwick and A. Shaw; *J. Org. Chem.*, 1989, **54**, 1218. 1010

89JOC1223 K. Tadano, J. Ishihara, H. Yamada and S. Ogawa; *J. Org. Chem.*, 1989, **54**, 1223. 828

89JOC1234 M. D. Bachi and E. Bosch; *J. Org. Chem.*, 1989, **54**, 1234. 938

89JOC1238 D. L. Boger and R. J. Wysocki, Jr.; *J. Org. Chem.*, 1989, **54**, 1238. 1130

89JOC1295 J. M. Klunder, T. Onami and K. B. Sharpless; *J. Org. Chem.*, 1989, **54**, 1295. 223

89JOC1320 B. P. Branchaud and M. S. Meier; *J. Org. Chem.*, 1989, **54**, 1320. 946

89JOC1375 G. A. Olah, A. Wu and O. Farooq; *J. Org. Chem.*, 1989, **54**, 1375. 693

89JOC1491 C. Nájera and M. Yus; *J. Org. Chem.*, 1989, **54**, 1491. 145

89JOC1647 P. Beak and D. A. Burg; *J. Org. Chem.*, 1989, **54**, 1647. 613

89JOC1671 E. Grovenstein, Jr., K. W. Black, S. C. Goel, R. L. Hughes, J. H. Northrop, D. L. Streeter and D. VanDerveer; *J. Org. Chem.*, 1989, **54**, 1671. 886

89JOC1757 M. Yamamoto, T. Takemori, S. Iwasa, S. Kohmoto and K. Yamada; *J. Org. Chem.*, 1989, **54**, 1757. 1090

89JOC1777 D. L. Boger and R. J. Mathvink; *J. Org. Chem.*, 1989, **54**, 1777. 933

89JOC1785 Y. Morita, M. Suzuki and R. Noyori; *J. Org. Chem.*, 1989, **54**, 1785. 132

89JOC1826 D. P. Curran, E. Bosch, J. Kaplan and M. Newcomb; *J. Org. Chem.*, 1989, **54**, 1826. 334

89JOC2027 L. Shi, W. Wang, Y. Wang and Y. Huang; *J. Org. Chem.*, 1989, **54**, 2027. 647

89JOC2117 J.-L. Giner, C. Margot and C. Djerassi; *J. Org. Chem.*, 1989, **54**, 2117. 113

89JOC2217 C. K. Chu, V. S. Bhadti, B. Doboszewski, Z. P. Gu, Y. Kosugi, K. C. Pullaiah and P. van Rogy; *J. Org. Chem.*, 1989, **54**, 2217. 585

89JOC2278 L. A. Paquette, D. T. DeRussy and J. C. Gallucci; *J. Org. Chem.*, 1989, **54**, 2278. 1098

89JOC2374 K. Umemura, H. Matsuyama, N. Watanabe, M. Kobayashi and N. Kamigata; *J. Org. Chem.*, 1989, **54**, 2374. 241

89JOC2391 A. Pecunioso and R. Menicagli; *J. Org. Chem.*, 1989, **54**, 2391. 1049

89JOC2492 C. Chatgilialoglu, D. Griller and M. Lesage; *J. Org. Chem.*, 1989, **54**, 2492. 936

89JOC2759 D. R. Deardorff, R. G. Linde, II, A. M. Martin and M. J. Shulman; *J. Org. Chem.*, 1989, **54**, 2759. 193

89JOC2893 E. W. Baxter, D. Labaree, S. Chao and P. S. Mariano; *J. Org. Chem.*, 1989, **54**, 2893. 835

89JOC3009 W. R. Roush and A. D. Palkowitz; *J. Org. Chem.*, 1989, **54**, 3009. 752

89JOC3114 G. A. Molander and S. W. Andrews; *J. Org. Chem.*, 1989, **54**, 3114. 216

89JOC3137 G. A. Kraus, S. Liras, T. O. Man and M. T. Molina; *J. Org. Chem.*, 1989, **54**, 3137. 827

89JOC3140 D. P. Curran and C.-T. Chang; *J. Org. Chem.*, 1989, **54**, 3140. 334, 337, 344, 936

89JOC3229 J. D. Hsi and M. Koreeda; *J. Org. Chem.*, 1989, **54**, 3229. 738, 740, 742

89JOC3239 T. L. Underiner and H. L. Goering; *J. Org. Chem.*, 1989, **54**, 3239. 187

89JOC3247 H. Xiong and R. D. Rieke; *J. Org. Chem.*, 1989, **54**, 3247. 116

89JOC3420 F. W. Hobbs, Jr.; *J. Org. Chem.*, 1989, **54**, 3420. 1010

89JOC3571 F.-D. Kopinke, G. Zimmermann and K. Anders; *J. Org. Chem.*, 1989, **54**, 3571. 925

89JOC3726 C. J. Elsevier and P. Vermeer; *J. Org. Chem.*, 1989, **54**, 3726. 966

89JOC3748 J. E. McMurry, T. Lectka and J. G. Rico; *J. Org. Chem.*, 1989, **54**, 3748. 688

89JOC3988 F. Kazmierczak and P. Helquist; *J. Org. Chem.*, 1989, **54**, 3988. 781

89JOC4055 T. Ibuka, N. Akimoto, M. Tanaka, S. Nishii, and Y. Yamamoto; *J. Org. Chem.*, 1989, **54**, 4055. 188

89JOC4072 W. H. Okamura, J. M. Aurrecoechea, R. A. Gibbs and A. W. Norman; *J. Org. Chem.*, 1989, **54**, 4072. 1039

89JOC4146 T. G. Back, M. V. Krishna and K. R. Muralidharan; *J. Org. Chem.*, 1989, **54**, 4146. 605

89JOC4267 J. A. Ragan, M. Nakatsuka, D. B. Smith, D. E. Uehling and S. L. Schreiber; *J. Org. Chem.*, 1989, **54**, 4267. 1079

89JOC4278 L. A. Paquette and G. Ladouceur; *J. Org. Chem.*, 1989, **54**, 4278. 826

89JOC4372 R. J. Mills, N. J. Taylor and V. Snieckus; *J. Org. Chem.*, 1989, **54**, 4372. 448

89JOC4391 N. X. Hu, Y. Aso, T. Otsubo and F. Ogura; *J. Org. Chem.*, 1989, **54**, 4391. 594, 611, 621

89JOC4453 G. Just and R. Singh; *J. Org. Chem.*, 1989, **54**, 4453. 1009

89JOC4474 K. M. Ho, C. H. Lam and T.-Y. Luh; *J. Org. Chem.*, 1989, **54**, 4474. 22, 23, 24, 436

89JOC4660 N. Tokitoh, N. Choi, M. Goto and W. Ando; *J. Org. Chem.*, 1989, **54**, 4660. 242

89JOC4840 M. Lourak, R. Vanderesse, Y. Fort and P. Caubère; *J. Org. Chem.*, 1989, **54**, 4840. 472

89JOC4998 A. Kamimura, H. Sasatani, T. Hashimoto and N. Ono; *J. Org. Chem.*, 1989, **54**, 4998. 252

89JOC5044 L. A. Paquette, J. P. Gilday and G. D. Maynard; *J. Org. Chem.*, 1989, **54**, 5044. 883

89JOC5054 L. A. Paquette and G. D. Maynard; *J. Org. Chem.*, 1989, **54**, 5054. 883

89JOC5308 M. Narisada, I. Horibe, F. Watanabe and K. Takeda; *J. Org. Chem.*, 1989, **54**, 5308. 428, 429

89JOC5324 A. Duréault, I. Tranchepain and J.-C. Depezay; *J. Org. Chem.*, 1989, **54**, 5324. 486, 1032

89JOC5350 J. K. Dickson, Jr., R. Tsang, J. M. Llera and B. Fraser-Reid; *J. Org. Chem.*, 1989, **54**, 5350. 828

89JOC5415 J. I. Luengo and M. Koreeda; *J. Org. Chem.*, 1989, **54**, 5415. 816

89JOC5424 O. D. Monera, M.-K. Chang and G. E. Means; *J. Org. Chem.*, 1989, **54**, 5424. 637

89JOC5648 X. K. Jiang, X.-Y. Li and K.-Y. Wang; *J. Org. Chem.*, 1989, **54**, 5648. 922

89JOC5811 L. W. Jenneskens, C. A. M. Hoefs and U. E. Wiersum; *J. Org. Chem.*, 1989, **54**, 5811. 776, 778

89JOC5828 G. K. Cook, W. J. Hornback, C. L. Jordan, J. H. McDonald, III and J. E. Munroe; *J. Org. Chem.*, 1989, **54**, 5828. 1027

89JOC5831 S. Hanessian, B. Thavonekham and B. DeHoff, *J. Org. Chem.*, 1989, **54**, 5831. 177, 178

89JOC5849	P. A. Grieco, E. Brandes, S. McCann and J. D. Clark; *J. Org. Chem.*, 1989, **54**, 5849.	826
89JOC5854	J. A. Marshall, E. D. Robinson and A. Zapata; *J. Org. Chem.*, 1989, **54**, 5854.	552
89JOC6022	A. R. Katritzky, S. Rachwal and B. Rachwal; *J. Org. Chem.*, 1989, **54**, 6022.	260
89JOC6064	H. C. Brown, C. D. Blue, D. J. Nelson and N. G. Bhat; *J. Org. Chem.*, 1989, **54**, 6064.	449
89JOC6101	W. G. Dauben, D. P. Bridon and B. A. Kowalczyk; *J. Org. Chem.*, 1989, **54**, 6101.	925, 935
89JOM(360)39	G. L. Larson; *J. Organomet. Chem.*, 1989, **360**, 39.	648
89JOM(364)245	H. Konishi, C. Matsumura, T. Okano and J. Kiji; *J. Organomet. Chem.*, 1989, **364**, 245.	1035
89JOM(366)87	Y. Huang, C. Chen and Y. Shen; *J. Organomet. Chem.*, 1989, **366**, 87.	648, 743
89JOM(369)C51	Y. Tsuji, T. Mukai, T. Kondo and Y. Watanabe; *J. Organomet. Chem.*, 1989, **369**, C51.	196
89JOM(371)279	O. G. Garkusha, I. A. Garbuzova, B. V. Lokshin and Z. V. Todres; *J. Organomet. Chem.*, 1989, **371**, 279.	890
89JOM(375)45	Yanchang Shen and Baozhen Yang; *J. Organomet. Chem.*, 1989, **375**, 45.	742
89JOU449	V. A. Roginskii; *J. Org. Chem. USSR*, 1989, **25**, 449.	921
89LA739	P. Binger, A. Brinkmann, P. Roefke and B. Schäfer; *Liebigs Ann. Chem.*, 1989, 739.	363
89LA2151	S. Scholz, H. Hofmeister, G. Neef, E. Ottow, C. Scheidges and R. Wiechert; *Liebigs Ann. Chem.* 1989, 2151.	558
B-89MI 101-01	R. C. Larock; "Comprehensive Organic Transformations," VCH, New York, 1989.	16, 21
B-89MI 104-01	E. J. Corey and X.-M. Cheng; in "The Logic of Chemical Synthesis," Wiley, New York, 1989, pp. 249–357.	111
B-89MI 104-02	R. M. Williams; in "Synthesis of Optically Active α-Amino Acids, Organic Chemistry Series," eds J. E. Baldwin and P. D. Magnus; Pergamon, Oxford, 1989, vol. 7.	166
B-89MI 108-01	I. Beranek and H. Fischer; "Free Radicals in Synthesis and Biology," ed. F. Minisci, Kluwer, Dordrecht, 1989, p. 303.	325
B-89MI 110-01	A. R. Bassindale and P. G. Taylor; "The Chemistry of Organosilicon Compounds," eds. S. Patai and Z. Rappoport, Wiley, Chichester, 1989, part 2, ch. 14, p. 893.	447, 450
B-89MI 112-01	R. C. Larock; "Comprehensive Organic Transformations," VCH, New York, 1989, pp. 233–238.	541
B-89MI 113-01	J. Gandler; in "The Chemistry of Double-bonded Functional Groups," ed. S. Patai, Wiley, Chichester, 1989, p. 733.	570
B-89MI 115-01	R. C. Larock; "Comprehensive Organic Transformations," VCH Publishers, New York, 1989, pp. 160-161.	687
B-89MI 118-01	R. M. Pollock, P. L. Bonds and C. L. Bevins; in "The Chemistry of Enones," eds. S. Patai and Z. Rappoport, Wiley, New York, 1989, vol. 1, p. 559.	797
B-89MI 119-01	A. C. Knipe; in "Organic Reaction Mechanisms," eds. A. C. Knipe and W. E. Watts, Wiley, Chichester, 1989, chap. 11, p. 335.	879
B-89MI 119-02	R. C. Larock; "Comprehensive Organic Transformations," VCH, New York, 1989, pp. 110–114.	882
B-89MI 119-03	C. Elschenbroich and A. Salzer; "Organometallics," VCH, New York, 1989.	885
B-89MI 119-04	G. V. Boyd; in "The Chemistry of Enones," eds. S. Patai and Z. Rappoport, Wiley, New York, 1989, part 1, pp. 281–315.	888
B-89MI 119-05	D. Duval and S. Géribaldi; in "The Chemistry of Enones," eds. S. Patai and Z. Rappoport, Wiley, New York, 1989, part 1, pp. 355–469.	888
B-89MI 119-06	R. A. Cox; in "Organic Reaction Mechanisms," eds. A. C. Knipe and W. E. Watts, Wiley, 1989, chap. 9, pp. 283–300.	919
B-89MI 119-07	I. Beranek and H. Fischer; in "Free Radicals in Synthesis and Biology," ed. F. Minisci, Kluwer, Dordrecht, 1989, p. 187.	927
B-89MI 119-08	I. Beranek and H. Fischer; in "Free Radicals in Synthesis and Biology," ed. F. Minisci, Kluwer, Dordrecht, 1989, p. 303.	921
89MI 119-09	G. M. Kramer and C. G. Scouten; *Adv. Carbocation Chem.*, 1989, **1**, 93–120.	906
89OM1593	S. L. Buchwald, R. B. Nielsen and J. C. Dewan; *Organometallics*, 1989, **8**, 1593.	397
89OR(37)57	I. Fleming, J. Dunoguès and R. Smithers; *Org. React.*, 1989, **37**, 57.	450, 451
89OS(67)60	A. I. Meyers, M. Boes and D. A. Dickman; *Org. Synth.*, 1989, **67**, 60.	149
89OS(67)125	G. L. Larson, I. Montes de Lopez-Cepero and L. Rodriguez Mieles; *Org. Synth.*, 1989, **67**, 125.	759
89OS(67)210	S. A. Miller and R. C. Gadwood; *Org. Synth.*, 1989, **67**, 210.	812
89S28	U. Azzenza, T. Denurra, E. Fenude, G. Melloni and G. Rassu; *Synthesis*, 1989, 28.	432
89S31	A. R. Katritzky, J. K. Gallos and K. Yannakopoulou; *Synthesis*, 1989, 31.	261, 1041
89S33	J. Barluenga, P. J. Campos and G. Canal; *Synthesis*, 1989, 33.	1040
89S42	X. Huang and H.-Z. Zhang; *Synthesis*, 1989, 42.	439
89S71	S. Blechert; *Synthesis*, 1989, 71.	414, 825, 836
89S217	J. B. Hendrickson and M. S. Hussoin; *Synthesis*, 1989, 217.	1071
89S323	A. R. Katritzky, L. Urogdi and A. Mayence; *Synthesis*, 1989, 323.	260, 261
89S472	W. Smit, S. O. Simonyan, V. A. Tarasov, G. S. Mikaelian, A. S. Gybin, I. I. Ibragimov, R. Caple, D. Froen and A. Kreager; *Synthesis*, 1989, 472.	530
89S477	H. Schick and I. Eichhorn; *Synthesis*, 1989, 477.	482
89S633	G. A. Tolstikov, M. S. Miftakhov, N. A. Danilova, Y. L. Vel'der and L. V. Spirikhin; *Synthesis*, 1989, 633.	491
89S747	A. R. Katritzky and K. Yannakopoulou; *Synthesis*, 1989, 747.	260, 261
89S883	D. Lenoir; *Synthesis*, 1989, 883.	687
89S898	K. C. Nicolaou, J. Y. Ramphal and Y. Abe; *Synthesis*, 1989, 898.	1134
89SC83	Y. Huang, Y. Shen and C. Chen; *Synth. Commun.*, 1989, **19**, 83.	648
89SC501	Y. Huang, C. Chen and Y. Shen; *Synth. Commun.*, 1989, **19**, 501.	648
89SC1705	D. Michelot; *Synth. Commun.*, 1989, **19**, 1705.	1055

89SC2109 T. Hayashi, A. Yamamoto and Y. Ito; *Synth. Commun.* 1989, **19**, 2109. 415, 826

89SC2199 T. Ogawa, K. Kusume, M. Tanaka, K. Hayami and H. Suzuki; *Synth. Commun.*, 1989, **19**, 2199. 469, 1005

89SC2415 J. T. Gupton, D. Krolikowski and M. Rusler; *Synth. Commun.*, 1989, **19**, 2415. 253

89SC2659 S. Wattanasin and F. G. Kathawala; *Synth. Commun.*, 1989, **19**, 2659. 767, 769

89SCI907 R. H. Grubbs and W. Tumas; *Science*, 1989, **243**, 907. 675

89SL1 H. Sakurai; *Synlett*, 1989, 1. 314

89SL64 B. H. Lipshutz, C. S. Ung and S. Sengupta; *Synlett.*, 1989, 64. 86

89T381 A. Alexakis and D. Jachiet; *Tetrahedron*, 1989, **45**, 381. 651

89T391 J. A. Marshall, J. D. Trometer and D. G. Cleary; *Tetrahedron*, 1989, **45**, 391. 226, 227

89T413 I. Fleming, M. Rowley, P. Cuadrado, A. M. González-Nogal and F. J. Pulido; *Tetrahedron*, 1989, **45**, 413. 495

89T435 T. Ibuka, M. Tanaka, H. Nemoto and Y. Yamamoto; *Tetrahedron*, 1989, **45**, 435. 227

89T455 G. Casy and R. J. K. Taylor; *Tetrahedron*, 1989, **45**, 455. 619

89T935 B. Giese, T. Linker and R. Muhn; *Tetrahedron*, 1989, **45**, 935. 330

89T951 W. P. Neumann, H. Hillgärtner, K. M. Baines, R. Dicke, K. Vorspohl, U. Kobs and U. Nussbeutel; *Tetrahedron*, 1989, **45**, 951. 924

89T979 D. R. Tueting, A. M. Echavarren and J. K. Stille; *Tetrahedron*, 1989, **45**, 979. 228, 479

89T1579 P. W. Rabideau; *Tetrahedron*, 1989, **45**, 1579. 95, 889

89T1859 D. S. Matteson; *Tetrahedron*, 1989, **45**, 1859. 660

89T2005 A. Krief, W. Dumont, M. Clarembeau, G. Bernard and E. Badaoui; *Tetrahedron*, 1989, **45**, 2005. 713

89T2023 A. Krief, W. Dumont, M. Clarembeau and E. Badaoui; *Tetrahedron*, 1989, **45**, 2023. 713

89T2649 M. Schlosser and S. Strunk; *Tetrahedron*, 1989, **45**, 2649. 397

89T3163 E. Schaumann, C. Friese and G. Adiwidjaja; *Tetrahedron*, 1989, **45**, 3163. 653

89T4017 C. F. Bernasconi; *Tetrahedron*, 1989, **45**, 4017. 888

89T4293 D. S. Brown, M. Bruno, R. J. Davenport and S. V. Ley; *Tetrahedron*, 1989, **45**, 4293. 1031

89T4913 R. M. Pollack; *Tetrahedron*, 1989, **45**, 4913. 130, 869

89T6197 A. Alexakis and D. Jachiet; *Tetrahedron*, 1989, **45**, 6197. 1024

89T6409 H. Künzer, G. Sauer and R. Wiechert; *Tetrahedron*, 1989, **45**, 6409. 555

89T6901 I. Ojima, N. Clos and C. Bastos; *Tetrahedron*, 1989, **45**, 6901. 76

89T7161 S. V. Ley, *et al.*; *Tetrahedron*, 1989, **45**, 7161. 697, 703

89T7643 P. C. B. Page, M. B. van Niel and J. C. Prodger; *Tetrahedron*, 1989, **45**, 7643. 165

89T7835 A. Johns, J. A. Murphy and M. S. Sherburn; *Tetrahedron*, 1989, **45**, 7835. 929

89TL57 A. De Mesmaeker, P. Hoffmann and B. Ernst; *Tetrahedron Lett.*, 1989, **30**, 57. 547

89TL159 W. Heinz, H.-J. Räder and K. Müllen; *Tetrahedron Lett.*, 1989, **30**, 159. 880

89TL173 C. Nájera and M. Yus; *Tetrahedron Lett.*, 1989, 173. 486, 1031

89TL179 P. Chabert, J. B. Ousset and C. Mioskowski; *Tetrahedon Lett.*, 1989, **30**, 179. 580

89TL225 K. V. Bhaskar and G. S. R. S. Rao; *Tetrahedron Lett.*, 1989, **30**, 225. 342, 1132

89TL229 G. J. P. H. Boons, G. A. van der Marel and J. H. van Boom; *Tetrahedron Lett.*, 1989, **30**, 229. 648

89TL243 V. Fiandanese, G. Marchese, F. Naso, L. Ronzini and D. Rotunno; *Tetrahedron Lett.*, 1989, 243. 484

89TL375 A. Yamamoto, Y. Ito and T. Hayashi; *Tetrahedron Lett.*, 1989, **30**, 375. 194

89TL477 K. M. Pietrusiewicz and M. Zablocka; *Tetrahedron Lett.*, 1989, **30**, 477. 158

89TL515 S. R. Angle and D. O. Arnaiz; *Tetrahedron Lett.*, 1989, **30**, 515 830, 831

89TL601 J. Elliott, D. Hall, and S. Warren; *Tetrahedron Lett.*, 1989, **30**, 601. 756

89TL617 J.-E. Bäckvall, J.-O. Vågberg and K. L. Granberg; *Tetrahedron Lett.*, 1989, **30**, 617. 193

89TL647 L. E. Overman and H. Wild; *Tetrahedron Lett.*, 1989, **30**, 647. 1098

89TL681 B. Giese, B. Kopping and C. Chatgilialoglu; *Tetrahedron Lett.*, 1989, **30**, 681. 331, 342, 936

89TL685 S. Bartel and F. Bohlmann; *Tetrahedron Lett.*, 1989, **30**, 685. 218, 221

89TL827 E. Lee, S. B. Ko, K. W. Jung and M. H. Chang; *Tetrahedron Lett.*, 1989, **30**, 827. 1130, 1132

89TL851 K. Tomioka, H. Fujita and K. Koga; *Tetrahedron Lett.*, 1989, **30**, 851. 199

89TL967 M. Shimizu and H. Yoshioka; *Tetrahedron Lett.*, 1989, **30**, 967. 650

89TL1073 K. Hartke, H. Gerber and U. Roesrath; *Tetrahedron Lett.*, 1989, **30**, 1073. 1047

89TL1197 P. Beak and W.-K. Lee; *Tetrahedron Lett.*, 1989, **30**, 1197. 151

89TL1209 J. V. Comasseto, H. M. C. Ferraz, C. A. Brandt and K. K. Gaeta; *Tetrahedron Lett.*, 1989, **30**, 1209. 243, 437

89TL1241 A. Cleve and F. Bohlmann; *Tetrahedron Lett.*, 1989, **30**, 1241. 769

89TL1265 K. Maruoka, H. Banno, K. Nonoshita and H. Yamamoto; *Tetrahedron Lett.*, 1989, **30**, 1265. 826

89TL1361 B. E. Maryanoff, A. B. Reitz, D. W. Graden and H. R. Almond, Jr.; *Tetrahedron Lett.*, 1989, **30**, 1361. 722

89TL1495 B. M. Trost and M. Acemoglu; *Tetrahedron Lett.*, 1989, **30**, 1495. 193, 288

89TL1575 F. Kido, S. C. Sinha, T. Abiko and A. Yoshikoshi; *Tetrahedron Lett.*, 1989, **30**, 1575. 822

89TL1613 F. A. Davis, M. C. Weismiller, G. S. Lal, B. C. Chen and R. M. Przeslawski; *Tetrahedron Lett.*, 1989, **30**, 1613. 147

89TL1641 J. Ichikawa, T. Sonoda and H. Kobayashi; *Tetrahedron Lett.*, 1989, **30**, 1641. 570

89TL1803 F. Chaigne, J.-P. Gotteland and M. Malacria; *Tetrahedron Lett.*, 1989, **30**, 1803. 194, 288

89TL2013 R. Hunter and G. D. Tomlinson; *Tetrahedron Lett.*, 1989, **30**, 2013. 280

89TL2153 B. Zwanenburg, L. Thijs and D. M. Edenberger; *Tetrahedron Lett.*, 1989, **30**, 2153. 1023

89TL2225 T. Jeffery; *Tetrahedron Lett.*, 1989, **30**, 2225. 121, 1002

89TL2403 T. Hiyama, Y. Hatanaka and K. Matsui; *Tetrahedron Lett.*, 1989, **30**, 2403. 1007

89TL2413	R. Tamura, S. Tamai and H. Suzuki; *Tetrahedron Lett.*, 1989, **30**, 2413.	251
89TL2469	S. E. Denmark, G. Rajendra and J. E. Marlin; *Tetrahedron Lett.*, 1989, **30**, 2469.	832
89TL2501	D. P. Curran and P. A. van Elburg; *Tetrahedron Lett.*, 1989, **30**, 2501.	348
89TL2549	J. Brocard, M. Mahmoudi, L. Pelinski and L. Maciejewsk; *Tetrahedron Lett.*, 1989, **30**, 2549.	816, 818
89TL2581	S. Cacchi, A. Arcadi and F. Marinelli; *Tetrahedron Lett.*, 1989, **30**, 2581.	1029
89TL2619	D. H. R. Barton and J. Cs. Jaszberenyi; *Tetrahedron Lett.*, 1989, **30**, 2619.	939
89TL2627	E. Vedejs and S. L. Dax; *Tetrahedron Lett.*, 1989, **30**, 2627.	524, 1140
89TL2733	M. Lesage, C. Chatgilialoglu and D. Griller; *Tetrahedron Lett.*, 1989, **30**, 2733.	5, 9, 10, 936
89TL2907	B. Giese and G. Thoma; *Tetrahedron Lett.*, 1989, **30**, 2907.	937
89TL3131	G. Opitz, T. Ehlis and K. Rieth; *Tetrahedron Lett.*, 1989, **30**, 3131.	619
89TL3155	Y. Ichinose, S.-I. Matsunaga, K. Fugami, K. Oshima and K. Utimoto; *Tetrahedron Lett.*, 1989, **30**, 3155.	333
89TL3159	Y. Takeyama, Y. Ichinose, K. Oshima and K. Utimoto; *Tetrahedron Lett.*, 1989, **30**, 3159.	1131
89TL3267	A. G. Sutherland and R. J. K. Taylor; *Tetrahedron Lett.*, 1989, **30**, 3267.	618
89TL3303	A. R. Katritzky, J. Jiang and L. Urogdi; *Tetrahedron Lett.*, 1989, **30**, 3303.	262, 1041
89TL3335	J. E. Baldwin and Y. Yamaguchi; *Tetrahedron Lett.*, 1989, **30**, 3335.	574
89TL3343	W. R. Bowman, B. A. Marples and N. A. Zaidi; *Tetrahedron Lett.*, 1989, **30**, 3343.	929
89TL3377	J. S. Sabol and R. J. Cregge; *Tetrahedron Lett.*, 1989, **30**, 3377.	828
89TL3487	R. C. Larock and S. K. Stolz-Dunn; *Tetrahedron Lett.*, 1989, **30**, 3487.	230
89TL3517	R. Oehrlein, R. Jeschke, B. Ernst and D. Bellus; *Tetrahedron Lett.*, 1989, **30**, 3517.	834
89TL3545	G. Cahiez and B. Laboue; *Tetrahedron Lett.*, 1989, **30**, 3545.	1045
89TL3581	S. J. Coote, S. G. Davies, D. Middlemiss and A. Naylor; *Tetrahedron Lett.*, 1989, **30**, 3581.	202
89TL3837	C. Nájera, B. Mancheño and M. Yus; *Tetrahedron Lett.*, 1989, **30**, 3837.	237
89TL3865	D. S. Middleton, N. S. Simpkins and N. K. Terrett; *Tetrahedron Lett.*, 1989, **30**, 3865.	244
89TL3927	J. M. Tour, P. V. Bedworth and R. Wu; *Tetrahedron Lett.*, 1989, **30**, 3927.	680
89TL3999	H. Kotsuki, I. Kadota and M. Ochi; *Tetrahedron Lett.*, 1989, **30**, 3999.	177
89TL4021	W. A. Smit, S. L. Kireev, O. M. Nefedov and V. A. Tarasov; *Tetrahedron Lett.*, 1989, **30**, 4021.	529
89TL4157	K. Kakinuma and H.-Y. Li; *Tetrahedron Lett.*, 1989, **30**, 4157.	552
89TL4271	N. Tokitoh, T. Suzuki and W. Ando; *Tetrahedron Lett.*, 1989, **30**, 4271.	983
89TL4581	P. M. Ayrey and S. Warren; *Tetrahedron Lett.*, 1989, **30**, 4581.	757
89TL4657	B. M. Trost and R. Braslau; *Tetrahedron Lett.*, 1989, **30**, 4657.	457
89TL4799	P. Knochel and M. C. P. Yeh; *Tetrahedron Lett.*, 1989, **30**, 4799.	1017
89TL4845	S. Hatakeyama, K. Osanai, H. Numata and S. Takano; *Tetrahedron Lett.*, 1989, **30**, 4845.	216, 748
89TL4951	M. Avignon-Tropis and J. R. Pougny; *Tetrahedron Lett.*, 1989, **30**, 4951.	1009
89TL4995	R. Nagata, H. Yamanaka, E. Okazaki and I. Saito; *Tetrahedron Lett.*, 1989, **30**, 4995.	993
89TL5017	F. J. Leeper and P. Padmanabhan; *Tetrahedron Lett.*, 1989, **30**, 5017.	1131
89TL5153	A. Suzuki, Y. Satoh, T. Tayano and S. Hara; *Tetrahedron Lett.*, 1989, **30**, 5153.	1104
89TL5455	J. S. Yadav, M. C. Chander and C. S. Rao; *Tetrahedron Lett.*, 1989, **30**, 5455.	1066
89TL5485	A. J. Clark and K. Jones; *Tetrahedron Lett.*, 1989, **30**, 5485.	346
89TL5643	A. Pelter, K. Smith, M. Rowlands and S. Elgendy; *Tetrahedron Lett.*, 1989, **30**, 5643.	765
89TL5647	A. Pelter, K. Smith, M. Rowlands and S. Elgendy; *Tetrahedron Lett.*, 1989, **30**, 5647.	765
89TL5693	J. A. Soderquist and B. Santiago; *Tetrahedron Lett.*, 1989, **30**, 5693.	652
89TL5747	A. G. Myers, N. S. Finney and E. Y. Kuo; *Tetrahedron Lett.*, 1989, **30**, 5747.	962
89TL5883	W. Oppolzer, T. H. Keller, M. Bedoya-Zurita and C. Stone; *Tetrahedron Lett.*, 1989, **30**, 5883.	195
89TL6027	J. R. Falck, V. Bolitt and C. Mioskowski; *Tetrahedron Lett.*, 1989, **30**, 6027.	1026
89TL6051	T. Tamao, K. Kobayashi and Y. Ito; *Tetrahedron Lett.*, 1989, 6051.	490
89TL6129	P. Dowd and S.-C. Choi; *Tetrahedron Lett.*, 1989, **30**, 6129.	245
89TL6271	J. S. Sabol and R. J. Cregge; *Tetrahedron Lett.*, 1989, **30**, 6271.	828
89TL6487	E. C. Stracker, W. Leong, J. A. Miller, T. M. Shoup and G. Zweifel; *Tetrahedron Lett.*, 1989, **30**, 6487.	449
89TL6577	A. Srikrishna and K. Krishnan; *Tetrahedron Lett.*, 1989, **30**, 6577.	828
89TL6829	K. T. Mead and H.-L. Yang; *Tetrahedron Lett.*, 1989, **30**, 6829.	282
89TL6963	A. B. Smith, III, K. J. Hale, L. M. Laakso, K. Chen and A. Riéra; *Tetrahedron Lett.*, 1989, **30**, 6963.	217
89TL7107	S. A. Bapuji, W. B. Motherwell and M. Shipman; *Tetrahedron Lett.*, 1989, **30**, 7107.	527
89TL7157	B. M. Trost and R. C. Holcomb; *Tetrahedron Lett.*, 1989, **30**, 7157.	523
89TL7353	S. M. Barros, J. V. Comasseto and J. Berriel; *Tetrahedron Lett.*, 1989, **30**, 7353.	245, 441
89ZOR654	S. N. Gaidamaka, O. A. Shantalii, A. N. Rozko and A. A. Grigor'ev; *Zh. Org. Khim.*, 1989, **25**, 654 (*Chem. Abstr.*, 1989, **111**, 173 703).	612

90AC59	Y. Osada, S. Koike, T. Fukushima, S. Ogasawara, T. Shikada and T. Ikaruja; *Appl. Catal.*, 1990, **59**, 59 (*Chem. Abstr.*, 1990, **112**, 220033j).	674
90ACR345	R. Noyori and H. Takaya; *Acc. Chem. Res.*, 1990, **23**, 345.	77
90ACR375	D. Gutman; *Acc. Chem. Res.*, 1990, **23**, 375.	321, 921
90ACS842	L. Jalander, L. Oksanen and A. Rosling; *Acta. Chem. Scand.*, 1990, **44**, 842.	482
90AG(E)609	L. A. Paquette; *Angew Chem., Int. Ed. Engl.*, 1990, **29**, 609.	806
90AG(E)801	T. Ibuka, H. Habashita, S. Furakoshi, N. Fujii, Y. Oguchi, T. Uyehara and Y. Yamamoto; *Angew. Chem., Int. Ed. Engl.*, 1990, **29**, 801.	188
90AG(E)977	G. Bringmann, R. Walter and R. Weirich; *Angew. Chem. Int. Ed. Engl.*, 1990, **29**, 977.	472

90AG(E)1371	H. Mayr; *Angew. Chem., Int. Ed. Engl.*, 1990, **29**, 1371.	294
90AG(E)1422	D. Hoppe, F. Hintze and P. Tebben; *Angew. Chem., Int. Ed. Engl.*, 1990, **29**, 1422.	138
90AG(E)1424	D. Hoppe, A. Carstens and T. Krämer; *Angew. Chem., Int. Ed. Engl.*, 1990, **29**, 1424.	138
90AG(E)1454	B. Boubia, A. Mann, F. D. Bellamy and C. Mioskowski; *Angew. Chem., Int. Ed. Engl.*, 1990, **29**, 1454.	738
90AJC1375	R. J. Dancer, R. K. Haynes, W. A. Loughlin and S. C. Vonwiller; *Aust. J. Chem.*, 1990, **43**, 1375.	486
90AJC1439	A. R. Carroll and W. C. Taylor; *Aust. J. Chem.*, 1990, **43**, 1439.	688
90APO(25)131	R. Sustmann and H. G. Korth; *Adv. Phy. Org. Chem.*, 1990, **25**, 131.	922
90BCJ1328	K. Sato, S. Inoue, N. Iwase and K. Honda; *Bull. Chem. Soc. Jpn.*, 1990, **63**, 1328.	839
90BCJ2124	S. Matsuoka, N. Yoneda, N. Miyaura, T. Fukuhara and A. Suzuki; *Bull. Chem. Soc. Jpn.*, 1990, **63**, 2124.	1009
90BSF401	L. Balas, D. Jhurry, L. Latxague, S. Grelier, Y. Morel, M. Hamdani, N. Ardoin and D. Astruc; *Bull. Chem. Soc. Fr.*, 1990, 401.	474
90BSF681	A. Krief, J. L. Laboureur, W. Dumont and D. Labar; *Bull. Chim. Soc. Fr.*, 1990, **127**, 681.	716
90CAR(201)185	J. Tandanier, C. M. Lee, D. Whittern and N. Wideburg; *Carbohydr. Res.*, 1990, **201**, 185.	769
90CB555	H. Priepke, R. Bruckner and K. Harms; *Chem Ber.*, 1990, **123**, 555.	552
90CB917	S. W. Scheuplein, A. Kusche, R. Bruckner and K. Harms; *Chem Ber.*, 1990, **123**, 917.	816, 817
90CB2191	U. Kobs and W. P. Neumann; *Chem. Ber.*, 1990, **123**, 2191.	1043
90CC84	M. E. Jung, I. D. Trifunovich, J. M. Gardiner and G. L. Clevenger; *J. Chem. Soc., Chem. Commun.*, 1990, 84.	582
90CC85	Y.-Z. Huang, L. Yi and C. Chen; *J. Chem. Soc., Chem. Commun.*, 1990, 85.	272, 743
90CC118	M. Ochiai, T. Ito, Y. Takaoka, Y. Masaki, M. Kunishima, S. Tani and Y. Nagao; *J. Chem. Soc., Chem. Commun.*, 1990, 118.	1016, 1145
90CC418	F. Kido, S. C. Sinha, T. Abiko, M. Watanabe and A. Yoshikoshi; *J. Chem. Soc., Chem. Commun.*, 1990, 418.	822
90CC587	J. Barluenga, F. Foubelo, R. González, F. J. Fañanas and M. Yus; *J. Chem. Soc., Chem. Commun.*, 1990, 587.	1090
90CC730	K. Yanada, R. Yanada and H. Meguri; *J. Chem. Soc., Chem. Commun.*, 1990, 730.	568
90CC819	T. G. Back and K. Yang; *J. Chem. Soc., Chem. Commun.*, 1990, 819.	436
90CC858	H. Fishcer, W. Bidell and J. Hofmann; *J. Chem. Soc., Chem. Commun.*, 1990, 858.	500
90CC934	D. Armesto, A. R. Agarrabeitia, W. M. Horspool and M. G. Gallego; *J. Chem. Soc., Chem. Commun.*, 1990, 934.	800
90CC1001	H. Nemoto, A. Satoh, M. Ando and K. Fukumoto; *J. Chem. Soc., Chem. Commun.*, 1990, 1001.	831
90CC1074	K. Orito, M. Ohto and H. Suginome; *J. Chem. Soc., Chem. Commun.*, 1990, 1074.	927
90CC1076	K. Orito, M. Ohto and H. Suginome; *J. Chem. Soc., Chem. Commun.*, 1990, 1076.	927
90CC1284	M. Hojo and S. Tanimoto; *J. Chem. Soc., Chem. Commun.*, 1990, 1284.	438
90CC1436	H. Ishibashi, M. Okano, H. Tamaki, K. Maruyama, T. Yakura and M. Ikeda; *J. Chem. Soc., Chem. Commun.*, 1990, 1436.	362
90CC1521	J. Barluenga, F. Foubelo, R. González, F. J. Fañanás and M Yus; *J. Chem. Soc., Chem. Commun.*, 1990, 1521.	1090
90CC1532	E. G. Nolen, A. Aliocco, J. Vitarius and K. McSorley; *J. Chem. Soc., Chem. Commun.*, 1990, 1532.	255
90CC1543	D. E. Cladingboel and P. J. Parsons; *J. Chem. Soc., Chem. Commun.*, 1990, 1543.	347
90CC1575	I. Fleming, I. T. Morgan and A. K. Sarkar; *J. Chem. Soc., Chem. Commun.*, 1990, 1575.	651, 943
90CC1589	E. W. Della, G. M. Elsey, N. J. Head and J. C. Walton; *J. Chem. Soc., Chem. Commun.*, 1990, 1589.	923
90CC1657	M. Murakata, M. Nakajima and K. Koga; *J. Chem. Soc., Chem. Commun.*, 1990, 1657.	135
90CE261	K. Ohe, S. Fukuzawa and S. Uemura; *Chem. Express*, 1990, **5**, 261 (*Chem. Abstr.*, 1990, **113**, 114 676).	611
90CJC1636	T. B. Lowinger and L. Weiler; *Can. J. Chem.*, 1990, **68**, 1636.	942
90CJC1698	J. P. Kutney, K. Piotrowska, Y.-H. Chen, K.-P. N. Cheng, Z. Gao and S. J. Rettig; *Can. J. Chem.*, 1990, **68**, 1698.	928
90CJC1714	F. G. Bordwell and J. A. Harrelson, Jr.; *Can. J. Chem.*, 1990, **68**, 1714.	922
90CJC1961	N. Nguyen, B. E. Harris, K. B. Clark and W. J. Leight; *Can. J. Chem.*, 1990, **68**, 1961.	799
90CL185	K. Ikenaga, S. Matsumoto, K. Kikukawa and T. Matsuda; *Chem. Lett.*, 1990, 185.	452, 490
90CL481	K. Sugita, A. Ohta, M. Onaka and Y. Izumi; *Chem. Lett.*, 1990, 481.	208
90CL593	T. Fujisawa, Y. Okumura and T. Sato; *Chem. Lett.*, 1990, 593.	183
90CL613	F. Kido, A. B. Kazi and A. Yoshikoshi; *Chem. Lett.*, 1990, 613	822
90CL1411	K. Narasaka, N. Saito, Y. Hayashi and H. Ichida; *Chem. Lett.*, 1990, 1411.	1140
90CL2069	Y. Murata and T. Nakai; *Chem. Lett.*, 1990, 2069.	823
90CL2149	M. Iyoda, M. Oda, Y. Kai, N. Kanehisa and N. Kasai; *Chem. Lett.*, 1990, 2149.	992
90CR1203	J. M. Schwab and B. S. Henderson; *Chem. Rev.*, 1990, **90**, 1203.	797
90CRV283	J. Barrau, J. Escudié and J. Satgé; *Chem. Rev.*, 1990, **90**, 283.	656
90H(31)1189	S. Mori and S. Takechi; *Heterocycles*, 1990, **31**, 1189.	173
90H(31)1359	J. T. Hunt, M. F. Malley and J. Z. Gougoutas; *Heterocycles*, 1990, **31**, 1359.	200
90H(31)1997	B. Alcaide, P. Areces, E. Borredon, C. Biurrun, J. P. Castells and J. Plumet; *Heterocycles*, 1990, **31**, 1997.	217
90HCA640	R. L. Snowden, R. Brauchli and M. Wuest; *Helv. Chim. Acta*, 1990, **73**, 640.	635
90IJC(B)390	G. K. Biswas, P. Bhattacharyya and D. Mal; *Indian J. Chem., Sect. B*, 1990, **29**, 390.	239
90IZV2665	N. A. Bumagin, L. I. Sukhomlinova, T. P. Tolstaya, A. N. Vanchikov and I. P. Beletskaya; *Izv. Akad. Nauk SSSR Ser. Khim.*, 1990, **39**, 2665.	486

90JA228	L. A. Paquette, M. P. Trova, J. Luo, A. E. Clough and L. B. Anderson; *J. Am. Chem. Soc.*, 1990, **112**, 228.	776
90JA277	L. A. Paquette, N. A. Pegg, D. Toops, G. D. Maynard and R. D. Rogers; *J. Am. Chem. Soc.*, 1990, **112**, 277.	1098
90JA316	K. Nonoshita, H. Banno, K. Maruoka and H. Yamamoto; *J. Am. Chem. Soc.*, 1990, **112**, 316.	415, 826
90JA408	B. M. Trost and S. A. King; *J. Am. Chem. Soc.*, 1990, **112**, 408.	194, 289
90JA697	S. Shambayati, J. F. Blake, S. G. Wierschke, W. L. Jorgensen and S. L. Schreiber; *J. Am. Chem. Soc.*, 1990, **112**, 697.	149
90JA878	E. J. Corey, C. M. Yu and D. H. Lee; *J. Am. Chem. Soc.*, 1990, **112**, 878.	977, 1055
90JA891	D. H. R. Barton and M. Ramesh; *J. Am. Chem. Soc.*, 1990, **112**, 891.	944
90JA896	V. Sniekus, J. C. Cuevas, C. P. Sloan, H. Liu and D. P. Curran; *J. Am. Chem. Soc.*, 1990, **112**, 896.	927
90JA1266	N. A. Porter, J. K. Kaplan and P. H. Dussault; *J. Am. Chem. Soc.*, 1990, **112**, 1266.	838
90JA1280	H. E. Zimmerman and P. A. Wang; *J. Am. Chem. Soc.*, 1990, **112**, 1280.	890
90JA1297	R. M. Moriarty, O. Prakash, R. K. Vaid and L. Zhao; *J. Am. Chem. Soc.*, 1990, **112**, 1297.	109
90JA1661	G. Stork, Y. Kobayashi, T. Suzuki and K. Zhao; *J. Am. Chem. Soc.*, 1990, **112**, 1661.	226
90JA1999	K. Honda, S. Inoue and K. Sato; *J. Am. Chem. Soc.*, 1990, **112**, 1999.	408, 823
90JA2034	J. E. Forbes and S. Z. Zard; *J. Am. Chem. Soc.*, 1990, **112**, 2034.	939, 946
90JA2264	L. S. Hegedus, M. A. Schwindt, S. De Lombaert and R. Imwinkelried; *J. Am. Chem. Soc.*, 1990, **112**, 2264.	165
90JA2759	M. A. Dombroski, S. A. Kates and B. B. Snider; *J. Am. Chem. Soc.*, 1990, **112**, 2759.	347
90JA2803	H. Alper and N. Hamel; *J. Am. Chem. Soc.*, 1990, **112**, 2803.	298
90JA2816	B. M. Trost and G. Kottirsch; *J. Am. Chem. Soc.*, 1990, **112**, 2816.	496, 1051
90JA3225	P. E. Eaton, C. Yang and Y. Xiong; *J. Am. Chem. Soc.*, 1990, **112**, 3225.	900
90JA3227	D. A. Hrovat and W. T. Borden; *J. Am. Chem. Soc.*, 1990, **112**, 3227.	900
90JA3228	R. M. Moriarty, S. M. Tuladhar, R. Penmasta and A. K. Awasthi; *J. Am. Chem. Soc.*, 1990, **112**, 3228.	900
90JA3252	H.-J. Kang and L. A. Paquette; *J. Am. Chem. Soc.*, 1990, **112**, 3252.	826
90JA3554	B. A. McCortney, B. M. Jacobson, M. Vreeke and E. S. Lewis; *J. Am. Chem. Soc.*, 1990, **112**, 3554.	107
90JA3717	A. G. Godfrey and B. Ganem; *J. Am. Chem. Soc.*, 1990, **112**, 3717.	637, 1077
90JA3945	L. E. Overman and A. F. Renaldo; *J. Am. Chem. Soc.*, 1990, **112**, 3945.	413
90JA4035	K. Mikami, K. Takahashi and T. Nakai; *J. Am. Chem. Soc.*, 1990, **112**, 4035.	826
90JA4054	L. T. Scott, M. J. Cooney and D. Johnels; *J. Am. Chem. Soc.*, 1990, **112**, 4054.	1019
90JA4063	B. H. Lipshutz, R. Crow, S. H. Dimock, E. L. Ellsworth, R. A. J. Smith and J. R. Behling; *J. Am. Chem. Soc.*, 1990, **112**, 4063.	207
90JA4072	E. Block and D. Putman; *J. Am. Chem. Soc.*, 1990, **112**, 4072.	707, 993
90JA4324	A. L. Casalnuovo and J. C. Calabrese; *J. Am. Chem. Soc.*, 1990, **112**, 4324.	1013
90JA4609	B. Lei and A. G. Fallis; *J. Am. Chem. Soc.*, 1990, **112**, 4609.	774
90JA4965	R. S. Jolly, G. Luedtke, D. Sheehan and T. Livinghouse; *J. Am. Chem. Soc.*, 1989, **112**, 4965.	523
90JA5244	T. Imamoto, T. Oshiki, T. Onozawa, T. Kusumoto and K. Sato; *J. Am. Chem. Soc.*, 1990, **112**, 5244.	158
90JA5562	L. A. Paquette, D. T. DeRussy, T. Vandenheste and R. D. Rogers; *J. Am. Chem. Soc.*, 1990, **112**, 5562.	1098
90JA5601	C. P. Jasperse and D. P. Curran; *J. Am. Chem. Soc.*, 1990, **112**, 5601.	344, 831
90JA6408	T. V. Rajanbabu, W. A. Nugent and M. S. Beattie; *J. Am. Chem. Soc.*, 1990, **112**, 6408.	587
90JA6416	L. M. Tolbert, Md. N. Islam, R. P. Johnson, P. M. Loiselle and W. C. Shakespeare; *J. Am. Chem. Soc.*, 1990, **112**, 6416.	499
90JA6422	G. A. Olah, G. K. Surya Prakash and P. Krishnamurti; *J. Am. Chem. Soc.*, 1990, **112**, 6422.	900
90JA6615	J.-E. Bäckvall, M. Sellén and B. Grant; *J. Am. Chem. Soc.*, 1990, **112**, 6615.	187
90JA6729	R. P. Bakale, M. A. Scialdone and C. R. Johnson; *J. Am. Chem. Soc.*, 1990, **112**, 6729.	668
90JA6738	D. P. Curran, W. Shen, J. Zhang and T. A. Heffner; *J. Am. Chem. Soc.*, 1990, **112**, 6738.	326, 328
90JA6740	N. A. Porter, E. Swann, J. Nally and A. T. McPhail; *J. Am. Chem. Soc.*, 1990, **112**, 6740.	328
90JA6842	P. White and R. Breslow; *J. Am. Chem. Soc.*, 1990, **112**, 6842.	927
90JA6950	N. S. Banait and W. P. Jencks; *J. Am. Chem. Soc.*, 1990, **112**, 6950.	623
90JA7407	M. Kageyama and T. Tamura; *J. Am. Chem. Soc.*, 1990, **112**, 7407.	697, 700, 703
90JA7431	P. Knochel; *J. Am. Chem. Soc.*, 1990, **112**, 7431.	163
90JA7682	E. W. Baxter, D. Labaree, H. L. Ammon and P. S. Mariano; *J. Am. Chem. Soc.*, 1990, **112**, 7682.	835
90JA7791	K. Maruoka, H. Banno, and H. Yamamoto; *J. Am. Chem. Soc.*, 1990, **112**, 7791.	825, 826
90JA8042	A. Alexakis, I. Marek, P. Manganey and J. F. Normant; *J. Am. Chem. Soc.*, 1990, **112**, 8042.	670, 967
90JA8084	D. Scarpetti and P. L. Fuchs; *J. Am. Chem. Soc.*, 1990, **112**, 8084.	618
90JA8135	T. S. Sorensen and S. M. Whitworth; *J. Am. Chem. Soc.*, 1990, **112**, 8135.	919
90JA8433	J. A. Duncan, R. T. Hendricks and K. S. Kwong; *J. Am. Chem. Soc.*, 1990, **112**, 8433.	575, 576
90JA8578	W. C. Shakespeare and R. P. Johnson; *J. Am. Chem. Soc.*, 1990, **112**, 8578.	1080
90JA8912	R. F. Childs, M. D. Kostyk, C. J. L. Lock and M. Mahendran; *J. Am. Chem. Soc.*, 1990, **112**, 8912.	894
90JA8991	C. S. Yannoni, G. G. Webb and P. C. Myhre; *J. Am. Chem. Soc.*, 1990, **112**, 8991.	906

90JA9284 L. A. Paquette, J. L. Romine, H.-S. Lin and J. Wright; *J. Am Chem. Soc.*, 1990, **112**, 9284. 558, 575, 584

90JA9330 G. C. M. Lee, B. Tobias, J. M. Holmes, D. A. Harcourt and M. E. Garst; *J. Am. Chem. Soc.*, 1990, **112**, 9330. 1014, 1029

90JA9388 J. Castro, H. Sörenson, A. Riera, C. Morin, A. Moyano, M. A. Pericàs and A. E. Greene; *J. Am. Chem. Soc.*, 1990, **112**, 9388. 530

90JA9401 D. P. Curran and C. M. Seong; *J. Am. Chem. Soc.*, 1990, **112**, 9401. 340

90JCI458 E. R. Laird and W. L. Jorgenson; *J. Chem. Inf. Comput. Sci.*, 1990, **30**, 458. 320

90JCR(S)388 Y. Liao, Y. Huang, L. Zhang and C. Chen; *J. Chem. Res. (S)*, 1990, 388. 647, 743

90JCS(P1)11 T. J. Grattan and J. S. Whitehurst; *J. Chem. Soc., Perkin Trans. 1*, 1990, 11. 826

90JCS(P1)428 W. Cabri, S. De. Bernardinis, F. Francalanci and S. Penco; *J Chem. Soc., Perkin Trans. 1*, 1990, 428. 433

90JCS(P1)937 S. Niwa and K. Soai; *J. Chem. Soc., Perkin Trans. 1*, 1990, 937. 1039

90JCS(P1)1226 M. Yoshitake, M. Yamamoto, S. Kohmoto and K. Yamada; *J. Chem. Soc., Perkin Trans. 1*, 1990, 1226. 216

90JCS(P1)1493 S. M. Roberts, C. Smith and R. J. Thomas; *J. Chem. Soc., Perkin Trans. 1*, 1990, 1493. 835

90JCS(P1)2185 M. P. S. Ishar, A. Wali and R. P. Ghandi; *J. Chem. Soc., Perkin Trans. 1*, 1990, 2185. 499

90JCS(P1)2207 M. Miura, H. Hashimoto, K. Itoh and M. Nomura; *J. Chem. Soc., Perkin Trans. 1*, 1990, 2207. 494

90JCS(P1)2352 M. J. Kelly, R. F. Newton, S. M. Roberts and J. F. Whitehead; *J. Chem. Soc., Perkin Trans. 1*, 1990, 2352. 173

90JCS(P1)2691 H. Bhandal, V. F. Patel, G. Pattenden and J. J. Russell; *J. Chem. Soc., Perkin Trans. 1*, 1990, 2691. 345, 946

90JCS(P1)2703 V. F. Patel and G. Pattenden; *J. Chem. Soc., Perkin Trans. 1*, 1990, 2703. 946

90JCS(P1)2709 H. Bhandal, A. R. Howell, V. F. Patel and G. Pattenden; *J. Chem. Soc., Perkin Trans. 1*, 1990, 2709. 335, 346, 495, 946

90JCS(P1)2851 D. M. X. Donnelly, J.-P. Finet, P. J. Guiry and R. M. Hutchinson; *J. Chem. Soc., Perkin Trans. 1*, 1990, 2851. 435

90JCS(P1)3017 T. Kataoka, K. Tsutsumi, K. Kano, K. Mori, M. Miyake, M. Yokota, H. Shimizu and M. Hori; *J. Chem. Soc., Perkin Trans. 1*, 1990, 3017. 943

90JCS(P2)2187 N. A. Al-Awadi; *J. Chem. Soc., Perkin Trans. 2*, 1990, 2187. 782

90JHC79 F. Minisci, E. Vismara and F. Fontana; *J. Heterocycl. Chem.*, 1990, **27**, 79. 934

90JHC503 N. Sato, A. Hayakawa and R. Takeuchi; *J. Heterocycl. Chem.*, 1990, **27**, 503. 1010

90JMOCL1 P. F. Dos Santos Filho and H. D. O. Pastore; *J. Mol. Catal.*, 1990, **58**, L1. 637

90JOC9 E. R. Laird and W. L. Jorgenson; *J. Org. Chem.*, 1990, **55**, 9. 320

90JOC58 F. G. Bordwell, M. J. Bausch, J. P. Cheng, T. H. Cripe, T. Y. Lynch and M. E. Mueller; *J. Org. Chem.*, 1990, **55**, 58. 922

90JOC63 R. F. Heck, W. J. Tao and S. Nesbitt; *J. Org. Chem.*, 1990, **55**, 63. 1009

90JOC227 J. A. Marshall, E. D. Robinson and J. Lebreton; *J. Org. Chem.*, 1990, **55**, 227. 818

90JOC324 E. J. Enholm, H. Satici and G. Prasad; *J. Org. Chem.*, 1990, **55**, 324. 225

90JOC393 S. M. Weinreb and J. Sisko; *J. Org. Chem.*, 1990, **55**, 393. 1042

90JOC408 R. Tamura, K.-I. Watabe, H. Katayama, H. Suzuki and Y. Yamamoto; *J. Org. Chem.*, 1990, **55**, 408. 257

90JOC414 A. Padwa, U. Chiacchio, Y. Garreau, J. M. Kassir, K. E. Krumpe and A. M. Schoffstall; *J. Org. Chem.*, 1990, **55**, 414. 550

90JOC441 M. Suzuki, Y. Morita and R. Noyori; *J. Org. Chem.*, 1990, **55**, 441. 1056, 1061

90JOC711 W. R. Dolbier, Jr. and S. L. Weaver; *J. Org. Chem.*, 1990, **55**, 711. 498

90JOC791 A. I. Meyers and S. Bienz; *J. Org. Chem.*, 1990, **55**, 791. 1001, 1134

90JOC991 J. M. Saa, M. Dopico, G. Martorell and A. Garcia-Raso; *J. Org. Chem.*, 1990, **55**, 991. 433

90JOC996 K. Okamoto, T. Kitagawa, K. Takeuchi, K. Komatsu, T. Kinoshita, S. Aonuma, M. Nagai and A. Miyabo; *J. Org. Chem.*, 1990, **55**, 996. 867

90JOC1019 G. D. Dzingeleski, G. Blotny and R. M. Pollack; *J. Org. Chem* , 1990, **55**, 1019. 877

90JOC1052 J. S. Lomas, D. Fain and S. Briand; *J. Org. Chem.*, 1990, **55**, 1052. 883

90JOC1127 B. M. Trost and C. A. Merlic; *J. Org. Chem.*, 1990, **55**, 1127. 238

90JOC1338 S. T. Bright, J. M. Coxon and P. J. Steel; *J. Org. Chem.*, 1990, **55**, 1338. 185

90JOC1416 F. E. Ziegler and Z.-I. Zheng; *J. Org. Chem.*, 1990, **55**, 1416. 176, 940

90JOC1421 Y.-D. Wu, K. N. Houk and J. A. Marshall; *J. Org. Chem.*, 1990, **55**, 1421. 816, 817

90JOC1506 M. Ohkita, T. Tsuji, M. Suzuki, M. Murakami and S. Nishida; *J. Org. Chem.*, 1990, **55**, 1506. 800

90JOC1670 E. J. Corey and S. W. Wright; *J. Org. Chem.*, 1990, **55**, 1670. 504

90JOC1705 O. Fujimura, K. Takai and K. Utimoto; *J. Org. Chem.*, 1990, **55**, 1705. 228

90JOC1786 D. L. J. Clive and R. J. Bergstra; *J. Org. Chem.*, 1990, **55**, 1786. 244

90JOC1874 L. L. Yeung, Y. C. Yip and T.-Y. Luh; *J. Org. Chem.*, 1990, **55** 1874. 991

90JOC1984 S. E. Denmark and L. K. Marble; *J. Org. Chem.*, 1990, **55**, 1984. 187

90JOC2016 M. B. Sassaman, G. K. S. Prakash and G. A. Olah; *J. Org. Chem.*, 1990, **55**, 2016. 208

90JOC2108 K. Tadano, M. Minami and S. Ogawa; *J. Org. Chem.*, 1990, **55**, 2108. 829

90JOC2128 D. H. Hua, S. N. Bharathi, P. D. Robinson and A. Tsujimoto; *J. Org. Chem.*, 1990, **55**, 2128. 694

90JOC2234 S. H. Pine, G. Shen, J. Bautista, C. Sutton, Jr., W. Yamada and L. Apodaca; *J. Org. Chem.*, 1990, **55**, 2234. 145

90JOC2294 S. Valverde, M. Bernabé, S. Garcia-Ochoa and A. M. Gómez; *J. Org. Chem.*, 1990, **55**, 2294. 187

90JOC2578 P. Beak and W. K. Lee; *J. Org. Chem.*, 1990, **55**, 2578. 151

90JOC2983	C. B. Ziegler, Jr.; *J. Org. Chem.*, 1990, **55**, 2983.	988
90JOC3007	E. Ciganek; *J. Org. Chem.*, 1990, **55**, 3007.	626
90JOC3147	E. K. Yau and J. K. Coward; *J. Org. Chem.*, 1990, **55**, 3147.	1074
90JOC3209	A. R. Katritzky, W.-Q. Fan and C. Fu; *J. Org. Chem.*, 1990, **55**, 3209.	261
90JOC3321	D. D. Tanner, D. J. Harrison, J. Chen, A. Kharrat, D. D. M. Wayner, D. Griller and D. J. McPhee; *J. Org. Chem.*, 1990, **55**, 3321.	946
90JOC3450	J. A. Marshall and E. D. Robinson; *J. Org. Chem.*, 1990, **55**, 3450.	957
90JOC3452	J. M. Tour, J. P. Cooper and S. L. Pendalwar; *J. Org. Chem.*, 1990, **55**, 3452.	456
90JOC3808	S. V. Luis, P. Ferrer and M. I. Burguete; *J. Org. Chem.*, 1990, **55**, 3808.	1081
90JOC3840	R. V. Williams, G. W. Kelley, J. Loebel, D. van der Helm and P. C. Bulman Page; *J. Org. Chem.*, 1990, **55**, 3840.	146
90JOC3961	Y. Shen, H. Liu, Y. Chen; *J. Org. Chem.*, 1990, **55**, 3961.	181
90JOC3967	G. H. Posner and C. M. Kinter; *J. Org. Chem.*, 1990, **55**, 3967.	788
90JOC4051	R. T. Jacobs, G. I. Feutrill and J. Meinwald; *J. Org. Chem.*, 1990, **55**, 4051.	776
90JOC4273	Z. Liu, B. Classon and B. Samuelsson; *J. Org. Chem.*, 1990, **55**, 4273.	587
90JOC4417	H. Kotsuki, I. Kadota and M. Ochi; *J. Org. Chem.*, 1990, **55**, 4417.	177
90JOC4614	G. B. Stone and L. S. Liebeskind; *J. Org. Chem.*, 1990, **55**, 4614.	836
90JOC4657	R. M. Williams, D. J. Aldous and S. C. Aldous; *J. Org. Chem.*, 1990, **55**, 4657.	1042
90JOC4853	J. Haruta, K. Nishi, S. Matsuda, S. Akai, Y. Tamura and Y. Kita; *J. Org. Chem.*, 1990, **55**, 4853.	1061, 1062
90JOC5045	T.-C. Wu, H. Xiong and R. D. Rieke; *J. Org. Chem.*, 1990, **55**, 5045.	115
90JOC5159	R. Ballini, M. Petrini and G. Rosini; *J. Org. Chem.*, 1990, **55**, 5159.	39
90JOC5359	L. S. Liebeskind and R. W. Fengl; *J. Org. Chem.*, 1990, **55**, 5359.	480, 1020
90JOC5386	U. Azzena, T. Denurra, G. Melloni and A. M. Piroddi; *J. Org. Chem.*, 1990, **55**, 5386.	432
90JOC5446	C. E. Tucker, S. A. Rao and P. Knochel; *J. Org. Chem.*, 1990, **55**, 5446.	486
90JOC5532	U. Azzena, T. Denurra, G. Melloni and A. M. Piroddi; *J. Org. Chem.*, 1990, **55**, 5532.	398
90JOC5553	S. Yamago and E. Nakamura; *J. Org. Chem.*, 1990, **55**, 5553.	360
90JOC5801	C. D. Weis and G. R. Newkome; *J. Org. Chem.*, 1990, **55**, 5801.	637, 638
90JOC5833	V. Farina, S. R. Baker, D. A. Benigni, S. I. Hauck and C. Sapino, Jr.; *J. Org. Chem.*, 1990, **55**, 5833.	1004
90JOC5923	N. Chatani, T. Sano, K. Ohe, Y. Kawasaki and S. Murai; *J. Org. Chem.*, 1990, **55**, 5923.	202
90JOC6061	G. A. Olah, R. Krishnamurti and G. K. Surya Prakash; *J. Org. Chem.*, 1990, **55**, 6061.	904
90JOC6194	S. Becker, Y. Fort and P. Caubère; *J. Org. Chem.*, 1990, **55**, 6194.	437, 438, 439
90JOC6236	C. Chen and D. J. Hart; *J. Org. Chem.*, 1990, **55**, 6236.	828
90JOC6356	S. Hara, K. Kishimura, A. Suzuki and R. S. Dhillon; *J. Org. Chem.*, 1990, **55**, 6356.	185
90JOM(381)155	K. Oka, R. Nakao, Y. Abe and T. Dohmaru; *J. Organometal. Chem.*, 1990, **381**, 155.	933
90JOM(382)121	M. T. Reetz and M. Sauerwald; *J. Organomet. Chem.*, 1990, **382**, 121.	202
90JOM(394)37	T. Gouyon, R. Sauvêtre and J.-F. Normant; *J. Organomet. Chem.*, 1990, **394**, 37.	653
90JOM(396)289	D. Damour, J. Pornet, B. Randrianoelina and L. Miginiac; *J. Organomet. Chem.*, 1990, **396**, 289.	978
90JOM(400)121	J. Satgé; *J. Organomet. Chem.*, 1990, **400**, 121.	656
B-90MI 108-01	W. Carruthers; "Cycloaddition Reactions in Organic Synthesis," Pergamon, Oxford, 1990.	359, 372
B-90MI 112-01	W. Carruthers; "Cycloaddition Reactions in Organic Synthesis," Pergamon, Oxford, 1990.	516, 519
B-90MI 114-01	Y. Huang, L. Shi, J. Yang, W. Xiao, S. Li and W. Wang; in "Heteroatom Chemistry," ed. E. Block, VCH Publishers, New York, 1990, pp. 189–206.	646, 647
B-90MI 119-01	Z. Rappoport (ed.); "The Chemistry of Enols," Wiley, Chichester, 1990.	869
B-90MI 119-02	J. R. Keefe and A. J. Kresge; in "The Chemistry of Enols," ed. Z. Rappoport, Wiley, Chichester, 1990, chap. 7.	869
B-90MI 119-03	B. Capon; in "The Chemistry of Enols," ed. Z. Rappoport, Wiley, Chichester, 1990, chap. 5.	869
B-90MI 119-04	J. Toullec; in "The Chemistry of Enols," ed. Z. Rappoport, Wiley, Chichester, 1990, chap. 6.	869
B-90MI 119-05	R. A. Cox; in "Organic Reaction Mechanisms," eds. A. C. Knipe and W. E. Watts, Wiley, New York, 1990, chap. 9, pp. 287–305.	891
B-90MI 119-06	M. Fieser and L. F. Fieser; "Reagents for Organic Synthesis," Wiley, New York, 1990, Vol. 15, p. 360.	898
B-90MI 121-01	E. Schaumann; in "The Chemistry of Double Bonded Functional Groups," ed. S. Patai, Wiley, New York, 1990, **2**, 1269.	1040
90OM1325	S. Inoue and Y. Sato; *Organometallics*, 1990, **9**, 1325.	291
90OM3086	M. B. Power, S. G. Bott, D. L. Clark, J. L. Atwood and A. R. Barron; *Organometallics*, 1990, **9**, 3086.	666
90OPP167	L. A. Paquette and J. P. Gilday; *Org. Prep. Proced. Int.*, 1990, **22**, 167–201.	885
90OR(38)1	D. J. Ager; *Org. React.*, 1990, **38**, 1.	160
90OR(38)225	M. J. Chapdelaine and M. Hulce; *Org. React.*, 1990, **38**, 225.	304
90OR(39)1	A. R. Chamberlin and S. H. Bloom; *Org. React.*, 1990, **39**, 1.	630, 633
90OS(68)32	R. L. Danheiser, S. Savariar and D. D. Cha; *Org. Synth.*, 1990, **68**, 32.	518
90OS(68)116	W. J. Scott, G. T. Crisp and J. K. Stille; *Org. Synth.*, 1990, **68**, 116.	480
90OS(68)130	N. Miyaura and A. Suzuki; *Org. Synth.*, 1990, **68**, 130.	489
90OS(68)138	S. Cacchi, E. Morera and G. Ortar; *Org. Synth.*, 1990, **68**, 138.	435
90OS(69)66	B. Giese and K. S. Gröninger; *Org. Synth.*, 1990, **69**, 66.	923
90OSC(7)226	M. J. Taschner, T. Rosen and C. H. Heathcock; *Org. Synth., Coll. Vol.*, 1990, **7**, 226.	505
90OSC(7)363	Z. G. Hajos and D. R. Parrish; *Org. Syn. Coll. Vol.*, 1990, **7**, 363.	578
90OSC(7)506	S. L. Taylor, T. R. Forbus and J. C. Martin; *Org. Syn. Coll. Vol.*, 1990, **7**, 506.	582

90PAC507 T. Asao; *Pure Appl. Chem.*, 1990, **62**, 507. 899, 909
90PAC531 K. Hafner; *Pure Appl. Chem.*, 1990, **62**, 531. 995
90PAC1597 P. A. Wender, R. Ternansky, M. de Long, S. Singh, A. Olivero and K. Rice; *Pure Appl. Chem.*, 1990, **62**, 1597. 364, 365
90PAC1867 B. Cazes; *Pure Appl. Chem.*, 1990, **62**, 1867. 496
90PAC1941 W. Oppolzer; *Pure Appl. Chem.*, 1990, **62**, 1941. 195
90PJC123 M. Olejnik, W. Jasiobedzki and M. Zieba; *Pol. J. Chem.*, 1990, **64**, 123. 995
90RTC87 K. Timmer, D. H. M. W. Thewissen, H. A. Meinema and E. J. Bulten; *Recl. Trav. Chim. Pays-Bas*, 1990, **109**, 87. 72, 88
90S1 S. Servi; *Synthesis*, 1990, 1. 77
90S89 T.-Y. Luh and Z. J. Ni; *Synthesis*, 1990, 89. 22, 23, 238, 436
90S115 K. Burger and N. Sewald; *Synthesis*, 1990, 115. 1042
90S259 T. Sato; *Synthesis*, 1990, 259. 667
90S285 T. Inghardt and T. Frejd; *Synthesis*, 1990, 285. 218
90S401 H. Kotsuki, N. Yoshimura, I. Kadota, Y. Ushio and M. Ochi; *Synthesis*, 1990, 401. 155
90S403 S. Dapperheld, M. Feldhues, H. Litterer, F. Sistig and P. Wegener; *Synthesis*, 1990, 403. 428
90S411 P. Serafinowski, *Synthesis*, 1990, 411. 585
90S448 U. Gerigk, M. Gerlach, W. P. Neumann, R. Vieler and V. Weintritt; *Synthesis*, 1990, 448. 331, 342
90S649 D. B. Berkowitz; *Synthesis*, 1990, 649. 3, 6
90S843 B. S. Bandodakar and G. Nagendrappa; *Synthesis*, 1990, 843. 574
90S1173 A. R. Katritzky, J. Borowiecka and W. Q. Fan; *Synthesis*, 1990, 1173. 261
90SC125 G. Bram, T. Fillebeen-Khan and N. Geraghty; *Synth. Commun.*, 1990, **10**, 125. 178
90SC423 P. Baeckström, L. Li, M. Wickramaratne and T. Norin; *Synth. Commun.*, 1990, **20**, 423. 828
90SC1293 P. G. Ciattini, E. Morera and G. Ortar; *Synth. Commun.*, 1990, **20**, 1293. 435
90SC1437 S. E. Drewes, N. D. Emslie, N. Karodia and G. Loizou; *Synth. Commun.*, 1990, **20**, 1437. 827
90SC1463 M. Cushman and J. Jurayi; *Synth. Commun.*, 1990, **20**, 1463. 583
90SC1681 J. V. Comasseto and J. Berriel; *Synth. Commun.*, 1990, **20**, 1681. 441
90SC2165 L. Brandsma, J. van Soolingen and H. Andringa; *Synth. Commun.*, 1990, **20**, 2165. 95, 889
90SC2181 M. J. Dabdoub, J. V. Comasseto, S. M. Barros and F. Moussa; *Synth. Commun.*, 1990, **20**, 2181. 999, 1035
90SC2349 H. Andringa, J. Hanekamp and L. Brandsma; *Synth. Commun.*, 1990, **20**, 2349. 464
90SC2733 L. Barajas, J. E. Hernández and S. Torres; *Synth. Commun.*, 1990, **20**, 2733. 1073, 1085
90SC3347 G. A. Poulton and C. A. Boulet; *Synth. Commun.*, 1990, **20**, 3347. 200, 255
90SC3477 B. S. Thyagarajan, R. Chandler and A. Santillan; *Synth. Commun.*, 1990, **20**, 3477. 515
90SL48 S. V. Ley, D. S. Brown and T. Hansson; *Syn. Lett.*, 1990, 48. 1031
90SL67 L. A. Paquette; *Synlett*, 1990, 67. 806
90SL97 P. Metz and C. Mues; *Synlett*, 1990, 97. 828
90SL119 B. H. Lipshutz; *Synlett.*, 1990, 119. 206
90SL166 A. Veit and B. Giese; *Synlett*, 1990, 166. 946
90SL211 R. A. Aitken and S. Seth; *Syn. Lett.*, 1990, 211. 1079
90SL212 R. A. Aitken and S. Seth; *Syn. Lett.*, 1990, 212. 1079
90SL219 C. Chatgilialoglu, A. Guerrini and G. Seconi; *Synlett*, 1990, 219. 936, 945
90SL224 M. Caruso, J. G. Knight and S. V. Ley; *Synlett*, 1990, 224. 196
90SL341 R. C. Larock and S. K. Stolz-Dunn; *Synlett*, 1990, 341. 230
90SL427 W. J. Youngs, D. Solooki, V. O. Kennedy and C. A. Tessier; *Syn. Lett.*, 1990, 427. 1010
90SL465 R. Schobert and U. Hohlein; *Synlett*, 1990, 465. 45
90SL547 G. M. Tombo, E. Didier and B. Loubinoux; *Syn. Lett.*, 1990, 547. 1039
90SL591 S. Kang, D. Lee and J. Lee; *Syn. Lett.*, 1990, 591. 1066
90SL701 A. Herunsalee, M. Isobe, Y. Fukuda and T. Goto; *Syn. Lett.*, 1990, 701. 1048
90SL705 D. Schummer and G. Hofle; *Synlett*, 1990, 705. 16, 17
90SL773 D. P. Curran, T. L. Fevig and M. J. Totleban; *Synlett.*, 1990, 773. 347
90T545 D. S. Middleton, N. S. Simkins and N. K. Terrett; *Tetrahedron*, 1990, **46**, 545. 943
90T987 A. R. Katritzky and P. A. Harris; *Tetrahedron*, 1990, **46**, 987. 260
90T1385 C. Thebtaranonth and Y. Thebtaranonth; *Tetrahedron*, 1990, **46**, 1385. 919
90T1791 M. Tramontini and L. Angiolini; *Tetrahedron*, 1990, **46**, 1791. 635
90T1859 J. Cossy, D. Belotti and J. P. Pete; *Tetrahedron*, 1990, **46**, 1859. 287, 362
90T1885 T. K. Sarkar, S. K. Ghosh and T. K. Satapathi; *Tetrahedron*, 1990, **46**, 1885. 593
90T2317 M. Newcomb, T. M. Deeb and D. J. Marquardt; *Tetrahedron*. 1990, **46**, 2317. 950
90T2329 M. Newcomb, T. M. Deeb and D. J. Marquardt; *Tetrahedron*. 1990, **46**, 2329. 950
90T2345 M. Newcomb, D. J. Marquardt and M. U. Kumar; *Tetrahedron*, 1990, **46**, 2345. 950
90T2371 M. Schultz, R. Kluge, L. Sivibi and B Kamm; *Tetrahedron*, 1990, **46**, 237. 556
90T2411 C. Margot, M. Rizzolio and M. Schlosser; *Tetrahedron*, 1990, **46**, 2411. 579
90T3061 A. C. Serra, C. M. M. da Silva Correa, M. A. M. S. A. Vieira and M. A. Gomes; *Tetrahedron*, 1990, **46**, 3061. 949
90T3155 S. Busato, O. Tinembart, Z.-D. Zhang and R. Scheffold; *Tetrahedron*, 1990, **46**, 3155. 946
90T3601 A. Srikrishna and G. Sundarababu; *Tetrahedron*, 1990, **46**, 3601. 826
90T4487 T. J. Sweeney and L. A. Paquette; *Tetrahedron*, 1990, **46**, 4487. 826
90T4503 A. Tackle and P. Kocieński, *Tetrahedron*, 1990, **46**, 4503. 224
90T4887 F. Kido, S. C. Sinha, T. Abiko, M. Watanabe and A. Yoshikoshi; *Tetrahedron*, 1990, **46**, 4887. 822
90T6501 C. J. Moody and R. J. Taylor; *Tetrahedron*, 1990, **46**, 6501. 400
90T6951 N. S. Simpkins; *Tetrahedron*, 1990, **46**, 6951. 144, 439
90T7033 J. S. Yadav, P. K. Deshpande and G. V. M. Sharma; *Tetrahedron*, 1990, **46**, 7033. 1066

90TL213 J.-B. Baudin, S. A. Julia, O. Ruel and Y. Wang; *Tetrahedron Lett.*, 1990, **31**, 213. 963
90TL227 K. Haraguchi, H. Tanaka and T. Miyasaka; *Tetrahedron Lett.*, 1990, **31**, 227. 243
90TL265 K. Takeda, K. Torii and H. Ogura; *Tetrahedron Lett.*, 1990, **31**, 265. 239
90TL357 P. Breuilles and D. Uguen; *Tetrahedron Lett.*, 1990, **31**, 357. 231
90TL365 Y. Kataoka, K. Takai, K. Oshima and K. Utimoto; *Tetrahedron Lett.*, 1990, **31**, 365. 457
90TL377 K. Maruoka, J. Sato, H. Banno and H. Yamamoto; *Tetrahedron Lett.*, 1990, **31**, 377. 826
90TL523 F. Tubery, D. S. Grierson and H. P. Husson; *Tetrahedron Lett.*, 1990, **31**, 523. 834
90TL597 J. Fiandor and S. Y. Tam; *Tetrahedron Lett.*, 1990, **31**, 597. 1089
90TL727 T. Tsunoda, O. Sasaki and S. Ito; *Tetrahedron Lett.*, 1990, **31**, 727. 835
90TL823 K. S. Feldman and A. K. K. Vong; *Tetrahedron Lett.*, 1990, **31**, 823. 231, 949
90TL869 K. Ritter; *Tetrahedron Lett.*, 1990, **31**, 869. 831
90TL879 J. d'Angelo and D. Desmaele; *Tetrahedron Lett.*, 1990, **31**, 879. 558
90TL1265 W. Oppolzer, T. H. Keller, D. L. Kuo and W. Pachinger; *Tetrahedron Lett.*, 1990, **31**, 1265. 195
90TL1653 F. A. Davis, P. Zhou and G. S. Lal; *Tetrahedron Lett.*, 1990, **31**, 1653. 147
90TL1675 M. Newcomb and M. U. Kumar; *Tetrahedron Lett.*, 1990, **31**, 1675. 950
90TL1679 D. M. Scott, A. T. McPhail and N. A. Porter; *Tetrahedron Lett.*, 1990, **31**, 1679. 326
90TL1841 E. V: Dehmlow and T. Stiehm; *Tetrahedron Lett.*, 1990, **31**, 1841. 990
90TL1865 Y. Langlois, L. Konopski, N. V. Bac, A. Chiaroni and C. Riche; *Tetrahedron Lett.*, 1990, **31**, 1865. 626
90TL1977 S. P. Tanis, M. C. McMills, T. A. Scahill and D. A. Kloosterman; *Tetrahedron Lett.*, 1990, **31**, 1977. 831
90TL2149 Y. Petit, C. Sanner and M. Larchevêque; *Tetrahedron Lett.*, 1990, **31**, 2149. 179
90TL2315 S. Takano, M. Suzuki, A. Kijima and K. Ogasawara; *Tetrahedron Lett.*, 1990, **31**, 2315. 339
90TL2397 J. F. Daeuble, C. McGettigan and J. M. Stryker; *Tetrahedron Lett.*, 1990, **31**, 2397. 458, 516
90TL2565 J. E. Forbes, C. Tailhan and S. Z. Zard; *Tetrahedron Lett.*, 1990, **31**, 2565. 938
90TL2623 A. Ishii, J. Nakayama, Y. Endo and M. Hoshino; *Tetrahedron Lett.*, 1990, **31**, 2623. 899
90TL2713 J. Kang, Y. R. Choi, B. J. Kim, J. U. Jeong, S. Lee, J. H. Lee and C. Pyun; *Tetrahedron Lett.*, 1990, **31**, 2713. 238
90TL2719 T. Hiyama and Y. Hatanaka; *Tetrahedron Lett.*, 1990, **31**, 2719. 490, 1028
90TL2879 J.-F. Normant, J.-C. Quirion, Y. Masuda and A. Alexakis; *Tetrahedron Lett.*, 1990, 2879.
90TL2901 S. Kim and Y. C. Kim; *Tetrahedron Lett.*, 1990, **31**, 2901. 446
90TL2957 J. Light and R. Breslow; *Tetrahedron Lett.*, 1990, **31**, 2957. 5, 936
90TL3241 W. G. Dauben, J. M. Cogen and V. Behar; *Tetrahedron Lett.*, 1990, **31**, 3241. 826
90TL3409 X. Fu and J. M. Cook; *Tetrahedron Lett.*, 1990, **31**, 3409. 776
90TL3641 S. A. Hitchcock and G. Pattenden; *Tetrahedron Lett.*, 1990, **31**, 3641. 343
90TL3699 D. J. Burton, G. A. Hartgraves and J. Hsu; *Tetrahedron Lett.*, 1990, **31**, 3699. 967
90TL3703 M.-H. Hung; *Tetrahedron Lett.*, 1990, **31**, 3703. 967
90TL3731 M. J. Robins, R. S. Vinayak and S. G. Wood; *Tetrahedron Lett.*, 1990, **31**, 3731. 1010
90TL3829 T. S. Lin, J. H. Yang, M.-C. Liu and J. L. Zhu; *Tetrahedron Lett.*, 1990, **31**, 3829. 585
90TL3991 D. H. R. Barton, D. O. Jang and J. Cs. Jaszberenyi; *Tetrahedron Lett.*, 1990, **31**, 3991. 938, 946
90TL4041 T. Mandai, M. Ueda, S. Hasegawa, M. Kawada and J. Tsuji; *Tetrahedron Lett.*, 1990, **31**, 4041. 826
90TL4101 J.-T. Lee and H. Alper; *Tetrahedron Lett.*, 1990, **31**, 4101. 13
90TL4147 A. Lubineau, J. Augé, N. Bellanger and S. Caillebourdin; *Tetrahedron Lett.*, 1990, **31**, 4147. 826
90TL4155 T. Hayashi; *Tetrahedron Lett.*, 1990, **31**, 4155. 836
90TL4203 R. N. Saičić and Z. Čeković; *Tetrahedron Lett.*, 1990, **31**, 4203. 231
90TL4301 S. J. Coutts, M. D. Wittman and J. Kallmerten; *Tetrahedron Lett.*, 1990, **31**, 4301. 816
90TL4393 E. Negishi, Y. Noda, F. Lamaty and E. J. Vawter; *Tetrahedron Lett.*, 1990, **31**, 4393. 1015
90TL4425 B. Kruse and R. Brückner; *Tetrahedron Lett.*, 1990, **31**, 4425. 816
90TL4433 R. Hirsenkorn, B. Haag-Zeino and R. R. Schmidt; *Tetrahedron Lett.*, 1990, **31**, 4433. 831
90TL4609 H. Kotsuki, I. Kadota and M. Ochi; *Tetrahedron Lett.*, 1990, **31**, 4609. 1021
90TL4681 D. H. R. Barton, D. O. Jang and J. Cs. Jaszberenyi; *Tetrahedron Lett.*, 1990, **31**, 4681. 17, 938, 946
90TL4859 F. Fotiadu, A. Archavlis and G. Buono; *Tetrahedron Lett.*, 1990, 4859. 499
90TL4931 G. V. M. Sharma and S. R. Vepachedu; *Tetrahedron Lett.*, 1990, **31**, 4931. 174
90TL5039 E. Lee, C. U. Hur and C. M. Park; *Tetrahedron Lett.*, 1990, **31**, 5039. 455
90TL5093 J. N. Kirwan, B. P. Roberts and C. R. Willis; *Tetrahedron Lett.*, 1990, **31**, 5093. 331
90TL5113 J. A. Soderquist and B. Santiago; *Tetrahedron Lett.*, 1990, **31**, 5113. 449
90TL5285 S. D. Burke, W. J. Porter, J. Rancourt and R. F. Kaltenbach; *Tetrahedron Lett.*, 1990, **37**, 5285. 831
90TL5449 J. R. McCarthy, D. P. Matthews, M. L. Edwards, D. M. Stemerick and E. T. Jarvi; *Tetrahedron Lett.*, 1990, **31**, 5449. 439
90TL5499 M. Isobe, Y. Hirose, K. Shimokawa, T. Nishikawa and T. Goto; *Tetrahedron Lett.*, 1990, **31**, 5499. 1048
90TL5525 S. E. de Laszlo, M. J. Ford, S. V. Ley and G. N. Maw; *Tetrahedron Lett.*, 1990, **31**, 5525. 614, 697, 700, 703
90TL5551 D. A. Singleton, K. M. Church and M. J. Lucero; *Tetrahedron Lett.*, 1990, **31**, 5551. 949
90TL5607 R. F. Cunico; *Tetrahedron Lett.*, 1990, **31**, 5607. 985
90TL5697 J. H. Byers and G. C. Lane; *Tetrahedron Lett.*, 1990, **31**, 5697. 243
90TL5749 J.-M. Gaudin and C. Morel; *Tetrahedron Lett.*, 1990, **31**, 5749. 504
90TL5897 Yi Liao and Yao-zeng Huang; *Tetrahedron Lett.*, 1990, **31**, 5897. 743
90TL6013 C. Chatgilialoglu, B. Giese and B. Kopping; *Tetrahedron Lett.*, 1990, **31**, 6013. 945
90TL6085 Z. Čekovič and R. N. Saičić; *Tetrahedron Lett.*, 1990, **31**, 6085. 231
90TL6155 R. Karaman, D. T. Kohlman and J. L. Fry; *Tetrahedron Lett.*, 1990, **31**, 6155. 136

90TL6179 B. Bennetau, M. Kremmp, J. Dunoguès and S. Ratton; *Tetrahedron Lett.*, 1990, **31**, 6179. 448

90TL6205 K. Fukumoto, H. Nemoto and M. Ando; *Tetrahedron Lett.*, 1990, **31**, 6205. 1082

90TL6209 S. Matsubara, H. Onishi and K. Utimoto; *Tetrahedron Lett.*, 1990, **31**, 9209. 208

90TL6291 C. J. Li and D. N. Harpp; *Tetrahedron Lett.*, 1990, **31**, 6291. 568

90TL6379 S. Matsubara, T. Kodama and K. Utimoto; *Tetrahedron Lett.*, 1990, **31**, 6379. 264

90TL6509 W. R. Roush, K. J. Moriarty and B. B. Brown; *Tetrahedron Lett.*, 1990, **31**, 6509. 489

90TL6669 Y. Watanabe, T. Yoneda, Y. Ueno and T. Toru; *Tetrahedron Lett.*, 1990, **31**, 6669. 243

90TL6799 N. A. Petasis and M. A. Patane; *Tetrahedron Lett.*, 1990, **31**, 6799. 826

90TL6979 J. Ezquerra, W. He and L. A. Paquette; *Tetrahedron Lett.*, 1990, **48**, 6979. 826

90TL6995 W. Oppolzer, T. N. Birkinshaw and G. Bernardinelli; *Tetrahedron Lett.*, 1990, **31**, 6995. 195

90TL7063 G. I. Nikishin and I. P. Kovalev; *Tetrahedron Lett.*, 1990, **31**, 7063. 1051

90TL7105 A. S. Kende and J. S. Mendoza; *Tetrahedron Lett.*, 1990, **31**, 7105. 615, 698

90TL7115 H. C. Brown and M. V. Rangaishenvi; *Tetrahedron Lett.*, 1990, **31**, 7115. 838

90TL7145 F. Khuong-Huu, T. Skrydstrup and M. Bénéchie; *Tetrahedron Lett.*, 1990, **31**, 7145. 1024

90TL7179 J. Tsuji, T. Mandai, T. Nakata, H. Murayama, H. Yamaoki, M. Ogawa and M. Kawada; *Tetrahedron Lett.*, 1990, **31**, 7179. 976, 1025

90TL7227 G. Lakshminarayana and S. Y. Mhaskar; *Tetrahedron Lett.*, 1990, **31**, 7227. 1019

90TL7257 B. H. Lipschutz, R. Keil and E. L. Ellsworth; *Tetrahedron Lett.*, 1990, **31**, 7257. 453

90TL7469 I. Saito, K. Yamaguchi, R. Nagata and E. Murahashi; *Tetrahedron Lett.*, 1990, **31**, 7469. 988

90TL7513 I. Paterson and A. N. Hulme; *Tetrahedron Lett.*, 1990, **31**, 7513. 831

90TL7627 S. Kim and J. M. Lee; *Tetrahedron Lett.*, 1990, **31**, 7627. 314, 1050

91ACR22 A. Padwa; *Acc. Chem. Res.*, 1991, **24**, 22. 366

91ACR139 T. V. RajanBabu; *Acc. Chem. Res.*, 1991, **24**, 139. 174, 320, 919

91ACR296 N. A. Porter, B. Giese and D. P. Curran; *Acc. Chem. Res.*, 1991, **24**, 296. 320, 326, 327, 328, 919

91AG(E)1173 K. Yamamoto, Y. Saitho, D. Iwaki and T. Ooka; *Angew. Chem., Int. Ed. Engl.*, 1991, **30**, 1173. 788

91AG(E)1372 R. Schwesinger, R. Link, G. Thiele, H. Rotter, D. Honert, H.-H. Limbach and F. Maennle; *Angew. Chem., Int. Ed. Engl.*, 1991, **30**, 1372. 574, 584

91AG(E)1387 K. C. Nicolaou and W. M. Dai; *Angew. Chem., Int. Ed. Engl.*, 1991, **30**, 1387. 1009

91AG(E)1465 U. Nubbemeyer, R. Oehrlein, J. Gonda, B. Ernst and D. Bellus; *Angew. Chem., Int. Ed. Engl.*, 1991, **30**, 1465. 834

91BCJ659 E. Weber, W. Seichter, R.-J. Wang and T. C. W. Mak; *Bull. Chem. Soc. Jpn.*, 1991, **64**, 659. 986

91BCJ2008 T. Umemoto and Y. Gotoh; *Bull. Chem. Soc. Jpn.*, 1991, **64**, 2008. 108

91CB247 D. Schinzer and M. Ruppelt; *Chem. Ber.*, 1991, **124**, 247. 979

91CB387 J. Hartung and B. Giese; *Chem. Ber.*, 1991, **124**, 387. 931

91CB2075 H. Hopf and H. Lipka; *Chem. Ber.*, 1991, **124**, 2075. 496

91CB2335 R. Aumann and B. Trentmann; *Chem. Ber.*, 1991, **124**, 2335. 500

91CC294 Ž. Čeković and R. Matović; *J. Chem. Soc., Chem. Commun.*, 1991, 294. 982

91CC354 J. H. Byers, T. G. Gleason and K. S. Knight; *J. Chem. Soc., Chem. Commun.*, 1991, 354. 244

91CC372 S. G. Manjunatha and S. Rajappa; *J. Chem. Soc., Chem. Commun.*, 1991, 372. 194

91CC535 A. S. C. Chan and J. P. Coleman; *J. Chem. Soc., Chem. Commun.*, 1991, 535. 14

91CC566 P. A. Morken, N. C. Baenziger, D. J. Burton, P. C. Bachand, C. R. Davis, S. D. Pedersen and S. W. Hansen; *J. Chem. Soc., Chem. Commun.*, 1991, 556. 990

91CC690 R. J. Ferrier, C.-K. Lee and T. A. Wood; *J. Chem. Soc., Chem. Commun.*, 1991, 690. 173

91CC948 S. Akiyama, S. Nakatsuji, K. Nomura, K. Matsuda and K. Nakashima; *J. Chem. Soc., Chem. Commun.*, 1991, 948. 1063

91CC1001 J. Barluenga, F. Foubelo, R. González, F. J. Fañanás and M. Yus; *J. Chem. Soc., Chem. Commun.*, 1991, 1001. 1090

91CC1290 L. Hill, S. P. Saberi, A. M. Z. Slawin, S. E. Thomas and D. J. Williams; *J. Chem. Soc., Chem. Commun.*, 1991, 1290. 987

91CC1399 G. L. Edwards, W. B. Motherwell, D. M. Powell and D. A. Sandham; *J. Chem. Soc., Chem. Commun.*, 1991, 1399. 133

91CC1579 J. W. Patterson and G. T. Huang; *J. Chem. Soc., Chem. Commun.*, 1991, 1579. 829

91CJC77 G. Berube and A. G. Fallis; *Can. J. Chem.*, 1991, **69**, 77. 810

91CL43 K. Tomooka, S.-Y. Wei, and T. Nakai; *Chem. Lett.*, 1991, 43. 816

91CL1193 N. Iwasawa, S. Hayakawa, K. Isobe and K. Narasaka; *Chem. Lett.*, 1991, 1193. 929

91CL1271 M. Isobe, T. Goto, T. Nishikawa and A. Ino; *Chem. Lett.*, 1991, 1271. 1009

91CL1691 H. Togo, M. Aoki, T. Kuramochi and M. Yokoyama; *Chem. Lett.*, 1991, 1691. 934

91CL1805 A. Ogawa, M. Sekiguchi, H. Shibuya, H. Kuniyasu, N. Takami, I. Ryu and N. Sonoda; *Chem. Lett.*, 1991, 1805. 1073

91CL2017 K. Sasaki, M. Sakai, Y. Sakakibara and K. Takagi; *Chem. Lett.*, 1991, 2017. 433

91CL2073 T. Katayama and M. Umeno; *Chem. Lett.*, 1991, 2073. 471

91COC(3)913 I. E. Markó; *Comp. Org. Chem.*, 1991, **3**, 913. 821, 823

91COC(5)785 R. K. Hill; *Comp. Org. Chem.*, 1991, **5**, 785. 806, 811

91COC(5)875 F. E. Ziegler; *Comp. Org. Chem.*, 1991, **5**, 875. 840

91COC(5)899 T. Hudlicky and J. W. Reed; *Comp. Org. Chem.*, 1991, **5**, 899. 801

91COC(5)999 J. J. Bronson and R. L. Danheiser; *Comp. Org. Chem.*, 1991, **5**, 999. 803

91COS(1)1 P. G. Williard; *Comp. Org. Synth.*, 1991, **1**, 1. 885

91COS(1)77 J. R. Hauske; *Comp. Org. Synth.*, 1991, **1**, 77. 885

91COS(1)107 B. H. Lipshutz; *Comp. Org. Synth.*, 1991, **1**, 107. 773, 778, 779, 780, 781, 782

91COS(1)139 C. Ferreri, G. Palumbo and R. Caputo; *Comp. Org. Synth.*, 1991, **1**, 139. 885

91COS(1)173	N. Saccomano; *Comp. Org. Synth.*, 1991, **1**, 173.	885
91COS(1)211	P. Knochel; *Comp. Org. Synth.*, 1991, **1**, 211.	885
91COS(1)231	T. Imamoto; *Comp. Org. Synth.*, 1991, **1**, 231.	885
91COS(1)251	G. A. Molander; *Comp. Org. Synth.*, 1991, **1**, 251.	885
91COS(1)355	R. A. Volkmann; *Comp. Org. Synth.*, 1991, **1**, 355.	1042
91COS(1)629	A. Krief; *Comp. Org. Synth.*, 1991, **1**, 629.	712, 713, 714, 716
91COS(1)729	S. E. Kelly; *Comp. Org. Synth.*, 1991, **1**, 729.	648
91COS(1)731	S. E. Kelly; *Comp. Org. Synth.*, 1991, **1**, 731.	758
91COS(1)749	S. E. Kelly; *Comp. Org. Synth.*, 1991, **1**, 749.	678
91COS(1)761	S. E. Kelly; *Comp. Org. Synth.*, 1991, **1**, 761.	744
91COS(1)792	S. E. Kelly; *Comp. Org. Synth.*, 1991, **1**, 792.	696, 702
91COS(1)807	S. E. Kelly; *Comp. Org. Synth.*, 1991, **1**, 807.	676, 677
91COS(2)81	H. Yamamoto; *Comp. Org. Synth.*, 1991, **2**, 81.	971, 1054
91COS(2)99	H. B. Mekelburger and C. S. Wilcox; *Comp. Org. Synth.*, 1991, **2**, 99.	873
91COS(2)133	C. H. Heathcock; *Comp. Org. Synth.*, 1991, **2**, 133.	883
91COS(2)181	C. H. Heathcock; *Comp. Org. Synth.*, 1991, **2**, 181.	879
91COS(2)234	C. H. Heathcock; *Comp. Org. Synth.*, 1991, **2**, 234.	884
91COS(2)551	K. Sonogashira; *Comp. Org. Synth.*, 1991, **2**, 551.	
91COS(3)1	D. Caine; *Comp. Org. Synth.*, 1991, **3**, 1.	130
91COS(3)85	A. Krief; *Comp. Org. Synth.*, 1991, **3**, 85.	140, 143, 145, 146, 164
91COS(3)193	D. R. Cheshire; *Comp. Org. Synth.*, 1991, **3**, 193.	135
91COS(3)207	J. M. Klunder and G. H. Posner; *Comp. Org. Synth.*, 1991, **3**, 207.	116, 118
91COS(3)241	D. W. Knight; *Comp. Org. Synth.*, 1991, **3**, 241.	279, 463
91COS(3)271	P. J. Garratt; *Comp. Org. Synth.*, 1991, **3**, 271.	111
91COS(3)293	G. A. Olah, R. Krishnamurti and G. K. S. Prakash; *Comp. Org. Synth.*, 1991, **3**, 293.	180, 212, 462
91COS(3)435	K. Tamao; *Comp. Org. Synth.*, 1991, **3**, 435.	465
91COS(3)468	K. Tamao; *Comp. Org. Synth.*, 1991, **3**, 468.	477
91COS(3)705	J. R. Hanson; *Comp. Org. Synth.*, 1991, **3**, 705.	380
91COS(3)721	B. Rickborn; *Comp. Org. Synth.*, 1991, **3**, 721.	384
91COS(3)733	B. Rickborn; *Comp. Org. Synth.*, 1991, **3**, 733.	386
91COS(3)777	D. J. Coveney; *Comp. Org. Synth.*, 1991, **3**, 777.	388
91COS(3)839	J. Mann; *Comp. Org. Synth.*, 1991, **3**, 839.	391
91COS(3)861	J. M. Clough; *Comp. Org. Synth.*, 1991, **3**, 861.	617, 704
91COS(3)887	G. B. Gill; *Comp. Org. Synth.*, 1991, **3**, 887.	393
91COS(3)913	I. E. Markó; *Comp. Org. Synth.*, 1991, **3**, 913.	233, 398, 407
91COS(3)975	J. A. Marshall; *Comp. Org. Synth.*, 1991, **3**, 975.	196, 233, 396, 404
91COS(4)1	M. E. Jung; *Comp. Org. Synth.*, 1991, **4**, 1.	300, 306, 889
91COS(4)69	V. J. Lee; *Comp. Org. Synth.*, 1991, **4**, 69.	300
91COS(4)139	V. J. Lee; *Comp. Org. Synth.*, 1991, **4**, 139.	312
91COS(4)169	J. A. Koslowski; *Comp. Org. Synth.*, 1991, **4**, 169.	311, 540
91COS(4)199	H.-G. Schmalz; *Comp. Org. Synth.*, 1991, **4**, 199.	305, 306
91COS(4)237	M. Hulce and M. J. Chapdelaine; *Comp. Org. Synth.*, 1991, **4**, 237.	304
91COS(4)451	R. K. Norris; *Comp. Org. Synth.*, 1991, **4**, 451.	474
91COS(4)483	S. V. Kessar; *Comp. Org. Synth.*, 1991, **4**, 483.	1081
91COS(4)585	S. A. Godleski; *Comp. Org. Synth.*, 1991, **4**, 585.	194
91COS(4)715	D. P. Curran; *Comp. Org. Synth.*, 1991, **4**, 715.	320, 329, 330, 331, 332, 334, 336, 546, 920, 923, 948
91COS(4)779	D. P. Curran; *Comp. Org. Synth.*, 1991, **4**, 779.	320, 336, 339, 340, 341, 342, 344, 347, 348, 546, 919
91COS(4)833	R. F. Heck; *Comp. Org. Synth.*, 1991, **4**, 833.	493
91COS(4)865	P. Knochel; *Comp. Org. Synth.*, 1991, **4**, 865.	207, 315, 538
91COS(4)872	P. Knochel; *Comp. Org. Synth.*, 1991, **4**, 872.	539
91COS(4)877	P. Knochel; *Comp. Org. Synth.*, 1991, **4**, 877.	539
91COS(4)879	P. Knochel; *Comp. Org. Synth.*, 1991, **4**, 879.	543
91COS(4)886	P. Knochel; *Comp. Org. Synth.*, 1991, **4**, 886.	542
91COS(4)888	P. Knochel; *Comp. Org. Synth.*, 1991, **4**, 888.	542
91COS(4)896	P. Knochel; *Comp. Org. Synth.*, 1991, **4**, 896.	539, 541
91COS(4)913	J. K. Stille; *Comp. Org. Synth.*, 1991, **4**, 913.	296
91COS(4)951	P. Helquist; *Comp. Org. Synth.*, 1991, **4**, 951.	298
91COS(4)999	V. Nair; *Comp. Org. Synth.*, 1991, **4**, 999.	550
91COS(4)1031	H. M. L. Davies; *Comp. Org. Synth.*, 1991, **4**, 1031.	299, 352, 353, 354, 495, 550
91COS(4)1069	A. Padwa; *Comp. Org. Synth.*, 1991, **4**, 1069.	365, 366, 368
91COS(4)1111	P. A. Wade; *Comp. Org. Synth.*, 1991, **4**, 1111.	368
91COS(5)1	B. B. Snider; *Comp. Org. Synth.*, 1991, **5**, 1.	524
91COS(5)63	J. E. Baldwin; *Comp. Org. Synth.*, 1991, **5**, 63.	354, 355
91COS(5)123	M. T. Crimmins; *Comp. Org. Synth.*, 1991, **5**, 123.	356, 357
91COS(5)239	R. D. Little; *Comp. Org. Synth.*, 1991, **5**, 239.	359
91COS(5)315	W. Oppolzer; *Comp. Org. Synth.*, 1991, **5**, 315.	373
91COS(5)451	D. L. Boger; *Comp. Org. Synth.*, 1991, **5**, 451.	375
91COS(5)513	W. R. Roush; *Comp. Org. Synth.*, 1991, **5**, 513.	374, 375
91COS(5)551	R. W. Sweger and A. W. Czarnik; *Comp. Org. Synth.*, 1991, **5**, 551.	785, 786, 1034
91COS(5)675	T. Durst and L. Breau; *Comp. Org. Synth.*, 1991, **5**, 675.	418
91COS(5)699	W. H. Okamura and A. R. de Lera; *Comp. Org. Synth.*, 1991, **5**, 699.	419
91COS(5)751	S. E. Denmark; *Comp. Org. Synth.*, 1991, **5**, 751.	421, 423
91COS(5)785	R. K. Hill; *Comp. Org. Synth.*, 1991, **5**, 785.	411

91COS(5)827	P. Wipf; *Comp. Org. Synth.*, 1991, **5**, 827.	414, 876
91COS(5)899	T. Hudlicky and J. W. Reed; *Comp. Org. Synth.*, 1991, **5**, 899.	402
91COS(5)999	J. J. Bronson and R. L. Danheiser; *Comp. Org. Synth.*, 1991, **5**, 999.	400
91COS(5)1115	R. H. Grubbs and S. H. Pine; *Comp. Org. Synth.*, 1991, **5**, 1115.	674
91COS(6)873	R. Brückner; *Comp. Org. Synth.*, 1991, **6**, 873.	404, 407
91COS(6)949	A. Krebs and J. Swienty-Busch; *Comp. Org. Synth.*, 1991, **6**, 949.	570
91COS(6)975	P. H. Kociénski; *Comp. Org. Synth.*, 1991, **6**, 975.	568
91COS(6)1000	P. Kocienski; *Comp. Org. Synth.*, 1991, **6**, 1000.	648
91COS(6)1011	P. C. Astles, S. V. Mortlock and E. J. Thomas; *Comp. Org. Synth.*, 1991, **6**, 1011.	595, 601, 624
91COS(8)327	R. O. Hutchins and M. K. Hutchins; *Comp. Org. Synth.*, 1991, **8**, 327.	887
91COS(8)417	S. Seigel; *Comp. Org. Synth.*, 1991, **8**, 417.	455, 456
91COS(8)443	H. Takaya and R. Noyori; *Comp. Org. Synth.*, 1991, **8**, 443.	456
91COS(8)471	D. J. Pasto; *Comp. Org. Synth.*, 1991, **8**, 471.	458
91COS(8)667	J. A. Labinger; *Comp. Org. Synth.*, 1991, **8**, 667.	515
91COS(8)703	K. Smith and A. Pelter; *Comp. Org. Synth.*, 1991, **8**, 703.	509
91COS(8)733	J. J. Eisch; *Comp. Org. Synth.*, 1991, **8**, 733.	511
91COS(8)835	P. Caubère and P. Coutrot; *Comp. Org. Synth.*, 1991, **8**, 835.	437, 444, 445, 454
91COS(8)850	P. Caubere and P. Coutrot; *Comp. Org. Synth.*, 1991, **8**, 850.	65, 68, 69
91COS(8)858	P. Caubere and P. Coutrot; *Comp. Org. Synth.*, 1991, **8**, 858.	53
91COS(8)895	M. Hudlický; *Comp. Org. Synth.*, 1991, **8**, 895.	425, 427, 429, 430
91CPB1659	S. Harusawa, H. Osaki, T. Kurokawa, H. Fujii, R. Yoneda and T. Kurihara; *Chem. Pharm. Bull.*, 1991, **39**, 1659.	836
91CRV197	M. I. Bruce; *Chem. Rev.*, 1991, **91**, 197.	1143, 1145
91CRV263	A. Padwa and S. F. Hornbuckle; *Chem. Rev.*, 1991, **91**, 263.	400, 820
91CRV375	M. Saunders and H. A. Jiménez-Vázquez; *Chem. Rev.*, 1991, **91**, 375.	891, 906, 918, 919
91CRV437	R. M. Hanson; *Chem. Rev.*, 1991, **91**, 437.	220
91CRV477	E. Lipczynska-Kochany; *Chem. Rev.*, 1991, **91**, 477.	935
91CRV1237	C. P. Jasperse, D. P. Curran and T. L. Fevig; *Chem. Rev.*, 1991, 1237.	320, 331, 336, 341, 342, 344, 348, 926, 935, 948
91G59	M. Tingoli, M. Tiecco, L. Testaferri and D. Chianelli; *Gazz. Chim. Ital.*, 1991, **121**, 59.	484
91H(32)2089	J. Kurita, T. Yamanaka and T. Tsuchiya; *Heterocycles*, 1991, **32**, 2089.	561
91HCA232	D. Muller, G. Umbricht, B. Weber and A. Pfaltz; *Helv. Chim. Acta*, 1991, **74**, 232.	194, 353
91HCA1135	B. Giese and G. Thoma; *Helv. Chim. Acta*, 1991, **74**, 1135.	937
91HCA1143	B. Giese and G. Thoma; *Helv. Chim. Acta*, 1991, **74**, 1143.	937
91HCA1924	A. R. Katritzky, S. Perumal, W. Kuzmierkiewicz, P. Lue and J. V. Greenhill; *Helv. Chim. Acta*, 1991, **74**, 1924.	260, 263
91JA631	S. H. Bertz, G. Dabbagh and A. M. Mujsce, *J. Am. Chem. Soc.*, 1991, **113**, 631.	118, 119
91JA694	A. G. Myers, P. M. Harrington and E. Y. Kuo; *J. Am. Chem. Soc.*, 1991, **113**, 694.	1038
91JA726	D. A. Evans, K. A. Woerpel, M. M. Hinman and M. M. Faul; *J. Am. Chem. Soc.*, 1991, **113**, 726.	353
91JA958	S. H. Bergens and B. Bosnich; *J. Am. Chem. Soc.*, 1991, **113**, 958.	133
91JA1055	J. Boivin, E. Fouquet and S. Z. Zard; *J. Am. Chem. Soc.*, 1991, **113**, 1055.	931
91JA1344	S. Eldin, R. M. Pollack and D. L. Whalen; *J. Am. Chem. Soc.*, 1991, **113**, 1344.	877
91JA1355	J. D. Enas, J. A. Palenzuela and W. H. Okamura; *J. Am. Chem. Soc.*, 1991, **113**, 1355.	795
91JA1364	T. Ohwada, N. Yamagata and K. Shudo; *J. Am. Chem. Soc.*, 1991, **113**, 1364.	492
91JA1440	E. Negishi, C. J. Rousset and K. Takagi; *J. Am. Chem. Soc.*, 1991, **113**, 1440.	1036
91JA1442	J. Bund, H.-J. Gais and I. Erdelmeier; *J. Am. Chem. Soc.*, 1991, **113**, 1442.	483
91JA1791	N. A. Porter, D. M. Scott, I. J. Rosentsein, B. Giese, A. Veit and H.-G. Zeitz; *J. Am. Chem. Soc.*, 1991, **113**, 1791.	326
91JA1850	B. M. Trost and M. K. Trost; *J. Am. Chem. Soc.*, 1991, **113**, 1850.	675
91JA2610	L. A. Paquette and H.-J. Kang; *J. Am. Chem. Soc.*, 1991, **113**, 2610.	826
91JA2762	C. M. G. Philippo, N. H. Vo and L. A. Paquette; *J. Am. Chem. Soc.*, 1991, **113**, 2762.	826
91JA3526	J. K. Whitesell and H. K. Yaser; *J. Am. Chem. Soc.*, 1991, **113**, 3526.	839
91JA3833	B. Zeng and R. M. Pollack; *J. Am. Chem. Soc.*, 1991, **113**, 3833.	877
91JA3873	J. D. Enas, G. Y. Shen and W. H. Okamura; *J. Am. Chem. Soc.*, 1991, **113**, 3873.	1029
91JA4026	E. J. Corey and D. H. Lee; *J. Am. Chem. Soc.*, 1991, **113**, 4026.	417, 831
91JA4712	R. D. Bach, J. W. Knight and M. L. Braden; *J. Am. Chem. Soc.*, 1991, **113**, 4712.	629
91JA4895	C. D. Poulter and D. S. Mautz; *J. Am. Chem. Soc.*, 1991, **113**, 4895.	840
91JA5079	A. H. Hoveyda and Z. Xu; *J. Am. Chem. Soc.*, 1991, **113**, 5079.	316
91JA5488	P. A. Grieco, J. D. Clark and C. T. Jagoe; *J. Am. Chem. Soc.*, 1991, **113**, 5488.	190, 402
91JA5735	S. A. Rao and P. Knochel; *J. Am. Chem. Soc.*, 1991, **113**, 5735.	123
91JA5751	F. E. Romesberg, J. H. Gilchrist, A. T. Harrison, D. J. Fuller and D. B. Collum; *J. Am. Chem. Soc.*, 1991, **113**, 5751.	131
91JA5893	A. Yanagisawa, S. Habaue and H. Yamamoto; *J. Am. Chem. Soc.*, 1991, **113**, 5893.	112
91JA5915	F. A. Gomez, S. E. Johnson and M. F. Hawthorne; *J. Am. Chem. Soc.*, 1991, **113**, 5915.	168
91JA5918	J. G. Stack, D. P. Curran, J. Rebek, Jr. and P. Ballester; *J. Am. Chem. Soc.*, 1991, **113**, 5918.	326
91JA6950	J. W. Bunting and J. P. Kanter; *J. Am. Chem. Soc.*, 1991, **113**, 6950.	629
91JA7002	N. A. Porter, R. Breyer, E. Swann, J. Nally, J. Pradhan, T. Allen and A. T. McPhail; *J. Am. Chem. Soc.*, 1991, **113**, 7002.	327, 328
91JA7350	B. M. Trost, T. A. Grese and D. M. T. Chan; *J. Am. Chem. Soc.*, 1991, **113**, 7350.	289, 362
91JA7363	B. M. Trost and T. A. Grese; *J. Am. Chem. Soc.*, 1991, **113**, 7363.	289
91JA7439	J. R. McCarthy, D. P. Matthews, D. M. Stemerick, E. W. Huber, P. Bey, B. J. Lippert, R. D. Snyder and P. S. Sunkara; *J. Am. Chem. Soc.*, 1991, **113**, 7439.	440

91JA7788 N. A. Porter, J. D. Bruhnke, W.-X. Wu, I. J. Rosenstein and R. A. Breyer; *J. Am. Chem. Soc.*, 1991, **113**, 7788. 326, 327

91JA7794 J. M. Hawkins and S. Loren; *J. Am. Chem. Soc.*, 1991, **113**, 7794. 374

91JA8821 J. R. I. Eubanks, L. B. Sims and A. Fry; *J. Am. Chem. Soc.*, 1991, **113**, 8821. 629

91JA8950 A. H. Hoveyda, Z. Xu, J. P. Morken and A. F. Houri; *J. Am. Chem. Soc.*, 1991, **113**, 8950. 316

91JA8955 A. Yanagisawa, S. Habaue and H. Yamamoto; *J. Am. Chem. Soc.*, 1991, **113**, 8955. 116

91JA8977 D. Wiedenfeld and R. Breslow; *J. Am. Chem. Soc.*, 1991, **113**, 8977. 927

91JA9007 B. M. Trost and M. C. Matelich; *J. Am. Chem. Soc.*, 1991, **113**, 9007. 287, 361

91JA9571 P. L. Hall, J. H. Gilchrist and D. B. Collum; *J. Am. Chem. Soc.*, 1991, **113**, 9571. 131

91JA9604 Y. Wakatsuki, H. Yamazaki, N. Kumegawa, T. Satoh and J. Y. Satoh; *J. Am. Chem. Soc.*, 1991, **113**, 9604. 992

91JA9667 J. A. Gerlt, J. W. Kozarich, G. L. Kenyon and P. G. Gassman; *J. Am. Chem. Soc.*, 1991, **113**, 9667. 869

91JA9682 R. K. Boeckman, Jr., S. K. Yoon and D. K. Heckendorn; *J. Am. Chem. Soc.*, 1991, **113**, 9682. 618, 707, 709

91JA9701 Y. Guindon, J. F. Lavalée, M. Llinas-Brunet, G. Horner and J. Rancourt; *J. Am. Chem. Soc.*, 1991, **113**, 9701. 327

91JA9888 C. E. Tucker and P. Knochel; *J. Am. Chem. Soc.*, 1991, **113**, 9888. 991

91JCR(S)1 C.-P. Chuang and T. H. J. Ngoi; *J. Chem. Res. (S)*, 1991, 1. 949

91JCR(S)62 M. A. Khan; *J. Chem. Res. (S)*, 1991, 62. 595

91JCR(S)172 T. Hayashi, Y. Okada and T. Inaba; *J. Chem. Res. (S)*, 1991, 172. 826

91JCR(S)216 C.-P. Chuang, S.-S. Hou and T. H. J. Ngoi; *J. Chem. Res. (S)*, 1991, 216. 950

91JCS(P1)103 S. J. Cole, J. N. Kirwan, B. P. Roberts and C. R. Willis; *J. Chem. Soc., Perkin Trans. 1*, 1991, 103. 5, 7, 9, 10, 16, 17, 331

91JCS(P1)429 V. J. Jephcote and E. J. Thomas; *J. Chem. Soc., Perkin Trans. 1*, 1991, 429. 603

91JCS(P1)561 S. Bauermeister, I. D. Gouws, H. F. Strauss and E. M. M. Venter; *J. Chem. Soc., Perkin Trans. 1*, 1991, 561. 828

91JCS(P1)656 C. Evans, R. McCague, S. M. Roberts and A. G. Sutherland; *J. Chem. Soc., Perkin Trans. 1*, 1991, 656. 4, 9

91JCS(P1)665 Y. Shen and Y. Xiang; *J. Chem. Soc., Perkin Trans. 1*, 1991, 665. 987

91JCS(P1)705 J. Cooper, D. W. Knight and P. T. Gallagher; *J. Chem. Soc., Perkin Trans. 1*, 1991, 705. 831

91JCS(P1)1173 S. Iwasa, M. Yamamoto, S. Kohmoto and K. Yamada; *J. Chem. Soc., Perkin Trans. 1*, 1991, 1173. 938

91JCS(P1)1255 M. F. Mahon, K. Molloy, C. A. Pittol, R. J. Pryce, S. M. Roberts, G. Ryback, V. Sik, J. O. Williams and J. A. Winders; *J. Chem. Soc., Perkin Trans. 1*, 1991, 1255. 522

91JCS(P1)1433 D. L. J. Clive, A. Khodabocus, P. G. Vernon, A. G. Angoh, L. Bordeleau, D. S. Middleton, C. Lowe and D. Kellner; *J. Chem. Soc., Perkin Trans. 1*, 1991, 1433. 942

91JCS(P1)1485 P. S. Brown, N. Greeves, A. B. McElroy and S. Warren; *J. Chem. Soc., Perkin Trans. 1*, 1991, 1485. 643

91JCS(P1)2089 A. S. R. Anjaneyulu and B. Isaa; *J. Chem. Soc., Perkin Trans. 1*, 1991, 2089. 827

91JCS(P1)2797 D. L. J. Clive and A. C. Joussef; *J. Chem. Soc., Perkin Trans. 1*, 1991, 2797. 943

91JCS(P1)2892 T. Kitamura, T. Tanaka, H. Taniguchi and P. J. Stang; *J. Chem. Soc., Perkin Trans. 1*, 1991, 2892. 1018

91JCS(P1)3199 J. R. Hwu and D. A. Anderson; *J. Chem. Soc., Perkin Trans. 1*, 1991, 3199. 834

91JCS(P1)3295 A. R. Katritzky, X. Zhao and I. V. Shcherbakova; *J. Chem. Soc., Perkin Trans. 1*, 1991, 3295. 262

91JCS(P2)1703 R. Taylor; *J. Chem. Soc., Perkin Trans. 2*, 1991, 1703. 777

91JHC289 T. Hatta, S. Mataka, M. Tashiro, H. Suzuki and K. Numano; *J. Heterocycl. Chem.*, 1991, **28**, 289. 232

91JMC2452 W. H. Okamura, B. Figadère, A. W. Norman, H. L. Henry, H. P. Koeffler and J. Y. Zhou; *J. Med. Chem.*, 1991, **34**, 2452. 1029

91JOC183 O. Piccolo, U. Azzena, G. Melloni, G. Delogu and E. Valoti; *J. Org. Chem.*, 1991, **56**, 183. 182, 475

91JOC261 C. Carfagna, A. Musco, G. Sallese, R. Santi and T. Fiorani; *J. Org. Chem.*, 1991, **56**, 261. 471, 483

91JOC321 B. B. Snider and A. J. Allentoff; *J. Org. Chem.*, 1991, **56**, 321. 810

91JOC353 J. T. Welch, J. S. Plummer and T. Chow; *J. Org. Chem.*, 1991, **56**, 353. 831

91JOC650 R. E. Ireland, P. Wipf and J. D. Armstrong, III; *J. Org. Chem.*, 1991, **56**, 650. 417, 830, 876

91JOC678 M. Ballestri, C. Chatgilialoglu, K. B. Clark, D. Griller, B. Giese and B. Kopping; *J. Org. Chem.*, 1991, **56**, 678. 5, 7, 9, 10, 17, 936, 943, 945

91JOC723 W. R. Cantrell and H. M. L. Davies; *J. Org. Chem.*, 1991, **56**, 723. 812

91JOC741 D. A. Evans, J. A. Gauchet-Prunet, E. M. Carreira and A. B. Charette; *J. Org. Chem.*, 1991, **56**, 741. 98

91JOC846 W. J. Bailey and A. Onopchenko; *J. Org. Chem.*, 1991, **56**, 846. 777

91JOC970 A. B. Cheikh, J. Chuche, N. Manisse, J. C. Pommelet, K.-P. Netsch, P. Lorenčak and C. Wentrup; *J. Org. Chem.*, 1991, **56**, 970. 835

91JOC997 T. Tsuji, M. Ohkita and S. Nishida; *J. Org. Chem.*, 1991, **56**, 997. 564

91JOC1003 S. E. Denmark and J. E. Marlin; *J. Org. Chem.*, 1991, **56**, 1003. 832

91JOC1014 R. V. Hoffman and H.-O. Kim; *J. Org. Chem.*, 1991, **56**, 1014. 840

91JOC1083 O. W. Gooding, C. C. Beard, D. Y. Jackson, D. L. Wren and G. F. Cooper; *J. Org. Chem.*, 1991, **56**, 1083. 967

91JOC1176 R. L. Danheiser and J. S. Nowick; *J. Org. Chem.*, 1991, **56**, 1176. 566, 681, 683

91JOC1185 D. S. Keegan, M. M. Midland, R. T. Werley and J. I. McLoughlin; *J. Org. Chem.*, 1991, **56**, 1185. 816, 818

91JOC1309 M. Newcomb and K. A. Weber; *J. Org. Chem.*, 1991, **56**, 1309. 950

91JOC1445 L. Zhu, R. M. Wehmeyer and R. D. Rieke; *J. Org. Chem.*, 1991, **56**, 1445. 123, 467

91JOC1500 H. C. Brown, V. K. Mahindroo, N. G. Bhat and B. Singaram; *J. Org. Chem.*, 1991, **56**, 1500. 1034

91JOC1543 B. Singaram, M. V. Rangaishenvi, H. C. Brown, C. T. Goralski and D. L. Hasha; *J. Org. Chem.*, 1991, **56**, 1543. 443

91JOC2045 P. P. Wickham, K. H. Hazen, H. Guo, G. Jones, K. H. Reuter and W. J. Scott; *J. Org. Chem.*, 1991, **56**, 2045. 1071

91JOC2089 T. Yamato, C. Hideshima, G. K. S. Prakash and G. A. Olah; *J. Org. Chem.*, 1991, **56**, 2089. 202

91JOC2091 R. L. Beard and A. I. Meyers; *J. Org. Chem.*, 1991, **56**, 2091 149

91JOC2179 A. S. Gozdz and P. Maslak; *J. Org. Chem.*, 1991, **56**, 2179. 949

91JOC2264 T. Hirao, T. Fujii, S.-i. Miyata and Y. Ohshiro; *J. Org. Chem.*, 1991, **56**, 2264. 931

91JOC2523 A. Padwa, K. E. Krumpe, Y. Gareau and U. Chiacchio; *J. Org. Chem.*, 1991, **56**, 2523. 550

91JOC2866 F. Fontana, F. Minisci, M. C. N. Barbosa and E. Vismara; *J. Org. Chem.*, 1991, **56**, 2866. 934

91JOC2911 W. L. Whipple and H. J. Reich; *J. Org. Chem.*, 1991, **56**, 2911. 1047

91JOC3205 T. Yoshino, S. Okamoto and F. Sato; *J. Org. Chem.*, 1991, **56**, 3205. 1049

91JOC3210 J. Sisko and S. M. Weinreb; *J. Org. Chem.*, 1991, **56**, 3210. 369

91JOC3465 T. Doi and T. Takahashi; *J. Org. Chem.*, 1991, **56**, 3465. 818

91JOC3572 R. E. Ireland, P. Wipf and J. Xiang; *J. Org. Chem.*, 1991, **56**, 3572. 417, 795, 830

91JOC3801 D. L. J. Clive and S. Daigneault; *J. Org. Chem.*, 1991, **56**, 3801. 928

91JOC3817 H. M. L. Davies, T. J. Clark and H. D. Smith; *J. Org. Chem.*, 1991, **56**, 3817. 812

91JOC3841 L. A. Paquette, D. Friedrich and R. D. Rogers; *J. Org. Chem.*, 1991, **56**, 3841. 826

91JOC3908 S. V. Kessar, P. Singh, N. P. Kaur, U. Chawla, K. Shukla, P. Aggarwal and D. Venugopal; *J. Org. Chem.*, 1991, **56**, 3908. 282

91JOC3912 P. J. Stang, M. D. Bachi, N. Bar-Ner, C. M. Crittell and B. L. Williamson; *J. Org. Chem.*, 1991, **56**, 3912. 1016

91JOC4091 M. Yamaguchi, A. Hayashi and T. Minami; *J. Org. Chem.*, 1991, **56**, 4091. 1038

91JOC4335 D. P. Curran, A. C. Abraham and H. Liu; *J. Org. Chem.*, 1991, **56**, 4335. 328, 927

91JOC4339 G. H. Posner and T. D. Nelson; *J. Org. Chem.*, 1991, **56**, 4339. 788

91JOC4370 T. Ibuka, H. Habashita, A. Otaka and N. Fujii; *J. Org. Chem.*, 1991, **56**, 4370. 188

91JOC4439 A. R. Katritzky, J. Pernak, W.-Q. Fan and F. Saczewski; *J. Org. Chem.*, 1991, **56**, 4439. 260

91JOC4718 G. D. Paderes, P. Metivier and W. L. Jorgensen; *J. Org. Chem.*, 1991, **56**, 4718. 427

91JOC5063 S. E. Denmark, H. Stadler, R. L. Dorow and J. H. Kim; *J. Org. Chem.*, 1991, **56**, 5063. 832

91JOC5696 H. M. L. Davies, E. Saikali and W. B. Young; *J. Org. Chem.*, 1991, **56**, 5696. 812

91JOC5782 W. Adam, R. Albert, L. Hasemann, V. O. Nava Salgado, B. Nestler, E.-M. Peters, K. Peters, F. Prechtl and H. Georg von Schnering; *J. Org. Chem.*, 1991, **56**, 5782. 785

91JOC5971 M. Gerlach, F. Jördens, H. Kuhn, W. P. Neumann and M. Peterseim; *J. Org. Chem.*, 1991, **56**, 5971. 936

91JOC5973 S.-Y. Wei, K. Tomooka and T. Nakai; *J. Org. Chem.*, 1991, **56**, 5973. 552

91JOC5974 H. P. Knoess, M. T. Furlong, M. J. Rozema and P. Knochel; *J. Org. Chem.*, 1991, **56**, 5974. 123

91JOC6206 C. Bonini, G. Righi and G. Sotgiu; *J. Org. Chem.*, 1991, **56**, 6206. 10

91JOC6447 D. L. J. Clive, C. Zhang, K. S. K. Murthy, W. D. Hayward and S. Daigneault; *J. Org. Chem.*, 1991, **56**, 6447. 688

91JOC6483 M. Miyashita, M. Hoshino and A. Yoshokoshi; *J. Org. Chem.*, 1991, **56**, 6483. 228

91JOC6873 J. W. Guiles and A. I. Meyers; *J. Org. Chem.*, 1991, **56**, 6873. 149

91JOC6981 G. H. Posner, R. D. Crouch, C. M. Kinter and J.-C. Carry; *J. Org. Chem*, 1991, **56**, 6981. 788

91JOC7177 S. Tsuboi, J. Sakamoto, T. Kawano, M. Utaka and A. Takeda; *J. Org. Chem.*, 1991, **56**, 7177. 571

91JOM(403)299 D. Mesnard and L. Miginiac; *J. Organomet. Chem.*, 1991, **403**, 299. 971

91JOM(405)1 Y. Dang and H. J. Geise; *J. Organomet. Chem.*, 1991, **405**, 1. 670

91JOM(417)289 C. Polizzi, C. Consoloni, L. Lardicci and A. M. Caporusso; *J. Organomet. Chem.*, 1991, **417**, 289. 466

91JOM(417)305 T. Mandai, H. Murayama, T. Nakata, H. Yamaoki, M. Ogawa, M. Kawada and J. Tsuji; *J. Organomet. Chem.*, 1991, **417**, 305. 976

91JPO449 I. Lee, C. K. Kim, B. H. Kong and B. C. Lee; *J. Phys. Org. Chem.*, 1991, **4**, 449. 130

91LA903 K. Hartke, H. Gerber and U. Roesrath; *Liebigs Ann. Chem.*, 1991, 903. 1047, 1048

B-91MI 105-01 M. Uemura; in "Advances in Metal–Organic Chemistry," ed. L. S. Liebeskind, JAI Press, Greenwich, CT, 1991, vol. 2, p. 195. 202

B-91MI 105-02 H. M. Colquhoun, D. J. Thompson and M. V. Twigg; "Carbonylation: Direct Synthesis of Carbonyl Compounds," Plenum Press, New York, 1991. 212

B-91MI 119-01 A. C. Knipe; in "Organic Reaction Mechanisms," eds. A. C. Knipe and W. E. Watts, Wiley, New York, 1991, chap. 11, p. 313–333. 845

B-91MI 119-02 K. H. Dötz and R. W. Hoffmann (eds.); "Organic Synthesis via Organometallics," (Proceedings of the Third Symposium in Marweg, July 11 to 14, 1990), Vieweg, Braunschweig, 1991. 885

B-91MI 119-03 W. P. Neumann; in "Chemistry and Technology of Silicon and Tin," eds. V. G. K. Das, N. S. Weng and M. Gielen, Oxford University Press, Oxford, 1991, p. 151. 924

B-91MI 121-01 A. M. Jones and S. P. Stanforth; in "Rodd's Chemistry of Carbon Compounds," 2nd edn., ed. M. Sainsbury, Elsevier, Amsterdam, 1991, suppl. 2, **1A/B**, 156. 998

91OM1620 V. V. Grushin and H. Alper; *Organometallics*, 1991, **10**, 1620. 427

91OM1771 M. Lazraq, C. Couret, J. Escudie, J. Satgé and M. Dräger; *Organometallics*, 1991, **10**, 1771. 766
91OR(40)1 N. E. Schore; *Org. React.*, 1991, **40**, 1. 528, 529
91OR(40)91 D. J. Pasto and R. T. Taylor; *Org. React.*, 1991, **40**, 91. 74
91PAC419 A. Suzuki; *Pure Appl. Chem.*, 1991, **63**, 419. 465, 473, 489
91RCR72 A. A. Kutyrev and V. V. Moskva; *Russ. Chem. Rev.* (*Engl. Transl.*), 1991, **60**, 72. 888
91RTC305 M. P. Doyle; *Recl. Trav. Chim. Pays-Bas*, 1991, **110**, 305. 353, 354
91S165 S. H. Pine, G. S. Shen and H. Hoang; *Synthesis*, 1991, 165. 769
91S235 S. Lin, H. Sheng and Y. Huang; *Synthesis*, 1991, 235. 1047
91S347 P. A. M. van der Klein, A. E. J. de Nooy, G. A. van der Marel and J. H. van Boom; *Synthesis*, 1991, 347. 185
91S423 R. Tamura, A. Kamimura and N. Ono; *Synthesis*, 1991, 423. 250
91S594 K. Mikami and T. Nakai; *Synthesis*, 1991, 594. 404, 407, 816
91S793 N. Petragnani and J. V. Comasseto; *Synthesis*, 1991, 793. 243, 595
91S827 J. M. Khurana, G. C. Maikap and P. Sahoo; *Synthesis*, 1991, 827. 568
91S897 N. Petragnani and J. V. Comasseto; *Synthesis*, 1991, 897. 243, 595, 611, 623
91S1151 R. Herges and C. Hoock; *Synthesis*, 1991, 1151. 982
91SC977 J. Wityak and J. B. Chan; *Synth. Commun.*, 1991, **21**, 977. 1019
91SC1455 V. Satyanarayana, C. P. Rao, G. L. D. Krupadanam and C. Srimannarayana; *Synth. Commun.*, 1991, **21**, 1455. 827
91SC2377 H. Hebri, E. Dunach and J. Perichon; *Synth. Commun.*, 1991, **21**, 2377. 427, 428
91SL63 D. P. Curran; *Synlett.*, 1991, 63. 320, 919
91SL238 Y. Ichikawa; *Synlett*, 1991, 238. 837
91SL251 A. Yanagisawa, Y. Noritake, N. Nomura and H. Yamamoto; *Synlett*, 1991, 251. 187
91SL412 D. J. Hart and R. Krishnamurthy; *Synlett*, 1991, 412. 243, 327
91SL423 M. Bulliard, H.-G. Zeitz and B. Giese; *Synlett.*, 1991, 423. 327
91SL425 M. Bulliard, H.-G. Zeitz and B. Giese; *Synlett.*, 1991, 425. 327
91SL435 D. H. R. Barton, D. O. Jang and J. Cs. Jaszberenyi; *Synlett*, 1991, 435. 938
91SL539 E. Nakamura; *Synlett*, 1991, 539. 122
91SL611 M. Julia, J. P. Férézou, R. Khourzon, A. Pancrazi and P. Robert; ***Synlett***, 1991, 611. 1071
91SL629 K. Mikami, O. Takahashi, K. Fujimoto and T. Nakai; *Synlett*, 1991, 629. 819
91SL651 M. Hirama, M. Tokuda and K. Fujiwara; *Syn. Lett.*, 1991, 651. 552
91SL684 H. C. Brown and A. M. Salunkhe; *Synlett*, 1991, 684. 486
91SL687 T. Ishiyama, N. Miyaura and A. Suzuki; *Synlett*, 1991, 687. 279
91SL697 T. Mandai, H. Kunitomi, K. Higashi, M. Kawada and J. Tsuji; *Synlett*, 1991, 697. 976
91SL764 A. G. M. Barrett, J. M. Hill, E. M. Wallace and J. A. Flygare; *Synlett*, 1991, 764. 648
91SL845 Y. Hatanaka and T. Hiyama; *Synlett*, 1991, 845. 490
91SL873 B. J. Banks, B. R. Fenner, V. F. Voss and M. J. Witty; *Synlett*, 1991, 873. 596
91SL919 M. Ohno, H. Tanaka, M. Komatsu and Y. Ohshiro; *Synlett*, 1991, 919. 284
91SR1 N. K. Gusarova, A. A. Tatarinova and L. M. Sinegouskaya; *Sulfur Reports*, 1991, 1. 441
91Txi W. S. Johnson; *Tetrahedron*, 1991, **47(41)**, xi. 295
91T173 B.-C. Chen, M. C. Weismiller, F. A. Davis, D. Boschelli, J. R. Empfield and A. B. Smith, III; *Tetrahedron*, 1991, **47**, 173. 147
91T343 E.-I. Negishi, T. Yoshida, A. Abramovitch, G. Lew and R. M. Williams; *Tetrahedron*, 1991, **47**, 343. 488, 983, 1033, 1034
91T519 G. V. M. Sharma and S. R. Vepachedu; *Tetrahedron*, 1991, **47**, 519. 174
91T1677 A. Alexakis, I. Marek, P. Mangeney and J.-F. Normant; *Tetrahedron*, 1991, **47**, 1677. 967
91T2237 J.-C. Wu, Z. Xi, C. Gioeli and J. Chattopadhyaya; *Tetrahedron*, 1991, **47**, 2237. 244
91T4223 C. M. Thompson and D. L. C. Green; *Tetrahedron*, 1991, **47**, 4223. 165, 879
91T4539 P. J. Stang, A. M. Arif and V. V. Zhdankin; *Tetrahedron*, 1991, **47**, 4539. 988
91T4847 P. Dowd and S.-C. Choi; *Tetrahedron*, 1991, **47**, 4847. 931
91T5425 C.-P. Chuang; *Tetrahedron*, 1991, **47**, 5425. 949
91T6275 P. Beslin and S. Perrio; *Tetrahedron*, 1991, **47**, 6275. 834
91T6813 J. C. Anderson, S. V. Ley, D. Santafianos and R. N. Sheppard; *Tetrahedron*, 1991, **47**, 6813. 780
91T7171 N. R. Curtis, A. B. Holmes and M. G. Looney; *Tetrahedron*, 1991, **47**, 7171. 829
91T8155 T. Schmidt, *Tetrahedron*, 1991, **47**, 8155. 582
91T8259 H. R. Sonawame, V. G. Naik, N. S. Bellur, V. G. Shah, P. C. Purohit, M. U. Kumar, D. G. Kulkarni and J. R. Ahuja; *Tetrahedron*, 1991, **47**, 8259. 802
91T8663 B. B. Snider and J. E. Merritt; *Tetrahedron*, 1991, **47**, 8663. 340
91T8855 M. Harmata and B. F. Herron; *Tetrahedron*, 1991, **47**, 8855. 622
91T8969 D. H. R. Barton, P. Blundell, J. Dorchak, D. O. Jang and J. Cs. Jaszberenyi; *Tetrahedron*, 1991, **47**, 8969. 938, 946
91T9503 D. Romo and A. I. Meyers; *Tetrahedron*, 1991, **47**, 9503. 134
91TA1 P. J. Cox and N. S. Simpkins; *Tetrahedron Asymmetry*, 1991, **2**, 1. 135, 842
91TA277 H. C. Brown, R. R. Iyer, V. H. Mahindroo and N. G. Bhat; *Tetrahedron Asymm.*, 1991, **2**, 277. 487
91TA643 K. Mikami, M. Terada, Y. Motoyama and T. Nakai; *Tetrahedron Asymmetry*, 1991, **2**, 643. 374
91TA843 H.-U. Blaser; *Tetrahedron Asymmetry*, 1991, **2**, 843. 76
91TA943 M. Lopp, T. Kanger, A. Müraus, T. Pehk and Ü. Lille; *Tetrahedron Asymmetry*, 1991, **2**, 943. 225
91TA1305 K. Narasaka, M. Saitou and N. Iwasawa; *Tetrahedron Asymmetry*, 1991, **2**, 1305. 375
91TL45 M. Newcomb and M. U. Kumar; *Tetrahedron Lett.*, 1991, **32**, 45. 949
91TL49 A. L. J. Beckwith and I. G. E. Davison; *Tetrahedron Lett.*, 1991, **32**, 49. 949
91TL53 M. E. Maier and B. Schöffling; *Tetrahedron Lett.*, 1991, **32**, 53. 1021

91TL179 M. A. Walters, C. S. McDonough, P. S. Brown, Jr. and A. B. Hoem; *Tetrahedron Lett.*, 1991, **32**, 179. 835

91TL199 J. Boivin, E. Crépon and S. Z. Zard; *Tetrahedron Lett.*, 1991, **32**, 199. 935

91TL243 D. C. Harrowven and G. Pattenden; *Tetrahedron Lett.*, 1991, **32**, 243. 345

91TL279 R. Grigg and J. Markandu; *Tetrahedron Lett.*, 1991, **32**, 279. 839

91TL319 S. W. Elmore and L. A. Paquette; *Tetrahedron Lett.*, 1991, **32**, 319. 810

91TL335 S. H. Tahir, M. M. Olmstead and M. J. Kurth; *Tetrahedron Lett.*, 1991, **32**, 335. 821

91TL417 A. Krief and P. Barbeaux; *Tetrahedron Lett.*, 1991, **32**, 417. 246

91TL687 R. Grigg, V. Loganathan, V. Santhakumar, V. Sridharan and A. Teasdale; *Tetrahedron Lett.*, 1991, 687. 494

91TL707 N. A. Porter, W.-X. Wu and A. T. McPhail; *Tetrahedron Lett*, 1991, **32**, 707. 326, 327

91TL843 D. E. Ward and B. F. Kaller; *Tetrahedron Lett.*, 1991, **32**, 843. 231

91TL1059 M. Fukumasa, K. Furuhashi, J. Umezawa, O. Takahashi and T. Hirai; *Tetrahedron Lett.*, 1991, **32**, 1059. 217

91TL1107 H. R. Sonawame, B. S. Nanjundiah, V. G. Shah, D. G. Kulkarni and J. R. Ahuja; *Tetrahedron Lett.*, 1991, **32**, 1107. 802

91TL1215 M. Kuroboshi and T. Hiyama; *Tetrahedron Lett.*, 1991, **32**, 1215. 574

91TL1459 K. Uneyama and H. Watanabe; *Tetrahedron Lett.*, 1991, **32**, 1459. 1047

91TL1691 C. L. Leone and A. R. Chamberlin; *Tetrahedron Lett.*, 1991, **32**, 1691. 43

91TL1707 D. M. S. Wheeler, T. Chamberlain, X. Fu, J. T. Pechacek, X. Peng and M. M. Wheeler; *Tetrahedron Lett.*, 1991, **32**, 1707. 1038, 1140

91TL1949 H. Künzer, M. Stahnke, G. Sauer and R. Wiechert; *Tetrahedron Lett.*, 1991, **32**, 1949. 24, 144

91TL2043 S. Kim, Y. G. Kim and J. H. Park; *Tetrahedron Lett.*, 1991, **32**, 2043. 629

91TL2169 A. Casarini, P. Dembech, G. Reginato, A. Ricci and G. Seconi; *Tetrahedron Lett.*, 1991, **32**, 2169. 1048

91TL2569 D. H. R. Barton, D. O. Jang and J. Cs Jaszberenyi; *Tetrahedron Lett.*, 1991, **32**, 2569. 587

91TL2703 D. H. R. Barton, J. C. Jaszberenyi and C. Tachdjian; *Tetrahedron Lett.*, 1991, **32**, 2703. 615, 700, 939

91TL2879 L.-H. Li, D. Wang and T. H. Chan; *Tetrahedron Lett.*, 1991, **32**, 2879. 161

91TL2913 Sunggak Kim and Yong Gil Kim; *Tetrahedron Lett.*, 1991, **32**, 2913. 739, 740

91TL3119 P. L. Bailey and R. F. W. Jackson; *Tetrahedron Lett.*, 1991, **32**, 3119. 622

91TL3171 C. Crévisy and J. Beau; *Tetrahedron Lett.*, 1991, **32**, 3171. 1038

91TL3209 I. G. Collado, R. Hernándex-Galán, G. M. Massanet, F. Rodriguez-Luis and J. Salvá; *Tetrahedron Lett.*, 1991, **32**, 3209. 831

91TL3357 G. Quinkert, M del Grosso, A. Bucher, J. W. Bats and G. Dürner; *Tetrahedron Lett.*, 1991, **32**, 3357. 558

91TL3397 T. Mandai, M. Ogawa, H. Yamaoki, T. Nakata, H. Murayama, M. Kawada and J. Tsuji; *Tetrahedron Lett.*, 1991, **32**, 3397. 976

91TL3405 S. Terashima, K. Nakatani, K. Arai and K. Yamada; *Tetrahedron Lett.*, 1991, **32**, 3405. 1029

91TL3551 T. Chou, S.-H. Hung, M.-L. Peng and S.-J. Lee; *Tetrahedron Lett.*, 1991, **32**, 3551. 24

91TL3605 M. A. Tius and N. K. Reddy; *Tetrahedron Lett.*, 1991, **32**, 3605. 1002

91TL3659 J. K. Crandall and T. A. Ayers; *Tetrahedron Lett.*, 1991, 3659. 497

91TL3679 M. E. Maier and T. Brandstetter; *Tetrahedron Lett.*, 1991, **32**, 3679. 554

91TL3739 P. P. M. A. Dols, L. Lacroix, A. J. H. Klunder and B. Zwanenburg; *Tetrahedron Lett.*, 1991, **32**, 3739. 582

91TL3755 T. R. Hoye and C. J. Dinsmore; *Tetrahedron Lett.*, 1991, **32**, 3755. 550

91TL3999 W.-B. Wang, L.-L. Shi, Z.-Q. Li and Y.-Z. Huang; *Tetrahedron Lett.*, 1991, **32**, 3999. 270

91TL4041 P. Molina, M. Alajarin and C. Lopez-Leonardo; *Tetrahedron Lett.*, 1991, **32**, 4041. 835

91TL4115 E. Widmer, P. Vogt and M. Schlageter; *Tetrahedron Lett.*, 1991, **32**, 4115. 1067

91TL4125 M. P. Crozet, L. Giraud, J.-F. Sabuco, P. Vanelle and M. Barreau; *Tetrahedron Lett.*, 1991, **32**, 4125. 156

91TL4199 B. Coates, D. Montgomery and P. J. Stevenson; *Tetrahedron Lett.*, 1991, **32**, 4199. 828

91TL4267 T. Yamasaki, J. T. Welch, J. S. Plummer and R. H. Gimi *Tetrahedron Lett.*, 1991, **32**, 4267. 828

91TL4311 K. Kudo, K. Saigo, Y. Hashimoto, H. Houchigai and M. Hasegawa; *Tetrahedron Lett.*, 1991, **32**, 4311. 189

91TL4453 E.-I. Negishi, Z. R. Owczarczyk and D. R. Swanson; *Tetrahedron Lett.*, 1991, 4453. 467

91TL4481 M. T. Reetz and E. H. Lauterbach; *Tetrahedron Lett.*, 1991, **32**, 4481. 839

91TL4591 A. N. Kasatkin, A. N. Kulak, G. A. Tolstikov and O. V. Shitikova; *Tetrahedron Lett.*, 1991, **32**, 4591. 801

91TL4619 Z. Wang, X.-J. Meng and G. W. Kabalka; *Tetrahedron Lett.*, 1991, **32**, 4619. 281

91TL4647 T. Oh, Z. Wrobel and S. M. Rubenstein; *Tetrahedron Lett.*, 1991, **32**, 4647. 404, 406, 819

91TL4709 M. J. Dickens and J.-L. Luche; *Tetrahedron Lett.*, 1991, **32**, 4709. 156

91TL4963 J. Katsuki and J. Inanaga; *Tetrahedron Lett.*, 1991, **32**, 4963. 16

91TL5137 H. Okamura, S. Kuroda, K. Tomita, S. Ikegami, Y. Sugimoto, S. Sakaguchi, T. Katsuki and M. Yamaguchi; *Tetrahedron Lett.*, 1991, **32**, 5137. 816

91TL5219 E. Piers and J. Y. Roberge; *Tetrahedron Lett.*, 1991, **32**, 5219 558

91TL5243 J. M. Nuss, B. H. Levine, R. A. Rennels and M. M. Heravi; *Tetrahedron Lett.*, 1991, **32**, 5243. 1015

91TL5321 J. Boivin, L. Elkaim, P. G. Ferro and S. Z. Zard; *Tetrahedron Lett.*, 1991, **32**, 5321. 1076

91TL5401 K. M. Bol and R. M. J. Liskamp; *Tetrahedron Lett.*, 1991, **32**, 5401. 817

91TL5501 A. I. Meyers, M. A. Gonzalez, V. Struzka, A. Akahane, J. Guiles and J. S. Warmus; *Tetrahedron Lett.*, 1991, **32**, 5501. 150

91TL5505 A. I. Meyers, J. Guiles, J. S. Warmus and M. A. Gonzalez; *Tetrahedron Lett.*, 1991, **32**, 5505. 150

91TL5509 A. I. Meyers, J. S. Warmus, M. A. Gonzalez, J. Guiles and A. Akahane; *Tetrahedron Lett.*, 1991, **32**, 5509. 150

91TL5525 P. Moreau, M. Al Neirabeyeh, G. Guillaumet and G. Coudert; *Tetrahedron Lett.*, 1991, **32**, 5525. 829

91TL5559 J. Inanaga, Y. Yokoyama, Y. Baba and M. Yamaguchi; *Tetrahedron Lett.*, 1991, **32**, 5559. 458, 516

91TL5709 E. D. Edstrom; *Tetrahedron Lett.*, 1991, **32**, 5709. 246

91TL5903 T. Nguyen and E.-I. Negishi; *Tetrahedron Lett.*, 1991, 5903. 481

91TL5969 I. Marek, J.-M. Lefrançois and J. F. Normant; *Tetrahedron Lett.*, 1991, **32**, 5969. 836

91TL6085 P. A. Magriotis, M. E. Scott and K. D. Kim; *Tetrahedron Lett.*, 1991, **32**, 6085. 1018

91TL6109 M. Alami and G. Linstrumelle; *Tetrahedron Lett.*, 1991, **32**, 6109. 1012

91TL6159 F. Kido, Y. Kawada, M. Kato and A. Yoshikoshi; *Tetrahedron Lett.*, 1991, **32**, 6159. 822

91TL6273 B. P. Branchaud and W. Detlefsen; *Tetrahedron Lett.*, 1991, 6273. 495

91TL6307 D. P. Curran and G. Thoma; *Tetrahedron Lett.*, 1991, **32**, 6307. 327

91TL6449 G. Ortar, P. G. Ciattini and E. Morera; *Tetrahedron Lett.*, 1991, **32**, 6449. 1030

91TL6473 P. G. Gassman and A. C. Lottes; *Tetrahedron Lett.*, 1991, 6473. 499

91TL6497 M. Harmata and R. J. Claassen, II; *Tetrahedron Lett.*, 1991, **32**, 6497. 146

91TL6541 J. Gore, D. Bouyssi and G. Balme; *Tetrahedron Lett.*, 1991, **32**, 6541. 1020

91TL6559 H. Togo, M. Aoki, T. Kuramochi and M. Yokoyama; *Tetrahedron Lett.*, 1991, **32**, 6559. 934

91TL6683 E.-I. Negishi and Z. Owczarczyk; *Tetrahedron Lett.*, 1991, 6683. 468

91TL6735 K. Suzuki, T. Matsumoto, T. Hosoya and M. Katsuki; *Tetrahedron Lett.*, 1991, **32**, 6735. 1071

91TL6749 S. Hara and A. Suzuki; *Tetrahedron Lett.*, 1991, **32**, 6749. 281

91TL6969 G. Jommi, F. Orsini, M. Resmini and M. Sisti; *Tetrahedron Lett.*, 1991, **32**, 6969. 812

91TL7187 D. H. R. Barton, D. O. Jang and J. Cs. Jaszberenyi; *Tetrahedron Lett.*, 1990, **32**, 7187. 17, 938

91TL7195 J. E. Macor and J. M. Wehner; *Tetrahedron Lett.*, 1991, **32**, 7195. 200

91TL7229 N. Krause and G. Handke; *Tetrahedron Lett.*, 1991, **32**, 7229. 960

91TL7233 M. M. Cid, U. Eggnauer, H. P. Weber and E. Pombo-Villar; *Tetrahedron Lett.*, 1991, **32**, 7233. 835

91TL7281 K. Fuji, F. Tanaka and M. Node; *Tetrahedron Lett.*, 1991, **32**, 7281. 134

91TL7373 R. E. Lowenthal and S. Masamune; *Tetrahedron Lett.*, 1991, **32**, 7373. 353

91TL7493 P. Galatsis and S. D. Millan; *Tetrahedron Lett.*, 1991, **32**, 7493. 929

91TL7557 K. Harano, H. Kiyonaga and T. Hisano; *Tetrahedron Lett.*, 1991, **32**, 7557. 836

91TL7601 I. Paterson, A. N. Hulme and D. J. Wallace; *Tetrahedron Lett.*, 1991, **32**, 7601. 831

91TL7683 T. Mandai, K. Ryoden, M. Kawada and J. Tsuji; *Tetrahedron Lett.*, 1991, **32**, 7683. 976

91ZOR488 K. A. Lukin, A. Y. Masunova, S. I. Kozhushkov, T. S. Kuznetsova, B. I. Ugrak, V. A. Piven' and N. S. Zefirov; *Zh. Org. Khim.*, 1991, **27**, 488. 500

91ZOR1137 N. V. Vasileva, V. F. Starichenko and V. A. Koptyug; *Zh. Org. Khim.*, 1991, **27**, 1137. 103

92ACR182 Y. Huang; *Acc. Chem. Res.*, 1992, **25**, 182. 647, 648, 743

92ACR188 C. Chatgilialoglu; *Acc. Chem. Res.*, 1992, 188. 4, 17, 331, 342, 919, 936, 938, 943, 945

92AG(E)90 K. Banert, H. Hückstädt and K. Vrobel; *Angew. Chem., Int. Ed. Engl.*, 1992, **31**, 90. 993

92AG(E)224 K. Nunn, P. Mosset, R. Grée, R. W. Saalfrank, K. Peters and H. G. von Schnering; *Angew. Chem., Int. Ed. Engl.*, 1992, **31**, 224. 966

92AG(E)274 P. J. Stang; *Angew. Chem., Int. Ed. Engl.*, 1992, **31**, 274. 108

92AG(E)647 R. Hoffmann and R. Brückner; *Angew. Chem., Int. Ed. Engl.*, 1992, **31**, 647. 405, 816

92AG(E)772 P. de Arms, C. G. Francisco and E. Suárez; *Angew. Chem., Int. Ed. Engl.*, 1992, **31**, 772. 931

92AG(E)1033 T. Arnold, B. Orschel and H.-U. Reissig; *Angew. Chem., Int. Ed. Engl.*, 1992, **31**, 1033. 955

92AG(E)1240 A. M. Boldi, J. Anthony, C. B. Knobler and F. Diederich; *Ang. Chem., Int. Ed. Engl.*, 1992, **31**, 1240. 999

92AG(E)1459 P. Knochel; *Angew. Chem., Int. Ed. Engl.*, 1992, **31**, 1459. 138

92AG(E)1505 J. Schwerdtfeger and D. Hoppe; *Angew. Chem., Int. Ed. Engl.*, 1992, **31**, 1505. 138

92AG(E)1611 R. Herges and C. Hoock; *Angew. Chem., Int. Ed. Engl.*, 1992, **31**, 1611. 982

92BCJ1841 S. Ikegami, H. Okamura, S. Kuroda, T. Katsuki and M. Yamaguchi; *Bull. Chem. Soc. Jpn.*, 1992, **65**, 1841. 404, 406, 819

92BCJ2366 M. Yamamoto, H. Munakata, T. Kishikawa, S. Kohmoto and K. Yamada; *Bull. Soc. Chem. Jpn.*, 1992, **65**, 2366. 582

92BCJ2863 S. Abe, N. Miyaura and A. Suzuki; *Bull. Chem. Soc. Jpn.*, 1992, **65**, 2863. 489

92BCJ2992 Y. Oshiro, S. Minakata, S. Itoh and M. Komatsu; *Bull. Chem. Soc. Jpn.*, 1992, **65**, 2992. 1010

92BCJ3487 T. Fujisawa, Y. Okumura, K. Morita and Y. Ukaji; *Bull. Chem. Soc. Jpn.*, 1992, **65**, 3487. 183

92BMC1619 S. S. Bhagwat, C. Boswell, C. Gude, N. Contardo, D. S. Cohen, J. Mathis, R. Dotson, W. Lee and S. Shetty; *Bioorg. Med. Chem. Lett.*, 1992, **2**, 1619. 1010

92BSF151 M. Laabassi and R. Grée; *Bull. Soc. Chim. Fr.*, 1992, **129**, 151. 963

92BSF171 I. Marek, A. Alexakis, P. Mangeney and J. F. Normant; *Bull. Soc. Chim. Fr.*, 1992, **129**, 171. 496

92BSF273 B. Friess, B. Cazes and J. Goré; *Bull. Soc. Chim. Fr.*, 1992, **129**, 273. 496

92CB149 T. Kauffmann and H. Kallweit; *Chem. Ber.*, 1992, **125**, 149. 692

92CB271 R. Brückner, S. W. Scheuplein, K. Harms and J. Suffert; *Chem. Ber.*, 1992, **125**, 271. 1029

92CB287 H. I. Tashtoush and R. Sustmann; *Chem. Ber.*, 1992, **125**, 287. 937

92CB1957 R. Hoffman and R. Brückner; *Chem. Ber.*, 1992, **125**, 1957. 398

92CB2667 H. Fischer, O. Podschadly, A. Früh, C. Troll, R. Stumpf and A. Schlageter; *Chem. Ber.*, 1992, **125**, 2667. 991

92CC13 M. C. P. Yeh and S. I. Tau; *J. Chem. Soc., Chem. Commun.*, 1992, 13. 291

92CC46 N. Komatsu, Y. Nishibayashi, T. Sugita and S. Uemura; *J. Chem. Soc., Chem. Commun.*, 1992, 46. 609, 983

92CC59 E. Baciocchi, C. Rol, G. C. Rosato and G. V. Sebastiani; *J. Chem. Soc., Chem. Commun.*, 1992, 59. 290

92CC319 M. J. Dunn and R. F. W. Jackson; *J. Chem. Soc., Chem. Commun.*, 1992, 319. 968

92CC350 K. I. Booker-Milburn, G. D. Heffernan and P. J. Parsons; *J. Chem. Soc., Chem. Commun.*, 1992, 350. 816

92CC613 J. C. López, A. M. Gómez and S. Vaverde; *J. Chem. Soc., Chem. Commun.*, 1992, 613. 942

92CC721 S. Ohira, K. Okai and T. Moritani; *J. Chem. Soc.; Chem. Commun.*, 1992, 721. 1078

92CC778 T. Kawase, S. Muro, H. Kurata and M. Oda; *J. Chem. Soc., Chem. Commun.*, 1992, 778. 990

92CC782 Y. Oshawa and T. Saji; *J. Chem. Soc., Chem. Commun.*, 1992, 782. 890

92CC1313 G. Pandey, G. Lakshmaiah and G. Kumaraswamy; *J. Chem. Soc. Chem. Commun.*, 1992, 1313. 285

92CC1477 S. T. Patel and J. M. Percy; *J. Chem. Soc., Chem. Commun.*. 1992, 1477. 570

92CL61 Y. Ukaji, M. Nishimura and T. Fujisawa; *Chem. Lett.*, 1992, 61. 352

92CL695 H. Fujishima, E. I. Takada, S. Hara and A. Suzuki; *Chem. Lett.*, 1992, 695. 1050

92CL1073 M. Hojo, M. Ohkuma and A. Hosomi; *Chem. Lett.*, 1992, 1073. 286

92CL1357 T. Kitamura, T. Takachi, S.-I. Soda, H. Kawasato and H. Taniguchi; *Chem. Lett.*, 1992, 1357. 985

92CL1725 H. Kuno, M. Shibagaki, K. Takahashi, I. Hondo and H. Matsushita; *Chem. Lett.*, 1992, 1725. 456

92CL2479 M. Yamaguchi, A. Hayashi and M. Hirama; *Chem. Lett.*, 1992, 2479. 1050

92CPL243 M. S. F. Lie Ken Jie and C. F. Wong; *Chem. Phys. Lipids.*, 1992, **61**, 243. 986

92CRV29 G. A. Molander; *Chem. Rev.*, 1992, **92**, 29. 937

92CRV209 V. I. Bregadze; *Chem. Rev.*, 1992, **92**, 209. 168

92CRV741 Y. Inoue; *Chem. Rev.*, 1992, **92**, 741. 357

92CRV771 B. E. Rossiter and N. M. Swingle; *Chem. Rev.*, 1992, **92**, 771. 305, 306

92CRV1007 H. B. Kagan and O. Riant; *Chem. Rev.*, 1992, **92**, 1007. 374, 375

92CSR105 J. C. Walton; *Chem. Soc. Rev.*, 1992, **21**, 105. 919, 923

92H(34)2247 E. J. Enholm and S. J. Jiang; *Heterocycles*, 1992, **34**, 2247. 179

92HCA1023 C. Fehr and O. Guntern; *Helv. Chim. Acta*, 1992, **75**, 1023. 582

92HCA1755 L. A. Paquette, S. W. Elmore, K. D. Combrink, E. R. Hickey and R. D. Rogers; *Helv. Chim. Acta*, 1992, **75**, 1755. 810

92JA375 E. J. Verner and T. Cohen; *J. Am. Chem. Soc.*, 1992, **114**, 375. 398, 405

92JA741 L. Chen, D. P. Dumas and C.-H. Wong; *J. Am. Chem. Soc.*, 1992, **114**, 741. 6

92JA1329 W. H. Pearson, D. P. Szura and M. J. Postich; *J. Am. Chem Soc.*, 1992, **114**, 1329. 372

92JA1491 G. S. Jones and W. J. Scott; *J. Am. Chem. Soc.*, 1992, **114**, 1491. 584

92JA2635 K. J. Shea, G. J. Stoddard, W. P. England and C. D. Haffner; *J. Am. Chem. Soc.*, 1992, **114**, 2635. 807

92JA3978 Q. Xie, E. Perez-Cordero and L. Echegoyen; *J. Am. Chem. Soc.*, 1992, **114**, 3978. 890

92JA3993 P. Magnus and J. Lacour; *J. Am. Chem. Soc.*, 1992, **114**, 3993. 263

92JA4051 A. Stolle, J. Ollivier, P. P. Piras, J. Salaün and A. de Meijere; *J. Am. Chem. Soc.*, 1992, **114**, 4051. 975

92JA4079 M. R. Sestrick, M. Miller and L. S. Hegedus; *J. Am. Chem. Soc.*, 1992, **114**, 4079. 987

92JA4128 X. Wang, S. D. Erickson, T. Limori and W. C. Still; *J. Am. Chem. Soc.*, 1992, **114**, 4128. 584

92JA4436 D. P. Curran and G. Thoma; *J. Am. Chem. Soc.*, 1992, **114**, 4436. 327

92JA4439 H.-J. Gais and G. Hellmann; *J. Am. Chem. Soc.*, 1992, **114**, 4439. 146

92JA4912 K. Durkin, D. C. Liotta, J. Rancourt, J. F. Lavalée, L. Boisvert and Y. Guindon; *J. Am. Chem. Soc.*, 1992, **114**, 4912. 327

92JA5018 L. A. Paquette and G. D. Maynard; *J. Am. Chem. Soc.*, 1992, **114**, 5018. 414

92JA5110 D. E. Stack, B. T. Dawson and R. D. Rieke; *J. Am. Chem. Soc.*, 1992, **114**, 5110. 121

92JA5432 B. M. Trost, J. B. Neilsen and K. Hoogsteen; *J. Am. Chem. Soc.*, 1992, **114**, 5432. 390

92JA5878 P. A. Wender and T. P. Mucciaro; *J. Am. Chem. Soc.*, 1992, **114**, 5878. 799

92JA5904 D. H. R. Barton, J. Cs. Jaszberenyi and E. A. Theodorakis; *J. Am. Chem. Soc.*, 1992, **114**, 5904. 944

92JA5959 M. A. Tius and M. A. Kerr; *J. Am. Chem. Soc.*, 1992. **114**, 5959. 556

92JA5998 J. K. Crandall, D. M. Coppert, T. Schuster and F. Lin; *J. Am. Chem. Soc.*, 1992, **114**, 5998. 983

92JA6050 D. P. Curran and M. J. Totleben; *J. Am. Chem. Soc.*, 1992, **114**, 6050. 347

92JA6330 P. R. Ashton, G. R. Brown, N. S. Isaacs, D. Giuffrida, F. H. Kohnke, J. P. Mathias, A. M. Z. Slawin, D. R. Smith, J. F. Stoddart and D. J. Williams; *J. Am. Chem. Soc.*, 1992, **114**, 6330. 571

92JA6556 E. Vedejs, A. R. Haight and W. O. Moss; *J. Am. Chem. Soc.*, 1992, **114**, 6556. 471

92JA6692 A. H. Hoveyda, J. P. Morken, A. F. Houri and Z. Xu; *J. Am. Chem. Soc.*, 1992, **114**, 6692. 316

92JA7007 J. G. Stack, D. P. Curran, S. V. Geib, J. Rebek, Jr. and P. Ballester; *J. Am. Chem. Soc.*, 1993, **114**, 7007. 326, 327, 328

92JA7375 L. A. Paquette, D. R. Sauer, D. G. Cleary, M. A. Kinsella, C. M. Blackwell and L. G. Anderson; *J. Am. Chem. Soc.*, 1992, **114**, 7375. 362, 555, 558

92JA7664 N. A. Porter, I. J. Rosenstein, R. A. Breyer, J. D. Bruhnke, W.-X. Wu and A. T. McPhail; *J. Am. Chem. Soc.*, 1992, **114**, 7664. 326, 327

92JA7903 B. M. Trost, S. Sharma and T. Schmidt; *J. Am. Chem. Soc.*, 1992, **114**, 7903.

92JA8053 W. F. Bailey, A. D. Khanolkar and K. V. Gavaskar; *J. Am. Chem. Soc.*, 1992, **114**, 8053. 316

92JA8483 A. I. Meyers, T. M. Sielecki, D. C. Crans, R. W. Marshman and T. H. Nguyen; *J. Am. Chem. Soc.*, 1992, **114**, 8483. 149

92JA8707 S. Ejiri, S. Yamago and E. Nakamura; *J. Am. Chem. Soc.*, 1992, **114**, 8707. 360

92JA8732 M. Shindo, K. Koga and K. Tomioka; *J. Am. Chem. Soc.*, 1992, **114**, 8732. 473

92JA8818 T. Ohwada; *J. Am. Chem. Soc.*, 1992, **114**, 8818. 584

92JA8869 K. Narasaka, Y. Hayashi, H. Shimadzu and S. Niihata; *J. Am. Chem. Soc.*, 1992, **114**, 8869. 358, 498

92JA9279 K. C. Nicolaou, A. Liu, Z. Zeng and S. McComb; *J. Am. Chem. Soc.*, 1992, **114**, 9279. 1038

92JA9419 A. B. Smith, III, J. R. Empfield, R. A. Rivero, H. A. Vaccaro, J. J.-W. Duan and M. M. Sulikowski; *J. Am. Chem. Soc.*, 1992, **114**, 9419. 573, 581, 584

92JA9492 J. A. Maguire, A. Petrillo and A. S. Goldman; *J. Am. Chem. Soc.*, 1992, **114**, 9492. 554

92JA9576 J. Bertran, I. Gallardo, M. Moreno and J.-M. Savéant; *J. Am. Chem. Soc.*, 1992, **114**, 9576. 107

92JA9673 J. D. White, M. P. Dillon and R. J. Butlin; *J. Am. Chem. Soc.*, 1992, **114**, 9673. 573

92JA10138 C.-H. Lin, T. Sugai, R. L. Halcomb, Y. Ichikawa and C.-H. Wong; *J. Am. Chem. Soc.*, 1992, **114**, 10138. 9

92JA10653 C. Kibayashi, S. Aoyagi and T. Wang; *J. Am. Chem. Soc.*, 1992, **114**, 10653. 1067

92JCR(S)309 A. K. Sarkar and N. Sarkar; *J. Chem. Res. (S)*, 1992, 309. 628

92JCS(P1)95 R. W. Carling, J. S. Clark, A. B. Holmes and D. Sartor; *J. Chem. Soc., Perkin Trans. 1*, 1992, 95. 177, 178

92JCS(P1)259 S. P. Saberi and S. E. Thomas; *J. Chem. Soc., Perkin Trans. 1*, 1992, 259. 987

92JCS(P1)311 A. Srikrishna and S. Nagaraju; *J. Chem. Soc., Perkin Trans. 1*, 1992, 311. 416

92JCS(P1)747 A. Pelter, K. Smith and K. D. Jones; *J. Chem. Soc., Perkin Trans. 1*, 1992, 747. 988

92JCS(P1)1313 N. J. G. Cox, S. D. Mills and G. Pattenden; *J. Chem. Soc., Perkin Trans. 1*, 1992, 1313. 340

92JCS(P1)1837 H. Suginome, Y. Nakayama and H. Senboku; *J. Chem. Soc., Perkin Trans. 1*, 1992, 1837. 931

92JCS(P1)1881 B. C. Ranu and A. R. Das; *J. Chem. Soc., Perkin Trans.* 1, 1992, 1881. 14

92JCS(P1)2155 D. Levin and S. Warren; *J. Chem. Soc., Perkin Trans. 1*, 1992, 2155. 643

92JCS(P1)2303 I. Coldham and S. Warren; *J. Chem. Soc., Perkin Trans. 1*, 1992, 2303. 599

92JCS(P1)2355 G. B. Gill, M. S. Hj. Idris and K. S. Kirollos; *J. Chem. Soc., Perkin Trans. 1*, 1992, 2355. 980

92JCS(P1)2565 M. S. Khana, O. V. Singh, C. P. Garg and R. P. Kapoor; *J. Chem. Soc., Perkin Trans. 1*, 1992, 2565. 556

92JCS(P1)2683 L. Crombie and A. D. Heavers; *J. Chem. Soc., Perkin Trans. 1*, 1992, 2683. 606

92JCS(P1)2727 S. Janardhanam and K. Rajagopalan; *J. Chem. Soc., Perkin Trans. 1*, 1992, 2727. 810

92JCS(P1)3165 C. Bokelmann, W. P. Neumann and M. Peterseim; *J. Chem. Soc., Perkin Trans. 1*, 1992, 3165. 936

92JCS(P1)3225 P. L. Burn, D. D. C. Bradley, R. H. Friend, D. A. Halliday, A. B. Holmes, R. W. Jackson and A. Kraft; *J. Chem. Soc., Perkin Trans. 1*, 1992, 3225. 623

92JCS(P1)3351 I. Fleming, S. Gil, A. K. Sarkar and T. Schmidlin; *J. Chem. Soc., Perkin Trans. 1*, 1992, 3351. 565

92JCS(P1)3407 P. M. Ayrey, M. A. Bolton, A. D. Buss, N. Greeves, D. Levin, P. Wallace and S. Warren; *J. Chem. Soc., Perkin Trans. 1*, 1992, 3407. 643

92JCS(P1)3419 P. J. Kocieński, M. Pritchard, S. N. Wadman, R. J. Whitby and C. L. Yeates; *J. Chem. Soc., Perkin Trans. 1*, 1992, 3419. 481

92JCS(P1)3431 P. A. Ashworth, N. J. Dixon, P. J. Kocieński and S. N. Wadman; *J. Chem. Soc., Perkin Trans. 1*, 1992, 3431. 481

92JMC241 A. Matsuda, M. Shinozaki, T. Yamaguchi, H. Homma, R. Nomoto, T. Miyasaka, Y. Watanabe and T. Abinu; *J. Med. Chem.*, 1992, **35**, 241. 1010

92JOC17 H. Nakahira, I. Ryu, M. Ikebe, Y. Oku, A. Ogawa, N. Kambe, N. Sonoda and S. Murai; *J. Org. Chem.*, 1992, **57**, 17. 666

92JOC50 A. Abad, C. Agulló, M. Arnó, A. C. Cuñat and R. J. Zaragozá; *J. Org. Chem.*, 1992, **57**, 50. 1082

92JOC428 K. Honda, S. Inoue and K. Sato; *J. Org. Chem.*, 1992, **57**, 428. 408

92JOC651 M. Avignon-Tropis, G. Linstrumelle, D. Guillerm, M. Fréchard-Ortuno, J. M. Berjeaud and J. R. Pougny; *J. Org. Chem.*, 1992, **57**, 651. 1064

92JOC678 J. M. Saá, G. Martorell and A. G. Raso; *J. Org. Chem.*, 1992, **57**, 678. 482, 1027

92JOC686 B. M. Trost and T. A. Grese; *J. Org. Chem.*, 1992, **57**, 686. 289, 362

92JOC714 G. Haufe, G. Alvernhe, D. Anker, A. Laurent and C. Saluzzo; *J. Org. Chem.*, 1992, **57**, 714. 780

92JOC774 L.-J. Zhang, Y.-Z. Huang, H.-X. Jiang, J. Duan-Hu and H. Liao; *J. Org. Chem.*, 1992, **57**, 774. 272

92JOC1024 Y. Yamamoto, Y. Chounan, M. Tanaka and T. Ibuka; *J. Org. Chem.*, 1992, **57**, 1024. 188

92JOC1146 D. J. Pasto, K. D. Sugi and D. E. Alonso; *J. Org. Chem.*, 1992, **57**, 1146. 498

92JOC1161 A. Padwa, D. J. Austin, M. Ishida, C. L. Muller, S. S. Murphree and P. E. Yeske; *J. Org. Chem.*, 1992, **57**, 1161. 470

92JOC1242 J. A. Marshall and X. Wang; *J. Org. Chem.*, 1992, **57**, 1242. 1057

92JOC1384 F. A. Gomez and M. F. Hawthorne; *J. Org. Chem.*, 1992, **57**, 1384. 168

92JOC1613 S. Valverde, J. C. López, A. M. Gómez and S. Garcia-Ochoa; *J. Org. Chem.*, 1992, **57**, 1613. 579

92JOC1615 Y. Kataoka, K. Takai, K. Oshima and K. Utimoto; *J. Org. Chem.*, 1992, **57**, 1615. 457

92JOC1618 J. J. Eisch, Z.-R. Liu and M. Singh; *J. Org. Chem.*, 1992, **57**, 1618. 14

92JOC1718 C. Cardellicchio, V. Fiandanese and F. Naso; *J. Org. Chem.* 1992, **57**, 1718. 470

92JOC2143 J. J. Eisch, Z.-R. Liu and M. P. Boleslawski; *J. Org. Chem.*, 1992, **57**, 2143. 13, 20

92JOC2160 R. Ballini, R. Castagnini and M. Petrini; *J. Org. Chem.*, 1992, **57**, 2160. 578

92JOC2188 H. Alper and V. V. Grushin; *J. Org. Chem.*, 1992, **57**, 2188. 1002

92JOC2270 J. D. White, J. C. Amiedo, S. Gut, S. Ohira and L. R. Jayasinghe; *J. Org. Chem.*, 1992, **57**, 2270. 582

92JOC2399 B. B. Snider and T. Kwon; *J. Org. Chem.*, 1992, **57**, 2399. 929

92JOC2427 C. Chatgilialoglu, M. Guerra, A. Guerrini, G. Seconi, K. B. Clark, D. Griller, J. Kanabus-Kaminska and J. A. Martinho-Simões; *J. Org. Chem.*, 1992, **57**, 2427. 5, 7, 9, 10, 17, 25

92JOC2856 J. A. Marsella, A. G. Gilicinski, A. M. Coughlin and G. P. Pez; *J. Org. Chem.*, 1992, **57**, 2856. 570

92JOC2941 J. C. Evans, C. T. Goralski and D. L. Hasha; *J. Org. Chem.*, 1992, **57**, 2941. 1037

92JOC2986 W. H. Pearson and J. M. Schkeryantz; *J. Org. Chem.*, 1992, **57**, 2986. 190

92JOC3145 S. Kusuda, Y. Watanabe, Y. Ueno and T. Toru; *J. Org. Chem.*, 1992, **57**, 3145 594, 668

92JOC3173 B. Fernández, J. A. M. Pérez, J. R. Granja, L. Castedo and A. Mouriño; *J. Org. Chem.*, 1992, **57**, 3173. 606

92JOC3387 J. A. Marshall and X. Wang; *J. Org. Chem.*, 1992, **57**, 3387. 1057

92JOC3405 C. Chatgilialoglu, A. Guerrini and M. Lucarini; *J. Org. Chem.*, 1992, **57**, 3405. 5, 7, 9, 10, 17, 25

92JOC3479 T. H. Eberlein, F. G. West and R. W. Tester; *J. Org. Chem.*, 1992, **57**, 3479. 400

92JOC3558 W. Cabri, I. Candiani, A. Bedeschi and R. Santi; *J. Org. Chem.*, 1992, **57**, 3558. 494

92JOC3615 B. B. Snider and K. Yang; *J. Org. Chem.*, 1992, **57**, 3615. 583

92JOC3757 J. Yu, J. R. Falck and C. Mioskowski; *J. Org. Chem.*, 1992, **57**, 3757. 183

92JOC3819 L. A. Castonguay, J. W. Guiles, A. K. Rappé and A. I. Meyers; *J. Org. Chem.*, 1992, **57**, 3819. 150

92JOC3994 B. Kopping, C. Chatgilialoglu, M. Zehdner and B. Giese; *J. Org. Chem.*, 1992, **57**, 3994. 327, 950

92JOC4070 K. Belyk, M. J. Rozema and P. Knochel; *J. Org. Chem.*, 1992, **57**, 4070. 974, 1056

92JOC4284 R. Hollis, L. Hughes, V. W. Bowry and K. U. Ingold; *J. Org. Chem.*, 1992, **57**, 4284. 340

92JOC4457 D. J. Hart and R. Krishnamurthy; *J. Org. Chem.*, 1992, **57**, 4457. 327, 328

92JOC4598 N. K. N. Yee and R. M. Coates; *J. Org. Chem.*, 1992, **57**, 4598. 24

92JOC4786 J. M. Tour, S. L. Pendalwar, C. M. Kafka and J. P. Cooper; *J. Org. Chem.*, 1992, **57**, 4786. 515

92JOC5060 S. L. Schreiber, D. Romo, D. D. Johnson, L. Plamondon and T. Miwa; *J. Org. Chem.*, 1992, **57**, 5060. 1067

92JOC5071 R. E. Ireland, T. K. Highsmith, L. D. Gegnas and J. L. Gleason; *J. Org. Chem.*, 1992, **57**, 5071. 1079

92JOC5121 X. Fu and J. M. Cook; *J. Org. Chem.*, 1992, **57**, 5121. 776

92JOC5208 M. A. Maestro, L. Castedo and A. Mouriño; *J. Org. Chem.*, 1992, **57**, 5208. 576

92JOC5557 K. Kim, J. Reibenspies and G. Sulikowski; *J. Org. Chem.*, 1992, **57**, 5557. 573

92JOC5805 T. Harada, T. Katsuhira and A. Oku; *J. Org. Chem.*, 1992, **57**, 5805. 431

92JOC5818 M. A. Brimble and G. M. Williams; *J. Org. Chem.*, 1992, **57**, 5818. 1045

92JOC6118 I. De Riggi, S. Gastaldi, J.-M. Surzur and M. P. Bertrand; *J. Org. Chem.*, 1992, **57**, 6118. 949

92JOC6344 K. Prasad, S. Ramaswamy and O. Repic; *J. Org. Chem.*, 1992, **57**, 6344. 1023

92JOC6803 M. D. Bachi, E. Bosch, D. Denenmark and D. Girsh; *J. Org. Chem.*, 1992, **57**, 6803. 938

92JOC7056 B. R. Henke, A. J. Kouklis and C. H. Heathcock; *J. Org. Chem.*, 1992, **57**, 7056. 369

92JOC7188 N. S. Barta, G. R. Cook, M. S. Landis and J. R. Stille; *J. Org. Chem.*, 1992, **57**, 7188. 835

92JOC7285 M. Ishizaki, O. Hoshino and Y. Iitaka; *J. Org. Chem.*, 1992, **57**, 7285. 584

92JOM(423)405 T. Mitsudo, M. Takagi, S.-W. Zhang and Y. Watanabe; *J. Organomet. Chem.*, 1992, **423**, 405. 493

92JOM(440)277 D. Mesnard and L. Miginiac; *J. Organomet. Chem.*, 1992, **440**, 277. 973

92JOM(441)1 H. J. M. Schoufs, A. Maercker and L. Brandsma; *J. Organomet. Chem.*, 1992, **441**, 1. 441

92JPC2151 J. L. Emdee, K. Brezinsky and I. Glassman; *J. Phys. Chem.*, 1992, 2151. 1127

92M587 H. Spreitzer; *Monatsch. Chem.*, 1992, **123**, 587. 564

92M939 E. Reimann and H. Benend; *Monatsch. Chem.*, 1992, **123**, 939. 560

B-92MI 104-01 J. March; in "Advanced Organic Chemistry: Reactions, Mechanisms and Structure," 4th edn., John Wiley, New York, 1992, ch. 10, p. 293. 107

B-92MI 104-02 M. Hudlický; in "Chemistry of Organic Fluorine Compounds," 2nd edn., Ellis Horwood, New York, 1992, p. 300. 107

B-92MI 104-03 C. H. Heathcock; in "Modern Synthetic Methods 1992," ed. R. Scheffold; Helvetica Chimica Acta, Basel, 1992, ch. 1, pp. 1–102. 130, 131

B-92MI 107-01 P. Perlmutter; "Conjugate Addition Reactions in Organic Synthesis," Pergamon, Oxford, 1992. 300

B-92MI 108-01 W. B. Motherwell and D. Crich; "Free Radical Chain Reactions in Organic Synthesis," Wiley, New York, 1992. 320, 330, 332, 342, 344

B-92MI 108-02 T.-L. Ho; "Tandem Organic Reactions," Wiley, New York, 1992. 347

B-92MI 108-03 J. March; "Advanced Organic Chemistry," Wiley, New York, 1992. 352

B-92MI 109-01 L. M. Harwood; "Polar Rearrangements," Oxford University Press, 1992. 377

B-92MI 112-01 W. B. Motherwell; "Free Radical Chain Reactions in Organic Synthesis," Academic Press, London, 1992. 546

B-92MI 112-02 R. Perlmutter; "Conjugate Addition Reactions in Organic Synthesis," Pergamon, Oxford, 1992, p. 339. 537

92MI 114-01 J. Kang, J. H. Kim and G. S. Jang; *Bull. Korean Chem. Soc.*, 1992, **13**, 192 (*Chem. Abstr.*, 1992, **117**, 25 600). 630

B-92MI 119-01 V. Snieckus; "Advances Carbanion Chemistry," Vol. 1, JAI Press, Connecticut, 1992. 845

B-92MI 119-02 J. March; "Advanced Organic Chemistry," 4th edn., Wiley, New York, 1992, p. 175. 552

B-92MI 119-03 T. L. Gilchrist; "Heterocyclic Chemistry," 2nd edn., Longman, Harlow, 1992, p. 132. 860, 861

B-92MI 119-04 J. March; "Advanced Organic Chemistry," 4th edn., Wiley, New York, 1992, p. 46. 861

B-92MI 119-05 J. March; "Advanced Organic Chemistry," 4th edn., Wiley, New York, 1992, p. 470. 872

B-92MI 119-06 J. March; "Advanced Organic Chemistry," 4th edn., Wiley, New York, 1992, p. 574. 883

B-92MI 119-07 J. March; "Advanced Organic Chemistry," 4th edn., Wiley, New York, 1992, p. 491. 884

B-92MI 119-08 J. March; "Advanced Organic Chemistry," 4th edn., Wiley, New York, 1992, p. 1080. 884

B-92MI 119-09	J. March; "Advanced Organic Chemistry," 4th edn., Wiley, New York, 1992, p. 633.	884
B-92MI 119-10	J. March; "Advanced Organic Chemistry," 4th edn., Wiley, New York, 1992, p. 182.	885
B-92MI 119-11	J. March; "Advanced Organic Chemistry," 4th edn., Wiley, New York, 1992, p. 1209.	887
B-92MI 119-12	J. March; "Advanced Organic Chemistry," 4th edn., Wiley, New York, 1992, p. 795.	889
B-92MI 119-13	J. March; "Advanced Organic Chemistry," 4th edn., Wiley, New York, 1992, p. 473.	889
B-92MI 119-14	L. M. Harwood; "Polar Rearrangements," Oxford University Press, Oxford, 1992, p. 13.	891
B-92MI 119-15	J. March; "Advanced Organic Chemistry," 4th edn., Wiley, New York, 1992, p. 320.	891, 904
B-92MI 119-16	D. H. R. Barton and S. I. Parekh; in "Half a Century of Free Radical Chemistry," Cambridge University Press, Cambridge, 1992.	919
B-92MI 119-17	W. B. Motherwell and D. Crich; "Best Synthetic Methods. Free Radical Chain Reactions in Organic Synthesis," Academic Press, London, 1992.	919, 927, 935, 938, 939
92MI 119-18	W. B. Motherwell; *Aldrichimica Acta*, 1992, **25**, 71.	919, 928
92OM493	H. Arzoumanian, F. Cochini, D. Nuel, J.-F. Petrignani and N. Rosas; *Organometallics*, 1992, **11**, 493.	976
92OR(41)1	T. Hudlicky, R. Fan, J. W. Reed and K. G. Gadamasetti; *Org. React.*, 1992, **41**, 1.	806, 811
92OR(41)135	B. H. Lipshutz and S. Segupta; *Org. React.*, 1992, **41**, 135.	116, 206, 311, 465, 476
92OR(42)1	P. Rabideau and Z. Marcinow; *Org. React.*, 1992, **42**, 1.	87
92PNA(89)915	G. A. Olah, S. Hamanaka, J. A. Wilkinson and J. A. Olah; *Proc. Natl. Acad. Sci. USA*, 1992, **89**, 915.	916
92PS(67)183	J. V. Commaseto; *Phosphorus Sulfur*, 1992, **67**, 183.	437, 438, 441
92PS(67)203	T. G. Back; *Phosphorus Sulfur*, 1992, **67**, 203.	441
92S151	B. M. Trost and M. C. Matelich; *Synthesis*, 1992, 151.	361
92S235	A. S. Demir and A. Jeganathan; *Synthesis*, 1992, 235.	560
92S248	P. Cintas; *Synthesis*, 1992, 248.	676
92S383	J. Sapi, Y. Grébille, J.-Y. Laronze and J. Lévy; *Synthesis*, 1992, 383.	628
92S461	B. P. Ernst and H. Laatsch; *Synthesis*, 1992, 461.	275
92S746	H. Yamanaka, T. Sakamoto, F. Shiga, A. Yasuhara, D. Uchiyama and Y. Kondo; *Synthesis*, 1992, 746.	1009
92S803	T. N. Mitchell; *Synthesis*, 1992, 803.	468
92S1035	B. B. Lohray; *Synthesis*, 1992, 1035.	185
92SC351	W. M. Owton, P. T. Gallagher and M. Brunaus; *Synth. Commun.*, 1992, **22**, 351.	254
92SC1953	M. T. Crimmins and D. Dedopoulou; *Synth. Commun.*, 1992, **22**, 1953.	482
92SL27	B. M. Trost and R. T. Matsuoka; *Synlett*, 1992, 27.	146
92SL57	F. Moulines, L. Djakovitch, J.-L. Fillaut and D. Astruc; *Synlett*, 1992, 57.	164
92SL107	R. D. Little, M. R. Masjedizadeh, K. D. Moeller and I. Dannecker-Doerig; *Synlett*, 1992, 107.	944
92SL197	P. J. Cox, A. Persad and N. S. Simpkins; *Synlett*, 1992, 197.	144
92SL199	P. J. Stang and B. L. Williamson; *Synlett*, 1992, 199.	1006
92SL207	T. Watanabe, N. Miyaura and A. Suzuki; *Synlett*, 1992, 207.	489
92SL266	F. Khuong-Huu and M. Bénéchie; *Synlett*, 1992, 266.	1024
92SL291	S. V. Ley and L. L. Yeung; *Synlett*, 1992, 291.	1023
92SL323	M. Miyazawa, S.-Z. Wang, H. Takeda and K. Yamamoto; *Synlett*, 1992, 323.	192
92SL442	B. Giese, W. Damm, M. Roth and M. Zehnder; *Synlett*, 1992, 442.	327
92SL635	H. Memmesheimer, U. Bergsträßer, J. Hoffmann, M. Baird and M. Regitz; *Synlett*, 1992, 635.	985
92SL715	J. P. Genet, E. Blart and M. Savignac; *Synlett*, 1992, 715.	469, 494, 1013
92SL721	S. Torii, H. Okumoto, M. A. Rashid and M. Mohri; *Synlett*, 1992, 721.	497
92SL743	T. Lübbers and H. J. Schäfer; *Synlett*, 1992, 743.	346
92SL766	D. Schinzer and J. Kabbara; *Synlett*, 1992, 766.	1012
92SL769	T. Ibuka and Y. Yamamoto; *Synlett*, 1992, 769.	188, 227
92SL801	W. P. Neumann and M. Peterseim; *Synlett*, 1992, 801.	17, 939
92SL809	K. I. Booker-Milburn; *Synlett*, 1992, 809.	929
92SL833	J. Ichikawa, S. Yonemaru and T. Minami; *Synlett*, 1992, 833.	429
92SL891	T. Sato, A. Kawase and T. Hirose; *Synlett*, 1992, 891.	164
92SL943	D. P. Curran, T. L. Fevig, C. P. Jasperse and M. J. Totleben; *Synlett*, 1992, 943.	347, 937
92T33	D. Caine and B. Stanhope; *Tetrahedron*, 1992, **48**, 33.	598
92T10265	M. Ishihara, Y. Masatsuga and K. Uneyama; *Tetrahedron*, 1992, **48**, 10265.	580
92T10315	S. Désert, P. Metzner and M. Ramdani; *Tetrahedron*, 1992, **48**, 10 315.	834
92T2515	D. Becker and H. J. E. Loewenthal; *Tetrahedron*, 1992, **48**, 2515.	622
92T2529	D. H. R. Barton; *Tetrahedron*, 1992, **48**, 2529.	935, 939, 946
92T2589	A. I. Meyers; *Tetrahedron*, 1992, **48**, 2589.	845, 860
92T3385	J. E. Baldwin, R. M. Adlington and R. Singh; *Tetrahedron*, 1992, **48**, 3385.	931
92T3413	J. E. Baldwin, R. M. Adlington and S. H. Ramcharitar; *Tetrahedron*, 1992, **48**, 3413.	340
92T3445	P. M. Esch, H. Hiemstra and W. N. Speckamp; *Tetrahedron*, 1992, **48**, 3445.	978
92T3749	K. Maruoka, J. Sato and H. Yamamoto; *Tetrahedron*, 1992, **48**, 3749.	387
92T4027	W. R. Bowman and P. J. Westlake; *Tetrahedron*, 1992, **48**, 4027.	931
92T4087	N. Kishi, T. Maeda, K. Mikami and T. Nakai; *Tetrahedron*, 1992, **48**, 4087.	816
92T5385	A. Padwa and K. E. Krumpe; *Tetrahedron*, 1992, **48**, 5385.	550
92T5709	M. Arai, B. H. Lipshutz and E. Nakamura; *Tetrahedron*, 1992, **48**, 5709.	124
92T6425	K. M. Bol and R. M. J. Liskamp; *Tetrahedron*, 1992, **48**, 6425.	406
92T6875	A. Clark and K. Jones; *Tetrahedron*, 1992, **48**, 6875.	345
92T6883	W. R. Bowman, D. S. Brown, C. A. Burns, B. A. Marples and N. A. Zaidi; *Tetrahedron*, 1992, **48**, 6883.	929

92T7083	D. H. R. Barton and M. Samadi; *Tetrahedron*, 1992, **48**, 7083.	333, 935
92T7091	D. H. R. Barton and C. Tachdjian; *Tetrahedron*, 1992, **48**, 7091.	332, 935
92T7109	D. H. R. Barton and C. Tachdjian; *Tetrahedron*, 1992, **35**, 7109.	700, 939
92T7121	D. H. R. Barton, P. Blundell and J. Cs. Jaszberenyi; *Tetrahedron*, 1992, **48**, 7121.	935
92T7383	L. A. Paquette and C. J. M. Stirling; *Tetrahedron*, 1992, **48**, 7383.	186
92T7435	D. H. R. Barton, J. Dorchak and J. Cs. Jaszberenyi; *Tetrahedron*, 1992, **48**, 7435.	939
92T8031	R. A. Batey, J. D. Harling and W. B. Motherwell; *Tetrahedron*, 1992, **48**, 8031.	928
92T8117	N. M. Ali, A. McKillop, M. B. Mitchell, R. A. Rebelo and P. J. Wallbank; *Tetrahedron*, 1992, **48**, 8117.	489
92T8963	T.-s. Chou, C.-W. Ko and T.-K. Yang; *Tetrahedron*, 1992, **48**, 8963.	145
92T9261	D. H. R. Barton, J. Camara, X. Cheng, S. D. Géro, J. Cs. Jaszberenyi and B. Quiclet-Sire; *Tetrahedron*, 1992, **48**, 9261.	944
92T9373	J. E. Baldwin, M. G. Moloney and A. F. Parsons; *Tetrahedron*, 1992, **48**, 9373.	946
92T9427	N. C. Ray, P. C. Raveendranath and T. A. Spencer; *Tetrahedron*, 1992, **48**, 9427.	14
92T9577	E. Erdik; *Tetrahedron*, 1992, **48**, 9577.	467
92TA853	S. Takano, T. Kamikubo, T. Sugihara and K. Ogasawara; *Tetrahedron Asymmetry*, 1992, **3**, 853.	596
92TA961	A. J. Walker; *Tetrahedron Asymmetry*, 1992, **3**, 961.	140, 143
92TA1089	C. G. Frost, J. Howarth and J. M. J. Williams; *Tetrahedron Asymmetry*; 1992, **3**, 1089.	192, 194
92TA1509	S.-K. Kang, S.-G. Kim and D.-G. Cho; *Tetrahedron Asymmetry*, 1992, **3**, 1509.	967
92TA1551	J. D. Brown; *Tetrahedron Asymmetry*, 1992, 1551.	389
92TL135	J. S. Yadav, M. C. Chander and K. K. Reddy; *Tetrahedron Lett.*, 1992, **33**, 135.	1002
92TL181	M. T. Crimmins, C. M. Dudek and A. W.-H. Cheung; *Tetrahedron Lett.*, 1992, **33**, 181.	928
92TL453	H. M. L. Davies and B. Hu; *Tetrahedron Lett.*, 1992, **33**, 453.	801
92TL461	H.-J. Gais and G. Bülow; *Tetrahedron Lett.*, 1992, 461.	485
92TL543	H. Imanieh, P. Quayle, M. Voaden, J. Conway and S. D. A. Street; *Tetrahedron Lett.*, 1992, **33**, 543.	160
92TL573	S. Hanessian, C. Girard and J. L. Chiara; *Tetrahedron Lett.*, 1992, **33**, 573.	587
92TL831	J. Barluenga, R. González and F. J. Fañanás; *Tetrahedron Lett.*, 1992, **33**, 831.	1090
92TL923	L. A. Paquette, R. Guevel and D. R. Sauer; *Tetrahedron Lett.*, 1992, **33**, 923.	813, 815
92TL1025	S. Knapp, A. B. J. Naughton and T. G. Murali Dhar; *Tetrahedron Lett.*, 1992, **33**, 1025.	782
92TL1057	J. Hatem, C. Henriet-Bernard, J. Grimaldi and R. Maurin; *Tetrahedron Lett.*, 1992, **33**, 1057.	979
92TL1295	H. Ito, T. Taguchi and Y. Hanzawa; *Tetrahedron Lett.*, 1992 **33**, 1295.	197
92TL1645	H. R. Sonawame, B. S. Nanjunidiah and G. M. Nazeruddin; *Tetrahedron Lett.*, 1992, **33**, 1645.	801
92TL1787	M. Ballestri, C. Chatgilialoglu, N. Cardi and A. Sommazzi; *Tetrahedron Lett.*, 1992, **33**, 1787.	933
92TL1855	C. W. Jefford, A. W. Sledeski, P. Lelandais and J. Boukouvalas; *Tetrahedron Lett.*, 1992, **33**, 1855.	133
92TL2299	J. S. Bajwa; *Tetrahedron Lett.*, 1992, **33**, 2299.	74
92TL2315	J. W. Grissom and T. L. Calkins; *Tetrahedron Lett.*, 1992, **33**, 2315.	1009
92TL2461	C. Chapuis; *Tetrahedron Lett.*, 1992, **33**, 2461.	789
92TL2589	Y. Sato, S. Watanabe and M. Shibasaki; *Tetrahedron Lett.*, 1992, 2589.	495
92TL2617	M. Lautens and R. K. Belter; *Tetrahedron Lett.*, 1992, **33**, 2617.	191
92TL2673	B. Giese, B. Carboni, T. Göbel, R. Muhn and F. Wetterich; *Tetrahedron Lett.*, 1992, **33**, 2673.	327
92TL2681	G. Linstrumelle, D. Chemin and M. Alami; *Tetrahedron Lett* , 1992, **33**, 2681.	1066
92TL2811	J. Gore, D. Bouyssi and G. Balme; *Tetrahedron Lett.*, 1992, **33**, 2811.	1020
92TL2851	A. Ali, D. C. Harrowven and G. Pattenden; *Tetrahedron Lett.*, 1992, **33**, 2851.	347
92TL3503	S. Torii, H. Okumoto, T. Kotani, S. Nakayasu and H. Ozaki; *Tetrahedron Lett.*, 1992, **33**, 3503.	1015
92TL3543	P. R. Jenkins, M. C. R. Symons, S. E. Booth and C. J. Swain; *Tetrahedron Lett.*, 1992, **33**, 3543.	338
92TL3563	J. Kant and V. Farina; *Tetrahedron Lett.*, 1992, 3563.	496
92TL3647	J. P. Quintard, I. Beaudet and J. L. Parrain; *Tetrahedron Lett.*, 1992, **33**, 3647.	1020
92TL3715	J. R. Hauske, P. Dorff , S. Julin , G. Martinelli and J. Bussolari; *Tetrahedron Lett.*, 1992, **33**, 3715.	1078
92TL3769	H. Ito, T. Nakamura, T. Taguchi and Y. Hanzawa; *Tetrahedron Lett.*, 1992, **33**, 3769.	197, 1057
92TL3899	C. A. M. Afonso, W. B. Motherwell, D. M. O'Shea and L. R. Roberts; *Tetrahedron Lett.*, 1992, **33**, 3899.	692
92TL4313	M. Mehmandoust, Y. Petit and M. Larchevêque; *Tetrahedron Lett.*, 1992, **33**, 4313.	837
92TL4401	P. A. Grieco and W. A. Carroll; *Tetrahedron Lett.*, 1992, **33**, 4401.	254
92TL4447	D. Dismaële and N. Champion; *Tetrahedron Lett.*, 1992, **33**, 4447.	980
92TL4563	D. S. Grierson, T. Skrydstrup, H. Audrain and G. Ulibarri; *Tetrahedron Lett.*, 1992, **33**, 4563.	818
92TL4725	T. G. Back and B. P. Dyck; *Tetrahedron Lett.*, 1992, **33**, 4725	609
92TL4735	P. A. Grieco, J. L. Collins and K. J. Henry, Jr.; *Tetrahedron Lett.*, 1992, **33**, 4735.	190, 402
92TL4815	P. G. Ciattini, E. Morera and G. Ortar; *Tetrahedron Lett.*, 1992, 4815.	489
92TL4843	S. A. Hitchcock and G. Pattenden; *Tetrahedron Lett.*, 1992, **31**, 4843.	347
92TL4985	R. F. de la Pradilla, M. B. Rubio, J. P. Marino and A. Viso; *Tetrahedron Lett.*, 1992, **33**, 4985.	600, 983
92TL4993	W. R. Bowman, D. N. Clark and R. J. Marmon; *Tetrahedron Lett.*, 1992, **33**, 4993.	950

92TL5017 D. H. R. Barton, C.-Y. Chern and J. C. Jaszberenyi; *Tetrahedron Lett.*, 1992, **33**, 5017. 333
92TL5261 J. Marco-Contelles, P. Ruiz, B. Sanchez and M. L. Jimenco; *Tetrahedron Lett.*, 1992, **33**, 5261. 175
92TL5311 J. K. Crandall and T. A. Ayers; *Tetrahedron Lett.*, 1992, 5311. 495
92TL5323 M. Paetow, H. Ahrens and D. Hoppe; *Tetrahedron Lett.*, 1992, **33**, 5323. 138
92TL5355 P. Pale, D. Grandjean and J. Chuche; *Tetrahedron Lett.*, 1992, **33**, 5355. 1018
92TL5363 M. Miura, K. Okuro, M. Furuune and M. Nomura; *Tetrahedron Lett.*, 1992, **33**, 5363. 1006
92TL5567 T. Fujisawa, B. I. Mobele and M. Shimizu; *Tetrahedron Lett.*, 1992, **33**, 5567. 21, 23
92TL5709 D. H. R. Barton, D. O. Jang and J. Cs. Jaszberenyi; *Tetrahedron Lett.*, 1992, **33**, 5709. 937, 938
92TL5757 T. Jeffery, S. Gueugnot and G. Linstrumelle; *Tetrahedron Lett.*, 1992, **33**, 5757. 121, 1002
92TL5795 K. Tomooka, T. Igarashi, M. Watanabe and T. Nakai; *Tetrahedron Lett.*, 1992, **33**, 5795. 405
92TL5913 J. L. Esker and M. Newcomb; *Tetrahedron Lett.*, 1992, **33**, 5913. 927, 950
92TL6187 W. B. Motherwell and D. A. Sandham; *Tetrahedron Lett.*, 1992, **33**, 6187. 133
92TL6193 J. S. Clark; *Tetrahedron Lett.*, 1992, **33**, 6193. 410, 820
92TL6367 Y. Kume and H. Ohta; *Tetrahedron Lett.*, 1992, **33**, 6367. 23
92TL6435 E. J. Corey and W.-C. Shieh; *Tetrahedron Lett.*, 1992, **33**, 6435. 116
92TL6461 A. De Nicola, J. Einhorn and J.-L. Luche; *Tetrahedron Lett.*, 1992, **33**, 6461. 166
92TL6607 E. Laborde; *Tetrahedron Lett.*, 1992, **33**, 6607. 285
92TL6629 D. H. R. Barton, D. O. Jang and J. Cs. Jaszberenyi; *Tetrahedron Lett.*, 1992, **33**, 6629. 939
92TL6673 E. Lee, D. S. Lee, Y. W. Choi and K. H. Lee; *Tetrahedron Lett.*, 1992, **33**, 6673. 4, 10
92TL6679 D. F. O'Keefe, M. C. Dannock and S. M. Marcuccio; *Tetrahedron Lett.*, 1992, 6679. 489
92TL6723 J. R. Falck and J. Yu; *Tetrahedron Lett.*, 1992, **33**, 6723. 183
92TL6783 T. Tanzawa and J. Schwartz; *Tetrahedron Lett.*, 1992, **33**, 6783. 568
92TL6935 H. M. L. Davies and N. J. S. Huby; *Tetrahedron Lett.*, 1992, **33**, 6935. 812
92TL7029 H. Tanaka, Y. Kameyama, S. Sumida and S. Torii; *Tetrahedron Lett.*, 1992, 7029. 496
92TL7181 T. Satoh, Y. Hayashi, Y. Mizu and K. Yamakawa; *Tetrahedron Lett.*, 1992, **33**, 7181. 598
92TL7307 W. Zhang and P. Dowd; *Tetrahedron Lett.*, 1992, **33**, 7307. 340
92TL7323 H. Von der Emde and R. Brückner; *Tetrahedron Lett.*, 1992, **33**, 7323. 406, 820
92TL7337 O. Cabon, M. Larchevêque, D. Buisson and R. Azerad; *Tetrahedron Lett.*, 1992, **33**, 7337. 8
92TL7429 C. Bonini, C. Giuliano, G. Righi and L. Rossi; *Tetrahedron Lett.*, 1992, **33**, 7429. 10
92TL7511 M. E. Maier and T. Brandstetter; *Tetrahedron Lett.*, 1992, **33**, 7511. 1038
92TL7521 K. Yamamoto, S. Yamada and K. Yamaguchi; *Tetrahedron Lett.*, 1992, **33**, 7521. 305
92TL7687 G. W. Kabalka, N. K. Reddy and C. Narayana; *Tetrahedron Lett.*, 1992, **33**, 7687. 168
92TL8017 D. Schinzer, J. Kabbara and K. Ringe; *Tetrahedron Lett.*, 1992, **33**, 8017. 979

93ACR219 K. U. Ingold, J. Lusztky and K. D. Raner; *Acc. Chem. Res.*, 1993, **23**, 219. 925
93ACR510 F. G. Bordwell and X.-M. Zhang; *Acc. Chem. Res.*, 1993, **26**, 510. 921
93ACS221 A. R. Martin and Y. Yang; *Acta Chem. Scand.*, 1993, **47**, 221. 489
93AG(E)547 O. Reiser; *Angew. Chem., Int. Ed. Engl.*, 1993, **32**, 547. 194
93AG(E)554 K. A. Hughes, P. G. Dopico, M. Sabat and M. G. Finn; *Angew. Chem., Int. Ed. Engl.*, 1993, **32**, 554. 991
93AG(E)566 P. von Matt and A. Pfaltz; *Angew. Chem., Int. Ed. Engl.*, 1993, **32**, 566. 194
93AG(E)1068 P. Wipf and S. Lim; *Angew. Chem., Int. Ed. Engl.*, 1993, **32**, 1068. 207
93AG(E)1187 J.-D. van Loon, P. Seiler and F. Diederich; *Angew. Chem., Int. Ed. Engl.*, 1993, **32**, 1187. 990
93AG(E)1315 M. Schäfer, N. Mahr, J. Wolf and H. Werner; *Angew. Chem., Int. Ed. Engl.*, 1993, **32**, 1315. 991
93AJC1321 R. F. C. Brown and R. P. McGeary; *Aust. J. Chem.*, 1993, **46**, 1321. 774
93BCJ819 N. Iwasawa, S. Hayakawa, M. Funahashi, K. Isobe and K. Narasako; *Bull. Chem. Soc. Jpn.*, 1993, **66**, 819. 929
93BCJ848 K. Kudo, Y. Hashimoto, H. Houchigai, M. Hasegama and K. Saigo; *Bull. Chem. Soc. Jpn.*, 1993, **66**, 848. 189
93BCJ987 Y. Wakatsuki, H. Yamazaki, N. Kumegawa and P. S. Johar; *Bull. Chem. Soc. Jpn.*, 1993, **66**, 987. 992
93BCJ2446 K. Sasaki, K. Nakao, Y. Kobayashi, M. Sakai, N. Uchino, Y. Sakakibari and K. Takagi; *Bull. Chem. Soc. Jpn.*, 1993, **66**, 2446. 468
93CB251 S. Arndt, G. Handke and N. Krause; *Chem. Ber.*, 1993, **126**, 251. 970
93CB261 N. Krause and S. Arndt; *Chem. Ber.*, 1993, **126**, 261. 970
93CB1077 M. Leise, H. Lang, W. Imhof and L. Zsolnai; *Chem. Ber.*, 1993, **126**, 1077. 956
93CB1171 U. Lüning and A. Kirsch; *Chem. Ber.*, 1993, **126**, 1171. 950
93CB1187 J. Hartung, B. Hertel and F. Trach; *Chem. Ber.*, 1993, **126**, 1187. 946
93CB1227 K. Brickmann and R. Brückner; *Chem. Ber.*, 1993, **126**, 1227. 406
93CB1447 T. Kauffmann, R. Sälker and K.-U. Voss; *Chem. Ber.*, 1993, **126**, 1447. 466
93CB1481 W. Adam, V. O. N. Salgudo, E.-M. Peters, K. Peters and H. G. von Schnering; *Chem. Ber.*, 1993, **126**, 1481. 566
93CB1623 E. Kloche, A. Matzcit, M. Gockeln and H. J. Schäfer; *Chem. Ber.*, 1993, **126**, 1623. 561
93CB1759 H. I. Tashtoush and R. Sustmann; *Chem. Ber.*, 1993, **126**, 1759. 937
93CC116 J. Vadecard, J.-C. Plaquevent, L. Duhamel and P. Duhamel; *J. Chem. Soc., Chem. Commun.*, 1993, 116. 573
93CC270 J. P. Dulcère, V. Agati and R. Faure; *J. Chem. Soc., Chem. Commun.*, 1993, 270. 955
93CC386 K. Tani, N. Ono, S. Okamoto and F. Sato; *J. Chem. Soc., Chem. Commun.*, 1993, 386. 457, 507
93CC424 D. Enders and S. Osborne; *J. Chem. Soc., Chem. Commun.*, 1993, 424. 106
93CC429 D. L. J. Clive and M. H. D. Postema; *J. Chem. Soc., Chem. Commun.*, 1993, 429. 244
93CC510 K. J. Henry and P. A. Grieco; *J. Chem. Soc., Chem. Commun.*, 1993, 510. 190
93CC614 S. Takano, M. Moriya and K. Ogasawara; *J. Chem. Soc., Chem. Commun.*, 1993, 614. 220

93CC617	A. D. Piscopio, N. Minowa, T. K. Chakraborty, K. Koide, P. Bertinato and K. C. Nicolaou; *J. Chem. Soc., Chem. Commun.*, 1993, 617.	106
93CC619	K. C. Nicolaou, P. Bertinato, A. D. Piscopio, T. K. Chakraborty and N. Minowa; *J. Chem. Soc., Chem. Commun.*, 1993, 619.	558, 828
93CC793	K. Shishido, K. Goto, A. Tsuda, Y. Takaishi and M. Shibuya; *J. Chem. Soc., Chem. Commun.*, 1993, 793.	827
93CC884	D. Milne and P. J. Murphy; *J. Chem. Soc., Chem. Commun.*, 1993, 884.	580, 842
93CC960	S. Augy-Dorey, D. H. Barton, S. D. Gero, B. Quiclet-Sire and I. Sagnard; *J. Chem. Soc., Chem. Commun.*, 1993, 960.	828
93CC1038	J.-P. Dulcère, V. Agati and J. Rodriguez; *J. Chem. Soc., Chem. Commun.*, 1993, 1038.	573
93CC1111	M. Penny and C. L. Willis; *J. Chem. Soc., Chem. Commun.*, 1993, 1111.	177
93CC1406	K. Toshima, K. Ohta, A. Ohtsuka, S. Matsumara and M. Nakata; *J. Chem. Soc., Chem. Commun.*, 1993, 1406.	956
93CC1595	H. Urabe, H. Inami and F. Sato; *J. Chem. Soc., Chem. Commun.*, 1993, 1595.	194
93CC1636	G. Pandey, K. S. S. P. Rao and B. B. V. S. Sekhar; *J. Chem. Soc., Chem. Commun.*, 1993, 1636.	943
93CC1682	J. Clayden and M. Julia; *J. Chem. Soc., Chem. Commun.*, 1993, 1682.	484
93CI(L)226	G. P. Miller; *Chem. Ind.* (*London*), 1993, 226.	890
93CL545	N. Iwasawa, M. Funahashi, S. Hayakawa and K. Narasaka; *Chem. Lett.*, 1993, 545.	929
93CL1397	K. Fujimoto, H. Sakai and T. Nakai; *Chem. Lett.*, 1993, 1397.	816
93CL1429	T. Moriya, N. Miyaura and A. Suzuki; *Chem. Lett.*, 1993, 1429.	489
93CM357	I. Kminek, J. Klimovic and P. N. Prasad; *Chem. Mater.*, 1993, **5**, 357.	986
93CPB55	H. Akita, I. Umezawa, M. Takano, C. Ohyama, H. Matsukura and T. Oishi; *Chem. Pharm. Bull.*, 1993, **41**, 55.	212
93CRV615	J. Cornelisse; *Chem. Rev.*, 1993, **93**, 615.	364, 365
93CRV1307	A. H. Hoveyda, D. A. Evans and G. C. Fu; *Chem. Rev.*, 1993, **93**, 1307.	299
93CRV1371	C. Chuit, R. J. P. Corriu, C. Reye and J. C. Young; *Chem. Rev.*, 1993, **93**, 1371.	490
93CRV2023	C. J. Li; *Chem. Rev.*, 1993, **93**, 2023.	826
93CRV2091	P. Dowd and W. Zhang; *Chem. Rev.*, 1993, **93**, 2091.	340, 919, 928, 929, 931
93CRV2117	P. Knochel and R. D. Singer; *Chem. Rev.*, 1993, **93**, 2117.	467
93CRV2779	R. J. Ferrier and S. Middleton; *Chem. Rev.*, 1993, **93**, 2779.	341
93CRV2933	M. Speranza; *Chem. Rev.*, 1993, **93**, 2933.	1142
93CSR347	D. C. Nonhebel; *Chem. Soc. Rev.*, 1993, **22**, 347.	919, 928
93H(35)349	J. E. Macor and J. M. Wehner; *Heterocycles*, 1993, **35**, 349.	200
93HCA418	M. Tarkoy, M. Bolli, B. Schweizer and C. Leumann; *Helv. Chim. Acta*, 1993, **76**, 481.	797
93HCA441	A. Veit, R. Lenz, M.-E. Seiler, M. Neuenbuger, M. Zehnder and B. Giese; *Helv. Chim. Acta*, 1993, **76**, 441.	327
93JA87	E. Lastra and L. S. Hegedus; *J. Am. Chem. Soc.*, 1993, **115**, 87.	165
93JA464	T. Toru, Y. Watanabe, M. Tsusaka and Y. Ueno; *J. Am. Chem. Soc.*, 1993, **115**, 464.	333
93JA1177	F. G. West and B. N. Naidu; *J. Am. Chem. Soc.*, 1993, **115**, 1177.	400
93JA2205	H. E. Zimmerman and P. A. Wang; *J. Am. Chem. Soc.*, 1993, **115**, 2205.	96
93JA2268	W. R. Roush and B. B. Brown; *J. Am. Chem. Soc.*, 1993, **115**, 2268.	788
93JA2581	F. E. Ziegler, C. A. Metcalf III, A. Nangia and G. Schulte; *J. Am. Chem. Soc.*, 1993, **115**, 2581.	176
93JA2970	J. D. White and M. S. Jensen; *J. Am. Chem. Soc.*, 1993, **115**, 2970.	216
93JA3328	S. Kim, G. H. Joe and J. Y. Do; *J. Am. Chem. Soc.*, 1993, **115**, 3328.	932
93JA3362	M. Yamaguchi, A. Hayashi and M. Hirama; *J. Am. Chem. Soc.*, 1993, **115**, 3362.	1051
93JA3776	A. Padwa, M. A. Filipkowski, M. Meske, S. H. Watterson and Z. Ni; *J. Am. Chem. Soc.*, 1993, **115**, 3776.	980
93JA3846	K. S. Feldman and C. M. Kraebel; *J. Am. Chem. Soc.*, 1993, **115**, 3846.	1077
93JA3855	B. Mudryk and T. Cohen; *J. Am. Chem. Soc.*, 1993, **115**, 3855.	236
93JA4413	D. H. McConville, J. R. Wolf and R. R. Schrock; *J. Am. Chem. Soc.*, 1993, **115**, 4413.	675
93JA5815	M. S. Congreve, A. B. Holmes, A. B. Hughes and M. G. Looney; *J. Am. Chem. Soc.*, 1993, **115**, 5815.	829
93JA5865	T. Mandai, J. Tsuji and Y. Tsujiguchi; *J. Am. Chem. Soc.*, 1993, **115**, 5865.	975
93JA6051	D. P. Curran and W. Shen; *J. Am. Chem. Soc.*, 1993, **115**, 6051.	339, 927
93JA6438	A. J. M. Caffyn and K. M. Nicholas; *J. Am. Chem. Soc.*, 1993, **115**, 6438.	200
93JA6609	Y. I. M. Nilsson, P. H. Andersson and J.-E. Bäckvall; *J. Am. Chem. Soc.*, 1993, **115**, 6609.	192
93JA6614	A. F. Houri, M. T. Didiuk, Z. Xu, N. R. Horan and A. H. Hoveyda; *J. Am. Chem. Soc.*, 1993, **115**, 6614.	316
93JA6652	A. I. Meyers and G. Milot; *J. Am. Chem. Soc.*, 1993, **115**, 6652.	151
93JA6997	J. P. Morken, M. T. Didiuk and A. H. Hoveyda; *J. Am. Chem. Soc.*, 1993, **115**, 6997.	316
93JA7033	N. Sakai, S. Mano, K. Nozaki and H. Takaya; *J. Am. Chem. Soc.*, 1993, **115**, 7033.	297
93JA7240	S. Hollenstein and T. Laube; *J. Am. Chem. Soc.*, 1993, **115**, 7240.	899
93JA8050	D. H. R. Barton, J. Cs. Jaszberenyi, E. A. Theodorakis and J. H. Reibenspies; *J. Am. Chem. Soc.*, 1993, **115**, 8050.	944
93JA8485	N. Suzuki, D. Y. Kondakov and T. Takahashi; *J. Am. Chem. Soc.*, 1993, **115**, 8485.	197
93JA8585	G. Thoma, D. P. Curran, S. V. Geib, B. Giese, W. Damm and F. Wetterich; *J. Am. Chem. Soc.*, 1993, **115**, 8585.	327
93JA8835	H. Ito, Y. Motoki, T. Taguchi and Y. Hanzawa; *J. Am. Chem Soc.*, 1993, **115**, 8835.	197
93JA8847	R. L. Funk, J. B. Stallman and J. A. Wos; *J. Am. Chem. Soc.*, 1993, **115**, 8847.	831
93JA9276	B. H. Lipshutz, K. Siegmann and E. Garcia; *J. Am. Chem. Soc.*, 1993, **115**, 9276.	473
93JA9421	B. M. Trost and Y. Shi; *J. Am. Chem. Soc.*, 1993, **115**, 9421.	494

93JA9806 B. K. Carpenter; *J. Am. Chem. Soc.*, 1993, **115**, 9806. 1127
93JA10400 R. A. Robinson, J. S. Clark and A. B. Holmes; *J. Am. Chem. Soc.*, 1993, **115**, 10 400. 829
93JA10464 T. Toru, Y. Watanabe, M. Tsusaka and Y. Ueno; *J. Am. Chem. Soc.*, 1993, **115**, 10 464. 946
93JA10671 B. A. Anderson, J. Bao, T. A. Brandvold, C. A. Challener, W. D. Wulff, Y.-C. Xu and A. L. Rheingold; *J. Am. Chem. Soc.*, 1993, **115**, 10 671. 551
93JCS(D)2629 P. D. Beer, O. Kocian, R. J. Mortimer and C. Ridgway; *J. Chem. Soc., Dalton Trans.*, 1993, 2629. 1137
93JCS(P1)9 S.-K. Kang, S.-G. Kim, D.-C. Park, J.-S. Lee, W.-J. Yoo and C. S. Pah; *J. Chem. Soc., Perkin Trans.* 1, 1993, 9. 15, 16
93JCS(P1)667 A. Srikrishna and K. Krishnan; *J. Chem. Soc., Perkin Trans. 1*, 1993, 667. 829
93JCS(P1)681 D. W. Jones and R. J. Marmon; *J. Chem. Soc., Perkin Trans. 1*, 1993, 681. 403
93JCS(P1)987 M. Ito, Y. Katsuka, Y. Yamano and K. Tsukida; *J. Chem. Soc., Perkin Trans. 1*, 1993, 987. 613
93JCS(P1)1275 I. Coldham; *J. Chem. Soc., Perkin Trans. 1*, 1993, 1275. 407
93JCS(P1)1849 J. Clayden, E. W. Collington, J. Elliot, S. J. Martin, A. B. McElroy, S. Warren and D. Waterson; *J. Chem. Soc., Perkin Trans. 1.*, 1993, 1849. 836
93JCS(P1)1943 S. Sengupta and S. Bhattacharya; *J. Chem. Soc., Perkin Trans. 1*, 1993, 1943. 486
93JCS(P1)2339 S. Ozaki, H. Matsushita and H. Ohmori; *J. Chem. Soc., Perkin Trans. 1*, 1993, 2339. 937
93JCS(P1)2417 H. Togo, M. Aoki, T. Kuramochi and M. Yokoyama; *J. Chem. Soc., Perkin Trans. 1*, 1993, 2417. 934
93JCS(P1)2913 J. Clayden and S. Warren; *J. Chem. Soc. Perkin Trans. 1.*, 1993, 2913. 836
93JCS(P1)3007 K. Kobayashi, H. Minakawa, H. Sakurai, S. Kujime and H. Suginome; *J. Chem. Soc., Perkin Trans. 1*, 1993, 3007. 929
93JCS(P1)3099 R. Armer, M. J. Begley, P. J. Cox, A. Persad and N. S. Simpkins; *J. Chem. Soc., Perkin Trans. 1*, 1993, 3099. 144
93JCS(P2)1617 S. M. Jeffery and C. J. M. Stirling; *J. Chem. Soc., Perkin Trans. 2*, 1993, 1617. 145
93JCS(P2)2163 S. M. Jeffery and C. J. M. Stirling; *J. Chem. Soc., Perkin Trans. 2*, 1993, 2163. 145
93JHC873 A. D. Woolhouse, G. J. Gainsford and D. R. Crump; *J. Heterocycl. Chem.*, 1993, **30**, 873. 784
93JOC11 E. Piers and J. Renaud; *J. Org. Chem.*, 1993, **58**, 11. 452
93JOC16 Y. Tsuji, N. Yamada and S. Tanaka; *J. Org. Chem.*, 1993, **58**, 16. 192
93JOC43 L. R. Collazo and F. S. Guziec Jr.; *J. Org. Chem.*, 1993, **58**, 43. 685, 686
93JOC249 C. Chatgilialoglu, C. Ferreri and M. Lucarini; *J. Org. Chem.*, 1993, **58**, 249. 17
93JOC322 R. L. Danheiser, Y. M. Choi, M. Menichincheri and E. J. Stoner; *J. Org. Chem.*, 1993, **58**, 322. 601, 781, 785
93JOC661 X. Fu and J. M. Cook; *J. Org. Chem.*, 1993, **58**, 661. 810
93JOC823 P. Beak and E. K. Yum; *J. Org. Chem.*, 1993, **58**, 823. 151
93JOC1083 A. Koch, C. Lamberth, F. Wetterich and B. Giese; *J. Org. Chem.*, 1993, **58**, 1083. 923
93JOC1109 P. Beak and W. K. Lee; *J. Org. Chem.*, 1993, **58**, 1109. 151
93JOC1207 T. Ibuka, T. Taga, H. Habashita, K. Nakai, H. Tamamura, N. Fujii, Y. Chounan, H. Nemoto and Y. Yamamoto; *J. Org. Chem.*, 1993, **58**, 1207. 189
93JOC1215 P. Galatsis, S. D. Millan and T. Faber; *J. Org. Chem.*, 1993, **58**, 1215. 929
93JOC1298 K. A. Reynolds, P. G. Dopico, M. J. Sundermann, K. A. Hughes and M. G. Finn; *J. Org. Chem.*, 1993, **58**, 1298. 991
93JOC1349 M. Tiecco, L. Testaferri, M. Tingoli and F. Marini; *J. Org. Chem.*, 1993, **58**, 1349. 782
93JOC1538 S.-I. Murahashi, Y. Imada, Y. Taniguchi and S. Higashiura; *J. Org. Chem.*, 1993, **58**, 1538. 192
93JOC1579 B. M. Trost and J. R. Parquette; *J. Org. Chem.*, 1993, **58**, 1579. 595, 598
93JOC1651 Y. W. Andemichael, Y. Huang and K. K. Wang; *J. Org. Chem.*, 1993, **58**, 1651. 497, 1057
93JOC1949 M.-Z. Deng, N.-S. Li and Y.-Z. Huang; *J. Org. Chem.*, 1993, **58**, 1949. 488
93JOC2115 S. A. Glover, Jr. and J. Warkentin; *J. Org. Chem.*, 1993, **58**, 2115. 339
93JOC2181 G. Solladié and C. Ziani-Chérif; *J. Org. Chem.*, 1993, **58**, 2181. 210
93JOC2186 B. Lei and A. G. Fallis; *J. Org. Chem.*, 1993, **58**, 2186. 774
93JOC2201 T. Oh-e, N. Miyaura and A. Suzuki; *J. Org. Chem.*, 1993, **58**, 2201. 480, 482, 489
93JOC2349 J. F. Devaux, I. Hanna, J. Y. Lallemand and T. Prangé; *J. Org. Chem.*, 1993, **58**, 2349. 810
93JOC2468 L. E. Overman, T. A. Grese and K. D. Hutchinson; *J. Org. Chem.*, 1993, **58**, 2468. 1039
93JOC2483 R. D. Rieke, D. E. Stack, B. T. Dawson and T.-C. Wu; *J. Org. Chem.*, 1993, **58**, 2483. 121
93JOC2492 R. D. Rieke, W. R. Klein and T.-C. Wu; *J. Org. Chem.*, 1993, **58**, 2492. 121
93JOC2501 E. N. Lawson, W. Kitching, C. H. L. Kennard and K. Y. Byriel; *J. Org. Chem.*, 1993, **58**, 2501. 572
93JOC2694 A. R. Sidduri, M. J. Rozema and P. Knochel; *J. Org. Chem.*, 1993, **58**, 2694. 1004
93JOC2894 J. Marco-Contelles, P. Ruiz-Fernández and B. Sánchez; *J. Org. Chem.*, 1993, **58**, 2894. 175
93JOC3038 A. Katritzky, J. Li and M. F. Gordeev; *J. Org. Chem.*, 1993, **58**, 3038. 956
93JOC3697 N. Komatsu, T. Murakami, Y. Nishibayashi, T. Sugita and S. Uemura; *J. Org. Chem.*, 1993, **58**, 3697. 983
93JOC3767 K. Burgess, K. K. Ho and C. Y. Ke; *J. Org. Chem.*, 1993, **58**, 3767. 185
93JOC4187 Y. Fukumoto, N. Chatani and S. Murai; *J. Org. Chem.*, 1993, **58**, 4187. 212
93JOC4237 A. H. Hoveyda and J. P. Morken; *J. Org. Chem.*, 1993, **58**, 4237. 316
93JOC4245 L. A. Paquette and M. A. Poupart; *J. Org. Chem.*, 1993, **58**, 4245. 810
93JOC4298 C. Aubert, J.-P. Gotteland and M. Malacria; *J. Org. Chem.*, 1993, **58**, 4298. 362
93JOC4376 A. R. Katritzky, L. Xie and W. Q. Fan; *J. Org. Chem.*, 1993, **58**, 4376. 262
93JOC4469 R. B. Bates and S. R. Taylor; *J. Org. Chem.*, 1993, **58**, 4469. 178
93JOC4589 J. C. McKew and M. J. Kurth; *J. Org. Chem.*, 1993, **58**, 4589. 831
93JOC4685 Y. Gimbert, A. Moradpour, G. Dive, D. Dehareng and K. Lahlil; *J. Org. Chem.*, 1993, **58**, 4685. 570
93JOC4866 K. Park, K. Yuan and W. J. Scott; *J. Org. Chem.*, 1993, **58**, 4866. 465

93JOC4890 F. A. Davis, P. Zhou and P. J. Carroll; *J. Org. Chem.*, 1993, **58**, 4890. 147

93JOC4897 T. Harada, T. Katsuhira, D. Hara, Y. Kotani, K. Maejima, R. Kaji and A. Oku; *J. Org. Chem.*, 1993, **58**, 4897. 431

93JOC4933 J. L. Esker and M. Newcomb; *J. Org. Chem.*, 1993, **58**, 4933 950

93JOC4952 L. A. Paquette and R. C. Thompson; *J. Org. Chem.*, 1993, **58**, 4952. 810

93JOC5028 S.-H. Chen, S. Huang, J. Kant, C. Fairchild, J. Wei and V. Farina; *J. Org. Chem.*, 1993, **58**, 5028. 939

93JOC5095 L. G. Beholz and J. R. Stille; *J. Org. Chem.*, 1993, **58**, 5095. 835

93JOC5163 T. Okano, T. Shimizu, K. Sumida and S. Eguchi; *J. Org. Chem.*, 1993, **58**, 5163. 569

93JOC5482 A. Balakumar, S. Janardhanam and K. Rajagopalan; *J. Org. Chem.*, 1993, **58**, 5482. 810

93JOC5574 W.-W. Weng and T.-Y. Luh; *J. Org. Chem.*, 1993, **58**, 5574. 1127

93JOC5702 J.-P. Dulcère, J. Crandall, R. Faure, M. Santelli, V. Agati, and M. N. Mihcubi; *J. Org. Chem.*, 1993, **58**, 5702. 955

93JOC5709 J.-P. Dulcère, M. N. Mihoubi and J. Rodriguez; *J. Org. Chem.*, 1993, **58**, 5709. 955

93JOC5873 R. L. Funk, T. A. Olmstead, M. Parvez and J. L. Stallman; *J. Org. Chem.*, 1993, **58**, 5873. 831

93JOC5892 J. Yu, H. S. Cho and J. R. Falck; *J. Org. Chem.*, 1993, **58**, 5892. 179, 183

93JOC5900 P. R. Brooks, R. Bishop, D. C. Craig, M. L. Scudder and J. A. Counter; *J. Org. Chem.*, 1993, **58**, 5900. 685, 686

93JOC5976 J. Barluenga, J. M. Montserrat and J. Florez; *J. Org. Chem.*, 1993, **58**, 5976. 466

93JOC6067 F. G. Bordwell, X.-M. Zhang and R. Filler; *J. Org. Chem.*, 1993, **58**, 6067. 921

93JOC6166 T. Katsuhira, T. Harada, K. Maejima, A. Osada and A. Oku; *J. Org. Chem.*, 1993, **58**, 6166. 198, 975

93JOC6174 D. B. Berkowitz, M. Eggen, Q. Shen and D. G. Sloss; *J. Org. Chem.*, 1993, **58**, 6174. 178

93JOC6217 B. B. Snider and B. A. McCarthy; *J. Org. Chem.*, 1993, **58**, 6217. 346

93JOC6255 J. C. Gilbert and R. D. Selliah; *J. Org. Chem.*, 1993, **58** 6255. 831

93JOC6538 A. I. Meyers and G. Milot; *J. Org. Chem.*, 1993, **58**, 6538. 151

93JOC6541 M. Hojo, H. Harada, J. Yoshizawa and A. Hosomi; *J. Org. Chem.*, 1993, **58**, 6541. 438

93JOC6716 E. Ghera, T. Yechezkel and A. Hassner; *J. Org. Chem.*, 1993, **58**, 6716. 360

93JOC6838 D. H. R. Barton, D. O. Jang and J. Cs. Jaszberenyi; *J. Org. Chem.*, 1993, **58**, 6838. 937, 938, 945

93JOC6851 K. S. Feldman, H. M. Berven, A. L. Romanelli and M. Parvez; *J. Org. Chem.*, 1993, **58**, 6851. 949

93JOC6857 D. L. J. Clive, H. W. Manning, T. L. B. Boivin and M. H. D. Postema; *J. Org. Chem.*, 1993, **58**, 6857. 942

93JOC7100 S. Eldin, D. L. Whalen and R. M. Pollack; *J. Org. Chem.*, 1993, **58**, 7100. 878

93JOC7180 J. A. Marshall and K. G. Pinney; *J. Org. Chem.*, 1993, **58**, 7180. 966

93JOC7228 B. B. Snider, N. H. Vo and B. M. Foxman; *J. Org. Chem.*, 1993, **58**, 7228. 929

93JOC7238 M. Bergdahl, M. Eriksson, M. Nilsson and T. Olsson; *J. Org. Chem.*, 1993, **58**, 7238. 311

93JOC7304 S. K. Taylor, J. A. Fried, Y. N. Grassl, A. E. Marolewski, E. A. Pelton, T.-J. Poel, D. S. Rezanka and M. R. Whittaker; *J. Org. Chem.*, 1993, **58**, 7304. 209

93JOC7507 L. Snyder and A. I. Meyers; *J. Org. Chem.*, 1993, **58**, 7507. 134

93JOC7610 E. Baciocchi and E. Muraglia; *J. Org. Chem.*, 1993, **58**, 7610. 926

93JOC7640 Q. Zhang, R. M. Mohan, L. Cook, S. Kazanis, D. Peisach, B. M. Foxman and B. B. Snider; *J. Org. Chem.*, 1993, **58**, 7640. 346, 926

93JOC7751 A. Srikrishna and K. Krishnan; *J. Org. Chem.*, 1993, **58** 7751. 826

93JOM(451)1 A. Llebaria and J. M. Moretó; *J. Organomet. Chem.*, 1993, **451**, 1. 106

93JOM(451)15 J. Tsuji and T. Mandai; *J. Organomet. Chem.*, 1993, **451**, 15. 975

93JOM(451)C1 D. J. Brondani, R. J. P. Corriu, S. E. Ayoubi, J. J. E. Moreau and M. W. C. Man; *J. Organomet. Chem.*, 1993, **451**, C1. 160

93JOM(460)127 S. Gronowitz, P. Björk, J. Malm and A.-B. Hörnfeldt; *J. Organomet. Chem.*, 1993, **460**, 127. 480

93LA521 N. Krause; *Liebigs Ann. Chem.*, 1993, 521. 970

93LA1133 K. Weinges, R. Braun, U. Huber-Patz and H. Irngartinger; *Liebigs Ann. Chem.*, 1993, 1133. 782

B-93MI 101-01 N. S. Simpkins; "Sulphones in Organic Synthesis," Pergamon Oxford, 1993. 24

B-93MI 104-01 N. S. Simpkins; in "Sulphones in Organic Synthesis, Tetrahedron Organic Chemistry Series," eds J. E. Baldwin and P. D. Magnus; Pergamon, Oxford, 1993, vol. 10, Chapter 3, pp. 100–182. 140, 146

B-93MI 105-01 N. S. Simpkins; "Sulphones in Organic Synthesis," Pergamon. Oxford, 1993. 230

93MI 107-01 C. Botteghi, S. Paganelli, A. Schionato and M. Marchetti; *Chirality*, 1991, **3**, 335. 297, 414

93MI 109-01 S. Pereira and M. Srebnik; *Aldrichimica Acta*, 1993, **26**, 17. 417

B-93MI 114-01 N. S. Simpkins; "Sulphones in Organic Synthesis," Pergamon. Oxford, 1993, ch 7. 612, 617

B-93MI 115-01 N. S. Simpkins; "Sulphones in Organic Synthesis," Pergamon Press, Oxford, 1993, pp. 273–284. 704

B-93MI 115-02 N. S. Simpkins; "Sulphones in Organic Synthesis," Pergamon Press, Oxford, 1993, pp. 254–273. 696

93MI 118-01 S. Pereira and M. Srebnik; *Aldrichimica Acta*, 1993, **26**, 17. 830

93MI 119-01 S. Pereira and M. Srebnik; *Aldrichimica Acta*, 1993, **26**, (1), 17. 876

93OM473 L.-B. Han, N. Kambe, A. Ogawa, I. Ryu and N. Sonoda; *Organometallics*, 1993, **12**, 473. 247

93OM1871 H. Arzoumanian, F. Cochini, D. Nuel and N. Rosas; *Organometallics*, 1993, **12**, 1871. 976

93OM3971 I.-Y. Wu, J.-H. Tsai, B.-C. Huang, S.-C. Chen and Y. C. Lin; *Organometallics*, 1993, **12**, 3971. 991

93OR(43)1 S. H. Pine; *Org. React.*, 1993, **43**, 1. 766

93OR(43)93 S. R. Wilson; *Org. React.*, 1993, **43**, 93. 400, 401, 803, 805

93OS(72)48 R. C. Sun and M. Okabe; *Org. Synth.*, 1993, **72**, 48. 16

93OS(72)199 J. Light and R. Breslow; *Org. Synth.*, 1993, **72**, 199. 936

93OS(72)216	J. R. McCarthy, D. P. Matthews and J. P. Paolini; *Org. Synth.*, 1993, **72**, 216.	440
93OSC(8)210	K. Narasaka; *Org. Syn. Coll. Vol.*, 1993, **8**, 210.	582
93PAC1153	D. P. Curran, J. Sisko, P. E. Yeske and H. Liu; *Pure Appl. Chem.*, 1993, **65**, 1153.	106
93PHC1	R. A. Aitken, I. Gosney and J. I. G. Cadogan; *Prog. Heterocycl. Chem.*, 1993, **5**, 1.	790
93S165	R. Zimmer; *Synthesis*, 1993, 165.	458, 955
93S250	A. I. Meyers, W. Schmidt and M. J. McKennon; *Synthesis*, 1993, 250.	473
93S263	A. Albini; *Synthesis*, 1993, 263.	782
93S441	A. Pommier and J. M. Pons; *Synthesis*, 1993, 441.	565, 681, 683, 785
93S460	Y. Tobe. K. Kakiuchi, K. Kobiro and Y. Odaira; *Synthesis*, 1993, 460.	561, 569
93S558	R. Gleiter, R. Merger, B. Treptow, W. Wittwer and G. Pflästerer; *Synthesis*, 1993, 558.	1001
93S615	J. Mulzer and M. Scharp; *Synthesis*, 1993, 615.	829
93S643	M. Sodeoka and M. Shibasaki; *Synthesis*, 1993, 643.	456
93S659	J.-L. Ripoll and Y. Vallee; *Synthesis*, 1993, 659.	772
93S729	J. Gonda, A. C. Helland, B. Ernst and D. Bellus; *Synthesis*, 1993, 729.	837
93S735	K. Ritter; *Synthesis*, 1993, 735.	433, 480
93S824	B. M. Trost and V. K. Chang; *Synthesis*, 1993, 824.	675
93S1092	B. B. Lohray and D. Enders; *Synthesis*, 1993, 1092.	186
93SC311	S. Janardhanam, P. Shanmugam and K. Rajagopalan; *Synth. Commun.*, 1993, **23**, 311.	810
93SC811	N. A. Abood; *Synth. Commun.*, 1993, **23**, 811.	210
93SL37	T. Takahashi and M. Nakazawa; *Synlett*, 1993, 37.	1023
93SL45	G. Cahiez and S. Marquais, *Synlett*, 1993, 45.	125, 465
93SL105	P. Rochet, J. M. Vatèle and J. Goré; *Synlett*, 1993, 105.	955, 971
93SL143	I. Ryu, M. Hasegawa, A. Kurihara, A. Ogawa, S. Tsunoi and N. Sonoda; *Synlett*, 1993, 143.	936
93SL171	B. Bennetau and J. Dunoguès; *Synlett*, 1993, 171.	447, 447
93SL189	R. Angell, P. J. Parsons and A. Naylor; *Synlett*, 1993, 189.	115
93SL217	M. Alami, B. Crousse, G. Linstrumelle, L. Mambu and M. Larchevêque; *Syn. Lett.*, 1993, 217.	1009
93SL219	R. F. W. Jackson, N. Wishart and M. J. Wythes; *Synlett*, 1993, 219.	968
93SL277	J. J. Eshelby, P. J. Crowley and P. J. Parsons; *Synlett*, 1993, 277.	227
93SL279	J. J. Eshelby, P. J. Crowley and P. J. Parsons; *Synlett*, 1993, 279.	227
93SL283	M. Harendza, K. Lessmann and W. P. Neumann; *Synlett.*, 1993, 283.	936
93SL286	M. Harendza, J. Junggebauer, K. Lessmann, W. P. Neumann and H. Tews; *Synlett*, 1993, 286.	925
93SL323	S. G. Davies and T. J. Donohoe; *Synlett*, 1993, 323.	202, 203
93SL413	M. Isobe, T. Tsukiyama and S. C. Peters; *Syn. Lett.*, 1993, 413.	1026
93SL499	M. J. Dunn, R. F. W. Jackson, J. Pietruszka, N. Wishart, D. Ellis and M. J. Wythes; *Synlett*, 1993, 499.	968
93SL527	K. Tomooka, K. Ishikawa and T. Nakai; *Synlett*, 1993, 527.	818
93SL568	A. Jutand and A. Mosleh; *Synlett*, 1993, 568.	483
93SL599	D. J. Aitken, F. Vergne, A. S. Phimmanao, D. Guillaume and H.-P. Husson; *Synlett*, 1993, 599.	224
93SL769	E. Tyrrell, P. Heshmati and L. Sarrazin; *Synlett*, 1993, 769.	199
93SL774	I. Paterson, K.-S. Yeung and J. B. Smaill; *Synlett*, 1993, 774.	640
93SL827	K. S. Feldman and H. M. Berven; *Synlett.*, 1993, 827.	949
93SL829	J. H. Rigby and A. C. Krueger; *Synlett*, 1993, 829.	145
93SL837	M. Knecht and W. Boland; *Synlett*, 1993, 837.	677
93SL839	J.-B. Baudin, M.-G. Commenil, S. A. Julia, L. Toupet and Y. Wang; *Synlett*, 1993, 839.	622
93SL875	H. R. Sonawane, N. S. Bellur, D. G. Kulkarni and J. R. Ahuja; *Synlett.*, 1993, 875.	801, 802
93SL931	P. J. Parsons and M. Stefinovic; *Synlett*, 1993, 931.	993
93SL1133	S. Casson and P. Kocienski; *Synlett*, 1993, 1133.	455
93T29	P. Knochel and C. J. Rao; *Tetrahedron*, 1993, **49**, 29.	468, 471
93T235	C. Dell'Erba, M. Novi, G. Petrillo and C. Tavani; *Tetrahedron*, 1993, **49**, 235.	486
93T537	T. Keumi, K. Matsuura, N. Nakayama, T. Tsubota, T. Morita, I. Takahashi and H. Kitajima; *Tetrahedron*, 1993, **49**, 537.	156
93T10175	M. Schlosser, O. Desponds, R. Lehmann, E. Moret and G. Rauchschwalbe; *Tetrahedron*, 1993, **49**, 10175.	113
93T1025	S. Y. Wei, K. Tomooka and T. Nakai; *Tetrahedron*, 1993, **49**, 1025.	810
93T1327	D. Guijarro, B. Mancheño and M. Yus; *Tetrahedron*, 1993, **49**, 1327.	136
93T1807	A. I. Meyers and R. H. Hutchings; *Tetrahedron*, 1993, **49**, 1807.	149
93T2093	G. Desimoni, G. Faita, S. Gatti Comini, P. P. Righetti and G. Tacconi; *Tetrahedron*, 1993, **49**, 2093.	826
93T3007	A. Pelter, G. F. Vaughan-Williams and R. M. Rosser; *Tetrahedron*, 1993, **49**, 3007.	212
93T3131	P. Beslin and S. Pierre; *Tetrahedron*, 1993, **49**, 3131.	834
93T4147	M. P. Trova and Y. Wang; *Tetrahedron*, 1993, **49**, 4147.	134
93T4789	J. W. Patterson; *Tetrahedron*, 1993, **49**, 4789.	829
93T4821	D. P. Curran and A. C. Abraham; *Tetrahedron*, 1993, **49**, 4821.	327, 328
93T4841	D. P. Curran and P. S. Ramamoorthy; *Tetrahedron*, 1993, **49**, 4841.	327, 328
93T5177	R. Grigg, V. Santhakumar, V. Sridharan, M. Thornton-Pett and A. W. Bridge; *Tetrahedron*, 1993, **49**, 5177.	837
93T5189	N. Garg, J. Plavec and J. Chattopadhyaya; *Tetrahedron*, 1993, **49**, 5189.	946
93T5203	H. M. L. Davies; *Tetrahedron*, 1993, **49**, 5203.	811
93T6419	D. P. G. Hamon, R. J. Massy-Westropp and P. Razzino; *Tetrahedron*, 1993, **49**, 6419.	328

93T6717 J. Bigorra, J. Font, C. O. de Echagüen and R. M. Ortuno; *Tetrahedron*, 1993, **49**, 6717. 579

93T6821 Z. Zhang-Lin, H. Yao-Zeng and S. Li-Lan; *Tetrahedron*, 1993, **49**, 6821. 766

93T7657 S. Harusawa, S. Takemura, H. Osaki, R. Yoneda and T. Kurihara; *Tetrahedron*, 1993, **49**, 7657. 836

93T7967 D. M. X. Donnelly, B. M. Fitzpatrick, B. A. O'Reilly and J.-P. Finet; *Tetrahedron*, 1993, **49**, 7967. 780

93T8433 M. Garcia, C. del Campo, E. F. Llama, J. M. Sanchez-Montero and J. V. Sinisterra; *Tetrahedron*, 1993, **49**, 8433. 11, 12

93T9079 N. Hossain, J. Plavec, C. Thibaudeau and J. Chattopadhyaya; *Tetrahedron*, 1993, **49**, 9079. 946

93T9479 T. Ibuka, K. Nakai, H. Habashita, K. Bessho, N. Fujii, Y. Chounan and Y. Yamamoto; *Tetrahedron*, 1993, **49**, 9479. 189

93T9775 K. Nunn, P. Mosset, R. Grée and R. W. Saalfrank; *Tetrahedron*, 1993, **49**, 9775. 966

93TA21 T. Honda, N. Kimura and M. Tsubuki; *Tetrahedron Asymmetry*, 1993, **4**, 21. 558

93TA399 B. Achmatowicz and J. Wicha; *Tetrahedron Asymmetry*, 1993, **4**, 399. 210

93TA625 D. W. Knight, N. Lewis, A. C. Share and D. Haigh; *Tetrahedron Asymmetry*, 1993, **4**, 625. 16

93TA879 A. K. Gupta and R. J. Kazlauskas; *Tetrahedron Asymmetry*, 1993, **4**, 879. 573

93TA1521 D. M. Rotstein and K. A. M. Walker; *Tetrahedron Asymmetry*, 1993, **4**, 1521. 223

93TA1795 S. Takano, Y. Sugihara and K. Ogasawara; *Tetrahedron Asymmetry*, 1993, **4**, 1795. 581

93TA1957 A. Q. Mi, Z. Y. Wang and Y. Z. Jiang; *Tetrahedron Asymmetry*, 1993, **4**, 1957. 135

93TL15 H. C. Brown, U. R. Khire and U. S. Racherla; *Tetrahedron Lett.*, 1993, **34**, 15. 1055

93TL63 A. B. Smith, III, R. C. Holcomb, M. C. Guzman, T. P. Keenan, P. A. Sprengeler and R. Hirschmann; *Tetrahedron Lett.*, 1993, **34**, 63. 166

93TL153 R. Grigg, P. Kennewell, A. Teasdale and V. Sridharan; *Tetrahedron Lett.*, 1993, **34**, 153. 546

93TL157 S. Brown, S. Clarkson, R. Grigg and V. Sridharan; *Tetrahedron Lett.*, 1993, **34**, 157. 546

93TL247 T. J. Sprules, J. D. Galpin and D. Macdonald; *Tetrahedron Lett.*, 1993, **34**, 247. 810

93TL297 R. Hoffmann, T. Rückert and R. Brückner; *Tetrahedron Lett.*, 1993, **34**, 297. 398

93TL415 J. Lee, M. Tsukazaki and V. Snieckus; *Tetrahedron Lett.*, 1993, **34**, 415. 429, 1089

93TL449 G. Majetich, Y. Zhang and G. Dreyer; *Tetrahedron Lett.*, 1993, **34**, 449. 957, 973

93TL607 T. Gillmann; *Tetrahedron Lett.*, 1993, **34**, 607. 497

93TL707 A. V. R. Rao, M. N. Bhanu and G. V. M. Sharma; *Tetrahedron Lett.*, 1993, **34**, 707. 1134

93TL749 J. E. Eshelman, J. L. Epps and J. Kallmerten; *Tetrahedron Lett.*, 1993, **34**, 749. 816, 819

93TL753 N. Sin and J. Kallmerten; *Tetrahedron Lett.*, 1993, **34**, 753. 816, 819

93TL867 J. A. Murphy and C. W. Patterson; *Tetrahedron Lett.*, 1993, **34**, 867. 929

93TL1011 Y. Gourdel, A. Ghanimi, P. Pellon and M. Le Corre; *Tetrahedron Lett.*, 1993, **34**, 1011. 158

93TL1103 C. L. Cywin and J. Kallmerten; *Tetrahedron Lett.*, 1993, **34**, 1103. 819

93TL1111 C. W. Jefford and J. Wang; *Tetrahedron Lett.*, 1993, **34**, 1111. 120

93TL1149 N. Ouvrard, J. Rodriguez and M. Santelli; *Tetrahedron Lett.*, 1993, **34**, 1149. 832

93TL1155 D. Grandjean and P. Pale; *Tetrahedron Lett.*, 1993, **34**, 1155. 431

93TL1197 O. Sieskind and P. Albrecht; *Tetrahedron Lett.*, 1993, **34**, 1197. 180

93TL1317 E. Kawashima, Y. Aoyama, T. Sekine, E. Nakamura, M. Kainosho, Y. Kyogoku and Y. Ishido; *Tetrahedron Lett.*, 1993, **34**, 1317. 925, 946

93TL1449 R. C. Hartley, S. Lamothe and T. H. Chan; *Tetrahedron Lett.*, 1993, **34**, 1449. 162

93TL1467 R. Keese and B. Bodenmann; *Tetrahedron Lett.*, 1993, **34**, 1467. 1038

93TL1593 J. E. Oliver, M. Schwarz, J. A. Klun, W. R. Lusby and R. M. Waters; *Tetrahedron Lett.*, 1993, **34**, 1593. 652

93TL1701 B. B. Snider, L. Armanetti and R. Baggio; *Tetrahedron Lett.*, 1993, **34**, 1701. 926

93TL1717 E. Hasegawa and D. P. Curran; *Tetrahedron Lett.*, 1993, **34**, 1717. 347, 937

93TL1721 N. A. Petasis and E. I. Bzowej; *Tetrahedron Lett.*, 1993, **34**, 1721. 826

93TL1769 J. Sprinz and G. Helmchen; *Tetrahedron Lett.*, 1993, **34**, 1769. 194

93TL1779 D. Bouvy, Z. Janousek, H. G. Viehe, B. Tinant and J.-P. Declercq; *Tetrahedron Lett.*, 1993, **34**, 1779. 995

93TL1851 B. D. Dorsey, K. J. Plzak and R. G. Ball; *Tetrahedron Lett.*, 1993, **34**, 1851. 829

93TL1917 J. Zindel, A. Zeeck, W. A. König and A. de Meijere; *Tetrahedron Lett.*, 1993, **34**, 1917. 166, 183

93TL1963 Y. Hasegawa, H. Kawasaki and K. Koga; *Tetrahedron Lett.*, 1993, **34**, 1963. 135

93TL2003 P. L. Ibáñaz and C. Nájera; *Tetrahedron Lett.*, 1993, **34**, 2003. 145, 1090

93TL2051 R. V. Hoffman and H. O. Kim; *Tetrahedron Lett.*, 1993, **34**, 2051. 179

93TL2091 R. C. Denis, J. Rancourt, E. Ghiro, F. Boutonnet and D. Gravel; *Tetrahedron Lett.*, 1993, **34**, 2091. 340, 1127

93TL2111 D. J. Brondani, R. J. P. Corriu, S. El Ayoubi, J. J. E. Moreau and M. W. C. Man; *Tetrahedron Lett.*, 1993, **34**, 2111. 160

93TL2135 I. Shimizu, T. Maruyama, T. Makuta and A. Yamamoto; *Tetrahedron Lett.*, 1993, **34**, 2135. 228

93TL2139 S. Torii, H. Okumoto, T. Tadokoro, A. Nishimura and M. A. Rashid; *Tetrahedron Lett.*, 1993, **34**, 2139. 1015

93TL2161 T. Mandai, T. Matsumoto, M. Kawada and J. Tsuji; *Tetrahedron Lett.*, 1993, **34**, 2161. 962

93TL2251 K. Araki and J. T. Welch; *Tetrahedron Lett.*, 1993, **34**, 2251. 830

93TL2271 B. M. Trost and C. J. Li; *Tetrahedron Lett.*, 1993, **34**, 2271. 1050

93TL2383 M. J. C. Buckle and I. Fleming; *Tetrahedron Lett.*, 1993, **34**, 2383. 1057

93TL2445 B. Giese, U. Hoffmann, M. Roth, A. Veit, C. Wyss, M. Zehnder and H. Zipse; *Tetrahedron Lett.*, 1993, **34**, 2445. 326, 327

93TL2581 P. A. Jacobi and W. Zheng; *Tetrahedron Lett.*, 1993, **34**, 2581. 199

93TL2597 Z. You and M. Koreeda; *Tetrahedron Lett.*, 1993, **34**, 2597. 816

93TL2733 D. H. R. Barton, S. I. Parekh and C.-L. Tse; *Tetrahedron Lett.*, 1993, **34**, 2733. 16, 940

93TL2911	C. W. Jefford and J. B. Wang; *Tetrahedron Lett.*, 1993, **34**, 2911.	120
93TL3021	A. G. Schultz and M. A. Holoboski; *Tetrahedron Lett.*, 1993, **34**, 3021.	556
93TL3041	C. C. Huval and D. A. Singleton; *Tetrahedron Lett.*, 1993, 3041.	470
93TL3063	D. E. Stack, W. R. Klein and R. D. Rieke; *Tetrahedron Lett.*, 1993, **34**, 3063.	207
93TL3129	D. Bouyssi, J. Goré, G. Balme, D. Louis and J. Wallach; *Tetrahedron Lett.*, 1993, **34**, 3129.	975
93TL3149	G. J. Dawson, C. G. Frost, J. M. J. Williams and S. J. Coote; *Tetrahedron Lett.*, 1993, **34**, 3149.	194
93TL3263	S.-M. H. Tabaei and C. V. Pittman, Jr.; *Tetrahedron Lett.*, 1993, **34**, 3263.	427
93TL3309	T. A. J. van der Heide, J. L. van der Baan, V. der Kimpe, F. Bickelhaupt and G. W. Klumpp; *Tetrahedron Lett.*, 1993, **34**, 3309.	194
93TL3523	L. A. Paquette, T. Z. Wang, S. Wang and C. M. P. Philippo; *Tetrahedron Lett.*, 1993, **34**, 3523.	826
93TL3539	Z. Li, S. Racha and E. Abushanab; *Tetrahedron Lett.*, 1993, **34**, 3539.	212
93TL3599	A. R. Hunt, S. K. Stewart and A. Whiting; *Tetrahedron Lett.*, 1993, **34**, 3599.	489
93TL3695	D. C. Chauret and J. M. Chong; *Tetrahedron Lett.*, 1993, **34**, 3695.	217
93TL3853	G. Linstrumelle and S. Gueugnot; *Tetrahedron Lett.*, 1993, **34**, 3853.	975, 1004, 1025
93TL3985	S. R. Angle and J. G. Breitenbucher; *Tetrahedron Lett.*, 1993, **34**, 3985.	831
93TL4121	M. F. Semmelhack and J. Gallagher; *Tetrahedron Lett.*, 1993, **34**, 4121.	111, 585
93TL4129	R. M. Moriarty, J. Kim and L. Guo; *Tetrahedron Lett.*, 1993, **34**, 4129.	109
93TL4177	R. Brückner, J. Suffert, A. Eggers and S. W. Scheuplein; *Tetrahedron Lett.*, 1993, **34**, 4177 and references therein.	1029
93TL4331	C. Larpent and G. Meignan; *Tetrahedron Lett.*, 1993, **34**, 4331.	446
93TL4385	J. S. Clark, S. A. Krowiak and L. J. Street; *Tetrahedron Lett.*, 1993, **34**, 4385.	196, 410
93TL4489	D. P. Curran, H. Qi, N. A. Porter, Q. Su and W.-X. Wu; *Tetrahedron Lett.*, 1993, **34**, 4489.	324
93TL4497	T. Zhou and J. R. Green; *Tetrahedron Lett.*, 1993, **34**, 4497.	196
93TL4505	F. E. Ziegler and P. G. Harran; *Tetrahedron Lett.*, 1993, **34**, 4505.	929, 935
93TL4541	G. H. Lee, E. B. Choi, E. Lee and C. S. Pak; *Tetrahedron Lett.*, 1993, **34**, 4541.	24, 439
93TL4591	M. Lautens and C. Gajda; *Tetrahedron Lett.*, 1993, **34**, 4591.	191
93TL4667	C. M. Moody and D. W. Young; *Tetrahedron Lett.*, 1993, 4667.	486
93TL4693	N. G. De Kimpe, M. A. Keppens and C. V. Stevens; *Tetrahedron Lett.*, 1993, **34**, 4693.	152
93TL4861	G. Pandey and G. Lakshmaiah; *Tetrahedron Lett.*, 1993, **34**, 4861.	370
93TL4865	A. Boto, C. Betancor, R. Hernández, M. S. Rodríguez and E. Suárez; *Tetrahedron Lett.*, 1993, **34**, 4865.	931
93TL5127	T. Kawabata, Y. Kiryu, Y. Sugiura and K. Fuji; *Tetrahedron Lett.*, 1993, **34**, 5127.	120
93TL5197	V. H. Rawal and H. M. Zhong; *Tetrahedron Lett.*, 1993, **34**, 5197.	929
93TL5213	K. Sakuma, J. H. Gilchrist, F. E. Romesberg, C. E. Cajthanu and D. B. Collum; *Tetradhedron Lett.*, 1993, **34**, 5213.	131
93TL5217	A. C. Bohnstedt, J. V. N. V. Prasad and D. H. Rich; *Tetrahedron Lett.*, 1993, **34**, 5217.	816, 817
93TL5251	F. H. Wartenberg, H. Junga and S. Blechert; *Tetrahedron Lett.*, 1993, **34**, 5251.	339, 341
93TL5257	W. Schwede, A. Cleve, E. Ottow and R. Wiechert; *Tetrahedron Lett.*, 1993, **34**, 5257.	1070
93TL5441	M. Schlosser and G.-F. Zhong; *Tetrahedron Lett.*, 1993, **34**, 5441.	113
93TL5523	M. Newcomb, J. H. Horner and H. Shahin; *Tetrahedron Lett.*, 1993, **34**, 5523.	950
93TL5601	T. Fujii, T. Hirao and Y. Oshiro; *Tetrahedron Lett.*, 1993, **34**, 5601.	290
93TL5621	R. A. Aitken, H. Hérion, A. Janosi, S. V. Raut, S. Seth, I. J. Shannon and F. C. Smith; *Tetrahedron Lett.*, 1993, **34**, 5621.	1079
93TL5885	B. Giese, W. Damm, F. Wetterich, H.-G. Zeitz, J. Rancourt and Y. Guindon; *Tetrahedron Lett.*, 1993, **34**, 5885.	327
93TL5911	F. Luo and R. Wang; *Tetrahedron Lett.*, 1993, **34**, 5911.	1048
93TL5923	M. Nakazawa, Y. Sakamoto and T. Takahashi; *Tetrahedron Lett.*, 1993, **34**, 5923.	818
93TL5943	M. Hojo, M. Ohkuma, N. Ishibashi and A. Hosomi; *Tetrahedron Lett.*, 1993, **34**, 5943.	366
93TL5995	E. J. Corey, M. C. Noe and W.-C. Shieh; *Tetrahedron Lett.*, 1993, **34**, 5995.	116
93TL6209	D. F. Taber, Y. Wang and S. J. Stachel; *Tetrahedron Lett.*, 1993, **34**, 6209.	944
93TL6403	M. Alami, F. Ferri and G. Linstrumelle; *Tetrahedron Lett.*, 1993, **34**, 6403.	469
93TL6505	D. H. R. Barton, C.-Y. Chern, J. C. Jaszberenyi and T. Shinada; *Tetrahedron Lett.*, 1993, **34**, 6505.	333
93TL6787	G. Pattenden and D. J. Schulz; *Tetrahedron Lett.*, 1993, **34**, 6787.	343, 931
93TL6877	J. L. Esker and M. Newcomb; *Tetrahedron Lett.*, 1993, **34**, 6877.	950
93TL7183	B. M. Trost, S. Sharma and T. Schmidt; *Tetrahedron Lett.*, 1993, **34**, 7183.	289
93TL7395	M. T. Reetz and H. Haning; *Tetrahedron Lett.*, 1993, **34**, 7395.	133
93TL7513	B. M. Trost and R. C. Bunt; *Tetrahedron Lett.*, 1993, **34**, 7513.	192
93TL7557	C. W. Jefford, J. B. Wang and Z.-H. Lu; *Tetrahedron Lett.*, 1993, **34**, 7557.	120
93TL7619	K. Yasui, Y. Goto, T. Yajima, Y. Taniseki, K. Fugami, A. Tanaka and Y. Tamura; *Tetrahedron Lett.*, 1993, **34**, 7619.	195
93TL7701	A. I. Meyers and L. J. Westrum; *Tetrahedron Lett.*, 1993, **34**, 7701.	134
93TL7819	D. J. Hart, R. Krishnamurthy, L. M. Pook and F. L. Seely; *Tetrahedron Lett.*, 1993, **34**, 7819.	924
93TL7849	P. V. Fish, A. R. Sudhakar and W. S. Johnson; *Tetrahedron Lett.*, 1993, **34**, 7849.	216, 979
93TL8047	M. H. Kress, B. F. Kaller and Y. Kishi; *Tetrahedron Lett.*, 1993, **34**, 8047.	196
93TL8051	D. H. R. Barton, J. Dorchak and J. Cs. Jaszberenyi; *Tetrahedron Lett.*, 1993, **34**, 8051.	443
93TL8245	B. Rondot, T. Durand, J. P. Girard, J. C. Rossi, L. Schio, S. P. Khanapure and J. Rokach; *Tetrahedron Lett.*, 1993, **34**, 8245.	940
93TL8263	K. Takahashi, T. Minami, Y. Ohara and T. Hiyama; *Tetrahedron Lett.*, 1993, **34**, 8263.	225
93TL8431	M. Schmittel and C. Whörle; *Tetrahedron Lett.*, 1993, **34**, 8431.	499

93TL8453 M. Kodama, S. Yoshio, S. Yamaguchi, Y. Fukuyama, H. Takayanagi, Y. Morinaka, S. Usui and Y. Fukazawa; *Tetrahedron Lett.*, 1993, **34**, 8453. 818

94AG(E)117 M. Wollenwerber, R. Pinkos, J. Leonhardt and H. Prinzbach; *Angew. Chem., Int. Ed. Engl.*, 1994, **33**, 117. 574

94AG(E)580 T. Mitsudo, H. Naruse, T. Kondo, Y. Ozaki and Y. Watanabe; *Angew. Chem., Int. Ed. Engl.*, 1994, **33**, 580. 518

94AG(E)683 H. Waldmann, E. Bläser, M. Jansen and H.-P. Letschert; *Angew. Chem., Int. Ed. Engl.*, 1994, **33**, 683. 370, 371

94CC41 A. J. Clark, D. I. Davies, K. Jones and C. Millbanks; *J. Chem. Soc., Chem. Commun.*, 1994, 41. 346

94CC251 R. S. Mali, P. K. Sandhu and A. Manekar-Tilve; *J. Chem. Soc., Chem. Commun.*. 1994, 251. 827

94CC789 F. Kido, K. Yamaji, S. C. Sinha, A. Yoshikoshi and M. Kato; *J. Chem. Soc., Chem. Commun.*, 1994, 789. 822

94CC889 A. Baramee, P. Charoenying, S. Rajviroongit, C. Thebtaranonth and Y. Thebtaranonth; *J. Chem. Soc., Chem. Commun.*, 1994, 889. 787

94CC901 N. Childa, K. Sughara and S. Ogawa; *J. Chem. Soc., Chem. Commun.*, 1994, 901. 828

94CC987 H. Nagano and Y. Kuno; *J. Chem. Soc., Chem. Commun.*, 1994, 987. 327

94CRV519 J. Iqbal, B. Bhatia and N. K. Nayyar; *Chem. Rev.*, 1994, 519. 320, 336, 345, 346, 925, 926, 937, 942, 946

94JA1597 R. A. Holton *et al.*; *J. Am. Chem. Soc.*, 1994, **116**, 1597. 842

94JA1599 R. A. Holton *et al.*; *J. Am. Chem. Soc.*, 1994, **116**, 1599. 842

94JA2223 H. Nishiyama, Y. Itoh, H. Matsumoto, S.-B. Park and K. Itoh; *J. Am. Chem. Soc.*, 1994, **116**, 2223. 299

94JA2902 A. S. Vellekoop and R. A. J. Smith; *J. Am. Chem. Soc.*, 1994, **116**, 2902. 313

94JA3123 J. P. Morken, M. T. Didiuk, M. S. Visser and A. H. Hoveyda; *J. Am. Chem. Soc.*, 1994, **116**, 3123. 316

94JA4131 K. Maruoka, H. Imoto, S. Saito and H. Yamamoto; *J. Am. Chem. Soc.*, 1994, **116**, 4131. 303

94JCS(P1)393 Y. Maeda, N. Shirai and Y. Sato; *J. Chem. Soc., Perkin Trans. 1.*, 1994, 393. 824

94JCS(P1)793 C. J. Baker, G. I. Fray, G. R. Geen and K. Ryan; *J. Chem. Soc., Perkin Trans. 1*, 1994, 793. 564

94JCS(P1)3115 M. W. Briscoe, R. D. Chambers, S. J. Mullins, T. Nakamura, J. F. S. Vaughan and F. G. Drakesmith; *J. Chem. Soc., Perkin Trans. 1*, 1994, 3115. 570

94JOC4 P. Garner and O. Dogan; *J. Org. Chem.*, 1994, **59**, 4. 370

94JOC281 S. Knapp and S. R. Nandan; *J. Org. Chem.*, 1994, **59**. 281. 582

94JOC311 C. K.-F. Chiu, S. V. Govindan and P. L. Fuchs; *J. Org. Chem.*, 1994, **59**, 311. 580, 781

94JOC332 S. D. Burke, A. D. Piscopio, M. E. Kort, M. A. Matulenko, M. H. Parker, D. M. Armistead and K. Shankaran; *J. Org. Chem.*, 1994, **59**, 332. 583

94JOC374 R. V. A. Orrū, J. B. P. A. Wijnberg, C. T. Bouwman and A. de Groot; *J. Org. Chem.*, 1994, **59**, 374. 584

94JOC394 K. H. Park and H. Rapoport; *J. Org. Chem.*, 1994, **59**, 394. 579, 584

94JOC406 K. Shishido, K. Goto, S. Miyoshi, Y. Takaishi and M. Shibuya; *J. Org. Chem.*. 1994, **59**, 406. 577

94JOC718 M. Journet and M. Malacria; *J. Org. Chem.*, 1994, **59**, 718. 340

94LA41 K. Mori and K. Uenishi; *Liebigs Ann. Chem.*, 1994, 41. 782

94MI 107-01 J. Leonard; *Contemp. Org. Synth.*, 1994, **1**, 387. 305, 306

B-94MI 110-01 S. Patai (ed.); "The Chemistry of Organic Arsenic, Antimony and Bismuth Compounds," Wiley, Chichester, 1994. 445

B-94MI 110-02 C. A. McAuliffe and A. G. Mackie; in "Chemistry of Arsenic, Antimony and Bismuth," ed. S. Patai, Wiley, New York, 1994, ch. 13, p. 527. 446

B-94MI 116-01 D. Lloyd and I. Gosney; in "The Chemistry of Organic Arsines, Antimony and Bismuth," ed. S. Patai, Wiley-Interscience, 1994, p. 657. 737

94OM94 C. E. Tucker, B. Greve, W. Klein and P. Knochel; *Organometallics*, 1994, **13**, 94. 991

94SL1 W. Smadja; *Synlett.*, 1994, 1. 320, 326, 327, 328, 919

94SL121 T. Takahashi, T. Okano, T. Harada, K. Imamura and H. Yamada; *Synlett*, 1994. 121. 166

94SL189 K. Voigt, U. Schick, F. E. Meyer and A. de Meijere; *Synlett*, 1994, 189. 494

94SL457 C. Darcel, C. Bruneau and P. H. Dixneuf; *Synlett*, 1994, 457. 962

94SL511 L. Tietze, H. Geissler, J. A. Gewert and U. Jakobi; *Synlett*, 1994, 511. 282

94T359 R. Grigg, P. Fretwell, C. Meerholtz and V. Sridharan; *Tetrahedron*, 1994, **50**, 359. 494

94T371 Y. Sato, S. Nukui, M. Sodeoka and M. Shibasaki; *Tetrahedron*. 1994, **50**, 371. 495

94T415 W. Oppolzer and C. Robyr; *Tetrahedron*, 1994, **50**, 415. 195

94T487 I. Shimizu and H. Ishii; *Tetrahedron*, 1994, **50**, 487. 561

94T639 D. H. R. Barton, S. I. Parekh, M. Tajbakhsh, E. A. Theodorakis and C.-L. Tse; *Tetrahedron*, 1994, **50**, 639. 777

94T655 M. Asaoka, K. Obuchi and H. Takei; *Tetrahedron*, 1994, **50**, 655. 558

94T1397 P. Magnus; *Tetrahedron*, 1994, **50**, 1397. 583

94T1539 R. O. Duthaler; *Tetrahedron*, 1994, **50**, 1539. 166

94T1651 J. C. Gilbert and R. D. Selliah; *Tetrahedron*, 1994, **50**, 1651. 831

94T3099 K. Hattori and H. Yamamoto; *Tetrahedron*, 1994, **50**, 3099. 830

94T3139 C. Yue, J. F. Nicolay, J. Royer and H. P. Husson; *Tetrahedron*. 1994, **50**, 3139. 104

94T3687 D. Goeppel, I. Munster and R. Bruckner; *Tetrahedron*, 1994, **50**, 3687. 816, 818

94T4519 M. P. Doyle, M. Y. Eismont, M. N. Protopopova and M. M. Y. Kwan; *Tetrahedron*, 1994, **50**, 4519. 353, 354

94T5957 W. F. Bailey and K. V. Gavaskar; *Tetrahedron*, 1994, **50**, 5957. 316

94TL15	J. A. De la Torre, M. Fernandez, D. Morgans, Jr., D. B. Smith, F. X. Talamas and A. Trejo; *Tetrahedron Lett.*, 1994, **35**, 15.	837
94TL363	G. G. Melikyan, R. C. Combs, J. Lamirand, M. Khan and K. M. Nicholas; *Tetrahedron Lett.*, 1994, **35**, 363.	199
94TL721	N. Hashimoto, T. Ishizuka and T. Kunieda; *Tetrahedron Lett.*, 1994, **35**, 721.	134
94TL1071	J. Leonard, D. Appleton and S. P. Fearnley; *Tetrahedron Lett.*, 1994, **35**, 1071.	791
94TL1161	S. Bienz, V. Enev and P. Huber; *Tetrahedron Lett.*, 1994, **35**, 1161.	985
94TL1255	S. Harusawa, N. Kase, R. Yoneda and T. Kurihara; *Tetrahedron Lett.*, 1994, **35**, 1255.	993
94TL1317	R. W. Jackson and K. J. Shea; *Tetrahedron Lett.*, 1994, **35**, 1317.	831
94TL1661	T. A. Engler, B. W. Draney and G. A. Gfesser; *Tetrahedron Lett.*, 1994, **35**, 1661.	827
94TL2145	T. Fukuyama, T. Li and G. Peng; *Tetrahedron Lett.*, 1994, **35**, 2145.	827
94TL2413	G. Pattenden, A. J. Smithies and D. S. Walter; *Tetrahedron Lett.*, 1994, **35**, 2413.	347
94TL2417	M. J. Begley, G. Pattenden, A. J. Smithies and D. S. Walter; *Tetrahedron Lett.*, 1994, **35**, 2417.	347
94TL2641	W. H. Pearson and E. P. Stevens; *Tetrahedron Lett.*, 1994, **35**, 2641.	371
94TL2675	T. E. Glass, P. A. Leber and P. L. Sandall; *Tetrahedron Lett.*, 1994, **35**, 2675.	784
94TL2787	J. Zhu, A. J. H. Klunder and B. Zwanenburg; *Tetrahedron Lett.*, 1994, **35**, 2787.	786
94TL4531	R. C. Denis and D. Gravel; *Tetrahedron Lett.*, 1994, **35**, 4531.	340

Subject Index

Every effort has been made to index as comprehensively as possible, and to standardize the terms used in the index in line with the IUPAC Recommendations. In view of the diverse nature of the terminology employed by the different authors, the reader is advised to search for related entries under the appropriate headings.

The index entries are presented in letter-by-letter alphabetical sequence. Compounds are normally indexed under the parent compound name, with the substituent component separated by a comma of inversion. An entry with a prefix/locant is filed after the same entry without any attachments, and in alphanumerical sequence. For example, 'dienes', '1,3-dienes', and '1-bromo-1,4-dienes' will be filed as:

Dienes
1,3-Dienes
1,4-Dienes, 1-bromo-

The Index is arranged in set-out style, with a maximum of three levels of heading. Location references refer to page number; major coverage of a subject is indicated by bold, elided page numbers: for example,

Halides, alkynyl, synthesis of, **1234–55**
  properties of, 345

See cross-references direct the user to the preferred term; for example,

Mercaptans *See* Thiols

See also cross-references provide the user with guideposts to terms of related interest, from the broader term to the narrower term, and appear at the end of the main heading to which they refer, e.g.

Thiones
  See also
    Thioketone